Hazardous Chemicals Desk Reference

Sixth Edition

Richard J. Lewis, Sr.

WILEY

A JOHN WILEY & SONS, INC., PUBLICATION

Library of Congress Cataloging-in-Publication Data:

Lewis, Richard J., Sr.
    Hazardous chemicals desk reference / Richard J. Lewis, Sr. — 6th ed.
        p. cm.
    Includes index.
    ISBN 978-0-470-18024-2 (cloth)
1. Hazardous substances—Handbooks, manuals, etc. I. Title.
    T55.3.H3L49 2008
    604.7—dc22   2008007188

Printed in the United States of America.

10 9 8 7 6 5 4 3 2 1

Hazardous Chemicals
Desk Reference

Dedicated to Grace Ross Lewis.
Her effort and advice made this edition possible.

Welcome to Ava Grace Herrmann,
the latest addition to the new generation to carry the torch.

Contents

Preface

This sixth edition of the *Hazardous Chemicals Desk Reference* again fills the need for a reference work of moderate size that serves the information needs of those who work with hazardous chemicals.

Over 5,800 entries are included. Over 500 new entries were added. Some entries present in previous editions were removed to make room for more relevant substances and maintain the book at a reasonable size. Two-thirds of the entries have been revised for this edition. Most of the new entries were selected because they are on the EPA TSCA Inventory. These are reported to be used in commerce in the United States. Emphasis was placed on adding and updating physical properties and updating all DOT Classifications.

The information was extracted from *Dangerous Properties of Industrial Materials, Eleventh Edition.* Citation to toxicity data and other less relevant information will be found in the above cited work. When cross-references to entries are not found in this book, they can be found in *Dangerous Properties of Industrial Materials, Eleventh Edition.*

Numerous synonyms have been added to assist in locating the many materials that are known under a variety of systematic and common names. The synonym cross-index contains the entry name as well as each synonym. This index should be consulted first to locate a material by name. Synonyms are given in English and also in other major languages such as French, German, Dutch, Polish, Japanese, and Italian.

Many additional physical and chemical properties have been added. Whenever available, physical descriptions, formulas, molecular weights, melting points, boiling points, explosion limits, flash points, densities, autoignition temperatures, and the like have been supplied.

A court order has vacated the OSHA Air Standards set in 1989 and contained in 29CFR 1910.1000. OSHA has decided to enforce only pre-1989 air standards. We have elected to include both the Transitional Limits that went into effect on December 31, 1992, and the Final Rule limits that went into effect September 1, 1989. These represent the current best judgment as to appropriate workplace air levels. While they may not be enforceable by OSHA, they are better guides than the OSHA Air Standards adopted in 1969.

The following classes of data are new or have been updated for all entries for which they apply.

1. ACGIH TLVs and BEIs reflect the latest recommendations and now include intended changes.

2. German MAK and BAT reflect the latest recommendations.

3. NTP 10th Annual Report On Carcinogens entries are identified.

4. DOT classifications were updated reflecting the HM-181 rule making.

5. CAS Registry numbers are provided for additional entries.

Each entry concludes with a safety profile, a textual summary of the hazards presented by the entry. The discussion of human exposures includes target organs and specific effects reported. Fire and explosion hazards are briefly summarized in terms of conditions of flammable or reactive hazard. Where feasible, fire-fighting materials and methods are discussed. Materials which are known to be incompatible with an entry are listed here.

Also included in the safety profile are comments on disaster hazards which serve to alert users of materials to the dangers that may be encountered on entering storage premises during a fire or other emergency. Although the presence of water, steam, acid fumes, or powerful vibrations can cause the decomposition of many materials into dangerous compounds, of particular concern are high temperatures (such as those resulting from a fire) since these can cause many otherwise mild chemicals to emit highly toxic gases or vapors such as NO_x, SO_x, acids, and so forth, or evolve vapors of antimony, arsenic, mercury, and the like.

Three cross-indices are provided as Appendices to permit rapid location of a material if either a Chemical Abstract Service (CAS) number, a synonym, or DOT Guide Number for the material is the point of entry.

Every effort has been made to include the most current and complete information. The author welcomes comments or corrections to the data presented.

Richard J. Lewis, Sr.

Acknowledgments

I extend thanks to Bob Esposito for his encouragement. My thanks to Melissa Yanuzzi for her expert professional advice and assistance in converting the manuscript to this volume.

Introduction

The list of potentially hazardous materials includes drugs, food additives, preservatives, ores, pesticides, dyes, detergents, lubricants, soaps, plastics, extracts from plant and animal sources, plants and animals that are toxic by contact or consumption, and industrial intermediates and waste products from production processes. Some of the information refers to materials of undefined composition. The chemicals included are assumed to exhibit the reported toxic effect in their pure state unless otherwise noted. However, even in the case of a supposedly "pure" chemical, there is usually some degree of uncertainty as to its exact composition and the impurities that may be present. This possibility must be considered in attempting to interpret the data presented because the toxic effects observed could in some cases be caused by a contaminant. Some radioactive materials are included but the effect reported is the chemically produced effect rather than the radiation effect.

For each entry the following data are provided when available: the DPIM code, hazard rating, entry name, CAS number, DOT number, molecular formula, molecular weight, line structural formula, a description of the material and physical properties, and synonyms. Following this are listed the toxicity data with references for reports of primary skin and eye irritation, mutation, reproductive, carcinogenic, and acute toxic dose data. The Consensus Reports section contains, where available, NTP 8th Annual Report on Carcinogens notation, IARC reviews, NTP Carcinogenesis Testing Program results, EPA Extremely Hazardous Substances List, the EPA Genetic Toxicology Program, and the Community Right-To-Know List. We also indicate the presence of the material in the update of the EPA TSCA inventory of chemicals in use in the United States. The next grouping consists of the U.S. Occupational Safety and Health Administration's (OSHA) permissible exposure levels, the American Conference of Governmental Industrial Hygienists' (ACGIH) Threshold Limit Values (TLVs), German Research Society's (MAK) values, National Institute for Occupational Safety and Health (NIOSH) recommended exposure levels, and U.S. Department of Transportation (DOT) classifications. Each entry concludes with a Safety Profile that discusses the toxic and other hazards of the entry. The Safety Profile concludes with the OSHA and NIOSH occupational analytical method, referenced by method name or number.

1. *DPIM Entry Code* identifies each entry by a unique code consisting of three letters and three numbers, for example, AAA123. The first letter of the entry code indicates the alphabetical position of the entry. Codes beginning with "A" are assigned to entries indexed with the A's. Each listing in the cross-indexes is referenced to its appropriate entry by the DPIM entry code.

2. *Entry Name* is the name of each material, selected, where possible, to be a commonly used designation.

3. *Hazard Rating* (**HR:**) is assigned to each material in the form of a number (1, 2, or 3) that briefly identifies the level of the toxicity or hazard. The letter "D" is used where the data available are insufficient to indicate a relative rating. In most cases a "D" rating is assigned when only in-vitro mutagenic or experimental reproductive data are available. Ratings are assigned on the basis of low (1), medium (2), or high (3) toxic, fire, explosive, or reactivity hazard.

The number "3" indicates an LD50 below 400 mg/kg or an LC50 below 100 ppm; or that the material is explosive, highly flammable, or highly reactive.

The number "2" indicates an LD50 of 400−4,000 mg/kg or an LC50 of 100−500 ppm; or that the material is flammable or reactive.

The number "1" indicates an LD50 of 4000−40,000 mg/kg or an LC50 of 500−5000 ppm; or that the material is combustible or has some reactivity hazard.

4. *Chemical Abstracts Service Registry Number* (*CAS:*) is a numeric designation assigned by the American Chemical Society's Chemical Abstracts Service and uniquely identifies a specific chemical compound. This entry allows one to conclusively identify a material regardless of the name or naming system used.

5. *DOT:* indicates a four-digit hazard code assigned by the U.S. Department of Transportation. This code is recognized internationally and is in agreement with the United Nations coding system. The code is used on transport documents, labels, and placards. It is also used to determine the regulations for shipping the material.

6. *Molecular Formula* (*mf:*) or *atomic formula* (*af:*) designates the elemental composition of the material and is structured according to the Hill System (see *Journal of the American Chemical Society*, 22(8): 478−494, 1900), in which carbon and hydrogen (if present) are listed first, followed by the other elemental symbols in alphabetical order. The formulas for compounds that do not contain carbon are ordered strictly alphabetically by element symbol. Compounds such as salts or those containing waters of hydration have molecular formulas incorporating the CAS dot-disconnect convention. In this convention, the components are listed individually and separated by a period. The individual components of the formula are given in order of decreasing carbon atom count, and the component ratios given. A lowercase "x" indicates that the ratio is unknown. A lower case "n" indicates a repeating, polymer-like structure. The formula is obtained from one of the cited references or a chemical reference text, or derived from the name of the material.

7. *Molecular Weight* (*mw:*) or *atomic weight* (*aw:*) is calculated from the molecular formula, using standard elemental molecular weights (carbon = 12.01).

8. *Structural Formula* is a line formula indicating the structure of a given material.

9. *Properties* (*PROP:*) are selected to be useful in evaluating the hazard of a material and designing its proper storage and use procedures. A definition of the material is included where necessary. The physical description of the material may refer to the form, color, and odor to aid in positive identification. When available, the boiling point, melting point, density, vapor pressure, vapor density, and refractive index are given. The flash point, autoignition temperature, and lower and upper explosive limits are included to aid in fire protection and control. An indication is given of the solubility or miscibility of the material in water and common solvents. Unless otherwise indicated, temperature is given in Celsius, pressure in millimeters of mercury. Levels identified as "IDLH:" indicate concentrations that meet the definition of "immediately dangerous to life or health concentrations" (IDLHs). These are defined according to the NIOSH Respirator Decision Logic (DHHS [NIOSH] Publication No. 87-108, NTIS Publication No. PB-91-151183). It is a situation "that poses a threat of exposure to airborne contaminants when that exposure is likely to cause death or

immediate or delayed permanent adverse health effects or prevent escape from such an environment."

10. *Synonyms* for the entry name are listed alphabetically. Synonyms include other chemical names, common or generic names, foreign names (with the language in parentheses), or codes. Some synonyms consist in whole or in part of registered trademarks. These trademarks are not identified as such. The reader is cautioned that some synonyms, particularly common names, may be ambiguous and refer to more than one material.

11. *Consensus Reports* lines supply additional information to enable the reader to make knowledgeable evaluations of potential chemical hazards. Two types of reviews are listed: (a) International Agency for Research on Cancer (IARC) monograph reviews, which are published by the United Nations World Health Organization (WHO); and (b) the National Toxicology Program (NTP).

a. Cancer Reviews. In the U.N. International Agency for Research on Cancer (IARC) monographs, information on suspected environmental carcinogens is examined, and summaries of available data with appropriate references are presented. Included in these reviews are synonyms, physical and chemical properties, uses and occurrence, and biological data relevant to the evaluation of carcinogenic risk to humans. The monographs in the series contain an evaluation of over 1200 materials.

The format of the IARC data line is as follows. The entry "IARC Cancer Review:" indicates that the carcinogenicity data pertaining to a compound have been reviewed by the IARC committee. The committee's conclusions are summarized in three words. The first word indicates whether the data pertain to humans or to animals. The next two words indicate the degree of carcinogenic risk as defined by IARC.

For experimental animals the evidence of carcinogenicity is assessed by IARC and judged to fall into one of four groups defined as follows:

(1) Sufficient Evidence of carcinogenicity is provided when there is an increased incidence of malignant tumors: (a) in multiple species or strains; (b) in multiple experiments (preferably with different routes of administration or using different dose levels); or (c) to an unusual degree with regard to the incidence, site, or type of tumor, or age at onset. Additional evidence may be provided by data on dose-response effects.

(2) Limited Evidence of carcinogenicity is available when the data suggest a carcinogenic effect but are limited because: (a) the studies involve a single species, strain, or experiment; (b) the experiments are restricted by inadequate dosage levels, inadequate duration of exposure to the agent, inadequate period of follow-up, poor survival, the use of too few animals, or inadequate reporting; or (c) the neoplasms produced often occur spontaneously and, in the past, have been difficult to classify as malignant by histological criteria alone (for example, lung adenomas and adenocarcinomas, and liver tumors in certain strains of mice).

(3) Inadequate Evidence is available when, because of major qualitative or quantitative limitations, the studies cannot be interpreted as showing either the presence or absence of a carcinogenic effect.

(4) No Evidence applies when several adequate studies are available that show that within the limitations of the tests used, the chemical is not carcinogenic.

It should be noted that the categories *Sufficient Evidence* and *Limited Evidence* refer only to the strength of the experimental evidence that these chemicals are carcinogenic and not to the extent of their carcinogenic activity nor to the mechanism involved. The classification of any chemical may change as new information becomes available.

The evidence for carcinogenicity from studies in humans is assessed by the IARC committees and judged to fall into one of four groups defined as follows:

(1) Sufficient Evidence of carcinogenicity indicates that there is a causal relationship between the exposure and human cancer.

(2) Limited Evidence of carcinogenicity indicates that a causal relationship is credible, but that alternative explanations, such as chance, bias, or confounding, could not adequately be excluded.

(3) Inadequate Evidence, which applies to both positive and negative evidence, indicates that one of two conditions prevailed: (a) there are few pertinent data; or (b) the available studies, while showing evidence of association, do not exclude chance, bias, or confounding.

(4) No Evidence applies when several adequate studies are available that do not show evidence of carcinogenicity.

This cancer review reflects only the conclusion of the IARC committee based on the data available for the committee's evaluation. Hence, for some substances there may be a disparity between the IARC determination and the information on the tumorigenic data lines (see paragraph 15). Also, some substances previously reviewed by IARC may be reexamined as additional data become available. These substances will contain multiple IARC review lines, each of which is referenced to the applicable IARC monograph volume.

An IARC entry indicates that some carcinogenicity data pertaining to a compound have been reviewed by the IARC committee. It indicates whether the data pertain to humans or to animals and whether the results of the determination are positive, suspected, indefinite, or negative, or whether there are no data.

This cancer review reflects only the conclusion of the IARC committee, based on the data available at the time of the committee's evaluation. Hence, for some materials there may be disagreement between the IARC determination and the tumorigenicity information in the toxicity data lines.

b. NTP Status. The notation "NTP 10th Annual Report on Carcinogens" indicated that the entry is listed on the report made to the U.S. Congress by the National Toxicology Program (NTP) as required by law. This listing implies that the entry is assumed to be a human carcinogen.

Another NTP notation indicates that the material has been tested by the NTP under its Carcinogenesis Testing Program. These entries are also identified as National Cancer Institute (NCI), which reported the studies before the NCI Carcinogenesis Testing Program was absorbed by NTP. To obtain additional information about NTP, the Carcinogenesis Testing Program, or the status of a particular material under test, contact the Toxicology Information and Scientific Evaluation Group, NTP/TRTP/NIEHS, Mail Drop 18-01, P.O. Box 12233, Research Triangle Park, NC 27709.

c. EPA Extremely Hazardous Substances List. This list was developed by the U.S. Environmental Protection Agency (EPA) as required by the Superfund Amendments and Reauthorization Act of 1986 (SARA). Title III, Section 304 requires notification by facilities of a release of certain extremely hazardous substances. These 402 substances were listed by the EPA in the *Federal Register* of November 17, 1986.

d. Community Right-To-Know List. This list was developed by the EPA as required by the Superfund Amendments and Reauthorization Act of 1986 (SARA). Title III, Sections 311−312 require manufacturing facilities to prepare Material Safety Data Sheets and notify local authorities of the presence of listed chemicals. Both specific chemicals and classes of chemicals are covered by these sections.

e. EPA Genetic Toxicology Program (GENE-TOX). This status line indicates that the material has had genetic effects reported in the literature during the period 1969−1979. The test protocol in the literature is evaluated by an EPA expert panel on mutations, and the positive or negative genetic effect of the substance is reported. To obtain additional information about this program, contact GENE-TOX Program, USEPA, 401 M Street, SW, TS796, Washington, DC 20460, telephone (202) 260-1513.

f. EPA TSCA Status Line. This line indicates that the material appears on the chemical inventory prepared by the Environmental Protection Agency in accordance with provisions of the Toxic Substances Control Act (TSCA). Materials reported in the inventory include those that are produced commercially in or are imported into this country. The reader should note, however, that materials already regulated by the EPA under FIFRA and by the Food and Drug Administration under the Food, Drug, and Cosmetic Act, as amended, are not included in the TSCA inventory. Similarly, alcohol, tobacco, and explosive materials are not regulated under TSCA. TSCA regulations should be consulted for an exact definition of reporting requirements. For additional information about TSCA, contact EPA, Office of Toxic Substances, Washington, DC 20402. Specific questions about the inventory can be directed to the EPA Office of Industry Assistance, telephone (800) 424-9065.

18. *Standards and Recommendations* section contains regulations by agencies of the U.S. government or recommendations by expert groups. "OSHA" refers to standards promulgated under Section 6 of the Occupational Safety and Health Act of 1970. "DOT" refers to materials regulated for shipment by the Department of Transportation. Because of frequent changes to and litigation of federal regulations, it is recommended that the reader contact the applicable agency for information about the current standards for a particular material. Omission of a material or regulatory notation from this edition does not imply any relief from regulatory responsibility.

a. OSHA Air Contaminant Standards. The values given are for the revised standards that were published in January 13, 1989 and were scheduled to take effect from September 1, 1989 through December 31, 1992. These are noted with the entry "OSHA PEL:" followed by "TWA" or "CL," meaning either time-weighted average or ceiling value, respectively, to which workers can be exposed for a normal 8-hour day, 40-hour work week without ill effects. For some materials, TWA, CL, and Pk (peak) values are given in the standard. In those cases, all three are listed. Finally, some entries may be followed by the designation "(skin)." This designation indicates that the compound may be absorbed by the skin and that, even though the air concentration may be below the standard, significant additional exposure through the skin may be possible.

b. ACGIH Threshold Limit Values. The American Conference of Governmental Industrial Hygienists (ACGIH) Threshold Limit Values are noted with the entry "ACGIH TLV:" followed by "TWA" or "CL," meaning either time-weighted average or ceiling value, respectively, to which workers can be exposed for a normal 8-hour day, 40-hour work week without ill effects. The notation "CL" indicates a ceiling limit that must not be exceeded. The notation "skin" indicates that the material penetrates intact skin, and skin contact should be avoided even though the TLV concentration is not exceeded. STEL indicates a short-term exposure limit, usually a 15-minute time-weighted average, which should not be exceeded. Biological Exposure Indices (*BEI:*) are, according to the ACGIH, set to provide a warning level ". . .of biological response to the chemical, or warning levels of that chemical or its metabolic product(s) in tissues, fluids, or exhaled air of exposed workers. . . ."

The latest annual TLV list is contained in the publication *Threshold Limit Values and Biological Exposure Indices.* This publication should be consulted for future trends in recommendations. The ACGIH TLVs are adopted in whole or in part by many countries and local administrative agencies throughout the world. As a result, these recommendations have a major effect on the control of workplace contaminant concentrations. The ACGIH may be contacted for additional information at Kemper Woods Center, 1330 Kemper Meadow Drive, Cincinnati, OH 45240.

c. DFG MAK. These lines contain the German Research Society's Maximum Allowable Concentration values. Those materials that are classified as to workplace hazard potential by the German Research Society are noted on this line. The MAK values are also revised annually and discussions of materials under consideration for MAK assignment are included in the annual publication together with the current values. *BAT:* indicates Biological Tolerance Value for a Working Material which is defined as, ". . .the maximum permissible quantity of a chemical compound, its metabolites, or any deviation from the norm of biological parameters induced by these substances in exposed humans." *TRK:* values are Technical Guiding Concentrations for workplace control of carcinogens. For additional information, write to Deutsche Forschungsgemeinschaft (German Research Society), Kennedyallee 40, D-5300 Bonn 2, Federal Republic of Germany. The publication *Maximum Concentrations at the Workplace and Biological Tolerance Values for Working Materials Report No. 34* can be obtained from VCH Publishers, Inc., 303 N.W. 12th Ave, Deerfield Beach, FL 33442-1788 or Verlag Chemie GmbH, Buchauslieferung, P.O. Box 1260/1280, D-6940 Weinheim, Federal Republic of Germany.

d. NIOSH REL. This line indicates that a NIOSH criteria document recommending a certain occupational exposure has been published for this compound or for a class of compounds to which this material belongs. These documents contain extensive data, analysis, and references. The more recent publications can be obtained from the National Institute for Occupational Safety and Health, U.S. Department of Health and Human Services, 4676 Columbia Pkwy., Cincinnati, OH 45226.

e. DOT Classification. This is the hazard classification according to the U.S. Department of Transportation (DOT) or the International Maritime Organization (IMO). This classification gives an indication of the hazards expected in transportation, and serves as a guide to the development of proper labels, placards, and shipping instructions. The basic hazard classes include compressed gases, flammables, oxidizers, corrosives, explosives, radioactive materials, and poisons. Although a material may be designated by only one hazard class, additional hazards may be indicated by adding labels or by using other means as directed by DOT. Many materials are regulated under general headings such as "pesticides" or "combustible liquids" as defined in the regulations. These are not noted here, as their specific concentration or properties must be known for proper classification. Special regulations may govern shipment by air. This information should serve *only as a guide*, because the regulation of transported materials is carefully controlled in most countries by federal and local agencies. Because there are frequent changes to regulations, it is recommended that the reader contact the applicable agency for information about the current standards for a particular material. United States transportation regulations are found in 40 CFR, Parts 100 to 189. Contact the U.S. Department of Transportation, Materials Transportation Bureau, Washington, DC 20590.

12. *Safety Profiles* are text summaries of the reported hazards of the entry. The word "experimental" indicates that the reported effects resulted from a controlled exposure of laboratory animals to the substance. Toxic effects reported include carcinogenic, reproductive, acute lethal, and human nonlethal effects, skin and eye irritation, and positive mutation study results.

Human effects are identified either by *human* or more specifically by *man, woman, child,* or *infant.* Specific symptoms or organ systems effects are reported when available.

Carcinogenicity potential is denoted by the words "confirmed," "suspected," or "questionable." The substance entries are grouped into three classes based on experimental evidence and the opinion of expert review groups. The OSHA, IARC, ACGIH, and DFG MAK decision schedules are not related or synchronized. Thus, an entry may have had a recent review by only one group. The most stringent classification of any regulation or expert group is taken as governing.

Class I--Confirmed Carcinogens

These substances are capable of causing cancer in exposed humans. An entry was assigned to this class if it had one or more of the following data items present:

a. an OSHA regulated carcinogen

b. an ACGIH assignment as a human or animal carcinogen

c. a DFG MAK assignment as a confirmed human or animal carcinogen

d. an IARC assignment of human or animal sufficient evidence of carcinogenicity, or higher

e. NTP 10th Annual Report on Carcinogens

Class II--Suspected Carcinogens

These substances may be capable of causing cancer in exposed humans. The evidence is suggestive, but not sufficient to convince expert review committees. Some entries have not yet had expert review, but contain experimental reports of carcinogenic activity. In particular, an entry is included if it has positive reports of carcinogenic endpoint in two species. As more studies are published, many Class II carcinogens will have their carcinogenicity confirmed. On the other hand, some will be judged noncarcinogenic in the future. An entry was assigned to this class if it had one or more of the following data items present:

a. an ACGIH assignment of suspected carcinogen

b. a DFG MAK assignment of suspected carcinogen

c. an IARC assignment of human or animal limited evidence

d. two animal studies reporting positive carcinogenic endpoint in different species

Class III--Questionable Carcinogens

For these entries there is minimal published evidence of possible carcinogenic activity. The reported endpoint is often neoplastic growth with no spread or invasion characteristic of carcinogenic pathology. An even weaker endpoint is that of equivocal tumorigenic agent (ETA). Reports are assigned this designation when the study was defective. The study may have lacked control animals, may have used a very small sample size, often may lack complete pathology reporting, or may suffer many other study design defects. Many of these studies were designed for other than carcinogenic evaluation, and the reported carcinogenic effect is a by-product of the study, not the goal. The data are presented because some of the substances studied may be carcinogens. There are insufficient data to affirm or deny the possibility. An entry was assigned to this class if it had one or more of the following data items present:

a. an IARC assignment of inadequate or no evidence

b. a single human report of carcinogenicity

c. a single experimental carcinogenic report, or duplicate reports in the same species

d. one or more experimental neoplastic or equivocal tumorigenic agent reports

Fire and explosion hazards are briefly summarized in terms of conditions of flammable or reactive hazard. Materials that are incompatible with the entry are listed here. Fire and explosion hazards are briefly summarized in terms of conditions of flammable or reactive hazard. Fire-fighting materials and methods are discussed where feasible. A material with a flash point of 100°F or less is considered dangerous; if the flash point is from 100 to 200°F, the flammability is considered moderate; if it is above 200°F, the flammability is considered low (the material is considered combustible).

Also included in the safety profile are disaster hazards comments, which serve to alert users of materials, safety professionals, researchers, supervisors, and firefighters to the dangers that may be encountered on entering storage premises during a fire or other emergency. Although the presence of water, steam, acid fumes, or powerful vibrations can cause many materials to decompose into dangerous compounds, we are particularly concerned with high temperatures (such as those resulting from a fire) because these can cause many otherwise inert chemicals to emit highly toxic gases or vapors such as NO_x, SO_x, acids, and so forth, or evolve vapors of antimony, arsenic, mercury, and the like.

Key to Abbreviations

abs – absolute
ACGIH – American Conference of Governmental Industrial Hygienists
af – atomic formula
alc – alcohol
alk – alkaline
amorph – amorphous
anhyd – anhydrous
approx – approximately
aq – aqueous
atm – atmosphere
autoign – autoignition
aw – atomic weight
BEI – ACGIH Biological Exposure Indexes
bp – boiling point
b range – boiling range
CAS – Chemical Abstracts Service
cc – cubic centimeter
CC – closed cup
CL – ceiling concentration
COC – Cleveland open cup
compd(s) – compound(s)
conc – concentration, concentrated
contg – containing
cryst – crystal(s), crystalline
d – density
D – day(s)
decomp – decomposition
deliq – deliquescent
dil – dilute
DOT – U.S. Department of Transportation
EPA – U.S. Environmental Protection Agency
eth – ether
(F) – Fahrenheit
FCC – Food Chemical Codex
FDA – U.S. Food and Drug Administration
flam – flammable
flash p – flash point
fp – freezing point
g – gram
glac – glacial
gran – granular, granules
H – hour(s)
HR: – hazard rating
htd – heated
htg – heating
hygr – hygroscopic
IARC – International Agency for Research on Cancer

immisc – immiscible
incomp – incompatible
insol – insoluble
IU – International Unit
kg – kilogram (one thousand grams)
L – liter
lel – lower explosive limit
liq – liquid
M – minute(s)
m^3 – cubic meter
mf – molecular formula
mg – milligram
misc – miscible
mL – milliliter
mm – millimeter
mod – moderately
mp – melting point
mppcf – million particles per cubic foot
mw – molecular weight
m, u – micro
mg – microgram
n – refractive index
ng – nanogram
NIOSH – National Institute for Occupational Safety and Health
nonflam – nonflammable
NTP – National Toxicology Program
OBS – obsolete
OC – open cup
org – organic
ORM – other regulated material (DOT)
OSHA – Occupational Safety and Health Administration
Pa – Pascals
PEL – permissible exposure level
pet – petroleum
pg – picogram (one trillionth of a gram)
Pk – peak concentration
pmole – picomole
powd – powder
ppb – parts per billion (v/v)
pph – parts per hundred (v/v)(percent)
ppm – parts per million (v/v)
ppt – parts per trillion (v/v)
prac – practically
prep – preparation
PROP – properties
refr – refractive

rhomb − rhombic
S, sec − second(s)
sl, slt − slight
sltly − slightly
sol − soluble
soln − solution
solv(s) − solvent(s)
spar − sparingly
spont − spontaneous(ly)
STEL − short-term exposure limit
subl − sublimes
TCC − Tag closed cup
tech − technical
temp − temperature
TLV − Threshold Limit Value
TOC − Tag open cup
TWA − time weighted average
uel − upper explosive limit
unk − unknown, unreported

ULC, ulc − Underwriters Laboratory Classification
USDA − U.S. Department of Agriculture
vac − vacuum
vap − vapor
vap d − vapor density
vap press − vapor pressure
visc − viscosity
vol − volume
W − week(s)
Y − year(s)
% − percent(age)
> − greater than
< − less than
<= − less than or equal to
=> − greater than or equal to

° − degrees of temperature in Celsius (centigrade)
°F − temperature in Fahrenheit

AAC250   CAS: 8021-27-0   HR: 1
ABIES ALBA OIL
PROP: Colorless to pale-yellow oil from the steam distillation of the crushed cones of *Abies Alba Mill* (FCTXAV 12,807,74).
SYNS: OIL OF ABIES ALBA □ OIL OF FUR □ OIL OF SILVER FIR □ OIL OF SILVER PINE □ SILVER FIR NEEDLE OIL □ SILVER FIR OIL □ SILVER PINE OIL □ TEMPLIN OIL
CONSENSUS REPORTS: Reported in EPA TSCA Inventory.
SAFETY PROFILE: Low toxicity by ingestion or skin contact. A skin irritant. When heated to decomposition it emits acrid smoke and irritating fumes.

AAG000   CAS: 105-57-7   HR: 3
ACETAL
DOT: UN 1088
mf: $C_6H_{14}O_2$ mw: 118.20
PROP: Colorless, volatile liquid; agreeable odor, nutty aftertaste. Mp: −100°, bp: 102.7°, flash p: −5°F (CC), lel: 1.65%, uel: 10.4%, d: 0.831, autoign temp: 446°F, vap press: 10 mm @ 8.0°, vap d: 4.08. Sltly sol in water; misc in alc and eth.
SYNS: ACETAAL (DUTCH) □ ACETAL DIETHY-LIQUE (FRENCH) □ ACETALE (ITALIAN) □ 1,1-DIAETHOXY-AETHAN (GERMAN) □ DIAETHYL-ACETAL (GERMAN) □ 1,1-DIETHOXY-ETHAAN (DUTCH) □ 1,1-DIETHOXYETHANE □ DIETHYL ACETAL □ 1,1-DIETOSSIETANO (ITALIAN) □ ETHYLIDENE DIETHYL ETHER □ USAF DO-45
CONSENSUS REPORTS: Reported in EPA TSCA Inventory.
DOT CLASSIFICATION: 3; Label: Flammable Liquid
SAFETY PROFILE: Moderately toxic by ingestion, inhalation, and intraperitoneal routes. A skin and eye irritant. A narcotic. Dangerous fire hazard when exposed to heat or flame; can react vigorously with oxidizing materials. Forms heat-sensitive explosive peroxides on contact with air. When heated to decomposition it emits acrid smoke and fumes. See also ETHERS and ALDEHYDES.

AAG250   CAS: 75-07-0   HR: 3
ACETALDEHYDE
DOT: UN 1089
mf: C_2H_4O mw: 44.06
PROP: Colorless, fuming liquid; pungent, fruity odor. Mp: −123.5°, bp: 20.8°, lel: 4.0%, uel: 57%, flash p: −36°F (CC), d: 0.804 @ 0°/20°, autoign temp: 347°F, vap d: 1.52. Misc in water, alc, and eth. IDLH 2000 ppm.
SYNS: ACETALDEHYD (GERMAN) □ ACETIC ALDEHYDE □ ALDEHYDE ACETIQUE (FRENCH) □ ALDEIDE ACETICA (ITALIAN) □ ETHANAL □ ETHYL ALDEHYDE □ FEMA No. 2003 □ NCI-C56326 □ OCTOWY ALDEHYD (POLISH) □ RCRA WASTE NUMBER U001
CONSENSUS REPORTS: NTP 10th Report on Carcinogens. IARC Cancer Review: Group 2B IMEMDT 7,77,87; Animal Sufficient Evidence IMEMDT 36,101,85; Human Inadequate Evidence IMEMDT 36,101,85. On Community Right-To-Know List. Reported in EPA TSCA Inventory. EPA Genetic Toxicology Program.
OSHA PEL: TWA 100 ppm; STEL 150 ppm
ACGIH TLV: CL 25, Confirmed Animal Carcinogen.
DFG MAK: 50 ppm (90 mg/m³), Suspected Carcinogen
DOT CLASSIFICATION: 3; Label: Flammable Liquid
SAFETY PROFILE: Confirmed carcinogen with experimental carcinogenic and tumorigenic data. Poison by intratracheal and intravenous routes. A human systemic

irritant by inhalation. An experimental routes. A human systemic irritant by inhalation. An experimental teratogen. Other experimental reproductive effects. A skin and severe eye irritant. A narcotic. Human mutation data reported. A common air contaminant. Highly flammable liquid. Mixtures of 30–60% of the vapor in air ignite above 100°. It can react violently with acid anhydrides, alcohols, ketones, phenols, NH_3, HCN, H_2S, halogens, P, isocyanates, strong alkalies, and amines. Reactions with cobalt chloride, mercury(II) chlorate, or mercury(II) perchlorate form violently in the presence of traces of metals or acids. Reaction with oxygen may lead to detonation. When heated to decomposition it emits acrid smoke and fumes.

AAG500   CAS: 75-39-8   HR: 2
ACETALDEHYDE AMMONIA
DOT: UN 1841
mf: $C_2H_4O \cdot H_3N$ mw: 61.10
PROP: White, crystalline solid. Bp: 110°, mp: 97°. Very sol in water, alc; sltly sol in eth.
SYNS: ACETALDEHYDE, AMINE SALT □ ALDEHYDE AMMONIA □ 1-AMINOETHANOL □ α-AMINOETHYL ALCOHOL □ ETHANOL, 1-AMINO-(8CI,9CI)
DOT CLASSIFICATION: 9; Label: CLASS 9
SAFETY PROFILE: It readily decomposes into acetaldehyde and ammonia when heated, causing the hazards of these substances. Moderate fire and explosion hazard when exposed to heat or flame. Can react with oxidizing materials. When heated to decomposition it emits toxic fumes of NH_3 and NO_x.

AAG850   CAS: 105-82-8   HR: 1
ACETALDEHYDE-DI-n-PROPYL
ACETAL
mf: $C_8H_{18}O_2$ mw: 146.26
SYNS: ACETALDEHYDE, DIPROPYL ACETAL □ 1,1-DIPROPOXYETHANE □ DIPROPYL ACETAL □ n-PROPYL ACETAL
CONSENSUS REPORTS: Reported in EPA TSCA Inventory.

SAFETY PROFILE: A skin irritant. When heated to decomposition it emits acrid smoke and irritating fumes.

AAH000   CAS: 16568-02-8   HR: 3
ACETALDEHYDE-N-METHYL-N-FOR-
MYLHYDRAZONE
mf: $C_4H_8N_2O$ mw: 100.14
SYNS: ACETALDEHYDE-N-FORMYL-N-METHYL-HYDRAZONE □ ETHYLIDENE GYROMITRIN □ GYROMITRIN □ N-METHYL-N-FORMYL HYDRA-ZONE of ACETALDEHYDE
CONSENSUS REPORTS: IARC Cancer Review: Group 3 IMEMDT 7,56,87; Animal Limited Evidence IMEMDT 7,391,87. EPA Genetic Toxicology Program.
SAFETY PROFILE: Poison via ingestion and possibly other routes. Questionable carcinogen with experimental carcinogenic and tumorigenic data. When heated to decomposition it emits toxic fumes of NO_x.

AAH250   CAS: 107-29-9   HR: 3
ACETALDEHYDE OXIME
DOT: UN 2332
mf: C_2H_5NO mw: 59.08
PROP: A water-sol, crystalline material; sol in alc, eth. Mp: (α) 46.5°, mp: (β) 12°, d: 0.966, bp: 114.5°, flash p: ≤72°F.
SYNS: ACETALDOXIME □ ALDOXIME □ ETHAN-AL OXIME □ ETHYLIDENEHYDROXYLAMINE □ USAF AM-5
CONSENSUS REPORTS: Reported in EPA TSCA Inventory.
DOT CLASSIFICATION: 3; Label: Flammable Liquid
SAFETY PROFILE: Poison via intraperitoneal route. Mutation data reported. A dangerous fire hazard with a flash point at room temperature. When heated to decomposition it emits toxic fumes of NO_x. See also ALDEHYDES.

AAH750   CAS: 107-89-1   HR: 3
ACETALDOL
DOT: UN 2839
mf: $C_4H_8O_2$ mw: 88.12
PROP: Clear, white-to-yellow syrupy liquid. Bp: 83° @ 20 mm, flash p: 150°F (OC), d: 1.11, autoign temp: 482°F, vap d: 3.04.

SYNS: ALDOL □ 3-BUTANOLAL □ 3-HYDROXY-
BUTANAL □ β-HYDROXYBUTYRALDEHYDE □ 3-
HYDROXYBUTYRALDEHYDE □ OXYBUTANAL □
OXYBUTYRIC ALDEHYDE

CONSENSUS REPORTS: Reported in EPA
TSCA Inventory.

DOT CLASSIFICATION: 6.1; Label: Poison

SAFETY PROFILE: Poison via skin contact.
Moderately toxic by ingestion. A skin and
eye irritant. A flammable liquid and fire
hazard when exposed to heat or flame; emits
crotonaldehyde and water when heated. See
CROTONALDEHYDE. Can react with
oxidizing materials.

AAI000   CAS: 60-35-5   HR: 3
ACETAMIDE
mf: C_2H_5NO mw: 59.08
PROP: Colorless crystals; mousy odor. Mp:
81°, bp: 221.2°, d: 1.159 @ 20°/4°, vap
press: 1 mm @ 65°. Decomp in hot water.
SYNS: ACETIC ACID AMIDE □ ACETIMIDIC ACID □
AMID KYSELINY OCTOVE (POLISH) □ ETHANAMIDE
□ METHANECARBOXAMIDE □ NCI-C02108
CONSENSUS REPORTS: IARC Cancer
Review: Group 2B IMEMDT 7,56,87;
Animal Sufficient Evidence IMEMDT
7,389,87. On Community Right-To-Know
List. Reported in EPA TSCA Inventory.
DFG MAK: Suspected Carcinogen
SAFETY PROFILE: Suspected carcinogen
with experimental carcinogenic and
neoplastigenic data. Moderately toxic by
intraperitoneal and possibly other routes. An
experimental teratogen. Other experimental
reproductive effects. Mutation data
reported. See also AMIDES. When heated
to decomposition it emits toxic fumes of
NO_x.

AAI250   CAS: 59-66-5   HR: 3
5-ACETAMIDE-1,3,4-THIADIAZOLE-2-
SULFONAMIDE
mf: $C_4H_6N_4O_3S_2$ mw: 222.26
SYNS: 2-ACETAMIDO-5-SULFONAMIDO-1,3,4-
THIADIAZOLE □ ACETAMIDOTHIADIAZ-
OLESULFONAMIDE □ ACETAMOX □ ACETAZOL-
AMID □ ACETAZOLAMIDE □ ACETAZOLEAMIDE □
ACETOZALAMIDE □ 2-ACETYLAMINO-1,3,4-
THIADIAZOLE-5-SULFONAMIDE □ N-(5-(AMINO-

SULFONYL)-1,3,4-THIADIAZOL-2-YL)ACETAMIDE □
CARBONIC ANHYDRASE INHIBITOR NO. 6063 □
CIDAMEX □ DEFILTRAN □ DEHYDRATIN □
DIACARB □ DIAKARB □ DIAMOX □ DIDOC □
DILURAN □ DIURAMID □ DIURETICUM-HOLZINGER
□ DIUTAZOL □ DONMOX □ EDEMOX □ EUMICTON
□ FONURIT □ GLAUPAX □ GLUPAX □ MUIRAMID □
NATRIONEX □ NEPHRAMIDE □ PHONURIT □ N-(5-
SULFAMOYL-1,3,4-THIADIAZOL-2-YL)ACET-AMIDE □
VETAMOX
CONSENSUS REPORTS: Reported in EPA
TSCA Inventory.
SAFETY PROFILE: Poison by
subcutaneous and intravenous routes.
Moderately toxic by intraperitoneal route.
Human systemic effects by ingestion:
dyspnea. An experimental teratogen by
many routes. Other experimental
reproductive effects. When heated to
decomposition it emits very toxic fumes of
NO_x and SO_x. A carbonic anhydrase
inhibitor and diuretic used to treat
glaucoma.

AAL750   CAS: 531-82-8   HR: 3
2-ACETAMIDO-4-(5-NITRO-2-FURYL)-
THIAZOLE
mf: $C_9H_7N_3O_4S$ mw: 253.25
SYNS: 2-ACETAMINO-4-(5-NITRO-2-FURYL)THIA-
ZOLE □ 2-ACETYLAMINO-4-(5-NITRO-2-FURYL)-
THIAZOLE □ N-(4-(5-NITRO-2-FURANYL)-2-
THIAZOLYL)ACETAMIDE □ N-(4-(5-NITRO-2-FURYL)-2-
THIAZOLYL)ACETAMIDE □ N-(4-(5-NITRO-2-
FURYL)THIAZOL-2-YL)ACETAMIDE
CONSENSUS REPORTS: IARC Cancer
Review: Group 2B IMEMDT 7,56,87;
Animal Sufficient Evidence IMEMDT
1,181,72; IMEMDT 7,185,74.
SAFETY PROFILE: Suspected carcinogen
with experimental carcinogenic,
tumorigenic, and neoplastigenic data.
Mutation data reported. When heated to
decomposition it emits very toxic fumes of
SO_x and NO_x.

AAQ250   CAS: 2832-40-8   HR: 2
ACETAMINE YELLOW CG
mf: $C_{15}H_{15}N_3O_2$ mw: 269.33
PROP: Brownish-yellow powder. Mp:
268°–270° Solubility in water: <0.1 mg/mL
@ 18°, 95% etoh: 1-5 mg/mL @ 22°

SYNS: ACETAMIDE, N-(4-((2-HYDROXY-5-METHYL-PHENYL)AZO)PHENYL)- □ 4-ACETAMIDO-2'-HYDRO-XY-5'-METHYLAZOBENZENE □ ACETATE FAST YELLOW G □ ACETOQUINONE LIGHT YELLOW □ ACETOQUINONE LIGHT YELLOW 4JLZ □ ALTCO SPERSE FAST YELLOW GFN NEW □ AMACEL YELLOW G □ ARTISIL DIRECT YELLOW G □ ARTISIL YELLOW G □ ARTISIL YELLOW 2GN □ CALCOSYN YELLOW GC □ CALCOSYN YELLOW GCN □ CELLITON DISCHARGE YELLOW GL □ CELLITON FAST YELLOW G □ CELLITON FAST YELLOW GA □ CELLITON FAST YELLOW GA-CF □ CELLITON YELLOW G □ CELUTATE YELLOW GH □ C.I. 11855 □ C.I. 3/11855 □ CIBACETE YELLOW GBA □ CIBACET YELLOW GBA □ CIBACET YELLOW 2GC □ C.I. DISPERSE YELLOW 3 □ CILLA FAST YELLOW G □ C.I. SOLVENT YELLOW 92 □ C.I. SOLVENT YELLOW 99 □ DIACELLITON FAST YELLOW G □ DISPERSE FAST YELLOW G □ DISPERSE YELLOW G □ DISPERSE YELLOW 3 □ DISPERSIVE YELLOW 3T □ DISPERSE YELLOW Z □ DISPERSOL FAST YELLOW G □ DISPERSOL PRINTING YELLOW G □ DISPERSOL YELLOW A-G □ DURGACET YELLOW G □ DUROSPERSE YELLOW G □ EASTONE YELLOW GN □ ESTEROQUINONE LIGHT YELLOW 4JL □ ESTONE YELLOW GN □ FENACET FAST YELLOW G □ FENACET YELLOW G □ GENACRON YELLOW G □ HISPACET FAST YELLOW G □ HISPERSE YELLOW G □ N-(4-((2-HYDROXY-5-METHYLPHEN-YL)AZO)PHENYL)ACETAMIDE □ 4'-((6-HYDROXY-m-TOLYL)AZO)ACETANILIDE □ INTERCHEM ACETATE YELLOW G □ INTERCHEM DISPERSE YELLOW GH □ INTRASPERSE YELLOW GBA □ INTRASPERSE YELLOW GBA EXTRA □ KAYALON FAST YELLOW G □ KAYASET YELLOW G □ KCA ACETATE FAST YELLOW G □ MICROSETILE YELLOW GR □ MIKETON FAST YELLOW G □ NACELAN FAST YELLOW CG □ NCI-C53781 □ NOVALON YELLOW 2GN □ NYLOQUINONE LIGHT YELLOW 4JL □ NYLOQUINONE YELLOW 4J □ OSTACET YELLOW P2G □ PALACET YELLOW GN □ PALANIL YELLOW G □ PAMACEL YELLOW G-3 □ PERLITON YELLOW G □ RELITON YELLOW C □ RESIREN YELLOW TG □ SAFARITONE YELLOW G □ SAMARON YELLOW PA3 □ SERINYL HOSIERY YELLOW GD □ SERIPLAS YELLOW GD □ SERISOL FAST YELLOW GD □ SETACYL YELLOW G □ SETACYL YELLOW 2GN □ SETACYL YELLOW P-2GL □ SILOTRAS YELLOW TSG □ SUPRACET FAST YELLOW G □ SYNTEN YELLOW 2G □ SYNTON YELLOW 2G □ TERASIL YELLOW GBA EXTRA □ TERASIL YELLOW 2GC □ TERTRANESE YELLOW N-2GL □ TULA-DISPERSE FAST YELLOW 2G □ VONTERYL YELLOW G □ VONTERYL YELLOW R □ YELLOW RELITON G □ YELLOW Z □ ZLUT DISPERZNI 3 □ ZLUT ROZPO-USTEDLOVA 77

CONSENSUS REPORTS: Community Right-To-Know List. Reported in EPA TSCA Inventory. IARC Cancer Review: Group 3 IMEMDT 48,149,90; Animal Inadequate Evidence IMEMDT 8,97,75; NTP Carcinogenesis Bioassay (feed); Clear Evidence: mouse, rat NTPTR* NTP-TR-222,82.

SAFETY PROFILE: Suspected carcinogen with experimental tumorigenic and carcinogenic data. Low toxicity by intraperitoneal route. An allergen. Mutation data reported. When heated to decomposition it emits toxic fumes of NO_x.

AAQ500   CAS: 103-84-4   HR: 3
ACETANILIDE
mf: C_8H_9NO mw: 135.18
PROP: White, shining, crystalline scales. Mp: 113.5°, bp: 305°, flash p: 345°F (OC), d: 1.2105 @ 4°/4°, autoign temp: 1004°F, vap press: 1 mm @ 114.0°, vap d: 4.65. Somewhat sol in water, alc, and eth.
SYNS: ACETAMIDE, N-PHENYL- □ ACETAMIDO-BENZENE □ ACETANIL □ ACETANILID □ ACETIC ACID ANILIDE □ ACETOANILIDE □ ACETYL-AMINOBENZENE □ ACETYLANILINE □ N-ACETYLANILINE □ AN □ ANILINE, N-ACETYL- □ ANTIFEBRIN □ PHENALGENE □ PHENALGIN □ N-PHENYLACETAMIDE □ USAF EK-3

CONSENSUS REPORTS: Reported in EPA TSCA Inventory. EPA Genetic Toxicology Program.

SAFETY PROFILE: A human poison by an unspecified route. Poison by ingestion and intravenous routes. Moderately toxic by intraperitoneal route. Human systemic effects by ingestion: hallucinations and distorted perceptions, sleepiness, constipation, cyanosis, respiratory stimulation, kidney damage, methemoglobinemia-carboxyhemoglobinemia, and decreased body temperature. Mutation data reported. When heated to decomposition it emits toxic fumes of NO_x. Combustible when exposed to heat or flame. See also ANILINE.

AAS250   CAS: 5421-48-7   HR: 3
(ACETATO)(DIETHOXYPHOSPHINYL)-
  MERCURY
mf: $C_6H_{13}HgO_5P$ mw: 396.75
PROP: IDLH 10 mg/m³ (as Hg).
SYN: (DIETHOXY-PHOSPHINYL)MERCURY ACETATE
CONSENSUS REPORTS: Mercury and its
compounds are on the Community Right-
To-Know List.
OSHA PEL: CL 0.1 mg(Hg)/m³ (skin)
ACGIH TLV: TWA 0.1 mg(Hg)/m³ (skin);
BEI: 35 μg/g creatinine total inorganic
mercury in urine preshift; 15 μg/g creatinine
total inorganic mercury in blood at end of
shift at end of workweek.
DFG MAK: Confirmed Animal Carcinogen
with Unknown Relevance to Humans
NIOSH REL: (Organomercury) TWA 0.01
mg(Hg)/m³
SAFETY PROFILE: Poison by
intraperitoneal route. See also MERCURY
COMPOUNDS. When heated to
decomposition it emits very toxic fumes of
Hg and PO_x.

AAS500   CAS: 21450-81-7   HR: 3
(ACETATO)(2,3,5,6-TETRAMETHYL-
  PHENYL)MERCURY
mf: $C_{12}H_{16}HgO_2$ mw: 392.87
PROP: IDLH 10 mg/m³ (as Hg).
SYN: (2,3,5,6-TETRAMETHYLPHENYL)MERCURY
ACETATE
CONSENSUS REPORTS: Mercury and its
compounds are on the Community Right-
To-Know List.
OSHA PEL: CL 0.1 mg(Hg)/m³ (skin)
ACGIH TLV: TWA 0.1 mg(Hg)/m³ (skin);
BEI: 35 μg/g creatinine total inorganic
mercury in urine preshift; 15 μg/g creatinine
total inorganic mercury in blood at end of
shift at end of workweek.
DFG MAK: Confirmed Animal Carcinogen
with Unknown Relevance to Humans
NIOSH REL: (Mercury, Aryl and Inorganic)
CL 0.1 mg/m³ (skin)
SAFETY PROFILE: Poison by intravenous
route. See also MERCURY
COMPOUNDS. When heated to
decomposition it emits toxic fumes of Hg.

AAT250   CAS: 64-19-7   HR: 3
ACETIC ACID
DOT: UN 2789/UN 2790
mf: $C_2H_4O_2$ mw: 60.06
PROP: Clear, colorless liquid; pungent odor.
Mp: 16.7°, bp: 118.1°, flash p: 109°F (CC),
lel: 5.4%, uel: 16.0% @ 212°F, d: 1.049 @
20°/4°, autoign temp: 869°F, vap press: 11.4
mm @ 20°, vap d: 2.07. Misc in water, alc,
and eth. IDLH 50 ppm.
SYNS: ACETIC ACID (aqueous solution) (DOT) □
ACETIC ACID, glacial or acetic acid solution, >80% acid, by
weight (UN 2790) (DOT) □ ACETIC ACID, GLACIAL □
ACETIC ACID solution, >10% but not >80% acid, by weight
(UN 2790) (DOT) □ ACIDE ACETIQUE (FRENCH) □
ACIDO ACETICO (ITALIAN) □ AZIJNZUUR (DUTCH) □
ESSIGSAEURE (GERMAN) □ ETHANOIC ACID □
ETHYLIC ACID □ FEMA No. 2006 □ GLACIAL ACETIC
ACID □ METHANE-CARBOXYLIC ACID □ OCTOWY
KWAS (POLISH) □ VINEGAR ACID
CONSENSUS REPORTS: Reported in EPA
TSCA Inventory.
OSHA PEL: TWA 10 ppm
ACGIH TLV: TWA 10 ppm; STEL 15 ppm
DFG MAK: 10 ppm (25 mg/m³)
DOT CLASSIFICATION: 8; Label: Corrosive
SAFETY PROFILE: A human poison by an
unspecified route. Moderately toxic by
various routes. A severe eye and skin
irritant. Can cause burns, lachrymation, and
conjunctivitis. Human systemic effects by
ingestion: changes in the esophagus,
ulceration, or bleeding from the small and
large intestines. Human systemic irritant
effects and mucous membrane irritant.
Experimental reproductive effects. Mutation
data reported. A common air contaminant.
A flammable liquid. A fire and explosion
hazard when exposed to heat or flame; can
react vigorously with oxidizing materials. To
fight fire, use CO_2, dry chemical, alcohol
foam, foam and mist. When heated to
decomposition it emits irritating fumes.
   Potentially explosive reaction with 5-
azidotetrazole, bromine pentafluoride,
chromium trioxide, hydrogen peroxide,
potassium permanganate, sodium peroxide,
and phosphorus trichloride. Potentially
violent reactions with acetaldehyde and
acetic anhydride. Ignites on contact with

potassium tert-butoxide. Incompatible with chromic acid, nitric acid, 2-amino-ethanol, NH_4NO_3, ClF_3, chlorosulfonic acid, $(O_3 +$ diallyl methyl carbinol), ethylenediamine, ethylene imine, $(HNO_3 + $ acetone), oleum, $HClO_4$, permanganates, $P(OCN)_3$, KOH, NaOH, xylene.

AAU000   CAS: 150-84-5   HR: 1
ACETIC ACID, CITRONELLYL ESTER
mf: $C_{12}H_{22}O_2$ mw: 198.34
PROP: Found in oils of Citronella Ceylon, geranium, and about 20 other oils (FCTXAV 11,1011,73). Colorless liquid; fruity odor. D: 0.883–0.893, refr index: 1.440–1.450, flash p: 212°F. Sol in alc and fixed oils; insol in glycerin, propylene glycol, and water @ 229°.
SYNS: ACETIC ACID-3,7-DIMETHYL-6-OCTEN-1-YL ESTER □ CITRONELLYL ACETATE (FCC) □ 2,6-DIMETHYL-2-OCTEN-8-OL ACETATE □ 3,7-DIMETHYL-6-OCTEN-1-YL ACETATE □ FEMA No. 2311
CONSENSUS REPORTS: Reported in EPA TSCA Inventory.
SAFETY PROFILE: Mildly toxic by ingestion. A human skin irritant. See also ESTERS. Combustible liquid. When heated to decomposition it emits acrid smoke and irritating fumes.

AAU250   CAS: 18461-55-7   HR: 3
ACETIC ACID-4,6-DINITRO-o-CRESYL ESTER
mf: $C_9H_8N_2O_6$ mw: 240.19
SYNS: 4,6-DINITRO-o-KRESYLESTER KYSELINY OCTOVE (CZECH) □ DNOK-ACETAT (CZECH)
NIOSH REL: (Dinitro ortho-Cresyl) TWA 0.2 mg/m³
SAFETY PROFILE: Poison by ingestion and intraperitoneal routes. A skin and severe eye irritant. When heated to decomposition it emits toxic fumes of NO_x.

AAW000   CAS: 56856-83-8   HR: 3
ACETIC ACID METHYLNITROSAMINO-METHYL ESTER
mf: $C_4H_8N_2O_3$ mw: 132.14
SYNS: α-ACETOXY DIMETHYLNITROSAMINE □ ACETOXYMETHYL-METHYL-NITROSAMIN (GERMAN) □ ACETOXYMETHYL METHYLNITRO-SAMINE □ N-α-

ACETOXYMETHYL-N-METHYL-NITROSAMINE □ 1-ACETOXY-N-NITROSODIMETH-YLAMINE □ AMMN □ ANN (GERMAN) □ DMN-OAC □ MAMN □ METHYL-(ACETOXYMETHYL)NITRO-SAMINE □ N-NITROSO-N-(ACETOXY)METHYL-N-METHYLAMINE □ N-NITROSO-N-METHYL-N-ACETOXYMETHYLAMINE
SAFETY PROFILE: Suspected carcinogen with experimental carcinogenic, neoplastigenic, and tumorigenic data. Poison by ingestion, subcutaneous, intravenous, and intraperitoneal routes. Experimental teratogenic data. Human mutation data reported. When heated to decomposition it emits toxic fumes of NO_x. See also NITROSAMINES, N-NITROSO COMPOUNDS, and ESTERS.

AAX175   CAS: 9003-22-9   HR: 1
ACETIC ACID, VINYL ESTER, POLY-MER with CHLOROETHYLENE
mf: $(C_4H_6O_2•C_2H_3Cl)_n$
PROP: White powder with bland odor. D: 1.4
SYNS: A 15 (polymer) □ ACETIC ACID ETHENYL ESTER POLYMER with CHLORETHENE (9CI) □ BAKELITE LP 70 □ BAKELITE VLFV □ BAKELITE VMCC □ BAKELITE VYNS □ BREON 351 □ CHLOROETHYLENEVINYL ACETATE POLYMER □ CORVIC 236581 □ DENKALAC 61 □ DIAMOND SHAMROCK 744 □ EXON 450 □ EXON 454 □ GEON 135 □ HOSTAFLEX VP 150 □ LEUCOVYL PA 1302 □ NORVINYL P 6 □ OPALON 400 □ PLIOVAC AO □ POLYVINYL CHLORIDE–POLYVINYL ACETATE □ PVC CORDO □ RHODOPAS 6000 □ SARPIFAN HP 1 □ SCONATEX □ SOLVIC 523KC □ SUMILIT PCX □ TENNUS 0565 □ TYGON □ VAGD □ VINNOL H 10/60 □ VINYL ACETATE–VINYL CHLORIDE COPOLYMER □ VINYL ACETATE–VINYL CHLO-RIDE POLYMER □ VINYL CHLORIDE–VINYL ACETATE POLYMER □ VINYLITE VYDR 21 □ VLVF □ VMCC □ VYNW
CONSENSUS REPORTS: IARC Cancer Review: Animal Limited Evidence IMEMDT 19,377,79. Reported in EPA TSCA Inventory.
SAFETY PROFILE: Suspected carcinogen with experimental tumorigenic data. When heated to decomposition it emits toxic fumes of HCl.

AAX250   CAS: 9003-20-7   HR: 1
ACETIC ACID VINYL ESTER
  POLYMERS
mf: $(C_4H_6O_2)_n$
PROP: Clear, water-white solid resin. Sol in
benzene, acetone; insol in water.
SYNS: ACETIC ACID ETHENYL ESTER
HOMOPOLYMER □ ASAHISOL 1527 □ ASB 516 □ AYAA
□ AYAF □ BAKELITE AYAA □ BAKELITE LP 90 □
BASCOREZ □ BOND CH 18 □ BOOKSAVER □ BORDEN
2123 □ CEVIAN A 678 □ D 50 □ DANFIRM □ DARATAK
□ DCA 70 □ DUVILAX BD 20 □ ELMER'S GLUE ALL □
EP 1463 □ FORMVAR 1285 □ GELVA CSV 16 □
GOHSENYL E 50 Y □ KURARE OM 100 □ LEMAC 1000 □
MERCKOGEN 6000 □ MOVINYL 114 □ NATIONAL 120-
1207 □ POLYVINYL ACETATE (FCC) □ PROTEX
(POLYMER) □ RHODOPAS M □ SOVIOL □ SP 60 ESTER
□ TOABOND 40H □ UCAR 130 □ VA 0112 □ VINAC B 7
□ VINYL ACETATE HOMOPOLYMER □ VINYL
ACETATE POLYMER □ VINYL ACETATE RESIN □
VINYL PRODUCTS R 10688 □ WINACET D
CONSENSUS REPORTS: IARC Cancer
Review: Animal Inadequate Evidence
IMEMDT 19,341,79. Reported in EPA
TSCA Inventory.
SAFETY PROFILE: Very low toxicity by
ingestion. Questionable carcinogen. When
heated to decomposition it emits acrid
smoke and irritating fumes. See also
ESTERS.

AAX500   CAS: 108-24-7   HR: 3
ACETIC ANHYDRIDE
DOT: UN 1715
mf: $C_4H_6O_3$ mw: 102.10
PROP: Colorless, very mobile, strongly
refractive liquid; very strong, irritating, acetic
odor. Mp: $-73.1°$, bp: $139.55°$, flash p:
$129°F$ (CC), d: 1.082 @ $20°/4°$, lel: 2.9%,
uel: 10.3%, autoign temp: $734°F$, vap press:
10 mm @ $36.0°$, vap d: 3.52. Sltly sol in
water; sol in org solvs. Decomp in hot water
and hot alc; misc in alc and eth. IDLH 200
ppm.
SYNS: ACETANHYDRIDE □ ACETIC ACID, ANHYDR-
IDE (9CI) □ ACETIC OXIDE □ ACETYL ANHYDRIDE □
ACETYL ETHER □ ACETYL OXIDE □ ANHYDRIDE
ACETIQUE (FRENCH) □ ANHY-DRID KYSELINY
OCTOVE □ ANIDRIDE ACETICA (ITALIAN) □
AZIJNZUURANHYDRIDE (DUTCH) □ ESSIGSAEURE-
ANHYDRID (GERMAN) □ ETHANOIC ANHYDRATE □
OCTOWY BEZWODNIK (POLISH)

CONSENSUS REPORTS: Reported in
EPA TSCA Inventory.
OSHA PEL: CL 5 ppm
ACGIH TLV: TWA 5 ppm.
DFG MAK: 5 ppm (20 mg/m³)
NIOSH REL: Acetic Anhydride: CL 5 ppm
DOT CLASSIFICATION: 8; Label: Corrosive
SAFETY PROFILE: Moderately toxic by
inhalation, ingestion, and skin contact. A
skin and severe eye irritant. A flammable
liquid. A fire and explosion hazard when
exposed to heat or flame. Potentially
explosive reactions with barium peroxide,
boric acid, chromium trioxide, 1,3-
diphenyltriazine, hydrochloric acid + water,
hypochlorous acid, nitric acid, perchloric
acid + water, peroxyacetic acid, potassium
permanganate, tetrafluoroboric acid, 4-
toluenesulfonic acid + water, and acetic acid
+ water. Reactions with ethanol + sodium
hydrogen sulfate, and hydrogen peroxide
form explosive products. Reactions with
ammonium nitrate +
hexamethylenetetrammonium acetate +
nitric acid form as products the military
explosives RDX and HMX. Reacts violently
with N-tert-butyl-phthalimic acid +
tetrafluoroboric acid, chromic acid, glycerol
+ phosphoryl chloride, and metal nitrates
(e.g., copper or sodium nitrates).
Incompatible with 2-aminoethanol, aniline,
chlorosulfonic acid, (CrO_3 + acetic acid),
ethylene-diamine, ethyleneimine, glycerol,
oleum, HF, permanganates, NaOH, Na_2O_2,
H_2SO_4, water, N_2O_2, (glycerol + phosphoryl
chloride). When heated to decomposition it
emits toxic fumes; can react vigorously with
oxidizing materials, will react violently on
contact with water or steam. Used in
production of drugs of abuse. To fight fire,
use CO_2, dry chemical, water mist, alcohol
foam. See also ANHYDRIDES.

AAX750   CAS: 93-29-8   HR: 2
ACETISOEUGENOL
mf: $C_{12}H_{14}O_3$ mw: 206.26

PROP: White crystals; clove odor. Flash p: 153°F. Sol in alc, chloroform, eth; insol in water.
SYNS: 4-ACETOXY-3-METHOXY-1-PROPENYL-BENZENE □ ACETYLISOEUGENOL □ FEMA No. 2470 □ ISOEUGENOL ACETATE □ ISOEUGENYL ACETATE (FCC) □ 2-METHOXY-4-PROPENYL-PHENYL ACETATE
CONSENSUS REPORTS: Reported in EPA TSCA Inventory.
SAFETY PROFILE: Moderately toxic by ingestion. Combustible liquid. When heated to decomposition it emits acrid smoke and irritating fumes.

AAY000   CAS: 102-01-2   HR: 2
ACETOACETANILIDE
mf: $C_{10}H_{11}NO_2$ mw: 177.22
PROP: White, crystalline solid. Mp: 86°, bp: decomp, flash p: 365°F (COC), d: 1.260 @ 20°, vap press: 0.01 mm @ 20°.
SYNS: AAN □ ACETANILIDE, 2-ACETYL- □ ACETOACETAMIDOBENZENE □ ACETO-ACETANILID □ ACETOACETIC ACID ANILIDE □ ACETOACETIC ANILIDE □ ((ACETOACETYL)-AMINO)BENZENE □ ACETOACETYLANILINE □ ACETYLACETANILIDE □ α-ACETYLACETANILIDE □ N-(ACETYLACETYL)-ANILINE □ ANILID KYSELINY ACETOCTOVE □ BUTANAMIDE, 3-OXO-N-PHENYL-(9CI) □ β-KETOBUTYRANILIDE □ 3-OXO-N-PHENYLBUT-ANAMIDE □ N-PHENYL-ACETOACETAMIDE □ USAF EK-1239
CONSENSUS REPORTS: Reported in EPA TSCA Inventory.
SAFETY PROFILE: Poison by intraperitoneal route. Moderately toxic by ingestion. A weak allergen. See also ACETANILIDE. Combustible when exposed to heat or flame. See ANILINE and CYANIDE for disaster hazard. When heated to decomposition it emits toxic NO_x fumes. To fight fire, use alcohol foam, water mist, CO_2, dry chemical.

ABA000   CAS: 93-68-5   HR: 2
ACETOACET-o-TOLUIDIDE
mf: $C_{11}H_{13}NO_2$ mw: 191.25
PROP: Crystals. Mp: 106°, bp: decomp, d: 1.300 @ 20°, vap press: 0.01 mm @ 20°, flash p: 320°F (COC).

SYNS: 2-ACETOACETYLAMINOTOLUENE □ ACETOACETYL-2-METHYLANILIDE □ 2'-METHYL-ACETOACETANILIDE
CONSENSUS REPORTS: Reported in EPA TSCA Inventory.
SAFETY PROFILE: Moderately toxic by ingestion. When heated to decomposition it emits toxic fumes of NO_x.

ABA500   CAS: 92-15-9   HR: 2
ACETOACETYL-o-ANISIDINE
mf: $C_{11}H_{13}NO_3$ mw: 207.25
PROP: Crystals. Mp: 86.6°, flash p: 325°F (OC), d: 1.132 @ 86.6°/20°, vap d: 7.0.
SYNS: o-ACETOACETANISIDE □ ACETOACET-o-ANISIDIN (CZECH) □ ACETOACETIC ACID-o-ANISIDIDE □ 2-ACETOACETYLAMINOANISOLE □ ACETOACETYL-o-ANISIDE □ ACETOACETYL-o-ANISINE □ o-METHOXYACETOACETANILIDE □ 2-METHOXYACETOACETANILIDE □ 2'-METHOXY-ACETOACETANILIDE
CONSENSUS REPORTS: Reported in EPA TSCA Inventory.
SAFETY PROFILE: A skin and eye irritant. When heated to decomposition it emits toxic fumes of NO_x. Combustible when exposed to heat or flame or oxidizing materials. To fight fire, use CO_2, mist, dry chemicals.

ABA750   CAS: 1271-55-2   HR: 3
ACETOFERROCENE
mf: $C_{12}H_{12}FeO$ mw: 228.09
PROP: Orange crystals from heptane. Mp: 85–86°.
SYNS: ACETYLFERROCENE □ 1-ACETYLFERROCENE □ FERROCENE, ACETYL- □ MONACETYLFERROCENE
CONSENSUS REPORTS: Reported in EPA TSCA Inventory.
DOT CLASSIFICATION: 3; Label: Flammable Liquid
SAFETY PROFILE: Poison by ocular and intravenous routes. A flammable liquid. When heated to decomposition it emits acrid smoke and irritating fumes.

ABB000   CAS: 968-81-0   HR: 3
ACETOHEXAMIDE
mf: $C_{15}H_{20}N_2O_4S$ mw: 324.43

A

PROP: Crystals from aq ethanol. Mp: 188–189°.
SYNS: 1-(p-ACETYLBENZENESULFONYL)-3-CYCLOHEXYLUREA □ 4-ACETYL-N-(((CYCLO-HEXYLAMINO)CARBONYL)-BENZENE-SULFONAMIDE □ CYCLAMIDE □ DIMELIN □ DIMELOR □ DYMELOR □ NCI-CO03247 □ ORDIMEL □ TSIKLAMID
CONSENSUS REPORTS: NCI Carcinogenesis Bioassay (feed); No Evidence: mouse, rat NCITR* NCI-CG-TR-50,78. Reported in EPA TSCA Inventory.
SAFETY PROFILE: Human reproductive effects by an unspecified route: stillbirth. Mildly toxic by ingestion. When heated to decomposition it emits very toxic fumes of SO_x and NO_x.

ABB500   CAS: 513-86-0   HR: 3
ACETOIN
DOT: UN 2621
mf: $C_4H_8O_2$ mw: 88.12
PROP: Sltly yellow liquid or crystalline solid; buttery odor. D: 1.016, bp: 147–148°, refr index: 1.417, mp: 15°, flash p: 106°F. Misc with water, alc, propylene glycol; insol in vegetable oil.
SYNS: ACETYL METHYL CARBINOL □ 2-BUTAN-OL-3-ONE □ 2,3-BUTANOLONE □ DIMETHYL-KETOL □ FEMA No. 2008 □ 3-HYDROXY-2-BUTANONE □ 1-HYDROXYETHYL METHYL KETONE □ γ-HYDROXY-β-OXOBUTANE
CONSENSUS REPORTS: Reported in EPA TSCA Inventory.
DOT CLASSIFICATION: 3; Label: Flammable Liquid
SAFETY PROFILE: Experimental reproductive effects. Mildly toxic by subcutaneous route. A moderate skin irritant. Flammable liquid. When heated to decomposition it emits acrid smoke and fumes. See also KETONES.

ABC000   CAS: 116-09-6   HR: 2
ACETOL (1)
mf: $C_3H_6O_2$ mw: 74.09
PROP: Colorless liquid. D: 1.084 @ 20°/4°, fp: −17° (approx), mp: −7°, bp: 145–146° decomp. Misc in water, alc, and eth.

SYNS: HYDROXYACETONE □ 1-HYDROXY-2-PROPANONE
CONSENSUS REPORTS: Reported in EPA TSCA Inventory.
SAFETY PROFILE: Moderately toxic by ingestion. Mutation data reported. An allergen. Implicated in aplastic anemia. A 10 gram dose may be fatal to an adult. Skin contact, inhalation, or ingestion can cause asthma, sneezing, irritation of eyes and nose, hives, and eczema. Combustible when exposed to heat or flame. When heated to decomposition it emits acrid smoke and fumes.

ABC475   CAS: 941-98-0   HR: 2
1'-ACETONAPHTHONE
mf: $C_{12}H_{10}O$ mw: 170.22
PROP: Crystals. Mp: 34°, bp: 302°, flash p: >230°F, d: 1.120.
SYNS: 1-ACETONAPHTHALENE □ α-ACETO-NAPHTHONE □ 1-ACETONAPHTHONE □ 1-ACETYLNAPHTHALENE □ ETHANONE, 1-(1-NAPHTHALENYL)-(9CI) □ METHYL α-NAPHTHYL KETONE □ METHYL 1-NAPHTHYL KETONE □ α-METHYL NAPHTHYL KETONE □ 1-(1-NAPHTH-ALENYL)ETHANONE □ α-NAPHTHYL METHYL KETONE □ 1-NAPHTHYL METHYL KETONE
CONSENSUS REPORTS: Reported in EPA TSCA Inventory.
DOT CLASSIFICATION: 3; Label: Flammable Liquid
SAFETY PROFILE: Moderately toxic by ingestion. A skin irritant. A combustible liquid. When heated to decomposition it emits acrid smoke and irritating fumes.

ABC500   CAS: 93-08-3   HR: 2
2'-ACETONAPHTHONE
mf: $C_{12}H_{10}O$ mw: 170.22
PROP: White needles; orange-blossom odor. Flash p: 264°F, mp: 56°, bp: 301–303°. Sol in fixed oils; sltly sol in propylene glycol; insol in glycerin.
SYNS: β-ACETONAPHTHALENE □ ACETONAPHTHONE □ β-ACETONAPHTHONE □ 2-ACETONAPHTHONE □ β-ACETYLNAPHTHALENE □ 2-ACETYLNAPHTHALENE □ FEMA No. 2723 □ METHYL-β-NAPHTHYL KETONE (FCC) □ METHYL-2-NAPHTHYL KETONE □ β-METHYL NAPHTHYL KETONE □ 1-(2-NAPHTHALENYL)ETHANONE □ β-

NAPHTHYL METHYL KETONE □ 2-NAPHTHYL
METHYL KETONE □ ORANGE CRYSTALS
CONSENSUS REPORTS: Reported in EPA
TSCA Inventory.
DOT CLASSIFICATION: 3; Label:
Flammable Liquid
SAFETY PROFILE: Moderately toxic by
ingestion. A human skin irritant. Flammable
liquid. When heated to decomposition it
emits acrid smoke and fumes.

ABC750   CAS: 67-64-1   HR: 3
ACETONE
DOT: UN 1090/UN 1091
mf: C_3H_6O mw: 58.09
PROP: Volatile, colorless liquid; fragrant
mintlike odor. Mp: −94.6°, bp: 56.2° @ 20
mm, refr index: 1.356, flash p: 0°F (CC), lel:
2.6%, uel: 12.8%, d: 0.7972 @ 15°, autoign
temp: (color) 869°F, vap press: 240 hPa @
20°, vap d: 2.00. Misc in water, alc, org
solvs, and ether. IDLH 2500 ppm [10%
LEL].
SYNS: ACETON (GERMAN, DUTCH, POLISH) □
ACETONE OILS (DOT) □ CHEVRON ACETONE □
DIMETHYLFORMALDEHYDE □ DIMETHYLKETAL □
DIMETHYL KETONE □ FEMA No. 3326 □ KETONE,
DIMETHYL □ KETONE PROPANE □ β-KETOPROPANE
□ METHYL KETONE □ PROPANONE □ 2-
PROPANONE □ PYROACETIC ACID □ PYROACETIC
ETHER □ RCRA WASTE NUMBER U002
CONSENSUS REPORTS: On Community
Right-To-Know List. Reported in EPA
TSCA Inventory.
OSHA PEL: TWA 750 ppm; STEL 1000
ppm
ACGIH TLV: TWA 500 ppm; STEL 750
ppm; Not Classifiable as a Human
Carcinogen; BEI: 50 mg/L acetone in urine
at end of shift.
DFG MAK: 500 ppm (1200 mg/m³)
NIOSH REL: (Ketones) 10H TWA 590
mg/m³
DOT CLASSIFICATION: 3; Label:
Flammable Liquid
SAFETY PROFILE: Moderately toxic by
various routes. A skin and severe eye
irritant. Human systemic effects by
inhalation: changes in EEG, changes in

carbohydrate metabolism, nasal effects,
conjunctiva irritation, respiratory system
effects, nausea and vomiting, and muscle
weakness. Human systemic effects by
ingestion: coma, kidney damage, and
metabolic changes. Narcotic in high
concentration. In industry, no injurious
effects have been reported other than skin
irritation resulting from its defatting action,
or headache from prolonged inhalation.
Experimental reproductive effects. A
common air contaminant. Highly flammable
liquid. Dangerous disaster hazard due to fire
and explosion hazard; can react vigorously
with oxidizing materials.

   Potentially explosive reaction with nitric
acid + sulfuric acid, bromine trifluoride,
nitrosyl chloride + platinum, nitrosyl
perchlorate, chromyl chloride, thiotrithiazyl
perchlorate, and (2,4,6-trichloro-1,3,5-
triazine + water). Reacts to form explosive
peroxide products with 2-methyl-1,3-
butadiene, hydrogen peroxide, and
peroxomonosulfuric acid. Ignites on contact
with activated carbon, chromium trioxide,
dioxygen difluoride + carbon dioxide, and
potassium-tert-butoxide. Reacts violently
with bromoform, chloroform + alkalies,
bromine, and sulfur dichloride.
Incompatible with CrO, (nitric + acetic
acid), NOCl, nitryl perchlorate,
permonosulfuric acid, NaOBr, (sulfuric acid
+ potassium dichromate), (thio-diglycol +
hydrogen peroxide), trichloromelamine, air,
HNO_3, chloroform, and H_2SO_4. To fight
fire, use CO_2, dry chemical, alcohol foam.
Used in production of drugs of abuse.

ABD000   CAS: 57-15-8   HR: 3
ACETONE CHLOROFORM
mf: $C_4H_7Cl_3O$ mw: 177.46
PROP: Hydrated crystals; camphor odor.
Mp: 97° (78° anhyd), bp: 167°.
SYNS: ANHYDROUS CHLOROBUTANOL □
CHLORBUTANOL □ CHLORBUTOL □ CHLORETONE
□ CHLOROBUTANOL □ CLORTRAN □ HCP □
METHAFORM □ SEDAFORM □ TRICHLORO-tert-BUTYL
ALCOHOL □ tert-TRICHLOROBUTYL ALCOHOL □

β,β,β-TRICHLORO-tert-BUTYL ALCOHOL □ 1,1,1-TRICHLORO-2-METHYL-2-PROPANOL
CONSENSUS REPORTS: Reported in EPA TSCA Inventory.
SAFETY PROFILE: Poison by ingestion. A narcotic. A skin and eye irritant. Mutation data reported. See also CHLORAL HYDRATE, which acts similarly. Dangerous; can react with oxidizing materials. Combustible when exposed to heat or flame. When heated to decomposition it emits toxic fumes of Cl⁻. See also PHOSGENE.

ABE000   HR: 3
ACETONE PEROXIDE
PROP: Shipped as a liquid or absorbed on cornstarch. The trimeric form is crystalline. Mp: 97°.
SAFETY PROFILE: Severe skin and eye irritant. Flammable by spontaneous chemical reaction; can react vigorously with reducing materials. The trimeric form is shock-sensitive and static-electricity-sensitive and may detonate.

ABE500   CAS: 75-05-8   HR: 3
ACETONITRILE
DOT: UN 1648
mf: C_2H_3N mw: 41.06
PROP: Colorless liquid; almond-ethereal, aromatic odor. Mp: −45°, bp: 81.1°, flash p: 42°F (COC), d: 0.7868 @ 20°/20°, vap d: 1.42, vap press: 100 mm @ 27°, lel: 4.4%, uel: 16%, autoign temp: 975°F. Misc in water, alc, and org solvs. Immisc in pet eth. IDLH 500 ppm.
SYNS: ACETONITRIL (GERMAN, DUTCH) □ CYANOMETHANE □ CYANURE de METHYL (FRENCH) □ ETHANENITRILE □ ETHYL NITRILE □ METHANECARBONITRILE □ METHANE, CYANO- □ METHYL CYANIDE □ METHYLKYANID □ NCI-C60822 □ RCRA WASTE NUMBER U003 □ USAF EK-488
CONSENSUS REPORTS: On Community Right-To-Know List. Reported in EPA TSCA Inventory.
OSHA PEL: TWA 40 ppm; STEL 60 ppm
ACGIH TLV: TWA 20 ppm skin; Not Classifiable as a Human Carcinogen.

DFG MAK: 40 ppm (70 mg/m³)
NIOSH REL: (Nitriles) TWA 34 mg/m³
DOT CLASSIFICATION: 3; Label: Flammable Liquid, Poison
SAFETY PROFILE: Poison by ingestion and intraperitoneal routes. Moderately toxic by several routes. An experimental teratogen. Other experimental reproductive effects. A skin and severe eye irritant. Human systemic effects by ingestion: convulsions, nausea or vomiting, and metabolic acidosis. Human respiratory system effects by inhalation. Mutation data reported. Dangerous fire hazard when exposed to heat, flame, or oxidizers. Explosion Hazard: See also CYANIDE and NITRILES. When heated to decomposition it emits highly toxic fumes of CN⁻ and NO$_x$. Potentially explosive reaction with lanthanide perchlorates and nitrogen-fluorine compounds. Exothermic reaction with sulfuric acid at 53°C. Will react with water, steam, acids to produce toxic and flammable vapors. Incompatible with oleum, chlorosulfonic acid, perchlorates, nitrating agents, indium, dinitrogen tetraoxide, N-fluoro compounds (e.g., perfluorourea + acetonitrile), HNO_3, SO_3. To fight fire, use foam, CO_2, dry chemical.

ABG750   CAS: 62-44-2   HR: 3
p-ACETOPHENETIDIDE
mf: $C_{10}H_{13}NO_2$ mw: 179.24
PROP: Solid. Mp: 137–138°, bp: 242–245°.
SYNS: 1-ACETAMIDO-4-ETHOXYBENZENE □ ACETO-p-PHENALIDE □ p-ACETOPHENETIDE □ ACETO-p-PHENETIDIDE □ ACETOPHENETIDIN □ ACETOPHENETIDINE □ ACETO-4-PHENETIDINE □ ACETOPHENETIN □ ACET-p-PHENALIDE □ ACET-p-PHENETIDIN □ ACETPHENETIDIN □ p-ACETPHENETIDIN □ ACETYLPHENETIDIN □ N-ACETYL-p-PHENETIDINE □ ACHROCIDIN □ ANAPAC □ APC □ ASA COMPOUND □ BROMO SELTZER □ BUFF-A-COMP □ CITRA-FORT □ CODEMPIRAL □ COMMOTIONAL □ CONTRADOL □ CORICIDIN □ CORIFORTE □ CORYBAN-D □ DAPRISAL □ DARVON COMPOUND □ DASIKON □ EMPIRIN COMPOUND □ p-ETHOXYACETANILIDE □ 4-ETHOXYACETANILIDE □ N-p-ETHOXYPHENYLACETAMIDE □ N-(4-ETHOXYPHENYL)ACETAMIDE □ FENACETINA □ FIORINAL □ MELABON □ PARACETOPHENETIDIN □ PERCOBARB □ PERCODAN □ p-PHENACETIN □ RCRA

WASTE NUMBER U187 □ SINUTAB □ TETRACYDIN □ XARIL □ ZACTIRIN COMPOUND
CONSENSUS REPORTS: NTP 10th Report on Carcinogens. IARC Cancer Review: Group 2A IMEMDT 7,310,87; Animal Inadequate Evidence IMEMDT 13,141,77; Human Limited Evidence IMEMDT 13,141,77; IMEMDT 24,135,80; Animal Limited Evidence IMEMDT 24,135,80; IMEMDT 24,135,80. Reported in EPA TSCA Inventory.
SAFETY PROFILE: Confirmed carcinogen producing tumors of the kidney and bladder. A human poison by an unspecified route. Poison by intravenous and possibly other routes. Moderately toxic by several routes. Human systemic effects by ingestion: cyanosis, liver damage, and methemo-globinemiacarboxyhemo-globinemia. Experimental teratogenic data. Other experimental reproductive effects. Mutation data reported. Chronic effects consist of weight loss, insomnia, shortness of breath, weakness, and often aplastic anemia. When heated to decomposition it emits toxic fumes of NO_x.

ABH000   CAS: 98-86-2   HR: 3
ACETOPHENONE
mf: C_8H_8O mw: 120.16
PROP: Colorless liquid or plates; sweet, pungent odor. Mp: 19.7°, bp: 202.3°, d: 1.026 @ 20°/4°, vap d: 4.14, vap press: 1 mm @ 15°, autoign temp: 1060°F. Very sol in propylene glycol and fixed oils; sol in alc, chloroform, and eth; sltly sol in water; insol in glycerin.
SYNS: ACETYLBENZENE □ BENZOYL METHIDE □ DYMEX □ FEMA No. 2009 □ HYPNONE □ KETONE METHYL PHENYL □ METHYL PHENYL KETONE □ 1-PHENYLETHANONE □ PHENYL METHYL KETONE □ USAF EK-496
CONSENSUS REPORTS: Reported in EPA TSCA Inventory.
ACGIH TLV: 10 ppm.
DOT CLASSIFICATION: 3; Label: Flammable Liquid
SAFETY PROFILE: Poison by intraperitoneal and subcutaneous routes.

Moderately toxic by ingestion. A skin and severe eye irritant. Mutation data reported. Narcotic in high concentration. A hypnotic. Flammable liquid. To fight fire, use foam, CO_2, dry chemical. When heated to decomposition it emits acrid smoke and fumes. See also KETONES.

ABH500   CAS: 61-00-7   HR: 3
ACETOPROMAZINE
mf: $C_{19}H_{22}N_2OS$ mw: 326.49
PROP: Orange oil. Bp: 208–210°@ 0.08 mm.
SYNS: ACEPROMAZINA □ ACEPROMAZINE □ ACEPROMIZINA □ ACETAZINE □ ACETHYL-PROMAZIN □ 3-ACETYL-10-(3-DIMETHYL-AMINOPROPYL)PHENOTHIAZINE □ ACETYL-PROMAZINE □ ANATRAN □ ANERGAN □ ATRAVET □ ATSETOZIN □ AY-57,062 □ AZEPRO-MAZINE □ 1522 CB □ 10-(3-DIMETHYL-AMINOPROPYL)PHENOTHI-AZINE-3-ETHYLONE □ 1-(10-(3-(DIMETHYLAMINO)-PROPYL)-10H-PHENOTHIAZIN-2-YL)ETHANONE □ 10-(3-DIMETHYLAMINOPROPYL)PHENOTHIAZIN-3-YLMETHYL KETONE □ LISERGAN □ NOTENQUIL □ NOTENSIL □ NOTESIL □ PLEGECYL □ PLEGICIN □ PLIVAPHEN □ SOPRINTIN □ SOPRONTIN □ SOPROTIN □ SV-1522 □ VETRANQUIL □ WY-1172
DOT CLASSIFICATION: 3; Label: Flammable Liquid
SAFETY PROFILE: Poison by ingestion, intravenous, and subcutaneous routes. A flammable liquid. When heated to decomposition it emits toxic fumes of SO_x and NO_x. See also KETONES. An animal tranquilizer.

ABI000   CAS: 350-03-8   HR: 3
3-ACETOPYRIDINE
mf: C_7H_7NO mw: 121.15
PROP: Liquid. Bp: 220°, bp: 106° @ 12 mm. Sol in water.
SYNS: β-ACETYLPYRIDINE □ 3-ACETYL-PYRIDINE □ METHYL PYRIDYL KETONE □ METHYL-β-PYRIDYL KETONE □ METHYL-3-PYRIDYL KETONE □ PYRIDINE, 3-ACETYL-
CONSENSUS REPORTS: Reported in EPA TSCA Inventory.
DOT CLASSIFICATION: 3; Label: Flammable Liquid
SAFETY PROFILE: Poison by ingestion. Moderately toxic by intraperitoneal route.

Mutation data reported. A flammable liquid. When heated to decomposition emits toxic fumes of NO_x. See also KETONES.

ABI500 CAS: 88-15-3 HR: 3
2-ACETOTHIENONE
mf: C_6H_6OS mw: 126.18
PROP: Liquid. D: 1.16 @ 24 mm, mp: 9°, bp: 213–214°.
SYNS: 2-ACETOTHIOPHENE □ 2-ACETYLTHIO-PHENE
CONSENSUS REPORTS: Reported in EPA TSCA Inventory.
DOT CLASSIFICATION: 3; Label: Flammable Liquid
SAFETY PROFILE: Poison by intraperitoneal route. A flammable liquid. When heated to decomposition emits toxic fumes of SO_x.

ABJ250 CAS: 103-89-9 HR: 2
p-ACETOTOLUIDIDE
mf: $C_9H_{11}NO$ mw: 149.21
PROP: Crystals from alc. Bp: 307°, fp: 335°F (CC), mp: 146°, d: 1.212, vap d: 5.14.
SYNS: p-ACETAMIDOTOLUENE □ p-ACETO-TOLUIDE □ 4-ACETOTOLUIDE □ 4-(ACETYL-AMINO)TOLUENE □ N-ACETYL-p-TOLUIDIDE □ ACETYL-p-TOLUIDINE □ p-METHYLACETANILIDE □ 4-METHYLACETANILIDE □ 4'-METHYLACETANILIDE
CONSENSUS REPORTS: Reported in EPA TSCA Inventory.
SAFETY PROFILE: Moderately toxic by ingestion. See also ACETANILIDE. Combustible. When heated to decomposition it emits toxic fumes of NO_x. To fight fire, use water, foam, CO_2, dry chemical.

ABO000 CAS: 60-31-1 HR: 3
2-ACETOXYETHYLTRIMETHYL-
AMMONIUM CHLORIDE
mf: $C_7H_{16}NO_2 \cdot Cl$ mw: 181.69
PROP: Deliquescent crystals, powder. Mp: 149–152°.
SYNS: ACECOLINE □ ACETYLCHOLINE CHLORIDE □ ACETYLCHOLINE HYDROCHLORIDE □ ACETYLCHOLINIUM CHLORIDE □ 2-(ACETYL-OXY)-N,N,N-TRIMETHYLETHANAMINIUM CHLORIDE □ ACH CHLORIDE □ ARTEROCOLINE □ CHOLINE

CHLORIDE ACETATE □ (2-HYDRO-XYETHYL)-TRIMETHYLAMMONIUM CHLORIDE ACETATE □ OVISOT □ TL 1505
CONSENSUS REPORTS: Reported in EPA TSCA Inventory.
SAFETY PROFILE: Poison by subcutaneous, intravenous, and parenteral routes. Moderately toxic by ingestion. When heated to decomposition it emits very toxic fumes of NO_x and Cl^-. A cholinergic agent. See also CHOLINE ACETATE (ESTER).

ABQ000 CAS: 6283-24-5 HR: 3
p-(ACETOXYMERCURI)ANILINE
mf: $C_8H_9HgNO_2$ mw: 351.77
PROP: Colorless prisms from ($CHCl_3$). Mp: 166–167°. Insol in water, Et_2O. Sltly sol in aqueous ethanol, $CHCl_3$.
SYNS: (ACETATO)(p-AMINOPHENYL)MERCURY □ p-AMINOPHENYLMERCURIC ACETATE
CONSENSUS REPORTS: Reported in EPA TSCA Inventory. On Community Right-To-Know List.
OSHA PEL: CL 0.1 mg(Hg)/m^3 (skin)
ACGIH TLV: TWA 0.1 mg(Hg)/m^3 (skin); BEI: 35 µg/g creatinine total inorganic mercury in urine preshift; 15 µg/g creatinine total inorganic mercury in blood at end of shift at end of workweek.
DFG MAK: Confirmed Animal Carcinogen with Unknown Relevance to Humans
NIOSH REL: (Mercury, Aryl and Inorganic) CL 0.1 mg/m^3 (skin)
SAFETY PROFILE: Poison by intravenous and intraperitoneal routes. See also MERCURY COMPOUNDS, ANILINE. When heated to decomposition it emits very toxic fumes of NO_x and Hg.

ABQ250 CAS: 54481-45-7 HR: 3
2-(ACETOXYMERCURI)-4-NITRO-
ANILINE
mf: $C_8H_8HgN_2O_4$ mw: 396.77
SYN: ACETATO(2-AMINO-5-NITROPHENYL)-MERCURY
OSHA PEL: CL 0.1 mg(Hg)/m^3 (skin)
ACGIH TLV: TWA 0.1 mg(Hg)/m^3 (skin); BEI: 35 µg/g creatinine total inorganic mercury in urine preshift; 15 µg/g creatinine

total inorganic mercury in blood at end of shift at end of workweek.

DFG MAK: Confirmed Animal Carcinogen with Unknown Relevance to Humans

NIOSH REL: (Mercury, Aryl and Inorganic) CL 0.1 mg/m³ (skin)

SAFETY PROFILE: Poison by intraperitoneal route. See also MERCURY COMPOUNDS, NITRO COMPOUNDS of AROMATIC HYDROCARBONS. When heated to decomposition it emits very toxic fumes of Hg and NO$_x$.

ABU000   CAS: 51-98-9   HR: 3
17-ACETOXY-19-NOR-17-α-PREGN-4-EN-20-YN-3-ONE

mf: C$_{22}$H$_{28}$O$_3$ mw: 340.50

PROP: Crystals from Me$_2$CO/hexane. Mp: 161–163°.

SYNS: 17-β-ACETOXY-19-NOR-17-α-PREGN-4-EN-20-YN-3-ONE □ (17-α)-17-(ACETYLOXY)-19-NORPREGN-4-EN-20-YN-3-ONE □ 17-ACETYLOXY-(17-α)-19-NORPREGN-4-ESTREN-17-β-OL-ACETATE-3-ONE □ 17-ENT □ 17-α-ETHINYL-19-NORTEST-OSTERONE ACETATE □ 17-α-ETHINYL-19-NORTESTOSTERONE-17-β-ACETATE □ 17-α-ETHYNYL-17-β-ACETOXY-19-NORANDROST-4-EN-3-ONE □ 17-α-ETHYNYL-17-HYDROXYESTR-4-EN-3-ONE ACETATE □ 17-α-ETHYNYL-19-NORTESTO-STERONE ACETATE □ 17-HYDROXY-19-NOR-17-α-PREGN-4-EN-20-YN-3-ONE ACETATE □ 17-β-HYDROXY-19-NOR-17-α-PREGN-4-EN-20-YN-3-ONE ACETATE □ NORETHINDRONE-17-ACETATE □ 19-NORETHISTERONE ACETATE □ 19-NORETHYNYLTESTOSTERONE ACETATE □ NORETHYSTERONE ACETATE □ NORLUTATE □ NORLUTINE ACETATE □ ORLUTATE

CONSENSUS REPORTS: IARC Cancer Review: Animal Limited Evidence IMEMDT 21,441,79; Animal Sufficient Evidence IMEMDT 6,179,74. EPA Genetic Toxicology Program.

SAFETY PROFILE: Suspected carcinogen with experimental tumorigenic data. Human reproductive effects by ingestion and implant routes: menstrual cycle changes, postpartum effects, and changes in fertility. A human teratogen by an unspecified route with developmental abnormalities of the urogenital system. Mutation data reported. When heated to decomposition it emits acrid smoke and irritating fumes. Used in the treatment of menstrual disorders and uterine bleeding.

ABU500   CAS: 62-38-4   HR: 3
ACETOXYPHENYLMERCURY

DOT: UN 1674

mf: C$_8$H$_8$HgO$_2$ mw: 336.75

PROP: Lustrous crystals. Mp: 149–152°. Sltly sol in water.

SYNS: ACETATE PHENYLMERCURIQUE (FRENCH) □ (ACETATO)PHENYLMERCURY □ ACETIC ACID, PHENYLMERCURY DERIV. □ (ACETOXYMERCURI)-BENZENE □ AGROSAN □ AGROSAND □ AGROSAN GN 5 □ ALGIMYCIN □ ANTIMUCIN WDR □ BENZENE, (ACETOXY-MERCURI)- □ BENZENE, (ACETOXY-MERCURIO)- □ BUFEN □ CEKUSIL □ CELMER □ CERESAN □ CERESAN UNIVERSAL □ CERESOL □ CONTRA CREME □ DYANACIDE □ FEMMA □ FENYL-MERCURIACETAT (CZECH) □ FMA □ FUNGITOX OR □ GALLOTOX □ HL-331 □ HONG KIEN □ HOSTAQUICK □ KWIKSAN □ LEYTOSAN □ LIQUIPHENE □ MERCURIPHENYL ACETATE □ MERCURY(II) ACETATE, PHENYL- □ MERCURY, ACETOXYPHENYL- □ MERGAMMA □ MERSOLITE □ MERSOLITE 8 □ METASOL 30 □ NORFORMS □ NYLMERATE □ OCTAN FENYLRTUTNATY (CZECH) □ PAMISAN □ PHENMAD □ PHENOMERCURIC ACETATE □ PHENYLMERCURI-ACETATE □ PHENYL MERCURIC ACETATE □ PHENYL-MERCURY ACETATE □ PHENYLQUECK-SILBERACETAT (GERMAN) □ PHIX □ PMA □ PMAC □ PMACETATE □ PMAL □ PMAS □ PURASAN-SC-10 □ PURATURF 10 □ QUICKSAN □ RCRA WASTE NUMBER P092 □ SANITIZED SPG □ SC-110 □ SCUTL □ SEEDTOX □ SHIMMEREX □ SPOR-KIL □ TAG □ TAG 331 □ TAG FUNGICIDE □ TAG HL 331 □ TRIGOSAN □ ZIARNIK

CONSENSUS REPORTS: IARC Cancer Review: Group 2B, Human Inadequate Evidence IMEMDT 58,239,93. EPA Extremely Hazardous Substances List. Reported in EPA TSCA Inventory. EPA Genetic Toxicology Program. Mercury and its compounds are on the Community Right-To-Know List.

OSHA PEL: CL 0.1 mg(Hg)/m³ (skin)

ACGIH TLV: TWA 0.1 mg(Hg)/m³ (skin); BEI: 35 μg/g creatinine total inorganic mercury in urine preshift; 15 μg/g creatinine total inorganic mercury in blood at end of shift at end of workweek.

DFG MAK: Confirmed Animal Carcinogen with Unknown Relevance to Humans

A

NIOSH REL: (Mercury, Aryl and Inorganic) CL 0.1 mg/m³ (skin)
DOT CLASSIFICATION: 6.1; Label: Poison
SAFETY PROFILE: Poison by ingestion, intravenous, intraperitoneal, subcutaneous, and possibly other routes. An experimental teratogen. Other experimental reproductive effects. Mutation data reported. See also MERCURY COMPOUNDS. When heated to decomposition it emits toxic fumes of Hg.

ABU800   CAS: 2114-33-2   HR: 1
2-ACETOXY-1-PHENYLPROPANE
mf: $C_{11}H_{14}O_2$ mw: 178.25
SYNS: ACETIC ACID, α-METHYL-PHENETHYL ESTER □ BENZYLMETHYLCARBINYL ACETATE □ METHYL-BENZYLCARBINYL ACETATE □ α-METHYL-β-PHENYLETHYL ACETATE □ 1-PHENYL-2-PROPANOL ACETATE
CONSENSUS REPORTS: Reported in EPA TSCA Inventory.
SAFETY PROFILE: A skin irritant. When heated to decomposition it emits acrid smoke and irritating fumes.

ABW600   CAS: 13121-71-6   HR: 3
ACETOXYTRICYCLOHEXYLSTANNANE
mf: $C_{20}H_{36}O_2Sn$ mw: 427.25
PROP: Rod-like crystals from aqueous ethanol. Mp: 61–63°.
SYN: STANNANE, ACETOXYTRICYCLOHEXYL-
OSHA PEL: TWA 0.1 mg(Sn)/m³
ACGIH TLV: TWA 0.1 mg(Sn)/m³ (skin)
SAFETY PROFILE: Poison by ingestion. When heated to decomposition it emits toxic fumes of Sn.

ABW750   CAS: 1907-13-7   HR: 3
ACETOXYTRIETHYLSTANNANE
mf: $C_8H_{18}O_2Sn$ mw: 264.95
SYNS: ACETOXYTRIETHYLTIN □ TRIAETHYLZINNACETAT (GERMAN) □ TRIETHYLTIN ACETATE
OSHA PEL: TWA 0.1 mg(Sn)/m³ (skin)
ACGIH TLV: TWA 0.1 mg(Sn)/m³; STEL 0.2 mg(Sn)/m³ (skin).
NIOSH REL: (Organotin Compounds) TWA 0.1 mg(Sn)/m³

SAFETY PROFILE: Poison by ingestion and intravenous routes. See also TIN COMPOUNDS. When heated to decomposition it emits acrid smoke and irritating Sn⁺ fumes.

ABX000   CAS: 2897-46-3   HR: 3
ACETOXYTRIHEXYLSTANNANE
mf: $C_{20}H_{42}O_2Sn$ mw: 433.31
SYNS: ACETOXYTRIHEXYLTIN □ TRIHEXYLTIN ACETATE □ TRI-N-HEXYLZINNACETAT (GERMAN)
OSHA PEL: TWA 0.1 mg(Sn)/m³ (skin)
ACGIH TLV: TWA 0.1 mg(Sn)/m³; STEL 0.2 mg(Sn)/m³ (skin).
NIOSH REL: (Organotin Compounds) TWA 0.1 mg(Sn)/m³
SAFETY PROFILE: Poison by skin contact and intravenous routes. Moderately toxic by ingestion. See also TIN COMPOUNDS. When heated to decomposition it emits acrid smoke and Sn⁺ fumes.

ABX250   CAS: 900-95-8   HR: 3
ACETOXYTRIPHENYLSTANNANE
mf: $C_{20}H_{18}O_2Sn$ mw: 409.07
PROP: White, crystalline solid. Mp: 120–123°. Practically insol.
SYNS: ACETATE de TRIPHENYL-ETAIN (FRENCH) □ ACETATO di STAGNO TRIFENILE (ITALIAN) □ ACETATOTRIPHENYLSTANNANE □ ACETOXY-TRIPHENYL-STANNAN (GERMAN) □ ACETOXY-TRIPHENYLSTANNANE □ ACETOXYTRIPHENYL-TIN □ (ACETYLOXY)TRIPHENYL-STANNANE (9CI) □ BATASAN □ BRESTAN □ ENT 25,208 □ FEN-OLOVO ACETATE □ FENTIN ACETAAT (DUTCH) □ FENTIN ACETAT (GERMAN) □ FENTIN ACETATE □ FENTINE ACETATE (FRENCH) □ FINTIN ACETATO (ITALIAN) □ GC 6936 □ HOE-2824 □ LIROMATIN □ LIROSTANOL □ PHENTIN ACETATE □ PHENTINOACETATE □ SUZU □ TINESTAN □ TINESTAN 60 WP □ TIN TRIPHENYL ACETATE □ TPTA □ TPZA □ TRIFENYLTINACETAAT (DUTCH) □ TRIPHENYLACETO STANNANE □ TRIPHENYL-TIN ACETATE □ TRIPHENYL-ZINNACETAT (GERMAN) □ TUBOTIN □ VP 1940
CONSENSUS REPORTS: EPA Extremely Hazardous Substances List. Reported in EPA TSCA Inventory.
OSHA PEL: TWA 0.1 mg(Sn)/m³ (skin)
ACGIH TLV: TWA 0.1 mg(Sn)/m³; STEL 0.2 mg(Sn)/m³ (skin).

NIOSH REL: (Organotin Compounds) TWA 0.1 mg(Sn)/m³
SAFETY PROFILE: Poison by ingestion, intraperitoneal, intravenous, and subcutaneous routes. Moderately toxic by skin contact. Questionable carcinogen with experimental neoplastigenic data. An experimental teratogen. Other experimental reproductive effects. A fungicide and algicide used as a wood preservative. When heated to decomposition it emits acrid smoke and Sn⁺ fumes. See also TIN COMPOUNDS.

ABX500   CAS: 97-44-9   HR: 3
ACETPHENARSINE
mf: $C_8H_{10}AsNO_5$ mw: 275.11
PROP: Crystalline material. Mp: 220–221°. Decomp @ 240–250°. Sltly sol in water.
SYNS: 3-ACETAMIDO-4-HYDROXY-PHENYL-ARSONIC ACID □ ACETARSOL □ ACETARSONE □ 3-ACETYLAMINO-4-HYDROXYPHENYLARSONIC ACID □ (3-(ACETYLAMINO)-4-HYDROXYPHENYL)-ARSONINE (9CI) □ N-ACETYL-4-HYDROXY-m-ARSANILIC ACID □ AMARSAN □ AMOEBAL □ ARSONIC ACID □ ARSPHEN □ DEVEGAN □ DISPARICIDA □ DYNARSAN □ EHRLICH 594 □ F 190 □ 190 F □ FOURNEAU 190 □ GINARSOL □ GOYL □ GYNOPLIX □ KHAROPHEN □ KUBARSOL □ LIMARSOL MALAGRIDE □ MEXYL □ MON-ARGAN □ NILACID □ ORALCID □ ORARSAN □ OSARSAL □ OSARSOLE □ OSVARSAN □ PALLICID □ PAROXYL □ SPIROCID □ SPIROZID □ STO-VARSAL □ STOVARSOL □ STOVARSOLAN □ SVC □ VAGISEPT □ VAGOFLOR
CONSENSUS REPORTS: Arsenic and its compounds are on the Community Right-To-Know List.
OSHA PEL: TWA 500 μg(As)/m³
ACGIH TLV: BEI: 35 μ (As)/L inorganic arsenic and methylated metabolites in urine
SAFETY PROFILE: Poison by ingestion and intravenous routes. Human systemic effects by ingestion: respiratory system, endocrine system, dermatitis, and fever. Human systemic effects by intravaginal route: hallucinations, distorted perceptions, convulsions, nausea or vomiting, decreased urine volume, and fever. Mutation data reported. See also ARSENIC COMPOUNDS. When heated to

decomposition it emits very toxic fumes of NO_x and As.

ABX750   CAS: 123-54-6   HR: 3
ACETYL ACETONE
DOT: UN 2310
mf: $C_5H_8O_2$ mw: 100.13
PROP: Colorless to sltly yellow liquid; pleasant odor. Mp: −23.2°, bp: 139° @ 746 mm, flash p: 105°F (OC), d: 0.952–0.962, refr index: 1.402, vap d: 3.45, autoign temp: 644°F. Misc in alc, ether, chloroform, acetone, glacial acetic acid, and propylene glycol; insol in glycerin and water.
SYNS: ACETOACETONE □ DIACETYLMETHANE □ FEMA No. 2841 □ PENTANEDIONE □ 2,4-PENTANEDIONE (FCC)
CONSENSUS REPORTS: Reported in EPA TSCA Inventory.
DOT CLASSIFICATION: 3; Label: Flammable Liquid
SAFETY PROFILE: Poison by ingestion and intraperitoneal routes. Moderately toxic by inhalation. A skin and severe eye irritant. Experimental reproductive effects. Mutation data reported. Flammable liquid when exposed to heat or flame. Incompatible with oxidizing materials. To fight fire, use alcohol foam, CO_2, dry chemical.

ABY000   CAS: 28322-02-3   HR: 3
4-ACETYLAMINOFLUORENE
mf: $C_{15}H_{13}NO$ mw: 223.29
SYNS: 4-ACETYLAMINOFLUOREN (GERMAN) □ N-FLUOREN-4-YLACETAMIDE □ N-4-FLUORENYLACETAMIDE
CONSENSUS REPORTS: EPA Genetic Toxicology Program.
SAFETY PROFILE: Poison by intraperitoneal route. Questionable carcinogen with experimental tumorigenic data. Experimental reproductive effects. Mutation data reported. When heated to decomposition it emits toxic fumes of NO_x.

ABY900   CAS: 140-40-9   HR: 3
2-ACETYLAMINO-5-NITROTHIAZOLE
mf: $C_5H_5N_3O_3S$ mw: 187.19

A

PROP: Needles from alc, elongated plates from acetic acid. The commercial product may be yellow. Mp: 264–265°. Sol in aq solns of NaOH and NH₃ with deep orange color.

SYNS: ACETAMIDE, N-(5-NITRO-2-THIAZOLYL)- □ ACETYL ENHEPTIN □ ACINITRAZOL □ ACINITR-AZOLE □ AMETOTERINA □ AMINITROZOL □ AMINITROZOLE □ CYZINE PREMIX □ ENHEPTIN A □ GYNOFON □ LAVOFLAGIN □ NITAZOL □ NITAZOLE □ NITHIAMIDE □ 5-NITRO-2-ACETILAMINOTIAZOLO □ N-(5-NITRO-2-THIAZOLYL)ACETAMIDE □ PLEOC-IDE □ TRICHLORAD □ TRICHOCID □ TRICHOMAN □ TRICHORAD □ TRICHORAL □ TRICOGEN □ TRI-COLAVAL □ TRICORAL □ TRICOSTERIL □ TRI-KOLAVAL □ TRITHEON

SAFETY PROFILE: Poison by ingestion. Mutation data reported. When heated to decomposition it emits toxic fumes of SO_x and NO_x.

ACA900   HR: D
ACETYLATED MONOGLYCERIDES
PROP: Esters of glycerin with acetic acid and edible fat-forming fatty acids. (FCC III) May be white to pale yellow liquids or solids; bland taste. Sol in alc, acetone; insol in water.

SYN: ACETYLATED MONO- and DIGLYCERIDES
SAFETY PROFILE: When heated to decomposition it emits acrid smoke and irritating fumes.

ACB250   CAS: 460-07-1   HR: 3
1-ACETYLAZIRIDINE
mf: C_4H_7NO mw: 85.12
SYN: ACETYLETHYLENEIMINE
CONSENSUS REPORTS: Reported in EPA TSCA Inventory. EPA Genetic Toxicology Program.
SAFETY PROFILE: Poison by intraperitoneal route. Questionable carcinogen with experimental tumorigenic and neoplastigenic data. When heated to decomposition it emits toxic fumes of NO_x.

ACC250   CAS: 644-31-5   HR: 3
ACETYL BENZOYL PEROXIDE (solid)
mf: $C_9H_8O_4$ mw: 180.17

PROP: White crystals. Mp: 36–37°, bp: 130° @ 19 mm. Sol in oils, alc, ether, and chloroform.
DOT CLASSIFICATION: Forbidden
SAFETY PROFILE: Poison by inhalation and ingestion. Severe irritant. A powerful oxidizing agent that is corrosive to the skin and mucous membranes. See also PEROXIDES, ORGANIC. Dangerous; shock or heat will cause detonation with evolution of toxic fumes; will react with water or steam to produce heat; can react vigorously with reducing materials. Flammable by spontaneous chemical reaction. To fight fire, use CO_2 or dry chemical. When heated to decomposition it emits acrid smoke and fumes.

ACD750   CAS: 506-96-7   HR: 3
ACETYL BROMIDE
DOT: UN 1716
mf: C_2H_3BrO mw: 122.96
PROP: Colorless, fuming liquid; turns yellow in air. Mp: −96.5°, bp: 76.7°, d: 1.52 @ 9.5°/4°. Decomp in water and alc; misc in benzene, ether, and chloroform.
CONSENSUS REPORTS: Reported in EPA TSCA Inventory.
DOT CLASSIFICATION: 8; Label: Corrosive
SAFETY PROFILE: Poison by ingestion, inhalation, skin contact, and intraperitoneal routes. See also HYDROBROMIC ACID and ACETIC ACID. Violent reaction on contact with water, steam, methanol, or ethanol produces toxic and reactive HBr. When heated to decomposition it emits highly corrosive and toxic fumes of carbonyl bromide and bromine. To fight fire, use dry chemical, CO_2.

ACF750   CAS: 75-36-5   HR: 3
ACETYL CHLORIDE
DOT: UN 1717
mf: C_2H_3ClO mw: 78.50
PROP: Colorless, pungent liquid. Fuming in air. Mp: −112°, bp: 51–52°, flash p: 40°F (CC), autoign temp: 734°F, d: 1.1051 @ 20°/4°, vap d: 2.70, lel: 5%. Decomp in

water and alc; misc in benzene, ether, and chloroform. Sol in Et_2O, C_6H_6.
SYNS: ACETIC ACID CHLORIDE □ ACETIC CHLORIDE □ ETHANOYL CHLORIDE □ RCRA WASTE NUMBER U006
CONSENSUS REPORTS: Reported in EPA TSCA Inventory.
DOT CLASSIFICATION: 3; Label: Flammable Liquid, Corrosive
SAFETY PROFILE: Poison by inhalation. Moderately toxic by ingestion. A human systemic irritant by inhalation. Violent hydrolysis reaction with water or steam produces heat, acetic acid, HCl, and other corrosive chlorides. May decompose during preparation. Dangerous fire hazard when exposed to heat or flame. Explosion hazard by spontaneous chemical reaction with dimethyl sulfoxide or ethanol. Also incompatible with PCl_3. When heated to decomposition it emits highly toxic fumes of phosgene and Cl⁻. To fight fire, use CO_2 or dry chemical. See also CHLORIDES.

ACH000   CAS: 616-91-1   HR: 3
N-ACETYL-l-CYSTEINE
mf: $C_5H_9NO_3S$ mw: 163.21
PROP: Crystals from water. Mp: 109–110°.
SYNS: l-α-ACETAMIDO-β-MERCAPTOPROPIONIC ACID □ ACETEIN □ ACETYLCYSTEINE □ N-ACETYLCYSTEINE □ N-ACETYL-N-CYSTEINE □ N-ACETYL-l-CYSTEINE (9CI) □ N-ACETYL-3-MERCAPTO-ALANINE □ AIRBRON □ BRONCHOL-YSIN □ FLUIMUCETIN □ FLUIMUCIL □ FLUMICIL □ INSPIR □ MERCAPTURIC ACID □ (R)-MERCAP-TURIC ACID □ MUCOLYTICUM □ MUCOLYTICUM LAPPE □ MUCOMYST □ MUCOSOLVIN □ NAC □ NAC-TB □ NSC-111180 □ PARVOLEX □ RESPAIRE
CONSENSUS REPORTS: Reported in EPA TSCA Inventory.
SAFETY PROFILE: Poison by intraperitoneal route. Moderately toxic by other routes. Mutation data reported. When heated to decomposition it emits very toxic fumes of NO_x and SO_x.

ACH090   CAS: 22439-58-3   HR: 3
2-ACETYLDIBENZOTHIOPHENE
PROP: A liquid.
mf: $C_{14}H_{10}OS$ mw: 226.30

SYNS: KETONE, 2-DIBENZOTHIENYL METHYL □ DIBENZOTHIEN-2-YL METHYL KETONE
DOT CLASSIFICATION: 3; Label: Flammable Liquid
SAFETY PROFILE: Poison by intravenous route. A flammable liquid. When heated to decomposition it emits toxic vapors of SO_x.

ACI500   CAS: 2386-25-6   HR: 3
3-ACETYL-2,4-DIMETHYL-PYRROLE
mf: $C_8H_{11}NO$ mw: 137.20
PROP: Solid. Mp: 137°.
SYN: 2,4-DIMETHYLPYRROL-3-YL METHYL KETONE
DOT CLASSIFICATION: 3; Label: Flammable Liquid
SAFETY PROFILE: Poison by intravenous and intraperitoneal routes. A flammable liquid. When heated to decomposition it emits toxic fumes of NO_x. See also KETONES.

ACI550   CAS: 1500-94-3   HR: 3
3-ACETYL-2,5-DIMETHYL-PYRROLE
mf: $C_8H_{11}NO$ mw: 137.20
SYNS: 2,5-DIMETHYLPYRROL-3-YL METHYL KETONE □ KETONE, 2,5-DIMETHYLPYRROL-3-YL METHYL
DOT CLASSIFICATION: 3; Label: Flammable Liquid
SAFETY PROFILE: A poison by intraperitoneal route. A flammable liquid. When heated to decomposition it emits toxic vapors of NO_x.

ACI640   CAS: 117568-24-8 HR: 3
10-ACETYLDITHRANOL
mf: $C_{16}H_{12}O_5$ mw: 284.28
SYNS: 10-(ACETYLOXY)-1,8-DIHYDROXY-9(10H)-ANTHRACENONE □ DITHRANOL, 10-ACETYL- □ 9(10H)-ANTHRACENONE, 1,8-DIHYDROXY-10-(1-OXOETHYL)- □ 9(10H)-ANTHRACENONE, 10-(ACETYLOXY)-1,8-DIHYDROXY-
SAFETY PROFILE: A poison by ingestion. A skin irritant. Mutation data reported. When heated to decomposition it emits acrid smoke and irritating vapors.

ACI750   CAS: 74-86-2   HR: 3
ACETYLENE
DOT: UN 1001

mf: C_2H_2 mw: 26.04
PROP: Colorless gas; garlic-like odor.
Flammable. Bp: −84.0° (subl), lel: 2.5%, uel:
82%, mp: −81.8°, flash p: 0°F (CC), d: 1.173
g/L @ 0°, autoign temp: 581°F, vap press:
40 atm @ 16.8°, vap d: 0.91, d: (liquid) 0.613
@ −80°, d: (solid) 0.730 @ −85°. Sltly sol in
water; mod sol in ethanol and acetic acid;
very sol in Me₂CO; almost misc in ether.
SYNS: ACETYLEN □ ACETYLENE, dissolved (DOT) □
ETHINE □ ETHYNE □ NARCYLEN
CONSENSUS REPORTS: Reported in EPA
TSCA Inventory.
OSHA PEL: CL 2500 ppm
ACGIH TLV: Simple asphyxiant
NIOSH REL: (Acetylene) 10H TWA no
exposure >2500 ppm
DOT CLASSIFICATION: Forbidden; DOT
Class 2.1; Label: Flammable Gas
SAFETY PROFILE: Mildly toxic by
inhalation. Human systemic effects by
inhalation: headache and dyspnea. Narcotic
in high concentration. In general industrial
practice, acetylene does not constitute a
serious toxic hazard. It is a very dangerous
fire hazard when exposed to heat, flame, or
oxidizers. Moderate explosion hazard when
exposed to heat or flame or by spontaneous
chemical reaction. At high pressures and
moderate temperatures, and in the absence
of air, acetylene has been known to
decompose explosively. Reacts with copper
to form the explosive copper acetylide.
Incompatible with brass, copper salts,
copper carbide, powdered Co, Hg, Hg salts,
K, Ag and Ag salts, RbH, CsH, halogens,
HNO_3, NaH, oxidants. Acetylene + halide
+ UV can explode. Molten K ignites in
C_2H_2 and then explodes. C_2H_2 reacts
vigorously with trifluoromethyl hypofluorite.
With O_2, C_2H_2 can detonate very
powerfully. See ACETYLIDES. When
ignited, it burns with an intensely hot flame;
can react vigorously with oxidizing
materials.

When mixed with O_2 in proportions of
40% or more, acetylene acts as a narcotic
and has been used in anesthesia. Acetylene
acts as a simple asphyxiant by diluting the
O_2 in the air to a level that will not support
life. However, the presence of impurities in
commercial acetylene may result in the
production of symptoms before an
asphyxiant concentration is reached. Thus:
10% in air produces a slight intoxication,
20% produces a staggering gait, 30%
produces general incoordination, 33% leads
to unconsciousness in 7 minutes, up to 80%
produces complete anesthesia, increased
blood pressure, narcosis, and stimulated
respiration.

Dizziness, headache, mild gastric
symptoms, and (in high concentration)
semi-asphyxia and brief loss of
consciousness have all been reported. See
ARGON for a discussion of simple
asphyxiants. To fight fire, use CO_2, water
spray, or dry chemical. Stop flow of gas.

ACJ125   HR: 3
ACETYLENE COMPOUNDS and
ALKYNES
SAFETY PROFILE: The carbon–carbon
triple bond is explosively unstable in many
acetylenic compounds. Both the lower
alkynes (i.e., propyne, butadyne, etc.) and
higher compounds may undergo explosive
decomposition. The presence of halogens
and heavy metal derivatives may increase
these explosive tendencies. See also
ACETYLENE, ACETYLIDES, and
specific compounds.

ACK000   CAS: 156-60-5   HR: 2
trans-ACETYLENE DICHLORIDE
mf: $C_2H_2Cl_2$ mw: 96.94
PROP: Colorless liquid; pleasant odor. Mp:
−50°, bp: 48°, flash p: 36°F, autoign temp:
860°F, lel: 9.7%, uel: 12.8%, d: 1.2743 @
25°/4°, vap press: 400 mm @ 30.8°, vap d:
3.34.
SYNS: *trans*-DICHLOROETHYLENE □ *trans*-1,2-
DICHLOROETHYLENE (MAK) □ RCRA WASTE
NUMBER U079
CONSENSUS REPORTS: Reported in EPA
TSCA Inventory.
DFG MAK: 200 ppm (800 mg/m³)

SAFETY PROFILE: Moderately toxic by ingestion. Mildly toxic by inhalation. Human systemic effects by inhalation: sleep, hallucinations, and distorted perceptions. Experimental reproductive effects. A skin and eye irritant. Mutation data reported. Exposure to high vapor concentration can cause nausea, vomiting, weakness, tremor, and cramps. Recovery is usually prompt following removal from exposure. Dermatitis may result from defatting action on skin. Dangerous fire hazard when exposed to heat, flame, or oxidizers. Moderate explosion hazard in the form of vapor when exposed to flame. Violent reaction with difluoromethylene dihypofluorite. Forms shock-sensitive explosive mixtures with dinitrogen tetraoxide. Reaction with solid caustic alkalies or their concentrated solutions produces chloracetylene gas that ignites spontaneously in air. Reacts violently with N_2O_4, KOH, Na, NaOH. Moderate explosion hazard in the form of vapor when exposed to flame. Can react vigorously with oxidizing materials. To fight fire, use water spray, foam, CO_2, dry chemical. When heated to decomposition it emits toxic fumes of Cl⁻. See also CHLORIDES; CHLORINATED HYDROCARBONS, ALIPHATIC; and ACETYLENE COMPOUNDS.

ACK250   CAS: 79-27-6   HR: 3
ACETYLENE TETRABROMIDE
mf: $C_2H_2Br_4$ mw: 345.68
PROP: Colorless to yellow liquid. Bp: 151° @ 54 mm, fp: −1°, d: 2.9638 @ 20°/4°, mp: 0.1°, autoign temp: 635°F. IDLH 8 ppm.
SYNS: MUTHMANN'S LIQUID □ TBE □ 1,1,2,2-TETRABROMAETHAN (GERMAN) □ TETRABRO-MOACETYLENE □ 1,1,2,2-TETRABROMOETANO (ITALIAN) □ S-TETRABROMOETHANE □ 1,1,2,2-TETRABROMOETHANE □ 1,1,2,2-TETRABRO-OMETHAAN (DUTCH)
CONSENSUS REPORTS: Reported in EPA TSCA Inventory. EPA Genetic Toxicology Program.
OSHA PEL: TWA 1 ppm

ACGIH TLV: TWA 1 ppm
DFG MAK: 1 ppm (14 mg/m³)
SAFETY PROFILE: Poison by inhalation, ingestion, and intraperitoneal routes. An eye and skin irritant and a narcotic. Questionable carcinogen with experimental neoplastigenic data. Mutation data reported. When heated it emits highly toxic fumes of carbonyl bromide and Br⁻. See also ACETYLENE COMPOUNDS and BROMIDES.

ACL750   CAS: 88-29-9   HR: 3
ACETYL ETHYL TETRAMETHYL
**   TETRALIN**
mf: $C_{18}H_{26}O$ mw: 258.44
PROP: White crystals.
SYNS: ACETYLETHYL TETRAMETHYLTETRALIN □ 6-ACETYL-1,1,4,4-TETRAMETHYL-7-ETHYL-1,2,3,4,-TETRALIN □ 7-ACETYL-1,1,4,4-TETRAMETHYL-1,2,3,4-TETRAHYDRONAPHTHALENE □ AETT □ ETHANONE-1-(3-ETHYL-5,6,7,8-TETRAHYDRO-5,5,8,8-TETRAMETHYL-2-NAPHTHALENYL)(9CI) □ 3'-ETHYL-5',6',7',8'-TETRAHYDRO-5',5',8'-TETRAMETH-YL-2'-ACETONAPHTHONE □ 1-(3-ETHYL-5,6,7,8-TETRAHYDRO-5,5,8,8-TETRAMETHYL-2-NAPHTH-ALENYL)-ETHANONE □ MUSK 36A □ POLYCYCLIC MUSK □ VERSALIDE
CONSENSUS REPORTS: Reported in EPA TSCA Inventory.
SAFETY PROFILE: Poison by ingestion. Moderately toxic by skin contact. A skin and eye irritant. Exposure causes blue coloration of internal organs and central nervous system effects, e.g., hyperexcitability, tremors, lack of coordination, hunched back, and loss of weight. It is slowly metabolized and excreted via feces. Symptoms persist for 90 days after exposure. Severity of symptoms seems proportional to length of exposure. It is freely absorbed via human skin. When heated to decomposition it emits acrid smoke and fumes.

ACM000   CAS: 557-99-3   HR: 3
ACETYL FLUORIDE
mf: C_2H_3FO mw: 62.05

PROP: Liquid or gas. D: 1.002 @ 15°/4°, mp: −60°, bp: 20.8°. Sltly sol in alc, ether, acetone, and benzene.
SYNS: FLUORID KYSELINY OCTOVE □ METHYL-CARBONYL FLUORIDE
CONSENSUS REPORTS: Reported in EPA TSCA Inventory.
OSHA PEL: TWA 2.5 mg(F)/m^3
ACGIH TLV: TWA 2.5 mg(F)/m^3; BEI: 3 mg/g creatinine of fluorides in urine prior to shift; 10 mg/g creatinine of fluorides in urine at end of shift.
NIOSH REL: (Fluorides, Inorganic) TWA 2.5 mg(F)/m^3
SAFETY PROFILE: Poison by inhalation. See also FLUORIDES. When heated to decomposition it emits toxic fumes of F⁻.

ACM750   CAS: 1068-57-1   HR: 3
ACETYL HYDRAZIDE
mf: $C_2H_6N_2O$ mw: 74.10
PROP: Needles from ethanol. Mp: 67°, bp: 127° @ 18 mm.
SYNS: ACETHYDRAZIDE □ ACETOHYDRAZIDE □ N-ACETYLHYDRAZINE □ ENT 61,241 □ ETHANEHYDRAZONIC ACID □ MONOACETYL-HYDRAZINE
CONSENSUS REPORTS: Reported in EPA TSCA Inventory.
SAFETY PROFILE: Poison by ingestion, subcutaneous, and intraperitoneal routes. Mutation data reported. Exposure can cause hemolysis and liver damage. See also PHENYLHYDRAZINE. When heated to decomposition it emits toxic fumes of NO$_x$.

ACN300   CAS: 39543-84-5   HR: 3
2-ACETYL-4-(2-HYDROXY-3-tert-BUTYLAMINOPROPOXY)BENZOFUR AN
mf: $C_{17}H_{23}NO_4$ mw: 305.41
PROP: A liquid.
SYNS: 1-(4-(3-((1,1-DIMETHYLETHYL)AMINO)-2-HYDROXYPROPOXY)-2-BENZOFURANYL)ETHA NONE □ ETHANONE, 1-(4-(3-((1,1-DIMETHYL-ETHYL)AMINO)-2-HYDROXYPROPOXY)-2-BENZOFURANYL)- □ KETONE, 4-(3-(tert-BUTYLAMINO)-2-HYDROXYPROP-OXY)-2-BENZOFURANYL METHYL
DOT CLASSIFICATION: 3; Label: Flammable Liquid

SAFETY PROFILE: A poison by intravenous route. A flammable liquid. When heated to decomposition it emits toxic vapors of NO$_x$.

ACN310   CAS: 39543-94-7   HR: 3
2-ACETYL-7-(2-HYDROXY-3-sec-BUTYLAMINO-PROPOXY)BENZO-FURAN
mf: $C_{17}H_{23}NO_4$ mw: 305.41
PROP: A liquid.
SYNS: ETHANONE, 1-(7-(2-HYDROXY-3-((1-METHYLPROPYL)AMINO)PROPOXY)-2-BENZOFURANYL)- □ KETONE, 7-(3-(sec-BUTYLAMINO)-2-HYDROXYPROPOXY)-2-BENZOFURANYL METHYL
DOT CLASSIFICATION: 3; Label: Flammable Liquid
SAFETY PROFILE: A poison by intravenous route. A flammable liquid. When heated to decomposition it emits toxic vapors of NO$_x$.

ACN320   CAS: 39543-80-1   HR: 3
2-ACETYL-7-(2-HYDROXY-3-tert-BUTYL-AMINOPROPOXY)-BENZOFURAN
mf: $C_{17}H_{23}NO_4$ mw: 305.41
PROP: A liquid.
SYNS: KETONE, 7-(3-(tert-BUTYLAMINO)-2-HYDROXYPROPOXY)-2-BENZOFURANYL METHYL □ 1-(7-(3-((1,1-DIMETHYLETHYL)AMINO)-2-HYDROXYPROPOXY)-2-BENZOFURANYL)ETHA NONE □ ETHANONE, 1-(7-(3-((1,1-DIMETHYL-ETHYL)AMINO)-2-HYDROXYPROPOXY)-2-BENZOFURANYL)-
DOT CLASSIFICATION: 3; Label: Flammable Liquid
SAFETY PROFILE: A poison by intravenous route. A flammable liquid. When heated to decomposition it emits toxic vapors of NO$_x$.

ACO000   HR: 3
ACETYLIDES
PROP: Colorless, flammable gases with odor similar to garlic. Can decompose spontaneously if pressure exceeds 15 PSIG.
SAFETY PROFILE: Severe explosion hazard when shocked or exposed to heat. Acetylides are very sensitive to shock, friction, and heat. They explode readily and

are among the few commercial explosives that contain no oxygen or nitrogen and therefore produce no gas. The explosion simply results from the large amount of heat instantaneously produced. Acetylides are used for detonating compositions, or in combination with lead azide in detonating rivets, where the acetylides reduce the flash point of the more insensitive azides. They are in a class with the fulminates and the azides as primary detonants. Because these materials are so sensitive to shock and temperature, they must be handled with extreme care. They must be kept cool, and should be kept wet if they are to be stored. (See FULMINATES for suggested precautions in storage and handling of acetylides.) Metal powders, such as finely divided Cu or Ag, should not be stored or kept with acetylene or acetylides because it is possible for these substances to react with these metal powders to form very sensitive acetylides, although they are not dangerous in themselves, can cause enough of a flash to ignite a possibly explosive mixture of gases and thus cause an explosion in a warehouse or storage area. Examples of commercially used acetylides are silver acetylide and copper acetylide. See also ACETYLENE. See also individual compounds.

ACO320   CAS: 16078-34-5   HR: 3
5-ACETYLINDOLINE
mf: $C_{10}H_{11}NO$ mw: 161.22
PROP: A liquid.
SYNS: INDOLINE, 5-ACETYL- □ KETONE, 5-INDOLINYL METHYL
DOT CLASSIFICATION: 3; Label: Flammable Liquid
SAFETY PROFILE: A poison by intravenous route. A flammable liquid. When heated to decomposition it emits toxic vapors of NO_x.

ACO500   CAS: 507-02-8   HR: 3
ACETYL IODIDE
DOT: UN 1898
mf: C_2H_3IO mw: 169.95

PROP: Brown, transparent, fuming liquid. Bp: 108°, d: 2.067 @ 20°/4°, decomp in water and alc; sol in ether.
CONSENSUS REPORTS: Reported in EPA TSCA Inventory.
ACGIH TLV: Proposed: (inhalable fraction) 0.1 mg/m³; Not Classifiable as a Human Carcinogen)
DOT CLASSIFICATION: 8; Label: Corrosive
SAFETY PROFILE: A toxic, corrosive material. Reacts with water or steam to produce toxic and corrosive fumes. Dangerous to use. When heated to decomposition it emits toxic fumes of I^-. See also IODIDES.

ACQ275   CAS: 65-82-7   HR: 3
N-ACETYL-l-METHIONINE
mf: $C_7H_{13}NO_3S$ mw:191.24
PROP: Colorless or white crystals or powder; odorless. Sol in water, alc, alkali and mineral acids; insol in ether.
SYNS: ACETYLMETHIONINE □ N-ACETYLMETHIONINE □ METHIONAMINE
CONSENSUS REPORTS: EPA TSCA Chemical Inventory, JUNE 1993
SAFETY PROFILE: Poison by intravenous route. When heated to decomposition emits toxic fumes of NO_x.

ACR400   CAS: 28895-91-2   HR: 3
ACETYLMETHYLNITROSOUREA
mf: $C_4H_7N_3O_3$ mw: 145.14
SYNS: ACETYL-METHYL-NITROSO-HARNSTOFF (GERMAN) □ N'-ACETYL-METHYLNITROSOUREA □ N-METHYL-N-NITROSO-N'-ACETYLUREA □ 1-METHYL-1-NITROSOACETYLUREA
SAFETY PROFILE: Poison by ingestion. Questionable carcinogen with experimental tumorigenic data. Mutation data reported. When heated to decomposition it emits toxic fumes of NO_x.

ACR500   CAS: 1053-74-3   HR: 3
3-ACETYL-10-(3'-N-METHYL-PIPERAZ-INO-N'-PROPYL)-PHENOTHIAZIN
mf: $C_{22}H_{27}N_3OS$ mw: 381.58
PROP: A liquid.
SYNS: KETONE, METHYL 10-(3-(4-METHYL-1-PIPERAZINYL)PROPYL)PHENOTHIAZIN-2-YL □

METHYL 10-(3-(4-METHYL-1-PIPERAZINYL)-
PROPYL)PHENOTHIAZIN-2-YL KETONE
· **DOT CLASSIFICATION:** 3; Label:
Flammable Liquid
SAFETY PROFILE: A poison by
intravenous route. Moderately toxic by
ingestion. A flammable liquid. When heated
to decomposition it emits toxic vapors of
SO_x.

ACT250   CAS: 5275-69-4   HR: 3
2-ACETYL-5-NITROFURAN
mf: $C_6H_5NO_4$ mw: 155.12
PROP: IDLH 2000 ppm.
SYN: (5-NITRO-2-FURYL) METHYL KETONE
CONSENSUS REPORTS: EPA Genetic
Toxicology Program.
DOT CLASSIFICATION: 3; Label:
Flammable Liquid
SAFETY PROFILE: Poison by
subcutaneous route. Moderately toxic by
ingestion. Mutation data reported. See also
KETONES. A flammable liquid. When
heated to decomposition it emits toxic
fumes of NO_x.

ARW200   CAS: 71751-41-2   HR: 3
AVERMECTIN B(SUB 1)
SYNS: AVID EC □ ABAMECTIN □ AFFIRM □
AGRIMEK □ AVERMECTIN B(SUB 1) TECHNICAL
GRADE □ AVOMEC □ MK 936 □ VERTIMEC □
ZEPHYR
SAFETY PROFILE: A poison by ingestion
and intracerebral routes. Moderately toxic by
inhalation and skin contact. When heated to
decomposition it emits acrid smoke and
irritating vapors.

ACV500   CAS: 110-22-5   HR: 3
ACETYL PEROXIDE
mf: $C_4H_6O_4$ mw: 118.04
PROP: Solid or colorless crystals or liquid
with very pungent odor. D: 1.18, mp: 30°,
bp: 63° @ 21 mm. Sltly sol in cold water,
decomp.
SYNS: ACETYL PEROXIDE, not >25% in solution (UN
2084) (DOT) □ ACETYL PEROXIDE, solid, or >25% in
solution (DOT) □ DIACETONE PEROXIDES, solid, or
>25% in solution (DOT) □ DIACETYL PEROXIDE (MAK)

CONSENSUS REPORTS: Reported in
EPA TSCA Inventory.
DFG MAK: Strong Skin Effects
DOT CLASSIFICATION: Forbidden
SAFETY PROFILE: Severe skin and eye
irritant. Questionable carcinogen with
experimental tumorigenic data. Dangerous
fire hazard by spontaneous chemical
reaction. A powerful oxidizing agent; can
cause ignition of organic materials on
contact. Severe explosion hazard when
shocked or exposed to heat. It may explode
spontaneously in storage and should be used
as soon as prepared. It will react with water
or steam to produce heat; can react
vigorously with reducing materials; emits
toxic fumes on contact with acid or acid
fumes. To fight fire, use CO_2, dry chemical.
   Storage and Handling: Must be kept below
27° and not warmed over 30°. Do not add
to hot materials. Do not add accelerator to
this material. Store in original container with
vented cap. Avoid bodily contact. This
material is nearly always stored and handled
as a 25% solution in an inert solvent. See
also ACETYL PEROXIDE 25% solution
(in dimethyl phthalate); and PEROXIDES,
ORGANIC.

ADA725   CAS: 50-78-2   HR: 3
ACETYLSALICYLIC ACID
mf: $C_9H_8O_4$ mw: 180.17
PROP: Colorless needles, crystals. Mp: 135°,
fp: 118°. Very sltly sol in alc, sol in benzene.
Solubility in water = 1% @ 37°, in ether =
5% @ 20°.
SYNS: AC 5230 □ ACENTERINE □ ACESAL □
ACETAL □ ACETICYL □ ACETILSALICILICO □
ACETILUM ACIDULATUM □ ACETISAL □ ACETOL □
ACETONYL □ ACETOPHEN □ ACETOSAL □
ACETOSALIC ACID □ ACETOSALIN □ o-ACETOXY-
BENZOIC ACID □ 2-ACETOXYBENZOIC ACID □
ACETYLIN □ 2-(ACETYLOXY)BENZOIC ACID □
ACETYLSAL □ ACETYLSALICYLSAEURE (GERMAN) □
ACIDE ACETYLSALICYLIQUE (FRENCH) □ ACIDO o-
ACETIL-BENZOICO (ITALIAN) □ ACIDO
ACETILSALICILICO (ITALIAN) □ ACIDUM
ACETYLSALICYLICUM □ ACIMETTEN □ ACISAL □
ACYLPYRIN □ ASA □ A.S.A. □ A.S.A. EMPIRIN □
ASAGRAN □ ASATARD □ ASPALON □ ASPERGUM □

ASPIRDROPS □ ASPIRIN □ ASPIRINE □ ASPRO □ ASTERIC □ BENASPIR □ BIALPIRINIA □ CAPRIN □ o-CARBOXYPHENYL ACETATE □ COLFARIT □ CONTRHEUMA RETARD □ CRYSTAR □ DELGESIC □ DOLEAN pH 8 □ DURAMAX □ ECM □ ECOTRIN □ EMPIRIN □ ENDYDOL □ ENTERICIN □ ENTEROPHEN □ ENTEROSARINE □ ENTROPHEN □ EXTREN □ GLOBOID □ HELICON □ IDRAGIN □ MEASURIN □ NEURONIKA □ NOVID □ POLOPIRYNA □ RHEUMIN TABLETTEN □ RHODINE □ SALACETIN □ SALCETOGEN □ SALETIN □ SOLPYRON □ XAXA
CONSENSUS REPORTS: EPA Genetic Toxicology Program. Reported in EPA TSCA Inventory.
OSHA PEL: TWA 5 mg/m³
ACGIH TLV: TWA 5 mg/m³
SAFETY PROFILE: Poison by ingestion, intraperitoneal, and possibly other routes. Human systemic effects by ingestion: acute pulmonary edema, body temperature increase, changes in kidney tubules, coma, constipation, dehydration, hematuria, hepatitis, nausea or vomiting, respiratory stimulation, somnolence, tinnitus, decreased urine volume. Implicated in aplastic anemia. A 10 gram dose to an adult may be fatal. A human teratogen. Human reproductive effects by ingestion and possibly other routes: menstrual cycle changes, parturition, various effects on newborn including Apgar score, developmental abnormalities of the cardiovascular and respiratory systems. Experimental animal reproductive effects. Human mutation data reported. An allergen; skin contact, inhalation, or ingestion can cause asthma, sneezing, irritation of eyes and nose, hives, and eczema. Combustible when exposed to heat or flame. When heated to decomposition it emits acrid smoke and fumes.

ADD400   CAS: 77-90-7   HR: D
ACETYL TRIBUTYL CITRATE
PROP: Clear viscous liquid. Mp: -80°, bp: 172°–174°, d: 1.05. Insol in water.
SAFETY PROFILE: When heated to decomposition it emits acrid smoke and irritating fumes.

ADD750   CAS: 77-89-4   HR: 2
ACETYL TRIETHYL CITRATE
mf: $C_{14}H_{22}O_8$ mw: 318.36
PROP: Bp: 197° @ 15 mm.
SYNS: ATEC □ CITRIC ACID, ACETYL TRIETHYL ESTER □ CITROFLEX A 2 □ 1,2,3-PROPANE-TRICARBOXYLIC ACID, 2-(ACETYLOXY)-, TRIETHYL ESTER (9CI) □ TRICARBALLYLIC ACID, β-ACETOXY-TRIBUTYL ESTER □ TRIETHYL ACETYLCITRATE □ TRIETHYL CITRATE, ACETATE □ TRIETHYLESTER KYSELINY ACETYLCITRONOVE
CONSENSUS REPORTS: Reported in EPA TSCA Inventory.
SAFETY PROFILE: Moderately toxic by intraperitoneal route. Mildly toxic by ingestion. See also ESTERS. When heated to decomposition it emits acrid smoke and fumes.

ADE050   CAS: 19005-95-9   HR: 3
3-ACETYL-2,4,5-TRIMETHYL-PYRROLE
mf: $C_9H_{13}NO$ mw: 151.23
PROP: A liquid.
SYNS: KETONE, METHYL 2,4,5-TRIMETHYL-PYRROL-3-YL □ METHYL 2,4,5-TRIMETHYLPYRROL-3-YL KETONE
DOT CLASSIFICATION: 3; Label: Flammable Liquid
SAFETY PROFILE: A poison by intraperitoneal route. A flammable liquid. When heated to decomposition it emits toxic vapors of NO_x.

ADE075   CAS: 1218-34-4   HR: 2
ACETYLTRYPTOPHAN
mf: $C_{13}H_{14}N_2O_3$ mw: 246.29
PROP: Solid. Mp: 190°.
SYNS: ACETYL-l-TRP □ ACETYL-l-TRYPTOPHAN □ N-ACETYL-l-TRYPTOPHAN □ N-ACETYLTRY-PTOPHAN □ (S)-N-ACETYLTRYPTOPHAN □ AC-TRY
CONSENSUS REPORTS: Reported in EPA TSCA Inventory.
SAFETY PROFILE: Moderately toxic by some routes. An experimental teratogen. Other experimental reproductive effects. When heated to decomposition it emits toxic fumes of NO_x.

ADF250   CAS: 12788-93-1   HR: 3
ACID BUTYL PHOSPHATE
DOT: UN 1718

mf: $C_4H_{10}O_4P$ mw: 153.1
PROP: Water-white liquid; sol in alc, acetone, and toluene; insol in water, petroleum, and naphtha. D: 1.120–1.125 @ 25°/40°, flash p: 230°F (COC).
SYNS: n-BUTYL ACID PHOSPHATE □ BUTYL PHOSPHORIC ACID
DOT CLASSIFICATION: 8; Label: Corrosive
SAFETY PROFILE: Toxic and corrosive. Combustible when exposed to heat or flame. When heated to decomposition it emits highly toxic fumes of PO_x. See also ESTERS and PHOSPHORIC ACID.

ADG250   CAS: 18472-87-2   HR: 3
ACID RED 92
mf: $C_{20}H_2Br_4Cl_4O_5 \cdot 2Na$ mw: 829.64
PROP: Orange-red crystals or powder. Sol in water and ethanol.
SYNS: AIZEN ACID PHLOXINE PB □ C.I. 45410 □ C.I. ACID RED 92 □ CYANOSIN □ CYANOSIN (ACID DYE) □ CYANOSINE □ D and C RED NO. 28 □ EOSIN BLUE □ EOSINE BLUE □ EOSINE BLUISH □ FOOD DYE RED No. 104 □ FOOD RED No. 104 □ JAPAN RED 104 □ ORIENT WATER PINK 2 □ PHLOXIN B □ PHLOXINE B □ PHLOXINE P □ RED 104 □ 11969 RED □ RED No. 104 □ 3427 VERI PUR PINK
CONSENSUS REPORTS: Reported in EPA TSCA Inventory. EPA Genetic Toxicology Program.
SAFETY PROFILE: Poison by intravenous route. An experimental teratogen. Other experimental reproductive effects. When heated to decomposition it emits very toxic fumes of Br^-, Cl^-, and Na_2O.

ADJ500   CAS: 260-94-6   HR: 3
ACRIDINE
DOT: UN 2713
mf: $C_{13}H_9N$ mw: 179.23
PROP: Small, colorless needles or prisms. Mp: 110.5°, bp: 346°, d: 1.005 @ 19.7°/4°, vap press: 1 mm @ 129.4°. Sltly sol in hot water; sol in alc, ether, and CS_2.
SYNS: 9-AZAANTHRACENE □ 10-AZAANTHRACENE □ BENZO(b)QUINOLINE □ 2,3-BENZOQUINOLINE □ DIBENZO(b,e)PYRIDINE
CONSENSUS REPORTS: Reported in EPA TSCA Inventory.
OSHA PEL: TWA 0.2 mg/m³

DOT CLASSIFICATION: 6.1; Label: KEEP AWAY FROM FOOD
SAFETY PROFILE: Poison by ingestion, subcutaneous, and intravenous routes. Mutation data reported. A skin, eye, and mucous membrane irritant. When heated to decomposition it emits toxic fumes of NO_x.

ADJ550   CAS: 7101-57-7   HR: 3
ACRIDINE-9-CARBOXAMIDE, N,N-DIETHYL-1,2,3,4-TETRAHYDRO-
mf: $C_{18}H_{22}N_2O$ mw: 282.42
PROP: A liquid.
SYNS: ACRIDINE-9-CARBOXAMIDE, 1,2,3,4-TETRAHYDRO-N,N-DIETHYL- □ KETONE, DIETHYLAMINO(1,2,3,4-TETRAHYDRO-9-ACRIDINYL)
DOT CLASSIFICATION: 3; Label: Flammable Liquid
SAFETY PROFILE: A poison by intraperitoneal route. A flammable liquid. When heated to decomposition it emits toxic vapors of NO_x.

ADQ600   CAS: 29023-85-6   HR: 3
4-(9-ACRIDINYL)-2-METHYL-3-THIOSEMICARBAZONE ACETONE
mf: $C_{18}H_{18}N_4S$ mw: 322.46
SAFETY PROFILE: A poison by ingestion. When heated to decomposition it emits toxic vapors of NO_x and SO_x.

ADR000   CAS: 107-02-8   HR: 3
ACROLEIN
DOT: UN 1092
mf: C_3H_4O mw: 56.07
PROP: Colorless or yellowish liquid; lachrymatory, disagreeable, choking odor. Mp: −87.7°, bp: 52.5°, flash p: <0°F, d: 0.841 @ 20°/4°, autoign temp: unstable (455°F), lel: 2.8%, uel: 31%, vap d: 1.94. Sol in water, alc, and ether. IDLH 2 ppm.
SYNS: ACQUINITE □ ACRALDEHYDE □ ACROLEINA (ITALIAN) □ ACROLEINE (DUTCH, FRENCH) □ ACRYLALDEHYD (GERMAN) □ ACRYLALDEHYDE □ ACRYLIC ALDEHYDE □ AKROLEIN (CZECH) □ AKROLEINA (POLISH) □ ALDEHYDE ACRYLIQUE (FRENCH) □ ALDEIDE ACRILICA (ITALIAN) □ ALLYL ALDEHYDE □ AQUALINE □ BIOCIDE □ CROLEAN □ ETHYLENE ALDEHYDE □ MAGNACIDE H □ NSC-8819 □

PROPENAL (CZECH) □ 2-PROPENAL □ PROP-2-EN-1-AL □ 2-PROPEN-1-ONE □ PROPYLENE ALDEHYDE □ RCRA WASTE NUMBER P003 □ SLIMICIDE

CONSENSUS REPORTS: IARC Cancer Review: Group 3 IMEMDT 7,78,87; Animal Inadequate Evidence IMEMDT 36,133,85; IMEMDT 19,479,79; Human Inadequate Evidence IMEMDT 36,133,85. Community Right-To-Know List. EPA Extremely Hazardous Substances List. Reported in EPA TSCA Inventory.

OSHA PEL: TWA 0.1 ppm; STEL 0.3 ppm

ACGIH TLV: STEL CL 0.1 ppm (skin); Not Classifiable as a Human Carcinogen

DFG MAK: Confirmed Animal Carcinogen with Unknown Relevance to Humans

DOT CLASSIFICATION: 6.1; Label: Poison, Flammable Liquid

SAFETY PROFILE: Human poison by inhalation and intradermal routes. Poison experimentally by most routes. Human systemic irritant and pulmonary system effects by inhalation include: lachrymation, delayed hypersensitivity with multiple organ involvement, and respiratory system damage. Severe eye and skin irritant. Experimental reproductive effects. Human mutation data reported. Questionable carcinogen. Dangerous fire hazard when exposed to heat, flame, or oxidizers. An explosion hazard. Incompatible with amines, SO_2, metal salts, oxidants, (light + heat). Violent polymerization reaction on contact with strong acid, strong base, weak acid conditions (e.g., nitrous fumes, sulfur dioxide, carbon dioxide), thiourea, or dimethylamine. When heated to decomposition it emits highly toxic fumes; can react vigorously with oxidizing materials. To fight fire, use CO_2, dry chemical, or alcohol foam.

ADR500   CAS: 100-73-2   HR: 2
ACROLEIN DIMER
DOT: UN 2607
mf: $C_6H_8O_2$ mw: 112.14
PROP: Liquid, sol in water. D: 1.0775 (20°), bp: 151.3°, fp: −100°, flash p: 118°F (OC).

SYNS: ACROLEIN DIMER, stabilized (DOT) □ 3,4-DIHYDRO-2H-PYRAN-2-CARBOXALDEHYDE □ 2,3-DIHYDRO-1,4-PYRAN-2-KARBOXALDEHYD □ 2-FORMYL-3,4-DIHYDRO-2H-PYRAN □ 5-HEXENAL, 2,6-EPOXY- □ PYRAN ALDEHYDE

DOT CLASSIFICATION: 3; Label: Flammable Liquid

SAFETY PROFILE: Mildly toxic by ingestion. A skin and severe eye irritant. A flammable liquid when exposed to heat, flame, or powerful oxidizing agents. To fight fire, use alcohol foam and multipurpose dry chemical. When heated to decomposition it emits acrid smoke and fumes.

ADS250   CAS: 79-06-1   HR: 3
ACRYLAMIDE
DOT: UN 2074
mf: C_3H_5NO mw: 71.09
PROP: White, crystalline solid. Leaflets from (C_6H_6). Mp: 84.5° ± 0.3°, bp: 125° @ 25 mm, d: 1.122 @ 30°, vap press: 1.6 mm @ 84.5°, vap d: 2.45. Very sol in water, alc, and ether. IDLH 60 mg/m³.

SYNS: ACRYLIC AMIDE □ AKRYLAMID (CZECH) □ AMID KYSELINY AKRYLOVE □ ETHYLENECARBOXAMIDE □ PROPENAMIDE □ 2-PROPENAMIDE □ RCRA WASTE NUMBER U007 □ VINYL AMIDE

CONSENSUS REPORTS: NTP 10th Report on Carcinogens. IARC Cancer Review: Group 2B IMEMDT 7,56,87; Animal Sufficient Evidence IMEMDT 39,41,86. EPA Extremely Hazardous Substances List. Community Right-To-Know List. Reported in EPA TSCA Inventory.

OSHA PEL: TWA 0.03 mg/m³ (skin)

ACGIH TLV: Animal Carcinogen, TWA 0.03 mg/m³ (skin)

DFG MAK: Animal Carcinogen, Suspected Human Carcinogen

NIOSH REL: TWA 0.3 mg/m³

DOT CLASSIFICATION: 6.1; Label: KEEP AWAY FROM FOOD

SAFETY PROFILE: Confirmed carcinogen with experimental carcinogenic and neoplastigenic data. Poison by ingestion, skin contact, and intraperitoneal routes. Experimental reproductive effects. Mutation data reported. A skin and eye irritant.

Intoxication from it has caused a peripheral neuropathy, erythema, and peeling palms. In industry, intoxication is mainly via dermal route, next via inhalation, and last via ingestion. Time of onset varied from 1–24 months to 8 years. Symptoms were, via dermal route, a numbness, tingling, and touch tenderness. In a couple of weeks, coldness of extremities; later, excessive sweating, bluish-red and peeling palms, marked fatigue and limb weakness. It is dangerous because it can be absorbed through the unbroken skin. From animal experiments it seems to be a central nervous system toxin. Adult rats fed an average of 30 mg/kg for 14 days were all partially paralyzed and had reduced their food consumption by 50 percent. Polymerizes violently at its melting point. When heated to decomposition it emits acrid fumes and NO_x.

ADS400   HR: D
ACRYLATE-ACRYLAMIDE RESINS
SAFETY PROFILE: When heated to decomposition it emits acrid smoke and irritating fumes.

ADS750   CAS: 79-10-7   HR: 3
ACRYLIC ACID
DOT: UN 2218
mf: $C_3H_4O_2$ mw: 72.07
PROP: Liquid with acrid odor. Misc in water, benzene, alc, chloroform, ether, and acetone. Mp: 13°, bp: 141° (polymerizes), d: 1.062, vap press: 10 mm @ 39.9°, flash p: 130°F (OC), vap d: 2.45.
SYNS: ACROLEIC ACID □ ACRYLIC ACID, GLACIAL □ ACRYLIC ACID, inhibited (DOT) □ ETHYLENECARBOXYLIC ACID □ GLACIAL ACRYLIC ACID □ KYSELINA AKRYLOVA □ PROPENE ACID □ PROPENOIC ACID □ 2-PROPENOIC ACID (9CI) □ RCRA WASTE NUMBER U008 □ VINYLFORMIC ACID
CONSENSUS REPORTS: IARC Cancer Review: Group 3 IMEMDT 7,56,87; Human Inadequate Evidence IMEMDT 19,47,79. Community Right-To-Know List. Reported in EPA TSCA Inventory.
OSHA PEL: TWA 10 ppm (skin)

ACGIH TLV: 2 ppm (skin); Not Classifiable as a Human Carcinogen
DOT CLASSIFICATION: 8; Label: Corrosive
SAFETY PROFILE: Poison by ingestion, skin contact, and intraperitoneal routes. An experimental teratogen. Other experimental reproductive effects. A severe skin and eye irritant. Questionable carcinogen with experimental carcinogenic and tumorigenic data. Corrosive. Flammable liquid. May undergo exothermic polymerization at room temperature. May become explosive if confined. A fire hazard when exposed to heat or flame.

ADT050   CAS: 17831-71-9   HR: 2
ACRYLIC ACID, DIESTER with
TETRAETHYLENE GLYCOL
mf: $C_{14}H_{22}O_7$ mw: 302.36
PROP: Photosensitive polyimide resin.
SYNS: ACRYLIC ACID, OXYBIS(ETHYLENEOXY-ETHYLENE) ESTER □ 2-PROPENOIC ACID, OXYBIS(2,1-ETHANEDIYLOXY-2,1-ETHANE-DIYL)ESTER □ TETRAETHYLENE GLYCOL DIACRYLATE
CONSENSUS REPORTS: Reported in EPA TSCA Inventory.
SAFETY PROFILE: Moderate skin and severe eye irritant. Questionable carcinogen with experimental tumorigenic data. Mutation data reported. When heated to decomposition it emits acrid smoke and irritating fumes.

ADX500   CAS: 107-13-1   HR: 3
ACRYLONITRILE
DOT: UN 1093
mf: C_3H_3N mw: 53.07
PROP: Colorless, mobile liquid; mild odor. Mp: −82°, bp: 77.3°, fp: −83°, flash p: 30°F (TCC), lel: 3.1%, uel: 17%, d: 0.806 @ 20°/4°, autoign temp: 898°F, vap press: 100 mm @ 22.8°, vap d: 1.83, flash p: (of 5% aq soln) <50°F. Sol in water. IDLH 85 ppm.
SYNS: ACRITET □ ACRYLNITRIL (GERMAN, DUTCH) □ ACRYLON □ ACRYLONITRILE, inhibited (DOT) □ ACRYLONITRILE MONOMER □ AKRYLONITRYL (POLISH) □ CARBACRYL □ CIANURO di VINILE (ITALIAN) □ CYANOETHYLENE □ CYANURE de VINYLE (FRENCH) □ ENT 54 □ FUMIGRAIN □

MILLER'S FUMIGRAIN □ NITRILE ACRILICO (ITALIAN) □ NITRILE ACRYLIQUE (FRENCH) □ PROPENENITRILE □ 2-PROPENENITRILE □ RCRA WASTE NUMBER U009 □ TL 314 □ VCN □ VENTOX □ VINYL CYANIDE □ VINYLKYANID

CONSENSUS REPORTS: NTP 10th Report on Carcinogens. IARC Cancer Review: Group 2A IMEMDT 7,79,87; Human Limited Evidence IMEMDT 19,73,79; Animal Limited Evidence IMEMDT 19,73,79. Community Right-To-Know List. EPA Extremely Hazardous Substances List. Reported in EPA TSCA Inventory.

OSHA PEL: TWA 2 ppm; CL 10 ppm/15M; Cancer Hazard

ACGIH TLV: TWA 2 ppm (skin), Confirmed Animal Carcinogen.

DFG MAK: DFG TRK: Animal Carcinogen, Suspected Human Carcinogen

NIOSH REL: TWA 1 ppm; CL 10 ppm/15M

DOT CLASSIFICATION: 3; Label: Flammable Liquid, Poison

SAFETY PROFILE: Confirmed human carcinogen with experimental carcinogenic, neoplastigenic, and tumorigenic data. Poison by inhalation, ingestion, skin contact, and other routes. Human systemic effects by inhalation and skin contact: conjunctiva irritation, somnolence, general anesthesia, cyanosis, and diarrhea. An experimental teratogen. Other experimental reproductive effects. Human mutation data reported. Dangerous fire hazard when exposed to heat, flame, or oxidizers. Moderate explosion hazard when exposed to flame. Can react vigorously with oxidizing materials (see also CYANIDE).

Acrylonitrile closely resembles hydrocyanic acid in its toxic action. By inhibiting the respiratory enzymes of tissue, it renders the tissue cells incapable of oxygen absorption. Poisoning is acute; there is little evidence of cumulative action on repeated exposure. Exposure to low concentration is followed by flushing of the face and increased salivation; further exposure results in irritation of the eyes and nose, photophobia, deepened respiration. If exposure continues, shallow respiration, nausea, vomiting, weakness, an oppressive feeling in the chest, and occasionally headache and diarrhea are other complaints. Several cases of mild jaundice accompanied by mild anemia and leucocytosis have been reported. Urinalysis is generally negative, except for an increase in bile pigment. Serum and bile thiocyanates are raised. See also HYDROCYANIC ACID. Unstable and easily oxidized. Explosive polymerization may occur on storage with silver nitrate. Potentially explosive reactions with benzyltrimethylammonium hydroxide + pyrrole, tetrahydrocarbazole + benzyltrimethylammonium hydroxide. Violent reactions with strong acids (e.g., nitric or sulfuric), strong bases, azoisobutyronitrile, dibenzoyl peroxide, di-tert-butylperoxide, or bromine. Incompatible with $AgNO_3$ and amines. To fight fire, use CO_2, dry chemical, or alcohol foam. When heated to decomposition it emits toxic fumes of NO_x and CN^-. See also NITRILES and CYANIDE.

ADY500   CAS: 9003-54-7   HR: 3
ACRYLONITRILE POLYMER with STYRENE
mf: $(C_8H_8 \cdot C_3H_3N)_x$
SYNS: ACRILAFIL □ ACRYLONITRILE-STYRENE COPOLYMER □ ACRYLONITRILE-STYRENE POLYMER □ ACRYLONITRILE-STYRENE RESIN □ ACS □ AS 61CL □ BAKELITE RMD 4511 □ CEVIAN HL □ DIALUX □ ESTYRENE AS □ KOSTIL □ LITAC □ LURAN □ LUSTRAN □ POLYSTYRENE-ACRYLONITRILE □ 2-PROPENENITRILE POLYMER with ETHENYLBENZENE □ REXENE 106 □ SANREX □ SN 20 □ STYREN-ACRYLONITRILEPOLYMER □ STYRENE-ACRYLONITRILE COPOLYMER □ TERULAN KP 2540 □ TYRIL

CONSENSUS REPORTS: IARC Cancer Review: Group 3 IMEMDT 7,56,87; Human No Adequate Data IMEMDT 19,73,79; Animal No Adequate Data IMEMDT 19,73,79. Reported in EPA TSCA Inventory. Cyanide and its compounds are on the Community Right-To-Know List.

SAFETY PROFILE: Moderately to highly toxic by ingestion. Questionable carcinogen.

See also NITRILES. When heated to decomposition it emits toxic fumes of NO_x and CN^-.

AEB750   CAS: 102488-99-3 HR: 3
ACTINOMYCIN L
SYN: ACTINOMYCIN 2104L
CONSENSUS REPORTS: IARC Cancer Review: Animal Sufficient Evidence IMEMDT 10,29,76.
SAFETY PROFILE: Confirmed carcinogen with experimental neoplastigenic data.

AEC000   CAS: 12623-78-8   HR: 3
ACTINOMYCIN S
SYN: ACTINOMYCIN 1048A
CONSENSUS REPORTS: IARC Cancer Review: Animal Limited Evidence IMEMDT 10,29,76; No Adequate Data IMEMDT 10,29,76.
SAFETY PROFILE: Suspected carcinogen with experimental neoplastigenic data.

AED250   CAS: 665-66-7   HR: 3
1-ADAMANTANAMINE
   HYDROCHLORIDE
mf: $C_{10}H_{17}N \cdot ClH$ mw: 187.74
PROP: Crystals from $EtOH/Et_2O$. Sol in water and ethanol; prac insol in Et_2O.
SYNS: ADAMANTANAMINE HYDROCHLORIDE □ ADAMANTINE HYDROCHLORIDE □ ADAMANTYL-AMINE HYDROCHLORIDE □ 1-ADAMANTYL-AMINE HYDROCHLORIDE □ AMANTADINE HYDRO-CHLORIDE □ AMAZOLON □ AMINO-ADAMANTANE HYDROCHLORIDE □ 1-AMINO-ADAMANTENE HYDROCHLORIDE □ EXP 105-1 □ MANTADAN □ NSC-83653 □ SYMMETREL □ TRICYCLO(3.3.1.1.$^{(3,7)}$)DECAN-1-AMINE, HYDROCHLORIDE (9CI) □ VIROFRAL
CONSENSUS REPORTS: Reported in EPA TSCA Inventory.
SAFETY PROFILE: Human poison by ingestion. Poison by ingestion, intraperitoneal, and intravenous routes. A human teratogen with developmental abnormalities of the circulatory system. Experimental reproductive effects. Human systemic effects by ingestion: distorted perceptions, euphoria, excitement, hallucinations. When heated to

decomposition it emits very toxic fumes of NO_x and HCl.

AEH000   CAS: 73-24-5   HR: 3
ADENINE
mf: $C_5H_5N_5$ mw: 135.15
PROP: Needles. Mp: 360–365° (anhyd) decomp.
SYNS: ADENINIMINE □ 6-AMINOPURINE □ 6-AMINO-1H-PURINE □ 6-AMINO-3H-PURINE □ 6-AMINO-9H-PURINE □ 1,6-DIHYDRO-6-IMINO-PURINE □ 3,6-DIHYDRO-6-IMINOPURINE □ LEUCO-4 □ 1H-PURIN-6-AMINE □ USAF CB-18 □ VITAMIN B4
CONSENSUS REPORTS: Reported in EPA TSCA Inventory.
SAFETY PROFILE: Poison by intraperitoneal route. Moderately toxic by ingestion. An experimental teratogen. Experimental reproductive effects. Mutation data reported. When heated to decomposition it emits toxic fumes of NO_x.

AEN000   CAS: 628-94-4   HR: 2
ADIPAMIDE
mf: $C_6H_{12}N_2O_2$ mw: 144.20
PROP: Crystals. Mp: 220°. Sol in alc.
SYNS: ADIPIC ACID DIAMIDE □ ADIPIC DIAMIDE □ 1,4-BUTANEDICARBOXAMIDE □ HEXANEDIAMIDE (9CI) □ NCI-C02095
CONSENSUS REPORTS: Reported in EPA TSCA Inventory.
SAFETY PROFILE: Moderately toxic by ingestion. Questionable carcinogen with experimental carcinogenic data. When heated to decomposition it emits toxic fumes of NO_x.

AEN250   CAS: 124-04-9   HR: 3
ADIPIC ACID
mf: $C_6H_{10}O_4$ mw: 146.16
PROP: White monoclinic prisms. Mp: 152°, flash p: 385°F (CC), d: 1.360 @ 25°/4°, vap press: 1 mm @ 159.5°, vap d: 5.04, autoign temp: 788°F, bp: 337.5°. Very sol in alc. Sol in acetone, water = 1.4% @ 15°; 0.6% @ 15° in ether.
SYNS: ACIFLOCTIN □ ACINETTEN □ ADILACTETTEN □ ADIPINIC ACID □ 1,4-BUTANEDICARBOXYLIC ACID □ FEMA No. 2011 □ 1,6-

HEXANEDIOIC ACID □ KYSELINA ADIPOVA (CZECH) □ MOLTEN ADIPIC ACID

CONSENSUS REPORTS: Reported in EPA TSCA Inventory.

ACGIH TLV: TWA 5 mg/^{3}

SAFETY PROFILE: Poison by intraperitoneal route. Moderately toxic by other routes. A severe eye irritant. Combustible when exposed to heat or flame; can react with oxidizing materials. When heated to decomposition it emits acrid smoke and fumes.

AEO000   CAS: 103-23-1   HR: 2
ADIPIC ACID BIS(2-ETHYLHEXYL)
    ESTER
mf: $C_{22}H_{42}O_4$ mw: 370.64

PROP: Liquid. D: 0.927 @ 20°/4°, bp: 181–185° @ 2 mm.

SYNS: ADIPOL 2EH □ BEHA □ BIS(2-ETHYL-HEXYL) ADIPATE □ BISOFLEX DOA □ DEHA □ DI-2-ETHYLHEXYL ADIPATE □ DIOCTYL ADIPATE □ DOA □ EFFEMOLL DOA □ ERGOPLAST AdDO □ FLEXOL A 26 □ HEXANEDIOIC ACID, BIS(2-ETHYL-HEXYL) ESTER □ HEXANEDIOIC ACID, DIOCTYL ESTER □ KODAFLEX DOA □ MONOPLEX DOA □ NCI-C54386 □ OCTYL ADIPATE □ PLASTOMOLL DOA □ PX-238 □ REOMOL DOA □ RUCOFLEX PLASTICIZER DOA □ SICOL 250 □ TRUFLEX DOA □ VESTINOL OA □ WICKENOL 158 □ WITAMOL 320

CONSENSUS REPORTS: IARC Cancer Review: Group 3 IMEMDT 7,56,87; Animal Limited Evidence IMEMDT 29,257,82. NTP Carcinogenesis Bioassay (feed); Clear Evidence: mouse NTPTR* NTP-TR-212,82; No Evidence: rat NTPTR* NTP-TR-212,82. Community Right-To-Know List. Reported in EPA TSCA Inventory.

SAFETY PROFILE: Moderately toxic by intravenous route. Mildly toxic by ingestion. Experimental reproductive effects. Mutation data reported. An eye and skin irritant. Questionable carcinogen with experimental carcinogenic data. See also ESTERS. When heated to decomposition it emits acrid smoke and irritating fumes.

AEO750   CAS: 105-99-7   HR: 2
ADIPIC ACID DIBUTYL ESTER
mf: $C_{14}H_{26}O_4$ mw: 258.40

PROP: Clear liquid. Mp: -20°, bp: 149°, d: 0.960. Flash Pt.: 155°. Insol in water.

SYNS: BUTYL ADIPATE □ DIBUTYL ADIPATE □ DI-N-BUTYL ADIPATE □ DIBUTYL ADIPINATE □ DIBUTYL HEXANEDIOATE □ EXPERIMENTAL TICK REPELLENT 3 □ HEXANEDIOIC ACID–DIBUTYL ESTER

CONSENSUS REPORTS: Reported in EPA TSCA Inventory.

SAFETY PROFILE: Mildly toxic by several routes. An experimental teratogen. Skin and eye irritant. See also ESTERS. A combustible liquid. When heated to decomposition it emits acrid smoke and irritating fumes.

AER250   CAS: 111-69-3   HR: 3
ADIPONITRILE
DOT: UN 2205
mf: $C_6H_8N_2$ mw: 108.16

PROP: Water-white liquid; practically odorless. Mp: 2.3°, bp: 295°, flash p: 199.4°F (OC), d: 0.965 @ 20°/4°, vap d: 3.73. Sol in EtOH, $CHCl_3$; insol in H_2O, Et_2O, CS_2.

SYNS: ADIPIC ACID DINITRILE □ ADIPIC ACID NITRILE □ ADIPODINITRILE □ 1,4-DICYANOBUTANE □ HEXANEDINITRILE □ HEXANEDIOIC ACID DINITRILE □ NITRILE ADIPICO (ITALIAN) □ TETRAMETHYLENE CYANIDE

CONSENSUS REPORTS: EPA Extremely Hazardous Substances List. Reported in EPA TSCA Inventory. Cyanide and its compounds are on the Community Right-To-Know List.

ACGIH TLV: TWA 2 ppm (skin)

NIOSH REL: TWA 18 mg/m^3

DOT CLASSIFICATION: 6.1; Label: KEEP AWAY FROM FOOD

SAFETY PROFILE: Poison by inhalation, ingestion, subcutaneous, and intraperitoneal routes. The nitrile group will behave as a cyanide when ingested or absorbed in the body. It produces disturbances of the respiration and circulation, irritation of the stomach and intestines, and loss of weight. Its low vapor pressure at room temperature makes exposure to harmful concentrations of its vapors unlikely if handled with reasonable care in well-ventilated areas.

Flammable when exposed to heat or flame. When heated to decomposition it emits toxic fumes of CN⁻. Can react with oxidizing materials. To fight fire, use foam, CO_2, dry chemical. See also HYDROCYANIC ACID and NITRILES.

AES750   CAS: 23214-92-8   HR: 3
ADRIAMYCIN
mf: $C_{27}H_{29}NO_{11} \cdot ClH$ mw: 543.57

PROP: Isolated from cultures of *Streptomyces peucetius var. Caesius.*

SYNS: ADM □ ADRIAMYCIN-HCl □ ADRIAMYCIN SEMIQUINONE □ ADRIBLASTINA □ DOXO-RUBICIN □ DX □ F.I 106 □ 14-HYDROXYDAUNO-MYCIN □ 14'-HYDROXYDAUNOMYCIN □ 14-HYDROXYDAUNORU-BICINE □ KW-125 □ NCI-C01514 □ NSC-123127

CONSENSUS REPORTS: NTP 10th Report on Carcinogens. IARC Cancer Review: Group 2A IMEMDT 7,82,87; Animal Sufficient Evidence IMEMDT 7,82,87; Human Inadequate Evidence IMEMDT 7,82,87.

SAFETY PROFILE: Confirmed carcinogen with experimental carcinogenic, neoplastigenic, and tumorigenic data. Poison by intraperitoneal, subcutaneous, parenteral, and intravenous routes. Human systemic effects by intravenous route: cardiac myopathy including infarction, nausea or vomiting, and effects on the hair. An experimental teratogen. Other experimental reproductive effects. Human mutation data reported. When heated to decomposition it emits very toxic fumes of NO_x and HCl.

AET600   CAS: 116425-35-5 HR: 3
AERUGIDIOL
mf: $C_{15}H_{22}O_3$ mw: 250.34

SYN: 6(1H)-AZULENONE, 2,3,3A,7,8,8A-HEXAHYDRO-1,3A-DIHYDROXY-1,4-DIMETHYL-7-(1-METHYLETH-YLIDENE)-, (1S,3AR,8AR)-

SAFETY PROFILE: A poison by ingestion. When heated to decomposition it emits acrid smoke and irritating vapors.

AET750   CAS: 1402-68-2   HR: 3
AFLATOXIN
PROP: Colorless to pale yellow crystals.

CONSENSUS REPORTS: NTP 10th Report on Carcinogens. IARC Cancer Review: Group 1 IMEMDT 7,83,87; Human Sufficient Evidence IMEMDT 7,83,87; Animal Sufficient Evidence IMEMDT 7,83,87

SAFETY PROFILE: Confirmed human carcinogen with experimental tumorigenic data. Human poison by ingestion. An experimental teratogen. Other experimental reproductive effects. Mutation data reported. See also various aflatoxins.

AEU250   CAS: 1162-65-8   HR: 3
AFLATOXIN B1
mf: $C_{17}H_{12}O_6$ mw: 312.29

PROP: A crystalline material. Mp: 268°.

SYNS: AFBI □ AFLATOXIN B

CONSENSUS REPORTS: IARC Cancer Review: Group 1 IMEMDT 7,83,87; Animal Sufficient Evidence IMEMDT 10,51,76; 1,145,72. EPA Genetic Toxicology Program.

SAFETY PROFILE: Confirmed human carcinogen with experimental tumorigenic, neoplastigenic, and carcinogenic data. Acute poison by ingestion, intraperitoneal, and possibly other routes. Experimental teratogenic and reproductive effects. Mutation data reported. When heated to decomposition it emits acrid smoke. See also various aflatoxins.

AEU750   CAS: 7220-81-7   HR: 3
AFLATOXIN B2
mf: $C_{17}H_{14}O_6$ mw: 314.31

PROP: Yellow crystals with blue fluorescence from MeOH.

SYN: DIHYDROAFLATOXIN B1

CONSENSUS REPORTS: IARC Cancer Review: Animal Sufficient Evidence IMEMDT 10,51,76; Animal Limited Evidence IMEMDT 1,145,72.

SAFETY PROFILE: Confirmed human carcinogen with experimental tumorigenic data. Poison by ingestion. Mutation data reported. When heated to decomposition it emits acrid smoke and fumes. See also various aflatoxins.

AEV000   CAS: 1165-39-5   HR: 3
AFLATOXIN G1
mf: $C_{17}H_{12}O_7$ mw: 328.29
PROP: Needles from MeOH exhibiting green fluorescence. Mp: 247–250°. Metabolite of *Aspergillus flavus link ex fries.*
CONSENSUS REPORTS: IARC Cancer Review: Animal Sufficient Evidence IMEMDT 10,51,76.
SAFETY PROFILE: Confirmed human carcinogen with experimental carcinogenic and neoplastigenic data. Poison by ingestion and intraperitoneal routes. Mutation data reported. When heated to decomposition it emits acrid smoke and irritating fumes. See also various aflatoxins.

AEV250   HR: 3
AFLATOXIN G1 mixed with AFLATOXIN B1
PROP: Metabolites of *Aspergillus flavus link ex fries*, Aflatoxin G1, 56.4%; Alfatoxin B1, 37.7%.
SAFETY PROFILE: Confirmed human carcinogen with experimental carcino-genic, neoplastigenic, and tumorigenic data. See also various aflatoxins.

AEW000   CAS: 6795-23-9   HR: 3
AFLATOXIN M1
mf: $C_{17}H_{12}O_7$ mw: 328.29
PROP: Crystals from MeOH exhibiting blue-violet fluorescence. Mp: 299° (decomp).
SYN: 4-HYDROXYAFLATOXIN B1
CONSENSUS REPORTS: IARC Cancer Review: Group 2B; Animal Sufficient Evidence IMEMDT 10,51,76; Animal Sufficient Evidence IMEMDT 56,245,93; Human Inadequate Evidence IMEMDT 56,245,93. EPA Genetic Toxicology Program.
SAFETY PROFILE: Confirmed carcinogen with experimental tumorigenic data. Poison by ingestion. Mutation data reported. When heated to decomposition it emits acrid smoke and irritating fumes. See also various aflatoxins.

AEW500   CAS: 29611-03-8   HR: 2
AFLATOXIN Ro
mf: $C_{17}H_{14}O_6$ mw: 314.31
PROP: Crystals from C_6H_6/hexane. Mp: 224–226°.
SYNS: AFL □ AFLATOXICOL □ AFLATOXICOL NATURAL EPIMER
SAFETY PROFILE: Suspected carcinogen with experimental carcinogenic data. Mutation data reported. When heated to decomposition it emits acrid smoke and irritating fumes. See also various aflatoxins.

AEX250   CAS: 9002-18-0   HR: 1
AGAR
PROP: Extracted from the red algae *Rhodopyceae*. Unground: in thin, translucent, membranous strips; ground: pale buff powder. Sol in boiling water; insol in cold water and org solvs.
SYNS: AGAR-AGAR □ AGAR AGAR FLAKE □ AGAR-AGAR GUM □ BENGAL GELATIN □ BENGAL ISINGLASS □ CEYLON ISINGLASS □ CHINESE ISINGLASS □ DIGENEA SIMPLEX MUCILAGE □ GELOSE □ JAPAN AGAR □ JAPAN ISINGLASS □ LAYOR CARANG □ NCI-C50475
CONSENSUS REPORTS: NTP Carcinogenesis Bioassay (feed); No Evidence: mouse, rat NTPTR* NTP-TR-230,82. Reported in EPA TSCA Inventory.
SAFETY PROFILE: Mildly toxic by ingestion. When heated to decomposition it emits acrid smoke and fumes.

AEX850   CAS: 644-06-4   HR: 2
AGERATOCHROMENE
mf: $C_{13}H_{16}O_3$ mw: 220.29
PROP: Pale-yellow needles from MeOH or oil. Mp: 49–50°, bp: 145–150° at 4 mm.
SYNS: 2H-1-BENZOPYRAN, 6,7-DIMETHOXY-2,2-DIMETHYL- □ 6,7-DIMETHOXY-2,2-DIMETHYL-2H-BENZO(b)PYRAN □ PRECOCENE 2 □ PRECOCENE II
CONSENSUS REPORTS: Reported in EPA TSCA Inventory.
SAFETY PROFILE: Questionable carcinogen with experimental carcinogenic data. Mutation data reported. When heated to decomposition it emits acrid smoke and irritating fumes.

-AFG250   **HR: 2**
AIR, refrigerated liquid
DOT: UN 1002/UN 1003
PROP: Bluish, mobile liquid. O_2 + N_2. Bp: −189° (liq); flash p: none; autoign temp: none.
SYNS: AIR, compressed (UN 1002) (DOT) □ AIR, refrigerated liquid (cryogenic liquid) (UN 1003) (DOT) □ AIR, refrigerated liquid (cryogenic liquid) non-pressurized (UN 1003) (DOT)
DOT CLASSIFICATION: 2.2; Label: Nonflammable Gas, Oxidizer (UN 1003); DOT Class: 2.2; Label: Nonflammable Gas (UN 1002); DOT Class: Nonflammable Gas; Label: Nonflammable Gas.
SAFETY PROFILE: Liquid air can cause tissue damage due to low temperature. Personnel exposed to compressed air may develop caisson disease (the bends, the chokes) if decompression is too rapid. Moderate explosion hazard when containers under pressure are shocked or exposed to heat or flame. Flammable materials, e.g., ethyl ether, hydrocarbons, or charcoal, which have been in contact with liquid air may explode very easily. Ordinary oxidation is greatly accelerated in compressed air. Moderately dangerous disaster hazard; can react vigorously with reducing materials.

AFH400   **CAS: 3011-89-0**   **HR: D**
AKLOMIDE
mf: $C_7H_5ClN_2O_3$ mw: 200.60
PROP: Crystals and gray scales from alc. Mp: 172°.
SYNS: AKLOMIX □ 2-CHLORO-4-NITROBENZAMIDE
SAFETY PROFILE: When heated to decomposition emits toxic fumes of Cl⁻ and NO_x.

AFH625   **CAS: 56-41-7**   **HR: D**
l-ALANINE
mf: $C_3H_7NO_2$ mw: 89.09
PROP: White crystalline powder; odorless with a sweet taste. Mp: 297° (decomp). Sol in water; insol in EtOH.
SYNS: ALANINE □ α-ALANINE □ l-(+)-ALANINE □ l-α-ALANINE □ (S)-ALANINE □ l-2-AMINO-PROPANOIC ACID □ (S)-2-AMINOPROPANOIC ACID □ α-AMINOPROPIONIC ACID

CONSENSUS REPORTS: Reported in EPA TSCA Inventory.
SAFETY PROFILE: Mutation data reported. When heated to decomposition it emits toxic fumes of NO_x.

AFI850   **CAS: 70536-17-3**   **HR: 3**
ALBUMIN MACRO AGGREGATES
SYNS: ALBUMIN □ MAA
SAFETY PROFILE: Poison by intravenous route. When heated to decomposition it emits acrid smoke and irritating fumes.

AFJ000   **HR: 3**
ALCOHOL, DENATURED
DOT: NA 1986/NA 1987
PROP: Liquid. Composed of alcohol and denaturants.
SYNS: DENATURED ALCOHOL (DOT) □ DENATURED SPIRITS
DOT CLASSIFICATION: 3; Label: Flammable Liquid (NA 1987); DOT Class: 3; Label: Flammable Liquid, Poison (NA 1986)
SAFETY PROFILE: Potentially poisonous by ingestion. Toxicity depends upon alcohols in question, generally ethanol with methanol as a denaturant. A flammable liquid and dangerous fire hazard; can react vigorously with oxidizing materials. Moderate explosion hazard. See ETHANOL, METHYL ALCOHOL, and n-PROPYL ALCOHOL.

AFJ500   **CAS: 52-01-7**   **HR: 3**
ALDACTAZIDE
mf: $C_{24}H_{32}O_4S$ mw: 416.62
PROP: Crystals. Mp: 134–135°.
SYNS: 7-α-ACETYLTHIO-3-OXO-17-α-PREGN-4-ENE-21,17-β-CARBOLACTONE □ 7-α-ACETYLTHIO-3-OXO-17-β-PREGN-4-ENE-21,17-β-CARBOLACTONE □ ALDACTIDE □ ALDACTONE □ ALDACTONE A □ 3-(3-KETO-7-α-ACETYLTHIO-17-β-HYDROXY-4-ANDROSTEN-17-α-YL)PROPIONIC ACID LACTONE □ OSIREN □ OSYROL □ 3'-(3-OXO-7-α-ACETYL-THIO-17-β-HYDROXYANDROST-4-EN-17-β-YL)PRO-PIONIC ACID LACTONE □ 17-α-PREGN-4-ENE-21-CARBOXYLIC ACID, 1-HYDROXY-7-α-MERCAPTO-3-OXO-α-LACTONE □ SC 9420 □ SC 15983 □ SPIRESIS □ SPIRIDON □ SPIROCTANIE □ SPIRO(17H-CYCLOPENTA(a)PHENANTHRENE-17,2'-(3'H)-FURAN) □

SPIRO(17H-CYCLOPENTA(a)PHENAN-THRENE-17,2'(5'H)FURAN), PREGN-4-ENE-21-CARBOXYLIC ACID DERIV. □ SPIROLACTONE □ SPIROLAKTON □ SPIROLANG □ SPIRONE □ SPIRONOLACTONE □ SPIRONOLACTONE A □ URACTONE □ VEROSPIRON □ VEROSPIRONE

CONSENSUS REPORTS: IARC Cancer Review: Group 3 IMEMDT 7,344,87; Animal Limited Evidence IMEMDT 24,259,80; Human Inadequate Evidence IMEMDT 24,259,80. Reported in EPA TSCA Inventory.

SAFETY PROFILE: Poison by intraperitoneal route. Human reproductive effects by ingestion and possibly other routes: men, impotence and breast development; women, menstrual cycle changes or disorders, changes in the breasts and lactation. An experimental teratogen. Other experimental reproductive effects. Other human systemic effects by ingestion: agranulocytosis, kidney tubule damage, increased urine volume, and changes in blood sodium and calcium levels. Questionable carcinogen. When heated to decomposition it emits toxic fumes of SO_x. Used to treat hypertension, edema of congestive heart failure, cirrhosis, and kidney failure.

AFJ700   CAS: 7779-41-1   HR: 1
ALDEHYDE C-10 DIMETHYLACETAL
mf: $C_{12}H_{26}O_2$ mw: 202.38
PROP: Fragrance ingredient.
SYNS: DECANAL, DIMETHYLACETAL □ DECYLALDEHYDE DMA □ 1,1-DIMETHOXY-DECANE □ 10,10-DIMETHOXYDECANE
CONSENSUS REPORTS: Reported in EPA TSCA Inventory.
SAFETY PROFILE: A skin irritant. When heated to decomposition it emits acrid smoke and irritating fumes.

AFJ800   HR: 2
ALDEHYDES
PROP: A class of chemicals with the general formula R•CHO, and characterized by an unsaturated carbonyl group (C=O).
SAFETY PROFILE: Aldehydes are widely used in many industrial processes. The US production of acetaldehyde in 1982 was 281,000 tons. The world production of acrolein in 1975 was 59,000 tons. Aldehydes occur in nature and are gaseous by-products of incomplete combustion of wood and coal, in exhaust from gasoline and diesel engines, industrial waste gases and fumes, tobacco smoke, and wood fires. Formaldehyde and acetaldehyde are carcinogens. Many of the aldehydes are mutagens. They are reactive compounds participating in oxidation, reduction, addition, and polymerization reactions. All the aldehydes possess anesthetic properties, but this is obscured by their highly irritating action on the eyes and mucous membranes of the respiratory tract. The lower aldehydes, very soluble in water, act chiefly on the eyes and tissues of the upper respiratory tract. The higher aldehydes, less soluble in water, tend to penetrate more deeply into the respiratory system and may affect the lungs. Some higher aldehydes and also the aromatic aldehydes may exhibit much lower toxicity. See also specific compounds.

AFK250   CAS: 309-00-2   HR: 3
ALDRIN
DOT: UN 2761/NA 2762
mf: $C_{12}H_8Cl_6$ mw: 364.90
PROP: Crystals. Mp: 104–105°. Sol in Me_2CO, C_6H_6; sltly sol in pet ether; almost insol in H_2O. IDLH 25 mg/m³.
SYNS: ALDREX □ ALDREX 30 □ ALDRIN, cast solid (DOT) □ ALDRINE (FRENCH) □ ALDRITE □ ALDROSOL □ ALTOX □ COMPOUND 118 □ DRINOX □ ENT 15,949 □ HEXACHLORO-HEXAHYDRO-endo-exo-DIMETHANONAPHTHAL-ENE □ 1,2,3,4,10,10-HEXACHLORO-1,4,4a,5,8,8a-HEXAHYDRO-1,4,5,8-DIMETHANONAPHTHALENE □ 1,2,3,4,10,10-HEXACHLORO-1,4,4a,5,8,8a-HEXAHYDRO-1,4-endo-exo-5,8-DIMETHANO-NAPHTHALENE □ 1,2,3,4,10,10-HEXACHLORO-1,4,4a,5,8,8a-HEXAHYDRO-exo-1,4,-endo-5,8-DIMETHANONAPHTHALENE □ HHDN □ NCI-C00044 □ OCTALENE □ RCRA WASTE NUMBER P004 □ SEEDRIN
CONSENSUS REPORTS: IARC Cancer Review: Group 3 IMEMDT 7,88,87; Human Inadequate Evidence IMEMDT 5,25,74; Animal Inadequate Evidence IMEMDT 5,25,74; NCI Carcinogenesis

A

Bioassay (feed); Clear Evidence: mouse NCITR* NCI-CG-TR-21,78; Inadequate Studies: rat NCITR* NCI-CG-TR-21,78. EPA Genetic Toxicology Program. EPA Extremely Hazardous Substances List. Community Right-To-Know List.
OSHA PEL: TWA 0.25 mg/m³ (skin)
ACGIH TLV: TWA (inhalable fraction) 0.05 mg/m³; (skin); Confirmed Animal Carcinogen with Unknown Revelance to Humans
DFG MAK: 0.25 mg/m³
NIOSH REL: (Aldrin) Reduce to lowest detectable level
DOT CLASSIFICATION: 6.1; Label: Poison
SAFETY PROFILE: Suspected carcinogen with experimental carcinogenic, neoplastigenic, and tumorigenic data. Poison by ingestion, skin contact, intravenous, and intraperitoneal routes. Human systemic effects by ingestion: excitement, tremors, and nausea or vomiting. An experimental teratogen. Other experimental reproductive effects. Continued acute exposure causes liver damage. Human mutation data reported. See also CHLORINATED HYDROCARBONS. When heated to decomposition it emits toxic fumes of Cl⁻.

AFK925   **HR: D**
ALGAE MEAL, DRIED
PROP: Mixture of algae cells from *Spongiococcum*, molasses, corn steep liquor, and a maximum of 0.3% ethoxyquin.
SAFETY PROFILE: When heated to decomposition it emits acrid smoke and irritating fumes.

AFL000   **CAS: 9005-32-7**   **HR: 2**
ALGINIC ACID
PROP: Extracted from brown seaweeds. White to yellow white fibrous powder; odorless and tasteless. Sol in alkaline solutions; insol in org solvs.
SYNS: KELACID □ LANDALGINE □ NORGINE □ PLOYMANNURONIC ACID □ SAZZIO
CONSENSUS REPORTS: Reported in EPA TSCA Inventory.

SAFETY PROFILE: Moderately toxic by intraperitoneal route. When heated to decomposition it emits acrid smoke and irritating fumes.

AFM250   **HR: 3**
ALIPHATIC and AROMATIC EPOXIDES
SAFETY PROFILE: Suspected carcinogen with experimental tumors of the skin, lung, and blood-forming tissues.

AFM500   **HR: 1**
ALKALIES
PROP: A term loosely applied to the hydroxides and carbonates of the alkali metals and alkaline earth metals, as well as the bicarbonate and hydroxide of ammonium. They can neutralize acids, change the color of indicators, and impart a soapy taste and feel to aq solns.
SAFETY PROFILE: Variable toxicity. As a group, they constitute the commonest causes of contact dermatitis. Systemically ammonia is most troublesome. See also AMMONIA. See also specific compound.

AFN250   **HR: 2**
ALKANES
PROP: All colorless neutral liquids with light odors. See also individual alkanes as listed (n-pentane, n-hexane, n-heptane, n-octane).
SAFETY PROFILE: Hexane can cause neuropathy with chronic exposure. Other alkanes or mixtures may have the same effect. Many are dangerous fire hazards when exposed to flame, heat, or oxidizers.

AFR250   **CAS: 584-79-2**   **HR: 3**
ALLETHRIN
mf: $C_{19}H_{26}O_3$ mw: 302.45
PROP: A viscous liquid.
SYNS: (+)-ALLELRETHONYL (+)-cis,trans-CHRYS-ANTHEMATE □ d-ALLETHRIN □ d-trans ALLE-THRIN □ ALLETHRIN I □ ALLEVIATE □ ALLYL CINERIN □ ALLYL HOMOLOG of CINERIN I □ d,l-2-ALLYL-4-HYDROXY-3-METHYL-2-CYCLOPENTEN-1-ONE-d,l-CHRYSANTHEMUM MONOCARBOXYLATE □ 3-ALLYL-4-KETO-2-METHYLCYCLOPENTENYL CHRYSAN-THEMUMMONOCARBOXYLATE □ 3-ALLYL-2-METHYL-4-OXO-2-CYCLOPENTEN-1-YL

CHRYSANTHEMATE □ dl-3-ALLYL-2-METHYL-4-OXOCYCLOPENT-2-ENYL dl-cis trans CHRYSANTHEMATE □ ALLYLRETHRONYL dl-cis-trans-CHRYSANTHEMATE □ BIOALLETHRIN □ BIOALTRINA □ CINERIN I ALLYL HOMOLOG □ DEPALLETHRIN □ ENT 17,510 □ EXTHRIN □ FDA 1446 □ FMC 249 □ NECARBOXYLIC ACID □ NIA 249 □ OMS 468 □ PALLETHRINE □ PYNAMIN □ PYNAMIN-FORTE □ PYRESIN □ PYRESYN □ SYNTHETIC PYRETHRINS
CONSENSUS REPORTS: Reported in EPA TSCA Inventory. EPA Genetic Toxicology Program.
SAFETY PROFILE: Poison by intravenous, intracerebral, and intraperitoneal routes. Moderately toxic by ingestion and skin contact. An allergen. An insecticide. It can cause liver and kidney damage by all routes of entry into the body. Lung congestion may occur due to exposure. Local contact may cause contact dermatitis. Inhalation may cause asthma, coughing, wheezing, running nose and eyes. Mutation data reported. See also ALLYL COMPOUNDS and ESTERS. Slight fire hazard. When heated to decomposition it emits acrid fumes.

AFS800   CAS: 14155-65-8   HR: 3
ALLO-GLAUCOTOXIGENIN
mf: $C_{23}H_{32}O_6$ mw: 404.55
SYNS: ALLOGLAUCOTOXIGENIN □ ALLOGLUCOTOXIGENIN □ 5-α-CARD-20(22)-ENOLIDE, 3-β,14,15-β-TRIHYDROXY-19-OXO- □ CARD-20(22)-ENOLIDE, 19-OXO-3,14,15-TRIHYDROXY-, (3-β,5-α,15-β)- □ (3-β,5-α,15-β)-19-OXO-3,14,15-TRIHYDROXYCARD-20(22)-ENOLIDE
SAFETY PROFILE: A poison by intravenous route. When heated to decomposition it emits acrid smoke and irritating vapors.

AFT750   CAS: 50-71-5   HR: 3
ALLOXAN
mf: $C_4H_2N_2O_4$ mw: 142.08
PROP: Orthorhombic crystals from AcOH or by subl. Mp: 256° (decomp), d: 1.70. Sol in water, alc, benzene, and acetone.
SYNS: MESOXALYLCARBAMIDE □ MESOXALYLUREA □ 2,4,5,6(1H,3H)-PYRIMIDINETETRONE □ 2,4,5,6-PYRIMIDINTETRON (CZECH) □ 2,4,5,6-TETRAOXOHEXAHYDROPYRIMIDINE

CONSENSUS REPORTS: Reported in EPA TSCA Inventory.
SAFETY PROFILE: Poison by intraperitoneal, intravenous, subcutaneous, and rectal routes. Moderately toxic by ingestion. An experimental teratogen. Other experimental reproductive effects. Mutation data reported. Produces diabetes in experimental animals. Decomposes in storage to release CO_2. Do not store in sealed container. Explodes when heated above 170°C. When heated to decomposition it emits toxic fumes of NO_x.

AFU750   CAS: 591-87-7   HR: 3
ALLYL ACETATE
DOT: UN 2333
mf: $C_5H_8O_2$ mw: 100.13
PROP: Liquid. Vap d: 3.45, bp: 103–104°, d: 0.928, flash p: 72°F. Insol in water.
SYNS: ACETIC ACID ALLYL ESTER □ ACETIC ACID-2-PROPENYL ESTER □ 3-ACETOXYPROPENE
CONSENSUS REPORTS: Reported in EPA TSCA Inventory.
DOT CLASSIFICATION: 3; Label: Flammable Liquid, Poison
SAFETY PROFILE: Poison by ingestion. Moderately toxic by inhalation and skin contact. A skin and eye irritant. When heated to decomposition it emits acrid smoke and irritating fumes. Dangerous fire hazard. See also ALLYL COMPOUNDS.

AFV500   CAS: 107-18-6   HR: 3
ALLYL ALCOHOL
DOT: UN 1098
mf: C_3H_6O mw: 58.09
PROP: Limpid liquid; pungent odor. Mp: −129°, fp: −50°, bp: 96–97°, lel: 2.5%, uel: 18%, flash p: 70°F (CC), d: 0.854 @ 20°/4°, autoign temp: 713°F, vap press: 10 mm @ 10.5°, vap d: 2.00. Misc in water, alc, and ether. IDLH 20 ppm.
SYNS: ALCOOL ALLILCO (ITALIAN) □ ALCOOL ALLYLIQUE (FRENCH) □ ALLILOWY ALKOHOL (POLISH) □ ALLYL AL □ ALLYLALKOHOL (GERMAN) □ ALLYLIC ALCOHOL □ 3-HYDROXY-PROPENE □ ORVINYLCARBINOL □ PROPENOL □ PROPEN-1-OL-3 □ 1-PROPEN-3-OL □ 2-PROPEN-1-OL □ PROPENYL

A

ALCOHOL □ 2-PROPENYL ALCOHOL □ RCRA WASTE NUMBER P005 □ SHELL UNDRAUTTED A □ VINYLCARBINOL □ WEED DRENCH

CONSENSUS REPORTS: EPA Extremely Hazardous Substances List. Reported in EPA TSCA Inventory.

OSHA PEL: TWA 2 ppm; STEL 4 ppm (skin)

ACGIH TLV: 0.5 ppm (skin); Not Classifiable as a Human Carcinogen

DFG MAK: Confirmed Animal Carcinogen with Unknown Relevance to Humans

DOT CLASSIFICATION: 6.1; Label: Poison, Flammable Liquid

SAFETY PROFILE: Suspected carcinogen. Poison by inhalation, ingestion, skin contact, subcutaneous, intraperitoneal, and possibly other routes. A skin, severe eye (human), and systemic irritant. Mutation data reported. Dangerous fire and explosion hazard when exposed to heat, flame, or oxidizers. Explosive or violent reaction with sulfuric acid, alkali + 2,4,6-trichloro-1,3,5-triazine, or 2,4,6-tris(bromoamino)-1,3,5-triazine. Reaction with carbon tetrachloride produces explosively unstable halogenated C_4 epoxides. Incompatible with chlorosulfonic acid, HNO_3, H_2SO_4, oleum, NaOH, diallyl phosphite, PCl_3, and tri-n-bromomelamine. When heated to decomposition it emits acrid smoke and fumes. To fight fire, use CO_2, alcohol foam, dry chemical. See also ALLYL COMPOUNDS.

AFW000   CAS: 107-11-9   HR: 3
ALLYLAMINE
DOT: UN 2334
mf: C_3H_7N mw: 57.11
PROP: Colorless liquid, burning taste, sharp ammonia odor. Bp: 56.5–58°, d: 0.761 @ 20°/4°, flash p: −20°F, autoign temp: 705°F, vap d: 2.00, lel: 2.2%, uel: 22%. Misc in water, alc, and ether.
SYNS: 3-AMINOPROPENE □ 3-AMINOPROPYLENE □ MONOALLYLAMINE □ 2-PROPENAMINE □ 2-PROPEN-1-AMINE

CONSENSUS REPORTS: EPA Extremely Hazardous Substances List. Reported in EPA TSCA Inventory.

DOT CLASSIFICATION: 6.1; Label: Poison, Flammable Liquid

SAFETY PROFILE: Poison by inhalation, ingestion, intraperitoneal, and skin contact routes. Human systemic effects by inhalation: lacrymation and lung effects. A systemic irritant. Mutation data reported. A severe eye and skin irritant. Extraordinary precautions against fumes are advised.

Dangerous fire and explosion hazard when exposed to heat, flame, or oxidizers. Highly reactive. When heated to decomposition it emits toxic fumes of NO_x. To fight fire, use alcohol foam, CO_2, dry chemical. See also ALLYL COMPOUNDS and AMINES.

AFW750   CAS: 140-67-0   HR: 2
p-ALLYLANISOLE
mf: $C_{10}H_{12}O$ mw: 148.22
PROP: Isolated from rind of *Persea Gratissima Garth,* and from Oil of Estragon; found in oils of Russian Anise, Basil, Fennel, Turpentine, and others (FCTXAV 14,601,76). Colorless to sltly yellow liquid; anise odor. Bp: 102° at 16 mm, d: 0.960–0.968, refr index: 1.519–1.524, flash p: 178°F. Sol in alc; insol in water.
SYNS: 4-ALLYL-1-METHOXYBENZENE □ CHAVICOL METHYL ETHER □ ESDRAGOL □ ISOANETHOLE □ p-METHOXYALLYLBENZENE □ 1-METHOXY-4-(2-PROPENYL)BENZENE □ METHYL CHAVICOL □ NCI-C60946 □ TARRAGON

CONSENSUS REPORTS: Reported in EPA TSCA Inventory.

SAFETY PROFILE: Moderate acute toxicity by many routes. A skin irritant. Questionable carcinogen with experimental carcinogenic and neoplastigenic data. Mutation data reported. Combustible liquid. When heated to decomposition it emits acrid smoke and irritating fumes. See also ALLYL COMPOUNDS. A spice used in foods, liqueurs, and perfumes.

AFY000   CAS: 106-95-6   HR: 3
ALLYL BROMIDE
DOT: UN 1099
mf: C_3H_5Br mw: 120.99
PROP: Colorless liquid; pungent odor. Mp:
$-119°$, bp: 71.3°, flash p: 30°F, d: 1.3980 @
20°/4°, autoign temp: 563°F, vap d: 4.17,
lel: 4.4%, uel: 7.3%. Insol in water.
SYNS: BROMALLYLENE □ 1-BROMO-2-PROPENE □
3-BROMOPROPENE □ 3-BROMOPROPYLENE
CONSENSUS REPORTS: Reported in EPA
TSCA Inventory.
DOT CLASSIFICATION: 3; Label:
Flammable Liquid, Poison
SAFETY PROFILE: Poison by ingestion and
intraperitoneal routes. Mildly toxic by
inhalation. Human mutation data reported.
See also ALLYL CHLORIDE and ALLYL
COMPOUNDS. Dangerous fire and
explosion hazard when exposed to heat,
flame, or oxidizers. When heated to
decomposition it emits toxic fumes of Br⁻.
To fight fire, use alcohol foam, water spray
or mist, CO_2, dry chemical.

AGA500   CAS: 123-68-2   HR: 3
ALLYL CAPROATE
mf: $C_9H_{16}O_2$ mw: 156.25
PROP: Bp: 186–188°. Insol in water; sol in
alc and ether.
SYNS: ALLYL HEXANOATE (FCC) □ FEMA No. 2032 □
2-PROPENYL-N-HEXANOATE
CONSENSUS REPORTS: Reported in EPA
TSCA Inventory.
SAFETY PROFILE: Poison by ingestion and
skin contact. Mutation data reported. An
irritant to human skin. When heated to
decomposition it emits acrid smoke and
irritating fumes. See also ALLYL
COMPOUNDS and ESTERS.

AGB250   CAS: 107-05-1   HR: 3
ALLYL CHLORIDE
DOT: UN 1100
mf: C_3H_5Cl mw: 76.53
PROP: Colorless liquid with pungent odor.
Mp: $-136.4°$, bp: 44.6°, d: 0.938 @ 20°/4°,
fp: $-134.5°$, flash p: $-25°F$, lel: 2.9%, uel:
11.2%, autoign temp: 905°F, vap d: 2.64.

Misc in org solvs. Sltly sol in water. IDLH
250 ppm.
SYNS: ALLILE (CLORURO di) (ITALIAN) □
ALLYLCHLORID (GERMAN) □ ALLYLE (CHLORURE d')
(FRENCH) □ CHLORALLYLENE □ CHLOROALLYLENE
□ 3-CHLOROPRENE □ 1-CHLORO PROPENE-2 □ 1-
CHLORO-2-PROPENE □ 3-CHLOROPROPENE □ 3-
CHLORO-1-PROPENE □ α-CHLOROPROPYLENE □ 3-
CHLOROPROPYLENE □ 3-CHLORO-1-PROPYLENE □
3-CHLORPROPEN (GERMAN) □ NCI-C04615 □ 2-
PROPENYL CHLORIDE
CONSENSUS REPORTS: IARC Cancer
Review: Group 3 IMEMDT 7,56,87; Animal
Inadequate Evidence IMEMDT 36,39,85;
NCI Carcinogenesis Bioassay (gavage); No
Evidence: rat NCITR* NCI-CG-TR-73,78;
Clear Evidence: mouse NCITR* NCI-CG-
TR-73,78. Reported in EPA TSCA
Inventory. EPA Genetic Toxicology
Program. Community Right-To-Know List.
OSHA PEL: TWA 1 ppm; STEL 2 ppm
ACGIH TLV: TWA 1 ppm; STEL 2 ppm;
Animal Carcinogen
DFG MAK: Confirmed Animal Carcinogen
with Unknown Relevance to Humans
NIOSH REL: TWA 1 ppm; CL 3 ppm/15M
DOT CLASSIFICATION: 3; Label:
Flammable Liquid, Poison
SAFETY PROFILE: Suspected carcinogen
with experimental tumorigenic data. Poison
by intraperitoneal and intravenous routes.
Moderately toxic by ingestion, inhalation,
and skin contact. Experimental teratogenic
and reproductive effects. A skin and eye
irritant. Human mutation data reported.
Chronic exposure may cause liver and
kidney damage. The vapors of allyl chloride
are quite irritating to the eyes, nose, and
throat. Contact of the liquid with the skin
may lead, in addition to local
vasoconstriction and numbness, to rapid
absorption and distribution through the
body. If remedial measures are not taken
promptly, such contact may result in burns
and internal injuries. Inhalation may cause
headache, dizziness, and in high
concentration, loss of consciousness;
however, even in low concentration, its odor
in most cases is irritating enough to give
warning of its presence. Concentration of

the vapors high enough to cause serious effects, including damage to the lungs, especially on repeated exposure, may not be intolerable. Consequently, the warning characteristics should never be disregarded. In general, precautions should be taken AT ALL TIMES to avoid spillage and accumulation of noticeable concentration of the vapors in the atmosphere. Acute exposure in experimental animals has resulted in marked inflammation of lungs, irritation of skin, and swelling of the kidneys. Chronically exposed animals have shown degenerative changes in the liver and kidneys. Reported human exposures have been principally cases of irritation of the eyes, skin, and respiratory tract, sometimes accompanied by aches and pains in the bones. Liver and kidney injury is possible. Dangerous fire and explosion hazard when exposed to heat, flame, or oxidizers. Vigorous or explosive reaction above $-70°C$ with alkyl aluminum chlorides (e.g., trichlorotriethyl dialuminum, ethyl aluminum dichloride, or diethyl aluminum chloride) + aromatic hydrocarbons (e.g., benzene or toluene). Violently exothermic polymerization reaction with Lewis acids (e.g., aluminum chloride, boron trifluoride, or sulfuric acid) and metals (e.g., aluminum, magnesium, zinc, or galvanized metals). Incompatible with HNO_3, ethylene imine, ethylenediamine, chlorosulfonic acid, oleum, NaOH. To fight fire, use CO_2, alcohol foam, dry chemical. See also CHLORINATED HYDROCARBONS; ALIPHATIC; ALLYL COMPOUNDS; and CHLORIDES.

Storage and Handling: Keep cool, away from heat sources. Maintain good ventilation. Work in a fume hood or with closed system if possible; otherwise, use adequate ventilation so that the odor of allyl chloride does not persist. If it should be necessary to enter an area in which the odor of allyl chloride is at all noticeable, use a gas mask equipped with an "organic vapor"

canister. Do not disregard the warning odor or eye irritation of allyl chloride.

A

AGB500   CAS: 2937-50-0   HR: 3
ALLYL CHLOROCARBONATE
DOT: UN 1722
mf: $C_4H_5ClO_2$ mw: 120.54
PROP: Liquid. Bp: 106–114°, flash p: 88°F (CC), d: 1.14, vap d: 4.2.
SYNS: ALLYL CHLOROFORMATE □ ALLYL CHLOROFORMATE (DOT) □ ALLYLESTER KYSELINY CHLORMRAVENCI □ CHLOROFORMIC ACID ALLYL ESTER
CONSENSUS REPORTS: Reported in EPA TSCA Inventory.
DOT CLASSIFICATION: 8; Label: Corrosive, Poison
SAFETY PROFILE: Poison by inhalation and ingestion. Corrosive. Dangerous when exposed to heat, open flame (or sparks), or powerful oxidizers. Can react with oxidizing materials. To fight fire, use alcohol foam, spray or mist, dry chemical. When heated to decomposition it emits toxic fumes of Cl⁻. See also ALLYL COMPOUNDS and ESTERS.

AGC000   CAS: 1866-31-5   HR: 2
ALLYL CINNAMATE
mf: $C_{12}H_{12}O_2$ mw: 188.24
PROP: Colorless to light-yellow liquid; cherry odor. D: 1.052 @ 25°/25°, bp: 150–152° @ 15 mm. Insol in water; sol in alc; very sol in ether.
SYNS: ALLYL-3-PHENYLACRYLATE □ PROPENYL CINNAMATE □ VINYL CARBINYL CINNAMATE
CONSENSUS REPORTS: Reported in EPA TSCA Inventory.
SAFETY PROFILE: Moderately toxic by ingestion. Human skin irritant. When heated to decomposition it emits acrid smoke and irritating fumes. See also ALLYL COMPOUNDS and ESTERS.

AGC500   CAS: 2705-87-5   HR: 3
ALLYL CYCLOHEXANEPROPIONATE
mf: $C_{12}H_{20}O_2$ mw: 196.32
PROP: Colorless liquid; pineapple odor. D: 0.945–0.950, refr index: 1.457–1.463, flash p:

212°F. Misc in alc, chloroform, ether; insol in glycerin and water.
SYNS: 3-ALLYLCYCLOHEXYL PROPIONATE □ ALLYL HEXAHYDROPHENYLPROPIONATE □ FEMA No. 2026
CONSENSUS REPORTS: Reported in EPA TSCA Inventory.
SAFETY PROFILE: Poison by ingestion. When heated to decomposition it emits acrid smoke and irritating fumes. Combustible liquid. See ALLYL COMPOUNDS and ESTERS.

AGE250   CAS: 93-15-2   HR: 3
4-ALLYL-1,2-DIMETHOXYBENZENE
mf: $C_{11}H_{14}O_2$ mw: 178.25
PROP: Colorless to pale-yellow liquid; clove, carnation odor. Bp: 255°, d: 1.032–1.036, refr index: 1.532, fp: −4°, flash p: 212°F. Sol in fixed oils; insol in glycerin and propylene glycol.
SYNS: 1-ALLYL-3,4-DIMETHOXYBENZENE □ 4-ALLYLVERATROLE □ 1,2-DIMETHOXY-4-ALLYL-BENZENE □ 1-(3,4-DIMETHOXYPHENYL)-2-PROPENE □ ENT 21,040 □ 1,3,4-EUGENOL METHYL ETHER □ EUGENYL METHYL ETHER □ FEMA No. 2475 □ METHYL EUGENOL (FCC) □ VERATROLE METHYL ETHER
CONSENSUS REPORTS: NTP 10th Report on Carcinogens. Reported in EPA TSCA Inventory.
SAFETY PROFILE: Confirmed carcinogen. Poison by intravenous route. Moderately toxic by ingestion and intraperitoneal routes. A skin irritant. Mutation data reported. Combustible liquid. When heated to decomposition it emits acrid smoke and irritating fumes. Some other alkenylbenzenes have carcinogenic activity. See also EUGENOL, ALLYL COMPOUNDS, and ETHERS.

AGH000   CAS: 1838-59-1   HR: 3
ALLYL FORMATE
DOT: UN 2336
mf: $C_4H_6O_2$ mw: 86.10
PROP: Liquid, sltly water-sol, sol in org solvs. D: 0.948 @ 18°/4°, bp: 83°, flash p: <−50°F.

SYNS: ALLYLESTER KYSELINY MRAVENCI □ FORMIC ACID, ALLYL ESTER
CONSENSUS REPORTS: Reported in EPA TSCA Inventory.
DOT CLASSIFICATION: 3; Label: Flammable Liquid, Poison
SAFETY PROFILE: Poison by ingestion. Moderately toxic by inhalation. Very flammable and reactive. Dangerous fire hazard. See also ALLYL COMPOUNDS and ESTERS. When heated to decomposition it yields irritating smoke and fumes.

AGH150   CAS: 106-92-3   HR: 3
ALLYL GLYCIDYL ETHER
DOT: UN 2219
mf: $C_6H_{10}O_2$ mw: 114.16
PROP: Bp: 153.9°, fp: −100° (forms glass), flash p: 135°F (OC), d: 0.9698 @ 20°/4°, vap press: 21.59 mm @ 60°, vap d: 3.94. IDLH 50 ppm.
SYNS: AGE □ ALLIL-GLICIDIL-ETERE (ITALIAN) □ 1-ALLILOSSI-2,3 EPOSSIPROPANO (ITALIAN) □ ALLYL-2,3-EPOXYPROPYL ETHER □ ALLYL-GLYCIDAETHER (GERMAN) □ 1-ALLYLOXY-2,3-EPOXY-PROPAAN (DUTCH) □ 1-ALLYLOXY-2,3-EPOXYPROPAN (GERMAN) □ 1-(ALLYLOXY)-2,3-EPOXYPROPANE □ NCI-C56666 □ OXYDE d'ALLYLE et de GLYCIDYLE (FRENCH) □ ((2-PROPENYLOXY)METHYL)OXIRANE
CONSENSUS REPORTS: Reported in EPA TSCA Inventory.
OSHA PEL: TWA 5 ppm; STEL 10 ppm
ACGIH TLV: 1 ppm; Not Classifiable as a Human Carcinogen
DFG MAK: Confirmed Animal Carcinogen, Suspected Human Carcinogen
NIOSH REL: (Glycidyl Ethers) CL 45 mg/m³/15M
DOT CLASSIFICATION: 3; Label: Flammable Liquid, Poison
SAFETY PROFILE: Confirmed animal carcinogen. Poison by ingestion. Moderately toxic by inhalation and skin contact. Mutation data reported. A severe skin and eye irritant. Can cause central nervous system depression and pulmonary edema. A flammable liquid when exposed to heat or flame; can react with oxidizing materials. To fight fire, use foam, CO_2, dry chemical.

A

When heated to decomposition it emits acrid smoke and irritating fumes. See also ALLYL COMPOUNDS.

AGH250   CAS: 142-19-8   HR: 2
ALLYL HEPTANOATE
mf: $C_{10}H_{18}O_2$ mw: 170.28
PROP: Colorless to pale-yellow liquid; fruity, sweet, pineapple odor. D: 0.880, refr index: 1.426, flash p: 154°F.
SYNS: ALLYL ENANTHATE □ ALLYL HEPTOATE □ ALLYL HEPTYLATE □ FEMA No. 2031 □ 2-PROPENYL HEPTANOATE
CONSENSUS REPORTS: Reported in EPA TSCA Inventory.
SAFETY PROFILE: Moderately toxic by ingestion and skin contact. A human skin irritant. See also ALLYL COMPOUNDS and ESTERS. Combustible liquid. When heated to decomposition it emits acrid smoke and irritating fumes.

AGI250   CAS: 556-56-9   HR: 3
ALLYL IODIDE
DOT: UN 1723
mf: C_3H_5I mw: 167.98
PROP: Yellow liquid; pungent odor. Mp: −99°, bp: 103.1°, d: 1.825 @ 20°/4°, vap d: 5.8.
SYNS: 3-IODOPROPENE □ 3-IODO-1-PROPENE □ 3-IODOPROPYLENE □ 1-PROPENE, 3-IODO-(9CI)
CONSENSUS REPORTS: Reported in EPA TSCA Inventory.
ACGIH TLV: Proposed: (inhalable fraction) 0.1 mg/m³; Not Classifiable as a Human Carcinogen)
DOT CLASSIFICATION: 3; Label: Flammable Liquid, Corrosive
SAFETY PROFILE: Poison by inhalation and ingestion. Mutation data reported. A powerful irritant. A flammable liquid. Incompatible with oxidizing materials. To fight fire, use water, foam, CO_2, dry chemical. When heated to decomposition it emits highly toxic fumes of I⁻. See also ALLYL COMPOUNDS and IODIDES.

AGI500   CAS: 79-78-7   HR: 2
ALLYL α-IONONE
mf: $C_{16}H_{24}O$ mw: 232.40
PROP: Colorless to yellow liquid; fruity, woody odor. D: 0.928–0.935, refr index: 1.503–1.507, flash p: 212°F. Sol in alc; insol in water @ 265°.
SYNS: FEMA No. 2033 □ 1-(2,6,6-TRIMETHYL-2-CYCLOHEXEN-1-YL)-1,6-HEPTADIEN-3-ONE
CONSENSUS REPORTS: Reported in EPA TSCA Inventory.
SAFETY PROFILE: A skin irritant. Combustible liquid. When heated to decomposition it emits acrid smoke and irritating fumes. See also ALLYL COMPOUNDS.

AGJ000   CAS: 1476-23-9   HR: 3
ALLYL ISOCYANATE
DOT: UN 2206/UN 2207/UN 2478/UN 3080
mf: C_4H_5NO mw: 83.10
PROP: Liquid. Bp: 85°.
SYN: ISOCYANIC ACID, ALLYL ESTER
CONSENSUS REPORTS: Reported in EPA TSCA Inventory.
DOT CLASSIFICATION: 6.1; Label: KEEP AWAY FROM FOOD (UN 2207); DOT Class: 6.1; Label: Poison (UN 2206); DOT Class: 6.1; Label: Poison, Flammable Liquid (UN 3080); DOT Class: 3; Label: Flammable Liquid, Poison (UN 2478)
SAFETY PROFILE: A poison. See also ALLYL COMPOUNDS and ESTERS. A flammable liquid. When heated to decomposition it emits toxic fumes of NO_x.

AGJ250   CAS: 57-06-7   HR: 3
ALLYL ISOTHIOCYANATE
DOT: UN 1545
mf: C_4H_5NS mw: 99.16
PROP: Colorless to pale-yellow liquid oil; irritating odor with mustard taste. Mp: −80°, d: 1.015 @ 15°/4°, bp: 150.7°, fp: −80°, flash p: 115°F, d: 1.013–1.016 @ 25°/25°, vap press: 10 mm @ 38.3°, vap d: 3.41, refr index: 1.527–1.531. Misc with alc, carbon disulfide, and ether.
SYNS: AITC □ ALLYL ISORHODANIDE □ ALLYL ISOSULFOCYANATE □ ALLYL ISOTHIOCYANATE, stabilized (DOT) □ ALLYL MUSTARD OIL □

ALLYLSENFOEL (GERMAN) □ ALLYL SEVENOLUM □ ALLYL THIOCARBONIMIDE □ ARTIFICIAL MUSTARD OIL □ CARBOSPOL □ FEMA No. 2034 □ ISOTHIO-CYANATE d'ALLYLE (FRENCH) □ 3-ISOTHIOCYANA-TO-1-PROPENE □ MUSTARD OIL □ NCI-C50464 □ OIL OF MUSTARD, artificial □ OLEUM SINAPIS VOLATILE □ 2-PROPENYL ISOTHIO-CYANATE □ REDSKIN □ SENF OEL (GERMAN) □ SYNTHETIC MUSTARD OIL □ VOLATILE OIL OF MUSTARD
CONSENSUS REPORTS: IARC Cancer Review: Group 3 IMEMDT 7,56,87; Animal Limited Evidence IMEMDT 36,55,85; NTP Carcinogenesis Bioassay (gavage); No Evidence: mouse NTPTR* NTP-TR-234,82; Clear Evidence: rat NTPTR* NTP-TR-234,82. Reported in EPA TSCA Inventory.
DOT CLASSIFICATION: 6.1; Label: Poison
SAFETY PROFILE: Suspected carcinogen with experimental neoplastigenic and tumorigenic data. Poison by ingestion, skin contact, intravenous, subcutaneous, and intraperitoneal routes. Experimental teratogenic and reproductive effects. An eye irritant. An allergen. May cause contact dermatitis. Mutation data reported. A flammable liquid. Highly reactive. When heated to decomposition (above 250°) or on contact with acid or acid fumes it emits highly toxic fumes of CN^-, SO_x, and NO_x. To fight fire, use foam, CO_2, dry chemical. See also ALLYL COMPOUNDS and ESTERS.

AGJ500   HR: 3
ALLYL MERCAPTAN
mf: C_3H_6S mw: 74.15
PROP: Water-white liquid with a strong garlic odor, darkens on standing. D: 0.925 @ 23°/4°, bp: 68°, flash p: 14°F.
SYN: 2-PROPENE-1-THIOL
SAFETY PROFILE: Poison by inhalation and ingestion. Strong irritant to skin and mucous membranes. When heated to decomposition it emits highly toxic fumes of SO_x. Very dangerous fire hazard. To fight fire, use water mist or spray, alcohol foam, CO_2, or dry chemical. See also ALLYL COMPOUNDS and MERCAPTANS.

AGM500   CAS: 4230-97-1   HR: 2
ALLYL OCTANOATE
mf: $C_{11}H_{20}O_2$ mw: 184.31
PROP: Colorless liquid; fruity odor. D: 0.8550.861, refr index: 1.425, flash p: 151°F. Sol in alc, fixed oils; sltly sol in propylene glycol; insol in glycerin and water @ 260°.
SYNS: ALLYL CAPRYLATE □ FEMA No. 2037 □ OCTANOIC ACID ALLYL ESTER □ OCTANOIC ACID-2-PROPENYL ESTER
CONSENSUS REPORTS: Reported in EPA TSCA Inventory.
SAFETY PROFILE: Moderately toxic by ingestion. A skin irritant. See also ALLYL COMPOUNDS and ESTERS. When heated to decomposition it emits acrid smoke and irritating fumes.

AGQ750   CAS: 7493-74-5   HR: 2
ALLYL PHENOXYACETATE
mf: $C_{11}H_{12}O_3$ mw: 192.23
PROP: Colorless to light-yellow liquid; heavy fruit odor.
SYN: ACETATE P.A.
CONSENSUS REPORTS: Reported in EPA TSCA Inventory.
SAFETY PROFILE: Moderately toxic by ingestion and skin contact. When heated to decomposition it emits acrid smoke and irritating fumes. See also ALLYL COMPOUNDS.

AGQ775   CAS: 21905-27-1   HR: 3
ALLYL PHENYL ARSINIC ACID
mf: $C_9H_{11}AsO_2$ mw: 226.12
SYNS: ALLYLHYDROXYPHENYLARSINE OXIDE □ ARSINE OXIDE, ALLYLHYDROXYPHENYL-
OSHA PEL: TWA 0.5 mg(As)/m³
SAFETY PROFILE: Poison by intravenous route. When heated to decomposition it emits toxic fumes of As.

AGR500   CAS: 2179-59-1   HR: 1
ALLYL PROPYL DISULFIDE
mf: $C_6H_{12}S_2$ mw: 148.30
PROP: Liquid with pungent odor. Bp: 66–69° @ 16 mm.
OSHA PEL: TWA 2 ppm; STEL 3 ppm

ACGIH TLV: TWA 2 ppm; STEL 3 ppm; (Proposed: 0.5 ppm (sensitizer))
DFG MAK: 2 ppm (12 mg/m³)
NIOSH REL: (Allyl Propyl Disulfide): TWA 2 ppm; STEL 3 ppm
SAFETY PROFILE: A powerful irritant. Moderately flammable by exposure to heat, flame, or oxidizers. When heated to decomposition it emits highly toxic SO_x. To fight fire, use foam, CO_2, dry chemical. See also ALLYL COMPOUNDS.

AGU250   CAS: 107-37-9   HR: 3
ALLYL TRICHLOROSILANE
DOT: UN 1724
mf: $C_3H_5Cl_3Si$ mw: 175.52
PROP: Colorless liquid; pungent, irritating odor. D: 1.222 @ 20°/4°, bp: 116° @ 750 mm, flash p: 95°F (COC).
SYNS: ALLYLTRICHLOROSILANE, stabilized (DOT) □ SILANE, TRICHLOROALLYL- □ TRICHLOROALLYLSILANE
CONSENSUS REPORTS: Reported in EPA TSCA Inventory.
DOT CLASSIFICATION: 8; Label: Corrosive, Flammable Liquid
SAFETY PROFILE: Poison by intravenous route. Corrosive. See ALLYL COMPOUNDS. When heated to decomposition it emits toxic Cl⁻. A dangerous fire hazard. To fight fire, use foam, mist, spray, dry chemical.

AGW750   CAS: 569-58-4   HR: 1
ALUMINON
mf: $C_{22}H_{23}N_3O_9$ mw: 473.48
PROP: Reddish-brown powder. Sol in H_2O; sltly sol in EtOH; insol in non-polar solvs.
SYNS: AMMONIUM AURINTRICARBOXYLATE □ AURINE-TRICARBOXYLATE d'AMMONIUM (FRENCH) □ AURINTRICARBOXYLIC ACID AMMONIUM SALT □ C.I. MORDANT VIOLET 39, TRIAMMONIUM SALT (8CI) □ LYSOFON □ TRIAMMONIUM AURINTRICARBOXYLATE
CONSENSUS REPORTS: Reported in EPA TSCA Inventory.
SAFETY PROFILE: Human reproductive effects by ingestion: changes in male fertility. Experimental reproductive effects. Mutation data reported. When heated to

decomposition it emits toxic fumes of NH_3 and NO_x.

AGX000   CAS: 7429-90-5   HR: 3
ALUMINUM
DOT: UN 1309/UN 1396/NA 9260
af: Al   aw: 26.98
PROP: Hard, strong, silvery-white ductile metal: in bulk form protected from oxidation in air by coherent Al_2O_3 coating. Mp: 660°, bp: 2494° @ 24 mm, d: 2.702, vap press: 1 mm @ 1284°. Sol in HCl, H_2SO_4, hot water, and alkalies.
SYNS: A 00 □ A 95 □ A 99 □ A 995 □ A 999 □ AA 1099 □ AA1199 □ AD 1 □ AD1M □ ADO □ AE □ ALAUN (GERMAN) □ ALLBRI ALUMINUM PASTE and POWDER □ ALUMINA FIBRE □ ALUMINIUM BRONZE □ ALUMINUM FLAKE □ ALUMINUM 27 □ ALUMINUM A00 □ ALUMINUM DEHYDRATED □ ALUMINUM METAL (OSHA) □ ALUMINUM, molten (NA 9260) (DOT) □ ALUMINUM POWDER □ ALUMINUM POWDER, coated (UN 1309) (DOT) □ ALUMINUM POWDER, uncoated (UN 1396) (DOT) □ ALUMINUM PYRO POWDERS (OSHA) □ ALUMINUM WELDING FUMES (OSHA) □ AO A1 □ AR2 □ AV00 □ AV000 □ C.I. 77000 □ EMANAY ATOMIZED ALUMINUM POWDER □ JISC 3108 □ JISC 3110 □ L16 □ METANA ALUMINUM PASTE □ NORAL ALUMINUM □ NORAL EXTRA FINE LINING GRADE □ NORAL INK GRADE ALUMINUM □ NORAL NON-LEAFING GRADE □ PAP-1
CONSENSUS REPORTS: Community Right-To-Know List (fume or dust). Reported in EPA TSCA Inventory.
OSHA PEL: Total Dust: TWA 15 mg/m³; Respirable Fraction: TWA 5 mg/m³; Pyro Powders and Welding Fumes: 5 mg/m³; Soluble Salts and Alkyls: 2 mg/m³
ACGIH TLV: (Proposed: Metal and Insoluble Compounds 1 mg/m³)
DFG MAK: 1.5 mg/m³; BAT: 200 µg/L in urine at end of shift
DOT CLASSIFICATION: 9; Label: CLASS 9 (NA 9260); DOT Class: 4.1; Label: Flammable Solid (UN 1309); DOT Class: 4.3; Label: Dangerous When Wet (UN 1396)
SAFETY PROFILE: Although aluminum is not generally regarded as an industrial poison, inhalation of finely divided powder has been reported to cause pulmonary fibrosis. It is a reactive metal and the

greatest industrial hazards are with chemical reactions. As with other metals the powder and dust are the most dangerous forms. Dust is moderately flammable and explosive by heat, flame, or chemical reaction with powerful oxidizers. To fight fire, use special mixtures of dry chemical.

Powdered aluminum undergoes the following dangerous interactions: explosive reaction after a delay period with $KClO_4$ + $Ba(NO_3)_2$ + KNO_3 + H_2O, also with $Ba(NO_3)_2$ + KNO_3 + sulfur + vegetable adhesives + H_2O. Mixtures with powdered AgCl, NH_4NO_3 or NH_4NO_3 + $Ca(NO_3)_2$ + formamide + H_2O are powerful explosives. Mixture with ammonium peroxodisulfate + water is explosive. Violent or explosive "thermite" reaction when heated with metal oxides, oxosalts (nitrates, sulfates), or sulfides, and with hot copper oxide worked with an iron or steel tool. Potentially explosive reaction with CCl_4 during ball milling operations. Many violent or explosive reactions with the following halocarbons have occurred in industry: bromomethane, bromotrifluoromethane, CCl_4, chlorodifluoromethane, chloroform, chloromethane, chloromethane + 2-methylpropane, dichlorodifluoromethane, 1,2-dichloroethane, dichloromethane, 1,2-dichloropropane, 1,2-difluorotetrafluoro-ethane, fluorotrichloroethane, hexa-chloroethane + alcohol, polytrifluoro-ethylene oils and greases, tetrachloro-ethylene, tetrafluoromethane, 1,1,1-trichloroethane, trichloroethylene, 1,1,2-trichlorotrifluoro-ethane, and trichlorotrifluoroethane-dichlorobenzene. Potentially explosive reaction with chloroform amidinium nitrate. Ignites on contact with vapors of $AsCl_3$, SCl_2, Se_2Cl_2, and PCl_5. Reacts violently on heating with Sb or As. Ignites on heating in $SbCl_3$ vapor. Ignites on contact with barium peroxide. Potentially violent reaction with sodium acetylide. Mixture with sodium peroxide may ignite or react violently. Spontaneously ignites in CS_2 vapor. Halogens: ignites in

chlorine gas, foil reacts vigorously with liquid Br_2, violent reaction with H_2O + I_2. Violent reaction with hydrochloric acid, hydro-fluoric acid, and hydrogen chloride gas. Violent reaction with disulfur dibromide. Violent reaction with the nonmetals phosphorus, sulfur, and selenium. Violent reaction or ignition with the interhalogens: bromine pentafluoride, chlorine fluoride, iodine chloride, iodine pentafluoride, and iodine heptafluoride. Burns when heated in CO_2. Ignites on contact with O_2, and mixtures with O_2 + H_2O ignite and react violently. Mixture with picric acid + water ignites after a delay period. Explosive reaction above 800°C with sodium sulfate. Violent reaction with sulfur when heated. Exothermic reaction with iron powder + water releases explosive hydrogen gas.

Aluminum powder also forms sensitive explosive mixtures with oxidants such as: liquid Cl_2 and other halogens, N_2O_4, tetranitromethane, bromates, iodates, $NaClO_3$, $KClO_3$, and other chlorates, $NaNO_3$, aqueous nitrates, $KClO_4$ and other perchlorate salts, nitryl fluoride, ammonium peroxodisulfate, sodium peroxide, zinc peroxide, and other peroxides, red phosphorus, and powdered polytetrafluoroethylene (PTFE).

Bulk aluminum may undergo the following dangerous interactions: exothermic reaction with butanol, methanol, 2-propanol, or other alcohols, sodium hydroxide to release explosive hydrogen gas. Reaction with diborane forms pyrophoric product. Ignition on contact with niobium oxide + sulfur. Explosive reaction with molten metal oxides, oxosalts (nitrates, sulfates), sulfides, and sodium carbonate. Reaction with arsenic trioxide + sodium arsenate + sodium hydroxide produces the toxic arsine gas. Violent reaction with chlorine trifluoride. Incandescent reaction with formic acid. Potentially violent alloy formation with palladium, platinum at mp of Al, 600°C. Vigorous dissolution reaction in methanol + carbon tetrachloride. Vigorous amalgamation reaction with mercury(II) salts + moisture. Violent reaction with molten silicon steels. Violent exothermic reaction above 600°C with sodium diuranate.

AGX250   **HR: 1**
ALUMINUM AMMONIUM SULFATE
mf: $Al_2(SO_4)_3(NH_4)_2SO_4 \cdot 24H_2O$ mw: 906
PROP: Colorless crystals; odorless with sweet taste. D: 1.645, mp: 94.5°, bp: loses 20 waters @ 120°. Sol in water, glycerin; insol in alc.
SAFETY PROFILE: Irritating if inhaled or ingested. Upon decomposition it emits toxic fumes of NO_x and SO_x.

AGX750   **CAS: 7727-15-3**   **HR: 2**
ALUMINUM BROMIDE
DOT: UN 1725/UN 2580
mf: $AlBr_3$ mw: 266.71
PROP: White to yellow-red lumps. Mp: 97.5°, bp: 263.3° @ 748 mm, d: 3.2, vap press: 1 mm @ 81.3°.
SYNS: ALUMINUM BROMIDE, anhydrous (UN 1725) (DOT) □ ALUMINUM BROMIDE, solution (UN 2580) (DOT) □ ALUMINUM TRIBROMIDE □ TRIBROMOALUMINUM
CONSENSUS REPORTS: Reported in EPA TSCA Inventory.
ACGIH TLV: TWA 2 mg(Al)/m³
DOT CLASSIFICATION: 8; Label: Corrosive
SAFETY PROFILE: A toxic, corrosive material. See also BROMIDES and ALUMINUM COMPOUNDS. Mixtures with sodium or potassium explode violently upon impact. When heated to decomposition it emits toxic fumes of Br^-. Do not add H_2O to anhydrous material. Hydrolysis can be violent.

AGY750   **CAS: 7446-70-0**   **HR: 3**
ALUMINUM CHLORIDE
DOT: UN 1726/UN 2581
mf: $AlCl_3$ mw: 133.33
PROP: White or colorless hexagonal deliquescent crystals or moisture sensitive plates. D: 2.44, mp: 192° @ 2.5 atm, bp:

subl @ 181°, vap press: 1 mm @ 100.0°. Violently sol in water; sol in alc and ether.
SYNS: ALLUMINIO(CLORURO DI) (ITALIAN) □ ALUMINUMCHLORID (GERMAN) □ ALUMINUM CHLORIDE (1:3) □ ALUMINUM CHLORIDE, anhydrous (DOT) □ ALUMINUM CHLORIDE, solution (DOT) □ ALUMINUM TRICHLORIDE □ CHLORURE d'ALUMINUM (FRENCH) □ PEARSALL □ TRICHLOROALUMINUM
CONSENSUS REPORTS: Reported in EPA TSCA Inventory.
ACGIH TLV: TWA 2 mg(Al)/m³
DOT CLASSIFICATION: 8; Label: Corrosive
SAFETY PROFILE: Moderately toxic by ingestion. Experimental teratogenic and reproductive effects. Mutation data reported. The dust is an irritant by ingestion, inhalation, and skin contact. Highly exothermic polymerization reactions with alkenes. Incompatible with nitrobenzenes or nitrobenzene + phenol. Highly exothermic reaction with water or steam produces toxic fumes of HCl. See also ALUMINUM COMPOUNDS, CHLORIDES, and HYDROCHLORIC ACID.

AHA000   **CAS: 12042-91-0**   **HR: 2**
ALUMINUM CHLORIDE HYDROXIDE
mf: $Al_2ClH_5O_5$ mw: 174.46
SYNS: ALUMINUM CHLORHYDRATE □ ALUMINUM CHLORHYDROL □ ALUMINUM CHLORHYDROXIDE □ ALUMINUM CHLOROHYDROXIDE □ ALUMINUM HYDROXIDE CHLORIDE □ ALUMINUM HYDROXY-CHLORIDE □ ASTRINGEN □ BASIC ALUMINUM CHLORATE □ CHLORHYDROL □ CHLORHYDROL, GRANULAR □ CHLORHYDROL, IMPALPABLE □ CHLOROHYDROL □ CHLOROPENTAHYDROXY-DIALUMINUM □ LOCRON EXTRA □ LOCRON FLAKES □ LOCRON POWDER □ LOCRON SOLUTION □ MACRO-SPHERICAL 95 □ MICRO DRY □ WICKENOL 303 □ WICKENOL 321 □ WICKENOL 323 □ WICKENOL 324
CONSENSUS REPORTS: Reported in EPA TSCA Inventory.
ACGIH TLV: TWA 2 mg(Al)/m³
SAFETY PROFILE: A mild human skin irritant. See also ALUMINUM COMPOUNDS, CHLORIDES, and SODIUM HYDROXIDE. When heated to decomposition it emits toxic fumes of Cl^-.

AHA250   CAS: 7047-84-9   HR: 1
ALUMINUM DEXTRAN

mf: $C_{18}H_{37}AlO_4$ mw: 344.48

PROP: Powder. A complex containing aluminum and dextran, a chain of molecular weight 2500, corresponding to a chain of 15 anhydroglucose units.

SYNS: ALUMINUM MONOSTEARATE □ ALUMINUM STEARATE (ACGIH) □ STEARIC ACID, ALUMINIUM SALT

CONSENSUS REPORTS: EPA TSCA Chemical Inventory.

ACGIH TLV: (Proposed: Metal and Insoluble Compounds 1 mg/m³)

SAFETY PROFILE: A nuisance dust. When heated to decomposition it emits acrid smoke and fumes. See also ALUMINUM COMPOUNDS.

AHB000   CAS: 7784-18-1   HR: 3
ALUMINUM FLUORIDE

mf: AlF_3 mw: 83.98

PROP: Solid, colorless crystals. Mp: 1291°, subl @ 1260°, d: 2.88, vap press: 1 mm @ 1238°, bp: 1537°. Sparingly sol in water; insol in org solvs.

SYNS: ALUMINUM FLUORURE (FRENCH) □ ALUMINUM TRIFLUORIDE □ FLUORID HLINITY (CZECH)

CONSENSUS REPORTS: Reported in EPA TSCA Inventory.

OSHA PEL: TWA 2.5 mg(F)/m³

ACGIH TLV: TWA 2.5 mg(F)/m³; BEI: 3 mg/g creatinine of fluorides in urine prior to shift; 10 mg/g creatinine of fluorides in urine at end of shift; TWA 2 mg(Al)/m³

NIOSH REL: (Fluorides, Inorganic) TWA 2.5 mg(F)/m³

SAFETY PROFILE: A poison by ingestion. Moderately toxic by subcutaneous route. A severe eye irritant. Violently impact-sensitive when in contact with Na and K. When heated to decomposition it emits toxic fumes of F⁻. See also FLUORIDES and ALUMINUM COMPOUNDS.

AHB500   CAS: 7784-21-6   HR: 3
ALUMINUM HYDRIDE

DOT: UN 2463

mf: AlH_3 mw: 30.01

PROP: Colorless powder or white amorph or microcrystal air- and moisture-sensitive solid. Usually made in THF or Et_2O and used *in situ*. Several solid forms exist. Solids $AlH_3 \bullet nEt_2O$, $n = 0.3–3$ obtained from Et_2O. Sol in Et_2O, THF.

SYNS: ALANE □ ALUMINUM TRIHYDRIDE □ α-ALUMINUM TRIHYDRIDE

CONSENSUS REPORTS: Reported in EPA TSCA Inventory.

ACGIH TLV: TWA 2 mg(Al)/m³

DOT CLASSIFICATION: 4.3; Label: Dangerous When Wet

SAFETY PROFILE: Hydrides of some metals (such as AsH_3) are extremely toxic. Dangerous fire hazard. An unstable material which is spontaneously flammable in air or O_2. Evolves explosive H_2 upon contact with moisture. Severe explosion hazard by chemical reaction wherein H_2 gas is produced, also in contact with methyl ethers contaminated by CO_2. Mixtures with tetrazole derivatives are explosive. Reacts with oxidizing materials. On contact with acid or acid fumes, it can emit toxic fumes. See also HYDRIDES and ALUMINUM COMPOUNDS.

AHC000   CAS: 21645-51-2   HR: 3
ALUMINUM HYDROXIDE

mf: AlH_3O_3 mw: 78.01

PROP: White, crystalline powder, balls, or granules. Solid from water. D: 2.42, mp: loses H_2O @ 300°. Practically insol in water; sol in mineral acids, alkalies, and caustic soda.

SYNS: AF 260 □ ALCOA 331 □ ALUMIGEL □ ALUMINA HYDRATE □ ALUMINA HYDRATED □ ALUMINA TRIHYDRATE □ α-ALUMINA TRIHYDRATE □ ALUMINIC ACID □ ALUMINUM HYDRATE □ ALUMINUM(III) HYDROXIDE □ ALUMINUM HYDROXIDE GEL □ ALUMINUM OXIDE HYDRATE □ ALUMINUM OXIDE TRIHYDRATE □ ALUMINUM TRIHYDRAT □ ALUMINUM TRIHYDROXIDE □ ALUSAL □ AMBEROL ST 140F □ AMPHOJEL □ BACO AF 260 □ BRITISH ALUMINUM AF 260 □ C.I. 77002 □ GHA 331 □ H 46 □ HIGILITE □ HYDRAL 705 □ LIQUIGEL □ PGA □ TRIHYDRATED ALUMINA

A

CONSENSUS REPORTS: Reported in EPA TSCA Inventory.
ACGIH TLV: TWA 2 mg(Al)/m^3
DFG MAK: 1.5 mg/m^3
SAFETY PROFILE: Poison by intraperitoneal route. Human systemic effects by ingestion: fever, osteomalacia, and gastrointestinal effects. When coprecipitated with bismuth hydroxide and reduced by H$_2$, it is violently flammable in air. Incompatible with chlorinated rubber.

AHC250   CAS: 24623-77-6   HR: 3
ALUMINUM HYDROXIDE OXIDE
mf: AlHO$_2$ mw: 59.99
PROP: White solid. Sol in hot NaOH.
SYNS: ALUMINUM METAHYDROXIDE □ HYDRATED ALUMINA
CONSENSUS REPORTS: Reported in EPA TSCA Inventory.
ACGIH TLV: TWA 2 mg(Al)/m^3
SAFETY PROFILE: Poison by intratracheal route. See also ALUMINUM COMPOUNDS.

AHC750   CAS: 18917-91-4   HR: D
ALUMINUM LACTATE
mf: C$_9$H$_{15}$AlO$_9$ mw: 294.22
SYNS: ALUMINUM, TRIS(2-HYDROXY-PROPANOATO-O^1),O^2- (9CI) □ ALUMINUM, TRIS(LACTATO)-
CONSENSUS REPORTS: Reported in EPA TSCA Inventory.
SAFETY PROFILE: An experimental teratogen. Other experimental reproductive effects. When heated to decomposition it emits acrid smoke and irritating fumes.

AHD250   HR: 3
ALUMINUM MAGNESIUM PHOSPHIDE
DOT: UN 1419
mf: Mg$_3$AlP$_3$ mw: 192.8
SYN: MAGNESIUM ALUMINUM PHOSPHIDE (DOT)
ACGIH TLV: TWA 2 mg(Al)/m^3
DOT CLASSIFICATION: 4.3; Label: Dangerous When Wet, Poison
SAFETY PROFILE: A poison. Dangerous fire hazard. Evolves spontaneously flammable PH$_3$ in contact with water. See also PHOSPHIDES, PHOSPHINE,

ALUMINUM COMPOUNDS, and MAGNESIUM COMPOUNDS.

AHD650   CAS: 1976-28-9   HR: D
ALUMINUM NICOTINATE
mf: C$_{18}$H$_{12}$AlN$_3$O$_6$ mw: 393.30
PROP: Solid.
SYNS: ALUNITINE □ MICALEX □ NICOTINIC ACID, ALUMINUM SALT □ 3-PYRIDINECARBOXYLIC ACID, ALUMINUM SALT □ TRIS(NICTINATO)ALUMINUM
SAFETY PROFILE: When heated to decomposition it emits toxic fumes of NO$_x$.

AHD750   CAS: 13473-90-0   HR: 3
ALUMINUM(III) NITRATE (1:3)
DOT: UN 1438
mf: N$_3$O$_9$•Al   mw: 213.01
PROP: White crystals or very hygroscopic solid. Bp: 50° @ 0.01 mm.
SYNS: ALUMINUM NITRATE (DOT) □ ALUMINUM TRINITRATE □ NITRIC ACID, ALUMINUM SALT □ NITRIC ACID, ALUMINUM(3+) SALT
CONSENSUS REPORTS: Reported in EPA TSCA Inventory.
ACGIH TLV: TWA 2 mg(Al)/m^3
DOT CLASSIFICATION: 5.1; Label: Oxidizer
SAFETY PROFILE: A poison. A severe eye and mild skin irritant. A powerful oxidizer. When heated to decomposition it emits toxic NO$_x$. See NITRATES and ALUMINUM COMPOUNDS. A nitrating agent.

AHE250   CAS: 1344-28-1   HR: 3
ALUMINUM OXIDE (2:3)
mf: Al$_2$O$_3$ mw: 101.96
PROP: White powder or solid. Mp: 2050°, bp: 2977°, d: 3.5–4.0, vap press: 1 mm @ 2158°. Sol in hot NaOH.
SYNS: A 1 (sorbent) □ A1-0109 P □ ABRAREX □ ACTIVATED ALUMINUM OXIDE □ ALCOA F 1 □ ALMITE □ ALON □ ALUMINA □ α-ALUMINA (OSHA) □ β-ALUMINA □ γ-ALUMINA □ ALUMINUM OXIDE □ α-ALUMINUM OXIDE □ β-ALUMINUM OXIDE □ γ-ALUMINUM OXIDE □ ALUMINUM SESQUIOXIDE □ ALUMITE □ ALUNDUM □ BROCKMANN, ALUMINUM OXIDE □ CAB-O-GRIP □ COMPALOX □ DIALUMINUM TRIOXIDE □ DISPAL □ DOTMENT 324 □ FASERTON □ G 2 (OXIDE) □ KHP 2 □ LUCALOX □ MICROGRIT WCA □ PS 1 □ RC 172DBM

CONSENSUS REPORTS: Community Right-To-Know List. Reported in EPA TSCA Inventory.
OSHA PEL: Total Dust: TWA 10 mg/m^3; (Proposed: 1 mg/m^3); Respirable Fraction: TWA 5 mg/m^3
ACGIH TLV: (Proposed: Metal and Insoluble Compounds 1 mg/m^3)
DFG MAK: Animal Carcinogen, Suspected Human Carcinogen; 1.5 mg/m^3
SAFETY PROFILE: Suspected carcinogen with experimental neoplastigenic and tumorigenic data by implantation. Inhalation of finely divided particles may cause lung damage (Shaver's disease). Exothermic reaction above 200°C with halocarbon vapors produces toxic HCl and phosgene. See also ALUMINUM COMPOUNDS.

AHE750   CAS: 20859-73-8   HR: 3
ALUMINUM PHOSPHIDE
DOT: UN 1397
mf: AlP   mw: 57.95
PROP: Dark-gray or dark-yellow moisture-sensitive crystals stable up to 10°. D: 2.85 @ 25°/4°, mp: >1000°.
SYNS: AIP □ AL-PHOS □ ALUMINUM FOSFIDE (DUTCH) □ ALUMINUM MONOPHOSPHIDE □ CELPHIDE □ CELPHOS □ DELICIA □ DETIA GAS EX-B □ FOSFURI di ALLUMINIO (ITALIAN) □ FUMITOXIN □ PHOSPHURES d'ALUMINUM (FRENCH) □ PHOSTOXIN □ QUICKPHOS □ RCRA WASTE NUMBER P006
CONSENSUS REPORTS: EPA Extremely Hazardous Substances List. Reported in EPA TSCA Inventory.
ACGIH TLV: TWA 2 mg(Al)/m^3
DOT CLASSIFICATION: 4.3; Label: Dangerous When Wet, Poison
SAFETY PROFILE: A human poison by inhalation and ingestion. Dangerous; in contact with water, steam, or alkali it slowly yields PH$_3$, which is spontaneously flammable in air. Explosive reaction on contact with mineral acids produces phosphine. When heated to decomposition it yields toxic PO$_x$. See also ALUMINUM COMPOUNDS, PHOSPHIDES, and PHOSPHINE.

AHG000   CAS: 11138-49-1   HR: 2
ALUMINUM SODIUM OXIDE
DOT: UN 1819/UN 2812
mf: NaAlO$_2$ mw: 82.0
PROP: White, hygroscopic powder. Mp: 1650°. Sol in water; insol in alc, strongly alkaline aqueous solutions.
SYNS: β-ALUMINA □ β"-ALUMINA □ J 242 □ NALCO 680 □ SODIUM ALUMINATE, solid (UN 2812) (DOT) □ SODIUM ALUMINATE, solution (UN 1819) (DOT) □ SODIUM ALUMINUM OXIDE □ SODIUM POLYALUMINATE
ACGIH TLV: TWA 2 mg(Al)/m^3
DOT CLASSIFICATION: 4.3; Label: Dangerous When Wet
SAFETY PROFILE: Moderate irritant to skin, eyes, and mucous membranes. A corrosive substance. When heated to decomposition it emits toxic fumes of Na$_2$O.

AHG500   HR: 2
ALUMINUM SODIUM SULFATE
mf: NaAl (SO$_4$)$_2$•12H$_2$O   mw: 458.29
PROP: Colorless crystals. Mp: 61°; d: 1.675. Anhydrous: sol in alc; sltly sol in water. Dodecahydrate: sol in water and alc.
SYNS: SODA ALUM □ SODIUM ALUMINUM SULFATE
SAFETY PROFILE: A weak sensitizer. A general-purpose food additive. Local contact may cause contact dermatitis. An irritant. See also SULFATES and ALUMINUM COMPOUNDS. When heated to decomposition it emits toxic fumes of SO$_x$ and Na$_2$O.

AHG750   CAS: 10043-01-3   HR: 2
ALUMINUM SULFATE (2:3)
mf: O$_{12}$S$_3$•2Al   mw: 342.14
PROP: White powder; sweet taste. Mp: decomp @ 770°, d: 2.71. Solubility in water 36.4% @ 20°.
SYNS: ALUM □ ALUMINUM ALUM □ ALUMINUM SULFATE □ ALUMINUM SULPHATE □ ALUMINUM TRISULFATE □ CAKE ALUM □ DIALUMINUM SULPHATE □ DIALUMINUM TRISULFATE □ SULFURIC ACID, ALUMINUM SALT (3:2)
CONSENSUS REPORTS: Reported in EPA TSCA Inventory. EPA FIFRA 1988

pesticide subject to registration or re-registration.

ACGIH TLV: TWA 2 mg(Al)/m^3
SAFETY PROFILE: Moderately toxic by ingestion and intraperitoneal routes. Experimental reproductive effects. Human mutation data reported. Hydrolyzes to form sulfuric acid, which irritates tissue, especially lungs. When heated to decomposition it emits toxic fumes of SO_x.

AHG875   CAS: 16962-07-5   HR: 3
ALUMINUM TETRAHYDROBORATE
DOT: UN 2870
mf: Al•3BH$_4$ mw: 71.53
SYNS: ALUMINUM BOROHYDRIDE (DOT) □ ALUMINUM BOROHYDRIDE in devices (DOT) □ ALUMINUM HYDROBORATE □ BORATE(1-), TETRAHYDRO-, ALUMINUM (3:1) (9CI)
ACGIH TLV: TWA 2 mg(Al)/m^3
DOT CLASSIFICATION: 4.2; Label: Spontaneously Combustible, Dangerous When Wet
SAFETY PROFILE: A poison. Spontaneously flammable in air. Explodes in oxygen with traces of water. Incompatible with alkenes and water. See also ALUMINUM COMPOUNDS and BORON COMPOUNDS.

AHJ750   CAS: 52645-53-1   HR: 3
AMBUSH
mf: C$_{21}$H$_{20}$Cl$_2$O$_3$ mw: 391.31
PROP: Crystals or pale-yellow viscous liquid. Bp: 220° @ 0.05 mm.
SYNS: AI3-29158 □ BW-21-Z □ ECTIBAN □ EXMIN □ FMC 33297 □ FMC 41655 □ ICI-PP 557 □ KESTREL (Pesticide) □ NDRC-143 □ NIA 33297 □ OUTFLANK □ OUTFLANK-STOCKADE □ PERMETHRIN (USDA) □ PERMETRIN (HUNGARIAN) □ PERMETRINA (PORT-UGUESE) □ 3-PHENOXYBENZYL (±)-3-(2,2-DICHLORO-VINYL)-2,2-DIMETHYLCYCLOPRO-PANECARBOXYL-ATE □ (3-PHENOXYPHENYL)-METHYL-3-(2,2-DICHLO-RETHENYL)-2,2-DIMETHYLCYCLOPROPANECARBO-XYLATE □ POUNCE □ PP 557 □ S-3151 □ SBP-1513 □ TALCORD □ WL 43479
SAFETY PROFILE: Poison by inhalation, intravenous, and intracerebral routes. Moderately toxic by ingestion. Experimental reproductive effects. Mutation data reported. A skin irritant. When heated to

decomposition it emits toxic fumes of Cl$^-$. See also ESTERS.

AHK500   CAS: 50-07-7   HR: 3
AMETYCIN
mf: C$_{15}$H$_{18}$N$_4$O$_5$ mw: 334.37
PROP: Deep blue-violet crystals. Sol in H$_2$O, MeOH, Me$_2$CO.
SYNS: 7-AMINO-9-α-METHOXYMITOSANE □ MIT-C □ MITO-C □ MITOCIN-C □ MITOMYCIN □ MITOMYCIN-C □ MITOMYCINUM □ MMC □ MUTAMYCIN □ MUTAMYCIN (MITOMYCIN for INJECTION) □ MYTOMYCIN □ NCI-C04706 □ NSC-26980 □ RCRA WASTE NUMBER U010
CONSENSUS REPORTS: IARC Cancer Review: Group 2B IMEMDT 7,56,87; Animal Sufficient Evidence IMEMDT 10,171,76; NCI Carcinogenesis Studies (ipr); Clear Evidence: rat RRCRBU 52,1,75; No Evidence: mouse RRCRBU 52,1,75. EPA Extremely Hazardous Substances List. EPA Genetic Toxicology Program. Reported in EPA TSCA Inventory.
SAFETY PROFILE: Confirmed carcinogen with experimental carcinogenic and neoplastigenic data. Poison by ingestion, subcutaneous, intravenous, and intraperitoneal routes. Human systemic effects by intravenous route: dyspnea and lung fibrosis, hemolysis with or without anemia, changes in tubules (including acute renal failure, acute tubular necrosis), normocytic anemia. Experimental teratogenic and reproductive effects. Human mutation data reported. When heated to decomposition it emits toxic fumes of NO_x. See also CARBAMATES and ESTERS.

AHL750   HR: D
AMIDES
PROP: Organic compounds containing the structural group $-CONH_2$, and closely related to the organic acids with the grouping $-COOH$. Common examples are: acetamide (CH_3CONH_2) and urea ($CO(NH_2)_2$).
SAFETY PROFILE: Most of the saturated amides have low toxicity, but the unsaturated and N-substituted amides are

irritants and may be absorbed via skin contact. Can cause injury to the liver, kidney, and brain.

AHO250   CAS: 57-67-0   HR: 2
N¹-AMIDINOSULFANILAMIDE

mf: $C_7H_{10}N_4O_2S$ mw: 214.27

PROP: Colorless, monoclinic crystals or needles from aqueous solns. Mp: 189–190°. Sltly sol in alc; insol in ether; solubility in water = 0.19% @ 37°.

SYNS: ABIGUANIL □ 4-AMINO-N-(AMINOIMINO-METHYL)BENZENESULFONAMIDE □ p-AMINO-BENZENESULFONYLGUANIDINE □ N-p-AMINOBENZENESULPHONYLGUANIDINE MONOHYDRATE □ 4-AMINO-N-(DIAMINOMETHYL-ENE)BENZENESULFONAMIDE □ ATERIAN □ BENZENESULFONAMIDE, 4-AMINO-N-(DIAMINOMETHYLENE)- □ N₁-(DIAMINO-METHYLENE)SULFANILAMIDE □ GANIDAN □ GUAMIDE □ GUANICIL □ GUANIDAN □ GUANIDINE, SULFANILYL- □ N₁-GUANYL-SULF-ANILAMIDE □ RESULFON □ RP 2275 □ RUOCID □ SHIGATOX □ SUGANYL □ SULFAGUANIDINE □ SULFAGUINE □ SULFANILGUANIDINE □ SULF-ANILYLGUANIDINE □ SULFOGUANIDINE □ SULFOGUENIL □ SULFOQUANIDINE □ SULGIN □ SULPHA-GUANIDINE

CONSENSUS REPORTS: Reported in EPA TSCA Inventory.

SAFETY PROFILE: Moderately toxic by intraperitoneal route. An experimental teratogen. Other experimental reproductive effects. See also SULFONATES. When heated to decomposition it emits very toxic fumes of SO_x and NO_x.

AHP750   HR: D
AMINES

PROP: A large group of organic compounds containing nitrogen and considered as derived from ammonia (NH_3) by replacement of one or more hydrogen atoms by an organic radical. See also specific compounds.

SAFETY PROFILE: Variable toxicity; some are poisons, some are only slightly toxic. Many are skin irritants and some are sensitizers. See also AROMATIC AMINES and AMINES, FATTY.

AHR240   CAS: 99-92-3   HR: 3
p-AMINOACETOPHENONE

mf: C_8H_9NO mw: 135.18

PROP: Crystalline. Mp: 106°, bp: 293–295°. Sol in C_6H_6, alc, hot acids, and ether. Insol in water.

SYNS: 4-ACETYLANILINE □ 4'-AMINOACETO-PHENONE □ p-AMINOACETYLBENZENE □ USAF EK-631

CONSENSUS REPORTS: Reported in EPA TSCA Inventory.

SAFETY PROFILE: Poison by ingestion and intraperitoneal routes. When heated to decomposition it emits toxic fumes of NO_x. See also AROMATIC AMINES.

AHR500   CAS: 99-03-6   HR: 2
3'-AMINOACETOPHENONE

mf: C_8H_9NO mw: 135.18

PROP: Yellow, oily liquid or pale-yellow plates from alc. Bp: 251° (slt decomp), mp: 97–99°. Insol in water; sol in alc and ether.

SYNS: m-ACETYLANILINE □ 3-ACETYLANILINE □ β-AMINOACETOPHENONE □ m-AMINOACETO-PHENONE □ m-AMINOACETYLBENZENE

CONSENSUS REPORTS: Reported in EPA TSCA Inventory.

SAFETY PROFILE: Moderately toxic by ingestion. Mildly toxic by skin contact. A skin and eye irritant. Mutation data reported. When heated to decomposition it emits toxic fumes of NO_x. See also AROMATIC AMINES.

AIA750   CAS: 82-45-1   HR: 2
1-AMINOANTHRAQUINONE

mf: $C_{14}H_9NO_2$ mw: 223.24

PROP: Orange to red needles. Mp: 256°, bp: sublimes. Insol in water; sol in HCl, alc, benzene, ether, and chloroform.

SYNS: 1-AMINO-9,10-ANTHRACENEDIONE □ 1-AMINOANTHRACHINON (CZECH) □ α-AMINOANTHRAQUINONE □ 1-AMINO-9,10-ANTHRAQUINONE □ α-ANTHRAQUINONYL-AMINE □ C.I. 37275 □ DIAZO FAST RED AL

CONSENSUS REPORTS: Reported in EPA TSCA Inventory.

SAFETY PROFILE: Moderately toxic by intraperitoneal route. An eye irritant. Questionable carcinogen with experimental tumorigenic data. Mutation data reported. When heated to decomposition it emits toxic NO_x. See also AMINES.

AIB000   CAS: 117-79-3   HR: 3
2-AMINOANTHRAQUINONE
mf: $C_{14}H_9NO_2$ mw: 223.24

PROP: Red needles from alc. Mp: 302–306°, bp: sublimes. Insol in water and ether; sol in alc and benzene.

SYNS: 2-AMINO-9,10-ANTHRACENEDIONE □ 2-AMINO-9,10-ANTHRAQUINONE □ β-AMINO-ANTHRAQUINONE □ β-ANTHRAQUINONYL-AMINE □ NCI-C01876

CONSENSUS REPORTS: NTP 10th Report on Carcinogens. IARC Cancer Review: Group 3 IMEMDT 7,56,87; Animal Limited Evidence IMEMDT 27,191,82. NCI Carcinogenesis Bioassay (feed); Clear Evidence: mouse, rat NCITR* NCI-CG-TR-144,78. Community Right-To-Know List. Reported in EPA TSCA Inventory.

SAFETY PROFILE: Confirmed carcinogen with experimental carcinogenic, neoplastigenic, and tumorigenic data. Moderately toxic via intraperitoneal route. Mutation data reported. When heated to decomposition it emits toxic NO_x. See also AMINES.

AIC250   CAS: 97-56-3   HR: 3
2-AMINO-5-AZOTOLUENE
mf: $C_{14}H_{15}N_3$ mw: 225.32

PROP: Yellow leaflets from alc. Mp: 100°.

SYNS: AAT □ o-AAT □ o-AMIDOAZOTOLUOL (GERMAN) □ AMINOAZOTOLUENE (indicator) □ o-AMINOAZOTOLUENE (MAK) □ 4'-AMINO-2,3'-AZOTOLUENE □ 4'-AMINO-2:3'-AZOTOLUENE □ o-AMINOAZOTOLUENO (SPANISH) □ o-AMINO-AZOTOLUOL □ 4-AMINO-2',3-DIMETHYL-AZOBENZENE □ 4'-AMINO-2,3'-DIMETHYL-AZOBENZENE □ o-AT □ BRASILAZINA OIL YELLOW R □ BUTTER YELLOW □ C.I. 11160 □ C.I. 11160B □ C.I. SOLVENT YELLOW 3 □ 2',3-DIMETHYL-4-AMINOAZOBENZENE □ FAST GARNET GBC BASE □ FAST OIL YELLOW □ FAST YELLOW AT □ FAST YELLOW B □ HIDACO OIL YELLOW □ 2-METHYL-4-((2-METHYLPHENYL)-AZO)BENZENAMINE □ OAAT □ OIL YELLOW □ OIL YELLOW 21 □ OIL YELLOW 2681 □ OIL YELLOW A □ OIL YELLOW AT □ OIL YELLOW C □ OIL YELLOW I □ OIL YELLOW 2R □ OIL YELLOW T

□ ORGANOL YELLOW 25 □ SOMALIA YELLOW R □ SUDAN YELLOW RRA □ o-TOLUENEAZO-o-TOLUIDINE □ o-TOLUOL-AZO-o-TOLUIDIN (GERMAN) □ 5-(o-TOLYLAZO)-2-AMINOTOLUENE □ 4-(o-TOLYLAZO)-o-TOLUIDINE □ TULABASE FAST GARNET GB □ TULABASE FAST GARNET GBC □ WAXAKOL YELLOW NL

CONSENSUS REPORTS: NTP 10th Report on Carcinogens. IARC Cancer Review: Group 2B IMEMDT 7,56,87; Animal Sufficient Evidence IMEMDT 8,61,75. Community Right-To-Know List. Reported in EPA TSCA Inventory. EPA Genetic Toxicology Program.

DFG MAK: Animal Carcinogen; Suspected Human Carcinogen

SAFETY PROFILE: Confirmed carcinogen with experimental carcinogenic, neoplastigenic, and tumorigenic data. Poison by ingestion. Moderately toxic by subcutaneous route. An experimental teratogen. Human mutation data reported. When heated to decomposition it emits toxic fumes of NO_x. See also AROMATIC AMINES.

AID500   CAS: 98-16-8   HR: 3
m-AMINOBENZAL FLUORIDE
DOT: UN 2948

mf: $C_7H_6F_3N$ mw: 161.14

PROP: Colorless liquid with aniline-like odor. Mp: 3°, bp: 189°, d: 1.303 @ 15.5°/15.5°, vap d: 5.56.

SYNS: m-AMINOBENZOTRIFLUORIDE □ 3-AMINOBENZOTRIFLUORIDE □ m-(TRIFLUO-ROMETHYL)ANILINE □ 3-(TRIFLUORO-METHYL)ANILINE □ 3-(TRIFLUOROMETHYL)-BENZENAMINE □ USAF MA-4

CONSENSUS REPORTS: EPA Extremely Hazardous Substances List. Reported in EPA TSCA Inventory.

DOT CLASSIFICATION: 6.1; Label: Poison

SAFETY PROFILE: Poison by inhalation, ingestion, and intraperitoneal routes. May be moderately toxic by other routes. See also AMINES and FLUORIDES. When heated to decomposition it emits very toxic fumes of F^- and NO_x.

AIF500   CAS: 137-07-5   HR: 3
2-AMINOBENZENETHIOL
mf: C_6H_7NS mw: 125.20
PROP: Needles or liquid. Mp: 23–26°, bp: 227.2°, flash p: 175°F, d: 1.168, vap d: 4.3.
SYNS: o-AMINOTHIOPHENOL □ 2-AMINO-THIOPHENOL □ o-MERCAPTOANILINE □ USAF EK-4376
CONSENSUS REPORTS: Reported in EPA TSCA Inventory.
SAFETY PROFILE: Poison by intraperitoneal route. Moderately toxic by ingestion. Moderately flammable. Can react with oxidizing materials. To fight fire, use water, foam, CO_2, mist or spray, dry chemical.

AIG000   CAS: 934-32-7   HR: 3
2-AMINOBENZIMIDAZOLE
mf: $C_7H_7N_3$ mw: 133.17
PROP: Aqueous leaflets or plates from water. Mp: 222–224°. Sol in water, alkalies, alc, acetone; very sltly sol in ether.
SYN: USAF EK-4037
CONSENSUS REPORTS: Reported in EPA TSCA Inventory.
SAFETY PROFILE: Poison by intravenous and intraperitoneal routes. Moderately toxic by ingestion. An experimental teratogen. Mutation data reported. When heated to decomposition it emits toxic NO_x. See also AROMATIC AMINES.

AIH000   CAS: 52329-60-9   HR: 3
2-AMINO-6-BENZIMIDAZOLYL
    PHENYLKETONE
mf: $C_{14}H_{11}N_3O$ mw: 237.28
SYNS: 2-AMINO-5-BENZOYLBENZIMIDAZOLE □ BENZIMIDAZOLE, 2-AMINO-5-BENZOYL- □ G-1029 □ KETONE, 2-AMINO-5-BENZIMIDAZOLYL PHENYL □ METHANONE, (2-AMINO-1H-BENZIMIDAZOL-5-YL)PHENYL- □ R 18986
DOT CLASSIFICATION: 3; Label: Flammable Liquid
SAFETY PROFILE: Acute poison by intraperitoneal route. Moderately toxic by ingestion. An experimental teratogen. See also KETONES. A flammable liquid. When heated to decomposition it emits very toxic fumes of NO_x.

AIH600   CAS: 150-13-0   HR: 2
p-AMINOBENZOIC ACID
mf: $C_7H_7NO_2$ mw: 137.15
PROP: Yellowish to red crystals or prisms. D: 1.374, mp: 187–188.5°. Sol in water, alkalies, EtOH, Et_2O, C_6H_6.
SYNS: ACIDO p-AMINOBENZOICO □ AMBEN □ AMINOBENZOIC ACID □ γ-AMINOBENZOIC ACID □ 4-AMINOBENZOIC ACID □ 1-AMINO-4-CARBOXY-BENZENE □ ANTICANITIC VITAMIN □ ANTI-CHROMOTRICHIA FACTOR □ BACL VITAMIN H1 □ BENZOIC ACID, 4-AMINO- □ p-CARBOXY-ANILINE □ 4-CARBOXYANILINE □ p-CARBOXY-PHENYLAMINE □ CHROMOTRICHIA FACTOR □ anti-CHROMOTRICHIA FACTOR □ KYSELINA p-AMINOBENZOOVA □ PABA □ PABANOL □ PARAMINOL □ PARANATE □ SUNBRELLA □ TRICHOCHROMOGENIC FACTOR □ VITAMIN BX □ VITAMIN H
CONSENSUS REPORTS: IARC Cancer Review: Group 3 IMEMDT 7,56,87; Animal Inadequate Evidence IMEMDT 16,249,78. Reported in EPA TSCA Inventory.
SAFETY PROFILE: Moderately toxic by ingestion and intravenous routes. Ingesting large doses can cause nausea, vomiting, skin rash, methemoglobinemia, and possibly toxic hepatitis. Experimental reproductive effects. Mutation data reported. Questionable carcinogen. Combustible. When heated to decomposition it emits toxic fumes of NO_x. A topical sunscreen.

AIR250   CAS: 1137-41-3   HR: 3
p-AMINOBENZOPHENONE
mf: $C_{13}H_{11}NO$ mw: 197.25
PROP: Leaflets from alc. Mp: 124°, bp: 246 @ 13 mm. Very sltly sol in cold water; very sol in alc.
SYN: USAF A-233
CONSENSUS REPORTS: Reported in EPA TSCA Inventory.
SAFETY PROFILE: Poison by intraperitoneal route. See also KETONES and AMINES. When heated to decomposition it emits toxic fumes of NO_x.

AIT250   CAS: 51-05-8   HR: 3
p-AMINOBENZOYLDIETHYLAMINO-
    ETHANOL HYDROCHLORIDE
mf: $C_{13}H_{20}N_2O_2 \cdot ClH$ mw: 272.81

PROP: Mono- or triclinic, six-sided plates. Mp: 153–156°. Sol in H_2O, EtOH; insol in Et_2O.

SYNS: ALLOCAINE □ p-AMINOBENZOIC ACID-2-DIETHYLAMINOETHYL ESTER, HYDROCHLORIDE □ 4-AMINOBENZOIC ACID 2-(DIETHYLAMINO)-ETHYL ESTER, HYDROCHLORIDE □ AMINOCAINE □ ANADOLOR □ ANESTHESOL □ ANESTIL □ ATOXICOCAINE □ BERNOCAINE □ CETAIN □ CHLOROCAINE □ DIETHYLAMINOETHANOL-4-AMINOBENZOATE HYDROCHLORIDE □ 2-DIETHYLAMINOETHYL-p-AMINOBENZOATE HYDROCHLORIDE □ DUGERASE □ ETHOCAINE □ IROCAINE □ ISOCAINE-ASID □ ISOCAINE-HEISLER □ JUVOCAINE □ KEROCAINE □ LACTOCAINE □ NAUCAINE □ NEOCAINE □ NOVOCAIN-CHLORHYDRAT (GERMAN) □ NOVOCAINE HYDROCHLORIDE □ NOVOCAIN HYDROCHLORID (GERMAN) □ PARACAIN □ PLANOCAINE □ PROCAINE HYDROCHLORIDE □ SCUROCAINE □ SEVICAINE □ SYNCAINE □ TOPOKAIN □ WESTOCAINE

CONSENSUS REPORTS: Reported in EPA TSCA Inventory. EPA Genetic Toxicology Program.

SAFETY PROFILE: Poison by ingestion, subcutaneous, intravenous, and intraperitoneal routes. Human systemic effects: acute renal failure. May have human reproductive effects. See also ESTERS. When heated to decomposition it emits very toxic fumes of HCl and NO_x. Used as a local anesthetic.

AIV500   CAS: 69-53-4   HR: 3
AMINOBENZYLPENICILLIN
mf: $C_{16}H_{19}N_3O_4S$ mw: 349.44
PROP: Solid. Mp: 199–202° (decomp).
SYNS: ACILLIN □ ADOBACILLIN □ ALPEN □ AMBLOSIN □ AMCILL □ AMFIPEN □ d-(−)-α-AMINOBENZYLPENICILLIN □ d-(−)-α-AMINOPENICILLIN □ 6-(d(−)-α-AMINOPHENYL-ACETAMIDO)PENICILLANIC ACID □ (AMINO-PHENYLMETHYL)-PENICILLIN □ AMIPENIX S □ AMPERIL □ AMPI-BOL □ AMPICILLIN (USDA) □ d-AMPICILLIN □ d-(−)-AMPICILLIN □ AMPICILLIN A □ AMPICILLIN ACID □ AMPICILLIN ANHYDRATE □ AMPICIN □ AMPIKEL □ AMPIMED □ AMPIPEN-IN □ AMPLISOM □ AMPLITAL □ AMPY-PENYL □ AUSTRAPEN □ AY-6108 □ BINOTAL □ BON-APICILLIN □ BRITACIL □ BRL □ BRL 1341 □ COPHARCILIN □ CYMBI □ DIVERCILLIN □ DOKTACILLIN □ GRAMPENIL □ GUICITRINA □ GUICITRINE □

LIFEAMPIL □ MARISILAN □ NSC-528986 □ NUVAPEN □ OMNIPEN □ P-50 □ PENBRISTOL □ PENBRITIN □ PENBRITIN PAEDIATRIC □ PENBRITIN SYRUP □ PENBROCK □ PENICLINE □ PENTREX □ PENTREXL □ PFIZERPEN A □ POLYCILLIN □ PONECIL □ PRINCIPEN □ QIDAMP □ RO-AMPEN □ SEMICILLIN □ SK-AMPICILLIN □ SYNPENIN □ TOKIOCILLIN □ TOLOMOL □ TOTACILLIN □ TOTALCICLINA □ TOTAPEN □ ULTRABION □ ULTRABRON □ VICCILLIN □ VICCILLIN S □ VICILLIN □ WY-5103

CONSENSUS REPORTS: EPA Genetic Toxicology Program. IARC Cancer Review: Group 3 IMEMDT 50,153,90; Animal Limited Evidence IMEMDT 50,153,90; Human Inadequate Evidence IMEMDT 50,153,90.

SAFETY PROFILE: Poison by intracerebral route. Moderately toxic by intraperitoneal route. Human systemic effects by ingestion: fever, agranulocytosis, and other blood effects. An experimental teratogen. Mutation data reported. Questionable carcinogen. When heated to decomposition it emits very toxic fumes of NO_x and SO_x.

AIX000   CAS: 1031-47-6   HR: 3
5-AMINO-1-BIS(DIMETHYLAMIDE)-PHOSPHORYL-3-PHENYL-1,2,4-TRIAZOLE
mf: $C_{12}H_{19}N_6OP$ mw: 294.34
PROP: Solid from alc (aq). Mp: 167–168°. Sol in most org solvs; spar sol in H_2O.
SYNS: 5-AMINO-1-BIS(DIMETHYLAMIDO)PHOSPHORYL-3-PHENYL-1,2,4-TRIAZOLE □ 5-AMINO-1-(BIS(DIMETHYL-AMINO)PHOSPHINYL)-3-PHENYL-1,2,4-TRIAZOLE □ 5-AMINO-3-FENIL-1-BIS(-DIMETILAMINO)-FOSFORIL-1,2,4-TRIAZOLO (ITALIAN) □ 5-AMINO-3-FENYL-1-BIS(DIMETHYL-AMINO)-FOSFORYL-1,2,4-TRIAZOOL (DUTCH) □ 5-AMINO-3-PHENYL-1-BIS(DIMETHYLAMINO)-PHOSPHORYL-1H-1,2,4-TRIAZOL (GERMAN) □ 5-AMINO-3-PHENYL-1-BIS(DIMETHYL-AMINO)-PHOSPHORYLE-1,2,4-TRIAZOLE (FRENCH) □ 5-AMINO-3-PHENYL-1,2,4-TRIAZOLE-1-YL-N,N,N',N'-TETRAMETHYLPHOSPH-ODIAMIDE □ 5-AMINO-3-PHENYL-1,2,4-TRIAZOLYL-1-BIS(DIMETHYLAMIDO)PHOSPHATE □ 5-AMINO-3-PHENYL-1,2,4-TRIAZOLYL-N,N,N'N'-TETRAMETHYL-PHOSPHONAMIDE □ p-(5-AMINO-3-PHENYL-1H-1,2,4-TRIAZOL-1-YL)-N,N,N'-TETRAMETHYL PHOSPHONIC DIAMIDE □ BIS(DIMETHYLAMINO)-3-AMINO-5-PHENYLTRIAZOLYL PHOSPHINE OXIDE □ ENT 27,223 □ NIAGARA 5943 □ 3-PHENYL-5-AMINO-1,2,4-TRIAZOLYL-(1)-(N,N'-TETRAMETHYL)

DIAMIDOPHOSPHONATE □ TRIAMIFOS (GERMAN, DUTCH, ITALIAN) □ TRIAMIPHOS □ TRIAMPHOS □ WEPSIN □ WEPSYN □ WEPSYN 155 □ WP 155
CONSENSUS REPORTS: EPA Extremely Hazardous Substances List.
SAFETY PROFILE: Poison by ingestion, skin contact, intraperitoneal, and possibly other routes. An experimental teratogen. Mutation data reported. When heated to decomposition it emits very toxic fumes of PO_x and NO_x.

AIX250   CAS: 56-18-8   HR: 3
AMINOBIS(PROPYLAMINE)
DOT: UN 2269
mf: $C_6H_{17}N_3$ mw: 131.26
PROP: Liquid. Bp: 100° @ 2 mm.
SYNS: AMINOBIS(PROPYLAMINE) □ BIS-(3-AMINOPROPYL)AMINE □ 3,3-DIAMINODIPROPYL-AMINE □ 3,3'-DIAMINODIPROPYLAMINE □ DIPROPYLENETRIAMINE □ IMINOBIS(PROPYL-AMINE) □ 3,3'-IMINOBIS(PROPYLAMINE) □ 1,3-PROPANEDIAMINE, N-(3-AMINOPROPYL)- □ PROPYLAMINE, 3,3'-IMINOBIS-
CONSENSUS REPORTS: Reported in EPA TSCA Inventory.
DOT CLASSIFICATION: 8; Label: Corrosive
SAFETY PROFILE: Poison by skin contact. Moderately toxic by ingestion. A skin and severe eye irritant. When heated to decomposition it emits toxic fumes of NO_x. An explosive.

AJD000   CAS: 60-32-2   HR: 2
6-AMINOCAPROIC ACID
mf: $C_6H_{13}NO_2$ mw: 131.20
PROP: Leaflets. Mp: 204–206°. Sol in water; insol in EtOH.
SYNS: ACEPRAMINE □ ACS □ AFIBRIN □ AMICAR □ AMINOCAPROIC ACID □ ω-AMINOCAPROIC ACID □ ε-AMINOCAPROIC ACID □ ω-AMINOHEXANOIC ACID □ AMINOKAPRON □ CAPRALENSE □ CAPRAMOL □ CAPROCID □ CAPROLISIN □ CL 10304 □ CY 116 □ EACA □ EACA KABI □ EACS □ EPSAMON □ EPSICAPRON □ HEMOCAPROL □ HEMOPAR □ HEPIN □ IPSILON □ 177 J.D. □ ε-LEUCINE □ ε-NORLEUCINE □ NSC-26154 □ RESPRAMIN
CONSENSUS REPORTS: Reported in EPA TSCA Inventory.
SAFETY PROFILE: Moderately toxic by intravenous route. Human systemic effects

by ingestion: changes in tubules (including acute renal failure, acute tubular necrosis), hematuria, and increased body temperature. Experimental reproductive effects. An eye irritant. When heated to decomposition it emits toxic fumes such as NO_x.

AJD750   CAS: 26148-68-5   HR: 3
AMINO-α-CARBOLINE
mf: $C_{11}H_9N_3$ mw: 183.2
PROP: Crystals from $CHCl_3$/hexane or EtOH. Mp: 202–203°.
SYNS: 2-AMINO-α-CARBOLINE □ 2-AMINO-9H-PYRIDO(2,3-B)INDOLE
CONSENSUS REPORTS: IARC Cancer Review: Group 2B IMEMDT 7,56,87; Animal Sufficient Evidence IMEMDT 40,245,86.
SAFETY PROFILE: Suspected carcinogen with experimental carcinogenic data. Human mutation data reported. When heated to decomposition it emits toxic fumes of NO_x.

AJK750   CAS: 1951-25-3   HR: 3
AMINODARONE
mf: $C_{25}H_{29}I_2NO_3$ mw: 645.35
SYNS: AMIODARONE □ 2-BUTYL-3-BENZOFURANYL p-((2-DIETHYLAMINO)ETHOXY)-m,m-DIIODOPHENYL KETONE □ 2-BUTYL-3-(3,5-DIIODO-4-(2-DIETHYLAMINOETHOXY)-BENZOYL)BENZOFURAN □ 2-N-BUTYL-3',5'-DIIODO-4'-N-DIETHYLAMINOETHOXY-3-BENZOYLBENZOFURAN □ L. 3428 □ LABAZ
DOT CLASSIFICATION: 3; Label: Flammable Liquid
SAFETY PROFILE: Poison by intravenous and intraperitoneal routes. Human systemic effects by ingestion: photosensitivity of the skin. A flammable liquid. When heated to decomposition it emits very toxic fumes of I^- and NO_x. A coronary vasodilator.

AJQ100   CAS: 109-55-7   HR: 3
1-AMINO-3-DIMETHYLAMINOPROPANE
mf: $C_5H_{14}N_2$ mw: 102.21
PROP: Colorless liquid. Mp: <−70°, bp: 132–135°, flash p: 100°F (OC), d: 0.8100 @ 30°, vap press: 10 mm @ 30°, vap d: 3.52.
SYNS: N,N-DIMETHYL-N-(3-AMINOPROPYL)-AMINE □ 3-(DIMETHYLAMINO)PROPYLAMINE □ N,N-

DIMETHYL-1,3-DIAMINOPROPANE □ N,N-DIMETHYL-1,3-PROPANEDIAMINE □ N,N-DIMETHYL-1,3-PROPYLENEDIAMINE

CONSENSUS REPORTS: Reported in EPA TSCA Inventory.

SAFETY PROFILE: Moderately toxic by ingestion and skin contact. A skin and eye irritant. Very flammable when exposed to heat, flame, or oxidizers. Reaction with 1,2-dichloroethane produces explosive acetylene gas. This and other amines ignite on contact with cellulose nitrate of high surface area. To fight fire, use alcohol foam, CO_2, dry chemical. When heated to decomposition it emits toxic fumes of NO_x. See also AMINES.

AJQ600   CAS: 77094-11-2   HR: 3
2-AMINO-3,4-DIMETHYLIMIDAZO(4,5-f)QUINOLINE
mf: $C_{12}H_{12}N_4$ mw: 212.28

PROP: Solid. Mp: 296–298° (sealed tube).

SYNS: 3,4-DIMETHYL-3H-IMIDAZO(4,5-f)QUINOLIN-2-AMINE □ MeIQ

CONSENSUS REPORTS: IARC Cancer Review: Group 2B IMEMDT 56,197,93; Animal Sufficient Evidence IMEMDT 56,197,93; Animal Inadequate Evidence IMEMDT 40,275,86; Human No Adequate Data IMEMDT 40,275,86; Human Inadequate Evidence IMEMDT 56,197,93.

SAFETY PROFILE: Confirmed carcinogen with experimental carcinogenic data. Mutation data reported. When heated to decomposition it emits toxic fumes of NO_x.

AJQ675   CAS: 77500-04-0   HR: 3
2-AMINO-3,8-DIMETHYLIMIDAZO(4,5-f)QUINOXALINE
mf: $C_{11}H_{11}N_5$ mw: 213.27

PROP: Crystals. Mp: 295–300° (sealed tube).

SYNS: 2-AMINO-3,8-DIMETHYL-3H-IMIDAZO(4,5-f)QUINOXALINE □ 3,8-DIMETHYL-3H-IMIDAZO(4,5-f)QUINOXALIN-2-AMINE

CONSENSUS REPORTS: IARC Cancer Review: Group 2B IMEMDT 56,211,93; Animal Sufficient Evidence IMEMDT 56,211,93; Animal Inadequate Evidence IMEMDT 40,283,86; Human No Adequate Data IMEMDT 40,283,86; Human Inadequate Evidence IMEMDT 56,211,93.

SAFETY PROFILE: Confirmed carcinogen with experimental carcinogenic data. Mutation data reported. When heated to decomposition it emits toxic fumes of NO_x.

AJR500   CAS: 68808-54-8   HR: 3
3-AMINO-1,4-DIMETHYL-5H-PYRIDO-(4,3-b)INDOLE ACETATE
mf: $C_{13}H_{13}N_3 \cdot C_2H_4O_2$ mw: 271.35

PROP: Pale-brown needles or small prisms from EtOAc. Mp: 252–262°.

SYNS: 1,4-DIMETHYL-5H-PYRIDO(4,3-b)INDOL-3-AMINE ACETATE □ 1,4-DIMETHYL-5H-PYRIDO(4,3-b)INDOL-3-AMINE MONOACETATE □ TRP-P-1 (ACETATE)

SAFETY PROFILE: Suspected carcinogen with experimental carcinogenic data. Mutation data reported. When heated to decomposition it emits toxic fumes of NO_x.

AJS100   CAS: 92-67-1   HR: 3
4-AMINODIPHENYL
mf: $C_{12}H_{11}N$ mw: 169.24

PROP: Leaflets or colorless crystals from alc (aq). Mp: 53°, bp: 302°, d: 1.160 @ 20°/20°, autoign temp: 842°F.

SYNS: p-AMINOBIPHENYL □ 4-AMINOBIPHENYL □ 4-AMINODIFENIL (SPANISH) □ p-AMINO-DIPHENYL □ BIPHENYLAMINE □ (1,1'-BIPHENYL)-4-AMINE □ p-BIPHENYLAMINE □ 4-BIPHENYL-AMINE □ PARAAMINODIPHENYL □ p-PHENYLANILINE □ XENYLAMIN (CZECH) □ XENYLAMINE

CONSENSUS REPORTS: NTP 10th Report on Carcinogens. IARC Cancer Review: Group 1 IMEMDT 7,91,87; Human Limited Evidence IMEMDT 1,74,72; Animal Sufficient Evidence IMEMDT 1,74,72; Human Sufficient Evidence IMEMDT 28,151,82. Reported in EPA TSCA Inventory. EPA Genetic Toxicology Program. Community Right-To-Know List.

OSHA PEL: Cancer Suspect Agent

ACGIH TLV: Confirmed Human Carcinogen

DFG MAK: Human Carcinogen

SAFETY PROFILE: Confirmed human carcinogen with experimental carcinogenic and tumorigenic data. Poison by ingestion and intraperitoneal routes. Human mutation data reported. An irritant. Effects resemble those of benzidine. See also BENZIDINE. Slight to moderate fire hazard when exposed to heat, flames (sparks), or powerful oxidizers. To fight fire, use water spray, mist, dry chemical. When heated to decomposition it emits toxic fumes of NO_x. See also AROMATIC AMINES.

AJS900   CAS: 2697-65-6   HR: 3
2-AMINO-ETHANESELENOL HYDROCHLORIDE
mf: $C_2H_7NSe \cdot ClH$ mw: 160.52
SYN: ETHANESELENOL, 2-AMINO-, HYDROCHLORIDE
OSHA PEL: TWA 0.2 mg(Se)/m³
ACGIH TLV: TWA 0.2 mg(Se)/m³
SAFETY PROFILE: Poison by ingestion. When heated to decomposition it emits toxic fumes of NO_x, Se, and HCl.

AJS950   CAS: 2697-60-1   HR: 3
2-AMINOETHANESELENOSULFURIC ACID
mf: $C_2H_7NO_3SSe$ mw: 204.12
SYN: SELENOSULFURIC ACID, 2-AMINOETHYL ESTER
OSHA PEL: TWA 0.2 mg(Se)/m³
ACGIH TLV: TWA 0.2 mg(Se)/m³
SAFETY PROFILE: Poison by intraperitoneal route. When heated to decomposition it emits toxic fumes of NO_x, SO_x, and Se.

AJT250   CAS: 60-23-1   HR: 3
2-AMINOETHANETHIOL
mf: C_2H_7NS mw: 77.16
PROP: Crystals from alc. Mp: 99–100°. Sol in MeOH, EtOH; spar sol in water.
SYNS: 2-AMINOETHYL MERCAPTAN □ BECAPTAN □ CISTEAMINA (ITALIAN) □ CYCTEINAMINE □ CYSTEAMIDE □ CYSTEAMINE □ DECARBOXYCYSTEINE □ LAMBRATEN □ MEA □ MECRAMINE □ MERCAMINE □ MERCAPTAMINE □ β-MERCAPTOETHYLAMINE □ (2-MERCAPTO-ETHYL)AMINE □ THIOETHANOLAMINE
CONSENSUS REPORTS: EPA Genetic Toxicology Program.
SAFETY PROFILE: Poison by intravenous, subcutaneous, and intraperitoneal routes.

Moderately toxic by ingestion. Experimental reproductive effects. Mutation data reported. When heated to decomposition it emits very toxic fumes of SO_x and NO_x.

AJU250   CAS: 929-06-6   HR: 2
2-AMINOETHOXYETHANOL
DOT: UN 3055
mf: $C_4H_{11}NO_2$ mw: 105.16
SYNS: 2-(2-AMINOETHOXY)ETHANOL □ DIGLYCOLAMINE
CONSENSUS REPORTS: Reported in EPA TSCA Inventory.
DOT CLASSIFICATION: 8; Label: Corrosive
SAFETY PROFILE: Moderately toxic by skin contact. Mildly toxic by ingestion. Severe eye and skin irritant. Corrosive and a powerful irritant. When heated to decomposition it emits toxic fumes of NO_x.

AJV000   CAS: 132-32-1   HR: 3
3-AMINO-9-ETHYLCARBAZOLE
mf: $C_{14}H_{14}N_2$ mw: 210.30
PROP: Solid. Mp: 98–100°. In cancer bioassay both free amine and hydrochloride salt were used. NCITR* NCI-CG-TR-93,78.
SYN: 3-AMINO-N-ETHYLCARBAZOLE
CONSENSUS REPORTS: Reported in EPA TSCA Inventory.
DFG MAK: Confirmed Animal Carcinogen with Unknown Relevance to Humans
SAFETY PROFILE: Suspected carcinogen with experimental carcinogenic data. Poison by ingestion and intraperitoneal routes. When heated to decomposition it emits toxic fumes of NO_x.

AJV250   CAS: 6109-97-3   HR: 3
3-AMINO-9-ETHYLCARBAZOLE-HYDROCHLORIDE
mf: $C_{14}H_{14}N_2 \cdot ClH$ mw: 246.76
PROP: In cancer bioassay both free amine and hydrochloride salt used. NCITR* NCI-CG-TR-93,78.
SYN: NCI-C03043
CONSENSUS REPORTS: NCI Carcinogenesis Bioassay Completed; Results Positive: mouse, rat NCITR* NCI-CG-TR-93,78.

A

SAFETY PROFILE: Suspected carcinogen with experimental carcinogenic data. Poison by ingestion. Mutation data reported. When heated to decomposition it emits very toxic fumes of NO_x and HCl.

AKB000   CAS: 140-31-8   HR: 3
N-AMINOETHYLPIPERAZINE
DOT: UN 2815
mf: $C_6H_{15}N_3$ mw: 129.24
PROP: Light-colored liquid. D: 0.9852 @ 20°/20°, mp: −19°, bp: 220.4°, flash p: 200°F (OC), vap d: 4.4.
SYNS: AMINOETHYLPIPERAZINE □ N-(β-AMINOETHYL)PIPERAZINE □ N-(2-AMINOETHYL)-PIPERAZINE □ 1-(2-AMINOETHYL)PIPERAZINE □ USAF DO-46
CONSENSUS REPORTS: Reported in EPA TSCA Inventory.
DOT CLASSIFICATION: 8; Label: Corrosive
SAFETY PROFILE: Poison by intraperitoneal route. Moderately toxic by ingestion and skin contact. Experimental reproductive effects. A skin and eye irritant. Mutation data reported. See also AMINES. Moderately flammable when exposed to heat, flame, sparks, or powerful oxidizers. To fight fire, use alcohol foam. When heated to decomposition it emits toxic fumes of NO_x.

AKE250   CAS: 116-85-8   HR: 3
1-AMINO-4-HYDROXYANTHRA-
    QUINONE
mf: $C_{14}H_9NO_3$ mw: 239.24
PROP: Red-violet powder or pink plates from C_6H_6, violet needles from pet ether. Mp: 207–208°. Sol in water, HCl, alc, ether, and benzene.
SYNS: 1A-4OA □ ACETATE FAST RED 2B □ ACETOQUINONE LIGHT GOOSEBERRY RL □ ACETYLON FAST PINK B □ AMACEL PINK B □ 1-AMINO-4-HYDROXY-9,10-ANTHRACENEDIONE □ 1-AMINO-4-OXYANTHRAQUINONE □ 9,10-ANTHRACENEDIONE, 1-AMINO-4-HYDROXY-(9CI) □ ANTHRAQUINONE, 1-AMINO-4-HYDROXY- □ ARTISIL DIRECT RED 3BP □ ARTISIL RED 3BP □ CALCOSYN PINK B □ CELANTHRENE RED 3BN □ CELLITON FAST PINK BA-CF □ CELLITON FAST PINK BN □ CELUTATE PINK B □ CELUTATE PINK BN □ CELUTATE PINK BY □ CERVEN DISPERZNI 15 □ CIBACETE RED 3B □

CIBACET RED 3B □ CIBACET RED E3B □ CILLA FAST PINK BN □ C.I. 60710 □ C.I. DISPERSE RED 15 □ C.I. SOLVENT RED 53 □ DIACELLITON FAST PINK B □ DISPERSE FAST PINK B □ DISPERSE RED 15 □ DISPERSE RED 25 □ DISPERSOL ORANGE D-G □ DURANOL RED 2B □ FENACET FAST PINK B □ 1-HYDROXY-4-AMINO-ANTHRAQUINONE □ 4-HYDROXY-1-ANTHRA-QUINONYLAMINE □ INTERCHEM ACETATE PINK BLF □ INTERCHEM HISPERSE PINK BH □ MICROSETILE PINK BN □ NACELAN PINK B □ NEOSETILE PINK BN □ ORACET RED 3B □ PARA M □ PERLITON PINK 3B □ SERISOL FAST RED 2B □ SETACYL PINK 3B □ SUPRACET BRILLIANT RED 2B
CONSENSUS REPORTS: Reported in EPA TSCA Inventory.
SAFETY PROFILE: Poison by intravenous route. Moderately toxic by intraperitoneal route. Mutation data reported. When heated to decomposition it emits toxic fumes of NO_x.

AKI900   CAS: 102516-61-0 HR: 3
3-(((3-AMINO-4-HYDROXYPHENYL)-
    PHENYLARSINO)THIO)ALANINE
mf: $C_{15}H_{17}AsN_2O_3S$ mw: 380.32
SYN: ALANINE, 3-(((3-AMINO-4-HYDROXY-PHENYL)PHENYLARSINO)THIO)-
OSHA PEL: TWA 0.5 mg(As)/m³
SAFETY PROFILE: Poison by intravenous route. When heated to decomposition it emits toxic fumes of NO_x, SO_x, and AS.

AKP750   CAS: 82-28-0   HR: 3
1-AMINO-2-METHYLANTHRAQUINONE
mf: $C_{15}H_{11}NO_2$ mw: 237.27
SYNS: ACETATE FAST ORANGE R □ ACETOQUINONE LIGHT ORANGE JL □ 1-AMINO-2-METHYL-9,10-ANTHRACENEDIONE □ ARTISIL ORANGE 3RP □ CELLITON ORANGE R □ C.I. 60700 □ C.I. DISPERSE ORANGE 11 □ CILLA ORANGE R □ DISPERSE ORANGE □ DURANOL ORANGE G □ 2-METHYL-1-ANTHRAQUINONYLAMINE □ MICROSETILE ORANGE RA □ NCI-C01901 □ NYLOQUINONE ORANGE JR □ PERLITON ORANGE 3R □ SERISOL ORANGE YL □ SUPRACET ORANGE R
CONSENSUS REPORTS: NTP 10th Report on Carcinogens. IARC Cancer Review: Group 3 IMEMDT 7,56,87; Animal Limited Evidence IMEMDT 27,199,82. NCI Carcinogenesis Bioassay (feed); Clear Evidence: mouse, rat NCITR* NCI-CG-

TR-111,78. Community Right-To-Know List. Reported in EPA TSCA Inventory. SAFETY PROFILE: Confirmed carcinogen with experimental carcinogenic, neoplastigenic, and tumorigenic data. Mutation data reported. When heated to decomposition it emits toxic fumes of NO_x.

AKS250   CAS: 67730-11-4   HR: 3
2-AMINO-6-METHYLDIPYRIDO(1,2-a:3',2'-d)IMIDAZOLE
mf: $C_{11}H_{10}N_4$ mw: 198.25
PROP: Yellow prisms from MeOH/EtOAc.
SYNS: GLU-P-I □ 6-ME-GLU-P-2 □ 6-METHYL DIPYRIDO(1,2-a:3',2'-d)IMIDAZOL-2-AMINE
CONSENSUS REPORTS: IARC Cancer Review: Group 2B IMEMDT 7,56,87; Animal Sufficient Evidence IMEMDT 40,223,86.
SAFETY PROFILE: Confirmed carcinogen with experimental carcinogenic data. Human mutation data reported. When heated to decomposition it emits toxic fumes of NO_x.

AKT600   CAS: 76180-96-6   HR: 3
2-AMINO-3-METHYLIMIDAZO(4,5-f)QUINOLINE
mf: $C_{11}H_{10}N_4$ mw: 198.25
PROP: Crystals from MeOH (aq).
CONSENSUS REPORTS: NTP 10th Report on Carcinogens. IARC Cancer Review: Group 2A IMEMDT 56,165,93; Animal Sufficient Evidence IMEMDT 40,261,86; Animal Sufficient Evidence IMEMDT 56,165,93; Human No Adequate Data IMEMDT 40,261,86; Human Inadequate Evidence IMEMDT 56,165,93.
SAFETY PROFILE: Confirmed carcinogen with experimental carcinogenic and tumorigenic data. Mutation data reported. When heated to decomposition it emits toxic fumes of NO_x.

AKZ200   CAS: 105650-23-5   HR: 3
2-AMINO-1-METHYL-6-PHENYLIMID-AZO(4,5-B)PYRIDINE
mf: $C_{13}H_{12}N_4$ mw: 224.29
SYNS: IMIDAZO(4,5-B)PYRIDINE, 2-AMINO-1-METHYL-6-PHENYL- □ 1H-IMIDAZO(4,5-B)PYRIDIN-2-AMINE, 1-METHYL-6-PHENYL-(9CI) □ 1H-IMIDAZO(4,5-B)PYRIDINE, 2-AMINO-1-METHYL-6-PHENYL- □ PHIP
CONSENSUS REPORTS: IARC Cancer Review: Group 2B IMEMDT 56,229,93; Animal Sufficient Evidence IMEMDT 56,229,93; Human Inadequate Evidence IMEMDT 56,229,93.
SAFETY PROFILE: Confirmed carcinogen with experimental carcinogenic data. Mutation data reported. When heated to decomposition it emits toxic vapors of NO_x.

ALD500   CAS: 62450-07-1   HR: 3
3-AMINO-1-METHYL-5H-PYRIDO(4,3-b)INDOLE
mf: $C_{12}H_{11}N_3$ mw: 197.26
SYNS: 3-AMINO-1-METHYL-γ-CARBOLINE □ 1-METHYL-3-AMINO-5H-PYRIDO(4,3-b)INDOLE □ TRP-P-2 □ TRYPTOPHAN P2
CONSENSUS REPORTS: IARC Cancer Review: Group 2B IMEMDT 7,56,87; Animal Sufficient Evidence IMEMDT 31,255,83. EPA Genetic Toxicology Program.
SAFETY PROFILE: Confirmed carcinogen with experimental carcinogenic and neoplastigenic data. Mutation data reported. When heated to decomposition it emits toxic fumes of NO_x.

ALD750   CAS: 68006-83-7   HR: 3
2-AMINO-3-METHYL-9H-PYRIDO(2,3-b)INDOLE
mf: $C_{12}H_{11}N_3$ mw: 197.2
PROP: Crystals from $CHCl_3$/hexane. Mp: 215–218°.
SYN: 2-AMINO-3-METHYL-α-CARBOLINE
CONSENSUS REPORTS: IARC Cancer Review: Group 2B IMEMDT 7,56,87; Animal Sufficient Evidence IMEMDT 40,253,86.
SAFETY PROFILE: Confirmed carcinogen with experimental carcinogenic data. Mutation data reported. When heated to decomposition it emits toxic fumes of NO_x.

ALF250   CAS: 127-79-7   HR: 2
4-AMINO-N-(4-METHYL-2-PYRIMIDIN-
   YL)-BENZENESULFONAMIDE
mf: $C_{11}H_{12}N_4O_2S$ mw: 264.33
PROP: Cream-colored powder, darkens on
exposure to light. Mp: 237° (decomp). Spar
sol in water.
SYNS: A-310 □ (p-AMINOBENZOLSULFONYL)-2-
AMINO-4-METHYLPYRIMIDIN (GERMAN) □
CREMOMERAZINE □ DEBENAL-M □ KELAMERAZINE
□ MEBACID □ MESULFA □ METHYLPYRIMAL □ N^1-(4-
METHYL-2-PYRIMIDIN-YL)SULFANILAMIDE □
METHYLSULFAZIN □ PERCOCCIDE □ PIRIMAL-M □
PYRALCID □ PYRIMAL M □ ROMEZIN □ RP 2632 □
2643-RP □ SEPTACIL □ SULFAMERADINE □
SULFAMERAZIN □ SULFAMETHYLDIAZINE □
SULPHAMERAZINE □ SUMEDINE □ VETA-MERAZINE
CONSENSUS REPORTS: Reported in EPA
TSCA Inventory.
SAFETY PROFILE: Moderately toxic by
intravenous and subcutaneous routes.
Experimental reproductive effects. When
heated to decomposition it emits very toxic
fumes of NO_x and SO_x.

ALL500   CAS: 99-56-9   HR: 2
2-AMINO-4-NITROANILINE
mf: $C_6H_7N_3O_2$ mw: 153.16
PROP: Dark-red needles from water. Mp:
201°. Sol in EtOH, Me_2CO, C_6H_6, $CHCl_3$;
mod sol in dil acids.
SYNS: C.I. 76020 □ 1,2-DIAMINO-4-NITROBENZENE
□ NCI-C03941 □ 4NDB □ 4-NITRO-1,2-BENZENEDI-
AMINE □ 4-NITRO-1,2-DIAMINOBENZENE □ p-NITRO-
o-PHENYLENE-DIAMINE □ 4-NITRO-o-PHENYLENE-
DIAMINE □ 4-NITRO-1,2-PHENYLENEDIAMINE □ 4-
NOPD
CONSENSUS REPORTS: IARC Cancer
Review: Group 3 IMEMDT 7,56,87; Animal
Inadequate Evidence IMEMDT 16,63,78.
NCI Carcinogenesis Bioassay (feed); No
Evidence: mouse, rat NCITR* NCI-CG-
TR-180,79. Reported in EPA TSCA
Inventory. EPA Genetic Toxicology
Program.
SAFETY PROFILE: Moderately toxic by
ingestion. An experimental teratogen. Other
experimental reproductive effects. Mutation
data reported. Questionable carcinogen.
When heated to decomposition it emits
toxic fumes of NO_x.

ALL750   CAS: 5307-14-2   HR: 3
4-AMINO-2-NITROANILINE
mf: $C_6H_7N_3O_2$ mw: 153.16
PROP: Black needles with strong green
reflection from water. Mp: 137°.
SYNS: C.I. 76070 □ C.I. OXIDATION BASE 22 □ 1,4-
DIAMINO-2-NITROBENZENE □ DURAFUR BROWN □
DURAFUR BROWN 2R □ DYE GS □ FOURAMIEN 2R □
FOURRINE 36 □ FOURRINE BROWN 2R □ NCI-C02222
□ 2NDB □ 2-NITRO-1,4-BENZENEDIAMINE □ 2-
NITRO-1,4-DIAMINOBENZENE □ NITRO-p-
PHENYLENEDIAMINE □ 2-NITRO-1,4-PHENYLENE-
DIAMINE □ o-NITRO-p-PHENYLENEDIAMINE (MAK)
□ 2-NITRO-p-PHENYLENEDIAMINE □ 2-NP □ 2-N-p-
PDA □ 2-NPPD □ OXIDATION BASE 22 □ URSOL
BROWN RR □ ZOBA BROWN RR
CONSENSUS REPORTS: IARC Cancer
Review: Group 3 IMEMDT 7,56,87; Animal
Inadequate Evidence IMEMDT 16,73,78.
NCI Carcinogenesis Bioassay (feed); No
Evidence: rat NCITR* NCI-CG-TR-169,79;
Clear Evidence: mouse NCITR* NCI-CG-
TR-169,79. Reported in EPA TSCA
Inventory. EPA Genetic Toxicology
Program.
DFG MAK: Confirmed Animal Carcinogen
with Unknown Relevance to Humans
SAFETY PROFILE: Suspected carcinogen
with experimental carcinogenic and
neoplastigenic data. Poison by
intraperitoneal route. Moderately toxic by
ingestion. An experimental teratogen. Other
experimental reproductive effects. Mutation
data reported. When heated to
decomposition it emits toxic fumes of NO_x.

ALO750   CAS: 2871-01-4   HR: 2
2-((4-AMINO-2-NITROPHENYL)AMINO)-
   ETHANOL
mf: $C_8H_{11}N_3O_3$ mw: 197.22
SYNS: ETHANOL, 2-((4-AMINO-2-NITROPHEN-
YL)AMINO)- □ HC RED NO. 3 □ 4-(2-HYDROXYETH-
YL)AMINO-3-NITROANILINE □ N^1-(2-HYDROXYETH-
YL)-2-NITRO-p-PHENYLENE-DIAMINE □ NCI-C54922
CONSENSUS REPORTS: IARC Cancer
Review: Group 3 IMEMDT 57,153,93;
Human Inadequate Evidence IMEMDT
57,153,93; Animal Inadequate Evidence
IMEMDT 57,153,93. NTP Carcinogenesis
Studies (gavage); Equivocal Evidence:

mouse NTPTR* NTP-TR-281,86; No
Evidence: rat NTPTR* NTP-TR-281,86.
Reported in EPA TSCA Inventory.
SAFETY PROFILE: Questionable
carcinogen with experimental carcinogenic
data. Mutation data reported. When heated
to decomposition it emits toxic fumes of
NO_x.

ALQ000   CAS: 121-66-4   HR: 3
2-AMINO-5-NITROTHIAZOLE
mf: $C_3H_3N_3O_2S$ mw: 145.15
PROP: Solid. Mp: 195–196°.
SYNS: AMINONITROTHIAZOLE □
AMINONITROTHIAZOLUM □ AMINZOL SOLUBLE □
ENHEPTIN □ ENTRAMIN □ NCI-C03065 □ NITRAMIN
□ NITRAMINE □ 5-NITRO-2-AMINOTHIAZOLE □
NITROMIN IDO □ 5-NITRO-2-THIAZOLYLAMINE □
USAF EK-6561
CONSENSUS REPORTS: IARC Cancer
Review: Group 3 IMEMDT 7,56,87; Animal
Limited Evidence IMEMDT 31,71,83. NCI
Carcinogenesis Bioassay (feed); No
Evidence: mouse NCITR* NCI-CG-TR-
53,78; Clear Evidence: rat NCITR* NCI-
CG-TR-53,78. Reported in EPA TSCA
Inventory.
SAFETY PROFILE: Poison by
intraperitoneal route. Experimental
reproductive effects. Questionable
carcinogen with experimental carcinogenic,
tumorigenic, and neoplastigenic data.
Mutation data reported. When heated to
decomposition it emits very toxic fumes of
NO_x and SO_x. Incompatible with HNO_3
and H_2SO_4. An antiprotozoal agent.

ALQ625   CAS: 58-60-6   HR: D
AMINONUCLEOSIDE PUROMYCIN
mf: $C_{12}H_{18}N_6O_3$ mw: 294.36
SYNS: AMINONUCLEOSIDE □ SAN □ SYTLOMYCIN
AMINONUCLEOSIDE
CONSENSUS REPORTS: Reported in EPA
TSCA Inventory.
SAFETY PROFILE: An experimental
teratogen. Other experimental reproductive
effects. Human mutation data reported.
When heated to decomposition it emits
toxic fumes of NO_x.

ALQ640   CAS: 92065-91-3   HR: 3
2-AMINO-2-OXOETHYL-2,2-DIMETHYL-
N-(((METHYLAMINO)CARBONYL)-
OXY)PROPANIMIDOTHIOATE
mf: $C_9H_{17}N_3O_3S$ mw: 247.35
SYN: PROPANIMIDOTHIOIC ACID, 2,2-DIMETHYL-N-
(((METHYLAMINO)CARBONYL)OXY)-, 2-AMINO-2-
OXOETHYL ESTER
SAFETY PROFILE: A poison by ingestion.
When heated to decomposition it emits
toxic vapors of NO_x and SO_x.

ALS990   CAS: 591-27-5   HR: 3
m-AMINOPHENOL
DOT: UN 2946
mf: C_6H_7NO mw: 109.14
PROP: Prisms from toluene. Mp: 123°. Sol
in water and alc; sltly sol in ether.
SYNS: m-AMINOFENOL (CZECH) □ 3-AMINO-1-
HYDROXYBENZENE □ m-AMINOPHENOL (DOT) □ 3-
AMINOPHENOL □ BASF URSOL EG □ C.I. 76545 □ C.I.
OXIDATION BASE 7 □ FOURAMINE EG □ FOURRINE
65 □ FOURRINE EG □ FURRO EG □ FUTRAMINE EG □
3-HYDROXYANILINE □ NAKO TEG □ PELAGOL EG □
RENAL EG □ TERTRAL EG □ URSOL EG □ ZOBA EG
CONSENSUS REPORTS: Reported in EPA
TSCA Inventory. EPA Genetic Toxicology
Program.
DOT CLASSIFICATION: 6.1; Label: KEEP
AWAY FROM FOOD
SAFETY PROFILE: Poison by ingestion,
subcutaneous, and intraperitoneal routes. An
experimental teratogen. Other experimental
reproductive effects. Mutation data
reported. A skin and eye irritant. When
heated to decomposition it emits toxic
fumes of NO_x.

ALT000   CAS: 95-55-6   HR: 3
2-AMINOPHENOL
DOT: UN 2512
mf: C_7H_7NO mw: 109.14
PROP: Colorless needles. Mp: 173°, bp:
subl. Sol in water and alc; very sol in ether.
SYNS: 2-AMINO-1-HYDROXYBENZENE □ o-
AMINOPHENOL □ BASF URSOL 3GA □ BENZOFUR GG
□ C.I. 76520 □ C.I. OXIDATION BASE 17 □ FOURAMINE
OP □ o-HYDROXYANILINE □ 2-HYDROXYANILINE □
NAKO YELLOW EGA □ PARADONE OLIVE GREEN B
□ PELAGOL 3GA □ PELAGOL GREY GG □ ZOBA 3GA

CONSENSUS REPORTS: Reported in EPA TSCA Inventory.
DOT CLASSIFICATION: 6.1; Label: KEEP AWAY FROM FOOD
SAFETY PROFILE: Poison by intraperitoneal and subcutaneous routes. Moderately toxic by ingestion. An experimental teratogen. Other experimental reproductive effects. An eye irritant. Mutation data reported. When heated to decomposition it emits toxic NO_x. See also AROMATIC AMINES.

ALT250   CAS: 123-30-8   HR: 3
4-AMINOPHENOL
DOT: UN 2512
mf: C_6H_7NO mw: 109.14
PROP: Colorless crystals or plates from water; sltly sol in water, alc, and ether; insol in chloroform. Mp: 189.6–190.2°, bp: 284° (decomp).
SYNS: ACTIVOL □ p-AMINOFENOL (CZECH) □ 4-AMINO-1-HYDROXYBENZENE □ p-AMINOPHENOL (DOT) □ BASF URSOL P BASE □ BENZOFUR P □ CERTINAL □ C.I. OXIDATION BASE 6A □ CITOL □ DURAFUR BROWN RB □ FOURAMINE P □ FOURRINE 84 □ FOURRINE P BASE □ FURRO P BASE □ p-HYDROXYANILINE □ 4-HYDROXYANILINE □ NAKO BROWN R □ PAP □ PARANOL □ PELAGOL GREY P BASE □ PELAGOL P BASE □ RENAL AC □ RODINAL □ TERTRAL P BASE □ URSOL P □ URSOL P BASE □ ZOBA BROWN P BASE
CONSENSUS REPORTS: Reported in EPA TSCA Inventory. EPA Genetic Toxicology Program.
DOT CLASSIFICATION: 6.1; Label: KEEP AWAY FROM FOOD
SAFETY PROFILE: Poison by ingestion, subcutaneous, and intraperitoneal routes. An experimental teratogen. Other experimental reproductive effects. An allergen and skin and eye irritant. Mutation data reported. Can cause contact dermatitis, bronchial asthma, and methemoglobinemia with cyanosis. When heated to decomposition it emits toxic fumes of NO_x.

ALV100   CAS: 73791-39-6   HR: 3
p-AMINOPHENYLARSINE OXIDE DIHYDRATE
mf: C_6H_6AsNO•$2H_2O$ mw: 219.09
SYNS: ANILINE, p-ARSENOSO-, DIHYDRATE □ 4-ARSENOSOANILINE, DIHYDRATE □ ARSINE, (p-AMINOPHENYL)OXO-, DIHYDRATE
OSHA PEL: TWA 0.5 mg(As)/m^3
ACGIH TLV: BEI: 35 μ (As)/L inorganic arsenic and methylated metabolites in urine
SAFETY PROFILE: Poison by intraperitoneal and intravenous routes. When heated to decomposition it emits toxic fumes of NO_x and As.

ALX120   CAS: 219959-86-1   HR: 3
AMINOPHENYLNORHARMAN
mf: $C_{17}H_{13}N_3$ mw: 259.31
SYNS: 9-(4'-AMINOPHENYL)-9H-PYRIDO(3,4-B)INDOLE □ BENZENAMINE, 4-(9H-PYRIDO(3,4-B)INDOL-9-YL)-
SAFETY PROFILE: A poison by ingestion. Questionable carcinogen with experimental neoplastigenic data reported. Mutation data reported. When heated to decomposition it emits toxic vapors of NO_x.

AMB500   CAS: 151-18-8   HR: 2
3-AMINOPROPIONITRILE
mf: $C_3H_6N_2$ mw: 70.11
PROP: Liquid; amine odor. Bp: 185°.
SYNS: β-AMINOPROPIONITRILE □ BAPN □ β-CYANOETHYLAMINE
CONSENSUS REPORTS: EPA Genetic Toxicology Program. Reported in EPA TSCA Inventory. Cyanide and its compounds are on the Community Right-To-Know List.
SAFETY PROFILE: Moderately toxic by intraperitoneal route. An experimental teratogen. Other experimental reproductive effects. Mutation data reported. Nitriles usually have cyanide-like effects. See also CYANIDE. Easily oxidized and unstable. A storage hazard; it polymerizes to an explosive yellow solid. When heated to decomposition it emits toxic fumes of CN^- and NO_x. For fire and explosion hazards see CYANIDE.

AMB750   CAS: 2079-89-2   HR: 2
β-AMINOPROPIONITRILE FUMARATE
mf: $C_3H_6N_2$•$2C_4H_4O_4$ mw: 302.27

SYNS: β-APN □ BAPN FUMARATE □ DI-β-AMINOPROPIONITRILE FUMARATE □ DI-BAPN FUMARATE

CONSENSUS REPORTS: Reported in EPA TSCA Inventory. Cyanide and its compounds are on the Community Right-To-Know List.

SAFETY PROFILE: Moderately toxic by ingestion and intraperitoneal routes. An experimental teratogen. When heated to decomposition it emits toxic fumes of NO_x and CN^-. See also NITRILES.

AMC000   CAS: 70-69-9   HR: 3
p-AMINOPROPIOPHENONE
mf: $C_9H_{11}NO$ mw: 149.21

PROP: Needles from water. Mp: 140°.

SYNS: 1-(4-AMINOPHENYL)-1-PROPANONE □ ETHYL-p-AMINOPHENYL KETONE □ PAPP □ PARAMINOPROPIOPHENONE □ USAF UCTL-1856

CONSENSUS REPORTS: Reported in EPA TSCA Inventory. EPA Extremely Hazardous Substances List.

DOT CLASSIFICATION: 3; Label: Flammable Liquid

SAFETY PROFILE: Poison by ingestion and intraperitoneal routes. Ingestion of large doses can cause cyanosis. A flammable liquid. When heated to decomposition it emits toxic fumes of NO_x.

AME500   CAS: 299-26-3   HR: 3
3-(2-AMINOPROPYL)INDOLE
mf: $C_{11}H_{14}N_2$ mw: 174.27

SYNS: INDOPAN □ α-METHYL-β-INDOLAETHYLAMINE (GERMAN) □ α-METHYL-β-INDOLEETHYLAMINE □ α-METHYLTRYPTAMINE

SAFETY PROFILE: Poison by ingestion and intraperitoneal routes. Moderately toxic by subcutaneous route. Human psychotropic effects by ingestion. An experimental teratogen. Other experimental reproductive effects. When heated to decomposition it emits toxic fumes of NO_x.

AMF250   CAS: 123-00-2   HR: 3
4-AMINOPROPYLMORPHOLINE
mf: $C_7H_{16}N_2O$ mw: 144.25

PROP: Liquid. Mp: −15°, bp: 224.7°, flash p: 220°F (OC), d: 0.9872 @ 20°/20°, vap press: 0.06 mm @ 20°, vap d: 4.97.

SYNS: N-(3-AMINOPROPYL)MORFOLIN □ N-(3-AMINOPROPYL)MORPHOLINE □ MORPHOLINE, N-AMINOPROPYL-

CONSENSUS REPORTS: Reported in EPA TSCA Inventory.

SAFETY PROFILE: A corrosive material. Moderately toxic by several routes. A severe skin and eye irritant. Combustible. Can react with oxidizing materials. To fight fire, use alcohol foam, dry chemical. When heated to decomposition it emits toxic fumes of NO_x.

AMG750   CAS: 54-62-6   HR: 3
AMINOPTERIDINE
mf: $C_{19}H_{20}N_8O_5$ mw: 440.47

PROP: Yellow needles, sol in sodium hydroxide soln.

SYNS: 4-AMINO-4-DEOXYPTEROYLGLUTAMATE □ 4-AMINO-PGA □ AMINOPTERIN □ 4-AMINOPTEROYLGLUTAMIC ACID □ APGA □ ENT 26,079 □ FOLIC ACID, 4-AMINO- □ NSC-739

CONSENSUS REPORTS: EPA Extremely Hazardous Substances List.

SAFETY PROFILE: Poison by ingestion and intraperitoneal routes. Human and experimental teratogenic data. Other experimental reproductive effects. Mutation data reported. Human systemic effects by ingestion: gastrointestinal. Questionable carcinogen with experimental tumorigenic data. When heated to decomposition it emits toxic fumes of NO_x.

AMI000   CAS: 504-29-0   HR: 3
2-AMINOPYRIDINE
DOT: UN 2671

mf: $C_5H_6N_2$ mw: 94.13

PROP: White powder or crystals from ligroin. Mp: 58.1, bp: 210.6°. Sol in water and ether; very sol in alc; sltly sol in ligroin. IDLH 5 ppm.

SYNS: o-AMINOPYRIDINE □ α-AMINOPYRIDINE □ AMINO-2-PYRIDINE □ α-PYRIDINAMINE □ α-PYRIDYLAMINE

CONSENSUS REPORTS: Reported in EPA TSCA Inventory.

OSHA PEL: TWA 0.5 ppm

A

ACGIH TLV: TWA 0.5 ppm
DFG MAK: 0.5 ppm (2 mg/m³)
DOT CLASSIFICATION: 6.1; Label: Poison
SAFETY PROFILE: Poison by ingestion, inhalation, subcutaneous, intravenous, and intraperitoneal routes. Toxic effects resemble strychnine poisoning. Human systemic effects by inhalation: somnolence, convulsions, and antipsychotic effects. Human central nervous system effects by inhalation. When heated to decomposition it emits highly toxic fumes of NO_x.

AMI250   CAS: 462-08-8   HR: 3
3-AMINOPYRIDINE
DOT: UN 2671
mf: $C_5H_6N_2$ mw: 94.13
PROP: Leaflets or crystals from benzene or ligroin. Mp: 64°, bp: 251°. Very sol in water, alc, ether; insol in ligroin.
SYNS: m-AMINOPYRIDINE (DOT) □ AMINO-3-PYRIDINE □ 3-PYRIDINAMINE □ 3-PYRIDYLAMINE
CONSENSUS REPORTS: Reported in EPA TSCA Inventory.
DOT CLASSIFICATION: 6.1; Label: Poison
SAFETY PROFILE: Poison by ingestion, intraperitoneal, subcutaneous, and intravenous routes. When heated to decomposition it emits toxic fumes of NO_x.

AMI500   CAS: 504-24-5   HR: 3
4-AMINOPYRIDINE
DOT: UN 2671
mf: $C_5H_6N_2$ mw: 94.13
PROP: Needles or crystals from benzene. Mp: 158°; sol in water; sltly sol in benzene and ether.
SYNS: AMINO-4-PYRIDINE □ γ-AMINOPYRIDINE □ p-AMINOPYRIDINE □ 4-AP □ AVITROL □ 4-PYRIDINAMINE □ 4-PYRIDYLAMINE □ RCRA WASTE NUMBER P008 □ VMI 10-3
CONSENSUS REPORTS: Reported in EPA TSCA Inventory. EPA Extremely Hazardous Substances List.
DOT CLASSIFICATION: 6.1; Label: Poison
SAFETY PROFILE: Poison by ingestion, subcutaneous, intravenous, and intraperitoneal routes. Human systemic effects by ingestion: hallucinations and distorted perceptions, dyspnea, nausea or vomiting. When heated to decomposition it emits toxic fumes of NO_x.

AMI600   CAS: 143621-35-6 HR: 3
3-AMINO-PYRIDINE-2-CARBOX-
ALDEHYDE
mf: $C_7H_9N_5S$ mw: 195.27
SYNS: 2-((3-AMINO-2-PYRIDINYL)METHYLENE)-HYDRAZINECARBOTHIOAMIDE □ 3-AMINOPYRIDINE-2-CARBOXALDEHYDE THIOSEMICARBAZONE □ HYDRAZINECARBO-THIOAMIDE, 2-((3-AMINO-2-PYRIDINYL)METHYLENE)- □ TRIAPINE
SAFETY PROFILE: A poison by intravenous route. When heated to decomposition it emits toxic vapors of NO_x and SO_x.

AMK500   CAS: 68-89-3   HR: 3
AMINOPYRINE SODIUM SULFONATE
mf: $C_{13}H_{17}N_3O_4S$•Na   mw: 334.38
PROP: Minute crystals. Sol in water.
SYNS: (ANTIPYRINYLMETHYLAMINO)METHANESULFONIC ACID SODIUM SALT □ METHYLAMINO-ANTIPYRINE SODIUM METHANESULFONATE □ 4-METHYLAMINO-1,5-DIMETHYL-2-PHENYL-3-PYRAZOLONE SODIUM METHANESULFONATE □ METHYLAMINOPHENYLDIMETHYLPYRAZOLONE METHANESULFONATE SODIUM □ 1-PHENYL-2,3-DIMETHYL-5-PYRAZOLONE-4-METHYLAMINO-METHANESULFONATESODIUM □ 1-PHENYL-2,3-DIMETHYLPYRAZOLONE-(5)-4-METHYLAMINO-METHANESULFONICACID SODIUM □ PHENYL DIMETHYL PYRAZOLON METHYL AMINOMETH-ANE SODIUM SULFONATE □ 4-SODIUM METHANE-SULFONATE METHYLAMINE-ANTIPYRINE □ SODIUM METHYLAMINO-ANTIPYRINE METHANESULFONATE □ SODIUM-4-METHYLAMINO-1,5-DIMETHYL-2-PHENYL-3-PYRAZOLONE 4-METHANESULFONATE □ SODIUM NORAMIDOPYRINE METHANESULFONATE □ SODIUM-1-PHENYL-2,3-DIMETHYL-4-METHYL-AMINOPYRAZOLON-N-METHANE-SULFONATE □ SODIUM-1-PHENYL-2,3-DIMETHYL-5-PYRAZOLONE-4-METHYLAMINO METHANE-SULFONATE □ SODIUM PHENYLDIMETHYL-PYRAZOLONMETHYLAMINO-METHANE SULFONATE
SAFETY PROFILE: Poison by subcutaneous route. Moderately toxic by several other routes. An experimental teratogen. Other experimental reproductive effects. Human mutation data reported. Questionable carcinogen with experimental neoplastigenic data. See also

SULFONATES. When heated to decomposition it emits very toxic fumes of NO_x, Na_2O, and SO_x.

AMN300   CAS: 7722-06-7   HR: 3
4-AMINO-1,2,5-SELENADIAZOLE-3-CARBOXAMIDE

mf: $C_3H_4N_4OSe$ mw: 191.07
SYNS: NSC-84963 □ 1,2,5-SELENADIAZOLE-3-CARBOXAMIDE, 4-AMINO-
OSHA PEL: TWA 0.2 mg(Se)/m³
ACGIH TLV: TWA 0.2 mg(Se)/m³
SAFETY PROFILE: Poison by intraperitoneal route. When heated to decomposition it emits toxic fumes of NO_x and Se.

AMQ500   CAS: 23757-42-8   HR: 3
4-AMINO-2,2,5,5-TETRAKIS(TRI-FLUOROMETHYL)-3-IMIDAZOLINE

mf: $C_7H_3F_{12}N_3$ mw: 357.13
PROP: Solid. Mp: 159.7–160.4°.
SYNS: 5-AMINO-2,2,4,4-TETRAKIS(TRIFLUORO-METHYL)IMIDAZOLIDINE □ EXP 338
SAFETY PROFILE: Poison by ingestion, intraperitoneal, and intravenous routes. An experimental teratogen. Other experimental reproductive effects. When heated to decomposition it emits very toxic fumes of F^- and NO_x.

AMR500   CAS: 26861-87-0   HR: 3
2-AMINO-1,3,4-THIADIAZOLEHYDRO-CHLORIDE

mf: $C_2H_3N_3S•ClH$ mw: 137.60
SYNS: 2-AMINO-1,3,4-THIADIAZOLE, MONOHYDROCHLORIDE □ ATDA HYDROCHLORIDE
CONSENSUS REPORTS: Reported in EPA TSCA Inventory.
SAFETY PROFILE: Poison by intraperitoneal route. An experimental teratogen. Other experimental reproductive effects. When heated to decomposition it emits very toxic fumes of HCl, SO_x, and NO_x.

AMT500   CAS: 139-13-9   HR: 3
AMINOTRIACETIC ACID

mf: $C_6H_9NO_6$ mw: 191.16

PROP: Prismatic crystals from water. Mp: 242° (decomp), bp: 167° @ 13 mm. Sltly sol in water.
SYNS: N,N-BIS(CARBOXYMETHYL)GLYCINE □ CHEL 300 □ COMPLEXON I □ GLYCINE, N,N-BIS(CARBOXYMETHYL)-(9CI) □ HAMPSHIRE NTA ACID □ KOMPLEXON I □ KYSELINA NITRILOTRIOCTOVA □ NCI-C02766 □ NITRILOTRIACETIC ACID □ NTA □ TITRIPLEX I □ TRIGLYCINE □ TRIGLYCOLLAMIC ACID □ TRILON A □ α-α',α"-TRIMETHYLAMINETRICARBOXYLIC ACID □ VERSENE NTA ACID
CONSENSUS REPORTS: NTP 10th Report on Carcinogens. IARC Cancer Review: Group 2B IMEMDT 48,181,90; Animal Sufficient Evidence IMEMDT 48,181,90; Human No Adequate Data IMEMDT 48,181,90. NCI Carcinogenesis Bioassay (feed); Clear Evidence: mouse, rat NCITR* NCI-CG-TR-6,77. Reported in EPA TSCA Inventory. Community Right-To-Know List.
SAFETY PROFILE: Confirmed carcinogen with experimental carcinogenic and neoplastigenic data. Poison by intraperitoneal route. Moderately toxic by ingestion. When heated to decomposition it emits toxic fumes of NO_x.

AMW000   CAS: 2432-99-7   HR: 3
11-AMINOUNDECANOIC ACID

mf: $C_{11}H_{23}NO_2$ mw: 201.35
PROP: Solid. Mp: 190–192°.
SYNS: AMINOUNDECANOIC ACID □ 11-AMINOUNDECYLIC ACID □ NCI-C50613
CONSENSUS REPORTS: IARC Cancer Review: Group 3 IMEMDT 7,56,87; Animal Limited Evidence IMEMDT 39,239,86. NTP Carcinogenesis Bioassay (feed): Clear Evidence: mouse, rat NTPTR* NTP-TR-216,82. Reported in EPA TSCA Inventory.
SAFETY PROFILE: Poison by ingestion. Questionable carcinogen with experimental carcinogenic and neoplastigenic data. Mutation data reported. When heated to decomposition it emits toxic fumes of NO_x.

A

AMY050   CAS: 61-82-5   HR: 3
AMITROLE
mf: $C_2H_4N_4$ mw: 84.10
PROP: Crystals from H_2O, EtOH, or
EtOAc. Mp: 159°.
SYNS: AMEROL □ AMINOTRIAZOLE □ 2-
AMINOTRIAZOLE □ 2-AMINO-1,3,4-TRIAZOLE □ 3-
AMINOTRIAZOLE □ 3-AMINO-s-TRIAZOLE □ 3-
AMINO-1,2,4-TRIAZOLE (ACGIH) □ 3-AMINO-1,2,4-
TRIAZOLE □ 3-AMINO-1H-1,2,4-TRIAZOLE □
AMINOTRIAZOLE (PLANT REGULATOR) □ AMINO
TRIAZOLE WEEDKILLER 90 □ AMINOTRIAZOL-
SPRITZPULVER □ AMITOL □ AMITRIL □ AMITRIL T.L.
□ AMITROL □ AMITROL 90 □ AMITROL-T □ AMIZOL
□ AMIZOL D □ AMIZOL DP NAU □ AMIZOL F □ AT □
ATA □ 3,A-T □ AT-90 □ AT LIQUID □ AZAPLANT □
AZAPLANT KOMBI □ AZOLAN □ AZOLE □
CAMPAPRIM A 1544 □ CYTROL □ CYTROL AMITROLE-
T □ CYTROLE □ DIUROL □ DIUROL 5030 □
DOMATOL □ DOMATOL 88 □ ELMASIL □ EMISOL □
EMISOL 50 □ EMISOL F □ ENT 25,445 □ FENAMINE □
FENAVAR □ HERBIDAL TOTAL □ HERBIZOLE □
KLEER-LOT □ ORGA-414 □ RADOXONE TL □
RAMIZOL □ RCRA WASTE NUMBER U011 □ SIMAZOL
□ SOLUTION CONCENTREE T271 □ TRIAZOLAMINE
□ 1H-1,2,4-TRIAZOL-3-AMINE □ USAF XR-22 □ VOROX
□ VOROX AA □ VOROX AS □ WEEDAR ADS □
WEEDAR AT □ WEEDAZIN □ WEEDAZIN ARGINIT □
WEEDAZOL □ WEEDAZOL GP2 □ WEEDAZOL SUPER
□ WEEDAZOL T □ WEEDAZOL TL □ WEEDEX
GRANULAT □ WEEDOCLOR □ X-ALL Liquid
CONSENSUS REPORTS: NTP 10th Report
on Carcinogens. IARC Cancer Review:
Group 2B IMEMDT 7,92,87; Human
Inadequate Evidence IMEMDT 41,293,86;
IMEMDT 7,31,74; Animal Sufficient
Evidence IMEMDT 7,31,74; IMEMDT
41,293,86. Reported in EPA TSCA
Inventory. EPA Genetic Toxicology
Program.
OSHA PEL: TWA 0.2 mg/m³
ACGIH TLV: TWA 0.2 mg/m³; Animal
Carcinogen
DFG MAK: 0.2 mg/m³ (as total dust);
Confirmed Animal Carcinogen with
Unknown Relevance to Humans
SAFETY PROFILE: Confirmed carcinogen
with experimental carcinogenic,
tumorigenic, and neoplastigenic data. Poison
by intraperitoneal route. Moderately toxic by
ingestion. An experimental teratogen. Other
experimental reproductive effects. Mutation

data reported. When heated to
decomposition it emits toxic fumes of
NO_x. An herbicide and plant growth
regulator.

AMY500   CAS: 7664-41-7   HR: 3
AMMONIA
DOT: UN 1005
mf: H_3N mw: 17.04
PROP: Colorless, alkaline, nonflammable
gas with extremely pungent odor; liquefied
by compression. Mp: −77.7°, bp: −33.35°,
lel: 16%, uel: 25%, d: 0.771 g/liter @ 0°,
0.817 g/liter @ −79°, autoign temp: 1204°F,
vap press: 10 atm @ 25.7°, vap d: 0.6. Very
sol in water; moderately sol in alc. IDLH
300 ppm.
SYNS: AM-FOL □ AMMONIA ANHYDROUS □
AMMONIA, anhydrous, liquefied (DOT) □ AMMONIAC
(FRENCH) □ AMMONIACA (ITALIAN) □ AMMONIA
GAS □ AMMONIAK (GERMAN) □ AMMONIA
SOLUTIONS, relative density <0.880 at 15 degrees C in water,
with >50% ammonia (DOT) □ AMONIAK (POLISH) □
ANHYDROUS AMMONIA □ NITRO-SIL □ R 717 □
SPIRIT of HARTSHORN
CONSENSUS REPORTS: EPA Extremely
Hazardous Substances List. Community
Right-To-Know List. Reported in EPA
TSCA Inventory.
OSHA PEL: TWA 35 ppm
ACGIH TLV: TWA 25 ppm; STEL 35 ppm
DFG MAK: 20 ppm (14 mg/m³)
NIOSH REL: CL 50 ppm
DOT CLASSIFICATION: 2.3; Label: Poison
Gas; DOT Class: 2.2; Label: Nonflammable
Gas
SAFETY PROFILE: A human poison by an
unspecified route. Poison experimentally by
inhalation. An eye, mucous membrane, and
systemic irritant by inhalation. Mutation data
reported. A common air contaminant.
Difficult to ignite. Explosion hazard when
exposed to flame or in a fire. NH_3 + air in a
fire can detonate. Potentially violent or
explosive reactions on contact with
interhalogens (e.g., bromine pentafluoride,
chlorine trifluoride), 1,2-dichloroethane
(with liquid NH_3), boron halides,
chloroformamideium nitrate, ethylene oxide
(polymerization reaction), magnesium

perchlorate, nitrogen trichloride, oxygen + platinum, or strong oxidants (e.g., potassium chlorate, nitryl chloride, chromyl chloride, dichlorine oxide, chromium trioxide, trioxygen difluoride, nitric acid, hydrogen peroxide, tetramethylammonium amide, thiocarbonyl azide thiocyanate, sulfinyl chloride, thiotriazyl chloride, ammonium peroxodisulfate, fluorine, nitrogen oxide, dinitrogen tetraoxide, and liquid oxygen). Forms sensitive explosive mixtures with air + hydrocarbons, 1-chloro-2,4-dinitrobenzene, 2- or 4-chloronitrobenzene (above 160°C/30 bar), ethanol + silver nitrate, germanium derivatives, stibine, and chlorine. Reactions with silver chloride, silver nitrate, silver azide, and silver oxide form the explosive silver nitride. Reactions with chlorine azide, bromine, iodine, iodine + potassium, heavy metals and their compounds (e.g., gold(III) chloride, mercury, and potassium thallium amide ammoniate), tellurium halides (e.g., tellurium tetrabromide and tellurium tetrachloride) and pentaborane(9) give explosive products. Incompatible in contact with Ag, acetaldehyde, acrolein, B, BI_3, halogens, $HClO_3$, ClO, chlorites, chlorosilane, (ethylene dichloride + liquid ammonia), Au, hexachloromelamine, (hydrazine + alkali metals), HBr, HOCl, $Mg(ClO_4)_2$, N_2O_4, NCl_3, NF_3, OF_2, P_2O_5, P_2O_3, picric acid, (K + AsH_3), (K + PH_3), (K + $NaNO_2$), potassium ferricyanide, potassium mercuric cyanide, (Na + CO), Sb, S, SCl_2, tellurium hydropentachloride, trichloromelamine, NO_2Cl, SbH_3, tetramethylammonium amide, $SOCl_2$, and thiotrithiazylchloride. Incandescent reaction when heated with calcium. Emits toxic fumes of NH_3 and NO_x when exposed to heat. To fight fire, stop flow of gas.

AMY700   CAS: 1407-03-0   HR: D
AMMONIATED GLYCYRRHIZIN
PROP: From roots of *Glycyrrhiza glabra*.
SYN: MONOAMMONIUM GLYCYRRHIZINATE

SAFETY PROFILE: When heated to decomposition it emits acrid smoke and irritating fumes.

ANA000   CAS: 631-61-8   HR: 3
AMMONIUM ACETATE
mf: $C_2H_4O_2 \cdot H_3N$ mw: 77.10
PROP: Crystals. Mp: 114°, d: 1.07.
SYN: ACETIC ACID, AMMONIUM SALT
CONSENSUS REPORTS: Reported in EPA TSCA Inventory.
SAFETY PROFILE: Poison by intravenous route. Moderately toxic by intraperitoneal route. When heated to decomposition it emits toxic fumes of NO_x and NH_3.

ANA750   CAS: 12164-94-2   HR: 3
AMMONIUM AZIDE
mf: H_4N_4 mw: 60.08
PROP: Colorless plates. Mp: 160°, bp: explodes, d: 1.346, vap press: 1 mm @ 59.2° (sublimes).
DOT CLASSIFICATION: Forbidden
SAFETY PROFILE: Poison by inhalation and ingestion. See also AZIDES. Moderately flammable. Unstable. Explosion hazard upon rapid heating.

ANB250   CAS: 1066-33-7   HR: 3
AMMONIUM BICARBONATE (1:1)
mf: $HCO_3 \cdot H_4N$ mw: 79.1
PROP: Hard, colorless to white crystals or solid; faint ammonia odor, stable at room temp, volatile. Decomp below mp. Mp: 107.5° (rapid heating), d: 1.586. Sol in water; insol in alc.
SYNS: ACID AMMONIUM CARBONATE □ AMMONIUM CARBONATE □ AMMONIUM HYDROGEN CARBONATE □ CARBONIC ACID, MONOAMMONIUM SALT □ MONOAMMONIUM CARBONATE
CONSENSUS REPORTS: Reported in EPA TSCA Inventory.
SAFETY PROFILE: Poison by intravenous route. When heated to decomposition it emits toxic fumes of NO_x and NH_3.

ANB500   CAS: 7789-09-5   HR: 3
AMMONIUM BICHROMATE
DOT: UN 1439

A

mf: $Cr_2H_8N_2O_7$ mw: 252.10
PROP: Bright red-orange crystals; air-stable monoclinic crystals. Mp: decomp, d: 2.936. Sol in water and alc. IDLH Ca [15 mg/m³ {as Cr(VI)}].
SYNS: AMMONIO (DICROMATO DI) (ITALIAN) □ AMMONIUMBICHROMAAT (DUTCH) □ AMMON-IUMDICHROMAAT (DUTCH) □ AMMONIUM-DICHROMAT (GERMAN) □ AMMONIUM DICHROMATE □ AMMONIUM DICHROMATE(VI) □ BICHROMATE d'AMMONIUM (FRENCH)
CONSENSUS REPORTS: Reported in EPA TSCA Inventory. Chromium and its compounds are on the Community Right-To-Know List.
OSHA PEL: CL 0.1 mg(CrO₃)/m³
ACGIH TLV: TWA 0.05 mg(Cr)/m³; Confirmed Human Carcinogen
NIOSH REL: (Chromium(VI)) TWA 25 μg(Cr(VI))/m³; CL 50 μg/m³/15M
DOT CLASSIFICATION: 5.1; Label: Oxidizer
SAFETY PROFILE: Confirmed human carcinogen. Poison by inhalation, ingestion, skin contact, and subcutaneous routes. See also CHROMIUM COMPOUNDS. An unstable oxidizer. Moderately flammable; reacts with reducing agents.

ANB600   CAS: 10192-30-0   HR: 2
AMMONIUM BISULFITE
mf: $H_3N \cdot H_2O_3S$ mw: 99.12
PROP: A solid.
SYNS: AMMONIUM HYDROGEN SULFITE □ AMMONIUM MONOSULFITE □ MONOAMMONIUM SULFITE □ SULFUROUS ACID, MONOAMMONIUM SALT
CONSENSUS REPORTS: Reported in EPA TSCA Inventory.
DOT CLASSIFICATION: 8; Label: Corrosive
SAFETY PROFILE: A corrosive solid. When heated to decomposition it emits toxic vapors of NH_4^-.

AND250   HR: 3
AMMONIUM CADMIUM CHLORIDE
mf: $4NH_4Cl \cdot CdCl_2$ mw: 397.3
PROP: Colorless, rhombic crystals. D: 2.01; sol in water.

CONSENSUS REPORTS: Cadmium and its compounds are on the Community Right-To-Know List.
OSHA PEL: TWA 5 μg(Cd)/m³
ACGIH TLV: TWA 0.002 mg(Cd)/m³ (respirable dust), Suspected Human Carcinogen); BEI: 5 μg/g creatinine in urine; 5 μg/L in blood
DFG MAK: DFG BAT: Blood: 1.5 μg/dL; Urine: 15 μg/dL; Suspected Carcinogen
NIOSH REL: (Cadmium) Reduce to lowest feasible level
SAFETY PROFILE: Confirmed human carcinogen. A poison. When heated to decomposition it emits toxic fumes of NH_3, NO_x, and Cl⁻. See also CADMIUM COMPOUNDS.

AND750   CAS: 1111-78-0   HR: 3
AMMONIUM CARBAMATE
mf: $CH_3NO_2 \cdot H_3N$ mw: 78.09
PROP: White, crystalline, rhombic powder; sol in water and alc; ammonia odor. Sublimes at 60°.
SYN: AMMONIUM AMINOFORMATE
CONSENSUS REPORTS: Reported in EPA TSCA Inventory.
SAFETY PROFILE: Poison by intravenous route. See also CARBAMATES.

ANE000   CAS: 506-87-6   HR: 3
AMMONIUM CARBONATE
mf: $(NH_4)_2CO_3$ mw: 96.11
PROP: Colorless crystals; strong odor of NH_3; sharp taste. Decomposes on standing to ammonium bicarbonate. Mp: 58°. Sltly sol in water.
SYNS: AMMONIUMCARBONAT (GERMAN) □ CARBONIC ACID, AMMONIUM SALT □ CARBONIC ACID, DIAMMONIUM SALT □ DIAMMONIUM CARBONATE
CONSENSUS REPORTS: Reported in EPA TSCA Inventory.
SAFETY PROFILE: Poison by subcutaneous and intravenous routes. When heated to decomposition it emits toxic fumes of NO_x and NH_3.

ANE250   CAS: 10192-29-7   HR: 3
AMMONIUM CHLORATE
mf: ClH_3NO_3 mw: 100.49
PROP: White, unstable, colorless crystals or needles. Very soluble in water.
SYN: CHLORIC ACID, AMMONIUM SALT
DOT CLASSIFICATION: Forbidden
SAFETY PROFILE: A powerful oxidizer. Moderately flammable due to spontaneous chemical reaction. Explosion hazard due to shock, chemical reaction, or exposure to heat. A storage hazard; it may explode at room temperature. Explodes when heated to 100°C. When contaminat-ed it is very sensitive. Solution in water may explode if heated or dried. When heated to decomposition it emits highly toxic fumes of Cl^- and NO_x. Incompat-ible with reducing materials; BrF_3; BrF_5.

ANE500   CAS: 12125-02-9   HR: 3
AMMONIUM CHLORIDE
mf: $H_4N \cdot Cl$ mw: 53.50
PROP: White, hygroscopic solid or crystals; salty taste. Bp: 520°, mp: 337.8°, d: 1.520, vap press: 1 mm @ 160.4° (sublimes). Sol in water, alc, and glycerin.
SYNS: AMCHLOR □ AMMONERIC □ AMMONIUMCHLORID (GERMAN) □ AMMONIUM MURIATE □ CHLORID AMONNY (CZECH) □ DARAMMON □ SAL AMMONIA □ SAL AMMONIAC □ SALAMMONITE □ SALMIAC
CONSENSUS REPORTS: Reported in EPA TSCA Inventory.
OSHA PEL: (Fume) TWA 10 mg/m³; STEL 20 mg/m³
ACGIH TLV: TWA 10 mg/m³; STEL 20 mg/m³; (Proposed: 1 mg/m³).
SAFETY PROFILE: Poison by subcutaneous, intravenous, and intramuscular routes. Moderately toxic by other routes. A severe eye irritant. Mutation data reported. Explosive reaction with potassium chlorate or bromine trifluoride. Violent reaction (ignition) with bromine pentafluoride, NH_4, NO_3, and IF_7. Reaction with hydrogen cyanide may give the explosive nitrogen trichloride. When heated to decomposition it emits very toxic fumes of NO_x, Cl^-, and NH_3.

ANF250   CAS: 16919-58-7   HR: 3
AMMONIUM CHLOROPLATINATE
mf: $Cl_6Pt \cdot 2H_4N$ mw: 443.89
PROP: Cubic, yellow crystals or solid. D: 3.065, mp: decomp. Aq solns slowly photoreduce with substitution. Sol in water. IDLH 4 mg/m³ (as Pt).
SYNS: AMMONIUM HEXACHLOROPLATIN-ATE(IV) □ AMMONIUM PLATINIC CHLORIDE □ DIAMMONIUM HEXACHLOROPLATINATE (2-) □ PLATINIC AMMONIUM CHLORIDE
CONSENSUS REPORTS: Reported in EPA TSCA Inventory.
OSHA PEL: TWA 0.002 mg(Pt)/m³
ACGIH TLV: TWA 0.002 mg(Pt)/m³
SAFETY PROFILE: Poison by inhalation and ingestion. Human pulmonary system effects by inhalation. See also PLATINUM COMPOUNDS. An explosively unstable compound. Incompatible with KOH (boiling with alkali yields a product which, after drying, will explode @ 205° or if mixed with combustibles). When heated to decomposition it emits very toxic fumes of Cl^-, NO_x, and NH_3.

ANF500   CAS: 7788-98-9   HR: 3
AMMONIUM CHROMATE
mf: $(NH_4)_2CrO_4$ mw: 152.10
PROP: Yellow, crystalline material. Mp: decomp @ 180°, d: 1.91 @ 12°. Sol in cold water. IDLH Ca [15 mg/m³ {as Cr(VI)}].
SYNS: AMMONIUM CHROMATE(VI) □ CHROMIC ACID, DIAMMONIUM SALT □ DIAMMONIUM CHROMATE □ NEUTRAL AMMONIUM CHROMATE
CONSENSUS REPORTS: Chromium and its compounds are on the Community Right-To-Know List.
OSHA PEL: CL 0.1 mg(CrO₃)/m³
ACGIH TLV: TWA 0.05 mg(Cr)/m³; Confirmed Human Carcinogen
NIOSH REL: (Chromium(VI)) TWA 25 µg(Cr(VI))/m³; CL 50 µg/m³/15M
SAFETY PROFILE: A poison. Mutation data reported. See also CHROMIUM COMPOUNDS. A powerful oxidizer. An

A

explosion hazard when shocked or heated. When heated to decomposition it emits toxic fumes of NH_3, CrO_3, and NO_x. Incompatible with reducing agents.

ANF750   HR: 3
AMMONIUM CHROMIC SULFATE
mf: $NH_4Cr(SO_4)_2 \cdot 12H_2O$ mw: 478.4
PROP: Green or violet crystals. Mp: 94° ($-9H_2O$ @ 94°), d: 1.720, water sol. IDLH Ca [15 mg/m³ {as Cr(VI)}].
CONSENSUS REPORTS: Chromium and its compounds are on the Community Right-To-Know List.
OSHA PEL: CL 0.1 mg(CrO_3)/m³
ACGIH TLV: TWA 0.05 mg(Cr)/m³; Confirmed Human Carcinogen
NIOSH REL: (Chromium(VI)) TWA 25 μg(Cr(VI))/m³; CL 50 μg/m³/15M
SAFETY PROFILE: A confirmed carcinogen. Poison. See also CHROMIUM COMPOUNDS and SULFATES. When heated to decomposition it emits toxic fumes of NH_3, NO_x, and SO_x.

ANF800   CAS: 7632-50-0   HR: 2
AMMONIUM CITRATE
mf: $C_6H_8O_7 \cdot xH_3N$ mw: 311.42
PROP: Granules or crystals. D: 1.48. Sol in water; sltly sol in alc.
SYNS: AMMONIUM CITRATE, DIBASIC (DOT) □ CITRIC ACID, AMMONIUM SALT □ DIAMMONIUM CITRATE
CONSENSUS REPORTS: Reported in EPA TSCA Inventory.
SAFETY PROFILE: Experimental poison by intravenous route. A skin and eye irritant. When heated to decomposition it emits acrid smoke and irritating fumes.

ANH000   CAS: 13826-83-0   HR: 3
AMMONIUM FLUOBORATE
mf: NH_4BF_4 mw: 104.86
PROP: White, colorless, rhombic crystals. D: 1.871 @ 15°, mp: sublimes. Sol in NH_4OH and water.
SYNS: AMMONIUM BOROFLUORIDE □ AMMONIUM FLUOROBORATE □ AMMONIUM

TETRAFLUOROBORATE □ AMMONIUM TETRAFLUOROBORATE(1-)
CONSENSUS REPORTS: Reported in EPA TSCA Inventory.
OSHA PEL: TWA 2.5 mg(F)/m³
ACGIH TLV: TWA 2.5 mg(F)/m³; BEI: 3 mg/g creatinine of fluorides in urine prior to shift; 10 mg/g creatinine of fluorides in urine at end of shift.
NIOSH REL: (Fluorides, Inorganic) TWA 2.5 mg(F)/m³
SAFETY PROFILE: A poison and strong irritant. See also FLUORIDES and BORON COMPOUNDS. When heated to decomposition it emits very toxic fumes of F^-, NO_x, and NH_3.

ANH250   CAS: 12125-01-8   HR: 3
AMMONIUM FLUORIDE
DOT: UN 2505
mf: $H_4N \cdot F$ mw: 37.05
PROP: White, colorless, deliquescent crystals. Mp: sublimes, d: 1.009 @ 25°. Very sol in water; sltly sol in alc.
SYNS: AMMONIUM FLUORURE (FRENCH) □ NEUTRAL AMMONIUM FLUORIDE
CONSENSUS REPORTS: Reported in EPA TSCA Inventory.
OSHA PEL: TWA 2.5 mg(F)/m³
ACGIH TLV: TWA 2.5 mg(F)/m³; BEI: 3 mg/g creatinine of fluorides in urine prior to shift; 10 mg/g creatinine of fluorides in urine at end of shift.
NIOSH REL: (fluorides, inorganic) TWA 2.5 mg(F)/m³
DOT CLASSIFICATION: 6.1; Label: KEEP AWAY FROM FOOD
SAFETY PROFILE: Poison by subcutaneous and intraperitoneal routes. See also FLUORIDES. When heated to decomposition it emits very toxic fumes of F^-, NO_x, and NH_3. Incompatible with ClF_3.

ANH300   CAS: 14874-86-3   HR: 3
AMMONIUM FLUOROBERYLLATE
mf: $BeF_4 \cdot 2H_4N$ mw: 121.11
SYNS: AMMONIUM TETRAFLUOROBERYLLATE □ BERYLLATE(2-), TETRAFLUORO-, DIAMMONIUM, (T-4)- □ BERYLLATE(2-), TETRAFLUORO-, DIAMMONIUM □ DIAMMONIUM BERYLLIUM-TETRAFLUORIDE □

DIAMMONIUM TETRA-FLUOROBERYLLATE □ (T-4)-TETRAFLUORO-BERYLLATE(2-) DIAMMONIUM
SAFETY PROFILE: A poison by ingestion, inhalation, and intraperitoneal routes. When heated to decomposition it emits toxic vapors of NH_4^-, Be, and F^-.

ANI250   CAS: 16962-40-6   HR: 3
AMMONIUM HEXAFLUOROTITANATE
mf: $F_6Ti \cdot H_4N_2$ mw: 193.96
PROP: White solid.
CONSENSUS REPORTS: Reported in EPA TSCA Inventory.
OSHA PEL: TWA 2.5 mg(F)/m³
ACGIH TLV: TWA 2.5 mg(F)/m³; BEI: 3 mg/g creatinine of fluorides in urine prior to shift; 10 mg/g creatinine of fluorides in urine at end of shift.
NIOSH REL: (Fluorides, Inorganic) TWA 2.5 mg(F)/m³
SAFETY PROFILE: Poison by intravenous route. See also FLUORIDES, AMMONIA, and TITANIUM COMPOUNDS. When heated to decomposition it emits very toxic fumes of F^- and NO_x.

ANI500   CAS: 13815-31-1   HR: 3
AMMONIUM HEXAFLUOROVANADATE
mf: $F_6H_{12}N_3V$ mw: 219.09
SYN: HEXAFLUORO VANADATE (3-) TRIAMMON-IUM SALT
OSHA PEL: TWA 2.5 mg(F)/m³
ACGIH TLV: TWA 2.5 mg(F)/m³; BEI: 3 mg/g creatinine of fluorides in urine prior to shift; 10 mg/g creatinine of fluorides in urine at end of shift; TWA 0.05 mg(V_2O_5)/m³
NIOSH REL: (Vanadium Compounds) CL 0.05 mg(V)/m³/15M
SAFETY PROFILE: Poison by intravenous route. See also FLUORIDES and VANADIUM COMPOUNDS. When heated to decomposition it emits very toxic NH_3, NO_x, VO_x, and fluorides.

ANJ000   CAS: 1341-49-7   HR: 3
AMMONIUM HYDROGEN FLUORIDE
DOT: UN 1727/UN 2817
mf: F_2H_5N mw: 57.06

PROP: White, colorless crystals. D: 1.51, mp: 126°, bp: 239°. Will etch glass. Very sol in water; sltly sol in alc.
SYNS: ACID AMMONIUM FLUORIDE □ AMMONIUM BIFLUORIDE □ AMMONIUM DIFLUORIDE □ AMMONIUM FLUORIDE comp. with HYDROGEN FLUORIDE (1:1) □ AMMONIUM HYDROFLUORIDE □ AMMONIUM HYDROGEN BIFLUORIDE □ AMMONIUM HYDROGEN DIFLUORIDE □ AMMONIUM HYDROGEN FLUORIDE, solid (UN 1727) (DOT) □ AMMONIUM HYDROGEN FLUORIDE, solution (UN 2817) (DOT)
CONSENSUS REPORTS: Reported in EPA TSCA Inventory.
OSHA PEL: TWA 2.5 mg(F)/m³
ACGIH TLV: TWA 2.5 mg(F)/m³; BEI: 3 mg/g creatinine of fluorides in urine prior to shift; 10 mg/g creatinine of fluorides in urine at end of shift.
NIOSH REL: (Fluorides, Inorganic) TWA 2.5 mg(F)/m³
DOT CLASSIFICATION: 8; Label: Corrosive (UN 1727); DOT Class: 8; Label: Corrosive, Poison (UN 2817)
SAFETY PROFILE: Caustic poison and strong irritant by all routes. See also HYDROFLUORIC ACID. When heated to decomposition it emits very toxic fumes of F^-, NO_x, and NH_3.

ANJ500   CAS: 7803-63-6   HR: 2
AMMONIUM HYDROGEN SULFATE
DOT: UN 2506
mf: NH_4HSO_4 mw: 115.11
PROP: White rhombic crystals; sol in water; insol in acetone. Mp: 146.9°, d: 1.78.
SYNS: ACID AMMONIUM SULFATE □ AMMONIUM ACID SULFATE □ AMMONIUM BISULFATE □ AMMONIUM MONOHYDROGEN SULFATE □ MONOAMMONIUM SULFATE □ SULFURIC ACID, MONOAMMONIUM SALT
CONSENSUS REPORTS: Reported in EPA TSCA Inventory.
DOT CLASSIFICATION: 8; Label: Corrosive
SAFETY PROFILE: Moderately toxic by ingestion. A corrosive. See also SULFATES. Dangerous; when heated to decomposition it emits highly toxic fumes of sulfuric acid and SO_x, NH_3, and NO_x.

A

ANJ750   CAS: 12124-99-1   HR: 3
AMMONIUM HYDROSULFIDE
mf: NH_4HS mw: 51.11
PROP: Powder or crystals. Mp: 118° (150 atm), d: 1.17, vap press: 400 mm @ 21.8°.
SYNS: AMMONIUM BISULFIDE □ AMMONIUM HYDROGEN SULFIDE □ AMMONIUM HYDROSULFIDE, solution (DOT) □ AMMONIUM MERCAPTAN □ AMMONIUM SULFHYDRATE □ MONOAMMONIUM SULFIDE □ SIRNIK AMONNY □ TRUE AMMONIUM SULFIDE
CONSENSUS REPORTS: Reported in EPA TSCA Inventory.
SAFETY PROFILE: Poison by ingestion, subcutaneous, and intravenous routes. Moderately toxic by skin contact. Pyroforic in air. See also SULFIDES. When heated to decomposition it emits very toxic fumes of SO_x, NO_x, and NH_3. Incompatible with zinc.

ANK250   CAS: 1336-21-6   HR: 3
AMMONIUM HYDROXIDE
DOT: NA 2672
mf: $H_4N•HO$ mw: 35.06
PROP: Clear, colorless liquid solution of ammonia; very pungent odor. D: 0.90, mp: −77°. Sol in water. Soln contains not more than 44% ammonia.
SYNS: AMMONIA AQUEOUS □ AMMONIA WATER 29% □ AMMONIA SOLUTIONS, with >10% but not >35% ammonia (UN 2672) (DOT) □ AMMONIA SOLUTIONS, with >35% but not >50% ammonia (UN 2073) (DOT) □ AQUA AMMONIA
CONSENSUS REPORTS: Reported in EPA TSCA Inventory.
NIOSH REL: (Ammonia) CL 50 ppm
DOT CLASSIFICATION: 8; Label: Corrosive (UN 2672); DOT Class: 2.2; Label: Nonflammable Gas (UN 2073)
SAFETY PROFILE: A human poison by ingestion. An experimental poison by inhalation and ingestion. A severe eye irritant. Human systemic irritant effects by ocular and inhalation routes. Mutation data reported. Incompatible with acrolein, nitromethane, acrylic acid, chlorosulfonic acid, dimethyl sulfate, halogens, (Au + aqua regia), HCl, HF, HNO_3, oleum, β-propiolactone, propylene oxide, $AgNO_3$, Ag_2O, (Ag_2O + C_2H_5OH), $AgMnO_4$, H_2SO_4. Dangerous; liquid can inflict burns. Use with adequate ventilation. When heated to decomposition it emits NH_3 and NO_x.

ANL100   CAS: 57267-78-4   HR: 1
AMMONIUM ISETHIONATE
mf: $C_2H_5O_4S•H_3N$ mw: 142.17
SYNS: ETHANESULFONIC ACID, 2-HYDROXY-, AMMONIUM SALT □ 2-HYDROXYETHANE-SULFONIC ACID AMMONIUM SALT
CONSENSUS REPORTS: Reported in EPA TSCA Inventory.
SAFETY PROFILE: An eye irritant. When heated to decomposition it emits toxic fumes of SO_x, NH_3, and NO_x.

ANM000   HR: 3
AMMONIUM MAGNESIUM CHROMATE
mf: $(NH_4)_2CrO_4•MgCrO_4•6H_2O$ mw: 400.5
PROP: Yellow crystals. Mp: decomp, d: 1.84. Very water-sol. IDLH Ca [15 mg/m³ {as Cr(VI)}].
CONSENSUS REPORTS: Chromium and its compounds are on the Community Right-To-Know List.
OSHA PEL: CL 0.1 mg(CrO_3)/m³
ACGIH TLV: TWA 0.05 mg(Cr)/m³; Confirmed Human Carcinogen
NIOSH REL: (Chromium(VI)) TWA 25 μg(Cr(VI))/m³; CL 50 μg/m³/15M
SAFETY PROFILE: A confirmed carcinogen. A poison. See also CHROMIUM COMPOUNDS and MAGNESIUM COMPOUNDS. Moderately flammable; can explode. Incompatible with reducing agents. When heated to decomposition it can emit toxic fumes of NH_3 and NO_x.

ANM500   CAS: 5421-46-5   HR: 3
AMMONIUM MERCAPTOACETATE
mf: $C_2H_3O_2S•H_3N$ mw: 108.15
PROP: Colorless liquid; strong skunk-like odor.

SYNS: AMMONIUM THIOGLYCOLATE □ AMMONIUM THIOGLYCOLLATE □ THIO-GLYCOLLIC ACID, AMMONIUM SALT □ USAF MO-2

CONSENSUS REPORTS: Reported in EPA TSCA Inventory.

SAFETY PROFILE: Poison by intravenous and intraperitoneal routes. An allergen; can cause contact dermatitis. Emits hydrogen sulfide. See also SULFIDES. When heated to decomposition it emits very toxic NO_x, SO_x, and NH_3.

ANM750   CAS: 13106-76-8   HR: 3
AMMONIUM MOLYBDATE

mf: $MoO_4 \cdot 2H_4N$ mw: 196.04

PROP: White solid. Sol in water. IDLH 1000 mg/m³ (as Mo).

SYNS: AMMONIUM PARAMOLYBDATE □ DIAMMONIUM MOLYBDATE □ MOLYBDIC ACID DIAMMONIUM SALT

CONSENSUS REPORTS: Reported in EPA TSCA Inventory.

OSHA PEL: TWA 5 mg(Mo)/m³

ACGIH TLV: TWA Soluble Compounds: TWA 0.5 mg(Mo)/m³ Confirmed Animal Carcinogen with Unknown Relevance to Humans

SAFETY PROFILE: Poison by ingestion and intraperitoneal routes. Moderately toxic by other routes. An irritant. See also MOLYBDENUM COMPOUNDS. When heated to decomposition it emits toxic fumes of NH_3 and NO_x.

ANN000   CAS: 6484-52-2   HR: 3
AMMONIUM(I) NITRATE(1:1)

DOT: UN 0222/UN 1942/UN 2426

mf: $HNO_3 \cdot H_3N$ mw: 80.06

PROP: Colorless crystals. Mp: 169.6°, d: 1.725 @ 25°, bp: decomp >210°. Solubility: 192/100 @ 20°.

SYNS: AMMONIUM NITRATE □ AMMONIUM NITRATE, liquid (hot concentrated solution) (UN 2426) (DOT) □ AMMONIUM NITRATE, with >0.2% combustible substances (UN 0222) (DOT) □ AMMONIUM NITRATE, with not >0.2% of combustible substances (UN 1942) (DOT) □ AMMONIUM SALTPETER □ HERCO PRILLS □ NITRIC ACID, AMMONIUM SALT □ VARIOFORM I

CONSENSUS REPORTS: Community Right-To-Know List. Reported in EPA TSCA Inventory.

DOT CLASSIFICATION: 5.1; Label: Oxidizer (UN 2426); DOT Class: EXPLOSIVE 1.1D; Label: EXPLOSIVE 1.1D (UN 0222); DOT Class: 5.1; Label: Oxidizer (UN 1942)

SAFETY PROFILE: A powerful oxidizer and an allergen. See also NITRATES. A relatively stable explosive that has, however, caused many industrial explosions. Violent or explosive spontaneous reactions with acetic anhydride + nitric acid, ammonium sulfate + potassium, copper iron(II) sulfide, sawdust, urea, barium nitrate, hot water, and ammonium chloride + water + zinc. Forms heat- or shock-sensitive explosive mixtures with acetic acid, aluminum + calcium nitrate + formamide (a blasting explosive), ammonia, charcoal + metal oxides (e.g., rust, copper oxide, zinc oxide above 80°C), chloride salts (e.g., ammonium chloride, calcium chloride, iron(III) chloride, and aluminum chloride), cyanoguanidine, fertilizers (e.g., super phosphate + organic materials above 90°C), hydrocarbon oils, powdered metals (e.g., aluminum, antimony, bismuth, cadmium, chromium, cobalt, copper, iron, lead, magnesium, manganese, nickel, tin, zinc, brass, stainless steel, titanium, and potassium), nonmetals (e.g., charcoal, and phosphorus), organic fuels (e.g., wax, oils, and stearates), potassium permanganate, sugar, sulfur, and trinitroanisole. Reaction with alkali metals (e.g., sodium) forms an explosive product. Ignites on contact with ammonium dichromate, potassium dichromate, potassium chromate, barium chloride, sodium chloride, potassium nitrate, and chromium(VI) salts. Can ignite when mixed with acetic acid. Use water in large amounts to fight fire. It is important that the mass of materials be kept cool and that burning be extinguished promptly. Ventilate well. May explode under confinement and high temperatures. When heated to decomposition it emits highly toxic fumes of

NO_x. Can react vigorously with reducing materials. Incompatible with, (NH_4Cl + heat), (C + heat), organic matter, P, NaOCl, $NaClO_4$. Occasional explosions in presence of oil, $(NH_4)_2SO_4$ with K or Na.

ANO500   CAS: 135-20-6   HR: 3
AMMONIUM-N-NITROSOPHENYL-HYDROXYLAMINE

mf: $C_6H_6N_2O_2 \cdot H_4N$ mw: 156.19

PROP: Needles from water. Mp: 163–164°. Sol in water and alc; insol in Et_2O.

SYNS: CUPFERRON □ N-HYDROXY-N-NITROSO-BENZENAMINE, AMMONIUM SALT □ KUPFERRON (CZECH) □ NCI-C03258 □ N-NITROSOFENYL-HYDROXYLAMIN AMONNY (CZECH) □ N-NITROSOPHENYLHYDROXYLAMIN AMMONIUM SALZ (GERMAN) □ N-NITROSOPHENYLHYDRO-XYLAMINE AMMONIUM SALT

CONSENSUS REPORTS: NTP 10th Report on Carcinogens. NCI Carcinogenesis Bioassay (feed); Clear Evidence: mouse, rat NCITR* NCI-CG-TR-100,78. Reported in EPA TSCA Inventory. Community Right-To-Know List.

SAFETY PROFILE: Confirmed carcinogen with experimental carcinogenic and tumorigenic data. Poison by intravenous route. An eye irritant. Solutions with thorium salts are unstable explosives above 15°C. Solutions with titanium or zirconium salts are unstable explosives above 40°C. When heated to decomposition it emits very toxic NH_3 and NO_x. See also N-NITROSO COMPOUNDS and AMINES.

ANO750   CAS: 1113-38-8   HR: 3
AMMONIUM OXALATE

mf: $C_2H_2O_4 \cdot 2H_3N$ mw: 124.12

PROP: Colorless crystals. Mp: decomp, d: 1.50. Sltly sol in water.

SYNS: ETHANEDIOIC ACID DIAMMONIUM SALT □ OXALIC ACID, DIAMMONIUM SALT

DOT CLASSIFICATION: 6.1; Label: KEEP AWAY FROM FOOD

CONSENSUS REPORTS: Reported in EPA TSCA Inventory.

SAFETY PROFILE: A poison. Can react violently with (NaOCl + ammonium acetate). When heated to decomposition it can emit toxic fumes of NH_3 and NO_x. See also OXALATES.

ANP000   HR: 3
AMMONIUM PENTA PEROXODI-CHROMATE

mf: $Cr_2H_8N_2O_{12}$ mw: 332.2

PROP: IDLH Ca [15 mg/m³ {as Cr(VI)}].

CONSENSUS REPORTS: Chromium and its compounds are on the Community Right-To-Know List.

OSHA PEL: CL 0.1 mg(CrO_3)/m³

ACGIH TLV: TWA 0.05 mg(Cr)/m³; Confirmed Human Carcinogen

NIOSH REL: (Chromium(VI)) TWA 25 μg(Cr(VI))/m³; CL 50 μg/m³/15M

SAFETY PROFILE: A confirmed carcinogen. An unstable compound. Detonation can be initiated by heat, friction, or impact. See also CHROMIUM COMPOUNDS. Explodes @ 50°. When heated to decomposition it emits toxic fumes of NO_x.

ANP625   CAS: 3825-26-1   HR: 3
AMMONIUM PERFLUOROOCTANOATE

mf: $C_8F_{15}O_2 \cdot H_4N$ mw: 431.13

PROP: Solid.

SYNS: AMMONIUM PENTADECAFLUORO-OCTANATE □ AMMONIUM PERFLUORO-CAPRILATE □ AMMONIUM PERFLUORO-CAPRYLATE □ APFO □ FC-143 □ PERFLUORO-AMMONIUM OCTANOATE

CONSENSUS REPORTS: Reported in EPA TSCA Inventory.

ACGIH TLV: 0.01 mg/m³; Animal Carcinogen

SAFETY PROFILE: Confirmed carcinogen. Poison by inhalation. Moderately toxic by ingestion. An eye and skin irritant. Experimental reproductive effects. When heated to decomposition it emits toxic fumes of F⁻ and NH_3.

ANQ750   HR: 3
AMMONIUM PEROXYCHROMATE

mf: $(NH_4)_3CrO_2$ mw: 234.1

PROP: Red-brown crystals. Mp: decomp @ 40°, bp: explodes @ 50°.

CONSENSUS REPORTS: Chromium and its compounds are on the Community Right-To-Know List. IDLH Ca [15 mg/m³ {as Cr(VI)}].
OSHA PEL: CL 0.1 mg(CrO₃)/m³
ACGIH TLV: TWA 0.05 mg(Cr)/m³; Confirmed Human Carcinogen
NIOSH REL: (Chromium(VI)) TWA 25 μg(Cr(VI))/m³; CL 50 μg/m³/15M
SAFETY PROFILE: A confirmed carcinogen. A poison. See also CHROMIUM COMPOUNDS. Moderately flammable by chemical reaction with reducing agents. A powerful oxidizer. Moderately explosive when heated. When heated to decomposition it emits toxic fumes of NOₓ and NH₃.

ANR000   CAS: 7727-54-0   HR: 3
AMMONIUM PERSULFATE
DOT: UN 1444
mf: O₈S₂•2H₄N   mw: 228.22
PROP: Colorless, white, monoclinic crystals. Mp: decomp @ 120°, d: 1.982. Stable as dry solid; decomposes in H₂O forming O₂.
SYNS: AMMONIUM PEROXYDISULFATE □ PERSULFATE d'AMMONIUM (FRENCH)
CONSENSUS REPORTS: Reported in EPA TSCA Inventory.
ACGIH TLV: TWA 0.1 mg/m³
DOT CLASSIFICATION: 5.1; Label: Oxidizer
SAFETY PROFILE: Poison by intravenous and intraperitoneal routes. Moderately toxic by ingestion. A powerful oxidizer that can react vigorously with reducing agents. Releases oxygen when heated. Mixtures with sodium peroxide are explosives sensitive to friction, heating above 75°C, or contact with CO₂ or water. Mixtures with (powdered aluminum + water) or (zinc + ammonia) are explosive. Violent reaction with iron or solutions of ammonia + silver salts. Solution with sulfuric acid is a strong oxidizing cleaning solution. When heated to decomposition it emits toxic fumes of SOₓ, NH₃, and NOₓ.

ANR500   CAS: 7783-28-0   HR: 2
AMMONIUM PHOSPHATE, DIBASIC

mf: H₆N₂•H₃O₄P   mw: 132.08
PROP: White crystals or powder; salty taste. D: 1.619, mp: 185° (decomp). Sol in water; insol in alc.
SYNS: AMMONIUM PHOSPHATE □ DIAMMON-IUM HYDROGEN PHOSPHATE □ DIBASIC AMMONIUM PHOSPHATE □ SECONDARY AMMONIUM PHOSPHATE
CONSENSUS REPORTS: Reported in EPA TSCA Inventory.
SAFETY PROFILE: Low to moderate toxicity. See also PHOSPHATES. When heated to decomposition it emits very toxic fumes of POₓ, NOₓ, and NH₃.

ANR750   CAS: 7772-76-1   HR: 2
AMMONIUM PHOSPHATE, MONOBASIC
mf: NH₄H₂PO₄   mw: 115
PROP: Brilliant-white crystals or powder. D: 1.803 @ 19°, mp: 190°. Sol in water.
SAFETY PROFILE: Incompatible with NaOCl.

ANS500   CAS: 131-74-8   HR: 3
AMMONIUM PICRATE
DOT: UN 0004/UN 1310
mf: C₆H₃N₃O₇•H₃N   mw: 246.16
PROP: Red or yellow, rhombic crystals. D: 1.719, mp: decomp, bp: explodes @ 423°. Solubility: 1.1/100 @ 20°.
SYNS: AMMONIUM CARBAZOATE □ AMMONIUM PICRATE, dry or wetted with <10% water, by weight (UN 0004) (DOT) □ AMMONIUM PICRATE, wetted with not <10% water, by weight (UN 1310) (DOT) □ AMMONIUM PICRONITRATE □ EXPLOSIVE D □ OBELINE PICRATE □ PHENOL, 2,4,6-TRINITRO-, AMMONIUM SALT (9CI) □ PICRATOL □ PICRIC ACID, AMMONIUM SALT □ RCRA WASTE NUMBER P009 □ 2,4,6-TRINITROPHENOL AMMONIUM SALT
CONSENSUS REPORTS: Reported in EPA TSCA Inventory.
DOT CLASSIFICATION: EXPLOSIVE 1.1D; Label: EXPLOSIVE 1.1D (UN 0004)
SAFETY PROFILE: An allergen. Moderately irritating to skin, eyes, and mucous membranes. Moderately flammable by spontaneous chemical reaction. A powerful oxidizer that reacts vigorously with reducing materials. Dangerous explosive when shocked or heated. The presence of trace

A

metals increases its heat sensitivity. See PICRIC ACID, NITRATES, and EXPLOSIVES, HIGH. When heated to decomposition it emits highly toxic fumes of NO_x.

ANT000   CAS: 9080-17-5   HR: 3
AMMONIUM POLYSULFIDE (solution)
DOT: UN 2818
SYNS: AMMONIUM POLYSULFIDE, solution (DOT) □ AMMONIUM SULFIDE (POLY-) □ AMMONIUM SULFIDE, solution, red □ AMMONIUM TRISULFIDE □ AP-S □ DIAMMONIUM TRISULFIDE
CONSENSUS REPORTS: Reported in EPA TSCA Inventory.
DOT CLASSIFICATION: 8; Label: Corrosive, Poison
SAFETY PROFILE: Poison by ingestion. Moderately toxic by skin contact. See also AMMONIUM HYDROSULFIDE. When heated to decomposition it emits very toxic fumes of NO_x, SO_x, and H_2S.

ANT100   HR: D
AMMONIUM POTASSIUM HYDROGEN
**   PHOSPHATE**
CONSENSUS REPORTS: Carcinogenic Determination: Indefinite IARC** 9,245,75.
SAFETY PROFILE: When heated to decomposition it emits acrid smoke and irritating fumes.

ANT300   CAS: 19441-09-9   HR: 3
AMMONIUM REINECKATE HYDRATE
mf: $C_4H_{10}N_7S_4 \cdot Cr \cdot H_2O$ mw: 354.47
SYN: CHROMATE(1-), DIAMMINETETRAKIS-(ISOTHIOCYANATO)-, AMMONIUM, HYDRATE
OSHA PEL: CL 0.1 mg(CrO_3)/m^3
SAFETY PROFILE: Poison by subcutaneous and intravenous routes. When heated to decomposition it emits toxic fumes of NO_x, SO_x, and CR.

ANU650   CAS: 7773-06-0   HR: 2
AMMONIUM SULFAMATE
mf: $H_2NO_3S \cdot H_4N$ mw: 114.14
PROP: Deliquescent, hygroscopic, crystalline material (white crystalline solid). Bp: 160° (decomp), mp: 131°. Sol in water,

liq NH_3, formamide, and glycerol. IDLH 1500 mg/m^3.
SYNS: AMCIDE □ AMICIDE □ AMMAT □ AMMATE □ AMMONIUM AMIDOSULFONATE □ AMMONIUM AMIDOSULPHATE □ AMMONIUM-SALZ der AMIDOSULFONSAEURE (GERMAN) □ AMMONIUM SULPHAMATE □ AMS □ IKURIN □ MONOAMMONIUM SULFAMATE □ SULFAMATE □ SULFAMIC ACID, MONOAMMONIUM SALT □ SULFAMINSAEURE (GERMAN)
CONSENSUS REPORTS: Reported in EPA TSCA Inventory.
OSHA PEL: TWA 10 mg/m^3; Respirable Fraction: 5 mg/m^3; (Proposed: 1 mg/m^3).
ACGIH TLV: TWA 10 mg/m^3; (Proposed: 1 mg/m^3).
DFG MAK: 15 mg/m^3
SAFETY PROFILE: Moderately toxic by ingestion and intraperitoneal routes. Somewhat explosive when heated or by spontaneous chemical reaction in a hot acid solution. A powerful oxidizer. When heated to decomposition it emits very toxic fumes of NH_3, NO_x, and SO_x. See also SULFONATES and SULFAMIC ACID.

ANU750   CAS: 7783-20-2   HR: 2
AMMONIUM SULFATE (2:1)
mf: $H_8N_2O_4S$ mw: 132.16
PROP: White rhombic crystals. Mp: >280° (decomp), d: 1.77. Sol in water; insol in alc.
SYNS: AMMONIUM SULPHATE □ DIAMMONIUM SULFATE □ SULFURIC ACID, DIAMMONIUM SALT
CONSENSUS REPORTS: Community Right-To-Know List. Reported in EPA TSCA Inventory.
SAFETY PROFILE: Moderately toxic by several routes. Human systemic effects by ingestion: hypermotility, diarrhea, nausea or vomiting. See also SULFATES. Incandescent reaction on heating with potassium chlorate. Reaction with sodium hypochlorite gives the unstable explosive nitrogen trichloride. Incompatible with (K + NH_4NO_3), KNO_2, (NaK + NH_4NO_3). When heated to decomposition it emits very toxic fumes of NO_x, NH_3, and SO_x.

ANV750   CAS: 13453-06-0   HR: D
AMMONIUM TELLURATE

mf: $(NH_4)_2TeO_4$ mw: 227.7
PROP: White powder. Mp: decomp, d: 3.01 @ 25°.
SYN: TELLURIC ACID, AMMONIUM SALT
CONSENSUS REPORTS: Reported in EPA TSCA Inventory.
OSHA PEL: TWA 0.1 mg(Te)/m³
ACGIH TLV: TWA 0.1 mg(Te)/m³
SAFETY PROFILE: Human mutation data reported. See also TELLURIUM COMPOUNDS. When heated to decomposition it emits very toxic fumes of Te, NO_x, and NH_3.

ANV800   CAS: 13820-41-2   HR: 3
AMMONIUM TETRACHLORO-PLATINATE

mf: $Cl_4Pt•2H_4N$ mw: 372.99
PROP: Red-brown solid or crystals. Mp: 140–150° (decomp), d: 2.936. Sol in water. IDLH 4 mg/m³ (as Pt).
SYNS: PLATINATE(2-), TETRACHLORO-, DIAMMONIUM □ TETRAMINE PLATINUM(II) CHLORIDE
CONSENSUS REPORTS: Reported in EPA TSCA Inventory.
OSHA PEL: TWA 0.002 mg(Pt)/m³
ACGIH TLV: TWA 0.002 mg(Pt)/m³
SAFETY PROFILE: Poison by intraperitoneal route. When heated to decomposition it emits toxic fumes of NH_3, Cl^-, and Pt.

ANW500   HR: 3
AMMONIUM TETRAPEROXO CHROMATE

mf: $CrH_{12}N_3O_8$ mw: 234.2
PROP: IDLH Ca [15 mg/m³ {as Cr(VI)}].
CONSENSUS REPORTS: Chromium and its compounds are on the Community Right-To-Know List.
OSHA PEL: CL 0.1 mg(CrO₃)/m³
ACGIH TLV: TWA 0.05 mg(Cr)/m³; Confirmed Human Carcinogen
NIOSH REL: (Chromium(VI)) TWA 25 μg(Cr(VI))/m³; CL 50 μg/m³/15M
SAFETY PROFILE: A confirmed carcinogen. A poison. Impact explodes @ 50° or in contact with H_2SO_4. See also

CHROMIUM COMPOUNDS. Incompatible with H_2SO_4. When heated to decomposition it emits toxic fumes of NO_x.

ANW750   CAS: 1762-95-4   HR: 3
AMMONIUM THIOCYANATE

mf: $CNS•H_4N$ mw: 76.13
PROP: Colorless solid or deliquescent crystals. Mp: 149.6°, bp: decomp @ 170°, d: 1.305. Very sol in H_2O, EtOH; sol in MeOH, Me_2CO; prac insol in $CHCl_3$ and EtOAc.
SYNS: AMMONIUM RHODANATE □ AMMONIUM RHODANIDE □ AMMONIUM SULFOCYANATE □ AMMONIUM SULFOCYANIDE □ AMTHIO □ RHODANID □ RHODANIDE □ TRANS-AID □ USAF EK-P-433 □ WEEDAZOL TL
CONSENSUS REPORTS: Reported in EPA TSCA Inventory. EPA Genetic Toxicology Program.
SAFETY PROFILE: Poison by ingestion and intraperitoneal routes. Human systemic effects by ingestion: hallucinations and distorted perceptions, nausea or vomiting, and other gastrointestinal effects. See also THIOCYANATES. When heated to decomposition it emits toxic fumes of NH_3, NO_x, SO_x, and CN^-. Incompatible with $KClO_3$ and mixtures with $Pb(NO_3)_2$.

ANY250   CAS: 7803-55-6   HR: 3
AMMONIUM VANADATE

DOT: UN 2859
mf: $O_3V•H_4N$ mw: 116.99
PROP: Colorless to yellow crystals or solid. Mp: 200° (decomp), d: 2.326.
SYNS: AMMONIUM METAVANADATE (DOT) □ RCRA WASTE NUMBER P119 □ VANADIC ACID, AMMONIUM SALT
CONSENSUS REPORTS: Reported in EPA TSCA Inventory. EPA Genetic Toxicology Program.
ACGIH TLV: TWA 0.05 mg(V_2O_5)/m³
NIOSH REL: (Vanadium Compounds) CL 0.05 mg(V)/m³/15M
DOT CLASSIFICATION: 6.1; Label: Poison
SAFETY PROFILE: Poison by ingestion, subcutaneous, intravenous, intratracheal, and intraperitoneal routes. Moderately toxic

A

by skin contact. An experimental teratogen. Other experimental reproductive effects. Mutation data reported. See also VANADIUM COMPOUNDS. When heated to decomposition it emits toxic fumes of NH_3, VO_x, and NO_x.

ANY750   HR: 3
AMMONIUM VANADO-ARSENATE
mf: $H_{40}N_{10}O_5 \cdot 3As_2O_5 \cdot 4O_4V_2$ mw: 1228.78
CONSENSUS REPORTS: Arsenic and its compounds are on the Community Right-To-Know List.
ACGIH TLV: TLV 0.01 mg/m³Confirmed Human Carcinogen; BEI: 35 μ (As)/L inorganic arsenic and methylated metabolites in urine
NIOSH REL: (Vanadium Compounds) CL 0.05 mg(V)/m³/15M; (Arsenic, Inorganic) CL 2 μg(As)/m³/15M
SAFETY PROFILE: Confirmed human carcinogen. Poison by subcutaneous and intravenous routes. See ARSENIC and VANADIUM COMPOUNDS. When heated to decomposition it emits very toxic NO_x, NH_3, and As.

AOA100   CAS: 61336-70-7   HR: 2
AMOXICILLIN TRIHYDRATE
mf: $C_{16}H_{19}N_3O_5S \cdot 3H_2O$ mw: 419.50
SYNS: α-AMINO-p-HYDROXYBENZYLPENICILLIN TRIHYDRATE □ (2S-(2-α,5-α,6-β(S*)))-6-((AMINO(4-HYDROXYPHENYL)ACETYL)AMINO)-3,3-DIMETHYL-7-OXO-4-THIA-1-AZABICYCLO-(3.2.0)HEPTANE-2-CARBOXYLIC ACID TRIHYDRATE □ BRL 2333 TRIHYDRATE
SAFETY PROFILE: Moderately toxic. An experimental teratogen. Other experimental reproductive effects. When heated to decomposition it emits toxic fumes of SO_x and NO_x.

AOA125   CAS: 61-19-8   HR: 1
AMP
mf: $C_{10}H_{14}N_5O_7P$ mw: 347.26
PROP: Solid. Mp: 196–200°.
SYNS: ADENOSINE-5'-MONOPHOSPHATE □ ADENOSINE-5-MONOPHOSPHORIC ACID □ ADENOSINE-5'-MONOPHOSPHORIC ACID □

ADENOSINE PHOSPHATE □ ADENOSINE-5'-PHOSPHATE □ ADENOSINE-5'-PHOSPHORIC ACID □ ADENOVITE □ ADENYL □ ADENYLIC ACID □ tert-ADENYLIC ACID □ A5MP □ 5-AMP □ 5'-AMP □ AMP (nucleotide) □ CARDIOMONE □ ERGADENYLIC ACID □ LYCEDAN □ MUSCLE ADENYLIC ACID □ MY-B-DEN □ MYOSTON □ NSC-20264 □ PHOSADEN □ PHOSPHADEN □ PHOSPHENTASIDE
CONSENSUS REPORTS: Reported in EPA TSCA Inventory.
SAFETY PROFILE: Slightly toxic by intraperitoneal route. Experimental reproductive effects. Human mutation data reported. When heated to decomposition it emits toxic fumes of PO_x and NO_x.

AOA500   CAS: 51-64-9   HR: 3
d-AMPHETAMINE
mf: $C_9H_{13}N$ mw: 135.23
PROP: Oil. Bp: 102° @ 16 mm.
SYNS: d-2-AMINO-1-PHENYLPROPANE □ (+)-AMPHETAMINE □ AMSUSTAIN □ DEPHADREN □ DEXAMPHETAMINE □ DEXEDRINE □ α-METHYLPHENETHYLAMINE, d-FORM □ d-1-PHENYL-2-AMINOPROPAN (GERMAN) □ d-1-PHENYL-2-AMINOPROPANE
CONSENSUS REPORTS: Reported in EPA TSCA Inventory.
SAFETY PROFILE: Poison by ingestion, subcutaneous, intravenous, and intraperitoneal routes. Experimental reproductive effects. Chronic exposure causes central nervous system damage and blood-pressure effects. When heated to decomposition it emits toxic NO_x. See other amphetamine entries.

AOC500   CAS: 1397-89-3   HR: 3
AMPHOTERICIN B
mf: $C_{47}H_{73}NO_{17}$ mw: 924.21
PROP: Deep-yellow prisms from DMF. Insol in H_2O.
SYNS: AMB □ AMPHOMORONAL □ AMPHOTERICIN beta □ AMPHOTERICINE B □ AMPHOZONE □ FUNGILIN □ FUNGISONE □ FUNGIZONE □ IAB □ IODOACETAMIDE □ MYSTECLIN-F □ NSC-527017 □ TEGOPEN
SAFETY PROFILE: Poison by intravenous and intraperitoneal routes. Human systemic effects by intravenous route: leukopenia, lung changes, and cardiac changes.

Experimental reproductive effects. Mutation data reported. When heated to decomposition it emits toxic fumes of NO_x.

AOD125   CAS: 7177-48-2   HR: 1
AMPICILLIN TRIHYDRATE
mf: $C_{16}H_{19}N_3O_4S \cdot 3H_2O$ mw: 403.50
SYNS: AMCAP □ AMCILL □ AMINOBENZYLPENICILLIN TRIHYDRATE □ α-AMINOBENZYLPENICILLIN TRIHYDRATE □ AMPERIL □ AMPICHEL □ AMPIKEL □ AMPINOVA □ AMPLIN □ ANCILLIN □ CYMBI □ DIVERCILLIN □ LIFEAMPIL □ MOREPEN □ NCI-C56086 □ PEN A □ PENSYN □ POLYCILLIN □ PRINCILLIN □ RO-AMPEN □ TRAFARBIOT □ UKOPEN □ VIDOPEN
SAFETY PROFILE: Mildly toxic by ingestion. An experimental teratogen. Other experimental reproductive effects. When heated to decomposition it emits toxic fumes of SO_x and NO_x.

AOD175   CAS: 121-25-5   HR: D
AMPROLIUM
mf: $C_{14}H_{19}CIN_4$ mw: 278.78
PROP: Crystals from methanol + ethanol. Decomp 248–249°. Sol in water, methanol, 95% ether; insol in isopropanol, butanol, dioxane, acetone, ethyl acetate, acetonitrile, isooctane.
SYNS: 1-[(4-AMINO-2-PROPYL-5-PYRIMIDINYL)-METHYL]-2-METHYLPYRIDINIUM CHLORIDE □ 1-(4-AMINO-2-n-PROPYL-5-PYRIMIDINYLMETHYL)-2-PICOLINIUM CHLORIDE □ CORID
SAFETY PROFILE: When heated to decomposition emits toxic fumes of Cl^-.

AOD725   CAS: 628-63-7   HR: 3
n-AMYL ACETATE
DOT: UN 1104
mf: $C_7H_{14}O_2$ mw: 130.21
PROP: Colorless liquid; pear- or banana-like odor. Mp: −78.5°, bp: 148° @ 737 mm, ULC: 55–60, lel: 1.1%, uel: 7.5%, flash p: 77°F (CC), d: 0.879 @ 20°/20°, autoign temp: 714°F, vap d: 4.5. Very sltly sol in water; misc in alc and ether. IDLH 1000 ppm.
SYNS: ACETATE d'AMYLE (FRENCH) □ ACETIC ACID, AMYL ESTER □ AMYL ACETATE (DOT) □ AMYL ACETIC ESTER □ AMYLAZETAT (GERMAN) □ AMYLESTER KYSELINY OCTOVE □ BIRNENOEL □ OCTAN AMYLU (POLISH) □ PEAR OIL □ PENT-ACETATE □ 1-PENTANOL ACETATE □ PENTYL ACETATE □ n-PENTYL ACETATE □ 1-PENTYL ACETATE □ PRIMARY AMYL ACETATE
CONSENSUS REPORTS: Reported in EPA TSCA Inventory.
OSHA PEL: TWA 100 ppm
ACGIH TLV: TWA 50 ppm; STEL 100 ppm
DFG MAK: 50 ppm
DOT CLASSIFICATION: 3; Label: Flammable Liquid
SAFETY PROFILE: Moderately toxic by intraperitoneal route. Human systemic effects by inhalation: conjunctiva irritation, headache, and somnolence. A human eye irritant. Apparently more toxic than butyl acetate. Chronic toxicity is of a low order. Dangerous fire hazard when exposed to heat or flame; can react with oxidizing materials. Moderately explosive in the form of vapor when exposed to flame. To fight fire, use alcohol foam, dry chemical. When heated to decomposition it emits acrid smoke and irritating fumes. See also ESTERS, AMYL ALCOHOL, and ACETIC ACID.

AOD735   CAS: 626-38-0   HR: 3
sec-AMYL ACETATE
DOT: UN 1104
mf: $C_7H_{14}O_2$ mw: 130.21
PROP: Colorless liquid. Bp: 120°, flash p: 73.4°F (CC), d: 0.862–0.866 @ 20°/20°, vap d: 4.48, lel: 1.1%, uel: 7.5%. Sltly sol in water; misc in alc and ether. IDLH 1000 ppm.
SYNS: 2-ACETOXYPENTANE □ sek.AMYLESTER KYSELINY OCTOVE □ 2-AMYLESTER KYSELINY OCTOVE □ 1-METHYLBUTYL ACETATE □ 2-PENTANOL, ACETATE □ 2-PENTYL ACETATE
CONSENSUS REPORTS: Reported in EPA TSCA Inventory.
OSHA PEL: TWA 125 ppm
ACGIH TLV: TWA 50 ppm; STEL 100 ppm
DFG MAK: 50 ppm
DOT CLASSIFICATION: 3; Label: Flammable Liquid
SAFETY PROFILE: Mildly toxic by inhalation. Human systemic effects by

inhalation: conjunctiva irritation. Dangerous fire hazard when exposed to heat or flame; can react with oxidizing materials. Moderately explosive in the form of vapor when exposed to heat or flame. To fight fire, use alcohol foam, dry chemical. When heated to decomposition it emits acrid smoke and irritating fumes.

AOD750   HR: 2
AMYL ACETATE (mixed isomers)
mf: $C_7H_{14}O_2$ mw: 130.21
PROP: Colorless liquid; pear-like odor. Mp: $-78.5°$, bp: 148° @ 737 mm, ULC: 55–60, lel: 1.1%, uel: 7.5%, flash p: 77°F (CC), d: 0.879 @ 20°/20°, autoign temp: 714°F, vap d: 4.5.
SYN: ACETIC ACID, AMYL ESTER
ACGIH TLV: TWA 50 ppm; STEL 100 ppm
DFG MAK: 100 ppm (525 mg/m³)
SAFETY PROFILE: A skin irritant. Mildly toxic by ingestion. Dangerous fire hazard; can react with oxidizing materials. Moderately explosive in the form of vapor when exposed to flame. To fight fire, use alcohol foam, dry chemical. When heated to decomposition it emits acrid smoke and irritating fumes.

AOE000   CAS: 71-41-0   HR: 3
AMYL ALCOHOL
mf: $C_5H_{12}O$ mw: 88.17
PROP: Clear liquid. Mp: $-79°$, bp: 137.8°, flash p: 91°F (CC), d: 0.8168 @ 20°/20°, ULC: 40, lel: 1.2%, uel: 10% @ 212°F, vap press: 1 mm @ 13.6°, 10 mm @ 44.9°, vap d: 3.04. Sol in water; misc in alc and ether.
SYNS: ALCOOL AMYLIQUE (FRENCH) □ N-AMYL ALCOHOL □ AMYL ALCOHOL, NORMAL □ N-AMYLALKOHOL (CZECH) □ N-BUTYLCARBINOL □ N-PENTANOL □ PENTANOL-1 □ PENTAN-1-OL □ PENTASOL □ PENTYL ALCOHOL □ PRIMARY AMYL ALCOHOL
CONSENSUS REPORTS: Reported in EPA TSCA Inventory.
SAFETY PROFILE: Moderately toxic by ingestion and skin contact. An eye and upper respiratory irritant by inhalation. A severe skin and eye irritant. Ingestion can

cause headache, nausea, vomiting, delirium, and methemoglobin formation. Mutation data reported. Extremely flammable if exposed to heat, flame, or powerful oxidizers. Moderately explosive when exposed to flame. Incompatible with oxidizing materials, hydrogen trisulfide. To fight fire, use alcohol foam, dry chemical.

AOE200   CAS: 598-74-3   HR: 3
iso-AMYLAMINE
DOT: UN 2733/UN 2734
mf: $C_5H_{13}N$ mw: 87.19
SYNS: 2-BUTANAMINE, 3-METHYL-(9CI) □ 1,2-DIMETHYLPROPANAMINE □ 1,2-DIMETHYL-PROPYLAMINE □ 3-METHYL-2-BUTANAMINE □ PROPYLAMINE, 1,2-DIMETHYL-
CONSENSUS REPORTS: Reported in EPA TSCA Inventory.
DOT CLASSIFICATION: 8; Label: Corrosive, Flammable Liquid (UN 2734); DOT Class: 3; Label: Flammable Liquid, Corrosive (UN 2733)
SAFETY PROFILE: Poison by intraperitoneal route. A flammable liquid. When heated to decomposition it emits toxic vapors of NO_x.

AOG000   CAS: 540-18-1   HR: 2
n-AMYL BUTYRATE
mf: $C_9H_{18}O_2$ mw: 158.27
PROP: Colorless liquid. D: 0.871, mp: $-73.2°$, bp: 186.4°. Sol in water, misc with alc and ether.
SYNS: AMYL BUTYRATE □ BUTANOIC ACID PENTYL ESTER □ PENTYL BUTYRATE
CONSENSUS REPORTS: Reported in EPA TSCA Inventory.
SAFETY PROFILE: Mildly toxic by ingestion. When heated to decomposition it emits acrid smoke and irritating fumes.

AOG500   CAS: 122-40-7   HR: 2
α-AMYL CINNAMALDEHYDE
mf: $C_{14}H_{18}O$ mw: 202.32
PROP: Pale-yellow oil or liquid; floral jasmine odor. D: 0.963, refr index: 1.554, bp: 174–175° @ 20 mm. Sol in fixed oils; insol in glycerin and propylene glycol.

SYNS: α-AMYL CINNAMIC ALDEHYDE □ α-AMYL-β-PHENYLACROLEIN □ FEMA No. 2061 □ JASMINALDEHYDE □ α-PENTYLCINNAM-ALDEHYDE

CONSENSUS REPORTS: Reported in EPA TSCA Inventory.

SAFETY PROFILE: Moderately toxic by ingestion. A severe skin irritant. See also ALDEHYDES. When heated to decomposition it emits acrid smoke and irritating fumes.

AOG600   HR: 1
AMYL CINNAMATE
mf: $C_{14}H_{18}O_2$ mw: 218.28

PROP: Colorless to pale-yellow liquid; slt cocoa odor. D: 0.992–0.997, refr index: 1.535, flash p: 212°F. Sol in fixed oils; sltly sol in propylene glycol; insol in glycerin @ 310°.

SYNS: FEMA No. 2063 □ ISOAMYL CINNAMATE □ ISOAMYL 3-PENTYL PROPENATE

SAFETY PROFILE: Combustible liquid. When heated to decomposition it emits acrid smoke and irritating fumes.

AOH100   CAS: 68527-78-6   HR: 1
AMYL CINNAMYLIDENE METHYL
ANTHRANILATE
mf: $C_{22}H_{25}NO_2$ mw: 335.48

SYNS: ANTHRANILIC ACID, N-(2-BENZYLID-ENEHEPTYLIDENE)-, METHYL ESTER □ METHYL N-(β-PENTYLCINNAMYLIDENE)ANTHRANILATE

CONSENSUS REPORTS: Reported in EPA TSCA Inventory.

SAFETY PROFILE: A skin irritant. When heated to decomposition it emits toxic fumes of NO_x.

AOI200   CAS: 692-95-5   HR: 3
AMYLDICHLORARSINE
mf: $C_5H_{11}AsCl_2$ mw: 216.98

SYNS: N-AMYLDICHLORARSINE □ ARSINE, AMYLDICHLORO- □ ARSINE, DICHLOROPENTYL- □ DICHLOROPENTYLARSINE □ PENTYL-DICHLOROARSINE

OSHA PEL: TWA 0.5 mg(As)/m³

SAFETY PROFILE: Poison by skin contact. Moderately toxic by inhalation. When heated to decomposition it emits toxic fumes of As and Cl⁻.

AOI800   CAS: 25377-72-4   HR: 3
n-AMYLENE PENTENE
DOT: UN 1108
mf: C_5H_{10} mw: 70.15

SYNS: AMYLENE □ PENTYLENE

CONSENSUS REPORTS: Reported in EPA TSCA Inventory.

DOT CLASSIFICATION: 3; Label: Flammable Liquid

SAFETY PROFILE: Moderately toxic. Very flammable; reacts with heat, flame, and oxidizing materials. To fight fire, use foam, CO_2, dry chemical.

AOJ000   HR: 3
AMYLENES, MIXED
DOT: UN 1106
mf: C_5H_{10} mw: 70.58

PROP: Water-white liquid. Bp: 32.2°, flash p: 0°F, d: 0.66 @ 20°.

CONSENSUS REPORTS: Reported in EPA TSCA Inventory.

DOT CLASSIFICATION: 3; Label: Flammable Liquid

SAFETY PROFILE: Moderately toxic. Very flammable; reacts with heat, flame, and oxidizing materials. To fight fire, use foam, CO_2, dry chemical.

AOJ500   CAS: 638-49-3   HR: 3
n-AMYL FORMATE
mf: $C_6H_{12}O_2$ mw: 116.18

PROP: Clear liquid. D: 0.902, 0.893 @ 15°/4°, mp: −73.5°, bp: 130.4°, flash p: 80°F. Very sltly sol in water; misc in alc and ether.

SYNS: AMYL FORMATE □ PENTYL FORMATE □ n-PENTYL FORMATE

CONSENSUS REPORTS: Reported in EPA TSCA Inventory.

SAFETY PROFILE: Very low toxicity by several routes. A skin irritant. See also ESTERS. Dangerously flammable; reacts vigorously with heat, flame, oxidizing materials. To fight fire, use foam, CO_2, dry chemical.

AOK750   CAS: 105-30-6   HR: 2
AMYL METHYL ALCOHOL

A

mf: $C_6H_{14}O$ mw: 102.20
PROP: Liquid. Bp: 130°, flash p: 114°F
(CC), d: 0.804, vap d: 3.52.
SYNS: 1,3-DIMETHYL BUTANOL □ ISOHEXYL
ALCOHOL □ ISOPROPYL DIMETHYL CARBINOL □
METHYLAMYL ALCOHOL □ METHYL ISOBUTYL
CARBINOL □ 2-METHYLPENTANOL-1 □ 2-METHYL-2-
PROPYLETHANOL
CONSENSUS REPORTS: Reported in EPA
TSCA Inventory.
SAFETY PROFILE: Moderately toxic by
ingestion and skin contact. A skin and
severe eye irritant. Human systemic irritant
by inhalation. A flammable liquid; can react
with oxidizing materials. To fight fire, use
CO_2, dry chemical. When heated to
decomposition it emits smoke and acrid
fumes.

AOL000   CAS: 13256-07-0   HR: 3
n-AMYL-N-METHYLNITROSAMINE
mf: $C_6H_{14}N_2O$ mw: 130.22
SYNS: AMN □ METHYLAMYLNITROSAMIN
(GERMAN) □ METHYLAMYLNITROSAMINE □ METH-
YL-N-AMYLNITROSAMINE □ N-METHYL-N-NITROSO-
PENTYLAMINE □ METHYL-N-PENTYL-NITROSAMINE
□ N-NITROSO-N-METHYL-N-AMYL-AMINE □ NITRO-
SOMETHYL-N-PENTYLAMINE
CONSENSUS REPORTS: EPA Genetic
Toxicology Program.
SAFETY PROFILE: Poison by ingestion,
subcutaneous, and intraperitoneal routes.
Suspected carcinogen with experimental
carcinogenic, neoplastigenic, and
tumorigenic data. Mutation data reported.
When heated to decomposition it emits
toxic NO_x. See also NITROSAMINES and
N-NITROSO COMPOUNDS.

AOL250   CAS: 1002-16-0   HR: 3
AMYL NITRATE
DOT: UN 1112
mf: $C_5H_{11}NO_3$ mw: 133.17
PROP: Liquid. Bp: 145°, flash p: 125°F
(OC), d: 0.99.
SYNS: AMYLESTER KYSELINY DUSICNE □ NITRATE
d'AMYLE (FRENCH)
DOT CLASSIFICATION: 3; Label:
Flammable Liquid

SAFETY PROFILE: Moderately toxic by
inhalation. A flammable liquid. An
oxidizing agent. When heated to
decomposition it emits toxic fumes of NO_x.

AOL500   CAS: 463-04-7   HR: 3
n-AMYL NITRITE
mf: $C_5H_{11}NO_2$ mw: 117.17
PROP: Clear, yellowish liquid; peculiar,
ethereal, fruity odor and pungent, aromatic
taste. Bp: 104°, d: 0.853 @ 20°/4°, autoign
temp: 408°F, vap d: 4.0.
SYNS: AMYL NITRITE (DOT) □ 1-NITROPENTANE □
NITROUS ACID, PENTYL ESTER □ PENTYL NITRITE
CONSENSUS REPORTS: Reported in EPA
TSCA Inventory.
SAFETY PROFILE: Moderately toxic by
inhalation and ingestion. Causes flushing of
skin, rapid pulse, headache, and fall in blood
pressure. Mutation data reported. See also
NITRITES and ESTERS. Flammable when
exposed to heat or flame or by spontaneous
chemical reaction. To fight fire, use alcohol
foam. An oxidizing material. Vapors
explode when heated. It will react with
oxidizing or reducing materials. When
heated to decomposition it emits toxic
fumes of NO_x.

AOM125   CAS: 9032-08-0   HR: D
AMYLOGLUCOSIDASE
PROP: A powder derived from *Rhizopus
niveus* with diatomaceous earth as a carrier.
SYNS: CARBOHYDRASE □ GLUCOMYLASE
SAFETY PROFILE: When heated to
decomposition it emits acrid smoke and
irritating fumes.

AOM250   CAS: 14938-35-3   HR: 2
4-n-AMYLPHENOL
mf: $C_{11}H_{16}O$ mw: 164.27
PROP: A liquid. Bp: 342°, vap d: 5.66, flash
p: 219°F (OC), d: 0.966.
SYN: p-PENTYLPHENOL
CONSENSUS REPORTS: Reported in EPA
TSCA Inventory.
DOT CLASSIFICATION: 6.1; Label: KEEP
AWAY FROM FOOD

SAFETY PROFILE: Questionable carcinogen with experimental neoplastigenic data. Moderately flammable. To fight fire, use foam, CO_2, dry chemical. When heated to decomposition it emits acrid smoke and irritating fumes.

AOM500   CAS: 26401-74-1   HR: 3
2-sec-AMYLPHENOL

mf: $C_{11}H_{16}O$ mw: 164.27

PROP: Clear, straw-colored liquid. D: 0.955–0.971 @ 30°/30°, bp: 235–250°, flash p: 200°F. Very sltly sol in water; sol in oils and org solvs.

SYN: o-(sec-PENTYL) PHENOL

CONSENSUS REPORTS: Reported in EPA TSCA Inventory.

SAFETY PROFILE: Poison by intravenous route. Questionable carcinogen with experimental neoplastigenic data by skin contact. Moderately flammable when exposed to heat or flame. To fight fire, use foam, fog, dry chemical, water mist or spray, multipurpose dry chemical. When heated to decomposition it emits acrid smoke and irritating fumes.

AON300   CAS: 131-18-0   HR: D
AMYL PHTHALATE

mf: $C_{18}H_{26}O_4$ mw: 306.44

SYNS: AMOIL □ 1,2-BENZENEDICARBOXYLIC ACID, DIPENTYL ESTER □ DIAMYL PHTHALATE □ DIPENTYL PHTHALATE □ DI-n-PENTYL-PHTHALATE □ DPP □ PHTHALIC ACID, DIPENTYL ESTER

CONSENSUS REPORTS: Reported in EPA TSCA Inventory.

SAFETY PROFILE: Experimental reproductive effects. When heated to decomposition it emits acrid smoke and irritating fumes.

AON350   CAS: 624-54-4   HR: 3
AMYL PROPIONATE

mf: $C_8H_{16}O_2$ mw: 144.24

PROP: Colorless liquid; fruity, apricot-pineapple odor. D: 0.866, refr index: 1.405–1.409, flash p: 106°F. Sol in alc, fixed oils; insol in glycerin, propylene glycol, water @ 160°.

SYNS: AMYL PROPANOATE □ n-AMYL PROPIONATE □ FEMA No. 2082 □ ISOAMYL PROPIONATE □ PENTYL PROPANOATE □ n-PENTYL PROPANOATE □ PENTYL PROPIONATE □ PROPANOIC ACID, PENTYL ESTER □ PROPIONIC ACID, PENTYL ESTER (6CI,7CI,8CI)

CONSENSUS REPORTS: Reported in EPA TSCA Inventory.

SAFETY PROFILE: Low toxicity by ingestion and skin contact. An eye irritant. A flammable liquid. When heated to decomposition it emits acrid smoke and irritating fumes.

AON750   CAS: 64-43-7   HR: 3
AMYTAL SODIUM

mf: $C_{11}H_{17}N_2O_3$•Na   mw: 248.29

SYNS: 5-ETHYL-5-ISOPENTYLBARBITURIC ACID SODIUM SALT □ 5-ETHYL-5-(3-METHYLBUTYL)-BARBITURIC ACID, SODIUM DERIVATIVE □ 5-ISOAMYL-5-ETHYLBARBITURIC ACID, SODIUM DERIVATIVE □ SODIUM AMYLOBARBITONE □ SODIUM ETHYLISOAMYLBARBITURATE □ SODIUM ISOAMYLETHYL BARBITURATE

SAFETY PROFILE: Poison by ingestion, subcutaneous, intravenous, and intraperitoneal routes. When heated to decomposition it emits toxic NO_x and Na_2O.

AOO760   CAS: 8015-64-3   HR: 2
ANGELICA OIL, root

PROP: Extracted from roots of *Angelica archangelica L.* A pale-yellow to amber liquid; pungent odor with bittersweet taste. Sol in fixed oils; sltly sol in mineral oil; insol in glycerin, propylene glycol.

SYNS: ANGELICA ROOT OIL □ ANGELIKA OEL □ OILS, ANGELICA ROOT

CONSENSUS REPORTS: Reported in EPA TSCA Inventory.

SAFETY PROFILE: Moderately toxic by ingestion. When heated to decomposition it emits acrid smoke and irritating fumes.

AOO790   HR: D
ANGELICA SEED OIL

PROP: Extracted from seeds of *Angelica archangelica L.* A light yellow liquid; sweet taste. Sol in fixed oils; sltly sol in mineral oil; insol in glycerin, propylene glycol.

SAFETY PROFILE: When heated to decomposition it emits acrid smoke and irritating fumes.

AOP500   HR: 2
ANHYDRIDES
PROP: Chemical compounds derived from acids by elimination of a molecule of water. Thus, sulfur trioxide (SO_3) is the anhydride of sulfuric acid (H_2SO_4); carbon dioxide (CO_2) is the anhydride of carbonic acid (H_2CO_3); phthalic acid ($C_6H_4(CO_2H)_2$) minus water gives phthalic anhydride ($C_6H_4(CO_2)O$). This term should not be confused with anhydrous, meaning without water.
SAFETY PROFILE: Anhydrides are acidic and react with bases in tissue. Thus, they tend to attack and irritate tissue.

AOQ000   CAS: 62-53-3   HR: 3
ANILINE
DOT: UN 1547
mf: C_6H_7N mw: 93.14
PROP: Colorless, oily liquid which darkens on exposure to light; characteristic odor. Mp: $-6°$, bp: 184.4°, lel: 1.3%, ULC: 20–25, flash p: 158°F (CC), fp: $-6.2°$, d: 1.02 @ 20°/4°, autoign temp: 1139°F, vap press: 1 mm @ 34.8°, vap d: 3.22. IDLH 100 ppm.
SYNS: AMINOBENZENE □ AMINOPHEN □ ANILIN (CZECH) □ ANILINA (ITALIAN, POLISH) □ ANILINE OIL □ BENZENAMINE □ BLUE OIL □ C.I. 76000 □ HUILE d'ANILINE (FRENCH) □ NCI-C03736 □ PHENYLAMINE
CONSENSUS REPORTS: IARC Cancer Review: Group 3 IMEMDT 7,99,87; Animal Inadequate Evidence IMEMDT 4,27,74; Human No Evidence IMEMDT 4,27,74. EPA Extremely Hazardous Substances List. Community Right-To-Know List. Reported in EPA TSCA Inventory.
OSHA PEL: TWA 2 ppm (skin)
ACGIH TLV: TWA 2 ppm (skin); Animal Carcinogen; BEI: 50 mg/g creatinine of total p-aminophenol in urine at end of shift or 1.5% of hemoglobin for methemoglobin in blood during or end of shift.

DFG MAK: 2 ppm (7.7 mg/m³), Confirmed Animal Carcinogen with Unknown Relevance to Humans; BAT: 1 mg/L in urine at end of shift
DOT CLASSIFICATION: 6.1; Label: Poison
SAFETY PROFILE: Suspected carcinogen with experimental neoplastigenic data. A human poison by an unspecified route. Poison experimentally by most routes including inhalation and ingestion. Experimental reproductive effects. A skin and severe eye irritant, and a mild sensitizer. In the body, aniline causes formation of methemoglobin, resulting in prolonged anoxemia and depression of the central nervous system; less acute exposure causes hemolysis of the red blood cells, followed by stimulation of the bone marrow. The liver may be affected with resulting jaundice. Long-term exposure to aniline dye manufacture has been associated with malignant bladder growths. A common air contaminant. A combustible liquid when exposed to heat or flame. To fight fire, use alcohol foam, CO_2, dry chemical. It can react vigorously with oxidizing materials. When heated to decomposition it emits highly toxic fumes of NO_x. Spontaneously explosive reactions occur with benzenediazonium-2-carboxylate, dibenzoyl peroxide, fluorine nitrate, nitrosyl perchlorate, red fuming nitric acid, peroxodisulfuric acid, and tetranitromethane. Violent reactions with boron trichloride, peroxyformic acid, diisopropyl peroxydicarbonate, fluorine, trichloronitromethane (145°C), acetic anhydride, chlorosulfonic acid, hexachloromelamine, ($HNO_3 + N_2O_4 + H_2SO_4$), (nitrobenzene + glycerin), oleum, (HCHO + $HClO_4$), perchromates, K_2O_2, β-propiolactone, $AgClO_4$, Na_2O_2, H_2SO_4, trichloromelamine, acids, peroxydisulfuric acid, FO_3Cl, diisopropyl peroxy-dicarbonate, n-haloimides, and trichloronitromethane. Ignites on contact with sodium peroxide + water. Forms heat- or shock-sensitive explosive mixtures with anilinium chloride

(detonates at 240°C/7.6 bar), nitromethane, hydrogen peroxide, 1-chloro-2,3-epoxypropane, and peroxomonosulfuric acid. Reactions with perchloryl fluoride, perchloric acid, and ozone form explosive products.

AOQ500   HR: 2
ANILINE DYES
SAFETY PROFILE: The finished dyes are generally very much less toxic than many of the intermediates occurring or used in the manufacture of the dyes. Some of the aniline dyes cause local irritating effects to the eyes, mucous membranes, and skin; the basic dyes are believed to be more irritating than the acid dyes. Allergic responses to aniline dyes have been known to occur. See also specific compounds. When heated to decomposition they emit toxic fumes of NO_x and possibly SO_x.

AOR500   CAS: 548-62-9   HR: 3
ANILINE VIOLET
mf: $C_{25}H_{30}N_3 \cdot Cl$ mw: 408.03
PROP: Dark green powder or bright blue-violet crystals. Mp: 215° (decomp). Sol in H_2O, EtOH, $CHCl_3$.
SYNS: ADERGON □ AIZEN CRYSTAL VIOLET □ AIZEN CRYSTAL VIOLET EXTRA PURE □ ANILINE VIOLET PYOKTANINE □ ATMONIL □ AVERMIN □ AXURIS □ BADIL □ BASIC VIOLET 3 □ BASIC VIOLET BN □ BISMUTH VIOLET □ BLAUES PYOKTANIN □ BRILLIANT VIOLET 5B □ CALCOZINE VIOLET C □ CALCOZINE VIOLET 6BN □ C.I. 42555 □ C.I. BASIC VIOLET 3 □ CRYSTAL VIOLET □ CRYSTAL VIOLET AO □ CRYSTAL VIOLET AON □ CRYSTAL VIOLET 6B □ CRYSTAL VIOLET 10B □ CRYSTAL VIOLET BASE □ CRYSTAL VIOLET 5BO □ CRYSTAL VIOLET 6BO □ CRYSTAL VIOLET BP □ CRYSTAL VIOLET BPC □ CRYSTAL VIOLET O □ CRYSTAL VIOLET CHLORIDE □ CRYSTAL VIOLET EXTRA PURE □ CRYSTAL VIOLET EXTRA PURE APN □ CRYSTAL VIOLET EXTRA PURE APNX □ CRYSTAL VIOLET FN □ CRYSTAL VIOLET HL2 □ CRYSTAL VIOLET PURE DSC □ CRYSTAL VIOLET PURE DSC BRILLIANT □ CRYSTAL VIOLET SS □ CRYSTAL VIOLET TECHNICAL □ CRYSTAL VIOLET USP □ GENTERSAL □ GENTIAN VIOLET □ GENTIANAVIOLETT □ GENTIAVERM □ GENTICID □ GENTIOLETTEN □ HECTOGRAPH VIOLET SR □ HECTO VIOLET R □ HEXAMETHYLPARA-OSANILINE CHLORIDE □ HEXAMETHYL-p-ROSANILINE CHLORIDE □ HEXAMETHYL-p-ROSANILINE HYDROCHLORIDE □ HEXAMETHYL VIOLET □ HIDACO BRILLIANT CRYSTAL VIOLET □ HIDACO CRYSTAL VIOLET □ KRISTALL-VIOLETT □ MEROXYL □ MEROXYLAN □ MEROXYLAN-WANDER □ MEROXYL-WANDER □ METHYLROS-ANILINCHLORID □ METHYLROSANILINE CHLOR-IDE □ METHYLROSANILINUM CHLORATUM □ METHYL VIOLET 10B □ METHYL VIOLET 10BD □ METHYL VIOLET 10BK □ METHYL VIOLET 10BN □ METHYL VIOLET 5BNO □ METHYL VIOLET 10BNS □ METHYL VIOLET 5BO □ METHYL VIOLET 10BO □ METHYLVIOLETT □ MITSUI CRYSTAL VIOLET □ NCI-C55969 □ OXIURAN □ OXYCOLOR □ OXYOZYL □ PAPER BLUE R □ PARAROSANILINE, N,N,N',N',N",N"-HEXAMETHYL-, CHLORIDE □ PLASTORESIN VIOLET 5BO □ PYOKTANIN □ PYOVERM □ VERMICID □ VIANIN □ VIOCID □ 12416 VIOLET □ VIOLET 6BN □ VIOLET 5BO □ VIOLET CP □ VIOLET GENCIANOVA □ VIOLET KRYSTALOVA □ VIOLET XXIII □ VIOLET ZASADITA 3
CONSENSUS REPORTS: Reported in EPA TSCA Inventory.
SAFETY PROFILE: Poison by ingestion, intravenous, and intraperitoneal routes. An experimental teratogen. Other experimental reproductive effects. A human skin irritant. Human mutation data reported. Questionable carcinogen with experimental carcinogenic data. When heated to decomposition it emits very toxic fumes of NO_x and Cl^-.

AOR640   CAS: 5410-78-6   HR: 3
4-ANILINODICHLOROARSINE,
HYDROCHLORIDE
mf: $C_6H_6AsCl_2N \cdot ClH$ mw: 274.41
SYNS: ANILINE, p-DICHLOROARSINO-, HYDROCHLORIDE □ ARSINE, (p-AMINOPHENYL)-DICHLORO-, HYDROCHLORIDE □ p-DICHLORO-ARSINOANILINE HYDROCHLORIDE
OSHA PEL: TWA 0.5 mg(As)/m³
SAFETY PROFILE: Poison by intravenous route. When heated to decomposition it emits toxic fumes of NO_x, As, Cl^-, and HCl.

AOR750   CAS: 122-98-5   HR: 3
2-ANILINOETHANOL
mf: $C_8H_{11}NO$ mw: 137.20
PROP: D: 1.1, bp: 268°, flash p: 305°F (OC). Sltly sol in water.

SYNS: N-(2-HYDROXYETHYL)PHENYLAMINE □ 2-
(PHENYLAMINO)ETHANOL □ PHENYL
ETHANOLAMINE □ N-PHENYLETHANOLAMINE
CONSENSUS REPORTS: Reported in EPA
TSCA Inventory.
SAFETY PROFILE: Poison by skin contact,
intraperitoneal, and intravenous routes.
Moderately toxic by ingestion. A skin and
severe eye irritant. Combustible when
exposed to heat or flame. To fight fire, use
dry chemical, water mist. When heated to
decomposition it emits toxic fumes of NO_x.

AOT525   CAS: 135-02-4   HR: 2
o-ANISALDEHYDE
PROP: Crystals. Mp: 37–39°, bp: 238°, d:
1.127, flash p: 244°F. Sol in water.
mf: $C_8H_8O_2$ mw: 136.16
SYNS: 2-ANISALDEHYDE □ BENZALDEHYDE, 2-
METHOXY-(9CI) □ o-METHOXYBENZALDEHYDE □ 2-
METHOXYBENZALDEHYDE □ 6-METHOXY-
BENZALDEHYDE □ 2-METHOXYBENZENE-
CARBOXALDEHYDE □ SALICYLALDEHYDE METHYL
ETHER
CONSENSUS REPORTS: Reported in EPA
TSCA Inventory.
SAFETY PROFILE: Moderately toxic by
ingestion. A skin irritant. Mutation data
reported. Combustible liquid. When heated
to decomposition it emits acrid smoke and
irritating fumes.

AOT530   CAS: 123-11-5   HR: 2
p-ANISALDEHYDE
mf: $C_8H_8O_2$ mw: 136.15
PROP: Colorless oil; hawthorn odor. D:
1.123 @ 20°/4°, refr index: 1.571–1.574,
mp: 0°, bp: 247–248°, flash p: 250°F. Misc
in alc, ether, fixed oils; sol in propylene
glycol; insol in glycerin and water.
SYNS: ANISIC ALDEHYDE □ FEMA No. 2670 □ 4-
METHOXYBENZALDEHYDE □ p-METHOXYBENZ-
ALDEHYDE (FCC)
CONSENSUS REPORTS: Reported in EPA
TSCA Inventory.
SAFETY PROFILE: Moderately toxic by
ingestion. A skin irritant. Mutation data
reported. Combustible liquid. When heated
to decomposition it emits acrid smoke and
irritating fumes.

AOU250   CAS: 8007-70-3   HR: 2
ANISE OIL
PROP: Consists of (80–90%) anethole and
small quantities of methyl chavicol, p-
methoxyacetophenone, and other materials.
Found in the dried ripe fruit of *Impinella
anisum L.* (FCTXAV 11,855,73). D:
0.978–0.988 @ 25°/25°.
SYNS: ANISEED OIL □ ANIS OEL (GERMAN) □ OIL
OF ANISE □ STAR ANISE OIL
CONSENSUS REPORTS: Reported in EPA
TSCA Inventory.
SAFETY PROFILE: Moderately toxic by
ingestion. A weak sensitizer. May cause
contact dermatitis. Mutation data reported.
Combustible liquid. When heated to
decomposition it emits acrid smoke and
irritating fumes.

AOV000   CAS: 94-30-4   HR: 2
p-ANISIC ACID, ETHYL ESTER
mf: $C_{10}H_{12}O_3$ mw: 180.21
PROP: Colorless liquid; fruity, anise odor.
D: 1.103 @ 25°/25°, refr index:
1.522–1.526, mp: 7–8°, bp: 263°, flash p:
212°F. Sol in alc and ether; sltly sol in water.
SYNS: ETHYL ANISATE □ ETHYL-p-ANISATE (FCC)
□ ETHYL-p-METHOXYBENZOATE □ ETHYL-4-
METHOXYBENZOATE □ FEMA No. 2420
CONSENSUS REPORTS: Reported in EPA
TSCA Inventory.
SAFETY PROFILE: Moderately toxic by
ingestion. See also ESTERS. Combustible
liquid. When heated to decomposition it
emits acrid smoke and irritating fumes.

AOV500   CAS: 3290-99-1   HR: 3
p-ANISIC ACID, HYDRAZIDE
mf: $C_8H_{10}N_2O_2$ mw: 166.20
SYNS: ANISIC ACID HYDRAZIDE □ ANISIC
HYDRAZIDE □ ANISOYLHYDRAZINE □ p-
ANISOYLHYDRAZINE □ p-METHOXYBENZOIC ACID
HYDRAZIDE □ 4-METHOXYBENZOIC ACID
HYDRAZIDE □ p-METHOXYBENZOIC HYDRAZIDE □
(p-METHOXYBENZOYL)HYDRAZINE □ 4-
METHOXYBENZOYLHYDRAZINE □ 4-METHOXY-
BENZOYL HYDRAZIDE
SAFETY PROFILE: Poison by intravenous
route. Questionable carcinogen with
experimental neoplastigenic data. When

heated to decomposition it emits toxic fumes of NO_x.

AOV900   CAS: 90-04-0   HR: 3
o-ANISIDINE
DOT: UN 2431
mf: C_7H_9NO mw: 123.17
PROP: Yellowish liquid. Mp: 5°, bp: 225°. Sol in acids; insol in H_2O; misc in EtOH, Et_2O, C_6H_6. IDLH 50 mg/m³.
SYNS: o-AMINOANISOLE □ 2-AMINOANISOLE □ 1-AMINO-2-METHOXYBENZENE □ 2-ANISIDINE □ o-ANISYLAMINE □ BENZENAMINE, 2-METHOXY-(9CI) □ 2-METHOXY-1-AMINOBENZENE □ o-METHOXYANILINE □ 2-METHOXYANILINE □ 2-METHOXYBENZENAMINE □ o-METHOXY-PHENYLAMINE
CONSENSUS REPORTS: IARC Cancer Review: Group 2B IMEMDT 7,56,87; Human Limited Evidence IMEMDT 27,63,82. EPA Genetic Toxicology Program. Reported in EPA TSCA Inventory. Community Right-To-Know List.
OSHA PEL: TWA 0.5 mg/m³
ACGIH TLV: TWA 0.5 mg/m³ (skin); Animal Carcinogen
DFG MAK: Animal Carcinogen, Suspected Human Carcinogen
SAFETY PROFILE: Confirmed carcinogen. Moderately toxic by ingestion. Mutation data reported. When heated to decomposition it emits toxic fumes of NO_x.

AOW000   CAS: 104-94-9   HR: 1
p-ANISIDINE
mf: C_7H_9NO mw: 123.16
PROP: Crystals, plates from aq soln. D: 1.089 @ 55°/55°, mp: 57°, bp: 246°, vap d: 4.28. Sol in alc, ether, and hot water (insol in cold water). IDLH 50 mg/m³.
SYNS: p-AMINOANISOLE □ 4-AMINOANISOLE □ 1-AMINO-4-METHOXYBENZENE □ 4-ANISIDINE □ p-ANISYLAMINE □ p-METHOXYANILINE □ 4-METHOXYANILINE □ 4-METHOXYBENZENAMINE □ 4-METHOXYBENZENEAMINE □ p-METHOXY-PHENYLAMINE
CONSENSUS REPORTS: IARC Cancer Review: Group 3 IMEMDT 7,56,87; Human Inadequate Evidence IMEMDT

27,63,82. Community Right-To-Know List. Reported in EPA TSCA Inventory.
OSHA PEL: TWA 0.5 mg/m³
ACGIH TLV: TWA 0.5 mg/m³ (skin); Not Classifiable as a Human Carcinogen
DFG MAK: 0.1 ppm (0.51 mg/m³)
SAFETY PROFILE: Moderately toxic by several routes. A mild sensitizer. May cause a contact dermatitis. Mutation data reported. Questionable carcinogen. See also ANILINE. When heated to decomposition it emits toxic fumes of NO_x.

AOX250   CAS: 134-29-2   HR: 3
o-ANISIDINE HYDROCHLORIDE
mf: $C_7H_9NO•ClH$ mw: 159.63
SYNS: o-AMINOANISOLE HYDROCHLORIDE □ 2-AMINOANISOLE HYDROCHLORIDE □ o-ANISYL-AMINE HYDROCHLORIDE □ BENZENAMINE, 2-METHOXY-, HYDROCHLORIDE (9CI) □ C.I. 37115 □ FAST RED BB BASE □ 2-METHOXY-1-AMINOBENZ-ENE HYDROCHLORIDE □ o-METHOXYANILINE HYDROCHLORIDE □ 2-METHOXYANILINE HYDROCHLORIDE □ 2-METHOXYBENZENAMINE HYDROCHLORIDE □ 2-METHOXYBENZENE-AMINE HYDROCHLORIDE □ o-METHOXYPHENYL-AMINE HYDROCHLORIDE □ NCI-C03747
CONSENSUS REPORTS: NTP 10th Report on Carcinogens. IARC Cancer Review: Group 2B IMEMDT 7,56,87; Animal Sufficient Evidence IMEMDT 27,63,82; Human No Adequate Data IMEMDT 27,63,82. NCI Carcinogenesis Bioassay (feed); Clear Evidence: mouse, rat NCITR* NCI-CG-TR-89,78. Community Right-To-Know List.
SAFETY PROFILE: Confirmed carcinogen with experimental carcinogenic, neoplastigenic, and tumorigenic data. Mutation data reported. When heated to decomposition it emits very toxic fumes of NO_x and HCl.

AOX500   CAS: 20265-97-8   HR: 1
p-ANISIDINE HYDROCHLORIDE
mf: $C_7H_9NO•ClH$ mw: 159.63
SYN: NCI-C03758
CONSENSUS REPORTS: IARC Cancer Review: Group 3 IMEMDT 7,56,87; Animal Inadequate Evidence IMEMDT 27,63,82;

NCI Carcinogenesis Bioassay (feed); No Evidence: mouse NCITR* NCI-CG-TR-116,78; Inadequate Studies: rat NCITR* NCI-CG-TR-116,78. Reported in EPA TSCA Inventory.

SAFETY PROFILE: Questionable carcinogen with experimental carcinogenic and tumorigenic data. Mutation data reported. When heated to decomposition it emits very toxic fumes of NO_x and HCl.

AOX750   CAS: 100-66-3   HR: 3
ANISOLE
DOT: UN 2222
mf: C_7H_8O mw: 108.15
PROP: Mobile liquid, clear straw color; phenol, anise odor. Vapor d: 3.72, mp: $-37.3°$, bp: 153.8°, flash p: 125°F (COC), d: 0.983–0.988, refr index: 1.513–1.518, vap press: 10 mm @ 42.2°, autoign temp: 887°F. Insol in water; sol in alc and ether.
SYNS: BENZENE, METHOXY □ ETHER, METHYL PHENYL □ FEMA No. 2097 □ METHOXYBENZENE □ METHYL PHENYL ETHER □ PHENYL METHYL ETHER
CONSENSUS REPORTS: Reported in EPA TSCA Inventory.
DOT CLASSIFICATION: 3; Label: Flammable Liquid
SAFETY PROFILE: Moderately toxic by ingestion and inhalation. A skin irritant. A flammable liquid. To fight fire, use foam, CO_2, dry chemical. When heated to decomposition it emits acrid fumes.

AOY250   CAS: 100-07-2   HR: 3
ANISOYL CHLORIDE
DOT: UN 1729
mf: $C_8H_7ClO_2$ mw: 170.60
PROP: Needle-like crystals. Mp: 22°, bp: 160–164° @ 35 mm. Insol in water; sol in ether and acetone.
SYNS: p-ANISYOL CHLORIDE □ BENZOYL CHLORIDE, METHOXY-(9CI) □ METHOXY-BENZOYL CHLORIDE
CONSENSUS REPORTS: Reported in EPA TSCA Inventory.
DOT CLASSIFICATION: 8; Label: Corrosive
SAFETY PROFILE: Corrosive to skin, eyes, mucous membranes, and other tissue.

Evolves HCl by hydrolysis. A storage hazard; can explode spontaneously at room temperature. When heated to decomposition it emits toxic fumes of Cl⁻ and may explode.

AOY270   CAS: 72432-14-5   HR: 3
N-ANISOYL-GABA
mf: $C_{12}H_{15}NO_4$ mw: 237.26
SYN: BUTANOIC ACID, 4-((4-METHOXY-BENZOYL)AMINO)-
SAFETY PROFILE: A poison by ingestion. When heated to decomposition it emits toxic vapors of NO_x.

AOY400   HR: 1
ANISYL ACETATE
mf: $C_{10}H_{12}O_3$ mw: 180.20
PROP: Colorless to slt yellow liquid; fruity, balsamic odor. D: 1.104, refr index: 1.511–1.516, flash p: 210°F. Sol in alc and most oils; insol in glycerin and propylene glycol.
SYNS: FEMA No. 2098 □ p-METHOXYBENZYL ACETATE
SAFETY PROFILE: Combustible liquid. When heated to decomposition it emits acrid smoke and irritating fumes.

APE100   HR: 2
ANNATTO EXTRACT
PROP: From solvent extraction of *Bixa orellana L.* seeds (JAPMA8 49,218,60). Yellow-red solutions or powder.
SYNS: ACHIOTE □ BIXA ORELLANA
SAFETY PROFILE: Moderately toxic by intraperitoneal route. Human systemic effects by skin contact. When heated to decomposition it emits acrid smoke and irritating fumes.

APE300   CAS: 60837-57-2   HR: D
ANOXOMER
PROP: A polymer consisting of 1,4-benzenediol, 2-(1,1-dimethylethyl)-polymer with diethylbenzene, 4-(1,1-dimethylethyl)phenol, 4-methoxyphenol, 4,4'(1-methylethylidene)bis(phenol) and 4-methylphenol.

SAFETY PROFILE: When heated to decomposition it emits acrid smoke and irritating fumes.

APG100   CAS: 613-13-8   HR: 2
2-ANTHRACENAMINE
mf: $C_{14}H_{11}N$ mw: 193.26
PROP: Yellow leaflets from alcohol. Mp: 238°, bp: subl @ 93° @ 9 mm. Insol in water; sltly sol in alc and ether.
SYNS: β-AMINOANTHRACENE □ 2-AMINOAN-THRACENE □ 2-ANTHRACYLAMINE □ 2-ANTHR-AMINE □ 2-ANTHRYLAMINE
SAFETY PROFILE: Moderately toxic by intraperitoneal route. Suspected carcinogen with experimental carcinogenic and tumorigenic data. An experimental teratogen. Mutation data reported. See also AMINES. When heated to decomposition it emits toxic fumes of NO_x.

APG500   CAS: 120-12-7   HR: 2
ANTHRACENE
mf: $C_{14}H_{10}$ mw: 178.24
PROP: Colorless crystals, monoclinic plates from EtOH, violet fluorescence when pure. Mp: 217°, lel: 0.6%, flash p: 250°F (CC), d: 1.24 @ 27°/4°, autoign temp: 1004°F, vap press: 1 mm @ 145.0° (subl), vap d: 6.15, bp: 339.9°. Insol in water. Solubility in alc @ 1.9/100 @ 20°; in ether 12.2/100 @ 20°.
SYNS: ANTHRACEN (GERMAN) □ ANTHRACIN □ GREEN OIL □ PARANAPHTHALENE □ TETRA OLIVE N2G
CONSENSUS REPORTS: IARC Cancer Review: Group 3 IMEMDT 7,56,87; Animal Inadequate Evidence IMEMDT 32,105,83; Human No Adequate Data IMEMDT 32,105,83. Reported in EPA TSCA Inventory. Community Right-To-Know List.
OSHA PEL: TWA 0.2 mg/m³
SAFETY PROFILE: Moderately toxic by intraperitoneal route. A skin irritant and allergen. Questionable carcinogen with experimental neoplastigenic and tumorigenic data. Mutation data reported. Combustible when exposed to heat, flame, or oxidizing materials. Moderately explosive when

exposed to flame, $Ca(OCl)_2$, chromic acid. To fight fire, use water, foam, CO_2, water spray or mist, dry chemical. Explodes on contact with fluorine.

API500   CAS: 118-92-3   HR: 2
ANTHRANILIC ACID
mf: $C_7H_7NO_2$ mw: 137.15
PROP: Needle-like crystals or leaflets. Mp: 144–148°, bp: subl, d: 1.412 @ 20°. Solubility: in water: 0.35/100 @ 14°, in 90% alc: 10.7/100 @ 10°, in ether: 16/100 @ 70°.
SYNS: o-AMIDOBENZOIC ACID □ o-AMINO-BENZOIC ACID □ 2-AMINOBENZOIC ACID □ 1-AMINO-2-CARBOXYBENZENE □ CARBOXY-ANILINE □ o-CARBOXYANILINE □ 2-CARBOXY-ANILINE □ NCI-C01730 □ VITAMIN L
CONSENSUS REPORTS: IARC Cancer Review: Group 3 IMEMDT 7,56,87; Animal Inadequate Evidence IMEMDT 16,265,78. NTP Carcinogenesis Bioassay (feed): No Evidence: mouse, rat NCITR* NCI-TR-36,78. Reported in EPA TSCA Inventory.
SAFETY PROFILE: Moderately toxic by ingestion and intraperitoneal routes. Experimental reproductive effects. Human mutation data reported. Questionable carcinogen with experimental tumorigenic data. Combustible. When heated to decomposition it emits toxic fumes of NO_x.

API750   CAS: 87-29-6   HR: 2
ANTHRANILIC ACID, CINNAMYL ESTER
mf: $C_{16}H_{15}NO_2$ mw: 253.32
PROP: Reddish-yellow powder; balsamic odor. Mp: 60°, flash p: 212°F. Sol in alc, chloroform, ether; insol in water.
SYNS: 2-AMINOBENZOIC ACID-3-PHENYL-2-PROPENYL ESTER □ CINNAMYL ALCOHOL ANTHRANILATE □ CINNAMYL-o-AMINO-BENZOATE □ CINNAMYL-2-AMINOBENZOATE □ CINNAMYL ANTHRANILATE (FCC) □ FEMA No. 2295 □ NCI-C03510 □ 3-PHENYL-2-PROPENYL-ANTHRANILATE □ 3-PHENYL-2-PROPEN-1-YL ANTHRANILATE
CONSENSUS REPORTS: IARC Cancer Review: Group 3 IMEMDT 7,56,87; Animal Limited Evidence IMEMDT 31,133,83; Animal Inadequate Evidence IMEMDT 16,287,78; NCI Carcinogenesis Bioassay

(feed); Clear Evidence: mouse, rat NCITR* NCI-CG-TR-196,80. Reported in EPA TSCA Inventory.

SAFETY PROFILE: Suspected carcinogen with experimental carcinogenic and neoplastigenic data. Mutation data reported. See also ESTERS. Combustible liquid. When heated to decomposition it emits toxic fumes of NO_x.

APJ250   CAS: 134-20-3   HR: 2
ANTHRANILIC ACID, METHYL ESTER
mf: $C_8H_9NO_2$ mw: 151.18
PROP: Crystals or plates from alc or colorless liquid; grape odor. D: 1.161–1.169, mp: 24–25°, bp: 225–230° @ 15 mm, flash p: 219°F. Very sol in water, propylene glycol, hot abs alc (23/100); insol in ether, chloroform, and glycerin.
SYNS: o-AMINOBENZOIC ACID METHYL ESTER □ 2-AMINOBENZOIC ACID METHYL ESTER □ o-CARBOMETHOXYANILINE □ 2-CARBOMETHO-XYANILINE □ FEMA No. 2682 □ 2-(METHOXY-CARBONYL)ANILINE □ METHYL o-AMINOBENZO-ATE □ METHYL 2-AMINOBENZOATE □ METHYL ANTHRANILATE (FCC) □ METHYLESTER KYSELINY ANTHRANILOVE □ NEROLI OIL, ARTIFICAL
CONSENSUS REPORTS: Reported in EPA TSCA Inventory.
SAFETY PROFILE: Moderately toxic by ingestion. Experimental reproductive effects. A skin irritant. See also ESTERS. Combustible liquid. When heated to decomposition it emits toxic fumes of NO_x.

APJ500   CAS: 133-18-6   HR: 1
ANTHRANILIC ACID, PHENETHYL
   ESTER
mf: $C_{15}H_{15}NO_2$ mw: 241.31
PROP: White to yellow crystals; grape odor.
SYNS: BENZOIC ACID, 2-AMINO-, 2-PHENYLETHYL ESTER □ BENZYLCARBINYL ANTHRANILATE □ β-PHENETHYL-o-AMINO-BENZOATE □ PHENETHYL ANTHRANILATE □ 2-PHENYLETHYL-o-AMINOBENZOATE □ PHENYL-ETHYL ANTHRANILATE □ 2-PHENYLETHYL ANTHRANILATE
CONSENSUS REPORTS: Reported in EPA TSCA Inventory.

SAFETY PROFILE: A skin irritant. See also ESTERS. When heated to decomposition it emits toxic fumes of NO_x.

APK250   CAS: 84-65-1   HR: 2
ANTHRAQUINONE
mf: $C_{14}H_8O_2$ mw: 208.22
PROP: Yellow rhombic crystals from $PhNO_2$ or AcOH. Mp: 286°, bp: 376.9°, flash p: 365°F (CC), d: 1.438, vap press: 1 mm @ 190.0°, vap d: 7.16. Sol in hot C_6H_6, toluene, $PhNO_2$; mod sol in EtOH; sltly sol in Et_2O. Insol in water.
SYNS: 9,10-ANTHRACENEDIONE □ ANTHRADIONE □ 9,10-ANTHRAQUINONE □ 9,10-DIOXOANTHRACENE
CONSENSUS REPORTS: Reported in EPA TSCA Inventory.
SAFETY PROFILE: Moderately toxic by intraperitoneal route. A mild allergen. Mutation data reported. Combustible when exposed to heat or flame. To fight fire, use water, foam, CO_2, water spray or mist, dry chemical. When heated to decomposition it emits acrid smoke and irritating fumes.

APK850   CAS: 131-14-6   HR: 1
2,6-ANTHRAQUINONYLDIAMINE
mf: $C_{14}H_{10}N_2O_2$ mw: 238.26
PROP: Reddish-brown prisms. Mp: 310–320°. Insol in $CHCl_3$.
SYNS: 9,10-ANTHRACENEDIONE, 2,6-DIAMINO- □ ANTHRAQUINONE, 2,6-DIAMINO- □ 2,6-DIAMINO-ANTHRACHINON □ 2,6-DIAMINOANTHRA-QUINONE □ 2,6-DIAMINO-9,10-ANTHRAQUINONE
CONSENSUS REPORTS: Reported in EPA TSCA Inventory.
SAFETY PROFILE: An eye irritant. Mutation data reported. When heated to decomposition it emits toxic fumes of NO_x.

AQB000   CAS: 31282-04-9   HR: 3
ANTIHELMYCIN
mf: $C_{20}H_{37}N_3O_{13}$ mw: 527.60
PROP: Amorphous. Mp: 160–180° (decomp).
SYNS: HYGROMIX-8 □ HYGROMYCIN B (USDA)

SAFETY PROFILE: Poison by intraperitoneal route. Mutation data reported. When heated to decomposition it emits toxic fumes of NO_x.

AQB750   CAS: 7440-36-0   HR: 3
ANTIMONY
DOT: UN 2871
af: Sb   aw: 121.75
PROP: Silvery or gray, lustrous metalloid. Mp: 630°, bp: 1635°, d: 6.684 @ 25°, vap press: 1 mm @ 886°. Insol in water; sol in hot concentrated H_2SO_4. IDLH 50 mg/m³ (as Sb).
SYNS: ANTIMONY BLACK □ ANTIMONY POWDER (DOT) □ ANTIMONY REGULUS □ ANTYMON (POLISH) □ C.I. 77050 □ STIBIUM
CONSENSUS REPORTS: Antimony and its compounds are on the Community Right-To-Know List. Reported in EPA TSCA Inventory.
OSHA PEL: TWA 0.5 mg(Sb)/m³
ACGIH TLV: TWA 0.5 mg(Sb)/m³
DFG MAK: 0.5 mg(Sb)/m³
NIOSH REL: TWA 0.5 mg(Sb)/m³
DOT CLASSIFICATION: 6.1; Label: KEEP AWAY FROM FOOD
SAFETY PROFILE: An experimental poison by intraperitoneal route. Questionable carcinogen with experimental carcinogenic data. Moderate fire and explosion hazard in the forms of dust and vapor when exposed to heat or flame. See also POWDERED METALS. When heated or on contact with acid it emits toxic fumes of SbH_3. Electrolysis of acid sulfides and stirred Sb halide yields explosive Sb. It can react violently with NH_4NO_3, halogens, BrN_3, BrF_3, $HClO_3$, ClO, ClF_3, HNO_3, KNO_3, $KMnO_4$, K_2O_2, $NaNO_3$, oxidants.

AQC500   CAS: 10025-91-9   HR: 3
ANTIMONY(III) CHLORIDE
DOT: UN 1733
mf: Cl_3Sb mw: 228.10
PROP: Colorless, rhombic, deliq, hygroscopic crystals which fume in the air. D: 3.06, mp: 73.4°, bp: 220°, vap press: 1 mm @ 49.2° (subl). Sol in cold EtOH, CS_2, Et_2O, CCl_4, and H_2O (small amounts); insol in quinoline, other org bases.
SYNS: ANTIMOINE (TRICHLORURE d') □ ANTIMONIO (TRICLORURO di) □ ANTIMONOUS CHLORIDE □ ANTIMONOUS CHLORIDE (DOT) □ ANTIMONTRICHLORID □ ANTIMONY BUTTER □ ANTIMONY CHLORIDE □ ANTIMONY CHLORIDE (DOT) □ ANTIMONY TRICHLORIDE □ ANTIMONY TRICHLORIDE, liquid (DOT) □ ANTIMONY TRICHLORIDE, solid (DOT) □ ANTIMONY TRICHLORIDE, solution (DOT) □ ANTIMO-ONTRICHL-ORIDE □ BUTTER of ANTIMONY □ CHLORID ANTIMONITY □ CHLORURE ANTIMONIEUX □ C.I. 77056 □ STIBINE, TRICHLORO- □ TRICHLOROSTIBINE □ TRICHLORURE d'ANTIMOINE
CONSENSUS REPORTS: Reported in EPA TSCA Inventory. Antimony and its compounds are on the Community Right-To-Know List.
OSHA PEL: TWA 500 µg(Sb)/m³
ACGIH TLV: TWA 0.5 mg(Sb)/m³
NIOSH REL: (Antimony) TWA 0.5 mg(Sb)/m³
DOT CLASSIFICATION: 8; Label: Corrosive
SAFETY PROFILE: Moderately toxic by ingestion. Human pulmonary system effects by inhalation. Corrosive by vigorous reaction with moisture, generating heat and hydrogen chloride gas (a strong irritant), which can cause pulmonary edema when inhaled. Systemic effects can be caused by the antimony. See also ANTIMONY COMPOUNDS. Experimental reproductive effects. Mutation data reported. When heated to decomposition it emits very toxic fumes of chlorine and antimony. It can react violently with aluminum, potassium, sodium.

AQD000   CAS: 7647-18-9   HR: 3
ANTIMONY(V) CHLORIDE
DOT: UN 1730/UN 1731
mf: Cl_5Sb mw: 299.01
PROP: Colorless or red-yellow oil or liquid; offensive odor. Mp: 4°, bp: 140°, d: 2.336, vap press: 1 mm @ 22.7°. Decomp in water; sol in HCl, HBr, CS_2, CCl_4, and $CHCl_3$.
SYNS: ANTIMONIC CHLORIDE □ ANTIMONIO (PENTACLORURO DI) (ITALIAN) □ ANTIMONPENT-ACHLORID (GERMAN) □ ANTIMONY PENTACHLOR-

IDE □ ANTIMONY PENTACHLORIDE (DOT) □ ANTIMONY PERCHLORIDE □ ANTIMOONPENTA-CHLORIDE (DUTCH) □ BUTTER of ANTIMONY □ PENTACHLOROANTIMONY □ PENTACHLORURE d'ANTIMOINE (FRENCH) □ PERCHLORURE d'ANTIMOINE (FRENCH)

CONSENSUS REPORTS: Reported in EPA TSCA Inventory. Antimony and its compounds are on the Community Right-To-Know List.

OSHA PEL: TWA 500 μg(Sb)/m³

ACGIH TLV: TWA 0.5 mg(Sb)/m³

NIOSH REL: (Antimony) TWA 0.5 mg(Sb)/m³

DOT CLASSIFICATION: 8; Label: Corrosive

SAFETY PROFILE: Poison by ingestion. Corrosive. Mutation data reported. See ANTIMONY COMPOUNDS and ANTIMONY(III) CHLORIDE. When heated to decomposition it emits very toxic fumes of Cl⁻ and Sb.

AQD500   **HR: 3**
ANTIMONY COMPOUNDS

CONSENSUS REPORTS: On Community Right-To-Know List.

SAFETY PROFILE: Most antimony compounds are poisons by ingestion, inhalation, and intraperitoneal routes. See also ANTIMONY. Locally antimony compounds irritate the skin and mucous membranes. Sb^{+++} and hot $HClO_3$ can form an explosive mixture.

AQE000   CAS: 7783-56-4   HR: 3
ANTIMONY(III) FLUORIDE (1:3)

DOT: NA 1549

mf: F_3Sb mw: 178.75

PROP: Colorless, rhombic, very deliq crystals. Mp: 292°, bp: 376° (subl), d: 4.379 @ 20.9°. Readily sol in water with part hydrolysis, sol in polar org solvs. Insol in C_6H_6, chlorobenzene, pet ether.

SYNS: ANTIMOINE FLUORURE (FRENCH) □ ANTIMONOUS FLUORIDE □ ANTIMONY TRIFLUORIDE □ ANTIMONY TRIFLUORIDE, solid or solution (DOT) □ STIBINE, TRIFLUORO- (9CI) □ TRIFLUOROANTIMONY □ TRIFLUOROSTIBINE

CONSENSUS REPORTS: Reported in EPA TSCA Inventory. Antimony and its

compounds are on the Community Right-To-Know List.

OSHA PEL: TWA 0.5 mg(Sb)/m³; TWA 2.5 mg(F)/m³

ACGIH TLV: TWA 2.5 mg(F)/m³; BEI: 3 mg/g creatinine of fluorides in urine prior to shift; 10 mg/g creatinine of fluorides in urine at end of shift; TWA 0.5 mg(Sb)/m³

NIOSH REL: TWA 0.5 mg(Sb)/m³

DOT CLASSIFICATION: 8; Label: Corrosive

SAFETY PROFILE: Poison by subcutaneous route. Corrosive to skin and eyes. See also FLUORIDES and ANTIMONY COMPOUNDS. When heated to decomposition it emits very toxic fumes of F⁻ and Sb.

AQE250   CAS: 58164-88-8   HR: 3
ANTIMONY LACTATE

DOT: UN 1550

mf: $C_9H_{15}O_9 \cdot Sb$ mw: 388.99

PROP: Tan mass, water-sol.

SYNS: ANTIMONY LACTATE, solid (DOT) □ LACTIC ACID, ANTIMONY SALT □ PROPANOIC ACID, 2-HYDROXY-, TRIANHYDRIDE with ANTIMONIC ACID (H_3SbO_3) (9CI)

CONSENSUS REPORTS: Reported in EPA TSCA Inventory. Antimony and its compounds are on the Community Right-To-Know List.

OSHA PEL: TWA 0.5 mg(Sb)/m³

ACGIH TLV: TWA 0.5 mg(Sb)/m³

NIOSH REL: (Antimony) TWA 0.5 mg(Sb)/m³

DOT CLASSIFICATION: 6.1; Label: KEEP AWAY FROM FOOD

SAFETY PROFILE: A poison. See also ANTIMONY COMPOUNDS. When heated to decomposition it emits toxic fumes of Sb.

AQE305   CAS: 77824-44-3   HR: 3
ANTIMONYL-2,4-DIHYDROXY
   PYRIMIDINE

mf: $C_8H_6N_4O_6Sb_2$ mw: 497.68

SYN: 2,4,10,12-TETRAOXA-6,16,17,18-TETRAAZA-3,11-DISTIBATRICYCLO(11.3.1.1^{5,9})OCTADECA-1(17),5,7,9(18),13,15-HEXAENE, 3,11-DIHYDROXY-

OSHA PEL: TWA 0.5 mg(Sb)/m³

ACGIH TLV: TWA 0.5 mg(Sb)/m³
NIOSH REL: 10H TWA 0.5 mg(Sb)/m³
SAFETY PROFILE: Poison by intraperitoneal route. When heated to decomposition it emits toxic fumes of NO_x and Sb.

AQE320   CAS: 77824-43-2   HR: 3
ANTIMONYL-7-FORMYL-8-HYDROXY-QUINOLINE-5-SULPHONATE
mf: $C_{20}H_{11}N_2O_{11}S_2Sb\cdot2Na$ mw: 687.18
SYN: 5-QUINOLINESULFONIC ACID, 8,8'-((HYDROXYSTIBYLENE)BIS(OXY))BIS(7-FORMYL)-, DISODIUM SALT
OSHA PEL: TWA 0.5 mg(Sb)/m³
ACGIH TLV: TWA 0.5 mg(Sb)/m³
NIOSH REL: 10H TWA 0.5 mg(Sb)/m³
SAFETY PROFILE: Poison by intraperitoneal route. When heated to decomposition it emits toxic fumes of NO_x, SO_x, and Sb.

AQF000   CAS: 1309-64-4   HR: 3
ANTIMONY OXIDE
mf: O_3Sb_2 mw: 291.50
PROP: White cubes. D: 5.2, mp: 650°, bp: 1550° (subl). Very sltly sol in water; sol in KOH and HCl.
SYNS: A 1530 □ A 1582 □ A 1588LP □ AMSPEC-KR □ ANTIMONIOUS OXIDE □ ANTIMONY(3+) OXIDE □ ANTIMONY PEROXIDE □ ANTIMONY SESQUIOXIDE □ ANTIMONY TRIOXIDE □ ANTIMONY WHITE □ ANTOX □ ANZON-TMS □ AP 50 □ BLUE STAR □ CHEMETRON FIRE SHIELD □ C.I. 77052 □ C.I. PIGMENT WHITE 11 □ DECHLOR-ANE A-O □ DIANTIMONY TRIOXIDE □ EXITELITE □ EXTREMA □ FLOWERS of ANTIMONY □ NCI-C55152 □ NYACOL A 1530 □ SENARMONTITE □ THERMOGUARD B □ THERMOGUARD S □ TIMONOX □ TWINKLING STAR □ VALENTINITE □ WEISSPIESSGLANZ □ WHITE STAR
CONSENSUS REPORTS: Reported in EPA TSCA Inventory. Antimony and its compounds are on the Community Right-To-Know List.
OSHA PEL: TWA 0.5 mg(Sb)/m³
ACGIH TLV: TWA 0.5 mg(Sb)/m³; Suspected Carcinogen
DFG MAK: Animal Carcinogen, Suspected Human Carcinogen
NIOSH REL: TWA 0.5 mg(Sb)/m³

SAFETY PROFILE: Confirmed carcinogen with experimental carcinogenic and neoplastigenic data. Poison by intravenous and subcutaneous routes. Moderately toxic by other routes. An experimental teratogen. Other experimental reproductive effects. Mutation data reported. See also ANTIMONY COMPOUNDS. When heated to decomposition it emits toxic Sb fumes. Incompatible with chlorinated rubber and heat of 216° and with BrF_3.

AQF250   CAS: 7783-70-2   HR: 3
ANTIMONY(V) PENTAFLUORIDE
DOT: UN 1732
mf: F_5Sb mw: 216.75
PROP: Oily, colorless liquid. Very reactive. Mp: 7.0°, bp: 149.5°, d: (liq) 2.99 @ 23°. Sol in water and KF.
SYNS: ANTIMONY FLUORIDE □ ANTIMONY(V) FLUORIDE □ ANTIMONY PENTAFLUORIDE (DOT) □ PENTAFLUOROANTIMONY
CONSENSUS REPORTS: Reported in EPA TSCA Inventory. Antimony and its compounds are on the Community Right-To-Know List. EPA Extremely Hazardous Substances List.
OSHA PEL: TWA 0.5 mg(Sb)/m³
ACGIH TLV: TWA 0.5 mg(Sb)/m³
NIOSH REL: (Antimony) TWA 0.5 mg(Sb)/m³
DOT CLASSIFICATION: 8; Label: Corrosive, Poison
SAFETY PROFILE: A poison by inhalation. A very reactive, corrosive liquid to skin, eyes, mucous membranes. See also FLUORIDES and ANTIMONY COMPOUNDS. Violent reaction with phosphates. When heated to decomposition it emits very toxic fumes of F⁻ and Sb.

AQF500   CAS: 1315-04-4   HR: 3
ANTIMONY PENTASULFIDE
mf: S_5Sb_2 mw: 403.80
PROP: Dark orange-yellow powder or solid. Mp: 75° (decomp), d: 4.120. Insol dilute aqueous acids; sol conc HCl.

A

SYNS: ANTIMONIAL SAFFRON □ ANTIMONIC SULFIDE □ ANTIMONY RED □ ANTIMONY SULFIDE □ C.I. 77061 □ GOLDEN ANTIMONY SULFIDE

CONSENSUS REPORTS: Reported in EPA TSCA Inventory. Antimony and its compounds are on the Community Right-To-Know List.

OSHA PEL: TWA 0.5 mg(Sb)/m^3

ACGIH TLV: TWA 0.5 mg(Sb)/m^3

NIOSH REL: (Antimony) TWA 0.5 mg(Sb)/m^3

SAFETY PROFILE: Moderately toxic by intraperitoneal route. See also ANTIMONY COMPOUNDS and SULFIDES. Flammable when exposed to heat or by chemical reaction with powerful oxidizers. Use water to fight fire. Moderately explosive when shocked or by spontaneous chemical reaction in contact with powerful oxidizers. When heated to decomposition or on contact with acid or acid fumes it emits highly toxic fumes of oxides of sulfur and antimony. Incompatible with water or steam to produce toxic and flammable vapors and with oxidizers, e.g., $Ag(ClO_3)_2$, $HClO_3$, ClO_2, $Mg(ClO_3)_2$, TlO, $Zn(ClO_3)_2$.

AQF750   CAS: 1314-60-9   HR: 2
ANTIMONY PENTOXIDE

mf: O_5Sb_2 mw: 323.50

PROP: Yellowish-white powder or deep yellow crystals. D: 3.78, mp: decomp @ 380°. Insol in water; sltly sol in warm KOH soln.

SYNS: ANTIMONIC "ACID" □ ANTIMONIC OXIDE □ ANTIMONY PENTAOXIDE □ DIANTIMONY PENTOXIDE □ STIBIC ANHYDRIDE

CONSENSUS REPORTS: Reported in EPA TSCA Inventory. Antimony and its compounds are on the Community Right-To-Know List.

OSHA PEL: TWA 500 μg(Sb)/m^3

ACGIH TLV: TWA 0.5 mg(Sb)/m^3

NIOSH REL: (Antimony) TWA 0.5 mg(Sb)/m^3

SAFETY PROFILE: Moderately toxic by intraperitoneal route. See also ANTIMONY COMPOUNDS.

AQG250   CAS: 28300-74-5   HR: 3
ANTIMONY POTASSIUM TARTRATE

DOT: UN 1551

mf: $C_8H_4O_{12}Sb_2•3H_2O•2K$ mw: 635.88

PROP: Colorless crystals to white powder. D: 2.607, mp: loses H_2O @ 100°.

SYNS: ANTIMONYL POTASSIUM TARTRATE □ EMETIQUE (FRENCH) □ ENT 50,434 □ POTASSIUM ANTIMONYL TARTRATE □ POTASSIUM ANTIMONYL-d-TARTRATE □ POTASSIUM ANTIMONY TARTRATE □ TARTAR EMETIC □ TARTARIZED ANTIMONY □ TARTRATE ANTIMONIO-POTASSIQUE (FRENCH) □ TARTRATED ANTIMONY

CONSENSUS REPORTS: Antimony and its compounds are on the Community Right-To-Know List.

OSHA PEL: TWA 0.5 mg(Sb)/m^3

ACGIH TLV: TWA 0.5 mg(Sb)/m^3

NIOSH REL: (Antimony) TWA 0.5 mg(Sb)/m^3

DOT CLASSIFICATION: 6.1; Label: KEEP AWAY FROM FOOD

SAFETY PROFILE: Human poison by intravenous route, producing liver and kidney changes, somnolence, dyspnea, and pupillary dilation. Poison by ingestion, subcutaneous, intravenous, intramuscular, and intraperitoneal routes. Large doses cause severe liver damage. Human mutation data reported. Used medicinally, the therapeutic dose is close to the toxic dose. Upon decomposition it emits toxic fumes of K_2O and Sb.

AQH000   CAS: 11071-15-1   HR: 3
I-ANTIMONY POTASSIUM TARTRATE

mf: $C_4H_5O_7Sb•K$ mw: 325.94

SYNS: POTASSIUM ANTIMONYL-l-TARTRATE □ l-TARTARIC ACID, ANTIMONY POTASSIUM SALT

CONSENSUS REPORTS: Reported in EPA TSCA Inventory. Antimony and its compounds are on the Community Right-To-Know List.

OSHA PEL: TWA 0.5 mg(Sb)/m^3

ACGIH TLV: TWA 0.5 mg(Sb)/m^3

NIOSH REL: (Antimony) TWA 0.5 mg(Sb)/m^3

SAFETY PROFILE: Poison by intraperitoneal route. See also ANTIMONY COMPOUNDS. When heated to decomposition it emits toxic fumes of Sb and K_2O.

AQH800   CAS: 16037-91-5   HR: 3
ANTIMONY SODIUM GLUCONATE

mf: $C_{12}H_{20}O_{17}Sb_2•3Na•9H_2O$ mw: 1048.91

SYNS: ESTIBOGLUCONATO SODICO □ d-GLUCONIC ACID, CYCLIC ESTER with ANTIMONIC ACID ($H_8Sb_2O_9$) (2:1),TRISODIUM SALT, NONA-HYDRATE □ d-GLUCONIC ACID, 2,4:2',4'-O-(OXYDISTIBYLIDYNE)BIS-, Sb,Sb'-DIOXIDE, TRISODIUM SALT, NONAHYDRATE □ MYOSTIBIN □ PENTOSTAM □ SODIUM STIBOGLUCONATE □ SOLUSTIBOSAN □ SOLUSTIN □ SOLUSURMIN □ SOLYUSURMIN □ STIBANATE □ STIBANOSE □ STIBATIN □ STIBINOL

OSHA PEL: TWA 0.5 mg(Sb)/m^3
ACGIH TLV: TWA 0.5 mg(Sb)/m^3
NIOSH REL: (Antimony) 10H TWA 0.5 mg(Sb)/m^3

SAFETY PROFILE: Poison by intraperitoneal route. When heated to decomposition it emits toxic fumes of Sb.

AQI750   CAS: 34521-09-0   HR: 3
ANTIMONY SODIUM TARTRATE

mf: $C_8H_4O_{12}Sb_2•2Na$ mw: 581.60

SYNS: ANTIMONY SODIUM OXIDE-l-(+)-TARTRATE □ NATRIUMANTIMONYLTARTRAT (GERMAN) □ SODIUM ANTIMONYL TARTRATE □ SODIUM ANTIMONY TARTRATE □ STIBNAL □ STIBUNAL

CONSENSUS REPORTS: Antimony and its compounds are on the Community Right-To-Know List.

OSHA PEL: TWA 0.5 mg(Sb)/m^3
ACGIH TLV: TWA 0.5 mg(Sb)/m^3
NIOSH REL: (Antimony) TWA 0.5 mg(Sb)/m^3

SAFETY PROFILE: Poison by subcutaneous, intravenous, and intraperitoneal routes. Human toxic effects by intravenous route. Human mutation data reported. See also ANTIMONY COMPOUNDS. When heated to decomposition it emits toxic fumes of Sb.

AQJ750   CAS: 6923-52-0   HR: 1
ANTIMONY TRIACETATE

mf: $C_6H_9O_6•Sb$ mw: 298.90

SYNS: ACETIC ACID, TRIANHYDRIDE with ANTIMONIC ACID □ ANTIMONY(III) ACETATE □ OCTAN ANTIMONITY (CZECH)

CONSENSUS REPORTS: Reported in EPA TSCA Inventory. Antimony and its compounds are on the Community Right-To-Know List.

OSHA PEL: TWA 0.5 mg(Sb)/m^3
ACGIH TLV: TWA 0.5 mg(Sb)/m^3
NIOSH REL: (Antimony) TWA 0.5 mg(Sb)/m^3

SAFETY PROFILE: Mildly toxic by ingestion. A skin and eye irritant. Mutation data reported. See also ANTIMONY COMPOUNDS. When heated to decomposition it emits acrid smoke and irritating fumes.

AQK000   CAS: 7789-61-9   HR: 3
ANTIMONY TRIBROMIDE

DOT: NA 1549

mf: Br_3Sb mw: 361.48

PROP: Yellow or white, deliquescent, crystalline mass becoming amber yellow when fused. Hygroscopic. Decomp by water. Mp: 96°, bp: 280°, d: 4.145, vap press: 1 mm @ 93.9°.

SYNS: ANTIMONY TRIBROMIDE □ ANTIMONY TRIBROMIDE, solid or solution (DOT) □ TRIBROMO-STIBINE

CONSENSUS REPORTS: Antimony and its compounds are on the Community Right-To-Know List. Reported in EPA TSCA Inventory.

OSHA PEL: TWA 0.5 mg(Sb)/m^3
ACGIH TLV: TWA 0.5 mg(Sb)/m^3
DOT CLASSIFICATION: 8; Label: Corrosive

SAFETY PROFILE: A poison. Corrosive to skin, eyes, and mucous membranes. Reaction with water liberates HBr and antimony trioxide. Can cause severe burns. See also ANTIMONY COMPOUNDS.

AQL500   CAS: 1345-04-6   HR: 3
ANTIMONY TRISULFIDE

mf: S_3Sb_2 mw: 339.68

PROP: Red-to-black crystals. Mp: 563°, d: 4.64. Sol in H_2SO_4; solubility in water: 0.002/100 @ 20° (decomp).

SYNS: ANTIMONOUS SULFIDE □ ANTIMONY GLANCE □ ANTIMONY ORANGE □ ANTIMONY SESQUISULFIDE □ ANTIMONY SULFIDE □ ANTIMONY TRISULFIDE COLLOID □ ANTIMONY VERMILION □ BLACK ANTIMONY □ C.I. 77060 □ C.I. PIGMENT RED 107 □ CRIMSON ANTIMONY □ DIANTIMONY TRISULFIDE □ LYMPHOSCAN □ NEEDLE ANTIMONY

CONSENSUS REPORTS: IARC Cancer Review: Group 3 IMEMDT 47,291,89; Animal Limited Evidence IMEMDT 47,291,89; Human Inadequate Evidence IMEMDT 47,291,89. Reported in EPA TSCA Inventory. Antimony and its compounds are on the Community Right-To-Know List.

OSHA PEL: TWA 500 μg(Sb)/m³
ACGIH TLV: TWA 0.5 mg(Sb)/m³
NIOSH REL: (Antimony) TWA 0.5 mg(Sb)/m³

SAFETY PROFILE: Poison by intraperitoneal route. Human blood and gastrointestinal system effects by inhalation. Questionable carcinogen. See also ANTIMONY COMPOUNDS and SULFIDES. Spontaneously flammable when exposed to strong oxidizers. Flammable when exposed to heat or flame. Moderately explosive by spontaneous reaction with chlorates, perchlorates, ClO, thallic oxide. When heated to decomposition or on contact with acid or acid fumes it emits highly toxic fumes of oxides of sulfur and antimony. Will react with water or steam to produce toxic and flammable vapors.

AQN000   CAS: 60-80-0   HR: 3
ANTIPYRINE
mf: $C_{11}H_{12}N_2O$ mw: 188.23
PROP: Fine, white crystals, leaflets, or scales. Mp: 113°, bp: 319° @ 174 mm, d: 1.19. Very sol in water and alc; sltly sol in ether.
SYNS: DIMETHYLOXYQUINAZINE □ 2,3-DIMETHYL-1-PHENYL-3-PYRAZOLIN-5-ONE □ 2,3-DIMETHYL-1-PHENYL-5-PYRAZOLONE □ OXYDIMETHYL-QUINAZINE □ PHENAZONE (pharmaceutical) □ 1-PHENYL-2,3-DIMETHYL-PYRAZOLE-5-ONE □ 1-PHENYL-2,3-DIMETHYL-5-PYRAZOLONE

CONSENSUS REPORTS: Reported in EPA TSCA Inventory.
SAFETY PROFILE: A human poison by an unspecified route. Moderately toxic via ingestion, subcutaneous, and intravenous routes. Questionable carcinogen with experimental tumorigenic data. Mutation data reported. When heated to decomposition it emits toxic fumes of NO_x.

AQN635   CAS: 86-88-4   HR: 3
ANTU
DOT: UN 1651
mf: $C_{11}H_{10}N_2S$ mw: 202.29
PROP: Crystals or prisms with bitter taste. Mp: 198°. Sltly sol in H_2O; sol in Me_2CO. IDLH 100 mg/m³.
SYNS: ALPHANAPHTYL THIOUREA □ ALPHANAPHTYL THIOUREE (FRENCH) □ ALRATO □ ANTURAT □ CHEMICAL 109 □ DIRAX □ KILL KANTZ □ KRYSID □ 1-NAFTIL-TIOUREA (ITALIAN) □ 1-NAFTYLTHIOUREUM (DUTCH) □ 1-NAPHT-HALENYLTHIOUREA □ α-NAPHTHALTHIOH-ARNSTOFF (GERMAN) □ α-NAPHTHOTHIOUREA □ α-NAPHTHYLTHIOCARBAMIDE □ 1-NAPHTHYL-THIOHARNSTOFF (GERMAN) □ α-NAPHTHYL-THIOUREA □ 1-NAPHTHYL THIOUREA (MAK) □ 1-(1-NAPHTHYL)-2-THIOUREA □ N-(1-NAPHTHYL)-2-THIOUREA □ α-NAPHTHYLTHIOUREA (DOT) □ 1-NAPHTHYL-THIOUREE (FRENCH) □ NAPHTOX □ RATTRACK □ RCRA WASTE NUMBER P072 □ SMEESANA □ U-5227 □ USAF EK-P-5976

CONSENSUS REPORTS: IARC Cancer Review: Animal Inadequate Evidence IMEMDT 30,347,83. Reported in EPA TSCA Inventory. EPA Extremely Hazardous Substances List. EPA Genetic Toxicology Program.
OSHA PEL: TWA 0.3 mg/m³
ACGIH TLV: TWA 0.3 mg/m³; Not Classifiable as a Human Carcinogen
DFG MAK: 0.3 mg/m³
DOT CLASSIFICATION: 6.1; Label: Poison
SAFETY PROFILE: Poison by ingestion and intraperitoneal routes. Moderately toxic to humans by an unspecified route. Questionable carcinogen with experimental tumorigenic data. Mutagenic data. A rodenticide used extensively. Death is caused by pulmonary edema. Chronic

toxicity has been known to cause dermatitis and a decrease in the white blood cells. When heated to decomposition it emits toxic fumes of NO_x and SO_x.

AQO300   CAS: 1107-26-2   HR: D
β-APO-8'-CAROTENAL

mf: $C_{30}H_{40}O$ mw: 416.65

PROP: Fine crystalline powder with dark metallic sheen or violet crystals. Mp: 139°. Sol in chloroform; sltly sol in acetone; insol in water.

SYNS: APO □ APOCAROTENAL

SAFETY PROFILE: When heated to decomposition it emits acrid smoke and irritating fumes.

AQP000   CAS: 1937-37-7   HR: 3
APOMINE BLACK GX

mf: $C_{34}H_{25}N_9O_7S_2•2Na$ mw: 781.78

SYNS: AHCO DIRECT BLACK GX □ AIREDALE BLACK ED □ AIZEN DIRECT DEEP BLACK EH □ AIZEN DIRECT DEEP BLACK GH □ AIZEN DIRECT DEEP BLACK RH □ AMANIL BLACK GL □ AMANIL BLACK WD □ ATLANTIC BLACK BD □ ATLANTIC BLACK C □ ATLANTIC BLACK E □ ATLANTIC BLACK EA □ ATLANTIC BLACK GAC □ ATLANTIC BLACK GG □ ATLANTIC BLACK GXCW □ ATLANTIC BLACK GXOO □ ATLANTIC BLACK SD □ ATUL DIRECT BLACK E □ AZINE DEEP BLACK EW □ AZOCARD BLACK EW □ AZOMINE BLACK EWO □ BELAMINE BLACK GX □ BENCIDAL BLACK E □ BENZAMIL BLACK E □ BENZO DEEP BLACK E □ BENZO LEATHER BLACK E □ BENZOFORM BLACK BCN-CF □ BLACK 2EMBL □ BLACK 4EMBL □ BRASILAMINA BLACK GN □ BRILLIANT CHROME LEATHER BLACK H □ CALCOMINE BLACK □ CALCOMINE BLACK EXL □ CARBIDE BLACK E □ CERN PRIMA 38 □ CHLORAMINE BLACK C □ CHLORAMINE BLACK EC □ CHLORAMINE BLACK ERT □ CHLORAMINE BLACK EX □ CHLORAMINE BLACK EXR □ CHLORAMINE BLACK XO □ CHLORAMINE CARBON BLACK S □ CHLORAMINE CARBON BLACK SJ □ CHLORAMINE CARBON BLACK SN □ CHLORAZOL BLACK E □ CHLORAZOL BLACK E (BIOLOGICAL STAIN) □ CHLORAZOL BLACK EA □ CHLORAZOL BLACK EN □ CHLORAZOL BURL BLACK E □ CHLORAZOL LEATHER BLACK ENP □ CHLORAZOL SILK BLACK G □ CHROME LEATHER BLACK E □ CHROME LEATHER BLACK EC □ CHROME LEATHER BLACK EM □ CHROME LEATHER BLACK G □ CHROME LEATHER BRILLIANT BLACK ER □ C.I. 30235 □ C.I. DIRECT BLACK 38 □ C.I. DIRECT BLACK 38, DISODIUM SALT □ COIR DEEP BLACK C □ COLUMBIA BLACK EP □ DIACOTTON DEEP BLACK □ DIACOTTON DEEP BLACK RX □ DIAMINE DEEP BLACK EC □ DIAMINE DIRECT BLACK E □ DIAPHTAMINE BLACK V □ DIAZINE BLACK E □ DIAZINE DIRECT BLACK E □ DIAZINE DIRECT BLACK G □ DIAZOL BLACK 2V □ DIPHENYL DEEP BLACK G □ DIRECT BLACK A □ DIRECT BLACK BRN □ DIRECT BLACK CX □ DIRECT BLACK CXR □ DIRECT BLACK E □ DIRECT BLACK EW □ DIRECT BLACK EX □ DIRECT BLACK FR □ DIRECT BLACK GAC □ DIRECT BLACK GW □ DIRECT BLACK GX □ DIRECT BLACK GXR □ DIRECT BLACK JET □ DIRECT BLACK META □ DIRECT BLACK METHYL □ DIRECT BLACK N □ DIRECT BLACK RX □ DIRECT BLACK SD □ DIRECT BLACK WS □ DIRECT BLACK Z □ DIRECT BLACK 3 □ DIRECT BLACK 38 □ DIRECT DEEP BLACK E □ DIRECT DEEP BLACK E EXTRA □ DIRECT DEEP BLACK EA-CF □ DIRECT DEEP BLACK EAC □ DIRECT DEEP BLACK EW □ DIRECT DEEP BLACK EX □ ENIANIL BLACK CN □ ERIE BLACK B □ ERIE BLACK BF □ ERIE BLACK GAC □ ERIE BLACK GXOO □ ERIE BLACK JET □ ERIE BLACK NUG □ ERIE BLACK RXOO □ ERIE BRILLIANT BLACK S □ ERIE FIBRE BLACK VP □ FENAMIN BLACK E □ FIBRE BLACK VF □ FIXANOL BLACK E □ FORMALINE BLACK C □ FORMIC BLACK BA □ FORMIC BLACK C □ FORMIC BLACK CW □ FORMIC BLACK MTG □ FORMIC BLACK TG □ HISPAMIN BLACK EF □ INTERCHEM DIRECT BLACK Z □ KAYAKU DIRECT DEEP BLACK EX □ KAYAKU DIRECT DEEP BLACK GX □ KAYAKU DIRECT DEEP BLACK S □ KAYAKU DIRECT LEATHER BLACK EX □ KAYAKU DIRECT SPECIAL BLACK AAX □ LURAZOL BLACK BA □ META BLACK □ MITSUI DIRECT BLACK EX □ MITSUI DIRECT BLACK GX □ NCI-C54557 □ NIPPON DEEP BLACK □ NIPPON DEEP BLACK GX □ PAPER BLACK BA □ PAPER BLACK T □ PAPER DEEP BLACK C □ PARAMINE BLACK B □ PARAMINE BLACK E □ PEERAMINE BLACK E □ PEERAMINE BLACK GXOO □ PHENAMINE BLACK BCN-CF □ PHENAMINE BLACK CL □ PHENAMINE BLACK E □ PHENAMINE BLACK E 200 □ PHENO BLACK EP □ PHENO BLACK SGN □ PONTAMINE BLACK E □ PONTAMINE BLACK EBN □ SANDOPEL BLACK EX □ SERISTAN BLACK B □ TELON FAST BLACK E □ TETRAZO DEEP BLACK G □ TERTRODIRECT BLACK E □ TETRODIRECT BLACK EFD □ UNION BLACK EM □ VONDACEL BLACK N

CONSENSUS REPORTS: NTP 10th Report on Carcinogens. IARC Cancer Review: Animal Sufficient Evidence IMEMDT 29,295,82; Human Limited Evidence IMEMDT 29,295,82. Reported in EPA TSCA Inventory. NTP Carcinogenesis Bioassay (feed): Clear Evidence: rat NCICTR* NCI-TR-108,78; No Evidence:

A

mouse NCICTR NCI-TR-108,78. On Community-Right-To-Know List.
SAFETY PROFILE: Confirmed carcinogen with carcinogenic and tumorigenic data. Moderately toxic by ingestion and inhalation. An eye irritant. Mutation data reported. When heated to decomposition it emits very toxic fumes of NO_x, Na_2O, and SO_2.

AQQ125   CAS: 17185-68-1   HR: 3
AQUA-1,2-DIAMINOPROPANEDIPER-
**   OXOCHROMIUM(IV) DIHYDRATE**
mf: $C_3H_{12}CrN_2O_5 \cdot 2H_2O$ mw: 244.16
PROP: IDLH Ca [15 mg/m^3 {as Cr(VI)}].
CONSENSUS REPORTS: Chromium and its compounds are on the Community Right-To-Know List.
OSHA PEL: CL 0.1 mg(CrO$_3$)/m^3
ACGIH TLV: TWA 0.05 mg(Cr)/m^3; Confirmed Human Carcinogen
NIOSH REL: (Chromium(VI)) TWA 25 μg(Cr(VI))/m^3; CL 50 μg/m^3/15M
SAFETY PROFILE: A confirmed carcinogen. May explode spontaneously at room temperature. Upon decomposition it emits toxic fumes of NO_x. See also CHROMIUM COMPOUNDS.

AQQ500   CAS: 9000-01-5   HR: 2
ARABIC GUM
mw: 240,000
PROP: Yellowish-amber lumps. A gum from the stems and branches of *Acacia senegal (L.) Willd.* or of *Acacia* (Fam. *Leguminosae*). Sol in water; insol in alc.
SYNS: ACACIA □ ACACIA DEALBATA GUM □ ACACIA GUM □ ACACIA SENEGAL □ ACACIA SYRUP □ AUSTRALIAN GUM □ GUM ARABIC □ GUM OVALINE □ GUM SENEGAL □ INDIAN GUM □ NCI-C50748 □ SENEGAL GUM □ STARSOL No. 1 □ WATTLE GUM
CONSENSUS REPORTS: NTP Carcinogenesis Bioassay (feed); No Evidence: mouse, rat NTPTR* NTP-TR-227,82. Reported in EPA TSCA Inventory.
SAFETY PROFILE: Very low toxicity by ingestion. Inhalation or ingestion has produced hives, eczema, and angiodema.

Experimental reproductive effects. A severe eye irritant. A weak allergen. Mutation data reported. Combustible. When heated to decomposition it emits acrid smoke.

AQQ900   CAS: 5536-17-4   HR: 3
9-β-d-ARABINO FURANOSYL ADENINE
mf: $C_{10}H_{13}N_5O_4$ mw: 267.28
PROP: Needles. Mp: 257–257.5°.
SYNS: ADENINE ARABINOSIDE □ ARABINOSYLADENINE □ β-d-ARABINOSYLADENINE □ 9-ARABINOSYLADENINE □ VIDARABIN □ VIDARABINE
CONSENSUS REPORTS: Reported in EPA TSCA Inventory.
SAFETY PROFILE: Poison by ingestion and intravenous routes. Moderately toxic by intraperitoneal route. An experimental teratogen. Other experimental reproductive effects. Human systemic effects by intravenous route: central nervous system, blood, and other effects. A skin and eye irritant. Human mutation data reported. When heated to decomposition it emits toxic fumes of NO_x.

AQR800   CAS: 9036-66-2   HR: D
ARABINOGALACTAN
PROP: Derived from water extraction of Western larch wood having galactose units and arabinose units in the ratio of approx. 6:1. Mp: >200° (decomp). Sol in water.
SYNS: (+)-ARABINOGALACTAN □ LARCH GUM □ POLYARABINOGALACTAN
SAFETY PROFILE: When heated to decomposition it emits acrid smoke and irritating fumes.

AQT500   CAS: 39300-45-3   HR: 3
ARATHANE
mf: $C_{18}H_{24}N_2O_6$ mw: 364.44
PROP: Liquid.
SYNS: CAPRYLDINITROPHENYL CROTONATE □ 2-CAPRYL-4,6-DINITROPHENYL CROTONATE □ CROTONATE de 2,4-DINITRO 6-(1-METHYL-HEPTYL)-PHENYLE (FRENCH) □ 4,6-DINITRO-2-CAPRYLPHENYL CROTONATE □ 4,6-DINITRO-2-(2-CAPRYL)PHENYL CROTONATE □ DINITRO(1-METHYLHEPTYL)PHENYL CROTONATE □ 2,4-DINITRO-6-(1-METHYLHEPTYL)-PHENYL CROTONATE □ 2,4-DINITRO-6-(2-OCTYL)-

PHENYL CROTONATE □ ENT 24,727 □ (6-(1-METHYL-HEPTYL)-2,4-DINITRO-FENYL)-CROTONAAT (DUTCH) □ (6-(1-METHYL-HEPTYL)-2,3-DINITRO-PHENYL)-CROTONAT (GERMAN) □ 2-(1-METHYLHEPTYL)-4,6-DINITROPHENYL CROTONATE □ (6-(1-METIL-EPITL)-2,4-DINITRO-FENIL)-CROTONATO (ITALIAN)

SAFETY PROFILE: Poison by ingestion and intravenous routes. An experimental teratogen. Other experimental reproductive effects. Mutation data reported. Questionable carcinogen with experimental neoplastigenic data. See NITRATES. When heated to decomposition it emits toxic fumes of NO_x.

AQT750   CAS: 63-75-2   HR: 3
ARECOLINE
mf: $C_8H_{13}NO_2$ mw: 155.22

PROP: Oily liquid. Bp: 94° @ 17 mm.
SYNS: ARECAIDINE METHYL ESTER □ ARECOLINE BASE □ METHYL-1,2,5,6-TETRA-HYDRO-1-METHYL-NICOTINATE □ N-METHYL-Δ-TETRAHYDRONICOTIN-IC ACID METHYL ESTER □ N-METHYLTETRAHYDRO-PYRIDINE-β-CARBO-XYLIC ACID METHYL ESTER □ 1,2,5,6-TETRA-HYDRO-1-METHYLNICOTINIC ACID, METHYL ESTER

CONSENSUS REPORTS: IARC Cancer Review: Animal Inadequate Evidence IMEMDT 37,141,85.

SAFETY PROFILE: Poison by subcutaneous and intraperitoneal routes. Moderately toxic by ingestion. Questionable carcinogen with experimental neoplastigenic data. It mimics the action of acetylcholine, a neurotransmitter, and is a parasympathetic nervous system stimulant. Its action on the central nervous system can cause tremors. Human mutation data reported. It is easily nitrosated to several nitrosamines. See also ESTERS and NITROSAMINES. It is the major alkaloid found in betel quid. Combustible, can react with oxidizing materials. When heated to decomposition it emits highly toxic fumes of NO_x.

AQV990   CAS: 101043-37-2 HR: 3
5-l-ARGININECYANOGINOSIN LA
mf: $C_{49}H_{74}N_{10}O_{12}$ mw: 995.33

SYNS: AKERSTOX □ CYANOGINOSIN LA, 5-l-ARGININE- □ CYANOGINOSIN LR □ MICROCYSTIN-A □ MICROCYSTIN-LR □ MICROCYSTIS AERUGINOSA

□ TOXIN □ TOXIN, BLUE GREEN ALGA, MICROCYSTIS AERUGINOSA □ TOXIN I (MICROCYSTIS AERUGINOSA) □ TOXIN-LR □ TOXIN T 17 (MICROCYSTIS AERUGINOSA)

SAFETY PROFILE: A poison by ingestion, inhalation, intraperitoneal, and intravenous route. Experimental reproductive effects. When heated to decomposition it emits toxic vapors of NO_x.

AQW000   CAS: 1119-34-2   HR: 2
l-ARGININE MONOHYDROCHLORIDE
mf: $C_6H_{14}N_4O_2$•ClH   mw: 210.70

PROP: White crystalline powder; odorless. Mp: 222–235° (decomp). Very sol in water; sltly sol in alc.
SYNS: ARGAMINE □ ARGININE HYDRO-CHLORIDE □ l-ARGININE HYDROCHLORIDE □ ARGININE MONOHYDROCHLORIDE □ ARGIVENE □ DETOX-ARGIN □ l-HYDROCHLORIDE ARGININE □ LEVARG-IN □ MINOPHAGEN A □ R-GENE

CONSENSUS REPORTS: Reported in EPA TSCA Inventory.

SAFETY PROFILE: Moderately toxic by intraperitoneal route. Mildly toxic by ingestion. An experimental teratogen. When heated to decomposition it emits very toxic fumes of NO_x and HCl.

AQW250   CAS: 7440-37-1   HR: 1
ARGON
af: Ar   aw: 39.94

DOT: UN 1006/UN 1951
PROP: Colorless, inert, odorless, tasteless, monatomic gas. Forms no true chemical compds. Forms clathrates with H_2O and hydroquinone. Forms complex with HBr. Mp: −189.2°, bp: −185.7°, d: 1.784 g/L @ 0°, 1.40 @ −186°, 1.65 @ −233°. Solubility in water 3.36 mL/100 g @ 20°.

CONSENSUS REPORTS: Reported in EPA TSCA Inventory.

DOT CLASSIFICATION: 2.2; Label: Nonflammable Gas

SAFETY PROFILE: A simple asphyxiant gas. As an inert gas, it has no specific inherent dangerous properties. Gases of this type have no specific toxicity effect, but they act by excluding O_2 from the lungs. The effect of simple asphyxiant gases is proportional to the extent to which they diminish the amount (partial pressure) of O_2 in the air that is breathed. The oxygen may be diminished to 75% of its normal percentage in air before appreciable symptoms develop, and this in turn requires the presence of a simple asphyxiant in a concentration of 33% in the mixture of air and gas. When the simple asphyxiant reaches a concentration of 50%, marked symptoms can be produced. A concentration of 75% is fatal in a matter of minutes. The first symptoms produced by simple asphyxiant gases such as argon are rapid respirations and air hunger. Mental alertness is diminished and muscular coordination is impaired. Later, judgment becomes faulty and all sensations are depressed. Emotional instability often results and fatigue occurs rapidly. As the asphyxia progresses, there may be nausea and vomiting, prostration, and loss of consciousness, and finally, convulsions, deep coma, and death.

AQX500   CAS: 76-25-5   HR: 3
ARISTOCORT ACETONIDE
mf: $C_{24}H_{31}O_6F$ mw: 434.55
SYNS: ACETOSPAN □ ARISTODERM □ ARISTOGEL □ 9-α-FLUORO-11-β,21-DIHYDROXY-16-α-ISOPROYLIDENEDIOXY-1,4-PREGNADIENE, 3,20-DIONE □ 9-α-FLUORO-16-HYDROXYPREDNISOL-ONE ACETONIDE □ 9-α-FLUORO-16-α-17-α-ISOPROPYLEDENE DIOXY PREDNISOLONE □ 9-α-FLUORO-16-α-17-α-ISOPROPYLIDENEDIOXY-Δ-1-HYDROCORTISONE □ FLUTONE □ KENACORT-A □ KENALOG □ TRAMACIN □ TRIAMCINCOLONE ACETONIDE □ TRIAMCINOLONE ACETONIDE □ TRIAMCINOLONE-16,17-ACETONIDE □ VETALOG
CONSENSUS REPORTS: Reported in EPA TSCA Inventory.
SAFETY PROFILE: Poison by subcutaneous and intraperitoneal routes. An experimental teratogen. Other experimental reproductive effects. Human mutation data reported. When heated to decomposition it emits acrid smoke and toxic fumes of F^-.

AQY250   CAS: 313-67-7   HR: 3

ARISTOLOCHINE
mf: $C_{17}H_{11}NO_7$ mw: 341.29
PROP: Crystals from DMF/EtOH, EtOH, or MeOH/Et$_2$O. Mp: 281–286° (decomp). From alcoholic extract of *Aristolochia indico* (CNCRA6 42,35,64).
SYNS: ARISTOLOCHIC ACID □ BIRTHWORT □ 8-METHOXY-6-NITROPHENANTHOL-(3,4-d)-1,3-DIOXOLE-5-CARBOXYLIC ACID □ NSC-50413
SAFETY PROFILE: Confirmed carcinogen. Poison by ingestion, intraperitoneal, and intravenous routes. Mutation data reported. When heated to decomposition it emits toxic fumes of NO_x.

  From "International Register of Potentially Toxic Chemicals: April 1982." Vol. 5 No. 1: The Ministry of Health of the Federal Republic of Germany has withdrawn from the national market drugs containing aristolochic acid. The decision resulted from the demonstration of a carcinogenic potential in a three-month ingestion toxicity study undertaken in rats. Aristolochic acid is claimed to promote phagocytosis and to have immunostimulant activity. A growth-inhibiting effect on experimentally induced tumors has been described, but this effect has not been shown to have any clinical relevance. Extracts of species of *Aristolochiacea* have traditionally been used as a bitter, and a broad range of therapeutic effects has been claimed.

AQY750   HR: 3
AROMATIC AMINES
PROP: Amines that contain one or more rings of unsaturated or cyclic HC, such as benzene. There are vast numbers of such amines. The term is largely due to the characteristic odor.
SAFETY PROFILE: Many of these aromatic amines are recognized as carcinogenic to the human bladder, ureter, and renal pelvis, intestines, lung, liver, and prostate. See also AMINES.

AQZ900   CAS: 585-54-6   HR: 3
ARSACETIN SODIUM SALT

mf: $C_8H_9AsNO_4 \cdot Na$ mw: 281.09
SYNS: ARSACETIN □ ARSANILIC ACID, N-ACETYL-, SODIUM SALT □ ARSONIC ACID, (4-(ACETYLAMINO)PHENYL)-, MONOSODIUM SALT (9CI) □ SODIUM ACETYLARSANILATE
OSHA PEL: TWA 0.5 mg(As)/m^3
ACGIH TLV: BEI: 35 μ (As)/L inorganic arsenic and methylated metabolites in urine
SAFETY PROFILE: Poison by intravenous route. Moderately toxic by subcutaneous route. When heated to decomposition it emits toxic fumes of NO$_x$ and As.

ARA100   CAS: 5687-22-9   HR: 3
9-ARSAFLUORENINIC ACID
mf: $C_{12}H_9AsO_2$ mw: 260.13
SYNS: ARSAFLUORINIC ACID □ 5H-BENZ(B)-ARSINDOLE, 5-HYDROXY-, 5-OXIDE □ DIBENZ-ARSENOLE, 5-HYDROXY-, 5-OXIDE □ 5H-DIBENZARSOLE, 5-HYDROXY-, 5-OXIDE □ DIBENZARSENOLIC ACID
SAFETY PROFILE: A poison by ingestion. When heated to decomposition it emits toxic vapors of As.

ARA250   CAS: 98-50-0   HR: 3
ARSANILIC ACID
mf: $C_6H_8AsNO_3$ mw: 217.06
PROP: Needles from aq solns. Mp: 232°, bp: decomp, $-H_2O$ @ 15°. Insol in Me$_2$CO, CHCl$_3$, C$_6$H$_6$; sltly sol in EtOH; sol in Et$_2$O, H$_2$O, conc acids, alkalies.
SYNS: p-AMINOBENZENEARSONIC ACID □ 4-AMINOBENZENEARSONIC ACID □ AMINO-PHENYLARSINE ACID □ p-AMINOPHENYLARSINE ACID □ p-AMINOPHENYLARSINIC ACID □ 4-AMINOPHENYLARSONIC ACID □ p-ANILINE-ARSONIC ACID □ ANTOXYLIC ACID □ p-ARSANIL-IC ACID □ 4-ARSANILIC ACID □ ATOXYLIC ACID
CONSENSUS REPORTS: IARC Cancer Review: Animal Inadequate Evidence IMEMDT 23,39,80. Reported in EPA TSCA Inventory. Arsenic and its compounds are on the Community Right-To-Know List.
OSHA PEL: TWA 0.5 mg(As)/m^3
ACGIH TLV: BEI: 35 μ (As)/L inorganic arsenic and methylated metabolites in urine
SAFETY PROFILE: Poison by intravenous and intraperitoneal routes. Moderately toxic by ingestion. Flammable, decomposes with heat to yield flammable vapors. When heated to decomposition or on contact with acid or acid fumes it emits highly toxic fumes of As and NO$_x$. See also ARSENIC COMPOUNDS and ANILINE.

ARA500   CAS: 127-85-5   HR: 3
ARSANILIC ACID, MONOSODIUM SALT
DOT: UN 2473
mf: $C_6H_7AsNO_3 \cdot Na$ mw: 239.05
PROP: Tetrahydrate: white, odorless, crystalline powder; faint salty taste. Sol in water; somewhat sol in alc.
SYNS: (4-AMINOPHENYL)ARSONIC ACID SODIUM SALT □ ANHYDROUS SODIUM ARSANILATE □ ARSAMIN □ ARSANILIC ACID SODIUM SALT □ ARSINOSOLVIN □ ARSONIC ACID, (4-AMINOPHEN-YL)-, MONOSODIUM SALT (9CI) □ ATOXYL □ MONOSODIUM (4-AMINOPHENYL)-ARSONATE □ NCI-C61176 □ NUARSOL □ PIGLET PRO-GEN V □ PRO-GEN SODIUM □ PROTOXYL □ SOAMIN □ SODIUM AMINARSONATE □ SODIUM p-AMINO-BENZENE-ARSONATE □ SODIUM AMINOPHENOL ARSONATE □ SODIUM p-AMINOPHENYLARSON-ATE □ SODIUM-ANALINE ARSONATE □ SODIUM ANILARSONATE □ SODIUM ARSANILATE □ SODIUM ARSANILATE (DOT) □ SODIUM p-ARSANILATE □ SODIUM ARSONILATE □ SONATE □ TRYPOXYL
CONSENSUS REPORTS: Arsenic and its compounds are on the Community Right-To-Know List.
OSHA PEL: TWA 0.5 mg(As)/m^3
ACGIH TLV: BEI: 35 μ (As)/L inorganic arsenic and methylated metabolites in urine
DOT CLASSIFICATION: 6.1; Label: KEEP AWAY FROM FOOD
SAFETY PROFILE: Poison by subcutaneous route. Can cause blindness. When heated to decomposition it emits very toxic fumes of As and NO$_x$.

ARA750   CAS: 7440-38-2   HR: 3
ARSENIC
DOT: UN 1558
af: As   aw: 74.92
PROP: Silvery to black, brittle, crystalline, or amorphous metalloid. Mp: 814° @ 36 atm, bp: subl @ 612°, d: black crystals 5.724 @ 14°, black amorphous 4.7, vap press: 1 mm

@ 372° (subl). Insol in water; sol in HNO_3. IDLH 5 mg/m³ (as As).

SYNS: ARSEN (GERMAN, POLISH) □ ARSENIC, metallic (DOT) □ ARSENIC BLACK □ ARSENIC-75 □ ARSENICALS □ COLLOIDAL ARSENIC □ GREY ARSENIC □ METALLIC ARSENIC

CONSENSUS REPORTS: IARC Cancer Review: Group 1 IMEMDT 7,100,87; Human Sufficient Evidence IMEMDT 23,39,80; Human Inadequate Evidence IMEMDT 2,48,73. Reported in EPA TSCA Inventory. Arsenic and its compounds are on the Community Right-To-Know List.

OSHA PEL: TWA 0.01 mg(As)/m³; Cancer Hazard

ACGIH TLV: TWA 0.01 mg/m³; Confirmed Human Carcinogen; BEI: 35 µ (As)/L inorganic arsenic and methylated metabolites in urine

DFG MAK: DFG TRK: 0.2 mg/m³ calculated as arsenic in that portion of dust that can possibly be inhaled

NIOSH REL: CL 2 µg(As)/m³

DOT CLASSIFICATION: 6.1; Label: Poison

SAFETY PROFILE: Confirmed human carcinogen producing liver tumors. Poison by subcutaneous, intramuscular, and intraperitoneal routes. Human systemic skin and gastrointestinal effects by ingestion. An experimental teratogen. Other experimental reproductive effects. Mutation data reported. Flammable in the form of dust when exposed to heat or flame or by chemical reaction with powerful oxidizers such as bromates, chlorates, iodates, peroxides, lithium, NCl_3, KNO_3, $KMnO_4$, Rb_2C_2, $AgNO_4$, NOCl, IF_5, CrO_3, ClF_3, ClO, BrF_3, BrF_5, BrN_3, RbC_3BCH, CsC_3BCH. Slightly explosive in the form of dust when exposed to flame. When heated or on contact with acid or acid fumes, it emits highly toxic fumes; can react vigorously on contact with oxidizing materials. Incompatible with bromine azide, dirubidium acetylide, halogens, palladium, zinc, platinum, NCl_3, $AgNO_3$, CrO_3, Na_2O_2, hexafluoroisopropylideneamino lithium.

ARB000   CAS: 10102-53-1   HR: 3

m-ARSENIC ACID

mf: $AsHO_3$ mw: 123.93

SYN: METAARSENIC ACID

CONSENSUS REPORTS: Reported in EPA TSCA Inventory. Arsenic and its compounds are on the Community Right-To-Know List.

OSHA PEL: TWA 0.01 mg(As)/m³; Cancer Hazard

ACGIH TLV: TWA 0.01 mg/m³; Confirmed Human Carcinogen; BEI: 35 µ (As)/L inorganic arsenic and methylated metabolites in urine

DFG MAK: Human Carcinogen

NIOSH REL: (Arsenic, Inorganic) CL 2 µg(As)/m³/15M

SAFETY PROFILE: Confirmed human carcinogen. When heated to decomposition it emits toxic fumes of arsenic. See also ARSENIC COMPOUNDS.

ARB250   CAS: 7778-39-4   HR: 3
o-ARSENIC ACID

DOT: UN 1553/UN 1554

mf: AsH_3O_4 mw: 141.95

SYNS: ACIDE ARSENIQUE LIQUIDE (FRENCH) □ ARSENATE □ ARSENIC ACID, liquid (DOT) □ ARSENIC ACID, solid (DOT) □ DESICCANT L-10 □ HI-YIELD DESICCANT H-10 □ ORTHOARSENIC ACID □ RCRA WASTE NUMBER P010 □ ZOTOX □ ZOTOX CRAB GRASS KILLER

CONSENSUS REPORTS: Reported in EPA TSCA Inventory. Arsenic and its compounds are on the Community Right-To-Know List.

OSHA PEL: TWA 0.01 mg(As)/m³; Cancer Hazard

ACGIH TLV: TWA 0.01 mg/m³; Confirmed Human Carcinogen; BEI: 35 µ (As)/L inorganic arsenic and methylated metabolites in urine

DFG MAK: Human Carcinogen

NIOSH REL: (Arsenic, Inorganic) CL 2 µg(As)/m³/15M

DOT CLASSIFICATION: 6.1; Label: Poison

SAFETY PROFILE: Confirmed human carcinogen. Poison by ingestion. An experimental teratogen. Human mutation data reported. When heated to decomposition it emits toxic fumes of arsenic. See also ARSENIC COMPOUNDS.

ARB750   CAS: 7778-44-1   HR: 3
ARSENIC ACID, CALCIUM SALT (2:3)
DOT: UN 1573
mf: $As_2O_8 \cdot 3Ca$ mw: 398.08
PROP: Colorless, amorphous powder. D: 3.620. Solubility in water: 0.013/100 @ 25°. IDLH 5 mg/m³ (as As).
SYNS: ARSENIATE de CALCIUM □ CALCIUM-ARSENAT □ CALCIUM ARSENATE (DOT) □ CALCIUM ORTHOARSENATE □ CHIP-CAL □ CHIP-CAL GRANULAR □ CUCUMBER DUST □ FENCAL □ FLAC □ KALO □ KALZIUMARSENIAT (GERMAN) □ KILMAG □ PENCAL □ SECURITY □ SPRACAL □ TRICALCIUM ARSENATE □ TRICALCIUMARSENAT (GERMAN) □ TURF-CAL
CONSENSUS REPORTS: IARC Cancer Review: Group 1 IMEMDT 7,100,87; Human Sufficient Evidence IMEMDT 23,39,80; Animal No Evidence IMEMDT 2,48,73; Animal Inadequate Evidence IMEMDT 23,39,80. Reported in EPA TSCA Inventory. Arsenic and its compounds are on the Community Right-To-Know List. EPA Extremely Hazardous Substances List.
OSHA PEL: TWA 0.01 mg(As)/m³; Cancer Hazard
ACGIH TLV: TWA 0.01 mg/m³; Confirmed Human Carcinogen; BEI: 35 μ (As)/L inorganic arsenic and methylated metabolites in urine
DFG MAK: Human Carcinogen
NIOSH REL: CL 2 μg(As)/m³/15M
DOT CLASSIFICATION: 6.1; Label: Poison
SAFETY PROFILE: Confirmed human carcinogen. Poison by ingestion. Moderately toxic by skin contact. When heated to decomposition it emits toxic fumes of arsenic.

ARC000   CAS: 7778-43-0   HR: 3
ARSENIC ACID, DISODIUM SALT
mf: $Na_2HAsO_4 \cdot 7H_2O$ mw: 312.01
PROP: Colorless white powder or solid, effloresces. D: 1.88, mp: $-7H_2O$ @ 130°,

bp: decomp @ 150°. Solubility in water: 61/100 @ 15°; sol in glycerin.
SYNS: DISODIUM ARSENATE □ DISODIUM ARSENIC ACID □ DISODIUM HYDROGEN ARSENATE □ DISODIUM HYDROGEN ORTHO-ARSENATE □ DISODIUM MONOHYDROGEN ARSENATE □ SODIUM ACID ARSENATE □ SODIUM ARSENATE □ SODIUM ARSENATE DIBASIC, anhydrous
CONSENSUS REPORTS: Reported in EPA TSCA Inventory. Arsenic and its compounds are on the Community Right-To-Know List.
OSHA PEL: TWA 0.5 mg(As)/m³; Cancer Hazard
ACGIH TLV: TWA 0.01 mg/m³; Confirmed Human Carcinogen; BEI: 35 μ (As)/L inorganic arsenic and methylated metabolites in urine
NIOSH REL: (Arsenic, Inorganic) CL 2 μg(As)/m³/15M
DFG MAK: Human Carcinogen
SAFETY PROFILE: Confirmed human carcinogen. Poison by intraperitoneal route. Human mutation data reported. When heated to decomposition it emits toxic fumes of arsenic. See ARSENIC COMPOUNDS.

ARC250   CAS: 10048-95-0   HR: 3
ARSENIC ACID, DISODIUM SALT, HEPTAHYDRATE
mf: $AsHO_4 \cdot 2Na \cdot 7H_2O$ mw: 427.05
PROP: Prisms. Mp: 40°. Very sol in water; sltly sol in EtOH.
SYNS: DISODIUM ARSENATE, HEPTAHYDRATE □ SODIUM ACID ARSENATE, HEPTAHYDRATE □ SODIUM ARSENATE, DIBASIC, HEPTAHYDRATE □ SODIUM ARSENATE HEPTAHYDRATE
CONSENSUS REPORTS: Arsenic and its compounds are on the Community Right-To-Know List.
OSHA PEL: TWA 0.01 mg(As)/m³; Cancer Hazard
ACGIH TLV: TWA 0.01 mg/m³; Confirmed Human Carcinogen; BEI: 35 μ (As)/L inorganic arsenic and methylated metabolites in urine
NIOSH REL: (Arsenic, Inorganic) CL 2 μg(As)/m³/15M
DFG MAK: Human Carcinogen

A

SAFETY PROFILE: Confirmed human carcinogen. Poison by subcutaneous route. An experimental teratogen. Other experimental reproductive effects. Human mutation data reported. See also ARSENIC COMPOUNDS. When heated to decomposition it emits toxic fumes of arsenic.

**ARC500   CAS: 7774-41-6   HR: 3
ARSENIC ACID, HEMIHYDRATE**
mf: $AsH_3O_4 \cdot 1/2H_2O$ mw: 150.96
PROP: White, translucent crystals. Mp: 35.5°, bp: $-H_2O$ @ 160°, d: 2.0–2.5.
SYNS: ARSENIC ACID, solid (DOT) □ ORTHOARSENIC ACID HEMIHYDRATE
CONSENSUS REPORTS: Arsenic and its compounds are on the Community Right-To-Know List.
OSHA PEL: TWA 0.01 mg(As); Cancer Hazard
ACGIH TLV: TWA 0.01 mg/m³; Confirmed Human Carcinogen; BEI: 35 μ (As)/L inorganic arsenic and methylated metabolites in urine
NIOSH REL: (Arsenic, Inorganic) CL 2 μg(As)/m³/15M
DFG MAK: Human Carcinogen
SAFETY PROFILE: Confirmed human carcinogen. Poison by intravenous route. When heated to decomposition it emits toxic fumes of arsenic. See also ARSENIC COMPOUNDS.

**ARC750   CAS: 7645-25-2   HR: 3
ARSENIC ACID, LEAD SALT**
DOT: UN 1617
mf: $AsH_3O_4 \cdot 7Pb$ mw: 1592.28
SYNS: ARSENIATE de PLOMB (FRENCH) □ LEAD ARSENATE
CONSENSUS REPORTS: Arsenic compounds and lead compounds are on the Community Right-To-Know List.
OSHA PEL: TWA 0.01 mg(As)/m³; Cancer Hazard
ACGIH TLV: TWA 0.01 mg/m³; Confirmed Human Carcinogen; BEI: 35 μ (As)/L inorganic arsenic and methylated metabolites in urine

NIOSH REL: (Lead, Inorganic): 10H TWA 0.10 mg(Pb)/m³; (Arsenic, Inorganic): CL 0.002 mg(As)/m³/15M
DFG MAK: Human Carcinogen
DOT CLASSIFICATION: 6.1; Label: Poison
SAFETY PROFILE: Confirmed human carcinogen. Poison by ingestion. See also LEAD COMPOUNDS and ARSENIC COMPOUNDS. When heated to decomposition it emits very toxic fumes of lead and arsenic.

**ARD000   CAS: 10103-50-1   HR: 3
ARSENIC ACID, MAGNESIUM SALT**
DOT: UN 1622
mf: $AsH_3O_4 \cdot 7Mg$ mw: 312.12
PROP: Monoclinic, white crystals. D: 2.60–2.61.
SYNS: ARSENIATE de MAGNESIUM (FRENCH) □ MAGNESIUM ARSENATE □ MAGNESIUM ARSENATE PHOSPHOR
CONSENSUS REPORTS: Reported in EPA TSCA Inventory. Arsenic and its compounds are on the Community Right-To-Know List.
OSHA PEL: TWA 0.01 mg(As)/m³; Cancer Hazard
ACGIH TLV: TWA 0.01 mg/m³; Confirmed Human Carcinogen; BEI: 35 μ (As)/L inorganic arsenic and methylated metabolites in urine
DFG MAK: Human Carcinogen
NIOSH REL: (Arsenic, Inorganic) CL 2 μg(As)/m³/15M
DOT CLASSIFICATION: 6.1; Label: Poison
SAFETY PROFILE: Confirmed human carcinogen. Poison by ingestion. When heated to decomposition it emits toxic fumes of arsenic. See also ARSENIC COMPOUNDS.

**ARD250   CAS: 7784-41-0   HR: 3
ARSENIC ACID, MONOPOTASSIUM
    SALT**
DOT: UN 1677
mf: $AsH_2O_4 \cdot K$ mw: 180.04
SYNS: MACQUER'S SALT □ MONOPOTASSIUM ARSENATE □ MONOPOTASSIUM DIHYDROGEN ARSENATE □ POTASSIUM ACID ARSENATE □

POTASSIUM ARSENATE □ POTASSIUM ARSENATE, MONOBASIC □ POTASSIUM DIHYDROGEN ARSENATE □ POTASSIUM HYDROGEN ARSENATE

CONSENSUS REPORTS: IARC Cancer Review: Human Sufficient Evidence IMEMDT 23,39,80. Reported in EPA TSCA Inventory. Arsenic and its compounds are on the Community Right-To-Know List.

OSHA PEL: TWA 0.01 mg(As)/m³; Cancer Hazard

ACGIH TLV: TWA 0.01 mg/m³; Confirmed Human Carcinogen; BEI: 35 μ (As)/L inorganic arsenic and methylated metabolites in urine

NIOSH REL: (Arsenic, Inorganic) CL 2 μg(As)/m³/15M

DOT CLASSIFICATION: 6.1; Label: Poison

SAFETY PROFILE: Confirmed human carcinogen. Mutation data reported. When heated to decomposition it emits toxic fumes of arsenic. See also ARSENIC COMPOUNDS.

ARD500   CAS: 15120-17-9   HR: 3
ARSENIC ACID, MONOSODIUM SALT
mf: $AsO_3 \cdot Na$ mw: 145.91
PROP: Needle-like fibrous hygroscopic crystals. Sol in water.
SYNS: ARSENIC ACID, SODIUM SALT (9CI) □ SODIUM ARSENATE □ SODIUM METAARSENATE □ SODIUM MONOHYDROGEN ARSENATE

CONSENSUS REPORTS: Arsenic and its compounds are on the Community Right-To-Know List.

OSHA PEL: TWA 0.5 mg(As)/m³; Cancer Hazard

ACGIH TLV: TWA 0.01 mg/m³; Confirmed Human Carcinogen; BEI: 35 μ (As)/L inorganic arsenic and methylated metabolites in urine

DFG MAK: Human Carcinogen

NIOSH REL: (Arsenic, Inorganic) CL 2 μg(As)/m³/15M

SAFETY PROFILE: Confirmed human carcinogen. A poison. Mutation data reported. See also ARSENIC COMPOUNDS. When heated to

decomposition it emits toxic fumes of arsenic.

ARD600   CAS: 10103-60-3   HR: 3
ARSENIC ACID, MONOSODIUM SALT
mf: $AsH_2O_4 \cdot Na$ mw: 163.93
PROP: Solid. Mp: 118°.
SYNS: MONOSODIUM ARSENATE □ SODIUM ARSENATE □ SODIUM DIHYDROGEN ARSENATE □ SODIUM DIHYDROGEN ORTHOARSENATE

CONSENSUS REPORTS: Arsenic and its compounds are on the Community Right-To-Know List.

OSHA PEL: TWA 0.5 mg(As)/m³; Cancer Hazard

ACGIH TLV: BEI: 35 μ (As)/L inorganic arsenic and methylated metabolites in urine

NIOSH REL: CL 2 μg(As)/m³/15M

DFG MAK: Human Carcinogen

SAFETY PROFILE: Confirmed human carcinogen. Poison by intravenous route. When heated to decomposition it emits toxic fumes of arsenic.

ARD750   CAS: 7631-89-2   HR: 3
ARSENIC ACID, SODIUM SALT
DOT: UN 1685
mf: $AsH_3O_4 \cdot 7Na$ mw: 202.94
SYNS: FATSCO ANT POISON □ SODIUM ARSENATE (DOT) □ SODIUM METAARSENATE □ SODIUM ORTHOARSENATE □ SWEENEY'S ANT-GO

CONSENSUS REPORTS: IARC Cancer Review: Human Sufficient Evidence IMEMDT 23,39,80; Animal Inadequate Evidence IMEMDT 2,48,73; IMEMDT 23,39,80. Reported in EPA TSCA Inventory. Arsenic and its compounds are on the Community Right-To-Know List. EPA Extremely Hazardous Substances List.

OSHA PEL: TWA 0.01 mg(As)/m³; Cancer Hazard

ACGIH TLV: TWA 0.01 mg/m³; Confirmed Human Carcinogen; BEI: 35 μ (As)/L inorganic arsenic and methylated metabolites in urine

NIOSH REL: CL 2 μg(As)/m³/15M

DOT CLASSIFICATION: 6.1; Label: Poison

SAFETY PROFILE: Confirmed human carcinogen with experimental tumorigenic

A

data. Poison by ingestion, intravenous, and intraperitoneal routes. An experimental teratogen. Other experimental reproductive effects. Mutation data reported. When heated to decomposition it emits toxic fumes of As and Na_2O. See also ARSENIC COMPOUNDS.

ARE000   CAS: 64070-83-3   HR: 3
ARSENIC(V) ACID, TRISODIUM SALT,
**   HEPTAHYDRATE (1:3:7)**
mf: $AsO_4•3Na•7H_2O$ mw: 334.03
SYN: TRISODIUM ARSENATE, HEPTAHYDRATE
CONSENSUS REPORTS: Arsenic and its compounds are on the Community Right-To-Know List.
OSHA PEL: TWA 0.01 mg(As)/m³; Cancer Hazard
ACGIH TLV: TWA 0.01 mg/m³; Confirmed Human Carcinogen; BEI: 35 µ (As)/L inorganic arsenic and methylated metabolites in urine
NIOSH REL: (Arsenic, Inorganic) CL 2 µg(As)/m³/15M
DFG MAK: Human Carcinogen
SAFETY PROFILE: Confirmed human carcinogen. Poison by intraperitoneal route. See also ARSENIC COMPOUNDS. When heated to decomposition it emits toxic fumes of arsenic.

ARE500   CAS: 8028-73-7   HR: 3
ARSENICAL DUST
DOT: UN 1562
SYNS: ARSENICAL FLUE DUST □ FLUE DUST, ARSENIC CONTAINING
CONSENSUS REPORTS: Reported in EPA TSCA Inventory. Arsenic and its compounds are on the Community Right-To-Know List.
OSHA PEL: TWA 0.5 mg(As)/m³
ACGIH TLV: BEI: 35 µ (As)/L inorganic arsenic and methylated metabolites in urine
NIOSH REL: CL 2 µg(As)/m³/15M
DOT CLASSIFICATION: 6.1; Label: Poison
SAFETY PROFILE: A poison. See also ARSENIC COMPOUNDS.

ARE750   CAS: 8028-73-7   HR: 3

ARSENICAL FLUE DUST
DOT: UN 1562
CONSENSUS REPORTS: Arsenic and its compounds are on the Community Right-To-Know List. Reported in EPA TSCA Inventory.
OSHA PEL: TWA 0.5 mg(As)/m³
DFG MAK: Human Carcinogen
DOT CLASSIFICATION: 6.1; Label: Poison
SAFETY PROFILE: Confirmed human carcinogen with experimental tumorigenic data. Poison by inhalation and ingestion. See also ARSENIC COMPOUNDS.

ARF250   CAS: 7784-33-0   HR: 3
ARSENIC(III) BROMIDE
DOT: UN 1555
mf: $AsBr_3$ mw: 314.65
PROP: Colorless, deliquescent, rhombic crystals. Mp: 32.8°, bp: 220.0°, vap press: 1 mm @ 41.8°, d: 3.3972 @ 25°, (liq) 3.3282.
SYNS: ARSENIC TRIBROMIDE □ ARSENOUS BROMIDE □ ARSENOUS TRIBROMIDE □ TRIBROMOARSINE
CONSENSUS REPORTS: Reported in EPA TSCA Inventory. Arsenic and its compounds are on the Community Right-To-Know List.
OSHA PEL: TWA 0.01 mg(As)/m³; Cancer Hazard
ACGIH TLV: TWA 0.01 mg/m³; Confirmed Human Carcinogen; BEI: 35 µ (As)/L inorganic arsenic and methylated metabolites in urine
NIOSH REL: (Arsenic, Inorganic) CL 2 µg(As)/m³/15M
DOT CLASSIFICATION: 6.1; Label: Poison
SAFETY PROFILE: Confirmed human carcinogen. A poison. See also ARSENIC COMPOUNDS and BROMIDES. When heated to decomposition it emits very toxic fumes of As and Br⁻.

ARF500   CAS: 7784-34-1   HR: 3
ARSENIC CHLORIDE
DOT: UN 1560
mf: $AsCl_3$ mw: 181.27
PROP: Colorless, oily liquid. Freezing to colorless crystals with pearly sheen. Fumes

in air. D: 2.15 @ 25°, mp: −16°, bp: 130°. Decomp in water and by UV light; misc in chloroform, CCl_4, ether, iodine, P, S, alkali iodides, oils and fats. Vap d: 6.25, vap press: 10 mm @ 23.5°.

SYNS: ARSENIC BUTTER □ ARSENIC(III) CHLORIDE □ ARSENIOUS CHLORIDE □ ARSENOUS CHLORIDE □ ARSENOUS TRICHLORIDE (9CI) □ CHLORURE d'ARSENIC (FRENCH) □ CHLORURE ARSENIEUX (FRENCH) □ FUMING LIQUID ARSENIC □ TRICHLOROARSINE □ TRICHLORURE d'ARSENIC (FRENCH)

CONSENSUS REPORTS: Reported in EPA TSCA Inventory. Arsenic and its compounds are on the Community Right-To-Know List. EPA Extremely Hazardous Substances List.

OSHA PEL: OSHA: Cancer Hazard

ACGIH TLV: TWA 0.01 mg/m³; Confirmed Human Carcinogen; BEI: 35 μ (As)/L inorganic arsenic and methylated metabolites in urine

NIOSH REL: (Arsenic, Inorganic) CL 2 μg(As)/m³/15M

DOT CLASSIFICATION: 6.1; Label: Poison

SAFETY PROFILE: Confirmed human carcinogen. A poison via inhalation. See also ARSENIC COMPOUNDS and CHLORIDES. Very poisonous; fumes in air. Mutation data reported. When heated to decomposition it emits very toxic fumes of As and Cl⁻. Highly reactive. Explodes with Na, K, and Al on impact.

ARF750   HR: 3
ARSENIC COMPOUNDS

SYN: ARSENICALS

CONSENSUS REPORTS: Arsenic and its compounds are on the Community Right-To-Know List.

OSHA PEL: Inorganic: TWA 0.01 mg(As)/m³; Cancer Hazard; Organic: TWA 0.5 mg(As)/m³

ACGIH TLV: Inorganic: TWA 0.01 mg/m³; Confirmed Human Carcinogen; BEI: 35 μ (As)/L inorganic arsenic and methylated metabolites in urine

NIOSH REL: CL 2 μg(As)/m³/15M

SAFETY PROFILE: Inorganic compounds are confirmed human carcinogens producing tumors of the mouth, esophagus, larynx, bladder, and paranasal sinus. Recognized carcinogens of the skin, lungs, and liver. Used as insecticides, herbicides, silvicides, defoliants, desiccants, and rodenticides. Poisoning from arsenic compounds may be acute or chronic. Acute poisoning usually results from swallowing arsenic compounds; chronic poisoning from either swallowing or inhaling. Acute allergic reactions to arsenic compounds used in medical therapy have been fairly common, the type and severity of reaction depending upon the compound. Inorganic arsenicals are more toxic than organics. Trivalent is more toxic than pentavalent. Acute arsenic poisoning (from ingestion) results in marked irritation of the stomach and intestines with nausea, vomiting, and diarrhea. In severe cases, the vomitus and stools are bloody and the patient goes into collapse and shock with weak, rapid pulse, cold sweats, coma, and death. Chronic arsenic poisoning, whether through ingestion or inhalation, may manifest itself in many different ways. There may be disturbances of the digestive system such as loss of appetite, cramps, nausea, constipation, or diarrhea. Liver damage may occur, resulting in jaundice. Disturbances of the blood, kidneys, and nervous system are not infrequent. Arsenic can cause a variety of skin abnormalities including itching, pigmentation, and even cancerous changes. A characteristic of arsenic poisoning is the great variety of sympt-oms that can be produced. Dangerous; when heated to decomposition, or when metallic arsenic contacts acids or acid fumes, or when water solutions of arsenicals are in contact with active metals such as Fe, Al, or Zn, highly toxic fumes of arsenic are emitted.

ARG500   HR: 3
ARSENIC HEMISELENIDE

mf: As_2Se mw: 228.78

A

CONSENSUS REPORTS: Arsenic compounds and its compounds as well as selenium and its compounds are on the Community Right-To-Know List.

OSHA PEL: TWA 0.01 mg(As)/m³; Cancer Hazard; TWA 0.2 mg(Se)/m³

ACGIH TLV: TWA 0.01 mg/m³; Confirmed Human Carcinogen; BEI: 35 μ (As)/L inorganic arsenic and methylated metabolites in urine; TWA 0.2 mg(Se)/m³

DFG MAK: DFG TRK: 0.2 mg/m³ calculated as arsenic in that portion of dust that can possibly be inhaled; 0.1 mg(Se)/m³

NIOSH REL: CL 2 μg(As)/m³

SAFETY PROFILE: Confirmed human carcinogen. When heated to decomposition it emits fumes of As and Se. Incompatible with oxidizing materials. When heated to decomposition it emits highly toxic fumes of Se and arsenic. See ARSENIC COMPOUNDS and SELENIUM COMPOUNDS.

ARG750   CAS: 7784-45-4   HR: 3
ARSENIC IODIDE
mf: AsI₃   mw: 455.62

PROP: Lustrous, orange-red, hexagonal crystals, leaves or platelets. Mp: 146°, bp: 403°, d: 4.38 @ 13°. Sol in H₂O, Et₂O, CS₂, xylene, dioxan; sltly sol in conc HCl.

SYNS: ARSENIC TRIIODIDE □ ARSENOUS IODIDE □ ARSENOUS TRIIODIDE (9CI) □ TRIIODOARSINE

CONSENSUS REPORTS: Reported in EPA TSCA Inventory. Arsenic and its compounds are on the Community Right-To-Know List.

OSHA PEL: TWA 0.01 mg(As)/m³; Cancer Hazard

ACGIH TLV: TWA 0.01 mg/m³; Confirmed Human Carcinogen; BEI: 35 μ (As)/L inorganic arsenic and methylated metabolites in urine; Proposed: (inhalable fraction) 0.1 mg/m³; Not Classifiable as a Human Carcinogen)

NIOSH REL: (Arsenic, Inorganic) CL 2 μg(As)/m³/15M

SAFETY PROFILE: Confirmed human carcinogen. A poison. See also ARSENIC

COMPOUNDS and IODIDES. Can form a shock-sensitive compound with sodium or potassium. When heated to decomposition it emits very toxic fumes of I⁻ and arsenic.

ARH250   HR: 3
ARSENIC PENTASULFIDE
mf: As₂S₅   mw: 310.2

PROP: Brownish-yellow, glassy, amorphous, highly refractive mass. Mp: 500° (subl).

CONSENSUS REPORTS: Arsenic and its compounds are on the Community Right-To-Know List.

OSHA PEL: TWA 0.01 mg(As)/m³

ACGIH TLV: TLV 0.01 mg/m³Confirmed Human Carcinogen; BEI: 35 μ (As)/L inorganic arsenic and methylated metabolites in urine

NIOSH REL: CL 2 μg(As)/m³/15M

SAFETY PROFILE: Confirmed human carcinogen. See also ARSENIC COMPOUNDS and SULFIDES. Flammable in the form of dust when exposed to heat or flame. Explosive when intimately mixed with powerful oxidizers, such as Cl₂, KNO₃, or chlorates. Will react with water and steam to produce toxic and flammable vapors. Incompatible with water, steam, and strong oxidizers.

ARH500   CAS: 1303-28-2   HR: 3
ARSENIC PENTOXIDE
DOT: UN 1559
mf: As₂O₅   mw: 229.84

PROP: White, amorphous, deliquescent solid. Mp: decomp @ 800°, d: 4.32. Sol in alc. Very sol in H₂O.

SYNS: ANHYDRIDE ARSENIQUE (FRENCH) □ ARSENIC ACID □ ARSENIC ACID ANHYDRIDE □ ARSENIC ANHYDRIDE □ ARSENIC OXIDE □ ARSENIC(V) OXIDE □ DIARSENIC PENTOXIDE □ RCRA WASTE NUMBER P011 □ ZOTOX

CONSENSUS REPORTS: IARC Cancer Review: Human Sufficient Evidence IMEMDT 23,39,80. Reported in EPA TSCA Inventory. Arsenic and its compounds are on the Community Right-

To-Know List. EPA Extremely Hazardous
Substances List.
OSHA PEL: OSHA: Cancer Hazard
ACGIH TLV: TWA 0.01 mg/m^3; Confirmed
Human Carcinogen; BEI: 35 μ (As)/L
inorganic arsenic and methylated
metabolites in urine
DFG MAK: Human Carcinogen
NIOSH REL: CL 2 μg(As)/m^3/15M
DOT CLASSIFICATION: 6.1; Label: Poison
SAFETY PROFILE: Confirmed human
carcinogen. Poison by ingestion and
intravenous routes. Experimental
reproductive effects. Mutation data
reported. Reacts vigorously with Rb_2C_2.
When heated to decomposition it emits
toxic fumes of arsenic. See also ARSENIC
COMPOUNDS.

**ARI000   CAS: 1303-33-9   HR: 3
ARSENIC SULFIDE**
DOT: NA 1557
mf: As_2S_3 mw: 246.04
PROP: Red needles or yellow in polycrystal
form. Bp: 707°, d: 3.43, mp: 327°. Insol in
water; sol in alkalies.
SYNS: ARSENIC SESQUISULFIDE □ ARSENIC
SULFIDE YELLOW □ ARSENIC SULPHIDE □ ARSENIC
TERSULPHIDE □ ARSENIC TRISULFIDE □ ARSENIC
TRISULFIDE (DOT) □ ARSENIC YELLOW □
ARSENIOUS SULPHIDE □ ARSENOUS SULFIDE □
AURIPIGMENT □ C.I. 77086 □ C.I. PIGMENT YELLOW
□ DIARSENIC TRISULFIDE □ DIARSENIC
TRISULPHIDE □ KING'S GOLD □ KING'S YELLOW □
ORPIMENT
CONSENSUS REPORTS: IARC Cancer
Review: Human Sufficient Evidence
IMEMDT 23,39,80. Reported in EPA
TSCA Inventory. Arsenic and its
compounds are on the Community Right-
To-Know List.
OSHA PEL: OSHA: Cancer Hazard
ACGIH TLV: TWA 0.01 mg/m^3; Confirmed
Human Carcinogen; BEI: 35 μ (As)/L
inorganic arsenic and methylated
metabolites in urine
NIOSH REL: (Arsenic, Inorganic) CL 2
μg(As)/m^3/15M
DOT CLASSIFICATION: 6.1; Label: Poison

SAFETY PROFILE: Confirmed human
carcinogen with experimental tumorigenic
data. A poison. Reacts violently with H_2O_2,
$(KNO_3 + S)$. When heated to
decomposition or on contact with acid or
acid fumes it emits highly toxic fumes of
SO_2, H_2S, and As. Reacts with water or
steam to emit toxic and flammable vapors.

**ARI250   CAS: 7784-35-2   HR: 3
ARSENIC TRIFLUORIDE**
mf: AsF_3 mw: 131.92
PROP: Very mobile, colorless liquid which
fumes in air and is readily hydrolyzed. D:
3.01, mp: −5.95°, bp: 63°, vap press: 100
mm @ 13.2°, 400 mm @ 41.5°. Insol in
water; sol in alc, benzene, and mercury.
SYNS: ARSENIC FLUORIDE □ ARSENOUS FLUORIDE
□ TRIFLUOROARSINE
CONSENSUS REPORTS: Reported in EPA
TSCA Inventory. Arsenic and its
compounds are on the Community Right-
To-Know List.
OSHA PEL: TWA 0.01 mg(As)/m^3; Cancer
Hazard
ACGIH TLV: TWA 0.01 mg/m^3; Confirmed
Human Carcinogen; BEI: 35 μ (As)/L
inorganic arsenic and methylated
metabolites in urine
NIOSH REL: (Arsenic, Inorganic) CL 2
μg(As)/m^3/15M
SAFETY PROFILE: Confirmed human
carcinogen. A poison by inhalation. Strong
reaction with P_2O_3. When heated to
decomposition it emits very toxic fumes of
As and F⁻. See also FLUORIDES and
ARSENIC COMPOUNDS.

**ARI750   CAS: 1327-53-3   HR: 3
ARSENIC TRIOXIDE**
DOT: UN 1561
mf: As_2O_3 mw: 197.84
PROP: Colorless, rhombic crystals (dimer,
claudetite), or white powder. D: 4.15, mp:
312°, bp: 460°. Solubility in water: 1.82/100
@ 20°; sol in alc. Cubes: Colorless. D: 3.865,
mp: 309°. Solubility in water: 1.2/100 @
20°.

A

SYNS: ACIDE ARSENIEUX □ ANHYDRIDE ARSENIEUX □ ARSENIC BLANC □ ARSENIC OXIDE □ ARSENIC(III) OXIDE □ ARSENIC SESQUIOXIDE □ ARSENICUM ALBUM □ ARSENIGEN SAURE □ ARSENIOUS ACID □ ARSENIOUS OXIDE □ ARSENIOUS TRIOXIDE □ ARSENITE □ ARSENOLITE □ ARSENOUS ACID □ ARSENOUS ACID ANHYDRIDE □ ARSENOUS ANHYDRIDE □ ARSENOUS OXIDE □ ARSENOUS OXIDE ANHYDRIDE □ ARSENTRIOXIDE □ ARSODENT □ CLAUDELITE □ CLAUDETITE □ CRUDE ARSENIC □ DIARSENIC TRIOXIDE □ RCRA WASTE NUMBER P012 □ WHITE ARSENIC

CONSENSUS REPORTS: IARC Cancer Review: Group 1 IMEMDT 7,100,87; Human Limited Evidence IMEMDT 2,48,73; Human Sufficient Evidence IMEMDT 23,39,80; Animal Inadequate Evidence IMEMDT 2,48,73; IMEMDT 23,39,80. Reported in EPA TSCA Inventory. Arsenic and its compounds are on the Community Right-To-Know List. EPA Extremely Hazardous Substances List.

OSHA PEL: TWA 0.01 mg(As)/m³; Cancer Hazard

ACGIH TLV: TWA 0.01 mg/m³; Confirmed Human Carcinogen; BEI: 35 μ (As)/L inorganic arsenic and methylated metabolites in urine

DFG MAK: Human Carcinogen

NIOSH REL: CL 2 μg(As)/m³/15M

DOT CLASSIFICATION: 6.1; Label: Poison

SAFETY PROFILE: Confirmed human carcinogen with experimental neoplastigenic and tumorigenic data. Poison by ingestion, subcutaneous, and intravenous routes. Human systemic effects by ingestion: sleep changes, muscle weakness, hypermotility, diarrhea, cardiac arrhythmias, coma, fatty degeneration of the liver, depressed renal function tests. An experimental teratogen. Other experimental reproductive effects. Mutation data reported. Reacts vigorously with Rb_2C_2, CIF_3, F_2, Hg, OF_2, $NaClO_3$. See also ARSENIC COMPOUNDS.

ARJ000   HR: 3
ARSENIC TRIOXIDE mixed with
**   SELENIUM DIOXIDE (1:1)**
mf: $AsO_3 \cdot O_2Se$ mw: 233.88

SYN: SELENIUM DIOXIDE mixed with ARSENIC TRIOXIDE (1:1)

CONSENSUS REPORTS: Arsenic and its compounds, as well as selenium and its compounds, are on the Community Right-To-Know List.

OSHA PEL: TWA 0.01 mg(As)/m³; Cancer Hazard

ACGIH TLV: TWA 0.01 mg/m³; Confirmed Human Carcinogen; BEI: 35 μ (As)/L inorganic arsenic and methylated metabolites in urine

DFG MAK: 0.1 mg(Se)/m³

NIOSH REL: (Arsenic, Inorganic): CL 0.002 mg(As)/m³/15M

SAFETY PROFILE: Confirmed human carcinogen with experimental tumorigenic data. See also ARSENIC COMPOUNDS and SELENIUM COMPOUNDS. When heated to decomposition it emits very toxic fumes of As and Se.

ARJ100   CAS: 56320-22-0   HR: 3
ARSENIC TRISULFIDE
DOT: NA 1557
mf: AsS_2 mw: 139.04
PROP: A solid.
SYNS: ARSENIC DISULFIDE □ ARSENIC SULFIDE (DOT)

OSHA PEL: OSHA: Cancer Hazard

DOT CLASSIFICATION: 6.1; Label: Poison

SAFETY PROFILE: Confirmed human carcinogen. A poison. When heated to decomposition it emits toxic vapors of As and SO_x.

ARJ250   HR: 3
ARSENIDES
CONSENSUS REPORTS: Arsenic and its compounds are on the Community Right-To-Know List.

SAFETY PROFILE: Compounds of arsenic and hydrogen or metals (i.e., transitional, alkaline earth, or rare-earth). These materials are dangerous because they readily emit very toxic arsine and arsenic fumes when exposed to heat, moisture, acids, and acid fumes.

ARJ500   CAS: 14060-38-9   HR: 3
ARSENIOUS ACID, SODIUM SALT
mf: $AsH_3O_3 \cdot 7Na$ mw: 286.88
PROP: Colorless or grayish-white powder.
D: 1.87.
SYNS: ARSENIOUS ACID, SODIUM SALT POLYMERS
□ ARSONIC ACID, SODIUM SALT (9CI) □ NATRIUM-
ARSENIT (GERMAN) □ SODIUM ORTHOARSENITE
CONSENSUS REPORTS: Arsenic and its
compounds are on the Community Right-
To-Know List.
OSHA PEL: TWA 0.01 mg(As)/m³; Cancer
Hazard
ACGIH TLV: TWA 0.01 mg/m³; Confirmed
Human Carcinogen; BEI: 35 μ (As)/L
inorganic arsenic and methylated
metabolites in urine
NIOSH REL: (Arsenic, Inorganic) CL 2
μg(As)/m³/15M
SAFETY PROFILE: Confirmed human
carcinogen. Poison by intraperitoneal and
subcutaneous routes. Moderately toxic by
ingestion. When heated to decomposition it
emits toxic fumes of arsenic.

ARJ750   CAS: 1303-18-0   HR: 3
ARSENOPYRITE
mf: AsFeS   mw: 162.83
SYNS: ARSENOMARCASITE □ MISPICKEL
CONSENSUS REPORTS: Arsenic and its
compounds are on the Community Right-
To-Know List.
OSHA PEL: TWA 0.01 mg(As)/m³
ACGIH TLV: BEI: 35 μ (As)/L inorganic
arsenic and methylated metabolites in urine
NIOSH REL: (Arsenic, Inorganic) CL 2
μg(As)/m³/15M
SAFETY PROFILE: Poison by intravenous
route. When heated to decomposition it
emits very toxic fumes of As and SO_x.

ARJ755   CAS: 1122-90-3   HR: 3
p-ARSENOSOANILINE
mf: C_6H_6AsNO mw: 183.05
SYN: ANILINE, p-ARSENOSO-
OSHA PEL: TWA 0.5 mg(As)/m³
SAFETY PROFILE: Poison by
intraperitoneal route. When heated to

decomposition it emits toxic fumes of NO_x
and As.

ARJ760   CAS: 4164-07-2   HR: 3
**p-ARSENOSO-N,N-BIS(2-CHLOROETH-
   YL)ANILINE**
mf: $C_{10}H_{12}AsCl_2NO$ mw: 308.05
SYN: ANILINE, p-ARSENOSO-N,N-BIS(2-
CHLOROETHYL)-
OSHA PEL: TWA 0.5 mg(As)/m³
SAFETY PROFILE: Poison by
intraperitoneal route. When heated to
decomposition it emits toxic fumes of NO_x,
As, and Cl^-.

ARJ770   CAS: 5185-80-8   HR: 3
**p-ARSENOSO-N,N-BIS(2-HYDROXY-
   ETHYL)ANILINE**
mf: $C_{10}H_{14}AsNO_3$ mw: 271.17
SYN: ANILINE, p-ARSENOSO-N,N-BIS(2-
HYDROXYETHYL)-
OSHA PEL: TWA 0.5 mg(As)/m³
SAFETY PROFILE: Poison by
intraperitoneal route. When heated to
decomposition it emits toxic fumes of NO_x
and As.

ARJ800   CAS: 4164-06-1   HR: 3
p-ARSENOSO-N,N-DIETHYLANILINE
mf: $C_{10}H_{14}AsNO$ mw: 239.17
SYN: ANILINE, p-ARSENOSO-N,N-DIETHYL-
OSHA PEL: TWA 0.5 mg(As)/m³
SAFETY PROFILE: Poison by
intraperitoneal route. When heated to
decomposition it emits toxic fumes of NO_x
and As.

ARJ900   CAS: 63951-03-1   HR: 3
ARSENOXIDE SODIUM
mf: $C_6H_5AsNO_2 \cdot Na$ mw: 221.03
SYN: PHENOL, 2-AMINO-4-ARSENOSO-, SODIUM
SALT
OSHA PEL: TWA 0.5 mg(As)/m³
ACGIH TLV: BEI: 35 μ (As)/L inorganic
arsenic and methylated metabolites in urine
SAFETY PROFILE: Poison by intravenous
route. When heated to decomposition it
emits toxic fumes of NO_x and As.

ARK250   CAS: 7784-42-1   HR: 3

A

ARSINE
DOT: UN 2188

mf: AsH$_3$ mw: 77.95

PROP: Thermally unstable, colorless; gas with mild garlic odor. D: 2.695 g/L, bp: −62.5°, vap d: 2.66, mp: −116°. Readily oxidized to As$_2$O$_3$. Very little tendency to protonate. Solubility in water: 28 mg/100 @ 20°. Sol in benzene and chloroform. IDLH 3 ppm.

SYNS: ARSENIC HYDRID □ ARSENIC HYDRIDE □ ARSENIC TRIHYDRIDE □ ARSENIURETTED HYDROGEN □ ARSENOUS HYDRIDE □ ARSENOWODOR (POLISH) □ ARSENWASSERSTOFF (GERMAN) □ HYDROGEN ARSENIDE

CONSENSUS REPORTS: IARC Cancer Review: Human Sufficient Evidence IMEMDT 23,39,80. Reported in EPA TSCA Inventory. Arsenic and its compounds are on the Community Right-To-Know List. EPA Extremely Hazardous Substances List.

OSHA PEL: TWA 0.05 ppm

ACGIH TLV: TWA 0.005 ppm

DFG MAK: 0.05 ppm (0.16 mg/m³)

NIOSH REL: (Arsine) CL 2 μg(As)/m³/15M

DOT CLASSIFICATION: 2.3; Label: Poison Gas, Flammable Gas

SAFETY PROFILE: Confirmed human carcinogen. Poison by inhalation. Human red blood cell, gastrointestinal system, central nervous system, and other systemic effects by inhalation. Flammable when exposed to flame. Moderately explosive when exposed to Cl$_2$, HNO$_3$, (K + NH$_3$), open flame, or powerful shock. Dangerous, more toxic than its oxidation product. When heated to decomposition it emits highly toxic fumes of arsenic. See also ARSENIC, ARSENIC COMPOUNDS, and HYDRIDES.

ARL250   CAS: 8022-37-5   HR: 2
ARTEMISIA OIL
PROP: Chief constituent is Thujone, and found in the plant *Artemisia absinthium* L. (FCTXAV 13,681,75).

SYNS: ABSINTHIUM □ ARTEMISIA OIL (WORMWOOD) □ OIL, ARTEMISIA

CONSENSUS REPORTS: Reported in EPA TSCA Inventory.

SAFETY PROFILE: Moderately toxic by ingestion. An allergen. Habitual users develop "absinthism" with tremors, vertigo, vomiting, and hallucinations. May cause a contact dermatitis. When heated to decomposition it emits acrid smoke and irritating fumes.

ARM000   CAS: 13425-94-0   HR: 3
ASALIN
mf: C$_{22}$H$_{33}$Cl$_2$N$_3$O$_4$ mw: 474.48

SYNS: N-ACETYL-SARCOLYSIL VALINE ETHYL ETHER □ ASALINE □ AZALINE □ ETHYL ESTER of N-ACETYL-dl-SARCOSYLYL-dl-VALINE

SAFETY PROFILE: Poison by ingestion, intramuscular, rectal, and intraperitoneal routes. An experimental teratogen. See also ESTERS.

ARM250   CAS: 1332-21-4   HR: 3
ASBESTOS
DOT: NA 2212

SYNS: AMIANTHUS □ AMOSITE (OBS.) □ AMPHIBOLE □ ASBEST (GERMAN) □ ASBESTOS FIBER □ FIBROUS GRUNERITE □ NCI-C08991 □ SERPENTINE

CONSENSUS REPORTS: NTP 10th Report on Carcinogens. IARC Cancer Review: Group 1 IMEMDT 7,106,87; Human Sufficient Evidence IMEMDT 2,17,73; IMEMDT 14,11,77; Animal Sufficient Evidence IMEMDT 2,17,73; IMEMDT 14,11,77. Reported in EPA TSCA Inventory. On Community Right-To-Know List. EPA Genetic Toxicology Program.

OSHA PEL: TWA 2 million fibers/m³; CL 10 million fibers/m³; Cancer and Lung Disease Hazard

ACGIH TLV: TWA 0.1 fibers/cc; Confirmed Human Carcinogen

DFG MAK: DFG TRK: (Fine dust particles that are able to reach the alveolar area of the lung) crocidolite: 0.05 × 10⁶ fibers/m³ (0.025 mg/m³) (definition of fiber: length greater than 5 μm; diameter less than 3 μm; length/diameter greater than 3:1, equivalent to 1 fiber/cc); chrysotile, amosite, anthophyllite, tremolite, actinolite: 1 × 10⁶

fibers/m³ (0.05 mg/m³), applicable when there is more than 2.5% asbestos in the dust; 2.0 mg/m³, applicable when there is less than or equal to 2.5 weight percent asbestos in fine dust
NIOSH REL: (asbestos): 0.1 fb/cc in a 400 L air sample
DOT CLASSIFICATION: 9; Label: CLASS 9
SAFETY PROFILE: Confirmed human carcinogen producing lung tumors. Experimental neoplastigenic and tumorigenic data. Human pulmonary system effects by inhalation. Usually at least 4 to 7 years of exposure are required before serious lung damage (fibrosis) results. Mutation data reported. A common air contaminant.

ARM260   CAS: 77536-66-4   HR: 3
ASBESTOS, ACTINOLITE
DOT: NA 2212
SYNS: ACTINOLITE ASBESTOS □ ASBESTOS (ACGIH)
CONSENSUS REPORTS: IARC Cancer Review: Group 1 IMEMDT 7,106,87; Animal Sufficient Evidence IMEMDT 14,11,77.
OSHA PEL: TWA 2 million fibers/m³; CL 10 million fibers/m³; Cancer Hazard
ACGIH TLV: TWA 0.1 fibers/cc; Confirmed Human Carcinogen
DFG MAK: DFG TRK: (Fine dust particles that are able to reach the alveolar area of the lung) 1×10^6 fibers/m³ (0.05 mg/m³), applicable when there is more than 2.5% asbestos in the dust
NIOSH REL: (asbestos): 0.1 fb/cc in a 400 L air sample
DOT CLASSIFICATION: 9; Label: CLASS 9
SAFETY PROFILE: Confirmed human carcinogen. See also other asbestos entries.

ARM262   CAS: 12172-73-5   HR: 3
ASBESTOS, AMOSITE
SYNS: AMOSITE ASBESTOS □ ASBESTOS (ACGIH) □ MYSORITE □ NCI-C60253A
CONSENSUS REPORTS: NTP 10th Report on Carcinogens. IARC Cancer Review: Group 1 IMEMDT 7,106,87; Animal Sufficient Evidence IMEMDT 2,17,73;

IMEMDT 14,11,77; Human Sufficient Evidence IMEMDT 2,17,73; IMEMDT 14,11,77. NTP Carcinogenesis Studies (feed); No Evidence: hamster NTPTR* NTP-TR-249,83. EPA Genetic Toxicology Program.
OSHA PEL: TWA 2 million fibers/m³; CL 10 million fibers/m³; Cancer Hazard
ACGIH TLV: TWA 0.1 fibers/cc; Confirmed Human Carcinogen
DFG MAK: DFG TRK: (Fine dust particles that are able to reach the alveolar area of the lung) 1×10^6 fibers/m³ (0.05 mg/m³), applicable when there is more than 2.5% asbestos in the dust
NIOSH REL: (asbestos): 0.1 fb/cc in a 400 L air sample
SAFETY PROFILE: Confirmed human carcinogen with experimental carcinogenic, neoplastigenic, and tumorigenic data. Mutation data reported.

ARM264   CAS: 77536-67-5   HR: 3
ASBESTOS, ANTHOPHYLITE
SYNS: ANTHOPHYLITE □ ASBESTOS (ACGIH) □ AZBOLEN ASBESTOS □ FERROANTHOPHYLLITE
CONSENSUS REPORTS: NTP 10th Report on Carcinogens. IARC Cancer Review: Animal Sufficient Evidence IMEMDT 2,17,73; IMEMDT 14,11,77; Human Sufficient Evidence IMEMDT 14,11,77. EPA Genetic Toxicology Program.
OSHA PEL: TWA 2 million fibers/m³; CL 10 million fibers/m³; Cancer Hazard
ACGIH TLV: TWA 0.1 fibers/cc; Confirmed Human Carcinogen
DFG MAK: DFG TRK: (Fine dust particles that are able to reach the alveolar area of the lung) 1×10^6 fibers/m³ (0.05 mg/m³), applicable when there is more than 2.5% asbestos in the dust
NIOSH REL: (asbestos): 0.1 fb/cc in a 400 L air sample
SAFETY PROFILE: Confirmed human carcinogen with experimental carcinogenic, neoplastigenic, and tumorigenic data. Mutation data reported.

A

ARM266   CAS: 17068-78-9   HR: 3
ASBESTOS, ANTHOPHYLLITE
SYNS: AZBOLEN ASBESTOS □ 16 F
CONSENSUS REPORTS: NTP 10th Report
on Carcinogens.
ACGIH TLV: TWA 0.1 fibers/cc; Confirmed
Human Carcinogen
SAFETY PROFILE: Confirmed carcinogen
with experimental tumorigenic data.
Mutation data reported.

ARM268   CAS: 12001-29-5   HR: 3
ASBESTOS, CHRYSOTILE
DOT: NA 2212
PROP: Silky white to green to brownish
fibers.
SYNS: 7-45 ASBESTOS □ ASBESTOS (ACGIH) □
AVIBEST C □ CALIDRIA RG 100 □ CALIDRIA RG 144 □
CALIDRIA RG 600 □ CASSIAR AK □ CHRYSOTILE
ASBESTOS □ HOOKER NO. 1 CHRYSOTILE ASBESTOS
□ K6-30 □ METAXITE □ NCI-C61223A □ PLASTIBEST 20
□ 5R04 □ RG 600 □ SERPENTINE □ SERPENTINE
CHRYSOTILE □ SYLODEX □ WHITE ASBESTOS □
WHITE ASBESTOS (chrysotile, actinolite, anthophyllite,
tremolite) (DOT)
CONSENSUS REPORTS: NTP 10th Report
on Carcinogens. IARC Cancer Review:
Human Sufficient Evidence IMEMDT
2,17,73; Animal Sufficient Evidence
IMEMDT 2,17,73. NTP Carcinogenesis
Studies (feed); Some Evidence: rat NTPTR*
NTP-TR-295,85. EPA Genetic Toxicology
Program.
OSHA PEL: TWA 2 million fibers/m³; CL
10 million fibers/m³; Cancer Hazard
ACGIH TLV: TWA 0.1 fibers/cc; Confirmed
Human Carcinogen
DFG MAK: DFG TRK: (Fine dust particles
that are able to reach the alveolar area of the
lung) 1×10^6 fibers/m³ (0.05 mg/m³),
applicable when there is more than 2.5%
asbestos in the dust
NIOSH REL: (asbestos): 0.1 fb/cc in a 400 L
air sample
DOT CLASSIFICATION: 9; Label: CLASS 9
SAFETY PROFILE: Confirmed human
carcinogen producing tumors of the lung.
Human mutation data reported. Poison by
intraperitoneal route. Human systemic

effects by inhalation: lung fibrosis,
dyspnea, and cough.

ARM275   CAS: 12001-28-4   HR: 3
ASBESTOS, CROCIDOLITE
DOT: NA 2212
mf: $ONa_2Fe_2O_{33}FeO_8SiO_2H_2O$ mw:
765.98
SYNS: AMORPHOUS CROCIDOLITE ASBESTOS □
ASBESTOS (ACGIH) □ BLUE ASBESTOS (DOT) □
BROWN ASBESTOS (DOT) □ CROCIDOLITE ASBESTOS
□ CROCIDOLITE (DOT) □ FIBROUS CROCIDOLITE
ASBESTOS □ KROKYDOLITH (GERMAN) □ NCI-C09007
CONSENSUS REPORTS: NTP 10th Report
on Carcinogens. IARC Cancer Review:
Animal Sufficient Evidence IMEMDT
14,11,77, IMEMDT 2,17,73; Human
Sufficient Evidence IMEMDT 14,11,77.
EPA Genetic Toxicology Program.
OSHA PEL: TWA 2 million fibers/m³; CL
10 million fibers/m³; Cancer Hazard
ACGIH TLV: TWA 0.1 fibers/cc; Confirmed
Human Carcinogen
DFG MAK: DFG TRK: (Fine dust particles
that are able to reach the alveolar area of the
lung) crocidolite: 0.05×10^6 fibers/m³
(0.025 mg/m³) (definition of fiber: length
greater than 5 μm; diameter less than 3 μm;
length/diameter greater than 3:1, equivalent
to 1 fiber/cc)
NIOSH REL: (asbestos): 0.1 fb/cc in a 400 L
air sample
DOT CLASSIFICATION: 9; Label: CLASS 9
SAFETY PROFILE: Confirmed human
carcinogen with experimental carcinogenic,
neoplastigenic, and tumorigenic data by
inhalation. Human mutation data reported.

ARM280   CAS: 77536-68-6   HR: 3
ASBESTOS, TREMOLITE
DOT: NA 2212
SYNS: ASBESTOS (ACGIH) □ FIBROUS TREMOLITE □
NCI-C08991 □ TREMOLITE ASBESTOS
CONSENSUS REPORTS: IARC Cancer
Review: Human Sufficient Evidence
IMEMDT 14,11,77; Animal Sufficient
Evidence IMEMDT 14,11,77.
OSHA PEL: TWA 2 million fibers/m³; CL
10 million fibers/m³; Cancer Hazard

ACGIH TLV: TWA 0.1 fibers/cc; Confirmed Human Carcinogen

DFG MAK: DFG TRK: (Fine dust particles that are able to reach the alveolar area of the lung) 1×10^6 fibers/m^3 (0.05 mg/m^3), applicable when there is more than 2.5% asbestos in the dust

NIOSH REL: (asbestos): 0.1 fb/cc in a 400 L air sample

DOT CLASSIFICATION: 9; Label: CLASS 9

SAFETY PROFILE: Confirmed human carcinogen with experimental tumorigenic and neoplastigenic data.

ARM500   CAS: 512-85-6   HR: 3
ASCARIDOLE

mf: $C_{10}H_{16}O_2$ mw: 168.26

PROP: Colorless unstable liquid. Mp: 3.3°, bp: 40° @ 2 mm, 115° @ 15 mm, d: 1.010 @ 20°/4°.

SYNS: ASCARIDOL □ ASCARIDOLE (organic peroxide) (DOT) □ ASCARISIN □ 2,3-DIOXABICYCLO(2.2.2)OCT-5-ENE, 1-ISOPROPYL-4-METHYL- □ 1-METHYL-4-(1-METHYLETHYL)-2,3-DIOXABICYCLO(2.2.2)OCT-5-ENE □ 1,4-PEROXIDO-p-MENTHENE-2

DOT CLASSIFICATION: Forbidden

SAFETY PROFILE: Poison by ingestion. Questionable carcinogen with experimental neoplastigenic and tumorigenic data. Flammable by spontaneous chemical reaction. An oxidizer. Explodes when heated >130° or when exposed to organic acids. Dangerous; heating emits toxic fumes and may explode; reacts with reducing materials. See also CHENOPODIUM OIL and PEROXIDES, ORGANIC.

ARN000   CAS: 50-81-7   HR: 2
I-ASCORBIC ACID

mf: $C_6H_8O_6$ mw: 176.14

PROP: White crystals. Mp: 192°. Sol in water; sltly sol in alc; insol in ether, chloroform, benzene, petroleum ether, fixed oils, and fats.

SYNS: ASCORBIC ACID □ l(+)-ASCORBIC ACID □ ASCORBUTINA □ CEVITAMIC ACID □ CEVITAMIN □ FEMA No. 2109 □ 3-KETO-l-GULOFURANOLACTONE □ l-3-KETOTHREOHEXURONIC ACID LACTONE □ NATRASCORB INJECTABLE □ NCI-C54808 □ 3-OXO-l-GULOFURANOLACTONE □ VITACIN □ VITAMIN C □

VITAMISIN □ VITASCORBOL □ XITIX □ l-XYLOASCORBIC ACID

CONSENSUS REPORTS: NTP Carcinogenesis Bioassay (feed); No Evidence: mouse, rat NTPTR* NTP-TR-247,83; NTPTR* NTP-TR-214,82. Reported in EPA TSCA Inventory.

SAFETY PROFILE: Moderately toxic by ingestion and intravenous routes. Human systemic effects by intravenous route: blood, changes in tubules (including acute renal failure, acute tubular necrosis). An experimental teratogen. Other experimental reproductive effects. Mutation data reported. When heated to decomposition it emits acrid smoke and irritating fumes.

ARN125   CAS: 134-03-2   HR: D
ASCORBIC ACID SODIUM SALT

mf: $C_6H_8O_6 \cdot Na$ mw: 199.13

PROP: Minute white to yellow crystals; odorless. Decomp at 218°. Freely sol in water; very sltly sol in alc; insol in chloroform, ether.

SYNS: l-ASCORBIC ACID SODIUM SALT □ ASCORBICIN □ ASCORBIN □ CEBITATE □ CENOLATE □ ISKIA-C □ MONOSODIUM ASCORBATE □ NATRASCORB □ NATRI-C □ SODASCORBATE □ SODIUM ASCORBATE (FCC) □ SODIUM-l-ASCORBATE □ VITAMIN C □ VITAMIN C SODIUM

CONSENSUS REPORTS: Reported in EPA TSCA Inventory.

SAFETY PROFILE: Human mutation data reported. When heated to decomposition it emits toxic fumes of Na_2O.

ARN150   HR: D
ASCORBYL PALMITATE

mf: $C_{22}H_{38}O_7$ mw: 414.54

PROP: White-to-yellowish powder; slt odor. Very sltly sol in water, vegetable oil.

SYN: PALMITOYL, l-ASCORBIC ACID

SAFETY PROFILE: When heated to decomposition it emits acrid smoke and irritating fumes.

ARN810   CAS: 70-47-3   HR: D
I-ASPARAGINE

mf: $C_4H_8N_2O_3 \cdot H_2O$ mw: 150.13

A

PROP: White crystalline powder or rhombic hemihedral crystals; sltly sweet taste. Mp: 234°. Sol in water; insol in alc, ether.
SYN: l-α-AMINOSUCCINAMIC ACID
SAFETY PROFILE: When heated to decomposition emits toxic fumes of NO_x.

ARN825   CAS: 22839-47-0   HR: 1
ASPARTAME
mf: $C_{14}H_{18}N_2O_5$ mw: 294.34
PROP: White crystalline powder from water or alc; odorless with a sweet taste. Mp: 190°. Sltly sol in water, alc.
SYNS: 3-AMINO-N-(α-CARBOXYPHENETHYL)SUCCINAMIC ACID N-METHYL ESTER, stereoisomer □ ASPARTYLPHENYLALANINE METHYL ESTER □ N-l-α-ASPARTYL-l-PHENYLALANINE 1-METHYL ESTER (9CI) □ CANDEREL □ DIPEPTIDE SWEETENER □ EQUAL □ METHYL ASPARTYLPHENYLALANATE □ 1-METHYL N-l-α-ASPARTYL-l-PHENYLALANINE □ NUTRASWEET □ SWEET DIPEPTIDE
SAFETY PROFILE: Human systemic effects by ingestion: allergic dermatitis. Experimental reproductive effects. When heated to decomposition it emits toxic fumes of NO_x.

ARN850   CAS: 56-84-8   HR: 1
l-ASPARTIC ACID
mf: $C_4H_7NO_4$ mw: 133.10
PROP: Colorless to white crystals or leaflets; acid taste. Mp: 270°. Sltly sol in water; insol in alc, ether.
SYNS: (S)-AMINOBUTANEDIOIC ACID □ l-AMINOSUCCINIC ACID □ ASPARAGIC ACID □ l-ASPARAGIC ACID □ ASPARAGINIC ACID □ l-ASPARAGINIC ACID □ ASPARTIC ACID □ (l)-ASPARTIC ACID □ l-(+)-ASPARTIC ACID □ (S)-ASPARTIC ACID
CONSENSUS REPORTS: Reported in EPA TSCA Inventory.
SAFETY PROFILE: Low toxicity by intraperitoneal route. When heated to decomposition emits toxic fumes of NO_x.

ARO500   CAS: 8052-42-4   HR: 3
ASPHALT
DOT: NA 1999

PROP: Black or dark-brown mass. Bp: <470°, flash p: 400°F (CC), d: 0.95−1.1, autoign temp: 905°F.
SYNS: ASPHALT, at or above its Fp (DOT) □ ASPHALT FUMES (ACGIH) □ ASPHALT, PETROLEUM □ ASPHALTUM □ BITUMEN (MAK) □ JUDEAN PITCH □ MINERAL PITCH □ PETROLEUM ASPHALT □ PETROLEUM BITUMEN □ PETROLEUM PITCH □ PETROLEUM ROOFING TAR □ ROAD ASPHALT (DOT) □ ROAD TAR (DOT)
CONSENSUS REPORTS: IARC Cancer Review: Group 3 IMEMDT 7,133,87; Human Inadequate Evidence IMEMDT 35,39,85. Reported in EPA TSCA Inventory.
ACGIH TLV: TWA 0.5 mg/m³; Not Classifiable as a Human Carcinogen
DFG MAK: Confirmed Animal Carcinogen with Unknown Relevance to Humans
NIOSH REL: (Asphalt Fumes) CL 5 mg/m³/15M
DOT CLASSIFICATION: 3; Label: Flammable Liquid
SAFETY PROFILE: Suspected carcinogen with experimental carcinogenic and tumorigenic data. A moderate irritant. May contain carcinogenic components. Combustible when exposed to heat or flame. To fight fire, use foam, CO_2, or dry chemical.

ARQ250   CAS: 83-89-6   HR: 3
ATABRINE
mf: $C_{23}H_{30}ClN_3O$ mw: 400.01
PROP: Bright yellow crystals. Mp: decomp @ 248°.
SYNS: ACRICHINE □ ACRINAMINE □ ACRIQUINE □ AKRICHIN □ ANTIMALARINA □ 6-CHLORO-9-((4-(DIETHYL AMINO)-1-METHYL BUTYL)AMINO)-2-METHOXYACRIDINE □ 3-CHLORO-7-METHOXY-9-(1-METHYL-4-DIETHYLAMINOBUTYLAMINO)ACRIDINE □ ERION □ HAFFKININE □ MEPACRINE □ 2-METHOXY-6-CHLORO-9-DIETHYLAMINOPENTYL-AMINOACRIDINE □ QUINACRINE
SAFETY PROFILE: Poison by intravenous and subcutaneous routes. Moderately toxic by ingestion. Mutation data reported. Experimental reproductive effects. Has been implicated in aplastic anemia. When heated to decomposition, it emits very toxic fumes of Cl^- and NO_x.

ARQ600   CAS: 479-20-9   HR: 3
ATRANORIN
mf: $C_{19}H_{18}O_8$ mw: 374.35
SYN: BENZOIC ACID, 3-FORMYL-2,4-DIHYDROXY-6-
METHYL-, 3-HYDROXY-4-(METHOXYCARBONYL)-2,5-
DIMETHYLPHENYL ESTER
SAFETY PROFILE: A poison by skin
contact. A skin irritant. When heated to
decomposition it emits acrid smoke and
irritating vapors.

ARQ725   CAS: 1912-24-9   HR: 3
ATRAZINE
mf: $C_8H_{14}ClN_5$ mw: 215.72
PROP: Crystals. Mp: 175−177°. Solubility at
25°: in water: 70 ppm; ether: 12,000 ppm;
chloroform: 52,000 ppm; methanol: 18,000
ppm.
SYNS: A 361 □ AATREX □ AATREX 4L □ AATREX
NINE-O □ AATREX 80W □ 2-AETHYLAMINO-4-CHLOR-
6-ISOPROPYLAMINO-1,3,5-TRIAZIN (GERMAN) □ 2-
AETHYLAMINO-4-ISOPROPYLAMINO-6-CHLOR-1,3,5-
TRIAZIN (GERMAN) □ AKTIKON □ AKTIKON PK □
AKTINIT A □ AKTINIT PK □ ARGEZIN □ ATAZINAX
□ ATRANEX □ ATRASINE □ ATRATOL A □ ATRAZIN
□ ATRED □ ATREX □ CANDEX □ CEKUZINA-T □ 2-
CHLORO-4-ETHYLAMINEISOPROPYLAMINE-s-
TRIAZINE □ 1-CHLORO-3-ETHYLAMINO-5-
ISOPROPYLAMINO-s-TRIAZINE □ 1-CHLORO-3-
ETHYLAMINO-5-ISOPROPYLAMINO-2,4,6-TRIAZINE □
2-CHLORO-4-ETHYLAMINO-6-ISOPROPYLAMINO-s-
TRIAZINE □ 2-CHLORO-4-ETHYLAMINO-6-
ISOPROPYLAMINO-1,3,5-TRIAZINE □ 6-CHLORO-N-
ETHYL-N'-(1-METHYLETHYL)-1,3,5-TRIAZINE-2,4-
DIAMINE (9CI) □ 2-CHLORO-4-(2-PROPYLAMINO)-6-
ETHYLAMINO-s-TRIAZINE □ CRISATRINA □
CRISAZINE □ CYAZIN □ FARMCO ATRAZINE □
FENAMIN □ FENAMINE □ FENATROL □ G 30027 □
GEIGY 30,027 □ GESAPRIM □ GESOPRIM □ GRIFFEX
□ HUNGAZIN □ HUNGAZIN PK □ INAKOR □
OLEOGESAPRIM □ PRIMATOL □ PRIMAZE □
RADAZIN □ RADIZINE □ SHELL ATRAZINE
HERBICIDE □ STRAZINE □ TRIAZINE A 1294 □
VECTAL □ VECTAL SC □ WEEDEX A □ WONUK □
ZEAZIN □ ZEAZINE
CONSENSUS REPORTS: EPA Genetic
Toxicology Program. Reported in EPA
TSCA Inventory.
OSHA PEL: TWA 5 mg/m³
ACGIH TLV: TWA 5 mg/m³; Not
Classifiable as a Human Carcinogen
DFG MAK: 2 mg/m³

SAFETY PROFILE: Poison by
intraperitoneal route. Moderately toxic by
ingestion. Mildly toxic by inhalation and skin
contact. An experimental teratogen. Other
experimental reproductive effects. Human
mutation data reported. A skin and severe
eye irritant. Questionable carcinogen with
experimental tumorigenic data. When heated
to decomposition it emits toxic fumes of Cl⁻
and NO_x.

ARR000   CAS: 51-55-8   HR: 3
ATROPINE
mf: $C_{17}H_{23}NO_3$ mw: 289.41
PROP: Solid. Mp: 116−117°. Colorless
crystalline alkaloid.
SYNS: ATROPIN (GERMAN) □ EYEULES □ dl-
HYOSCYAMINE □ 2-PHENYLHYDRACRYLIC ACID-3-α-
TROPANYL ESTER □ β-PHENYL-γ-OXYPROPION-
SAEURE-TROPYL-ESTER (GERMAN) □ 1-α-H,5-α-H-
TROPAN-3-α-OL (±)-TROPATE (ESTER) □ dl-TROPAN-
YL-2-HYDROXY-1-PHENYLPROPIONATE □ TROPIC
ACID, ESTER with TROPINE □ TROPIC ACID-3-α-
TROPANYL ESTER □ TROPINE TROPATE □ dl-
TROPYLTROPATE □ (±)-TROPYL TROPATE
CONSENSUS REPORTS: Reported in EPA
TSCA Inventory.
SAFETY PROFILE: Poison by ingestion,
subcutaneous, intravenous, and
intraperitoneal routes. Human systemic
effects by ingestion and intramuscular
routes: visual field changes, mydriasis
(pupillary dilation), and muscle weakness.
An experimental teratogen. Other
experimental reproductive effects. An
alkaloid. When heated to decomposition it
emits toxic fumes of NO_x.

ARR500   CAS: 55-48-1   HR: 3
ATROPINE SULFATE (2:1)
mf: $C_{34}H_{46}N_2O_6 \bullet H_2O_4S$ mw: 676.90
PROP: Crystals. Mp: 190−194°.
SYNS: ATROPIN SIRAN (CZECH) □ ATROPINSULFAT
(GERMAN) □ SULFATE d'ATROPINE (FRENCH) □ 1-α-
H,5-α-H-TROPAN-3-α-OL (±)-TROPATE (ESTER),
SULFATE (2:1) SALT □ dl-TROPANYL-2-HYDROXY-1-
PHENYLPROPIONATE SULFATE □ TROPINTRAN
CONSENSUS REPORTS: Reported in EPA
TSCA Inventory.

SAFETY PROFILE: Poison by subcutaneous, intravenous, and intraperitoneal routes. Moderately toxic by ingestion. Human (child) pulmonary system effects by ingestion. Human systemic effects: decreased body temperature, cardiac arrhythmias. An experimental teratogen. Other experimental reproductive effects. See also ATROPINE. When heated to decomposition it emits very toxic fumes of NO_x and SO_x.

ART250 CAS: 12192-57-3 HR: 3
1-AUROTHIO-d-GLUCOPYRANOSE
mf: $C_6H_{11}O_5S•Au$ mw: 392.20

PROP: Yellow crystals from EtOH (aq). Sol in water, insol in org solvs.

SYNS: AUREOTAN □ AUROMYOSE □ AUROTAN □ AUROTHIOGLUCOSE □ AURUMINE □ AUTHRON □ BRENOL □ (d-GLUCOPYRANOSYLTHIO)GOLD □ (1-d-GLUCOSYLTHIO)GOLD □ GLYSANOL B □ GOLD THIOGLUCOSE □ GTG □ ORONOL □ ROMOSOL □ SOLGANAL □ SOLGANAL B □ (1-THIO-d-GLUCO-PYRANOSATO)GOLD □ 1-THIO-GLUCOPYRANOSE, MONOGOLD(1+) SALT □ THIOGLUCOSE d'OR (FRENCH)

CONSENSUS REPORTS: IARC Cancer Review: Group 1 IMEMDT 7,56,87; Animal Limited Evidence IMEMDT 13,39,77.

SAFETY PROFILE: Confirmed carcinogen with experimental carcinogenic and neoplastigenic data. A deadly human poison by an unspecified route. An experimental poison by intramuscular route. Moderately toxic by subcutaneous and intravenous routes. Human systemic effects: nausea or vomiting, cholestatic jaundice, and eye effects. An experimental teratogen. Other experimental reproductive effects. See also GOLD COMPOUNDS. When heated to decomposition it emits very toxic fumes of SO_x. Used to treat rheumatoid arthritis.

ARW250 CAS: 75-80-9 HR: 3
AVERTIN
mf: $C_2H_3Br_3O$ mw: 282.78

PROP: Crystals, needles, or prisms with ethereal odor and aromatic taste. Mp: 80°, bp: 92–93° @ 10 mm. Sltly water-sol; sol in alc and org solvs.

SYNS: BROMETHOL □ ETHOBROM □ NARCOLAN □ NARKOLAN □ RENARCOL □ TRIBROMETHANOL □ 2,2,2-TRIBROMOETHANOL □ TRIBROMOETHYL ALCOHOL □ 2,2,2-TRIBROMOETHYL ALCOHOL

CONSENSUS REPORTS: Reported in EPA TSCA Inventory.

SAFETY PROFILE: Poison by intravenous and intraperitoneal routes. Moderately toxic by ingestion and other routes. Dangerous when heated; see also BROMIDES.

ARY000 CAS: 320-67-2 HR: 3
AZACYTIDINE
mf: $C_8H_{12}N_4O_5$ mw: 244.24

PROP: Solid. Mp: 232–234° (decomp).

SYNS: 5-AC □ 5-ACZ □ 4-AMINO-1-β-d-RIBOFURANOSYL-d-TRIAZIN-2(1H)-ONE □ 4-AMINO-1-β-d-RIBOFURANOSYL-1,3,5-TRIAZIN-2(1H)-ONE □ ANTIBIOTIC U 18496 □ AZACITIDINE □ 5-AZACYTIDINE □ 5'-AZACYTIDINE □ LADAKAMYCIN □ MYLOSAR □ NCI-C01569 □ NSC-102816 □ U 18496

CONSENSUS REPORTS: NTP 10th Report on Carcinogens. IARC Cancer Review: Group 3 IMEMDT 7,56,87; Animal Limited Evidence IMEMDT 26,37,81; NCI Carcinogenesis Bioassay (ipr); Inadequate Studies: rat NCITR* NCI-CG-TR-42,78; Clear Evidence: mouse NCITR* NCI-CG-TR-42,78. EPA Genetic Toxicology Program.

SAFETY PROFILE: Confirmed carcinogen with experimental carcinogenic, neoplastigenic, tumorigenic data. Poison by ingestion, intravenous, and intraperitoneal routes. Human systemic effects by intravenous route: nausea, vomiting and diarrhea, reduction in white cell count (luekopenia and agranulocytosis). An experimental teratogen. Other experimental reproductive effects. Human mutation data reported. A skin irritant. When heated to decomposition it emits toxic fumes of NO_x.

ASA500 CAS: 115-02-6 HR: 3
AZASERINE
mf: $C_5H_7N_3O_4$ mw: 173.15

PROP: Light-yellow needles from EtOH (aq). Mp: 146–162° (decomp). Produced by the strain *Streptomyces fragilis* (85ERAY 2,1249,78).

SYNS: AZASERIN ☐ l-AZASERINE ☐ AZS ☐ CI-337 ☐ CL 337 ☐ CN-15,757 ☐ DIAZOACETATE (ESTER)-l-SERINE ☐ l-DIAZOACETATE (ESTER) SERINE ☐ DIAZO-ACETIC ACID ESTER with SERINE ☐ o-DIAZOACETYL-l-SERINE ☐ NSC-742 ☐ P-165 ☐ RCRA WASTE NUMBER U015 ☐ l-SERINE DIAZOACETATE ☐ l-SERINE DIAZOACETATE (ester)

CONSENSUS REPORTS: IARC Cancer Review: Group 2B IMEMDT 7,56,87; Animal Limited Evidence IMEMDT 10,73,76. EPA Genetic Toxicology Program.

SAFETY PROFILE: Suspected carcinogen with experimental carcinogenic, neoplastigenic, and tumorigenic data. Poison by ingestion, intraperitoneal, and subcutaneous routes. An experimental teratogen. Other experimental reproductive effects. Human mutation data reported. When heated to decomposition it emits toxic fumes of NO_x.

ASA600   CAS: 214899-21-5 HR: 3
AZASPIRACID
mf: $C_{47}H_{71}NO_{12}$ mw: 842.08

SAFETY PROFILE: A poison by ingestion. When heated to decomposition it emits toxic vapors of NO_x.

ASB250   CAS: 446-86-6   HR: 3
AZATHIOPRINE
mf: $C_9H_7N_7O_2S$ mw: 277.29

PROP: Pale-yellow crystals from Me_2CO (aq). Mp: 243–244° (decomp).

SYNS: AZANIN ☐ AZATIOPRIN ☐ AZOTHIOPRINE ☐ BW 57-322 ☐ CCUCOL ☐ IMURAN ☐ IMUREK ☐ IMUREL ☐ METHYLNITROIMIDAZOLYLMERCAPTOPURINE ☐ 6-(1'-METHYL-4'-NITRO-5'-IMIDAZOLYL)-MERCAPTO-PURINE ☐ 6-(METHYL-p-NITRO-5-IMIDAZOLYL)-THIOPURINE ☐ 6-(1-METHYL-p-NITRO-5-IMIDAZOL-YL)-THIOPURINE ☐ 6-((1-METHYL-4-NITROIMIDAZOL-5-YL)THIO)PURINE ☐ 6-(1-METHYL-4-NITROIMIDAZ-OL-5-YLTHIO)PURINE ☐ 6-((1-METHYL-4-NITRO-1H-IMIDAZOL-5-YL)THIO)-1H-PURINE ☐ NCI-C03474 ☐ NSC-39084 ☐ RORASUL

CONSENSUS REPORTS: NTP 10th Report on Carcinogens. IARC Cancer Review: Group 1 IMEMDT 7,119,87; Human Sufficient Evidence IMEMDT 26,47,81;

Animal Limited Evidence IMEMDT 26,47,81. NCI Carcinogenesis Studies (ipr); No Evidence: rat CANCAR 40,1935,77; Clear Evidence: mouse CANCAR 40,1935,77. EPA Genetic Toxicology Program.

SAFETY PROFILE: Confirmed human carcinogen producing bladder tumors and leukemia. Poison by subcutaneous, intradermal, and intraperitoneal routes. Moderately toxic by ingestion. Human systemic effects: liver changes, hypermotility, diarrhea, nausea or vomiting, increased body temperature, BP lowering, decreased urine volume or anuria, normocytic anemia, bone marrow changes. An experimental teratogen. Other experimental reproductive effects. Human mutation data reported. When heated to decomposition it emits very toxic fumes of NO_x and SO_x. An immunosuppressant.

ASH500   CAS: 86-50-0   HR: 3
AZINPHOS METHYL
mf: $C_{10}H_{12}N_3O_3PS_2$ mw: 317.34

PROP: Crystals or brown, waxy solid. D: 1.44, mp: 74°. Very sltly sol in water, very sol in $CHCl_3$, toluene. IDLH 10 mg/m³.

SYNS: AZINFOS-METHYL (DUTCH) ☐ AZINPHOS METHYL, liquid (DOT) ☐ AZINPHOS-METILE (ITALIAN) ☐ BAY 9027 ☐ BAYER 17147 ☐ BENZOTRIAZINE derivative of a METHYL DITHIOPHOSPHATE ☐ BENZOTRIAZINEDITHIOPHOSPHORIC ACID DIMETHOXY ESTER ☐ CARFENE ☐ COTNION METHYL ☐ CRYSTHION 2L ☐ CRYSTHYON ☐ DBD ☐ S-(3,4-DIHYDRO-4-OXO-BENZO(α)(1,2,3)TRIAZIN-3-YLMETHYL)-O,O-DIMETHYL PHOSPHORODITHIOATE ☐ S-(3,4-DIHYDRO-4-OXO-1,2,3-BENZOTRIAZIN-3-YLMETHYL)-O,O-DIMETHYL PHOSPHORODITHIOATE ☐ O,O-DIMETHYL-S-(BENZAZIMINOMETHYL) DITHIOPHOSPHATE ☐ O,O-DIMETHYL-S-(1,2,3-BENZOTRIAZINYL-4-KETO)METHYL PHOSPHORO-DITHIOATE ☐ O,O-DIMETHYL-S-(3,4-DIHYDRO-4-KETO-1,2,3-BENZOTRIAZINYL-3-METHYL) DITHIOPHOSPHATE ☐ DIMETHYLDITHIO-PHOSPHORIC ACID N-METHYLBENZAZIMIDE ESTER ☐ O,O-DIMETHYL-S-(4-OXO-3H-1,2,3-BENZOTRIZIANE-3-METHYL)PHOSPHORODITHIOATE ☐ O,O-DIMETH-YL-S-(4-OXOBENZOTRIAZINO-3-METHYL)PHOS-PHORO-DITHIOATE ☐ O,O-DIMETHYL-S-(4-OXO-1,2,3-BENZOTRIAZINO(3)-METHYL) THIOTHIONOPHOSPH-ATE ☐ O,O-DIMETHYL-S-((4-OXO-3H-1,2,3-BENZOTRIAZIN-3-YL)-METHYL)-DITHIOFOSFAAT

(DUTCH) □ O,O-DIMETHYL-S-((4-OXO-3H-1,2,3-BENZOTRIAZIN-3-YL)-METHYL)-DITHIOPHOSPHAT (GERMAN) □ O,O-DIMETHYL-S-4-OXO-1,2,3-BENZOTRIAZIN-3(4H)-YLMETHYL PHOSPHORODI-THIOATE □ O,O-DIMETIL-S-((4-OXO-3H-1,2,3-BENZOTRIAZIN-3-IL)-METIL)-DITIOFOSFATO (ITALIAN) □ ENT 23,233 □ GOTHNION □ GUSATHION □ GUTHION (DOT) □ 3-(MERCAPTOMETHYL)-1,2,3-BENZOTRIAZIN-4(3H)-ONE-O,O-DIMETHYL PHOSPHORODITHIOATE □ 3-(MERCAPTOMETHYL)-1,2,3-BENZOTRIAZIN-4(3H)-ONE-O,O-DIMETHYL PHOSPHORODITHIOATE-S-ESTER □ METHYL-AZINPHOS □ N-METHYLBENZAZIMIDE, DIMETHYLDITHIOPHOSPHORIC ACID ESTER □ METHYL GUTHION □ METILTRIAZOTION □ NCI-C00066

CONSENSUS REPORTS: NCI Carcinogenesis Bioassay (feed); Inadequate Studies: rat NCITR* NCI-CG-TR-69,78; No Evidence: mouse NCITR* NCI-CG-TR-69,78. EPA Genetic Toxicology Program. EPA Extremely Hazardous Substances List.

OSHA PEL: TWA 0.2 mg/m³ (skin)

ACGIH TLV: TWA 0.2 mg/m³ (skin, sensitizer); Not Classifiable as a Human Carcinogen

DFG MAK: 0.2 mg/m³

SAFETY PROFILE: Poison by inhalation, ingestion, skin contact, intravenous, and intraperitoneal routes. An experimental teratogen. Other experimental reproductive effects. Human mutation data reported. Questionable carcinogen with experimental tumorigenic data. See also PARATHION and ESTERS. When heated to decomposition it emits very toxic fumes of PO_x, SO_x, and NO_x.

ASI000 CAS: 1072-52-2 HR: 3
1-AZIRIDINE ETHANOL
mf: C_4H_9NO mw: 87.14

PROP: Liquid. Bp: 154–156°.

SYNS: 2-(1-AZIRIDINYL)ETHANOL □ β-HYDROXY-1-ETHYLAZIRIDINE □ N-(β-HYDROXYETHYL)-AZIRIDINE □ 2-HYDROXY-1-ETHYLAZIRIDINE □ N-(2-HYDROXYETHYL)AZIRIDINE □ N-HYDROXYETH-YL ETHYLENE IMINE □ N-(2-HYDROXYETH-YL)ETHYLENIMINE □ 1-(2-HYDROXYETHYL)-ETHYLENIMINE

CONSENSUS REPORTS: IARC Cancer Review: Group 3 IMEMDT 7,56,87; Animal

Limited Evidence IMEMDT 9,47,75. Reported in EPA TSCA Inventory.

SAFETY PROFILE: Poison by ingestion, skin contact, and intravenous routes. A skin and eye irritant. Questionable carcinogen with experimental neoplastigenic data. Mutation data reported. When heated to decomposition it emits toxic fumes of NO_x.

ASI300 CAS: 4638-44-2 HR: 3
1-AZIRIDINYL m-(BIS(2-CHLOROETH-YL)AMINO)PHENYL KETONE
mf: $C_{13}H_{16}Cl_2N_2O$ mw: 287.21

PROP: A liquid.

DOT CLASSIFICATION: 3; Label: Flammable Liquid

SAFETY PROFILE: A poison by an unspecified route. A flammable liquid. When heated to decomposition it emits toxic vapors of NO_x and Cl⁻.

ASL250 CAS: 103-33-3 HR: 2
AZOBENZENE
mf: $C_{12}H_{10}N_2$ mw: 182.23

PROP: Orange, monoclinic crystals. Mp: 68°, bp: 297°, d: 1.203 @ 20°/4°, vap press: 1 mm @ 103.5°. Insol in water. Solubility in alc = 4.2/100 @ 20° in ether (ligroin) = 12/100 @ 20°.

SYNS: AZOBENZEEN (DUTCH) □ AZOBENZIDE □ AZOBENZOL □ AZOBISBENZENE □ AZODIBENZENE □ AZODIBENZENEAZOFUME □ BENZENEAZO-BENZENE □ DIAZOBENZENE □ DIPHENYLDIAZENE □ 1,2-DIPHENYLDIAZENE □ DIPHENYLDIIMIDE □ ENT 14,611 □ NCI-C02926 □ USAF EK-704

CONSENSUS REPORTS: IARC Cancer Review: Group 3 IMEMDT 7,56,87; Animal Limited Evidence IMEMDT 8,75,75; NCI Carcinogenesis Bioassay (feed); Clear Evidence: rat NCITR* NCI-CG-TR-154,79; No Evidence: mouse NCITR* NCI-CG-TR-154,79. Reported in EPA TSCA Inventory.

SAFETY PROFILE: Moderately toxic by ingestion and possibly other routes. Questionable carcinogen with experimental carcinogenic, neoplastigenic, and tumorigenic data. When heated to decomposition it emits toxic fumes of NO_x.

ASL750   CAS: 78-67-1   HR: 3
AZOBISISOBUTYLONITRILE
DOT: UN 2952
mf: $C_8H_{12}N_4$ mw: 164.24
PROP: Crystals from EtOH. Mp: 107°
(decomp).
SYNS: ACETO AZIB □ AIBN □ α,α'-AZOBISISO-
BUTYLONITRILE □ AZOBISISOBUTYRONITRILE □ 2,2'-
AZOBIS(ISOBUTYRONITRILE) □ 2,2'-AZOBIS(2-
METHYLPROPIONITRILE) □ AZODIISOBUTYRO-
NITRILE □ α,α'-AZODIISOBUTYRONITRILE □ 2,2'-
AZODIISOBUTYRONITRILE □ AZODIISOBUTYRO-
NITRILE (DOT) □ 2,2'-DICYANO-2,2'-AZOPROPANE □
POLY-ZOLE AZDN □ POROFOR 57 □ VAZO 64
CONSENSUS REPORTS: Cyanide and its
compounds are on the Community Right-
To-Know List. Reported in EPA TSCA
Inventory.
DOT CLASSIFICATION: 4.1; Label:
Flammable Solid, Explosive
SAFETY PROFILE: Poison by
intraperitoneal route. Moderately toxic by
ingestion. Easily oxidized, unstable. Violent
exothermic decomposition when heated.
Solution in acetone may decompose
explosively. Explodes when heated with
heptane. When heated to decomposition it
emits toxic fumes of NO_x and CN^-. See also
NITRILES. A free-radical generator.

ASM300   CAS: 123-77-3   HR: 3
AZODICARBONAMIDE
mf: $C_2H_4N_4O_2$ mw: 116.08
PROP: Yellow to orange-red crystalline
powder. Mp: above 180° (decomp). Sltly sol
in dimethyl sulfoxide; insol in water, org
solvs.
SAFETY PROFILE: Flammable solid. When
heated to decomposition it emits toxic
fumes of NO_x.

ASO750   CAS: 495-48-7   HR: 3
AZOXYBENZENE

mf: $C_{12}H_{10}N_2O$ mw: 198.23
PROP: Yellow, rhombic crystals. D: 1.248
@ 20°/20°, mp: 36°, bp: decomp. Insol in
water; solubility in alc = 11.4/100 @ 15°,
solubility in ether (ligroin) = 43.5/100 @
15°.
SYNS: AZOBENZENE OXIDE □ AZOSSIBENZENE
(ITALIAN) □ AZOXYBENZEEN (DUTCH) □
AZOXYBENZIDE □ AZOXYBENZOL (GERMAN) □
AZOXYDIBENZENE □ ORDINARY AZOXYBENZENE
CONSENSUS REPORTS: Reported in EPA
TSCA Inventory.
SAFETY PROFILE: Poison by
subcutaneous route. Moderately toxic by
ingestion, skin contact, and other routes. A
skin and eye irritant. Mutation data reported.
Combustible. When heated to
decomposition it emits toxic fumes of NO_x.

ASP000   CAS: 16301-26-1   HR: 3
AZOXYETHANE
mf: $C_4H_{10}N_2O$ mw: 102.16
PROP: Bp: 46°.
SYNS: AZOXYAETHAN (GERMAN) □
DIETHYLDIAZENE-1-OXIDE
SAFETY PROFILE: Poison by
subcutaneous and intravenous routes. An
experimental teratogen. Questionable
carcinogen with experimental carcinogenic
and tumorigenic data. When heated to
decomposition it emits toxic fumes of NO_x.

ASP250   CAS: 25843-45-2   HR: 3
AZOXYMETHANE
mf: $C_2H_6N_2O$ mw: 74.10
PROP: Oil. Bp: 98°.
SYN: AOM
SAFETY PROFILE: Suspected carcinogen
with experimental carcinogenic and
tumorigenic data. Poison by subcutaneous
route. An experimental teratogen. Mutation
data reported. When heated to
decomposition it emits toxic fumes of NO_x

B

BAC000   CAS: 9014-01-1   HR: 3
BACILLUS SUBTILIS CARLSBERG
PROP: A commercial raw proteolytic
enzyme used in laundry detergents
(FCTXAV 7,581,69).
SYNS: ALCALASE □ ALK-ENZYME □ BACILLO-
PEPTIDASE A □ BACILLOPEPTIDASE B □ BIOPRASE □
COLISTINASE □ E.C. 3.4.4.16 □ E.C. 3.4.21.14 □
MAXATASE □ NAGARSE □ SUBTILISIN (9CI, ACGIH) □
SUBTILISIN CARLSBURG □ SUBTILISIN NOVO □
SUBTILOPEPTIDASE A □ SUBTILOPEPTIDASE B □
SUBTILOPEPTIDASE BPN' □ SUBTILOPEPTIDASE C □
THERMOASE PC-10
CONSENSUS REPORTS: Reported in EPA
TSCA Inventory.
ACGIH TLV: CL 0.00006 mg/m³
SAFETY PROFILE: Moderately toxic by
ingestion. An eye irritant. When heated to
decomposition it emits toxic fumes of NO_x.

BAC250   CAS: 1405-87-4   HR: 3
BACITRACIN
PROP: White to pale-buff, hygroscopic
powder; odorless or slt odor. Freely sol in
water, alc, methanol, and glacial acetic acid;
insol in acetone, chloroform, and ether.
When heated to decomposition it emits
acrid smoke and irritating fumes.
SYNS: AYFIVIN □ BACIGUENT □ BACI-JEL □
BACILIQUIN □ BACITEK OINTMENT □ FORTRACIN □
PARENTRACIN □ PENITRACIN □ TOPITRACIN □
USAF CB-7 □ ZUTRACIN
CONSENSUS REPORTS: Reported in EPA
TSCA Inventory.
SAFETY PROFILE: A poison by
intraperitoneal and intravenous routes.
Moderately toxic by ingestion and
subcutaneous routes. Mutation data
reported.

BAC260   CAS: 55852-84-1   HR: 1
BACITRACIN METHYLENE DISALICYL-
    ATE

PROP: White to brownish-gray powder.
Disagreeable odor. Sol in water, pyridine,
ethanol; less sol in acetone, ether, chloro-
form, pentane, benzene.
SYNS: BMD □ FORTRACIN (BACITRACIN-MD) □
KEMITRACIN 10 □ MD BACITRACIN
CONSENSUS REPORTS: Reported in EPA
TSCA Inventory.
SAFETY PROFILE: Low oral toxicity.
When heated to decomposition it emits
acrid smoke and irritating fumes.

BAD400   HR: D
BAKER'S YEAST EXTRACT
PROP: From ruptured cells of *Saccharomyces
cerevisiae*. Liquid, paste or powder. Water sol.
SYNS: AUTOLYZED YEAST EXTRACT □ BAKER'S
YEAST GLYCAN
SAFETY PROFILE: When heated to
decomposition it emits acrid smoke and
irritating fumes.

BAD750   CAS: 59-52-9   HR: 3
BAL
mf: $C_3H_8OS_2$ mw: 124.23
PROP: Viscous, oily liquid; pungent odor.
Bp: 140° @ 40 mm, vap d: 4.3, d: 1.2385 @
25°/4°.
SYNS: BRITISH ANTILEWISITE □ DICAPTOL □
DIMERCAPROL PROPANOL □ DIMERCAPTOL □ 2,3-
DIMERCAPTOL-1-PROPANOL □ DIMERCAPTOPRO-
PANOL □ 2,3-DIMERCAPTOPROPANOL □ 2,3-
DIMERCAPTOPROPAN-1-OL □ DITHIOGLYCEROL □
1,2-DITHIOGLYCEROL □ 2,3-DITHIOPROPANOL □
SULFACTIN □ USAF ME-1
CONSENSUS REPORTS: EPA Genetic
Toxicology Program. Reported in EPA
TSCA Inventory.
SAFETY PROFILE: Poison via ingestion,
intramuscular, parenteral, intraperitoneal,
and intravenous routes. Experimental
teratogenic effects. Human systemic effects
by intramuscular route: hemorrhage and

dermatitis. Human blood and systemic skin effects by intramuscular route. It causes redness and swelling when applied locally to the skin, but does not produce blisters or ulcers. Intensely irritating to eyes and mucous membranes. Systemic symptoms are caused by injection. When heated to decomposition, it emits toxic fumes of SO_x. Used as an antidote to arsenic, gold, and mercury poisoning.

BAE750   **HR: 1**
BALSAM of PERU
PROP: Dark-brown, viscid liquid; vanilla odor. Sol in fixed oils; sltly sol in propylene glycol; insol in glycerin. Extracted from *Myroxylon pereirae Klotzsch.*
SYNS: BALSAM PERU OIL (FCC) ◻ PERUVIAN BALSAM
SAFETY PROFILE: A mild allergen. Combustible when heated. When heated to decomposition it emits acrid smoke and irritating fumes.

BAG250   CAS: 144-02-5   HR: 3
BARBITAL SODIUM
mf: $C_8H_{12}N_2O_3 \cdot Na$ mw: 207.21
PROP: Bitter crystals or powder.
SYNS: BARBITAL Na ◻ BARBITAL SOLUBLE ◻ BARBITONE SODIUM ◻ DIETHYLBARBITURATE MONOSODIUM ◻ 5,5-DIETHYLBARBITURIC ACID SODIUM deriv. ◻ DIETHYLMALONYLUREA SODIUM ◻ EMBINAL ◻ MEDINAL ◻ NATRINAL ◻ NATRIUM-BARBITALS (GERMAN) ◻ NERVOSETON ◻ 2,4,6(1H,3H,5H)-PYRIMIDINETRIONE, 5,5-DIETHYL-, MONOSODIUM SALT (9CI) ◻ SODIUM BARBITAL ◻ SODIUM BARBITONE ◻ SODIUM DIETHYL-BARBITURATE ◻ SODIUM-5,5-DIETHYLBARBITURATE ◻ SODIUM ETHYLBARBITAL ◻ SODIUM MALONYL-UREA ◻ SODIUM VERONAL ◻ SOLUBLE BARBITAL ◻ SOPRINAL ◻ THYALONE ◻ VERONAL SODIUM
SAFETY PROFILE: Poison by ingestion, subcutaneous, intravenous, and intraperitoneal routes. Large doses cause marked depression (sometimes preceded by excitation), prolonged coma, and death. Experimental teratogenic and reproductive effects. Allergic skin reactions may occur on contact. Implicated in development of aplastic anemia. Questionable carcinogin

with experimental tumorigenic and neoplastigenic data. A truly habit-forming drug. Other experimental reproductive effects. Mutation data reported. Combustible. When heated to decomposition it emits toxic fumes of NO_x and Na_2O. See also BARBITURATES.

BAH250   CAS: 7440-39-3   HR: 3
BARIUM
DOT: UN 1400
af: Ba   aw: 137.36
PROP: Silver-white, sltly lustrous, somewhat malleable metal. Mp: 727°, bp: 1640°, d: 3.5 @ 20°, vap press: 10 mm @ 1049°. Dissolves in H_2O forming $Ba(OH)_2$ solns. Solution in $NH_3(l)$ blue-black soln.
CONSENSUS REPORTS: Reported in EPA TSCA Inventory. Community Right-To-Know List.
OSHA PEL: TWA 0.5 mg(Ba)/m³
ACGIH TLV: TWA 0.5 mg(Ba)/m³; Not Classifiable as a Human Carcinogen
DFG MAK: 0.5 mg(Ba)/m³
DOT CLASSIFICATION: 4.3; Label: Dangerous When Wet
SAFETY PROFILE: Water and stomach acids solubilize barium salts and can cause poisoning. Symptoms are vomiting, colic, diarrhea, slow irregular pulse, transient hypertension, and convulsive tremors and muscular paralysis. Death may occur in a few hours to a few days. Half-life of barium in bone has been estimated at 50 days. Dust is dangerous and explosive when exposed to heat, flame, or chemical reaction. Violent or explosive reaction with water, CCl_4, fluorotrichloromethane, trichloroethylene, and C_2Cl_4. Incompatible with acids, $C_2Cl_3F_3$, $C_2H_2FCl_3$, C_2HCl_3 and water, 1,1,2-trichlorotrifluoroethane, and fluorotrichloroethane. The powder may ignite or explode in air or other oxidizing gases. See also BARIUM COMPOUNDS.

BAH500   CAS: 543-80-6   HR: 3
BARIUM ACETATE
mf: $C_4H_6O_4 \cdot Ba$ mw: 255.44

PROP: White or colorless crystals. Decomp on heating with $BaCO_3$ formation. Very sol in H_2O.
SYNS: ACETIC ACID, BARIUM SALT □ BARIUM DIACETATE □ OCTAN BARNATY (CZECH)
CONSENSUS REPORTS: Reported in EPA TSCA Inventory. Barium and its compounds are on the Community Right-To-Know List.
OSHA PEL: TWA 0.5 mg(Ba)/m³
ACGIH TLV: TWA 0.5 mg(Ba)/m³; Not Classifiable as a Human Carcinogen
DFG MAK: 0.5 mg(Ba)/m³
SAFETY PROFILE: Poison via ingestion, intravenous, and subcutaneous routes. When heated to decomposition it emits acrid smoke and fumes. See also BARIUM COMPOUNDS.

BAI000   CAS: 18810-58-7   HR: 3
BARIUM AZIDE
DOT: UN 0224/UN 1571
mf: BaN_6 mw: 221.40
PROP: Monoclinic prisms or crystals, decomp on heating with loss of N_2 at about 12°. Mp: evolves N_2 at about 120°, bp: explodes, d: 2.936. Very sol in H_2O; sltly sol in EtOH; insol in Et_2O.
SYNS: BARIUM AZIDE, dry or wetted with <50% water, by weight (UN 0224) (DOT) □ BARIUM AZIDE, wetted with not <50% water, by weight (UN 1571) (DOT)
CONSENSUS REPORTS: Reported in EPA TSCA Inventory. Barium and its compounds are on the Community Right-To-Know List.
OSHA PEL: TWA 0.5 mg(Ba)/m³
ACGIH TLV: TWA 0.5 mg(Ba)/m³; Not Classifiable as a Human Carcinogen
DFG MAK: 0.5 mg(Ba)/m³
DOT CLASSIFICATION: EXPLOSIVE 1.1A; Label: EXPLOSIVE 1.1A, Poison (UN 0224); DOT Class: 4.1; Label: Flammable Solid, Poison (UN 1571)
SAFETY PROFILE: A poison. Moderate explosion hazard when shocked or heated to 275°. Spontaneously flammable in air. Very unstable. When heated to decomposition it emits toxic fumes of NO_x. See also

BARIUM COMPOUNDS (soluble) and AZIDES.

BAI750   CAS: 13967-90-3   HR: 3
BARIUM BROMATE
DOT: UN 2719
mf: $Ba(BrO_3)_2 \cdot H_2O$ mw: 411.21
PROP: White or colorless crystals or crystalline powder. Decomp on heating with O_2 evolution and $BaBr_2$ formation. Mp: decomp @ 260°, d: 3.99 @ 18°.
CONSENSUS REPORTS: Barium and its compounds are on the Community Right-To-Know List. EPA TSCA Chemical Inventory.
OSHA PEL: TWA 0.5 mg(Ba)/m³
ACGIH TLV: TWA 0.5 mg(Ba)/m³; Not Classifiable as a Human Carcinogen
DFG MAK: 0.5 mg(Ba)/m³
DOT CLASSIFICATION: 5.1; Label: Oxidizer, Poison
SAFETY PROFILE: Very toxic. Fire hazard by chemical reaction with easily oxidized materials. Explodes at 300°. Mixtures with sulfur are unstable storage hazards; igniting immediately at 91°C and after a 2–11 day delay period at room temperature. Incompatible with Al, As, C, Cu, metal sulfides, organic matter, P, and reducing materials. When heated to decomposition it emits toxic fumes of Br⁻. See also BARIUM COMPOUNDS (soluble) and BROMINE.

BAI800   CAS: 1191-79-3   HR: 3
BARIUM CADMIUM STEARATE
mf: $C_{72}H_{140}O_8 \cdot Ba \cdot Cd$ mw: 1383.86
SYNS: CADMIUM BARIUM STEARATE □ OCTA-DECANOIC ACID, BARIUM CADMIUM SALT (4:1:1) (9CI) □ STEARIC ACID, BARIUM CADMIUM SALT (4:1:1)
CONSENSUS REPORTS: Reported in EPA TSCA Inventory.
OSHA PEL: TWA 5 μg(Cd)/m³
ACGIH TLV: TWA 0.01 mg(Cd)/m³; Suspected Carcinogen
NIOSH REL: TWA reduce to lowest feasible level
SAFETY PROFILE: Confirmed human carcinogen. Moderately toxic by ingestion.

When heated to decomposition it emits toxic fumes of Ba and Cd.

BAJ250   CAS: 513-77-9   HR: 3
BARIUM CARBONATE (1:1)
mf: $CO_3 \cdot Ba$ mw: 197.35
PROP: White orthorhombic powder or crystals, becomes hexagonal at 8° and cubic at 976°. Decomp on heating with CO_2 loss. Mp: 1740° @ 90 atm, bp: decomp, d: 4.43. Dissolves in acids to form corresponding Ba salts. Practically insol in H_2O; insol in alc EtOH.
SYNS: BARIUM CARBONATE □ CARBONIC ACID, BARIUM SALT (1:1) □ C.I. 77099 □ C.I. PIGMENT WHITE 10
CONSENSUS REPORTS: Reported in EPA TSCA Inventory. Barium and its compounds are on the Community Right-To-Know List.
OSHA PEL: TWA 0.5 mg(Ba)/m^3
ACGIH TLV: TWA 0.5 mg(Ba)/m^3; Not Classifiable as a Human Carcinogen
DFG MAK: 0.5 mg(Ba)/m^3
SAFETY PROFILE: Poison by ingestion, intravenous, and intraperitoneal routes. Human systemic effects by ingestion: stomach ulcers, muscle weakness, paresthesias and paralysis, hypermotility, diarrhea, nausea or vomiting, lung changes. Experimental reproductive effects. Incompatible with BrF_3 and 2-furanpercarboxylic acid. See also BARIUM COMPOUNDS (soluble).

BAJ500   CAS: 13477-00-4   HR: 3
BARIUM CHLORATE
DOT: UN 1445
mf: $Cl_2O_6 \cdot Ba$ mw: 304.24
PROP: Colorless prisms or white powder. Mp: loses H_2O @ 414°, d: 3.18.
SYN: CHLORIC ACID, BARIUM SALT
CONSENSUS REPORTS: Reported in EPA TSCA Inventory. Barium and its compounds are on the Community Right-To-Know List.
OSHA PEL: TWA 0.5 mg(Ba)/m^3
ACGIH TLV: TWA 0.5 mg(Ba)/m^3; Not Classifiable as a Human Carcinogen

DFG MAK: 0.5 mg(Ba)/m^3
DOT CLASSIFICATION: 5.1; Label: Oxidizer, Poison
SAFETY PROFILE: A poison. For fire and explosion hazards, see CHLORATES. Incompatible with Al, As, C, charcoal, Cu, MnO_2, metal sulfides, S_4N_4, organic matter, P, S. See also BARIUM COMPOUNDS (soluble).

BAK000   CAS: 10361-37-2   HR: 3
BARIUM CHLORIDE
mf: $BaCl_2$ mw: 208.24
PROP: Colorless, deliquescent, orthorhombic, flat crystals. Undergoes orthorhombic to cubic phase transition at 9°. Mp: transition @ 925° to cubic crystals, bp: 1560°, d: 3.856 @ 24°. Very sol in H_2O; practically insol in EtOH. IDLH 50 mg/m^3 (as Ba).
SYNS: BARIUM DICHLORIDE □ NCI-C61074 □ SBa 0108E
CONSENSUS REPORTS: Reported in EPA TSCA Inventory. Barium and its compounds are on the Community Right-To-Know List. EPA Genetic Toxicology Program.
OSHA PEL: TWA 0.5 mg(Ba)/m^3
ACGIH TLV: TWA 0.5 mg(Ba)/m^3; Not Classifiable as a Human Carcinogen
DFG MAK: 0.5 mg(Ba)/m^3
SAFETY PROFILE: A poison by ingestion, subcutaneous, intravenous, and intraperitoneal routes. Inhalation absorption of barium chloride equals 60–80%; oral absorption equals 10–30%. Experimental reproductive effects. Mutation data reported. See also BARIUM COMPOUNDS (soluble). When heated to decomposition it emits toxic fumes of Cl⁻.

BAK250   CAS: 10294-40-3   HR: 3
BARIUM CHROMATE(VI)
mf: $Ba \cdot CrO_4$ mw: 255.36
PROP: Heavy, pale-yellow, crystalline powder; darkens on heating. D: 4.498 @ 15°. Sol in strong acids; insol in org solvents.

B

SYNS: BARIUM CHROMATE (1:1) □ BARIUM CHROMATE OXIDE □ BARYTA YELLOW □ CHROMIC ACID, BARIUM SALT (1:1) □ C.I. 77103 □ C.I. PIGMENT YELLOW 31 □ LEMON CHROME □ LEMON YELLOW □ PERMANENT YELLOW □ STEINBUHL YELLOW □ ULTRAMARINE YELLOW

CONSENSUS REPORTS: IARC Cancer Review: Group 1 IMEMDT 7,165,87; Animal Inadequate Evidence IMEMDT 2,100,73; Human Sufficient Evidence IMEMDT 23,205,80. Reported in EPA TSCA Inventory. Barium and its compounds are on the Community Right-To-Know List.

OSHA PEL: TWA 0.1 mg $(C_3O_3)m^3$; 0.5 mg(Ba)/m^3

ACGIH TLV: TWA 0.5 mg(Ba)/m^3; Not Classifiable as a Human Carcinogen; 0.05 mg(Cr)/m^3; Confirmed Human Carcinogen

DFG MAK: 0.5 mg(Ba)/m^3

NIOSH REL: TWA 0.001 mg(Cr(VI))/m^3

SAFETY PROFILE: Confirmed human carcinogen. A poison. Mutation data reported. Reacts vigorously with reducing materials. See also BARIUM COMPOUNDS (soluble) and CHROMIUM COMPOUNDS. Used in pyrotechnics and as an explosive initiator.

BAK500   HR: 3
BARIUM COMPOUNDS (soluble)

CONSENSUS REPORTS: Barium and its compounds are on the Community Right-To-Know List.

OSHA PEL: Soluble Compounds: TWA 0.5 mg(Ba)/m^3

ACGIH TLV: Soluble Compounds:TWA 0.5 mg/m^3

DFG MAK: Soluble Compounds: 0.5 mg/m^3

DOT CLASSIFICATION: 6.1; Label: Poison

SAFETY PROFILE: The chromate is a human carcinogen. The soluble barium salts, such as the chloride and sulfide, are poisonous when ingested. The insoluble sulfate used in radiography is not acutely toxic. See also BARIUM SULFATE. Few cases of industrial systemic poisoning have been reported, but one investigator describes a fatal case of poisoning attributed

to barium oxide, the symptoms being severe abdominal pain with vomiting, dyspnea, rapid pulse, paralysis of the arm and leg, and eventually cyanosis and death. The same investigator produced paralysis in animals with barium oxide and carbonate. The usual result of exposure to the sulfide, oxide, and carbonate is irritation of the eyes, nose, and throat, and of the skin, producing dermatitis. The salts mentioned are somewhat caustic.

BAK750   CAS: 542-62-1   HR: 3
BARIUM CYANIDE
DOT: UN 1565
mf: C_2BaN_2 mw: 189.38
PROP: White, crystalline powder.
SYNS: BARIUM CYANIDE, solid (DOT) □ BARIUM DICYANIDE □ RCRA WASTE NUMBER P013

CONSENSUS REPORTS: Reported in EPA TSCA Inventory. Cyanide and its compounds, as well as barium and its compounds, are on the Community Right-To-Know List.

OSHA PEL: TWA 0.5 mg(Ba)/m^3

ACGIH TLV: TWA 0.5 mg(Ba)/m^3; Not Classifiable as a Human Carcinogen

DFG MAK: 0.5 mg(Ba)/m^3

DOT CLASSIFICATION: 6.1; Label: Poison

SAFETY PROFILE: A deadly poison. See also CYANIDE and BARIUM COMPOUNDS (soluble). When heated to decomposition it emits toxic fumes of CN^-.

BAL275   CAS: 6332-68-9   HR: 3
BARIUM DIBENZYLPHOSPHATE
mf: $C_{28}H_{28}BaO_8P_2$ mw: 691.84

OSHA PEL: TWA 0.5 mg(Ba)/m^3

ACGIH TLV: TWA 0.5 mg(Ba)/m^3; Not Classifiable as a Human Carcinogen

DFG MAK: 0.5 mg(Ba)/m^3

SAFETY PROFILE: Poison by intravenous route. When heated to decomposition it emits toxic fumes of PO_x and Ba.

BAM000   CAS: 7787-32-8   HR: 3
BARIUM FLUORIDE
mf: BaF_2 mw: 175.34

PROP: White, colorless powder or cubic crystals. Mp: 1368°, bp: 2137°, d: 4.89. Sltly sol in H$_2$O.
SYN: BARYUM FLUORURE (FRENCH)
CONSENSUS REPORTS: Reported in EPA TSCA Inventory. Barium and its compounds are on the Community Right-To-Know List.
OSHA PEL: TWA 0.5 mg(Ba)/m^3; 2.5 mg(F)/m^3
ACGIH TLV: TWA 0.5 mg(Ba)/m^3; Not Classifiable as a Human Carcinogen; TWA 2.5 mg(F)/m^3; BEI: 3 mg/g creatinine of fluorides in urine prior to shift; 10 mg/g creatinine of fluorides in urine at end of shift.
DFG MAK: 0.5 mg(Ba)/m^3
NIOSH REL: (Fluorides, Inorganic) TWA 2.5 mg(F)/m^3
SAFETY PROFILE: A poison by ingestion and intraperitoneal routes. Moderately toxic by subcutaneous route. An experimental teratogen. See also FLUORIDES and BARIUM COMPOUNDS (soluble). When heated to decomposition it emits toxic fumes of F$^-$.

BAN250   CAS: 10022-31-8   HR: 3
BARIUM(II) NITRATE (1:2)
DOT: UN 1446
mf: N$_2$O$_6$•Ba   mw: 261.36
PROP: Lustrous, colorless, cubic crystals. Mp: 592°, bp: decomp, d: 3.24 @ 23°. Decomp on heating with evolution of NO$_2$ and O$_2$ and formation of BaO. Insol in EtOH. IDLH 50 mg/m^3 (as Ba).
SYNS: BARIUM DINITRATE □ BARIUM NITRATE (DOT) □ DUSICNAN BARNATY (CZECH) □ NITRATE de BARYUM (FRENCH) □ NITRIC ACID, BARIUM SALT
CONSENSUS REPORTS: Reported in EPA TSCA Inventory. Barium and its compounds are on the Community Right-To-Know List.
OSHA PEL: TWA 0.5 mg(Ba)/m^3
ACGIH TLV: TWA 0.5 mg(Ba)/m^3; Not Classifiable as a Human Carcinogen
DFG MAK: 0.5 mg(Ba)/m^3

DOT CLASSIFICATION: 5.1; Label: Oxidizer, Poison
SAFETY PROFILE: A poison by ingestion, subcutaneous, parenteral, and intravenous routes. An irritant to skin and eyes. When heated to decomposition it emits very toxic fumes of NO$_x$. An oxidizer. Mixtures with finely divided aluminum-magnesium alloys are easily ignitable and extremely sensitive to friction or impact. Such mixtures are used in chemical photoflash applications. Incompatible with (Mg + BaO$_2$ + Zn), Al, and Mg alloys. When heated to decomposition it emits toxic fumes of NO$_x$. See also BARIUM COMPOUNDS (soluble) and NITRATES.

BAO000   CAS: 1304-28-5   HR: 3
BARIUM OXIDE
DOT: UN 1884
mf: BaO   mw: 153.34
PROP: White to yellowish-white powder or cubic crystals; moisture-sensitive. Mp: 1913°, bp: 2000° (approx), d: 5.72. Mod sol in EtOH; insol in Me$_2$CO.
SYNS: BARIUM MONOXIDE □ BARIUM PROTOXIDE □ BARYTA □ CALCINED BARYTA □ OXYDE de BARYUM (FRENCH)
CONSENSUS REPORTS: Reported in EPA TSCA Inventory. Barium and its compounds are on the Community Right-To-Know List.
OSHA PEL: TWA 0.5 mg(Ba)/m^3
ACGIH TLV: TWA 0.5 mg(Ba)/m^3; Not Classifiable as a Human Carcinogen
DFG MAK: 0.5 mg(Ba)/m^3
DOT CLASSIFICATION: 6.1; Label: KEEP AWAY FROM FOOD
SAFETY PROFILE: A poison via subcutaneous route. See also BARIUM COMPOUNDS (soluble). Combustible by spontaneous chemical reaction; produces heat on contact with water or steam. Reacts with H$_2$O, Ba(OH)$_2$. Incompatible with H$_2$S, hydroxylamine, N$_2$O$_4$, triuranium octaoxide, SO$_3$.

B

BAO250   CAS: 1304-29-6   HR: 3
BARIUM PEROXIDE
DOT: UN 1449
mf: BaO_2 mw: 169.34
PROP: Pale, grayish-white powder. Mp: 450°, bp: loses O_2 @ 800°, d: 4.96. Decomp on heating to BaO and O_2. Dissolves in water with formation of H_2O_2.
SYNS: BARIO (PEROSSIDO di) (ITALIAN) □ BARIUM BINOXIDE □ BARIUM DIOXIDE □ BARIUMPEROXID (GERMAN) □ BARIUMPEROXYDE (DUTCH) □ BARIUM SUPEROXIDE □ DIOXYDE de BARYUM (FRENCH) □ PEROXYDE de BARYUM (FRENCH)
CONSENSUS REPORTS: Reported in EPA TSCA Inventory. Barium and its compounds are on the Community Right-To-Know List.
OSHA PEL: TWA 0.5 mg(Ba)/m³
ACGIH TLV: TWA 0.5 mg(Ba)/m³; Not Classifiable as a Human Carcinogen
DFG MAK: 0.5 mg(Ba)/m³
DOT CLASSIFICATION: 5.1; Label: Oxidizer, Poison
SAFETY PROFILE: A poison via subcutaneous route. A powerful oxidizer. Explodes on contact with acetic anhydride. Ignites when mixed with calcium-silicon alloys, powdered aluminum, powdered magnesium, water + organic compounds. Mixtures with propane react violently when heated. The powder ignites when heated to 265°C with selenium. Wood ignites with friction from the peroxide. Incompatible with H_2S, water, peroxyformic acid, hydroxylamine solution, mixture of (Mg + Zn + Ba(NO$_3$)$_2$), and organic matter. See also BARIUM COMPOUNDS (soluble) and PEROXIDES, INORGANIC.

BAO300   CAS: 50864-67-0   HR: 3
BARIUM POLYSULFIDE
SYNS: BARIUMPOLYSULFID □ BARIUM SULFIDE □ SOLABAR □ SOLBAR
OSHA PEL: TWA 0.5 mg(Ba)/m³
ACGIH TLV: TWA 0.5 mg(Ba)/m³; Not Classifiable as a Human Carcinogen
DFG MAK: 0.5 mg(Ba)/m³
SAFETY PROFILE: Poison by ingestion. Human systemic effects by ingestion: flaccid

paralysis without anesthesia, muscle weakness, and dyspnea. When heated to decomposition it emits toxic fumes of SO_x and Ba.

BAO750   CAS: 17125-80-3   HR: 3
BARIUM SILICOFLUORIDE
mf: $F_6Si•Ba$ mw: 279.43
PROP: White or colorless rhombohedral crystalline powder. D: 4.29 @ 21°/4°, mp: 300° (decomp). Decomp on heating to form SiF_4 and BaF_2. Sltly sol in H_2O. Insol in EtOH.
SYNS: BARIUM FLUOROSILICATE □ BARIUM FLUOSILICATE □ BARIUM HEXAFLUOROSILICATE □ BARIUM HEXAFLUOROSILICATE(2-) □ BARIUMSILICOFLUORID □ BARIUM SILICON FLUORIDE □ SILICATE(2-), HEXAFLUORO-, BARIUM □ SILICATE(2-), HEXAFLUORO-, BARIUM (1:1) (9CI) □ SILICON FLUORIDE BARIUM SALT
CONSENSUS REPORTS: Reported in EPA TSCA Inventory. Barium and its compounds are on the Community Right-To-Know List.
OSHA PEL: TWA 0.5 mg(Ba)/m³; TWA 2.5 mg(F)/m³
ACGIH TLV: TWA 0.5 mg(Ba)/m³; Not Classifiable as a Human Carcinogen; TWA 2.5 mg(F)/m³; BEI: 3 mg/g creatinine of fluorides in urine prior to shift; 10 mg/g creatinine of fluorides in urine at end of shift.
DFG MAK: 0.5 mg(Ba)/m³
NIOSH REL: (Fluorides, Inorganic) TWA 2.5 mg(F)/m³
DOT CLASSIFICATION: 6.1; Label: KEEP AWAY FROM FOOD
SAFETY PROFILE: A poison by ingestion. When heated to decomposition it emits toxic fumes of F⁻. See also BARIUM COMPOUNDS (soluble).

BAO900   CAS: 20236-55-9   HR: 3
BARIUM STYPHNATE
DOT: NA 0473
SYNS: 1,3-BENZENEDIOL, 2,4,6-TRINITRO-, BARIUM SALT, HYDRATE (2:1:1) □ RESORCINOL, 2,4,6-TRINITRO-, BARIUM SALT, HYDRATE (2:1:1)
DOT CLASSIFICATION: Explosive 1.1A; Label: Explosive 1.1A

SAFETY PROFILE: An explosive. When heated to decomposition it emits toxic vapors of NO_x and fumes of Ba.

BAP000   CAS: 7727-43-7   HR: 2
BARIUM SULFATE
mf: $O_4S \cdot Ba$ mw: 233.40
PROP: White, heavy, orthorhombic, odorless powder or crystals. Undergoes orthorhombic to monoclinic phase transition at 11°. D: 4.50 @ 15°, mp: 1580°. Sltly sol in H_2O. Insol in water or dilute acids.
SYNS: ACTYBARYTE □ ARTIFICIAL BARITE □ ARTIFICIAL HEAVY SPAR □ BAKONTAL □ BARIDOL □ BARITE □ BARITOP □ BAROSPERSE □ BAROTRAST □ BARYTA WHITE □ BARYTES □ BAYRITES □ BLANC FIXE □ C.I. 77120 □ C.I. PIGMENT WHITE 21 □ CITOBARYUM □ COLONATRAST □ ENAMEL WHITE □ ESOPHOTRAST □ EWEISS □ E-Z-PAQUE □ FINEMEAL □ LACTOBARYT □ LIQUIBARINE □ MACROPAQUE □ NEOBAR □ ORATRAST □ PERMANENT WHITE □ PRECIPITATED BARIUM SULPHATE □ RAYBAR □ REDI-FLOW □ SOLBAR □ SULFURIC ACID, BARIUM SALT (1:1) □ SUPRAMIKE □ TRAVAD □ UNIBARYT
CONSENSUS REPORTS: Reported in EPA TSCA Inventory. Barium and its compounds are on the Community Right-To-Know List.
OSHA PEL: Total Dust: TWA 10 mg/m³; Respirable Fraction: 5 mg/m³
ACGIH TLV: TWA (nuisance particulate) 10 mg/m³ of total dust (when toxic impurities are not present, e.g., quartz <1%)
SAFETY PROFILE: Questionable carcinogen with experimental tumorigenic data. Mutation data reported. A relatively insoluble salt used as an opaque medium in radiography. Soluble impurities can lead to toxic reactions. Heating with aluminum can produce an explosion. Incompatible with aluminum and potassium. When heated to decomposition it emits toxic fumes of SO_x.

BAP250   CAS: 21109-95-5   HR: 3
BARIUM SULFIDE
mf: BaS   mw: 169.4
PROP: Cubic, colorless crystals. Moisture-sensitive. D: 4.25 @ 15°, mp: 1200°.

CONSENSUS REPORTS: Barium and its compounds are on the Community Right-To-Know List.
SAFETY PROFILE: A poison. Flammable by spontaneous chemical reaction, air, moisture, or acid fumes may cause it to ignite. For explosion and disaster hazards, see SULFIDES. To fight fire, use CO_2, dry chemical. Reacts violently with phosphorus(V) oxide. Mixtures with lead dioxide, potassium chlorate, or potassium nitrite explode when heated. Incompatible with Cl_2O, $Ca(NO_3)_2$, $Sr(NO_3)_2$, $Ca(ClO_3)_2$, $Sr(ClO_3)_2$, $(ClO_3)_2$. See also BARIUM COMPOUNDS (soluble) and SULFIDES.

BAP750   CAS: 12009-21-1   HR: 2
BARIUM ZIRCONIUM(IV) OXIDE
mf: $O_4Zr_4 \cdot Ba$ mw: 566.22
PROP: Light gray-buff powder or white powder. D: 5.52, mp: 2510°. Insol in water and alkalies; sltly sol in acid. IDLH 50 mg/m³ (as Zr).
SYNS: BARIUM ZIRCONATE □ BARIUM ZIRCONIUM OXIDE □ BARIUM ZIRCONIUM TRIOXIDE □ ZIRCONATE, BARIUM (1:1)
CONSENSUS REPORTS: Reported in EPA TSCA Inventory. Barium and its compounds are on the Community Right-To-Know List.
OSHA PEL: TWA 5 mg(Zr)/m³; STEL 10 mg(Zr)/m³
ACGIH TLV: TWA 5 mg(Zr)/m³; STEL 10 mg(Zr)/m³
DFG MAK: 1 mg(Zr)/m³
SAFETY PROFILE: Moderately toxic by ingestion and intraperitoneal routes. Inhalation produces interstitial pneumonitis. See also ZIRCONIUM COMPOUNDS and BARIUM COMPOUNDS.

BAR250   CAS: 8015-73-4   HR: 2
BASIL OIL
PROP: Contains about 55% methyl chavicol and 35% of alcohols calculated as lenatoal and other compounds found in the leaves of *Ocimum resilium L.* (FCTXAR 11,855,73). A pale-yellow liquid; floral, spicy odor. Sol in

fixed oils and propylene glycol; insol in glycerin.

SYNS: BASIL OIL, EUROPEAN TYPE (FCC) □ BASIL OIL, SWEET □ OCIMUM BASILICUM OIL □ OIL OF BASIL □ OILS, BASIL

CONSENSUS REPORTS: Reported in EPA TSCA Inventory.

SAFETY PROFILE: Moderately toxic by ingestion. A skin and eye irritant. When heated to decomposition it emits acrid smoke and irritating fumes.

BAT500   **HR: 2**
BAY OIL

PROP: Consists mainly of eugenol and chavicol (55–65%), major portion of balance consists of terpenes (α-pinene, myrcene, and dipentene), small quantities of citrol, nerol, cineol, and other terpenoids have also been found (FCTXAV 11,855,73). Yellow or brown liquid; aromatic odor, pungent, spicy taste. Sol in alc and glacial acetic acid.

SYNS: BAY LEAF OIL □ BOIS d'INDE □ LAUREL LEAF OIL □ MYRCIA OIL □ MYRICIA OIL □ OIL OF BAY □ OIL OF MYRCIA

CONSENSUS REPORTS: Reported in EPA TSCA Inventory.

SAFETY PROFILE: Moderately toxic by ingestion. When heated to decomposition it emits acrid smoke.

BAT750   CAS: 14816-18-3   HR: 3
BAYTHION

mf: $C_{12}H_{15}N_2O_3PS$ mw: 298.32

PROP: Liquid. D: 1.176° @ 20 mm, fp: 5–6°, bp: 102° @ 0.01 mm (decomp).

SYNS: B 77488 □ BAY 5621 □ BAY 77488 □ BAYRE 77488 □ BENZOYL CYANIDE-o-(DIETHOXY-PHOSPHINOTHIOYL)OXIME □ O,O-DIAETHYL-o-(α-CYANBENZYLIDEN-AMINO)-THIONPHOSPHAT (GERMAN) □ O,O-DIAETHYL-o-(α-CYANO-BENZYLIDENAMINO)-MONOTHIOPHOSPHAT (GERMAN) □ α-(((DIETHOXYPHOSPHINO-THIOYL)OXY)IMINO)BENZENEACETONITRILE □ (DIETHOXY-THIOPHOSPHORYLOXYIMINO)-PHENYL ACETONITRILE □ O,O-DIETHYL PHOSPHORO-THIOATE, o-ESTER with PHENYLGLYOXYLONITRILE OXIME □ ENT 27,488 □ 4-ETHOXY-7-PHENYL-3,5-DIOXA-6-AZA-4-PHOSPHAOCT-6-ENE-8-NITRILE-4-SULFIDE □ PHENYLGLYOXYLONITRILE OXIME-O,O-DIETHYL PHOSPHOROTHIOATE □ PHOXIME □ PHOXIN □ SEBACIL □ VALEXONE □ VOLATON

CONSENSUS REPORTS: Cyanide and its compounds are on the Community Right-To-Know List.

SAFETY PROFILE: Poison by ingestion. An experimental teratogen. When heated to decomposition it emits very toxic fumes of CN⁻, NO_x, PO_x, and SO_x. See also NITRILES.

BAT850   CAS: 8021-39-4   HR: 3
BEECHWOOD CRESOATE

PROP: Yellowish, greasy, liquid with smokey odor and sharp burned taste. Relatively sol in water.

SYNS: CRESOATE, WOOD □ RCRA WASTE NUMBER U051

CONSENSUS REPORTS: NTP 10th Report on Carcinogens. Reported in EPA TSCA Inventory.

SAFETY PROFILE: Confirmed carcinogen. When heated to decomposition it emits acrid smoke and irritating fumes.

BAU000   CAS: 8012-89-3   HR: 1
BEESWAX

PROP: Yellow to brownish-yellow, soft to brittle wax. Mp: 62–65°, d: 0.95–0.96. Sol in chloroform, ether, fixed oils; sltly sol in alc.

SYNS: BEESWAX, WHITE □ BEESWAX, YELLOW

SAFETY PROFILE: A mild allergen. Combustible when heated.

BAU750   CAS: 147-24-0   HR: 3
BENADRYL HYDROCHLORIDE

mf: $C_{17}H_{21}NO•ClH$ mw: 291.85

PROP: Crystals from EtOH/Et_2O. Mp: 161–162°. Sol in H_2O.

SYNS: AMBENYL □ BAX □ BENA □ BENADRYL □ BENDYLATE □ BENOCTEN □ BENZEHIST □ BENZ-HYDRAMINE HYDROCHLORIDE □ 2-(BENZHYDRYL-OXY)-N,N-DIMETHYLETHYL-AMINEHYDRO-CHLORIDE □ DABYLEN □ DIFENHYDRAMINE HYDROCHLORIDE □ DIMETHYLAMINE BENZHYDRYL ESTER HYDROCHLORIDE □ β-DIMETHYLAMINOETHYL BENZHYDRYL ESTER HYDROCHLORIDE □ DIPHENYLHYDRAMINE HYDROCHLORIDE □ 2-(DIPHENYLMETHOXY)-N,N-DIMETHYL-ETHANAMINE HYDROCHLORIDE □ 2-DIPHENYLMETHOXY-N,N-DIMETHYLETHYLAMINE HYDROCHLORIDE □ DOLESTAN □ ELDADRYL □ FELBEN □ FENYLHIST □ HALBMOND □ α-HYDR-

OXYDIPHENYLMETHANE-β-DIMETHYLAMINOETHYL ETHER HYDROCHLORIDE □ NCI-C56075 □ ROHYDRA □ SK-DIPHENHYDRAMINE □ VALDRENE □ WEHYDRYL

CONSENSUS REPORTS: Reported in NTP Carcinogenesis Studies (feed); Equivocal Evidence: rat NTPTR* NTP-TR-355,89; (feed); No Evidence: mouse NTPTR* NTP-TR-355,89.

SAFETY PROFILE: Poison by ingestion, subcutaneous, intravenous, and intraperitoneal routes. Human systemic effects by ingestion or skin contact: arrhythmias, ataxia, blood pressure elevation, convulsions, distorted perceptions, eye effects, and hallucinations. Experimental teratogenic and reproductive effects. Questionable carcinogen with experimental tumorigenic data. When heated to decomposition it emits very toxic fumes of NO_x and HCl. See also ESTERS and ETHERS.

BAV575   CAS: 17804-35-2   HR: 3
BENOMYL
mf: $C_{14}H_{18}N_4O_3$ mw: 290.36
PROP: Very sltly sol in H_2O; sol in $CHCl_3$; less sol in other org solvents.
SYNS: ARILATE □ BBC □ BENLATE 50 □ BENOMYL 50W □ BNM □ 1-(BUTYLCARBAMOYL)-2-BENZIMID-AZOLECARBAMIC ACID, METHYL ESTER □ 1-(BUTYLCARBAMOYL)-2-BENZIMIDAZOL-METHYL-CARBAMAT (GERMAN) □ 1-(N-BUTYLCARBAMOYL)-2-(METHOXY-CARBOXAMIDO)-BENZIMIDAZOL (GERMAN) □ DU PONT 1991 □ FUNDASOL □ FUNGICIDE 1991 □ MBC □ METHYL-1-(BUTYL-CARBAMOYL)-2-BENZIMIDAZOLYLCARBAMATE □ TERSAN 1991
CONSENSUS REPORTS: Reported in EPA TSCA Inventory. EPA Genetic Toxicology Program.
OSHA PEL: Total Dust: TWA 10 mg/m³; Respirable Fraction: 5 mg/m³
ACGIH TLV: TWA 10 mg/m³; Not Classifiable as a Human Carcinogen; (Proposed: 1 mg/m³; Not Classifiable as a Human Carcinogen)
SAFETY PROFILE: Poison by ingestion. Mildly toxic by inhalation. Experimental teratogenic and reproductive effects. Human mutation data reported. A human skin irritant. When heated to decomposition it emits toxic fumes of NO_x. See also CARBAMATES.

BAV750   CAS: 1302-78-9   HR: 1
BENTONITE
PROP: A clay containing appreciable amounts of the clay mineral montmorillonite; light yellow or green, cream, pink, gray to black solid. Insol in water and common org solvs.
SYNS: ALBAGEL PREMIUM USP 4444 □ BENTONITE 2073 □ BENTONITE MAGMA □ HI-JEL □ IMVITE I.G.B.A. □ MAGBOND □ MONTMORILLONITE □ PANTHER CREEK BENTONITE □ SOUTHERN BENTONITE □ TIXOTON □ VOLCLAY □ VOLCLAY BENTONITE BC □ WILKINITE
CONSENSUS REPORTS: Reported in EPA TSCA Inventory.
SAFETY PROFILE: Poison by intravenous route causing blood clotting. Questionable carcinogen with experimental tumorigenic data.

BAW250   CAS: 205-99-2   HR: 3
BENZ(e)ACEPHENANTHRYLENE
mf: $C_{20}H_{12}$ mw: 252.32
PROP: Needles from C_6H_6 or EtOH. Mp: 168°.
SYNS: 3,4-BENZ(e)ACEPHENANTHRYLENE □ 2,3-BENZFLUORANTHENE □ 3,4-BENZFLUORANTHENE □ BENZO(b)FLUORANTHENE □ BENZO(e)FLUOR-ANTHENE □ 2,3-BENZOFLUORANTHENE □ 3,4-BENZOFLUORANTHENE □ 2,3-BENZOFLUOR-ANTHRENE □ B(b)F
CONSENSUS REPORTS: NTP 10th Report on Carcinogens. IARC Cancer Review: Group 2B IMEMDT 7,56,87; Animal Sufficient Evidence IMEMDT 32,147,83; IMEMDT 3,69,73. EPA Genetic Toxicology Program.
ACGIH TLV: Suspected Carcinogen
SAFETY PROFILE: Confirmed carcinogen with experimental carcinogenic and tumorigenic data. Mutation data reported. When heated to decomposition it emits acrid smoke and irritating fumes.

B

BAY300   CAS: 98-87-3   HR: 3
BENZAL CHLORIDE
DOT: UN 1886
mf: $C_7H_6Cl_2$ mw: 161.03
PROP: Very refractive liquid. Mp: $-16°$, bp: $214°$, d: 1.29.
SYNS: BENZYL DICHLORIDE □ BENZYLENE CHLORIDE □ BENZYLIDENE CHLORIDE □ BENZYLIDENE CHLORIDE (DOT) □ CHLOROBENZAL □ CHLORURE de BENZYLIDENE □ (DICHLORO-METHYL)BENZENE □ α-α-DICHLOROTOLUENE □ RCRA WASTE NUMBER U017 □ TOLUENE, α-α-DICHLORO-
CONSENSUS REPORTS: IARC Cancer Review: Human Inadequate Evidence IMEMDT 29,65,82; Animal Limited Evidence IMEMDT 29,65,82. Reported in EPA TSCA Inventory. EPA Genetic Toxicology Program. EPA Extremely Hazardous Substances List. Community Right-To-Know List.
DFG MAK: Confirmed Human Carcinogen
DOT CLASSIFICATION: 6.1; Label: Poison
SAFETY PROFILE: Confirmed carcinogen with experimental carcinogenic and neoplastigenic data. Poison by inhalation. Moderately toxic by ingestion. A strong irritant and lachrymator. Causes central nervous system depression. Mutation data reported. When heated to decomposition it emits toxic fumes of Cl^-. See also CHLORINATED HYDROCARBONS, AROMATIC.

BAY500   CAS: 100-52-7   HR: 3
BENZALDEHYDE
mf: C_7H_6O mw: 106.13
PROP: Colorless liquid; burning taste with bitter almond odor. Mp: $-26°$, bp: $179°$, fp: $-56.9°$ (to $-55°$), flash p: $148°F$, d: 1.041, autoign temp: $377°F$, vap press: 1 mm @ $26.2°$, vap d: 3.65, refr index: 1.544. Sltly sol in water; misc in alc, ether, oils.
SYNS: ALMOND ARTIFICIAL ESSENTIAL OIL □ ARTIFICIAL ALMOND OIL □ BENZENECARB-ALDEHYDE □ BENZENECARBONAL □ BENZOIC ALDEHYDE □ FEMA No. 2127 □ NCI-C56133
CONSENSUS REPORTS: NTP Carcinogenesis Studies (gavage); Some Evidence mouse; NTP-TR-378,90; No Evidence: rat NTP-TR-378,90. EPA Genetic Toxicology Program. Reported in EPA TSCA Inventory.
SAFETY PROFILE: Poison by ingestion and intraperitoneal routes. Moderately toxic by subcutaneous route. An allergen. Acts as a feeble local anesthetic. Local contact may cause contact dermatitis. Causes central nervous system depression in small doses and convulsions in larger doses. A skin irritant. Questionable carcinogen with experimental tumorigenic data. Mutation data reported. Combustible liquid. To fight fire, use water (may be used as a blanket), alcohol, foam, dry chemical. A strong reducing agent. Reacts violently with peroxyformic acid and other oxidizers. See also ALDEHYDES.

BAY750   CAS: 633-03-4   HR: 3
BENZALDEHYDE GREEN
mf: $C_{27}H_{33}N_2 \cdot HO_4S$ mw: 482.69
PROP: Bright green crystals or powder. Mp: $210°$ (decomp). Sol in H_2O, EtOH, and $CHCl_3$.
SYNS: ADC BRILLIANT GREEN CRYSTALS □ AIZEN DIAMOND GREEN GH □ ANILINE GREEN □ ASTRA DIAMOND GREEN GX □ AVON GREEN A-4379 □ BASIC BRIGHT GREEN □ BRILLIANT GREEN SULFATE □ CALCOZINE BRILLIANT GREEN G □ C.I. 42040 □ C.I. BASIC GREEN 1, SULFATE (1:1) □ DEORLENE GREEN JJO □ DIAMOND GREEN G □ EMERALD GREEN □ ETHYL GREEN □ FAST GREEN JJO □ HIDACO BRILLIANT GREEN □ MALACHITE GREEN G □ MITSUI BRILLIANT GREEN G □ TERTROPHENE BRILLIANT GREEN G □ TOKYO ANILINE BRILLIANT GREEN
CONSENSUS REPORTS: Reported in EPA TSCA Inventory.
SAFETY PROFILE: Poison by ingestion, intraperitoneal, and intravenous routes. A mild human skin irritant. Mutation data reported. See also ALDEHYDES and SULFATES. When heated to decomposition it emits very toxic fumes of NO_x, NH_3, and SO_x.

BBA000   CAS: 1708-39-0   HR: 2
BENZAL GLYCERYL ACETAL
mf: $C_{10}H_{12}O_3$ mw: 180.22

PROP: Colorless to pale-yellow liquid; mild almond odor. D: 1.183–1.193, refr index: 1.535–1.541, flash p: 165°F.
SYNS: BENZALDEHYDE GLYCERYL ACETAL (FCC) □ BENZYLIDENE GLYCEROL □ BUTYL PHENYL ACETATE □ FEMA No. 2209 □ 2-PHENYL-m-DIOXAN-5-OL
SAFETY PROFILE: Moderately toxic by ingestion and intraperitoneal routes. Mildly toxic by skin contact. Combustible liquid. When heated to decomposition it emits acrid smoke and irritating fumes.

BBC250   CAS: 56-55-3   HR: 3
BENZ(a)ANTHRACENE
mf: $C_{18}H_{12}$ mw: 228.30
PROP: Colorless leaflets or plates from EtOH/AcOH. Mp: 160°, bp: 400°.
SYNS: BA □ BENZANTHRACENE □ 1,2-BENZANTHRACENE □ 1,2-BENZ(a)ANTHRACENE □ 1,2-BENZANTHRAZEN (GERMAN) □ BENZANTHRENE □ 1,2-BENZANTHRENE □ BENZOANTHRACENE □ BENZO(a)ANTHRACENE □ 1,2-BENZOANTHRACENE □ BENZO(a)PHENANTHRENE □ BENZO(b)PHEN-ANTHRENE □ 2,3-BENZOPHENANTHRENE □ 2,3-BENZPHENANTHRENE □ NAPHTHANTHRACENE □ RCRA WASTE NUMBER U018 □ TETRAPHENE
CONSENSUS REPORTS: NTP 10th Report on Carcinogens. IARC Cancer Review: Group 2A IMEMDT 7,56,87; Animal Sufficient Evidence IMEMDT 32,135,83; IMEMDT 3,45,73. EPA Genetic Toxicology Program. Reported in EPA TSCA Inventory.
ACGIH TLV: Suspected Human Carcinogen
SAFETY PROFILE: Confirmed carcinogen with experimental carcinogenic, neoplastigenic, and tumorigenic data by skin contact and other routes. Poison by intravenous route. Human mutation data reported. It is found in oils, waxes, smoke, food, drugs. When heated to decomposition it emits acrid smoke and irritating fumes.

BBJ500   CAS: 1477-19-6   HR: 3
BENZARONE
mf: $C_{17}H_{14}O_3$ mw: 266.31
PROP: Solid. Mp: 126–127°.
SYNS: BENZOFURAN, (2-ETHYL-3-(4'-HYDROXY-BENZOYL)) □ 2-ETHYL-3-BENZOFURANYL p-

HYDROXYPHENYL KETONE □ 2-ETHYL-3-(p-HYDROXYBENZOYL)BENZOFURAN □ 2-ETHYL-4'-HYDROXY-3-BENZOYLBENZOFURAN □ ETHYL-2 (HYDROXY-4 BENZOYL)-3 BENZOFURANNE □ FRAGIVIX
CONSENSUS REPORTS: Reported in EPA TSCA Inventory.
DOT CLASSIFICATION: 3; Label: Flammable Liquid
SAFETY PROFILE: Poison by intraperitoneal route. An experimental teratogen. Other experimental reproductive effects. A flammable liquid. When heated to decomposition it emits acrid and irritating smoke and fumes. See also KETONES.

BBJ750   CAS: 59-97-2   HR: 3
BENZAZOLINE HYDROCHLORIDE
mf: $C_{10}H_{12}N_2 \cdot ClH$ mw: 196.70
PROP: Solid. Mp: 171–172°.
SYNS: ARTERODY □ BENZYLIMIDAZOLINE HYDROCHLORIDE □ 2-BENZYL-2-IMIDAZOLINE MONOHYDROCHLORIDE □ IMIDALINE HYDROCHLORIDE □ PRISCOL □ PRISCOLINE HYDROCHLORIDE □ TOLAVAD □ TOLAZOLINE CHLORIDE □ TOLAZOLINE HYDROCHLORIDE □ TOLPAL
CONSENSUS REPORTS: Reported in EPA TSCA Inventory.
SAFETY PROFILE: Poison by ingestion, intravenous, and intraperitoneal routes. Human systemic effects by intravenous route: change in heart rate, sweating, ulceration or bleeding from duodeum, ulceration or bleeding from small intestine, unspecified vascular effects. When heated to decomposition it emits very toxic fumes of NO_x and HCl.

BBK000   CAS: 300-62-9   HR: 3
BENZEDRINE
mf: $C_9H_{13}N$ mw: 135.23
PROP: Liquid or oil. Bp: 203°, flash p: <212°F (OC), d: 0.931, vap d: 4.65. Sltly sol in H_2O.
SYNS: ACTEDRON □ ADIPAN □ ALLODENE □ dl-AMPHETAMINE □ ANOREXIDE □ (±)-BENZEDRINE □ dl-BENZEDRINE □ DEOXYNOREPHEDRINE □ (±)-DESOXYNOREPHEDRINE □ racemic-DESOXYNORE-PHEDRINE □ ELASTONON □ ISOAMYCIN □ ISOMYN □ MECODRIN □ α-METHYLBENZENEETHANEAMINE

□ dl-α-METHYLPHENETHYLAMINE □ (±)-α-METHYLPHENETHYLAMINE □ NOREPHEDRANE □ NOVYDRINE □ ORTEDRINE □ PHENEDRINE □ dl-1-PHENYL-2-AMINOPROPANE □ PROFAMINA □ PROPISAMINE □ PSYCHEDRINE □ RAPHETAMINE □ SIMPATEDRIN □ SYMPAMINE □ SYMPATEDRINE □ WECKAMINE

CONSENSUS REPORTS: Reported in EPA TSCA Inventory. EPA Extremely Hazardous Substances List.

SAFETY PROFILE: A deadly human poison by an unspecified route. An experimental poison by ingestion, subcutaneous, intraperitoneal, and intravenous routes. Experimental reproductive effects. Mutation data reported. A central nervous system stimulant. Overdoses cause hyperactivity, restlessness, insomnia, rapid pulse, rise in blood pressure, dilated pupils, dryness of the throat. Combustible when exposed to heat, flame, or oxidizers. When heated to decomposition it emits toxic fumes of NO_x. To fight fire, use CO_2, dry chemical, alcohol foam, water mist, fog. See other benzedrine entries.

BBK500   CAS: 51-63-8   HR: 3
d-BENZEDRINE SULFATE
mf: $C_{18}H_{26}N_2 \cdot H_2O_4S$ mw: 368.54
PROP: Plates.
SYNS: ACEDRON □ ADJUDETS □ ADRIXINE □ AFATIN □ ALBEMAP □ AMDEX □ d-AMFETASUL □ AMITRENE □ AMPHAETEX □ AMPHEDRINE □ AMPHEREX □ (+)-AMPHETAMINE SULFATE □ d-AMPHETAMINE SULFATE □ AMSUSTAIN □ APETAIN □ ARDEX □ BETAFEDRINA □ BETAFEDRINE □ d-BETAPHEDRINE □ CARRTIME □ CRADEX □ DADEX □ DADOX d-CITRAMINE □ DELLIPSOIDS □ DEPHADREN □ DESOXYN □ DEXAIME □ DEXALINE □ DEXALME □ DEXAMED □ DEXAMINE □ DEXAMPHAMINE □ DEXAMPHETAMINE □ DEXAMPHETAMINE SULFATE □ DEXAMYL □ DEXEDRINA □ DEXEDRINE SULFATE □ DEXIES □ DEXTROAMPHETAMINE SULFATE □ DEXTRO-α-METHYLPHENETHYLAMINE SULFATE □ DEXTRO-1-PHENYL-2-AMINOPROPANE SULFATE □ DEXTRO-β-PHENYLISOPROPYLAMINE SULFATE □ FASTBALLS □ HEARTS □ (S)-α-METHYL-BENZENEETHANAMINE SULFATE (2:1) □ d-α-METHYLPHENETHYLAMINE SULFATE □ ORANGES □ PELLCAFS □ PELLCAP □ PELLCAPS □ PERKE □ PHENOPROMIN □ d-1-PHENYL-2-AMINOPROPANE SULFATE □ d-β-

PHENYLISOPROPYLAMINE SULFATE □ PHETADEX □ PSYCHODRINE □ REVIDEX □ SIMPAMINA-D □ SYMPAMINA-D □ TEMPODEX □ TUPHETAMINE □ TYDEX □ ZAMINE

SAFETY PROFILE: Poison by ingestion, intraperitoneal, subcutaneous, and intravenous routes. A human teratogen that causes developmental abnormalities of the central nervous system. Experimental reproductive effects including other teratogenic effects. A habit-forming stimulant. When heated to decomposition it emits very toxic fumes of SO_x and NO_x. See also other benzidrine compounds and SULFATES.

BBL000   CAS: 142-04-1   HR: 3
BENZENAMINE HYDROCHLORIDE
DOT: UN 1548
mf: $C_6H_7N \cdot ClH$ mw: 129.60
PROP: Crystals. Vap d: 4.46, d: 1.22, mp: 198°, bp: 245°, flash p: 380°F (OC).
SYNS: ANILINE CHLORIDE □ ANILINE HYDROCHLORIDE (DOT) □ "ANILINE SALT" □ ANILINIUM CHLORIDE □ CHLORHYDRATE d'ANILINE (FRENCH) □ CHLORID ANILINU (CZECH) □ NCI-C03736 □ PHENYLAMINE HYDROCHLORIDE □ SUL ANILINOVA (CZECH) □ USAF EK-442

CONSENSUS REPORTS: IARC Cancer Review: Animal Limited Evidence IMEMDT 27,39,82. NCI Carcinogenesis Bioassay Completed; Results Positive: rat NCITR* NCI-CG-TR-130,78; Results Negative: mouse NCITR* NCI-CG-TR-130,78. Reported in EPA TSCA Inventory. EPA Genetic Toxicology Program.

DOT CLASSIFICATION: 6.1; Label: KEEP AWAY FROM FOOD

SAFETY PROFILE: Suspected carcinogen with experimental carcinogenic and tumorigenic data. Poison by intraperitoneal route. Moderately toxic by ingestion. Experimental teratogenic effects. Human mutation data reported. A skin and eye irritant. Combustible when exposed to heat or flame. When heated to decomposition or on contact with acid or acid fumes, it emits highly toxic fumes of aniline and chlorine compounds. Reacts explosively with aniline at 240°C/7.6 bar. Can react vigorously with oxidizing materials. To fight fire, use water, CO_2, water mist or spray, dry chemical. See also ANILINE.

BBL250   CAS: 71-43-2   HR: 3
BENZENE

DOT: UN 1114

mf: C_6H_6 mw: 78.12

PROP: Clear, colorless liquid. Mp: 5.51°, bp: 80.093–80.094°, flash p: 12°F (CC), d: 0.8794 @ 20°, autoign temp: 1044°F, lel: 1.4%, uel: 8.0%, vap press: 100 mm @ 26.1°, vap d: 2.77, ULC: 95–100. Very sltly sol in H_2O; misc in most org solvs. IDLH 500 ppm.

SYNS: (6)ANNULENE □ BENZEEN (DUTCH) □ BENZEN (POLISH) □ BENZIN (OBS.) □ BENZINE (OBS.) □ BENZOL (DOT) □ BENZOLE □ BENZOLENE □ BENZOLO (ITALIAN) □ BICARBURET of HYDROGEN □ CARBON OIL □ COAL NAPHTHA □ CYCLO-HEXATRIENE □ FENZEN (CZECH) □ MINERAL NAPHTHA □ MOTOR BENZOL □ NCI-C55276 □ NITRATION BENZENE □ PHENE □ PHENYL HYDRIDE □ PYROBENZOL □ PYROBENZOLE □ RCRA WASTE NUMBER U019

CONSENSUS REPORTS: NTP 10th Report on Carcinogens. IARC Cancer Review: Group 1 IMEMDT 7,120,87; Human Limited Evidence IMEMDT 7,203,74; Animal Inadequate Evidence IMEMDT 7,203,74; IARC Cancer Review: Animal Limited Evidence IMEMDT 29,93,82; Human Sufficient Evidence IMEMDT 29,93,82. NTP Carcinogenesis Studies (gavage); Clear Evidence: mouse, rat NTPTR* NTP-TR-289,86. EPA Genetic Toxicology Program. Reported in EPA TSCA Inventory. On Community Right-To-Know List.

OSHA PEL: TWA 1 ppm; STEL 5 ppm; Pk 5 ppm/15M/8H; Cancer Hazard

ACGIH TLV: TWA 0.5 ppm; STEL 2.5 ppm (skin); Confirmed Human Carcinogen; BEI: 25 μ/g creatinine of Sphenylmercapturic acid in urine at end of shift

DFG MAK: DFG TRK: Human Carcinogen

NIOSH REL: TWA 0.32 mg/m³; CL 3.2 mg/m³/15M

DOT CLASSIFICATION: 3; Label: Flammable Liquid

SAFETY PROFILE: Confirmed human carcinogen producing myeloid leukemia, Hodgkin's disease, and lymphomas by inhalation. Experimental carcinogenic, neoplastigenic, and tumorigenic data. A human poison by inhalation. An experimental poison by skin contact, intraperitoneal, intravenous, and possibly other routes. Moderately toxic by ingestion and subcutaneous routes. A severe eye and moderate skin irritant. Human systemic effects by inhalation and ingestion: blood changes, increased body temperature. Experimental teratogenic and reproductive effects. Human mutation data reported. A narcotic. In industry, inhalation is the primary route of chronic benzene poisoning. Poisoning by skin contact has been reported. Recent (1987) research indicates that effects are seen at less than 1 ppm. Exposures needed to be reduced to 0.1 ppm before no toxic effects were observed. Elimination is chiefly through the lungs. A common air contaminant.

A dangerous fire hazard when exposed to heat or flame. Explodes on contact with diborane, bromine pentafluoride, permanganic acid, peroxomonosulfuric acid, and peroxodisulfuric acid. Forms sensitive, explosive mixtures with iodine pentafluoride, silver perchlorate, nitryl perchlorate, nitric acid, liquid oxygen, ozone, and arsenic pentafluoride + potassium methoxide (explodes above 30°C). Ignites on contact with sodium peroxide + water, dioxygenyl tetrafluoroborate, iodine heptafluoride, and dioxygen difluoride. Vigorous or incandescent reaction with hydrogen + Raney nickel (above 210°C), uranium hexafluoride, and bromine trifluoride. Can react vigorously with oxidizing materials, such as Cl_2, CrO_3, O_2, $NClO_4$, O_3, perchlorates, ($AlCl_3$ + $FClO_4$), (H_2SO_4 + permanganates), K_2O_2, ($AgClO_4$ + acetic acid), Na_2O_2. Moderate explosion hazard when exposed to heat or flame. Use with adequate ventilation. To fight fire, use foam, CO_2, dry chemical.

Poisoning occurs most commonly via inhalation of the vapor, although benzene can penetrate the skin and cause poisoning. Locally, benzene has a comparatively strong irritating effect, producing erythema and burning, and, in more severe cases, edema and even blistering. Exposure to high concentrations of the vapor (3000 ppm or higher) may result from failure of equipment or spillage. Such exposure, while rare in industry, may cause acute poisoning, characterized by the narcotic action of benzene on the central nervous system. The anesthetic action of benzene is similar to that of other anesthetic gases, consisting of a preliminary stage of excitation followed by depression and, if exposure is continued, death through respiratory failure. The chronic, rather than the acute, form of benzene poisoning is important in industry. It is a recognized leukemogen. There is no specific blood picture occurring in cases of chronic benzol poisoning. The bone marrow may be hypoplastic, normal, or hyperplastic, the changes reflected in the peripheral blood. Anemia, leucopenia, macrocytosis, reticulocytosis, thrombocytopenia, high color index, and prolonged bleeding time may be present. Cases of myeloid leukemia have been reported. For the worker, repeated blood examinations are necessary, including hemoglobin determinations, white and red cell counts, and differential smears. Where a worker shows a progressive drop in either red or white cells, or where the white count remains below <5000/mm³ or the red count remains below 4.0 million/mm³, on two successive monthly examinations, the worker should be immediately removed from benzene exposure. Elimination is chiefly through the lungs, when fresh air is breathed. The portion that is absorbed is oxidized, and the oxidation products are combined with sulfuric and glycuronic acids and eliminated in the urine. This may be used as a diagnostic sign. Benzene has a definite cumulative action, and exposure to a relatively high concentration is not serious from the point of view of causing damage to the blood-forming system, provided the exposure is not repeated. In acute poisoning, the worker becomes confused and dizzy, complains of tightening of the leg muscles and of pressure over the forehead, then passes into a stage of excitement. If allowed to remain exposed, he quickly becomes stupefied and lapses into coma. In nonfatal cases, recovery is usually complete with no permanent disability. In chronic poisoning the onset is slow, with the symptoms vague; fatigue, headache, dizziness, nausea and loss of appetite, loss of weight, and weakness are common complaints in early cases. Later, pallor, nosebleeds, bleeding gums, menorrhagia, petechiae, and purpura may develop. There is great individual variation in the signs and symptoms of chronic benzene poisoning.

BBL500   CAS: 122-78-1   HR: 2
BENZENEACETALDEHYDE
mf: C_8H_8O mw: 120.16

PROP: Oily, colorless liquid that polymerizes and grows more viscous on standing; odor similar to lilac and hyacinth. Has been crystallized, mp: 33–34°, d: (25/25) 1.023–1.030, refr index: 1.525–1.545, bp: (10) 78°, n: (20/D) 1.524–1.528, flash p: 154°F. Sltly sol in water; sol in alc, ether, and propylene glycol. One part is sol in two parts of 80% alc forming a clear solution.

SYNS: FEMA No. 2874 □ HYACINTHIN □ PAA □ PHENYLACETALDEHYDE (FCC) □ PHENYLACETIC ALDEHYDE □ PHENYLETHANAL □ α-TOLU-ALDEHYDE □ α-TOLUIC ALDEHYDE

CONSENSUS REPORTS: Reported in EPA TSCA Inventory.

SAFETY PROFILE: Moderately toxic by ingestion. Human skin irritant. Combustible liquid. When heated to decomposition it emits acrid smoke and irritating fumes. See also ALDEHYDES.

BBL750   CAS: 98-05-5   HR: 3
BENZENEARSONIC ACID
mf: $C_6H_7AsO_3$ mw: 202.05
PROP: Colorless crystals from water. D:
1.760, mp: 160° decomp. Sol in water.
SYNS: PHENYL ARSENIC ACID □ PHENYLARSONIC
ACID
CONSENSUS REPORTS: Reported in EPA
TSCA Inventory. EPA Extremely
Hazardous Substances List. Arsenic and its
compounds are on the Community Right-
To-Know List.
OSHA PEL: TWA 0.5 mg/(As)m³
ACGIH TLV: BEI: 35 μ (As)/L inorganic
arsenic and methylated metabolites in urine
SAFETY PROFILE: A deadly poison by
ingestion and intravenous routes. See also
ARSENIC COMPOUNDS. When heated
to decomposition it emits toxic fumes of As.

BBP000   CAS: 123-61-5   HR: 3
BENZENE-1,3-DIISOCYANATE
mf: $C_8H_4N_2O_2$ mw: 160.14
PROP: Crystals. Mp: 51–55°, bp: 102–104°
@ 8 mm.
SYNS: BENZENE-1,3-DIISOCYANATE □ BENZENE,
1,3-DIISOCYANATO- □ 1,3-DIISOCYANATOBENZENE
□ NACCONATE 400 □ m-PHENYLENE DIISOCYANATE
□ m-PHENYLENE ISOCYANATE
CONSENSUS REPORTS: Reported in EPA
TSCA Inventory. Cyanide and its
compounds are on the Community Right-
To-Know List.
NIOSH REL: TWA (Diisocyanates) 0.005
ppm; CL 0.02 ppm/10M
SAFETY PROFILE: A sensitizer at very low
concentrations. Deadly poison by
intravenous route. When heated to
decomposition it emits toxic fumes of NO_x
and CN^-. See also ESTERS.

BBP250   CAS: 623-26-7   HR: 2
p-BENZENEDINITRILE
mf: $C_8H_4N_2$ mw: 128.14
PROP: Crystals. Mp: 222°, vap d: 4.42.
SYNS: 4-CYANOBENZONITRILE □ p-DICYANO-
BENZENE □ 1,4-DICYANOBENZENE □ NITRIL
KYSELINY TEREFTALOVE (CZECH) □ p-PDN □ p-
PHTHALODINITRILE □ TEREFTALODINITRIL
(CZECH) □ TEREPHTHALONITRILE

CONSENSUS REPORTS: Reported in EPA
TSCA Inventory. Cyanide and its
compounds are on the Community Right-
To-Know List.
SAFETY PROFILE: Moderately toxic by
ingestion and intraperitoneal routes. An eye
irritant. When heated to decomposition it
emits toxic fumes of CN^- and NO_x. See also
NITRILES.

BBP750   CAS: 608-73-1   HR: 3
BENZENE HEXACHLORIDE
mf: $C_6H_6Cl_6$ mw: 290.82
PROP: Technical grade contains 68.7% α-
BHC, 6.5% β-BHC, and 13.5% γ-BHC
(JPFCD2 14,305,79). White, crystalline
powder. Mp: 113°, vap press: 0.0317 mm @
20°.
SYNS: BHC (USDA) □ COMPOUND-666 □ DBH □ ENT
8,601 □ GAMMEXANE □ HCCH □ HEXA □
HEXACHLOR □ HEXACHLORAN □ HEXACHLORO-
CYCLOHEXANE □ 1,2,3,4,5,6-HEXACHLORO-
CYCLOHEXANE □ HEXYLAN □ JACUTIN □ LATKA
666
CONSENSUS REPORTS: NTP 10th Report
on Carcinogens. IARC Cancer Review:
Animal Sufficient Evidence IMEMDT
5,47,74.
ACGIH TLV: TWA 0.5 mg/m³ (skin)
SAFETY PROFILE: Confirmed carcinogen
with experimental carcinogenic,
neoplastigenic, and tumorigenic data by
ingestion and skin contact. Poison by
ingestion, skin contact, and subcutaneous
routes. Human systemic effects by
inhalation: headache, nausea or vomiting,
and fever. Implicated in aplastic anemia.
Experimental reproductive effects. Mutation
data reported. Lindane is more toxic than
DDT or dieldrin. Potentially violent reaction
with dimethylformamide + iron. When
heated to decomposition it emits highly
toxic fumes of phosgene, HCl, and Cl^-. See
other benzenehexachloride entries.
    A toxic organochlorine that is persistent
in the environment and accumulates in
mammalian tissue. For cattle, the oral LD50
<= 100 mg/kg. The various isomers have
different actions; the γ (lindane) and α

isomers are central nervous system stimulants, the principal symptom being convulsions. The β and Δ isomers are central nervous system depressants. The use of thermal vaporizers with lindane has caused acute poisoning by inhalation.

The dangerous acute dose of the technical mixture has been estimated at about 30 g and the dangerous dose of lindane at about 7 to 15 g. However, as already mentioned, a single dose of 45 mg (or approximately 0.65 mg/kg) of lindane caused convulsions. Lindane shows a marked difference in toxicity to different species. Its toxic effect on laboratory animals compares favorably with that of DDT, but for several domestic animals, notably calves, lindane is more toxic than DDT or dieldrin. On a chronic systemic basis the α, β and γ isomers are experimental carcinogens. Has been implicated in aplastic anemia.

Dermatitis and perhaps other manifestations based on sensitivity represent a sort of chronic, though probably not systemic intoxication, which has been observed in humans.

The signs and symptoms of confirmed acute poisoning in humans have paralleled those in experimental animals. These signs and symptoms are: excitation, hyperirritability, loss of equilibrium, clonic-tonic convulsions, and later depression.

There is some evidence that the pulmonary edema and vascular collapse may be of neurogenic origin also. The symptoms in animals systemically poisoned by the γ-isomer alone are essentially similar to those caused by mixtures, although the onset may be earlier. Workers acutely exposed to high air concentrations of lindane and its decomposition products show headache, nausea, and irritation of eyes, nose, and throat.

In rare instances, urticaria has followed exposure to lindane vapor. Unlike the signs and symptoms already mentioned, this allergic manifestation occurs only in susceptible individuals, and usually only after a period of sensitization.

BBQ000   CAS: 319-84-6   HR: 3
BENZENE HEXACHLORIDE-α-isomer
mf: $C_6H_6Cl_6$ mw: 290.82
PROP: Solid. Mp: 158°.
SYNS: α-BENZENEHEXACHLORIDE □ α-BHC □ ENT 9,232 □ α-HCH □ α-HEXACHLORANE □ HEXA-CHLORCYCLOHEXAN (GERMAN) □ α-HEXACHLORO-CYCLOHEXANE □ α-1,2,3,4,5,6-HEXACHLOROCYCLO-HEXANE (MAK) □ 1-α,2-α,3-β,4-α,5-β,6-β-HEXA-CHLOROCYCLOHEXANE □ α-LINDANE
CONSENSUS REPORTS: NTP 10th Report on Carcinogens. IARC Cancer Review: Animal Sufficient Evidence IMEMDT 20,195,79; IMEMDT 5,47,74. EPA Genetic Toxicology Program. Reported in EPA TSCA Inventory.
DFG MAK: 0.5 mg/m³
SAFETY PROFILE: Confirmed carcinogen with experimental carcinogenic, tumorigenic, and neoplastigenic data. Poison by ingestion. Mutation data reported. When heated to decomposition it emits toxic fumes of Cl⁻. See also BENZENE HEXACHLORIDE and other benzenehexachloride entries.

BBQ500   CAS: 58-89-9   HR: 3
BENZENE HEXACHLORIDE-γ-isomer
mf: $C_6H_6Cl_6$ mw: 290.82
PROP: Solid. Mp: 112.5°.
SYNS: AALINDAN □ AFICIDE □ AGRISOL G-20 □ AGROCIDE □ AGRONEXIT □ AMEISENATOD □ AMEISENMITTEL MERCK □ APARSIN □ APHTIRIA □ APLIDAL □ ARBITEX □ BBH □ BEN-HEX □ BENTOX 10 □ γ-BENZENE HEXACHLORIDE □ BEXOL □ BHC □ γ-BHC □ CELANEX □ CHLORESENE □ CODECHINE □ DBH □ DETMOL-EXTRAKT □ DETOX 25 □ DEVORAN □ DOL GRANULE □ DRILL TOX-SPEZIAL AGLUKON □ ENT 7,796 □ ENTOMOXAN □ EXAGAMA □ FORLIN □ GALLOGAMA □ GAMACID □ GAMAPHEX □ GAMENE □ GAMISO □ GAMMA-COL □ GAMMAHEXA □ GAMMAHEXANE □ GAMMALIN □ GAMMOPAZ □ HCCH □ HCH □ γ-HCH □ HECLOTOX □ HEXA-CHLORAN □ γ-HEXACHLORAN □ γ-HEXACHLORANE □ γ-HEXACHLOROBENZENE □ 1-α,2-α,3-β,4-α,5-α,6-β-HEXACHLOROCYCLOHEXANE □ γ-HEXACHLORO-CYCLOHEXANE (MAK) □ 1,2,3,4,5,6-HEXACHLORO-CYCLOHEXANE, γ-ISOMER □ HEXATOX □ HEXICIDE

□ HGI □ INEXIT □ ISOTOX □ JACUTIN □ KOKOTINE □ KWELL □ LENDINE □ LENTOX □ LIDENAL □ LINDAGRAIN □ LINDANE (ACGIH, DOT, USDA) □ LINTOX □ MILBOL 49 □ MSZYCOL □ NCI-C00204 □ NEO-SCABICIDOL □ NEXIT □ NOVIGAM □ OVADZIAK □ PEDRACZAK □ QUELLADA □ RCRA WASTE NUMBER U129 □ SANG gamma □ STREUNEX □ TAP 85 □ VITON

CONSENSUS REPORTS: NTP 10th Report on Carcinogens. IARC Cancer Review: Animal Sufficient Evidence IMEMDT 5,47,74; IMEMDT 20,195,79. NCI Carcinogenesis Bioassay (feed); No Evidence: mouse, rat NCITR* NCI-CG-TR-14,77. EPA Extremely Hazardous Substances List. EPA Genetic Toxicology Program. Community Right-To-Know List. Reported in EPA TSCA Inventory. EPA Extremely Hazardous Substances List.

OSHA PEL: TWA 0.5 mg/m^3 (skin)
ACGIH TLV: TWA 0.5 mg/m^3 (skin)
DFG MAK: 0.1 mg/m^3; Not Classifiable as a Human Carcinoge

SAFETY PROFILE: Confirmed carcinogen with experimental carcinogenic and neoplastigenic data. A human systemic poison by ingestion. Also a poison by ingestion, skin contact, intraperitoneal, intravenous, and intramuscular routes. Human systemic effects by ingestion: convulsions, dyspnea, and cyanosis. Experimental teratogenic and reproductive effects. Mutation data reported. See also BENZENE HEXACHLORIDE and other benzene hexachloride entries. When heated to decomposition it emits toxic fumes of Cl$^-$, HCl, and phosgene.

BBQ750   **HR: 3**
BENZENEHEXACHLORIDE (mixed isomers)
mf: $C_6H_6Cl_6$ mw: 290.82
PROP: Technical BHC contains about 64% α, 10% β, 13% γ, 9% Δ, and 1% ε isomers of 1,2,3,4,5,6-hexachlorocyclohexane (IARC** 5,47,74).
SYNS: BENZAHEX □ BENZEX □ DOL □ DOLMIX □ FBHC □ FHCH □ 1,2,3,4,5,6-HEXACHLOROCYCLO-HEXANE (mixture of isomers) □ HEXYCLAN □ KOTOL □ SOPROCIDE □ TECHNICAL BHC □ TECHNICAL HCH

CONSENSUS REPORTS: IARC Cancer Review: Animal Sufficient Evidence IMEMDT 5,47,74; IMEMDT 20,195,79.
SAFETY PROFILE: Confirmed carcinogen with experimental tumorigenic and neoplastigenic data. Poison by inhalation and ingestion. Human systemic effects by an unspecified route: convulsions. Potentially dangerous reaction with DMF in presence of Fe, also CCl$_4$. When heated to decomposition it emits highly toxic fumes of Cl$^-$, HCl, and phosgene. See also BENZENE HEXACHLORIDE and other benzenehexachloride entries.

BBR000   CAS: 319-85-7   HR: 3
trans-α-BENZENEHEXACHLORIDE
mf: $C_6H_6Cl_6$ mw: 290.82
PROP: Solid. Mp: 297°.
SYNS: β-BENZENEHEXACHLORIDE □ β-BHC □ ENT 9,233 □ β-HCH □ β-HEXACHLOROBENZENE □ β-HEXACHLOROCYCLOHEXANE □ 1-α,2-β,3-α,4-β,5-α,6-β-HEXACHLOROCYCLOHEXANE □ β-1,2,3,4,5,6-HEXACHLOROCYCLOHEXANE (MAK) □ β-ISOMER □ β-LINDANE
CONSENSUS REPORTS: NTP 10th Report on Carcinogens. IARC Cancer Review: Animal Sufficient Evidence IMEMDT 5,47,74; Animal Limited Evidence IMEMDT 20,195,79. Reported in EPA TSCA Inventory.
DFG MAK: 0.5 mg/m^3
SAFETY PROFILE: Confirmed carcinogen with experimental neoplastigenic data. Mildly toxic by ingestion. When heated to decomposition it emits very toxic fumes of Cl$^-$, HCl, and phosgene. See also BENZENE HEXACHLORIDE and other benzenehexachloride entries.

BBS250   CAS: 98-11-3   HR: 3
BENZENESULFONIC ACID
mf: $C_6H_6O_3S$ mw: 158.18
PROP: Deliquescent plates or tablets. Mp: 43–44°.
SYN: PHENYLSULFONIC ACID
CONSENSUS REPORTS: Reported in EPA TSCA Inventory.

B

SAFETY PROFILE: Poison by ingestion, skin contact, and probably inhalation. A severe skin and eye irritant. See also SULFATES and SULFONATES.

BBS300   CAS: 80-17-1   HR: 3
BENZENESULFONIC HYDRAZIDE
DOT: UN 2970
mf: $C_6H_8N_2O_2S$ mw: 172.22
SYNS: BENZENESULFOHYDRAZIDE □ BENZENE-SULFONIC ACID, HYDRAZIDE □ BENZENE-SULFONOHYDRAZIDE □ BENZENE-SULFONYL HYDRAZIDE □ BENZENESULFONYL HYDRAZINE □ BENZENE SULPHONOHYDRAZIDE □ CELOGEN BSH □ ChKhZ 9 □ GENITRON BSH □ HYDRAZIDE BSG □ NITROPORE OBSH □ PHENYLSULFOHYDRAZIDE □ PHENYLSULFONYL HYDRAZIDE □ PHENYLSULFO-NYLHYDRAZINE □ POROFOR BSH □ POROFOR-BSH-PULVER □ POROFOR ChKhZ 9
CONSENSUS REPORTS: Reported in EPA TSCA Inventory.
DOT CLASSIFICATION: 4.1; Label: Flammable Solid
SAFETY PROFILE: Poison by ingestion. A flammable solid. When heated to decomposition it emits toxic vapors of NO_x and SO_x.

BBS750   CAS: 98-09-9   HR: 3
BENZENESULFONYL CHLORIDE
DOT: UN 2225
mf: $C_6H_5ClO_2S$ mw: 176.62
PROP: Liquid. D: 1.384 @ 15°/15°, mp: 14.5°, bp: 251–252°.
SYNS: BENZENE SULFONCHLORIDE □ BENZENE-SULFONIC (ACID) CHLORIDE □ BENZENE SULPHONYL CHLORIDE (DOT) □ BENZENO-SULFO-CHLOREK (POLISH) □ BENZENOSULPHOCHLORIDE □ BSC-REFINE D □ PHENYL-SULFONYL CHLORIDE □ RCRA WASTE NUMBER U020
CONSENSUS REPORTS: Reported in EPA TSCA Inventory.
DOT CLASSIFICATION: 8; Label: Corrosive
SAFETY PROFILE: Poison by intraperitoneal route. A dangerous storage hazard. It may explode in a sealed bottle. Explosive reaction with dimethyl sulfoxide. Reacts vigorously with methyl formamide. When heated to decomposition it emits

toxic fumes of Cl^- and SO_x. See also SULFONATES.

BBT250   CAS: 368-43-4   HR: 3
BENZENESULPHONYL FLUORIDE
mf: $C_6H_5FO_2S$ mw: 160.17
PROP: Clear liquid. Bp: 209°, fp: −5°, flash p: 196°F, d: 1.329, vap press: 8 mm @ 80°, vap d: 5.52.
CONSENSUS REPORTS: Reported in EPA TSCA Inventory.
SAFETY PROFILE: A poison by intraperitoneal route. Slightly irritating to skin. Flammable when exposed to heat or flame. It can react vigorously with oxidizing materials. To fight fire, use water, foam, CO_2, water spray or mist, dry chemical. When heated to decomposition it emits toxic fumes of F^- and SO_x. See also FLUORIDES and SULFATES.

BBU250   CAS: 533-73-3   HR: 3
1,2,4-BENZENETRIOL
mf: $C_6H_6O_3$ mw: 126.12
PROP: Plates from Et_2O. Mp: 140.5° (subl). Sol in water.
SYNS: HYDROXYHYDROQUINONE □ HYDROXY-QUINOL □ OXYHYDROCHINON (GERMAN) □ OXYHYDROQUINONE □ 1,2,4-TRIHYDROXYBENZENE
CONSENSUS REPORTS: EPA Genetic Toxicology Program. Reported in EPA TSCA Inventory.
SAFETY PROFILE: Poison by subcutaneous and intraperitoneal routes. Human mutation data reported. When heated to decomposition it emits acrid smoke and irritating fumes.

BBV250   CAS: 613-94-5   HR: 3
BENZHYDRAZIDE
mf: $C_7H_8N_2O$ mw: 136.17
PROP: Crystals from water. Mp: 112.5°. Sol in water, acids, EtOH, C_6H_6, and Me_2CO.
SYNS: BENZOHYDRAZIDE □ BENZOHYDRAZINE □ BENZOIC HYDRAZIDE □ BENZOYL HYDRAZIDE
CONSENSUS REPORTS: Reported in EPA TSCA Inventory.
SAFETY PROFILE: Poison by subcutaneous route. Questionable

carcinogen with experimental carcinogenic and neoplastigenic data. Violent reaction with benzeneseleninic acid. When heated to decomposition it emits toxic fumes of NO_x.

BBV500   CAS: 58-73-1   HR: 3
BENZHYDRYL
mf: $C_{17}H_{21}NO$ mw: 255.39
PROP: Oil. Bp: 163–167° @ 3 mm.
SYNS: ALERYL □ ALLEDRYL □ ALLERGAN B □ ALLERGEVAL □ ALLERGICAL □ ALLERGIN □ ALLERGINA □ ALLERGIVAL □ AMIDRYL □ ANTISTOMINUM □ ANTOMIN □ AUTOMIN □ BAGAODRYL □ BARAMINE □ BENA □ BENACHLOR □ BENADON □ BENADRIN □ BENADRYL □ BEN-ALLERGIN □ BENAPON □ BENODIN □ BENODINE □ BENYLAN □ BENZANTINE □ BENZHYDRAMINE □ BENZHYDRAMINUM □ BENZHYDRIL □ o-BENZHYDRYLDIMETHYLAMINOETHANOL □ 2-(BENZHYDRYLOXY)-N,N-DIMETHYLETHYLAMINE □ 2-(BENZOHYDRYLOXY)-N,N-DIMETHYLETHYLAMINE □ BETRAMIN □ DABYLEN □ DEBENDRIN □ DERMISTINE □ DERMODRIN □ DESENTOL □ DIABENYL □ DIABYLEN □ DIBONDRIN □ DIFEDRYL □ DIFENHYDRAMIN □ DIFENIDRAMINA (ITALIAN) □ DIHIDRAL □ DIMEDROL □ DIMEDRYL □ β-DIMETHYLAMINO-AETHYL-BENZHYDRYL-AETHER (GERMAN) □ β-DIMETHYLAMINOETHANOL DIPHENYLMETHYL ETHER □ α-(2-DIMETHYLAMINO-ETHOXY)DIPHENYLMETHANE □ β-DIMETHYL-AMINOETHYLBENZHYDRYLETHER □ DIPHANTINE □ DIPHENYLHYDRAMINE □ 2-(DIPHENYLMETH-OXY)-N,N-DIMETHYLETHYLAMINE □ DRYISTAN □ DRYLISTAN □ DYLAMON □ ETANAUTINE □ HISTAXIN □ HYADRINE □ IBIODRAL □ MEDIDRYL □ MEPHADRYL □ NAUSEN □ PROBEDRYL □ RESTAMIN □ RESTAMINE □ RIGIDIL □ RIGIDYL □ S51 □ SYNTEDRIL □ SYNTODRIL □ VENA
CONSENSUS REPORTS: Reported in EPA TSCA Inventory.
SAFETY PROFILE: Deadly human poison by an unspecified route. Poison by ingestion, intravenous, intraperitoneal, and subcutaneous routes. Experimental reproductive effects. Human systemic effects by ingestion: somnolence, alteration of operant conditioning, changes in psychophysiological tests. Human mutation data reported. When heated to decomposition it emits toxic fumes of NO_x. See also ETHERS.

BBW500   CAS: 132-69-4   HR: 3
BENZIDAMINE HYDROCHLORIDE
mf: $C_{19}H_{23}N_3O•ClH$ mw: 345.91
PROP: Mp: 160°. Very sol in H_2O.
SYNS: AF 864 □ BENALGIN □ BENZINDAMINE HYDROCHLORIDE □ BENZYDAMINE HYDRO-CHLORIDE □ 1-BENZYL-3-γ-DIMETHYL-AMINOPROPOXY-1H-INDAZOLE HYDROCHLORIDE □ 1-BENZYL-3-(3-(DIMETHYLAMINO)PROPOXY)-1H-INDAZOLE HYDROCHLORIDE □ BENZYRIN □ DIFFLAM □ N,N-DIMETHYL-3((1-PHENYLMETHYL)-1H-INDAZOL-3-YL)OXY-1-PROPANAMINE HYDRO-CHLORIDE □ DORINAMIN □ ENZAMIN □ EPIROTIN □ IMOTRYL □ INDOLIN □ RIRILIM □ RIRIPEN □ SALYZORON □ TAMAS □ TANTUM □ VERAX
SAFETY PROFILE: Poison by intraperitoneal, subcutaneous, and intravenous routes. Moderately toxic by ingestion. An experimental teratogen. Other experimental animal reproductive effects. An eye irritant. A nonsteroidal anti-inflammatory analgesic. When heated to decomposition it emits very toxic fumes of HCl and NO_x.

BBX000   CAS: 92-87-5   HR: 3
BENZIDINE
DOT: UN 1885
mf: $C_{12}H_{12}N_2$ mw: 184.26
PROP: Grayish-yellow, crystalline powder; white or sltly reddish crystals, powder, or leaf from water or alc. Mp: 127.5–128.7° @ 740 mm, bp: 401.7°, d: 1.250 @ 20°/4°.
SYNS: BENZIDIN (CZECH) □ BENZIDINA (ITALIAN) □ BENZYDYNA (POLISH) □ p,p-BIANILINE □ 4,4'-BIANILINE □ (1,1'-BIPHENYL)-4,4'-DIAMINE (9CI) □ 4,4'-BIPHENYLDIAMINE □ 4,4'-BIPHENYLENEDIAMINE □ C.I. 37225 □ C.I. AZOIC DIAZO COMPONENT 112 □ p,p'-DIAMINOBIPHENYL □ 4,4'-DIAMINOBIPHENYL □ 4,4'-DIAMINO-1,1'-BIPHENYL □ p-DIAMINODIPHENYL □ 4,4'-DIAMINODIPHENYL □ p,p'-DIANILINE □ 4,4'-DIPHENYLENEDIAMINE □ FAST CORINTH BASE B □ NCI-C03361 □ RCRA WASTE NUMBER U021
CONSENSUS REPORTS: NTP 10th Report on Carcinogens. IARC Cancer Review: Human Limited Evidence IMEMDT 1,80,72; Human Sufficient Evidence IMEMDT 29,149,82; Animal Sufficient Evidence IMEMDT 1,80,72; IMEMDT 29,149,82. EPA Genetic Toxicology

Program. Community Right-To-Know List. Reported in EPA TSCA Inventory.
OSHA PEL: OSHA: Cancer Suspect Agent
ACGIH TLV: Confirmed Human Carcinogen
DFG MAK: Human Carcinogen
DOT CLASSIFICATION: 6.1; Label: Poison
SAFETY PROFILE: Confirmed human carcinogen producing bladder tumors. Experimental carcinogenic and tumorigenic data. Poison by ingestion and intraperitoneal routes. Human mutation data reported. Can cause damage to blood, including hemolysis and bone marrow depression. On ingestion causes nausea and vomiting, which may be followed by liver and kidney damage. Any exposure is considered extremely hazardous. When heated to decomposition it emits highly toxic fumes of NO_x. See also AROMATIC AMINES.

BBX750   CAS: 531-85-1   HR: 2
BENZIDINE HYDROCHLORIDE
mf: $C_{12}H_{12}N_2 \cdot 2ClH$ mw: 257.18
PROP: Leaflets. Sol in H_2O.
SYNS: (1,1'-BIPHENYL)-4,4'-DIAMINE, DIHYDRO-CHLORIDE □ DIHIDROCLORURO de BENZIDINA (SPANISH)
CONSENSUS REPORTS: Reported in EPA TSCA Inventory. EPA Genetic Toxicology Program.
SAFETY PROFILE: Suspected carcinogen with experimental carcinogenic and tumorigenic data. Human mutation data reported. When heated to decomposition it emits very toxic fumes of HCl and NO_x.

BBY000   CAS: 531-86-2   HR: 3
BENZIDINE SULFATE
mf: $C_{12}H_{12}N_2 \cdot H_2O_4S$ mw: 282.34
PROP: Hair dye.
SYN: (1,1'-BIPHENYL)-4,4'-DIAMINE SULFATE (1:1)
OSHA PEL: OSHA: Carcinogen
SAFETY PROFILE: Confirmed human carcinogen with experimental carcinogenic data. See also BENZIDINE and SULFATES. When heated to decomposition it emits toxic fumes of SO_x and NO_x.

BBY300   HR: 2
BENZIDINE SULPHATE and HYDRA-ZINE-BENZENE
mf: $C_6H_8N_2 \cdot C_{12}H_{12}N_2 \cdot H_2O_4S$ mw: 390.50
SYN: HYDRAZINE-BENZENE and BENZIDINE SULFATE
SAFETY PROFILE: Suspected carcinogen with experimental carcinogenic data. When heated to decomposition it emits toxic fumes of NO_x and SO_x.

BCA000   CAS: 57-37-4   HR: 3
BENZILIC ACID-β-DIETHYLAMINO-ETHYL ESTER HYDROCHLORIDE
mf: $C_{20}H_{25}NO_3 \cdot ClH$ mw: 363.92
PROP: Crystals from Me_2CO. Mp: 177–178°. Sol in H_2O; insol in Et_2O.
SYNS: ACTOZINE □ AMIOYL □ AMISYL □ AMITAKON □ AMIZIL HYDROCHLORIDE □ ARCADINE □ AY-5406 □ BENACTIZINE HYDRO-CHLORIDE □ BENACTYZIN (CZECH) □ BENACTY-ZINE CHLORIDE □ BENACTYZINE HYDROCHLORIDE □ BENAKTIN □ BENZILATE DU DIETHYLAMINO-ETHANOL CHLORHYDRATE (FRENCH) □ CAFRON □ CEDAD □ CEVANOL □ DESTENDO □ β-DIETHYL-AMINOETHYL BENZILATE HYDRO-CHLORIDE □ 2-DIETHYLAMINOETHYL BENZILATE HYDRO-CHLORIDE □ 2-DIETHYLAMINO-ETHYL DIPHENYL-GLYCOLATE HYDROCHLORIDE □ 2-(DIFENYL-HYDROXYACETOXY)ETHYL-DIETHYLAMMONIUM-CHLORID (CZECH) □ DIPHENYL-GLYCOLLIC ACID-2-(DIETHYLAMINO)ETHYL ESTER HYDROCHLORIDE □ FOBEX □ IBIOTYZIL □ KATRON □ LEUCIDIL □ NERVACTON □ NERVATIL □ NEURAKTIL □ NEUROBENZIL □ NEUROLEPTONE □ NUTINAL □ PARASAN □ PARPON □ PHOBEX □ PROCALM □ STOIKON □ SUAVITIL □ TRANQUILLIN □ VALLADAN □ WIN 5606
CONSENSUS REPORTS: Reported in EPA TSCA Inventory.
SAFETY PROFILE: Poison by ingestion, intraperitoneal, subcutaneous, intradermal, and intravenous routes. Human systemic effects by ingestion of very small amounts: toxic psychosis. Experimental reproductive effects. When heated to decomposition it emits very toxic fumes of NO_x and HCl.

BCA300   CAS: 14090-77-8   HR: 3
α-BENZIL MONOXIME
mf: $C_{14}H_{11}NO_2$ mw: 225.26

SYNS: α-BENZIL MONOOXIME □ BENZIL, MONOXIME □ BENZIL, MONOOXIME □ BENZIL, β-MONOXIME □ BENZIL, OXIME □ 1,2-DIPHENYL-ETHANEDIONE MONOOXIME □ ETHANEDIONE, DIPHENYL-, MONOOXIME

SAFETY PROFILE: A poison by intravenous route. Moderately toxic by ingestion. When heated to decomposition it emits toxic vapors of NO_x.

BCB750   CAS: 51-17-2   HR: 3
BENZIMIDAZOLE

mf: $C_7H_6N_2$ mw: 118.15

PROP: Tabular crystals or plates. Mp: 170.5°, bp: >360°. Sol in alc; sparingly sol in water.

SYNS: 3-AZAINDOLE □ AZINDOLE □ o-BENZIMIDAZOLE □ 1H-BENZIMIDAZOLE (9CI) □ BENZIMINAZOLE □ 1,3-BENZODIAZOLE □ BENZOIMIDAZOLE □ BZI □ 1,3-DIAZAINDENE □ N,N'-METHENYL-o-PHENYLENEDIAMINE □ NSC-759

CONSENSUS REPORTS: Reported in EPA TSCA Inventory.

SAFETY PROFILE: Poison by intravenous and intraperitoneal routes. Moderately toxic by ingestion. Mutation data reported. When heated to decomposition it emits highly toxic fumes of NO_x.

BCC500   CAS: 583-39-1   HR: 3
2-BENZIMIDAZOLETHIOL

mf: $C_7H_6N_2S$ mw: 150.21

PROP: Plates from alc (aq). Mp: 298°. Sol in EtOH; sltly sol in H_2O.

SYNS: ANTIEGENE MB □ ANTIOXIDANT MB (CZECH) □ AOMB □ ASM MB □ 2-MERCAPTO-BENZIMIDAZOLE □ MERCAPTOBENZOIMIDAZOLE □ 2-MERCAPTOBENZOIMIDAZOLE □ MERKAPTO-BENZIMIDAZOL (CZECH) □ NCI-C60980 □ o-PHENYLENETHIOUREA □ USAF EK-6540 □ USAF XF-21

CONSENSUS REPORTS: Reported in EPA TSCA Inventory.

SAFETY PROFILE: Poison by intraperitoneal and intravenous routes. Moderately toxic by ingestion. Skin and eye irritant. When heated to decomposition it emits toxic fumes of SO_x and NO_x. See also MERCAPTANS.

BCE500   CAS: 81-07-2   HR: 3
1,2-BENZISOTHIAZOL-3(2H)-ONE-1,1-DIOXIDE

mf: $C_7H_5NO_3S$ mw: 183.19

PROP: White crystals or powder from water; odorless with sweet taste. Mp: 224° (decomp), bp: subl. Sol in water, alc, chloroform, and ether.

SYNS: ANHYDRO-o-SULFAMINEBENZOIC ACID □ 3-BENZISOTHIAZOLINONE-1,1-DIOXIDE □ o-BENZOIC SULPHIMIDE □ o-BENZOSULFIMIDE □ BENZO-SULPHIMIDE □ BENZO-2-SULPHIMIDE □ o-BENZOYL SULFIMIDE □ o-BENZOYL SULPHIMIDE □ 1,2-DIHYDRO-2-KETOBENZISOSULFONAZOLE □ 1,2-DIHYDRO-2-KETOBENZISOSULPHONAZOLE □ 2,3-DIHYDRO-3-OXOBENZISOSULFONAZOLE □ 2,3-DIHYDRO-3-OXOBENZISOSULPHONAZOLE □ GARANTOSE □ GLUCID □ GLUSIDE □ HERMESETAS □ 3-HYDROXYBENZISOTHIAZOL-S,S-DIOXIDE □ INSOLUBLE SACCHARINE □ KANDISET □ NATREEN □ RCRA WASTE NUMBER U202 □ SACARINA □ SACCAHARIMIDE □ SACCHARINA □ SACCHARIN ACID □ SACCHARINE □ SACCHARINOL □ SACCHARINOSE □ SACCHAROL □ SAXIN □ SUCRE EDULCOR □ SUCRETTE □ o-SULFOBENZIMIDE □ o-SULFOBENZOIC ACID IMIDE □ 2-SULPHOBENZOIC IMIDE □ SYKOSE □ SYNCAL □ ZAHARINA

CONSENSUS REPORTS: IARC Cancer Review: Group 2B IMEMDT 7,334,87; Human Inadequate Evidence IMEMDT 22,111,80; Animal Sufficient Evidence IMEMDT 22,111,80. EPA Genetic Toxicology Program. Reported in EPA TSCA Inventory. Community Right-To-Know List.

SAFETY PROFILE: Confirmed carcinogen with experimental neoplastigenic and tumorigenic data. Mild acute toxicity by ingestion. Experimental teratogenic and reproductive effects. Mutation data reported. When heated to decomposition it emits toxic NO_x and SO_x.

BCH750   CAS: 10085-81-1   HR: 3
BENZOCTAMINE HYDROCHLORIDE

mf: $C_{18}H_{19}N \cdot ClH$ mw: 285.84

PROP: Solid. Mp: 320–322°.

SYNS: BA 30,803 □ 1-METHYLAMINOMETHYLDIBENZO(b,c)BICYCLO(2,2,2)OCTADIENE HYDROCHLORIDE □ N-METHYLETHANO-ANTHRACENE-9-(10H)-METHYLAMINE HYDROCHLORIDE □ TACITIN

B

SAFETY PROFILE: Poison by intravenous route. Moderately toxic by ingestion. Experimental teratogenic effects. A sedative and muscle relaxant. When heated to decomposition it emits very toxic fumes of NO_x and HCl^-.

BCI500   CAS: 135-87-5   HR: 3
BENZODIOXANE HYDROCHLORIDE
mf: $C_{14}H_{19}NO_2 \cdot ClH$ mw: 269.80
PROP: Solid. Mp: 232–236°.
SYNS: BENODAINE HYDROCHLORIDE □ 1-(1,4-BENZODIOXAN-2-YLMETHYL)PIPERIDINE-HYDROCHLORIDE □ F 933 □ FOURNEAU 933 □ 2-PIPERIDINOMETHYL-1,4-BENZODIOXAN HYDROCHLORIDE □ 2-(1-PIPERIDYLMETHYL)-1,4-BENZODIOXAN HYDROCHLORIDE □ PIPEROXANE HYDROCHLORIDE
SAFETY PROFILE: Poison by intraperitoneal and intravenous routes. Moderately toxic by ingestion and subcutaneous routes. Experimental reproductive effects. When heated to decomposition it emits very toxic fumes of NO_x and HCl.

BCJ000   CAS: 5208-87-7   HR: 3
1,3-BENZODIOXOLE-5-(2-PROPEN-1-OL)
mf: $C_{10}H_{10}O_3$ mw: 178.20
SYNS: 1'-HYDROXYSAFROLE □ 1,2-METHYLENEDIOXY-4-(1-HYDROXYALLYL)BENZENE □ α-VINYLPIPERONYL ALCOHOL
SAFETY PROFILE: Suspected carcinogen with experimental carcinogenic, neoplastigenic, and tumorigenic data. Human mutation data reported. When heated to decomposition it emits acrid smoke and irritating fumes.

BCJ005   CAS: 22791-33-9   HR: 3
1,3-BENZODIOXOL-4-OL, 2,2-DIMETHYL-, ACETYLMETHYL-CARBAMATE
mf: $C_{13}H_{15}NO_5$ mw: 265.29
SYNS: ACETYLMETHYLCARBAMIC ACID 2,2-DIMETHYL-1,3-BENZODIOXOL-4-YL ESTER □ CARBAMIC ACID, ACETYLMETHYL-, 2,2-DIMETHYL-1,3-BENZODIOXOL-4-YL ESTER □ CARBAMIC ACID, ACETYLMETHYL-, 2,3-(ISOPROPYLIDENE-DIOXY)PHENYL ESTER □ NC-6897, ACETYL DERIVATIVE

SAFETY PROFILE: A poison by ingestion. When heated to decomposition it emits toxic vapors of NO_x.

BCJ150   CAS: 54531-52-1   HR: 3
BENZODOL
mf: $(C_6H_7AsO_4 \cdot CH_2O)_n$
PROP: Sol in water, alc, and NaOH.
SYNS: ARSONIC ACID, (4-HYDROXYPHENYL)-, polymer with FORMALDEHYDE □ (4-HYDROXY-PHENYL)ARSONIC ACID polymer with FORMALDEHYDE □ POLYBENZARSOL
OSHA PEL: TWA 0.5 mg(As)/m³
ACGIH TLV: BEI: 35 μ (As)/L inorganic arsenic and methylated metabolites in urine
SAFETY PROFILE: Poison by intraperitoneal route. Low toxicity by ingestion. When heated to decomposition it emits toxic fumes of As.

BCJ250   CAS: 205-82-3   HR: 3
BENZO(j)FLUORANTHENE
mf: $C_{20}H_{12}$ mw: 252.32
PROP: Yellow crystals from EtOH. Mp: 165°, bp: 240–260° @ 2 mm.
SYNS: 10,11-BENZFLUORANTHENE □ BENZ(j)-FLUOROANTHRENE □ BENZO(1)FLUORANTHENE □ 7,8-BENZOFLUORANTHENE □ B(j)F □ DIBENZO-(a,jk)FLUORENE
CONSENSUS REPORTS: NTP 10th Report on Carcinogens. IARC Cancer Review: Group 2B IMEMDT 7,56,87; Animal Limited Evidence IMEMDT 3,82,73; Animal Sufficient Evidence IMEMDT 32,155,83.
SAFETY PROFILE: Confirmed carcinogen with experimental carcinogenic, neoplastigenic, and tumorigenic data. Mutation data reported. When heated to decomposition it emits acrid smoke and irritating fumes.

BCJ280   CAS: 207-08-9   HR: 3
BENZO(k)FLUORANTHENE
mf: $C_{20}H_{12}$ mw: 252.32
PROP: Yellow prisms from C_6H_6 or AcOH. Mp: 217°, bp: 480°.
SYNS: 8,9-BENZOFLUORANTHENE □ 11,12-BENZOFLUORANTHENE □ 11,12-

BENZO(k)FLUORANTHENE □ 2,3,1',8'-BINAPHTHYLENE □ DIBENZO(b,jk)FLUORENE

CONSENSUS REPORTS: NTP 10th Report on Carcinogens. IARC Cancer Review: Group 2B IMEMDT 7,56,87; Animal Sufficient Evidence IMEMDT 32,163,83; Human No Adequate Data IMEMDT 32,163,83.

SAFETY PROFILE: Confirmed carcinogen with experimental tumorigenic data. Mutation data reported. When heated to decomposition it emits acrid smoke and irritating fumes.

BCK250   CAS: 271-89-6   HR: 2
BENZOFURAN
mf: C_8H_6O mw: 118.14
PROP: Liquid. D: 1.078° @ 15°/15°, bp: 166.5–168° @ 735 mm.
SYNS: BENZO(b)FURAN □ 2,3-BENZOFURAN □ BENZOFURFURAN □ COUMARONE □ NCI-C56166 □ 1-OXINDENE

CONSENSUS REPORTS: Reported in EPA TSCA Inventory. NTP Carcinogenesis Studies (gavage): Clear Evidence: mouse NTPTR* NTP-TR-370,89; (gavage): Some Evidence: rat NTPTR* NTP-TR-370,89. EPA TSCA Chemical Inventory.

SAFETY PROFILE: Confirmed carcinogen with experimental carcinogenic data reported. Moderately toxic by intraperitoneal route. Mutation data reported. When heated to decomposition it emits acrid smoke and fumes.

BCL750   CAS: 65-85-0   HR: 2
BENZOIC ACID
mf: $C_7H_6O_2$ mw: 122.13
PROP: White crystalline powder, leaflets, or needles from water. Mp: 122°, bp: 249°, flash p: 250°F (CC), d: 1.316, autoign temp: 1060°F, vap press: 1 mm @ 96.0° (sublimes), vap d: 4.21. Very sltly sol in water; sol in alc, ether, chloroform, and fixed oils.
SYNS: ACIDE BENZOIQUE (FRENCH) □ BENZENE-CARBOXYLIC ACID □ BENZENEFORMIC ACID □ BENZENEMETHANOIC ACID □ BENZOATE □ BENZOESAEURE (GERMAN) □ BENZOIC ACID (DOT)

□ CARBOXYBENZENE □ DRACYLIC ACID □ KYSELINA BENZOOVA (CZECH) □ PHENYL CARBOXYLIC ACID □ PHENYLFORMIC ACID □ RETARDER BA □ RETARDEX □ SALVO LIQUID □ SALVO POWDER □ TENN-PLAS

CONSENSUS REPORTS: Reported in EPA TSCA Inventory. EPA Genetic Toxicology Program.

SAFETY PROFILE: Moderately toxic by ingestion, subcutaneous, and intraperitoneal routes. A severe eye irritant. A human skin and severe eye irritant. Mutation data reported. Combustible when exposed to heat or flame; can react with oxidizing materials. The powder burns rapidly in oxygen. To fight fire, use water, CO_2, water spray or mist, dry chemical. When heated to decomposition it emits acrid smoke and irritating fumes.

BCM000   CAS: 120-51-4   HR: 2
BENZOIC ACID, BENZYL ESTER
mf: $C_{14}H_{12}O_2$ mw: 212.26
PROP: Leaflets found in Peru and tolu balsams, in ylang-ylang, and in about 20 other essential oils (FCTXAV 11,1011,73). Colorless oily liquid; slt aromatic odor. Mp: 21°, bp: 324°, flash p: 298°F (CC), d: 1.116, refr index: 1.568, vap d: 7.3, autoign temp: 898°F. Misc with alc, chloroform, ether; insol in glycerin, water.
SYNS: ASCABIN □ ASCABIOL □ BENYLATE □ BENZOIC ACID, PHENYLMETHYL ESTER □ BENZYL ALCOHOL BENZOIC ESTER □ BENZYL BENZENE-CARBOXYLATE □ BENZYL BENZOATE (FCC) □ BENZYLETS □ BENZYL PHENYLFORMATE □ COLEBENZ □ FEMA No. 2138 □ NOVOSCABIN □ PERUSCABIN □ SCABANCA □ VANZOATE □ VENZONATE

CONSENSUS REPORTS: Reported in EPA TSCA Inventory.

SAFETY PROFILE: Moderately toxic by ingestion and skin contact. Combustible liquid. Can react with oxidizing materials. To fight fire, use CO_2, water spray or mist, dry chemical. When heated to decomposition it emits acrid and irritating fumes and smoke. See also ESTERS.

BCP250   CAS: 119-53-9   HR: 3
BENZOIN
mf: $C_{14}H_{12}O_2$ mw: 212.26
PROP: Externally reddish yellow, internally milky white tree resin with agreeable vanilla-like odor.
SYNS: ACETOPHENONE, 2-HYDROXY-2-PHENYL- □ BENZOYLPHENYLCARBINOL □ BITTER ALMOND OIL CAMPHOR □ ETHANONE, 2-HYDROXY-1,2-DIPHENYL- □ FENYL-α-HYDROXYBENZYLKETON □ α-HYDR-OXYBENZYL PHENYL KETONE □ α-HYDROXY-α-PHENYLACETOPHENONE □ 2-HYDROXY-2-PHENYL-ACETOPHENONE □ KETONE, α-HYDROXYBENZYL PHENYL □ NCI-C50011 □ WY-42956
CONSENSUS REPORTS: NCI Carcinogenesis Bioassay (feed); No Evidence: mouse, rat NCITR* NCI-CG-TR-204,80. Reported in EPA TSCA Inventory.
DOT CLASSIFICATION: 3; Label: Flammable Liquid
SAFETY PROFILE: Slightly toxic by ingestion and skin contact. Mutation data reported. A flammable liquid. When heated to decomposition it emits acrid smoke and irritating fumes. See also KETONES.

BCP690   CAS: 190133-94-9 HR: 3
5H-BENZO(d)NAPHTH(2,1-B)AZEPIN-12-OL, 11-CHLORO-6,6A,7,8,9,13b-HEXAHYDRO-7-METHYL-, HYDROCHLORIDE, (6as,13br)-
mf: $C_{19}H_{20}ClNO \cdot ClH$ mw: 350.29
SYN: (–)-SCH 39166 HYDROCHLORIDE
SAFETY PROFILE: A poison by intramuscular route. When heated to decomposition it emits toxic vapors of NO_x, HCl, and Cl^-.

BCQ250   CAS: 100-47-0   HR: 3
BENZONITRILE
DOT: UN 2224
mf: C_7H_5N mw: 103.13
PROP: Transparent, colorless oil; almond-like odor. D: 1.246 @ 20°/4°, bp: 191°, mp: −12.8°.
SYNS: BENZENENITRILE □ BENZOIC ACID NITRILE □ BENZONITRILE (DOT) □ CYANOBENZENE □ FENYLKYANID □ PHENYL CYANIDE

CONSENSUS REPORTS: Reported in EPA TSCA Inventory. Cyanide and its compounds are on the Community Right-To-Know List.
DOT CLASSIFICATION: 6.1; Label: Poison
SAFETY PROFILE: Poison by intraperitoneal and subcutaneous routes. Moderately toxic by ingestion, inhalation, and skin contact. See also NITRILES. A skin irritant. Combustible liquid. When heated to decomposition it emits toxic fumes of CN^- and NO_x.

BCQ500   CAS: 189-55-9   HR: 3
BENZO(rst)PENTAPHENE
mf: $C_{24}H_{14}$ mw: 302.38
PROP: Green-yellow needles from toluene. Mp: 280–282°, bp: 275° @ 0.05 mm (subl).
SYNS: DB(a,i)P □ DIBENZO(a,i)PYRENE □ DIBENZO(b,h)PYRENE □ 1,2,7,8-DIBENZOPYRENE □ 3,4:9,10-DIBENZOPYRENE □ DIBENZ(a,i)PYRENE □ 1,2:7,8-DIBENZPYRENE □ 3,4:9,10-DIBENZPYRENE □ RCRA WASTE NUMBER U064
CONSENSUS REPORTS: NTP 10th Report on Carcinogens. IARC Cancer Review: Group 2B IMEMDT 7,56,87; Animal Sufficient Evidence IMEMDT 3,215,73; IMEMDT 32,337,83. EPA Genetic Toxicology Program.
SAFETY PROFILE: Confirmed carcinogen with experimental neoplastigenic and tumorigenic data. Mutation data reported. When heated to decomposition it emits acrid smoke and irritating fumes.

BCS250   CAS: 119-61-9   HR: 3
BENZOPHENONE
mf: $C_{13}H_{10}O$ mw: 182.23
PROP: Rhombic prisms (stable form), monoclinic prisms (labile form), white crystals; persistent rose-like odor. Mp (α): 49°, mp (β): 26°, mp (γ): 47°, bp: 305.4°, d (α): 1.0976 @ 50°/50°, d (β): 1.108 @ 23°/40°, vap press: 1 mm @ 108.2. Sol in fixed oils; sltly sol in propylene glycol; insol in glycerin.
SYNS: BENZOYLBENZENE □ DIPHENYL KETONE □ DIPHENYLMETHANONE □ FEMA No. 2134 □ α-OXODIPHENYLMETHANE □ PHENYL KETONE

CONSENSUS REPORTS: Reported in EPA TSCA Inventory.

DOT CLASSIFICATION: 3; Label: Flammable Liquid

SAFETY PROFILE: Moderately toxic by ingestion and intraperitoneal routes. Combustible when heated. Incompatible with oxidizers. When heated to decomposition it emits acrid and irritating fumes. See also KETONES.

BCS750   CAS: 50-32-8   HR: 3
BENZO(a)PYRENE

mf: $C_{20}H_{12}$ mw: 252.32

PROP: Pale-yellow crystals. Mp: 177°, bp: 312° @ 10 mm. Insol in water; sol in benzene, toluene, and xylene.

SYNS: BENZO(d,e,f)CHRYSENE □ 3,4-BENZOPIRENE (ITALIAN) □ 3,4-BENZOPYRENE □ 6,7-BENZOPYRENE □ BENZ(a)PYRENE □ 3,4-BENZPYREN (GERMAN) □ 3,4-BENZ(a)PYRENE □ 3,4-BENZYPYRENE □ B(a)P □ RCRA WASTE NUMBER U022

CONSENSUS REPORTS: NTP 10th Report on Carcinogens. IARC Cancer Review: Group 2A IMEMDT 7,56,87; Animal Sufficient Evidence IMEMDT 32,211,83; IMEMDT 3,91,73. Reported in EPA TSCA Inventory.

OSHA PEL: TWA 0.2 mg/m³

SAFETY PROFILE: Confirmed carcinogen with experimental carcinogenic, neoplastigenic, and tumorigenic data. A poison via subcutaneous, intraperitoneal, and intrarenal routes. Experimental teratogenic and reproductive effects. Human mutation data reported. A skin irritant. A common air contaminant of water, food, and smoke. When heated to decomposition it emits acrid smoke and fumes. See other benzopyrenes.

BCV250   CAS: 21247-98-3   HR: 3
BENZO(a)PYRENE-6-METHANOL

mf: $C_{21}H_{14}O$ mw: 282.35

PROP: Pale-yellow crystals from C_6H_6. Mp: 270–271°.

SYN: 6-HYDROXYMETHYLBENZO(a)PYRENE

CONSENSUS REPORTS: EPA Genetic Toxicology Program.

SAFETY PROFILE: Suspected carcinogen with experimental carcinogenic, neoplastigenic, and tumorigenic data. Mutation data reported. When heated to decomposition it emits acrid smoke and fumes.

BDC250   CAS: 583-63-1   HR: 3
o-BENZOQUINONE

mf: $C_6H_4O_2$ mw: 108.10

PROP: Solid. Mp: 60–70° (decomp).

SYNS: 1,2-BENZOQUINONE □ BENZOQUINONE (DOT) □ 3,5-CYCLOHEXADIENE-1,2-DIONE □ o-QUINONE

SAFETY PROFILE: A poison. Mutation data reported. When heated to decomposition it emits acrid smoke and irritating fumes.

BDE750   CAS: 120-78-5   HR: 3
BENZOTHIAZOLE DISULFIDE

mf: $C_{14}H_8N_2S_4$ mw: 332.48

PROP: Cream to pale-yellow powder. Mp: 186°, d: 1.5.

SYNS: ALTAX □ BENZOTHIAZOLYL DISULFIDE □ 2-BENZOTHIAZOLYL DISULFIDE □ BIS(BENZOTHIAZ-OLYL)DISULFIDE □ BIS(2-BENZOTHIAZYL) DISULFIDE □ DI-2-BENZOTHIAZOLYLDISULFIDE □ DIBENZOTHIAZYL DISULFIDE □ 2,2'-DIBENZO-THIAZYLDISULFIDE □ DIBENZOYL-THIAZYL DISULFIDE □ DIBENZTHIAZYL DISULFIDE □ 2,2'-DITHIOBIS(BENZOTHIAZOLE) □ DWUSIARCZEK DWUBENZOTIAZYLU (POLISH) □ MBTS □ MBTS RUBBER ACCELERATOR □ 2-MERCAPTOBENZO-THIAZOLEDISULFIDE □ 2-MERCAPTOBENZO-THIAZYLDISULFIDE □ ROYAL MBTS □ THIOFIDE □ USAF B-33 □ USAF CY-5 □ USAF EK-5432 □ VULKACIT DM □ VULKACIT DM/MGC

CONSENSUS REPORTS: Reported in EPA TSCA Inventory.

SAFETY PROFILE: Poison by intravenous and intraperitoneal routes. Slightly toxic by ingestion. Experimental teratogenic and reproductive effects. Questionable carcinogen with experimental tumorigenic data. Mutation data reported. When heated to decomposition it emits very toxic fumes of SO_x and NO_x. See also SULFIDES.

B

BDF000   CAS: 149-30-4   HR: 3
2-BENZOTHIAZOLETHIOL
mf: $C_7H_5NS_2$ mw: 167.25
PROP: Light-yellow powder or needles from
MeOH (aq). Mp: 177–179°. Sltly sol in
EtOH, Et_2O, and AcOH; insol in H_2O; sol
in alkalies.
SYNS: BENZOTHIAZOLE-2-THIONE □ 2(3H)-
BENZOTHIAZOLETHIONE □ 2-BENZOTHIAZOLYL
MERCAPTAN □ CAPTAX □ KAPTAX □ MBT □
MERCAPTOBENZOTHIAZOLE □ 2-MERCAPTO-
BENZOTHIAZOLE □ 2-MERKAPTOBENZOTIAZOL □
2-MERKAPTOBENZTHIAZOL □ NCI-C56519 □ PENNAC
MBT POWDER □ ROKON □ ROTAX □ SULFADENE □
USAF GY-3 □ USAF XR-29 □ VULKACIT MERCAPTO
CONSENSUS REPORTS: NTP
Carcinogenesis Studies (gavage); Some
Evidence: rat NTPTR* NTP-TR-332,88:
(gavage); Equivocal Evidence: mouse
NTPTR* NTP-TR-332,88. Reported in
EPA TSCA Inventory.
SAFETY PROFILE: Suspected carcinogen
with experimental carcinogenic and
tumorigenic data. Poison by ingestion and
intraperitoneal routes. Experimental
teratogenic and reproductive effects.
Mutation data reported. Incompatible with
oxidizers. When heated to decomposition or
on contact with acids or acid fumes it emits
toxic SO_x and NO_x. See also
MERCAPTANS.

BDG000   CAS: 102-77-2   HR: 3
2-BENZOTHIAZOLYL-N-
**   MORPHOLINOSULFIDE**
mf: $C_{11}H_{12}N_2OS_2$ mw: 252.37
SYNS: AMAX □ 2-BENZOTHIAZOLYLSULFENYL
MORPHOLINE □ 4-(2-BENZOTHIAZOLYLTHIO)-
MORPHOLINE □ 2-(MORPHOLINOTHIO)-
BENZOTHIAZOLE □ MORPHOLINYLMERCAPTO-
BENZOTHIAZOLE □ 2-(4-MORPHOLINYLTHIO)-
BENZOTHIAZOLE □ N-(OXYDIETHYLENE)-
BENZOTHIAZOLE-2-SULFEN-AMIDE □ SANTOCURE
MOR □ SULFENAMIDE M □ USAF CY-7 □ VULCAFOR
BSM
CONSENSUS REPORTS: Reported in EPA
TSCA Inventory.
SAFETY PROFILE: Poison by
intraperitoneal route. Moderately toxic by
ingestion. Questionable carcinogen with
experimental neoplastigenic data.

Experimental teratogenic effects. An eye
irritant. Mutation data reported. See also
MERCAPTANS and SULFIDES. When
heated to decomposition it emits very toxic
fumes of NO_x and SO_x.

BDH000   CAS: 90-16-4   HR: 3
1,2,3-BENZOTRIAZIN-4(1H)-ONE
mf: $C_7H_5N_3O$ mw: 147.15
PROP: Needles from cyclohexane.
SYNS: BENZAZIMIDE □ BENZAZIMIDONE □
BENZOKETCTRIAZINE □ 3H-1,2,3-BENZOTRIAZIN-4-
ONE □ 4-KETOBENZOTRIAZINE □ USAF MA-2
CONSENSUS REPORTS: Reported in EPA
TSCA Inventory.
SAFETY PROFILE: Poison by
intraperitoneal route. Experimental
reproductive effects. When heated to
decomposition it emits toxic fumes of NO_x.

BDH250   CAS: 95-14-7   HR: 3
1H-BENZOTRIAZOLE
mf: $C_6H_5N_3$ mw: 119.14
PROP: Needles from C_6H_6. Mp: 100°. Sol in
C_6H_6.
SYNS: 1,2-AMINOZOPHENYLENE □ AZIMIDO-
BENZENE □ AZIMINOBENZENE □ BENZENE
AZIMIDE □ BENZISOTRIAZOLE □ 1,2,3-BENZO-
TRIAZOLE □ COBRATEC #99 □ 2,3-DIAZAINDOLE □
NCI-C03521 □ NSC-3058 □ 1,2,3-TRIAZAINDENE □ U-
6233
CONSENSUS REPORTS: NCI
Carcinogenesis Bioassay (feed); Inadequate
Studies: mouse, rat NCITR* NCI-CG-TR-
88,78. Reported in EPA TSCA Inventory.
SAFETY PROFILE: Poison by intravenous
route. Moderately toxic by ingestion and
intraperitoneal routes. Questionable
carcinogen with experimental tumorigenic
data. Mutation data reported. May detonate
at 220°C or during vacuum distillation.
When heated to decomposition it emits
toxic fumes of NO_x.

BDH500   CAS: 98-08-8   HR: 3
BENZOTRIFLUORIDE
DOT: UN 2338
mf: $C_7H_5F_3$ mw: 146.12

PROP: Water-white liquid; aromatic odor. Mp: $-29.1°$, bp: $98-99°$ @ 725 mm, flash p: $54°F$ (CC), d: 1.197 @ $15.5°/15.5°$, vap d: 5.04, vap press: 11 mm @ $0°$.

SYNS: BENZENYL FLUORIDE □ BENZYLIDYNE FLUORIDE □ PHENYLFLUOROFORM □ (TRIFLUORO-METHYL)BENZENE □ α,α,α-TRIFLUOROTOLUENE □ ω-TRIFLUOROTOLUENE □ USAF MA-16

CONSENSUS REPORTS: Reported in EPA TSCA Inventory.

DOT CLASSIFICATION: 3; Label: Flammable Liquid

SAFETY PROFILE: Poison by intraperitoneal route. Moderately toxic by subcutaneous route. See also FLUORIDES. Dangerous fire hazard. To fight fire, use water, foam, CO_2, spray mist, dry chemical. When heated to decomposition it emits toxic fumes of F^-. Incompatible with oxidizing materials.

BDJ250   CAS: 2310-17-0   HR: 3
S-((3-BENZOXAZOLINYL-6-CHLORO-2-OXO)METHYL) O,O-DIETHYLPHOS-PHORODITHIOATE

mf: $C_{12}H_{15}ClNO_4PS_2$ mw: 367.82

PROP: Crystals with garlic odor. Mp: $47.5-48.0°$. Insol in H_2O and hydrocarbons.

SYNS: AZOFENE □ BENZOPHOSPHATE □ BENZPHOS □ CHIPMAN 11974 □ S-(6-CHLORO-3-(MERCAPTOMETHYL)-2-BENZOXAZOLINONE)-O,O-DIETHYL PHOSPHORODITHIOATE □ 3-(6-CHLORO-2-OXOBENZOXAZOLIN-3-YL)METHYL-O,O-DIETHYL PHOSPHOROTHIOLOTHIONATE □ O,O-DIAETHYL-S-(6-CHLOR-2-OXO-BEN(b)-1,3-OXALIN-3-YL)-METHYL-DIT HIOPHOSPHAT (GERMAN) □ O,O-DIETHYL-S-((6-CHLOOR-2-OXO-BENZOXAZOLIN-3-YL)-METHYL)-DITHIO FOSFAAT (DUTCH) □ O,O-DIETHYL-S-(6-CHLOROBENZOXAZOLINYL-3-METHYL)DITHIOPH-OSPHATE □ O,O-DIETHYL-S-((6-CHLORO-2-OXOBENZOXAZOLIN-3-YL)METHYL) PHOSPHORO-DITHIOATE □ O,O-DIETHYL-S-(6-CHLORO-2-OXO-BENZOXAZOLIN-3-YL)METHYL-PHOSPHORO THIOLOTHIONATE □ 3-DIETHYLDITHIOPHOS-PHORYLMETHYL-6-CHLOROBENZOXAZOLONE-2 □ O,O-DIETIL-S-((6-CLORO-2-OXO-BENZOSSAZOLIN-3-IL)-METIL)-DITIOFOSFATO (ITALIAN) □ ENT 27,163 □ FOZALON □ NIA-9241 □ NIAGARA 9241 □ NPH-1091 □ PHASOLON □ PHOSALON □ PHOSALONE □ PHOZALON □ RHODIA RP 11974 □ RUBITOX □ ZOLON □ ZOLONE □ ZOLONE PM □ ZOOLON

CONSENSUS REPORTS: EPA: Farm Worker Field Reentry FEREAC 39,16888,74.

SAFETY PROFILE: Poison by ingestion, skin contact, and possibly other routes. A cholinesterase inhibitor. See also PARATHION. When heated to decomposition it emits very toxic fumes of Cl^-, NO_x, PO_x, and SO_x.

BDL750   CAS: 582-61-6   HR: 3
BENZOYL AZIDE

mf: $C_7H_5N_3O$ mw: 147.14

SYNS: BENZAZIDE □ BENZOIC ACID AZIDE

DOT CLASSIFICATION: Forbidden

SAFETY PROFILE: May explode when heated above $120°C$. See also AZIDES.

BDL860   CAS: 611-95-0   HR: 3
p-BENZOYLBENZOIC ACID

mf: $C_{14}H_{10}O_3$ mw: 226.23

SYN: BENZOIC ACID, 4-BENZOYL-

SAFETY PROFILE: A poison by intravenous route. When heated to decomposition it emits acrid smoke and irritating vapors.

BDM500   CAS: 98-88-4   HR: 3
BENZOYL CHLORIDE

DOT: UN 1736

mf: C_7H_5ClO mw: 140.57

PROP: Colorless, fuming, pungent liquid; decomposes in water. Fp: $-1°$, mp: $-0.5°$, bp: $197°$, flash p: $162°F$ (CC), d: 1.22 @ $15°/15°$, vap press: 1 mm @ $32.1°$, vap d: 4.88.

SYNS: BENZENECARBONYL CHLORIDE □ BENZOIC ACID, CHLORIDE □ BENZOYL CHLORIDE (DOT) □ α-CHLOROBENZALDEHYDE

CONSENSUS REPORTS: IARC Cancer Review: Group 3 IMEMDT 7,56,87; Human Inadequate Evidence IMEMDT 29,83,82; Animal Inadequate Evidence IMEMDT 29,83,82. Community Right-To-Know List. Reported in EPA TSCA Inventory. EPA Genetic Toxicology Program.

DFG MAK: Confirmed Human Carcinogen

DOT CLASSIFICATION: 8; Label: Corrosive

B

SAFETY PROFILE: Confirmed carcinogen with experimental tumorigenic data by skin contact. Human systemic effects by inhalation: unspecified effects on olfaction and respiratory systems. Corrosive effects on the skin, eyes, and mucous membranes by inhalation. Flammable when exposed to heat or flame. Will react with water or steam to produce heat and toxic and corrosive fumes. Violent or explosive reaction with dimethyl sulfoxide, and aluminum chloride + naphthalene. To fight fire, use alcohol foam, CO_2, dry chemical. Incompatible with dimethyl sulfoxide, $(NaN_3 + KOH)$, water, steam, and oxidizers. When heated to decomposition it emits toxic fumes of Cl^-. See also CHLORIDES and ALDEHYDES.

BDN600 CAS: 110690-43-2 HR: 3
3-(3-(6-BENZOYLOXY-3-CYANO-2-PYRIDYLOXYCARBONYL)BENZOYL)-1-ETHOXYMETHYL-5-FLUOROURACIL

mf: $C_{28}H_{19}FN_4O_8$ mw: 558.51
SYNS: BENZOIC ACID, 3-((3-(ETHOXYMETHYL)-5-FLUORO-3,6-DIHYDRO-2,6-DIOXO-1(2H)-PYRIMIDINYL)CARB ONYL)-, □ BOF-A2 □ EMITEFUR
SAFETY PROFILE: A poison by ingestion and subcutaneous routes. Experimental reproductive effects. Mutation data reported. When heated to decomposition it emits toxic vapors of NO_x and F^-.

BDR750 CAS: 4342-36-3 HR: 3
BENZOYLOXYTRIBUTYLSTANNANE

mf: $C_{19}H_{32}O_2Sn$ mw: 411.20
SYNS: TRIBUTYLTIN BENZOATE □ TRI-N-BUTYL-ZINN BENZOATE (GERMAN)
CONSENSUS REPORTS: Reported in EPA TSCA Inventory.
OSHA PEL: TWA 0.1 mg(Sn)/m³ (skin)
ACGIH TLV: TWA 0.1 mg(Sn)/m³; STEL 0.2 mg(Sn)/m³ (skin).
DFG MAK: 0.0021 ppm (0.05 mg/m³)
NIOSH REL: (Organotin Compounds) TWA 0.1 mg(Sn)/m³
SAFETY PROFILE: Poison by ingestion and intravenous routes. Moderately toxic by subcutaneous route. See also TIN

COMPOUNDS. When heated to decomposition it emits acrid smoke and irritating fumes.

BDS000 CAS: 94-36-0 HR: 3
BENZOYL PEROXIDE

mf: $C_{14}H_{10}O_4$ mw: 242.24
PROP: White, granular, tasteless, odorless powder or prisms. Mp: 106–108.6° (decomp), bp: decomposes explosively, autoign temp: 176°F. Sol in benzene, acetone, chloroform; sltly sol in alc; insol in water. IDLH 1500 mg/m³.
SYNS: ACETOXYL □ ACNEGEL □ AZTEC BPO □ BENOXYL □ BENZAC □ BENZAKNEW □ BENZOIC ACID, PEROXIDE □ BENZOPEROXIDE □ BENZOYL □ BENZOYLPEROXID (GERMAN) □ BENZOYLPEROXY-DE (DUTCH) □ BENZOYL SUPEROXIDE □ BZF-60 □ CADET □ CADOX □ CLEARASIL BENZOYL PEROXIDE LOTION □ CLEARASIL BP ACNE TREATMENT □ CUTICURA ACNE CREAM □ DEBROXIDE □ DIBENZOYLPEROXID (GERMAN) □ DIBENZOYL PEROXIDE (MAK) □ DIBENZOYLPEROXYDE (DUTCH) □ DIPHENYLGLYOXAL PEROXIDE □ DRY AND CLEAR □ EPI-CLEAR □ FOSTEX □ GAROX □ INCIDOL □ LOROXIDE □ LUCIDOL □ LUPERCO □ LUPEROX FL □ NAYPER B and BO □ NOROX BZP-250 □ NOVADELOX □ OXY-5 □ OXY-10 □ OXYLITE □ OXY WASH □ PANOXYL □ PEROSSIDO di BENZOILE (ITALIAN) □ PEROXYDE de BENZOYLE (FRENCH) □ PERSADOX □ QUINOLOR COMPOUND □ SULFOXYL □ SUPEROX □ THERADERM □ TOPEX □ VANOXIDE □ XERAC
CONSENSUS REPORTS: IARC Cancer Review: Group 3 IMEMDT 7,56,87; Animal Inadequate Evidence IMEMDT 36,267,85; Human Inadequate Evidence IMEMDT 36,267,85. Reported in EPA TSCA Inventory. EPA Genetic Toxicology Program. Community Right-To-Know List.
OSHA PEL: TWA 5 mg/m³
ACGIH TLV: TWA 5 mg/m³; Not Classifiable as a Human Carcinogen
DFG MAK: 5 mg/m³; Weak allergin and skin irritant
NIOSH REL: (Benzoyl Peroxide) TWA 5 mg/m³
SAFETY PROFILE: Poison by intraperitoneal route. Can cause dermatitis, asthmatic effects, testicular atrophy, and vasodilation. An allergen and eye irritant.

Human mutation data reported. Questionable carcinogen with experimental tumorigenic data. Moderate fire hazard by spontaneous chemical reaction in contact with reducing agents. It ignites readily and burns rapidly. A powerful oxidizer. Dangerous explosion hazard; may explode spontaneously when heated to above melting point, or when overheated under confinement. It is moderately sensitive to heat, shock, friction, or contact with combustible materials. Explosive decomposition above the mp (103°) forms flammable products.

Explosive or violent reaction on contact with N,N-dimethylaniline, aniline, dimethyl sulfide, lithium tetrahydroaluminate, and N-bromosuccinimide + 4-toluic acid. Mixture with carbon tetrachloride + ethylene explodes at elevated temperatures and pressures. Reacts violently in contact with various organic or inorganic acids, alcohols, amines, metallic naphthenates, as well as with polymerization accelerators, e.g., dimethylaniline, and (CCl_4 + C_2H_4). Violent reaction with charcoal when heated above 50°. Decomposition produces dense white smoke of benzoic acid, phenyl benzoate, terphenyls, biphenyls, benzene, and carbon dioxide. Vigorous reaction leading to ignition with methylmethacrylate, and vinyl acetate + ethyl acetate. To fight fire, use water spray, foam. All precautions must be taken to guard against fire and explosion hazards. Keep in a cool place, out of the direct rays of the sun, away from sparks, open flames, and other sources of heat, avoid shock, rough handling, friction from grinding, etc. Isolated storage is required; keep away from possible contact with acids, alcohols, ethers, or other reducing agents or polymerization catalysts such as dimethylaniline. Complete instructions on storage and handling available from manufacturer. See also PEROXIDES.

BDU500   CAS: 22071-15-4   HR: 3
2-(m-BENZOYLPHENYL)PROPIONIC

ACID
mf: $C_{16}H_{14}O_3$ mw: 254.30
SYNS: ALRHEUMAT □ ALRHEUMUM □ m-BENZOYLHYDRATROPIC ACID □ 3-BENZOYLHYDRATROPIC ACID □ 2-(3-BENZOYL-PHENYL)PROPIONIC ACID □ CAPISTEN □ FASTUM □ ISO-K □ KEFENID □ KETOPROFEN □ KETOPRON □ LERTUS □ MEPROFEN □ ORUDIS □ ORUVAIL □ PROFENID □ 19583 RP
CONSENSUS REPORTS: Reported in EPA TSCA Inventory.
SAFETY PROFILE: Poison by ingestion, subcutaneous, intravenous, rectal, and intraperitoneal routes. Human systemic effects by an unspecified route: headache, nausea or vomiting, and degenerative changes in the brain, changes in kidney tubules. An experimental teratogen. Other experimental reproductive effects. When heated to decomposition it emits acrid smoke and irritating fumes. An anti-inflammatory and analgesic agent.

BDX000   CAS: 140-11-4   HR: 3
BENZYL ACETATE
mf: $C_9H_{10}O_2$ mw: 150.19
PROP: Colorless liquid; sweet, floral fruity odor. Mp: −51.5°, bp: 134° @ 102 mm, flash p: 216°F (CC), d: 1.06, autoign temp: 862°F, vap press: 1 mm @ 45°, vap d: 5.1, refr index: 1.501. Sol in alc, most fixed oils, propylene glycol; insol in glycerin and water @ 214°.
SYNS: ACETIC ACID BENZYL ESTER □ ACETIC ACID PHENYLMETHYL ESTER □ α-ACETOXYTOLUENE □ BENZYL ETHANOATE □ FEMA No. 2135 □ NCI-C06508
CONSENSUS REPORTS: IARC Cancer Review: Group 3 IMEMDT 7,56,87; Animal Limited Evidence IMEMDT 40,109,86. NTP Carcinogenesis Studies (gavage); Some Evidence: mouse, rat NTPTR* NTP-TR-250,86. Reported in EPA TSCA Inventory.
ACGIH TLV: TWA 10 ppm; Not Classifiable as a Human Carcinogen
SAFETY PROFILE: A poison by inhalation. Moderately toxic by ingestion and subcutaneous routes. Human systemic effects by inhalation: an antipsychotic, unspecified respiratory and urinary system

effects. Questionable carcinogen with experimental tumorigenic data. Combustible liquid. To fight fire, use alcohol foam, CO_2. When heated to decomposition it emits irritating fumes. See also ESTERS.

BDX500   CAS: 100-51-6   HR: 3
BENZYL ALCOHOL
mf: C_7H_8O mw: 108.15
PROP: Found in jasmine, hyacinth, ylang-ylang oils, and at least two dozen other essential oils (FCTXAV 11,1011,73). Water-white liquid; faint, aromatic odor, sharp burning taste. Mp: −15.3°, bp: 205.3°, flash p: 213°F (CC), d: 1.050, autoign temp: 817°F, vap press: 1 mm @ 58.0°, vap d: 3.72, refr index: 1.540. Misc with alc, chloroform, ether, and water @ 206°(decomp). Moderately sol in water.
SYNS: BENZAL ALCOHOL □ BENZENECARBINOL □ BENZENEMETHANOL □ BENZOYL ALCOHOL □ FEMA No. 2137 □ HYDROXYTOLUENE □ α-HYDROXY-TOLUENE □ NCI-C06111 □ PHENOLCARBINOL □ PHENYLCARBINOL □ PHENYLMETHANOL □ PHENYLMETHYL ALCOHOL □ α-TOLUENOL
CONSENSUS REPORTS: EPA Genetic Toxicology Program. Reported in EPA TSCA Inventory.
SAFETY PROFILE: Poison by ingestion, intraperitoneal, intravenous, and parenteral routes. Moderately toxic by inhalation, skin contact, and subcutaneous routes. A moderate skin and severe eye irritant. Mutation data reported. Combustible liquid. Mixtures with sulfuric acid decompose explosively at 180°. Exothermic polymerization is catalyzed by HBr + iron when heated above 100°. To fight fire, use alcohol foam, CO_2, dry chemical. When heated to decomposition it emits acrid smoke and fumes. See also ALCOHOLS.

BDY669   CAS: 61-33-6   HR: 3
BENZYL-6-AMINOPENICILLINIC ACID
mf: $C_{16}H_{18}N_2O_4S$ mw: 334.42
PROP: Crystals.
SYNS: ABBOCILLIN □ (5R,6R)-BENXYLPENICILLIN □ BENZOPENICILLIN □ BENZYLPENICILLIN □ BENZYLPENICILLIN G □ BENZYLPENICILLINIC ACID

□ CILLORAL □ CILOPEN □ COMPOCILLIN G □ COSMOPEN □ DROPCILLIN □ FREE BENZYL-PENICILLIN □ GALOFAK □ GELACILLIN □ LIQUACILLIN □ PENICILLIN G □ PHENYLACET-AMIDOPENICILLANIC ACID □ (PHENYLMETHYL) PENICILLINIC ACID □ PRADUPEN □ SPECILLINE G
CONSENSUS REPORTS: EPA Genetic Toxicology Program.
SAFETY PROFILE: Poison by ingestion, intravenous, intracerebral, intraspinal, subcutaneous, and possibly other routes. Human (child) systemic effects by parenteral route: changes in cochlear (inner ear) structure or function, convulsions, and dyspnea. Questionable carcinogen with experimental tumorigenic data. Mutation data reported. When heated to decomposition it emits very toxic fumes of NO_x and SO_x. See other penicillin entries.

BEC000   CAS: 100-39-0   HR: 2
BENZYL BROMIDE
DOT: UN 1737
mf: C_7H_7Br mw: 171.05
PROP: Clear, refractive liquid; pleasant odor, lachrymator, insol in water. Mp: −4.0°, bp: 198°, d: 1.438 @ 22°/0°, vap d: 5.8.
SYNS: (BROMOMETHYL)BENZENE □ p-(BROMO-METHYL)NITROBENZENE □ BROMOPHENYL-METHANE □ ω-BROMOTOLUENE □ α-BROMOTOLUENE (DOT)
CONSENSUS REPORTS: Reported in EPA TSCA Inventory.
DOT CLASSIFICATION: 6.1; Label: Poison, Corrosive
SAFETY PROFILE: Intensely irritating and corrosive to skin, eyes, and mucous membranes. Large doses cause central nervous system depression. Mutation data reported. Reaction with molecular sieve produces toxic hydrogen bromide gas. See also BROMIDES.

BEC500   CAS: 85-68-7   HR: 2
BENZYL BUTYL PHTHALATE
mf: $C_{19}H_{20}O_4$ mw: 312.39
PROP: Clear, oily liquid. Mp: <−35°, bp: 370°, flash p: 390°F, d: 1.116 @ 25°/25°, vap d: 10.8.

SYNS: BBP □ 1,2-BENZENEDICARBOXYLIC ACID, BUTYL PHENYLMETHYL ESTER □ BUTYL BENZYL PHTHALATE □ n-BUTYL BENZYL PHTHALATE □ NCI-C54375 □ PALATINOL BB □ SANTICIZER 160 □ SICOL 160 □ UNIMOLL BB

CONSENSUS REPORTS: IARC Cancer Review: Group 3 IMEMDT 7,56,87; Animal Inadequate Evidence IMEMDT 29,193,82; NTP Carcinogenesis Bioassay (feed); No Evidence: mouse NTPTR* NTP-TR-213,82; Clear Evidence: rat NTPTR* NTP-TR-213,82. Reported in EPA TSCA Inventory. Community Right-To-Know List.

SAFETY PROFILE: Questionable carcinogen with experimental carcinogenic data. Moderately toxic by ingestion, skin contact, and intraperitoneal routes. Experimental reproductive effects. See also ESTERS. Combustible when exposed to heat or flame; can react with oxidizers. To fight fire, use spray or mist, CO_2, dry chemical. When heated to decomposition it emits acrid smoke and irritating fumes.

BED000   CAS: 103-37-7   HR: 2
BENZYL n-BUTYRATE

mf: $C_{11}H_{14}O_2$ mw: 178.25

PROP: Colorless liquid; floral plum-like odor. D: 1.006, refr index: 1.492, flash p: 212°F. Sol in fixed oils; insol in glycerin, propylene glycol, water @ 239°.

SYNS: BENZYL n-BUTANOATE □ FEMA No. 2140

CONSENSUS REPORTS: Reported in EPA TSCA Inventory.

SAFETY PROFILE: Moderately toxic by ingestion. See also ESTERS. Combustible liquid. When heated to decomposition it emits acrid smoke and irritating fumes.

BEE375   CAS: 100-44-7   HR: 3
BENZYL CHLORIDE

DOT: UN 1738

mf: C_7H_7Cl mw: 126.59

PROP: Colorless liquid, very refractive; irritating, unpleasant odor. Mp: −48°, bp: 99° @ 62 mm, lel: 1.1%, flash p: 153°F, d: 1.11 @ 4°/4°, autoign temp: 1085°F, vap d: 4.36. IDLH 10 ppm.

SYNS: BENZILE (CLORURO di) (ITALIAN) □ BENZYLCHLORID (GERMAN) □ BENZYLE (CHLORURE de) (FRENCH) □ CHLOROMETHYL-BENZENE □ CHLOROPHENYLMETHANE □ α-CHLOROTOLUENE □ ω-CHLOROTOLUENE □ α-CHLORTOLUOL (GERMAN) □ CHLORURE de BENZYLE (FRENCH) □ NCI-C06360 □ RCRA WASTE NUMBER P028 □ TOLYL CHLORIDE

CONSENSUS REPORTS: IARC Cancer Review: Animal Limited Evidence IMEMDT 29,49,82; Animal Sufficient Evidence IMEMDT 11,217,76; Human Inadequate Evidence IMEMDT 29,49,82. EPA Genetic Toxicology Program. Community Right-To-Know List. Reported in EPA TSCA Inventory. EPA Extremely Hazardous Substances List.

OSHA PEL: TWA 1 ppm

ACGIH TLV: TWA 1 ppm; Animal Carcinogen

DFG MAK: Confirmed Human Carcinogen

NIOSH REL: (Benzyl Chloride) CL 5 mg/m^3/15M

DOT CLASSIFICATION: 6.1; Label: Poison, Corrosive

SAFETY PROFILE: Confirmed carcinogen with experimental carcinogenic and tumorigenic data. Poison by inhalation. Moderately toxic by ingestion and subcutaneous routes. Experimental reproductive effects. Human mutation data reported. A corrosive irritant to skin, eyes, and mucous membranes. Flammable and moderately explosive when exposed to heat or flame. Can react vigorously with oxidizing materials. May explode during distillation. The decomposition rate can reach explosive violence in presence of metals such as iron. Catalytic impurities (e.g., aluminum, iron, rust) or sodium acetate + pyridine + iron (at 115°C) may cause violent polymerization reactions. Will react with water or steam to produce toxic and corrosive fumes. Incompatible with dimethyl sulfoxide. Used in production of drugs of abuse. When heated to decomposition it emits toxic fumes of Cl^-. See also CHLORINATED HYDROCARBONS, AROMATIC.

B

BEF500   CAS: 501-53-1   HR: 3
BENZYL CHLOROFORMATE
DOT: UN 1739
mf: $C_8H_7ClO_2$ mw: 170.60
PROP: Colorless to pale-yellow liquid or oil; odor of phosgene. Mp: 0°, bp: 103° @ 20 mm.
SYNS: BENZYLCARBONYL CHLORIDE □ BENZYL CHLOROCARBONATE (DOT) □ BENZYL CHLORO-FORMATE (DOT) □ BENZYLOXYCARBONYL CHLORIDE □ BZCF □ CARBOBENZOXY CHLORIDE □ CARBOBENZYLOXY CHLORIDE □ CHLOROFORMIC ACID BENZYL ESTER
CONSENSUS REPORTS: Reported in EPA TSCA Inventory.
DOT CLASSIFICATION: 8; Label: Corrosive
SAFETY PROFILE: Poison by ingestion and inhalation routes. A powerful corrosive irritant. Thermally unstable. Will react with water or steam to produce toxic and corrosive fumes and heat. Iron salts catalyze the explosive decomposition of the ester. When heated to decomposition it emits toxic fumes of Cl⁻ and phosgene. See also PHOSGENE, ESTERS, and CHLORIDES.

BEG750   CAS: 103-41-3   HR: 2
BENZYL CINNAMATE
mf: $C_{16}H_{14}O_2$ mw: 238.30
PROP: Found in balsams of Peru, tolu, styrax, copaiba, and others (FCTXAV 11,1011,73). White crystals; aromatic odor. Mp: 39°, bp: 350.0°, vap press: 1 mm @ 173.8°, flash p: 212°F. Sol in fixed oils; insol in glycerin and propylene glycol.
SYNS: BENZYL ALCOHOL CINNAMIC ESTER □ BENZYL γ-PHENYLACRYLATE □ CINNAMEIN □ trans-CINNAMIC ACID BENZYL ESTER □ FEMA No. 2142 □ 3-PHENYL-2-PROPENOIC ACID PHENYLMETHYL ESTER (9CI)
CONSENSUS REPORTS: Reported in EPA TSCA Inventory.
SAFETY PROFILE: Moderately toxic by ingestion. A mild allergen and skin irritant. Combustible liquid. See also ESTERS. When heated to decomposition it emits acrid smoke and irritating fumes.

BEL525   CAS: 7083-24-1   HR: 3
BENZYL 1-(2-(DIMETHYLAMINO)-
    PROPYL)PYRROL-2-YL, CITRATE
    KETONE
mf: $C_{17}H_{22}N_2O \cdot C_6H_8O_7$ mw: 462.55
SYN: 1-(2-(DIMETHYLAMINO)PROPYL)-2-(PHENYLACETYL)PYRROLE CITRATE
DOT CLASSIFICATION: 3; Label: Flammable Liquid
SAFETY PROFILE: A poison by intravenous route. A flammable liquid. When heated to decomposition it emits acrid smoke and irritating vapors.

BEL550   CAS: 73747-22-5   HR: 3
BENZYLDIMETHYLAMMONIUM
    HEXAFLUOROARSENATE
mf: $C_9H_{13}N \cdot AsF_6H$ mw: 325.16
SYNS: BENZYLAMINE, N,N-DIMETHYL-, HEXA-FLUOROARSENATE (1-) □ N,N-DIMETHYLBENZYL-AMINE HEXAFLUOROARSENATE
OSHA PEL: TWA 0.5 mg(As)/m³
SAFETY PROFILE: Poison by intravenous route. When heated to decomposition it emits toxic fumes of NO_x, F⁻, and As.

BEL850   CAS: 10094-34-5   HR: 1
BENZYL DIMETHYLCARBINYL n-
    BUTYRATE
mf: $C_{14}H_{20}O_2$ mw: 220.34
PROP: Colorless liquid with plum aroma. Mp: >70°.
SYNS: BENZYL DIMETHYLCARBINYL BUTYRATE □ BUTYRIC ACID, α-α-DIMETHYLPHENETHYL ESTER □ DIMETHYLBENZYLCARBINYL BUTYRATE □ α-α-DIMETHYLPHENETHYL BUTYRATE
CONSENSUS REPORTS: Reported in EPA TSCA Inventory.
SAFETY PROFILE: Low oral toxicity. A skin irritant. When heated to decomposition it emits acrid smoke and irritating fumes.

BEM000   CAS: 139-07-1   HR: 3
BENZYLDIMETHYLDODECYLAMMONIU
    M CHLORIDE
mf: $C_{21}H_{38}N \cdot Cl$ mw: 340.05
PROP: Solid. Mp: 31–32°.
SYN: DODECYL DIMETHYL BENZYLAMMONIUM CHLORIDE
SAFETY PROFILE: Poison by ingestion and intraperitoneal routes. A skin and eye

irritant. When heated to decomposition it emits very toxic fumes of NO_x, NH_3, and Cl^-.

BEM750   CAS: 525-02-0   HR: 3
1-BENZYL-2,5-DIMETHYL SEROTONIN HYDROCHLORIDE
mf: $C_{19}H_{22}N_2O \cdot ClH$ mw: 330.89

PROP: Solid. Mp: 230–231°.

SYNS: 3-(2-AMINOETHYL)-1-BENZYL-5-METHOXY-2-METHYLINDOLE HYDROCHLORIDE □ BAS □ BENANSERIN HYDROCHLORIDE □ BENZYL ANTISEROTONIN □ 1-BENZYL-2-METHYL-3-(2-AMINOETHYL)-5-METHOXYINDOLE HYDROCHLORIDE □ 1-BENZYL-2-METHYL-5-METHOXY-TRYPTAMINE HYDROCHLORIDE □ SEROTONIN BENZYL ANALOG □ WOOLLEY'S ANTISEROTONIN

SAFETY PROFILE: Poison by intraperitoneal route. A serotonin antagonist that causes psychotropic effects in humans. Experimental reproductive effects. When heated to decomposition it emits very toxic fumes of HCl and NO_x.

BEN000   CAS: 121-54-0   HR: 3
BENZYLDIMETHYL(2-(2-(p-(1,1,3,3-TETRAMETHYLBUTYL)PHENOXY)ETHOXY)ETHYL) AMMONIUM CHLORIDE
mf: $C_{27}H_{42}NO_2 \cdot Cl$ mw: 448.15

PROP: Colorless crystals or plates. Mp: 164–166°. Very sol in H_2O; sol in Me_2CO, EtOH, and $CHCl_3$.

SYNS: ANTI-GERM 77 □ ANTISEPTOL □ BENZETHONIUM CHLORIDE □ BENZETONIUM CHLORIDE □ BENZYLDIMETHYL-p-(1,1,3,3-TETRAMETHYLBUTYL)PHENOXYETHOXY-ETHYLAMMONIUM CHLORIDE □ BZT □ DIAPP □ DIISOBUTYLPHENOXYETHOXYETHYLDIMETHYL BENZYL AMMONIUM CHLORIDE □ DISILYN □ HYAMINE □ HYAMINE 1622 □ NCI-C61494 □ p-tert-OCTYLPHENOXYETHOXYETHYLDIMETHYLBENZYL AMMONIUM CHLORIDE □ PHEMERIDE □ PHEMEROL CHLORIDE □ PHEMITHYN □ POLYMINE D □ QUATRACHLOR □ SOLAMINE

CONSENSUS REPORTS: Reported in EPA TSCA Inventory.

SAFETY PROFILE: Poison by ingestion, subcutaneous, intraperitoneal, and intravenous routes. A severe eye irritant. Questionable carcinogen with experimental neoplastigenic data. Mutation data reported.

When heated to decomposition it emits very toxic fumes of Cl^-, NH_3, and NO_x. A topical anti-infective agent.

BEO250   CAS: 103-50-4   HR: 2
BENZYL ETHER
mf: $C_{14}H_{14}O$ mw: 198.28

PROP: Colorless to pale-yellow liquid. Mp: 5°, bp: 182–183° @ 22 mm, flash p: 275°F (CC), d: 1.056, vap d: 6.84, refr index: 1.557.

SYNS: BENZYL OXIDE (CZECH) □ DIBENZYLETHER (CZECH) □ FEMA No. 2371

CONSENSUS REPORTS: Reported in EPA TSCA Inventory.

SAFETY PROFILE: Moderately toxic by ingestion. Vapors are probably narcotic in high concentration. A skin and eye irritant. Combustible when exposed to heat or flame; can react with oxidizing materials. Moderate explosion hazard by spontaneous chemical reaction. To fight fire, use CO_2, dry chemical. See also ETHERS.

BEP250   CAS: 104-57-4   HR: 2
BENZYL FORMATE
mf: $C_8H_8O_2$ mw: 136.16

PROP: Colorless liquid with powerful fruity, spicy odor. D: 1.083–1.092, bp: 202°. Insol in water.

SYNS: BENZYL ALCOHOL FORMATE □ BENZYL METHANOATE

CONSENSUS REPORTS: Reported in EPA TSCA Inventory.

SAFETY PROFILE: Moderately toxic by ingestion and skin contact. Mutation data reported. Probably narcotic in high concentrations. See also ESTERS. When heated to decomposition it emits acrid, irritating fumes.

BEP500   CAS: 10453-86-8   HR: 3
5-BENZYL-3-FURYL METHYL(±)-cis,trans-CHRYSANTHEMATE
mf: $C_{22}H_{26}O_3$ mw: 338.48

SYNS: BENZOFUROLINE □ BENZYLFUROLINE □ (5-BENZYL-3-FURYL) METHYL-2,2-DIMETHYL-3-(2-METHYLPROPENYL)-CYCLOPROPANECARBOXYLATE □ CHRYSON □ CHRYSRON □ DIMETHYL-3-(2-METHYL-1-PROPENYL)CYCLOPROPANE-CARBOXYLATE □ ENT 27,474 □ FMC 17370 □ FOR-SYN

□ NIA 17170 □ NRDC 104 □ NSC-195022 □ OMS-1206 □ PREMGARD □ PYNOSECT □ PYRETHERM □ RESMETHRIN □ RESMETRINA (PORTUGUESE) □ SBP-1382 □ S.B. PENICK 1382 □ SYNTHRIN

CONSENSUS REPORTS: EPA FIFRA 1988 pesticide subject to registration or re-registration. EPA Genetic Toxicology Program.

SAFETY PROFILE: Poison by ingestion, and intravenous routes. Moderately toxic by inhalation and skin contact. When heated to decomposition it emits acrid and irritating fumes. See also ESTERS.

BES300   CAS: 52098-16-5   HR: 3
1-BENZYL-2-INDOLYL HYDROXY-
** METHYL KETONE**
mf: $C_{17}H_{15}NO_2$ mw: 265.33
SYNS: 1-BENZYL-2-(HYDROXYACETYL)INDOLE □ KETONE, 1-BENZYL-2-INDOLYL HYDROXYMETHYL-
DOT CLASSIFICATION: 3; Label: Flammable Liquid
SAFETY PROFILE: A poison by intraperitoneal route. A flammable liquid. When heated to decomposition it emits toxic vapors of NO_x.

BEU250   CAS: 622-78-6   HR: 3
BENZYL-ISOTHIOCYANATE
mf: C_8H_7NS mw: 149.22
PROP: Orange-red, crystalline solid. Mp: 41°, bp: 230°, d: 1.125
SYNS: BENZYL MUSTARD OIL □ BENZYLSENFOEL (GERMAN) □ ISOTHIOCYANIC ACID BENZYL ESTER
CONSENSUS REPORTS: Reported in EPA TSCA Inventory.
SAFETY PROFILE: Poison by intraperitoneal and subcutaneous routes. Intensely irritating. Mutation data reported. Moderate fire hazard via heat, flame, and oxidizers. To fight fire, use water, spray, foam, dry chemical. When heated to decomposition it emits very toxic NO_x and SO_x. See also ESTERS and THIOCYANATES.

BEU500   CAS: 538-28-3   HR: 3
BENZYLISOTHIOUREA HYDRO-
** CHLORIDE**

mf: $C_8H_{10}N_2S \cdot ClH$ mw: 202.72
PROP: Dimorphic. Mp: 146–148°.
SYNS: BENZYLISOTHIOURONIUM CHLORIDE □ 2-BENZYLISOTHIOURONIUM CHLORIDE □ BENZYL THIOPSEUDOUREA HYDROCHLORIDE □ 2-BENZYL-2-THIO-PSEUDOUREA HYDROCHLORIDE □ BENZYL-THIURONIUM CHLORIDE □ S-BENZYLTHIURONIUM CHLORIDE □ BTKH □ ISOTHIOURONIUM CHLORIDE, BENZYL □ 2-THIO-2-BENZYL-PSEUDOUREA HYDROCHLORIDE □ TL 944 □ USAF EK-2124
CONSENSUS REPORTS: Reported in EPA TSCA Inventory.
SAFETY PROFILE: Poison by ingestion, intraperitoneal, subcutaneous, and intravenous routes. When heated to decomposition it emits very toxic fumes of HCl, SO_x, and NO_x.

BFC200   CAS: 101670-78-4 HR: 3
5-BENZYLOXY-3-ISONIPECOTOYLIND-
** OLE**
mf: $C_{21}H_{21}N_2O_2$ mw: 333.44
SYNS: INDOLE, 5-BENZYLOXY-3-ISONIPECOTOYL-□ KETONE, 5-BENZYLOXY-3-INDOLYL 4-PIPERIDYL
DOT CLASSIFICATION: 3; Label: Flammable Liquid
SAFETY PROFILE: A poison by intravenous route. A flammable liquid. When heated to decomposition it emits toxic vapors of NO_x.

BFC750   CAS: 1538-09-6   HR: 2
BENZYLPENCILLINDIBENZYLETHYLEN
** EDIAMINE SALT**
mf: $C_{32}H_{36}N_4O_8S_2 \cdot C_{16}H_{20}N_2$ mw: 909.22
PROP: Crystals. Mp: 123–124°.
SYNS: BEACILLIN □ BEN-P □ BENZACILLIN □ BENZATHINE BENZYLPENICILLIN □ BENZATHINE PENICILLIN □ BENZATHINE PENICILLIN G □ BENZETHACIL □ BENZYLPENICILLIN BENZATHINE □ BICA-PENICILLIN □ BICILLIN □ CEPACILINA □ CEPACILLINA □ CILLENTA □ DBED DIPENCILLIN G □ DBED PENICILLIN □ DEBECILLIN □ DEBECYLINA □ DIAMINE DIPENICILLIN G □ DIAMINOCILLIAN □ DIBENCIL □ DIBENCILLIN □ N,N'-DIBENZYLETHYL-ENEDIAMINE BIS(BENZYL PENICILLIN) □ DIBENZYL-ETHYLENEDIAMINE-DI-PENICILLIN G □ N,N'-DIBENZYLETHYLENEDIAMINE, compounded with PENICILLIN G (1:2) □ DIPO-SAFT □ DURABIOTIC □ DURA-PENITA □ DUROPENIN □ EXTENCILLINE □ EXTENICILLINE □ LENTOCILLIN □ LENTOPENIL □

LEOMYPEN ☐ LONGACILIAN ☐ LONGICIL ☐ LPG ☐ MEGACILLIN SUSPENSION ☐ MOLDAMIN ☐ NCI-C56100 ☐ NEOLIN ☐ PENADUR ☐ PENADUR L-A ☐ PENDEPON ☐ PEN-DI-BEN ☐ PENDITAN ☐ PENDURAN ☐ PENICILLIN G, compounded with N,N'-DIBENZYLETHYLENEDIAMINE (2:1) ☐ PENICILLIN G SALT of N,N'-DIBENZYLETHYLENEDIAMINE ☐ PENIDURAL ☐ PENIDURE ☐ PENILENTE ☐ PERMAPEN ☐ RETARPEN ☐ TARDOCILLIN ☐ VETARCILLIN ☐ VICIN ☐ WYCILLINA

CONSENSUS REPORTS: Reported in EPA TSCA Inventory.

SAFETY PROFILE: Moderately toxic by ingestion and intraperitoneal routes. Experimental reproductive effects. When heated to decomposition it emits very toxic fumes of NO_x and SO_x. See other penicillin entries.

BFD250   CAS: 69-57-8   HR: 3
BENZYL PENICILLINIC ACID SODIUM SALT

mf: $C_{16}H_{17}N_2O_4S•Na$ mw: 356.40

PROP: Needles from butanol (aq). Mp: 215° (decomp). Sol in H_2O and MeOH.

SYNS: AMERICAN PENICILLIN ☐ BENZYL-PENICILLIN SODIUM ☐ CRYSTAPEN ☐ MYCOFARM ☐ NOVOCILLIN ☐ PEN-A-BRASIVE ☐ PENICILLIN-G, MONOSODIUM SALT ☐ PENICILLIN G, SODIUM ☐ PENICILLIN G, SODIUM SALT ☐ PENILARYN ☐ PENZYLPENICILLIN SODIUM SALT ☐ SODIUM BENZYLPENICILLIN ☐ SODIUM BENZYLPENICILLIN G ☐ SODIUM BENZYLPENICILLINATE ☐ SODIUM PENICILLIN ☐ SODIUM PENICILLIN G ☐ SODIUM PENICILLIN II ☐ VETICILLIN

CONSENSUS REPORTS: EPA Genetic Toxicology Program.

SAFETY PROFILE: Poison by intracerebral, parenteral, and intramuscular routes. Moderately toxic via intravenous route. Mildly toxic by ingestion. Experimental teratogenic and reproductive effects. Questionable carcinogen with experimental tumorigenic data. When heated to decomposition it emits very toxic fumes of NO_x, Na_2O, and SO_x. An antibiotic. See other penicillin entries.

BFD400   HR: 1
BENZYL PHENYLACETATE

mf: $C_{15}H_{14}O_2$ mw: 226.27

PROP: Colorless liquid; sweet, floral odor with honey undertone. D: 1.095–1.099, refr index: 1.553–1.558, flash p: 212°F. Sol in alc, chloroform, ether.

SYN: FEMA No. 2149

SAFETY PROFILE: Combustible liquid. When heated to decomposition it emits acrid smoke and irritating fumes.

BFD760   CAS: 3762-27-4   HR: 3
BENZYLPHOSPHONIC ACID DIBUTYL ESTER

mf: $C_{15}H_{25}O_3P$ mw: 284.37

SYNS: DI-N-BUTYL BENZYLPHOSPHONATE ☐ PHOSPHONIC ACID, BENZYL-, DIBUTYL ESTER ☐ PHOSPHONIC ACID, (PHENYLMETHYL)-, DIBUTYL ESTER

SAFETY PROFILE: A poison by ingestion. When heated to decomposition it emits toxic vapors of PO_x.

BFD800   HR: D
BENZYL PROPIONATE

mf: $C_{10}H_{12}O_2$ mw: 164.20

PROP: Colorless liquid; sweet, floral fruity odor. D: 1.028–1.032, refr index: 1.496–1.500. Sol alc, most oils; sltly sol in propylene glycol; insol in glycerin, water.

SYN: FEMA No. 2150

SAFETY PROFILE: When heated to decomposition it emits acrid smoke and irritating fumes.

BFJ750   CAS: 118-58-1   HR: 2
BENZYL SALICYLATE

mf: $C_{14}H_{12}O_3$ mw: 228.26

PROP: Thick colorless liquid; pleasant odor. Bp: 208° @ 26 mm, d: 1.175 @ 20°, refr index: 1.579. Sol in fixed oils; insol in glycerin and propylene glycol.

SYNS: BENZYL-o-HYDROXYBENZOATE ☐ FEMA No. 2151

CONSENSUS REPORTS: Reported in EPA TSCA Inventory.

SAFETY PROFILE: Moderately toxic by ingestion. See also BENZYL ALCOHOL, SALICYLIC ACID, and ESTERS. Combustible when exposed to heat or flame. When heated to decomposition it emits acrid smoke and irritating fumes. Incompatible with oxidizing materials.

BFL000   CAS: 3012-37-1   HR: 3
BENZYL THIOCYANATE
mf: C_8H_7NS mw: 149.22
PROP: Orange-red crystals or solid. Mp: 41–42°, bp: 230°, d: 1.125.
SYNS: BENZYL MUSTARD OIL □ PHENYLMETHYL ESTER THIOCYANIC ACID (9CI) □ SOLVAT 14 □ α-THIOCYANATOTOLUENE □ TROPEOLIN
CONSENSUS REPORTS: Reported in EPA TSCA Inventory.
SAFETY PROFILE: Poison by subcutaneous and intraperitoneal routes. See also THIOCYANATES. When heated to decomposition it emits very toxic fumes of NO_x, SO_x, and CN^-.

BFL250   CAS: 98-07-7   HR: 3
BENZYL TRICHLORIDE
DOT: UN 2226
mf: $C_7H_5Cl_3$ mw: 195.47
PROP: Clear, colorless to yellowish liquid; penetrating odor. Mp: −5°, bp: 221°, d: 1.38 @ 15.5°/15.5°, vap d: 6.77.
SYNS: BENZENYL CHLORIDE □ BENZENYL TRICHLORIDE □ BENZOIC TRICHLORIDE □ BENZOTRICHLORIDE (DOT, MAK) □ BENZYLIDYNE CHLORIDE □ CHLORURE de BENZENYLE (FRENCH) □ PHENYL CHLOROFORM □ PHENYLTRICHLORO-METHANE □ RCRA WASTE NUMBER U023 □ TOLUENE TRICHLORIDE □ TRICHLOORMETHYL-BENZEEN (DUTCH) □ TRICHLORMETHYLBENZOL (GERMAN) □ TRICHLOROMETHYLBENZENE □ 1-(TRICHLOROMETHYL)BENZENE □ TRICLORO-METILBENZENE (ITALIAN) □ TRICHLOROPHENYL-METHANE □ α,α,α-TRICHLOROTOLUENE □ ω,ω,ω-TRICHLOROTOLUENE □ TRICLOROTOLUENE (ITALIAN)
CONSENSUS REPORTS: NTP 10th Report on Carcinogens. IARC Cancer Review: Human Limited Evidence IMEMDT 29,73,82; Animal Sufficient Evidence IMEMDT 29,73,82. EPA Genetic Toxicology Program. EPA Extremely Hazardous Substances List. Reported in EPA TSCA Inventory.
ACGIH TLV: CL 0.1 (skin); Suspected Human Carcinogen

DFG MAK: Confirmed Human Carcinogen
DOT CLASSIFICATION: 8; Label: Corrosive
SAFETY PROFILE: Confirmed carcinogen with experimental carcinogenic data by skin contact and neoplastigenic data by inhalation. Experimental poison by inhalation. Corrosive to the skin, eyes, and mucous membranes. Large doses can cause central nervous system depression. Mutation data reported. When heated to decomposition it emits toxic fumes of Cl^-. See also CHLORINATED HYDROCARBONS, AROMATIC.

BFM750   CAS: 4525-46-6   HR: 3
BENZYL TRIMETHYL AMMONIUM IODIDE
mf: $C_{10}H_{16}N \cdot I$ mw: 277.17
PROP: Solid. Mp: 181–182°.
SYNS: BENZYLDIMETHYLAMINE METHIODIDE □ PHENMETHYL TRIMETHYLAMMONIUM IODIDE
CONSENSUS REPORTS: Reported in EPA TSCA Inventory.
ACGIH TLV: Proposed: (inhalable fraction) 0.1 mg/m³; Not Classifiable as a Human Carcinogen)
SAFETY PROFILE: Poison by intraperitoneal and intravenous routes. See also IODIDES. When heated to decomposition it emits very toxic fumes of NO_x, NH_3, and I^-.

BFO000   CAS: 8007-75-8   HR: 1
BERGAMOT OIL rectified
PROP: Yellow-green liquid; agreeable odor. *Composition:* 1-linalyl acetate, 1-linalool, d-limonene, dipentene, bergaptene. By rectification of bergamot oil expressed, under vacuum, to remove completely the furocoumarins and other related nonvolatile residues; found in the fruit of citrus *Bergamia risso et poiteau* (Fam. *Rutaceae*) (FCTXAV 11,1011,73). D: 0.875–0.880 @ 25°/25°. Misc with alc, glacial acetic acid; sol in fixed oils; insol in glycerin, propylene glycol.
SYNS: BERGAMOTTE OEL (GERMAN) □ OIL OF BERGAMOT, coldpressed □ OIL OF BERGAMOT, rectified
CONSENSUS REPORTS: Reported in EPA TSCA Inventory.

SAFETY PROFILE: Mildly toxic by ingestion. A mild skin irritant and allergen. Combustible. When heated to decomposition it emits acrid smoke and irritating fumes.

BFO250   CAS: 12161-82-9   HR: 3
BERTRANDITE
mf: $H_{10}O_9Si_2•H_2O•Be_4$ mw: 264.34
PROP: Colorless, pale-yellow, orthorhombic crystals.
SYN: BERYLLIUM SILICATE HYDRATE
CONSENSUS REPORTS: IARC Cancer Review: Group 1 IMEMDT 58,41,93; Human Sufficient Evidence IMEMDT 58,41,93; Animal Sufficient Evidence IMEMDT 1,17,72; Animal Sufficient Evidence IMEMDT 23,143,80; Animal Sufficient Evidence IMEMDT 58,41,93. Reported in EPA TSCA Inventory. Beryllium and its compounds are on the Community Right-To-Know List.
OSHA PEL: TWA 0.002 mg(Be)/m³; STEL 0.005 mg(Be)/m³/30M; CL 0.025 mg(Be)/m³
ACGIH TLV: TWA 0.002 mg(Be)/m³; Confirmed Human Carcinogen; (Proposed: TWA 0.00005 mg(Be)/m³; STEL 0.0002 mg(Be)/m³ (sensitizer) (skin); Confirmed Human Carcinogen)
NIOSH REL: (Beryllium) CL not to exceed 0.0005 mg(Be)/m³
SAFETY PROFILE: Confirmed carcinogen. See also BERYLLIUM and BERYLLIUM COMPOUNDS. When heated to decomposition it emits very toxic fumes of BeO.

BFO500   CAS: 1302-52-9   HR: 3
BERYL
mf: $Al_2O_{18}Si_6•3Be$ mw: 537.53
PROP: Colorless, white, blue-green, green-yellow, yellow, or blue crystals. D: 2.63–2.91.
SYNS: BERYLLIUM ALUMINOSILICATE □ BERYLLIUM ALUMINUM SILICATE □ BERYL ORE
CONSENSUS REPORTS: NTP 10th Report on Carcinogens. IARC Cancer Review:

Group 1 IMEMDT 58,41,93; Human Sufficient Evidence IMEMDT 58,41,93; Animal Sufficient Evidence IMEMDT 1,17,72; Animal Sufficient Evidence IMEMDT 23,143,80; Animal Sufficient Evidence IMEMDT 58,41,93. Reported in EPA TSCA Inventory. Beryllium and its compounds are on the Community Right-To-Know List.
OSHA PEL: TWA 0.002 mg(Be)/m³; STEL 0.005 mg(Be)/m³/30M; CL 0.025 mg(Be)/m³
ACGIH TLV: TWA 0.002 mg(Be)/m³; Confirmed Human Carcinogen; (Proposed: TWA 0.00005 mg(Be)/m³; STEL 0.0002 mg(Be)/m³ (sensitizer) (skin); Confirmed Human Carcinogen)
NIOSH REL: (Beryllium) CL not to exceed 0.0005 mg(Be)/m³
SAFETY PROFILE: Confirmed carcinogen with experimental carcinogenic, neoplastigenic, and tumorigenic data. See also BERYLLIUM COMPOUNDS and SILICATES. When heated to decomposition it emits toxic fumes of BeO.

BFO750   CAS: 7440-41-7   HR: 3
BERYLLIUM
DOT: UN 1966/UN 1567
af: Be   aw: 9.01
PROP: A silvery-white, relatively soft, lustrous metal, ductile at red heat. Unreactive to H_2O and air; dissolves vigorously in dil acids. Be reacts with aq alkalies or H_2. Mp: 1287–1292°, bp: 2970°, d: 1.85. IDLH 4 mg/m³ (as Be).
SYNS: BERYLLIUM-9 □ BERYLLIUM COMPOUNDS, n.o.s. (UN 1566) (DOT) □ BERYLLIUM, powder (UN 1567) (DOT) □ GLUCINIUM □ GLUCINUM □ RCRA WASTE NUMBER P015
CONSENSUS REPORTS: NTP 10th Report on Carcinogens. IARC Cancer Review: Group 1 IMEMDT 58,41,93; Human Sufficient Evidence IMEMDT 58,41,93; Animal Sufficient Evidence IMEMDT 1,17,72; Animal Sufficient Evidence IMEMDT 23,143,80; Animal Sufficient Evidence IMEMDT 58,41,93. Beryllium and its compounds are on the Community

Right-To-Know List. Reported in EPA TSCA Inventory.
OSHA PEL: TWA 0.002 mg(Be)/m³; STEL 0.005 mg(Be)/m³/30M; CL 0.025 mg(Be)/m³
ACGIH TLV: TWA 0.002 mg(Be)/m³; Confirmed Human Carcinogen; (Proposed: TWA 0.00005 mg(Be)/m³; STEL 0.0002 mg(Be)/m³ (sensitizer) (skin); Confirmed Human Carcinogen)
DFG MAK: DFG TRK: Animal Carcinogen, Suspected Human Carcinogen. Grinding of beryllium metal and alloys: 0.005 mg/m³ calculated as beryllium in that portion of dust that can possibly be inhaled; other beryllium compounds: 0.002 mg/m³ calculated as beryllium in that portion of dust that can possibly be inhaled
NIOSH REL: CL not to exceed 0.0005 mg(Be)/m³
DOT CLASSIFICATION: 6.1; Label: Poison (UN 1566); DOT Class: 6.1; Label: Poison, Flammable Solid (UN 1567)
SAFETY PROFILE: Confirmed carcinogen with experimental carcinogenic, neoplastigenic, and tumorigenic data. A deadly poison by intravenous route. Human systemic effects by inhalation: lung fibrosis, dyspnea, and weight loss. Human mutation data reported. See also BERYLLIUM COMPOUNDS. A moderate fire hazard in the form of dust or powder, or when exposed to flame or by spontaneous chemical reaction. Slight explosion hazard in the form of powder or dust. Incompatible with halocarbons. Reacts incandescently with fluorine or chlorine. Mixtures of the powder with CCl₄ or trichloroethylene will flash or spark on impact. When heated to decomposition in air it emits very toxic fumes of BeO. Reacts with Li and P.

BFP000   CAS: 543-81-7   HR: 3
BERYLLIUM ACETATE
mf: $C_4H_6O_4 \cdot Be$ mw: 127.11
PROP: Plates. Mp: decomp @ 300°.
SYN: BERYLLIUM ACETATE, NORMAL

CONSENSUS REPORTS: IARC Cancer Review: Group 1 IMEMDT 58,41,93; Human Sufficient Evidence IMEMDT 58,41,93; Animal Sufficient Evidence IMEMDT 1,17,72; Animal Sufficient Evidence IMEMDT 23,143,80; Animal Sufficient Evidence IMEMDT 58,41,93. Beryllium and its compounds are on the Community Right-To-Know List.
OSHA PEL: TWA 0.002 mg(Be)/m³; STEL 0.005 mg(Be)/m³/30M; CL 0.025 mg(Be)/m³
ACGIH TLV: TWA 0.002 mg(Be)/m³; Confirmed Human Carcinogen; (Proposed: TWA 0.00005 mg(Be)/m³; STEL 0.0002 mg(Be)/m³ (sensitizer) (skin); Confirmed Human Carcinogen)
DFG MAK: Animal Carcinogen, Suspected Human Carcinogen
NIOSH REL: (Beryllium) CL not to exceed 0.0005 mg(Be)/m³
SAFETY PROFILE: Confirmed carcinogen. Poison by intraperitoneal route. See also BERYLLIUM COMPOUNDS. When heated to decomposition it emits toxic fumes of BeO.

BFP250   CAS: 12770-50-2   HR: 3
BERYLLIUM ALUMINUM ALLOY
PROP: Alloy is 62% beryllium and 38% aluminum (ENVRAL 21,63,80).
SYNS: ALUMINUM ALLOY, Al,Be □ ALUMINUM BERYLLIUM ALLOY
CONSENSUS REPORTS: NTP 10th Report on Carcinogens. IARC Cancer Review: Group 1 IMEMDT 58,41,93; Human Sufficient Evidence IMEMDT 58,41,93; Animal Sufficient Evidence IMEMDT 1,17,72; Animal Sufficient Evidence IMEMDT 23,143,80; Animal Sufficient Evidence IMEMDT 58,41,93. Beryllium and its compounds are on the Community Right-To-Know List.
OSHA PEL: TWA 0.002 mg(Be)/m³; STEL 0.005 mg(Be)/m³/30M; CL 0.025 mg(Be)/m³
ACGIH TLV: TWA 0.002 mg(Be)/m³; Confirmed Human Carcinogen; (Proposed:

TWA 0.00005 mg(Be)/m³; STEL 0.0002 mg(Be)/m³ (sensitizer) (skin); Confirmed Human Carcinogen)
DFG MAK: Animal Carcinogen, Suspected Human Carcinogen
NIOSH REL: (Beryllium) CL not to exceed 0.0005 mg(Be)/m³
SAFETY PROFILE: Confirmed carcinogen with experimental carcinogenic and tumorigenic data. See also BERYLLIUM COMPOUNDS. When heated to decomposition it emits very toxic BeO.

BFP500   CAS: 66104-24-3   HR: 3
BERYLLIUM CARBONATE
mf: $C_2H_2Be_3O_8$ mw: 181.07
PROP: White powder. Decomposes >200°. Insol in water.
SYNS: BERYLLIUM CARBONATE, BASIC □ BERYLLIUMOXIDE CARBONATE □ BIS(CARBONATO(2-))DIHYDROXYTRIBERYLLIUM
CONSENSUS REPORTS: NTP 10th Report on Carcinogens. IARC Cancer Review: Group 1 IMEMDT 58,41,93; Human Sufficient Evidence IMEMDT 58,41,93; Animal Sufficient Evidence IMEMDT 1,17,72; Animal Sufficient Evidence IMEMDT 23,143,80; Animal Sufficient Evidence IMEMDT 58,41,93. Reported in EPA TSCA Inventory. Beryllium and its compounds are on the Community Right-To-Know List.
OSHA PEL: TWA 0.002 mg(Be)/m³; STEL 0.005 mg(Be)/m³/30M; CL 0.025 mg(Be)/m³
ACGIH TLV: TWA 0.002 mg(Be)/m³; Confirmed Human Carcinogen; (Proposed: TWA 0.00005 mg(Be)/m³; STEL 0.0002 mg(Be)/m³ (sensitizer) (skin); Confirmed Human Carcinogen)
DFG MAK: Animal Carcinogen, Suspected Human Carcinogen
NIOSH REL: CL not to exceed 0.0005 mg(Be)/m³
SAFETY PROFILE: Confirmed carcinogen. See also BERYLLIUM COMPOUNDS. When heated to decomposition it emits toxic BeO dust.

BFP750   CAS: 13106-47-3   HR: 3
BERYLLIUM CARBONATE (1:1)
mf: CO₃•Be   mw: 69.02
PROP: Insol in water. Decomposes in hot water.
SYN: CARBONIC ACID BERYLLIUM SALT (1:1)
CONSENSUS REPORTS: IARC Cancer Review: Group 1 IMEMDT 58,41,93; Human Sufficient Evidence IMEMDT 58,41,93; Animal Sufficient Evidence IMEMDT 1,17,72; Animal Sufficient Evidence IMEMDT 23,143,80; Animal Sufficient Evidence IMEMDT 58,41,93. Reported in EPA TSCA Inventory. Beryllium and its compounds are on the Community Right-To-Know List.
OSHA PEL: TWA 0.002 mg(Be)/m³; STEL 0.005 mg(Be)/m³/30M; CL 0.025 mg(Be)/m³
ACGIH TLV: TWA 0.002 mg(Be)/m³; Confirmed Human Carcinogen; (Proposed: TWA 0.00005 mg(Be)/m³; STEL 0.0002 mg(Be)/m³ (sensitizer) (skin); Confirmed Human Carcinogen)
DFG MAK: 50 ppm (90 mg/m³)
NIOSH REL: (Beryllium) CL not to exceed 0.0005 mg(Be)/m³
SAFETY PROFILE: Confirmed carcinogen. Poison by intraperitoneal route. See also BERYLLIUM COMPOUNDS. When heated to decomposition it emits highly toxic fumes of BeO.

BFQ000   CAS: 7787-47-5   HR: 3
BERYLLIUM CHLORIDE
mf: BeCl₂   mw: 79.91
PROP: Colorless, deliquescent needles, or orthorhombic crystals. Undergoes transition to high temp orthorhombic polymorph at 4°. Mp: 415°, bp: 520°, d: 1.899 @ 25°, vap press: 1 mm @ 291° (subl). Very sol in H₂O, EtOH, Et₂O, or Py; sltly sol in C₆H₆, CHCl₃, CS₂; insol in NH₃ or Me₂CO.
SYN: BERYLLIUM DICHLORIDE
CONSENSUS REPORTS: NTP 10th Report on Carcinogens. IARC Cancer Review: Group 1 IMEMDT 58,41,93; Human Sufficient Evidence IMEMDT 58,41,93;

Animal Sufficient Evidence IMEMDT 1,17,72; Animal Sufficient Evidence IMEMDT 23,143,80; Animal Sufficient Evidence IMEMDT 58,41,93. EPA Genetic Toxicology Program. Reported in EPA TSCA Inventory. Beryllium and its compounds are on the Community Right-To-Know List.

OSHA PEL: TWA 0.002 mg(Be)/m³; STEL 0.005 mg(Be)/m³/30M; CL 0.025 mg(Be)/m³

ACGIH TLV: TWA 0.002 mg(Be)/m³; Confirmed Human Carcinogen; (Proposed: TWA 0.00005 mg(Be)/m³; STEL 0.0002 mg(Be)/m³ (sensitizer) (skin); Confirmed Human Carcinogen)

DFG MAK: Animal Carcinogen, Suspected Human Carcinogen

NIOSH REL: (Beryllium) CL not to exceed 0.0005 mg(Be)/m³

SAFETY PROFILE: Confirmed carcinogen with experimental tumorigenic data. Poison by ingestion and intraperitoneal routes. An experimental teratogen. Other experimental reproductive effects. Mutation data reported. When heated to decomposition it emits very toxic fumes of BeO and Cl⁻. See also BERYLLIUM COMPOUNDS and CHLORIDES.

BFQ500   **HR: 3**
BERYLLIUM COMPOUNDS
PROP: Pure beryllium is a hard, brittle, silvery metal. Mp: 1278°, bp: 2970°, d: 1.85. Beryllium oxide: white powder. Mp: 2530°, d: 3.0. Beryllium chloride: white to faintly yellow powder, deliquescent. Mp: 399°, bp: 482°. Beryllium fluoride: glassy, hygroscopic solid. Mp: 545°, d: 2.0. Beryllium nitrate: white to slightly yellow crystals. Mp: 60°). Beryllium sulfate: Mp: 550°.

CONSENSUS REPORTS: IARC Cancer Review: Group 1 IMEMDT 58,41,93; Human Sufficient Evidence IMEMDT 58,41,93; Animal Sufficient Evidence IMEMDT 1,17,72; Animal Sufficient Evidence IMEMDT 23,143,80; Animal Sufficient Evidence IMEMDT 58,41,93. Beryllium and its compounds are on the Community Right-To-Know List.

OSHA PEL: TWA 0.002 mg(Be)/m³; STEL 0.005 mg(Be)/m³/30M; CL 0.025 mg(Be)/m³

ACGIH TLV: TWA 0.002 mg(Be)/m³; Confirmed Human Carcinogen; (Proposed: TWA 0.00005 mg(Be)/m³; STEL 0.0002 mg(Be)/m³ (sensitizer) (skin); Confirmed Human Carcinogen)

DFG MAK: DFG TRK: Animal Carcinogen, Suspected Human Carcinogen. Grinding of beryllium metal and alloys: 0.005 mg/m³ calculated as beryllium in that portion of dust that can possibly be inhaled; other beryllium compounds: 0.002 mg/m³ calculated as beryllium in that portion of dust that can possibly be inhaled

SAFETY PROFILE: Confirmed carcinogens. Beryllium compounds can enter the body through inhalation of dusts and fumes and may act locally on the skin. Even alloys of low beryllium content have been shown to be dangerous. In industry, inhalation of the dust can cause severe lung damage with symptoms appearing within months. Effects have been reported in persons living near processing plants and in families of beryllium workers. The fluoride, ammonium fluoride, sulfate, oxide, and hydroxide occur during extraction from beryllium ore. Exposure to the oxide may occur in processing of beryllium alloys and beryllium ceramics.

   The extraction of Be from its ore is attended by exposure to acid salts of the metal, particularly the fluoride (BeF₂), the ammonium fluoride and sulfate (BeSO₄), and also to beryllium oxide (BeO), and hydroxide [Be(OH)₂]. Exposure to the oxide also occurs in the casting of beryllium alloys and in operations with beryllia ceramics. In the manufacture of fluorescent powders, lamps, and sign tubes, there may be exposure to beryllium carbonate and to more complex salts, such as ZnMnBe silicate. Exposure to beryllium compounds encountered in the extraction of the metal or its oxide from the ore, particularly the halide salts, has been attended, in certain individuals, by the development of dermatitis of an edematous and papulovesicular type, chronic skin ulcers, rhinitis, nasopharyngitis, epistaxis, bronchitis, and in severe cases, by the development of an acute pneumonitis, with cough, scanty sputum, low-grade fever, rales, dyspnea, and substernal pain. Radiographs show diffuse haziness throughout both lungs, followed by the appearance of soft, ill-defined opacities. The condition occurs while the worker is exposed, sometimes within 1 or 2 months of starting work, and recovery occurs within 2 months, as a rule, though radiographic changes sometimes persist for longer periods. Occasionally, recovery may not occur and lung fibrosis results. In severe cases of pneumonitis, the patient may die. Necropsies have revealed diffuse pulmonary edema, hemorrhagic extravasation, large numbers of plasma cells, and a relative absence of polymorphonuclear infiltration. On the basis of experimental work with animals, certain investigators are of the opinion that the acute upper and lower respiratory effects are due chiefly to the acid radical present in the dust or fume, but this view has little support. A delayed form of lung disease, characterized by the occurrence of granulomatous areas in the lung tissue, has been reported in workers manufacturing fluorescent powders, lamps, and sign tubes, casting beryllium master alloys, and producing beryllium from beryl ore. Symptoms can start during exposure, but they may be delayed up to 5 years or more after the last exposure. The commonest symptoms are coughing, shortness of breath, loss of appetite, loss of weight, and fatigue. Rales are usually present in the bases and axillae, and the red cell count is frequently elevated. Cyanosis is common, and the pulse and respiratory rates are often increased. Radiographically, three stages of the disease are described: (1) a diffuse, uniform granular shadowing extending throughout both lung fields; (2) a diffuse reticular pattern on the granular background; and (3) the appearance of distinct nodules scattered through the lungs, with some enlargement and blurring of the hilar shadows. The intensity of the shadowing is usually greater in the middle third of the lung fields. The prognosis is poor. Clinical improvement may occur gradually over a period of several years, but there appears to be little tendency for the radiographic shadowing to clear. In certain cases, the disease has progressed gradually for some months or years, with death resulting from respiratory and cardiac failure. In several instances necropsies have shown the presence of a diffuse fibrosis with coarse strands of hyalinized collagen between the alveoli and, in some places, replacing them. The hyalinized areas contain granulomatous foci, the alveolar walls are thickened and fibrosed, the blood vessels being engorged and dilated. In some cases the hilar lymph nodes show granulomatous change and fibrosis. Granulomatous change has also been noted in the liver and hyaline fibrosis in the spleen. Two cases of delayed lung disease coming to autopsy have presented papular lesions on the dorsum of the hands; on the biopsy these showed "sarcoid-like" lesions with central necrosis.

Several cases have been reported in which localized granulomatous lesions developed following penetrating wounds caused by splinters of glass from broken fluorescent light tubes. Several weeks or months following the accident, swellings were noted in the injured areas and excision revealed granulomatous tumors, which in one case was shown to contain beryllium.

There is no specific treatment, but temporary remissions have been produced by ACTH and cortisone.

BFQ750   CAS: 12010-12-7   HR: 3
BERYLLIUM COMPOUND with NIOBIUM
(12:1)
mf: $Be_{12}Nb$ mw: 201.03
CONSENSUS REPORTS: IARC Cancer
Review: Group 1 IMEMDT 58,41,93;
Human Sufficient Evidence IMEMDT
58,41,93; Animal Sufficient Evidence
IMEMDT 1,17,72; Animal Sufficient
Evidence IMEMDT 23,143,80; Animal
Sufficient Evidence IMEMDT 58,41,93.
Beryllium and its compounds are on the
Community Right-To-Know List.
OSHA PEL: TWA 0.002 mg(Be)/m^3; STEL
0.005 mg(Be)/m^3/30M; CL 0.025
mg(Be)/m^3
ACGIH TLV: TWA 0.002 mg(Be)/m^3;
Confirmed Human Carcinogen; (Proposed:
TWA 0.00005 mg(Be)/m^3; STEL 0.0002
mg(Be)/m^3 (sensitizer) (skin); Confirmed
Human Carcinogen)
NIOSH REL: (Beryllium) CL not to exceed
0.005 mg(Be)/m^3
SAFETY PROFILE: Confirmed carcinogen
with experimental tumorigenic data. When
heated to decomposition in air it emits very
toxic fumes of BeO. See also BERYLLIUM
COMPOUNDS and NIOBIUM.

BFR000   CAS: 12232-67-6   HR: 3
BERYLLIUM COMPOUND with
TITANIUM (12:1)
mf: $Be_{12}Ti$ mw: 156.02
SYN: TITANIUM compounded with BERYLLIUM (1:12)
CONSENSUS REPORTS: IARC Cancer
Review: Group 1 IMEMDT 58,41,93;
Human Sufficient Evidence IMEMDT
58,41,93; Animal Sufficient Evidence
IMEMDT 1,17,72; Animal Sufficient
Evidence IMEMDT 23,143,80; Animal
Sufficient Evidence IMEMDT 58,41,93.
Beryllium and its compounds are on the
Community Right-To-Know List.
OSHA PEL: TWA 0.002 mg(Be)/m^3; STEL
0.005 mg(Be)/m^3/30M; CL 0.025
mg(Be)/m^3
ACGIH TLV: TWA 0.002 mg(Be)/m^3;
Confirmed Human Carcinogen; (Proposed:

TWA 0.00005 mg(Be)/m^3; STEL 0.0002
mg(Be)/m^3 (sensitizer) (skin); Confirmed
Human Carcinogen)
NIOSH REL: (Beryllium) CL not to exceed
0.0005 mg(Be)/m^3
SAFETY PROFILE: Confirmed carcinogen
with experimental tumorigenic data. See also
BERYLLIUM COMPOUNDS and
TITANIUM COMPOUNDS. When heated
to decomposition it emits very toxic fumes
of BeO.

BFR250   CAS: 12400-16-7   HR: 3
BERYLLIUM COMPOUND with
VANADIUM (12:1)
mf: $Be_{12}V$ mw: 159.06
SYN: VANADIUM compounded with BERYLLIUM (1:12)
CONSENSUS REPORTS: IARC Cancer
Review: Group 1 IMEMDT 58,41,93;
Human Sufficient Evidence IMEMDT
58,41,93; Animal Sufficient Evidence
IMEMDT 1,17,72; Animal Sufficient
Evidence IMEMDT 23,143,80; Animal
Sufficient Evidence IMEMDT 58,41,93.
Beryllium and its compounds are on the
Community Right-To-Know List.
OSHA PEL: TWA 0.002 mg(Be)/m^3; STEL
0.005 mg(Be)/m^3/30M; CL 0.025
mg(Be)/m^3
ACGIH TLV: TWA 0.002 mg(Be)/m^3;
Confirmed Human Carcinogen; (Proposed:
TWA 0.00005 mg(Be)/m^3; STEL 0.0002
mg(Be)/m^3 (sensitizer) (skin); Confirmed
Human Carcinogen)
NIOSH REL: (Beryllium) CL not to exceed
0.0005 mg(Be)/m^3; (REL to Vanadium) 1.0
mg(V)/m^3
SAFETY PROFILE: Confirmed carcinogen
with experimental tumorigenic data. See also
BERYLLIUM COMPOUNDS and
VANADIUM COMPOUNDS. When
heated to decomposition it emits very toxic
fumes of BeO and VO_x.

BFR500   CAS: 7787-49-7   HR: 3
BERYLLIUM FLUORIDE
mf: BeF_2 mw: 47.01

B

PROP: Amorphous, colorless, hexagonal crystals. Undergoes transition from low temp quartz to high-temp quartz structure types at 2°. Readily forms glass. Mp: 552°, d: 1.986 @ 25°. Subl @ 8°. Very sol in H_2O; sltly sol in EtOH.

SYN: BERYLLIUM DIFLUORIDE

CONSENSUS REPORTS: NTP 10th Report on Carcinogens. IARC Cancer Review: Group 1 IMEMDT 58,41,93; Human Sufficient Evidence IMEMDT 58,41,93; Animal Sufficient Evidence IMEMDT 1,17,72; Animal Sufficient Evidence IMEMDT 23,143,80; Animal Sufficient Evidence IMEMDT 58,41,93. Beryllium and its compounds are on the Community Right-To-Know List. Reported in EPA TSCA Inventory.

OSHA PEL: TWA 0.002 mg(Be)/m³; STEL 0.005 mg(Be)/m³/30M; CL 0.025 mg(Be)/m³

ACGIH TLV: TWA 0.002 mg(Be)/m³; Confirmed Human Carcinogen; (Proposed: TWA 0.00005 mg(Be)/m³; STEL 0.0002 mg(Be)/m³ (sensitizer) (skin); Confirmed Human Carcinogen); TWA 2.5 mg(F)/m³; BEI: 3 mg/g creatinine of fluorides in urine prior to shift; 10 mg/g creatinine of fluorides in urine at end of shift.

NIOSH REL: (Beryllium) CL not to exceed 0.0005 mg(Be)/m³

SAFETY PROFILE: Confirmed carcinogen with experimental carcinogenic and tumorigenic data by inhalation. Poison by ingestion, subcutaneous, intravenous, and intraperitoneal routes. See also BERYLLIUM COMPOUNDS and FLUORIDES. Incompatible with Mg. When heated to decomposition, it emits very toxic fumes of BeO and F⁻.

BFR750   CAS: 7787-52-2   HR: 3
BERYLLIUM HYDRIDE
mf: BeH_2 mw: 11.03
PROP: White solid.
CONSENSUS REPORTS: IARC Cancer Review: Group 1 IMEMDT 58,41,93; Human Sufficient Evidence IMEMDT 58,41,93; Animal Sufficient Evidence IMEMDT 1,17,72; Animal Sufficient Evidence IMEMDT 23,143,80; Animal Sufficient Evidence IMEMDT 58,41,93. Beryllium and its compounds are on the Community Right-To-Know List.

OSHA PEL: TWA 0.002 mg(Be)/m³; STEL 0.005 mg(Be)/m³/30M; CL 0.025 mg(Be)/m³

ACGIH TLV: TWA 0.002 mg(Be)/m³; Confirmed Human Carcinogen; (Proposed: TWA 0.00005 mg(Be)/m³; STEL 0.0002 mg(Be)/m³ (sensitizer) (skin); Confirmed Human Carcinogen); TWA 2.5 mg(F)/m³; BEI: 3 mg/g creatinine of fluorides in urine prior to shift; 10 mg/g creatinine of fluorides in urine at end of shift.

NIOSH REL: (Beryllium) CL not to exceed 0.0005 mg(Be)/m³

SAFETY PROFILE: Confirmed carcinogen. A dangerous fire hazard. When heated to 220°C it liberates explosive hydrogen gas. Reacts violently with methanol, water, and dilute acids. When heated to decomposition it emits toxic fumes of BeO. See BERYLLIUM COMPOUNDS and HYDRIDES.

BFS000   CAS: 13598-15-7   HR: 3
BERYLLIUM HYDROGEN PHOSPHATE (1:1)
mf: $BeHO_4P$ mw: 104.99
SYNS: BERYLLIUM PHOSPHATE □ PHOSPHORIC ACID, BERYLLIUM SALT (1:1) □ PHOSPHOROUS ACID, BERYLLIUM SALT

CONSENSUS REPORTS: NTP 10th Report on Carcinogens. IARC Cancer Review: Group 1 IMEMDT 58,41,93; Human Sufficient Evidence IMEMDT 58,41,93; Animal Sufficient Evidence IMEMDT 1,17,72; Animal Sufficient Evidence IMEMDT 23,143,80; Animal Sufficient Evidence IMEMDT 58,41,93. Beryllium and its compounds are on the Community Right-To-Know List.

OSHA PEL: TWA 0.002 mg(Be)/m³; STEL 0.005 mg(Be)/m³/30M; CL 0.025 mg(Be)/m³

B

ACGIH TLV: TWA 0.002 mg(Be)/m³; Confirmed Human Carcinogen; (Proposed: TWA 0.00005 mg(Be)/m³; STEL 0.0002 mg(Be)/m³ (sensitizer) (skin); Confirmed Human Carcinogen)
NIOSH REL: (Beryllium) CL not to exceed 0.0005 mg(Be)/m³
SAFETY PROFILE: Confirmed carcinogen with experimental carcinogenic and tumorigenic data. Poison by intravenous route. See also BERYLLIUM COMPOUNDS and PHOSPHATES. When heated to decomposition it emits very toxic fumes of BeO and PO$_x$.

**BFS250   CAS: 13327-32-7   HR: 3
BERYLLIUM HYDROXIDE**
mf: H$_2$O$_2$•Be   mw: 43.03
PROP: Colorless, orthorhombic, amorphous powder or crystals. Decomp on heating with H$_2$O loss forming BeO. Mp: decomp @ 138°. Practically insol in H$_2$O.
SYNS: BERYLLIUM DIHYDROXIDE □ BERYLLIUM HYDRATE
CONSENSUS REPORTS: NTP 10th Report on Carcinogens. IARC Cancer Review: Group 1 IMEMDT 58,41,93; Human Sufficient Evidence IMEMDT 58,41,93; Animal Sufficient Evidence IMEMDT 1,17,72; Animal Sufficient Evidence IMEMDT 23,143,80; Animal Sufficient Evidence IMEMDT 58,41,93. Beryllium and its compounds are on the Community Right-To-Know List. Reported in EPA TSCA Inventory.
OSHA PEL: TWA 0.002 mg(Be)/m³; STEL 0.005 mg(Be)/m³/30M; CL 0.025 mg(Be)/m³
ACGIH TLV: TWA 0.002 mg(Be)/m³; Confirmed Human Carcinogen; (Proposed: TWA 0.00005 mg(Be)/m³; STEL 0.0002 mg(Be)/m³ (sensitizer) (skin); Confirmed Human Carcinogen)
NIOSH REL: (Beryllium) CL not to exceed 0.0005 mg(Be)/m³
SAFETY PROFILE: Confirmed carcinogen with experimental carcinogenic and tumorigenic data. Poison by intravenous

route. See also BERYLLIUM COMPOUNDS. When heated to decomposition it emits very toxic fumes of BeO.

**BFS750   HR: 3
BERYLLIUM MANGANESE ZINC
    SILICATE**
mf: BeMnO$_4$SiZn   mw: 221.41
SYNS: MANGANESE ZINC BERYLLIUM SILICATE □ ZINC MANGANESE BERYLLIUM SILICATE
CONSENSUS REPORTS: IARC Cancer Review: Group 1 IMEMDT 58,41,93; Human Sufficient Evidence IMEMDT 58,41,93; Animal Sufficient Evidence IMEMDT 1,17,72; Animal Sufficient Evidence IMEMDT 23,143,80; Animal Sufficient Evidence IMEMDT 58,41,93. Beryllium, manganese, zinc, and their compounds are on the Community Right-To-Know List.
OSHA PEL: TWA 0.002 mg(Be)/m³; STEL 0.005 mg(Be)/m³/30M; CL 0.025 mg(Be)/m³
ACGIH TLV: TWA 0.002 mg(Be)/m³; Confirmed Human Carcinogen; (Proposed: TWA 0.00005 mg(Be)/m³; STEL 0.0002 mg(Be)/m³ (sensitizer) (skin); Confirmed Human Carcinogen)
NIOSH REL: (Beryllium) CL not to exceed 0.0005 mg(Be)/m³
SAFETY PROFILE: Confirmed carcinogen with experimental tumorigenic data. When heated to decomposition it emits very toxic fumes of BeO and ZnO. See also BERYLLIUM COMPOUNDS, MANGANESE COMPOUNDS, and ZINC COMPOUNDS.

**BFT000   CAS: 13597-99-4   HR: 3
BERYLLIUM NITRATE**
DOT: UN 2464
mf: BeN$_2$O$_6$ mw: 133.03
PROP: Deliquescent, white, amorphous solid or white-yellowish crystals. Mp: 60°, bp: decomp @ 100–200°.
SYNS: BERYLLIUM DINITRATE □ NITRIC ACID, BERYLLIUM SALT

CONSENSUS REPORTS: IARC Cancer Review: Group 1 IMEMDT 58,41,93; Human Sufficient Evidence IMEMDT 58,41,93; Animal Sufficient Evidence IMEMDT 1,17,72; Animal Sufficient Evidence IMEMDT 23,143,80; Animal Sufficient Evidence IMEMDT 58,41,93. Beryllium and its compounds are on the Community Right-To-Know List. Reported in EPA TSCA Inventory.
OSHA PEL: TWA 0.002 mg(Be)/m³; STEL 0.005 mg(Be)/m³/30M; CL 0.025 mg(Be)/m³
ACGIH TLV: TWA 0.002 mg(Be)/m³; Confirmed Human Carcinogen; (Proposed: TWA 0.00005 mg(Be)/m³; STEL 0.0002 mg(Be)/m³ (sensitizer) (skin); Confirmed Human Carcinogen)
NIOSH REL: CL not to exceed 0.0005 mg(Be)/m³
DOT CLASSIFICATION: 5.1; Label: Oxidizer, Poison
SAFETY PROFILE: Confirmed carcinogen. Poison by intraperitoneal, intravenous, and subcutaneous routes. Experimental reproductive effects. When heated to decomposition it emits very toxic fumes of BeO and NO$_x$. See also BERYLLIUM COMPOUNDS and NITRATES.

BFT250   CAS: 1304-56-9   HR: 3
BERYLLIUM OXIDE
mf: BeO   mw: 25.01
PROP: White, amorphous powder or white, hexagonal crystals; piezoelectric and pyroelectric. Undergoes hexagonal to tetragonal transition at 21°. Mp: 2507°, bp: 3900° (approx), d: 3.025. Dissolves in conc H$_2$SO$_4$ and in fused KOH. Sltly sol in H$_2$O.
SYNS: BERYLLIA □ BERYLLIUM MONOXIDE □ THERMALOX
CONSENSUS REPORTS: NTP 10th Report on Carcinogens. IARC Cancer Review: Group 1 IMEMDT 58,41,93; Human Sufficient Evidence IMEMDT 58,41,93; Animal Sufficient Evidence IMEMDT 1,17,72; Animal Sufficient Evidence IMEMDT 23,143,80; Animal Sufficient

Evidence IMEMDT 58,41,93. Beryllium and its compounds are on the Community Right-To-Know List. Reported in EPA TSCA Inventory.
OSHA PEL: TWA 0.002 mg(Be)/m³; STEL 0.005 mg(Be)/m³/30M; CL 0.025 mg(Be)/m³
ACGIH TLV: TWA 0.002 mg(Be)/m³; Confirmed Human Carcinogen; (Proposed: TWA 0.00005 mg(Be)/m³; STEL 0.0002 mg(Be)/m³ (sensitizer) (skin); Confirmed Human Carcinogen)
NIOSH REL: (Beryllium) CL not to exceed 0.0005 mg(Be)/m³
SAFETY PROFILE: Confirmed carcinogen with experimental tumorigenic data. Experimental teratogenic data. Other experimental reproductive effects. See also BERYLLIUM COMPOUNDS. Incompatible with (Mg + heat). When heated to decomposition it emits very toxic fumes of BeO.

BFT500   CAS: 19049-40-2   HR: 3
BERYLLIUM OXYACETATE
mf: C$_{12}$H$_{18}$Be$_4$O$_{13}$ mw: 406.34
PROP: Colorless cubic crystals from CHCl$_3$. Undergoes cubic to orthorhombic transition at 1°. Mp: 284°, bp: 331°. Sol in CHCl$_3$, AcOH; sltly sol in EtOH and Et$_2$O.
SYNS: BERYLLIUM ACETATE, BASIC □ BERYLLIUM OXIDE ACETATE □ HEXAKIS(μ-ACETATO-O:O')-μ⁴-OXOTETRABERYLLIUM □ HEXAKIS(μ-ACETATO)-μ⁴-OXOTETRABERYLLIUM
CONSENSUS REPORTS: IARC Cancer Review: Group 1 IMEMDT 58,41,93; Human Sufficient Evidence IMEMDT 58,41,93; Animal Sufficient Evidence IMEMDT 1,17,72; Animal Sufficient Evidence IMEMDT 23,143,80; Animal Sufficient Evidence IMEMDT 58,41,93. Beryllium and its compounds are on the Community Right-To-Know List.
OSHA PEL: TWA 0.002 mg(Be)/m³; STEL 0.005 mg(Be)/m³/30M; CL 0.025 mg(Be)/m³
ACGIH TLV: TWA 0.002 mg(Be)/m³; Confirmed Human Carcinogen; (Proposed:

B

TWA 0.00005 mg(Be)/m³; STEL 0.0002 mg(Be)/m³ (sensitizer) (skin); Confirmed Human Carcinogen)
NIOSH REL: (Beryllium) CL not to exceed 0.0005 mg(Be)/m³
SAFETY PROFILE: Confirmed carcinogen. See BERYLLIUM COMPOUNDS. When heated to decomposition it emits toxic fumes of BeO.

BFT750   CAS: 63990-88-5   HR: 3
BERYLLIUM OXYFLUORIDE
mf: BeF_2O_2 mw: 79.01
CONSENSUS REPORTS: IARC Cancer Review: Group 1 IMEMDT 58,41,93; Human Sufficient Evidence IMEMDT 58,41,93; Animal Sufficient Evidence IMEMDT 1,17,72; Animal Sufficient Evidence IMEMDT 23,143,80; Animal Sufficient Evidence IMEMDT 58,41,93. Beryllium and its compounds are on the Community Right-To-Know List.
OSHA PEL: TWA 0.002 mg(Be)/m³; STEL 0.005 mg(Be)/m³/30M; CL 0.025 mg(Be)/m³
ACGIH TLV: TWA 0.002 mg(Be)/m³; Confirmed Human Carcinogen; (Proposed: TWA 0.00005 mg(Be)/m³; STEL 0.0002 mg(Be)/m³ (sensitizer) (skin); Confirmed Human Carcinogen)
NIOSH REL: (Beryllium) CL not to exceed 0.0005 mg(Be)/m³
SAFETY PROFILE: Confirmed carcinogen. Poison by ingestion, subcutaneous, intravenous, and intraperitoneal routes. See also BERYLLIUM COMPOUNDS and FLUORIDES. When heated to decomposition it emits very toxic fumes of BeO and F⁻.

BFU000   CAS: 13597-95-0   HR: 3
BERYLLIUM PERCHLORATE
mf: $Be(ClO_4)_2$ mw: 207.91
PROP: Very hygroscopic crystals, sol in water: 148.6 g/100 mL.
CONSENSUS REPORTS: IARC Cancer Review: Group 1 IMEMDT 58,41,93; Human Sufficient Evidence IMEMDT

58,41,93; Animal Sufficient Evidence IMEMDT 1,17,72; Animal Sufficient Evidence IMEMDT 23,143,80; Animal Sufficient Evidence IMEMDT 58,41,93.
OSHA PEL: TWA 0.002 mg(Be)/m³; STEL 0.005 mg(Be)/m³/30M; CL 0.025 mg(Be)/m³
ACGIH TLV: TWA 0.002 mg(Be)/m³; Confirmed Human Carcinogen; (Proposed: TWA 0.00005 mg(Be)/m³; STEL 0.0002 mg(Be)/m³ (sensitizer) (skin); Confirmed Human Carcinogen)
NIOSH REL: CL not to exceed 0.0005 mg(Be)/m³
SAFETY PROFILE: Confirmed carcinogen. A powerful oxidant used in propellant and igniter systems. When heated to decomposition it emits toxic fumes of Cl⁻ and BeO. See also BERYLLIUM COMPOUNDS and PERCHLORATES.

BFU250   CAS: 13510-49-1   HR: 3
BERYLLIUM SULFATE (1:1)
mf: $O_4S•Be$ mw: 105.07
PROP: Colorless, tetragonal, hygroscopic crystals. Undergoes polymorphic transitions at 5° and 6°. On further heating dissoc without melting to BeO, SO_3, SO_2, and O_2 crystals. Mp: 550–600° (decomp), d: 2.443. Insol in H_2O.
SYN: SULFURIC ACID, BERYLLIUM SALT (1:1)
CONSENSUS REPORTS: NTP 10th Report on Carcinogens. IARC Cancer Review: Group 1 IMEMDT 58,41,93; Human Sufficient Evidence IMEMDT 58,41,93; Animal Sufficient Evidence IMEMDT 1,17,72; Animal Sufficient Evidence IMEMDT 23,143,80; Animal Sufficient Evidence IMEMDT 58,41,93. Beryllium and its compounds are on the Community Right-To-Know List. Reported in EPA TSCA Inventory.
OSHA PEL: TWA 0.002 mg(Be)/m³; STEL 0.005 mg(Be)/m³/30M; CL 0.025 mg(Be)/m³
ACGIH TLV: TWA 0.002 mg(Be)/m³; Confirmed Human Carcinogen; (Proposed: TWA 0.00005 mg(Be)/m³; STEL 0.0002

mg(Be)/m³ (sensitizer) (skin); Confirmed Human Carcinogen)
NIOSH REL: (Beryllium) CL not to exceed 0.0005 mg(Be)/m³
SAFETY PROFILE: Confirmed carcinogen with experimental tumorigenic data. Acute poison by inhalation, ingestion, intraperitoneal, subcutaneous, intravenous, and intratracheal routes. See also BERYLLIUM COMPOUNDS and SULFATES. Mutation data reported. When heated to decomposition it emits very toxic fumes of SO_x and BeO.

BFU500   CAS: 7787-56-6   HR: 3
BERYLLIUM SULFATE TETRA-
   HYDRATE (1:1:4)
mf: $O_4S \cdot Be \cdot 4H_2O$ mw: 177.15
PROP: Colorless, tetragonal crystals. Decomp on heating with H_2O loss. Very sol in H_2O.
SYNS: BERYLLIUM SULPHATE TETRAHYDRATE □ SULFURIC ACID, BERYLLIUM SALT (1:1), TETRAHYDRATE
CONSENSUS REPORTS: NTP 10th Report on Carcinogens. IARC Cancer Review: Group 1 IMEMDT 58,41,93; Human Sufficient Evidence IMEMDT 58,41,93; Animal Sufficient Evidence IMEMDT 1,17,72; Animal Sufficient Evidence IMEMDT 23,143,80; Animal Sufficient Evidence IMEMDT 58,41,93. Beryllium and its compounds are on the Community Right-To-Know List.
OSHA PEL: TWA 0.002 mg(Be)/m³; STEL 0.005 mg(Be)/m³/30M; CL 0.025 mg(Be)/m³
ACGIH TLV: TWA 0.002 mg(Be)/m³; Confirmed Human Carcinogen; (Proposed: TWA 0.00005 mg(Be)/m³; STEL 0.0002 mg(Be)/m³ (sensitizer) (skin); Confirmed Human Carcinogen)
NIOSH REL: CL not to exceed 0.0005 mg(Be)/m³
SAFETY PROFILE: Confirmed carcinogen with experimental carcinogenic data by inhalation. Deadly poison by subcutaneous and intravenous routes. Human mutation data reported. See also BERYLLIUM COMPOUNDS and SULFATES. When heated to decomposition it emits very toxic fumes of BeO and SO_x.

BFU750   HR: 3
BERYLLIUM TETRAHYDROBORATE
mf: B_2BeH_8 mw: 38.70
CONSENSUS REPORTS: IARC Cancer Review: Group 1 IMEMDT 58,41,93; Human Sufficient Evidence IMEMDT 58,41,93; Animal Sufficient Evidence IMEMDT 1,17,72; Animal Sufficient Evidence IMEMDT 23,143,80; Animal Sufficient Evidence IMEMDT 58,41,93. Beryllium and its compounds are on the Community Right-To-Know List.
OSHA PEL: TWA 0.002 mg(Be)/m³; STEL 0.005 mg(Be)/m³/30M; CL 0.025 mg(Be)/m³
ACGIH TLV: TWA 0.002 mg(Be)/m³; Confirmed Human Carcinogen; (Proposed: TWA 0.00005 mg(Be)/m³; STEL 0.0002 mg(Be)/m³ (sensitizer) (skin); Confirmed Human Carcinogen)
NIOSH REL: CL not to exceed 0.0005 mg(Be)/m³
SAFETY PROFILE: Confirmed carcinogen. Ignites and then explodes in air or on contact with water. Upon decomposition it emits toxic fumes of BeO and BO_x. See also BERYLLIUM COMPOUNDS and BORON COMPOUNDS.

BFV000   HR: 3
BERYLLIUM TETRAHYDROBORATE-
   TRIMETHYLAMINE
mf: $C_3H_{17}B_2BeN$ mw: 97.78
CONSENSUS REPORTS: IARC Cancer Review: Group 1 IMEMDT 58,41,93; Human Sufficient Evidence IMEMDT 58,41,93; Animal Sufficient Evidence IMEMDT 1,17,72; Animal Sufficient Evidence IMEMDT 23,143,80; Animal Sufficient Evidence IMEMDT 58,41,93. Beryllium and its compounds are on the Community Right-To-Know List.

B

OSHA PEL: TWA 0.002 mg(Be)/m³; STEL 0.005 mg(Be)/m³/30M; CL 0.025 mg(Be)/m³
ACGIH TLV: TWA 0.002 mg(Be)/m³; Confirmed Human Carcinogen; (Proposed: TWA 0.00005 mg(Be)/m³; STEL 0.0002 mg(Be)/m³ (sensitizer) (skin); Confirmed Human Carcinogen)
NIOSH REL: CL not to exceed 0.0005 mg(Be)/m³
SAFETY PROFILE: Confirmed carcinogen. It will ignite in contact with air or water. When heated to decomposition it emits toxic fumes of BeO, BO$_x$, and NO$_x$. See also BERYLLIUM COMPOUNDS and BORON COMPOUNDS.

BFV250 CAS: 39413-47-3 HR: 3
BERYLLIUM ZINC SILICATE
mf: O$_2$Si•Zn•Be   mw: 134.47
SYN: ZINC BERYLLIUM SILICATE
CONSENSUS REPORTS: NTP 10th Report on Carcinogens. IARC Cancer Review: Group 1 IMEMDT 58,41,93; Human Sufficient Evidence IMEMDT 58,41,93; Animal Sufficient Evidence IMEMDT 1,17,72; Animal Sufficient Evidence IMEMDT 23,143,80; Animal Sufficient Evidence IMEMDT 58,41,93. Beryllium and its compounds, as well as zinc and its compounds, are on the Community Right-To-Know List.
OSHA PEL: TWA 0.002 mg(Be)/m³; STEL 0.005 mg(Be)/m³/30M; CL 0.025 mg(Be)/m³
ACGIH TLV: TWA 0.002 mg(Be)/m³; Confirmed Human Carcinogen; (Proposed: TWA 0.00005 mg(Be)/m³; STEL 0.0002 mg(Be)/m³ (sensitizer) (skin); Confirmed Human Carcinogen)
NIOSH REL: (Beryllium) CL not to exceed 0.0005 mg(Be)/m³
SAFETY PROFILE: Confirmed carcinogen with experimental tumorigenic data. When heated to decomposition it emits toxic fumes of BeO and ZnO. See also BERYLLIUM COMPOUNDS, ZINC COMPOUNDS, and SILICATES.

BFW000 HR: 3
BETEL NUT
PROP: Mottled brown with fawn color. Extract of 50 g sun-dried betel nut in 100 mL boiling water (IJCNAW 17,469,76).
SYNS: ARECA CATECHU □ ARECA CATECHU Linn., fruit extract □ ARECA CATECHU Linn., nut extract □ BN □ PINANG □ POOGIPHALAM, nut extract □ SUPARI, nut extract
CONSENSUS REPORTS: IARC Cancer Review: Animal Limited Evidence IMEMDT 37,141,85.
SAFETY PROFILE: Suspected carcinogen with experimental carcinogenic and neoplastigenic data. Moderately toxic by intraperitoneal route. Experimental teratogenic and reproductive effects. When heated to decomposition it emits toxic fumes of NO$_x$. See also ARECA NUT, other betel entries, and SMOKELESS TOBACCO.

BFW125 HR: 2
BETEL QUID EXTRACT
CONSENSUS REPORTS: IARC Cancer Review: Human Inadequate Evidence IMEMDT 37,141,85; Animal Limited Evidence IMEMDT 37,141,85.
SAFETY PROFILE: Suspected carcinogen with experimental carcinogenic and tumorigenic data by skin contact. Human mutation data reported. See other betel entries.

BFW135 HR: 3
BETEL TOBACCO EXTRACT
SYN: JAFFNA TOBACCO
CONSENSUS REPORTS: IARC Cancer Review: Human Sufficient Evidence IMEMDT 37,141,85; Animal Limited Evidence IMEMDT 37,141,85.
SAFETY PROFILE: Confirmed human carcinogen. Human mutation data reported. See also SMOKELESS TOBACCO and other betel entries.

BFW500 CAS: 319-86-8 HR: 2
Δ-BHC
mf: C$_6$H$_6$Cl$_6$ mw: 290.82

PROP: Solid. Mp: 129–132°.
SYNS: Δ-BENZENEHEXACHLORIDE □ ENT 9,234 □
1-α,2-α,3-α,4-β,5-α,6-β-HEXACHLOROCYCLOHEXANE □
Δ-HEXACHLOROCYCLOHEXANE □ Δ-1,2,3,4,5,6-
HEXACHLOROCYCLOHEXANE □ Δ-LINDANE
CONSENSUS REPORTS: Reported in EPA
TSCA Inventory.
SAFETY PROFILE: Moderately toxic by
ingestion. When heated to decomposition it
emits toxic fumes of Cl⁻. See also
CHLORINATED HYDROCARBONS,
ALIPHATIC.

BFW750   CAS: 128-37-0   HR: 2
BHT (food grade)
mf: $C_{15}H_{24}O$ mw: 220.39
PROP: White, crystalline solid; faint
characteristic odor. Bp: 265°, fp: 68°, flash
p: 260°F (TOC), d: 1.048 @ 20°/4°, vap d:
7.6, mp: 71°. Sol in alc; insol in water and
propylene glycol.
SYNS: ADVASTAB 401 □ AGIDOL □ ANTIOXIDANT
DBPC □ ANTIOXIDANT 29 □ AO 29 □ AO 4K □ 2,6-
BIS(1,1-DIMETHYLETHYL)-4-METHYLPHENOL □ BUKS
□ BUTYLATED HYDROXYTOLUENE □ BUTYLHYDRO-
XYTOLUENE □ CAO 1 □ CAO 3 □ CATALIN CAO-3 □
CHEMANOX 11 □ DBMP □ DBPC (technical grade) □
DIBUTYLATED HYDROXYTOLUENE □ 2,6-DI-tert-
BUTYL-p-CRESOL (OSHA, ACGIH) □ 2,6-DI-tert-BUTYL-
1-HYDROXY-4-METHYLBENZENE □ 3,5-DI-tert-BUTYL-
4-HYDROXYTOLUENE □ 2,6-DI-terc.BUTYL-p-KRESOL
(CZECH) □ 2,6-DI-tert-BUTYL-p-METHYLPHENOL □ 2,6-
DI-tert-BUTYL-4-METHYLPHENOL □ FEMA No. 2184 □
4-HYDROXY-3,5-DI-tert-BUTYLTOLUENE □ IMPRUVOL
□ IONOL □ IONOL (antioxidant) □ 4-METHYL-2,6-DI-
terc.BUTYLFENOL (CZECH) □ METHYL-DI-tert-
BUTYLPHENOL □ 4-METHYL-2,6-DI-tert-BUTYLPHEN-
OL □ NCI-C03598 □ NONOX TBC □ PARABAR 441 □
SUSTANE □ TENOX BHT □ TOPANOL □ VANLUBE
PCX
CONSENSUS REPORTS: IARC Cancer
Review: Group 3 IMEMDT 7,56,87; Animal
Limited Evidence IMEMDT 40,161,86.
NCI Carcinogenesis Bioassay Completed;
(feed): No Evidence: mouse, rat NCITR*
NCI-CG-TR-150,79. Reported in EPA
TSCA Inventory. EPA Genetic Toxicology
Program.
OSHA PEL: TLV 10 mg/m³
ACGIH TLV: TLV 10 mg/m³; Not
Classifiable as a Human Carcinogen

SAFETY PROFILE: Poison by
intraperitoneal and intravenous routes.
Moderately toxic by ingestion. An
experimental teratogen. Other experimental
reproductive effects. A human skin irritant.
A skin and eye irritant. Questionable
carcinogen with experimental carcinogenic
and neoplastigenic data. Combustible when
exposed to heat or flame. It can react with
oxidizing materials. To fight fire, use CO_2,
dry chemical. When heated to
decomposition it emits acrid smoke and
fumes.

BFX000   CAS: 613-35-4   HR: 3
4',4'''-BIACETANILIDE
mf: $C_{16}H_{16}N_2O_2$ mw: 268.34
PROP: Solid. Mp: 317°.
SYNS: N,N'-(1,1'-BIPHENYL)-4,4'-DIYLBIS-ACETAMIDE
4',4'''-BIACETANILIDE □ N,N'-4,4'-BIPHENYLYLENE-
BISACETAMIDE □ 4,4'-DIACETYLAMINOBIPHENYL □
N,N'-DIACETYL BENZIDINE □ 4,4'-DIACETYLBENZ-
IDINE
CONSENSUS REPORTS: IARC Cancer
Review: Group 2B IMEMDT 7,56,87;
Animal Sufficient Evidence IMEMDT
16,293,78. Reported in EPA TSCA
Inventory.
SAFETY PROFILE: Suspected carcinogen
with experimental carcinogenic,
neoplastigenic, and tumorigenic data.
Mutation data reported. When heated to
decomposition it emits toxic fumes of NO_x.

BFX250   CAS: 2130-56-5   HR: 2
5,5'-BIANTHRANILIC ACID
mf: $C_{14}H_{12}N_2O_4$ mw: 272.28
PROP: Needles. Mp: 300°.
SYNS: 3,3'-BENZIDINEDICARBOXYLIC ACID □ 4,4'-
DIAMINOBIPHENYL-3,3'-DICARBOXYLIC ACID □ 4,4'-
DIAMINO-3,3'-BIPHENYLDICARBOXYLIC ACID □ 3,3'-
DICARBOXYBENZIDINE □ KWAS BENZYDYNO-
DWUKAROKSYLOWY (POLISH)
CONSENSUS REPORTS: Reported in EPA
TSCA Inventory.
SAFETY PROFILE: Questionable
carcinogen with experimental tumorigenic
data. Mutation data reported. When heated
to decomposition it emits toxic fumes of
NO_x.

B

BFX500   CAS: 103-29-7   HR: 3
BIBENZYL
mf: $C_{14}H_{14}$ mw: 182.28
PROP: Flash p: 264°F, autoign temp: 896°F, d: 1.0, vap d: 6.29, bp: 284°, mp: 52°.
SYNS: DIBENZYL □ 1,2-DIPHENYLETHANE
CONSENSUS REPORTS: Reported in EPA TSCA Inventory.
SAFETY PROFILE: Poison by intravenous route. Moderately toxic by intraperitoneal route. Combustible. To fight fire, use water, spray, mist, alcohol foam, dry chemical. When heated to decomposition it emits acrid smoke and fumes.

BGA750   CAS: 1464-53-5   HR: 3
1,1′-BI(ETHYLENE OXIDE)
mf: $C_4H_6O_2$ mw: 86.10
PROP: Colorless liquid. Bp: 142°, mp: 19°, d: 1.113 @ 18°/4°.
SYNS: BIOXIRANE □ 2,2′-BIOXIRANE □ BUTADIEN DIOXYD (GERMAN) □ BUTADIENE DIEPOXIDE □ 1,3-BUTADIENE DIEPOXIDE □ BUTADIENE DIOXIDE □ BUTANE DIEPOXIDE □ DEB □ DIEPOXYBUTANE □ 2,4-DIEPOXYBUTANE □ 1,2:3,4-DIEPOXYBUTANE □ DIOXYBUTADIENE □ ENT 26,592 □ ERYTHRITOL ANHYDRIDE □ RCRA WASTE NUMBER U085
CONSENSUS REPORTS: NTP 10th Report on Carcinogens. EPA Extremely Hazardous Substances List. EPA Genetic Toxicology Program. Community Right-To-Know List. Reported in EPA TSCA Inventory.
SAFETY PROFILE: Confirmed carcinogen with experimental tumorigenic data. Poison by ingestion, inhalation, skin contact, and intraperitoneal routes. Human mutation data reported. A severe skin and eye irritant. When heated to decomposition it emits acrid smoke and irritating fumes.

BGC250   CAS: 69382-20-3   HR: 3
BINDON ETHYL ETHER
mf: $C_{20}H_{14}O_3$ mw: 302.34
SYNS: BINDON ATHYLATHER □ 2-(3-ETHOXY-1-INDANYLIDENE)-1,3-DINDANDIONE
SAFETY PROFILE: Poison by intraperitoneal route. An experimental teratogen. Other experimental reproductive effects. When heated to decomposition it

emits acrid smoke and irritating fumes. See also ETHERS.

BGD088   CAS: 205943-18-6 HR: 3
BIOREX
mf: $C_{22}H_{19}Cl_2NO_3 \cdot C_{19}H_{30}O_5 \cdot C_8H_{10} \cdot C_5H_9NO \cdot C_4H_6O_3$ mw: 1062.24
SYN: CYCLOPROPANECARBOXYLIC ACID, 3-(2,2-DICHLOROETHENYL)-2,2-DIMETHYL-, CYANO(3-PHENOXYPHENYL) METHYL ESTER, MIXED WITH ACETIC ACID ANHYDRIDE, 5-((2-(2-BUTOXYETHOXY)ETHOXY)METHYL)-6-PROPYL-1,3-BENZODIOXOLE, DIMETHYLBENZENE AND 1-METHYL 2-PYRROLIDINONE
SAFETY PROFILE: A poison by ingestion. When heated to decomposition it emits toxic vapors of NO_x and Cl^-.

BGD100   CAS: 58-85-5   HR: D
BIOTIN
mf: $C_{10}H_{16}N_2O_3S$ mw: 244.31
PROP: White crystalline powder or fine long needles. Mp: 232–233°. Sltly sol in water, alc; insol in common org solvs.
SYNS: BIOEPIDERM □ BIOS II □ (+)-BIOTIN □ d-BIOTIN □ D-BIOTIN □ D-(+)-BIOTIN □ COENZYME R □ FACTOR S □ FACTOR S (VITAMIN) □ VITAMIN B7 □ VITAMIN H
CONSENSUS REPORTS: EPA TSCA Chemical Inventory, JUNE 1993.
SAFETY PROFILE: Experimental reproductive effects. When heated to decomposition it emits toxic fumes of NO_x, SO_x.

BGE000   CAS: 92-52-4   HR: 3
BIPHENYL
mf: $C_{12}H_{10}$ mw: 154.22
PROP: Monoclinic, white scales, with a pleasant odor. Mp: 71°, bp: 255°, flash p: 235°F (CC), d: 0.991 @ 75°/4°, autoign temp: 1004°F, vap d: 5.31, lel: 0.6% @ 232°, uel: 5.8% @ 331°F. IDLH 100 mg/m³.
SYNS: BIBENZENE □ 1,1′-BIPHENYL □ DIPHENYL (OSHA) □ LEMONENE □ PHENADOR-X □ PHENYLBENZENE □ PHPH □ XENENE
CONSENSUS REPORTS: EPA Genetic Toxicology Program. Reported in EPA TSCA Inventory. Community Right-To-Know List.
OSHA PEL: TWA 0.2 ppm

ACGIH TLV: TWA 0.2 ppm
DFG MAK: 0.16 ppm (1 mg/m^3)
SAFETY PROFILE: Poison by intravenous route. Moderately toxic by ingestion. A powerful irritant by inhalation in humans. Human systemic effects by inhalation of very small amounts: flaccid paralysis, nausea or vomiting, and other unspecified gastrointestinal effects. Questionable carcinogen with experimental tumorigenic and neoplastigenic data. Mutation data reported. Combustible when exposed to heat or flame; can react with oxidizing materials. To fight fire, use CO_2, dry chemical, water spray, mist, fog. When heated to decomposition it emits acrid smoke and fumes.

BGJ250   CAS: 90-43-7   HR: 3
2-BIPHENYLOL
mf: $C_{12}H_{10}O$ mw: 170.22
PROP: Needles from pet ether. Mp: 56°, bp: 275°.
SYNS: o-BIPHENYLOL □ (1,1'-BIPHENYL)-2-OL □ o-DIPHENYLOL □ DOWCIDE 1 □ DOWCIDE 1 ANTIMICROBIAL □ 2-HYDROXYBIFENYL (CZECH) □ o-HYDROXYBIPHENYL □ 2-HYDROXYBIPHENYL □ o-HYDROXYDIPHENYL □ 2-HYDROXYDIPHENYL □ KIWI LUSTR 277 □ NCI-C50351 □ OPP □ ORTHO-HYDROXYDIPHENYL □ ORTHOPHENYL-PHENOL □ ORTHOXENOL □ o-PHENYLPHENOL □ 2-PHENYL-PHENOL □ PREVENTOL O EXTRA □ REMOL TRF □ TETROSIN OE □ TORSITE □ TUMESCAL OPE □ USAF EK-2219 □ o-XENOL
CONSENSUS REPORTS: IARC Cancer Review: Group 3 IMEMDT 7,56,87; Animal Inadequate Evidence IMEMDT 30,329,83; NTP Carcinogenesis Studies (dermal); No Evidence: mouse NTPTR* NTP-TR-301,86. Reported in EPA TSCA Inventory. On Community Right-To-Know List.
SAFETY PROFILE: A poison by intraperitoneal route. Moderately toxic by ingestion and possibly other routes. An experimental teratogen. Other experimental reproductive effects. Human mutation data reported. Severe eye and moderate skin irritant. Questionable carcinogen with experimental carcinogenic data. When

heated to decomposition it emits acrid smoke and irritating fumes.

BGJ500   CAS: 92-69-3   HR: 3
4-BIPHENYLOL
mf: $C_{12}H_{10}O$ mw: 170.22
PROP: Needles or plates from EtOH (aq). Mp: 164–165°, bp: 305–308°.
SYNS: p-HYDROXYBIPHENYL □ 4-HYDROXYBIPHENYL □ p-HYDROXYDIPHENYL □ 4-HYDROXYDIPHENYL □ PARAXENOL □ p-PHENYLPHENOL □ 4-PHENYLPHENOL
CONSENSUS REPORTS: Reported in EPA TSCA Inventory.
SAFETY PROFILE: Acute poison by intraperitoneal route. Questionable carcinogen with experimental carcinogenic and tumorigenic data. When heated to decomposition it emits acrid, irritating fumes.

BGJ750   CAS: 132-27-4   HR: 3
2-BIPHENYLOL, SODIUM SALT
mf: $C_{12}H_9O$•Na   mw: 192.20
SYNS: BACTROL □ (1,1'-BIPHENYL)-2-OL, SODIUM SALT □ D.C.S. □ DORVICIDE A □ DOWICIDE □ DOWICIDE A □ DOWICIDE A & A FLAKES □ DOWIZID A □ 2-HYDROXYBIPHENYL SODIUM SALT □ 2-HYDROXYDIPHENYL SODIUM □ 2-HYDROXY-DIPHENYL, SODIUM SALT □ MIL-DU-RID □ MYSTOX WFA □ NATRIPHENE □ OPP-Na □ OPP-SODIUM □ ORPHENOL □ PHENOL, o-PHENYL-, SODIUM deriv. □ o-PHENYLPHENOL, SODIUM SALT □ 2-PHENYL-PHENOL SODIUM SALT □ PREVENTOL-ON □ PREVENTOL ON & ON EXTRA □ SODIUM 2-BIPHENYLOLATE □ SODIUM (1,1'-BIPHENYL)-2-OLATE □ SODIUM, (2-BIPHENYLYLOXY)- □ SODIUM 2-HYDROXYDIPHENYL □ SODIUM ORTHO PHENYLPHENATE □ SODIUM o-PHENYLPHENATE □ SODIUM 2-PHENYLPHENATE □ SODIUM o-PHENYLPHENOL □ SODIUM o-PHENYLPHENOLATE □ SODIUM o-PHENYLPHENOXIDE □ SOPP □ STOPMOLD B □ TOPANE
CONSENSUS REPORTS: IARC Cancer Review: Group 2B IMEMDT 7,56,87; Animal Limited Evidence IMEMDT 30,329,83. Reported in EPA TSCA Inventory.
SAFETY PROFILE: Suspected carcinogen with experimental carcinogenic, neoplastigenic, and tumorigenic data.

Moderately toxic by ingestion. Experimental teratogenic and reproductive effects. A human skin irritant. A severe skin irritant to experimental animals. When heated to decomposition it emits toxic fumes of Na_2O. See also 2-BIPHENYLOL.

BGK500   CAS: 91-95-2   HR: 3
3,3′,4,4′-BIPHENYLTETRAMINE
mf: $C_{12}H_{14}N_4$ mw: 214.30
PROP: Crystals from MeOH. Mp: 178–179°.
SYNS: 3,3′-DIAMINOBENZIDENE □ 3,3′,4,4′-DIPHENYLTETRAMINE □ 3,3′,4,4′-TETRAAMINO-BIPHENYL
CONSENSUS REPORTS: Reported in EPA TSCA Inventory.
DFG MAK: Confirmed Animal Carcinogen with Unknown Relevance to Humans
SAFETY PROFILE: Suspected carcinogen with experimental tumorigenic data. Moderately toxic by ingestion. Mutation data reported. When heated to decomposition it emits toxic fumes of NO_x. See also AROMATIC AMINES.

BGK750   CAS: 7411-49-6   HR: 3
3,3′,4,4′-BIPHENYLTETRAMINE
**   TETRAHYDROCHLORIDE**
mf: $C_{12}H_{14}N_4 \cdot 4ClH$ mw: 360.14
PROP: Crystals. Sol in acids.
SYNS: 3,3′-DIAMINOBENZIDINE TETRAHYDRO-CHLORIDE □ 3,3′,4,4′-TETRAAMINOBIPHENYL TETRAHYDROCHLORIDE
CONSENSUS REPORTS: Reported in EPA TSCA Inventory.
DFG MAK: Confirmed Animal Carcinogen with Unknown Relevance to Humans
SAFETY PROFILE: Suspected carcinogen with experimental neoplastigenic and tumorigenic data. Poison by intraperitoneal route. When heated to decomposition it emits very toxic fumes of HCl and NO_x. See also AROMATIC AMINES.

BGM100   CAS: 37940-57-1   HR: 3
4-BIPHENYLYL ETHYLKETONE
mf: $C_{15}H_{14}O$ mw: 210.29

SYNS: KETONE, 4-BIPHENYL ETHYL □ 4-PHENYLPROPIOPHENONE □ PROPIOPHENONE, 4′-PHENYL-
DOT CLASSIFICATION: 3; Label: Flammable Liquid
SAFETY PROFILE: A poison by intravenous route. A flammable liquid. When heated to decomposition it emits acrid smoke and irritating vapors.

BGO500   CAS: 366-18-7   HR: 3
2,2′-BIPYRIDINE
mf: $C_{10}H_8N_2$ mw: 156.20
PROP: White crystals or prisms from pet ether. Mp: 69.7°, bp: 272–273°. Sol in H_2O, EtOH, Et_2O, C_6H_6, $CHCl_3$, and dilute acids.
SYNS: BIPYRIDINE □ α,α′-BIPYRIDINE □ α,α′-BIPYRIDYL □ 2,2′-BIPYRIDYL □ 2,2′-BYPYRIDIN □ CI-588 □ α,α′-DIPYRIDYL □ 2,2′-DIPYRIDYL
CONSENSUS REPORTS: Reported in EPA TSCA Inventory.
SAFETY PROFILE: Poison by ingestion, subcutaneous, and intraperitoneal routes. Experimental teratogenic data. Questionable carcinogen with experimental tumorigenic data. Mutation data reported. When heated to decomposition it emits toxic fumes of NO_x.

BGO750   CAS: 8001-88-5   HR: 2
BIRCH TAR OIL
PROP: Brown liquid; leather-like odor. D: 0.886–0.950. Found in the tar of the bark and wood of *Betula pendula Roth* (Fam. *Betulaceae*) and prepared by steam distillation of the tar obtained by dry distillation of the bark and wood (FCTXAV 11,1011,73). Sol in fixed oils; insol in glycerin, mineral oil, and propylene glycol.
SYN: BIRCH TAR OIL, RECTIFIED (FCC)
CONSENSUS REPORTS: Reported in EPA TSCA Inventory.
SAFETY PROFILE: A skin irritant. Moderately irritating to eyes and mucous membranes. A mild allergen. Combustible when exposed to heat or flame; can react with oxidizing materials.

BGP250　CAS: 304-28-9　HR: 3
2,7-BIS(ACETAMIDO)FLUORENE
mf: $C_{17}H_{16}N_2O_2$　mw: 280.35
SYNS: 2,7-DIACETAMIDOFLUORENE □ 2,7-DIACETYLAMINOFLUORENE □ 2,7-FAA □ N,N'-FLUOREN-2,7-YLBISACETAMIDE □ 2,7-FLUORENYLBISACETAMIDE □ N,N'-FLUOREN-2,7-YLENEBISACETAMIDE □ N,N'-2,7-FLUORENYL-ENEBISACETAMIDE □ N,N'-(FLUOREN-2,7-YLENE)BIS(ACETYLAMINE) □ N,N'-2,7-FLUORENYL-ENEDIACETAMIDE
CONSENSUS REPORTS: EPA Genetic Toxicology Program.
SAFETY PROFILE: Suspected carcinogen with experimental carcinogenic, neoplastigenic, and tumorigenic data. Mutation data reported. When heated to decomposition it emits toxic fumes of NO_x.

BGQ000　CAS: 5967-09-9　HR: 3
BIS(ACETOXYDIBUTYLSTANNANE)
　　OXIDE
mf: $C_{20}H_{42}O_5Sn_2$　mw: 600.00
SYNS: BIS(DIBUTYLACETOXYTIN)OXIDE □ DIACETOXYTETRABUTYLDISTANNOXANE
CONSENSUS REPORTS: Reported in EPA TSCA Inventory.
OSHA PEL: TWA 0.1 mg(Sn)/m³ (skin)
ACGIH TLV: TWA 0.1 mg(Sn)/m³; STEL 0.2 mg(Sn)/m³ (skin).
NIOSH REL: (Organotin Compounds) TWA 0.1 mg(Sn)/m³
SAFETY PROFILE: Poison by intravenous route. See also TIN COMPOUNDS. When heated to decomposition it emits acrid smoke and irritating fumes.

BGT000　CAS: 28434-86-8　HR: 3
BIS(4-AMINO-3-CHLOROPHENYL)
　　ETHER
mf: $C_{12}H_{10}Cl_2N_2O$　mw: 269.14
SYNS: 3,3'-DICHLOR-4,4'-DIAMINO-DIPHENYL-AETHER (GERMAN) □ 3,3'-DICHLORO-4,4'-DIAMINODIPHENYL ETHER □ 4,4'-OXYBIS(2-CHLOROANILINE) □ 4,4'-OXYBIS(2-CHLORO-BENZENAMINE)
CONSENSUS REPORTS: IARC Cancer Review: Group 2B IMEMDT 7,56,87; Animal Sufficient Evidence IMEMDT 16,309,78.

SAFETY PROFILE: Suspected carcinogen with experimental carcinogenic data. Mutation data reported. When heated to decomposition it emits toxic fumes of Cl^- and NO_x. See also ETHERS.

BGT250　CAS: 314-13-6　HR: 3
4,4'-BIS(1-AMINO-8-HYDROXY-2,4-
　　DISULFO-7-NAPHTHYLAZO)-3,3'-
　　BITOLYL, TETRASODIUM SALT
mf: $C_{34}H_{24}N_6O_{14}S_4 \cdot 4Na$　mw: 960.84
PROP: Blue crystals with brown/green luster. Sol in H_2O, EtOH, acid, and alkalies.
SYNS: 4,4'-BIS(7-(1-AMINO-8-HYDROXY-2,4-DISULFO)-NAPHTHYLAZO)-3,3'-BITOLYL, TETRASODIUM SALT □ 4,4'-BIS(1-AMINO-8-HYDROXY-2,4-DISULPHO-7-NAPHTHYLAZO)-3,3'-BITOLYL, TETRASODIUM SALT □ BLEKIT EVANSA (POLISH) □ CHLORAZOL SKY BLUE FF □ C.I. 23860 □ C.I. DIRECT BLUE 53 □ DIAMINE SKY BLUE FF □ DIAZOBLEU □ DIAZOL PURE BLUE FF □ DYE EVANS BLUE □ EB □ EVABLIN □ EVANS BLUE DYE □ GEIGY-BLAU 536 □ T 1824
CONSENSUS REPORTS: IARC Cancer Review: Group 3 IMEMDT 7,56,87; Animal Limited Evidence IMEMDT 8,151,75. Reported in EPA TSCA Inventory. EPA Genetic Toxicology Program.
SAFETY PROFILE: Poison by intraperitoneal route. Moderately toxic by intravenous route. An experimental teratogen. Other experimental reproductive effects. Questionable carcinogen with experimental tumorigenic data. Mutation data reported. When heated to decomposition it emits very toxic fumes of SO_x, Na_2O, and NO_x.

BGU600　CAS: 7300-34-7　HR: 3
1,4-BIS(3-AMINOPROPOXY)BUTANE
mf: $C_{12}H_{24}N_2O_2$　mw: 228.38
SYNS: 1,4-BIS(γ-AMINOPROPOXY)BUTANE □ 1,4-BUTANEDIOL BIS(3-AMINOPROPYL) ETHER □ α,OMEGA-DIAMINO-4,9-DIOXADODECANE □ 1,12-DIAMINO-4,9-DIOXADODECANE □ 4,9-DIOXA-1,12-DIAMINODODECANE □ 4,9-DIOXADODECANE-1,12-DIAMINE □ 1-PROPANAMINE, 3,3'-(1,4-BUTANEDIYL-BIS(OXY))BIS- □ PROPYLAMINE, 3,3'-(TETRAMETHYL-ENEDIOXY)BIS- □ 3,3'-(TETRAMETHYLENE-DIOXY)BIS(PROPYLAMINE) □ 3,3'-(TETRAMETHYL-ENEDIOXY)DI(PROPANAMINE)

SAFETY PROFILE: A poison by skin contact. Moderately toxic by ingestion and inhalation. When heated to decomposition it emits toxic vapors of NO_x.

BGU750   CAS: 105-83-9   HR: 3
BIS(γ-AMINOPROPYL)METHYLAMINE
mf: $C_7H_{19}N_3$ mw: 145.29
PROP: Liquid, completely miscible in water. D: 0.9307 @ 20°/20°, bp: 240.6°, fp: −29.6°, flash p: 220°F.
SYNS: BIS(ω-AMINOPROPYL)METHYLAMINE □ BIS(3-AMINOPROPYL)METHYLAMINE □ N,N-BIS(γ-AMINOPROPYL)METHYLAMINE □ N,N-BIS(3-AMINOPROPYL)METHYLAMINE □ 3,7'-DIAMINO-N-METHYLDIPROPYLAMINE □ METHYLBIS(3-AMINOPROPYL)AMINE
CONSENSUS REPORTS: Reported in EPA TSCA Inventory.
SAFETY PROFILE: Poison by inhalation and skin contact. Moderately toxic by ingestion. A skin and severe eye irritant. See also AMINES. Combustible when exposed to heat or flame. To fight fire, use foam, fog, dry chemical. When heated to decomposition it emits toxic fumes of NO_x.

BGV000   CAS: 7209-38-3   HR: 3
1,4-BIS(AMINOPROPYL)PIPERAZINE
mf: $C_{10}H_{24}N_4$ mw: 200.38
SYN: BIS(AMINOPROPYL)PIPERAZINE (DOT)
CONSENSUS REPORTS: Reported in EPA TSCA Inventory.
SAFETY PROFILE: Poison by intravenous route. A corrosive material and a powerful irritant to skin, eyes, and mucous membranes. When heated to decomposition it emits toxic fumes of NO_x.

BGW100   CAS: 3426-43-5   HR: 3
4,4'-BIS(((4-ANILINO-6-METHOXY-s-TRIAZIN-2-YL)AMINO)-2,2'-STILBENEDISULFONIC ACID) DISODIUM SALT
mf: $C_{34}H_{28}N_{10}O_8S_2 \cdot 2Na$ mw: 814.82
SYN: DISODIUM-4,4'-BIS((4-ANILINO-6-METHOXY-s-TRIAZIN-2-YL)AMINO)STILBENE-2,2'-DISULFONATE
CONSENSUS REPORTS: Reported in EPA TSCA Inventory.

SAFETY PROFILE: Poison by intraperitoneal route. An eye irritant. When heated to decomposition it emits toxic fumes of SO_x and Cl^-.

BGY700   CAS: 1271-54-1   HR: 3
BIS-BENZENE CHROMIUM
mf: $C_{12}H_{12}Cr$ mw: 208.24
PROP: Air-sensitive brown-black crystals. Mp: 284–285°. Sol in C_6H_6; sltly sol in Et_2O. IDLH 250 mg/m³ [as Cr(II)].
SYNS: CHROMIUM, BIS(BENZENE)-(8CI) □ CHROMIUM, BIS(eta⁶)-BENZENE)-(9CI) □ CHROMIUM(II), DIPHENYL- □ DIBENZENE-CHROMIUM □ DIPHENYLCHROMIUM
OSHA PEL: TWA 0.5 mg(Cr)/m³
ACGIH TLV: TWA 0.5 mg(Cr)/m³; Not Classifiable as a Carcinogen
SAFETY PROFILE: Poison by intravenous route. When heated to decomposition it emits toxic fumes of Cr.

BGY720   CAS: 12089-29-1   HR: 3
BIS(BENZENE)CHROMIUM IODIDE
mf: $C_{12}H_{12}Cr \cdot I$ mw: 335.14
PROP: Light-sensitive, air-stable yellow solid. Sol in H_2O and EtOH. IDLH 25 mg/m³ [as Cr(III)].
SYNS: BIS(BENZENE)CHROMIUM(1+)IODIDE □ CHROMIUM(1+), BIS(BENZENE)-, IODIDE (8CI) □ (CHROMIUM(1+), BIS(eta⁶)-BENZENE)-, IODIDE (9CI) □ CHROMIUM, BIS(BENZENE)IODO- □ CHROMIUM(III), DIPHENYL-, IODIDE □ DIBENZENECHROMIUM IODIDE □ DIPHENYLCHROMIUM(III) IODIDE
OSHA PEL: TWA 0.5 mg(Cr)/m³; Not Classifiable as a Carcinogen
ACGIH TLV: TWA 0.5 mg(Cr)/m³; Not Classifiable as a Carcinogen; Proposed: (inhalable fraction) 0.1 mg/m³; Not Classifiable as a Human Carcinogen)
SAFETY PROFILE: Poison by intravenous route. When heated to decomposition it emits toxic fumes of Cr and I^-.

BHA750   CAS: 155-04-4   HR: 3
BIS(2-BENZOTHIAZOLYLTHIO)ZINC
mf: $C_{14}H_8N_2S_4 \cdot Zn$ mw: 397.85
SYNS: 2-BENZOTHIAZOLETHIOL, ZINC SALT (2:1) □ BIS(MERCAPTOBENZOTHIAZOLATO)ZINC □ HERMAT

Zn-MBT □ 2-MERCAPTOBENZOTHIAZOLE ZINC SALT □ OXAF □ PENNAC ZT □ TISPERSE MB-58 □ USAF GY-7 □ VULKACIT ZM □ ZENITE □ ZENITE SPECIAL □ ZETAX □ ZINC-2-BENZOTHIAZOLETHIOLATE □ ZINC BENZOTHIAZOLYL MERCAPTIDE □ ZINC BENZOTHIAZOL-2-YLTHIOLATE □ ZINC BENZO-THIAZYL-2-MERCAPTIDE □ ZINC MERCAPTO-BENZOTHIAZOLATE □ ZINC-2-MERCAPTO-BENZOTHIAZOLE □ ZINC MERCAPTOBENZO-THIAZOLE SALT □ ZMBT □ ZnMB

CONSENSUS REPORTS: Reported in EPA TSCA Inventory. Zinc compounds are on the Community Right-To-Know List.

SAFETY PROFILE: Poison by intraperitoneal route. Moderately toxic by ingestion and subcutaneous routes. Questionable carcinogen with experimental carcinogenic data. When heated to decomposition it emits very toxic fumes of SO_x, NO_x, and ZnO. See also ZINC COMPOUNDS and MERCAPTANS.

BHB000   CAS: 64092-23-5   HR: 3
BIS(2-BENZOYLBENZOATO)BIS(3-(1-
METHYL-2-
PYRROLIDINYL)PYRIDINE) NICKEL
TRIHYDRATE

mf: $C_{48}H_{46}N_4NiO_6 \cdot 3H_2O$ mw: 887.75
PROP: IDLH Ca [10 mg/m³ (as Ni)].
SYN: NICOTINE, COMPOUND, with NICKEL(II)-o-BENZOYL BENZOATE TRIHYDRATE (2:1)

CONSENSUS REPORTS: NTP 10th Report on Carcinogens. Nickel and its compounds are on the Community Right-To-Know List.

OSHA PEL: TWA 0.1 mg(Ni)/m³
ACGIH TLV: TWA 0.1 mg(Ni)/m³; Not Classifiable as a Carcinogen
NIOSH REL: (Inorganic Nickel) TWA 0.015 mg(Ni)/m³

SAFETY PROFILE: Confirmed Human Carcinogen. Poison by ingestion and intraperitoneal routes. See also NICKEL COMPOUNDS and NICOTINE. When heated to decomposition it emits toxic fumes of NO_x.

BHB950   CAS: 4028-32-4   HR: 1
4,4'-BIS((4-BIS((2-HYDROXYETHYL)-
AMINO)-6-CHLORO-s-TRIAZIN-2-
YL)AMINO)-2,2'-STILBENEDI-

SULFONIC ACID, DISODIUM SALT

mf: $C_{28}H_{30}Cl_2N_{10}O_{10}S_2 \cdot 2Na$ mw: 847.68

CONSENSUS REPORTS: Reported in EPA TSCA Inventory.

SAFETY PROFILE: An eye irritant. When heated to decomposition it emits toxic fumes of NO_x, SO_x, and Cl⁻.

BHK250   CAS: 15546-16-4   HR: 3
BIS(BUTOXYMALEOYLOXY)DIBUTYL-
STANNANE

mf: $C_{24}H_{40}O_8Sn$ mw: 575.33
SYNS: DI-N-BUTYLTIN DI(MONOBUTYL)MALEATE □ DI-N-BUTYL-ZINN-DI(MONOBUTYL)MALEINAT (GERMAN)

CONSENSUS REPORTS: Reported in EPA TSCA Inventory.

OSHA PEL: TWA 0.1 mg(Sn)/m³ (skin)
ACGIH TLV: TWA 0.1 mg(Sn)/m³; STEL 0.2 mg(Sn)/m³ (skin).
NIOSH REL: (Organotin Compounds) TWA 0.1 mg(Sn)/m³

SAFETY PROFILE: Poison by ingestion. See also TIN COMPOUNDS. When heated to decomposition it emits acrid smoke and irritating fumes.

BHK500   CAS: 29575-02-8   HR: 2
BIS(BUTOXYMALEOYLOXY)DIOCTYLS
TANNANE

mf: $C_{32}H_{56}O_8Sn$ mw: 687.57
SYNS: DI-N-OCTYLTIN BIS(BUTYL MALEATE) □ DI-N-OCTYLTIN DIMONOBUTYLMALEATE □ DI-N-OCTYLZINN-DIMONOBUTYLMALEINAT (GERMAN)

CONSENSUS REPORTS: Reported in EPA TSCA Inventory.

OSHA PEL: TWA 0.1 mg(Sn)/m³ (skin)
ACGIH TLV: TWA 0.1 mg(Sn)/m³; STEL 0.2 mg(Sn)/m³ (skin).
NIOSH REL: (Organotin Compounds) TWA 0.1 mg(Sn)/m³

SAFETY PROFILE: Moderately toxic by ingestion. See also TIN COMPOUNDS. When heated to decomposition it emits acrid smoke and irritating fumes.

B

BHL100   CAS: 25155-25-3   HR: 1
α-α'-BIS(tert-BUTYLPEROXY)-
DIISOPROPYLBENZENE
mf: $C_{20}H_{34}O_4$ mw: 338.54
SYNS: PEROXIDE, (PHENYLENEBIS(1-METHYL-
ETHYLIDENE))BIS(1,1-DIMETHYLETHYL)- □
PEROXIDE, (PHENYLENEDIISOPROPYLIDENE)BIS(tert-
BUTYL)- □ (PHENYLENEDIISOPROPYLIDENE)BIS(tert-
BUTYLPEROXIDE) □ VUL-CUP □ VUL-CUP 40KE □
VUL-CUP R
CONSENSUS REPORTS: Reported in EPA
TSCA Inventory.
SAFETY PROFILE: A skin irritant. When
heated to decomposition it emits acrid
smoke and irritating fumes.

BHL800   CAS: 58705-49-0   HR: 3
2,3-BIS(CARBOMETHOXYMERCAPTO)-
QUINOXALINE
mf: $C_{12}H_{10}N_2O_4S_2$ mw: 310.36
SYNS: AI3-25722 □ CARBONIC ACID, DITHIODI-, o,o'-
DIMETHYL S,S'-(2,3-QUINOXALINEDIYL) ESTER □
CARBONOTHIOIC ACID, S,S'-2,3-QUINOXALINEDIYL
o,o'-DIMETHYL ESTER □ CARBONIC ACID, THIO-, o,o'-
DIMETHYL S,S'-2,3-QUINOXALINEDIYL ESTER □
CARBONIC ACID, THIO-, o-METHYL ESTER, S,S-
DIESTER WITH 2,3-QUINOXALINEDITHIOL □ o,o'-
DIMETHYL S,S'-2,3-QUINOXALINEDIYL
THIOCARBONATE □ SAS 2185
SAFETY PROFILE: A poison by
intravenous route. When heated to
decomposition it emits toxic vapors of NO_x
and SO_x.

BHM000   CAS: 111-17-1   HR: 3
BIS(2-CARBOXYETHYL) SULFIDE
mf: $C_6H_{10}O_4S$ mw: 178.22
PROP: Very sol in alc, hot water, acetate;
sltly sol in water. Mp: 134°.
SYNS: DIETHYL SULFIDE-2,2'-DICARBOXYLIC ACID
□ KYSELINA-β,β'-THIODIPROPIONOVA (CZECH) □
TDPA □ 2-(2,3,5,6-TETRAMETHYLPHENOXY)-
PROPIONIC ACID □ 4-THIAHEPTANEDIOIC ACID □
THIODIPROPIONIC ACID □ β,β'-THIODIPROPIONIC
ACID □ 3,3'-THIODIPROPIONIC ACID □ TYOX A
CONSENSUS REPORTS: Reported in EPA
TSCA Inventory.
SAFETY PROFILE: A poison by
intraperitoneal and intravenous routes.
Moderately toxic by ingestion. A skin and
eye irritant. When heated to decomposition

it emits toxic fumes of SO_x. See also
SULFIDES.

BHM750   CAS: 94-17-7   HR: 2
BIS(p-CHLOROBENZOYL) PEROXIDE
mf: $C_{14}H_8Cl_2O_4$ mw: 311.12
PROP: A white, granular material. Insol in
water; sol in org solvs.
SYNS: CADPX PS □ p-CHLOROBENZOYL PEROXIDE
(DOT) □ p,p'-DICHLOROBENZOYL PEROXIDE □ DI-(4-
CHLOROBENZOYL) PEROXIDE
CONSENSUS REPORTS: Reported in EPA
TSCA Inventory.
SAFETY PROFILE: Moderately toxic by
intraperitoneal route. Probably an irritant to
skin and mucous membranes. Dangerous
fire hazard; a powerful oxidizer. Store in a
cool place away from fire hazards, sparks,
open flames, and out of the direct rays of
the sun. Dangerous explosion hazard; this
material may explode by heat (over 38°) or
contamination. Any contaminant that acts as
an accelerator to the polymerization or
decomposition of this material can cause an
explosion. Heat or contact with certain
fumes or mists can cause it to explode. To
fight small fires, use CO_2 or foam
extinguishers. Water spray or mist may also
be used. Dry chemical is effective. When
heated to decomposition it emits toxic
fumes of Cl⁻. See also PEROXIDES,
ORGANIC.

BHN000   CAS: 366-93-8   HR: 3
trans-N,N'-BIS(2-CHLOROBENZYL)-1,4-
CYCLOHEXANEBIS(METHYLAMINE)
DIHYDROCHLORIDE
mf: $C_{22}H_{28}Cl_2N_2$•2ClH   mw: 464.34
SYNS: AY 9944 □ trans-1,4-BIS(2-DICHLOROBENZYL-
AMINOETHYL)CYCLOHEXANE DICHLORHYDRATE
(FRENCH) □ trans-N,N'-(1,4-CYCLOHEXYLENE-
DIMETHYLENE)BIS(2-CHLOROBENZYLAMINE)
DIHYDROCHLORIDE
SAFETY PROFILE: Poison by ingestion.
Experimental teratogenic and reproductive
effects. Inhibits cholesterol synthesis. When
heated to decomposition it emits very toxic
fumes of NO_x and Cl⁻.

BHN750   CAS: 334-22-5   HR: 3
BIS-β-CHLOROETHYLAMINE
mf: $C_4H_9Cl_2N$ mw: 142.04
SYNS: N,N-BIS-(β-CHLORAETHYL)-AMIN (GERMAN) □ NH-LOST □ NOR-NITROGEN MUSTARD □ NSC-10873
SAFETY PROFILE: Poison by intraperitoneal, subcutaneous, and intravenous routes. Experimental reproductive effects. Human mutation data reported. When heated to decomposition it emits very toxic NO_x and Cl^-.

BHO250   CAS: 821-48-7   HR: 3
BIS(2-CHLOROETHYL)AMINE
HYDROCHLORIDE
mf: $C_4H_9Cl_2N \cdot ClH$ mw: 178.50
SYNS: BIS(β-CHLOROETHYL)AMINE HYDRO-CHLORIDE □ N,N-BIS(2-CHLOROETHYL)AMINE HYDROCHLORIDE □ BIS(2-CHLOROETHYL)-AMMONIUM CHLORIDE □ 2-CHLORO-N-(2-CHLOROETHYL)ETHANAMINE HYDROCHLORIDE □ β,β'-DICHLORODIETHYLAMINE HYDROCHLORIDE □ 2,2'-DICHLORODIETHYLAMINE HYDROCHLORIDE □ DI-2-CHLOROETHYLAMINE HYDROCHLORIDE □ LEO 72a □ NC 26 □ NOR-HN2 □ NOR-HN2 HYDROCHLOR-IDE □ NOR-LOST HYDROCHLORID (GERMAN) □ NORNITROGEN MUSTARD HYDROCHLORIDE □ NSC-10873 □ SK 555 □ TL 161
CONSENSUS REPORTS: EPA Genetic Toxicology Program. Reported in EPA TSCA Inventory.
SAFETY PROFILE: A poison by inhalation, intraperitoneal, intramuscular, and subcutaneous routes. An experimental teratogen. Human mutation data reported. When heated to decomposition it emits toxic fumes of NH_3, NO_x, and Cl^-.

BHO500   CAS: 1215-16-3   HR: 3
4'-(BIS(2-CHLOROETHYL)AMINO)-
ACETANILIDE
mf: $C_{12}H_{16}Cl_2N_2O$ mw: 275.20
SYNS: p-ACETYLAMINOPHENYL DERIVATIVE of NITROGEN MUSTARD □ LONIN 3
SAFETY PROFILE: Poison via intraperitoneal route. An experimental teratogen. Mutation data reported. When heated to decomposition it emits very toxic fumes of Cl^- and NO_x.

BHP150   CAS: 24813-03-4   HR: 3
1-((BIS(2-CHLOROETHYL)AMINO)-
BENZOYL)PIPERIDINE
mf: $C_{16}H_{22}Cl_2N_2O$ mw: 329.30
SYNS: KETONE, m-(BIS(2-CHLOROETHYL)AMINO)PHENYL PIPERIDINO □ PIPERIDINE, 1-(m-(BIS(2-CHLOROETHYL)-AMINO)BENZOYL)-
DOT CLASSIFICATION: 3; Label: Flammable Liquid
SAFETY PROFILE: A poison by an unspecified route. A flammable liquid. When heated to decomposition it emits toxic vapors of NO_x and Cl^-.

BHP750   CAS: 1492-93-9   HR: 3
4'-(BIS(2-CHLOROETHYL)AMINO)-2-
FLUORO ACETANILIDE
mf: $C_{12}H_{15}Cl_2FN_2O$ mw: 293.19
SYN: p-FLUOROACETYLAMINOPHENYL DERIVATIVE of NITROGEN MUSTARD
SAFETY PROFILE: Poison by intraperitoneal route. An experimental teratogen. When heated to decomposition it emits very toxic fumes of Cl^-, F^-, and NO_x.

BHQ760   CAS: 21447-86-9   HR: 3
1-(3-(BIS(2-CHLOROETHYL)AMINO-4-
METHYLBENZOYL)AZIRIDINE)
mf: $C_{14}H_{18}Cl_2N_2O$ mw: 301.24
SYNS: AZIRIDINE, 1-(3-(BIS(2-CHLOROETHYL)-AMINO-p-TOLUOYL))- □ KETONE, 1-AZIRIDINYL 3-(BIS(2-CHLOROETHYL)AMINO)-p-TOLYL
DOT CLASSIFICATION: 3; Label: Flammable Liquid
SAFETY PROFILE: A poison by an unspecified route. A flammable liquid. When heated to decomposition it emits toxic vapors of NO_x and Cl^-.

BHR400   CAS: 21447-39-2   HR: 3
1-(3-(BIS(2-CHLOROETHYL)AMINO)-4-
METHYLBENZOYL)MORPHOLINE
mf: $C_{16}H_{22}Cl_2N_2O_2$ mw: 345.30
SYNS: KETONE, 3-(BIS(2-CHLOROETHYL)AMINO)-p-TOLYL MORPHOLINO- □ MORPHOLINE, 4-(3-(BIS(2-CHLOROETHYL)AMINO)-p-TOLUOYL)-
DOT CLASSIFICATION: 3; Label: Flammable Liquid
SAFETY PROFILE: A poison by an unspecified route. A flammable liquid.

When heated to decomposition it emits toxic vapors of NO_x and Cl^-.

BHT750   CAS: 531-76-0   HR: 3
dl-3-(p-(BIS(2-CHLOROETHYL)AMINO)-
PHENYL)ALANINE
mf: $C_{13}H_{18}Cl_2N_2O_2$ mw: 305.23
PROP: Needles from MeOH. Mp: 180–181°.
SYNS: 3-(p-(BIS(2-CHLOROETHYL)AMINO)PHENYL)ALANINE □ 4-(BIS(2-CHLOROETHYL)AMINO)-dl-PHENYLALANINE □ CB-3307 □ p-DI-(2-CHLORAETHYL)-AMINO-dl-PHENYL-ALANIN (GERMAN) □ p-DI(2-CHLOROETHYL)AMINO-dl-PHENYLALANINE □ MERFALAN □ MERPHALAN □ o-MERPHALAN □ NCI-C04944 □ NSC-14210 □ PHENYLALANIN-LOST (GERMAN) □ dl-PHENYLALAN-INE MUSTARD □ SAKOLYSIN (GERMAN) □ SARCOCLORIN □ dl-SARCOLYSIN □ dl-SARCOLYSINE
CONSENSUS REPORTS: IARC Cancer Review: Group 2B IMEMDT 7,56,87; Animal Limited Evidence IMEMDT 9,167,75; NCI Carcinogenesis Studies (ipr); Clear Evidence: mouse CANCAR 40,1935,77; No Evidence: rat CANCAR 40,1935,77.
SAFETY PROFILE: Suspected carcinogen with experimental tumorigenic data. A poison by ingestion, intraperitoneal, intravenous, and intracerebral routes. An experimental teratogen. Other experimental reproductive effects. Mutation data reported. An antineoplastic agent. When heated to decomposition it emits very toxic fumes of Cl^- and NO_x.

BHW300   CAS: 5185-77-3   HR: 3
2-(p-BIS(2-CHLOROETHYL)AMINO-
PHENYL)-1,3,2-DITHIARSENOLANE
mf: $C_{12}H_{16}AsCl_2NS_2$ mw: 384.23
SYN: 1,3,2-DITHIARSENOLANE, 2-(p-BIS(2-CHLOROETHYL)AMINOPHENYL)-
OSHA PEL: TWA 0.5 mg(As)/m³
SAFETY PROFILE: Poison by intraperitoneal route. When heated to decomposition it emits toxic fumes of NO_x, SO_x, As, and Cl^-.

BHX300   CAS: 4587-15-9   HR: 3
m-(BIS(2-CHLOROETHYL)AMINO)-

PHENYL MORPHOLINO KETONE
mf: $C_{15}H_{20}Cl_2N_2O_2$ mw: 331.27
DOT CLASSIFICATION: 3; Label: Flammable Liquid
SAFETY PROFILE: A poison by an unreported route. A flammable liquid. When heated to decomposition it emits toxic vapors of NO_x and Cl^-.

BIA100   CAS: 21447-87-0   HR: 3
3-(BIS(2-CHLOROETHYL)AMINO)-p-
TOLYL PIPERIDYL KETONE
mf: $C_{17}H_{24}Cl_2N_2O$ mw: 343.33
SYNS: 1-(3-(BIS(2-CHLOROETHYL)AMINO)-p-TOLUOYL)PIPERIDINE □ KETONE, 3-(BIS(2-CHLOROETHYL)AMINO)-p-TOLYL PIPERIDYL- □ PIPERIDINE, 1-(3-(BIS(2-CHLOROETHYL)AMINO)-p-TOLUOYL)-
DOT CLASSIFICATION: 3; Label: Flammable Liquid
SAFETY PROFILE: A poison by an unreported route. A flammable liquid. When heated to decomposition it emits toxic vapors of NO_x and Cl^-.

BIA250   CAS: 66-75-1   HR: 3
5-(BIS(2-
CHLOROETHYL)AMINO)URACIL
mf: $C_8H_{11}Cl_2N_3O_2$ mw: 252.12
PROP: Crystals from MeOH (aq). Mp: 206° (decomp).
SYNS: AMINOURACIL MUSTARD □ 5-(BIS(2-CHLOROETHYL)AMINO)-2,4(1H,3H)PYRIMIDINEDIONE □ 5-N,N-BIS(2-CHLOROETHYL)AMINOURACIL □ CB-4835 □ CHLORETHAMINACIL □ DEMETHYLDOPAN □ DESMETHYLDOPAN □ 5-(DI-(β-CHLOROETHYL)-AMINO)URACIL □ 5-(DI-2-CHLOROETHYL)AMINO-URACIL □ 2,6-DIHYDROXY-5-BIS(2-CHLOROETHYL)-AMINOPYRAMIDINE □ ENT 50,439 □ NCI-C04820 □ NORDOPAN □ NSC-34462 □ RCRA WASTE NUMBER U237 □ SK-19849 □ U-8344 □ URACILLOST □ URACILMOSTAZA □ URACIL MUSTARD □ URAMUSTIN □ URAMUSTINE
CONSENSUS REPORTS: IARC Cancer Review: Group 2B IMEMDT 7,370,87; Animal Sufficient Evidence IMEMDT 9,235,75; NCI Carcinogenesis Studies (ipr); Clear Evidence: mouse, rat RRCRBU 52,1,75. EPA Genetic Toxicology Program.
SAFETY PROFILE: Suspected carcinogen with experimental carcinogenic and

neoplastigenic data. A deadly poison by ingestion and intraperitoneal routes. Mutation data reported. When heated to decomposition it emits very toxic fumes of Cl^- and NO_x.

BIA300   CAS: 5185-71-7   HR: 3
N,N-BIS(2-CHLOROETHYL)-p-ARS-ANILIC ACID

mf: $C_{10}H_{14}AsCl_2NO_3$ mw: 342.07

SYN: p-ARSANILIC ACID, N,N-BIS(2-CHLOROETHYL)-

OSHA PEL: TWA 0.5 mg(As)/m³

SAFETY PROFILE: Poison by intraperitoneal route. When heated to decomposition it emits toxic fumes of NO_x, As, and Cl^-.

BID250   CAS: 538-07-8   HR: 3
BIS(2-CHLOROETHYL)ETHYLAMINE

mf: $C_6H_{13}Cl_2N$ mw: 170.10

SYNS: 2,2'-DICHLOROTRIETHYLAMINE □ ETHYLBIS(β-CHLOROETHYL)AMINE □ ETHYLBIS(2-CHLOROETHYL)AMINE □ ETHYL-S □ HN1 □ TL 329 □ TL 1149

CONSENSUS REPORTS: Reported in EPA TSCA Inventory. EPA Extremely Hazardous Substances List.

SAFETY PROFILE: Deadly poison by inhalation, skin contact, ingestion, intravenous, subcutaneous, and intraperitoneal routes. When heated to decomposition it emits very toxic fumes of Cl^- and NO_x.

BID750   CAS: 111-91-1   HR: 3
BIS(β-CHLOROETHYL)FORMAL

mf: $C_5H_{10}Cl_2O_2$ mw: 173.05

PROP: Liquid. Bp: 217.5°, flash p: 230°F (OC), d: 1.23, vap d: 5.9.

SYNS: BIS(2-CHLOROETHOXY)METHANE □ BIS(2-CHLOROETHYL)FORMAL □ DICHLOROETHYL FORMAL □ DI-2-CHLOROETHYL FORMAL □ FORMALDEHYDE BIS(β-CHLOROETHYL) ACETAL □ 1,1'-(METHYLENEBIS(OXY)BIS(2-CHLOROETHANE)) □ RCRA WASTE NUMBER U024

CONSENSUS REPORTS: Reported in EPA TSCA Inventory.

SAFETY PROFILE: Poison by ingestion, inhalation, and skin contact. A skin and eye irritant. Combustible when exposed to heat or flame. Incompatible with oxidizers. To fight fire, use alcohol foam, foam, CO_2, dry chemical. When heated to decomposition it emits toxic fumes of Cl^-. See also CHLORIDES.

BID800   CAS: 29023-83-4   HR: 3
9-(2,2-BIS(2-CHLOROETHYL)HYDRAZINO)ACRIDINE MONOHYDROCHLORIDE

mf: $C_{17}H_{17}Cl_2N_3 \bullet ClH$ mw: 370.73

SYNS: ACRIDINE, 9-(2,2-BIS(2-CHLOROETHYL)-HYDRAZINO)-, MONOHYDROCHLORIDE □ 9-(2',2'-BIS-β-CHLORO-ETHYL-HYDRAZINO)ACRIDINE HYDROCHLORIDE

SAFETY PROFILE: A poison by ingestion. When heated to decomposition it emits toxic vapors of NO_x, HCl, and Cl^-.

BIE250   CAS: 51-75-2   HR: 3
BIS(β-CHLOROETHYL)METHYLAMINE

mf: $C_5H_{11}Cl_2N$ mw: 156.07

PROP: Dark liquid. Mp: 1° @ 10 mm, bp: 86–87° @ 11 mm, d: 1.09 @ 25°, vap press: 0.17 mm @ 25°, vap d: 5.9. Sltly sol in water.

SYNS: BIS(2-CHLOROETHYL)METHYLAMINE □ N,N-BIS(2-CHLOROETHYL)METHYLAMINE □ CARYOLYSIN □ CHLORMETHINE □ CLORAMIN □ DICHLORAMINE □ DICHLOREN (GERMAN) □ β,β'-DICHLORODIETHYL-N-METHYLAMINE □ DI(2-CHLOROETHYL)METHYL-AMINE □ 2,2'-DICHLORO-N-METHYLDIETHYLAMINE □ EMBICHIN □ ENT 25,294 □ HN2 □ MBA □ MECHLORETHAMINE □ N-METHYL-BIS-CHLORAETHYLAMIN (GERMAN) □ METHYLBIS(β-CHLOROETHYL)AMINE □ N-METHYL-BIS(β-CHLOROETHYL)AMINE □ N-METHYL-BIS(2-CHLOROETHYL)AMINE (MAK) □ N-METHYL-2,2'-DICHLORODIETHYLAMINE □ METHYLDI(2-CHLOROETHYL)AMINE □ N-METHYL-LOST □ MUSTARGEN □ MUSTINE □ MUTAGEN □ NITROGEN MUSTARD □ N-LOST (GERMAN) □ NSC-762 □ TL 146

CONSENSUS REPORTS: EPA Genetic Toxicology Program. Reported in EPA TSCA Inventory. EPA Extremely Hazardous Substances List. Community Right-To-Know List.

DFG MAK: Human Carcinogen

SAFETY PROFILE: Confirmed human carcinogen producing skin tumors by skin contact. Experimental carcinogenic, tumorigenic, and neoplastigenic data. A deadly poison by inhalation, ingestion, skin contact, and most other routes. Experimental teratogenic and reproductive effects. A powerful skin and eye irritant. Human mutation data reported. It has been used as a blistering agent in chemical warfare. When heated to decomposition it emits very toxic fumes of Cl⁻ and NOₓ.

BIE500   CAS: 55-86-7   HR: 3
BIS(2-CHLOROETHYL)METHYLAMINE HYDROCHLORIDE
mf: $C_5H_{11}Cl_2N \cdot ClH$ mw: 192.53
PROP: Leaflets from Me_2CO or $CHCl_3$. Mp: 119°.
SYNS: ANTIMIT □ AZOTOYPERITE □ C 6866 □ CAROLYSINE □ CARYOLYSINE □ CARYOLYSINE HYDROCHLORIDE □ CHLORAMIN □ CHLORAMINE □ CHLORAMIN HYDROCHLORIDE □ CHLORETHAMINE □ CHLORETHAZINE □ CHLORMETHINE HYDRO-CHLORIDE □ CHLORMETHINUM □ 2-CHLORO-N-(2-CHLOROETHYL)-N-METHYLETHANAMINE HYDROCHLORIDE □ DEMA □ DICHLOREN □ DICHLOREN HYDROCHLORIDE □ β,β'-DICHLORO-DIETHYL-N-METHYLAMINE HYDROCHLORIDE □ DI(2-CHLOROETHYL)METHYLAMINE HYDRO-CHLORIDE □ 1,5-DICHLORO-3-METHYL-3-AZAPENTANE HYDROCHLORIDE □ 2,2'-DICHLORO-N-METHYLDIETHYLAMINE HYDROCHLORIDE □ DIMITAN □ EMBECHINE □ EMBICHIN □ EMBICHIN HYDROCHLORIDE □ EMBIKHINE □ ERASOL □ ERASOL HYDROCHLORIDE □ ERASOL-IDO □ HN2.HCl □ HN2 HYDROCHLORIDE □ KLORAMIN □ N-LOST □ MBA HYDROCHLORIDE □ MEBICHLORAMINE □ MECHLORETHAMINE HYDROCHLORIDE □ MERCHLORETHANAMINE □ METHYLBIS(β-CHLOROETHYL)AMINE HYDRO-CHLORIDE □ N-METHYL-BIS-β-CHLORETHYLAMINE HYDROCHLORIDE □ METHYLBIS(2-CHLORO-ETHYL)AMINE HYDROCHLORIDE □ N-METHYLBIS(2-CHLOROETHYL)AMINE HYDROCHLORIDE □ N-METHYL-2,2'-DICHLORODIETHYLAMINE HYDRO-CHLORIDE □ N-METHYL-DI-2-CHLOROETHYLAMINE HYDROCHLORIDE □ METHYLDI(β-CHLORO-ETHYL)AMINE HYDROCHLORIDE □ METHYLDI(2-CHLOROETHYL)AMINE HYDROCHLORIDE □ MITOXINE □ N-MUSTARD (GERMAN) □ MUSTARGEN □ MUSTARGEN HYDROCHLORIDE □ MUSTINE HYDROCHLOR □ MUSTINE HYDROCHLORIDE □ NCI-C56382 □ NITOL □ NITOL "TAKEDA" □ NITROGEN MUSTARD HYDROCHLORIDE □ NITROGRANULOGEN □ NITROGRANULOGEN HYDROCHLORIDE □ NSC-762 □ NSC-762 HYDROCHLORIDE □ PLIVA □ STICKSTOFFLOST □ ZAGREB

CONSENSUS REPORTS: NTP 10th Report on Carcinogens. IARC Cancer Review: Group 2A IMEMDT 7,269,87; Animal Sufficient Evidence IMEMDT 9,193,75. EPA Genetic Toxicology Program.
SAFETY PROFILE: Confirmed carcinogen with experimental carcinogenic, neoplastigenic, and tumorigenic data. Deadly poison by ingestion, intravenous, subcutaneous, intraperitoneal, and parenteral routes. Experimental teratogenic and reproductive effects. Human systemic effects by intravenous route: nausea or vomiting, reduction in the number of white blood cells and blood platelets. Other experimental reproductive effects. Human mutation data reported.

BIF250   CAS: 494-03-1   HR: 3
N,N-BIS(2-CHLOROETHYL)-2-NAPHTH-YLAMINE
mf: $C_{14}H_{15}Cl_2N$ mw: 268.20
PROP: Platelets from pet ether. Mp: 54–56°, bp: 210° @ 5 mm.
SYNS: 2-BIS(2-CHLOROETHYL)AMINONAPHTHALENE □ BIS(2-CHLOROETHYL)-β-NAPHTHYLAMINE □ CHLOR-NAFTINA □ CHLORNAPHAZIN □ CHLORNAPHTHIN □ CHLORONAFTINA □ CHLORONAPHTHINE □ CLORNAPHAZINE □ DICHLOROETHYL-β-NAPHTHYLAMINE □ DI(2-CHLOROETHYL)-β-NAPHTHYLAMINE □ N,N-DI(2-CHLOROETHYL)-β-NAPHTHYLAMINE □ 2-N,N-DI(2-CHLOROETHYL)-NAPHTHYLAMINE □ ERYSAN □ NAPHTHYLAMINE MUSTARD □ β-NAPHTHYL-BIS-(β-CHLOROETHYL)-AMINE □ 2-NAPHTHYLBIS(2-CHLOROETHYL)AMINE □ β-NAPHTHYL-DI-(2-CHLOROETHYL)AMINE □ NSC-62209 □ R48 □ RCRA WASTE NUMBER U026
CONSENSUS REPORTS: IARC Cancer Review: Group 1 IMEMDT 7,130,87; Animal Sufficient Evidence IMEMDT 4,119,74; Human Sufficient Evidence IMEMDT 4,119,74. EPA Genetic Toxicology Program.
SAFETY PROFILE: Confirmed human carcinogen producing bladder tumors. Human and experimental carcinogenic data. Moderately toxic by intraperitoneal route.

When heated to decomposition it emits very toxic fumes of Cl⁻ and NOₓ.

BIF750　CAS: 154-93-8　HR: 3
N,N'-BIS(2-CHLOROETHYL)-N-NITRO-
**　SOUREA**
mf: $C_5H_9Cl_2N_3O_2$　　mw: 214.07
PROP: Light-yellow powder. Mp: 30–32°.
SYNS: BCNU □ BiCNU □ BISCHLORO-ETHYLNITROSOUREA □ BIS(2-CHLOROETHYL)-NITROSOUREA □ 1,3-BIS(β-CHLOROETHYL)-1-NITROSOUREA □ 1,3-BIS-(2-CHLOROETHYL)-1-NITROSOUREA □ CARMUBRIS □ CARMUSTIN □ CARMUSTINE □ FDA 0345 □ NCI-C04773 □ NITRUMON □ NSC-409962 □ SK 27702 □ SRI 1720
CONSENSUS REPORTS: NTP 10th Report on Carcinogens. IARC Cancer Review: Group 2A IMEMDT 7,150,87; Human Limited Evidence IMEMDT 26,79,81; Animal Sufficient Evidence IMEMDT 26,79,81. NCI Carcinogenesis Studies (ipr); Some Evidence: rat CANCAR 40,1935,77; Clear Evidence: mouse CANCAR 40,1935,77. EPA Genetic Toxicology Program.
SAFETY PROFILE: Confirmed carcinogen with experimental carcinogenic and tumorigenic data. A human poison by parenteral route. An experimental poison by ingestion, intravenous, intraperitoneal, parenteral, and subcutaneous routes. Human systemic effects by parenteral, intravenous, and possibly other routes: nausea or vomiting, reduced white blood cell and blood platelet counts, bone marrow damage, and potentially fatal respiratory system effects including lung fibrosis, dyspnea, and cyanosis. Experimental teratogenic and reproductive effects. Human mutation data reported. When heated to decomposition it emits very toxic fumes of Cl⁻ and NOₓ. See also N-NITROSO COMPOUNDS.

BIH250　CAS: 505-60-2　HR: 3
BIS(2-CHLOROETHYL)SULFIDE
mf: $C_4H_8Cl_2S$　　mw: 159.08
PROP: Colorless (if pure), to light-yellow, oily liquid. Mp: 13–14°, bp: 215–217°, flash p: 221°F, d: 1.2741 @ 20°/4°, vap d: 5.4, vap press: 0.09 mm @ 30°.
SYNS: BIS(β-CHLOROETHYL)SULFIDE □ BIS(2-CHLOROETHYL)SULPHIDE □ 1-CHLORO-2-(β-CHLOROETHYLTHIO)ETHANE □ β,β-DICHLOR-ETHYL-SULPHIDE □ 2,2'-DICHLORODIETHYL SULFIDE □ DI-2-CHLOROETHYL SULFIDE □ β,β'-DICHLOROETHYL SULFIDE □ 2,2'-DICHLOROETHYL SULPHIDE (MAK) □ DISTILLED MUSTARD □ KAMPSTOFF "LOST" □ MUSTARD GAS □ MUSTARD HD □ MUSTARD VAPOR □ SCHWEFEL-LOST □ S-LOST □ S MUSTARD □ SULFUR MUSTARD □ SULFUR MUSTARD GAS □ SULPHUR MUSTARD GAS □ 1,1'-THIOBIS(2-CHLOROETHANE) □ YELLOW CROSS LIQUID □ YPERITE
CONSENSUS REPORTS: NTP 10th Report on Carcinogens. IARC Cancer Review: Group 1 IMEMDT 7,259,87; Animal Sufficient Evidence IMEMDT 9,181,75; Human Limited Evidence IMEMDT 9,181,75. EPA Extremely Hazardous Substances List. Community Right-To-Know List. EPA Genetic Toxicology Program. Reported in EPA TSCA Inventory.
DFG MAK: Human Carcinogen
SAFETY PROFILE: Confirmed human carcinogen with experimental carcinogenic, neoplastigenic, and tumorigenic data. A human poison by inhalation and subcutaneous routes. An experimental poison by inhalation, skin contact, subcutaneous, and intravenous routes. An experimental teratogen. A severe human skin and eye irritant. Human mutation data reported. A military blistering gas. Strongly affects the skin, eyes, lungs, and gastric system. Pulmonary lesions are often fatal. It penetrates the skin deeply and injures blood vessels. Minute amounts can cause inflammation. Secondary infections are common. Combustible when exposed to heat or flame; can be ignited by a large explosive charge. It will react with water or steam to produce toxic and corrosive fumes. Vigorous reaction with oxidizing materials. Incompatible with bleaching powder. To fight fire, use water, foam, CO_2, dry chemical. Dangerous; when heated to decomposition or on contact with acid or

acid fumes it emits highly toxic fumes of SO_x and Cl^-. See also SULFIDES and CHLORIDES.

BII250   CAS: 108-60-1   HR: 3
BIS(2-CHLOROISOPROPYL) ETHER
DOT: UN 2490
mf: $C_6H_{12}Cl_2O$ mw: 171.08
PROP: Colorless liquid. Bp: 187.8°, fp: >−20°, flash p: 185°F (OC), d: 1.11 @ 25°/25°, vap d: 6.0, vap press: 0.10 mm @ 20°.
SYNS: BIS(β-CHLOROISOPROPYL)ETHER □ BIS(2-CHLORO-1-METHYLETHYL) ETHER □ BIS(1-CHLORO-2-PROPYL) ETHER □ (2-CHLORO-1-METHYLETHYL) ETHER □ DCIP □ DCIP (nematocide) □ DICHLORODI-ISOPROPYL ETHER □ β,β'-DICHLORODIISOPROPYL ETHER □ DICHLOROISOPROPYL ETHER □ 2,2'-DICHLOROISOPROPYL ETHER □ DICHLORO-ISOPROPYL ETHER (DOT) □ NCI-C50044 □ NEMAMORT □ 2,2'-OXYBIS(1-CHLOROPROPANE) □ PROPANE, 2,2'-OXYBIS(1-CHLORO)- □ RCRA WASTE NUMBER U027
CONSENSUS REPORTS: IARC Cancer Review: Group 3 IMEMDT 7,56,87; Animal Limited Evidence IMEMDT 41,149,86. NCI Carcinogenesis Bioassay (gavage); No Evidence: rat NCITR* NCI-CG-TR-191,79. Community Right-To-Know List. Reported in EPA TSCA Inventory.
DOT CLASSIFICATION: 6.1; Label: Poison
SAFETY PROFILE: Poison by ingestion. Moderately toxic by skin contact and inhalation. An eye irritant. Questionable carcinogen. Mutation data reported. A corrosive material. Moderate fire hazard when exposed to heat, flame, or powerful oxidizers. Incompatible with oxidizing materials. To fight fire, use water to blanket fire; foam, CO_2, dry chemical. When heated to decomposition it emits highly toxic fumes of Cl^-. See also ETHERS.

BIJ250   CAS: 13483-18-6   HR: 2
BIS-1,2-(CHLOROMETHOXY)ETHANE
mf: $C_4H_8Cl_2O_2$ mw: 159.02
PROP: Viscous liquid. Bp: 99–100° @ 22 mm, d: 1.2879 @ 14°/15°.
SYN: ETHYLENE GLYCOL BIS(CHLOROMETHYL)-ETHER

CONSENSUS REPORTS: IARC Cancer Review: Group 3 IMEMDT 7,56,87; Animal Sufficient Evidence IMEMDT 15,31,77. Reported in EPA TSCA Inventory. Glycol ethers are on the Community Right-To-Know List.
SAFETY PROFILE: Questionable carcinogen with experimental neoplastigenic data. See also GLYCOL ETHERS. When heated to decomposition it emits toxic fumes of Cl^-.

BIK000   CAS: 542-88-1   HR: 3
BIS(CHLOROMETHYL) ETHER
DOT: UN 2249
mf: $C_2H_4Cl_2O$ mw: 114.96
PROP: Volatile liquid. Bp: 105°, d: 1.315 @ 20°, vap d: 4.0, flash p: <19°, fp: −41.5°.
SYNS: BCME □ BIS-CME □ CHLORO(CHLORO-METHOXY)METHANE □ DICHLORDIMETHYLAETHER (GERMAN) □ sym-DICHLORODIMETHYL ETHER (DOT) □ sym-DICHLOROMETHYL ETHER □ DIMETHYL-1,1'-DICHLOROETHER □ OXYBIS(CHLOROMETHANE) □ RCRA WASTE NUMBER P016
CONSENSUS REPORTS: NTP 10th Report on Carcinogens. IARC Cancer Review: Group 1 IMEMDT 7,131,87; Animal Sufficient Evidence IMEMDT 4,231,74; Human Sufficient Evidence IMEMDT 4,231,74. Community Right-To-Know List. EPA Extremely Hazardous Substances List. Reported in EPA TSCA Inventory.
OSHA PEL: OSHA: Cancer Suspect Agent
ACGIH TLV: TWA 0.001 ppm; Confirmed Human Carcinogen
DFG MAK: Human Carcinogen
DOT CLASSIFICATION: 6.1; Label: Poison
SAFETY PROFILE: Confirmed human carcinogen with experimental carcinogenic, neoplastigenic, and tumorigenic data. Poison by inhalation, ingestion, and skin contact. Human systemic effects by inhalation: irritation of the conjunctiva, unspecified nasal and respiratory effects. Human mutation data reported. A dangerous fire hazard. When heated to decomposition it emits very toxic fumes of Cl^-. See also ETHERS.

BIK250   CAS: 534-07-6   HR: 3
BIS(CHLOROMETHYL)KETONE
DOT: UN 2649
mf: C$_3$H$_4$Cl$_2$O   mw: 126.97
PROP: Crystals. Mp: 45°, bp: 173°, d:
1.3826 @ 46°/4°, vap d: 4.38. Sol in water.
SYNS: sym-DICHLOROACETONE □ α,α'-
DICHLOROACETONE □ α,γ-DICHLOROACETONE □
1,3-DICHLOROACETONE □ 1,3-DICHLOROACETONE
(DOT) □ 1,3-DICHLORO-2-PROPANONE
CONSENSUS REPORTS: EPA Genetic
Toxicology Program. Reported in EPA
TSCA Inventory. EPA Extremely
Hazardous Substances List.
DOT CLASSIFICATION: 6.1; Label: Poison
SAFETY PROFILE: Poison by inhalation.
Mutation data reported. A systemic irritant
by ingestion and inhalation routes. See also
KETONES. Dangerous; when heated to
decomposition it emits highly toxic fumes of
Cl⁻.

BIM250   CAS: 55-56-1   HR: 3
1,6-BIS(5-(p-CHLOROPHENYL)BI-
    GUANIDINO)HEXANE
mf: C$_{22}$H$_{30}$Cl$_2$N$_{10}$ mw: 505.52
PROP: Solid. Mp: 134°.
SYNS: 1,6-BIS(p-
CHLOROPHENYLDIGUANIDO)HEXANE □ CHLO-
RHEXIDIN (CZECH) □ CHLORHEXIDINE □ 1,6-DI(4'-
CHLOROPHENYLDIGUANIDO)HEXANE □ 1,1'-
HEXAMETHYLENEBIS(5-(p-CHLOROPHENYL)-
BIGUANIDE) □ HIBITANE □ NOLVASAN □
ROTERSEPT □ STERIDO
SAFETY PROFILE: Poison by
intraperitoneal and intravenous routes.
Mildly toxic by ingestion. Experimental
reproductive effects. A human skin irritant.
Mutation data reported. When heated to
decomposition it emits very toxic fumes of
Cl⁻ and NO$_x$.

BIM500   CAS: 72-54-8   HR: 3
1,1-BIS(4-CHLOROPHENYL)-2,2-
    DICHLOROETHANE
mf: C$_{14}$H$_{10}$Cl$_4$ mw: 320.04
PROP: Crystalline solid from pet ether. Mp:
111°, vap d: 11.
SYNS: 1,1-BIS(p-CHLOROPHENYL)-2,2-DICHLORO-
ETHANE □ 2,2-BIS(p-CHLOROPHENYL)-1,1-DICHLORO-

ETHANE □ 2,2-BIS(4-CHLOROPHENYL)-1,1-DICHLORO-
ETHANE □ DDD □ p,p'-DDD □ 1,1-DICHLOOR-2,2-
BIS(4-CHLOOR FENYL)-ETHAAN (DUTCH) □ 1,1-
DICHLOR-2,2-BIS(4-CHLOR-PHENYL)-AETHAN
(GERMAN) □ 1,1-DICHLORO-2,2-BIS(p-
CHLOROPHENYL)ETHANE □ 1,1-DICHLORO-2,2-BIS(p-
CHLOROPHENYL)ETHANE (DOT) □ 1,1-DICHLORO-
2,2-BIS(4-CHLOROPHENYL)-ETHANE (FRENCH) □ 1,1-
DICHLORO-2,2-BIS(PARACHLOROPHENYL)ETHANE
(DOT) □ 1,1-DICHLORO-2,2-DI(4-CHLOROPHENYL)-
ETHANE □ DICHLORODIPHENYL DICHLORO-
ETHANE □ p,p'-DICHLORODIPHENYLDICHLORO-
ETHANE □ 1,1-DICLORO-2,2-BIS(4-CLORO-FENIL)-
ETANO (ITALIAN) □ DILENE □ ENT 4,225 □ ME-1700
□ NCI-C00475 □ RCRA WASTE NUMBER U060 □
RHOTHANE □ RHOTHANE D-3 □ ROTHANE □ p,p'-
TDE □ TDE (DOT) □ TETRACHLORODIPHENYL-
ETHANE
CONSENSUS REPORTS: IARC Cancer
Review: Animal Sufficient Evidence
IMEMDT 5,83,74. NCI Carcinogenesis
Bioassay (feed); Clear Evidence: rat NCITR*
NCI-CG-TR-131,78; No Evidence: mouse
NCITR* NCI-CG-TR-131,78. EPA Genetic
Toxicology Program.
SAFETY PROFILE: Confirmed carcinogen
with experimental carcinogenic,
neoplastigenic, and tumorigenic data. Poison
by ingestion. Moderately toxic by skin
contact. Mutation data reported. An
insecticide. When heated to decomposition
it emits toxic fumes of Cl⁻. See also DDT.

BIM750   CAS: 72-55-9   HR: 3
2,2-BIS(p-CHLOROPHENYL)-1,1-DI-
    CHLOROETHYLENE
mf: C$_{14}$H$_8$Cl$_4$ mw: 318.02
SYNS: DDE □ p,p'-DDE □ DDT DEHYDROCHLORIDE
□ 1,1-DICHLORO-2,2-BIS(p-CHLOROPHENYL)-ETHYL-
ENE □ p,p'-DICHLORODIPHENYLDICHLORO-
ETHYLENE □ 1,1'-(DICHLOROETHENYLIDENE)BIS(4-
CHLOROBENZENE) □ NCI-C00555
CONSENSUS REPORTS: IARC Cancer
Review: Animal Limited Evidence
IMEMDT 5,83,74. NCI Carcinogenesis
Bioassay (feed); Clear Evidence: mouse
NCITR* NCI-CG-TR-131,78; No
Evidence: rat NCITR* NCI-CG-TR-131,78.
EPA Genetic Toxicology Program.
SAFETY PROFILE: Suspected carcinogen
with experimental carcinogenic and
neoplastigenic data. Poison by ingestion.

Experimental reproductive effects. Mutation data reported. An insecticide. When heated to decomposition it emits very toxic fumes of Cl⁻. See also CHLORINATED HYDROCARBONS, ALIPHATIC.

BIO750   CAS: 115-32-2   HR: 3
1,1-BIS(p-CHLOROPHENYL)-2,2,2-
    TRICHLOROETHANOL
mf: $C_{14}H_9Cl_5O$ mw: 370.48
PROP: Solid. Mp: 78.5°. Material used in cancer bioassay was 40–60% pure NCITR* NCI-CG-TR-90,78.
SYNS: ACARIN □ 1,1-BIS(CHLOROPHENYL)-2,2,2-TRICHLOROETHANOL □ 1,1-BIS(4-CHLOROPHENYL)-2,2,2-TRICHLOROETHANOL □ CARBAX □ CEKUDIFOL □ 4-CHLORO-α-(4-CHLOROPHENYL)-α-(TRICHLORO-METHYL)BENZENEMETHANOL □ CPCA □ DECOFOL □ DICHLOROKELTHANE □ DI-(p-CHLOROPHENYL)-TRICHLOROMETHYLCARBINOL □ 4,4'-DICHLORO-α-(TRICHLOROMETHYL)BENZHYDROL □ DICOFOL □ DTMC □ ENT 23,648 □ FW 293 □ HIFOL □ KELTANE □ p,p'-KELTHANE □ KELTHANE (DOT) □ KELTHANE DUST BASE □ KELTHANETHANOL □ MILBOL □ MITIGAN □ NCI-C00486 □ 2,2,2-TRICHLOOR-1,1-BIS(4-CHLOOR FENYL)-ETHANOL (DUTCH) □ 1,1,1-TRICHLOR-2,2-BIS(4-CHLORPHENYL)-AETHANOL (GERMAN) □ 2,2,2-TRICHLOR-1,1-BIS(4-CHLOR-PHENYL)-AETHANOL (GERMAN) □ 2,2,2-TRICHLORO-1,1-BIS(4-CHLOROPHENYL)-ETHANOL (FRENCH) □ 2,2,2-TRICHLORO-1,1-BIS(4-CLORO-FENIL)-ETANOLO (ITALIAN) □ 2,2,2-TRICHLORO-1,1-DI-(4-CHLOROPHENYL)ETHANOL
CONSENSUS REPORTS: IARC Cancer Review: Group 3 IMEMDT 7,56,87; Animal Limited Evidence IMEMDT 30,87,83. NCI Carcinogenesis Bioassay (feed); Clear Evidence: mouse NCITR* NCI-CG-TR-90,78; No Evidence: rat NCITR* NCI-CG-TR-90,78. Community Right-To-Know List.
SAFETY PROFILE: Poison by ingestion and skin contact. Moderately toxic by intraperitoneal route. Human mutation data reported. Questionable carcinogen with experimental carcinogenic data. An experimental teratogen. Other experimental reproductive effects. When heated to decomposition it emits toxic fumes of Cl⁻.

BIQ250   CAS: 40334-69-8   HR: 3
BIS(2-CHLOROVINYL)CHLOROARSINE

mf: $C_4H_4AsCl_3$ mw: 233.35
SYNS: DICHLOROVINYLARSINE CHLORIDE □ DICHLOROVINYLCHLOROARSINE (DOT) □ L-2 □ LEWISITE II
CONSENSUS REPORTS: Arsenic and its compounds are on the Community Right-To-Know List.
OSHA PEL: TWA 0.5 mg(As)/m³
DOT CLASSIFICATION: Forbidden
SAFETY PROFILE: A poison by skin contact and subcutaneous routes. When heated to decomposition it emits very toxic fumes of As and Cl⁻. See also ARSENIC COMPOUNDS.

BIQ500   CAS: 111-94-4   HR: 3
BIS(β-CYANOETHYL)AMINE
mf: $C_6H_9N_3$ mw: 123.18
PROP: Liquid. Mp: −5.5°, bp: 135° @ 1 mm, d: 1.463 @ 25°, vap d: 3.3.
SYNS: BBCE □ BIS-(2-CYANOETHYL)AMINE □ N,N-BIS(2-CYANOETHYL)AMINE □ 2-CYANO-N-(2-CYANO-ETHYL)ETHANAMINE □ DI(2-CIANOETIL)AMMINA (ITALIAN) □ 2,2'-DICYANODIETHYLAMINE □ DI-(2-CYANOETHYL)AMINE □ IDPN □ 3,3'-IMINOBIS-PROPANENITRILE □ IMINO-β,β'-DIPROPIONITRILE □ β,β-IMINODIPROPIONITRILE □ β,β'-IMINODIPROPIO-NITRILE □ 3,3'-IMINODIPROPIONITRILE □ 2341 I.S. □ USAF A-8564
CONSENSUS REPORTS: Reported in EPA TSCA Inventory. Cyanide and its compounds are on the Community Right-To-Know List.
SAFETY PROFILE: A poison by intraperitoneal route. Moderately toxic by ingestion and skin contact. Experimental teratogenic and reproductive effects. A skin and severe eye irritant. A storage hazard, may explode in a sealed container. When heated to decomposition it emits toxic fumes of NOₓ and CN⁻. See also NITRILES and AMINES.

BIQ660   CAS: 63942-43-8   HR: 3
N,N'-BIS(2-CYANO-2-METHYLPROPION-
    ALDEHYDEO-(N-METHYL-
    CARBAMOYL)OXIME)SULFIDE
mf: $C_{14}H_{20}N_6O_4S$ mw: 368.46
SYN: 5,11-DIOXA-9-THIA-4,7,9,12-TETRAAZAPENTADECA-3,12-DIENEDINITRILE, 6,10-DIOXO-2,2,7,9,14,14-HEXAMETHYL-

SAFETY PROFILE: A poison by ingestion. When heated to decomposition it emits toxic vapors of NO_x and SO_x.

BIS500   CAS: 3465-75-6   HR: 3
BIS(DECANOYLOXY)DI-n-BUTYL-STANNANE
mf: $C_{28}H_{56}O_4Sn$ mw: 575.53
SYN: BIS(DECANOYLOXY)DI-N-BUTYLTIN
OSHA PEL: TWA 0.1 mg(Sn)/m³ (skin)
ACGIH TLV: TWA 0.1 mg(Sn)/m³; STEL 0.2 mg(Sn)/m³ (skin).
NIOSH REL: (Organotin Compounds) TWA 0.1 mg(Sn)/m³
SAFETY PROFILE: Poison by ingestion. A severe skin and eye irritant. See also TIN COMPOUNDS. When heated to decomposition it emits acrid and irritant fumes.

BIV900   CAS: 73926-85-9   HR: 3
BIS(DIBUTYLAMMONIUM)HEXACHLOROSTANNATE
mf: $C_{16}H_{40}N_2 \cdot Cl_6Sn$ mw: 591.97
SYNS: AMMONIUMYL, DIBUTYL-, HEXACHLORO-STANNATE(2-) (2:1) □ DIBUTYLAMINE, HEXACHLOROSTANNANE (2:1)
OSHA PEL: TWA 2 mg(Sn)/m³
ACGIH TLV: TWA 2 mg(Sn)/m³
SAFETY PROFILE: Poison by intravenous route. When heated to decomposition it emits toxic fumes of NO_x, Sn, and Cl⁻.

BIW750   CAS: 13927-77-0   HR: 3
BIS(DIBUTYLDITHIOCARBAMATO)NICKEL
mf: $C_{18}H_{36}N_2S_4 \cdot Ni$ mw: 467.51
PROP: Green crystals from C_6H_6/EtOH. Mp: 91°. Sol in C_6H_6, Me_2CO.
SYNS: DIBUTYLDITHIOCARBAMIC ACID, NICKEL SALT □ NICKEL DIBUTYLDITHIOCARBAMATE □ UV CHEK AM 104 □ VANGUARD N
CONSENSUS REPORTS: NTP 10th Report on Carcinogens. Reported in EPA TSCA Inventory. Nickel and its compounds are on the Community Right-To-Know List.
SAFETY PROFILE: Low toxicity by ingestion. Confirmed human carcinogen with experimental tumorigenic data. See also

NICKEL COMPOUNDS and CARBAMATES. When heated to decomposition it emits very toxic fumes of SO_x and NO_x.

BIX000   CAS: 136-23-2   HR: 3
BIS(DIBUTYLDITHIOCARBAMATO)ZINC
mf: $C_{18}H_{38}N_2S_4Zn$ mw: 476.19
PROP: White powder. Mp: 104–108°, d: 1.24 @ 20°/20°.
SYNS: ACETO ZDBD □ BUTAZATE □ BUTAZATE 50-D □ BUTYL ZIMATE □ BUTYL ZIRAM □ DIBUTYL-DITHIO-CARBAMIC ACID ZINC COMPLEX □ DIBUTYLDITHIOCARBAMIC ACID ZINC SALT □ USAF GY-5 □ VULCACURE □ VULKACIT LDB/C □ ZINC-BIBUTYLDITHIOCARBAMATE □ ZINC-DIBUTYLDI-THIOCARBAMATE □ ZINC-N,N-DIBUTYLDI-THIOCARBAMATE
CONSENSUS REPORTS: Reported in EPA TSCA Inventory. Zinc and its compounds are on the Community Right-To-Know List.
SAFETY PROFILE: Poison by intraperitoneal route. Questionable carcinogen with experimental tumorigenic data. When heated to decomposition it emits very toxic fumes of NO_x, ZnO, and SO_x. See also ZINC COMPOUNDS and CARBAMATES.

BIX500   CAS: 15442-77-0   HR: 3
BIS(3,4-DICHLOROBENZOATO)NICKEL
mf: $C_{14}H_6Cl_4NiO_4$ mw: 438.71
CONSENSUS REPORTS: NTP 10th Report on Carcinogens. Nickel and its compounds are on the Community Right-To-Know List.
OSHA PEL: TWA 0.1 mg (Ni)/m³
ACGIH TLV: TWA 0.2 mg(Ni)/m³; Human Carcinogen)
NIOSH REL: (Inorganic Nickel) TWA 0.015 mg(Ni)/m³
SAFETY PROFILE: Confirmed human carcinogen. Poison by intravenous route. See also NICKEL COMPOUNDS and CHLORIDES. When heated to decomposition it emits toxic fumes of Cl⁻.

B

BJK560   CAS: 30947-30-9   HR: 3
((3,5-BIS(1,1-DIMETHYLETHYL)-4-HYDROXYPHENYL)METHYL)PHOSPHONIC ACID, MONOETHYL ESTER, NICKEL(2+) SALT (2:1)
mf: $C_{34}H_{56}O_8P_2 \cdot Ni$ mw: 713.55
SYNS: IRGASTAB 2002 □ IRGASTAB 2002 HT
CONSENSUS REPORTS: NTP 10th Report on Carcinogens.
SAFETY PROFILE: Confirmed human carcinogen. Moderately toxic by ingestion. When heated to decomposition it emits toxic vapors of PO_x and Ni.

BIX750   CAS: 133-14-2   HR: 3
BIS(2,4-DICHLOROBENZOYL)-PEROXIDE
mf: $C_{14}H_6Cl_4O_4$ mw: 380.00
SYNS: CADOX TS □ CADOX TS 40,50 □ DI-2,4-DICHLOROBENZOYL PEROXIDE, >75% with water (DOT) □ LUPERCO CST
CONSENSUS REPORTS: Reported in EPA TSCA Inventory.
DOT CLASSIFICATION: Forbidden
SAFETY PROFILE: Poison by intraperitoneal route. Explosion Hazard: Pure compound is extremely shock sensitive and decomposes rapidly @ 80°. When heated to decomposition it emits toxic fumes of Cl^-. See also PEROXIDES, ORGANIC; and ESTERS.

BJA200   CAS: 90466-79-8   HR: 3
BIS(2,2-DIETHOXYETHYL)DISELENIDE
mf: $C_{12}H_{26}O_4Se_2$ mw: 392.30
SYN: DISELENIDE, BIS(2,2-DIETHOXYETHYL)-
OSHA PEL: TWA 0.2 mg(Se)/m^3
ACGIH TLV: TWA 0.2 mg(Se)/m^3
SAFETY PROFILE: Poison by intravenous route. When heated to decomposition it emits toxic fumes of Se.

BJB500   CAS: 14239-68-0   HR: 3
BIS(DIETHYLDITHIOCARBAMATO)CADMIUM
mf: $C_{10}H_{20}CdN_2S_4$ mw: 408.96
SYNS: CADMIUM DIETHYL DITHIOCARBAMATE □ ETHYL CADMATE □ ETHYL TUADS

CONSENSUS REPORTS: Reported in EPA TSCA Inventory. Cadmium and its compounds are on the Community Right-To-Know List.
OSHA PEL: TWA 5 μg(Cd)/m^3
ACGIH TLV: TWA 0.002 mg(Cd)/m^3 (respirable dust), Suspected Human Carcinogen); BEI: 5 μg/g creatinine in urine; 5 μg/L in blood
DFG MAK: DFG BAT: Blood 1.5 μg/dL; Urine 15 μg/dL, Suspected Carcinogen
NIOSH REL: (Cadmium) Reduce to lowest feasible level
SAFETY PROFILE: Confirmed human carcinogen with experimental tumorigenic data. Mutation data reported. When heated to decomposition it emits very toxic fumes of NO_x and SO_x. See also CADMIUM COMPOUNDS and CARBAMATES.

BJB750   CAS: 14239-51-1   HR: 3
BIS(DIETHYLDITHIOCARBAMATO)MERCURY
mf: $C_{10}H_{20}HgN_2S_4$ mw: 497.15
PROP: Yellow crystals from Me_2CO. Mp: 127–130°. IDLH 10 mg/m^3 (as Hg).
CONSENSUS REPORTS: Mercury and its compounds are on the Community Right-To-Know List.
OSHA PEL: CL 0.1 mg(Hg)/m^3 (skin)
ACGIH TLV: TWA 0.1 mg(Hg)/m^3 (skin); BEI: 35 μg/g creatinine total inorganic mercury in urine preshift; 15 μg/g creatinine total inorganic mercury in blood at end of shift at end of workweek.
DFG MAK: Confirmed Animal Carcinogen with Unknown Relevance to Humans
NIOSH REL: (Organomercury): TWA 0.01 mg/m^3; STEL 0.03 mg/m^3 (skin)
SAFETY PROFILE: Poison by intravenous and intraperitoneal routes. See also MERCURY COMPOUNDS and CARBAMATES. When heated to decomposition it emits very toxic fumes of NO_x, SO_x, and Hg.

BJC000   CAS: 14324-55-1   HR: 3
BIS(DIETHYLDITHIOCARBAMATO)ZINC
mf: $C_{10}H_{22}N_2S_4 \cdot Zn$ mw: 363.95
PROP: White powder. D: 1.47 @ 20°/20°.
SYNS: DIETHYLDITHIOCARBAMIC ACID ZINC SALT □ ETHAZATE □ ETHYL CYMATE □ ETHYL ZIMATE □ ETHYL ZIRUM □ VULCACURE □ VULKACIT LDA □ ZINC DIETHYLDITHIOCARBAMATE □ ZINC-N,N-DIETHYLDITHIOCARBAMATE
CONSENSUS REPORTS: Reported in EPA TSCA Inventory. Zinc and its compounds are on the Community Right-To-Know List.
SAFETY PROFILE: Poison by intraperitoneal route. Moderately toxic by ingestion and subcutaneous routes. Severe irritant to eyes, nose, and throat. Questionable carcinogen with experimental carcinogenic and tumorigenic data. Mutation data reported. When heated to decomposition it emits very toxic fumes of NO_x and SO_x. See also ZINC COMPOUNDS and CARBAMATES.

BJE550   CAS: 2162-74-5   HR: 3
BIS(2,6-DIISOPROPYLPHENYL)-CARBODIIMIDE
mf: $C_{25}H_{34}N_2$ mw: 362.61
SYNS: BENZENAMINE, N,N'-METHANETETRAYL-BIS(2,6-BIS(1-METHYLETHYL)- □ CARBO D □ CARBODIIMIDE, BIS(2,6-DIISOPROPYLPHENYL)- □ N,N'-METHANETETRAYLBIS(2,6-BIS(1-METHYLETHYL)BENZENAMINE □ STABOXOL 1
SAFETY PROFILE: A poison by ingestion, intraperitoneal, and inhalation. When heated to decomposition it emits toxic vapors of NO_x.

BJE750   CAS: 115-26-4   HR: 3
BIS(DIMETHYLAMIDO)FLUORO PHOSPHATE
mf: $C_4H_{12}FN_2OP$ mw: 154.15
PROP: Liquid. Misc in H_2O and most org solvs. D: 1.115 mm @ 20°, bp: 67° @ 4 mm.
SYNS: BFP □ BFPO □ BIS(DIMETHYLAMIDO)-PHOSPHORYL FLUORIDE □ BIS(DIMETHYLAMINO)-FLUOROPHOSPHATE □ BISDIMETHYLAMINO-FLUOROPHOSPHINE OXIDE □ CR 409 □ DIFO □ DIMEFOX □ DMF □ ENT 19,109 □ FLUOPHOSPHORIC ACID DI(DIMETHYLAMIDE) □ FLUORURE de N,N,N',N'-TETRAMETHYLE PHOSPHORO-DIAMIDE (FRENCH) □

HANANE □ PESTOX IV □ PESTOX XIV □ PESTOX 14 □ T-2002 □ TERRA-SYSTAM □ TERRA-SYTAM □ TERRASYTUM □ N,N,N',N'-TETRAMETHYL-DIAMIDO-FOSFORZUUR-FLUORIDE (DUTCH) □ TETRAMETHYL-DIAMIDOPHOSPHORIC FLUORIDE □ N,N,N',N'-TETRAMETHYL-DIAMIDO-PHOSPHORSAEURE-FLUORID (GERMAN) □ TETRAMETHYLPHOSPHORO-DIAMIDIC FLUORIDE □ N,N,N,N-TETRAMETHYL-PHOSPHORODIAMIDIC FLUORIDE □ N,N,N',N'-TETRAMETIL-FOSFORODIAMMIDO-FLUORURO (ITALIAN) □ TETRA SYTAM □ TL 792 □ WACKER S 14/10
CONSENSUS REPORTS: EPA Extremely Hazardous Substances List.
OSHA PEL: TWA 2.5 mg(F)/m^3
ACGIH TLV: TWA 2.5 mg(F)/m^3; BEI: 3 mg/g creatinine of fluorides in urine prior to shift; 10 mg/g creatinine of fluorides in urine at end of shift.
NIOSH REL: (Fluorides, Inorganic) TWA 2.5 mg(F)/m^3
SAFETY PROFILE: Poison by ingestion, skin contact, intraperitoneal, subcutaneous, and intravenous routes. When heated to decomposition it emits very toxic fumes of F^-, NO_x, and PO_x.

BJF000   CAS: 494-38-2   HR: 3
3,6-BIS(DIMETHYLAMINO)ACRIDINE
mf: $C_{17}H_{19}N_3$ mw: 265.39
PROP: Yellow needles from EtOH. Mp: 180–181°. Sol in EtOH and Me_2CO.
SYNS: ACRIDINE ORANGE □ ACRIDINE ORANGE FREE BASE □ BASIC ORANGE 3RN □ 2,8-BISDIMETHYLAMINOACRIDINE □ BRILLIANT ACRIDINE ORANGE E □ C.I. 46005 □ C.I. No. 46005:1 □ C.I. BASIC ORANGE 14 □ C.I. SOLVENT ORANGE 15 □ 3,6-DI(DIMETHYLAMINO)ACRIDINE □ EUCHRYSINE □ RHODULINE ORANGE □ SOLVENT ORANGE 15 □ N,N,N'-TETRAMETHYL-3,6-ACRIDINEDIAMINE □ WAXOLINE ORANGE A
CONSENSUS REPORTS: IARC Cancer Review: Group 3 IMEMDT 7,56,87; Animal Inadequate Evidence IMEMDT 16,145,78.
SAFETY PROFILE: Poison by subcutaneous route. Questionable carcinogen with experimental tumorigenic and carcinogenic data. Mutation data reported. When heated to decomposition it emits toxic fumes of NO_x.

BJH750   CAS: 3033-62-3   HR: 3
BIS(2-DIMETHYLAMINOETHYL) ETHER
mf: $C_8H_{20}N_2O$ mw: 160.30

PROP: Bp: 180–182°.

SYN: DMAEE □ NIAX CATALYST AL

CONSENSUS REPORTS: Reported in EPA TSCA Inventory.

ACGIH TLV: TWA 0.05 ppm; STEL 0.15 ppm (skin))

SAFETY PROFILE: Poison by skin contact. Moderately toxic by ingestion. Experimental reproductive effects. A severe skin and eye irritant. See also ETHERS. When heated to decomposition it emits toxic fumes of NO_x.

BJI250   CAS: 61-73-4   HR: 3
3,7-BIS(DIMETHYL AMINO)PHENAZA THIONIUM CHLORIDE
mf: $C_{16}H_{18}N_3S \cdot Cl$ mw: 319.88

PROP: Dark bronze-green crystals with bronze luster. Sol in H_2O and EtOH.

SYNS: AIZEN METHYLENE BLUE BH □ BASIC BLUE 9 □ 3,7-BIS(DIMETHYLAMINO)PHENOTHIAZIN-5-IUM CHLORIDE □ CALCOZINE BLUE ZF □ CHROMOSMON □ C.I. 52015 (CZECH) □ C.I. BASIC BLUE 9 □ D&C BLUE NUMBER 1 □ EXTERNAL BLUE 1 □ HIDACO METHYLENE BLUE SALT FREE □ LEATHER PURE BLUE HB □ METHYLENE BLUE □ METHYLENE BLUE A □ METHYLENE BLUE BB □ METHYLENE BLUE BB ZINC FREE □ METHYLENE BLUE CHLORIDE □ METHYLENE BLUE CHLORIDE (biological stain) □ METHYLENE BLUE D □ METHYLENE BLUE (medicinal) □ METHYLENE BLUE I (medicinal) □ METHYLENE BLUE NF (medicinal) □ METHYLENE BLUE POLYCHROME □ METHYLENE BLUE USP (medicinal) □ METHYLENE BLUE USP XII (medicinal) □ METHYLENIUM CERULEUM □ METHYLTHIONINE CHLORIDE □ METHYLTHIONIUM CHLORIDE □ MITSUI METHYLENE BLUE □ MODR METHYLENOVA (CZECH) □ SANDOCRYL BLUE BRL □ SCHULTZ No. 1038 □ SWISS BLUE □ TETRAMETHYLTHIONINE CHLORIDE □ YAMAMOTO METHYLENE BLUE B

CONSENSUS REPORTS: EPA Genetic Toxicology Program. Reported in EPA TSCA Inventory.

SAFETY PROFILE: Poison by ingestion, intraperitoneal, intravenous, and subcutaneous routes. Human systemic effects: cyanosis, blood changes. Experimental reproductive effects. Mutation data reported. When heated to

decomposition it emits very toxic fumes of NO_x, SO_x, and Cl^-.

BJK500   CAS: 137-30-4   HR: 3
BIS(DIMETHYLDITHIOCARBAMATO)-ZINC
mf: $C_6H_{12}N_2S_4 \cdot Zn$ mw: 305.81

PROP: White powder. Mp: 248–250°, d: 1.65 @ 20°/20°.

SYNS: AAPROTECT □ AAVOLEX □ AAZIRA □ ACCELERATOR L □ ACETO ZDED □ ACETO ZDMD □ ALCOBAM ZM □ AMYL ZIMATE □ ANTENE □ BIS(DIMETHYLCARBAMODITHIOATO-S,S')ZINC □ BIS(DIMETHYLDITHIOCARBAMATE de ZINC) (FRENCH) □ BIS(N,N-DIMETIL-DITIOCARBAMMATO) DI ZINCO (ITALIAN) □ CARBAMIC ACID, DIMETHYLDITHIO-, ZINC SALT (2:1) □ CARBAZINC □ CIRAM □ CORONA COROZATE □ COROZATE □ CUMAN □ CUMAN L □ CYMATE □ DIMETHYLCARBAMODITHIOIC ACID, ZINC COMPLEX □ DIMETHYLCARBAMODITHIOIC ACID, ZINC SALT □ DIMETHYLDITHIOCARBAMATE ZINC SALT □ DIMETHYLDITHIOCARBAMIC ACID, ZINC SALT □ DRUPINA 90 □ ENT 988 □ EPTAC 1 □ FUCLASIN □ FUCLASIN ULTRA □ FUKLASIN □ FUNGOSTOP □ HERMAT ZDM □ HEXAZIR □ KARBAM WHITE □ METHASAN □ METHAZATE □ METHYL ZIMATE □ METHYL ZINEB □ METHYL ZIRAM □ MEXENE □ MEZENE □ MILBAM □ MILBAN □ MOLURAME □ MYCRONIL □ NCI-C50442 □ ORCHARD BRAND ZIRAM □ POMARSOL Z FORTE □ PRODARAM □ RHODIACID □ SOXINAL PZ □ SOXINOL PZ □ TRICARBAMIX Z □ TSIMAT □ TSIRAM (RUSSIAN) □ USAF P-2 □ VANCIDE MZ-96 □ ZERLATE □ ZIMATE □ ZIMATE METHYL □ ZINC BIS(DIMETHYLDITHIOCARBAMATE) □ ZINC BIS(DIMETHYLDITHIOCARBAMOYL)DISULPHIDE □ ZINC DIMETHYLDITHIOCARBAMATE □ ZINC N,N-DIMETHYLDITHIOCARBAMATE □ ZINCMATE □ ZINK-BIS(N,N-DIMETHYL-DITHIOCARBAMAAT) (DUTCH) □ ZINK-BIS(N,N-DIMETHYL-DITHIOCARBAMAT) (GERMAN) □ ZINKCARBAMATE □ ZINK-(N,N-DIMETHYL-DITHIOCARBAMAT) (GERMAN) □ ZIRAM □ ZIRAMVIS □ ZIRASAN □ ZIRBERK □ ZIREX 90 □ ZIRIDE □ ZIRTHANE □ ZITOX

CONSENSUS REPORTS: IARC Cancer Review: Group 3 IMEMDT 7,56,87; Animal Inadequate Evidence IMEMDT 12,259,76; NTP Carcinogenesis Bioassay (feed); Clear Evidence: mouse, rat NTPTR* NTP-TR-238,83. EPA Genetic Toxicology Program. Reported in EPA TSCA Inventory. Zinc and its compounds are on the Community Right-To-Know List.

SAFETY PROFILE: Poison by ingestion, intraperitoneal, and intravenous routes. Moderately toxic by inhalation. Questionable carcinogen with experimental carcinogenic and tumorigenic data. An experimental teratogen. Other experimental reproductive effects. Human mutation data reported. See also ZINC COMPOUNDS and CARBAMATES. Severe irritant to eyes, nose, and throat. When heated to decomposition it emits very toxic fumes of NO_x and SO_x.

BJK600   CAS: 80387-97-9   HR: D
(((3,5-BIS(1,1-DIMETHYLETHYL)-4-HYDROXYPHENYL)METHYL)THIO)A CETIC ACID 2-ETHYLHEXYL ESTER

mf: $C_{25}H_{42}O_3S$ mw: 422.73

SYN: ACETIC ACID, (((3,5-BIS(1,1-DIMETHYLETHYL)-4-HYDROXYPHENYL)METHYL)THIO)-, 2-ETHYLHEXYL ESTER

CONSENSUS REPORTS: Reported in EPA TSCA Inventory.

SAFETY PROFILE: An experimental teratogen. Other experimental reproductive effects. When heated to decomposition it emits toxic fumes of SO_x.

BJK780   CAS: 32613-12-0   HR: 3
1,1'-BIS(DIMETHYLOCTOXYSILYL)-FERROCENE

mf: $C_{30}H_{54}FeO_2Si_2$ mw: 558.87

SYNS: 1,1'-BIS(DIMETHYL(OCTYLOXY)SILYL)FERROCENE □ FERROCENE, 1,1'-BIS(DIMETHYL(OCTYLOXY)SILYL)-

SAFETY PROFILE: A poison by inhalation. Low toxicity by ingestion. When heated to decomposition it emits acrid smoke and irritating vapors.

BJL600   CAS: 97-74-5   HR: 3
BIS(DIMETHYLTHIOCARBAMOYL)-SULFIDE

mf: $C_6H_{12}N_2S_3$ mw: 208.38

PROP: Yellow crystals from EtOH. Mp: 104°. Very sol in EtOH, $CHCl_3$; sltly sol in cold Et_2O.

SYNS: ACETO TMTM □ BIS(DIMETHYLTHIO-CARBAMYL) MONOSULFIDE □ CARBAMIC ACID, DIMETHYLDITHIO-, ANHYDROSULFIDE □ MONEX □

MONO-THIURAD □ MONOTHIURAM □ PENNAC MS □ TETRAMETHYLTHIURAMMONIUM SULFIDE □ TETRAMETHYLTHIURAM MONOSULFIDE □ TETRAMETHYLTHIURAM SULFIDE □ TETRAMETHYL-TRITHIO CARBAMIC ANHYDRIDE □ 1,1'-THIOBIS(N,N-DIMETHYLTHIO)FORMAMIDE □ THIONEX □ THIONEX RUBBER ACCELERATOR □ TMTM □ TMTMS □ UNADS □ USAF B-32 □ USAF EK-P-6255 □ VULKACIT THIURAM MS/C

CONSENSUS REPORTS: Reported in EPA TSCA Inventory.

SAFETY PROFILE: Poison by ingestion and intraperitoneal routes. Questionable carcinogen with experimental tumorigenic data. Mutation data reported. An experimental teratogen. When heated to decomposition it emits very toxic fumes of NO_x and SO_x. See also SULFIDES.

BJM700   CAS: 38998-91-3   HR: 3
BIS(1,3-DITHIOCYANATO-1,1,3,3-TETRABUTYLDISTANNOXANE)

mf: $C_{36}H_{72}N_4O_2S_4Sn_4$ mw: 1196.12

SYNS: DISTANNOXANE, BIS(1,3-DITHIOCYANATO-1,1,3,3-TETRABUTYL)- □ DI-μ-(THIOCYANATODI-n-BUTYLSTANNYLOXO)BIS(THIOCYANATODI-n-BUTYLTIN)

OSHA PEL: TWA 0.1 mg(Sn)/m³
ACGIH TLV: TWA 0.1 mg(Sn)/m³; STEL 0.2 mg/m³ (skin)
NIOSH REL: 10H TWA 0.1 mg(Sn)/m³

SAFETY PROFILE: Poison by intravenous route. When heated to decomposition it emits toxic fumes of NO_x, SO_x, and Sn.

BJN250   CAS: 2386-90-5   HR: 2
BIS(2,3-EPOXYCYCLOPENTYL) ETHER

mf: $C_{10}H_{14}O_3$ mw: 182.24

SYNS: EP-205 □ ERR 4205 □ 2,2'-OXYBIS-6-OXABICYCLO-(3.1.0)HEXANE

CONSENSUS REPORTS: EPA Genetic Toxicology Program. Reported in EPA TSCA Inventory.

SAFETY PROFILE: Moderately toxic by ingestion. A systemic irritant by skin contact and ingestion. Experimental reproductive effects. Questionable carcinogen with experimental carcinogenic and neoplastigenic data. See also ETHERS. When heated to decomposition it emits acrid smoke and irritating fumes.

B

BJO225   CAS: 109-44-4   HR: 2
BIS(2-ETHOXYETHYL) ADIPATE
mf: $C_{14}H_{26}O_6$ mw: 290.40
SYNS: ADIPIC ACID, BIS(2-ETHOXYETHYL) ESTER □ DIETHOXY ETHYL ADIPATE □ HEXANOIC ACID, BIS(2-ETHOXYETHYL) ESTER
CONSENSUS REPORTS: Reported in EPA TSCA Inventory.
SAFETY PROFILE: A severe skin irritant. When heated to decomposition it emits acrid smoke and irritating fumes.

BJP000   CAS: 122-34-9   HR: 3
2,4-BIS(ETHYLAMINO)-6-CHLORO-s-TRIAZINE
mf: $C_7H_{12}ClN_5$ mw: 201.69
PROP: Crystals. Mp: 228–229°.
SYNS: AKTINIT S □ AQUAZINE □ BATAZINA □ 2,4-BIS(AETHYLAMINO)-6-CHLOR-1,3,5-TRIAZIN (GERMAN) □ BITEMOL □ BITEMOL S 50 □ CAT (herbicide) □ CDT □ CEKUSAN □ CEKUZINA-S □ CET □ 1-CHLORO-3,5-BISETHYLAMINO-2,4,6-TRIAZINE □ 2-CHLORO-4,6-BIS(ETHYLAMINO)-s-TRIAZINE □ 2-CHLORO-4,6-BIS(ETHYLAMINO)-1,3,5-TRIAZINE □ FRAMED □ GEIGY 27,692 □ GESARAN □ GESATOP □ HERBAZIN □ HERBEX □ HERBOXY □ HUNGAZIN DT □ PREMAZINE □ PRIMATOL S □ RADOCON □ RADOKOR □ SIMANEX □ SIMAZIN □ SIMAZINE (USDA) □ SIMAZINE 80W □ TAFAZINE □ TAPHAZINE □ ZEAPUR
CONSENSUS REPORTS: EPA Genetic Toxicology Program. Reported in EPA TSCA Inventory.
SAFETY PROFILE: Poison by intravenous route. Moderately toxic by ingestion. Questionable carcinogen with experimental tumorigenic data. An experimental teratogen. Other experimental reproductive effects. A skin and eye irritant. Mutation data reported. May cause weight loss and reduced red blood cell count. When heated to decomposition it emits very toxic fumes of Cl^- and NO_x.

BJP425   CAS: 29471-80-5   HR: 3
BIS(ETHYLENEDIAMINE)(MERCURICTETRATHIOCYANATO)COPPER
mf: $(C_4H_{16}CuN_4 \cdot C_4HgN_4S_4)_x$
PROP: IDLH 10 mg/m³ (as Hg).
SYNS: COPPER, BIS(ETHYLENEDIAMINE)(MERCURICTETRATHIOCYA NATO)- □ COPPER(2+), BIS(ETHYLENEDIAMINE)-, TETRAKIS(THIOCYANATO)MERCURATE(2-), POLYMERS
ACGIH TLV: TWA 0.1 mg(Hg)/m³ (skin); BEI: 35 μg/g creatinine total inorganic mercury in urine preshift; 15 μg/g creatinine total inorganic mercury in blood at end of shift at end of workweek.
DFG MAK: Confirmed Animal Carcinogen with Unknown Relevance to Humans
NIOSH REL: (Organomercury): TWA 0.01 mg/m³; STEL 0.03 mg/m³ (skin)
SAFETY PROFILE: Poison by intraperitoneal route. When heated to decomposition it emits toxic fumes of NO_x, SO_x, Hg, and Cl^-.

BJP450   CAS: 1192-75-2   HR: 3
BISETHYLENEUREA
mf: $C_5H_8N_2O$ mw: 112.15
SYNS: AZIRIDINE, 1,1'-CARBONYLBIS- □ BIS(1-AZIRIDINYL)KETONE □ CARBONYLBIS(AZIRIDINE) □ CARBONYLBIS(1-AZIRIDINE) □ DIETHYLENEUREA □ N,N'-DIETHYLENEUREA
DOT CLASSIFICATION: 3; Label: Flammable Liquid
SAFETY PROFILE: A poison by intraperitoneal route. A flammable liquid. When heated to decomposition it emits toxic vapors of NO_x.

BJQ250   CAS: 2781-10-4   HR: 3
BIS(2-ETHYLHEXANOYLOXY)DIBUTYL STANNANE
mf: $C_{24}H_{48}O_4Sn$ mw: 519.41
SYNS: DIBUTYLBIS((2-ETHYLHEXANOYL)OXY)-STANNANE □ DIBUTYLBIS((2-ETHYL-1-OXOHEXYL)OXY)-STANNANE (9CI) □ DIBUTYLTIN BIS(α-ETHYLHEXANOATE) □ DIBUTYLTIN BIS(2-ETHYLHEXANOATE) □ DIBUTYLTIN DI(2-ETHYLHEXANOATE) □ DI-n-BUTYLTIN DI-2-ETHYLHEXANOATE □ DIBUTYLTIN DI(2-ETHYLHEXOATE)
CONSENSUS REPORTS: Reported in EPA TSCA Inventory.
OSHA PEL: TWA 0.1 mg(Sn)/m³ (skin)
ACGIH TLV: TWA 0.1 mg(Sn)/m³; STEL 0.2 mg(Sn)/m³ (skin).
NIOSH REL: (Organotin Compounds) TWA 0.1 mg(Sn)/m³

SAFETY PROFILE: Poison by ingestion and intravenous routes. See also TIN COMPOUNDS. When heated to decomposition it emits acrid smoke and irritating fumes.

BJR750   CAS: 298-07-7   HR: 3
BIS(2-ETHYLHEXYL) PHOSPHATE
mf: $C_{16}H_{35}O_4P$ mw: 322.48
PROP: Viscous liquid. D: 0.975 @ 25 mm, bp: 155° @ 0.015 mm. Sol in C_6H_6, hexane, and 4-methyl-2-pentanone; sltly sol in H_2O.
SYNS: BIS(2-ETHYLHEXYL)HYDROGEN PHOSPHATE □ BIS(2-ETHYLHEXYL)ORTHOPHOSPHORIC ACID □ BIS(2-ETHYLHEXYL)PHOSPHORIC ACID □ DEHPA EXTRACTANT □ DI(2-ETHYLHEXYL)PHOSPHATE □ DI-2(ETHYLHEXYL)PHOSPHORIC ACID □ DI-(2-ETHYLHEXYL)PHOSPHORIC ACID (DOT) □ 2-ETHYL-1-HEXANOL HYDROGEN PHOSPHATE □ HDEHP □ KYSELINA DI-(2-ETHYLHEXYL)FOSFORECNA
CONSENSUS REPORTS: Reported in EPA TSCA Inventory.
SAFETY PROFILE: Poison by intraperitoneal route. A corrosive material. A severe eye and skin irritant. When heated to decomposition it emits toxic fumes of PO_x.

BJS250   CAS: 122-62-3   HR: 2
BIS(2-ETHYLHEXYL) SEBACATE
mf: $C_{26}H_{50}O_4$ mw: 426.76
PROP: Light, clear liquid; mild odor. Mp: −48°, fp: −55°, bp: 256° @ 5 mm, flash p: 410°F, d: 0.914 @ 20°/4°, vap d: 14.7.
SYNS: BISOFLEX DOS □ DECANEDIOIC ACID, BIS(2-ETHYLHEXYL) ESTER □ DI(2-ETHYLHEXYL)SEBACATE □ DIOCTYL SEBACATE □ DOS □ 2-ETHYLHEXYL SEBACATE □ MONOPLEX DOS □ OCTOIL S □ OCTYL SEBACATE □ PX 438 □ STALFLEX DOS □ UNIFLEX DOS
CONSENSUS REPORTS: Reported in EPA TSCA Inventory.
SAFETY PROFILE: Moderately toxic by ingestion and intravenous routes. See also ESTERS. Combustible when exposed to heat or flame; can react with oxidizing materials. To fight fire, use foam, CO_2, dry chemical. When heated to decomposition it emits acrid and irritating fumes.

BJT250   CAS: 2440-45-1   HR: 3
BIS(ETHYLMERCURI) PHOSPHATE
mf: $C_4H_{11}Hg_2O_4P$ mw: 555.30
PROP: Solid. IDLH 10 mg/m³ (as Hg).
SYNS: ETHYLMERCURIC PHOSPHATE □ ETHYLMERCURY PHOSPHATE □ LIGNASAN FUNGICIDE □ LIGNASAN-X □ NEW IMPROVED CERESAN □ NEW IMPROVED GRANOSAN
CONSENSUS REPORTS: Mercury and its compounds are on the Community Right-To-Know List.
OSHA PEL: TWA 0.01 mg(Hg)/m³; STEL 0.03 mg/m³ (skin)
ACGIH TLV: TWA 0.01 mg(Hg)/m³; BEI: 35 μg/g creatinine total inorganic mercury in urine preshift; 15 μg/g creatinine total inorganic mercury in blood at end of shift at end of workweek.
DFG MAK: Confirmed Animal Carcinogen with Unknown Relevance to Humans
SAFETY PROFILE: Poison by ingestion and subcutaneous routes. See also MERCURY COMPOUNDS, ORGANIC. An experimental teratogen. When heated to decomposition it emits very toxic fumes of Hg and PO_x.

BJT800   CAS: 18771-38-5   HR: 3
BIS(ETHYLTHIO)METHYLENE
MALONONITRILE
mf: $C_8H_{10}N_2S_2$ mw: 198.32
SYNS: CP 26890 □ (BIS(ETHYLTHIO)METHYLENE)PROPANEDINITRILE □ MALONONITRILE, CARBONYL-, DIETHYL MERCAPT-OLE □ PROPANEDINITRILE, (BIS(ETHYLTHIO)-METHYLENE)-
SAFETY PROFILE: A poison by ingestion and skin contact. A severe eye irritant. When heated to decomposition it emits toxic vapors of NO_x and SO_x.

BJU000   CAS: 502-55-6   HR: 3
BIS(ETHYLXANTHOGEN) DISULFIDE
mf: $C_6H_{10}O_2S_4$ mw: 242.40
PROP: Yellow needles. Mp: 28–32°.
SYNS: AULIGEN □ BEK □ BEXIDE □ BEXT □ BIETHYLXANTHOGENTRISULFIDE □ BIS(ETHYL-XANTHIC)DISULFIDE □ DEX □ DIETHYLDITHIO BIS(THIONOFORMATE) □ DIETHYL DIXANTHOGEN □ DIETHYL XANTHOGENATE □ DIETHYLXANTH-

B

OGEN DISULFIDE □ DITHIOBIS(THIOFORMIC ACID)-o,o-DIETHYL ESTER □ DIXANTHOGEN □ ETHYL XANTHOGEN DISULFIDE □ EXD □ K PREPARATION □ THIOPEROXYDICARBONIC ACID DIETHYL ESTER

CONSENSUS REPORTS: Reported in EPA TSCA Inventory.

SAFETY PROFILE: Poison by ingestion and intraperitoneal routes. Moderately toxic by skin contact and possibly other routes. See also ESTERS and SULFIDES. When heated to decomposition it emits highly toxic fumes of SO_x.

BJW800   CAS: 4387-13-7   HR: 3
BIS(FORMYLMETHYL) MERCURY
mf: $C_4H_6HgO_2$ mw: 286.69
PROP: Crystals from EtOH. Mp: 92–94°.
IDLH 10 mg/m³ (as Hg).
SYN: MERCURIDIACETALDEHYDE
OSHA PEL: CL 0.1 mg(Hg)/m³ (skin)
ACGIH TLV: TWA 0.01 mg(Hg)/m³; BEI: 35 µg/g creatinine total inorganic mercury in urine preshift; 15 µg/g creatinine total inorganic mercury in blood at end of shift at end of workweek.
DFG MAK: Confirmed Animal Carcinogen with Unknown Relevance to Humans
NIOSH REL: (Organomercury): TWA 0.01 mg/m³; STEL 0.03 mg/m³ (skin)
SAFETY PROFILE: Poison by intravenous route. When heated to decomposition it emits toxic fumes of Hg.

BJW825   CAS: 5188-42-1   HR: 3
BIS(GUANIDINIUM) CHROMATE
mf: $C_2H_{10}N_6$•CrH_2O_4 mw: 236.20
PROP: IDLH Ca [15 mg/m³ {as Cr(VI)}].
SYN: BIGUANIDINE, CHROMATE
OSHA PEL: CL 0.1 mg(CrO₃)/m³
ACGIH TLV: TWA 0.05 mg(Cr)/m³;
Confirmed Human Carcinogen
NIOSH REL: (Chromium(VI)) TWA 0.025 mg/m³; CL 0.05 mg/15M
SAFETY PROFILE: A confirmed carcinogen. Poison by intravenous route. When heated to decomposition it emits toxic fumes of NO_x and Cr.

BJX800   CAS: 63270-67-7   HR: 3
BIS(l-HISTIDINATO)MANGANESE TETRAHYDRATE
mf: $C_{12}H_{16}MnN_6O_4$•$4H_2O$ mw: 435.36
SYNS: MANGANESE, BIS(l-HISTIDINATO)-, TETRAHYDRATE □ MANGANESE, BIS(l-HISTIDINATO-N,O)-, TETRAHYDRATE
OSHA PEL: CL 5 mg(Mn)/m³
ACGIH TLV: TWA 5 mg(Mn)/m³
SAFETY PROFILE: Poison by an unspecified route. When heated to decomposition it emits toxic fumes of NO_x and Mn.

BJZ000   CAS: 66-76-2   HR: 3
BISHYDROXYCOUMARIN
mf: $C_{19}H_{12}O_6$ mw: 336.31
PROP: Very small crystals from cyclohexanone with a slight pleasant odor and bitter taste. Mp: 288–289°. Sol in alkali.
SYNS: ACADYL □ ACAVYL □ ANTITROMBOSIN □ BARACOUMIN □ BHC □ BIS(4-HYDROXYCOUMARIN-3-YL)METHANE □ CUMA □ CUMID □ DICOUMARIN □ DICOUMAROL □ DICUMAN □ DICUMARINE □ DI-(4-HYDROXY-3-COUMARINYL)METHANE □ DI-4-HYDROXY-3,3'-METHYLENEDICOUMARIN □ DUFALONE □ KUMORAN □ MELITOXIN □ 3,3'-METHYLEEN-BIS(4-HYDROXY-CUMARINE) (DUTCH) □ 3,3'-METHYLEN-BIS(4-HYDROXY-CUMARIN) (GERMAN) □ 3,3'-METHYLENEBIS(4-HYDROXY-1,2-BENZOPYR-ONE) □ 3,3'-METHYLENEBIS(4-HYDROXYCOUMARIN) □ 3,3'-METHYLENE-BIS(4-HYDROXYCOUMARINE) (FRENCH) □ 3,3'-METILEN-BIS(4-IDROSSI-CUMARINA) (ITALIAN) □ TEMPARIN □ TROMBOSAN
CONSENSUS REPORTS: Reported in EPA TSCA Inventory.
SAFETY PROFILE: Poison by ingestion, subcutaneous, intravenous, and intraperitoneal routes. An experimental teratogen. Human reproductive effects by ingestion and possibly other routes: fetal death, unspecified developmental abnormalities, stillbirth, and unspecified neonatal effects. An anticoagulant. Excessive doses can cause hemorrhages. When heated to decomposition it emits acrid smoke and fumes. See also WARFARIN.

BKA000   CAS: 548-00-5   HR: 3
BIS(4-HYDROXY-3-COUMARIN) ACETIC ACID ETHYL ESTER
mf: $C_{22}H_{16}O_8$ mw: 408.38
PROP: Amorphous or crystalline from Me$_2$CO. Mp: 151° (amorphous), mp: 173° (crystalline).
SYNS: BIS-3,3'-(4-HYDROXYCOUMARINYL)ACETIC ACID ETHYL ESTER □ BIS-(4-HYDROXY-3-COUMARIN-YL)ETHYL ACETATE □ BIS(4-HYDROXY-2-OXO-2H-1-BENZOPYRAN-3-YL)ACETIC ACID ETHYL ESTER □ BOEA □ B.O.E.A. □ 3,3'-(CARBOXYMETHYLENE)BIS(4-HYDROXYCOUMARIN) ETHYL ESTER □ DICUMACYL □ ETHYL BISCOUMACETATE □ ETHYL BIS(4-HYDROXYCOUMARINYL)ACETATE □ ETHYL BIS(4-HYDROXY-3-COUMARINYL)ACETATE □ ETHYLDICOUMAROL □ ETHYLDICOUMAROL ACETATE □ ETHYL-4,4'-DIHYDROXYDICOUMARINYL-3,3'-ACETATE □ NEODICOUMARIN □ NEODI-COUMAROL □ NEODICUMARINUM □ PELENTAN □ STABILENE □ TROMBARIN □ TROMBIL □ TROMBOLYSAN □ TROMEXAN □ TROMEXAN ETHYL ACETATE
SAFETY PROFILE: Poison by intraperitoneal route. Moderately toxic by ingestion and subcutaneous routes. An experimental teratogen. Human reproductive effects by ingestion: developmental abnormalities of the cardiovascular system, stillbirth, and unspecified neonatal effects. An anticoagulant. See also WARFARIN and ESTERS. When heated to decomposition it emits acrid and irritating fumes.

BKE500   CAS: 120-40-1   HR: 2
N,N-BIS(2-HYDROXYETHYL)DODECAN AMIDE
mf: $C_{16}H_{33}NO_3$ mw: 287.50
PROP: Solid. Mp: 36°
SYNS: BIS(2-HYDROXYETHYL)LAURAMIDE □ N,N-BIS(HYDROXYETHYL)LAURAMIDE □ N,N-BIS(β-HYDROXYETHYL)LAURAMIDE □ N,N-BIS(2-HYDROXYETHYL)LAURAMIDE □ CLINDROL 101CG □ CLINDROL SUPERAMIDE 100L □ COCO DIETHANOL-AMIDE □ COCONUT OIL AMIDE of DIETHANOL-AMINE □ COMPERLAN LD □ CONDENSATE PL □ CRILLON L.D.E. □ DIETHANOLLAURAMIDE □ N,N-DIETHANOLLAURAMIDE □ N,N-DIETHANOLLAURIC ACID AMIDE □ EMID 6511 □ EMID 6541 □ ETHYLAN MLD □ HETAMIDE ML □ LAURAMIDE DEA □ LAURIC ACID DIETHANOLAMIDE □ LAURIC DIETHANOL-

AMIDE □ LAUROYL DIETHANOLAMIDE □ LAURYL DIETHANOLAMIDE □ LDA □ LDE □ MONAMID 150-LW □ NCI-C55323 □ NINOL AA62 □ NINOL AA-62 EXTRA □ NINOL 4821 □ ONYXOL 345 □ REWOMID DLMS □ RICHAMIDE 6310 □ ROLAMID CD □ STANDAMIDD LD □ STEINAMID DL 203 S □ SUPER AMIDE L-9A □ SYNOTOL L-60 □ UNAMIDE J-56 □ VARAMID ML 1
CONSENSUS REPORTS: Reported in EPA TSCA Inventory.
SAFETY PROFILE: Moderately toxic by ingestion. When heated to decomposition it emits toxic fumes of NO$_x$. See also AMIDES.

BKF250   CAS: 2784-94-3   HR: 3
N',N'-BIS(2-HYDROXYETHYL)-N-METH-YL-2-NITRO-p-PHENYLENEDIAMINE
mf: $C_{11}H_{17}N_3O_4$ mw: 255.31
SYNS: HC BLUE 1 □ NCI-C04159
CONSENSUS REPORTS: IARC Cancer Review: Group 2B IMEMDT 57,129,93; Animal Sufficient Evidence IMEMDT 57,129,93; Human Inadequate Evidence IMEMDT 57,129,93. NTP Carcinogenesis Studies (feed); Some Evidence: rat NTPTR* NTP-TR-271,85; (feed); Clear Evidence: mouse NTPTR* NTP-TR-271,85. Reported in EPA TSCA Inventory.
SAFETY PROFILE: Suspected carcinogen with experimental carcinogenic data. Mutation data reported. See also AMINES. When heated to decomposition it emits toxic fumes of NO$_x$.

BKH500   CAS: 794-93-4   HR: 3
BIS(HYDROXYMETHYL)FURATRIZINE
mf: $C_{11}H_{11}N_5O_5$ mw: 293.27
PROP: Yellow crystals. Mp: 161° (decomp).
SYNS: 3-BIS(HYDROXYMETHYL)AMINO-6-(5-NITRO-2-FURYLETHENYL)-1,2,4-TRIAZINE □ DHNT □ 3-DI(HYDROXYMETHYL)AMINO-6-(5-NITRO-2-FURYL-ETHENYL)-1,2,4-TRIAZINE □ 3-DI(HYDROXYMETHYL)-AMINO-6-(2-(5-NITRO-2-FURYL)VINYL)-1,2,4-TRIAZINE □ DIHYDROXYMETHYL FURATRIZINE □ FURATONE □ FURATONE-S □ N-(6-(5-NITROFURFURYLIDENE-METHYL)-1,2,4-TRIAZIN-3-YL)IMINODIMETHANOL □ 6-(5-NITRO-2-FURYLVINYL)-3-(DIHYDROXYDIMETH-YLAMINO)-1,2,4-TRIAZENE □ N-(6-(2-(5-NITRO-2-FURYL)VINYL)-1,2,4-TRIAZIN-3-YL)IMINODIMETHAN-OL □ ((6-(2-(5-NITRO-2-FURYL)VINYL)-as-TRIAZIN-3-YL)IMINO)DIMETHANOL □ PANFURAN-S

B

CONSENSUS REPORTS: IARC Cancer Review: Group 3 IMEMDT 7,56,87; Animal Inadequate Evidence IMEMDT 24,77,80; Human No Adequate Data IMEMDT 24,77,80.
SAFETY PROFILE: Suspected carcinogen with experimental carcinogenic and tumorigenic data. Moderately toxic by ingestion, intraperitoneal, and subcutaneous routes. Mutation data reported. An antibacterial agent. When heated to decomposition it emits toxic fumes of NO_x.

BKJ275   CAS: 15702-65-5   HR: 3
BIS(8-HYDROXYQUINOLINE-5-
**   SULFONIC ACID) MANGANESE(II)**
mf: $C_{18}H_{12}N_2O_8S_2 \cdot Mn$ mw: 503.38
SYNS: BIS(5-SULFO-8-QUINOLINOLATO-N^1,O^8) MANGANESE(II) □ MANGANESE, BIS(5-SULFO-8-QUINOLINOLATO)-
OSHA PEL: CL 5 mg(Mn)/m^3
ACGIH TLV: TWA 5 mg(Mn)/m^3
SAFETY PROFILE: Poison by intravenous route. When heated to decomposition it emits toxic fumes of NO_x, SO_x, and Mn.

BKJ500   CAS: 73816-43-0   HR: 3
BIS(3-INDOLEMETHYLENEMORPH-
**   OLINIUM)HEXA-CHLOROSTANNATE**
mf: $C_{26}H_{30}N_4O_2 \cdot Cl_6Sn$ mw: 761.99
SYN: MORPHOLINIUM, (3-INDOLYLMETHYLENE)-, HEXACHLOROSTANNATE(2-) (2:1)
OSHA PEL: TWA 2 mg(Sn)/m^3
ACGIH TLV: TWA 2 mg(Sn)/m^3
SAFETY PROFILE: Poison by intravenous route. When heated to decomposition it emits toxic fumes of NO_x, Sn, and Cl⁻.

BKK250   CAS: 25168-24-5   HR: 2
BIS(ISOOCTYLOXYCARBONYLMETHYL
**   THIO)DIBUTYL STANNANE**
mf: $C_{28}H_{56}O_4S_2Sn$ mw: 639.65
SYNS: BIS(2-ETHYLHEXYLOXYCARBONYLMETHYLTHIO)DIBUTYL STANNANE □ DIBUTYL-TIN BIS(ISOOCTYLTHIO-GLYCOLLATE) □ DIBUTYLZINN-S,S'-BIS(ISOOCTYL-THIOGLYCOLAT) (GERMAN)
CONSENSUS REPORTS: Reported in EPA TSCA Inventory.
OSHA PEL: TWA 0.1 mg(Sn)/m^3 (skin)

ACGIH TLV: TWA 0.1 mg(Sn)/m^3; STEL 0.2 mg(Sn)/m^3 (skin).
NIOSH REL: (Organotin Compounds) TWA 0.1 mg(Sn)/m^3
SAFETY PROFILE: Moderately toxic by ingestion. See also TIN COMPOUNDS. When heated to decomposition it emits toxic fumes of SO_x.

BKK500   CAS: 26636-01-1   HR: 2
BIS(ISOOCTYLOXYCARBONYLMETHYL
**   THIO)DIMETHYLSTANNANE**
mf: $C_{22}H_{44}O_4S_2Sn$ mw: 555.47
SYNS: BIS(2-ETHYLHEXYLOXYCARBONYLMETHYLTHIO)DIMETHYLSTANNANE □ DIMETHYL-TIN BIS(ISOOCTYLTHIO-GLYCOLLATE) □ DIMETHYLZINN-S,S'-BIS(ISOOCTYL--THIOGLYCOLAT) (GERMAN)
CONSENSUS REPORTS: Reported in EPA TSCA Inventory.
OSHA PEL: TWA 0.1 mg(Sn)/m^3 (skin)
ACGIH TLV: TWA 0.1 mg(Sn)/m^3; STEL 0.2 mg(Sn)/m^3 (skin).
NIOSH REL: (Organotin Compounds) TWA 0.1 mg(Sn)/m^3
SAFETY PROFILE: Moderately toxic by ingestion. See also TIN COMPOUNDS. When heated to decomposition it emits toxic fumes of SO_x.

BKK750   CAS: 26401-97-8   HR: 2
BIS(ISOOCTYLOXYCARBONYLMETHYL
**   THIO)DIOCTYL STANNANE**
mf: $C_{36}H_{72}O_4S_2Sn$ mw: 751.89
SYNS: ADVASTAB 17 MO □ BIS(MERCAPTOACETATE)DIOCTYL-TIN BIS(ISOOCT-YL) ESTER □ DIISOOCTYL ((DIOCTYLSTANNYLENE)-DITHIO)DIACETATE □ DIOCTYLTIN BIS(ISOOCTYL MERCAPTOACETATE) □ DIOCTYLTIN-S,S'-BIS(ISOOCTYL MERCAPTOACETATE) □ DIOCTYLTIN BIS(ISOOCTYL THIOGLYCOLATE) □ DIOCTYL-TIN BIS(ISOOCTYLTHIOGLYCOLLATE) □ DI-n-OCTYLTIN DIISOOCTYL THIOGLYCOLATE □ DI-n-OCTYL-ZINN-DI-ISOOCTYLTHIOGLYKOLAT (GERMAN) □ DOTG □ THERMOLITE 831
CONSENSUS REPORTS: Reported in EPA TSCA Inventory.
OSHA PEL: TWA 0.1 mg(Sn)/m^3 (skin)
ACGIH TLV: TWA 0.1 mg(Sn)/m^3; STEL 0.2 mg(Sn)/m^3 (skin).

DFG MAK: 0.1 mg(Sn)/m³ calculated as total dust
NIOSH REL: (Organotin Compounds) TWA 0.1 mg(Sn)/m³
SAFETY PROFILE: Moderately toxic by ingestion and skin contact. An experimental teratogen. See also TIN COMPOUNDS and MERCAPTANS. When heated to decomposition it emits toxic fumes of SO_x.

BKL000   CAS: 33568-99-9   HR: 2
BIS(ISOOCTYLOXYMALEOYLOXY)DIOCTYLSTANNANE
mf: $C_{40}H_{72}O_8Sn$ mw: 799.81
SYNS: (Z,Z)-BIS((3-CARBOXYACRYLOYL)OXY)DIOCTYL-STANNANE DIISOOCTYL ESTER (8CI) □ (Z,Z)-4,4'-((DIOCTYL-STANNYLENE)BIS(OXY))BIS(4-OXO-2-BUTANOIC ACID) DIISOOCTYL ESTER □ DIOCTYLTINBIS(ISOOCTYL MALEATE)
CONSENSUS REPORTS: Reported in EPA TSCA Inventory.
OSHA PEL: TWA 0.1 mg(Sn)/m³ (skin)
ACGIH TLV: TWA 0.1 mg(Sn)/m³; STEL 0.2 mg(Sn)/m³ (skin).
DFG MAK: 0.1 mg(Sn)/m³ calculated as total dust
NIOSH REL: (Organotin Compounds) TWA 0.1 mg(Sn)/m³
SAFETY PROFILE: Moderately toxic by ingestion. See also TIN COMPOUNDS. When heated to decomposition it emits acrid smoke and irritating fumes.

BKL250   CAS: 7287-19-6   HR: 2
2,4-BIS(ISOPROPYLAMINO)-6-METHYL-MERCAPTO-s-TRIAZINE
mf: $C_{10}H_{19}N_5S$ mw: 241.40
PROP: Solid. Mp: 118–120°. Very sltly sol in H_2O.
SYNS: 4,6-BIS(ISOPROPYLAMINO)-2-METHYL-MERCAPTO-s-TRIAZINE □ 2,4-BIS(ISOPROPYLAMINO)-6-METHYLTHIO-s-TRIAZINE □ 2,4-BIS(ISOPROPYL-AMINO)-6-METHYLTHIO-1,3,5-TRIAZINE □ N,N'-BIS(1-METHYLETHYL)-6-METHYL-THIO-1,3,5-TRIAZINE-2,4-DIAMINE □ CAPAROL □ G 34161 □ GESAGARD □ MERKAZIN □ 2-METHYLMERCAPTO-4,6-BIS(ISOPROP-YLAMINO)-s-TRIAZINE □ 2-METHYLTHIO-4,6-BIS(ISOPROPYLAMINO)-s-TRIAZINE □ POLISIN □ PRIMATOL Q □ PROMETREX □ PROMETRIN □

PROMETRYN □ PROMETRYNE (USDA) □ SELEKTIN □ SESAGARD
SAFETY PROFILE: Moderately toxic by ingestion. Experimental reproductive effects. An eye irritant. Mutation data reported. An herbicide. When heated to decomposition it emits very toxic fumes of NO_x and SO_x. See also MERCAPTANS.

BKL750   CAS: 3006-93-7   HR: 3
1,3-BISMALEIMIDO BENZENE
mf: $C_{14}H_8N_2O_4$ mw: 268.24
SYNS: 1,3-DIMALEIMIDOBENZENE □ HVA 2 □ HVA-2 CURING AGENT □ M-PHDM □ N,N'-(m-PHENYL-ENE)BISMALEIMIDE □ 1,1'-(m-PHENYLENE)BIS-1H-PYROLE-2,5-DIONE (9CI) □ N,N'-(m-PHENYLENE-DIMALEIMIDE)
CONSENSUS REPORTS: Reported in EPA TSCA Inventory.
SAFETY PROFILE: Poison by ingestion, inhalation, and intraperitoneal routes. When heated to decomposition it emits toxic fumes of NO_x.

BKM500   CAS: 1187-00-4   HR: 3
BIS(METHANE SULFONYL)-d-MANNIT-OL
mf: $C_8H_{18}O_{10}S_2$ mw: 338.38
SYNS: 1,6-BIS-o-METHYLSULFONYL-d-MANNITOL □ CB 2511 □ 1,6-DIMESYL-d-MANNITOL □ 1,6-DIMETHANESULFONATE-d-MANNITOL □ 1,6-DIMETHANE-SULFONOXY-d-MANNITOL □ 1,6-DIMETHANESULPHONOXY-1,6-DIDEOXY-d-MANNITOL □ DMM □ d-MANNITOL BUSULFAN □ MANNITOL MYLERAN □ MANNOGRANOL □ MM □ NSC-37538
CONSENSUS REPORTS: EPA Genetic Toxicology Program.
SAFETY PROFILE: Poison by intravenous route. Moderately toxic by intraperitoneal route. Mildly toxic by ingestion. Questionable carcinogen with experimental neoplastigenic data. Mutation data reported. When heated to decomposition it emits toxic fumes of SO_x.

BKO250   CAS: 15546-11-9   HR: 3
BIS(METHOXYMALEOYLOXY)DIBUTYL STANNANE
mf: $C_{18}H_{28}O_8Sn$ mw: 491.15

B

SYNS: DIBUTYLBIS((3-CARBOXYACRYLOYL)OXY)-STANNANE DIMETHYL ESTER (Z,Z) (8CI) □ DIBUTYLTIN BIS(METHYL MALEATE) □ DIBUTYLTIN BIS(MONOMETHYL MALEATE) □ DIBUTYLTIN METHYL MALEATE □ 6,6-DIBUTYL-4,8,11-TRIOXO-5,7,12-TRIOXA-6-STANNATRIDECA-2,9-DIENOIC ACID METHYL ESTER □ DI-n-BUTYLZINN-DIMONO-METHYLMALEINAT (GERMAN) □ STAN-GUARD 156

CONSENSUS REPORTS: Reported in EPA TSCA Inventory.

OSHA PEL: TWA 0.1 mg(Sn)/m³ (skin)

ACGIH TLV: TWA 0.1 mg(Sn)/m³; STEL 0.2 mg(Sn)/m³ (skin).

NIOSH REL: (Organotin Compounds) TWA 0.1 mg(Sn)/m³

SAFETY PROFILE: Poison by ingestion. See also TIN COMPOUNDS. When heated to decomposition it emits acrid smoke and irritating fumes.

BKS640   CAS: 81877-66-9   HR: 3
1,2:5,6-BIS-o-(1-METHYLETHYLIDENE)-α-d-GLUCOFURANOSE, ((((2-(DI-METHYLAMINO)-2-OXO-1-(METHYLTHIO)ETHYLIDENE)AMINO)OXY)CARBONYL)METHYLAMIDO-SULFITE

mf: $C_{19}H_{31}N_3O_{10}S_2$ mw: 525.65

SAFETY PROFILE: A poison by ingestion. When heated to decomposition it emits toxic vapors of NO_x and SO_x.

BKS810   CAS: 3810-81-9   HR: 3
BIS(METHYLMERCURIC)SULFATE

mf: $C_2H_6Hg_2O_4S$ mw: 527.32

PROP: Platelets from water. Sltly sol in EtOH. Mp: 255° (decomp).

SYNS: ARETAN-NIEUW □ B 4992 □ BIS-(METHYL-MERCURY)-SULFATE □ BIS-(METHYLMERKURI)-SULFAT □ CERESAN UNIVERSAL-FEUCHTBEIZE □ CEREWET □ COMPOUND-4992 □ MERCURY, SULFATOBIS(METHYL- □ METHYLMERCURIC SULFATE □ SULFURIC ACID, BIS(METHYLMERCURY) SALT

OSHA PEL: TWA 0.01 mg(Hg)/m³; STEL 0.03 mg/m³ (skin)

ACGIH TLV: TWA 0.01 mg(Hg)/m³; BEI: 35 µg/g creatinine total inorganic mercury in urine preshift; 15 µg/g creatinine total inorganic mercury in blood at end of shift at end of workweek.

DFG MAK: Confirmed Animal Carcinogen with Unknown Relevance to Humans

SAFETY PROFILE: Poison by ingestion. When heated to decomposition it emits toxic fumes of SO_x and Hg.

BKT300   CAS: 63942-42-7   HR: 3
N,N'-BIS(2-METHYLSULFONYL-2-METHYLPROPIONALDEHYDE-o-(N-METHYLCARBAMOYL)OXIME)-SULFIDE

mf: $C_{14}H_{26}N_4O_8S_3$ mw: 474.62

SYN: PROPANAL, 2-METHYL-2-(METHYLSULFONYL)-o,o'-(THIOBIS((METHYLIMINO)CARBONYL))DIOXIME

SAFETY PROFILE: A poison by ingestion. When heated to decomposition it emits toxic vapors of NO_x and SO_x.

BKU120   CAS: 63942-44-9   HR: 3
N,N'-BIS(1-METHYLTHIO-1-(N,N-DI-METHYLCARBONYL)FORM-ALDEHYDE-o-(N-METHYLCARB-AMOYL)OXIME)SULFIDE

mf: $C_{14}H_{24}N_6O_6S_3$ mw: 468.62

SYN: ETHANIMIDOTHIOIC ACID, N,N'-(THIOBIS((METHYLAMINO)CARBONYLOXY))BIS(2-(DIMETHYLAMINO)-2-OXO-, DIMETHYL ESTER

SAFETY PROFILE: A poison by ingestion. When heated to decomposition it emits toxic vapors of NO_x and SO_x.

BKU500   CAS: 103-34-4   HR: 3
N,N'-BISMORPHOLINE DISULFIDE

mf: $C_8H_{16}N_2O_2S_2$ mw: 236.38

PROP: Tan to gray powder or crystals. Mp: 124–125°, d: 1.36 @ 25°.

SYNS: ACCEL R □ BISMORPHOLINO DISULFIDE □ DIMORPHOLINE DISULFIDE □ DIMORPHOLINO DISULFIDE □ DITHIOBISMORPHOLINE □ 4,4'-DITHIOBIS(MORPHOLINE) □ N,N-DITHIODI-MORPHOLINE □ 4,4'-DITHIODIMORPHOLINE □ 4,4'-DITHIOMORPHOLINE □ MORPHOLINE DISULFIDE □ MORPHOLINODISULFIDE □ SULFASAN □ SULFASAN R POWDER □ USAF B-17 □ USAF EK-T-6645

CONSENSUS REPORTS: Reported in EPA TSCA Inventory.

SAFETY PROFILE: Poison by intraperitoneal and intravenous routes. Moderately toxic by ingestion. Mutation data reported. See also MORPHOLINE. When heated to decomposition it emits very toxic fumes of NO_x and SO_x.

BKU750   CAS: 7440-69-9   HR: 3
BISMUTH
af: Bi   aw: 208.98
PROP: Hexagonal silver-white or reddish metallic crystals. Mp: 271.3°, bp: 1420–1560°, d: 9.80, vap press: 1 mm @ 1021°.
SYN: BISMUTH-209
CONSENSUS REPORTS: Reported in EPA TSCA Inventory.
SAFETY PROFILE: Poisonous to humans. See also BISMUTH COMPOUNDS. Flammable when exposed to flame. Reaction with [$Bi(OH)_3$ + $Al(OH)_3$], coprecipitated and H_2 reduced produces a spontaneously flammable product. Moderately dangerous, can react with acid or acid fumes to emit toxic fumes. Incompatible with Al, BrF_3, acids, NOF, NH_4NO_3, $HClO_3$, Cl_2, IF_5, HNO_3, $HClO_4$.

BKV750   HR: 3
BISMUTH COMPOUNDS
SAFETY PROFILE: Bismuth and its salts can cause kidney damage, although the degree of such damage is usually mild. Large doses can be fatal. Industrially it is considered one of the less toxic of the heavy metals, although intoxication has occurred from its use in medicine. The similarity between the pharmacologic and toxic behavior of lead and bismuth has been pointed out in the literature. Like lead, bismuth may be liberated from tissue deposits during periods of acidosis. Serious and sometimes fatal poisoning may occur from the injection of large doses into closed cavities and from extensive application to burns. Death of animals from bismuth nephritis following injections of soluble salts occurs within several hours to 24 days, the time being generally inversely proportional to the dose, and it appears to be in the order of 5–10 times higher than the dose by slow intravenous injection for rabbits. It is stated

that the administration of bismuth should be stopped when gingivitis appears, for otherwise serious ulcerative stomatitis is likely to result. Other toxic results may develop, such as malaise, albuminuria, diarrhea, skin reactions, and sometimes serious exodermatitis. Industrial bismuth poisoning has not been reported, although bismuth absorbed in industrial cases may complicate a diagnosis of plumbism, since the dark line in the gums, which is often present in lead poisoning, is also produced by bismuth. All bismuth compounds do not have equal toxicity. See also individual entries.
  Treatment and Antidotes: Personnel showing some of the symptoms noted above, which might indicate that they were absorbing too much bismuth into the body, should be removed from exposure as soon as possible. Get medical advice. Personnel should be cautioned against careless handling of these materials.

BKW000   CAS: 21260-46-8   HR: 1
BISMUTH DIMETHYL DITHIO-
CARBAMATE
mf: $C_9H_{18}N_3S_6$•Bi   mw: 569.64
SYNS: BISMATE □ TRIS(DIMETHYLDITHIO-CARBAMATO)BISMUTH
CONSENSUS REPORTS: Reported in EPA TSCA Inventory.
SAFETY PROFILE: Low toxicity by ingestion. Questionable carcinogen with experimental tumorigenic data. See also BISMUTH COMPOUNDS and CARBAMATES. When heated to decomposition it emits very toxic fumes of SO_x and NO_x.

BKW250   CAS: 10361-44-1   HR: 3
BISMUTH NITRATE
mf: BiN_3O_9 mw: 395.01
PROP: Triclinic, colorless, sltly hygroscopic crystals. Bp: $-5H_2O$ @ 80°, d: 2.83, mp: 30° (decomp).
SYN: NITRIC ACID, BISMUTH(3+) SALT
CONSENSUS REPORTS: Reported in EPA TSCA Inventory.

B

SAFETY PROFILE: Poison by intravenous route. Moderately toxic by intraperitoneal route. Experimental reproductive effects. When heated to decomposition it emits toxic fumes of Bi and NO_x. See also BISMUTH COMPOUNDS and NITRATES.

BKY000   CAS: 1304-82-1   HR: 3
BISMUTH TELLURIDE
mf: Bi_2Te_3 mw: 800.76
PROP: Gray crystals or solid. D: 7.7.
SYNS: BISMUTH SESQUITELLURIDE □ BISMUTH TELLURIDE, UNDOPED
CONSENSUS REPORTS: Reported in EPA TSCA Inventory.
OSHA PEL: Total Dust: TWA 0.1 mg(Te)/m³; Respirable Fraction: TWA 5 mg/m³; Se doped: 5 mg/m³
ACGIH TLV: TWA 10 mg/m³; Not Classifiable as a Human Carcinogen; Se doped: 5 mg/m³; Not Classifiable as a Human Carcinogen
SAFETY PROFILE: Moderate fire hazard by spontaneous chemical reaction with powerful oxidizers. Reacts with moisture to evolve a toxic gas. Slight explosion hazard by chemical reaction with powerful oxidizers; reacts with moisture. When heated to decomposition it emits toxic fumes of Te. See also BISMUTH COMPOUNDS and TELLURIUM COMPOUNDS.

BLA000   CAS: 13826-66-9   HR: 2
BIS(NITRATO-O)OXOZIRCONIUM
mf: N_2O_7Zr mw: 231.24
PROP: IDLH 50 mg/m³ (as Zr).
SYN: ZIRCONYL NITRATE
CONSENSUS REPORTS: Reported in EPA TSCA Inventory.
OSHA PEL: TWA 5 mg(Zr)/m³; STEL 10 mg(Zr)/m³
ACGIH TLV: TWA 5 mg(Zr)/m³; STEL 10 mg(Zr)/m³
DFG MAK: 1 mg(Zr)/m³
SAFETY PROFILE: Moderately toxic by ingestion and intraperitoneal routes. See also

ZIRCONIUM COMPOUNDS and NITRATES. When heated to decomposition it emits toxic fumes of NO_x.

BLC000   CAS: 868-18-8   HR: 2
BISODIUM TARTRATE
mf: $C_4H_4O_6 \cdot 2Na$ mw: 194.06
PROP: Transparent crystals; colorless and odorless. Sol in water.
SYNS: 2,3-DIHYDROXY-(R-(R*,R*))-BUTANEDIOIC ACID DISODIUM SALT (9CI) □ DISODIUM TARTRATE □ DISODIUM l-(+)-TARTRATE □ SODIUM TARTRATE (FCC) □ SODIUM l-(+)-TARTRATE
CONSENSUS REPORTS: Reported in EPA TSCA Inventory.
SAFETY PROFILE: Moderately toxic by ingestion. When heated to decomposition it emits acrid smoke and irritating fumes.

BLC250   CAS: 10380-28-6   HR: 3
BIS(8-OXYQUINOLINE)COPPER
mf: $C_{18}H_{12}CuN_2O_2$ mw: 351.86
PROP: Yellow-green powder or crystals. Insol in H_2O and common org solvs.
SYNS: BIOQUIN □ BIOQUIN 1 □ BIS(8-QUINOLINATO)COPPER □ BIS(8-QUINOLINOL-ATO)COPPER □ BIS(8-QUINOLINOLATO-N^1,O^8)-COPPER □ CELLU-QUIN □ COPPER-8 □ COPPER HYDROXYQUINOLATE □ COPPER-8-HYDROXY-QUINOLATE □ COPPER-8-HYDROXYQUINOLINATE □ COPPER-8-HYDROXYQUINOLINE □ COPPER OXINATE □ COPPER (2+) OXINATE □ COPPER OXINE □ COPPER OXYQUINOLATE □ COPPER OXYQUINOL-INE □ COPPER QUINOLATE □ COPPER-8-QUINOLATE □ COPPER-8-QUINOLINOL □ COPPER QUINOLINOL-ATE □ COPPER-8-QUINOLINOLATE □ CUNILATE □ CUNILATE 2472 □ CUPRIC-8-HYDROXYQUINOLATE □ CUPRIC-8-QUINOLINOLATE □ DOKIRIN □ FRUITDO □ 8-HYDROXYQUINOLINE COPPER COMPLEX □ MILMER □ OXIME COPPER □ OXINE COPPER □ OXINE CUIVRE □ OXYQUINOLINOLEATE de CUIVRE (FRENCH) □ QUINONDO
CONSENSUS REPORTS: IARC Cancer Review: Group 3 IMEMDT 7,56,87; Animal Inadequate Evidence IMEMDT 15,103,77. Reported in EPA TSCA Inventory. Copper and its compounds are on the Community Right-To-Know List. EPA FIFRA 1988 pesticide subject to registration or re-registration.

SAFETY PROFILE: Poison by intraperitoneal route. Moderately toxic by ingestion and inhalation. Questionable carcinogen with experimental tumorigenic data. Mutation data reported. See also COPPER COMPOUNDS. When heated to decomposition it emits toxic fumes of NO_x.

BLD000   CAS: 42310-84-9   HR: 3
BISPENTAFLUOROSULFUR OXIDE
mf: $F_{10}OS_2$ mw: 270.12
PROP: Colorless liquid. Mp: $-118°$, bp: $31°$.
SYN: SULFUR FLUORIDE OXIDE
OSHA PEL: TWA 2.5 mg(F)/m^3
ACGIH TLV: TWA 2.5 mg(F)/m^3; BEI: 3 mg/g creatinine of fluorides in urine prior to shift; 10 mg/g creatinine of fluorides in urine at end of shift.
NIOSH REL: TWA 2.5 mg(F)/m^3
SAFETY PROFILE: Poison by inhalation. See also FLUORIDES. When heated to decomposition it emits very toxic fumes of F^- and SO_x.

BLD500   CAS: 80-05-7   HR: 3
BISPHENOL A
mf: $C_{15}H_{16}O_2$ mw: 228.31
PROP: White flakes; mild phenolic odor. Mp: $156-157°$, bp: $250-252°$ @ 13 mm. Insol in water; sol in alcohol and dilute alkalies; sltly sol in CCl_4.
SYNS: BISFEROL A (GERMAN) □ 2,2-BIS-4'-HYDROXYFENYLPROPAN (CZECH) □ BIS(4-HYDROXYPHENYL) DIMETHYLMETHANE □ BIS(4-HYDROXYPHENYL)PROPANE □ 2,2-BIS(p-HYDROXYPHENYL)PROPANE □ 2,2-BIS(4-HYDROXYPHENYL)PROPANE □ DIAN □ p,p'-DIHYDROXYDIPHENYLDIMETHYLMETHANE □ 4,4'-DIHYDROXYDIPHENYLDIMETHYLMETHANE □ p,p'-DIHYDROXYDIPHENYLPROPANE □ 2,2-(4,4'-DIHYDROXYDIPHENYL)PROPANE □ 4,4'-DIHYDROXYDIPHENYLPROPANE □ 4,4'-DIHYDROXYDIPHENYL-2,2-PROPANE □ 4,4'-DIHYDROXY-2,2-DIPHENYLPROPANE □ β-DI-p-HYDROXYPHENYLPROPANE □ 2,2-DI(4-HYDROXYPHENYL)PROPANE □ DIMETHYL BIS(p-HYDROXYPHENYL)METHANE □ DIMETHYLMETHYL-ENE-p,p'-DIPHENOL □ 2,2-DI(4-PHENYLOL)PROPANE □ p,p'-ISOPROPYLIDENE-BISPHENOL □ 4,4'-ISOPROPYLIDENEBISPHENOL □ p,p'-ISOPROPYLIDENEDIPHENOL □ NCI-C50635

CONSENSUS REPORTS: NTP Carcinogenesis Bioassay (feed); Inadequate Studies: mouse, rat NTPTR* NTP-TR-215,82. Community Right-To-Know List. Reported in EPA TSCA Inventory.
DFG MAK: 5 ppm
SAFETY PROFILE: Poison by intraperitoneal route. Moderately toxic by ingestion, inhalation, and skin contact. Experimental teratogenic and reproductive effects. A skin and eye irritant. When heated to decomposition it emits acrid and irritating fumes.

BLD750   CAS: 1675-54-3   HR: 3
BISPHENOL A DIGLYCIDYL ETHER
mf: $C_{21}H_{24}O_4$ mw: 340.45
SYNS: 2,2-BIS(4-(2,3-EPOXYPROPYLOXY)PHENYL)PROPANE □ BIS(4-GLYCIDYLOXYPHENYL)DIMETHYLAMETHANE □ 2,2-BIS(p-GLYCIDYLOXYPHENYL)PROPANE □ BIS(4-HYDROXYPHENYL)DIMETHYLMETHANE DIGLYCIDYL ETHER □ 2,2-BIS(p-HYDROXYPHENYL)-PROPANE, DIGLYCIDYL ETHER □ 2,2-BIS(4-HYDROXYPHENYL)PROPANE, DIGLYCIDYL ETHER □ D.E.R. 332 □ DIGLYCIDYL BISPHENOL A ETHER □ DIGLYCIDYL ETHER of 2,2-BIS(p-HYDROXYPHENYL)-PROPANE □ DIGLYCIDYL ETHER of 2,2-BIS(4-HYDROXYPHENYL)PROPANE □ DIGLYCIDYL ETHER of BISPHENOL A □ DIGLYCIDYL ETHER of 4,4'-ISOPROPYLIDENEDIPHENOL □ 4,4'-DIHYDROXYDI-PHENYLDIMETHYLMETHANE DIGLYCIDYL ETHER □ p,p'-DIHYDROXYDIPHENYLDIMETHYLMETHANE DIGLYCIDYL ETHER □ EPI-REZ 508 □ EPI-REZ 510 □ EPON 828 □ EPOXIDE A □ ERL-2774 □ 4,4'-ISOPROPYLIDENEDIPHENOL DIGLYCIDYL ETHER □ 2,2'-((1-METHYLETHYLIDENE)BIS(4,1-PHENYLENEOXYMETHYLENE))BISOXIRANE
CONSENSUS REPORTS: EPA Genetic Toxicology Program. Reported in EPA TSCA Inventory.
DFG MAK: Confirmed Animal Carcinogen with Unknown Relevance to Humans
SAFETY PROFILE: Suspected carcinogen. Poison by skin contact. Mildly toxic by ingestion. Mutation data reported. A skin and severe eye irritant. Experimental reproductive effects. Questionable carcinogen with experimental carcinogenic and tumorigenic data. See also ETHERS. When heated to decomposition it emits acrid and irritating fumes.

BLE500   CAS: 74-31-7   HR: 3
1,4-BIS(PHENYL AMINO)BENZENE
mf: $C_{18}H_{16}N_2$ mw: 260.36
PROP: Gray crystals or solid. D: 1.20, mp: 147°, vap d: 9.0.
SYNS: AGERITE □ AGERITEDPPD □ N,N'-DIFENYL-p-FENYLENDIAMIN (CZECH) □ DIPHENYL-p-PHENYLENEDIAMINE □ N,N'-DIPHENYL-p-PHENYLENEDIAMINE □ DPPD □ FLEXAMINE G □ JZF □ NONOX DPPD □ p-PHENYLAMINODI-PHENYLAMINE □ 4-PHENYLAMINODIPHENYLAMINE □ USAF GY-2
CONSENSUS REPORTS: Reported in EPA TSCA Inventory.
SAFETY PROFILE: Poison by intraperitoneal route. Moderately toxic by ingestion. A weak allergen. Experimental teratogenic and reproductive effects. An eye irritant. Questionable carcinogen with experimental tumorigenic data. Mutation data reported. Combustible when exposed to heat or flame; can react with oxidizing materials. When heated to decomposition it emits toxic fumes of NO_x.

BLK000   CAS: 128-80-3   HR: 2
1,4-BIS(p-TOLYLAMINO)ANTHRA-QUINONE
mf: $C_{28}H_{22}N_2O_2$ mw: 418.52
PROP: Dark green crystals or powder. Sol in C_6H_6 or acids; sltly sol in Me_2CO; insol in H_2O and EtOH.
SYNS: ALIZARINE CYANINE GREEN BASE □ AMAPLAST GREEN OZ □ ARLOSOL GREEN B □ BIS-1,4-p-TOLYLAMINOANTHRCHINON (CZECH) □ C-GREEN 10 □ C.I. 61565 □ C.I. SOLVENT GREEN 3 □ CYANINE GREEN G BASE □ D&C GREEN No. 6 □ 1,4-DI-p-TOLUIDINOANTHRAQUINONE □ FAT SOLUBLE GREEN ANTHRAQUINONE □ 11091 GREEN □ GREEN No. 2 □ MICRO-LEX GREEN 5B □ NITRO FAST GREEN GB □ ORGANOL FAST GREEN J □ QUINIZARINE GREEN BASE □ SUDAN GREEN 4B □ TOYO ORIENTAL OIL BLUE G □ WAXOLINE GREEN
CONSENSUS REPORTS: Reported in EPA TSCA Inventory.
SAFETY PROFILE: Moderately toxic by ingestion. An eye irritant. When heated to decomposition it emits toxic fumes of NO_x.

BLL750   CAS: 56-35-9   HR: 3
BIS(TRIBUTYL TIN)OXIDE
mf: $C_{24}H_{54}OSn_2$ mw: 596.16
PROP: Air-sensitive liquid. D: 1.17 @ 20°/4°, bp: 220–230° @ 10 mm.
SYNS: BIOMET TBTO □ BIS-(TRI-N-BUTYLCIN)OXID (CZECH) □ BIS(TRIBUTYLOXIDE) of TIN □ BIS(TRIBUTYLSTANNYL)OXIDE □ BIS(TRI-N-BUTYLZINN)-OXYD (GERMAN) □ BTO □ BUTINOX □ C-Sn-9 □ ENT 24,979 □ HEXABUTYLDISTANNOXANE □ HEXABUTYLDITIN □ KYSLICNIK TRI-N-BUTYLCINICITY (CZECH) □ L.S. 3394 □ OTBE (FRENCH) □ OXYBIS(TRIBUTYLTIN) □ OXYDE de TRIBUTYLETAIN □ TBOT □ TBTO □ TRI-n-BUTYL-STANNANE OXIDE □ TRIBUTYLTIN OXIDE
CONSENSUS REPORTS: Reported in EPA TSCA Inventory.
OSHA PEL: TWA 0.1 mg(Sn)/m³ (skin)
ACGIH TLV: TWA 0.1 mg(Sn)/m³; STEL 0.2 mg(Sn)/m³ (skin).
DFG MAK: 0.0021 ppm (0.05 mg/m³)
NIOSH REL: (Organotin Compounds) TWA 0.1 mg(Sn)/m³
SAFETY PROFILE: A poison by ingestion, intraperitoneal, and intravenous routes. Moderately toxic by skin contact. An experimental teratogen. Other experimental reproductive effects. Questionable carcinogen with experimental carcinogenic data. Mutation data reported. A severe eye irritant. See also TIN COMPOUNDS. When heated to decomposition it emits acrid and irritating fumes.

BLN500   CAS: 57-52-3   HR: 3
BIS(TRIETHYL TIN) SULFATE
mf: $C_{12}H_{30}O_4SSn_2$ mw: 507.86
SYNS: TRIAETHYLZINNSULFAT (GERMAN) □ TRIETHYLHYDROXY-STANNANE SULFATE (2:1) (8CI) □ TRIETHYLHYDROXYTIN SULFATE □ TRIETHYLTIN SULPHATE
OSHA PEL: TWA 0.1 mg(Sn)/m³ (skin)
ACGIH TLV: TWA 0.1 mg(Sn)/m³; STEL 0.2 mg(Sn)/m³ (skin).
NIOSH REL: (Organotin Compounds) TWA 0.1 mg(Sn)/m³
SAFETY PROFILE: Poison by ingestion, intraperitoneal, subcutaneous, intravenous, and parenteral routes. See also TIN COMPOUNDS and SULFATES. When

heated to decomposition it emits toxic fumes of SO_x.

BLQ525   CAS: 21259-75-6   HR: 3
BIS(TRIFLUOROMETHYLTHIO)-
**   MERCURY**
mf: $C_2F_6HgS_2$ mw: 402.73
SYN: MERCURY, BIS(TRIFLUOROMETHYLTHIO)-
ACGIH TLV: TWA 0.01. STEL 0.03 mg/m³ (skin)
NIOSH REL: (MERCURY, ORGANO) TWA 0.01 mg/m³. STEL 0.03 mg/m³ (Sk)
SAFETY PROFILE: A poison by intravenous route. When heated to decomposition it emits toxic vapors of SO_x and Hg.

BLS250   CAS: 14264-16-5   HR: 3
BIS(TRIPHENYLPHOSPHINE)DI-
**   CHLORONICKEL**
mf: $C_{24}H_{54}P_2 \cdot Cl_2Ni$ mw: 534.33
SYNS: BIS(TRI-N-BUTYLPHOSPHINE)DICHLORONICKEL □ TRIBUTYL-PHOSPHINE compounded with NICKELCHLORIDE (2:1)
CONSENSUS REPORTS: NTP 10th Report on Carcinogens. Reported in EPA TSCA Inventory. Nickel and its compounds are on the Community Right-To-Know List.
OSHA PEL: TWA 0.1 mg (Ni)/m³
ACGIH TLV: TWA 0.2 mg(Ni)/m³; Human Carcinogen)
SAFETY PROFILE: Confirmed human carcinogen. Poison by intravenous route. See also NICKEL COMPOUNDS. When heated to decomposition it emits very toxic fumes of Cl^- and PO_x.

BLS750   CAS: 1624-02-8   HR: 2
BIS(TRIPHENYL SILYL)CHROMATE
mf: $C_{36}H_{30}CrO_4Si_2$ mw: 634.84
PROP: IDLH Ca [15 mg/m³ {as Cr(VI)}].
SYN: CHROMIC ACID, BIS(TRIPHENYLSILYL) ESTER
CONSENSUS REPORTS: Reported in EPA TSCA Inventory. Chromium and its compounds are on the Community Right-To-Know List.
OSHA PEL: CL 0.1 mg(CrO₃)/m³
ACGIH TLV: TWA 0.05 mg(CrO₃)/m³

NIOSH REL: (Chromium(VI)): TWA 0.025 mg(Cr(VI))/m³; CL 0.05/15M
SAFETY PROFILE: Moderately toxic by ingestion and skin contact. See also CHROMIUM COMPOUNDS and ESTERS. When heated to decomposition it emits toxic fumes of CrO_3 particulates.

BLS900   CAS: 73940-87-1   HR: 3
BIS(TRIPHENYLTIN)ACETYLENEDICAR
**   BOXYLATE**
mf: $C_{40}H_{30}O_4Sn_2$ mw: 812.08
SYNS: ETHYNYLENEBIS(CARBONYLOXY)BIS(TRIPHENYLSTANNANE) □ STANNANE, ETHYNYLENEBIS(CARBONYL-OXY)BIS(TRIPHENYL-
OSHA PEL: TWA 0.1 mg(Sn)/m³
ACGIH TLV: TWA 0.1 mg(Sn)/m³; STEL 0.2 mg/m³ (skin)
NIOSH REL: (Organotin compound): 10H TWA 0.1 mg(Sn)/m³
SAFETY PROFILE: Poison by intravenous route. When heated to decomposition it emits toxic fumes of Sn.

BLT300   CAS: 1067-29-4   HR: 3
BIS(TRIPROPYLTIN)OXIDE
mf: $C_{18}H_{42}OSn_2$ mw: 511.98
PROP: Air-sensitive liquid. Bp: 154.5° @ 3.5 mm.
SYNS: DISTANNOXANE, 1,1,1,3,3,3-HEXAPROPYL- □ 1,1,1,3,3,3-HEXAPROPYLDISTANNOXANE
CONSENSUS REPORTS: Reported in EPA TSCA Inventory.
OSHA PEL: TWA 0.1 mg(Sn)/m³
ACGIH TLV: TWA 0.1 mg(Sn)/m³; STEL 0.2 mg/m³ (skin)
NIOSH REL: (Organotin Compound): 10H TWA 0.1 mg(Sn)/m³
SAFETY PROFILE: Poison by intravenous route. When heated to decomposition it emits toxic fumes of Sn.

BLT775   CAS: 38402-95-8   HR: 3
BIS(TRIS(p-DIMETHYLAMINOPHENYL)-
**   PHOSPHINE OXIDE)STANNIC**
**   CHLORIDE COMPLEX**
mf: $C_{48}H_{60}N_6O_2P_2 \cdot Cl_4Sn$ mw: 1075.57
SYN: PHOSPHINE OXIDE, TRIS(p-DIMETHYLAMINO-PHENYL)-, compounded with STANNIC CHLORIDE (2:1)

OSHA PEL: TWA 2 mg(Sn)/m³
ACGIH TLV: TWA 2 mg(Sn)/m³
SAFETY PROFILE: Poison by intravenous route. When heated to decomposition it emits toxic fumes of NO_x, PO_x, Sn, and Cl⁻.

BLU000   CAS: 13356-08-6   HR: 2
BIS(TRIS(β,β-DIMETHYLPHENETHYL)-TIN)OXIDE
mf: $C_{60}H_{78}OSn_2$ mw: 1052.76
PROP: Crystals or powder. Sol in $CHCl_3$, C_6H_6. Sltly sol in Me_2CO; insol in H_2O.
SYNS: BENDEX □ BIS(TRIS(2-METHYL-2-PHENYLPROPYL)TIN)OXIDE □ DI(TRI-(2,2-DIMETHYL-2-PHENYLETHYL)TIN)OXIDE □ ENT 27,738 □ FENBUTATIN OXIDE □ HEXAKIS(β,β-DIMETHYL-PHENETHYL)DISTANNOXANE □ HEXAKIS(2-METHYL-2-PHENYLPROPYL)DISTANNOXANE □ SD 14114 □ SHELL SD-14114 □ TORQUE □ VENDEX
OSHA PEL: TWA 0.1 mg(Sn)/m³ (skin)
ACGIH TLV: TWA 0.1 mg(Sn)/m³; STEL 0.2 mg(Sn)/m³ (skin).
NIOSH REL: (Organotin Compounds) TWA 0.1 mg(Sn)/m³
SAFETY PROFILE: Moderately toxic by ingestion and skin contact. See also TIN COMPOUNDS. When heated to decomposition it emits acrid smoke and irritating fumes.

BLV250   CAS: 13394-86-0   HR: 3
(m,o'-BITOLYL)-4-AMINE
mf: $C_{14}H_{15}N$ mw: 197.30
PROP: Oil. Bp: 201° @ 15 mm.
SYNS: 2',3-DIMETHYL-4-AMINOBIPHENYL □ 3,2'-DIMETHYL-4-AMINOBIPHENYL □ 3,2'-DIMETHYL-4-AMINODIPHENYL □ 3,2'-DIMETHYL-4-BIPHENYLAMINE □ 3,2'-DMAB
CONSENSUS REPORTS: EPA Genetic Toxicology Program.
SAFETY PROFILE: Suspected carcinogen with experimental carcinogenic and tumorigenic data. Moderately toxic by intraperitoneal route. Experimental reproductive effects. Mutation data reported. When heated to decomposition it emits toxic fumes of NO_x. See also AROMATIC AMINES.

BLV500   CAS: 8013-76-1   HR: 3
BITTER ALMOND OIL
PROP: Volatile oil from dried ripe kernels of bitter almonds or from other kernels containing amygdalin, such as apricots, cherries, plums, and especially peaches. Colorless liquid; strong almond odor. Bp: 179°, d: 1.045–1.070 @ 15°. Sltly sol in water; sol in fixed oils and propylene glycol; insol in glycerin.
SYNS: ALMOND OIL BITTER, FFPA (FCC) □ OIL, BITTER ALMOND
CONSENSUS REPORTS: Reported in EPA TSCA Inventory.
SAFETY PROFILE: A human poison by ingestion. Moderately toxic by skin contact. A skin irritant. When heated to decomposition it emits toxic fumes of CN⁻.

BLW250   CAS: 8006-82-4   HR: 1
BLACK PEPPER OIL
PROP: From steam distillation of dried fruit of *Piper nigrum L.* (Fam. *Piperaceae*). Main constituents include α- and β-pinene, β-caryophyllene, l-limonene, d-hydrocarveol, piperidine, and piperrine (FCTXAV 16,637,78). A colorless to greenish liquid; odor and taste of pepper. Sol in fixed oils, mineral oil, propylene glycol; sltly sol in glycerin.
CONSENSUS REPORTS: Reported in EPA TSCA Inventory.
SAFETY PROFILE: A moderate skin irritant. Mutation data reported. When heated to decomposition it emits acrid smoke and irritating fumes.

BLW750   CAS: 21725-46-2   HR: 3
BLADEX
mf: $C_9H_{13}ClN_6$ mw: 240.73
PROP: A white, crystalline material. Mp: 167°.
SYNS: BLADEX 80WP □ 2-CHLORO-4-(1-CYANO-1-METHYLETHYLAMINO)-6-ETHYLAMINO-1,3,5-TRIAZINE □ 2-CHLORO-4-ETHYLAMINO-6-(1-CYANO-1-METHYL)ETHYLAMINO-s-TRIAZINE □ 2-(4-CHLORO-6-ETHYLAMINO-s-TRIAZINE-2-YLAMINO)-2-METHYL-PROPIONITRILE □ 2-(4-CHLORO-6-ETHYLAMINO-1,3,5-TRIAZINE-2-YLAMINO)-2-METHYLPROPIONITRILE □ 2-((4-CHLORO-6-(ETHYLAMINO)-1,3,5-TRIAZIN-2-

YL)AMINO)-2-METHYL-PROPANENITRILE □ 2-((4-CHLORO-6-(ETHYLAMINO)-s-TRIAZIN-2-YL)AMINO)-2-METHYLPROPIONITRILE □ CYANAZINE □ DW3418 □ FORTROL □ PAYZE □ SD 15418 □ WL 19805

CONSENSUS REPORTS: EPA Genetic Toxicology Program. Cyanide and its compounds are on the Community Right-To-Know List.

SAFETY PROFILE: Poison by ingestion and intraperitoneal routes. Moderately toxic by skin contact. An experimental teratogen. Mutation data reported. See also NITRILES. An herbicide. When heated to decomposition it emits very toxic fumes of Cl^-, NO_x, and CN^-.

BLY000   CAS: 11056-06-7   HR: 3
BLEOMYCIN

PROP: A group of related glycopeptide antibiotics isolated from *Streptomyces verticillus*.
SYNS: BLENOXANE □ BLEO □ BLEOCIN □ BLM

CONSENSUS REPORTS: IARC Cancer Review: Group 2B IMEMDT 7,134,87; Human Inadequate Evidence IMEMDT 26,97,81. EPA Genetic Toxicology Program.

SAFETY PROFILE: A human poison by intravenous route; moderately toxic to humans by intramuscular route. Poison experimentally by intravenous and intraperitoneal routes. Human systemic effects by ingestion and intramuscular routes: dyspnea and fibrosing alveolitis (lung). Experimental reproductive effects. An eye irritant. Human mutation data reported. When heated to decomposition it emits toxic fumes of NO_x. See other bleomycin entries.

BMA550   HR: D
BOIS de ROSE OIL

PROP: From steam distillation of chipped wood of *Aniba rosaeodora* var. *amazonica* Ducke, (Fam. Lauraceae). Colorless to pale yellow liquid; slt pleasant floral odor. Sol in fixed oils, propylene glycol, mineral oil; sltly sol in glycerin.
SYN: LIGNALOE OIL

SAFETY PROFILE: When heated to decomposition it emits acrid smoke and irritating fumes.

BMA750   CAS: 8001-85-2   HR: 2
BONE OIL

PROP: Product of destructive distillation of bones in preparation of bone charcoal containing nitrogenous compounds such as pyridine, aniline, methylamine, and pyrrole (27ZTAP 3,25,69).
SYNS: ANIMAL OIL □ DIPPEL'S OIL □ OIL OF HARTSHORN

CONSENSUS REPORTS: Reported in EPA TSCA Inventory.

SAFETY PROFILE: Moderately toxic by ingestion. When heated to decomposition it emits toxic fumes of NO_x.

BMC000   CAS: 10043-35-3   HR: 3
BORIC ACID

mf: BH_3O_3 mw: 61.84

PROP: White crystals, powder, or pearly scales. Mp: 171° (decomp), loses 1.5 H_2O @ 300°, d: 1.435 @ 15°.
SYNS: BORACIC ACID □ BOROFAX □ BORSAEURE (GERMAN) □ NCI-C56417 □ ORTHOBORIC ACID □ THREE ELEPHANT

CONSENSUS REPORTS: Reported in EPA TSCA Inventory.

SAFETY PROFILE: A human poison by ingestion and possibly other routes. Moderately toxic by skin contact and subcutaneous routes in humans. Poison experimentally by inhalation and subcutaneous routes. Moderately toxic experimentally by intraperitoneal and intravenous routes. Human systemic effects: anorexia, changes in kidney tubules, nausea or vomiting, wakefulness. Ingestion or absorption by other routes may also cause diarrhea, abdominal cramps, erythematous lesions on skin and mucous membranes, circulatory collapse, tachycardia, cyanosis, delirium, convulsions, and coma. Death has occurred from ingestion of less than 5 g in infants, and from 5 to 20 g in adults. Chronic exposure may result in borism (dry skin, eruptions, and gastrointestinal

B

disturbances). Experimental reproductive effects. Mutation data reported. A human skin irritant. See also BORON COMPOUNDS. Incompatible with K, $(CH_3CO)_2O$.

BMC250   CAS: 34099-73-5   HR: 3
BORIC ACID, ETHYL ESTER
DOT: UN 1176
mf: $C_2H_7BO_3$ mw: 89.90
PROP: Colorless liquid, mild odor, decomp in water. Bp: 120°, flash p: 52°F (CC), d: 0.864 @ 26.5°, vap d: 5.04.
SYN: ETHYL BORATE (DOT)
DOT CLASSIFICATION: 3; Label: Flammable Liquid
SAFETY PROFILE: A severe eye irritant. See also BORON COMPOUNDS and ESTERS. Dangerous fire hazard when exposed to heat or flame; will react with water or steam to produce flammable vapors. Incompatible with oxidizers, heat, and open flame. To fight fire, use CO_2, dry chemical.

BMD000   CAS: 507-70-0   HR: 3
BORNEOL
DOT: UN 1312
mf: $C_{10}H_{18}O$ mw: 154.28
PROP: Hexagonal crystals; peppery odor and burning taste. Mp: 208°, bp: 212°, flash p: 150°F, d: 1.01 @ 20°/4°, vap d: 5.31.
SYNS: BAROS CAMPHOR □ BHIMSAIM CAMPHOR □ BICYCLO(2.2.1)HEPTAN-2-OL, 1,7,7-TRIMETHYL-, endo- (9CI) □ 2-BORNANOL, endo- □ BORNEO CAMPHOR □ trans-BORNEOL □ BORNEOL (DOT) □ BORNYL ALCOHOL □ CAMPHANE, 2-HYDROXY- □ 2-CAMPHANOL □ CAMPHOL □ DRYOBALANOPS CAMPHOR □ 2-HYDROXYCAMPHANE □ MALAYAN CAMPHOR □ SUMATRA CAMPHOR □ endo-1,7,7-TRIMETHYL-BICYCLO(2.2.1)HEPTAN-2-OL
CONSENSUS REPORTS: Reported in EPA TSCA Inventory.
DOT CLASSIFICATION: 4.1; Label: Flammable Solid
SAFETY PROFILE: Moderately toxic by ingestion. Mutation data reported. A mild irritant. Flammable when exposed to heat or flame; can react with oxidizing materials. To fight fire, use water, CO_2, water spray, dry

chemical. When heated to decomposition it emits acrid smoke and fumes.

BMD100   CAS: 76-49-3   HR: 1
BORNYL ACETATE
mf: $C_{12}H_{20}O_2$ mw: 196.29
PROP: Colorless liquid or white crystalline solid; sweet, piney odor. D: 0.981–0.985, refr index: 1.462, flash p: 192°F. Sol in alc, fixed oils; sltly sol in water; insol in glycerin, propylene glycol @ 226°.
SYNS: l-BORNYL ACETATE □ FEMA No. 2159
SAFETY PROFILE: Combustible liquid. When heated to decomposition it emits acrid smoke and irritating fumes.

BMD300   CAS: 464-41-5   HR: 3
2-BORNYL CHLORIDE
mf: $C_{10}H_{17}Cl$ mw: 172.72
SYNS: BORNANE, 2-CHLORO-, endo- □ 2-CHLORO-CAMPHANE □ BICYCLO(2.2.1)HEPTANE, 2-CHLORO-1,7,7-TRIMETHYL-, endo- □ BORNYL CHLORIDE □ TERPENE HYDROCHLORIDE □ TURPENTINE CAMPHOR □ endo-2-CHLORO-1,7,7-TRIMETHYL-BICYCLO(2.2.1)HEPTANE
SAFETY PROFILE: A poison by inhalation. Moderately toxic by ingestion. When heated to decomposition it emits toxic vapors of Cl^-.

BMD500   CAS: 7440-42-8   HR: 3
BORON
af: B   aw: 10.81
PROP: Monoclinic crystals, yellow or brown amorphous powder. Mp: 2190°, bp: 3660°, d: 3.33 @ 20°.
CONSENSUS REPORTS: Reported in EPA TSCA Inventory.
SAFETY PROFILE: A poison by ingestion. See also BORON COMPOUNDS. A relatively inert metal except in the form of powder or when exposed to highly oxidizing agents. Amorphous boron is very reactive, sometimes violently. Flammable in the form of dust when exposed to air, or by chemical reaction. An explosion hazard in the form of dust, which ignites on contact with air. Reacts with NaOH at 5°, Na_2CO_3 at 8°. Reacts explosively when ground with lead

fluoride or silver fluoride. Ignites in contact with gaseous chlorine or fluorine at room temperature. Incompatible with NH_3, Br_2, BrF_3, Cs_2C_2, Cl_2, CuO, HIO_3, PbO_2, HNO_3, NO, NOF, N_2O, $KClO_3$, KNO_3, Rb_2C_2, S, BrF_5, IF_5, metal fluorides, interhalogens, nitryl fluoride (FNO_2), OF_2, KNO_2, NO_x, Na_2O_2, PbO, air. See also POWDERED METALS.

BME500   HR: 3
BORON COMPOUNDS
SAFETY PROFILE: Very toxic and therefore considered an industrial poison. Used in medicine as sodium borate, boric acid, or borax, which is a common cleanser. Fatal poisoning of children has been caused by the accidental substitution of boric acid for powdered milk. The medical literature reveals instances of accidental poisoning due to boric acid, ingestion of borates or boric acid, and, presumably, absorption of boric acid from wounds and burns. The fatal dose of orally ingested boric acid for an adult is somewhat greater than 15 to 20 g and, for an infant, 5 to 6 g. Boron is one of a group of elements, such as Pb, Mn, As, that affects the central nervous system. Boron poisoning causes depression of the circulation, persistent vomiting, and diarrhea, followed by profound shock and coma. The temperature becomes subnormal and a scarlatina-form rash may cover the entire body. Containers of boric acid should be plainly labeled and should differ radically from those that contain powdered milk, particularly in institutions such as hospitals.

BMG000   CAS: 1303-86-2   HR: 2
BORON OXIDE
mf: B_2O_3 mw: 69.62
PROP: Vitreous or colorless. Two crystalline forms. Bp: 2250°, mp: 450° (approx), d: 2.46. IDLH 2000 mg/m³.
SYNS: BORIC ANHYDRIDE □ BORON SESQUIOXIDE □ BORON TRIOXIDE □ FUSED BORIC ACID
CONSENSUS REPORTS: Reported in EPA TSCA Inventory.

OSHA PEL: Total Dust: TWA 10 mg/m³; Respirable Fraction: TWA 5 mg/m³
ACGIH TLV: TWA 10 mg/m³
DFG MAK: 15 mg/m³
SAFETY PROFILE: Moderately toxic by ingestion and intraperitoneal routes. An eye and skin irritant. A pesticide. Mixed with CaO and put into fused $CaCl_2$, the mixture incandesces. See also BORON COMPOUNDS.

BMG400   CAS: 10294-33-4   HR: 3
BORON TRIBROMIDE
DOT: UN 2692
mf: BBr_3 mw: 250.54
PROP: Colorless, fuming liquid. Very moisture-sensitive. Mp: −46°, bp: 91.3°, d: 2.650 @ 0°, vap press: 40 mm @ 14.0°, 100 mm @ 33.5°. Sol in CCl_4, SO_2 (l), SCl_2; mod sol in methylcyclohexane.
SYNS: BORON BROMIDE □ TRONA
CONSENSUS REPORTS: Reported in EPA TSCA Inventory.
OSHA PEL: CL 1 ppm
ACGIH TLV: CL 1 ppm
DOT CLASSIFICATION: 8; Label: Corrosive, Poison
SAFETY PROFILE: A poison. Corrosive. A skin, eye, and mucous membrane irritant. Dangerous; may explode when heated. This and other boron halides react with water or steam to produce toxic and corrosive fumes and may explode. Incompatible with K; Na. When heated to decomposition it emits toxic fumes of Br⁻. See also BORON COMPOUNDS and HYDROBROMIC ACID.

BMG500   CAS: 10294-34-5   HR: 3
BORON TRICHLORIDE
DOT: UN 1741
mf: BCl_3 mw: 117.16
PROP: Colorless gas, fuming liquid. Pungent, irritating odor. Very easily hydrolyzed. Mp: −107°, bp: 12.5°, d: 1.349 @ 11°/4°, vap press: 1 atm @ 12.7°, vap d: 4.03.

B

SYNS: BORON CHLORIDE □ CHLORURE de BORE (FRENCH)

CONSENSUS REPORTS: Reported in EPA TSCA Inventory. EPA Extremely Hazardous Substances List.

DOT CLASSIFICATION: 2.3; Label: Poison Gas, Corrosive

SAFETY PROFILE: A poison by inhalation. A corrosive and severe irritant to skin, eyes, and mucous membranes. Reacts with water or steam to produce heat, toxic and corrosive fumes. Violent reaction with aniline or phosphine. Incompatible with hexafluorisopropylidene amino lithium, NO_2, grease, organic matter, O_2. When heated to decomposition it emits toxic fumes of Cl^-. See also BORON COMPOUNDS and HYDROCHLORIC ACID.

BMG700   CAS: 7637-07-2   HR: 3
BORON TRIFLUORIDE
DOT: UN 1008
mf: BF_3 mw: 67.81
PROP: Colorless nonflammable gas; pungent, irritating odor. Mp: $-128.4°$, bp: $-100.0°$, d: 2.99 g/L. Sol in H_2O, and org solvs, e.g., alcohols, ethers (forming adducts). IDLH 25 ppm.
SYNS: BORON FLUORIDE □ FLUORURE de BORE (FRENCH)
CONSENSUS REPORTS: Reported in EPA TSCA Inventory. EPA Extremely Hazardous Substances List.
OSHA PEL: CL 1 ppm
ACGIH TLV: CL 1 ppm
DFG MAK: 1 ppm (3 mg/m³)
NIOSH REL: (Boron Trifluoride) No Exposure Limit
DOT CLASSIFICATION: 2.3; Label: Poison Gas
SAFETY PROFILE: A poison by inhalation. A strong irritant. See also BORON COMPOUNDS and FLUORIDES. A nonflammable gas. Dangerous; when heated to decomposition or upon contact with water or steam, will produce toxic and corrosive fumes of F^-. Incompatible with

alkali metals, alkaline earth metals (except Mg), alkyl nitrates, and CaO.

BMG750   CAS: 7578-36-1   HR: 3
BORON TRIFLUORIDE–ACETIC ACID COMPLEX
DOT: UN 1742
SYNS: ACETIC ACID, compd. with BORON FLUORIDE (BF3) (8CI) □ BORON FLUORIDE, compd. with ACETIC ACID
CONSENSUS REPORTS: Reported in EPA TSCA Inventory.
DOT CLASSIFICATION: 8; Label: Corrosive
SAFETY PROFILE: A very corrosive material. When heated to decomposition it emits very toxic fumes of F^-, B oxides. See BORON COMPOUNDS, ACETIC ACID, and FLUORIDES.

BMG800   CAS: 13319-75-0   HR: 2
BORON TRIFLUORIDE DIHYDRATE
DOT: UN 2851
mf: $BF_3 \cdot 2H_2O$ mw: 103.85
SYNS: BORANE, TRIFLUORO-, DIHYDRATE □ BORON FLUORIDE DIHYDRATE □ BORON TRIFLUORIDE DIHYDRATE (DOT)
DOT CLASSIFICATION: 8; Label: Corrosive
SAFETY PROFILE: Moderately toxic by inhalation. A corrosive irritant. When heated to decomposition it emits toxic vapors of B and F^-.

BMH000   CAS: 353-42-4   HR: 3
BORON TRIFLUORIDE-DIMETHYL ETHER
DOT: UN 2965
mf: $C_2H_6O \cdot BF_3$ mw: 113.89
PROP: Moisture-sensitive liquid. D: 1.239, mp: $-14°$, bp: 126–127°.
SYNS: BORON TRIFLUORIDE DIMETHYL ETHERATE (DOT) □ FLUORID BORITY-DIMETHYLETHER (1:1)
CONSENSUS REPORTS: EPA Extremely Hazardous Substances List. Reported in EPA TSCA Inventory.
DOT CLASSIFICATION: 4.3; Label: Dangerous When Wet, Corrosive, Flammable Liquid
SAFETY PROFILE: Poison by inhalation. Corrosive. Flammable liquid. When heated

to decomposition it emits toxic fumes of F⁻.
See also ETHERS and BORON
COMPOUNDS.

BMK290   CAS: 27098-03-9   HR: 3
BOTRYODIPLODIN
mf: $C_7H_{12}O_3$ mw: 144.19
PROP: Crystals from Et$_2$O. Mp: 50–52°.
SYNS: (−)-BOTRYODIPLODIN □ 2-HYDROXY-3-
METHYL-4-ACETYLTETRAHYDROFURANE □ METHYL
TETRAHYDRO-5-HYDROXY-4-METHYL-3-FURYL KET-
ONE □ 1-(TETRAHYDRO-5-HYDROXY-4-METHYL-3-
FURANYL)-ETHANONE (9CI)
DOT CLASSIFICATION: 3; Label:
Flammable Liquid
SAFETY PROFILE: A poison. Human
mutation data reported. A flammable liquid.
When heated to decomposition it emits
acrid smoke and fumes. See also
KETONES.

BML000   HR: 3
BRACKEN FERN, DRIED
SYNS: 1-CYCLOHEXENE-1-CARBOXYLIC ACID, 3,4,5
□ S. EGRELTRI ATUNUN (TURKISH) □ PTERIDIUM
AQUILINUM □ PTERIS AQUALINA
CONSENSUS REPORTS: IARC Cancer
Review: Human Inadequate Evidence
IMEMDT 40,47,86; Animal Sufficient
Evidence IMEMDT 40,47,86.
SAFETY PROFILE: Confirmed carcinogen
with experimental carcinogenic,
neoplastigenic, and tumorigenic data.
Experimental teratogenic and reproductive
effects. Mutation data reported.

BMM500   CAS: 2580-78-1   HR: 2
BRILLIANT BLUE R
mf: $C_{22}H_{16}N_2O_{11}S_3 \cdot 2Na$ mw: 626.56
SYNS: CAVALITE BRILLIANT BLUE R □ C.I. 61200 □
C.I. REACTIVE BLUE 19 □ C.I. REACTIVE BLUE 19,
DISODIUM SALT □ REACTIVE BLUE 19 □ REMALAN
BRILLIANT BLUE R □ REMAZOL BRILLIANT BLUE R
CONSENSUS REPORTS: Reported in EPA
TSCA Inventory.
SAFETY PROFILE: Questionable
carcinogen with experimental tumorigenic
data. When heated to decomposition it
emits very toxic fumes of Na$_2$O, NO$_x$, and
SO$_x$. See also SULFONATES.

BMM575   CAS: 59803-98-4   HR: 3
BRIMONIDINE
mf: $C_{11}H_{10}BrN_5$ mw: 292.17
SYNS: 5-BROMO-6-(2-IMIDAZOLIN-2-
YLAMINO)QUINOXALINE □ LK 14304-18 □ 6-
QUINOXALINAMINE, 5-BROMO-N-(4,5-DIHYDRO-1H-
IMIDAZOL-2-YL)- □ UK 14304 □ UK14304
SAFETY PROFILE: A poison by ingestion.
When heated to decomposition it emits
toxic vapors of NO$_x$ and Br⁻.

BMM650   CAS: 314-40-9   HR: 2
BROMACIL
mf: $C_9H_{13}BrN_2O_2$ mw: 261.15
PROP: Crystals or solid from EtOH (aq).
Sltly sol in H$_2$O; mod sol in strong aq bases
from Me$_2$CO, MeCN, and EtOH.
SYNS: BOREA □ BROMAZIL □ 5-BROMO-3-sec-
BUTYL-6-METHYLURACIL □ 5-BROMO-6-METHYL-3-(1-
METHYLPROPYL)-2,4(1H,3H)-PYRIMIDINEDIONE □ 5-
BROMO-6-METHYL-3-(1-METHYLPROPYL)URACIL □ 3-
sek.BUTYL-5-BROM-6-METHYLURACIL (GERMAN) □
CYNOGAN □ DU PONT HERBICIDE 976 □ EEREX
GRANULAR WEED KILLER □ EEREX WATER
SOLUBLE CONCENTRATE WEED KILLER □ HERBI-
CIDE 976 □ HYVAR □ HYVAREX □ HYVAR X □
HYVAR X BROMACIL □ HYVAR X WEED KILLER □
KROVAR II □ NALKIL □ URAGAN □ URAGON □
UROX B WATER SOLUBLE CONCENTRATE WEED
KILLER □ UROX HX GRANULAR WEED KILLER
CONSENSUS REPORTS: EPA Genetic
Toxicology Program.
OSHA PEL: TWA 10 ppm
ACGIH TLV: TWA 10 mg/m³; Animal
Carcinogen
SAFETY PROFILE: Moderately toxic by
ingestion. An experimental teratogen.
Mutation data reported. Questionable
carcinogen. An herbicide. When heated to
decomposition it emits very toxic fumes of
Br⁻ and NO$_x$.

BMN500   HR: 3
BROMATES
SAFETY PROFILE: Generally considered to
be more toxic than chlorates; cause central
nervous system paralysis. They may form
methemoglobin, but less actively than
chlorates. See also specific compounds.
Flammable in the form of gas, vapor, or
dust by chemical reaction with (powdered

B

metals + acids); Al; As; CaH_2; C; Cu; powdered metals; metal sulfides; organic matter; PH_4I; P; SrH; S; (H_2SO_4 + metals). When heated to decomposition they emit toxic fumes of Br^-; can react with reducing materials.

BMO000   CAS: 9001-00-7   HR: 3
BROMELAIN
PROP: From pineapples *Ananas comosus* and *Ananas bracteatus L.* White to tan amorphous powder. Sol in water; insol in alc, chloroform, ether.
SYNS: ANANASE □ BROMELAINS □ BROMELIN □ E.C. 3.4.4.24 □ EXTRANASE □ INFLAMEN □ PLANT PROTEASE CONCENTRATE □ TRAUMANASE
CONSENSUS REPORTS: Reported in EPA TSCA Inventory.
SAFETY PROFILE: A poison via intraperitoneal and intravenous routes. When heated to decomposition it emits acrid smoke and fumes.

BMO750   HR: 3
BROMIDES
SAFETY PROFILE: The most common inorganic bromides are Na, K, NH_4, Ca, and Mg bromides. Methyl and ethyl bromides are among the most common organic bromides. The inorganic bromides produce depression, emaciation, and, in severe cases, psychosis and mental deterioration. Bromide rashes (bromoderma), especially of the face and resembling acne and furunculosis, often occur when bromide inhalation or administration is prolonged. Organic bromides, such as methyl bromide and ethyl bromide, are volatile liquids of relatively high toxicity. See also specific compounds. When strongly heated they emit highly toxic fumes of Br^-.

BMO825   CAS: 68952-98-7   HR: D
BROMINATED VEGETABLE (SOYBEAN)
OIL
PROP: Pale-yellow to dark-brown viscous, oily liquid; bland or fruity odor and bland

taste. Sol in alc, chloroform, ether, hexane, fixed oils; insol in water.
SYN: VEGETABLE (SOYBEAN) OIL, brominated
SAFETY PROFILE: Experimental reproductive effects. When heated to decomposition it emits toxic fumes of Br^-.

BMP000   CAS: 7726-95-6   HR: 3
BROMINE
DOT: UN 1744
mf: Br_2 mw: 159.82
PROP: Rhombic crystals or dark red-brown liquid with a strong disagreeable pungent odor. Strong oxidant. Fp: $-7.3°$, bp: $59.5°$, d: 2.928 @ $59°$, 3.12 @ $20°$, vap press: 175 mm @ $21°$, 1 atm @ $58.2°$, vap d: 5.5. Sol in H_2O. Misc in most org solvs, although it may react. IDLH 3 ppm.
SYNS: BROM (GERMAN) □ BROME (FRENCH) □ BROMINE, solution (DOT) □ BROMO (ITALIAN) □ BROOM (DUTCH)
CONSENSUS REPORTS: EPA Extremely Hazardous Substances List. Reported in EPA TSCA Inventory. EPA Genetic Toxicology Program.
OSHA PEL: TWA 0.1 ppm; STEL 0.3 ppm
ACGIH TLV: TWA 0.1 ppm; STEL 0.2 ppm
DFG MAK: 0.1 ppm (0.66 mg/m^3)
DOT CLASSIFICATION: 8; Label: Corrosive, Poison
SAFETY PROFILE: A human poison by ingestion and moderately toxic by inhalation. A poison by ingestion and inhalation experimentally. Corrosive. The action of bromine is essentially the same as that of chlorine, irritating the mucous membranes of the eyes and upper respiratory tract. Severe exposure may result in pulmonary edema. Usually, however, the irritant qualities of the chemical force the worker to leave the exposure area before serious poisoning can result. Chronic exposure is similar to the therapeutic ingestion of excessive bromides. See also BROMIDES. Regular physical examinations should be made of people who work with bromine or bromides. Flammable in the form of liquid or vapor by spontaneous chemical reaction with reducing materials. A

very powerful oxidizer. Highly dangerous; when heated it emits highly toxic fumes; will react with water or steam to produce toxic and corrosive fumes. Reacts explosively with diethylzinc, germane, disilane, dimethylformamide, hydrogen, isobutyrophenone, metal azides (particularly silver or sodium azide), potassium, silane and homologs, praseodymium, antimony, trimethylamine, ammonia. Mixtures with lithium or sodium are shock-sensitive explosives. Ignition on contact with germanium, mono- or di-alkali metal acetylides, trialkyl boranes, copper acetylides. Violent reaction with carbonyl compounds (aldehydes, ketones, carboxylic acids), diethyl ether, phosphine, natural rubber, aluminum, mercury, titanium. Vigorous reaction with methanol and other alcohols, tetrahydrofuran, mixtures of ethanol and phosphorus. Incompatible with acetaldehyde, C_2H_2, acrylonitrile, NH_3, Sb, B, Ca_3N_2, Cs_2O, Cs_2C_2, CsC_2H, ClF_3C_2, CuH_2, dimethyl formamide, ethyl phosphine, F_2, Fe_2C, isobutyrophenone, Li_2C_2, Li_2Si_2, Mg_3P_2, $Ni(Co)_4$, NI_3, olefins, OF_2, O_3, P, PO_x, Rb_2C_2, RbC_2H, Na_2C_2, NaC_2H, Sr_3P, Sn, UC_2, ZrC2, reducing materials.

**BMP250   CAS: 13973-87-0   HR: 3
BROMINE AZIDE**
mf: BrN_3 mw: 121.93
PROP: Crystals or red liquid. Mp: 45°, bp: explodes.
SYNS: BROMINE NITRIDE □ NITROGEN BROMIDE
DOT CLASSIFICATION: Forbidden
SAFETY PROFILE: A poison. Can explode spontaneously. The solid, liquid and vapor are shock-sensitive explosives. Concentrated solutions in organic solvents may explode. Moderate fire hazard in the form of vapor by chemical reaction. A powerful oxidant. Moderately explosive when exposed to heat. The liquid explodes on contact with arsenic, sodium, silver foil, or phosphorus. Incompatible with Sb, ethyl ether, Ag, metals. When heated to decomposition it emits highly toxic fumes of Br^- and explodes. See also BROMINE and AZIDES.

**BMQ000   CAS: 7789-30-2   HR: 3
BROMINE PENTAFLUORIDE**
DOT: UN 1745
mf: BrF_5 mw: 174.91
PROP: Colorless fuming liquid. An extremely vigorous fluorinating agent. Mp: −60.5°, bp: 40.5°, d: 2.466 @ 25°, vap d: 6.05.
OSHA PEL: TWA 0.1 ppm; TWA 2.5 mg(F)/m³
ACGIH TLV: TWA 0.1 ppm; TWA 2.5 mg(F)/m³; BEI: 3 mg/g creatinine of fluorides in urine prior to shift; 10 mg/g creatinine of fluorides in urine at end of shift.
NIOSH REL: (Inorganic Fluorides) TWA 2.5 mg(F)/m³
DOT CLASSIFICATION: 5.1; Label: Oxidizer, Poison, Corrosive
SAFETY PROFILE: A poisonous, corrosive, and extremely reactive gas. It is a powerful oxidizer. Will react with water or steam to produce toxic and corrosive fumes. The liquefied gas reacts violently with many organic compounds and some inorganic compounds. Explodes or ignites on contact with hydrogen-containing materials (e.g., acetic acid, ammonia, benzene, ethanol, hydrogen, hydrogen sulfide, methane, cork, grease, paper, wax, chloromethane). Reacts violently and may ignite on contact with acids, halogens, nonmetals, metal halides, metals, oxides, concentrated nitric or sulfuric acids, aluminum powder, ammonium chloride, antimony, arsenic, arsenic pentoxide, barium, bismuth, boron powder, boron trioxide, calcium oxide, carbon monoxide, charcoal, chlorine, chromium, chromium trioxide, cobalt powder, iodine, iodine pentoxide, iridium powder, iron powder, lithium powder, manganese, magnesium oxide, molybdenum, molybdenum trioxide, nickel powder, red phosphorus, phosphorus pentoxide,

potassium iodide, rhodium powder, selenium, sulfur, sulfur dioxide, tellurium, tungsten, tungsten trioxide, water, zinc. When heated to decomposition it emits very toxic fumes of F⁻ and Br⁻. See also BROMINE and FLUORIDES.

BMQ325   CAS: 7787-71-5   HR: 3
BROMINE TRIFLUORIDE
DOT: UN 1746
mf: BrF₃   mw: 136.91
PROP: Colorless, fuming liquid. Mp: 8.8°, bp: 127°, d: 2.84.
CONSENSUS REPORTS: Reported in EPA TSCA Inventory.
OSHA PEL: TWA 2.5 mg(F)/m³
ACGIH TLV: TWA 2.5 mg(F)/m³; BEI: 3 mg/g creatinine of fluorides in urine prior to shift; 10 mg/g creatinine of fluorides in urine at end of shift.
NIOSH REL: (Inorganic Fluorides) TWA 2.5 mg(F)/m³
DOT CLASSIFICATION: 5.1; Label: Oxidizer, Poison, Corrosive
SAFETY PROFILE: Poisonous and corrosive. Very reactive, a powerful oxidizer. Explosive or violent reaction with organic materials, water, acetone, ammonium halides, antimony, antimony trichloride oxide, arsenic, benzene, boron, bromine, carbon, carbon monoxide, carbon tetrachloride, carbon tetraiodide, chloromethane, cobalt, ether, halogens, iodine, powdered molybdenum, niobium, 2-pentanone, phosphorus, potassium hexachloroplatinate, pyridine, silicon, silicone grease, sulfur, tantalum, tin dichloride, titanium, toluene, vanadium, uranium, uranium hexafluoride. Incompatible with Sb_2O_3, $BaCl_2$, Bi_2O_5, $CdCl_2$, $CaCl_2$, $CsCl$, $LiCl$, $MnIO_3$, metals, Nb_2O_5, $PtBr_4$, $PtCl_4$, (Pt + KFO), KBr, KCl, KI, $RhBr_4$, RbCl, AgCl, NaBr, NaCl, NaI, Ta_2O_5, Sn, W, UO_x, rubber, plastics. The product of reaction with pyridine ignites when dry. When heated to decomposition it emits toxic fumes of F⁻ and Br⁻. Very dangerous. See also BROMINE PENTAFLUORIDE, FLUORIDES, and BROMINE.

BMQ800   CAS: 61288-13-9   HR: 3
BROMKAL 80
CONSENSUS REPORTS: NTP 10th Report on Carcinogens, 2000: Reasonably anticipated to be human carcinogen
SAFETY PROFILE: Confirmed carcinogen.

BMR750   CAS: 79-08-3   HR: 3
α-BROMOACETIC ACID
DOT: UN 1938
mf: $C_2H_3O_2Br$ mw: 138.04
PROP: Hygroscopic crystals, sol in water and alc. D: 1.93, mp: 50°, bp: 208°. Sol in H_2O and EtOH.
SYNS: ACIDE BROMACETIQUE (FRENCH) □ BROMOACETIC ACID □ BROMOACETIC ACID, solid or solution (DOT) □ BROMOETHANOIC ACID □ α-BROMOETHANOIC ACID □ KYSELINA BROMOCTOVA □ MONOBROMESSIGSAEURE (GERMAN) □ MONOBROMOACETIC ACID □ TO NTU
CONSENSUS REPORTS: Reported in EPA TSCA Inventory.
DOT CLASSIFICATION: 8; Label: Corrosive
SAFETY PROFILE: Poison by ingestion, intraperitoneal, and intravenous routes. Irritating and corrosive to skin and mucous membranes. Mutation data reported. When heated to decomposition it emits toxic fumes of Br⁻. See also BROMIDES.

BMS300   CAS: 143-84-0   HR: 3
1-BROMOACETYL-α-α-DIPHENYL-4-
PIPERIDINEMETHANOL
mf: $C_{20}H_{22}BrNO_2$ mw: 388.34
SYNS: KETONE, BROMOMETHYL 4-(DIPHENYLHYDROXYMETHYL)PIPERIDINO □ 4-PIPERIDINEMETHANOL, 1-BROMOACETYL-α-α-DIPHENYL-
DOT CLASSIFICATION: 3; Label: Flammable Liquid
SAFETY PROFILE: A poison by intraperitoneal route. A flammable liquid. When heated to decomposition it emits toxic vapors of NO_x and Br⁻.

BMT150   CAS: 101652-13-5 HR: 3
3-BROMOALLYL ISOCYANATE
DOT: UN 2206/UN 2207/UN 2478/UN 3080
mf: C_4H_4BrNO mw: 162.00
SYN: ISOCYANIC ACID, 3-BROMOALLYL ESTER
DOT CLASSIFICATION: 6.1; Label: KEEP AWAY FROM FOOD (UN2207); 6.1; Label: Poison (UN2206); 6.1; Label: Poison, Flammable Liquid (UN3080); 3; Label: Flammable Liquid, Poison (UN2478)
SAFETY PROFILE: A poison by intraperitoneal route. A flammable liquid. When heated to decomposition it emits toxic vapors of NO_x and Br^-.

BMW250   CAS: 5798-79-8   HR: 3
BROMOBENZYLNITRILE
mf: C_8H_6BrN mw: 196.06
PROP: Pure: Yellowish-white crystals. Tech: brown, oily liquid with pungent odor of sour fruit; mp: 29°, bp: 242°, fp: 25.5°, flash p: none, d: 1.5160 @ 20°, vap d: 6.8, vap press: 0.011 mm @ 20°.
SYNS: BBC □ BBN □ BROMBENZYL CYANIDE □ α-BROMOBENZYL CYANIDE □ α-BROMOBENZYL-NITRILE □ α-BROMOPHENYLACETONITRILE □ α-BROMO-α-TOLUNITRILE □ CA □ CAMITE
CONSENSUS REPORTS: Cyanide and its compounds are on the Community Right-To-Know List.
SAFETY PROFILE: Poison by ingestion. Moderately toxic to humans by inhalation. When heated to decomposition it emits very toxic fumes of NO_x, Br^-, and CN^-. See also NITRILES.

BMX500   CAS: 109-65-9   HR: 3
1-BROMOBUTANE
DOT: UN 1126
mf: C_4H_9Br mw: 137.04
PROP: Colorless to pale straw-colored liquid. Mp: −112.3°, bp: 101.6°, flash p: 65°F (OC), d: 1.276 @ 20°/8°, autoign temp: 509°F, vap d: 4.72, lel: 2.8% @ 212°F, uel: 6.6% @ 212°F.
SYNS: BUTYL BROMIDE (DOT) □ n-BUTYL BROMIDE (DOT)

CONSENSUS REPORTS: EPA Genetic Toxicology Program. Reported in EPA TSCA Inventory.
DOT CLASSIFICATION: 3; Label: Flammable Liquid
SAFETY PROFILE: Moderately toxic by intraperitoneal route. Mildly toxic by inhalation. Dangerous fire hazard when exposed to heat, flame, or oxidizers. Violent reaction with bromobenzene + sodium above 30°C. Can react with oxidizing materials. To fight fire, use CO_2, dry chemical, mist or spray. See also BROMIDES.

BMX750   CAS: 78-76-2   HR: 2
2-BROMOBUTANE
DOT: UN 2339
mf: C_4H_9Br mw: 137.04
PROP: Colorless liquid. Fp: <−50°, bp: 91.4°, flash p: 70°F, d: 1.257 @ 25°/25°.
SYNS: sec-BUTYL BROMIDE □ METHYLETHYL-BROMOMETHANE
CONSENSUS REPORTS: EPA Genetic Toxicology Program. Reported in EPA TSCA Inventory.
DOT CLASSIFICATION: 3; Label: Flammable Liquid
SAFETY PROFILE: Narcotic in high concentrations. Questionable carcinogen with experimental neoplastigenic data. See also BROMIDES and CHLORINATED HYDROCARBONS, ALIPHATIC. Flammable liquid. Dangerous fire hazard when exposed to heat or flame. When heated to decomposition it emits toxic fumes of Br^-; can react with oxidizing materials. To fight fire, use water, spray or mist, foam, CO_2, dry chemical.

BNA250   CAS: 353-59-3   HR: 1
BROMOCHLORODIFLUOROMETHANE
DOT: UN 1974
mf: $CBrClF_2$ mw: 165.37
PROP: Colorless gas. Fp: −160.5°, bp: −4°.
SYNS: CHLORODIFLUOROBROMOMETHANE (DOT) □ CHLORODIFLUOROMONOBROMOMETHANE □ FLUGEX 12B1 □ FLUOROCARBON 1211 □ FREON 12B1 □ HALON 1211 □ R12B1 (DOT)

B

CONSENSUS REPORTS: Reported in EPA TSCA Inventory.

DOT CLASSIFICATION: 2.2; Label: Nonflammable Gas

SAFETY PROFILE: Mutation data reported. An asphyxiant. See also ARGON for description of inert gas asphyxiants. When heated to decomposition it emits very toxic fumes of Br⁻, Cl⁻, and F⁻.

BNA350   CAS: 122322-22-9 HR: 3
2-(3-BROMO-4-CHLOROPHENYL)-4-
** CHLORO-5-((6-CHLORO-3-**
** PYRIDINYL)METHOXY)-3(2H)-**
** PYRIDAZINONE**
mf: $C_{16}H_9BrCl_3N_3O_2$ mw: 461.54
SYN: 3(2H)-PYRIDAZINONE, 2-(3-BROMO-4-CHLOROPHENYL)-4-CHLORO-5-((6-CHLORO-3-PYRIDINYL)METHOXY)-
SAFETY PROFILE: A poison by ingestion. When heated to decomposition it emits toxic vapors of NO_x, Br⁻, and Cl⁻.

BNA750   CAS: 41198-08-7 HR: 3
O-(4-BROMO-2-CHLOROPHENYL)-O-
** ETHYL-S-PROPYL**
** PHOSPHOROTHIOATE**
mf: $C_{11}H_{15}BrClO_3PS$ mw: 373.65
SYNS: CGA 15324 □ CURACRON □ POLYCRON □ PROFENOFOS □ SELECRON
SAFETY PROFILE: Poison by ingestion and skin contact. When heated to decomposition it emits very toxic SO_x, PO_x, Br⁻, and Cl⁻. See also ESTERS.

BNA825   CAS: 109-70-6   HR: 2
1-BROMO-3-CHLOROPROPANE
DOT: UN 2688
mf: C_3H_6BrCl mw: 157.45
PROP: Bp: 142–143°.
SYNS: 3-BROMOPROPYL CHLORIDE □ 1,3-CHBP □ ω-CHLOROBROMOPROPANE □ 1-CHLORO-3-BROMOPROPANE (DOT) □ 3-CHLOROPROPYL BROMIDE □ TRIMETHYLENE BROMIDE CHLORIDE □ TRIMETHYLENE CHLOROBROMIDE
CONSENSUS REPORTS: Reported in EPA TSCA Inventory.
DOT CLASSIFICATION: 6.1; Label: KEEP AWAY FROM FOOD

SAFETY PROFILE: Moderately toxic by ingestion. When heated to decomposition it emits toxic fumes of Cl⁻ and Br⁻. See also CHLORINATED HYDROCARBONS, ALIPHATIC; and BROMIDES.

BNA880   CAS: 3737-00-6   HR: 3
3-BROMO-1-CHLOROPROPENE
mf: C_3H_4BrCl mw: 155.42
SYN: 1-PROPENE, 3-BROMO-1-CHLORO-
SAFETY PROFILE: A poison by ingestion. Low toxicity by inhalation. Human systemic effects. When heated to decomposition it emits toxic vapors of Br⁻ and Cl⁻.

BNC750   CAS: 59-14-3   HR: 2
5-BROMO-2'-DEOXYURIDINE
mf: $C_9H_{11}BrN_2O_5$ mw: 307.13
PROP: Solid. Mp: 187–189°.
SYNS: BDU □ 5-BDU □ BROMODEOXYURIDINE □ 5-BROMODEOXYURIDINE □ 5-BROMO-2-DEOXYURIDINE □ 5-BROMODESOXYURIDINE □ BROMOURACIL DEOXYRIBOSIDE □ 5-BROMOURACIL DEOXYRIBOSIDE □ 5-BROMOURACIL-2-DEOXYRIBOSIDE □ BROXURIDINE □ BRUDR □ BUDR □ 5-BUDR
CONSENSUS REPORTS: Reported in EPA TSCA Inventory. EPA Genetic Toxicology Program.
SAFETY PROFILE: Moderately toxic by subcutaneous, intravenous, intraperitoneal, and possibly other routes. Mildly toxic by ingestion. Experimental teratogenic and reproductive effects. Human mutation data reported. When heated to decomposition it emits very toxic fumes of Br⁻ and NO_x.

BNC800   CAS: 27312-17-0 HR: 1
2-BROMO-1,5-DIAMINO-4,8-DIHYDRO-
** XYANTHRAQUINONE**
mf: $C_{14}H_9BrN_2O_4$ mw: 349.16
SYNS: ANTHRAQUINONE, 2-BROMO-1,5-DIAMINO-4,8-DIHYDROXY- □ MODR OSTACETOVA LR
CONSENSUS REPORTS: Reported in EPA TSCA Inventory.
SAFETY PROFILE: An eye irritant. When heated to decomposition it emits toxic fumes of NO_x and Br⁻.

BND500 CAS: 75-27-4 HR: 3
BROMODICHLOROMETHANE
mf: CHBrCl$_2$ mw: 163.83
PROP: Colorless liquid. Mp: −57.1°, bp: 88.4–88.6°, d: 1.971 @ 25°/25°.
SYNS: BDCM □ DICHLOROBROMOMETHANE □ NCI-C55243
CONSENSUS REPORTS: NTP 10th Report on Carcinogens. NTP Carcinogenesis Studies (gavage); Clear Evidence: rat, mouse NTPTR* NTP-TR-321,87. EPA Genetic Toxicology Program. Community Right-To-Know List. Reported in EPA TSCA Inventory.
SAFETY PROFILE: Confirmed carcinogen with experimental carcinogenic data. Moderately toxic by ingestion. Human mutation data reported. When heated to decomposition it emits very toxic fumes of Br⁻ and Cl⁻. See also CHLORINATED HYDROCARBONS, ALIPHATIC; and BROMIDES.

BNG750 CAS: 776-74-9 HR: 3
BROMODIPHENYLMETHANE
DOT: UN 1770
mf: C$_{13}$H$_{11}$Br mw: 247.15
PROP: Solid. Mp: 45°, bp: 184° @ 20 mm. Decomp in hot water; sol in alc; very sol in benzene.
SYN: DIPHENYLMETHYL BROMIDE (DOT)
CONSENSUS REPORTS: Reported in EPA TSCA Inventory.
DOT CLASSIFICATION: 8; Label: Corrosive
SAFETY PROFILE: A corrosive poison. When heated to decomposition it emits toxic fumes of Br⁻. See also BROMIDES.

BNH000 CAS: 776-74-9 HR: 3
BROMODIPHENYLMETHANE (solution)
DOT: UN 1770
mf: C$_{13}$H$_{11}$Br mw: 247.15
SYN: DIPHENYL METHYL BROMIDE, solution (DOT)
CONSENSUS REPORTS: Reported in EPA TSCA Inventory.
DOT CLASSIFICATION: 8; Label: Corrosive
SAFETY PROFILE: A corrosive, irritating liquid. When heated to decomposition it emits toxic fumes of Br⁻. See also

BROMODIPHENYLMETHANE and BROMIDES.

BNH500 CAS: 17372-87-1 HR: 3
BROMOEOSINE
mf: C$_{20}$H$_8$Br$_4$O$_5$•2Na mw: 693.90
PROP: Red crystals with bluish tinge, or brownish-red powder. Sol in H$_2$O; sltly sol in EtOH; insol in Et$_2$O.
SYNS: AIZEN EOSINE GH □ BROMO ACID □ BROMOFLUORESCEIC ACID □ BROMO FLUORESCEIN □ BRONZE BROMO □ CERTIQUAL EOSINE □ C.I. 45380 □ D&C RED No. 22 □ DISODIUM EOSIN □ EOSINE □ EOSINE SODIUM SALT □ EOSINE YELLOWISH □ EOSIN GELBLICH (GERMAN) □ FENAZO EOSINE XG □ HIDACID DIBROMO FLUORESCEIN □ IRGALITE BRONZE RED CL □ PHLOXINE TONER B □ PHLOX RED TONER X-1354 □ PURE EOSINE YY □ 11445 RED □ SODIUM EOSINATE □ SYMULER EOSIN TONER □ 2,4,5,7-TETRABROMO-9-o-CARBOXYPHENYL-6-HYDROXY-3-ISOXANTHONE, DISODIUM SALT □ 2,4,5,7-TETRABROMO-3,6-FLUORANDIOL □ TETRABROMOFLUORESCEIN □ 2',4',5',7'-TETRABROMOFLUORESCEIN DISODIUM SALT □ TETRABROMOFLUORESCEIN S □ TETRABROMOFLUORESCEIN SOLUBLE □ 2-(2,4,5,7-TETRABROMO-6-HYDROXY-3-OXO-3H-XANTHENE-9-YL)BENZOIC ACID, DISODIUM SALT □ TOYO EOSINE G □ 1903 YELLOW PINK
CONSENSUS REPORTS: IARC Cancer Review: Animal Inadequate Evidence IMEMDT 15,183,77. EPA Genetic Toxicology Program. Reported in EPA TSCA Inventory.
SAFETY PROFILE: Poison by intravenous and intraperitoneal routes. Moderately toxic by ingestion. Questionable carcinogen with experimental tumorigenic data. When heated to decomposition it emits very toxic fumes of Br⁻ and Na$_2$O. See also BROMIDES.

BNI000 CAS: 3132-64-7 HR: 3
3-BROMO-1,2-EPOXYPROPANE
DOT: UN 2558
mf: C$_3$H$_5$BrO mw: 136.99
PROP: Flash p: <22°.
SYNS: EPIBROMHYDRIN □ EPIBROMOHYDRIN (DOT) □ EPIBROMOHYDRINE
CONSENSUS REPORTS: EPA Genetic Toxicology Program. Reported in EPA TSCA Inventory.

B

DOT CLASSIFICATION: 6.1; Label: Poison
SAFETY PROFILE: Poison by intraperitoneal route. Human mutation data reported. A dangerous fire hazard when exposed to heat or flame. When heated to decomposition it emits toxic fumes of Br⁻. See also BROMIDES.

BNI500   CAS: 540-51-2   HR: 3
2-BROMO ETHANOL
mf: C_2H_5BrO mw: 124.98
PROP: D: 1.79 @ 0°/4°, bp: 149–150°.
SYNS: BE □ BROMOETHANOL □ ETHYLENEBROMOHYDRIN □ GLYCOL BROMOHYDRIN
CONSENSUS REPORTS: EPA Genetic Toxicology Program. Reported in EPA TSCA Inventory.
SAFETY PROFILE: Poison by intraperitoneal route. Questionable carcinogen with experimental neoplastigenic and tumorigenic data. Mutation data reported. When heated to decomposition it emits toxic fumes of Br⁻. See also BROMIDES.

BNK000   CAS: 77-65-6   HR: 3
2-BROMO-2-ETHYLBUTYRYLUREA
mf: $C_7H_{13}BrN_2O_2$ mw: 237.13
PROP: Solid. Mp: 116–119°.
SYNS: ADALIN □ ADDISOMNOL □ N-(AMINOCARBONYL)-2-BROMO-2-ETHYLBUTANAMIDE □ BROMACETOCARBAMIDE □ BROMADAL □ BROMADEL □ BROMODIETHYLACETYLCARBAMIDE □ BROMODIETHYLACETYLUREA □ (α-BROMO-α-ETHYLBUTYRYL)CARBAMIDE □ (α-BROMO-α-ETHYLBUTYRYL)UREA □ 1-BROMO-ETHYL-BUTYRYL-UREA □ 2-BROMO-2-ETHYLBUTYRLUREA □ CARBOMAL □ DIACID □ DORMITURIN □ FYDALIN □ HOGGAR □ KARBROMAL □ KARTRYL □ NCI-C03805 □ NENESIN □ NYCTAL □ PARKOSED □ PELIDORM □ PIANADALIN □ PLANADALIN □ TILDIN □ URADAL
CONSENSUS REPORTS: NCI Carcinogenesis Bioassay (feed); No Evidence: mouse, rat NCITR* NCI-CG-TR-173,79. Reported in EPA TSCA Inventory.
SAFETY PROFILE: Poison by ingestion, subcutaneous, and possibly other routes. Moderately toxic via intravenous and intraperitoneal routes. Mutation data reported. A sedative, hypnotic, and central nervous system depressant. When heated to decomposition it emits very toxic fumes of NO_x and Br⁻.

BNL000   CAS: 75-25-2   HR: 3
BROMOFORM
DOT: UN 2515
mf: $CHBr_3$ mw: 252.75
PROP: Colorless heavy liquid or hexagonal crystals. Mp: 6–7°, bp: 149°, flash p: none, d: 2.887 @ 20°/4°. IDLH 850 ppm.
SYNS: BROMOFORME (FRENCH) □ BROMOFORMIO (ITALIAN) □ METHENYL TRIBROMIDE □ NCI-C55130 □ RCRA WASTE NUMBER U225 □ TRIBROMMETHAAN (DUTCH) □ TRIBROMMETHAN (GERMAN) □ TRIBRO-MOMETAN (ITALIAN) □ TRIBROMOMETHANE
CONSENSUS REPORTS: Reported in EPA TSCA Inventory. Community Right-To-Know List.
OSHA PEL: TWA 0.5 ppm (skin)
ACGIH TLV: TWA 0.5 ppm (skin); Animal Carcinogen
DFG MAK: Confirmed Animal Carcinogen with Unknown Relevance to Humans
DOT CLASSIFICATION: 6.1; Label: KEEP AWAY FROM FOOD
SAFETY PROFILE: Suspected carcinogen with experimental neoplastigenic data. A human poison by ingestion. Moderately toxic by intraperitoneal and subcutaneous routes. Human mutation data reported. A lachrymator. It can damage the liver to a serious degree and cause death. It has anesthetic properties similar to those of chloroform, but is not sufficiently volatile for inhalation purposes and is far too toxic for human use. As a sedative and antitussive its medicinal application has resulted in numerous poisonings. Inhalation of small amounts causes irritation, provoking the flow of tears and saliva, and reddening of the face. Abuse can lead to addiction and serious consequences. Explosive reaction with crown ethers or potassium hydroxide. Violent reaction with acetone or bases. Incompatible with Li or NaK alloys. When

heated to decomposition it emits highly toxic fumes of Br⁻. See also BROMIDES.

BNL275   CAS: 23483-74-1   HR: 3
BROMO(2-HYDROXYETHYL)MERCURY AMMONIA SALT
SYNS: 2-(BROMOMERCURI) ETHANOL-AMMONIA (1:0.8 moles) compound □ MERCURY, BROMO(2-HYDROXYETHYL)-, compound with AMMONIA (1:0.8 moles)
OSHA PEL: TWA 0.01 mg(Hg)/m³; CL 0.03 mg(Hg)/m³ (skin)
ACGIH TLV: TWA 0.01 mg(Hg)/m³; BEI: 35 μg/g creatinine total inorganic mercury in urine preshift; 15 μg/g creatinine total inorganic mercury in blood at end of shift at end of workweek.
DFG MAK: Confirmed Animal Carcinogen with Unknown Relevance to Humans
SAFETY PROFILE: Poison by intravenous route. When heated to decomposition it emits toxic fumes of NH_3, Hg, and Br⁻.

BNM250   CAS: 478-84-2   HR: 3
2-BROMO-d-LYSERGIC ACID DIETHYLAMIDE
mf: $C_{20}H_{26}BrN_3O$ mw: 404.40
SYNS: BOL □ BOL-148 □ d-2-BROM-DIETHYLAMIDE of LYSERGIC ACID □ BROM LSD □ BROMLYSER-GAMIDE □ 2-BROM-d-LYSERGIC ACID DIETHYL-AMINE □ 2-BROMO-9,10-DIDEHYDRO-N,N-DIETHYL-6-METHYLERGOLINE-8-β-CARBOXAMIDE □ BROMOLYSERGIDE □ 9,10-DIDEHYDRO-N,N-DIETHYL-2-BROMO-6-METHYLERGOLINE-8-β-CARBOXAMIDE □ USAF SZ-1
SAFETY PROFILE: Poison by intraperitoneal and intravenous routes. Experimental teratogenic and reproductive effects. Human systemic effects by ingestion: dilation of the arteries or veins. Many lysergic acid derivatives have central nervous system effects. When heated to decomposition it emits very toxic fumes such as Br⁻ and NO_x. See other lysergic acid derivatives.

BNM750   HR: 3
BROMOMETHANE mixed with DIBRO-MOETHANE
DOT: UN 1647

SYN: METHYL BROMIDE and ETHYLENE DIBROMIDE MIXTURE, liquid (DOT)
DOT CLASSIFICATION: 6.1; Label: Poison
SAFETY PROFILE: A poison. See also BROMIDES. When heated to decomposition it emits toxic fumes of Br⁻.

BNN550   CAS: 128-93-8   HR: 1
1-BROMO-4-(METHYLAMINO)ANTHRA-QUINONE
mf: $C_{15}H_{10}BrNO_2$ mw: 316.17
PROP: Red-brown needles from Py. Mp: 195–196°.
SYNS: 9,10-ANTHRACENEDIONE, 1-BROMO-4-(METHYLAMINO)- □ ANTHRAQUINONE, 1-BROMO-4-(METHYLAMINO)- □ 1-METHYLAMINO-4-BROMANTHRACHINON □ 1-(METHYLAMINO)-4-BROMOANTHRAQUINONE
CONSENSUS REPORTS: Reported in EPA TSCA Inventory.
SAFETY PROFILE: An eye irritant. When heated to decomposition it emits toxic fumes of NO_x and Br⁻.

BNP250   CAS: 107-82-4   HR: 3
1-BROMO-3-METHYL BUTANE
DOT: UN 2341
mf: $C_5H_{11}Br$ mw: 151.05
PROP: Colorless liquid. D: 1.210, mp: −112°, bp: 120–121°, flash p: 21°. Sltly sol in water; misc with alc and ether.
SYNS: ISOAMYL BROMIDE □ ISOPENTYL BROMIDE □ 3-METHYLBUTYL BROMIDE
CONSENSUS REPORTS: EPA Genetic Toxicology Program. EPA TSCA Inventory.
DOT CLASSIFICATION: 3; Label: Flammable Liquid
SAFETY PROFILE: Moderately toxic by intraperitoneal route. Flammable liquid. Dangerous fire hazard when exposed to heat or flame. When heated to decomposition it emits toxic fumes of Br⁻. See also BROMIDES.

BNP750   CAS: 496-67-3   HR: 2
2-BROMO-3-METHYLBUTYRYLUREA
mf: $C_6H_{11}BrN_2O_2$ mw: 223.10
SYNS: ABROVAL □ ALLUVAL □ ALURAL □ N-(AMINOCARBONYL)-2-BROMO-3-METHYLBUTANAM-IDE □ BROMARAL □ BROMCARBAMIDE □ BROMISO-

VAL □ BROMISOVALERYLUREA □ α-BROMISO-
VALERYLUREA □ BROMISOVALUM □ BROMIZOVAL
□ BROMOCARBAMIDE □ α-BROMO-β-DIMETHYL-
PROPANOYLUREA □ α-BROMOISOVALERIC ACID
UREIDE □ α-BROMOISOVALEROYLUREA □ (α-
BROMOISOVALERYL)UREA □ BROMOVAL □
BROMOVALEROCARBAMIDE □ BROMOVALERYL-
UREA □ BROMOXIL □ BROMURAL □ BROMUVAN □
BROMVALERYLUREA □ BROMVALETONE □
BROMVALETONUM □ BROMVALUREA □ BROMYL □
BROVALIN □ BROVALUREA □ BROVARIN □ BVU □
CALMOTIN □ DIAGRABROMYL □ DIBROLUUR □
DORMIGENE □ ISOBROMYL □ ISOVAL □ MONO-
BROMOISOVALERYLUREA □ 2-MONOBROMOISO-
VALERYLUREA □ PIVADORM □ PIVADORN □
SOMNUROL □ UPIOL □ UVALERAL

CONSENSUS REPORTS: Reported in EPA
TSCA Inventory.
SAFETY PROFILE: Moderately toxic by
ingestion. Human systemic effects by
ingestion: nausea or vomiting, and coma. A
sedative and hypnotic agent. When heated
to decomposition it emits very toxic fumes
of Br⁻ and NOₓ.

BNR750   CAS: 78-77-3   HR: 2
1-BROMO-2-METHYLPROPANE
mf: C₄H₉Br   mw: 137.04
PROP: Liquid. Flash p: 22°C, d: 1.253 @
20°/4°, fp: −117.4°, bp: 90.5–91°.
SYNS: 1-BUTYL BROMIDE □ iso-BUTYL BROMIDE □
ISOBUTYL BROMIDE
CONSENSUS REPORTS: EPA Genetic
Toxicology Program. Reported in EPA
TSCA Inventory.
DOT CLASSIFICATION: 3; Label:
Flammable Liquid
SAFETY PROFILE: Questionable
carcinogen with experimental neoplastigenic
data. Moderately toxic by intraperitoneal
route. A dangerous fire hazard when
exposed to heat or flame. When heated to
decomposition it emits toxic fumes of Br⁻.
See also BROMIDES.

BNT250   CAS: 52-51-7   HR: 3
2-BROMO-2-NITRO-1,3-PROPANEDIOL
mf: C₃H₆BrNO₄   mw: 200.01
PROP: Crystals. Mp: 130–133°. Very sol in
H₂O.

SYNS: 2-BROMO-2-NITROPANE-1,3-DIOL □ 2-
BROMO-2-NITROPROPAN-1,3-DIOL □ β-BROMO-β-
NITROTRIMETHYLENEGLYCOL □ BRONOCOT □
BRONOPOL □ BRONOSOL
CONSENSUS REPORTS: Reported in
EPA TSCA Inventory.
SAFETY PROFILE: Poison by ingestion,
subcutaneous, intravenous, and
intraperitoneal routes. Moderately toxic by
skin contact. An eye and human skin irritant.
An antiseptic. When heated to
decomposition it emits very toxic fumes of
NOₓ and Br⁻.

BNU500   CAS: 107-81-3   HR: 3
2-BROMOPENTANE
DOT: UN 2343
mf: C₅H₁₁Br   mw: 151.07
PROP: Colorless to yellow liquid; strong
odor. Bp: 120°, fp: <−30°, d: 1.211 @
25°/25°, flash p: 90°F.
CONSENSUS REPORTS: Reported in EPA
TSCA Inventory.
DOT CLASSIFICATION: 3; Label:
Flammable Liquid
SAFETY PROFILE: Poison by
intraperitoneal route. Mildly toxic by
inhalation. A local irritant and narcotic in
high concentration. Ingestion can cause liver
damage. A dangerous fire hazard when
exposed to heat or flame. When heated to
decomposition it emits toxic fumes of Br⁻.
See also BROMIDES and
CHLORINATED HYDROCARBONS,
ALIPHATIC.

BNV750   CAS: 16532-79-9   HR: 3
4-BROMOPHENYLACETONITRILE
mf: C₈H₆BrN   mw: 196.06
SYNS: 4-BROMOBENZENEACETONITRILE □ p-
BROMOBENZYL CYANIDE □ 4-BROMOBENZYL-
CYANIDE □ p-BROMOPHENYLACETONITRILE □ 2-(4-
BROMOPHENYL)ACETONITRILE
CONSENSUS REPORTS: Reported in EPA
TSCA Inventory. Cyanide and its
compounds are on the Community Right-
To-Know List.
SAFETY PROFILE: Poison by intravenous
route. See also BROMIDES and

NITRILES. When heated to decomposition it emits very toxic fumes of Br⁻, NO$_x$, and CN⁻.

BNV752   CAS: 301644-27-9 HR: 3
4-((3-((4-BROMOPHENYL)AMINO)-4,5-DIHYDRO-2H-BENZ(gNDAZOL-2-YL)ACETYL)MORPHOLINE

mf: $C_{23}H_{23}BrN_4O_2$ mw: 467.37

SAFETY PROFILE: A poison by ingestion. When heated to decomposition it emits toxic vapors of NO$_x$ and Br⁻.

BNV800   CAS: 107359-69-3 HR: 3
2-(4-BROMOPHENYL)-4-CHLORO-5-((4-CHLOROPHENYL)METHOXY)-3(2H)-PYRIDAZINONE

mf: $C_{17}H_{11}BrCl_2N_2O_2$ mw: 426.11

SYN: 3(2H)-PYRIDAZINONE, 2-(4-BROMOPHENYL)-4-CHLORO-5-((4-CHLOROPHENYL)METHOXY)-

SAFETY PROFILE: A poison by ingestion. When heated to decomposition it emits toxic vapors of NO$_x$, Br⁻, and Cl⁻.

BNX035   CAS: 107359-76-2 HR: 3
5-((4-BROMOPHENYL)METHOXY)-4-CHLORO-2-(4-CHLORO-2-FLUORO-PHENYL)-3(2H)-PYRIDAZINONE

mf: $C_{17}H_{10}BrCl_2FN_2O_2$ mw: 444.10

SYN: 3(2H)-PYRIDAZINONE, 5-((4-BROMOPHENYL)-METHOXY)-4-CHLORO-2-(4-CHLORO-2-FLUOROPHEN-YL)-

SAFETY PROFILE: A poison by ingestion. When heated to decomposition it emits toxic vapors of NO$_x$, F⁻, Br⁻, and Cl⁻.

BNX330   CAS: 315706-76-4 HR: 3
3-(4-BROMOPHENYL)-N-(4-PROPYL-CYCLOHEXYL)-2-PROPENAMIDE

mf: $C_{18}H_{24}BrNO$ mw: 350.30

SAFETY PROFILE: A poison by ingestion. When heated to decomposition it emits toxic vapors of NO$_x$ and Br⁻.

BNX750   CAS: 106-94-5   HR: 3
1-BROMOPROPANE

mf: C_3H_7Br mw: 123.01

PROP: Liquid. Mp: −110°, bp: 71°, d: 1.35 @ 20°/4°, autoign temp: 914°F, flash p: <22°, lel: 4.6%.

SYNS: 1-BROMOPROPANE (DOT) □ PROPYL BROMIDE

CONSENSUS REPORTS: Reported in EPA TSCA Inventory.

SAFETY PROFILE: Moderately toxic by ingestion and intraperitoneal routes. Mildly toxic by inhalation. Experimental reproductive effects. Mutation data reported. Dangerous fire hazard when heated or exposed to flame or oxidizers. To fight fire, use water, foam, CO$_2$, dry chemical. When heated to decomposition it emits toxic fumes of Br⁻. See also BROMIDES.

BNY000   CAS: 75-26-3   HR: 3
2-BROMOPROPANE

DOT: UN 2344

mf: C_3H_7Br mw: 122.98

PROP: Liquid. Flash p: <14°, d: 1.31 @ 20°/4°, fp: −89°, bp: 59.35°.

SYN: ISOPROPYL BROMIDE

CONSENSUS REPORTS: Reported in EPA TSCA Inventory.

DOT CLASSIFICATION: 3; Label: Flammable Liquid

SAFETY PROFILE: Moderately toxic by intraperitoneal route. A very flammable liquid and dangerous fire hazard. When heated to decomposition it emits toxic fumes of Br⁻. See also BROMIDES.

BNZ000   CAS: 598-31-2   HR: 3
BROMO-2-PROPANONE

DOT: UN 1569

mf: C_3H_5BrO mw: 136.99

PROP: Liquid which turns violet rapidly. D: 1.634°, fp: −36.5°, bp: 136.5° @ 725 mm.

SYNS: ACETONYL BROMIDE □ ACETYL METHYL BROMIDE □ BROMOACETONE □ BROMOACETONE (DOT) □ BROMOACETONE, liquid (DOT) □ BROMO-METHYL METHYL KETONE □ 1-BROMO-2-PROPAN-ONE □ MONOBROMOACETONE □ RCRA WASTE NUMBER P017

DOT CLASSIFICATION: 6.1; Label: Poison

SAFETY PROFILE: A poisonous gas. Moderately toxic to humans by inhalation. When heated to decomposition it emits toxic fumes of Br⁻. See also BROMIDES.

BOB250   CAS: 590-92-1   HR: 2
3-BROMOPROPIONIC ACID
mf: $C_3H_5BrO_2$ mw: 152.99
PROP: Plates. D: 1.485, mp: 62.5°, bp: 140–142° @ 45 mm.
SYN: β-BROMOPROPIONIC ACID
CONSENSUS REPORTS: Reported in EPA TSCA Inventory.
SAFETY PROFILE: Moderately toxic by intraperitoneal route. Questionable carcinogen with experimental tumorigenic data. Mutation data reported. When heated to decomposition it emits toxic fumes of Br^-. See also BROMIDES.

BOC600   CAS: 122322-20-7 HR: 3
5-((6-BROMO-3-PYRIDINYL)METHOXY)-
**   4-CHLORO-2-(4-CHLOROPHENYL)-**
**   3(2H)-PYRIDAZINONE**
mf: $C_{16}H_{10}BrCl_2N_3O_2$ mw: 427.10
SYN: 3(2H)-PYRIDAZINONE, 5-((6-BROMO-3-PYRIDINYL)METHOXY)-4-CHLORO-2-(4-CHLORO-PHENYL)-
SAFETY PROFILE: A poison by ingestion. When heated to decomposition it emits toxic vapors of NO_x, Br^-, and Cl^-.

BOE750   CAS: 13465-73-1   HR: 3
BROMOSILANE
mf: BrH_3Si mw: 111.02
SYN: SILYL BROMIDE
DOT CLASSIFICATION: Forbidden
SAFETY PROFILE: Ignites spontaneously upon exposure to air. When heated to decomposition it emits toxic fumes of Br^-.

BOF500   CAS: 128-08-5   HR: 3
N-BROMOSUCCINIMIDE
mf: $C_4H_4BrNO_2$ mw: 178.00
PROP: White to pale-buff, fine, orthorhombic, crystalline powder with faint odor of bromine. Mp: 173–175°, d: 2.098. Sol in Me_2CO; sltly sol in AcOH, H_2O, and CCl_4; prac insol in hexane.
SYNS: 1-BROMO-2,5-PYRROLIDINEDIONE □ N-BROMOSUCCIMIDE □ SUCCINBROMIMIDE □ SUCCINIBROMIMIDE
CONSENSUS REPORTS: Reported in EPA TSCA Inventory.

SAFETY PROFILE: Poison by intraperitoneal route. An irritating poison to skin, eyes, and mucous membranes. Reacts explosively with aniline, diallyl sulfide, and hydrazine hydrate. Explosive reaction with propiononitrile after heating to 105°C for 24 hours. Violent reaction with dibenzoyl peroxide + 4-toluic acid. When heated to decomposition it emits toxic fumes of Br^- and NO_x. See also BROMIDES and NITROGEN MONOXIDE.

BOG255   CAS: 106-38-7   HR: 2
p-BROMOTOLUENE
mf: C_7H_7Br mw: 171.05
SYNS: PARABROMOTOLUENE □ TOLUENE, p-BROMO-
CONSENSUS REPORTS: Reported in EPA TSCA Inventory.
SAFETY PROFILE: Moderately toxic by inhalation and intraperitoneal routes. When heated to decomposition it emits toxic vapors of Br^-.

BOH750   CAS: 75-62-7   HR: 3
BROMOTRICHLOROMETHANE
mf: $CBrCl_3$ mw: 198.27
PROP: Colorless liquid. Fp: −5.8°, bp: 104.2°, d: 2.01 @ 20°/4°.
CONSENSUS REPORTS: Reported in EPA TSCA Inventory.
SAFETY PROFILE: Poison by ingestion and intraperitoneal routes. Narcotic in high concentration. Mutation data reported. See also CHLOROFORM. Incompatible with ethylene. When heated to decomposition it emits very toxic fumes of Cl^- and Br^-.

BOI750   CAS: 2767-54-6   HR: 3
BROMOTRIETHYLSTANNANE
mf: $C_6H_{15}BrSn$ mw: 285.81
PROP: Colorless liquid. D: 1.630, mp: −13.5°, bp: 221.° Sol in org solvs.
SYNS: TRIETHYLSTANNIUM BROMIDE □ TRIETHYL TIN BROMIDE
CONSENSUS REPORTS: Reported in EPA TSCA Inventory.
OSHA PEL: TWA 0.1 mg(Sn)/m^3 (skin)

B

ACGIH TLV: TWA 0.1 mg(Sn)/m³; STEL 0.2 mg(Sn)/m³ (skin).
NIOSH REL: (Organotin Compounds) TWA 0.1 mg(Sn)/m³
SAFETY PROFILE: Moderately toxic by inhalation. Experimental reproductive effects. See also TIN COMPOUNDS and BROMIDES. When heated to decomposition it emits toxic fumes of Br⁻.

BOJ000   CAS: 598-73-2   HR: 3
BROMO TRIFLUOROETHYLENE
mf: BrF_3C_2 mw: 160.94
PROP: Gas. Bp: −2.5°.
SYNS: BROMOTRIFLUOROETHENE □ TRIFLUORO-BROMOETHYLENE □ TRIFLUOROVINYLBROMIDE
CONSENSUS REPORTS: Reported in EPA TSCA Inventory.
SAFETY PROFILE: A poison. A flammable gas. Ignites spontaneously in air. Incompatible with powerful oxidizers, O_2. When heated to decomposition it emits highly toxic fumes of Br⁻, F⁻, and $COCF_2$.

BOL000   CAS: 51-20-7   HR: 2
5-BROMOURACIL
mf: $C_4H_3BrN_2O_2$ mw: 191.00
PROP: Prisms from H_2O. Mp: 293°.
CONSENSUS REPORTS: Reported in EPA TSCA Inventory. EPA Genetic Toxicology Program.
SAFETY PROFILE: Moderately toxic by intraperitoneal route. Experimental reproductive effects. Mutation data reported. When heated to decomposition it emits very toxic fumes of Br⁻ and NO_x.

BOL750   CAS: 357-57-3   HR: 3
BRUCINE
DOT: UN 1570
mf: $C_{23}H_{26}N_2O_4$ mw: 394.51
PROP: Crystals, powder, or monoclinic prisms. Mp: 105° (hydrate), mp: 78° (anhydrate). Sol in EtOH and $CHCl_3$; sltly sol in C_6H_6 and Et_2O. An alkaloid extracted from *Strychnos* seeds (WQCHM* 4,-,74).
SYNS: BRUCINA (ITALIAN) □ (−)-BRUCINE □ BRUCINE (DOT) □ 2,3-DIMETHOXYSTRYCHNIDIN-10-ONE □ DIMETHOXY STRYCHNINE (DOT) □ 2,3-

DIMETHOXYSTRYCHNINE □ 10,11-DIMETHY-STRYCHNINE □ STRYCHNIDIN-10-ONE, 2,3-DIMETHOXY-(9CI) □ STRYCHNINE, 2,3-DIMETHOXY-□ RCRA WASTE NUMBER P018
CONSENSUS REPORTS: Reported in EPA TSCA Inventory.
DOT CLASSIFICATION: 6.1; Label: Poison
SAFETY PROFILE: A poison by subcutaneous, intravenous, and intraperitoneal routes. An alkaloid-like strychnine, but one-sixth as toxic. When heated to decomposition it emits toxic fumes of NO_x. See also STRYCHNINE.

BOM250   CAS: 129-74-8   HR: 3
BUCLIZINE DIHYDROCHLORIDE
mf: $C_{28}H_{33}ClN_2$•2ClH   mw: 506.00
PROP: Crystals. Mp: 265–266°.
SYNS: BUCLODIN □ 1-(p-tert-BUTYLBENZYL)-4-(p-CHLORODIPHENYLMETHYL)PIPERAZINE DIHYDRO-CHLORIDE □ 1-(p-tert-BUTYLBENZYL-4-p-CHLORO-α-PHENYLBENZYL)PIPERAZINE DIHYDROCHLORIDE □ 1-(p-CHLOROBENZHYDRYL)-4-(p-tert-BUTYLBENZYL)-DIETHYLENEDIAMINE DIHYDROCHLORIDE □ 1-p-CHLOROBENZHYDRYL-4-p-(tert)-BUTYLBENZYL-PIPERAZINE DIHYDROCHLORIDE □ HISTABUTYZINE DIHYDROCHLORIDE □ LONGIFENE □ SOFTRAN □ UCB 4445 □ VIBAZINE
CONSENSUS REPORTS: Reported in EPA TSCA Inventory.
SAFETY PROFILE: Poison by intraperitoneal route. An experimental teratogen. When heated to decomposition it emits very toxic fumes of NO_x and HCl.

BOO632   CAS: 5486-03-3   HR: D
BUQUINOLATE
mf: $C_{20}H_{27}NO_5$ mw: 361.42
PROP: Crystals. Mp: 288–291°.
SYNS: BONAID □ ETHYL-6,7-DIISOBUTOXY-4-HYDROXYQUINOLINE-3-CARBOXYLATE □ 4-HYDROXY-6,7-BIS(2-METHYLPROPOXY)-3-QUINOLINECARBOXYLIC ACID ETHER ESTER □ 4-HYDROXY-6,7-DIISOBUTOXY-3-QUINOLINECARBOXYLIC ACID ETHYL ESTER
SAFETY PROFILE: When heated to decomposition it emits acrid smoke and irritating fumes.

BOP100   CAS: 25339-57-5   HR: 3
BUTADIENE
DOT: UN 1010

mf: C$_4$H$_6$ mw: 54.10
SYNS: BUTADIENES, inhibited (DOT) □ PLIOLITE
DOT CLASSIFICATION: 2.1; Label:
Flammable Gas
SAFETY PROFILE: A flammable gas. When
heated to decomposition it emits acrid
smoke and irritating vapors.

BOP500   CAS: 106-99-0   HR: 3
1,3-BUTADIENE
mf: C$_4$H$_6$ mw: 54.10
PROP: Colorless gas; mild aromatic odor.
Very reactive. Bp: −2.6°, mp: −113°, fp:
−108.9°, flash p: −105°F, lel: 2.0%, uel:
11.5%, d: 0.621 @ 20°/4°, autoign temp:
788°F, vap d: 1.87, vap press: 1840 mm @
21°. IDLH 2000 ppm [10%LEL].
SYNS: BIETHYLENE □ BIVINYL □ BUTADIEEN
(DUTCH) □ BUTA-1,3-DIEEN (DUTCH) □ BUTADIEN
(POLISH) □ BUTA-1,3-DIEN (GERMAN) □ BUTA-1,3-
DIENE □ α-γ-BUTADIENE □ DIVINYL □ ERYTHRENE
□ NCI-C50602 □ PYRROLYLENE □ VINYLETHYLENE
CONSENSUS REPORTS: NTP 10th Report
on Carcinogens. IARC Cancer Review:
Group 2A IMEMDT 54,237,92; Animal
Sufficient Evidence IMEMDT 39,155,86;
IARC Cancer Review: Animal Sufficient
Evidence IMEMDT 54,237,92; Human
Limited Evidence IMEMDT 54,237,92;
Human Inadequate Evidence IMEMDT
39,155,86; NTP Carcinogenesis Studies
(inhalation); Clear Evidence: mouse
NTPTR* NTP-TR-288,84. Reported in
EPA TSCA Inventory. Community Right-
To-Know List.
OSHA PEL: TWA 1 ppm; STEL 5 ppm
ACGIH TLV: TWA 2 ppm; Suspected
Human Carcinogen
DFG MAK: Confirmed Human Carcinogen
NIOSH REL: Reduce to lowest feasible level
SAFETY PROFILE: Confirmed carcinogen
with experimental carcinogenic and
neoplastigenic data. An experimental
teratogen. Mutation data reported.
Inhalation of high concentrations can cause
unconsciousness and death. Human
systemic effects by inhalation: cough,
hallucinations, distorted perceptions,
changes in the visual field and other
unspecified eye effects. The vapors are
irritating to eyes and mucous membranes. If
spilled on skin or clothing, it can cause
burns or frostbite (due to rapid
vaporization). Chronic systemic poisoning
in humans has not been reported.
 Dangerous fire hazard when exposed to
heat, flame, or powerful oxidizers. Upon
exposure to air it forms explosive peroxides
sensitive to heat, shock, or heating above
27°C. May decompose explosively when
heated above 200°C/1.0 kbar. Explodes on
contact with aluminum tetrahydroborate.
Potentially explosive reaction with NO$_x$ +
O$_2$, ethanol + iodine + mercury oxide (at
35°C), ClO$_2$, crotonaldehyde (above 180°C),
buten-3-yne (with heat and pressure).
Reaction with sodium nitrite forms a
spontaneously flammable product.
Exothermic reaction with boron trifluoride
etherate + phenol. To fight fire, stop flow of
gas. When heated to decomposition it emits
acrid smoke and fumes.

BOP750   CAS: 30031-64-2   HR: 3
l-BUTADIENE DIEPOXIDE
mf: C$_4$H$_6$O$_2$ mw: 86.10
PROP: Solid or liquid. Mp: 24–25.6°, bp:
144.5–145°.
SYNS: (S-(R*,R*))-2,2'-BIOXIRANE □ l-DIEPOXY-
BUTANE □ (2S,3S)-DIEPOXYBUTANE □ 1-1,2:3,4-
DIEPOXYBUTANE □ (2S,3S)-1,2:3,4-DIEPOXYBUTANE □
NSC-32606
CONSENSUS REPORTS: IARC Cancer
Review: Group 2B IMEMDT 7,56,87;
Animal Sufficient Evidence IMEMDT
11,115,76. EPA Genetic Toxicology
Program.
SAFETY PROFILE: Suspected carcinogen
with experimental neoplastigenic data.
Poison by intraperitoneal route. Mutation
data reported. When heated to
decomposition it emits acrid and irritating
fumes.

BOR500   CAS: 106-97-8   HR: 3
BUTANE
DOT: UN 1011
mf: C$_4$H$_{10}$ mw: 58.14

PROP: Colorless gas; faint disagreeable odor. Bp: $-0.5°$, fp: $-135°$, lel: 1.9%, uel: 8.5%, flash p: $-76°F$ (CC), d: 0.599, autoign temp: 761°F, vap press: 2 atm @ 18.8°, vap d: 2.046. Sltly sol in H_2O; mod sol in Et_2O and $CHCl_3$.

SYNS: n-BUTANE (DOT) □ BUTANE MIXTURES (DOT) □ BUTANEN (DUTCH) □ BUTANI (ITALIAN) □ DIETHYL □ METHYLETHYLMETHANE

CONSENSUS REPORTS: Reported in EPA TSCA Inventory.

OSHA PEL: TWA 800 ppm

ACGIH TLV: TWA 800 ppm

DFG MAK: 1000 ppm (2400 mg/m³)

DOT CLASSIFICATION: 2.1; Label: Flammable Gas

SAFETY PROFILE: Mildly toxic by inhalation. Causes drowsiness. An asphyxiant. Very dangerous fire hazard when exposed to heat, flame, or oxidizers. Highly explosive when exposed to flame, or when mixed with [$Ni(CO)_4 + O_2$]. To fight fire, stop flow of gas. When heated to decomposition it emits acrid smoke and fumes.

BOS000   CAS: 110-60-1   HR: 3
1,4-BUTANEDIAMINE

mf: $C_4H_{12}N_2$ mw: 88.18

PROP: Crystals with strong odor. Mp: 27–28°, bp: 158–159°.

SYNS: BUTYLENEDIAMINE □ 1,4-BUTYLENE-DIAMINE □ 1,4-DIAMINOBUTANE □ PUTRESCIN □ PUTRESCINE □ TETRAMETHYLENEDIAMINE □ 1,4-TETRAMETHYLENEDIAMINE

CONSENSUS REPORTS: Reported in EPA TSCA Inventory.

SAFETY PROFILE: Poison by subcutaneous, intravenous, and rectal routes. Moderately toxic by ingestion. An experimental teratogen. Human mutation data reported. When heated to decomposition it emits toxic fumes of NO_x. See also 1,3-BUTANEDIAMINE and AMINES.

BOS250   CAS: 584-03-2   HR: 2
1,2-BUTANEDIOL

mf: $C_4H_{10}O_2$ mw: 90.14

PROP: D: 1.0, vap d: 3.1, bp: 194°, flash p: 194°F.

SYN: 1,2-BUTYLENE GLYCOL

CONSENSUS REPORTS: Reported in EPA TSCA Inventory.

SAFETY PROFILE: Moderately toxic by ingestion. Combustible when exposed to heat or flame. To fight fire, use alcohol foam. When heated to decomposition it emits acrid and irritating fumes.

BOS500   CAS: 107-88-0   HR: 1
1,3-BUTANEDIOL

mf: $C_4H_{10}O_2$ mw: 90.14

PROP: Viscous liquid. Bp: 207.5°, fp: $<-50°$, flash p: 250°F, d: 1.006 @ 20°/20°, autoign temp: 741°F, vap press: 0.06 mm @ 20°, vap d: 3.2.

SYNS: 1,3-BUTANDIOL (GERMAN) □ BUTANE-1,3-DIOL □ β-BUTYLENE GLYCOL □ 1,3-BUTYLENE GLYCOL (FCC) □ 1,3-DIHYDROXYBUTANE □ METHYLTRIMETHYLENE GLYCOL

CONSENSUS REPORTS: Reported in EPA TSCA Inventory.

SAFETY PROFILE: Mildly toxic by ingestion and subcutaneous routes. A skin and eye irritant. See also ETHYLENE GLYCOL. Experimental reproductive effects. Combustible when exposed to heat or flame. Incompatible with oxidizing materials. To fight fire, use foam, alcohol foam, CO_2, dry chemical. When heated to decomposition it emits acrid smoke and irritating fumes.

BOS750   CAS: 110-63-4   HR: 3
1,4-BUTANEDIOL

mf: $C_4H_{10}O_2$ mw: 90.14

PROP: Nearly odorless, colorless, viscous liquid or crystals; to needles on chilling. Mp: 16°, bp: 230°, flash p: 250°F (OC), d: 1.02 @ 20°, vap d: 3.1.

SYNS: BUTANE-1,4-DIOL □ 1,4-BUTYLENE GLYCOL □ 1,4-DIHYDROXYBUTANE □ 1,4-TETRAMETHYLENE GLYCOL

CONSENSUS REPORTS: Reported in EPA TSCA Inventory.

SAFETY PROFILE: A human poison by an unspecified route. Moderately toxic by ingestion and intraperitoneal routes. Human systemic effects: altered sleep time. Combustible when exposed to heat or flame. To fight fire, use alcohol foam, mist, foam, CO_2, dry chemical. Incompatible with oxidizing materials. When heated to decomposition it emits acrid smoke and fumes.

BOT000   CAS: 513-85-9   HR: 1
2,3-BUTANEDIOL
mf: $C_4H_{10}O_2$ mw: 90.14
PROP: Colorless liquid or solid. Bp: 180°, fp: 19°, flash p: 185°F (TOC), d: 1.0095 @ 20°/20°, autoign temp: 756°F, vap press: 0.17 mm @ 20°, vap d: 3.1.
SYNS: 2,3-BUTYLENE GLYCOL □ 2,3-DIHYDROXYBUTANE □ DIMETHYLENE GLYCOL
CONSENSUS REPORTS: Reported in EPA TSCA Inventory.
SAFETY PROFILE: Mildly toxic by ingestion. See also ETHYLENE GLYCOL. Flammable when exposed to heat or flame. Incompatible with oxidizing materials. To fight fire, use alcohol foam, CO_2, dry chemical. When heated to decomposition it emits acrid smoke and fumes.

BOT250   CAS: 55-98-1   HR: 3
1,4-BUTANEDIOL DIMETHYL
**   SULFONATE**
mf: $C_6H_{14}O_6S_2$ mw: 246.32
PROP: White crystals or needles. Mp: 116°.
SYNS: 1,4-BIS(METHANESULFONOXY)BUTANE □ (1,4-BIS(METHANESULFONYLOXY)BUTANE) □ BISULFAN □ BISULPHANE □ 1,4-BUTANEDIOL DIMETHANESULPHONATE □ BUZULFAN □ C.B. 2041 □ CITOSULFAN □ 1,4-DIMESYLOXYBUTANE □ 1,4-DIMETHANESULFONOXYBUTANE □ 1,4-DI(METHANESULFONYLOXY)BUTANE □ 1,4-DIMETHANESULPHONYLOXYBUTANE □ 1,4-DIMETHYLSULFONOXYBUTANE □ GT41 □ GT 2041 □ LEUCOSULFAN □ MABLIN □ METHANESULFONIC ACID TETRAMETHYLENE ESTER □ MIELUCIN □ MISULBAN □ MITOSTAN □ MYELOLEUKON □ MYLERAN □ NCI-C01592 □ NSC-750 □ SULPHABUTIN □ TETRAMETHYLENE BIS(METHANESULFONATE) □ TETRAMETHYLENE DIMETHANE SULFONATE □ X 149
CONSENSUS REPORTS: NTP 10th Report on Carcinogens. IARC Cancer Review:

Group 1 IMEMDT 7,137,87; Animal Inadequate Evidence IMEMDT 4,247,74; Human Inadequate Evidence IMEMDT 4,247,74. EPA Genetic Toxicology Program.
SAFETY PROFILE: Confirmed carcinogen producing leukemia, kidney, and uterine tumors. Experimental neoplastigenic and tumorigenic data. Poison by ingestion, subcutaneous, intraperitoneal, intravenous, and possibly other routes. Ingestion by pregnant women can cause cancer of the reproductive system of the fetus including the uterus. Human teratogenic effects by ingestion and possibly other routes include developmental abnormalities of the eye, ear, craniofacial area including the nose and tongue, gastrointestinal system, endocrine system, urogenital system, and other unspecified areas. Other human reproductive effects by ingestion and possibly other routes include: impotence, changes in the uterus, cervix, and vagina, and menstrual-cycle disorders. Experimental reproductive effects. Human systemic effects by ingestion: general arteriolar or venous dilation of the eye, changes in structure or function of salivary glands. When heated to decomposition it emits toxic fumes of SO_x. See also SULFONATES.

BOT500   CAS: 431-03-8   HR: 3
2,3-BUTANEDIONE
DOT: UN 2346
mf: $C_4H_6O_2$ mw: 86.10
PROP: Greenish-yellow liquid; strong odor. Bp: 88°, flash p: 80°F, d: 0.9904 @ 15°/15°, refr index: 1.393–1.397, vap d: 3.00. Misc in alc, fixed oils, propylene glycol; sol in glycerin, alc, water.
SYNS: BIACETYL □ BUTADIONE □ BUTANEDIONE (DOT) □ DIACETYL (FCC) □ 2,3-DIKETOBUTANE □ DIMETHYL DIKETONE □ DIMETHYLGLYOXAL □ GLYOXAL, DIMETHYL- □ FEMA No. 2370
CONSENSUS REPORTS: Reported in EPA TSCA Inventory.

DOT CLASSIFICATION: 3; Label:
Flammable Liquid
SAFETY PROFILE: A poison by ingestion
and intraperitoneal routes. A skin irritant.
Human inhalation hazard in popcorn
manufacture. Human mutation data
reported. Flammable liquid. Dangerous fire
hazard when exposed to heat or flame. To
fight fire, use alcohol foam, CO_2, dry
chemical. When heated to decomposition it
emits acrid smoke and fumes. See also
KETONES.

BOU250   CAS: 1633-83-6   HR: 3
BUTANE SULTONE
mf: $C_4H_8O_3S$ mw: 136.18
PROP: Liquid. D: 1.33 @ 20°/4°, mp:
12.5–14.5°, bp: 134–136° @ 4 mm.
SYNS: BUTANESULFONE □ Δ-BUTANE SULTONE □
1,4-BUTANESULTONE (MAK) □ 1,4-BUTYLENE
SULFONE □ Δ-VALEROSULTONE
CONSENSUS REPORTS: EPA Genetic
Toxicology Program. Reported in EPA
TSCA Inventory.
DFG MAK: Animal Carcinogen, Suspected
Human Carcinogen
SAFETY PROFILE: Suspected carcinogen
with experimental tumorigenic data. Poison
by subcutaneous, intravenous, and
intraperitoneal routes. Moderately toxic by
ingestion. Experimental reproductive
effects. Human mutation data reported. See
also SULFONATES. When heated to
decomposition it emits toxic fumes of SO_x.

BOV000   CAS: 96-48-0   HR: 2
4-BUTANOLIDE
mf: $C_4H_6O_2$ mw: 86.10
PROP: Colorless liquid; mild caramel odor.
Mp: −44°, bp: 203–204°, flash p: 209°F
(OC), d: 1.441 @ 0°, refr index: 1.434–1.454
@ 25°, vap d: 3.0. Misc in H_2O.
SYNS: γ-6480 □ γ-BL □ BLO □ BLON □ BUTYRIC
ACID LACTONE □ α-BUTYROLACTONE □ γ-
BUTYROLACTONE (FCC) □ BUTYRYL LACTONE □ 4-
DEOXYTETRONIC ACID □ DIHYDRO-2(3H)-FURAN-
ONE □ FEMA No. 3291 □ 4-HYDROXYBUTANOIC ACID
LACTONE □ γ-HYDROXYBUTYRIC ACID CYCLIC
ESTER □ 4-HYDROXYBUTYRIC ACID γ-LACTONE □ γ-

HYDROXYBUTYROLACTONE □ NCI-C55878 □
TETRAHYDRO-2-FURANONE
CONSENSUS REPORTS: IARC Cancer
Review: Group 3 IMEMDT 7,56,87; Animal
No Evidence IMEMDT 11,231,76. EPA
Genetic Toxicology Program. Reported in
EPA TSCA Inventory.
SAFETY PROFILE: Moderately toxic by
ingestion, intravenous, and intraperitoneal
routes. An experimental teratogen. Other
experimental reproductive effects.
Questionable carcinogen with experimental
tumorigenic data by skin contact. Mutation
data reported. Less acutely toxic than β-
propiolactone. Combustible when exposed
to heat or flame; can react with oxidizing
materials. To fight fire, use foam, alcohol
foam, CO_2, dry chemical. Potentially
explosive reaction with butanol + 2,4-
dichlorophenol + sodium hydroxide. When
heated to decomposition it emits acrid and
irritating fumes.

BOW250   CAS: 25167-67-3   HR: 3
1-BUTENE
mf: C_4H_8 mw: 56.11
PROP: A colorless, flammable gas; sltly
aromatic odor. Bp: −6.3°, fp: −185.3°, lel:
1.6%, uel: 9.3%, flash p: −80° (−112°F), d:
0.668 @ 0°/1°, vap d: 1.93, vap press: 3480
mm @ 21°, autoign temp: 723°F.
SYNS: BUTYLENE □ α-BUTYLENE
CONSENSUS REPORTS: Reported in EPA
TSCA Inventory.
ACGIH TLV: (Proposed: TWA 250 ppm.)
SAFETY PROFILE: A simple asphyxiant.
Very dangerous fire hazard when exposed to
heat, flame, or oxidizers. To fight fire, stop
flow of gas. Moderately explosive when
exposed to flame. Mixtures with aluminum
tetrahydroborate explode after an induction
period. When heated to decomposition it
emits acrid smoke and fumes.

BOX750   CAS: 106-88-7   HR: 3
1-BUTENE OXIDE
DOT: UN 3022
mf: C_4H_8O mw: 72.12

B

PROP: Colorless liquid. D: 0.8312 @ 20°/20°, bp: 63°, flash p: 5°F, lel: 1.5%, uel: 18.3%. Sol in water; misc with most org solvs.
SYNS: BUTYLENE OXIDE □ 1,2-BUTYLENE OXIDE □ 1,2-BUTYLENE OXIDE, stabilized (DOT) □ EPOXY-BUTANE □ 1,2-EPOXYBUTANE □ ETHYLENE OXIDE, ETHYL- □ ETHYL ETHYLENE OXIDE □ ETHYLOXIR-ANE □ NCI-C55527
CONSENSUS REPORTS: NTP Carcinogenesis Studies (inhalation); Clear Evidence: rat; No Evidence: mouse NTPTR* NTP-TR-329,88. Community Right-To-Know List. EPA Genetic Toxicology Program. Reported in EPA TSCA Inventory.
DFG MAK: Animal Carcinogen, Suspected Human Carcinogen
SAFETY PROFILE: Confirmed carcinogen with experimental carcinogenic data. Moderately toxic by ingestion and skin contact. Mildly toxic by inhalation. Experimental reproductive effects. Mutation data reported. Dangerous fire hazard when exposed to heat, flame, or powerful oxidizers. To fight fire, use dry chemical, water spray, mist or fog, alcohol foam. When heated to decomposition it emits acrid smoke and fumes.

BOY000   CAS: 6117-91-5   HR: 2
2-BUTEN-1-OL
mf: C_4H_8O mw: 72.12
PROP: Colorless liquid. Mp: <30°, bp: 118°, flash p: 92°F, d: 0.8726 @ 0°/4°, vap d: 2.49.
SYNS: 2-BUTENOL □ 2-BUTENYL ALCOHOL □ CROTONYL ALCOHOL □ CROTYL ALCOHOL
CONSENSUS REPORTS: Reported in EPA TSCA Inventory.
SAFETY PROFILE: Moderately toxic by ingestion and skin contact. Mutation data reported. Dangerous fire hazard when exposed to heat or flame; can react with oxidizing materials. To fight fire, use alcohol foam, CO_2, dry chemical. When heated to decomposition it emits acrid smoke and fumes. See also ALCOHOLS.

BOY500   CAS: 78-94-4   HR: 3
3-BUTEN-2-ONE
DOT: UN 1251
mf: C_4H_6O mw: 70.10
PROP: Colorless liquid; powerfully irritating odor. Bp: 81.4°, flash p: 20°F (CC), d: 0.8393 @ 25°/4°, vap d: 2.41.
SYNS: ACETYL ETHYLENE □ 3-BUTENE-2-ONE □ METHYLENE ACETONE □ METHYL-VINYL-CETONE (FRENCH) □ METHYLVINYLKETON (GERMAN) □ METHYL VINYL KETONE □ γ-OXO-α-BUTYLENE □ VINYL METHYL KETONE
CONSENSUS REPORTS: Reported in EPA TSCA Inventory. EPA Extremely Hazardous Substances List.
ACGIH TLV: STEL CL 0.2 ppm (skin, sensitizer)
DOT CLASSIFICATION: 3; Label: Flammable Liquid
SAFETY PROFILE: Poison by ingestion, inhalation, and intraperitoneal routes. A severe irritant to skin, eyes, and mucous membranes. A lachrymator. Mutation data reported. See also KETONES. Dangerous fire hazard when exposed to heat, flame, or oxidizers. To fight fire, use CO_2, dry chemical. When heated to decomposition it emits acrid smoke and fumes.

BPJ850   CAS: 111-76-2   HR: 3
2-BUTOXYETHANOL
DOT: UN 2369
mf: $C_6H_{14}O_2$ mw: 118.20
PROP: Clear, mobile liquid; pleasant odor. Fp: −74.8°, bp: 171–172°, flash p: 160°F (COC), d: 0.9012 @ 20°/20°, vap press: 300 mm @ 140°. IDLH 700 ppm.
SYNS: BUCS □ BUTOKSYETYLOWY ALKOHOL (POLISH) □ 2-BUTOSSI-ETANOLO (ITALIAN) □ 2-BUTOXY-AETHANOL (GERMAN) □ BUTOXYETHANOL □ n-BUTOXYETHANOL □ 2-BUTOXY-1-ETHANOL □ BUTYL CELLOSOLVE □ o-BUTYL ETHYLENE GLYCOL □ BUTYL GLYCOL □ BUTYLGLYCOL (FRENCH, GERMAN) □ BUTYL OXITOL □ DOWANOL EB □ EGBE □ EKTASOLVE EB □ ETHYLENE GLYCOL-n-BUTYL ETHER □ ETHYLENE GLYCOL MONOBUTYL ETHER (MAK, DOT) □ GAFCOL EB □ GLYCOL BUTYL ETHER □ GLYCOL ETHER EB □ GLYCOL ETHER EB ACETATE □ GLYCOL MONOBUTYL ETHER □ JEFFERSOL EB □ MONOBUTYL GLYCOL ETHER □ 3-OXA-1-HEPTANOL □ POLY-SOLV EB

CONSENSUS REPORTS: Reported in EPA TSCA Inventory. Glycol ethers are on the Community Right-To-Know List.

OSHA PEL: TWA 25 ppm (skin)

ACGIH TLV: 20 ppm (skin); Confirmed Animal Carcinogen.

DFG MAK: 20 ppm (98 mg/m^3)

DOT CLASSIFICATION: 6.1; Label: KEEP AWAY FROM FOOD

SAFETY PROFILE: Poison by ingestion, skin contact, intraperitoneal, and intravenous routes. Moderately toxic via inhalation and subcutaneous routes. Human systemic effects by inhalation: nausea or vomiting, headache, unspecified eye effects. Experimental teratogenic and reproductive effects. A skin irritant. Combustible liquid when exposed to heat or flame. To fight fire, use foam, CO_2, dry chemical. Incompatible with oxidizing materials, heat, and flame. When heated to decomposition it emits acrid smoke and irritating fumes.

BPK250   CAS: 78-51-3   HR: 3
2-BUTOXYETHANOL PHOSPHATE

mf: $C_{18}H_{39}O_7P$ mw: 398.54

PROP: Light-colored liquid; butyl-like odor. Mp: $-70°$, bp: 200–230° @ 4 mm, flash p: 435°F, d: 1.02 @ 20°/20°, vap press: 0.03 mm @ 150°, vap d: 13.8.

SYNS: KP 140 □ KRONITEX KP-140 □ PHOSFLEX T-BEP □ TBEP □ TRI(2-BUTOXYETHANOL PHOSPHATE) □ TRIBUTOXYETHYL PHOSPHATE □ TRI(2-BUTOXY-ETHYL) PHOSPHATE □ TRIBUTYL CELLOSOLVE PHOSPHATE □ TRIS(2-BUTOXYETHYL) ESTER PHOSPHORIC ACID □ TRIS(2-BUTOXYETHYL) PHOSPHATE

CONSENSUS REPORTS: Reported in EPA TSCA Inventory.

SAFETY PROFILE: A poison by intravenous route. Moderately toxic by ingestion. A skin and eye irritant. Combustible when exposed to heat or flame. Dangerous; see also PHOSPHATES; can react with oxidizing materials. To fight fire, use water, foam, CO_2, dry chemical. When heated to decomposition it emits toxic fumes of PO_x.

BPL500   CAS: 124-16-3   HR: 2
1-BUTOXY ETHOXY-2-PROPANOL

mf: $C_9H_{20}O_3$ mw: 176.29

PROP: D: 0.9310 @ 20°/20°, bp: 230.3°, fp: $-90°$, flash p: 250°F (OC). Sol in water.

SYN: 1-(2-BUTOXYETHOXY)-2-PROPANOL

CONSENSUS REPORTS: Reported in EPA TSCA Inventory.

SAFETY PROFILE: Moderately toxic by ingestion and skin contact. A skin and eye irritant. Combustible when exposed to heat or flame. To fight fire, use alcohol foam, dry chemical, spray, or mist. When heated to decomposition it emits acrid and irritating fumes.

BPM000   CAS: 112-07-2   HR: 3
2-BUTOXYETHYL ACETATE

mf: $C_8H_{16}O_3$ mw: 160.24

PROP: Colorless liquid; fruity odor. Bp: 192.3°, d: 0.9424 @ 20°/20°, fp: $-63.5°$, flash p: 190°F. Sol in hydrocarbons and org solvs; insol in water.

SYNS: 2-BUTOXYETHANOL ACETATE □ 2-BUTOXYETHYL ESTER ACETIC ACID □ BUTYL CELLOSOLVE ACETATE □ EKTASOLVE EB ACETATE □ ETHYLENE GLYCOL MONOBUTYL ETHER ACETATE (MAK) □ GLYCOL MONOBUTYL ETHER ACETATE

CONSENSUS REPORTS: Reported in EPA TSCA Inventory. Glycol ethers are on the Community Right-To-Know List.

ACGIH TLV: 20 ppm; Confirmed Animal Carcinogen

DFG MAK: 20 ppm (130 mg/m^3)

SAFETY PROFILE: Moderately toxic by ingestion and skin contact. Mild skin irritant. Flammable when exposed to heat, flame, or oxidizers. To fight fire, use alcohol foam. When heated to decomposition it emits acrid smoke and irritating fumes. See also ESTERS.

BPU750   CAS: 123-86-4   HR: 3
n-BUTYL ACETATE

DOT: UN 1123

mf: $C_6H_{12}O_2$ mw: 116.18

PROP: Colorless liquid; strong fruity odor. Fp: $-77°$, bp: 126°, ULC: 50–60, lel: 1.4%, uel: 7.5%, flash p: 72°F, d: 0.88 @ 20°/20°, refr index: 1.393–1.396, autoign temp: 797°F, vap press: 15 mm @ 25°. Misc with alc, ether, and propylene glycol. Sol in EtOH, Et₂CO, and Me₂CO; insol in H_2O. IDLH 1700 ppm [10%LEL].

SYNS: ACETATE de BUTYLE (FRENCH) □ ACETIC ACID n-BUTYL ESTER □ BUTILE (ACETATI di) (ITALIAN) □ BUTYLACETAT (GERMAN) □ BUTYL ACETATE □ 1-BUTYL ACETATE □ BUTYLACETATEN (DUTCH) □ BUTYLE (ACETATE de) (FRENCH) □ BUTYL ETHANOATE □ FEMA No. 2174 □ OCTAN n-BUTYLU (POLISH)

CONSENSUS REPORTS: Reported in EPA TSCA Inventory.

OSHA PEL: TWA 150 ppm; STEL 200 ppm

ACGIH TLV: Proposed: 150 ppm; STEL 200 ppm

DFG MAK: 100 ppm (480 mg/m³)

DOT CLASSIFICATION: 3; Label: Flammable Liquid

SAFETY PROFILE: Moderately toxic by intraperitoneal route. Mildly toxic by inhalation and ingestion. An experimental teratogen. A skin and severe eye irritant. Human systemic effects by inhalation: conjunctiva irritation, unspecified nasal and respiratory system effects. A mild allergen. High concentrations are irritating to eyes and respiratory tract and cause narcosis. Evidence of chronic systemic toxicity is inconclusive. Flammable liquid. Moderately explosive when exposed to flame. Ignites on contact with potassium tert-butoxide. To fight fire, use alcohol foam, CO_2, dry chemical. When heated to decomposition it emits acrid and irritating fumes. See also ESTERS.

BPV000   CAS: 105-46-4   HR: 3
sec-BUTYL ACETATE
DOT: UN 1123
mf: $C_6H_{12}O_2$ mw: 116.18
PROP: Colorless liquid; mild odor. Bp: 112°, flash p: 18°, d: 0.862–0.866 @ 20°/20°, vap d: 4.00, lel: 1.3%, uel: 7.5%. IDLH 1700 ppm [10%LEL].
SYNS: ACETATE de BUTYLE SECONDAIRE (FRENCH) □ ACETIC ACID-2-BUTOXY ESTER □ ACETIC ACID-1-

METHYLPROPYL ESTER (9CI) □ 2-BUTANOL ACETATE □ sec-BUTYL ACETATE □ 2-BUTYL ACETATE □ sec-BUTYL ALCOHOL ACETATE

CONSENSUS REPORTS: Reported in EPA TSCA Inventory.

OSHA PEL: TWA 200 ppm

ACGIH TLV: TWA 200 ppm

DFG MAK: 100 ppm (480 mg/m³)

DOT CLASSIFICATION: 3; Label: Flammable Liquid

SAFETY PROFILE: An irritant and allergen. See also ESTERS. Flammable liquid. To fight fire, use alcohol foam, CO_2, dry chemical. When heated to decomposition it emits acrid and irritating fumes.

BPV100   CAS: 540-88-5   HR: 3
tert-BUTYL ACETATE
DOT: UN 1123
mf: $C_6H_{12}O_2$ mw: 116.18
PROP: Liquid. Bp: 97–98°. IDLH 1500 ppm [10%LEL].
SYNS: ACETIC ACID-tert-BUTYL ESTER □ ACETIC ACID-1,1-DIMETHYLETHYL ESTER □ TEXACO LEAD APPRECIATOR □ TLA

CONSENSUS REPORTS: Reported in EPA TSCA Inventory.

OSHA PEL: TWA 200 ppm

ACGIH TLV: TWA 200 ppm

DFG MAK: 100 ppm (480 mg/m³)

DOT CLASSIFICATION: 3; Label: Flammable Liquid

SAFETY PROFILE: Poison by inhalation and ingestion. Flammable. To fight fire, use alcohol foam, CO_2, dry chemical. When heated to decomposition it emits acrid smoke and irritating fumes.

BPV250   CAS: 591-60-6   HR: 1
BUTYL ACETOACETATE
mf: $C_8H_{14}O_3$ mw: 158.22
PROP: Bp: 214°, flash p: 185°F, d: 0.96, vap d: 5.55.
SYNS: ACETOACETIC ACID BUTYL ESTER □ 3-OXO-BUTANOIC ACID BUTYL ESTER

CONSENSUS REPORTS: Reported in EPA TSCA Inventory.

SAFETY PROFILE: Mildly toxic by ingestion. A skin and eye irritant. See also

ESTERS. Flammable. To fight fire, use alcohol foam, CO_2, dry chemical. When heated to decomposition it emits acrid and irritating fumes.

BPW100   CAS: 141-32-2   HR: 3
n-BUTYL ACRYLATE
DOT: UN 2348
mf: $C_7H_{12}O_2$ mw: 128.19
PROP: Water-white, extremely reactive monomer. Bp: 69° @ 50 mm, fp: −64.6°, flash p: 120°F (OC), d: 0.89 @ 25°/25°, vap press: 10 mm @ 35.5°, vap d: 4.42.
SYNS: ACRYLIC ACID BUTYL ESTER □ ACRYLIC ACID n-BUTYL ESTER (MAK) □ BUTYL ACRYLATE □ BUTYLACRYLATE, INHIBITED (DOT) □ BUTYL-2-PROPENOATE
CONSENSUS REPORTS: IARC Cancer Review: Group 3 IMEMDT 7,56,87; Animal Inadequate Evidence IMEMDT 39,67,86. Reported in EPA TSCA Inventory. Community Right-To-Know List.
OSHA PEL: TWA 10 ppm
ACGIH TLV: TWA 2 ppm (sensitizer); Not Classifiable as a Carcinogen
DFG MAK: 2 ppm (11 mg/m³)
DOT CLASSIFICATION: 3; Label: Flammable Liquid
SAFETY PROFILE: Moderately toxic by ingestion, inhalation, skin contact, and intraperitoneal routes. Experimental reproductive effects. A skin and eye irritant. Questionable carcinogen. A flammable liquid when exposed to heat or flame. To fight fire, use foam, CO_2, dry chemical. Incompatible with oxidizing materials. When heated to decomposition it emits acrid and irritating fumes. See also ESTERS.

BPW500   CAS: 71-36-3   HR: 3
n-BUTYL ALCOHOL
mf: $C_4H_{10}O$ mw: 74.14
PROP: Colorless liquid; vinous odor. Bp: 117.4°, ULC: 40, lel: 1.4%, uel: 11.2%, fp: −90°, flash p: 95–100°F, d: 0.80978 @ 20°/4°, autoign temp: 689°F, vap press: 5.5 mm @ 20°, vap d: 2.55. Misc in alc, ether, and org solvs. Mod sol in water. IDLH 1400 ppm [10%LEL].

SYNS: ALCOOL BUTYLIQUE (FRENCH) □ BUTANOL (FRENCH) □ n-BUTANOL □ BUTAN-1-OL □ 1-BUTANOL □ BUTANOL (DOT) □ BUTANOLEN (DUTCH) □ BUTANOLO (ITALIAN) □ BUTYL ALCOHOL (DOT) □ BUTYL HYDROXIDE □ BUTYLOWY ALKOHOL (POLISH) □ BUTYRIC or NORMAL PRIMARY BUTYL ALCOHOL □ CCS 203 □ FEMA No. 2178 □ 1-HYDROXYBUTANE □ METHYLOLPROPANE □ PROPYLCARBINOL □ PROPYLMETHANOL □ RCRA WASTE NUMBER U031
CONSENSUS REPORTS: Community Right-To-Know List. EPA Genetic Toxicology Program. Reported in EPA TSCA Inventory.
OSHA PEL: CL 50 ppm (skin)
ACGIH TLV: TWA 20 ppm
DFG MAK: 100 ppm (310 mg/m³)
SAFETY PROFILE: A poison by intravenous route. Moderately toxic by skin contact, ingestion, subcutaneous, and intraperitoneal routes. Human systemic effects by inhalation: conjunctiva irritation, unspecified respiratory system effects, and nasal effects. Experimental reproductive effects. A severe skin and eye irritant. Though animal experiments have shown the butyl alcohols to possess toxic properties, they have produced few cases of poisoning in industry, probably because of their low volatility. The use of normal butyl alcohol is reported to have resulted in irritation of the eyes, with corneal inflammation, slight headache and dizziness, slight irritation of the nose and throat, and dermatitis about the fingernails and along the side of the fingers. Keratitis has also been reported. Mutation data reported. See also ALCOHOLS. Flammable liquid. Moderately explosive when exposed to flame. Incompatible with Al, chromium trioxide, oxidizing materials. To fight fire, use water spray, alcohol foam, CO_2, dry chemical. When heated to decomposition it emits acrid smoke and fumes.

BPW750   CAS: 78-92-2   HR: 3
sec-BUTYL ALCOHOL
mf: $C_4H_{10}O$ mw: 74.14

B

PROP: Colorless liquid. Mp: −89°, bp: 99.5°, flash p: 14°, d: 0.808 @ 20°/4°, autoign temp: 763°F, vap press: 10 mm @ 20°, vap d: 2.55, lel: 1.7% @ 212°F, uel: 9.8% @ 212°F. IDLH 2000 ppm.
SYNS: ALCOOL BUTYLIQUE SECONDAIRE (FRENCH) □ sec-BUTANOL (DOT) □ BUTAN-2-OL □ 2-BUTANOL □ BUTANOL SECONDAIRE (FRENCH) □ 2-BUTYL ALCOHOL □ BUTYLENE HYDRATE □ CCS 301 □ ETHYLMETHYL CARBINOL □ 2-HYDROXYBUTANE □ METHYLETHYLCARBINOL □ S.B.A.
CONSENSUS REPORTS: Community Right-To-Know List. Reported in EPA TSCA Inventory.
OSHA PEL: TWA 100 ppm
ACGIH TLV: TWA 100 ppm
DFG MAK: 100 ppm (310 mg/m³)
SAFETY PROFILE: Poison by intravenous and intraperitoneal routes. Mildly toxic by ingestion. Experimental reproductive effects. A skin and eye irritant. See also n-BUTYL ALCOHOL and ALCOHOLS. Dangerous fire hazard when exposed to heat or flame. Auto-oxidizes to an explosive peroxide. Ignites on contact with chromium trioxide. To fight fire, use water spray, alcohol foam, CO₂, dry chemical. Incompatible with oxidizing materials. When heated to decomposition it emits acrid smoke and fumes.

BPX000   CAS: 75-65-0   HR: 3
tert-BUTYL ALCOHOL
mf: C₄H₁₀O   mw: 74.14
PROP: Colorless liquid or rhombic prisms or plates with camphoraceous odor. Mp: 25.5°, bp: 82.8°, flash p: 50°F (CC), d: 0.781 @ 25°/4°, autoign temp: 896°F, vap press: 40 mm @ 24.5°, vap d: 2.55, lel: 2.4%, uel: 8.0%. Misc in H₂O. IDLH 1600 ppm.
SYNS: ALCOOL BUTYLIQUE TERTIAIRE (FRENCH) □ tert-BUTANOL □ BUTANOL TERTIAIRE (FRENCH) □ tert-BUTYL HYDROXIDE □ 1,1-DIMETHYLETHANOL □ 2-METHYL-2-PROPANOL □ NCI-C55367 □ TRIMETHYL-CARBINOL
CONSENSUS REPORTS: Community Right-To-Know List. Reported in EPA TSCA Inventory. EPA Genetic Toxicology Program.
OSHA PEL: TWA 100 ppm; STEL 150 ppm

ACGIH TLV: TWA 100 ppm; Not Classifiable as a Human Carcinogen
DFG MAK: 100 ppm (310 mg/m³)
SAFETY PROFILE: Moderately toxic by ingestion, intravenous, and intraperitoneal routes. An experimental teratogen. Other experimental reproductive effects. Dangerous fire hazard when exposed to heat or flame. Moderately explosive in the form of vapor when exposed to flame. Ignites on contact with potassium-sodium alloys. To fight fire, use alcohol foam, CO₂, dry chemical. Incompatible with oxidizing materials, H₂O₂. See also n-BUTYL ALCOHOL and ALCOHOLS.

BPX750   CAS: 109-73-9   HR: 3
n-BUTYLAMINE
DOT: UN 1125
mf: C₄H₁₁N   mw: 73.16
PROP: Liquid; ammonia-like odor. Mp: −50°, bp: 78°, flash p: 10°F (OC), 10°F (CC), d: 0.74–0.76 @ 20°/20°, autoign temp: 594°F, vap d: 2.52, lel: 1.7%, uel: 9.8%. IDLH 300 ppm.
SYNS: 1-AMINO-BUTAAN (DUTCH) □ 1-AMINOBUTAN (GERMAN) □ 1-AMINOBUTANE □ 1-BUTANAMINE □ n-BUTILAMINA (ITALIAN) □ n-BUTYLAMIN (GERMAN) □ BUTYLAMINE (OSHA) □ MONOBUTILAMINA □ MONOBUTYLAMINE □ MONO-n-BUTYLAMINE □ NORVALAMINE
CONSENSUS REPORTS: Reported in EPA TSCA Inventory.
OSHA PEL: CL 5 ppm (skin)
ACGIH TLV: CL 5 ppm
DFG MAK: 5 ppm (15 mg/m³)
DOT CLASSIFICATION: 3; Label: Flammable Liquid; DOT Class: 3; Label: Flammable Liquid, Corrosive
SAFETY PROFILE: Poison by ingestion, skin contact, and intravenous routes. Moderately toxic by inhalation, intraperitoneal, and parenteral routes. A corrosive and severe skin irritant. Questionable carcinogen with experimental tumorigenic data. Mutation data reported. A flammable liquid and dangerous fire hazard when exposed to heat, flame, or oxidizing materials. To fight fire, use alcohol foam,

CO_2, dry chemical. Explodes on contact with perchloryl fluoride. When heated to decomposition it emits toxic fumes of NO_x. See also AMINES.

BPY000   CAS: 13952-84-6   HR: 3
sec-BUTYLAMINE
DOT: UN 2733/UN 2734
mf: $C_4H_{11}N$ mw: 73.16
PROP: Liquid. Mp: −104°, bp: 63°, flash p: 15°F, d: 0.724 @ 20°.
SYNS: 2-AB □ 2-AMINOBUTANE □ BUTAFUME □ 2-BUTANAMINE □ DECCOTANE □ FRUCOTE □ 1-METHYLPROPYLAMINE □ TUTANE
CONSENSUS REPORTS: Reported in EPA TSCA Inventory.
DFG MAK: 5 ppm (15 mg/m³)
DOT CLASSIFICATION: 8; Label: Corrosive, Flammable Liquid (UN 2734); DOT Class: 3; Label: Flammable Liquid, Corrosive (UN 2733)
SAFETY PROFILE: A poison by ingestion. A powerful irritant. Moderately toxic by skin contact. Dangerous fire hazard when exposed to heat or flame. To fight fire, use alcohol foam, water spray or mist, dry chemical. Incompatible with oxidizing materials. When heated to decomposition it emits toxic fumes of NO_x. A fungicide.

BPY250   CAS: 75-64-9   HR: 3
tert-BUTYLAMINE
DOT: UN 2733/UN 2734
mf: $C_4H_{11}N$ mw: 73.16
PROP: Colorless liquid. Mp: −67.5°, bp: 46.4°, fp: −72.65°, d: 0.700 @ 15°, lel: 1.7% @ 212°F, uel: 8.9% @ 212°F, vap d: 2.5, autoign temp: 716°F.
SYNS: 2-AMINOISOBUTANE □ 2-AMINO-2-METHYLPROPANE □ BUTYLAMINE, tertiary □ 1,1-DIMETHYLETHYLAMINE □ TRIMETHYLAMINO-METHANE
CONSENSUS REPORTS: Reported in EPA TSCA Inventory.
DFG MAK: 5 ppm (15 mg/m³)
DOT CLASSIFICATION: 8; Label: Corrosive, Flammable Liquid (UN 2734); DOT Class: 3; Label: Flammable Liquid, Corrosive (UN 2733)

SAFETY PROFILE: Poison by ingestion. Moderately toxic to humans by inhalation. A corrosive liquid. See also n-BUTYLAMINE and AMINES. Very dangerous fire hazard when exposed to heat or flame. Very exothermic reaction with 2,2-dibromo-1,3-dimethylcyclopropanoic acid. To fight fire, use alcohol foam. When heated to decomposition it emits toxic fumes of NO_x.

BQC000   CAS: 111-75-1   HR: 2
2-BUTYLAMINOETHANOL
mf: $C_6H_{15}NO$ mw: 117.22
PROP: Liquid. Bp: 200°, flash p: 170°F (OC), d: 0.89, vap d: 4.03. Sol in H_2O.
SYN: 2-n-BUYTLAMINOETHANOL
CONSENSUS REPORTS: Reported in EPA TSCA Inventory.
SAFETY PROFILE: Moderately toxic by ingestion and intraperitoneal routes. A skin and severe eye irritant. See also AMINES. Combustible when exposed to heat or flame. To fight fire, use alcohol foam, foam, CO_2, dry chemical. Incompatible with oxidizing materials. When heated to decomposition it emits toxic fumes of NO_x.

BQD250   CAS: 3775-90-4   HR: 3
tert-BUTYL AMINO ETHYL METHACRYLATE
mf: $C_{10}H_{19}NO_2$ mw: 185.30
PROP: Liquid. Bp: 100–105°, d: 0.914, flash p: 205°F (OC).
SYNS: AGEFLEX FM-4 □ 2-(tert-BUTYLAMINO)ETHYL METHACRYLATE
CONSENSUS REPORTS: Reported in EPA TSCA Inventory.
SAFETY PROFILE: Poison by intraperitoneal route. See also ESTERS and AMINES. Combustible when exposed to heat or flame. To fight fire, use alcohol foam, water spray or mist, dry chemical. When heated to decomposition it emits toxic fumes of NO_x.

BQH850   CAS: 1126-78-9   HR: 3
N-BUTYLANILINE
DOT: UN 2738
mf: $C_{10}H_{15}N$ mw: 149.26

PROP: Liquid. D: 0.936 @ 20°/4°, bp: 249°. Sol in acids, EtOH, C_6H_6, $CHCl_3$; insol in H_2O.

SYNS: BENZENAMINE, N-BUTYL-(9CI) □ N-(n-BUTYL)ANILINE □ N-n-BUTYLANILINE (DOT) □ N-BUTYLBENZENAMINE (9CI) □ 4-(PHENYLAMINO)-BUTANE

CONSENSUS REPORTS: Reported in EPA TSCA Inventory.

DOT CLASSIFICATION: 6.1; Label: Poison

SAFETY PROFILE: Poison by an unspecified route. Moderately toxic by skin contact and ingestion. A skin and eye irritant. When heated to decomposition it emits toxic fumes of NO_x. See also ANILINE DYES.

BQI000   CAS: 25013-16-5   HR: 3
BUTYLATED HYDROXYANISOLE
mf: $C_{11}H_{16}O_2$ mw: 180.27

PROP: White waxy solid; faint characteristic odor. Mp: 104–105°. Sol in alc and propylene glycol; insol in water.

SYNS: ANTRANCINE 12 □ BHA (FCC) □ BUTYL-HYDROXYANISOLE □ tert-BUTYLHYDROXYANISOLE □ tert-BUTYL-4-HYDROXYANISOLE □ 2(3)-tert-BUTYL-4-HYDROXYANISOLE □ BUTYLOHYDROKSYANIZOL (POLISH) □ EMBANOX □ FEMA No. 2183 □ NIPANTIOX 1-F □ PREMERGE PLUS □ SUSTANE □ SUSTANE 1-F □ TENOX BHA □ VERTAC

CONSENSUS REPORTS: NTP 10th Report on Carcinogens. IARC Cancer Review: Group 2B IMEMDT 7,56,87; Animal Sufficient Evidence IMEMDT 40,123,86. Reported in EPA TSCA Inventory. EPA Genetic Toxicology Program.

SAFETY PROFILE: Confirmed carcinogen with experimental carcinogenic, neoplastigenic, and tumorigenic data. Moderately toxic by ingestion and intraperitoneal routes. Experimental reproductive effects. Mutation data reported. When heated to decomposition it emits acrid and irritating fumes.

BQI050   HR: D
BUTYLATED HYDROXYMETHYL-PHENOL
mf: $C_{15}H_{24}O_2$ mw: 236.35

PROP: White crystalline powder. Mp: 140–141°. Sol in alc; insol in water, propylene glycol.

SYN: 4-HYDROXYMETHYL-2,6-DI-tert-BUTYLPHENOL

SAFETY PROFILE: When heated to decomposition it emits acrid smoke and irritating fumes.

BQI250   CAS: 1070-19-5   HR: 3
tert-BUTYL AZIDOFORMATE
mf: $C_5H_9N_3O_2$ mw: 143.17

PROP: Bp: 73–74° @ 70 mm.

SYNS: t-BUTOXYCARBONYL AZIDE □ tert-BUTOXYCARBONYL AZIDE (DOT) □ tert-BUTYLOXY-CARBONYL AZIDE □ CARBONAZIDIC ACID, 1,1-DIMETHYLETHYL ESTER □ FORMIC ACID, AZIDO-, tert-BUTYL ESTER

CONSENSUS REPORTS: Reported in EPA TSCA Inventory.

DOT CLASSIFICATION: Forbidden

SAFETY PROFILE: An unstable shock- and heat-sensitive explosive. It may explode above 100°C and ignites at 143°C. When heated to decomposition it emits toxic fumes of NO_x. See also AZIDES.

BQI750   CAS: 104-51-8   HR: 1
n-BUTYLBENZENE
mf: $C_{10}H_{14}$ mw: 134.24

PROP: Colorless liquid. Mp: −81.2°, bp: 182.1°, d: 0.875 @ 13°/4°, vap press: 1 mm @ 22.7°, autoign temp: 774°F, lel: 0.8%, uel: 5.8%, vap d: 4.6.

SYN: 1-PHENYLBUTANE

CONSENSUS REPORTS: Reported in EPA TSCA Inventory.

DOT CLASSIFICATION: 3; Label: Flammable Liquid

SAFETY PROFILE: Mildly toxic by ingestion. Flammable when exposed to heat or flame. To fight fire, use alcohol foam, CO_2, dry chemical. Incompatible with oxidizing materials. When heated to decomposition it emits acrid and irritating fumes.

BQJ000   CAS: 135-98-8   HR: 3
sec-BUTYLBENZENE
mf: $C_{10}H_{14}$ mw: 134.24
PROP: Colorless liquid. Mp: −82.7°, bp: 173.5°, fp: −75.8°, flash p: 126°F (TOC), d: 0.8621 @ 20°, vap press: 1 mm @ 18.6°, vap d: 4.62, autoign temp: 788°F, lel: 0.8%, uel: 6.9%.
SYN: 2-PHENYLBUTANE
CONSENSUS REPORTS: Reported in EPA TSCA Inventory.
DOT CLASSIFICATION: 3; Label: Flammable Liquid
SAFETY PROFILE: Moderately toxic by ingestion. A skin and eye irritant. Flammable liquid when exposed to heat or flame. To fight fire, use foam, CO_2, dry chemical, water spray or mist. Incompatible with oxidizing materials. When heated to decomposition it emits acrid smoke and fumes.

BQJ250   CAS: 98-06-6   HR: 3
tert-BUTYLBENZENE
mf: $C_{10}H_{14}$ mw: 134.24
PROP: Colorless liquid. Bp: 170–171°, fp: −58°, flash p: 140°F (TOC), d: 0.8665 @ 20°, vap press: 1 mm @ 13.0°, vap d: 4.62, autoign temp: 842°F, lel: 0.7% @ 212°F, uel: 5.7% @ 212°F.
SYNS: 2-METHYL-2-PHENYLPROPANE □ PSEUDO-BUTYLBENZENE □ TRIMETHYLPHENYLMETHANE
CONSENSUS REPORTS: Reported in EPA TSCA Inventory.
DOT CLASSIFICATION: 3; Label: Flammable Liquid
SAFETY PROFILE: Mildly toxic by ingestion. Flammable liquid when exposed to heat or flame. To fight fire, use foam, CO_2, dry chemical, water spray, fog, mist. Incompatible with oxidizing materials. When heated to decomposition it emits acrid smoke and fumes.

BQJ350   CAS: 122-43-0   HR: 1
BUTYLBENZENEACETATE
mf: $C_{12}H_{16}O_2$ mw: 192.28
SYNS: ACETIC ACID, PHENYL-, BUTYL ESTER □ BENZENEACETIC ACID, BUTYL ESTER (9CI) □ BUTYL PHENYLACETATE □ n-BUTYL PHENYLACETATE □ PHENYLETHANOIC ACID BUTYL ESTER
CONSENSUS REPORTS: Reported in EPA TSCA Inventory.
SAFETY PROFILE: Low toxicity by ingestion. A skin irritant. When heated to decomposition it emits acrid smoke and irritating fumes.

BQJ500   CAS: 583-03-9   HR: 3
α-BUTYLBENZENEMETHANOL
mf: $C_{11}H_{16}O$ mw: 164.27
SYNS: α-BUTYLBENZYL ALCOHOL □ FENIPENTOL □ 1-HYDROXY-1-PHENYLPENTANE □ PANCORAL □ PC 1 □ PH BC □ PHENYLBUTYLCARBINOL □ 1-PHENYL-1-HYDROXYPENTANE □ PHENYLPENTAN-OL □ 1-PHENYLPENTANOL
CONSENSUS REPORTS: Reported in EPA TSCA Inventory.
SAFETY PROFILE: Poison by intraperitoneal route. Moderately toxic by ingestion and subcutaneous routes. An experimental teratogen. Other experimental reproductive effects. Stimulates the production of bile by the liver. When heated to decomposition it emits acrid smoke and irritating fumes.

BQK250   CAS: 136-60-7   HR: 2
BUTYL BENZOATE
mf: $C_{11}H_{14}O_2$ mw: 178.25
PROP: Liquid. Mp: −21.5°, bp: 248–249°, flash p: 225°F (OC), d: 1.01 @ 15°/15°, vap press: <0.01 mm @ 20°, vap d: 6.15.
SYNS: ANTHRAPOLE AZ □ BENZOIC ACID-n-BUTYL ESTER □ n-BUTYL BENZOATE □ DAI CARI XBN
CONSENSUS REPORTS: Reported in EPA TSCA Inventory.
SAFETY PROFILE: Moderately toxic by skin contact. Mildly toxic by ingestion. Severe skin irritant and moderate eye irritant. Combustible when exposed to heat or flame; can react with oxidizing materials. To fight fire, use CO_2, dry chemical, water mist, fog, spray. When heated to decomposition it emits acrid and irritating fumes. See also ESTERS.

B

BQK500   CAS: 98-73-7   HR: 2
p-tert-BUTYL BENZOIC ACID
mf: $C_{11}H_{14}O_2$ mw: 178.25
PROP: Colorless, fine, crystalline powder.
Mp: 163–164.4°, d: 1.142 @ 20°/4°.
SYN: TBBA
CONSENSUS REPORTS: Reported in EPA
TSCA Inventory.
SAFETY PROFILE: Moderately toxic by
ingestion. Experimental reproductive
effects. An irritant. Combustible when
exposed to heat or flame. Incompatible with
oxidizing materials. To fight fire, use foam,
CO_2, dry chemical. When heated to
decomposition it emits acrid smoke and
irritating fumes.

BQM000   CAS: 102-79-4   HR: 2
N-BUTYL-N,N-BIS(HYDROXY ETHYL)-
   AMINE
mf: $C_8H_{19}NO_2$ mw: 161.28
PROP: Liquid. Mp: −70°, bp: 273–275°,
flash p: 245°F (OC), d: 0.97, vap d: 5.55.
SYNS: N-BUTYLDIETHANOLAMINE □ N-BUTYL-2,2′-
IMINODIETHANOL
CONSENSUS REPORTS: Reported in EPA
TSCA Inventory.
SAFETY PROFILE: Mildly toxic via
ingestion. A skin and severe eye irritant.
Combustible when exposed to heat or
flame. To fight fire, use alcohol foam, foam,
CO_2, dry chemical. Incompatible with
oxidizing materials. When heated to
decomposition it emits toxic fumes of NO_x.
See also AMINES.

BQM250   CAS: 507-19-7   HR: 2
tert-BUTYL BROMIDE
mf: C_4H_9Br mw: 137.04
PROP: Colorless liquid. Mp: −20°, bp:
72.8°, fp: −16.3°, d: 1.20 @ 15°/4°.
SYNS: 2-BROMOISOBUTANE □ 2-BROMO-2-
METHYLPROPANE (DOT) □ TRIMETHYLBROMO-
METHANE
CONSENSUS REPORTS: EPA Genetic
Toxicology Program. Reported in EPA
TSCA Inventory.
DOT CLASSIFICATION: 3; Label:
Flammable Liquid

SAFETY PROFILE: Moderately toxic by
intraperitoneal route. Questionable
carcinogen with experimental
neoplastigenic data. When heated to
decomposition it emits toxic fumes of Br^-.
See also BROMIDES.

BQM500   CAS: 109-21-7   HR: 3
n-BUTYL n-BUTANOATE
mf: $C_8H_{16}O_2$ mw: 144.24
PROP: Colorless liquid; pineapple odor. Bp:
166°, flash p: 128°F (OC), d: 0.67–0.871,
refr index: 1.405, vap d: 5.0. Misc with alc,
ether, vegetable oils; sltly sol in propylene
glycol, water.
SYNS: BUTYL BUTYRATE (FCC) □ n-BUTYL
BUTYRATE □ n-BUTYL n-BUTYRATE □ FEMA No. 2186
CONSENSUS REPORTS: Reported in EPA
TSCA Inventory.
SAFETY PROFILE: Moderately toxic via
intraperitoneal route. Mildly toxic by
ingestion. Moderately irritating to eyes, skin,
and mucous membranes by inhalation.
Narcotic in high concentrations. Flammable
liquid. To fight fire, use alcohol foam, foam,
CO_2, dry chemical. Incompatible with
oxidizing materials. When heated to
decomposition it emits acrid and irritating
fumes.

BQP000   CAS: 7492-70-8   HR: 1
BUTYL BUTYROLACTATE
mf: $C_{11}H_{20}O_4$ mw: 216.28
PROP: Colorless liquid; butter, creamlike
odor. D: 0.970, refr index: 1.420, flash p:
212°F. Misc with alc, fixed oils; sol in
propylene glycol; insol in water.
SYNS: BUTANOIC ACID-2-BUTOXY-1-METHYL-2-
OXOETHYL ESTER (9CI) □ BUTYL BUTYRYL LACTATE
□ BUTYRIC ACID ESTER with BUTYL LACTATE □
FEMA No. 2190 □ LACTIC ACID, BUTYL ESTER,
BUTYRATE
CONSENSUS REPORTS: Reported in EPA
TSCA Inventory.
SAFETY PROFILE: A skin irritant. See also
ESTERS. Combustible liquid. When heated
to decomposition it emits acrid smoke and
irritating fumes.

BQP250   CAS: 592-35-8   HR: 3
BUTYL CARBAMATE
mf: $C_5H_{11}NO_2$ mw: 117.17
SYNS: CARBAMIC ACID, BUTYL ESTER ◻ USAF EL-101 ◻ USAF FO-1
CONSENSUS REPORTS: Reported in EPA TSCA Inventory.
SAFETY PROFILE: A poison via intraperitoneal route. Moderately toxic via subcutaneous route. Experimental teratogenic effects. Questionable carcinogen with experimental neoplastigenic data. Mutation data reported. See also CARBAMATES. When heated to decomposition it emits toxic fumes of NO_x.

BQP500   CAS: 124-17-4   HR: 2
BUTYL CARBITOL ACETATE
mf: $C_{10}H_{20}O_4$ mw: 204.30
PROP: Colorless liquid. Fp: $-32.2°$, bp: 247°, flash p: 240°F (OC), d: 0.981 @ 20°/20°, autoign temp: 570°F, vap press: 0.01 mm @ 20°.
SYNS: 2-(2-BUTOXYETHOXY)ETHANOL ACETATE ◻ 2-(2-BUTOXYETHOXY)ETHYL ACETATE ◻ DIETHYLENE GLYCOL BUTYL ETHER ACETATE ◻ DIGLYCOL MONOBUTYL ETHER ACETATE ◻ EKTASOLVE DB ACETATE ◻ GLYCOL ETHER DB ACETATE
CONSENSUS REPORTS: Reported in EPA TSCA Inventory. Glycol ethers are on the Community Right-To-Know List.
SAFETY PROFILE: Moderately toxic by ingestion. Mild skin and eye irritant. Combustible when exposed to heat or flame. To fight fire, use foam, CO_2, dry chemical. Incompatible with oxidizing materials, heat, flame. When heated to decomposition it emits acrid smoke and irritating fumes.

BQP750   CAS: 85-70-1   HR: 2
BUTYL CARBOBUTOXYMETHYL
PHTHALATE
mf: $C_{18}H_{24}O_6$ mw: 336.42
SYNS: BUTYL PHTHALATE BUTYL GLYCOLATE ◻ BUTYL PHTHALYL BUTYL GLYCOLATE ◻ DIBUTYL-o-(o-CARBOXYBENZOYL) GLYCOLATE ◻ DIBUTYL-o-CARBOXYBENZOYLOXYACETATE ◻ SANTICIZIER B-16

CONSENSUS REPORTS: Reported in EPA TSCA Inventory.
SAFETY PROFILE: Mildly toxic via intraperitoneal route. Experimental teratogenic and reproductive effects. Mutation data reported. An eye irritant. When heated to decomposition it emits acrid and irritating fumes.

BQQ250   CAS: 38252-74-3   HR: 3
N-BUTYL-(3-CARBOXY PROPYL)-
NITROSAMINE
mf: $C_8H_{16}N_2O_3$ mw: 188.26
SYNS: BCPN ◻ 4-(BUTYLNITROSOAMINO)BUTANOIC ACID ◻ N-NITROSO-N-BUTYL-N-(3-CARBOXYPROPYL)AMINE
CONSENSUS REPORTS: NTP 10th Report on Carcinogens. IARC Cancer Review: Animal Limited Evidence IMEMDT 17,51,78. EPA Genetic Toxicology Program.
SAFETY PROFILE: Confirmed carcinogen with experimental carcinogenic and tumorigenic data. Mutation data reported. When heated to decomposition it emits toxic fumes of NO_x. See also NITROSAMINES.

BQQ750   CAS: 109-69-3   HR: 3
n-BUTYL CHLORIDE
mf: C_4H_9Cl mw: 92.58
PROP: Colorless liquid. Mp: $-123.1°$, bp: 78.5°, lel: 1.9%, uel: 10.1%, flash p: 15°F (OC), d: 0.892 @ 15°, autoign temp: 860°F, fp: $-123.1°$, vap d: 3.20.
SYNS: BUTYL CHLORIDE (DOT) ◻ 1-CHLOROBUTANE (DOT) ◻ CHLORURE de BUTYLE (FRENCH) ◻ NCI-C06155 ◻ N-PROPYLCARBINYL CHLORIDE
CONSENSUS REPORTS: NTP Carcinogenesis Studies (gavage); No Evidence: mouse, rat NTPTR* NTP-TR-312,86. EPA Genetic Toxicology Program. Reported in EPA TSCA Inventory.
SAFETY PROFILE: Moderately toxic by ingestion. Mutation data reported. See CHLORINATED HYDROCARBONS, ALIPHATIC. Skin and eye irritant. Dangerous fire hazard when exposed to heat or flame. Moderately explosive when

exposed to flame. When heated to decomposition it emits highly toxic fumes of phosgene and Cl⁻. To fight fire, use foam, CO_2, dry chemical. Incompatible with oxidizing materials.

BQR000   CAS: 507-20-0   HR: 3
tert-BUTYL CHLORIDE
mf: C_4H_9Cl mw: 92.58
PROP: Liquid. Flash p: 32°F, d: 0.87, vap d: 3.2, bp: 51°, fp: −27.1°.
SYNS: 2-CHLOROISOBUTANE □ 2-CHLORO-2-METHYLPROPANE □ TRIMETHYLCHLOROMETHANE
CONSENSUS REPORTS: Reported in EPA TSCA Inventory.
SAFETY PROFILE: Questionable carcinogen with experimental neoplastigenic data. Dangerous fire hazard when exposed to heat, flame (sparks), and oxidizers. To fight fire, use water, spray, fog, alcohol foam, dry chemical. When heated to decomposition it emits toxic fumes of Cl⁻. See also CHLORINATED HYDROCARBONS, ALIPHATIC.

BQR100   CAS: 107-59-5   HR: 3
tert-BUTYL CHLOROACETATE
mf: $C_6H_{11}ClO_2$ mw: 150.62
SYNS: ACETIC ACID, CHLORO-, tert-BUTYL ESTER □ ACETIC ACID, CHLORO-, 1,1-DIMETHYLETHYL ESTER □ CHLOROACETIC ACID tert-BUTYL ESTER □ 1,1-DIMETHYLETHYL CHLOROACETATE
SAFETY PROFILE: A poison by ingestion. Moderately toxic by inhalation skin contact. S severe skiin and moderate eye irritant. When heated to decomposition it emits toxic vapors of Cl⁻.

BQV000   CAS: 1189-85-1   HR: 3
tert-BUTYL CHROMATE
mf: $C_8H_{18}CrO_4$ mw: 230.26
PROP: Red crystals from pet ether. IDLH 15 mg/m³ {as Cr(VI)}.
SYN: CHROMIC ACID, DI-tert-BUTYL ESTER
CONSENSUS REPORTS: Chromium and its compounds are on the Community Right-To-Know List.
OSHA PEL: CL 0.1 mg(CrO₃)/m³ (skin)
ACGIH TLV: CL 0.1 mg(CrO₃)/m³ (skin)

NIOSH REL: (Chromium(VI)) CL 0.001 Mg(Cr(VI))/m³
SAFETY PROFILE: A very flammable mixture. When heated to decomposition it emits acrid and irritating fumes. See CHROMIUM COMPOUNDS and ESTERS.

BQV750   CAS: 2409-55-4   HR: 3
2-tert-BUTYL-p-CRESOL
mf: $C_{11}H_{16}O$ mw: 164.27
PROP: Clear liquid, sol in org solvs and aqueous potassium hydroxide. Fp: 23.1°, bp: 118–119° @ 14 mm, d: 0.922, flash p: 116°F.
SYNS: 2-tert-BUTYL-p-KRESOL (CZECH) □ 2-tert-BUTYL-4-METHYLPHENOL
CONSENSUS REPORTS: Reported in EPA TSCA Inventory.
SAFETY PROFILE: A poison by intraperitoneal and intravenous routes. Moderately toxic by ingestion and skin contact. Questionable carcinogen with experimental neoplastigenic data. A severe skin and eye irritant. Mutation data reported. Flammable liquid when exposed to heat, flame, or oxidizers. To fight fire, use alcohol foam, foam, water spray, fog, dry chemical. When heated to decomposition it emits acrid and irritating fumes.

BQY300   CAS: 684-82-2   HR: 3
sec-BUTYLDICHLOROARSINE
mf: $C_4H_9AsCl_2$ mw: 202.95
SYNS: ARSINE, sec-BUTYLDICHLORO- □ ARSONOUS DICHLORIDE, (1-METHYLPROPYL)-(9CI) □ sec-BUTYL-DICHLORARSINE □ DICHLORO(1-METHYLPROPYL)-ARSINE
OSHA PEL: TWA 0.5 mg(As)/m³
SAFETY PROFILE: Poison by inhalation. When heated to decomposition it emits toxic fumes of As and Cl⁻.

BRA550   CAS: 17563-48-3   HR: 3
n-BUTYLDIETHYLTIN IODIDE
mf: $C_8H_{19}ISn$ mw: 360.86
SYN: STANNANE, BUTYLDIETHYLIODO-
OSHA PEL: TWA 0.1 mg(Sn)/m³

ACGIH TLV: TWA 0.1 mg(Sn)/m³; STEL 0.2 mg/m³ (skin)
NIOSH REL: (Organotin compound): 10H TWA 0.1 mg(Sn)/m³
SAFETY PROFILE: Poison by intravenous route. When heated to decomposition it emits toxic fumes of Sn and I⁻.

BRF500   CAS: 50-33-9   HR: 3
4-BUTYL-1,2-DIPHENYL-3,5-DIOXO
  PYRAZOLIDINE
mf: $C_{19}H_{20}N_2O_2$ mw: 308.41
PROP: Crystals from EtOH. Mp: 105.5–106.5°.
SYNS: ALINDOR □ ALQOVERIN □ ANERVAL □ ANTADOL □ ANUSPIRAMIN □ ARTIZIN □ ARTRIZONE □ ARTROPAN □ AZDID □ AZOLID □ BENZONE □ BETAZED □ BUSONE □ BUTACOMPREN □ BUTACOTE □ BUTALAN □ BUTALGINA □ BUTALIDON □ BUTAPIRAZOL □ BUTAPYRAZOLE □ BUTARECBON □ BUTARTRINA □ BUTAZINA □ BUTAZONA □ BUTAZONE □ BUTIDIONA □ BUTONE □ BUTOZ □ 4-BUTYL-1,2-DIPHENYLPYRAZOLIDINE-3,5-DIONE □ BUTYLPYRIN □ BUVETZONE □ BUZON □ DIGIBUTINA □ DIOSSIDONE □ 3,5-DIOXO-1,2-DIPHENYL-4-N-BUTYLPYRAZOLIDENE □ DIOZOL □ DIPHEBUZOL □ DIPHENYLBUTAZONE □ 1,2-DIPHENYL-4-BUTYL-3,5-DIOXOPYRAZOLIDINE □ ELMEDAL □ EQUI BUTE □ ERIBUTAZONE □ ESTEVE □ FENARTIL □ FENIBUTAZONA □ FENIBUTOL □ FENILBUTINE □ FENILIDINA □ FENOTONE □ FENYLBUTAZON □ FLEXAZONE □ INTALBUT □ IPSOFLAME □ LINGEL □ MALGESIC □ MEPHABUTAZONE □ MERIZONE □ NADOZONE □ NCI-C56531 □ NOVOPHENYL □ PHEBUZIN □ PHENBUTAZOL □ PHENOPYRINE □ PHENYLBUTAZ-ON (GERMAN) □ PHENYLBUTAZONE □ PIRARREUM-OL "B" □ PRAECIRHEUMIN □ PYRAZOLIDIN □ REUDO □ REUMASYL □ REUMAZOL □ REUPOLAR □ RUBATONE □ SCANBUTAZONE □ SHIGRODIN □ TAZONE □ TEVCODYNE □ THERAZONE □ TODALGIL □ UZONE □ WESCOZONE □ ZOLAPHEN □ ZOLIDINUM □ ZORANE
CONSENSUS REPORTS: IARC Cancer Review: Group 3 IMEMDT 7,316,87; Human Inadequate Evidence IMEMDT 13,183,77. EPA Genetic Toxicology Program. Reported in EPA TSCA Inventory.
SAFETY PROFILE: Suspected human carcinogen producing leukemia. A human poison by parenteral route. An experimental poison by ingestion, intraperitoneal, subcutaneous, intravenous, and intramuscular routes. Human systemic effects by ingestion and possibly other routes: fever, blood pressure increase, other unspecified vascular effects, damage to kidney tubules and glomeruli, decreased urine volume, blood in the urine, reduction in the number of white blood cells, and agranulocytosis. Experimental teratogenic and reproductive effects. Human mutation data reported. An eye irritant. An anti-inflammatory agent. When heated to decomposition it emits toxic fumes of NO_x.

BRF550   CAS: 20333-40-8   HR: 3
BUTYL DISELENIDE
mf: $C_8H_{18}Se_2$ mw: 272.18
SYNS: DIBUTYL DISELENIDE □ DI-n-BUTYL-DISELENIDE □ DIBUTYLDISELENIUM □ DISELEN-IDE, DIBUTYL-(9CI)
OSHA PEL: TWA 0.2 mg(Se)/m³
ACGIH TLV: TWA 0.2 mg(Se)/m³
SAFETY PROFILE: Poison by intravenous route. When heated to decomposition it emits toxic fumes of Se.

BRH750   CAS: 142-96-1   HR: 3
n-BUTYL ETHER
DOT: UN 1149
mf: $C_8H_{18}O$ mw: 130.26
PROP: Colorless liquid. Mp: −98°, bp: 142°, flash p: 77°F, d: 0.784 @ 0°/4°, autoign temp: 382°F, vap d: 4.48, lel: 1.5%, uel: 7.6%.
SYNS: 1-BUTOXYBUTANE □ BUTYL ETHER (DOT) □ DI-n-BUTYL ETHER (DOT) □ DIBUTYL OXIDE □ ETHER BUTYLIQUE (FRENCH) □ 1,1'-OXYBIS(BUTANE)
CONSENSUS REPORTS: Reported in EPA TSCA Inventory.
DOT CLASSIFICATION: 3; Label: Flammable Liquid
SAFETY PROFILE: Mildly toxic by inhalation, ingestion, and skin contact. Human systemic effects by inhalation: conjunctiva irritation and unspecified nasal effects. An experimental skin and human eye irritant. See also ETHERS. Dangerous fire hazard when exposed to heat, flame, or

B

oxidizers. Incompatible with NCl₃ and oxidizing materials. To fight fire, use alcohol foam, dry chemical. When heated to decomposition it emits acrid smoke and fumes.

BRH760   CAS: 6863-58-7   HR: 3
sec-BUTYL ETHER
mf: C₈H₁₈O   mw: 130.26
SYNS: BIS(2-BUTYL)ETHER □ BUTANE, 2,2'-OXYBIS-(9CI) □ DI-sec-BUTYL ETHER □ 2,2'-OXYBISBUTANE
CONSENSUS REPORTS: Reported in EPA TSCA Inventory.
DOT CLASSIFICATION: 3; Label: Flammable Liquid
SAFETY PROFILE: Poison by inhalation. A flammable liquid. When heated to decomposition it emits acrid smoke and irritating vapors.

BRI000   CAS: 123-05-7   HR: 3
BUTYL ETHYL ACETALDEHYDE
mf: C₈H₁₆O   mw: 128.24
PROP: Bp: 163.4°, flash p: 125°F (OC), autoign temp: 387°F, d: 0.8205, vap press: 1.8 mm @ 20°, vap d: 4.42.
SYNS: ETHYLBUTYLACETALDEHYDE □ α-ETHYL-CAPROALDEHYDE □ 2-ETHYLHEXALDEHYDE □ ETHYLHEXALDEHYDE (DOT) □ 2-ETHYLHEXANAL □ β-PROPYL-α-ETHYLACROLEIN
CONSENSUS REPORTS: Reported in EPA TSCA Inventory.
SAFETY PROFILE: Moderately toxic by ingestion and intraperitoneal routes. Mildly toxic by inhalation and skin contact. An eye and severe skin irritant. See also ALDEHYDES. Dangerous fire hazard; spontaneously flammable in air. To fight fire, use foam, CO₂, dry chemical, water spray, mist, fog. Incompatible with oxidizing materials. When heated to decomposition it emits acrid and irritating fumes.

BRI250   CAS: 149-57-5   HR: 2
BUTYL ETHYL ACETIC ACID
mf: C₈H₁₆O₂   mw: 144.24
PROP: Flash p: 260°F (OC), bp: 225–228°.
SYNS: α-ETHYLCAPROIC ACID □ 2-ETHYL-HEXANOIC ACID □ 2-ETHYLHEXOIC ACID

CONSENSUS REPORTS: Reported in EPA TSCA Inventory.
ACGIH TLV: TWA 5 mg/m³
SAFETY PROFILE: Moderately toxic by ingestion and skin contact. An experimental teratogen. A skin and severe eye irritant. Combustible when exposed to heat or flame. When heated to decomposition, it emits acrid and irritating fumes.

BRJ750   CAS: 2425-74-3   HR: 3
tert-BUTYL FORMAMIDE
mf: C₅H₁₁NO   mw: 101.17
CONSENSUS REPORTS: Reported in EPA TSCA Inventory.
SAFETY PROFILE: A poison by intravenous route. When heated to decomposition it emits toxic fumes of NOₓ.

BRK000   CAS: 592-84-7   HR: 3
n-BUTYL FORMATE
DOT: UN 1128
mf: C₅H₁₀O₂   mw: 102.15
PROP: Colorless liquid. Mp: −90°, bp: 106.0°, flash p: 64°F (CC), d: 0.911, autoign temp: 612°F, vap press: 40 mm @ 31.6°, vap d: 3.52, lel: 1.7%, uel: 8%.
SYNS: BUTYLESTER KYSELINY MRAVENCI □ BUTYL FORMATE (DOT)
CONSENSUS REPORTS: Reported in EPA TSCA Inventory.
DOT CLASSIFICATION: 3; Label: Flammable Liquid
SAFETY PROFILE: Moderately toxic by ingestion. Mildly toxic by inhalation. Human systemic effects by inhalation: muscle contractions and spasticity, conjunctiva irritation, and unspecified respiratory changes. An irritant and narcotic in high concentrations. See also ESTERS, n-BUTYL ALCOHOL, and FORMIC ACID. Dangerous fire hazard when exposed to heat or flame. To fight fire, use alcohol foam, foam, CO₂, dry chemical. Incompatible with oxidizing materials. When heated to decomposition it emits acrid and irritating fumes.

BRK750   CAS: 2426-08-6   HR: 3
n-BUTYL GLYCIDYL ETHER
mf: $C_7H_{14}O_2$ mw: 130.21

PROP: Bp: 327°. IDLH 250 ppm.

SYNS: AGEFLEX BGE ☐ BGE ☐ BGE (OSHA) ☐ BUTYL GLYCIDYL ETHER ☐ 2,3-EPOXYPROPYL BUTYL ETHER ☐ ETHER, BUTYL 2,3-EPOXYPROPYL ☐ ETHER, BUTYL GLYCIDYL ☐ GLYCIDYL BUTYL ETHER ☐ TK 10408

CONSENSUS REPORTS: Reported in EPA TSCA Inventory.

OSHA PEL: TWA 25 ppm

ACGIH TLV: TWA 25 ppm

DFG MAK: Confirmed Animal Carcinogen with Unknown Relevance to Humans

NIOSH REL: (Glycidyl Ethers) CL 30 mg/m³/15M

SAFETY PROFILE: Suspected Carcinogen. Moderately toxic by ingestion, skin contact, and intraperitoneal routes. Mildly toxic by inhalation. An experimental teratogen. Mutation data reported. A skin and severe eye irritant. See also ETHERS. When heated to decomposition it emits acrid and irritating fumes.

BRK800   CAS: 7665-72-7   HR: 3
t-BUTYL GLYCIDYL ETHER
mf: $C_7H_{14}O_2$ mw: 130.21

SYNS: t-BGE ☐ 1,1-DIMETHYLETHYL GLYCIDYL ETHER ☐ OXIRANE, ((1,1-DIMETHYLETHOXY)-METHYL)- ☐ PROPANE, 1-tert-BUTOXY-2,3-EPOXY-

CONSENSUS REPORTS: Reported in EPA TSCA Inventory.

DFG MAK: Confirmed Animal Carcinogen with Unknown Relevance to Humans

SAFETY PROFILE: Suspected carcinogen. Moderately toxic by ingestion. Mutation data reported. When heated to decomposition it emits acrid smoke and irritating vapors.

BRK900   CAS: 626-82-4   HR: 1
BUTYL HEXANOATE
mf: $C_{10}H_{20}O_2$ mw: 172.30

SYNS: BUTYL CAPROATE ☐ n-BUTYL HEXANOATE ☐ HEXANOIC ACID, BUTYL ESTER

CONSENSUS REPORTS: Reported in EPA TSCA Inventory.

SAFETY PROFILE: Low toxicity by ingestion and skin contact. A skin irritant. When heated to decomposition it emits acrid smoke and irritating fumes.

BRM250   CAS: 75-91-2   HR: 2
tert-BUTYLHYDROPEROXIDE
mf: $C_4H_{10}O_2$ mw: 90.14

PROP: Water-white liquid. Flash p: 80°F or above, fp: −35°, d: 0.860, mp: −8°, bp: 40° @ 23 mm, vap d: 2.07. Sltly sol in water; very sol in esters and alc.

SYNS: terc.BUTYLHYDROPEROXID (CZECH) ☐ CADOX TBH ☐ 1,1-DIMETHYLETHYL HYDRO-PEROXIDE ☐ HYDROPEROXYDE de BUTYLE TERTIAIRE (FRENCH) ☐ 2-HYDROPEROXY-2-METHYLPROPANE ☐ PERBUTYL H ☐ TBHP-70 ☐ TRIGONOX A-75 (CZECH)

CONSENSUS REPORTS: EPA Genetic Toxicology Program. Reported in EPA TSCA Inventory.

DFG MAK: Moderate skin effects

DOT CLASSIFICATION: Forbidden

SAFETY PROFILE: Moderately toxic by ingestion and inhalation. A severe skin and eye irritant. Mutation data reported. At highest dosage levels, symptoms noted were severe depression, incoordination, and cyanosis. Death was due to respiratory arrest. Very dangerous fire hazard when exposed to heat or flame, or by spontaneous chemical reaction such as with reducing materials. Moderately explosive; may explode during distillation. Violent reaction with traces of acid. Concentrated solutions may ignite spontaneously on contact with molecular sieve. Mixtures with transition metal salts may react vigorously and release oxygen. Forms an unstable solution with 1,2-dichloroethane. To fight fire, use alcohol foam, CO_2, dry chemical. When heated to decomposition it emits acrid smoke and fumes. See also PEROXIDES, ORGANIC.

BRM500   CAS: 1948-33-0   HR: 3
tert-BUTYLHYDROQUINONE
mf: $C_{10}H_{14}O_2$ mw: 166.24

B

PROP: White crystalline solid; characteristic odor. Mp: 126.5–128.5°. Sol in alc, ether; insol in water.
SYNS: MONO-tert-BUTYLHYDROQUINONE □ MTBHQ □ SUSTANE □ TBHQ (FCC) □ TENOX TBHQ
CONSENSUS REPORTS: Reported in EPA TSCA Inventory.
SAFETY PROFILE: Poison by intraperitoneal route. Moderately toxic by ingestion and inhalation. Mutation data reported. When heated to decomposition it emits acrid smoke and irritating fumes.

BRQ100   CAS: 551-08-6   HR: 2
3-BUTYLIDENE PHTHALIDE
mf: $C_{12}H_{12}O_2$ mw: 188.24
PROP: Needles from $CHCl_3$. Mp: 82–83°.
SYNS: BUTYLIDENE PHTHALIDE □ n-BUTYLIDENE PHTHALIDE □ 1(3H)-ISOBENZOFURANONE, 3-BUTYLIDENE-(9CI) □ PHTHALIDE, 3-BUTYLIDENE-
CONSENSUS REPORTS: Reported in EPA TSCA Inventory.
SAFETY PROFILE: Moderately toxic by ingestion. A skin irritant. When heated to decomposition it emits acrid smoke and irritating fumes.

BRQ350   HR: 3
BUTYL ISOBUTYRATE
mf: $C_8H_{16}O_2$ mw: 44.44
PROP: Colorless liquid; apple-pineapple odor. D: 0.859–0.864, refr index: 1.401, flash p: 113°F. Misc with alc, ether, fixed oils; insol in glycerin, propylene glycol, water @ 166°.
SYN: FEMA No. 2188
SAFETY PROFILE: Flammable liquid. When heated to decomposition it emits acrid smoke and irritating fumes.

BRQ500   CAS: 111-36-4   HR: 3
n-BUTYL ISOCYANATE
DOT: UN 2485
mf: C_5H_9NO mw: 99.15
PROP: Colorless liquid. Bp: 115°, d: 0.880 @ 20°/4°.
SYNS: BIC □ ISOCYANIC ACID, BUTYL ESTER
CONSENSUS REPORTS: Reported in EPA TSCA Inventory.

DOT CLASSIFICATION: 3; Label: Flammable Liquid, Poison; DOT Class: 6.1; Label: Poison; DOT Class: 6.1; Label: Poison, Flammable Liquid; DOT Class: 3; Label: Flammable Liquid, Poison
SAFETY PROFILE: A poison by ingestion and intravenous routes. Mildly toxic by inhalation. A powerful irritant to eyes, skin, and mucous membranes. A flammable liquid. See also CYANATES and NITROGEN MONOXIDE.

BRQ800   CAS: 73791-40-9   HR: 3
BUTYL(ISOPROPYL)ARSINIC ACID
mf: $C_7H_{17}AsO_2$ mw: 208.16
SYNS: ARSINE OXIDE, BUTYLHYDROXYISOPROPYL- □ BUTYLHYDROXYISOPROPYLARSINE OXIDE
OSHA PEL: TWA 0.5 mg(As)/m^3
SAFETY PROFILE: Poison by intravenous route. When heated to decomposition it emits toxic fumes of As.

BRR600   CAS: 138-22-7   HR: 3
n-BUTYL LACTATE
mf: $C_7H_{14}O_3$ mw: 146.21
PROP: Liquid. Sltly sol in water; misc in alc and ether. Mp: −43°, bp: 188°, flash p: 160°F (OC), d: 0.968, autoign temp: 720°F, vap d: 5.04, vap press: 0.4 mm @ 20°.
SYNS: BUTYL α-HYDROXYPROPIONATE □ BUTYL LACTATE □ 2-HYDROXYPROPANOIC ACID, BUTYL ESTER □ LACTIC ACID, BUTYL ESTER
CONSENSUS REPORTS: Reported in EPA TSCA Inventory.
OSHA PEL: TWA 5 ppm
ACGIH TLV: TWA 5 ppm
SAFETY PROFILE: Poison by intraperitoneal route. A skin irritant. Toxic concentration in air for humans is about 4 ppm. Flammable when exposed to heat or flame; can react with oxidizing materials. To fight fire, use alcohol foam, foam, CO_2, dry chemical. When heated to decomposition it emits acrid smoke and irritating fumes. See also ESTERS, n-BUTYL ALCOHOL, and LACTIC ACID.

BRR800   CAS: 5606-24-6   HR: 3
N-BUTYLMELAMINE
mf: $C_7H_{14}N_6$ mw: 182.27
SYNS: 2,6-DIAMINO-4-BUTYLAMINO-s-TRIAZINE □
MELAMINE, BUTYL- □ s-TRIAZINE, 2,6-DIAMINO-4-
BUTYLAMINO- □ 1,3,5-TRIAZINE-2,4,6-TRIAMINE, N-
BUTYL-
SAFETY PROFILE: A poison by
intravenous route. When heated to
decomposition it emits toxic vapors of NO_x.

BRR900   CAS: 109-79-5   HR: 3
n-BUTYL MERCAPTAN
DOT: UN 2347
mf: $C_4H_{10}S$ mw: 90.20
PROP: Colorless liquid; skunk-like odor.
Mp: −116°, bp: 98°, d: 0.8365 @ 25°/4°,
flash p: 35°F, vap d: 3.1. IDLH 500 ppm.
SYNS: BUTANETHIOL (OSHA) □ BUTYL
MERCAPTAN □ n-BUTYL MERCAPTAN (ACGIH,DOT) □
NCI-C60866
CONSENSUS REPORTS: Reported in EPA
TSCA Inventory.
OSHA PEL: TWA 0.5 ppm
ACGIH TLV: TWA 0.5 ppm
DFG MAK: 0.5 ppm (1.9 mg/m³)
NIOSH REL: (n-Alkane Mono Thiols) CL
0.5 ppm/15M
DOT CLASSIFICATION: 3; Label:
Flammable Liquid
SAFETY PROFILE: Poison by
intraperitoneal route. Moderately toxic by
ingestion. An eye irritant. Dangerous fire
hazard by exposure to heat, flame, sparks, or
powerful oxidizers. Reacts violently with
HNO_3. Incompatible with acids, acid fumes,
oxidizing materials, heat, flame, and sparks.
To fight fire, use alcohol foam. When
heated to decomposition it emits toxic SO_x.
See also MERCAPTANS.

BRT000   CAS: 532-34-3   HR: 2
n-BUTYL MESITYL OXIDE OXALATE
mf: $C_{12}H_{18}O_4$ mw: 226.30
PROP: Yellow to pale-red liquid. Bp:
256–270°, d: 1.052–1.060 @ 20°/4°, flash p:
315°F.
SYNS: BMOO □ BUTOPYRONOXYL □ BUTYL-3,4-
DIHYDRO-2,2-DIMETHYL-4-OXO-2H-PYRAN-6-
CARBOXYLATE □ n-BUTYL ESTER of 3,4-DIHYDRO-2,2-
DIMETHYL-4-OXO-2H-PYRAN-6-CARBOXYLIC ACID □
n-BUTYLMESITYLOXID OXALATE □ 2-CARBO-n-
BUTOXY-6,6-DIMETHYL-5,6-DIHYDRO-1,4-PYRONE □
3,4-DIHYDRO-2,2-DIMETHYL-4-OXO-2H-PYRAN-6-
CARBOXYLIC ACID-n-BUTYL ESTER □ DIHDYROPYR-
ONE □ α,α-DIMETHYL-α'-CARBOBUTOXY-DIHYDRO-
γ-PYRONE □ 2,2-DIMETHYL-6-CARBOBUTOXY-2,3-
DIHYDRO-4-PYRONE □ ENT 9 □ INDALONE
CONSENSUS REPORTS: Reported in EPA
TSCA Inventory.
SAFETY PROFILE: Moderately toxic by
ingestion. Produces liver necrosis in
experimental animals. A mild skin irritant.
See also OXALATES and ESTERS.
Combustible when exposed to heat or
flame. When heated to decomposition it
emits acrid fumes.

BRU500   CAS: 83-66-9   HR: 3
6-tert-BUTYL-3-METHYL-2,4-DINITRO
    ANISOLE
mf: $C_{12}H_{16}N_2O_5$ mw: 268.30
SYNS: 2,6-DINITRO-3-METHOXY-4-tert-
BUTYLTOLUENE □ MUSK AMBRETTE
CONSENSUS REPORTS: Reported in EPA
TSCA Inventory.
SAFETY PROFILE: A poison by ingestion.
Mutation data reported. A skin irritant.
When heated to decomposition it emits
toxic fumes of NO_x. See also AROMATIC
AMINES.

BRU780   CAS: 628-28-4   HR: 3
BUTYL METHYL ETHER (DOT)
DOT: UN 2350
mf: $C_5H_{12}O$ mw: 88.17
SYNS: BUTANE, 1-METHOXY-(9CI) □ ETHER, BUTYL
METHYL □ α-METHOXYBUTANE □ 1-
METHOXYBUTANE □ METHYL BUTYL ETHER □
METHYL n-BUTYL ETHER
DOT CLASSIFICATION: 3; Label:
Flammable Liquid
SAFETY PROFILE: Poison by inhalation. A
flammable liquid. When heated to
decomposition it emits acrid smoke and
irritating vapors.

BRV500   CAS: 544-16-1   HR: 3
n-BUTYL NITRITE
mf: $C_4H_9NO_2$ mw: 103.14

PROP: Oily liquid, characteristic odor, misc in alc and ether. Bp: 78°, d: 0.9114 @ 0°/4°, vap d: 3.5, flash p: 10°.

SYNS: BUTYL NITRITE (DOT) □ NBN □ NCI-C56553 □ NITROUS ACID-n-BUTYL ESTER

CONSENSUS REPORTS: Reported in EPA TSCA Inventory.

SAFETY PROFILE: A poison by ingestion and intraperitoneal routes. Mildly toxic by inhalation. An irritant. Human systemic effects by ingestion: methemoglobinemia-carboxyhemoglobinemia. Resembles amyl nitrite in causing fall in blood pressure, headache, pulse throbbing, and weakness. Mutation data reported. Flammable when exposed to heat or flame or by spontaneous chemical reaction. When heated to decomposition it emits toxic fumes of NO_x. See also NITRITES, n-BUTYL ALCOHOL, and ESTERS.

BRV750   CAS: 924-43-6   HR: 3
sec-BUTYL NITRITE

mf: $C_4H_9NO_2$ mw: 103.14

PROP: Liquid. Bp: 68°, d: 0.8981 @ 0°/4°, vap d: 3.5.

SYNS: NITROUS ACID-sec-BUTYL ESTER □ NITROUS ACID-1-METHYL PROPYL ESTER

CONSENSUS REPORTS: Reported in EPA TSCA Inventory.

SAFETY PROFILE: Moderately toxic by ingestion, inhalation, and intraperitoneal routes. Mutation data reported. Flammable when exposed to heat or flame or by spontaneous chemical reaction. An oxidizer. Potentially explosive. To fight fire, use water, spray, foam, dry chemical. When heated to decomposition it emits toxic fumes of NO_x. See also n-BUTYL NITRITE, NITRITES, and ESTERS.

BRY500   CAS: 924-16-3   HR: 3
n-BUTYL-N-NITROSO-1-BUTAMINE

mf: $C_8H_{18}N_2O$ mw: 158.28

PROP: Pale-yellow liquid. Bp: 235°.

SYNS: DBN □ DBNA □ DI-n-BUTYLNITROSAMIN (GERMAN) □ DIBUTYLNITROSOAMINE □ DI-n-BUTYLNITROSAMINE □ N,N-DI-n-BUTYLNITROS-AMINE □ N,N-DIBUTYLNITROSOAMINE □ NDBA □

N-NITROSODIBUTYLAMINE □ N-NITROSODI-n-BUTYLAMINE (MAK) □ RCRA WASTE NUMBER U172

CONSENSUS REPORTS: NTP 10th Report on Carcinogens. IARC Cancer Review: Group 2B IMEMDT 7,56,87; Animal Suþ⁻ cient Evidence IMEMDT 28,151,82; IMEMDT 17,51,78; IMEMDT 4,197,74; Human Limited Evidence IMEMDT 17,51,78. Community Right-To-Know List. EPA Genetic Toxicology Program. Reported in EPA TSCA Inventory.

DFG MAK: Animal Carcinogen, Suspected Human Carcinogen

SAFETY PROFILE: Confirmed carcinogen with experimental carcinogenic, tumorigenic, and neoplastigenic data. Moderately toxic by ingestion, subcutaneous, and intraperitoneal routes. Experimental teratogenic effects. Human mutation data reported. When heated to decomposition it emits toxic fumes of NO_x. See also NITROSAMINES.

BSA250   CAS: 869-01-2   HR: 3
n-BUTYLNITROSOUREA

mf: $C_5H_{11}N_3O_2$ mw: 145.19

PROP: Solid. Mp: 82.5–84°.

SYNS: BNU □ BUTYLNITROSOHARNSTOFF (GERMAN) □ N-n-BUTYL-N-NITROSOUREA □ 1-BUTYL-1-NITROSOUREA □ N-NITROSOBUTYLUREA

CONSENSUS REPORTS: EPA Genetic Toxicology Program.

SAFETY PROFILE: Suspected carcinogen with experimental carcinogenic and tumorigenic data. A poison by ingestion. Moderately toxic by subcutaneous route. Experimental teratogenic and reproductive effects. Mutation data reported. When heated to decomposition it emits toxic fumes of NO_x. See also NITROSAMINES.

BSA500   CAS: 3913-02-8   HR: 1
2-BUTYL-1-OCTANOL

mf: $C_{12}H_{26}O$ mw: 186.38

PROP: Liquid. Mp: −80°, flash p: 230°F (OC), bp: 253.3°, d: 0.8355 @ 20°/20°, vap d: 6.42.

SYN: 2-BUTYLOCTYL ALCOHOL

CONSENSUS REPORTS: Reported in EPA TSCA Inventory.

SAFETY PROFILE: Mildly toxic by ingestion. A skin and eye irritant. See also ALCOHOLS. Combustible when exposed to heat or flame. Incompatible with oxidizing materials. To fight fire, use CO_2, dry chemical. When heated to decomposition it emits acrid and irritating fumes.

BSC250　CAS: 107-71-1　HR: 3
tert-BUTYL PERACETATE
mf: $C_6H_{12}O_3$　mw: 132.18
PROP: Clear, colorless, benzene solution; insol in water; sol in org solvs. D: 0.923, vap press: 50 mm @ 26°, flash p: <80°F (COC).
SYNS: t-BUTYL PERACETATE □ t-BUTYL PEROXYACETATE □ tert-BUTYL PEROXYACETATE, >76% in solution (DOT) □ ETHANEPEROXOIC ACID, 1,1-DIMETHYLETHYL ESTER □ LUPERSOL 70 □ TRIGONOX F-C50
CONSENSUS REPORTS: Reported in EPA TSCA Inventory.
DFG MAK: Moderate skin irritant
DOT CLASSIFICATION: Forbidden
SAFETY PROFILE: Moderately toxic by ingestion. Mildly toxic by inhalation. Moderate skin and eye irritant. A shock- and heat-sensitive explosive. Dangerous fire hazard when exposed to heat, flame, reducing agents. To fight fire, use dry chemical, alcohol foam, spray and mist. When heated to decomposition it emits acrid smoke and fumes. See also PEROXIDES, ORGANIC; and ESTERS.

BSC500　CAS: 614-45-9　HR: 3
tert-BUTYL PERBENZOATE
mf: $C_{11}H_{14}O_3$　mw: 194.25
PROP: Colorless to slightly yellow liquid; mild aromatic odor. Bp: 112° (decomp), flash p: 19°, fp: 8°, vap press: 0.33 mm @ 50°, d: 1.0. Insol in water; sol in org solvs.
SYNS: terc.BUTYLESTER KYSELINY PEROXY-BENZOOVE (CZECH) □ terc.BUTYLPER-BENZOAN (CZECH) □ t-BUTYL PERBENZOATE □ t-BUTYL PEROXY BENZOATE □ ESPEROX 10 □ NOVOX □ PERBENZOATE de BUTYLE TERTIAIRE (FRENCH) □ TRIGONOX C

CONSENSUS REPORTS: Reported in EPA TSCA Inventory.

SAFETY PROFILE: Moderately toxic by ingestion. A skin and eye irritant. Questionable carcinogen with experimental tumorigenic data. Mutation data reported. See also PEROXIDES, ORGANIC. Potentially explosive when heated above 115°C. Explosive reaction on contact with organic matter or copper(I) bromide + limonene. When heated to decomposition it emits acrid smoke and fumes.

BSD000　CAS: 19910-65-7　HR: 2
sec-BUTYL PEROXYDICARBONATE
mf: $C_{10}H_{18}O_6$　mw: 234.28
SYNS: DI-sec-BUTYL PEROXYDICARBONATE □ DI-sec-BUTYL PEROXYDICARBONATE, not more than 52% in solution (DOT) □ DI-sec-BUTYL PEROXYDICARBONATE, technically pure (DOT)
CONSENSUS REPORTS: Reported in EPA TSCA Inventory.
SAFETY PROFILE: Moderately toxic by skin contact. See also PEROXIDES, ORGANIC. When heated to decomposition it emits acrid smoke and irritating fumes.

BSD250　CAS: 927-07-1　HR: 3
tert-BUTYL PEROXYPIVALATE
mf: $C_9H_{18}O_3$　mw: 174.27
PROP: Colorless liquid. D: 0.854 @ 25°/25°, fp: <19°, flash p: >155°F (OC), rapid decomp @ 21°. Insol in water and ethylene glycol; sol in most org solvs.
SYNS: t-BUTYL PEROXYPIVALATE □ tert-BUTYL PERPIVALATE □ tert-BUTYL TRIMETHYLPEROXY-ACETATE □ ESPEROX 31M □ LUPERSOL 11 □ TRIGONOX 25/75 □ TRIGONOX 25-C75
CONSENSUS REPORTS: Reported in EPA TSCA Inventory.
SAFETY PROFILE: Mildly toxic by ingestion. Moderately flammable by heat, flame (sparks), oxidizers. Can explode on heating. To fight fire, use water, fog, mist, alcohol foam, dry chemical. When heated to decomposition it emits acrid smoke and fumes. See also PEROXIDES, ORGANIC.

BSE000   CAS: 89-72-5   HR: 3
o-sec-BUTYLPHENOL
mf: $C_{10}H_{14}O$ mw: 150.24
PROP: Colorless liquid. Bp: 226–228° @ 25 mm, fp: 12°, flash p: 225°F, d: 0.981 @ 25°/25°.
SYN: 2-sec.-BUTYLFENOL (CZECH)
CONSENSUS REPORTS: Reported in EPA TSCA Inventory.
OSHA PEL: TWA 5 ppm (skin)
ACGIH TLV: TWA 5 ppm (skin)
SAFETY PROFILE: A poison by intraperitoneal and intravenous routes. Moderately toxic by ingestion and skin contact. A severe skin and eye irritant. Combustible when exposed to heat or flame. To fight fire, use foam, spray, CO_2, dry chemical. When heated to decomposition it emits acrid and irritating fumes. See also PHENOL and other butyl phenols.

BSE250   CAS: 99-71-8   HR: 3
p-sec-BUTYLPHENOL
mf: $(CH_3CHC_2H_5)C_6H_4OH$ mw: 150.2
PROP: Nearly white flakes. Bp: 135.4–136.5° @ 25 mm, fp: 51°, flash p: 240°F, d: 0.963 @ 60°/60°, mp: 60°.
SYN: 4-sec BUTYL PHENOL
CONSENSUS REPORTS: Reported in EPA TSCA Inventory.
SAFETY PROFILE: Poison by intravenous and intraperitoneal routes. Moderately toxic by ingestion. Combustible when exposed to heat or flame. When heated to decomposition it emits toxic fumes. To fight fire, use foam, CO_2, dry chemical. Incompatible with oxidizing materials. See also PHENOL and other butyl phenols.

BSE450   CAS: 1638-22-8   HR: 2
4-n-BUTYLPHENOL
mf: $C_{10}H_{14}O$ mw: 150.24
PROP: Solid or liquid. D: 0.976 @ 22°/4°, mp: 22°, bp: 248°.
CONSENSUS REPORTS: Reported in EPA TSCA Inventory.

SAFETY PROFILE: A poison. Questionable carcinogen with experimental tumorigenic data. When heated to decomposition it emits acrid smoke and irritating fumes. See also PHENOL and other butyl phenols.

BSE500   CAS: 98-54-4   HR: 3
4-t-BUTYLPHENOL
mf: $C_{10}H_{14}O$ mw: 150.24
PROP: Crystals, needles, or practically white flakes. Mp: 99°, bp: 236–238°, d: 0.9081 @ 114°/4°, vap press: 1 mm @ 70.0°, vap d: 5.1.
SYNS: p-tert-BUTYLFENOL (CZECH) □ BUTYLPHEN □ p-tert-BUTYLPHENOL (MAK) □ 4-(1,1-DIMETHYL-ETHYL)PHENOL □ 1-HYDROXY-4-tert-BUTYLBENZ-ENE □ UCAR BUTYLPHENOL 4-T
CONSENSUS REPORTS: Reported in EPA TSCA Inventory.
DFG MAK: 0.08 ppm (0.5 mg/m³)
SAFETY PROFILE: Poison by intraperitoneal route. Moderately toxic by skin contact and ingestion. A skin and severe eye irritant. Questionable carcinogen with experimental neoplastigenic data. Combustible when exposed to heat or flame; can react with oxidizing materials. To fight fire, use foam, CO_2, dry chemical. When heated to decomposition it emits acrid and irritating fumes. See also PHENOL and other butyl phenols.

BSF750   CAS: 1126-79-0   HR: 2
BUTYL PHENYL ETHER
mf: $C_{10}H_{14}O$ mw: 150.24
PROP: Liquid. Flash p: 180°F (OC), d: 0.9, vap d: 5.2, mp: −19°, bp: 210°.
SYN: BUTOXYPHENYL
CONSENSUS REPORTS: Reported in EPA TSCA Inventory.
SAFETY PROFILE: Moderately toxic by ingestion. See also ETHERS. When heated to decomposition it emits acrid and irritating fumes.

BSH100   CAS: 87-18-3   HR: 2
p-tert-BUTYLPHENYL SALICYLATE
mf: $C_{17}H_{18}O_3$ mw: 270.13

SYNS: BENZOIC ACID, 2-HYDROXY-, 4-(1,1-DIMETHYLETHYL)PHENYL ESTER □ p-terc.BUTYL-FENYLESTER KYSELINY SALICYLOVE
CONSENSUS REPORTS: Reported in EPA TSCA Inventory.
SAFETY PROFILE: Moderately toxic by ingestion. When heated to decomposition it emits acrid smoke and irritating fumes.

BSH250   CAS: 78-48-8   HR: 3
BUTYL PHOSPHOROTRITHIOATE
mf: $C_{12}H_{27}OPS_3$ mw: 314.54
PROP: Liquid. Bp: 167–170° @ 1 mm, d: 1.06 @ 20 mm. Insol in water; sol in aliphatic, aromatic, and chlorinated hydrocarbons.
SYNS: B-1,776 □ BUTIFOS □ BUTIPHOS □ CHEMAGRO 1,776 □ CHEMAGRO B-1776 □ DEF □ DEF DEFOLIANT □ DE-GREEN □ E-Z-OFF D □ FOS-FALL "A" □ ORTHO PHOSPHATE DEFOLIANT □ S,S,S-TRIBUTYL PHOSPHOROTRITHIOATE □ S,S,S-TRIBUTYL TRITHIOPHOSPHATE
CONSENSUS REPORTS: Reported in EPA TSCA Inventory.
SAFETY PROFILE: A poison by ingestion, skin contact, and intraperitoneal routes. Experimental reproductive effects. Animal experiments show an anti-cholinesterase effect. When heated to decomposition it emits toxic fumes of PO_x and SO_x. See also PARATHION, PHOSPHATES, ESTERS, and SULFATES.

BSI000   CAS: 536-69-6   HR: 3
5-BUTYL PICOLINIC ACID
mf: $C_{10}H_{13}NO_2$ mw: 179.24
PROP: Plates from pet ether. Mp: 108–109°.
SYNS: 5-BUTYL-2-PYRIDINECARBOXYLIC ACID □ FUSARIC ACID □ FUSARINIC ACID
CONSENSUS REPORTS: EPA Genetic Toxicology Program. Reported in EPA TSCA Inventory.
SAFETY PROFILE: A poison by ingestion, intraperitoneal, subcutaneous, and intravenous routes. When heated to decomposition it emits toxic fumes of NO_x.

BSJ500   CAS: 590-01-2   HR: 3
BUTYL PROPANOATE
DOT: UN 1914

mf: $C_7H_{14}O_2$ mw: 130.2
PROP: Water-white liquid; apple-like odor. Mp: −89.6°, bp: 145.4°, flash p: 90°F, d: 0.893 @ 0°/0°, autoign temp: 800°F, vap d: 4.49.
SYNS: BUTYL PROPIONATE □ n-BUTYL PROPIONATE □ PROPANOIC ACID BUTYLESTER (9CI)
CONSENSUS REPORTS: Reported in EPA TSCA Inventory.
DOT CLASSIFICATION: 3; Label: Flammable Liquid
SAFETY PROFILE: Mildly toxic by ingestion. A skin irritant. Dangerously flammable when exposed to heat or flame. To fight fire, use foam, CO_2, dry chemical. Incompatible with oxidizing materials. See also ESTERS, n-BUTYL ALCOHOL, and PROPIONIC ACID.

BSK000   CAS: 98-29-3   HR: 3
4-tert-BUTYLPYROCATECHOL
mf: $C_{10}H_{14}O_2$ mw: 166.24
PROP: Crystals. Fp: 52°, bp: 285°, flash p: 265°F, d: 1.049 @ 60°/25°.
SYNS: 4-tert-BUTYLCATECHOL □ p-tert-BUTYLPYROCATECHOL □ 4-tert-BUTYLPYROKATECHIN (CZECH) □ 4-(1,1-DIMETHYLETHYL)-1,2-BENZENEDIOL □ SYNOX TBC
CONSENSUS REPORTS: Reported in EPA TSCA Inventory.
SAFETY PROFILE: A poison by intravenous route. Moderately toxic by ingestion and skin absorption. A severe skin and eye irritant. Mutation data reported. Combustible when exposed to heat or flame. To fight fire, use CO_2, dry chemical, fog, mist. When heated to decomposition it emits acrid and irritating fumes.

BSL500   CAS: 2273-43-0   HR: 3
BUTYL STANNOIC ACID
mf: $C_4H_{10}O_2Sn$ mw: 208.83
PROP: White infusible solid. Sol in Me_2CO.
SYN: BUTYLHYDROXYOXOSTANNANE
CONSENSUS REPORTS: Reported in EPA TSCA Inventory.
OSHA PEL: TWA 0.1 mg(Sn)/m³ (skin)
ACGIH TLV: TWA 0.1 mg(Sn)/m³; STEL 0.2 mg(Sn)/m³ (skin).

NIOSH REL: (Organotin Compounds) TWA 0.1 mg(Sn)/m^3
SAFETY PROFILE: A poison by intravenous route. See also TIN COMPOUNDS. When heated to decomposition it emits acrid smoke and irritating fumes.

BSM000   CAS: 339-43-5   HR: 3
1-BUTYL-3-SULFANILYL UREA
mf: $C_{11}H_{17}N_3O_3S$ mw: 271.37
PROP: Solid. Mp: 144–145°.
SYNS: ALENTIN □ N-(4-AMINOBENZENESULFONYL)-N'-BUTYLUREA □ 4-AMINO-N-((BUTYLAMINO)CARBONYL)BENZENE-SULFONAMIDE □ AMINOPHENUROBUTANE □ BUCARBAN □ BUCROL □ BUKARBAN □ BURCOL □ BUTISULFINA □ N'-(BUTYLCARBAMOYL)SULFANIL-AMIDE □ N^1-(BUTYLCARBAMOYL)SULFANILAMIDE □ N-BUTYLSULFANILYLUREA □ CARBUTAMID □ CARBUTAMIDE □ CICLORAL □ DIABORAL □ EMED-AN □ GLUCIDORAL □ GLUCOFREN □ GLYBUTAMIDE □ INBUTON □ INVENOL □ NADISAN □ NADIZAN □ NORBORAL □ ORANIL □ ORANYL □ ORASULIN □ N^1-SULFANILYL-N^2-BUTYLCARBAMIDE □ N^1-SULFANIL-YL-N^2-BUTYLUREA □ N-SULFANILYL-N'BUTYLUREE (FRENCH) □ U 6987
CONSENSUS REPORTS: Reported in EPA TSCA Inventory.
SAFETY PROFILE: A poison by intraperitoneal route. Moderately toxic by ingestion and subcutaneous routes. An experimental teratogen. Other experimental reproductive effects. When heated to decomposition it emits very toxic fumes of NO_x and SO_x.

BSN500   CAS: 628-83-1   HR: 3
n-BUTYL THIOCYANATE
mf: C_5H_9NS mw: 115.21
SYNS: n-BUTYL RHODANATE □ BUTYRHODANID (GERMAN) □ 1-THIOCYANOBUTANE
CONSENSUS REPORTS: Reported in EPA TSCA Inventory.
SAFETY PROFILE: A poison by ingestion and subcutaneous routes. When heated to decomposition it emits very toxic fumes of NO_x and SO_x. See also THIOCYANATES.

BSO200   CAS: 70303-47-8   HR: 3
(BUTYLTHIO)TRIOCTYLSTANNANE
mf: $C_{28}H_{60}SSn$ mw: 547.63
SYNS: STANNANE, (BUTYLTHIO)TRIOCTYL- □ TRIOCTYL(BUTYLTHIO)STANNANE
OSHA PEL: TWA 0.1 mg(Sn)/m^3
ACGIH TLV: TWA 0.1 mg(Sn)/m^3; STEL 0.2 mg/m^3 (skin)
NIOSH REL: (Organotin compound): 10H TWA 0.1 mg(Sn)/m^3
SAFETY PROFILE: Poison by intraperitoneal route. When heated to decomposition it emits toxic fumes of SO_x and Sn.

BSO500   CAS: 1516-32-1   HR: 3
n-BUTYL THIOUREA
mf: $C_5H_{12}N_2S$ mw: 132.25
SYN: USAF D-5
CONSENSUS REPORTS: Reported in EPA TSCA Inventory.
SAFETY PROFILE: A poison by intraperitoneal route. When heated to decomposition it emits very toxic fumes of NO_x and SO_x.

BSP250   CAS: 5593-70-4   HR: 3
BUTYL TITANATE
mf: $C_{16}H_{36}O_4 \cdot Ti$ mw: 340.42
PROP: Colorless to light-yellow liquid or oil with the odor of butanol. Mp: −55°, bp: 155° @ 1 mm, d: 0.993 @ 25°/4°, flash p: 170°F, vap d: 11.5.
SYN: TETRABUTYLTITANATE (CZECH)
CONSENSUS REPORTS: Reported in EPA TSCA Inventory.
ACGIH TLV: Animal Carcinogen, Suspected Human Carcinogen
SAFETY PROFILE: Suspected carcinogen. A poison by intravenous route. Moderately toxic by ingestion. See n-BUTYL ALCOHOL and TITANIUM COMPOUNDS. Flammable when exposed to heat or flame. To fight fire, use water, spray, foam, dry chemical. Incompatible with oxidizing materials. When heated to decomposition it emits acrid and irritating fumes.

BSP500   CAS: 98-51-1   HR: 2
p-tert-BUTYLTOLUENE
mf: $C_{11}H_{16}$ mw: 148.27
PROP: Colorless liquid. D: 0.861 @ 20°/4°,
mp: −54°, bp: 189–192°. IDLH 100 ppm.
SYNS: p-METHYL-tert-BUTYLBENZENE □ 1-METHYL-
4-tert-BUTYLBENZENE □ TBT
CONSENSUS REPORTS: Reported in EPA
TSCA Inventory.
OSHA PEL: TWA 10 ppm; STEL 20 ppm
ACGIH TLV: TWA 1 ppm
DFG MAK: 10 ppm (60 mg/m³)
SAFETY PROFILE: Moderately toxic by
inhalation and ingestion. A skin and human
eye irritant. Human systemic effects by
inhalation: nausea or vomiting, conjunctiva
irritation, unspecified effects on the sense of
taste. Inhalation of vapors causes irritation
of lungs and depression of central nervous
system. Prolonged exposure may result in
damage to liver and kidneys. Flammable
when exposed to heat or flame.
Incompatible with oxidizing materials.
When heated to decomposition it emits
acrid smoke and fumes.

BSQ000   CAS: 64-77-7   HR: 2
1-BUTYL-3-(p-TOLYL SULFONYL)UREA
mf: $C_{12}H_{18}N_2O_3S$ mw: 270.38
PROP: Crystals. Mp: 128.5–129.5°. Insol in
H_2O; sol in $CHCl_3$, dil acids, and alkalies.
SYNS: AGLICID □ ARKOZAL □ ARTOSIN □
ARTOZIN □ BUTAMID □ N-((BUTYLAMINO)-
CARBONYL)-4-METHYLBENZENESULFONAMIDE □ 1-
BUTYL-3-(p-METHYLPHENYLSULFONYL)UREA □ n-
BUTYL-N'-p-TOLUENESULFONYLUREA □ N-n-BUTYL-
N'-TOSYLUREA □ 1-BUTYL-3-TOSYLUREA □ BZ 55 □ D
860 □ DIABEN □ DIABETAMID □ DIABETOL □
DIABUTON □ DOLIPOL □ DRABET □ HLS 831 □
IPOGLICONE □ MOBENOL □ NCI-CO1763 □ ORABET
□ ORALIN □ OREZAN □ ORINASE □ ORINAZ □
OTERBEN □ RASTINON □ SK-TOLBUTAMIDE □ N-
(SULFONYL-p-METHYLBENZENE)-N'-N-BUTYLUREA □
TOLBUSAL □ TOLBUTAMID □ TOLBUTAMIDE □ 1-p-
TOLUENESULFONYL-3-BUTYLUREA □ TOLUINA □
TOLUMID □ TOLUVAN □ N-(p-TOLYLSULFONYL)-N'-
BUTYLCARBAMIDE □ 3-(p-TOLYL-4-SULFONYL)-1-
BUTYLUREA □ TOLYLSULFONYLBUTYLUREA □
WILLBUTAMIDE
CONSENSUS REPORTS: NCI
Carcinogenesis Bioassay (feed); No

Evidence: mouse, rat NCITR* NCI-CG-
TR-31,77. Reported in EPA TSCA
Inventory. EPA Genetic Toxicology
Program.
SAFETY PROFILE: Moderately toxic by
ingestion and several other routes. A human
teratogen. Human reproductive effects by
ingestion and possibly other routes:
stillbirth, developmental abnormalities of
the cardiovascular (circulatory) system and
urogenital system, and unspecified neonatal
effects. Human systemic effects by
ingestion: nausea or vomiting, hypoglycemia.
Other experimental teratogenic and
reproductive effects. Mutation data
reported. Implicated in aplastic anemia.
When heated to decomposition it emits very
toxic fumes of NO_x and SO_x.

BSR000   CAS: 7521-80-4   HR: 3
BUTYLTRICHLOROSILANE
DOT: UN 1747
mf: $C_4H_9Cl_3Si$ mw: 191.57
PROP: Colorless liquid. Vap d: 6.4, flash p:
130°F (OC), d: 1.16 @ 20°/4°, bp:
148–149°.
SYN: TRICHLOROBUTYLSILANE
CONSENSUS REPORTS: Reported in EPA
TSCA Inventory.
DOT CLASSIFICATION: 8; Label: Corrosive
SAFETY PROFILE: A corrosive poison. See
also CHLOROSILANE. Flammable liquid
when exposed to heat, flame (sparks), or
oxidizers. To fight fire, use water to blanket
fire, fog, mist, dry chemical, alcohol foam.
Reacts with water or steam to produce heat
and toxic and corrosive fumes. When heated
to decomposition it emits highly toxic fumes
of Cl^-.

BSR250   CAS: 1118-46-3   HR: 2
BUTYL TRICHLORO STANNANE
mf: $C_4H_9Cl_3Sn$ mw: 282.17
PROP: Liquid. D: 0.85 @ 20°/4°, bp: 93°
@ 10 mm.
SYN: CHLORID-N-BUTYLCINICITY (CZECH)
CONSENSUS REPORTS: Reported in EPA
TSCA Inventory.

OSHA PEL: TWA 0.1 mg(Sn)/m³ (skin)
ACGIH TLV: TWA 0.1 mg(Sn)/m³; STEL
0.2 mg(Sn)/m³ (skin).
NIOSH REL: (Organotin Compounds) TWA
0.1 mg(Sn)/m³
SAFETY PROFILE: Moderately toxic by
ingestion. A severe skin and eye irritant.
Mutation data reported. See also TIN
COMPOUNDS. When heated to
decomposition it emits toxic fumes of Cl⁻.

BSS000   CAS: 25852-70-4   HR: 2
**BUTYLTRIS(ISOOCTYLOXYCARBONYL
   METHYLTHIO)STANNANE**
mf: $C_{34}H_{66}O_6S_3Sn$ mw: 785.87
SYN: BUTYLTRIS(2-ETHYLHEXYLOXYCARBONYL-
METHYLTHIO)STANNANE
CONSENSUS REPORTS: Reported in EPA
TSCA Inventory.
OSHA PEL: TWA 0.1 mg(Sn)/m³ (skin)
ACGIH TLV: TWA 0.1 mg(Sn)/m³; STEL
0.2 mg(Sn)/m³ (skin).
NIOSH REL: (Organotin Compounds) TWA
0.1 mg(Sn)/m³
SAFETY PROFILE: Moderately toxic by
ingestion. See also TIN COMPOUNDS.
When heated to decomposition it emits
toxic fumes of SO_x.

BSS100   CAS: 109-42-2   HR: 1
BUTYL 10-UNDECENOATE
mf: $C_{15}H_{28}O_2$ mw: 240.43
SYNS: BUTYL UNDECYLENATE □ 10-UNDECENOIC
ACID, BUTYL ESTER
CONSENSUS REPORTS: Reported in EPA
TSCA Inventory.
SAFETY PROFILE: Mildly toxic by
ingestion. A skin irritant. When heated to
decomposition it emits acrid smoke and
irritating fumes.

BSS250   CAS: 592-31-4   HR: 2
N-BUTYLUREA
mf: $C_5H_{12}N_2O$ mw: 116.19
PROP: Needles from C_6H_6. Mp: 96°.
SYN: NCI-CO2131
CONSENSUS REPORTS: Reported in EPA
TSCA Inventory. EPA Genetic Toxicology
Program.

SAFETY PROFILE: Moderately toxic by
parenteral route. Mutation data reported.
When heated to decomposition it emits
toxic fumes of NO_x.

BSS500   HR: 3
**1-BUTYLUREA and SODIUM NITRITE
   (2:1)**
SAFETY PROFILE: Suspected carcinogen
with experimental carcinogenic,
neoplastigenic, and tumorigenic data. An
experimental teratogen. When heated to
decomposition it emits toxic fumes of NO_x.
See also NITRITES.

BST500   CAS: 110-65-6   HR: 3
2-BUTYNE-1,4-DIOL
DOT: UN 2716
mf: $C_4H_6O_2$ mw: 86.10
PROP: Plates from EtOAc or C_6H_6. Mp:
57–57°, bp: 145° @ 15 mm. Very sol in
H_2O, EtOH; sltly sol in $CHCl_3$.
SYN: 1,4-BUTYNEDIOL (DOT)
CONSENSUS REPORTS: Reported in EPA
TSCA Inventory.
DOT CLASSIFICATION: 6.1; Label: KEEP
AWAY FROM FOOD
SAFETY PROFILE: A poison by ingestion.
A skin sensitizer upon long or repeated
contact. Moderately explosive. When heated
to decomposition it emits acrid smoke and
fumes and may explode. Explosive reaction
with traces of alkalies, alkali earth
hydroxides, halide salts, strong acids,
mercury salts + strong acids. See also
ACETYLENE COMPOUNDS.

BST900   CAS: 1606-83-3   HR: D
**1,1'-(2-BUTYNYLENEDIOXY)BIS(3-
   CHLORO)-2-PROPANOL)**
mf: $C_{10}H_{16}Cl_2O_4$ mw: 271.16
SYNS: 2-PROPANOL, 1,1'-(2-BUTYNYLENEDIOXY)-
BIS(3-CHLORO- □ U 27,151
CONSENSUS REPORTS: Reported in EPA
TSCA Inventory.
SAFETY PROFILE: Experimental
reproductive effects. When heated to
decomposition it emits toxic fumes of Cl⁻.

BSU250   CAS: 123-72-8   HR: 3
n-BUTYRALDEHYDE
DOT: UN 1129
mf: C_4H_8O mw: 72.12
PROP: Colorless, mobile liquid; pungent, nutty odor. Mp: $-100°$, bp: $74.7°$, flash p: $20°F$ (CC), $(-6°)$, d: 0.7988 @ $25°$, autoign temp: $446°F$, lel: 2.5%, uel: 12.5%, vap d: 2.5. Sol in water; misc with ether @ $74.8°$.
SYNS: ALDEHYDE BUTYRIQUE (FRENCH) □ ALDEIDE BUTIRRICA (ITALIAN) □ BUTAL □ BUTALDEHYDE □ BUTALYDE □ BUTANAL □ n-BUTANAL (CZECH) □ n-BUTYL ALDEHYDE □ BUTYRAL □ BUTYRALDEHYD (GERMAN) □ BUTYRALDEHYDE (CZECH) □ BUTYRIC ALDEHYDE □ FEMA No. 2219 □ NCI-C56291
CONSENSUS REPORTS: Community Right-To-Know List. Reported in EPA TSCA Inventory.
DOT CLASSIFICATION: 3; Label: Flammable Liquid
SAFETY PROFILE: Moderately toxic by ingestion, inhalation, skin contact, intraperitoneal, and subcutaneous routes. Severe skin and eye irritant. Human immunological effects by inhalation: delayed hypersensitivity. See also ALDEHYDES. Highly flammable liquid. To fight fire, use foam, CO_2, dry chemical. Incompatible with oxidizing materials. Reacts vigorously with chlorosulfonic acid, HNO_3, oleum, H_2SO_4. When heated to decomposition it emits acrid smoke and fumes.

BSU500   CAS: 110-69-0   HR: 3
n-BUTYRALDEHYDE OXIME
DOT: UN 2840
mf: C_4H_9NO mw: 87.14
PROP: Liquid. Mp: $-29.5°$, bp: $152°$, flash p: $136°F$ (CC), d: 0.923, vap d: 3.01.
SYNS: BUTANAL OXIME □ BUTYRALDOXIME (DOT) □ N-BUTYRALDOXIME □ SKINO #1 □ TROYKYD ANTI-SKIN BTO □ USAF AM-6
CONSENSUS REPORTS: Reported in EPA TSCA Inventory.
DOT CLASSIFICATION: 3; Label: Flammable Liquid
SAFETY PROFILE: A poison by intraperitoneal route. Mutation data

reported. Flammable liquid when exposed to heat or flame. To fight fire, use alcohol foam, dry chemical. Highly explosive. Can explode during vacuum distillation. Incompatible with oxidizing materials, metallic impurities. When heated to decomposition it emits toxic fumes of NO_x. See also ALDEHYDES.

BSW000   CAS: 107-92-6   HR: 2
n-BUTYRIC ACID
DOT: UN 2820
mf: $C_4H_8O_2$ mw: 88.12
PROP: Colorless liquid; strong, rancid-butter odor. Mp: $-7.9°$, bp: $163.5°$, flash p: $161°F$, d: 0.9590 @ $20°/20°$, refr index: 1.397, autoign temp: $846°F$, vap press: 0.43 mm @ $20°$, vap d: 3.04, lel: 2.0%, uel: 10.0%. Misc in H_2O, EtOH, Et_2O.
SYNS: BUTANOIC ACID □ BUTTERSAEURE (GERMAN) □ ETHYLACETIC ACID □ FEMA No. 2221 □ 1-PROPANECARBOXYLIC ACID □ PROPYLFORMIC ACID
CONSENSUS REPORTS: Reported in EPA TSCA Inventory.
DOT CLASSIFICATION: 8; Label: Corrosive
SAFETY PROFILE: Moderately toxic by ingestion, skin contact, subcutaneous, intraperitoneal, and intravenous routes. Human mutation data reported. Severe skin and eye irritant. A corrosive material. Combustible liquid. Could react with oxidizing materials. Incandescent reaction with chromium trioxide above $100°$. To fight fire, use alcohol foam, CO_2, dry chemical. When heated to decomposition it emits acrid smoke and irritating fumes.

BSW500   CAS: 539-90-2   HR: 1
BUTYRIC ACID ISOBUTYL ESTER
mf: $C_8H_{16}O_2$ mw: 144.24
PROP: Colorless liquid; apple-pineapple odor. D: 0.858–0863, refr index: 1.402. Sol in alc, fixed oils; sltly sol in water; insol in glycerin.
SYNS: FEMA No. 2187 □ ISOBUTYL BUTANOATE □ ISOBUTYL BUTYRATE (FCC) □ 2-METHYLPROPYL BUTYRATE

CONSENSUS REPORTS: Reported in EPA TSCA Inventory.

SAFETY PROFILE: Mildly toxic by ingestion and intraduodenal routes. A skin irritant. See also ESTERS. When heated to decomposition it emits acrid smoke and irritating fumes.

BSW550   CAS: 106-31-0   HR: 1
BUTYRIC ANHYDRIDE
DOT: UN 2739
mf: $C_8H_{14}O_3$ mw: 158.22
SYNS: ANHYDRID KYSELINY MASELNE □ BUTANO-IC ACID, ANHYDRIDE (9CI) □ BUTANOIC ANHYDRIDE □ BUTYRANHYDRID □ BUTYRIC ACID ANHYDRIDE □ n-BUTYRIC ACID ANHYDRIDE □ n-BUTYRIC ANHYDR-IDE □ BUTYRYL OXIDE
CONSENSUS REPORTS: Reported in EPA TSCA Inventory.
DOT CLASSIFICATION: 8; Label: Corrosive
SAFETY PROFILE: Mildly toxic by ingestion. A corrosive liquid. When heated to decomposition it emits acrid smoke and irritating vapors.

BSX000   CAS: 3068-88-0   HR: 3
β-BUTYROLACTONE
mf: $C_4H_6O_2$ mw: 86.10
SYNS: 3-HYDROXYBUTANOIC ACID-β-LACTONE □ HYDROXYBUTYRIC ACID LACTONE □ 3-HYDROXYBUTYRIC ACID LACTONE □ 4-METHYL-2-OXETANONE
CONSENSUS REPORTS: IARC Cancer Review: Group 2B IMEMDT 7,56,87; Animal Sufficient Evidence IMEMDT 11,225,76. Reported in EPA TSCA Inventory.
SAFETY PROFILE: Suspected carcinogen with experimental carcinogenic, neoplastigenic, and tumorigenic data. Mildly toxic by ingestion. A moderate skin irritant. Mutation data reported. When heated to

decomposition it emits acrid and irritating fumes. See also 4-BUTYROLACTONE.

BSX250   CAS: 109-74-0   HR: 3
BUTYRONITRILE
DOT: UN 2411
mf: C_4H_7N mw: 69.12
PROP: Colorless liquid. D: 0.796 @ 15°, mp: −112.6°, bp: 117°, flash p: 79°F (OC). Sltly sol in water; sol in alc and ether.
SYNS: BUTANENITRILE □ n-BUTANENITRILE □ BUTYRIC ACID NITRILE □ BUTYRONITRILE (DOT) □ 1-CYANOPROPANE □ PROPYL CYANIDE
CONSENSUS REPORTS: Reported in EPA TSCA Inventory. Cyanide and its compounds are on the Community Right-To-Know List.
NIOSH REL: (Nitriles) TWA 22 mg/m^3
DOT CLASSIFICATION: 3; Label: Flammable Liquid, Poison
SAFETY PROFILE: A poison by ingestion, skin contact, intraperitoneal, and subcutaneous routes. Moderately toxic by inhalation. Experimental reproductive data. A skin irritant. Dangerous fire hazard when exposed to heat, flame, or oxidizers. To fight fire, use alcohol foam. When heated to decomposition it emits toxic fumes of NO_x and CN^-.

BSY400   CAS: 75464-11-8   HR: 3
10-BUTYRYLDITHRANOL
mf: $C_{18}H_{16}O_4$ mw: 296.34
SYNS: 9(10H)-ANTHRACENONE, 1,8-DIHYDROXY-10-(1-OXOBUTYL)- □ 10-BUTYRYL DITHRANOL □ BUTANTRONE □ BUTYRYL DITHRANOL □ DITHRANOL, 10-BUTYRYL-
SAFETY PROFILE: A poison by ingestion. A mild skin irritant. Mutation data reported. When heated to decomposition it emits acrid smoke and irritating vapors.

C

CAD000   CAS: 7440-43-9   HR: 3
CADMIUM
af: Cd   aw: 112.40
PROP: Hexagonal, ductile crystals or soft, silver-white, lustrous, malleable metal. Tarnishes in air, particularly moist air. Mp: 321°, bp: 767°, d: 8.642, vap press: 1 mm @ 394°. Sol in dil acids (H_2 evolved). IDLH 9 mg/m³ (as Cd).
SYNS: C.I. 77180 □ COLLOIDAL CADMIUM □ KADMIUM (GERMAN)
CONSENSUS REPORTS: NTP 10th Report on Carcinogens. IARC Cancer Review: Group 1 IMEMDT 58,119,93; Animal Sufficient Evidence IMEMDT 2,74,73; Animal Sufficient Evidence IMEMDT 11,39,76; Human Sufficient Evidence IMEMDT 58,119,93; Human Limited Evidence IMEMDT 7,139,87; Animal Limited Evidence IMEMDT 58,119,93. Cadmium and its compounds are on the Community Right-To-Know List. Reported in EPA TSCA Inventory. EPA Genetic Toxicology Program.
OSHA PEL: TWA 5 μg(Cd)/m³
ACGIH TLV: TWA 0.01 mg(Cd)/m³ (metal), Suspected Human Carcinogen; TWA 0.002 mg(Cd)/m³ (respirable dust), Suspected Human Carcinogen); BEI: 5 μg/g creatinine in urine; 5 μg/L in blood
DFG MAK: DFG BAT: Blood 1.5 μg/dL; Urine 15 μg/dL. MAK: Animal Carcinogen, Suspected Human Carcinogen
NIOSH REL: (Cadmium) Reduce to lowest feasible level
SAFETY PROFILE: Confirmed human carcinogen with experimental carcinogenic, tumorigenic, and neoplastigenic data. A human poison by inhalation and possibly other routes. Poison experimentally by ingestion, inhalation, intraperitoneal, subcutaneous, and intravenous routes. In humans inhalation causes an excess of protein in the urine. Experimental teratogenic and reproductive effects. Mutation data reported. The dust ignites spontaneously in air and is flammable and explosive when exposed to heat, flame, or by chemical reaction with oxidizing agents, metals, HN_3, Zn, Se, and Te. Explodes on contact with hydrazoic acid. Violent or explosive reaction when heated with ammonium nitrate. Vigorous reaction when heated with nitryl fluoride. When heated to a high temperature it emits toxic fumes of Cd. See also CADMIUM COMPOUNDS.

CAD250   CAS: 543-90-8   HR: 3
CADMIUM(II) ACETATE
mf: $C_2H_4O_2$•1/2Cd   mw: 116.25
PROP: Monoclinic, colorless crystals; odor of acetic acid. Mp: 256°, bp: decomp, d: 2.341.
SYNS: ACETIC ACID, CADMIUM SALT □ BIS(ACETOXY)CADMIUM □ CADMIUM ACETATE (DOT) □ CADMIUM DIACETATE □ C.I. 77185
CONSENSUS REPORTS: IARC Cancer Review: Group 1 IMEMDT 58,119,93; Human Sufficient Evidence IMEMDT 58,119,93; Animal Sufficient Evidence IMEMDT 58,119,93. Reported in EPA TSCA Inventory. EPA Genetic Toxicology Program. Cadmium and its compounds are on the Community Right-To-Know List.
OSHA PEL: TWA 5 μg(Cd)/m³
ACGIH TLV: TWA 0.002 mg(Cd)/m³ (respirable dust), Suspected Human Carcinogen); BEI: 5 μg/g creatinine in urine; 5 μg/L in blood
NIOSH REL: (Cadmium) Reduce to lowest feasible level
SAFETY PROFILE: Confirmed human carcinogen. Poison by intraperitoneal route.

An experimental teratogen. Other experimental reproductive effects. Human mutation data reported. When heated to decomposition it emits toxic fumes of Cd. See also CADMIUM COMPOUNDS.

CAD275   CAS: 5743-04-4   HR: 3
CADMIUM ACETATE DIHYDRATE
mf: $C_4H_6O_4 \cdot Cd \cdot 2H_2O$ mw: 266.54
PROP: Crystals, becoming anhydrous at 130°; slt acetic acid odor. D: 2.01, 2.341 (anhydrous), mp: 255° (anhydrous). Sol in water and alc.
SYNS: ACETIC ACID, CADMIUM SALT, DIHYDRATE □ CADMIUM DIACETATE DIHYDRATE
OSHA PEL: TWA 5 μg(Cd)/m³
ACGIH TLV: TWA 0.01 mg(Cd)/m³; Suspected Carcinogen
SAFETY PROFILE: Confirmed human carcinogen. Mutation data reported. When heated to decomposition it emits toxic fumes of Cd.

CAD325   CAS: 22750-53-4   HR: 3
CADMIUM AMIDE
mf: CdH_4N_2 mw: 144.46
PROP: White solid, which turns brown in air.
SYN: CADMIUM DIAMIDE
CONSENSUS REPORTS: Cadmium compounds are on the Community Right-To-Know List.
OSHA PEL: TWA 5 μg(Cd)/m³
ACGIH TLV: TWA 0.002 mg(Cd)/m³ (respirable dust), Suspected Human Carcinogen); BEI: 5 μg/g creatinine in urine; 5 μg/L in blood
NIOSH REL: (Cadmium) Reduce to lowest feasible level
SAFETY PROFILE: Confirmed human carcinogen. May explode if heated. Reacts violently with water. When heated to decomposition it emits toxic fumes of Cd and NO$_x$. See also CADMIUM COMPOUNDS and AMIDES.

CAD350   CAS: 14215-29-3   HR: 3
CADMIUM AZIDE

mf: CdN_6 mw: 196.45
PROP: White crystals.
SYN: CADMIUM DIAZIDE
CONSENSUS REPORTS: Cadmium compounds are on the Community Right-To-Know List.
OSHA PEL: TWA 5 μg(Cd)/m³
ACGIH TLV: TWA 0.002 mg(Cd)/m³ (respirable dust), Suspected Human Carcinogen); BEI: 5 μg/g creatinine in urine; 5 μg/L in blood
NIOSH REL: (Cadmium) Reduce to lowest feasible level
SAFETY PROFILE: Confirmed human carcinogen. The dry solid is an unstable heat- and friction-sensitive explosive. When heated to decomposition it emits toxic fumes of NO$_x$ and Cd. See also CADMIUM COMPOUNDS and AZIDES.

CAD500   CAS: 7495-93-4   HR: 3
CADMIUM BIS(2-ETHYLHEXYL) PHOSPHITE
mf: $C_{32}H_{68}O_6P_2 \cdot Cd$ mw: 723.34
SYN: BIS(2-ETHYLHEXYL) ESTER PHOSPHOROUS ACID CADMIUM SALT
CONSENSUS REPORTS: Cadmium and its compounds are on the Community Right-To-Know List.
OSHA PEL: TWA 5 μg(Cd)/m³
ACGIH TLV: TWA 0.002 mg(Cd)/m³ (respirable dust), Suspected Human Carcinogen); BEI: 5 μg/g creatinine in urine; 5 μg/L in blood
NIOSH REL: (Cadmium) Reduce to lowest feasible level
SAFETY PROFILE: Confirmed human carcinogen. Poison by intraperitoneal route. When heated to decomposition it emits toxic fumes of PO$_x$ and Cd. See also CADMIUM COMPOUNDS.

CAD550   CAS: 20246-69-9   HR: 3
CADMIUM BIS(PENTYLDITHIO-CARBAMATE)
mf: $C_{12}H_{24}N_2S_4 \cdot Cd$ mw: 437.02
SYNS: CADMIUM BIS(N-AMYLDITHIOCARBAMATE) □ CADMIUM, BIS(PENTYLDITHIOCARBAMATO)- □ CADMIUM DIAMYL DITHIOCARBAMATE

CONSENSUS REPORTS: IARC Cancer Review: Group 1 IMEMDT 58,119,93; Human Sufficient Evidence IMEMDT 58,119,93; Animal Sufficient Evidence IMEMDT 58,119,93.
SAFETY PROFILE: Confirmed carcinogen. Low toxicity by ingestion and intraperitoneal routes. When heated to decomposition it emits toxic vapors of NO_x, Cl^-, and cadmium.

CAD600   CAS: 7789-42-6   HR: 3
CADMIUM BROMIDE
mf: Br_2Cd mw: 272.22
PROP: Pearly or colorless hexagonal crystals; hygroscopic. Mp: 570°, bp: 863°, d: 5.192. Sol in water, alc, and Me_2CO; moderately sol in acetone.
SYN: CADMIUM DIBROMIDE
CONSENSUS REPORTS: Reported in EPA TSCA Inventory.
OSHA PEL: TWA 5 µg(Cd)/m³
ACGIH TLV: TWA 0.01 mg(Cd)/m³; Suspected Carcinogen
SAFETY PROFILE: Confirmed human carcinogen. When heated to decomposition it emits toxic fumes of Cd and Br^-.

CAD750   CAS: 2191-10-8   HR: 3
CADMIUM CAPRYLATE
mf: $C_{16}H_{30}O_4$•Cd   mw: 398.86
SYN: OCTANOIC ACID, CADMIUM SALT (2:1)
CONSENSUS REPORTS: Reported in EPA TSCA Inventory. Cadmium and its compounds are on the Community Right-To-Know List.
OSHA PEL: TWA 5 µg(Cd)/m³
ACGIH TLV: TWA 0.002 mg(Cd)/m³ (respirable dust), Suspected Human Carcinogen); BEI: 5 µg/g creatinine in urine; 5 µg/L in blood
NIOSH REL: (Cadmium) Reduce to lowest feasible level
SAFETY PROFILE: Confirmed human carcinogen. Poison by ingestion and intratracheal routes. When heated to decomposition it emits toxic fumes of Cd. See also CADMIUM COMPOUNDS.

CAD800   CAS: 513-78-0   HR: 3
CADMIUM CARBONATE
mf: CO_3•Cd   mw: 172.41
PROP: Powder. D: 4.258.
SYNS: CADMIUM MONOCARBONATE □ CARBONIC ACID, CADMIUM SALT □ CHEMCARB □ KALCIT □ MIKROKALCIT □ SUPERMIKROKALCIT
CONSENSUS REPORTS: IARC Cancer Review: Group 1 IMEMDT 58,119,93; Human Sufficient Evidence IMEMDT 58,119,93; Animal Sufficient Evidence IMEMDT 58,119,93; Reported in EPA TSCA Inventory.
OSHA PEL: TWA 5 µg(Cd)/m³
ACGIH TLV: TWA 0.01 mg(Cd)/m³; Suspected Carcinogen
NIOSH REL: (Cadmium, dust and fume) lowest feasible concentration
SAFETY PROFILE: Confirmed human carcinogen. Poison by ingestion. Mutation data reported. When heated to decomposition it emits toxic fumes of cadmium.

CAD900   HR: 3
CADMIUM CDTA
SYNS: ACETIC ACID, (1,2-CYCLOHEXYLIDENEDINITRILO)TETRA-, CADMIUM COMPLEX, TRANS- □ ZYKLOHEXANDIAMINE-TETRAESSIGSAEURE KADMIUMKOMPLEXE
CONSENSUS REPORTS: IARC Cancer Review: Group 1 IMEMDT 58,119,93; Human Sufficient Evidence IMEMDT 58,119,93; Animal Sufficient Evidence IMEMDT 58,119,93.
SAFETY PROFILE: Confirmed carcinogen. A poison by intravenous route. When heated to decomposition it emits toxic vapors of NO_x and cadmium.

CAE000   HR: 3
CADMIUM CHLORATE
mf: $CdCl_2O_6$ mw: 279.31
PROP: Colorless, deliquescent prisms. Mp: 80°, d: 2.28 @ 18°.
CONSENSUS REPORTS: Cadmium and its compounds are on the Community Right-To-Know List.
OSHA PEL: TWA 5 µg(Cd)/m³

ACGIH TLV: TWA 0.002 mg(Cd)/m³ (respirable dust), Suspected Human Carcinogen); BEI: 5 µg/g creatinine in urine; 5 µg/L in blood

NIOSH REL: (Cadmium) Reduce to lowest feasible level

SAFETY PROFILE: Confirmed human carcinogen. A powerful oxidizing agent. Flammable by chemical reaction with reducing agents. Moderate explosion hazard when shocked or exposed to heat. Violent or explosive reaction with sulfides (e.g., copper(II) sulfide (explodes); antimony(II) sulfide; arsenic(III) sulfide; tin(II) sulfide; tin(IV) sulfide). When heated to decomposition it emits toxic fumes of Cd and Cl⁻. See also CHLORATES.

CAE250   CAS: 10108-64-2   HR: 3
CADMIUM CHLORIDE
mf: CdCl₂   mw: 183.30

PROP: Hexagonal, colorless crystals. Mp: 568°, bp: 969.6°, d: 4.047 @ 25°, vap press: 10 mm @ 656°. Sol in H₂O: sltly sol in EtOH.

SYNS: CADDY □ CADMIUM DICHLORIDE □ KADMIUMCHLORID (GERMAN) □ VI-CAD

CONSENSUS REPORTS: NTP 10th Report on Carcinogens. IARC Cancer Review: Group 1 IMEMDT 58,119,93; Animal Sufficient Evidence IMEMDT 2,74,73; Animal Sufficient Evidence IMEMDT 11,39,76; Human Sufficient Evidence IMEMDT 58,119,93; IARC Cancer Review: Animal Sufficient Evidence IMEMDT 58,119,93; EPA Genetic Toxicology Program. Cadmium and its compounds are on the Community Right-To-Know List. Reported in EPA TSCA Inventory.

OSHA PEL: TWA 5 µg(Cd)/m³

ACGIH TLV: TWA 0.002 mg(Cd)/m³ (respirable dust), Suspected Human Carcinogen); BEI: 5 µg/g creatinine in urine; 5 µg/L in blood

DFG MAK: Animal Carcinogen, Suspected Human Carcinogen

NIOSH REL: (Cadmium) Reduce to lowest feasible level

SAFETY PROFILE: Confirmed human carcinogen with experimental carcinogenic and tumorigenic data. Poison by ingestion, inhalation, skin contact, intraperitoneal, subcutaneous, intravenous, and possibly other routes. Human systemic effects by ingestion: blood pressure, acute pulmonary edema, hypermotility, diarrhea. Experimental teratogenic and reproductive effects. Human mutation data reported. Reacts violently with BrF₃ and K. When heated to decomposition it emits very toxic fumes of Cd and Cl⁻. See also CADMIUM COMPOUNDS and CHLORIDES.

CAE375   CAS: 72589-96-9   HR: 3
CADMIUM CHLORIDE, DIHYDRATE
mf: CdCl₂•2H₂O   mw: 219.34

PROP: White crystals.

CONSENSUS REPORTS: Cadmium and its compounds are on the Community Right-To-Know List.

OSHA PEL: TWA 5 µg(Cd)/m³

ACGIH TLV: TWA 0.002 mg(Cd)/m³ (respirable dust), Suspected Human Carcinogen); BEI: 5 µg/g creatinine in urine; 5 µg/L in blood

DFG MAK: Animal Carcinogen, Suspected Human Carcinogen

NIOSH REL: (Cadmium) Reduce to lowest feasible level

SAFETY PROFILE: Confirmed human carcinogen with experimental tumorigenic data. An experimental teratogen. Other experimental reproductive effects. When heated to decomposition it emits toxic fumes of Cl⁻ and Cd. See also CADMIUM CHLORIDE, CADMIUM COMPOUNDS, and CHLORIDES.

CAE425   CAS: 7790-78-5   HR: 3
CADMIUM CHLORIDE, HYDRATE (2:5)
mf: CdCl₂•5/2H₂O   mw: 228.35

PROP: Crystals.

CONSENSUS REPORTS: Cadmium and its compounds are on the Community Right-To-Know List.

OSHA PEL: TWA 5 µg(Cd)/m³

ACGIH TLV: TWA 0.002 mg(Cd)/m³ (respirable dust), Suspected Human Carcinogen); BEI: 5 µg/g creatinine in urine; 5 µg/L in blood
DFG MAK: Animal Carcinogen, Suspected Human Carcinogen
NIOSH REL: (Cadmium) Reduce to lowest feasible level
SAFETY PROFILE: Confirmed human carcinogen. Poison by ingestion and intraperitoneal routes. Experimental reproductive effects. Human mutation data reported. When heated to decomposition it emits toxic fumes of Cl⁻ and Cd. See also CADMIUM CHLORIDE, CADMIUM COMPOUNDS, and CHLORIDES.

CAE500   CAS: 35658-65-2   HR: 3
CADMIUM CHLORIDE, MONOHYDRATE
mf: CdCl₂•H₂O   mw: 201.32
PROP: White powder.
CONSENSUS REPORTS: Cadmium and its compounds are on the Community Right-To-Know List.
OSHA PEL: TWA 5 µg(Cd)/m³
ACGIH TLV: TWA 0.002 mg(Cd)/m³ (respirable dust), Suspected Human Carcinogen); BEI: 5 µg/g creatinine in urine; 5 µg/L in blood
DFG MAK: Animal Carcinogen, Suspected Human Carcinogen
NIOSH REL: (Cadmium) Reduce to lowest feasible level
SAFETY PROFILE: Confirmed human carcinogen with experimental carcinogenic and tumorigenic data. Experimental reproductive effects. When heated to decomposition it emits very toxic fumes of Cd and Cl⁻. See also CADMIUM CHLORIDE, CADMIUM COMPOUNDS, and CHLORIDES.

CAE750   HR: 3
CADMIUM COMPOUNDS
CONSENSUS REPORTS: IARC Cancer Review: Group 1 IMEMDT 58,119,93; Animal Sufficient Evidence IMEMDT 58,119,93; Human Sufficient Evidence

IMEMDT 58,119,93. Cadmium and its compounds are on the Community Right-To-Know List.
OSHA PEL: TWA 5 µg(Cd)/m³
ACGIH TLV: TWA 0.002 mg(Cd)/m³ (respirable dust), Suspected Human Carcinogen); BEI: 5 µg/g creatinine in urine; 5 µg/L in blood
DFG MAK: DFG BAT: Blood 1.5 µg/dL; Urine 15 µg/dL. MAK: Suspected Carcinogen
NIOSH REL: (Cadmium, dust and fume) Reduce to lowest feasible level
SAFETY PROFILE: Confirmed human carcinogens producing lung tumors. Poison by ingestion. The irritating and emetic action is so violent, however, that little of the cadmium has time to be absorbed and fatal poisoning rarely ensues. Experimental carcinogens and teratogens. Cases of human poisoning have been reported from ingestion of food or beverages prepared or stored in cadmium-plated containers. Inhalation of fumes or dusts affects the respiratory tract and the kidneys. Brief exposure to high concentrations may result in pulmonary edema and death. Fatal concentrations may be breathed without sufficient discomfort to warn a worker to leave the exposure site. Cadmium oxide fumes can cause metal fume fever resembling that caused by zinc oxide fumes. When heated to decomposition cadmium compounds emit toxic fumes of Cd.

CAE800   HR: 3
CADMIUM DIACETATE MONOHYDRATE
mf: C₄H₆O₄•Cd•H₂O   mw: 248.52
SYN: CADMIUM(II) ACETATE, MONOHYDRATE
CONSENSUS REPORTS: IARC Cancer Review: Group 1 IMEMDT 58,119,93; Human Sufficient Evidence IMEMDT 58,119,93; Animal Sufficient Evidence IMEMDT 58,119,93.
ACGIH TLV: TWA 0.002 mg(Cd)/m³ (respirable dust), Suspected Human Carcinogen); BEI: 5 µg/g creatinine in urine; 5 µg/L in blood

SAFETY PROFILE: Confirmed carcinogen. Mutation data reported. When heated to decomposition it emits toxic vapors of cadmium.

CAF500   HR: 3
CADMIUM DICYANIDE
mf: C_2CdN_2 mw: 164.44
CONSENSUS REPORTS: Cadmium and its compounds and cyanide and its compounds are on the Community Right-To-Know List.
OSHA PEL: TWA 5 μg(Cd)/m³
ACGIH TLV: TWA 0.002 mg(Cd)/m³ (respirable dust), Suspected Human Carcinogen); BEI: 5 μg/g creatinine in urine; 5 μg/L in blood
NIOSH REL: (Cadmium) Reduce to lowest feasible level
SAFETY PROFILE: Confirmed human carcinogen. A poison. Incompatible with magnesium. When heated to decomposition it emits toxic fumes of Cd and CN⁻. See also CADMIUM COMPOUNDS and CYANIDE.

CAF750   CAS: 15954-91-3   HR: 3
CADMIUM(II) EDTA COMPLEX
SYN: (ETHYLENEDINITRILO)TETRAACETIC ACID CADMIUM(II) COMPLEX
CONSENSUS REPORTS: Cadmium and its compounds are on the Community Right-To-Know List.
OSHA PEL: TWA 5 μg(Cd)/m³
ACGIH TLV: TWA 0.002 mg(Cd)/m³ (respirable dust), Suspected Human Carcinogen); BEI: 5 μg/g creatinine in urine; 5 μg/L in blood
NIOSH REL: (Cadmium) Reduce to lowest feasible level
SAFETY PROFILE: Confirmed human carcinogen. Poison by intraperitoneal and intravenous routes. When heated to decomposition it emits toxic fumes of NO_x and Cd.

CAG000   CAS: 14486-19-2   HR: 3
CADMIUM FLUOBORATE
mf: B_2CdF_8 mw: 286.02

SYNS: BORATE(1-), TETRAFLUORO-, CADMIUM (2:1) (9CI) □ CADMIUM FLUOROBORATE □ CADMIUM TETRAFLUOROBORATE (7CI) □ TL 1026
CONSENSUS REPORTS: Reported in EPA TSCA Inventory. Cadmium and its compounds are on the Community Right-To-Know List.
OSHA PEL: TWA 5 μg(Cd)/m³
ACGIH TLV: TWA 0.002 mg(Cd)/m³ (respirable dust), Suspected Human Carcinogen); BEI: 5 μg/g creatinine in urine; 5 μg/L in blood
NIOSH REL: (Cadmium) Reduce to lowest feasible level
SAFETY PROFILE: Suspected human carcinogen. Poison by ingestion and inhalation. When heated to decomposition it emits very toxic fumes of Cd and F⁻. See TETRAFLUOROBORATE.

CAG250   CAS: 7790-79-6   HR: 3
CADMIUM FLUORIDE
mf: CdF_2 mw: 150.40
PROP: Cubic, white, non-hygroscopic, non-volatile crystals. Mp: 1078°, bp: 1748°, d: 6.64, vap press: 1 mm @ 1112°. Sltly sol in H_2O.
SYN: CADMIUM FLUORURE (FRENCH)
CONSENSUS REPORTS: Reported in EPA TSCA Inventory. Cadmium and its compounds are on the Community Right-To-Know List.
OSHA PEL: TWA 5 μg(Cd)/m³
ACGIH TLV: TWA 0.002 mg(Cd)/m³ (respirable dust), Suspected Human Carcinogen); BEI: 5 μg/g creatinine in urine; 5 μg/L in blood
NIOSH REL: (Cadmium) Reduce to lowest feasible level
SAFETY PROFILE: Confirmed human carcinogen. Poison by subcutaneous route. Violent reaction with K. When heated to decomposition it emits very toxic fumes of Cd and F⁻. See also FLUORIDES and CADMIUM COMPOUNDS.

CAG500   CAS: 17010-21-8   HR: 3
CADMIUM FLUOSILICATE
mf: CdF$_6$Si   mw: 254.49
PROP: Hexagonal, colorless crystals.
SYNS: CADMIUM FLUOROSILICATE □ CADMIUM
HEXAFLUOROSILICATE (7CI) □ CADMIUM SILICON
FLUORIDE □ SILICATE(2-), HEXAFLUORO-, CADMIUM
(8CI,9CI) □ TL 1070
CONSENSUS REPORTS: Cadmium and its
compounds are on the Community Right-
To-Know List.
OSHA PEL: TWA 5 µg(Cd)/m^3
ACGIH TLV: TWA 0.002 mg(Cd)/m^3
(respirable dust), Suspected Human
Carcinogen); BEI: 5 µg/g creatinine in urine;
5 µg/L in blood
NIOSH REL: (Cadmium) Reduce to lowest
feasible level
DOT CLASSIFICATION: 6.1; Label: KEEP
AWAY FROM FOOD
SAFETY PROFILE: Confirmed human
carcinogen. Poison by ingestion and
inhalation. When heated to decomposition it
emits very toxic fumes of Cd and F$^-$.

CAG600   CAS: 22537-48-0   HR: 3
CADMIUM, ION (Cd^{2+})
mf: Cd   mw: 112.40
SYNS: CADMIUM(2Cd^{2+}) □ CADMIUM CATION □
CADMIUM ION
CONSENSUS REPORTS: IARC Cancer
Review: Group 1 IMEMDT 58,119,93;
Human Sufficient Evidence IMEMDT
58,119,93; Animal Sufficient Evidence
IMEMDT 58,119,93.
ACGIH TLV: TWA 0.002 mg(Cd)/m^3
(respirable dust), Suspected Human
Carcinogen); BEI: 5 µg/g creatinine in urine;
5 µg/L in blood
SAFETY PROFILE: Confirmed carcinogen.
A poison by intravenous route. Mutation
data reported.

CAG750   CAS: 16039-55-7   HR: 3
CADMIUM LACTATE
mf: C$_6$H$_{10}$O$_6$•Cd   mw: 290.56
PROP: Needles.
SYN: LACTIC ACID, CADMIUM SALT

CONSENSUS REPORTS: Cadmium and its
compounds are on the Community Right-
To-Know List.
OSHA PEL: TWA 5 µg(Cd)/m^3
ACGIH TLV: TWA 0.002 mg(Cd)/m^3
(respirable dust), Suspected Human
Carcinogen); BEI: 5 µg/g creatinine in
urine; 5 µg/L in blood
NIOSH REL: (Cadmium) Reduce to lowest
feasible level
SAFETY PROFILE: Confirmed human
carcinogen. A poison. When heated to
decomposition it emits toxic fumes of Cd.
See also CADMIUM COMPOUNDS.

CAG775   CAS: 2605-44-9   HR: 3
CADMIUM LAURATE
mf: C$_{24}$H$_{46}$O$_4$•Cd   mw: 511.10
SYNS: CADMIUM DILAURATE □ CADMIUM
DODECANOATE □ DODECANOIC ACID, CADMIUM
SALT (9CI) □ LAURIC ACID, CADMIUM SALT (2:1)
CONSENSUS REPORTS: Reported in EPA
TSCA Inventory.
OSHA PEL: TWA 5 µg(Cd)/m^3
ACGIH TLV: TWA 0.002 mg(Cd)/m^3
(respirable dust), Suspected Human
Carcinogen); BEI: 5 µg/g creatinine in urine;
5 µg/L in blood
SAFETY PROFILE: Confirmed human
carcinogen. Moderately toxic by ingestion.
When heated to decomposition it emits
toxic fumes of Cd.

CAH000   CAS: 10325-94-7   HR: 3
CADMIUM NITRATE
mf: CdN$_2$O$_6$ mw: 236.42
PROP: Strongly hygroscopic, white,
prismatic needles. Mp: 350–360°. Very sol in
H$_2$O; sol in EtOAc.
SYNS: CADMIUM DINITRATE □ CADMIUM(II)
NITRATE □ NITRIC ACID, CADMIUM SALT
CONSENSUS REPORTS: IARC Cancer
Review: Group 1 IMEMDT 58,119,93;
Human Sufficient Evidence IMEMDT
58,119,93; Animal Sufficient Evidence
IMEMDT 58,119,93. Reported in EPA
TSCA Inventory. EPA Genetic Toxicology
Program. Cadmium and its compounds are
on the Community Right-To-Know List.

OSHA PEL: TWA 5 μg(Cd)/m³
ACGIH TLV: TWA 0.002 mg(Cd)/m³
(respirable dust), Suspected Human
Carcinogen); BEI: 5 μg/g creatinine in urine;
5 μg/L in blood
NIOSH REL: (Cadmium) Reduce to lowest
feasible level
SAFETY PROFILE: Confirmed human
carcinogen. Poison by ingestion and possibly
other routes. Moderately toxic by inhalation.
Mutation data reported. When heated to
decomposition it emits very toxic fumes of
Cd and NO$_x$. See also CADMIUM
COMPOUNDS and NITRATES.

CAH250   CAS: 10022-68-1   HR: 3
CADMIUM(II) NITRATE
TETRAHYDRATE (1:2:4)
mf: N$_2$O$_6$•Cd•4H$_2$O   mw: 308.50
PROP: Crystals in H$_2$O. Mp: 59.4°.
SYNS: DUSICNAN KADEMNATY (CZECH) □ NITRIC
ACID, CADMIUM SALT, TETRAHYDRATE
CONSENSUS REPORTS: Cadmium and its
compounds are on the Community Right-
To-Know List.
OSHA PEL: TWA 5 μg(Cd)/m³
ACGIH TLV: TWA 0.002 mg(Cd)/m³
(respirable dust), Suspected Human
Carcinogen); BEI: 5 μg/g creatinine in urine;
5 μg/L in blood
NIOSH REL: (Cadmium) Reduce to lowest
feasible level
SAFETY PROFILE: Confirmed human
carcinogen. Poison by ingestion. A severe
skin and moderate eye irritant. Mutation
data reported. See also CADMIUM
COMPOUNDS, CADMIUM NITRATE,
and NITRATES. When heated to
decomposition it emits very toxic fumes of
Cd and NO$_x$.

CAH500   CAS: 1306-19-0   HR: 3
CADMIUM OXIDE
mf: CdO   mw: 128.40
PROP: (1) Amorphous, brown powder; (2)
cubic, brown crystals. Changes color on
heating. Mp (1): <1426°, mp (2): decomp @
950°, bp: 1559°, d (1): 6.95, d (2): 8.15, vap

press: 1 mm @ 1000°. Subl at 7°. Insol in
H$_2$O; sol in acids and NH$_3$. IDLH 9 mg/m³
(as Cd).
SYNS: CADMIUM MONOXIDE □ KADMU TLENEK
(POLISH) □ NCI-C02551
CONSENSUS REPORTS: NTP 10th Report
on Carcinogens. IARC Cancer Review:
Group 1 IMEMDT 58,119,93; Animal
Sufficient Evidence IMEMDT 2,74,73;
Animal Sufficient Evidence IMEMDT
11,39,76; Animal Sufficient Evidence
IMEMDT 58,119,93; Human Limited
Evidence IMEMDT 11,39,76; Human
Sufficient Evidence IMEMDT 58,119,93.
Reported in EPA TSCA Inventory. EPA
Extremely Hazardous Substances List.
Cadmium and its compounds are on the
Community Right-To-Know List.
OSHA PEL: TWA 5 μg(Cd)/m³
ACGIH TLV: TWA 0.002 mg(Cd)/m³
(respirable dust), Suspected Human
Carcinogen); BEI: 5 μg/g creatinine in urine;
5 μg/L in blood
DFG MAK: Animal Carcinogen, Suspected
Human Carcinogen
NIOSH REL: (Cadmium) Reduce to lowest
feasible level
SAFETY PROFILE: Confirmed human
carcinogen with experimental neoplastigenic
data. Poison by ingestion, inhalation, and
intraperitoneal routes. An experimental
teratogen. Other experimental reproductive
effects. Human systemic effects by
inhalation include: change in the sense of
smell, change in heart rate, blood pressure
increase, an excess of protein in the urine,
and other kidney or bladder changes.
Mixtures with magnesium explode when
heated. When heated to decomposition it
emits toxic fumes of Cd. See also
CADMIUM COMPOUNDS.

CAH750   HR: 3
CADMIUM OXIDE FUME
mf: CdO   mw: 128.40
SYN: CADMIUM FUME
CONSENSUS REPORTS: IARC Cancer
Review: Group 1 IMEMDT 58,119,93;

Human Sufficient Evidence IMEMDT 58,119,93; Animal Sufficient Evidence IMEMDT 58,119,93. Reported in EPA TSCA Inventory. Cadmium and its compounds are on the Community Right-To-Know List.
OSHA PEL: TWA 5 μg(Cd)/m³
ACGIH TLV: TWA 0.002 mg(Cd)/m³ (respirable dust), Suspected Human Carcinogen); BEI: 5 μg/g creatinine in urine; 5 μg/L in blood
NIOSH REL: (Cadmium, dust and fume) Reduce to lowest feasible level
SAFETY PROFILE: Confirmed human carcinogen. Moderately toxic to humans by inhalation. Human pulmonary system effects by inhalation, including: coughing, difficult breathing, and cyanosis. A strong irritant via inhalation. When heated to decomposition it emits toxic fumes of Cd. See also CADMIUM OXIDE and CADMIUM COMPOUNDS.

CAI000   CAS: 13477-17-3   HR: 3
CADMIUM PHOSPHATE
mf: $Cd_3O_8P_2 \cdot 4H_2O$ mw: 599.22
PROP: Amorphous or colorless crystals. Mp: 1180°.
SYN: TL 1182
CONSENSUS REPORTS: Reported in EPA TSCA Inventory. Cadmium and its compounds are on the Community Right-To-Know List.
OSHA PEL: TWA 5 μg(Cd)/m³
ACGIH TLV: TWA 0.002 mg(Cd)/m³ (respirable dust), Suspected Human Carcinogen); BEI: 5 μg/g creatinine in urine; 5 μg/L in blood
NIOSH REL: (Cadmium) Reduce to lowest feasible level
SAFETY PROFILE: Confirmed human carcinogen. Poison by inhalation. When heated to decomposition it emits toxic fumes of Cd and PO_x. See CADMIUM COMPOUNDS and PHOSPHATES.

CAI125   CAS: 12014-28-7   HR: 3
CADMIUM PHOSPHIDE
mf: Cd_3P_2 mw: 399.18
PROP: Gray needles or platelets.
CONSENSUS REPORTS: Cadmium compounds are on the Community Right-To-Know List.
OSHA PEL: TWA 5 μg(Cd)/m³
ACGIH TLV: TWA 0.002 mg(Cd)/m³ (respirable dust), Suspected Human Carcinogen); BEI: 5 μg/g creatinine in urine; 5 μg/L in blood
DFG MAK: DFG BAT: Blood 1.5 μg/dL; Urine 15 μg/dL. MAK: Suspected Carcinogen
NIOSH REL: (Cadmium) Reduce to lowest feasible level
SAFETY PROFILE: Confirmed human carcinogen. Explosive reaction with concentrated nitric acid. When heated to decomposition it emits toxic fumes of PO_x and Cd. See also CADMIUM COMPOUNDS and PHOSPHIDES.

CAI250   HR: 3
CADMIUM PROPIONATE
mf: $C_6H_{10}CdO_5$ mw: 258.55
CONSENSUS REPORTS: Cadmium and its compounds are on the Community Right-To-Know List.
OSHA PEL: TWA 5 μg(Cd)/m³
ACGIH TLV: TWA 0.002 mg(Cd)/m³ (respirable dust), Suspected Human Carcinogen); BEI: 5 μg/g creatinine in urine; 5 μg/L in blood
DFG MAK: DFG BAT: Blood 1.5 μg/dL; Urine 15 μg/dL. MAK: Suspected Carcinogen
NIOSH REL: (Cadmium) Reduce to lowest feasible level
SAFETY PROFILE: Confirmed human carcinogen. The salt has exploded. Incompatible with 3-pentanone vapor. When heated to decomposition it emits toxic fumes of Cd. See also CADMIUM COMPOUNDS.

CAI350   CAS: 18897-36-4   HR: 3
CADMIUM 2-PYRIDINETHIONE
mf: $C_{10}H_8CdN_2O_2S_2$ mw: 364.72

SYNS: CADMIUM, BIS(1-HYDROXY-2(1H)-PYRIDINETHIONATO)- □ CADMIUM PT □ CdPT
OSHA PEL: TWA 5 µg(Cd)/m³
ACGIH TLV: TWA 0.002 mg(Cd)/m³ (respirable dust), Suspected Human Carcinogen); BEI: 5 µg/g creatinine in urine; 5 µg/L in blood
NIOSH REL: (Cadmium) TWA reduce to lowest feasible level
SAFETY PROFILE: Confirmed human carcinogen. Poison by ingestion and intravenous routes. When heated to decomposition it emits toxic fumes of NO_x, SO_x, and Cd.

CAI400   CAS: 19010-79-8   HR: 3
CADMIUM SALICYLATE
mf: $C_{14}H_{10}CdO_6$ mw: 386.64
PROP: Monohydrate small needles or plates. Mp: 242°. Sltly sol in cold water, methanol, eth; very sol in boiling water.
SYNS: BIS(2-HYDROXYBENZOATO-O¹O²-), (T-4)-CADMIUM (9CI) □ CADMIUM, BIS(SALICYLATO)-
OSHA PEL: TWA 5 µg(Cd)/m³
ACGIH TLV: TWA 0.002 mg(Cd)/m³ (respirable dust), Suspected Human Carcinogen); BEI: 5 µg/g creatinine in urine; 5 µg/L in blood
SAFETY PROFILE: Confirmed human carcinogen. Poison by ingestion. When heated to decomposition it emits toxic fumes of Cd.

CAI500   HR: 3
CADMIUM SELENIDE
mf: CdSe   mw: 191.36
PROP: Preparative hazard.
CONSENSUS REPORTS: Cadmium and its compounds as well as selenium and its compounds are on the Community Right-To-Know List.
OSHA PEL: TWA 5 µg(Cd)/m³
ACGIH TLV: TWA 0.002 mg(Cd)/m³ (respirable dust), Suspected Human Carcinogen); BEI: 5 µg/g creatinine in urine; 5 µg/L in blood

DFG MAK: DFG BAT: Blood 1.5 µg/dL; Urine 15 µg/dL. MAK: Suspected Carcinogen; 0.1 mg(Se)/m³
NIOSH REL: (Cadmium) Reduce to lowest feasible level
SAFETY PROFILE: Confirmed human carcinogen. Selenium compounds are considered to be poisons. When heated to decomposition it emits toxic fumes of Cd and Se. See also CADMIUM COMPOUNDS and SELENIUM COMPOUNDS.

CAI750   CAS: 141-00-4   HR: 3
CADMIUM SUCCINATE
mf: $C_4H_4O_4$•Cd   mw: 228.48
SYNS: CADMINATE □ SUCCINIC ACID, CADMIUM SALT (1:1)
CONSENSUS REPORTS: Reported in EPA TSCA Inventory. Cadmium and its compounds are on the Community Right-To-Know List.
OSHA PEL: TWA 5 µg(Cd)/m³
ACGIH TLV: TWA 0.002 mg(Cd)/m³ (respirable dust), Suspected Human Carcinogen); BEI: 5 µg/g creatinine in urine; 5 µg/L in blood
DFG MAK: DFG BAT: Blood 1.5 µg/dL; Urine 15 µg/dL. MAK: Suspected Carcinogen
NIOSH REL: (Cadmium) Reduce to lowest feasible level
SAFETY PROFILE: Confirmed human carcinogen. Poison by ingestion and intraperitoneal routes. When heated to decomposition it emits toxic fumes of Cd. See also CADMIUM COMPOUNDS.

CAJ000   CAS: 10124-36-4   HR: 3
CADMIUM SULFATE (1:1)
mf: O_4S•Cd   mw: 208.46
PROP: Rhombic, white crystals or prisms. Mp: 1000°, d: 4.691. Sol in H_2O; very sltly sol in MeOH, EtOH, and EtOAc.
SYNS: CADMIUM SULFATE □ CADMIUM SULPHATE □ SULFURIC ACID, CADMIUM(2+) SALT □ SULPHURIC ACID, CADMIUM SALT (1:1)
CONSENSUS REPORTS: NTP 10th Report on Carcinogens. IARC Cancer Review:

Group 1 IMEMDT 58,119,93; Animal Sufficient Evidence IMEMDT 2,74,73; Animal Sufficient Evidence IMEMDT 11,39,76; Animal Sufficient Evidence IMEMDT 58,119,93; Human Sufficient Evidence IMEMDT 58,119,93. Reported in EPA TSCA Inventory. EPA Genetic Toxicology Program. Cadmium and its compounds are on the Community Right-To-Know List.

OSHA PEL: TWA 5 μg(Cd)/m^3

ACGIH TLV: TWA 0.002 mg(Cd)/m^3 (respirable dust), Suspected Human Carcinogen); BEI: 5 μg/g creatinine in urine; 5 μg/L in blood

DFG MAK: Animal Carcinogen, Suspected Human Carcinogen

NIOSH REL: (Cadmium) Reduce to lowest feasible level

SAFETY PROFILE: Confirmed human carcinogen with experimental carcinogenic data. Poison by ingestion, subcutaneous, and intraperitoneal routes. Experimental teratogenic and reproductive effects. Mutation data reported. See also CADMIUM COMPOUNDS and SULFATES. When heated to decomposition it emits very toxic fumes of Cd and SO$_x$.

CAJ250   CAS: 7790-84-3   HR: 3
CADMIUM SULFATE (1:1) HYDRATE (3:8)

mf: O$_4$S•Cd•8/3H$_2$O   mw: 256.51

PROP: Crystals from aq soln @ 174°.

SYNS: CADMIUM SULFATE OCTAHYDRATE □ SULFURIC ACID, CADMIUM SALT, HYDRATE

CONSENSUS REPORTS: IARC Cancer Review: Animal Sufficient Evidence IMEMDT 2,74,73. Cadmium and its compounds are on the Community Right-To-Know List.

OSHA PEL: TWA 5 μg(Cd)/m^3

ACGIH TLV: TWA 0.002 mg(Cd)/m^3 (respirable dust), Suspected Human Carcinogen); BEI: 5 μg/g creatinine in urine; 5 μg/L in blood

NIOSH REL: (Cadmium) Reduce to lowest feasible level

SAFETY PROFILE: Confirmed human carcinogen with experimental tumorigenic and neoplastigenic data. Experimental teratogenic and reproductive effects. Mutation data reported. When heated to decomposition it emits very toxic fumes of Cd and SO$_x$. See also CADMIUM SULFATE, CADMIUM COMPOUNDS, and SULFATES.

CAJ500   CAS: 13477-21-9   HR: 3
CADMIUM SULFATE TETRAHYDRATE

mf: O$_4$S•Cd•4H$_2$O   mw: 280.54

SYN: SULFURIC ACID, CADMIUM SALT, TETRAHYDRATE

CONSENSUS REPORTS: Cadmium and its compounds are on the Community Right-To-Know List.

OSHA PEL: TWA 5 μg(Cd)/m^3

ACGIH TLV: TWA 0.002 mg(Cd)/m^3 (respirable dust), Suspected Human Carcinogen); BEI: 5 μg/g creatinine in urine; 5 μg/L in blood

NIOSH REL: (Cadmium) Reduce to lowest feasible level

SAFETY PROFILE: Confirmed human carcinogen with experimental neoplastigenic data. When heated to decomposition it emits very toxic fumes of Cd and SO$_x$. See also CADMIUM COMPOUNDS.

CAJ750   CAS: 1306-23-6   HR: 3
CADMIUM SULFIDE

mf: CdS   mw: 144.46

PROP: Hexagonal, lemon-yellow to orange crystals. Mp: 1750° @ 100 atm, bp: subl in N$_2$, subl @ 9°, d: 4.82. Sltly sol in H$_2$O.

SYNS: AURORA YELLOW □ CADMIUM GOLDEN 366 □ CADMIUM LEMON YELLOW 527 □ CADMIUM MONOSULFIDE □ CADMIUM ORANGE □ CADMIUM PRIMROSE 819 □ CADMIUM SULPHIDE □ CADMIUM YELLOW □ CADMIUM YELLOW 000 □ CADMIUM YELLOW 892 □ CADMIUM YELLOW CONC. DEEP □ CADMIUM YELLOW CONC. GOLDEN □ CADMIUM YELLOW CONC. LEMON □ CADMIUM YELLOW CONC. PRIMROSE □ CADMIUM YELLOW 10G CONC. □ CADMIUM YELLOW OZ DARK □ CADMIUM YELLOW PRIMROSE 47-4100 □ CADMOPUR GOLDEN YELLOW N

□ CADMOPUR YELLOW □ CAPSEBON □ C.I. 77199 □ C.I. PIGMENT ORANGE 20 □ C.I. PIGMENT YELLOW 37 □ FERRO LEMON YELLOW □ FERRO ORANGE YELLOW □ FERRO YELLOW □ GREENOCKITE □ NCI-C02711

CONSENSUS REPORTS: NTP 10th Report on Carcinogens. IARC Cancer Review: Group 1 IMEMDT 58,119,93; Animal Sufficient Evidence IMEMDT 2,74,73; Animal Sufficient Evidence IMEMDT 11,39,76; Animal Sufficient Evidence IMEMDT 58,119,93; Human Sufficient Evidence IMEMDT 58,119,93. EPA Genetic Toxicology Program. Cadmium and its compounds are on the Community Right-To-Know List. Reported in EPA TSCA Inventory.

OSHA PEL: TWA 5 μg(Cd)/m³
ACGIH TLV: TWA 0.002 mg(Cd)/m³ (respirable dust), Suspected Human Carcinogen); BEI: 5 μg/g creatinine in urine; 5 μg/L in blood
DFG MAK: Animal Carcinogen, Suspected Human Carcinogen
NIOSH REL: (Cadmium) Reduce to lowest feasible level
SAFETY PROFILE: Confirmed human carcinogen with experimental carcinogenic and tumorigenic data. Moderately toxic by ingestion and inhalation. Human mutation data reported. When heated to decomposition it emits very toxic fumes of Cd and SO_x. See also CADMIUM COMPOUNDS and SULFIDES.

CAJ760   **HR: 3**
CADMIUM SULFIDE (AMORPHOUS)
CONSENSUS REPORTS: IARC Cancer Review: Group 1 IMEMDT 58,119,93; Human Sufficient Evidence IMEMDT 58,119,93; Animal Sufficient Evidence IMEMDT 58,119,93.
SAFETY PROFILE: Confirmed carcinogen. Mutation data reported. When heated to decomposition it emits toxic vapors of cadmium.

CAJ770   **HR: 3**
CADMIUM SULFIDE mixed with ZINC

SULFIDE (5:95)
SYN: K-82
CONSENSUS REPORTS: IARC Cancer Review: Group 1 IMEMDT 58,119,93; Human Sufficient Evidence IMEMDT 58,119,93; Animal Sufficient Evidence IMEMDT 58,119,93.
ACGIH TLV: TWA 0.002 mg(Cd)/m³ (respirable dust), Suspected Human Carcinogen); BEI: 5 μg/g creatinine in urine; 5 μg/L in blood
NIOSH REL: (Cadmium, dust and fume) Reduce to lowest feasible level
SAFETY PROFILE: Confirmed carcinogen. When heated to decomposition it emits toxic vapors of SO_x, cadmium, and zinc.

CAJ772   CAS: 63661-05-2   HR: 3
CADMIUM SULFIDE mixed with ZINC
**   SULFIDE (8:92)**
SYN: K-83
CONSENSUS REPORTS: IARC Cancer Review: Group 1 IMEMDT 58,119,93; Human Sufficient Evidence IMEMDT 58,119,93; Animal Sufficient Evidence IMEMDT 58,119,93.
ACGIH TLV: TWA 0.002 mg(Cd)/m³ (respirable dust), Suspected Human Carcinogen); BEI: 5 μg/g creatinine in urine; 5 μg/L in blood
NIOSH REL: (Cadmium, dust and fume) Reduce to lowest feasible level
SAFETY PROFILE: Confirmed human carcinogen. When heated to decomposition it emits toxic vapors of SO_x, cadmium, and zinc.

CAK000   **HR: 3**
CADMIUM THERMOVACUUM AEROSOL
mf: Cd   mw: 112.40
SYN: AEROSOL of THERMOVACUUM CADMIUM
CONSENSUS REPORTS: IARC Cancer Review: Group 1 IMEMDT 58,119,93; Human Sufficient Evidence IMEMDT 58,119,93; Animal Sufficient Evidence IMEMDT 58,119,93. Cadmium and its compounds are on the Community Right-To-Know List.

OSHA PEL: TWA 5 μg(Cd)/m³
ACGIH TLV: TWA 0.002 mg(Cd)/m³
(respirable dust), Suspected Human
Carcinogen); BEI: 5 μg/g creatinine in urine;
5 μg/L in blood
NIOSH REL: (Cadmium, dust and fume)
Reduce to lowest feasible level
SAFETY PROFILE: Confirmed human
carcinogen. Moderately toxic by an
unspecified route. When heated to
decomposition it emits very toxic fumes of
Cd. See also CADMIUM and CADMIUM
COMPOUNDS.

CAK250   CAS: 73419-42-8   HR: 3
CADMIUM-THIONEINE
mf: $C_{18}H_{30}N_6O_4S_2$•Cd   mw: 571.06
PROP: Cadmium(II) is bound to the protein
thioneine from rat or rabbit liver (BCPCA6
26,25,77).
CONSENSUS REPORTS: Cadmium and its
compounds are on the Community Right-
To-Know List.
OSHA PEL: TWA 5 μg(Cd)/m³
ACGIH TLV: TWA 0.002 mg(Cd)/m³
(respirable dust), Suspected Human
Carcinogen); BEI: 5 μg/g creatinine in urine;
5 μg/L in blood
NIOSH REL: (Cadmium) Reduce to lowest
feasible level
SAFETY PROFILE: Confirmed human
carcinogen. Deadly poison by intravenous
route. When heated to decomposition it
emits very toxic fumes of NO_x, SO_x, and
Cd. See also CADMIUM COMPOUNDS.

CAK285   CAS: 17650-98-5   HR: 3
CAERULEIN
mf: $C_{58}H_{73}N_{13}O_{21}S_2$ mw: 1352.42
SYN: CERULEIN
SAFETY PROFILE: A poison by
subcutaneous route. When heated to
decomposition it emits toxic vapors of NO_x
and SO_x.

CAK500   CAS: 58-08-2   HR: 3
CAFFEINE
mf: $C_8H_{10}N_4O_2$ mw: 194.22

PROP: White, fleecy masses; odorless with
bitter taste. Mp: 235° (anhyd). Sol in water,
alc, chloroform, ether.
SYNS: CAFFEIN □ COFFEIN (GERMAN) □ COFFEINE
□ 3,7-DIHYDRO-1,3,7-TRIMETHYL-1H-PURINE-2,6-
DIONE □ ELDIATRIC C □ FEMA No. 2224 □
GUARANINE □ KOFFEIN (GERMAN) □ METHYL-
THEOBROMIDE □ 1-METHYLTHEOBROMINE □ 7-
METHYLTHEOPHYLLINE □ NCI-C02733 □ NO-DOZ □
ORGANEX □ THEIN □ THEINE □ 1,3,7-TRIMETHYL-
2,6-DIOXOPURINE □ 1,3,7-TRIMETHYLXANTHINE
CONSENSUS REPORTS: Reported in EPA
TSCA Inventory. EPA Genetic Toxicology
Program.
SAFETY PROFILE: A human poison by
ingestion. An experimental poison by
ingestion, subcutaneous, intraperitoneal,
intramuscular, rectal, and intravenous
routes. Human systemic effects: ataxia,
blood pressure elevation, change in heart
rate, changes in tubules, convulsions or
effect on seizure threshold, diarrhea,
distorted perceptions, hallucinations,
hypermotility, muscle contraction,
musculoskeletal tumors, nausea or vomiting,
toxic psychosis, tremors. A human teratogen
causing developmental abnormalities of the
craniofacial and musculoskeletal systems,
pregnancy termination (abortion), and
stillbirth. Human maternal effects include an
unspecified effect on labor or childbirth.
Human mutation data reported. An
experimental teratogen. Other experimental
reproductive effects. Questionable
carcinogen with experimental carcinogenic
data. Large doses (above 1.0 g) cause
palpitation, excitement, insomnia, dizziness,
headache, and vomiting. Continued
excessive use of caffeine in tea or coffee
may lead to digestive disturbances,
constipation, palpitations, shortness of
breath, and depressed mental states. It is
also implicated in cardiac disorders under
those conditions. When heated to
decomposition it emits toxic fumes of NO_x.

CAL750   CAS: 62-54-4   HR: 3
CALCIUM ACETATE
mf: $C_4H_6O_4$•Ca   mw: 158.18

PROP: Fine, white, hygroscopic, bulky powder. Very sol in water; sltly sol in alc.
SYNS: ACETATE of LIME □ BROWN ACETATE □ CALCIUM DIACETATE □ GRAY ACETATE □ LIME ACETATE □ LIME PYROLIGNITE □ SORBO-CALCIAN □ SORBO-CALCION □ TELTOZAN □ VINEGAR SALTS
CONSENSUS REPORTS: Reported in EPA TSCA Inventory.
SAFETY PROFILE: Poison by intravenous and intraperitoneal routes. Mutation data reported. See also CALCIUM COMPOUNDS. When heated to decomposition it emits acrid smoke and fumes.

CAM000   CAS: 5902-95-4   HR: 3
CALCIUM ACID METHYL ARSONATE

mf: $C_2H_8As_2O_6 \cdot Ca$ mw: 318.02
SYNS: CALAR □ CALCIUM ACID METHANE-ARSONATE □ CALCIUM HYDROGEN METHANE-ARSONATE □ CALCIUM METHANEARSONATE □ CAMA □ SUPER CRAB-E-RAD-CALAR □ SUPER DAL-E-RAD □ SUPER DAL-E-RAD-CALAR □ USAF AN-11
CONSENSUS REPORTS: Arsenic and its compounds are on the Community Right-To-Know List.
OSHA PEL: TWA 0.5 mg(As)/m³
ACGIH TLV: BEI: 35 μ (As)/L inorganic arsenic and methylated metabolites in urine
SAFETY PROFILE: Moderately toxic by intraperitoneal and possibly other routes. Arsenic compounds are considered to be poisons. An herbicide. When heated to decomposition it emits toxic fumes of As. See also ARSENIC COMPOUNDS and CALCIUM COMPOUNDS.

CAM200   CAS: 9005-35-0   HR: 3
CALCIUM ALGINATE

mf: $[(C_6H_7O_6)_2Ca]_n$ mw: 195.16
PROP: White to yellow, granular powder. Insol in water, org solvs.
SYNS: ALGIN □ CA 33 □ CALGINATE □ COMBINACE □ KALTOSTAT
CONSENSUS REPORTS: Reported in EPA TSCA Inventory.
SAFETY PROFILE: Poison by intravenous route. Moderately toxic by intraperitoneal

route. When heated to decomposition it emits acrid smoke and irritating fumes.

CAM222   CAS: 10103-62-5   HR: 3
CALCIUM ARSENATE

mf: $AsH_3O_4 \cdot xCa$ mw: 422.51
SYN: ARSENIC ACID, CALCIUM SALT
CONSENSUS REPORTS: NTP 10th Report on Carcinogens, 2000:known to be human carcinogen
SAFETY PROFILE: Confirmed human carcinogen. When heated to decomposition it emits toxic vapors of NO_x.

CAM300   CAS: 15194-98-6   HR: 3
CALCIUM ARSENITE

DOT: NA 1574
mf: $AsO_4 \cdot Ca$ mw: 179.00
SYNS: ARSENENOUS ACID, CALCIUM SALT (2:1) □ CALCIUM ARSENITE, solid (DOT) □ PROTARS
DOT CLASSIFICATION: 6.1; Label: Poison
SAFETY PROFILE: A poison. When heated to decomposition it emits toxic vapors of As.

CAM500   CAS: 27152-57-4   HR: 3
CALCIUM ARSENITE

DOT: NA 1574
mf: $As_2O_6 \cdot 3Ca$ mw: 366.08
PROP: White, granular powder.
SYNS: ARSENIOUS ACID, CALCIUM SALT □ CALCIUM ARSENITE, solid (DOT) □ MONOCALCIUM ARSENITE
CONSENSUS REPORTS: Arsenic and its compounds are on the Community Right-To-Know List.
OSHA PEL: OSHA: Cancer Hazard
ACGIH TLV: TWA 0.01 mg/m³; Confirmed Human Carcinogen; BEI: 35 μ (As)/L inorganic arsenic and methylated metabolites in urine; BEI: 10 μg/g creatinine in urine; 10 μg/L in blood
NIOSH REL: (Inorganic Arsenic) CL 0.002 mg(As)/m³/15M
DOT CLASSIFICATION: 6.1; Label: Poison
SAFETY PROFILE: Confirmed carcinogen. A poison by inhalation and ingestion. When heated to decomposition it emits toxic fumes of As. See also ARSENIC

COMPOUNDS and CALCIUM COMPOUNDS.

CAM600   CAS: 5743-27-1   HR: D
CALCIUM ASCORBATE
mf: $C_{12}H_{14}CaO_{12}\cdot 2H_2O$ mw: 426.35
PROP: White crystalline powder; odorless. Sol in water; sltly sol in alc; insol in ether.
SAFETY PROFILE: When heated to decomposition it emits acrid smoke and irritating fumes.

CAM750   CAS: 6485-34-3   HR: 3
CALCIUM-o-BENZOSULFIMIDE
mf: $C_{14}H_{10}N_2O_6S_2\cdot Ca$ mw: 406.46
PROP: White, crystalline powder; odorless or faint aromatic odor; sol in water.
SYNS: 1,2-BENZISOTHIAZOL-3(2H)-ONE-1,1-DIOXIDE, CALCIUM SALT □ CALCIUM-o-BENZOSULPHIMIDE □ CALCIUM-2-BENZO-SULPHIMIDE □ CALCIUM SACCHARIN □ CALCIUM SACCHARINA □ CALCIUM SACCHARINATE □ DARAMIN □ SACCHARIN CALCIUM □ SULPHO-BENZOIC IMIDE CALCIUM SALT
CONSENSUS REPORTS: Reported in EPA TSCA Inventory.
SAFETY PROFILE: Mutagenic data reported. When heated to decomposition it emits toxic fumes of SO_x and NO_x.

CAN000   CAS: 13780-03-5   HR: 3
CALCIUM BISULFITE
DOT: UN 1923
PROP: Colorless or sltly yellowish liquid; strong sulfur dioxide odor. D: 1.06.
SYNS: CALCIUM DITHIONITE (DOT) □ CALCIUM HYDROSULFITE (DOT) □ SULFUROUS ACID, CALCIUM SALT (2:1) (8CI,9CI)
CONSENSUS REPORTS: Reported in EPA TSCA Inventory.
DOT CLASSIFICATION: 4.2; Label: Spontaneously Combustible
SAFETY PROFILE: A poison via ingestion. Strong irritant via skin and eye contact, ingestion, and inhalation. Spontaneously combustible. When heated to decomposition it emits toxic fumes of SO_x. See also SULFITES and SULFUROUS ACID.

CAN400   HR: 1
CALCIUM BROMATE
mf: $Ca(BrO_3)_2\cdot H_2O$ mw: 313.90
PROP: White crystalline powder. Very sol in water.
SAFETY PROFILE: A nuisance dust.

CAN750   CAS: 75-20-7   HR: 3
CALCIUM CARBIDE
DOT: UN 1402
mf: C_2Ca mw: 64.10
PROP: Rhombic, moisture-sensitive, gray crystals. Mp: approx 2300°, d: 2.222.
SYNS: ACETYLENOGEN □ CALCIUM ACETYLIDE □ CALCIUM DICARBIDE
CONSENSUS REPORTS: Reported in EPA TSCA Inventory.
DOT CLASSIFICATION: 4.3; Label: Dangerous When Wet
SAFETY PROFILE: Reaction on contact with moisture forms explosive acetylene gas. Flammable on contact with moisture, acid or acid fumes; evolves heat or flammable vapors. Moderate explosion hazard. Incandescent reaction with Cl_2 (245°C), Br_2 (350°C), I_2 (305°C), HCl gas + heat, PbF_2, Mg + heat. Incompatible with Se, (KOH + Cl_2), $AgNO_3$, Na_2O_2, $SnCl_2$, S, water. Mixtures with iron(III) chloride, iron(III) oxide, tin(II) chloride are easily ignited and burn fiercely. Vigorous reaction with methanol after an induction period. Addition to silver nitrate solutions precipitates the dangerously explosive silver acetylide. Copper salt solutions behave similarly. See also CALCIUM HYDROXIDE and ACETYLENE.

CAO000   CAS: 1317-65-3   HR: 1
CALCIUM CARBONATE
mf: $CO_3\cdot Ca$ mw: 100.09
PROP: White microcrystalline powder. Mp: 825° (α), 1339° (β) @ 102.5 atm, d: 2.7–2.95. Found in nature as the minerals limestone, marble, aragonite, calcite, and vaterite. Odorless, tasteless powder or crystals. Two crystalline forms are of commercial importance: aragonite, orthorhombic, mp:

825° (decomp), d: 2.83, formed at temperatures above 30°; calcite, hexagonal-rhombohedral, mp: 1339° (102.5 atm), d: 2.711, formed at temperatures below 30°. At about 825° it decomposes into CaO and CO_2. Practically insol in water, alc; sol in dilute acids.

SYNS: AGRICULTURAL LIMESTONE □ AGSTONE □ ARAGONITE □ ATOMIT □ BELL MINE PULVERIZED LIMESTONE □ CALCITE □ CARBONIC ACID, CALCIUM SALT (1:1) □ CHALK □ DOLOMITE □ FRANKLIN □ LIMESTONE (FCC) □ LITHOGRAPHIC STONE □ MARBLE □ NATURAL CALCIUM CARBONATE □ PORTLAND STONE □ SOHNHOFEN STONE □ VATERITE

CONSENSUS REPORTS: Reported in EPA TSCA Inventory.

OSHA PEL: Total Dust: 15 mg/m³; Respirable Fraction: 5 mg/m³

ACGIH TLV: TWA (nuisance particulate) 10 mg/m³ of total dust (when toxic impurities are not present, e.g., quartz <1%)

SAFETY PROFILE: A nuisance dust. An eye and skin irritant. Ignites on contact with F_2. Incompatible with acids, ammonium salts, $(Mg + H_2)$. Calcium carbonate is a common air contaminant. See also CALCIUM COMPOUNDS.

CAO250   CAS: 9049-05-2   HR: 2
CALCIUM CARRAGHEENATE

PROP: A mixture of highly sulfated polygalactosides. It is extracted from seaweed (FAONAU 53A,398,74).

SYNS: ALGIN GUM □ CALCIUM CARAGEENIN □ CALCIUM CARRAGEENAN □ CARRAGEENAN, CALCIUM(II) SALT □ VISCARIN 402

CONSENSUS REPORTS: Reported in EPA TSCA Inventory.

SAFETY PROFILE: Moderately toxic by ingestion. Experimental reproductive effects. When heated to decomposition it emits toxic fumes of SO_x. See also CALCIUM COMPOUNDS.

CAO500   CAS: 10137-74-3   HR: 2
CALCIUM CHLORATE

DOT: UN 1452/UN 2429
mf: $Cl_2O_6 \cdot Ca$ mw: 206.98

PROP: Monoclinic, yellowish-white, deliquescent crystals. Mp: 340° (loses H_2O @ >100°), d: 2.711. Very sol in H_2O.

SYNS: CALCIUM CHLORATE, aqueous solution (DOT) □ CHLORATE de CALCIUM (FRENCH)

DOT CLASSIFICATION: 5.1; Label: Oxidizer

SAFETY PROFILE: Moderately toxic by ingestion and intraperitoneal routes. A powerful oxidant. Incompatible with Al, As, C, Cu, charcoal, MnO_2, metal sulfides, S, dibasic organic acids, organic matter, P. When heated to decomposition it emits toxic fumes of Cl⁻. See also CHLORATES for fire, disaster, and explosion hazards.

CAO750   CAS: 10043-52-4   HR: 2
CALCIUM CHLORIDE

mf: $CaCl_2$ mw: 110.98

PROP: Cubic, colorless, deliq crystals. Mp: 782°, bp: >1600°, d: 2.512 @ 25°. Very sol in H_2O; sol in EtOH, Me_2CO, and AcOH.

SYNS: CALCIUM CHLORIDE, anhydrous □ CALPLUS □ CALTAC □ DOWFLAKE □ LIQUIDOW □ PELADOW □ SNOMELT □ SUPERFLAKE ANHYDROUS

CONSENSUS REPORTS: Reported in EPA TSCA Inventory. EPA Genetic Toxicology Program. EPA FIFRA 1988 pesticide subject to registration or re-registration.

SAFETY PROFILE: Moderately toxic by ingestion. Poison by intravenous, intramuscular, intraperitoneal, and subcutaneous routes. Human systemic effects: dermatitis, changes in calcium. Questionable carcinogen with experimental tumorigenic data. Mutation data reported. Reacts violently with $(B_2O_3 + CaO)$, BrF_3. Reaction with zinc releases explosive hydrogen gas. Catalyzes exothermic polymerization of methyl vinyl ether. Exothermic reaction with water. When heated to decomposition it emits toxic fumes of Cl⁻. See also CALCIUM COMPOUNDS and CHLORIDES.

CAP000   CAS: 14674-72-7   HR: 3
CALCIUM CHLORITE

DOT: UN 1453
mf: $CaCl_2O_4$ mw: 174.98
PROP: White solid.

DOT CLASSIFICATION: 5.1; Label: Oxidizer
SAFETY PROFILE: A strong oxidizer. Ignites on contact with potassium thiocyanate. Reaction with Cl_2 yields explosive ClO_2. When heated to decomposition it emits toxic fumes of Cl^-. See also CHLORITES and CALCIUM COMPOUNDS.

CAP500   CAS: 13765-19-0   HR: 3
CALCIUM CHROMATE
mf: $CrO_4•Ca$ mw: 156.08
PROP: Monoclinic prisms; yellow colored crystals. Sltly sol in H_2O; insol in EtOH and Me_2CO. IDLH Ca [15 mg/m³ {as Cr(VI)}].
SYNS: CALCIUM CHROMATE (VI) □ CALCIUM CHROME YELLOW □ CALCIUM CHROMIUM OXIDE (CaCrO4) □ CALCIUM MONOCHROMATE □ CHROMIC ACID, CALCIUM SALT (1:1) □ C.I. 77223 □ C.I. PIGMENT YELLOW 33 □ GELBIN □ RCRA WASTE NUMBER U032 □ YELLOW ULTRAMARINE
CONSENSUS REPORTS: IARC Cancer Review: Group 1 IMEMDT 49,49,90; Human Sufficient Evidence IMEMDT 23,205,80; Animal Sufficient Evidence IMEMDT 23,205,80. Reported in EPA TSCA Inventory. EPA Genetic Toxicology Program. Chromium and its compounds are on the Community Right-To-Know List.
OSHA PEL: CL 0.1 mg(CrO₃)/m³
ACGIH TLV: TWA 0.001 mg(Cr)/m³; Suspected Human Carcinogen
DFG MAK: DFG TRK: 0.1 mg/m³ calculated as CrO_3 in that portion of dust that can possibly be inhaled; 0.2 mg/m³ arc-welding by hand; others 0.1 mg/m³. Animal Carcinogen, Suspected Human Carcinogen
NIOSH REL: (Chromium(VI)) TWA 0.001 mg(Cr(VI))/m³
SAFETY PROFILE: Suspected human carcinogen with experimental carcinogenic, neoplastigenic, and tumorigenic data. Experimental reproductive effects. Mutation data reported. A powerful oxidizer. Mixture with boron burns violently if ignited. See also CHROMIUM COMPOUNDS and CALCIUM COMPOUNDS.

CAP750   CAS: 10060-08-9   HR: 3
CALCIUM CHROMATE(VI) DIHYDRATE
mf: $CrO_4•Ca•2H_2O$ mw: 192.12
PROP: IDLH Ca [15 mg/m³ {as Cr(VI)}].
SYNS: CALCIUM CHROME YELLOW □ CHROMIC ACID, CALCIUM SALT (1:1), DIHYDRATE □ C.I. 77223 □ C.I. PIGMENT YELLOW 33 □ GELBIN YELLOW ULTRAMARINE □ PIGMENT YELLOW 33 □ STEINBUHL YELLOW
CONSENSUS REPORTS: IARC Cancer Review: Animal Sufficient Evidence IMEMDT 2,100,72. Chromium and its compounds are on the Community Right-To-Know List.
OSHA PEL: CL 0.1 mg(CrO³)/m³
ACGIH TLV: TWA 0.05 mg(Cr)/m³; Confirmed Human Carcinogen
NIOSH REL: (Chromium(VI)) TWA 0.001 mg(Cr(VI))/m³
SAFETY PROFILE: Confirmed human carcinogen with experimental tumorigenic and carcinogenic data. Poison by ingestion and implant routes. Mutation data reported. A powerful oxidizer. See also CHROMIUM COMPOUNDS and CALCIUM COMPOUNDS.

CAQ000   HR: 1
CALCIUM COMPOUNDS
SAFETY PROFILE: The fumes evolved by burning calcium in air are composed of calcium oxide (quicklime), which is an irritant to the skin, eyes, and mucous membranes. Generally speaking, calcium compounds should be considered toxic only when they contain toxic components (such as arsenic, etc.) or as calcium oxide or hydroxide. Calcium compounds are common air contaminants.

CAQ250   CAS: 156-62-7   HR: 3
CALCIUM CYANAMIDE
DOT: UN 1403
mf: $CN_2•Ca$ mw: 80.11
PROP: Hexagonal, rhombohedral, colorless, moisture-sensitive crystals. Mp: 1300°, subl @ >1500°. Decomposes in water. Compound not hydrated; compound

contains more than 0.1% calcium (FEREAC 41,15972,76).

SYNS: AERO-CYANAMID □ AERO CYANAMID GRANULAR □ AERO CYANAMID SPECIAL GRADE □ ALZODEF □ CALCIUM CARBIMIDE □ CALCIUM CYANAMID □ CCC □ CYANAMIDE □ CYANAMIDE CALCIQUE (FRENCH) □ CYANAMIDE, CALCIUM SALT (1:1) □ CYANAMID GRANULAR □ CYANAMID SPECIAL GRADE □ CY-L 500 □ LIME-NITROGEN (DOT) □ NCI-C02937 □ NITROGEN LIME □ NITROLIME □ USAF CY-2

CONSENSUS REPORTS: NCI Carcinogenesis Bioassay (feed); No Evidence: mouse, rat NCITR* NCI-CG-TR-163,79. Community Right-To-Know List. Reported in EPA TSCA Inventory.

OSHA PEL: TWA 0.5 mg/m^3

ACGIH TLV: TWA 0.5 mg/m^3; Not Classifiable as a Human Carcinogen

DFG MAK: 1 mg/m^3

DOT CLASSIFICATION: 4.3; Label: Dangerous When Wet

SAFETY PROFILE: Poison by ingestion, inhalation, skin contact, intravenous, and intraperitoneal routes. Moderately toxic to humans by ingestion. Questionable carcinogen with experimental tumorigenic data. Mutation data reported. The fatal dose, by ingestion, is probably around 20 to 30 g for an adult. It does not have a cyanide effect. Calcium cyanamide is not believed to have a cumulative action. Flammable. Reaction with water forms the explosive acetylene gas. When heated to decomposition it emits toxic fumes of NO_x and CN^-. See also CALCIUM COMPOUNDS, AMIDES, and CYANIDE.

CAQ500   CAS: 592-01-8   HR: 3
CALCIUM CYANIDE

DOT: UN 1575

mf: C_2CaN_2 mw: 92.12

PROP: Rhombohedral crystals or white powder. Mp: decomp >350°.

SYNS: CALCID □ CALCIUM CYANIDE MIXTURE, solid (DOT) □ CALCYAN □ CALCYANIDE □ CYANOGAS □ CYANURE de CALCIUM (FRENCH) □ RCRA WASTE NUMBER P021

CONSENSUS REPORTS: Cyanide and its compounds are on the Community Right-To-Know List. Reported in EPA TSCA Inventory.

OSHA PEL: TWA 5 mg(CN)/m^3

ACGIH TLV: CL 5 mg(CN)/m^3 (skin)

DFG MAK: 5 mg/m^3

NIOSH REL: (Cyanide) CL 5 mg(CN)/m^3/10M

DOT CLASSIFICATION: 6.1; Label: Poison

SAFETY PROFILE: A deadly poison by ingestion and probably other routes. When heated to decomposition it emits toxic fumes of NO_x and CN^-. See also CALCIUM COMPOUNDS and CYANIDE.

CAR000   CAS: 139-06-0   HR: 3
CALCIUM CYCLOHEXYLSULPHAMATE

mf: $C_{12}H_{24}N_2O_6S_2$•Ca   mw: 396.58

PROP: White, crystalline powder; almost odorless; freely sol in water; practically insol in alc, benzene, chloroform, and ether.

SYNS: CALCIUM CYCLAMATE □ CALCIUM CYCLO-HEXANESULFAMATE □ CALCIUM CYCLOHEXANE SULPHAMATE □ CALCIUM CYCLOHEXYLSULFAMATE □ CYCLAMATE CALCIUM □ CYCLAMATE, CALCIUM SALT □ CYCLAN □ CYCLOHEXANESULFAMIC ACID, CALCIUM SALT □ CYCLOHEXYLSULPHAMIC ACID, CALCIUM SALT □ CYLAN □ DIETIL □ KALZIUMZY-KLAMATE (GERMAN) □ SUCARYL CALCIUM

CONSENSUS REPORTS: IARC Cancer Review: Group 3 IMEMDT 7,178,87; Animal Limited Evidence IMEMDT 22,55,80; Human Inadequate Evidence IMEMDT 22,55,80. Reported in EPA TSCA Inventory. EPA Genetic Toxicology Program.

SAFETY PROFILE: Poison by ingestion and intravenous routes. Experimental reproductive effects. Questionable carcinogen with experimental tumorigenic and neoplastigenic data. Human mutation data reported. When heated to decomposition it emits very toxic fumes of SO_x and NO_x. See also CALCIUM COMPOUNDS.

CAR775 **HR: D**
CALCIUM DISODIUM EDTA
mf: $C_{10}H_{12}CaN_2Na_2O_8 \cdot 2H_2O$ mw: 410.30
PROP: White crystalline powder; hygroscopic with a faint salt taste. Sol in water.
SYNS: CALCIUM DISODIUM EDETATE □ CALCIUM DISODIUM ETHYLENEDIAMINETETRAACETATE □ CALCIUM DISODIUM (ETHYLENEDINITRILO)-TETRAACETATE
SAFETY PROFILE: When heated to decomposition it emits toxic fumes of NO_x.

CAS000 **CAS: 7789-75-5** **HR: 2**
CALCIUM FLUORIDE
mf: CaF_2 mw: 78.08
PROP: Hygroscopic, cubic, colorless crystals; luminous with heat. Mp: 1418°, d: 3.180. Practically insol in H_2O; insol in Me_2CO; sol in acids.
SYNS: ACID-SPAR □ CALCIUM DIFLUORIDE □ FLUORITE □ FLUORSPAR □ IRTRAN 3 □ LIPARITE □ MET-SPAR
CONSENSUS REPORTS: Reported in EPA TSCA Inventory.
OSHA PEL: TWA 2.5 mg(F)/m³
ACGIH TLV: TWA 2.5 mg(F)/m³; BEI: 3 mg/g creatinine of fluorides in urine prior to shift; 10 mg/g creatinine of fluorides in urine at end of shift.
NIOSH REL: (Inorganic Fluorides) TWA 2.5 mg(F)/m³
SAFETY PROFILE: Moderately toxic by intraperitoneal route. Mildly toxic by ingestion. An experimental teratogen. Other experimental reproductive effects. Mutation data reported. See also FLUORIDES and CALCIUM COMPOUNDS. When heated to decomposition it emits toxic fumes of F^-.

CAS250 **CAS: 544-17-2** **HR: 3**
CALCIUM FORMATE
mf: $C_2H_2O_4 \cdot Ca$ mw: 130.12
PROP: Colorless, orthorhombic crystals. Also exists in several other polymorphic forms. Very sol in H_2O; insol in EtOH.
SYNS: FORMIC ACID, CALCIUM SALT □ MRAVENCAN VAPENATY (CZECH)

CONSENSUS REPORTS: Reported in EPA TSCA Inventory.
SAFETY PROFILE: Poison by intravenous route. Moderately toxic by ingestion. An eye irritant. When heated to decomposition it emits acrid smoke and fumes. See also CALCIUM COMPOUNDS.

CAS750 **CAS: 299-28-5** **HR: 2**
CALCIUM GLUCONATE
mf: $C_{12}H_{22}O_{14} \cdot Ca$ mw: 430.42
PROP: White, fluffy powder or granules; odorless and tasteless. Sol in hot water; less sol in cold water; insol in alc, acetic acid, and other org solvs. Mp: loses H_2O @ 120°.
SYNS: CALCICOL □ CALCIOFON □ CALCIPUR □ CALCIUM d-GLUCONATE □ CALCIUM HEXA-GLUCONATE □ CALGLUCOL □ CALGLUCON □ DRAGOCAL □ EBUCIN □ GLUCAL □ GLUCOBIOGEN □ GLUCONATE de CALCIUM (FRENCH) □ GLUCONATO di CALCIO □ d-GLUCONIC ACID, CALCIUM SALT (2:1) (9CI) □ KALPREN □ NOVOCAL
CONSENSUS REPORTS: Reported in EPA TSCA Inventory.
SAFETY PROFILE: Moderately toxic by subcutaneous, intraperitoneal, and intravenous routes. Human systemic effects in infants by intramuscular route: dermatitis and fever. When heated to decomposition it emits acrid smoke and fumes. See also CALCIUM COMPOUNDS.

CAS825 **HR: 1**
CALCIUM HEXAMETAPHOSPHATE
SAFETY PROFILE: A nuisance dust.

CAT225 **CAS: 1305-62-0** **HR: 2**
CALCIUM HYDROXIDE
mf: CaH_2O_2 mw: 74.10
PROP: Rhombic, trigonal, colorless crystals or white power; sltly bitter taste. Mp: loses H_2O @ 580°, bp: decomp, d: 2.343. Sltly sol in water and glycerin; insol in alc.
SYNS: BELL MINE □ BIOCALC □ CALCIUM DIHYDROXIDE □ CALCIUM HYDRATE □ CALCIUM HYDROXIDE (ACGIH, OSHA) □ CALVIT □ CARBOXIDE □ HYDRATED LIME □ KALKHYDRATE □ KEMIKAL □ LIMBUX □ LIME MILK □ LIME WATER □ MILK OF LIME □ SLAKED LIME

CONSENSUS REPORTS: Reported in EPA TSCA Inventory.
OSHA PEL: TWA 5 mg/m³
ACGIH TLV: TWA 5 mg/m³
SAFETY PROFILE: Mildly toxic by ingestion. A severe eye irritant. A skin, mucous membrane, and respiratory system irritant. Mutation data reported. Causes dermatitis. Dust is considered to be a significant industrial hazard. A common air contaminant. Violent reaction with maleic anhydride, nitroethane, nitromethane, nitroparaffins, nitropropane, phosphorus. Reaction with polychlorinated phenols + potassium nitrate forms extremely toxic products. See also CALCIUM COMPOUNDS.

CAT500   CAS: 7789-80-2   HR: 1
CALCIUM IODATE
mf: $Ca(IO_3)_2 \cdot H_2O$ mw: 407.90
PROP: White powder or colorless monoclinic crystals. Decomposes on heating. Sltly sol in water; insol in alc.
SAFETY PROFILE: A nuisance dust.

CAT650   CAS: 5001-51-4   HR: D
CALCIUM LACTOBIONATE
mf: $C_{24}H_{42}CaO_{24}$ mw: 754.66
PROP: White powder. Mp: 120° (decomp). Sol in water; insol in alc, ether.
SYN: CALCIUM 4-(β-d-GALACTOSIDO)-d-GLUCONATE
SAFETY PROFILE: When heated to decomposition it emits acrid smoke and irritating fumes.

CAT750   CAS: 7789-82-4   HR: 3
CALCIUM MOLYBDATE
mf: $MoO_4 \cdot Ca$ mw: 200.02
PROP: White crystals. An electrical conductor. Mp: 965° (decomp). Insol in H_2O. IDLH 1000 mg/m³ (as Mo).
SYNS: CALCIUM MOLYBDENUM OXIDE (CaMoO4) □ MOLYBDATE, CALCIUM □ MOLYBDIC ACID (H2MoO4), CALCIUM SALT (1:1)
CONSENSUS REPORTS: Reported in EPA TSCA Inventory.

OSHA PEL: TWA Total Dust: 10 mg/m³; Respirable Fraction: 5 mg/m³
ACGIH TLV: Insoluble Compounds: inhalable fraction, 10 mg(Mo)/m³, 3 mg(Mo)/m³, respirable fraction.
SAFETY PROFILE: Poison by intraperitoneal route. See also MOLYBDENUM and CALCIUM COMPOUNDS.

CAU000   CAS: 10124-37-5   HR: 3
CALCIUM(II) NITRATE (1:2)
DOT: UN 1454
mf: $N_2O_6 \cdot Ca$ mw: 164.10
PROP: Hygroscopic, colorless, cubic crystals. Mp: 561°. Decomposes on heating. Very sol in H_2O and EtOH; sol in MeOH and Me_2CO; insol in Et_2O.
SYNS: CALCIUM DINITRATE □ CALCIUM NITRATE (DOT) □ CALCIUM SALTPETER □ NITRIC ACID, CALCIUM SALT (8CI,9CI) □ NORGE SALTPETER □ NORWAY SALTPETER □ NORWEGIAN SALTPETER □ SYNFAT 1006
CONSENSUS REPORTS: Reported in EPA TSCA Inventory.
DOT CLASSIFICATION: 5.1; Label: Oxidizer
SAFETY PROFILE: A poison by ingestion. An irritant. A strong oxidant. Forms powerfully explosive mixtures with aluminum + ammonium nitrate + formamide + water, ammonium nitrate + hydrocarbon oils, ammonium nitrate + water-soluble fuels, and organic materials. When heated to decomposition it emits toxic fumes of NO_x. See also NITRATES and CALCIUM COMPOUNDS.

CAU300   CAS: 142-17-6   HR: D
CALCIUM OLEATE
mf: $C_{36}H_{66}CaO_4$ mw: 602.97
PROP: Pale yellow transparent solid.
SYNS: 9-OCTADECENOIC ACID CALCIUM SALT □ OLEIC ACID CALCIUM SALT
SAFETY PROFILE: When heated to decomposition it emits acrid smoke and irritating fumes.

CAU500   **CAS: 1305-78-8**   **HR: 3**
CALCIUM OXIDE
DOT: UN 1910
mf: CaO   mw: 56.08
PROP: Cubic, colorless, white crystals. Mp: 2580°, d: 3.37, bp: 2850°. Sol in water and glycerin; insol in alc. IDLH 25 mg/m^3.
SYNS: AIRLOCK □ BELL CML(E) □ BURNT LIME □ CALCIA □ CALOXOL CP2 □ CALOXOL W3 □ CALX □ CALXYL □ CML 21 □ CML 31 □ DESICAL P □ LIME □ LIME, BURNED □ LIME, UNSLAKED (DOT) □ OXYDE de CALCIUM (FRENCH) □ QUICKLIME (DOT) □ RHENOSORB C □ RHENOSORB F □ WAPNIOWY TLENEK (POLISH)
CONSENSUS REPORTS: Reported in EPA TSCA Inventory.
OSHA PEL: TWA 5 mg/m^3
ACGIH TLV: TWA 2 mg/m^3
DFG MAK: 5 mg/m^3
DOT CLASSIFICATION: 8; Label: Corrosive
SAFETY PROFILE: A caustic and irritating material. See also CALCIUM COMPOUNDS. A common air contaminant. A powerful caustic to living tissue. The powdered oxide may react explosively with water. Mixtures with ethanol may ignite if heated and thus can cause an air-vapor explosion. Violent reaction with (B$_2$O$_3$ + CaCl$_2$) interhalogens (e.g., BF$_3$, ClF$_3$), F$_2$, HF, P$_2$O$_5$ + heat, water. Incandescent reaction with liquid HF. Incompatible with phosphorus(V) oxide.

CAU750   **CAS: 137-08-6**   **HR: 2**
CALCIUM-d-PANTOTHENATE
mf: C$_{19}$H$_{34}$N$_2$O$_{10}$•Ca   mw: 490.63
PROP: White, sltly hygroscopic powder; odorless; bitter taste; crystals from MeOH. Mp: 195–196° (decomp). Sol in water and glycerin; insol in alc, chloroform, and ether.
SYNS: CALCIUM d(+)-N-(α,γ-DIHYDROXY-β,β-DIMETHYLBUTYRYL)-β-ALANINATE □ CALCIUM PANTHOTHENATE (FCC) □ CALCIUM PANTOTHEN-ATE □ d-CALCIUM PANTOTHENATE □ CALPANATE □ DEXTRO CALCIUM PANTOTHENATE □ N-(2,4-DIHYDROXY-3,3-DIMETHYLBUTYRYL)-β-ALANINE CALCIUM □ PANCAL □ PANTHOJECT □ PANTHOLIN □ PANTOTHENATE CALCIUM □ PANTOTHENIC ACID, CALCIUM SALT □ (+)-PANTOTHENIC ACID, CALCIUM SALT □ VITAMIN B-5

CONSENSUS REPORTS: Reported in EPA TSCA Inventory.
SAFETY PROFILE: Moderately toxic by intraperitoneal, subcutaneous, and intravenous routes. Mildly toxic by ingestion. A vitamin. See also CALCIUM COMPOUNDS. When heated to decomposition it emits toxic fumes of NO$_x$.

CAV250   **CAS: 10118-76-0**   **HR: 3**
CALCIUM PERMANGANATE
DOT: UN 1456
mf: Mn$_2$O$_8$•Ca   mw: 277.96
PROP: Violet, deliquescent crystals. Mp: decomp, d: 2.4.
SYNS: ACERDOL □ KALIUMPERMANGANAT (GERMAN) □ PERMANGANIC ACID(HMnO4), CALCIUM SALT (8CI,9CI)
CONSENSUS REPORTS: Manganese and its compounds are on the Community Right-To-Know List.
OSHA PEL: CL 5 mg(Mn)/m^3
ACGIH TLV: TWA 0.03 mg(Mn)/m^3
DOT CLASSIFICATION: 5.1; Label: Oxidizer
SAFETY PROFILE: Poison by intravenous route. See also CALCIUM COMPOUNDS, MANGANESE COMPOUNDS, and PERMANGANATES. A strong oxidant. May explode on contact with acetic acid or acetic anhydride. Ignites on contact with cellulose. Incompatible with hydrogen peroxide.

CAV500   **CAS: 1305-79-9**   **HR: 3**
CALCIUM PEROXIDE
DOT: UN 1457
mf: CaO$_2$ mw: 72.08
PROP: Yellow crystals or powder or white crystals, decomposes in air. Mp: decomp @ 275°. Insol in water; sol in acids, forming hydrogen peroxide.
SYNS: CALCIUM DIOXIDE □ CALCIUM SUPEROXIDE
CONSENSUS REPORTS: Reported in EPA TSCA Inventory.
DOT CLASSIFICATION: 5.1; Label: Oxidizer
SAFETY PROFILE: Irritating in concentrated form. Will react with moisture to form slaked lime. Flammable if hot and

mixed with finely divided combustible material. Mixtures with oxidizable materials can also be ignited by grinding and are explosion hazards. A strong alkali. An oxidizer. Mixtures with polysulfide polymers may ignite. See also CALCIUM COMPOUNDS, CALCIUM HYDROXIDE, and PEROXIDES, INORGANIC.

CAW100   CAS: 7757-93-9   HR: 1
CALCIUM PHOSPHATE, DIBASIC
mf: $CaHPO_4 \cdot 2H_2O$ mw: 172.09
PROP: White powder or crystals. Sol in dilute acid; insol in water, alc.
SYN: DICALCIUM PHOSPHATE
SAFETY PROFILE: Skin and eye irritant. A nuisance dust.

CAW250   CAS: 1305-99-3   HR: 3
CALCIUM PHOSPHIDE
DOT: UN 1360
mf: Ca_3P_2 mw: 182.18
PROP: Red-brown crystals. Mp: >1600°, d: 2.238 @ 25°. Insol in EtOH, Et_2O, and C_6H_6.
SYNS: CALCIUM PHOTOPHOR □ PHOTOPHOR
CONSENSUS REPORTS: Reported in EPA TSCA Inventory.
DOT CLASSIFICATION: 4.3; Label: Dangerous When Wet, Poison
SAFETY PROFILE: Highly toxic due to phosphide, which in presence of moisture emits phosphine. The phosphine may ignite spontaneously in air. Incandescent reaction with oxygen at 300°C. Incompatible with dichlorine oxide. When heated to decomposition it emits toxic fumes of PO_x. See also CALCIUM COMPOUNDS and PHOSPHIDES.

CAW400   CAS: 4075-81-4   HR: 2
CALCIUM PROPIONATE
mf: $C_6H_{10}CaO_4$ mw: 186.22
PROP: White crystals; faint odor of propionic acid. Sol in water.
SYNS: BIOBAN-C □ CALCIUM DIPROPIONATE □ CALCIUM PROPIONATE □ PROPANOIC ACID, CALCIUM SALT (9CI)

SAFETY PROFILE: Moderately toxic by ingestion. When heated to decomposition it emits acrid smoke and irritating fumes.

CAW500   CAS: 9007-13-0   HR: 2
CALCIUM RESINATE
DOT: UN 1313/UN 1314
mf: $Ca(C_{44}H_{62}O_4)_2$ mw: 1349.50
PROP: Yellowish-white, amorphous powder or lumps.
SYNS: CALCIUM RESINATE (UN 1313) (DOT) □ CALCIUM RESINATE, fused (UN 1314) (DOT) □ LIMED ROSIN □ RESIN ACIDS and ROSIN ACIDS, CALCIUM SALTS □ URAPRINT 62-126
CONSENSUS REPORTS: Reported in EPA TSCA Inventory.
DOT CLASSIFICATION: 4.1; Label: Flammable Solid
SAFETY PROFILE: Flammable solid when heated; can react with oxidizing materials. When heated to decomposition it emits acrid smoke and fumes. See also CALCIUM COMPOUNDS.

CAW525   HR: D
CALCIUM RICINOLEATE
SAFETY PROFILE: When heated to decomposition it emits acrid smoke and irritating fumes.

CAW850   CAS: 1344-95-2   HR: 1
CALCIUM SILICATE
PROP: Varying proportions of CaO and SiO_2. White powder. Insol in water.
SYNS: CALCIUM HYDROSILICATE □ CALCIUM MONOSILICATE □ CALCIUM POLYSILICATE □ CALCIUM SILICATE, synthetic nonfibrous (ACGIH) □ CALFLO E □ CALSIL □ CS LAFARGE □ FLORITE R □ MARIMET 45 □ MICROCAL 160 □ MICROCAL ET □ MICRO-CEL □ MICRO-CEL A □ MICRO-CEL B □ MICRO-CEL C □ MICRO-CEL E □ MICRO-CEL T □ MICRO-CEL T26 □ MICRO-CEL T38 □ MICRO-CEL T41 □ PROMAXON P60 □ SILENE EF □ SILMOS T □ SOLEX □ STABINEX NW 7PS □ STARLEX L □ SW 400 □ TOYOFINE A
OSHA PEL: Total Dust: 15 mg/m³; Respirable Fraction: 5 mg/m³
ACGIH TLV: TWA (nuisance particulate) 10 mg/m³ of total dust (when toxic impurities

are not present, e.g., quartz <1%); Not Classifiable as a Human Carcinogen
SAFETY PROFILE: A nuisance dust.

CAX250   CAS: 16925-39-6   HR: 3
CALCIUM SILICOFLUORIDE
mf: CaF_6Si mw: 182.17
PROP: White, crystalline powder. Hydrolyzes in H_2O forming complex mixtures of products. Decomp on heating with formation of CaF_2 and SiF_4 at 225°. Sltly sol in EtOH.
SYNS: CALCIUM FLUOROSILICATE □ CALCIUM FLUOSILICATE □ CALCIUM HEXAFLUOROSILICATE □ SILICATE(2-), HEXAFLUORO-, CALCIUM (1:1) (9CI)
CONSENSUS REPORTS: Reported in EPA TSCA Inventory.
OSHA PEL: TWA 2.5 mg(F)/m^3
ACGIH TLV: TWA 2.5 mg(F)/m^3; BEI: 3 mg/g creatinine of fluorides in urine prior to shift; 10 mg/g creatinine of fluorides in urine at end of shift.
NIOSH REL: (Inorganic Fluorides) TWA 2.5 mg(F)/m^3
DOT CLASSIFICATION: 6.1; Label: KEEP AWAY FROM FOOD
SAFETY PROFILE: Poison by ingestion and subcutaneous routes. See also CALCIUM COMPOUNDS. When heated to decomposition it emits toxic fumes of F^-.

CAX260   CAS: 23209-59-8   HR: 3
CALCIUM SODIUM METAPHOSPHATE
mf: $HO_3P•Ca•Na$ mw: 143.05
SYN: METAPHOSPHORIC ACID, CALCIUM SODIUM SALT
CONSENSUS REPORTS: Reported in EPA TSCA Inventory.
DFG MAK: Confirmed Animal Carcinogen with Unknown Relevance to Humans
SAFETY PROFILE: Suspected carcinogen with experimental tumorigenic data. When heated to decomposition it emits toxic fumes of PO_x.

CAX275   HR: D
CALCIUM SORBATE
PROP: Solid. Sltly sol in water.

SAFETY PROFILE: When heated to decomposition it emits acrid smoke and irritating fumes.

CAX350   CAS: 1592-23-0   HR: 1
CALCIUM STEARATE
PROP: Variable proportions of calcium stearate and calcium palmitate. Fine white powder; slt characteristic odor. Insol in water, alc, ether.
SYNS: AQUACAL □ CALCIUM DISTEARATE □ CALSTAR □ FLEXICHEM □ FLEXICHEM CS □ G 339 S □ NOPCOTE C 104 □ OCTADECANOIC ACID, CALCIUM SALT □ STAVINOR 30 □ SYNPRO STEARATE □ WITCO G 339S
CONSENSUS REPORTS: Reported in EPA TSCA Inventory.
ACGIH TLV: TWA 10 mg/m^3, total dust
SAFETY PROFILE: A nuisance dust. When heated to decomposition it emits acrid smoke and irritating fumes.

CAX500   CAS: 7778-18-9   HR: 1
CALCIUM SULFATE
mf: $CaSO_4$ mw: 136.14
PROP: Pure anhydrous, colorless or white powder or odorless crystals. D: 2.964, mp: 1570°. Dissolves in acids. Sltly sol in H_2O.
SYNS: ANHYDROUS CALCIUM SULFATE □ CRYSALBA □ DRIERITE □ GIBS □ PLASTER of PARIS □ THIOLITE
OSHA PEL: Total Dust: 15 mg/m^3; Respirable Fraction: 5 mg/m^3
ACGIH TLV: TWA (nuisance particulate) 10 mg/m^3 of total dust (when toxic impurities are not present, e.g., quartz <1%)
DFG MAK: 6 mg/m^3
SAFETY PROFILE: A nuisance dust. Reacts violently with aluminum when heated. Mixtures with diazomethane react exothermically and eventually explode. Mixtures with phosphorus ignite at high temperatures. When heated to decomposition it emits toxic fumes of SO_x. See also CALCIUM COMPOUNDS and SULFATES.

CAX750   CAS: 10101-41-4   HR: 1
CALCIUM(II) SULFATE DIHYDRATE
(1:1:2)
mf: $O_4S \cdot Ca \cdot 2H_2O$ mw: 172.18
PROP: Colorless, monoclinic, hygroscopic crystals. D: 2.32, mp: 128°, bp: 163°. Sltly sol in water.
SYNS: ALABASTER ☐ ANNALINE ☐ C.I. 77231 ☐ C.I. PIGMENT WHITE 25 ☐ GYPSUM ☐ GYPSUM STONE ☐ LAND PLASTER ☐ LIGHT SPAR ☐ MAGNESIA WHITE ☐ MINERAL WHITE ☐ NATIVE CALCIUM SULFATE ☐ PRECIPITATED CALCIUM SULFATE ☐ SATINITE ☐ SATIN SPAR ☐ SULFURIC ACID, CALCIUM(2+) SALT, DIHYDRATE ☐ TERRA ALBA
OSHA PEL: Total Dust: 15 mg/m³; Respirable Fraction: 5 mg/m³
ACGIH TLV: TWA (nuisance particulate) 10 mg/m³ of total dust (when toxic impurities are not present, e.g., quartz <1%)
SAFETY PROFILE: Human systemic effects by inhalation: fibrosing alveolitis (growth of fibrous tissue in the lung), unspecified respiratory system effects, and unspecified effects on the nose. Questionable carcinogen with experimental carcinogenic data. Long considered a nuisance dust (depending on silica content). When heated to decomposition it emits toxic fumes of SO_x. See also CALCIUM SULFATE, CALCIUM COMPOUNDS, and SULFATES.

CAY250   CAS: 2092-16-2   HR: 3
CALCIUM THIOCYANATE
mf: $C_2N_2S_2 \cdot Ca$ mw: 156.24
PROP: White, deliquescent crystals. Very sol in H_2O; sol in MeOH and EtOH.
SYNS: CALCIUM DITHIOCYANATE ☐ CALCIUM RHODANID (GERMAN) ☐ CALCIUMRHODANID ☐ CALCIUM SULFOCYANATE ☐ THIOCYAN
CONSENSUS REPORTS: Reported in EPA TSCA Inventory.
SAFETY PROFILE: Poison by ingestion and intravenous routes. See also THIOCYANATES and CALCIUM COMPOUNDS. When heated to decomposition it emits toxic fumes of NO_x and SO_x.

CAY500   CAS: 12111-24-9   HR: 2
CALCIUM TRISODIUM DIETHYLENE
TRIAMINE PENTAACETATE
mf: $C_{14}H_{18}N_3O_{10} \cdot CaNa_3$ mw: 497.40
SYNS: Ba 2797 ☐ CALCIUM CHEL-330 ☐ CALCIUM-DTPA ☐ CALCIUM TRISODIUM CHEL 330 ☐ CALCIUM TRISODIUM DTPA ☐ CALCIUM TRISODIUM PENTETATE ☐ CALCIUM TRISODIUM SALT of DIETHYLENETRIAMINEPENTAACETIC ACID ☐ DIETHYLENETRIAMINE PENTAACETIC ACID, CALCIUM TRISODIUM SALT ☐ DITRIPENTAT ☐ DTPA CALCIUM TRISODIUM SALT ☐ PENTACIN ☐ PENTACINE ☐ PENTETATE TRISODIUM CALCIUM ☐ PENTHAMIL
CONSENSUS REPORTS: Reported in EPA TSCA Inventory.
SAFETY PROFILE: Moderately toxic by intravenous and intraperitoneal routes. Experimental teratogenic and reproductive effects. Mutation data reported. When heated to decomposition it emits toxic fumes of Na_2O and NO_x. See also CALCIUM COMPOUNDS.

CAY710   CAS: 68955-53-3   HR: 3
C12-14-tert-ALKYL AMINES
SYN: PRIMENE 81-R
SAFETY PROFILE: A poison by ingestion. Moderately toxic by skin contact. Low toxicity by inhalation. A severe skin irritant. When heated to decomposition it emits toxic vapors of NO_x.

CAY950   CAS: 8065-83-6   HR: 3
CALO-CLOR
mf: $Cl_2Hg_2 \cdot Cl_2Hg$ mw: 743.57
ACGIH TLV: TWA 0.1 mg(Hg)/m³ (skin); BEI: 35 µg/g creatinine total inorganic mercury in urine preshift; 15 µg/g creatinine total inorganic mercury in blood at end of shift at end of workweek.
NIOSH REL: (Mercury, Aryl and Inorganic) CL 0.1 mg/m³ (skin)
SAFETY PROFILE: Poison by ingestion. When heated to decomposition it emits toxic fumes of Hg and Cl⁻.

CBA500   CAS: 79-92-5   HR: 1
CAMPHENE
mf: $C_{10}H_{16}$ mw: 136.26

PROP: Colorless cubic crystals; oily odor. Mp: 50–51°, bp: 159°, d: 0.842 @ 54°/4°, refr index: 1.452 @ 55°. Sol in alc; misc in fixed oils; insol in water.
SYNS: BICYCLO(2.2.1)HEPTANE, 2,2-DIMETHYL-3-METHYLENE-(9CI) □ FEMA No. 2229
CONSENSUS REPORTS: Reported in EPA TSCA Inventory.
SAFETY PROFILE: Mutation data reported. Combustible; yields flammable vapors when heated and can react with oxidizing materials. To fight fire, use water spray, foam, fog, CO2. When heated to decomposition it emits acrid smoke and irritating fumes.

CBA750   CAS: 76-22-2   HR: 3
CAMPHOR
DOT: UN 2717
mf: $C_{10}H_{16}O$ mw: 152.26
PROP: White, transparent, crystalline masses; penetrating odor; pungent, aromatic taste. Mp: 180°, bp: 204°, lel: 0.6%, uel: 3.5%, flash p: 150°F (CC), d: 0.992 @ 25°/4°, autoign temp: 871°F, vap d: 5.24. IDLH 200 mg/m³.
SYNS: 2-BORNANONE □ 2-CAMPHANONE □ CAMPHOR-natural □ CAMPHOR, synthetic (ACGIH, DOT) □ FORMOSA CAMPHOR □ GUM CAMPHOR □ HUILE de CAMPHRE (FRENCH) □ JAPAN CAMPHOR □ KAMPFER (GERMAN) □ 2-KETO-1,7,7-TRIMETHYLNORCAMPH-ANE □ LAUREL CAMPHOR □ MATRICARIA CAMPHOR □ 2-OXOBORNANE □ 1,7,7-TRIMETHYLBICYCLO(2.2.1)-2-HEPTANONE □ 1,7,7-TRIMETHYLNORCAMPHOR
CONSENSUS REPORTS: Reported in EPA TSCA Inventory.
OSHA PEL: TWA 2 mg/m³
ACGIH TLV: TWA 2 ppm; STEL 3 ppm; Not Classifiable as a Human Carcinogen
DFG MAK: 2 ppm (13 mg/m³)
DOT CLASSIFICATION: 4.2; Label: Flammable Solid
SAFETY PROFILE: A human poison by ingestion and possibly other routes. An experimental poison by inhalation, subcutaneous, and intraperitoneal routes. A local irritant. Ingestion causes nausea, vomiting, dizziness, excitation, and convulsions. Mutation data reported. Used as a topical anti-infective and anti-itching agent. Flammable liquid when exposed to heat or flame; can react with oxidizing materials. Vapor is explosive when exposed to heat or flame or CrO3. To fight fire, use foam, carbon dioxide, dry chemical. See also KETONES and other camphor entries.

CBB500   CAS: 8008-51-3   HR: 3
CAMPHOR OIL
DOT: UN 1130
PROP: Colorless or yellowish, oily, fragrant liquid. Bp: 175–200°, flash p: 117°F (CC), d: 0.875–0.900 @ 20°/20°. Insol in water; sol in chloroform, ether, oils, and in approx 3 vols alc. Found in the trees and bark of *Cinnamomum carphora sieb* (Fam. *Lauraceae*) and prepared by fractional distillation of crude camphor oil after the camphor has been crystallized out; a white, viscous liquid with cineole as the principal ingredient along with monoterpenes (FCTXAV 11,1011,73).
SYNS: CAMPHOR OIL, RECTIFIED □ CAMPHOR OIL WHITE □ CAMPHOR OIL YELLOW □ FORMOSA CAMPHOR OIL □ FORMOSE OIL OF CAMPHOR □ JAPANESE CAMPHOR OIL □ JAPANESE OIL OF CAMPHOR □ LIGHT CAMPHOR OIL □ LIGHT OIL OF CAMPHOR □ LIQUID CAMPHOR □ OIL CAMPHOR SASSAFRASSY □ OIL OF CAMPHOR RECTIFIED □ OIL OF CAMPHOR WHITE □ WHITE CAMPHOR OIL □ WHITE OIL OF CAMPHOR
CONSENSUS REPORTS: Reported in EPA TSCA Inventory.
DOT CLASSIFICATION: 3; Label: Flammable Liquid
SAFETY PROFILE: A human poison by ingestion. Human systemic effects by ingestion: convulsions, tremors, and unspecified respiratory system effects. A skin irritant. Flammable liquid when exposed to heat or flame; can react with oxidizing materials. To fight fire, use foam, CO2, dry chemical, mist, fog. See also SAFROL and CAMPHOR.

CBC100   HR: D
CANANGA OIL
PROP: From flowers of the tree *Cananga odorata* f. et Thoms., (Fam. Anonaceae). Yellow liquid; harsh floral odor. Sol in fixed oils, mineral oil; insol in glycerin, propylene glycol.
SAFETY PROFILE: When heated to decomposition it emits acrid smoke and irritating fumes.

CBC400   HR: D
CANDIDA LIPOLYTICA
PROP: Derived from *Candida lipolytica* Fam. Cryptococcaceae.
SAFETY PROFILE: When heated to decomposition it emits acrid smoke and irritating fumes.

CBE800   CAS: 514-78-3   HR: D
CANTHAXANTHIN
mf: $C_{40}H_{52}O_2$ mw: 564.80
PROP: Purple or dark crystalline powder. Mp: 218°. Sol in chloroform; very sltly sol in acetone; insol in water.
SYNS: CANTHA □ β-CAROTENE-4,4'-DIONE □ 4,4'-DIKETO-β-CAROTENE
SAFETY PROFILE: When heated to decomposition it emits acrid smoke and irritating fumes.

CBF250   CAS: 302-22-7   HR: 3
CAP
mf: $C_{23}H_{29}ClO_4$ mw: 404.97
PROP: Crystals from Me_2CO/Et_2O. Mp: 211–212°.
SYNS: 17-ACETOXY-6-CHLORO-6-DEHYDROPROG-ESTERONE □ 17-α-ACETOXY-6-CHLORO-6-DEHYDROPROGESTERONE □ 17-α-ACETOXY-6-CHLORO-6,7-DEHYDROPROGESTERONE □ 17-α-ACETOXY-6-CHLOROPREGNA-4,6-DIENE-3,20-DIONE □ 17-α-ACETOXY-6-CHLORO-4,6-PREGNADIENE-3,20-DIONE □ 17-(ACETYLOXY)-6-CHLOROPREGNA-4,6-DIENE-3,20-DIONE □ CHLORMADINON ACETATE □ CHLORMADINONE ACETATE □ CHLORMADINONU (POLISH) □ 6-CHLORO-17-α-ACETOXY-4,6-PREGNADI-ENE-3,20-DIONE □ Δ^6-6-CHLORO-17-α-ACETOXY-PROGESTERONE □ 6-CHLORO-Δ^6-17-ACETOXY-PROGESTERONE □ 6-CHLORO-Δ^6-(17-α)ACETOXY-PROGESTERONE □ 6-CHLORO-Δ^6-DEHYDRO-17-

ACETOXYPROGESTERONE □ 6-CHLORO-6-DEHYDRO-17-α-ACETOXYPROGESTERONE □ 6-CHLORO-6-DEHYDRO-17-α-HYDROXYPROGESTER-ONE ACETATE □ 6-CHLORO-17-α-HYDROXYPREGNA-4,6-DIENE-3,20-DIONE ACETATE □ 6-CHLORO-17-α-HYDROXY-Δ^6-PROGESTERONE ACETATE □ CHLOROMADINONE ACETATE □ 6-CHLORO-$\Delta^{4,6}$-PREGNADIENE-17-α-OL-3,20-DIONE-17-ACETATE □ 6-CHLORO-PREGNA-4,6-DIEN-17-α-OL-3,20-DIONE ACETATE □ CLORDION □ CMA □ C-QUENS □ 6-DEHYDRO-6-CHLORO-17-α-ACETOXYPROGESTERONE □ LORMIN □ LUTINYL □ NSC-92338 □ RS 1280 □ SKEDULE □ ST 155
CONSENSUS REPORTS: IARC Cancer Review: Animal Limited Evidence IMEMDT 21,365,79; Animal Sufficient Evidence IMEMDT 6,149,74.
SAFETY PROFILE: Suspected carcinogen with experimental carcinogenic and tumorigenic data. Moderately toxic by intraperitoneal route. Human maternal and reproductive effects by ingestion, intramuscular, and possibly other routes: ovary, uterus, cervix, vagina, and fallopian tube changes; menstrual cycle changes or disorders; changes in fertility; and other unspecified female effects. A human teratogen that causes developmental abnormalities of the endocrine system in the fetus. Experimental teratogenic and reproductive effects. An oral contraceptive. When heated to decomposition it emits toxic fumes of Cl^-.

CBF550   CAS: 485-50-7   HR: 3
l-CAPNOIDINE
mf: $C_{20}H_{17}NO_6$ mw: 367.38
SYNS: (–)-ADLUMIDINE □ l-ADLUMIDINE □ CAPNOIDINE □ (–)-CAPNOIDINE □ FURO(3,4-E)-1,3-BENZODIOXOL-8(6H)-ONE, 6-(5,6,7,8-TETRAHYDRO-6-METHYL-1,3-DIOXOLO(4,5-G) ISOQUINOLIN-5-YL)-, (R-(R*,R*))-
SAFETY PROFILE: A poison by intravenous route. When heated to decomposition it emits toxic vapors of NO_x.

CBF700   CAS: 105-60-2   HR: 2
CAPROLACTAM
mf: $C_6H_{11}NO$ mw: 113.18

PROP: White crystals or leaflets from ligroin. Mp: 69°, bp: 139° @ 12 mm, vap press: 6 mm @ 120°.

SYNS: AMINOCAPROIC LACTAM □ 6-AMINOHEXAN-OIC ACID CYCLIC LACTAM □ 2-AZACYCLOHEPTAN-ONE □ 6-CAPROLACTAM □ ω-CAPROLACTAM (MAK) □ CAPROLATTAME (FRENCH) □ CYCLOHEXANONE ISO-OXIME □ EPSYLON KAPROLAKTAM (POLISH) □ HEXAHYDRO-2-AZEPINONE □ HEXAHYDRO-2H-AZEPIN-2-ONE □ 6-HEXANELACTAM □ HEXANONE ISOXIME □ HEXANONISOXIM (GERMAN) □ 1,6-HEXOLACTAM □ e-KAPROLAKTAM (CZECH) □ 2-KETOHEXAMETHYLENIMINE □ NCI-C50646 □ 2-OXOHEXAMETHYLENIMINE □ 2-PERHYDROAZEPIN-ONE

CONSENSUS REPORTS: IARC Cancer Review: Group 4 IMEMDT 7,56,87; Animal No Evidence IMEMDT 39,247,86. NCI Carcinogenesis Studies (feed); No Evidence: mouse, rat NTPTR* NTP-TR-214,82. Reported in EPA TSCA Inventory.

OSHA PEL: Dust: 1 mg/m³; STEL 3 mg/m³; Vapor: 5 ppm; STEL 10 ppm

ACGIH TLV: TWA (aerosol and vapor) 5 mg/m³; Not Suspected as a Human Carcinogen

DFG MAK: 5 mg/m³

NIOSH REL: (Caprolactam, dust) TWA 1 mg/m³; STEL 3 mg/m³; (Caprolactam, vapor) TWA 0.22 ppm; STEL 0.66 ppm

SAFETY PROFILE: Moderately toxic by ingestion, skin contact, intraperitoneal, and subcutaneous routes. Human systemic effects by inhalation: nose and throat irritation, cough. Experimental reproductive effects. A skin and eye irritant. Potentially explosive reaction with acetic acid + dinitrogen trioxide. When heated to decomposition it emits toxic fumes of NO_x.

CBF760   CAS: 465-42-9   HR: 3
CAPSANTHIN

mf: $C_{40}H_{56}O_3$ mw: 584.88

SYN: β,KAPPA-CAROTEN-6'-ONE, 3,3'-DIHYDROXY-, (3R,3'S,5'R)-

SAFETY PROFILE: A poison by skin contact. When heated to decomposition it emits acrid smoke and irritating vapors.

CBF800   CAS: 2425-06-1   HR: 3
CAPTAFOL

mf: $C_{10}H_9Cl_4NO_2S$ mw: 349.06

PROP: Crystals. Mp: 160–161°.

SYNS: CAPTOFOL □ DIFOLATAN □ DIFOSAN □ FOLCID □ ORTHO 5865 □ SANSPOR □ SULFONIMIDE □ SULPHEIMIDE □ N-(1,1,2,2-TETRACHLORA-ETHYLTHIO)CYCLOHEX-4-EN-1,4-DIACARBOXIMID (GERMAN) □ N-(1,1,2,2-TETRACHLORAETHYLTHIO)-TETRAHYDROPHTHALAMID (GERMAN) □ N-1,1,2,2-TETRACHLOROETHYLMERCAPTO-4-CYCLOHEXENE-1,2-CARBOXIMIDE □ N-((1,1,2,2-TETRACHLORO-ETHYL)SULFENYL)-cis-4-CYCLOHEXENE-1,2-DICARBOXIMIDE □ N-(1,1,2,2-TETRACHLORO-ETHYLTHIO)-4-CYCLOHEXENE-1,2-DICARBOXIMIDE

CONSENSUS REPORTS: IARC Cancer Review: Group 2A IMEMDT 53,353,91; Animal Sufficient Evidence IMEMDT 53,353,91; Human No Available Data IMEMDT 53,353,91. EPA Genetic Toxicology Program.

OSHA PEL: TWA 0.1 mg/m³

ACGIH TLV: TWA 0.1 mg/m³; Not Classifiable as a Human Carcinogen

SAFETY PROFILE: Confirmed carcinogen with experimental carcinogenic data. Poison by intraperitoneal route. Moderately toxic by ingestion. An experimental teratogen. Other experimental reproductive effects. Mutation data reported. A fungicide. When heated to decomposition it emits very toxic fumes of Cl^-, NO_x, and SO_x.

CBG000   CAS: 133-06-2   HR: 3
CAPTAN

mf: $C_9H_8Cl_3NO_2S$ mw: 300.59

PROP: Odorless crystals from CCl_4 or C_6H_6. Mp: 172–173°, d: 1.745. Practically insol in water; sol in benzene, alcohol, and chloroform.

SYNS: AACAPTAN □ AGROSOL S □ AGROX 2-WAY and 3-WAY □ AMERCIDE □ BANGTON □ BEAN SEED PROTECTANT □ CAPTAF □ CAPTANCAPTENEET 26,538 □ CAPTANE □ CAPTAN-STREPTOMYCIN 7.5-0.1 POTATO SEED PIECE PROTECTANT □ CAPTEX □ ENT 26,538 □ ESSO FUNGICIDE 406 □ FLIT 406 □ FUNGUS BAN TYPE II □ GLYODEX 3722 □ GRANOX PPM □ GUSTAFSON CAPTAN 30-DD □ HEXACAP □ KAPTAN □ LE CAPTANE (FRENCH) □ MALIPUR □ MERPAN □ MICRO-CHECK 12 □ NCI-C00077 □ NERACID □ ORTHOCIDE □ OSOCIDE □ SR406 □

STAUFFER CAPTAN □ 3a,4,7,7a-TETRAHYDRO-N-(TRICHLOROMETHANESULPHENYL)PHTHALIMIDE □ 3a,4,7,7a-TETRAHYDRO-2-((TRICHLOROMETHYL)THIO)-1H-ISOINDOLE-1,3(2H)-DIONE □ 1,2,3,6-TETRAHYDRO-N-(TRICHLOROMETHYLTHIO)PHTHALIMIDE □ N-(TRICHLOR-METHYLTHIO)-PHTHALIMID (GERMAN) □ N-TRICHLOROMETHYLMERCAPTO-4-CYCLOHEXENE-1,2-DICARBOXIMIDE □ N-(TRICHLOROMETHYL-MERCAPTO)-Δ⁴-TETRAHYDROPHTHALIMIDE □ N-TRICHLOROMETHYLTHIOCYCLOHEX-4-ENE-1,2-DICARBOXIMIDE □ N-TRICHLOROMETHYLTHIO-cis-Δ⁴-CYCLOHEXENE-1,2-DICARBOXIMIDE □ N-((TRICHLOROMETHYL)THIO)-4-CYCLOHEXENE-1,2-DICARBOXIMIDE □ TRICHLOROMETHYLTHIO-1,2,5,6-TETRAHYDROPHTHALAMIDE □ N-((TRICHLORO-METHYL)THIO)TETRAHYDROPHTHALIMIDE □ N-TRICHLOROMETHYLTHIO-3A,4,7,7A-TETRAHYDROPH-THALIMIDE □ VANCIDE 89 □ VANGARD K □ VANICIDE □ VONDCAPTAN

CONSENSUS REPORTS: IARC Cancer Review: Group 3 IMEMDT 7,56,87; Animal Limited Evidence IMEMDT 30,295,83. NCI Carcinogenesis Bioassay (feed); Clear Evidence: mouse NCITR* NCI-CG-TR-15,77; No Evidence: rat NCITR* NCI-CG-TR-15,77. EPA Genetic Toxicology Program. Community Right-To-Know List. Reported in EPA TSCA Inventory.
OSHA PEL: TWA 5 mg/m³
ACGIH TLV: TWA 5 mg/m³ (Sensitizer); Confirmed Animal Carcinogen with Unknown Revelance to Humans
SAFETY PROFILE: Poison by intraperitoneal route. Moderately toxic to humans by ingestion. Moderately toxic experimentally by ingestion and inhalation routes. Experimental teratogenic and reproductive effects. Questionable carcinogen with experimental tumorigenic and neoplastigenic data. Human mutation data reported. When heated to decomposition it emits toxic fumes of Cl⁻, SOₓ, and NOₓ.

CBG125   CAS: 8028-89-5   HR: D
CARAMEL
PROP: Dark-brown to black liquid or solid; burnt-sugar odor, pleasant bitter taste. Sol in water (colloidal).
SYN: CARAMEL COLOR

CONSENSUS REPORTS: Reported in EPA TSCA Inventory.
SAFETY PROFILE: Mutation data reported. When heated to decomposition it emits acrid smoke and irritating fumes.

CBG500   CAS: 8000-42-8   HR: 2
CARAWAY OIL
PROP: The main constituent of caraway oil is 1-carvone; found in the fruits of *Carum carvi* L. (Fam. *Umbelliferae*) (FCTXAV 11,1011,73). Colorless liquid; odor and taste of caraway.
SYNS: KUEMMEL OIL (GERMAN) □ OIL OF CARAWAY
CONSENSUS REPORTS: Reported in EPA TSCA Inventory.
SAFETY PROFILE: Moderately toxic by ingestion and skin contact. A skin irritant. Mutation data reported. When heated to decomposition it emits acrid smoke and irritating fumes. See also l(−)-CARVONE.

CBH750   HR: 3
CARBAMATES
PROP: Compounds based upon carbamic acid, NH₂COOH. Used only in the form of its numerous salts and derivatives.
SAFETY PROFILE: Many carbamates are poisons or moderately toxic, and some are carcinogenic, teratogenic, or mutagenic. They are used as insecticides, fungicides, herbicides, and as accelerators in the vulcanization of rubber. There is little data on persistence or breakdown in the environment.

The N-alkylcarbamates and thiocarbamates can react with nitrite under mildly acid conditions to form N-nitroso compounds. Nitrite is found in soils, in human saliva, and in cured meats. N-nitrosodimethylamine is formed by soil microorganisms from thiram. Other N-nitroso compounds could similarly be formed from other carbamate pesticides. However, the extent of the reaction of carbamates and nitrite in humans is not known. The N-nitrosodialkylamines formed from dialkylthiocarbamate pesticides and nitrite are potent animal carcinogens and mutagens. The N-nitroso derivatives of several N-alkylcarbamates produce cancers in experimental animals at small doses.

Carbaryl, semicarbazide hydrochloride, n-propyl carbamate, Maneb, Zineb, Ferbam, and Thiram are experimental teratogens.

Many of the carbamates have central nervous system effects. Carbaryl and Zectran are acetylcholinesterase inhibitors.

Ethylenethiourea, which produces thyroid carcinomas in rats and liver cell tumors in mice by ingestion, is formed from ethylenebisdithiocarbamates such as Maneb and Zineb by metabolic processes and cooking.

See also individual compounds, NITROSAMINES, and N-NITROSO COMPOUNDS.

CBI250   CAS: 120-02-5   HR: 3
4-CARBAMIDOPHENYL BIS(CARBOXY-
METHYLTHIO)ARSENITE
mf: $C_{11}H_{13}AsN_2O_5S_2$ mw: 392.30
SYNS: 2,2'-((4-((AMINOCARBONYL)AMINO)PHENYL)ARSINIDENE)BIS(THIO)BISACETIC ACID □ BIS(CARBOXYMETHYL-MERCAPTO)(p-UREIDOPHENYL)ARSINE □ BIS(CARBOXYMETHYLTHIO)(p-UREIDOPHENYL)-ARSINE □ (p-CARBAMOYLAMINO)PHENYLARSINO-BIS(2-THIO-ACETIC ACID) □ CC 914 □ C.C. No. 914 □ MERCAPTOACETIC ACID, DIESTER with DITHIO-p-UREIDOBENZENEARSONOUS ACID □ PHENYL UREA-p-DI(CARBOXYMETHYL) THIOARSENITE □ THIOCARBARSONE □ (p-UREIDOPHENYLARSYLENEDITHIO)DIACETIC ACID
CONSENSUS REPORTS: Arsenic and its compounds are on the Community Right-To-Know List.
OSHA PEL: TWA 0.5 mg(As)/m³
ACGIH TLV: BEI: 35 μ (As)/L inorganic arsenic and methylated metabolites in urine
SAFETY PROFILE: Poison by intraperitoneal and intravenous routes. Moderately toxic by ingestion. See also ARSENIC COMPOUNDS, MERCAPTANS, and ESTERS. When heated to decomposition it emits very toxic fumes of As and SO_x.

CBJ000   CAS: 121-59-5   HR: 3
N-CARBAMOYLARSANILIC ACID
mf: $C_7H_9AsN_2O_4$ mw: 260.10
PROP: White, nearly odorless powder or needles from water. Sltly acid taste. Mp: 174°. Sltly sol in cold H_2O and org solvs; sol in hot H_2O.
SYNS: AMABEVAN □ AMEBAN □ AMEBARSONE □ AMIBIARSON □ AMINARSON □ AMINARSONE □ AMINOARSON □ (4-((AMINOCARBONYL)AMINO)-PHENYL)ARSONIC ACID □ ARSAMBIDE □ p-ARSONOPHENYLUREA □ p-CARBAMIDOBENZENE-ARSONIC ACID □ CARBAMINOPHENYL-p-ARSONIC ACID □ p-CARBAMINO PHENYL ARSONIC ACID □ 4-CARBAMYLAMINOPHENYLARSONIC ACID □ N-CARBAMYL ARSANILIC ACID □ CARBARSONE (USDA) □ CARBASONE □ FENARSONE □ HISTOCARB □ LEUCARSONE □ p-UREIDOBENZENEARSONIC ACID □ 4-UREIDO-1-PHENYLARSONIC ACID
CONSENSUS REPORTS: Arsenic and its compounds are on the Community Right-To-Know List.
OSHA PEL: TWA 0.5 mg(As)/m³
ACGIH TLV: BEI: 35 μ (As)/L inorganic arsenic and methylated metabolites in urine
SAFETY PROFILE: Poison by ingestion. Moderately toxic by intraperitoneal route. Questionable carcinogen with experimental tumorigenic data. See also ARSENIC COMPOUNDS. When heated to decomposition it emits very toxic fumes of As and NO_x.

CBJ750   CAS: 618-25-7   HR: 3
N-(CARBAMOYLMETHYL)ARSANILIC
ACID
mf: $C_8H_{11}AsN_2O_4$ mw: 274.13
PROP: White, crystalline powder.
SYNS: (4-((2-AMINO-2-OXOETHYL)AMINO)-PHENYL)ARSONIC ACID □ 4-ARSONOPHENYL-GLYCINAMIDE □ p-((CARBAMOYLMETHYL)AMINO)-BENZENEARSONIC ACID □ SODIUM-N-PHENYL-GLYCINAMIDE-p-ARSONATE □ TRYPARSAMIDE
CONSENSUS REPORTS: Arsenic and its compounds are on the Community Right-To-Know List.
OSHA PEL: TWA 0.5 mg(As)/m³
ACGIH TLV: BEI: 35 μ (As)/L inorganic arsenic and methylated metabolites in urine

SAFETY PROFILE: Poison by ingestion and intramuscular route. Moderately toxic by intravenous route. See also ARSENIC COMPOUNDS. When heated to decomposition it emits very toxic fumes of As and NO_x.

CBM000   CAS: 122-42-9   HR: 3
CARBANILIC ACID ISOPROPYL ESTER
mf: $C_{10}H_{13}NO_2$ mw: 179.24

PROP: A white, crystalline solid; sol in acetone and benzene. Mp: 90°.

SYNS: BAN-HOE □ BEET-KLEEN □ CHEM-HOE □ IFC □ IPPC □ ISOPROPIL-N-FENIL-CARBAMMATO (ITALIAN) □ ISOPROPYL CARBANILATE □ ISOPROPYL CARBANILIC ACID ESTER □ ISOPROPYL-N-FENYL-CARBAMAAT (DUTCH) □ ISOPROPYL-N-PHENYL-CARBAMAT (GERMAN) □ ISOPROPYL PHENYL-CARBAMATE □ ISOPROPYL-N-PHENYLCARBAMATE □ o-ISOPROPYL-N-PHENYL CARBAMATE □ ISOPROPYL-N-PHENYLURETHAN (GERMAN) □ ORTHO GRASS KILLER □ N-PHENYLCARBAMATE d'ISOPROPYLE (FRENCH) □ PHENYLCARBAMIC ACID-1-METHYLETH-YL ESTER □ N-PHENYL ISOPROPYL CARBAMATE □ PREMALOX □ PROFAM □ PROPHAM □ TRIHERBIDE □ TRIHERBIDE-IPC □ TUBERIT □ TUBERITE □ USAF D-9 □ Y 2

CONSENSUS REPORTS: IARC Cancer Review: Group 3 IMEMDT 7,56,87; Animal Inadequate Evidence IMEMDT 12,189,76. Reported in EPA TSCA Inventory. EPA Genetic Toxicology Program.

SAFETY PROFILE: Poison by intraperitoneal route. Moderately toxic to humans by ingestion. Moderately toxic experimentally by ingestion and possibly other routes. An experimental teratogen. Human mutation data reported. Questionable carcinogen with experimental neoplastigenic data. An herbicide. When heated to decomposition it emits toxic fumes of NO_x. See also CARBAMATES.

CBM500   CAS: 116-06-3   HR: 3
CARBANOLATE
mf: $C_7H_{14}N_2O_2S$ mw: 190.29

PROP: A solid material or crystals. Mp: 98–100°. Sltly sol in water.

SYNS: ALDECARB □ ALDICARB (USDA) □ ALDICARBE (FRENCH) □ AMBUSH □ ENT 27,093 □ 2-METHYL-2-(METHYLTHIO)PROPANAL-O-((METHYL-AMINO)CARBONYL)OXIME □ 2-METHYL-2-(METHYL-THIO)PROPIONALDEHYDE-O-(METHYLCARBAMOYL)-OXIME □ 2-METHYL-2-(METHYLTHIO)PROPION-ALDEHYDE OXIME □ 2-METHYL-2-METHYLTHIO-PROPIONALDEHYD-O-(N-METHYL-CARBAMOYL)-OXIM (GERMAN) □ 2-METIL-2-TIOMETIL-PROPION-ALDEID-O-(N-METIL-CARBAMOIL)-OSSIMA (ITALIAN) □ NCI-C08640 □ OMS-771 □ RCRA WASTE NUMBER P070 □ TEMIC □ TEMIK □ TEMIK G10 □ UC-21149

CONSENSUS REPORTS: IARC Cancer Review: Group 3 IMEMDT 53,93,91; Animal Inadequate Evidence IMEMDT 53,93,91; Human No Available Data IMEMDT 53,93,91. NCI Carcinogenesis Bioassay (feed); No Evidence: mouse, rat NCITR* NCI-CG-TR-136,79. Reported in EPA TSCA Inventory. EPA Extremely Hazardous Substances List.

SAFETY PROFILE: Deadly poison by ingestion, skin contact, subcutaneous, and possibly other routes. Human mutation data reported. Questionable carcinogen. A powerful systemic poison. In 1985 over 150 people in California exhibited toxic effects from eating watermelons contaminated with aldicarb. When heated to decomposition it emits very toxic fumes of NO_x and SO_x.

CBM750   CAS: 63-25-2   HR: 3
CARBARYL
mf: $C_{12}H_{11}NO_2$ mw: 201.24

PROP: White crystals. Mp: 142°, d: 1.232 @ 20°/20°. IDLH 100 mg/m³.

SYNS: ARILAT □ ARILATE □ ARYLAM □ ATOXAN □ BERCEMA NMC50 □ BUG MASTER □ CAPROLIN □ CARBAMINE □ CARBARYL (ACGIH,DOT,OSHA) □ CARBATOX □ CARBATOX-60 □ CARBATOX-75 □ CARBAVUR □ CARBOMATE □ CARPOLIN □ CARYLDERM □ CEKUBARYL □ COMPOUND 7744 □ CRAG SEVIN □ CRUNCH □ DENAPON □ DEVICARB □ DICARBAM □ DYNA-CARBYL □ ENT 23,969 □ EXPERIMENTAL INSECTICIDE 7744 □ GAMONIL □ GERMAIN'S □ HEXAVIN □ KARBARYL (POLISH) □ KARBASPRAY □ KARBATOX □ KARBATOX 75 □ KARBATOX ZAWIESINOWY □ KARBOSEP □ LATKA 7744 □ MENAPHTAM □ N-METHYLCARBAMATE de 1-NAPHTYLE □ METHYLCARBAMATE-1-NAPHTHALEN-OL □ METHYLCARBAMATE-1-NAPHTHOL □ N-METHYLCARBAMATE de 1-NAPHTYLE (FRENCH) □ METHYLCARBAMIC ACID-1-NAPHTHYL ESTER □ N-METHYL-1-NAFTYL-CARBAMAAT (DUTCH) □ N-METHYL-1-NAPHTHYL-CARBAMAT (GERMAN) □ N-

METHYL-α-NAPHTHYLCARBAMATE □ N-METHYL-1-NAPHTHYL CARBAMATE □ N-METHYL-α-NAPHTHYL-URETHAN □ N-METIL-1-NAFTIL-CARBAMMATO (ITALIAN) □ MONSUR □ MUGAN □ MURVIN □ NAC □ 1-NAFTYLESTER KYSELINY METHYLKARBAMINOVE □ α-NAFTYL-N-METHYLKARBAMAT □ 1-NAPHTHALEN8OL, METHYLCARBAMATE (9CI) □ α-NAPHTHALENYL METHYLCARBAMATE □ 1-NAPHTHALENYL METHYLCARBAMATE □ 1-NAPHTHOL N-METHYL-CARBAMATE □ α-NAPHTHYL METHYLCARBAMATE □ α-NAPHTHYL N-METHYLCARBAMATE □ 1-NAPHTHYL METHYLCARBAMATE □ 1-NAPHTHYL N-METHYL-CARBAMATE □ 1-NAPHTHYL-N-METHYL-KARBAMAT □ NMC 50 □ OLTITOX □ OMS-29 □ PANAM □ POMEX □ PROSEVOR 85 □ RAVYON □ RYLAM □ SAVIT □ SEFFEIN □ SEPTENE □ SEVIMOL □ SEVIN □ SEVIN 4 □ SEVIN (OSHA) □ SEWIN □ SOK □ TERCYL □ TOXAN □ TRICARNAM □ UC 7744 □ UNION CARBIDE 7,744 □ VETOX □ VIOXAN

CONSENSUS REPORTS: IARC Cancer Review: Group 3 IMEMDT 7,56,87; Animal Inadequate Evidence IMEMDT 12,37,76. Community Right-To-Know List.

OSHA PEL: TWA 5 mg/m^3

ACGIH TLV: TWA 5 mg/m^3; Not Classifiable as a Human Carcinogen; (Proposed: 1 mg/m^3; Sen; Confirmed Animal Carcinogen with Unknown Revelance to Humans

DFG MAK: 5 mg/m^3

NIOSH REL: (Carbaryl) TWA 5 mg/m^3

SAFETY PROFILE: Poison by ingestion, intravenous, intraperitoneal, and possibly other routes. Human systemic effects by ingestion: sensory change involving peripheral nerves and muscle weakness. Experimental teratogenic and reproductive effects. Questionable carcinogen with experimental carcinogenic and tumorigenic data. Human mutation data reported. An eye and severe skin irritant. Absorbed by all routes, although skin absorption is slow. No accumulation in tissue. Symptoms include blurred vision, headache, stomachache, vomiting. Symptoms similar to but less severe than those due to parathion. A reversible cholinesterase inhibitor. See also CARBAMATES and ESTERS. When heated to decomposition it emits toxic fumes of NO$_x$.

CBN000   CAS: 86-74-8   HR: 3
CARBAZOLE
mf: C$_{12}$H$_9$N   mw: 167.22
PROP: White crystals or plates from xylene. Mp: 244.8°, bp: 354.7°, d: 1.10 @ 18°/4°, vap press: 400 mm @ 323.0°. Sltly sol in most org solvs; sol in hot EtOH.
SYNS: 9-AZAFLUORENE □ 9H-CARBAZOLE □ DIBENZOPYRROLE □ DIBENZO(b,d)PYRROLE □ DIPHENYLENEIMINE □ DIPHENYLENIMIDE □ DIPHENYLENIMINE □ USAF EK-600
CONSENSUS REPORTS: IARC Cancer Review: Group 3 IMEMDT 7,56,87; Animal Limited Evidence IMEMDT 32,239,83. Reported in EPA TSCA Inventory.
SAFETY PROFILE: Poison by intraperitoneal route. Questionable carcinogen. Moderately toxic by ingestion. Mutation data reported. A pesticide. When heated to decomposition it emits toxic fumes of NO$_x$.

CBN100   CAS: 86-72-6   HR: 1
4-(3-CARBAZOLYLAMINO)PHENOL
mf: C$_{18}$H$_{14}$N$_2$O   mw: 274.34
SYNS: CARBAZOLE, 3-(p-HYDROXYANILINO)- □ 3-(4'-HYDROXYFENYL)AMINOKARBAZOL □ PHENOL, 4-(3-CARBAZOLYLAMINO)- □ R-BASE
CONSENSUS REPORTS: Reported in EPA TSCA Inventory.
SAFETY PROFILE: A skin and eye irritant. When heated to decomposition it emits toxic fumes of NO$_x$.

CBN375   HR: 3
CARBENDAZIM and SODIUM NITRITE (5:1)
SYNS: METHYL-2-BENZIMIDAZOLE CARBAMATE and SODIUM NITRITE □ SODIUM NITRITE and CARBENDAZIM (1:5) □ SODIUM NITRITE and METHYL-2-BENZIMIDAZOLE CARBAMATE
SAFETY PROFILE: Suspected carcinogen with experimental carcinogenic data. An experimental teratogen. When heated to decomposition it emits toxic fumes of Na$_2$O and NO$_x$. See also NITRITES and CARBAMATES.

CBQ750   CAS: 112-15-2   HR: 2
CARBITOL ACETATE

mf: $C_8H_{16}O_4$ mw: 176.24

PROP: Liquid. Bp: 217.4°, fp: −25°, flash p: 230°F (OC), d: 1.0114 @ 20°/20°, vap press: 0.05 mm @ 20°, vap d: 6.07.

SYNS: DIETHYLENE GLYCOL MONOETHYL ETHER ACETATE □ DIGLYCOL MONOETHYL ETHER ACETATE □ EKTASOLVE de ACETATE □ 2-(2-ETHOXYETHOXY)ETHANOL ACETATE □ GLYCOL ETHER de ACETATE

CONSENSUS REPORTS: Reported in EPA TSCA Inventory. Glycol ether compounds are on the Community Right-To-Know List.

SAFETY PROFILE: Moderately toxic by ingestion. A skin and eye irritant. See also GLYCOL ETHERS. Combustible when exposed to heat; can react with oxidizing materials. To fight fire, use alcohol foam, water, CO_2, dry chemical. When heated to decomposition it emits acrid smoke and fumes.

CBR000   CAS: 111-90-0   HR: 2
CARBITOL CELLOSOLVE

mf: $C_6H_{14}O_3$ mw: 134.20

PROP: Very hygroscopic, colorless liquid; mild pleasant odor. Bp: 201.9°, flash p: 201°F (OC), d: 0.986 @ 25°/4°, vap d: 4.62. Misc in water.

SYNS: APV □ CARBITOL □ CARBITOL SOLVENT □ DIETHYLENE GLYCOL ETHYL ETHER □ DIETHYLENE GLYCOL MONOETHYL ETHER □ DIGLYCOL MONOETHYL ETHER □ DIOXITOL □ DOWANOL □ DOWANOL DE □ ETHOXY DIGLYCOL □ 2-(2-ETHOXYETHOXY)ETHANOL □ ETHYL CARBITOL □ ETHYL DIETHYLENE GLYCOL □ ETHYLENE DIGLYCOL MONOETHYL ETHER □ LOSUNGSMITTEL APV □ MONOETHYL ETHER of DIETHYLENE GLYCOL □ POLY-SOLV □ SOLVOSOL

CONSENSUS REPORTS: Reported in EPA TSCA Inventory. Glycol ether compounds are on the Community Right-To-Know List.

SAFETY PROFILE: Moderately toxic by ingestion, intravenous, intraperitoneal, and possibly other routes. Mildly toxic by skin contact. A skin and eye irritant. Experimental reproductive effects. Mutation data reported. Combustible when exposed to heat; can react with oxidizing materials.

To fight fire, use alcohol foam, CO_2, dry chemical. When heated to decomposition it emits acrid smoke and irritating fumes.

CBR125   CAS: 1138-80-3   HR: 3
CARBOBENZOXYLGLYCINE

mf: $C_{10}H_{11}NO_4$ mw: 209.22

PROP: A solid. Mp: 119–120°.

SYNS: BENZYLOXYCARBONYLGLYCINE □ N-BENZYLOXYCARBONYLGLYCINE □ N-CARBOBENZOYLGLYCINE □ CARBOBENZOYL GLYCINE □ CARBOBENZYLOXYGLYCINE □ N-CARBOBENZYLOXYGLYCINE □ (CBZ)GLY □ Z-GLY

CONSENSUS REPORTS: Reported in EPA TSCA Inventory.

SAFETY PROFILE: Poison by intravaginal route. Experimental reproductive effects. When heated to decomposition it emits toxic fumes of NO_x. See also ESTERS.

CBR250   CAS: 1722-62-9   HR: 3
CARBOCAINE HYDROCHLORIDE

mf: $C_{15}H_{22}N_2O\bullet ClH$ mw: 282.85

PROP: A solid. Mp: 262–264°. Sol in H_2O.

SYNS: CHLOROCAIN □ N-(2,6-DIMETHYLPHENYL)-1-METHYL-2-PIPERIDINECARBOXAMIDE-MONO-HYDROCHLORIDE □ MEAVERIN □ MEPIVACAINE HYDROCHLORIDE □ dl-MEPIVACAINE HYDRO-CHLORIDE □ MEPIVASTESIN □ 1-METHYL-2',6'-PIPECOLOXYLIDIDE HYDROCHLORIDE □ dl-1-METHYL-2',6'-PIPECOLOXYLIDIDE HYDROCHLORIDE □ (1-METHYL-dl-PIPERIDINE-2-CARBOXYLIC ACID)-2,6-DIMETHYLANILIDE HYDROCHLORIDE □ SCANDICAIN

CONSENSUS REPORTS: Reported in EPA TSCA Inventory.

SAFETY PROFILE: Poison by subcutaneous, intravenous, subcutaneous, and implant routes. Experimental reproductive effects. An anesthetic. When heated to decomposition it emits very toxic fumes of HCl and NO_x.

CBR300   CAS: 60731-46-6   HR: 3
CARBOCALCITONIN

mf: $C_{148}H_{244}N_{42}O_{47}$ mw: 3364.34

SYNS: (AMINOSUBERIC ACID 1,7)-EEL CALCITONIN □ 1,7-DICARBACALCITONIN (sal), 1-BUTANOIC ACID-26-1-ASPARTIC ACID-27-1-VALINE-29-1-ALANINE- □ 1,7-DICARBACALCITONIN (EEL), 1-BUTANOIC ACID- □ ELCATONIN □ HC-58

C

SAFETY PROFILE: A poison by ingestion, subcutaneous, intravenous, and intramuscular routes. When heated to decomposition it emits acrid smoke and irritating vapors.

CBR675   CAS: 638-23-3   HR: 2
CARBOCISTEINE
mf: $C_5H_9NO_4S$ mw: 179.21
PROP: l-Form: Mp: 204–207°. dl-Form: Spherical aggregates of needles.
SYNS: CARBOCIT □ CARBOCYSTEINE □ S-(CARBOXYMETHYL)CYSTEINE □ l-CARBOXY-METHYLCYSTEINE □ 3-(CARBOXYMETHYLTHIO)-ALANINE □ l-3-((CARBOXYMETHYL)THIO)ALANINE □ FLUIFORT □ L.J. 206 □ LOVISCOL □ MUCICLAR □ MUCOCIS □ MUCODYNE □ MUCOLASE □ MUCOLEX □ MUCOPRONT □ PECTOX □ PULMOCLASE □ REOMUCIL □ RHINATHIOL □ RINATIOL □ THIODRIL □ TRANSBRONCHIN
CONSENSUS REPORTS: Reported in EPA TSCA Inventory.
SAFETY PROFILE: Moderately toxic by intraperitoneal route. Mildly toxic by ingestion. Experimental reproductive effects. When heated to decomposition it emits toxic fumes of SO_x and NO_x.

CBS275   CAS: 1563-66-2   HR: 3
CARBOFURAN
mf: $C_{12}H_{15}NO_3$ mw: 221.28
PROP: White, crystalline solid; odorless. Mp: 150–152°, d: 1.180 @ 20°/20°, vap press: 2×10^{-5} mm @ 33°. Sltly sol in water.
SYNS: BAY 70143 □ CURATERR □ D 1221 □ 2,3-DIHYDRO-2,2-DIMETHYLBENZOFURANYL-7-N-METHYLCARBAMATE □ 2,3-DIHYDRO-2,2-DIMETHYL-7-BENZOFURANYL METHYLCARBAMATE □ 2,2-DIMETHYL-7-COUMARANYL-N-METHYLCARBAMATE □ 2,2-DIMETHYL-2,3-DIHYDROBENZOFURAN-7-YL ESTER, METHYLCARBAMIC ACID □ 2,2-DIMETHYL-2,3-DIHYDRO-7-BENZOFURANYL-N-METHYLCARBAMATE □ ENT 27,164 □ FMC 10242 □ FURADAN □ FURODAN □ METHYL CARBAMIC ACID 2,3-DIHYDRO-2,2-DIMETHYL-7-BENZOFURANYL ESTER □ NIA 10242 □ NIAGRA 10242 □ YALTOX
CONSENSUS REPORTS: EPA Extremely Hazardous Substances List. Reported in EPA TSCA Inventory. EPA Genetic Toxicology Program.
OSHA PEL: TWA 0.1 mg/m³

ACGIH TLV: TWA 0.1 mg/m³; Not Classifiable as a Human Carcinogen
SAFETY PROFILE: Poison by inhalation, ingestion, skin contact, and intravenous routes. Experimental teratogenic and reproductive effects. Human mutation data reported. When heated to decomposition it emits toxic fumes of NO_x. See also CARBAMATES.

CBS400   HR: D
CARBOHYDRASE, ASPERGILLUS
PROP: From fermentation of *Aspergillus oryzae* var. Tan amorphous powder or liquid. Sol in water.
SAFETY PROFILE: When heated to decomposition it emits acrid smoke and irritating fumes.

CBS410   HR: D
CARBOHYDRASE and PROTEASE, mixed
PROP: From controlled fermentation of *Bacillus licheniformus* var. Brown amorphous powders or liquid. Sol in water; insol in alc, chloroform, ether.
SAFETY PROFILE: When heated to decomposition it emits acrid smoke and irritating fumes.

CBS500   CAS: 497-18-7   HR: 3
CARBOHYDRAZIDE
mf: CH_6N_4O mw: 90.11
PROP: Crystals from EtOH (aq). Mp: 153–154°. Sol in H_2O; practically insol in org solvs.
SYNS: 4-AMINOSEMICARBAZIDE □ CARBAZIC ACID HYDRAZIDE □ CARBAZIDE □ CARBODIHYDRAZIDE □ CARBONIC ACID DIHYDRAZIDE □ CARBONIC DIHYDRAZIDE □ CARBONOHYDRAZIDE □ CARBONYLDIHYDRAZINE □ 1,3-DIAMINOUREA
DOT CLASSIFICATION: Forbidden
CONSENSUS REPORTS: Reported in EPA TSCA Inventory.
SAFETY PROFILE: Poison by intravenous and intraperitoneal routes. Reacts with nitrous acid to form the explosive carbonic diazide. When heated to decomposition it emits toxic fumes of NO_x.

CBS750   CAS: 63042-08-0   HR: 2
4'-CARBOMETHOXY-2,3'-DIMETHYL-AZOBENZENE

mf: $C_{16}H_{16}N_2O_3$ mw: 284.34

SYNS: 4'-CARBOMETHOXY-2,3'-DIMETHYLAZOBENZOL □ CARBONIC ACID METHYL-4-(o-TOLYLAZO)-o-TOLYL ESTER □ 2,3'-DIMETHYLAZOBENZENE-4'-METHYLCARBONATE

SAFETY PROFILE: Questionable carconigen with experimental tumorigenic data. When heated to decomposition it emits toxic fumes of NO_x.

CBT250   CAS: 4564-87-8   HR: 3
CARBOMYCIN

mf: $C_{42}H_{67}NO_{16}$ mw: 842.10

PROP: Laths from MeOH (aq). Mp: 210–214° (decomp).

SYNS: CARBOMYCIN A □ DELTAMYCIN A □ 9-DEOXY-12,13-EPOXY-9-OXOLEUCOMYCIN V 3-ACETATE 4^B-(3-METHYLBUTANOATE) □ M-4209 □ MAGNAMYCIN □ MAGNAMYCIN A

SAFETY PROFILE: Poison by subcutaneous route. Moderately toxic by intravenous and intramuscular routes. When heated to decomposition it emits toxic fumes of NO_x.

CBT500   CAS: 7440-44-0   HR: 1
CARBON

DOT: UN 1361/UN 1362

af: C   aw: 12.01

PROP: Black crystals, powder or diamond form. Mp: 3652–3697° (subl), bp: approx 4200°, d (amorph): 1.8–2.1, d (graphite): 2.25, d (diamond): 3.51, vap press: 1 mm @ 3586°. IDLH 1250 mg/m³.

SYNS: ACTICARBONE □ ACTIVATED CARBON □ AG 3 □ AG 5 □ AG 3 (ADSORBENT) □ AG 5 (ADSORBENT) □ AK (ADSORBENT) □ ANTHRASORB □ AR 3 □ ART 2 □ AU 3 □ BAU □ BG 6080 □ BLACK LEAD □ CARBON, activated (DOT) □ CARBON-12 □ CARBON, animal or vegetable origin (DOT) □ CARBOPOL EXTRA □ CARBOPOL M □ CARBOPOL Z 4 □ CARBOPOL Z EXTRA □ CARBOSIEVE □ CARBOSORBIT R □ CECARBON □ CF 8 □ CF 8 (CARBON) □ C.I. 77265 □ C.I. PIGMENT BLACK 10 □ CLF II □ CMB 50 □ CMB 200 □ COKE POWDER □ COLUMBIA LCK □ CONDUCTEX □ CUZ 3 □ CWN 2 □ DARCO □ FILTRASORB □ FILTRASORB 200 □ FILTRASORB 400 □ GRAPHITE □ GRAPHITE SYNTHETIC (ACGIH,OSHA) □ GROSAFE □

HYDRODARCO □ IRGALITE 1104 □ JADO □ K 257 □ MA 100 (CARBON) □ NORIT □ NUCHAR □ OU-B □ PELIKAN C 11/1431a □ PLUMBAGO □ SKG □ SKT □ SKT (ADSORBENT) □ SU 2000 □ SUCHAR 681 □ SUPERSORBON IV □ SUPERSORBON S 1 □ U 02 □ WATERCARB □ WITCARB 940 □ XE 340 □ XF 4175L

CONSENSUS REPORTS: Reported in EPA TSCA Inventory.

OSHA PEL: (Natural graphite) TWA 2.5 mg/m³; (Synthetic graphite) TWA Total Dust: 10 mg/m³; Respirable Fraction: 5 mg/m³

ACGIH TLV: TWA 2 mg/m³ (respirable dust)

DFG MAK: 1.5 mg/m³

DOT CLASSIFICATION: 4.2; Label: Spontaneously Combustible

SAFETY PROFILE: Moderately toxic by intravenous route. Experimental reproductive effects. It can cause a dust irritation, particularly to the eyes and mucous membranes. See also CARBON BLACK, SOOT. Combustible when exposed to heat. Dust is explosive when exposed to heat or flame or oxides, peroxides, oxosalts, halogens, interhalogens, O_2, (NH_4NO_3 + heat), (NH_4ClO_4 @ 240°), bromates, $Ca(OCl)_2$, chlorates, (Cl_2 + $Cr(OCl)_2$), ClO, iodates, IO_5, $Pb(NO_3)_2$, $HgNO_3$, HNO_3, (oils + air), (K + air), Na_2S, $Zn(NO_3)_2$. Incompatible with air, metals, oxidants, unsaturated oils.

CBT750   CAS: 1333-86-4   HR: 1
CARBON BLACK

PROP: A generic term applied to a family of high-purity colloidal carbons commercially produced by carefully controlled pyrolysis of gaseous or liquid hydrocarbons. Carbon blacks, including commercial colloidal carbons such as furnace blacks, lampblacks and acetylene blacks, usually contain less than several tenths percent of extractable organic matter and less than one percent ash. IDLH 1750 mg/m³.

SYNS: ACETYLENE BLACK □ ARO □ AROFLOW □ AROGEN □ AROMEX □ AROTONE □ AROVEL □ ARROW □ ATLANTIC □ BLACK PEARLS □ CANCARB □ CARBODIS □ CARBOLAC □ CARBOLAC 1 □

CARBOMET □ CARBON BLACK, ACETYLENE □ CARBON BLACK BV and V □ CARBON BLACK, CHANNEL □ CARBON BLACK, FURNACE □ CARBON BLACK, LAMP □ CARBON BLACK, THERMAL □ CHANNEL BLACK □ C.I. 77266 □ C.I. PIGMENT BLACK 6 □ C.I. PIGMENT BLACK 7 □ CK3 □ COLLOCARB □ COLUMBIA CARBON □ CONDUCTEX □ CONTINENTAL □ CONTINEX □ CORAX □ CORAX P □ CROFLEX □ CROLAC □ DEGUSSA □ DELUSSA BLACK FW □ DIXIE □ DIXIECELL □ DIXIEDENSED □ DIXITHERM □ DUREX □ EAGLE GERMANTOWN □ ELF □ ELFTEX □ ESSEX □ EXCELSIOR □ EXPLOSION ACETYLENE BLACK □ EXPLOSION BLACK □ FARBRUSS □ FECTO □ FLAMRUSS □ FURNAL □ FURNEX □ FURNEX N 765 □ GAS-FURNACE BLACK □ GASTEX □ HUBER □ HUMENEGRO □ IMPINGEMENT BLACK □ KETJENBLACK EC □ KOSMINK □ KOSMOBIL □ KOSMOLAK □ KOSMOS □ KOSMOTHERM □ KOSMOVAR □ MAGECOL □ METANEX □ MICRONEX □ MIIKE 20 □ MODULEX □ MOGUL □ MOGUL L □ MOLACCO □ MONARCH □ NEO-SPECTRA □ NEO-SPECTRA II □ NEOTEX □ OIL-FURNACE BLACK □ P-33 □ P68 □ P1250 □ PEERLESS □ PELLETEX □ PHILBLACK □ PHILBLACK N 550 □ PHILBLACK N 765 □ PHILBLACK O □ PIGMENT BLACK 7 □ PRINTEX □ PRINTEX 60 □ RAVEN □ RAVEN 30 □ RAVEN 420 □ RAVEN 500 □ RAVEN 8000 □ REBONEX □ REGAL □ REGAL 99 □ REGAL 300 □ REGAL 330 □ REGAL 600 □ REGAL 400R □ REGAL SRF □ REGENT □ ROYAL SPECTRA □ SEVACARB □ SEVAL □ SHAWINIGAN ACETYLENE BLACK □ SHELL CARBON □ SPECIAL BLACK 1V & V □ SPECIAL SCHWARZ □ SPHERON □ SPHERON 6 □ STATEX □ STATEX N 550 □ STERLING □ STERLING N 765 □ STERLING NS □ STERLING SO 1 □ SUPERBA □ SUPER-CARBOVAR □ SUPER-SPECTRA □ TEXAS □ THERMA-ATOMIC BLACK □ THERMAL ACETYLENE BLACK □ THERMATOMIC □ THERMAX □ THERMBLACK □ TINOLITE □ TM 30 □ TORCH BRAND □ TRIANGLE □ UCET □ UKARB □ UNITED □ VELVETEX □ VULCAN □ WITCO □ WITCOBLAK NO. 100 □ WYEX

CONSENSUS REPORTS: IARC Cancer Review: Group 3 IMEMDT 7,142,87; Human Inadequate Evidence IMEMDT 33,35,84; Animal Inadequate Evidence IMEMDT 33,35,84.
OSHA PEL: TWA 3.5 mg/m³
ACGIH TLV: TWA 3.5 mg/m³; Not Classifiable as a Human Carcinogen
NIOSH REL: (Carbon Black) TWA 3.5 mg/m³
SAFETY PROFILE: Mildly toxic by ingestion, inhalation, and skin contact.

Questionable carcinogen. Mutation data reported. See also CARBON. A nuisance dust in high concentrations. While it is true that the tiny particulates of carbon black contain some molecules of carcinogenic materials, the carcinogens are apparently held tightly and are not eluted by hot or cold water, gastric juices, or blood plasma.

CBU250   CAS: 124-38-9   HR: 1
CARBON DIOXIDE
DOT: UN 1013/UN 1845/UN 2187
mf: CO_2 mw: 44.01
PROP: Colorless, odorless gas. Mp: 57° (sublimes @ −78.5°), vap d: 1.53 @ 78.2°. Sltly sol in water, forming H_2CO_3. IDLH 40,000 ppm.
SYNS: ANHYDRIDE CARBONIQUE (FRENCH) □ CARBON DIOXIDE, refrigerated liquid (UN 2187) (DOT) □ CARBON DIOXIDE, solid (UN 1845) (DOT) □ CARBONIC ACID ANHYDRIDE □ CARBONIC ACID GAS □ CARBONIC ANHYDRIDE □ CARBON OXIDE □ DRY ICE □ DRY ICE (UN 1845) (DOT) □ KHLADON 744 □ KOHLENDIOXYD (GERMAN) □ KOHLENSAEURE (GERMAN) □ R 744
CONSENSUS REPORTS: Reported in EPA TSCA Inventory.
OSHA PEL: TWA 10,000 ppm; STEL 30,000 ppm
ACGIH TLV: TWA 5000 ppm; STEL 30,000 ppm
DFG MAK: 5000 ppm (9100 mg/m³)
NIOSH REL: (Carbon Dioxide) TWA 10,000 ppm; CL 30,000 ppm/10M
DOT CLASSIFICATION: 2.2; Label: Nonflammable Gas; DOT Class: 9; Label: None (UN 1845)
SAFETY PROFILE: An asphyxiant. See discussion of simple asphyxiants under ARGON. Experimental teratogenic and reproductive effects. Contact of solid carbon dioxide snow with the skin can cause burns. Dusts of magnesium, zirconium, titanium, and some magnesium-aluminum alloys ignite and then explode in CO_2 atmospheres. Dusts of aluminum, chromium, and manganese ignite and then explode when heated in CO_2. Several bulk metals will burn in CO_2. Reacts vigorously

with (Al + Na₂O₂), Cs₂O, Mg(C₂H₅)₂, Li, (Mg + Na₂O₂), K, KHC, Na, Na₂C₂, NaK, Ti. CO_2 fire extinguishers can produce highly incendiary sparks of 5–15 mJ at 10–20 kV by electrostatic discharge. Incompatible with acrylaldehyde, aziridine, metal acetylides, sodium peroxide.

CBV000   CAS: 53569-62-3   HR: 2
CARBON DIOXIDE mixed with NITROUS OXIDE
DOT: UN 1015
mf: $CO_2 \cdot N_2O$ mw: 88.03
SYNS: CARBON DIOXIDE, mixture with NITROGEN OXIDE (N₂O) □ CARBON DIOXIDE–NITROUS OXIDE mixture (DOT)
NIOSH REL: (Carbon Dioxide) TWA 10,000 ppm; CL 30,000 ppm/10M; (N₂O as Anesthetic Agent) TWA 25 ppm/1H
DOT CLASSIFICATION: 2.2; Label: Nonflammable Gas
SAFETY PROFILE: See components as listed. An anesthetic mixture. Combustible. An oxidizing mixture. Can react with reducing materials.

CBV250   CAS: 8063-77-2   HR: 1
CARBON DIOXIDE mixed with OXYGEN
DOT: UN 1014
SYNS: CARBOGEN (8CI) □ CARBON DIOXIDE-OXYGEN mixture (DOT)
NIOSH REL: (Carbon Dioxide) TWA 10,000 ppm; CL 30,000 ppm/10M
DOT CLASSIFICATION: 2.2; Label: Nonflammable Gas
SAFETY PROFILE: Possible asphyxiant.

CBV500   CAS: 75-15-0   HR: 3
CARBON DISULFIDE
DOT: UN 1131
mf: CS_2 mw: 76.13
PROP: Highly refracting, clear, colorless liquid; nearly odorless when pure. Mp: −111.6°, d: 1.293 @ 0°/4°, bp: 46.5°, lel: 1.3%, uel: 50%, flash p: −22°F (CC), autoign temp: 257°F, vap press: 400 mm @ 28°, vap d: 2.64. Misc in EtOH, Et₂O, and C₆H₆; sltly sol in H₂O. IDLH 500 ppm.

SYNS: CARBON BISULFIDE (DOT) □ CARBON BISULPHIDE □ CARBON DISULPHIDE □ CARBONE (SUFURE de) (FRENCH) □ CARBONIO (SOLFURO di) (ITALIAN) □ CARBON SULFIDE □ CARBON SULPHIDE (DOT) □ DITHIOCARBONIC ANHYDRIDE □ KOHLENDISULFID (SCHWEFELKOHLENSTOFF) (GERMAN) □ KOOLSTOFDISULFIDE (ZWAVEL-KOOLSTOF) (DUTCH) □ NCI-C04591 □ RCRA WASTE NUMBER P022 □ SCHWEFELKOHLENSTOFF (GERMAN) □ SOLFURO di CARBONIO (ITALIAN) □ SULPHOCARBONIC ANHYDRIDE □ WEEVILTOX □ WEGLA DWUSIARCZEK (POLISH)
CONSENSUS REPORTS: Reported in EPA TSCA Inventory. EPA Genetic Toxicology Program. Community Right-To-Know List. EPA Extremely Hazardous Substances List.
OSHA PEL: TWA 4 ppm; STEL 12 ppm (skin)
ACGIH TLV: TWA 10 ppm (skin); BEI: 5 mg (2-thiothiazolidine-4-carboxylic acid (TTCA))/g creatinine in urine
DFG MAK: 5 ppm (16 mg/m³); BAT: 8 mg/L of 4-thio-4-thiazolidine carboxylic acid (TTCA) at end of shift
NIOSH REL: (Carbon Disulfide) TWA 1 ppm; CL 10 ppm/15M
DOT CLASSIFICATION: 3; Label: Flammable Liquid, Poison
SAFETY PROFILE: A human poison by unspecified route. Mildly toxic to humans by inhalation. An experimental poison by intraperitoneal route. Human reproductive effects on spermatogenesis by inhalation. Experimental teratogenic and reproductive effects. Human mutation data reported. The main toxic effect is on the central nervous system, acting as a narcotic and anesthetic in acute poisoning with death following from respiratory failure. In chronic poisoning, the effect on the nervous system is one of central and peripheral damage, which may be permanent if the damage has been severe.

Flammable liquid. A dangerous fire hazard when exposed to heat, flame, sparks, friction, or oxidizing materials. Severe explosion hazard when exposed to heat or flame. Ignition and potentially explosive reaction when heated in contact with rust or iron. Mixtures with sodium or potassium-sodium alloys are powerful, shock-sensitive explosives. Explodes on contact with permanganic acid. Potentially explosive reaction with nitrogen oxide, chlorine (catalyzed by iron). Mixtures with dinitrogen tetraoxide are heat-, spark-, and shock-sensitive explosives. Reacts with metal azides to produce shock- and heat-sensitive, explosive metal azidodithioformates. Aluminum powder ignites in CS_2 vapor. The vapor ignites on contact with fluorine. Reacts violently with azides, CsN_3, ClO, ethylamine diamine, ethylene imine, $Pb(N_3)_2$, LiN_3, (H_2SO_4 + permanganates), KN_3, RbN_3, NaN_3, phenylcopper-triphenylphosphine complexes. Incompatible with air, metals, oxidants. To fight fire, use water, CO_2, dry chemical, fog, mist. When heated to decomposition it emits highly toxic fumes of SO_x.

CBW400   CAS: 2463-45-8   HR: 3
CARBONIC ACID, CYCLIC 3-CHLORO-
    PROPYLENE ESTER
mf: $C_4H_5ClO_3$ mw: 136.54
SYN: 1,3-DIOXOLAN-2-ONE, 4-(CHLOROMETHYL)-
CONSENSUS REPORTS: Reported in EPA TSCA Inventory.
SAFETY PROFILE: Poison by ingestion. Experimental reproductive effects. When heated to decomposition it emits toxic fumes of Cl⁻.

CBW750   CAS: 630-08-0   HR: 3
CARBON MONOXIDE
DOT: UN 1016/NA 9202
mf: CO   mw: 28.01
PROP: Colorless, odorless, tasteless gas. Mp: −213°, bp: −190°, lel: 12.5%, uel: 74.2%, d: (gas) 1.250 g/L @ 0°, (liquid) 0.793, autoign temp: 1128°F. Very sltly sol in H_2O; sol in AcOH, MeOH, and EtOH. IDLH 1200 ppm.
SYNS: CARBONE (OXYDE de) (FRENCH) □ CARBONIC OXIDE □ CARBONIO (OSSIDO di) (ITALIAN) □ CARBON MONOXIDE (ACGIH,OSHA) □ CARBON MONOXIDE (UN 1016) (DOT) □ CARBON MONOXIDE, refrigerated liquid (cryogenic liquid) (NA 9202) (DOT) □ CARBON OXIDE (CO) □ EXHAUST GAS □ FLUE GAS □ KOHLENMONOXID (GERMAN) □

KOHLENOXYD (GERMAN) □ KOOLMONOXYDE (DUTCH) □ OXYDE de CARBONE (FRENCH) □ WEGLA TLENEK (POLISH)
CONSENSUS REPORTS: Reported in EPA TSCA Inventory.
OSHA PEL: TWA 35 ppm; CL 200 ppm
ACGIH TLV: 25 ppm; BEI: 3% of hemoglobin indicating carboxyhemoglobin in blood at end of shift; 20 ppm CO in end-exhaled air at end of shift.
DFG MAK: 30 ppm (35 mg/m³); BAT: 5% carboxyhemoglobin in blood at end of shift
NIOSH REL: (Carbon Monoxide) TWA 35 ppm; CL 200 ppm
DOT CLASSIFICATION: 2.3; Label: Poison Gas, Flammable Gas
SAFETY PROFILE: Mildly toxic by inhalation in humans but has caused many fatalities. Experimental teratogenic and reproductive effects. Human systemic effects by inhalation: changes in psychophysiological tests and methemoglobinemia-carboxyhemoglobinemia. Can cause asphyxiation by preventing hemoglobin from binding oxygen. After removal from exposure, the half-life of elimination from the blood is one hour. Chronic exposure effects can occur at lower concentrations. A common air contaminant. Acute cases of poisoning resulting from brief exposures to high concentrations seldom result in any permanent disability if recovery takes place. Chronic effects as the result of repeated exposure to lower concentrations have been described, particularly in the Scandinavian literature. Auditory disturbances and contraction of the visual fields have been demonstrated. Glycosuria does occur, and heart irregularities have been reported. Other workers have found that where the poisoning has been relatively long and severe, cerebral congestion and edema may occur, resulting in long-lasting mental or nervous damage. Repeated exposure to low concentration of the gas, up to 100 ppm in air, is generally believed to cause no signs of poisoning or permanent damage.

Industrially, sequelae are rare, as exposure, though often severe, is usually brief. It is a common air contaminant.

A dangerous fire hazard when exposed to flame. Severe explosion hazard when exposed to heat or flame. Violent or explosive reaction on contact with bromine trifluoride, bromine pentafluoride, chlorine dioxide, or peroxodisulfuryl difluoride. Mixture of liquid CO with liquid O_2 is explosive. Reacts with sodium or potassium to form explosive products sensitive to shock, heat, or contact with water. Mixture with copper powder + copper(II) perchlorate + water forms an explosive complex. Mixture of liquid CO with liquid dinitrogen oxide is a rocket propellant combination. Ignites on warming with iodine heptafluoride. Ignites on contact with cesium oxide + water. Potentially explosive reaction with iron(III) oxide between 0° and 150°C. Exothermic reaction with CIF_3, (Li + H_2O), NF_3, OF_2, (K + O_2), Ag_2O, (Na + NH_3). To fight fire, stop flow of gas.

CBX109   CAS: 1885-14-9   HR: 3
CARBONOCHLORIDIC ACID PHENYL
**   ESTER**
DOT: UN 2746
mf: $C_7H_5ClO_2$ mw: 156.57
PROP: Bp: 68–71° @ 9 mm.
SYNS: CHLOROFORMIC ACID PHENYL ESTER □ FENYLESTER KYSELINY CHLORMRAVENCI (CZECH) □ PHENYL CHLOROCARBONATE □ PHENYL CHLOROFORMATE □ PHENYLCHLOROFORMATE (DOT)
CONSENSUS REPORTS: Reported in EPA TSCA Inventory.
DOT CLASSIFICATION: 6.1; Label: Poison, Corrosive
SAFETY PROFILE: Poison by inhalation. Moderately toxic by ingestion and skin contact. A corrosive skin and eye irritant. See also ESTERS. When heated to decomposition it emits toxic fumes of Cl^-.

CBX750   CAS: 558-13-4   HR: 3
CARBON TETRABROMIDE
DOT: UN 2516

mf: CBr_4 mw: 331.65
PROP: Colorless, monoclinic tablets. Mp: (α) 48.4°, (β) 90.1°, bp: 102° @ 50 mm, d: 2.961 @ 99.5°/4°, vap press: 40 mm @ 96.3°. Sol in EtOH, Et_2O, and $CHCl_3$; insol in H_2O.
SYNS: BROMID UHLICITY □ CARBON BROMIDE □ METHANE, TETRABROMIDE □ METHANE, TETRABROMO- □ TETRABROMIDE METHANE □ TETRABROMOMETHANE
CONSENSUS REPORTS: Reported in EPA TSCA Inventory.
OSHA PEL: TWA 0.1 ppm; STEL 0.3 ppm
ACGIH TLV: TWA 0.1 ppm; STEL 0.3 ppm
DOT CLASSIFICATION: 6.1; Label: KEEP AWAY FROM FOOD
SAFETY PROFILE: Poison by subcutaneous and intravenous routes. Narcotic in high concentration. Mixture with Li particles is an impact-sensitive explosive. Explodes on contact with hexacyclohexyldilead. When heated to decomposition it emits toxic fumes of Br^-. See also CHLORINATED HYDROCARBONS, ALIPHATIC.

CBY000   CAS: 56-23-5   HR: 3
CARBON TETRACHLORIDE
DOT: UN 1846
mf: CCl_4 mw: 153.81
PROP: Colorless liquid; heavy, ethereal odor. Mp: −22.6°, bp: 76.8°, flash p: none, d: 1.632 @ 0°/4°, vap press: 100 mm @ 23.0°. Sol in EtOH and Et_2O; practically insol in H_2O. IDLH 200 ppm.
SYNS: BENZINOFORM □ CARBONA □ CARBON CHLORIDE □ CARBON TET □ CZTEROCHLOREK WEGLA (POLISH) □ ENT 4,705 □ FASCIOLIN □ FLUKOIDS □ METHANE TETRACHLORIDE □ NECATORINA □ NECATORINE □ PERCHLORO- METHANE □ R 10 □ RCRA WASTE NUMBER U211 □ TETRACHLOORKOOLSTOF (DUTCH) □ TETRACHLO- ORMETAAN □ TETRACHLORKOHLENSTOFF, (GERMAN) □ TETRACHLORMETHAN (GERMAN) □ TETRACHLOROCARBON □ TETRACHLOROMETHANE □ TETRACHLORURE de CARBONE (FRENCH) □ TETRACLOROMETANO (ITALIAN) □ TETRACLORURO di CARBONIO (ITALIAN) □ TETRAFINOL □ TETRAFORM □ TETRASOL □ UNIVERM □ VERMOESTRICID

c

CONSENSUS REPORTS: NTP 10th Report on Carcinogens. IARC Cancer Review: Group 2B IMEMDT 7,143,87; Animal Sufficient Evidence IMEMDT 20,371,79; IMEMDT 1,53,72; Human Inadequate Evidence IMEMDT 1,53,72; Human Limited Evidence IMEMDT 20,371,79. Community Right-To-Know List. EPA Genetic Toxicology Program. Reported in EPA TSCA Inventory.
OSHA PEL: TWA 2 ppm
ACGIH TLV: TWA 5 ppm; STEL 10 (skin); Suspected Human Carcinogen
DFG MAK: 10 ppm (64 mg/m³); BEI: 1.6 mL/m³ in alveolar air 1 hour after exposure; Suspected Carcinogen
NIOSH REL: (Carbon Tetrachloride) CL 2 ppm/60M
DOT CLASSIFICATION: 6.1; Label: Poison
SAFETY PROFILE: Confirmed carcinogen with experimental carcinogenic, neoplastigenic, and tumorigenic data. A human poison by ingestion and possibly other routes. Poison by subcutaneous and intravenous routes. Mildly toxic by inhalation. Human systemic effects by inhalation and ingestion: nausea or vomiting, pupillary constriction, coma, antipsychotic effects, tremors, somnolence, anorexia, unspecified respiratory system and gastrointestinal system effects. Experimental teratogenic and reproductive effects. An eye and skin irritant. Damages liver, kidneys, and lungs. Mutation data reported. A narcotic. Individual susceptibility varies widely. Contact dermatitis can result from skin contact.

Carbon tetrachloride has a narcotic action resembling that of chloroform, though not as strong. Following exposure to high concentrations, the victim may become unconscious, and, if exposure is not terminated, death can follow from respiratory failure. The aftereffects following recovery from narcosis are more serious than those of delayed chloroform poisoning, usually taking the form of damage to the kidneys, liver, and lungs. Exposure to lower concentrations, insufficient to produce unconsciousness, usually results in severe gastrointestinal upset and may progress to serious kidney and hepatic damage. The kidney lesion is an acute nephrosis; the liver involvement consists of an acute degeneration of the central portions of the lobules. When recovery takes place, there may be no permanent disability. Marked variation in individual susceptibility to carbon tetrachloride exists; some persons appear to be unaffected by exposures that seriously poison their fellow workers. Alcoholism and previous liver and kidney damage seem to render the individual more susceptible. Concentrations on the order of 1000 to 1500 ppm are sufficient to cause symptoms if exposure continues for several hours. Repeated daily exposure to such concentration may result in poisoning.

Though the common form of poisoning following industrial exposure is usually one of gastrointestinal upset, which may be followed by renal damage, other cases have been reported in which the central nervous system has been affected, resulting in the production of polyneuritis, narrowing of the visual fields, and other neurological changes. Prolonged exposure to small amounts of carbon tetrachloride has also been reported as causing cirrhosis of the liver.

Locally, a dermatitis may be produced following long or repeated contact with the liquid. The skin oils are removed and the skin becomes red, cracked, and dry. The effect of carbon tetrachloride on the eyes either as a vapor or as a liquid, is one of irritation with lachrymation and burning.

Industrial poisoning is usually acute with malaise, headache, nausea, dizziness, and confusion, which may be followed by stupor and sometimes loss of consciousness. Symptoms of liver and kidney damage may follow later with development of dark urine, sometimes jaundice and liver enlargement, followed by scanty urine, albuminuria, and renal casts; uremia may develop and cause death. Where exposure has been less acute, the symptoms are usually headache, dizziness, nausea, vomiting, epigastric distress, loss of appetite, and fatigue. Visual disturbances (blind spots, spots before the eyes, a visual "haze," and restriction of the visual fields), secondary anemia, and occasionally a slight jaundice may occur. Dermatitis may be noticed on the exposed parts.

Forms impact-sensitive explosive mixtures with particulates of many metals, e.g., aluminum (when ball milled or heated to 152° in a closed container), barium (bulk metal also reacts violently), beryllium, potassium (200 times more shock sensitive than mercury fulminate), potassium-sodium alloy (more sensitive than potassium), lithium, sodium, zinc (burns readily). Also forms explosive mixtures with chlorine trifluoride, calcium hypochlorite (heat-sensitive), calcium disilicide (friction- and pressure- sensitive), triethyldialuminum trichloride (heat- sensitive), decaborane(14) (impact-sensitive), dinitrogen tetraoxide. Violent or explosive reaction on contact with fluorine. Forms explosive mixtures with ethylene between 25° and 105° and between 30 and 80 bar. Potentially explosive reaction on contact with boranes. 9:1 mixtures of methanol and CCl_4 react exothermically with aluminum, magnesium, or zinc. Potentially dangerous reaction with dimethyl formamide, 1,2,3,4,5,6-hexachlorocyclohexane, or dimethylacetamide when iron is present as a catalyst. CCl_4 has caused explosions when used as a fire extinguisher on wax and uranium fires. Incompatible with aluminum trichloride, dibenzoyl peroxide, potassium-tert-butoxide. Vigorous exothermic reaction with allyl alcohol, $Al(C_2H_5)_3$, (benzoyl peroxide + C_2H_4), BrF_3, diborane, disilane, liquid O_2, Pu, ($AgClO_4$ + HCl), potassium-tert-butoxide, tetraethylenepentamine, tetrasilane, trisilane, Zr. When heated to decomposition it emits toxic fumes of Cl^- and phosgene. It has been banned from household use by the FDA. See also

CHLORINATED HYDROCARBONS, ALIPHATIC.

CBY250   CAS: 75-73-0   HR: 2
CARBON TETRAFLUORIDE
DOT: UN 1982
mf: CF_4 mw: 88.01
PROP: Colorless gas. Mp: −184°, bp: −127.7°, d: 1.96 @ −184°. Sltly sol in H_2O.
SYNS: ARCTON 0 □ CARBON FLUORIDE □ F 14 □ FC 14 □ FREON 14 □ HALOCARBON 14 □ HALON 14 □ METHANE, TETRAFLUORO- □ PERFLUOROMETHANE □ R 14 □ R14 (DOT) □ REFRIGERANT 14 □ REFRIGERANT R 14 □ R 14 (REFRIGERANT) □ TETRAFLUOROCARBON □ TETRAFLUOROMETHANE □ TETRAFLUOROMETHANE (DOT)
CONSENSUS REPORTS: Reported in EPA TSCA Inventory.
DOT CLASSIFICATION: 2.2; Label: Nonflammable Gas
SAFETY PROFILE: Mildly toxic by inhalation. Less chronically toxic than carbon tetrachloride. Violent reaction with Al. When heated to decomposition it emits toxic fumes of F^-. See also FLUORIDES.

CBY750   CAS: 75-46-7   HR: 2
CARBON TRIFLUORIDE
DOT: UN 1984/UN 3136
mf: CHF_3 mw: 70.02
PROP: Colorless, odorless gas. Mp: −163°, bp: −82.2°, d: 1.52 (liquid) @ −100°. Sol in water.
SYNS: ARCTON □ CARBON TRIFLUORIDE □ FLUOROFORM □ FLUORYL □ FREON 23 □ FREON F-23 □ GENETRON 23 □ HALOCARBON 23 □ METHYL TRIFLUORIDE □ R 23 □ TRIFLUOROMETHANE □ TRIFLUOROMETHANE, refrigerated, liquid (UN 3136) (DOT) □ TRIFLUOROMETHANE (UN 1984) (DOT)
CONSENSUS REPORTS: EPA Genetic Toxicology Program. Reported in EPA TSCA Inventory.
DOT CLASSIFICATION: 2.2; Label: Nonflammable Gas
SAFETY PROFILE: Narcotic in high concentration. A mild respiratory irritant. Mutation data reported. See also FLUORIDES. When heated to decomposition it emits toxic fumes of F^-.

CCA500   CAS: 353-50-4   HR: 3
CARBONYL FLUORIDE
DOT: UN 2417
mf: CF_2O mw: 66.01
PROP: Colorless gas; pungent; hygroscopic.
Readily hydrolyzes to CO_2 and HF. Mp:
−114°, bp: −83°, d: 1.139 @ −114°.
SYNS: CARBON DIFLUORIDE OXIDE □ CARBON
FLUORIDE OXIDE □ CARBONIC DIFLUORIDE □
CARBON OXYFLUORIDE □ CARBONYL DIFLUORIDE
□ DIFLUOROFORMALDEHYDE □ FLUOPHOSGENE □
FLUOROFORMYL FLUORIDE □ FLUOROPHOSGENE □
RCRA WASTE NUMBER U033
CONSENSUS REPORTS: Reported in EPA
TSCA Inventory.
OSHA PEL: TWA 2 ppm; STEL 5 ppm
ACGIH TLV: TWA 2 ppm; STEL 5 ppm
DOT CLASSIFICATION: 2.3; Label: Poison
Gas
SAFETY PROFILE: A poison. Moderately
toxic by inhalation. A powerful irritant.
Hydrolyzes instantly to form HF on contact
with moisture. See also CARBONYLS,
HYDROFLUORIC ACID, and
FLUORINE. Incompatible with
hexafluoroisopropylideneamino-lithium.
When heated to decomposition it emits
toxic fumes of CO and F⁻. See CARBON
MONOXIDE for fire and explosion hazard.

CCB609   HR: 3
CARBONYLS
PROP: The (CO) group with a metal (M).
They may exist as dimeric acetylene
derivatives (MOC□COM) or as salts of
hexahydroxybenzene.
SAFETY PROFILE: Most carbonyls are
highly toxic. The toxicity of carbonyls
depends in part, but not always entirely, on
their ready decomposition, which releases
carbon monoxide. Symptoms are due in part
to carbon monoxide and in part to the direct
irritating action of the carbonyl. See specific
carbonyl in question. Many carbonyl metals
ignite spontaneously in air, some with a
delay period. Others are moderate fire and
explosion hazards when exposed to heat or
flame. Carbonyls of alkali metals are
potentially explosive. Hypergolic reaction

with dinitrogen tetraoxide. They react with
water or steam to produce toxic and
flammable vapors; can react vigorously with
oxidizing materials. When heated to
decomposition they emit highly toxic
fumes of carbon monoxide. See also
CARBON MONOXIDE and
POWDERED METALS.

CCC000   CAS: 463-58-1   HR: 3
CARBONYL SULFIDE
DOT: UN 2204
mf: COS   mw: 60.07
PROP: Gas or liquid. Hydrolyzed by water.
Mp: −138.2°, bp: 50.2°, lel: 12%, uel: 28.5%,
d: liq 1.24 @ −87°, vap d: 2.1, d: 1.19 @ 50
mm. Very sltly sol in water, alc, and toluene.
SYNS: CARBON OXIDE SULFIDE □ CARBON
OXYSULFIDE □ CARBONYL SULFIDE-³²S □
OXYCARBON SULFIDE
CONSENSUS REPORTS: Community
Right-To-Know List. Reported in EPA
TSCA Inventory.
DOT CLASSIFICATION: 2.3; Label: Poison
Gas, Flammable Gas
SAFETY PROFILE: Poison by
intraperitoneal route. Mildly toxic by
inhalation. Narcotic in high concentration.
An irritant. May liberate highly toxic
hydrogen sulfide upon decomposition. A
very dangerous fire hazard and moderate
explosion hazard when exposed to heat or
flame. Can react vigorously with oxidizing
materials. To fight fire, stop flow of gas or
use CO_2, dry chemical, or water spray.
When heated to decomposition it emits
toxic fumes of CO. See also CARBONYLS
and SULFIDES.

CCC500   CAS: 5234-68-4   HR: 3
CARBOXINE
mf: $C_{12}H_{13}NO_2S$ mw: 235.32
PROP: Solid. Sltly sol in H_2O; sol in C_6H_6,
EtOH, and MeOH. Very sol in Me_2CO.
SYNS: 5-CARBOXANILIDO-2,3-DIHYDRO-6-METHYL-
1,4-OXATHIIN □ CARBOXIN (USDA) □ D 735 □ DCMO
□ 2,3-DIHYDRO-5-CARBOXANILIDO-6-METHYL-1,4-
OXATHIIN □ 5,6-DIHYDRO-2-METHYL-3-CARBOX-
ANILIDO-1,4-OXATHIIN (GERMAN) □ 2,3-DIHYDRO-6-

METHYL-1,4-OXATHIIN-5-CARBOXANILIDE □ 5,6-DIHYDRO-2-METHYL-1,4-OXATHIIN-3-CARBOX-ANILIDE □ 5,6-DIHYDRO-2-METHYL-N-PHENYL-1,4-OXATHIIN-3-CARBOXAMIDE □ F 735 □ FLO PRO V SEED PROTECTANT □ VITAVAX

SAFETY PROFILE: Poison by ingestion. Moderately toxic by skin contact and possibly other routes. Mutation data reported. When heated to decomposition it emits very toxic fumes of NO_x and SO_x.

CCE500   CAS: 493-52-7   HR: 2
2-CARBOXY-4'-(DIMETHYLAMINO)-
AZOBENZENE

mf: $C_{15}H_{15}N_3O_2$ mw: 269.33
PROP: Shiny violet crystals.
SYNS: C.I. 13020 □ C.I. ACID RED 2 □ p-(DIMETHYLAMINO)AZOBENZENE-o-CARBOXYLIC ACID □ 4'-DIMETHYLAMINOAZOBENZENE-2-CARBOXYLIC ACID □ o-((p-(DIMETHYLAMINO)-PHENYL)AZO)BENZOIC ACID □ 2-((4-DIMETHYL-AMINO)PHENYLAZO)BENZOIC ACID □ METHYL RED
CONSENSUS REPORTS: IARC Cancer Review: Group 3 IMEMDT 7,56,87; Animal Inadequate Evidence IMEMDT 8,161,75. Reported in EPA TSCA Inventory. EPA Genetic Toxicology Program.
SAFETY PROFILE: Questionable carcinogen with experimental tumorigenic data. Mutation data reported. When heated to decomposition it emits toxic fumes of NO_x.

CCF125   CAS: 12758-40-6   HR: 2
CARBOXYETHYLGERMANIUM
SESQUIOXIDE

mf: $C_6H_{10}Ge_2O_7$ mw: 339.34
SYNS: BIS-β-CARBOXYETHYLGERMANIUM SESQUIOXIDE □ 2-CARBOXYETHYLGERMASE-SQUIOXANE □ 3,3'-(DIOXODIGERMOXANYLENE) DIPROPANOIC ACID □ DIPROPANOIC ACID GERMANIUM SESQUIOXIDE □ Ge 132 □ GERMANATE-(2-), BIS(2-CARBOXYLATOETHYL)TRIOXODI-, DIHYDROGEN (9CI) □ 3,3'-(GERMANOIC ANHYDRIDE) DIPROPANOIC ACID
CONSENSUS REPORTS: Reported in EPA TSCA Inventory.
SAFETY PROFILE: Moderately toxic by intravenous route. Experimental reproductive effects. See also GERMANIUM COMPOUNDS.

CCG500   CAS: 36568-91-9   HR: 3
(4-(CARBOXY METHOXY)-3-CHLORO-
PHENYL)(5,5-DIETHYL-2,4,6(1H,3H,-
5H)-PYRIMIDINETRIONATO)-O²-
MERCURY, MONOSODIUM SALT

mf: $C_{16}H_{18}ClHgN_2O_6•Na$ mw: 593.39
PROP: IDLH 10 mg/m³ (as Hg).
SYNS: MERBAPHEN □ NOVASUROL
CONSENSUS REPORTS: Mercury and its compounds are on the Community Right-To-Know List.
OSHA PEL: CL 0.1 mg(Hg)/m³ (skin)
ACGIH TLV: TWA 0.1 mg(Hg)/m³ (skin); BEI: 35 μg/g creatinine total inorganic mercury in urine preshift; 15 μg/g creatinine total inorganic mercury in blood at end of shift at end of workweek.
DFG MAK: Confirmed Animal Carcinogen with Unknown Relevance to Humans
NIOSH REL: (Mercury, Aryl and Inorganic) CL 0.1 mg/m³ (skin)
SAFETY PROFILE: Poison by intravenous route. See also MERCURY COMPOUNDS. When heated to decomposition it emits very toxic fumes of Cl^-, NO_x, and Hg.

CCH300   CAS: 22041-28-7   HR: 3
1-CARBOXYMETHYL-1-METHYL-
PYRROLIDINIUM IODIDE METHYL
ESTER

mf: $C_8H_{16}NO_2•I$ mw: 285.15
SYN: PYRROLIDINIUM, 1-CARBOXYMETHYL-1-METHYL-, IODIDE, METHYL ESTER
SAFETY PROFILE: A poison by intravenous route. When heated to decomposition it emits toxic vapors of NO_x and I^-.

CCI550   CAS: 1197-16-6   HR: 3
p-CARBOXY PHENYLARSENOXIDE

mf: $C_7H_5AsO_3$ mw: 212.04
SYNS: ARSINE, OXO(4-CARBOXY)PHENYL- □ BENZOIC ACID, 4-ARSENOSO-
OSHA PEL: TWA 0.5 mg(As)/m³
SAFETY PROFILE: Poison by intravenous route. When heated to decomposition it emits toxic fumes of As.

CCJ625   **CAS: 8000-66-6**   **HR: 1**
CARDAMON OIL
PROP: From the seed of *Elettaria cardamomun* (L.) Maton (Fam. *Zingiberazeae*). Colorless liquid; aromatic penetrating odor of cardamom, pungent taste. Misc with alc.
SYNS: CARDAMON □ OIL OF CARDAMON
CONSENSUS REPORTS: Reported in EPA TSCA Inventory.
SAFETY PROFILE: Mildly toxic by ingestion. Mutation data reported. When heated to decomposition it emits acrid smoke and fumes.

CCK000   **CAS: 3599-32-4**   **HR: 3**
CARDIO-GREEN
mf: $C_{43}H_{48}N_2O_6S_2 \cdot Na$ mw: 776.04
PROP: Green powder. Mp: 243–245° (decomp).
SYNS: ICG □ INDOCYANINE GREEN □ IR 125 □ UJOVIRIDIN □ WOFAVERDIN
CONSENSUS REPORTS: Reported in EPA TSCA Inventory.
SAFETY PROFILE: Poison by intravenous route. When heated to decomposition it emits very toxic fumes of SO_x, Na_2O, and NO_x.

CCK125   **CAS: 87-33-2**   **HR: 2**
CARDIS
DOT: UN 2907
mf: $C_6H_8N_2O_8$ mw: 236.16
PROP: Hard, colorless crystals. Mp: 71°. Sparingly sol in water. Freely sol in org solvs, such as acetone, alc, and ether.
SYNS: ASTRIDINE □ CARDIO □ CARVANIL □ CARVASIN □ CEDOCARD □ CLAODICAL □ COROSORBIDE □ COROVLISS □ 1,4:3,6-DIANHYDRO-SORBITOL-2,5-DINITRATE □ DINITROSORBIDE □ DISORLON □ DURANITRAT □ EURECOR □ FLINDIX □ GLENTONIN-RETARD □ HARRICAL □ IBD □ ISDIN □ ISO-BID □ ISOKET □ ISOMACK □ ISO-PUREN □ ISORBID □ ISORDIL □ ISORDIL TEMBIDS □ ISOSORBIDE DINITRATE □ ISOSTENASE □ ISOTRATE □ KORODIL □ LANGORAN □ LASERDIL □ MAYCOR □ MONOCLAIR □ MYOREXON □ NITROSORBID □ NITROSORBIDE □ NITROSORBON □ NOSIM □ RESOIDAN □ RIFLOC RETARD □ RIGEDAL □ SORBANGIL □ SORBID □ SORBIDE NITRATE □ SORBIDILAT □ SORBIDINITRATE □ SORBISLO □ SORBITRATE □ SORBONIT □ SORQUAD □ SORQUAT □ VASCARDIN □ VASORBATE □ VASOTRATE
CONSENSUS REPORTS: Reported in EPA TSCA Inventory.
DOT CLASSIFICATION: 4.1; Label: Flammable Solid
SAFETY PROFILE: Moderately toxic by ingestion, intraperitoneal, intramuscular, and subcutaneous routes. Experimental reproductive effects. Mutation data reported. A flammable solid. When heated to decomposition it emits toxic fumes of NO_x. A coronary vasodilator. See also NITRATES.

CCK590   **CAS: 1390-65-4**   **HR: D**
CARMINE
mf: $C_{22}H_{20}O_{13}$ mw: 492.39
PROP: An aqueous extract of cochineal obtained from the dried female insects *Dactylopius coccus costa (Cossus cacti* L.). Bright red crystals from water. Decomp @ 250°. Sol in water, alc, ether; insol in benzene, chloroform. Carmine is the aluminum or calcium-aluminum lake on aluminum hydroxide substrate of carminic acid.
SYNS: B ROSE LIQUID □ CARMINIC ACID
SAFETY PROFILE: When heated to decomposition it emits acrid smoke and irritating fumes.

CCK665   **CAS: 305-84-0**   **HR: 1**
CARNOSINE
mf: $C_9H_{14}N_4O_3$ mw: 226.27
PROP: Needles. Mp: 246–250° (decomp).
SYNS: β-ALANYL-l-HISTIDINE □ l-CARNOSINE □ l-HISTIDINE, N-β-ALANYL- □ IGNOTINE □ KARNOZZN □ N-2-M
CONSENSUS REPORTS: Reported in EPA TSCA Inventory.
SAFETY PROFILE: Mildly toxic by intraperitoneal route. An experimental teratogen. Other experimental reproductive effects. When heated to decomposition it emits toxic fumes of NO_x.

CCK685   **CAS: 7235-40-7**   **HR: D**
β-CAROTENE
mf: $C_{40}H_{56}$ mw: 536.88

PROP: Deep purple prisms from C_6H_6/MeOH; or red crystals from pet ether; or crystalline powder. Mp: 183° (sealed tube). Sol in carbon disulfide, benzene, chloroform; sltly sol in ether, hexane, veg. oil; insol in water.

SYN: CAROTENE

SAFETY PROFILE: When heated to decomposition it emits acrid smoke and irritating fumes.

CCL250   CAS: 9000-07-1   HR: 2
CARRAGEEN

PROP: A sulfated polysaccharide. Dried plant of seaweed *Chondrus crispus, Chondrus ocellatus, Eucheuma cottonil, Eucheuma spinosum, Gigartina acicularis, Gigartina pistillata, Gigartina radula, Gigartina stellata.* Yellow-white when powdered. Sol in water @ 80°; insol in org solvs. Dried, bleached *Chondrus crispus* containing salts of sulfated polygalactose esters.

SYNS: 3,6-ANHYDRO-d-GALACTAN □ AUBYGEL GS □ AUBYGUM DM □ BURTONITE-V-40-E □ CARASTAY □ CARASTAY G □ CARRAGEENAN (FCC) □ CARRA-GEENAN GUM □ CARRAGHEANIN □ CARRAGHEEN □ CARRAGHEENAN □ CHONDRUS □ CHONDRUS EXTRACT □ COLLOID 775 □ COREINE □ EUCHEUMA SPINOSUM GUM □ FLANOGEN ELA □ GALOZONE □ GELCARIN □ GELCARIN HMR □ GELOZONE □ GENU □ GENUGEL □ GENUGEL CJ □ GENUGOL RLV □ GENUVISCO J □ GUM CARRAGEENAN □ GUM CHON 2 □ GUM CHROND □ IRISH GUM □ IRISH MOSS EXTRACT □ IRISH MOSS GELOSE □ KILLEEN □ LYGOMME CDS □ PEARLPUSS □ PELLUGEL □ PENCOGEL □ PIG-WRACK □ SATIAGEL GS 350 □ SATIAGUM 3 □ SATIAGUM STANDARD □ SEAKEM CARRAGEENIN □ SEATREM □ SELF ROCK MOSS □ VISCARIN

CONSENSUS REPORTS: IARC Cancer Review: Group 3 IMEMDT 7,56,87; Animal Limited Evidence IMEMDT 10,181,76. Reported in EPA TSCA Inventory.

SAFETY PROFILE: Poison by intravenous route. Questionable carcinogen with experimental neoplastigenic and tumorigenic data. When heated to decomposition it emits acrid smoke and fumes.

CCL500   HR: 3
CARRAGEENAN, DEGRADED

PROP: Carrageenan derived from *Eucheuma spinosum*, degraded by acid hydrolysis; average molecular weight 20,000–40,000 (CALEDQ 4,171,78).

CONSENSUS REPORTS: IARC Cancer Review: Animal Sufficient Evidence IMEMDT 31,79,83.

SAFETY PROFILE: Confirmed carcinogen with experimental carcinogenic, neoplastigenic, and tumorigenic data. See also CARRAGEEN. When heated to decomposition it emits toxic fumes of SO_x.

CCL750   CAS: 8015-88-1   HR: 1
CARROT SEED OIL

PROP: Distilled from the seeds of *Daucus carota L.* (Fam. *Umbelliferae*) (FCTXAV 14,659,76). Light-yellow to amber liquid; aromatic odor. Sol in fixed oils, mineral oil; insol in glycerin, propylene glycol.

SYNS: DAUCUS OIL □ OILS, CARROT

CONSENSUS REPORTS: Reported in EPA TSCA Inventory.

SAFETY PROFILE: A skin irritant. When heated to decomposition it emits acrid smoke and irritating fumes.

CCM000   CAS: 499-75-2   HR: 3
CARVACROL

mf: $C_{10}H_{14}O$ mw: 150.24

PROP: Colorless to pale-yellow liquid; spicy thymol odor. D: 0.974–0.980, mp: 3.5°, bp: 237–238°, refr index: 1.521–1.526, flash p: 212°F. Sol in alc, ether; insol in water.

SYNS: 2-p-CYMENOL □ FEMA No. 2245 □ 2-HYDROXY-p-CYMENE □ ISOPROPYL-o-CRESOL □ 5-ISOPROPYL-2-METHYLPHENOL □ ISOTHYMOL □ 2-METHYL-5-ISOPROPYLPHENOL □ o-THYMOL

CONSENSUS REPORTS: Reported in EPA TSCA Inventory.

SAFETY PROFILE: Poison by ingestion, intravenous, and subcutaneous routes. Moderately toxic by skin contact. A severe skin irritant. Combustible liquid. When heated to decomposition it emits acrid smoke and irritating fumes.

CCM100   CAS: 2244-16-8   HR: 3
d-CARVONE
mf: $C_{10}H_{14}O$ mw: 150.24
PROP: Colorless liquid or oil; caraway odor.
D: 0.956–0.960, bp: 230°, refr index:
1.96–1.499. Sol in propylene glycol, fixed
oils; misc in alc; insol in glycerin.
SYNS: (+)-CARVONE □ d(+)-CARVONE □ (S)-
CARVONE □ (S)-(+)-CARVONE □ FEMA No. 2249 □ d-p-
MENTHA-6,8,(9)-DIEN-2-ONE □ d-1-METHYL-4-
ISOPROPENYL-6-CYCLOHEXEN-2-ONE □ (S)-2-
METHYL-5-(1-METHYLETHENYL)-2-CYCLOHEXEN-1-
ONE
CONSENSUS REPORTS: Reported in EPA
TSCA Inventory.
SAFETY PROFILE: Poison by ingestion and
skin contact. A skin irritant. When heated to
decomposition it emits acrid smoke and
irritating fumes.

CCM120   CAS: 6485-40-1   HR: 3
l(−)-CARVONE
mf: $C_{10}H_{14}O$ mw: 150.22
PROP: Colorless liquid or oil; spearmint
odor. D: 0.956–0.960, bp: 230–231°, refr
index: 1.495–1.499. Sol in propylene glycol,
fixed oils; misc in alc; insol in glycerin.
SYNS: (−)-CARVONE □ 1-CARVONE □ (R)-CARVONE
□ FEMA No. 2249 □ 1-6,8(9)-p-MENTHADIEN-2-ONE □
(R)-(−)-p-MENTHA-6,8-DIEN-2-ONE □ 1-1-METHYL-4-
ISOPROPENYL-6-CYCLOHEXEN-2-ONE □ (R)-2-
METHYL-5-(1-METHYLETHENYL)-2-CYCLOHEXEN-1-
ONE (9CI)
SAFETY PROFILE: Poison by intravenous
route. Moderately toxic by ingestion. When
heated to decomposition it emits acrid
smoke and irritating fumes.

CCN000   CAS: 87-44-5   HR: 1
CARYOPHYLLENE
mf: $C_{15}H_{26}$ mw: 206.41
PROP: Colorless to sltly yellow oily liquid;
clove odor. Found in oil of clove, cinnamon
leaves, and copaiba balsam, and in minor
quantities in various other essential oils,
especially lavender; prepared by isolation
from clove leaf oil, clove stem oil, cinnamon
leaf oil, or pine oil fractions (FCTXAV
11,1011,73). D: 0.897–0.910, refr index:

1.498–1.504, bp: 118–119° @ 9.7 mm, flash
p: 206°F. Sol in alc, ether; insol in water.
SYNS: β-CARYOPHYLLENE (FCC) □ FEMA No. 2252 □
8-METHYLENE-4,11,11-(TRIMETHYL)BICYCLO-
(7.2.0)UNDEC-4-ENE
CONSENSUS REPORTS: Reported in
EPA TSCA Inventory.
SAFETY PROFILE: A skin irritant.
Combustible liquid. When heated to
decomposition it emits acrid smoke and
irritating fumes.

CCN100   CAS: 1139-30-6   HR: 1
β-CARYOPHYLLENE EPOXIDE
mf: $C_{15}H_{24}O$ mw: 220.39
PROP: Crystals. Mp: 63.5–64°.
SYNS: CARYOPHYLLENE EPOXIDE □
CARYOPHYLLENE OXIDE □ (−)-CARYOPHYLLENE
OXIDE □ β-CARYOPHYLLENE OXIDE □ EPOXY-
CARYOPHYLLENE □ (−)-EPOXYDIHYDROCARY-
OPHYLLENE □ 5-OXATRICYCLO(8.2.0.0^{4,6})DODECANE,
4,12,12-TRIMETHYL-9-METHYLENE-, (1R,4R,6R,10S)- □
4,11,11-TRIMETHYL-8-METHYLENE-5OXATRI-CYCLO-
(8.2.0.0(4,6))DODECANE
CONSENSUS REPORTS: Reported in EPA
TSCA Inventory.
SAFETY PROFILE: Low toxicity by
ingestion and skin contact. A skin irritant.
When heated to decomposition it emits
acrid smoke and irritating fumes.

CCO500   HR: D
CASCARILLA OIL
PROP: From steam distillation of bark of
Croton *cascarilla* Benn. or *Croton eluteria Benn.*
(Fam. Euphorbiaceae). Light yellow to
brown liquid; spicy odor. Sol in fixed oils;
insol in glycerin, propylene glycol.
SYN: SWEETWOOD BARK OIL
SAFETY PROFILE: When heated to
decomposition it emits acrid smoke and
irritating fumes.

CCO750   CAS: 8007-80-5   HR: 3
CASSIA OIL
PROP: Chief constituent is cinnamic
aldehyde, found in the leaves and twigs of
Cinnamomum cassia blume (FCTXAV
13,91,75). Yellow liquid; cinnamon odor,
spicy burning taste. Sol in fixed oils,

propylene glycol; insol in glycerin, mineral oil.

SYNS: ARTIFICIAL CINNAMON OIL □ CINNAMON BARK OIL □ CINNAMON BARK OIL, CEYLON TYPE (FCC) □ CINNAMON OIL □ KASSIA OEL (GERMAN) □ OIL OF CASSIA □ OIL OF CHINESE CINNAMON □ OIL OF CINNAMON □ OIL OF CINNAMON, CEYLON □ OILS, CINNAMON

CONSENSUS REPORTS: Reported in EPA TSCA Inventory.

SAFETY PROFILE: Poison by skin contact. Moderately toxic by ingestion and intraperitoneal routes. A human skin irritant. Mutation data reported. See also CINNAMALDEHYDE and ALDEHYDES. When heated to decomposition it emits acrid smoke and irritating fumes.

CCP000   HR: 3
CASTOR BEAN
DOT: UN 2969

PROP: An annual may grow higher than 15 feet. The large, lobed leaves may be 3 feet across. The spiny seed pods grow in clusters and contain plump seeds that are white with brown or black mottling. The seeds have a pleasant taste.

SYNS: AFRICAN COFFEE TREE □ CASTOR BEANS (DOT) □ CASTOR FLAKE (DOT) □ CASTOR MEAL (DOT) □ CASTOR OIL PLANT □ CASTOR POMACE (DOT) □ HIGUERETA (CUBA, PUERTO RICO) □ HIGUERILLA (MEXICO) □ KOLI (HAWAII) □ LA'AU-'AILA (HAWAII) □ MAN'S MOTHERWORT □ MEXICO WEED □ PA'AILA (HAWAII) □ PALMA CHRISTI (HAITI) □ RICIN (HAITI) □ RICINO (PUERTO RICO) □ RICINUS COMMUNIS □ STEADFAST □ WONDER TREE

DOT CLASSIFICATION: 9; Label: None

SAFETY PROFILE: Deadly poison by ingestion in humans. The seeds contain the deadly poison ricin, a plant lectin (toxalbumin) which inhibits protein synthesis in the intestinal wall. Ingestion of the seeds can cause after a delay period of several hours: nausea, vomiting, diarrhea, and intestinal dysfunction. There may be massive fluid and electrolyte loss. Ingestion of as few as 2 seeds could be fatal. A potent allergen. When heated to decomposition it emits toxic fumes of NO_x. See also RICIN.

CCP250   CAS: 8001-79-4   HR: 1
CASTOR OIL

PROP: From seeds of *Ricinus communis L.* (Fam. *Euphorbiaceae*). A colorless to pale-yellow, viscous liquid; bland taste, characteristic odor. Mp: −12°, bp: 313°, flash p: 445°F (CC), d: 0.96, autoign temp: 840°F. Sol in alc; misc in abs alc, glacial acetic acid, chloroform, and ether.

SYNS: AROMATIC CASTOR OIL □ CASTOR OIL AROMATIC □ COSMETOL □ CRYSTAL O □ GOLD BOND □ NCI-C55163 □ NEOLOID □ OIL OF PALMA CHRISTI □ PHORBYOL □ RICINUS OIL □ RICIRUS OIL □ TANGANTANGAN OIL

CONSENSUS REPORTS: Reported in EPA TSCA Inventory.

SAFETY PROFILE: An allergen. A human skin and eye irritant. Combustible when exposed to heat. Spontaneous heating may occur. To fight fire, use CO_2, dry chemical, fog, mist. See also CASTOR BEAN.

CCP850   CAS: 120-80-9   HR: 3
CATECHOL
mf: $C_6H_6O_2$ mw: 110.12

PROP: Colorless crystals or needles from water. Mp: 105°, bp: 240°, flash p: 261°F (CC), d: 1.341 @ 15°, vap press: 10 mm @ 118.3°, vap d: 3.79. Sol in water, chloroform, and benzene; very sol in alc and ether.

SYNS: o-BENZENEDIOL □ 1,2-BENZENEDIOL □ CATECHIN □ C.I. 76500 □ C.I. OXIDATION BASE 26 □ o-DIHYDROXYBENZENE □ 1,2-DIHYDROXYBENZENE □ o-DIOXYBENZENE □ o-DIPHENOL □ DURAFUR DEVELOPER C □ FOURAMINE PCH □ FOURRINE 68 □ o-HYDROQUINONE □ o-HYDROXYPHENOL □ 2-HYDROXYPHENOL □ NCI-C55856 □ OXYPHENIC ACID □ PELAGOL GREY C □ o-PHENYLENEDIOL □ PYROCATECHIN □ PYROCATECHINIC ACID □ PYROCATECHOL □ PYROCATECHUIC ACID

CONSENSUS REPORTS: IARC Cancer Review: Group 3 IMEMDT 7,56,87; Animal Inadequate Evidence IMEMDT 15,155,77. Reported in EPA TSCA Inventory. EPA Genetic Toxicology Program.

OSHA PEL: TWA 5 ppm (skin)

ACGIH TLV: TWA 5 ppm (skin); Animal Carcinogen

SAFETY PROFILE: Poison by ingestion, subcutaneous, intraperitoneal, intravenous, and parenteral routes. Moderately toxic by skin contact. Experimental reproductive effects. Can cause dermatitis on skin contact. An allergen. Human mutation data reported. Questionable carcinogen. Systemic effects similar to those of phenol. Combustible when exposed to heat or flame; can react vigorously with oxidizing materials. Hypergolic reaction with concentrated nitric acid. To fight fire, use water, CO_2, dry chemical. When heated to decomposition it emits acrid smoke and irritating fumes. See also PHENOL.

CCP900   CAS: 2050-46-6   HR: 1
CATECHOL DIETHYL ETHER
mf: $C_{10}H_{14}O_2$ mw: 166.24
PROP: Crystals. Mp: 43–45°, bp: 219°, d: 1.0.
SYNS: BENZENE, o-DIETHOXY- □ BENZENE, 1,2-DIETHOXY-(9CI) □ o-DIETHOXYBENZENE □ 1,2-DIETHOXYBENZENE
CONSENSUS REPORTS: Reported in EPA TSCA Inventory.
SAFETY PROFILE: An eye irritant. When heated to decomposition it emits acrid smoke and irritating fumes.

CCQ500   CAS: 8007-20-3   HR: 2
CEDAR LEAF OIL
PROP: Constituent is d-α-thujone, found in leaves of *Thuja occidentalis L.* (Fam. *Cupressaaceae*) (FCTXAV 12,807,74). Yellowish, volatile oil; strong sage odor. D: 0.910–0.920. Sol in fixed oils, mineral oil, propylene glycol; insol in glycerin.
SYNS: OIL OF ARBOR VITAE □ OIL OF CEDAR LEAF □ OILS, CEDAR LEAF □ OIL THUJA □ OIL OF THUJA □ OIL OF WHITE CEDAR □ THUJA OIL □ WHITE CEDAR OIL
CONSENSUS REPORTS: Reported in EPA TSCA Inventory.
SAFETY PROFILE: Moderately toxic by ingestion and skin contact. A skin irritant. Ingestion of large quantities causes hypertension, bradycardia, tachypnea, convulsions, death. When heated to decomposition it emits acrid smoke and fumes. See also ARTEMISIA OIL.

CCR510   CAS: 29597-36-2   HR: 1
CEDR-8-ENE EPOXIDE
SYNS: ANDRANE □ CEDRANE, 8,9-EPOXIDE
CONSENSUS REPORTS: Reported in EPA TSCA Inventory.
SAFETY PROFILE: Very low toxicity by ingestion and skin contact. A skin irritant. When heated to decomposition it emits acrid smoke and irritating fumes.

CCR524   CAS: 39900-38-4   HR: 1
CEDROL FORMATE
mf: $C_{16}H_{26}O_2$ mw: 250.42
PROP: Cosmetic chemical.'
SYNS: CEDRYL FORMATE □ 1H-3-α-7-METHANO-AZULEN-6-OL, OCTAHYDRO-3,6,8,8-TETRAMETHYL-, FORMATE, (3R-(3-α-3a-β,6-α-7-β,8aα-))-
CONSENSUS REPORTS: Reported in EPA TSCA Inventory.
SAFETY PROFILE: Very low toxicity by ingestion and skin contact. A skin irritant. When heated to decomposition it emits acrid smoke and irritating fumes.

CCR525   CAS: 67874-81-1   HR: 1
CEDROL METHYL ETHER
mf: $C_{16}H_{28}O$ mw: 236.44
PROP: Colorless to pale yellow liquid. D: 0.97.
SYNS: CEDRAMBER □ 1H-3a,7-METHANOAZULENE, OCTAHYDRO-6-METHOXY-3,6,8,8-TETRAMETHYL-,(3R-(3-α-3a-β, 6-α-7-β,8aα-))- □ METHYL CEDRYL ETHER
CONSENSUS REPORTS: Reported in EPA TSCA Inventory.
SAFETY PROFILE: A skin irritant. When heated to decomposition it emits acrid smoke and irritating fumes.

CCS575   HR: D
CEFTIOFUR
PROP: Powder.
SAFETY PROFILE: When heated to decomposition it emits acrid smoke and irritating fumes.

CCT250   CAS: 9005-81-6   HR: 2
CELLOPHANE
mf: $(C_6H_{10}O_5)_n$
SYN: VISKING CELLOPHANE
CONSENSUS REPORTS: Reported in EPA TSCA Inventory.
SAFETY PROFILE: Questionable carcinogen with experimental tumorigenic data by implant. See also POLYMERS. When heated to decomposition it emits acrid smoke and irritating fumes.

CCU050   CAS: 9004-38-0   HR: D
CELLULOSE ACETATE MONOPHTHALATE
PROP: White powder. Insoluble in water.
SYNS: ACETYL PHTHALYL CELLULOSE □ CAP-WAKO □ CELLACETATE □ CELLULOSE, ACETATE HYDROGEN 1,2-BENZENEDICARBOXYLATE (9CI) □ CELLULOSE, ACETATE PHTHALATE □ CELLULOSE ACETOPHTHALATE □ CELLULOSE ACETYLPHTHAL-ATE
CONSENSUS REPORTS: Reported in EPA TSCA Inventory.
SAFETY PROFILE: An experimental teratogen. When heated to decomposition it emits acrid smoke and irritating fumes.

CCU100   HR: 1
CELLULOSE, MICROCRYSTALLINE
PROP: Fine white crystalline powder from treatment of α-cellulose with mineral acids. Insol in water, most org solvs.
SYN: CELLULOSE GEL
SAFETY PROFILE: A nuisance dust. When heated to decomposition it emits acrid smoke and irritating fumes.

CCU150   CAS: 9004-34-6   HR: 1
CELLULOSE, POWDERED
PROP: Fine white fibrous particles from treatment of bleached cellulose from wood or cotton. Insol in water and most org solvs.
SYNS: ABICEL □ β-AMYLOSE □ ARBOCEL □ ARBOCEL BC 200 □ ARBOCELL B 600/30 □ AVICEL □ AVICEL 101 □ AVICEL 102 □ AVICEL PH 101 □ AVICEL PH 105 □ CELLEX MX □ α-CELLULOSE □ CELLULOSE 248 □ CELLULOSE (ACGIH,OSHA) □ CELLULOSE CRYSTALLINE □ CELUFI □ CEPO □ CEPO CFM □ CEPO S 20 □ CEPO S 40 □ CHROMEDIA CC 31 □

CHROMEDIA CF 11 □ CUPRICELLULOSE □ ELCEMA F 150 □ ELCEMA G 250 □ ELCEMA P 050 □ ELCEMA P 100 □ FRESENIUS D 6 □ HEWETEN 10 □ HYDROXY-CELLULOSE □ KINGCOT □ LA 01 □ MN-CELLULOSE □ ONOZUKA P 500 □ PYROCELLULOSE □ RAYOPH-ANE □ RAYWEB Q □ REXCEL □ SIGMACELL □ SOLKA-FIL □ SOLKA-FLOC □ SOLKA-FLOC BW □ SOLKA-FLOC BW 20 □ SOLKA-FLOC BW 100 □ SOLKA-FLOC BW 200 □ SOLKA-FLOC BW 2030 □ SPARTOSE OM-22 □ SULFITE CELLULOSE □ TOMOFAN □ TUNICIN □ WHATMAN CC-31
OSHA PEL: Total Dust: 15 mg/m³; Respirable Fraction: 5 mg/m³
ACGIH TLV: TWA (nuisance particulate) 10 mg/m³ of total dust (when toxic impurities are not present, e.g., quartz <1%)
SAFETY PROFILE: A nuisance dust. When heated to decomposition it emits acrid smoke and irritating fumes.

CCU250   CAS: 9004-70-0   HR: 3
CELLULOSE TETRANITRATE
DOT: UN 0340/UN 0341/UN 0342/UN 0343/UN 2059/UN 2555/UN 2556/UN 2557
mf: $C_{12}H_{16}(ONO_2)_4O_6$ mw: 504.3
PROP: White, amorphous solid. D: 1.66, flash p: 55°F.
SYNS: AS □ C 2018 □ CA 80-15 □ CELEX □ CELLOIDIN □ CELLULOSE NITRATE □ CELLULOSE, NITRATE (9CI) □ COLLODION □ COLLODION COTTON □ COLLODION WOOL □ COLLOXYLIN □ CORIAL EM FINISH F □ E 1440 □ FLEXIBLE COLLODION □ FM-NTS □ GUNCOTTON □ HX 3/5 □ KODAK LR 115 □ LR 115 □ NITROCELLULOSE, dry or wetted with <25% water (or alcohol), by weight (UN 0340) (DOT) □ NITROCELLULOSE, plasticized with not <18% plasticizing substance, by weight (UN 0343) (DOT) □ NITROCELLULOSE, solution, flammable with not >12.6% nitrogen, by weight (UN 2059) (DOT) □ NITROCELLULOSE, unmodified or plasticized with <18% plasticizing substance (UN 0341) (DOT) □ NITROCELLULOSE, wetted with not <25% alcohol, by weight (UN 0342) (DOT) □ NITROCELLULOSE with alcohol not <25% alcohol by weight, and not >12.6% nitrogen (UN 2556) (DOT) □ NITROCELLULOSE with plasticizing not <18% plasticizing substance, by weight (UN 2557) (DOT) □ NITROCELLULOSE with water not <25% water, by weight (UN 2555) (DOT) □ NITROCELLULOSE E950 □ NITROCOTTON □ NITRON □ NITRON (NITROCELLULOSE) □ NIXON N/C □ NTs 62 □ NTs 218 □ NTs 222 □ NTs 539 □ NTs 542 □ PARLODION □ PYRALIN □ PYROXYLIN □ RF 10 □ RS □ R.S. NITROCELLULOSE □ SOLUBLE GUN COTTON □ SS □ SYNPOR □ TSAPOLAK 964 □ XYLOIDIN

CONSENSUS REPORTS: Reported in EPA TSCA Inventory.

DOT CLASSIFICATION: EXPLOSIVE 1.1D; Label: EXPLOSIVE 1.1D (UN 0340, UN 0341); DOT Class: EXPLOSIVE 1.3C; Label: EXPLOSIVE 1.3C (UN 0343, UN 0342); DOT Class: 3; Label: Flammable Liquid (UN 2059); DOT Class: 4.1; Label: Flammable Solid (UN 2556, UN 2557, UN 2555)

SAFETY PROFILE: Very low oral toxicity. Flammable solid. Highly dangerous fire hazard in the dry state when exposed to heat, flame, or powerful oxidizers. When wet with 35% of denatured ethanol it is about as hazardous as ethanol alone or gasoline. Dry cellulose tetranitrate burns rapidly with intense heat and ignites easily. Moderately dangerous explosion hazard. To fight fire, use copious volumes of water; alcohol foam. CO_2 is effective in extinguishing fires of nitrocellulose solvents. See also EXPLOSIVES, HIGH.

CCW250   HR: 3
CEMENT (rubber)
PROP: Flash p: 50°F or less.
SYNS: CEMENT, RUBBER □ RUBBER CEMENT
SAFETY PROFILE: May contain benzene or other toxic solvents. See specific constituent. Dangerous fire hazard when exposed to heat or flame; can react with oxidizing materials.

CCX000   CAS: 123-03-5   HR: 3
CEPACOL CHLORIDE
mf: $C_{21}H_{38}N \cdot Cl$ mw: 340.05
PROP: A solid. Mp: 87–88°. Sol in water.
SYNS: ACETOQUAT CPC □ AKTIVEX □ AMMONYX CPC □ BIOSEPT □ CEEPRYN □ CEEPRYN CHLORIDE □ CEPRIM □ CETAMIUM □ CETYLPYRIDINIUM CHLORIDE □ N-CETYLPYRIDINIUM CHLORIDE □ 1-CETYLPYRIDINIUM CHLORIDE □ DOBENDAN □ HEXADECYLPYRIDINIUM CHLORIDE □ n-HEXADE-CYLPYRIDINIUM CHLORIDE □ 1-HEXADECYLPY-RIDINIUM CHLORIDE □ INTEXSAN CPC □ PRISTACIN □ PYRISEPT □ QUATERNARIO CPC
CONSENSUS REPORTS: Reported in EPA TSCA Inventory.

SAFETY PROFILE: Poison by ingestion, intraperitoneal, subcutaneous, and intravenous routes. Moderately toxic by skin contact. A skin and eye irritant. When heated to decomposition it emits very toxic fumes of NO_x and Cl^-.

CCX500   CAS: 21593-23-7   HR: 2
CEPHAPIRIN
mf: $C_{17}H_{17}N_3O_6S_2$ mw: 423.49
PROP: Crystals from Me_2CO (aq). Mp: 155°.
SYNS: CEFAPIRIN (GERMAN) □ 3-(HYDROXY-METHYL)-8-OXO-7-(2-(4-PYRIDYLTHIO)ACETAMIDO)-5-THIA-1-AZABICYCLO(4.2.0)OCT-2-ENE-2-CARBOXYLIC ACID, ACETATE (ESTER)
SAFETY PROFILE: Moderately toxic by intraperitoneal route. Human systemic effects by intravenous route: jaundice. Experimental reproductive effects. When heated to decomposition it emits very toxic fumes of NO_x and SO_x.

CCY250   CAS: 7440-45-1   HR: 3
CERIUM
af: Ce   aw: 140.13
PROP: Malleable gray metal, forms lustrous crystals that tarnish in air. Cubic or hexagonal, steel-gray crystals. Mp: 804°, bp: 3433°, d: (cubic form): 6.90, hexagonal form: 6.75. Reacts with moist air readily and with H_2O (slow in cold), acids, and alkalies.
CONSENSUS REPORTS: Reported in EPA TSCA Inventory.
SAFETY PROFILE: Cerium resembles aluminum in its pharmacological action as well as in its chemical properties. The insoluble salts such as the oxalates are stated to be nontoxic even in large doses. It is used to prevent vomiting in pregnancy. The average dose is from 0.05 to 0.5 g.

The effect on the central nervous system of the rare-earth metals following inhalation may preclude welding operations with these materials to any large extent. Cerium is stated to produce polycythemia but is useless in the treatment of anemia owing to its toxic effects. The salts of cerium increase the blood coagulation rate. See also RARE

EARTHS. A strong reducing agent. Moderate fire hazard; ignites spontaneously in air at 150–180°. Moderate explosion hazard in the form of dust when exposed to flame. The metal or its alloys spark with friction. Many alloys are pyrophoric in air. See also IRON DUST. Explosive reaction with zinc. Very exothermic reaction with antimony or bismuth. Ignites when heated in atmospheres of CO_2 + N_2, Cl_2, or Br_2. Violent reaction when heated with phosphorus (400°C), silicon (1400°C).

CCY500   CAS: 537-00-8   HR: 3
CERIUM ACETATE
mf: $C_6H_9O_6$•Ce   mw: 317.27
PROP: White powder. Sol in water.
SYNS: CERIUM TRIACETATE □ CEROUS ACETATE
CONSENSUS REPORTS: Reported in EPA TSCA Inventory.
SAFETY PROFILE: Human central nervous system effects. See also CERIUM COMPOUNDS. When heated to decomposition it emits acrid and irritating fumes.

CCY750   CAS: 7790-86-5   HR: 3
CERIUM CHLORIDE
mf: $CeCl_3$ mw: 246.47
PROP: Colorless or white solid or deliquescent crystals. Mp: 722°, bp: 1705°, d: 3.92. Sol in water and THF.
SYNS: CERIUM(III) CHLORIDE □ CERIUM TRICHLORIDE □ CEROUS CHLORIDE
CONSENSUS REPORTS: Reported in EPA TSCA Inventory. EPA Genetic Toxicology Program.
SAFETY PROFILE: Poison by intravenous, intraperitoneal, and subcutaneous routes. Moderately toxic by ingestion. See also CERIUM COMPOUNDS. When heated to decomposition it emits toxic fumes of Cl^-.

CCZ000   CAS: 512-24-3   HR: 3
CERIUM CITRATE
mf: $C_6H_8O_7$•Ce   mw: 332.26
SYNS: CERIUM(III) CITRATE □ CEROUS CITRATE □ 2-HYDROXY-1,2,3-PROPANETRISCARBOXYLIC ACID CERIUM(3+) SALT (1:1) (9CI)

SAFETY PROFILE: Poison by intraperitoneal route. Experimental reproductive effects. See also CERIUM COMPOUNDS. When heated to decomposition it emits acrid and irritating fumes.

CDA250   HR: 2
CERIUM COMPOUNDS
PROP: Compounds of cerium and the other rare-earth elements are generally of low toxicity. The greatest exposures are likely to be during manufacture of cerium. Exposed workers have experienced sensitivity to heat, itching, and skin lesions. Large doses to experimental animals have caused writhing, ataxia (loss of muscle coordination), labored respiration, sedation, hypotension, and death by cardiovascular collapse. The chloride, bromide, nitrate, bromate, and perchlorate salts are water soluble and thus are more likely to cause systemic effects when ingested. The sulfates, iodides, and iodates are less water soluble. Oxides, oxalates, sulfides, carbonates, fluorides, and phosphates are insoluble. The salts of cerium increase the blood coagulation rate. Cerium tartrate has been found to produce a direct injurious action on the hearts of small animals. Cerium oxalate has been used to suppress motion sickness and to suppress vomiting during pregnancy (by ingestion of 1 g/24 hr). The toxicity of cerium compounds may be taken to be that of cerium, except when the anion has a toxicity of its own. See also CERIUM and RARE EARTHS.

CDA750   CAS: 7758-88-5   HR: 1
CERIUM FLUORIDE
mf: CeF_3 mw: 197.12
PROP: White, hexagonal crystals or solid. D: 6.16, mp: 1460°, bp: 2300°. Insol in water; sol in H_2SO_4.
SYNS: CERIUM FLUORURE (FRENCH) □ CERIUM TRIFLUORIDE □ CEROUS FLUORIDE
CONSENSUS REPORTS: Reported in EPA TSCA Inventory.

OSHA PEL: TWA 2.5 mg(F)/m³
ACGIH TLV: TWA 2.5 mg(F)/m³; BEI: 3 mg/g creatinine of fluorides in urine prior to shift; 10 mg/g creatinine of fluorides in urine at end of shift.
NIOSH REL: (Inorganic Fluorides) TWA 2.5 mg(F)/m³
SAFETY PROFILE: Low toxicity by ingestion. See FLUORIDES and CERIUM COMPOUNDS. When heated to decomposition it emits toxic fumes of F⁻.

CDB000   CAS: 10108-73-3   HR: 3
CERIUM(III) NITRATE
mf: N₃O₉•Ce   mw: 326.15
SYNS: CERIUM NITRATE □ CERIUM(3+) NITRATE □ CERIUM TRINITRATE □ CEROUS NITRATE □ DUSICNAN CERITY (CZECH) □ NITRIC ACID, CERIUM(3+) SALT (8CI, 9CI)
CONSENSUS REPORTS: Reported in EPA TSCA Inventory.
SAFETY PROFILE: Poison by intravenous and intraperitoneal routes. Moderately toxic by ingestion. Experimental reproductive effects. See also CERIUM COMPOUNDS and NITRATES. When heated to decomposition it emits toxic fumes of NOₓ.

CDB400   CAS: 13590-82-4   HR: D
CERIUM(IV) SULFATE
mf: O₈S₂•Ce   mw: 332.24
PROP: Yellow crystals or powder. Sol in H₂O.
SYNS: CERIC DISULFATE □ CERIC SULFATE □ CERIC SULPHATE □ CERIUM DISULFATE □ CERIUM SULFATE □ CERIUM(4+) SULFATE □ SULFURIC ACID, CERIUM SALT (2:1)
CONSENSUS REPORTS: Reported in EPA TSCA Inventory.
SAFETY PROFILE: Experimental reproductive effects. When heated to decomposition it emits acrid smoke and irritating fumes.

CDC000   CAS: 7440-46-2   HR: 3
CESIUM
DOT: UN 1407
af: Cs   aw: 132.91

PROP: Bright, shiny, hexagonal crystals; silver-white, ductile metal; or possibly a silvery liquid. Golden when ultra pure. Spontaneously ignites in the atmosphere forming cesium oxides, carbonates and hydroxide. Mp: 28.5°, bp: 668°, d: 1.873, vap press: 1 mm @ 279°. Reacts violently with H₂O forming CsOH and dihydrogen.
SYN: CESIUM-133
CONSENSUS REPORTS: Reported in EPA TSCA Inventory.
DOT CLASSIFICATION: 4.3; Label: Dangerous When Wet
SAFETY PROFILE: Moderately toxic by intraperitoneal route. Cesium is quite similar to potassium in its elemental state. It has been shown, however, to have pronounced physiological action in experimentation with animals. Hyper-irritability, including marked spasms, has been shown to follow the administration of cesium in amounts equal to the potassium content of the diet. It has been found that replacing the potassium in the diet of rats with cesium caused death after 10–17 days. Ignites spontaneously in air. Violent reaction with water, moisture, or steam releases hydrogen gas which explodes. Violent reaction with acids, halogens, and other oxidizing materials. Incandescent reaction with nonmetals (e.g., sulfur, phosphorus). See also SODIUM.

CDC375   CAS: 61136-62-7   HR: 3
CESIUM ARSENATE
mf: AsO₄•3Cs   mw: 537.65
SYN: ARSENIC ACID, TRICESIUM SALT
CONSENSUS REPORTS: Arsenic and its compounds are on the Community Right-To-Know List.
OSHA PEL: OSHA: Cancer Hazard
SAFETY PROFILE: Poison by ingestion. Experimental teratogenic effects by inhalation. When heated to decomposition it emits toxic fumes of As. See also ARSENIC COMPOUNDS and CESIUM.

CDC500   CAS: 7787-69-1   HR: 2
CESIUM BROMIDE
mf: BrCs   mw: 212.82
PROP: Deliquescent colorless cubic crystals.
Mp: 636°, bp: 1300°. Very sol in water.
CONSENSUS REPORTS: Reported in EPA
TSCA Inventory.
SAFETY PROFILE: Moderately toxic by
intraperitoneal route. See also CESIUM and
BROMIDES. When heated to
decomposition it emits toxic fumes of Br⁻.

CDC750   CAS: 534-17-8   HR: 2
CESIUM CARBONATE
mf: $CO_3 \cdot 2Cs$ mw: 325.83
PROP: Deliquescent colorless monoclinic
crystals. Very sol in H_2O; sol in EtOH and
Et_2O.
SYNS: CARBONIC ACID, DICESIUM SALT □
DICESIUM CARBONATE
CONSENSUS REPORTS: EPA Genetic
Toxicology Program. Reported in EPA
TSCA Inventory.
SAFETY PROFILE: Moderately toxic by
ingestion. Mutation data reported. When
heated to decomposition it emits acrid
smoke and fumes. See also CESIUM.

CDD000   CAS: 7647-17-8   HR: 2
CESIUM CHLORIDE
mf: ClCs   mw: 168.36
PROP: Deliquescent cubic crystals.
Undergoes transition to high temp
polymorph at 4°. D: 3.99, mp: 646°, bp:
1209°. Very sol in H_2O, MeOH, and EtOH;
insol in Me_2CO.
SYNS: CESIUM MONOCHLORIDE □ DICESIUM
DICHLORIDE □ TRICESIUM TRICHLORIDE
CONSENSUS REPORTS: Reported in EPA
TSCA Inventory. EPA Genetic Toxicology
Program.
SAFETY PROFILE: Moderately toxic by
ingestion and intraperitoneal routes.
Experimental reproductive effects. Mutation
data reported. Reacts violently with BF_3. See
also CESIUM. When heated to
decomposition it emits toxic fumes of Cl⁻.

CDD500   CAS: 13400-13-0   HR: 3
CESIUM FLUORIDE
mf: CsF   mw: 151.91
PROP: Deliquescent colorless cubic crystals.
Mp: 703°, bp: 1251°. Very sol in H_2O and
MeOH; insol in Py.
SYNS: CESIUM MONOFLUORIDE □ DICESIUM
DIFLUORIDE □ TRICESIUM TRIFLUORIDE
CONSENSUS REPORTS: Reported in EPA
TSCA Inventory.
OSHA PEL: TWA 2.5 mg(F)m³
ACGIH TLV: TWA 2.5 mg(F)/m³; BEI: 3
mg/g creatinine of fluorides in urine prior to
shift; 10 mg/g creatinine of fluorides in
urine at end of shift.
NIOSH REL: (Inorganic Fluorides) TWA 2.5
mg(F)/m³
SAFETY PROFILE: A poison. Incompatible
with benzenediazonium tetrafluoroborate
and difluoroamine. When heated to
decomposition it emits toxic fumes of F⁻.

CDD750   CAS: 21351-79-1   HR: 3
CESIUM HYDROXIDE
DOT: UN 2681/UN 2682
mf: CsHO   mw: 149.92
PROP: Colorless to yellowish, very
deliquescent crystals. Undergoes transition
from orthorhombic to cubic at 2°. Mp:
315°, d: 3.675. Very sol in H_2O and EtOH.
SYNS: CAESIUM HYDROXIDE, solid (UN 2682) (DOT) □
CAESIUM HYDROXIDE, solution (UN 2681) (DOT) □
CESIUM HYDRATE □ CESIUM HYDROXIDE (ACGIH,
OSHA) □ CESIUM HYDROXIDE DIMER
CONSENSUS REPORTS: Reported in EPA
TSCA Inventory.
OSHA PEL: TWA 2 mg/m³
ACGIH TLV: TWA 2 mg/m³
DOT CLASSIFICATION: 8; Label: Corrosive
SAFETY PROFILE: Poison by
intraperitoneal route. Moderately toxic by
ingestion. A powerful caustic. A corrosive
skin and eye irritant. See also CESIUM.

CDE000   CAS: 7789-17-5   HR: 2
CESIUM IODIDE
mf: CsI   mw: 259.81

PROP: Deliquescent colorless orthorhombic crystals. Mp: 626°, bp: 1280°. Very sol in H_2O; sol in EtOH.
SYNS: CESIUM MONOIODIDE □ DICESIUM DIIODIDE □ TRICESIUM TRIIODIDE
CONSENSUS REPORTS: Reported in EPA TSCA Inventory.
ACGIH TLV: Proposed: (inhalable fraction) 0.1 mg/m³; Not Classifiable as a Human Carcinogen)
SAFETY PROFILE: Moderately toxic by ingestion and intraperitoneal routes. See also CESIUM and IODIDES. When heated to decomposition, it emits toxic fumes of I⁻.

CDE250   CAS: 7789-18-6   HR: 2
CESIUM(I) NITRATE (1:1)
DOT: UN 1451
mf: $NO_3 \cdot Cs$ mw: 194.92
PROP: Colorless, hexagonal or cubic, glittering crystalline powder. Undergoes hexagonal to cubic transition at 1°. Piezoelectric. Mp: 414°, bp: decomp, d: 3.685, 2.71 @ 500° (liq). Very sol in H_2O; sol in Me_2CO; sltly sol in EtOH.
SYNS: CESIUM NITRATE (DOT) □ NITRIC ACID, CESIUM SALT
CONSENSUS REPORTS: Reported in EPA TSCA Inventory. EPA Genetic Toxicology Program.
DOT CLASSIFICATION: 5.1; Label: Oxidizer
SAFETY PROFILE: Moderately toxic by ingestion and intraperitoneal routes. Mutation data reported. When heated to decomposition it emits toxic fumes of NO_x. See also CESIUM and NITRATES.

CDH000   CAS: 94-41-7   HR: 3
CHALCONE
mf: $C_{15}H_{12}O$ mw: 208.27
SYNS: 2-BENZALACETOPHENONE □ 1-BENZOYL-1-PHENYLETHENE □ β-BENZOYLSTYRENE □ 2-BENZYLIDENEACETOPHENONE □ CINNAMO-PHENONE □ 1,3-DIPHENYL-1-PROPEN-3-ONE □ 3-PHENYLACRYLOPHENONE □ β-PHENYL-ACRYLO-PHENONE □ 1-PHENYL-2-BENZOYLETHYLENE □ PHENYL STYRYL KETONE
CONSENSUS REPORTS: Reported in EPA TSCA Inventory.

DOT CLASSIFICATION: 3; Label: Flammable Liquid
SAFETY PROFILE: Poison by intravenous route. See also KETONES. When heated to decomposition it emits acrid smoke and irritating fumes.

CDH500   CAS: 8002-66-2   HR: 1
CHAMOMILE OIL
PROP: By steam distillation of the flowers and stalks of *Matrilaria chamomilla* L. (FCTXAV 12,807,74).
Blue–yellowish–brown liquid; strong odor and bitter aromatic taste. Composed of amyl and butyl esters of angelic, tiglic acids, and butyric acid. D: 0.905–0.915 @ 15°/15°. Sol in fixed oils, propylene glycol; insol in mineral oil, glycerin.
SYNS: BLUE CHAMOMILE OIL □ CAMOMILE OIL GERMAN □ CHAMOMILE-GERMAN OIL □ GERMAN CHAMOMILE OIL □ HUNGARIAN CHAMOMILE OIL □ KAMILLENOEL □ OILS, CHAMOMILE, GERMAN
CONSENSUS REPORTS: Reported in EPA TSCA Inventory.
SAFETY PROFILE: Low toxicity by ingestion and skin contact. A mild allergen. A skin irritant. See also ESTERS. When heated to decomposition it emits acrid and irritating fumes.

CDH750   CAS: 8015-92-7   HR: 1
CHAMOMILE OIL (ROMAN)
PROP: Obtained by the steam distillation of the dried flowers of *Anthemis nobilis* L. (FCTXAV 12,807,74). Blue liquid, turning brownish-yellow; strong aromatic odor. Composition: Amyl and butyl esters of angelic and tiglic acids, butyric acid, etc. D: 0.905–0.915 @ 15°/15°. Sol in fixed oils, mineral oil, propylene glycol; insol in glycerin.
SYN: CAMOMILE OIL, ENGLISH TYPE (FCC)
CONSENSUS REPORTS: Reported in EPA TSCA Inventory.
SAFETY PROFILE: Low toxicity by ingestion and skin contact. A mild allergen. A skin irritant. See also ESTERS. Combustible when heated. When heated to

decomposition it emits acrid smoke and irritating fumes.

**CDI000   CAS: 64365-11-3   HR: 1
CHARCOAL, ACTIVATED (DOT)**
DOT: NA 1361
af: C   aw: 12.01
PROP: Black porous solid, coarse granules or powder. Insol in water, org solvs.
SYNS: ACTIVATED CARBON □ CARBON, ACTIVATED □ CARBORAFFIN □ CARBORAFINE □ KARBORAFIN □ NUCHAR 722
CONSENSUS REPORTS: Reported in EPA TSCA Inventory.
DOT CLASSIFICATION: 4.2; Label: Spontaneously Combustible
SAFETY PROFILE: It can cause a dust irritation, particularly to the eyes and mucous membranes. Combustible when exposed to heat. Dust is flammable and explosive when exposed to heat or flame or oxides.

**CDI250   CAS: 16291-96-6   HR: 2
CHARCOAL (BRIQUETTES)**
DOT: NA 1361
af: C   aw: 12.01
PROP: Black amorphous solid.
Composition: carbon + impurities. Mp: >3500°, bp: 4200°, d: 3.51.
SYNS: CHARCOAL □ CHARCOAL SCREENINGS (DOT) □ CHARCOAL WOOD (DOT)
CONSENSUS REPORTS: Reported in EPA TSCA Inventory.
DOT CLASSIFICATION: 4.2; Label: Spontaneously Combustible
SAFETY PROFILE: Carbon itself has no toxic action, but it contains impurities that may be toxic. Fire hazard: reacts with liquid air, $Ba(ClO_3)_2$, BrF_5, ClO, $Ca(ClO_3)_2$, ClF_2, F_2, H_2O_2, $Mg(ClO_3)_2$, $(O_2 + wood)$, perchlorates, peroxides, $(P + air)$, K + $KClO_3$, KNO_3, RuO_4, $AgNO_3$, $NaClO_3$, $(AgCl + NaO_2)$, S, $(S + NaNO_3)$, $Zn(ClO_3)_2$. Heats spontaneously, particularly when wet, freshly calcined, or tightly packed, and it can ignite and burn. Slight explosion hazard when exposed to heat or flame. To fight fire, use water, mist, foam, or dry

chemical. When heated to decomposition it emits acrid smoke and fumes.

**CDJ000   HR: 3
CHARCOAL (SHELL)**
DOT: NA 1361
SYN: CHARCOAL, SHELL (DOT)
DOT CLASSIFICATION: 4.2; Label: Spontaneously Combustible
SAFETY PROFILE: A flammable solid. See also CHARCOAL (BRIQUETTES).

**CDM250   CAS: 1401-55-4   HR: 3
CHESTNUT TANNIN**
SYNS: CASTANEA SATIVA MILL TANNIN □ TANNIN from CHESTNUT
SAFETY PROFILE: Poison by subcutaneous, intramuscular, intravenous, and intraperitoneal routes. Questionable carcinogen with experimental tumorigenic data. See also TANNIN. When heated to decomposition it emits acrid and irritating fumes.

**CDN200   CAS: 78-95-5   HR: 3
CHLORACETONE**
DOT: UN 1695
mf: C_3H_5ClO mw: 92.53
PROP: Colorless, lachrymatory liquid with pungent odor. Mp: −44.5°, bp: 119°, d: 1.162.
SYNS: ACETONYL CHLORIDE □ A-STOFF □ CHLORACETONE □ CHLOROACETONE □ CHLORO-ACETONE, stabilized (DOT) □ CHLOROPROPANONE □ 1-CHLORO-2-PROPANONE □ MONOCHLORACETONE □ MONOCHLOROACETONE □ MONOCHLORO-ACETONE, inhibited (DOT) □ MONOCHLOROACETONE, stabilized (DOT) □ MONOCHLOROACETONE, unstabilized (DOT) □ TONITE
CONSENSUS REPORTS: Reported in EPA TSCA Inventory.
ACGIH TLV: CL 1 ppm (skin)
DOT CLASSIFICATION: 6.1; Label: Poison (UN 1695); DOT Class: Forbidden
SAFETY PROFILE: Poison by inhalation, ingestion, and skin contact. Mutation data reported. A lachrymator poison gas. See also CHLORINATED HYDROCARBONS, ALIPHATIC; ACETONE. Flammable when exposed to heat or flame, or oxidizers.

Old material can explode. When heated to decomposition it emits highly toxic fumes.

CDN500   CAS: 107-14-2   HR: 3
CHLORACETONITRILE
DOT: UN 2668
mf: C_2H_2ClN mw: 75.50
PROP: Fuming liquid. D: 1.193 @ 20°, bp: 126–127°.
SYNS: CHLOROACETONITRILE (DOT) □ α-CHLOROACETONITRILE □ 2-CHLOROACETONITRILE □ CHLOROMETHYL CYANIDE □ MONOCHLORO-ACETONITRILE □ MONOCHLOROMETHYL CYANIDE □ USAF KF-5
CONSENSUS REPORTS: Reported in EPA TSCA Inventory. Cyanide and its compounds are on the Community Right-To-Know List.
DOT CLASSIFICATION: 6.1; Label: Poison
SAFETY PROFILE: Poison by ingestion, skin contact, and intraperitoneal routes. Moderately toxic by inhalation. A skin irritant. Human mutation data reported. Questionable carcinogen with experimental tumorigenic data. Flammable liquid. See also NITRILES. When heated to decomposition it emits very toxic fumes of Cl⁻, NO_x, and CN⁻.

CDN505   CAS: 1341-24-8   HR: 3
CHLORACETOPHENONE
mf: C_8H_7ClO mw: 154.60
SYN: ETHANONE, 1-PHENYL-, MONOCHLORO DERIV.
SAFETY PROFILE: A poison by intravenous route. Low toxicity by inhalation. A mild eye irritant. When heated to decomposition it emits toxic vapors of Cl⁻.

CDN550   CAS: 75-87-6   HR: 3
CHLORAL
DOT: UN 2075
mf: C_2HCl_3O mw: 147.38
SYNS: ACETALDEHYDE, TRICHLORO-(9CI) □ ANHYDROUS CHLORAL □ CHLORAL, anhydrous, inhibited (DOT) □ CLORALIO □ GRASEX □ RCRA WASTE NUMBER U034 □ TRICHLOROACETALDEHYDE □ 2,2,2-TRICHLOROACETALDEHYDE □ TRICHLORO-ETHANAL

CONSENSUS REPORTS: Reported in EPA TSCA Inventory.
DOT CLASSIFICATION: 6.1; Label: Poison
SAFETY PROFILE: A poison. Mutation data reported.

CDO000   CAS: 302-17-0   HR: 3
CHLORAL HYDRATE
mf: $C_2HCl_3O \cdot H_2O$ mw: 165.40
PROP: Transparent, colorless crystals; aromatic, penetrating, sltly acrid odor and sltly bitter, caustic taste. Mp: 52°, bp: 97.5°, d: 1.9.
SYNS: AQUACHLORAL □ Bi 3411 □ CHLORALDURAT □ DORMAL □ FELSULES □ HYDRAL □ HYDRAL de CHLORAL □ KESSODRATE □ LORINAL □ NOCTEC □ NORTEC □ NYCOTON □ PHALDRONE □ RECTULES □ SK-CHLORAL HYDRATE □ SOMNI SED □ SOMNOS □ SONTEC □ TOSYL □ TRAWOTOX □ TRICHLORACETALDEHYD-HYDRAT (GERMAN) □ TRICHLOROACETALDEHYDE HYDRATE □ TRICHLOROACETALDEHYDE MONOHYDRATE □ 2,2,2-TRICHLORO-1,1-ETHANEDIOL
CONSENSUS REPORTS: Reported in EPA TSCA Inventory. EPA Genetic Toxicology Program.
SAFETY PROFILE: A human poison by ingestion and possibly other routes. Poison experimentally by ingestion, intravenous, and rectal routes. Moderately toxic by subcutaneous, parenteral, and intraperitoneal routes. Experimental reproductive effects. Human systemic effects by ingestion: general anesthetic, cardiac arrhythmias, blood pressure depression, eye effects, coma, pulse rate increase, arrhythmias. Human mutation data reported. Questionable carcinogen with experimental carcinogenic and tumorigenic data by skin contact. A sedative, anesthetic, and narcotic. Combustible when exposed to heat or flame. When heated to decomposition it emits toxic fumes of Cl⁻.

CDO250   CAS: 95-06-7   HR: 3
2-CHLORALLYL
   DIETHYLDITHIOCARBAMATE
mf: $C_8H_{14}ClNS_2$ mw: 223.80

PROP: Amber liquid or oil. Bp: 129° @ 1 mm, d: 1088 @ 25°. Very sltly sol in H_2O; sol in most org solvs.
SYNS: CDEC □ CHLORALLYL DIETHYLDITHIO-CARBAMATE □ 2-CHLOROALLYL DIETHYLDITHIO-CARBAMATE □ 2-CHLOROALLYL-N,N-DIETHYLDITHIOCARBAMATE □ 2-CHLORO-2-PROPENE-1-THIOL DIETHYLDITHIOCARBAMATE □ 2-CHLORO-2-PROPENYL DIETHYLCARBAMO-DITHIOATE □ CP 4572 □ DIETHYLCARBAMO-DITHIOIC ACID 2-CHLORO-2-PROPENYL ESTER □ DIETHYLDITHIOCARBAMIC ACID-2-CHLOROALLYL ESTER □ NCI-C00453 □ SULFALLATE □ THIOALLATE □ VEGADEX □ VEGADEX SUPER
CONSENSUS REPORTS: NTP 10th Report on Carcinogens. IARC Cancer Review: Group 2B IMEMDT 7,56,87; Animal Sufficient Evidence IMEMDT 30,283,83. NCI Carcinogenesis Bioassay (feed); Clear Evidence: mouse, rat NCITR* NCI-CG-TR-115,78. EPA Genetic Toxicology Program.
SAFETY PROFILE: Confirmed carcinogen with experimental carcinogenic data. Moderately toxic by ingestion and skin contact. Mutation data reported. An herbicide. When heated to decomposition it emits very toxic fumes of Cl⁻, NO_x, and SO_x. See also ALLYL COMPOUNDS, CARBAMATES, and ESTERS.

CDO500   CAS: 305-03-3   HR: 3
CHLORAMBUCIL
mf: $C_{14}H_{19}Cl_2NO_2$ mw: 304.24
PROP: Flattened needles from pet ether. Mp: 64–66°. Sol in Et_2O.
SYNS: AMBOCHLORIN □ AMBOCLORIN □ 4-(BIS(2-CHLOROETHYL)AMINO)BENZENEBUTANOIC ACID □ γ-(p-BIS(2-CHLOROETHYL)AMINOPHENYL)BUTYRIC ACID □ 4-(p-BIS(β-CHLOROETHYL)AMINOPHENYL)-BUTYRIC ACID □ 4-(p-(BIS(2-CHLOROETHYL)AMINO)-PHENYL)BUTYRIC ACID □ CB 1348 □ CHLORAMINO-PHEN □ CHLORAMINOPHENE □ CHLOROAMBUCIL □ CHLOROBUTIN □ CHLOROBUTINE □ N,N-DI-2-CHLOROETHYL-γ-p-AMINOPHENYLBUTYRIC ACID □ p-N,N-DI-(β-CHLOROETHYL)AMINOPHENYL BUTYRIC ACID □ p-(N,N-DI-2-CHLOROETHYL)AMINOPHENYL BUTYRIC ACID □ γ-(p-DI(2-CHLOROETHYL)AMINO-PHENYL)BUTYRIC ACID □ ECLORIL □ ELCORIL □ LEUKERAN □ LEUKERSAN □ LEUKORAN □ LINFOLIZIN □ LINFOLYSIN □ NCI-C03485 □ NSC-3088

□ PHENYLBUTYRIC ACID NITROGEN MUSTARD □ RCRA WASTE NUMBER U035
CONSENSUS REPORTS: NTP 10th Report on Carcinogens. IARC Cancer Review: Group 1 IMEMDT 7,144,87; Human Inadequate Evidence IMEMDT 9,125,75; Human Limited Evidence IMEMDT 26,115,81; Animal Limited Evidence IMEMDT 26,115,81; Animal Sufficient Evidence IMEMDT 9,125,75. EPA Genetic Toxicology Program.
SAFETY PROFILE: Confirmed carcinogen producing leukemia. Experimental carcinogenic and neoplastigenic data. Poison by ingestion, intravenous, intraperitoneal, and subcutaneous routes. Human systemic effects by ingestion: convulsions, cough, dyspnea, and interstitial fibrosis. Human reproductive effects by ingestion and possibly other routes: changes in spermatogenesis, menstrual cycle changes or disorders, and teratogenic effects of the fetal urogenital system. Experimental teratogenic and reproductive effects. Human mutation data reported. An anti-neoplastic agent. When heated to decomposition it emits very toxic fumes of Cl⁻ and NO_x.

CDP000   CAS: 127-65-1   HR: 3
CHLORAMINE T
mf: $C_7H_8ClNO_2S•Na$ mw: 228.66
PROP: Faintly yellow crystals or powder (as Na salt). Mp: 167–170° (as Na salt). Mod sol in H_2O; insol in C_6H_6, $CHCl_3$; decomp in EtOH (Na salt).
SYNS: ACTI-CHLORE □ AKTIVIN □ ANEXOL □ BENZENESULFONAMIDE, N-CHLORO-4-METHYL-, SODIUM SALT (9CI) □ BERKENDYL □ CHLORALONE □ CHLORASAN □ CHLORASEPTINE □ CHLORAZAN □ CHLORAZENE □ CHLORAZONE □ CHLOROZONE □ CHLORSEPTOL □ CLORINA □ CLOROSAN □ DESINFECT □ EUCLORINA □ GANSIL □ GYNE-CLORINA □ HALAMID □ HELIOGEN □ KLORAMIN □ KLORAMINE-T □ MULTICHLOR □ SODIUM CHLORAMINE T □ SODIUM p-TOLUENESULFONYL-CHLORAMIDE □ SODIUM TOSYLCHLORAMIDE □ TAMPULES □ TOCHLORINE □ TOLAMINE □ TOSYLCHLORAMIDE SODIUM
CONSENSUS REPORTS: Reported in EPA TSCA Inventory.

SAFETY PROFILE: Poison by parenteral and intravenous routes. Human mutagenic data reported. When heated to decomposition it emits toxic fumes of Cl⁻, SO$_x$, Na$_2$O, and NO$_x$. See also SULFONATES and CHLORIDES.

CDP250   CAS: 56-75-7   HR: 3
CHLORAMPHENICOL
mf: C$_{11}$H$_{12}$Cl$_2$N$_2$O$_5$ mw: 323.15
PROP: Pale-yellow needles (H$_2$O or 1,2-dichloroethane) or crystals. Mp: 151°. Sltly sol in water. Sol in EtOH, EtOAc, and Me$_2$CO; insol in C$_6$H$_6$ and pet ether.
SYNS: ALFICETYN □ AMBOFEN □ AMPHENICOL □ AMPHICOL □ AMSECLOR □ ANACETIN □ AQUAMYCETIN □ AUSTRACIL □ AUSTRACOL □ BIOCETIN □ BIOPHENICOL □ CAF □ CAM □ CAP □ CATILAN □ CHEMICETIN □ CHEMICETINA □ CHLOMIN □ CHLOMYCOL □ CHLORAMEX □ CHLORAMFICIN □ CHLORAMFILIN □ d-CHLOR-AMPHENICOL □ d-threo-CHLORAMPHENICOL □ CHLORAMSAAR □ CHLORASOL □ CHLORA-TABS □ CHLORICOL □ CHLORNITROMYCIN □ CHLOROCAPS □ CHLOROCID □ CHLOROCIDIN C TETRAN □ CHLOROCOL □ CHLOROJECT L □ CHLOROMAX □ CHLOROMYCETIN □ CHLORONITRIN □ CHLOROPTIC □ CHLOROVULES □ CIDOCETINE □ CIPLAMYCETIN □ CLORAMIDINA □ CLOROAMFEN-ICOLO (ITALIAN) □ CLOROMISAN □ CLOROSINTEX □ COMYCETIN □ CPH □ CYLPHENICOL □ DESPHEN □ DETREOMYCINE □ DEXTROMYCETIN □ d-(−)-threo-2-DICHLOROACETAMIDO-1-p-NITROPHENYL-1,3-PROPANEDIOL □ d-threo-N-DICHLOROACETYL-1-p-NITROPHENYL-2-AMINO-1,3-PROPANEDIOL □ d-(−)-2,2-DICHLORO-N-(β-HYDROXY-α-(HYDROXYMETHYL)-p-NITROPHENYLETHYL)ACETAMIDE □ d-(−)-threo-2,2-DICHLORO-N-(β-HYDROXY-α-(HYDROXYMETHYL))-p-NITROPHENETHYLACETAMIDE □ d-threo-N-(1,1'-DIHYDROXY-1-p-NITROPHENYLISOPROPYL)-DICHLOROACETAMIDE □ DOCTAMICINA □ ECONOCHLOR □ EMBACETIN □ EMETREN □ ENICOL □ ENTEROMYCETIN □ ERBAPLAST □ ERTILEN □ FARMICETINA □ FENICOL □ GLOBENIC-OL □ GLOROUS □ HALOMYCETIN □ HORTFENICOL □ I 337A □ INTRAMYCETIN □ ISMICETINA □ ISOPHENICOL □ ISOPTO FENICOL □ KAMAVER □ KEMICETINE □ LEUKOMYAN □ LEVOMYCETIN □ LOROMISIN □ MASTIPHEN □ MEDIAMYCETINE □ MICOCHLORINE □ MICROCETINA □ MYCHEL □ MYCINOL □ NCI-C55709 □ d-(−)-threo-1-p-NITROPHEN-YL-2-DICHLORACETAMIDO-1,3-PROPANEDIOL □ d-threo-1-(p-NITROPHENYL)-2-(DICHLOROACETYL-AMINO)-1,3-PROPANEDIOL □ NORIMYCIN V □

NOVOCHLOROCAP □ NOVOMYCETIN □ NOVOPHENICOL □ NSC-3069 □ OFTALENT □ OLEOMYCETIN □ OPTHOCHLOR □ OTOPHEN □ PANTOVERNIL □ PARAXIN □ PETNAMYCETIN □ QUEMICETINA □ RIVOMYCIN □ ROMPHENIL □ SEPTICOL □ SINTOMICETINA □ STANOMYCETIN □ SYNTHOMYCINE □ TEVCOCIN □ TIFOMYCINE □ TREOMICETINA □ U-6062 □ UNIMYCETIN □ VETICOL
CONSENSUS REPORTS: NTP 10th Report on Carcinogens. IARC Cancer Review: Group 2B IMEMDT 7,145,87; Human Limited Evidence IMEMDT 10,85,76. Reported in EPA TSCA Inventory. EPA Genetic Toxicology Program.
SAFETY PROFILE: Confirmed human carcinogen producing leukemia, aplastic anemia, and other bone marrow changes. Experimental tumorigenic data. Poison by intravenous and subcutaneous routes. Moderately toxic by ingestion and intraperitoneal routes. Human systemic effects by an unknown route: changes in plasma or blood volume, unspecified liver effects, and hemorrhaging. Experimental teratogenic and reproductive effects. Human mutation data reported. An antibiotic. When heated to decomposition it emits very toxic fumes of NO$_x$ and Cl⁻. See also other chloramphenicol entries.

CDP700   CAS: 530-43-8   HR: 2
CHLORAMPHENICOL PALMITATE
mf: C$_{27}$H$_{42}$Cl$_2$N$_2$O$_6$ mw: 561.61
PROP: Crystals from C$_6$H$_6$. Mp: 90°.
SYNS: CAP-P □ CAP-PALMITATE □ CHLORPHENICOL MONOPALMITATE □ DETREOPAL □ α-ESTER PALMITIC ACID with D-threo-(−)-2,2-DICHLORO-N-(β-HYDROXY-α-(HYDROXY-METHYL)-p-NITROPHENETHYL)ACETAMIDE
SAFETY PROFILE: Moderately toxic by oral route. An experimental teratogen. Other experimental reproductive effects. An antibiotic. When heated to decomposition it emits very toxic fumes of NO$_x$ and Cl⁻. See also other chloramphenicol entries.

CDQ000   HR: 3
CHLORATES
PROP: Chlorates are a combination of a metal or hydrogen and $^-ClO_3$ monovalent radical. They are crystalline and somewhat deliquescent.

SAFETY PROFILE: The principal toxic effects of chlorates are the production of methemoglobin in the blood and destruction of red blood corpuscles. The latter may lead to irritation of the kidneys. Damage to heart muscle has been reported.

Dangerous fire hazard in contact with flammable matter. When contaminated with oxidizable materials, they are particularly sensitive to friction, heat, and shock. They are powerful oxidizing agents and can undergo violent reactions with reducing materials. Dangerous explosion hazard when shocked, exposed to heat, or rubbed, particularly when contaminated with sugar, charcoal, shellac, sulfur, starch, sawdust, sulfuric acid, ammonium compounds, cyanides, phosphorous or antimony sulfide, Al, (metals + acids), As_2S_3, CaH_2, MnO_2, metal sulfides, organic acids, powdered metals, Hg_3P_4, PHI_4, SCN, (S + Cu), Se, NaH_2PO_2, SrH, SO_2. Chlorates when mixed with combustible materials may form explosive mixtures. For instance, potassium chlorate, when mixed with sulfur or with other combustible substances explodes on friction. Pure chlorates which have been spilled on the floor, or mixed with small amounts of impurities, become very sensitive to shock and friction. Water is considered the best agent for fighting fires involving chlorates. When heated to decomposition they can emit toxic fumes of Cl^- and explode.

CDR250   CAS: 14362-31-3   HR: 3
CHLORCYCLIZINE HYDROCHLORIDE
mf: $C_{18}H_{21}ClN_2 \cdot xClH$ mw: 556.08

SYNS: AH-289 HYDROCHLORIDE □ CHLORCYCLIZINIUM CHLORIDE □ 1-(p-CHLOROBENZHYDRYL)-4-METHYLPIPERAZINE HYDROCHLORIDE □ DIPARALENE HYDROCHLORIDE □ ERAMIDE

SAFETY PROFILE: Poison by ingestion, subcutaneous, intravenous, and intraperitoneal routes. An experimental teratogen. Other experimental reproductive effects. When heated to decomposition it emits very toxic fumes of HCl and NO_x.

CDR575   CAS: 5566-34-7   HR: 3
trans-CHLORDAN
mf: $C_{10}H_6Cl_8$ mw: 409.76

SYNS: γ-CHLORDAN □ γ(trans)-CHLORDANE □ 4,7-METHANOINDAN, 2,2,4,5,6,7,8,8-OCTACHLORO-3a,4,7,7a-TETRAHYDRO- □ 4,7-METHANO-1H-INDENE, 2,2,4,5,6,7,8,8-OCTACHLORO-2,3,3a,4,7,7a-HEXAHYDRO-(9CI) □ 2,2,4,5,6,7,8,8-OCTACHLORO-3a,4,7,7a-TETRA-HYDRO-4,7-METHANOINDAN

CONSENSUS REPORTS: IARC Cancer Review: Group 2B IMEMDT 53,115,91; Animal Sufficient Evidence IMEMDT 53,115,91; Human Inadequate Evidence IMEMDT 53,115,91.

SAFETY PROFILE: Confirmed carcinogen. Moderately toxic by ingestion. Experimental reproductive effects. When heated to decomposition it emits toxic fumes of Cl^-.

CDR750   CAS: 57-74-9   HR: 3
CHLORDANE
mf: $C_{10}H_6Cl_8$ mw: 409.76

PROP: Colorless to amber; odorless, viscous liquid. Bp: 175°, d: 1.57–1.63 @ 15.5°/15.5°. IDLH 100 mg/m^3.

SYNS: ASPON-CHLORDANE □ BELT □ CD 68 □ CHLOORDAAN (DUTCH) □ CHLORDAN □ γ-CHLO-RDAN □ CHLORDANE, liquid (DOT) □ CHLORINDAN □ CHLOR KIL □ CHLORODANE □ CHLORTOX □ CLORDAN (ITALIAN) □ CLORODANE □ CORTILAN-NEU □ DICHLOROCHLORDENE □ DOWCHLOR □ ENT 9,932 □ ENT 25,552-X □ HCS 3260 □ KYPCHLOR □ M 140 □ M 410 □ NCI-C00099 □ NIRAN □ 1,2,4,5,6,7,8,8-OCTACHLOOR-3a,4,7,7a-TETRAHYDRO-4,7-endo-METHANO-INDAAN (DUTCH) □ OCTACHLOR □ OCTACHLORODIHYDRODICYCLOPENTADIENE □ 1,2,4,5,6,7,8,8-OCTACHLORO-2,3,3a,4,7,7a-HEXAHYDRO-4,7-METHANOINDENE □ 1,2,4,5,6,7,8,8-OCTACHLORO-2,3,3a,4,7,7a-HEXAHYDRO-4,7-METHANO-1H-INDENE □ 1,2,4,5,6,7,8,8-OCTACHLORO-3a,4,7,7a-HEXAHYDRO-4,7-METHYLENE INDANE □ OCTACHLORO-4,7-METHANOHYDROINDANE □ OCTACHLORO-4,7-METHANOTETRAHYDROINDANE □ 1,2,4,5,6,7,8,8-OCTACHLORO-4,7-METHANO-3a,4,7,7a-TETRAHYDRO-INDANE □ 1,2,4,5,6,7,8,8-OCTACHLORO-3a,4,7,7a-TETRAHYDRO-4,7-METHANOINDAN □ 1,2,4,5,6,7,8,8-

OCTACHLORO-3a,4,7,7a-TETRAHYDRO-4,7-
METHANOINDANE ☐ 1,2,4,5,6,7,10,10-OCTACHLORO-
4,7,8,9-TETRAHYDRO-4,7-METHYLENEINDANE ☐
1,2,4,5,6,7,8,8-OCTACHLOR-3a,4,7,7a-TETRAHYDRO-4,7-
endo-METHANO-INDAN (GERMAN) ☐ OCTA-KLOR ☐
OKTATERR ☐ ORTHO-KLOR ☐ 1,2,4,5,6,7,8,8-
OTTOCHLORO-3A,4,7,7A-TETRAIDRO-4,7-endo-
METANO-INDANO (ITALIAN) ☐ RCRA WASTE
NUMBER U036 ☐ SD 5532 ☐ SHELL SD-5532 ☐
SYNKLOR ☐ TAT CHLOR 4 ☐ TOPICHLOR 20 ☐
TOPICLOR ☐ TOPICLOR 20 ☐ TOXICHLOR ☐
VELSICOL 1068

CONSENSUS REPORTS: IARC Cancer
Review: Group 2B IMEMDT 53,115,91;
Animal Sufficient Evidence IMEMDT
20,45,79; Animal Limited Evidence IARC
Monographs, Supplement. IMEMDT
7,146,87; Animal Sufficient Evidence
IMEMDT 53,115,91; Human Inadequate
Evidence IMEMDT 20,45,79; Human
Inadequate Evidence IMEMDT 53,115,91.
NCI Carcinogenesis Bioassay (feed); Clear
Evidence: mouse NCITR* NCI-CG-TR-
8,77; No Evidence: rat NCITR* NCI-CG-
TR-8,77. EPA Genetic Toxicology Program.
Community Right-To-Know List. EPA
Extremely Hazardous Substances List.

OSHA PEL: TWA 0.5 mg/m³ (skin)
ACGIH TLV: TWA 0.5 mg/m³ (skin);
Animal Carcinogen
DFG MAK: 0.5 mg/m³; Confirmed Animal
Carcinogen with Unknown Relevance to
Humans

SAFETY PROFILE: Confirmed carcinogen
with experimental carcinogenic data. Poison
to humans by ingestion and possibly other
routes. An experimental poison by ingestion,
inhalation, intravenous, and intraperitoneal
routes. Moderately toxic by skin contact.
Human systemic effects by ingestion or skin
contact: tremors, convulsions, excitement,
ataxia (loss of muscle coordination), and
gastritis. Experimental teratogenic and
reproductive effects. Human mutation data
reported. Combustible liquid. It is no longer
permitted for use as a termiticide in homes.

A central nervous system stimulant whose
exact mode of action is unknown, but it may
involve microsomal enzyme stimulation.
Animals poisoned by this and related
compounds show an extremely marked loss
of appetite and neurological symptoms. The
fatal dose to humans is unknown. It has
been estimated to be between 6 and 60 g
(0.2 and 2 ounces). One person receiving
an accidental skin application of 25%
solution (amounting to something over 30
g of technical chlordane) developed
symptoms within about 40 minutes and
died, apparently of respiratory failure, before
medical attention was obtained. In two
patients, death followed exposure to low
ingestion doses of chlordane (2–4 g). On
microscopic examination, both patients
showed severe chronic fatty degeneration of
the liver, characteristic of chronic
alcoholism. Although these two fatalities
cannot be attributed exclusively to
chlordane, they are entirely consistent with
previous observations that the toxicity of
other chlorinated hydrocarbons is much
enhanced in the presence of chronic liver
damage. The dangerous chronic dose in
humans is unknown.

Experimental animals exposed to repeated
small doses exhibit hyperexcitability,
tremors, and convulsions, and those that
survive long enough show marked anorexia
and loss of weight. Symptoms in animals
frequently occur within an hour of the
administration of a large dose, but death
often is delayed for several days depending
on the dosage and route of administration.
In any event, symptoms are of longer
duration with chlordane than with DDT
under similar conditions.

Laboratory analyses on poisoned animals
are essentially normal, except that the
insecticide is found in tissues by means of
bioassay. A method for specific, quantitative
chemical analysis for chlordane is now
available using small amounts of
subcutaneous fat. Chronically poisoned
animals show degenerative changes in the
liver and kidney tubules.

When heated to decomposition chlordane
emits toxic fumes of Cl⁻.

CDR760   CAS: 12789-03-6   HR: 3
CHLORDANE
SYNS: CHLORDANE, TECHNICAL □ 4,7-
METHANOINDEN, 1,2,4,5,6,7,8,9-OCTACHLORO-
3A,4,7,7A-TETRAHYDRO- □ RCRA WASTE NUMBER
U036
CONSENSUS REPORTS: IARC Cancer
Review: Group 2B IMEMDT 53,115,91;
Animal Sufficient Evidence IMEMDT
53,115,91; Human Inadequate Evidence
IMEMDT 53,115,91.
SAFETY PROFILE: Suspected carcinogen.
A poison by ingestion. Moderately toxic by
skin contact. Experimental reproductive
effects. Mutation data reported.

CDS000   CAS: 115-28-6   HR: 3
CHLORENDIC ACID
mf: $C_9H_4Cl_6O_4$ mw: 388.83
SYNS: HET ACID □ 1,4,5,6,7,7-HEXACHLORO-5-
NORBORNENE-2,3-DICARBOXYLIC ACID □ KYSELINA
3,6-ENDOMETHYLEN-3,4,5,6,7,7-HEXACHLOR-Δ^4-
TETRAHYDROFTALOVA (CZECH) □ KYSELINA HET
(CZECH) □ NCI-C55072
CONSENSUS REPORTS: NTP 10th Report
on Carcinogens. NTP Carcinogenesis
Studies (feed); Clear Evidence: mouse, rat
NTPTR* NTP-TR-304,87. Reported in
EPA TSCA Inventory.
SAFETY PROFILE: Confirmed carcinogen
with experimental carcinogenic data. A
severe eye and mild skin irritant. When
heated to decomposition it emits toxic
fumes of Cl⁻.

CDS125   CAS: 16672-87-0   HR: 2
CHLORETHEPHON
mf: $C_2H_6ClO_3P$ mw: 144.50
PROP: Very hygroscopic needles from
benzene. Mp: 74–75°. Freely sol in water,
methanol, acetone, ethylene glycol,
propylene glycol; sltly sol in benzene,
toluene; practically insol in pet ether.
SYNS: AMCHEM 68-250 □ BROMOFLOR □
CAMPOSAN □ CEP □ 2-CEPA □ CEPHA □ CEPHA 10LS
□ 2-CHLORAETHYL-PHOSPHONSAEURE (GERMAN) □
2-CHLORETHYLPHOSPHONIC ACID □ 2-
CHLOROETHANEPHOSPHONIC ACID □ ETHEFON □
ETHEL □ ETHEPHON □ ETHEVERSE □ ETHREL □
FLORDIMEX □ FLOREL □ G 996 □ KAMPOSAN □
ROLL-FRUCT □ TOMATHREL

CONSENSUS REPORTS: EPA Genetic
Toxicology Program.
SAFETY PROFILE: Moderately toxic by
ingestion. Mildly toxic by skin contact. A
plant growth regulator. Caution: Spray
formulations are quite acidic, about pH 1.0.
May be irritating to exposed skin and eyes,
or if inhaled. When heated to
decomposition it emits toxic fumes of Cl⁻
and PO_x.

CDS750   CAS: 470-90-6   HR: 3
CHLORFENVINFOS
mf: $C_{12}H_{14}Cl_3O_4P$ mw: 359.58
PROP: Amber liquid. Mp: −23°, bp:
124–126° @ 0.008 mm.
SYNS: APACHLOR □ BIRLANE □ C-10015 □ CFV □
CGA 26351 □ CHLOFENVINPHOS □ O-2-CHLOOR-1-
(2,4-DICHLOOR-FENYL)-VINYL-O,O-DIETHYLFOSFAAT
(DUTCH) □ O-2-CHLOR-1-(2,4-DICHLOR-PHENYL)-
VINYL-O,O-DIAETHYLPHOSPHAT (GERMAN) □
CHLORFENVINFOS □ CHLORFENVINPHOS □ 2-
CHLORO-1-(2,4-DICHLOROPHENYL)VINYL DIETHYL
PHOSPHATE □ β-2-CHLORO-1-(2',4'-DICHLOROPHEN-
YL) VINYL DIETHYLPHOSPHATE □ CHLOROFEN-
VINPHOS □ CHLORPHENVINFOS □ CHLORPHEN-
VINPHOS □ O-2-CLORO-1-(2,4-DICLORO-FENIL)-
VINYL-O,O-DIETILFOSFATO (ITALIAN) □ COMPOUND
4072 □ CVP □ DERMATON □ O,O-DIAETHYL-O-1-(4,5-
DICHLORPHENYL)-2-CHLOR-VINYL-PHOSPHAT
(GERMAN) □ 2,4-DICHLORO-α-
(CHLOROMETHYLENE)BENZYL ALCOHOL DIETHYL
PHOSPHATE □ O,O-DIETHYL-O-(2-CHLORO-1-(2',4'-
DICHLOROPHENYL)VINYL) PHOSPHATE □ ENT 24,969
□ GC 4072 □ OMS 1328 □ PHOSPHATE de O,O-
DIETHYLE et de O-2-CHLORO-1-(2,4-DICHLORO-
PHENYL) VINYLE (FRENCH) □ SAPECRON □ SHELL
4072 □ STELADONE □ SUPONA □ SUPONE □ UNITOX
□ VINYLPHATE
CONSENSUS REPORTS: EPA Extremely
Hazardous Substances List.
SAFETY PROFILE: Poison by ingestion,
skin contact, intraperitoneal, subcutaneous,
and intravenous routes. Human systemic
effects by skin contact: unspecified blood
system effects. Mutation data reported. A
cholinesterase inhibitor. An insecticide. See
also PARATHION. When heated to
decomposition it emits very toxic fumes of
Cl⁻ and PO_x.

CDT750   CAS: 96-24-2   HR: 3
CHLORHYDRIN
DOT: UN 2689
mf: $C_3H_7ClO_2$ mw: 110.55
PROP: Colorless liquid. Bp: 213° decomp, d: 1.326.
SYNS: α-CHLORHYDRIN □ CHLORODEOXY-GLYCEROL □ 1-CHLORO-2,3-DIHYDROXYPROPANE □ 3-CHLORO-1,2-DIHYDROXYPROPANE □ α-CHLOROHYDRIN □ 1-CHLOROPROPANE-2,3-DIOL □ 1-CHLORO-2,3-PROPANEDIOL □ 3-CHLOROPROPANE-1,2-DIOL □ 3-CHLORO-1,2-PROPANEDIOL □ 3-CHLOROPROPYLENE GYLCOL □ β,β'-DIHYDROXY-ISOPROPYL CHLORIDE □ 2,3-DIHYDROXYPROPYL CHLORIDE □ EPIBLOC □ GLYCERIN-α-MONOCHLOR-HYDRIN □ GLYCEROL CHLOROHYDRIN □ GLYCEROL-α-CHLOROHYDRIN □ GLYCEROL-α-MONOCHLOROHYDRIN (DOT) □ GLYCERYL-α-CHLOROHYDRIN □ MONOCHLORHYDRIN □ MONOCHLOROHYDRIN □ α-MONOCHLOROHYDRIN □ U-5897
CONSENSUS REPORTS: Reported in EPA TSCA Inventory. EPA Genetic Toxicology Program.
DOT CLASSIFICATION: 6.1; Label: KEEP AWAY FROM FOOD
SAFETY PROFILE: Poison by ingestion and intraperitoneal routes. Moderately toxic by inhalation. Experimental reproductive effects. A severe eye irritant. Questionable carcinogen with experimental tumorigenic data. Mutation data reported. A chemosterilant for rodents. Combustible when exposed to heat or flame. Reaction with perchloric acid forms a sensitive explosive product more powerful than glyceryl nitrate. When heated to decomposition it emits toxic fumes of Cl^-.

CDU000   CAS: 7790-93-4   HR: 3
CHLORIC ACID
DOT: UN 2626
mf: $ClHO_3$ mw: 84.46
PROP: Colorless solution. Fairly stable in cold H_2O up to 30%. Strong oxidant, stable as alkali metal salts. Mp: $<-20°$, bp: decomp @ 40°, d: 1.282 @ 14.2°.
SYN: CHLORIC ACID, solution, containing not more than 10% acid (DOT)

CONSENSUS REPORTS: Reported in EPA TSCA Inventory.
DOT CLASSIFICATION: 5.1; Label: Oxidizer
SAFETY PROFILE: A poison. A strong irritant by ingestion and inhalation. Dangerous fire hazard; ignites organic matter upon contact. A very powerful oxidizing agent. Violent or explosive reaction with oxidizable materials. Aqueous solutions decompose explosively during evaporation. Solutions greater than 40% are unstable. Reacts violently with NH_3, Sb, Sb_2S_3, As_2S_3, Bi, CuS, PHI_4, SnS_2, SnS. Reaction with cellulose causes ignition after a delay period. Dangerous reaction with metal sulfides and metal chlorides (e.g., incandescent reaction with antimony trisulfide, arsenic trisulfide, tin(II)sulfide, tin(IV) sulfide, explosion on contact with copper sulfide). Reaction with metals (e.g., antimony, bismuth, iron) forms explosive products. When heated to decomposition it emits toxic fumes of Cl^-. See also CHLORATES and CHLORINE.

CDU250   HR: D
CHLORIDES
SAFETY PROFILE: Varies widely. Sodium chloride (table salt) has very low toxicity, while carbonyl chloride (phosgene) is lethal in small doses. Therefore, see specific entries. When heated to decomposition or on contact with acids or acid fumes, they evolve highly toxic chloride fumes. Some organic chlorides decompose to yield phosgene.

CDV100   CAS: 8001-35-2   HR: 3
CHLORINATED CAMPHENE
mf: $C_{10}H_{10}Cl_8$ mw: 413.80
PROP: Yellow, waxy solid; pleasant piney odor. Mp: 65–90°. Almost insol in water; very sol in aromatic hydrocarbons. IDLH 200 mg/m³.
SYNS: AGRICIDE MAGGOT KILLER (F) □ ALLTEX □ ALLTOX □ ATTAC 6 □ ATTAC 6-3 □ CAMPHECHLOR □ CAMPHOCHLOR □ CAMPHOCLOR □ CAMPHOFENE HUILEUX □ CHEM-PHENE □ CHLOROCAMPHENE □ CLOR CHEM T-590 □ COMPOUND 3956 □ CRESTOXO

CRISTOXO 90 □ ENT 9,735 □ ESTONOX □ FASCO-TERPENE □ GENIPHENE □ GY-PHENE □ HERCULES 3956 □ HERCULES TOXAPHENE □ KAMFOCHLOR □ M 5055 □ MELIPAX □ MOTOX □ NCI-C00259 □ OCTACHLOROCAMPHENE □ PCC □ PENPHENE □ PHENACIDE □ PHENATOX □ POLYCHLORCAMPH-ENE □ POLYCHLORINATED CAMPHENES □ POLYCHLOROCAMPHENE □ RCRA WASTE NUMBER P123 □ STROBANE-T-90 □ SYNTHETIC 3956 □ TOXADUST □ TOXAFEEN (DUTCH) □ TOXAKIL □ TOXAPHEN (GERMAN) □ TOXAPHENE □ TOXON 63 □ TOXYPHEN □ VERTAC 90% □ VERTAC TOXAPHENE 90

CONSENSUS REPORTS: NTP 10th Report on Carcinogens. IARC Cancer Review: Group 2B IMEMDT 7,56,87; Human Limited Evidence IMEMDT 20,327,79; Animal Sufficient Evidence IMEMDT 20,327,79. NCI Carcinogenesis Bioassay (feed); Clear Evidence: mouse, rat NCITR* NCI-CG-TR-37,79. EPA Extremely Hazardous Substances List.

OSHA PEL: TWA 0.5 mg/m³; STEL 1 mg/m³ (skin)

ACGIH TLV: TWA 0.5 mg/m³; STEL 1 mg/m³ (skin); Animal Carcinogen

DFG MAK: Animal Carcinogen, Suspected Human Carcinogen

SAFETY PROFILE: Confirmed carcinogen with experimental carcinogenic and tumorigenic data. Human poison by ingestion and possibly other routes. Experimental poison by ingestion, intraperitoneal, and possibly other routes. Moderately toxic experimentally by inhalation and skin contact. Human systemic effects by ingestion and skin contact: somnolence, convulsions or effect on seizure threshold, coma, and allergic skin dermatitis. A skin irritant; absorbed through the skin. Experimental teratogenic and reproductive effects. Human mutation data reported. Liver injury has been reported. Lethal amounts of toxaphene can enter the body through the mouth, lungs, and skin. Systemic absorption of the insecticide is increased by the presence of digestible oils, and liquid preparations of the insecticide, which penetrate the skin more readily than do dusts and wettable powders.

A toxic mixture of organochlorine pesticides stored to some extent in body fat. It resembles chlordane and, to some extent, camphor in its physiological action. It causes diffuse stimulation of the brain and spinal cord resulting in generalized convulsions of a tonic or clonic character. Death usually results from respiratory failure. Detoxification appears to occur in the liver. The lethal ingestion dose for humans is estimated to be 2–7 g, a toxicity of about four times that of DDT. At least seven human deaths have been reported due to toxaphene, all in children. Two families have been made ill by eating vegetables containing a large residue of toxaphene. When heated to decomposition it emits toxic fumes of Cl⁻.

CDV175   CAS: 31242-93-0   HR: 2
CHLORINATED DIPHENYL OXIDE

mf: $C_{12}H_4Cl_6O$ mw: 376.86

PROP: Light-yellow, very viscous liquid. Bp: 230–260° @ 8 mm, d: 1.60 @ 20°/60°, autoign temp: 1148°F, vap d: 13.0. IDLH 5 mg/m³.

SYNS: BENZENE, 1,1'-OXYBIS-, HEXACHLORO derivatives (9CI) □ ETHER, HEXACHLOROPHENYL □ HEXACHLORODIPHENYL ETHER □ HEXACHLO-RODIPHENYL OXIDE □ PHENYL ETHER, HEXACHLORO derivative (8CI) □ TRICHLORO DIPHENYL ETHER □ TRICHLORO DIPHENYL OXIDE

OSHA PEL: TWA 0.5 mg/m³
ACGIH TLV: TWA 0.5 mg/m³
DFG MAK: 0.5 mg/m³

SAFETY PROFILE: Moderately toxic by ingestion and probably by inhalation. Combustible when exposed to heat, flame, or oxidizing materials. To fight fire, use water spray, fog, foam, dry chemical, CO_2. When heated to decomposition it emits toxic fumes of Cl⁻. See also ETHERS and ALDRIN (a closely related compound).

CDV250   HR: 2
CHLORINATED HYDROCARBONS, ALIPHATIC

SYNS: ALIPHATIC CHLORINATED HYDROCARBONS □ CHLORINATED HC, ALIPHATIC

SAFETY PROFILE: Suspected carcinogen with experimental tumors of the liver, lung, skin, and blood-forming tissues. The substitution of a chlorine (or other halogen) atom for a hydrogen greatly increases the anesthetic action of the aliphatic hydrocarbons and increases the range of their systemic effects. In many cases, the chlorine derivative is quite toxic. In general, the unsaturated chlorine derivatives are more narcotic but less toxic than the saturated derivatives. In the saturated group, the narcotic effect is proportional to the number of chlorine atoms. This relationship is not true for toxicity.

In dealing with these chlorinated hydrocarbons, it must be remembered that a toxic action may result from repeated exposure to concentrations that are too low to produce a narcotic effect, and that, consequently, are too low to give warning of danger. Individual susceptibility varies widely. Certain workmen may be seriously affected by concentrations that seem to have no effect on fellow employees at the same exposure.

In general reactivity decreases with greater substitution of halogen for hydrogen atoms. Halogenated (i.e., fluorine-, chlorine-, or bromine-containing) acetylene compounds are unstable and should be treated as explosives. Lightly substituted haloalkanes are highly flammable and can react with divalent light metals to form dangerously reactive products. Lightly substituted haloalkenes are highly flammable, peroxidizable, and may polymerize violently. When heated to decomposition they emit highly toxic fumes of phosgene. They may react violently with Al, liquid O_2, K, and Na.

CDV500 HR: 3
CHLORINATED HYDROCARBONS, AROMATIC
SYN: CHLORINATED HC AROMATIC
SAFETY PROFILE: In most instances, it is difficult to predict the toxicity of these compounds. However, in the case of most aromatic chlorine compounds, their toxicity is usually no greater, and frequently is less, than that of the corresponding aromatic hydrocarbons, with the notable exception of naphthalene and the various biphenyls. They can react with oxidizing materials. React violently with Al, liquid O_2, K, or Na. When heated to decomposition they emit toxic fumes of Cl^-.

CDV575 HR: 3
CHLORINATED NAPHTHALENES
SAFETY PROFILE: Questionable carcinogens that can cause tumors of the liver. Severe irritants by ingestion, inhalation, and skin contact. The action of the chlorinated naphthalenes on the body is quite similar to that of the chlorinated biphenyls, the chief effects being the production of chloracne of the skin, and systemically an acute yellow atrophy of the liver. When heated to decomposition they emit toxic fumes of Cl^-.

CDV625 CAS: 56641-03-3 HR: 3
CHLORINATED POLYETHER POLYURETHAN
PROP: Polymer formed from toluene diisocyanate and 1,4-butanediol and cured with 4,4'-methylenebis(o-chloroaniline) (CNREA8 36,3973,76).
mf: $(C_{13}H_{12}Cl_2N_2 \bullet C_9H_6N_2O_2 \bullet (C_4H_8O)_n H_2O)_x$
SYNS: OSTAMER □ POLYURETHANE Y-238 □ Y-238
CONSENSUS REPORTS: IARC Cancer Review: Animal Sufficient Evidence IMEMDT 19,303,79.
SAFETY PROFILE: Confirmed carcinogen with experimental tumorigenic data. When heated to decomposition it emits toxic fumes of Cl^- and NO_x.

CDV750 CAS: 7782-50-5 HR: 3
CHLORINE
DOT: UN 1017
mf: Cl_2 mw: 70.90
PROP: Greenish-yellow gas, liquid, or rhombic crystals. Mp: $-101°$, bp: $-34.9°$, d:

(liquid) 1.47 @ 0° (3.65 atm), vap press:
4800 mm @ 20°, vap d: 2.49. Sol in water.
IDLH 10 ppm.

SYNS: BERTHOLITE □ CHLOOR (DUTCH) □ CHLOR
(GERMAN) □ CHLORE (FRENCH) □ CHLORINE MOL.
□ CLORO (ITALIAN) □ MOLECULAR CHLORINE

CONSENSUS REPORTS: Reported in EPA
TSCA Inventory. Community Right-To-
Know List. EPA Extremely Hazardous
Substances List.

OSHA PEL: TWA 0.5 ppm; STEL 1 ppm
ACGIH TLV: TWA 0.5 ppm; STEL 1 ppm;
Not Classifiable as a Human Carcinogen
DFG MAK: 0.5 ppm (1.5 mg/m³)
NIOSH REL: (Chlorine) CL 0.5 ppm/15M
DOT CLASSIFICATION: 2.3; Label: Poison
Gas

SAFETY PROFILE: Moderately toxic to
humans by inhalation. Very irritating by
inhalation. Human mutation data reported.
Human respiratory system effects by
inhalation: changes in the trachea or
bronchi, emphysema, chronic pulmonary
edema or congestion. A strong irritant to
eyes and mucous membranes. Questionable
carcinogen.

Chlorine is extremely irritating to the
mucous membranes of the eyes and the
respiratory tract at 3 ppm. Combines with
moisture to form HCl. Both these
substances, if present in quantity, cause
inflammation of the tissues with which they
come in contact. A concentration of 3.5
ppm produces a detectable odor; 15 ppm
causes immediate irritation of the throat.
Concentrations of 50 ppm are dangerous for
even short exposures; 1000 ppm may be
fatal, even when exposure is brief. Because
of its intensely irritating properties, severe
industrial exposure seldom occurs, as the
worker is forced to leave the exposure area
before he can be seriously affected. In cases
where this is impossible, the initial irritation
of the eyes and mucous membranes of the
nose and throat is followed by coughing, a
feeling of suffocation, and, later, pain and a
feeling of constriction in the chest. If
exposure has been severe, pulmonary edema

may follow, with rales being heard over the
chest. It is a common air contaminant.

Explodes on contact with acetylene + heat
or UV light, air + ethylene, molten
aluminum, ammonia, amidosulfuric acid,
antimony trichloride + tetramethyl silane (at
100°), benzene + light, biuret, bromine
pentafluoride + heat, tert-butanol, butyl
rubber + naphtha, carbon disulfide + iron
catalyst, chlorinated pyridine + iron powder,
3-chloropropyne, cobalt(II) chloride +
methanol, diborane, dibutyl phthalate (at
118°), dichloro(methyl)arsine (in a sealed
container), diethyl ether, dimethyl
phosphoramidiate, dioxygen difluoride,
disilyl oxide, 4,4'-dithiodimorpholine, ethane
over activated carbon (at 350°), fluorine +
sparks, gasoline, glycerol (above 70° in a
sealed container), hexachlorodisilane (above
300°), hydrocarbon oils or waxes, iron(III)
chloride + monomers (e.g., styrene),
methane over mercury oxide, methanol,
methanol + tetrapyridine cobalt(II) chloride,
naphtha + sodium hydroxide, nitrogen
triiodide, oxygen difluoride, white
phosphorus (in liquid Cl_2), phosphorus
compounds, polypropylene + zinc oxide,
propane (at 300°), silicones when heated in a
sealed container [e.g., polydimethyl siloxane
(above 88°), polymethyl trifluoropropyl-
siloxane (above 68°)], stibine, synthetic
rubber (in liquid Cl_2), tetraselenium
tetranitride, trimethyl thionophosphate.
Explosive products are formed on reaction
with alkylthiouronium salts, amidosulfuric
acid, acidic ammonium chloride solutions,
aziridine, bis(2,4-dinitrophenyl)disulfide,
cyanuric acid, phenyl magnesium bromide.
Mixtures with ethylene are explosives
initiated by light, heat, or by the presence of
mercury, mercury oxide, silver oxide, lead
oxide (at 100°). Mixtures with hydrogen are
explosives initiated by sparks, light, heating
to over 280°, or the presence of yellow
mercuric oxide or nitrogen trichloride.
Mixtures with hydrogen and other gases
(e.g., air, hydrogen chloride, oxygen) are also
explosive.

Ignition or explosive reaction with metals (e.g., aluminum, antimony powder, bismuth powder, brass, calcium powder, copper, germanium, iron, manganese, potassium, tin, vanadium powder). Reaction with some metals requires moist Cl_2 or heat. Ignites with diethyl zinc (on contact), polyisobutylene (at 130°), metal acetylides, metal carbides, metal hydrides (e.g., potassium hydride, sodium hydride, copper hydride), metal phosphides (e.g., copper(II) phosphide), methane + oxygen, hydrazine, hydroxylamine, calcium nitride, nonmetals (e.g., boron, active carbon, silicon, phosphorus), nonmetal hydrides (e.g., arsine, phosphine, silane), steel (above 200° or as low as 50° when impurities are present), sulfides (e.g., arsenic disulfide, boron trisulfide, mercuric sulfide), trialkyl boranes.

Violent reaction with alcohols, N-aryl sulfinamides, dimethyl formamide, polychlorobiphenyl, sodium hydroxide, hydrochloric acid + dinitroanilines. Incandescent reaction when warmed with cesium oxide (above 150°), tellurium, arsenic, tungsten dioxide. Potentially dangerous reaction with hydrocarbons + Lewis acids releases toxic and reactive HCl gas.

Can react to cause fires or explosions upon contact with turpentine, illuminating gas, polypropylene, rubber, sulfamic acid, $As_2(CH_3)_4$, UC_2, acetaldehyde, alcohols, alkylisothiourea salts, alkyl phosphines, Al, Sb, As, AsS_2, AsH_3, Ba_3P_2, C_6H_6, Bi, B, BPI_2, B_2S_3, brass, BrF_5, Ca, $(CaC_2 + KOH)$, $Ca(ClO_2)_2$, Ca_3N_2, Ca_3P_2, C, CS_2, Cs, $CsHC_2$, Co_2O, Cs_3N, $(C + Cr(OCl)_2)$, CuH_2, CuC_2, dialklyl phosphines, diborane, dibutyl phthalate, $Zn(C_2H_5)_2$, C_2H_6, C_2H_4, ethylene imine, $C_2H_5PH_2$, F_2, Ge, glycerol, $(NH_2)_2$, $(H_2O + KOH)$, I_2, hydroxylamine, Fe, FeC_2, Li, Li_2C_2, Li_6C_2, Mg, Mg_2P_3, Mn, Mn_3P_2, HgO, HgS, Hg, Hg_3P_2, CH_4, Nb, NI_3, OF_2, H_2SiO, $(OF_2 + Cu)$, PH_3, P, $P(SNC)_3$, P_2O_3, PCB's, K, KHC_2, KH, Ru, $RuHC_2$, Si, SiH_2, Ag_2O, Na, $NaHC_2$, Na_2C_2, SnF_2, SbH_3, Sr_3P, Te, Th, Sn, WO_2, U, V, Zn, ZrC_2.

CDW000   CAS: 13973-88-1   HR: 3
CHLORINE AZIDE
mf: ClN_3 mw: 77.48
PROP: An explosive gas. Bp: 15°. Sol in org solvs.
SYN: NITROGEN CHLORIDE
DOT CLASSIFICATION: Forbidden
SAFETY PROFILE: Strong irritant by inhalation. An extremely unstable explosive. Reacts with liquid ammonia to form an explosive liquid. Explosive reaction with 1,3-butadiene, C_2H_6, C_2H_4, CH_4, C_3H_8, phopshorus, silver azide, sodium. Reacts with water or steam to produce toxic and corrosive fumes of HCl. Has been used as an initiator in chemical gas lasers. When heated to decomposition it emits toxic fumes of Cl^- and NO_x. See also CHLORINE and AZIDES.

CDW450   CAS: 10049-04-4   HR: 3
CHLORINE DIOXIDE
mf: ClO_2 mw: 67.45
PROP: Red-yellow or orange-green gas or orange-red crystals. Unstable in light, stable in the dark. Mp: −59°, bp: 11°, d: 3.09 g/L @ 11°. Insol in water. IDLH 5 ppm. IDLH 4 mg/m^3 (as Pt).
SYNS: ALCIDE □ ANTHIUM DIOXIDE □ CHLORINE DIOXIDE, not hydrated (DOT) □ CHLORINE OXIDE □ CHLORINE(IV) OXIDE □ CHLORINE PEROXIDE □ CHLOROPEROXYL □ CHLORYL RADICAL □ DOXCIDE 50
CONSENSUS REPORTS: Reported in EPA TSCA Inventory. Community Right-To-Know List.
OSHA PEL: TWA 0.1 ppm; STEL 0.3 ppm
ACGIH TLV: TWA 0.1 ppm; STEL 0.3 ppm
DFG MAK: 0.1 ppm (0.28 mg/m^3)
DOT CLASSIFICATION: Forbidden
SAFETY PROFILE: Moderately toxic by inhalation. Experimental reproductive effects. Mutation data reported. An eye irritant. A powerful explosive sensitive to spark, impact, sunlight, or heating rapidly to 100°C. A powerful oxidizer. Concentrations of greater than 10% in air are explosive. Explodes on mixing with carbon monoxide, hydrocarbons (e.g., butadiene, ethane,

ethylene, methane, propane), fluoramines (e.g., difluoramine, trifluoramine). Mixtures with hydrogen explode with sparking or contact with platinum. Explodes on contact with mercury, potassium hydroxide, phosphorus pentachloride + chlorine. Ignites or explodes on contact with non-metals (e.g., phosphorus, sulfur, sugar). Reacts violently with F_2, NHF_2. Reacts with water or steam to produce toxic and corrosive fumes of HCl. When heated to decomposition it emits toxic fumes of Cl^-. See also CHLORINE.

CDX250   CAS: 13637-63-3   HR: 3
CHLORINE PENTAFLUORIDE
DOT: UN 2548
mf: ClF_5 mw: 130.45
PROP: Colorless gas, extremely vigorous fluorinating agent. D: 2.105 @ $-80°$, mp: $-103°$, bp: $-13.1°$.
SYNS: CHLORINE FLUORIDE (ClF5) □ CHLORINE PENTAFLUORIDE (DOT)
OSHA PEL: TWA 2.5 mg(F)/m³
ACGIH TLV: TWA 2.5 mg(F)/m³; BEI: 3 mg/g creatinine of fluorides in urine prior to shift; 10 mg/g creatinine of fluorides in urine at end of shift.
NIOSH REL: (Inorganic Fluorides) TWA 2.5 mg(F)/m³
DOT CLASSIFICATION: 2.3; Label: Poison Gas, Oxidizer, Corrosive
SAFETY PROFILE: Poison by inhalation. A corrosive material. Vigorous reaction in contact with water or anhydrous nitric acid. Violent reaction on contact with metals. When heated to decomposition it emits very toxic fumes of Cl^- and F^-. See also CHLORINE, FLUORINE, FLUORIDES, and CHLORINE TRIFLUORIDE.

CDX750   CAS: 7790-91-2   HR: 3
CHLORINE TRIFLUORIDE
DOT: UN 1749
mf: ClF_3 mw: 92.45
PROP: Colorless gas to yellow liquid; sweet odor. One of the most reactive chemical

compds known. Mp: $-83°$, bp: $11.8°$, d: 1.77 @ 13°. IDLH 20 ppm.
SYNS: CHLORINE FLUORIDE □ CHLOROTRIFLUORIDE □ TRIFLUORURE de CHLORE (FRENCH)
CONSENSUS REPORTS: Reported in EPA TSCA Inventory.
OSHA PEL: CL 0.1 ppm
ACGIH TLV: CL 0.1 ppm
DFG MAK: 0.1 ppm (0.38 mg/m³)
DOT CLASSIFICATION: 2.3; Label: Poison Gas, Oxidizer, Corrosive
SAFETY PROFILE: Human poison by inhalation. An eye irritant. See also FLUORIDES, CHLORINE, and FLUORINE. Spontaneously flammable. A powerful oxidant which may react violently with oxidizable materials. A rocket propellant.

Explosive reaction with water, bis(trifluoromethyl)sulfide or -disulfide, polychlorotrifluoroethylene, trifluoromethanesulfenyl chloride, and other hydrogen- containing materials (e.g., ammonia, coal gas, hydrogen, hydrogen sulfide, methane, acetic acid, benzene, ether, cotton, paper, wood). Forms shock-sensitive explosive mixtures with highly chlorinated compounds (e.g., carbon tetrachloride), nitroaryl compounds (e.g., trinitrotoluene, hexanitrobiphenyl, hexanitrodiphenyl amine, hexanitrodiphenyl sulfide, hexanitrodiphenyl ether). Reaction with ammonium fluoride or ammonium hydrogen fluoride forms explosive gaseous products.

Ignition on contact with boron-containing materials, iodine, finely divided refractory materials (e.g., asbestos, glass wool, sand, tungsten carbide), fluorinated polymers (with flowing trifluoride).

Violent reaction with acids (e.g., nitric or sulfuric), chromium trioxide, ruthenium, selenium tetrafluoride (above 106°C), metals, metal oxides, metal salts, nonmetals, nonmetal salts, organic matter, glass wool, acetic acid, Al, Sb, As, Cu, Ir, Fe, Pb, Mg, Mo, Os, P, K, Rh, Se, Si, Ag, Na, S, Te, Sn, W, Zn, oxides, CO, graphite, HgI_2, HNO_3,

K_2CO_3, KI, rubber, AgN_3, $AgNO_3$, NaOH, V_2P_5, WO_3. Incompatible with fuels, nitro compounds. When heated to decomposition or in reaction with water or steam it emits toxic fumes of F^- and Cl^-.

CDY250   **HR: 3**
CHLORITES
SAFETY PROFILE: Many chlorite salts are heat- and impact-sensitive explosives. The metal salts are powerful oxidants. They are much less stable than the analogous chlorates. React violently with NH_3, organic matter, or metals. See individual chlorites.

CDY500   **CAS: 107-20-0**   **HR: 3**
CHLOROACETALDEHYDE
DOT: UN 2232
mf: C_2H_3ClO mw: 78.50
PROP: Clear, colorless liquid; pungent odor. Bp: 90.0–100.1° (40% soln), fp: −16.3° (40% soln), flash p: 190°F, d: 1.19 @ 25°/25° (40% soln), vap press: 100 mm @ 45° (40% soln). IDLH 45 ppm.
SYNS: CHLOROACETALDEHYDE MONOMER □ 2-CHLOROACETALDEHYDE □ 2-CHLOROETHANAL □ 2-CHLORO-1-ETHANAL □ MONOCHLORO-ACETALDEHYDE □ RCRA WASTE NUMBER P023
CONSENSUS REPORTS: Reported in EPA TSCA Inventory. EPA Genetic Toxicology Program.
OSHA PEL: CL 1 ppm
ACGIH TLV: CL 1 ppm
DFG MAK: Confirmed Animal Carcinogen with Unknown Relevance to Humans
DOT CLASSIFICATION: 6.1; Label: Poison
SAFETY PROFILE: Suspected carcinogen. Poison by ingestion, skin contact, and intraperitoneal routes. Mutation data reported. Combustible when exposed to heat or flame. Reacts with oxidizing materials. To fight fire, use water, foam, CO_2, dry chemical. When heated to decomposition it emits toxic fumes of Cl^-. See also ALDEHYDES and CHLORIDES.

CDY850   **CAS: 79-07-2**   **HR: 3**
2-CHLOROACETAMIDE

mf: C_2H_4ClNO mw: 93.52
PROP: Crystals. Mp: 120°, bp: 225° (decomp). Sol in H_2O, EtOH; sltly sol in Et_2O.
SYNS: CHLORACETAMID (GERMAN) □ CHLORO-ACETAMIDE □ α-CHLOROACETAMIDE □ 2-CHLOROETHANAMIDE □ USAF DO-29
CONSENSUS REPORTS: Reported in EPA TSCA Inventory.
SAFETY PROFILE: Poison by ingestion, intravenous, and intraperitoneal routes. Mutation data reported. When heated to decomposition it emits very toxic Cl^- and NO_x. See also N-CHLOROACETAMIDE.

CEA000   **CAS: 79-11-8**   **HR: 3**
CHLOROACETIC ACID
DOT: UN 1750/UN 1751
mf: $C_2H_3ClO_2$ mw: 94.50
PROP: Colorless crystals in three forms. Mp: (α) 61.3°, (β) 56.2°, (τ) 52.5°, bp: 189°, flash p: 259°F, d: 1.58 @ 20°/20°, vap d: 3.26. Very sol in H_2O; sol in org solvs.
SYNS: ACIDE CHLORACETIQUE (FRENCH) □ ACIDE MONOCHLORACETIQUE (FRENCH) □ ACIDOMONOCLOROACETICO (ITALIAN) □ CHLORACETIC ACID □ CHLOROACETIC ACID, liquid (UN 1750) (DOT) □ CHLOROACETIC ACID, solid (UN 1751) (DOT) □ α-CHLOROACETIC ACID □ CHLOROETHANOIC ACID □ KYSELINA CHLOROCTOVA □ MCA □ MONOCHLOORAZIJNZUUR (DUTCH) □ MONOCHLORACETIC ACID □ MONOCHLORESSIGSAEURE (GERMAN) □ MONOCHLOROACETIC ACID □ MONOCHLORO-ETHANOIC ACID □ NCI-C60231
CONSENSUS REPORTS: Reported in EPA TSCA Inventory. EPA Genetic Toxicology Program. EPA Extremely Hazardous Substances List. Community Right-To-Know List.
DOT CLASSIFICATION: 8; Label: Corrosive, Poison (UN 1750)
SAFETY PROFILE: Poison by ingestion, inhalation, subcutaneous, and intravenous routes. A corrosive skin, eye, and mucous membrane irritant. Questionable carcinogen with experimental tumorigenic data. Mutation data reported. Combustible liquid when exposed to heat or flame. To fight fire, use water spray, fog, mist, dry chemical,

foam. When heated to decomposition it emits toxic fumes of Cl⁻. See also CHLORIDES.

CEA750   CAS: 532-27-4   HR: 3
α-CHLOROACETOPHENONE
DOT: UN 1697
mf: C_8H_7ClO mw: 154.60
PROP: Leaflets from pet ether. Mp: 54°, bp: 139–141° @ 14 mm. IDLH 15 mg/m³.
SYNS: CAF □ CAP □ CHEMICAL MACE □ CHLORACETOPHENONE □ ω-CHLOROACETOPHEN-ONE □ 1-CHLOROACETOPHENONE □ CHLORO-ACETOPHENONE, liquid or solid (DOT) □ CHLORO-METHYL PHENYL KETONE □ 2-CHLORO-1-PHENYL-ETHANONE □ CN □ ETHANONE, 2-CHLORO-1-PHENYL- □ MACE (lachrymator) □ NCI-C55107 □ PHENACYL CHLORIDE □ PHENYLCHLORO-METHYLKETONE
CONSENSUS REPORTS: Reported in EPA TSCA Inventory. Community Right-To-Know List.
OSHA PEL: TWA 0.05 ppm
ACGIH TLV: TWA 0.05 ppm; Not Classifiable as a Human Carcinogen
DOT CLASSIFICATION: 6.1; Label: Poison
SAFETY PROFILE: A human poison by inhalation. An experimental poison by ingestion, inhalation, intraperitoneal, and intravenous routes. Human systemic effects by inhalation: lachrymation, conjunctiva irritation, and unspecified eye effects, cough, and dyspnea. A severe eye and moderate skin irritant. Questionable carcinogen with experimental neoplastigenic data by skin contact. A riot control agent. When heated to decomposition it emits toxic fumes of Cl⁻. See also KETONES.

CEB250   CAS: 99-91-2   HR: 3
p-CHLOROACETOPHENONE
mf: C_8H_7ClO mw: 154.60
PROP: Pale straw-colored liquid or white crystals; fragrant, non-persistent odor. Mp: 20°, bp: 236°, fp: 59°, d: 1.19 @ 20°, vap press: 0.012 mm @ 0°, vap d: 5.2, flash p: 244°F.
SYNS: 4-CHLOROACETOPHENONE □ 4'-CHLOROACETOPHENONE □ 1-(4-CHLORO-PHENYL)ETHANONE □ USAF DO-1

CONSENSUS REPORTS: Reported in EPA TSCA Inventory.
SAFETY PROFILE: Poison by intraperitoneal route. Moderately toxic by ingestion. A powerful irritant and lachrymator. Human systemic effects by inhalation: lachrymation and unspecified effects on the eye and sense of smell. Combustible when exposed to heat or flame. To fight fire, use water, foam, alcohol foam, dry chemical. When heated to decomposition or on contact with water or steam it emits toxic fumes of Cl⁻.

CEC000   CAS: 140-49-8   HR: 2
4'-CHLOROACETYL ACETANILIDE
mf: $C_{10}H_{10}ClNO_2$ mw: 211.66
SYNS: p-ACETAMIDOPHENACYL CHLORIDE □ p-(ACETYLAMINO)PHENACYL CHLORIDE □ 4'-(CHLOROACETYL)ACETANILIDE □ NCI-C03770
CONSENSUS REPORTS: NCI Carcinogenesis Bioassay (feed); No Evidence: mouse, rat NCITR* NCI- CG-TR-177,79. Reported in EPA TSCA Inventory.
SAFETY PROFILE: Moderately toxic by ingestion. Mutation data reported. When heated to decomposition it emits very toxic fumes of Cl⁻ and NOₓ. See also CHLORIDES.

CEC100   CAS: 1132-20-3   HR: 3
N-(CHLOROACETYL)-3-AZABICYCLO-
(3.2.1)NONANE
mf: $C_{10}H_{16}ClNO$ mw: 201.72
SYNS: 3-AZABICYCLO(3.2.2)NONANE, 3-(CHLOROACETYL)- □ KETONE, 3-AZABI-CYCLO(3.2.2)NONYL CHLOROMETHYL
DOT CLASSIFICATION: 3; Label: Flammable Liquid
SAFETY PROFILE: A poison by intravenous route. A flammable liquid. When heated to decomposition it emits toxic vapors of NOₓ and Cl⁻.

CEC250   CAS: 79-04-9   HR: 3
CHLOROACETYL CHLORIDE
DOT: UN 1752
mf: $C_2H_2Cl_2O$ mw: 112.94

PROP: Water-white or sltly yellow liquid. Bp: 108–110°, fp: −22.5°, flash p: none, d: 1.495 @ 0°.
SYNS: CHLORACETYL CHLORIDE □ CHLORID KYSELINY CHLOROCTOVE □ CHLOROACETIC ACID CHLORIDE □ CHLOROACETIC CHLORIDE □ CHLORURE de CHLORACETYLE (FRENCH) □ MONOCHLOROACETYL CHLORIDE
CONSENSUS REPORTS: Reported in EPA TSCA Inventory.
OSHA PEL: TWA 0.05 ppm
ACGIH TLV: TWA 0.05 ppm; STEL 0.15 ppm
DOT CLASSIFICATION: 8; Label: Corrosive, Poison
SAFETY PROFILE: Poison by ingestion and intravenous routes. Mildly toxic by inhalation. Corrosive. A lachrymator. When heated to decomposition it emits toxic fumes of Cl⁻.

CEC300   CAS: 143-85-1   HR: 3
1-CHLOROACETYL-α-α-DIPHENYL-4-PIPERIDINEMETHANOL
mf: $C_{20}H_{22}ClNO_2$ mw: 343.88
SYNS: KETONE, CHLOROMETHYL 4-(DIPHENYL-HYDROXYMETHYL)PIPERIDINO □ 4-PIPERIDINEME-THANOL, 1-CHLOROACETYL-α-α-DIPHENYL-
DOT CLASSIFICATION: 3; Label: Flammable Liquid
SAFETY PROFILE: A poison by intraperitoneal route. A flammable liquid. When heated to decomposition it emits toxic vapors of NO_x and Cl⁻.

CEC700   CAS: 24040-34-4   HR: 3
2-CHLOROACETYLFLUORENE
mf: $C_{15}H_{11}ClO$ mw: 242.71
SYNS: ETHANONE, 2-CHLORO-1-(9H-FLUOREN-2-YL)- □ FLUORENE, 2-(CHLOROACETYL)- □ KETONE, CHLOROMETHYL 2-FLUORENYL
DOT CLASSIFICATION: 3; Label: Flammable Liquid
SAFETY PROFILE: A poison by intraperitoneal route. A flammable liquid. When heated to decomposition it emits toxic vapors of Cl⁻.

CEE750   CAS: 920-37-6   HR: 3
2-CHLOROACRYLONITRILE

mf: C_4H_2ClN mw: 87.51
PROP: Liquid. Bp: 88°, flash p: 46.4°F.
SYNS: CHLOROACRYLONITRILE □ α-CHLOROACRYLONITRILE
CONSENSUS REPORTS: Reported in EPA TSCA Inventory. Cyanide and its compounds are on the Community Right-To-Know List.
DFG MAK: Confirmed Animal Carcinogen with Unknown Relevance to Humans
SAFETY PROFILE: Suspected carcinogen. Poison by intravenous route. A powerful irritant. A dangerous fire hazard when exposed to heat or flame. To fight fire, use water, dry chemical, CO_2, foam. When heated to decomposition it emits very toxic fumes of Cl⁻, NO_x, and CN⁻. See also NITRILES.

CEG550   CAS: 4080-31-3   HR: 3
1-(3-CHLOROALLYL)-3,5,7-TRIAZA-1-AZONIAADAMANTANE CHLORIDE
mf: $C_9H_{16}ClN_4$•Cl   mw: 251.19
SYNS: DOWCO 184 □ DOWICIDE Q □ DOWICIL 75 □ DOWICIL 100 □ QUATERNIUM 15 □ 3,5,7-TRIAZA-1-AZONIAADAMANTANE, 1-(3-CHLOROALLYL)-, CHLORIDE
CONSENSUS REPORTS: Reported in EPA TSCA Inventory.
SAFETY PROFILE: Poison by ingestion. Moderately toxic by skin contact. Experimental teratogenic effects. A skin irritant. Mutation data reported. When heated to decomposition it emits toxic fumes of NO_x and Cl⁻.

CEG600   CAS: 615-66-7   HR: 1
3-CHLORO-4-AMINOANILINE
mf: $C_6H_7ClN_2$ mw: 142.60
PROP: Needles. Mp: 64°.
SYNS: 3-CHLOR-p-FENYLENDIAMIN (CZECH) □ 2-CHLORO-1,4-BENZENEDIAMINE □ o-CHLORO-p-PHENYLENEDIAMINE □ 2-CHLORO-p-PHENYLENE-DIAMINE □ C.I. 76065 □ URSOL BROWN O
CONSENSUS REPORTS: Reported in EPA TSCA Inventory.
SAFETY PROFILE: An experimental teratogen. Other experimental reproductive effects. An eye irritant. Decomposes explosively at 165°C/33 mbar. When heated

to decomposition it emits toxic fumes of Cl⁻ and NOₓ. See also AROMATIC AMINES.

CEH250   CAS: 95-85-2   HR: 3
p-CHLORO-o-AMINOPHENOL
DOT: UN 2673
mf: C_6H_6ClNO mw: 143.58
PROP: Plates from H_2O. Mp: 140–141°.
SYN: 2-AMINO-4-CHLOROPHENOL (DOT)
CONSENSUS REPORTS: Reported in EPA TSCA Inventory.
DOT CLASSIFICATION: 6.1; Label: Poison
SAFETY PROFILE: A poison. Moderately toxic by ingestion. Mutation data reported. When heated to decomposition it emits very toxic fumes of Cl⁻ and NOₓ. See also AROMATIC AMINES and CHLORIDES.

CEH670   CAS: 95-51-2   HR: 3
2-CHLOROANILINE
mf: C_6H_6ClN mw: 127.58
PROP: Liquid. Mp: −1.94°, bp: 208.84°, d: 1.21 @ 20°/4°. Practically insol in water; sol in most org solvs, also in acids.
SYNS: 1-AMINO-2-CHLOROBENZENE □ o-CHLO-RANILINE □ o-CHLOROANILINE □ o-CHLOROANIL-INE, liquid □ o-CHLOROANILINE, solid □ 2-CHLORO-BENZENAMINE (9CI) □ FAST YELLOW GC BASE
CONSENSUS REPORTS: EPA Genetic Toxicology Program. Reported in EPA TSCA Inventory.
SAFETY PROFILE: Poison by skin contact, ingestion, and subcutaneous routes. Mutation data reported. When heated to decomposition it emits toxic fumes of Cl⁻ and NOₓ. See also ANILINE DYES.

CEH675   CAS: 108-42-9   HR: 3
3-CHLOROANILINE
mf: C_6H_6ClN mw: 127.58
PROP: Liquid. Bp: 230.5°, mp: −10.4°, d: 1.2225 @ 15°/15°. Practically insol in water; sol in most common org solvs.
SYNS: m-AMINOCHLOROBENZENE □ 1-AMINO-3-CHLOROBENZENE □ 3-CHLOORANILINEN (DUTCH) □ m-CHLORANILINE □ m-CHLOROANILINE □ 3-CHLOROANILINE (ITALIAN) □ m-CHLOROANILINE, liquid □ m-CHLOROANILINE, solid □ 3-CHLORO-BENZENAMINE □ m-CHLOROPHENYLAMINE □ 3-

CHLOROPHENYLAMINE □ FAST ORANGE GC BASE □ ORANGE GC BASE
CONSENSUS REPORTS: EPA Genetic Toxicology Program. Reported in EPA TSCA Inventory.
SAFETY PROFILE: Poison by ingestion, skin contact, subcutaneous, and intravenous routes. Mutation data reported. When heated to decomposition it emits toxic fumes of Cl⁻ and NOₓ. See also ANILINE DYES.

CEH680   CAS: 106-47-8   HR: 3
4-CHLOROANILINE
mf: C_6H_6ClN mw: 127.58
PROP: Orthorhombic crystals from alc or pet ether; needles from toluene. Mp: 70–71°, bp: 232°, d: 1.169. Sol in hot water; freely sol in alc, ether, acetone, carbon disulfide.
SYNS: 1-AMINO-4-CHLOROBENZENE □ 4-CHLORANILIN (CZECH) □ p-CHLORANILINE □ p-CHLOROANILINE □ p-CHLOROANILINE, liquid □ p-CHLOROANILINE, solid □ 4-CHLOROBENZENAMINE □ 4-CHLOROBENZENEAMINE □ 4-CHLOROPHENYL-AMINE □ NCI-C02039 □ RCRA WASTE NUMBER P024
CONSENSUS REPORTS: EPA Genetic Toxicology Program. Reported in EPA TSCA Inventory.
DFG MAK: Animal Carcinogen, Suspected Human Carcinogen
SAFETY PROFILE: Confirmed carcinogen with experimental neoplastigenic and tumorigenic data. Poison by ingestion, inhalation, skin contact, subcutaneous, and intravenous routes. A skin and severe eye irritant. Mutation data reported. When heated to decomposition it emits toxic fumes of Cl⁻ and NOₓ. See also ANILINE DYES.

CEI000   CAS: 82-44-0   HR: 3
1-CHLOROANTHRAQUINONE
mf: $C_{14}H_8ClO_2$ mw: 243.67
PROP: Yellow needles from EtOH. Mp: 162°.
SYNS: 1-CHLORANTHRACHINON (CZECH) □ 1-CHLORO-9,10-ANTHRACENEDIONE □ α-CHLOROANTHRAQUINONE □ 1-CHLORO-9,10-ANTHRAQUINONE □ α-MONOCHLORO-ANTHRAQUINONE

CONSENSUS REPORTS: Reported in EPA TSCA Inventory.

SAFETY PROFILE: Poison by intravenous route. An eye irritant. When heated to decomposition it emits very toxic fumes of Cl^- and NO_x.

CEI500   CAS: 89-98-5   HR: 3
o-CHLOROBENZALDEHYDE

mf: C_7H_5ClO mw: 140.57

PROP: Liquid or needles with strong odor. Fp: 11°, bp: 213–214°.

SYNS: o-CHLOORBENZALDEHYDE (DUTCH) □ 2-CHLOORBENZALDEHYDE (DUTCH) □ 2-CHLORBENZ-ALDEHYD (GERMAN) □ 2-CHLOROBENZALDEHYDE □ o-CHLOROBENZENECARBOXALDEHYDE □ 2-CLOROBENZALDEIDE (ITALIAN) □ USAF M-7

CONSENSUS REPORTS: Reported in EPA TSCA Inventory.

SAFETY PROFILE: Poison by intraperitoneal and intravenous routes. When heated to decomposition it emits toxic fumes of Cl^-. See also ALDEHYDES and CHLORIDES.

CEJ125   CAS: 108-90-7   HR: 3
CHLOROBENZENE

DOT: UN 1134

mf: C_6H_5Cl mw: 112.56

PROP: Clear, colorless liquid with faint odor. Bp: 131.7°, lel: 1.3%, uel: 7.1% @ 150°, mp: −45°, flash p: 85°F (CC), d: 1.11 @ 20°/4°, autoign temp: 1180°F, vap press: 10 mm @ 22.2°, vap d: 3.88. IDLH 1000 ppm.

SYNS: BENZENE CHLORIDE □ CHLOORBENZEEN (DUTCH) □ CHLORBENZENE □ CHLORBENZOL □ CHLOROBENZEN (POLISH) □ CHLOROBENZOL (DOT) □ CLOROBENZENE (ITALIAN) □ MCB □ MONOCHLO-ORBENZEEN (DUTCH) □ MONOCHLORBENZENE □ MONOCHLORBENZOL (GERMAN) □ MONOCHLORO-BENZENE □ MONOCLOROBENZENE (ITALIAN) □ NCI-C54886 □ PHENYL CHLORIDE □ RCRA WASTE NUMBER U037

CONSENSUS REPORTS: NTP Carcinogenesis Studies (gavage); Some Evidence: rat NTPTR* NTP-TR-261,85; No Evidence: mouse NTPTR* NTP-TR-261,85. Reported in EPA TSCA Inventory. Community Right-To-Know List.

OSHA PEL: TWA 75 ppm

ACGIH TLV: TWA 10 ppm; Animal Carcinogen; BEI: 150 mg/g creatinine of total 4-chlorocatechol in urine at end of shift; 25 mg/g creatinine of total p-chloropohenol in urine at end of shift

DFG MAK: 10 ppm (47 mg/m³)

DOT CLASSIFICATION: 3; Label: Flammable Liquid

SAFETY PROFILE: Suspected carcinogen. Moderately toxic by ingestion and intraperitoneal routes. Experimental teratogenic and reproductive effects. Mutation data reported. Strong narcotic with slight irritant qualities. Dichlorobenzols are strongly narcotic. Little is known of the effects of repeated exposures at lower concentrations, but it may cause kidney and liver damage. The industrial illnesses reported may possibly be due to nitrobenzol. Dangerous fire hazard when exposed to heat or flame. Moderate explosion hazard when exposed to heat or flame. Potentially explosive reaction with powdered sodium or phosphorus trichloride + sodium. Violent reaction with $AgClO_4$. Reacts vigorously with oxidizers. See also CHLORINATED HYDROCARBONS, AROMATIC. To fight fire, use foam, CO_2, dry chemical, water to blanket fire. Associated with EPA Superfund sites.

CEM000   CAS: 873-32-5   HR: 3
o-CHLOROBENZONITRILE

mf: C_7H_4ClN mw: 137.57

PROP: Crystals. Mp: 42–43°, bp: 232°.

SYNS: o-CHLORBENZONITRIL (CZECH) □ NITRIL KYSELINY-o-CHLORBENZOOVE (CZECH)

CONSENSUS REPORTS: Reported in EPA TSCA Inventory. Cyanide and its compounds are on the Community Right-To-Know List.

SAFETY PROFILE: Poison by intraperitoneal route. Moderately toxic by ingestion. An eye irritant. When heated to decomposition or on contact with water, steam, acid, or acid fumes it emits toxic fumes of Cl^- and CN^-. See also NITRILES.

CEM825 CAS: 98-56-6 HR: 1
p-CHLOROBENZOTRIFLUORIDE
mf: $C_7H_4ClF_3$ mw: 180.56
PROP: Bp: 135–136°.
SYNS: (p-CHLOROPHENYL)TRIFLUOROMETHANE □
p-CHLOROTRIFLUOROMETHYLBENZENE □ 4-
CHLOROTRIFLUOROMETHYLBENZENE □ 1-CHLORO-
4-(TRIMETHYL)-BENZENE (9CI) □ α,α,α-TRIFLUORO-4-
CHLOROTOLUENE □ p-(TRIFLUOROMETHYL)-
CHLOROBENZENE □ p-TRIFLUOROMETHYLPHENYL
CHLORIDE
CONSENSUS REPORTS: Reported in EPA
TSCA Inventory.
SAFETY PROFILE: Mildly toxic by
ingestion and inhalation. Human mutation
data reported. Flammable. Strongly
exothermic reaction with sodium
dimethylsulfinate. When heated to
decomposition it emits toxic fumes of F−
and Cl−. See also CHLORINATED
HYDROCARBONS, AROMATIC; and
FLUORIDES.

CEO100 CAS: 14415-66-8 HR: 3
2-(p-CHLOROBENZOYL)-1-(2-MORPHO-
** LINOETHYL)PYRROLE MONO-**
** HYDROCHLORIDE**
mf: $C_{17}H_{19}ClN_2O_2 \cdot ClH$ mw: 355.29
SYN: KETONE, p-CHLOROPHENYL 1-(2-
MORPHOLINOETHYL)PYRROL-2-YL,
HYDROCHLORIDE
DOT CLASSIFICATION: 3; Label:
Flammable Liquid
SAFETY PROFILE: A poison by
intravenous route. A flammable liquid.
When heated to decomposition it emits
toxic vapors of NO_x, HCl, and Cl−.

CEQ600 CAS: 2698-41-1 HR: 3
o-CHLOROBENZYLIDENE MALONO-
** NITRILE**
mf: $C_{10}H_5ClN_2$ mw: 188.62
PROP: White crystals. Mp: 95°, bp: 313°.
IDLH 2 mg/m³.
SYNS: o-CHLOROBENZAL MALONONITRILE □ 2-
CHLOROBENZAL MALONONITRILE □ o-CHLORO-
BENZYLIDENE MALONITRILE □ 2-CHLORO-
BENZYLIDENE MALONITRILE □ 2-CHLOROBMN □
CS □ β,β-DICYANO-o-CHLOROSTYRENE □ NCI-C55118
□ PROPANEDINITRILE((2-CHLOROPHENYL)-
METHYLENE) □ USAF KF-11

CONSENSUS REPORTS: Reported in EPA
TSCA Inventory. Cyanide and its
compounds are on the Community Right-
To-Know List.
OSHA PEL: CL 0.05 ppm (skin)
ACGIH TLV: CL 0.05 ppm (skin); Not
Classifiable as a Human Carcinogen
SAFETY PROFILE: Poison by ingestion,
intraperitoneal, and intravenous routes.
Moderately toxic by inhalation. Human
systemic effects by inhalation: conjunctiva
irritation, cough, and unspecified respiratory
system effects. A human skin and eye
irritant. Human exposure data suggest
relatively low systemic toxicity, but intense
irritation of eyes, skin, and mucous
membranes. Mutation data reported. A tear
gas used for riot control. When heated to
decomposition it emits very toxic fumes of
Cl−, NO_x, and CN−. See also NITRILES.

CES650 CAS: 74-97-5 HR: 2
CHLOROBROMOMETHANE
DOT: UN 1887
mf: CH_2BrCl mw: 129.39
PROP: Clear, colorless liquid; sweet odor.
Fp: −88°, bp: 67.8°, flash p: none, d: 1.930
@ 25°/25°, vap d: 4.46. Sol in organic solvs;
insol in water. IDLH 2000 ppm.
SYNS: BROMOCHLOROMETHANE □ HALON 1011 □
METHYLENE CHLOROBROMIDE □ MIL-B-4394-B □
MONO-CHLORO-MONO-BROMO-METHANE
CONSENSUS REPORTS: Reported in EPA
TSCA Inventory.
OSHA PEL: TWA 200 ppm
ACGIH TLV: TWA 200 ppm
DFG MAK: 200 ppm (1100 mg/m³)
DOT CLASSIFICATION: 6.1; Label: KEEP
AWAY FROM FOOD
SAFETY PROFILE: Mildly toxic by
ingestion and inhalation. Mutation data
reported. This material has a narcotic action
of moderate intensity, although of
prolonged duration. Animals exposed for
several weeks to 1000 ppm had blood
bromide levels as high as 350 mg/100 g.
Therefore, until further data are available, it
should be considered at least as toxic as

carbon tetrachloride and more than minimal exposure to its vapors should be avoided. Dangerous; when heated to decomposition it emits highly toxic fumes of Br⁻ and Cl⁻. See also BROMIDES and CHLORINATED HYDROCARBONS, ALIPHATIC.

CET000   CAS: 73926-87-1   HR: 3
trans-CHLORO(2-(3-BROMOPROPION-
AMIDO)CYCLOHEXYL)MERCURY
mf: $C_9H_{15}BrClHgNO$ mw: 469.20
PROP: IDLH 10 mg/m³ (as Hg).
SYNS: 3-BROMO-N-(2-CHLOROMERCURICYCLOHEX-YL)PROPIONAMIDE □ MERCURY (E)-CHLORO(2-(3-BROMOPROPIONAMIDO)-CYCLOHEXYL)
CONSENSUS REPORTS: Mercury and its compounds are on the Community Right-To-Know List.
OSHA PEL: TWA 0.01 mg(Hg)/m³; STEL 0.03 mg/m³ (skin)
ACGIH TLV: TWA 0.1 mg(Hg)/m³ (skin); BEI: 35 µg/g creatinine total inorganic mercury in urine preshift; 15 µg/g creatinine total inorganic mercury in blood at end of shift at end of workweek.
DFG MAK: Confirmed Animal Carcinogen with Unknown Relevance to Humans
NIOSH REL: (Organomercury): TWA 0.01 mg/m³; STEL 0.03 mg/m³ (skin)
SAFETY PROFILE: Poison by intravenous route. See also MERCURY COMPOUNDS. When heated to decomposition it emits very toxic fumes of Br⁻, Cl⁻, NOₓ, and Hg.

CEU250   CAS: 78-86-4   HR: 3
2-CHLOROBUTANE
mf: C_4H_9Cl mw: 92.58
PROP: Flash p: 14°F, d: 0.87, vap d: 3.2, bp: 68.50°.
SYN: sec-BUTYL CHLORIDE
CONSENSUS REPORTS: Reported in EPA TSCA Inventory.
SAFETY PROFILE: Mildly toxic by ingestion, inhalation, and skin contact. Questionable carcinogen with experimental neoplastigenic data. Dangerous fire hazard when exposed to heat, open flame (sparks),

or oxidizers. To fight fire, use water, water spray, fog, mist, dry chemical, alcohol foam. When heated to decomposition it emits toxic fumes of Cl⁻. See also CHLORINATED HYDROCARBONS, ALIPHATIC.

CEU500   CAS: 928-51-8   HR: 2
4-CHLORO-1-BUTANOL
mf: C_4H_9ClO mw: 108.58
PROP: D: 1.088 @ 20°/4°, bp: 84–85° @ 16 mm.
SYNS: 4-CHLORBUTAN-1-OL (GERMAN) □ 4-CHLORO-1-BUTANE-OL □ 4-CHLOROBUTANOL □ TETRAMETHYLENE CHLOROHYDRIN
CONSENSUS REPORTS: Reported in EPA TSCA Inventory.
SAFETY PROFILE: Moderately toxic by ingestion. Questionable carcinogen with experimental neoplastigenic data. Mutation data reported. When heated to decomposition it emits toxic fumes of Cl⁻. See also CHLORIDES and ALCOHOLS.

CEV840   CAS: 76706-99-5   HR: 3
S-(4-CHLORO-2-BUTYNYL) DIPHENYL-
PHOSPHINOTHIOATE
mf: $C_{16}H_{14}ClOPS$ mw: 320.78
SYNS: 2-BUTYNE, 1-CHLORO-4-MERCAPTO-, S-ESTER WITH DIPHENYLPHOSPHINOTHIOATE □ PHOS-PHINOTHIOIC ACID, DIPHENYL-, S-(4-CHLORO-2-BUTYNYL) ESTER
SAFETY PROFILE: A poison by subcutaneous route. When heated to decomposition it emits toxic vapors of POₓ, SOₓ, and Cl⁻.

CEX800   CAS: 79622-59-6   HR: 2
3-CHLORO-N-(5-CHLORO-2,6-DINITRO-
4-TRIFLUOROMETHYLPHENYL)-5-
TRIFLUOROMETHYL-2-PYRIDIN-
AMINE
mf: $C_{13}H_4Cl_2F_6N_4O_4$ mw: 465.11
SYNS: 3-CHLORO-N-(3-CHLORO-5-TRIFLUORO-METHYL-2-PYRIDYL)-α,α,α-TRIFLUORO-2,6-DINITRO-p- □ TOLUIDINE □ ASC 66825 □ ASC 67178 □ FLUAZINAM □ FROWNCIDE □ IKF 1216 □ PP 192 □ 2-PYRIDINAMINE, 3-CHLORO-N-(3-CHLORO-2,6-DINITRO-4-(TRIFLUOROMETHYL)PHENYL)-5-(TRIFLUOROMETHYL)- □ SHIRLAN (ZENECA) □ SHIRLAN FLOW

SAFETY PROFILE: Moderately toxic by ingestion and inhalation. Experimental reproductive effects. When heated to decomposition it emits toxic vapors of NO_x, F^-, and Cl^-.

CFA500   CAS: 126-85-2   HR: 3
2-CHLORO-N-(2-CHLOROETHYL)-N-METHYLETHANAMINE-N-OXIDE

mf: $C_5H_{11}Cl_2NO$ mw: 172.07

SYNS: 2,2'-DICHLORO-N-METHYLDIETHYLAMINE-N-OXIDE □ DIETHYLAMINE, 2,2'-DICHLORO-N-METH-YL-, OXIDE □ HN2 AMINE OXIDE □ HN2 OXIDE MUSTARD □ MBAO □ MECHLORETHAMINE OXIDE □ METHYL-BIS(β-CHLOROETHYL)AMINE OXIDE □ METHYLBIS(β-CHLOROETHYL)AMINE N-OXIDE □ N-METHYL-DI-2-CHLOROETHYLAMINE-N-OXIDE □ MITOMEN □ MITOMIN □ NITROGEN MUSTARD OXIDE □ NITROGEN MUSTARD-N-OXIDE □ NITROMIN □ NMO □ N-OXYD-LOST □ N-OXYD-MUSTARD □ NSC-10107 □ OXY-NH2

SAFETY PROFILE: Poison by intravenous and intraperitoneal routes. Experimental reproductive effects. Questionable carcinogen with experimental carcinogenic and tumorigenic data. Mutation data reported. When heated to decomposition it emits toxic fumes of Cl^- and NO_x.

CFA750   CAS: 302-70-5   HR: 3
2-CHLORO-N-(2-CHLOROETHYL)-N-METHYLETHANAMINE-N-OXIDE HYDROCHLORIDE

mf: $C_5H_{11}Cl_2NO \cdot ClH$ mw: 208.53

PROP: A solid. Mp: 109–110°.

SYNS: CHLORMETHINE-N-OXIDE HYDROCHLOR-IDE □ 2,2'-DICHLORO-N-METHYLDIETHYLAMINE N-OXIDE HYDROCHLORIDE □ HN2 OXIDE HYDRO-CHLORIDE □ MBAO HYDROCHLORIDE □ MECHLO-RETHAMINE OXIDE HYDROCHLORIDE □ METHYL-BIS-(β-CHLORAETHYL)-AMIN-N-OXYD-HYDRO-CHLORID (GERMAN) □ METHYLBIS(β-CHLOROETH-YL)AMINE-N-OXIDE HYDROCHLORIDE □ N-METHYLBIS(2-CHLOROETHYL)AMINE-N-OXIDE HYDROCHLORIDE □ N-METHYL-2,2'-DICHLORO-DIETHYLAMINE-N-OXIDE HYDROCHLORIDE □ METHYLDI(2-CHLOROETHYL)AMINE-N-OXIDE HYDROCHLORIDE □ MITOMEN □ MUSTRON □ NITROGEN MUSTARD OXIDE □ NITROGEN MUSTARD-N-OXIDE □ NITROGEN MUSTARD-N-OXIDE HYDROCHLORIDE □ NITROMIM □ NITROMIN HYDROCHLORIDE □ N-OXYD-LOST □ NSC-10107 □

OSSIAMINA □ OSSICHLORIN □ OXYAMINE □ SK-598 □ XA 2

CONSENSUS REPORTS: IARC Cancer Review: Group 2B IMEMDT 7,56,87; Animal Sufficient Evidence IMEMDT 9,209,75; EPA Genetic Toxicology Program.

SAFETY PROFILE: Suspected carcinogen with experimental tumorigenic data. Poison by ingestion, subcutaneous, intravenous, and intraperitoneal routes. Mutation data reported. An antineoplastic agent. When heated to decomposition it emits toxic fumes of Cl^- and NO_x.

CFA800   CAS: 107359-74-0 HR: 3
4-CHLORO-2-(4-CHLORO-2-FLUORO-PHENYL)-5-((4-CHLOROPHENYL)-METHOXY)-3(2H)-PYRIDAZINONE

mf: $C_{17}H_{10}Cl_3FN_2O_2$ mw: 399.64

SYN: 3(2H)-PYRIDAZINONE, 4-CHLORO-2-(4-CHLORO-2-FLUOROPHENYL)-5-((4-CHLOROPHENYL)-METHOXY)-

SAFETY PROFILE: A poison by ingestion. When heated to decomposition it emits toxic vapors of NO_x, F^-, and Cl^-.

CFC600   CAS: 122322-19-4 HR: 3
4-CHLORO-2-(4-CHLOROPHENYL)-5-((6-IODO-3-PYRIDINYL)METHOXY)-3(2H)-PYRIDAZINONE

mf: $C_{16}H_{10}Cl_2IN_3O_2$ mw: 474.09

SYN: 3(2H)-PYRIDAZINONE, 4-CHLORO-2-(4-CHLOROPHENYL)-5-((6-IODO-3-PYRIDINYL)-METHOXY)-

SAFETY PROFILE: A poison by ingestion. When heated to decomposition it emits toxic vapors of NO_x, I^-, and Cl^-.

CFD990   CAS: 59-50-7   HR: 3
4-CHLORO-m-CRESOL

mf: C_7H_7ClO mw: 142.59

PROP: Odorless crystals (when pure) from pet ether. Mp: 55.5°, bp: 235°. Somewhat sol in water; very sol in org solvs.

SYNS: APTAL □ BAKTOL □ BAKTOLAN □ CAND-ASEPTIC □ p-CHLOR-m-CRESOL □ CHLOROCRESOL □ p-CHLOROCRESOL □ p-CHLORO-m-CRESOL □ 6-CHLORO-m-CRESOL □ 2-CHLORO-HYDROXY-TOLUENE □ 6-CHLORO-3-HYDROXYTOLUENE □ 4-CHLORO-3-METHYLPHENOL □ 3-METHYL-4-CHLOROPHENOL □ OTTAFACT □ PARMETOL □

PAROL □ PCMC □ PREVENTOL CMK □ RASCHIT □ RASEN-ANICON □ RCRA WASTE NUMBER U039

CONSENSUS REPORTS: Reported in EPA TSCA Inventory. Chlorophenol compounds are on the Community Right-To-Know List.

SAFETY PROFILE: Poison by intravenous, subcutaneous, and intraperitoneal routes. Moderately toxic by ingestion. An allergen. Mutation data reported. Incompatible with sodium hydroxide. When heated to decomposition it emits toxic fumes of Cl⁻ and phosgene. See also CRESOL and CHLOROPHENOLS.

CFK000   CAS: 93-71-0   HR: 3
2-CHLORO-N,N-DIALLYLACETAMIDE
mf: $C_8H_{12}ClNO$ mw: 173.66

PROP: Amber liquid. Bp: 74° @ 0.3 mm. Sltly sol in water; sol in alc, hexane, and xylene.

SYNS: ALIDOCHLOR □ ALLIDOCHLOR □ CDAA □ CDAAT □ α-CHLORO-N,N-DIALLYLACETAMIDE □ 2-CHLORO-N,N-DI-2-PROPENYLACETAMIDE □ CP 6,343 □ DIALLYLCHLOROACETAMIDE □ N,N-DIALLYL-CHLOROACETAMIDE □ N,N-DIALLYL-α-CHLORO-ACETAMIDE □ N,N-DIALLYL-2-CHLOROACETAMIDE □ NCI-CO4035 □ RADOX □ RANDOX □ RANTOX T

CONSENSUS REPORTS: Reported in EPA TSCA Inventory.

SAFETY PROFILE: Poison by skin contact. Moderately toxic by ingestion. An herbicide. When heated to decomposition it emits very toxic fumes of Cl⁻ and NOₓ. See also ALLYL COMPOUNDS.

CFK125   CAS: 95-83-0   HR: 3
4-CHLORO-1,2-DIAMINOBENZENE
mf: $C_6H_7ClN_2$ mw: 142.60

PROP: Leaflets from H_2O. Mp: 76°.

SYNS: p-CHLORO-o-PHENYLENEDIAMINE □ 4-CHLORO-o-PHENYLENEDIAMINE □ 4-CHLORO-1,2-PHENYLENEDIAMINE □ 4-Cl-o-PD □ 1,2-DIAMINO-4-CHLOROBENZENE □ 3,4-DIAMINOCHLOROBENZENE □ 3,4-DIAMINO-1-CHLOROBENZENE □ NCI-C03292 □ URSOL OLIVE 6G

CONSENSUS REPORTS: NTP 10th Report on Carcinogens. IARC Cancer Review: Group 2B IMEMDT 7,56,87; Human Limited Evidence IMEMDT 27,81,82; Animal Sufficient Evidence IMEMDT 27,81,82. NCI Carcinogenesis Bioassay (feed); Clear Evidence: mouse, rat NCITR* NCI-CG-TR-63,78. Reported in EPA TSCA Inventory.

SAFETY PROFILE: Confirmed carcinogen with experimental carcinogenic and neoplastigenic data. Human mutation data reported. When heated to decomposition it emits toxic fumes of Cl⁻ and NOₓ. See also AROMATIC AMINES.

CFK500   CAS: 124-48-1   HR: 2
CHLORODIBROMOMETHANE
mf: $CHBr_2Cl$ mw: 208.29

PROP: Colorless to pale-yellow, heavy liquid. Bp: 121–122°, fp: −22°, d: 2.440 @ 25°/25°.

SYNS: CDBM □ DIBROMOCHLOROMETHANE □ NCI-C55254

CONSENSUS REPORTS: IARC Cancer Review: Group 3 IMEMDT 52,243,91; Animal Limited Evidence IMEMDT 52,243,91; Human Inadequate Evidence IMEMDT 52,243,91. NTP Carcinogenesis Studies (gavage); Some Evidence: mouse NTPTR* NTP-TR-282,86; No Evidence: rat NTPTR* NTP-TR-282,85. Reported in EPA TSCA Inventory.

SAFETY PROFILE: Moderately toxic by ingestion. Questionable carcinogen with experimental carcinogenic data. Human mutation data reported. Compounds of this type are generally irritating and narcotic. See also BROMOFORM and CHLOROFORM. When heated to decomposition it emits toxic fumes of Cl⁻ and Br⁻.

CFX000   CAS: 15972-60-8   HR: 2
2-CHLORO-2',6'-DIETHYL-N-(METHOXY-METHYL)ACETANILIDE
mf: $C_{14}H_{20}ClNO_2$ mw: 269.80

PROP: Crystals. Sltly sol in H_2O; sol in Me_2CO, C_6H_6, EtOH, and EtOAc.

SYNS: ALACHLOR (USDA) □ ALANEX □ ALOCHLOR □ CHLORESSIGSAEURE-N-(METHOXYMETHYL)-2,6-DIAETHYLANILID (GERMAN) □ 2-CHLORO-N-(2,6-DIETHYLPHENYL)-N-(METHOXYMETHYL)ACETAM-IDE □ CP 50144 □ LASSO □ LAZO □ METACHLOR □ METHACHLOR □ PILLARZO

CONSENSUS REPORTS: EPA Genetic Toxicology Program.
ACGIH TLV: inhalable fraction 1 mg/m³; (sensitizer); Confirmed Animal Carcinogen with Unknown Revelance to Humans
SAFETY PROFILE: Moderately toxic by ingestion, skin contact, and possibly other routes. Questionable carcinogen with experimental carcinogenic data. Human mutation data reported. When heated to decomposition it emits very toxic fumes of Cl⁻ and NO$_x$.

CFX250   CAS: 75-68-3   HR: 1
1-CHLORO-1,1-DIFLUOROETHANE
DOT: UN 2517
mf: C$_2$H$_3$ClF$_2$ mw: 100.50
PROP: Gas. Mp: −131°, bp: −9.5°, d: 1.19, lel: 9.0%, uel: 14.8%. Insol in water.
SYNS: CFC 142b □ CHLORODIFLUOROETHANES (DOT) □ CHLOROETHYLIDENE FLUORIDE □ α-CHLOROETHYLIDENE FLUORIDE □ DIFLUORO-CHLOROETHANES (DOT) □ 1,1-DIFLUORO-1-CHLOROETHANE □ FC142b □ FLUOROCARBON FC142b □ FREON 142 □ FREON 142b □ GENETRON 101 □ GENETRON 142b □ GENTRON 142B □ HYDROCHLOROFLUOROCARBON 142b □ R142B (DOT)
CONSENSUS REPORTS: Reported in EPA TSCA Inventory.
DFG MAK: 1000 ppm (4200 mg/m³)
DOT CLASSIFICATION: 2.1; Label: Flammable Gas
SAFETY PROFILE: Very mildly toxic by inhalation. Mutation data reported. A very dangerous fire hazard when exposed to heat, flame, or oxidizing materials. To fight fire, stop flow of gas. Can react vigorously with oxidizing materials. When heated to decomposition it emits toxic fumes of F⁻ and Cl⁻.

CFX500   CAS: 75-45-6   HR: 1
CHLORODIFLUOROMETHANE
DOT: UN 1018
mf: CHClF$_2$ mw: 86.47
PROP: Gas. D: 1.49 @ 69°/4°, mp: −146°, bp: −40.8°, fp: −160°, autoign temp: 1170°F. Sltly sol in water.

SYNS: ALGEON 22 □ ALGOFRENE 22 □ ALGOFRENE TYPE 6 □ ARCTON 4 □ ARCTON 22 □ CFC 22 □ CHLORODIFLUOROMETHANE □ CHLORODIFLUOROMETHANE (ACGIH,DOT,OSHA) □ DAIFLON 22 □ DIFLUOROCHLOROMETHANE □ DIFLUOROMONOCHLOROMETHANE □ DYMEL 22 □ ELECTRO-CF 22 □ ESKIMON 22 □ F 22 □ FC 22 □ FLUGENE 22 □ FLUOROCARBON-22 □ FORANE 22 □ FREON □ FREON 22 □ FRIGEN □ FRIGEN 22 □ GENETRON 22 □ HALTRON 22 □ ISCEON 22 □ ISOTRON 22 □ KHALADON 22 □ MONO-CHLORODIFLUOROMETHANE □ PROPELLANT 22 □ R-22 □ R22 (DOT) □ REFRIGERANT 22 □ UCON 22 □ UCON 22/HALOCARBON 22
CONSENSUS REPORTS: IARC Cancer Review: Group 3 IMEMDT 7,149,87; Human Inadequate Evidence IMEMDT 41,237,86; Animal Limited Evidence IMEMDT 41,237,86. Reported in EPA TSCA Inventory. EPA Genetic Toxicology Program.
OSHA PEL: TWA 1000 ppm
ACGIH TLV: TWA 1000 ppm; Not Classifiable as a Human Carcinogen
DFG MAK: 500 ppm (1800 mg/m³)
DOT CLASSIFICATION: 2.2; Label: Nonflammable Gas
SAFETY PROFILE: Mildly toxic by inhalation. Experimental reproductive effects. Mutation data reported. An asphyxiant in high concentrations. At elevated pressures, 50% mixtures with air are combustible although ignition is difficult. When heated to decomposition it emits toxic fumes of F⁻ and Cl⁻. See also CHLORINATED HYDROCARBONS, ALIPHATIC; and FLUORIDES.

CFY000   CAS: 58-93-5   HR: 3
6-CHLORO-3,4-DIHYDRO-2H-1,2,4-BENZOTHIADIAZINE-7-SULFON-AMIDE- 1,1-DIOXIDE
mf: C$_7$H$_8$ClN$_3$O$_4$S$_2$ mw: 297.75
PROP: A solid. Mp: 273–275°.
SYNS: AQUARILLS □ AQUARIUS □ BREMIL □ 6-CHLORO-3,4-DIHYDRO-7-SULFAMOYL-2H-1,2,4-BENZOTHIADIAZINE-1,1-DIOXIDE □ 6-CHLORO-7-SULFAMOYL-3,4-DIHYDRO-2H-1,2,4-BENZOTHIADI-AZINE-1,1-DIOXIDE □ CHLOROSULTHIADIL □ CHLORSULFONAMIDO DIHYDROBENZOTHIADI-AZINE DIOXIDE □ CHLORZIDE □ CIDREX □ DICHLOROSAL □ DICHLOTIAZID □ DICHLOTRIDE □

DICLOTRIDE □ 3,4-DIHYDRO-6-CHLORO-7-SULFAMYL-1,2,4-BENZOTHIADIAZINE-1,1-DIOXIDE □ DIHYDRO-CHLOROTHIAZID □ DIHYDROCHLOROTHIAZIDE □ 3,4-DIHYDROCHLOROTHIAZIDE □ DIHYDROXY-CHLOROTHIAZIDUM □ DIREMA □ DISALUNIL □ DRENOL □ DYAZIDE □ ESIDREX □ ESIDRIX □ FLUVIN □ HCTZ □ HCZ □ HIDRIL □ HIDRO-CHLORTIAZID □ HIDRORONOL □ HIDROTIAZIDA □ HYDRO-AQUIL □ HYDROCHLORTHIAZID □ HYDRODIURETIC □ HYDRO-DIURIL □ HYDRO-SALURIC □ HYDROTHIDE □ HYPOTHIAZIDE □ IDROTIAZIDE □ IVAUGAN □ JEN-DIRIL □ MASCHITT □ MEGADIURIL □ NCI-C55925 □ NEFRIX □ NEO-CODEMA □ NEOFLUMEN □ ORETIC □ PANURIN □ RO-HYDRAZIDE □ SU 5879 □ THIARETIC □ THIURETIC □ THLARETIC □ URODIAZIN □ VETIDREX □ ZIDE

CONSENSUS REPORTS: Reported in EPA TSCA Inventory. EPA Genetic Toxicology Program.

SAFETY PROFILE: Poison by intraperitoneal and intravenous routes. Moderately toxic by ingestion and subcutaneous routes. Human systemic effects by ingestion: sodium level changes, chlorine level changes, acute pulmonary edema, nausea or vomiting. Experimental reproductive effects. Questionable carcinogen with experimental tumorigenic data. Mutation data reported. A diuretic. When heated to decomposition it emits very toxic fumes of SO_x, Cl⁻, and NO_x.

CGB500   CAS: 1779-25-5   HR: 1
CHLORO DIISOBUTYL ALUMINUM
mf: $C_8H_{18}AlCl$ mw: 176.69
SYNS: ALLUMINIO DIISOBUTIL-MONOCLORURO (ITALIAN) □ BIS(ISOBUTYL)ALUMINUM CHLORIDE □ CHLOROBIS(2-METHYLPROPYL)ALUMINUM □ DIISOBUTYLALUMINUM CHLORIDE □ DIISOBUTYL-ALUMINUM MONOCHLORIDE □ DIISOBUTYL-CHLOROALUMINUM

CONSENSUS REPORTS: Reported in EPA TSCA Inventory.

ACGIH TLV: TWA 2 mg(Al)/m³

SAFETY PROFILE: Mildly toxic by inhalation. See also ALUMINUM COMPOUNDS and CHLORIDES. Ignites spontaneously in air. When heated to decomposition it emits toxic fumes of Cl⁻.

CGH800   CAS: 29023-82-3   HR: 3
2-CHLORO-9-(2,2-DIMETHYLHY-DRAZINO)ACRIDINE
mf: $C_{15}H_{14}ClN_3$ mw: 271.77
SYN: ACRIDINE, 2-CHLORO-9-(2,2-DIMETHYL-HYDRAZINO)-
SAFETY PROFILE: A poison by ingestion. When heated to decomposition it emits toxic vapors of NO_x and Cl⁻.

CGJ280   CAS: 17256-39-2   HR: 3
1-CHLORO-N,N-DIMETHYL-2-PROP-ANAMINE HYDROCHLORIDE
mf: $C_5H_{12}ClN•ClH$ mw: 158.09
SYNS: (β-CHLOROISOPROPYL)DIMETHYLAMINE HYDROCHLORIDE □ ETHYLAMINE, 2-CHLORO-N,N-TRIMETHYL-, HYDROCHLORIDE □ 2-PROPANAMINE, 1-CHLORO-N,N-DIMETHYL-, HYDROCHLORIDE
SAFETY PROFILE: A poison by ingestion. Moderately toxic by skin contact. A severe eye irritant. When heated to decomposition it emits toxic vapors of NO_x, HCl, and Cl⁻.

CGL750   CAS: 25567-67-3   HR: 3
CHLORODINITROBENZENE
DOT: UN 1577
mf: $C_6H_3ClN_2O_4$ mw: 202.56
SYNS: CHLORODINITROBENZENE (mixed isomers) (DOT) □ DINITROCHLOROBENZENE □ DINITROCHLOROBENZENE (DOT)
DOT CLASSIFICATION: 6.1; Label: Poison
SAFETY PROFILE: A poison by ingestion. Mutation data reported. Potentially explosive. When heated to decomposition it emits very toxic fumes of Cl⁻ and NO_x. See also other chloro-dinitrobenzenes.

CGM000   CAS: 97-00-7   HR: 3
1-CHLORO-2,4-DINITROBENZENE
mf: $C_6H_3ClN_2O_4$ mw: 202.56
PROP: Yellow rhombic crystals from Et₂O; insol in water. Mp(α): 51°, mp(β): 43°, mp(γ): 27°, bp: 315° (sltly decomp), lel: 2.0%, uel: 22%, flash p: 382°F (CC), d(α): 1.687 @ 22°, d(β): 1.680 @ 20°/4°, vap d: 6.98.
SYNS: 1-CHLOOR-2,4-DINITROBENZEEN (DUTCH) □ 1-CHLOR-2,4-DINITROBENZENE □ 4-CHLORO-1,3-DINITROBENZENE □ 6-CHLORO-1,3-DINITRO-BENZENE □ 1-CHLORO-2,4-DINITROBENZOL (GERMAN) □ 1-CLORO-2,4-DINITROBENZENE (ITALIAN) □ 2,4-DINITROCHLOROBENZENE □ 1,3-

DINITRO-4-CHLOROBENZENE □ 2,4-DINITRO-1-CHLOROBENZENE □ DINITROCHLOROBENZOL □ DNCB

CONSENSUS REPORTS: Reported in EPA TSCA Inventory.

SAFETY PROFILE: Poison by skin contact and intraperitoneal routes. Moderately toxic by ingestion. A severe human skin and eye irritant. Acts as a primary irritant as well as a sensitizer of skin. An allergen. Mutation data reported. Combustible when exposed to heat or flame. A moderate explosion hazard when exposed to flame, sparks, heated to 150°, or when shocked in a sealed container. Explosive reaction with ammonia at 170°C/40 bar. To fight fire, use CO_2, dry chemical. Reacts violently with hydrazine sulfate or hydrazine hydrate. See also NITRO COMPOUNDS of AROMATIC HYDROCARBONS.

CGN000   CAS: 712-48-1   HR: 3
CHLORODIPHENYLARSINE
DOT: UN 1699
mf: $C_{12}H_{10}AsCl$ mw: 264.59
PROP: Colorless crystals when pure, technical product is dark-brown liquid. Mp: 44°, bp: 333° (decomp), d: 1.333 @ 40° (solid), 1.358 @ 45° (liquid), vap press: 0.00049 mm @ 20°, vap d: 9.15. Insol in H_2O; sol in org solvs.
SYNS: BLUE CROSS □ CLARK I □ DA □ DIPHENYL-ARSINOUS CHLORIDE □ DIPHENYLCHLOORARSINE (DUTCH) □ DIPHENYLCHLOROARSINE (DOT) □ SNEEZING GAS

CONSENSUS REPORTS: Arsenic and its compounds are on the Community Right-To-Know List.

OSHA PEL: TWA 0.5 mg(As)/m³

DOT CLASSIFICATION: 6.1; Label: Poison

SAFETY PROFILE: A human poison by inhalation. Poison experimentally by inhalation and skin contact. A powerfully irritating military poison. Exposure yields cold-like symptoms, plus headache, vomiting and nausea. A nonpersistent gas. Decontamination is by use of chlorine or caustic soda in confined spaces. When heated to decomposition it emits toxic fumes of As and Cl⁻. See also ARSENIC COMPOUNDS.

CGO500   CAS: 115-96-8   HR: 3
2-CHLOROETHANOL PHOSPHATE
mf: $C_6H_{12}Cl_3O_4P$ mw: 285.50
PROP: Flash p: 421°F (COC), boiling range: 210–220° @ 20 mm, d: 1.425 @ 20°/20°, autoign temp: 1115°F, vap press: 0.5 mm @ 145°.
SYNS: CELLUFLEX □ FYROL CEF □ NCI-C60128 □ NIAX FLAME RETARDANT 3 CF □ TRICHLORETHYL PHOSPHATE □ TRI-β-CHLOROETHYL PHOSPHATE □ TRI(2-CHLOROETHYL)PHOSPHATE □ TRIS(2-CHLOROETHYL)ESTER PHOSPHORIC ACID □ TRIS(β-CHLOROETHYL) PHOSPHATE □ TRIS(2-CHLOROETH-YL) PHOSPHATE

CONSENSUS REPORTS: IARC Cancer Review: Group 3 IMEMDT 48,109,90; Animal Inadequate Evidence IMEMDT 48,109,90. Reported in EPA TSCA Inventory. EPA Genetic Toxicology Program.

SAFETY PROFILE: Poison by intraperitoneal route. Moderately toxic by ingestion. Experimental reproductive effects. Questionable carcinogen with experimental tumorigenic data. A skin and eye irritant. Combustible when exposed to heat or flame. When heated to decomposition it emits very toxic fumes of PO_x and Cl⁻. See also PHOSPHATES, CHLORIDES, and ESTERS.

CGU199   CAS: 627-11-2   HR: 3
2-CHLOROETHYL CHLOROFORMATE
mf: $C_3H_4Cl_2O_2$ mw: 142.97
PROP: Bp: 155.7°, d: 1.3847, insoluble in water.
SYNS: 2-CHLORETHYLESTER KYSELINY CHLORMRAVENCI □ CHLOROFORMIC ACID 2-CHLOROETHYL ESTER □ FORMIC ACID, CHLORO-, 2-CHLOROETHYL ESTER □ TL 207

CONSENSUS REPORTS: EPA Extremely Hazardous Substances List. EPA Genetic Toxicology Program. Reported in EPA TSCA Inventory.

DOT CLASSIFICATION: 6.1; Label: Poison, Corrosive

SAFETY PROFILE: Poison by inhalation. May also cause fatalities by ingestion or skin contact. May cause burns to the eyes and skin. When heated to decomposition it emits toxic fumes of Cl⁻.

CGV250   CAS: 13010-47-4   HR: 3
1-(2-CHLOROETHYL)-3-CYCLOHEXYL-
1-NITROSOUREA

mf: $C_9H_{16}ClN_3O_2$ mw: 233.73

PROP: Yellow powder. Mp: 90°.

SYNS: BELUSTINE □ CCNU □ CECENU □ CEENU □ CHLOROETHYLCYCLOHEXYLNITROSOUREA □ N-(2-CHLOROETHYL)-N'-CYCLOHEXYL-N-NITROSOUREA □ (CHLORO-2-ETHYL)-1-CYCLOHEXYL-3-NITROSOUREA □ CINU □ (CLORO-2-ETIL)-1-CICLOESIL-3-NITROSO-UREA (ITALIAN) □ ICIG 1109 □ LOMUSTINE □ NCI-C04740 □ NSC-79037 □ RB 1509 □ SRI 2200

CONSENSUS REPORTS: NTP 10th Report on Carcinogens. IARC Cancer Review: Group 2A IMEMDT 7,150,87; Human Limited Evidence IMEMDT 26,137,81; Animal Sufficient Evidence IMEMDT 26,137,81. NCI Carcinogenesis Studies (ipr); Clear Evidence: mouse CANCAR 40,1935,77; No Evidence: rat CANCAR 40,1935,77. EPA Genetic Toxicology Program.

SAFETY PROFILE: Confirmed carcinogen with experimental carcinogenic and tumorigenic data. Poison by ingestion, intraperitoneal, subcutaneous, intravenous, and possibly other routes. Human systemic effects by ingestion: anorexia, nausea or vomiting, leukopenia (decrease in the white blood cell count), and thrombocytopenia (decrease in the number of blood platelets). Experimental teratogenic and reproductive effects. Human mutation data reported. When heated to decomposition it emits very toxic fumes of Cl⁻ and NO$_x$. See also N-NITROSO COMPOUNDS.

CGV600   CAS: 4261-68-1   HR: 3
2-CHLOROETHYLDIISOPROPYLAMINE
HYDROCHLORIDE

mf: $C_8H_{18}ClN\cdot ClH$ mw: 200.18

SYNS: (β-CHLOROETHYL)DIISOPROPYLAMINE HYDROCHLORIDE □ N-(CHLOROETHYL)-DIISOPROPYLAMINE HYDROCHLORIDE □ N-(2-CHLOROETHYL)-N-(1-METHYLETHYL)-2-PROPANAMINE HYDROCHLORIDE □ 2-(DIISOPROPYLAMINO)ETHYL CHLORIDE HYDROCHLORIDE □ 2-PROPANAMINE, N-(2-CHLOROETHYL)-N-(1-METHYLETHYL)-, HYDROCHLORIDE □ TRIETHYLAMINE, 2"-CHLORO-1,1'-DIMETHYL-, HYDROCHLORIDE

SAFETY PROFILE: A poison by ingestion, inhalation, and skin contact. A severe skin and eye irritant. When heated to decomposition it emits toxic vapors of NO$_x$, HCl, and Cl⁻.

CGW000   CAS: 107-99-3   HR: 3
N-(2-CHLOROETHYL)DIMETHYLAMINE

mf: $C_4H_{10}ClN$ mw: 107.60

PROP: Liquid. Vap d: 3.72.

SYNS: CHLORO(DIMETHYLAMINO)ETHANE □ β-CHLOROETHYLDIMETHYLAMINE □ (2-CHLORO-ETHYL)DIMETHYLAMINE □ DIMETHYLAMINOETHYL CHLORIDE □ β-(DIMETHYLAMINO)ETHYL CHLORIDE □ 2-DIMETHYLAMINOETHYLCHLORIDE □ DIMETH-YL(2-CHLOROETHYL)AMINE □ HN 1 □ NITROGEN HALF MUSTARD

CONSENSUS REPORTS: EPA Genetic Toxicology Program. Reported in EPA TSCA Inventory.

SAFETY PROFILE: Poison by an unspecified route. A systemic irritant. Mutation data reported. When heated to decomposition it emits highly toxic fumes of Cl⁻ and NO$_x$.

CGY750   CAS: 693-07-2   HR: 3
CHLOROETHYL ETHYL SULFIDE

mf: C_4H_9ClS mw: 124.64

PROP: Liquid with penetrating odor. D: 1.07 @ 25°/4°, bp: 156.5°.

SYNS: 2-CHLOROETHYL ETHYL SULFIDE □ 2-CHLOROETHYL ETHYL THIOETHER □ 1-CHLORO-2-(ETHYLTHIO)ETHANE □ ETHYL-β-CHLOROETHYL SULFIDE □ ETHYL-2-CHLOROETHYL SULFIDE □ β-ETHYLMERKAPTOETHYLCHLORID (CZECH) □ 2-(ETHYLTHIO)CHLOROETHANE □ 2-ETHYLTHIO-ETHYL CHLORIDE □ HALF-MUSTARD GAS □ h-MG

CONSENSUS REPORTS: Reported in EPA TSCA Inventory. EPA Genetic Toxicology Program.

SAFETY PROFILE: Poison by ingestion and subcutaneous routes. Mutation data reported. A severe skin and eye irritant. See also ETHERS and SULFIDES. When heated to decomposition it emits very toxic fumes of Cl⁻ and SO_x.

CHC500   CAS: 107-27-7   HR: 3
CHLOROETHYL MERCURY
mf: C_2H_5ClHg mw: 265.11
PROP: Silvery, iridescent leaflets from EtOH. Mp: 196–198°. Sltly sol in Et_2O and EtOH; sol in $CHCl_3$; insol in H_2O. IDLH 10 mg/m³ (as Hg).
SYNS: CERESAN □ EMC □ ETHYLMERCURIC CHLORIDE □ ETHYLMERCURY CHLORIDE □ GANOZAN □ GRANOSAN
CONSENSUS REPORTS: Mercury and its compounds are on the Community Right-To-Know List.
OSHA PEL: TWA 0.01 mg(Hg)/m³; STEL 0.03 mg/m³ (skin)
ACGIH TLV: TWA 0.01 mg(Hg)/m³; BEI: 35 µg/g creatinine total inorganic mercury in urine preshift; 15 µg/g creatinine total inorganic mercury in blood at end of shift at end of workweek.
DFG MAK: Confirmed Animal Carcinogen with Unknown Relevance to Humans
NIOSH REL: (Organomercury) TWA 0.01 mg/m³; STEL 0.03 mg/m³ (skin)
SAFETY PROFILE: Poison by ingestion, inhalation, skin contact, and intraperitoneal routes. An experimental teratogen. Other experimental reproductive effects. Human mutation data reported. See also MERCURY COMPOUNDS, ORGANIC. When heated to decomposition it emits very toxic fumes of Cl⁻ and Hg.

CHC750   CAS: 3570-58-9   HR: 3
2-CHLOROETHYL METHANE-SULFONATE
mf: $C_3H_7ClO_3S$ mw: 158.61
PROP: Bp: 125–126° @ 9 mm.
SYNS: CB 1506 □ β-CHLOROETHYLMETHANESULFONATE □ CHLORO-ETHYL METHANESULPHONATE □ CHLORO-METHANE SULFONATE d'ETHYLE (FRENCH) □ METHANESULFONIC ACID CHLOROETHYL ESTER □ NSC-18016
CONSENSUS REPORTS: EPA Genetic Toxicology Program.

SAFETY PROFILE: Poison by ingestion, intravenous, and intraperitoneal routes. Experimental reproductive effects. Mutation data reported. See also SULFONATES. When heated to decomposition it emits very toxic fumes of Cl⁻ and SO_x.

CHD250   CAS: 13909-09-6   HR: 3
1-(2-CHLOROETHYL)-3-(4-METHYL-CYCLOHEXYL)-1-NITROSOUREA
mf: $C_{10}H_{18}ClN_3O_2$ mw: 247.76
PROP: Mp: 64° (decomp).
SYNS: N-(2-CHLOROETHYL)-N'-(trans-4-METHYL-CYCLOHEXYL)-N-NITROSOUREA □ 1-(2-CHLORO-ETHYL)-3-(trans-4-METHYLCYCLOHEXYL)-1-NITROSOUREA □ ICIG 1110 □ ME-CCNU □ METHYL-CCNU □ trans-METHYL-CCNU □ METHYL-LOMUSTINE □ NCI-C04955 □ NSC-95441 □ SEMUSTINE
CONSENSUS REPORTS: NTP 10th Report on Carcinogens. IARC Cancer Review: Group 1 IMEMDT 7,150,87; Animal Limited Evidence 7,150,87; Human Sufficient Evidence IMEMDT 7,150,87. NCI Carcinogenesis Studies (ipr); Clear Evidence: rat CANCAR 40,1935,77; No Evidence: mouse CANCAR 40,1935,77.
SAFETY PROFILE: Confirmed human carcinogen producing leukemia. Experimental carcinogenic and tumorigenic data. Poison by ingestion, intraperitoneal, intravenous, and possibly other routes. Mutation data reported. Human systemic effects by ingestion: nausea or vomiting, damage to kidney tubules and glomeruli, and hematuria (blood in the urine). When heated to decomposition it emits very toxic fumes of Cl⁻ and NO_x. See also N-NITROSO COMPOUNDS.

CHE000   CAS: 3647-69-6   HR: D
4-(2-CHLOROETHYL)MORPHOLINE HYDROCHLORIDE
mf: $C_6H_{12}ClNO•ClH$ mw: 186.10
CONSENSUS REPORTS: Reported in EPA TSCA Inventory.
SAFETY PROFILE: Experimental reproductive effects. Mutation data reported. When heated to decomposition it emits very toxic fumes of Cl⁻ and NO_x.

C

CHF500   CAS: 6296-45-3   HR: 2
2-CHLOROETHYL-N-
NITROSOURETHANE
mf: $C_5H_9ClN_2O_3$ mw: 180.61
SYNS: N-(2-CHLOROETHYL)-N-
NITROSOETHYLCARBAMATE □ N-(β-CHLOROETHYL)-
N-NITROSOURETHAN □ ETHYL-N-(β-CHLORO-
ETHYL)-N-NITROSOCARBAMATE □ TL 154
SAFETY PROFILE: Poison by inhalation,
ingestion, and intraperitoneal routes.
Questionable carcinogen with experimental
tumorigenic data. Mutation data reported.
Many N-nitroso compounds are
carcinogens. See also CARBAMATES and
N-NITROSO COMPOUNDS. When
heated to decomposition it emits very toxic
fumes of Cl⁻ and NO_x.

CHG000   CAS: 113-18-8   HR: 3
1-CHLORO-3-ETHYL-1-PENTEN-4-YN-3-
OL
mf: C_7H_9ClO mw: 144.61
PROP: Liquid with pungent, aromatic odor,
slowly darkening on exposure to light and
air. D: 1.065–1.070 @ 25°/4°, bp: 173–174°.
Misc most org solvents; immisc in H_2O.
SYNS: A 71 □ AETHYL-CHLORVYNOL □ ALVINOL □
ARVYNOL □ β-CHLOROVINYL ETHYLETHYNYL
CARBINOL □ 3-(β-CHLOROVINYL)-1-PENTYN-3-OL □
ETCHLORVINOLO □ ETHCHLOROVYNOL □
ETHCHLORVINYL □ ETHCHLORVYNOL □
ETHOCHLORVYNOL □ ETHYL-β-CHLORO-
VINYLETHYNYL CARBINOL □ ETHYLCHLORVYNOL
□ NORMONSON □ NORMOSAN □ NORMOSON □
NOSTEL □ PLACIDIL □ PLACIDYL □ ROERIDORM □
SERENIL □ SERENSIL
SAFETY PROFILE: Poison by ingestion,
subcutaneous, intraperitoneal, and
intravenous routes. Human systemic effects
by ingestion: general anesthesia and
thrombocytopenia (reduction in the number
of blood platelets). Human effects on
newborn by an unspecified route: drug
dependency and Apgar score (condition of
newborn). Experimental teratogenic and
reproductive effects. When heated to
decomposition it emits toxic fumes of Cl⁻.

CHI250   CAS: 110-75-8   HR: 3
2-CHLOROETHYL VINYL ETHER

mf: C_4H_7ClO mw: 106.56
PROP: Liquid. Mp: −70.3°, bp: 108°, d: 1.05
@ 20°/4°, flash p: 80°F (OC).
SYNS: 2-CHLORETHYL VINYL ETHER □ (2-
CHLOROETHOXY)ETHENE □ RCRA WASTE
NUMBER U042 □ VINYL-β-CHLOROETHYL ETHER □
VINYL-2-CHLOROETHYL ETHER
CONSENSUS REPORTS: Reported in
EPA TSCA Inventory.
SAFETY PROFILE: Poison by ingestion.
Moderately toxic by inhalation and skin
contact. A severe eye and skin irritant. See
also ETHERS. Dangerous fire hazard when
exposed to heat, flame, or oxidizers.
Potentially explosive. May form dangerous
peroxides on exposure to air. To fight fire,
use alcohol foam, dry chemical. When
heated to decomposition it emits toxic
fumes of Cl⁻. See also CHLORIDES and
ETHERS.

CHI900   CAS: 593-70-4   HR: 3
CHLOROFLUOROMETHANE
mf: CH_2ClF mw: 68.48
PROP: Gas. Bp: −9°.
SYNS: CFC 31 □ FC 31 □ FREON 31 □
MONOCHLOROMONOFLUOROMETHANE □ R 31 □ R
31 (refrigerant)
CONSENSUS REPORTS: IARC Cancer
Review: Group 3 IMEMDT 7,56,87; Animal
Limited Evidence IMEMDT 41,229,86.
DFG MAK: Animal Carcinogen, Suspected
Human Carcinogen
SAFETY PROFILE: Confirmed carcinogen
with experimental carcinogenic data.
Moderately toxic by inhalation. Mutation
data reported. When heated to
decomposition it emits very toxic fumes of
Cl⁻ and F⁻. See also CHLORINATED
HYDROCARBONS, ALIPHATIC; and
FLUORIDES.

CHJ500   CAS: 67-66-3   HR: 3
CHLOROFORM
DOT: UN 1888
mf: $CHCl_3$ mw: 119.37
PROP: Colorless liquid; heavy, ethereal
odor. Mp: −63.2°, bp: 61.3°, flash p: none,
d: 1.481 @ 25°/4°, vap press: 100 mm @

10.4°, vap d: 4.12. Sltly sol in H_2O. IDLH 500 ppm.

SYNS: CHLOROFORME (FRENCH) □ CLOROFORMIO (ITALIAN) □ FORMYL TRICHLORIDE □ METHANE TRICHLORIDE □ METHENYL TRICHLORIDE □ METHYL TRICHLORIDE □ NCI-C02686 □ R 20 (refrigerant) □ RCRA WASTE NUMBER U044 □ TCM □ TRICHLOORMETHAAN (DUTCH) □ TRICHLOR-METHAN (CZECH) □ TRICHLOROFORM □ TRICHLOROMETHANE □ TRICLOROMETANO (ITALIAN)

CONSENSUS REPORTS: NTP 10th Report on Carcinogens. IARC Cancer Review: Group 2B IMEMDT 7,152,87; Animal Limited Evidence IMEMDT 1,61,72; Human Limited Evidence IMEMDT 20,401,79; Animal Sufficient Evidence IMEMDT 20,401,79. NCI Carcinogenesis Bioassay (gavage); Clear Evidence: mouse, rat NCITR* NCI-CG-TR,1976. EPA Genetic Toxicology Program. EPA Extremely Hazardous Substances List. Community Right-To-Know List. Reported in EPA TSCA Inventory.

OSHA PEL: TWA 2 ppm

ACGIH TLV: TWA 10 ppm; Suspected Human Carcinogen; Animal Carcinogen

DFG MAK: 10 ppm (50 mg/m³); Confirmed Animal Carcinogen with Unknown Relevance to Humans)

NIOSH REL: (Waste Anesthetic Gases and Vapors) CL 2 ppm/1H; (Chloroform) CL 2 ppm/60M

DOT CLASSIFICATION: 6.1; Label: Poison

SAFETY PROFILE: Confirmed carcinogen with experimental carcinogenic, neoplastigenic, and tumorigenic data. A human poison by ingestion and inhalation. An experimental poison by ingestion and intravenous routes. Moderately toxic experimentally by intraperitoneal and subcutaneous routes. Human systemic effects by inhalation: hallucinations and distorted perceptions, nausea, vomiting, and other unspecified gastrointestinal effects. Human mutation data reported. Experimental teratogenic and reproductive effects.

Inhalation of the concentrated vapor causes dilation of the pupils with reduced reaction to light, as well as reduced intraocular pressure (experimental). In the initial stages there is a feeling of warmth of the face and body, then an irritation of the mucous membranes, conjunctiva, and skin; followed by excitation, loss of reflexes, sensation, and consciousness. Prolonged inhalation will bring on paralysis accompanied by cardiac-respiratory failure and finally death.

Chloroform has been widely used as an anesthetic. However, due to its toxic effects, this use is being abandoned. Concentrations of 68,000–82,000 ppm in air can kill most animals in a few minutes. 14,000 ppm may cause death after an exposure of from 30 to 60 minutes. 5000–6000 ppm can be tolerated by animals for 1 hour without serious disturbances. The maximum concentration tolerated for several hours or for prolonged exposure with slight symptoms is 2000–2500 ppm. Prolonged administration as an anesthetic may lead to such serious effects as profound toxemia and damage to the liver, heart, and kidneys. Experimental prolonged but light anesthesia in dogs produces a typical hepatitis.

Explosive reaction with sodium + methanol or sodium methoxide + methanol. Mixtures with sodium or potassium are impact-sensitive explosives. Reacts violently with acetone + alkali (e.g., sodium hydroxide, potassium hydroxide, or calcium hydroxide), Al, disilane, Li, Mg, methanol + alkali, nitrogen tetroxide, perchloric acid + phosphorus pentoxide, potassium-tert-butoxide, sodium methylate, NaK. Incompatible with dinitrogen tetraoxide, fluorine, metals, or triisopropylphosphine. Nonflammable. When heated to decomposition it emits toxic fumes of Cl^-.

See also CHLORINATED HYDROCARBONS, ALIPHATIC.

CHJ750 CAS: 54-31-9 HR: 3
4-CHLORO-N-FURFURYL-5-
SULFAMOYLANTHRANILIC ACID
mf: $C_{12}H_{11}ClN_2O_5S$ mw: 330.76
PROP: Crystals from EtOH (aq). Mp: 206°.
SYNS: AISEMIDE □ ALUZINE □ 5-
(AMINOSULFURANYL)-4-CHLORO-2-((2-
FURNAYLMETHYL)AMINO)BENZOIC ACID □
BERONALD □ CHLOR-N-(2-FURYLMETHYL)-5-
SULFAMYLANTHRANILSAEURE (GERMAN) □ 4-
CHLORO-N-(2-FURYLMETHYL)-5-SULFAMOYL-
ANTHRANILIC ACID □ DESDEMIN □ DIURAL □
DRYPTAL □ ERROLON □ EUTENSIN □ FRUSEMIDE □
FRUSEMIN □ FRUSID □ FULSIX □ FULUVAMIDE □
FURANTHRIL □ FURANTHRYL □ FURANTRIL □
FURESIS □ FUROSEDON □ FUROSEMID □
FUROSEMIDE □ FUROSEMIDE "MITA" □ FURSEMID □
FURSEMIDE □ FUSID □ HYDRO-RAPID □ KATLEX □
LASEX □ LASIX □ LB 502 □ LOWPSTRON □
MACASIROOL □ NCI-C55936 □ NICOROL □ PREFEMIN
□ PROFEMIN □ RADONNA □ ROSEMIDE □ SALIX □
SEGURIL □ TRANSIT □ TROFURIT □ UREX □
UROSEMIDE
CONSENSUS REPORTS: IARC Cancer
Review: Group 3 IMEMDT 50,277,90;
Human Inadequate Evidence IMEMDT
50,277,90; Animal Inadequate Evidence
IMEMDT 50,277,90. EPA Genetic
Toxicology Program.
SAFETY PROFILE: Poison by intravenous
route. Moderately toxic by ingestion and
intraperitoneal routes. Human systemic
effects by intravenous route: change in the
sensitivity of the ear to sound, tinnitus,
unspecified effects on the heart, constriction
of the arteries, a decrease in urine volume,
interstitial nephritis, metabolic alkalosis,
pulse rate decrease, fall in blood pressure.
Ingestion can damage the liver.
Experimental teratogenic and reproductive
effects. Questionable carcinogen with
experimental carcinogenic effects. Human
mutation data reported. When heated to
decomposition it emits very toxic fumes of
Cl^-, NO_x, and SO_x.

CHL250 CAS: 13654-91-6 HR: 3
CHLOROHEXYL ISOCYANATE
DOT: UN 2206/UN 2207/UN 2478/UN
3080
mf: $C_7H_{10}ClNO$ mw: 159.63

SYNS: 6-CHLORHEXYLISOKYANAT □ ISOCYANIC
ACID, 6-CHLOROHEXYL ESTER
DOT CLASSIFICATION: 6.1; Label: KEEP
AWAY FROM FOOD (UN 2207); DOT
Class: 6.1; Label: Poison (UN 2206); DOT
Class: 6.1; Label: Poison, Flammable
Liquid (UN 3080); DOT Class: 3; Label:
Flammable Liquid, Poison (UN 2478)
SAFETY PROFILE: Poison by inhalation.
See also THIOCYANATES and ESTERS.
A flammable liquid. When heated to
decomposition it emits very toxic fumes of
Cl^- and NO_x.

CHM000 CAS: 615-67-8 HR: 3
CHLOROHYDROQUINONE
mf: $C_6H_5ClO_2$ mw: 144.56
PROP: Leaflets from $CHCl_3$; needles from
toluene. Mp: 106°, bp: 263°.
CONSENSUS REPORTS: Reported in EPA
TSCA Inventory.
SAFETY PROFILE: Poison by ingestion and
intraperitoneal routes. Moderately toxic by
skin contact. Experimental reproductive
effects. When heated to decomposition it
emits toxic fumes of Cl^-. See also
CHLORIDES.

CHP250 CAS: 303-47-9 HR: 3
(−)-N-((5-CHLORO-8-HYDROXY-3-
METHYL-1-OXO-7-ISOCHROMAN-
YL)CARBONYL)-3-PHENYLALANINE
mf: $C_{29}H_{18}ClNO_6$ mw: 403.84
PROP: Crystals from xylene. Mp: 169°.
SYNS: (R)N-((5-CHLORO-3,4-DIHYDRO-8-HYDROXY-3-
METHYL-1-OXO-1H-2-BENZOPYRAN-7-YL))PHENYL-
ALANINE □ NCI-C56586 □ OCHRATOXIN A
CONSENSUS REPORTS: NTP 10th Report
on Carcinogens. IARC Cancer Review:
Group 2B IMEMDT 56,489,93; Animal
Limited Evidence IMEMDT 31,191,83;
Animal Sufficient Evidence IMEMDT
56,489,93; Human Inadequate Evidence
IMEMDT 31,191,83; Human Inadequate
Evidence IMEMDT 56,489,93.
SAFETY PROFILE: Confirmed carcinogen
with carcinogenic and neoplastigenic data.
Poison by ingestion, intraperitoneal,
intravenous, and subcutaneous routes.

Experimental teratogenic and reproductive effects. Mutation data reported. When heated to decomposition it emits very toxic fumes of Cl⁻ and NO_x.

CHP500   CAS: 5160-02-1   HR: 2
5-CHLORO-2-((2-HYDROXY-1-NAPHTH-YL)AZO)-p-TOLUENE SULFONIC ACID, BARIUM SALT
mf: $C_{17}H_{12}ClN_2O_4S \cdot 1/2Ba$ mw: 444.49
SYNS: BRIGHT RED □ BRILLIANT RED □ BRILLIANT SCARLET □ BRILLIANT TONER Z □ BRONZE RED RO □ BRONZE SCARLET □ 5-CHLORO-2-((2-HYDROXY-1-NAPHTHALENYL)AZO)-4-METHYLBENZENE SULFONIC ACID, BARIUM SALT (2:1) □ 5-CHLORO-2-((2-HYDROXY-1-NAPHTHALENYL)AZO)-4-METHYL-BENZENE SULPHONIC ACID, BARIUM SALT □ 1-(4-CHLORO-o-SULFO-5-TOLYLAZO)-2-NAPHTHOL,-BARIUM SALT □ C.I. PIGMENT RED □ COSMETIC CORAL RED KO BLUISH □ DAINICHI LAKE RED C □ D&C RED No. 9 □ DESERT RED □ ELJON LAKE RED C □ HAMILTON RED □ HELIO RED TONER LCLL □ IRGALITE RED CBN □ ISOL LAKE RED LCS 12527 □ LAKE RED C □ LATEXOL SCARLET R □ LD RUBBER RED 16913 □ LUTETIA RED CLN □ MICROTEX LAKE RED CR □ MOHICAN RED A-8008 □ NCI-C53792 □ No. 3 CONC. SCARLET □ PARIDINE RED LCL □ PIGMENT RED CD □ POTOMAC RED □ RECOLITE RED LAKE C □ 1860 RED □ RED SCARLET □ SANYO LAKE RED C □ SEGNALE RED LC □ SICO LAKE RED 2L □ SUPEROL RED C RT-265 □ SYMULER LAKE RED C □ TERMO-SOLIDO RED LCG □ TEXAN RED TONER D □ TONER LAKE RED C □ TRANSPARENT BRONZE SCARLET □ VULCAFIX SCARLET R □ VULCAN RED LC □ VULCOL FAST RED L □ WAYNE RED X-2486
CONSENSUS REPORTS: IARC Cancer Review: Group 3 IMEMDT 57,203,93; Animal Inadequate Evidence IMEMDT 8,107,75; Human No Adequate Data IMEMDT 8,107,75; Human Inadequate Evidence IMEMDT 57,203,93. NTP Carcinogenesis Bioassay (feed); Clear Evidence: rat NTPTR* NTP-TR-225,82; No Evidence: mouse NTPTR* NTP-TR-225,82. Reported in EPA TSCA Inventory.
SAFETY PROFILE: Questionable carcinogen with experimental carcinogenic and tumorigenic data. Mutation data reported. When heated to decomposition it emits very toxic fumes of SO_x, NO_x, and Cl⁻. See also SULFONATES.

CHR500   CAS: 130-26-7   HR: 3
5-CHLORO-7-IODO-8-QUINOLINOL
mf: C_9H_5ClINO mw: 305.50
PROP: Brownish-yellow powder, darkens when exposed to light. Sltly sol in Et_2O.
SYNS: ALCHLOQUIN □ AMEBIL □ AMOENOL □ BACTOL □ BARQUINOL □ BUDOFORM □ CHINOFORM □ 5-CHLOR-7-JOD-8-HYDROXY-CHINOLIN (GERMAN) □ 5-CHLORO-8-HYDROXY-7-IODOQUINOLINE □ 5-CHLORO-7-IODO-8-HYDROXYQUINOLINE □ CHLOROIODOQUINE □ CLIOQUINOL □ CLIQUINOL □ ECZECIDIN □ EMAFORM □ ENTERO-BIO FORM □ ENTEROQUINOL □ ENTEROSEPTOL □ ENTERO-VIOFORM □ ENTEROZOL □ ENTERUM LOCORTEN □ ENTROKIN □ HI-ENTEROL □ HYDRIODIDE-ENTROL □ IODENTEROL □ IODOCHLORHYDROXYQUINOL □ IODOCHLORHYDROXYQUINOLINE □ 7-IODO-5-CHLORO-8-HYDROXYQUINOLINE □ 7-IODO-5-CHLOROXINE □ IODOENTEROL □ NIOFORM □ QUINAMBICIDE □ ROMETIN □ VIOFORM □ VIOFORM N.N.R.
CONSENSUS REPORTS: Reported in EPA TSCA Inventory. EPA Genetic Toxicology Program.
SAFETY PROFILE: Poison by ingestion. Moderately toxic by intraperitoneal route. Human systemic effects by ingestion: change in central nervous system electrical function, optic nerve damage, and changes in vision. Experimental teratogenic and reproductive effects. Human mutation data reported. When heated to decomposition it emits very toxic fumes of Cl⁻, I⁻, and NO_x.

CHS250   CAS: 4288-84-0   HR: D
1-CHLORO-2-ISOPROPOXY-2-PROPANOL
mf: $C_6H_{13}ClO_2$ mw: 152.64
SYNS: 1-CHLORO-3-(PENTYLOXY)-2-PROPANOL □ 2-PROPANOL, 1-CHLORO-3-ISOPROPOXY- □ U 25,352
CONSENSUS REPORTS: Reported in EPA TSCA Inventory.
SAFETY PROFILE: Experimental reproductive effects. When heated to decomposition it emits toxic fumes of Cl⁻.

CHU500   CAS: 59-85-8   HR: 3
p-CHLOROMERCURIC BENZOIC ACID
mf: $C_7H_5ClHgO_2$ mw: 357.16
PROP: A solid. Mp: 273°.

SYNS: (p-CARBOXYPHENYL)CHLOROMERCURY □ p-
(CHLOROMERCURI)BENZOIC ACID □ USAF D-3
CONSENSUS REPORTS: Reported in EPA
TSCA Inventory. Mercury and its
compounds are on the Community Right-
To-Know List.
OSHA PEL: CL 0.1 mg(Hg)/m³ (skin)
ACGIH TLV: TWA 0.1 mg(Hg)/m³ (skin);
BEI: 35 µg/g creatinine total inorganic
mercury in urine preshift; 15 µg/g creatinine
total inorganic mercury in blood at end of
shift at end of workweek.
DFG MAK: Confirmed Animal Carcinogen
with Unknown Relevance to Humans
NIOSH REL: (Mercury, Aryl and Inorganic)
CL 0.1 mg/m³ (skin)
SAFETY PROFILE: Poison by
intraperitoneal route. See also MERCURY
COMPOUNDS. When heated to
decomposition it emits very toxic fumes of
Cl⁻ and Hg.

**CHW675   CAS: 90-03-9   HR: 3
o-CHLOROMERCURIPHENOL**
mf: C₆H₅ClHgO   mw: 329.15
PROP: Crystals from H₂O. Mp: 152.5°.
IDLH 10 mg/m³ (as Hg).
SYNS: CHLORO(o-HYDROXYPHENYL)MERCURY □ o-
HYDROXYFENYLMERKURICHLORID □ o-HYDROXY-
PHENYLMERCURIC CHLORIDE □ MERCUFENOL
CHLORIDE □ MERCURY, CHLORO(2-HYDROXY-
PHENYL)- □ MYRINGACAINE DROPS □ PHENOL, o-
(CHLOROMERCURI)- □ SALICRESIN FLUID □ U-7743 □
USPULUM
CONSENSUS REPORTS: Reported in EPA
TSCA Inventory. Mercury and its
compounds as well as chlorophenol
compounds are on the Community Right-
To-Know List.
OSHA PEL: CL 0.1 mg(Hg)/m³ (skin)
ACGIH TLV: TWA 0.1 mg(Hg)/m³ (skin);
BEI: 35 µg/g creatinine total inorganic
mercury in urine preshift; 15 µg/g creatinine
total inorganic mercury in blood at end of
shift at end of workweek.
DFG MAK: Confirmed Animal Carcinogen
with Unknown Relevance to Humans
NIOSH REL: (Mercury, Aryl and Inorganic)
CL 0.1 mg/m³ (skin)

SAFETY PROFILE: Poison by ingestion,
subcutaneous, intravenous, and
intraperitoneal routes. An antiseptic. See
also MERCURY COMPOUNDS,
ORGANIC; and CHLOROPHENOLS.
When heated to decomposition it emits
toxic fumes of Cl⁻ and Hg.

**CHW750   CAS: 623-07-4   HR: 3
p-CHLOROMERCURIPHENOL**
mf: C₆H₅ClHgO   mw: 329.15
PROP: Plates from Me₂CO. Mp: 226–227°.
IDLH 10 mg/m³ (as Hg).
SYNS: CHLORO(p-HYDROXYPHENYL)MERCURY □ p-
(CHLOROMERCURI)PHENOL
CONSENSUS REPORTS: Reported in EPA
TSCA Inventory. Mercury and its
compounds as well as chlorophenol
compounds are on the Community Right-
To-Know List.
OSHA PEL: CL 0.1 mg(Hg)/m³ (skin)
ACGIH TLV: TWA 0.1 mg(Hg)/m³ (skin);
BEI: 35 µg/g creatinine total inorganic
mercury in urine preshift; 15 µg/g creatinine
total inorganic mercury in blood at end of
shift at end of workweek.
DFG MAK: Confirmed Animal Carcinogen
with Unknown Relevance to Humans
NIOSH REL: (Mercury, Aryl and Inorganic)
CL 0.1 mg/m³ (skin)
SAFETY PROFILE: Poison by
intraperitoneal route. When heated to
decomposition it emits very toxic fumes of
Cl⁻ and Hg. See also MERCURY
COMPOUNDS, ORGANIC; and
CHLOROPHENOLS.

**CIF775   CAS: 60177-39-1   HR: D
CHLOROMETHYLATED AMINATED
      STRYENE-DIVINYLBENZENE RESIN**
SAFETY PROFILE: When heated to
decomposition it emits toxic fumes of Cl⁻
and NO₂.

**CIG250   CAS: 6325-54-8   HR: 3
7-CHLOROMETHYL BENZ(a)ANTHRA-
      CENE**
mf: C₁₉H₁₃Cl   mw: 276.77
SYN: ICR 451

SAFETY PROFILE: Poison by intravenous route. Questionable carcinogen with experimental neoplastigenic data. Mutation data reported. When heated to decomposition it emits toxic Cl⁻. See also CHLORINATED HYDROCARBONS, AROMATIC.

CIK750   CAS: 321-54-0   HR: 3
3-CHLORO-4-METHYL-7-COUMARINYL DIETHYLPHOSPHATE
mf: $C_{14}H_{16}ClO_6P$ mw: 346.72
SYNS: COROXON □ COUMAPHOS-O-ANALOG □ COUMAPHOS OXYGEN ANALOG (USDA) □ O,O-DI(2-CHLOROETHYL)-7-(3-CHLORO-4-METHYLCOUMARIN-YL)PHOSPHATE □ O,O-DIETHYL-O-(3-CHLORO-4-METHYLCOUMARIN-7-YL)PHOSPHATE □ DIETHYL-3-CHLORO-4-METHYL-7-COUMARINYL PHOSPHATE □ PHOSPHORIC ACID, DIETHYL ESTER, with 3-CHLORO-7-HYDROXY-4-METHYLCOUMARIN
SAFETY PROFILE: Deadly poison by ingestion. When heated to decomposition it emits very toxic fumes of PO_x and Cl⁻. See also ESTERS and PHOSPHATES.

CIL800   CAS: 4362-40-7   HR: 3
4-(CHLOROMETHYL)-2,2-DIMETHYL-1,3-DIOXOLANE
mf: $C_6H_{11}ClO_2$ mw: 150.62
CONSENSUS REPORTS: Reported in EPA TSCA Inventory.
SAFETY PROFILE: Poison by ingestion. Experimental reproductive effects. When heated to decomposition it emits toxic fumes of Cl⁻.

CIM000   CAS: 3188-13-4   HR: 3
CHLOROMETHYL ETHYL ETHER
mf: C_3H_7ClO mw: 94.54
PROP: Flash p: <−2.2°F, bp: 83°.
SYNS: CHLOROMETHOXY ETHANE □ ETHOXY CHLOROMETHANE □ ETHOXY METHYL CHLORIDE
CONSENSUS REPORTS: Reported in EPA TSCA Inventory.
SAFETY PROFILE: A poison by inhalation and ingestion. A very dangerous fire and explosion hazard when exposed to heat or flame. See also ETHERS.

CIN750   CAS: 13345-62-5   HR: 3

7-CHLOROMETHYL-12-METHYL BENZ-(a)ANTHRACENE
mf: $C_{20}H_{15}Cl$ mw: 290.80
SYN: IRC 453
CONSENSUS REPORTS: EPA Genetic Toxicology Program.
SAFETY PROFILE: Poison by intravenous route. Questionable carcinogen with experimental neoplastigenic data. Mutation data reported. When heated to decomposition it emits toxic Cl⁻. See also CHLORINATED HYDROCARBONS, AROMATIC.

CIO250   CAS: 107-30-2   HR: 3
CHLOROMETHYL METHYL ETHER
DOT: UN 1239
mf: C_2H_5ClO mw: 80.52
PROP: Liquid. Flash p: <73.4°F, d: 1.070 @ 25 mm, bp: 59.5°.
SYNS: CHLORDIMETHYLETHER (CZECH) □ CMME □ DIMETHYLCHLOROETHER □ ETHER METHYLIQUE MONOCHLORE (FRENCH) □ METHYLCHLOROMETH-YL ETHER (DOT) □ METHYL CHLOROMETHYL ETHER, anhydrous (DOT) □ MONOCHLORODIMETHYL ETHER (MAK) □ RCRA WASTE NUMBER U046
CONSENSUS REPORTS: NTP 10th Report on Carcinogens. IARC Cancer Review: Group 1 IMEMDT 7,131,87; Animal Sufficient Evidence IMEMDT 4,239,74; Human Limited Evidence IMEMDT 4,239,74. EPA Genetic Toxicology Program. Reported in EPA TSCA Inventory. Community Right-To-Know List. EPA Extremely Hazardous Substances List.
OSHA PEL: OSHA: Cancer Suspect Agent
ACGIH TLV: Suspected Human Carcinogen
DFG MAK: Human Carcinogen
NIOSH REL: (Methyl Chloromethyl Ether) TWA use 29 CFR 1910.1006
DOT CLASSIFICATION: 6.1; Label: Poison, Flammable Liquid
SAFETY PROFILE: Confirmed human carcinogen with experimental carcinogenic, tumorigenic, and neoplastigenic data. Poison by inhalation. Moderately toxic by ingestion. Human mutation data reported. A very dangerous fire hazard when exposed to heat or flame. To fight fire, use alcohol foam,

water, CO_2, or dry chemical. Reaction with divalent metals forms a very reactive product. When heated to decomposition it emits toxic fumes of Cl^-. See also ETHERS and CHLORINATED HYDROCARBONS, ALIPHATIC.

CIQ500   CAS: 16339-16-5   HR: 3
2-CHLORO-N-METHYL-N-NITROSO-
      ETHYLAMINE
mf: $C_3H_7ClN_2O$ mw: 122.57
SYNS: 2-CHLORO-2-METHYL-N-NITROSO-
ETHANAMINE □ METHYL-2-CHLORAETHYL-
NITROSAMIN (GERMAN) □ METHYL(2-
CHLOROETHYL)NITROSAMINE □ N-
NITROSOMETHYL-2-CHLOROETHYLAMINE
SAFETY PROFILE: Poison by ingestion, intravenous, and possibly other routes. Questionable carcinogen with experimental tumorigenic data. Many nitrosamine compounds are carcinogens. When heated to decomposition it emits very toxic fumes of Cl^- and NO_x. See also NITROSAMINES.

CIR250   CAS: 94-74-6   HR: 3
(4-CHLORO-2-METHYLPHENOXY)-
      ACETIC ACID
mf: $C_9H_9ClO_3$ mw: 200.63
PROP: Crystals from C_6H_6 or toluene. Mp: 120°.
SYNS: AGRITOX □ AGROXONE □ ANICON KOMBI □ ANICON M □ BH MCPA □ BORDERMASTER □ BROMINAL M & PLUS □ B-SELEKTONON M □ CHIPTOX □ 4-CHLORO-o-CRESOXYACETIC ACID □ 4-CHLORO-o-TOLOXYACETIC ACID □ ((4-CHLORO-o-TOLYL)OXY)ACETIC ACID □ CHWASTOX □ CORNOX-M □ DED-WEED □ DICOPUR-M □ DICOTEX □ DOW MCP AMINE WEED KILLER □ EMCEPAN □ EMPAL □ HEDAPUR M 52 □ HERBICIDE M □ HORMOTUHO □ 4K-2M □ KILSEM □ KREZONE □ LEGUMEX DB □ LEUNA M □ LEYSPRAY □ LINORMONE □ M 40 □ 2M-4C □ MCP □ MCPA □ MEPHANAC □ METAXON □ METHOXONE □ 2-METHYL-4-CHLOROPHENOXY-ACETIC ACID □ 2-METHYL-4-CHLORPHENOXY-ESSIGSAEURE (GERMAN) □ 2M-4KH □ NETAZOL □ OKULTIN M □ PHENOXYLENE SUPER □ RAPHONE □ RAZOL DOCK KILLER □ RHOMENE □ RHONOX □ SEPPIC MMD □ SHAMROX □ SOVIET TECHNICAL HERBICIDE 2M-4C □ TRASAN □ U 46 M-FLUID □ USTINEX □ VACATE □ VERDONE □ VESAKONTUHO

MCPA □ WEEDAR MCPA CONCENTRATE □ WEEDONE MCPA ESTER □ WEED-RHAP □ ZELAN
CONSENSUS REPORTS: IARC Cancer Review: Group 2B IMEMDT 7,156,87; Animal Inadequate Evidence IMEMDT 30,255,83; Human Inadequate Evidence IMEMDT 30,255,83; Human Limited Evidence IMEMDT 41,357,86. Reported in EPA TSCA Inventory. EPA Genetic Toxicology Program.
SAFETY PROFILE: Suspected carcinogen. Poison by subcutaneous and intravenous routes. Moderately toxic by ingestion. Human systemic effects by ingestion: blood pressure decrease and coma. Experimental teratogenic and reproductive effects. Mutation data reported. An herbicide. When heated to decomposition it emits toxic fumes of Cl^-.

CIR500   CAS: 93-65-2   HR: 3
4-CHLORO-2-METHYLPHENOXY-α-
      PROPIONIC ACID
mf: $C_{10}H_{11}ClO_3$ mw: 214.66
SYNS: ACIDE 2-(4-CHLORO-2-METHYL-PHENOXY)PROPIONIQUE (FRENCH) □ ACIDO 2-(4-CLORO-2-METIL-FENOSSI)-PROPIONICO (ITALIAN) □ BH MECOPROP □ CHIPCO TURF HERBICIDE MCPP □ 2-(4-CHLOOR-2-METHYL-FENOXY)-PROPIONZUUR (DUTCH) □ 2-(4-CHLOR-2-METHYL-PHENOXY)-PROPIONSAEURE (GERMAN) □ (+)-α-(4-CHLORO-2-METHYLPHENOXY) PROPIONIC ACID □ 2-(4-CHLORO-2-METHYLPHENOXY)PROPIONIC ACID □ 2-(4-CHLOROPHENOXY-2-METHYL)PROPIONIC ACID □ 2-(p-CHLORO-o-TOLYLOXY)PROPIONIC ACID □ CMPP □ COMPITOX □ FBC CMPP □ HEDONAL MCPP □ ISO-CORNOX □ KILPROP □ LIRANOX □ 2M-4CP □ MCPP □ 2-MCPP □ MCPP 2,4-D □ MCPP-D-4 □ MCPP-K-4 □ MECOMEC □ MECOPEOP □ MECOPER □ MECOPEX □ MECOPROP □ MECOTURF □ MECPROP □ MEPRO □ METHOXONE □ α-(2-METHYL-4-CHLOROPHENOXY)-PROPIONIC ACID □ 2-(2-METHYL-4-CHLOROPHENO-XY)PROPIONIC ACID □ 2-METHYL-4-CHLOROPHENO-XY-α-PROPIONIC ACID □ 2-(2-METHYL-4-CHLOR-PHENOXY)-PROPIONSAEURE (GERMAN) □ 2M 4KHP □ N.B. MECOPROP □ PROPAL □ PROPONEX-PLUS □ RANKOTEX □ RD 4593 □ RUNCATEX □ U 46 □ U 46 KV-ESTER □ U 46 KV-FLUID □ VI-PAR □ VI-PEX
CONSENSUS REPORTS: IARC Cancer Review: Group 2B IMEMDT 7,156,87; Human Limited Evidence IMEMDT 41,357,86. EPA Genetic Toxicology

Program. Reported in EPA TSCA Inventory.

SAFETY PROFILE: Suspected carcinogen. Poison by ingestion. Moderately toxic by skin contact and intraperitoneal routes. Experimental teratogenic and reproductive effects. Mutation data reported. An herbicide. When heated to decomposition it emits toxic fumes of Cl⁻.

CIR750　　CAS: 22316-47-8　　HR: 3
7-CHLORO-1-METHYL-5-PHENYL-1H-
1,5-BENZODIAZEPINE-2,4(3H,5H)-
DIONE
mf: $C_{16}H_{13}ClN_2O_2$　　mw: 300.76
PROP: A solid. Mp: 180–182°.
SYNS: CHLOREPIN □ CLOBAZAM □ CLOREPIN □ FRISIUM □ H-4723 □ HR 376 □ LM-2717 □ 1-PHENYL-5-METHYL-8-CHLORO-1,2,4,5-TETRAHYDRO-2,4-DIOXO-3H-1,5-BENZODIAZEPINE □ RU-4723 □ URBANYL
SAFETY PROFILE: Poison by ingestion and intraperitoneal routes. Moderately toxic by subcutaneous route. Human systemic effects by ingestion: wakefulness, withdrawal, nausea and vomiting. An experimental teratogen. Other experimental reproductive effects. A tranquilizer. When heated to decomposition it emits very toxic fumes of NO_x and Cl⁻. See also DIAZEPAM.

CIS750　　CAS: 2058-52-8　　HR: 3
2-CHLORO-11-(4-METHYLPIPERAZ-
INO)DIBENZO(b,f)(1,4)THIAZEPINE
mf: $C_{18}H_{18}ClN_3S$　　mw: 343.90
PROP: Crystals from Et_2O/pet ether. Mp: 118–120°.
SYNS: 2-CHLORO-11-(4-METHYL-1-PIPERAZINYL)-DIBENZO(b,f)(1,4)THIAZEPINE □ DIBENZO-THIAZEPINE
SAFETY PROFILE: Poison by ingestion. An experimental teratogen. Other experimental reproductive effects. When heated to decomposition it emits very toxic fumes of Cl⁻, NO_x, and SO_x.

CIU750　　CAS: 563-47-3　　HR: 3
3-CHLORO-2-METHYLPROPENE
DOT: UN 2554
mf: C_4H_7Cl　　mw: 90.56

PROP: Colorless, volatile liquid; disagreeable odor. Bp: 72.17°, lel: 2.3%, uel: 9.3%, fp: <−80°, d: 0.9257 @ 20°/4°, vap press: 101.7 mm @ 20°, vap d: 3.12, flash p: −10°. Misc in alc and ether.
SYNS: 3-CHLOR-2-METHYL-PROP-1-EN (GERMAN) □ γ-CHLOROISOBUTYLENE □ 3-CHLORO-2-METHYL-1-PROPENE □ CHLORURE de METHALLYLE (FRENCH) □ 3-CLORO-2-METIL-PROP-1-ENE (ITALIAN) □ CLORURO di METALLILE (ITALIAN) □ ISOBUTENYL CHLORIDE □ METHALLYL CHLORIDE □ α-METHALLYL CHLORIDE □ 2-METHYL-ALLYLCHLORID (GERMAN) □ β-METHYLALLYL CHLORIDE □ 2-METHYLALLYL CHLORIDE □ METHYL ALLYL CHLORIDE (DOT) □ NCI-C54820
CONSENSUS REPORTS: NTP 10th Report on Carcinogens. NTP Carcinogenesis Studies (gavage); Clear Evidence: mouse, rat NTPTR* NTP-TR-300,86. Reported in EPA TSCA Inventory.
DFG MAK: Confirmed Animal Carcinogen with Unknown Relevance to Humans
DOT CLASSIFICATION: 3; Label: Flammable Liquid
SAFETY PROFILE: Confirmed carcinogen with experimental carcinogenic, neoplastigenic, and tumorigenic data. Experimental reproductive effects. An irritant. Human mutation data reported. Very dangerous fire hazard when exposed to heat, flame, or oxidizers. Moderately explosive when exposed to heat or flame. Can react vigorously with oxidizing materials. To fight fire, use alcohol foam, CO_2, dry chemical. When heated to decomposition it emits toxic fumes of Cl⁻. See also CHLORINATED HYDROCARBONS, ALIPHATIC; and ALLYL COMPOUNDS.

CIV000　　CAS: 6959-48-4　　HR: 3
3-(CHLOROMETHYL) PYRIDINE
HYDROCHLORIDE
mf: C_6H_6ClN•ClH　　mw: 164.04
SYN: NCI-C03838
CONSENSUS REPORTS: NCI Carcinogenesis Bioassay (gavage); Clear Evidence: mouse, rat NCITR* NCI-CG-TR-95,78. EPA Genetic Toxicology Program.

SAFETY PROFILE: Suspected carcinogen with experimental carcinogenic data. Poison by ingestion. Mutation data reported. When heated to decomposition it emits very toxic fumes of NO_x and Cl^-.

CJA185   CAS: 635-22-3   HR: 3
4-CHLORO-3-NITROANILINE
mf: $C_6H_5ClN_2O_2$ mw: 172.58
PROP: Yellow needles from H_2O. Mp: 103°.
SYN: ANILINE, 4-CHLORO-3-NITRO-
CONSENSUS REPORTS: Reported in EPA TSCA Inventory.
SAFETY PROFILE: Poison by ingestion and intraperitoneal routes. Experimental reproductive effects. Skin and eye irritant. When heated to decomposition it emits toxic fumes of NO_x and Cl^-.

CJA950   CAS: 25167-93-5   HR: 3
CHLORONITROBENZENE
DOT: UN 1578
mf: $C_6H_4ClNO_2$ mw: 157.56
SYNS: CHLORONITROBENZENE, ortho, liquid (DOT) □ MONONITROCHLOROBENZENE □ NITROCHLOROBENZENE
DOT CLASSIFICATION: 6.1; Label: Poison
SAFETY PROFILE: A poison. Mutation data reported. When heated to decomposition it emits toxic fumes of Cl^- and NO_x. See also other chloronitrobenzene entries and NITRO COMPOUNDS of AROMATIC HYDROCARBONS.

CJB250   CAS: 121-73-3   HR: 3
1-CHLORO-3-NITROBENZENE
DOT: UN 1578
mf: $C_6H_4ClNO_2$ mw: 157.56
PROP: Pale-yellow crystals or prisms. Mp: 46°, flash p: 103°, bp: 236°, d: 1.534 @ 20°/4°.
SYNS: CHLORO-m-NITROBENZENE □ m-CHLORONITROBENZENE □ m-CHLORONITRO-BENZENE (DOT) □ m-NITROCHLOROBENZENE □ m-NITROCHLOROBENZENE, solid (DOT)
CONSENSUS REPORTS: Reported in EPA TSCA Inventory.
DOT CLASSIFICATION: 6.1; Label: Poison

SAFETY PROFILE: Poison by ingestion and inhalation. It forms methemoglobin in the body and gives rise to cyanosis and blood changes. Its effects are cumulative and analogous to those of nitrobenzene. The para compound is thought to be somewhat less toxic than the ortho compound. Chemically, it is probably converted in the body to chloroaniline, which is also poisonous. In industry, it is the dust of this material that is most often the source of intoxication. Flammable liquid and dangerous fire hazard when exposed to heat or flame. It can react with oxidizing materials. When heated to decomposition it emits toxic fumes of Cl^-, NO_x, and phosgene. See also other chloronitrobenzene entries and NITRO COMPOUNDS of AROMATIC HYDROCARBONS.

CJB750   CAS: 88-73-3   HR: 3
CHLORO-o-NITROBENZENE
DOT: UN 1578
mf: $C_6H_4ClNO_2$ mw: 157.56
PROP: Yellow crystals or needles. Mp: 32–33°, bp: 245–246°, d: 1.348, flash p: 123°.
SYNS: o-CHLORONITROBENZENE □ o-CHLORO-NITROBENZENE, liquid (DOT) □ 1-CHLORO-2-NITROBENZENE □ 2-CHLORONITROBENZENE □ 2-CHLORO-1-NITROBENZENE □ o-NITROCHLO-ROBENZENE □ ONCB
CONSENSUS REPORTS: Reported in EPA TSCA Inventory.
DFG MAK: Confirmed Animal Carcinogen with Unknown Relevance to Humans
DOT CLASSIFICATION: 6.1; Label: Poison
SAFETY PROFILE: Suspected carcinogen with experimental carcinogenic and neoplastigenic data. Poison by ingestion, skin contact, and probably inhalation. Combustible when exposed to heat or flame. To fight fire, use water, foam. Potentially explosive reaction with ammonia at 160°C/30 bar. When heated to decomposition it emits toxic fumes of Cl^-, NO_x, and phosgene. See also other chloronitrobenzene entries and NITRO

COMPOUNDS of AROMATIC HYDROCARBONS.

CJD600   CAS: 135-12-6   HR: 1
4-CHLORO-2-NITROPHENYL p-CHLOROPHENYL ETHER
mf: $C_{12}H_7Cl_2NO_3$ mw: 284.10

SYNS: BENZENE, 4-CHLORO-1-(4-CHLORO-PHENOXY)-2-NITRO- □ 4,4'-DICHLOR-2-NITRO-DIFENYLETHER □ ETHER, 4-CHLOROPHENYL (4'-CHLORO-2'-NITRO)PHENYL

CONSENSUS REPORTS: Reported in EPA TSCA Inventory.

SAFETY PROFILE: A skin and eye irritant. When heated to decomposition it emits toxic fumes of NO_x and Cl^-.

CJE000   CAS: 600-25-9   HR: 3
1-CHLORO-1-NITROPROPANE
mf: $C_3H_6ClNO_2$ mw: 123.55

PROP: Liquid. Bp: 139.5°, flash p: 144°F (OC), d: 1.209 @ 20°/20°, vap d: 4.26. IDLH 100 ppm.

SYN: CHLORONITROPROPANE

OSHA PEL: TWA 2 ppm

ACGIH TLV: TWA 2 ppm

DFG MAK: 20 ppm (100 mg/m³)

SAFETY PROFILE: Poison by ingestion, subcutaneous, and possibly other routes. Moderately toxic by inhalation. Causes injury to kidneys, liver, and cardiovascular system. Mutation data reported. Flammable liquid when exposed to heat, flame (sparks), and oxidizers. Moderately explosive when exposed to heat. To fight fire, use alcohol foam, water, CO_2, or dry chemical. When heated to decomposition it emits toxic fumes of Cl^- and NO_x. See also other chloropropane entries and CHLORIDES.

CJG800   CAS: 121-86-8   HR: 1
2-CHLORO-4-NITROTOLUENE
mf: $C_7H_6ClNO_2$ mw: 171.59

PROP: Needles. Fp: 62°, mp: 65°.

SYN: TOLUENE, 2-CHLORO-4-NITRO-

CONSENSUS REPORTS: Reported in EPA TSCA Inventory.

SAFETY PROFILE: A skin and eye irritant. When heated to decomposition it emits toxic fumes of NO_x and Cl^-.

CJI500   CAS: 76-15-3   HR: 1
CHLOROPENTAFLUOROETHANE
DOT: UN 1020

mf: C_2ClF_5 mw: 154.47

PROP: Colorless gas. Bp: −37.7°, mp: −106°, d: 1.5678 @ −42°. Insol in water; sol in alc and ether.

SYNS: F-115 □ FLUOROCARBON-115 □ FREON 115 □ GENETRON 115 □ HALOCARBON 115 □ MONOCHLOROPENTAFLUOROETHANE (DOT)

CONSENSUS REPORTS: Reported in EPA TSCA Inventory.

OSHA PEL: TWA 1000 ppm

ACGIH TLV: TWA 1000 ppm

DOT CLASSIFICATION: 2.2; Label: Nonflammable Gas

SAFETY PROFILE: Mildly toxic by inhalation. A nonflammable gas. When heated to decomposition it emits toxic fumes of F^- and Cl^-.

CJJ000   CAS: 3691-35-8   HR: 3
CHLOROPHACINONE
mf: $C_{23}H_{15}ClO_3$ mw: 374.83

SYNS: AFNOR □ CAID □ CHLOORFACINON (DUTCH) □ 2(2-(4-CHLOOR-FENYL-2-FENYL)-ACETYL)-INDAAN-1,3-DION (DUTCH) □ CHLORFACINON (GERMAN) □ 2-(α-p-CHLOROPHENYLACETYL)-INDANE-1,3-DIONE □ 2-((p-CHLOROPHENYL)PHENYL-ACETYL)-1,3-INDANDIONE □ 2(2-(4-CHLOROPHENYL)-2-PHENYLACETYL)INDAN-1,3-DIONE □ 2-((4-CHLOROPHENYL)PHENYLACETYL)-1H-INDENE-1,3,(2H)-DIONE □ CHLORPHACINON (ITALIAN) □ 2(2-(4-CHLOR-PHENYL-2-PHENYL)ACETYL)INDAN-1,3-DION (GERMAN) □ ((4-CHLORPHENYL)-1-PHENYL)-ACETYL-1,3-INDANDION (GERMAN) □ 1-(4-CHLORPHENYL)-1-PHENYL-ACETYL-INDAN-1,3-DION (GERMAN) □ 2(2-(4-CLORO-FENIL-2-FENIL)-ACETIL)INDAN-1,3-DIONE (ITALIAN) □ DELTA □ DRAT □ LIPHADIONE □ LM 91 □ MICROZUL □ MURIOL □ 2-(2-PHENYL-2-(4-CHLOROPHENYL)ACET-YL)-1,3-INDANDIONE □ QUICK □ RAMUCIDE □ RANAC □ RATOMET □ RAVIAC □ ROZOL □ TOPITOX

CONSENSUS REPORTS: EPA Extremely Hazardous Substances List.

SAFETY PROFILE: Poison by ingestion and skin contact. Human systemic effects by ingestion: vascular changes. A pesticide.

When heated to decomposition it emits toxic fumes of Cl⁻.

CJJ250   CAS: 6164-98-3   HR: 3
CHLOROPHENAMIDINE

mf: $C_{10}H_{13}ClN_2$ mw: 196.70

PROP: Crystals. Mp: 32°, bp: 163–165° @ 14 mm.

SYNS: ACARON □ BERMAT □ C 8514 □ CARZOL □ CDM □ CHLORDIMEFORM □ CHLORFENAMIDINE □ N'-(4-CHLORO-2-METHYLPHENYL)-N,N-DIMETHYLMETHANIMIDAMIDE □ CHLORO-PHENAMADIN □ N'-(4-CHLORO-o-TOLYL)-N,N-DIMETHYLFORMAMIDINE □ CHLORPHENAMIDINE □ N'-(4-CHLOR-o-TOLYL)-N,N-DIMETHYL-FORMAMIDIN (GERMAN) □ CIBA 8514 □ N,N-DIMETHYL-N'-(2-METHYL-4-CHLOROPHENYL)-FORMAMIDINE □ N,N-DIMETHYL-N'-(2-METHYL-4-CHLORPHENYL)-FORMADIN (GERMAN) □ ENT 27,335 □ ENT 27,567 □ EP-333 □ FUNDAL □ FUNDAL 500 □ FUNDEX □ GALECRON □ N'-(2-METHYL-4-CHLOROPHENYL)-N,N-DIMETHYLFORMAMIDINE □ N'-(2-METHYL-4-CHLORPHENYL)-FORMAMIDIN-HYDROCHLORID (GERMAN) □ NSC-190935 □ RS 141 □ SCHERING 36268 □ SN 36268 □ SPANON □ SPANONE

CONSENSUS REPORTS: IARC Cancer Review: Group 3 IMEMDT 7,56,87. EPA Genetic Toxicology Program.

SAFETY PROFILE: Poison by ingestion, skin contact, and intraperitoneal routes. Experimental reproductive effects. Human mutation data reported. An eye irritant. Questionable carcinogen with experimental carcinogenic data. When heated to decomposition it emits very toxic fumes of NO_x and Cl⁻.

CJK250   CAS: 95-57-8   HR: 3
2-CHLOROPHENOL

mf: C_6H_5ClO mw: 128.56

PROP: Light-amber liquid. Mp: 9°, fp: 7°, bp: 174.9°, d: 1.263 @ 20°/4°, flash p: 147°F, vap press: 1 mm @ 12.1°. Sltly water sol; very sol in alc, ether, and alkali.

SYNS: o-CHLOROPHENOL □ o-CHLOROPHENOL, liquid □ o-CHLOROPHENOL, solid □ o-CHLORPHENOL (GERMAN) □ RCRA WASTE NUMBER U048

CONSENSUS REPORTS: Reported in EPA TSCA Inventory. Chlorophenol compounds are on the Community Right-To-Know List.

SAFETY PROFILE: Poison by ingestion, intraperitoneal, and intravenous routes. Experimental reproductive effects. Questionable carcinogen with experimental tumorigenic data. Mutation data reported. Flammable liquid when exposed to heat, flame, or oxidizers. To fight fire, use alcohol foam. When heated to decomposition it emits toxic fumes of Cl⁻. See also CHLOROPHENOLS and CHLORIDES.

CJK500   CAS: 108-43-0   HR: 3
3-CHLOROPHENOL

mf: C_6H_5ClO mw: 128.56

PROP: Crystals or needles from pet ether. Mp: 33°, bp: 210–214°, d: 1.245 @ 45°/4°, vap press: 1 mm @ 44.2°, flash p: >112°.

SYN: m-CHLOROPHENOL

CONSENSUS REPORTS: Reported in EPA TSCA Inventory. Chlorophenol compounds are on the Community Right-To-Know List.

SAFETY PROFILE: Poison by intraperitoneal route. Moderately toxic by ingestion and subcutaneous routes. Questionable carcinogen with experimental tumorigenic data by skin contact. Mutation data reported. Flammable or combustible liquid. When heated to decomposition it emits toxic fumes of Cl⁻. See also CHLOROPHENOLS.

CJK750   CAS: 106-48-9   HR: 3
4-CHLOROPHENOL

DOT: UN 2020/UN 2021

mf: C_6H_5ClO mw: 128.56

PROP: Needle-like, white to straw-colored crystals; unpleasant odor. Flash p: 250°F, d: 1.246 @ 60°/25°, vap press: 1 mm @ 49.8°, mp: 43.5°, d: 1.246 @ 60°/25° (þ&□ -form), mp: 34.2° (β-form) mp: 43.5° (γ-form), bp: 220°. Sltly water sol; very sol in alc, chloroform, and ether.

SYNS: p-CHLORFENOL (CZECH) □ p-CHLOROPHENOL □ PARACHLOROPHENOL □ PHENOL, 4-CHLORO-

CONSENSUS REPORTS: Reported in EPA TSCA Inventory. Chlorophenol compounds are on the Community Right-To-Know List.

SAFETY PROFILE: Poison by inhalation and intraperitoneal routes. Moderately toxic by ingestion, skin contact, and subcutaneous routes. A severe skin and eye irritant. Human systemic effects by inhalation: excitement, irritability. Mutation data reported. Combustible when exposed to heat or flame. To fight fire, use water, spray, mist, fog, foam, dry chemical. When heated to decomposition it emits toxic fumes of Cl⁻. See also CHLOROPHENOLS and CHLORIDES.

CJL000   HR: 3
CHLOROPHENOLS
CONSENSUS REPORTS: Chlorophenol compounds are on the Community Right-To-Know List.
SAFETY PROFILE: Many are suspected experimental carcinogens. Most are strong eye and skin irritants. They are systemic irritants by inhalation, ingestion, and skin contact. Generally mutagenic.

Trichlorophenols are generally poisons and may be carcinogens. They may contain 2,3,7,8-tetrachlorodibenzo-p-dioxin (TCDD) as a contaminant. Some trichlorophenols are used as herbicides (e.g., 2,4,5-T and silvex). Human exposure may cause chloracne, liver dysfunction, muscle weakness, and prophyria.

Pentachlorophenol is a poison by several routes. Human exposure causes increased respiration, fever, tachycardia, muscle weakness, and cardiac failure. Many toxic effects are due to impurities in commercial-grade material. A teratogen and mutagen. Pentachlorophenol and 2,4,6-trichlorophenol may interfere with mitochondrial oxidative phosphorylation. When heated to decomposition they emit toxic fumes of Cl⁻. See also specific compounds, PHENOL, and CHLORIDES.

CJM250   CAS: 58-39-9   HR: 3
4-(3-(2-CHLOROPHENOTHIAZIN-10-
YL)PROPYL)-1-PIPERAZINE-
ETHANOL

mf: $C_{21}H_{26}ClN_3OS$ mw: 404.01
PROP: Crystals from EtOAc. Mp: 100–101°, bp: 278–281° @ 1 mm.
SYNS: 2-CHLORO-10-3-(1-(2-HYDROXYETHYL)-4-PIPERAZINYL)PROPYL PHENOTHIAZINE □ DECENTAN □ ETAPERAZIN □ ETAPERAZINE □ ETHAPERAZINE □ FENTAZIN □ 1-(2-HYDROXYETHYL)-4-(3-(2-CHLORO-10-PHENOTHIAZINYL)PROPYL)PIPERAZINE □ γ-(4-(β-HYDROXYETHYL)PIPERAZIN-1-YL)PROPYL-2-CHLOROPHENOTHIAZINE □ 1',1-(2-IDROSSIETIL)4-(3-(2-CLORO-10-FENOTIAZIL)PROPILPIPERAZINA (ITALIAN) □ PERFENAZINA (ITALIAN) □ PERPHENAZIN □ PERPHENAZINE □ TRIFARON □ TRILAFON
SAFETY PROFILE: Poison by ingestion, intravenous, subcutaneous, intraperitoneal, and intramuscular routes. Human systemic effects by intramuscular route: muscle spasms. Experimental teratogenic and reproductive effects. Human mutation data reported. When heated to decomposition it emits very toxic fumes of SO_x, NO_x, and Cl⁻.

CJM750   CAS: 2865-70-5   HR: 3
10-CHLOROPHENOXARSINE
mf: $C_{12}H_8AsClO$ mw: 278.57
PROP: Colorless prisms from CHCl₃/octane. Mp: 122°. Very sol in C_6H_6, Me₂CO, CHCl₃; mod sol in EtOH, Et₂O, and AcOH.
SYNS: 10-CHLORO-10H-PHENOXARSINE □ 10-CHLOROPHENOXYARSINE □ DID 95
CONSENSUS REPORTS: Reported in EPA TSCA Inventory. Arsenic and its compounds are on the Community Right-To-Know List.
OSHA PEL: TWA 0.5 mg(As)/m³
SAFETY PROFILE: Poison by ingestion. See also ARSINE and ARSENIC COMPOUNDS. When heated to decomposition it emits very toxic fumes of As and Cl⁻.

CJN000   CAS: 122-88-3   HR: 2
p-CHLOROPHENOXYACETIC ACID
mf: $C_8H_7ClO_3$ mw: 186.60
PROP: Needles or prisms from H₂O. Mp: 159°.

SYNS: (4-CHLOROPHENOXY)ACETIC ACID □ 4-CP □ CPA □ MARKS 4-CPA □ PCPA □ SURE-SET □ TOMATO FIX CONCENTRATE □ TOMATO HOLD □ TOMATOTONE

CONSENSUS REPORTS: Reported in EPA TSCA Inventory.

SAFETY PROFILE: Moderately toxic by ingestion and intraperitoneal routes. Human mutation data reported. When heated to decomposition it emits toxic fumes of Cl⁻.

CJO250   CAS: 43121-43-3   HR: 3
1-(4-CHLOROPHENOXY)-3,3-DIMETH-YL-1-(1,2,4-TRIAZOL-1-YL)-2-BUTAN-2-ONE

mf: $C_{14}H_{16}ClN_3O_2$ mw: 293.78

PROP: A solid. Mp: 82–83°.

SYNS: AMIRAL □ BAY 6681 F □ BAYLETON □ BAY-MEB-6447 □ 1-((tert-BUTYLCARBONYL-4-CHLOROPHENOXY)METHYL)-1H-1,2,4-TRIAZOLE □ 1-(4-CHLOROPHENOXY)-3,3-DIMETHYL-1-(1H-1,2,4-TRIAZOL-1-YL)-2-BUTANONE □ MEB 6447 □ TRIADIMEFON

SAFETY PROFILE: Poison by ingestion. Mutation data reported. When heated to decomposition it emits very toxic fumes of Cl⁻ and NOₓ. See also KETONES.

CJR200   CAS: 20265-96-7   HR: 3
4-CHLOROPHENYLAMINE HYDROCHLORIDE

mf: $C_6H_6ClN \cdot ClH$ mw: 164.04

SYNS: 1-AMINO-4-CHLOROBENZENE HYDROCHLORIDE □ ANILINE, p-CHLORO-, HYDROCHLORIDE □ BENZENAMINE, 4-CHLORO-, HYDROCHLORIDE □ p-CHLOROANILINE HYDROCHLORIDE □ 4-CHLOROANILINE HYDROCHLORIDE □ p-CHLOROANILINIUM CHLORIDE □ 4-CHLOROBENZENAMINE HYDROCHLORIDE □ p-CHLOROPHENYLAMINE HYDROCHLORIDE

CONSENSUS REPORTS: NTP Carcinogenesis Studies (gavage): Clear Evidence: rat NTPTR* NTP-TR-351,89; (gavage); Some Evidence: mouse NTPTR* NTP-TR-351,89. Reported in EPA TSCA Inventory.

SAFETY PROFILE: Suspected carcinogen with carcinogenic data. Mutation data reported. When heated to decomposition it emits toxic vapors of NOₓ, HCl, and Cl⁻.

CJR210   CAS: 301644-25-7   HR: 3
4-((3-((4-CHLOROPHENYL)AMINO)-4,5-DIHYDRO-2H-BENZ(G)INDAZOL-2-YL)ACETYL)MORPHOLINE

mf: $C_{23}H_{23}ClN_4O_2$ mw: 422.91

SAFETY PROFILE: A poison by ingestion. When heated to decomposition it emits toxic vapors of NOₓ and Cl⁻.

CJR220   CAS: 301644-24-6   HR: 3
3-((4-CHLOROPHENYL)AMINO)-4,5-DIHYDRO-N-(PHENYLMETHYL)-2H-BENZ(g)INDAZOLE-2-ACETAMIDE

mf: $C_{26}H_{23}ClN_4O$ mw: 442.95

SAFETY PROFILE: A poison by ingestion. When heated to decomposition it emits toxic vapors of NOₓ and Cl⁻.

CJR550   CAS: 3574-96-7   HR: 3
2-(o-CHLOROPHENYL)BENZIMID-AZOLE

mf: $C_{13}H_9ClN_2$ mw: 228.69

SYNS: 1H-BENZIMIDAZOLE, 2-(2-CHLOROPHENYL)- □ 2-(2-CHLOROPHENYL)-1H-BENZIMIDAZOLE □ G 572

SAFETY PROFILE: A poison by ingestion. When heated to decomposition it emits toxic vapors of NOₓ and Cl⁻.

CJR909   CAS: 68-88-2   HR: 3
1-(p-CHLORO-α-PHENYLBENZYL)-4-(2-((2-HYDROXYETHOXY)ETHYL)-PIPERAZINE)

mf: $C_{21}H_{27}ClN_2O_2$ mw: 374.95

SYNS: ATARA □ ATARAX □ ATARAXOID □ ATARAZOID □ ATAZINA □ ATERAX □ 1-(p-CHLOROBENZHYDRYL)-4-(2-(2-HYDROXYETHOXY)-ETHYL)DIETHYLENEDIAMINE □ 1-(p-CHLOROBENZ-HYDRYL)-4-(2-(2-HYDROXYETHOXY)ETHYL)-PIPERAZINE □ N-(4-CHLOROBENZHYDRYL)-N'-(HYDROXYETHOXYETHYL)PIPERAZINE □ 1-(p-CHLORODIPHENYLMETHYL)-4-(2-(2-HYDROXYETHO-XY)ETHYL)PIPERAZINE □ 2-(2-(4-(p-CHLORO-α-PHENYLBENZYL)-1-PIPERAZINYL)ETHOXY)ETHANOL □ DEINAIT □ EQUIPOISE □ FENAROL □ HYCHOTINE □ HYDROXINE □ HYDROXYCINE □ HYDROXYZINE □ IDROSSIZINA □ NEO-CALMA □ NEUROZINA □ NP 212 □ PAMAZONE □ PARENTERAL □ PAXISTIL □ PLACIDOL □ PLAXIDOL □ TRAN-Q □ TRAQUIZINE □ UCB 492 □ U.CB 4492 □ VESPARAZ-WIRKSTOFF

SAFETY PROFILE: Poison by intravenous and intraperitoneal routes. Moderately toxic by ingestion. Experimental teratogenic and reproductive effects. When heated to decomposition it emits very toxic fumes of Cl⁻ and NO$_x$.

CJT800 CAS: 128758-36-1 HR: 3
2-(4-CHLOROPHENYL)-5-((4-CHLORO-PHENYL)METHOXY)-4-IODO-3(2H)-PYRIDAZINONE

mf: $C_{17}H_{11}Cl_2IN_2O_2$ mw: 473.10

SYN: 3(2H)-PYRIDAZINONE, 2-(4-CHLOROPHENYL)-5-((4-CHLOROPHENYL)METHOXY)-4-IODO-

SAFETY PROFILE: A poison by ingestion. When heated to decomposition it emits toxic vapors of NO$_x$, I⁻, and Cl⁻.

CJV250 CAS: 35367-38-5 HR: 2
1-(4-CHLOROPHENYL)-3-(2,6-DIFLUO-ROBENZOYL)UREA

mf: $C_{14}H_9ClF_2N_2O_2$ mw: 310.70

PROP: Crystals. Mp: 230–232°.

SYNS: N-((((4-CHLOROPHENYL)AMINO)CARBONYL)-2,6-DIFLUOROBENZAMIDE □ DIFLUBENZURON □ DIFLURON □ DIMILIN □ DU 112307 □ ENT 29,054 □ OMS 1804 □ PDD 6040I □ PH 60-40 □ PHILIPS-DUPHAR PH 60-40 □ TH 6040 □ THOMPSON-HAYWARD TH6040

CONSENSUS REPORTS: Reported in EPA TSCA Inventory. EPA Genetic Toxicology Program.

SAFETY PROFILE: Moderately toxic by skin contact. Mildly toxic by ingestion. Mutation data reported. When heated to decomposition it emits very toxic fumes of Cl⁻, F⁻, and NO$_x$.

CJX750 CAS: 150-68-5 HR: 2
3-(p-CHLOROPHENYL)-1,1-DIMETHYL-UREA

mf: $C_9H_{11}ClN_2O$ mw: 198.67

PROP: Crystals or thin rectangular prisms from MeOH; slight odor. Mp: 170.5–171.51°, vap press: 0.002 mm @ 100°. Sltly sol in water and hydrocarbons.

SYNS: 3-(4-CHLOOR-FENYL)-1,1-DIMETHYLUREUM (DUTCH) □ CHLORFENIDIM □ N-(p-CHLOROPHEN-YL)-N',N'-DIMETHYLUREA □ N'-(4-CHLOROPHENYL)-N,N-DIMETHYLUREA □ 1-(p-CHLOROPHENYL)-3,3-DIMETHYLUREA □ 3-(4-CHLOROPHENYL)-1,1-DIMETHYLUREA □ 1-(4-CHLORO PHENYL)-3,3-DIMETHYLUREE (FRENCH) □ 3-(4-CHLOR-PHENYL)-1,1-DIMETHYL-HARNSTOFF (GERMAN) □ 3-(4-CLORO-FENIL)-1,1-DIMETIL-UREA (ITALIAN) □ CMU □ N,N-DIMETHYL-N'-(4-CHLOROPHENYL)UREA □ 1,1-DIMETHYL-3-(p-CHLOROPHENYL)UREA □ HERBICIDES, MONURON □ KARMEX MONURON HERBICIDE □ KARMEX W. MONURON HERBICIDE □ LIROBETAREX □ MONUREX □ MONURON □ MONUROX □ MONURUON □ MONUURON □ NCI-C02846 □ TELVAR □ TELVAR MONURON WEED-KILLER □ USAF P-8 □ USAF XR-41

CONSENSUS REPORTS: IARC Cancer Review: Group 3 IMEMDT 7,56,87; Animal Sufficient Evidence IMEMDT 12,167,76. Reported in EPA TSCA Inventory. EPA FIFRA 1988 pesticide subject to registration or re-registration. EPA Genetic Toxicology Program.

SAFETY PROFILE: Moderately toxic by ingestion, intraperitoneal, and possibly other routes. Experimental teratogenic and reproductive effects. Questionable carcinogen with experimental carcinogenic data. Mutation data reported. An herbicide. When heated to decomposition it emits very toxic fumes of NO$_x$ and Cl⁻.

CJY120 CAS: 5131-60-2 HR: 3
4-CHLORO-m-PHENYLENEDIAMINE

mf: $C_6H_7ClN_2$ mw: 142.60

PROP: Needles. Mp: 91°.

SYNS: C.I. 76027 □ 4-CHLORO-1,3-BENZENEDIAMINE □ 1-CHLORO-2,4-DIAMINOBENZENE □ 4-CHLORO-PHENE-1,3-DIAMINE □ 4-CHLOROPHENYL-ENE-1,3-DIAMINE □ 4-CHLORO-1,3-PHENYLENE-DIAMINE □ 4-Cl-m-PD □ NCI-C03305

CONSENSUS REPORTS: IARC Cancer Review: Group 3 IMEMDT 7,56,87; Animal Inadequate Evidence IMEMDT 27,81,82. NCI Carcinogenesis Bioassay (feed); Clear Evidence: mouse, rat NCITR* NCI-CG-TR-85,78. Reported in EPA TSCA Inventory.

SAFETY PROFILE: Suspected carcinogen with experimental carcinogenic, neoplastigenic, and tumorigenic data. Experimental reproductive effects. Mutation data reported. When heated to decomposition it emits toxic fumes of Cl⁻ and NO$_x$. See also AROMATIC AMINES.

CKA030   CAS: 90035-12-4   HR: 3
3-(3-(4-(2-(4-CHLOROPHENYL)ETHYL)-
   PHENYL)-1,2,3,4-TETRAHYDRO-1-
   NAPHTHALENYL)-4-HYDROXY2H-1-
   BENZOPYRAN-2-ONE
mf: $C_{33}H_{27}ClO_3$ mw: 507.05
SAFETY PROFILE: A poison by ingestion.
When heated to decomposition it emits
toxic vapors of Cl⁻.

CKA575   CAS: 21905-40-8   HR: 3
(m-CHLOROPHENYL)HYDROXY(β-
   HYDROXYPHENETHYL)ARSINE
   OXIDE
mf: $C_{14}H_{14}AsClO_3$ mw: 340.65
SYNS: ARSINE OXIDE, (m-CHLOROPHENYL)-
HYDROXY(β-HYDROXYPHENETHYL)- □ 2-PHENYL-2-
HYDROXYETHYL, m-CHLOROPHENYL ARSINIC ACID
OSHA PEL: TWA 0.5 mg(As)/m³
SAFETY PROFILE: Poison by intravenous
route. When heated to decomposition it
emits toxic fumes of As and Cl⁻.

CKA580   CAS: 24671-21-4   HR: 3
4-(4-(p-CHLOROPHENYL)-4-HYDROXY-
   PIPERIDINO)-4'-(DIMETHYLAMINO)-
   BUTYROPHENO NE
mf: $C_{23}H_{29}ClN_2O_2$ mw: 400.99
SYNS: BUTYROPHENONE, 4-(4-(p-CHLOROPHENYL)-
4-HYDROXYPIPERIDINO)-4'-(DIMETHYLAMINO)- □ 4-
(DIMETHYLAMINO)PHENYL-3-(4-(4-CHLOROPHENYL)-
4-HYDROXYPIPERIDINO)-PROPYL KETONE
DOT CLASSIFICATION: 3; Label:
Flammable Liquid
SAFETY PROFILE: A poison by
intravenous route. A flammable liquid.
When heated to decomposition it emits
toxic vapors of NOₓ and Cl⁻.

CKA750   CAS: 2909-38-8   HR: 3
m-CHLOROPHENYL ISOCYANATE
DOT: UN 2206/UN 2207/UN 2478/UN
3080
mf: C_7H_4ClNO mw: 153.57
PROP: Water-white liquid with irritating
odors. Mp: −4°, bp: 101° @ 30 mm. Sol in
org solvs.
SYNS: m-CHLORFENYLISOKYANAT □ ISOCYANIC
ACID-m-CHLOROPHENYL ESTER

CONSENSUS REPORTS: Reported in EPA
TSCA Inventory.
DOT CLASSIFICATION: 6.1; Label: KEEP
AWAY FROM FOOD (UN 2207); DOT
Class: 6.1; Label: Poison (UN 2206); DOT
Class: 6.1; Label: Poison, Flammable
Liquid (UN 3080); DOT Class: 3; Label:
Flammable Liquid, Poison (UN 2478)
SAFETY PROFILE: Poison by inhalation
and ingestion. A flammable liquid when
exposed to heat or flame. When heated to
decomposition it emits toxic fumes of Cl⁻,
CN⁻, and NOₓ. See also ESTERS and
THIOCYANATES.

CKB000   CAS: 104-12-1   HR: 3
p-CHLOROPHENYL ISOCYANATE
DOT: UN 2206/UN 2207/UN 2478/UN
3080
mf: C_7H_4ClNO mw: 153.57
PROP: White solid or crystals. Mp: 31°, bp:
204°. Sol in org solvs.
SYNS: p-CHLORFENYLISOKYANAT (CZECH) □
ISOCYANIC ACID-p-CHLOROPHENYL ESTER □ PCPI
CONSENSUS REPORTS: Reported in EPA
TSCA Inventory.
DOT CLASSIFICATION: 6.1; Label: KEEP
AWAY FROM FOOD (UN 2207); DOT
Class: 6.1; Label: Poison (UN 2206); DOT
Class: 6.1; Label: Poison, Flammable Liquid
(UN 3080); DOT Class: 3; Label:
Flammable Liquid, Poison (UN 2478)
SAFETY PROFILE: Poison by ingestion and
inhalation. Unspecified human systemic
effects. A severe eye and moderate skin
irritant. A flammable liquid when exposed to
heat or flame. Dangerous, can explode on
distillation. When heated to decomposition
it emits toxic fumes of Cl⁻, CN⁻, and NOₓ.

CKB250   CAS: 500-92-5   HR: 3
1-(p-CHLOROPHENYL)-5-ISOPROPYL-
   BIGUANIDE
mf: $C_{11}H_{16}ClN_5$ mw: 253.77
PROP: White powder or rectangular plates
from toluene. Mp: 130–131°.
SYNS: BIGUMAL □ CHLORGUANIDE □
CHLOROGUANIDE □ PALUDRINE □ PROGUANIL

SAFETY PROFILE: Poison by ingestion, intravenous, and intraperitoneal routes. Experimental reproductive effects. When heated to decomposition it emits toxic fumes of Cl⁻ and NOₓ.

CKD500   CAS: 1746-81-2   HR: 2
3-(4-CHLOROPHENYL)-1-METHOXY-1-
**   METHYLUREA**
mf: $C_9H_{11}ClN_2O_2$ mw: 214.67
PROP: A solid. Mp: 76–78°.
SYNS: AFESIN □ ARESIN □ AREZIN □ AREZINE □ ARRESIN □ N-(4-CHLOROPHENYL)-N'-METHOXY-N-METHYLUREA □ N'-(4-CHLOROPHENYL)-N-METHOXY-N-METHYLUREA □ 3-(4-CHLORPHENYL)-1-METHOXY-1-METHYLHARNSTOFF (GERMAN) □ HOE 2747 □ MONOLINURON □ PREMALIN
CONSENSUS REPORTS: Reported in EPA TSCA Inventory.
SAFETY PROFILE: Moderately toxic by ingestion. Experimental teratogenic and reproductive effects. When heated to decomposition it emits very toxic fumes of Cl⁻ and NOₓ.

CKD750   CAS: 1867-66-9   HR: 3
2-(o-CHLOROPHENYL)-2-(METHYL-
**   AMINO)CYCLOHEXANONE**
**   HYDROCHLORIDE**
mf: $C_{13}H_{16}ClNO•ClH$ mw: 274.21
SYNS: CI 581 □ CL 369 □ CN-52,372-2 □ KETAJECT □ KETALAR □ KETAMINE □ KETAMINE HYDROCHLORIDE □ KETANEST □ KETASET □ KETAVET □ KETOLAR □ VETALAR
CONSENSUS REPORTS: EPA Genetic Toxicology Program.
SAFETY PROFILE: Poison by intramuscular, intraperitoneal, and intravenous routes. Moderately toxic by ingestion. Human systemic effects by intravenous and possibly other routes: analgesia, coma, hallucinations and distorted perceptions, dyspnea. An experimental teratogen. An anesthetic. When heated to decomposition it emits very toxic fumes of Cl⁻ and NOₓ.

CKD800   CAS: 73791-42-1   HR: 3
(4-CHLOROPHENYL)METHYLARSINIC

ACID
mf: $C_7H_8AsClO_2$ mw: 234.52
SYNS: ARSINE OXIDE, (p-CHLOROPHENYL)-HYDROXYMETHYL- □ (p-CHLOROPHENYL)HYDRO-XYMETHYLARSINE OXIDE
OSHA PEL: TWA 0.5 mg(As)/m³
SAFETY PROFILE: Poison by intravenous route. When heated to decomposition it emits toxic fumes of As and Cl⁻.

CKF000   CAS: 3942-54-9   HR: 3
o-CHLOROPHENYL METHYL-
**   CARBAMATE**
mf: $C_8H_8ClNO_2$ mw: 185.62
SYNS: 2-CHLOROPHENYL-N-METHYLCARBAMATE □ CPMC □ ETROFOL □ HOPCIDE
CONSENSUS REPORTS: EPA Genetic Toxicology Program. Reported in EPA TSCA Inventory.
SAFETY PROFILE: A poison by ingestion and possibly other routes. An insecticide. See also CARBAMATES. When heated to decomposition it emits very toxic fumes of Cl⁻ and NOₓ.

CKJ100   CAS: 17766-66-4   HR: 3
4-(p-CHLOROPHENYL)PIPERAZINYL
**   3,4,5-TRIMETHYOXYPHENYL**
**   KETONE**
mf: $C_{20}H_{23}ClN_2O_4$ mw: 390.90
SYNS: 1-(p-CHLOROPHENYL)-4-(3,4,5-TRIMETHOXYBENZOYL)PIPERAZINE □ KETONE, 4-(p-CHLOROPHENYL)PIPERAZINYL 3,4,5-TRIMETHOXY-PHENYL □ PIPERAZINE, 1-(p-CHLOROPHENYL)-4-(3,4,5-TRIMETHOXYBENZOYL)-
DOT CLASSIFICATION: 3; Label: Flammable Liquid
SAFETY PROFILE: A poison by intraperitoneal route. A flammable liquid. When heated to decomposition it emits toxic vapors of NOₓ and Cl⁻.

CKK050   CAS: 23597-98-0   HR: 3
5-(p-CHLOROPHENYL)-2,3,5,6-TETRA-
**   HYDROIMIDAZO(1,2-c)QUINAZOLIN**
mf: $C_{16}H_{14}ClN_3$ mw: 283.78
SYNS: AW-15'1129 □ IMIDAZO(1,2-c)QUINAZOLINE, 5-(4-CHLOROPHENYL)-2,3,5,6-TETRAHYDRO- □ IMID-AZO(1,2-c)QUINAZOLINE, 2,3,5,6-TETRAHYDRO-5-(p-CHLOROPHENYL)-

SAFETY PROFILE: A poison by ingestion. Mutation data reported. When heated to decomposition it emits toxic vapors of NO_x and Cl^-.

CKM000   CAS: 116-29-0   HR: 2
p-CHLOROPHENYL-2,4,5-TRICHLORO-
**   PHENYL SULFONE**
mf: $C_{12}H_6Cl_4O_2S$ mw: 356.04
PROP: Crystals. Mp: 148–149°. Nearly water-insol.
SYNS: AKARITOX ☐ AREDION ☐ 4-CHLOROPHEN-YL-2,4,5-TRICHLOROPHENYL SULFONE ☐ p-CHLOROPHENYL-2,4,5-TRICHLOROPHENYL SULPHONE ☐ DUPHAR ☐ ENT 23,737 ☐ FMC 5488 ☐ MITION ☐ NIA 5488 ☐ POLACARITOX ☐ ROZTOZOL ☐ SULFONE-2,4,4',5-TETRACHLORODIPHENYL ☐ TEDION ☐ TEDION V-18 ☐ 2,4,4',5-TETRACHLOOR-DIFENYL-SULFON (DUTCH) ☐ 2,4,4',5-TETRACHLOR-DIPHENYL-SULFON (GERMAN) ☐ 2,4,4',5-TETRACHLO-RODIPHENYL SULFONE ☐ 2,4,5,4'-TETRACHLORO-DIPHENYLSULPHONE ☐ 2,4,4',5-TETRACLORO-DIFENIL-SOLFONE (ITALIAN) ☐ TETRADICHLONE ☐ TETRADIFON ☐ TETRADIPHON ☐ TETRAFIDON ☐ 1,2,4-TRICHLORO-5-((4-CHLOROPHENYL)SULFONYL)-BENZENE ☐ V-18
CONSENSUS REPORTS: EPA Genetic Toxicology Program.
SAFETY PROFILE: Moderately toxic by ingestion. Mildly toxic by skin contact. An experimental teratogen. Used to control worms in crops. When heated to decomposition it emits highly toxic fumes of Cl^- and SO_x.

CKM250   CAS: 26571-79-9   HR: 3
CHLOROPHENYLTRICHLOROSILANE
DOT: UN 1753
mf: $C_6H_4Cl_4Si$ mw: 245.99
PROP: Colorless to pale-yellow liquid, readily hydrolyzed by moisture with the liberation of HCl (a mixture of 3 isomers). Bp: 230°, d: 1.439 @ 25°/25°, flash p: 255°F (COC).
DOT CLASSIFICATION: 8; Label: Corrosive
SAFETY PROFILE: A poison irritant by ingestion and inhalation. A corrosive irritant to the skin, eyes, and mucous membranes. Combustible when exposed to heat or flame. In contact with water it readily hydrolyzes to HCl and evolves heat. When heated to decomposition it emits toxic fumes of Cl^-. See also CHLOROSILANES.

CKN000   CAS: 1406-65-1   HR: 3
CHLOROPHYLL
PROP: Dark-green liquid.
SYNS: CHLOROFOLIN ☐ CHLOROFYL ☐ CHLOROPHYLLS ☐ C.I. 1956 ☐ DEODOPHYLL ☐ E 140 ☐ L-GRUEN No. 1 ☐ L-GRUEN No. 1 (GERMAN)
SAFETY PROFILE: Poison by intravenous and intraperitoneal routes. When heated to decomposition it emits toxic fumes of NO_x.

CKN500   CAS: 76-06-2   HR: 3
CHLOROPICRIN
DOT: UN 1580/UN 1583
mf: CCl_3NO_2 mw: 164.37
PROP: Sltly oily, colorless liquid. D: 1.692 @ 0.4°, fp: −69°, mp: −64°, bp: 112.3° @ 766 mm, vap press: 40 mm @ 33.80, vap d: 6.69. Sol in water, alc, and ether. IDLH 2 ppm.
SYNS: ACQUINITE ☐ CHLOORPIKRINE (DUTCH) ☐ CHLOR-O-PIC ☐ CHLOROPICRIN, liquid (DOT) ☐ CHLOROPICRIN, ABSORBED (DOT) ☐ CHLOROPICRINE (FRENCH) ☐ CHLORPIKRIN (GERMAN) ☐ CLOROPICRINA (ITALIAN) ☐ DOJYOPICRIN ☐ DOLOCHLOR ☐ LARVACIDE ☐ MICROLYSIN ☐ NCI-C00533 ☐ NITROCHLOROFORM ☐ NITROTRICHLOROMETHANE ☐ PIC-CLOR ☐ PICFUME ☐ PICRIDE ☐ PROFUME A ☐ PS ☐ TRICHLOORNITROMETHAAN (DUTCH) ☐ TRICHLORNITROMETHAN (GERMAN) ☐ TRICHLORONITROMETHANE ☐ TRI-CLOR ☐ TRICLORO-NITRO-METANO (ITALIAN)
CONSENSUS REPORTS: NCI Carcinogenesis Bioassay (gavage); No Evidence: mouse NCITR* NCI-GC-TR-65,78. Reported in EPA TSCA Inventory.
OSHA PEL: TWA 0.1 ppm
ACGIH TLV: TWA 0.1 ppm; Not Classifiable as a Human Carcinogen
DFG MAK: 0.1 ppm (0.68 mg/m³)
DOT CLASSIFICATION: 6.1; Label: Poison (UN 1580); DOT Class: 6.1; Label: Poison, KEEP AWAY FROM FOOD (UN 1583)
SAFETY PROFILE: Poison by ingestion, intravenous, and intraperitoneal routes. Moderately toxic by inhalation. Human

systemic effects by inhalation: lachrymation, conjunctiva irritation, and pulmonary changes. Mutation data reported. A powerful irritant that affects all body surfaces. It causes lachrymation, vomiting, bronchitis, pulmonary edema, irritation to gastrointestinal and respiratory tracts. Questionable carcinogen with experimental tumorigenic data. An additional toxic effect is its reaction with SH-groups in hemoglobin, thus interfering with oxygen transport. Photochemical transformation of chloropicrin into phosgene (carboxy chloride, $COCl_2$) has been reported. A concentration of 1 ppm causes a smarting pain in the eyes and therefore in itself constitutes a good warning of exposure. Inhalation causes vomiting, probably due to swallowing saliva in which small amounts of chloropicrin have dissolved. Its primary lethal effect is to produce lung injury and it is a difficult gas to protect oneself against because it is chemically inert and does not react with the usual chemicals used in gas masks. Four ppm is sufficient to render a worker unfit for action and 20 ppm, when breathed from 1 to 2 minutes, causes definite bronchial or pulmonary lesions. Industrially it is used as a warning agent in commercial fumigants. It is more toxic than chlorine but less so than phosgene.

Above a critical volume it can be shocked into detonation. Mixtures with 3-bromopropyne are shock- and heat-sensitive explosives. Violent reaction with aniline + heat, alcoholic sodium hydroxide, sodium methoxide, and propargyl bromide. When heated to decomposition it emits very toxic fumes of Cl^- and NO_x.

Used for insect and rodent control in grain elevators and bins and as a soil fumigant and fungicide. See also NITRO COMPOUNDS.

CKO750   CAS: 16941-12-1   HR: 3
CHLOROPLATINIC ACID
DOT: UN 2507
mf: $Cl_6Pt•2H$ mw: 409.81

PROP: Reddish-brownish-yellow deliquescent solid or crystalline mass. D: 2.431, mp: 60°. Easily sol in water and alc. IDLH 4 mg/m³ (as Pt).
SYNS: CHLOROPLATINIC(IV) ACID □ DIHYDROGEN HEXACHLOROPLATINATE □ DIHYDROGEN HEXACHLOROPLATINATE(2-) □ HEXACHLORO-PLATINIC ACID □ HEXACHLOROPLATINIC(IV) ACID □ HEXACHLOROPLATINIC(4+) ACID, HYDROGEN- □ HYDROGEN HEXACHLOROPLATINATE(4+) □ PLATINIC CHLORIDE
CONSENSUS REPORTS: Reported in EPA TSCA Inventory. EPA Genetic Toxicology Program.
OSHA PEL: TWA 0.002 mg(Pt)/m³
ACGIH TLV: TWA 0.002 mg(Pt)/m³
DOT CLASSIFICATION: 8; Label: Corrosive
SAFETY PROFILE: Poison by intravenous and intraperitoneal routes. Mutation data reported. See PLATINUM COMPOUNDS and CHLORIDES. Incompatible with BrF_3. When heated to decomposition it emits toxic fumes of Cl^-.

CKP500   CAS: 69-09-0   HR: 3
CHLOROPROMAZINE HYDRO-CHLORIDE
mf: $C_{17}H_{19}ClN_2S•ClH$ mw: 355.35
SYNS: AMINAZIN MONOHYDROCHLORIDE □ AMPLIACTIL MONOHYDROCHLORIDE □ CHLORACTIL □ CHLORAZIN □ 2-CHLORO-10-(3-DIMETHYLAMINOPROPYL) PHENOTHIAZINE MONOHYDROCHLORIDE □ CHLOROPROMAZINE MONOHYDROCHLORIDE □ CPZ □ 10-(3-DIMETHYLAMINOPROPYL)-2-CHLOROPHENO-THIAZINE MONOHYDROCHLORIDE □ HEBANIL □ HIBANIL □ HIBERNAL □ HYBERNAL □ KLOR-PROMAN □ KLORPROMEX □ LARGACTIL MONOHYDROCHLORIDE □ LARGAKTYL □ MEGAPHEN □ NCI-CO5210 □ NEURAZINE □ NORCOZINE □ PHENOTHIAZINE HYDROCHLORIDE □ PLEGOMAZIN □ PROMACID □ PROMAPAR □ PROPAPHEN □ PROPAPHENIN HYDROCHLORIDE □ PSYCHOZINE □ 4560 RP HYDROCHLORIDE □ SONAZINE □ TAROCTYL □ THORAZINE □ THORAZINE HYDROCHLORIDE □ TORAZINA □ TRANZINE □ UNITENSEN
CONSENSUS REPORTS: Reported in EPA TSCA Inventory.
SAFETY PROFILE: Poison by ingestion, intraperitoneal, intravenous, and subcutaneous routes. An experimental

teratogen. Experimental reproductive effects. An anti-emetic and antipsychotic drug. Human systemic effects: anorexia (human), excitement, gastrointestinal changes, irritability, pulse rate increase, respiratory stimulation, rigidity, somnolence, sweating. Mutation data reported. When heated to decomposition it emits very toxic fumes of Cl^-, NO_x, and SO_x.

CKP600   CAS: 627-30-5   HR: 2
3-CHLOROPROPANOL
DOT: UN 2849
mf: C_3H_7ClO mw: 94.55
SYNS: 3-CHLORPROPAN-1-OL □ 1-PROPANOL, 3-CHLORO- □ TRIMETHYLENE CHLOROHYDRIN
CONSENSUS REPORTS: Reported in EPA TSCA Inventory.
DOT CLASSIFICATION: 6.1; Label: KEEP AWAY FROM FOOD
SAFETY PROFILE: Moderately toxic by ingestion. Mutation data reported.

CKP750   CAS: 540-54-5   HR: 3
1-CHLOROPROPANE
DOT: UN 1278
mf: C_3H_7Cl mw: 78.55
PROP: Colorless liquid, chloroform-like odor. Mp: −122.8°, d: 0.897 @ 15°/4°, bp: 46.60°, lel: 2.6%, uel: 11.1%, flash p: <0°F, d: 0.890, vap d: 2.71, autoign temp: 968°F.
SYN: N-PROPYL CHLORIDE
CONSENSUS REPORTS: Reported in EPA TSCA Inventory.
DOT CLASSIFICATION: 3; Label: Flammable Liquid
SAFETY PROFILE: A moderately poisonous irritant to skin, eyes, and mucous membranes. Narcotic in high concentrations. Flammable liquid and dangerous fire hazard when exposed to heat, flame, or oxidizers. Moderately explosive when exposed to flame. Keep away from heat and open flame; can react vigorously with oxidizing materials. To fight fire, use CO_2, dry chemical. When heated to decomposition it emits toxic fumes of Cl^-.

See also CHLORINATED HYDROCARBONS, ALIPHATIC.

CKQ000   CAS: 75-29-6   HR: 3
2-CHLOROPROPANE
DOT: UN 2356
mf: C_3H_7Cl mw: 78.54
PROP: Flash p: −25.6°F, lel: 2.8%, uel: 10.7%, d: 0.868 @ 15°, mp: −117°, bp: 34.8°.
SYN: ISOPROPYL CHLORIDE
DOT CLASSIFICATION: Flammable Liquid; Label: Flammable Liquid
SAFETY PROFILE: Mutation data reported. A flammable liquid. A very dangerous fire hazard when exposed to heat, flame, or oxidizers. When heated to decomposition it emits toxic fumes of Cl^-. See also 1-CHLOROPROPANE.

CKQ750   CAS: 15121-11-6   HR: 3
CHLOROPROPANEDIOL CYCLIC SULFITE
mf: $C_3H_5ClO_3S$ mw: 156.69
SYN: 1-CHLOROMETHYLETHYLENE GLYCOL CYCLIC SULFITE
CONSENSUS REPORTS: Reported in EPA TSCA Inventory.
SAFETY PROFILE: Poison by intravenous route. See also SULFITES. Experimental reproductive effects. When heated to decomposition it emits very toxic fumes of Cl^- and SO_x.

CKR500   CAS: 78-89-7   HR: 3
2-CHLORO-1-PROPANOL
DOT: UN 2611
mf: C_3H_7ClO mw: 94.55
PROP: Colorless liquid; mild non-residual odor. Bp: 133.5°, flash p: 125°F (CC), d: 1.103 @ 20°, vap d: 3.26.
SYNS: 2-CHLOROPROPANOL □ 2-CHLOROPROPYL ALCOHOL □ PROPYLENECHLOROHYDRIN
CONSENSUS REPORTS: Reported in EPA TSCA Inventory.
ACGIH TLV: TWA 1 ppm (skin); Not Classifiable as a Human Carcinogen
DOT CLASSIFICATION: 6.1; Label: Poison

SAFETY PROFILE: Poison by ingestion. Moderately toxic by inhalation and skin contact. A skin and severe eye irritant. Flammable liquid when exposed to heat, flame, or powerful oxidizers. To fight fire, use alcohol foam, CO_2, dry chemical. When heated to decomposition it emits toxic fumes of Cl^-.

CKS000   CAS: 557-98-2   HR: 3
2-CHLORO-1-PROPENE
DOT: UN 2456
mf: C_3H_5Cl mw: 76.53
PROP: Colorless liquid or gas. Bp: 22.65°, fp: −137.4°, d: 0.918 @ 9°, flash p: −4°, lel: 4.5%, uel: 16%, mp: −138.6°.
SYN: 2-CHLOROPROPENE (DOT)
CONSENSUS REPORTS: Reported in EPA TSCA Inventory.
DOT CLASSIFICATION: 3; Label: Flammable Liquid
SAFETY PROFILE: Mildly toxic by inhalation. Mutation data reported. Very dangerous fire hazard when exposed to heat, flame, sparks, or powerful oxidizers. To fight fire, use water, spray, mist, fog, dry chemical, alcohol foam. When heated to decomposition it emits toxic fumes of Cl^-. See also CHLORIDES.

CKS750   CAS: 598-78-7   HR: 3
α-CHLOROPROPIONIC ACID
DOT: UN 2511
mf: $C_3H_5ClO_2$ mw: 108.53
PROP: Sol in water. D: 1.260–1.268 @ 20°, bp: 183–187°, flash p: 225°F.
CONSENSUS REPORTS: Reported in EPA TSCA Inventory.
ACGIH TLV: TWA 0.1 ppm (skin)
DOT CLASSIFICATION: 8; Label: Corrosive
SAFETY PROFILE: Poison by skin contact. A corrosive. Combustible when exposed to heat or flame. To fight fire, use water, foam, alcohol foam. When heated to decomposition it emits toxic fumes of Cl^-. See also 3-CHLOROPROPIONIC ACID.

CKT000   CAS: 17639-93-9   HR: 3
2-CHLOROPROPIONIC ACID METHYL ESTER
DOT: UN 2933
mf: $C_4H_7ClO_2$ mw: 122.56
SYN: METHYL-2-CHLOROPROPIONATE (DOT)
CONSENSUS REPORTS: Reported in EPA TSCA Inventory.
DOT CLASSIFICATION: 3; Label: Flammable Liquid
SAFETY PROFILE: Poison by intraperitoneal route. See also ESTERS. A flammable liquid when exposed to heat or flame. When heated to decomposition it emits toxic fumes of Cl^-.

CKT500   CAS: 6285-05-8   HR: 3
p-CHLOROPROPIOPHENONE
mf: C_9H_9ClO mw: 168.63
PROP: A solid. Mp: 37–38°, bp: 134–137° @ 31 mm.
SYN: USAF EK-5296
CONSENSUS REPORTS: Reported in EPA TSCA Inventory.
SAFETY PROFILE: Poison by intraperitoneal and intravenous routes. When heated to decomposition it emits toxic fumes of Cl^-.

CKW000   CAS: 109-09-1   HR: 3
2-CHLOROPYRIDINE
DOT: UN 2822
mf: C_5H_4ClN mw: 113.55
PROP: Colorless, oily liquid or crystals. Mp: 65°, bp: 170°, d: 1.205 @ 15°, vap press: 1 mm @ 13.3°, vap d: 3.93.
SYNS: o-CHLOROPYRIDINE □ α-CHLOROPYRIDINE
CONSENSUS REPORTS: Reported in EPA TSCA Inventory.
DOT CLASSIFICATION: 6.1; Label: Poison
SAFETY PROFILE: Poison by ingestion, inhalation, skin contact, and intraperitoneal routes. Combustible when exposed to heat or flame. Can react with oxidizing materials. When heated to decomposition it emits very toxic fumes of Cl^-, NO_x, and phosgene.

CKW330   CAS: 122322-24-1   HR: 3
5-((6-CHLORO-3-PYRIDINYL)METHO-
XY)-2-(3,4-DICHLOROPHENYL)-4-
IODO-3(2H)-PYRIDAZINONE
mf: $C_{16}H_9Cl_3IN_3O_2$ mw: 508.53
SYN: 3(2H)-PYRIDAZINONE, 5-((6-CHLORO-3-
PYRIDINYL)METHOXY)-2-(3,4-DICHLOROPHENYL)-4-
IODO-
SAFETY PROFILE: A poison by ingestion.
When heated to decomposition it emits
toxic vapors of NO_x, I^-, and Cl^-.

CLD750   CAS: 95-88-5   HR: 3
4-CHLORORESORCINOL
mf: $C_6H_5ClO_2$ mw: 144.56
PROP: Crystals from C_6H_6. Mp: 89°, bp:
147°.
CONSENSUS REPORTS: Reported in EPA
TSCA Inventory. EPA Genetic Toxicology
Program. Chlorophenols are on the
Community Right-To-Know List.
SAFETY PROFILE: Poison by ingestion and
intraperitoneal routes. Experimental
reproductive effects. An eye irritant. A hair
dye component. When heated to
decomposition it emits toxic fumes of Cl^-.
See also CHLOROPHENOLS and
RESORCINOL.

CLE250   HR: 3
CHLOROSILANES
PROP: Compounds of Si, Cl, and H where
the total number of atoms of Cl and H add
up to 4. SiH_xCl_{4-x}.
SAFETY PROFILE: Poison by ingestion and
inhalation, and a poisonous irritant to skin,
eyes, and mucous membranes. Toxicity is
based on HCl which is formed upon
hydrolysis of a chlorosilane. Self-ignites in
air. With a little ammonia, it forms a self-
igniting product. They react with water or
steam to produce heat and toxic and
corrosive fumes of HCl. When heated to
decomposition they emit highly toxic fumes
of Cl^-.

CLE750   CAS: 2039-87-4   HR: 1
o-CHLOROSTYRENE
mf: C_8H_7Cl mw: 138.60

PROP: A solid. Mp: −63.15°, bp: 188.6°, d:
1.100 @ 20°/4°.
CONSENSUS REPORTS: Reported in EPA
TSCA Inventory.
OSHA PEL: TWA 50 ppm; STEL 75 ppm
ACGIH TLV: TWA 50 ppm; STEL 75 ppm
SAFETY PROFILE: A skin and eye irritant.
When heated to decomposition it emits
toxic fumes of Cl^-. See also
CHLORINATED HYDROCARBONS,
AROMATIC.

CLG500   CAS: 7790-94-5   HR: 3
CHLOROSULFURIC ACID
DOT: UN 1754/UN 2240
mf: $ClHO_3S$ mw: 116.52
PROP: Strong acid; clear to cloudy or
colorless to pale-yellow liquid; sharp odor.
Fumes in moist air. Mp: −80°, bp: 155°, d:
1.77 @ 28°, vap press: 1 mm @ 32°, vap d:
4.02. Sol in $CHCl_3$, CH_2Cl_2, and Py; insol in
CS_2 and CCl_4.
SYNS: CHLOROSULFONIC ACID □ CHLOROSULFON-
IC ACID (with or without sulfur trioxide) (UN 1754) (DOT) □
CHROMOSULFURIC ACID (UN 2240) (DOT) □
MONOCHLOROSULFURIC ACID □ SULFONIC ACID,
MONOCHLORIDE □ SULFURIC CHLOROHYDRIN
CONSENSUS REPORTS: Reported in EPA
TSCA Inventory.
DOT CLASSIFICATION: 8; Label: Corrosive,
Poison (UN 1754); DOT Class: 8; Label:
Corrosive (UN 2240)
SAFETY PROFILE: A poison irritant. See
also SULFURIC ACID. Chlorosulfonic acid
is corrosive, can cause severe acid burns and
is very irritating to the eyes, lungs, and
mucous membranes. It can cause acute toxic
effects either in the liquid or vapor state.
Inhalation of concentrated vapor may cause
loss of consciousness with serious damage
to lung tissue. Contact of liquid with the
eyes can cause severe burns if the liquid is
not immediately and completely removed. It
also causes severe skin burns due to its
highly corrosive action. Upon ingestion it
will irritate the mouth, esophagus, and
stomach to a serious degree and on contact
with skin cause dermatitis. It may cause
conjunctivitis even in the vapor form. If

spilled on a person, remove all contaminated clothing, wash contaminated skin with copious amounts of water, followed by a baking soda solution. Irrigate eyes with warm water for 15 minutes. Consult a physician.

Stored drums should be vented two times per month to control the hydrogen pressure, which is produced by the action of acid on the drum metal. Decomposes explosively on contact with water, alcohol, or acids. Explosive reaction with phosphorus. Violent reaction with silver nitrate. Potentially violent reaction with sulfuric acid or diphenyl ether. Incompatible with acetic acid, acetic anhydride, acetonitrile, acrolein, acrylic acid, acrylonitrile, allyl alcohol, allyl chloride, 2-amino ethanol, ammonium hydroxide, aniline, n-butyraldehyde, creosote oil, cresol, cumene, dichloroethyl ether, diethylene glycol monomethyl ether, diisobutylene, diisopropylether, epichloro hydrin, ethyl acetate, ethyl acrylate, ethylene chlorohydrin, ethylene cyanohydrin, ethylene diamine, ethylene glycol, ethylene glycol monoethyl ether acetate, ethylene imine, glyoxal, HCl, HF, H_2O_2, isoprene, mesityl oxide, metal powders, methyl ethyl ketone, HNO_3, 2-nitropropane, β-propiolactone, propylene oxide, pyridene, NaOH, sulfolane, styrene monomer, vinyl acetate, vinylidene chloride, water, organic matter, combustibles. Dangerous. To fight fire, avoid water, use dry chemicals. When heated to decomposition it emits toxic fumes of Cl⁻ and SO_x. See SULFURIC ACID, HYDROCHLORIC ACID, and SULFONATES.

CLH750   CAS: 58-94-6   HR: 2
CHLOROTHIAZIDE
mf: $C_7H_6ClN_3O_4S_2$ mw: 295.73
PROP: Mp: 342.5°–343° (decomp). Sol in alkali.
SYNS: ALURENE □ CHLORIAZID □ 6-CHLORO-2H-1,2,4-BENZOTHIADIAZINE-7-SULFONAMIDE-1,1-DIOXIDE □ 6-CHLORO-7-SULFAMOYL-2H-1,2,4-BENZOTHIADIAZINE-1,1-DIOXIDE □ CHLORO-THIAZID □ CHLORSAL □ CHLORTHIAZIDE □

CHLORURIT □ CHLOTRIDE □ CLOTRIDE □ CT □ DIURESAL □ DIURIL □ DIURILIX □ DIURITE □ DIUTRID □ FLUMEN □ MINZIL □ NEO-DEMA □ SALISAN □ SALUNIL □ SALURETIL □ SALURIC □ SK-CHLOROTHIAZIDE □ THIAZIDE □ URINEX □ WARDUZIDE □ YADALAN
CONSENSUS REPORTS: Reported in EPA TSCA Inventory. EPA Genetic Toxicology Program.
SAFETY PROFILE: Moderately toxic by intraperitoneal and intravenous routes. Mildly toxic by ingestion. Experimental reproductive effects. Has been implicated in aplastic anemia. When heated to decomposition it emits very toxic fumes of SO_x, NO_x, and Cl⁻.

CLH800   CAS: 7672-94-8   HR: 3
6-CHLORO-2-THIO-2H-1,3-BENZO-XAZINE-2,4(3H)-DIONE
mf: $C_8H_4ClNO_2S$ mw: 213.64
SYN: 2H-1,3-BENZOXAZINE-2,4(3H)-DIONE, 6-CHLORO-2-THIO-
SAFETY PROFILE: A poison by ingestion. When heated to decomposition it emits toxic vapors of NO_x, SO_x, and Cl⁻.

CLH820   CAS: 41219-30-1   HR: 3
6-CHLORO-4-THIOCHROMANYL o,o-DIMETHYL DITHIOPHOSPHATE
mf: $C_{11}H_{14}ClO_2PS_3$ mw: 340.85
SYNS: S-(6-CHLORO-3,4-DIHYDRO-2H-1-BENZO-THIOPYRAN-4-YL) o,o-DIMETHYL PHOSPHORO-DITHIOATE □ PHOSPHORODITHIOIC ACID, S-(6-CHLORO-3,4-DIHYDRO-2H-1-BENZOTHIOPYRAN-4-YL) o,o-DIMETHYL ESTER
SAFETY PROFILE: A poison by ingestion. When heated to decomposition it emits toxic vapors of PO_x, SO_x, and Cl⁻.

CLJ750   CAS: 2812-73-9   HR: 3
CHLOROTHIOFORMIC ACID ETHYL ESTER
DOT: UN 2826
mf: C_3H_5ClOS mw: 124.59
PROP: Bp: 52–55° @ 40 mm.
SYN: ETHYL CHLOROTHIOFORMATE (DOT)
DOT CLASSIFICATION: 8; Label: Corrosive, Poison
SAFETY PROFILE: Probably a poison by inhalation and ingestion. A corrosive irritant

to skin, eyes, and mucous membranes. See also ESTERS and CHLORIDES. Flammable when exposed to heat or flame. When heated to decomposition it emits very toxic fumes of Cl⁻ and SO_x.

CLK100   CAS: 95-49-8   HR: 3
o-CHLOROTOLUENE
DOT: UN 2238
mf: C_7H_7Cl mw: 126.59
PROP: Liquid. Mp: −34°, bp: 159°, d: 1.08 @ 20°/4°. Volatile with steam. Sltly sol in water; freely sol in alc, benzene, chloroform, ether.
SYNS: 2-CHLORO-1-METHYLBENZENE (9CI) □ 2-CHLOROTOLUENE □ HALSO 99 □ 1-METHYL-2-CHLOROBENZENE □ 2-METHYLCHLOROBENZENE □ o-TOLYL CHLORIDE
CONSENSUS REPORTS: Reported in EPA TSCA Inventory.
OSHA PEL: TWA 50 ppm
ACGIH TLV: TWA 50 ppm
DOT CLASSIFICATION: 3; Label: Flammable Liquid
SAFETY PROFILE: Moderately toxic by unspecified routes. Flammable when exposed to heat or flame. When heated to decomposition it emits toxic fumes of Cl⁻. See also TOLYL CHLORIDE and CHLORINATED HYDROCARBONS, AROMATIC.

CLK130   CAS: 25168-05-2   HR: 3
CHLOROTOLUENES
DOT: UN 2238
mf: C_7H_7Cl mw: 126.59
SYNS: BENZENE, CHLOROMETHYL-(9CI) □ CHLOROMETHYLBENZENE □ CHLOROTOLUENE □ ar-CHLOROTOLUENE □ TOLUENE, ar-CHLORO-
CONSENSUS REPORTS: Reported in EPA TSCA Inventory.
DOT CLASSIFICATION: 3; Label: Flammable Liquid
SAFETY PROFILE: Low toxicity by inhalation. A flammable liquid. When heated to decomposition it emits toxic vapors of Cl⁻.

CLK200   CAS: 87-60-5   HR: 3

3-CHLORO-o-TOLUIDINE
mf: C_7H_8ClN mw: 141.61
PROP: Bp: 245°.
SYNS: 1-AMINO-2-CHLORO-6-METHYLBENZENE □ 1-AMINO-3-CHLORO-2-METHYLBENZENE □ 2-AMINO-6-CHLOROTOLUENE □ AZOIC DIAZO COMPONENT 46 □ 3-CHLORO-2-METHYLANILINE □ 3-CHLOR-2-TOLUIDIN (CZECH) □ FAST SCARLET TR BASE □ SCARLET TR BASE
CONSENSUS REPORTS: Reported in EPA TSCA Inventory.
DOT CLASSIFICATION: 6.1; Label: KEEP AWAY FROM FOOD
SAFETY PROFILE: Poison by ingestion. Mutation data reported. When heated to decomposition it emits toxic fumes of Cl⁻ and NO_x. See also other chloro toluidine entries.

CLK210   CAS: 615-65-6   HR: 3
2-CHLORO-p-TOLUIDINE
mf: C_7H_8ClN mw: 141.61
PROP: Liquid. D: 1.151 @ 20°, mp: 7°, bp: 219° @ 732 mm.
SYNS: BENZENAMINE, 2-CHLORO-4-METHYL- □ 2-CHLORO-4-METHYLANILINE □ 2-CHLOR-4-TOLUIDIN (CZECH) □ 4-METHYL-2-CHLOROANILINE
CONSENSUS REPORTS: Reported in EPA TSCA Inventory.
DOT CLASSIFICATION: 6.1; Label: KEEP AWAY FROM FOOD
SAFETY PROFILE: Poison by ingestion. A severe eye and skin irritant. Mutation data reported. When heated to decomposition it emits toxic fumes of Cl⁻ and NO_x. See also other chloro toluidine entries.

CLK215   CAS: 95-74-9   HR: 3
3-CHLORO-p-TOLUIDINE
mf: C_7H_8ClN mw: 141.61
PROP: A solid or liquid. Mp: 26°, bp: 237–238.5°.
SYNS: 1-AMINO-3-CHLORO-4-METHYLBENZENE □ 4-AMINO-2-CHLOROTOLUENE □ 2-CHLORO-4-AMINOTOLUENE □ 3-CHLORO-4-METHYLANILINE □ CPT □ DKC 1347 □ DRC 1339 □ NCI-C02040
CONSENSUS REPORTS: Reported in EPA TSCA Inventory. NCI Carcinogenogenesis Bioassay (Feed); Results Negative: Mouse, Rat NCITR* NCI-CG-TR-145,78

DOT CLASSIFICATION: 6.1; Label: KEEP AWAY FROM FOOD

SAFETY PROFILE: Poison by ingestion, intravenous, and intraperitoneal routes. Mutation data reported. When heated to decomposition it emits toxic fumes of Cl⁻ and NO$_x$. See also other chloro toluidine entries.

CLK220   CAS: 95-69-2   HR: 3
4-CHLORO-o-TOLUIDINE

mf: C$_7$H$_8$ClN   mw: 141.61

PROP: Leaflets from EtOH. Mp: 29–30°, bp: 236–238° @ 730 mm.

SYNS: AMARTHOL FAST RED TR BASE □ 2-AMINO-5-CHLOROTOLUENE □ AZOENE FAST RED TR BASE □ AZOGENE FAST RED TR □ AZOIC DIAZO COMPONENT 11 BASE □ BRENTAMINE FAST RED TR BASE □ 5-CHLORO-2-AMINOTOLUENE □ 4-CHLORO-2-METHYLANILINE □ 4-CHLORO-6-METHYLANILINE □ 4-CHLORO-2-METHYLBENZENEAMINE □ 4-CHLORO-2-TOLUIDINE □ DAITO RED BASE TR □ DEVAL RED K □ DEVAL RED TR □ DIAZO FAST RED TRA □ FAST RED BASE TR □ FAST RED 5CT BASE □ FAST RED TR □ FAST RED TR11 □ FAST RED TR BASE □ FAST RED TRO BASE □ KAKO RED TR BASE □ KAMBAMINE RED TR □ 2-METHYL-4-CHLOROANILINE □ MITSUI RED TR BASE □ RED BASE CIBA IX □ RED BASE IRGA IX □ RED BASE NTR □ RED TR BASE □ SANYO FAST RED TR BASE □ TULABASE FAST RED TR

CONSENSUS REPORTS: NTP 10th Report on Carcinogens. IARC Cancer Review: Group 2A IMEMDT 7,56,87; Human Inadequate Evidence IMEMDT 16,277,78; Animal Sufficient Evidence IMEMDT 30,61,83. Reported in EPA TSCA Inventory.

DFG MAK: Human Carcinogen

DOT CLASSIFICATION: 6.1; Label: KEEP AWAY FROM FOOD

SAFETY PROFILE: Confirmed carcinogen. Poison by ingestion and subcutaneous routes. Human mutation data reported. In the presence of copper(II) chloride catalyst decomposition occurs above 239°C. When heated to decomposition it emits toxic fumes of Cl⁻ and NO$_x$. See also other chloro toluidine entries.

CLK225   CAS: 95-79-4   HR: 3
5-CHLORO-o-TOLUIDINE

mf: C$_7$H$_8$ClN   mw: 141.61

PROP: Solid. Bp: 237° @ 722 mm, mp: 21–22°.

SYNS: ACCO FAST RED KB BASE □ 1-AMINO-3-CHLORO-6-METHYLBENZENE □ 2-AMINO-4-CHLOROTOLUENE □ ANSIBASE RED KB □ AZOENE FAST RED KB BASE □ AZOIC DIAZO COMPONENT 32 □ 4-CHLORO-2-AMINOTOLUENE □ 3-CHLORO-6-METHYLANILINE □ 5-CHLORO-2-METHYLANILINE □ FAST RED KB AMINE □ FAST RED KB BASE □ FAST RED KB SALT □ FAST RED KB SALT SUPRA □ FAST RED KBS SALT □ GENAZO RED KB SOLN □ HILTONIL FAST RED KB BASE □ LAKE RED KB BASE □ METROGEN RED FORMER KB SOLN □ NAPHTHOSOL FAST RED KB BASE □ NCI-C02051 □ PHARMAZOID RED KB □ RED KB BASE □ SPECTROLENE RED KB □ STABLE RED KB BASE

CONSENSUS REPORTS: NTP Carcinogenesis Bioassay (feed): Clear Evidence: mouse NCITR* NCI-TR-187,79; (feed): Inadequate Studies: rat NCITR* NCI-TR-187,79. Reported in EPA TSCA Inventory. EPA Genetic Toxicology Program.

DFG MAK: Confirmed Animal Carcinogen with Unknown Relevance to Humans

DOT CLASSIFICATION: 6.1; Label: KEEP AWAY FROM FOOD

SAFETY PROFILE: Suspected carcinogen with experimental carcinogenic and tumorigenic data. Moderately toxic by ingestion. When heated to decomposition it emits very toxic fumes of Cl⁻ and NO$_x$. See also AROMATIC AMINES.

CLK227   CAS: 87-63-8   HR: 3
6-CHLORO-o-TOLUIDINE

mf: C$_7$H$_8$ClN   mw: 141.61

SYNS: 2-AMINO-3-CHLOROTOLUENE □ 3-CHLORO-2-AMINOTOLUENE □ 6-CHLORO-2-METHYLANILINE □ 6-CHLORO-2-TOLUIDINE □ o-TOLUIDINE, 6-CHLORO-

CONSENSUS REPORTS: Reported in EPA TSCA Inventory.

DOT CLASSIFICATION: 6.1; Label: KEEP AWAY FROM FOOD

SAFETY PROFILE: Poison by subcutaneous route. Mutation data reported.

When heated to decomposition it emits toxic vapors of NO_x and Cl⁻.

CLK235   CAS: 3165-93-3   HR: 3
4-CHLORO-2-TOLUIDINE HYDROCHLORIDE

DOT: UN 1579

mf: $C_7H_8ClN \cdot ClH$ mw: 178.07

SYNS: AMARTHOL FAST RED TR BASE □ AMARTHOL FAST RED TR SALT □ 2-AMINO-5-CHLOROTOLUENE HYDROCHLORIDE □ AZANIL RED SALT TRD □ AZOENE FAST RED TR SALT □ AZOGENE FAST RED TR □ AZOIC DIAZO COMPONENT 11 BASE □ BRENTAMINE FAST RED TR SALT □ CHLORHYDRATE de 4-CHLOROORTHO-TOLUIDINE (FRENCH) □ 5-CHLORO-2-AMINO-TOLUENE HYDROCHLORIDE □ 4-CHLORO-2-METHYLANILINE HYDROCHLORIDE □ 4-CHLORO-6-METHYLANILINE HYDROCHLORIDE □ 4-CHLORO-2-METHYLBENZENE-AMINE HYDROCHLORIDE □ 4-CHLORO-o-TOLUIDINE HYDROCHLORIDE □ 4-CHLORO-o-TOLUIDINE HYDROCHLORIDE (DOT) □ C.I. 37085 □ C.I. AZOIC DIAZO COMPONENT 11 □ DAITO RED SALT TR □ DEVOL RED K □ DEVOL RED TA SALT □ DEVOL RED TR □ DIAZO FAST RED TR □ DIAZO FAST RED TRA □ FAST RED 5CT SALT □ FAST RED SALT TR □ FAST RED SALT TRA □ FAST RED SALT TRN □ FAST RED TR SALT □ HINDASOL RED TR SALT □ KROMON GREEN B □ 2-METHYL-4-CHLOROANILINE HYDROCHLORIDE □ NATASOL FAST RED TR SALT □ NCI-C02368 □ NEUTROSEL RED TRVA □ OFNA-PERL SALT RRA □ RCRA WASTE NUMBER U049 □ RED BASE CIBA IX □ RED BASE IRGA IX □ RED SALT CIBA IX □ RED SALT IRGA IX □ RED TRS SALT □ SANYO FAST RED SALT TR

CONSENSUS REPORTS: NTP 10th Report on Carcinogens. IARC Cancer Review: Group 2A IMEMDT 48,123,90; Animal Inadequate Evidence, Human Inadequate Evidence IMEMDT 16,277,78. NCI Carcinogenesis Bioassay (feed); Clear Evidence: mouse; No Evidence: rat NCITR* NCI-CG-TR-165,79. Reported in EPA TSCA Inventory.

DOT CLASSIFICATION: 6.1; Label: KEEP AWAY FROM FOOD

SAFETY PROFILE: Confirmed carcinogen with experimental carcinogenic data. Moderately toxic by intraperitoneal route. When heated to decomposition it emits toxic fumes of NO_x and Cl⁻. See also other chloro toluidine entries.

CLO750   CAS: 569-57-3   HR: 3
CHLOROTRIANISENE

mf: $C_{23}H_{21}ClO_3$ mw: 380.89

PROP: Crystals from MeOH. Mp: 114–116°.

SYNS: ANISENE □ CHLORESTROLO □ 1,1',1"-(1-CHLORO-1-ETHENYL-2-YLIDENE)-TRIS(4-METHOXYBENZENE) □ CHLOROTRIANIZEN □ CHLOROTRISIN □ CHLOROTRIS(p-METHOXY-PHENYL)ETHYLENE □ CHLORTRIANISEN □ CLORESTROLO □ CLOROTRISIN □ CTA □ HORMONISENE □ KHLORTRIANIZEN □ MERBENTUL □ METACE □ NSC-10108 □ RIANIL □ TACE □ TACE-FN □ TRI-p-ANISYLCHLOROETHYLENE □ TRIS(p-METHOXYPHENYL)CHLOROETHYLENE

CONSENSUS REPORTS: IARC Cancer Review: Animal Inadequate Evidence IMEMDT 21,139,79; Human Limited Evidence IMEMDT 21,139,79.

SAFETY PROFILE: Suspected human carcinogen with experimental tumorigenic data. Human reproductive effects by ingestion: changes in fertility. Used in cancer treatment. When heated to decomposition it emits very toxic fumes of Cl⁻.

CLP500   CAS: 1461-22-9   HR: 3
CHLOROTRIBUTYLSTANNANE

mf: $C_{12}H_{27}ClSn$ mw: 325.53

PROP: Liquid. D: 1.2105 @ 20°, bp: 171–173° @ 25 mm.

SYNS: CHLORID TRI-n-BUTYLCINICITY (CZECH) □ TRIBUTYLCHLOROSTANNANE □ TRI-n-BUTYLTIN CHLORIDE □ TRI-n-BUTYLZINN-CHLORID (GERMAN)

CONSENSUS REPORTS: Reported in EPA TSCA Inventory.

OSHA PEL: TWA 0.1 mg(Sn)/m³ (skin)

ACGIH TLV: TWA 0.1 mg(Sn)/m³; STEL 0.2 mg(Sn)/m³ (skin).

DFG MAK: 0.0021 ppm (0.05 mg/m³)

NIOSH REL: (Organotin Compounds) TWA 0.1 mg(Sn)/m³

SAFETY PROFILE: Poison by ingestion and skin contact. A severe eye irritant. Mutation data reported. Tributyl tin compounds are extremely toxic to marine life. See also TIN COMPOUNDS. When heated to decomposition it emits toxic fumes of Cl⁻.

CLP750   CAS: 1929-82-4   HR: 3
2-CHLORO-6-(TRICHLOROMETHYL)-
**   PYRIDINE**
mf: $C_6H_3Cl_4N$ mw: 230.90
PROP: Crystals. Mp: 62–63°. Very sltly sol in H_2O.
SYNS: DOWCO-163 □ NITRAPYRIN (ACGIH) □ N-SERVE NITROGEN STABILIZER
CONSENSUS REPORTS: NCI Carcinogenesis Studies (ipr); Clear Evidence: mouse, rat RRCRBU 52,1,75. Reported in EPA TSCA Inventory.
OSHA PEL: Total Dust: 15 mg/m³; Respirable Fraction: 5 mg/m³
ACGIH TLV: TWA 10 mg/m³; STEL 20 mg/m³; Not Classifiable as a Human Carcinogen
SAFETY PROFILE: Poison by ingestion. Moderately toxic by skin contact. Experimental reproductive effects. Mutation data reported. When heated to decomposition it emits very toxic fumes of Cl⁻ and NO_x.

CLQ500   CAS: 15529-90-5   HR: 3
CHLORO(TRIETHYLPHOSPHINE)GOLD
mf: $C_6H_{15}AuClP$ mw: 350.60
PROP: Crystals from EtOH. Mp: 78°, bp: 210° @ 0.03 mm. Sol in $CHCl_3$ and EtOH.
SYNS: SK&F 36914 □ TRIETHYLPHOSPHINEAUROUS CHLORIDE
SAFETY PROFILE: Poison by ingestion. Experimental teratogenic and reproductive effects. Human mutation data reported. When heated to decomposition it emits very toxic fumes of Cl⁻ and PO_x. See also PHOSPHINE and GOLD.

CLQ750   CAS: 79-38-9   HR: 3
CHLOROTRIFLUOROETHYLENE
DOT: UN 1082
mf: C_2ClF_3 mw: 116.47
PROP: A gas. Fp: −157.5°, bp: −26.2°, flash p: −18°F, lel: 24%, uel: 40.3%.
SYNS: 1-CHLORO-1,2,2-TRIFLUOROETHYLENE □ 2-CHLORO-1,1,2-TRIFLUOROETHYLENE □ CHLOR-TRIFLUORAETHYLEN (GERMAN) □ CTFE □ DAIFLON □ FLUOROPLAST 3 □ GENETRON 1113 □ MONO-CHLOROTRIFLUOROETHYLENE □ TRIFLUORO-CHLOROETHYLENE (DOT) □ 1,1,2-TRIFLUORO-2-CHLOROETHYLENE □ TRIFLUOROMONO-CHLOROETHYLENE □ TRIFLUOROVINYL CHLORIDE □ TRITHENE
CONSENSUS REPORTS: Reported in EPA TSCA Inventory.
DOT CLASSIFICATION: 2.1; Label: Flammable Gas
SAFETY PROFILE: Poison by ingestion and intraperitoneal routes. Moderately toxic by inhalation. Very dangerous fire hazard when exposed to heat, flames (sparks), or oxidizers. To fight fire, stop flow of gas. Violent reaction when mixed with (Br_2 + O_2) or (ClF_3 + water). Potentially explosive polymerization reaction with ethylene. Incompatible with 1,1-dichloroethylene; oxygen. When heated to decomposition it emits toxic fumes of F⁻ and Cl⁻. See also CHLORINATED HYDROCARBONS, ALIPHATIC; and FLUORIDES.

CLR250   CAS: 75-72-9   HR: 1
CHLOROTRIFLUOROMETHANE
DOT: UN 1022
mf: $CClF_3$ mw: 104.46
PROP: Colorless gas; ethereal odor. Mp: −181°, bp: −81.4°, fp: −181°.
SYNS: ARCTON 3 □ F 13 □ FREON 13 □ GENETRON 13 □ HALOCARBON 13/UCON 13 □ MONOCHLOROTRIFLUOROMETHANE (DOT) □ R 13 □ R13 (DOT) □ TRIFLUOROCHLOROMETHANE (DOT) □ TRIFLUOROMETHYL CHLORIDE □ TRIFLUORO-MONOCHLOROCARBON
CONSENSUS REPORTS: Reported in EPA TSCA Inventory.
DFG MAK: 1000 ppm (4300 mg/m³)
DOT CLASSIFICATION: 2.2; Label: Nonflammable Gas
SAFETY PROFILE: A mild irritant. Narcotic in high concentrations. Reacts violently with Al. When heated to decomposition it emits highly toxic fumes of F⁻ and Cl⁻.

CLT000   CAS: 1066-45-1   HR: 3
CHLOROTRIMETHYLSTANNANE
mf: C_3H_9ClSn mw: 199.26
PROP: Colorless needles. Mp: 42°, bp: 154–156°.

SYNS: CHLOROTRIMETHYLTIN □ TRIMETHYL-CHLOROSTANNANE □ TRIMETHYLCHLOROTIN □ TRIMETHYLSTANNYL CHLORIDE □ TRIMETHYLTIN CHLORIDE

CONSENSUS REPORTS: EPA Extremely Hazardous Substances List. Reported in EPA TSCA Inventory.

OSHA PEL: TWA 0.1 mg(Sn)/m³ (skin)

ACGIH TLV: TWA 0.1 mg(Sn)/m³; STEL 0.2 mg(Sn)/m³ (skin).

NIOSH REL: (Organotin Compounds) TWA 0.1 mg(Sn)/m³

SAFETY PROFILE: A deadly poison by intravenous route. Experimental reproductive effects. See also TIN COMPOUNDS. When heated to decomposition it emits toxic fumes of Cl⁻.

CLU000 CAS: 639-58-7 HR: 3
CHLOROTRIPHENYLSTANNANE

mf: C₁₈H₁₅ClSn   mw: 385.47

PROP: Colorless crystals from alc. Mp: 106°, bp: 240° @ 13.5 mm. Insol in water; sol in org solvs.

SYNS: AQUATIN □ BRESTANOL □ CHLORO-TRIPHENYLTIN □ FENTIN CHLORIDE □ GC 8993 □ GENERAL CHEMICALS 8993 □ HOE 2872 □ LS 4442 □ TINMATE □ TPTC □ TRIPHENYLCHLOROSTANNANE □ TRIPHENYLCHLOROTIN □ TRIPHENYLTIN CHLORIDE

CONSENSUS REPORTS: Reported in EPA TSCA Inventory. EPA Extremely Hazardous Substances List.

OSHA PEL: TWA 0.1 mg(Sn)/m³ (skin)

ACGIH TLV: TWA 0.1 mg(Sn)/m³; STEL 0.2 mg(Sn)/m³ (skin).

NIOSH REL: (Organotin Compounds) TWA 0.1 mg(Sn)/m³

SAFETY PROFILE: Poison by ingestion, intraperitoneal, and intravenous routes. Experimental reproductive effects. Mutation data reported. An insect chemosterilant. See also TIN COMPOUNDS. When heated to decomposition it emits toxic fumes of Cl⁻.

CLU250 CAS: 2279-76-7 HR: 3
CHLOROTRIPROPYLSTANNANE

mf: C₉H₂₁ClSn   mw: 283.44

PROP: Colorless liquid. D: 1.2678 @ 28°, mp: −23.5°. Sol in org solvs.

SYNS: TRIPROPYLTIN CHLORIDE □ TRI-n-PROPYLTIN CHLORIDE

OSHA PEL: TWA 0.1 mg(Sn)/m³ (skin)

ACGIH TLV: TWA 0.1 mg(Sn)/m³; STEL 0.2 mg(Sn)/m³ (skin).

NIOSH REL: (Organotin Compounds) TWA 0.1 mg(Sn)/m³

SAFETY PROFILE: Poison by intravenous route. See also TIN COMPOUNDS. When heated to decomposition it emits toxic fumes of Cl⁻.

CLU500 CAS: 10008-90-9 HR: 3
CHLORO(TRIVINYL)STANNANE

mf: C₆H₉ClSn   mw: 235.29

PROP: Colorless liquid. Bp: 59–60° @ 6 mm.

SYN: TRIVINYLTIN CHLORIDE

OSHA PEL: TWA 0.1 mg(Sn)/m³ (skin)

ACGIH TLV: TWA 0.1 mg(Sn)/m³; STEL 0.2 mg(Sn)/m³ (skin).

NIOSH REL: TWA (Organotin Compounds) 0.1 mg(Sn)/m³

SAFETY PROFILE: Poison by intravenous route. See also TIN COMPOUNDS and CHLORIDES. When heated to decomposition it emits toxic fumes of Cl⁻.

CLV000 CAS: 541-25-3 HR: 3
CHLOROVINYLARSINE DICHLORIDE

mf: C₂H₂AsCl₃   mw: 207.31

PROP: Liquid; faint odor of geranium. Bp: 190° decomp, fp: −13°, d: 1.888 @ 20°/4°, vap press: 0.4 mm @ 20°, vap d: 7.15.

SYNS: (2-CHLOROETHENYL) ARSONOUS DICHLORIDE □ β-CHLOROVINYLBICHLOROARSINE □ 2-CHLOROVINYLDICHLOROARSINE □ (2-CHLOROVINYL)DICHLOROARSINE □ DICHLORO(2-CHLOROVINYL)ARSINE □ LEWISITE □ LEWISITE (ARSENIC COMPOUND)

CONSENSUS REPORTS: Reported in EPA TSCA Inventory. EPA Genetic Toxicology Program. EPA Extremely Hazardous Substances List. Arsenic and its compounds are on the Community Right-To-Know List.

SAFETY PROFILE: A human poison by inhalation. Poison experimentally by inhalation, skin contact, subcutaneous, intraperitoneal, and intravenous routes. An experimental teratogen. A blistering-type military poison. Lewisite is absorbed through skin; as little as 2 mL on the skin can cause death. Has a delayed action similar to distilled mustard gas. This gas exhibits a systemic poisoning effect on humans. When heated to decomposition it emits toxic fumes of Cl⁻ and As. See also ARSENIC COMPOUNDS.

CLW000   CAS: 88-04-0   HR: 3
4-CHLORO-3,5-XYLENOL
mf: C_8H_9ClO mw: 156.62
PROP: Crystals or prisms from (C_6H_6); phenolic odor. Mp: 115.5°, bp: 246°. Sltly water-sol.
SYNS: BENZYTOL □ 4-CHLORO-3,5-DIMETHYL-PHENOL □ CHLORO-XYLENOL □ p-CHLORO-m-XYLENOL □ DESSON □ DETTOL □ ESPADOL □ HUSEPT EXTRA □ OTTASEPT □ OTTASEPT EXTRA □ PCMX □ RBA 777
CONSENSUS REPORTS: Reported in EPA TSCA Inventory. Chlorophenols are on the Community Right-To-Know List.
SAFETY PROFILE: Poison by intraperitoneal route. Moderately toxic by ingestion. An experimental teratogen. An eye irritant. An antimicrobial agent. See also CHLOROPHENOLS and CHLORINATED HYDROCARBONS, AROMATIC. When heated to decomposition it emits toxic fumes of Cl⁻.

CLW250   CAS: 50892-23-4   HR: 3
(4-CHLORO-6-(2,3-XYLIDINO)-2-PY-RIMIDINYLTHIO)ACETIC ACID
mf: $C_{14}H_{14}ClN_3O_2S$ mw: 323.82
PROP: Crystals from EtOAc. Mp: 150–153°.
SYNS: ((4-CHLORO-6-((2,3-DIMETHYLPHENYL)-AMINO)-2-PYRIMIDINYL)THIO)ACETIC ACID □ WY-14,643
SAFETY PROFILE: Moderately toxic by ingestion. Suspected carcinogen with experimental carcinogenic and tumorigenic data. Mutation data reported. When heated to decomposition it emits very toxic fumes of Cl⁻, NOₓ, and SOₓ.

CLW500   CAS: 65089-17-0   HR: 3
2-((4-CHLORO-6-(2,3-XYLIDINO)-2-PYRIMIDINYL)THIO)-N-(2-HYDROXYETHYL)ACETAMIDE
mf: $C_{16}H_{19}ClN_4O_2S$ mw: 366.90
PROP: Crystals from Me₂CO. Mp: 144–146°.
SYNS: BR-931 □ PIRINIXIL
SAFETY PROFILE: Suspected carcinogen with experimental carcinogenic data. Mutation data reported. When heated to decomposition it emits very toxic fumes of Cl⁻, NOₓ, and SOₓ.

CLX000   CAS: 54749-90-5   HR: 3
CHLOROZOTOCIN
mf: $C_9H_{16}ClN_3O_7$ mw: 313.73
SYNS: 1-(2-CHLOROETHYL)-3-(d-GLUCOPYRANOS-2-YL)-1-NITROSOUREA □ 2-((((2-CHLOROETHYL)NITRO-SOAMINO)CARBONYL)AMINO)-2-DEOXY-d-GLUCOPYRANOSE □ 2-((((2-CHLOROETHYL)NITROSO-AMINO)CARBONYL)AMINO)-2-DEOXY-d-GLUCOSE □ 2-(3-(2-CHLOROETHYL)-3-NITROSOUREIDO)-2-DEOXY-d-GLUCOSOPYRANOSE □ 2-(3-(2-CHLOROETHYL)-3-NITROSOUREIDO)-d-GLUCO-PYRANOSE □ CHLZ □ CZT □ DCNU □ NSC-178248 □ NSC-D 254157
CONSENSUS REPORTS: NTP 10th Report on Carcinogens. IARC Cancer Review: Group 2A IMEMDT 50,65,90; Animal Sufficient Evidence IMEMDT 50,65,90; Human No Adequate Data IMEMDT 50,65,90. EPA Genetic Toxicology Program.
SAFETY PROFILE: Confirmed carcinogen with experimental carcinogenic and tumorigenic data. Poison by subcutaneous, intravenous, and intraperitoneal routes. Human systemic effects by intravenous route: anorexia, leukopenia, nausea or vomiting, thrombocytopenia. Mutation data reported. When heated to decomposition it emits very toxic fumes of Cl⁻ and NOₓ. See also NITROSAMINES.

CMA100   CAS: 2921-88-2   HR: 3
CHLORPYRIFOS
mf: $C_9H_{11}Cl_3NO_3PS$ mw: 350.59
PROP: Crystals with mild mercaptan odor. Mp: 42–43.5°. Very sltly sol in H_2O; sol in most org solvs.

SYNS: BRODAN ☐ CHLORPYRIFOS-ETHYL ☐ CHLORPYRIPHOS ☐ CHLORPYRIPHOS-ETHYL ☐ DETMOL U.A. ☐ O,O-DIAETHYL-O-3,5,6-TRICHLOR-2-PYRIDYLMONOTHIOPHOSPHAT ☐ O,O-DIETHYL O-3,5,6-TRICHLORO-2-PYRIDYL PHOSPHOROTHIOATE ☐ DOWCO 179 ☐ DURSBAN ☐ DURSBAN F ☐ ENT 27,311 ☐ ERADEX ☐ ETHION, dry ☐ LORSBAN ☐ OMS-0971 ☐ PIRIDANE ☐ 2-PYRIDINOL, 3,5,6-TRICHLORO-, O-ESTER with O,O-DIETHYL PHOSPHOROTHIOATE ☐ PYRINEX ☐ STIPEND

CONSENSUS REPORTS: EPA Genetic Toxicology Program.

OSHA PEL: TWA 0.2 mg/m³ (skin)

ACGIH TLV: TWA 0.1 mg/m³ (skin); Not Classifiable as a Human Carcinogen

SAFETY PROFILE: Poison by ingestion, intraperitoneal, skin contact, and inhalation routes. Human systemic effects by ingestion: paresthesia, muscle weakness, coma. Experimental reproductive effects: developmental toxicity. Mutation data reported. When heated to decomposition it emits very toxic fumes of Cl⁻, NO$_x$, PO$_x$, and SO$_x$.

CMA250   CAS: 5598-13-0   HR: 2
CHLORPYRIFOS-METHYL

mf: C₇H₇Cl₃NO₃PS   mw: 322.53

PROP: A solid. Mp: 44.5–45.5°. Sltly sol in H₂O; very sol in org solvs.

SYNS: O,O-DIMETHYL-O-(3,5,6-TRICHLORO-2-PYRIDYL)PHOSPHOROTHIOATE ☐ DOWCO 217 ☐ DURSBAN METHYL ☐ ENT 27,520 ☐ METHYL CHLORPYRIFOS ☐ METHYL DURSBAN ☐ NOLTRAN ☐ NSC-60380 ☐ OMS-1155 ☐ RELDAN ☐ ZERTELL

SAFETY PROFILE: Moderately toxic by ingestion, intraperitoneal, and skin contact routes. A skin irritant. Experimental reproductive effects. When heated to decomposition it emits very toxic fumes of Cl⁻, NO$_x$, PO$_x$, and SO$_x$. A pesticide.

CMA750   CAS: 57-62-5   HR: 3
CHLORTETRACYCLINE

mf: C₂₂H₂₃ClN₂O₈   mw: 478.92

PROP: Golden-yellow crystals. Mp: 168–169°. Sltly sol in water; very sol in aq soln pH 7.65; freely sol in the "cellosolves," dioxane, "Carbitol"; sol in methanol, ethanol, butanol, acetone, ethyl acetate, and benzene; insol in ether and pet ether.

SYNS: ACRONIZE ☐ AUREOCINA ☐ AUREOMYCIN ☐ AUREOMYCIN A-377 ☐ AUREOMYKOIN ☐ BIO-MITSIN ☐ BIOMYCIN ☐ 7-CHLORO-4-(DIMETHYL-AMINO)-1,4,4a,5,5a,6,11,12a-OCTAHYDRO-2-NAPHTH-ACENECARBOXAMIDE ☐ 7-CHLOROTETRACYCLINE ☐ CHRYSOMYKINE ☐ CTC ☐ DUOMYCIN ☐ FLAMYCIN

CONSENSUS REPORTS: Reported in EPA TSCA Inventory.

SAFETY PROFILE: Poison by intravenous and intraperitoneal routes. Moderately toxic by ingestion. Experimental reproductive effects. When heated to decomposition it emits toxic fumes of Cl⁻ and NO$_x$. See also TETRACYCLINE.

CMC750   CAS: 67-97-0   HR: 3
CHOLECALCIFEROL

mf: C₂₇H₄₄O   mw: 384.71

PROP: White crystals; odorless. Mp: 87–88°. Insol in water; sol in alc, chloroform, and fatty oils.

SYNS: COLECALCIFEROL ☐ 7-DEHYDROCHO-LESTROL, ACTIVATED ☐ DELSTEROL ☐ DEPARAL ☐ D3-VIGANTOL ☐ OLEOVITAMIN D3 ☐ RICKETON ☐ 9,10-SECOCHOLESTA-5,7,10(19)-TRIEN-3-β-OL ☐ TRIVITAN ☐ VIGORSAN ☐ VITAMIN D3 ☐ VITINC DAN-DEE-3

CONSENSUS REPORTS: Reported in EPA TSCA Inventory.

SAFETY PROFILE: Poison by ingestion. An experimental teratogen. When heated to decomposition it emits acrid smoke and irritating fumes.

CMD750   CAS: 57-88-5   HR: 2
CHOLESTEROL

mf: C₂₇H₄₆O   mw: 386.73

PROP: White or faint yellow, pearly leaflets from aq alc. Mp: 148.5° (anhyd), bp: 360° (decomp).

SYNS: CHOLEST-5-EN-3-β-OL ☐ Δ⁵-CHOLESTEN-3-β-OL ☐ 5-CHOLESTEN-3-β-OL ☐ 5:6-CHOLESTEN-3-β-OL ☐ CHOLESTERIN ☐ CHOLESTEROL BASE H ☐ CHOLESTERYL ALCOHOL ☐ CHOLESTRIN ☐ CHOLESTROL ☐ CORDULAN ☐ DUSOLINE ☐ DUSORAN ☐ DYTHOL ☐ HYDROCERIN ☐ 3-β-HYDROXYCHOLEST-5-ENE ☐ KATHRO ☐ LANOL ☐

NIMCO CHOLESTEROL BASE H □ PROVITAMIN D □ SUPER HARTOLAN □ TEGOLAN

CONSENSUS REPORTS: IARC Cancer Review: Group 3 IMEMDT 7,161,87; Human Inadequate Evidence IMEMDT 31,95,83; Animal Inadequate Evidence IMEMDT 10,99,76. Reported in EPA TSCA Inventory. EPA Genetic Toxicology Program.

SAFETY PROFILE: Experimental teratogenic and reproductive effects. Questionable carcinogen with experimental carcinogenic and tumorigenic data. Mutation data reported. Used in pharmaceutical and dermal preparations as an emulsifying agent. When heated to decomposition it emits acrid smoke and irritating fumes.

CME250   CAS: 3546-10-9   HR: 3
CHOLESTERYL-p-BIS(2-CHLOROETH-YL)AMINO PHENYLACETATE
mf: C₃₉H₅₉Cl₂NO₂   mw: 644.89
SYNS: (p-(BIS(2-CHLOROETHYL)AMINO)PHENYL)ACETATE CHOLESTEROL □ (p-(BIS(2-CHLOROETHYL)-AMINO)PHENYL)ACETIC ACID CHOLESTEROL ESTER □ (4-(BIS(2-CHLOROETHYL)AMINO)PHENYL)ACETIC ACID CHOLESTERYL ESTER □ 5-CHOLESTEN-3-β-OL 3-(p-(BIS(2-CHLOROETHYL)AMINO)PHENYL)ACETATE □ FENESTERIN □ FENESTRIN □ NCI-C01558 □ NSC-104469 □ PHENESTERINE □ PHENESTRIN

CONSENSUS REPORTS: NCI Carcinogenesis Bioassay (gavage); Clear Evidence: mouse, rat NCITR* NCI-CG-TR-60,78.

SAFETY PROFILE: Suspected carcinogen with experimental carcinogenic and neoplastigenic data. When heated to decomposition it emits very toxic fumes of Cl⁻ and NOₓ.

CME400   CAS: 11041-12-6   HR: 2
CHOLESTYRAMINE
SYNS: CHOLESTYRAMINE CHLORIDE □ CHOLESTYRAMINE RESIN □ COLESTYRAMIN □ CUEMID □ QUANTALAN □ QUESTRAN

CONSENSUS REPORTS: Reported in EPA TSCA Inventory.

SAFETY PROFILE: Low toxicity by ingestion. Questionable human carcinogen producing colon tumors. An experimental teratogen. Other experimental reproductive effects. Toxic effects by ingestion: acidosis and nosebleeds. When heated to decomposition it emits acrid smoke and irritating fumes.

CMF300   HR: D
CHOLINE BITARTRATE
mf: C₉H₁₉NO₇   mw: 253.25
PROP: White crystalline powder; acetic taste. Sol in water; sltly sol in alc; insol in ether, chloroform, benzene.
SYN: (2-HYDROXYETHYL)TRIMETHYLAMMONIUM BITARTRATE
SAFETY PROFILE: When heated to decomposition emits toxic fumes of NOₓ.

CMF400   CAS: 999-81-5   HR: 3
CHOLINE DICHLORIDE
mf: C₅H₁₃ClN•Cl   mw: 158.09
PROP: Crystals. Mp: 245° (decomp). Very sol in H₂O.
SYNS: ANTYWYLEGACZ □ CCC PLANT GROWTH REGULANT □ CE CE CE □ 2-CHLORAETHYL-TRIMETHYLAMMONIUMCHLORID □ CHLORCHO-LINCHLORID □ CHLORCHOLINE CHLORIDE □ CHLORMEQUAT □ CHLORMEQUAT CHLORIDE □ CHLOROCHOLINE CHLORIDE □ (β-CHLOROETH-YL)TRIMETHYLAMMONIUM CHLORIDE □ (2-CHLOROETHYL)TRIMETHYLAMMONIUM CHLORIDE □ 2-CHLORO-N,N,N-TRIMETHYLETHANAMINIUM CHLORIDE □ 60-CS-16 □ CYCLOCEL □ CYCOCEL □ CYCOCEL-EXTRA □ CYCOGAN □ CYCOGAN EXTRA □ CYOCEL □ EI 38,555 □ ETHANAMINIUM, 2-CHLORO-N,N,N-TRIMETHYL-, CHLORIDE (9CI) □ HICO CCC □ HORMOCEL-2CCC □ INCRECEL □ LIHOCIN □ NCI-C02960 □ RETACEL □ STABILAN □ TRIMETHYL-β-CHLORETHYL-AMMONIUMCHLORID □ TUR

CONSENSUS REPORTS: NTP Carcinogenesis Bioassay (feed); No Evidence: mouse, rat NCITR* NCI-TR-158,79. EPA Extremely Hazardous Substances List. Reported in EPA TSCA Inventory.

SAFETY PROFILE: Human poison by ingestion and intravenous routes. Moderately toxic by skin contact. Human systemic effects: respiratory depression. Questionable carcinogen with experimental neoplastigenic data. Mutation data reported.

When heated to decomposition it emits toxic fumes of Cl⁻.

CMF750   CAS: 67-48-1   HR: 3
CHOLINE HYDROCHLORIDE

mf: $C_5H_{14}NO \cdot Cl$ mw: 139.65

PROP: Colorless to white, deliquescent, hygroscopic crystals; slt odor of trimethylamine. Sol in water and alc.

SYNS: BIOCOLINA □ CHLORIDE de CHOLINE (FRENCH) □ CHOLINE CHLORHYDRATE □ CHOLINE CHLORIDE (FCC) □ CHOLINIUM CHLORIDE □ HEPACHOLINE □ (2-HYDROXYETHYL)TRIMETHYL-AMMONIUM CHLORIDE □ LIPOTRIL

CONSENSUS REPORTS: Reported in EPA TSCA Inventory.

SAFETY PROFILE: Poison by intraperitoneal and intravenous routes. Moderately toxic experimentally by ingestion and subcutaneous routes. Mutation data reported. A lipotropic agent which induces the reduction in fats contained in the liver. When heated to decomposition it emits toxic fumes of Cl⁻, SO_x, and NO_x. See also CHOLINE.

CMG800   CAS: 15005-90-0   HR: 3
CHROMALUM HEXAHYDRATE

mf: $Cr_2O_{12}S_3 \cdot 6H_2O$ mw: 500.30

PROP: IDLH 25 mg/m³ [as Cr(III)].

SYN: CHROMIUM(III) SULFATE, HEXAHYDRATE (2:3:6)

OSHA PEL: TWA 0.5 mg(Cr)/m³

ACGIH TLV: TWA 0.5 mg(Cr)/m³; Not Classifiable as a Carcinogen.

SAFETY PROFILE: Poison by intravenous route. When heated to decomposition it emits toxic fumes of SO_x and Cr⁻.

CMH000   CAS: 1066-30-4   HR: 3
CHROMIC ACETATE

mf: $C_6H_9O_6 \cdot Cr$ mw: 229.15

PROP: Gray-green powder or bluish-green pasty mass. IDLH 25 mg/m³ [as Cr(III)].

SYNS: CHROMIC ACETATE(III) □ CHROMIUM ACETATE □ CHROMIUM(III) ACETATE □ CHROMIUM TRIACETATE

CONSENSUS REPORTS: IARC Cancer Review: Group 3 IMEMDT 7,165,87; Animal Inadequate Evidence IMEMDT

2,100,73; IMEMDT 23,205,80. Chromium and its compounds are on the Community Right-To-Know List. Reported in EPA TSCA Inventory.

OSHA PEL: TWA 0.5 mg(Cr)/m³

ACGIH TLV: TWA 0.5 mg(Cr)/m³; Not Classifiable as a Carcinogen

SAFETY PROFILE: Questionable carcinogen with experimental tumorigenic data. Moderately toxic by intravenous route. Human mutation data reported. See also CHROMIUM COMPOUNDS. When heated to decomposition it emits acrid smoke and irritating fumes.

CMH250   CAS: 7738-94-5   HR: 3
CHROMIC(VI) ACID

mf: CrH_2O_4 mw: 118.02

PROP: Found in solution. IDLH Ca [15 mg/m³ {as Cr(VI)}].

SYNS: ACIDE CHROMIQUE (FRENCH) □ CHROMIC ACID

CONSENSUS REPORTS: Reported in EPA TSCA Inventory. Chromium and its compounds are on the Community Right-To-Know List.

OSHA PEL: CL 0.1 mg(CrO₃)/m³

ACGIH TLV: TWA 0.05 mg(Cr)/m³, Confirmed Human Carcinogen

DFG MAK: Animal Carcinogen, Suspected Human Carcinogen

NIOSH REL: (Chromium(VI)) TWA 0.025 mg(Cr(VI))/m³; CL 0.05/15M

SAFETY PROFILE: Confirmed human carcinogen. Poison by subcutaneous route. Mutation data reported. A powerful oxidizer. A powerful irritant of skin, eyes, and mucous membranes. Can cause a dermatitis, bronchoasthma, "chrome holes," damage to the eyes. Dangerously reactive. Incompatible with acetic acid, acetic anhydride, tetrahydronaphthalene, acetone, alcohols, alkali metals, ammonia, arsenic, bromine penta fluoride, butyric acid, n,n-dimethylformamide, hydrogen sulfide, peroxyformic acid, phosphorus, potassium hexacyanoferrate, pyridine, selenium, sodium, sulfur, and many other materials. See also CHROMIUM COMPOUNDS.

CMH260   CAS: 1308-14-1   HR: 3
CHROMIC(III) ACID
mf: CrH_3O_3 mw: 103.03
PROP: Green flocculent solid or crystals. Readily hydrolyzed. Practically insol in H_2O; sol in mineral acids. IDLH 25 mg/m^3 [as Cr(III)].
SYNS: CHROMIC (III) HYDROXIDE □ CHROMIUM(III) HYDROXIDE □ CHROMIUM TRIHYDROXIDE
CONSENSUS REPORTS: IARC Cancer Review: Group 3 IMEMDT 49,49,90; Human Inadequate Evidence IMEMDT 49,49,90; Animal Inadequate Evidence IMEMDT 49,49,90. Reported in EPA TSCA Inventory.
OSHA PEL: CL 0.1 mg(CrO$_3$)/m^3
ACGIH TLV: TWA 0.05 mg(Cr)/m^3; Confirmed Human Carcinogen
NIOSH REL: (Chromium(VI)) TWA 0.001 mg(Cr)/m^3
SAFETY PROFILE: A confirmed carcinogen. A poison. A powerful oxidizer. A powerful irritant of skin, eyes, and mucous membranes.

CMI250   CAS: 24613-89-6   HR: 3
CHROMIC CHROMATE
mf: $Cr_3O_{12}•2Cr$ mw: 452.00
SYNS: CHROMIC ACID, CHROMIUM(3+) SALT (3:2) □ CHROMIUM CHROMATE (MAK)
CONSENSUS REPORTS: IARC Cancer Review: Animal Sufficient Evidence IMEMDT 2,100,73. Reported in EPA TSCA Inventory. Chromium and its compounds are on the Community Right-To-Know List.
OSHA PEL: CL 0.1 mg(CrO$_3$)/m^3
ACGIH TLV: TWA 0.05 mg(Cr)/m^3; Confirmed Human Carcinogen
DFG MAK: Animal Carcinogen, Suspected Human Carcinogen
NIOSH REL: (Chromium(VI)) TWA 0.001 mg(Cr(VI))/m^3
SAFETY PROFILE: Confirmed carcinogen with experimental carcinogenic and

neoplastigenic data. Very powerful oxidizer. See also CHROMIUM COMPOUNDS.

CMI500   CAS: 1308-31-2   HR: 3
CHROMITE (mineral)
mf: Cr_2FeO_4 mw: 223.85
PROP: Black or brown-black cubic crystals. Relatively insol in acids.
SYNS: CHROME ORE □ CHROMITE □ CHROMITE ORE □ IRON CHROMITE
CONSENSUS REPORTS: NTP 10th Report on Carcinogens. IARC Cancer Review: Group 3 IMEMDT 7,165,87; Animal Inadequate Evidence IMEMDT 23,205,80. Chromium and its compounds are on the Community Right-To-Know List.
OSHA PEL: TWA 0.5 mg(Cr)/m^3
ACGIH TLV: TWA 0.05 mg/m^3 (ore processing); Confirmed Human Carcinogen (ore processing)
SAFETY PROFILE: Confirmed human carcinogen during ore processing. Human mutation data reported. See also CHROMIUM COMPOUNDS and IRON.

CMI750   CAS: 7440-47-3   HR: 3
CHROMIUM
af: Cr   aw: 52.00
PROP: Hard, ductile, blue-white metal. Resists oxidation in air. Bp: 26° @ 2690 mm. More reactive to acids than Mo or W and can be rendered passive. Rapidly attacked by fused NaOH + KNO$_3$ or KClO$_4$. IDLH 250 mg/m^3 (as Cr).
SYNS: CHROME □ CHROMIUM METAL (OSHA)
CONSENSUS REPORTS: IARC Cancer Review: Group 3 IMEMDT 7,165,87; Animal Inadequate Evidence IMEMDT 23,205,80. Chromium and its compounds are on the Community Right-To-Know List. Reported in EPA TSCA Inventory.
OSHA PEL: TWA 1 mg/m^3
ACGIH TLV: TWA 0.5 (Cr)mg/m^3; Not Classifiable as a Carcinogen
SAFETY PROFILE: Confirmed human carcinogen with experimental tumorigenic data. Powder will explode spontaneously in air. Ignites and is potentially explosive in

atmospheres of carbon dioxide. Violent or explosive reaction when heated with ammonium nitrate. May ignite or react violently with bromine pentafluoride. Incandescent reaction with nitrogen oxide or sulfur dioxide. Incompatible with oxidants. See also CHROMIUM COMPOUNDS.

CMJ000   CAS: 628-52-4   HR: 3
CHROMIUM ACETATE HYDRATE
mf: $C_4H_6O_4 \cdot Cr \cdot H_2O$ mw: 188.12
PROP: Air-sensitive dark red crystals. Stable in air for short period. Sltly sol in H_2O and EtOH; sol in hot H_2O. IDLH 250 mg/m^3 [as Cr(II)].
SYNS: ACETIC ACID, CHROMIUM (2+) SALT (8CI, 9CI) □ CHROMIUM(2+) ACETATE □ CHROMIUM(II) ACETATE □ CHROMIUM DIACETATE □ CHROMOUS ACETATE □ CHROMOUS ACETATE MONOHYDRATE
CONSENSUS REPORTS: Reported in EPA TSCA Inventory. Chromium and its compounds are on the Community Right-To-Know List.
OSHA PEL: TWA 0.5 mg(Cr)/m^3
ACGIH TLV: TWA 0.5 mg(Cr)/m^3; Not Classifiable as a Carcinogen
SAFETY PROFILE: Mildly toxic by ingestion. The anhydrous acetate ignites spontaneously in air. See also CHROMIUM COMPOUNDS. When heated to decomposition it emits acrid smoke and irritating fumes.

CMJ100   CAS: 29689-14-3   HR: 1
CHROMIUM CARBONATE
mf: $CH_2O_3 \cdot xCr$ mw: 426.03
SYNS: BASIC CHROMIUM CARBONATE □ CARBONIC ACID, CHROMIUM SALT
CONSENSUS REPORTS: Reported in EPA TSCA Inventory.
SAFETY PROFILE: When heated to decomposition it emits toxic fumes of Cr.

CMJ250   CAS: 10025-73-7   HR: 3
CHROMIUM CHLORIDE
mf: Cl_3Cr mw: 158.35

PROP: Red-violet flaky crystals. Mp: 1152°, bp: 1300° (subl). Insol in cold H_2O; sltly sol in hot H_2O. IDLH 25 mg/m^3 [as Cr(III)].
SYNS: CHROMIC CHLORIDE □ CHROMIUM(III) CHLORIDE (1:3) □ CHROMIUM CHLORIDE, anhydrous □ CHROMIUM TRICHLORIDE □ C.I. 77295 □ PURATRON-IC CHROMIUM CHLORIDE □ TRICHLOROCHROMIUM
CONSENSUS REPORTS: IARC Cancer Review: Group 3 IMEMDT 49,49,90; Human Inadequate Evidence IMEMDT 49,49,90; Animal Inadequate Evidence IMEMDT 49,49,90. Reported in EPA TSCA Inventory. EPA Genetic Toxicology Program. Chromium and its compounds are on the Community Right-To-Know List. EPA Extremely Hazardous Substances List.
OSHA PEL: TWA 0.5 mg(Cr)/m^3
ACGIH TLV: TWA 0.5 mg(Cr)/m^3; Not Classifiable as a Carcinogen
SAFETY PROFILE: Poison by skin contact, inhalation, and intraperitoneal routes. Experimental teratogenic and reproductive effects. Human mutation data reported. Questionable carcinogen. Reacts violently with lithium under nitrogen atmosphere. When heated to decomposition it emits toxic fumes of Cl$^-$.

CMJ500   HR: 3
CHROMIUM COMPOUNDS
CONSENSUS REPORTS: Chromium and its compounds are on the Community Right-To-Know List.
SAFETY PROFILE: Chromate salts are suspected human carcinogens producing tumors of the lungs, nasal cavity, and paranasal sinus. Chromic acid and its salts have a corrosive action on the skin and mucous membranes. The lesions are confined to the exposed parts, affecting chiefly the skin of the hands and forearms and the mucous membranes of the nasal septum. The characteristic lesion is a deep, penetrating ulcer, which, for the most part, does not tend to suppurate, and which is slow in healing. Small ulcers, about the size of a matchhead, may be found, chiefly around the base of the nails, on the

knuckles, dorsum of the hands and forearms. These ulcers tend to be clean and progress slowly. They are frequently painless, even though quite deep. They heal slowly and leave scars. On the mucous membranes of the nasal septum, the ulcers are usually accompanied by purulent discharge and crusting. If exposure continues, perforation of the nasal septum may result but produces no deformity of the nose. Hexavalent compounds are more toxic than the trivalent. Eczematous dermatitis due to trivalent chromium compounds has been reported.

CMJ560   CAS: 7788-97-8   HR: 3
CHROMIUM(III) FLUORIDE
DOT: UN 1756/UN 1757
mf: CrF_3 mw: 109.00
PROP: IDLH 25 mg/m³ [as Cr(III)].
SYNS: CHROME FLUORURE □ CHROMIC FLUORIDE □ CHROMIC FLUORIDE, solid (UN1756) (DOT) □ CHROMIC FLUORIDE, solution (UN1757) (DOT) □ CHROMIC TRIFLUORIDE □ CHROMIUM TRIFLUORIDE
CONSENSUS REPORTS: Reported in EPA TSCA Inventory.
OSHA PEL: 8H TWA 0.5 mg(Cr)/m³; TWA 2.5 mg(F)/m³
ACGIH TLV: TWA 0.5 mg(Cr)/m³; TWA 2.5 mg(F)/m³
NIOSH REL: (Fluorides, Inorganic) TWA 2.5 mg(F)/m³
DOT CLASSIFICATION: 8; Label: Corrosive
SAFETY PROFILE: A poison by ingestion. A corrosive. When heated to decomposition it emits toxic vapors of Cr and F⁻.

CMJ600   CAS: 13548-38-4   HR: 3
CHROMIUM(III) NITRATE
DOT: UN 2720
mf: CrN_3O_9 mw: 238.03
PROP: Very deliquescent, pale green powder. Non-volatile. Sol in H_2O, EtOAc, MeCN, and DMSO; insol in C_6H_6, CCl_4, and $CHCl_3$. IDLH 25 mg/m³ [as Cr(III)].
SYNS: CHROMIC NITRATE □ CHROMIUM NITRATE □ CHROMIUM (3+) NITRATE □ CHROMIUM NITRATE (DOT) □ CHROMIUM TRINITRATE □ NITRIC ACID, CHROMIUM (3+) SALT

CONSENSUS REPORTS: IARC Cancer Review: Group 3 IMEMDT 49,49,90; Human Inadequate Evidence IMEMDT 49,49,90; Animal Inadequate Evidence IMEMDT 49,49,90. Reported in EPA TSCA Inventory.
OSHA PEL: TWA 0.5 mg(Cr)/m³
ACGIH TLV: TWA 0.5 mg(Cr)/m³; Not Classifiable as a Carcinogen
DOT CLASSIFICATION: 5.1; Label: Oxidizer
SAFETY PROFILE: Poison by intraperitoneal route. Moderately toxic by subcutaneous and ingestion routes. Mutation data reported. Questionable carcinogen. When heated to decomposition it emits toxic fumes of NO_x and Cr.

CMJ900   CAS: 1308-38-9   HR: 3
CHROMIUM(III) OXIDE (2:3)
mf: Cr_2O_3 mw: 152.00
PROP: Green crystals. Mp: 2275°. IDLH 25 mg/m³ [as Cr(III)].
SYNS: ANADOMIS GREEN □ ANIDRIDE CROMIQUE (FRENCH) □ CASALIS GREEN □ CHROME GREEN □ CHROME OCHER □ CHROME OXIDE □ CHROME OXIDE GREEN □ CHROMIA □ CHROMIC ACID □ CHROMIC ACID GREEN □ CHROMIC OXIDE □ CHROMIUM OXIDE □ CHROMIUM(III) OXIDE □ CHROMIUM(3+) OXIDE □ CHROMIUM SESQUIOXIDE □ CHROMIUM(3+) TRIOXIDE □ C.I. 77288 □ C.I. No. 77278 □ C.I. PIGMENT GREEN 17 □ DICHROMIUM TRIOXIDE □ 11661 GREEN □ GREEN CHROME OXIDE □ GREEN CHROMIC OXIDE □ GREEN CINNABAR □ GREEN ROUGE □ GUIGNER'S GREEN □ LEAF GREEN □ LEVANOX GREEN GA □ OIL GREEN □ OXIDE of CHROMIUM □ ULTRAMARINE GREEN
CONSENSUS REPORTS: NTP 10th Report on Carcinogens. IARC Cancer Review: Group 3 IMEMDT 7,165,87; Animal Inadequate Evidence IMEMDT 23,205,80. Reported in EPA TSCA Inventory. Chromium and its compounds are on the Community Right-To-Know List.
OSHA PEL: TWA 0.5 mg(Cr)/m³
ACGIH TLV: TWA 0.5 mg(Cr)/m³; Not Classifiable as a Carcinogen
DFG MAK: Suspected Carcinogen
SAFETY PROFILE: Confirmed carcinogen with experimental tumorigenic data. Mutation data reported. Probably a severe

eye, skin, and mucous membrane irritant. A powerful oxidizer. Reacts violently with CLF₃. See also CHROMIUM COMPOUNDS.

CMK000   CAS: 1333-82-0   HR: 3
CHROMIUM(VI) OXIDE (1:3)
DOT: UN 1463/NA 1463/UN 1755
mf: CrO₃   mw: 100.00
PROP: Dark orange-red, rhombic, deliquescent crystals. D: 2.70, mp: 190°, bp: decomp, sol: 61.7 g/100 cc @ 0°, 67.45 g/100 cc @ 100°. Very sol in H₂O; sol in H₂SO₄ and org solvs. IDLH 15 mg/m³ {as Cr(VI)}.
SYNS: ANHYDRIDE CHROMIQUE (FRENCH) □ ANIDRIDE CROMICA (ITALIAN) □ CHROME (TRIOXYDE de) (FRENCH) □ CHROMIC ACID □ CHROMIC(VI) ACID □ CHROMIC ACID, solid (NA 1463) (DOT) □ CHROMIC ACID, solution (UN 1755) (DOT) □ CHROMIC ANHYDRIDE □ CHROMIC TRIOXIDE □ CHROMIUM OXIDE □ CHROMIUM(VI) OXIDE □ CHROMIUM TRIOXIDE □ CHROMIUM(6+) TRIOXIDE □ CHROMIUM TRIOXIDE, anhydrous (DOT) □ CHROMIUM TRIOXIDE, anhydrous (UN 1463) (DOT) □ CHROMO (TRIOSSIDO di) (ITALIAN) □ CHROMS-AEUREANHYDRID (GERMAN) □ CHROMTRIOXID (GERMAN) □ CHROOMTRIOXYDE (DUTCH) □ CHROOMZUURANHYDRIDE (DUTCH) □ MONO-CHROMIUM OXIDE □ MONOCHROMIUM TRIOXIDE □ PURATRONIC CHROMIUM TRIOXIDE
CONSENSUS REPORTS: IARC Cancer Review: Group 1 IMEMDT 7,165,87; Animal Sufficient Evidence IMEMDT 23,205,80. EPA Genetic Toxicology Program. Chromium and its compounds are on the Community Right-To-Know List. Reported in EPA TSCA Inventory.
OSHA PEL: CL 0.1 mg(CrO₃)/m³
ACGIH TLV: TWA 0.05 mg(Cr)/m³; Confirmed Human Carcinogen
DFG MAK: 0.1 mg/m³, Suspected Carcinogen
NIOSH REL: (Chromium(VI)) TWA 0.025 mg(Cr(VI))/m³; CL 0.05/15M
DOT CLASSIFICATION: 5.1; Label: Oxidizer, Corrosive (NA 1463, UN 1463); DOT Class: 8; Label: Corrosive (UN 1755)
SAFETY PROFILE: Confirmed human carcinogen producing nasal and lung tumors. Experimental carcinogenic and tumorigenic data. Poison by ingestion, intraperitoneal, and subcutaneous routes. Experimental teratogenic and reproductive effects. Human mutation data reported. Corrosive. Probably a severe eye, skin, and mucous membrane irritant. See also CHROMIUM COMPOUNDS.

A powerful oxidizer. Explosive reaction with acetaldehyde, acetic acid + heat, acetic anhydride + heat, benzaldehyde, benzene, benzylthylaniline, butyraldehyde, 1,3-dimethylhexahydropyrimidone, diethyl ether, ethylacetate, isopropylacetate, methyl dioxane, pelargonic acid, pentyl acetate, phosphorus + heat, propionaldehyde, and other organic materials or solvents. Forms a friction- and heat-sensitive explosive mixture with potassium hexacyanoferrate. Ignites on contact with alcohols, acetic anhydride + tetrahydronaphthalene, acetone, butanol, chromium(II) sulfide, cyclohexanol, dimethyl formamide, ethanol, ethylene glycol, methanol, 2-propanol, pyridine. Violent reaction with acetic anhydride + 3-methylphenol (above 75°C), acetylene, bromine pentafluoride, glycerol, hexamethylphosphoramide, peroxyformic acid, selenium, sodium amide. Incandescent reaction with alkali metals (e.g., sodium, potassium), ammonia, arsenic, butyric acid (above 100°C), chlorine trifluoride, hydrogen sulfide + heat, sodium + heat, and sulfur. Incompatible with N,N-dimethylformamide.

CMK300   CAS: 7789-04-0   HR: 1
CHROMIUM PHOSPHATE
mf: Cr•H₃O₄P   mw: 150.00
SYNS: ARNAUDON'S GREEN □ ARNAUDON'S GREEN (HEMIHEPTAHYDRATE) □ CHROMIC PHOSPHATE □ CHROMIUM MONOPHOSPHATE □ CHROMIUM ORTHOPHOSPHATE □ PHOSPHORIC ACID CHROMIUM (III) SALT □ PHOSPHORIC ACID, CHROMIUM(3+) SALT (1:1) □ PLESSY'S GREEN (HEMIHEPTAHYDRATE)
CONSENSUS REPORTS: IARC Cancer Review: Group 3 IMEMDT 49,49,90; Human Inadequate Evidence IMEMDT

49,49,90; IARC Cancer Review: Animal Inadequate Evidence IMEMDT 49,49,90. Reported in EPA TSCA Inventory.
OSHA PEL: TWA 0.5 mg(Cr)/m^3
ACGIH TLV: TWA 0.5 mg(Cr)/m^3; Not Classifiable as a Carcinogen
SAFETY PROFILE: When heated to decomposition it emits toxic fumes of PO$_x$ and Cr.

CMK400   CAS: 37224-57-0   HR: 2
CHROMIUM POTASSIUM ZINC OXIDE
SYNS: POTASSIUM ZINC CHROMATE □ ZINC POTASSIUM CHROMATE
CONSENSUS REPORTS: IARC Cancer Review: Human Sufficient Evidence IMEMDT 23,205,80; Animal Sufficient Evidence IMEMDT 2,100,73. Chromium and its compounds, as well as zinc and its compounds, are on the Community Right-To-Know List.
OSHA PEL: CL 0.1 mg(CrO$_3$)/m^3
ACGIH TLV: TWA 0.01 mg(Cr)/m^3; Confirmed Human Carcinogen
DFG MAK: Human Carcinogen
NIOSH REL: (Chromium (VI)) TWA 0.001 mg(Cr(VI))/m^3
SAFETY PROFILE: Confirmed carcinogen with experimental carcinogenic data. When heated to decomposition it emits toxic fumes of Cr$^-$ and Zn$^-$.

CMK425   CAS: 10031-37-5   HR: 3
CHROMIUM SULFATE, PENTADECA-
HYDRATE
mf: O$_{12}$S$_3$•2Cr•15H$_2$O   mw: 662.38
SYNS: SULFURIC ACID, CHROMIUM(3+) SALT (3:2), PENTADECAHYDRATE □ WOOL MORDANT
ACGIH TLV: TWA 0.5 mg(Cr)/m^3; Not Classifiable as a Carcinogen
SAFETY PROFILE: Poison by intraperitoneal route. When heated to decomposition it emits toxic fumes of SO$_x$ and Cr.

CMK500   CAS: 15930-94-6   HR: 3
CHROMIUM(6+)ZINC OXIDE HYDRATE
(1:2:6:1)
mf: CrO$_4$•H$_2$O$_2$•Zn$_2$•H$_2$O   mw: 298.78

SYNS: BUTTERCUP YELLOW □ CHROMIC ACID, ZINC SALT (1:2) □ ZINC CHROMATE HYDROXIDE □ ZINC CHROMATE(VI) HYDROXIDE □ ZINC HYDROXYCHROMATE □ ZINC YELLOW
CONSENSUS REPORTS: IARC Cancer Review: Human Sufficient Evidence IMEMDT 23,205,80; Animal Sufficient Evidence IMEMDT 2,100,73. Chromium and its compounds, as well as zinc and its compounds, are on the Community Right-To-Know List.
OSHA PEL: CL 0.1 mg(CrO$_3$)/m^3
ACGIH TLV: TWA 0.01 mg(Cr)/m^3; Confirmed Human Carcinogen
DFG MAK: Human Carcinogen
NIOSH REL: (Chromium (VI)) TWA 0.001 mg(Cr(VI))/m^3
SAFETY PROFILE: Confirmed human carcinogen. Mutation data reported. When heated to decomposition it emits toxic fumes of ZnO. See also CHROMIUM and ZINC COMPOUNDS.

CML125   CAS: 14977-61-8   HR: 3
CHROMYL CHLORIDE
DOT: UN 1758
mf: Cl$_2$CrO$_2$ mw: 154.90
PROP: Dark-red liquid; yellow-red vapor; musty burning odor. Readily hydrol to HCl and CrO$_3$. Fumes in air. Sol in org solvs and inorganic acid halides. Mp: −96.5°, bp: 115.7°, d: 1.9145 @ 25°/4°, vap press: 20 mm @ 20°. IDLH Ca [15 mg/m^3 {as Cr(VI)}].
SYNS: CHLORURE de CHROMYLE (FRENCH) □ CHROMIC OXYCHLORIDE □ CHROMIUM CHLORIDE OXIDE □ CHROMIUM DICHLORIDE DIOXIDE □ CHROMIUM DIOXIDE DICHLORIDE □ CHROMIUM(VI) DIOXYCHLORIDE □ CHROMIUM OXYCHLORIDE □ CHROMOXYCHLORID (GERMAN) □ CHROMYLCHLOR-ID (GERMAN) □ CHROOMOXYLCHLORIDE (DUTCH) □ CROMILE, CLORURO di (ITALIAN) □ CROMO, OSSICLO-RURO di (ITALIAN) □ DICHLORODIOXOCHROMIUM □ DIOXODICHLOROCHROMIUM □ OXYCHLORURE CHROMIQUE (FRENCH)
CONSENSUS REPORTS: Reported in EPA TSCA Inventory. Chromium and its compounds are on the Community Right-To-Know List.
OSHA PEL: TWA 0.5 mg(Cr)/m^3

ACGIH TLV: TWA 0.025 ppm
DFG MAK: Suspected Carcinogen
NIOSH REL: (Chromium(VI)) TWA 0.001 mg(Cr(VI))/m³
DOT CLASSIFICATION: 8; Label: Corrosive
SAFETY PROFILE: Suspected carcinogen. Probably a poison by various routes. Mutation data reported. Corrosive. A strong irritant. Hydrolyzes to form chromic and hydrochloric acids. A strong oxidizer and chlorinating agent. Violent reaction with water. Reacts violently with alcohol, ether, acetone, turpentine. Ignites or explodes on contact with nonmetal halides (e.g., disulfur dichloride, phosphorus trichloride, and phosphorus tribromide), nonmetal hydrides (e.g., hydrogen sulfide and hydrogen phosphide), flowers of sulfur, moist phosphorus, sodium azide, and urea. During preparation can violently explode. Incompatible with ammonia, disulfur dichloride, organic solvents, phosphorus, phosphorus trichloride, sodium azide, and sulfur. When heated to decomposition it emits toxic fumes of Cl⁻. See also CHROMIUM COMPOUNDS.

CML810   CAS: 218-01-9   HR: 3
CHRYSENE
mf: $C_{18}H_{12}$ mw: 228.30
PROP: Plates from C_6H_6 or AcOH with reddish-violet fluorescence. Occurs in coal tar. Is formed during distillation of coal, in very small amount during distillation or pyrolysis of many fats and oils. Orthorhombic bipyramidal plates from benzene. D: 1.274, mp: 255–256°. Sublimes easily in vacuum, bp: 448°. Sltly sol in alc, ether, carbon disulfide, and glacial acetic acid; moderately sol in boiling benzene; insol in water. Chrysene is generally only sltly sol in cold org solvs, but fairly sol in these solvents when hot, including glacial acetic acid.
SYNS: BENZO(a)PHENANTHRENE □ 1,2-BENZO-PHENANTHRENE □ BENZ(a)PHENANTHRENE □ 1,2-BENZPHENANTHRENE □ 1,2,5,6-DIBENZO-NAPHTHALENE □ RCRA WASTE NUMBER U050

CONSENSUS REPORTS: IARC Cancer Review: Group 3 IMEMDT 7,56,87; Animal Limited Evidence IMEMDT 32,247,83; Human No Adequate Data IMEMDT 32,247,83. EPA Genetic Toxicology Program. Reported in EPA TSCA Inventory.
OSHA PEL: 0.2 mg/m³
ACGIH TLV: Animal Carcinogen
DFG MAK: Animal Carcinogen, Suspected Human Carcinogen
NIOSH REL: (Chrysene) To be controlled as a carcinogen
SAFETY PROFILE: Confirmed carcinogen with experimental carcinogenic, neoplastigenic, and tumorigenic data by skin contact. Human mutation data reported. When heated to decomposition it emits acrid smoke and fumes.

CMM330   CAS: 6459-94-5   HR: 3
C.I. ACID RED 114, DISODIUM SALT
mf: $C_{37}H_{28}N_4O_{10}S_3 \cdot 2Na$ mw: 830.85
SYNS: ACID LEATHER RED BG □ ACID RED 114 □ AMACID MILLING RED PRS □ BENZYL FAST RED BG □ BENZYL RED BR □ CERVEN KYSELA 114 □ C.I. 23635 □ C.I. ACID RED 114 □ ELCACID MILLING FAST RED RS □ ERIONYL RED RS □ FENAFOR RED PB □ FOLAN RED B □ INTRAZONE RED BR □ KAYANOL MILLING RED RS □ LEATHER FAST RED B □ LEVANOL RED GG □ MIDLON RED PRS □ MILLING FAST RED B □ MILLING RED B □ MILLING RED BB □ MILLING RED SWB □ 1,3-NAPHTHALENEDISULFONIC ACID, 8-((3,3'-DIMETHYL-4'-((4-(((4-METHYLPHENYL)SULFON-YL)OXY) PHENYL)AZO)(1,1'-BIPHENYL)-4-YL)AZO)-7-HYDROXY-, DISODIUM SALT □ NCI C61096 □ POLAR RED RS □ SANDOLAN RED N-RS □ SELLA FAST RED RS □ SULPHONOL FAST RED R □ SULPHONOL RED R □ SUMINOL MILLING RED RS □ SUPRANOL FAST RED 3G □ SUPRANOL FAST RED GG □ SUPRANOL RED PBX-CF □ SUPRANOL RED R □ TELON FAST RED GG □ TETRACID MILLING RED B □ TETRACID MILLING RED G □ VONDAMOL FAST RED RS
CONSENSUS REPORTS: IARC Cancer Review: Group 2B IMEMDT 57,247,93; Animal Sufficient Evidence IMEMDT 57,247,93; Human Inadequate Evidence IMEMDT 57,247,93. NTP Carcinogenesis Studies (drinking); Clear Evidence: rat NTPTR* NTP-TR-405,91. Reported in EPA TSCA Inventory.

SAFETY PROFILE: Confirmed carcinogen with experimental carcinogenic data. Mutation data reported. When heated to decomposition it emits toxic vapors of NO_x and SO_x.

CMN750　　CAS: 2610-05-1　　HR: D
C.I. DIRECT BLUE 1, TETRASODIUM SALT

mf: $C_{34}H_{28}N_6O_{16}S_4 \cdot 4Na$　　mw: 996.88

SYNS: AIREDALE BLUE FFD □ AMANIL SKY BLUE 6B □ ATLANTIC RESIN FAST BLUE □ BELAMINE SKY BLUE FF □ CALCODUR RESIN FAST BLUE □ CHICAGO BLUE 6B □ CHLORAZOL SKY BLUE FF □ CHROME LEATHER SKY BLUE □ C.I. 24410 □ C.I. DIRECT BLUE 1 □ DIACOTTON SKY BLUE 6B □ DIPHENYL BRILLIANT BLUE FF □ DIRECT BRILLIANT BLUE FF □ ENIANIL BRILLIANT BLUE FF □ FENAMIN SKY BLUE 3F □ HISPAMIN SKY BLUE 6B □ KAYAKU DIRECT SKY BLUE 6B □ LUMICREASE BLUE 4GL □ NAPHTAMINE SKY BLUE DD □ NCI-C61109 □ PONTAMINE SKY BLUE □ PYRAZOL FAST BRILLIANT BLUE VP □ SHIKISO DIRECT SKY BLUE 6B □ TERTRODIRECT BLUE FF □ VONDACEL BLUE FF

CONSENSUS REPORTS: Reported in EPA TSCA Inventory.

SAFETY PROFILE: An experimental teratogen. Other experimental reproductive effects. Mutation data reported. When heated to decomposition it emits very toxic fumes of NO_x, Na_2O, and SO_x.

CMO000　　CAS: 2602-46-2　　HR: 3
C.I. DIRECT BLUE 6, TETRASODIUM SALT

mf: $C_{32}H_{20}N_6O_{14}S_4 \cdot 4Na$　　mw: 932.78

PROP: A dye.

SYNS: AIREDALE BLUE 2BD □ AIZEN DIRECT BLUE 2BH □ AMANIL BLUE 2BX □ ATLANTIC BLUE 2B □ ATUL DIRECT BLUE 2B □ AZOCARD BLUE 2B □ AZOMINE BLUE 2B □ BELAMINE BLUE 2B □ BENCIDAL BLUE 2B □ BENZANIL BLUE 2B □ BENZO BLUE GS □ BLUE 2B □ BRASILAMINA BLUE 2B □ CALCOMINE BLUE 2B □ CHLORAMINE BLUE 2B □ CHLORAZOL BLUE B □ CHROME LEATHER BLUE 2B □ C.I. 22610 □ CRESOTINE BLUE 2B □ DIACOTTON BLUE BB □ DIAMINE BLUE 2B □ DIAPHTAMINE BLUE BB □ DIAZINE BLUE 2B □ DIAZOL BLUE 2B □ DIPHENYL BLUE 2B □ DIRECT BLUE 6 □ ENIANIL BLUE 2BN □ FENAMIN BLUE 2B □ FIXANOL BLUE 2B □ HISPAMIN BLUE 2B □ INDIGO BLUE 2B □ KAYAKU DIRECT □ MITSUI DIRECT BLUE 2BN □ NAPHTAMINE

BLUE 2B □ NB2B □ NCI-C54579 □ NIAGARA BLUE 2B □ NIPPON BLUE BB □ PARAMINE BLUE 2B □ PHENAMINE BLUE BB □ PHENO BLUE 2B □ PONTAMINE BLUE BB □ SODIUM DIPHENYL-4,4'-BIS-AZO-2''-8''-AMINO-1''-NAPHTHOL-3'',6 '' DISULPHONATE □ TERTRODIRECT BLUE 2B □ VONDACEL BLUE 2B

CONSENSUS REPORTS: NTP 10th Report on Carcinogens. IARC Cancer Review: Animal Sufficient Evidence IMEMDT 29,311,82; Human Inadequate Evidence IMEMDT 29,311,82. NCI Carcinogenesis Bioassay (feed); Clear Evidence: rat NCITR* NCI-CG-TR-108,78; No Evidence: mouse NCITR* NCI-CG-TR-108,78. Reported in EPA TSCA Inventory. Community Right-To-Know List.

SAFETY PROFILE: Confirmed carcinogen with experimental carcinogenic, neoplastigenic, and tumorigenic data. Experimental teratogenic and reproductive effects. Mutation data reported. When heated to decomposition it emits very toxic fumes of NO_x, Na_2O, and SO_x.

CMO250　　CAS: 72-57-1　　HR: 3
C.I. DIRECT BLUE 14, TETRASODIUM SALT

mf: $C_{34}H_{28}N_6O_{14}S_4 \cdot 4Na$　　mw: 964.88

PROP: Dark blue crystals or powder. Sol in H_2O and acids.

SYNS: AMANIL SKY BLUE □ AMIDINE BLUE 4B □ AZIDINE BLUE 3B □ AZZURRO DIRETTO 3B □ BENCIDAL BLUE 3B □ BENZAMINE BLUE □ BENZO BLUE □ BLEU DIAMINE □ BLUE EMB □ BRASILAMINA BLUE 3B □ CENTRALINE BLUE 3B □ CHLORAMINE BLUE □ CHLORAZOL BLUE 3B □ CHROME LEATHER BLUE 3B □ C.I. 23850 □ C.I. DIRECT BLUE 14 □ CONGOBLAU 3B □ CONGO BLUE □ CRESOTINE BLUE 3B □ DIAMINE BLUE 3B □ DIANILBLAU □ DIANIL BLUE □ DIAZINE BLUE 3B □ DIPHENYL BLUE 3B □ DIRECT BLUE 14 □ HISPAMIN BLUE 3BX □ NAPHTAMINE BLUE 2B □ NAPHTHYLAMINE BLUE □ NCI-C61289 □ NIAGARA BLUE □ ORION BLUE 3B □ PARAMINE BLUE 3B □ PARKIBLEU □ PARKIPAN □ PONTAMINE BLUE 3BX □ PYRAZOL BLUE 3B □ PYROTROPBLAU □ RCRA WASTE NUMBER U236 □ RENOLBLAU 3B □ SODIUM DITOLYLDIAZOBIS-8-AMINO-1-NAPHTHOL-3,6-DISULFONATE □ SODIUM DITOLYLDIAZOBIS-8-AMINO-1-NAPHTHOL-3,6-DISULPHONATE □ TB □ TRIANOL DIRECT BLUE 3B □ TRIPAN BLUE □

TRYPANBLAU (GERMAN) □ TRYPAN BLUE □ TRYPAN BLUE SODIUM SALT

CONSENSUS REPORTS: IARC Cancer Review: Group 2B IMEMDT 7,56,87; Animal Sufficient Evidence IMEMDT 8,267,75. EPA Genetic Toxicology Program. Reported in EPA TSCA Inventory.

SAFETY PROFILE: Suspected carcinogen with experimental carcinogenic, neoplastigenic, and tumorigenic data. Poison by intraperitoneal, intravenous, and subcutaneous routes. Experimental teratogenic and reproductive effects. Mutation data reported. When heated to decomposition it emits very toxic fumes of NO_x, Na_2O, and SO_x.

CMO500 CAS: 2429-74-5 HR: 3
C.I. DIRECT BLUE 15, TETRASODIUM SALT

mf: $C_{34}H_{28}N_6O_{16}S_4 \cdot 4Na$ mw: 996.88

SYNS: AIREDALE BLUE D □ AIZEN DIRECT SKY BLUE 5BH □ AMANIL SKY BLUE □ ATLANTIC SKY BLUE A □ ATUL DIRECT SKY BLUE □ AZINE SKY BLUE 5B □ BELAMINE SKY BLUE A □ BENZANIL SKY BLUE □ BENZO SKY BLUE S □ BENZO SKY BLUE A-CF □ CHLORAMINE SKY BLUE A □ CHLORAMINE SKY BLUE 4B □ CHROME LEATHER PURE BLUE □ C.I. 24400 □ C.I. DIRECT BLUE 15 □ CRESOTINE PURE BLUE □ DIACOTTON SKY BLUE 5B □ DIAMINE SKY BLUE CI □ DIAPHTAMINE PURE BLUE □ DIAZOL PURE BLUE 4B □ DIPHENYL BRILLIANT BLUE □ DIPHENYL SKY BLUE 6B □ DIRECT BLUE 10G □ DIRECT BLUE 15 □ DIRECT BLUE HH □ DIRECT PURE BLUE □ DIRECT PURE BLUE M □ DIRECT SKY BLUE A □ ENIANIL PURE BLUE AN □ FENAMIN SKY BLUE □ HISPAMIN SKY BLUE 3B □ KAYAKU DIRECT SKY BLUE 5B □ MITSUI DIRECT SKY BLUE 5B □ MODR PRIMA 15 □ NAPHTAMINE BLUE 10G □ NCI C61290 □ NIAGARA BLUE 4B □ NIAGARA SKY BLUE □ NIPPON DIRECT SKY BLUE □ NITTO DIRECT SKY BLUE 5B □ PHENAMINE SKY BLUE A □ PONTAMINE SKY BLUE 5BX □ PONTACYL SKY BLUE 4BX □ SHIKISO DIRECT SKY BLUE 5B □ SKY BLUE 4B □ SKY BLUE 5B □ TERTRODIRECT BLUE F □ VONDACEL BLUE HH

CONSENSUS REPORTS: IARC Cancer Review: Group 2B IMEMDT 57,235,93; Animal Sufficient Evidence IMEMDT 57,235,93; Human Inadequate Evidence IMEMDT 57,235,93. Reported in EPA TSCA Inventory.

SAFETY PROFILE: A confirmed carcinogen with experimental carcinogenic data reported. An experimental teratogen. Other experimental reproductive effects. Mutation data reported. When heated to decomposition it emits very toxic fumes of NO_x, Na_2O, and SO_x.

CMO750 CAS: 16071-86-6 HR: 3
C.I. DIRECT BROWN

mf: $C_{31}H_{20}N_6O_9S \cdot Cu \cdot 2Na$ mw: 762.15

SYNS: AIZEN PRIMULA BROWN BRLH □ AMANIL SUPRA BROWN LBL □ ATLANTIC RESIN FAST BROWN BRL □ BENZAMIL SUPRA BROWN BRLL □ CALCODUR BROWN BRL □ CHLORAMINE FAST BROWN BRL □ CHROME LEATHER BROWN BRLL □ C.I. 30145 □ DERMA FAST BROWN W-GL □ DIPHENYL FAST BROWN BRL □ DIRECT BROWN 95 □ NCI-C54568 □ SATURN BROWN LBR □ SOLAR BROWN PL □ TETRAMINE FAST BROWN BRS

CONSENSUS REPORTS: IARC Cancer Review: Animal Limited Evidence IMEMDT 29,321,82; Human Limited Evidence IMEMDT 29,321,82; NCI Carcinogenesis Bioassay (feed); Clear Evidence: rat NCITR* NCI-CG-TR-108,78; No Evidence: mouse NCITR* NCI-CG-TR-108,78. Reported in EPA TSCA Inventory. Community Right-To-Know List.

SAFETY PROFILE: Low toxicity by ingestion. Suspected carcinogen with experimental carcinogenic and neoplastigenic data. Mutation data reported. When heated to decomposition it emits very toxic fumes of Na_2O, SO_x, and NO_x.

CMP800 HR: 3
CIGARETTE REFINED TAR

SYNS: CIGARETTE TAR □ COLOMBIAN BLACK TOBACCO CIGARETTE REFINED TAR □ TAR, from tobacco □ TOBACCO REFINED TAR □ TOBACCO TAR □ U.S. BLENDED LIGHT TOBACCO CIGARETTE REFINED TAR

DOT CLASSIFICATION: 3; Label: Flammable Liquid

SAFETY PROFILE: Suspected carcinogen with experimental carcinogenic data. Experimental reproductive effects. Mutation data reported. See also TOBACCO and NICOTINE.

CMP969   CAS: 104-55-2   HR: 3
CINNAMALDEHYDE
mf: C_9H_8O mw: 132.17

PROP: Found in Ceylon and Chinese cinnamon oils. Yellowish, oily liquid; strong odor of cinnamon. D: 1.048–1.052, mp: −7.5°, bp: 246.0° (some decomp), d: 1.048–1.052 @ 25°/25°, refr index: 1.619–1.623, flash p: 248°F. Very sltly sol in water; misc with alc, ether, chloroform, fixed oils.

SYNS: BENZYLIDENEACETALDEHYDE □ CASSIA ALDEHYDE □ CINNAMAL □ CINNAMYL ALDEHYDE □ CINNIMIC ALDEHYDE □ FEMA No. 2286 □ NCI-C56111 □ PHENYLACROLEIN □ 3-PHENYLACROLEIN □ 3-PHENYLPROPENAL □ 3-PHENYL-2-PROPENAL □ ZIMTALDEHYDE

CONSENSUS REPORTS: Reported in EPA TSCA Inventory.

SAFETY PROFILE: Poison by intravenous and parenteral routes. Moderately toxic by ingestion and intraperitoneal routes. A severe human skin irritant. Mutation data reported. Combustible liquid. May ignite after a delay period in contact with NaOH. When heated to decomposition it emits acrid smoke and fumes. See also ALDEHYDES.

CMP975   CAS: 621-82-9   HR: 3
CINNAMIC ACID
mf: $C_9H_8O_2$ mw: 148.17

PROP: Occurs free and partly esterified in storax, balsam Peru or Tolu, oil of cinnamon, coca leaves. White monoclinic crystals; honey floral odor. D: (4°/4°) 1.2475, mp: 133°, bp: 300°, flash p: 212°F. One gram dissolves in about 2000 mL water at 25° (more sol in hot water), in 6 mL alc, 5 mL methanol, 12 mL chloroform. Freely sol in benzene, ether, acetone, glacial acetic acid, carbon disulfide, fixed oils.

SYNS: FEMA No. 2288 □ PHENYLACRYLIC ACID □ tert-β-PHENYLACRYLIC ACID □ 3-PHENYLACRYLIC ACID □ 3-PHENYLPROPENOIC ACID □ 3-PHENYL-2-PROPENOIC ACID □ ZIMTSAEURE (GERMAN)

CONSENSUS REPORTS: Reported in EPA TSCA Inventory.

SAFETY PROFILE: Poison by intravenous and intraperitoneal routes. Moderately toxic by ingestion. A skin irritant. Combustible liquid. When heated to decomposition it emits acrid smoke and fumes.

CMQ510   CAS: 8015-91-6   HR: 2
CINNAMON LEAF OIL
PROP: Extracted by steam distillation of leaves from *Cinnamomum zeylanicum* Nees. Light to dark brown liquid; spicy cinnamon, clove odor and taste. Sol in fixed oils, propylene glycol, mineral oil; insol in glycerin.

SYNS: CEYLON CINNAMON BARK OIL □ CEYLON-ZIMT OEL □ CINNAMON LEAF OIL, CEYLON □ CINNAMON OIL, CEYLON □ OIL OF CINNAMON, CEYLON

CONSENSUS REPORTS: Reported in EPA TSCA Inventory.

SAFETY PROFILE: Moderately toxic by ingestion. Low toxicity by skin contact. Mutation data reported. When heated to decomposition it emits acrid smoke and irritating fumes.

CMQ730   CAS: 103-54-8   HR: 2
CINNAMYL ACETATE
mf: $C_{11}H_{12}O_2$ mw: 176.23

PROP: Colorless liquid; sweet floral odor. D: 1.047–1.051, refr index: 1.539–1.543, flash p: 244°F. Misc with chloroform, ether, fixed oils; insol in glycerin, water @ 264°.

SYNS: ACETIC ACID, CINNAMYL ESTER □ FEMA No. 2293 □ γ-PHENYLALLYL ACETATE □ 3-PHENYL-2-PROPEN-1-YL ACETATE

CONSENSUS REPORTS: Reported in EPA TSCA Inventory.

SAFETY PROFILE: Moderately toxic by ingestion and intraperitoneal routes. A skin irritant. Combustible liquid. When heated to decomposition it emits acrid smoke and fumes. See also ALLYL COMPOUNDS.

CMQ740   CAS: 104-54-1   HR: 2
CINNAMYL ALCOHOL
mf: $C_9H_{10}O$ mw: 134.19

C

PROP: Occurs (in the esterified form) in storax and in balsam Peru, cinnamon leaves, hyacinth oil. Needles or crystalline mass; odor of hyacinth. Mp: 33°, d: 1.0397, bp: 250.0°, n: (20/D) 1.58190. Sol in water, glycerol, and propylene glycol; freely sol in alc, ether, other common org solvs.

SYNS: CINNAMIC ALCOHOL □ CINNAMYL ALCOHOL, SYNTHETIC □ FEMA No. 2294 □ γ-PHENYLALLYL ALCOHOL □ 3-PHENYLALLYL ALCOHOL □ 3-PHENYL-2-PROPEN-1-OL □ STYRONE □ STYRYL CARBINOL

CONSENSUS REPORTS: Reported in EPA TSCA Inventory.

SAFETY PROFILE: Moderately toxic by ingestion. A skin irritant. Mutation data reported. When heated to decomposition it emits acrid smoke and fumes. See also ALCOHOLS and ALLYL COMPOUNDS.

CMQ800   CAS: 103-61-7   HR: 1
CINNAMYL BUTYRATE
mf: $C_{13}H_{16}O_2$ mw: 204.29
SYNS: BUTYNOIC ACID, 3-PHENYL-2-PROPENYL ESTER □ BUTYRIC ACID, CINNAMYL ESTER □ PHENYLPROPENYL n-BUTYRATE

CONSENSUS REPORTS: Reported in EPA TSCA Inventory.

SAFETY PROFILE: A skin irritant. When heated to decomposition it emits acrid smoke and irritating fumes.

CMR500   CAS: 104-65-4   HR: 2
CINNAMYL FORMATE
mf: $C_{10}H_{10}O_2$ mw: 162.20
PROP: Colorless liquid; balsamic odor. D: 1.077–1.082, refr index: 1.550–1.556, flash p: 212°F. Misc with alc, chloroform, ether, fixed oils; insol in water @ 250°.

SYNS: CINNAMYL ALCOHOL, FORMATE □ CINNAMYL METHANOATE □ FEMA No. 2299 □ FORMIC ACID, CINNAMYL ESTER □ 3-PHENYL-2-PROPEN-1-YL FORMATE

CONSENSUS REPORTS: Reported in EPA TSCA Inventory.

SAFETY PROFILE: Moderately toxic by ingestion. See also ESTERS. Combustible liquid. When heated to decomposition it emits acrid smoke and irritating fumes.

CMR800   HR: 2
CINNAMYL ISOVALERATE
mf: $C_{14}H_{18}O_2$ mw: 218.30
PROP: Colorless to sltly yellow liquid; spicy, floral, fruity odor. D: 0.991–0.996, refr index: 1.518–1.524, flash p: 212°F. Misc in alc, chloroform, ether, most oils; insol in glycerin, propylene glycol, and water @ 313°.

SYN: FEMA No. 2302

SAFETY PROFILE: Combustible liquid. When heated to decomposition it emits acrid smoke and irritating fumes.

CMR850   CAS: 103-56-0   HR: 2
CINNAMYL PROPIONATE
mf: $C_{12}H_{14}O_2$ mw: 190.24
PROP: Colorless to pale-yellow liquid; spicy, fruity, balsamic odor. D: 1.029–1.033, refr index: 1.523–1.537, flash p: 212°F. Misc in alc, chloroform, ether, most oils; insol in glycerin, propylene glycol, and water @ 289°.

SYNS: FEMA No. 2301 □ 3-PHENYL-2-PROPENYL PROPIONATE □ 3-PHENYL-2-PROPEN-1-YL PROPIONATE □ PROPIONIC ACID, CINNAMYL ESTER

CONSENSUS REPORTS: Reported in EPA TSCA Inventory.

SAFETY PROFILE: Moderately toxic by ingestion. A skin irritant. Combustible liquid. When heated to decomposition it emits acrid smoke and irritating fumes.

CMS212   CAS: 62865-26-3   HR: 3
C.I. PIGMENT YELLOW 35
SYNS: B-3-Zh □ CADMIUM GOLDEN □ CADMIUM LEMON □ CADMIUM PRIMROSE □ CADMIUM SULFIDE mixed with ZINC SULFIDE (1:1) □ C.I. 77205

OSHA PEL: TWA 5 µg(Cd)/m³
ACGIH TLV: TWA 0.01 mg(Cd)/m³; Suspected Carcinogen
NIOSH REL: (Cadmium) TWA reduce to lowest feasible level
SAFETY PROFILE: Confirmed human carcinogen. When heated to decomposition it emits toxic fumes of Cd.

CMS238   CAS: 1229-55-6   HR: 1
C.I. SOLVENT RED

mf: $C_{17}H_{14}N_2O_2$ mw: 278.33
SYNS: BRILLIANT FAT SCARLET R □ CERES RED G □ CERES RED G 102 □ C.I. 12150 □ C.I. FOOD RED 16 □ C RED 2 □ FAT RED BG □ FAT RED G □ FAT RED RS □ FAT SOLUBLE RED S □ FOOD RED 16 □ LACQUER RED V 2G □ 2-NAPHTHALENOL, 1-((2-METHOXY-PHENYL)AZO)-(9CI) □ OIL PINK □ OIL RED □ OIL RED 113 □ OIL RED OG □ OIL SCARLET 389 □ OIL SOLUBLE RED S □ OIL VERMILION □ OIL VERMILION LP □ OLEAL RED G □ ORGANOL VERMILION □ ORIENT OIL RED OG □ PLASTORESIN RED FR □ RESINOL RED G □ SICO FAT RED BG NEW □ SILOTRAS RED TG □ SOLVENT RED 1 □ SOMALIA RED PG □ SUDAN R □ SUDAN RED 290 □ SUDAN RED G (6CI)
CONSENSUS REPORTS: Reported in EPA TSCA Inventory.
SAFETY PROFILE: An eye irritant. When heated to decomposition it emits toxic fumes of NO_x.

CMS324   CAS: 66408-78-4   HR: 1
CITRAL ETHYLENE GLYCOL ACETAL
mf: $C_{12}H_{20}O_2$ mw: 196.32
SYNS: CITRACETAL □ 1,3-DIOXOLANE, 2-(2,6-DIMETHYL-1,5-HEPTADIENYL)-
CONSENSUS REPORTS: Reported in EPA TSCA Inventory.
SAFETY PROFILE: A skin irritant. When heated to decomposition it emits acrid smoke and irritating fumes.

CMS750   CAS: 77-92-9   HR: 3
CITRIC ACID
mf: $C_6H_8O_7$ mw: 192.14
PROP: Colorless, odorless crystals (crystals are monoclinic holohedra and crystallize from hot conc aq soln); acid taste. Mp: 135° (monohydrate), mp: 153° (anhydrous), bp: decomp, d: 1.665, flash p: 212°F. Very sol in H_2O and EtOH; mod sol in Et_2O.
SYNS: ACILETTEN □ CITRETTEN □ CITRIC ACID, anhydrous □ CITRO □ FEMA No. 2306 □ 2-HYDROXY-1,2,3-PROPANETRICARBOXYLIC ACID □ β-HYDROXYTRICARBALLYLIC ACID □ KYSELINA CITRONOVA (CZECH)
CONSENSUS REPORTS: Reported in EPA TSCA Inventory.
SAFETY PROFILE: Poison by intravenous route. Moderately toxic by subcutaneous and intraperitoneal routes. Mildly toxic by

ingestion. A severe eye and moderate skin irritant. An irritating organic acid, some allergenic properties. Combustible liquid. Potentially explosive reaction with metal nitrates. When heated to decomposition it emits acrid smoke and fumes.

CMS845   CAS: 106-23-0   HR: 2
CITRONELLAL
mf: $C_{10}H_{18}O$ mw: 154.25
PROP: Colorless to sltly yellow liquid; intense lemon-citronella-rose odor. D: 0.850–0.860, refr index: 1.446–1.456, flash p: 170°F. Sol in alc, most oils; sltly sol in propylene glycol; insol in glycerin and water.
SYNS: 3,7-DIMETHYL-6-OCTENAL □ FEMA No. 2307
SAFETY PROFILE: Combustible liquid. When heated to decomposition it emits acrid smoke and irritating fumes.

CMS850   CAS: 107-75-5   HR: 1
CITRONELLAL HYDRATE
mf: $C_{10}H_{20}O_2$ mw: 172.30
PROP: Colorless liquid; sweet, floral, lily odor. D: 0.918–0.923, refr index: 1.447–1.450, flash p: 212°F. Sol in fixed oils and propylene glycol; insol in glycerin.
SYNS: CYCLALIA □ CYCLOSIA □ 3,7-DIMETHYL-7-HYDROXYOCTANAL □ FEMA No. 2583 □ FIXOL □ HYDROXYCITRONELLAL (FCC) □ 7-HYDROXY-CITRONELLAL □ 7-HYDROXY-3,7-DIMETHYL OCTANAL □ 7-HYDROXY-3,7-DIMETHYLOCTAN-1-AL □ LAURINE □ LILYL ALDEHYDE □ MUSUET SYNTHETIC □ MUSUETTINE PRINCIPLE □ PHIXIA
CONSENSUS REPORTS: Reported in EPA TSCA Inventory.
SAFETY PROFILE: A skin irritant. Combustible liquid. When heated to decomposition it emits acrid smoke and irritating fumes. See also ALDEHYDES.

CMT050   CAS: 2436-90-0   HR: 1
CITRONELLENE
mf: $C_{10}H_{18}$ mw: 138.28
PROP: Bp: 154–155°, d: 0.757 @ 20°/4°.
SYNS: DIHYDROMYRCENE □ 3,7-DIMETHYL-1,6-OCTADIENE □ 1,6-OCTADIENE, 3,7-DIMETHYL-
CONSENSUS REPORTS: Reported in EPA TSCA Inventory.

SAFETY PROFILE: A skin irritant. When heated to decomposition it emits acrid smoke and irritating fumes.

CMT250   CAS: 106-22-9   HR: 3
CITRONELLOL
mf: $C_{10}H_{20}O$ mw: 156.30
PROP: Colorless oily liquid; rose odor. D: 0.850–0.860, refr index: 1.454–1.462, flash p: 215°F. Sol in fixed oils, propylene glycol; sltly sol in water; insol in glycerin @ 225°.
SYNS: CEPHROL □ 2,6-DIMETHYL-2-OCTEN-8-OL □ 3,7-DIMETHYL-6-OCTEN-1-OL □ FEMA No. 2309 □ FEMA No. 2980 □ RHODINOL □ RODINOL
CONSENSUS REPORTS: Reported in EPA TSCA Inventory.
SAFETY PROFILE: Poison by intravenous route. Moderately toxic by ingestion, skin contact, and intramuscular routes. A severe skin irritant. A combustible liquid. When heated to decomposition it emits acrid smoke and irritating fumes. See also ALCOHOLS.

CMT600   HR: 2
CITRONELLYL BUTYRATE
mf: $C_{14}H_{26}O_2$ mw: 226.36
PROP: Colorless liquid; strong, fruity-rosy odor. D: 0.873–0.883, refr index: 1.444–1.448, flash p: 212°F. Misc in alc, ether, chloroform, most oils; insol in water @ 245°.
SYNS: 3,7-DIMETHYL-6-OCTEN-1-YL BUTYRATE □ FEMA No. 2312
SAFETY PROFILE: Combustible liquid. When heated to decomposition it emits acrid smoke and irritating fumes.

CMT750   CAS: 105-85-1   HR: 1
CITRONELLYL FORMATE
mf: $C_{11}H_{20}O_2$ mw: 184.31
PROP: Colorless liquid; strong, fruity odor. D: 0.890–0.93, refr index: 1.443–1.452, flash p: 197°F. Sol in alc, fixed oils; sltly sol in propylene glycol; insol in glycerin, water @ 235°.
SYNS: 3,7-DIMETHYL-6-OCTEN-1-OL FORMATE □ 2,6-DIMETHYL-2-OCTEN-8-YL FORMATE □ 3,7-DIMETHYL-6-OCTEN-1-YL FORMATE □ FEMA No. 2314

□ FORMIC ACID, CITRONELLYL ESTER □ FORMIC ACID-3,7-DIMETHYL-6-OCTEN-1-YL ESTER
CONSENSUS REPORTS: Reported in EPA TSCA Inventory.
SAFETY PROFILE: Mildly toxic by ingestion. A human skin irritant. Combustible liquid. When heated to decomposition it emits acrid smoke and irritating fumes. See also ESTERS and FORMIC ACID.

CMT900   HR: 2
CITRONELLYL ISOBUTYRATE
mf: $C_{14}H_{26}O_2$ mw: 226.36
PROP: Colorless liquid; rosy-fruity odor. D: 0.870–0.880, refr index: 1.440–1.448, flash p: 212°F. Misc in alc, chloroform, ether, most oils; insol in water @ 249°.
SYNS: 3,7-DIMETHYL-6-OCTEN-1-YL ISOBUTYRATE □ FEMA No. 2313
SAFETY PROFILE: Combustible liquid. When heated to decomposition it emits acrid smoke and irritating fumes.

CMU050   CAS: 139-70-8   HR: 1
CITRONELLYL PHENYLACETATE
mf: $C_{18}H_{26}O_2$ mw: 274.44
SYNS: ACETIC ACID, PHENYL-, 3,7-DIMETHYL-6-OCTENYL ESTER □ BENZENEACETIC ACID, 3,7-DIMETHYL-6-OCTENYL ESTER (9CI) □ 3,7-DIMETHYL-6-OCTEN-1-YL PHENYLACETATE
CONSENSUS REPORTS: Reported in EPA TSCA Inventory.
SAFETY PROFILE: A skin irritant. When heated to decomposition it emits acrid smoke and irritating fumes.

CMU100   HR: 2
CITRONELLYL PROPIONATE
mf: $C_{13}H_{24}O_2$ mw: 212.33
PROP: Colorless liquid; fruity-rosy odor. D: 0.877–0.886, refr index: 1.443–1.449, flash p: 212°F. Misc in alc, most oils; insol in water @ 242°.
SYN: FEMA No. 2316
SAFETY PROFILE: Combustible liquid. When heated to decomposition it emits acrid smoke and irritating fumes.

CMU475   CAS: 2278-50-4   HR: 1
C.I. VAT BLACK 8
mf: $C_{45}H_{19}N_3O_4$ mw: 665.67
SYNS: BENZADONE GREY M □ 1H-BENZ(6,7)INDAZ-
OLO(2,3,4-fgh)NAPHTH(2'',3'':6',7')INDOLO(3',2':5,6)ANTHR
A(2,1,9-mna) ACRIDINE-5,8,13,25-TETRAONE □ CALED-
ON GREY M □ CERN KYPOVA 8 □ C.I. 71000 □ IND-
ANTHREN GREY M □ INDANTHREN GREY MG □
MIKETHRENE GREY M □ MIKETHRENE GREY MG □
NIHONTHRENE GREY M □ OSTANTHREN GREY M □
PARADONE GREY M □ PARADONE GREY MG □ SED
OSTANTHRENOVA M □ VAT GRAY S □ VAT GREY S
CONSENSUS REPORTS: Reported in EPA
TSCA Inventory.
SAFETY PROFILE: Mildly toxic by
intraperitoneal and ingestion routes. An eye
irritant.When heated to decomposition it
emits toxic fumes of NO_x.

CMU850   CAS: 542-46-1   HR: 1
cis-CIVETONE
mf: $C_{17}H_{30}O$ mw: 250.47
PROP: Crystals; strong musty odor. Mp:
31–32°, bp: 342°.
SYNS: CIVETONE □ 9-CYCLOHEPTADECEN-1-ONE
□ 9-CYCLOHEPTADECEN-1-ONE, (Z)-(8CI,9CI)
CONSENSUS REPORTS: Reported in EPA
TSCA Inventory.
SAFETY PROFILE: A skin irritant. When
heated to decomposition it emits acrid
smoke and irritating fumes.

CMU900   HR: D
CLARY OIL
PROP: From steam distillation of flowering
tops and leaves of *Salvia sclarea L.* (Fam.
Labiatae). Pale yellow liquid; herbaceous
odor. Sol in fixed oils, mineral oil; insol in
glycerin, propylene glycol.
SYNS: CLARY SAGE OIL □ OIL OF MUSCATEL
SAFETY PROFILE: When heated to
decomposition it emits acrid smoke and
irritating fumes.

CMX850   CAS: 2971-90-6   HR: 1
CLOPIDOL
mf: $C_7H_7Cl_2NO$ mw: 192.05
PROP: Insol in H_2O.
SYNS: COCCIDIOSTAT C □ COYDEN □ 3,5-
DICHLORO-2,6-DIMETHYL-4-PYRIDINOL □ LERBEK □

METHYLCHLOROPINDOL □ METHYLCHLORPINDOL
□ METILCLORPINDOL
OSHA PEL: Total Dust: 15 mg/m³;
Respirable Fraction: 5 mg/m³
ACGIH TLV: TWA 10 mg/m³; Not
Classifiable as a Human Carcinogen
SAFETY PROFILE: A nuisance dust. When
heated to decomposition it emits very toxic
fumes of Cl⁻ and NO_x.

CMX880   CAS: 40665-92-7   HR: D
CLOPROSTENOL
mf: $C_{22}H_{29}ClO_6$ mw: 424.96
SYNS: ESTRUMATE □ ICI 80996 □ racemic-ICI 80,996 □
I.C.I. LTD. COMPOUND NUMBER 80996
SAFETY PROFILE: Experimental
reproductive effects. When heated to
decomposition it emits toxic fumes of Cl⁻.

CMX010   HR: D
CLORSULON
SAFETY PROFILE: When heated to
decomposition emits toxic fumes of Cl⁻.

CMY500   CAS: 8015-97-2   HR: 2
CLOVE LEAF OIL MADAGASCAR
PROP: From steam distillation of leaves of
Eugenis caryophyllata Thunberg (*Eugenia
aromatica* L. Baill.) (Fam. *Myrtaceae*). Pale-
yellow liquid. Refr index: 1.527–1.538 @
20°. Sol in propylene glycol, fixed oils; insol
in glycerin, mineral oil.
SYNS: CLOVE LEAF OIL □ OILS, CLOVE LEAF
CONSENSUS REPORTS: Reported in EPA
TSCA Inventory.
SAFETY PROFILE: Moderately toxic by
ingestion and skin contact. A severe skin
irritant. When heated to decomposition it
emits acrid smoke and fumes.

CMY750   HR: 3
**COAL CONVERSION MATERIALS, SRC-
    II HEAVY DISTILLATE**
SYN: SRC-II HEAVY DISTILLATE
SAFETY PROFILE: Suspected carcinogen
with experimental carcinogenic data by skin
contact.

CMY760   **HR: 3**
COAL DUST
PROP: Black powder or dust.
SYNS: ANTHRACITE PARTICLES □ COAL FACINGS □ COAL, GROUND BITUMINOUS (DOT) □ COAL-MILLED □ COAL SLAG-MILLED □ SEA COAL
OSHA PEL: Respirable Quartz Fraction less than 5% SiO_2: TWA 2 mg/m³; Respirable Quartz Fraction greater than or equal to 5% SiO_2: 0.1 mg/m³
ACGIH TLV: Bituminous: TWA 0.9 mg/m³; Not Classifiable as a Human Carcinogen; Anthracite: 0.4 mg/m³; Not Classifiable as a Human Carcinogen
ACGIH TLV: Confirmed Animal Carcinogen with Unknown Relevance to Humans
SAFETY PROFILE: Suspected carcinogen with experimental tumorigenic data. Variable toxicity depending upon SiO_2 content. See also SILICA. Moderately flammable when exposed to heat, flame, or chemical reaction with oxidizers. Slightly explosive when exposed to flame.

CMY800   **CAS: 8007-45-2**   **HR: 3**
COAL TAR
SYNS: CARBO-CORT □ COAL TAR, AEROSOL □ COAL TAR SOLUTION USP □ CRUDE COAL TAR □ ESTAR □ IMPERVOTAR □ LAV □ LAVATAR □ PICIS CARBONIS □ PIXALBOL □ PIX CARBONIS □ PIX LITHANTHRACIS □ POLYTAR BATH □ SUPERTAH □ SYNTAR □ TAR □ TAR, COAL □ ZETAR
CONSENSUS REPORTS: NTP 10th Report on Carcinogens. IARC Cancer Review: Group 1 IMEMDT 7,175,87; Animal Sufficient Evidence IMEMDT 34,65,84; IMEMDT 35,83,85; IMEMDT 3,22,73; Human Sufficient Evidence IMEMDT 34,65,84; IMEMDT 3,22,73; Human Limited Evidence IMEMDT 35,83,85. Reported in EPA TSCA Inventory.
OSHA PEL: TWA 0.2 mg/m³; Carcinogen
DFG MAK: Human Carcinogen
NIOSH REL: (Coal Tar Products) TWA 0.1 mg/m³
DOT CLASSIFICATION: 3; Label: Flammable Liquid
SAFETY PROFILE: Confirmed human carcinogen with experimental carcinogenic and tumorigenic data. Mutation data reported. A human and experimental skin irritant. A flammable liquid. When heated to decomposition it emits acrid smoke and irritating fumes.

CMY805   **CAS: 8007-45-2**   **HR: 2**
COAL TAR, AEROSOL
OSHA PEL: TWA 0.2 mg/m³
NIOSH REL: TWA 0.1 mg/m³ CHE fraction
CONSENSUS REPORTS: Reported in EPA TSCA Inventory.
SAFETY PROFILE: Questionable carcinogen with experimental carcinogenic data. When heated to decomposition it emits acrid smoke and irritating fumes.

CMY825   **CAS: 8001-58-9**   **HR: 3**
COAL TAR CREOSOTE
SYNS: AWPA #1 □ BRICK OIL □ COAL TAR OIL □ COAL TAR OIL (DOT) □ CREOSOTE □ CREOSOTE, from COAL TAR □ CREOSOTE OIL □ CREOSOTE P1 □ CREOSOTUM □ CRESYLIC CREOSOTE □ HEAVY OIL □ LIQUID PITCH OIL □ NAPHTHALENE OIL □ PRESERV-O-SOTE □ RCRA WASTE NUMBER U051 □ TAR OIL □ WASH OIL
CONSENSUS REPORTS: NTP 10th Report on Carcinogens. IARC Cancer Review: Group 2A IMEMDT 7,177,87; Animal Sufficient Evidence, Human Limited Evidence IMEMDT 35,83,85; Animal Sufficient Evidence IMEMDT 3,22,73. Reported in EPA TSCA Inventory.
NIOSH REL: (Coal Tar Products) TWA 0.1 mg/m³ CHE fraction
DOT CLASSIFICATION: 3; Label: Flammable Liquid
SAFETY PROFILE: Confirmed carcinogen with experimental carcinogenic data. Poison by ingestion. Experimental reproductive effects. Mutation data reported. A flammable liquid. When heated to decomposition it emits acrid smoke and fumes.

CMY900   **CAS: 65996-92-1**   **HR: 3**
COAL TAR DISTILLATE
DOT: UN 1136

SYNS: COAL TAR DISTILLATES □ COAL TAR
DISTILLATES, flammable (DOT) □ DISTILLATES (COAL
TAR)
CONSENSUS REPORTS: Reported in NTP
10th Report on Carcinogens. Reported in
EPA TSCA Inventory.
DOT CLASSIFICATION: 3; Label:
Flammable Liquid
SAFETY PROFILE: A carcinogen. A
flammable liquid. When heated to
decomposition it emits acrid smoke and
irritating vapors.

CMY920   HR: 3
COAL TAR DYE
CONSENSUS REPORTS: Reported in NTP
10th Report on Carcinogens.
SAFETY PROFILE: Confirmed carcinogen.
When heated to decomposition it emits
acrid smoke and irritating vapors.

CMZ100   CAS: 65996-93-2   HR: 3
COAL TAR PITCH VOLATILES
PROP: IDLH 80 mg/m^3.
SYNS: PITCH □ PITCH, COAL TAR
CONSENSUS REPORTS: IARC Cancer
Review: Group 1 IMEMDT 7,174,87;
Animal Sufficient Evidence, Human
Sufficient Evidence IMEMDT 35,83,85;
Human Sufficient Evidence IMEMDT
3,22,73. Reported in EPA TSCA Inventory.
OSHA PEL: TWA 0.2 mg/m^3; Carcinogen
ACGIH TLV: TWA 0.2 mg/m^3 (volatile),
Confirmed Human Carcinogen
NIOSH REL: (Coal Tar Products) TWA 0.1
mg/m^3 CHE fraction
DOT CLASSIFICATION: 3; Label:
Flammable Liquid
SAFETY PROFILE: Confirmed carcinogen
with experimental carcinogenic and
neoplastigenic data by skin contact. When
heated to decomposition it emits acrid
smoke and fumes.

CNA250   CAS: 7440-48-4   HR: 3
COBALT
af: Co   aw: 58.93
PROP: Gray, hard, magnetic, lustrous,
ductile, somewhat malleable, silvery-blue

metal. Mp: 1495°, bp: 28° @ 3100 mm, d:
8.92, Brinell hardness: 125, latent heat of
fusion: 62 cal/g, latent heat of vaporization:
1500 cal/g, specific heat (15–100°): 0.1056
cal/g/°C. Exists in two allotropic forms. At
room temperature, the hexagonal form is
more stable than the cubic form; both forms
can exist at room temperature. Stable in air
or toward water at ordinary temperatures.
Readily sol in dil HNO$_3$; very slowly
attacked by HCl or cold H$_2$SO$_4$. The
hydrated salts of cobalt are red, and the sol
salts form red solns that become blue on
adding conc HCl. IDLH 20 mg/m^3 (as Co).
SYNS: AQUACAT □ C.I. 77320 □ COBALT-59 □
KOBALT (GERMAN, POLISH) □ NCI-C60311 □ SUPER
COBALT
CONSENSUS REPORTS: Reported in EPA
TSCA Inventory. Cobalt and its compounds
are on the Community Right-To-Know List.
OSHA PEL: TWA 0.05 mg/m^3
ACGIH TLV: (metal, dust, and fume) TWA
0.02 mg(Co)/m^3; Animal Carcinogen
DFG MAK: DFG TRK: Animal Carcinogen,
Suspected Human Carcinogen
NIOSH REL: (Cobalt) Insufficient evidence
for recommending limit
SAFETY PROFILE: Confirmed carcinogen
with experimental neoplastigenic and
tumorigenic data. Poison by intravenous,
intratracheal, and intraperitoneal routes.
Moderately toxic by ingestion. Inhalation of
the dust may cause pulmonary damage. The
powder may cause dermatitis. Ingestion of
soluble salts produces nausea and vomiting
by local irritation. Powdered cobalt ignites
spontaneously in air. Flammable when
exposed to heat or flame. Explosive reaction
with hydrazinium nitrate, ammonium nitrate
+ heat, and 1,3,4,7-tetramethylisoindole (at
390°C). Ignites on contact with bromine
pentafluoride. Incandescent reaction with
acetylene or nitryl fluoride. See also
COBALT COMPOUNDS.

CNA500   CAS: 6147-53-1   HR: 3
COBALT ACETATE TETRAHYDRATE
mf: C$_4$H$_6$O$_4$•Co•4H$_2$O   mw: 249.11

SYNS: ACETIC ACID, COBALT(2+) SALT, TETRA-HYDRATE □ COBALT DIACETATE TETRAHYDRATE □ COBALTOUS ACETATE TETRAHYDRATE □ OCTAN KOBALTNATY (CZECH)

CONSENSUS REPORTS: Cobalt and its compounds are on the Community Right-To-Know List. IARC Cancer Review: Group 2B IMEMDT 52,363,91; Human Inadequate Evidence IMEMDT 52,363,91.

SAFETY PROFILE: Moderately toxic by ingestion. Questionable carcinogen. A skin and eye irritant. Human mutation data reported. See also COBALT COMPOUNDS. When heated to decomposition it emits acrid smoke and irritating fumes.

CNB500   CAS: 10210-68-1   HR: 3
COBALT CARBONYL

mf: $C_8Co_2O_8$ mw: 341.94

PROP: Air-sensitive orange-red platelets or crystals. D: 1.87, mp: 51°, decomp above 52°. Decomp on exposure to air. Insol in water; sol in org solvs.

SYNS: COBALT, DI-MU-CARBONYLHEXACARBONYLDI-, (CO-CO) □ COBALT OCTACARBONYL □ COBALT TETRACARBONYL □ COBALT TETRACARBONYL DIMER □ DI-MU-CARBON-YLHEXACARBONYLDICOBALT □ DICOBALT CARBONYL □ DICOBALT OCTACARBONYL □ OCTACARBONYLDICOBALT

CONSENSUS REPORTS: IARC Cancer Review: Group 2B IMEMDT 52,363,91; Human Inadequate Evidence IMEMDT 52,363,91. Reported in EPA TSCA Inventory. Cobalt and its compounds are on the Community Right-To-Know List. EPA Extremely Hazardous Substances List.

OSHA PEL: TWA 0.1 mg(Co)/m³

ACGIH TLV: TWA 0.1 mg(Co)/m³

SAFETY PROFILE: Poison by inhalation and intraperitoneal routes. Questionable carcinogen. Decomposes in air to form a product that ignites spontaneously in air. When heated to decomposition it emits acrid smoke and fumes. See also CARBONYLS and COBALT COMPOUNDS.

CNB599   CAS: 7646-79-9   HR: 3
COBALT(II) CHLORIDE

mf: Cl_2Co mw: 129.83

PROP: Pale blue hygroscopic powder or crystals. Mp: 735°, bp: 1049°, d: 3.348. Sol in H_2O and polar org solvs.

SYNS: COBALT DICHLORIDE □ COBALT MURIATE □ COBALTOUS CHLORIDE □ COBALTOUS DICHLORIDE □ KOBALT CHLORID (GERMAN)

CONSENSUS REPORTS: IARC Cancer Review: Group 2B IMEMDT 52,363,91; Animal Limited Evidence IMEMDT 52,363,91; Human Inadequate Evidence IMEMDT 52,363,91. Reported in EPA TSCA Inventory. EPA Genetic Toxicology Program. Cobalt and its compounds are on the Community Right-To-Know List.

SAFETY PROFILE: Suspected carcinogen with experimental carcinogenic data. Poison experimentally by ingestion, skin contact, intraperitoneal, intravenous, and subcutaneous routes. Moderately toxic to humans by ingestion. Human systemic effects by ingestion: anorexia, goiter (increased thyroid size), and weight loss. Experimental teratogenic and reproductive effects. Human mutation data reported. Incompatible with metals (e.g., sodium and potassium). See also COBALT. When heated to decomposition it emits toxic fumes of Cl⁻.

CNB850   HR: 3
COBALT COMPOUNDS

CONSENSUS REPORTS: Cobalt and its compounds are on the Community Right-To-Know List.

DFG MAK: DFG TRK: 0.5 mg/m³ calculated as cobalt in that portion of dust that can possibly be inhaled in the production of cobalt powder and catalysts; hard metal (tungsten carbide) and magnet production (processing of powder, machine pressing, and mechanical processing of unsintered articles); other cobalt alloys and compounds: 0.1 mg/m³ calculated as cobalt in that portion of dust that can possibly be inhaled. Animal Carcinogen, Suspected Human Carcinogen.

SAFETY PROFILE: Confirmed carcinogen with experimental neoplastigenic and tumorigenic data. Cobalt has a low toxicity by ingestion. Ingestion of soluble salts produces nausea and vomiting by local irritation. In animals, administration of cobalt salts produces an increase in the total red cell mass of the blood. In humans, a single case of poisoning with liver and kidney damage has been attributed to cobalt. Locally, cobalt has been shown to produce dermatitis and investigators have been able to demonstrate a hypersensitivity of the skin to cobalt. There have been reports of hematologic, digestive, and pulmonary changes in humans. See also specific compounds.

CNC000   CAS: 71-48-7   HR: 3
COBALT DIACETATE
mf: $C_4H_6O_4\bullet Co$ mw: 177.03
PROP: Light-pink crystals. Very sol in H_2O; sol in EtOH.
SYNS: ACETIC ACID, COBALT(2+) SALT □ COBALT ACETATE □ COBALT(2+) ACETATE □ COBALT(II) ACETATE □ COBALTOUS DIACETATE
CONSENSUS REPORTS: IARC Cancer Review: Group 2B IMEMDT 52,363,91; Human Inadequate Evidence IMEMDT 52,363,91. Cobalt and its compounds are on the Community Right-To-Know List. Reported in EPA TSCA Inventory. EPA Genetic Toxicology Program.
NIOSH REL: (Cobalt Metal, Dust and Fume) TWA 0.05 mg/m³
SAFETY PROFILE: Poison by intravenous route. Moderately toxic by ingestion. Questionable carcinogen. Mutation data reported. See also COBALT COMPOUNDS. When heated to decomposition it emits acrid smoke and irritating fumes.

CNC100   CAS: 10026-17-2   HR: 3
COBALT(II) FLUORIDE
mf: CoF_2 mw: 96.93

PROP: Small light-brown or reddish-pink, hexagonal crystals. Bp: 1400°. Reacts with water. Sltly sol in water.
SYNS: COBALT DIFLUORIDE □ COBALTOUS FLUORIDE
CONSENSUS REPORTS: Reported in EPA TSCA Inventory.
OSHA PEL: TWA 2.5 mg(F)/m³
ACGIH TLV: TWA 2.5 mg(F)/m³; BEI: 3 mg/g creatinine of fluorides in urine prior to shift; 10 mg/g creatinine of fluorides in urine at end of shift.
NIOSH REL: (Fluorides, Inorganic): 10H TWA 2.5 mg(F)/m³
SAFETY PROFILE: Poison by ingestion. When heated to decomposition it emits toxic fumes of Co and F⁻.

CNC230   CAS: 16842-03-8   HR: 3
COBALT HYDROCARBONYL
mf: C_4HCoO_4 mw: 171.98
PROP: Light-yellow liquid or gas. Unstable, but distillable in current of CO. Mp: −33°, bp: 10°. Sol org solvs; spar sol in H_2O.
CONSENSUS REPORTS: Cobalt and its compounds are on the Community Right-To-Know List.
OSHA PEL: TWA 0.1 mg(Co)/m³
ACGIH TLV: TWA 0.1 mg(Co)/m³
SAFETY PROFILE: Poison by inhalation. See also COBALT COMPOUNDS.

CNC233   CAS: 1307-86-4   HR: 2
COBALT HYDROXIDE
mf: CoH_3O_3 mw: 109.96
SYNS: COBALT(III) HYDROXIDE □ COBALTIC HYDROXIDE □ COBALT TRIHYDROXIDE
CONSENSUS REPORTS: IARC Cancer Review: Group 2B IMEMDT 52,363,1991; Human Inadequate Evidence IMEMDT 52,363,1991.
SAFETY PROFILE: Mutation data reported. Questionable carcinogen with experimental data reported. When heated to decomposition it emits toxic vapors of Co.

CNC245   **HR: D**
COBALT LINOLEATE
SAFETY PROFILE: When heated to
decomposition it emits acrid smoke and
irritating fumes.

CNC250   CAS: 13762-14-6   HR: D
COBALT MOLYBDATE
mf: $CoMoO_4$ mw: 218.87
PROP: Shiny black crystals or olive-green
powder. Mp: 1040°. There is a high-pressure
form with the wolframite structure. IDLH
1000 mg/m^3 (as Mo).
SYNS: COBALT(2+) MOLYBDATE □ COBALT
MOLYBDENUM OXIDE □ COBALTOUS MOLYBDATE
□ HT-400 E 1/8'
CONSENSUS REPORTS: Reported in EPA
TSCA Inventory. EPA Genetic Toxicology
Program. Cobalt and its compounds are on
the Community Right-To-Know List.
OSHA PEL: TWA 5 mg(Mo)/m^3
ACGIH TLV: TWA Soluble Compounds:
TWA 0.5 mg(Mo)/m^3 Confirmed Animal
Carcinogen with Unknown Relevance to
Humans
NIOSH REL: (Cobalt) Insufficient evidence
for recommending limit
SAFETY PROFILE: Mutation data reported.
See also COBALT COMPOUNDS and
MOLYBDENUM COMPOUNDS.

CNC500   CAS: 10141-05-6   HR: 3
COBALT(II) NITRATE
mf: CoN_2O_6 mw: 182.95
PROP: Mp: 55°, d: 1.87.
SYNS: COBALT BIS(NITRATE) □ COBALT DINITRATE
□ COBALT NITRATE □ COBALT(2+) NITRATE □
COBALTOUS NITRATE □ NITRIC ACID, COBALT(2+)
SALT
CONSENSUS REPORTS: IARC Cancer
Review: Group 2B; Human Inadequate
Evidence IMEMDT 52,363,91. Reported in
EPA TSCA Inventory. Cobalt and its
compounds are on the Community Right-
To-Know List.
SAFETY PROFILE: Poison by ingestion,
intramuscular, and subcutaneous routes.
Experimental reproductive effects.
Questionable carcinogen with experimental

tumorigenic data. Used in animal feed.
Explosive reaction with ammonium
hexacyanoferrate(II) at 220°C. Potentially
explosive reaction with carbon. When
heated to decomposition it emits toxic
fumes of NO$_x$. See also COBALT
COMPOUNDS and NITRATES.

CND125   CAS: 1307-96-6   HR: 3
COBALT(II) OXIDE
mf: CoO   mw: 74.93
PROP: Powder, cubic, hexagonal crystals or
solid. Color varies from olive green to red,
depending on the particle size, but the
commercial material is usually dark gray and
contains about 76% Co. The color of
anhydrous CoO depends upon the degree of
dispersion. It may be yellow, gray, brown,
reddish, bluish, or black. Mp: about 1935°,
d: 5.7 to 6.7. Practically insol in water or alc;
sol in acids or alkalies. Easily reduced to Co
by C or CO. Reacts at high temperatures
with silica, alumina, and zinc oxide to form
pigments.
SYNS: C.I. 77322 □ C.I. PIGMENT BLACK 13 □
COBALT BLACK □ COBALT MONOOXIDE □ COBALT
MONOXIDE □ COBALTOUS OXIDE □ COBALT OXIDE
□ COBALT(2+) OXIDE □ MONOCOBALT OXIDE □
ZAFFRE
CONSENSUS REPORTS: IARC Cancer
Review: Group 2B IMEMDT 52,363,91;
Animal Sufficient Evidence IMEMDT
52,363,91; Human Inadequate Evidence
IMEMDT 52,363,91. Cobalt and its
compounds are on the Community Right-
To-Know List. Reported in EPA TSCA
Inventory.
SAFETY PROFILE: Suspected carcinogen
with experimental carcinogenic and
tumorigenic data. Poison by ingestion,
subcutaneous, and intratracheal routes.
Moderately toxic by intramuscular route.
Violent reaction with hydrogen peroxide.
See also COBALT. Note: The commercial
oxides are usually not definite chemical
compounds but mixtures of the cobalt
oxides.

**CND825   CAS: 1308-04-9   HR: 2
COBALT(III) OXIDE**
mf: Co₂O₃   mw: 165.86
SYNS: C.I. 77323 □ COBALTIC OXIDE □ COBALT
OXIDE (8CI,9CI) □ COBALT(3+) OXIDE □ COBALT
PEROXIDE □ COBALT SESQIOXIDE □ COBALT
SESQUIOXIDE □ COBALT TRIOXIDE □ DICOBALT
OXIDE □ DICOBALT TRIOXIDE
CONSENSUS REPORTS: IARC Cancer
Review: Group 2B IMEMDT 52,363,91;
Human Inadequate Evidence IMEMDT
52,363,91. Reported in EPA TSCA
Inventory. Cobalt compounds are on the
Community Right-To-Know List.
SAFETY PROFILE: Moderately toxic by
intraperitoneal and subcutaneous routes.
Questionable carcinogen. Violent reaction
with hydrogen peroxide. The oxide increases
the sensitivity of nitroalkanes (e.g.,
nitromethane, nitroethane, and 1-
nitropropane) to heat or detonation. See
also COBALT COMPOUNDS.

**CND900   CAS: 13478-33-6   HR: 3
COBALT(II) PERCHLORATE, HEXAHY-
   DRATE**
mf: Cl₂O₈•Co•6H₂O   mw: 365.95
SYNS: COBALT DIPERCHLORATE HEXAHYDRATE □
COBALTOUS PERCHLORATE, HEXAHYDRATE □
COBALT PERCHLORATE HEXAHYDRATE □
PERCHLORIC ACID, COBALT(II) SALT, HEXAHYDRATE
DOT CLASSIFICATION: 5.1; Label: Oxidizer
SAFETY PROFILE: A poison by
intraperitoneal route. An oxidizer. When
heated to decomposition it emits toxic
vapors of Co and Cl⁻.

**CNE000   CAS: 68956-82-1   HR: 3
COBALT RESINATE, precipitated**
DOT: UN 1318
mf: Co(C₄₄H₆₂O₄)₂   mw: 1368.81
PROP: Brown-red powder.
CONSENSUS REPORTS: Reported in EPA
TSCA Inventory. Cobalt and its compounds
are on the Community Right-To-Know List.
DOT CLASSIFICATION: 4.1; Label:
Flammable Solid
SAFETY PROFILE: A dangerous fire hazard
when exposed to heat, flame, oxidizers, or

air. Ignites spontaneously in air. See also
COBALT COMPOUNDS. When heated to
decomposition it emits acrid smoke and
irritating fumes.

**CNE125   CAS: 10124-43-3   HR: 3
COBALT(II) SULFATE (1:1)**
mf: O₄S•Co   mw: 154.99
PROP: Red to lavender or dark bluish solid
or dimorphic, orthorhombic crystals. D:
3.71. Stable to 708°. Dissolves slowly in
boiling water.
SYNS: COBALTOUS SULFATE □ COBALT SULFATE □
COBALT SULFATE (1:1) □ COBALT (2+) SULFATE □
COBALT(II) SULPHATE □ SULFURIC ACID, COBALT(2+)
SALT (1:1)
CONSENSUS REPORTS: IARC Cancer
Review: Group 2B IMEMDT 52,363,91;
Human Inadequate Evidence IMEMDT
52,363,91. Cobalt and its compounds are on
the Community Right-To-Know List. EPA
Genetic Toxicology Program. Reported in
EPA TSCA Inventory.
SAFETY PROFILE: Poison by intravenous
and intraperitoneal routes. Moderately toxic
by ingestion. Questionable carcinogen.
When heated to decomposition it emits
toxic fumes of SOₓ. See also COBALT
COMPOUNDS.

**CNE200   CAS: 1317-42-6   HR: 3
COBALT(II) SULFIDE**
mf: CoS   mw: 90.99
PROP: Exists in two forms. α-CoS: black,
amorphous powder. Sol in HCl. β-CoS: gray
powder or reddish-silver octahedral crystals.
Mp: above 1100°, d: 5.45. Practically insol in
water; sol in acids.
SYNS: COBALT MONOSULFIDE □ COBALTOUS
SULFIDE □ COBALT SULFIDE □ COBALT SULFIDE
(amorphous)
CONSENSUS REPORTS: IARC Cancer
Review: Group 2B IMEMDT 52,363,91;
Animal Limited Evidence IMEMDT
52,363,91; Human Inadequate Evidence
IMEMDT 52,363,91. Reported in EPA
TSCA Inventory. Cobalt and its compounds
are on the Community Right-To-Know List.

C

SAFETY PROFILE: Suspected carcinogen with experimental tumorigenic data. Mutation data reported. If dried at 300°C it ignites spontaneously in air. See also COBALT COMPOUNDS and SULFIDES.

CNE240   **HR: D**
COBALT TALLATE
SAFETY PROFILE: When heated to decomposition it emits acrid smoke and irritating fumes.

CNE750   **CAS: 50-36-2**   **HR: 3**
COCAINE
mf: $C_{17}H_{21}NO_4$ mw: 303.39
PROP: Colorless to white crystals or prisms from alc. Mp: 98°, bp: 187–188°. Volatile, especially above 90°. Soluble in alcohol, chloroform, ether, oil turpentine, olive oil, liquid petrolatum, acetone, ethyl acetate, carbon disulfide. Sparingly soluble in water.
SYNS: BENZOYLMETHYLECGONINE □ BERNICE □ BERNIES □ BURESE □ 2-β-CARBOMETHOXY-3-β-BENZOXYTROPANE □ "C" CARRIE □ CECIL □ CHOLLY □ (−)-COCAINE □ β-COCAINE □ l-COCAINE □ COKE □ CORINE □ ECGONINE, METHYL ESTER, BENZOATE (ESTER) □ ERITROXILINA □ ERYTROXY-LIN □ GIRL □ GOLD DUST □ HAPPY DUST □ 3-β-HYDROXY-1-α-H,5-α-H-TROPANE-2-β-CARBOXYLIC ACID METHYL ESTER, BENZOATE □ KOKAIN □ KOKAN □ KOKAYEEN □ METHYL-3-β-HYDROXY-1-α-H,5-α-H-TROPANE-2-β-CARBOXYLATE BENZOATE (ESTER) □ NEUROCAINE □ STAR DUST □ 2-β-TROPANECARBOXYLIC ACID, 3-β-HYDROXY-, METH-YL ESTER, BENZOATE (ESTER) □ 3-TROPANYLBEN-ZOATE-2-CARBOXYLIC ACID METHYL ESTER
DOT CLASSIFICATION: Forbidden
SAFETY PROFILE: A human poison by ingestion and possibly other routes. Poison experimentally by ingestion, intraperitoneal, intravenous, subcutaneous, and parenteral routes. Human central nervous system effects by ingestion and possibly other routes: general anesthesia, hallucinations or distorted perceptions, and convulsions. An eye irritant. A widely abused, controlled substance. Abuse leads to habituation or addiction. In medicine, it is used as a local narcotic anesthetic applied topically to mucous membranes. The free base is soluble

in fats and thus is used for ointments and oily solutions. For water-soluble applications, the sulfate or hydrochloride is used. See also ESTERS. When heated to decomposition it emits highly toxic fumes.

CNF175   **CAS: 61789-30-8**   **HR: D**
COCOA FATTY ACIDS, POTASSIUM
     SALTS
SYN: SOAP
CONSENSUS REPORTS: Reported in EPA TSCA Inventory.
SAFETY PROFILE: An experimental teratogen. Other experimental reproductive effects. When heated to decomposition it emits acrid smoke and irritating fumes.

CNF250   **CAS: 104-61-0**   **HR: 2**
COCONUT ALDEHYDE
mf: $C_9H_{16}O_2$ mw: 156.25
PROP: Colorless to sltly yellow liquid; coconut odor. D: 0.958–0.966, refr index: 1.446–1.450, flash p: 212°F. Sol in alc, fixed oils, propylene glycol; insol in water.
SYNS: ALDEHYDE C-18 □ γ-N-AMYLBUTYROLACTONE □ FEMA No. 2781 □ 4-HYDROXYNONANOIC ACID, γ-LACTONE □ γ-NONALACTONE (FCC) □ 1,4-NONALOLIDE □ PRUNOLIDE
CONSENSUS REPORTS: Reported in EPA TSCA Inventory.
SAFETY PROFILE: Moderately toxic by ingestion. A skin irritant. Mutation data reported. Combustible liquid. When heated to decomposition it emits acrid smoke and irritating fumes. See also ALDEHYDES.

CNF330   **CAS: 68603-42-9**   **HR: 3**
COCONUT OIL ACID DIETHANOLAMINE
SYNS: AMIDES, COCO, N,N-BIS(HYDROXYETHYL) □ CLINDROL 200CGN □ CLINDROL 202CGN □ CLINDR-OL SUPERAMIDE 100CG □ COCAMIDE DEA □ COCO-NUT DIETHANOLAMIDE □ COCONUT DIETHANOL-AMINE □ COCONUT OIL ACID DIETHANOLAMINE □ COCONUT OIL ACID DIETHANOLAMINE CONDENS-ATE □ ETHYLAN LD □ N,N-BIS(2-HYDROXYETH-YL)COCOAMIDE □ N,N-BIS(2-HYDROXYETHYL)-COCONUT FATTY ACID AMIDE □ N,N-BIS(2-HYDRO-XYETHYL)COCONUT OIL AMIDE □ NCI-C55312 □ NINOL 2012E

CONSENSUS REPORTS: Reported in EPA TSCA Inventory. EPA FIFRA 1988 pesticide subject to registration or registration. NCI Carcinogenesis Studies (derm). clear evidence: mouse; (derm). equivocal evidence: rat

SAFETY PROFILE: A poison by ingestion. A moderate skin irritant. Questionable carcinogen with experimental Carcinogenic data reported. When heated to decomposition it emits acrid smoke and irritating vapors.

CNG929   HR: 3
COKE OVEN EMISSIONS
SYN: COKE OVEN EMISSIONS (OSHA)
CONSENSUS REPORTS: Reported in NTP 10th Report on Carcinogens.
OSHA PEL: OSHA: Cancer Hazard
NIOSH REL: 0.5–0.7 mg/m^3 (total particulates)
SAFETY PROFILE: Confirmed carcinogen. Mutation data reported. When heated to decomposition it emits acrid smoke and irritating vapors.

CNG938   CAS: 64-86-8   HR: 3
COLCHICINE
mf: $C_{22}H_{25}NO_6$ mw: 399.48
PROP: Pale-yellow scales or powder. Mp: 155–157°. Crystals from ethyl acetate, pale-yellow needles. One gram dissolves in 22 mL water, 220 mL ether, 100 mL benzene; freely sol in alcohol or chloroform; practically insol in pet ether.
SYNS: 7-ACETAMIDO-6,7-DIHYDRO-1,2,3,10-TETRAMETHOXY-BENZO(a)HEPTALEN-9(5H)-ONE □ N-ACETYL TRIMETHYLCOLCHICINIC ACID METHYL ETHER □ COLCHICIN (GERMAN) □ COLCHICINA (ITALIAN) □ 7-α-H-COLCHICINE □ COLCHINEOS □ COLCHISOL □ COLCIN □ COLSALOID □ CONDYLON □ NSC-757 □ N-(5,6,7,9-TETRAHYDRO-1,2,3,10-TETRAMETHOXY-9-OXOBENZO(α)HEPTALEN-7-YL)-ACETAMIDE
CONSENSUS REPORTS: EPA Extremely Hazardous Substances List. EPA Genetic Toxicology Program. Reported in EPA TSCA Inventory.
SAFETY PROFILE: A human poison by ingestion and intravenous routes. Poison experimentally by most routes. Human systemic effects: aplastic anemia, blood pressure depression, body temperature decrease, changes in kidney tubules, dyspnea, flaccid paralysis without anesthesia, gastrointestinal effects, kidney damage and hemorrhaging, muscle contraction or spasticity, muscle weakness, nausea or vomiting, respiratory stimulation, and somnolence. An experimental teratogen. Experimental reproductive effects. A severe eye irritant. Human mutation data reported. Inhibits the formation of microtubules and thus impairs cell division. When heated to decomposition it emits toxic fumes of NO_x.

CNH792   CAS: 8001-61-4   HR: 2
COPAIBA OIL
PROP: From steam distillation of South American *Copaifera L.* (Fam. *Leguminosae*) balsam. Colorless to yellow liquid; characteristic odor, aromatic, slightly bitter taste. D: 0.880–0.907, refr index: 1.493–1.500 @ 20°. Sol in alc, fixed oils, mineral oil.
SYNS: BALSAM CAPTIVI □ BALSAMS, COPAIBA □ COPAIBA BALSAM □ COPAIBA OLEORESIN □ JESUIT'S BALSAM
CONSENSUS REPORTS: Reported in EPA TSCA Inventory.
SAFETY PROFILE: Mildly toxic by ingestion. Large doses cause vomiting and diarrhea. Can also cause dermatitis and kidney damage. When heated to decomposition it emits acrid smoke and irritating fumes.

CNI000   CAS: 7440-50-8   HR: 2
COPPER
af: Cu   aw: 63.54
PROP: Reddish, malleable and ductile metal. Slowly weathers to green patina. Mp: 1083°, bp: 25° @ 2595 mm, d: 8.92, vap press: 1 mm @ 1628°. IDLH 100 mg/m^3 (as Cu).
SYNS: ALLBRI NATURAL COPPER □ ANAC 110 □ ARWOOD COPPER □ BRONZE POWDER □ CDA 101 □ CDA 102 □ CDA 110 □ CDA 122 □ C.I. 77400 □ C.I. PIGMENT METAL 2 □ COPPER-AIRBORNE □ COPPER BRONZE □ COPPER-MILLED □ COPPER SLAG-

AIRBORNE □ COPPER SLAG-MILLED □ 1721 GOLD □ GOLD BRONZE □ KAFAR COPPER □ M1 (COPPER) □ M2 (COPPER) □ OFHC Cu □ RANEY COPPER

CONSENSUS REPORTS: Reported in EPA TSCA Inventory. Copper and its compounds are on the Community Right-To-Know List.

OSHA PEL: TWA (dust, mist) 1 mg(Cu)/m³; (fume and respirable particles) 0.1 mg/m³

ACGIH TLV: TWA (fume) 0.2 mg/m³; (dust, mist) 1 mg(Cu)/m³

DFG MAK: (dust) 1 mg/m³; (fume) 0.1 mg/m³

SAFETY PROFILE: Toxic by inhalation. Questionable carcinogen with experimental tumorigenic data. Experimental teratogenic and reproductive effects. Human systemic effects by ingestion: nausea and vomiting. See also COPPER COMPOUNDS. Liquid copper explodes on contact with water. Potentially explosive reaction with acetylenic compounds, 3-bromopropyne, ethylene oxide, lead azide, and ammonium nitrate. Ignites on contact with chlorine, chlorine trifluoride, fluorine (above 121°), and hydrazinium nitrate (above 70°). Reacts violently with C_2H_2, bromates, chlorates, iodates, $(Cl_2 + OF_2)$, dimethyl sulfoxide + trichloroacetic acid, ethylene oxide, H_2O_2, hydrazine mononitrate, hydrazoic acid, H_2S + air, $Pb(N_3)_2$, K_2O_2, NaN_3, Na_2O_2, sulfuric acid. Incandescent reaction with potassium dioxide. Incompatible with 1-bromo-2-propyne.

CNI250   CAS: 142-71-2   HR: 3
COPPER ACETATE
mf: $C_4H_6O_4 \cdot Cu$ mw: 181.64
PROP: Greenish-blue powder or small crystals.
SYNS: ACETATE de CUIVRE (FRENCH) □ ACETIC ACID, CUPRIC SALT □ COPPER(2+) ACETATE □ COPPER(II) ACETATE □ COPPER DIACETATE □ COPPER(2+) DIACETATE □ CRYSTALLIZED VERDIGRIS □ CRYSTALS of VENUS □ CUPRIC ACETATE □ CUPRIC DIACETATE □ NEUTRAL VERDIGRIS □ OCTAN MEDNATY (CZECH)
CONSENSUS REPORTS: Reported in EPA TSCA Inventory. Copper and its

compounds are on the Community Right-To-Know List.
ACGIH TLV: TWA 1 mg(Cu)/m³
SAFETY PROFILE: Poison by subcutaneous and intraperitoneal routes. Moderately toxic by ingestion. Experimental reproductive effects. When heated to decomposition it emits acrid smoke and irritating fumes. See also COPPER COMPOUNDS.

CNI500   CAS: 12540-13-5   HR: 3
COPPER ACETYLIDE
mf: C_2Cu mw: 87.56
PROP: A black or brown solid.
SYN: COPPER CARBIDE
CONSENSUS REPORTS: Copper and its compounds are on the Community Right-To-Know List.
ACGIH TLV: TWA 1 mg(Cu)/m³
DOT CLASSIFICATION: Forbidden
SAFETY PROFILE: Ignites and then explodes when heated to 100°C. Much more sensitive to impact, friction, and heat than copper(I) acetylide (the red-brown form). See also COPPER COMPOUNDS and ACETYLIDES.

CNI600   CAS: 11133-98-5   HR: 3
COPPER ALLOY, Cu, Be
SYNS: BERYLLIUM-COPPER ALLOY □ COPPER-BERYLLIUM ALLOY
CONSENSUS REPORTS: IARC Cancer Review: Group 1 IMEMDT 58,41,93; Human Sufficient Evidence IMEMDT 58,41,93; Animal Sufficient Evidence IMEMDT 1,17,72; Animal Sufficient Evidence IMEMDT 23,143,80; Animal Sufficient Evidence IMEMDT 58,41,93. Copper and its compounds, as well as beryllium and its compounds, are on the Community Right-To-Know List.
OSHA PEL: TWA 0.002 mg(Be)/m³; STEL 0.005 mg(Be)/m³/30M; CL 0.025 mg(Be)/m³
ACGIH TLV: TWA 0.002 mg(Be)/m³; Confirmed Human Carcinogen; (Proposed: TWA 0.00005 mg(Be)/m³; STEL 0.0002

mg(Be)/m³ (sensitizer) (skin); Confirmed Human Carcinogen)

SAFETY PROFILE: Confirmed carcinogen. Cases of berylliosis have been reported from exposure to so-called low beryllium alloys. Human systemic effects by inhalation: dyspnea, fibrosing alveolitis, weight loss, or decreased weight gain. See also BERYLLIUM COMPOUNDS and COPPER COMPOUNDS. When heated to decomposition it emits very toxic fumes of BeO.

CNJ900   CAS: 26506-47-8   HR: 2
COPPER CHLORATE
DOT: UN 2721

mf: HClO₃•xCu   mw: 529.24

SYNS: CHLORIC ACID, COPPER SALT □ COPPER CHLORATE (DOT)

ACGIH TLV: TWA (fume) 0.2 mg/m³; (dust, mist) 1 mg(Cu)/m³

DOT CLASSIFICATION: 5.1; Label: Oxidizer

SAFETY PROFILE: An oxidizer. When heated to decomposition it emits toxic fumes of Cl⁻.

CNJ950   CAS: 1344-67-8   HR: 3
COPPER CHLORIDE
DOT: UN 2802

SYNS: COPPER CHLORIDE (DOT) □ KIRTICOPPER

CONSENSUS REPORTS: Reported in EPA TSCA Inventory.

ACGIH TLV: TWA (fume) 0.2 mg/m³; (dust, mist) 1 mg(Cu)/m³

DOT CLASSIFICATION: 8; Label: Corrosive

SAFETY PROFILE: A corrosive. A skin and eye irritant. When heated to decomposition it emits toxic vapors of Cu and Cl⁻.

CNK500   CAS: 7447-39-4   HR: 3
COPPER(II) CHLORIDE (1:2)
DOT: UN 2802

mf: Cl₂Cu   mw: 134.44

PROP: Yellowish-brown, hygroscopic powder. Mp: 498°, d: 3.054.

SYNS: COPPER BICHLORIDE □ COPPER(2+) CHLORIDE □ COPPER(II) CHLORIDE □ CUPRIC CHLORIDE □ CUPRIC DICHLORIDE

CONSENSUS REPORTS: Copper and its compounds are on the Community Right-To-Know List. Reported in EPA TSCA Inventory.

ACGIH TLV: TWA (fume) 0.2 mg/m³; (dust, mist) 1 mg(Cu)/m³

DOT CLASSIFICATION: 8; Label: Corrosive

SAFETY PROFILE: Poison by intravenous and intraperitoneal routes. Experimental reproductive effects. Mutation data reported. See also COPPER COMPOUNDS and CHLORIDES. Can react violently with K and Na. When heated to decomposition it emits toxic fumes of Cl⁻.

CNK625   CAS: 866-82-0   HR: 2
COPPER(I) CITRATE
mf: C₆H₄O₇•2Cu   mw: 315.18

SYNS: CITRIC ACID, COPPER(2+) SALT (8CI) □ COPPER CITRATE □ COPPER SALT 2-HYDROXY-1,2,3-PROPANETRICARBOXYLIC ACID (1:2) □ CUPRIC CITRATE □ CUPROCITROL □ 2-HYDROXY-1,2,3-PROPANETRICARBOXYLIC ACID COPPER(2+) SALT (1:2) (9CI)

CONSENSUS REPORTS: Copper and its compounds are on the Community Right-To-Know List. Reported in EPA TSCA Inventory.

SAFETY PROFILE: Moderately toxic by ingestion. Experimental reproductive effects. An experimental teratogen. See also COPPER COMPOUNDS.

CNK700   CAS: 55158-44-6   HR: 3
COPPER-COBALT-BERYLLIUM
SYNS: BERYLLIUM-COPPER-COBALT ALLOY □ COPPER ALLOY, Cu, Be, Co

CONSENSUS REPORTS: IARC Cancer Review: Group 1 IMEMDT 58,41,93; Human Sufficient Evidence IMEMDT 58,41,93; Animal Sufficient Evidence IMEMDT 1,17,72; Animal Sufficient Evidence IMEMDT 23,143,80; Animal Sufficient Evidence IMEMDT 58,41,93. Copper, cobalt, and beryllium and their compounds are on the Community Right-To-Know List.

OSHA PEL: TWA 0.002 mg(Be)/m³; STEL 0.005 mg(Be)/m³/30M; CL 0.025 mg(Be)/m³
ACGIH TLV: TWA 0.002 mg(Be)/m³; Confirmed Human Carcinogen; (Proposed: TWA 0.00005 mg(Be)/m³; STEL 0.0002 mg(Be)/m³ (sensitizer) (skin); Confirmed Human Carcinogen)
NIOSH REL: (Beryllium) CL not to exceed 0.0005 mg(Be)/m³; (Cobalt) Insufficient evidence for recommending limit
SAFETY PROFILE: Confirmed carcinogen. See COPPER, BERYLLIUM, and COBALT COMPOUNDS. When heated to decomposition it emits very toxic fumes of BeO.

CNK750   HR: 3
COPPER COMPOUNDS
CONSENSUS REPORTS: Copper and its compounds are on the Community Right-To-Know List.
ACGIH TLV: Inorganic: TWA (fume) 0.2 mg/m³; (dust, mist) 1 mg(Cu)/m³
SAFETY PROFILE: As the sublimed oxide, copper may be responsible for one form of metal fume fever. In animals, inhalation of copper dust has caused hemolysis of the red blood cells, deposition of hemofuscin in the liver and pancreas, and injury to the lung cells; injection of the dust has caused cirrhosis of the liver and pancreas, and a condition closely resembling hemochromatosis, or bronzed diabetes. However, considerable trial exposure to copper compounds has not resulted in such disease. As regards local effect, copper chloride and sulfate have been reported as causing irritation of the skin and conjunctiva, possibly on an allergic basis. Cuprous oxide is irritating to the eyes and upper respiratory tract. Discoloration of the skin is often seen in persons handling copper, but this does not indicate any actual injury. There is an excess of cancer cases in the copper smelting industry. In humans the ingestion of a large quantity of copper sulfate has caused vomiting, gastric pain, dizziness, exhaustion, anemia, cramps, convulsions, shock, coma, and death. Symptoms attributed to damage to the nervous system and kidney have been recorded, jaundice has been observed, and, in some cases, the liver has been enlarged. Deaths have been reported to have occurred following the ingestion of as little as 27 g of the salt, while other victims have recovered after having taken up to 120 g. Many copper-containing compounds are used as fungicides. Many copper salts form highly unstable acetylides. Those formed in basic solutions from (Cu⁺ salts + C₂H₂) are less stable than those formed from Cu⁺⁺ salts. Copper salts + hydrazine react strongly, and with nitromethane these salts are explosive.

CNL000   CAS: 544-92-3   HR: 3
COPPER CYANIDE
DOT: UN 1587
mf: CCuN   mw: 89.56
PROP: Monoclinic, white prisms; white-cream powder; dark-green orthorhombic crystals; dark red monoclinic crystals. Mp: 474° in N₂, bp: decomp, d: 2.92. Sol in NH₃ (aq); insol in H₂O and alcohols.
SYNS: COPPER(I) CYANIDE □ CUPRICIN □ CUPROUS CYANIDE □ RCRA WASTE NUMBER P029
CONSENSUS REPORTS: Reported in EPA TSCA Inventory. Cyanide and its compounds, as well as copper and its compounds, are on the Community Right-To-Know List.
ACGIH TLV: TWA 1 mg(Cu)/m³
DOT CLASSIFICATION: 6.1; Label: Poison
SAFETY PROFILE: A poison. Reacts violently with magnesium. When heated to decomposition it emits very toxic CN⁻ and NOₓ. See also CYANIDE and COPPER COMPOUNDS.

CNL250   CAS: 14763-77-0   HR: 3
COPPER(II) CYANIDE
mf: C₂CuN₂   mw: 115.58
PROP: Yellowish-green powder. Mp: decomp before melting.

SYNS: COPPER CYANAMIDE □ COPPER CYANIDE (DOT) □ CUPRIC CYANIDE (DOT) □ CYANURE de CUIVRE (FRENCH)

CONSENSUS REPORTS: Copper and its compounds, as well as cyanide and its compounds, are on the Community Right-To-Know List.

ACGIH TLV: TWA 1 mg(Cu)/m³

DOT CLASSIFICATION: 6.1; Label: Poison, KEEP AWAY FROM FOOD

SAFETY PROFILE: Poison by intraperitoneal route. See also CYANIDE and COPPER COMPOUNDS. Incompatible with magnesium. When heated to decomposition it emits toxic fumes of NO$_x$ and CN⁻.

CNM100   CAS: 527-09-3   HR: D
COPPER GLUCONATE

mf: C$_{12}$H$_{22}$CuO$_{14}$ mw: 453.84

PROP: Fine light blue powder. Sol in water; sltly sol in alc.

SAFETY PROFILE: When heated to decomposition it emits acrid smoke and irritating fumes.

CNM750   CAS: 3251-23-8   HR: 2
COPPER(II) NITRATE

mf: CuN$_2$O$_6$ mw: 187.55

PROP: Large, blue-green, deliquescent, orthorhombic crystals. D: 2.047, sublimes at 150–225°, mp: 255–256°. Sol in water, ethyl acetate, dioxane; dissolves in and reacts vigorously with ether.

SYNS: COPPER DINITRATE □ COPPER(2+) NITRATE □ CUPRIC DINITRATE □ CUPRIC NITRATE (DOT)

CONSENSUS REPORTS: Copper and its compounds are on the Community Right-To-Know List. Reported in EPA TSCA Inventory.

ACGIH TLV: TWA (fume) 0.2 mg/m³; (dust, mist) 1 mg(Cu)/m³

SAFETY PROFILE: Moderately toxic by ingestion. A severe eye and skin irritant. Potentially explosive reaction above 220°C with ammonium or potassium hexacyanoferrate(II). Reaction with ammonia + potassium amide gives explosive product. Violent reaction with acetic anhydride. May ignite on prolonged contact with paper. Concentrated solutions may ignite in contact with tin or aluminum foil. Used as a fungicide, herbicide, and as a catalyst component in solid rocket fuel. When heated to decomposition it emits toxic fumes of NO$_x$. See also COPPER COMPOUNDS and NITRATES.

CNN500   CAS: 10290-12-7   HR: 3
COPPER ORTHOARSENITE

DOT: UN 1586

mf: AsCuHO$_3$ mw: 187.47

PROP: Yellowish-green powder. Mp: decomp.

SYNS: ACID COPPER ARSENITE □ AIR-FLO GREEN □ ARSONIC ACID, COPPER(2+) SALT (1:1) (9CI) □ COPPER ARSENITE, solid (DOT) □ CUPRIC ARSENITE □ CUPRIC GREEN □ SCHEELES GREEN □ SCHEELE'S MINERAL □ SWEDISH GREEN

CONSENSUS REPORTS: Arsenic and its compounds, as well as copper and its compounds, are on the Community Right-To-Know List.

OSHA PEL: OSHA: Cancer Hazard

ACGIH TLV: TWA 0.01 mg/m³; Confirmed Human Carcinogen; BEI: 35 μ (As)/L inorganic arsenic and methylated metabolites in urine; TWA (fume) 0.2 mg/m³; (dust, mist) 1 mg(Cu)/m³

NIOSH REL: (Arsenic, Inorganic) CL 0.002 mg(As)/m³/15M

DOT CLASSIFICATION: 6.1; Label: Poison

SAFETY PROFILE: Confirmed human carcinogen. Poison. When heated to decomposition it emits toxic fumes of As. See also ARSENIC COMPOUNDS and COPPER COMPOUNDS.

CNO000   CAS: 1317-39-1   HR: 2
COPPER(I) OXIDE

mf: Cu$_2$O   mw: 143.08

PROP: Octahedral, cubic, red or yellow crystals or powder. Mp: 1235°, bp: loses O$_2$ @ 1800° (decomp), d: 6.0. Insol in water.

SYNS: BROWN COPPER OXIDE □ C.I. 77402 □ COPOX □ COPPER NORDOX □ COPPER SARDEX □ CUPPER OXIDE (RUSSIAN) □ CUPROUS OXIDE □ DICOPPER MONOXIDE □ FUNGIMAR □

KUPFEROXYDUL (GERMAN) □ OLEOCUIVRE □ OLEO NORDOX □ RED COPPER OXIDE □ YELLOW CUPROCIDE

CONSENSUS REPORTS: Reported in EPA TSCA Inventory. Copper and its compounds are on the Community Right-To-Know List.

ACGIH TLV: TWA (fume) 0.2 mg/m³; (dust, mist) 1 mg(Cu)/m³

SAFETY PROFILE: Moderately toxic by ingestion. Experimental reproductive effects. A fungicide. Violent, potentially explosive reaction with concentrated peroxyformic acid. Violent reaction when heated with aluminum. See also COPPER COMPOUNDS.

CNO500　CAS: 17031-32-2　HR: 3
COPPER(II) PERCHLORATE, DIHYDR-ATE

mf: $Cl_2O_8 \cdot Cu \cdot 2H_2O$　mw: 298.48

SYNS: CUPRIC DIPERCHLORATE TETRAHYDRATE □ PERCHLORIC ACID, COPPER(II) SALT, DIHYDRATE (8CI, 9CI)

CONSENSUS REPORTS: Copper and its compounds are on the Community Right-To-Know List.

ACGIH TLV: TWA (fume) 0.2 mg/m³; (dust, mist) 1 mg(Cu)/m³

DOT CLASSIFICATION: 5.1; Label: Oxidizer

SAFETY PROFILE: Poison by intraperitoneal route. See also COPPER COMPOUNDS and COPPER(II) PERCHLORATE. An oxidizer. When heated to decomposition it emits toxic fumes of Cl⁻.

CNP250　CAS: 7758-98-7　HR: 3
COPPER(II) SULFATE (1:1)

mf: $O_4S \cdot Cu$　mw: 159.60

PROP: Blue or white rhombic crystals or crystalline granules or hygroscopic powder. D: 2.284, mp: 200°. Very sol in H_2O; sol in MeOH or glycerol; sltly sol in EtOH.

SYNS: BCS COPPER FUNGICIDE □ BLUE COPPER □ BLUE STONE □ BLUE VITRIOL □ COPPER MONO-SULFATE □ COPPER SULFATE □ CP BASIC SULFATE □ CUPRIC SULFATE □ KUPFERSULFAT (GERMAN) □ ROMAN VITRIOL □ SULFATE de CUIVRE (FRENCH) □

SULFURIC ACID, COPPER(2+) SALT (1:1) □ TNCS 53 □ TRIANGLE

CONSENSUS REPORTS: Copper and its compounds are on the Community Right-To-Know List. Reported in EPA TSCA Inventory. EPA Genetic Toxicology Program.

ACGIH TLV: TWA (fume) 0.2 mg/m³; (dust, mist) 1 mg(Cu)/m³

SAFETY PROFILE: A human poison by ingestion. An experimental poison by ingestion, subcutaneous, parenteral, intravenous, and intraperitoneal routes. Human systemic effects by ingestion: gastritis, diarrhea, nausea or vomiting, damage to kidney tubules, and hemolysis. Questionable carcinogen with experimental tumorigenic data. An experimental teratogen. Other experimental reproductive effects. Mutation data reported. Reacts violently with hydroxylamine, magnesium. See also COPPER COMPOUNDS and SULFATES. When heated to decomposition it emits toxic fumes of SO_x.

CNR000　CAS: 8001-31-8　HR: 3
COPRA (OIL)

DOT: UN 1363

PROP: From the kernel of the fruit of the coconut palm *Cocos nucifera*. Fatty solid or liquid; sweet, nutty taste. Mp: 21–27°.

SYNS: COCONUT BUTTER □ COCONUT MEAL PELLETS, containing 6–13% moisture and no more than 10% residual fat (DOT) □ COCONUT OIL (FCC) □ COCONUT PALM OIL □ COPRA (DOT) □ COPRA PELLETS (DOT) □ FREE COCONUT OIL

CONSENSUS REPORTS: Reported in EPA TSCA Inventory.

DOT CLASSIFICATION: 4.2; Label: Spontaneously Combustible

SAFETY PROFILE: Flammable solid when exposed to heat or flame. May spontaneously heat and ignite if stored wet and hot.

CNR100　CAS: 3486-66-6　HR: 3
COPTISINE

mf: $C_{19}H_{14}NO_4$　mw: 320.32

SYN: BIS(1,3)BENZODIOXOLO(5,6-A:4',5'-G)QUINOL-IZINIUM, 6,7-DIHYDRO-

SAFETY PROFILE: A poison by ingestion. When heated to decomposition it emits toxic vapors of NO_x.

CNR735   CAS: 8008-52-4   HR: 2
CORIANDER OIL
PROP: From steam distillation of ripe fruit of *Coriandrum sativum* L. (Fam. *Umbelliferae*). Colorless liquid; characteristic odor and taste. D: 0.863–0.875, refr index: 1.462 @ 20°.
SYNS: OIL OF CORIANDER □ OILS, CORIANDER
CONSENSUS REPORTS: Reported in EPA TSCA Inventory.
SAFETY PROFILE: Moderately toxic by ingestion. Mutation data reported. A skin irritant. When heated to decomposition it emits acrid smoke and fumes.

CNR850   HR: D
CORN ENDOSPERM OIL
PROP: Reddish-brown liquid.
SAFETY PROFILE: When heated to decomposition it emits acrid smoke and irritating fumes.

CNS000   CAS: 8001-30-7   HR: 1
CORN OIL
PROP: Light-yellow, clear, oily liquid; faint characteristic odor. Mp: −10°, flash p: 490°F (CC), d: 0.92, autoign temp: 740°F. From wet milling of *Zea mays* (85DIA2 2,70,77).
CONSENSUS REPORTS: Reported in EPA TSCA Inventory.
SAFETY PROFILE: Human skin irritant. An experimental teratogen. May be an allergen. Combustible liquid when exposed to heat or flame. Dangerous spontaneous heating may occur during storage if leaks impregnate rags, waste, etc. To fight fire, use CO_2, dry chemical.

CNS100   HR: D
CORN SILK and CORN SILK EXTRACT
PROP: Extracted from the fresh styles and stigmas of *Zea mays* L.

SAFETY PROFILE: When heated to decomposition it emits acrid smoke and irritating fumes.

CNS750   CAS: 50-23-7   HR: 3
CORTISOL
mf: $C_{21}H_{30}N_{40}O_5S$ mw: 954.97
PROP: Striated blocks from 2-propanol. Mp: 220° (decomp).
SYNS: AEROSEB-HC □ ANTI-INFLAMMATORY HORMONE □ BARSEB HC □ CETACORT □ COBADEX □ COMPOUND F □ CORT-DOME □ CORTISOL ALCOHOL □ CORTISPRAY □ DERMACORT □ EF CORLIN □ GENACORT □ HC □ HEB-CORT □ HIDRO-COLISONA □ 11-β-HYDROCORTISONE □ HYDRO-CORTISONE FREE ALCOHOL □ HYDROCORTISYL □ HYDROCORTONE □ 17-HYDROXYCORTICOSTERONE □ 11-β-HYDROXYCORTISONE □ HYTONE LOTION □ KENDALL'S COMPOUND F □ NSC-10483 □ OPTEF □ PERMICORT □ 4-PREGNENE-11-β,17-α,21-TRIOL 3,20-DIONE □ REICHSTEIN'S SUBSTANCE M □ SCHEROS-ON F □ 11-β,17,21-TRIHYDROXYPREGN-4-ENE-3,20-DIONE □ 11-β,17-α-21-TRIHYDROXY-4-PREGNENE-3,20-DIONE
CONSENSUS REPORTS: Reported in EPA TSCA Inventory.
SAFETY PROFILE: Poison by intraperitoneal route. Moderately toxic by subcutaneous route. An experimental teratogen. Other experimental reproductive effects. Mutation data reported. A steroid. When heated to decomposition it emits very toxic fumes of SO_x and NO_x. See also CORTISONE.

CNT250   HR: 3
CORUNDUM FUME
PROP: Half finely divided alumina, half silica (JPBAA7 69,81,55).
SAFETY PROFILE: Poison by intratracheal route. See also ALUMINUM OXIDE (2:3) and SILICA.

CNT750   HR: 2
COTTON DUST
PROP: IDLH 100 mg/m³.
OSHA PEL: TWA 1 mg/m³ (raw dust); 0.2 mg/m³ (yarn manufacturing); 0.75 mg/m³ (slashing and weaving); 0.5 mg/m³ (other operations)

ACGIH TLV: TWA 0.2 mg/m³ (raw dust)
DFG MAK: 1.5 mg/m³ (raw cotton)
NIOSH REL: (Cotton Dust) CL 0.200 mg/m³ lint-free

SAFETY PROFILE: Human pulmonary effects. Causes a mild febrile condition of the lungs resembling metal fume fever. Coarser grades of cotton contain more dust than the finer varieties, and therefore constitute a greater hazard. It is considered an inert dust and indeed it is, within the meaning of the term. However, it can cause some illness, due to the allergens or fungi in the cotton or on the dust. Workers in processing rooms may develop conjunctivitis or blepharitis from the burned products of the gassing of the double yarn. It is a mild allergen. Inhalation may produce bronchial asthma, sneezing, and eczema in sensitized persons. Moderate fire and explosion hazard when exposed to heat or flame; can react with oxidizing materials.

CNT950   HR: D
COTTONSEED, MODIFIED PRODUCTS

PROP: Extracted from decorticated, partially defatted, cooked, ground cotton-seed kernels.
SAFETY PROFILE: When heated to decomposition it emits acrid smoke and irritating fumes.

CNU000   CAS: 8001-29-4   HR: 2
COTTONSEED OIL (unhydrogenated)

PROP: Oily, pale-yellow, nearly odorless liquid from seeds of species of *Gossypium hirsutum*. Flash p: 486°F (CC), fp: 0–5°, d: 0.915–0.921 @ 25°/25°, autoign temp: 650°F.
SYNS: DEODORIZED WINTERIZED COTTONSEED OIL □ NCI-C50168
SAFETY PROFILE: Questionable carcinogen with experimental tumorigenic data. Experimental teratogenic effects. An allergen. Combustible liquid when exposed to heat or flame. However, if allowed to impregnate rags or oily waste, it can become

a dangerous hazard due to spontaneous heating. To fight fire, use CO_2, dry chemical.

CNU750   CAS: 56-72-4   HR: 3
COUMAPHOS

mf: $C_{14}H_{16}ClO_5PS$ mw: 362.78
PROP: A solid. Mp: 95°.
SYNS: AGRIDIP □ ASUNTHOL □ BAYER 21/199 □ BAYMIX 50 □ 3-CHLORO-7-HYDROXY-4-METHYL-COUMARIN-O,O-DIETHYL PHOSPHOROTHIOATE □ 3-CHLORO-7-HYDROXY-4-METHYL-COUMARIN-O-ESTER with O,O-DIETHYL PHOSPHOROTHIOATE □ O-3-CHLORO-4-METHYL-7-COUMARINYL-O,O-DIETHYL PHOSPHOROTHIOATE □ 3-CHLORO-4-METHYL-7-COUMARINYL DIETHYL PHOSPHOROTHIOATE □ 3-CHLORO-4-METHYL-7-HYDROXYCOUMARIN DIETHYL THIOPHOSPHORIC ACID ESTER □ 3-CHLORO-4-METHYLUMBELLIFERONE-O-ESTER with O,O-DIETHYL PHOSPHOROTHIOATE □ CUMAFOS (DUTCH) □ O,O-DIAETHYL-O-(3-CHLOR-4-METHYL-CUMARIN-7-YL)-MONOTHIOPHOSPHAT (GERMAN) □ O,O-DIETHYL-O-(3-CHLOOR-4-METHYL-CUMARIN-7-YL)MONOTHIOFOSFAAT (DUTCH) □ O,O-DIETHYL-O-(3-CHLORO-4-METHYL-7-COUMARINYL)PHOSPHORO-THIOATE □ O,O-DIETHYL-O-(3-CHLORO-4-METHYLCOUMARINYL-7) THIOPHOSPHATE □ O,O-DIETHYL-O-(3-CHLORO-4-METHYL-2-OXO-2H-BENZOPYRAN-7-YL)PHOSPHOROTHIOATE □ O,O-DIETHYL-3-CHLORO-4-METHYL-7-UMBELLIFERONE THIOPHOSPHATE □ O,O-DIETHYL-O-(3-CHLORO-4-METHYLUMBELLIFERYL)PHOSPHOROTHIOATE □ DIETHYL-3-CHLORO-4-METHYLUMBELLIFERYL THIONOPHOSPHATE □ DIETHYL THIOPHOSPHORIC ACIDESTER of 3-CHLORO-4-METHYL-7-HYDROXY-COUMARIN □ O,O-DIETIL-O-(3-CLORO-4-METIL-CUMARIN-7-IL-MONOTIOFOSFATO) (ITALIAN) □ DIOLICE □ ENT 17,956 □ MELDONE □ NCI-C08662 □ THIOPHOSPHATE de O,O-DIETHYLE et de O-(3-CHLORO-4-METHYL-7-COUMARINYLE) (FRENCH) □ UMBETHION
CONSENSUS REPORTS: NCI Carcinogenesis Bioassay (feed); No Evidence: mouse, rat NCITR* NCI-CG-TR-96,79. EPA Extremely Hazardous Substances List.
SAFETY PROFILE: Poison by ingestion, skin contact, inhalation, and intraperitoneal routes. Mutation data reported. When heated to decomposition, it emits very toxic fumes of SO_x, PO_x, and Cl⁻. See also COUMARIN.

CNU825   CAS: 7400-08-0   HR: 2
4-COUMARIC ACID

mf: $C_9H_8O_3$ mw: 164.17

PROP: Needles. Mp: 210–213°. Crystallizes in anhydrous form from conc hot aq soln, but as monohydrate from dilute aq soln on slow cooling. Sltly sol in cold water; sol in hot water, alc, and ether. Practically insol in benzene and ligroin.
SYNS: p-COUMARIC ACID □ p-CUMARIC ACID □ p-HYDROXYCINNAMIC ACID □ 4-HYDROXYCINNAMIC ACID □ 4'-HYDROXYCINNAMIC ACID □ p-HYDROXY-PHENYLACRYLIC ACID □ β-(4-HYDROXYPHENYL)-ACRYLIC ACID □ 3-(4-HYDROXYPHENYL)-2-PROP-ENOIC ACID
CONSENSUS REPORTS: Reported in EPA TSCA Inventory.
SAFETY PROFILE: Moderately toxic by intraperitoneal route. Experimental reproductive effects. When heated to decomposition it emits acrid smoke and fumes. See also COUMARIN.

CNV000   CAS: 91-64-5   HR: 3
COUMARIN
mf: $C_9H_6O_2$ mw: 146.15
PROP: Rhombic crystals; fragrant, pleasant odor; burning taste. Mp: 70°, bp: 291.0°, vap press: 1 mm @ 106.0°. Sol in EtOH, hot H_2O, and alkalies.
SYNS: 2H-1-BENZOPYRAN-2-ONE □ 1,2-BENZO-PYRONE □ cis-o-COUMARINIC ACID LACTONE □ COUMARINIC ANHYDRIDE □ o-HYDROXYCINNAMIC ACID LACTONE □ o-HYDROXYZIMTSAEURE-LACTON (GERMAN) □ NCI-C60297 □ 2-OXO-1,2-BENZOPYRAN □ RATTEX □ TONKA BEAN CAMPHOR
CONSENSUS REPORTS: IARC Cancer Review: Group 3 IMEMDT 7,56,87; Animal Limited Evidence IMEMDT 10,113,76. EPA Genetic Toxicology Program. Reported in EPA TSCA Inventory.
SAFETY PROFILE: Poison by ingestion, intraperitoneal, and subcutaneous routes. Questionable carcinogen with experimental tumorigenic data. Experimental teratogenic effects. Mutation data reported. Combustible when exposed to heat or flame. When heated to decomposition it emits acrid smoke and fumes. See also KETONES and ANHYDRIDES.

CNW500   CAS: 1319-77-3   HR: 3
CRESOL
DOT: UN 2022
mf: C_7H_8O mw: 108.15
PROP: Mixture of isomeric cresols obtained from coal tar, colorless or yellowish to brown-yellow or pinkish liquid; phenolic odor. Mp: 10.9–35.5°, bp: 191–203°, flash p: 178°F, d: 1.030–1.038 @ 25°/25°, vap press: 1 mm @ 38–53°, vap d: 3.72.
SYNS: ACIDE CRESYLIQUE (FRENCH) □ BACILLOL □ CRESOLI (ITALIAN) □ CRESYLIC ACID □ HYDROXYTOLUOLE (GERMAN) □ KRESOLE (GERMAN) □ KRESOLEN (DUTCH) □ KREZOL (POLISH) □ RCRA WASTE NUMBER U052 □ TEKRESOL □ ar-TOLUENOL □ TRICRESOL
CONSENSUS REPORTS: Community Right-To-Know List.
OSHA PEL: TWA 5 ppm (skin)
ACGIH TLV: TWA 5 ppm
DFG MAK: (all isomers) 5 ppm (22 mg/m³)
NIOSH REL: (Cresol) TWA 10 mg/m³
DOT CLASSIFICATION: 6.1; Label: Poison
SAFETY PROFILE: A poison by ingestion. Moderately toxic by skin contact. Corrosive to skin and mucous membranes. Systemic poisoning has rarely been reported, but it is possible that absorption may result in damage to the kidneys, liver, and nervous system. The main hazard accompanying its use in industry lies in severe chemical burns and dermatitis. Flammable when exposed to heat or flame; can react vigorously with oxidizing materials. Slightly explosive in the form of vapor when exposed to heat or flame. Explosive Range: 1.35% @ 300°F. Reacts violently with HNO_3, oleum, or chlorosulfonic acid. When heated to decomposition it emits highly toxic and irritating fumes. To fight fire, use foam, CO_2, dry chemical. See also other cresol entries and PHENOL.

CNW750   CAS: 108-39-4   HR: 3
m-CRESOL
DOT: UN 2076
mf: C_7H_8O mw: 108.15
PROP: Colorless to yellowish liquid; phenolic odor. Mp: 10.9°, bp: 202.8°, lel: 1.1% @ 302°F, flash p: 202°F, d: 1.04 @

22°/4°, autoign temp: 1038°F, vap press: 1 mm @ 52.0°, vap d: 3.72. IDLH 250 ppm.
SYNS: 3-CRESOL □ m-CRESYLIC ACID □ 1-HYDROXY-3-METHYLBENZENE □ m-HYDROXY-TOLUENE □ m-KRESOL □ m-METHYLPHENOL □ 3-METHYLPHENOL □ m-OXYTOLUENE □ RCRA WASTE NUMBER U052 □ m-TOLUOL
CONSENSUS REPORTS: Community Right-To-Know List. Reported in EPA TSCA Inventory. EPA Genetic Toxicology Program.
OSHA PEL: TWA 5 ppm (skin)
ACGIH TLV: TWA 5 ppm
NIOSH REL: (Cresol) TWA 10 mg/m³
DOT CLASSIFICATION: 6.1; Label: Poison
SAFETY PROFILE: Poison by ingestion, intravenous, intraperitoneal, and subcutaneous routes. Moderately toxic by skin contact. Severe eye and skin irritant. An experimental teratogen. Human mutation data reported. Questionable carcinogen with experimental neoplastigenic data. Flammable when exposed to heat or flame. Moderately explosive in the form of vapor when exposed to heat or flame. See also other cresol entries and PHENOL.

CNX000   CAS: 95-48-7   HR: 3
o-CRESOL
DOT: UN 2076
mf: C₇H₈O   mw: 108.15
PROP: Crystals or liquid darkening with exposure to air and light. Mp: 30°, bp: 191°, flash p: 178°F, d: 1.05 @ 20°/4°, autoign temp: 1110°F, vap press: 1 mm @ 38.2°, vap d: 3.72, lel: 1.4% @ 300°F. IDLH 250 ppm.
SYNS: 2-CRESOL □ o-CRESYLIC ACID □ 1-HYDROXY-2-METHYLBENZENE □ o-HYDROXYTOLUENE □ o-KRESOL (GERMAN) □ o-METHYLPHENOL □ 2-METHYLPHENOL □ ORTHOCRESOL □ o-OXYTOLUENE □ RCRA WASTE NUMBER U052 □ o-TOLUOL
CONSENSUS REPORTS: EPA Extremely Hazardous Substances List. Community Right-To-Know List. EPA Genetic Toxicology Program. Reported in EPA TSCA Inventory.
OSHA PEL: TWA 5 ppm (skin)

ACGIH TLV: TWA 5 ppm
NIOSH REL: (Cresol) TWA 10 mg/m³
DOT CLASSIFICATION: 6.1; Label: Poison
SAFETY PROFILE: Poison by ingestion, inhalation, subcutaneous, intravenous, and intraperitoneal routes. Moderately toxic by skin contact. A severe eye and skin irritant. Human mutation data reported. Questionable carcinogen with experimental neoplastigenic data. Flammable when exposed to heat, flame, or oxidants. To fight fire, water may be used to blanket fire; foam, fog, mist, dry chemical. See also other cresol entries and PHENOL.

CNX250   CAS: 106-44-5   HR: 3
p-CRESOL
DOT: UN 2076
mf: C₇H₈O   mw: 108.15
PROP: Found in a score of essential oils, including ylang-ylang and oil of jasmine (FCTXAV 12,385,74). Crystals; phenolic odor. Mp: 35.5°, bp: 201.8°, lel: 1.1% @ 302°F, flash p: 202°F, d: 1.0341 @ 20°/4°, autoign temp: 1038°F, vap press: 1 mm @ 53.0°, vap d: 3.72. IDLH 250 ppm.
SYNS: 4-CRESOL □ p-CRESYLIC ACID □ 1-HYDROXY-4-METHYLBENZENE □ p-HYDROXYTOLUENE □ 4-HYDROXYTOLUENE □ p-KRESOL □ 1-METHYL-4-HYDROXYBENZENE □ p-METHYLPHENOL □ 4-METHYLPHENOL □ p-OXYTOLUENE □ PARAMETHYL PHENOL □ RCRA WASTE NUMBER U052 □ p-TOLUOL □ p-TOLYL ALCOHOL
CONSENSUS REPORTS: Community Right-To-Know List. Reported in EPA TSCA Inventory. EPA Genetic Toxicology Program.
OSHA PEL: TWA 5 ppm (skin)
ACGIH TLV: TWA 5 ppm
NIOSH REL: (Cresol) TWA 10 mg/m³
DOT CLASSIFICATION: 6.1; Label: Poison
SAFETY PROFILE: Poison by ingestion, skin contact, subcutaneous, intravenous, and intraperitoneal routes. A severe skin and eye irritant. Questionable carcinogen with experimental neoplastigenic data by itself and with 7,12-dimethyl benz(a)anthracene. Combustible when exposed to heat or flame. Moderately explosive in the form of

vapor when exposed to heat or flame. To fight fire, use CO_2, dry chemical, alcohol foam. See also other cresol entries and PHENOL.

COB250   CAS: 4170-30-3   HR: 3
CROTONALDEHYDE
DOT: UN 1143
mf: C_4H_6O mw: 70.09
PROP: Water-white, mobile liquid; pungent suffocating odor. Bp: 104°, fp: −76.0°, lel: 2.1%, uel: 15.5%, flash p: 55°F, d: 0.853 @ 20°/20°, vap d: 2.41, autoign temp: 405°F. IDLH 50 ppm.
SYNS: 2-BUTENAL □ CROTONALDEHYDE, stabilized (DOT) □ CROTONIC ALDEHYDE □ KROTON-ALDEHYD (CZECH) □ β-METHYLACROLEIN □ RCRA WASTE NUMBER U053
CONSENSUS REPORTS: EPA Extremely Hazardous Substances List. Reported in EPA TSCA Inventory.
OSHA PEL: TWA 2 ppm
ACGIH TLV: STEL CL 0.3 ppm (skin); Animal Carcinogen
DFG MAK: Suspected Carcinogen
DOT CLASSIFICATION: 3; Label: Flammable Liquid, Poison
SAFETY PROFILE: Suspected carcinogen with experimental carcinogenic data. Poison by ingestion and inhalation. Mutation data reported. An eye, skin, and mucous membrane irritant. A lachrymating material that can cause corneal burns and is very dangerous to the eyes. Caution: Keep away from heat and open flame. Keep container closed. Use with adequate ventilation. Extremely irritating to eyes, skin, mucous membranes. When necessary, the lachrymatory effect of the vapors may be counteracted by ammonia fumes. Dangerous fire hazard when exposed to heat or flame; can react with oxidizing materials. To fight fire, use alcohol foam, CO_2, dry chemical. Reacts violently with 1,3-butadiene. Violent hypergolic reaction with concentrated nitric acid. When heated to decomposition it emits acrid smoke and fumes. See also ALDEHYDES.

COB260   CAS: 123-73-9   HR: 3
(E)-CROTONALDEHYDE
mf: C_4H_6O mw: 70.10
PROP: Water-white, mobile liquid; pungent, suffocating odor. Mp: −69°, bp: 102.2°, fp: −76.0°, lel: 2.1%, uel: 15.5%, flash p: 55°F, d: 0.853 @ 20°/20°, vap d: 2.41, autoign temp: 450°F. Moderately sol in H_2O.
SYNS: ALDEHYDE CROTONIQUE (FRENCH) □ trans-2-BUTENAL □ (E)-2-BUTENAL □ CROTONAL □ CROTONALDEHYDE □ CROTONIC ALDEHYDE □ 1,2-ETHANEDIOL DIPROPANOATE (9CI) □ ETHYLENE GLYCOL DIPROPIONATE (8CI) □ ETHYLENE PROPIONATE □ β-METHYL ACROLEIN □ NCI-C56279 □ PROPYLENE ALDEHYDE □ RCRA WASTE NUMBER U053 □ TOPANEL
CONSENSUS REPORTS: EPA Extremely Hazardous Substances List. Reported in EPA TSCA Inventory.
OSHA PEL: TWA 2 ppm
ACGIH TLV: TWA 2 ppm
DFG MAK: Confirmed Animal Carcinogen with Unknown Relevance to Humans
SAFETY PROFILE: Suspected carcinogen. A poison by ingestion, subcutaneous, and intraperitoneal routes. Mutation data reported. A lachrymating material that is very dangerous to the eyes. Human respiratory system irritant by inhalation. Can cause corneal burns and is irritating to the skin. In case of contact, immediately flush the skin or eyes with water for at least 15 minutes and get medical attention. See also ALDEHYDES. Dangerous fire hazard when exposed to heat or flame. To fight fire, use alcohol foam, CO_2, dry chemical. Incompatible with 1,3-butadiene and oxidizing materials. When heated to decomposition it emits acrid smoke and fumes.

COB500   CAS: 3724-65-0   HR: 3
CROTONIC ACID
DOT: UN 2823
mf: $C_4H_6O_2$ mw: 86.10
PROP: Colorless, needle-like crystals. Bp: 185°, mp: 72°, flash p: 190°F (COC), d: 1.018 @ 15°/4°, vap press: 0.19 mm @ 20°, vap d: 2.97.

SYNS: α-BUTENOIC ACID □ 2-BUTENOIC ACID □ CROTONIC ACID, solid □ α-CROTONIC ACID □ 3-METHYLACRYLIC ACID □ β-METHYLACRYLIC ACID

CONSENSUS REPORTS: Reported in EPA TSCA Inventory.

DOT CLASSIFICATION: 8; Label: Corrosive

SAFETY PROFILE: Poison by intraperitoneal route. Moderately toxic by ingestion, skin contact, and subcutaneous routes. A powerful corrosive and irritant. Flammable when exposed to heat or flame; can react with oxidizing materials. To fight fire, use alcohol foam, CO_2, dry chemical. When heated to decomposition it emits acrid smoke and irritating fumes.

COB750   CAS: 623-70-1   HR: 3
α-CROTONIC ACID ETHYL ESTER
DOT: UN 1862

mf: $C_6H_{10}O_2$ mw: 114.16

PROP: Colorless, monoclinic prisms or water-white liquid; pungent odor. Mp: 45° (solid), bp: 209° (solid), 137° (liquid), flash p: 36.0°F, d: 0.9207 @ 20°/20°, vap d: 3.93.

SYNS: 2-BUTENOIC ACID, ETHYL ESTER, (E)-(9CI) □ trans-2-BUTENOIC ACID ETHYL ESTER □ CROTONATE d'ETHYLE (FRENCH) □ ETHYL (E)-2-BUTENOATE □ ETHYLCROTONATE □ ETHYL (E)-CROTONATE □ ETHYL trans-CROTONATE □ ETHYL CROTONATE (DOT)

CONSENSUS REPORTS: Reported in EPA TSCA Inventory.

DOT CLASSIFICATION: 3; Label: Flammable Liquid

SAFETY PROFILE: Moderately toxic by ingestion and probably by inhalation. A skin, mucous membrane, and severe eye irritant. Very dangerous fire hazard when exposed to heat or flame; can react vigorously with oxidizing materials. To fight fire, use foam, CO_2, or dry chemical. See also ESTERS. When heated to decomposition it emits acrid smoke and fumes.

COB900   CAS: 623-68-7   HR: 2
CROTONIC ANHYDRIDE
mf: $C_8H_{10}O_3$ mw: 154.18

SYNS: ANHYDRID KYSELINY KROTONOVE □ 2-BUTENOIC ACID, ANHYDRIDE (9CI) □ CROTONIC ACID ANHYDRIDE

CONSENSUS REPORTS: Reported in EPA TSCA Inventory.

SAFETY PROFILE: Moderately toxic by ingestion. A skin irritant. When heated to decomposition it emits acrid smoke and irritating fumes.

COC500   CAS: 503-17-3   HR: 3
CROTONYLENE
DOT: UN 1144

mf: CH_3CCCH_3 mw: 54.09

PROP: Liquid. Mp: −32.8°, bp: 27°, flash p: <−4°F, lel: 1.4%, d: 0.688 @ 25°, vap d: 1.91.

SYNS: 2-BUTYNE □ DIMETHYLACETYLENE

CONSENSUS REPORTS: Reported in EPA TSCA Inventory.

DOT CLASSIFICATION: 3; Label: Flammable Liquid

SAFETY PROFILE: A simple asphyxiant. Very dangerous fire hazard when exposed to heat or flame; can react with oxidizing materials. Moderately explosive in the form of vapor when exposed to heat or flame. To fight fire, use foam, CO_2, dry chemicals. See also ACETYLENE COMPOUNDS and ARGON (for a description of simple asphyxiants).

COD750   CAS: 68308-34-9   HR: 3
CRUDE SHALE OILS
DOT: UN 1288

SYNS: BLUE OIL □ GREEN OIL □ RAW SHALE OIL □ SHALE OIL (DOT) □ UNFINISHED LUBRICATING OIL

CONSENSUS REPORTS: IARC Cancer Review: Group 1 IMEMDT 7,339,87; Human Sufficient Evidence IMEMDT 35,161,85; Animal Limited Evidence IMEMDT 35,161,85; Animal Sufficient Evidence IMEMDT 3,22,73.

DOT CLASSIFICATION: 3; Label: Flammable Liquid

SAFETY PROFILE: Confirmed human carcinogen with experimental carcinogenic, neoplastigenic, and tumorigenic data. Mildly toxic by ingestion, skin contact, and intraperitoneal routes. A skin irritant. Experimental reproductive effects. Mutation data reported. Flammable when exposed to heat and flame. When heated to decomposition it emits acrid smoke and fumes.

COE000   CAS: 1309-32-6   HR: 3
CRYPTOHALITE
DOT: UN 2854
mf: $F_6Si•2H_4N$ mw: 178.19
PROP: Mp: subl, d: 2.01.
SYNS: AMMONIUM FLUOROSILICATE (DOT) □ AMMONIUM FLUOSILICATE □ AMMONIUM HEXA-FLUOROSILICATE □ AMMONIUM SILICOFLUORIDE □ AMMONIUM SILICON FLUORIDE □ DIAMMONIUM FLUOSILICATE □ DIAMMONIUM HEXAFLUOROSILIC-ATE(2-) □ DIAMMONIUM SILICON HEXAFLUORIDE □ FLUOSILICATE de AMMONIUM (FRENCH)
OSHA PEL: TWA 2.5 mg(F)/m³
ACGIH TLV: TWA 2.5 mg(F)/m³; BEI: 3 mg/g creatinine of fluorides in urine prior to shift; 10 mg/g creatinine of fluorides in urine at end of shift.
NIOSH REL: (Fluorides, Inorganic) TWA 2.5 mg(F)/m³
SAFETY PROFILE: Poison by ingestion and subcutaneous routes. See also HEXAFLUOROSILICATE (2-) DIHYDROGEN and FLUORIDES. When heated to decomposition it emits very toxic fumes of F⁻, NH_3, and NO_x.

COE175   CAS: 8007-87-2   HR: 1
CUBEB OIL
PROP: From steam distillation of mature, unripe fruit of *piper cubeba L.* (Fam. *Piperaceae*). Colorless to light green liquid; spicy odor, slt acrid taste. D: 0.898–0.928, refr index: 1.492–1.502 @ 20°. Sol in fixed oils, mineral oil; insol in glycerin, propylene glycol.
SYNS: OIL OF CUBEB □ OILS, CUBEB
CONSENSUS REPORTS: Reported in EPA TSCA Inventory.
SAFETY PROFILE: A skin irritant. When heated to decomposition it emits acrid smoke and irritating fumes.

COE500   CAS: 122-03-2   HR: 2
CUMALDEHYDE

mf: $C_{10}H_{12}O$ mw: 148.22
PROP: Found in at least 50 essential oils, such as cumin, eucalyptus species, cinnamon, boldo, and rue, and as main constituent of oil of *Pectis papposa harn* and *gray* (FCTXAV 12,385,74). Colorless to pale-yellow liquid; pungent odor of cumin. D: 0.976–0.980, refr index: 1.529–1.534, flash p: 199°F. Sol in alc, ether; insol in water.
SYNS: p-CUMIC ALDEHYDE □ CUMINALDEHYDE □ CUMINIC ALDEHYDE (FCC) □ CUMINYL ALDEHYDE □ FEMA No. 2341 □ p-ISOPROPYLBENZALDEHYDE □ 4-ISOPROPYLBENZALDEHYDE □ p-ISOPROPYLBEN-ZENECARBOXALDEHYDE □ 4-(1-METHYLETHYL)-BENZALDEHYDE (9CI)
CONSENSUS REPORTS: Reported in EPA TSCA Inventory.
SAFETY PROFILE: Moderately toxic by ingestion and skin contact. A skin irritant. Combustible liquid. When heated to decomposition it emits acrid smoke and irritating fumes. See also ALDEHYDES.

COE750   CAS: 98-82-8   HR: 3
CUMENE
DOT: UN 1918
mf: C_9H_{12} mw: 120.21
PROP: Colorless liquid. Mp: −96.0°, bp: 152°, flash p: 111°F, d: 0.864 @ 20°/4°, vap press: 10 mm @ 38.3°, autoign temp: 795°F, lel: 0.9%, uel: 6.5%, vap d: 4.1. IDLH 900 ppm [10%LEL].
SYNS: BENZENE ISOPROPYL □ CUM □ CUMEEN (DUTCH) □ 2-FENILPROPANO (ITALIAN) □ 2-FENYL-PROPAAN (DUTCH) □ ISOPROPILBENZENE (ITALIAN) □ ISOPROPYLBENZEEN (DUTCH) □ ISOPROPYL BENZENE □ ISOPROPYLBENZOL □ ISOPROPYL-BENZOL (GERMAN) □ 2-PHENYLPROPANE □ RCRA WASTE NUMBER U055
CONSENSUS REPORTS: Community Right-To-Know List. Reported in EPA TSCA Inventory. EPA Genetic Toxicology Program.
OSHA PEL: TWA 50 ppm (skin)
ACGIH TLV: TWA 50 ppm
DFG MAK: 50 ppm (250 mg/m³)
DOT CLASSIFICATION: 3; Label: Flammable Liquid
SAFETY PROFILE: Moderately toxic by ingestion. Mildly toxic by inhalation and skin

c

contact. Human systemic effects by inhalation: an antipsychotic, unspecified changes in the sense of smell and respiratory system. An eye and skin irritant. Potential narcotic action. Central nervous system depressant. There is no apparent difference between the toxicity of natural cumene and that derived from petroleum. See also BENZENE and TOLUENE. Flammable liquid when exposed to heat or flame; can react with oxidizing materials. Violent reaction with HNO_3, oleum, chlorosulfonic acid. To fight fire, use foam, CO_2, dry chemical.

COF000   CAS: 93-53-8   HR: 2
CUMENE ALDEHYDE
mf: $C_9H_{10}O$ mw: 134.19
PROP: Colorless liquid; floral odor. D: 0.998–1.006, refr index: 1.515–1.520, flash p: 156°F. Sol in fixed oils; sltly sol in propylene glycol; insol in glycerin.
SYNS: FEMA No. 2886 □ α-FORMYLETHYLBENZENE □ HYACINTHAL □ HYDRATROP ALDEHYDE □ HYDRATROPIC ALDEHYDE □ α-METHYL PHENYL-ACETALDEHYDE □ α-METHYL-α-TOLUIC ALDEHYDE □ 2-PHENYLPROPANAL □ α-PHENYLPROPION-ALDEHYDE □ 2-PHENYLPROPIONALDEHYDE (FCC)
CONSENSUS REPORTS: Reported in EPA TSCA Inventory.
SAFETY PROFILE: Moderately toxic by ingestion. Combustible liquid. When heated to decomposition it emits acrid smoke and irritating fumes. See also ALDEHYDES.

COF325   CAS: 8014-13-9   HR: 2
CUMIN OIL
PROP: From steam distillation of *Cuminum cyminum L.* Light-yellow to brown liquid; strong odor. D: 0.905–0.925, refr index: 1.501 @ 20°. Sol in fixed oils, mineral oil; very sol in glycerin, propylene glycol.
SYNS: CUMMIN □ OILS, CUMIN
CONSENSUS REPORTS: Reported in EPA TSCA Inventory.
SAFETY PROFILE: Moderately toxic by ingestion and skin contact. A skin irritant. Mutation data reported. When heated to

decomposition it emits acrid smoke and irritating fumes.

COF350   CAS: 479-13-0   HR: D
CUMOESTEROL
mf: $C_{15}H_8O_5$ mw: 268.23
PROP: Crystals. Mp: 385° (sublimes @ 325°). Sltly sol in alkaline water, methanol, chloroform, ether, carbon tet, and benzene. Insol in acidified water and pet ether.
SYNS: 6H-BENZOFURO(3,2-c)(1)BENZOPYRAN-6-ONE, 3,9-DIHDYROXY- □ COUMESTROL □ CUMOSTROL
CONSENSUS REPORTS: Reported in EPA TSCA Inventory.
SAFETY PROFILE: An experimental teratogen. Other experimental reproductive effects. Mutation data reported. When heated to decomposition it emits acrid smoke and irritating fumes.

COF500   CAS: 12002-03-8   HR: 3
CUPRIC ACETOARSENITE
mf: $C_4H_6As_6Cu_4O_{16}$ mw: 1013.78
PROP: Emerald-green powder.
SYNS: (ACETATO)(TRIMETAARSENITO)DICOPPER □ ACETOARSENITE de CUIVRE (FRENCH) □ BASLE GREEN □ C.I. 77410 □ C.I. PIGMENT GREEN 21 (9CI) □ COPPER ACETOARSENITE (DOT) □ COPPER ACETOARSENITE, solid (DOT) □ EMERALD GREEN □ ENT 884 □ FRENCH GREEN □ GENUINE PARIS GREEN □ IMPERIAL GREEN □ KING'S GREEN □ MEADOW GREEN □ MINERAL GREEN □ MITIS GREEN □ MOSS GREEN □ MOUNTAIN GREEN □ NEUWIED GREEN □ NEW GREEN □ ORTHO P-G BAIT □ PARIS GREEN □ PARROT GREEN □ PATENT GREEN □ POWDER GREEN □ SCHWEINFURTER-GRUEN □ SCHWEINFURT GREEN □ SOWBUG & CUTWORM BAIT □ SWEDISH GREEN □ VIENNA GREEN
CONSENSUS REPORTS: EPA Extremely Hazardous Substances List. Arsenic and its compounds, as well as copper and its compounds, are on the Community Right-To-Know List.
OSHA PEL: TWA 0.5 mg(As)/m³
DOT CLASSIFICATION: 6.1; Label: Poison
SAFETY PROFILE: Poison by ingestion and possibly other routes. An insecticide. When heated to decomposition it emits very toxic fumes of As. See also ARSENIC

COMPOUNDS and COPPER
COMPOUNDS.

COF675   CAS: 370-81-0   HR: D
CUPRIZONE
mf: $C_{14}H_{22}N_4O_2$ mw: 278.40
SYNS: BISCYCLOHEXANONE OXALDIHYDRAZONE
□ CUPRIZANE □ ETHANEDIOIC ACID BIS(CYCLOHE-
XYLIDENE HYDRAZIDE)
CONSENSUS REPORTS: Reported in EPA
TSCA Inventory.
SAFETY PROFILE: Experimental
reproductive effects. Mutation data
reported. When heated to decomposition it
emits toxic fumes of NO_x. See also
OXALATES.

COG000   CAS: 8024-37-1   HR: 2
CURCUMIN
SYNS: C.I. 75300 □ CURCUMA OIL □ CURCUMINE □
NCI-C60015 □ TURMERIC OIL □ TURMERIC
OLEORESIN
CONSENSUS REPORTS: Reported in EPA
TSCA Inventory.
SAFETY PROFILE: Moderately toxic by
intraperitoneal route. A skin irritant.
Mutation data reported. When heated to
decomposition it emits acrid smoke and
irritating fumes.

COH000   HR: 3
CUTTING OILS
SAFETY PROFILE: Often carcinogenic.
The cause of "cutting oil" dermatitis is
generally due to an insoluble oil. However it
can occasionally be caused by a soluble oil.
Many have looked for a causative factor
other than the oil itself. Bacteria have
frequently been blamed, although insoluble
oils are usually sterile, while the soluble oils
may contain bacteria. The metal slivers that
occur in these oils after use have also been
blamed as well as the sulfur, chlorine, and
inhibitors that they contain. The oil itself
can plug the pores and form boils.
Combustible when exposed to heat or
flame. See also MINERAL OIL.

COH500   CAS: 420-04-2   HR: 3
CYANAMIDE
mf: CH_2N_2 mw: 42.05
PROP: Deliquescent crystals. Mp: 45°, bp:
260°, flash p: 285°F, d: 1.282, vap d: 1.45.
SYNS: AMIDOCYANOGEN □ CARBAMONITRILE □
CARBIMIDE □ CYANOAMINE □ CYANOGENAMIDE □
CYANOGEN NITRIDE □ HYDROGEN CYANAMIDE □
USAF EK-1995
CONSENSUS REPORTS: Reported in EPA
TSCA Inventory. Cyanide and its
compounds are on the Community Right-
To-Know List.
OSHA PEL: TWA 2 mg/m³
ACGIH TLV: TWA 2 mg/m³
SAFETY PROFILE: Poison by ingestion,
inhalation, and intraperitoneal routes.
Moderately toxic by skin contact.
Experimental reproductive effects.
Combustible when exposed to heat or
flame. To fight fire, use CO_2, dry chemical.
Thermally unstable. Contact with moisture
(water), acids, or alkalies may cause a violent
reaction above 40°. Concentrated aqueous
solutions may undergo explosive
polymerization. Mixture with 1,2-
phenylenediamine salts may cause explosive
polymerization. When heated to
decomposition or on contact with acid or
acid fumes, it emits toxic fumes of CN^- and
NO_x. See also CYANIDE and AMIDES.

COI250   CAS: 917-61-3   HR: 3
CYANIC ACID, SODIUM SALT
DOT: UN 2207/UN 2478/UN 3080
mf: CNO•Na   mw: 65.01
SYNS: CYANSAN □ SAN-CYAN □ SODIUM
ISOCYANATE □ WEECON □ ZASSOL
CONSENSUS REPORTS: Cyanide and its
compounds are on the Community Right-
To-Know List. Reported in EPA TSCA
Inventory.
DOT CLASSIFICATION: 6.1; Label: KEEP
AWAY FROM FOOD (UN 2207); DOT
Class: 6.1; Label: Poison (UN 2206); DOT
Class: 6.1; Label: Poison, Flammable Liquid
(UN 3080); DOT Class: 3; Label:
Flammable Liquid, Poison (UN 2478)

SAFETY PROFILE: Poison by ingestion, intraperitoneal, and intramuscular routes. Human systemic effects by ingestion: weight loss, changes in the visual field, and other eye effects. See also CYANATES. When heated to decomposition it emits very toxic fumes of CN^- and Na_2O.

COI500   CAS: 57-12-5   HR: 3
CYANIDE
DOT: UN 1935
mf: CN^- mw: 26.02
SYNS: CARBON NITRIDE ION (CN^{1-}) □ CYANIDE-(CN^{1-}) □ CYANIDE, dry (UN 1588) □ CYANIDE ANION □ CYANIDE ION □ CYANIDE(CN^{1-}) ION □ CYANIDE SOLUTIONS (DOT) □ CYANURE □ HYDROCYANIC ACID, ION(CN^{1-}) □ ISOCYANIDE □ RCRA WASTE NUMBER P030

CONSENSUS REPORTS: Cyanide and its compounds are on the Community Right-To-Know List.
OSHA PEL: TWA 5 mg(CN)/m³
ACGIH TLV: CL 5 mg/m³ (skin)
DFG MAK: 5 mg/m³
NIOSH REL: (Cyanide) TWA CL 5 mg/m³/10M
DOT CLASSIFICATION: 6.1; Label: Poison, KEEP AWAY FROM FOOD
SAFETY PROFILE: Very poisonous by most routes. Cyanide directly stimulates the chemoreceptors of the carotid and aortic bodies with a resultant hyperpnea (increase in the depth and rate of respiration). Cardiac irregularities are often noted, but the heart invariably outlasts the respirations. Death is due to respiratory arrest of central origin. It can occur within seconds or minutes of the inhalation of high concentrations of HCN gas. Because of slower absorption, death may be more delayed after the ingestion of cyanide salts, but the critical events still occur within the first hour. Two other sources of cyanide have been responsible for human poisoning: the naturally occurring amygdalin and the drug nitroprusside.

Amygdalin is a cyanogenic glycoside found in apricot, peach, and similar fruit pits and in sweet almonds (Sayre and Kaymakcalan, 1941). It is a chemical combination of glucose, benzaldehyde, and cyanide from which the latter can be released by the action of β-glucosidase or emulsion. Although these enzymes are not found in mammalian tissues, the human intestinal microflora appears to possess these or similar enzymes capable of effecting cyanide release resulting in human poisoning. For this reason, amygdalin may be as much as 40 times more toxic by the oral route as compared with intravenous injection. Amygdalin is the major ingredient of Laetrile, and this alleged anticancer drug has also been responsible for human cyanide poisoning.

An ethical drug that may also cause cyanide poisoning in overdose is the potent vascular smooth-muscle relaxant sodium nitroprusside. Although nitroprusside is related chemically to ferricyanide, unlike the latter it penetrates into erythrocytes and reacts with hemoglobin to release its cyanide (Smith and Kruszyna, 1974). Fortunately, the therapeutic margin for nitroprusside appears to be quite large.

Cyanide is commonly found in certain rat and pest poisons, silver and metal polishes, photographic solutions, and fumigating products. Compounds such as potassium cyanide can also be readily purchased from chemical stores. Cyanide is readily absorbed from all routes, including the skin, mucous membranes, and by inhalation, although alkali salts of cyanide are toxic only when ingested. Death may occur with ingestion of even small amounts of sodium or potassium cyanide and can occur within minutes or hours depending on route of exposure. Inhalation of toxic fumes represents a potentially rapidly fatal type of exposure. A blood cyanide level of greater than 0.2 μg/mL is considered toxic. Lethal cases have usually had levels above 1 μg/mL.

Clinically, cyanide poisoning is reported to produce a bitter, almond odor on the breath of the patient; however, only a small proportion of the population is genetically able to discern this characteristic odor. Typically, cyanide has a bitter, burning taste, and, following poisoning, symptoms of salivation, nausea without vomiting, anxiety, confusion, vertigo, giddiness, lower-jaw stiffness, convulsions, opisthotonos, paralysis, coma, cardiac arrhythmias, and transient respiratory stimulation followed by respiratory failure may occur. Bradycardia is a common finding, but in most cases heartbeat usually outlasts respiration (*Wexler et al.*, 1947). A prolonged expiratory phase is considered to be characteristic of cyanide poisoning. (Casarett and Doull's, Toxicology, The Basic Science of Poisons 2nd ed. Doull, Klaassen and Amdur (eds), Macmillan Pub. Co. Inc. New York, NY).

The volatile cyanides resemble HCN physiologically, inhibiting tissue oxidation and causing death through asphyxia. Cyanogen is probably as toxic as HCN; the nitriles are generally considered somewhat less toxic, probably because of their lower volatility. The non-volatile cyanide salts appear to be relatively nontoxic systemically, so long as they are not ingested and care is taken to prevent the formation of HCN. Workers, such as electroplaters and picklers, who are daily exposed to cyanide solutions may develop a "cyanide" rash, characterized by itching, and by macular, papular, and vesicular eruptions. Frequently there is secondary infection. Exposure to small amounts of cyanide compounds over long periods of time is reported to cause loss of appetite, headache, weakness, nausea, dizziness, and symptoms of irritation of the upper respiratory tract and eyes. See also specific cyanide compounds.

Flammable by chemical reaction with heat, moisture, acid. Many cyanides rather easily evolve HCN, a flammable gas that is highly toxic. Carbon dioxide from the air is sufficiently acidic to liberate HCN from cyanide solutions. Reaction with hypochlorite solutions may be violent at pH 10.0–10.3. Cyanides explode if melted with nitrites or chlorates at about 450°. Violent reaction with F_2, Mg, nitrates, HNO_3, nitrites. Metal cyanides are easily oxidized and may be thermally unstable. N-cyano derivatives may be reactive or unstable. Many organic nitriles can be very reactive under the right conditions. When heated to decomposition or on contact with acid, acid fumes, water, or steam, cyanides emit toxic and flammable vapors of CN^-. See also HYDROCYANIC ACID.

CON750   **HR: D**
CYANO(4-FLUORO-3-PHENOXYPHEN-YL)-METHYL-3-(2,2-DICHLORO-ETHENYL)-2,2-DIMETHYLCYCLO-PROPANE-CARBOXYLATE
SAFETY PROFILE: When heated to decomposition emits toxic fumes of CN^-, F^-, and Cl^-.

COO000   CAS: 460-19-5   HR: 3
CYANOGEN
DOT: UN 1026
mf: C_2N_2 mw: 52.04
PROP: Colorless gas; pungent odor. Mp: −34.4°, bp: −21.0°, d: 0.866 @ 17°/4°, lel: 6.6%, uel: 32%, vap d: 1.8.
SYNS: CARBON NITRIDE □ CYANOGENE (FRENCH) □ CYANOGEN GAS (DOT) □ DICYANOGEN □ ETHANEDINITRILE □ NITRILOACETONITRILE □ OXALIC ACID DINITRILE □ OXALONITRILE □ OXALYL CYANIDE □ PRUSSITE □ RCRA WASTE NUMBER P031
CONSENSUS REPORTS: Reported in EPA TSCA Inventory. Cyanide and its compounds are on the Community Right-To-Know List.
OSHA PEL: TWA 10 ppm
ACGIH TLV: TWA 10 ppm
DFG MAK: 10 ppm (22 mg/m³)
DOT CLASSIFICATION: 2.3; Label: Poison Gas, Flammable Gas
SAFETY PROFILE: A poison by subcutaneous and possibly other routes. Moderately toxic by inhalation. Human systemic effects by inhalation: damage to the olfactory nerves and irritation of the conjunctiva. A systemic irritant by inhalation and subcutaneous routes. A human eye

C

irritant. Very dangerous fire hazard when exposed to heat, flames (sparks), or oxidizers. To fight fire, stop flow of gas. Potentially explosive reaction with powerful oxidants (e.g., dichlorine oxide, fluorine, oxygen, ozone). When heated to decomposition or on contact with acid, acid fumes, water, or steam will react to produce highly toxic fumes of NO_x and CN^-. See also other cyanogen entries and CYANIDE.

COO500   CAS: 506-68-3   HR: 3
CYANOGEN BROMIDE
DOT: UN 1889
mf: CBrN   mw: 105.93
PROP: Colorless needles. Mp: 52°, bp: 61.6°, d: 2.015 @ 20°/4°, vap press: 100 mm @ 22.6°.
SYNS: BROMINE CYANIDE □ BROMOCYAN □ BROMOCYANOGEN □ BROMURE de CYANOGEN (FRENCH) □ CAMPILIT □ CYANOBROMIDE □ CYANOGEN MONOBROMIDE □ RCRA WASTE NUMBER U246 □ TL 822
CONSENSUS REPORTS: Cyanide and its compounds are on the Community Right-To-Know List. EPA Extremely Hazardous Substances List. Reported in EPA TSCA Inventory.
DOT CLASSIFICATION: 6.1; Label: Poison, Corrosive
SAFETY PROFILE: A human and experimental poison by inhalation. Corrosive. When heated to decomposition it emits very toxic fumes of CN^- and Br^-. Possibly unstable. See also other cyanogen entries; CYANIDE; and BROMIDES.

COO750   CAS: 506-77-4   HR: 3
CYANOGEN CHLORIDE
DOT: UN 1589
mf: CClN   mw: 61.47
PROP: Colorless liquid or gas; lachrymatory and irritating odor. Mp: −6.5°, bp: 13.1°, d: 1.218 @ 4°/4°, vap press: 1010 mm @ 20°, vap d: 1.98.
SYNS: CHLORCYAN □ CHLORINE CYANIDE □ CHLOROCYAN □ CHLOROCYANIDE □ CHLORO-CYANOGEN □ CHLORURE de CYANOGENE □ CYANOGEN CHLORIDE (ACGIH,OSHA) □ CYANOGEN

CHLORIDE, inhibited (DOT) □ RCRA WASTE NUMBER P033
CONSENSUS REPORTS: Cyanide and its compounds are on the Community Right-To-Know List. Reported in EPA TSCA Inventory.
OSHA PEL: CL 0.3 ppm
ACGIH TLV: CL 0.3 ppm
DOT CLASSIFICATION: 2.3; Label: Poison Gas, Flammable Gas
SAFETY PROFILE: Poison by ingestion, subcutaneous, and possibly other routes. Toxic by inhalation. Human systemic effects by inhalation: lachrymation, conjunctiva irritation, and chronic pulmonary edema or congestion. A primary irritant. A severe human eye irritant. An insecticide. Flammable when exposed to heat or flame. When heated to decomposition or on contact with water or steam, it will react to produce highly toxic and corrosive fumes of Cl^-, CN^-, and NO_x. See also other cyanogen entries, CYANIDE, and CHLORIDES.

COP000   CAS: 506-78-5   HR: 3
CYANOGEN IODIDE
mf: CIN   mw: 152.92
PROP: Colorless solid. Mp: 146.5°, vap press: 1 mm @ 25.2°.
SYNS: IODINE CYANIDE □ JODCYAN
CONSENSUS REPORTS: Cyanide and its compounds are on the Community Right-To-Know List. EPA Extremely Hazardous Substances List. Reported in EPA TSCA Inventory.
ACGIH TLV: Proposed: (inhalable fraction) 0.1 mg/m^3; Not Classifiable as a Human Carcinogen)
DOT CLASSIFICATION: 6.1; Label: Poison, KEEP AWAY FROM FOOD
SAFETY PROFILE: A poison by ingestion and subcutaneous routes. Violent reaction with P. See other cyanogen entries; CYANIDE and IODIDES. When heated to decomposition it emits very toxic fumes of NO_x, CN^-, and I^-.

COP525   CAS: 17380-21-1   HR: 3
5-CYANO-3-INDOLYL ISOPROPYL
**   KETONE**
mf: $C_{13}H_{12}N_2O$ mw: 212.27
SYNS: INDOLE-5-CARBONITRILE, 3-(2-METHYL-PROPIONYL)- □ 3-(2-METHYLPROPIONYL)-5-INDOLECARBONITRILE
DOT CLASSIFICATION: 3; Label:
Flammable Liquid
SAFETY PROFILE: A poison by
intravenous route. A flammable liquid.
When heated to decomposition it emits
toxic vapors of NO_x.

COP550   CAS: 17380-19-7   HR: 3
5-CYANO-3-INDOLYLMETHYL KETONE
mf: $C_{11}H_8N_2O$ mw: 184.21
SYNS: 3-ACETYLINDOLE-5-CARBONITRILE □ INDO-LE-5-CARBONITRILE, 3-ACETYL-
DOT CLASSIFICATION: 3; Label:
Flammable Liquid
SAFETY PROFILE: A poison by
intravenous route. A flammable liquid.
When heated to decomposition it emits
toxic vapors of NO_x.

COQ390   HR: D
(+)CYANO(3-PHENOXYPHENYL)-
**   METHYL(±)-1-(DIFLUOROMETHO-**
**   XY)-α-(1-METHYLETHYL)BENZENE-**
**   ACETATE**
SYN: FLUCYTHRINATE
SAFETY PROFILE: When heated to
decomposition emits toxic fumes of CN^-,
F^-.

COR500   CAS: 1067-99-8   HR: 2
(3-CYANOPROPYL)DIETHOXY-
**   (METHYL) SILANE**
mf: $C_9H_{19}NO_2Si$ mw: 201.38
SYNS: BUTYRONITRILE, 4-(DIETHOXYMETHYLSIL-YL)- □ DIETHOXY-3-KYANPROPYL-METHYLSILAN □ SILANE, (3-CYANOPROPYL)DIETHOXY(METHYL)-
CONSENSUS REPORTS: Reported in EPA
TSCA Inventory.
SAFETY PROFILE: Moderately toxic by
ingestion. A skin irritant. When heated to
decomposition it emits toxic fumes of NO_x
and Si.

COR800   CAS: 1067-47-6   HR: 1
(3-CYANOPROPYL) TRIETHOXYSILANE
mf: $C_{10}H_{21}NO_3Si$ mw: 231.41
SYNS: BUTYRONITRILE, 4-(TRIETHOXYSILYL)- □ SILANE, (3-CYANOPROPYL)TRIETHOXY- □ SILANE, TRIETHOXY(3-CYANOPROPYL)- □ TRIETHOXY-3-KYANPROPYLSILAN
CONSENSUS REPORTS: Reported in EPA
TSCA Inventory.
SAFETY PROFILE: Mildly toxic by
ingestion. A skin irritant. When heated to
decomposition it emits toxic fumes of NO_x
and Si.

COU000   CAS: 14901-08-7   HR: 3
CYCASIN
mf: $C_8H_{16}N_2O_7$ mw: 252.26
PROP: A solid. Mp: 154° (decomp)
SYNS: CYCAS REVOLUTA GLUCOSIDE □ CYKAZINE □ β-d-GLUCOSYLOXYAZOXYMETHANE □ METHYL-AZOXYMETHANOL GLUCOSIDE □ METHYLAZOXY-METHANOL-β-d-GLUCOSIDE □ (METHYL-ONN-AZOXY)METHYL-β-d-GLUCOPYRANOSIDE
CONSENSUS REPORTS: EPA Genetic
Toxicology Program. IARC Cancer Review:
Group 2B IMEMDT 7,56,87; Human
Inadequate Evidence IMEMDT 10,121,76;
Animal Sufficient Evidence IMEMDT
10,121,76; IMEMDT 1,157,72.
SAFETY PROFILE: Confirmed carcinogen
with experimental carcinogenic and
tumorigenic data. A poison by ingestion. An
experimental teratogen. Mutation data
reported. When heated to decomposition it
emits toxic fumes of NO_x.

COU500   CAS: 103-95-7   HR: 2
CYCLAMEN ALDEHYDE
mf: $C_{13}H_{18}O$ mw: 190.31
PROP: Colorless liquid; strong, floral odor.
D: 0.946–0.952, refr index: 1.503–1.508. Sol
in fixed oils; insol in propylene glycol,
glycerin.
SYNS: ALDEHYDE B □ CYCLAMAL □ FEMA No. 2743 □ p-ISOPROPYL-α-METHYLHYDROCINNAMIC ALDEHYDE □ p-ISOPROPYL-α-METHYLPHENYLPROP-YL ALDEHYDE □ α-METHYL-p-ISOPROPYLHYDRO-CINNAMALDEHYDE □ 2-METHYL-3-(p-ISOPROPYL-PHENYL)PROPIONALDEHYDE

CONSENSUS REPORTS: Reported in EPA TSCA Inventory.

SAFETY PROFILE: Moderately toxic by ingestion. A human skin irritant. See also ALDEHYDES. When heated to decomposition it emits acrid smoke and irritating fumes.

COU510   CAS: 7149-24-8   HR: 1
CYCLAMEN ALDEHYDE DIETHYL
**   ACETAL**
mf: $C_{17}H_{28}O_2$ mw: 264.45

SYNS: HYDROCINNAMALDEHYDE, p-ISOPROPYL-α-METHYL-, DIETHYL ACETAL □ α-METHYL-p-ISOPROPYL HYDROCINNAMIC ALDEHYDE DIETHYL ACETAL

CONSENSUS REPORTS: Reported in EPA TSCA Inventory.

SAFETY PROFILE: A skin irritant. When heated to decomposition it emits acrid smoke and irritating fumes.

COW780   CAS: 72117-72-7   HR: 1
α-CYCLOCITRYLIDENE-4-METHYL-
**   BUTAN-3-ONE**
mf: $C_{15}H_{24}O$ mw: 220.39

SYNS: DIMETHYLIONONE □ 1,3-DIMETHYL-α-IONONE □ 1-PENTEN-3-ONE, 1-(2,6,6-TRIMETHYL-2-CYCLOHEXEN-1-YL)-2-METHYL-

CONSENSUS REPORTS: Reported in EPA TSCA Inventory.

SAFETY PROFILE: A skin irritant. When heated to decomposition it emits acrid smoke and irritating fumes.

COY000   CAS: 544-25-2   HR: 3
1,3,5-CYCLOHEPTATRIENE
DOT: UN 2603
mf: C_7H_8 mw: 92.14

PROP: A liquid. D: 0.888 @ 18.5°/4°, bp: 117° @ 749 mm, flash p: 39.2°F.

SYNS: CYCLOHEPTATRIENE (DOT) □ TROPILIDENE □ TROPILIDIN

DOT CLASSIFICATION: 3; Label: Flammable Liquid, Poison

SAFETY PROFILE: Poison by ingestion and skin contact. Mutation data reported. A very dangerous fire hazard when exposed to heat, flame, or oxidizers. Potentially violent reaction with nitrogen monoxide. When

heated to decomposition it emits acrid smoke and fumes.

COY100   CAS: 12125-77-8   HR: 3
CYCLOHEPTATRIENE MOLYBDENUM
**   TRICARBONYL**
mf: $C_{10}H_8MoO_3$ mw: 272.12

PROP: Red hexagonal prisms from hexane. Mp: 95° decomp. IDLH 1000 mg/m³ (as Mo).

PROP: Mp: 100–101° decomp.

SYN: MOLYBDENUM, TRICARBONYL(1,3,5-CYCLOHEPTATRIENE)-

OSHA PEL: TWA Total Dust: 10 mg/m³; Respirable Fraction: 5 mg/m³

ACGIH TLV: Insoluble Compounds: inhalable fraction, 10 mg(Mo)/m³, 3 mg(Mo)/m³, respirable fraction.

SAFETY PROFILE: Poison by intravenous route. A skin and eye irritant. When heated to decomposition it emits toxic fumes of Mo.

CPA775   CAS: 4998-76-9   HR: 3
CYCLOHEXANAMINE HYDROCHLOR-
**   IDE**
mf: $C_6H_{13}N \cdot ClH$ mw: 135.66

SYNS: AMINOCYCLOHEXANE HYDROCHLORIDE □ AMINOHEXAHYDROBENZENE HYDROCHLORIDE □ HEXAHYDROANILINE HYDROCHLORIDE

CONSENSUS REPORTS: EPA Genetic Toxicology Program. Reported in EPA TSCA Inventory.

SAFETY PROFILE: Poison by intraperitoneal route. Moderately toxic by ingestion. An experimental teratogen. Other experimental reproductive effects. Mutation data reported. When heated to decomposition it emits toxic fumes of NO_x and HCl.

CPB000   CAS: 110-82-7   HR: 3
CYCLOHEXANE
DOT: UN 1145
mf: C_6H_{12} mw: 84.18

PROP: Colorless, mobile liquid; pungent odor. Mp: 6.5°, bp: 81°, fp: 4.6°, flash: p: 1.4°F, ULC: 90–95, lel: 1.3%, uel: 8.4%, d: 0.7791 @ 20°/4°, autoign temp: 473°F, vap

press: 100 mm @ 60.8°, vap d: 2.90. Prac insol in H_2O; sol in MeOH; misc in most org solvs. IDLH 1300 ppm [10%LEL].
SYNS: CICLOESANO (ITALIAN) □ CYCLOHEXAAN (DUTCH) □ CYCLOHEXAN (GERMAN) □ CYKLO-HEKSAN (POLISH) □ HEXAHYDROBENZENE □ HEXAMETHYLENE □ HEXANAPHTHENE □ RCRA WASTE NUMBER U056
CONSENSUS REPORTS: Community Right-To-Know List.
OSHA PEL: TWA 300 ppm
ACGIH TLV: TWA 100 ppm
DFG MAK: 200 ppm (720 mg/m³)
DOT CLASSIFICATION: 3; Label: Flammable Liquid
SAFETY PROFILE: Poison by intravenous route. Moderately toxic by ingestion. A systemic irritant by inhalation and ingestion. A skin irritant. Mutation data reported. Flammable liquid. Dangerous fire hazard when exposed to heat or flame; can react with oxidizing materials. Moderate explosion hazard in the form of vapor when exposed to flame. When mixed hot with liquid dinitrogen tetraoxide an explosion can result. To fight fire, use foam, CO_2, dry chemical, spray, fog. When heated to decomposition it emits acrid smoke and fumes.

CPB050   CAS: 1122-56-1   HR: 1
CYCLOHEXANECARBOXAMIDE
mf: $C_7H_{13}NO$ mw: 127.21
PROP: Prisms from H_2O. Mp: 186–188°. Hygroscopic. Very sol in EtOH, Et_2O.
SYNS: CYCLOHEXAMETHYLENE CARBAMIDE □ CYCLOHEXANAMIDE □ CYCLOHEXANEFORMAMIDE □ CYCLOHEXYLCARBOXAMIDE □ CYCLOHEXYL CARBOXYAMIDE □ HEXAHYDROBENZOIC ACID AMIDE
CONSENSUS REPORTS: Reported in EPA TSCA Inventory.
SAFETY PROFILE: A skin irritant. When heated to decomposition it emits toxic fumes of NO_x.

CPB625   CAS: 1569-69-3   HR: 3
CYCLOHEXANETHIOL
DOT: UN 3054
mf: $C_6H_{12}S$ mw: 116.24

PROP: Oil. D: 0.991, bp: 158–160°. Sol in EtOH, $CHCl_3$; insol in H_2O.
SYNS: CYCLOHEXYL MERCAPTAN (DOT) □ CYKLOHEXANTHIOL □ CYKLOHEXYLMERKAPTAN (CZECH)
CONSENSUS REPORTS: Reported in EPA TSCA Inventory.
NIOSH REL: (Cyclohexanethiol) CL 0.5 ppm/15M
DOT CLASSIFICATION: 3; Label: Flammable Liquid
SAFETY PROFILE: An eye and severe skin irritant. When heated to decomposition it emits toxic fumes of SO_x. See also MERCAPTANS.

CPB750   CAS: 108-93-0   HR: 3
CYCLOHEXANOL
mf: $C_6H_{12}O$ mw: 100.16
PROP: Colorless needles or viscous liquid; hygroscopic, camphor-like odor. Mp: 24°, bp: 161.5°, flash p: 154°F (CC), d: 0.9449 @ 25°/4°, vap press: 1 mm @ 21.0°, vap d: 3.45, autoign temp: 572°F. Sol in EtOH, Et_2O; mod sol in H_2O; misc in nonpolar solvents. IDLH 400 ppm.
SYNS: ADRONAL □ ANOL □ CICLOESANOLO (ITALIAN) □ CYCLOHEXYL ALCOHOL □ CYKLOHEKSANOL (POLISH) □ HEXAHYDROPHENOL □ HEXALIN □ HYDRALIN □ HYDROPHENOL □ HYDROXYCYCLOHEXANE □ NAXOL
CONSENSUS REPORTS: EPA Genetic Toxicology Program. Reported in EPA TSCA Inventory.
OSHA PEL: TWA 50 ppm (skin)
ACGIH TLV: TWA 50 ppm (skin)
DFG MAK: 50 ppm (210 mg/m³)
SAFETY PROFILE: Poison by intravenous route. Moderately toxic by ingestion, subcutaneous, and intramuscular routes. Mildly toxic by skin contact. Human systemic effects by inhalation: conjunctiva irritation and changes in the olfactory and respiratory systems. Has caused damage to kidneys, liver, and blood vessels in experimental animals. Experimental reproductive effects. Human mutation data reported. A severe eye irritant. Narcotic-like action. Flammable when exposed to heat or

flame; can react with oxidizing materials. Ignites on contact with chromium trioxide. Violent reaction with HNO_3. Incompatible with oxidants. To fight fire, use alcohol foam, foam, CO_2, dry chemical. When heated to decomposition it emits acrid smoke and fumes. See also ALCOHOLS.

CPC000   CAS: 108-94-1   HR: 3
CYCLOHEXANONE
DOT: UN 1915
mf: $C_6H_{10}O$ mw: 98.16
PROP: Colorless oily liquid; acetone-like odor. Mp: −45.0°, bp: 155°, ULC: 35–40, lel: 1.1% @ 100°, flash p: 111°F, d: 0.9478 @ 20°/4°, autoign temp: 788°F, vap press: 10 mm @ 38.7°, vap d: 3.4. Mod sol in H_2O. IDLH 700 ppm.
SYNS: CICLOESANONE (ITALIAN) □ CYCLOHEXAN-ON (DUTCH) □ CYKLOHEKSANON (POLISH) □ HEXAN-ON □ KETOHEXAMETHYLENE □ NADONE □ NCI-C55005 □ PIMELIC KETONE □ RCRA WASTE NUMBER U057 □ SEXTONE
CONSENSUS REPORTS: Reported in EPA TSCA Inventory.
OSHA PEL: TWA 25 ppm (skin)
ACGIH TLV: TWA 20 ppm, STEL 50 ppm (skin); Confirmed Animal Carcinogen
DFG MAK: Confirmed Animal Carcinogen with Unknown Relevance to Humans
NIOSH REL: (Ketone (Cyclohexanone)) TWA 100 mg/m³
DOT CLASSIFICATION: 3; Label: Flammable Liquid
SAFETY PROFILE: Suspected carcinogen. Moderately toxic by ingestion, inhalation, subcutaneous, intravenous, and intraperitoneal routes. A skin and severe eye irritant. Human systemic effects by inhalation: changes in the sense of smell, conjunctiva irritation, and unspecified respiratory system changes. Human irritant by inhalation. Mild narcotic properties have also been ascribed to it. Human mutation data reported. Experimental reproductive effects. Flammable liquid when exposed to heat or flame; can react vigorously with oxidizing materials. Slight explosion hazard in its vapor form, when exposed to flame.

Explosive reaction with nitric acid at 75°C. Reaction with hydrogen peroxide + nitric acid forms an explosive peroxide. To fight fire, use alcohol foam, dry chemical, or CO_2. When heated to decomposition it emits acrid smoke and irritating fumes. See also KETONES and CYCLOHEXANE.

CPC300   CAS: 78-18-2   HR: 2
CYCLOHEXANONE PEROXIDE
mf: $C_{12}H_{22}O_5$ mw: 246.34
SYNS: CYCLOHEXANOL, 1-((1-HYDROPERO-XYCYCLOHEXYL)DIOXY)- □ 1-HYDROPEROXY-CYCLOHEXYL-1-HYDROXYCYCLOHEXYL PEROXIDE □ 1-HYDROXY-1-HYDROPEROXYDICYCLOHEXYL PEROXIDE □ 1-HYDROXY-1'-HYDROPEROXYDI-CYCLOHEXYL PEROXIDE □ PEROXIDE, 1-HYDRO-PEROXYCYCLOHEXYL 1-HYDROXYCYCLOHEXYL
CONSENSUS REPORTS: Reported in EPA TSCA Inventory.
DFG MAK: Strong Skin Effects
SAFETY PROFILE: Slightly toxic by parenteral route. A severe eye and skin irritant. When heated to decomposition it emits acrid smoke and irritating vapors.

CPC579   CAS: 110-83-8   HR: 3
CYCLOHEXENE
DOT: UN 2256
mf: C_6H_{10} mw: 82.15
PROP: Colorless liquid. Bp: 83°, fp: −103.7°, flash p: <21.2°F, d: 0.8102 @ 20°/4°, vap press: 160 mm @ 38°, autoign temp: 590°F, vap d: 2.8, lel: 1.2%. IDLH 2000 ppm.
SYNS: BENZENETETRAHYDRIDE □ CYKLOHEKSEN (POLISH) □ 1,2,3,4-TETRAHYDROBENZENE
CONSENSUS REPORTS: Reported in EPA TSCA Inventory.
OSHA PEL: 300 ppm
ACGIH TLV: 300 ppm
DFG MAK: 300 ppm (1000 mg/m³)
DOT CLASSIFICATION: 3; Label: Flammable Liquid
SAFETY PROFILE: Moderately toxic by inhalation and ingestion. A very dangerous fire hazard when exposed to flame; can react with oxidizers. Dangerous; keep away from

heat and open flame. To fight fire, use foam, CO_2, dry chemical.

CPD000   CAS: 286-20-4   HR: 3
CYCLOHEXENE OXIDE

mf: $C_6H_{10}O$ mw: 98.16

PROP: Clear liquid. Bp: 130–131°, flash p: 81°F, d: 0.9678 @ 25°/4°, vap d: 3.5.

SYNS: CCHO □ CYCLOHEXANE OXIDE □ CYCLOHEXENE EPOXIDE □ CYCLOHEXENE-1-OXIDE □ 1,2-CYCLOHEXENE OXIDE □ CYCLOHEXYLENE OXIDE □ 1,2-EPOXYCYCLOHEXANE □ 7-OXABI-CYCLO(4.1.0)HEPTANE □ TETRAMETHYL-ENEOXIRANE

CONSENSUS REPORTS: Reported in EPA TSCA Inventory. EPA Genetic Toxicology Program.

SAFETY PROFILE: Moderately toxic by ingestion, skin contact, intraperitoneal, and intramuscular routes. Mildly toxic by inhalation. Questionable carcinogen with experimental tumorigenic data. Mutation data reported. A flammable liquid and dangerous fire hazard when exposed to heat or flame. When heated to decomposition it emits acrid smoke and irritant fumes.

CPD250   CAS: 930-68-7   HR: 3
2-CYCLOHEXEN-1-ONE

mf: C_6H_8O mw: 96.14

PROP: A liquid. Bp: 169–171°.

SYN: CYCLOHEXENONE

CONSENSUS REPORTS: Reported in EPA TSCA Inventory.

SAFETY PROFILE: A poison by ingestion, inhalation, intraperitoneal, and skin contact routes. Mutation data reported. When heated to decomposition it emits acrid smoke and irritant fumes. See also KETONES.

CPD750   CAS: 100-40-3   HR: 3
CYCLOHEXENYLETHYLENE

mf: C_8H_{12} mw: 108.20

PROP: Liquid. Bp: 128°, fp: −109°, flash p: 60°F (TOC), d: 0.832 @ 20°/4°, autoign temp: 517°F, vap press: 25.8 mm @ 38°, vap d: 3.76.

SYNS: BUTADIENE DIMER □ 4-ETHENYL-1-CYCLOHEXENE □ NCI-C54999 □ 1,2,3,4-TETRA-HYDROSTYRENE □ 1-VINYLCYCLOHEXENE-3 □ 1-VINYLCYCLOHEX-3-ENE □ 4-VINYLCYCLOHEXENE □ 4-VINYLCYCLOHEXENE-1 □ 4-VINYL-1-CYCLOHEXENE

CONSENSUS REPORTS: IARC Cancer Review: Group 3 IMEMDT 7,56,87; Animal Inadequate Evidence IMEMDT 11,277,76; Animal Limited Evidence IMEMDT 39,181,86. NTP Carcinogenesis Studies (gavage); Clear Evidence: mouse NTPTR* NTP-TR-303,86; Inadequate Studies: rat NTPTR* NTP-TR-303,86. Reported in EPA TSCA Inventory.

ACGIH TLV: TWA 0.1 ppm (skin); Animal Carcinogen

DFG MAK: Animal Carcinogen, Suspected Human Carcinogen

SAFETY PROFILE: Confirmed carcinogen with experimental carcinogenic, neoplastigenic, and tumorigenic data. Moderately toxic by ingestion and inhalation. Mildly toxic by skin contact. Experimental reproductive effects. Dangerous fire hazard when exposed to heat, flame, or oxidizers. Can react with oxidizers. To fight fire, use foam, CO_2, dry chemical.

CPE500   CAS: 10137-69-6   HR: 2
CYCLOHEXENYL TRICHLOROSILANE

DOT: UN 1762

mf: $C_6H_9Cl_3Si$ mw: 215.59

PROP: Colorless, fuming liquid; HCl odor. Bp: 202°, d: 1.263 @ 25°/25°, flash p: 200°F (COC).

SYNS: CYCLOHEXENE, 4-(TRICHLOROSILYL)- □ CYCLOHEXENYLTRICHLOROSILANE (DOT) □ TRICHLORO-3-CYCLOHEXENYLSILANE

CONSENSUS REPORTS: Reported in EPA TSCA Inventory.

DOT CLASSIFICATION: 8; Label: Corrosive

SAFETY PROFILE: Moderately toxic by ingestion and skin contact. An eye and severe skin irritant. A corrosive material. It fumes in moist air, releasing HCl. Combustible when exposed to heat or flame. When heated to decomposition it emits toxic fumes of Cl⁻. See also CHLOROSILANES.

CPF000   CAS: 622-45-7   HR: 3
CYCLOHEXYL ACETATE
DOT: UN 2243
mf: $C_8H_{14}O_2$ mw: 142.22
PROP: Pale-yellow liquid; fruity odor. Bp: 177°, d: 0.996, vap d: 4.9, flash p: 136°F, autoign temp: 633°F.
SYNS: CYCLOHEXANOL ACETATE □ CYCLO-HEXANOLAZETAT (GERMAN) □ CYCLOHEXANYL ACETATE □ CYCLOHEXYLESTER KYSELINY OCTOVE
CONSENSUS REPORTS: Reported in EPA TSCA Inventory.
DOT CLASSIFICATION: 3; Label: Flammable Liquid
SAFETY PROFILE: Moderately toxic by subcutaneous route. Mildly toxic by ingestion and skin contact. Human systemic effects by inhalation: conjunctiva irritation and unspecified respiratory system changes. A systemic irritant to humans. A skin and eye irritant. Flammable liquid when exposed to heat or flame. When heated to decomposition it emits acrid smoke and irritating fumes.

CPF500   CAS: 108-91-8   HR: 3
CYCLOHEXYLAMINE
DOT: UN 2357
mf: $C_6H_{13}N$ mw: 99.20
PROP: Liquid; strong, fishy odor. Mp: −17.7°, bp: 134.5°, flash p: 69.8°F, d: 0.865 @ 25°/25°, autoign temp: 560°F, vap d: 3.42. Misc in H_2O, org solvs.
SYNS: AMINOCYCLOHEXANE □ AMINO-HEXAHYDROBENZENE □ CHA □ CYCLOHEX-ANAMINE □ HEXAHYDROANILINE □ HEXA-HYDROBENZENAMINE
CONSENSUS REPORTS: IARC Cancer Review: Group 3 IMEMDT 7,178,87; Animal Limited Evidence IMEMDT 7,178,87. EPA Extremely Hazardous Substances List. EPA Genetic Toxicology Program. Reported in EPA TSCA Inventory.
OSHA PEL: TWA 10 ppm

ACGIH TLV: TWA 10 ppm; Not Classifiable as a Human Carcinogen
DFG MAK: 10 ppm (41 mg/m³)
DOT CLASSIFICATION: 8; Label: Corrosive, Flammable Liquid
SAFETY PROFILE: A poison by ingestion, skin contact, and intraperitoneal routes. Experimental teratogenic and reproductive effects. A severe human skin irritant. Can cause dermatitis and convulsions. Human mutation data reported. Questionable carcinogen. Flammable liquid. Dangerous fire hazard when exposed to heat, flame, or oxidizers. To fight fire, use alcohol foam, CO_2, dry chemical. When heated to decomposition it emits toxic fumes of NO_x.

CPI250   CAS: 95-33-0   HR: 3
N-CYCLOHEXYL-2-BENZOTHIAZOLE-
SULFENAMIDE
mf: $C_{13}H_{16}N_2S_2$ mw: 264.43
PROP: Light tan or buff powder. Mp: 103–104°, d: 1.27 @ 25°.
SYNS: ACCELERATOR CZ □ ACCICURE HBS □ BENZOTHIAZYL-2-CYCLOHEXYLSULFENAMIDE □ CBS □ CONAC A □ CONAC S □ CURAX □ N-CYCLO-HEXYL-2-BENZOTHIAZOLESULFENAMIDE □ N-CYCLOHEXYL-2-BENZOTHIAZYLSULFENAMIDE □ DELAC S □ DURAX □ EKAGOM CBS □ NOCCELER CZ □ PENNAC CBS □ RHODIFAX 16 □ ROYAL CBTS □ SANCELER CM-PO □ SANTOCURE □ SANTOCURE VULCANIZATION ACCELERATOR □ SOXINOL CZ □ SULFENAMIDE TS □ SULFENAX □ SULFENAX CB □ SULFENAX CB 30 □ SULFENAX CB/K □ THIOHEXAM □ VULCAFOR CBS □ VULCAFOR HBS □ VULKACIT C □ VULKACIT CZ □ VULKACIT CZ/C □ VULKACIT CZ/K
CONSENSUS REPORTS: Reported in EPA TSCA Inventory.
SAFETY PROFILE: A poison by intravenous route. Questionable carcinogen with experimental tumorigenic data. An experimental teratogen. Experimental reproductive effects. When heated to decomposition it emits toxic fumes of SO_x and NO_x.

CPI375   CAS: 32921-23-6   HR: 3
4-(CYCLOHEXYLCARBONYL)PYRIDINE
mf: $C_{12}H_{15}NO$ mw: 189.28

SYNS: CYCLOHEXYL 4-PYRIDYL KETONE □ KETONE, CYCLOHEXYL 4-PYRIDYL □ PYRIDINE, 4-(CYCLOHEXYLCARBONYL)-

DOT CLASSIFICATION: 3; Label: Flammable Liquid

SAFETY PROFILE: A poison by intraperitoneal route. A flammable liquid. When heated to decomposition it emits toxic vapors of NO_x.

CPN500   CAS: 3173-53-3   HR: 3
CYCLOHEXYL ISOCYANATE
DOT: UN 2488

mf: $C_7H_{11}NO$ mw: 125.19

PROP: Oil. Mp: 168–170°.

SYNS: CYCLOHEXANE, ISOCYANATO-(9CI) □ ISOCYANATOCYCLOHEXANE □ ISOCYANIC ACID, CYCLOHEXYL ESTER □ NSC-87419

CONSENSUS REPORTS: Reported in EPA TSCA Inventory.

DOT CLASSIFICATION: 6.1; Label: Poison; DOT Class: 6.1; Label: Poison, Flammable Liquid; DOT Class: 3; Label: Flammable Liquid, Poison

SAFETY PROFILE: Poison by intravenous and intraperitoneal routes. Mutation data reported. A flammable liquid when exposed to heat or flame. When heated to decomposition it emits toxic fumes of NO_x. See also CYANATES and ESTERS.

CPQ275   CAS: 6837-24-7   HR: 3
1-CYCLOHEXYL-2-PYRROLIDINONE
mf: $C_{10}H_{17}NO$ mw: 167.28

SYNS: N-CYCLOHEXYLPYRROLIDINONE □ N-CYCLOHEXYLPYRROLIDONE □ 2-PYRROLIDINONE, 1-CYCLOHEXYL-

CONSENSUS REPORTS: Reported in EPA TSCA Inventory.

SAFETY PROFILE: Poison by ingestion. Moderately toxic by inhalation and skin contact. A severe skin and eye irritant. When heated to decomposition it emits toxic fumes of NO_x.

CPQ625   CAS: 100-88-9   HR: 3
N-CYCLOHEXYLSULPHAMIC ACID
mf: $C_6H_{13}NO_3S$ mw: 179.26

PROP: Crystals; sweet-sour taste. Mp: 169–170°. Fairly strong acid. Very sparingly soluble in water. Slowly hydrolyzed by hot water.

SYNS: CYCLAMATE □ CYCLAMIC ACID □ CYCLOHEXANESULPHAMIC ACID □ CYCLOHEXYLAMIDOSULPHURIC ACID □ CYCLOHEXYLAMINESULPHONIC ACID □ CYCLOHEXYLSULFAMIC ACID (9CI) □ CYCLOHEXYLSULPHAMIC ACID □ HEXAMIC ACID □ SUCARYL □ SUCARYL ACID

CONSENSUS REPORTS: Reported in EPA TSCA Inventory.

SAFETY PROFILE: Suspected human carcinogen producing bladder tumors. Poison by intravenous route. Mildly toxic by ingestion. When heated to decomposition it emits toxic fumes of SO_x and NO_x.

CPR250   CAS: 98-12-4   HR: 3
CYCLOHEXYLTRICHLOROSILANE
DOT: UN 1763

mf: $C_6H_{11}Cl_3Si$ mw: 217.61

PROP: A liquid. Bp: 198.6–200°.

SYNS: CYCLOHEXANE, 1-(TRICHLOROSILYL)- □ SILANE, TRICHLOROCYCLOHEXYL-

CONSENSUS REPORTS: Reported in EPA TSCA Inventory.

DOT CLASSIFICATION: 8; Label: Corrosive

SAFETY PROFILE: A highly toxic and corrosive material. When heated to decomposition it emits toxic fumes of Cl^-. See also CHLOROSILANES.

CPR800   CAS: 121-82-4   HR: 3
CYCLONITE
DOT: UN 0072/UN 0118/UN 0391/UN 0483

mf: $C_3H_6N_6O_6$ mw: 222.15

PROP: White, crystalline powder. Mp: 202°.

SYNS: CYCLONITE, desensitized (UN 0483) (DOT) □ CYCLONITE, wetted (UN 0072) (DOT) □ CYCLOTRIMETHYLENENITRAMINE □ CYCLOTRIMETHYLENETRINITRAMINE □ CYCLOTRIMETHYLENE-TRINITRAMINE, desensitized (UN 0483) (DOT) □ CYCLOTRIMETHYLENETRINITRAMINE, wetted (UN 0072) (DOT) □ CYKLONIT □ ESAIDRO-1,3,5-TRINITRO-1,3,5-TRIAZINA (ITALIAN) □ HEKSOGEN (POLISH) □ HEXAHYDRO-1,3,5-TRINITRO-1,3,5-TRIAZIN (GERMAN) □ HEXAHYDRO-1,3,5-TRINITRO-1,3,5-TRIAZINE □ HEXOGEEN (DUTCH) □ HEXOGEN □ HEXOGEN, desensitized (UN 0483) (DOT) □ HEXOGEN (Explosive) □

HEXOGEN 5W □ HEXOGEN, wetted (UN 0072) (DOT) □ HEXOLITE □ HEXOLITE, dry or wetted with <15% water, by weight (UN 0118) (DOT) □ PBX(AF) 108 □ PBXW 108(E) □ RDX □ RDX, desensitized (UN 0483) (DOT) □ RDX and HMX MIXTURES, desensitized with not <10% phlegmatizer by weight (UN 0391) (DOT) □ RDX and HMX MIXTURES, wetted with not <15% water by weight (UN 0391) (DOT) □ RDX, wetted with not <15% water by weight (UN 0072) (DOT) □ T4 □ 1,3,5-TRIAZINE, HEXAHYDRO-1,3,5-TRINITRO-(9CI) □ TRIMETHYLEENTRINITRAMINE (DUTCH) □ TRIMETHYLENETRINITRAMINE □ sym-TRIMETHYL-ENETRINITRAMINE □ TRINITROCYCLOTRIMETHYL-ENE TRIAMINE □ 1,3,5-TRINITRO-1,3,5-TRIAZA-CYCLOHEXANE
CONSENSUS REPORTS: Reported in EPA TSCA Inventory.
OSHA PEL: TWA 1.5 mg/m³ (skin)
ACGIH TLV: TWA 0.5 mg/m³ (skin); Not Classifiable as a Human Carcinogen
DOT CLASSIFICATION: EXPLOSIVE 1.1D; Label: EXPLOSIVE 1.1D
SAFETY PROFILE: Poison by ingestion, intraperitoneal, and intravenous routes. An experimental teratogen. Other experimental reproductive effects. A corrosive irritant to skin, eyes, and mucous membranes. Cases of epileptiform convulsions have been reported from exposure. It is one of the most powerful high explosives in use today. Has more shattering power than TNT and is often mixed with TNT as a bursting charge for aerial bombs, mines, and torpedoes. It is easily initiated by mercury fulminate, which may be used as a booster. When heated to decomposition it emits toxic fumes of NO_x. See also AMINES, NITRATES, and EXPLOSIVES, HIGH.

CPR840 CAS: 12245-39-5 HR: 3
(1,5-CYCLOOCTADIENE)(2,4-PENTA-NEDIONATO)RHODIUM
mf: $C_{13}H_{19}O_2Rh$ mw: 310.23
PROP: Yellow crystals from pet ether. Mp: 125–128° decomp. Sol in hexane, $CHCl_3$. IDLH 100 mg/m³ (as Rh).
SYNS: ACETYLACETONATE-1,5-CYCLOOCTADIENE RHODIUM □ RHODIUM, ((1,2,5,6-eta)-1,5-CYCLOOCTA-DIENE)(2,4-PENTANEDIONATO-O,O')- □ RHOD-IUM, (1,5-CYCLOOCTADIENE)(2,4-PENTANEDIONATO)-
OSHA PEL: TWA 0.1 mg(Rh)/m³
ACGIH TLV: TWA 1 mg(Rh)/m³

SAFETY PROFILE: Poison by intraperitoneal route. When heated to decomposition it emits toxic fumes of Rh.

CPS000 CAS: 115-25-3 HR: 1
CYCLOOCTAFLUOROBUTANE
DOT: UN 1976
mf: C_4F_8 mw: 200.03
PROP: Colorless, odorless gas. Mp: −41.4°, bp: −6.04°, d (liquid): 1.513 @ −70°F.
SYNS: FC-C 318 □ FREON C-318 □ HALOCARBON C-138 □ OCTAFLUOROCYCLOBUTANE (DOT) □ PERFLUOROCYCLOBUTANE □ PROPELLANT C318 □ R-C 318
CONSENSUS REPORTS: EPA Genetic Toxicology Program. Reported in EPA TSCA Inventory.
DOT CLASSIFICATION: 2.2; Label: Nonflammable Gas
SAFETY PROFILE: Mildly toxic by ingestion and inhalation. Can cause slight transient effects at high concentrations. No anesthesia or central nervous system effects. Nonflammable gas. Mutation data reported. When heated to decomposition it emits highly toxic fumes of F^-.

CPU500 CAS: 542-92-7 HR: 3
1,3-CYCLOPENTADIENE
mf: C_5H_6 mw: 66.11
PROP: Colorless liquid. Mp: −85°, bp: 41–42°, d: 0.80475 @ 19°/4°, flash p: 77°F. Misc in EtOH and C_6H_6. IDLH 750 ppm.
SYNS: CYCLOPENTADIENE □ PENTOLE □ PYROPENTYLENE □ R-PENTINE
CONSENSUS REPORTS: Reported in EPA TSCA Inventory.
OSHA PEL: TWA 75 ppm
ACGIH TLV: TWA 75 ppm
DFG MAK: 75 ppm (210 mg/m³)
SAFETY PROFILE: Low toxicity by ingestion. A dangerous fire hazard when exposed to heat or flame; can react with oxidizing materials. Moderate explosion hazard in the form of gas when exposed to heat or by chemical reaction. It decomposes violently at high temperatures and pressures. Dimerization is highly exothermic. Explosive reaction with fuming nitric acid,

dinitrogen tetroxide, sulfuric acid. Reaction with nitrogen oxide + oxygen forms an explosive product. Reaction with oxygen forms a flame-sensitive explosive product. Ignites on contact with oxygen + ozone. Reacts vigorously on contact with potassium hydroxide. Incompatible with oxides of nitrogen, sulfuric acid. When heated to decomposition it emits acrid smoke and fumes.

CPV000   CAS: 12079-65-1   HR: 3
CYCLOPENTADIENYLMANGANESE TRICARBONYL
mf: $C_8H_5MnO_3$ mw: 204.07
PROP: Pale-yellow crystals with camphoraceous odor. Mp: 76.8–77.1°.
SYNS: MANGANESE CYCLOPENTADIENYL TRICARBONYL □ MCT
CONSENSUS REPORTS: Reported in EPA TSCA Inventory. Manganese and its compounds are on the Community Right-To-Know List.
OSHA PEL: TWA 0.1 mg(Mn)/m³ (skin)
ACGIH TLV: TWA 0.1 mg(Mn)/m³
SAFETY PROFILE: A poison by ingestion, inhalation, intraperitoneal, and intravenous routes. A mild narcotic which can damage kidneys. When heated to decomposition it emits acrid smoke and irritating fumes. See also MANGANESE COMPOUNDS and CARBON MONOXIDE.

CPV750   CAS: 287-92-3   HR: 3
CYCLOPENTANE
DOT: UN 1146
mf: C_5H_{10} mw: 70.15
PROP: Colorless liquid. Bp: 49.3°, fp: −93.7°, flash p: 19.4°F, autoign temp: 716°F, d: 0.745 @ 20°/4°, vap press: 400 mm @ 31.0°, vap d: 2.42.
SYN: PENTAMETHYLENE
CONSENSUS REPORTS: Reported in EPA TSCA Inventory.
OSHA PEL: TWA 600 ppm
ACGIH TLV: TWA 600 ppm
DOT CLASSIFICATION: 3; Label: Flammable Liquid

SAFETY PROFILE: Mildly toxic by ingestion and inhalation. High concentrations have narcotic action. A very dangerous fire hazard when exposed to heat or flame; can react with oxidizers. To fight fire, use foam, CO_2, dry chemical. When heated to decomposition it emits acrid smoke and fumes.

CPW300   CAS: 1679-07-8   HR: 3
CYCLOPENTANETHIOL
DOT: UN 1228/UN 3071
mf: $C_5H_{10}S$ mw: 102.21
SYNS: CYCLOPENTYL MERCAPTAN □ MERCAPTOCYCLOPENTANE
CONSENSUS REPORTS: Reported in EPA TSCA Inventory.
DOT CLASSIFICATION: 3; Label: Flammable Liquid, Poison (UN1228); DOT Class: 6.1; Label: Poison, Flammable Liquid (UN3071)
SAFETY PROFILE: A human poison. A flammable liquid. When heated to decomposition it emits toxic vapors of SO_x.

CPW500   CAS: 120-92-3   HR: 3
CYCLOPENTANONE
DOT: UN 2245
mf: C_5H_8O mw: 84.13
PROP: Liquid with a pleasant odor. Mp: −58.2°, bp: 130.6°, flash p: 79°F, d: 0.9509 @ 18°/4°, vap d: 2.3. Sparingly sol in H_2O.
SYNS: ADIPIC KETONE □ DUMASIN □ KETOCYCLOPENTANE □ KETOPENTAMETHYLENE
CONSENSUS REPORTS: Reported in EPA TSCA Inventory.
DOT CLASSIFICATION: 3; Label: Flammable Liquid
SAFETY PROFILE: Moderately toxic by intraperitoneal and subcutaneous routes. A skin and severe eye irritant. Dangerous fire hazard when exposed to heat or flame; can react with oxidizers. To fight fire, use alcohol foam, foam, CO_2, dry chemical. Potentially explosive reaction with hydrogen peroxide + nitric acid. When heated to decomposition it emits acrid smoke and fumes. See also KETONES.

c

CPX750   CAS: 142-29-0   HR: 3
CYCLOPENTENE
DOT: UN 2246
mf: C_5H_8 mw: 68.13
PROP: A liquid. Mp: −135.3°, bp: 44.242°,
flash p: −20°F, d: 0.77199 @ 20°.
CONSENSUS REPORTS: Reported in EPA
TSCA Inventory.
DOT CLASSIFICATION: 3; Label:
Flammable Liquid
SAFETY PROFILE: Moderately toxic by
ingestion and skin contact. A very
dangerous fire hazard when exposed to
flame or heat; can react with oxidizing
materials. Keep away from heat and open
flame. To fight fire, use foam, CO_2, dry
chemical.

CQC500   CAS: 6055-19-2   HR: 3
CYCLOPHOSPHAMIDE HYDRATE
mf: $C_7H_{15}Cl_2N_2O_2P \cdot H_2O$ mw: 279.13
SYNS: 1-BIS(2-CHLOROETHYL)AMINO-1-OXA-2-AZA-
5-OXAPHOSPHORIDINE MONOHYDRATE □ 2-(BIS(2-
CHLOROETHYL)AMINO)-1-OXA-3-AZA-2-
PHOSPHOCYCLOHEXANE 2-OXIDE MONOHYDRATE
□ (BIS(CHLORO-2-ETHYL)AMINO)-2-TETRAHYDRO-
3,4,5,6-OXAZAPHOSPHORINE-1,3,2-OXIDE-2-
MONOHYDRATE □ BIS(2-CHLOROETHYL)PHOSPHOR-
AMIDE CYCLIC PROPANOLAMIDE ESTER MONO-
HYDRATE □ N,N-BIS(β-CHLOROETHYL)-N',O-
PROPYLENEPHOSPHORIC ACID ESTER AMINE
MONOHYDRATE □ N,N-BIS(2-CHLOROETHYL)-
TETRAHYDRO-2H-1,3,2-OXAPHOSPHORIN-2-AMINE-2-
OXIDE MONOHYDRATE □ N,N-BIS(β-CHLOROETH-
YL)-N',O-TRIMETHYLENEPHOSPHORIC ACID ESTER
DIAMIDE MONOHYDRATE □ CB-4564 □ CLAFEN □
CYCLIC N',O-PROPYLENE ESTER of N,N-BIS(2-
CHLOROETHYL)PHOSPHORODIAMIDIC ACID MONO-
HYDRATE □ CYCLOPHOSPHAMIDE MONOHYDRATE
□ CYCLOPHOSPHAMIDUM □ CYCLOPHOSPHAN □
CYCLOPHOSPHANE □ CYCLOPHOSPHANUM □ CYTO-
PHOSPHAN □ CYTOXAN □ 2-(DI(2-CHLOROETHYL)-
AMINO)-1-OXA-3-AZA-2-PHOSPHACYCLOHEXANE-2-
OXIDE MONOHYDRATE □ N,N-DI-(2-CHLOROETHYL)-
AMINO-N,O-PROPYLENE PHOSPHORIC ACID ESTER
DIAMIDE MONOHYDRATE □ ENDOXANA □ ENDOX-
AN-ASTA □ ENDOXAN MONOHYDRATE □ ENDOXAN
R □ ENDUXAN □ GENOXAL □ MITOXAN □ NSC-
26271 □ PROCYTOX □ SEMDOXAN □ SENDOXAN □
SENDUXAN
CONSENSUS REPORTS: IARC Cancer
Review: Human Sufficient Evidence
IMEMDT 26,165,81; Human Limited
Evidence IMEMDT 9,135,75; Animal

Sufficient Evidence IMEMDT 9,135,75;
Animal Sufficient Evidence IMEMDT
26,165,81.
SAFETY PROFILE: Confirmed human
carcinogen. Poison by ingestion and
intravenous routes. Experimental
reproductive effects. Mutation data
reported. When heated to decomposition it
emits toxic fumes of Cl^-, PO_x, and NO_x.

CQC650   CAS: 50-18-0   HR: 3
CYCLOPHOSPHORAMIDE
mf: $C_7H_{15}Cl_2N_2O_2P$ mw: 261.11
PROP: Crystals. Mp: 41–45°. Water-sol; sltly
sol in org solvs.
SYNS: ASTA □ ASTA B518 □ B 518 □ N,N-BIS-(β-
CHLORAETHYL)-N',O-PROPYLEN-PHOSPHORSAEURE-
ESTER-DIAMID (GERMAN) □ 2-(BIS(2-CHLOROETHYL)-
AMINO)-2H-1,3,2-OXAAZAPHOSPHORINE 2-OXIDE □
N,N-BIS(2-CHLOROETHYL)-N'-(3-HYDROXYPROPYL)-
PHOSPHORODIAMIDIC ACID intramol. ESTER □ BIS(2-
CHLOROETHYL)PHOSPHORAMIDE-CYCLIC PROP-
ANOLAMIDE ESTER □ N,N-BIS(2-CHLOROETHYL)-
N',O-PROPYLENEPHOSPHORIC ACID ESTER DIAMIDE
□ N,N-BIS(2-CHLOROETHYL)TETRAHYDRO-2H-1,3,2-
OXAPHOSPHORIN-2-AMINE-2-OXIDE □ N,N-BIS(β-
CHLOROETHYL)-N',O-TRIMETHYLENEPHOSPHORIC
ACID ESTER DIAMIDE □ CB 4564 □ CLAFEN □
CLAPHENE □ CP □ CPA □ CTX □ CY □ CYCLOPHOS-
PHAMIDE □ CYCLOPHOSPHAMIDUM □ CYCLO-
PHOSPHAN □ CYCLOSTIN □ CYTOPHOSPHAN □
CYTOXAN □ N,N-DI-(2-CHLOROETHYL)-N,o-PROP-
YLENE-PHOSPHORIC ACID ESTER DIAMIDE □
ENDOXAN □ ENDOXANAL □ GENOXAL □
HEXADRIN □ MITOXAN □ NCI-C04900 □ NEOSAR □
NSC-26271 □ 2-H-1,3,2-OXAZAPHOSPHORINANE □
PROCYTOX □ RCRA WASTE NUMBER U058 □
SEMDOXAN □ SENDUXAN □ SK 20501 □ ZYKLO-
PHOSPHAMID (GERMAN)
CONSENSUS REPORTS: NTP 10th Report
on Carcinogens. IARC Cancer Review:
Group 1 IMEMDT 7,182,87; Human
Sufficient Evidence IMEMDT 26,165,81;
Animal Sufficient Evidence IMEMDT
26,165,81; IMEMDT 9,135,75; Human
Limited Evidence IMEMDT 9,135,75. NCI
Carcinogenesis Studies (ipr); Clear Evidence:
mouse, rat RRCRBU 52,1,75. EPA Genetic
Toxicology Program.
SAFETY PROFILE: Confirmed human
carcinogen producing leukemia, Hodgkin's
disease, gastrointestinal and bladder tumors.

Experimental carcinogenic, neoplastigenic, and teratogenic data. A human poison by ingestion and many other routes. Human systemic effects: kidney changes (hepatic dysfunction), leukopenia (reduced white blood cell count), nausea and alopecia (loss of hair), liver changes, agranulocytosis. Human reproductive and teratogenic effects by multiple routes: spermatogenesis, testicular changes, epididymis and sperm duct changes, menstrual cycle changes, fetal developmental abnormalities of the craniofacial area, musculoskeletal and cardiovascular systems. Experimental reproductive effects. Human mutation data reported. A powerful skin irritant. Used as an immunosuppressive agent in nonmalignant diseases. When heated to decomposition it emits highly toxic fumes of PO_x, NO_x, and Cl^-.

CQD750   CAS: 75-19-4   HR: 3
CYCLOPROPANE
DOT: UN 1027
mf: C_3H_6 mw: 42.09
PROP: Colorless gas with ethereal odor. Mp: $-126.6°$, bp: $-33.5°$, lel: 2.4%, uel: 10.4%, d: 1.879 g/L @ 0°, autoign temp: 932°F. Mod sol in H_2O; very sol in EtOH and Et_2O. A minor constituent of MAPP gas.
SYNS: CYCLOPROPANE, liquefied (DOT) □ TRIMETHYLENE
CONSENSUS REPORTS: IARC Cancer Review: Animal No Adequate Data IMEMDT 7,93,87. Reported in EPA TSCA Inventory.
DOT CLASSIFICATION: 2.1; Label: Flammable Gas
SAFETY PROFILE: Mutation data reported. Questionable carcinogen. High concentrations are narcotic. Human reproductive effects. Very dangerous fire hazard when exposed to heat or flame; can react with oxidizing materials. Explosion Hazard: Moderate in the form of vapor when exposed to heat or flame. To fight fire, stop flow of gas, then use CO_2, dry chemical, or water spray. When heated to decomposition it emits acrid smoke and fumes.

CQH100   CAS: 59865-13-3   HR: 3
CYCLOSPORIN A
mf: $C_{62}H_{111}N_{11}O_{12}$ mw: 1202.84
PROP: Needles from Me_2CO. Mp: 148–151°.
SYNS: ANTIBIOTIC S 7481F1 □ CICLOSPORIN □ CYCLOSPORIN □ CYCLOSPORINE □ CYCLOSPORINE A □ OL 27-400 □ S 7481F1 □ SANDIMMUN □ SANDIMMUNE
CONSENSUS REPORTS: NTP 10th Report on Carcinogens.
SAFETY PROFILE: Confirmed carcinogen producing Hodgkin's disease. Experimental reproductive effects. Poison by intraperitoneal and intravenous routes. Moderately toxic by ingestion. Human systemic effects by ingestion: increased body temperature, cyanosis. Mutation data reported. When heated to decomposition it emits toxic fumes of NO_x.

CQH250   CAS: 2691-41-0   HR: 3
CYCLOTETRAMETHYLENE TETRANI-TRAMINE
DOT: UN 0226/UN 0484
mf: $C_4H_8N_8O_8$ mw: 296.20
PROP: A solid. Mp: 286°.
SYNS: CYCLOTETRAMETHYLENETETRANITRAMINE, desensitized (UN 0483) (DOT) □ CYCLOTETRAMETHYL-ENETETRANITRAMINE (dry or unphlegmatized) (DOT) □ CYCLOTETRAMETHYLENETETRANITRAMINE, wetted (UN 0226) (DOT) □ HMX □ HMX (dry or unphlegmatized) (DOT) □ HMX, wetted (UN 0226) (DOT) □ beta HMY □ HW 4 □ LX 14-0 □ OCTOGEN □ OCTOGEN, desensitized (UN 0483) (DOT) □ OCTOGEN, wetted with not <15% water, by weight (UN 0226) (DOT) □ OKTOGEN □ TETRAMETHYLENETETRANITRAMINE
CONSENSUS REPORTS: Reported in EPA TSCA Inventory.
DOT CLASSIFICATION: Forbidden (dry or unphlegmatized); DOT Class: EXPLOSIVE 1.1D; Label: EXPLOSIVE 1.1D
SAFETY PROFILE: A poison by ingestion and intravenous routes. An explosive. Decomposes violently at 279°C. When heated to decomposition it emits toxic

fumes of NO_x. See also EXPLOSIVES, HIGH.

CQH650   CAS: 13121-70-5   HR: 3
CYHEXATIN
mf: $C_{18}H_{34}OSn$ mw: 385.21
PROP: White solid. Very sparingly sol in H_2O; sparingly sol in Me_2CO and MeOH; sol in $CHCl_3$. IDLH 80 mg/m³.
SYNS: DOWCO-213 □ ENT 27,395-X □ M 3180 □ PLICTRAN □ PLYCTRAN □ TCTH □ TRICYCLOHEXYLHYDROXYSTANNANE □ TRICYCLOHEXYLHYDROXYTIN □ TRICYCLOHEXYLTIN HYDROXIDE □ TRICYCLO-HEXYLZINNHYDROXID (GERMAN)
OSHA PEL: TWA 0.1 mg(Sn)/m³; TWA 5 mg/m³
ACGIH TLV: TWA 5 mg/m³; TWA 0.1 mg(Sn)/m³; STEL 0.2 mg/m³ (skin); Not Classifiable as a Human Carcinogen
NIOSH REL: (Organotin Compounds) TWA 0.1 mg(Sn)/m³
SAFETY PROFILE: Poison by ingestion, inhalation, and intraperitoneal routes. Moderately toxic by skin contact. Experimental reproductive effects. When heated to decomposition it emits acrid smoke and irritating fumes. See also TIN COMPOUNDS.

CQI000   CAS: 99-87-6   HR: 3
p-CYMENE
mf: $C_{10}H_{14}$ mw: 134.24
PROP: Colorless to pale-yellow liquid; odorless. Mp: −68.2°, bp: 176°, lel: 0.7% @ 100°, ULC: 30–35, flash p: 117°F (CC), d: 0.853, refr index: 1.489, autoign temp: 817°F, vap d: 4.62, vap press: 1 mm @ 17.3°, flash p (technical): 127°F, uel (technical): 5.6%. Found in nearly 100 volatile oils, including lemongrass, sage, thyme, coriander, star anise, and cinnamon (FCTXAV 12,385,74). Sol in alc, ether, acetone, benzene.
SYNS: CAMPHOGEN □ CYMENE □ CYMOL □ DOLCYMENE □ FEMA No. 2356 □ 4-ISOPROPYL-1-METHYLBENZENE □ p-ISOPROPYLTOLUENE □ p-METHYL-CUMENE □ p-METHYLISOPROPYL BENZENE □ 1-METHYL-4-ISOPROPYLBENZENE □ PARACYMENE □ PARACYMOL

CONSENSUS REPORTS: Reported in EPA TSCA Inventory.
SAFETY PROFILE: Mildly toxic by ingestion. Humans sustain central nervous system effects at low doses. A skin irritant. Flammable liquid. Explosion Hazard: Slight in the form of vapor. To fight fire, use foam, CO_2, dry chemical. When heated to decomposition it emits acrid smoke and fumes.

CQJ750   CAS: 56-17-7   HR: 3
CYSTAMINE DIHYDROCHLORIDE
mf: $C_4H_{12}N_2S_2 \cdot 2ClH$ mw: 225.22
PROP: Prisms from EtOH. Mp: 212°.
SYNS: AED □ 2-AMINOETHYL DISULFIDE DIHYDROCHLORIDE □ 2,2'-DITHIO-BIS-(ETHYLAMINE) DIHYDROCHLORIDE □ USAF CB-34
CONSENSUS REPORTS: Reported in EPA TSCA Inventory.
SAFETY PROFILE: A poison by subcutaneous and intraperitoneal routes. Experimental reproductive effects. When heated to decomposition it emits very toxic fumes of HCl, SO_x, and NO_x. See also SULFIDES.

CQK000   CAS: 52-90-4   HR: 2
l-CYSTEINE
mf: $C_3H_7NO_2S$ mw: 121.17
PROP: A solid. Mp: 178°. Sol in H_2O, AcOH, and NH_3. An amino acid derived from cystine, occurring naturally in the l-form, which will be considered here. Colorless crystals; sol in water, ammonium hydroxide, and acetic acid; insol in ether, acetone, benzene, carbon disulfide, and carbon tetrachloride.
SYNS: CYSTEIN □ CYSTEINE □ l-(+)-CYSTEINE □ HALF-CYSTEINE □ HALF-CYSTINE □ β-MERCAPTOALANINE □ THIOSERINE
CONSENSUS REPORTS: Reported in EPA TSCA Inventory. EPA Genetic Toxicology Program.
SAFETY PROFILE: Moderately toxic by ingestion, intraperitoneal, and subcutaneous routes. Experimental reproductive effects. Human mutation data reported. When

heated to decomposition it emits very toxic fumes of SO$_x$ and NO$_x$.

CQK250   CAS: 52-89-1   HR: 2
l-CYSTEINE HYDROCHLORIDE
mf: C$_3$H$_7$NO$_2$S•ClH   mw: 157.63
PROP: White crystalline powder; characteristic acetic taste. Mp: 175° (decomp). Sol in water, alc.
SYNS: CYSTEINE CHLORHYDRATE □ CYSTEINE HYDROCHLORIDE □ l-CYSTEINE HYDROCHLORIDE □ l-CYSTEINE MONOHYDROCHLORIDE (FCC)
CONSENSUS REPORTS: Reported in EPA TSCA Inventory.
SAFETY PROFILE: Moderately toxic by intraperitoneal, intravenous, and possibly other routes. Mutation data reported. When heated to decomposition it emits very toxic fumes of NO$_x$, SO$_x$, and Cl$^-$.

CQK325   CAS: 56-89-3   HR: 1
l-CYSTINE
mf: C$_6$H$_{12}$N$_2$O$_4$S$_2$ mw: 240.30
PROP: Plates or prisms. Mp: 258–261° (decomp) (sealed tube). Sol in hot H$_2$O, mineral acids, and aq alkali. Naturally occurring levorotatory form. Colorless to white hexagonal tablets from water. Decomp 260–261°. d-Cystine: Crystals. Sltly sol in water. dl-Cystine, the synthetic racemic form: Crystals. Sltly sol in water. meso-Cystine, the internally compensated form: Crystals. Sltly sol in water.
SYNS: CYSTEINE DISULFIDE □ CYSTIN □ (–)-CYSTINE □ CYSTINE ACID □ DICYSTEINE □ β,β'-DITHIODIALANINE □ GELUCYSTINE □ OXIDIZED l-CYSTEINE
SAFETY PROFILE: Low toxicity by ingestion. When heated to decomposition it emits toxic fumes of PO$_x$ and SO$_x$.

CQL250   CAS: 115-93-5   HR: 3
CYTHIOATE
mf: C$_8$H$_{12}$NO$_5$PS$_2$ mw: 297.30
PROP: A solid. Insol in H$_2$O; sol in C$_6$H$_6$, Me$_2$CO, Et$_2$O, and EtOH.
SYNS: AC 26,691 □ AMERICAN CL-26691 □ AMERICAN CYANAMID CL-26,691 □ O-(4-(AMINOSULFONYL)PHEN-YL) O,O-DIMETHYL PHOSPHOROTHIOATE □ BENZENE-SULFONAMIDE, p-HYDROXY-, O-ESTER with

O,O-DIMETHYL PHOSPHOROTHIOATE □ CL 26691 □ CYFLEE □ O,O-DIMETHYL O-p-SULFAMOYLPHENYL PHOSPHOROTHIOATE □ ENT 25,640 □ PHOSPHORO-THIOIC ACID, O-(4-(AMINOSULFONYL)PHENYL) O,O-DIMETHYL ESTER (9CI) □ PROBAN
CONSENSUS REPORTS: Reported in EPA TSCA Inventory.
SAFETY PROFILE: Poison by ingestion. An insecticide. When heated to decomposition it emits very toxic fumes of NO$_x$, PO$_x$, and SO$_x$.

CQM325   CAS: 9007-43-6   HR: D
CYTOCHROME C
PROP: Reduced form crystallizes as separate needles; oxidized form as rosettes. Mol wt about 13,000. Cytochrome c2: Needles changing to squares. Mol wt about 13,000. Cytochrome c3: Needles. Mol wt 11,300.
SYNS: CROMOCI □ CYTOREST □ FERRICYTO-CHROME C □ FERROCYTOCHROME C □ HEMATIN-PROTEIN □ HORSE-CYTOCHROME C □ HORSE HEART CYTOCHROME C □ LANDRAX □ MYOHEMA-TIN □ NITROSYLFERRICYTOCHROME C
CONSENSUS REPORTS: Reported in EPA TSCA Inventory.
SAFETY PROFILE: Experimental reproductive effects. Mutation data reported. When heated to decomposition it emits toxic fumes of NO$_x$.

CQN000   CAS: 4465-94-5   HR: 3
CYTOXAL ALCOHOL
mf: C$_7$H$_{17}$Cl$_2$N$_2$O$_3$P•C$_6$H$_{13}$N   mw: 378.33
SYNS: 2-(BIS(2-CHLOROETHYL)AMINO)TETRA-HYDROOXAZAPHOSPHORINE CYCLOHEXYLAMINE SALT □ N,N-BIS(2-CHLOROETHYL)-N'-(3-HYDROXY-PROPYL)PHOSPHORODIAMIDATE, CYCLOHEXYL-AMMONIUM SALT □ N,N-BIS(2-CHLOROETHYL)-N'-3-PHOSPHORODIAMIDIC ACID HYDROXYLPROPYL-CYCLOHEXYLAMINE SALT □ CYTOXYL ALCOHOL CYCLOHEXYLAMMONIUM SALT □ NCI-C04922 □ NSC-52695
CONSENSUS REPORTS: NCI Carcinogenesis Studies (ipr); Clear Evidence: mouse, rat RRCRBU 52,1,75.
SAFETY PROFILE: Suspected carcinogen with experimental carcinogenic and tumorigenic data. Poison by intravenous route. Moderately toxic by ingestion and

subcutaneous routes. Experimental teratogenic and reproductive effects. Human mutation data reported. When heated to decomposition it emits very toxic fumes of NO_x, NH_3, PO_x, and Cl^-. See also ALCOHOLS.

D

DAB600 CAS: 4342-03-4 HR: 3
DACARBAZINE
mf: $C_6H_{10}N_6O$ mw: 182.22

PROP: Ivory microcrystals. Mp: 250–255° (decomp).

SYNS: DETICENE □ DIC □ (DIMETHYLTRIAZENO)-IMIDAZOLECARBOXAMIDE □ 4-(DIMETHYL-TRIAZENO)-IMIDAZOLE-5-CARBOXAMIDE □ 4-(3,3-DIMETHYL-1-TRIAZENO)IMIDAZOLE-5-CARBOX-AMIDE □ 4-(5)-(3,3-DIMETHYL-1-TRIAZENO)IMIDAZ-OLE-5(4)-CARBOXAMIDE □ 5-(DIMETHYLTRIAZENO)-IMIDAZOLE-4-CARBOXAMIDE □ 5-(3,3-DIMETHYL-TRIAZENO)IMIDAZOLE-4-CARBOXAMIDE □ 5-(3,3-DIMETHYL-1-TRIAZENO)IMIDAZOLE-4-CARBOX-AMIDE □ 5-(3,3-DIMETHYL-1-TRIAZENYL)-1H-IMIDAZOLE-4-CARBO-XAMIDE □ DTIC □ DTIC-DOME □ NCI-C04717 □ NSC-45388

CONSENSUS REPORTS: NTP 10th Report on Carcinogens. IARC Cancer Review: Group 2B IMEMDT 7,184,87; Human Limited Evidence IMEMDT 26,203,81; Animal Sufficient Evidence IMEMDT 26,203,81. NCI Carcinogenesis Studies (ipr); Clear Evidence: mouse, rat RRCRBU 52,1,75. EPA Genetic Toxicology Program.

SAFETY PROFILE: Confirmed carcinogen with experimental carcinogenic and tumorigenic data. Poison by intraperitoneal and parenteral routes. Moderately toxic by ingestion and intravenous routes. Experimental teratogenic effects. Human systemic effects by intravenous route: nausea or vomiting, leukopenia (reduced white blood cell count), and changes in dehydrogenase enzymatic activity. Mutation data reported. When heated to decomposition it emits toxic fumes of NO_x.

DAB630 CAS: 73245-91-7 HR: 3
DACTIMICIN SULFATE
mf: $C_{18}H_{36}N_6O_6 \cdot 2H_2O_4S \cdot H_2O$ mw: 614.72

SYNS: ANTIBIOTIC SF 2052 SULFATE □ l-CHIRO-INOSITOL, 4-AMINO-1,4-DIDEOXY-3-o-(2,6-DIAMINO-2,3,4,6,7-PENTADEOXY-β-l-LYXO-HEPTOPYRANOSYL)-6-o-METHYL-1-(2-(FORMIMIDOYLAMINO)-N-METHYL-ACETAMIDO)-, SULFATE (1:2),HYDRATE □ SF-2052 SULFATE

SAFETY PROFILE: A poison by intravenous and intramuscular route. Moderately toxic by subcutaneous route. When heated to decomposition it emits toxic vapors of NO_x and SO_x.

DAB879 CAS: 469-62-5 HR: 3
DARVON
mf: $C_{22}H_{29}NO_2$ mw: 339.52

PROP: Crystals from pet ether. Mp: 75–76°.

SYNS: DEXTROPROPOXYPHENE □ α-(+)-4-DIMETHYLAMINO-1,2-DIPHENYL-3-METHYL-2-BUTANOL PROPIONATE ESTER □ DOLENE □ DOLOXENE □ (+)-PROPOXYPHENE □ d-PROPOXYPHENE □ PROXAGESIC □ SK 65

SAFETY PROFILE: Poison by ingestion, intraperitoneal, subcutaneous, and intravenous routes. Human systemic effects by ingestion: change in cardiac rate, respiratory depression, and coma. When heated to decomposition it emits toxic fumes of NO_x.

DAC000 CAS: 20830-81-3 HR: 3
DAUNOMYCIN
mf: $C_{27}H_{29}NO_{10}$ mw: 527.57

PROP: Thin, red needles. Mp: 190° (decomp). Isolated from cultures of a *Streptomyces* (CNREA8 32,1029,72).

SYNS: ACETYLADRIAMYCIN □ CERUBIDIN □ DAUNAMYCIN □ DAUNORUBICIN □ DAUNORUBIC-INE □ DM □ FI6339 □ LEUKAEMOMYCIN C □ NCI-C04693 □ NSC-82151 □ RCRA WASTE NUMBER U059 □ RP 13057 □ 13,057 R.P. □ RUBIDOMYCIN □ RUBI-DOMYCINE □ RUBOMYCIN C □ RUBOMYCIN C 1 □ STREPTOMYCES PEUCETIUS

CONSENSUS REPORTS: IARC Cancer Review: Group 2B IMEMDT 7,56,87; Animal Sufficient Evidence IMEMDT

10,145,76. NCI Carcinogenesis Studies (ipr); Clear Evidence: rat CANCAR 40,1935,77; No Evidence: mouse CANCAR 40,1935,77. EPA Genetic Toxicology Program.

SAFETY PROFILE: Confirmed carcinogen with experimental carcinogenic, neoplastigenic, and tumorigenic data. Human poison by ingestion. Experimental poison by subcutaneous, intravenous, and intraperitoneal routes. Experimental teratogenic and reproductive effects. Human mutation data reported. When heated to decomposition it emits toxic fumes of NO_x. See also DAUNOMYCIN HYDROCHLORIDE.

DAC450   CAS: 12011-76-6   HR: 3
DAWSONITE

mf: $CH_2AlO_5 \cdot Na$ mw: 144.00

SYN: CRYSTALLINE DEHYDROXY SODIUM ALUMINUM, CARBONATE

DFG MAK: Animal Carcinogen, Suspected Human Carcinogen

SAFETY PROFILE: Suspected carcinogen with experimental neoplastigenic data.

DAD200   CAS: 50-29-3   HR: 3
DDT

mf: $C_{14}H_9Cl_5$ mw: 354.48

PROP: Colorless crystals or white to sltly off-white powder. Odorless or with slight aromatic odor. Mp: 108.5–109°. IDLH 500 mg/m³.

SYNS: AGRITAN □ ANOFEX □ ARKOTINE □ AZOTOX □ α,α-BIS(p-CHLOROPHENYL)-β,β,β-TRICHLOROETHANE □ 1,1-BIS-(p-CHLOROPHENYL)-2,2,2-TRICHLOROETHANE □ 2,2-BIS(p-CHLOROPHEN-YL)-1,1,1-TRICHLOROETHANE □ BOSAN SUPRA □ BOVIDERMOL □ CHLOROPHENOTHAN □ CHLORO-PHEN-OTHANE □ CHLOROPHENOTOXUM □ CITOX □ CLOFENO-TANE □ p,p'-DDT □ DEDELO □ DEOVAL □ DETOX □ DETOXAN □ DIBOVAN □ DICHLORODI-PHENYLTRICHLORO-ETHANE □ DICHLORODIPHEN-YLTRICHLOROETHANE (DOT) □ p,p'-DICHLORO-DIPHENYLTRICHLOROETHANE □ 4,4'-DICHLORODI-PHENYLTRICHLOROETHANE □ DICOPHANE □ DIDIGAM □ DIDIMAC □ DIPHENYLTRICHLORO-ETHANE □ DODAT □ DYKOL □ ENT 1,506 □ ESTONATE □ GENITOX □ GESAFID □ GESAPON □ GESAREX □ GESAROL □ GUESAPON □ GUESAROL □ GYRON □ HAVERO-EXTRA □ HILDIT □ IVORAN □

IXODEX □ KOPSOL □ MICRO DDT 75 □ MUTOXIN □ NCI-C00464 □ NEOCID □ PARA-CHLOROCIDUM □ PEB1 □ PENTACHLORIN □ PENTECH □ PPZEIDAN □ R50 □ RCRA WASTE NUMBER U061 □ RUKSEAM □ SANTOBANE □ TECH DDT □ 1,1,1-TRICHLOOR-2,2-BIS(4-CHLOOR FENYL)-ETHAAN (DUTCH) □ 1,1,1-TRICHLOR-2,2-BIS(4-CHLOR-PHENYL)-AETHAN (GERMAN) □ TRICHLOROBIS (4-CHLOROPHENYL) ETHANE □ 1,1,1-TRICHLORO-2,2-DI(4-CHLOROPHEN-YL)ETHANE □ 1,1'-(2,2,2-TRICHLOROETHYLIDENE)-BIS(4-CHLOROBENZENE) □ 1,1,1-TRICLORO-2,2-BIS(4-CLORO-FENIL)-ETANO (ITALIAN) □ ZEIDANE □ ZERDANE

CONSENSUS REPORTS: NTP 10th Report on Carcinogens. IARC Cancer Review: Group 2B IMEMDT 7,186,87; Animal Sufficient Evidence IMEMDT 5,83,74; Human Inadequate Evidence IMEMDT 5,83,74. NCI Carcinogenesis Bioassay (feed); No Evidence: mouse, rat NCITR* NCI-CG-TR-131,78. Reported in EPA TSCA Inventory. EPA Genetic Toxicology Program.

OSHA PEL: TWA 1 mg/m³ (skin)

ACGIH TLV: TWA 1 mg/m³; Animal Carcinogen

NIOSH REL: (DDT) TWA 0.5 mg/m³; avoid skin contact

DFG MAK: 1 mg/m³

SAFETY PROFILE: Confirmed carcinogen with experimental carcinogenic, neoplastigenic, tumorigenic, and teratogenic data. Human poison by ingestion. Experimental poison by ingestion, skin contact, subcutaneous, intravenous, and intraperitoneal routes. Experimental reproductive effects. Human systemic effects by ingestion: anesthetic, convulsions, headache, analgesia, cardiac arrhythmias, nausea or vomiting, sweating, and unspecified pulmonary changes. Human mutation data reported. An insecticide. When heated to decomposition it emits toxic fumes of Cl^-. See also CHLORINAT-ED HYDROCARBONS, AROMATIC.

A dose of 20 g has proved highly dangerous though not fatal to a human. This dose was taken by 5 persons who vomited an unknown portion of the material and even so recovered only incompletely after 5 weeks. Smaller doses produced less important symptoms with relatively rapid recovery. Experimental ingestion of 1.5 g resulted in great discomfort and moderate neurological changes including paresthesia, tremor, moderate ataxia, exaggeration of part of the reflexes, headache, and fatigue. Vomiting followed only after 11 hours. Recovery was complete on the following day. The fatal dose of DDT for humans is not known. Judging from the literature, no one has ever been killed by DDT in the absence of other insecticides and/or a variety of toxic solvents. However, these common solvent formulations are highly fatal when taken in small doses, partly because of the toxicity of the solvent, and perhaps because of the increased absorbability of the DDT; several fatal cases in humans have been reported. Little is known of the hazard of chronic DDT poisoning. Human volunteers have ingested up to 35 mg/day for 21 months with no ill effects.

DDT and some of its degradation products, particularly DDE, are stored in fat. This storage effect leads to a concentration of DDT at higher levels of the food chain. DDT stored in the fat is at least largely inactive since a greater total dose may be stored in an experimental animal than is sufficient as a lethal dose for that same animal if given at one time. A study based on 75 human cases reported an average of 5.3 ppm of DDT stored in the fat. A higher content of DDT and its derivatives (up to 434 ppm of DDE and 648 ppm of DDT) was found in workers who had very extensive exposure. Without exception, the samples were taken from persons who were either asymptomatic or suffering from some disease completely unrelated to DDT. Careful hospital examination of workers who had been very extensively exposed and who had volunteered for examination revealed no abnormality that could be attributed to DDT. Much higher levels have been found in humans than have been observed in the fat of experimental animals that were apparently asymptomatic. DDT stored in the fat is eliminated only very gradually when further dosage is discontinued. However, weight loss can speed the release of this stored DDT (and DDE) into the blood. After a single dose, the secretion of DDT in the milk and its excretion in the urine reach their height within a day or two and continue at a lower level thereafter.

DAE400   CAS: 17702-41-9   HR: 3
DECABORANE
DOT: UN 1868
mf: $B_{10}H_{14}$ mw: 122.24
PROP: Colorless needles or crystals. Solid by sublimation. Mp: 99.6°, d: 0.94 (solid), d: 0.78 (liquid @ 100°), bp: 213°, vap press: 19 mm @ 100°. Sol in CS_2. IDLH 15 mg/m^3.
SYN: DECABORANE(14)
CONSENSUS REPORTS: Reported in EPA TSCA Inventory. EPA Extremely Hazardous Substances List.
OSHA PEL: TWA 0.05 ppm; STEL 0.15 ppm (skin)
ACGIH TLV: TWA 0.05 ppm; STEL 0.15 ppm (skin)
DFG MAK: 0.05 ppm (0.25 mg/m^3)
DOT CLASSIFICATION: 4.1; Label: Flammable Solid, Poison
SAFETY PROFILE: Poison by inhalation, ingestion, skin contact, and intraperitoneal routes. Ignites in O_2 at 100°C. Forms impact-sensitive explosive mixtures with ethers (e.g., dioxane) and halocarbons (e.g., carbon tetrachloride). Incompatible with dimethyl sulfoxide. When heated to decomposition it emits toxic fumes of boron oxides. See also BORON COMPOUNDS and BORANES.

DAE450   CAS: 25152-84-5   HR: 2
trans,trans-2,4-DECADIENAL
mf: $C_{10}H_{16}O$ mw: 152.23
PROP: Yellow liquid; chicken fat odor. D: 0.806–0.876, refr index: 1.514–1.516, bp:

58–61° @ 0.05 mm, flash p: 212°F. Sol in alc, fixed oils; insol in water @ 104°.
SYNS: (E,E)-2,4-DECADIENAL □ (2E,4E)-DECADIENAL □ (2E,4E)-2,4-DECADIENAL □ FEMA No. 3135 □ HEPTENYL ACROLEIN
SAFETY PROFILE: Combustible liquid. When heated to decomposition it emits acrid smoke and irritating fumes.

DAE525   CAS: 2392-39-4   HR: 3
DECADRON PHOSPHATE
mf: $C_{22}H_{28}FO_8P•2Na$ mw: 516.45
PROP: Crystals. Mp: 233–235°.
SYNS: DEXACORT □ DEXADRESON □ DEXAGRO □ DEXAMETHASONE DISODIUM PHOSPHATE □ DEXAMETHASONE SODIUM PHOSPHATE □ DEXAMETHAZONE SODIUM PHOSPHATE □ DISODIUM DEXAMETHASONE PHOSPHATE □ 9-FLUORO-11-β,17,21-TRIHYDROXY-16-α-METHYL-PREGNA-1,4-DIENE-3,20-DIONE-21-(DIHYDROGEN PHOSPHATE) DISODIUM SALT □ MEGACORT □ SODIUM DEXAMETHASONE PHOSPHATE □ SOLDESAM □ SOLU-DECADRON □ SPERSADOX □ TURBINAIRE
CONSENSUS REPORTS: Reported in EPA TSCA Inventory.
SAFETY PROFILE: Poison by intravenous route. Moderately toxic by ingestion and intraperitoneal routes. Human systemic effects by intravenous route: peritonitis, central nervous system, and gastrointestinal changes. An experimental teratogen. Other experimental reproductive effects. When heated to decomposition it emits toxic fumes of F^-, PO_x, and Na_2O.

DAE600   CAS: 15652-38-7   HR: 3
DECAFENTIN
mf: $C_{28}H_{36}P•C_{18}H_{15}BrClSn$ mw: 868.99
SYNS: A-36 □ CELA A-36 □ DECYLTRIPHENYLPHOSPHONIUM BROMOCHLORO-TRIPHENYLSTANNATE □ (DECYL-TRIPHENYL-PHOSPHONIUM)-TRIPHENYL-BROM-CHLOR-STANNAT (GERMAN) □ STANNOPLUS □ STANNORAM □ STANNPLOUS
OSHA PEL: TWA 0.1 mg(Sn)/m³ (skin)
ACGIH TLV: TWA 0.1 mg(Sn)/m³; STEL 0.2 mg(Sn)/m³ (skin).
NIOSH REL: (Organotin Compounds) TWA 0.1 mg(Sn)/m³
SAFETY PROFILE: Poison by skin contact. Moderately toxic by ingestion. When heated

to decomposition it emits very toxic fumes of PO_x, Br^-, and Cl^-. A pesticide. See also TIN COMPOUNDS.

DAE800   CAS: 91-17-8   HR: 3
DECAHYDRONAPHTHALENE
DOT: UN 1147
mf: $C_{10}H_{18}$ mw: 138.28
PROP: Water-white liquid with slight menthol odor. Mp: (cis) −43.3°, mp: (trans) −30.7°, bp: (cis) 195.6°, bp: (trans) 187.3°, flash p: 136°F, (CC), autoign temp: 482°F, vap press: (cis) 1 mm @ 22.5°, (trans) 10 mm @ 47.2°, d: (cis) 0.8963 @ 20°/4°, vap d: 4.76, lel: 0.7% @ 212°F, uel: 4.9% @ 212°F.
SYNS: BICYCLO(4.4.0)DECANE □ DEC □ DECALIN □ DECALIN (DOT) □ DECALIN SOLVENT □ DE-KALIN □ DEKALINA (POLISH) □ NAPHTHALANE □ NAPHTH-ANE □ PERHYDRONAPHTHALENE
CONSENSUS REPORTS: Reported in EPA TSCA Inventory.
DOT CLASSIFICATION: 3; Label: Flammable Liquid
SAFETY PROFILE: Moderately toxic by inhalation and ingestion. Questionable carcinogen with experimental carcinogenic and neoplastigenic data. Mildly toxic by skin contact. Human systemic effects by inhalation: conjunctiva irritation, unspecified olfactory and pulmonary system changes. Can cause kidney damage. Mutation data reported. A skin and eye irritant. Flammable liquid when exposed to heat or flame, can react with oxidizing materials. To fight fire, use foam, CO_2, dry chemical. When heated to decomposition it emits acrid smoke and fumes.

DAF100   CAS: 10519-11-6   HR: 1
DECAHYDRO-β-NAPHTHYL ACETATE
mf: $C_{12}H_{20}O_2$ mw: 196.32
SYN: 2-NAPHTHOL, DECAHYDRO-, ACETATE
CONSENSUS REPORTS: Reported in EPA TSCA Inventory.
SAFETY PROFILE: A skin irritant. When heated to decomposition it emits acrid smoke and irritating fumes.

DAF150   CAS: 10519-12-7   HR: 1
DECAHYDRO-β-NAPHTHYL FORMATE
mf: $C_{11}H_{18}O_2$ mw: 182.29
SYNS: DECALINYL FORMATE □ 2-NAPHTHALENOL, DECAHYDRO-, FORMATE □ 2-NAPHTHOL, DECAHYDRO-, FORMATE □ SANTALOZONE
CONSENSUS REPORTS: Reported in EPA TSCA Inventory.
SAFETY PROFILE: A skin irritant. When heated to decomposition it emits acrid smoke and irritating fumes.

DAF200   CAS: 705-86-2   HR: 1
Δ-DECALACTONE
mf: $C_{10}H_{18}O_2$ mw: 170.28
PROP: Colorless liquid; coconut, fruity odor, butterlike on dilution. Refr index: 1.456–1.459. Very sol in alc and propylene glycol; insol in water @ 281°.
SYNS: AMYL-Δ-VALEROLACTONE □ DECANOLIDE-1,5 □ FEMA No. 2361
CONSENSUS REPORTS: Reported in EPA TSCA Inventory.
SAFETY PROFILE: A skin and eye irritant. When heated to decomposition it emits acrid smoke and irritating fumes.

DAG000   CAS: 112-31-2   HR: 2
1-DECANAL
mf: $C_{10}H_{20}O$ mw: 156.30
PROP: Colorless to light-yellow liquid; floral, fatty odor. D: 0.830 @ 15°/4°, bp: 208°, flash p: 185°F, mp: −5° (approx). Found in over 50 sources including citrus oils, citronella, and lemongrass (FCTXAV 11,477,73). Sol in 80% alc, fixed oils, volatile oils, and mineral oils; insol in water and glycerol.
SYNS: ALDEHYDE C-10 □ C-10 ALDEHYDE □ CAPRALDEHYDE □ CAPRIC ALDEHYDE □ CAPRIN-ALDEHYDE □ CAPRINIC ALDEHYDE □ DEC-ALDEHYDE □ n-DECALDEHYDE □ DECANAL □ n-DECANAL □ DECANALDEHYDE □ DECYL ALDEHYDE □ n-DECYL ALDEHYDE □ 1-DECYL ALDEHYDE □ DECYLIC ALDEHYDE □ FEMA No. 2362
CONSENSUS REPORTS: Reported in EPA TSCA Inventory.
SAFETY PROFILE: Moderately toxic by ingestion. Mildly toxic by skin contact. A severe skin irritant. Mutation data reported.

Combustible liquid. When heated to decomposition it emits acrid smoke and irritating fumes. See also ALDEHYDES.

DAG200   CAS: 112-31-2   HR: 2
1-DECANAL (mixed isomers)
SYN: FEMA No. 2362
CONSENSUS REPORTS: Reported in EPA TSCA Inventory.
SAFETY PROFILE: Moderately toxic by ingestion. A severe skin irritant. See also 1-DECANAL. When heated to decomposition it emits acrid smoke and fumes.

DAG400   CAS: 124-18-5   HR: 3
DECANE
DOT: UN 2247
mf: $C_{10}H_{22}$ mw: 142.29
PROP: Liquid. Mp: −30°, bp: 174°, lel: 0.8%, uel: 5.4%, flash p: 115°F (CC), d: 0.730 @ 20°/4°, autoign temp: 410°F, vap press: 1 mm @ 16.5°, vap d: 4.90.
SYN: n-DECANE (DOT)
CONSENSUS REPORTS: Reported in EPA TSCA Inventory.
DOT CLASSIFICATION: 3; Label: Flammable Liquid
SAFETY PROFILE: Questionable carcinogen with experimental tumorigenic data. A simple asphyxiant. Narcotic in high concentrations. Flammable liquid when exposed to heat or flame. Can react with oxidizing materials. Moderately explosive in its vapor form. To fight fire, use foam, CO_2, dry chemical. Emitted from modern building materials (CENEAR 69,22,91). See also ARGON for discussion of asphyxiants.

DAG600   CAS: 2016-57-1   HR: 3
1-DECANEAMINE
DOT: UN 2733/UN 2734
mf: $C_{10}H_{23}N$ mw: 157.34
PROP: Liquid. Mp: 15.0°, bp: 218.0–218.5° @ 747 mm, flash p: 210°F, d: 0.79 @ 20°, vap d: 5.5.
SYNS: 1-AMINODECANE □ DECYLAMINE

CONSENSUS REPORTS: Reported in EPA TSCA Inventory.

DOT CLASSIFICATION: 8; Label: Corrosive, Flammable Liquid (UN 2734); DOT Class: 3; Label: Flammable Liquid, Corrosive (UN 2733)

SAFETY PROFILE: Poison by ingestion and skin contact. A skin irritant. Flammable when exposed to heat or flame; can react with oxidizing materials. To fight fire, use alcohol foam, foam, dry chemical. When heated to decomposition it emits toxic fumes of NO_x. See AMINES and AMINES, FATTY.

DAH400   CAS: 334-48-5   HR: 3
DECANOIC ACID
mf: $C_{10}H_{20}O_2$ mw: 172.30

PROP: White crystals or needles with an unpleasant odor. D: 0.8858 @ 40°/4°, bp: 270°, mp: 31.5°. Sol in EtOH, Et_2O, Me_2CO, C_6H_6, $CHCl_3$, and alkalies.

SYNS: CAPRIC ACID □ n-CAPRIC ACID □ CAPRINIC ACID □ CAPRYNIC ACID □ n-DECANOIC ACID □ n-DECOIC ACID □ DECYLIC ACID □ n-DECYLIC ACID □ HEXACID 1095 □ NEO-FAT 10 □ 1-NONANECARBO-XYLIC ACID

CONSENSUS REPORTS: Reported in EPA TSCA Inventory.

SAFETY PROFILE: Poison by intravenous route. Mutation data reported. A moderate skin irritant. When heated to decomposition it emits acrid smoke and irritating fumes.

DAH450   CAS: 10024-58-5   HR: 1
DECANOIC ACID, DIESTER with
TRIETHYLENE GLYCOL (mixed
isomers)
mf: $C_{26}H_{50}O_6$ mw: 458.76

SYN: DIDECANOYLTRIETHYLENE GLYCOL ESTER (mixed isomers)

CONSENSUS REPORTS: Reported in EPA TSCA Inventory.

SAFETY PROFILE: Mildly toxic by ingestion and skin contact. When heated to decomposition it emits acrid smoke and irritating fumes.

DAI350   CAS: 3913-71-1   HR: 2
2-DECENAL
mf: $C_{10}H_{18}O$ mw: 154.28

PROP: Sltly yellow liquid; orange odor. D: 0.836–0.846, refr index: 1.452–1.457. Sol in alc, fixed oils; insol in water.

SYNS: trans-2-DECEN-1-AL □ DECENALDEHYDE □ FEMA No. 2366

CONSENSUS REPORTS: Reported in EPA TSCA Inventory.

SAFETY PROFILE: Moderately toxic by skin contact. Mildly toxic by ingestion. A severe skin irritant. When heated to decomposition it emits acrid smoke and fumes. See also ALDEHYDES.

DAI360   CAS: 21662-09-9   HR: 1
cis-4-DECENAL
mf: $C_{10}H_{18}O$ mw: 154.28

PROP: Colorless to sltly yellow liquid; fatty, orangelike odor. D: 0.847, refr index: 1.442–1.444. Sol in alc, fixed oils; insol in water.

SYNS: cis-4-DECEN-1-AL (FCC) □ FEMA No. 3264

CONSENSUS REPORTS: Reported in EPA TSCA Inventory.

SAFETY PROFILE: Low toxicity by ingestion and skin contact. A skin irritant. When heated to decomposition it emits acrid smoke and irritating fumes.

DAI450   CAS: 50816-18-7   HR: 1
9-DECENYL ACETATE
mf: $C_{12}H_{22}O_2$ mw: 198.34

PROP: Bp: 95–97° @ 0.09 mm.

SYNS: ACETIC ACID, 9-DECENYL ESTER □ 9-DECEN-1-OL, ACETATE □ DECENYL ACETATE

CONSENSUS REPORTS: Reported in EPA TSCA Inventory.

SAFETY PROFILE: Low toxicity by ingestion and skin contact. A skin irritant. When heated to decomposition it emits acrid smoke and irritating fumes.

DAI495   CAS: 18507-89-6   HR: D
DECOQUINATE
mf: $C_{24}H_{35}NO_5$ mw: 417.53

PROP: Crystals. Mp: 86–87°.

SYNS: DECCOX ▢ 6-DECYLOXY-7-ETHOXY-4-HYDROXY-3-QUINOLINECARBOXYLID ACID ETHYL ESTER ▢ ETHYL 6-(N-DECYLOXY)-7-ETHOXY-4-HYDROXYQUINOLINE-3-CARBOXYLATE

SAFETY PROFILE: When heated to decomposition it emits acrid smoke and irritating fumes.

DAI600   CAS: 112-30-1   HR: 2
DECYL ALCOHOL
mf: $C_{10}H_{22}O$ mw: 158.32

PROP: Found in sweet orange and a few other essential oils (FCTXAV 11,95,73). Colorless, viscous, refractive liquid; floral fruity odor. Mp: 7°, fp: 7°, bp: 232–239° @ 700 mm, flash p: 180°F (OC), d: 0.8297 @ 20°/4°, refr index: 1.435–1.439, vap press: 1 mm @ 69.5°, vap d: 5.3. Sol in alc, ether, mineral oil, propylene glycol, fixed oils; insol in glycerin water @ 233°.

SYNS: AGENT 504 ▢ ALCOHOL C-10 ▢ ANTAK ▢ C 10 ALCOHOL ▢ CAPRIC ALCOHOL ▢ CAPRINIC ALCOHOL ▢ DECANAL DIMETHYL ACETAL ▢ DECANOL ▢ n-DECANOL ▢ 1-DECANOL (FCC) ▢ n-DECATYL ALCOHOL ▢ n-DECYL ALCOHOL ▢ DECYLIC ALCOHOL ▢ DYTOL S-91 ▢ EPAL 10 ▢ FEMA No. 2365 ▢ LOROL 22 ▢ NONYLCARBINOL ▢ PRIMARY DECYL ALCOHOL ▢ ROYALTAC ▢ SIPOL L10

CONSENSUS REPORTS: Reported in EPA TSCA Inventory.

SAFETY PROFILE: Moderately toxic by skin contact. Mildly toxic by ingestion and inhalation. A severe human skin and eye irritant. Experimental reproductive effects. Questionable carcinogen with experimental tumorigenic data. Combustible when exposed to heat or flame; can react with oxidizing materials. To fight fire, use foam, CO_2, dry chemical. When heated to decomposition it emits acrid smoke and irritating fumes. See also ALCOHOLS.

DAJ000   CAS: 1322-98-1   HR: 3
DECYL BENZENE SODIUM SULFON-ATE
mf: $C_{16}H_{25}O_3S \cdot Na$ mw: 320.46

SYNS: SODIUM DECYLBENZENESULFONAMIDE ▢ SODIUM DECYLBENZENESULFONATE

CONSENSUS REPORTS: Reported in EPA TSCA Inventory.

SAFETY PROFILE: Poison by intravenous route. Moderately toxic by ingestion. A severe eye irritant. When heated to decomposition it emits toxic fumes of SO_x. See also SULFONATES.

DAJ450   CAS: 6285-34-3   HR: 3
4-DECYLOXY-2-HYDROXYPHENYL 4-DECYLOXYPHENYL KETONE
mf: $C_{33}H_{50}O_4$ mw: 510.83

DOT CLASSIFICATION: 3; Label: Flammable Liquid

SAFETY PROFILE: A poison by intravenous route. A flammable liquid. When heated to decomposition it emits acrid smoke and irritating vapors.

DAL400   CAS: 23107-12-2   HR: 3
DEHYDRORETRONECINE
mf: $C_8H_{11}NO_2$ mw: 153.20

SYNS: 3,8-DIDEHYDRORETRONECINE ▢ (R)-2,3-DIHYDRO-1-HYDROXY-1H-PYRROLIZINE-7-METHANOL

CONSENSUS REPORTS: IARC Cancer Review: Animal Sufficient Evidence IMEMDT 10,333,76.

SAFETY PROFILE: Confirmed carcinogen with experimental carcinogenic and neoplastigenic data. Poison by intraperitoneal route. Human mutation data reported. When heated to decomposition it emits toxic fumes of NO_x.

DAL600   CAS: 84-17-3   HR: 3
DEHYDROSTILBESTROL
mf: $C_{18}H_{18}O_2$ mw: 266.36

PROP: A solid. Mp: 121–122°.

SYNS: 3,4-BIS(p-HYDROXYPHENYL)-2,4-HEXADIENE ▢ 3,4-BIS(4-HYDROXYPHENYL)-2,4-HEXADIENE ▢ CYCLADIENE ▢ DIENESTROL ▢ DIENOESTROL ▢ β-DIENOESTROL ▢ DIENOL ▢ 4,4'-(1,2-DIETHYLIDENE-1,2-ETHANEDIYL)BISPHENOL ▢ p,p'-(DIETHYLI-DENEETHYLENE)DIPHENOL ▢ 4,4'-(DIETHYLIDENE-ETHYLENE)DIPHENOL ▢ DINOVEX ▢ DI(p-OXYPHENYL)-2,4-HEXADIENE ▢ DV ▢ ESTRAGARD ▢ ESTRODIENOL ▢ ESTRORAL ▢ FOLLIDIENE ▢ FOLLORMON ▢ GYNEFOLLIN ▢ HORMOFEMIN ▢ 4,4'-HYDROXY-γ,Δ-DIPHENYL-β,Δ-HEXADIENE ▢ ISODIENESTROL ▢ OESTRASID ▢ OESTRODIENE ▢ OESTRODIENOL ▢ OESTRORAL ▢ PARA-DIEN ▢

RESTROL □ RETALON □ SEXADIEN □ SYNESTROL □ TESERENE □ WILLNESTROL

CONSENSUS REPORTS: IARC Cancer Review: Animal Inadequate Evidence IMEMDT 21,161,79; Human Limited Evidence IMEMDT 21,161,79.

SAFETY PROFILE: Suspected human carcinogen. Human mutation data reported. Experimental reproductive effects. Used as a drug for the treatment of postmenopausal symptoms. When heated to decomposition it emits acrid smoke and irritating fumes.

DAM700   CAS: 50-13-5   HR: 3
DEMEROL HYDROCHLORIDE
mf: $C_{15}H_{21}NO_2 \cdot ClH$ mw: 283.83

PROP: Minute crystals. Mp: 186–189°. Insol in C_6H_6 and Et_2O.

SYNS: ALGIL □ ALODAN (GEROT) □ ANTIDUROL □ CENTRALGIN □ CHLORBICYCLENE (FRENCH) □ DEMEROL □ DISPADOL □ DOLANTAL □ DOLANTIN □ DOLANTIN HYDROCHLORIDE □ DOLANTOL □ DOLAREN □ DOLARGAN □ DOLCONTRAL □ DOLENAL □ DOLENOL □ DOLESTINE □ DOLIN □ DOLOGAL □ DOLONEURINE □ DOLOPETHIN □ DOLOSAL □ DOLVANOL □ ENDOLAT □ ETHYL-1-METHYL-4-PHENYLISONIPECOTATE HYDRO-CHLORIDE □ ETHYL-1-METHYL-4-PHENYL-PIPERIDINE-4-CARBOXYLATE HYDROCHLORIDE □ ETHYL-1-METHYL-4-PHENYLPIPERIDYL-4-CARBOXYLATE HYDROCHLORIDE □ ISONIPECAINE HYDROCHLORIDE □ LIDOL □ LYDOL □ MEFEDINA □ MEPADIN □ MEPERIDINE HYDROCHLORIDE □ MEPHEDINE □ 1-METHYL-4-CARBETHOXY-4-PHENYLPIPERIDINE HYDROCHLORIDE □ N-METHYL-4-PHENYL-4-CARBETHOXYPIPERIDINE HYDROCHLORIDE □ 1-METHYL-4-PHENYL-4-CARBOETHOXYPIPERIDINE HYDROCHLORIDE □ 1-METHYL-4-PHENYLISONIPECOTIC ACID ETHYL ESTER HYDROCHLORIDE □ OPERIDINE □ PANTALGINE □ PENTANTIN □ PENTANTIN HYDROCHLORIDE □ PETHIDINE CHLORIDE □ PETIDIN □ PIRIDOSAL □ S 140 □ SAUTERALGYL □ SPASMEDAL □ SPASMODOLIN □ SYNELAUDINE □ WY 554

SAFETY PROFILE: Poison by ingestion, subcutaneous, intravenous, and intraperitoneal routes. Moderately toxic by parenteral route. Experimental teratogenic effects. Mutation data reported. An analgesic. When heated to decomposition it emits very toxic fumes of HCl and NO_x.

DAN100   CAS: 57852-57-0   HR: 3
4-DEMETHOXYDAUNORUBICIN HYDROCHLORIDE
mf: $C_{26}H_{27}NO_9 \cdot ClH$ mw: 534.00

SYN: DAUNOMYCIN, 4-DEMETHOXY-, HYDROCHLORIDE

SAFETY PROFILE: A poison by ingestion, subcutaneous, intravenous, and intraperitoneal routes. When heated to decomposition it emits toxic vapors of NO_x, HCl, and Cl^-.

DAO800   CAS: 867-27-6   HR: 3
DEMETON-O-METHYL
mf: $C_6H_{15}O_3PS_2$ mw: 230.30

PROP: Characterisitic odor. D: 1.2. Sol in water at 20°.

SYNS: BAY 15203 □ DEMETON-O-METILE (ITALIAN) □ O,O-DIMETHYL-O-(2-AETHYLTHIO-AETHYL) MONOTHIOPHOSPHAT (GERMAN) □ O,O-DIMETHYL-O-ETHYLMERCAPTOETHYL THIOPHOSPHATE □ O,O-DIMETHYL 2-ETHYLMERCAPTOETHYL THIOPHOSPH-ATE, THIONO ISOMER □ O,O-DIMETHYL-O-(2-ETHYL-THIO-ETHYL)-MONOTHIOFOSFAAT (DUTCH) □ O,O-DIMETHYL-O-2-(ETHYLTHIO)ETHYL PHOSPHORO-THIOATE □ O,O-DIMETIL-O-(2-ETILTIO-ETIL)-MONOTIOFOSFATO (ITALIAN) □ ENT 18,862 □ β-ETHYLMERCAPTOETHYL DIMETHYL THIONO-PHOSPHATE □ O-(2-(ETHYLTHIO)ETHYL)-O,O-DIMETHYL PHOSPHOROTHIOATE □ 2-(ETHYLTHIO)-ETHYL DIMETHYL PHOSPHOROTHIONATE □ METHYL-DEMETON-O □ O-METHYLDEMETON □ METHYLCISTOX □ METHYLSYSTOX □ METHYL-MERCAPTOPHOS □ THIOPHOSPHATE de O,O-DIMETHYLE et de O-2-ETHYLTHIO-ETHYLE (FRENCH)

SAFETY PROFILE: Poison by ingestion, skin contact, inhalation, and intravenous routes. Experimental teratogenic and reproductive effects. When heated to decomposition it emits very toxic fumes of PO_x and SO_x. See also DEMETON-O + DEMETON-S and other DEMETON entries.

DAP000   CAS: 301-12-2   HR: 3
DEMETON-O-METHYL SULFOXIDE
mf: $C_6H_{15}O_3PS_2$ mw: 230.30

PROP: Yellow liquid. D: 1.289 @ 20°/4°, bp: 106° @ 0.01 mm. Sol in H_2O, most org solvs; insol in pet ether.

SYNS: BAY 21097 □ DEMETON-S-METHYL-SULFOXID (GERMAN) □ DEMETON-S-METHYL SULFOXIDE □

DEMETON-METHYL SULPHOXIDE □ O,O-DIMETHYL-S-(2-AETHYLSULFINYL-AETHYL)-THIOLPHOSPHAT (GERMAN) □ O,O-DIMETHYL-S-(2-ETHTHIONYL-ETHYL) PHOSPHOROTHIOATE □ DIMETHYL-S-(2-ETHTHIONYLETHYL) THIOPHOSPHATE □ O,O-DIMETHYL-S-(2-ETHYLSULFINYL-ETHYL)-MONO-THIOFOSFAAT (DUTCH) □ O,O-DIMETHYL-S-(2-(ETHYLSULFINYL)ETHYL) PHOSPHOROTHIOATE □ O,O-DIMETHYL-S-(2-ETHYLSULFINYL)ETHYL THIOPHOSPHATE □ O,O-DIMETHYL-S-ETHYL-SULPHINYLETHYL PHOSPHOROTHIOLATE □ O,O-DIMETHYL-S-(3-OXO-3-THIA-PENTYL)-MONOTHIO-PHOSPHAT (GERMAN) □ O,O-DIMETIL-S-(2-ETIL-SOLFINIL-ETIL)-MONOTIOFOSFATO (ITALIAN) □ ENT 24,964 □ S-(2-(ETHYLSULFINYL)ETHYL)-O,O-DIMETH-YL PHOSPHOROTHIOATE □ ISOMETHYLSYSTOX SULFOXIDE □ METAISOSYSTOXSULFOXIDE □ METASYSTEMOX □ METASYSTOX-R □ METHYL DEMETON-O-SULFOXIDE □ METILMERCAPTO-FOSOKSID □ OXYDEMETONMETHYL □ OXYDE-METON-METILE (ITALIAN) □ R 2170 □ THIO-PHOSPHATE de O,O-DIMETHYLE et de S-2-ETHYL-SULFINYLETHYLE (FRENCH)

CONSENSUS REPORTS: EPA Genetic Toxicology Program.

SAFETY PROFILE: Poison by ingestion, skin contact, intravenous, and intraperitoneal routes. Human mutation data reported. When heated to decomposition it emits very toxic fumes of PO_x and SO_x. See also other demeton entries.

DAP200   CAS: 126-75-0   HR: 3
DEMETON-S

mf: $C_8H_{19}O_3PS_2$ mw: 258.36

PROP: Colorless oily liquid. Bp: 128° @ 0.1KPa. Sol in water.

SYNS: O,O-DIAETHYL-S-(2-AETHYLTHIO-AETHYL)-MONOTHIOPHOSPHAT (GERMAN) □ DIAETHYL-THIOPHOSPHORSAEUREESTER des AETHYLTHIO-GLYKOL (GERMAN) □ DIETHYL-S-(2-ETHIOETHYL)-THIOPHOSPHATE □ O,O-DIETHYL-S-(2-ETHTHIO-ETHYL)PHOSPHOROTHIOATE □ O,O-DIETHYL-S-ETHYL-2-ETHYLMERCAPTOPHOSPHOROTHIOLATE □ O,O-DIETHYL-S-(2-ETHYLTHIO-ETHYL)-MONO-THIOFOSFAAT (DUTCH) □ O,O-DIETHYL-S-2-(ETHYLTHIO)ETHYL PHOSPHOROTHIOATE □ O,O-DIETHYL-S-(2-(ETHYLTHIO)ETHYL) PHOSPHORO-THIOLATE (USDA) □ O,O-DIETIL-S-(2-ETILTIO-ETIL)-MONOTIOFOSFATO (ITALIAN) □ O,O-DIETYL-S-2-ETYLMERKAPTOETYLTIOFOSFAT (CZECH) □ 2-(ETHYLTHIO)-ETHANETHIOL S-ESTER with O,O-DIETHYL PHOSPHOROTHIOATE □ ISODEMETON □ IZOSYSTOX (CZECH) □ PO-SYSTOX □ THIOLDE-METON □ THIOL SYSTOX □ THIOPHOSPHATE de O,O-DIETHYLE et de S-(2-ETHYLTHIO-ETHYLE) (FRENCH)

SAFETY PROFILE: Poison by ingestion, intraperitoneal, and subcutaneous routes. When heated to decomposition it emits very toxic fumes of PO_x and SO_x. See also DEMETON-O + DEMETON-S and other demeton entries.

DAP400   CAS: 919-86-8   HR: 3
DEMETON-S-METHYL

mf: $C_6H_{15}O_3PS_2$ mw: 230.30

PROP: Pale-yellow oil. Bp: 89° @ 0.15 mm, d: 1.21 @ 20°/4°. Sltly sol in H_2O.

SYNS: BAY 18436 □ BAYER 25/154 □ DEMETON-S-METILE (ITALIAN) □ O,O-DIMETHYL-S-(2-AETHYL-THIO-AETHYL)-MONOTHIOPHOSPHAT (GERMAN) □ O,O-DIMETHYL-S-(2-ETHTHIOETHYL)PHOSPHORO-THIOATE □ DIMETHYL-S-(2-ETHTHIOETHYL)THIO-PHOSPHATE □ O,O-DIMETHYL-S-ETHYLMERCAPTO-ETHYL THIOPHOSPHATE □ O,O-DIMETHYL-S-ETHYLMERCAPTOETHYL THIOPHOSPHATE, THIOLO ISOMER □ O,O-DIMETHYL-S-(2-ETHYLTHIO-ETHYL)-MONOTHIOFOSFAAT (DUTCH) □ O,O-DIMETHYL-S-(2-(ETHYLTHIO)ETHYL)PHOSPHOROTHIOATE □ O,O-DIMETHYL-S-(3-THIA-PENTYL)-MONOTHIOPHOSPHAT (GERMAN) □ O,O-DIMETIL-S-(2-ETILITIO-ETIL)-MONOTIOFOSFATO (ITALIAN) □ DURATOX □ S-(2-(ETHYLTHIO)ETHYL)-O,O-DIMETHYL PHOSPHORO-THIOATE □ S-(2-(ETHYLTHIO)ETHYL)DIMETHYL PHOSPHOROTHIOLATE □ S-(2-(ETHYLTHIO)ETHYL)-O,O-DIMETHYL THIOPHOSPHATE □ ISOMETASYSTOX □ ISOMETHYLSYSTOX □ METAISOSEPTOX □ METAISOSYSTOX □ METASYSTOX FORTE □ METHYL DEMETON THIOESTER □ METHYL ISOSYSTOX □ METHYL-MERCAPTOFOS TEOLOVY □ THIOPHOS-PHATE de O,O-DIMETHYLE et de S-2-ETHYLTHIO-ETHYLE (FRENCH)

CONSENSUS REPORTS: Reported in EPA TSCA Inventory. EPA Extremely Hazardous Substances List.

ACGIH TLV: TWA 0.05 ppm (skin)(sensitizer); Not Classifiable as a Human Carcinogen

SAFETY PROFILE: Poison by ingestion, inhalation, skin contact, intraperitoneal, and intravenous routes. Mutation data reported. An insecticide. When heated to decomposition it emits very toxic fumes of PO_x and SO_x. See also DEMETON-O + DEMETON-S and other demeton entries.

DAP600   CAS: 17040-19-6   HR: 3
DEMETON-S-METHYL-SULPHONE
mf: $C_6H_{15}O_5PS_2$ mw: 262.30
PROP: Crystals. Mp: 60°, bp: 120° @ 0.03 mm.
SYNS: BAYER 20315 ◻ DEMETON-S-METHYLSULFON (GERMAN) ◻ DEMETON-S-METHYLSULFONE ◻ O,O-DIMETHYL-S-(2-AETHYLSULFONYL-AETHYL)-THIOLPHOSPHAT (GERMAN) ◻ O,O-DIMETHYL-S-(2-ETHSULFONYLETHYL)PHOSPHOROTHIOATE ◻ DIMETHYL-S-(2-ETHSULFONYLETHYL)THIOPHOS-PHATE ◻ O,O-DIMETHYL-S-ETHYL-2-SULFONYLETH-YL PHOSPHOROTHIOLATE ◻ O,O-DIMETHYL-S-ETHYLSULPHONYLETHYL PHOSPHOROTHIOLATE ◻ DIOXYDEMETON-S-METHYL ◻ E 158 ◻ ISOMETA-SYSTOX SULFONE ◻ ISOMETHYLSYSTOX SULFONE ◻ M 3/158 ◻ METAISOSYSTOX-SOLFON 20 315
SAFETY PROFILE: Poison by ingestion, inhalation, intraperitoneal, and intravenous routes. Moderately toxic by skin contact. Mutation data reported. An insecticide. When heated to decomposition it emits very toxic fumes of PO_x and SO_x. See also DEMETON-O + DEMETON-S and other demeton entries.

DAQ400   CAS: 83-44-3   HR: 3
DEOXYCHOLATIC ACID
mf: $C_{24}H_{40}O_4$ mw: 392.64
PROP: A white crystalline powder from EtOH. Mp: 187–189°. Sol in alc, acetone; sltly sol in ether and chloroform; insol in water.
SYNS: CHOLEIC ACID ◻ CHOLEREBIC ◻ CHOLOREBIC ◻ DEGALOL ◻ DEOXYCHOLIC ACID (FCC) ◻ 7-α-DEOXYCHOLIC ACID ◻ DESOXYCHOLIC ACID ◻ DESOXYCHOLSAEURE (GERMAN) ◻ 3,12-DIHYDROXYCHOLANIC ACID ◻ 3-α,12-α-DIHYDRO-XYCHOLANIC ACID ◻ 3-α,12-α-DIHYDROXY-5-β-CHOLANOIC ACID ◻ 3-α,12-α-DIHYDROXY-5-β-CHOLAN-24-OIC ACID ◻ 3-α,12-α-DIHYDROXY-CHOLANSAEURE (GERMAN) ◻ DROXOLAN ◻ 17-β-(1-METHYL-3-CARBOXYPROPYL)-ETIOCHOLANE-3-α,12-α-DIOL ◻ PYROCHOL ◻ SEPTOCHOL
CONSENSUS REPORTS: Reported in EPA TSCA Inventory.
SAFETY PROFILE: Poison by intraperitoneal route. Moderately toxic by ingestion and intravenous routes. Questionable carcinogen with experimental tumorigenic data. Experimental reproductive effects. Mutation data reported. When heated to decomposition it emits acrid smoke and irritating fumes.

DAQ850   CAS: 951-77-9   HR: D
DEOXYCYTIDINE
mf: $C_9H_{13}N_3O_4$ mw: 227.25
PROP: A solid. Mp: 207–210°.
SYNS: CYTOSINE DEOXYRIBOSIDE ◻ dCYD ◻ 2'-DEOXYCYTIDINE ◻ DEOXYRIBONUCLEOSIDE CYTOSINE ◻ DEOXYRIBOSE CYTIDINE ◻ DESOXYCYTIDIN (GERMAN)
CONSENSUS REPORTS: EPA Genetic Toxicology Program. Reported in EPA TSCA Inventory.
SAFETY PROFILE: Experimental reproductive effects. Mutation data reported. When heated to decomposition it emits toxic fumes of NO_x.

DAR400   CAS: 50-91-9   HR: 3
2'-DEOXY-5-FLUOROURIDINE
mf: $C_9H_{11}FN_2O_5$ mw: 246.22
PROP: Crystals from butyl acetate. Mp: 150–151°.
SYNS: DEOXYFLUOROURIDINE ◻ 1-β-d-2'-DEOXYRIBOFURANOSYL-5-FLUOROURACIL ◻ FDUR ◻ FLOXURIDIN ◻ FLOXURIDINE ◻ FLUORODEOXY-URIDINE ◻ β-5-FLUORO-2'-DEOXYURIDINE ◻ 5-FLUORODEOXYURIDINE ◻ 5-FLUORO-2-DEOXY-URIDINE ◻ 5-FLUORO-2'-DEOXYURIDINE ◻ 5-FLUOROURACIL DEOXYRIBOSIDE ◻ 5-FLUORO-URACIL-2'-DEOXYRIBOSIDE ◻ FLUORURIDINE DEOXYRIBOSE ◻ FUDR ◻ 5-FUDR ◻ NSC-27640 ◻ RO 5-0360
CONSENSUS REPORTS: EPA Genetic Toxicology Program.
SAFETY PROFILE: Poison by ingestion. Moderately toxic by intraperitoneal route. An experimental teratogen. Other experimental reproductive effects. Human systemic effects: hypermotility, diarrhea, nausea, vomiting and other gastrointestinal effects, allergic dermatitis, and bone marrow changes. Human mutation data reported. When heated to decomposition it emits very toxic fumes of F^- and NO_x.

DAR600   CAS: 154-17-6   HR: 3
2-DEOXYGLUCOSE
mf: $C_6H_{12}O_5$ mw: 164.18

PROP: Crystals from acetone or butanone. Mp: 142–144°. α-Form: Crystals from isopropanol. Mp: 134–136°.
SYNS: 2-DEOXY-d-ARABINO-HEXOSE ◻ 2-DEOXY-3-ARABINO-HEXOSE ◻ d-2-DEOXYGLUCOSE ◻ 2-DEOXY-d-GLUCOSE ◻ 2-DEOXY-d-GLUCOSE (FRENCH) ◻ 2-DG ◻ NSC-15193
CONSENSUS REPORTS: Reported in EPA TSCA Inventory.
SAFETY PROFILE: Poison by subcutaneous route. Moderately toxic by intraperitoneal route. An experimental teratogen. Other experimental reproductive effects. When heated to decomposition it emits acrid smoke and fumes.

DAS000   CAS: 54-42-2   HR: 2
2'-DEOXY-5-IODOURIDINE
mf: $C_9H_{11}IN_2O_5$ mw: 354.12
PROP: Crystals from H_2O. Mp: 160°.
SYNS: ALLERGAN 211 ◻ DENDRID ◻ 1-(2-DEOXY-β-d-RIBOFURANOSYL)-5-IODOURACIL ◻ 1-β-d-2'-DEOXYRIBOFURANOSYL-5-IODOURACIL ◻ EMANIL ◻ HERPESIL ◻ HERPIDU ◻ HERPLEX ◻ HERPLEX LIQUIFILM ◻ IDEXUR ◻ IDOXENE ◻ IDOXURIDIN ◻ IDOXURIDINE ◻ IDU ◻ IDUCHER ◻ IDULEA ◻ IDUOCULOS ◻ IDUR ◻ IDURIDIN ◻ 5-IODODE-OXYURIDINE ◻ 5-IODO-2'-DEOXYURIDINE ◻ 5-IODOURACIL DEOXYRIBOSIDE ◻ IUDR ◻ 5-IUDR ◻ JODDEOXIURIDIN ◻ KERECID ◻ NSC-39661 ◻ OPHTHALMADINE ◻ SK&F 14287 ◻ STOXIL ◻ SYNMIOL
CONSENSUS REPORTS: Reported in EPA TSCA Inventory. EPA Genetic Toxicology Program.
SAFETY PROFILE: Moderately toxic by intraperitoneal route. Experimental teratogenic and reproductive effects. Questionable carcinogen with experimental carcinogenic data. Human mutation data reported. When heated to decomposition it emits very toxic fumes of I⁻ and NO_x.

DAZ115   CAS: 313-06-4   HR: D
DEPOFEMIN
mf: $C_{26}H_{36}O_3$ mw: 396.62
PROP: Crystals from benzene + pet ether. Mp: 151–152°. Sol in ether, methanol, benzene, chloroform, peanut oil, cottonseed oil, corn oil, and sesame oil. The limit of solubility in the oils is about 400 mg/mL.
SYNS: DEPOESTRADIOL ◻ DEPOESTRADIOL CYPIONATE ◻ ECP ◻ ESTRADEP ◻ ESTRADIOL CYCLOPENTYLPROPIONATE ◻ ESTRADIOL-17-CYCLOPENTYLPROPIONATE ◻ ESTRADIOL-17-β-CYCLOPENTYLPROPIONATE ◻ ESTRADIOL-CYPIONATE ◻ ESTRADIOL-17-CYPIONATE ◻ ESTRADIOL-17-β-CYPIONATE ◻ (17-β)-ESTRA-1,3,5(10)-TRIENE-3,17-DIOL 17-CYCLOPENTANEPROPANOATE (9CI)
CONSENSUS REPORTS: Reported in EPA TSCA Inventory.
SAFETY PROFILE: An experimental teratogen. Other experimental reproductive effects. A steroid. When heated to decomposition it emits acrid smoke and fumes.

DBA800   CAS: 300-42-5   HR: 3
DESOXYEPHEDRINE HYDRO-CHLORIDE
mf: $C_{10}H_{15}N•ClH$ mw: 185.72
PROP: Crystals from $EtOH/Et_2O$. Mp: 135–136°.
SYNS: A 884 ◻ AMDRAM ◻ AMEDRINE ◻ AMPHEDROXY ◻ AMPHEDROXYN ◻ APAMINE ◻ BOMBITA ◻ C 6379 ◻ CORVITIN ◻ DAROPERVAMIN ◻ DEA OXO-5 ◻ DEOFED ◻ DEOXYEPHEDRINE ◻ DEPOXIN ◻ DESAMINE ◻ DESFEDRIN ◻ DESOSSIEFEDRINA ◻ DES-OXA-D ◻ DESOXEDRINE ◻ DESOXIN ◻ DESOXO-5 ◻ DESOXYFED ◻ DESOXYN ◻ DESOXYPHED ◻ DESTIM ◻ DETREX ◻ DEXOPHRINE ◻ DEXOVAL ◻ N,α-DIMETHYL-PHENETHYLAMINE HYDROCHLORIDE ◻ DOPIDRIN ◻ DOXEPHRIN ◻ DOXYFED ◻ DRINALFA ◻ EFFROX-INE ◻ ESTIMULEX ◻ EUPHODRIN ◻ FENYPRIN ◻ GEROBIT ◻ GEROVIT ◻ HEROPON ◻ ISOPHEN ◻ KEMODRIN ◻ LANAZINE ◻ MADRINE ◻ METAMFET-AMINA ◻ METAMPHETAMIN ◻ METANFETAMINA ◻ METHAMPHETAMINE HYDROCHLORIDE ◻ METHEDRINE ◻ METHEDRINE HYDROCHLORIDE ◻ METHOXYN ◻ METHYLAMPHETAMINE HYDROCHLORIDE ◻ METHYLBENZEDRIN ◻ METHYLISOMIN ◻ N-METHYL-β-PHENYLISOPROPYL-AMINHYDROCHLORID (GERMAN) ◻ METHYL-PROPAMINE ◻ NEODRINE ◻ NEOPHARMEDRINE ◻ NORODIN ◻ OXYDRENE ◻ OXYFED ◻ PERVITIN ◻ PHILOPON ◻ PREMODRIN ◻ SEMOXYDRINE ◻ SPEED ◻ STIMULEX ◻ TONEDRIN ◻ VONEDRINE
SAFETY PROFILE: Poison by ingestion, intravenous, intraperitoneal, subcutaneous, and intramuscular routes. Human systemic

effects by ingestion: altered sleep patterns, anorexia, and change in heart rate. Experimental reproductive effects. An eye irritant. A powerful central nervous system stimulant. When heated to decomposition it emits very toxic fumes of NO_x and HCl. See also BENZEDRINE.

DBB200   CAS: 125-33-7   HR: 3
2-DESOXYPHENOBARBITAL
mf: $C_{12}H_{14}N_2O_2$ mw: 218.28
PROP: White crystalline powder; odorless with slty bitter taste. Mp: 281°–282°. Sol in water <1 mg/mL @ 19°.
SYNS: 5-AETHYL-5-PHENYL-HEXAHYDRO-PYRIMIDIN-4,6-DION (GERMAN) □ CYRAL □ 2-DEOXYPHENOBARBITAL □ DESOXYPHENO-BARBITONE □ 5-ETHYLDIHYDRO-5-PHENYL-4,6(1H,5H)-PYRIMIDINEDIONE □ 5-ETHYLHEXA-HYDRO-4,6-DIOXO-5-PHENYLPHYIMIDINE □ 5-ETHYLHEXAHYDRO-5-PHENYLPYRIMIDINE-4,6-DIONE □ 5-ETHYL-5-PHENYLHEXAHYDROPYRIMID-INE-4,6-DIONE □ HEXADIONA □ HEXAMIDINE □ HEXAMIDINE (the antispasmodic) □ LEPIMIDIN □ LEPSIRAL □ MAJSOLIN □ MIDONE □ MILEPSIN □ MISODINE □ MISOLYNE □ MIZODIN □ MIZOLIN □ MYLEPSIN □ MYLEPSINUM □ MYSEDON □ MYSOLINE □ NCI-C56360 □ 5-PHENYL-5-ETHYL-HEXAHYDROPYRIMIDINE-4,6-DIONE □ PRILEPSIN □ PRIMACIONE □ PRIMACLONE □ PRIMACONE □ PRIMAKTON □ PRIMIDON □ PRIMIDONE □ PRYSOLINE □ PYRIMIDONE MEDI-PETS □ ROE 101 □ SERTAN
SAFETY PROFILE: Poison by ingestion and intraperitoneal routes. Human teratogenic effects include developmental abnormalities of the craniofacial area, skin and skin appendages, and cardiovascular system. Human reproductive effects: effects on newborn, including unusual growth statistics, drug dependence, physical and other neonatal changes. Experimental teratogenic and reproductive effects. Human mutation data reported. An addictive drug. When heated to decomposition it emits toxic fumes of NO_x. See also BARBITURATES.

DBC400   CAS: 1177-87-3   HR: D
DEXAMETHASONE ACETATE
mf: $C_{24}H_{31}FO_6$ mw: 434.55

SYNS: DECADRON-LA □ DECTAN □ 9-FLUORO-11-β,17,21-TRIHYDROXY-16-α-METHYLPREGNA-1,4-DIENE-3,20-DIONE ACETATE
CONSENSUS REPORTS: Reported in EPA TSCA Inventory.
SAFETY PROFILE: Experimental teratogenic and reproductive effects. A steroid. When heated to decomposition it emits toxic fumes of F^-.

DBC800   CAS: 9004-54-0   HR: 2
DEXTRAN 1
PROP: A linear water-sol polymer of average molecular weight 200,000 (ARPAAQ 67,589,59).
CONSENSUS REPORTS: Reported in EPA TSCA Inventory.
SAFETY PROFILE: Questionable carcinogen with experimental neoplastigenic, tumorigenic, and teratogenic data. Other experimental reproductive effects. When heated to decomposition it emits acrid smoke and fumes. See also other dextrans.

DBD000   CAS: 9004-54-0   HR: 2
DEXTRAN 2
PROP: A linear, water-sol polymer of average molecular weight 100,000 (ARPAAQ 67,589,59).
CONSENSUS REPORTS: Reported in EPA TSCA Inventory.
SAFETY PROFILE: Questionable carcinogen with experimental neoplastigenic data. When heated to decomposition it emits acrid smoke and fumes. See also other DEXTRANS.

DBD200   CAS: 9004-54-0   HR: 2
DEXTRAN 5
PROP: A highly branched, water-sol polymer (ARPAAQ 67,589,59).
CONSENSUS REPORTS: Reported in EPA TSCA Inventory.
SAFETY PROFILE: Questionable carcinogen with experimental neoplastigenic data. When heated to decomposition it emits acrid smoke and fumes. See also other dextrans.

DBD400   **CAS: 9004-54-0**   **HR: 3**
DEXTRAN 10
PROP: A branched, water-sol polymer of
average molecular weight 89,400 (ARPAAQ
67,589,59).
CONSENSUS REPORTS: Reported in EPA
TSCA Inventory.
SAFETY PROFILE: Suspected carcinogen
with experimental carcinogenic data. When
heated to decomposition it emits acrid
smoke and fumes. See also other
DEXTRANS.

DBD600   **CAS: 9004-54-0**   **HR: 2**
DEXTRAN 11
PROP: A highly branched, water-sol
polymer of average molecular weight 71,400
(ARPAAQ 67,589,59).
CONSENSUS REPORTS: Reported in EPA
TSCA Inventory.
SAFETY PROFILE: Questionable
carcinogen with experimental neoplastigenic
and tumorigenic data. When heated to
decomposition it emits acrid smoke and
fumes. See also other dextrans.

DBD700   **CAS: 9004-54-0**   **HR: 2**
DEXTRAN 70
PROP: Chemical and physical properties of
the dextrans vary with the methods of
production. Native dextrans usually have
high molecular weight; lower-molecular-
weight clinical dextrans are usually prepared
by depolymerization of native dextrans or by
synthesis. All dextrans are composed
exclusively of α-d-glucopyranosyl units,
differing only in degree of branching and
chain length.
SYNS: DEXTRAN □ DEXTRAVEN □ EXPANDEX □
GENTRAN □ HEMODEX □ INTRADEX □ MACROSE □
ONKOTIN □ PLAVOLEX □ POLYGLUCIN □ PROMIT
CONSENSUS REPORTS: Reported in EPA
TSCA Inventory.
SAFETY PROFILE: Human systemic effects
by intravenous route: dermatitis and changes
in the vascular and cardiac systems.
Experimental reproductive effects. An

experimental teratogen. See also various
dextrans.

DBD800   **CAS: 9004-53-9**   **HR: 1**
DEXTRINS
mf: $(C_6H_{10}O_5)_n \cdot xH_2O$
PROP: An intermediate product formed by
the hydrolysis of starches. It describes a
class of substances. Yellow or white
powder or granules. Sol in boiling water to
give a gummy solution; insol in alc and
ether; forms colloids.
SYNS: ARTIFICIAL GUM □ DEXTRANS □ STARCH
GUM □ TAPIOCA □ VEGETABLE GUM
SAFETY PROFILE: Mildly toxic by
intravenous route. When heated to
decomposition it emits acrid smoke and
irritating fumes.

DBF200   **CAS: 642-65-9**   **HR: 3**
2-DIACETAMIDOFLUORENE
mf: $C_{17}H_{15}NO_2$ mw: 265.33
SYNS: N-ACETYL-N-9H-FLUOREN-2-YL-ACETAMIDE
□ N-DIACETYL-2-AMINOFLUORENE □ N,N-
DIACETYL-2-AMINOFLUORENE □ 2-DIACETYL-
AMINOFLUORENE □ N,N-DIACETYL-2-FLUOREN-
AMINE □ F-diAA □ 2-FLUORENYLDIACETAMIDE □ N-
FLUOREN-2-YLDIACETAMIDE □ N-2-FLUORENYLDI-
ACETAMIDE
SAFETY PROFILE: Suspected carcinogen
with experimental carcinogenic,
neoplastigenic, and tumorigenic data. When
heated to decomposition it emits toxic
fumes of NO_x.

DBF600   **CAS: 102-62-5**   **HR: 2**
1,3-DIACETIN
mf: $C_7H_{12}O_5$ mw: 176.19
PROP: Crystals. D: 1.178 @ 15°/15°, bp:
280°, mp: 40°.
SYNS: 1,3-DIACETATE GLYCEROL □ 1,2-DIACETATE
1,2,3-PROPANETRIOL □ DIACETIN □ 1,2-DI-ACETIN □
2,3-DIACETIN □ DIACETYL GLYCERINE □
DIGLYCERIDE ACETIC ACID □ GLYCEROL
DIACETATE □ GLYCERYL-1,3-DIACETATE □
(HYDROXYMETHYL)ETHYLENE ACETATE
CONSENSUS REPORTS: Reported in EPA
TSCA Inventory.
SAFETY PROFILE: Moderately toxic by
subcutaneous and intravenous routes. Mildly

toxic by ingestion. When heated to decomposition it emits acrid smoke and irritating fumes.

DBF750   CAS: 123-42-2   HR: 3
DIACETONE ALCOHOL
DOT: UN 1148

mf: $C_6H_{12}O_2$ mw: 116.18

PROP: Liquid; oily; faint pleasant odor. Mp: −47 to −54°, bp: 164°, flash p: 148°F, d: 0.9306 @ 25°/4°, autoign temp: 1118°F, vap d: 4.00, vap press: 1.1 mm @ 20°, lel: 1.8%, uel: 6.9%, flash p: (acetone free) 136°F. Sol in water. IDLH 1800 ppm [10%LEL].

SYNS: DIACETONALCOHOL (DUTCH) □ DIACETON-ALCOOL (ITALIAN) □ DIACETONALKOHOL (GER-MAN) □ DIACETONE □ DIACETONE-ALCOOL (FRENCH) □ DIKETONE ALCOHOL □ 4-HYDROXY-2-KETO-4-METHYLPENTANE □ 4-HYDROXY-4-METHYL-PENTAN-2-ON (GERMAN, DUTCH) □ 4-HYDROXY-4-METHYLPENTANONE-2 □ 4-HYDROXY-4-METHYL PENTAN-2-ONE □ 4-HYDROXY-4-METHYL-2-PENTANONE □ 4-IDROSSI-4-METIL-PENTAN-2-ONE (ITALIAN) □ 2-METHYL-2-PENTANOL-4-ONE □ PYRANTON □ TYRANTON

CONSENSUS REPORTS: Reported in EPA TSCA Inventory.

OSHA PEL: TWA 50 ppm

ACGIH TLV: TWA 50 ppm

DFG MAK: 50 ppm (240 mg/m³)

NIOSH REL: (Ketones) TWA 240 mg/m³

DOT CLASSIFICATION: 3; Label: Flammable Liquid

SAFETY PROFILE: Moderately toxic by ingestion and intraperitoneal routes. Mildly toxic by skin contact. Human systemic effects by inhalation: headache, nausea or vomiting, eye and pulmonary changes. A skin, mucous membrane, and severe eye irritant. Can cause anemia and damage to liver and kidneys. Narcotic in high concentration. Flammable liquid when exposed to heat or flame; can react with oxidizing materials. Explosive in the form of vapor when exposed to heat or flame. To fight fire, use alcohol foam, foam, CO_2, dry chemical. When heated to decomposition it emits acrid smoke and irritating fumes. See also KETONES.

DBF800   CAS: 1067-33-0   HR: 3
DIACETOXYDIBUTYL STANNANE
mf: $C_{12}H_{24}O_4Sn$ mw: 351.05

PROP: Clear, colorless oily liquid with a slight acetic acid odor. Bp: 144.5–145.5° @ 10 mm (decomp), mp: 8.5–10°, fp: 5–10°, flash p: 290°F (OC), d: 1.31 @ 25°, vap d: 12.1. Sol in org solvs.

SYNS: BA 2726 □ BIS(ACETYLOXY)DIBUTYLSTANNANE □ DIACET-OXYBUTYLTIN □ DIACETOXYDIBUTYLTIN □ DIBUTYL TIN DIACETATE □ FOMREZ SUL-3 □ NCI-C02028 □ T 1 (Catalyst)

CONSENSUS REPORTS: NCI Carcinogenesis Bioassay (feed); Inadequate Studies: mouse, rat NCITR* NCI-CG-TR-183,79. Reported in EPA TSCA Inventory.

OSHA PEL: TWA 0.1 mg(Sn)/m³ (skin)

ACGIH TLV: TWA 0.1 mg(Sn)/m³; STEL 0.2 mg(Sn)/m³ (skin).

NIOSH REL: (Organotin Compounds) TWA 0.1 mg(Sn)/m³

SAFETY PROFILE: Poison by ingestion and intravenous routes. Experimental reproductive effects. Combustible when exposed to heat or flame; can react with oxidizing materials. To fight fire, use water, foam, CO_2, dry chemical. When heated to decomposition it emits acrid smoke and irritating fumes. See also TIN COMPOUNDS.

DBH000   CAS: 95-45-4   HR: 3
DIACETYL DIOXIME
mf: $C_4H_8N_2O_2$ mw: 116.14

PROP: Triclinic crystals from EtOH (aq). Mp: 238–240°. Sol in EtOH, Me_2CO, Et_2O, and alk soln; very sltly sol in H_2O and $CHCl_3$.

SYNS: 2,3-DIISONITROSOBUTANE □ DIMETHYL-GLYOXIME

CONSENSUS REPORTS: Reported in EPA TSCA Inventory.

SAFETY PROFILE: Poison by ingestion. Mutation data reported. When heated to decomposition it emits toxic fumes of NO_x.

DBH700 **HR: D**
DIACETYL TARTARIC ACID ESTERS of MONO- and DIGLYCERIDES
PROP: Vary from sticky, viscous liquid to waxy solid; faint acid odor. Sol in oil, methanol, acetone, acetic acid, water.
SAFETY PROFILE: When heated to decomposition it emits acrid smoke and irritating fumes.

DBI099 **CAS: 10311-84-9** **HR: 3**
DIALIFOR
mf: $C_{14}H_{17}NO_4PS_2$ mw: 393.86
PROP: Crystals from toluene/hexane. Mp: 67–69°.
SYNS: S-(2-CHLORO-1-(1,3-DIHYDRO-1,3-DIOXO-2H-ISOINDOL-2-YL)ETHYL)-O,O-DIETHYL PHOS-PHORO-DITHIOATE □ S-(2-CHLORO-1-PHTHALIMIDOETHYL)-O,O-DIETHYL PHOSPHORODITHIOATE □ O,O-DIETH-YL-S-(2-CHLORO-1-PHTHALIMIDOETHYL)PHOS-PHORODITHIOATE □ ENT 27,320 □ HERCULES 14503 □ PHOSPHORODITHIOIC ACID-S-(2-CHLORO-1-(1,3-DIHYDRO-1,3-DIOXO-2H-ISOINDOL-2-YL))ETHYL-O,O-DIETHYL ESTER □ PHOSPHORODITHIOIC ACID-S-((2-CHLORO-1-PHTHALIMIDOETHYL)-O,)-DIETHYL ESTER □ TORAK
SAFETY PROFILE: Poison by ingestion and skin contact. An experimental teratogen. Other experimental reproductive effects. When heated to decomposition it emits toxic fumes of SO_x, PO_x, and NO_x.

DBI200 **CAS: 2303-16-4** **HR: 3**
DIALLATE
mf: $C_{10}H_{17}Cl_2NOS$ mw: 270.24
PROP: Brown liquid. Bp: 150° @ 9 mm, mp: 25–30°. Sltly sol in water; sol in org solvs.
SYNS: AVADEX □ CP 15,336 □ DATC □ 2,3-DCDT □ DIALLAAT (DUTCH) □ DIALLAT (GERMAN) □ 2,3-DICHLORALLYL-N,N-(DIISOPROPYL)-THIOCARBAMAT (GERMAN) □ S-(2,3-DICHLOR-ALLYL)-N,N-DIISOPROPYL-MONOTHIOCARBAMAAT (DUTCH) □ DICHLOROALLYL DIISOPROPYLTHIOCARBAMATE □ S-2,3-DICHLOROALLYL DIISOPROPYLTHIOCARBAM-ATE □ 2,3-DICHLOROALLYL-N,N-DIISOPROPYLTHIOL-CARBAMATE □ 2,3-DICHLORO-2-PROPENE-1-THIOL DIISOPROPYLCARBAMATE □ S-(2,3-DICHLORO-2-PROPENYL)ESTER, BIS(1-METHYLETHYL) CARBAMO-THIOIC ACID □ S-(2,3-DICLORO-ALLIL)-N,N-DIISO-PROPIL-MONOTIOCARBAMMATO (ITALIAN) □ DI-ISOPROPYL-THIOLOCARBAMATE de S-(2,3-DICHLORO-

ALLYLE) (FRENCH) □ PYRADEX □ RCRA WASTE NUMBER U062
CONSENSUS REPORTS: IARC Cancer Review: Group 3 IMEMDT 7,56,87; Animal Limited Evidence IMEMDT 30,235,83; Animal Sufficient Evidence IMEMDT 12,69,76. EPA Genetic Toxicology Program. Community Right-To-Know List. EPA Extremely Hazardous Substances List.
SAFETY PROFILE: Poison by ingestion. Moderately toxic by skin contact. Questionable carcinogen with experimental carcinogenic and tumorigenic data. Mutation data reported. When heated to decomposition it emits very toxic fumes of Cl^-, NO_x, and SO_x. See also CARBAMATES and ALLYL COMPOUNDS.

DBI600 **CAS: 124-02-7** **HR: 3**
DIALLYLAMINE
DOT: UN 2359
mf: $C_6H_{11}N$ mw: 97.18
PROP: Liquid, sol in water. D: 0.7889 @ 20°, bp: 112°, fp: −100°, flash p: 69.8°F.
SYNS: DI-2-PROPENYLAMINE □ N-2-PROPENYL-2-PROPEN-1-AMINE
CONSENSUS REPORTS: Reported in EPA TSCA Inventory.
DOT CLASSIFICATION: 3; Label: Flammable Liquid
SAFETY PROFILE: Poison by ingestion, skin contact, and intraperitoneal routes. Moderately toxic by inhalation. Human systemic effects by inhalation route: eye lachrymation and changes in the trachea or bronchi. A skin and severe eye irritant. A dangerous fire hazard when exposed to heat or flame. When heated to decomposition it emits toxic fumes of NO_x. See also AMINES and ALLYL COMPOUNDS.

DBJ400 **CAS: 17381-88-3** **HR: 3**
DIALLYLDIBROMO STANNANE
mf: $C_6H_{10}Br_2Sn$ mw: 360.67
SYN: DIALLYLTIN DIBROMIDE
OSHA PEL: TWA 0.1 mg(Sn)/m³ (skin)

ACGIH TLV: TWA 0.1 mg(Sn)/m³; STEL 0.2 mg(Sn)/m³ (skin).
NIOSH REL: (Organotin Compounds) TWA 0.1 mg(Sn)/m³
SAFETY PROFILE: Poison by intravenous route. When heated to decomposition it emits toxic fumes of Br⁻. See also ALLYL COMPOUNDS, BROMIDES, and TIN COMPOUNDS.

DBK000   CAS: 557-40-4   HR: 3
DIALLYL ETHER
DOT: UN 2360
mf: $C_6H_{10}O$ mw: 98.16
PROP: Liquid; odor of radishes. Bp: 94.3°, d: 0.805, vap d: 3.38, flash p: 20°F (OC).
SYNS: ALLYLETHER □ 3,3'-OXYBIS(1-PROPENE) □ PROPENYL ETHER
CONSENSUS REPORTS: Reported in EPA TSCA Inventory.
DOT CLASSIFICATION: 3; Label: Flammable Liquid, Poison
SAFETY PROFILE: Poison by ingestion. Moderately toxic by skin contact. A skin and eye irritant. A dangerous fire hazard when exposed to heat, flame, or oxidizing materials. To fight fire, use alcohol foam. Reacts with air to form explosive peroxides. Violent explosions have occurred during distillation. When heated to decomposition it emits acrid smoke and fumes. See also ALLYL COMPOUNDS and ETHERS.

DBK200   CAS: 999-21-3   HR: 3
DIALLYL MALEATE
mf: $C_{10}H_{12}O_4$ mw: 196.22
PROP: Liquid. Vap d: 6.6.
SYNS: MALEIC ACID, DIALLYL ESTER □ SIPOMER DAM
CONSENSUS REPORTS: Reported in EPA TSCA Inventory.
SAFETY PROFILE: Poison by ingestion and intraperitoneal routes. Moderately toxic by skin contact. A skin and eye irritant. When heated to decomposition it emits acrid smoke and irritating fumes. See also ALLYL COMPOUNDS and ESTERS.

DBL200   CAS: 131-17-9   HR: 3
DIALLYL PHTHALATE
mf: $C_{14}H_{14}O_4$ mw: 246.28
PROP: Nearly colorless, oily liquid. Bp: 157°, flash p: 330°F, d: 1.120 @ 20°/20°, vap d: 8.3.
SYNS: DAPON 35 □ DAPON R □ DI-2-PROPENYL ESTER, 1,2-BENZENEDICARBOXYLIC ACID □ NCI-C50657 □ PHTHALIC ACID, DIALLYL ESTER □ o-PHTHALIC ACID, DIALLYL ESTER
CONSENSUS REPORTS: NTP Carcinogenesis Studies (gavage); Equivocal Evidence: rat NTPTR* NTP-TR-284,85. Reported in EPA TSCA Inventory.
SAFETY PROFILE: Suspected carcinogen with experimental carcinogenic data. Moderately toxic by ingestion, skin contact, intraperitoneal, and subcutaneous routes. An eye irritant. Mutation data reported. Combustible when exposed to heat or flame; can react with oxidizing materials. To fight fire, use CO_2 or dry chemical. When heated to decomposition it emits acrid smoke and irritating fumes. See also ALLYL COMPOUNDS and ESTERS.

DBL300   CAS: 91297-11-9   HR: 3
DIALLYL SELENIDE
mf: $C_6H_{10}Se$ mw: 184.11
SYN: SELENIDE, DIALLYL-
OSHA PEL: TWA 0.2 mg(Se)/m³
ACGIH TLV: TWA 0.2 mg(Se)/m³
SAFETY PROFILE: Poison by intravenous route. When heated to decomposition it emits toxic fumes of Se.

DBM800   CAS: 59-33-6   HR: 3
DIAMINIDE MALEATE
mf: $C_{17}H_{23}N_3O \cdot C_4H_4O_4$ mw: 401.51
PROP: A solid. Mp: 100–101°. Very sol in H_2O.
SYNS: AH □ ANISOPYRADAMINE □ ANTHISAN MALEATE □ ANTIHIST □ N-DIMETHYLAMINOETHYL-N-p-METHOXY-α-AMINOPYRIDINE MALEATE □ 2-((2-(DIMETHYLAMINO)ETHYL)(p-METHOXYBENZYL)-AMINO)PYRIDINE BIMALEATE □ 2-((2-(DIMETHYL-AMINO)ETHYL)(p-METHOXYBENZYL)AMINO)-PYRIDINE MALEATE □ N,N-DIMETHYL-N'-(4-METHOXYBENZYL)-N'-(2-PYRIDYL)ETHYLENEDI-AMINE MALEATE □ HISTATEX □ MEPYRAMINE MALEATE □ N-p-METHOXYBENZYL-N'-N'-DIMETHYL-

N-α-PYRIDYLETHYLENEDIAMINE MALEATE □ MINIHIST □ NEOANTERGAN MALEATE □ PARAMAL □ PARAMINYL MALEATE □ PYMAFED □ PYRA MALEATE □ PYRANILAMINE MALEATE □ PYRANINYL □ PYRANISAMINE MALEATE □ PYRILAMINE MALEATE □ RENSTAMIN □ 2786 R.P. MALEATE □ STANGEN MALEATE □ STATOMIN MALEATE □ THYLOGEN MALEATE

CONSENSUS REPORTS: Reported in EPA TSCA Inventory.

SAFETY PROFILE: A human poison by ingestion. An experimental poison by ingestion, subcutaneous, intravenous, and intraperitoneal routes. Experimental reproductive effects. Questionable carcinogen with experimental tumorigenic data. Mutation data reported. An antihistamine. When heated to decomposition it emits toxic fumes of NO_x.

DBN600   CAS: 92-62-6   HR: 3
3,6-DIAMINOACRIDINIUM
mf: $C_{13}H_{11}N_3$ mw: 209.27
PROP: Yellow crystals. Mp: 284–286°.
SYNS: 3,6-ACRIDINEDIAMINE □ 2,8-DIAMINO-ACRIDINE □ 3,6-DIAMINOACRIDINE □ 2,8-DIAMINO-ACRIDINIUM □ 3,7-DIAMINO-5-AZAANTHRACENE □ ISOFLAV BASE □ PROFLAVIN □ PROFLAVINE □ PROFOLIOL □ PROFORMIPHEN □ PROFUNDOL □ PROFURA □ PROGARMED □ PRO-GEN □ PROGESIC

CONSENSUS REPORTS: IARC Cancer Review: Animal Inadequate Evidence IMEMDT 24,195,80. Reported in EPA TSCA Inventory. EPA Genetic Toxicology Program.

SAFETY PROFILE: Poison by intravenous, intraperitoneal, and subcutaneous routes. Questionable carcinogen. Mutation data reported. When heated to decomposition it emits toxic fumes of NO_x. See also other diaminoacridine entries.

DBO000   CAS: 615-05-4   HR: 3
2,4-DIAMINOANISOLE
mf: $C_7H_{10}N_2O$ mw: 138.19
PROP: Needles. Mp: 68°.
SYNS: C.I. 76050 □ C.I. OXIDATION BASE 12 □ 2,4-DAA □ 2,4-DIAMINEANISOLE □ 2,4-DIAMINOANISOL □ 2,4-DIAMINOANISOLE BASE □ m-DIAMINOANISOLE 1,3-DIAMINO-4-METHOXYBENZENE □ 2,4-DIAMINO-1-

METHOXYBENZENE □ FURRO L □ 4-METHOXY-1,3-BENZENEDIAMINE □ p-METHOXY-m-PHENYLENEDI-AMINE □ 4-METHOXY-m-PHENYLENEDIAMINE □ 4-MMPD □ PELAGOL DA □ PELAGOL GREY L □ PELAGOL L

CONSENSUS REPORTS: IARC Cancer Review: Group 2B IMEMDT 7,56,87; Human Limited Evidence IMEMDT 27,103,82. EPA Genetic Toxicology Program. Reported in EPA TSCA Inventory.

DFG MAK: Animal Carcinogen, Suspected Human Carcinogen

NIOSH REL: (2,4-diaminoanisole) Reduce to lowest feasible level

SAFETY PROFILE: Confirmed carcinogen. Poison by intraperitoneal route. Moderately toxic by ingestion. Experimental reproductive effects. Human mutation data reported. A skin irritant. When heated to decomposition it emits toxic fumes of NO_x. See also other diaminoanisole entries.

DBO400   CAS: 39156-41-7   HR: 3
2,4-DIAMINOANISOLE SULPHATE
mf: $C_7H_{10}N_2O \bullet H_2O_4S$ mw: 236.27
PROP: Off white to violet powder. Sol in water and ethanol.
SYNS: ANISOLE, 2,4-DIAMINO-, HYDROGEN SULFATE □ ANISOLE, 2,4-DIAMINO-, SULFATE □ 1,3-BENZENEDIAMINE, 4-METHOXY, SULFATE (1:1) (9CI) □ C.I. 76051 □ C.I. OXIDATION BASE 12A □ 2,4-DAA SULFATE □ 2,4-DIAMINOANISOLE SULPHATE □ 2,4-DIAMINO-ANISOL SULPHATE □ 2,4-DIAMINO-1-METHOXYBENZENE □ 1,3-DIAMINO-4-METHOXY-BENZENE SULPHATE □ 2,4-DIAMINO-1-METHOXY-BENZENE SULPHATE □ 2,4-DIAMINOSOLE SULPHATE □ DURAFUR BROWN MN □ FOURAMINE BA □ FOURRINE 76 □ FOURRINE SLA □ FURRO SLA □ 4-METHOXY-1,3-BENZENEDIAMINE SULFATE □ 4-METHOXY-1,3-BENZENEDIAMINE SULFATE (1:1) □ 4-METHOXY-1,3-BENZENEDIAMINE SULPHATE □ 4-METHOXY-m-PHENYLENEDIAMINE SULFATE □ p-METHOXY-m-PHENYLENEDIAMINE SULPHATE □ 4-METHOXY-m-PHENYLENEDIAMINE SULPHATE □ 4-MMPD SULPHATE □ NAKO TSA □ NCI-C01989 □ OXIDATION BASE 12A □ PELAGOL BA □ PELAGOL GREY □ PELAGOL GREY SLA □ PELAGOL SLA □ RENAL SLA □ URSOL SLA □ ZOBA SLE

CONSENSUS REPORTS: NTP 10th Report on Carcinogens. IARC Cancer Review: Animal Sufficient Evidence IMEMDT

27,103,82; Animal Inadequate Evidence IMEMDT 16,51,78. NCI Carcinogenesis Bioassay (feed); Clear Evidence: mouse, rat NCITR* NCI-CG-TR-84,78. Reported in EPA TSCA Inventory. EPA Genetic Toxicology Program. Community Right-To-Know List.
SAFETY PROFILE: Confirmed carcinogen with experimental carcinogenic, neoplastigenic, and tumorigenic data. Poison by intraperitoneal route. Mutation data reported. When heated to decomposition it emits very toxic fumes of NO_x and SO_x. See also other diaminoanisole entries.

**DBP909   CAS: 145-49-3   HR: 3
1,5-DIAMINOANTHRARUFIN**
mf: $C_{14}H_{10}N_2O_4$ mw: 270.26
PROP: Dark red crystals or powder. Sol concentration in H_2SO_4; sltly sol in H_2O and EtOH.
SYNS: 4,8-DIAMINOANTHRARUFIN □ 1,5-DIAMINO-4,8-DIHYDROXY-9,10-ANTHRACENEDIONE □ 1,5-DIAMINO-4,8-DIHYDROXYANTHRAQUINONE □ leuco-1,5-DIAMINO-4,8-DIHYDROXYANTHRAQUINONE □ 4,8-DIAMINO-1,5-DIHYDROXYANTHRAQUINONE □ 1,5-DIHYDROXY-4,8-DIAMINOANTHRACHINON (CZECH) □ 1,5-DIHYDROXY-4,8-DIAMINOANTHRA-QUINONE
CONSENSUS REPORTS: Reported in EPA TSCA Inventory.
SAFETY PROFILE: Poison by intravenous route. An eye irritant. Mutation data reported. When heated to decomposition it emits toxic fumes of NO_x.

**DBT200   CAS: 92-26-2   HR: 3
3,6-DIAMINO-2,7-DIMETHYLACRIDINE**
mf: $C_{15}H_{15}N_3$ mw: 237.33
PROP: Crystals from $PhNH_2$. Mp: 325°.
SYNS: ACRIDINE YELLOW BASE □ 2,8-DIAMINO-3,7-DIMETHYLACRIDINE
CONSENSUS REPORTS: Reported in EPA TSCA Inventory.
SAFETY PROFILE: Poison by subcutaneous route. Mutation data reported. When heated to decomposition it emits toxic fumes of NO_x.

**DBU800   CAS: 13426-91-0   HR: 2
1,2-DIAMINOETHANE COPPER
     COMPLEX**
DOT: UN 1761
mf: $C_2H_{10}N_2 \cdot xCu$ mw: 506.92
PROP: Herbicide.
SYNS: COPPER-ETHYLENEDIAMINE COMPLEX □ CUPRIETHYLENE DIAMINE □ CUPRIETHYLENEDIAMINE, solution (DOT) □ ETHANE, 1,2-DIAMINO-, COPPER COMPLEX □ KOMEEN □ KOPLEX AQUATIC HERBICIDE
CONSENSUS REPORTS: Copper and its compounds are on the Community Right-To-Know List.
DOT CLASSIFICATION: 8; Label: Corrosive, Poison
SAFETY PROFILE: Moderately toxic by ingestion. An irritating and corrosive material to the skin, eyes, and mucous membranes. When heated to decomposition it emits toxic fumes of NO_x. See also COPPER COMPOUNDS.

**DBV400   CAS: 1239-45-8   HR: 3
2,7-DIAMINO-10-ETHYL-9-PHENYL-
     PHENANTHRIDINIUM BROMIDE**
mf: $C_{21}H_{20}N_3 \cdot Br$ mw: 394.35
PROP: Dark-red crystals from EtOH. Mp: 238–240°.
SYNS: 3,8-DIAMINO-5-ETHYL-6-PHENYLPHENAN-THRIDINIUM BROMIDE □ 2,7-DIAMINO-9-PHENYL-10-ETHYLPHENANTHRIDINIUM BROMIDE □ 2,7-DI-AMINO-9-PHENYLPHENANTHRIDINE ETHOBROMIDE □ DROMILAC □ ETHIDIUM BROMIDE □ HOMIDIUM BROMIDE □ RD 1572
CONSENSUS REPORTS: EPA Genetic Toxicology Program.
SAFETY PROFILE: Poison by intraperitoneal and subcutaneous routes. Human mutation data reported. When heated to decomposition it emits very toxic fumes of NO_x and Br^-. See also BROMIDES.

**DBX400   CAS: 8048-52-0   HR: 3
3,6-DIAMINO-10-METHYLACRIDINIUM
     CHLORIDE with 3,6-ACRIDINE-
     DIAMINE**
mf: $C_{14}H_{14}N_3 \cdot Cl \cdot C_{13}H_{11}N_3$ mw: 469.03

SYNS: ACRIFLAVIN □ ACRIFLAVINE mixture with PROFLAVINE □ ACRIFLAVINIUM CHLORIDE □ ACRIFLAVINIUM CHLORIDUM □ ACRIFLAVON □ ANGIFLAN □ ASSIFLAVINE □ AVLON □ BIAL-FLAVINA □ BIOACRIDIN □ BOVOFLAVIN □ BURNOL □ BUROFLAVIN □ CHOLIFLAVIN □ CHROMO-FLAVINE □ DIACRID □ 3,6-DIAMINOACRIDINE mixture with 3,6-DIAMINO-10-METHYLACRIDINIUM CHLORIDE □ 2,8-DIAMINO-10-METHYLACRIDINIUM CHLORIDE mixture with 2,8-DIAMINOACRIDINE □ EUFLAVINE □ FLAVACRIDINUM HYDROCHLORICUM □ FLAVINE □ FLAVIOFORM □ FLAVIPIN □ FLAVISEPT □ GLYCO-FLAVINE □ GONACRINE □ ISRAVIN □ MEDIFLAVIN □ NEUTRAL ACRIFLAVINE □ PANFLAVIN □ PANTONSILETTEN □ TRACHOSEPT □ TRIPLA-ETILO □ TRYPAFLAVINE □ VETAFLAVIN □ XANTH-ACRIDINUM □ ZORIFLAVIN

CONSENSUS REPORTS: IARC Cancer Review: Group 3 IMEMDT 7,56,87; Animal Inadequate Evidence IMEMDT 13,31,77. EPA Genetic Toxicology Program.
SAFETY PROFILE: Poison by subcutaneous route. Questionable carcinogen with experimental tumorigenic data. Human mutation data reported. A topical antiseptic used in the treatment of gonorrhea. When heated to decomposition it emits very toxic fumes of NO_x and Cl^-. See also 3,6-DIAMINO-10-METHYLACRIDINIUM CHLORIDE.

DBY700   CAS: 82-33-7   HR: 2
1,4-DIAMINO-5-NITRO ANTHRA-QUINONE
mf: $C_{14}H_9N_3O_4$ mw: 283.26
PROP: Violet needles from EtOH.
SYNS: 9,10-ANTHRACENEDIONE, 1,4-DIAMINO-5-NITRO-(9CI) □ CELLITON FAST VIOLET B □ CELLIT-ON FAST VIOLET BA-CF □ CELLITON VIOLET B □ C.I. 62030 □ CIBACET BRILLIANT VIOLET 3B □ C.I. DISPERSE VIOLET 8 □ CILLA FAST VIOLET B □ DIACELLITON FAST VIOLET B □ 1,4-DIAMINO-5-NITRO-9,10-ANTHRACENEDIONE □ DIANIX FAST VIOLET B □ DISPERSE VIOLET 2S □ DURANOL BRILLIANT BLUE VIOLET BR □ DURANOL BRILLIANT VIOLET BR □ FENACET FAST VIOLET B □ KAYALON FAST VIOLET BR □ MIKETON FAST VIOLET B □ NITROCRESOLAMINE □ PALANIL VIOLET 3B □ PERLITON VIOLET B □ SAMARON BRILLIANT VIOLET B □ SERISOL FAST VIOLET B □ SUPRACET FAST VIOLET B □ TERASIL BRILLIANT VIOLET 3B □ VIOLET 2S □ VONTERYL VIOLET 2B

CONSENSUS REPORTS: Reported in EPA TSCA Inventory.
SAFETY PROFILE: Poison by intravenous route. Questionable carcinogen with experimental tumorigenic data. Mutation data reported. When heated to decomposition it emits toxic fumes of NO_x.

DCC800   CAS: 141-86-6   HR: 3
2,6-DIAMINOPYRIDINE
mf: $C_5H_7N_3$ mw: 109.15
PROP: Crystals or leaflets. Mp: 121.5°, bp: 148–150° @ 5 mm.
CONSENSUS REPORTS: Reported in EPA TSCA Inventory.
SAFETY PROFILE: Poison by intravenous and intraperitoneal routes. Mutation data reported. When heated to decomposition it emits toxic fumes of NO_x.

DCD050   CAS: 25007-79-8   HR: 3
2,6-DIAMINO-4-(2-PYRIDYL)-s-TRIAZINE
mf: $C_8H_8N_6$ mw: 188.22
SYNS: AI 3-51408 □ s-TRIAZINE, 2,6-DIAMINO-4-(2-PYRIDYL)- □ 1,3,5-TRIAZINE-2,4-DIAMINE, 6-(2-PYRIDINYL)-
SAFETY PROFILE: A poison by intravenous route. When heated to decomposition it emits toxic vapors of NO_x.

DCE000   CAS: 636-23-7   HR: 3
2,4-DIAMINOTOLUENE DIHYDRO-CHLORIDE
mf: $C_7H_{10}N_2 \cdot 2ClH$ mw: 195.11
SYN: METATOLYLENEDIAMINE DIHYDRO-CHLORIDE
CONSENSUS REPORTS: Reported in EPA TSCA Inventory.
SAFETY PROFILE: Poison by intraperitoneal route. Moderately toxic by ingestion. Questionable carcinogen with experimental neoplastigenic data. When heated to decomposition it emits very toxic fumes of NO_x and HCl.

DCE200   CAS: 615-45-2   HR: 3
2,5-DIAMINOTOLUENE DIHYDRO-CHLORIDE

mf: $C_7H_{10}N_2 \cdot 2ClH$ mw: 195.11
SYNS: 2-METHYL-1,4-BENZENEDIAMINE
DIHYDROCHLORIDE □ p-TOLUENEDIAMINE
DIHYDROCHLORIDE
CONSENSUS REPORTS: Reported in EPA
TSCA Inventory.
SAFETY PROFILE: Poison by ingestion.
Experimental teratogenic effects. Mutation
data reported. When heated to
decomposition it emits very toxic fumes of
NO_x and HCl. See also CHLORIDES.

DCE600   CAS: 615-50-9   HR: 3
2,5-DIAMINOTOLUENE SULFATE
mf: $C_7H_{10}N_2 \cdot H_2O_4S$ mw: 220.27
PROP: Pale purple powder. Mp: > 300°.
SYNS: C.I. 76043 □ p-DIAMINOTOLUENE SULFATE □
2,5-DIAMINOTOLUENE SULPHATE □ 2-METHYL-1,4-
BENZENEDIAMINE SULFATE □ 2-METHYL-p-
PHENYLENEDIAMINE SULPHATE □ NCI-C01832 □ p-
TOLUENEDIAMINE SULFATE □ 2,5-TOLUENEDI-
AMINE SULFATE □ TOLUENE-2,5-DIAMINE, SULFATE
(1:1) (8CI) □ p-TOLUENEDIAMINE SULPHATE □
TOLUENE-2,5-DIAMINE SULPHATE □ p-TOLUYLENE-
DIAMINE SULPHATE □ TOLUYLENE-2,5-DIAMINE
SULPHATE □ p-TOLYLENEDIAMINE SULPHATE
CONSENSUS REPORTS: IARC Cancer
Review: Animal Indefinite Evidence
IMEMDT 16,97,78. NCI Carcinogenesis
Bioassay Completed; Results Indefinite:
mouse, rat NCITR* NCI-CG-TR-126,78.
Reported in EPA TSCA Inventory.
SAFETY PROFILE: Poison by ingestion and
intraperitoneal routes. Mutation data
reported. When heated to decomposition it
emits very toxic fumes of NO_x and SO_x. See
also SULFATES.

DCF200   CAS: 1455-77-2   HR: 2
3,5-DIAMINO-s-TRIAZOLE
mf: $C_2H_5N_5$ mw: 99.12
PROP: Prisms from H_2O. Mp: 206°.
SYNS: GUANAZOLE □ NCI-C04819 □ NSC-1895
CONSENSUS REPORTS: NCI
Carcinogenesis Studies (ipr); Equivocal
Evidence: rat CANCAR 40,1935,77; No
Evidence: mouse CANCAR 40,1935,77.
Reported in EPA TSCA Inventory.
SAFETY PROFILE: Human systemic effects
by intravenous route: leukopenia (reduced

white blood cell count) and thrombocytopenia (reduced blood platelet count).
Human mutation data reported.
Questionable carcinogen with experimental
tumorigenic data. When heated to
decomposition it emits toxic fumes of NO_x.

DCG000   HR: 3
DIAMMINEMALONATO PLATINUM (II)
mf: $C_3H_8N_2O_4Pt$ mw: 331.22
PROP: IDLH 4 mg/m^3 (as Pt).
CONSENSUS REPORTS: Reported in EPA
TSCA Inventory.
SAFETY PROFILE: Poison by
intraperitoneal route. Mutation data
reported. When heated to decomposition it
emits toxic fumes of NO_x. See also
PLATINUM COMPOUNDS.

DCG800   CAS: 7784-44-3   HR: 3
DIAMMONIUM HYDROGEN ARSENATE
DOT: UN 1546
mf: $AsH_3O_4 \cdot 2H_3N$ mw: 176.03
PROP: White powder or plate-like colorless
crystals or prisms. Mp: decomp to yield
NH_3.
SYNS: AMMONIUM ACID ARSENATE □ AMMONIUM
ARSENATE, solid (DOT) □ DIAMMONIUM ARSENATE □
DIAMMONIUM MONOHYDROGEN ARSENATE □
DIBASIC AMMONIUM ARSENATE □ SECONDARY
AMMONIUM ARSENATE
CONSENSUS REPORTS: Arsenic and its
compounds are on the Community Right-
To-Know List.
OSHA PEL: OSHA: Cancer Hazard
ACGIH TLV: TWA 0.01 mg/m^3; Confirmed
Human Carcinogen; BEI: 35 μ (As)/L
inorganic arsenic and methylated
metabolites in urine
NIOSH REL: (Inorganic Arsenic) CL 0.002
mg(As)/m^3/15M
DOT CLASSIFICATION: 6.1; Label: Poison
SAFETY PROFILE: Confirmed human
carcinogen. A poison. When heated to
decomposition it emits very toxic fumes of
As, NO_x, and NH_3. See also ARSENIC.

DCH000   CAS: 3164-29-2   HR: 3
DIAMMONIUM TARTRATE
mf: $C_4H_6O_6 \cdot 2H_3N$ mw: 184.18
PROP: White crystals. D: 1.60. Sol in water.
SYNS: AMMONIUM TARTRATE (DOT) □
AMMONIUM-d-TARTRATE □ 2,3-
DIHYDROXYBUTANEDIOIC ACID, DIAMMONIUM
SALT □ l-TARTARIC ACID, AMMONIUM SALT □
TARTARIC ACID, DIAMMONIUM SALT
CONSENSUS REPORTS: Reported in EPA
TSCA Inventory.
SAFETY PROFILE: Poison by intravenous
route. Moderately toxic by subcutaneous
route. When heated to decomposition it
emits very toxic fumes of NH_3 and NO_x.

DCH200   CAS: 2050-92-2   HR: 3
DIAMYL AMINE
DOT: UN 2841
mf: $C_{10}H_{23}N$ mw: 157.34
PROP: Water-white liquid. Bp: 210–211°,
flash p: 124°F, d: 0.777 @ 20°/20°, vap d:
5.42.
SYNS: DI-n-AMYLAMINE (DOT) □ DIPENTYLAMINE
□ PENTYL PENTYLAMINE
CONSENSUS REPORTS: Reported in EPA
TSCA Inventory.
DOT CLASSIFICATION: 6.1; Label: KEEP
AWAY FROM FOOD
SAFETY PROFILE: Poison by inhalation,
ingestion, and skin contact. A severe skin
irritant. See also AMINES. Flammable
liquid when exposed to heat or flame; can
react with oxidizing materials. To fight fire,
use alcohol foam, foam, CO_2, dry chemical.
When heated to decomposition it emits
toxic fumes of NO_x.

DCI000   CAS: 120-95-6   HR: 3
DI-tert-AMYLPHENOL
mf: $C_{16}H_{26}O$ mw: 234.42
PROP: Molten form. Mp: 23°.
SYNS: 2,4-DI-tert-AMYLPHENOL □ 2,4-DI-tert-
PENTYLPHENOL □ PHENOL, 2,4-DI-tert-PENTYL- □
PRODOX 156
CONSENSUS REPORTS: Reported in EPA
TSCA Inventory.
DOT CLASSIFICATION: 6.1; Label: KEEP
AWAY FROM FOOD

SAFETY PROFILE: Poison by ingestion. An
eye irritant. When heated to decomposition
it emits acrid smoke and irritating fumes.
See also PHENOL.

DCJ200   CAS: 119-90-4   HR: 3
o-DIANISIDINE
mf: $C_{14}H_{16}N_2O_2$ mw: 244.32
PROP: Colorless leaflets or crystals. Mp:
137–138°, flash p: 403°F, vap d: 8.5. Sol in
C_6H_6 and AcOH; sltly sol in H_2O.
SYNS: ACETAMINE DIAZO BLACK RD □ AMACEL
DEVELOPED NAVY SD □ AZOENE FAST BLUE BASE □
AZOFIX BLUE B SALT □ AZOGNE FAST BLUE B □
BLUE BN BALSE □ BRENTAMINE FAST BLUE B BASE □
CELLITAZOL B □ C.I. 24110 □ C.I. AZOIC DIAZO
COMPONENT 48 □ CIBACETE DIAZO NAVY BLUE 2B
□ C.I. DISPERSE BLACK 6 □ DIACELLITON FAST GREY
G □ DIACEL NAVY DC □ o-DIANISIDIN (CZECH,
GERMAN) □ o-DIANISIDINA (ITALIAN) □
O,O'DIANISIDINE □ 3,3'-DIANISIDINE □ DIATO BLUE
BASE B □ 3,3'-DIMETHOXYBENZIDIN (CZECH) □ 3,3'-
DIMETHOXYBENZIDINE □ 3,3'-DIMETOSSIBENZO-
DINA (ITALIAN) □ FAST BLUE B BASE □ HILTONIL
FAST BLUE B BASE □ HILTOSAL FAST BLUE B SALT □
HINDASOL BLUE B SALT □ KAKO BLUE B SALT □
KAYAKU BLUE B BASE □ LAKE BLUE B BASE □
MEISEI TERYL DIAZO BLUE HR □ MITSUI BLUE B
BASE □ NAPHTHANIL BLUE B BASE □ NEUTROSEL
NAVY BN □ RCRA WASTE NUMBER U091 □ SANYO
FAST BLUE SALT B □ SETACYL DIAZO NAVY R □
SPECTROLENE BLUE B
CONSENSUS REPORTS: NTP 10th Report
on Carcinogens. IARC Cancer Review:
Group 2B IMEMDT 7,198,87; Animal
Sufficient Evidence IMEMDT 4,41,74. EPA
Genetic Toxicology Program. Community
Right-To-Know List. Reported in EPA
TSCA Inventory.
DFG MAK: Animal Carcinogen, Suspected
Human Carcinogen
NIOSH REL: (o-Dianisidine-Based Dyes)
Reduce to lowest feasible concentration
SAFETY PROFILE: Confirmed carcinogen
with experimental tumorigenic data.
Moderately toxic by ingestion. Mutation data
reported. Combustible when exposed to
heat or flame. When heated to
decomposition it emits toxic fumes of NO_x.

D

DCJ400   CAS: 91-93-0   HR: 3
DIANISIDINE DIISOCYANATE

mf: $C_{16}H_{12}N_2O_4$ mw: 296.30

SYNS: 4,4'-DIISOCYANATO-3,3'-DIMETHOXY-1,1'-BIPHENYL ☐ 3,3'-DIMETHOXYBENZIDINE-4,4'-DIISOCYANATE ☐ 3,3'-DIMETHOXY-4,4'-BIPHENYL-ENE DIISOCYANATE ☐ NCI-C02175

CONSENSUS REPORTS: IARC Cancer Review: Group 3 IMEMDT 7,56,87; Animal Limited Evidence IMEMDT 39,279,86. NCI Carcinogenesis Bioassay (feed); No Evidence: mouse NCITR* NCI-CG-TR-128,79; Clear Evidence: rat NCITR* NCI-CG-TR-128,79.

NIOSH REL: (Diisocyanates) TWA 0.005 ppm; CL 0.02 ppm/10M

SAFETY PROFILE: Poison by intravenous route. A strong sensitizer. Questionable carcinogen with experimental carcinogenic data. When heated to decomposition it emits toxic fumes of NO_x. See also CYANATES.

DCJ800   CAS: 61790-53-2   HR: 1
DIATOMACEOUS EARTH

PROP: Composed of skeletons of small aquatic plants related to algae and contains as much as 88% amorphous silica (DTLVS* 4,120,80). White to buff-colored solid. Insol in water; sol in hydrofluoric acid.

SYNS: AMORPHOUS SILICA ☐ CELITE ☐ D.E. ☐ DIATOMACEOUS EARTH, NATURAL ☐ DIATOMACEOUS SILICA ☐ DIATOMITE ☐ INFUSORIAL EARTH ☐ KIESELGUHR ☐ SILICA, AMORPHOUS-DIATOMACEOUS EARTH (UNCALCINED) (ACGIH)

CONSENSUS REPORTS: IARC Cancer Review: Group 3 IMEMDT 7,341,87; Animal Inadequate Evidence IMEMDT 42,39,87; Human Inadequate Evidence IMEMDT 42,39,87. Reported in EPA TSCA Inventory.

OSHA PEL: TWA 6 mg/m³

ACGIH TLV: TWA (nuisance particulate) 10 mg/m³ of total dust (when toxic impurities are not present, e.g., quartz <1%)

DFG MAK: 4 mg/m³ as fine dust

SAFETY PROFILE: A nuisance dust that may cause fibrosis of the lungs. Roasting or calcining at high temperatures produces cristobalite and tridymite, thus increasing the fibrogenicity of the material. A questionable carcinogen.

DCK700   CAS: 283-66-9   HR: 3
1,6-DIAZA-3,4,8,9,12,13-HEXAOXABICYCLO(4.4.4)TETRADECANE

mf: $C_6H_{12}N_2O_6$ mw: 208.17

PROP: Crystals.

SYN: HEXAMETHYLENETRIPEROXYDIAMINE

DOT CLASSIFICATION: Forbidden

SAFETY PROFILE: The dry material is a powerful explosive that is heat- and shock-sensitive. Explodes on contact with bromine or sulfuric acid. When heated to decomposition it emits toxic fumes of NO_x. See also PEROXIDES.

DCK759   CAS: 439-14-5   HR: 3
DIAZEPAM

mf: $C_{16}H_{13}ClN_2O$ mw: 284.76

PROP: Plates or crystals from Me_2CO/pet ether. Mp: 125–126°.

SYNS: ALBORAL ☐ AMIPROL ☐ ANSIOLISINA ☐ APAURIN ☐ APOZEPAM ☐ ATENSINE ☐ ATILEN ☐ BIALZEPAM ☐ CALMOCITENE ☐ CERCINE ☐ 7-CHLORO-1,3-DIHYDRO-1-METHYL-5-PHENYL-2H-1,4-BENZODIAZEPIN-2-ONE ☐ 7-CHLORO-1-METHYL-5-3H-1,4-BENZODIAZEPIN-2(1H)-ONE ☐ 7-CHLORO-1-METHYL-2-OXO-5-PHENYL-3H-1,4-BENZODIAZEPINE ☐ 7-CHLORO-1-METHYL-5-PHENYL-2H-1,4-BENZODI-AZEPIN-2-ONE ☐ 7-CHLORO-1-METHYL-5-PHENYL-1,3-DIHYDRO-2H-1,4-BENZODIAZEPIN-2-ONE ☐ CONDI-TION ☐ DIACEPAN ☐ DIAPAM ☐ DIAZETARD ☐ DIENPAX ☐ DIPAM ☐ DOMALIUM ☐ DUKSEN ☐ E-PAM ☐ ERIDAN ☐ FAUSTAN ☐ FRUSTAN ☐ GIHITAN ☐ KIATRIUM ☐ LEMBROL ☐ LEVIUM ☐ LIBERETAS ☐ METHYL DIAZEPINONE ☐ 1-METHYL-5-PHENYL-7-CHLORO-1,3-DIHYDRO-2H-1,4-BENZODIAZEPIN-2-ONE ☐ MOROSAN ☐ NSC-77518 ☐ PACITRAN ☐ PAXATE ☐ PLIDAN ☐ QUETINIL ☐ QUIATRIL ☐ RELAMINAL ☐ RELANIUM ☐ RENBORIN ☐ SAROMET ☐ SEDIPAM ☐ SEDUXEN ☐ SERENACK ☐ SERENZIN ☐ SETONIL ☐ SONACON ☐ STESOLID ☐ TENSOPAM ☐ TRANIMUL ☐ TRANQUIRIT ☐ UMBRIUM ☐ UNISEDIL ☐ VALEO ☐ VALITRAN ☐ VALIUM ☐ VATRAN ☐ VIVAL ☐ ZIPAN

CONSENSUS REPORTS: IARC Cancer Review: Group 3 IMEMDT 7,189,87; Animal Inadequate Evidence IMEMDT 7,189,87; Human Inadequate Evidence

IMEMDT 7,189,87. Reported in EPA TSCA Inventory. EPA Genetic Toxicology Program.

SAFETY PROFILE: Poison by ingestion, parenteral, subcutaneous, intravenous, and intraperitoneal routes. Moderately toxic by skin contact. Questionable carcinogen with experimental tumorigenic data. Human systemic effects: dermatitis, effect on inflammation or mediation of inflammation, change in cardiac rate, somnolence, respiratory depression, and other respiratory changes, visual field changes, diplopia (double vision), change in motor activity, muscle contraction or spasticity, ataxia (loss of muscle coordination), an antipsychotic and general anesthetic. Human reproductive effects by ingestion and intravenous routes causing developmental abnormalities of the fetal cardiovascular (circulatory) system and postnatal effects. Experimental teratogenic and reproductive effects. Human mutation data reported. An allergen. A drug for the treatment of anxiety. When heated to decomposition it emits very toxic fumes of Cl⁻ and NO$_x$.

DCL125   CAS: 2294-47-5   HR: 3
1,4-DIAZIDOBENZENE
mf: $C_6H_4N_6$ mw: 160.16
PROP: Yellow crystals from Et$_2$O. Mp: 83°.
SYNS: BENZENE, 1,4-DIAZIDO- □ p-DIAZIDO-BENZENE (DOT) □ 1,4-DIAZIDOBENZENE □ p-PHENYLENE DIAZIDE
DOT CLASSIFICATION: Forbidden
SAFETY PROFILE: Explodes violently when heated. When heated to decomposition it emits toxic fumes of NO$_x$. See also AZIDES.

DCL600   CAS: 629-13-0   HR: 3
1,2-DIAZIDOETHANE
mf: $C_2H_4N_6$ mw: 112.09
PROP: Bp: 54–55° @ 11 mm.
SYN: ETHANE, 1,2-DIAZIDO-
DOT CLASSIFICATION: Forbidden
SAFETY PROFILE: Explodes on heating or on contact with sulfuric acid. Upon

decomposition it emits toxic fumes of NO$_x$. See also AZIDES.

DCM750   CAS: 333-41-5   HR: 3
DIAZINON
mf: $C_{12}H_{21}N_2O_3PS$ mw: 304.38
PROP: Liquid with faint ester-like odor. Bp: 84° @ 0.002 mm, d: 1.116 @ 20°/4°. Miscible in org solvs.
SYNS: ALFA-TOX □ BASUDIN □ BASUDIN 10 G □ BAZUDEN □ DAZZEL □ O,O-DIAETHYL-O-(2-ISOPROPYL-4-METHYL-PYRIMIDIN-6-YL)-MONOTHIOPHOSPHAT (GERMAN) □ O,O-DIAETHYL-O-(2-ISOPROPYL-4-METHYL)-6-PYRIMIDYL-THIONOPHOSPHAT (GERMAN) □ DIANON □ DIATERR-FOS □ DIAZAJET □ DIAZATOL □ DIAZIDE □ DIAZINONE □ DIAZITOL □ DIAZOL □ O,O-DIETHYL-O-(2-ISOPROPYL-4-METHYL-PYRIMIDIN-6-YL)MONOTHIOFOSFAAT (DUTCH) □ O,O-DIETHYL-O-(2-ISOPROPYL-4-METHYL-6-PYRIMIDINYL)PHOS-PHOROTHIOATE □ O,O-DIETHYL-O-(2-ISOPROPYL-6-METHYL-4-PYRIMIDINYL) PHOSPHOROTHIOATE □ DIETHYL 4-(2-ISOPROPYL-6-METHYLPYRIMIDINYL)-PHOSPHOROTHIONATE □ O,O-DIETHYL-O-(2-ISOPROPYL-4-METHYL-6-PYRIMIDYL)PHOSPHORO-THIOATE □ O,O-DIETHYL-O-(2-ISOPROPYL-4-METHYL-6-PYRIMIDYL) THIONOPHOSPHATE □ O,O-DIETHYL-2-ISOPROPYL-4-METHYLPYRIMIDYL-6-THIOPHOSPHATE □ O,O-DIETHYL-O-6-METHYL-2-ISOPROPYL-4-PYRIMIDINYL PHOSPHOROTHIOATE □ O,O-DIETIL-O-(2-ISOPROPIL-4-METIL-PIRIMIDIN-6-IL)-MONOTIOFOSFATO (ITALIAN) □ DIMPYLATE □ DIPOFENE □ DIZINON □ DYZOL □ ENT 19,507 □ G 301 □ G-24480 □ GARDENTOX □ GEIGY 24480 □ O-2-ISOPROPYL-4-METHYLPYRIMIDYL-O,O-DIETHYL PHOSPHOROTHIOATE □ ISOPROPYLMETHYL-PYRIMIDYL DIETHYL THIOPHOSPHATE □ KAYAZINON □ KAYAZOL □ NCI-C08673 □ NEDCIDOL □ NEOCIDOL □ NIPSAN □ NUCIDOL □ SAROLEX □ SPECTRACIDE □ THIOPHOSPHATE de O,O-DIETHYLE et de o-2-ISOPROPYL-4-METHYL-6-PYRIMIDYLE (FRENCH)
CONSENSUS REPORTS: NCI Carcinogenesis Bioassay (feed); No Evidence: mouse, rat NCITR* NCI-CG-TR-137,79. Reported in EPA TSCA Inventory. EPA Genetic Toxicology Program.
OSHA PEL: TWA 0.1 mg/m³ (skin)
ACGIH TLV: TWA 0.01 mg/m³ (skin); Not Classifiable as a Human Carcinogen
DFG MAK: 1 mg/m³
SAFETY PROFILE: Poison by ingestion, skin contact, subcutaneous, intravenous, and

intraperitoneal routes. Mildly toxic by inhalation. Human systemic effects by ingestion: changes in motor activity, muscle weakness, and sweating. Experimental teratogenic and reproductive effects. A skin and severe eye irritant. Human mutation data reported. When heated to decomposition it emits very toxic fumes of NO_x, PO_x, and SO_x.

DCN800   CAS: 623-73-4   HR: 3
DIAZOACETIC ESTER
mf: $C_4H_6N_2O_2$ mw: 114.12
PROP: Yellow oil with pungent odor. Mp: −22°, bp: 141° @ 720 mm.
SYNS: DAAE □ DIAZOACETIC ACID, ETHYL ESTER □ DIAZOESSIGSAEURE-AETHYLESTER (GERMAN) □ EDA □ ETHOXYCARBONYLDIAZOMETHANE □ ETHYL DIAZOACETATE
SAFETY PROFILE: Poison by ingestion and intravenous routes. Questionable carcinogen with experimental carcinogenic and tumorigenic data. Can explode. Explodes on contact with tris(dimethylamino) antimony. When heated to decomposition it emits toxic fumes of NO_x. See also ESTERS.

DCO800   CAS: 820-75-7   HR: 3
N-(DIAZOACETYL)GLYCINE HYDRA-ZINE
mf: $C_4H_7N_5O_2$ mw: 157.16
SYNS: N-DIAZOACETILGLICINA-IDRAZIDE (ITALIAN) □ DIAZOACETYLGLYCINE HYDRAZIDE □ N-DIAZOACETYL GLYCYLHYDRAZIDE □ NSC-58404
CONSENSUS REPORTS: EPA Genetic Toxicology Program.
SAFETY PROFILE: Poison by intravenous route. Moderately toxic by ingestion, intraperitoneal, and subcutaneous routes. Questionable carcinogen with experimental carcinogenic and neoplastigenic data. Mutation data reported. When heated to decomposition it emits toxic fumes including NO_x.

DCP800   CAS: 334-88-3   HR: 3
DIAZOMETHANE
mf: CH_2N_2 mw: 42.05

PROP: Yellow gas at ordinary temp which forms yellow solns in ethereal solvs. Mp: −145°, bp: −23°, d: 1.45. IDLH 2 ppm.
SYNS: AZIMETHYLENE □ DIAZIRINE
CONSENSUS REPORTS: IARC Cancer Review: Group 3 IMEMDT 7,56,87; Animal Sufficient Evidence IMEMDT 7,223,74. EPA Genetic Toxicology Program. Community Right-To-Know List.
OSHA PEL: TWA 0.2 ppm
ACGIH TLV: TWA 0.2 ppm; Suspected Human Carcinogen
DFG MAK: Animal Carcinogen, Suspected Human Carcinogen
SAFETY PROFILE: Confirmed carcinogen with experimental tumorigenic data. A poisonous irritant by inhalation. A powerful allergen. It can cause pulmonary edema and frequently causes hypersensitivity leading to asthmatic symptoms. Mutation data reported. Highly explosive when shocked, exposed to heat, or by chemical reaction. Undiluted liquid or gas may explode on contact with alkali metals, rough surfaces, heat (100°C), high-intensity light, or shock. When heated to decomposition or on contact with acid or acid fumes it emits highly toxic fumes of NO_x. Incompatible with alkali metals; calcium sulfate.

DCQ600   CAS: 2435-76-9   HR: 3
DIAZOURACIL
mf: $C_4H_2N_4O_2$ mw: 138.10
PROP: Crystals from H_2O. Mp: 198° (decomp).
SYNS: 5-DIAZOPYRIMIDINE-2,4(3H)-DIONE □ 5-DIAZO-2,4(1H,3H)-PYRIMIDINEDIONE □ 5-DIAZOURACIL □ 2,4-DIOSSI-5-DIAZOPIRIMIDINA (ITALIAN) □ 2,6-DIOXO-5-DIAZOPYRIMIDINE □ DU □ NSC-23519 □ (1,2,3)OXADIAZOLO(5,4-d)PYRIMIDIN-5(4H)-ONE
CONSENSUS REPORTS: Reported in EPA TSCA Inventory.
SAFETY PROFILE: Poison by subcutaneous and intraperitoneal routes. Mutation data reported. When heated to decomposition it emits toxic fumes of NO_x.

D

DCQ800 CAS: 34493-98-6 HR: 3
DIBEKACIN
mf: $C_{18}H_{37}N_5O_8$ mw: 451.60
PROP: A solid.
SYNS: DEBECACIN □ DIDEOXYKANAMYCIN B □ 3',4'-DIDEOXYKANAMYCIN B □ DKB □ ORBICIN
CONSENSUS REPORTS: EPA Genetic Toxicology Program.
SAFETY PROFILE: Poison by intraperitoneal, subcutaneous, intramuscular, and intravenous routes. Moderately toxic by ingestion. Experimental teratogenic and reproductive effects. An antibacterial agent. When heated to decomposition it emits toxic fumes of NO_x.

DCR400 CAS: 203-20-3 HR: 3
DIBENZ(a,j)ACEANTHRYLENE
mf: $C_{24}H_{14}$ mw: 302.38
PROP: Bright-yellow needles from C_6H_6. Mp: 181–181.3°.
SYN: 15,16-BENZDEHYDROCHOLANTHRENE
SAFETY PROFILE: Poison by intravenous route. Questionable carcinogen with experimental tumorigenic data. When heated to decomposition it emits acrid smoke and irritating fumes.

DCS200 CAS: 1977-10-2 HR: 3
DIBENZACEPIN
mf: $C_{18}H_{18}ClN_3O$ mw: 327.84
PROP: Pale-yellow cryst from pet ether. Mp: 109–110°.
SYNS: 2-CHLORO-11-(4-METHYL-1-PIPERAZINYL)-DIBENZO(b,f)(1,4)OXAZEPINE □ 2-CHLORO-11-(4-METHYL-1-PIPERAZINYL)-DIBENZO(b,f)(1,4)OXO-AZEPINE □ CL-62362 □ CL-71563 □ CLOXAZEPINE □ DIBENZOAZEPINE □ HF3170 □ LOXAPINE □ LW 3170 □ OXILAPINE □ S-805 □ SUM 3170
SAFETY PROFILE: Poison by ingestion, intraperitoneal, subcutaneous, and intravenous routes. Experimental teratogenic and reproductive effects. A tranquilizer. Many dibenz-azepine compounds have central nervous system effects. When heated to decomposition it emits very toxic fumes of Cl^- and NO_x.

DCS400 CAS: 226-36-8 HR: 3

DIBENZ(a,h)ACRIDINE
mf: $C_{21}H_{13}N$ mw: 279.35
PROP: Yellow crystals. Mp: 228°.
SYNS: 7-AZADIBENZ(a,h)ANTHRACENE □ DB(a,h)AC □ DIBENZ(a,d)ACRIDINE □ 1,2,5,6-DIBENZACRIDINE □ 1,2,5,6-DIBENZOACRIDINE □ 1,2,5,6-DINAPHTHACRIDINE
CONSENSUS REPORTS: NTP 10th Report on Carcinogens. IARC Cancer Review: Group 2B IMEMDT 7,56,87; Animal Sufficient Evidence IMEMDT 32,277,83; IMEMDT 3,247,73. EPA Genetic Toxicology Program.
SAFETY PROFILE: Confirmed carcinogen with experimental carcinogenic and tumorigenic data. Mutation data reported. When heated to decomposition it emits toxic fumes of NO_x. See also ANTHRACENE.

DCS600 CAS: 224-42-0 HR: 3
DIBENZ(a,j)ACRIDINE
mf: $C_{21}H_{13}N$ mw: 279.35
PROP: Yellow crystals. Mp: 216–217°.
SYNS: 7-AZADIBENZ(a,j)ANTHRACENE □ DB(a,j)AC □ DIBENZ(a,f)ACRIDINE □ 1,2,7,8-DIBENZACRIDINE □ 3,4,5,6-DIBENZACRIDINE □ DIBENZO(a,j)ACRIDINE □ 3,4,6,7-DINAPHTHACRIDINE
CONSENSUS REPORTS: NTP 10th Report on Carcinogens. IARC Cancer Review: Group 2B IMEMDT 7,56,87; Animal Sufficient Evidence IMEMDT 32,283,83; IMEMDT 3,254,73. EPA Genetic Toxicology Program.
SAFETY PROFILE: Confirmed carcinogen with experimental carcinogenic and tumorigenic data. Experimental reproductive effects. Mutation data reported. When heated to decomposition it emits toxic fumes of NO_x. See also ANTHRACENE.

DCS821 CAS: 125276-72-4 HR: 2
DIBENZ(a,h)ACRIDINE 3,4-DIOL-1,2-EPOXIDE
mf: $C_{21}H_{15}NO_3$ mw: 329.37
SYNS: BENZ(h)OXIRENO(5,6)BENZ(1,2-A)ACRIDINE-2,3-DIOL, 1A-α,2,3,13C-TETRAHYDRO-, (1A-α,2-β,3-α,13C-α)-(+–)- □ (+/–)-3α,4β-DIHYDROXY-1α,2α-EPOXY-1,2,3,4-TETRAHYDRODIBEN-Z(a,h)ACRIDINE

SAFETY PROFILE: Questionable carcinogen with experimental carcinogenic data reported. Mutation data reported. When heated to decomposition it emits toxic vapors of NO_x.

DCT050   CAS: 51-50-3   HR: 3
DIBENZAMINE
mf: $C_{16}H_{18}ClN$ mw: 259.80
SYNS: N-(2-CHLOROETHYL)DIBENZYLAMINE □ DIBENZYL CHLORETHYLAMINE □ N,N-DIBENZYL-β-CHLOROETHYLAMINE □ SYMPATHOLYTIN
SAFETY PROFILE: Poison by intravenous and intraperitoneal routes. Moderately toxic by ingestion and subcutaneous routes. Experimental reproductive effects. Can cause leukopenia (reduced white blood cell count). Mutation data reported. When heated to decomposition it emits very toxic fumes of Cl⁻ and NO_x. See also AROMATIC AMINES.

DCT400   CAS: 53-70-3   HR: 3
DIBENZ(a,h)ANTHRACENE
mf: $C_{22}H_{14}$ mw: 278.36
PROP: Silvery leaflets from AcOH. Mp: 266–267°.
SYNS: 1,2:5,6-BENZANTHRACENE □ DBA □ DB(a,h)A □ 1,2,5,6-DBA □ 1,2,5,6-DIBENZANTHRACEEN (DUTCH) □ 1,2,5,6-DIBENZANTHRACENE □ 1,2:5,6-DIBENZ(a)-ANTHRACENE □ DIBENZO(a,h)ANTHRACENE □ 1,2:5,6-DIBENZOANTHRACENE □ RCRA WASTE NUMBER U063
CONSENSUS REPORTS: NTP 10th Report on Carcinogens. IARC Cancer Review: Group 2A IMEMDT 7,56,87; Animal Sufficient Evidence IMEMDT 32,299,83; IMEMDT 3,178,73. EPA Genetic Toxicology Program. Reported in EPA TSCA Inventory.
SAFETY PROFILE: Confirmed carcinogen with experimental carcinogenic, tumorigenic, and neoplastigenic data. Poison by intravenous route. Human mutation data reported. When heated to decomposition it emits acrid smoke and irritating fumes.

DCY000   CAS: 194-59-2   HR: 3
7H-DIBENZO(c,g)CARBAZOLE

mf: $C_{20}H_{13}N$ mw: 267.34
PROP: Needles or crystals from alc. Mp: 158°.
SYNS: 7-AZA-7H-DIBENZO(c,g)FLUORENE □ 7H-DB(c,g)C □ 3,4,5,6-DIBENZCARBAZOL □ 3,4,5,6-DIBENZCARBAZOLE □ 3,4,5,6-DIBENZOCARBAZOLE □ 3,4,5,6-DINAPHTHACARBAZOLE
CONSENSUS REPORTS: NTP 10th Report on Carcinogens. IARC Cancer Review: Group 2B IMEMDT 7,56,87; Animal Sufficient Evidence IMEMDT 32,315,83; IMEMDT 3,260,73.
SAFETY PROFILE: Confirmed carcinogen with experimental carcinogenic, neoplastigenic, and tumorigenic data. Poison by intraperitoneal route. Mutation data reported. When heated to decomposition it emits toxic fumes of NO_x.

DCY200   CAS: 189-64-0   HR: 3
DIBENZO(b,def)CHRYSENE
mf: $C_{24}H_{14}$ mw: 302.38
PROP: Golden-orange plates from trichlorobenzene. Mp: 315°.
SYNS: BD(a,h)P □ DIBENZO(a,h)PYRENE □ 1,2,6,7-DIBENZOPYRENE □ 3,4,8,9-DIBENZOPYRENE □ 3,4,8,9-DIBENZPYRENE
CONSENSUS REPORTS: NTP 10th Report on Carcinogens. IARC Cancer Review: Group 2B IMEMDT 7,56,87; Animal Sufficient Evidence IMEMDT 32,331,83; IMEMDT 3,207,73.
SAFETY PROFILE: Confirmed carcinogen with experimental carcinogenic and tumorigenic data. When heated to decomposition it emits acrid smoke and irritating fumes. Mutation data reported.

DCY400   CAS: 191-30-0   HR: 3
DIBENZO(def,p)CHRYSENE
mf: $C_{24}H_{14}$ mw: 302.38
PROP: Pale-yellow plates from C_6H_6/EtOH. Mp: 164–165°
SYNS: BA 51-090462 □ DB(a,l)P □ DIBENZO(a,d)-PYRENE □ DIBENZO(a,l)PYRENE □ 1,2:3,4-DIBENZOPYRENE □ 1,2,9,10-DIBENZOPYRENE □ 2,3:4,5-DIBENZOPYRENE □ 1,2,3,4-DIBENZPYRENE □ 4,5,6,7-DIBENZPYRENE
CONSENSUS REPORTS: NTP 10th Report on Carcinogens. IARC Cancer Review:

Group 2B IMEMDT 7,56,87; Animal Sufficient Evidence IMEMDT 32,343,83; Animal Limited Evidence IMEMDT 3,224,73.

SAFETY PROFILE: Confirmed carcinogen with experimental tumorigenic data. When heated to decomposition it emits acrid smoke and irritating fumes.

DDC810   CAS: 153857-27-3 HR: 2
(+–)-DIBENZO(a,l)PYRENE-11,12-DI-HYDRODIOL

mf: $C_{24}H_{16}O_2$ mw: 336.39

SYN: DIBENZO(def,p)CHRYSENE-11,12-DIOL, 11,12-DIHYDRO-, (11S,12S)-REL-

SAFETY PROFILE: Questionable carcinogen with experimental carcinogenic data reported. When heated to decomposition it emits acrid smoke and irritating vapors.

DDE200   CAS: 257-07-8   HR: 3
DIBENZ(b,f)(1,4)OXAZEPINE

mf: $C_{13}H_9NO$ mw: 195.23

SYNS: CR □ EA 3547

CONSENSUS REPORTS: Reported in EPA TSCA Inventory.

SAFETY PROFILE: Poison by intraperitoneal and intravenous routes. Moderately toxic by ingestion and inhalation. Experimental teratogenic and reproductive effects. A human skin and eye irritant. Questionable carcinogen with experimental carcinogenic and tumorigenic data. When heated to decomposition it emits toxic fumes of NO_x.

DDE300   CAS: 2743-38-6   HR: 1
DIBENZOYLTARTARIC ACID

mf: $C_{18}H_{14}O_8$ mw: 358.32

PROP: Crystals. Mp: 89–92°.

SYNS: BUTANEDIOIC ACID, 2,3-BIS(BENZOYLOXY)-, (R-(R*,R*))- □ TARTARIC ACID, DIBENZOATE

CONSENSUS REPORTS: Reported in EPA TSCA Inventory.

SAFETY PROFILE: An eye irritant. When heated to decomposition it emits acrid smoke and irritating fumes.

DDG600   CAS: 73926-81-5   HR: 3
DIBENZYLETHYLSULFONIUM IODIDE MERCURIC IODIDE

PROP: IDLH 10 mg/m³ (as Hg).

SYN: DIBENZYLETHYLSULFONIUM IODIDE with MERCURY IODIDE (1:1)

CONSENSUS REPORTS: Mercury and its compounds are on the Community Right-To-Know List.

ACGIH TLV: Proposed: (inhalable fraction) 0.1 mg/m³; Not Classifiable as a Human Carcinogen)

DFG MAK: Confirmed Animal Carcinogen with Unknown Relevance to Humans

NIOSH REL: TWA 0.05 mg(Hg)/m³

SAFETY PROFILE: Poison by intravenous route. See also IODIDES and MERCURY IODIDE. When heated to decomposition it emits very toxic fumes of Hg, I⁻ and SO_x.

DDG800   CAS: 63-92-3   HR: 3
DIBENZYLINE HYDROCHLORIDE

mf: $C_{18}H_{22}ClNO•ClH$ mw: 340.32

PROP: Crystals from EtOH/Et₂O. Mp: 137.5–140°.

SYNS: 688A □ BENSYLYT □ 2-(N-BENZYL-2-CHLOROETHYLAMINO)-1-PHENOXYPROPANE HYDROCHLORIDE □ BENZYL(2-CHLOROETHYL)(1-METHYL-2-PHENOXYETHYL)AMINE HYDRO-CHLORIDE □ N-BENZYL-N-PHENOXYISOPROPYL-β-CHLORETHYLAMINE HYDROCHLORIDE □ BENZYLYT □ BLOCADREN □ N-(2-CHLOROETHYL)-N-(1-METHYL-2-PHENOXYETHYL)BENZENE-METHANAMINE HYDROCHLORIDE □ N-(2-CHLORO-ETHYL)-N-(1-METHYL-2-PHENOXYETHYL)BENZYL-AMINE HYDROCHLORIDE □ DIBENZYLENE □ DIBENZYLIN □ DIBENZYRAN □ FENOXYBENZAMIN □ NCI-C01661 □ PHENOXYBENZAMIDE HYDROCHLORIDE □ N-PHENOXYISOPROPYL-N-BENZYL-β-CHLOROETHYLAMINE HYDROCHLORIDE □ N-2-PHENOXYISOPROPYL-N-BENZYL-CHLOROETHYLAMINE HYDROCHLORIDE □ SKF 688A

CONSENSUS REPORTS: NTP 10th Report on Carcinogens. IARC Cancer Review: Group 2B IMEMDT 7,56,87; Animal Sufficient Evidence IMEMDT 24,185,80. NCI Carcinogenesis Bioassay Completed; Results Positive: mouse, rat NCITR* NCI-CG-TR-72,78.

SAFETY PROFILE: Confirmed carcinogen with experimental carcinogenic and teratogenic data. Poison by intraperitoneal, intravenous, and subcutaneous routes. Human systemic effects by ingestion: changes in tubules, including acute renal failure, acute tubular necrosis. Moderately toxic by ingestion. Other experimental reproductive effects. Mutation data reported. A long-acting adrenergic blocker. When heated to decomposition it emits very toxic fumes of NO_x and Cl^-.

DDH000   CAS: 780-24-5   HR: 3
DIBENZYLMERCURY
mf: $C_{14}H_{14}Hg$ mw: 382.87
PROP: Colorless crystals or needles from alc. Mp: 111°. Insol in Et_2O, pet ether; sltly sol in EtOH and C_6H_6; sol in org solvs. IDLH 10 mg/m³ (as Hg).
OSHA PEL: TWA 0.01 mg(Hg)/m³; STEL 0.03 mg/m³ (skin)
ACGIH TLV: TWA 0.01 mg(Hg)/m³; BEI: 35 µg/g creatinine total inorganic mercury in urine preshift; 15 µg/g creatinine total inorganic mercury in blood at end of shift at end of workweek.
DFG MAK: Confirmed Animal Carcinogen with Unknown Relevance to Humans
NIOSH REL: (Mercury, Aryl and Inorganic) CL 0.1 mg/m³ (skin)
SAFETY PROFILE: Poison by intravenous route. See also MERCURY COMPOUNDS, ORGANIC. When heated to decomposition it emits toxic fumes of Hg.

DDI450   CAS: 19287-45-7   HR: 3
DIBORANE
DOT: UN 1911/NA 1911
mf: B_2H_6 mw: 27.68
PROP: Colorless air- and moisture-sensitive gas; sickly-sweet odor. Mp: −165.5°, bp: −92.5°, d: 0.447 (liquid @ −112°), 0.577 (solid @ −183°), vap press: 224 mm @ −112°, autoign temp: 38–52°, lel: 0.9%, uel: 98%, flash p: −90°F. Sol in THF as BH_3-THF complex. IDLH 15 ppm.
SYNS: BOROETHANE □ BORON HYDRIDE □ DIBORANE MIXTURES (NA 1911) □ DIBORON HEXAHYDRIDE

CONSENSUS REPORTS: Reported in EPA TSCA Inventory. EPA Extremely Hazardous Substances List.
OSHA PEL: TWA 0.1 ppm
ACGIH TLV: TWA 0.1 ppm
DFG MAK: 0.1 ppm (0.1 mg/m³)
DOT CLASSIFICATION: 2.1; Label: Flammable Gas (NA 1911); DOT Class: 2.3; Label: Poison Gas, Flammable Gas
SAFETY PROFILE: Poison by inhalation. An irritant to skin, eyes, and mucous membranes comparable to chlorine, fluorine, arsine, and phosgene. The liquid causes local inflammation, blisters, redness, and swelling. Injuries to central nervous system, liver, and kidneys have also been produced in experimental animals. Similar observations have been reported in humans, resulting at times in a reaction resembling metal fume fever. Human exposure to pentaborane has produced signs of severe central nervous system irritation such as drowsiness, dizziness, visual disturbances, muscle twitching, and in severe cases, painful muscle spasm. Dangerously flammable when exposed to heat or flame or by chemical reaction. On contact with moisture, hydrogen is usually evolved. Highly explosive when exposed to heat or flame. Explosive reaction with air, tetravinyllead, O_2 above 165°C, octanol oxime + sodium hydroxide, benzene vapor, HNO_3Cl_2. Violent reaction with halocarbon liquids. Other boron hydrides evolve H_2 upon contact with moisture or can propagate a flame rapidly enough to cause an explosion. Heat can cause these materials to decompose violently or at least to evolve H_2. They also react with water or steam to evolve hydrogen. Reaction with Al or Li forms complex hydrides that may ignite spontaneously in air. Powerful oxidizing agents, such as chlorine gas, etc., can react violently with boron hydrides. Pentaborane (stable) is spontaneously flammable in air. See also BORANES and HYDRIDES.

DDJ000   CAS: 10318-26-0   HR: 3
DIBROMODULCITOL
mf: $C_6H_{12}Br_2O_4$ mw: 308.00
PROP: A solid. Mp: 187–188°.
SYNS: DBD □ 1,6-DIBROMODIDEOXYDULCITOL □
1,6-DIBROMO-1,6-DIDEOXYDULCITOL □ 1,6-
DIBROMO-1,6-DIDEOXYGALACTITOL □ 1,6-
DIBROMO-1,6-DIDEOXY-d-GALACTITOL □
DIBROMODULCITOL □ 1,6-DIBROMODULCITOL □
ELOBROMOL □ GALACTICOL □ MITOLAC □
MITOLACTOL □ NCI-C04795 □ NSC-104800
CONSENSUS REPORTS: NCI
Carcinogenesis Bioassay Completed; Results
Positive: mouse, rat (RRCRBU 52,1,75).
SAFETY PROFILE: Poison by ingestion.
Moderately toxic by intraperitoneal route.
Questionable carcinogen with experimental
carcinogenic, neoplastigenic, and
tumorigenic data. Human mutation data
reported. An anti-cancer agent taken orally.
When heated to decomposition it emits very
toxic fumes of Br⁻.

DDJ400   CAS: 3252-43-5   HR: 3
DIBROMOACETONITRILE
mf: C_2HBr_2N mw: 198.86
PROP: Water disinfection agent.
CONSENSUS REPORTS: Cyanide and its
compounds are on the Community Right-
To-Know List. Reported in EPA TSCA
Inventory.
SAFETY PROFILE: Poison by intravenous
route. Questionable carcinogen with
experimental carcinogenic data.
Experimental reproductive effects. Human
mutation data reported. See also NITRILES
and BROMIDES. When heated to
decomposition it emits very toxic fumes of
NO_x, Br⁻, and CN⁻.

DDJ800   CAS: 624-61-3   HR: 3
DIBROMOACETYLENE
mf: C_2Br_2 mw: 183.83
PROP: Heavy liquid with unpleasant odor.
Mp: −25°, bp: explodes, d: 2 (approx), vap
d: 6.35.
DOT CLASSIFICATION: Forbidden
SAFETY PROFILE: Ignites spontaneously
in air. Explodes when heated. When heated
to decomposition it emits toxic fumes of
Br⁻. See also ACETYLENE
COMPOUNDS.

DDJ900   CAS: 26249-12-7   HR: 3
DIBROMOBENZENE
DOT: UN 2711
mf: $C_6H_4Br_2$ mw: 235.92
SYNS: BENZENE, DIBROMO-
DOT CLASSIFICATION: 3; Label:
Flammable Liquid
CONSENSUS REPORTS: Reported in EPA
TSCA Inventory.
SAFETY PROFILE: Moderately toxic by
intraperitoneal route. A flammable liquid.
When heated to decomposition it emits
toxic vapors of Br⁻.

DDK600   CAS: 6305-43-7   HR: 3
2,2'-DIBROMOBIACETYL
mf: $C_4H_4Br_2O_2$ mw: 243.90
PROP: Crystals from $CHCl_3$. Mp: 116–117°.
SYN: α,α'-DIBROMOBIACETYL
CONSENSUS REPORTS: Reported in EPA
TSCA Inventory.
SAFETY PROFILE: Poison by intravenous
and intraperitoneal routes. When heated to
decomposition it emits toxic fumes of Br⁻.
See also BROMIDES.

DDL800   CAS: 96-12-8   HR: 3
1,2-DIBROMO-3-CHLOROPROPANE
DOT: UN 2872
mf: $C_3H_5Br_2Cl$ mw: 236.35
PROP: Bp: 196°, flash p: 170°F (TOC).
SYNS: BBC 12 □ 1-CHLORO-2,3-DIBROMOPROPANE
□ 3-CHLORO-1,2-DIBROMOPROPANE □ DBCP □
DIBROMCHLORPROPAN (GERMAN) □ 1,2-DIBROM-3-
CHLOR-PROPAN (GERMAN) □ DIBROMOCHLOROPRO-
PANE □ 1,2-DIBROMO-3-CLORO-PROPANO (ITALIAN)
□ 1,2-DIBROOM-3-CHLOORPROPAAN (DUTCH) □
FUMAGON □ FUMAZONE □ NCI-C00500 □ NEMA-
BROM □ NEMAFUME □ NEMAGON □ NEMAGONE □
NEMAGON SOIL FUMIGANT □ NEMANAX □ NEMA-
PAZ □ NEMASET □ NEMATOCIDE □ NEMATOX □
NEMAZON □ OS 1897 □ OXY DBCP □ RCRA WASTE
NUMBER U066 □ SD 1897
CONSENSUS REPORTS: NTP 10th Report
on Carcinogens. IARC Cancer Review:
Group 2B IMEMDT 7,191,87; Animal

Sufficient Evidence IMEMDT 15,139,77; Human Limited Evidence IMEMDT 20,83,79; Animal Sufficient Evidence IMEMDT 20,83,79. NCI Carcinogenesis Bioassay Completed; Results Positive: mouse, rat NCITR* NCI-CG-TR-28,78. EPA Genetic Toxicology Program. Community Right-To-Know List. Reported in EPA TSCA Inventory.
OSHA PEL: TWA 0.001 ppm; Cancer Hazard
DFG MAK: Animal Carcinogen, Suspected Human Carcinogen
NIOSH REL: (Dibromochloropropane) CL 0.01 ppm/30M
DOT CLASSIFICATION: 6.1; Label: KEEP AWAY FROM FOOD
SAFETY PROFILE: Confirmed human carcinogen with experimental carcinogenic and teratogenic data. Poison by ingestion, inhalation, and subcutaneous routes. Moderately toxic by skin contact. An eye and severe skin irritant. Narcotic in high concentrations. Has been implicated in causing human sterility in male factory workers. Human mutation data reported. A soil fumigant. Combustible. When heated to decomposition it emits toxic fumes of Cl^- and Br^-. See also CHLORIDES and BROMIDES.

DDM000   CAS: 10222-01-2   HR: 3
α,α-DIBROMO-α-CYANOACETAMIDE
mf: $C_3H_2Br_2N_2O$ mw: 241.89
SYNS: DBNPA □ DIBROMOCYANOACETAMIDE □ 2,2-DIBROMO-3-NITRILOPROPIONAMIDE
CONSENSUS REPORTS: Cyanide and its compounds are on the Community Right-To-Know List. EPA FIFRA 1988 pesticide subject to registration or re-registration. Reported in EPA TSCA Inventory.
SAFETY PROFILE: Poison by ingestion and intravenous routes. A severe skin and eye irritant. When heated to decomposition it emits very toxic fumes of Br^- and NO_x. See also NITRILES.

DDM400   CAS: 996-08-7   HR: 3
DIBROMODIBUTYLSTANNANE
mf: $C_8H_{18}Br_2Sn$ mw: 392.77
PROP: Mp: 20°.
SYNS: DIBROMODIBUTYLTIN □ DIBUTYL TIN DIBROMIDE
OSHA PEL: TWA 0.1 mg(Sn)/m³ (skin)
ACGIH TLV: TWA 0.1 mg(Sn)/m³; STEL 0.2 mg(Sn)/m³ (skin).
NIOSH REL: (Organotin Compounds) TWA 0.1 mg(Sn)/m³
SAFETY PROFILE: Poison by ingestion. Moderately toxic by skin contact. See also TIN COMPOUNDS. When heated to decomposition it emits toxic fumes of Br^-.

DDM500   CAS: 35691-65-7   HR: D
1,2-DIBROMO-2,4-DICYANOBUTANE
mf: $C_6H_6Br_2N_2$ mw: 265.96
SYNS: 2-BROMO-2-(BROMOMETHYL)GLUTARONITRILE □ GLUTARO-NITRILE, 2-BROMO-2-(BROMOMETHYL)- □ METHYL-DIBROMOGLUTARONITRILE □ PENTANEDINITRILE, 2-BROMO-2-(BROMOMETHYL)-
CONSENSUS REPORTS: Reported in EPA TSCA Inventory.
SAFETY PROFILE: Experimental reproductive effects. When heated to decomposition it emits toxic fumes of NO_x and Br^-.

DDM820   CAS: 23611-68-9   HR: 3
6,8-DIBROMO-DIHYDRO-1,3-BENZOX-
    AZINE-2-THIONE-4-ONE
mf: $C_8H_3Br_2NO_2S$ mw: 337.00
SYNS: 2H-1,3-BENZOXAZINE-2,4(3H)-DIONE, 6,8-DIBROMO-2-THIO- □ 6,8-DIBROMO-2-THIO-2H-1,3-BENZOXAZINE-2,4(3H)-DIONE
SAFETY PROFILE: A poison by ingestion. When heated to decomposition it emits toxic vapors of NO_x, SO_x, and Br^-.

DDO200   CAS: 596-03-2   HR: D
DIBROMOFLUORESCEIN
mf: $C_{20}H_{10}Br_2O_5$ mw: 490.12
PROP: Cosmetic colorant.
SYNS: ACID ORANGE 11 □ C.I. 45370:1 □ C.I. SOLVENT RED 72 □ D&C ORANGE NO. 5 □ 4',5'-DIBROMOFLUORORESCEIN □ FLUORESCEIN, 4',5'-DIBROMO- □ SOLVENT RED 72 □ SPIRO(ISOBENZO-

FURAN-1(3H),9'-(9H)XANTHEN)-3-ONE, 4',5'-DIBROMO-3',6'-DIHYDROXY- (9CI)
CONSENSUS REPORTS: Reported in EPA TSCA Inventory.
SAFETY PROFILE: Experimental reproductive effects. Mutation data reported. When heated to decomposition it emits toxic fumes of Br⁻.

DDP000   CAS: 1689-84-5   HR: 3
3,5-DIBROMO-4-HYDROXYBENZO-NITRILE
mf: $C_7H_3Br_2NO$ mw: 276.93
PROP: Needles. Mp: 187°.
SYNS: BRITTOX □ BROMINAL □ BROMINEX □ BROMINIL □ BROMOXYNIL □ BROXYNIL □ BUCTRIL □ BUCTRIL INDUSTRIAL □ BUTILCHLOROFOS □ CHIPCO BUCTRIL □ CHIPCO CRAB-KLEEN □ 2,6-DIBROMO-4-CYANOPHENOL □ 3,5-DIBROMO-4-HYDROXYPHENYLCYANIDE □ ENT 20,852 □ 4-HYDROXY-3,5-DIBROMOBENZONITRILE □ MB 10064 □ ME4 BROMINAL □ NU-LAWN WEEDER □ OXYTRIL M
CONSENSUS REPORTS: Cyanide and its compounds are on the Community Right-To-Know List.
SAFETY PROFILE: Poison by ingestion and intravenous routes. An herbicide. When heated to decomposition it emits highly toxic fumes of NO_x, CN⁻, and Br⁻. See also NITRILES.

DDP200   CAS: 3562-84-3   HR: 3
3,5-DIBROMO-4-HYDROXYPHENYL-2-ETHYL-3-BENZOFURANYL KETONE
mf: $C_{17}H_{12}Br_2O_3$ mw: 424.11
PROP: Yellow prisms. Mp: 151°.
SYNS: BENZBROMARON □ BENZBROMARONE □ DESURIC □ 3-(3,5-DIBROMO-4-HYDROXYBENZOYL)-2-ETHYLBENZOFURAN □ (3,5-DIBROMO-4-HYDROXYPHENYL)(2-ETHYL-3-BENZOFURANYL)METHANONE □ EXURATE □ L2214 □ MINURIC □ MJ 10061 □ URICOVAC
CONSENSUS REPORTS: Reported in EPA TSCA Inventory.
DOT CLASSIFICATION: 3; Label: Flammable Liquid
SAFETY PROFILE: Poison by ingestion, intravenous, and intraperitoneal routes. Moderately toxic by subcutaneous route. Experimental teratogenic and reproductive effects. A uricosuric agent which promotes the excretion of uric acid in the urine. A flammable liquid. When heated to decomposition it emits toxic fumes of Br⁻. See also KETONES.

DDP800   CAS: 74-95-3   HR: 3
DIBROMOMETHANE
DOT: UN 2664
mf: CH_2Br_2 mw: 173.85
PROP: Colorless, heavy liquid. Fp: −52.7°, bp: 95.6–97.4°, d: 2.485 @ 25°/25°, vap d: 6.05. Sltly sol in water.
SYNS: METHYLENE BROMIDE □ METHYLENE DIBROMIDE □ RCRA WASTE NUMBER U068
CONSENSUS REPORTS: Community Right-To-Know List. Reported in EPA TSCA Inventory.
DOT CLASSIFICATION: 6.1; Label: KEEP AWAY FROM FOOD
SAFETY PROFILE: A poison. Moderately toxic by subcutaneous route. Mildly toxic by inhalation. Mutation data reported. Mixtures with potassium explode on light impact. When heated to decomposition it emits toxic fumes of Br⁻. See also BROMIDES.

DDR200   CAS: 696-24-2   HR: 3
DIBROMOPHENYLARSINE
mf: $C_6H_5AsBr_2$ mw: 311.85
PROP: Bp: 285° (decomp), d: 2.103.
SYNS: PHENYLARSONOUS DIBROMIDE □ PHENYLDIBROMOARSINE
CONSENSUS REPORTS: Arsenic and its compounds are on the Community Right-To-Know List.
OSHA PEL: TWA 0.5 mg(As)/m³
SAFETY PROFILE: Poison by skin contact and intravenous routes. See also ARSENIC COMPOUNDS. When heated to decomposition it emits very toxic fumes of As and Br⁻.

DDT800   CAS: 111-92-2   HR: 3
n-DIBUTYLAMINE
DOT: UN 2248
mf: $C_8H_{19}N$ mw: 129.28
PROP: A liquid. Mp: −61.9°; bp: 159°, flash p: 125°F (OC), d: 0.76, vap d: 4.46, vap

press: 2 mm @ 20°. Sol in water and alcohol.

SYNS: n-BUTYL-1-BUTANAMINE □ DI-n-BUTYLAMINE □ DI(n-BUTYL)AMINE (DOT)

CONSENSUS REPORTS: Reported in EPA TSCA Inventory. EPA Genetic Toxicology Program.

DOT CLASSIFICATION: 8; Label: Corrosive, Flammable Liquid

SAFETY PROFILE: Poison by ingestion and subcutaneous routes. Moderately toxic by skin contact and inhalation. Corrosive. A severe skin and eye irritant. Mutation data reported. Flammable liquid when exposed to heat or flame; can react with oxidizing materials. To fight fire, use alcohol foam, foam, CO_2, dry chemical. Exothermic reaction with cellulose nitrate does not proceed to ignition. When heated to decomposition it emits toxic fumes of NO_x.

DDU600   CAS: 102-81-8   HR: 3
2-N-DIBUTYLAMINOETHANOL

DOT: UN 2873

mf: $C_{10}H_{23}NO$ mw: 173.34

PROP: Liquid. Bp: 222°, flash p: 220°F (OC), d: 0.85, vap d: 6.0.

SYNS: BU2AE □ DIBUTYLAMINOETHANOL □ 2-DIBUTYLAMINOETHANOL □ 2-DI-n-BUTYL-AMINOETHANOL □ N,N-DI-n-BUTYLAMINOETHANOL (DOT) □ β-N-DIBUTYLAMINOETHYL ALCOHOL □ N,N-DIBUTYLETHANOLAMINE □ N,N-DIBUTYL-N-(2-HYDROXYETHYL)AMINE

CONSENSUS REPORTS: Reported in EPA TSCA Inventory.

OSHA PEL: TWA 2 ppm

ACGIH TLV: TWA 0.5 ppm (skin)

DOT CLASSIFICATION: 6.1; Label: KEEP AWAY FROM FOOD

SAFETY PROFILE: Poison by intraperitoneal route. Moderately toxic by ingestion and skin contact. A severe eye and skin irritant. Combustible; can react with oxidizing materials. To fight fire, use CO_2, dry chemical. When heated to decomposition it emits toxic fumes of NO_x. See also AMINES and ALCOHOLS.

DDV225   CAS: 7128-68-9   HR: 3
DI-n-BUTYLAMMONIUM HEXA-FLUOROARSENATE

mf: $C_8H_{19}N \cdot AsF_6H$ mw: 319.21

SYN: DIBUTYLAMINE, HEXAFLUOROARSENATE(1-)

OSHA PEL: TWA 0.5 mg(As)/m^3

SAFETY PROFILE: Poison by intravenous route. When heated to decomposition it emits toxic fumes of NO_x, As, and F⁻.

DDV250   CAS: 2850-61-5   HR: 3
DIBUTYLARSINIC ACID

mf: $C_8H_{19}AsO_2$ mw: 222.19

PROP: Prisms from H_2O or crystals from Me_2CO. Mp: 138°. Sol in EtOH; sltly sol in H_2O.

SYNS: ARSINE OXIDE, DIBUTYLHYDROXY- □ ARSINIC ACID, DIBUTYL-(9CI)

OSHA PEL: TWA 0.5 mg(As)/m^3

SAFETY PROFILE: Poison by intravenous route. When heated to decomposition it emits toxic fumes of As.

DDV600   CAS: 77-58-7   HR: 3
DIBUTYLBIS(LAUROYLOXY)-STANNANE

mf: $C_{32}H_{64}O_4Sn$ mw: 631.65

PROP: Pale-yellow liquid to colorless solid (when pure). Mp: 23°, bp: non-distillable @ 10 mm, flash p: 455°F (OC), d: 1.066 @ 20°/20°, vap d: 21.8.

SYNS: BIS(DODECANOYLOXY)DI-n-BUTYL-STANNANE □ BIS(LAUROYLOXY)DIBUTYLSTANNANE □ BIS(LAUROYLOXY)DI(n-BUTYL)STANNANE □ BUTYNORATE □ DBTL □ DIBUTYLBIS(LAUROYLOXY)TIN □ DI-n-BUTYLTIN DI(DODECANOATE) □ DIBUTYLTIN DILAURATE (USDA) □ DIBUTYLTIN LAURATE □ DIBUTYL-ZINN-DILAURAT (GERMAN) □ FOMREZ SUL-4 □ LAUDRAN DI-n-BUTYLCINICITY (CZECH) □ LAURIC ACID, DIBUTYLSTANNYLENE derivative □ LAURIC ACID, DIBUTYLSTANNYLENE SALT □ STABILIZER D-22 □ THERM CHEK 820 □ TIN DIBUTYL DILAURATE □ TINOSTAT

CONSENSUS REPORTS: Reported in EPA TSCA Inventory.

OSHA PEL: TWA 0.1 mg(Sn)/m^3 (skin)

ACGIH TLV: TWA 0.1 mg(Sn)/m^3; STEL 0.2 mg(Sn)/m^3 (skin).

NIOSH REL: (Organotin Compounds) TWA 0.1 mg(Sn)/m^3

SAFETY PROFILE: Poison by ingestion and intraperitoneal routes. A skin and eye irritant. Avoid the vapor produced by heating. Combustible when exposed to heat or flame; reacts with oxidizers. When heated to decomposition it emits acrid smoke and fumes. See also TIN COMPOUNDS.

DDV800   CAS: 78-46-6   HR: 3
DIBUTYL BUTANEPHOSPHONATE
mf: $C_{12}H_{27}O_3P$ mw: 250.36
PROP: Colorless liquid with mild pleasant odor. Bp: 160–162° @ 20 mm, flash p: 311° (COC), d: 8.62.
SYNS: DIBUTYL BUTYLPHOSPHONATE □ NSC-2666
CONSENSUS REPORTS: Reported in EPA TSCA Inventory.
SAFETY PROFILE: Poison by intraperitoneal and intravenous routes. Combustible when exposed to heat or flame. It can react vigorously with oxidizing materials. To fight fire, use foam, CO_2, or dry chemical. When heated to decomposition it emits toxic fumes of PO_x.

DDW200   CAS: 112-73-2   HR: 2
DIBUTYL CARBITOL
mf: $C_{12}H_{26}O_3$ mw: 218.38
PROP: Practically colorless liquid, characteristic odor, sltly sol in water. D: 0.8853 @ 20°/20°, bp: 256°, fp: −60.2°, flash p: 245°F (OC).
SYNS: BIS(BUTOXYETHYL) ETHER □ BIS(2-BUTOXYETHYL) ETHER □ BUTYL DIGLYME □ 2,2'-DIBUTOXYETHYL ETHER □ DIETHYLENEGLYCOL DIBUTYL ETHER □ DIETHYLENEGLYCOL DI-n-BUTYL ETHER □ 1,1'-(OXYBIS(2,1-ETHANEDIYLOXY))-BISBUTANE □ 5,8,11-TRIOXAPENTADECANE
CONSENSUS REPORTS: Reported in EPA TSCA Inventory. Glycol ether compounds are on the Community Right-To-Know List.
SAFETY PROFILE: Moderately toxic by ingestion. Mildly toxic by skin contact. Experimental reproductive effects. A skin and eye irritant. See also GLYCOL ETHERS. Combustible when exposed to heat or flame. To fight fire, use foam or alcohol foam. When heated to

decomposition it emits acrid smoke and irritating fumes.

DDY000   CAS: 4593-81-1   HR: 3
DIBUTYLDICHLOROGERMANE
mf: $C_8H_{18}Cl_2Ge$ mw: 257.75
SYNS: DI-n-BUTYLGERMANEDICHLORIDE □ DICHLORODIBUTYLGERMANE
CONSENSUS REPORTS: Reported in EPA TSCA Inventory.
SAFETY PROFILE: Poison by ingestion and intraperitoneal routes. Mutation data reported. See also GERMANIUM COMPOUNDS. When heated to decomposition it emits very toxic fumes of Cl^-.

DDY200   CAS: 683-18-1   HR: 3
DIBUTYLDICHLOROSTANNANE
mf: $C_8H_{18}Cl_2Sn$ mw: 303.85
PROP: White, crystalline solid. Mp: 43°, bp: 135° @ 10 mm, flash p: 335°F (OC), d: 1.36 @ 50°, vap press: 2 mm @ 100°, vap d: 10.5.
SYNS: CHLORID DI-n-BUTYLCINICITY (CZECH) □ D.B.T.C. □ DIBUTYLDICHLOROTIN □ DIBUTYLTIN CHLORIDE □ DIBUTYLTIN DICHLORIDE □ DI-n-BUTYLTIN DICHLORIDE □ DI-n-BUTYL-ZINN-DICHLORID (GERMAN) □ DICHLORODIBUTYL-STANNANE □ DICHLORODIBUTYLTIN
CONSENSUS REPORTS: Reported in EPA TSCA Inventory.
OSHA PEL: TWA 0.1 mg(Sn)/m³ (skin)
ACGIH TLV: TWA 0.1 mg(Sn)/m³; STEL 0.2 mg(Sn)/m³ (skin).
NIOSH REL: (Organotin Compounds) TWA 0.1 mg(Sn)/m³
SAFETY PROFILE: Poison by ingestion, intravenous, and intraperitoneal routes. Moderately toxic by skin contact. A severe skin and eye irritant. Experimental reproductive effects. Mutation data reported. See also TIN COMPOUNDS. Combustible when exposed to heat or flame. A dangerous material; emits highly toxic fumes of HCl; will react with water or steam to produce heat and toxic fumes; can react vigorously with oxidizing materials. To

fight fire, use water, foam, CO_2, dry chemical.

DDY600   CAS: 10584-98-2   HR: 2
DIBUTYLDI(2-ETHYLHEXYLOXYCAR-BONYLMETHYLTHIO)STANNANE

mf: $C_{28}H_{56}O_4S_2Sn$ mw: 639.65

SYNS: BIS(2-ETHYLHEXYLOXYCARBONYLMETHYLTHIO)DIBUTYL STANNANE □ BIS(2-ETHYLHEXYLTHIOGLYCOLATE)-DIBUTYLTIN □ DI-n-BUTYLTIN DI-2-ETHYLHEXYL-THIOGLYCOLATE □ DI-n-BUTYL-ZINN DI-2-AETHYLHEXYL THIOGLYKOLAT (GERMAN)

CONSENSUS REPORTS: Reported in EPA TSCA Inventory.

OSHA PEL: TWA 0.1 mg(Sn)/m³ (skin)

ACGIH TLV: TWA 0.1 mg(Sn)/m³; STEL 0.2 mg(Sn)/m³ (skin).

NIOSH REL: (Organotin Compounds) TWA 0.1 mg(Sn)/m³

SAFETY PROFILE: Moderately toxic by ingestion. See also TIN COMPOUNDS. When heated to decomposition it emits toxic fumes of SO_x.

DDY800   CAS: 563-25-7   HR: 3
DIBUTYLDIFLUOROSTANNANE

mf: $C_8H_{18}F_2Sn$ mw: 270.95

SYN: DIBUTYLTIN DIFLUORIDE

CONSENSUS REPORTS: Reported in EPA TSCA Inventory.

OSHA PEL: TWA 0.1 mg(Sn)/m³ (skin)

ACGIH TLV: TWA 0.1 mg(Sn)/m³; STEL 0.2 mg(Sn)/m³ (skin).

NIOSH REL: (Organotin Compounds) TWA 0.1 mg(Sn)/m³

SAFETY PROFILE: Poison by ingestion. See also TIN COMPOUNDS and FLUORIDES. When heated to decomposition it emits toxic fumes of F^-.

DDZ000   CAS: 7392-96-3   HR: 3
DIBUTYL(DIFORMYLOXY)STANNANE

mf: $C_{10}H_{20}O_4Sn$ mw: 322.99

SYNS: DI-n-BUTYLTIN DIFORMATE □ MRAVENCAN DI-n-BUTYLCINICITY (CZECH)

OSHA PEL: TWA 0.1 mg(Sn)/m³ (skin)

ACGIH TLV: TWA 0.1 mg(Sn)/m³; STEL 0.2 mg(Sn)/m³ (skin).

NIOSH REL: (Organotin Compounds) TWA 0.1 mg(Sn)/m³

SAFETY PROFILE: Poison by ingestion. A severe skin and eye irritant. When heated to decomposition it emits acrid and irritating fumes. See also TIN COMPOUNDS.

DEA600   CAS: 3465-74-5   HR: 3
DIBUTYLDIPENTANOYLOXY-STANNANE

mf: $C_{18}H_{36}O_4Sn$ mw: 435.23

SYNS: DI-n-BUTYLTIN DIPENTANOATE □ DI(PENTANOYLOXY)DIBUTYLSTANNANE □ VALERAN DI-n-BUTYLCINICITY (CZECH)

OSHA PEL: TWA 0.1 mg(Sn)/m³ (skin)

ACGIH TLV: TWA 0.1 mg(Sn)/m³; STEL 0.2 mg(Sn)/m³ (skin).

NIOSH REL: (Organotin Compounds) TWA 0.1 mg(Sn)/m³

SAFETY PROFILE: Poison by ingestion. A severe skin and eye irritant. See also TIN COMPOUNDS. When heated to decomposition it emits acrid and irritating fumes.

DEC000   CAS: 625-22-9   HR: 3
DIBUTYL ESTER SULFURIC ACID

mf: $C_8H_{18}O_4S$ mw: 210.32

PROP: Liquid with sharp odor. Bp: 115–116° @ 6 mm.

SYNS: DI-n-BUTYLSULFAT (GERMAN) □ DIBUTYL SULFATE

SAFETY PROFILE: Poison by ingestion. Mildly toxic by subcutaneous route. Questionable carcinogen with experimental tumorigenic data. See also ESTERS and SULFATES. When heated to decomposition it emits toxic fumes of SO_x.

DEC100   CAS: 4062-60-6   HR: 3
N,N-DI-tert-BUTYLETHYLENEDIAMINE

mf: $C_{10}H_{24}N_2$ mw: 172.36

SYNS: AR 81242 □ N,N'-BIS(1,1-DIMETHYLETHYL)-1,2-ETHANEDIAMINE □ 1,2-ETHANEDIAMINE, N,N'-BIS(1,1-DIMETHYLETHYL)-

SAFETY PROFILE: A poison by ingestion. Low toxicity by ingestion. A mild skin irritant. When heated to decomposition it emits toxic vapors of NO_x.

D

DEC400 **CAS: 761-65-9** **HR: 3**
N,N-DI-n-BUTYLFORMAMIDE
mf: $C_9H_{19}NO$ mw: 157.29
PROP: Liquid. Bp: 235°.
SYNS: DBF □ DIBUTYLAMID KYSELINY MRAVENCI □ N,N-DI-n-BUTYLFORMAMIDE
CONSENSUS REPORTS: Reported in EPA TSCA Inventory.
SAFETY PROFILE: Poison by intraperitoneal route. An experimental teratogen. When heated to decomposition it emits toxic fumes of NO_x.

DEC600 **CAS: 105-75-9** **HR: 3**
DIBUTYL FUMARATE
mf: $C_{12}H_{20}O_4$ mw: 228.32
PROP: Colorless, clear, mobile liquid; typical odor. Bp: 285.1°, fp: −19°, flash p: 300°F (OC), d: 0.986 @ 20°/20°, vap d: 7.88.
SYNS: DIBUTYLESTER KYSELINY FUMAROVE □ FUMARIC ACID, DIBUTYL ESTER
CONSENSUS REPORTS: Reported in EPA TSCA Inventory.
SAFETY PROFILE: Poison by intraperitoneal route. Mildly toxic by ingestion and skin contact. An eye, skin, and mucous membrane irritant. Combustible when exposed to heat or flame; can react with oxidizing materials. To fight fire, use foam, CO_2, dry chemical. When heated to decomposition it emits acrid smoke and fumes.

DED600 **CAS: 105-76-0** **HR: 3**
DIBUTYL MALEATE
mf: $C_{12}H_{20}O_4$ mw: 228.32
PROP: Liquid. Mp: −85° (sets to a glass), bp: 281°, flash p: 285°F (OC), d: 0.9964 @ 20°/20°, vap d: 7.9.
SYNS: 2-BUTENEDIOIC ACID, DIBUTYL ESTER □ DBM □ MALEIC ACID, DIBUTYL ESTER □ RC COMONOMER DBM □ STAFLEX DBM
CONSENSUS REPORTS: Reported in EPA TSCA Inventory.
SAFETY PROFILE: Poison by intraperitoneal route. Moderately toxic by ingestion. Mildly toxic by skin contact. An eye and skin irritant. See also ESTERS and n-BUTYL ALCOHOL. Combustible when exposed to heat or flame; can react with oxidizing materials. To fight fire, use foam, CO_2, dry chemical, alcohol foam. When heated to decomposition it emits acrid smoke and irritating fumes.

DEE000 **CAS: 629-35-6** **HR: 3**
DIBUTYLMERCURY
mf: $C_8H_{18}Hg$ mw: 314.85
PROP: Liquid. Bp: 105° @ 10 mm, d: 1.779, vap d: 10.8. Sol in most org solvs; insol in H_2O. IDLH 10 mg/m³ (as Hg).
SYN: DIBUTYLRTUT
CONSENSUS REPORTS: Mercury and its compounds are on the Community Right-To-Know List. Reported in EPA TSCA Inventory.
OSHA PEL: TWA 0.01 mg(Hg)/m³; STEL 0.03 mg/m³ (skin)
ACGIH TLV: TWA 0.01 mg(Hg)/m³; BEI: 35 µg/g creatinine total inorganic mercury in urine preshift; 15 µg/g creatinine total inorganic mercury in blood at end of shift at end of workweek.
DFG MAK: Confirmed Animal Carcinogen with Unknown Relevance to Humans
NIOSH REL: (Mercury, Organo) TWA 0.01 mg/m³; STEL 0.03 mg/m³ (skin)
SAFETY PROFILE: Poison by intraperitoneal route. See also MERCURY COMPOUNDS, ORGANIC. Flammable when exposed to heat or flame. Can react vigorously with oxidizing materials. When heated to decomposition or on contact with acid or acid fumes it emits highly toxic fumes of mercury.

DEE200 **CAS: 691-88-3** **HR: 3**
DI-sec-BUTYLMERCURY
mf: $C_8H_{18}Hg$ mw: 314.85
PROP: Liquid; unstable in air and light. Bp: 93–96° @ 18 mm. IDLH 10 mg/m³ (as Hg).
CONSENSUS REPORTS: Mercury and its compounds are on the Community Right-To-Know List.
OSHA PEL: TWA 0.01 mg(Hg)/m³; STEL 0.03 mg/m³ (skin)

ACGIH TLV: TWA 0.01 mg(Hg)/m³; BEI: 35 µg/g creatinine total inorganic mercury in urine preshift; 15 µg/g creatinine total inorganic mercury in blood at end of shift at end of workweek.
DFG MAK: Confirmed Animal Carcinogen with Unknown Relevance to Humans
NIOSH REL: (Mercury, Organo) TWA 0.01 mg/m³; STEL 0.03 mg/m³ (skin)
SAFETY PROFILE: Poison by intraperitoneal route. See also MERCURY COMPOUNDS, ORGANIC. When heated to decomposition it emits toxic fumes of Hg.

DEF150   CAS: 27371-95-5   HR: 3
2,2-DIBUTYL-1,3,2-OXATHIA-STANNOLANE

mf: $C_{10}H_{22}OSSn$ mw: 309.07
SYN: 1,3,2-OXATHIASTANNOLANE, 2,2-DIBUTYL-
OSHA PEL: TWA 0.1 mg(Sn)/m³ (skin)
ACGIH TLV: TWA 0.1 mg(Sn)/m³; STEL 0.2 mg/m³ (skin)
NIOSH REL: (Organotin Compounds): 10H TWA 0.1 mg(Sn)/m³
SAFETY PROFILE: Poison by ingestion and intravenous routes. Moderately toxic by skin contact. When heated to decomposition it emits toxic fumes of SO_x and Sn.

DEF200   CAS: 78-20-6   HR: 2
2,2-DIBUTYL-1,3,2-OXATHIASTANNOL-ANE-5-OXIDE

mf: $C_{10}H_{20}O_2SSn$ mw: 323.05
SYNS: DIBUTYL(THIOACETOXY)STANNANE □ DI-n-BUTYLZINN THIOGLYKOLAT (GERMAN)
CONSENSUS REPORTS: Reported in EPA TSCA Inventory.
OSHA PEL: TWA 0.1 mg(Sn)/m³ (skin)
ACGIH TLV: TWA 0.1 mg(Sn)/m³; STEL 0.2 mg(Sn)/m³ (skin).
NIOSH REL: (Organotin Compounds) TWA 0.1 mg(Sn)/m³
SAFETY PROFILE: Moderately toxic by ingestion. See also TIN COMPOUNDS. When heated to decomposition it emits toxic fumes of SO_x.

DEF400   CAS: 818-08-6   HR: 3
DIBUTYLOXOSTANNANE

mf: $C_8H_{18}OSn$ mw: 248.95
PROP: White, amorphous powder or polymeric infusible solid. Mp: decomp without melting, bulk density: 0.5, vap d: 8.6.
SYNS: DBOT □ DIBUTYLOXIDE of TIN □ DIBUTYLOXOTIN □ DIBUTYLSTANNANE OXIDE □ DIBUTYLTIN OXIDE □ DI-n-BUTYLTIN OXIDE □ DI-n-BUTYL-ZINN-OXYD (GERMAN) □ KYSLICNIK DI-n-BUTYLCINICITY (CZECH)
CONSENSUS REPORTS: Reported in EPA TSCA Inventory.
OSHA PEL: TWA 0.1 mg(Sn)/m³ (skin)
ACGIH TLV: TWA 0.1 mg(Sn)/m³; STEL 0.2 mg(Sn)/m³ (skin).
NIOSH REL: (Organotin Compounds) TWA 0.1 mg(Sn)/m³
SAFETY PROFILE: Poison by ingestion and intraperitoneal routes. A skin and eye irritant. Flammable when exposed to flame; can react with oxidizing materials. To fight fire, use dry chemical, fog, CO_2. When heated to decomposition it emits acrid smoke and irritating fumes. See also TIN COMPOUNDS.

DEG200   CAS: 101-96-2   HR: 3
N,N'-DI-sec-BUTYL-p-PHENYLENEDI-AMINE

mf: $C_{14}H_{24}N_2$ mw: 220.40
PROP: Liquid. Mp: 17.8°, flash p: 285°F (OC), d: 0.94–0.95 @ 24°/24°.
SYN: TENAMENE 2
CONSENSUS REPORTS: Reported in EPA TSCA Inventory.
SAFETY PROFILE: Poison by ingestion. Moderately toxic by inhalation and skin contact. Corrosive to skin. A mild allergen. Symptoms of exposure are sweating, flushing, shortness of breath, and slow pulse. Combustible when exposed to heat or flame; can react with oxidizing materials. To fight fire, use foam, CO_2, dry chemical. When heated to decomposition it emits toxic fumes of NO_x. See also AMINES.

D

DEG600   CAS: 2528-36-1   HR: 2
DIBUTYL PHENYL PHOSPHATE
mf: $C_{14}H_{23}O_4P$ mw: 286.34
SYN: PHOSPHORIC ACID, DIBUTYL PHENYL ESTER
CONSENSUS REPORTS: Reported in EPA
TSCA Inventory.
ACGIH TLV: TWA 0.3 ppm (skin)
SAFETY PROFILE: Moderately toxic by
ingestion. Experimental reproductive
effects. When heated to decomposition it
emits toxic fumes of PO_x.

DEG700   CAS: 107-66-4   HR: 2
DIBUTYL PHOSPHATE
mf: $C_8H_{19}PO_4$ mw: 210.2
PROP: Pale-amber liquid or oil. Bp:
135–138° @ 0.05 mm, decomp >100°. Sol
in butanol and CCl_4. IDLH 30 ppm.
SYNS: DIBUTYL ACID PHOSPHATE □ DIBUTYL
HYDROGEN PHOSPHATE □ DIBUTYL PHOSPHATE □
DI-n-BUTYL PHOSPHATE
CONSENSUS REPORTS: Reported in EPA
TSCA Inventory.
OSHA PEL: TWA 1 ppm; STEL 2 ppm
ACGIH TLV: TWA 1 ppm; STEL 2 ppm
SAFETY PROFILE: Moderately toxic by
ingestion. When heated to decomposition it
emits toxic fumes of PO_x. See also
PHOSPHATES.

DEH200   CAS: 84-74-2   HR: 3
DIBUTYL PHTHALATE
mf: $C_{16}H_{22}O_4$ mw: 278.38
PROP: Oily liquid; mild odor. Mp: −35°, bp:
340°, flash p: 315°F (CC), d: 1.047–1.049 @
20°/20°, autoign temp: 757°F, vap d: 9.58.
IDLH 4000 mg/m³.
SYNS: BENZENE-o-DICARBOXYLIC ACID DI-n-
BUTYL ESTER □ o-BENZENEDICARBOXYLIC ACID,
DIBUTYL ESTER □ n-BUTYL PHTHALATE (DOT) □
CELLUFLEX DPB □ DBP □ DIBUTYL-1,2-
BENZENEDICARBOXYLATE □ DI-n-BUTYL
PHTHALATE □ ELAOL □ HEXAPLAS M/B □
PALATINOL C □ POLYCIZER DBP □ PX 104 □ RCRA
WASTE NUMBER U069 □ STAFLEX DBP □ WITCIZER
300
CONSENSUS REPORTS: On the
Community Right-To-Know List. EPA
Genetic Toxicology Program. Reported in
EPA TSCA Inventory.

OSHA PEL: TWA 5 mg/m³
ACGIH TLV: TWA 5 mg/m³
SAFETY PROFILE: Moderately toxic by
intraperitoneal and intravenous routes.
Mildly toxic by ingestion. Human systemic
eye effects by ingestion, hallucinations,
distorted perceptions, nausea or vomiting,
and kidney, ureter, or bladder changes.
Experimental teratogenic and reproductive
effects. Mutation data reported.
Combustible when exposed to heat or
flame; can react with oxidizing materials.
Violent reaction with Cl_2. Incompatible with
chlorine. To fight fire, use CO_2, dry
chemical. When heated to decomposition it
emits acrid smoke and fumes. See also
ESTERS, PHTHALIC ACID, and n-
BUTYL ALCOHOL.

DEH600   CAS: 109-43-3   HR: 1
DIBUTYL SEBACATE
mf: $C_{18}H_{34}O_4$ mw: 314.52
PROP: Clear liquid. Bp: 180° @ 3 mm, fp:
−11°, flash p: 353°F (COC), d: 0.936 @
20°/20°, vap d: 10.8.
SYNS: BIS(n-BUTYL)SEBACATE □ DECANEDIOIC
ACID, DIBUTYL ESTER □ DI-n-BUTYL SEBACATE □
KODAFLEX DBS □ MONOPLEX DBS □ POLYCIZER
DBS □ PX 404 □ SEBACIC ACID, DIBUTYL ESTER □
STAFLEX DBS
CONSENSUS REPORTS: Reported in EPA
TSCA Inventory.
SAFETY PROFILE: Mildly toxic by
ingestion. Experimental reproductive
effects. Combustible liquid when exposed to
heat or flame; can react with oxidizing
materials. To fight fire, use CO_2, dry
chemical. When heated to decomposition it
emits acrid smoke and fumes. See also
ESTERS and n-BUTYL ALCOHOL.

DEH650   CAS: 7399-02-2   HR: 3
**2,2'-((DIBUTYLSTANNYLENE)BIS-
(THIO))BISACETIC ACID DINONYL
ESTER**
mf: $C_{30}H_{60}O_4S_2Sn$ mw: 667.71
SYNS: ACETIC ACID, 2,2'-
((DIBUTYLSTANNYLENE)BIS(THIO))BIS-, DINONYL
ESTER □ ACETIC ACID, ((DIBUTYLSTANNYLENE)-
DITHIO)DI-, DINONYL ESTER (8CI) □ MELLITE 131 □

8-OXA-3,5-DITHIA-4-STANNAHEPTADECANOIC ACID,
4,4-DIBUTYL-7-OXO-, NONYLESTER (9CI)
OSHA PEL:8H TWA 0.1 mg(Sn)/m³ (skin)
ACGIH TLV: TWA 0.1 mg(Sn)/m³; STEL
0.2 mg/m³ (skin)
NIOSH REL: (Organotin Compounds):
TWA 0.1 mg(Sn)/m³
SAFETY PROFILE: Poison by ingestion.
When heated to decomposition it emits
toxic fumes of SO$_x$ and Sn.

**DEI200   CAS: 4253-22-9   HR: 3
DIBUTYLTHIOXOSTANNANE**
mf: C$_8$H$_{18}$SSn   mw: 265.01
SYNS: DIBUTYLTIN SULFIDE □ TIN DIBUTYL
MERCAPTIDE
CONSENSUS REPORTS: Reported in EPA
TSCA Inventory.
OSHA PEL: TWA 0.1 mg(Sn)/m³ (skin)
ACGIH TLV: TWA 0.1 mg(Sn)/m³; STEL
0.2 mg(Sn)/m³ (skin).
NIOSH REL: (Organotin Compounds) TWA
0.1 mg(Sn)/m³
SAFETY PROFILE: Poison by ingestion.
Mutation data reported. See also TIN
COMPOUNDS and SULFIDES. When
heated to decomposition it emits toxic
fumes of SO$_x$.

**DEJ000   CAS: 13323-62-1   HR: 3
DIBUTYLTIN DIOLEATE**
mf: C$_{44}$H$_{54}$O$_4$Sn   mw: 765.67
SYNS: BIS(OLEOYLOXY)DIBUTYLSTANNANE □ CN
447 □ DIBUTYLBIS(OLEOYLOXY)STANNANE □
DIBUTYLBIS((1-OXO-9-OCTADECENYL)OXY)-
STANNANE (Z,Z)
CONSENSUS REPORTS: Reported in EPA
TSCA Inventory.
OSHA PEL: TWA 0.1 mg(Sn)/m³ (skin)
ACGIH TLV: TWA 0.1 mg(Sn)/m³; STEL
0.2 mg(Sn)/m³ (skin).
NIOSH REL: (Organotin Compounds) TWA
0.1 mg(Sn)/m³
SAFETY PROFILE: Poison by intravenous
route. See also TIN COMPOUNDS. When
heated to decomposition it emits acrid
smoke and irritating fumes.

**DEJ200   CAS: 78-06-8   HR: 3
DIBUTYLTIN MERCAPTOPROPIONATE**

mf: C$_{11}$H$_{22}$O$_2$SSn   mw: 337.08
SYNS: 2,2-DIBUTYLDIHYDRO-6H-1,3,2-
OXATHIASTANNIN-6-ONE □ 2,2-DIBUTYL-1-OXA-2-
STANNA-3-THIACYCLOHEXAN-6-ONE □ DIBUTYLTIN-
O,S-MERCAPTOPROPIONATE □ DIBUTYLTIN-S,O-3-
MERCAPTOPROPIONATE □ DIBUTYLTIN-S,O-β-
MERCAPTOPROPIONATE □ DIBUTYL(3-MERCAPTO-
PROPIONATO(2-))TIN □ MERCAPTO-PROPIONIC ACID,
DIBUTYLTIN SALT
CONSENSUS REPORTS: Reported in EPA
TSCA Inventory.
OSHA PEL: TWA 0.1 mg(Sn)/m³ (skin)
ACGIH TLV: TWA 0.1 mg(Sn)/m³; STEL
0.2 mg(Sn)/m³ (skin).
NIOSH REL: (Organotin Compounds) TWA
0.1 mg(Sn)/m³
SAFETY PROFILE: Poison by intravenous
route. See also TIN COMPOUNDS and
MERCAPTANS. When heated to
decomposition it emits toxic fumes of SO$_x$.

**DEL000   CAS: 79-43-6   HR: 2
DICHLORACETIC ACID**
DOT: UN 1764
mf: C$_2$H$_2$Cl$_2$O$_2$ mw: 128.94
PROP: Colorless, corrosive liquid; pungent
odor. Mp (a): 10°, (b): −4°, bp: 194°, d:
1.5634 @ 20°/4°, vap press: 1 mm @ 44.0°,
vap d: 4.45.
SYNS: BICHLORACETIC ACID □ DCA □
DICHLORETHANOIC ACID □ 2,2-DICHLOROACETIC
ACID □ DICHLOROETHANOIC ACID □ KYSELINA
DICHLOROCTOVA □ URNER'S LIQUID
CONSENSUS REPORTS: Reported in EPA
TSCA Inventory.
DOT CLASSIFICATION: 8; Label: Corrosive
SAFETY PROFILE: Moderately toxic by
skin contact and ingestion. It is corrosive to
the skin, eyes, and mucous membranes.
Questionable carcinogen with experimental
tumorigenic data. Will react with water or
steam to produce toxic and corrosive fumes.
When heated to decomposition it emits
toxic fumes of Cl⁻. See also CHLORIDES.

**DEL400   CAS: 3589-22-8   HR: 3
DICHLORIMIPRAMINE**
mf: C$_{19}$H$_{22}$Cl$_2$N$_2$ mw: 349.33
SYNS: 5H-DIBENZ(b,f)AZEPINE,-3,7-DICHLORO-5-(3-
(DIMETHYLAMINO)PROPYL)-10,11-DIHYDRO- □ 5H-

DIBENZ(b,f)AZEPINE,-10,11-DIHYDRO-3,7-DICHLORO-5-(3-(DIMETHYLAMINO)PROPYL)- □ 5H-DIBENZ(b,f)-AZEPINE, 3,7-DICHLORO-10,11-DIHYDRO-5-(3-(DIMETHYLAMINO)PROPYL)- □ G 28364

SAFETY PROFILE: A poison by intravenous route. When heated to decomposition it emits toxic vapors of NO_x and Cl^-.

DEM800 CAS: 116-54-1 HR: 3
DICHLOROACETIC ACID METHYL ESTER
DOT: UN 2299
mf: $C_3H_4Cl_2O_2$ mw: 142.97
PROP: Colorless liquid; ethereal odor. Bp: 143.0°, d: 1.3809 @ 19.2°/19.2°, vap d: 4.93.
SYNS: METHYL DICHLOROACETATE (DOT) □ METHYL DICHLOROETHANOATE
CONSENSUS REPORTS: Reported in EPA TSCA Inventory.
DOT CLASSIFICATION: 6.1; Label: KEEP AWAY FROM FOOD
SAFETY PROFILE: Poisonous irritant to the skin, eyes, and mucous membranes. Hydrolyzes upon contact with moisture to form a product corrosive to tissue. See also DICHLOROACETIC ACID and ESTERS. Dangerous; when heated to decomposition it emits highly toxic fumes of phosgene and Cl^-.

DEN400 CAS: 79-36-7 HR: 2
DICHLOROACETYL CHLORIDE
DOT: UN 1765
mf: C_2HCl_3O mw: 147.38
PROP: Fuming liquid, acrid odor, misc in ether. D: 1.5315 @ 16°/4°, bp: 108°, flash p: 151°F, vap d: 5.8.
SYNS: CHLORID KYSELINY DICHLOROCTOVE □ CHLORURE de DICHLORACETYLE (FRENCH) □ DICHLORACETYL CHLORIDE □ α,α-DICHLOROACETYL CHLORIDE □ 2,2-DICHLOROACETYL CHLORIDE □ DICHLOROACETYL CHLORIDE (DOT) □ DICHLORO-ETHANOYL CHLORIDE
CONSENSUS REPORTS: Reported in EPA TSCA Inventory.
DOT CLASSIFICATION: 8; Label: Corrosive
SAFETY PROFILE: Questionable carcinogen with experimental tumorigenic

data. Moderately toxic by ingestion, inhalation, and skin contact. Corrosive to the skin, eyes, and mucous membranes. Combustible when exposed to heat or flame. When heated to decomposition it emits toxic fumes of Cl^-. See also CHLORIDES.

DEN600 CAS: 7572-29-4 HR: 3
DICHLOROACETYLENE
mf: C_2Cl_2 mw: 94.92
PROP: Volatile liquid. Mp: −66 to −64°, bp: 33°.
SYNS: DICHLOROETHYNE □ ETHYNE, DICHLORO-(9CI)
CONSENSUS REPORTS: IARC Cancer Review: Group 3 IMEMDT 7,56,87; Animal Limited Evidence IMEMDT 39,369,86.
OSHA PEL: CL 0.1 ppm
ACGIH TLV: CL 0.1 ppm
DFG MAK: Animal Carcinogen, Suspected Human Carcinogen
DOT CLASSIFICATION: Forbidden
SAFETY PROFILE: Confirmed carcinogen with experimental carcinogenic data. Poison by inhalation. Central nervous system effects. Can be formed by thermal decomposition (>70°) from trichloroethylene. Symptoms include a disabling nausea and intense jaw pain. Strong explosive when shocked or exposed to heat or air. Can react vigorously with oxidizing materials. When heated to decomposition or on contact with acid or acid fumes it emits highly toxic fumes of Cl^-. See also ACETYLENE COMPOUNDS and CHLORINATED HYDROCARBONS, ALIPHATIC.

DEN880 CAS: 27695-60-9 HR: 3
2-((4-(DICHLOROACETYL)PHENYL)-AMINO)-2-HYDROXY-1-(4-PHENOXY-PHENYL)ETHANONE
mf: $C_{22}H_{17}Cl_2NO_4$ mw: 430.30
SYNS: ETHANONE, 2-((4-(DICHLOROACETYL)-PHENYL)AMINO)-2-HYDROXY-1-(4-PHENOXY-PHENYL)- □ KETONE, 2-((4-(DICHLOROACETYL)-PHENYL)AMINO)-2-HYDROXY-1-(4-PHENOXYPHEN-YL)-

DOT CLASSIFICATION: 3; Label: Flammable Liquid

SAFETY PROFILE: A poison by intraperitoneal route. A flammable liquid. When heated to decomposition it emits toxic vapors of NO_x and Cl^-.

DEO700   CAS: 82-46-2   HR: 1
1,5-DICHLORO-9,10-ANTHRAQUINONE
mf: $C_{14}H_6Cl_2O_2$ mw: 277.10
PROP: Yellow needles from AcOH. Mp: 245°.
SYNS: 9,10-ANTHRACENEDIONE, 1,5-DICHLORO- □ 1,5-DICHLORANTHRACHINON □ 1,5-DICHLORO-ANTHRAQUINONE
CONSENSUS REPORTS: Reported in EPA TSCA Inventory.
SAFETY PROFILE: An eye irritant. When heated to decomposition it emits toxic fumes of Cl^-.

DEO750   CAS: 82-43-9   HR: 1
1,8-DICHLORO-9,10-ANTHRAQUINONE
mf: $C_{14}H_6Cl_2O_2$ mw: 277.10
PROP: Pale-yellow needles from $PhNO_2$. Mp: 201–202°.
SYNS: 9,10-ANTHRACENEDIONE, 1,8-DICHLORO- □ 1,8-DICHLORANTHRACHINON □ 1,8-DICHLORO-ANTHRAQUINONE
CONSENSUS REPORTS: Reported in EPA TSCA Inventory.
SAFETY PROFILE: An eye irritant. When heated to decomposition it emits toxic fumes of Cl^-.

DEP599   CAS: 541-73-1   HR: 2
m-DICHLOROBENZENE
mf: $C_6H_4Cl_2$ mw: 147.00
PROP: Liquid. D: 1.288 @ 20°/4°, fp: −26.25°, bp: 173°.
SYN: 1,3-DICHLOROBENZENE
CONSENSUS REPORTS: Reported in EPA TSCA Inventory. Community Right-To-Know List.
SAFETY PROFILE: Moderately toxic by intraperitoneal route. Mutation data reported. When heated to decomposition it emits toxic fumes of Cl^-. See also o-DICHLOROBENZENE and p-DICHLOROBENZENE.

DEP600   CAS: 95-50-1   HR: 3
o-DICHLOROBENZENE
DOT: UN 1591
mf: $C_6H_4Cl_2$ mw: 147.00
PROP: Clear liquid. Mp: −17.5°, bp: 180.5°, fp: −22°, flash p: 151°F, d: 1.307 @ 20°/20°, vap d: 5.05, autoign temp: 1198°F, lel: 2.2%, uel: 9.2%. IDLH 200 ppm.
SYNS: BENZENE, 1,2-DICHLORO- □ CHLOROBEN □ CHLORODEN □ CLOROBEN □ DCB □ o-DICHLOR BENZOL □ o-DICHLOROBENZENE □ 1,2-DICHLOROBENZENE □ o-DICHLOROBENZENE (ACGIH,OSHA) □ o-DICHLOROBENZENE (DOT) □ DILANTIN DB □ DILATIN DB □ DIZENE □ DOWTHERM E □ NCI-C54944 □ ODB □ ODCB □ ORTHODICHLOROBENZENE □ ORTHODI-CHLOROBENZOL □ SPECIAL TERMITE FLUID □ TERMITKIL
CONSENSUS REPORTS: IARC Cancer Review: Group 3 IMEMDT 7,192,87; Animal Inadequate Evidence IMEMDT 7,231,74; IMEMDT 29,213,82; Human Inadequate Evidence IMEMDT 7,231,74; IMEMDT 29,213,82. Reported in EPA TSCA Inventory. Community Right-To-Know List.
OSHA PEL: CL 50 ppm
ACGIH TLV: TWA 25 ppm; STEL 50 ppm; Not Classifiable as a Human Carcinogen
DFG MAK: 50 ppm (300 mg/m³)
DOT CLASSIFICATION: 6.1; Label: KEEP AWAY FROM FOOD
SAFETY PROFILE: Poison by ingestion and intravenous routes. Moderately toxic by inhalation and intraperitoneal routes. An experimental teratogen. Other experimental reproductive effects. An eye, skin, and mucous membrane irritant. Causes liver and kidney injury. Questionable carcinogen. Mutation data reported. A pesticide. Flammable when exposed to heat or flame. Can react vigorously with oxidizing materials. To fight fire, use water, foam, CO_2, or dry chemical. Slow reaction with aluminum may lead to explosion during storage in a sealed aluminum container.

When heated to decomposition it emits toxic fumes of Cl⁻. See also CHLOROBENZENE and CHLORINATED HYDROCARBONS, AROMATIC.

DEP800   CAS: 106-46-7   HR: 3
p-DICHLOROBENZENE
DOT: UN 1592
mf: $C_6H_4Cl_2$ mw: 147.00
PROP: White crystals or leaflets with strong penetrating odor. Mp: 54°, bp: 174°, flash p: 150°F (CC), d: 1.4581 @ 20.5°/4°, vap press: 10 mm @ 54.8°, vap d: 5.08. IDLH 150 ppm.
SYNS: p-CHLOROPHENYL CHLORIDE □ p-DICHLOORBENZEEN (DUTCH) □ 1,4-DICHLOOR-BENZEEN (DUTCH) □ p-DICHLORBENZOL (GERMAN) □ 1,4-DICHLOR-BENZOL (GERMAN) □ DI-CHLO-RICIDE □ 1,4-DICHLOROBENZENE (MAK) □ DICHLOROBENZENE, PARA, solid (DOT) □ p-DICHLOROBENZOL □ p-DICLOROBENZENE (ITALIAN) □ 1,4-DICLOROBENZENE (ITALIAN) □ EVOLA □ NCI-C54955 □ PARACIDE □ PARA CRYSTALS □ PARADI □ PARADICHLORBENZOL (GERMAN) □ PARADICHLOROBENZENE □ PARADICHLORO-BENZOL □ PARADOW □ PARAMOTH □ PARA-NUGGETS □ PARAZENE □ PDB □ PDCB □ PERSIA-PERAZOL □ RCRA WASTE NUMBER U070 □ RCRA WASTE NUMBER U071 □ RCRA WASTE NUMBER U072 □ SANTOCHLOR
CONSENSUS REPORTS: NTP 10th Report on Carcinogens. IARC Cancer Review: Group 2B IMEMDT 7,192,87; Animal Inadequate Evidence IMEMDT 7,231,74; IMEMDT 29,213,82. Human Inadequate Evidence IMEMDT 7,231,74; Reported in EPA TSCA Inventory. EPA Genetic Toxicology Program. Community Right-To-Know List.
OSHA PEL: TWA 75 ppm; STEL 110 ppm
ACGIH TLV: TWA 10 ppm, Confirmed Animal Carcinogen.
DFG MAK: 50 ppm (300 mg/m³)
DOT CLASSIFICATION: 6.1; Label: KEEP AWAY FROM FOOD
NIOSH REL: (p-Dichlorobenzene): (1.7 ppm LOQ)
SAFETY PROFILE: Confirmed carcinogen with experimental carcinogenic data. An experimental teratogen. A human poison by an unspecified route. Moderately toxic to humans by ingestion. Moderately toxic experimentally by ingestion, subcutaneous, and intraperitoneal routes. Other experimental reproductive effects. Human systemic effects by ingestion: unspecified changes in the eyes, lungs, thorax and respiration, and decreased motility or constipation. Can cause liver injury in humans. A human eye irritant. Mutation data reported. A fumigant. Flammable liquid when exposed to heat, flame, or oxidizers. Dangerous; can react vigorously with oxidizing materials. To fight fire, use water, foam, CO_2, dry chemical. When heated to decomposition it emits toxic fumes of Cl⁻. See also CHLORINATED HYDROCARBONS, AROMATIC.

DEQ600   CAS: 91-94-1   HR: 3
3',3'-DICHLOROBENZIDINE
mf: $C_{12}H_{10}Cl_2N_2$ mw: 253.14
PROP: Crystals or needles from alc. Mp: 133°. Insol in water; sol in alc, benzene, and glacial acetic acid.
SYNS: C.I. 23060 □ CURITHANE C126 □ DCB □ 4,4'-DIAMINO-3,3'-DICHLOROBIPHENYL □ 4,4'-DIAMINO-3,3'-DICHLORODIPHENYL □ 3,3'-DICHLORBENZIDIN (CZECH) □ 3,3'-DICHLOROBENZIDINA (SPANISH) □ DICHLOROBENZIDINE □ o,o'-DICHLOROBENZIDINE □ 3,3'-DICHLOROBENZIDINE □ DICHLOROBENZID-INE BASE □ 3,3'-DICHLOROBIPHENYL-4,4'-DIAMINE □ 3,3'-DICHLORO-4,4'-BIPHENYLDIAMINE □ 3,3'-DICHLORO-4,4'-DIAMINOBIPHENYL □ 3,3'-DICHLORO-4,4'-DIAMINO(1,1-BIPHENYL) □ RCRA WASTE NUMBER U073
CONSENSUS REPORTS: NTP 10th Report on Carcinogens. IARC Cancer Review: Group 2B IMEMDT 7,193,87; Human Inadequate Evidence IMEMDT 29,239,82; Animal Sufficient Evidence IMEMDT 29,239,82; IMEMDT 4,49,74. Reported in EPA TSCA Inventory. Community Right-To-Know List. EPA Genetic Toxicology Program.
OSHA PEL: Cancer Suspect Agent
ACGIH TLV: Animal Carcinogen
DFG MAK: DFG TRK: Animal Carcinogen, Suspected Human Carcinogen

NIOSH REL: (Benzidine-based Dye) Reduce to lowest feasible level
SAFETY PROFILE: Confirmed carcinogen with experimental carcinogenic and tumorigenic data. Human mutation data reported. When heated to decomposition it emits very toxic fumes of Cl⁻ and NO$_x$.

DEQ800   CAS: 612-83-9   HR: 3
3,3'-DICHLOROBENZIDINE DIHYDRO- CHLORIDE
mf: $C_{12}H_{10}Cl_2N_2 \cdot 2ClH$ mw: 326.06
SYN: 3,3'-DICHLORO-(1,1'-BIPHENYL)-4,4'-DIAMINE DIHYDROCHLORIDE
CONSENSUS REPORTS: NTP 10th Report on Carcinogens. Reported in EPA TSCA Inventory.
OSHA PEL: Cancer Suspect Agent
SAFETY PROFILE: Confirmed carcinogen. Moderately toxic by ingestion. Mutation data reported. When heated to decomposition it emits very toxic fumes of Cl⁻ and NO$_x$.

DER000   CAS: 510-15-6   HR: 3
4,4'-DICHLOROBENZILIC ACID ETHYL ESTER
mf: $C_{16}H_{14}Cl_2O_3$ mw: 325.20
PROP: Viscous liquid, sometimes yellow, sltly sol in water. Bp: 146–148°, vap press: 2.2×10^{-6} mm @ 20°.
SYNS: ACAR □ ACARABEN 4E □ AKAR □ BENZILAN □ BENZ-o-CHLOR □ CHLORBENZILATE □ CHLORO-BENZYLATE □ COMPOUND 338 □ 4,4'-DICHLORBEN-ZILSAEUREAETHYLESTER (GERMAN) □ 4,4'-DICHLOROBENZILATE □ ENT 18,596 □ ETHYL 4-CHLORO-α-(4-CHLOROPHENYL)-α-HYDROXYBENZ-ENEACETATE □ ETHYL-p,p'-DICHLOROBENZILATE □ ETHYL-4,4'-DICHLOROBENZILATE □ ETHYL-4,4'-DICHLORODIPHENYL GLYCOLLATE □ ETHYL-4,4'-DICHLOROPHENYL GLYCOLLATE □ ETHYL ESTER of 4,4'-DICHLOROBENZILIC ACID □ ETHYL-2-HYDROXY-2,2-BIS(4-CHLOROPHENYL)ACETATE □ FOLBEX □ FOLBEX SMOKE-STRIPS □ G 338 □ G 23992 □ GEIGY 338 □ KOP MITE □ NCI-C00408 □ NCI-C60413 □ RCRA WASTE NUMBER U038
CONSENSUS REPORTS: IARC Cancer Review: Group 3 IMEMDT 7,56,87; Animal Limited Evidence IMEMDT 30,73,83; Animal Sufficient Evidence IMEMDT 5,75,74. NCI Carcinogenesis Bioassay Completed; Results Positive: mouse

NCITR* NCI-CG-TR-75,78. NCI Carcinogenesis Bioassay Completed; Results Indefinite: rat NCITR* NCI-CG-TR-75,78. Community Right-To-Know List. Reported in EPA TSCA Inventory.
SAFETY PROFILE: Suspected carcinogen with experimental carcinogenic, neoplastigenic, and tumorigenic data. Moderately toxic by ingestion. A skin and eye irritant. A pesticide. When heated to decomposition it emits toxic fumes of Cl⁻.

DER600   CAS: 51-44-5   HR: 3
3,4-DICHLOROBENZOIC ACID
mf: $C_7H_4Cl_2O_2$ mw: 191.01
PROP: Needles from EtOH (aq) or C_6H_6. Mp: 208–209°.
SYNS: SYNSTIGMINE □ SYNTOSTIGMIN □ VAGOSTIGMIN
CONSENSUS REPORTS: Reported in EPA TSCA Inventory.
SAFETY PROFILE: Poison by subcutaneous route. When heated to decomposition it emits toxic fumes of Cl⁻.

DEU100   CAS: 70134-26-8   HR: 3
DICHLOROBIS(2-CHLOROCYCLOHEX-YL)SELENIUM
mf: $C_{12}H_{20}Cl_4Se$ mw: 385.08
SYN: SELENIUM, DICHLOROBIS(2-CHLORO-CYCLOHEXYL)-
OSHA PEL: TWA 0.2 mg(Se)/m³
ACGIH TLV: TWA 0.2 mg(Se)/m³
SAFETY PROFILE: Poison by intravenous route. When heated to decomposition it emits toxic fumes of Se and Cl⁻.

DEU115   CAS: 18252-65-8   HR: 3
cis-DICHLOROBIS(DIMETHYLSELENIDE)PLATINUM(II)
mf: $C_4H_{12}Cl_2PtSe_2$ mw: 484.07
PROP: IDLH 4 mg/m³ (as Pt).
SYNS: NSC-271675 □ PLATINUM(II), BIS(METHYL SELENIDE)DICHLORO-, cis- □ PLATINUM, DICHLORO-BIS(METHYL SELENIDE)-, cis- □ PLATINUM, DICHLOROBIS(SELENOBIS(METHANE))-(SP-4-2)
OSHA PEL: TWA 0.2 mg(Se)/m³
ACGIH TLV: TWA 0.2 mg(Se)/m³; TWA 0.002 mg(Pt)/m³

SAFETY PROFILE: Poison by intravenous route. When heated to decomposition it emits toxic fumes of Se, Pt, and Cl⁻.

DEU125 CAS: 74037-18-6 HR: 3
DICHLOROBIS(2-ETHOXYCYCLOHEX-YL)SELENIUM
mf: $C_{16}H_{30}Cl_2O_2Se$ mw: 404.32
SYN: SELENIUM, DICHLOROBIS(2-ETHOXY-CYCLOHEXYL)-
OSHA PEL: TWA 0.2 mg(Se)/m³
ACGIH TLV: TWA 0.2 mg(Se)/m³
SAFETY PROFILE: Poison by intravenous route. When heated to decomposition it emits toxic fumes of Se and Cl⁻.

DEU200 CAS: 38780-42-6 HR: 3
cis-DICHLOROBIS(PYRROLIDINE)-PLATINUM(II)
mf: $C_8H_{18}Cl_2N_2Pt$ mw: 408.27
PROP: IDLH 4 mg/m³ (as Pt).
SYN: cis-DIPYRROLIDINEDICHLOROPLATINUM(II)
SAFETY PROFILE: Poison by intraperitoneal route. Questionable carcinogen with experimental tumorigenic data. Mutation data reported. See also PLATINUM COMPOUNDS. When heated to decomposition it emits very toxic fumes of Cl⁻ and NOₓ.

DEV000 CAS: 764-41-0 HR: 3
1,4-DICHLORO-2-BUTENE
mf: $C_4H_6Cl_2$ mw: 125.00
PROP: Colorless liquid. Mp: 1–3°, bp: 156°, d: 1.183 @ 25°/4°.
SYNS: DCB □ 1,4-DCB □ 1,4-DICHLOROBUTENE-2 (MAK) □ RCRA WASTE NUMBER U074
CONSENSUS REPORTS: Reported in EPA TSCA Inventory. EPA Genetic Toxicology Program.
ACGIH TLV: Animal Carcinogen, Suspected Human Carcinogen
DFG MAK: Animal Carcinogen, Suspected Human Carcinogen
SAFETY PROFILE: Confirmed carcinogen with experimental carcinogenic and neoplastigenic data. Poison by ingestion, inhalation, and intravenous routes. Moderately toxic by skin contact. An experimental teratogen. Other experimental reproductive effects. Mutation data reported. A severe skin and eye irritant. When heated to decomposition it emits toxic fumes of Cl⁻. See also CHLORINATED HYDROCARBONS, ALIPHATIC.

DEV100 CAS: 760-23-6 HR: 2
3,4-DICHLORO-1-BUTENE
mf: $C_4H_6Cl_2$ mw: 125.00
SYN: 1-BUTENE, 3,4-DICHLORO-
CONSENSUS REPORTS: Reported in EPA TSCA Inventory.
SAFETY PROFILE: Moderately toxic by ingestion. Experimental reproductive effects. Mutation data reported. When heated to decomposition it emits toxic fumes of Cl⁻.

DEV200 CAS: 11069-19-5 HR: 3
DICHLOROBUTENE
DOT: NA 2920
mf: $C_4H_6Cl_2$ mw: 125.00
SYNS: BUTENE, DICHLORO- □ DICHLORO-BUTYLENE
DOT CLASSIFICATION: 8; Label: Corrosive, Flammable Liquid
SAFETY PROFILE: A flammable liquid. When heated to decomposition it emits toxic vapors of Cl⁻.

DEV400 CAS: 821-10-3 HR: 3
1,4-DICHLORO-2-BUTYNE
mf: $C_4H_4Cl_2$ mw: 122.98
PROP: D: 1.26° @ 20°/4°, bp: 165–168°.
SYN: 1,4-DICHLOROBUTYNE
CONSENSUS REPORTS: Reported in EPA TSCA Inventory.
SAFETY PROFILE: Poison by intravenous route. When heated to decomposition it emits toxic fumes of Cl⁻. Probably a dangerous fire and explosion hazard. See also ACETYLENE COMPOUNDS and CHLORINATED HYDROCARBONS, ALIPHATIC.

D

DEW000   CAS: 333-25-5   HR: 3
DICHLORO(2-CHLOROVINYL)ARSINE OXIDE

SYN: LEWISITE I OXIDE

CONSENSUS REPORTS: Arsenic and its compounds are on the Community Right-To-Know List.

OSHA PEL: TWA 0.5 mg(As)/m³

SAFETY PROFILE: Poison by ingestion, intravenous, and subcutaneous routes. See also ARSENIC COMPOUNDS. When heated to decomposition it emits very toxic fumes of Cl⁻ and As. See also CHLOROVINYLARSINE DICHLORIDE.

DEX000   CAS: 14913-33-8   HR: 3
trans-DICHLORODIAMMINE-PLATINUM(II)

mf: $C_{12}H_6N_2Pt$ mw: 300.07

PROP: Pale yellow crystals. Mp: 270° (decomp). Less sol in H_2O than cis-form; sol in DMF and DMSO. IDLH 4 mg/m³ (as Pt).

SYNS: trans-DIAMMINEDICHLOROPLATINUM(II) □ trans-PLATINUM(II)DIAMMINEDICHLORIDE

CONSENSUS REPORTS: EPA Genetic Toxicology Program.

SAFETY PROFILE: Poison by intraperitoneal route. Questionable carcinogen with experimental tumorigenic data. Human mutation data reported. See also PLATINUM COMPOUNDS. When heated to decomposition it emits toxic fumes of NO_x and Cl⁻.

DEX200   CAS: 43047-99-0   HR: 3
DICHLORODIBENZOFURAN

mf: $C_{12}H_6Cl_2O$ mw: 237.08

SYN: DIBENZOFURAN, DICHLORO-

SAFETY PROFILE: A poison by ingestion. When heated to decomposition it emits toxic vapors of Cl⁻.

DEY800   CAS: 1719-53-5   HR: 3
DICHLORODIETHYLSILANE

DOT: UN 1767

mf: $C_4H_{10}Cl_2Si$ mw: 157.13

PROP: Liquid. Mp: −96°, bp: 131.0°, d: 1.05, vap d: 5.41, flash p: 75.2°F.

SYN: DIETHYLDICHLOROSILANE (DOT)

CONSENSUS REPORTS: Reported in EPA TSCA Inventory.

DOT CLASSIFICATION: 8; Label: Corrosive, Flammable Liquid

SAFETY PROFILE: Poison by intraperitoneal route. Moderately toxic by ingestion. Corrosive to tissue. Dangerous fire hazard when exposed to heat, flame, or oxidizers. Can react vigorously with oxidizing materials. To fight fire, use foam, CO_2, dry chemical. When heated to decomposition or in reaction with water or steam it emits toxic and corrosive fumes of Cl⁻. See also CHLOROSILANES.

DEZ000   CAS: 866-55-7   HR: 3
DICHLORODIETHYLSTANNANE

mf: $C_4H_{10}Cl_2Sn$ mw: 247.73

PROP: Water-white crystals. Mp: 85°, bp: 277°.

SYNS: DIAETHYLZINNDICHLORID (GERMAN) □ DICHLORODIETHYLTIN □ DIETHYLDICHLORO-STANNANE □ DIETHYLSTANNYL DICHLORIDE □ DIETHYLTIN CHLORIDE □ DIETHYLTIN DICHLORIDE

OSHA PEL: TWA 0.1 mg(Sn)/m³ (skin)

ACGIH TLV: TWA 0.1 mg(Sn)/m³; STEL 0.2 mg(Sn)/m³ (skin).

NIOSH REL: (Organotin Compounds) TWA 0.1 mg(Sn)/m³

SAFETY PROFILE: Poison by ingestion and intravenous routes. See also TIN COMPOUNDS and CHLORIDES. When heated to decomposition it emits toxic fumes of Cl⁻.

DFA000   CAS: 1649-08-7   HR: 1
1,2-DICHLORO-1,1-DIFLUOROETHANE

mf: $C_2H_2Cl_2F_2$ mw: 134.94

PROP: A liquid. Fp: −101°, bp: 45–47°, d: 1.416 @ 20°/4°.

CONSENSUS REPORTS: Reported in EPA TSCA Inventory.

SAFETY PROFILE: Experimental reproductive effects. Mildly toxic by inhalation. When heated to decomposition it emits very toxic fumes of Cl⁻ and F⁻.

**DFA600   CAS: 75-71-8   HR: 1
DICHLORODIFLUOROMETHANE**
DOT: UN 1028
mf: CCl_2F_2 mw: 120.91
PROP: Colorless, almost odorless gas. Mp: −158°, bp: −29°, vap press: 5 atm @ 16.1°. IDLH 15,000 ppm.
SYNS: ALGOFRENE TYPE 2 □ ARCTON 6 □ DIFLUORODICHLOROMETHANE □ DWUCHLORO-DWUFLUOROMETAN (POLISH) □ ELECTRO-CF 12 □ ESKIMON 12 □ F 12 □ FC 12 □ FLUOROCARBON-12 □ FREON 12 □ FREON F-12 □ FRIGEN 12 □ GENETRON 12 □ HALON □ ISCEON 122 □ ISOTRON 12 □ KAISER CHEMICALS 12 □ LEDON 12 □ PROPELLANT 12 □ RCRA WASTE NUMBER U075 □ R12 (DOT) □ REFRIGERANT 12 □ UCON 12 □ UCON 12/HALOCARBON 12
CONSENSUS REPORTS: Reported in EPA TSCA Inventory. EPA Genetic Toxicology Program.
OSHA PEL: TWA 1000 ppm
ACGIH TLV: TWA 1000 ppm; Not Classifiable as a Human Carcinogen
DFG MAK: 1000 ppm (5000 mg/m³)
DOT CLASSIFICATION: 2.2; Label: Nonflammable Gas
SAFETY PROFILE: Human systemic effects by inhalation: conjunctiva irritation, fibrosing alveolitis, and liver changes. Narcotic in high concentrations. Nonflammable gas. Can react violently with Al. When heated to decomposition it emits highly toxic fumes of phosgene, Cl⁻, and F⁻.

**DFB400   CAS: 56275-41-3 HR: 1
DICHLORODIFLUOROMETHANE with
      1,1-DIFLUOROETHANE**
DOT: UN 1954
mf: $C_2H_4F_2 \cdot CCl_2F_2$ mw: 186.97
SYNS: DICHLORODIFLUOROMETHANE and DIFLUOROETHANE AZEOTROPIC MIXTURE (DOT) □ FREON 500 □ R500 (DOT) □ UCON 500/HALOCARBON 500
DOT CLASSIFICATION: 2.2; Label: Nonflammable Gas
SAFETY PROFILE: A simple asphyxiant. See also components as listed. When heated

to decomposition it emits very toxic fumes of Cl⁻ and F⁻.

**DFC200   CAS: 2767-41-1   HR: 3
DICHLORODIHEXYLSTANNANE**
mf: $C_{12}H_{26}Cl_2Sn$ mw: 359.97
SYN: DIHEXYLTIN DICHLORIDE
OSHA PEL: TWA 0.1 mg(Sn)/m³ (skin)
ACGIH TLV: TWA 0.1 mg(Sn)/m³; STEL 0.2 mg(Sn)/m³ (skin).
NIOSH REL: (Organotin Compounds) TWA 0.1 mg(Sn)/m³
SAFETY PROFILE: Poison by ingestion and intravenous routes. See also TIN COMPOUNDS. When heated to decomposition it emits toxic fumes of Cl⁻.

**DFC300   CAS: 23611-67-8 HR: 3
6,8-DICHLORO-DIHYDRO-1,3-BENZOX-
      AZINE-2-THIONE-4-ONE**
mf: $C_8H_3Cl_2NO_2S$ mw: 248.08
SYNS: 2H-1,3-BENZOXAZINE-2,4(3H)-DIONE, 6,8-DICHLORO-2-THIO- □ 6,8-DICHLORO-2-THIO-2H-1,3-BENZOXAZINE-2,4(3H)-DIONE
SAFETY PROFILE: A poison by ingestion. When heated to decomposition it emits toxic vapors of NO_x, SO_x, and Cl⁻.

**DFC800   CAS: 33770-60-4 HR: 3
(2,5-DICHLORO-3,6-DIHYDROXY-p-
      BENZOQUINOLATO)MERCURY**
mf: $C_6Cl_2HgO_4$ mw: 407.55
PROP: IDLH 10 mg/m³ (as Hg).
SYNS: 2,5-DICHLORO-3,6-DIHYDROXY-p-BENZOQUINONE, MERCURY SALT □ (2,5-DICHLORO-3,6-DIHYDROXY-p-BENZOQUINONE), MERCURY SALT
CONSENSUS REPORTS: Mercury and its compounds are on the Community Right-To-Know List.
OSHA PEL: CL 0.1 mg(Hg)/m³ (skin)
ACGIH TLV: TWA 0.1 mg(Hg)/m³ (skin); BEI: 35 µg/g creatinine total inorganic mercury in urine preshift; 15 µg/g creatinine total inorganic mercury in blood at end of shift at end of workweek.
DFG MAK: Confirmed Animal Carcinogen with Unknown Relevance to Humans
NIOSH REL: (Mercury, Aryl and Inorganic) CL 0.1 mg/m³ (skin)

SAFETY PROFILE: Poison by intravenous route. See also MERCURY COMPOUNDS. When heated to decomposition it emits very toxic fumes of Cl⁻ and Hg.

DFE200   CAS: 118-52-5   HR: 2
1,3-DICHLORO-5,5-DIMETHYL HYDANTOIN

mf: $C_5H_6Cl_2N_2O_2$ mw: 197.03

PROP: Crystals, liberates chlorine on contact with hot water; prisms from $CHCl_3$. Mp: 132°. Subl @ 100°; conflagrates @ 212°; d: 1.5 @ 20°, vap d: 6.8. Sol in H_2O; mod in sol AcOH and EtOH. IDLH 5 mg/m³.

SYNS: DACTIN □ DAKTIN □ DANTOIN □ DCA □ DICHLORANTIN □ DICHLORODIMETHYLHYDANTOIN □ 1,3-DICHLORO-5,5-DIMETHYL-2,4-IMIDAZOLIDINEDIONE □ 1,3-DICHLORO-5,5'-METHYLHYDANTOIN □ HALANE □ HYDAN □ HYDAN (antiseptic) □ NCI-C03054 □ OMCHLOR

CONSENSUS REPORTS: Reported in EPA TSCA Inventory.

OSHA PEL: TWA 0.2 mg/m³; STEL 0.4 mg/m³

ACGIH TLV: TWA 0.2 mg/m³; STEL 0.4 mg/m³

SAFETY PROFILE: Moderately toxic by ingestion. Mildly toxic by inhalation. A severe skin irritant. Mutation data reported. Avoid excessive contact because of effects of active chlorine on skin. Some of the hydantoins are central nervous system depressants. Mixtures with xylene may explode. Will react with water or steam to produce toxic and corrosive fumes. When heated to decomposition it emits toxic fumes of Cl⁻ and NO_x. See also CHLORIDES.

DFE259   CAS: 75-78-5   HR: 3
DICHLORODIMETHYLSILANE

DOT: UN 1162

mf: $C_2H_6Cl_2Si$ mw: 129.06

PROP: Liquid. D: 1.06 @ 20°/4°, mp: −16°.

SYNS: DIMETHYLDICHLOROSILANE (DOT) □ DIMETHYL-DICHLORSILAN

CONSENSUS REPORTS: EPA Extremely Hazardous Substances List. Reported in EPA TSCA Inventory.

DOT CLASSIFICATION: 3; Label: Flammable Liquid, Corrosive

SAFETY PROFILE: Poison by ingestion and intraperitoneal routes. Moderately toxic by inhalation. A skin and severe eye irritant. Violent reaction on contact with water. When heated to decomposition it emits toxic fumes of Cl⁻. See also CHLOROSILANES.

DFF000   CAS: 80-10-4   HR: 3
DICHLORO DIPHENYLSILANE

DOT: UN 1769

mf: $C_{12}H_{10}Cl_2Si$ mw: 253.21

PROP: Colorless liquid. Mp: −22°, bp: 303°, d: 1.19 @ 20°, vap d: 8.45.

SYNS: DICHLOR-DIFENYLSILAN □ DIPHENYL DICHLOROSILANE (DOT)

CONSENSUS REPORTS: Reported in EPA TSCA Inventory.

DOT CLASSIFICATION: 8; Label: Corrosive

SAFETY PROFILE: A poison irritant to skin, eyes, and mucous membranes. See also CHLOROSILANES. Can react vigorously with oxidizing materials. When heated to decomposition or on contact with acid or acid fumes it emits toxic fumes of Cl⁻.

DFF400   CAS: 867-36-7   HR: 3
DICHLORODIPROPYLSTANNANE

mf: $C_6H_{14}Cl_2Sn$ mw: 275.79

PROP: Colorless crystals. Mp: 82.5–83°, bp: 118–121° @ 10 mm.

SYNS: DICHLORODIPROPYLTIN □ DIPROPYLTIN CHLORIDE □ DIPROPYLTIN DICHLORIDE □ DI-n-PROPYLTIN DICHLORIDE

OSHA PEL: TWA 0.1 mg(Sn)/m³ (skin)

ACGIH TLV: TWA 0.1 mg(Sn)/m³; STEL 0.2 mg(Sn)/m³ (skin).

NIOSH REL: (Organotin Compounds) TWA 0.1 mg(Sn)/m³

SAFETY PROFILE: Poison by ingestion. See also TIN COMPOUNDS and CHLORIDES. When heated to decomposition it emits toxic fumes of Cl⁻.

DFF809   CAS: 75-34-3   HR: 3
1,1-DICHLOROETHANE
DOT: UN 2362
mf: C₂H₄Cl₂   mw: 98.96
PROP: Colorless liquid; aromatic, ethereal odor; hot, saccharine taste. Mp: −97.7°, lel: 5.6%, fp: −98°, bp: 57.3°, flash p: 22°F (TOC), d: 1.174 @ 20°/4°, vap press: 230 mm @ 25°, vap d: 3.44, autoign temp: 856°F. IDLH 3000 ppm.
SYNS: AETHYLIDENCHLORID (GERMAN) □ CHLORINATED HYDROCHLORIC ETHER □ CHLORURE d'ETHYLIDENE (FRENCH) □ CLORURO di ETILIDENE (ITALIAN) □ 1,1-DICHLOORETHAAN (DUTCH) □ 1,1-DICHLORAETHAN (GERMAN) □ 1,1-DICLOROETANO (ITALIAN) □ ETHYLIDENE CHLORIDE □ ETHYLIDENE DICHLORIDE □ NCI-C04535 □ RCRA WASTE NUMBER U076
CONSENSUS REPORTS: NCI Carcinogenesis Bioassay (gavage); Inadequate Studies: mouse, rat NCITR* NCI-CG-TR-66,78. Reported in EPA TSCA Inventory.
OSHA PEL: TWA 100 ppm
ACGIH TLV: TWA 100 ppm; Not Classifiable as a Human Carcinogen
DFG MAK: 100 ppm (410 mg/m³)
NIOSH REL: (1,1-Dichloroethane) Handle with caution
DOT CLASSIFICATION: 3; Label: Flammable Liquid
SAFETY PROFILE: Moderately toxic by ingestion. Experimental teratogenic effects. Questionable carcinogen with experimental tumorigenic data. Liver damage reported in experimental animals. A very dangerous fire hazard and moderate explosion hazard when exposed to heat or flame; can react vigorously with oxidizing materials. To fight fire, use alcohol foam, water, foam, CO₂, dry chemical. When heated to decomposition it emits highly toxic fumes of phosgene and Cl⁻.

DFH200   CAS: 598-14-1   HR: 3
DICHLOROETHYLARSINE
DOT: UN 1892
mf: C₂H₅AsCl₂   mw: 174.89

PROP: Colorless liquid; fruity, biting, irritating odor. Mp: −65°, bp: 156° decomp, d: 1.742 @ 14°, vap press: 2.29 mm @ 21.5°, vap d: 6.03. Sol in H₂O; misc in EtOH and C₆H₆.
SYNS: ARSENIC DICHLOROETHANE □ ARSONOUS DICHLORIDE, ETHYL-(9CI) □ DICK (GERMAN) □ ED □ ETHYLARSONOUS DICHLORIDE □ ETHYLI-DICHLORARSINE □ ETHYLIDICHLOROARSINE (DOT) □ TL 214
CONSENSUS REPORTS: Arsenic and its compounds are on the Community Right-To-Know List.
OSHA PEL: TWA 0.5 mg(As)/m³
DOT CLASSIFICATION: 6.1; Label: Poison
SAFETY PROFILE: A human poison by inhalation. Experimentally, a deadly poison by inhalation and subcutaneous routes, and probably by ingestion. A severe irritant. A military poison gas. Can react with oxidizing materials. Will react with water or steam to produce toxic and corrosive fumes. Dangerous; on contact with acid or acid fumes it emits highly toxic fumes of Cl⁻, As, and phosgene. See also ARSENIC COMPOUNDS.

DFH600   CAS: 321-55-1   HR: 2
O,O-DI(2-CHLOROETHYL)-O-(3-CHLORO-4-METHYLCOUMARIN-7-YL) PHOSPHATE
mf: C₁₄H₁₄Cl₃O₆P   mw: 415.60
PROP: Crystals from EtOH. Mp: 91°.
SYNS: O,O-BIS(2-CHLOROETHYL)-O-(3-CHLORO-4-METHYL-7-COUMARINYL) PHOSPHATE □ 2-CHLOROETHANOL HYDROGEN PHOSPHATE ESTER with 3-CHLORO-7-HYDROXY-4-METHYLCOUMARIN □ 2-CHLOROETHANOL PHOSPHATE DIESTER ESTER with 3-CHLORO-7-HYDROXY-4-METHYLCOUMARIN □ 3-CHLORO-7-HYDROXY-4-METHYLCOUMARIN BIS(2-CHLOROETHYL)PHOSPHATE □ 3-CHLORO-4-METHYL-UMBELLIFERONE BIS(2-CHLOROETHYL)PHOSPHATE □ DI-(2-CHLOROETHYL)-3-CHLORO-4-METHYL-COUMARIN-7-YL PHOSPHATE □ DI-(2-CHLORO-ETHYL)-3-CHLORO-4-METHYL-7-COUMARINYL PHOSPHATE □ EUSTIDIL □ GALLOXON □ GALOX-ANE □ 96H60 □ HALOXON □ HELMIRANE □ HELMIR-ON □ HELMIRONE □ LOXON □ LUXON □ LXON
SAFETY PROFILE: Moderately toxic by ingestion and intraperitoneal routes. Human mutation data reported. When heated to decomposition it emits very toxic fumes of

PO$_x$ and Cl$^-$. See also other coumarin entries.

DFH800   CAS: 25323-30-2   HR: 3
DICHLOROETHYLENE
DOT: UN 1150
mf: C$_2$H$_2$Cl$_2$ mw: 96.94
DOT CLASSIFICATION: 3; Label: Flammable Liquid
SAFETY PROFILE: Moderately toxic by ingestion. Mildly toxic by inhalation. Flammable when exposed to heat or flame. When heated to decomposition it emits toxic fumes of Cl$^-$. See also VINYLIDENE CHLORIDE.

DFI200   CAS: 156-59-2   HR: 1
cis-DICHLOROETHYLENE
mf: C$_2$H$_2$Cl$_2$ mw: 96.94
PROP: Colorless liquid; pleasant odor. Mp: −80.5°, bp: 59°, lel: 9.7%, uel: 12.8%, flash p: 39°F, d: 1.291 @ 15°/4°, vap press: 400 mm @ 41.0°, vap d: 3.34.
SYN: 1,2-DICHLOROETHYLENE
CONSENSUS REPORTS: Reported in EPA TSCA Inventory.
DFG MAK: 200 ppm (800 mg/m^3)
SAFETY PROFILE: Mildly toxic by ingestion and inhalation. In high concentration it is irritating and narcotic. Has produced liver and kidney injury in experimental animals. Mutation data reported. Sometimes thought to be nonflammable, however, it is a dangerous fire hazard when exposed to heat or flame. Reaction with solid caustic alkalies or their concentrated solutions produces chloracetylene gas, which ignites spontaneously in air. Reacts violently with N$_2$O$_4$, KOH, Na, NaOH. Moderate explosion hazard in the form of vapor when exposed to flame. Can react vigorously with oxidizing materials. To fight fire, use water spray, foam, CO$_2$, dry chemical. When heated to decomposition it emits toxic fumes of Cl$^-$. See also VINYLIDENE CHLORIDE and CHLORINATED HYDROCARBONS, ALIPHATIC.

DFI210   CAS: 540-59-0   HR: 3
1,2-DICHLOROETHYLENE
mf: C$_2$H$_2$Cl$_2$ mw: 96.94
PROP: Liquid with ethereal odor. Bp: 55°. IDLH 1000 ppm.
SYNS: ACETYLENE DICHLORIDE □ 1,2-DICHLOR-AETHEN (GERMAN) □ sym-DICHLOROETHYLENE □ DICHLORO-1,2-ETHYLENE (FRENCH) □ DIOFORM □ NCI-C56031
CONSENSUS REPORTS: Reported in EPA TSCA Inventory. Community Right-To-Know List.
OSHA PEL: TWA 200 ppm
ACGIH TLV: TWA 200 ppm
DFG MAK: 200 ppm (800 mg/m^3)
SAFETY PROFILE: Poison by inhalation. Moderately toxic by ingestion. A skin irritant. When heated to decomposition it emits highly toxic fumes of Cl$^-$. See also ACETYLENE COMPOUNDS, and CHLORINATED HYDROCARBONS, ALIPHATIC.

DFJ050   CAS: 111-44-4   HR: 3
DICHLOROETHYL ETHER
DOT: UN 1916
mf: C$_4$H$_8$Cl$_2$O   mw: 143.02
PROP: Colorless, stable liquid. Bp: 178.5°, fp: −51.9°, flash p: 131°F (CC), d: 1.2220 @ 20°/20°, autoign temp: 696°F, vap press: 0.7 mm @ 20°, vap d: 4.93. Misc in Et$_2$O, MeOH, and C$_6$H$_6$. IDLH 100 ppm.
SYNS: BIS(β-CHLOROETHYL) ETHER □ BIS(2-CHLOROETHYL) ETHER □ CHLOREX □ 1-CHLORO-2-(β-CHLOROETHOXY)ETHANE □ CHLOROETHYL ETHER □ CLOREX □ DCEE □ 2,2'-DICHLOOR-ETHYLETHER (DUTCH) □ 2,2'-DICHLOR-DIAETHYL-AETHER (GERMAN) □ 2,2'-DICHLORETHYL ETHER □ β,β-DICHLORODIETHYL ETHER □ DICHLOROETHER □ DI(β-CHLOROETHYL)ETHER □ β,β'-DICHLOROETH-YL ETHER □ sym-DICHLOROETHYL ETHER □ 2,2'-DICHLOROETHYL ETHER (MAK) □ DICHLOROETHYL OXIDE □ 2,2'-DICLOROETILETERE (ITALIAN) □ DWUCHLORODWUETYLOWY ETER (POLISH) □ ENT 4,504 □ ETHER DICHLORE (FRENCH) □ 1,1'-OXYBIS(2-CHLORO)ETHANE □ OXYDE de CHLORETHYLE (FRENCH) □ RCRA WASTE NUMBER U025
CONSENSUS REPORTS: IARC Cancer Review: Group 3 IMEMDT 7,56,87; Animal Sufficient Evidence IMEMDT 9,117,75. Reported in EPA TSCA Inventory. On

Community Right-To-Know List. On EPA Extremely Hazardous Substances List.

OSHA PEL: TWA 5 ppm; STEL 10 ppm (skin)

ACGIH TLV: TWA 5 ppm; STEL 10 ppm (skin); Not Classifiable as a Human Carcinogen

DFG MAK: 10 ppm (59 mg/m³)

DOT CLASSIFICATION: 6.1; Label: Poison, Flammable Liquid

SAFETY PROFILE: A poison by ingestion, skin contact, and inhalation. A skin, eye, and mucous membrane irritant. Questionable carcinogen with experimental carcinogenic and tumorigenic data. Mutation data reported. Exposure to 1000 ppm for 30 to 60 minutes may result in death within days. The odor is easily detectable at 35 ppm which causes only slight irritation. Flammable liquid when exposed to heat, flame, or oxidants. Dangerous explosion hazard; reacts vigorously with oleum, chlorosulfonic acid. Reacts with water or steam to evolve toxic and corrosive fumes. Can react vigorously with oxidizing materials. To fight fire, use water, foam, mist, fog, spray, dry chemical. When heated to decomposition it emits toxic fumes of Cl⁻. See also ETHERS.

DFJ400   CAS: 20198-77-0   HR: 3
2,3-DICHLORO-N-ETHYLMALEINIMIDE
mf: C₆H₅Cl₂NO₂   mw: 194.02
SYN: N-ETHYL-DICHLOROMALEINIMIDE
SAFETY PROFILE: Poison by intraperitoneal and intravenous routes. An experimental teratogen. Experimental reproductive effects. When heated to decomposition it emits very toxic fumes of Cl⁻ and NOₓ.

DFJ800   CAS: 1125-27-5   HR: 3
DICHLOROETHYLPHENYLSILANE
DOT: UN 2435
mf: C₈H₁₀Cl₂Si   mw: 205.17
PROP: Liquid.
SYN: ETHYL PHENYL DICHLOROSILANE (DOT)

CONSENSUS REPORTS: Reported in EPA TSCA Inventory.

DOT CLASSIFICATION: 8; Label: Corrosive

SAFETY PROFILE: Poison by ingestion and inhalation. A poison irritant to skin, eyes, and mucous membranes. Corrosive. Will react with water or steam to produce toxic and corrosive fumes. Can react with oxidizing materials. When heated to decomposition it emits toxic fumes of Cl⁻ and phenol. See also CHLOROSILANES.

DFK000   CAS: 1789-58-8   HR: 3
DICHLOROETHYLSILANE
DOT: UN 1183
mf: C₂H₆Cl₂Si   mw: 129.07
PROP: Liquid. Vap d: 4.45, bp: 74–75°, flash p: <73.4°F.
SYN: ETHYL DICHLOROSILANE (DOT)
CONSENSUS REPORTS: Reported in EPA TSCA Inventory.
DOT CLASSIFICATION: 4.3; Label: Danger When Wet, Corrosive, Flammable Liquid
SAFETY PROFILE: Poison by ingestion and inhalation. A severe irritant to skin, eyes, and mucous membranes. Corrosive. Dangerous fire hazard if exposed to heat, open flames, or powerful oxidizers. Will react with water or steam to produce heat and toxic and corrosive fumes. To fight fire, use foam, dry chemical, mist, spray. When heated to decomposition it emits toxic fumes of Cl⁻ and phosgene. See also CHLOROSILANES.

DFK600   CAS: 97-17-6   HR: 3
DICHLOROFENTHION
mf: C₁₀H₁₃Cl₂O₃PS   mw: 315.16
PROP: A liquid. A nonvolatile, residual organic phosphate nematocide and insecticide. Bp: 126–131° @ 0.2 mm, d: 1.3. Insol in water; sol in most org solvs.
SYNS: BROMEX □ O,O-DIAETHYL-O-2,4-DICHLOR-PHENYL-MONOTHIOPHOSPHAT (GERMAN) □ O,O-DIAETHYL-O-2,4-DICHLORPHENYL-THIONO-PHOSPHAT (GERMAN) □ DICHLOFENTHION □ DICHLOFENTION □ 2,4-DICHLORO-PHENOL-O-ESTER with O,O-DIETHYL PHOSPHOROTHIOATE □ O-2,4-DICHLOROPHENYL-O,O-DIETHYL PHOSPHOROTHIO-

ATE □ 2,4-DICHLORO-PHENYL DIETHYL PHOS-PHOROTHIONATE □ O,O-DIETHYL-O-(2,4-DICHLOOR-FENYL)-MONOTHIOFOSFAAT (DUTCH) □ O,O-DIETHYL-O-(2,4-DICHLOROPHENYL) PHOSPHORO-THIOATE □ DIETHYL 2,4-DICHLOROPHENYL PHOSPHOROTHIONATE □ O,O-DIETHYL-O-2,4-DICHLOROPHENYL THIOPHOSPHATE □ O,O-DIETIL-O-(2,4-DICLORO-FENIL)-MONOTIOFOSFATO (ITALIAN) □ ECP □ ENT 17,470 □ HEXA-NEMA □ MOBILAWN □ NEMACIDE □ THIOPHOSPHATE de O-2,4-DICHLOROPHENYLE et de O,O-DIETHYLE (FRENCH) □ TRI-VC 13 □ VC13 NEMACIDE

SAFETY PROFILE: Poison by ingestion and skin contact. A very toxic insecticide. Mutation data reported. See also ESTERS and PARATHION. When heated to decomposition it emits very toxic fumes of PO_x, SO_x, and Cl^-.

DFL000   CAS: 75-43-4   HR: 1
DICHLOROFLUOROMETHANE
DOT: UN 1029
mf: $CHCl_2F$ mw: 102.92
PROP: Heavy, colorless gas. Mp: −135°, bp: 8.9°, d: 1.48, vap press: 2 atm @ 28.4°, vap d: 3.82. IDLH 5000 ppm.
SYNS: ALGOFRENE TYPE 5 □ ARCTON 7 □ DICHLOROMONOFLUOROMETHANE (OSHA, DOT) □ DWUCHLOROFLUOROMETAN (POLISH) □ FC-21 □ FLUORODICHLOROMETHANE □ FREON 21 □ GENETRON 21 □ R21 (DOT)
CONSENSUS REPORTS: Reported in EPA TSCA Inventory.
OSHA PEL: TWA 10 ppm
ACGIH TLV: TWA 10 ppm
DFG MAK: 10 ppm (43 mg/m³)
DOT CLASSIFICATION: 2.2; Label: Nonflammable Gas
SAFETY PROFILE: Mildly toxic by inhalation. Experimental reproductive effects. When heated to decomposition it emits very toxic fumes of Cl^- and F^-.

DFN800   CAS: 1193-54-0   HR: 3
DICHLOROMALEIMIDE
mf: $C_4HCl_2NO_2$ mw: 165.96
PROP: A solid. Mp: 175°.
SYNS: DICHLOROMALEINIMIDE □ 3,4-DICHLORO-2,5-PYRROLIDINEDIONE
SAFETY PROFILE: Poison by intraperitoneal route. Experimental

teratogenic and reproductive effects. When heated to decomposition it emits very toxic fumes of Cl^- and NO_x.

DFP200   CAS: 593-89-5   HR: 3
DICHLOROMETHYLARSINE
DOT: NA 1556
mf: CH_3AsCl_2 mw: 160.86
PROP: Colorless liquid. Bp: 89–91° @ 200 mm, fp: −59°, flash p: >221°F, d: 1.84 @ 20°/4°, vap press: 10 mm @ 24.3°, vap d: 5.40, mp: −42.5°.
SYNS: ARSONOUS DICHLORIDE, METHYL-(9CI) □ METHYLARSINE DICHLORIDE □ METHYLARSONOUS DICHLORIDE □ METHYLDICHLORARSINE □ METHYLDICHLOROARSINE (DOT) □ TL 294
CONSENSUS REPORTS: Arsenic and its compounds are on the Community Right-To-Know List.
OSHA PEL: TWA 0.5 mg/(As)/m³
DOT CLASSIFICATION: 6.1; Label: Poison
SAFETY PROFILE: Poison irritant to skin, eyes, and mucous membranes and poison by ingestion and inhalation. A blistering type of military poison. It is rapidly detoxified in the body. A moderately persistent gas. Combustible when exposed to heat or flame. To fight fire, use water, foam, CO_2, dry chemical. Explosive reaction with chlorine. Can react vigorously with oxidizing materials. Dangerous; when heated to decomposition or on contact with acid or acid fumes it emits highly toxic fumes of Cl^- and As. See also CHLOROVINYLARSINE DICHLORIDE and ARSENIC COMPOUNDS.

DFP800   CAS: 1123-61-1   HR: 3
DICHLORO-N-METHYLMALEIMIDE
mf: $C_5H_3Cl_2NO_2$ mw: 179.99
SYNS: 2,3-DICHLORO-N-METHYLMALEIMIDE □ N-METHYLDICHLOROMALEINIMIDE
SAFETY PROFILE: Poison by intraperitoneal and intravenous routes. An experimental teratogen. Other experimental reproductive effects. When heated to decomposition it emits very toxic fumes of Cl^- and NO_x.

DFQ800   CAS: 149-74-6   HR: 3
DICHLOROMETHYLPHENYLSILANE
DOT: UN 2437
mf: C$_7$H$_8$Cl$_2$Si   mw: 191.14
PROP: Colorless liquid. D: 1.18 @ 20°/4°, bp: 205°.
SYNS: METHYLPHENYLDICHLOROSILANE (DOT) □ PHENYLMETHYLDICHLOROSILANE
CONSENSUS REPORTS: Reported in EPA TSCA Inventory. EPA Extremely Hazardous Substances List.
DOT CLASSIFICATION: 8; Label: Corrosive
SAFETY PROFILE: Poison by inhalation, subcutaneous, and intraperitoneal routes. Corrosive to eyes, skin, and mucous membranes. Flammable liquid. When heated to decomposition it emits toxic fumes of Cl⁻. See also CHLOROSILANES.

DFS000   CAS: 75-54-7   HR: 3
DICHLOROMETHYLSILANE
DOT: UN 1242
mf: CH$_4$Cl$_2$Si   mw: 115.04
PROP: Colorless liquid; acrid hydrochloric acid-like odor. Bp: 41°, d: 1.10 @ 20/4°, mp: −93°, flash p: −26°F. Sol in benzene, ether, and heptane.
SYNS: METHYL DICHLOROSILANE (DOT) □ METHYL-DICHLORSILAN (CZECH)
CONSENSUS REPORTS: Reported in EPA TSCA Inventory.
DOT CLASSIFICATION: 4.3; Label: Danger When Wet, Corrosive, Flammable Liquid
SAFETY PROFILE: Moderately toxic by inhalation. Corrosive. A severe irritant to skin, eyes, and mucous membranes. Ignites spontaneously in air. A very dangerous fire hazard when exposed to heat or flame. Forms impact-sensitive explosive mixtures with potassium permanganate, lead(II) oxide, lead(IV) oxide, copper oxide, silver oxide. To fight fire, use water, foam, CO$_2$, mist. When heated to decomposition it emits toxic fumes of Cl⁻. See also CHLOROSILANE.

DFT000   CAS: 117-80-6   HR: 3
2,3-DICHLORO-1,4-NAPHTHOQUINONE
mf: C$_{10}$H$_4$Cl$_2$O$_2$ mw: 227.04

PROP: Golden-yellow crystals or needles from alc. Mp: 194–195°, vap d: 7.8. Insol in water; moderately sol in org solvs.
SYNS: ALGISTAT □ COMPOUND 604 □ DICHLONE (DOT) □ 2,3-DICHLOR-1,4-NAPHTHOCHINON (GERMAN) □ 2,3-DICHLORO-1,4-NAPHTHALENE-DIONE □ 2,3-DICHLORO-1,4-NAPHTHAQUINONE □ DICHLORONAPHTHOQUINONE □ 2,3-DICHLORO-NAPHTHOQUINONE □ 2,3-DICHLORO-α-NAPHTHO-QUINONE □ 2,3-DICHLORONAPHTHOQUINONE-1,4 □ ENT 3,776 □ PHYGON □ PHYGON PASTE □ PHYGON SEED PROTECTANT □ PHYGON XL □ QUINTAR □ QUINTAR 540F □ SANQUINON □ UNIROYAL □ USR 604 □ U.S. RUBBER 604
CONSENSUS REPORTS: Reported in EPA TSCA Inventory.
SAFETY PROFILE: Poison by ingestion and intraperitoneal routes. Mildly toxic by skin contact. A skin, eye, and mucous membrane irritant. Large doses can cause central nervous system depression. Questionable carcinogen with experimental carcinogenic and neoplastigenic data. A fungicide and algicide. When heated to decomposition it emits toxic fumes of Cl⁻. See also CHLORIDES.

DFT800   CAS: 1836-75-5   HR: 3
2,4-DICHLORO-4'-NITRODIPHENYL ETHER
mf: C$_{12}$H$_7$Cl$_2$NO$_3$ mw: 284.10
PROP: Crystals or solid. Mp: 70–71°. Very sltly sol in H$_2$O.
SYNS: 2',4'-DICHLORO-4-NITROBIPHENYL ETHER □ 2,4-DICHLORO-1-(4-NITROPHENOXY)BENZENE □ 4-(2,4-DICHLOROPHENOXY)NITROBENZENE □ 2,4-DICHLOROPHENYL-p-NITROPHENYL ETHER □ 2,4-DICHLOROPHENYL-4-NITROPHENYL ETHER □ 2,4,-DICHLORPHENYL-4-NITROPHENYLAETHER (GERMAN) □ FW 925 □ MEZOTOX □ NCI-C00420 □ NICLOFEN □ NIP □ NITOFEN □ NITRAFEN □ NITRAPHEN □ NITROCHLOR □ 4'-NITRO-2,4-DICHLORODIPHENYL ETHER □ NITROFEN □ NITROFENE (FRENCH) □ NITROPHEN □ NITROPHENE □ PREPARATION 125 □ TOK □ TOK-2 □ TOK E □ TOK E-25 □ TOK E 40 □ TOKKORN □ TOK WP-50 □ TRIZILIN
CONSENSUS REPORTS: NTP 10th Report on Carcinogens. IARC Cancer Review: Group 2B IMEMDT 7,56,87; Animal Sufficient Evidence IMEMDT 30,271,83.

NCI Carcinogenesis Bioassay (feed); No Evidence: rat NCITR* NCI-CG-TR-184,79; Clear Evidence: mouse, rat NCITR* NCI-CG-TR-26,78; Clear Evidence: mouse NCITR* NCI-CG-TR-184,79. EPA Genetic Toxicology Program. Community Right-To-Know List. Reported in EPA TSCA Inventory.
SAFETY PROFILE: Confirmed carcinogen with experimental carcinogenic data. Poison by ingestion. Moderately toxic by inhalation and possibly other routes. Experimental teratogenic and reproductive effects. A skin and severe eye irritant. Mutation data reported. A broad-spectrum herbicide. See also NITRO COMPOUNDS of AROMATIC HYDROCARBONS and ETHERS. When heated to decomposition it emits very toxic fumes of Cl⁻ and NOₓ.

DFU000   CAS: 594-72-9   HR: 3
1,1-DICHLORO-1-NITROETHANE
DOT: UN 2650
mf: $C_2H_3Cl_2NO_2$ mw: 143.96
PROP: Liquid. Bp: 124°, flash p: 168°F(OC), d: 1.4153 @ 20°/20°, vap d: 4.97. IDLH 25 ppm.
SYNS: 1,1-DICHLOOR-1-NITROETHAAN (DUTCH) □ 1,1-DICHLOR-1-NITROAETHAN (GERMAN) □ DICHLORONITROETHANE □ 1,1-DICLORO-1-NITROETANO (ITALIAN) □ ETHIDE
OSHA PEL: TWA 2 ppm
ACGIH TLV: TWA 2 ppm
DFG MAK: 10 ppm (60 mg/m³)
DOT CLASSIFICATION: 6.1; Label: Poison
SAFETY PROFILE: Poison by ingestion and intraperitoneal routes. Moderately toxic by inhalation. A strong irritant. Inhalation causes pulmonary edema. A fumigant for produce. Flammable when exposed to heat, flame, or oxidizers. Can react vigorously with oxidizing materials. To fight fire, use water, CO_2, dry chemical. When heated to decomposition it emits highly toxic fumes of Cl⁻ and NOₓ.

DFV400   CAS: 50-65-7   HR: 3
2',5-DICHLORO-4'-

NITROSALICYLANILIDE
mf: $C_{13}H_8Cl_2N_2O_4$ mw: 327.13
PROP: Pale-yellow crystals. Mp: 225–230°. Sltly sol in EtOH, CHCl₃, and Et₂O.
SYNS: BAY 2353 □ BAYER 73 □ BAYER 2353 □ BAYLUSCID □ CHEMAGRO 2353 □ 5-CHLORO-N-(2-CHLORO-4-NITROPHENYL)-2-HYDROXYBENZAMIDE □ 5-CHLORO-2'-CHLORO-4'-NITROSALICYLANILIDE □ 2-CHLORO-4-NITROPHENYLAMIDE-6-CHLOROSALICYLIC ACID □ N-(2-CHLORO-4-NITROPHENYL)-5-CHLOROSALICYLAMIDE □ CLONITRALID □ 2',5-DICHLOR-4'-NITRO-SALIZYLSAEUREANILID (GERMAN) □ DICHLOSALE □ ENT 25,823 □ FENASAL □ HL 2447 □ 2-HYDROXY-5-CHLORO-N-(2-CHLORO-4-NITROPHENYL)BENZAMIDE □ IOMESAN □ IOMEZAN □ NICLOSAMIDE □ PHENASAL □ VERMITIN □ YOMESAN
SAFETY PROFILE: Poison by intravenous and intraperitoneal routes. Moderately toxic by ingestion. Experimental reproductive effects. Human mutation data reported. When heated to decomposition it emits very toxic fumes of Cl⁻ and NOₓ.

DFW730   CAS: 153436-22-7 HR: 3
4,6-DICHLORO-3-((1E)-3-OXO-3-(PHENYLAMINO)-1-PROPENYL)-1H-INDOLE-2-CARBOXYLIC ACID,
mf: $C_{18}H_{12}Cl_2N_2O_3$ mw: 375.21
SAFETY PROFILE: A poison by intravenous route. When heated to decomposition it emits toxic vapors of NOₓ and Cl⁻.

DFX000   CAS: 30586-10-8   HR: 3
DICHLOROPENTANE
DOT: UN 1152
mf: $C_5H_{10}Cl_2$ mw: 141.05
PROP: Clear, light-yellow liquid. Bp: 130°, flash p: 106°F (OC), vap d: 4.86, d: 1.06–1.08 @ 20°.
SYN: DICHLOROPENTANES (DOT)
DOT CLASSIFICATION: 3; Label: Flammable Liquid
SAFETY PROFILE: Flammable liquid when exposed to heat or flame. Can react vigorously with oxidizing materials. To fight fire, use water, foam, CO_2, dry chemical. When heated to decomposition it emits highly toxic fumes of Cl⁻ and phosgene. See also 1,5-DICHLOROPENTANE, and

CHLORINATED HYDROCARBONS, ALIPHATIC.

DFX400   CAS: 536-29-8   HR: 3
DICHLOROPHENARSINE HYDROCHLORIDE

mf: $C_6H_6AsCl_2NO•ClH$ mw: 290.41

PROP: White hygroscopic powder from EtOH. Mp: 146–148°. Sol in H_2O with hydrolysis.

SYNS: 2-AMINO-4-DICHLOROARSINOPHENOL HYDROCHLORIDE □ (3-AMINO-4-HYDROXYPHEN-YL)ARSONOUS DICHLORIDE MONOHYDROCHLOR-IDE □ 3-AMINO-4-HYDROXYPHENYL DICHLORARS-INE HYDROCHLORIDE □ (3-AMINO-4-HYDROXY-PHENYL)DICHLOROARSINE HYDROCHLORIDE □ ARSECLOR □ CHLORARSOL □ CHLORASEN □ CLORARSEN □ DICHLOROMAPHARSEN □ FILARSEN □ FONTARSOL □ HALARSOL □ R.P. 2591

CONSENSUS REPORTS: Arsenic and its compounds, as well as chlorophenol compounds, are on the Community Right-To-Know List.

OSHA PEL: TWA 0.5 mg(As)/m³

ACGIH TLV: BEI: 35 μ (As)/L inorganic arsenic and methylated metabolites in urine

SAFETY PROFILE: Poison by intravenous, intraperitoneal, and possibly other routes. Moderately toxic by ingestion. Human systemic effects by parenteral route: hypermotility, diarrhea, nausea, vomiting. See also ARSENIC COMPOUNDS and CHLOROPHENOLS. When heated to decomposition it emits very toxic fumes of As, NO_x, and Cl^-.

DFX800   CAS: 120-83-2   HR: 3
2,4-DICHLOROPHENOL

mf: $C_6H_4Cl_2O$ mw: 163.00

PROP: Colorless crystals or needles. Mp: 45°, bp: 210°, flash p: 237°F, d: 1.383 @ 60°/25°, vap d: 5.62, vap press: 1 mm @ 53.0°.

SYNS: DCP □ 2,4-DCP □ NCI-C55345 □ RCRA WASTE NUMBER U081

CONSENSUS REPORTS: IARC Cancer Review: Human Limited Evidence IMEMDT 41,319,86. Reported in EPA TSCA Inventory. EPA Genetic Toxicology Program. Community Right-To-Know List.

SAFETY PROFILE: Suspected carcinogen with experimental carcinogenic and teratogenic data. Poison by intraperitoneal route. Moderately toxic by ingestion and subcutaneous routes. An experimental teratogen. Mutation data reported. Combustible when exposed to heat or flame. Can react vigorously with oxidizing materials. To fight fire, use alcohol foam, foam, CO_2, dry chemical. When heated to decomposition, or on contact with acid or acid fumes, it emits highly toxic fumes of Cl^-. See also CHLOROPHENOLS.

DFY000   CAS: 87-65-0   HR: 3
2,6-DICHLOROPHENOL

mf: $C_6H_4Cl_2O$ mw: 163.00

PROP: Needles from pet ether. Mp: 66–67°, bp: 219–220°.

SYNS: 2,6-DICHLORFENOL (CZECH) □ RCRA WASTE NUMBER U082

CONSENSUS REPORTS: Reported in EPA TSCA Inventory. EPA Genetic Toxicology Program. Chlorophenol compounds are on the Community Right-To-Know List.

SAFETY PROFILE: Poison by intraperitoneal route. Moderately toxic by ingestion. A severe skin and eye irritant. When heated to decomposition it emits toxic fumes of Cl^-. See also CHLOROPHENOLS.

DFY400   CAS: 97-16-5   HR: 3
2,4-DICHLOROPHENOL BENZENE-SULFONATE

mf: $C_{12}H_8Cl_2O_3S$ mw: 303.16

SYNS: COMPOUND 923 □ 2,4-DICHLOROPHENYL BENZENESULFONATE □ 2,4-DICHLOROPHENYL BENZENESULPHONATE □ 2,4-DICHLOROPHENYL ESTER of BENZENESULFONIC ACID □ 2,4-DICHLOROPHENYL ESTER BENZENESULPHONIC ACID □ DPBS □ EM 923 □ GENITE □ GENITOL

CONSENSUS REPORTS: Chlorophenol compounds are on the Community Right-To-Know List.

SAFETY PROFILE: Poison by intravenous route. Moderately toxic by ingestion and possibly other routes. Questionable carcinogen with experimental carcinogenic and tumorigenic data. An irritant. A pesticide. See also CHLOROPHENOLS. When heated to decomposition it emits very toxic fumes of Cl⁻ and SO$_x$.

DGA425   CAS: 73791-41-0   HR: 3
3-(2,4-DICHLOROPHENOXY)-2-HYDRO-XYPROPYL-o-CHLOROPHENYL ARSINIC ACID

mf: $C_{15}H_{14}AsCl_3O_4$ mw: 439.56

SYNS: ARSINE OXIDE, (o-CHLOROPHENYL)(3-(2,4-DICHLOROPHENOXY)-2-HYDROXYPROPYL)-HYDROXY- □ (o-CHLOROPHENYL)(3-(2,4-DICHLOROPHENOXY)-2-HYDROXYPROPYL)-HYDROXYARSINEOXIDE

OSHA PEL: TWA 0.5 mg(As)/m³

SAFETY PROFILE: Poison by intravenous route. When heated to decomposition it emits toxic fumes of As and Cl⁻.

DGB000   CAS: 120-36-5   HR: 3
2-(2,4-DICHLOROPHENOXY) PROPIONIC ACID

mf: $C_9H_8Cl_2O_3$ mw: 235.07

SYNS: ACIDE-2-(2,4-DICHLORO-PHENOXY) PROPIONIQUE (FRENCH) □ ACIDO-2-(2,4-DICLORO-FENOSSI)-PROPIONICO (ITALIAN) □ CORNOX RD □ CORNOX RK □ DESORMONE □ 2-(2,4-DICHLOOR-FENOXY)-PROPIONZUUR (DUTCH) □ α-(2,4-DICHLOROPHENOXY) PROPIONIC ACID □ DICHLOROPROP □ 2-(2,4-DICHLOR-PHENOXY)-PROPIONSAEURE (GERMAN) □ DICHLORPROP □ 2,4-DP □ 2-(2,4-DP) □ HEDONAL □ HEDONAL DP □ HORMATOX □ KILDIP □ POLYCLENE □ POLYMONE □ POLYTOX □ RD 406 □ SERITOX 50 □ U46 □ U46 DP-FLUID □ VISKO-RHAP □ WEEDONE DP □ WEEDONE 170

CONSENSUS REPORTS: IARC Cancer Review: Group 2B IMEMDT 7,156,87; Human Limited Evidence IMEMDT 41,357,86. Reported in EPA TSCA Inventory.

SAFETY PROFILE: Suspected carcinogen. Poison by ingestion. Moderately toxic by skin contact. An experimental teratogen. Other experimental reproductive effects. Mutation data reported. A fumigant. When heated to decomposition it emits toxic fumes of Cl⁻.

DGB600   CAS: 696-28-6   HR: 3

DICHLOROPHENYLARSINE

mf: $C_6H_5AsCl_2$ mw: 222.93

PROP: Colorless gas or liquid, changes to yellow. Bp: 127–129° @ 13 mm, fp: −15.6°, d: 1.654 @ 20°, vap press: 0.021 mm @ 20°, vap d: 7.7.

SYNS: ARSONOUS DICHLORIDE, PHENYL-(9CI) □ DICHLOR-FENYLARSIN □ FDA □ FENILDICLORO-ARSINA (ITALIAN) □ FENYLDICHLORARSIN □ PHENYLARSINEDICHLORIDE □ PHENYLARSONOUS DICHLORIDE □ PHENYL DICHLORARSINE □ PHENYLDICHLOROARSINE □ RCRA WASTE NUMBER P036 □ TL 69

CONSENSUS REPORTS: Arsenic and its compounds are on the Community Right-To-Know List. Reported in EPA TSCA Inventory. EPA Extremely Hazardous Substances List.

OSHA PEL: TWA 0.5 mg(As)/m³

SAFETY PROFILE: Poison by inhalation, ingestion, skin contact, and intravenous routes. See also ARSENIC. A lachrymator type of military poison gas. When exposed to heat, water, or steam it reacts to produce corrosive fumes of Cl⁻. When heated to decomposition it emits highly toxic fumes of arsenic.

DGD600   CAS: 330-55-2   HR: 3
3-(3,4-DICHLOROPHENYL)-1-METHO-XYMETHYLUREA

mf: $C_9H_{10}Cl_2N_2O_2$ mw: 249.11

PROP: Solid. Mp: 93–94°. Sltly sol in water; partially sol in acetone and alc.

SYNS: 3-(3,4-DICHLOOR-FENYL)-1-METHOXY-1-METHYLUREUM (DUTCH) □ 3-(3,4-DICHLORO-FENIL)-1-METOSSI-1-METIL-UREA (ITALIAN) □ 3-(3,4-DICHLOROPHENYL)-1-METHOXY-1-METHYLUREA □ N'-(3,4-DICHLOROPHENYL)-N-METHOXY-N-METHYLUREA □ 1-(3,4-DICHLOROPHENYL)3-METHOXY-3-METHYLUREE (FRENCH) □ N-(3,4-DICHLOROPHENYL)-N'-METHYL-N'-METHOXYUREA □ 3-(3,4-DICHLOR-PHENYL)-1-METHOXY-1-METHYL-HARNSTOFF (GERMAN) □ 3-(4,5-DICHLORPHENYL)-1-METHOXY-1-METHYLHARNSTOFF (GERMAN) □ DU PONT 326 □ DU PONT HERBICIDE 326 □ GARNITAN □ HERBICIDE 326 □ HOE 2810 □ LINEX 4L □ LINOROX □ LINUREX □ LINURON □ LINURON (herbicide) □ LOREX □ LOROX □ LOROX LINURON WEED KILLER □ METHOXYDIURON □ 1-METHOXY-1-METHYL-3-(3,4-DICHLOROPHENYL)UREA □ PREMALIN □ SARCLEX □ SCARCLEX □ SINURON

CONSENSUS REPORTS: Reported in EPA TSCA Inventory. EPA Genetic Toxicology Program.
SAFETY PROFILE: Poison by inhalation. Moderately toxic by ingestion. Mutation data reported. A selective herbicide used in farming. When heated to decomposition it emits very toxic fumes of Cl⁻ and NO$_x$. See also 3-(p-CHLOROPHENYL)-1,1-DIMETHYLUREA.

DGE400   CAS: 644-97-3   HR: 3
DICHLOROPHENYLPHOSPHINE
DOT: UN 2798
mf: $C_6H_5Cl_2P$ mw: 178.98
PROP: Pungent, odorous liquid. D: 1.33 @ 20°/4°, bp: 99–101° @ 5 mm.
SYNS: PHENYLDICHLOROPHOSPHINE □ PHENYL-PHOSPHINE DICHLORIDE □ PHENYLPHOSPHONOUS ACID DICHLORIDE □ PHENYLPHOSPHONOUS DICHLORIDE □ PHENYL PHOSPHORUS DICHLORIDE □ PHENYL PHOSPHORUS DICHLORIDE (DOT)
CONSENSUS REPORTS: Reported in EPA TSCA Inventory.
DOT CLASSIFICATION: 8; Label: Corrosive
SAFETY PROFILE: A poison irritant to skin, eyes, and mucous membranes and poison by ingestion and inhalation. When heated to decomposition it emits very toxic fumes of Cl⁻ and PO$_x$. See also PHOSPHINE.

DGF100   CAS: 50673-11-5   HR: 3
((2,5-DICHLOROPHENYL)THIO)-
**   METHYLCARBAMIC ACID, 2,3-**
**   DIHYDRO-2,2-DIMETHYL-7-**
**   BENZOFURANYL ESTER**
mf: $C_{18}H_{17}Cl_2NO_3S$ mw: 398.32
SAFETY PROFILE: A poison by ingestion. When heated to decomposition it emits toxic vapors of NO$_x$, SO$_x$, and Cl⁻.

DGF200   CAS: 27137-85-5   HR: 3
(DICHLOROPHENYL)TRICHLORO-
**   SILANE**
DOT: UN 1766
mf: $C_6H_3Cl_5Si$ mw: 280.43

PROP: Straw-colored liquid, sol in benzene and perchloroethylene (mixture of isomers). D: 1.562, bp: 260°, flash p: 286°F.
SYNS: DICHLOROPHENYLTRICHLOROSILANE (DOT) □ TRICHLORO(DICHLOROPHENYL)SILANE
CONSENSUS REPORTS: Reported in EPA TSCA Inventory. EPA Extremely Hazardous Substances List.
DOT CLASSIFICATION: 8; Label: Corrosive
SAFETY PROFILE: Poison by ingestion, inhalation, subcutaneous, and intraperitoneal routes. Corrosive to the eyes, skin, and mucous membranes. On contact with moisture it releases corrosive HCl. Combustible when exposed to heat or flame. When heated to decomposition it emits toxic fumes of Cl⁻. See also CHLOROSILANES.

DGF800   CAS: 142-28-9   HR: 2
1,3-DICHLOROPROPANE
mf: $C_3H_6Cl_2$ mw: 112.99
PROP: Colorless liquid. Bp: 120.4°, d: 1.201 @ 15°, vap d: 3.90, flash p: 69.8°F.
SYN: TRIMETHYLENE DICHLORIDE
CONSENSUS REPORTS: Reported in EPA TSCA Inventory.
SAFETY PROFILE: Moderately toxic by ingestion. Mutation data reported. A very dangerous fire hazard when exposed to heat or flame. When heated to decomposition it emits highly toxic fumes of Cl⁻ and phosgene. See also CHLORINATED HYDROCARBONS, ALIPHATIC; and PROPYLENE DICHLORIDE.

DGG000   CAS: 8003-19-8   HR: 3
DICHLOROPROPANE-DICHLORO-
**   PROPENE MIXTURE**
mf: $C_3H_6Cl_2 \cdot C_3H_4Cl_2$ mw: 223.96
PROP: D-D Soil fumigant consists of chlorinated C$_3$ hydrocarbons (100%), 1,3-dichloropropene, 3,3-dichloropropene, 1,2-dichloropropane, 2,3-dichloropropene, and related C$_3$ chlorinated hydrocarbons (SHELL*).
SYNS: D-D □ DD MIXTURE □ DD SOIL FUMIGANT □ 1,3-DICHLOROPROPENE and 1,2-

DICHLOROPROPANE MIXTURE □ DICHLORPROPAN-DICHLORPROPENGEMISCH (GERMAN) □ DOWFUME N □ ENT 8,420 □ NEMAFENE □ TELONE □ VIDDEN D

SAFETY PROFILE: Poison by ingestion and inhalation. Moderately toxic by skin contact. Severe skin and eye irritant. Mutation data reported. A fumigant. When heated to decomposition it emits toxic fumes of Cl⁻. See also PROPYLENE DICHLORIDE; and CHLORINATED HYDROCARBONS, ALIPHATIC.

DGG400   CAS: 96-23-1   HR: 3
1,3-DICHLORO-2-PROPANOL
DOT: UN 2750

mf: $C_3H_6Cl_2O$ mw: 128.99

PROP: Colorless liquid; ether-like odor. Bp: 174°, d: 1.367 @ 20°/4°, vap press: 1 mm @ 28.0°, vap d: 4.45, flash p: 165°F (OC), mp: −4°. Sol in H_2O and Et_2O.

SYNS: DICHLOROHYDRIN □ α-DICHLOROHYDRIN □ sym-DICHLOROISOPROPYL ALCOHOL □ 1,3-DICHLOROPROPANOL-2 (DOT) □ GLYCEROL α,γ-DICHLOROHYDRIN □ sym-GLYCEROL DICHLORO-HYDRIN □ U 25,354

CONSENSUS REPORTS: Reported in EPA TSCA Inventory. EPA Genetic Toxicology Program.

DFG MAK: Animal Carcinogen, Suspected Human Carcinogen

DOT CLASSIFICATION: 6.1; Label: Poison

SAFETY PROFILE: Suspected carcinogen. Poison by ingestion and inhalation. Moderately toxic by skin contact. Human mutation data reported. A skin irritant. Action may be similar to that of carbon tetrachloride, but more irritating to mucous membranes. Flammable when exposed to heat, flame, or oxidizers. To fight fire, use alcohol foam, dry chemical, fog, mist, or spray. Dangerous; when heated to decomposition it emits highly toxic fumes of Cl⁻ and phosgene.

DGG450   CAS: 616-23-9   HR: 3
2,3-DICHLOROPROPANOL
mf: $C_3H_6Cl_2O$ mw: 128.99

SYNS: 1,2-DICHLOROPROPANOL-3 □ 2,3-DICHLORO-1-PROPANOL □ 1,2-DICHLORO-3-PROPANOL □ GLYCEROL-α,β-DICHLOROHYDRIN

CONSENSUS REPORTS: Reported in EPA TSCA Inventory.

SAFETY PROFILE: Poison by ingestion and skin contact. Moderately toxic by inhalation. A skin and severe eye irritant. Mutation data reported. When heated to decomposition it emits toxic fumes of Cl⁻. See also CHLORINATED HYDROCARBONS, AROMATIC.

DGG700   CAS: 26952-23-8   HR: 3
DICHLOROPROPENE
DOT: UN 2047

mf: $C_3H_4Cl_2$ mw: 110.97

SYNS: DICHLOROPROPYLENE □ 1-PROPENE, DICHLORO-

CONSENSUS REPORTS: Reported in EPA TSCA Inventory.

DOT CLASSIFICATION: 3; Label: Flammable Liquid

SAFETY PROFILE: A flammable liquid. When heated to decomposition it emits toxic vapors of Cl⁻.

DGG950   CAS: 542-75-6   HR: 3
1,3-DICHLOROPROPENE
DOT: UN 2047

mf: $C_3H_4Cl_2$ mw: 110.97

PROP: Liquid. Bp: 103–110°, flash p: 95°F, d: 1.22, vap d: 3.8.

SYNS: α-CHLOROALLYL CHLORIDE □ γ-CHLOROALLYL CHLORIDE □ DICHLOROPROPENE (DOT) □ 1,3-DICHLOROPROPYENE-1 □ DICHLORO-PROPYLENE □ α,γ-DICHLOROPROPYLENE □ 1,3-DICHLOROPROPYLENE □ NCI-C03985 □ RCRA WASTE NUMBER U084 □ TELONE □ TELONE II SOIL FUMIGANT □ VIDDEN D

CONSENSUS REPORTS: NTP 10th Report on Carcinogens. IARC Cancer Review: Group 2B IMEMDT 7,195,87; Human Inadequate Evidence IMEMDT 41,113,86; Animal Sufficient Evidence IMEMDT 41,113,86. NTP Carcinogenesis Studies (gavage); Clear Evidence: mouse, rat NTPTR* NTP-TR-269,86. Reported in EPA TSCA Inventory. EPA Genetic

Toxicology Program. Community Right-To-Know List.

OSHA PEL: TWA 1 ppm (skin)

ACGIH TLV: TWA 1 ppm (skin); Not Classifiable as a Human Carcinogen

DFG MAK: Animal Carcinogen, Suspected Human Carcinogen

DOT CLASSIFICATION: 3; Label: Flammable Liquid

SAFETY PROFILE: Confirmed carcinogen with experimental carcinogenic data. Poison by ingestion and intraperitoneal routes. Moderately toxic by skin contact. Mildly toxic by inhalation. A strong irritant. Mutation data reported. A pesticide. A flammable liquid and dangerous fire hazard when exposed to heat, flame, or oxidizers. Reacts vigorously with oxidizing materials. To fight fire, use water, foam, CO_2, dry chemical. When heated to decomposition it emits toxic fumes of Cl^-. See also ALLYL COMPOUNDS and CHLORIDES.

DGH200   CAS: 10061-01-5   HR: 3
cis-1,3-DICHLOROPROPENE

mf: $C_3H_4Cl_2$ mw: 110.97

PROP: Flash p: 69.8°F (21°C), bp: 104.3°.

SYNS: (Z)-1,3-DICHLOROPROPENE □ cis-1,3-DICHLOROPROPYLENE

CONSENSUS REPORTS: EPA Genetic Toxicology Program.

DFG MAK: Animal Carcinogen, Suspected Human Carcinogen

SAFETY PROFILE: Confirmed carcinogen with experimental neoplastigenic data. Human mutation data reported. A dangerous fire hazard when exposed to heat, flame, or oxidizers. When heated to decomposition it emits toxic fumes of Cl^-. See also CHLORINATED HYDROCARBONS, ALIPHATIC.

DGH400   CAS: 78-88-6   HR: 3
2,3-DICHLOROPROPENE

mf: $C_3H_4Cl_2$ mw: 110.97

PROP: Flash p: 50°F, bp: 94°.

SYNS: 2,3-DICHLORO-1-PROPENE □ 2,3-DICHLORO-PROPYLENE □ NSC-60520

CONSENSUS REPORTS: Reported in EPA TSCA Inventory. EPA Genetic Toxicology Program.

SAFETY PROFILE: Poison by ingestion. Moderately toxic by inhalation and skin contact. Human mutation data reported. A severe skin irritant. A very dangerous fire hazard when exposed to heat, flame, or oxidizers. When heated to decomposition it emits toxic fumes of Cl^-. See also CHLORINATED HYDROCARBONS, ALIPHATIC.

DGI000   CAS: 709-98-8   HR: 3
DICHLOROPROPIONANILIDE

mf: $C_9H_9Cl_2NO$ mw: 218.09

PROP: Light-brown solid (pure); liquid (technical grade). Mp (pure): 85–89°, bp (technical grade): 91–95°.

SYNS: BAY 30130 □ CHEM RICE □ CRYSTAL PROPANIL-4 □ DCPA □ N-(3,4-DICHLOROPHENYL)PROPANAMIDE □ N-(3,4-DICHLOROPHENYL)PROPIONAMIDE □ 3,4-DICHLOROPROPIONANILIDE □ 3',4'-DICHLOROPRO-PIONANILIDE □ DIPRAM □ DPA □ FARMCO PROPANIL □ FW 734 □ GRASCIDE □ HERBAX TECHNICAL □ MONTROSE PROPANIL □ PROPANEX □ PROPANID □ PROPANIDE □ PROPANIL □ PROPIONIC ACID-3,4-DICHLOROANILIDE □ PROP-JOB □ RISELECT □ ROGUE □ ROSANIL □ S 10165 □ STAM □ STAM F 34 □ STAM LV 10 □ STAM M-4 □ STAMPEDE □ STAMPEDE 3E □ STAM SUPERNOX □ STREL □ SUPERNOX □ SURCOPUR □ SURPUR □ VERTAC

CONSENSUS REPORTS: EPA Genetic Toxicology Program.

SAFETY PROFILE: Poison by ingestion. Moderately toxic by an unspecified route. Mildly toxic by skin contact. Mutation data reported. When heated to decomposition it emits very toxic fumes of Cl^- and NO_x.

DGI400   CAS: 75-99-0   HR: 2
2,2-DICHLOROPROPIONIC ACID

mf: $C_3H_4Cl_2O_2$ mw: 142.97

PROP: White to tan powder. D: 1.39 @ 22.6°/4°, bp: 185–190°. Sol in water.

SYNS: BASFAPON □ BASFAPON B □ BASFAP-ON/BASFAPON N □ BASINEX □ BH DALAPON □ CRISAPON □ DALAPON (USDA) □ DALAPON 85 □ DED-WEED □ DEVIPON □ α-DICHLOROPROPIONIC

ACID □ α,α-DICHLOROPROPIONIC ACID □ DOWPON □ DOWPON M □ GRAMEVIN □ KENAPON □ LIROPON □ PROPROP □ RADAPON □ REVENGE □ UNIPON

CONSENSUS REPORTS: EPA Genetic Toxicology Program. Reported in EPA TSCA Inventory.

OSHA PEL: TWA 1 ppm

ACGIH TLV: TWA 5 ppm, Not Classifiable as a Human Carcinogen

DFG MAK: 1 ppm (5.9 mg/m³)

SAFETY PROFILE: A corrosive with low toxicity by skin contact. A skin irritant. Mutation data reported. When heated to decomposition it emits toxic fumes of Cl⁻.

DGI600   CAS: 127-20-8   HR: 2
α,α-DICHLOROPROPIONIC ACID SODIUM SALT

mf: $C_3H_3Cl_2O_2 \cdot Na$ mw: 164.95

SYNS: BASFAPON B □ DALAPON □ DALAPON SODIUM □ DALAPON SODIUM SALT □ 2,2-DICHLOROPROPIONIC ACID, SODIUM SALT □ DOWPON □ 2,2-DPA □ GRAMEVIN □ NATRIUMSALZ DER 2,2-DICHLORPROPIONSAEURE □ RADAPON □ SODIUM DALAPON □ SODIUM-α,α-DICHLOROPROP-IONATE □ SODIUM-2,2-DICHLOROPROPIONATE □ UNIPON

CONSENSUS REPORTS: Reported in EPA TSCA Inventory. EPA Genetic Toxicology Program.

DFG MAK: 1 ppm (6 mg/m³)

SAFETY PROFILE: Moderately toxic by ingestion. Mutation data reported. When heated to decomposition it emits toxic fumes of Na_2O and Cl⁻.

DGI630   CAS: 315706-69-5 HR: 3
2,4-DICHLORO-N-(4-PROPYLCYCLO-HEXYL)BENZAMIDE

mf: $C_{16}H_{21}Cl_2NO$ mw: 314.25

SAFETY PROFILE: A poison by ingestion. When heated to decomposition it emits toxic vapors of NO_x and Cl⁻.

DGI700   CAS: 26952-23-8   HR: 3
DICHLOROPROPYLENE

DOT: UN 2047

mf: $C_3H_4Cl_2$ mw: 110.97

SYN: 1-PROPENE, DICHLORO-

CONSENSUS REPORTS: Reported in EPA TSCA Inventory.

DOT CLASSIFICATION: 3; Label: Flammable Liquid

SAFETY PROFILE: A flammable liquid. When heated to decomposition it emits toxic vapors of Cl⁻.

DGK200   CAS: 320-72-9   HR: 3
3,5-DICHLOROSALICYLIC ACID

mf: $C_7H_4Cl_2O_3$ mw: 207.01

PROP: Crystals from EtOH (aq). Mp: 219–220° (subl).

SYN: USAF DO-68

CONSENSUS REPORTS: Reported in EPA TSCA Inventory.

SAFETY PROFILE: Poison by intraperitoneal route. When heated to decomposition it emits toxic fumes of Cl⁻.

DGK300   CAS: 4109-96-0   HR: 3
DICHLOROSILANE

mf: Cl_2H_2Si mw: 101.01

CONSENSUS REPORTS: Reported in EPA TSCA Inventory.

PROP: A gas. Mp: −122°, bp: 8.3°.

SYNS: CHLOROSILANE □ DICHLOROSILANE □ SILICON CHLORIDE HYDRIDE

SAFETY PROFILE: Moderately toxic by inhalation. Ignites spontaneously in air. Confined mixtures with air are spontaneously explosive. When heated to decomposition it emits toxic fumes of Cl⁻. See also CHLOROSILANES.

DGL600   CAS: 1320-37-2   HR: 1
DICHLOROTETRAFLUOROETHANE

DOT: UN 1958

mf: $C_2Cl_2F_4$ mw: 170.92

PROP: Colorless gas. Bp: 3.5°.

SYNS: DWUCHLOROCZTEROFLUOROETAN (POLISH) □ R114 (DOT) □ TETRAFLUORO-DICHLOROETHANE

CONSENSUS REPORTS: Reported in EPA TSCA Inventory.

OSHA PEL: TWA 1000 ppm

ACGIH TLV: TWA 1000 ppm

DOT CLASSIFICATION: 2.2; Label: Nonflammable Gas

D

SAFETY PROFILE: A mildly toxic irritant; narcotic in high concentrations. An asphyxiant. Reacts violently with alcohol. When heated to decomposition it emits toxic fumes of F⁻ and Cl⁻.

DGM600   CAS: 1918-13-4   HR: 3
2,6-DICHLOROTHIOBENZAMIDE
mf: $C_7H_5Cl_2NS$ mw: 206.09
PROP: A solid. Mp: 152°. Sltly sol in H_2O.
SYNS: CHLOROTHIAMIDE □ DCBN □ 2,6-DICHLOROBENZENECARBOTHIOAMIDE □ SD 7961 □ WL-5792
CONSENSUS REPORTS: EPA Genetic Toxicology Program.
SAFETY PROFILE: Poison by ingestion and intraperitoneal route. Moderately toxic by skin contact. Mutation data reported. An herbicide. When heated to decomposition it emits very toxic fumes of Cl⁻, NO_x, and SO_x.

DGN200   CAS: 2782-57-2   HR: 2
1,3-DICHLORO-s-TRIAZINE-2,4,6-
(1H,3H,5H)-TRIONE
DOT: UN 2465
mf: $C_3H_2Cl_2N_3O_3$ mw: 198.98
PROP: White crystals; chlorine odor. Mp: 226–226.7°. Moderately sol in water.
SYNS: ACL 70 □ CDB 60 □ DICHLOROCYANURIC ACID □ DICHLOROISOCYANURATE □ DICHLORO-ISOCYANURIC ACID □ DICHLOROISO-CYANURIC ACID, dry or dichloroisocyanuric acid salts (DOT) □ FI CLOR 71 □ HILITE 60 □ ISOCYANURIC ACID, DICHLORO- □ ISOCYANURIC DICHLORIDE □ ORCED □ KYSELINA DICHLORISOKYANUROVA (CZECH) □ TROCLOSENE
CONSENSUS REPORTS: Reported in EPA TSCA Inventory.
DOT CLASSIFICATION: 5.1; Label: Oxidizer
SAFETY PROFILE: Moderately toxic by ingestion. Human systemic effects by ingestion: ulceration or bleeding from stomach. Autopsy findings include gastrointestinal tract irritation, tissue edema, liver and kidney congestion. A severe eye and skin irritant. When heated to decomposition it emits chlorides and carbon monoxide.

DGO800   CAS: 594-31-0   HR: 3
DICHLOROTRIPHENYLANTIMONY
mf: $C_{18}H_{15}Cl_2Sb$ mw: 423.98
PROP: Crystals from MeOH or EtOH/$CHCl_3$. Mp: 143°.
SYNS: ANTIMONY TRIPHENYLDICHLORIDE □ DICHLOROTRIPHENYLSTIBINE □ TRIPHENYL-ANTIMONY DICHLORIDE
CONSENSUS REPORTS: Reported in EPA TSCA Inventory. Antimony and its compounds are on the Community Right-To-Know List.
OSHA PEL: TWA 0.5 mg(Sb)/m³
ACGIH TLV: TWA 0.5 mg(Sb)/m³
NIOSH REL: TWA 0.5 mg/m³
SAFETY PROFILE: Poison by ingestion. See also ANTIMONY COMPOUNDS. When heated to decomposition it emits very toxic fumes of Cl⁻ and Sb.

DGP400   CAS: 612-12-4   HR: 3
α,α'-DICHLORO-o-XYLENE
mf: $C_8H_8Cl_2$ mw: 175.06
PROP: Crystals from pet ether. D: 1.39 @ 0°, mp: 55°, bp: 239–241°.
SYNS: 1,2-BIS(CHLOROMETHYL)BENZENE □ o-XYLYLENE DICHLORIDE
CONSENSUS REPORTS: Reported in EPA TSCA Inventory.
SAFETY PROFILE: Poison by intravenous route. Mutation data reported. See also CHLORINATED HYDROCARBONS, AROMATIC. When heated to decomposition it emits toxic Cl⁻.

DGP900   CAS: 62-73-7   HR: 3
DICHLORVOS
mf: $C_4H_7Cl_2O_4P$ mw: 220.98
PROP: Liquid with aromatic odor. Bp: 120° @ 14 mm, bp: 88° @ 3 mm. Sltly sol in water and glycerin; misc with aromatic and chlorinated hydrocarbon solvents and alc. IDLH 100 mg/m³.
SYNS: APAVAP □ ASTROBOT □ ATGARD □ BAY 19149 □ BENFOS □ BIBESOL □ BREVINYL □ CANOGARD □ CEKUSAN □ CHLORVINPHOS □ CYANOPHOS □ CYPONA □ DDVF □ DDVP □ DEDEVAP □ DERIBAN □ DERRIBANTE □ DEVIKOL □ (2,2-DICHLOOR-VINYL)-DIMETHYL-FOSFAAT (DUTCH) □ DICHLOORVO (DUTCH) □ DICHLORFOS (POLISH) □

2,2-DICHLOROETHENOL DIMETHYL PHOSPHATE □ 2,2-DICHLOROETHENYL DIMETHYL PHOSPHATE □ 2,2-DICHLOROETHENYL PHOSPHORIC ACID DIMETHYL ESTER □ DICHLOROPHOS □ DICHLOROVAS □ 2,2-DICHLOROVINYL ALCOHOL, DIMETHYL PHOSPHATE □ 2,2-DICHLOROVINYL DIMETHYL PHOSPHATE □ 2,2-DICHLOROVINYL DIMETHYL PHOSPHORIC ACID ESTER □ DICHLO-ROVOS □ DICHLORPHOS □ (2,2-DICHLOR-VINYL)-DIMETHYL-PHOSPHAT (GERMAN) □ O-(2,2-DICHLORVINYL)-O,O-DIMETHYLPHOSPHAT (GERMAN) □ (2,2-DICLORO-VINIL)DIMETILFOSFATO (ITALIAN) □ DIMETHYL-2,2-DICHLOROETHENYL PHOSPHATE □ DIMETHYL DICHLOROVINYL PHOSPHATE □ O,O-DIMETHYL DICHLOROVINYL PHOSPHATE □ DIMETHYL-2,2-DICHLOROVINYL PHOSPHATE □ O,O-DIMETHYL-O-2,2-DICHLORO-VINYL PHOSPHATE □ O,O-DIMETHYL-O-(2,2-DICHLOR-VINYL)-PHOSPHAT (GERMAN) □ DIVIPAN □ DQUIGARD □ DUO-KILL □ DURAVOS □ ENT 20,738 □ EQUIGEL □ ESTROSEL □ ESTROSOL □ FECAMA □ FLY-DIE □ FLY FIGHTER □ HERKAL □ KRECALVIN □ LINDAN □ MAFU □ MARVEX □ MOPARI □ NCI-C00113 □ NERKOL □ NOGOS □ NO-PEST □ NO-PEST STRIP □ NSC-6738 □ NUVA □ OKO □ OMS 14 □ PHOSPHATE de DIMETHYLE et de 2,2-DICHLOROVINYLE (FRENCH) □ PHOSPHORIC ACID-2,2-DICHLOROETHENYL DIMETHYL ESTER □ PHOSVIT □ SD-1750 □ SZKLARNIAK □ TAP 9VP □ TASK □ TASK TABS □ TENAC □ TETRAVOS □ VAPONA □ VAPONITE □ VERDICAN □ VERDIPOR □ VINYLOFOS □ VINYLOPHOS

CONSENSUS REPORTS: IARC Cancer Review: Group 2B IMEMDT 53,267,91; Animal Sufficient Evidence IMEMDT 53,267,91; Animal Inadequate Evidence IMEMDT 20,97,79; Human No Adequate Data IMEMDT 20,97,79; Human Inadequate Evidence IMEMDT 53,267,91. NCI Carcinogenesis Bioassay (feed); No Evidence: mouse, rat NCITR* NCI-CG-TR-10,77. EPA Genetic Toxicology Program. Community Right-To-Know List. EPA Extremely Hazardous Substances List.

OSHA PEL: TWA 1 mg/m^3 (skin)

ACGIH TLV: TWA 0.1 ppm (skin, sensitizer); Not Classifiable as a Human Carcinogen

DFG MAK: 0.1 ppm (1 mg/m^3)

SAFETY PROFILE: Confirmed carcinogen with carcinogenic and tumorigenic data. Poison by ingestion, inhalation, skin contact, subcutaneous, intravenous, and intraperitoneal routes. Experimental teratogenic and reproductive effects. Human mutation data reported. A cholinesterase inhibitor, it is used in flea (pest) collars for pets. No neurotoxicity has been observed. It is very rapidly metabolized and excreted. When heated to decomposition it emits very toxic fumes of Cl$^-$ and PO$_x$. See also PARATHION.

DGQ875   CAS: 141-66-2   HR: 3
DICROTOPHOS
mf: C$_8$H$_{16}$NO$_5$P   mw: 237.22

SYNS: BIDIRL □ BIDRIN □ C 709 □ CARBICRON □ CIBA 709 □ DIAPADRIN □ DICROTOFOS (DUTCH) □ 3-(DIMETHOXYPHOSPHINYLOXY)-N,N-DIMETHYL-cis-CROTONAMIDE □ 3-(DIMETHOXYPHOSPHINYLOXY)-N,N-DIMETHYLISOCROTONAMIDE □ 3-(DIMETHYL-AMINO)-1-METHYL-3-OXO-1-PROPENYL DIMETHYL PHOSPHATE □ cis-2-DIMETHYLCARBAMOYL-1-METHYLVINYL DIMETHYLPHOSPHATE □ O,O-DIMETHYL-O-(2-DIMETHYL-CARBAMOYL-1-METHYL-VINYL)PHOSPHAT (GERMAN) □ O,O-DIMETHYL-O-(N,N-DIMETHYLCARBAMOYL-1-METHYLVINYL) PHOSPHATE □ O,O-DIMETHYL-O-(1,4-DIMETHYL-3-OXO-4-AZA-PENT-1-ENYL)FOSFAAT (DUTCH) □ O,O-DIMETHYL-O-(1,4-DIMETHYL-3-OXO-4-AZA-PENT-1-ENYL)PHOSPHATE □ DIMETHYLPHOSPHATE ESTER with 3-HYDROXY-N,N-DIMETHYL-cis-CROTONAMIDE □ DIMETHYL PHOSPHATE of 3-HYDROXY-N,N-DIMETHYL-cis-CROTONAMIDE □ O,O-DIMETIL-O-(1,4-DIMETIL-3-OXO-4-AZA-PENT-1-ENIL)-FOSFATO (ITALIAN) □ EKTAFOS □ ENT 24,482 □ 3-HYDROXY-DIMETHYL CROTONAMIDE DIMETHYL PHOSPHATE □ 3-HYDROXY-N,N-DIMETHYL-cis-CROTONAMIDE DIMETHYL PHOSPHATE □ PHOSPHATE de DIMETHYLE et de 2-DIMETHYLCARBAMOYL-1-METHYL VINYLE (FRENCH) □ SD 3562 □ SHELL SD-3562

CONSENSUS REPORTS: EPA Farm Worker Reentry (39 FR 16888,74). EPA Genetic Toxicology Program. EPA Extremely Hazardous Substances List.

OSHA PEL: TWA 0.25 mg/m^3 (skin)

ACGIH TLV: TWA 0.05 mg/m^3 (skin); Not Classifiable as a Human Carcinogen

SAFETY PROFILE: Poison by ingestion, inhalation, skin contact, subcutaneous, intravenous, and intraperitoneal routes. Mutation data reported. Used to control the coffee borer and certain economically important pests of cotton. When heated to decomposition it emits very toxic fumes of NO_x and PO_x. See also ESTERS.

DGR200   CAS: 12001-89-7   HR: 3
DICUMENE CHROMIUM

mf: $C_{18}H_{24} \cdot Cr$ mw: 292.42

SYNS: BIS(CUMENE)CHROMIUM □ BIS(pi-CUMENE)CHROMIUM □ BIS(ISOPROPYL-BENZENE)CHROMIUM □ DICUMENYLCHROMIUM

CONSENSUS REPORTS: Chromium and its compounds are on the Community Right-To-Know List.

OSHA PEL: TWA 1 mg(Cr)/m³

ACGIH TLV: TWA 0.05 mg(Cr)/m³; Confirmed Human Carcinogen

NIOSH REL: (Chromium(VI)) CL 1 µg(Cr(VI))/m³

SAFETY PROFILE: A confirmed carcinogen. Poison by skin contact and intravenous routes. Moderately toxic by ingestion. A skin and eye irritant. See also CHROMIUM COMPOUNDS. When heated to decomposition it emits acrid and irritating fumes.

DGR600   CAS: 80-43-3   HR: 1
DI-α-CUMYL PEROXIDE

mf: $C_{18}H_{22}O_2$ mw: 270.40

SYNS: ACTIVE DICUMYL PEROXIDE □ BIS(α,α-DIMETHYLBENZYL)PEROXIDE □ CUMENE PEROXIDE □ CUMYL PEROXIDE □ DICUMYL PEROXIDE (DOT) □ DI-CUP □ DI-CUP 40 KF □ DI-CUPR □ DIISOPROPYLBENZENE PEROXIDE □ ISOPROPYLBENZENE PEROXIDE □ LUPERCO □ LUPEROX □ LUPEROX 500R □ LUPEROX 500T □ VAROX DCP-R □ VAROX DCP-T

CONSENSUS REPORTS: Reported in EPA TSCA Inventory.

SAFETY PROFILE: Mildly toxic by ingestion. See also PEROXIDES. When heated to decomposition it emits acrid smoke and irritating fumes.

DGT500   CAS: 849-99-0   HR: 1
DICYCLOHEXYL ADIPATE

mf: $C_{18}H_{30}O_4$ mw: 310.48

SYNS: ERGOPLAST ADC □ HEXANEDIOIC ACID, DICYCLOHEXYL ESTER (9CI)

CONSENSUS REPORTS: Reported in EPA TSCA Inventory.

SAFETY PROFILE: Mildly toxic by ingestion and intraperitoneal routes. Experimental teratogenic and reproductive effects. When heated to decomposition it emits acrid smoke and fumes.

DGT600   CAS: 101-83-7   HR: 3
N,N-DICYCLOHEXYLAMINE

DOT: UN 2565

mf: $C_{12}H_{23}N$ mw: 181.36

PROP: Liquid; fishy odor. Mp: 20°, fp: −0.1°, bp: 256°, flash p: >210°F (OC), d: 0.910, vap d: 6.27.

SYNS: N-CYCLOHEXYLCYCLOHEXANAMINE □ DCHA □ DICHA □ DICYCLOHEXYLAMINE (DOT) □ DICYKLOHEXYLAMIN (CZECH) □ DODECAHYDRODIPHENYLAMINE

CONSENSUS REPORTS: IARC Cancer Review: Group 3 IMEMDT 7,178,87; Animal Inadequate Evidence IMEMDT 22,55,80. Reported in EPA TSCA Inventory.

DOT CLASSIFICATION: 8; Label: Corrosive

SAFETY PROFILE: Poison by ingestion and subcutaneous routes. Corrosive. A severe skin and eye irritant. Questionable carcinogen with experimental tumorigenic data. Human mutation data reported. Combustible when exposed to heat or flame; can react with oxidizing materials. To fight fire, use alcohol foam, CO_2, dry chemical. When heated to decomposition it emits toxic fumes of NO_x. See also CYCLOHEXYLAMINE.

DGU200   CAS: 3129-91-7   HR: 3
DICYCLOHEXYLAMINE NITRITE

DOT: UN 2687

mf: $C_{12}H_{23}N \cdot HNO_2$ mw: 228.38

SYNS: DECHAN □ DICHAN (CZECH) □ DICYCLO-HEXYLAMINONITRITE □ DICYCLOHEXYL-AMMONIUM NITRITE □ DICYKLOHEXYLAMIN NITRIT (CZECH) □ DICYNIT (CZECH) □ DODECA-HYDROPHENYLAMINE NITRITE □ DUSITAN DICYKLOHEXYLAMINU (CZECH)

CONSENSUS REPORTS: Reported in EPA TSCA Inventory.

DOT CLASSIFICATION: 4.1; Label: Flammable Solid

SAFETY PROFILE: Poison by ingestion and subcutaneous routes. Questionable carcinogen with experimental tumorigenic data. A flammable liquid. When heated to decomposition it emits very toxic fumes of HNO_2 and NO_x. See also NITRITES.

DGV100   CAS: 16069-36-6   HR: 3
DICYCLOHEXYL-18-CROWN-6
mf: $C_{20}H_{36}O_6$ mw: 372.56
PROP: Colorless or pale-yellow wax. Mp: 38–54°, bp: 344°.
SYNS: DICYCLOHEXANO-18-CROWN-6 □ EICOSAHYDRO DIBENZO(b,k)(1,4,7,10,13,16)-HEXAOXACYCLOOCTADECIN
CONSENSUS REPORTS: Reported in EPA TSCA Inventory.
SAFETY PROFILE: Poison by ingestion, skin contact, and intraperitoneal routes. An experimental teratogen. An eye and skin irritant. When heated to decomposition it emits acrid smoke and fumes.

DGV600   CAS: 119-60-8   HR: 3
DICYCLOHEXYL KETONE
mf: $C_{13}H_{22}O$ mw: 194.35
PROP: Bp: 145° @ 15 mm.
CONSENSUS REPORTS: Reported in EPA TSCA Inventory.
DOT CLASSIFICATION: 3; Label: Flammable Liquid
SAFETY PROFILE: Poison by intravenous route. See also KETONES. A flammable liquid. When heated to decomposition it emits acrid and irritating fumes.

DGV900   CAS: 22771-17-1   HR: 3
DICYCLOHEXYLTIN OXIDE
mf: $C_{12}H_{22}OSn$ mw: 301.03
SYN: STANNANE, DICYCLOHEXYLOXO-
OSHA PEL: TWA 0.1 mg(Sn)/m³ (skin)
ACGIH TLV: TWA 0.1 mg(Sn)/m³ (skin)
SAFETY PROFILE: Poison by ingestion. When heated to decomposition it emits toxic fumes of Sn.

DGW000   CAS: 77-73-6   HR: 3
DICYCLOPENTADIENE
DOT: UN 2048

mf: $C_{10}H_{12}$ mw: 132.22
PROP: Colorless crystals. Mp: 32.9°, bp: 166.6°, d: 0.976 @ 35°, vap press: 10 mm @ 47.6°, vap d: 4.55, flash p: 90°F (OC).
SYNS: BICYCLOPENTADIENE □ BISCYCLOPENTA-DIENE □ 1,3-CYCLOPENTADIENE, DIMER □ DICYKLOPENTADIEN (CZECH) □ DIMER CYKLOPENTADIENU (CZECH) □ 3a,4,7,7a-TETRAHYDRO-4,7-METHANOINDENE
CONSENSUS REPORTS: Reported in EPA TSCA Inventory. EPA Genetic Toxicology Program.
OSHA PEL: TWA 5 ppm
ACGIH TLV: TWA 5 ppm
DFG MAK: 0.5 ppm (2.7 mg/m³)
DOT CLASSIFICATION: 3; Label: Flammable Liquid
SAFETY PROFILE: Poison by ingestion and intraperitoneal routes. Moderately toxic by inhalation. Mildly toxic by skin contact. A severe skin and moderate eye irritant. Dangerous fire hazard when exposed to heat or flame; can react with oxidizing materials. To fight fire, use alcohol foam. When heated to decomposition it emits acrid smoke and fumes.

DGW200   CAS: 1271-19-8   HR: 3
DICYCLOPENTADIENYLDICHLOROTIT
ANIUM
mf: $C_{10}H_{10}Cl_2Ti$ mw: 249.00
PROP: Bright-red acicular crystals from toluene. Mp: 289°.
SYNS: DICHLOROBIS(ETA⁵)-2,4-CYCLOPENTADIEN-1-YL-TITANIUM (9CI) □ DICHLORODICYCLOPENTA-DIENYLTITANIUM □ DICHLORODI-pi-CYCLOPENTA-DIENYLTITANIUM □ DICHLOROTITANOCENE □ DICYCLOPENTADIENYLTITANIUMDICHLORIDE □ NCI-C04502 □ TITANIUM FERROCENE □ TITANOCENE □ TITANOCENE, DICHLORIDE
CONSENSUS REPORTS: Reported in EPA TSCA Inventory. EPA Genetic Toxicology Program.
SAFETY PROFILE: Poison by intravenous and intraperitoneal routes. Questionable carcinogen with experimental neoplastigenic, tumorigenic, and teratogenic data. Mutation data reported. See also TITANIUM COMPOUNDS. When heated to decomposition it emits toxic fumes of Cl⁻.

DHB400   **CAS: 60-57-1**   **HR: 3**
DIELDRIN
DOT: UN 2761
mf: $C_{12}H_8Cl_6O$ mw: 380.90
PROP: White crystals; odorless. Mp: 176–177°, vap d: 13.2. Insol in water; sol in common org solvs. IDLH 50 mg/m³.
SYNS: ALVIT □ COMPOUND 497 □ DIELDREX □ DIELDRINE (FRENCH) □ DIELDRITE □ ENT 16,225 □ HEOD □ HEXACHLOROEPOXYOCTAHYDRO-endo,exo-DIMETHANONAPHTHALENE □ 3,4,5,6,9,9-HEXA-CHLORO-1a,2,2a,3,6,6a,7,7a-OCTAHYDRO-2,7:3,6-DIMETHANONAPHTH(2,3-b)OXIRENE □ ILLOXOL □ INSECTICIDE No. 497 □ NCI-C00124 □ OCTALOX □ PANORAM D-31 □ QUINTOX □ RCRA WASTE NUMBER P037
CONSENSUS REPORTS: IARC Cancer Review: Group 3 IMEMDT 7,196,87; Human Inadequate Evidence IMEMDT 5,125,74; Animal Sufficient Evidence IMEMDT 5,125,74. NCI Carcinogenesis Bioassay (feed); Clear Evidence: mouse NCITR* NCI-CG-TR-21,78; No Evidence: rat NCITR* NCI-CG-TR-22,78; Inadequate Studies: rat NCITR* NCI-CG-TR-21,78.
OSHA PEL: TWA 0.25 mg/m³ (skin)
ACGIH TLV: TWA 0.25 mg/m³ (skin); Not Classifiable as a Human Carcinogen
DFG MAK: 0.25 mg/m³
NIOSH REL: (Dieldrin) Lowest reliable detectable level
DOT CLASSIFICATION: 6.1; Label: Poison
SAFETY PROFILE: A human poison by ingestion and possibly other routes. Poison experimentally by inhalation, ingestion, skin contact, intravenous, and intraperitoneal routes. Experimental teratogenic and reproductive data. Absorbed readily through the skin and by other routes. It is a central nervous system stimulant. Questionable carcinogen with experimental carcinogenic, neoplastigenic, and tumorigenic data. Human mutation data reported. An insecticide. Dieldrin is considerably more toxic than DDT by ingestion and skin contact. Dieldrin or its derivatives may accumulate in the body from chronic low dosages. When heated to decomposition it emits toxic fumes of Cl⁻. See also ALDRIN.

DHB800   **CAS: 564-00-1**   **HR: 3**
meso-1,2,3,4-DIEPOXYBUTANE
mf: $C_4H_6O_2$ mw: 86.10
PROP: A liquid. Mp: −19°, bp: 140°.
SYNS: (R*,S*)-2,2'-BIOXIRANE □ 1,2:3,4-DIANHYDROERYTHRITOL □ meso-DIEPOXYBUTANE □ (R*,S*)-DIEPOXYBUTANE □ ERYTHRITOL ANHYDRIDE
CONSENSUS REPORTS: IARC Cancer Review: Group 2B IMEMDT 7,56,87; Animal Sufficient Evidence IMEMDT 11,115,76.
SAFETY PROFILE: Confirmed carcinogen with experimental carcinogenic, neoplastigenic, and tumorigenic data. Poison by skin contact. Mutation data reported. When heated to decomposition it emits acrid smoke and irritating fumes.

DHE485   **HR: 3**
DIESEL EXHAUST
SAFETY PROFILE: Suspected carcinogen with experimental carcinogenic data. When heated to decomposition it emits acrid smoke and irritating fumes.

DHE750   **HR: 2**
DIESEL FUEL MARINE
SYNS: DFM □ DIESEL FUEL NO. 4 □ DISTILLATE FUEL, MARINE, PETROLEUM DERIV. □ MARINE DIESEL FUEL □ NCI-C54795 □ PETROLEUM DERIVED DISTILLATE FUEL, MARINE
CONSENSUS REPORTS: IARC Cancer Review: Group 2B IMEMDT 45,219,89; Animal Limited Evidence IMEMDT 45,219,89. Reported in NTP Carcinogenesis Studies (dermal); Equivocal Evidence: mouse NTPTR* NTP-TR-310,86
ACGIH TLV: TWA 100 mg/m³ (skin); Confirmed Animal Carcinogen with Unknown Revelance to Humans
SAFETY PROFILE: Suspected carcinogen. Low toxicity by ingestion and skin contact. A skin irritant. Experimental reproductive effects. When heated to decomposition it emits acrid smoke and irritating vapors.

D

DHE800   CAS: 68476-30-2   HR: 3
DIESEL FUEL MARINE
DOT: NA 1993
PROP: Brown, slightly viscous liquid. Flash
p: 100°F, d: <1, autoign temp: 494°F.
SYNS: API No. 2 FUEL OIL □ FUEL OIL NO. 2 (DOT) □
GAS OIL □ GAS OIL (DOT) □ HOME HEATING OIL No.
2 □ #2 HOME HEATING OILS □ NUMBER 2 BURNER
FUEL □ NUMBER 2 FUEL OIL
CONSENSUS REPORTS: Reported in EPA
TSCA Inventory. IARC Cancer Review:
Group 3 IMEMDT 45,239,89; IARC Cancer
Review: Animal Limited Evidence
IMEMDT 45,239,89
ACGIH TLV: TWA 100 mg/m³ (skin);
Confirmed Animal Carcinogen with
Unknown Revelance to Humans
DOT CLASSIFICATION: 3; Label:
Flammable Liquid
SAFETY PROFILE: Mildly toxic by
ingestion. A moderate skin and eye irritant.
Questionable carcinogen with experimental
carcinogenic data. A flammable liquid and
dangerous fire hazard when exposed to heat,
flame, or oxidizers. To fight fire, use foam,
CO_2, dry chemical. When heated to
decomposition it emits acrid smoke and
irritating fumes. See also MINERAL OIL
and KEROSENE.

DHE850   CAS: 68476-34-6   HR: 2
DIESEL FUEL NO. 2
SYNS: COMMERCIAL DIESEL FUEL NO. 2 □ FUELS,
DIESEL, NO. 2 □ NO. 2 DIESEL FUEL □ PRIMIER
DIESEL FUEL
SAFETY PROFILE: Questionable
carcinogen with experimental carcinogenic
data reported. When heated to
decomposition it emits acrid smoke and
irritating vapors.

DHE900   CAS: 68334-30-5   HR: 2
DIESEL FUELS
DOT: NA 1993
SYNS: AUTOMOTIVE DIESEL OIL □ DIESEL FUEL
(DOT) □ DIESEL OIL (PETROLEUM) □ DIESEL OILS □
DIESEL TEST FUEL □ FUELS, DIESEL □ OLEJ
NAPEDOWY III
CONSENSUS REPORTS: IARC Cancer
Review: Group 3 IMEMDT 45,219,89;

Human Inadequate Evidence IMEMDT
45,219,89. Reported in EPA TSCA
Inventory.
ACGIH TLV: TWA 100 mg/m³ (skin);
Confirmed Animal Carcinogen with
Unknown Revelance to Humans
DOT CLASSIFICATION: 3; Label: None
SAFETY PROFILE: Low toxicity by
ingestion. A skin irritant. Questionable
carcinogen. When heated to decomposition
it emits acrid smoke and irritating vapors.

DHF000   CAS: 111-42-2   HR: 3
DIETHANOLAMINE
mf: $C_4H_{11}NO_2$ mw: 105.16
PROP: A faintly colored, viscous liquid or
deliquescent prisms. Mp: 28°, bp: 270°
(decomp), flash p: 305°F (OC), d: 1.0919 @
30°/20°, autoign temp: 1224°F, vap press: 5
mm @ 138°, vap d: 3.65. Very sol in water.
SYNS: BIS(2-HYDROXYETHYL)AMINE □ DEA □
DIAETHANOLAMIN (GERMAN) □ DIETHANOLAMIN
(CZECH) □ DIETHYLOLAMINE □ 2,2'-DIHYDROXY-
DIETHYLAMINE □ DI(2-HYDROXYETHYL)AMINE □
DIOLAMINE □ 2,2'-IMINOBISETHANOL □ 2,2'-
IMINODIETHANOL □ NCI-C55174
CONSENSUS REPORTS: Community
Right-To-Know List. Reported in EPA
TSCA Inventory.
OSHA PEL: TWA 3 ppm
ACGIH TLV: TWA 0.46 ppm (skin)
SAFETY PROFILE: Poison by
intraperitoneal route. Moderately toxic by
ingestion and subcutaneous routes. Mildly
toxic by skin contact. A severe eye and mild
skin irritant. Experimental reproductive
effects. Combustible when exposed to heat
or flame; can react with oxidizing materials.
To fight fire, use alcohol foam, water, CO_2,
dry chemical. When heated to
decomposition it emits toxic fumes such as
NO_x. See also AMINES.

DHF200   CAS: 5716-15-4   HR: 2
DIETHANOLAMMONIUM MALEIC
**   HYDRAZIDE**
mf: $C_4H_{11}NO_2 \cdot C_4H_4N_2O_2$ mw: 217.26
SYNS: 6-HYDROXY-3-(2H)-PYRIDAZINONE
DIETHANOLAMINE □ 2,2'-IMINODI-ETHANOL with 1,2-

DIHYDRO-3,6-PYRIDAZINEDIONE (1:1) □ MALEIC HYDRAZIDE DIETHANOLAMINE SALT □ MAZIDE 30 □ NCI-C54660 □ SLO-GRO

CONSENSUS REPORTS: Reported in EPA TSCA Inventory.

SAFETY PROFILE: Moderately toxic by ingestion. Questionable carcinogen with experimental tumorigenic data. Mutation data reported. When heated to decomposition it emits toxic fumes of NO_x and NH_3.

DHG000   CAS: 78-62-6   HR: 3
DIETHOXYDIMETHYLSILANE
DOT: UN 2380
mf: $C_6H_{16}O_2Si$ mw: 148.31
PROP: A liquid. Bp: 114°, d: 0.86, vap press: 10 mm @ 13.3°, vap d: 5.1, flash p: <73.4°F.
SYNS: DIMETHYL-DIETHOXYSILAN (CZECH) □ DIMETHYLDIETHOXYSILANE (DOT)
CONSENSUS REPORTS: Reported in EPA TSCA Inventory.
DOT CLASSIFICATION: 3; Label: Flammable Liquid
SAFETY PROFILE: Mildly toxic by inhalation and ingestion. A skin and eye irritant. A dangerous fire hazard when exposed to heat, flame, or oxidizers. When heated to decomposition it emits acrid smoke and irritating fumes.

DHH400   CAS: 950-10-7   HR: 3
2-(DIETHOXYPHOSPHINYLIMINO)-4-
METHYL-1,3-DITHIOLANE
mf: $C_8H_{16}NO_3PS_2$ mw: 269.34
SYNS: AC 47470 □ AMERICAN CYANAMID CL-47470 □ CL-47,470 □ CYCLIC PROPYLENE (DIETHOXYPHOS-PHINYL)DITHIOIMIDOCARBONATE □ CYTROLANE □ p,p-DIETHYL CYCLIC PROPYLENE ESTER of PHOSPHONODITHIOIMIDOCARBONIC ACID □ DIETHYL (4-METHYL-1,3-DITHIOLAN-2-YLIDENE)-PHOSPHOROAMIDATE □ EI-47470 □ ENT 25,991 □ MEPHOSFOLAN □ (4-METHYL-1,3-DITHIOLAN-2-YLIDENE)PHOSPHORAMIDIC ACID, DIETHYL ESTER
CONSENSUS REPORTS: EPA Extremely Hazardous Substances List. Reported in EPA TSCA Inventory. EPA Genetic Toxicology Program.
SAFETY PROFILE: Poison by ingestion and skin contact. When heated to decomposition it emits very toxic fumes of NO_x, PO_x, and SO_x.

DHH800   CAS: 3054-95-3   HR: 3
3,3-DIETHOXYPROPENE
DOT: UN 2374
mf: $C_7H_{14}O_2$ mw: 130.21
PROP: Flash p: <73.4°.
SYNS: ACROLEIN ACETAL □ ACRYLALDEHYDE DIETHYL ACETAL □ 3,3-DIETHOXY-1-PROPENE □ PROPENAL DIETHYL ACETAL □ 1-PROPENE, 3,3-DIETHOXY-(9CI)
DOT CLASSIFICATION: 3; Label: Flammable Liquid
SAFETY PROFILE: Mutation data reported. A flammable liquid. A dangerous fire hazard when exposed to heat, flame, or oxidizers. When heated to decomposition it emits acrid smoke and fumes.

DHI000   CAS: 97-96-1   HR: 3
DIETHYL ACETALDEHYDE
DOT: UN 1178
mf: $C_6H_{12}O$ mw: 100.18
PROP: Colorless liquid; pungent odor. Fp: −89°, bp: 117°, flash p: 70°F (OC), d: 0.808–0.814, vap press: 13.7 mm @ 20°, vap d: 3.45, lel: 1.2%, uel: 7.7%. Misc in alc, ether; sltly sol in water.
SYNS: ALDEHYDE-2-ETHYLBUTYRIQUE (FRENCH) □ 2-ETHYLBUTANAL □ ETHYL BUTYRALDEHYDE □ α-ETHYLBUTYRALDEHYDE □ ETHYL BUTYR-ALDEHYDE (DOT) □ 2-ETHYLBUTYRALDEHYDE (DOT,FCC) □ 2-ETHYLBUTYRIC ALDEHYDE □ FEMA No. 2426
CONSENSUS REPORTS: Reported in EPA TSCA Inventory.
DOT CLASSIFICATION: 3; Label: Flammable Liquid
SAFETY PROFILE: Moderately toxic by ingestion. Mildly toxic by inhalation. A skin irritant. Flammable liquid. Can react vigorously with oxidizing materials. To fight fire, use alcohol foam, CO_2, dry chemical. When heated to decomposition it emits acrid smoke and fumes. See also ALDEHYDES.

DHI200   CAS: 685-91-6   HR: 3
N,N-DIETHYLACETAMIDE

mf: $C_6H_{13}NO$ mw: 115.20
PROP: Liquid. Mp: <65°, bp: 180°, flash p: 170°F, d: 0.92, vap d: 4.0.
CONSENSUS REPORTS: Reported in EPA TSCA Inventory.
SAFETY PROFILE: Moderately toxic by ingestion and intraperitoneal routes. Questionable carcinogen with experimental tumorigenic data. Flammable when exposed to heat or flame. To fight fire, use foam, mist, CO_2, dry chemical. When heated to decomposition it emits toxic fumes of NO_x.

DHI400   CAS: 88-09-5   HR: 3
DIETHYLACETIC ACID
mf: $C_6H_{12}O_2$ mw: 116.18
PROP: Colorless, volatile liquid; rancid odor. Fp −15°, mp: −93°, bp: 190°, flash p: 78°F (CC), d: 0.917, vap press: 10 mm @ 15.3°, vap d: 4.0, autoign temp: 865°F. Misc in alc, ether, water.
SYNS: 2-ETHYL BUTANOIC ACID □ α-ETHYLBUTYRIC ACID □ 2-ETHYLBUTYRIC ACID (FCC) □ FEMA No. 2429 □ 3-PENTANECARBOXYLIC ACID
CONSENSUS REPORTS: Reported in EPA TSCA Inventory.
SAFETY PROFILE: Moderately toxic by ingestion and skin contact. An irritant to skin and mucous membranes. A severe eye irritant. See also ESTERS. Narcotic in high concentrations. Flammable liquid. To fight fire, use CO_2, dry chemical, alcohol foam. When heated to decomposition it emits acrid smoke and fumes.

DHJ200   CAS: 109-89-7   HR: 3
DIETHYLAMINE
DOT: UN 1154
mf: $C_4H_{11}N$ mw: 73.16
PROP: Colorless liquid; ammonia-like odor. Mp: −38.9°, bp: 55.5°, flash p: −0.4°F, d: 0.711 @ 18°/4°, fp: −50°, autoign temp: 594°F, vap press: 400 mm @ 38.0°, vap d: 2.53, lel: 1.8%, uel: 10.1%. IDLH 200 ppm.
SYNS: 2-AMINOPENTANE □ DIAETHYLAMIN (GERMAN) □ N,N-DIETHYLAMINE □ DIETILAMINA (ITALIAN) □ DWUETYLOAMINA (POLISH) □ N-ETHYL-ETHANAMINE

CONSENSUS REPORTS: Reported in EPA TSCA Inventory.
OSHA PEL: TWA 10 ppm; STEL 25 ppm
ACGIH TLV: TWA 5 ppm; STEL 15 ppm; Not Classifiable as a Carcinogen
DFG MAK: 5 ppm (15 mg/m³)
DOT CLASSIFICATION: 3; Label: Flammable Liquid
SAFETY PROFILE: Moderately toxic by ingestion, inhalation, and skin contact. A skin and severe eye irritant. Exposure to strong vapor can cause severe cough and chest pains. Contact with liquid can damage eyes, possibly permanently; contact with skin causes necrosis and vesciculation. A very dangerous fire hazard when exposed to heat, flame, or oxidizers. To fight fire, use alcohol foam, CO_2, dry chemical. Explodes on contact with dicyanofurazan. Violent reaction with sulfuric acid. Ignites on contact with cellulose nitrate of sufficiently high surface area. When heated to decomposition it emits toxic fumes of NO_x. See also AMINES.

DHJ600   CAS: 3010-02-4   HR: 3
N,N-DIETHYLAMINOACETONITRILE
mf: $C_6H_{12}N_2$ mw: 112.20
SYNS: (DIETHYLAMINO)ACETONITRILE □ N,N-DIETHYLGLYCINONITRILE □ NITRIL KISELINY DIETHYLAMINOOCTOVE (CZECH)
CONSENSUS REPORTS: Cyanide and its compounds are on the Community Right-To-Know List. Reported in EPA TSCA Inventory.
SAFETY PROFILE: Poison by skin contact. Moderately toxic by inhalation. A skin and severe eye irritant. When heated to decomposition it emits toxic fumes of NO_x. See also NITRILES.

DHK400   CAS: 137-58-6   HR: 3
2-(DIETHYLAMINO)-2′,6′-ACETOXYLID-
IDE
mf: $C_{14}H_{22}N_2O$ mw: 234.38
PROP: Needles from C_6H_6 or EtOH. Mp: 68–69°, bp: 180–182° @ 4 mm.
SYNS: ANESTACON □ DIETHYLAMINOACETO-2,6-XYLIDIDE □ α-DIETHYLAMINOACETO-2,6-XYLIDIDE

□ α-DIETHYLAMINO-2,6-ACETOXYLIDIDE □ DIETH-YLAMINOACET-2,6-XYLIDIDE □ α-DIETHYLAMINO-2,6-DIMETHYLACETANILIDE □ ω-DIETHYLAMINO-2,6-DIMETHYLACETANILIDE □ α-DIETILAMINO-2,6-DIMETILACETANILIDE (ITALIAN) □ DUNCAINE □ GRAVOCAIN □ ISICAINA □ LEOSTESIN □ LIDA-MANTLE □ LIDOCAINE □ LIGNOCAINE □ MARICA-INE □ RUCAINA □ SOLCAIN □ XILOCAINA (ITALIAN) □ XYCIANE □ XYLESTESIN □ XYLOCAIN □ XYLOCITIN □ XYLOTOX

CONSENSUS REPORTS: Reported in EPA TSCA Inventory.

SAFETY PROFILE: Poison by ingestion, intravenous, intraperitoneal, and subcutaneous routes. Human systemic effects: blood pressure lowering, changes in heart rate, coma, convulsions, distorted perceptions, dyspnea, excitement, hallucinations, muscle contraction or spasticity, pulse rate, respiratory depression, toxic psychosis. An experimental teratogen. Other experimental reproductive effects. A local anesthetic. Mutation data reported. When heated to decomposition it emits toxic fumes of NO_x.

DHK600   CAS: 73-78-9   HR: 3
2-(DIETHYLAMINO)-2',6'-ACETOXYLID-IDE HYDROCHLORIDE
mf: $C_{14}H_{22}N_2O•ClH$ mw: 270.84
PROP: A solid. Mp: 127–129°. Sol in water.
SYNS: ANESTACON HYDROCHLORIDE □ 2-(DIETHYLAMINO)-2',6'-ACETOXYLIDIDE MONO-HYDROCHLORIDE □ α-DIETHYLAMINO-2,5-ACETOXYLIDINE HYDROCHLORIDE □ ω-DIETHYLAMINO-2,6-DIMETHYLACETANILIDE HYDROCHLORIDE □ 2-(DIETHYLAMINO)-N-(2,6-DIMETHYLPHENYL)ACETAMIDE MONOHYDRO-CHLORIDE □ DUNCAINE HYDROCHLORIDE □ GRAVOCAIN HYDROCHLORIDE □ ISICAINE HYDROCHLORIDE □ LEOSTESIN HYDROCHLORIDE □ LIDOCAINE HYDROCHLORIDE □ LIDOTHESIN HYDROCHLORIDE □ LIGNOCAINE HYDRO-CHLORIDE □ RUCAINA HYDROCHLORIDE □ S 202 □ XYCAINE HYDROCHLORIDE □ XYLESTESIN HYDROCHLORIDE □ XYLOCAINE HYDROCHLORIDE □ XYLOCARD □ XYLOCITIN HYDROCHLORIDE □ XYLONEURAL □ XYLOTOX HYDROCHLORIDE

CONSENSUS REPORTS: Reported in EPA TSCA Inventory. EPA Genetic Toxicology Program.

SAFETY PROFILE: Poison by ingestion, intraperitoneal, intravenous, subcutaneous, intramuscular, and intratracheal routes. Human systemic effects: somnolence, respiratory depression, low blood pressure, cardiomyopathy including infarction, pulse rate increase. An experimental teratogen. Other experimental reproductive effects. A skin and eye irritant. An anesthetic. When heated to decomposition it emits very toxic fumes of NO_x and HCl.

DHM500   CAS: 2869-83-2   HR: 2
3-(DIETHYLAMINO)-7-((p-(DIMETHYL-AMINO)PHENYL)AZO)-5-PHENYL-PHENAZINIUM CHLORIDE
mf: $C_{30}H_{31}N_6•Cl$ mw: 511.12
PROP: Dark green crystals or powder.
SYNS: C.I. 11050 □ JANUS GREEN B □ JANUS GREEN V

CONSENSUS REPORTS: Reported in EPA TSCA Inventory.

SAFETY PROFILE: Questionable carcinogen with experimental tumorigenic data. Mutation data reported. When heated to decomposition it emits toxic fumes of NO_x.

DHO500   CAS: 100-37-8   HR: 3
2-DIETHYLAMINOETHANOL
DOT: UN 2686
mf: $C_6H_{15}NO$ mw: 117.22
PROP: Colorless, hygroscopic liquid. Bp: 162°, flash p: 140°F (OC), d: 0.8851 @ 20°/20°, vap press: 1.4 mm @ 20°, vap d: 4.03. Sol in water. IDLH 100 ppm.
SYNS: DEAE □ DIAETHYLAMINOAETHANOL (GERMAN) □ DIETHYLAMINOETHANOL □ β-DIETHYLAMINOETHANOL □ N-DIETHYLAMINO-ETHANOL □ 2-(DIETHYLAMINO)ETHANOL □ 2-N-DIETHYLAMINOETHANOL □ DIETHYLAMINO-ETHANOL (DOT) □ β-DIETHYLAMINOETHYL ALCOHOL □ DIETHYLETHANOLAMINE □ N,N-DIETHYLETHANOLAMINE □ N,N-DIETHYL-N-(β-HYDROXYETHYL)AMINE □ 2-HYDROXYTRIETH-YLAMINE

CONSENSUS REPORTS: Reported in EPA TSCA Inventory.
OSHA PEL: TWA 10 ppm (skin)
ACGIH TLV: TWA 10 ppm (skin)

DFG MAK: 5 ppm (24 mg/m³)
DOT CLASSIFICATION: 3; Label:
Flammable Liquid
SAFETY PROFILE: Poison by
intraperitoneal and intravenous routes.
Moderately toxic by ingestion, skin contact,
subcutaneous, and intramuscular routes.
Human systemic effects by inhalation:
nausea or vomiting. A skin and severe eye
irritant. Flammable liquid when exposed to
heat or flame; can react with oxidizing
materials. To fight fire, use alcohol foam,
CO_2, dry chemical. When heated to
decomposition it emits toxic fumes of NO_x.
See also AMINES.

DHP200   CAS: 6376-26-7   HR: 3
o-(DIETHYLAMINOETHOXY)BENZ-
ANILIDE
mf: $C_{19}H_{24}N_2O_2$ mw: 312.45
PROP: A solid. Mp: 44°.
SYNS: o-DIAETHYLAMINOAETHOXY-BENZANILID
(GERMAN) □ 2-(2-
(DIETHYLAMINO)ETHOXY)BENZANILIDE
SAFETY PROFILE: Poison by ingestion,
subcutaneous, and intravenous routes. An
experimental teratogen. Other experimental
reproductive effects. When heated to
decomposition it emits toxic fumes of NO_x.

DHQ100   CAS: 140-82-9   HR: 3
2-(2-(DIETHYLAMINO)ETHOXY)-
ETHANOL
mf: $C_8H_{19}NO_2$ mw: 161.28
SYNS: DIETHYLAMINOETHOXYETHANOL □ 2-(β-
(DIETHYLAMINO)ETHOXY)ETHANOL □ ETHANOL, 2-
(2-(DIETHYLAMINO)ETHOXY)-
SAFETY PROFILE: A poison by ingestion
and skin contact. A severe eye and mild skin
irritant. When heated to decomposition it
emits toxic vapors of NO_x.

DHQ800   CAS: 17822-74-1   HR: 3
2-(2-(DIETHYLAMINO)ETHOXY)-3-
METHYLBENZANILIDE
mf: $C_{20}H_{25}N_2O_2$ mw: 325.47
SAFETY PROFILE: Poison by
subcutaneous route. An experimental

teratogen. When heated to decomposition it
emits toxic fumes of NO_x.

DHU000   CAS: 479-50-5   HR: 3
1-(2'-DIETHYLAMINO)ETHYLAMINO-4-
METHYLTHIOXANTHENONE
mf: $C_{20}H_{24}N_2OS$ mw: 340.52
PROP: Yellow crystals from EtOH (aq).
Mp: 64–65°.
SYNS: 1-((2-(DIETHYLAMINO)ETHYL)AMINO)-4-
METHYL-9H-THIOXANTHEN-9-ONE □ LUCANTHON
□ LUCANTHONE □ MIRACIL D □ NILODIN
CONSENSUS REPORTS: EPA Genetic
Toxicology Program.
SAFETY PROFILE: Poison by intravenous
route. Human mutation data reported.
When heated to decomposition it emits very
toxic fumes of NO_x and SO_x.

DHZ050   CAS: 13406-60-5   HR: 3
1-(2-(DIETHYLAMINO)ETHYL)-2-(p-
ETHOXYBENZYL)-5-BENZIMIDAZOL-
YL METHYL KETONE
mf: $C_{24}H_{31}N_3O_2$ mw: 393.58
SYN: KETONE, 1-(2-(DIETHYLAMINO)ETHYL)-2-(p-
ETHOXYBENZYL)-5-BENZIMIDAZOLYL METHYL
DOT CLASSIFICATION: 3; Label:
Flammable Liquid
SAFETY PROFILE: A poison by
intraperitoneal route. A flammable liquid.
When heated to decomposition it emits
toxic vapors of NO_x.

DIE350   CAS: 15451-93-1   HR: 3
1-(2-(DIETHYLAMINO)ETHYL)-2-p-
PHENETIDINO-5-BENZIMIDAZOLYL
METHYL KETONE
mf: $C_{23}H_{30}N_4O_2$ mw: 394.57
SYN: KETONE, 1-(2-(DIETHYLAMINO)ETHYL)-2-p-
PHENETIDINO-5-BENZIMIDAZOLYL METHYL
DOT CLASSIFICATION: 3; Label:
Flammable Liquid
SAFETY PROFILE: A poison by
subcutaneous route. A flammable liquid.
When heated to decomposition it emits
toxic vapors of NO_x.

DIN800 CAS: 29232-93-7 HR: 2
2-DIETHYLAMINO-6-METHYLPYRIMID-
IN-4-YL DIMETHYL PHOSPHORO-
THIONATE

mf: $C_{11}H_{20}N_3O_3PS$ mw: 305.37

PROP: Straw-colored oil. D: 1.157 @ 30 mm. Very sltly sol in H_2O. Misc in most org solvs.

SYNS: ACTELIC □ ACTELLIC □ ACTELLIFOG □ BLEX □ O-(2-DIETHYLAMINO-6-METHYLPYRIMIDIN-4-YL)-O,O-DIMETHYL PHOSPHOROTHIOATE □ O-(2-(DIETHYLAMINO)-6-METHYL-4-PYRIMIDINYL)-O,O-DIMETHYL PHOSPHOROTHIOATE □ ENT 27,699GC □ METHYL PIRIMIPHOS □ PIRIMIFOS-METHYL □ PLANT PROTECTION PP511 □ PP511 □ PYRIMIDINE PHOSPHATE □ PYRIMIPHOS METHYL □ SILOSAN

SAFETY PROFILE: Moderately toxic by ingestion. Mutation data reported. When heated to decomposition it emits very toxic fumes of NO_x, PO_x, and SO_x.

DIO200 CAS: 15421-84-8 HR: 3
7-DIETHYLAMINO-5-METHYL-s-TRIA-
ZOLO(1,5-a)PYRIMIDINE

mf: $N_5C_{10}H_{15}$ mw: 205.30

PROP: Bitter-tasting powder. Mp: 98–99.4°.

SYNS: AR 12008 □ N,N-DIETHYL-5-METHYL-(1,2,4)TRIAZOLO(1,5-a)PYRIMIDINE-7-AMINE □ 5-METHYL-7-DIETHYLAMINO-s-TRIAZOLO-(1,5-a)PYRIMIDINE □ ROCORNAL □ TRAPIDIL □ TRAPYMIN

SAFETY PROFILE: Poison by ingestion, intraperitoneal, subcutaneous, and intravenous routes. An experimental teratogen. Other experimental reproductive effects. A coronary vasodilator. When heated to decomposition it emits toxic fumes of NO_x. See also AMINES.

DIP100 CAS: 5185-78-4 HR: 3
2-(p-(DIETHYLAMINOPHENYL))-1,3,2-
DITHIARSENOLANE

mf: $C_{12}H_{18}AsNS_2$ mw: 315.35

SYN: 1,3,2-DITHIARSENOLANE, 2-(p-(DIETHYLAMINO)PHENYL)-

OSHA PEL: TWA 0.5 mg(As)/m³

SAFETY PROFILE: Poison by intraperitoneal route. When heated to decomposition it emits toxic fumes of NO_x, SO_x, and As.

DIR000 CAS: 522-00-9 HR: 3
10-(2-DIETHYLAMINOPROPYL)-
PHENOTHIAZINE

mf: $C_{19}H_{24}N_2S$ mw: 312.51

PROP: Crystals. Mp: 53–55°.

SYNS: AETHOPROPROPAZIN □ ATHAPROPAZINE □ ATHOPROPAZIN □ DIBUTIL □ 10-(2-DIETHYLAMINO-2-METHYLETHYL)PHENOTHIAZINE □ 2-DIETHYLAMINO-1-PROPYL-N-DIBENZOPARATHI-AZINE □ N,N-DIETHYL-α-METHYL-10H-PHENOTHI-AZINE-10-ETHANAMINE □ ETHOPROPAZINE □ ETOPROPEZINA □ FEMPROPAZINE □ FENPROPAZINA □ ISOTAZIN □ ISOTHAZIN □ ISOTHIAZINE □ LYSIVANE □ PARCIDOL □ PARDIDOL □ PARDISOL □ PARFEZINE □ PARKIN □ PARKISOL □ PARPHEZEIN □ PARSIDOL □ PARSITAN □ PHENOPROPAZINE □ PHENOPROZINE □ PRODICTAZIN □ PRODIERAZINE □ PROFENAMINA (ITALIAN) □ PROFENAMINUM □ ROCHIPEL □ ROCIPEL □ RODIPAL □ RP 3356 □ SC 2538 □ SKF 2538 □ TOMIL □ W 483

SAFETY PROFILE: Poison by ingestion, subcutaneous, and intravenous routes. An anticholinergic agent used to treat Parkinson's disease. When heated to decomposition it emits very toxic fumes of NO_x and SO_x.

DIR100 CAS: 67465-66-1 HR: 3
10-(2-(DIETHYLAMINO)PROPYL)-10H-
PYRIDO(3,2-B)(1,4)BENZOTHIAZINE

mf: $C_{18}H_{23}N_3S$ mw: 313.50

SYNS: 10H-PYRIDO(3,2-B)(1,4)BENZOTHIAZINE, 10-(2-(DIETHYLAMINO)PROPYL)- □ D 212

SAFETY PROFILE: A poison by ingestion and intraperitoneal routes. When heated to decomposition it emits toxic vapors of NO_x and SO_x.

DIS700 CAS: 91-66-7 HR: 2
N,N-DIETHYLANILINE

DOT: UN 2432

mf: $C_{10}H_{15}N$ mw: 149.26

PROP: Colorless to yellow liquid. D: (25°/4°) 0.9302, bp: 215–216°, mp: −38°, n: (24/D) 1.5394. Volatile with steam. Sltly sol in alc, chloroform, ether. One gram dissolves in 70 mL water at 12°. Sol in acids; sltly sol in EtOH, Et_2O, $CHCl_3$.

SYNS: BENZENAMINE, N,N-DIETHYL-(9CI) □ DEA □ DIAETHYLANILIN (GERMAN) □ N,N-DIETHYL-AMINOBENZENE □ N,N-DIETHYLANILIN (CZECH) □ DIETHYLANILINE □ N,N-DIETHYLBENZENAMINE □ DIETHYLPHENYLAMINE
CONSENSUS REPORTS: Reported in EPA TSCA Inventory.
DOT CLASSIFICATION: 6.1; Label: KEEP AWAY FROM FOOD
SAFETY PROFILE: Moderately toxic by ingestion and intraperitoneal routes. When heated to decomposition it emits toxic fumes of NO_x. See also ANILINE DYES.

DIS775   CAS: 5185-76-2   HR: 3
N,N-DIETHYL-p-ARSANILIC ACID
mf: $C_{10}H_{16}AsNO_3$ mw: 273.19
SYN: p-ARSANILIC ACID, N,N-DIETHYL-
OSHA PEL: TWA 0.5 mg(As)/m³
SAFETY PROFILE: Poison by intraperitoneal route. When heated to decomposition it emits toxic fumes of NO_x and As.

DIS850   CAS: 4964-27-6   HR: 3
DIETHYL ARSINIC ACID
mf: $C_4H_{11}AsO_2$ mw: 166.07
SYNS: ARSINE OXIDE, DIETHYLHYDROXY- □ DIETHYLHYDROXY ARSINE OXIDE
OSHA PEL: TWA 0.5 mg(As)/m³
SAFETY PROFILE: Poison by intravenous route. When heated to decomposition it emits toxic fumes of As.

DIU000   CAS: 25340-17-4   HR: 3
DIETHYL BENZENE
DOT: UN 2049
mf: $C_{10}H_{14}$ mw: 134.24
PROP: Colorless, mobile liquid. Bp: 183.8°, flash p: 134°F, d: 0.868 @ 25°/25°, autoign temp: 743–842°F, vap press: 1 mm @ 20.7°, vap d: 4.62.
SYNS: DIETHYLBENZENE (DOT) □ DIETHYLBENZOL
CONSENSUS REPORTS: Reported in EPA TSCA Inventory.
DOT CLASSIFICATION: 3; Label: Flammable Liquid
SAFETY PROFILE: Mildly toxic by ingestion. A skin and eye irritant. Flammable liquid when exposed to heat or flame; can react with oxidizing materials. To fight fire, use CO_2, dry chemical. When heated to decomposition it emits acrid smoke and fumes. See also ETHYL BENZENE.

DIU200   CAS: 141-93-5   HR: 2
m-DIETHYLBENZENE
mf: $C_{10}H_{14}$ mw: 134.24
PROP: A solid. Bp: 181–182°.
CONSENSUS REPORTS: Reported in EPA TSCA Inventory.
SAFETY PROFILE: Moderately toxic by ingestion. When heated to decomposition it emits acrid and irritating fumes. See also DIETHYLBENZENE.

DIV000   CAS: 542-63-2   HR: 3
DIETHYLBERYLLIUM
mf: $C_4H_{10}Be$ mw: 67.13
PROP: Colorless liquid. Mp: −13°, bp: 63° @ 3 mm, vap d: 2.3.
CONSENSUS REPORTS: IARC Cancer Review: Group 1 IMEMDT 58,41,93; Human Sufficient Evidence IMEMDT 58,41,93; Animal Sufficient Evidence IMEMDT 1,17,72; Animal Sufficient Evidence IMEMDT 23,143,80; Animal Sufficient Evidence IMEMDT 58,41,93. Beryllium and its compounds are on the Community Right-To-Know List.
OSHA PEL: TWA 0.002 mg(Be)/m³; STEL 0.005 mg(Be)/m³/30M; CL 0.025 mg(Be)/m³
ACGIH TLV: TWA 0.002 mg(Be)/m³; Confirmed Human Carcinogen; (Proposed: TWA 0.00005 mg(Be)/m³; STEL 0.0002 mg(Be)/m³ (sensitizer) (skin); Confirmed Human Carcinogen)
SAFETY PROFILE: Confirmed human carcinogen. Very poisonous. Dangerous fire hazard when exposed to heat or flame. Spontaneously flammable in air. Can react vigorously with oxidizing materials. To fight fire, use special extinguishing agents, dry chemical. Explodes on contact with water. Upon decomposition it emits poisonous

fumes of BeO. See also BERYLLIUM COMPOUNDS.

DIV600   CAS: 2641-56-7   HR: 3
DIETHYLBIS(OCTANOYLOXY)-
**   STANNANE**
mf: $C_{20}H_{40}O_4Sn$ mw: 463.29
SYNS: DIETHYLBIS(1-OXOOCTYL)OXY)STANNANE □ DIETHYLTIN DICAPRYLATE □ DIETHYLTIN DIOCTANOATE
OSHA PEL: TWA 0.1 mg(Sn)/m³ (skin)
ACGIH TLV: TWA 0.1 mg(Sn)/m³; STEL 0.2 mg(Sn)/m³ (skin).
NIOSH REL: (Organotin Compounds) TWA 0.1 mg(Sn)/m³
SAFETY PROFILE: Poison by ingestion. See also TIN COMPOUNDS. When heated to decomposition it emits acrid and irritating fumes.

DIV800   HR: 3
DIETHYLCADMIUM
mf: $C_4H_{10}Cd$ mw: 170.5
PROP: An oil; decomp by moisture. D: 1.6562, mp: −21°, bp: 64°.
CONSENSUS REPORTS: Cadmium and its compounds are on the Community Right-To-Know List.
OSHA PEL: TWA 5 μg(Cd)/m³
ACGIH TLV: TWA 0.002 mg(Cd)/m³ (respirable dust), Suspected Human Carcinogen); BEI: 5 μg/g creatinine in urine; 5 μg/L in blood
DFG MAK: DFG BAT: Blood 1.5 μg/dL; Urine 15 μg/dL, Suspected Carcinogen
NIOSH REL: (Cadmium) Reduce to lowest feasible level
SAFETY PROFILE: Confirmed human carcinogen. A poison. A dangerous fire and explosion hazard. Explodes when heated rapidly to 130°C. On exposure to air it forms white fumes that turn brown and explode. The vapor explodes when heated to 180°C. When heated to decomposition it emits highly toxic fumes of cadmium. See also CADMIUM COMPOUNDS.

DIW000   CAS: 90-89-1   HR: 3
DIETHYLCARBAMAZINE

mf: $C_{10}H_{21}N_3O$ mw: 199.34
PROP: Crystals. Mp: 47–49°, bp: 108.5–111°.
SYNS: BITIRAZINE □ CARBAMAZINE □ CARBILAZINE □ CARICIDE □ CATACIDE □ CYPIP □ 1-DIETHYLCARBAMOYL-4-METHYLPIPERAZINE □ 1-DIETHYLCARBAMYL-4-METHYLPIPERZINE □ N,N-DIETHYL-4-METHYL-1-PIPERAZINECARBOXAMIDE □ DITRAZINE BASE □ ETHODRYL □ 84L □ 1-METHYL-4-DIETHYLCARBAMYLPIPERAZINE □ NOTEZINE □ RP 3799 □ SPATONIN
CONSENSUS REPORTS: Reported in EPA TSCA Inventory.
SAFETY PROFILE: Poison by intraperitoneal route. Human systemic effects by ingestion: allergic dermatitis. An experimental teratogen. Mutation data reported. When heated to decomposition it emits toxic fumes of NO$_x$ and HCl. An additive permitted in the food and drinking water of animals and/or for the treatment of food-producing animals.

DIW200   CAS: 1642-54-2   HR: 3
DIETHYLCARBAMAZINE ACID CITRATE
mf: $C_{10}H_{21}N_3O \cdot C_6H_8O_7$ mw: 391.48
PROP: A solid. Mp: 137–138.5°.
SYNS: BANOCIDE □ CARICIDE □ CARITROL □ DICAROCIDE □ DIETHYLCARBAMAZANE CITRATE □ DIETHYLCARBAMAZINE CITRATE □ DIETHYLCARBAMAZINE HYDROGEN CITRATE □ 1-DIETHYLCARBAMOYL-4-METHYLPIPERAZINE DIHYDROGEN CITRATE □ N,N-DIETHYL-4-METHYL-1-PIPERAZINE CARBOXAMIDE CITRATE □ N,N-DIETHYL-4-METHYL-1-PIPERAZINECARBOXAMIDE DIHYDROGEN CITRATE □ N,N-DIETHYL-4-METHYL-1-PIPERAZINECARBOXAMIDE-2-HYDROXY-1,2,3-PROPANETRICARBOXYLATE □ DITRAZIN □ DITRAZIN CITRATE □ DITRAZINE □ DITRAZINE CITRATE □ ETHODRYL CITRATE □ FRANOCIDE □ FRANOZAN □ HETRAZAN □ LOXURAN □ 1-METHYL-4-DIETHYLCARBAMOYLPIPERAZINE CITRATE
CONSENSUS REPORTS: Reported in EPA TSCA Inventory.
SAFETY PROFILE: Poison by inhalation and intravenous routes. Moderately toxic by ingestion, subcutaneous, and intraperitoneal routes. An experimental teratogen. When heated to decomposition it emits toxic fumes of NO$_x$.

D

DIW400   CAS: 88-10-8   HR: 2
DIETHYLCARBAMOYL CHLORIDE
mf: $C_5H_{10}ClNO$ mw: 135.61
PROP: Liquid. Mp: $-44°$, bp: $186°$, vap d: 4.1.
SYNS: DIETHYLCARBAMIC CHLORIDE □ DIETHYL-CARBAMIDOYL CHLORIDE □ N,N-DIETHYL-CARBAMOYL CHLORIDE □ DIETHYLCARBAMYL CHLORIDE
CONSENSUS REPORTS: Reported in EPA TSCA Inventory.
DFG MAK: Confirmed Animal Carcinogen with Unknown Relevance to Humans
SAFETY PROFILE: Suspected carcinogen with experimental carcinogenic data. Moderately toxic by intraperitoneal route. Mutation data reported. Reacts with water or steam to produce toxic and corrosive fumes. When heated to decomposition it emits highly toxic fumes of Cl^- and NO_x. See also CARBAMATES and CHLORIDES.

DIW800   CAS: 112-36-7   HR: 3
DIETHYL CARBITOL
mf: $C_8H_{18}O_3$ mw: 162.26
PROP: Colorless liquid; sol in water and hydrocarbons. D: 0.9082 @ $20°/20°$, bp: $189°$, fp: $-44°$, flash p: $180°F$ (OC), vap d: 5.6.
SYNS: BIS(2-ETHOXYETHYL)ETHER □ DIETHYLENE GLYCOL DIETHYL ETHER □ 1-ETHOXY-2-(β-ETHOXY-ETHOXY)ETHANE □ ETHYL DIGLYME □ 3,6,9-TRIOXAUNDECANE
CONSENSUS REPORTS: Reported in EPA TSCA Inventory. Glycol ether compounds are on the Community Right-To-Know List.
SAFETY PROFILE: Moderately toxic by ingestion. An experimental teratogen. Other experimental reproductive effects. An eye irritant. Flammable when exposed to heat or flame. When heated to decomposition it emits acrid smoke and irritating fumes. See also GLYCOL ETHERS.

DIX200   CAS: 105-58-8   HR: 3
DIETHYL CARBONATE
DOT: UN 2366
mf: $C_5H_{10}O_3$ mw: 118.15

PROP: Colorless liquid; mild odor. Mp: $43°$, bp: $125.8°$, flash p: $77°F$ (OC), d: 0.975 @ $20°/4°$, vap press: 10 mm @ $23.8°$, vap d: 4.07.
SYNS: DEC □ DIAETHYLCARBONAT (GERMAN) □ DIETHYL CARBONATE (DOT) □ ETHOXYFORMIC ANHYDRIDE □ ETHYL CARBONATE □ EUFIN □ NCI-C60899
CONSENSUS REPORTS: Reported in EPA TSCA Inventory.
DOT CLASSIFICATION: 3; Label: Flammable Liquid
SAFETY PROFILE: Mildly toxic by subcutaneous route. Questionable carcinogen with experimental tumorigenic and teratogenic data. A dangerous fire hazard when exposed to heat or flame; can react with oxidizing materials. To fight fire, use foam, CO_2, dry chemical. When heated to decomposition it emits acrid smoke and fumes. See also ANHYDRIDES.

DIY000   CAS: 814-49-3   HR: 3
DIETHYL CHLOROPHOSPHATE
mf: $C_4H_{10}ClO_3P$ mw: 172.56
PROP: Water-white liquid with irritating, unpleasant odor. Bp: $88–90°$ @ 15 mm, d: 1.21 @ $20°/4°$, vap d: 5.94.
SYNS: CHLOROPHOSPHORIC ACID DIETHYL ESTER □ DIETHOXYPHOSPHORUS OXYCHLORIDE
CONSENSUS REPORTS: EPA Extremely Hazardous Substances List. Reported in EPA TSCA Inventory.
SAFETY PROFILE: Deadly poison by skin contact. Poison by ingestion. A cholinesterase inhibitor. See also PARATHION. Trace HCl catalyzes a hazardous reaction during the preparation of diethyl phosphate from diethyl ·chlorophosphate. When heated to decomposition it emits very toxic fumes of Cl^- and PO_x.

DIY800   CAS: 104-78-9   HR: 3
N,N-DIETHYL-1,3-DIAMINOPROPANE
DOT: UN 2684
mf: $C_7H_{18}N_2$ mw: 130.27
PROP: Liquid. Bp: $165–170°$, flash p: $138°F$ (OC), d: 0.82, vap d: 4.48.

SYNS: 1-AMINO-3-(DIETHYLAMINO)PROPANE □ N,N-DIETHYLAMINOPROPYLAMINE □ N-(3-DIETHYLAMINOPROPYL)AMINE □ 3-(DIETHYL-AMINO)PROPYLAMINE (DOT) □ DIETHYLAMINO-TRIMETHYLENAMINE

CONSENSUS REPORTS: Reported in EPA TSCA Inventory.

DOT CLASSIFICATION: 8; Label: Corrosive, Flammable Liquid

SAFETY PROFILE: Moderately toxic by ingestion and skin contact. Corrosive to the eyes, skin, and mucous membranes. A sensitizer. See also AMINES. Flammable liquid when exposed to heat or flame; can react with oxidizing materials. To fight fire, use foam, CO_2, dry chemical. When heated to decomposition it emits toxic fumes of NO_x.

DIZ100   CAS: 1609-47-8   HR: 3
DIETHYL DICARBONATE

mf: $C_6H_{10}O_5$ mw: 162.16

PROP: Viscous liquid; fruity odor. Bp: 58.5–62° @ 3 mm, d: 1.12, visc (20°): 1.97 cp. Soluble in alc, esters, ketones, and hydrocarbons.

SYNS: BAYCOVIN □ DEPC □ DICARBONIC ACID DIETHYL ESTER □ DIETHYL ESTER of PYROCARBONIC ACID □ DIETHYL OXYDIFORMATE □ DIETHYL PYROCARBONATE □ DIETHYL PYRO-CARBONIC ACID □ DKD □ ETHYL PYROCARBONATE □ OXYDIFORMIC ACID DIETHYL ESTER □ PIREF □ PYROCARBONATE d'ETHYLE (FRENCH) □ PYRO-CARBONIC ACID, DIETHYL ESTER □ PYROKOHL-ENSAEURE DIAETHYL ESTER (GERMAN)

CONSENSUS REPORTS: Reported in EPA TSCA Inventory.

SAFETY PROFILE: Poison by ingestion, inhalation, and intraperitoneal routes. Concentrated DEPC is irritating to eyes, mucous membranes, and skin. When heated to decomposition it emits acrid smoke and fumes. See also ESTERS.

DJA325   CAS: 10161-85-0   HR: 3
O,O-DIETHYL Se-(2-DIETHYLAMINO-ETHYL)PHOSPHOROSELENOATE

mf: $C_{10}H_{24}NO_3PSe$ mw: 316.28

SYN: PHOSPHOROSELENOIC ACID, Se-(2-(DIETHYLAMINO)ETHYL) O,O-DIETHYL ESTER

OSHA PEL: TWA 0.2 mg(Se)/m³

ACGIH TLV: TWA 0.2 mg(Se)/m³

SAFETY PROFILE: Poison by subcutaneous route. When heated to decomposition it emits toxic fumes of NO_x, PO_x, and Se.

DJA330   CAS: 18048-06-1   HR: 3
DIETHYL (2-(DIETHOXYMETHYLSILYL)-ETHYL)PHOSPHONATE

mf: $C_{11}H_{27}O_5PSi$ mw: 298.44

SYNS: DIETHYL (2-(METHYLDIETHOXYSILYL)-ETHYL)PHOSPHONATE □ PHOSPHONIC ACID, (2-(DIETHOXYMETHYLSILYL)ETHYL)-, DIETHYL ESTER □ SILANE 40-43

SAFETY PROFILE: A poison by ingestion and skin contact. A moderate skin and severe eye irritant. When heated to decomposition it emits toxic vapors of PO_x.

DJB000   CAS: 2767-55-7   HR: 3
DIETHYLDIIODOSTANNANE

mf: $C_4H_{10}I_2Sn$ mw: 430.63

PROP: Very sltly sol white crystals or white needles. Mp: 44°, bp: 240–245° (decomp).

SYNS: DIETHYLSTANNIUM DIIODIDE □ DIETHYLSTANNIUMDIJODID (GERMAN) □ DIETHYLTIN DIIODIDE □ TIN, DIETHYL-, DIIODIDE

OSHA PEL: TWA 0.1 mg(Sn)/m³ (skin)

ACGIH TLV: TWA 0.1 mg(Sn)/m³; STEL 0.2 mg(Sn)/m³ (skin).

NIOSH REL: (Organotin Compounds) TWA 0.1 mg(Sn)/m³

SAFETY PROFILE: Poison by ingestion and intraperitoneal routes. See also TIN COMPOUNDS and IODIDES. When heated to decomposition it emits toxic fumes of I⁻.

DJB460   CAS: 23125-28-2   HR: 3
DIETHYL 2,6-DIMETHYL-4(2-PYRIDYL)-1,4-DIHYDRO-3,5-PYRIDINE-DICARBOXYLATE

mf: $C_{18}H_{22}N_2O_4$ mw: 330.42

SYNS: BAY 1518 □ (2,4'-BIPYRIDINE)-3',5'-DICARBOXYLIC ACID, 1',4'-DIHYDRO-2',6'-DIMETHYL-,-DIETHYL ESTER □ ISOCINCHOMERONIC ACID, 1,4-DIHYDRO-2,6-DIMETHYL-4-(2-PYRIDYL)-, DIETHYL ESTER □ 3,5-PYRIDINEDICARBOXYLIC ACID, 1,4-DIHYDRO-2,6-DIMETHYL-4-(2-PYRIDYL)-, DIETHYL ESTER

SAFETY PROFILE: A poison by intravenous route. When heated to decomposition it emits toxic vapors of NO$_x$.

DJC000   CAS: 72-56-0   HR: 3
DIETHYLDIPHENYL DICHLOROETHANE
mf: C$_{18}$H$_{20}$Cl$_2$ mw: 307.28
PROP: Crystals. Mp: 60–61°.
SYNS: 1,1-BIS(p-ETHYLPHENYL)-2,2-DICHLORO-ETHANE □ 2,2-BIS(p-ETHYLPHENYL)-1,1-DICHLO-ROETHANE □ α,α-DICHLORO-2,2-BIS(p-ETHYLPHEN-YL)ETHANE □ 1,1-DICHLORO-2,2-BIS(p-ETHYLPHEN-YL)ETHANE □ 1,1-DICHLORO-2,2-BIS(4-ETHYLPHEN-YL)ETHANE □ 2,2-DICHLORO-1,1-BIS(p-ETHYLPHEN-YL)ETHANE □ DI(p-ETHYLPHENYL)DICHLOROETH-ANE □ ETHYLAN □ p,p-ETHYL DDD □ p,p'-ETHYL-DDD □ NCI-C02868 □ PERTHANE □ Q-137
CONSENSUS REPORTS: NCI Carcinogenesis Bioassay (feed); Clear Evidence: mouse NCITR* NCI-CG-TR-156,79; No Evidence: rat NCITR* NCI-CG-TR-156,79.
SAFETY PROFILE: Poison by intravenous route. Mildly toxic by ingestion. Questionable carcinogen with experimental carcinogenic and tumorigenic data. Experimental teratogenic and reproductive effects. Mutation data reported. A pesticide. When heated to decomposition it emits toxic fumes of Cl$^-$. See also CHLORINATED HYDROCARBONS, ALIPHATIC.

DJC400   CAS: 85-98-3   HR: 3
1,3-DIETHYL-1,3-DIPHENYLUREA
mf: C$_{17}$H$_{20}$N$_2$O   mw: 268.39
PROP: Colorless crystals. Mp: 73°, d: 1.12, bp: 326°, flash p: 302°F (CC), vap d: 9.3.
SYNS: BIS(N-ETHYL-N-PHENYL)UREA □ N,N-DIETHYLCARBANILIDE □ N,N'-DIETHYL-N,N'-DIPHENYLUREA □ sym-DIETHYLDIPHENYLUREA □ USAF EK-1047
CONSENSUS REPORTS: Reported in EPA TSCA Inventory.
SAFETY PROFILE: Poison by intraperitoneal route. Moderately toxic by ingestion. Combustible when exposed to heat or flame. Probably a slight explosion hazard, although it is a component of smokeless explosive mixtures. When heated to decomposition it burns and emits very toxic fumes of NO$_x$. To fight fire, use dry chemical, CO$_2$, spray or mist. An explosion regulator.

DJD400   CAS: 136-92-5   HR: 3
DIETHYLDITHIOCARBAMIC ACID SELENIUM(II) SALT
mf: C$_{10}$H$_{20}$N$_2$S$_4$•Se   mw: 375.52
SYNS: ETHYL SELENAC □ SELENIUM DIETHYLDITHIOCARBAMATE
CONSENSUS REPORTS: Selenium and its compounds are on the Community Right-To-Know List.
OSHA PEL: TWA 0.2 mg(Se)/m^3
ACGIH TLV: TWA 0.2 mg(Se)/m^3
DFG MAK: 0.1 mg(Se)/m^3
SAFETY PROFILE: Poison by ingestion and intraperitoneal routes. See also SELENIUM COMPOUNDS and CARBAMATES. When heated to decomposition it emits very toxic fumes of NO$_x$, SO$_x$, and Se.

DJD600   CAS: 111-46-6   HR: 2
DIETHYLENE GLYCOL
mf: C$_4$H$_{10}$O$_3$ mw: 106.14
PROP: Clear, colorless, practically odorless, syrupy liquid. Fp: −8°, mp: −6.5°, bp: 133° @ 14 mm, flash p: 255°F, d: 1.1184 @ 20°/20°, autoign temp: 444°F, vap press: 1 mm @ 91.8°, vap d: 3.66. Sol in water.
SYNS: BIS(2-HYDROXYETHYL) ETHER □ BRECOLANE NDG □ CARBITOL □ DEACTIVATOR E □ DEACTIVATOR H □ DEG □ DICOL □ DIGLYCOL □ DIHYDROXYDIETHYL ETHER □ β,β'-DIHYDROXY-DIETHYL ETHER □ 2,2'-DIHYDROXYETHYL ETHER □ DISSOLVANT APV □ ETHYLENE DIGLYCOL □ GLYCOL ETHER □ GLYCOL ETHYL ETHER □ 3-OXAPENTANE-1,5-DIOL □ 3-OXA-1,5-PENTANEDIOL □ 2,2'-OXYBISETHANOL □ 2,2'-OXYDIETHANOL □ TL4N
CONSENSUS REPORTS: Reported in EPA TSCA Inventory. Glycol ether compounds are on the Community Right-To-Know List.
DFG MAK: 10 ppm
SAFETY PROFILE: Moderately toxic to humans by ingestion. Poison experimentally by inhalation. Moderately toxic by ingestion and intravenous routes. Questionable carcinogen with experimental carcinogenic,

tumorigenic, and teratogenic data. An eye and human skin irritant. Combustible when exposed to heat or flame; can react with oxidizing materials. To fight fire, use alcohol foam, water, CO_2, dry chemical. Mixtures with sodium hydroxide decompose exothermically when heated to 230°C and release explosive hydrogen gas. When heated to decomposition it emits acrid smoke and irritating fumes. See also GLYCOL ETHERS.

DJD700   CAS: 13988-26-6   HR: 2
DIETHYLENE GLYCOL BISPHTHALATE
mf: $C_{12}H_{12}O_5$ mw: 236.24

SYNS: 2,5,8-BENZOTRIOXACYCLOUNDECIN-1,9-DIONE, 3,4,6,7-TETRAHYDRO-(9CI) □ HOWFLEX GBP

CONSENSUS REPORTS: Reported in EPA TSCA Inventory.

SAFETY PROFILE: Questionable carcinogen with experimental tumorigenic data. When heated to decomposition it emits acrid smoke and irritating fumes.

DJE400   CAS: 693-21-0   HR: 3
DIETHYLENE GLYCOL DINITRATE
DOT: UN 0075

mf: $C_4H_8N_2O_7$ mw: 196.14

PROP: Liquid. Vap d: 6.76.

SYNS: BIS(HYDROXYAETHYL)-AETHER-DINITRAT (GERMAN) □ DIETHYLENE GLYCOL DINITRATE, containing at least 25% phlegmatizer (DOT) □ DIETHYLENGLYKOLDINITRATE (CZECH) □ DIGLYCOLDINITRAAT (DUTCH) □ DIGLYCOL (DINITRATE de) (FRENCH) □ DIGLYKOLDINITRAT (GERMAN) □ DI(HYDROXYETHYL) ETHER DINITRATE □ DINITRATE de DIETHYLENE-GLYCOL (FRENCH) □ DINITRODIGLICOL (ITALIAN) □ DINITRODIGLYKOL (CZECH)

CONSENSUS REPORTS: Reported in EPA TSCA Inventory. Glycol ether compounds are on the Community Right-To-Know List.

DOT CLASSIFICATION: Forbidden; DOT Class: EXPLOSIVE 1.1D; Label: EXPLOSIVE 1.1D (UN 0075)

SAFETY PROFILE: Moderately toxic by ingestion. Ingestion of this compound can cause a drop in blood pressure and cardiac disturbances. A dangerous fire hazard when exposed to heat or flame; can react

vigorously with oxidizing or reducing materials. A dangerous explosive sensitive to heat, shock, and vibration. Used in low-freezing dynamites and some permissible explosives. Upon decomposition it emits toxic fumes of NO_x. See also GLYCOL ETHERS, NITRATES, and EXPLOSIVES, HIGH.

DJF200   CAS: 112-34-5   HR: 2
DIETHYLENE GLYCOL MONOBUTYL ETHER
mf: $C_8H_{18}O_3$ mw: 162.26

PROP: Colorless liquid. Mp: −68.1°, bp: 230.6°, flash p: 172°F, d: 0.9553 @ 20°/4°, autoign temp: 442°F, vap press: 0.02 mm @ 20°, vap d: 5.58.

SYNS: BUCB □ BUTOXYDIETHYLENE GLYCOL □ BUTOXYDIGLYCOL □ 2-(2-BUTOXYETHOXY)ETHANOL □ BUTYL CARBITOL □ o-BUTYL DIETHYLENE GLYCOL □ BUTYL DIOXITOL □ DIETHYLENE GLYCOL-n-BUTYL ETHER □ DIGLYCOL MONOBUTYL ETHER □ DOWANOL DB □ EKTASOLVE DB □ GLYCOL ETHER DB □ JEFFERSOL DB □ POLY-SOLV DB

CONSENSUS REPORTS: Reported in EPA TSCA Inventory. Glycol ether compounds are on the Community Right-To-Know List.

DFG MAK: 100 mg/m³

SAFETY PROFILE: Moderately toxic by ingestion and intraperitoneal routes. Mildly toxic by skin contact. A severe eye irritant. Combustible when exposed to heat or flame; can react with oxidizing materials. To fight fire, use alcohol foam, CO_2, or dry chemical. When heated to decomposition it emits acrid smoke and irritating fumes. See also GLYCOL ETHERS.

DJG000   CAS: 111-77-3   HR: 2
DIETHYLENE GLYCOL MONOMETHYL ETHER
mf: $C_5H_{12}O_3$ mw: 120.17

PROP: Hygroscopic, water-white liquid. Mp: −70°, bp: 194.2°, flash p: 200°F (OC), d: 1.0354 @ 20°/4°, vap press: 0.2 mm @ 20°, vap d: 4.14.

SYNS: DIETHYLENE GLYCOL METHYL ETHER □ DIGLYCOL MONOMETHYL ETHER □ DOWANOL DM □ ETHYLENE DIGLYCOL MONOMETHYL ETHER □

MECB □ METHOXYDIGLYCOL □ 2-(2-METHOXY-ETHOXY)ETHANOL □ β-METHOXY-β'-HYDROXY-DIETHYL ETHER □ METHYL CARBITOL □ POLY-SOLV DM

CONSENSUS REPORTS: Reported in EPA TSCA Inventory. Glycol ether compounds are on the Community Right-To-Know List.

SAFETY PROFILE: Moderately toxic by skin contact and intraperitoneal routes. Mildly toxic by ingestion. An experimental teratogen. Other experimental reproductive effects. An eye irritant. Combustible when exposed to heat or flame; can react with oxidizing materials. Reacts violently with $Ca(OCl)_2$, chlorosulfonic acid, and oleum. To fight fire, use dry chemical, alcohol foam, water spray or mist, CO_2. When heated to decomposition it emits acrid smoke and irritating fumes. See also GLYCOL ETHERS.

DJG600   CAS: 111-40-0   HR: 3
DIETHYLENETRIAMINE
DOT: UN 2079
mf: $C_4H_{13}N_3$ mw: 103.20
PROP: Yellow, viscous liquid; mild ammonia-like odor. Mp: −39°, bp: 207°, flash p: 215°F (OC), d: 0.9586 @ 20°/20°, autoign temp: 750°F, vap press: 0.22 mm @ 20°, vap d: 3.48. Misc in H_2O and EtOH.
SYNS: AMINOETHYLETHANEDIAMINE □ N-(2-AMINOETHYL)ETHYLENEDIAMINE □ 3-AZAPENT-ANE-1,5-DIAMINE □ BIS(β-AMINOETHYL)AMINE □ BIS(2-AMINOETHYL)AMINE □ D.E.H. 20 □ DETA □ 2,2'-DIAMINODIETHYLAMINE □ 2,2'-IMINOBISETHYL-AMINE

CONSENSUS REPORTS: Reported in EPA TSCA Inventory.
OSHA PEL: TWA 1 ppm
ACGIH TLV: TWA 1 ppm (skin)
DOT CLASSIFICATION: 8; Label: Corrosive
SAFETY PROFILE: Poison by skin contact and intraperitoneal routes. Moderately toxic by ingestion. Corrosive. A severe skin and eye irritant. High concentration of vapors causes irritation of respiratory tract, nausea, and vomiting. Repeated exposures can cause asthma and sensitization of skin. Combustible when exposed to heat or flame; can react with oxidizing materials. Mixture with

nitromethane is a shock-sensitive explosive. Ignites on contact with cellulose nitrate of high surface area. To fight fire, use alcohol foam. When heated to decomposition it emits toxic fumes of NO_x. See also AMINES.

DJG800   CAS: 67-43-6   HR: 2
(DIETHYLENETRINITRILO)PENTA-ACETIC ACID
mf: $C_{14}H_{23}N_3O_{10}$ mw: 393.40
PROP: Crystals from H_2O. Mp: 219–220°. Sol in alkalies and H_2O.
SYNS:
((CARBOXYMETHYLIMINO)BIS(ETHYLENENITRILO))TETRAACETIC ACID □ CHEL 330 □ CHEL 330 ACID □ CHEL DTPA □ DIETHYLENETRIAMINEPENTAACETIC ACID □ 1,1,4,7,7-DIETHYLENETRIAMINEPENTA-ACETIC ACID □ DTPA □ HAMP-EX ACID □ MONAQUEST □ PENTHAMIL □ PERMA KLEER □ 3,6,9-TRIS(CARBOXYMETHYL)-3,6,9-TRIAZAUNDE-CANEDIOIC ACID

CONSENSUS REPORTS: Reported in EPA TSCA Inventory.
SAFETY PROFILE: Moderately toxic by intraperitoneal route. When heated to decomposition it emits toxic fumes of NO_x.

DJI400   CAS: 100-36-7   HR: 3
N,N-DIETHYLETHYLENEDIAMINE
DOT: UN 2685
mf: $C_6H_{16}N_2$ mw: 116.24
PROP: Liquid. Bp: 149–150°, flash p: 115°F (OC), d: 0.82 @ 20°/20°, vap d: 4.00.
SYNS: N,N-DIETHYL-1,2-ETHANEDIAMINE □ USAF AM-1

CONSENSUS REPORTS: Reported in EPA TSCA Inventory.
DOT CLASSIFICATION: 8; Label: Corrosive, Flammable Liquid
SAFETY PROFILE: Poison by intraperitoneal route. Moderately toxic by ingestion and skin contact. A skin and severe eye irritant. Flammable liquid when exposed to heat or flame; can react with oxidizing materials. To fight fire, use alcohol foam, CO_2, dry chemical. When heated to decomposition it emits toxic fumes of NO_x. See also AMINES.

D

DJJ393   CAS: 32522-68-2   HR: 3
o,o-DIETHYL S-(2-(ETHYLTHIO)-6-
    METHYL-4-PYRIMIDINYL)
    PHOSPHORODITHIOATE
mf: $C_{11}H_{20}N_2O_2PS_3$ mw: 339.48
SYNS: PHOSPHORODITHIOIC ACID, o,o-DIETHYL S-
(2-(ETHYLTHIO)-6-METHYL-4-PYRIMIDINYL) ESTER □
HERCULES 6937
SAFETY PROFILE: A poison by ingestion
and subcutaneous routes. When heated to
decomposition it emits toxic vapors of NO_x,
SO_x and PO_x.

DJJ850   CAS: 26645-10-3   HR: 3
DIETHYL GOLD BROMIDE
mf: $C_4H_{10}AuBr$ mw: 335.02
SYNS: BROMODIETHYLGOLD □ DIETHYLGOLD
BROMIDE (DOT)
DOT CLASSIFICATION: Forbidden
SAFETY PROFILE: Explodes at 70°C.
When heated to decomposition it emits
toxic fumes of Br^-. See also GOLD
COMPOUNDS and BROMIDES.

DJK800   CAS: 16111-62-9   HR: 2
DI(2-ETHYLHEXYL) PEROXYDI-
    CARBONATE
mf: $C_{18}H_{34}O_6$ mw: 346.52
SYNS: PEROXYDICARBONIC ACID, BIS(2-ETHYL-
HEXYL) ESTER □ PEROXYDICARBONIC ACID, DI(2-
ETHYLHEXYL) ESTER
CONSENSUS REPORTS: Reported in EPA
TSCA Inventory.
SAFETY PROFILE: Moderately toxic by
ingestion. When heated to decomposition it
emits acrid smoke and irritating fumes. See
also PEROXIDES, ORGANIC.

DJL000   CAS: 577-11-7   HR: 3
DI-(2-ETHYLHEXYL) SODIUM SULFO-
    SUCCINATE
mf: $C_{20}H_{38}O_7S \cdot Na$ mw: 445.63
PROP: White, waxlike, plastic solid; octyl
alcohol odor. Sol in hexane, glycerin, alc;
sltly sol in water and org solvs.
SYNS: AEROSOL GPG □ ALCOPOL O □ ALPHASOL
OT □ BEROL 478 □ BIS(ETHYLHEXYL) ESTER of
SODIUM SULFOSUCCINIC ACID □ BIS(2-ETHYLHEX-
YL)SODIUM SULFOSUCCINATE □ BIS(2-ETHYLHEX-
YL)-S-SODIUM SULFOSUCCINATE □ 1,4-BIS(2-
ETHYLHEXYL) SODIUM SULFOSUCCINATE □ 1,4-BIS(2-
ETHYLHEXYL)SULFOBUTANEDIOIC ACID ESTER,
SODIUM SALT □ CELANOL DOS 75 □ CLESTOL □
COLACE □ COMPLEMIX □ CONSTONATE □ COPROL
□ DEFILIN □ DIOCTLYN □ DIOCTYLAL □ DIOCTYL
ESTER of SODIUM SULFOSUCCINATE □ DIOCTYL
ESTER of SODIUM SULFOSUCCINIC ACID □ DIOCTYL-
MEDO FORTE □ DIOCTYL SODIUM SULFOSUCCINATE
(FCC) □ DIOCTYL SULFOSUCCINATE SODIUM SALT
□ DIOMEDICONE □ DIOSUCCIN □ DIOTILAN □
DIOVAC □ DOCUSATE SODIUM □ DOXINATE □
DOXOL □ DSS □ DULSIVAC □ DUOSOL □ 2-
ETHYLHEXYL SULFOSUCCINATE SODIUM □
HUMIFEN WT 27G □ KONLAX □ KOSATE □
LAXINATE □ MANOXAL OT □ MERVAMINE □
MODANE SOFT □ MOLATOC □ MOLCER □ MOLOFAC
□ MONAWET MD 70E □ NEKAL WT-27 □ NEVAX □
NIKKOL OTP 70 □ NORVAL □ OBSTON □ RAPISOL □
REGUTOL □ REQUTOL □ REVAC □ SANMORIN OT 70
□ SBO □ SOBITAL □ SODIUM BIS(2-ETHYLHEXYL)
SULFOSUCCINATE □ SODIUM DI-(2-ETHYLHEXYL)
SULFOSUCCINATE □ SODIUM DIOCTYL SULFO-
SUCCINATE □ SODIUM DIOCTYL SULPHOSUCCINATE
□ SODIUM-2-ETHYLHEXYLSULFOSUCCINATE □
SODIUM SULFODI-(2-ETHYLHEXYL)SULFOSUCCINATE
□ SOFTIL □ SOLIWAX □ SOLUSOL-75% □ SOLUSOL-
100% □ SULFIMEL DOS □ TEX WET 1001 □ TRITON
GR-5 □ VATSOL OT □ VELMOL □ WAXSOL □ WETAID
SR
CONSENSUS REPORTS: Reported in EPA
TSCA Inventory.
SAFETY PROFILE: Poison by intravenous
route. Moderately toxic by ingestion and
intraperitoneal routes. A skin and severe eye
irritant. See also ESTERS. When heated to
decomposition it emits toxic fumes of SO_x
and Na_2O.

DJL400   CAS: 1615-80-1   HR: 3
1,2-DIETHYLHYDRAZINE
mf: $C_4H_{12}N_2$ mw: 88.18
PROP: Bp: 86°, d: 0.797 @ 26°. Sol in alc
and ether.
SYNS: 1,2-DIAETHYLHYDRAZINE (GERMAN) □ N-N'-
DIETHYLHYDRAZINE □ sym-DIETHYLHYDRAZINE □
HYDRAZOETHANE □ HYDROAZOETHANE □ RCRA
WASTE NUMBER U086 □ SDEH
CONSENSUS REPORTS: IARC Cancer
Review: Group 2B IMEMDT 7,56,87;
Animal Sufficient Evidence IMEMDT
4,153,74.
SAFETY PROFILE: Confirmed carcinogen
with experimental carcinogenic,
tumorigenic, and teratogenic data. It is also a

transplacental carcinogen. Mutation data reported. When heated to decomposition it emits toxic fumes of NO_x. See also HYDRAZINE.

DJN000   CAS: 3710-84-7   HR: 3
DIETHYLHYDROXYLAMINE
mf: $C_4H_{11}NO$ mw: 89.16
PROP: A liquid. D: 0.867 @ 20°/0°, mp: −8°, bp: 133°. Sol in water.
SYNS: DEHA □ N,N-DIETHYLHYDROXYLAMINE
CONSENSUS REPORTS: Reported in EPA TSCA Inventory. EPA Genetic Toxicology Program.
SAFETY PROFILE: Poison by skin contact. Moderately toxic by ingestion and intraperitoneal routes. Experimental reproductive effects. Human mutation data reported. When heated to decomposition it emits toxic fumes of NO_x. See also AMINES.

DJN750   CAS: 96-22-0   HR: 3
DIETHYL KETONE
DOT: UN 1156
mf: $C_5H_{10}O$ mw: 86.15
PROP: Colorless, mobile liquid; acetone-like odor. Mp: −42°, bp: 101°, flash p: 55°F, d: 0.8159 @ 19°/4°, vap d: 2.96, autoign temp: 842°F, lel: 1.6%. Mod sol in water; misc in alc and ether.
SYNS: DEK □ DIETHYLCETONE (FRENCH) □ DIMETHYLACETONE □ METACETONE □ METHACETONE □ PENTANONE-3 □ 3-PENTANONE □ PROPIONE
CONSENSUS REPORTS: Reported in EPA TSCA Inventory.
OSHA PEL: TWA 200 ppm
ACGIH TLV: TWA 200 ppm; STEL 300 ppm
DOT CLASSIFICATION: 3; Label: Flammable Liquid
SAFETY PROFILE: Moderately toxic by ingestion, intraperitoneal, and intravenous routes. A skin and eye irritant. Mutation data reported. Dangerous fire hazard when exposed to heat or flame; can react vigorously with oxidizing materials. To fight fire, use alcohol foam, foam, CO_2, dry

chemical. Reacts with hydrogen peroxide + nitric acid to form a shock- and heat-sensitive explosive peroxide. When heated to decomposition it emits acrid smoke and irritating fumes. See also KETONES.

DJO000   CAS: 50-37-3   HR: 3
N,N-DIETHYLLYSERGAMIDE
mf: $C_{20}H_{25}N_3O$ mw: 323.48
SYNS: ACID □ CUBES □ DELYSID □ 9,10-DIDE-HYDRO-N,N-DIETHYL-6-METHYL-ERGOLINE-8-β-CARBOXAMIDE □ HEAVENLY BLUE □ LSD □ d-LSD □ LSD-25 □ LYSERGAMID □ LYSERGAURE DIETHYL-AMID □ d-LYSERGIC ACID DIETHYLAMIDE □ LYSERGIC ACID DIETHYLAMIDE-25 □ LYSERGIDE □ LYSERGSAEUREDIAETHYLAMID □ PEARLY GATES □ ROYAL BLUE □ WEDDING BELLS
CONSENSUS REPORTS: EPA Genetic Toxicology Program.
SAFETY PROFILE: Poison by ingestion, subcutaneous, intraperitoneal, and intravenous routes. Mutation data reported. Human systemic effects by ingestion and intramuscular routes: euphoria, hallucinations, distorted perceptions, excitement, anorexia, nausea and vomiting. An experimental teratogen. Other experimental reproductive effects. Mutation data reported. A much-abused hallucinogen. A federally regulated substance. When heated to decomposition it emits toxic fumes of NO_x.

DJO100   CAS: 557-18-6   HR: 3
DIETHYL MAGNESIUM
mf: $C_4H_{10}Mg$ mw: 82.43
PROP: A solid. Mp: 176° (decomp).
CONSENSUS REPORTS: Reported in EPA TSCA Inventory.
SAFETY PROFILE: Ignites on contact with moist air, water, or carbon dioxide. See also MAGNESIUM COMPOUNDS.

DJO400   CAS: 627-44-1   HR: 3
DIETHYL MERCURY
mf: $C_4H_{10}Hg$ mw: 258.73
PROP: Colorless liquid; hazel-like odor. Bp: 159°, d: 2.43 @ 20°. Sol in Et_2O; less sol in EtOH.

CONSENSUS REPORTS: Reported in EPA TSCA Inventory. Mercury and its compounds are on the Community Right-To-Know List.

OSHA PEL: TWA 0.01 mg(Hg)/m³; STEL 0.03 mg/m³ (skin)

ACGIH TLV: TWA 0.01 mg(Hg)/m³; BEI: 35 μg/g creatinine total inorganic mercury in urine preshift; 15 μg/g creatinine total inorganic mercury in blood at end of shift at end of workweek.

DFG MAK: Confirmed Animal Carcinogen with Unknown Relevance to Humans

SAFETY PROFILE: A deadly human poison by inhalation. Poison by ingestion and intraperitoneal routes. An experimental teratogen. See also MERCURY COMPOUNDS, ORGANIC. Flammable when exposed to heat or flame; can react with oxidizing materials. When heated to decomposition or on contact with acid or acid fumes it emits highly toxic fumes of Hg.

DJP520   CAS: 22936-17-0   HR: 3
o,o-DIETHYL o-(4-(1-(((((1-METHYL-ETHYL)AMINO)CARBONYL)OXY)IMI NO)ETHYL) PHENYL) PHOSPHORO-THIOIC ACID ESTER

mf: $C_{16}H_{25}N_2O_5PS$ mw: 388.46

SYN: ACETOPHENONE, 4'-HYDROXY-, o-(ISOPROPYLCARBAMOYL)OXIME, o-ESTER WITH o,o-DIETHYL PHOSPHOROTHIO

SAFETY PROFILE: A poison by ingestion. When heated to decomposition it emits toxic vapors of NO_x, SO_x, and PO_x.

DJP600   CAS: 50285-72-8   HR: 3
1,1-DIETHYL-3-METHYL-3-NITROSO-UREA

mf: $C_6H_{13}N_3O_2$ mw: 159.22

SYNS: NITROSO-1,1-DIETHYL-3-METHYLUREA □ NITROSOMETHYLDIAETHYLHARNSTOFF □ NITROSOMETHYLDIETHYLUREA □ 1-NITROSO-1-METHYL-3,3-DIETHYLUREA

SAFETY PROFILE: Poison by subcutaneous route. Questionable carcinogen with experimental carcinogenic and tumorigenic data. Mutation data reported. When heated to decomposition it emits toxic fumes of NO_x. See also N-NITROSO COMPOUNDS.

DJQ300   CAS: 7310-87-4   HR: 3
3,3'-DIETHYL-9-METHYLSELENO-CARBOCYANINE IODIDE

mf: $C_{22}H_{23}N_2Se_2 \bullet I$ mw: 600.29

SYN: BENZOSELENAZOLIUM, 3-ETHYL-2-(3-(3-ETHYL-2-BENZOSELENAZOLINYLIDENE)-2-METHYLPROPENYL)-, IODIDE

OSHA PEL: TWA 0.2 mg(Se)/m³

ACGIH TLV: TWA 0.2 mg(Se)/m³; Proposed: (inhalable fraction) 0.1 mg/m³; Not Classifiable as a Human Carcinogen)

SAFETY PROFILE: Poison by intravenous route. When heated to decomposition it emits toxic fumes of NO_x, Se, and I⁻.

DJS200   CAS: 59-26-7   HR: 3
N,N-DIETHYLNICOTINAMIDE

mf: $C_{10}H_{14}N_2O$ mw: 178.26

PROP: Pale yellow oil or crystals with sltly bitter taste. Mp: 24–26°, bp: 296–300° (decomp).

SYNS: ANACARDONE □ ANACORDONE □ ASTROCAR □ BETAPYRIMIDUM □ CAMPHOZONE □ CARBAMIDAL □ CARDAMINE □ CARDIAGEN □ CARDIAMID □ CARDIAMINA □ CARDIAMINE □ CARDIMON □ CITOCOR □ CORACON □ CORAETHAMIDE □ CORAETHAMIDUM □ CORALEPT □ CORAMINE □ CORAVITA □ CORAZONE □ CORDIAMID □ CORDIAMIN □ CORDIAMINE □ CORDITON □ CORDYNIL □ COREDIOL □ CORESPIN □ CORETHAMIDE □ CORETONE □ CORMED □ CORMID □ CORMOTYL □ CORNOTONE □ COROTONIN □ COROVIT □ CORVITAN □ CORVITOL □ CORVITONE □ CORYWAS □ DANAMINE □ DIAETHYL-NICOTINAMID (GERMAN) □ DIETHYL-NICOTAMIDE □ N,N-DIETHYL-3-PYRIDINE-CARBOXAMIDE □ DIETILAMIDE-CARBOPIRIDINA □ DINACORYL □ DYNACORYL □ DYNAMICARDE □ ELITONE □ EUCORAN □ HANSACOR □ INICARDIO □ KARDIAMID □ KARDONYL □ KORDIAMIN □ LEPTAMIN □ MEDIAMID □ NIAMINE □ NICAMIDE □ NICETAMIDE □ NICETHAMIDE □ NICOR □ NICORDAMIN □ NICORINE □ NICORYL □ NICOTINIC ACID DIETHYLAMIDE □ NIKARDIN □ NIKETAMID □ NIKETHAROL □ NIKETHYL □ NIKETILAMID □ NIKORIN □ NIQUETAMIDA □ NISETAMIDE □ PERCORAL □ PROCARDINE □ PROCORMAN □ PYRICAROYL □ PYRIDINE-3-CARBOXYDIETHYL-AMIDE □ PYRIDINE-3-CARBOXYLIC ACID DIETHYLAMIDE □ REFORMIN □ REHORMIN □

SALVACARD □ SALVACORIN □ SOLYACORD □ STELLAMINE □ STIMINOL □ STIMULIN □ TONOCARD □ TONOCOR □ VASAZOL □ VENTRAMINE

SAFETY PROFILE: Poison by ingestion, intravenous, intraperitoneal, and subcutaneous routes. When heated to decomposition it emits toxic fumes of NO_x.

DJT200   CAS: 95-92-1   HR: 3
DIETHYL OXALATE
DOT: UN 2525
mf: $C_6H_{10}O_4$ mw: 146.16
PROP: Colorless, oily, aromatic liquid; decomp in water. Mp: −40.6°, bp: 185.4°, flash p: 168°F (OC), d: 1.079 @ 20°/4°, vap d: 5.04. Sltly sol in H_2O.
SYNS: DIETHYL ETHANEDIOATE □ ETHYL OXALATE □ ETHYL OXALATE (DOT) □ OXALIC ACID, DIETHYL ESTER
CONSENSUS REPORTS: Reported in EPA TSCA Inventory.
DOT CLASSIFICATION: 6.1; Label: KEEP AWAY FROM FOOD
SAFETY PROFILE: Poison by ingestion. Flammable liquid when exposed to heat or flame; can react with oxidizing materials. To fight fire, use foam, CO_2, dry chemical. When heated to decomposition it emits acrid smoke and fumes. See also OXALATES and ESTERS.

DJT400   CAS: 702-54-5   HR: 3
5,5-DIETHYL-1,3-OXAZIN-2,4-DIONE
mf: $C_8H_{13}NO_3$ mw: 171.22
PROP: Crystals from Et_2O. Mp: 97–98°.
SYNS: DIETADIONE (ITALIAN) □ DIETHADION □ DIETHADIONE □ 5,5-DIETHYLDIHYDRO-2H-1,3-OXAZINE-2,4(3H)-DIONE □ 5,5-DIETHYL-1,3-OXAZINE-2,4-DIONE □ 5,5-DIETHYLTETRAHYDRO-2H-1,3-OXAZINE-2,4(3H)-DIONE □ 5,5-DIETILDIIDRO-1,3-OSSAZIN-2,4-DIONE (ITALIAN) □ DIETROXINE □ DIHYDRO-5,5-DIETHYL-2H-1,3-OXAZINE-2,4(3H)-DIONE □ DIIDRO-5,5-DIETIL-2H-1,3-OSSAZIN-2,4(3H)-DIONE (ITALIAN) □ DIOXONE □ L 1811 □ LEDOSTEN □ LEPTON □ PERSISTEN □ TOCE □ TOCEN
CONSENSUS REPORTS: Reported in EPA TSCA Inventory.
SAFETY PROFILE: Poison by ingestion, intravenous, intraperitoneal, subcutaneous, and intramuscular routes. An analeptic

(central nervous system stimulant). When heated to decomposition it emits toxic fumes of NO_x.

DJT800   CAS: 514-73-8   HR: 3
3,3'-DIETHYLPENTAMETHINE-
   THIACYANINE IODIDE
mf: $C_{23}H_{24}N_2S_2 \cdot I$ mw: 519.51
PROP: Green needles from MeOH. Mp: 248° (decomp).
SYNS: ABMINTHIC □ ANELMID □ ANGUIFUGAN □ COMPOUND 01748 □ DEJO □ DELVEX □ DIETHYL-THIADICARBOCYANINE IODIDE □ 3,3'-DIETHYL-THIADICARBOCYANINE IODIDE □ DILO-MBRIN □ DITHIAZANINE IODIDE □ DITHIAZANIN IODIDE □ DITHIAZININE □ EASTMAN 7663 □ 3-ETHYL-2-(5-(3-ETHYL-2-BENZOTHIAZOLIN-YLIDENE)-1,3-PENTA-DIENYL)BENZOTHIAZOLIUM IODIDE □ L-01748 □ NETOCYD □ NK 136 □ OMNI-PASSIN □ PARTEL □ TELMICID □ TELMID □ TELMIDE □ VERCIDON
CONSENSUS REPORTS: EPA Extremely Hazardous Substances List. Reported in EPA TSCA Inventory.
ACGIH TLV: Proposed: (inhalable fraction) 0.1 mg/m³; Not Classifiable as a Human Carcinogen)
SAFETY PROFILE: Poison by ingestion, intraperitoneal, and intravenous routes. When heated to decomposition it emits very toxic fumes of I^-, SO_x, and NO_x. See also IODIDES.

DJU200   CAS: 512-48-1   HR: 3
2,2-DIETHYL-4-PENTENAMIDE
mf: $C_9H_{17}NO$ mw: 155.27
PROP: A solid. Mp: 75–76°.
SYNS: DIAETHYLALLYLACETAMIDE (GERMAN) □ EPINOVAL □ NOVONAL
SAFETY PROFILE: Poison to humans by ingestion. An experimental poison by ingestion, rectal, and intraperitoneal routes. Human systemic effects by ingestion: muscle spasms, cardiac arrhythmias, and respiratory depression. When heated to decomposition it emits toxic fumes of NO_x.

DJU600   CAS: 14666-78-5   HR: 3
DIETHYL PEROXYDICARBONATE
mf: $C_6H_{10}O_6$ mw: 178.14

SYNS: DIETHYL PEROXYDICARBONATE, >27% in solution (DOT) ☐ DIETHYL PEROXYDIFORMATE ☐ ETHYL PEROXYCARBONATE ☐ PEROXYDICARBONIC ACID, DIETHYL ESTER

DOT CLASSIFICATION: Forbidden

SAFETY PROFILE: The impure material is a powerful explosive extremely sensitive to heat or impact. When heated to decomposition it emits acrid smoke and fumes. See also PEROXIDES.

DJV200   CAS: 93-05-0   HR: 3
DIETHYL-p-PHENYLENEDIAMINE
mf: $C_{10}H_{16}N_2$ mw: 164.28
PROP: A liquid. Bp: 260–262°.
SYNS: p-AMINODIETHYLANILINE ☐ p-(DIETHYL-AMINO)ANILINE ☐ 4-(DIETHYLAMINO)ANILINE ☐ N,N'-DIETHYL-p-FENYLENDIAMIN ☐ N,N'-DIETHYL-p-PHENYLENEDIAMINE ☐ DIETHYL-PARAPHENYL-ENEDIAMINE ☐ DPD

CONSENSUS REPORTS: Reported in EPA TSCA Inventory.

SAFETY PROFILE: Poison by ingestion, skin contact, subcutaneous, and intravenous routes. Human systemic skin effects by skin contact: hemorrhage, allergic dermatitis, and primary irritation. Mutation data reported. When heated to decomposition it emits toxic fumes of NO_x. See also AMINES.

DJV800   CAS: 64036-46-0   HR: 3
DIETHYL PHENYLTIN ACETATE
mf: $C_{12}H_{18}O_2Sn$ mw: 312.99
SYN: ACETOXYDIETHYLPHENYLSTANNANE
OSHA PEL: TWA 0.1 mg(Sn)/m³ (skin)
ACGIH TLV: TWA 0.1 mg(Sn)/m³; STEL 0.2 mg(Sn)/m³ (skin).
NIOSH REL: (Organotin Compounds) TWA 0.1 mg(Sn)/m³
SAFETY PROFILE: Poison by ingestion. See also TIN COMPOUNDS. When heated to decomposition it emits acrid and irritating fumes.

DJW600   CAS: 2524-04-1   HR: 3
O,O-DIETHYLPHOSPHORO-CHLOR-IDOTHIOATE
DOT: UN 2751
mf: $C_4H_{10}ClO_2PS$ mw: 188.62

PROP: A liquid. D: 1.202 @ 20°/4°, bp: 94–96° @ 20 mm.
SYNS: CHLORO-PHOSPHONOTHIOIC ACID-O,O-DIETHYL ESTER ☐ DIETHYLCHLOROTHIO-PHOSPHATE ☐ DIETHYLCHLORTHIOFOSFAT (CZECH) ☐ DIETHYLTHIOPHOSPHORYL CHLORIDE (DOT)
CONSENSUS REPORTS: Reported in EPA TSCA Inventory.
DOT CLASSIFICATION: 8; Label: Corrosive
SAFETY PROFILE: Poison by inhalation and skin contact. Moderately toxic by ingestion. Corrosive. Probably a severe eye and skin irritant. See also ESTERS. When heated to decomposition it emits very toxic fumes of Cl⁻, PO_x, and SO_x.

DJW890   CAS: 22941-96-4   HR: 3
4-(o-(o,o-DIETHYLPHOSPHOROTHIO-YL))BENZALDOXIMINO-N-METHYL-CARBAMATE
mf: $C_{13}H_{19}N_2O_5PS$ mw: 346.37
SYNS: PHOSPHOROTHIOIC ACID, o,o-DIETHYL o-(4-(((((METHYLAMINO)CARBONYL)OXY)IMINO)METHYL)PHENYL)ESTER ☐ R 14487 ☐ STAUFFER R 14487
SAFETY PROFILE: A poison by ingestion. When heated to decomposition it emits toxic vapors of NO_x, PO_x, and SO_x.

DJX000   CAS: 84-66-2   HR: 3
DIETHYL PHTHALATE
mf: $C_{12}H_{14}O_4$ mw: 222.26
PROP: Clear, colorless liquid. Mp: −0.3°, bp: 298°, flash p: 325°F (OC), d: 1.110, vap d: 7.66.
SYNS: ANOZOL ☐ 1,2-BENZENEDICARBOXYLIC ACID, DIETHYL ESTER ☐ DIETHYL-o-PHTHALATE ☐ ESTOL 1550 ☐ ETHYL PHTHALATE ☐ NCI-C60048 ☐ NEANTINE ☐ PALATINOL A ☐ PHTHALIC ACID, DIETHYL ESTER ☐ PHTHALOL ☐ PHTHAL-SAEUREDIAETHYLESTER (GERMAN) ☐ PLACIDOL E ☐ RCRA WASTE NUMBER U088 ☐ SOLVANOL
CONSENSUS REPORTS: Reported in EPA TSCA Inventory.
OSHA PEL: TWA 5 mg/m³
ACGIH TLV: TWA 5 mg/m³; Not Classifiable as a Carcinogen
SAFETY PROFILE: Poison by intravenous route. Moderately toxic by ingestion, subcutaneous, and intraperitoneal routes.

Human systemic effects by inhalation: lachrymation, respiratory obstruction, and other unspecified respiratory system effects. An eye irritant and systemic irritant by inhalation. An experimental teratogen. Other experimental reproductive effects. Narcotic in high concentrations. Combustible when exposed to heat or flame. To fight fire, use water spray, mist, foam. When heated to decomposition it emits acrid smoke and irritating fumes.

DJY000   CAS: 21600-43-1   HR: 3
3,3-DIETHYL-1-(m-PYRIDYL)TRIAZENE
mf: $C_9H_{14}N_4$ mw: 178.27

SYNS: PYDT □ 1-(PYRIDYL-3-)-3,3-DIAETHYL-TRIAZEN (GERMAN) □ m-PYRIDYL-DIETHYL-TRIAZENE □ 1-PYRIDYL-3,3-DIETHYLTRIAZENE □ 1-(PYRIDYL-3)-3,3-DIETHYLTRIAZENE □ 1-(3-PYRIDYL)-3,3-DIETHYLTRIAZENE

SAFETY PROFILE: Poison by ingestion and subcutaneous routes. Questionable carcinogen with experimental carcinogenic, neoplastigenic, tumorigenic, and teratogenic data. Human mutation data reported. A transplacental carcinogen. When heated to decomposition it emits toxic fumes of NO_x.

DJY200   CAS: 13593-03-8   HR: 3
O,O-DIETHYL-O-2-QUINOXALYLTHIO-PHOSPHATE
mf: $C_{12}H_{15}N_2O_3PS$ mw: 298.32

PROP: Crystals. Mp: 31–32°. Bp: 142° (decomp) @ 0.0003 mm.

SYNS: BAY 5821 □ BAY 77049 □ BAYRUSIL □ CHINALPHOS □ O,O-DIAETHYL-O-(CHINOXALYL-(2))-MONOTHIOPHOSPHAT (GERMAN) □ DIETHQUIN-ALPHION □ DIETHQUINALPHIONE □ O,O-DIETHYL-O-(2-CHINOXALYL)PHOSPHOROTHIOATE □ O,O-DIETHYL-O-QUINOXALIN-2-YL PHOSPHOROTHIOATE □ O,O-DIETHYL-O-(2-QUINOXALINYL) PHOSPHORO-THIOATE □ O,O-DIETHYL-O-(2-QUINOXALYL) PHOSPHOROTHIOATE □ EKALUX □ ENT 27,394 □ NSC-190986 □ QUINALPHOS □ SAN 6538 I □ SANDOZ 6538 □ SPENCER S-6538 □ SRA 7312 □ WIE OBEN

SAFETY PROFILE: Poison by ingestion, inhalation, skin contact, parenteral, and intraperitoneal routes. Experimental reproductive effects. Mutation data reported. An insecticide. When heated to

decomposition it emits very toxic fumes of NO_x, PO_x, and SO_x.

DJY600   CAS: 110-40-7   HR: 1
DIETHYL SEBACATE
mf: $C_{14}H_{26}O_4$ mw: 258.40

PROP: Colorless to sltly yellow liquid; faint fruity odor. D: 0.960–0.965, refr index: 1.435. Misc with alc, ether, other org solvs, fixed oils; insol in water @ 302°.

SYNS: DIETHYL DECANEDIOATE □ DIETHYL-1,10-DECANEDIOATE □ ETHYL SEBACATE □ FEMA No. 2376 □ SEBACIC ACID, DIETHYL ESTER

CONSENSUS REPORTS: Reported in EPA TSCA Inventory.

SAFETY PROFILE: Mildly toxic by ingestion. A skin irritant. See also ESTERS. When heated to decomposition it emits acrid smoke and irritating fumes.

DJY800   CAS: 5117-17-9   HR: 3
N,N-DIETHYLSELENOUREA
mf: $C_5H_{12}N_2Se$ mw: 179.15

SYNS: 1,1-DIETHYL-2-SELENOUREA □ USAF B-100

CONSENSUS REPORTS: Reported in EPA TSCA Inventory. Community Right-To-Know List.

OSHA PEL: TWA 0.2 mg(Se)/m^3
ACGIH TLV: TWA 0.2 mg(Se)/m^3
DFG MAK: 0.1 mg(Se)/m^3

SAFETY PROFILE: Poison by intraperitoneal route. See also SELENIUM COMPOUNDS. When heated to decomposition it emits very toxic fumes of NO_x and Se.

DKA600   CAS: 56-53-1   HR: 3
DIETHYLSTILBESTEROL
mf: $C_{18}H_{20}O_2$ mw: 268.38

PROP: Small crystals or plates from EtOAc or C_6H_6. Mp: 171–172°.

SYNS: ACNESTROL □ AGOSTILBEN □ ANTIGESTIL □ BIO-DES □ 3,4-BIS(p-HYDROXYPHENYL)-3-HEXENE □ BUFON □ CLIMATERINE □ COMESTROL □ COMESTROL ESTROBENE □ CYREN □ DAWE'S DESTROL □ DEB □ DES (synthetic estrogen) □ DESMA □ DESTROL □ DIASTYL □ DIBESTROL □ DICORVIN □ DI-ESTRYL □ trans-4,4'-(1,2-DIETHYL-1,2-ETHENEDI-YL)BISPHENOL □ 4,4'-(1,2-DIETHYL-1,2-ETHENEDIYL)-BIS-PHENOL □ α,α'-DIETHYLSTIL-BENEDIOL □ α,α'-

DIETHYL-(E)-4,4'-STILBENEDIOL □ α,α'-DIETHYL-4,4'-STILBENEDIOL □ trans-α,α'-DIETHYL-4,4'-STILBENEDIOL □ 2,2'-DIETHYL-4,4'-STILBENEDIOL □ trans-DIETHYLSTILBESTEROL □ DIETHYLSTILBESTROL □ trans-DIETHYLSTILBESTROL □ DIETHYLSTILBOESTEROL □ trans-DIETHYLSTILBOESTEROL □ DIETILESTILBESTROL (SPANISH) □ 4,4'-DIHYDROXY-DIETHYLSTILBENE □ 4,4'-DIHYDROXY-α,β-DIETHYL-STILBENE □ 3,4'(4,4'-DIHYDROXYPHENYL)HEX-3-ENE □ DISTILBENE □ DOMESTROL □ DYESTROL □ ESTILBEN □ ESTRIL □ ESTROBENE □ ESTROGEN □ ESTROMENIN □ ESTROSYN □ FOLLIDIENE □ FONATOL □ GRAFESTROL □ GYNOPHARM □ HIBESTROL □ IDROESTRIL □ ISCOVESCO □ MAKAROL □ MENOSTILBEEN □ MICREST □ MICROEST □ MILESTROL □ NEO-OESTRANOL 1 □ NSC-3070 □ OEKOLP □ OESTROGENINE □ OESTROL VETAG □ OESTROMENIN □ OESTROMENSIL □ OESTROMENSYL □ OESTROMIENIN □ OESTROMON □ PABESTROL □ PALESTROL □ PERCUTATRINE OESTROGENIQUE ISCOVESCO □ PROTECTONA □ RCRA WASTE NUMBER U089 □ RUMESTROL 1 □ RUMESTROL 2 □ SEDESTRAN □ SERRAL □ SEXOCRETIN □ SIBOL □ SINTESTROL □ STIBILIUM □ STIL □ STILBESTROL □ STILBESTRONE □ STILBETIN □ STILBOEFRAL □ STILBOESTROFORM □ STILBOESTROL □ STILBOFOLLIN □ STILBOL □ STILKAP □ STIL-ROL □ SYNESTRIN □ SYNTHO-ESTRIN □ SYNTHOFOLIN □ SYNTOFOLIN □ TAMPOVAGAN STILBOESTROL □ TYLOSTERONE □ VAGESTROL

CONSENSUS REPORTS: NTP 10th Report on Carcinogens. IARC Cancer Review: Group 1 IMEMDT 7,273,87; Human Limited Evidence IMEMDT 6,55,74; IMEMDT 21,173,79; Animal Sufficient Evidence IMEMDT 21,173,79; IMEMDT 6,55,74. EPA Genetic Toxicology Program. Reported in EPA TSCA Inventory.

SAFETY PROFILE: Confirmed carcinogen producing skin, liver, and lung tumors in exposed humans as well as uterine and other reproductive system tumors in the female offspring of exposed women. Experimental carcinogenic, neoplastigenic, tumorigenic, and teratogenic data. A transplacental carcinogen. A human teratogen by many routes. Poison by intraperitoneal and subcutaneous routes. It causes glandular system effects by skin contact. Human reproductive effects by ingestion: abnormal spermatogenesis; changes in testes, epididymis, and sperm duct; menstrual cycle changes or disorders; changes in female fertility; unspecified maternal effects; developmental abnormalities of the fetal urogenital system; germ cell effects in offspring; and delayed effects in newborn. Implicated in male impotence and enlargement of male breasts. Other experimental reproductive effects. Mutation data reported. When heated to decomposition it emits acrid smoke and fumes. See also ETHINYL ESTRADIOL.

DKB000   CAS: 130-80-3   HR: 3
DIETHYLSTILBESTROL DIPROPION-ATE

mf: $C_{24}H_{28}O_4$ mw: 380.52

PROP: Crystals or plates from MeOH. Mp: 104°.

SYNS: CLINESTROL □ CYREN B □ DESD □ DIBESTIL □ trans-4,4'-(1,2-DIETHYL-1,2-ETHENEDI-YL)BISPHENOL DIPROPIONATE □ α,α'-DIETHYL-4,4'-STILBENEDIOL, DIPROPIONATE □ α,α'-DIETHYL-4,4'-STILBENEDIOL trans-DIPROPIONATE □ trans-α,α'-DIETHYL-4,4'-STILBENEDIOL DIPROPIONATE □ α,α'-DIETHYL-4,4'-STILBENEDIOL DIPROPIONYL ESTER □ DIETHYLSTILBENE DIPROPIONATE □ DIETHYLSTILBESTEROL DIPROPIONATE □ DIETHYLSTILBESTROL PROPIONATE □ DIHYDROXYDIETHYLSTILBENE DIPROPIONATE □ 4,4'-DIHYDROXY-α,β-DIETHYLSTILBENE DIPROPIONATE □ DIPROPIONATO de ESTILBENE (SPANISH) □ p,p'-DIPROPIONOXY-trans-α,β-DIETHYLSTILBENE □ DISTILBENE □ ESTILBEN □ ESTILBIN □ ESTROBEN □ ESTROBENE □ ESTROGENIN □ ESTROSTILBEN □ EUVESTIN □ GYNOLETT □ HORFEMINE □ NEO-OESTRANOL II □ OESTROGYNAEDRON □ ORESTOL □ PABESTROL □ SINCICLAN □ STILBESTROL DIETHYL DIPROPIONATE □ STILBESTROL DIPROPIONATE □ STILBESTROL PROPIONATE □ STILBESTRONATE □ STILBOESTROL DIPROPIONATE □ STILBOFAX □ STILRONATE □ SYNESTRIN □ SYNOESTRON □ SYNTESTRIN □ SYNTESTRINE □ WILLESTROL

CONSENSUS REPORTS: IARC Cancer Review: Animal Sufficient Evidence IMEMDT 21,173,79. EPA Genetic Toxicology Program.

SAFETY PROFILE: Confirmed carcinogen with experimental tumorigenic data. An experimental teratogen. Other experimental reproductive effects. Human mutagenic data. When heated to decomposition it emits acrid smoke and irritating fumes. See also DIETHYLSTILBESTEROL.

DKB110   CAS: 64-67-5   HR: 3
DIETHYL SULFATE
DOT: UN 1594
mf: $C_4H_{10}O_4S$ mw: 154.20
PROP: Colorless, oily liquid; faint ethereal odor. Mp: $-25°$, bp: 209.5° (decomp to ethyl ether), flash p: 220°F (CC), d: 1.18 @ 18°/0°, autoign temp: 817°F, vap press: 1 mm @ 47.0°, vap d: 5.31. Insol in water; decomp by hot water; misc with alc and ether. Insol in water.
SYNS: DIAETHYLSULFAT (GERMAN) □ DIETHYLESTER KYSELINY SIROVE □ DIETHYL ESTER SULFURIC ACID □ DIETHYL TETRAOXOSULFATE □ DS □ ETHYL SULFATE
CONSENSUS REPORTS: NTP 10th Report on Carcinogens. IARC Cancer Review: Group 2A IMEMDT 7,198,87; Animal Sufficient Evidence IMEMDT 4,277,74. EPA Genetic Toxicology Program. Community Right-To-Know List. Reported in EPA TSCA Inventory.
DFG MAK: DFG TRK: Animal Carcinogen, Human Suspected Carcinogen
DOT CLASSIFICATION: 6.1; Label: Poison
SAFETY PROFILE: Confirmed carcinogen with experimental carcinogenic and tumorigenic data. Poison by inhalation and subcutaneous routes. Moderately toxic by ingestion and skin contact. A severe skin irritant. An experimental teratogen. Mutation data reported. Combustible when exposed to heat or flame; can react with oxidizing materials. Moisture causes liberation of H_2SO_4. Violent reaction with potassium tert-butoxide. Reacts violently with 3,8-dinitro-6-phenylphenanthridine + water. Reaction with iron + water forms explosive hydrogen gas. To fight fire, use alcohol foam, H_2O foam, CO_2, dry chemicals. When heated to decomposition it emits toxic fumes of SO_x. See also SULFATES.

DKB170   CAS: 3513-92-6   HR: 3
o,o-DIETHYL o-TETRAHYDROFURFUR-YL ESTER PHOSPHOROTHIOIC ACID
mf: $C_9H_{19}O_4PS$ mw: 254.31
SYNS: BAYER 19564 □ ENT 23,444 □ PHOSPHORO-THIOIC ACID, o,o-DIETHYL o-((TETRAHYDRO-2-FURANYL)METHYL) ESTER
SAFETY PROFILE: A poison by ingestion. When heated to decomposition it emits toxic vapors of SO_x and PO_x.

DKC400   CAS: 105-55-5   HR: 3
1,3-DIETHYLTHIOUREA
mf: $C_5H_{12}N_2S$ mw: 132.25
PROP: Crystals. Mp: 77°. Sol in H_2O and EtOH.
SYNS: N,N'-DIETHYLTHIOCARBAMIDE □ N,N'-DIETHYLTHIOUREA □ 1,3-DIETHYL-2-THIOUREA □ NCI-C03816 □ PENNZONE E □ THIATE H □ U 15030 □ USAF EK-1803
CONSENSUS REPORTS: NCI Carcinogenesis Bioassay (feed); Clear Evidence: rat NCITR* NCI-CG-TR-149,79; No Evidence: mouse NCITR* NCI-CG-TR-149,79. Reported in EPA TSCA Inventory. EPA Genetic Toxicology Program.
SAFETY PROFILE: Poison by ingestion. Moderately toxic by intraperitoneal route. Questionable carcinogen with experimental carcinogenic data. Mutation data reported. When heated to decomposition it emits very toxic fumes of NO_x and SO_x.

DKC800   CAS: 134-62-3   HR: 3
DIETHYL-m-TOLUAMIDE
mf: $C_{12}H_{17}NO$ mw: 191.30
PROP: A liquid, sol in water, alc, and ether. Bp: 160° @ 19 mm, d: 0.996 @ 20°/4°.
SYNS: AI 3-22542 □ AUTAN □ BAKER'S ANTIFOL □ CHEMFORM □ DEET □ DELPHENE □ m-DELPHENE □ DET □ m-DET □ m-DETA □ DETAMIDE □ DIELTAMID □ N,N-DIETHYL-3-METHYLBENZAMIDE □ DIETHYLTOLUAMIDE □ N,N-DIETHYL-m-TOL-UAMIDE □ ENT 20,218 □ ENT 22,542 □ FLYPEL □ METADELPHENE □ 3-METHYL-N,N-DIETHYLBENZ-AMIDE □ MGK DIETHYLTOLUAMIDE □ NAUGATUCK DET □ OFF □ REPEL □ REPPER-DET □ REPUDIN-SPECIAL □ m-TOLUIC ACID DIETHYLAMIDE

D

CONSENSUS REPORTS: Reported in EPA TSCA Inventory.
SAFETY PROFILE: Poison by intravenous route. Moderately toxic by ingestion and skin contact. Human systemic effects: coma, convulsions, dermatitis, mydriasis (pupillary dilation), nausea or vomiting, stiffness. An eye and skin irritant. Experimental reproductive effects by skin contact. Mutation data reported. Can cause central nervous system disturbances. A pesticide. DEET is the active ingredient in most commercial insect repellents. When heated to decomposition it emits toxic fumes of NO$_x$.

DKD200   CAS: 63980-20-1   HR: 3
DIETHYL TRIAZENE
mf: C$_4$H$_{11}$N$_3$ mw: 101.18
PROP: A liquid. Bp: 84° @ 145 mm.
SYNS: DIETHYL-TRIAZENE □ N,N'-DIETHYL-TRIAZENE □ 1,3-DIETHYLTRIAZENE □ 1,3-DIETHYL-1-TRIAZENE □ 1,3-DIETHYLTRIAZINE □ 1-TRIAZENE, 1,3-DIETHYL-(9CI)
SAFETY PROFILE: Poison by subcutaneous route. Questionable carcinogen with experimental carcinogenic and tumorigenic data. An experimental teratogen. Mutation data reported. When heated to decomposition it emits toxic fumes of NO$_x$.

DKE600   CAS: 557-20-0   HR: 3
DIETHYLZINC
DOT: UN 1366
mf: C$_4$H$_{10}$Zn   mw: 123.51
PROP: Liquid. Mp: −28°, bp: 118°, d: 1.187 @ 18°.
SYNS: ZINC ETHIDE □ ZINC ETHYL (DOT)
CONSENSUS REPORTS: Reported in EPA TSCA Inventory. Zinc and its compounds are on the Community Right-To-Know List.
DOT CLASSIFICATION: 4.2; Label: Spontaneously Combustible
SAFETY PROFILE: Presumed to be a poison. Ignites spontaneously in air. Dangerously flammable by spontaneous chemical reaction in air, or with oxidizing materials. A dangerous explosion hazard.

Explosive reaction with alkenes + diiodomethane, sulfur dioxide. Reacts violently with bromine, water, nitro compounds. Ignites on contact with air, ozone, methanol, or hydrazine. Reacts violently with nonmetal halides (e.g., arsenic trichloride or phosphorus trichloride) to produce pyrophoric triethyl arsine or triethyl phosphine. To fight fire, do not use water, foam, or halogenated extinguishing agents. Use dry materials, such as graphite, sand, etc. When heated to decomposition it emits toxic fumes of ZnO. See also ZINC COMPOUNDS.

DKG100   CAS: 368-68-3   HR: 3
3,4-DIFLUOROBENZENEARSONIC ACID
mf: C$_6$H$_5$AsF$_2$O$_3$ mw: 238.03
SYN: BENZENEARSONIC ACID, 3,4-DIFLUORO-
OSHA PEL: TWA 0.5 mg(As)/m^3
SAFETY PROFILE: Poison by intraperitoneal route. When heated to decomposition it emits toxic fumes of As and F$^-$.

DKG850   CAS: 75-61-6   HR: 1
DIFLUORODIBROMOMETHANE
DOT: UN 1941
mf: CBr$_2$F$_2$ mw: 209.83
PROP: Colorless, heavy liquid. Fp: −141°, bp: 23° @ 24.5 mm, d: 2.288 @ 15°/4°, mp: −110° (−1°). Insol in water. IDLH 2000 ppm.
SYNS: DIBROMODIFLUOROMETHANE □ FREON 12-B2 □ HALON 1202 □ R12B2 (DOT)
CONSENSUS REPORTS: Reported in EPA TSCA Inventory.
OSHA PEL: TWA 100 ppm
ACGIH TLV: TWA 100 ppm
DFG MAK: 100 ppm (870 mg/m^3)
DOT CLASSIFICATION: 9; Label: None
SAFETY PROFILE: Mildly toxic by inhalation. A non-flammable liquid. When heated to decomposition it emits very toxic fumes of Br$^-$ and F$^-$.

DKH200   CAS: 3582-17-0   HR: 3
DIFLUORODIMETHYLSTANNANE
mf: $C_2H_6F_2Sn$ mw: 186.77
PROP: White crystals. Bp: decomp @ <360°. Sol in water.
SYNS: DIMETHYLTIN DIFLUORIDE □ DIMETHYL-TIN FLUORIDE
CONSENSUS REPORTS: Reported in EPA TSCA Inventory.
OSHA PEL: TWA 0.1 mg(Sn)/m³ (skin)
ACGIH TLV: TWA 0.1 mg(Sn)/m³; STEL 0.2 mg(Sn)/m³ (skin).
NIOSH REL: (Organotin Compounds) TWA 0.1 mg(Sn)/m³
SAFETY PROFILE: Poison by intravenous route. See also TIN COMPOUNDS and FLUORIDES. When heated to decomposition it emits toxic fumes of F⁻.

DKI400   CAS: 368-97-8   HR: 3
DIFLUOROPHENYLARSINE
mf: $C_6H_5AsF_2$ mw: 190.03
SYN: PHENYLDIFLUOROARSINE
CONSENSUS REPORTS: Arsenic and its compounds are on the Community Right-To-Know List.
OSHA PEL: TWA 0.5 mg(As)/m³
SAFETY PROFILE: Poison by skin contact and intravenous routes. See also FLUORIDES and ARSENIC COMPOUNDS. When heated to decomposition it emits very toxic fumes of As and F⁻.

DKI600   CAS: 22494-42-4   HR: 3
5-(2,4-DIFLUOROPHENYL)SALICYLIC ACID
mf: $C_{13}H_8F_2O_3$ mw: 250.21
PROP: Crystals from C_6H_6/pet ether. Mp: 210–211°.
SYNS: DIFLUNISAL □ 2',4'-DIFLUORO-4-HYDROXY-(1,1'-BIPHENYL)-3-CARBOXYLIC ACID □ 2',4'-DIFLUORO-4-HYDROXY-3-BIPHENYLCARBOXYLIC ACID □ 2',4'-DIFLUORO-4-HYDROXY-(1',1-DIPHENYL)-3-CARBOXYLIC ACID □ DOLOBID □ DOLOBIL □ DOLOBIS □ FLOVACIL □ FLUNIGET □ 2-(HYDROXY)-5-(2,4-DIFLUOROPHENYL)BENZOIC ACID □ MK 647
SAFETY PROFILE: Poison by ingestion, subcutaneous, and intraperitoneal routes. Human systemic effects by ingestion: tolerance, and cholestatic jaundice (due to the stoppage of the flow of bile), agranulocytosis, increased body temperature. An experimental teratogen. Other experimental reproductive effects. An analgesic and anti-inflammatory agent. When heated to decomposition it emits toxic fumes of F⁻. See also FLUORIDES.

DKL400   CAS: 11024-24-1   HR: 3
DIGITONIN
mf: $C_{56}H_{92}O_{29}$ mw: 1229.48
PROP: Crystals from EtOH. Mp: 235–240°.
SYN: DIGITIN
CONSENSUS REPORTS: Reported in EPA TSCA Inventory.
SAFETY PROFILE: Poison by ingestion, intravenous, intraperitoneal, and subcutaneous routes. Mutation data reported. See also DIGITALIS. When heated to decomposition it emits acrid smoke and irritating fumes.

DKL800   CAS: 71-63-6   HR: 3
DIGITOXIN
mf: $C_{41}H_{64}O_{13}$ mw: 765.05
PROP: A solid. Mp: 256–257°.
SYNS: ACEDOXIN □ ASTHENTHILO □ CARDIDIGIN □ CARDIGIN □ CARDITOXIN □ CRISTAPURAT □ CRYSTALLINE DIGITALIN □ CRYSTODIGIN □ DIGILONG □ DIGIMED □ DIGIMERCK □ DIGISIDIN □ DIGITALIN □ DIGITALINE (FRENCH) □ DIGIT-ALINE CRISTALLISEE □ DIGITALINE NATIVELLE □ DIGITALINUM VERUM □ DIGITOPHYLLIN □ DIGITOXIGENIN-TRIDIGITOXOSID (GERMAN) □ DIGITOXIGENIN TRIDIGITOXOSIDE □ DITAVEN □ GLUCODIGIN □ LANATOXIN □ MONO-GLYCO-COARD □ MYODIGIN □ PURODIGIN □ PURPURID □ TARDIGAL □ TRI-DIGITOXOSIDE (GERMAN) □ UNIDIGIN
CONSENSUS REPORTS: Reported in EPA TSCA Inventory. EPA Extremely Hazardous Substances List.
SAFETY PROFILE: A deadly poison by most routes. Human systemic effects: arrhythmias, cardiomyopathy, EKG changes, nausea or vomiting, paresthesia, pulse rate increase, thrombocytopenia. Human reproductive effects by ingestion: reduced viability of newborn. An eye

irritant. When heated to decomposition it emits acrid smoke and irritating fumes. See also DIGITALIS.

DKM200   CAS: 2238-07-5   HR: 3
DIGLYCIDYL ETHER
mf: $C_6H_{10}O_3$ mw: 130.16
PROP: Liquid. D: 1.126 @ 25°/4°, bp: 98–99° @ 11 mm. IDLH 10 ppm.
SYNS: BIS(2,3-EPOXYPROPYL)ETHER □ DGE □ DI(2,3-EPOXYPROPYL) ETHER
CONSENSUS REPORTS: EPA Extremely Hazardous Substances List. Reported in EPA TSCA Inventory. EPA Genetic Toxicology Program.
OSHA PEL: TWA 0.1 ppm
ACGIH TLV: TWA 0.01 ppm; Not Classifiable as a Human Carcinogen
DFG MAK: 0.1 ppm (0.54 mg/m³); Confirmed Animal Carcinogen with Unknown Relevance to Humans
NIOSH REL: (Glycidyl Ethers) CL 1 mg/m³/15M
SAFETY PROFILE: Suspected carcinogen with experimental tumorigenic data. Poison by ingestion, inhalation, and intravenous routes. Moderately toxic by skin contact. A severe eye and skin irritant. Mutation data reported. Chronic exposure can cause bone marrow depression. When heated to decomposition it emits acrid smoke and fumes. See also ETHERS.

DKN400   CAS: 20830-75-5   HR: 3
DIGOXIN
mf: $C_{41}H_{64}O_{14}$ mw: 781.05
PROP: White, crystalline powder. Crystals from alc (aq). Mp: 265° (decomp). Glycoside isolated from *Digitalis lanata* (JPETAB 52,1,34).
SYNS: CHLOROFORMIC DIGITALIN □ DIGACIN □ DIGITALIS GLYCOSIDE □ DIGOXIGENIN-TRIDIGITOXOSID (GERMAN) □ DIGOXINE □ HOMOLLE'S DIGITALIN □ LANICOR □ LANOXIN □ ROUGOXIN □ SK-DIGOXIN
CONSENSUS REPORTS: EPA Extremely Hazardous Substances List. Reported in EPA TSCA Inventory. EPA Genetic Toxicology Program.

SAFETY PROFILE: A deadly poison by most routes. Human systemic effects by ingestion: anorexia, cardiac arrhythmias, nausea and vomiting, visual field changes, pulse rate decrease, fall in blood pressure. An experimental teratogen. When heated to decomposition it emits acrid and irritating fumes. See also DIGITALIS.

DKO000   CAS: 51622-02-7   HR: 3
DIHEPTYLMERCURY
mf: $C_{14}H_{30}Hg$ mw: 399.03
PROP: IDLH 10 mg/m³ (as Hg).
CONSENSUS REPORTS: Mercury and its compounds are on the Community Right-To-Know List. OSHA PEL: TWA 0.01 mg(Hg)/m³; STEL 0.03 mg/m³ (skin)
ACGIH TLV: TWA 0.01 mg(Hg)/m³; BEI: 35 µg/g creatinine total inorganic mercury in urine preshift; 15 µg/g creatinine total inorganic mercury in blood at end of shift at end of workweek.
DFG MAK: Confirmed Animal Carcinogen with Unknown Relevance to Humans
NIOSH REL: (Mercury, Organo) TWA 0.01 mg/m³; STEL 0.03 mg/m³ (skin)
SAFETY PROFILE: Poison by intraperitoneal route. See also MERCURY COMPOUNDS, ORGANIC. When heated to decomposition it emits toxic fumes of Hg.

DKO600   CAS: 143-16-8   HR: 3
DIHEXYLAMINE
mf: $C_{12}H_{27}N$ mw: 185.40
PROP: Liquid or crystals. Mp: 193–195°, flash p: 220°F (OC), d: 0.78, vap d: 6.38.
SYN: DI-N-HEXYLAMINE
CONSENSUS REPORTS: Reported in EPA TSCA Inventory.
SAFETY PROFILE: Poison by ingestion, skin contact, and intravenous routes. A skin and severe eye irritant. Flammable when exposed to heat or flame; can react with oxidizing materials. To fight fire, use CO_2, dry chemical. When heated to decomposition it emits toxic fumes of NO_x. See also AMINES.

DKP600 CAS: 84-75-3 HR: 1
DI-n-HEXYL PHTHALATE
mf: $C_{20}H_{30}O_4$ mw: 334.50
PROP: Liquid. Mp: $-58°$, bp: 210° @ 5 mm, flash p: 350°F, d: 0.995 @ 20°/20°, vap d: 11.5.
SYNS: 1,2-BENZENEDICARBOXYLIC ACID DIHEXYL ESTER □ DIHEXYL PHTHALATE □ PHTHALIC ACID DIHEXYL ESTER
CONSENSUS REPORTS: Reported in EPA TSCA Inventory.
SAFETY PROFILE: Very mildly toxic by ingestion and skin contact. Experimental reproductive effects. A skin and eye irritant. Combustible when exposed to heat or flame; can react with oxidizing materials. To fight fire, use foam, CO_2, dry chemical. See also PHTHALIC ACID and ESTERS.

DKQ000 CAS: 57-41-0 HR: 3
DIHYDANTOIN
mf: $C_{15}H_{12}N_2O_2$ mw: 252.29
PROP: A solid. Mp: 295–298°.
SYNS: ALEVIATIN □ ANTISACER □ AURANILE □ CAUSOIN □ CITRULLAMON □ COMITAL □ CONVUL □ DANTEN □ DANTINAL □ DANTOINAL KLINOS □ DANTOINE □ DENYL □ DIDAN-TDC-250 □ DIFENILHIDANTOINA (SPANISH) □ DIFENIN □ DIFHYDAN □ DIHYCON □ DI-HYDAN □ DILANTIN □ DILANTINE □ DINTOIN □ DIPHANTOIN □ DIPHEDAL □ DIPHENINE □ DIPHENTOIN □ DIPHENYLAN □ DIPHENYLHYDANTOIN □ 5,5-DIPHENYLHYDANTOIN □ DIPHENYLHYDANTOINE (FRENCH) □ 5,5-DIPHENYLIMIDAZOLIDIN-2,4-DIONE □ 5,5-DIPHENYL-2,4-IMIDAZOLIDINEDIONE □ DI-PHETINE □ DITOINATE □ DPH □ EKKO CAPSULES □ ELEPSINDON □ ENKELFEL □ EPAMIN □ EPANUTIN □ EPASMIR "5" □ EPDANTOINE SIMPLE □ EPELIN □ EPIFENYL □ EPIHYDAN □ EPILAN □ EPILANTIN □ EPINAT □ EPISED □ EPTAL □ EPTOIN □ FENANTOIN □ FENIDANTOIN "S" □ FENYLEPSIN □ FENYTOINE □ GEROT-EPILAN-D □ HIDAN □ HIDANTILO □ HIDANTINA SENOSIAN □ HIDANTINA VITORIA □ HIDANTOMIN □ HYDANTAL □ HYDANTOIN □ ICTALIS SIMPLE □ IDANTOIN □ KESSODANTEN □ LABOPAL □ LEHYDAN □ LEPITOIN □ LEPSIN □ MINETOIN □ NCI-C55765 □ NEOS-HIDANTOINA □ NOVANTOINA □ OM-HYDANTOINE □ OXYLAN □ PHANANTIN □ PHENATOINE □ RITMENAL □ SACERIL □ SANEPIL □ SILANTIN □ SODANTON □ SOLANTIN □ SYLANTOIC □ TACOSAL □ THILOPHENYL □ TOIN UNICELLES □ ZENTRONAL □ ZENTROPIL

CONSENSUS REPORTS: NTP 10th Report on Carcinogens. IARC Cancer Review: Group 2B IMEMDT 7,319,87; Human Limited Evidence IMEMDT 13,201,77; Animal Sufficient Evidence IMEMDT 13,201,77. EPA Genetic Toxicology Program.
SAFETY PROFILE: Confirmed carcinogen producing lymphoma, Hodgkin's disease, tumors of the skin and appendages. Experimental carcinogenic and tumorigenic data. A human poison by ingestion. Poison experimentally by ingestion, subcutaneous, intravenous, and intraperitoneal routes. Moderately toxic by an unspecified route. Experimental teratogenic and reproductive effects. Human systemic effects by ingestion: dermatitis, change in motor activity (specific assay), ataxia (loss of muscle coordination), degenerative brain changes, encephalitis, hallucinations, distorted perceptions, irritability, and jaundice. Human teratogenic effects by ingestion: developmental abnormalities of the central nervous system, cardiovascular (circulatory) system, musculoskeletal system, craniofacial area, skin and skin appendages, eye, ear, other developmental abnormalities. Effects on newborn include abnormal growth statistics (e.g., reduced weight gain), physical abnormalities, other postnatal measures or effects, and delayed effects. Human mutation data reported. A drug for the treatment of grand mal and psychomotor seizures. When heated to decomposition it emits toxic fumes of NO_x.

DKV150 CAS: 619-01-2 HR: 2
DIHYDROCARVEOL
mf: $C_{10}H_{18}O$ mw: 154.28
PROP: Colorless, oily liquid; spearmint odor. D: 0.921–0.926, refr index: 1.477–1.481, flash p: 153°F. Sol in alc, fixed oils; insol in water.
SYNS: 1,6-DIHYDROCARVEOL □ FEMA No. 2379 □ 8-p-MENTHEN-2-OL □ 6-METHYL-3-ISOPROPYLCYCLOHEXANOL

CONSENSUS REPORTS: Reported in EPA TSCA Inventory.

SAFETY PROFILE: A moderate skin and eye irritant. A combustible liquid. When heated to decomposition it emits acrid smoke and irritating fumes.

DKV160   CAS: 20777-49-5   HR: 1
DIHYDROCARVEYL ACETATE
mf: $C_{12}H_{20}O_2$ mw: 196.32
SYNS: CYCLOHEXANOL, 2-METHYL-5-(1-METHYLETHENYL)-, ACETATE,(1-α-2-β,5α)- (9CI) □ DIHYDROCARVEOL ACETATE □ DIHYDROCARVYL ACETATE □ p-MENTH-8-EN-2-OL, ACETATE □ p-MENTH-8-EN-2-YL ACETATE □ 2-METHYL-5-(1-METHYLETHENYL)CYCLOHEXYL ACETATE
CONSENSUS REPORTS: Reported in EPA TSCA Inventory.
SAFETY PROFILE: A skin irritant. When heated to decomposition it emits acrid smoke and irritating fumes.

DKV175   CAS: 7764-50-3   HR: 2
d-DIHYDROCARVONE
mf: $C_{10}H_{16}O$ mw: 152.26
PROP: Colorless liquid; spearmint-like odor. D: 0.923–0.928, refr index: 1.470–1.474. Sol in alc, fixed oils; insol in water.
SYNS: FEMA No. 3565 □ p-MENTH-8-EN-2-ONE □ 8-p-MENTHEN-2-ONE □ d-2-METHYL-5-(1-METHYLENENYL)-CYCLOHEXANONE
CONSENSUS REPORTS: Reported in EPA TSCA Inventory.
SAFETY PROFILE: Moderately toxic by subcutaneous route. A skin irritant. When heated to decomposition it emits acrid smoke and irritating fumes.

DKW800   CAS: 125-28-0   HR: 3
DIHYDROCODEINE
mf: $C_{18}H_{23}NO_3$ mw: 301.42
PROP: A solid. Mp: 55°.
SYNS: CODHYDRINE □ COHYDRIN □ DEHACODIN □ DF 118 □ DIDRATE □ DIHYDRIN □ 7,8-DIHYDROCODEINE □ DIHYDRONEOPINE □ DROCODE □ HYDROCODIN □ 6-HYDROXY-3-METHOXY-N-METHYL-4,5-EPOXYMORPHINAN □ NADEINE □ NOVICODIN □ PARACODIN □ PARACODINE □ PARZONE □ RAPACODIN

SAFETY PROFILE: Poison by ingestion, intravenous, and subcutaneous routes. Human systemic effects by ingestion and subcutaneous routes: somnolence, miosis (pupillary constriction), and respiratory depression. An analgesic. Can cause drug dependency with repeated doses. When heated to decomposition it emits toxic fumes of NO_x. See also CODEINE.

DLH630   CAS: 113-52-0   HR: 3
10,11-DIHYDRO-5-(3-(DIMETHYL-
    AMINO)PROPYL)-5H-DIBENZ-
    (b,f)AZEPINE HYDROCHLORIDE
mf: $C_{19}H_{24}N_2 \cdot ClH$ mw: 316.91
PROP: Crystals from Me₂CO. Mp: 174–175°.
SYNS: ANTIDEPRIN HYDROCHLORIDE □ BERKOMINE □ CENSTIM □ CENSTIN □ CHIMO-REPTIN □ CHRYTEMIN □ CO CAP IMIPRAMINE 25 □ DEPRINOL □ 10,11-DIHYDRO-N,N-DIMETHYL-5H-DIBENZ(b,f)AZEPINE-5-PROPANAMINE MONO-HYDROCHLORIDE □ 5-(3-DIMETHYLAMINO-PROPYL)-10,11-DIHYDRO-5H-DIBENZ(b,f)AZEPINE HYDROCHLORIDE □ N-(3-DIMETHYLAMINOPROPYL)IMINODIBENZYL HYDROCHLORIDE □ N-(γ-DIMETILAMINOPROPIL)-IMINODIBENZILE CLORIDRATO (ITALIAN) □ DIMIPRESSIN □ DYNA-ZINA □ EFURANOL □ EUPRAMIN □ FEINALMIN □ G 22150 □ G 22355 □ IA-PRAM □ IMAVATE □ IMIDOBENZYLE □ IMIDOL □ IMILANYLE □ IMIPRAMINA (ITALIAN) □ IMIPRAMINE □ IMIPRAMINE HYDROCHLORIDE □ IMIPRAMINE MONOHYDROCHLORIDE □ IMIPRIN □ IMP HYDROCHLORIDE □ INTALPRAM □ IPROGEN □ IRAMIL □ JANIMINE □ LOFEPRAMINE □ MELIPRAM □ MELIPRAMINE □ MELIPRAMINE HYDROCHLORIDE □ MELIPRAMIN HYDROCHLORIDE □ NSC-114900 □ PERSAMINE □ PERTOFRAM □ PRESAMINE □ PROMIBEN □ PYRLEUGAN □ SK-PRAMINE □ SK-PRAMINE HYDROCHLORIDE □ SURPLIX □ TEPERINE □ TIMOLET □ TOFRANIL □ TOFRANILE
CONSENSUS REPORTS: Reported in EPA TSCA Inventory.
SAFETY PROFILE: Human poison by ingestion. An experimental poison by ingestion, intravenous, subcutaneous, and intraperitoneal routes. An experimental teratogen. Human systemic effects by ingestion: sleep, somnolence, convulsions, muscle contraction or spasticity, coma,

blood pressure decrease, dyspnea (difficulty in breathing), paresthesia (abnormal sensations), and kidney changes. Experimental reproductive effects. Mutation data reported. Used in the treatment of depression. When heated to decomposition it emits very toxic fumes of NO_x and HCl. See also DIAZEPAM.

DLH830   CAS: 50673-08-0   HR: 3
2,3-DIHYDRO-2,2-DIMETHYL-7-BENZO-FURANYL (BUTYLTHIO)METHYL-CARBAMATE
mf: $C_{16}H_{23}NO_3S$ mw: 309.46
SYN: CARBAMIC ACID, (BUTYLTHIO)METHYL-, 2,3-DIHYDRO-2,2-DIMETHYL-7-BENZOFURANYL ESTER
SAFETY PROFILE: A poison by ingestion. When heated to decomposition it emits toxic vapors of NO_x and SO_x.

DLH850   CAS: 50802-69-2   HR: 3
2,3-DIHYDRO-2,2-DIMETHYL-7-BENZO-FURANYL((4-FLUOROPHENYL)-THIO)METHYLCARBAMATE
mf: $C_{18}H_{18}FNO_3S$ mw: 347.43
SYN: CARBAMIC ACID, ((4-FLUOROPHENYL)THIO)METHYL-, 2,3-DIHYDRO-2,2-DIMETHYL-7-BENZOFURANYL ESTER
SAFETY PROFILE: A poison by ingestion. When heated to decomposition it emits toxic vapors of NO_x, SO_x, and F^-.

DLH870   CAS: 50802-70-5   HR: 3
2,3-DIHYDRO-2,2-DIMETHYL-7-BENZO-FURANYLMETHYL((4-NITROPHEN-YL)THIO)CARBAMATE
mf: $C_{18}H_{18}N_2O_5S$ mw: 374.44
SYN: CARBAMIC ACID, METHYL((4-NITROPHENYL)-THIO)-, 2,3-DIHYDRO-2,2-DIMETHYL-7-BENZOFURANYL ESTER
SAFETY PROFILE: A poison by ingestion. When heated to decomposition it emits toxic vapors of NO_x and SO_x.

DLH890   CAS: 50673-06-8   HR: 3
2,3-DIHYDRO-2,2-DIMETHYL-7-BENZO-FURANYLMETHYL(1,1,2,2-TETRA-CHLOROETHYL)THIOCARBAMATE
mf: $C_{14}H_{15}Cl_4NO_3S$ mw: 419.16

SYN: CARBAMIC ACID, METHYL((1,1,2,2-TETRACHLOROETHYL)THIO)-, 2,3-DIHYDRO-2,2-DIMETHYL-7-BENZOFURANYL ESTER
SAFETY PROFILE: A poison by ingestion. When heated to decomposition it emits toxic vapors of NO_x, SO_x, and Cl^-.

DLI630   CAS: 326800-80-0 HR: 3
2,5-DIHYDRO-1,2-DIMETHYL-3-(2-NAPHTHALENYL)-1H-PYRROLE, REL-(2R,3R)-2,3-DIHYDROXY-BUTANEDIOATE (1:1)
mf: $C_{16}H_{17}N \cdot C_4H_6O_6$ mw: 373.41
SAFETY PROFILE: A poison by subcutaneous route. When heated to decomposition it emits toxic vapors of NO_x.

DLO400   CAS: 18497-13-7   HR: 3
DIHYDROGEN HEXACHLOROPLATIN-ATE HEXAHYDRATE
mf: $Cl_6Pt \cdot 2H \cdot 6H_2O$ mw: 521.97
SYN: PLATINATE(2-), HEXACHLORO-, DIHYDROGEN, HEXAHYDRATE
OSHA PEL: TWA 0.002 mg(Pt)/m³
ACGIH TLV: TWA 0.002 mg(Pt)/m³
SAFETY PROFILE: Poison by intraperitoneal route. When heated to decomposition it emits toxic fumes of Pt and Cl^-.

DLQ800   CAS: 21842-58-0   HR: 3
12,β,13,α-DIHYDROJERVINE
mf: $C_{26}H_{39}NO_3$ mw: 413.66
SAFETY PROFILE: Poison by ingestion. Experimental teratogenic and reproductive effects. When heated to decomposition it emits toxic fumes of NO_x.

DLS600   CAS: 58-28-6   HR: 3
10,11-DIHYDRO-5-(3-(METHYLAMINO)-PROPYL)-5H-DIBENZ(b,f)AZEPINE HYDROCHLORIDE
mf: $C_{18}H_{22}N_2 \cdot ClH$ mw: 302.88
PROP: A solid. Mp: 214–218°.
SYNS: DESIPRAMINE HYDROCHLORIDE □ DESMETHYLIMIPRAMINE HYDROCHLORIDE □ DIMETHYLIMIPRAMINE HYDROCHLORIDE □ DMI HYDROCHLORIDE □ EX 4355 □ G 35020 □ GMI □ IMIPRAMINEDEMETHYL HYDROCHLORIDE □ IRENE □ JB 8181 □ N-(γ-METHYLAMINOPROPYL)IMINODI-BENZYL HYDROCHLORIDE □ NORPRAMIN □

NORTIMIL □ NSC-114901 □ PERTOFRAN □ PERTOFRANE □ RMI9,384A
SAFETY PROFILE: Poison by ingestion, intraperitoneal, subcutaneous, and intravenous routes. Human systemic effects by ingestion: decreased urine volume, sodium level changes, chlorine level changes. An experimental teratogen. Other experimental reproductive effects. Mutation data reported. When heated to decomposition it emits very toxic fumes of NO_x and HCl.

DLU650 CAS: 301644-18-8 HR: 3
4,5-DIHYDRO-N-(1-METHYLETHYL)-3-(PHENYLAMINO)-2H-BENZ(g)INDAZOLE-2-ACETAMIDE
mf: $C_{22}H_{24}N_4O$ mw: 360.46
SAFETY PROFILE: A poison by ingestion. When heated to decomposition it emits toxic vapors of NO_x.

DLY000 CAS: 146-22-5 HR: 3
1,3-DIHYDRO-7-NITRO-5-PHENYL-2H-1,4-BENZODIAZEPIN-2-ONE
mf: $C_{15}H_{11}N_3O_3$ mw: 281.29
PROP: Yellow crystals from EtOH. Mp: 224–226°.
SYNS: BENZALIN □ CALSMIN □ EATAN □ EPIBENZALIN □ EPINELBON □ EUNOCTIN □ HIPNAX □ HIPSAL □ LA 1 □ MOGADAN □ NELBON □ NEOZEPAM □ NEUCHLONIC □ NITRADOS □ NITRAZEPAM □ NITRENPAX □ 7-NITRO-5-PHENYL-2,3-DIHYDRO-1H-1,4-BENZODIAZEPIN-2-ONE □ NSC-58775 □ PAXISYN □ PELSON □ RADEDORM □ RELACT □ RO 4-5360 □ RO 5-3059 □ SOMNASED □ SOMNIBEL □ SOMNITE □ SONEBON □ SONNOLIN □ SUREM □ UNISOMNIA
CONSENSUS REPORTS: EPA Genetic Toxicology Program.
SAFETY PROFILE: Poison by intraperitoneal and intravenous routes. Moderately toxic by ingestion. Experimental reproductive effects. Mutation data reported. An anticonvulsant and hypnotic agent. When heated to decomposition it emits toxic fumes of NO_x. See also DIAZEPAM.

DMA500 CAS: 301644-21-3 HR: 3
4-((4,5-DIHYDRO-3-(PHENYLAMINO)-2H-BENZ(G)INDAZOL-2-YL)ACETYL)MORPHOLINE
mf: $C_{23}H_{24}N_4O_2$ mw: 388.47
SAFETY PROFILE: A poison by ingestion. When heated to decomposition it emits toxic vapors of NO_x.

DMC000 CAS: 68-94-0 HR: 2
1,7-DIHYDRO-6H-PURIN-6-ONE
mf: $C_5H_4N_4O$ mw: 136.13
PROP: Needles.
SYNS: HYPOXANTHINE □ 9H-PURIN-6-OL □ PURIN-6(3H)-ONE □ 6(1H)-PURINONE
CONSENSUS REPORTS: Reported in EPA TSCA Inventory.
SAFETY PROFILE: Moderately toxic by intraperitoneal route. An experimental teratogen. When heated to decomposition it emits toxic fumes of NO_x.

DMC200 CAS: 110-87-2 HR: 3
DIHYDROPYRAN
mf: C_5H_8O mw: 84.13
PROP: Colorless, mobile liquid; ethereal odor. Bp: 86–87°, flash p: 0°F, d: 0.922 @ 19°/15, vap d: 2.90.
SYNS: Δ^2-DIHYDROPYRAN □ 3,4-DIHYDROPYRAN □ 2H-3,4-DIHYDROPYRAN
CONSENSUS REPORTS: Reported in EPA TSCA Inventory.
SAFETY PROFILE: A flammable and very dangerous fire hazard when exposed to heat or flame; can react vigorously with oxidizing materials. Keep away from heat and open flame. To fight fire, use alcohol foam, CO_2, or dry chemical. When heated to decomposition it emits acrid smoke and irritating fumes.

DMC600 CAS: 123-33-1 HR: 2
1,2-DIHYDROPYRIDAZINE-3,6-DIONE
mf: $C_4H_4N_2O_2$ mw: 112.10
PROP: Crystals. Mp: >300°. Sol in water and alc.
SYNS: BURTOLIN □ CHEMFORM □ DE-CUT □ DE-SPROUT □ 1,2-DIHYDRO-3,6-PYRADIZINEDIONE □ 1,2-DIHYDRO-3,6-PYRIDAZINEDIONE □ DREXEL-SUPER

P ☐ ENT 18,870 ☐ FAIR 30 ☐ FAIR PS ☐ HYDRAZID KYSELINY MALEINOVE ☐ 6-HYDROXY-3(2H)-PYRIDAZINONE ☐ KMH ☐ MAH ☐ MAINTAIN 3 ☐ MALAZIDE ☐ MALEIC ACID HYDRAZIDE ☐ MALEIC HYDRAZIDE ☐ MALEIC HYDRAZIDE 30% ☐ MALEIC HYDRAZINE ☐ MALEIN 30 ☐ MALEINSAEUREHYDR-AZID ☐ N,N-MALEOYLHYDRAZINE ☐ MALZID ☐ MH ☐ MH 30 ☐ MH-40 ☐ MH 36 BAYER ☐ RCRA WASTE NUMBER U148 ☐ REGULOX ☐ REGULOX W ☐ REGULOX 50 W ☐ RETARD ☐ ROYAL MH-30 ☐ ROYAL SLO-GRO ☐ SLO-GRO ☐ SPROUT/OFF ☐ SPROUT-STOP ☐ STUNTMAN ☐ SUCKER-STUFF ☐ SUPER-DE-SPROUT ☐ SUPER SPROUT STOP ☐ SUPER SUCKER-STUFF ☐ SUPER SUCKER-STUFF HC ☐ 1,2,3,6-TETRAHYDRO-3,6-DIOXOPYRIDAZINE ☐ VONDAL-HYDE ☐ VONDRAX

CONSENSUS REPORTS: IARC Cancer Review: Group 3 IMEMDT 7,56,87; Animal Inadequate Evidence IMEMDT 4,173,74. Reported in EPA TSCA Inventory.

SAFETY PROFILE: Moderately toxic by ingestion. Questionable carcinogen with experimental tumorigenic data. Mutation data reported. Can cause chronic liver damage and acute central nervous system effects. When heated to decomposition it emits highly toxic fumes of NO_x. See also HYDRAZINE.

DMD600   CAS: 94-58-6   HR: 3
DIHYDROSAFROLE
mf: $C_{10}H_{12}O_2$ mw: 164.22
PROP: An oily liquid. Bp: 228°, d: 1.0695 @ 20°.
SYNS: 1,2-(METHYLENEDIOXY)-4-PROPYLBENZENE ☐ 5-PROPYL-1,3-BENZODIOXOLE ☐ 4-PROPYL-1,2-METHYLENEDIOXYBENZENE ☐ RCRA WASTE NUMBER U090
CONSENSUS REPORTS: IARC Cancer Review: Group 2B IMEMDT 7,56,87; Animal Sufficient Evidence IMEMDT 10,231,76; Animal Limited Evidence IMEMDT 1,169,72. Reported in EPA TSCA Inventory. EPA Genetic Toxicology Program.
SAFETY PROFILE: Confirmed carcinogen with experimental carcinogenic data. Moderately toxic by ingestion and intraperitoneal routes. A skin irritant. When heated to decomposition it emits acrid smoke and irritating fumes.

DME000   CAS: 128-46-1   HR: 3
DIHYDROSTREPTOMYCIN
mf: $C_{21}H_{41}N_7O_{12}$ mw: 583.69
SYNS: DHMS ☐ DST
CONSENSUS REPORTS: EPA Genetic Toxicology Program.
SAFETY PROFILE: Poison by intravenous and intramuscular routes. Moderately toxic by subcutaneous and intraperitoneal routes. Human teratogenic effects by unspecified route: developmental abnormalities of the eye and ear. An experimental teratogen. Other experimental reproductive effects. Mutation data reported. A derivative of streptomycin; has anesthetic properties. When heated to decomposition it emits toxic fumes of NO_x.

DMG400   CAS: 89-84-9   HR: 2
2′,4′-DIHYDROXYACETOPHENONE
mf: $C_8H_8O_3$ mw: 152.16
PROP: Leaflets or needles. Mp: 147°.
SYNS: 4-ACETYLRESORCINOL ☐ 2,4-DIHYDROXY-ACETOPHENONE ☐ 1-(2,4-DIHYDROXYPHENYL)ETH-ANONE ☐ RESACETOPHENONE ☐ β-RESACETO-PHENONE ☐ RESOACETOPHENONE
CONSENSUS REPORTS: Reported in EPA TSCA Inventory.
SAFETY PROFILE: Moderately toxic by ingestion. Experimental reproductive effects. A severe eye irritant. When heated to decomposition it emits acrid smoke and irritating fumes.

DMH000   CAS: 81-64-1   HR: 3
1,4-DIHYDROXYANTHRAQUINONE
mf: $C_{14}H_8O_4$ mw: 240.22
PROP: Red crystals from alc. Mp: 194°, bp: 200–202°, vap press: 1 mm @ 196.7°, vap d: 8.3.
SYNS: 1,4-DIHYDROXYANTHRACHINON (CZECH) ☐ 1,4-DIHYDROXY-9,10-ANTHRAQUINONE ☐ 1,4-DIOXYANTHRAQUINONE (RUSSIAN) ☐ QUINIZARIN
CONSENSUS REPORTS: Reported in EPA TSCA Inventory.
SAFETY PROFILE: Poison by intravenous route. Moderately toxic by intraperitoneal route. Mutation data reported. An eye irritant. A weak allergen. When heated to

decomposition it emits acrid smoke and irritating fumes.

DMH400   CAS: 117-10-2   HR: 3
1,8-DIHYDROXYANTHRAQUINONE
mf: $C_{14}H_8O_4$ mw: 240.22
PROP: Reddish-yellow needles or leaflets. Mp: 193°, vap d: 8.3. Sol in alc, alkalies.
SYNS: ALTAN □ ANTRAPUROL □ CHRYSAZIN □ DANTHRON □ DANTRON □ DIAQUONE □ 1,8-DIHYDROXY-9,10-ANTHRACENEDIONE □ 1,8-DIHYDROXYANTHRACHINON (CZECH) □ DIONONE □ DORBANE □ DORBANEX □ DUOLAX □ ISTIN □ LAXANORM □ LAXANTHREEN □ LAXIPUR □ LAXIPURIN □ LTAN □ MODANE □ USAF ND-59 □ ZWITSALAX
CONSENSUS REPORTS: NTP 10th Report on Carcinogens. Reported in EPA TSCA Inventory.
SAFETY PROFILE: Confirmed carcinogen with experimental carcinogenic data. Moderately toxic by intraperitoneal route. An eye irritant. Questionable carcinogen with experimental carcinogenic and neoplastigenic data. Human mutation data reported. A laxative. When heated to decomposition it emits acrid smoke and irritating fumes.

DMS410   CAS: 93780-95-1   HR: 2
(1R,2S,3S,4R)-3,4-DIHYDROXY-1,2-
    EPOXY-1,2,3,4-TETRAHYDRO-
    DIBENZ(c,h)ACRIDINE
mf: $C_{21}H_{15}NO_3$ mw: 329.37
SYNS: BENZ(c)OXIRENO(5,6)BENZ(1,2-H)ACRIDINE-2,3-DIOL, 1A,2,3,13C-TETRAHYDRO-, (1AS-(1A-α,2-β,3-α,13C-α))- □ (+)-(1R,2S,3S,4R)-3,4-DIHYDROXY-1,2-EPOXY-1,2,3,4-TETRAHYDRODIBENZ(c,h)ACRIDINE
SAFETY PROFILE: Questionable carcinogen with experimental carcinogenic data reported. Mutation data reported. When heated to decomposition it emits toxic vapors of NO_x.

DMI400   CAS: 2373-98-0   HR: 3
3,3′-DIHYDROXYBENZIDINE
mf: $C_{12}H_{12}N_2O_2$ mw: 216.26
PROP: Plates from Me_2CO. Mp: 160°.
SYNS: 6,6′-DIAMINO-m,m′-BIPHENOL □ 4,4′-DIAMINO-3,3′-BIPHENYLDIOL □ 3,3′-

DIOXYBENZIDINE □ 3,3′-DWUOKSYBENZYDYNA (POLISH)
SAFETY PROFILE: Suspected carcinogen with experimental carcinogenic, neoplastigenic, and tumorigenic data. Mutation data reported. When heated to decomposition it emits toxic fumes of NO_x.

DMI600   CAS: 131-56-6   HR: 3
2,4-DIHYDROXYBENZOPHENONE
mf: $C_{13}H_{10}O_3$ mw: 214.23
PROP: Needles from H_2O. Mp: 142.6–144.6°. Sol in concentrations of H_2SO_4.
SYNS: 2,4-DIHYDROXYBENZOFENON (CZECH) □ EASTMAN INHIBITOR DHPB □ QUINSORB 010 □ SYNTASE 100 □ UF 1 □ USAF DO-28 □ USAF ND-54 □ UVINUL 400
CONSENSUS REPORTS: Reported in EPA TSCA Inventory.
SAFETY PROFILE: Poison by intravenous and intraperitoneal routes. Mildly toxic by ingestion. An eye irritant. When heated to decomposition it emits acrid smoke and irritating fumes.

DMV600   CAS: 7683-59-2   HR: 3
3,4-DIHYDROXY-α-((ISOPROPYL-
    AMINO)-METHYL)BENZYL
    ALCOHOL
mf: $C_{11}H_{17}NO_3$ mw: 211.29
SYNS: A 21 □ ALEUDRIN □ ALUDRINE □ ASIPRENOL □ ASMALAR □ ASSIPRENOL □ BELLASTHMAN □ BRONKEPHRINE □ DIHYDROXYPHENYLETHANOLISOPROPYLAMINE □ 1-(3,4-DIHYDROXYPHENYL)-2-ISOPROPYLAMINO-ETHANOL □ EPINEPHRINE ISOPROPYL HOMOLOG □ 4-(1-HYDROXY-2-((1-METHYLETHYL)AMINO)ETHYL)-1,2-BENZENEDIOL □ IPA □ ISONORENE □ ISOPREN-ALINE □ ISOPROPYDRIN □ ISOPROPYLADRENALINE □ ISOPROPYLAMINOMETHYL-3,4-DIHYDROXY-PHENYL CARBINOL □ α-(ISOPROPYLAMINO-METHYL)PROTOCATECHUYL ALCOHOL □ ISOPROPYLARTERENOL □ N-ISOPROPYL-β-DIHYDROXYPHENYL-β-HYDROXYETHYLAMINE □ ISOPROPYL NORADRENALINE □ N-ISOPROPYL-NORADRENALINE □ 1-ISOPROPYLNORADRENALINE □ ISOPROTERENOL □ l-ISOPROTERENOL □ ISORENIN □ ISUPREL □ ISUPREN □ LOMUPREN □ NEODRENAL □ NEO-EPININE □ NORISODRINE □

NOVODRIN □ PROTERNOL □ RESPIFRAL □ SAVENTRINE □ VAPO-N-ISO □ WIN 5162

SAFETY PROFILE: Poison by ingestion, subcutaneous, intravenous, and intraperitoneal routes. An experimental teratogen. Other experimental reproductive effects. Human systemic effects by intramuscular route: increased pulse and cardiac rate. A bronchodilator. Mutation data reported. When heated to decomposition it emits toxic fumes of NO_x.

**DMW000   CAS: 7361-61-7   HR: 3
5,6-DIHYDRO-2-(2,6-XYLIDINO)-4H-1,3-
  THIAZINE**

mf: $C_{12}H_{16}N_2S$ mw: 220.36

PROP: Crystals from C_6H_6/pet ether. Sol in dil acids, C_6H_6, Me_2CO, and $CHCl_3$; insol in H_2O.

SYNS: BAY 1470 □ BAY VA 1470 □ N-(5,6-DIHYDRO-4H-1,3-THIAZINYL)-2,6-XYLIDINE □ 2-(2,6-DIMETHYLANILINO)-5,6-DIHYDRO-4H-1,3-THIAZINE □ 2-(2,6-DIMETHYLPHENYLAMINO)-4H-5,6-DIHYDRO-1,3-THIAZINE □ N-(2,6-DIMETHYLPHENYL)-5,6-DIHYDRO-4H-1,3-THIAZIN-2-AMINE □ N-(2,6-DIMETHYLPHENYL)-5,6-DIHYDRO-4H-1,3-THIAZINE-2-AMINE (9CI) □ ROMPUN □ WH 7286 □ XYLAZINE (USDA) □ XYLZIN

SAFETY PROFILE: Poison by ingestion, subcutaneous, and intravenous routes. Human systemic effects: change in motor activity, fall in blood pressure, miosis, pleural thickening, pulse rate decrease, somnolence. When heated to decomposition it emits very toxic fumes of NO_x and SO_x.

**DMX200   CAS: 2318-18-5   HR: 3
2,12-DIHYDROXY-4-METHYL-11,16-
  DIOXOSENECIONANIUM**

mf: $C_{19}H_{28}NO_6$ mw: 366.48

PROP: Bevelled plates from EtOAc or Me_2CO. Mp: 196.5–197.5°.

SYNS: trans-15-ETHYLIDENE-12-β-HYDROXY-4,12-α,13-β-TRIMETHYL 8-OXO-4,8 SECOSENEC-1-ENINE □ 12-HYDROXY-4-METHYL-4,8-SECOSENECIONAN-8,11,16-TRIONE □ NSC-89945 □ RENARDIN □ RENARDINE □ SENKIRKIN □ SENKIRKINE

CONSENSUS REPORTS: IARC Cancer Review: Group 3 IMEMDT 7,56,87; Animal Limited Evidence IMEMDT 31,231,83; Animal Inadequate Evidence IMEMDT 10,327,76.

SAFETY PROFILE: Poison by ingestion and intraperitoneal routes. Questionable carcinogen with experimental neoplastigenic data. Mutation data reported. When heated to decomposition it emits toxic fumes of NO_x.

**DNA200   CAS: 59-92-7   HR: 3
l-DIHYDROXYPHENYL-l-ALANINE**

mf: $C_9H_{11}NO_4$ mw: 197.21

PROP: Prisms or needles from H_2O + SO_2; plates from EtOH (aq). Mp: 285.5° (decomp).

SYNS: 2-AMINO-3-(3,4-DIHYDROXYPHENYL)PROPANOIC ACID □ BENDOPA □ BIODOPA □ BROCADOPA □ CEREPAP □ CIDANDOPA □ DA □ DEADOPA □ DIHYDROXY-l-PHENYLALANINE □ (−)-3-(3,4-DIHYDROXYPHENYL)-l-ALANINE □ β-(3,4-DIHYDROXYPHENYL)-α-ALANINE □ l-α-DIHYDROXYPHENYLALANINE □ l-β-(3,4-DIHYDROXYPHENYL)ALANINE □ l-3,4-DIHYDROXY-PHENYL-α-ALANINE □ β-(3,4-DIHYDROXYPHENYL)-l-ALANINE □ 3-(3,4-DIHYDROXYPHENYL)-l-ALANINE □ 3,4-DIHYDROXYPHENYLALANINE □ (−)-3,4-DI-HYDROXYPHENYLALANINE □ 3,4-DIHYDRO-XYPHENYL-l-ALANINE □ 3,4-DIHYDROXY-l-PHENYLALANINE □ l-3,4-DIHYDROXYPHENYL-ALANINE □ (−)-DOPA □ l-DOPA □ DOPAFLEX □ DOPAL □ DOPARKINE □ DOPASOL □ DOPRIN □ ELDOPAL □ EURODOPA □ HELFO DOPA □ l-o-HYDROXYTYROSINE □ 3-HYDROXY-l-TYROSINE □ INSULAMINA □ LARODOPA □ MAIPEDOPA □ PARDA □ RO 4-6316 □ SOBIODOPA □ VELDOPA

CONSENSUS REPORTS: Reported in EPA TSCA Inventory.

SAFETY PROFILE: Poison by ingestion. Moderately toxic by intravenous and intraperitoneal routes. Human systemic effects by ingestion: somnolence, hallucinations and distorted perceptions, toxic psychosis, motor activity changes, ataxia, dyspnea. Experimental teratogenic and reproductive effects. Questionable human carcinogen producing skin tumors. Human mutation data reported. An anticholinergic agent used as an anti-Parkinsonian drug. When heated to decomposition it emits toxic fumes of NO_x.

DNA600   CAS: 13055-82-8   HR: 3
7-(3-(2-(3,5-DIHYDROXYPHENYL-2-
HYDROXY-ETHYLAMINO)PROPYL))-
THEOPHYLLINE HYDROCHLORIDE
mf: $C_{18}H_{23}N_5O_5 \cdot ClH$ mw: 425.92
PROP: Crystals. Mp: 249–250°.
SYNS: BRONCHODIL □ BRONCHOSPASMIN □
REPROTEROL HYDROCHLORIDE □ 7-(3-((β,3,5-
TRIHYDROXYPHENETHYL)AMINO)PROPYL)THEOPHY
LLINE MONOHYDROCHLORIDE □ W-2946M
SAFETY PROFILE: Poison by intravenous
route. Experimental reproductive effects.
When heated to decomposition it emits very
toxic fumes of HCl and NO_x. See also
THEOPHYLLINE.

DNA800   CAS: 555-30-6   HR: 3
l-(−)-3-(3,4-DIHYDROXYPHENYL)-2-
METHYLALANINE
mf: $C_{10}H_{13}NO_4$ mw: 211.24
PROP: Crystals. Sol in isopropanol, EtOH,
and H_2O.
SYNS: ALDOMET □ ALDOMETIL □ ALDOMIN □
ALPHA MEDOPA □ AMD □ BAYER 1440 L □
BAYPRESOL □ l(−)-β-(3,4-DIHYDROXYPHENYL)-α-
METHYLALANINE □ DOPAMET □ DOPEGYT □
DOPTAEC □ 3-HYDROXY-α-METHYL-l-TYROSINE □
HYPERPAX □ l-(α-MD) □ MEDOMET □ MEDOPREN □
METHOPLAIN □ α-METHYL-l-3,4-DIHYDROXY-
PHENYLALANINE □ α-METHYL-β-(3,4-DIHYDROXY-
PHENYL)-l-ALANINE □ l-α-METHYL-3,4-DIHYDROXY-
PHENYLALANINE □ l-(−)-α-METHYL-β-(3,4-DIHYDRO-
XYPHENYL)ALANINE □ METHYLDOPA □ α-METHYL-
l-DOPA □ l-α-METHYLDOPA □ MK. B51 □ MK 351 □
NCI-C55721 □ NR.C 2294 □ PRESINOL □ PRESOLISIN □
SEDOMETIL □ SEMBRINA
CONSENSUS REPORTS: Reported in EPA
TSCA Inventory. EPA Genetic Toxicology
Program.
SAFETY PROFILE: Poison by
intraperitoneal route. Moderately toxic by
ingestion and intravenous routes. Human
systemic effects by ingestion: fasciculations,
hallucinations, distorted perceptions,
tremors, allergic dermatitis, necrotic
gastrointestinal changes. An experimental
teratogen. Human reproductive effects:
menstrual cycle changes or disorders, effects
on newborn including abnormal neonatal
measures and growth statistics, biochemical
and metabolic changes. Experimental

reproductive effects. Mutation data
reported. When heated to decomposition it
emits toxic fumes of NO_x.

DNA850   CAS: 27953-64-6   HR: 3
17-α,21-DIHYDROXY-14-α-PREGN-4-
ENE-3,20-DIONE 21-IODOACETATE
mf: $C_{23}H_{31}IO_5$ mw: 514.44
SYN: 14-α-PREGN-4-ENE-3,20-DIONE, 17-α,21-
DIHYDROXY-, 21-IODOACETATE
SAFETY PROFILE: A poison by
intravenous route. When heated to
decomposition it emits toxic vapors of I^-.

DNB000   CAS: 2589-47-1   HR: 3
17R,21-α-DIHYDROXY-4-PROPYLAJ-
MALANIUM HYDROGEN TARTRATE
mf: $C_{23}H_{32}N_2O_2 \cdot C_4H_6O_6$ mw: 518.67
PROP: Crystals from EtOH/Et₂O. Mp:
149–152°.
SYNS: GT-1012 □ NEO-GILURYTMAL □ NPA □
PRAJMALINE BITARTRATE □ PRAJMALINE
HYDROGEN TARTRATE □ N-PROPYLAJMALINE
BITARTRATE □ N-PROPYLAJMALINE HYDROGEN
TARTRATE □ N-PROPYLAJMALINIUM BITARTRATE □
N-PROPYLAJMALINIUMHYDROGENTARTRAT
(GERMAN) □ N^4-PROPYLAJMALINIUM HYDROGEN
TARTRATE
SAFETY PROFILE: Poison by ingestion and
intravenous routes. An experimental
teratogen. Human systemic effects by
ingestion: hallucinations and distorted
perceptions. Experimental reproductive
effects. An antiarrhythmic agent. When
heated to decomposition it emits toxic
fumes of NO_x.

DNB200   CAS: 53609-64-6   HR: 3
DI(2-HYDROXY-n-PROPYL)AMINE
mf: $C_6H_{14}N_2O_3$ mw: 162.22
SYNS: BHP □ N-BIS(2-
HYDROXYPROPYL)NITROSAMINE □ 2,2'-BISHYDRO-
XYPROPYLNITROSAMINE □ DHPN □ 2,2'-DIHYDRO-
XYDI-n-PROPYLNITROSOAMINE □ N,N-DI-(2-
HYDROXYPROPYL)NITROSAMINE □ DIISOPROPA-
NOLNITROSAMINE □ DIPN □ N-NITROSOBIS(2-
HYDROXYPROPYL)AMINE □ N-NITROSO-N,N-DI(2-
HYDROXYPROPYL)AMINE □ N-NITROSO-1,1'-
IMINODI-2-PROPANOL □ 1,1'-NITROSOIMINODI-2-
PROPANOL

SAFETY PROFILE: Suspected carcinogen with experimental carcinogenic, neoplastigenic, tumorigenic, and teratogenic data. Moderately toxic by subcutaneous route. Human mutation data reported. When heated to decomposition it emits toxic fumes of NO_x. See also NITROSAMINES.

DNE000   CAS: 488-17-5   HR: 3
2,3-DIHYDROXYTOLUENE
mf: $C_7H_8O_2$ mw: 124.15
SYNS: 3-METHYL-1,2-BENZENEDIOL □ 3-METHYL-CATECHOL □ 3-METHYLPYROCATECHOL □ 2,3-TOLUENEDIOL
CONSENSUS REPORTS: Reported in EPA TSCA Inventory.
SAFETY PROFILE: Poison by intravenous route. When heated to decomposition it emits acrid smoke and fumes.

DNE400   CAS: 3468-11-9   HR: 2
1,3-DIIMINOISOINDOLINE
mf: $C_8H_7N_3$ mw: 145.18
PROP: Yellow crystals. Mp: 199° (decomp). Sol in alcohols, acids; sltly sol in H_2O.
SYNS: AFASTOGEN BLUE 5040 □ 1,3-DIIMINOISOINDOLIN (CZECH) □ FASTOGEN BLUE FP-3100 □ FASTOGEN BLUE SH-100 □ MODR FRALOSTANOVA 3G (CZECH) □ PHTHALIMIDIMIDE □ PHTHALOCYANINE BLUE 01206 □ PHTHALOGEN
CONSENSUS REPORTS: Reported in EPA TSCA Inventory.
SAFETY PROFILE: A severe eye and skin irritant. Questionable carcinogen with experimental carcinogenic and tumorigenic data. When heated to decomposition it emits toxic fumes of NO_x.

DNE500   CAS: 624-74-8   HR: 3
DIIODOACETYLENE
mf: C_2I_2 mw: 277.83
PROP: Needles from ligroin. Mp: 76.0–76.5°
SYNS: DIIODOETHYNE □ ETHYNE, DIIODO-
DOT CLASSIFICATION: Forbidden
SAFETY PROFILE: An explosive sensitive to impact, crushing, or heating to 84°C. When heated to decomposition it emits

toxic fumes of I^-. See also IODIDES and ACETYLENE COMPOUNDS.

DNF450   CAS: 4662-17-3   HR: 3
3,5-DIIODO-4-HYDROXYPHENYL 2,5-DIMETHYL-3-FURYL KETONE
mf: $C_{13}H_{10}I_2O_3$ mw: 468.03
SYNS: DB 136 □ KETONE, 3,5-DIIODO-4-HYDROXYPHENYL 2,5-DIMETHYL-3-FURYL
DOT CLASSIFICATION: 3; Label: Flammable Liquid
SAFETY PROFILE: A poison by intraperitoneal and intravenous routes. A flammable liquid. When heated to decomposition it emits toxic vapors of I^-.

DNF500   CAS: 4568-82-5   HR: 3
3,5-DIIODO-4-HYDROXYPHENYL 2-FURYL KETONE
mf: $C_{11}H_6I_2O_3$ mw: 439.97
SYNS: DB 134 □ DIIODO-3,3 HYDROXY-4 BENZOYL 2 FURANNE □ KETONE, 3,5-DIIODO-4-HYDROXY-PHENYL 2-FURYL
DOT CLASSIFICATION: 3; Label: Flammable Liquid
SAFETY PROFILE: A poison by intraperitoneal route. A flammable liquid. When heated to decomposition it emits toxic vapors of F^-.

DNF600   CAS: 83-73-8   HR: 3
DIIODOHYDROXYQUIN
mf: $C_9H_5I_2NO$ mw: 396.95
PROP: Crystals from xylene.
SYNS: DIIODOHYDROXYQUIN □ DIIODOHYDROXYQUINOLINE □ 5,7-DIIODO-8-HYDROXYQUINOLINE □ 5,7-DIIODO-OXINE □ DIIODOQUIN □ 5,7-DIIODO-8-QUINOLINOL □ DINOLEINE □ DIODOQUIN □ DIODOXYLIN □ DI-QUINOL □ DIREXIODE □ DISOQUIN □ DYODIN □ EMBEQUIN □ ENTEROSEPT □ FLORAQUIN □ FLUORAQUIN □ 8-HYDROXY-5,7-DIIODOQUINOLINE □ IODOQUINOL □ IOQUIN SUSPENSION □ LANODOXIN □ MOEBIQUIN □ QUINADOME □ SEARLEQUIN □ SEBAQUIN □ SS 578 □ YODOXIN □ ZOAQUIN
CONSENSUS REPORTS: Reported in EPA TSCA Inventory.
SAFETY PROFILE: Poison by ingestion and intravenous routes. Human systemic effects by ingestion: eye effects. Mutation data

reported. When heated to decomposition it emits very toxic fumes of I⁻ and NO_x.

DNG000   CAS: 305-85-1   HR: 3
2,6-DIIODO-4-NITROPHENOL
mf: $C_6H_3I_2NO_3$ mw: 390.90
PROP: Light-yellow feathery crystals from AcOH. Mp: 157°.
SYNS: ANCYLOL □ DIISOPHENOL □ DISOFEN □ DISOPHENOL □ DNP
CONSENSUS REPORTS: Reported in EPA TSCA Inventory.
SAFETY PROFILE: Poison by ingestion, intraperitoneal, subcutaneous, intravenous, and parenteral routes. An anthelmintic. When heated to decomposition it emits very toxic fumes of I⁻ and NO_x. See also NITRO COMPOUNDS of AROMATIC HYDROCARBONS.

DNH125   CAS: 141-04-8   HR: 2
DIISOBUTYL ADIPATE
mf: $C_{14}H_{26}O_4$ mw: 258.40
SYNS: DIBA □ FTAFLEX DIBA □ ISOBUTYL ADIPATE
CONSENSUS REPORTS: Reported in EPA TSCA Inventory.
SAFETY PROFILE: Moderately toxic by intraperitoneal route. Mildly toxic by ingestion. Experimental teratogenic effects. When heated to decomposition it emits acrid smoke and fumes.

DNH400   CAS: 110-96-3   HR: 3
DIISOBUTYLAMINE
DOT: UN 2361
mf: $C_8H_{19}N$ mw: 129.28
PROP: Water-white liquid; amine odor. Fp: −77°, mp: −70°, bp: 139°, flash p: 69.8°F, d: 0.745 @ 20°/4°, vap press: 10 mm @ 30.6°, vap d: 4.46. Sltly sol in water.
SYN: 2-METHYL-N-(2-METHYLPROPYL)-1-PROPANAMINE
CONSENSUS REPORTS: Reported in EPA TSCA Inventory.
DOT CLASSIFICATION: 3; Label: Flammable Liquid
SAFETY PROFILE: Poison by ingestion. A dangerous fire hazard when exposed to heat or flame; can react vigorously with oxidizing materials. To fight fire, use alcohol foam, CO_2, dry chemical. When heated to decomposition it emits toxic fumes of NO_x.

DNH800   CAS: 108-82-7   HR: 2
DIISOBUTYL CARBINOL
mf: $C_9H_{20}O$ mw: 144.29
PROP: Colorless liquid. Fp: −65°, bp: 179°, flash p: 165°F, d: 0.8121 @ 20°/20°, vap press: 0.3 mm @ 20°, vap d: 4.98, lel: 0.8% @ 212°F, uel: 6.1% @ 212°F.
SYNS: 2,6-DIMETHYL HEPTANOL-4 □ 2,6-DIMETHYL-4-HEPTANOL □ sec-NONYL ALCOHOL
CONSENSUS REPORTS: Reported in EPA TSCA Inventory.
SAFETY PROFILE: Moderately toxic by ingestion and intraperitoneal routes. Mildly toxic by skin contact. A powerful systemic irritant by inhalation. A skin and eye irritant. Can cause central nervous system and liver damage when ingested. Combustible when exposed to heat or flame; can react with oxidizing materials. To fight fire, use alcohol foam, foam, CO_2, dry chemical. When heated to decomposition it emits acrid smoke and fumes.

DNI600   CAS: 1191-15-7   HR: 3
DIISOBUTYLHYDROALUMINUM
mf: $C_8H_{19}Al$ mw: 142.25
PROP: Colorless pyrophoric liquid. Fp: −80°, bp: 140° @ 4 mm, d: 0.798. Misc in hydrocarbon solvents. Sol in Et_2O, C_6H_6, toluene, and cyclohexane.
SYNS: AL-ALCHILI (ITALIAN) □ AL-DIISOBUTYL □ BIS(ISOBUTYL)HYDROALUMINUM □ DIISOBUTYL-ALUMINIUM HYDRIDE □ DIISOBUTYLALUMINUM HYDRIDE □ HYDROBIS(2-METHYLPROPYL)-ALUMINUM □ HYDRODIISOBUTYLALUMINUM
CONSENSUS REPORTS: Reported in EPA TSCA Inventory.
ACGIH TLV: TWA 2 mg(Al)/m³
SAFETY PROFILE: Mildly toxic by inhalation. Dangerous fire hazard; ignites spontaneously in air. To fight fire, do not use water, foam, or halogenated extinguishing agents. See also HYDRIDES and ALUMINUM COMPOUNDS.

D

DNI800   CAS: 108-83-8   HR: 3
DIISOBUTYL KETONE
DOT: UN 1157
mf: $C_9H_{18}O$ mw: 142.27
PROP: Liquid. Bp: 166°, flash p: 140°F, d: 0.81, vap d: 4.9, lel: 0.8% @ 212°F, uel: 6.2% @ 212°F. IDLH 500 ppm.
SYNS: DIISOBUTILCHETONE (ITALIAN) □ DI-ISOBUTYLCETONE (FRENCH) □ DIISOBUTYLKETON (DUTCH, GERMAN) □ s-DIISOPROPYLACETONE □ 2,6-DIMETHYL-HEPTAN-4-ON (DUTCH, GERMAN) □ 2,6-DIMETHYLHEPTAN-4-ONE □ 2,6-DIMETHYL-4-HEPTANONE □ 2,6-DIMETIL-EPTAN-4-ONE (ITALIAN) □ ISOBUTYL KETONE □ ISOVALERONE □ VALERONE
CONSENSUS REPORTS: Reported in EPA TSCA Inventory.
OSHA PEL: TWA 25 ppm
ACGIH TLV: TWA 25 ppm
DFG MAK: 50 ppm (290 mg/m³)
NIOSH REL: (Ketones) TWA 140 mg/m³
DOT CLASSIFICATION: 3; Label: Flammable Liquid
SAFETY PROFILE: Moderately toxic by ingestion and inhalation. Mildly toxic by skin contact. Human systemic effects by inhalation: headache, nausea or vomiting, and unspecified eye effects. An eye and skin irritant. Narcotic in high concentrations. Flammable liquid when exposed to heat or flame; can react with oxidizing materials. To fight fire, use CO_2, dry chemical, water spray, mist or fog. When heated to decomposition it emits acrid smoke and fumes. See also KETONES.

DNJ000   CAS: 61947-30-6   HR: 3
DIISOBUTYLOXOSTANNANE
mf: $C_8H_{18}OSn$ mw: 248.95
SYNS: DIISOBUTYLTIN OXIDE □ KYSLICNIK DIISOBUTYLCINICITY (CZECH)
OSHA PEL: TWA 0.1 mg(Sn)/m³ (skin)
ACGIH TLV: TWA 0.1 mg(Sn)/m³; STEL 0.2 mg(Sn)/m³ (skin).
NIOSH REL: (Organotin Compounds) TWA 0.1 mg(Sn)/m³
SAFETY PROFILE: Poison by ingestion. An eye and severe skin irritant. When heated to decomposition it emits acrid smoke and fumes. See also TIN COMPOUNDS.

DNJ400   CAS: 84-69-5   HR: 2
DIISOBUTYL PHTHALATE
mf: $C_{16}H_{22}O_4$ mw: 278.38
PROP: Liquid. Mp: −64°, flash p: 385°F, d: 1.039–1.043, vap d: 9.59.
SYNS: DIBP □ DIISOBUTYLESTER KYSELINY FTALOVE □ HATCOL DIBP □ HEXAPLAS M/1B □ KODAFLEX DIBP □ PALATINOL IC
CONSENSUS REPORTS: Reported in EPA TSCA Inventory.
SAFETY PROFILE: Moderately toxic by intraperitoneal route. Mildly toxic by ingestion and skin contact. Experimental teratogenic and reproductive effects. Combustible when exposed to heat or flame. To fight fire, use foam, CO_2, dry chemical. When heated to decomposition it emits acrid smoke and fumes.

DNJ600   CAS: 3437-84-1   HR: 3
DIISOBUTYRYL PEROXIDE
mf: $C_8H_{14}O_4$ mw: 174.20
CONSENSUS REPORTS: Reported in EPA TSCA Inventory.
SAFETY PROFILE: May explode when dried at room temperature. When heated to decomposition it emits acrid smoke and fumes. See also PEROXIDES.

DNJ800   CAS: 822-06-0   HR: 3
1,6-DIISOCYANATOHEXANE
DOT: UN 2281
mf: $C_8H_{12}N_2O_2$ mw: 168.22
PROP: Oil. D: 1.053 @ 20°/4°, bp: 121–122° @ 9 mm.
SYNS: DESMODUR H □ DESMODUR N □ HEXA-METHYLENDIISOKYANAT □ HEXAMETHYLENE DIISOCYANATE □ HEXAMETHYLENE DIISOCYAN-ATE (DOT) □ HEXAMETHYLENE-1,6-DIISOCYANATE □ 1,6-HEXAMETHYLENE DIISOCYANATE □ 1,6-HEXANEDIOL DIISOCYANATE □ HMDI □ ISOCYANIC ACID, DIESTER with 1,6-HEXANEDIOL □ ISOCYANIC ACID, HEXAMETHYLENE ESTER □ METYLENO-BIS-FENYLOIZOCYJANIAN □ SZESCIOMETYLEN-ODWUIZOCYJANIAN □ TL 78
CONSENSUS REPORTS: Reported in EPA TSCA Inventory.
ACGIH TLV: TWA 0.005 ppm
DFG MAK: 0.005 ppm (0.035 mg/m³)

NIOSH REL: (Diisocyanates) TWA 0.005 ppm; CL 0.02 ppm/10M
DOT CLASSIFICATION: 6.1; Label: Poison
SAFETY PROFILE: Poison by inhalation and intravenous routes. Moderately toxic by ingestion and skin contact. Potentially explosive reaction with alcohols + base. When heated to decomposition it emits toxic fumes of NO_x. See also CYANATES.

**DNK800   CAS: 27215-10-7   HR: 2
DIISOOCTYL ACID PHOSPHATE**
DOT: UN 1902
mf: $C_{16}H_{35}O_4P$ mw: 322.48
PROP: A corrosive liquid.
SYN: DIISOOCTYL PHOSPHATE (DOT)
CONSENSUS REPORTS: Reported in EPA TSCA Inventory.
DOT CLASSIFICATION: 8; Label: Corrosive
SAFETY PROFILE: Moderately toxic by irritation to skin, eyes, and mucous membranes. A corrosive compound. When heated to decomposition it emits toxic fumes of PO_x. See also PHOSPHATES.

**DNM200   CAS: 108-18-9   HR: 3
DIISOPROPYLAMINE**
DOT: UN 1158
mf: $C_6H_{15}N$ mw: 101.22
PROP: Colorless liquid. Bp: 83–84°, flash p: 19.4°F, d: 0.722 @ 220.0°, vap d: 3.5. IDLH 200 ppm.
SYNS: DIPA □ N-(1-METHYLETHYL)-2-PROPANAMINE □ 2-PROPANAMINE, N-(1-METHYLETHYL)-
CONSENSUS REPORTS: Reported in EPA TSCA Inventory.
OSHA PEL: TWA 5 ppm (skin)
ACGIH TLV: TWA 5 ppm (skin)
DOT CLASSIFICATION: 3; Label: Flammable Liquid
SAFETY PROFILE: Moderately toxic by ingestion and subcutaneous routes. Mildly toxic by inhalation. Mutation data reported. A skin and severe eye irritant. Inhalation of fumes can cause pulmonary edema. A very dangerous fire hazard when exposed to heat or flame; can react vigorously with oxidizing materials. To fight fire, use alcohol foam,

foam, CO_2, dry chemical. When heated to decomposition it emits toxic fumes of NO_x. See also AMINES.

**DNO200   CAS: 15721-33-2   HR: 3
DIISOPROPYLBERYLLIUM**
mf: $C_6H_{14}Be$ mw: 95.19
CONSENSUS REPORTS: IARC Cancer Review: Group 1 IMEMDT 58,41,93; Human Sufficient Evidence IMEMDT 58,41,93; Animal Sufficient Evidence IMEMDT 1,17,72; Animal Sufficient Evidence IMEMDT 23,143,80; Animal Sufficient Evidence IMEMDT 58,41,93. Beryllium and its compounds are on the Community Right-To-Know List.
OSHA PEL: TWA 0.002 mg(Be)/m³; STEL 0.005 mg(Be)/m³/30M; CL 0.025 mg(Be)/m³
ACGIH TLV: TWA 0.002 mg(Be)/m³; Confirmed Human Carcinogen; (Proposed: TWA 0.00005 mg(Be)/m³; STEL 0.0002 mg(Be)/m³ (sensitizer) (skin); Confirmed Human Carcinogen)
DFG MAK: DFG TRK: 0.002 mg(Be)/m³; Animal Carcinogen, Suspected Human Carcinogen
SAFETY PROFILE: Confirmed human carcinogen. Explosive reaction on contact with water. When heated to decomposition it emits toxic fumes of BeO. See also BERYLLIUM COMPOUNDS.

**DNP000   CAS: 96-80-0   HR: 2
N,N-DIISOPROPYL ETHANOLAMINE**
mf: $C_8H_{19}NO$ mw: 145.28
SYNS: 2-DIISOPROPYLAMINOETHANOL □ DIISOPROPYL ETHANOLAMINE
CONSENSUS REPORTS: Reported in EPA TSCA Inventory.
SAFETY PROFILE: Moderately toxic by ingestion and skin contact. A skin and severe eye irritant.

**DNQ800   CAS: 1071-39-2   HR: 3
DIISOPROPYLMERCURY**
mf: $C_6H_{14}Hg$ mw: 286.79

PROP: A liquid. Bp: 63° @ 10 mm, d: 2.00 @ 20°/4°, vap d: 9.9. IDLH 10 mg/m³ (as Hg).

CONSENSUS REPORTS: Mercury and its compounds are on the Community Right-To-Know List.

OSHA PEL: TWA 0.01 mg(Hg)/m³; STEL 0.03 mg/m³ (skin)

ACGIH TLV: TWA 0.01 mg(Hg)/m³; BEI: 35 µg/g creatinine total inorganic mercury in urine preshift; 15 µg/g creatinine total inorganic mercury in blood at end of shift at end of workweek.

DFG MAK: Confirmed Animal Carcinogen with Unknown Relevance to Humans

NIOSH REL: (Mercury, Organo) TWA 0.01 mg/m³; STEL 0.03 mg/m³ (skin)

SAFETY PROFILE: Poison by intraperitoneal route. Mercury compounds are poisons. When heated to decomposition it emits toxic fumes of Hg. See also MERCURY COMPOUNDS, ORGANIC.

DNR200   CAS: 23668-76-0   HR: 3
DIISOPROPYLOXOSTANNANE

mf: $C_6H_{14}OSn$ mw: 220.89

PROP: Solid. Insol in water.

SYNS: DIISOPROPYLTIN OXIDE □ KYSLICNIK DIISOPROPYLCINICITY (CZECH)

OSHA PEL: TWA 0.1 mg(Sn)/m³ (skin)

ACGIH TLV: TWA 0.1 mg(Sn)/m³; STEL 0.2 mg(Sn)/m³ (skin).

NIOSH REL: (Organotin Compounds) TWA 0.1 mg(Sn)/m³

SAFETY PROFILE: Poison by ingestion. An eye and severe skin irritant. When heated to decomposition it emits acrid smoke and irritating fumes. See also TIN COMPOUNDS.

DNR400   CAS: 105-64-6   HR: 3
DIISOPROPYL PERDICARBONATE

mf: $C_8H_{14}O_6$ mw: 206.22

PROP: Colorless, crystalline solid. Rapid decomp @ 63°F, mp: 8–10°, d: 1.080 @ 15.5°/4°. Almost insol in water; miscible with aliphatic and aromatic hydrocarbons, esters, ethers, and chlorinated hydrocarbons.

SYNS: DIISOPROPYL PEROXYDICARBONATE □ ISOPROPYL PERCARBONATE □ ISOPROPYL PEROXYDICARBONATE □ PEROXYDICARBONATE d'ISOPROPYLE □ PEROXYDICARBONIC ACID, BIS(1-METHYLETHYL) ESTER

CONSENSUS REPORTS: Reported in EPA TSCA Inventory.

SAFETY PROFILE: Moderately toxic by ingestion and skin contact. A severe eye irritant. Very dangerous fire hazard. Dangerously unstable above 10°C. An impact- and heat-sensitive explosive. Solutions may spontaneously explode (the hazard increases with concentration). Storage in sealed containers may be dangerous. Explodes on contact with amines or potassium iodide. May explode on contact with organic matter. When heated to decomposition it emits acrid smoke and fumes. See also PEROXIDES, ORGANIC.

DNR800   CAS: 2078-54-8   HR: 3
2,6-DIISOPROPYLPHENOL

mf: $C_{12}H_{18}O$ mw: 178.30

PROP: A colorless liquid or solid. Fp: 17.9°, mp: 19°, bp: 136° @ 30 mm, flash p: 235°F (CC), d: 0.955 @ 20°/4°.

SYNS: 2,6-BIS(1-METHYLETHYL)PHENOL □ DIPRIVAN □ ICI 35868 □ PHENOL, 2,6-BIS(1-METHYLETHYL)-(9CI) □ PROPOFOL

CONSENSUS REPORTS: Reported in EPA TSCA Inventory.

SAFETY PROFILE: Poison by intravenous and intraperitoneal routes. Experimental reproductive effects. Combustible when exposed to heat or flame; can react with oxidizing materials. To fight fire, use foam, CO_2, dry chemical. When heated to decomposition it emits acrid smoke and fumes. See also PHENOL.

DNS000   CAS: 26762-93-6   HR: 2
DIISOPROPYLPHENYLHYDROPEROXID
E (solution)

DOT: UN 2171

mf: $C_{12}H_{19}O_2$ mw: 195.30

PROP: Colorless to pale-yellow liquid.

SYN: DIISOPROPYLBENZENE HYDROPEROXIDE, not more than 72% in solution (DOT)

CONSENSUS REPORTS: Reported in EPA TSCA Inventory.

DOT CLASSIFICATION: Forbidden

SAFETY PROFILE: Questionable carcinogen with experimental tumorigenic data. A powerful oxidizer. When heated to decomposition it emits acrid smoke and fumes. See also PEROXIDES, ORGANIC.

DNT000   CAS: 38802-82-3   HR: 3
DIISOPROPYLTIN DICHLORIDE
mf: $C_6H_{14}Cl_2Sn$ mw: 275.79
PROP: Colorless crystals. Sol in water. Mp: 84°
SYN: DICHLORODIISOPROPYLSTANNANE
OSHA PEL: TWA 0.1 mg(Sn)/m³ (skin)
ACGIH TLV: TWA 0.1 mg(Sn)/m³; STEL 0.2 mg(Sn)/m³ (skin).
NIOSH REL: (Organotin Compounds) TWA 0.1 mg(Sn)/m³
SAFETY PROFILE: Poison by intravenous route. When heated to decomposition it emits toxic fumes of Cl⁻. See also TIN COMPOUNDS and CHLORIDES.

DNU000   CAS: 630-93-3   HR: 3
DILANTIN
mf: $C_{15}H_{11}N_2O_2•Na$ mw: 274.27
PROP: Hygroscopic crystals with bitter, soapy taste.
SYNS: ALEPSIN □ ANTILEPSIN □ ANTISACER □ AURANILE □ CITRULLAMON □ DANTEN □ DANTOIN □ DENYL □ DENYLSODIUM □ DERIZENE □ DIFENIN □ DIFETOIN □ DIFHYDAN □ DI-HYDAN □ DIHYDANTOIN □ DILANTIN SODIUM □ DI-LEN □ DINTOINA □ DIPHANTOINE SODIUM □ DIPHEDAN □ DIPHENATE □ DIPHENIN □ DIPHENINE SODIUM □ DIPHENTOIN □ DIPHENYLAN SODIUM □ DIPHENYLHYDANTOIN SODIUM □ 5,5-DIPHENYL-HYDANTOIN SODIUM □ 5,5-DIPHENYL-2,4-IMIDAZOLIDINE-DIONE, MONOSODIUM SALT □ DI-PHETINE □ DITOIN □ DIVULSAN □ DPH □ ENKEFAL □ EPAMIN □ EPANUTIN □ EPELIN □ EPIFENYL □ EPIHYDAN □ EPILAN-D □ EPILANTIN □ EPINAT □ EPTOIN □ FENANTOIN □ FENITOIN □ FENYTOINE □ HYDANTIN SODIUM □ HYDANTOIN SODIUM □ IDANTOIL □ IDANTOINAL □ LEPITOIN □ LEPITOIN SODIUM □ MINETOIN □ NOVANTOINA □ NOVODIPHENYL □ OM-HYDANTOINE SODIUM □ PHENYTOIN SODIUM □ SACERIL □ SDPH □ SODANTON □ SODIUM DIPHENYLHYDANTOIN □

SODIUM DIPHENYL HYDANTOINATE □ SODIUM-5,5-DIPHENYLHYDANTOINATE □ SODIUM-5,5-DIPHENYL-2,4-IMIDAZOLIDINEDIONE □ SOLANTOIN □ SOLANTYL □ SOLUBLE PHENYTOIN □ SYLANTOIC □ TACOSAL □ THILOPHENYT □ ZENTROPIL

CONSENSUS REPORTS: IARC Cancer Review: Animal Sufficient Evidence IMEMDT 13,201,77. Reported in EPA TSCA Inventory.

SAFETY PROFILE: Confirmed carcinogen. Experimental teratogen. Other experimental reproductive effects. Poison by ingestion, subcutaneous, intravenous, and intraperitoneal routes. Human systemic effects by ingestion: anorexia, respiratory depression, nausea or vomiting, hemorrhage, dermatitis, and endocrine effects. Mutation data reported. An anticonvulsant and cardiac depressant used for the treatment of grand mal and psychomotor seizures. When heated to decomposition it emits very toxic fumes of NO_x and Na_2O.

DNU300   CAS: 71-68-1   HR: 3
DILAUDID
mf: $C_{17}H_{19}NO_3•ClH$ mw: 321.83
PROP: A solid. Mp: 305–315° (decomp).
SYNS: DIHYDROMORPHINONE HYDROCHLORIDE □ DILAUDID HYDROCHLORIDE □ 4,5-α-EPOXY-3-HYDROXY-17-METHYLMORPHINAN-6-ONE HYDROCHLORIDE □ HYDROMORPHONE HYDROCHLORIDE □ HYMORPHAN
SAFETY PROFILE: Poison by subcutaneous and intravenous routes. Experimental teratogenic effects. A powerful analgesic. When heated to decomposition it emits very toxic fumes of NO_x and HCl. See also MORPHINE.

DNU330   CAS: 32093-26-8   HR: 3
DILITHIUM N-ACETYL-I-ASPARTATE
mf: $C_6H_7NO_5•2Li$ mw: 187.01
SYNS: l-ASPARTIC ACID, N-ACETYL-, DILITHIUM SALT □ AKF-94
SAFETY PROFILE: A poison by intracerebral route. When heated to decomposition it emits toxic vapors of NO_x and Li.

DNU390   **HR: D**
DILL SEED OIL, AMERICAN TYPE
PROP: From steam distillation of the salks, leaves, and seeds of *Anethum graveolens* L. Yellowish liquid. D: 0.884–0.900, refr index: 1.480 @ 20°. Sol in propylene glycol; insol in glycerin.
SYNS: DILL OIL □ DILL HERB OIL, AMERICAN TYPE
SAFETY PROFILE: When heated to decomposition it emits acrid smoke and irritating fumes.

DNU400   CAS: 8006-75-5   HR: 1
DILL SEED OIL, EUROPEAN TYPE
PROP: From steam distillation of the dried ripe fruit of *Anethum graveolens* L. (Fam. *Umbelliferae*). Yellowish liquid; caraway odor and taste. D: 0.890–0.915, refr index: 1.4836 @ 20°. Sol in fixed oils, mineral oil, and propylene glycol; insol in glycerin.
SYNS: DILL FRUIT OIL □ DILL HERB OIL □ DILL OIL □ DILL SEED OIL □ DILL WEED OIL
CONSENSUS REPORTS: Reported in EPA TSCA Inventory.
SAFETY PROFILE: Mildly toxic by ingestion. A skin irritant. Mutation data reported. When heated to decomposition it emits acrid smoke and fumes.

DNW000   CAS: 63869-15-8   HR: 3
DIMERCUROUS METHANE ARSONATE
mf: $CH_3AsO_3 \cdot 2Hg$ mw: 539.14
SYN: METHANEARSONIC ACID DIMERCURY SALT
CONSENSUS REPORTS: Arsenic and its compounds, as well as mercury and its compounds, are on the Community Right-To-Know List.
OSHA PEL: TWA 0.5 mg(As)/m³; CL 0.1 mg(Hg)/m³ (skin)
ACGIH TLV: BEI: 35 μ (As)/L inorganic arsenic and methylated metabolites in urine; TWA 0.1 mg(Hg)/m³ (skin); BEI: 35 μg/g creatinine total inorganic mercury in urine preshift; 15 μg/g creatinine total inorganic mercury in blood at end of shift at end of workweek.
DFG MAK: Confirmed Animal Carcinogen with Unknown Relevance to Humans

NIOSH REL: (Mercury, Organo) TWA 0.01 mg/m³; STEL 0.03 mg/m³ (skin)
SAFETY PROFILE: Poison by intraperitoneal route. When heated to decomposition it emits very toxic fumes of As and Hg. See also MERCURY COMPOUNDS and ARSENIC COMPOUNDS.

DNX500   CAS: 8015-19-8   HR: 3
DIMETHISTERONE and ETHINYL
**   ESTRADIOL**
mf: $C_{23}H_{32}O_2 \cdot C_{20}H_{24}O_2$ mw: 636.99
SYNS: ETHINYL ESTRADIOL and DIMETHISTERONE □ ORACON □ OVIN □ SECROVIN
SAFETY PROFILE: Suspected human carcinogen producing uterine tumors. Human reproductive effects by ingestion: abnormalities of the uterus, cervix, and vagina. A steroid. When heated to decomposition it emits acrid smoke and irritating fumes.

DNZ200   CAS: 86-51-1   HR: 3
2,3-DIMETHOXYBENZALDEHYDE
mf: $C_9H_{10}O_3$ mw: 166.18
SAFETY PROFILE: A poison by ingestion. When heated to decomposition it emits acrid smoke and irritating vapors.

DOA800   CAS: 20325-40-0   HR: 3
3,3'-DIMETHOXYBENZIDINE DIHYDRO-
**   CHLORIDE**
mf: $C_{14}H_{16}N_2O_2 \cdot 2ClH$ mw: 317.24
SYNS: C.I. DISPERSE BLACK 6 DIHYDROCHLORIDE □ o-DIANISIDINE DIHYDROCHLORIDE □ 3,3-DIMETHOXY-(1,1'-BIPHENYL)-4,4'-DIAMINE DIHYDROCHLORIDE
CONSENSUS REPORTS: NTP Carcinogenesis Studies (Gavage); Clear Evidence: rat NCITR* NTP-TR-372,90. Reported in EPA TSCA Inventory.
NIOSH REL: (Benzidine-Based Dye) Reduce to lowest feasible level
SAFETY PROFILE: Confirmed carcinogen with experimental carcinogenic and tumorigenic data. Experimental reproductive data. Mutation data reported.

When heated to decomposition it emits very toxic fumes of NO_x and HCl.

DOE200   CAS: 120-20-7   HR: 3
3,4-DIMETHOXYDOPAMINE
mf: $C_{10}H_{15}NO_2$ mw: 181.26
PROP: Crystals from C_6H_6/pet ether. Mp: 124°, bp: 188° @ 15 mm, d: 1.08 @ 28°/4°, vap d: 6.25.
SYNS: DIMETHYOXYDOPAMINE □ 3,4-DIMETHOXY-PHENETHYLAMINE □ 3,4-DIMETHOXY-β-PHENETH-YLAMINE □ DIMETHOXYPHENYL-ETHYLAMINE □ 3,4-DIMETHOXYPHENYLETHYLAMINE □ 3,4-DIMETH-OXY-β-PHENYLETHYL-AMINE □ β-(3,4-DIMETHOXY-PHENYL)ETHYLAMINE □ 2-(3,4-DIMETHOXYPHEN-YL)ETHYLAMINE □ 3,4-DIMETHOXYPHENYLETHYL-AMINE (base) □ DIMETHYLMESCALINE □ DIMPEA □ DMPE □ DMPEA □ HOMOVERATRYLAMINE
CONSENSUS REPORTS: Reported in EPA TSCA Inventory.
SAFETY PROFILE: Poison by intravenous and intraperitoneal routes. When heated to decomposition it emits toxic fumes of NO_x.

DOE600   CAS: 110-71-4   HR: 3
1,2-DIMETHOXYETHANE
DOT: UN 2252
mf: $C_4H_{10}O_2$ mw: 90.14
PROP: Liquid; sharp, ethereal odor. D: 0.86877, mp: −58°, bp: 82–83°, n: (24/D) 1.3739, flash p: 4.5°C (40°F). Miscible with water and alc; sol in hydrocarbon solvents.
SYNS: DIMETHOXYETHANE □ α,β-DIMETHO-XYETHANE □ 1,2-DIMETHOXYETHANE (DOT) □ DIMETHYLCELLOSOLVE □ 2,5-DIOXAHEXANE □ EGDME □ ETHYLENE DIMETHYL ETHER □ ETHYLENE GLYCOL DIMETHYL ETHER □ GLYCOL DIMETHYL ETHER □ GLYME □ MONOETHYLENE GLYCOL DIMETHYL ETHER □ MONOGLYME
CONSENSUS REPORTS: Reported in EPA TSCA Inventory. Glycol ether compounds are on the Community Right-To-Know List.
DOT CLASSIFICATION: 3; Label: Flammable Liquid
SAFETY PROFILE: An experimental teratogen. Other experimental reproductive effects. Readily forms an explosive peroxide. A very dangerous fire hazard when exposed to heat, flame, or oxidizers. Mixture with lithium tetrahydroaluminate may ignite or explode if heated. When heated to decomposition it emits acrid smoke and fumes. See also GLYCOL ETHERS.

DOF400   CAS: 117-82-8   HR: 3
DIMETHOXY ETHYL PHTHALATE
mf: $C_{14}H_{18}O_6$ mw: 282.32
PROP: Light-colored, clear liquid; mild aromatic odor. Mp: −40° (forms gel), bp: 190–210° @ 4 mm, flash p: 360°F, d: 1.171 @ 20°/20°, vap press: 0.3 mm @ 150°, vap d: 9.75.
SYNS: 1,2-BENZENEDICARBOXYLIC ACID BI(2-METHOXYETHYL)ESTER (9CI) □ BIS(METHOXY-ETHYL) PHTHALATE □ BIS(2-METHOXYETHYL) PHTHALATE □ DI(2-METHOXYETHYL)PHTHALATE □ DMEP □ KESSCOFLEX MCP □ 2-METHOXYETHYL PHTHALATE □ PHTHALIC ACID BIS(2-METHOXYETHYL) ESTER
CONSENSUS REPORTS: EPA Genetic Toxicology Program. Reported in EPA TSCA Inventory.
SAFETY PROFILE: Moderately toxic by ingestion and intraperitoneal routes. Mildly toxic by inhalation. Experimental teratogenic and reproductive effects. A skin and eye irritant. Mutation data reported. A pesticide. Combustible when exposed to heat or flame; can react with oxidizing materials. To fight fire, use water, foam, CO_2, dry chemical. When heated to decomposition it emits acrid smoke and irritating fumes.

DOG700   CAS: 1125-88-8   HR: 2
DIMETHOXYMETHYLBENZENE
mf: $C_9H_{12}O_2$ mw: 152.21
SYNS: BENZALDEHYDE, DIMETHYL ACETAL □ DIMETHOXYPHENYLMETHANE
CONSENSUS REPORTS: Reported in EPA TSCA Inventory.
SAFETY PROFILE: Moderately toxic by ingestion. A skin irritant. When heated to decomposition it emits acrid smoke and irritating fumes.

DOJ200   CAS: 91-10-1   HR: 3
2,6-DIMETHOXYPHENOL
mf: $C_8H_{10}O_3$ mw: 154.18

PROP: A solid. Mp: 55–56°, bp: 262–267°. Sltly sol in H_2O.

SYNS: ALDRICH □ 1,3-DIMETHYL PYROGALLATE □ PYROGALLOL DIMETHYLETHER □ PYROGALLOL-1,3-DIMETHYL ETHER □ SYRINGOL

CONSENSUS REPORTS: Reported in EPA TSCA Inventory.

SAFETY PROFILE: Poison by intravenous route. Moderately toxic by ingestion. When heated to decomposition it emits acrid smoke and irritating fumes. See also ETHERS.

DOK200   CAS: 6358-53-8   HR: 3
1-((2,5-DIMETHOXYPHENYL)AZO)-2-NAPHTHOL

mf: $C_{18}H_{16}N_2O_3$ mw: 308.36

PROP: Mp: 156°. Sltly water-sol; mod sol in alc.

SYNS: C.I. 12156 □ C.I. SOLVENT RED 80 □ CITRUS RED No. 2 □ 2,5-DIMETHOXYBENZENEAZO-β-NAPHTHOL □ 1-((2,5-DIMETHOXYPHENYL)AZO)-2-NAPHTHALENOL □ 2,5-DIMETHOXY-1-(PHENYLAZO)-2-NAPHTHOL □ 1-(1-(2,5-DIMETHOXYPHENYL)AZO)-2-NAPHTHOL □ 1-(2,5-DIMETHYLOXYPHENYLAZO)-2-NAPHTHOL

CONSENSUS REPORTS: IARC Cancer Review: Group 2B IMEMDT 7,56,87; Animal Sufficient Evidence IMEMDT 8,101,75. EPA Genetic Toxicology Program.

SAFETY PROFILE: Confirmed carcinogen with experimental carcinogenic data. Mutation data reported. When heated to decomposition it emits toxic fumes of NO_x.

DOM100   CAS: 24991-55-7   HR: D
DIMETHOXY POLYETHYLENE GLYCOL

mf: $(C_2H_4O)_n \cdot C_2H_6O$

SYNS: GLYCOLS, POLYETHYLENE, DIMETHYL ETHER □ GLYME-23 □ α-ω-METHOXYPOLY-(ETHYLENE OXIDE) □ POLYETHYLENE GLYCOL DIMETHYL ETHER □ POLY(OXY-1,2-ETHANEDIYL), α-METHYL-ω-METHOXY-(9CI) □ POLYOXYETHYLENE DIMETHYL ETHER □ SELEXOL

CONSENSUS REPORTS: Reported in EPA TSCA Inventory.

SAFETY PROFILE: Experimental reproductive effects. When heated to decomposition it emits acrid smoke and irritating fumes.

DON400   CAS: 23435-31-6   HR: 3
2',5'-DIMETHOXYSTILBENAMINE

mf: $C_{16}H_{17}NO_2$ mw: 255.34

SYNS: (trans)-2,5-DIMETHOXY-4'-AMINOSTILBENE □ 4-(2,5-DIMETHOXYPHENETHYL)ANILINE □ 4-(2-(2,5-DIMETHOXYPHENYL)ETHYL)BENZENAMINE □ 4-(2,5-DIMETHOXY)STILBENAMINE □ 2,5-DIMETHOXY-4'-STILBENAMINE

SAFETY PROFILE: Suspected carcinogen with experimental carcinogenic data. When heated to decomposition it emits toxic fumes of NO_x.

DOO600   CAS: 534-15-6   HR: 3
DIMETHYLACETAL

DOT: UN 2377

mf: $C_4H_{10}O_2$ mw: 90.14

PROP: Colorless liquid; strong aromatic odor. Bp: 64.5°, flash p: 34°F, d: 0.848 @ 25°, vap d: 3.1.

SYNS: ACETALDEHYDE DIMETHYL ACETAL □ 1,1-DIMETHOXYETHANE (DOT) □ DIMETHYL ALDEHYDE □ ETHYLIDENE DIMETHYL ETHER □ METHYL FORMYL

CONSENSUS REPORTS: Glycol ether compounds are on the Community Right-To-Know List. Reported in EPA TSCA Inventory.

DOT CLASSIFICATION: 3; Label: Flammable Liquid

SAFETY PROFILE: Mildly toxic by inhalation, ingestion, and skin contact. A skin and eye irritant. A very dangerous fire hazard when exposed to heat, flame, or oxidizers. When exposed to heat or flame it can react vigorously with oxidizing materials. To fight fire, use foam, CO_2, dry chemical. When heated to decomposition it emits acrid smoke and irritating fumes. See also GLYCOL ETHERS.

DOO800   CAS: 127-19-5   HR: 2
N,N-DIMETHYLACETAMIDE

mf: C_4H_9NO mw: 87.14

PROP: Colorless oily liquid; weak fishy odor. Mp: −20°, bp: 165°, d: 0.943 @ 20°/4°, vap d: 3.01, vap press: 1.3 mm @ 25°, flash p: 171°F (TOC), lel: 1.8%, uel: 11.5% @ 740 mm and 160°. Misc in water. IDLH 300 ppm.

SYNS: ACETDIMETHYLAMIDE □ ACETIC ACID DIMETHYLAMIDE □ DIMETHYLACETAMIDE □ DIMETHYLACETONE AMIDE □ DIMETHYLAMIDE ACETATE □ DMA □ DMAC □ NSC-3138 □ U-5954

CONSENSUS REPORTS: Reported in EPA TSCA Inventory.

OSHA PEL: TWA 10 ppm (skin)

ACGIH TLV: TWA 10 ppm (skin); Not Classifiable as a Human Carcinogen; BEI: 30 mg/g creatinine of N-methylacetamide in urine at end of shift

DFG MAK: 10 ppm (36 mg/m^3)

SAFETY PROFILE: Moderately toxic by skin contact, inhalation, intravenous, and intraperitoneal routes. Mildly toxic by ingestion. Experimental teratogenic and reproductive effects. A skin and eye irritant. Less toxic than dimethylformamide. Mutation data reported. Combustible when exposed to heat and flame. A moderate explosion hazard. Violent reaction with halogenated compounds (e.g., carbon tetrachloride, hexachlorocyclohexane) when heated above 90°C. Iron powder catalyzes the reaction so that it initiates at 71°C. When heated to decomposition it emits toxic fumes of NO$_x$.

DOP200   CAS: 13265-60-6   HR: 3
O,O-DIMETHYL-S-(2-(ACETYLAMINO)-ETHYL) DITHIOPHOSPHATE

mf: C$_6$H$_{14}$NO$_3$PS$_2$ mw: 243.30

SYNS: S-(2-(ACETYLAMINO)ETHYL)-O,O-DIMETHYL PHOSPHORODITHIOATE □ AMIPHOS □ CP 49674 □ DAEP □ O,O-DIMETHYL-S-(2-ACETAMIDOETHYL) ESTER PHOSPHORODITHIOIC ACID □ O,O-DIMETH-YL-S-(2-ACETYLAMINOETHYL) PHOSPHORODITHIO-ATE □ N-((O,O-DIMETHYLPHOSPHORODITHIOYL)-ETHYL)ACETAMIDE □ ENT 27,346 □ MONSANTO CP-49674 □ NSC-190945 □ PHOSPHORODITHIOIC ACID, O,O-DIMETHYL ESTER, S-ESTER with N-(2-MERCAPTOETHYL)ACETAMIDE

SAFETY PROFILE: Poison by ingestion, skin contact, intraperitoneal, and subcutaneous routes. Experimental teratogenic and reproductive effects. When heated to decomposition it emits very toxic NO$_x$, PO$_x$, and SO$_x$. See also MERCAPTANS.

DOP600   CAS: 30560-19-1   HR: 3
O,S-DIMETHYLACETYLPHOS-PHOROAMIDOTHIOATE

mf: C$_4$H$_{10}$NO$_3$PS   mw: 183.18

SYNS: ACEPHAT (GERMAN) □ ACEPHATE □ ACETYLPHOSPHORAMIDOTHIOIC ACID-O,S-DIMETHYL ESTER □ CHEVRON RE 12,420 □ ENT 27,822 □ ORTHENE □ ORTHENE-755 □ ORTHO 12420 □ ORTRAN □ ORTRIL □ RE 12420 □ 75 SP

CONSENSUS REPORTS: EPA Genetic Toxicology Program.

SAFETY PROFILE: Poison by ingestion. Moderately toxic by skin contact and inhalation. Human mutation data reported. When heated to decomposition it emits very toxic fumes of NO$_x$, PO$_x$, and SO$_x$. See also ESTERS.

DOQ300   CAS: 627-93-0   HR: 2
DIMETHYL ADIPATE

mf: C$_8$H$_{14}$O$_4$ mw: 174.22

PROP: A liquid. Fp: 0°, mp: 8°.

SYNS: DIMETHYL HEXANEDIOATE □ METHYL ADIPATE

CONSENSUS REPORTS: Reported in EPA TSCA Inventory.

SAFETY PROFILE: Moderately toxic by intraperitoneal route. Experimental teratogenic and reproductive effects. When heated to decomposition it emits acrid smoke and irritating fumes.

DOQ350   CAS: 1191-16-8   HR: 1
3,3-DIMETHYLALLYL ACETATE

mf: C$_7$H$_{12}$O$_2$ mw: 128.19

PROP: Oil. Bp: 54–56° @ 32 mm.

SYNS: 2-BUTEN-1-OL, 3-METHYL-, ACETATE □ DIMETHYLALLYL ACETATE □ γ,γ-DIMETHYLALLYL ACETATE □ ISOPENT-2-ENYL ACETATE □ 3-METHYL-2-BUTENYL ACETATE □ PRENYL ACETATE

CONSENSUS REPORTS: Reported in EPA TSCA Inventory.

SAFETY PROFILE: Mildly toxic by ingestion. A skin irritant. When heated to decomposition it emits acrid smoke and irritating fumes.

DOQ400   CAS: 359-83-1   HR: 3
2-(3,3-DIMETHYLALLYL)CYCLAZOCINE

mf: C$_{19}$H$_{27}$NO   mw: 285.47

SYNS: 2-DIMETHYLALLYL-5,9-DIMETHYL-2'-HYDRO-XYBENZOMORPHAN □ 2-(3,3-DIMETHYLALLYL)-2',2'-HYDROXY-5,9-DIMETHYL-6,7-BENZOMORPHAN □ FORTALGESIC □ FORTALIN □ FORTRAL □ 1,2,3,4,5,6-HEXAHYDRO-6,11-DIMETHYL-3-(3-METHYL-2-BUTENYL)-2,6-METHANO-3-BENZAZOCINE □ 2'-HYDROXY-5,9-DIMETHYL-2-(3,3-DIMETHYLALLYL)-6,7-BENZOMORPHAN □ dl-2'-HYDROXY-5,9-DIMETHYL-2-(3,3-DIMETHYLALLYL)-6,7-BENZOMORPHAN □ II-C-2 □ KF-1820 □ LITICON □ 3-(3-METHYL-2-BUTENYL)-1,2,3,4,5,6-HEXAHYDRO-6,11-DIMETHYL-2,6-METHANO-3-BENZAZOCIN-8-OL □ NIH 7958 □ NSC-107430 □ PENTAGIN □ PENTAZOCINE □ SOSIGON □ TALWAN □ TALWIN □ WIN 20228

SAFETY PROFILE: Poison by ingestion, subcutaneous, intramuscular, intraperitoneal, and intravenous routes. Experimental reproductive effects. Human systemic effects by intramuscular and intravenous routes: wakefulness, euphoria, hallucinations or distorted perceptions, tremors, convulsions, excitement, motor activity changes, muscle weakness, analgesia, withdrawal, parasympathomimetic effects, nausea or vomiting, and dermititis. Can cause drug dependency and other central nervous system effects. An analgesic. When heated to decomposition it emits toxic fumes of NO_x. See also ALLYL COMPOUNDS.

DOQ800   CAS: 124-40-3   HR: 3
DIMETHYLAMINE
DOT: UN 1032/UN 1160
mf: C_2H_7N mw: 45.10
PROP: Gas. D: 0.680 @ 0°/4°, mp: −96°, bp: 7°. Very sol in water. IDLH 500 ppm.
SYNS: DIMETHYLAMINE, anhydrous (DOT) □ DIMETHYLAMINE, aqueous solution (DOT) □ DIMETHYLAMINE, solution (DOT) □ DMA □ N-METHYLMETHANAMINE □ RCRA WASTE NUMBER U092
CONSENSUS REPORTS: EPA Genetic Toxicology Program. Reported in EPA TSCA Inventory.
OSHA PEL: TWA 10 ppm
ACGIH TLV: TWA 5 ppm; STEL 15 ppm; Not Classifiable as a Human Carcinogen
DFG MAK: 2 ppm (3.7 mg/m³)

DOT CLASSIFICATION: 2.1; Label: Flammable Gas (UN 1032); DOT Class: 3; Label: Flammable Liquid (UN 1160)
SAFETY PROFILE: Poison by ingestion. Moderately toxic by inhalation and intravenous routes. Mutation data reported. An eye irritant. Corrosive to the eyes, skin, and mucous membranes. A flammable gas. When heated to decomposition it emits toxic fumes of NO_x. Incompatible with acrylaldehyde, fluorine, and maleic anhydride

DOR200   CAS: 74-94-2   HR: 3
DIMETHYLAMINE BORANE
mf: $C_2H_7N \cdot BH_3$ mw: 58.94
PROP: Solid. Mp: 37°, bp: 49° @ 0.01 mm.
SYNS: BORANE with DIMETHYLAMINE (1:1) □ DMAB □ N-METHYLMETHANAMINE with BORANE (1:1)
CONSENSUS REPORTS: Reported in EPA TSCA Inventory.
SAFETY PROFILE: Poison by ingestion, intraperitoneal, and intravenous routes. A skin and eye irritant. Mutation data reported. When heated to decomposition it emits toxic fumes of NO_x. See also DIMETHYLAMINE and BORANE.

DOR500   CAS: 25988-97-0   HR: D
DIMETHYLAMINE-EPICHLOROHYDRIN
COPOLYMER
SYN: EPICHLOROHYDRIN-DIMETHYLAMINE COPOLYMER
SAFETY PROFILE: When heated to decomposition it emits acrid smoke and irritating fumes.

DOS000   CAS: 315-18-4   HR: 3
4-(DIMETHYLAMINE)-3,5-XYLYL-N-
METHYLCARBAMATE
mf: $C_{12}H_{18}N_2O_2$ mw: 222.32
PROP: Crystals. Mp: 85°, vap press: <0.1 mm @ 139°.
SYNS: 4-(DIMETHYLAMINO)-3,5-DIMETHYLPHENOL METHYLCARBAMATE (ESTER) □ 4-(DIMETHYL-AMINO)-3,5-DIMETHYLPHENYL ESTER, METHYLCARBAMIC ACID □ 4-(DIMETHYLAMINO)-3,5-DIMETHYLPHENYL-N-METHYLCARBAMATE □ 4-(DIMETHYLAMINO)-3,5-XYLENOL METHYLCARBAM-ATE (ESTER) □ 4-(DIMETHYLAMINO)-3,5-XYLYL ESTER

METHYLCARBAMIC ACID □ 4-DIMETHYLAMINO-3,5-XYLYL METHYLCARBAMATE □ 4-DIMETHYLAMINO-3,5-XYLYL-N-METHYLCARBAMATE □ 4-(N,N-DIMETHYL-AMINO)-3,5-XYLYL N-METHYLCARBAMATE □ DOWCO 139 □ ENT 25,766 □ METHYL-4-DIMETHYL-AMINO-3,5-XYLYL CARBAMATE □ METHYL-4-DIMETHYLAMINO-3,5-XYLYL ESTER of CARBAMIC ACID □ MEXACARBATE (DOT) □ NCI-C00544 □ OMS-47 □ ZACTRAN □ ZECTANE □ ZECTRAN □ ZEXTRAN

CONSENSUS REPORTS: IARC Cancer Review: Group 3 IMEMDT 7,56,87; Animal Inadequate Evidence IMEMDT 12,237,76. NCI Carcinogenesis Bioassay (feed); No Evidence: mouse, rat NCITR* NCI-CG-TR-147,78. EPA Extremely Hazardous Substances List.

SAFETY PROFILE: Poison by ingestion, skin contact, and intraperitoneal routes. Experimental teratogenic effects. Questionable carcinogen with experimental neoplastigenic data. When heated to decomposition it emits toxic fumes of NO_x. See also ESTERS and CARBAMATES.

DOS200   CAS: 926-64-7   HR: 3
DIMETHYLAMINOACETONITRILE
DOT: UN 2378
mf: $C_4H_8N_2$ mw: 84.14
PROP: Flash p: <73.4°F.
SYNS: N-(CYANOMETHYL)DIMETHYLAMINE □ 2-DIMETHYLAMINOACETONITRILE (DOT) □ N,N-DIMETHYLGLYCINONITRILE

CONSENSUS REPORTS: Cyanide and its compounds are on the Community Right-To-Know List.

DOT CLASSIFICATION: 3; Label: Flammable Liquid, Poison
SAFETY PROFILE: Poison by ingestion, skin contact, and ocular routes. Moderately toxic by inhalation. A skin and eye irritant. A dangerous fire hazard when exposed to heat or flame. When heated to decomposition it emits toxic fumes of NO_x and CN^-. See also NITRILES.

DOS400   CAS: 143563-20-6 HR: 3
N'-DIMETHYLAMINOACETYL-PARTRIC-IN A DIMETHYLAMINOETHYLAMIDE DIASPARTAT

mf: $C_7H_7NO_4 \cdot 1/2C_{67}H_{103}N_5O_{19}$ mw: 810.43
SYNS: SPA-S-753 □ l-ASPARTIC ACID, COMPD. WITH 18-DECARBOXY-40-DEMETHYL-3,7-DIDEOXO-N3')-((DIMETHYLAMINO) ACETYL)-18-(((2-(DIMETHYL-AMINO)ETHYL)AMINO)CARBONYL)-3,7-DIHYDROXY-N47)-METHYL-5-OXOCANDICIDIN D, CYCLIC 15,19-HEMIACETAL (2:1)
SAFETY PROFILE: A poison by intravenous route. When heated to decomposition it emits toxic vapors of NO_x.

DOT000   CAS: 58-15-1   HR: 3
DIMETHYLAMINOANTIPYRINE
mf: $C_{13}H_{17}N_3O$ mw: 231.33
PROP: Colorless leaflets, somewhat water-sol; crystals from toluene. Mp: 107–109°. Sol in H_2O, EtOH, $CHCl_3$, Et_2O, toluene, and acids.
SYNS: AMIDAZOPHEN □ AMIDOFEBRIN □ AMIDO-PHEN □ AMIDOPHENAZONE □ AMIDOPYRAZOLINE □ AMIDOPYRIN □ AMINOFENAZONE (ITALIAN) □ AMINOPHENAZONE □ AMINOPYRINE □ ANAFE-BRINA □ BRUFANEUXOL □ DAP □ DEREUMA □ DIMAPYRIN □ DIMETHYLAMINO-ANALGESINE □ 4-(DIMETHYLAMINO)ANTIPYRINE □ DIMETHYLAMINOAZOPHENE □ 4-(DIMETHYL-AMINO)-1,2-DIHYDRO-1,5-DIMETHYL-2-PHENYL-3H-PYRAZOL-3-ONE □ 4-DIMETHYLAMINO-2,3-DIMETHYL-1-PHENYL-3-PYRAZOLIN-5-ONE □ 4-DIMETHYLAMINO-2,3-DIMETHYL-1-PHENYL-5-PYRAZOLONE □ DIMETHYLAMINOPHENAZON (GERMAN) □ DIMETHYLAMINOPHENAZONE □ 4-DIMETHYLAMINOPHENAZONE □ DIMETHYLAMINO-PHENYLDIMETHYLPYRAZOLIN □ 4-DIMETHYL-AMINO-1-PHENYL-2,3-DIMETHYLPYRAZOLONE □ 3-keto-1,5-DIMETHYL-4-DIMETHYLAMINO-2-PHENYL-2,3-DIHYDROPYRAZOLE □ 1,5-DIMETHYL-4-DIMETHYLAMINO-2-PHENYL-3-PYRAZOLONE □ 2,3-DIMETHYL-4-DIMETHYLAMINO-1-PHENYL-5-PYRAZOLONE □ DIPIRIN □ DIPYRIN □ FEBRININA □ FEBRON □ ITAMIDONE □ MAMALLET-A □ NETSUSARIN □ NOVAMIDON □ 1-PHENYL-2,3-DIMETHYL-4-DIMETHYLAMINOPYRAZOLONE-5 □ 1-PHENYL-2,3-DIMETHYL-4-DIMETHYLAMINOPYRAZ-OL-5-ONE □ PIRAMIDON □ PIRIDOL □ PIROMIDINA □ POLINALIN □ PYRADONE □ PYRAMIDON □ PYRAMIDONE

CONSENSUS REPORTS: Reported in EPA TSCA Inventory. EPA Genetic Toxicology Program.
SAFETY PROFILE: Human poison by unspecified route. Experimental poison by ingestion, subcutaneous, intramuscular,

D

intravenous, and intraperitoneal routes. Moderately toxic by parenteral route. Experimental teratogenic and reproductive effects. Questionable carcinogen when mixed with $NaNO_2$ (1:1). Mutation data reported. Can cause bone marrow depression resulting in leucopenia. Has been implicated in development of aplastic anemia. A tranquilizer. When heated to decomposition it emits toxic fumes of NO_x.

DOT300   CAS: 60-11-7   HR: 3
4-DIMETHYLAMINOAZOBENZENE
mf: $C_{14}H_{15}N_3$ mw: 225.32

PROP: Yellow, crystalline tablets; yellow leaflets from EtOH. Mp: 115°. Sol in EtOH, Me_2CO, and C_6H_6; insol in H_2O.

SYNS: ATUL FAST YELLOW R □ BENZENEAZODIMETHYLANILINE □ BRILLIANT FAST YELLOW □ BUTTER YELLOW □ CERASINE YELLOW GG □ C.I. 11020 □ C.I. SOLVENT YELLOW 2 □ DAB □ p-DIMETHYLAMINOAZOBENZEN (CZECH) □ DIMETHYLAMINOAZOBENZENE □ N,N-DIMETHYL-p-AMINOAZOBENZENE □ N,N-DIMETHYL-4-AMINOAZOBENZENE □ p-DIMETHYLAMINOAZOBENZENE □ 4-(N,N-DIMETHYLAMINO)AZOBENZENE □ DIMETHYL-AMINOAZOBENZOL □ p-DIMETHYLAMINO-AZOBENZOL (GERMAN) □ 4-DIMETHYLAMINOAZO-BENZOL □ 4-DIMETHYLAMINOPHENYLAZO-BENZENE □ N,N-DIMETHYL-p-AZOANILINE □ N,N-DIMETHYL-p-PHENYLAZOANILINE □ N,N-DIMETHYL-4-(PHENYLAZO)BENZAMINE □ N,N-DIMETHYL-4-(PHENYLAZO)BENZENAMINE □ DIMETHYL YELLOW □ DIMETHYL YELLOW-N,N-DIMETHYLANILINE □ DMAB □ ENIAL YELLOW 2G □ FAST OIL YELLOW B □ FAT YELLOW □ GRASAL BRILLIANT YELLOW □ JAUNE de BEURRE (FRENCH) □ METHYL YELLOW □ OIL YELLOW □ OLEAL YELLOW 2G □ ORGANOL YELLOW ADM □ ORIENT OIL YELLOW GG □ P.D.A.B. □ PETROL YELLOW WT □ RCRA WASTE NUMBER U093 □ RESINOL YELLOW GR □ RESOFORM YELLOW GGA □ SILOTRAS YELLOW T2G □ SOMALIA YELLOW A □ STEAR YELLOW JB □ SUDAN YELLOW □ TOYO OIL YELLOW G □ USAF EK-338 □ WAXOLINE YELLOW AD □ YELLOW G SOLUBLE in GREASE □ ZLUT MASELNA (CZECH)

CONSENSUS REPORTS: NTP 10th Report on Carcinogens. IARC Cancer Review: Group 2B IMEMDT 7,56,87; Animal Sufficient Evidence IMEMDT 8,125,75. EPA Genetic Toxicology Program.

Community Right-To-Know List. Reported in EPA TSCA Inventory.

OSHA PEL: Cancer Suspect Agent

NIOSH REL: (4-Dimethylaminoazobenzene) TWA use 29 CFR 1910.1015

SAFETY PROFILE: Confirmed carcinogen with experimental carcinogenic, neoplastigenic, and tumorigenic data. Poison by ingestion and intraperitoneal routes. Experimental teratogenic and reproductive effects. Human mutation data reported. When heated to decomposition it emits toxic fumes of NO_x.

DOT800   CAS: 536-17-4   HR: 3
p-DIMETHYLAMINOBENZ-ALRHO-DANINE
mf: $C_{12}H_{12}N_2OS_2$ mw: 264.38

PROP: Red crystals or powder. Mp: 285–288° (decomp). Sltly sol in Me_2CO, EtOH, and $CHCl_3$; insol in H_2O.

SYNS: p-(DIMETHYLAMINO)BENZAL-5-RHODANINE □ 5-(p-DIMETHYLAMINOBENZAL)RHODANINE □ 5-(p-DIMETHYLAMINOBENZOYLIDENE)RHODANINE □ p-DIMETHYLAMINOBENZYLIDENE RHODAMINE □ USAF PD-20

CONSENSUS REPORTS: Reported in EPA TSCA Inventory.

SAFETY PROFILE: Poison by intraperitoneal route. When heated to decomposition it emits very toxic fumes of NO_x and SO_x.

DOU600   CAS: 140-56-7   HR: 3
p-DIMETHYLAMINOBENZENEDIAZO-SODIUM SULPHONATE
mf: $C_8H_{10}N_3O_3S \cdot Na$ mw: 251.26

PROP: Yellow-brown crystals. Mod sol in water; sol in DMF.

SYNS: BAYER 5072 □ DAPA □ DAS □ DEKSONAL □ DEXON □ p-DIMETHYLAMINOBENZENE DIAZO SODIUM SULFONATE □ p-(DIMETHYLAMINO)BENZ-ENEDIAZOSULFONATE □ p-DIMETHYLAMINOBENZ-ENEDIAZOSULFONIC ACID, SODIUM SALT □ 4-DIMETHYLAMINOBENZENEDIAZOSULFONIC ACID, SODIUM SALT □ p-(DIMETHYLAMINO)BENZENEDI-AZOSULPHONATE □ p-(DIMETHYLAMINO)BENZENE-DIAZOSULPHONIC ACID, SODIUM SALT □ 4-DIMETHYLAMINOBENZENEDIAZOSULPHONIC ACID, SODIUM SALT □ p-DIMETHYLAMINOBENZOLDIA-ZOSULFONAT (NATRIUMSALZ) (GERMAN) □ (4-

(DIMETHYLAMINO)PHENYL)DIAZENESULFONIC ACID, SODIUM SALT □ 4-((DIMETHYLAMINO)PHENYL)DIAZENESULFONIC ACID, SODIUM SALT □ p-(DIMETHYLAMINO)-PHENYLDIAZO-NATRIUMSULFONAT (GERMAN) □ N,N-DIMETHYL-p-ANILINEDIAZOSULFONIC ACID SODIUM SALT □ FENAMINOSULF □ GOLD ORANGE MP □ LESAN □ NCI-C03010 □ SODIUM-p-(DIMETHYLAMINO)BENZENEDIAZOSULFONATE □ SODIUM-4-(DIMETHYLAMINO)BENZENEDIAZOSULFONATE □ SODIUM-p-(DIMETHYLAMINO)BENZENEDIAZOSULPHONATE □ SODIUM-4-(DIMETHYLAMINO)BENZENEDIAZOSULPHONATE □ SODIUM-(4-(DIMETHYLAMINO)PHENYL)DIAZENESULFONATE □ TROPAEOLIN D

CONSENSUS REPORTS: IARC Cancer Review: Group 3 IMEMDT 7,56,87; Animal Inadequate Evidence IMEMDT 8,147,75. NCI Carcinogenesis Bioassay (feed); No Evidence: mouse, rat NCITR* NCI-CG-TR-101,78. EPA Genetic Toxicology Program.

SAFETY PROFILE: Poison by ingestion, intravenous, and intraperitoneal routes. Experimental teratogenic effects. Human mutation data reported. Questionable carcinogen. A fungicide. When heated to decomposition it emits very toxic fumes of NO_x, Na_2O, and SO_x.

DOU700   CAS: 6843-30-7   HR: 3
5-DIMETHYLAMINO-3-BENZOYLINDOLE
mf: $C_{17}H_{16}N_2O$ mw: 264.35
SYN: KETONE, 5-DIMETHYLAMINO-3-INDOLYL PHENYL
DOT CLASSIFICATION: 3; Label: Flammable Liquid
SAFETY PROFILE: A poison by intravenous route. A flammable liquid. When heated to decomposition it emits toxic vapors of NO_x.

DOV870   CAS: 90293-54-2   HR: 3
8-((DIMETHYLAMINO)CARBONYL)-5-OXO-2,4,9-TRIMETHYL-6,11-DIOXA-3-THIA-2,4,7,10-TETRAAZADODECA-7,9-DIENOIC ACID, 2,3-DIHYDRO-2,2-DIMETHYL-7-BENZOFURANYL ESTER
mf: $C_{21}H_{29}N_5O_7S$ mw: 495.61

SAFETY PROFILE: A poison by ingestion. When heated to decomposition it emits toxic vapors of NO_x and SO_x.

DOY800   CAS: 108-01-0   HR: 3
N-DIMETHYLAMINOETHANOL
DOT: UN 2051
mf: $C_4H_{11}NO$ mw: 89.16
PROP: A liquid. Bp: 135°, flash p: 105°F (OC), d: 0.8866 @ 20°/4°, vap d: 3.03.
SYNS: DEANOL □ DIMETHYLAETHANOLAMIN (GERMAN) □ DIMETHYLAMINOAETHANOL (GERMAN) □ DIMETHYLAMINOETHANOL □ β-DIMETHYLAMINOETHANOL □ N,N-DIMETHYLAMINOETHANOL □ 2-(DIMETHYLAMINO)ETHANOL □ β-DIMETHYLAMINOETHYL ALCOHOL □ DIMETHYLETHANOLAMINE □ N,N-DIMETHYLETHANOLAMINE □ DIMETHYLETHANOLAMINE (DOT) □ N,N-DIMETHYL-2-HYDROXYETHYLAMINE □ N,N-DIMETHYL-N-(2-HYDROXYETHYL)AMINE □ DMAE □ β-HYDROXYETHYLDIMETHYLAMINE

CONSENSUS REPORTS: Reported in EPA TSCA Inventory.
DOT CLASSIFICATION: 3; Label: Flammable Liquid
SAFETY PROFILE: Moderately toxic by ingestion, inhalation, skin contact, intraperitoneal, and subcutaneous routes. A skin and severe eye irritant. Used medically as a central nervous system stimulant. Flammable liquid when exposed to heat or flame; can react vigorously with oxidizing materials. Ignites spontaneously in contact with cellulose nitrate of high surface area. To fight fire, use alcohol foam, foam, CO_2, dry chemical. When heated to decomposition it emits toxic fumes of NO_x.

DPG600   CAS: 2867-47-2   HR: 3
DIMETHYLAMINOETHYL METH-ACRYLATE
DOT: UN 2522
mf: $C_8H_{15}NO_2$ mw: 157.24
PROP: Liquid, sol in water and org solvs. D: 0.933 @ 25°, bp: 182–190°, flash p: 165°F (TOC), vap d: 5.4.
SYNS: AGEFLEX FM-1 □ 2-(DIMETHYLAMINO)ETHANOL METHACRYLATE □ 2-(DIMETHYLAMINO)ETHYL ESTER METHACRYLIC ACID □ N,N-DIMETHYLAMINOETHYL METH-

ACRYLATE □ β-DIMETHYLAMINOETHYL METH-
ACRYLATE □ 2-(DIMETHYLAMINO)ETHYL METH-
ACRYLATE □ USAF RH-3

CONSENSUS REPORTS: Reported in EPA TSCA Inventory.

DOT CLASSIFICATION: 6.1; Label: Poison

SAFETY PROFILE: Poison by intraperitoneal route. Moderately toxic by ingestion and inhalation. A skin, eye, and mucous membrane irritant. A powerful lachrymator. Flammable when exposed to sparks, heat, open flame, or oxidizers. To fight fire, use alcohol foam, dry chemical, spray. When heated to decomposition it emits toxic fumes of NO_x. See also ESTERS.

DPH600   CAS: 4985-15-3   HR: 3
5-DIMETHYLAMINOETHYLOXYIMINO-5H-DIBENZO(a,d)CYCLOHEPTA-1,4-DIENE HYDROCHLORIDE

mf: $C_{19}H_{22}N_2O \cdot ClH$ mw: 330.89

PROP: A solid. Mp: 185–187°.

SYNS: AGEDAL □ BAY 1521 □ 5-(DIMETHYL-AMINOAETHYL-OXYIMINO)-5H-DIBENZO(a,d)-CYCLOHEPTA-1,4-DIENHYDROCHLORID (GERMAN) □ 5-(DIMETHYLAMINOOXYIMINO)-5H-DIBENZO(a,b)-CYCLOHEPTA-1,4-DIENE HYDROCHLORIDE □ 5-(DIMETILAMINOETILOSIMINO-5H-DIBENZO(a,d)-CICLOEPTA-1,4-DIENE) CLORIDRATO (ITALIAN) □ NOGEDAL □ NOXIPTILINE HYDROCHLORIDE □ NOXIPTILIN HYDROCHLORID (GERMAN) □ NOXIPTYLINE HYDROCHLORIDE

SAFETY PROFILE: Poison by ingestion, subcutaneous, intravenous, intramuscular, and intraperitoneal routes. Experimental reproductive effects. When heated to decomposition it emits very toxic fumes of NO_x and HCl.

DPI750   CAS: 96811-96-0   HR: 3
2-((2-(DIMETHYLAMINO)ETHYL)-(SELENOPHENE-2-YLMETHYL)-AMINO)PYRIDINE

mf: $C_{14}H_{19}N_3Se$ mw: 308.32

SYN: PYRIDINE, 2-((2-(DIMETHYLAMINO)-ETHYL)(SELENOPHENE-2-YLMETHYL)AMINO)-

OSHA PEL: TWA 0.2 mg(Se)/m³

ACGIH TLV: TWA 0.2 mg(Se)/m³

SAFETY PROFILE: Poison by parenteral route. When heated to decomposition it emits toxic fumes of NO_x and Se.

DPL000   CAS: 55738-54-0   HR: 3
trans-2-((DIMETHYLAMINO)METHYL-IMINO)-5-(2-(5-NITRO-2-FURYL)-VINYL)-1,3,4-OXADIAZOLE

mf: $C_{11}H_{12}N_5O_4$ mw: 277.27

CONSENSUS REPORTS: IARC Cancer Review: Group 2B IMEMDT 7,56,87; Animal Limited Evidence IMEMDT 7,147,74. EPA Genetic Toxicology Program.

SAFETY PROFILE: Confirmed carcinogen with experimental carcinogenic data. Mutation data reported. When heated to decomposition it emits toxic fumes of NO_x.

DPL300   CAS: 81862-00-2   HR: 3
2-(DIMETHYLAMINO)-N-(((METHYL(((2-PHENYL-1,3-DIOXAN-5-YL)METHO-XY) SULFINYL)AMINO)CARBONYL)-OXY)-2-OXO-ETHANIMIDOTHIOIC ACID, METHYL ESTER

mf: $C_{18}H_{25}N_3O_7S_2$ mw: 459.58

SAFETY PROFILE: A poison by ingestion. When heated to decomposition it emits toxic vapors of NO_x and SO_x.

DPN200   CAS: 605-65-2   HR: 3
5-(DIMETHYLAMINO)-1-NAPHTHAL-ENESULFONYL CHLORIDE

mf: $C_{12}H_{12}ClNO_2$ mw: 237.70

PROP: Crystals from Me_2CO (aq). Mp: 69°.

SYNS: 1-CHLOROSULFONYL-5-DIMETHYL-AMINONAPHTHALENE □ DANSYL □ DANSYL CHLORIDE □ DIMETHYLAMINONAPHTHALENE-SULFONYL CHLORIDE □ 1-DIMETHYLAMINO-NAPHTHALENE-5-SULFONYL CHLORIDE □ 1-(DIMETHYLAMINO)-5-NAPHTHALENESULFONYL-CHLORIDE □ 5-DIMETHYLAMINONAPHTHYL-5-SULFONYL CHLORIDE

CONSENSUS REPORTS: Reported in EPA TSCA Inventory.

SAFETY PROFILE: Poison by intravenous route. When heated to decomposition it emits very toxic fumes of Cl^- and NO_x.

DPO200   CAS: 539-17-3   HR: 3
4-(p-DIMETHYLAMINOPHENYLAZO)-
ANILINE
mf: $C_{14}H_{16}N_4$ mw: 240.34
SYNS: ACETILE DIAZO BLACK N □ ADAB □ p-
AMINOBENZENEAZODIMETHYLANILINE □ 4-
AMINO-DAB □ 4-AMINO-4'-DIMETHYLAMINO-
AZOBENZENE □ 4'-AMINO-N,N-DIMETHYL-4-
AMINOAZOBENZENE □ 4-((4-AMINOPHENYL)AZO)-
N,N-DIMETHYLBENZENAMINE □ C.I. 11025 □ C.I.
DISPERSE BLACK 3 □ DIAZO NERO MICROSETILE G □
INTERCHEM ACETATE DEVELOPED BLACK □ MEISEI
TERYL DIAZO BLACK CR □ MICROSETILE DIAZO
BLACK G □ SUPRACET DIAZO BLACK A
CONSENSUS REPORTS: Reported in EPA
TSCA Inventory.
SAFETY PROFILE: Poison by
intraperitoneal route. Mutation data
reported. When heated to decomposition it
emits toxic fumes of NO_x.

DPO275   CAS: 73688-85-4   HR: 3
4-(p-DIMETHYLAMINOPHENYLAZO)-
BENZENEARSONIC ACID
HYDROCHLORIDE
mf: $C_{14}H_{16}AsN_3O_3 \cdot ClH$ mw: 385.71
PROP: Mp: 203° decomposes.
SYN: BENZENEARSONIC ACID, 4-(p-
DIMETHYLAMINOPHENYLAZO)-, HYDROCHLORIDE
OSHA PEL: TWA 0.5 mg(As)/m³
SAFETY PROFILE: Poison by intravenous
route. When heated to decomposition it
emits toxic fumes of NO_x, As, and HCl.

DPQ200   CAS: 33804-48-7   HR: 3
4-((p-(DIMETHYLAMINO)PHENYL)AZO)-
N-METHYLACETANILIDE
mf: $C_{17}H_{20}N_4O$ mw: 296.41
SYNS: N'-ACETYL-N'-METHYL-4'-AMINO-N,N-
DIMETHYL-4-AMINOAZOBENZENE □ 4-(N-ACETYL-N-
METHYL)AMINO-4'-(N',N'-DIMETHYLAMINO)-AZO-
BENZENE □ N',N'-DIMETHYL-4'-AMINO-N-ACETYL-N-
MONOMETHYL-4-AMINOAZOBENZENE □ N-(4-((4-
(DIMETHYLAMINO)PHENYL)AZO)PHENYL)-N-
METHYLACETAMIDE
SAFETY PROFILE: Poison by
intraperitoneal route. Questionable
carcinogen with experimental tumorigenic
data. Mutation data reported. When heated
to decomposition it emits toxic fumes of
NO_x.

DPS700   CAS: 102504-71-2   HR: 3
4-DIMETHYLAMINOPHENYL-2-((4-
PHENYL-1,2,5,6-TETRAHYDRO-1-
PYRIDYL)ETHYL) KETONE
mf: $C_{22}H_{26}N_2O$ mw: 334.50
SYNS: 4'-(DIMETHYLAMINO)-3-(4-PHENYL-1,2,5,6-
TETRAHYDRO-1-PYRIDYL)PROPIOPHENONE □
PROPIOPHENONE, 4'-(DIMETHYLAMINO)-3-(4-
PHENYL-1,2,3,6-TETRAHYDRO-1-PYRIDYL)-
DOT CLASSIFICATION: 3; Label:
Flammable Liquid
SAFETY PROFILE: A poison by
intravenous route. A flammable liquid.
When heated to decomposition it emits
toxic vapors of NO_x.

DPT300   CAS: 24955-83-7   HR: 3
5-DIMETHYLAMINO-3-
PIPERIDINOACETYLINDOLE
mf: $C_{13}H_{16}N_2O$ mw: 216.31
SYNS: INDOLE, 5-(DIMETHYLAMINO)-3-
(PIPERIDINOACETYL)- □ 3-(3-DIMETHYL-
AMINOPROPIONYL)INDOLE □ 1-PROPANONE, 3-
(DIMETHYLAMINO)-1-(1H-INDOL-3-YL)- □ 1-
PROPANONE, 3-(DIMETHYLAMINO)-1-INDOL-3-YL-
SAFETY PROFILE: A poison by
intravenous route. When heated to
decomposition it emits toxic vapors of NO_x.

DPU000   CAS: 1738-25-6   HR: 3
3-(DIMETHYLAMINO)PROPIONITRILE
mf: $C_5H_{10}N_2$ mw: 98.17
PROP: Liquid. Mp: −43°, bp: 170°, d:
0.8617, vap d: 3.35, flash p: 145°F.
SYN: β-DIMETHYLAMINOPROPIONITRILE
CONSENSUS REPORTS: Reported in EPA
TSCA Inventory. Cyanide and its
compounds are on the Community Right-
To-Know List.
SAFETY PROFILE: Poison by intravenous
route. Moderately toxic by ingestion and
skin contact. A skin and eye irritant.
Flammable liquid when exposed to heat,
flame, or oxidizers; can react with oxidizing
materials. To fight fire, use foam, CO_2, dry
chemical. When heated to decomposition it
emits highly toxic fumes of NO_x and CN^-.
See also NITRILES.

DPW630   CAS: 296269-51-7 HR: 3
α-(3-(DIMETHYLAMINO)PROPYL)-1,5-DIPHENYL-1H-PYRAZOLE-4-METHANOL

mf: $C_{21}H_{25}N_3O$ mw: 335.45

SAFETY PROFILE: A poison by ingestion. When heated to decomposition it emits toxic vapors of NO_x.

DPY700   CAS: 296269-49-3 HR: 3
α-(3-(DIMETHYLAMINO)PROPYL)-5-(1-METHYLETHYL)-1-PHENYL-1H-PYRAZOLE-4-METHANOL

mf: $C_{18}H_{27}N_3O$ mw: 301.43

SAFETY PROFILE: A poison by ingestion. When heated to decomposition it emits toxic vapors of NO_x.

DQA400   CAS: 60-87-7   HR: 3
10-(2-(DIMETHYLAMINO)PROPYL)-PHENOTHIAZINE

mf: $C_{17}H_{20}N_2S$ mw: 284.45

SYNS: A-91033 □ APROBIT □ ATOSIL □ AVOMINE □ DIMAPP □ DIMETHYLAMINO-ISOPROPYL-PHENTHIAZIN (GERMAN) □ (2-DIMETHYLAMINO-2-METHYL)ETHYL-N-DIBENZOPARATHIAZINE □ N-(2'-DIMETHYLAMINO-2'-METHYL)ETHYLPHENO-THIAZINE □ 10-(2-(DIMETHYLAMINO)-2-METHYLETHYL)PHENOTHIAZINE □ N-DIMETHYL-AMINO-2-METHYLETHYL THIODIPHENYLAMINE □ (DIMETHYLAMINO-2-PROPYL-10-PHENOTHIAZINE HYDROCHLORIDE (FRENCH) □ DIPRAZINE □ DIPROZIN □ FARGAN □ FENAZIL □ FENERGAN □ FENETAZINA □ HIBERNA □ HISTARGAN □ IERGIGAN □ ISOPHENERGAN □ ISOPROMETHAZINE □ LERCIGAN □ LERGIGAN □ LILLY 1516 □ LILLY 01516 □ NCI-C60673 □ PHARGAN □ PHENERGAN □ PHENSEDYL □ PILPOPHEN □ PIPOLPHEN □ PROAZ-AIMINE □ PROAZAMINE □ PROCIT □ PROMA-ZINAMIDE □ PROMETASIN □ PROMETAZIN □ PROMETHIAZINE □ PROMEZATHINE □ PROREX □ PROTAZINE □ PROTHAZIN □ PROVIGAN □ PYRETHIA □ PYRETHIAZINE □ ROMERGAN □ 3277 RP □ 3389 R.P. □ 4182 R.P. □ SKF 1498 □ SYNALGOS □ TANIDIL □ THIERGAN □ VALLERGINE □ WY 509

SAFETY PROFILE: Poison by ingestion, intravenous, intramuscular, intraperitoneal, and subcutaneous routes. Human systemic effects by ingestion: pupillary dilation, wakefulness, hallucinations, and distorted perceptions. An experimental teratogen. Other experimental reproductive effects.

Human mutation data reported. A severe eye irritant. When heated to decomposition it emits very toxic fumes of NO_x and SO_x.

DQA710   CAS: 13065-64-0 HR: 3
10-(3-(DIMETHYLAMINO)PROPYL)-PHENOTHIAZIN-2-YL MORPHOLINO-METHYL KETONE

mf: $C_{23}H_{29}N_3O_2S$ mw: 411.61

SYN: PHENOTHIAZINE, 10-(3-(DIMETHYLAMINO)-PROPYL)-2-(MORPHOLINOACETYL)-

DOT CLASSIFICATION: 3; Label: Flammable Liquid

SAFETY PROFILE: A poison by intraperitoneal route. A flammable liquid. When heated to decomposition it emits toxic vapors of SO_x and NO_x.

DQA720   CAS: 296269-48-2 HR: 3
α-(3-(DIMETHYLAMINO)PROPYL)-1-PHENYL-5-PROPYL-1H-PYRAZOLE-4-METHANOL

mf: $C_{18}H_{27}N_3O$ mw: 301.43

SAFETY PROFILE: A poison by ingestion. When heated to decomposition it emits toxic vapors of NO_x.

DQB400   CAS: 5683-33-0   HR: 3
2-DIMETHYLAMINOPYRIDINE

mf: $C_7H_{10}N_2$ mw: 122.19

SYNS: DIMETHYLAMINO-2 PYRIDINE □ PYRIDINE, 2-DIMETHYLAMINO- □ 2-PYRIDINAMIDE, N,N-DIMETHYL-

SAFETY PROFILE: A poison by intravenous route. When heated to decomposition it emits toxic vapors of NO_x.

DQC400   CAS: 1628-58-6   HR: 2
2-(p-(DIMETHYLAMINO)STYRYL)-BENZO-THIAZOLE

mf: $C_{17}H_{16}N_2S$ mw: 280.41

SYN: 2-(4-DIMETHYLAMINOSTYRYL)BENZO-THIAZOLE

CONSENSUS REPORTS: Reported in EPA TSCA Inventory.

SAFETY PROFILE: Experimental reproductive effects. Questionable carcinogen with experimental neoplastigenic data. When heated to decomposition it emits very toxic fumes of NO_x and SO_x.

DQD000   CAS: 897-55-2   HR: 3
4-(4-(DIMETHYLAMINO)STYRYL)-
**   QUINOLINE**
mf: $C_{19}H_{18}N_2$ mw: 274.39
SYNS: 2-(4-N,N-
DIMETHYLAMINOSTYRYL)QUINOLINE □ 4-(p-
(DIMETHYLAMINO)STYRYL)QUINOLINE
CONSENSUS REPORTS: Reported in EPA
TSCA Inventory.
SAFETY PROFILE: Poison by intravenous
route. Experimental reproductive effects.
Questionable carcinogen with experimental
neoplastigenic data. When heated to
decomposition it emits toxic fumes of NO_x.

DQD400   CAS: 1596-84-5   HR: 3
DIMETHYLAMINOSUCCINAMIC ACID
mf: $C_6H_{12}N_2O_3$ mw: 160.20
PROP: A solid. Mp: 154–155°.
SYNS: ALAR □ ALAR-85 □ AMINOZIDE □ B 995 □
BERNSTEINSAEURE-2,2-DIMETHYLHYDRAZID
(GERMAN) □ B-NINE □ BUTANEDIOIC ACID
MONO(2,2-DIMETHYLHYDRAZIDE) □ DAMINOZIDE
(USDA) □ DIMAS □ N-DIMETHYL AMINO-β-
CARBAMYL PROPIONIC ACID □ N-
(DIMETHYLAMINO)SUCCINAMIC ACID □ N-
DIMETHYLAMINO-SUCCINAMIDSAEURE (GERMAN) □
DMASA □ DMSA □ KYLAR □ NCI-C03827 □ SADH □
SUCCINIC ACID-2,2-DIMETHYLHYDRAZIDE □
SUCCINIC-1,1-DIMETHYL HYDRAZIDE
CONSENSUS REPORTS: EPA Genetic
Toxicology Program. NCI Carcinogenesis
Bioassay (feed); Clear Evidence: mouse, rat
NCITR* NCI-CG-TR-83,78.
SAFETY PROFILE: Suspected carcinogen
with experimental carcinogenic and
tumorigenic data. Moderately toxic by
ingestion and intraperitoneal routes. When
heated to decomposition it emits toxic
fumes of NO_x.

DQF800   CAS: 121-69-7   HR: 3
N,N-DIMETHYLANILINE
DOT: UN 2253
mf: $C_8H_{11}N$ mw: 121.20
PROP: Yellowish-brown oily liquid. Mp:
2.5°, bp: 193.1°, flash p: 145°F (CC), d:
0.9557 @ 20°/4°, ULC: 20–25, autoign
temp: 700°F, vap press: 1 mm @ 29.5°, vap
d: 4.17. IDLH 100 ppm.

SYNS: BENZENAMINE, N,N,-DIMETHYL-(9CI) □
(DIMETHYLAMINO)BENZENE □ N-DIMETHYL-
ANILINE (OSHA) □ N,N-DIMETHYLBENZENEAMINE
□ DIMETHYLPHENYLAMINE □ N,N-DIMETHYL-
PHENYLAMINE □ DWUMETYLOANILINA (POLISH) □
NCI-C56428 □ VERSNELLER NL 63/10
CONSENSUS REPORTS: Reported in EPA
TSCA Inventory. Community Right-To-
Know List.
OSHA PEL: TWA 5 ppm; STEL 10 ppm
(skin)
ACGIH TLV: TWA 5 ppm; STEL 10 ppm
(skin); Not Classifiable as a Human
Carcinogen
DFG MAK: 5 ppm (25 mg/m³); Confirmed
Animal Carcinogen with Unknown
Relevance to Humans
DOT CLASSIFICATION: 6.1; Label: Poison
SAFETY PROFILE: Suspected carcinogen
with equivocal tumorigenic data. Human
poison by ingestion. Moderately toxic by
inhalation and skin contact. A skin irritant.
Human systemic effects by ingestion: nausea
or vomiting. Physiological action is similar
to, but less toxic than, aniline. A central
nervous system depressant. Mutation data
reported. Flammable liquid when exposed to
heat, flame, or oxidizers. Explodes on
contact with benzoyl peroxide or
diisopropyl peroxydicarbonate. To fight fire,
use foam, CO_2, dry chemical. When heated
to decomposition it emits highly toxic fumes
of aniline and NO_x. See also ANILINE.

DQG700   CAS: 13367-92-5   HR: 3
DIMETHYL ARSINIC SULFIDE
mf: $C_2H_6As_2S_2$ mw: 244.04
SYNS: ARSINE SULFIDE, DIMETHYLDI- □
DIMETHYLDIARSINE SULFIDE
OSHA PEL: TWA 0.5 mg(As)/m³
SAFETY PROFILE: Poison by intravenous
route. When heated to decomposition it
emits toxic fumes of SO_x and As.

DQJ200   CAS: 57-97-6   HR: 3
DIMETHYLBENZANTHRACENE
mf: $C_{20}H_{16}$ mw: 256.36
PROP: Leaflets from Me_2CO/EtOH. Mp:
122–123°.

SYNS: DBA □ DIMETHYLBENZ(a)ANTHRACENE □ 7,12-DIMETHYLBENZANTHRACENE □ 7,12-DIMETHY-LBENZ(a)ANTHRACENE □ 9,10-DIMETHYL-BENZA-NTHRACENE □ 9,10-DIMETHYLBENZ(a)-ANTHRAC-ENE □ 9,10-DIMETHYL-1,2-BENZANTHRAC-ENE □ 9,10-DIMETHYL-1,2-BENZANTHRAZEN (GERMAN) □ DIMETHYL-BENZANTHRENE □ 7,12-DIMETHYL--BENZO(a)ANTHRACENE □ 1,4-DIMETHYL-2,3-BENZPHEN-ANTHRENE □ DMBA □ 7,12-DMBA □ NCI-C03918 □ RCRA WASTE NUMBER U094

CONSENSUS REPORTS: Reported in EPA TSCA Inventory. EPA Genetic Toxicology Program.

SAFETY PROFILE: Suspected carcinogen with experimental carcinogenic, neoplastigenic, tumorigenic, and teratogenic data. A transplacental carcinogen. Poison by ingestion, intravenous, subcutaneous, intraperitoneal, and intratracheal routes. Other experimental reproductive effects. Human mutation data reported. A skin irritant. When heated to decomposition it emits acrid smoke and irritating fumes.

DQM000 CAS: 612-82-8 HR: 3
3,3'-DIMETHYLBENZIDINE DIHYDRO-
CHLORIDE

mf: $C_{14}H_{16}N_2 \cdot 2ClH$ mw: 285.24

SYNS: 4,4'-DIAMINO-3,3'-DIMETHYLBIPHENYL DIHYDROCHLORIDE □ 3,3'-DIMETHYLBIPHENYL-4,4'-BIPHENYLDIAMINE DIHYDROCHLORIDE □ 2,3'-DIMETHYLBIPHENYL-4,4'-DIAMINE DIHYDRO-CHLORIDE □ o-TOLIDINE DIHYDROCHLORIDE

CONSENSUS REPORTS: Reported in NTP Carcinogenesis Studies (Drinking); Clear Evidence: rat NTPTR* NTP-TR-390,91. Reported in EPA TSCA Inventory.

SAFETY PROFILE: Suspected carcinogen with experimental carcinogenic data. Mutation data reported. When heated to decomposition it emits very toxic fumes of HCl and NO_x.

DQM600 CAS: 22781-23-3 HR: 3
2,2-DIMETHYL-1,3-BENZODIOX-4-OL
METHYLCARBAMATE

mf: $C_{11}H_{13}NO_4$ mw: 223.25

PROP: Crystals. Mp: 129–130°. Very sltly sol in H_2O.

SYNS: BENCARBATE □ BENDIOCARB □ BICAM ULV □ 2,2-DIMETHYL-1,3-BENZDIOXOL-4-YL-N-METHYL-CARBAMATE □ 2,2-DIMETHYLBENZO-1,3-DIOXOL-4-YL METHYLCARBAMATE □ 2,2-DIMETHYL-4-(N-METHYLAMINOCARBOXYLATO)-1,3-BENZODIOXOLE □ 2,2-DIMETHYL-4-(N-METHYLCARBAMATO)-1,3-BENZODIOXOLE □ DYCARB □ FICAM □ GARVOX □ 2,3-ISOPROPYLIDENEDIOXYPHENYL METHYL-CARBAMATE □ MC6897 □ METHYLCARBAMIC ACID-2,3-(ISOPROPYLIDENEDIOXY)PHENYL ESTER □ MULTAMAT □ NIOMIL □ ROTATE □ TATTOO □ TURCAM

CONSENSUS REPORTS: EPA Genetic Toxicology Program.

SAFETY PROFILE: Poison by ingestion. Moderately toxic by skin contact. When heated to decomposition it emits toxic fumes of NO_x. See also CARBAMATES.

DQO650 CAS: 2626-34-8 HR: 3
5,6-DIMETHYL-2,1,3-BENZOSELENO-
DIAZOLE

mf: $C_8H_8N_2Se$ mw: 211.14

SYN: 2,1,3-BENZOSELENADIAZOLE, 5,6-DIMETHYL-

OSHA PEL: TWA 0.2 mg(Se)/m^3

ACGIH TLV: TWA 0.2 mg(Se)/m^3

SAFETY PROFILE: Poison by intravenous route. When heated to decomposition it emits toxic fumes of NO_x and Se.

DQP800 CAS: 103-83-3 HR: 3
N,N-DIMETHYLBENZYLAMINE

DOT: UN 2619

mf: $C_9H_{13}N$ mw: 135.23

PROP: Corrosive liquid. Bp: 181°.

SYNS: ARALDITE ACCELERATOR 062 □ BDMA □ BENZYLDIMETHYLAMINE □ BENZYL-N,N-DIMETHYLAMINE □ N-BENZYLDIMETHYLAMINE □ N,N-DIMETHYLBENZENEMETHANAMINE □ N-(PHENYLMETHYL)DIMETHYLAMINE □ SUMINE 2015

CONSENSUS REPORTS: Reported in EPA TSCA Inventory.

DOT CLASSIFICATION: 8; Label: Corrosive

SAFETY PROFILE: Poison by ingestion. Moderately toxic by inhalation and skin contact. Corrosive. A severe eye and skin irritant. Flammable when exposed to heat or flame. When heated to decomposition it emits toxic fumes of NO_x.

DQQ200   CAS: 100-86-7   HR: 2
DIMETHYL BENZYL CARBINOL
mf: $C_{10}H_{14}O$ mw: 150.24
PROP: Needles or white crystalline solid; floral odor. D: 0.972–0.977, mp: 24°, bp: 214–216°, flash p: 198°F. Sol in fixed oils, mineral oil, propylene glycol; insol in glycerin.
SYNS: BENZYL DIMETHYL CARBINOL □ α,α-DIMETHYLPHENETHYL ALCOHOL □ 1,1-DIMETHYL-2-PHENYLETHANOL □ DMBC □ FEMA No. 2393
CONSENSUS REPORTS: Reported in EPA TSCA Inventory.
SAFETY PROFILE: Moderately toxic by ingestion. Combustible liquid. When heated to decomposition it emits acrid smoke and irritating fumes.

DQQ375   HR: 2
DIMETHYL BENZYL CARBINYL ACETATE
mf: $C_{12}H_{16}O_2$ mw: 192.26
PROP: Colorless liquid to solid at room temp; floral, fruity odor. D: 0.995–1.002, refr index: 1.490–1.495, flash p: 212°F. Sol in fixed oils; sltly sol in propylene glycol; insol in water.
SYNS: α,α-DIMETHYLPHENETHYL ACETATE □ FEMA No. 2392
SAFETY PROFILE: Combustible liquid. When heated to decomposition it emits acrid smoke and irritating fumes.

DQQ380   HR: 2
DIMETHYL BENZYL CARBINYL BUTYRATE
mf: $C_{14}H_{20}O_2$ mw: 220.31
PROP: Colorless liquid; prunelike odor. D: 0.960–0.981, refr index: 1.473–1.493 @ 25°, flash p: 151°F. Sol in alc, fixed oils; insol in water, propylene glycol.
SYNS: α,α-DIMETHYLPHENRTHYL BUTYRATE □ FEMA No. 2394
SAFETY PROFILE: Combustible liquid. Use in accordance with good manufacturing practice.

DQR200   CAS: 506-63-8   HR: 3
DIMETHYL BERYLLIUM
mf: C_2H_6Be mw: 39.09
PROP: White needles or crystals. Bp: sublimes @ 200°.
CONSENSUS REPORTS: IARC Cancer Review: Group 1 IMEMDT 58,41,93; Human Sufficient Evidence IMEMDT 58,41,93; Animal Sufficient Evidence IMEMDT 1,17,72; Animal Sufficient Evidence IMEMDT 23,143,80; Animal Sufficient Evidence IMEMDT 58,41,93. Beryllium and its compounds are on the Community Right-To-Know List.
OSHA PEL: TWA 0.002 mg(Be)/m³; STEL 0.005 mg(Be)/m³/30M; CL 0.025 mg(Be)/m³
ACGIH TLV: TWA 0.002 mg(Be)/m³; Confirmed Human Carcinogen; (Proposed: TWA 0.00005 mg(Be)/m³; STEL 0.0002 mg(Be)/m³ (sensitizer) (skin); Confirmed Human Carcinogen)
SAFETY PROFILE: Confirmed human carcinogen. A poison. Flammable when exposed to heat or flame; can react with oxidizing materials. Explosive reaction on contact with water. Ignites on contact with moist air or carbon dioxide. Upon decomposition it emits highly toxic fumes of BeO. See also BERYLLIUM COMPOUNDS.

DQR289   HR: 3
DIMETHYLBERYLLIUM-1,2-DIME-THOXYETHANE
mf: $C_2H_6Be \cdot C_4H_{10}O_2$ mw: 129.21
CONSENSUS REPORTS: IARC Cancer Review: Group 1 IMEMDT 58,41,93; Human Sufficient Evidence IMEMDT 58,41,93; Animal Sufficient Evidence IMEMDT 1,17,72; Animal Sufficient Evidence IMEMDT 23,143,80; Animal Sufficient Evidence IMEMDT 58,41,93. Beryllium and its compounds are on the Community Right-To-Know List.
OSHA PEL: TWA 0.002 mg(Be)/m³; STEL 0.005 mg(Be)/m³/30M; CL 0.025 mg(Be)/m³
ACGIH TLV: TWA 0.002 mg(Be)/m³; Confirmed Human Carcinogen; (Proposed:

TWA 0.00005 mg(Be)/m³; STEL 0.0002 mg(Be)/m³ (sensitizer) (skin); Confirmed Human Carcinogen)
SAFETY PROFILE: Confirmed human carcinogen. Ignites spontaneously in air. Upon decomposition it emits highly toxic fumes of BeO. See also BERYLLIUM COMPOUNDS.

DQR600   CAS: 657-24-9   HR: 3
1,1-DIMETHYLBIGUANIDE
mf: $C_4H_{11}N_5$ mw: 129.20
PROP: IDLH 3 mg/m³.
SYNS: N,N-DIMETHYLBIGUANIDE □ N,N-DIMETHYLDIGUANIDE □ FLUMAMINE □ GLUCOPHAGE □ GLUCOPHAGE LA 6023 □ GLUEOPHOGE □ LA 6023 □ MELBIN □ METFORMIN □ NNDG
SAFETY PROFILE: Poison by subcutaneous and intraperitoneal routes. Mildly toxic by parenteral route. Experimental teratogenic effects. Mutation data reported. When heated to decomposition it emits toxic fumes of NO_x.

DQS000   CAS: 91-97-4   HR: 3
3,3'-DIMETHYL-4,4'-BIPHENYLENE
**   DIISOCYANATE**
mf: $C_{16}H_{12}N_2O_2$ mw: 264.30
PROP: Crystals from chlorobenzene. Mp: 70°.
SYNS: 4,4'-DIISOCYANATO-3,3'-DIMETHYL-1,1'-BIPHENYL □ ISOCYANIC ACID, 3,3'-DIMETHYL-4,4'-BIPHENYLENE ESTER
CONSENSUS REPORTS: Reported in EPA TSCA Inventory.
NIOSH REL: TWA (Diisocyanates) 0.005 ppm; CL 0.02 ppm/10M
SAFETY PROFILE: Poison by intravenous route. When heated to decomposition it emits toxic fumes of NO_x. See also CYANATES and ESTERS.

DQT200   CAS: 75-83-2   HR: 3
2,2-DIMETHYLBUTANE
mf: C_6H_{14} mw: 86.20
PROP: Liquid. Bp: 49.7°, mp: −98.2°, flash p: −54°F, fp: −101.9°, d: 0.649, autoign temp: 797°F, vap press: 400 mm @ 31.0°, vap d: 3.00, lel: 1.2%, uel: 7.0%.
SYN: NEOHEXANE (DOT)
CONSENSUS REPORTS: Reported in EPA TSCA Inventory.
OSHA PEL: TWA 500 ppm; STEL 1000 ppm
ACGIH TLV: TWA 500 ppm; STEL 1000 ppm
DFG MAK: 200 ppm (720 mg/m³)
NIOSH REL: (Alkanes) TWA 350 mg/m³
SAFETY PROFILE: Probably an irritant and narcotic in high concentration. A very dangerous fire and explosion hazard when exposed to heat or flame; can react vigorously with oxidizing materials. Keep away from heat or open flame. To fight fire, use foam, CO_2, dry chemical. When heated to decomposition it emits acrid smoke and irritating fumes.

DQT400   CAS: 79-29-8   HR: 3
2,3-DIMETHYLBUTANE
DOT: UN 2457
mf: C_6H_{14} mw: 86.20
PROP: Liquid. Mp: −135°, bp: 58.0°, flash p: −20°F, d: 0.662 @ 20°/4°, autoign temp: 788°F, vap press: 400 mm @ 39.0°, vap d: 3.0, lel: 1.2%, uel: 7.0%.
CONSENSUS REPORTS: Reported in EPA TSCA Inventory.
OSHA PEL: TWA 500 ppm; STEL 1000 ppm
ACGIH TLV: TWA 500 ppm; STEL 1000 ppm
DFG MAK: 200 ppm (720 mg/m³)
NIOSH REL: TWA (Alkanes) 350 mg/m³
DOT CLASSIFICATION: 3; Label: Flammable Liquid
SAFETY PROFILE: Probably an irritant and narcotic in high concentration. A very dangerous fire and explosion hazard when exposed to heat or flame; can react vigorously with oxidizing materials. Keep away from heat and open flame. To fight fire, use foam, CO_2, dry chemical. When heated to decomposition it emits acrid smoke and irritating fumes.

D

DQU600   CAS: 108-09-8   HR: 3
1,3-DIMETHYL BUTYLAMINE
DOT: UN 2379
mf: $C_6H_{15}N$ mw: 101.22
PROP: A liquid. Bp: 106–109°, flash p: 55°F
(OC), d: 0.750 @ 20°/20°.
SYN: 1,3-DIMETHYLBUTYLAMINE (DOT)
CONSENSUS REPORTS: Reported in EPA
TSCA Inventory.
DOT CLASSIFICATION: 3; Label:
Flammable Liquid
SAFETY PROFILE: Poison by intravenous
route. Moderately toxic by ingestion and
skin contact. Mildly toxic by inhalation. A
dangerous fire and explosion hazard when
exposed to heat or flame; can react
vigorously with oxidizing materials. To fight
fire, use foam, CO_2, dry chemical. When
heated to decomposition it emits toxic
fumes of NO_x. See also AMINES.

DQW800   CAS: 506-82-1   HR: 3
DIMETHYLCADMIUM
mf: C_2H_6Cd mw: 142.47
PROP: Oil or liquid; decomp by water; foul
odor. D: 1.984, mp: −2.4°, bp: 106°.
CONSENSUS REPORTS: Cadmium and its
compounds are on the Community Right-
To-Know List.
OSHA PEL: TWA 5 μg(Cd)/m³
ACGIH TLV: TWA 0.002 mg(Cd)/m³
(respirable dust), Suspected Human
Carcinogen); BEI: 5 μg/g creatinine in urine;
5 μg/L in blood
DFG MAK: DFG BAT: Blood 1.5 μg/dL;
Urine 15 μg/dL; Suspected Carcinogen
NIOSH REL: (Cadmium) Reduce to lowest
feasible level
SAFETY PROFILE: Confirmed human
carcinogen. Contact with air produces the
friction-sensitive explosive dimethyl
cadmium peroxide. Explodes when heated
above 150°C. Ignition may occur on contact
with air if the surface area is large. See also
CADMIUM COMPOUNDS.

DQY950   CAS: 79-44-7   HR: 3
DIMETHYLCARBAMOYL CHLORIDE

DOT: UN 2262
mf: C_3H_6ClNO mw: 107.55
PROP: Liquid. Mp: −33°, bp: 165–167°, d:
1.678 @ 20°/4°, vap d: 3.73.
SYNS: CARBAMIC ACID, DIMETHYL-(9CI) □
CARBAMYL CHLORIDE, N,N-DIMETHYL- □ CHLORID
KYSELINY DIMETHYLKARBAMINOVE □
CHLOROFORMIC ACID DIMETHYLAMIDE □ DDC □
DIMETHYLAMID KYSELINY CHLORMRAVENCI □
(DIMETHYLAMINO)CARBONYL CHLORIDE □ N,N-
DIMETHYLAMINOCARBONYL CHLORIDE □
DIMETHYLCARBAMIC ACID CHLORIDE □ N,N-
DIMETHYLCARBAMIC ACID CHLORIDE □
DIMETHYLCARBAMIC CHLORIDE □ DIMETHYL-
CARBAMIDOYL CHLORIDE □ N,N-DIMETHYL-
CARBAMIDOYL CHLORIDE □ N,N-DIMETHYL-
CARBAMOYL CHLORIDE □ DIMETHYL CARBAMOYL
CHLORIDE (ACGIH,DOT) □ DIMETHYLCARBAMYL
CHLORIDE □ N,N-DIMETHYLCARBAMYL CHLORIDE
□ DIMETHYLCHLOROFORMAMIDE □ DIMETHYL-
KARBAMOYLCHLORID □ DMCC □ RCRA WASTE
NUMBER U097 □ TL 389
CONSENSUS REPORTS: NTP 10th Report
on Carcinogens. IARC Cancer Review:
Group 2A IMEMDT 7,199,87; Animal
Sufficient Evidence IMEMDT 12,77,76;
Human Inadequate Evidence IMEMDT
12,77,76. EPA Genetic Toxicology Program.
Community Right-To-Know List. Reported
in EPA TSCA Inventory.
ACGIH TLV: TWA 0.005 ppm (skin);
Suspected Human Carcinogen
DFG MAK: Animal Carcinogen, Suspected
Human Carcinogen
DOT CLASSIFICATION: 8; Label: Corrosive
SAFETY PROFILE: Confirmed carcinogen
with experimental carcinogenic,
neoplastigenic, and tumorigenic data. Poison
by intraperitoneal route. Moderately toxic by
inhalation and ingestion. Human mutation
data reported. Can cause skin and papillary
tumors by skin contact, and squamous cell
carcinoma by inhalation. Will react with
water or steam to produce toxic and
corrosive fumes. A powerful lachrymator.
When heated to decomposition it emits very
toxic fumes of Cl⁻ and NO_x. See also
CHLORIDES.

DRC000   CAS: 4584-46-7   HR: 3

DIMETHYL(2-CHLOROETHYL)AMINE HYDROCHLORIDE

mf: $C_4H_{10}ClN \cdot ClH$ mw: 144.06

SYNS: 2-CHLORO-N,N-DIMETHYLETHYLAMINE HYDROCHLORIDE □ DIMETHYL-β-CHLORO-ETHYLAMINE HYDROCHLORIDE

CONSENSUS REPORTS: Reported in EPA TSCA Inventory.

SAFETY PROFILE: Poison by intraperitoneal and subcutaneous routes. Questionable carcinogen with experimental neoplastigenic data. Mutation data reported. When heated to decomposition it emits very toxic fumes of Cl⁻ and NO$_x$.

DRF600   CAS: 1467-79-4   HR: 3
DIMETHYLCYANAMIDE

mf: $C_3H_6N_2$ mw: 70.11

PROP: Colorless, mobile liquid. Mp: −41.0°, bp: 162–163°, flash p: 160°F (TCC), d: 0.8767 @ 30°, vap press: 40 mm @ 80°, vap d: 2.55.

CONSENSUS REPORTS: Reported in EPA TSCA Inventory.

SAFETY PROFILE: Poison by ingestion, skin contact, and intraperitoneal routes. Moderately toxic by inhalation. Flammable when exposed to heat, flame, or oxidizers. Can react with oxidizing materials. To fight fire, use foam, CO_2, or dry chemical. When heated to decomposition or in reaction with water or steam it produces toxic fumes of NO$_x$ and CN⁻ and flammable vapors. See also CYANIDE.

DRF709   CAS: 98-94-2   HR: 3
N,N-DIMETHYLCYCLOHEXANAMINE

DOT: UN 2264

mf: $C_8H_{17}N$ mw: 127.26

SYNS: CYCLOHEXYLDIMETHYLAMINE □ N-CYCLOHEXYLDIMETHYLAMINE □ (DIMETHYL-AMINO)CYCLOHEXANE □ N,N-DIMETHYLAMINO-CYCLOHEXANE □ DIMETHYLCYCLOHEXYLAMINE □ N,N-DIMETHYLCYCLOHEXYLAMINE (DOT) □ POLYCAT 8

CONSENSUS REPORTS: Reported in EPA TSCA Inventory.

DOT CLASSIFICATION: 8; Label: Corrosive

SAFETY PROFILE: Poison by ingestion. Moderately toxic by inhalation. When

heated to decomposition it emits toxic fumes of NO$_x$.

DRI500   HR: D
DIMETHYL DIALKYL AMMONIUM CHLORIDE

SAFETY PROFILE: When heated to decomposition it emits acrid smoke and irritating fumes.

DRJ800   CAS: 78-63-7   HR: 2
2,5-DIMETHYL-2,5-DI(tert-BUTYLPEROXY)HEXANE

mf: $C_{16}H_{34}O_4$ mw: 290.50

PROP: Colorless to light-yellow liquid. D: 0.85, fp: 8°, flash p: >180°F (COC), bp: 250°. Insol in water; sol in many orgsolvs.

SYNS: 2,5-DIMETHYL-2,5-DI(t-BUTYLPEROXY)-HEXANE □ PEROXIDE, (1,1,4,4-TETRAMETHYL-1,4-BUTANEDIYL)BIS(1,1-DIMETHYLETHYL) □ PEROXIDE, (1,1,4,4-TETRAMETHYLTETRAMETHYLENE)BIS(tert-BUTYL) □ TRIGONOX 101-101/45 □ VAROX

CONSENSUS REPORTS: Reported in EPA TSCA Inventory.

SAFETY PROFILE: Moderately toxic by intraperitoneal route. Combustible when exposed to heat, flames, or reducing agents. To fight fire, use water spray, foam, dry chemical. When heated to decomposition it emits acrid smoke and irritating fumes. Used in the polymerization of styrene and in cross-linking of various grades of polyethylene. See also PEROXIDES, ORGANIC.

DRL100   CAS: 50539-74-7   HR: 3
2,2-DIMETHYL-2,3-DIHYDROBENZO-FURAN-7-YL-N-(4-BROMOPHENYL-THIO)-N-METHYLCARBAMATE

mf: $C_{18}H_{18}BrNO_3S$ mw: 408.34

SYN: CARBAMIC ACID, ((p-BROMOPHENYL)THIO)-METHYL-, 2,3-DIHYDRO-2,2-DIMETHYL-7-BENZO-FURANYL ESTER

SAFETY PROFILE: A poison by ingestion. When heated to decomposition it emits toxic vapors of NO$_x$, SO$_x$, and Br⁻.

DRL425   CAS: 90293-48-4   HR: 3
N,5-DIMETHYL-4-((DIMETHYLAMINO)-CARBONYL)-N-((4-(1,1-DIMETHYL-

ETHYL)PHENYL)THIO)-2,7-DIOXA-3,6-DIAZAOCTA-3,5-DIENAMIDE

mf: $C_{19}H_{28}N_4O_4S$ mw: 408.57

SAFETY PROFILE: A poison by ingestion. When heated to decomposition it emits toxic vapors of NO_x and SO_x.

DRM000   CAS: 3759-07-7   HR: 3

9,9-DIMETHYL-10-DIMETHYLAMINO-PROPYLACRIDAN HYDROGEN TARTRATE

mf: $C_{20}H_{26}N_2 \cdot C_4H_4O_6$ mw: 442.56

PROP: A solid. Mp: 155–156°.

SYNS: DIMETACRINE BITARTRATE □ DIMETACRIN HYDROGENTARTRATE □ DIMETHACRINE TARTRATE □ 10-(3-(DIMETHYLAMINO)PROPYL)-9,9-DIMETHYL-ACRIDAN TARTRATE (1:1) □ 9,9-DIMETHYL-10-(3-DIMETHYLAMINO)PROPYLACRIDINE TARTRATE □ ISOTONIL □ ISTONYL □ MIROISTONIL □ MO 709 □ SD 709 □ ((R-R*,R*))-N,N,9,9-TETRAMETHYL-10(9H)-ACRIDINEPROPANAMINE-2,3-DIHYDROXY-BUTANEDIOATE (1:1)

SAFETY PROFILE: Poison by ingestion, intravenous, and intraperitoneal routes. Moderately toxic by subcutaneous route. Experimental teratogenic and reproductive effects. When heated to decomposition it emits toxic fumes of NO_x.

DRM110   CAS: 3215-89-2   HR: 3

3,3-DIMETHYL-4-(DIMETHYLAMINO)-4-(o-TOLYL)BUTYL o-TOLYL KETONE

mf: $C_{23}H_{31}NO$ mw: 337.55

SYN: KETONE, 3,3-DIMETHYL-4-(DIMETHYLAMINO)-4-(o-TOLYL)BUTYL o-TOLYL

DOT CLASSIFICATION: 3; Label: Flammable Liquid

SAFETY PROFILE: A poison by ingestion. A flammable liquid. When heated to decomposition it emits toxic vapors of NO_x.

DRO400   CAS: 133-55-1   HR: D

N,N′-DIMETHYL-N,N′-DINITROSO-TEREPHTHALAMIDE

DOT: UN 2973

mf: $C_{10}H_{10}N_4O_4$ mw: 250.24

SYNS: 1,4-BENZENEDICARBOXAMIDE, N,N′-DIMETHYL-N,N′-DINITROSO-(9CI) □ N,N′-DIMETHYL-N,N′-DINITROSO-1,4-BENZENEDICARBOXAMIDE □ N,N′-DINITROSO-N,N′-DIMETHYLTEREPHTAL-SAUREAMID □ N,N′-DINITROSO-N,N′-DIMETHYL-TEREPHTHALAMIDE, not >72% as a paste (DOT)

CONSENSUS REPORTS: EPA Genetic Toxicology Program. Reported in EPA TSCA Inventory.

DOT CLASSIFICATION: 4.1; Label: Flammable Solid, EXPLOSIVE

SAFETY PROFILE: Mutation data reported. Many N-nitroso compounds are carcinogens. A flammable solid and explosive. When heated to decomposition it emits toxic fumes of NO_x. See also N-NITROSO COMPOUNDS.

DRQ400   CAS: 624-92-0   HR: 3

DIMETHYL DISULFIDE

DOT: UN 2381

mf: $C_2H_6S_2$ mw: 94.20

PROP: A liquid. Flash p: 44.6°F, bp: 109.7°, d: 1.057 @ 16°/4°, vap press: 28.6 mm @ 25°, vap d: 3.24.

CONSENSUS REPORTS: Reported in EPA TSCA Inventory. EPA Extremely Hazardous Substances List.

ACGIH TLV: 0.5 ppm (skin)

DOT CLASSIFICATION: 3; Label: Flammable Liquid

SAFETY PROFILE: Poison by inhalation. A very dangerous fire hazard when exposed to heat, flame, or oxidizers. Can react vigorously with oxidizing materials. See also SULFIDES.

DRR200   CAS: 2540-82-1   HR: 3

O,O-DIMETHYL DITHIOPHOSPHORYL-ACETIC ACID-N-METHYL-N-FORMYLAMIDE

mf: $C_6H_{12}NO_4PS_2$ mw: 257.28

PROP: Yellow viscous oil or crystal mass. D: 1.361 @ 20°/4°, mp: 25–26°. Slt sol in H_2O; misc in most org solvs.

SYNS: AFLIX □ ANTHIO □ ANTIO □ CP 53926 □ O,O-DIMETHYL-S-(N-FORMYL-N-METHYLCARBAMOYL-METHYL) PHOSPHORODITHIOATE □ O,O-DIMETHYL-S-(3-METHYL-2,4-DIOXO-3-AZA-BUTYL)-DITHIO-FOSFAAT (DUTCH) □ O,O-DIMETHYL-S-(3-METHYL-2,4-DIOXO-3-AZA-BUTYL)-DITHIOPHOSPHAT (GERMAN) □ O,O-DIMETHYL-S-(N-METHYL-N-FORMYL-CARB-AMOYLMETHYL)-DITHIOPHOSPHAT □ O,O-DIMETH-YL-S-(N-METHYL-N-FORMYLCARBAMOYL-METHYL)-PHOSPHORODITHIOATE □ O,O-DIMETHYL PHOS-PHORODITHIOATE N-FORMYL-2-MERCAPTO-N-METHYLACETAMIDE-S-ESTER □ O,O-DIMETIL-S-(N-

FORMIL-N-METIL-CARBAMOIL-METIL)-DITIOFOSF-ATO (ITALIAN) □ ENT 27,257 □ FORMOTHION □ S-(2-(FORMYLMETHYLAMINO)-2-OXOETHYL)-O,O-DIMETHYLPHOSPHORODITHIOATE □ N-FORMYL-N-METHYLCARBAMOYLMETHYL-O,O-DIMETHYL PHOSPHORODITHIOATE □ S-(N-FORMYL-N-METHYL-CARBAMOYLMETHYL)-O,O-DIMETHYL PHOSPHORO-DITHIOATE □ S-(N-FORMYL-N-METHYLCARBAMOYL-METHYL) DIMETHYL PHOSPHOROTHIOLOTHIONATE □ S 6900 □ SAN 244 I □ SAN 6913 I □ SAN 7107 I □ SPENCER S-6900 □ VEL 4284

CONSENSUS REPORTS: EPA Extremely Hazardous Substances List.

SAFETY PROFILE: Poison by ingestion, inhalation, skin contact, and intravenous routes. Mutation data reported. When heated to decomposition it emits very toxic fumes of NO_x, PO_x, and SO_x. See also ESTERS.

DRS800   CAS: 120-08-1   HR: 3
6,7-DIMETHYLESCULETIN
mf: $C_{11}H_{10}O_4$ mw: 206.21
PROP: Needles from H_2O. Mp: 144°.
SYNS: AESCULETIN DIMETHYL ETHER □ 6,7-DIMETHOXYBENZOPYRAN-2-ONE □ 6,7-DIMETHOXYCOUMARIN □ ESCOPARONE □ ESCULETIN DIMETHYL ETHER □ SCOPARON □ SCOPARONE

SAFETY PROFILE: Poison by ingestion and intraperitoneal routes. Experimental reproductive effects. An antihypertensive agent. When heated to decomposition it emits acrid smoke and irritating fumes.

DRV300   CAS: 71108-04-8   HR: 3
5,5-DIMETHYL-2-(ETHYLIMINO)-1,3-DITHIOLAN-4-ONE-o-((METHYL-AMINO)CARBONYL)OXIME
mf: $C_9H_{15}N_3O_2S_2$ mw: 261.39
SYN: 1,3-DITHIOLAN-4-ONE, 5,5-DIMETHYL-2-(ETHYLIMINO)-, o-((METHYLAMINO)-CARBONYL)OXIME

SAFETY PROFILE: A poison by ingestion. When heated to decomposition it emits toxic vapors of NO_x and SO_x.

DSQ810   CAS: 302542-44-5 HR: 3
3,5-DIMETHYL-N-(2-METHYLPHENYL)-4-NITRO-1H-PYRAZOLE-1-ACETAMIDE
mf: $C_{14}H_{16}N_4O_3$ mw: 288.31

SAFETY PROFILE: A poison by subcutaneous and ocular routes. Moderately toxic by ingestion. When heated to decomposition it emits toxic vapors of NO_x.

DSQ830   CAS: 302542-50-3 HR: 3
3,5-DIMETHYL-N-(3-METHYLPHENYL)-1H-PYRAZOLE-1-ACETAMIDE
mf: $C_{14}H_{17}N_3O$ mw: 243.31
SAFETY PROFILE: A poison by subcutaneous and ocular routes. Moderately toxic by ingestion. When heated to decomposition it emits toxic vapors of NO_x.

DSB000   CAS: 68-12-2   HR: 3
DIMETHYLFORMAMIDE
DOT: UN 2265
mf: C_3H_7NO mw: 73.11
PROP: Colorless, mobile liquid; fishy or faint amine odor. Mp: −61°, bp: 152.8°, lel: 2.2% @ 100°, uel: 15.2% @ 100°, flash p: 136°, d: 0.945 @ 22.4°/4°, autoign temp: 833°F, vap press: 3.7 mm @ 25°, vap d: 2.51. Misc in H_2O, EtOH, Et_2O, C_6H_6, and $CHCl_3$. IDLH 500 ppm.
SYNS: DIMETHYLFORMAMID (GERMAN) □ N,N-DIMETHYL FORMAMIDE □ N,N-DIMETHYL-FORMAMIDE (DOT) □ DIMETILFORMAMIDE (ITALIAN) □ DIMETYLFORMAMIDU (CZECH) □ DMF □ DMFA □ DWUMETHYLOFORMAMID (POLISH) □ N-FORMYLDIMETHYLAMINE □ NCI-C60913 □ NSC-5356 □ U-4224

CONSENSUS REPORTS: IARC Cancer Review: Group 2B IMEMDT 47,171,89; Human Limited Evidence IMEMDT 47,171,89; Animal Inadequate Evidence IMEMDT 47,171,89. EPA Genetic Toxicology Program. Reported in EPA TSCA Inventory.
OSHA PEL: TWA 10 ppm (skin)
ACGIH TLV: TWA 10 ppm (skin); Not Classifiable as a Human Carcinogen; BEI: 40 mg/L N-methylforamide in urine at end of shift
DFG MAK: 10 ppm (30 mg/m³)
DOT CLASSIFICATION: 3; Label: Flammable Liquid
SAFETY PROFILE: Suspected carcinogen. Moderately toxic by ingestion, intravenous,

subcutaneous, intramuscular, and intraperitoneal routes. Mildly toxic by skin contact and inhalation. Experimental teratogenic and reproductive effects. A skin and severe eye irritant. Human mutation data reported. Flammable liquid when exposed to heat or flame; can react with oxidizing materials. Explosion hazard when exposed to flame. Explosive reaction with bromine, potassium permanganate, triethylaluminum + heat. Forms explosive mixtures with lithium azide (shock-sensitive above 200°C), uranium perchlorate. Ignition on contact with chromium trioxide. Violent reaction with chlorine, sodium hydroborate + heat, diisocyanatomethane, carbon tetrachloride + iron, 1,2,3,4,5,6-hexachlorocyclohexane + iron. Vigorous exothermic reaction with magnesium nitrate, sodium + heat, sodium hydride + heat, sulfinyl chloride + traces of iron or zinc, 2,4,6-trichloro-1,3,5-triazine (with gas evolution), and many other materials. Avoid contact with halogenated hydrocarbons, inorganic and organic nitrates, (2,5-dimethyl pyrrole + P(OCl)$_3$), C_6Cl_6, methylene diisocyanates, P_2O_3. To fight fire, use foam, CO_2, dry chemical. When heated to decomposition it emits toxic fumes of NO$_x$.

DSB200   CAS: 533-74-4   HR: 3
DIMETHYLFORMOCARBOTHIALDINE
mf: $C_5H_{10}N_2S_2$ mw: 162.29
PROP: Crystals from Me$_2$CO/hexane. Mp: 106°. Sol in alc.
SYNS: BASAMID □ BASAMID G □ BASAMID-GRANULAR □ BASAMID P □ BASAMID-PUDER □ CARBOTHIALDIN □ CARBOTHIALDINE □ CRAG 974 □ CRAG FUNGICIDE 974 □ CRAG NEMACIDE □ CRAG 85W □ DAZOMET □ 3,5-DIMETHYLPERHYDRO-1,3,5-THIADIAZIN-2-THION (CZECH, GERMAN) □ 3,5-DIMETHYLTETRAHYDRO-1,3,5-THIADIAZINE-2-THIONE □ 3,5-DIMETHYLTETRAHYDRO-1,3,5-2H-THIADIAZINE-2-THIONE □ 3,5-DIMETHYL-1,2,3,5-TETRAHYDRO-1,3,5-THIADIAZINETHIONE-2 □ 3,5-DIMETHYLTETRAHYDRO-2H-1,3,5-THIADIAZINE-2-THIONE □ 3,5-DIMETHYL-1,3,5-2H-TETRAHYDRO-THIADIAZINE-2-THIONE □ 3,5-DIMETHYL-2-THIONOTETRAHYDRO-1,3,5-THIADIAZINE □ 3,5-DIMETIL-PERIDRO-1,3,5-THIADIAZIN-2-TIONE (ITALIAN) □ DMTT □ FENNOSAN B 100 □ MICOFUME

□ MYLON (CZECH) □ MYLONE □ MYLONE 85 □ N 521 □ NALCON 243 □ NEFUSAN □ PREZERVIT □ STAUFFER N 521 □ TETRAHYDRO-2H-3,5-DIMETHYL-1,3,5-THIADIAZINE-2-THIONE □ TETRAHYDRO-3,5-DIMETHYL-2H-1,3,5-THIADIAZINE-2-THIONE □ THIAZON □ THIAZONE □ 2-THIO-3,5-DIMETHYL-TETRAHYDRO-1,3,5-THIADIAZINE □ TIAZON □ TROYSAN 142 □ UCC 974
CONSENSUS REPORTS: Reported in EPA TSCA Inventory.
SAFETY PROFILE: Poison by ingestion and intraperitoneal routes. Moderately toxic by skin contact and subcutaneous routes. A severe eye irritant. A mild primary skin irritant and sensitizer. When heated to decomposition it emits very toxic fumes of NO$_x$ and SO$_x$.

DSC100   CAS: 4568-81-4   HR: 3
2,5-DIMETHYL-3-FURYL p-HYDROXY-PHENYL KETONE
mf: $C_{13}H_{12}O_3$ mw: 216.25
SYNS: DB 135 □ DIMETHYL-2,5 (HYDROXY 4 BENZOYL) 3 FURANNE □ KETONE, 2,5-DIMETHYL-3-FURYL p-HYDROXYPHENYL
DOT CLASSIFICATION: 3; Label: Flammable Liquid
SAFETY PROFILE: A poison by intravenous route. Moderately toxic by intraperitoneal route. A flammable liquid. When heated to decomposition it emits acrid smoke and irritating vapors.

DSD775   CAS: 106-72-9   HR: 1
2,6-DIMETHYL-5-HEPTENAL
mf: $C_9H_{16}O$ mw: 140.23
PROP: Pale-yellow liquid or oil; melon odor. D: 0.852–0.858, refr index: 1.443–1.448
SYN: FEMA No. 2497
SAFETY PROFILE: Skin and eye irritant. When heated to decomposition it emits acrid smoke and irritating fumes.

DSE800   HR: 3
DIMETHYLHEXANE DIHYDRO-PEROXIDE (dry)
PROP: Fine, white crystals; insol in hydrocarbons; sltly sol in water, esters, and glycerin; sol in other org solvs. Mp: 104°.

SYNS: 2,5-DIMETHYL-2,5-DIHYDROPEROXY-HEXANE, >82% with water (DOT) □ HEXANE, 2,5-DIMETHYL-, 2,5-DIHYDROPEROXIDE

CONSENSUS REPORTS: Reported in EPA TSCA Inventory.

DOT CLASSIFICATION: Forbidden

SAFETY PROFILE: A reactive peroxide. When heated to decomposition it emits acrid smoke and fumes. See also PEROXIDES, ORGANIC.

DSF400   CAS: 57-14-7   HR: 3
1,1-DIMETHYLHYDRAZINE

DOT: UN 1163

mf: $C_2H_8N_2$ mw: 60.12

PROP: Colorless liquid; ammonia-like odor. Hygroscopic, water-misc. Mp: $-58°$, bp: 63.3°, flash p: 5°F, d: 0.791 @ 22°, vap press: 157 mm @ 25°, vap d: 1.94, autoign temp: 480°F, lel: 2%, uel: 95%. Sol in H_2O and EtOH. IDLH 15 ppm.

SYNS: DIMAZINE □ DIMETHYLHYDRAZINE □ asym-DIMETHYLHYDRAZINE □ N,N-DIMETHYLHYDRAZINE □ uns-DIMETHYLHYDRAZINE □ unsym-DIMETHYLHYDRAZINE □ 1,1-DIMETHYL-HYDRAZINE (GERMAN) □ DIMETHYLHYDRAZINE, unsymmetrical (DOT) □ DMH □ NIESYMETRYCZNA DWU METYLOHYDRAZYNA (POLISH) □ RCRA WASTE NUMBER U098 □ UDMH (DOT)

CONSENSUS REPORTS: NTP 10th Report on Carcinogens. IARC Cancer Review: Group 2B IMEMDT 7,56,87; Animal Sufficient Evidence IMEMDT 4,137,74. EPA Genetic Toxicology Program. Community Right-To-Know List. EPA Extremely Hazardous Substances List. Reported in EPA TSCA Inventory.

OSHA PEL: TWA 0.5 ppm (skin)

ACGIH TLV: TWA 0.01 ppm (skin), Confirmed Animal Carcinogen.

DFG MAK: Animal Carcinogen, Suspected Human Carcinogen

NIOSH REL: (Hydrazines) CL 0.15 mg/m³/2H

DOT CLASSIFICATION: 6.1; Label: Poison, Flammable Liquid, Corrosive

SAFETY PROFILE: Confirmed carcinogen with experimental carcinogenic, tumorigenic, and teratogenic data. Other experimental reproductive effects. Poison by ingestion, intraperitoneal, intravenous, and intracerebral routes. Moderately toxic by inhalation and skin contact. Human mutation data reported. A plant growth control agent. Corrosive. A powerful reducing agent. A dangerous fire hazard. It is hypergolic with many oxidants (e.g., dinitrogen tetroxide, hydrogen peroxide, and nitric acid). Dangerous when exposed to heat, flame, or oxidizers; can react vigorously with oxidizing materials such as air, fuming HNO_3, $(HNO_3 + N_2O_4)$, NO. A high-energy propellant for liquid-fueled rockets. To fight fire, use alcohol foam, CO_2, dry chemical. When heated to decomposition it emits highly toxic fumes of NO_x. See also HYDRAZINE.

DSF600   CAS: 540-73-8   HR: 3
1,2-DIMETHYLHYDRAZINE

DOT: UN 2382

mf: $C_2H_8N_2$ mw: 60.12

PROP: Clear, colorless, flammable, hygroscopic, fuming liquid; fishy ammonia odor. Flash p: <73.4°F, bp: 81°, mp: $-9°$, d: 0.8274 @ 20°/4°. Sol in H_2O, EtOH, etc.

SYNS: 1,2-DIMETHYLHYDRAZIN (GERMAN) □ DIMETHYLHYDRAZINE, symmetrical (DOT) □ N,N'-DIMETHYLHYDRAZINE □ sym-DIMETHYL-HYDRAZINE □ 1,2-DIMETHYL-HYDRAZINE □ DMH □ HYDRAZOMETHANE □ RCRA WASTE NUMBER U099 □ SDMH □ SYMETRYCZNA DWUMETYLOHYDRAZYNA (POLISH)

CONSENSUS REPORTS: IARC Cancer Review: Group 2B IMEMDT 7,56,87; Animal Sufficient Evidence IMEMDT 4,145,74. EPA Genetic Toxicology Program.

DFG MAK: Animal Carcinogen, Suspected Human Carcinogen

DOT CLASSIFICATION: 3; Label: Flammable Liquid, Poison

SAFETY PROFILE: Confirmed carcinogen with experimental carcinogenic, neoplastigenic, tumorigenic, and teratogenic data. Poison by ingestion, intraperitoneal, intravenous, subcutaneous, and intramuscular routes. Moderately toxic by inhalation. Human mutation data reported. A very dangerous fire hazard when exposed to heat, flame, or oxidizers. A high-energy propellant for liquid-fueled rockets. When heated to decomposition it emits toxic fumes of NO$_x$.

DSF800 CAS: 306-37-6 HR: 3
1,2-DIMETHYLHYDRAZINE DIHYDRO-CHLORIDE
mf: C$_2$H$_8$N$_2$•2ClH   mw: 133.04
PROP: A solid. Mp: 170°.
SYNS: N,N'-DIMETHYLHYDRAZINE DIHYDRO-CHLORIDE □ sym-DIMETHYLHYDRAZINE DIHYDRO-CHLORIDE □ DMH
CONSENSUS REPORTS: Reported in EPA TSCA Inventory. EPA Genetic Toxicology Program.
SAFETY PROFILE: Suspected carcinogen with experimental carcinogenic, neoplastigenic, and tumorigenic data. Poison by ingestion and subcutaneous routes. Experimental reproductive effects. Mutation data reported. A rocket fuel. When heated to decomposition it emits very toxic fumes of HCl and NO$_x$.

DSG000 CAS: 593-82-8 HR: 3
1,1-DIMETHYLHYDRAZINE HYDRO-CHLORIDE
mf: C$_2$H$_8$N$_2$•ClH   mw: 96.58
PROP: Hygroscopic crystals from EtOH. Mp: 83°.
NIOSH REL: (Hydrazines) CL 0.15 mg/m^3/2H
SAFETY PROFILE: Poison by ingestion, intraperitoneal, and intravenous routes. Questionable carcinogen with experimental tumorigenic data. Mutation data reported. When heated to decomposition it emits very toxic fumes of HCl and NO$_x$.

DSG200 CAS: 56400-60-3 HR: 3
1,2-DIMETHYLHYDRAZINE HYDRO-CHLORIDE
mf: C$_2$H$_8$N$_2$•ClH   mw: 96.58
SYNS: sym-DIMETHYLHYDRAZINE HYDRO-CHLORIDE □ DMH
SAFETY PROFILE: Poison by ingestion, intraperitoneal, subcutaneous, and intravenous routes. Questionable carcinogen

with experimental carcinogenic, neoplastigenic, and tumorigenic data. Mutation data reported. When heated to decomposition it emits very toxic fumes of HCl and NO$_x$. See also 1,1-DIMETHYL HYDRAZINE.

DSG330 CAS: 1086-34-6 HR: 3
9-(2,2-DIMETHYLHYDRAZINO)ACRID-INE MONOHYDROCHLORIDE
mf: C$_{15}$H$_{15}$N$_3$•ClH   mw: 273.79
SYN: ACRIDINE, 9-(2,2-DIMETHYLHYDRAZINO)-, MONOHYDROCHLORIDE
SAFETY PROFILE: A poison by ingestion. When heated to decomposition it emits toxic vapors of NO$_x$, HCl, and Cl$^-$.

DSG400 CAS: 26049-69-4 HR: 3
2-(2,2-DIMETHYLHYDRAZINO)-4-(5-NITRO-2-FURYL)THIAZOLE
mf: C$_9$H$_{10}$N$_4$O$_3$S   mw: 254.29
SYN: DMNT
CONSENSUS REPORTS: EPA Genetic Toxicology Program.
SAFETY PROFILE: Suspected carcinogen with experimental carcinogenic data. Mutation data reported. When heated to decomposition it emits very toxic fumes of NO$_x$ and SO$_x$.

DSG600 CAS: 868-85-9 HR: 3
DIMETHYL HYDROGEN PHOSPHITE
mf: C$_2$H$_7$O$_3$P   mw: 110.06
PROP: D: 1.20 @ 20°/4°, bp: 56.5° @ 8 mm.
SYNS: BIS(HYDROXYMETHYL)PHOSPHINE OXIDE □ DIMETHOXYPHOSPHINE OXIDE □ DIMETHYL ACID PHOSPHITE □ DIMETHYLESTER KYSELINY FOSFORITE (CZECH) □ DIMETHYLFOSFIT □ DIMETHYLFOSFONAT □ DIMETHYLHYDRO-GENPHOSPHITE □ DIMETHYL PHOSPHITE □ DIMETHYL PHOSPHONATE □ DIMETHYL PHOSPHOROUS ACID □ HYDROGEN DIMETHYL PHOSPHITE □ METHYL PHOSPHONATE □ NCI-C54773 □ PHOSPHOROUS ACID DIMETHYL ESTER
CONSENSUS REPORTS: IARC Cancer Review: Group 3 IMEMDT 48,85,90; Animal Limited Evidence IMEMDT 48,85,90. NTP Carcinogenesis Studies (gavage); No Evidence: mouse NTPTR*

NTP-TR-287,85; Clear Evidence: rat NTPTR* NTP-TR-287,85. Reported in EPA TSCA Inventory.

DFG MAK: Confirmed Animal Carcinogen with Unknown Relevance to Humans

SAFETY PROFILE: Suspected carcinogen with experimental carcinogenic data. Moderately toxic by ingestion and skin contact. A skin and eye irritant. Mutation data reported. When heated to decomposition it emits toxic fumes of PO$_x$.

DSK200   CAS: 119-38-0   HR: 3
DIMETHYL-5-(1-ISOPROPYL-3-METHYL-PYRAZOLYL)CARBAMATE

mf: C$_{10}$H$_{17}$N$_3$O$_2$ mw: 211.30

SYNS: DIMETHYLCARBAMATE-d'l-ISOPROPYL-3-METHYL-5-PYRAZOLYLE (FRENCH) □ DIMETHYL-CARBAMIC ACID 3-METHYL-1-(1-METHYLETHYL)-1H-PYRAZOL-5-YL ESTER □ ENT 19,060 □ GEIGY G-23611 □ ISOLAN □ ISOLANE (FRENCH) □ (1-ISOPROPIL-3-METIL-1H-PIRAZOL-5-IL)-N,N-DIMETIL-CARBAMMATO (ITALIAN) □ (1-ISOPROPYL-3-METHYL-1H-PYRAZOL-5-YL)-N,N-DIMETHYLCARBAMAAT (DUTCH) □ (1-ISOPROPYL-3-METHYL-1H-PYRAZOL-5-YL)-N,N-DIMETHYL-CARBAMAT (GERMAN) □ ISOPROPYL-METHYLPYRAZOLYL DIMETHYLCARBAMATE □ 1-ISOPROPYL-3-METHYL-5-PYRAZOLYL DIMETHYL-CARBAMATE □ 1-ISOPROPYL-3-METHYLPYRAZOLYL-(5)-DIMETHYLCARBAMATE □ 5-METHYL-2-ISOPROP-YL-3-PYRAZOLYL DIMETHYLCARBAMATE □ PRIMIN □ SAOLAN

CONSENSUS REPORTS: EPA Extremely Hazardous Substances List.

SAFETY PROFILE: Poison by ingestion, skin contact, and intraperitoneal routes. Questionable carcinogen with experimental tumorigenic data. Mutation data reported. An insecticide. When heated to decomposition it emits toxic fumes of NO$_x$. See also CARBAMATES.

DSK300   CAS: 74038-78-1   HR: 3
4,4-DIMETHYL-1-ISOPROPYL-2-NONYL-2-IMIDAZOLINE

mf: C$_{17}$H$_{34}$N$_2$ mw: 266.53

SYN: 2-IMIDAZOLINE, 4,4-DIMETHYL-1-ISOPROPYL-2-NONYL-

SAFETY PROFILE: A poison by intravenous route. When heated to decomposition it emits toxic vapors of NO$_x$.

DSK900   CAS: 534-13-4   HR: 2
1,3-DIMETHYLISOTHIOUREA

mf: C$_3$H$_8$N$_2$S   mw: 104.19

PROP: Colorless, exceedingly deliquescent crystals. Mp: 52°. Very sol in water, alc, acetone; sparingly sol in benzene, ether, carbon disulfide; very sltly sol in pet ether.

SYNS: DIMETHYLTHIOCARBAMIDE □ N,N'-DIMETHYLTHIOCARBAMIDE □ sym-DIMETHYL-THIOUREA □ 1,3-DIMETHYLTHIOUREA

CONSENSUS REPORTS: Reported in EPA TSCA Inventory.

SAFETY PROFILE: Moderately toxic by ingestion. Experimental teratogenic and reproductive effects. When heated to decomposition it emits toxic fumes of SO$_x$ and NO$_x$.

DSL600   CAS: 2999-74-8   HR: 3
DIMETHYLMAGNESIUM

mf: C$_2$H$_6$Mg   mw: 54.38

PROP: A solid. Stable to 2°.

CONSENSUS REPORTS: Reported in EPA TSCA Inventory.

SAFETY PROFILE: The solid and its solution in ether ignite on contact with water. The powder ignites on contact with moist air. When heated to decomposition it emits irritating fumes of MgO. See also MAGNESIUM COMPOUNDS.

DSM000   CAS: 766-39-2   HR: 2
α,β-DIMETHYLMALEIC ANHYDRIDE

mf: C$_6$H$_6$O$_3$ mw: 126.12

PROP: Pearly plates or leaflets. Mp: 96°, bp: 223°.

SYN: DIMETHYLMALEIC ANHYDRIDE

CONSENSUS REPORTS: Reported in EPA TSCA Inventory.

SAFETY PROFILE: Questionable carcinogen with experimental tumorigenic data. When heated to decomposition it emits acrid smoke and irritating fumes. See also ANHYDRIDES.

DSM450   CAS: 593-74-8   HR: 3
DIMETHYL MERCURY

mf: C$_2$H$_6$Hg   mw: 230.67

PROP: Volatile, colorless liquid with faint sweet odor. D: 3.1874 @ 20°/4°, bp: 92° @ 761 mm. Insoluble in water; very sol in alc and ether.

SYN: MERCURY, DIMETHYL

CONSENSUS REPORTS: IARC Cancer Review: Group 2B IMEMDT 58,239,93; Human Inadequate Evidence IMEMDT 58,239,93. Reported in EPA TSCA Inventory.

OSHA PEL: TWA 0.01 mg(Hg)/m³; CL 0.03 mg(Hg)/m³ (skin)

ACGIH TLV: TWA 0.01 mg(Hg)/m³; BEI: 35 µg/g creatinine total inorganic mercury in urine preshift; 15 µg/g creatinine total inorganic mercury in blood at end of shift at end of workweek.

DFG MAK: Confirmed Animal Carcinogen with Unknown Relevance to Humans

SAFETY PROFILE: Suspected carcinogen. Highly toxic. Mutation data reported. Easily flammable. When heated to decomposition it emits toxic fumes of Hg.

DSN600   CAS: 7203-92-1   HR: 3
3,3-DIMETHYL-1-p-METHOXYPHENYL-
**   TRIAZENE**

mf: $C_9H_{13}N_3O$ mw: 179.25

SYNS: 1-p-METHOXYFENYL-3,3-DIMETHYLTRIAZEN (CZECH) □ 1-(p-METHOXYPHENYL)-3,3-DIMETHYL-TRIAZENE □ 1-(4-METHYLOXYPHENYL)-3,3-DIMETHYLTRIAZINE

CONSENSUS REPORTS: EPA Genetic Toxicology Program.

SAFETY PROFILE: Poison by ingestion. Moderately toxic by subcutaneous route. Questionable carcinogen with experimental carcinogenic data. Mutation data reported. When heated to decomposition it emits toxic fumes of NO_x.

DSO000   CAS: 950-37-8   HR: 3
O,O-DIMETHYL-S-(5-METHOXY-1,3,4-
**   THIADIAZOLINYL-3-METHYL)**
**   DITHIOPHOSPHATE**

mf: $C_6H_{11}N_2O_4PS_3$ mw: 302.34

PROP: Crystals from MeOH. Mp: 39–40°. Very sltly sol in H_2O.

SYNS: CIBA-GEIGY GS 13005 □ S-(2,3-DIHYDRO-5-METHOXY-2-OXO-1,3,4-THIADIAZOL-3-METHYL) □ (O,O-DIMETHYL)-S-(-2-METHOXY-Δ²-1,3,4-THIA-DIAZOLIN-5-ON-4-YLMETHYL)DITHIOPHOSPHATE DIMETHYL PHOSPHOROTHIOLOTHIONATE □ O,O-DIMETHYL-S-(2-METHOXY-1,3,4-THIADIAZOL-5-(4H)-ONYL-(4)-METHYL)-DITHIOPHOSPHAT (GERMAN) □ O,O-DIMETHYL-S-(2-METHOXY-1,3,4-THIADIAZOL-5(4H)-ONYL-(4)-METHYL) PHOSPHORODITHIOATE □ O,O-DIMETHYL-S-(2-METHOXY-1,3,4 (4H)-THIODIAZ-OL-5-ON-4-YL)-METHYL)DITHIOFOSFAAT (DUTCH) □ O,O-DIMETIL-S-((2-METOSSI-1,3,4-(4H)-TIADIZAOL-5-ON-4-IL)-METIL)-DITIFOSFATO (ITALIAN) □ DMTP (JAPAN) □ ENT 27,193 □ FISONS NC 2964 □ GEIGY 13005 □ METHIDATHION □ S-((5-METHOXY-2-OXO-1,3,4-THIADIAZOL-3(2H)-YL)METHYL)-O,O-DIMETHYL PHOSPHORODITHIOATE □ SOMONIL □ SURPRACIDE □ ULTRACIDE

CONSENSUS REPORTS: EPA Extremely Hazardous Substances List.

SAFETY PROFILE: Poison by ingestion and skin contact. Moderately toxic by inhalation. Human mutation data reported. Human systemic effects: coma, lachrymation, miosis. A severe eye irritant. An insecticide. When heated to decomposition it emits very toxic fumes of NO_x, PO_x, and SO_x.

DSO200   CAS: 23422-53-9   HR: 3
N,N-DIMETHYL-N'-(((METHYLAMINO)-
**   CARBONYL)OXY)PHENYLMETHANI**
**   MIDAMIDE MONOHYDROCHLORIDE**

mf: $C_{11}H_{15}N_3O_2 \cdot ClH$ mw: 257.75

PROP: Powder. Very sol in H_2O; sol in MeOH; sltly sol in Me_2CO, $CHCl_3$, and hexane.

SYNS: CARZOL SP □ DICARZOL □ m-(((DIMETHYL-AMINO)METHYLENE)AMINO)PHENYLMETHYL CARBAMATE,HYDROCHLORIDE □ 3-DIMETHYL-AMINOMETHYLENEIMINOPHENYL-N-METHYL-CARBAMATE, HYDROCHLORIDE □ ENT 27,566 □ EP-332 □ FORMETANATE HYDROCHLORIDE □ MORTON EP332 □ NOR-AM EP 332 □ SCHERING 36056 □ SN 36056

CONSENSUS REPORTS: Reported in EPA TSCA Inventory. EPA Extremely Hazardous Substances List.

SAFETY PROFILE: Poison by ingestion and intraperitoneal routes. Mildly toxic by skin contact. When heated to decomposition it emits very toxic fumes of NO_x and HCl.

DSP400   CAS: 60-51-5   HR: 3
O,O-DIMETHYL METHYLCARBAMOYL-

METHYL PHOSPHORODITHIOATE

mf: $C_5H_{12}NO_3PS_2$ mw: 229.27

PROP: Crystals from Et_2O or toluene/hexane. Mp: 51–52°. Sol in H_2O, alcohols, $CHCl_3$, C_6H_6, and ketones.

SYNS: AC-12682 □ AMERICAN CYANAMID 12880 □ BI-58 □ CEKUTHOATE □ CL 12880 □ CYGON □ CYGON INSECTICIDE □ DAPHENE □ DE-FEND □ DEMOS-L40 □ DEVIGON □ DIMATE 267 □ DIMETATE □ DIME-THOAAT (DUTCH) □ DIMETHOAT (GERMAN) □ DIMETHOATE (USDA) □ DIMETHOAT TECHNISCH 95% □ DIMETHOGEN □ O,O-DIMETHYLDITHIO-PHOSPHORYLACETIC ACID-N-MONOMETHYLAMIDE SALT □ O,O-DIMETHYL-DITHIOPHOSPHORYL-ESSIGSAEURE MONOMETHYLAMID (GERMAN) □ O,O-DIMETHYL-S-(2-(METHYLAMINO)-2-OXOETHYL) PHOSPHORODITHIOATE □ O,O-DIMETHYL-S-(N-METHYL-CARBAMOYL)-METHYL-DITHIOFOSFAAT (DUTCH) □ (O,O-DIMETHYL-S-(N-METHYL-CARBAMOYL-METHYL)-DITHIOPHOSPHAT) (GERMAN) □ O,O-DIMETHYL-S-(N-METHYLCARBAMOYLMETHYL) DITHIOPHOSPHATE □ O,O-DIMETHYL-S-(N-METHYL-CARBAMOYLMETHYL) PHOSPHORODITHIOATE □ O,O-DIMETHYL-S-(N-METHYLCARBAMYLMETHYL) THIOTHIONOPHOSPHATE □ O,O-DIMETHYL-S-(N-MONOMETHYL)-CARBAMYL METHYLDITHIOPHO-SPHATE □ O,O-DIMETHYL-S-(2-OXO-3-AZA-BUTYL)-DITHIOPHOSPHAT (GERMAN) □ O,O-DIMETIL-S-(N-METIL-CARBAMOIL-METIL)-DITIOFOSFATO (ITALIAN) □ DIMETON □ DIMEVUR □ DITHIOPHOSPHATE de O,O-DIMETHYLE et de S(-N-METHYLCARBAMOYL-METHYLE) (FRENCH) □ EI-12880 □ ENT 24,650 □ EXPERIMENTAL INSECTICIDE 12,880 □ FERKETHION □ FORTION NM □ FOSFAMID □ FOSFOTOX □ FOSTION MM □ L-395 □ LURGO □ S-METHYL-CARBAMOYLMETHYL-O,O-DIMETHYL PHOSPHORO-DITHIOATE □ N-MONOMETHYLAMIDE of O,O-DIMETHYLDITHIOPHOSPHORYLACETIC ACID □ NC-262 □ NCI-C00135 □ PERFECTHION □ PHOSPHAMID □ PHOSPHORODITHIOIC ACID-O,O-DIMETHYL-S-(2-(METHYLAMINO)-2-OXOETHYL) ESTER □ RACUSAN □ RCRA WASTE NUMBER P044 □ REBELATE □ ROGODIAL □ ROGOR □ ROXION U.A. □ SINORATOX □ TRIMETION

CONSENSUS REPORTS: NCI Carcinogenesis Bioassay (feed); No Evidence: mouse, rat NCITR* NCI-CG-TR-4,77. Reported in EPA TSCA Inventory. EPA Genetic Toxicology Program. EPA Extremely Hazardous Substances List.

SAFETY PROFILE: A deadly human poison. Poison by ingestion, skin contact, intraperitoneal, and subcutaneous routes. Moderately toxic by intravenous route. Human systemic effects: coma, dyspnea, fasciculations. Questionable carcinogen with experimental carcinogenic data. Experimental teratogenic and reproductive effects. Human mutation data reported. When heated to decomposition it emits very toxic fumes of NO_x, PO_x, and SO_x. See also ESTERS.

DSP600   CAS: 23135-22-0   HR: 3
N',N'-DIMETHYL-N-((METHYLCAR-BAMOYL)OXY)-1-METHYLTHIOOX-AMIMIDIC ACID

mf: $C_7H_{13}N_3O_3S$ mw: 219.29

PROP: Solid in H_2O. Mp: 100–102°. Sol in H_2O, Me_2CO, EtOH, and MeOH.

SYNS: D-1410 □ 2-(DIMETHYLAMINO)-N-(((METHYLAMINO)CARBONYL)OXY)-2-OXOETH-ANIMIDOTHIOIC ACID METHYL ESTER □ 2-DIMETHYLAMINO-1-(METHYLTHIO)GLYOXAL-o-METHYLCARBAMOYLMONOXIME □ N,N-DIMETHYL-α-METHYLCARBAMOYLOXYIMINO-α-(METHYLTHIO)-ACETAMIDE □ N',N'-DIMETHYL-N-((METHYLCARB-AMOYL)OXY)-1-THIOOXAMIMIDIC ACID METHYL ESTER □ DPX 1410 □ INSECTICIDE-NEMATICIDE 1410 □ METHYL-2-(DIMETHYLAMINO)-N-(((METHYLAMINO)CARBONYL)OXY)-2-OXO-ETHANIMIDOTHIOATE □ METHYL-1-(DIMETHYL-CARBAMOYL)-N-(METHYLCARBAMOYLOXY)THIO-FORMIMIDATE □ S-METHYL-1-(DIMETHYLCARBAMO-YL)-N-((METHYLCARBAMO-YL)OXY)THIOFORMIMID-ATE □ METHYL-N',N'-DIMETHYL-N-((METHYL-CARBAMOYL)OXY)-1-THIOOXAMIMIDATE □ OXAMYL □ THIOXAMYL □ VYDATE □ VYDATE L INSECTIC-IDE/NEMATICIDE □ VYDATE L OXAMYL INSECTIC-8IDE/NEMATOCIDE

CONSENSUS REPORTS: EPA Extremely Hazardous Substances List.

SAFETY PROFILE: Poison by ingestion, skin contact, and inhalation. Experimental reproductive effects. Moderately toxic by skin contact. When heated to decomposition it emits very toxic fumes of NO_x and SO_x.

DSP710   CAS: 71108-06-0   HR: 3
5,5-DIMETHYL-2-((1-METHYLETHYL)-IMINO)1,3-DITHIOLAN-4-ONE, o-((METHYL((TRICHLOROMETHYL)THIO)AMINO)CARBONYL)OXIME

mf: $C_{11}H_{16}Cl_3N_3O_2S_3$ mw: 424.83

SAFETY PROFILE: A poison by ingestion. When heated to decomposition it emits toxic vapors of NO_x, SO_x, and Cl^-.

DSQ000   CAS: 122-14-5   HR: 3
DIMETHYL-3-METHYL-4-NITRO-
  PHENYLPHOSPHOROTHIONATE
mf: $C_9H_{12}NO_5PS$ mw: 277.25
PROP: Yellow oil. D: 1.323 @ 25°/4°, bp:
140–145° @ 0.1 mm. Insol in H_2O; sltly sol
in ligroin.
SYNS: ACCOTHION □ ACEOTHION □ AGRIA 1050 □
AGRIYA 1050 □ AGROTHION □ AMERICAN CYAN-
AMID CL-47,300 □ ARBOGAL □ BAY 41831 □ BAYER
41831 □ BAYER S 5660 □ CEKUTROTHION □ CL 47300
□ CP 47114 □ CYFEN □ CYTEL □ CYTEN □ O,O-
DIMETHYL-O-(3-METHYL-4-NITROFENYL)-
MONOTHIOFOSFAAT (DUTCH) □ O,O-DIMETHYL-O-
(3-METHYL-4-NITRO-PHENYL)-MONOTHIOPHOSPHAT
(GERMAN) □ O,O-DIMETHYL-O-(3-METHYL-4-
NITROPHENYL) PHOSPHOROTHIOATE □ O,O-
DIMETHYL-O-(3-METHYL-4-NITROPHENYL)
THIOPHOSPHATE □ O,O-DIMETHYL-O-(3-METHYL)
PHOSPHOROTHIOATE □ O,O-DIMETHYL-O-(4-NITRO-
3-METHYLPHENYL)THIOPHOSPHATE □ O,O-
DIMETHYL-O-4-NITRO-m-TOLYL PHOSPHOROTHIO-
ATE □ O,O-DIMETIL-O-(3-METIL-4-NITRO-FENIL)-
MONOTIOFOSFATO (ITALIAN) □ EI 47300 □ ENT
25,715 □ FALITHION □ FENITOX □ FENITROTHION □
FENITROTION (HUNGARIAN) □ FOLETHION □ H-35-F
87 (BVM) □ 8057HC □ KOTION □ MEP (Pesticide) □
METATHIONE □ METATION □ METHYLNITROPHOS
□ MONSANTO CP 47114 □ NITROPHOS □ NOVATH-
ION □ NUVANOL □ OLEOSUMIFENE □ OMS 43 □
OVADOFOS □ PENNWALT C-4852 □ PHENITROTHION
□ S 112A □ S 5660 □ SUMITHIAN □ THIOPHOSPHATE
de O,O-DIMETHYLE et de O-(3-METHYL-4-
NITROPHENYLE) (FRENCH) □ VERTHION
CONSENSUS REPORTS: EPA Genetic
Toxicology Program.
SAFETY PROFILE: Poison by ingestion,
inhalation, intravenous, and intraperitoneal
routes. Moderately toxic by skin contact,
intratracheal, and subcutaneous routes.
Human systemic effects: coma, diarrhea,
dyspnea, gastrointestinal changes,
hypermotility, nausea or vomiting,
respiratory depression. Mutation data
reported. When heated to decomposition it
emits very toxic fumes of NO_x, PO_x, and
SO_x.

DSQ840   CAS: 302542-60-5 HR: 3
3,5-DIMETHYL-N-(4-METHYLPHENYL)-
  1H-PYRAZOLE-1-ACETAMIDE
mf: $C_{14}H_{17}N_3O$ mw: 243.31

SAFETY PROFILE: A poison by
subcutaneous and ocular routes. Moderately
toxic by ingestion. When heated to
decomposition it emits toxic vapors of NO_x.

DSR200   CAS: 20241-03-6   HR: 3
3,3-DIMETHYL-1-(m-METHYLPHEN-
  YL)TRIAZENE
mf: $C_9H_{13}N_3$ mw: 163.25
SYNS: 3,3-DIMETHYL-1-(m-TOLYL)TRIAZENE □ 1-(m-
METHYLPHENYL)-3,3-DIMETHYLTRIAZENE □ 1-(3-
METHYLPHENYL)-3,3-DIMETHYLTRIAZENE
SAFETY PROFILE: Poison by ingestion and
intraperitoneal routes. Moderately toxic by
subcutaneous route. Questionable
carcinogen with experimental carcinogenic
data. Mutation data reported. When heated
to decomposition it emits toxic fumes of
NO_x.

DSR400   CAS: 756-79-6   HR: 2
DIMETHYL METHYLPHOSPHONATE
mf: $C_3H_9O_3P$ mw: 124.09
PROP: Pleasant-smelling liquid. Bp: 66–68°
@ 10 mm.
SYNS: DMMP □ METHYLPHOSPHONIC ACID
DIMETHYL ESTER □ NCI-C56762
CONSENSUS REPORTS: Reported in EPA
TSCA Inventory.
SAFETY PROFILE: Moderately toxic by
intravenous route. Experimental
reproductive effects. Questionable
carcinogen with experimental carcinogenic
data. Mutation data reported. An
experimental nerve gas stimulant. A flame
retardant. When heated to decomposition it
emits toxic fumes of PO_x.

DST000   CAS: 2032-65-7   HR: 3
3,5-DIMETHYL-4-METHYLTHIOPHENYL-
  N-METHYLCARBAMATE
mf: $C_{11}H_{15}NO_2S$ mw: 225.33
PROP: Crystals or powder. Mp: 117–118°.
Sol in most org solvs; pract insol in H_2O.
SYNS: BAY 9026 □ BAYER 37344 □ 3,5-DIMETHYL-4-
(METHYLTHIO)PHENOL METHYLCARBAMATE □ 3,5-
DIMETHYL-4-METHYL-THIOPHENYL-N-CARBAMAT
(GERMAN) □ DRAZA □ ENT 25,726 □ H 321 □
MERCAPTODIMETHUR (DOT) □ MESUROL □
METHIOCARB □ METHYL CARBAMIC ACID-4-

(METHYLTHIO)-3,5-XYLYL ESTER □ 4-METHYL-MERCAPTO-3,5-DIMETHYLPHENYL N-METHYL-CARBAMATE □ 4-METHYLMERCAPTO-3,5-XYLYL METHYLCARBAMATE □ 4-METHYLTHIO-3,5-DIMETHYLPHENYL METHYLCARBAMATE □ 4-(METHYLTHIO)-3,5-XYLENOL METHYLCARBAMATE □ 4-(METHYLTHIO)-3,5-XYLYL METHYLCARBAMATE □ METMERCAPTURON □ OMS-93

CONSENSUS REPORTS: EPA Extremely Hazardous Substances List.

SAFETY PROFILE: Poison by ingestion, skin contact, and intraperitoneal routes. Used as an insecticide, molluscicide, and bird repellent. When heated to decomposition it emits very toxic fumes of NO_x and SO_x. See also ESTERS and CARBAMATES.

DSU300 CAS: 302959-32-6 HR: 3
1,2-DIMETHYL-3-(2-NAPHTHALENYL)-(2R,3S)-REL-3-PYRROLIDINOL DROCHLORIDE

mf: $C_{16}H_{19}NO \cdot ClH$ mw: 277.79

SAFETY PROFILE: A poison by subcutaneous route. When heated to decomposition it emits toxic vapors of NO_x, HCl, and Cl^-.

DSU400 CAS: 86-56-6 HR: 3
N,N-DIMETHYL-1-NAPHTHYLAMINE

mf: $C_{12}H_{13}N$ mw: 171.26

PROP: A liquid. D: 1.052 @ 4°/4°, bp: 272–274°.

SYNS: 1-DIMETHYLAMINONAPHTHALENE □ DIMETHYL-α-NAPHTHYLAMINE □ α-DIMETHYL-NAPHTHYLAMINE □ N,N-DIMETHYL-α-NAPHTHYL-AMINE

CONSENSUS REPORTS: Reported in EPA TSCA Inventory.

SAFETY PROFILE: Poison by intraperitoneal route. Moderately toxic by ingestion. Mutation data reported. When heated to decomposition it emits toxic fumes of NO_x.

DSV200 CAS: 4164-28-7 HR: 3
DIMETHYLNITRAMINE

mf: $C_2H_6N_2O_2$ mw: 90.10

PROP: Needles from ligroin. Mp: 58°, bp: 187°. Sol in H_2O.

SYNS: DIMETHYLNITRAMIN (GERMAN) □ DIMETHYLNITROAMINE □ DMNM □ DMNO □ N-NITRODIMETHYLAMINE □ N-NITRO-DMA

SAFETY PROFILE: Poison by intraperitoneal route. Moderately toxic by ingestion. Questionable carcinogen with experimental tumorigenic data. Mutation data reported. When heated to decomposition it emits toxic fumes of NO_x.

DSX400 CAS: 7227-92-1 HR: 3
3,3-DIMETHYL-1-(p-NITROPHEN-YL)TRIAZENE

mf: $C_8H_{10}N_4O_2$ mw: 194.22

SYNS: 1-p-NITROFENYL-3,3-DIMETHYLTRIAZEN (CZECH) □ 1-(p-NITROPHENYL-3,3-DIMETHYL-TRIAZEN (GERMAN) □ 1-(p-NITROPHENYL)-3,3-DIMETHYL-TRIAZENE □ 1-(4-NITROPHENYL)-3,3-DIMETHYLTRIAZENE

SAFETY PROFILE: Poison by subcutaneous route. Moderately toxic by ingestion. Questionable carcinogen with experimental neoplastigenic and tumorigenic data. Human mutation data reported. When heated to decomposition it emits toxic fumes of NO_x.

DSY600 CAS: 138-89-6 HR: 3
N,N-DIMETHYL-p-NITROSOANILINE

DOT: UN 1369

mf: $C_8H_{10}N_2O$ mw: 150.20

PROP: Green plates from Et_2O. Mp: 92.5–93.5°. Sol in EtOH, Et_2O; sltly sol in H_2O.

SYNS: ACCELERINE □ p-(DIMETHYLAMINO)-NITROSOBENZENE □ 4-(DIMETHYLAMINO)NITROSO-BENZENE □ DIMETHYL-p-NITROSOANILINE (DOT) □ N,N-DIMETHYL-4-NITROSOBENZENAMINE □ DIMETHYL(p-NITROSOPHENYL)AMINE □ NCI-C01821 □ NDMA □ p-NITROSO-N,N-DIMETHYLANILINE □ 4-NITROSODIMETHYLANILINE □ p-NITROSODIMETH-YLANILINE (DOT) □ PARANITROSODIMETHYL-ANILIDE □ ULTRA BRILLIANT BLUE P

CONSENSUS REPORTS: Reported in EPA TSCA Inventory.

DOT CLASSIFICATION: 4.2; Label: Spontaneously Combustible

SAFETY PROFILE: Poison by ingestion. Mutation data reported. Questionable carcinogen with experimental tumorigenic data. Flammable when exposed to heat, flame, or oxidizers. Violent reaction with acetic anhydride + acetic acid. When heated to decomposition it emits toxic fumes of NO_x.

DTA000   CAS: 1456-28-6   HR: 3
2,6-DIMETHYLNITROSOMORPHOLINE
mf: $C_6H_{12}N_2O_2$ mw: 144.20
SYNS: DIMETHYLNITROSOMORPHOLINE □ 2,6-DIMETHYL-N-NITROSOMORPHOLINE □ DMNM □ Me2NMOR □ NITROSO-2,6-DIMETHYLMORPHOLINE □ N-NITROSO-2,6-DIMETHYLMORPHOLINE
CONSENSUS REPORTS: EPA Genetic Toxicology Program.
SAFETY PROFILE: Suspected carcinogen with experimental carcinogenic, tumorigenic, and neoplastigenic data. Poison by ingestion and subcutaneous routes. Mutation data reported. Used as a model carcinogenic and carcinogenic metabolite. When heated to decomposition it emits toxic fumes of NO_x. See also N-NITROSO COMPOUNDS.

DTB200   CAS: 13256-32-1   HR: 3
1,3-DIMETHYLNITROSOUREA
mf: $C_3H_7N_3O_2$ mw: 117.13
SYNS: DIMETHYLNITROSOHARNSTOFF (GERMAN) □ N,N'-DIMETHYLNITROSOUREA □ 1,3-DIMETHYL-N-NITROSOUREA □ NITROSODIMETHYLUREA □ N-NITROSODIMETHYLUREA
CONSENSUS REPORTS: EPA Genetic Toxicology Program.
SAFETY PROFILE: Suspected carcinogen with experimental carcinogenic, neoplastigenic, tumorigenic, and teratogenic data. Poison by ingestion and intravenous routes. Mutation data reported. When heated to decomposition it emits toxic fumes of NO_x. See also N-NITROSO COMPOUNDS.

DTC600   CAS: 122-19-0   HR: 3
DIMETHYLOCTADECYLBENZYLAMMO NIUM CHLORIDE
mf: $C_{27}H_{50}N \bullet Cl$ mw: 424.23
SYNS: AMMONYX 4 □ AMMONYX CA SPECIAL □ ARQUAD DM18B-90 □ BARQUAT SB-25 □ BENZYLDIMETHYLSTEARYLAMMONIUM CHLORIDE

□ BENZYLSTEARYLDIMETHYLAMMONIUM CHLORIDE □ CARSOQUAT SDQ-25 □ DEHYQUART STC-25 □ DIMETHYLBENZYLOCTADECYLAMMONIUM CHLORIDE □ INTEXAN SB-85 □ J SOFT C 4 □ KATAMINE AB □ NISSAN CATION S2-100 □ N-OCTADECYL-N-BENZYL-N,N-DIMETHYLAMMONIUM-CHLORIDE □ OCTADECYLDIMETHYLBENZYL-AMMONIUM CHLORIDE □ ORTHOSAN MB □ QUATERNOL 1 □ STEARALKONIUM CHLORIDE □ STEARYLDIMETHYLBENZYLAMMONIUM CHLORIDE □ STEBAC □ TALLOW BENZYL DIMETHYLAMMON-IUM CHLORIDE □ TRITON X-40 □ VARISOFT SDC
CONSENSUS REPORTS: Reported in EPA TSCA Inventory.
SAFETY PROFILE: Poison by intraperitoneal route. Moderately toxic by ingestion. A human skin irritant and severe experimental eye irritant. When heated to decomposition it emits very toxic fumes of NO_x, NH_3, and Cl^-.

DTC800   CAS: 5392-40-5   HR: 2
3,7-DIMETHYL-2,6-OCTADIENAL
mf: $C_{10}H_{16}O$ mw: 152.26
PROP: Mobile, pale-yellow liquid; strong lemon odor. D: 0.891–0.897 @ 15°, refr index: 1.486–1.490, flash p: 198°F. Sol in 5 volumes of 60% alc; sol in all proportions of benzyl benzoate, diethyl phthalate, glycerin, propylene glycol, mineral oil, fixed oils, and 95% alc; insol in water.
SYNS: BUTOBEN □ BUTYL p-HYDROXYBENZOATE □ CITRAL (FCC) □ FEMA No. 2203 □ NCI-C56348 □ NERAL
CONSENSUS REPORTS: Reported in EPA TSCA Inventory.
SAFETY PROFILE: Moderately toxic by intraperitoneal route. Mildly toxic by ingestion. Experimental reproductive effects. A severe human and experimental skin irritant. Mutation data reported. Combustible liquid. When heated to decomposition it emits acrid smoke and irritating fumes.

DTD000   CAS: 106-24-1   HR: 3
3,7-DIMETHYL-(E)-2,6-OCTADIEN-1-OL
mf: $C_{10}H_{18}O$ mw: 154.28
PROP: Colorless to pale-yellow, oily liquid; pleasant floral odor. D: 0.870–0.890 @ 15°,

refr index: 1.469–1.478, mp: 15°, bp: 230°, flash p: 214°F. Sol in fixed oils, propylene glycol; sltly sol in water; insol in glycerin @ 230°.

SYNS: 2,6-DIMETHYL-trans-2,6-OCTADIEN-8-OL □ 3,7-DIMETHYL-trans-2,6-OCTADIEN-1-OL □ FEMA No. 2507 □ GERANIOL (FCC) □ GERANIOL ALCOHOL □ GERANIOL EXTRA □ GERANYL ALCOHOL □ GUANIOL □ LEMONOL

CONSENSUS REPORTS: Reported in EPA TSCA Inventory.

SAFETY PROFILE: Poison by intravenous route. Moderately toxic by ingestion, subcutaneous, and intramuscular routes. A severe human skin irritant. Combustible liquid. When heated to decomposition it emits acrid smoke and irritating fumes.

DTD200   CAS: 106-25-2   HR: 2
2-cis-3,7-DIMETHYL-2,6-OCTADIEN-1-OL

mf: $C_{10}H_{18}O$ mw: 154.28

PROP: Colorless oily liquid; sweet, rose odor. D: 0.875–0.880, refr index: 1.467–1.478, bp: 225–226°. Sol in alc, chloroform, ether, water @ 227°.

SYNS: 3,7-DIMETHYL-(Z)-2,6-OCTADIEN-1-OL □ FEMA No. 2770 □ NEROL (FCC)

CONSENSUS REPORTS: Reported in EPA TSCA Inventory.

SAFETY PROFILE: Moderately toxic by intramuscular route. Mildly toxic by ingestion. A skin irritant. When heated to decomposition it emits acrid smoke and irritating fumes.

DTD800   CAS: 105-87-3   HR: 1
trans-3,7-DIMETHYL-2,6-OCTADIEN-1-OL ACETATE

mf: $C_{12}H_{20}O_2$ mw: 196.32

PROP: Colorless, sweet, clear, oily liquid; odor of lavender. D: 0.907–0.918 @ 15°, refr index: 1.458–1.464, bp: 130–132° @ 16 mm, flash p: 219°F. Sol in alc, fixed oils, ether; sltly sol in propylene glycol; insol in water and glycerol.

SYNS: ACETIC ACID GERANIOL ESTER □ 3,7-DIMETHYL-2-trans-6-OCTADIENYL ACETATE □ trans-3,7-DIMETHYL-2,6-OCTADIEN-1-YL ACETATE □ trans-2,6-DIMETHYL-2,6-OCTADIEN-8-YL ETHANOATE □ FEMA No. 2509 □ GERANIOL ACETATE □ GERANYL ACETATE (FCC) □ NCI-C54728

CONSENSUS REPORTS: NTP Carcinogenesis Studies (gavage); No Evidence: mouse, rat NTPTR* NTP-TR-252,87. Reported in EPA TSCA Inventory.

SAFETY PROFILE: Mildly toxic by ingestion. A human skin irritant. Mutation data reported. Combustible liquid. When heated to decomposition it emits acrid smoke and irritating fumes. See also ESTERS.

DTE600   CAS: 106-21-8   HR: 2
DIMETHYLOCTANOL

mf: $C_{10}H_{22}O$ mw: 158.32

PROP: Colorless liquid; sweet, rose odor. D: 0.26–0.842, refr index: 1.435. Sol in fixed oils, propylene glycol; insol in glycerin.

SYNS: DIHYDROCITRONELLOL □ 2,6-DIMETHYL-8-OCTANOL □ 3,7-DIMETHYL-1-OCTANOL (FCC) □ FEMA No. 2391 □ GERANIOL TETRAHYDRIDE □ PELARGOL □ PERHYDROGERANIOL □ TETRAHYDROGERANIOL

CONSENSUS REPORTS: Reported in EPA TSCA Inventory.

SAFETY PROFILE: Moderately toxic by skin contact. A skin irritant. When heated to decomposition it emits acrid smoke and irritating fumes.

DTF400   CAS: 141-25-3   HR: 2
2,6-DIMETHYL-1-OCTEN-8-OL

mf: $C_{10}H_{20}O$ mw: 156.30

PROP: Flash p: 212°F.

SYNS: α-CITRONELLOL □ 3,7-DIMETHYL-7-OCTEN-1-OL □ FEMA No. 2981 □ RHODINOL (FCC)

CONSENSUS REPORTS: Reported in EPA TSCA Inventory.

SAFETY PROFILE: Moderately toxic by intramuscular route. Combustible liquid. When heated to decomposition it emits acrid smoke and irritating fumes.

DTF820   CAS: 5538-94-3   HR: 3
N,N-DIMETHYL-N-OCTYL-1-OCTANAMINIUM CHLORIDE

mf: $C_{18}H_{40}N \cdot Cl$ mw: 306.04

SYNS: AMMONIUM, DIMETHYLDIOCTYL-, CHLORIDE □ DIMETHYLDIOCTYLAMMONIUM CHLORIDE □ DIOCTYLDIMETHYLAMMONIUM CHLORIDE □ DODIGEN 2617 □ HOE-S 2617 □ 1-OCTANAMINIUM, N,N-DIMETHYL-N-OCTYL-, CHLORIDE □ QUERTON 28CL □ RC 5626

CONSENSUS REPORTS: EPA FIFRA 1988 pesticide subject to registration or re-registration.

CONSENSUS REPORTS: Reported in EPA TSCA Inventory.

SAFETY PROFILE: A poison by ingestion. When heated to decomposition it emits toxic vapors of NO_x and Cl^-.

DTH000   CAS: 1955-45-9   HR: 3
3,3-DIMETHYL-2-OXETHANONE

mf: $C_5H_8O_2$ mw: 100.13

SYNS: 3,3-DIMETHYL-2-OXETANONE □ DIMETHYL PROPIOLACTONE □ 3,3-DIMETHYL-β-PROPIOL-ACTONE □ NCI-C04126 □ PIVALIC ACID LACTONE □ PIVALOLACTONE

CONSENSUS REPORTS: NCI Carcinogenesis Bioassay (gavage); No Evidence: mouse NCITR* NCI-CG-TR-140,78; Clear Evidence: rat NCITR* NCI-CG-TR-140,78. Reported in EPA TSCA Inventory.

SAFETY PROFILE: Poison by ingestion. Questionable carcinogen with experimental carcinogenic and tumorigenic data. Mutation data reported. When heated to decomposition it emits acrid smoke and irritating fumes.

DTH400   CAS: 2273-45-2   HR: 3
DIMETHYLOXOSTANNANE

mf: C_2H_6OSn mw: 164.77

PROP: White powder. Insol in water.

SYN: DIMETHYLTIN OXIDE

CONSENSUS REPORTS: Reported in EPA TSCA Inventory.

OSHA PEL: TWA 0.1 mg(Sn)/m³ (skin)

ACGIH TLV: TWA 0.1 mg(Sn)/m³; STEL 0.2 mg(Sn)/m³ (skin).

NIOSH REL: (Organotin Compounds) TWA 0.1 mg(Sn)/m³

SAFETY PROFILE: Poison by intravenous route. When heated to decomposition it emits acrid smoke and irritating fumes. See also TIN COMPOUNDS.

DTJ400   CAS: 122-09-8   HR: 3
α,α-DIMETHYLPHENETHYLAMINE

mf: $C_{10}H_{15}N$ mw: 149.26

SYNS: α,α-DIMETHYLBENZEETHANAMINE □ 1,1-DIMETHYL-2-PHENYLETHANAMINE □ α,α-DIMETHYL-β-PHENYLETHYLAMINE □ DUROMINE □ LIPOPILL □ LONAMIN □ MG 18370 □ MG 18570 □ MIRAPRONT □ PHENTERMINE □ 2-PHENYL-tert-BUTYLAMINE □ RCRA WASTE NUMBER P046 □ WILPO

CONSENSUS REPORTS: Reported in EPA TSCA Inventory.

SAFETY PROFILE: Poison by ingestion, intravenous, and intraperitoneal routes. Human systemic effects by ingestion: sympathomimetic. Mutation data reported. When heated to decomposition it emits toxic fumes of NO_x.

DTK600   CAS: 2747-31-1   HR: 3
N,N-DIMETHYL-p-PHENYLAZOANILINE-N-OXIDE

mf: $C_{14}H_{15}N_3O$ mw: 241.32

SYNS: DAB-N-OXIDE □ 4-DIMETHYLAMINOAZOBENZENE AMINE-N-OXIDE □ N,N-DIMETHYLAMINOAZOBENZENE-N-OXIDE

SAFETY PROFILE: Poison by intraperitoneal route. Moderately toxic by ingestion. Questionable carcinogen with experimental tumorigenic data. When heated to decomposition it emits toxic fumes of NO_x.

DTM600   CAS: 154-99-4   HR: 3
o,p-DIMETHYL-β-PHENYLETHYLHY-DRAZINE DIHYDROGEN SULFATE

SYNS: β-(2,4-DIMETHYLPHENYL)ETHYLHYDRAZINE DIHYDROGEN SULPHATE □ LON 41

SAFETY PROFILE: Poison by ingestion and subcutaneous routes. Experimental reproductive effects. When heated to decomposition it emits very toxic fumes of SO_x and NO_x.

DTP000   CAS: 7227-91-0   HR: 3
3,3-DIMETHYL-1-PHENYLTRIAZENE

mf: $C_8H_{11}N_3$ mw: 149.22

SYNS: 3,3-DIMETHYL-1-PHENYL-1-TRIAZENE □ DMPT □ 1-FENYL-3,3-DIMETHYLTRIAZIN □ NSC-3094 □ PDMT □ PDT □ 1-PHENYL-3,3-DIMETHYLTRIAZENE □ PHENYLDIMETHYLTRIAZINE □ X 119
CONSENSUS REPORTS: EPA Genetic Toxicology Program.
SAFETY PROFILE: Poison by ingestion and intraperitoneal routes. Questionable carcinogen with experimental carcinogenic and tumorigenic data. Experimental teratogenic effects. Human mutation data reported. Decomposes explosively on attempted distillation at atmospheric pressure. When heated to decomposition it emits toxic fumes of NO_x.

DTQ400   CAS: 10265-92-6   HR: 3
O,S-DIMETHYL PHOSPHORAMIDO-THIOATE
mf: $C_2H_8NO_2PS$ mw: 141.14
PROP: Crystals. Mp: 40°. Sltly water-sol; sol in alc.
SYNS: ACEPHATE-MET □ BAY 71628 □ BAYER 71628 □ CHEVRON 9006 □ CHEVRON ORTHO 9006 □ O,S-DIMETHYL ESTER AMIDE of AMIDOTHIOATE □ ENT 27,396 □ HAMIDOP □ METAMIDOFOS ESTRELLA □ METHAMIDOPHOS □ MONITOR □ MTD □ NSC-190987 □ ORTHO 9006 □ PILLARON □ SRA 5172 □ TAHMABON □ TAMARON □ THIOPHOSPHOR-SAEURE-O,S-DIMETHYLESTERAMID (GERMAN)
CONSENSUS REPORTS: EPA Extremely Hazardous Substances List.
SAFETY PROFILE: Poison by ingestion, inhalation, skin contact, subcutaneous, and intraperitoneal routes. Human systemic effects by ingestion: fasciculations, pupillary constriction, and sweating. A cholinesterase inhibitor type of insecticide. When heated to decomposition it emits very toxic fumes of NO_x, PO_x, and SO_x. See also PARATHION.

DTQ600   CAS: 2524-03-0   HR: 3
O,O-DIMETHYLPHOSPHOROCHLORI-DOTHIOATE
mf: $C_2H_6ClO_2PS$ mw: 160.56
PROP: A liquid. D: 1.326, bp: 68° @ 12 mm.
SYNS: DIMETHYL CHLOROTHIOPHOSPHATE (DOT) □ DIMETHYLCHLORTHIOFOSAT (CZECH) □ O,O-

DIMETHYLESTER KYSELINY CHLORTHIOFOSFORECNE (CZECH) □ DIMETHYL PHOSPHORO-CHLORIDOTHIOATE (DOT) □ METHYL PCT □ PHOSPHOROCHLORIDOTHIOIC ACID-O,O-DIMETHYL ESTER
CONSENSUS REPORTS: Reported in EPA TSCA Inventory. EPA Extremely Hazardous Substances List.
SAFETY PROFILE: Poison by inhalation. Moderately toxic by ingestion and skin contact. Corrosive. When heated to decomposition it emits very toxic fumes of Cl^-, PO_x, and SO_x.

DTR200   CAS: 131-11-3   HR: 2
DIMETHYL PHTHALATE
mf: $C_{10}H_{10}O_4$ mw: 194.20
PROP: Colorless, odorless liquid. Mp: 0°, bp: 282.4°, flash p: 295°F (CC), d: 1.189 @ 25°/25°, autoign temp: 1032°F, vap d: 6.69, vap press: 1 mm @ 100.3°. IDLH 2000 mg/m³.
SYNS: AVOLIN □ 1,2-BENZENEDICARBOXYLIC ACID DIMETHYL ESTER □ DIMETHYL-1,2-BENZENEDICARBOXYLATE □ DIMETHYL BENZENEORTHODICARBOXYLATE □ DMP □ ENT 262 □ FERMINE □ METHYL PHTHALATE □ MIPAX □ NTM □ PALATINOL M □ PHTHALIC ACID METHYL ESTER □ PHTHALSAEUREDIMETHYLESTER (GERMAN) □ RCRA WASTE NUMBER U102 □ SOLVANOM □ SOLVARONE
CONSENSUS REPORTS: Reported in EPA TSCA Inventory. Community Right-To-Know List.
OSHA PEL: TWA 5 mg/m³
ACGIH TLV: TWA 5 mg/m³
SAFETY PROFILE: Moderately toxic by ingestion and intraperitoneal routes. Mildly toxic by inhalation. Experimental teratogenic and reproductive effects. Mutation data reported. An eye irritant. A pesticide and insect repellent. Combustible when exposed to heat or flame; can react with oxidizing materials. To fight fire, use CO_2, dry chemical. When heated to decomposition it emits acrid smoke and irritating fumes. See also ESTERS.

DTR850   **HR: 2**
DIMETHYLPOLYSILOXANE
mf: $[(CH_3)_2SiO{-}]$
PROP: Clear, colorless, viscous liquid. D: 0.96, refr index: 1.400. Sol in hydrocarbon solvents; insol in water.
SYNS: DIMETHYL SILICONE □ POLYDIMETHYL-SILOXANE
SAFETY PROFILE: Combustible liquid. When heated to decomposition it emits acrid smoke and irritating fumes.

DTS400   **CAS: 3282-30-2**   **HR: 3**
2,2-DIMETHYLPROPANOYL CHLORIDE
DOT: UN 2438
mf: C_5H_9ClO mw: 120.59
PROP: Bp: 105–106°.
SYNS: 2,2-DIMETHYLPROPIONYL CHLORIDE □ NEOPANTANOYL CHLORIDE □ PIVALIC ACID CHLORIDE □ PIVALOLYL CHLORIDE □ PIVALOYL CHLORIDE □ PIVALYL CHLORIDE □ TRIMETHYL ACETYL CHLORIDE (DOT)
CONSENSUS REPORTS: Reported in EPA TSCA Inventory.
DOT CLASSIFICATION: 8; Label: Corrosive, Flammable Liquid, Poison
SAFETY PROFILE: A corrosive irritant to skin, eyes, and mucous membranes. The liquid is flammable when exposed to heat, flame, or oxidizers. When heated to decomposition it emits toxic fumes of Cl⁻.

DTU400   **CAS: 5910-89-4**   **HR: 3**
2,3-DIMETHYLPYRAZINE
mf: $C_6H_8N_2$ mw: 108.16
PROP: Colorless liquid; nutty cocoa odor. D: 1.000–1.022 @ 20°, refr index: 1.506–1.509, flash p: 147°F (OC), d: 0.99, vap d: 3.72, bp: 156–158°. Misc with water, org solvs. Sol in water and org solvs.
SYNS: 2,3-DIMETHYL-1,4-DIAZINE □ FEMA No. 3271
CONSENSUS REPORTS: Reported in EPA TSCA Inventory.
SAFETY PROFILE: Moderately toxic by ingestion and intraperitoneal routes. Flammable liquid when exposed to heat, sparks, or flame. When heated to decomposition it emits toxic fumes of NO_x.

DTU600   **CAS: 123-32-0**   **HR: 3**
2,5-DIMETHYLPYRAZINE
mf: $C_6H_8N_2$ mw: 108.16
PROP: Colorless liquid; potato taste. D: 0.980–1.000, refr index: 1.497–1.501, flash p: 147°F (OC), d: 0.99, vap d: 3.72, bp: 155°, mp: 15°. Misc with water, org solvs. Sol in H_2O, EtOH, and Et_2O.
SYNS: 2,5-DIMETHYL-1,4-DIAZINE □ FEMA No. 3272
CONSENSUS REPORTS: Reported in EPA TSCA Inventory.
SAFETY PROFILE: Moderately toxic by ingestion and intraperitoneal routes. Mutation data reported. Flammable liquid when exposed to heat, open flame, spark, oxidizers. To fight fire, use water spray, mist, dry chemical, CO_2, foam. When heated to decomposition it emits toxic fumes of NO_x.

DTU800   **CAS: 108-50-9**   **HR: 2**
2,6-DIMETHYLPYRAZINE
mf: $C_6H_8N_2$ mw: 108.16
PROP: Prisms or white to yellow crystals; nutty, coffee odor. Mp: 48°, d: 0.965 @ 50°, bp: 155.6°. Sol in H_2O, EtOH, and Et_2O.
SYN: FEMA No. 3273
CONSENSUS REPORTS: Reported in EPA TSCA Inventory.
SAFETY PROFILE: Moderately toxic by ingestion and intraperitoneal routes. Mutation data reported. When heated to decomposition it emits toxic fumes of NO_x.

DTV200   **CAS: 21600-42-0**   **HR: 3**
(3,3-DIMETHYL-1-(m-PYRIDYL-N-OXIDE))TRIAZENE
mf: $C_7H_{10}N_4O$ mw: 166.21
SYNS: 3-(3',3'-DIMETHYLTRIAZENO)-PYRIDIN-N-OXID (GERMAN) □ 3-(3',3'-DIMETHYLTRIAZ-ENO)PYRIDINE-N-OXIDE □ PYNDT □ 1-(PYRIDYL-3-N-OXID)-3,3-DIMETHYL-TRIAZEN (GERMAN) □ 1-(PYRIDYL-3-N-OXIDE)-3,3-DIMETHYLTRIAZENE
CONSENSUS REPORTS: EPA Genetic Toxicology Program.
SAFETY PROFILE: Poison by intravenous and subcutaneous routes. Questionable carcinogen with experimental carcinogenic and tumorigenic data. Human mutation data

reported. When heated to decomposition it emits toxic fumes of NO_x.

DTV300   CAS: 625-84-3   HR: D
2,5-DIMETHYLPYRROLE
mf: C_6H_9N mw: 95.15
PROP: Colorless to yellow oily liquid. D: 0.935–0.945 @ 20°/4°, refr index: 1.503-1.506, bp: 165° @ 760 mm. Very sol in alc, and ether; very sltly sol in water.
SYN: FEMA No. 7071
SAFETY PROFILE: When heated to decomposition emits toxic fumes of NO_x.

DTV330   CAS: 22041-39-0   HR: 3
N,N-DIMETHYL-3-(PYRROLIDIN-1-YL)PROPIONAMIDE
mf: $C_9H_{18}N_2O$ mw: 170.29
SYN: PROPIONAMIDE, N,N-DIMETHYL-3-(PYRROLIDIN-1-YL)-
SAFETY PROFILE: A poison by intravenous route. When heated to decomposition it emits toxic vapors of NO_x.

DUB800   CAS: 1145-73-9   HR: 3
N,N-DIMETHYL-4-STILBENAMINE
mf: $C_{16}H_{17}N$ mw: 223.34
SYNS: 4-DIMETHYLAMINOSTILBEN (GERMAN) □ N,N-DIMETHYL-4-AMINOSTILBENE □ N,N-DIMETHYL-p-STYRYLANILINE □ STILBENYL-N,N-DIMETHYLAMINE
SAFETY PROFILE: Poison by ingestion and intraperitoneal routes. Questionable carcinogen with experimental carcinogenic and tumorigenic data. Mutation data reported. When heated to decomposition it emits toxic fumes of NO_x.

DUC000   CAS: 838-95-9   HR: 3
(E)-N,N-DIMETHYL-4-STILBENAMINE
mf: $C_{16}H_{17}N$ mw: 223.34
SYNS: trans-p-(DIMETHYLAMINO)STILBENE □ trans-4-DIMETHYLAMINOSTILBENE □ 4-DIMETHYLAMINO-trans-STILBENE □ (E)-N,N,-DIMETHYL-4-(2-PHENYL-ETHENYL)BENZENAMINE □ trans-N,N-DIMETHYL-4-STILBENAMINE
CONSENSUS REPORTS: EPA Genetic Toxicology Program.
SAFETY PROFILE: Poison by ingestion and subcutaneous routes. Questionable carcinogen with experimental carcinogenic and tumorigenic data. Mutation data reported. When heated to decomposition it emits toxic fumes of NO_x.

DUD100   CAS: 77-78-1   HR: 3
DIMETHYL SULFATE
DOT: UN 1595
mf: $C_2H_6O_4S$ mw: 126.14
PROP: Colorless, odorless liquid. Mp: −31.8°, fp: −27°, bp: 188° (decomp), flash p: 182°F (OC), d: 1.332 @ 15°, vap d: 4.35, autoign temp: 370°F. Sltly sol in H_2O, hexane, EtOH, C_6H_6; sol in Et_2O and Me_2CO. IDLH 7 ppm.
SYNS: DIMETHYLESTER KYSELINY SIROVE (CZECH) □ DIMETHYL MONOSULFATE □ DIMETHYLSULFAAT (DUTCH) □ DIMETHYLSULFAT (CZECH) □ DIMETIL-SOLFATO (ITALIAN) □ DMS □ DMS(METHYL SULFATE) □ DWUMETYLOWY SIARCZAN (POLISH) □ METHYLE (SULFATE de) (FRENCH) □ METHYL SULFATE (DOT) □ RCRA WASTE NUMBER U103 □ SULFATE de METHYLE (FRENCH) □ SULFATE DIMETHYLIQUE (FRENCH) □ SULFURIC ACID, DIMETHYL ESTER
CONSENSUS REPORTS: NTP 10th Report on Carcinogens. IARC Cancer Review: Group 2A IMEMDT 7,200,87; Animal Sufficient Evidence IMEMDT 4,271,74; Human Inadequate Evidence IMEMDT 4,271,74. EPA Genetic Toxicology Program. Community Right-To-Know List. EPA Extremely Hazardous Substances List. Reported in EPA TSCA Inventory.
OSHA PEL: TWA 0.1 ppm (skin)
ACGIH TLV: TWA 0.1 ppm (skin); Animal Carcinogen
DFG MAK: DFG TRK: Production: 0.02 ppm; Use: 0.04 ppm; Animal Carcinogen, Suspected Human Carcinogen
DOT CLASSIFICATION: 3; Label: Poison, Corrosive
SAFETY PROFILE: Confirmed carcinogen with experimental carcinogenic, tumorigenic, and teratogenic data. Human poison by inhalation. Experimental poison by ingestion, inhalation, intravenous, and subcutaneous routes. Other experimental reproductive effects. Human mutation data

reported. A corrosive irritant to skin, eyes, and mucous membranes. There is no odor or initial irritation to give warning of exposure. On brief, mild exposures, conjunctivitis, catarrhal inflammation of the mucous membranes of the nose, throat, larynx, and trachea, and possibly some reddening of the skin develop after the latent period. With longer, heavier exposures, the cornea shows clouding, the irritation changes to the nasopharynx are more marked, and after 6 to 8 hours pulmonary edema may develop. Death may occur in 3 or 4 days. The liver and kidneys are frequently damaged. Spilling of the liquid on the skin can cause ulceration and local necrosis. In patients surviving severe exposure, there may be serious injury of the liver and kidneys, with suppression of urine, jaundice, albuminuria, and hematuria appearing. Death, resulting from the kidney or liver damage, may be delayed for several weeks. Flammable when exposed to heat, flame, or oxidizers. Can react with oxidizing materials. Violent reaction with NH_4OH and NaN_3. To fight fire, use water, foam, CO_2, dry chemical. When heated to decomposition it emits toxic fumes of SO_x. See also SULFATES.

DUD800   CAS: 67-68-5   HR: 2
DIMETHYL SULFOXIDE
mf: C_2H_6OS mw: 78.14
PROP: Clear, water-white, hygroscopic liquid; garlic-onion-oyster odor. Mp: 18.5°, bp: 189°, flash p: 203°F (OC), d: 1.100 @ 20°, vap press: 0.37 mm @ 20°, lel: 3.0%, uel: 43%, autoign temp: 574°F (301°C). Misc in H_2O and org solvs.
SYNS: A 10846 □ DELTAN □ DEMASORB □ DEMAVET □ DEMESO □ DEMSODROX □ DERMASORB □ DIMETHYL SULPHOXIDE □ DIMEXIDE □ DIPIRARTRIL-TROPICO □ DMS-70 □ DMS-90 □ DMSO □ DOLICUR □ DOLIGUR □ DOMOSO □ DROMISOL □ DURASORB □ GAMASOL 90 □ HYADUR □ INFILTRINA □ M 176 □ METHYLSULFIN-YLMETHANE □ METHYL SULFOXIDE □ NSC-763 □ RIMSO-50 □ SOMIPRONT □ SQ 9453 □ SULFINYLBIS(METHANE) □ SYNTEXAN □ TOPSYM

CONSENSUS REPORTS: Reported in EPA TSCA Inventory. EPA Genetic Toxicology Program.
SAFETY PROFILE: Slightly toxic by ingestion. Moderately toxic by intravenous and intraperitoneal routes. Human systemic effects by intravenous route: nausea or vomiting and jaundice. Experimental teratogenic and reproductive effects. A skin and eye irritant. Questionable carcinogen with experimental tumorigenic data. Human mutation data reported. Can cause an anaphylactic reaction. Corneal opacity reported only in rabbits, dogs, and pigs. It freely penetrates the skin and may carry dissolved chemicals with it into the body. Combustible when exposed to heat or flame; can react with oxidizing materials. To fight fire, use water, foam, alcohol foam, CO_2, dry chemical. Violent or explosive reaction with many acyl, aryl, and nonmetal halides (e.g., acetyl chloride, benzenesulfonyl chloride, bromobenzoyl acetanilide, cyanuric chloride, iodine pentafluoride, $Mg(ClO_4)_2$, CH_3Br, NIO_4, oxalyl chloride, P_2O_3, phosphorus trichloride, phosphoryl chloride, silver fluoride, silver difluoride, sodium hydride, sulfur dichloride, disulfur dichloride, sulfuryl chloride, tetrachlorosilane, thionyl chloride). Violent or explosive reaction with boron compounds (e.g., borane, nonahydrononaborate(2-) ion), 4(4'-bromobenzoyl)acetanilide, carbonyl diisothiocyanate, dinitrogen tetraoxide, hexachlorocyclotriphosphazine, copper + trichloroacetic acid, metal alkoxides (e.g., potassium tert-butoxide, sodium isopropoxide), trifluoroacetic acid anhydride. Incompatible with magnesium perchlorate, metal oxosalts, perchloric acid, periodic acid, sulfur trioxide. Forms powerfully explosive mixtures with metal salts of oxoacids (e.g., aluminum perchlorate, sodium perchlorate, iron(III) nitrate). When heated to decomposition it emits toxic fumes of SO_x.

DUE000   CAS: 120-61-6   HR: 2
DIMETHYL TEREPHTHALATE
mf: $C_{10}H_{10}O_4$ mw: 194.20
PROP: Crystals from Et_2O. Mp: 141–142°,
bp: 284°.
SYNS: 1,4-BENZENE DICARBOXYLIC ACID
DIMETHYL ESTER (9CI) □ DIMETHYL-1,4-BENZENE
DICARBOXYLATE □ METHYL-4-CARBOMETHOXY
BENZOATE □ NCI-C50055 □ TEREPHTHALIC ACID
METHYL ESTER
CONSENSUS REPORTS: NTP
Carcinogenesis Bioassay (feed) Equivocal
Evidence: mouse NCITR* NCI-TR-121,79;
(feed) No Evidence rat NCITR* NCI-TR-
121,79. Reported in EPA TSCA Inventory.
SAFETY PROFILE: Moderately toxic by
intraperitoneal route. Mildly toxic by
ingestion. An eye irritant. Mutation data
reported. When heated to decomposition it
emits acrid smoke and irritating fumes.

DUG425   CAS: 2530-10-1   HR: 3
DIMETHYLTHIENYLCETONE
mf: $C_8H_{10}OS$ mw: 154.24
SYNS: ETHANONE, 1-(2,5-DIMETHYL-3-THIENYL)- □
KETONE, 2,5-DIMETHYL-3-THIENYL METHYL
DOT CLASSIFICATION: 3; Label:
Flammable Liquid
SAFETY PROFILE: A poison by
intraperitoneal route. A flammable liquid.
When heated to decomposition it emits
toxic vapors of SO_x

DUG550   CAS: 66637-35-2   HR: 3
3,4-DIMETHYL-2,5-THIOMORPHOLINE-
   DIONE, 2-(o-((METHYL((TRICHLORO-
   METHYL)THIO)AMINO)CARBONYL)O
   XIME)
mf: $C_9H_{12}Cl_3N_3O_3S_2$ mw: 380.71
SAFETY PROFILE: A poison by ingestion.
When heated to decomposition it emits
toxic vapors of NO_x, SO_x, and Cl^-.

DUG800   CAS: 2767-47-7   HR: 3
DIMETHYLTIN DIBROMIDE
mf: $C_2H_6Br_2Sn$ mw: 308.59
PROP: Colorless or white crystals. Mp: 76°,
bp: 208–213°. Sol in water and org solvs.
SYN: DIBROMODIMETHYL STANNANE
OSHA PEL: TWA 0.1 mg(Sn)/m³ (skin)

ACGIH TLV: TWA 0.1 mg(Sn)/m³; STEL
0.2 mg(Sn)/m³ (skin).
NIOSH REL: (Organotin Compounds) TWA
0.1 mg(Sn)/m³
SAFETY PROFILE: Poison by intravenous
route. When heated to decomposition it
emits toxic fumes of Br^-. See also TIN
COMPOUNDS and BROMIDES.

DUH600   CAS: 55-80-1   HR: 2
N,N-DIMETHYL-p-((m-TOLYL)AZO)-
   ANILINE
mf: $C_{15}H_{17}N_3$ mw: 239.35
SYNS: 4-(N,N-DIMETHYLAMINO)-3'-
METHYLAZOBENZENE □ N,N-DIMETHYL-p-(3'-
METHYLPHENYLAZO)ANILINE □ N,N-DIMETHYL-4-
((3-METHYLPHENYL)AZO)BENZENAMINE □ MDAB □
3'-MDAB □ 3'-METHYLBUTTERGELB (GERMAN) □ 3'-
METHYL-DAB □ 3'-METHYL-4-DIMETHYLAMINOAZO-
BENZEN (CZECH) □ M'-METHYL-p-DIMETHYLAMINO-
AZOBENZENE □ 3'-METHYL-4-DIMETHYLAMINO-
AZOBENZENE □ 3'-METHYL-N,N-DIMETHYL-4-
AMINOAZOBENZENE □ 3'-METHYLDIMETHYL-
AMINOAZOBENZOL (GERMAN)
CONSENSUS REPORTS: Reported in EPA
TSCA Inventory. EPA Genetic Toxicology
Program.
SAFETY PROFILE: Moderately toxic by
ingestion. An experimental teratogen.
Questionable carcinogen with experimental
carcinogenic, neoplastigenic, and
tumorigenic data. Mutation data reported.
When heated to decomposition it emits
toxic fumes of NO_x.

DUK800   CAS: 2164-17-2   HR: 2
1,1-DIMETHYL-3-(α,α,α-TRIFLUORO-m-
   TOLYL) UREA
mf: $C_{10}H_{11}F_3N_2O$ mw: 232.23
PROP: Crystals. Mp: 163–164.5°. Sol in
most org solvs; very sltly sol in H_2O.
SYNS: C 2059 □ CIBA 2059 □ COTORAN □ COTORAN
MULTI 50WP □ COTTONEX □ N,N-DIMETHYL-N'-(3-
TRIFLUOROMETHYLPHENYL)UREA □ 1,1-DIMETHYL-
3-(3-TRIFLUOROMETHYLPHENYL)UREA □
FLUOMETURON □ HERBICIDE C-2059 □ LANEX □
NCI-C08695 □ PAKHTARAN □ 3-(5-TRIFLUORMETHYL-
PHENYL)-,1-DIMETHYLHARNSTOFF (GERMAN) □ N-
(m-TRIFLUOROMETHYLPHENYL)-N',N'-DIMETHYL-
UREA □ N-(3-TRIFLUOROMETHYLPHENYL)-N'-N'-
DIMETHYLUREA □ 3-(m-TRIFLUOROMETHYL-
PHENYL)-1,1-DIMETHYLUREA

CONSENSUS REPORTS: EPA Genetic Toxicology Program. IARC Cancer Review: Group 3 IMEMDT 7,56,87; Animal Inadequate Evidence IMEMDT 30,245,83. NCI Carcinogenesis Bioassay (feed); No Evidence: rat NCITR* NCI-CG-TR-195,80; Equivocal Evidence: mouse NCITR* NCI-CG-TR-195,80. Reported in EPA TSCA Inventory.
SAFETY PROFILE: Moderately toxic by ingestion and intraperitoneal routes. Questionable carcinogen with experimental carcinogenic data. Mutation data reported. When heated to decomposition it emits very toxic fumes of F^- and NO_x.

DUM200   CAS: 96-31-1   HR: 2
1,3-DIMETHYLUREA
mf: $C_3H_8N_2O$ mw: 88.13
PROP: Colorless rhombic crystals. D: 1.14, mp: 106°, bp: 270°. Sol in water and alc.
SYNS: N,N'-DIMETHYLHARNSTOFF (GERMAN) □ N,N'-DIMETHYLUREA □ sym-DIMETHYLUREA □ SYMMETRIC DIMETHYLUREA
CONSENSUS REPORTS: Reported in EPA TSCA Inventory.
SAFETY PROFILE: Moderately toxic by intraperitoneal route. Experimental teratogenic and reproductive effects. Human mutation data reported. When heated to decomposition it emits toxic fumes of NO_x.

DUO500   CAS: 69853-15-2   HR: 3
DIMORPHOLINIUM HEXACHLORO-
    STANNATE
mf: $C_8H_{10}Cl_6N_2O_2Sn$ mw: 497.59
SYN: MORPHOLINIUM, HEXACHLOROSTANNATE(2-) (2:1)
OSHA PEL: TWA 2 mg(Sn)/m³
ACGIH TLV: TWA 2 mg(Sn)/m³
SAFETY PROFILE: Poison by intravenous route. When heated to decomposition it emits toxic fumes of NO_x, Sn, and Cl^-.

DUP300   CAS: 148-01-6   HR: 3
DINITOLMIDE
mf: $C_8H_7N_3O_5$ mw: 225.18
PROP: Yellowish solid or needles from EtOH (aq). Mp: 181°. Very sltly sol in water;

sol in acetone, acetonitrile, and dimethyl formamide.
SYNS: COCCIDINE A □ COCCIDOT □ DINITOLMID □ 3,5-DINITRO-o-TOLUAMIDE □ D.O.T. □ 2-METHYL-3,5-DINITROBENZAMIDE □ ZOALENE □ ZOAMIX
OSHA PEL: TWA 5 mg/m³
ACGIH TLV: 1 mg/m³; Not Classifiable as a Human Carcinogen
SAFETY PROFILE: Poison by intravenous route. Moderately toxic by ingestion. Mutation data reported. A strong exothermic reaction above 248°C has caused industrial explosions. When heated to decomposition it emits toxic fumes of NO_x. See also NITRO COMPOUNDS of AROMATIC HYDROCARBONS.

DUP600   CAS: 97-02-9   HR: 3
2,4-DINITROANILINE
mf: $C_6H_5N_3O_4$ mw: 183.14
PROP: Yellow, needle-like crystals. Mp: 188°, flash p: 435°F (CC), d: 1.615, vap d: 6.31. Insol in water.
SYNS: 2,4-DINITRANILINE □ 2,4-DINITROANILIN (GERMAN) □ 2,4-DINITROANILINA (ITALIAN) □ 2,4-DINITROBENZENAMIME □ DNA □ NCI-C60753
CONSENSUS REPORTS: Reported in EPA TSCA Inventory.
SAFETY PROFILE: Poison by ingestion and intraperitoneal routes. Experimental teratogenic and reproductive effects. Mutation data reported. An eye irritant. Combustible and explosive when exposed to heat or flame; can react with oxidizing materials. To fight fire, use CO_2, dry chemical. Mixtures with charcoal ignite at 350°C. Vigorous reaction with chlorine + hydrochloric acid evolves gases. When heated to decomposition it emits highly toxic fumes of NO_x.

DUP800   CAS: 119-27-7   HR: 3
2,4-DINITROANISOL
mf: $C_7H_6N_2O_5$ mw: 198.15
PROP: Colorless to yellow crystals from alc. Mp: 83°, bp: sublimes, d: 1.341 @ 20°/4°, vap d: 6.83.

SYNS: α-DINITROANISOLE □ 2,4-DINITROANISOLE □ 2,4-DINITROPHENYLMETHYL ETHER □ 1-METHOXY-2,4-DINITROBENZENE

CONSENSUS REPORTS: Reported in EPA TSCA Inventory.

SAFETY PROFILE: Poison by ingestion. Mutation data reported. When heated to decomposition it emits toxic fumes of NO_x. See also NITRO COMPOUNDS of AROMATIC HYDROCARBONS and NITRATES.

DUQ150   **HR: D**
3,5-DINITROBENZAMIDE
SAFETY PROFILE: When heated to decomposition emits toxic fumes of NO_x.

DUQ180   CAS: 25154-54-5   HR: 3
DINITROBENZENE
DOT: UN 1597
mf: $C_6H_4N_2O_4$ mw: 168.12
SYNS: DINITROBENZENE, solution (DOT) □ DINITROBENZOL, solid (DOT)
OSHA PEL: TWA 1 mg/m³ (skin)
ACGIH TLV: TWA 0.15 ppm (skin)
DFG MAK: Confirmed Animal Carcinogen with Unknown Relevance to Humans
DOT CLASSIFICATION: 6.1; Label: Poison
SAFETY PROFILE: Suspected carcinogen. A poison. When heated to decomposition it emits toxic fumes of NO_x. See also o-DINITROBENZENE.

DUQ200   CAS: 99-65-0   HR: 3
m-DINITROBENZENE
DOT: UN 1597
mf: $C_6H_4N_2O_4$ mw: 168.12
PROP: Yellowish crystals from alc. Mp: 89°, bp: 291°.
SYNS: BINITROBENZENE □ 1,3-DINITROBENZENE □ 2,4-DINITROBENZENE □ 1,3-DINITROBENZOL □ DWUNITROBENZEN (POLISH)
CONSENSUS REPORTS: Reported in EPA TSCA Inventory. EPA Genetic Toxicology Program.
OSHA PEL: TWA 1 mg/m³ (skin)
ACGIH TLV: TWA 0.15 ppm (skin)
DFG MAK: Confirmed Animal Carcinogen with Unknown Relevance to Humans

DOT CLASSIFICATION: 6.1; Label: Poison
SAFETY PROFILE: Suspected carcinogen. Human poison by ingestion. Experimental poison by ingestion, intraperitoneal, and intravenous routes. Human systemic effects by skin contact: cyanosis and motor activity changes. Experimental reproductive effects. An eye irritant. Mutation data reported. Mixture with nitric acid is a high explosive. Mixture with tetranitromethane is a high explosive very sensitive to sparks. When heated to decomposition it emits toxic fumes of NO_x. See also o- and p-DINITROBENZENE.

DUQ400   CAS: 528-29-0   HR: 3
o-DINITROBENZENE
DOT: UN 1597
mf: $C_6H_4N_2O_4$ mw: 168.12
PROP: Colorless needles or plates from alc. Mp: 118°, bp: 319°, flash p: 302°F (CC), d: 1.571 @ 0°/4°, vap d: 5.79. Sol in EtOH and $CHCl_3$; sltly sol in H_2O. IDLH 50 mg/m³.
SYN: 1,2-DINITROBENZENE
OSHA PEL: TWA 1 mg/m³ (skin)
ACGIH TLV: TWA 0.15 ppm (skin)
DFG MAK: Confirmed Animal Carcinogen with Unknown Relevance to Humans
DOT CLASSIFICATION: 6.1; Label: Poison
SAFETY PROFILE: Suspected carcinogen. Poison by inhalation and ingestion. Moderately toxic by skin contact. Can cause liver, kidney, and central nervous system injury. Combustible when exposed to heat or flame; can react vigorously with oxidizing materials. A severe explosion hazard when shocked or exposed to heat or flame. It is used in bursting charges and to fill artillery shells. Mixtures with nitric acid are highly explosive. To fight fire, use water, CO_2, dry chemical. Dangerous; when heated to decomposition it emits highly toxic fumes of NO_x and explodes. See also m- and p-DINITROBENZENE and NITRO COMPOUNDS of AROMATIC HYDROCARBONS.

DUQ600   **CAS: 100-25-4**   **HR: 3**
p-DINITROBENZENE
DOT: UN 1597
mf: $C_6H_4N_2O_4$ mw: 168.12
PROP: White crystals, needles or prisms
from alc. Mp: 173°, bp: 299°. Volatile with
steam. IDLH 50 mg/m³.
SYN: DITHANE A-4
CONSENSUS REPORTS: Reported in EPA
TSCA Inventory.
OSHA PEL: TWA 1 mg/m³ (skin)
ACGIH TLV: TWA 0.15 ppm (skin)
DFG MAK: Confirmed Animal Carcinogen
with Unknown Relevance to Humans
DOT CLASSIFICATION: 6.1; Label: Poison
SAFETY PROFILE: Suspected carcinogen.
Poison by ingestion. Mutation data reported.
Mixture with nitric acid is a high explosive.
When heated to decomposition it emits
toxic fumes of NO_x. See also o- and m-
DINITROBENZENE.

DUR425   **CAS: 2218-96-4**   **HR: 3**
2,4-DINITROBENZENETHIOL
mf: $C_6H_4N_2O_4S$ mw: 200.18
SYN: BENZENETHIOL, 2,4-DINITRO-
SAFETY PROFILE: A poison by ingestion.
When heated to decomposition it emits
toxic vapors of NO_x and SO_x.

DUR800   **CAS: 87-31-0**   **HR: 3**
5,7-DINITRO-1,2,3-BENZOXADIAZOLE
DOT: UN 0074
mf: $C_6H_2N_4O_5$ mw: 210.12
SYNS: DDNP □ DIAZO □ 2-DIAZO-4,6-DINITRO-
BENZENE-1-OXIDE □ DIAZODINITROPHENOL (dry)
(DOT) □ DIAZODINITROPHENOL, wetted with not <40%
H_2O or mixture of alcohol & H_2O (UN 0074) (DOT) □
INITIATING EXPLOSIVE DIAZODINITROPHENOL
(DOT)
DOT CLASSIFICATION: Forbidden; DOT
Class: EXPLOSIVE 1.1A; Label:
EXPLOSIVE 1.1A (UN 0074)
SAFETY PROFILE: An explosive. When
heated to decomposition it emits toxic
fumes of NO_x. See also NITRO
COMPOUNDS of AROMATIC
HYDROCARBONS, and EXPLOSIVES,
HIGH.

DUS000   **CAS: 1528-74-1**   **HR: 2**
4,4'-DINITROBIPHENYL
mf: $C_{12}H_8N_2O_4$ mw: 244.22
PROP: Needles from EtOH or toluene. Mp:
239–239.5°.
SYN: 4,4'-DINITROBIFENYL (CZECH)
CONSENSUS REPORTS: Reported in
EPA TSCA Inventory.
SAFETY PROFILE: An eye irritant.
Questionable carcinogen with experimental
tumorigenic data. Mutation data reported.
When heated to decomposition it emits
toxic fumes of NO_x.

DUS600   **CAS: 2401-85-6**   **HR: 3**
2,4-DINITRO-1-CHLORO-NAPHTHAL-
ENE
mf: $C_{10}H_5ClN_2O_4$ mw: 252.62
PROP: Yellow needles from C_6H_6 or AcOH
(aq). Mp: 146.5°.
SYN: 1-CHLORO-2,4-DINITRONAPHTHALENE
CONSENSUS REPORTS: Reported in EPA
TSCA Inventory.
SAFETY PROFILE: Poison by unspecified
route. Questionable carcinogen with
experimental carcinogenic and
neoplastigenic data. When heated to
decomposition it emits very toxic fumes of
Cl^- and NO_x. See also 2,4-
DINITROANILINE.

DUS700   **CAS: 534-52-1**   **HR: 3**
DINITRO-o-CRESOL
mf: $C_7H_6N_2O_5$ mw: 198.15
PROP: Yellow, prismatic crystals from alc.
Mp: 85.8°, vap d: 6.82. IDLH 5 mg/m³.
SYNS: ANTINONIN □ ARBOROL □ CAPSINE □
CHEMSECT DNOC □ DEGRASSAN □ DEKRYSIL □
DETAL □ DINITROCRESOL □ 2,4-DINITRO-o-CRESOL
□ 4,6-DINITRO-o-CRESOL □ 4,6-DINITRO-o-CRESOLO
(ITALIAN) □ DINITRODENDTROXAL □ 3,5-DINITRO-
2-HYDROXYTOLUENE □ 4,6-DINITRO-o-KRESOL
(CZECH) □ 4,6-DINITROKRESOL (DUTCH) □
DINITROL □ DINITROMETHYL CYCLOHEXYLTRIEN-
OL □ 2,4-DINITRO-6-METHYLPHENOL □ DINOC □
DINURANIA □ DITROSOL □ DN-DRY MIX No. 2 □
DNOK (CZECH) □ DWUNITRO-o-KREZOL (POLISH) □
EFFUSAN □ ELGETOL □ ELIPOL □ ENT 154 □
EXTRAR □ HEDOLIT □ K III □ KRENITE (OBS.) □
KRESAMONE □ KREZOTOL 50 □ LE DINITROCRES-

OL-4,6 (FRENCH) □ LIPAN □ 2-METHYL-4,6-
DINITROPHENOL □ NITRADOR □ NITROFAN □
PROKARBOL □ RAFEX □ RAPHATOX □ RCRA WASTE
NUMBER P047 □ SANDOLIN □ SELINON □ SINOX □
TRIFOCIDE □ TRIFRINA □ WINTERWASH □
ZAHLREICHE BEZEICHNUNGEN (GERMAN)
CONSENSUS REPORTS: Reported in EPA
TSCA Inventory. EPA Genetic Toxicology
Program. Community Right-To-Know List.
EPA Extremely Hazardous Substances List.
OSHA PEL: TWA 0.2 mg/m³ (skin)
ACGIH TLV: TWA 0.2 mg/m³ (skin)
DFG MAK: 0.2 mg/m³
NIOSH REL: (Dinitro-o-Cresol) TWA 0.2
mg/m³
DOT CLASSIFICATION: 6.1; Label: Poison
SAFETY PROFILE: Human poison by
unspecified route. Experimental poison by
ingestion, inhalation, skin contact,
intraperitoneal, and intravenous routes.
Human systemic effects by ingestion and
inhalation: somnolence, headache, abnormal
brain recordings from specific areas of the
central nervous system, cardiac and
gastrointestinal changes. Mutation data
reported. An eye and skin irritant. Less toxic
than the para form, but is still highly toxic.
A pesticide. See also NITRO
COMPOUNDS of AROMATIC
HYDROCARBONS and other dinitrocresol
entries.

DUT000   **CAS: 497-56-3**   **HR: 3**
3,5-DINITRO-o-CRESOL
mf: $C_7H_6N_2O_5$ mw: 198.15
PROP: Yellow prisms from EtOH. Mp: 85°.
NIOSH REL: (Dinitro-Ortho-Cresol) TWA
0.2 mg/m³
SAFETY PROFILE: Poison by
subcutaneous route. When heated to
decomposition it emits toxic fumes of NO_x.
See also NITRO COMPOUNDS of
AROMATIC HYDROCARBONS and
other dinitrocresol entries.

DUT600   **CAS: 609-93-8**   **HR: 3**
2,6-DINITRO-p-CRESOL
mf: $C_7H_6N_2O_5$ mw: 198.15

PROP: Yellow needles from pet ether. Mp:
84°.
SYNS: DINITRO-p-CRESOL □ DNPC □ VICTORIA
ORANGE □ VICTORIA YELLOW
CONSENSUS REPORTS: Reported in EPA
TSCA Inventory.
SAFETY PROFILE: Poison by
intraperitoneal route. Mutation data
reported. When heated to decomposition it
emits toxic fumes of NO_x. See also other
dinitrocresol entries.

DUU600   **CAS: 2312-76-7**   **HR: 3**
4,6-DINITRO-o-CRESOL SODIUM SALT
mf: $C_7H_5N_2O_5$•Na   mw: 220.13
PROP: Brilliant, orange-yellow dye.
SYNS: CORODINOC □ CRESOTOL □ DINITRO-o-
CRESOL SODIUM SALT □ 3,5-DINITRO-o-CRESOL
SODIUM SALT □ 2,4-DINITRO-6-METHYLPHENOL
SODIUM SALT □ DINOC □ DNOC SODIUM SALT □
DYNOSOL □ EK 54 □ ELGETOL □ KRENITE (OBS.) □
KREZONITE □ 2-METHYL-4,6-DINITROPHENOL
SODIUM SALT □ SINOX □ SODIUM-4,6-DINITRO-o-
CRESOXIDE □ SODIUM SALT of 4,6-DINITRO-o-
CRESOL
CONSENSUS REPORTS: Reported in EPA
TSCA Inventory.
NIOSH REL: (Dinitro-Ortho-Cresol) TWA
0.2 mg/m³
SAFETY PROFILE: Poison by ingestion,
skin contact, and subcutaneous routes.
Flammable. A pesticide. When heated to
decomposition it emits toxic fumes of
Na_2O. See also other dinitrocresol entries.

DUV600   **CAS: 1582-09-8**   **HR: 2**
2,6-DINITRO-N,N-DIPROPYL-4-(TRI-
    FLUOROMETHYL)BENZENAMINE
mf: $C_{13}H_{16}F_3N_3O_4$ mw: 335.32
PROP: A solid. Mp: 48.5–49°. Very sol in
Me_2CO and xylene; very sltly sol in H_2O.
Technical product contains 84–88 ppm
dipropylnitrosoamine NCITR* NCI-CG-
TR-34,78.
SYNS: AGREFLAN □ AGRIFLAN 24 □ CRISALIN □
DIGERMIN □ 2,6-DINITRO-N,N-DI-N-PROPYL-α,α,α-
TRIFLURO-p-TOLUIDINE □ 2,6-DINITRO-4-TRIFLUOR-
METHYL-N,N-DIPROPYLANILIN (GERMAN) □ 4-(DI-N-
PROPYLAMINO)-3,5-DINITRO-1-TRIFLUOROMETHYL-
BENZENE □ N,N-DI-N-PROPYL-2,6-DINITRO-4-
TRIFLUOROMETHYLANILINE □ N,N-DIPROPYL-4-

TRIFLUOROMETHYL-2,6-DINITROANILINE □
ELANCOLAN □ L-36352 □ LILLY 36,352 □ M.T.F. □ NCI-
C00442 □ NITRAN □ OLITREF □ SUPER-TREFLAN □
SU SEGURO CARPIDOR □ SYNFLORAN □
TREFANOCIDE □ TREFICON □ TREFLAM □ TREFLAN
□ TREFLANOCIDE ELANCOLAN □ TRI-4 □ TRIFLOR-
AN □ TRIFLUORALIN (USDA) □ α,α,α-TRIFLUORO-2,6-
DINITRO-N,N-DIPROPYL-p-TOLUIDINE □ 4-(TRIFLU-
OROMETHYL)-2,6-DINITRO-N,N-DIPROPYLANILINE □
TRIFLURALIN □ TRIFLURALINA 600 □ TRIFLURALINE
□ TRIFUREX □ TRIKEPIN □ TRIM □ TRISTAR

CONSENSUS REPORTS: NCI
Carcinogenesis Bioassay (feed); Clear
Evidence: mouse NCITR* NCI-CG-TR-
34,78; No Evidence: rat NCITR* NCI-CG-
TR-34,78. EPA Genetic Toxicology
Program. Community Right-To-Know List.
SAFETY PROFILE: Moderately toxic by
ingestion and intraperitoneal routes.
Experimental teratogenic and reproductive
effects. Questionable carcinogen with
experimental carcinogenic and tumorigenic
data. Human mutation data reported. When
heated to decomposition it emits very toxic
fumes of F⁻ and NO$_x$. See also
FLUORIDES.

DUV710   CAS: 600-40-8   HR: 3
1,1-DINITROETHANE
mf: C$_2$H$_4$N$_2$O$_4$ mw: 120.08
SYNS: 1,1-DINITROETHANE (dry) (DOT) □ ETHANE,
1,1-DINITRO-
DOT CLASSIFICATION: Forbidden
SAFETY PROFILE: A poison by
intraperitoneal route. An unstable solid
forbidden for transport. When heated to
decomposition it emits toxic vapors of NO$_x$.

DUW400   CAS: 70-34-8   HR: 3
2,4-DINITRO-1-FLUOROBENZENE
mf: C$_6$H$_3$FN$_2$O$_4$ mw: 186.11
PROP: Crystals or oil. Crystals mp: 27°; oil
mp: 12°, bp: 137° @ 20 mm. Sol in ether,
benzene, and propylene glycol.
SYNS: 2,4-DINITROFLUOROBENZENE □ 2,4-DNFB □
1-FLUORO-2,4-DINITROBENZENE □ 1,2,4-
FLUORODINITROBENZENE
CONSENSUS REPORTS: Reported in EPA
TSCA Inventory. EPA Genetic Toxicology
Program.

SAFETY PROFILE: Poison by ingestion,
skin contact, and subcutaneous routes. A
powerful irritant and vesicant. Mutation data
reported. Solutions in ether may explode
when evaporated. When heated to
decomposition it emits highly toxic fumes of
NO$_x$ and F⁻. See also NITRO
COMPOUNDS of AROMATIC
HYDROCARBONS and FLUORIDES.

DUX650   CAS: 606-37-1   HR: 3
1,3-DINITRONAPHTHALENE
mf: C$_{10}$H$_6$N$_2$O$_4$ mw: 218.18
SYN: NAPHTHALENE, 1,3-DINITRO-
DFG MAK: Confirmed Animal Carcinogen
with Unknown Relevance to Humans
SAFETY PROFILE: Suspected carcinogen.
Mutation data reported. When heated to
decomposition it emits toxic vapors of NO$_x$.

DUX700   CAS: 605-71-0   HR: 3
1,5-DINITRONAPHTHALENE
mf: C$_{10}$H$_6$N$_2$O$_4$ mw: 218.17
PROP: Needles from AcOH or Me$_2$CO.
Mp: 218°. Sol in hot C$_6$H$_6$ and Py.
SYN: 1,5-DINITRONAPHTHALENE
DFG MAK: Confirmed Animal Carcinogen
with Unknown Relevance to Humans
SAFETY PROFILE: A suspected
carcinogen. Mutation data reported.
Mixtures with sulfur or sulfuric acid (used in
commercial reactions) may explode if heated
to 120°C. Initiation temperature depends on
the quality of the dinitronaphthalene. When
heated to decomposition it emits toxic
fumes of NO$_x$. See also NITRO
COMPOUNDS of AROMATIC
HYDROCARBONS.

DUX710   CAS: 602-38-0   HR: 3
1,8-DINITRONAPHTHALENE
mf: C$_{10}$H$_6$N$_2$O$_4$ mw: 218.18
SYN: NAPHTHALENE, 1,8-DINITRO-
CONSENSUS REPORTS: Reported in EPA
TSCA Inventory.
DFG MAK: Confirmed Animal Carcinogen
with Unknown Relevance to Humans

SAFETY PROFILE: Suspected carcinogen. Mutation data reported. When heated to decomposition it emits toxic vapors of NO_x.

DUX800   CAS: 605-69-6   HR: 3
2,4-DINITRO-1-NAPHTHOL
mf: $C_{10}H_6N_2O_5$ mw: 234.18
PROP: Yellow needles or leaflets from EtOH or $CHCl_3$. Mp: 140°, vap d: 8.08. Sltly sol in Et_2O, EtOH, C_6H_6, and H_2O.
SYNS: C.I. 10315 □ 2,4-DINITRO-1-NAFTOL □ 2-4 DINITRO-α-NAPHTOL □ 2-4 DINITRO-α-NAPHTOL (FRENCH) □ GOLDEN YELLOW □ MANCHESTER YELLOW □ MARITUS YELLOW □ NAPHTHOL YELLOW □ NAPHTHYLENE YELLOW □ SAFFRON YELLOW □ ZLUT MARCIOVA □ ZLUT NAFTOLOVA
CONSENSUS REPORTS: Reported in EPA TSCA Inventory.
SAFETY PROFILE: Poison by subcutaneous, intramuscular, intravenous, and intraperitoneal routes. Human reproductive effects by skin contact: skin toxicity. Mutation data reported. For fire, disaster, and explosion hazards, see NITRATES.

DUY600   CAS: 25550-58-7   HR: 3
DINITROPHENOL
DOT: UN 0076/UN 1320/UN 1599
mf: $C_6H_4N_2O_5$ mw: 184.12
SYNS: DINITROPHENOL □ DINITROPHENOL, dry or wetted with <15% water, by weight (UN 0076) (DOT) □ DINITROPHENOL, wetted with not <15% water, by weight (UN 1320) (DOT) □ DINITROPHENOL SOLUTIONS (UN 1599) (DOT)
DOT CLASSIFICATION: EXPLOSIVE 1.1D; Label: EXPLOSIVE 1.1D, Poison (UN 076); DOT Class: 4.1; Label: Flammable Solid, Poison (UN 1320); DOT Class: 6.1; Label: Poison (UN 1599)
SAFETY PROFILE: Poison by ingestion and subcutaneous routes. An explosive and flammable solid. When heated to decomposition it emits toxic fumes of NO_x. See also NITRO COMPOUNDS of AROMATIC HYDROCARBONS.

DUZ000   CAS: 51-28-5   HR: 3
2,4-DINITROPHENOL
mf: $C_6H_4N_2O_5$ mw: 184.12

PROP: Yellow crystals or plates from water. Mp: 113°, d: 1.683 @ 24°, vap d: 6.35. Sol in EtOH, Me_2CO, and C_6H_6; sltly sol in H_2O.
SYNS: ALDIFEN □ CHEMOX PE □ 2,4-DINITROFENOL (DUTCH) □ DINITROFENOLO (ITALIAN) □ α-DINITROPHENOL □ 2,4-DNP □ FENOXYL CARBON N □ 1-HYDROXY-2,4-DINITROBENZENE □ MAROXOL-50 □ NITRO KLEENUP □ NSC-1532 □ RCRA WASTE NUMBER P048 □ SOLFO BLACK B □ SOLFO BLACK BB □ SOLFO BLACK 2B SUPRA □ SOLFO BLACK G □ SOLFO BLACK SB □ TERTROSULPHUR BLACK PB □ TERTROSULPHUR PBR
CONSENSUS REPORTS: Reported in EPA TSCA Inventory. EPA Genetic Toxicology Program.
SAFETY PROFILE: A deadly human poison by ingestion. An experimental poison by ingestion, inhalation, intravenous, intraperitoneal, subcutaneous, and intramuscular routes. Moderately toxic by skin contact. Experimental teratogenic and reproductive effects. Human systemic effects: body temperature increase, change in heart rate, coma. A skin irritant. Mutation data reported. Phytotoxic. A pesticide. An explosive. Forms explosive salts with alkalies and ammonia. When heated to decomposition it emits toxic fumes of NO_x. See also NITRO COMPOUNDS of AROMATIC HYDROCARBONS.

DVD400   CAS: 75321-20-9   HR: 3
1,3-DINITROPYRENE
mf: $C_{16}H_8N_2O_4$ mw: 292.26
PROP: Light-brown needles from C_6H_6/MeOH. Mp: 274–276°.
SYN: DINITROPYRENE
CONSENSUS REPORTS: IARC Cancer Review: Group 3 IMEMDT 46,201,89; Animal Limited Evidence IMEMDT 46,201,89; Human No Adequate Data IMEMDT 46,201,89.
DFG MAK: Confirmed Animal Carcinogen with Unknown Relevance to Humans
SAFETY PROFILE: Suspected carcinogen with experimental carcinogenic data. Mutation data reported. When heated to decomposition it emits toxic fumes of NO_x.

DVD600　　CAS: 42397-64-8　HR: 3
1,6-DINITROPYRENE
mf: $C_{16}H_8N_2O_4$　　mw: 292.26
PROP: Light-brown needles from C_6H_6/MeOH.
SYN: DINITROPYRENE
CONSENSUS REPORTS: NTP 10th Report on Carcinogens. IARC Cancer Review: Group 2B IMEMDT 46,215,89; Animal Sufficient Evidence IMEMDT 46,215,89; Human No Adequate Data IMEMDT 46,215,89.
DFG MAK: Confirmed Animal Carcinogen with Unknown Relevance to Humans
SAFETY PROFILE: Confirmed carcinogen with experimental carcinogenic data. Human mutation data reported. When heated to decomposition it emits toxic fumes of NO_x.

DVD800　　CAS: 42397-65-9　HR: 3
1,8-DINITROPYRENE
mf: $C_{16}H_8N_2O_4$　　mw: 292.26
PROP: Light-brown needles from C_6H_6/MeOH.
SYN: DINITROPYRENE
CONSENSUS REPORTS: NTP 10th Report on Carcinogens. IARC Cancer Review: Group 2B IMEMDT 46,231,89; Animal Sufficient Evidence IMEMDT 46,231,89; Human No Adequate Data IMEMDT 46,231,89.
DFG MAK: Confirmed Animal Carcinogen with Unknown Relevance to Humans
SAFETY PROFILE: Confirmed carcinogen with experimental carcinogenic data. Human mutation data reported. When heated to decomposition it emits toxic fumes of NO_x.

DVD900　　CAS: 117929-15-4　HR: 2
2,7-DINITROPYRENE
mf: $C_{16}H_8N_2O_4$　　mw: 292.26
SYN: PYRENE, 2,7-DINITRO-
DFG MAK: Confirmed Animal Carcinogen with Unknown Relevance to Humans
SAFETY PROFILE: Suspected carcinogen. Mutation data reported. When heated to decomposition it emits toxic vapors of NO_x.

DVF200　　CAS: 140-79-4　HR: 3
DINITROSOPIPERAZINE
mf: $C_4H_8N_4O_2$　　mw: 144.16
PROP: White or cream colored crystals. Mp: 158°, vap d: 4.97.
SYNS: DINITROSOPIPERAZIN (GERMAN) □ N,N'-DINITROSOPIPERAZINE □ 1,4-DINITROSO-PIPERAZINE □ DNPZ □ NSC-339 □ USAF DO-36
CONSENSUS REPORTS: EPA Genetic Toxicology Program. Reported in EPA TSCA Inventory.
SAFETY PROFILE: Suspected carcinogen with experimental carcinogenic, neoplastigenic, tumorigenic, and teratogenic data. Poison by ingestion, subcutaneous, and intraperitoneal routes. Experimental reproductive effects. Human mutation data reported. When heated to decomposition it emits toxic fumes of NO_x. See also N-NITROSO COMPOUNDS.

DVF300　　CAS: 118-02-5　HR: 3
2,4-DINITROSO-m-RESORCINOL
mf: $C_6H_4N_2O_4$　　mw: 168.12
SYNS: BENZENE-1,3-DIOL, 2,4-DINITROSO- □ 2,4-DINITRORESORCINOL (heavy metal salts of) (dry) (DOT) □ RESORCINOL, 2,4-DINITROSO-
DOT CLASSIFICATION: Forbidden
SAFETY PROFILE: A poison by intraperitoneal route. An unstable substance forbidden for transport. When heated to decomposition it emits toxic vapors of NO_x.

DVF400　　CAS: 101-25-7　HR: 3
3,7-DINITROSO-1,3,5,7-TETRAAZABI-CYCLO[3.3.1]NONANE
DOT: UN 2972
mf: $C_5H_{10}N_6O_2$　　mw: 186.18
SYNS: ACETO DNPT 40 □ ACETO DNPT 80 □ ACETO DNPT 100 □ CHKHZ 18 □ DINITROSOPENTAMETH-YLENETETRAMINE □ N,N-DINITROSOPENTAMETH-YLENETETRAMINE □ N^1,N^3-DINITROSOPENTA-METHYLENETETRAMINE □ 3,4-DI-N-NITROSO-PENTAMETHYLENETETRAMINE □ 3,7-DI-N-NITROSOPENTAMETHYLENETETRAMINE □ DNPMT □ DNPT □ 1,5-METHYLENE-3,7-DINITROSO-1,3,5,7-TETRAAZACYCLOOCTAINE □ 1,5-METHYLENE-3,7-DINITROSO-1,3,5,7-TETRAAZACYCLOOCTANE □ POROFOR CHKHC-18 □ POROPHOR B □ UNICEL-ND □ UNICEL NDX □ VULCACEL B-40 □ VULCACEL BN

CONSENSUS REPORTS: IARC Cancer Review: Group 3 IMEMDT 7,56,87; Animal No Evidence IMEMDT 11,241,76. Reported in EPA TSCA Inventory. EPA Genetic Toxicology Program.
DOT CLASSIFICATION: 4.1; Label: Flammable Solid, EXPLOSIVE
SAFETY PROFILE: Poison by intravenous, intraperitoneal, and subcutaneous routes. Moderately toxic by ingestion. Questionable carcinogen. Mutation data reported. Can ignite when handled and burns very rapidly. Many N-nitroso compounds are carcinogens. A blowing agent. When heated to decomposition it emits toxic fumes of NO_x. See also N-NITROSO COMPOUNDS.

DVG600   CAS: 25321-14-6   HR: 3
DINITROTOLUENE
DOT: UN 2038
mf: $C_7H_6N_2O_4$ mw: 182.15
PROP: IDLH 50 mg/m³.
SYNS: BENZENE, METHYLDINITRO- □ DINITRO-PHENYLMETHANE □ DINITROTOLUENES, liquid or solid (DOT) □ METHYLDINITROBENZENE □ TOLUENE, ar,ar-DINITRO-
CONSENSUS REPORTS: Reported in EPA TSCA Inventory. EPA Genetic Toxicology Program.
OSHA PEL: TWA 1.5 mg/m³ (skin)
ACGIH TLV: TWA 0.2 mg/m³ (skin); Animal Carcinogen
DFG MAK: Animal Carcinogen, Suspected Human Carcinogen
NIOSH REL: (Dinitrotoluene) Reduce to lowest level
DOT CLASSIFICATION: 6.1; Label: Poison
SAFETY PROFILE: Confirmed carcinogen with experimental tumorigenic and teratogenic data. A poison. Experimental reproductive effects. Mutation data reported. Flammable. When heated to decomposition it emits toxic fumes of NO_x. See also 2,4-DINITROTOLUENE.

DVG800   CAS: 602-01-7   HR: 2
2,3-DINITROTOLUENE
mf: $C_7H_6N_2O_4$ mw: 182.15

PROP: Needles from pet ether. Mp: 63°.
SYNS: 2,3-DNT □ 1-METHYL-2,3-DINITRO-BENZENE (9CI)
CONSENSUS REPORTS: Reported in EPA TSCA Inventory.
OSHA PEL: TWA 1.5 mg/m³ (skin)
NIOSH REL: (Dinitroluene) Reduce to lowest level
SAFETY PROFILE: Moderately toxic by ingestion. Mutation data reported. A skin irritant. When heated to decomposition it emits toxic fumes of NO_x. See also 2,4-DINITROTOLUENE.

DVH000   CAS: 121-14-2   HR: 3
2,4-DINITROTOLUENE
mf: $C_7H_6N_2O_4$ mw: 182.15
PROP: Yellow needles from CS_2. Mp: 69.5°, bp: 300°, d: 1.521 @ 15°, vap d: 6.27, flash p: 404°F.
SYNS: 2,4-DINITROTOLUOL □ DNT □ 2,4-DNT □ 1-METHYL-2,4-DINITROBENZENE □ NCI-C01865 □ RCRA WASTE NUMBER U105
CONSENSUS REPORTS: NCI Carcinogenesis Bioassay (feed); No Evidence: mouse NCITR* NCI-CG-TR-54,78; Some Evidence: rat NCITR* NCI-CG-TR-54,78. Reported in EPA TSCA Inventory.
OSHA PEL: TWA 1.5 mg/m³ (skin)
NIOSH REL: (Dinitroluene) Reduce to lowest level
SAFETY PROFILE: Suspected carcinogen with experimental carcinogenic and neoplastigenic data. Poison by ingestion and subcutaneous routes. Experimental reproductive effects. A skin irritant. Mutation data reported. An irritant and an allergen. Can cause anemia, methemoglobinemia, cyanosis, and liver damage. Combustible when exposed to heat or flame; can react with oxidizing materials. To fight fire, use water spray or mist, dry chemical. Decomposes when heated to 250°C. There are instances of explosion during manufacture or storage. Mixture with nitric acid is a high explosive. Mixture with sodium carbonate can decompose with significant pressure increase at 210°C.

Mixtures with other alkalies may have the same effect. Ignites on contact with sodium oxide. When heated to decomposition it emits toxic fumes of NO_x.

DVH200   CAS: 619-15-8   HR: 2
2,5-DINITROTOLUENE
mf: $C_7H_6N_2O_4$ mw: 182.15
PROP: Needles from EtOH. Mp: 52.5°.
SYNS: 2,5-DNT □ 2-METHYL-1,4-DINITROBENZENE
CONSENSUS REPORTS: Reported in EPA TSCA Inventory.
OSHA PEL: TWA 1.5 mg/m³ (skin)
NIOSH REL: (Dinitrotoluene) Reduce to lowest level
SAFETY PROFILE: Moderately toxic by ingestion. Mutation data reported. A skin irritant. When heated to decomposition it emits toxic fumes of NO_x. See also 2,4-DINITROTOLUENE.

DVH400   CAS: 606-20-2   HR: 3
2,6-DINITROTOLUENE
mf: $C_7H_6N_2O_4$ mw: 182.15
PROP: Needles from EtOH. Mp: 66°.
SYNS: 2,6-DNT □ 2-METHYL-1,3-DINITROBENZENE □ RCRA WASTE NUMBER U106
CONSENSUS REPORTS: Reported in EPA TSCA Inventory.
OSHA PEL: TWA 1.5 mg/m³ (skin)
NIOSH REL: (Dinitrotoluene) Reduce to lowest level
SAFETY PROFILE: Poison by ingestion. A skin irritant. Questionable carcinogen with experimental tumorigenic data. Mutation data reported. When heated to decomposition it emits toxic fumes of NO_x. See also 2,4-DINITROTOLUENE.

DVH600   CAS: 610-39-9   HR: 2
3,4-DINITROTOLUENE
mf: $C_7H_6N_2O_4$ mw: 182.15
PROP: A solid. Mp: 61°.
SYNS: 3,4-DNT □ 4-METHYL-1,2-DINITROBENZENE
CONSENSUS REPORTS: Reported in EPA TSCA Inventory.
OSHA PEL: TWA 1.5 mg/m³ (skin)

NIOSH REL: (Dinitrotoluene) Reduce to lowest level
SAFETY PROFILE: Moderately toxic by ingestion. Mutation data reported. A skin irritant. When heated to decomposition it emits toxic fumes of NO_x. See also 2,4-DINITROTOLUENE.

DVJ200   CAS: 363-24-6   HR: 3
DINOPROSTONE
mf: $C_{20}H_{32}O_5$ mw: 352.52
PROP: Crystals. Mp: 66–68°.
SYNS: (5Z,11-α,13E,15S)-11,15-DIHYDROXY-9-OXOPROSTA-5,13-DIEN-1-OIC ACID □ 7-(3-HYDROXY-2-(3-HYDROXY-1-OCTENYL)-5-OXOCYCLOPENTYL)-5-HEPTENOIC ACID □ PGE2 □ PROSTAGLANDIN E2 □ (−)-PROSTAGLANDIN E2 □ (15S)-PROSTAGLANDIN E2 □ PROSTIN E2 □ U-12062
SAFETY PROFILE: Poison by subcutaneous and intravenous routes. Moderately toxic by ingestion and intraperitoneal routes. An experimental teratogen. Human reproductive effects by intravenous, intraplacental, and intravaginal routes: changes in the uterus, cervix and vagina; termination of pregnancy; and changes in fertility. Experimental reproductive effects. Mutation data reported. When heated to decomposition it emits acrid smoke and irritating fumes.

DVJ800   CAS: 3648-18-8   HR: 3
DIOCTYLDI(LAUROYLOXY)STANNANE
mf: $C_{40}H_{80}O_4Sn$ mw: 743.89
SYNS: BIS(DODECANOLOXY)DIOCTYLSTANNANE □ BIS(LAUROYLOXY)DIOCTYLSTANNANE □ DIDO-DECANOYLOXYDIOCTYLSTANNANE □ DIOCTYL-BIS(LAUROYLOXY)STANNANE □ DIOCTYLDIDO-DECANOYLOXYSTANNANE □ DIOCTYLTIN DILAURATE □ DI-n-OCTYLTIN DILAURATE □ DI-n-OCTYL-ZINN DILAURAT (GERMAN)
CONSENSUS REPORTS: Reported in EPA TSCA Inventory.
OSHA PEL: TWA 0.1 mg(Sn)/m³ (skin)
ACGIH TLV: TWA 0.1 mg(Sn)/m³; STEL 0.2 mg(Sn)/m³ (skin).
NIOSH REL: (Organotin Compounds) TWA 0.1 mg(Sn)/m³
SAFETY PROFILE: Poison by intraperitoneal route. Mildly toxic by

ingestion. When heated to decomposition it emits acrid smoke and irritating fumes. See also TIN COMPOUNDS.

DVK200   CAS: 16091-18-2   HR: 2
2,2-DIOCTYL-1,3,2-DIOXASTANNEPIN-4,7-DIONE
mf: $C_{20}H_{36}O_4Sn$ mw: 459.25
SYNS: DIOCTYLSTANNYLENE MALEATE □ DIOCTYLTIN MALEATE □ DI-n-OCTYLTIN MALEATE □ DI-n-OCTYLZINN MALEINAT □ ESTABEX U 18 □ LIV 1176 □ MELLITE 825 □ STANN OMF □ THERMOLITE 813 □ TVS 8105
CONSENSUS REPORTS: Reported in EPA TSCA Inventory.
OSHA PEL: TWA 0.1 mg(Sn)/m³ (skin)
ACGIH TLV: TWA 0.1 mg(Sn)/m³; STEL 0.2 mg(Sn)/m³ (skin).
DFG MAK: 0.1 mg(Sn)/m³ calculated as total dust
NIOSH REL: (Organotin Compounds) TWA 0.1 mg(Sn)/m³
SAFETY PROFILE: Moderately toxic by ingestion. When heated to decomposition it emits acrid smoke and irritating fumes. See also TIN COMPOUNDS.

DVK600   CAS: 141-02-6   HR: 3
DIOCTYL FUMARATE
mf: $C_{20}H_{36}O_4$ mw: 340.56
PROP: Clear, mobile liquid; mild odor. Bp: 211–220°, flash p: 365°F (COC), d: 0.942 @ 20°/20°.
SYNS: BIS(2-ETHYLHEXYL) FUMARATE □ 2-BUTENEDIOIC ACID BIS(2-ETHYLHEXYL) ESTER □ DI(2-ETHYLHEXYL) FUMARATE □ DOF □ 2-ETHYLHEXYL FUMARATE □ RC COMONOMER DOF
CONSENSUS REPORTS: Reported in EPA TSCA Inventory.
SAFETY PROFILE: Poison by intraperitoneal route. An eye and severe skin irritant. Combustible when exposed to heat or flame; can react with oxidizing materials. To fight fire, use foam, CO₂, dry chemical. See also ESTERS and FUMARIC ACID.

DVL200   CAS: 15535-79-2   HR: 2
2,2-DIOCTYL-1,3,2-OXATHIASTANNOLANE-5-OXIDE
mf: $C_{18}H_{36}O_2SSn$ mw: 435.29
SYNS: DIOCTYLTHIOACETOXYSTANNANE □ DIOCTYLTIN THIOGLYCOLATE □ DI-n-OCTYLTIN THIOGLYCOLATE □ DI-n-OCTYL-ZINN THIOGLYKOLAT (GERMAN)
CONSENSUS REPORTS: Reported in EPA TSCA Inventory.
OSHA PEL: TWA 0.1 mg(Sn)/m³ (skin)
ACGIH TLV: TWA 0.1 mg(Sn)/m³; STEL 0.2 mg(Sn)/m³ (skin).
NIOSH REL: (Organotin Compounds) TWA 0.1 mg(Sn)/m³
SAFETY PROFILE: Moderately toxic by ingestion. When heated to decomposition it emits toxic fumes of SO_x. See also TIN COMPOUNDS.

DVL400   CAS: 870-08-6   HR: 2
DIOCTYLOXOSTANNANE
mf: $C_{16}H_{34}OSn$ mw: 361.19
SYNS: DIOCTYLTIN OXIDE □ DI-n-OCTYLTIN OXIDE □ DI-n-OCTYL-ZINN OXYD (GERMAN) □ OXODIOCTYLSTANNANE
CONSENSUS REPORTS: Reported in EPA TSCA Inventory.
OSHA PEL: TWA 0.1 mg(Sn)/m³ (skin)
ACGIH TLV: TWA 0.1 mg(Sn)/m³; STEL 0.2 mg(Sn)/m³ (skin).
DFG MAK: 0.1 mg(Sn)/m³ calculated as total dust
NIOSH REL: (Organotin Compounds) TWA 0.1 mg(Sn)/m³
SAFETY PROFILE: Moderately toxic by ingestion. When heated to decomposition it emits acrid smoke and irritating fumes. See also TIN COMPOUNDS.

DVL600   CAS: 117-84-0   HR: 2
n-DIOCTYL PHTHALATE
mf: $C_{24}H_{38}O_4$ mw: 390.62
SYNS: o-BENZENEDICARBOXYLIC ACID DIOCTYL ESTER □ 1,2-BENZENEDICARBOXYLIC ACID DIOCTYL ESTER □ CELLUFLEX DOP □ DINOPOL NOP □ DIOCTYL-o-BENZENEDICARBOXYLATE □ DIOCTYL PHTHALATE □ DNOP □ OCTYL PHTHALATE □ n-OCTYL PHTHALATE □ PX-138 □ RCRA WASTE NUMBER U107 □ VINICIZER 85
CONSENSUS REPORTS: Reported in EPA TSCA Inventory.

SAFETY PROFILE: Mildly toxic by ingestion. Experimental teratogenic and reproductive effects. A skin and severe eye irritant. Used as a plasticizer. When heated to decomposition it emits acrid smoke and irritating fumes. See also ESTERS.

DVL700   CAS: 117-81-7   HR: 3
DI-sec-OCTYL PHTHALATE
mf: $C_{24}H_{38}O_4$ mw: 390.62
PROP: A liquid. D: 0.986 @ 20°, mp: −46°, bp: 231° @ 5 mm. IDLH 5000 mg/m³.
SYNS: BEHP □ BIS(2-ETHYLHEXYL-)-1,2-BENZENE-DICARBOXYLATE □ BIS(2-ETHYL-HEXYL)PHTHALATE □ BISOFLEX 81 □ BISOFLEX DOP □ COMPOUND 889 □ DAF 68 □ DEHP □ DI(2-ETHYLHEXYL)ORTHO-PHTHALATE □ DI(2-ETHYLHEXYL)PHTHALATE □ DIOCTYL PHTHALATE □ DOP □ ERGOPLAST FDO □ ETHYLHEXYL PHTHALATE □ 2-ETHYLHEXYL PHTHALATE □ EVIPLAST 80 □ EVIPLAST 81 □ FLEXIMEL □ FLEXOL DOP □ FLEXOL PLASTICIZER DOP □ GOOD-RITE GP 264 □ HATCOL DOP □ HERCOFLEX 260 □ KODAFLEX DOP □ MOLLAN O □ NCI-C52733 □ NUOPLAZ DOP □ OCTOIL □ OCTYL PHTHALATE □ PALATINOL AH □ PHTHALIC ACID DIOCTYL ESTER □ PITTSBURGH PX-138 □ PLATINOL AH □ PLATINOL DOP □ RC PLASTICIZER DOP □ RCRA WASTE NUMBER U028 □ REOMOL DOP □ REOMOL D 79P □ SICOL 150 □ STAFLEX DOP □ TRUFLEX DOP □ VESTINOL AH □ VINICIZER 80 □ WITCIZER 312
CONSENSUS REPORTS: NTP 10th Report on Carcinogens. IARC Cancer Review: Group 2B IMEMDT 7,56,87; Human Inadequate Evidence IMEMDT 29,269,82; Animal Sufficient Evidence IMEMDT 29,269,82. NTP Carcinogenesis Bioassay (feed); Clear Evidence: mouse, rat NTPTR* NTP-TR-217,82. EPA Genetic Toxicology Program. Reported in EPA TSCA Inventory. Community Right-To-Know List.
OSHA PEL: TWA 5 mg/m³; STEL 10 mg/m³
ACGIH TLV: TWA 5 mg/m³; Confirmed Animal Carcinogen with Unknown Revelance to Humans
DFG MAK: 10 mg/m³
NIOSH REL: (DEHP) Reduce to lowest feasible level

SAFETY PROFILE: Confirmed carcinogen with experimental carcinogenic and tumorigenic data. Experimental teratogenic data. Other experimental reproductive effects. Poison by intravenous route. Human systemic effects by ingestion: gastrointestinal tract effects. A mild skin and eye irritant. When heated to decomposition it emits acrid smoke.

DVL800   CAS: 69226-45-5   HR: 3
DIOCTYL(1,2-PROPYLENEDIOXYBIS-
(MALEOYLDIOXY))STANNANE
mf: $C_{27}H_{42}O_8Sn$ mw: 613.38
SYNS: DI-n-OCTYLTIN DI(1,2-PROPYLENEGLYCOL-MALEATE) □ DI-n-OCTYL-ZINN-DI-(1,2-PROPYLENGLYKOLMALEINAT)(GERMAN)
OSHA PEL: TWA 0.1 mg(Sn)/m³ (skin)
ACGIH TLV: TWA 0.1 mg(Sn)/m³; STEL 0.2 mg(Sn)/m³ (skin).
NIOSH REL: (Organotin Compounds) TWA 0.1 mg(Sn)/m³
SAFETY PROFILE: Poison by intraperitoneal route. Mildly toxic by ingestion. When heated to decomposition it emits acrid smoke and irritating fumes. See also TIN COMPOUNDS.

DVM000   CAS: 3572-47-2   HR: 3
DIOCTYLTHIOXOSTANNANE
mf: $C_{16}H_{34}SSn$ mw: 377.25
SYN: DI-n-OCTYLTIN SULFIDE
CONSENSUS REPORTS: Reported in EPA TSCA Inventory.
OSHA PEL: TWA 0.1 mg(Sn)/m³ (skin)
ACGIH TLV: TWA 0.1 mg(Sn)/m³; STEL 0.2 mg(Sn)/m³ (skin).
NIOSH REL: (Organotin Compounds) TWA 0.1 mg(Sn)/m³
SAFETY PROFILE: Poison by intravenous route. When heated to decomposition it emits toxic fumes of SO_x. See also SULFIDES and TIN COMPOUNDS.

DVM400   CAS: 22205-30-7   HR: 2
DI-n-OCTYLTIN BIS(DODECYL
MERCAPTIDE)
mf: $C_{44}H_{88}O_4S_2Sn$ mw: 864.13

SYN: BIS(MERCAPTO)DIOCTYLTIN BIS(DODECYL) ESTER

CONSENSUS REPORTS: Reported in EPA TSCA Inventory.

OSHA PEL: TWA 0.1 mg(Sn)/m³ (skin)

ACGIH TLV: TWA 0.1 mg(Sn)/m³; STEL 0.2 mg(Sn)/m³ (skin).

NIOSH REL: (Organotin Compounds) TWA 0.1 mg(Sn)/m³

SAFETY PROFILE: Moderately toxic by ingestion. When heated to decomposition it emits toxic fumes of SO$_x$. See also TIN COMPOUNDS and SULFIDES.

DVM600   CAS: 10039-33-5   HR: 2
DI-n-OCTYLTIN BIS(2-ETHYLHEXYL MALEATE)

mf: C$_{40}$H$_{72}$O$_8$Sn   mw: 799.81

SYNS: BIS(HYDROGEN MALEATO)DIOCTYLTIN BIS(2-ETHYLHEXYL) ESTER ☐ DI-n-OCTYL-ZINN-BIS(2-AETHYLHEXYLMALEINAT) (GERMAN)

CONSENSUS REPORTS: Reported in EPA TSCA Inventory.

OSHA PEL: TWA 0.1 mg(Sn)/m³ (skin)

ACGIH TLV: TWA 0.1 mg(Sn)/m³; STEL 0.2 mg(Sn)/m³ (skin).

NIOSH REL: (Organotin Compounds) TWA 0.1 mg(Sn)/m³

SAFETY PROFILE: Moderately toxic by ingestion. When heated to decomposition it emits acrid smoke and irritating fumes. See also TIN COMPOUNDS.

DVM800   CAS: 15571-58-1   HR: 2
DI-n-OCTYLTIN BIS(2-ETHYLHEXYL) MERCAPTOACETATE

mf: C$_{36}$H$_{72}$O$_4$S$_2$Sn   mw: 751.89

SYNS: BIS(2-ETHYLHEXYLTHIOGLYCOLATE)DIOCTYLTIN ☐ BIS(MERCAPTOACETATE)DIOCTYLTIN BIS(2-ETHYLHEXYL) ESTER ☐ 10-ETHYL-4,4-DIOCTYL-7-OXO-8-OXA-3,5-DITHIA-4-STANNATETRADECANOIC ACID-2-ETHYLHEXYL ESTER ☐ DI-N-OCTYLTIN-2-ETHYLHEXYLDIMERCAPTOETHANOATE ☐ DI-N-OCTYLTIN-THIOGLYCOLIC ACID 2-ETHYLHEXYL ESTER ☐ OTS 11

CONSENSUS REPORTS: Reported in EPA TSCA Inventory.

OSHA PEL: TWA 0.1 mg(Sn)/m³ (skin)

ACGIH TLV: TWA 0.1 mg(Sn)/m³; STEL 0.2 mg(Sn)/m³ (skin).

DFG MAK: 0.1 mg(Sn)/m³ calculated as total dust

NIOSH REL: (Organotin Compounds) TWA 0.1 mg(Sn)/m³

SAFETY PROFILE: Moderately toxic by ingestion. When heated to decomposition it emits toxic fumes of SO$_x$. See also TIN COMPOUNDS and ESTERS.

DVN800   CAS: 3033-29-2   HR: 3
DI-n-OCTYLTIN β-MERCAPTOPROPIONATE

mf: C$_{19}$H$_{38}$O$_2$SSn   mw: 449.32

SYNS: DIHYDRO-2,2-DIOCTYL-6H-1,3,2-OXATHIASTANNIN-6-ONE ☐ DIOCTYLTIN-β-MERCAPTOPROPIONATE ☐ DI-n-OCTYL-ZINN β-MERCAPTOPROPIONAT (GERMAN)

CONSENSUS REPORTS: Reported in EPA TSCA Inventory.

OSHA PEL: TWA 0.1 mg(Sn)/m³ (skin)

ACGIH TLV: TWA 0.1 mg(Sn)/m³; STEL 0.2 mg(Sn)/m³ (skin).

NIOSH REL: (Organotin Compounds) TWA 0.1 mg(Sn)/m³

SAFETY PROFILE: Poison by intraperitoneal route. Moderately toxic by ingestion. When heated to decomposition it emits toxic fumes of SO$_x$. See also TIN COMPOUNDS and MERCAPTANS.

DVN909   CAS: 3594-15-8   HR: 3
DIOCTYLTIN-3,3'-THIODIPROPIONATE

mf: C$_{22}$H$_{42}$O$_4$SSn   mw: 521.39

SYN: 2,2-DIOCTYL-1,3-DIOXA-2-STANNA-7-THIADECAN-4,10-DIONE

OSHA PEL: TWA 0.1 mg(Sn)/m³ (skin)

ACGIH TLV: TWA 0.1 mg(Sn)/m³; STEL 0.2 mg(Sn)/m³ (skin).

NIOSH REL: (Organotin Compounds) TWA 0.1 mg(Sn)/m³

SAFETY PROFILE: Poison by intravenous route. When heated to decomposition it emits toxic fumes of SO$_x$. See also TIN COMPOUNDS.

DVO920   CAS: 99591-73-8   HR: 3
1,5,2,4-DIOXADITHIEPANE-2,2,4,4-TETRAOXIDE

mf: C$_3$H$_6$O$_6$S$_2$ mw: 202.21

SYNS: 1,5,2,4-DIOXADITHIEPANE, 2,2,4,4-
TETRAOXIDE □ CYCLIC SOSO □ CYCLIC-SOSO □
CYCLODISONE □ NSC-348948
SAFETY PROFILE: A poison by ingestion
and intraperitoneal routes. Mutation data
reported. When heated to decomposition it
emits toxic vapors of SO$_x$.

DVQ000   CAS: 123-91-1   HR: 3
DIOXANE
DOT: UN 1165
mf: C$_4$H$_8$O$_2$ mw: 88.11
PROP: Colorless liquid with pleasant odor.
Mp: 12°, fp: 11°, bp: 101.1°, lel: 2.0%, uel:
22.2%, flash p: 54°F (CC), d: 1.0353 @
20°/4°, autoign temp: 356°F, vap press: 40
mm @ 25.2°, vap d: 3.03. Sol in EtOH and
C$_6$H$_6$. IDLH 500 ppm.
SYNS: DIETHYLENE DIOXIDE □ 1,4-DIETHYLENE
DIOXIDE □ DIETHYLENE ETHER □ DI(ETHYLENE
OXIDE) □ DIOKAN □ DIOKSAN (POLISH) □
DIOSSANO-1,4 (ITALIAN) □ DIOXAAN-1,4 (DUTCH) □
p-DIOXAN (CZECH) □ DIOXAN-1,4 (GERMAN) □ p-
DIOXANE □ 1,4-DIOXANE (MAK) □ DIOXANNE
(FRENCH) □ DIOXYETHYLENE ETHER □ GLYCOL
ETHYLENE ETHER □ NCI-C03689 □ RCRA WASTE
NUMBER U108 □ TETRAHYDRO-p-DIOXIN □
TETRAHYDRO-1,4-DIOXIN
CONSENSUS REPORTS: NTP 10th Report
on Carcinogens. IARC Cancer Review:
Group 2B IMEMDT 7,201,87; Animal
Sufficient Evidence IMEMDT 11,247,76.
NCI Carcinogenesis Bioassay (oral); Clear
Evidence: mouse, rat NCITR* NCI-CG-
TR-80,78. EPA Genetic Toxicology
Program. Glycol ether compounds are on
the Community Right-To-Know List.
Reported in EPA TSCA Inventory.
OSHA PEL: TWA 25 ppm (skin)
ACGIH TLV: TWA 20 ppm (skin);
Confirmed Animal Carcinogen with
Unknown Revelance to Humans
DFG MAK: 20 ppm (73 mg/m³); Not
Classifiable as a Human Carcinogen
NIOSH REL: CL (Dioxane) 1 ppm/30M
DOT CLASSIFICATION: 3; Label:
Flammable Liquid
SAFETY PROFILE: Confirmed carcinogen
with experimental carcinogenic,
neoplastigenic, tumorigenic, and teratogenic

data. Poison by intraperitoneal route.
Moderately toxic by ingestion and
inhalation. Mildly toxic by skin contact.
Human systemic effects by inhalation:
lachrymation, conjunctiva irritation,
convulsions, high blood pressure,
unspecified respiratory and gastrointestinal
system effects. Mutation data reported. An
eye and skin irritant. The irritant effects
probably provide sufficient warning, in acute
exposures, to enable a worker to leave
exposure before being seriously affected.
Repeated exposure to low concentrations
has resulted in human fatalities, the organs
chiefly affected being the liver and kidneys.
   A very dangerous fire and explosion
hazard when exposed to heat or flame; can
react vigorously with oxidizing materials.
Violent reaction with (H$_2$ + Raney Ni),
AgClO$_4$. Can form dangerous peroxides
when exposed to air. Potentially explosive
reaction with nitric acid + perchloric acid,
Raney nickel catalyst (above 210°C). Forms
explosive mixtures with decaborane (impact-
sensitive), triethynylaluminum (sensitive to
heating or drying). Violent reaction with
sulfur trioxide. Incompatible with sulfur
trioxide. To fight fire, use alcohol foam,
CO$_2$, dry chemical. When heated to
decomposition it emits acrid smoke and
irritating fumes. See also GLYCOL
ETHERS.

DVQ709   CAS: 78-34-2   HR: 3
DIOXATHION
mf: C$_{12}$H$_{26}$O$_6$P$_2$S$_4$ mw: 456.56
PROP: Nonvolatile, stable solid or brown
liquid (tech grade). D: 1.257 @ 26°/4°, mp:
−20°, bp: 60–68° @ 0.5 mm.
Nonflammable. Insol in water.
SYNS: BIS(DITHIOPHOSPHATE de O,O-DIETHYLE) de
S,S'-(1,4-DIOXANNE-2,3-DIYLE) (FRENCH) □ DELNAV □
1,4-DIOSSAN-2,3-DIYL-BIS(O,O-DIETIL-DITIOFOSFATO)
(ITALIAN) □ 1,4-DIOXAAN-2,3-DIYL-BIS(O,O-DIETHYL-
DITHIOFOSFAAT) (DUTCH) □ 2,3-p-DIOXANDITHIOL
S,S-BIS(O,O-DIETHYL PHOSPHORODITHIOATE) □ 1,4-
DIOXAN-2,3-DIYL-BIS(O,O-DIAETHYL-DITHIO-
PHOSPHAT) (GERMAN) □ 1,4-DIOXAN-2,3-DIYL-
BIS(O,O-DIETHYLPHOSPHOROTHIOLOTHIONATE) □
1,4-DIOXAN-2,3-DIYL-O,O,O',O'-TETRAETHYL

DI(PHOSPHORODITHIOATE) □ 2,3-p-DIOXANE-S,S-BIS(O,O-DIETHYLPHOSPHORODITHIOATE) □ p-DIOXANE-2,3-DITHIOL-S,S-DIESTER with O,O-DIETHYL PHOSPHORODITHIOATE □ p-DIOXANE-2,3-DIYL ETHYL PHOSPHORODITHIOATE □ ENT 22,897 □ NCI-C00395 □ PHOSPHORODITHIOIC ACID-S,S'-1,4-DIOXANE-2,3-DIYL-O,O,O',O'-TETRAETHYL ESTER

CONSENSUS REPORTS: NCI Carcinogenesis Bioassay (feed); No Evidence: mouse, rat NCITR* NCI-CG-TR-125,78. EPA Extremely Hazardous Substances List.

OSHA PEL: TWA 0.2 mg/m³ (skin)

ACGIH TLV: TLV: TWA 0.1 mg/m³ (skin); Not Classifiable as a Human Carcinogen

SAFETY PROFILE: Poison by ingestion, inhalation, skin contact, and intraperitoneal routes. Mutation data reported. A cholinesterase inhibitor. When heated to decomposition it emits very toxic fumes of PO_x and SO_x. See also PARATHION.

DVR200   CAS: 105-11-3   HR: 2
DIOXIME-p-BENZOQUINONE

mf: $C_6H_6N_2O_2$ mw: 138.14

PROP: Pale-yellow crystals. Mp: 240° (decomp).

SYNS: ACTOR Q □ 1,4-BENZOQUINONE DIOXINE □ 2,5-CYCLOHEXADIENE-1,4-DIONE DIOXIME □ DIBENZO PQD □ DIOXIME-1,4-CYCLOHEXA-DIENEDIONE □ DIOXIME-2,5-CYCLOHEXADIENE-1,4-DIONE □ G-M-F □ NCI-C03850 □ PQD □ QDO □ QUINONE DIOXIME □ p-QUINONE DIOXIME □ p-QUINONE OXIME

CONSENSUS REPORTS: IARC Cancer Review: Group 3 IMEMDT 7,56,87; Animal Limited Evidence IMEMDT 29,185,82. NCI Carcinogenesis Bioassay (feed); Clear Evidence: rat NCITR* NCI-CG-TR-179,79; No Evidence: mouse NCITR* NCI-CG-TR-179,79. Reported in EPA TSCA Inventory.

SAFETY PROFILE: Moderately toxic by ingestion. Questionable carcinogen with experimental neoplastigenic and tumorigenic data. Mutation data reported. When heated to decomposition it emits toxic fumes of NO_x.

DVR600   CAS: 100-79-8   HR: 1
DIOXOLAN

mf: $C_6H_{12}O_3$ mw: 132.18

PROP: Water-white liquid. Mp: −26.4°, bp: 75°, flash p: 35°F (OC), d: 1.065, vap press: 70 mm @ 20°, vap d: 2.6.

SYNS: CYCLIC (HYDROXYMETHYL)ETHYLENE ACETAL ACETONE □ 2,2-DIMETHYL-1,3-DIOXOLANE-4-METHANOL □ 2,2-DIMETHYL-5-HYDROXYMETHYL-1,3-DIOXOLANE □ 2,2-DIMETHYL-4-OXYMETHYL-1,3-DIOXOLANE □ DIOXOLANE (DOT) □ GIE □ GLYCEROLACETONE □ GLYCEROL DIMETHYL-KETAL □ 4-HYDROXYMETHYL-2,2-DIMETHYL-1,3-DIOXOLANE □ ISOPROPYLIDENE GLYCEROL □ 1,2-o-ISOPROPYLIDENE GLYCEROL □ SOLKETAL

CONSENSUS REPORTS: Reported in EPA TSCA Inventory.

SAFETY PROFILE: A poison by intravenous route. An eye irritant. Mutation data reported. A very dangerous fire hazard when exposed to heat or flame; can react vigorously with oxidizing materials. To fight fire, use alcohol foam, CO_2, dry chemical. When heated to decomposition it emits acrid smoke and fumes.

DVR800   CAS: 646-06-0   HR: 2
1,3-DIOXOLANE

DOT: UN 1166

mf: $C_3H_6O_2$ mw: 74.09

PROP: A liquid. D: 1.066 @ 15°/4°, fp: −95°, bp: 78° @ 750 mm, flash p: 35.6°F. Misc in water.

SYNS: 1,3-DIOXACYCLOPENTANE □ 1,3-DIOXOLAN □ ETHYLAENE GLYCOL FORMAL □ FORMAL GLYCOL □ GLYCOL FORMAL

CONSENSUS REPORTS: Reported in EPA TSCA Inventory.

ACGIH TLV: TWA 20 ppm

DOT CLASSIFICATION: 3; Label: Flammable Liquid

SAFETY PROFILE: Moderately toxic by ingestion and intraperitoneal routes. Mildly toxic by skin contact and inhalation. A skin and severe eye irritant. Mutation data reported. A very dangerous fire hazard when exposed to heat or flame; can react with oxidizers. Used in lithium batteries. Potentially explosive reaction with lithium perchlorate. When heated to decomposition it emits acrid smoke and irritating fumes.

DVV200   CAS: 1118-42-9   HR: 3
DIPENTYLTIN DICHLORIDE
mf: $C_{10}H_{22}Cl_2Sn$ mw: 331.91
SYN: DICHLORODIPENTYLSTANNANE
OSHA PEL: TWA 0.1 mg(Sn)/m³ (skin)
ACGIH TLV: TWA 0.1 mg(Sn)/m³; STEL 0.2 mg(Sn)/m³ (skin).
NIOSH REL: (Organotin Compounds) TWA 0.1 mg(Sn)/m³
SAFETY PROFILE: Poison by intravenous route. When heated to decomposition it emits toxic fumes of Cl⁻. See also TIN COMPOUNDS and CHLORIDES.

DVV600   CAS: 82-66-6   HR: 3
DIPHENADIONE
mf: $C_{23}H_{16}O_3$ mw: 340.39
PROP: Pale-yellow crystals from alc. Mp: 147°. Sol in Me_2CO and AcOH; sltly sol in C_6H_6.
SYNS: DIDANDIN □ DIPAXIN □ DIPHACIN □ DIPHACINONE □ DIPHENACIN □ 2-DIPHENYL-ACETYL-1,3-DIKETOHYDRINDENE □ 2-(DIPHENYL-ACETYL)INDAN-1,3-DIONE □ 2-DIPHENYLACETYL-1,3-INDANDIONE □ 2-(DIPHENYLACETYL)-1H-INDENE-1,3(2H)-DIONE □ PID □ PROMAR □ RAMIK □ RATINDAN 1 □ U 1363
CONSENSUS REPORTS: EPA Extremely Hazardous Substances List.
SAFETY PROFILE: Poison by ingestion. Inhibits blood clotting, leading to hemorrhages. Action similar to coumadin (warfarin). A pesticide used in rodent control. When heated to decomposition it emits acrid smoke and irritating fumes.

DVW100   CAS: 82-21-3   HR: 1
1,5-DIPHENOXYANTHRAQUINONE
mf: $C_{26}H_{16}O_4$ mw: 392.42
SYNS: 9,10-ANTHRACENEDIONE, 1,5-DIPHENOXY- □ 1,5-DIFENOXYANTHRACHINON
CONSENSUS REPORTS: Reported in EPA TSCA Inventory.
SAFETY PROFILE: An eye irritant. When heated to decomposition it emits acrid smoke and irritating fumes.

DVX200   CAS: 86-29-3   HR: 3
DIPHENYLACETONITRILE
mf: $C_{14}H_{11}N$ mw: 193.26
PROP: A solid. Mp: 75–76°, bp: 181° @ 12 mm.
SYNS: BENZYHDRYLCYANIDE □ α-CYANO-DIPHENYLMETHANE □ DIPAN □ DIPHENATRILE □ DIPHENYL-α-CYANOMETHANE □ DIPHENYLMETH-YLCYANIDE □ α-PHENYLBENZYLCYANIDE □ α-PHENYLPHENYLACETONITRILE □ USAF KF-13
CONSENSUS REPORTS: Reported in EPA TSCA Inventory. Cyanide and its compounds are on the Community Right-To-Know List.
SAFETY PROFILE: Poison by ingestion, intraperitoneal, and intravenous routes. Moderately toxic by subcutaneous route. Questionable carcinogen with experimental carcinogenic and tumorigenic data. When heated to decomposition it emits toxic fumes of NO_x and CN⁻. See also NITRILES.

DVX600   CAS: 2510-95-4   HR: 3
2,3-DIPHENYLACRYLONITRILE
mf: $C_{15}H_{11}N$ mw: 205.27
SYNS: BENZAL-(BENZYL-CYANID) (GERMAN) □ BENZYLIDENEPHENYLACETONITRILE □ α-CYANO-STILBENE □ α,β-DIPHENYLACRYLONITRILE □ F 2387 □ α-PHENYLCINNAMONITRILE □ α-(PHENYLMETH-YLENE)BENZENEACETONITRILE □ α-STILBENE-CARBONITRILE □ USAF A-9789
CONSENSUS REPORTS: Reported in EPA TSCA Inventory. Cyanide and its compounds are on the Community Right-To-Know List.
SAFETY PROFILE: Poison by intraperitoneal route. When heated to decomposition it emits toxic fumes of NO_x and CN⁻. See also NITRILES.

DVX800   CAS: 122-39-4   HR: 3
DIPHENYLAMINE
mf: $C_{12}H_{11}N$ mw: 169.24
PROP: Crystals; floral odor. Mp: 52.9°, bp: 302.0°, flash p: 307°F (CC), d: 1.16, autoign temp: 1173°F, vap press: 1 mm @ 108.3°, vap d: 5.82. Sol in benzene, ether, and carbon disulfide; insol in water.

SYNS: ANILINOBENZENE □ BIG DIPPER □ C.I. 10355 □ DFA □ N,N-DIPHENYLAMINE □ DPA □ NO SCALD □ N-PHENYLANILINE □ N-PHENYLBENEZENAMINE □ SCALDIP

CONSENSUS REPORTS: Reported in EPA TSCA Inventory. EPA Genetic Toxicology Program.

OSHA PEL: TWA 10 mg/m^3

ACGIH TLV: TWA 10 mg/m^3; Not Classifiable as a Human Carcinogen

SAFETY PROFILE: Poison by ingestion. Experimental teratogenic effects. Action similar to aniline but less severe. Combustible when exposed to heat or flame. Can react violently with hexachloromelamine or trichloromelamine. Can react with oxidizing materials. To fight fire, use CO$_2$, dry chemical. When heated to decomposition it emits highly toxic fumes of NO$_x$. See also ANILINE, AMINES, and AROMATIC AMINES.

DVY100   CAS: 6217-24-9   HR: 3
DIPHENYLARSINOUS ACID

mf: C$_{12}$H$_{11}$AsO   mw: 246.15

SYNS: ARSINE, DIPHENYLHYDROXY- □ ARSINE, HYDROXYDIPHENYL- □ DIPHENYLHYDROXYARSINE

OSHA PEL: TWA 0.5 mg(As)/m^3

SAFETY PROFILE: Poison by intravenous route. When heated to decomposition it emits toxic fumes of As.

DVZ000   CAS: 102-09-0   HR: 2
DIPHENYL CARBONATE

mf: C$_{13}$H$_{10}$O$_3$ mw: 214.23

PROP: Needles. Mp: 78°, bp: 306°.

SYNS: CARBONIC ACID, DIPHENYL ESTER □ PHENYL CARBONATE

CONSENSUS REPORTS: Reported in EPA TSCA Inventory.

SAFETY PROFILE: Questionable carcinogen with experimental neoplastigenic and tumorigenic data. When heated to decomposition it emits acrid smoke and irritating fumes.

DWB800   CAS: 1241-94-7   HR: 3
DIPHENYL-2-ETHYLHEXYL PHOSPHATE

mf: C$_{20}$H$_{27}$O$_4$P   mw: 362.44

SYNS: 2-ETHYL-1-HEXANOL ESTER with DIPHENYL PHOSPHATE □ 2-ETHYLHEXYL DIPHENYL ESTER PHOSPHORIC ACID □ 2-ETHYLHEXYL DIPHENYLPHOSPHATE □ SANTICIZER 141 (MONSANTO)

CONSENSUS REPORTS: Reported in EPA TSCA Inventory.

SAFETY PROFILE: Poison by intravenous route. Moderately toxic by intraperitoneal route. When heated to decomposition it emits toxic fumes of PO$_x$.

DWC600   CAS: 102-06-7   HR: 3
DIPHENYLGUANIDINE

mf: C$_{13}$H$_{13}$N$_3$ mw: 211.29

PROP: White powder or needles from alc. Mp: 150°, d: 1.115 @ 25°. Sol in Et$_2$O, CHCl$_3$, and dil acids; sltly sol in H$_2$O.

SYNS: N,N'-DIPHENYLGUANIDINE □ 1,3-DIPHENYLGUANIDINE □ DPG □ DPG ACCELERATOR □ DWUFENYLOGUANIDYNA (POLISH) □ MELANILINE □ NCI-C60924 □ USAF B-19 □ USAF EK-1270 □ VULCACID D □ VULKACIT D/C □ VULKAZIT

CONSENSUS REPORTS: Reported in EPA TSCA Inventory.

SAFETY PROFILE: Poison by ingestion and intraperitoneal routes. Experimental teratogenic and reproductive effects. Mutation data reported. When heated to decomposition it emits toxic fumes of NO$_x$.

DWC650   CAS: 4657-20-9   HR: D
2,6-DIPHENYL-2,4,6,6,8,8-HEXAMETH-YLCYCLOTETRASILOXANE

mf: C$_{18}$H$_{28}$O$_4$Si$_4$ mw: 420.82

SYN: CYCLOTETRASILOXANE, 2,6-DIPHENYL-2,4,4,6,8,8-HEXAMETHYL-

CONSENSUS REPORTS: Reported in EPA TSCA Inventory.

SAFETY PROFILE: Experimental reproductive effects. When heated to decomposition it emits acrid smoke and irritating fumes.

DWD800   CAS: 587-85-9   HR: 3
DIPHENYLMERCURY

mf: C$_{12}$H$_{10}$Hg   mw: 354.81

PROP: White crystals or needles from alc. D: 2.318, mp: 124.5–125° (sublimes), bp: 204° @ 10.5 mm. Insol in water.
CONSENSUS REPORTS: Reported in EPA TSCA Inventory. Mercury and its compounds are on the Community Right-To-Know List.
OSHA PEL: CL 0.1 mg(Hg)/m³ (skin)
ACGIH TLV: TWA 0.1 mg(Hg)/m³ (skin); BEI: 35 μg/g creatinine total inorganic mercury in urine preshift; 15 μg/g creatinine total inorganic mercury in blood at end of shift at end of workweek.
DFG MAK: Confirmed Animal Carcinogen with Unknown Relevance to Humans
NIOSH REL: (Mercury, Aryl and Inorganic) CL 0.1 mg/m³ (skin)
SAFETY PROFILE: Poison by intraperitoneal route. Moderately toxic by ingestion. Incompatible with nonmetal oxides. When heated to decomposition it emits toxic fumes of Hg. See also MERCURY COMPOUNDS.

DWH550   CAS: 778-25-6   HR: D
DIPHENYLMETHYLSILANOL
mf: C₁₃H₁₄OSi   mw: 214.36
PROP: Crystals. Mp: 165–166°, bp: 130–134° @ 2 mm.
SYN: SILANOL, DIPHENYLMETHYL-
CONSENSUS REPORTS: Reported in EPA TSCA Inventory.
SAFETY PROFILE: Experimental reproductive effects. When heated to decomposition it emits acrid smoke and irritating fumes.

DWI000   CAS: 86-30-6   HR: 3
DIPHENYLNITROSAMINE
mf: C₁₂H₁₀N₂O   mw: 198.24
PROP: Yellow plates or green crystals. Mp: 66.5°.
SYNS: CURETARD A □ DELAC J □ DIPHENYLNITROSAMIN (GERMAN) □ DIPHENYL N-NITROSOAMINE □ N,N-DIPHENYLNITROSAMINE □ NAUGARD TJB □ NCI-C02880 □ NDPA □ NDPhA □ N-NITROSODIFENYLAMIN (CZECH) □ NITROSO-DIPHENYLAMINE □ N-NITROSODIPHENYLAMINE □ N-NITROSO-N-PHENYLANILINE □ NITROUS

DIPHENYLAMIDE □ REDAX □ RETARDER J □ TJB □ VULCALENT A □ VULCATARD □ VULKALENT A (CZECH) □ VULTROL
CONSENSUS REPORTS: IARC Cancer Review: Group 3 IMEMDT 7,56,87; Animal Limited Evidence IMEMDT 27,213,82. NCI Carcinogenesis Bioassay (feed); Clear Evidence: rat NCITR* NCI-CG-TR-164,79; No Evidence: mouse NCITR* NCI-CG-TR-164,79. Reported in EPA TSCA Inventory. EPA Genetic Toxicology Program. Community Right-To-Know List.
SAFETY PROFILE: Moderately toxic by ingestion. An eye irritant. Questionable carcinogen with experimental carcinogenic and tumorigenic data. Human mutation data reported. Dangerous fire hazard when exposed to heat, flame, or oxidizing materials. Can react vigorously with oxidizing materials. When heated to decomposition it emits highly toxic fumes of NOₓ. See also NITROSAMINES.

DWL400   CAS: 10087-89-5   HR: 3
1,1-DIPHENYL-2-PROPYNYL-N-CYCLOHEXYLCARBAMATE
mf: C₂₂H₂₃NO₂   mw: 333.46
PROP: A solid. Mp: 160–161°.
SYNS: 1,1-DIPHENYL-2-PROPYN-1-OL CYCLOHEXANECARBAMATE □ 1,1-DIPHENYL-2-PROPYNYL ESTER CYCLOHEXANECARBAMIC ACID □ ENPROMATE
CONSENSUS REPORTS: EPA Genetic Toxicology Program.
SAFETY PROFILE: Poison by intraperitoneal route. Moderately toxic by ingestion. Questionable carcinogen with experimental tumorigenic data. Mutation data reported. When heated to decomposition it emits toxic fumes of NOₓ. See also CARBAMATES.

DWN150   CAS: 56-33-7   HR: D
1,3-DIPHENYL-1,1,3,3-TETRAMETH-YLDISILOXANE
mf: C₁₆H₂₂OSi₂   mw: 286.56
PROP: Bp: 110–111° @ 1 mm.
SYN: DISILOXANE, 1,3-DIPHENYL-1,1,3,3-TETRAMETHYL-

CONSENSUS REPORTS: Reported in EPA TSCA Inventory.

SAFETY PROFILE: Experimental reproductive effects. When heated to decomposition it emits acrid smoke and irritating fumes.

DWN200   CAS: 60-10-6   HR: 3
DIPHENYLTHIOCARBAZONE
mf: $C_{13}H_{12}N_4S$ mw: 256.35

PROP: Bluish-black crystalline powder from alc (aq). Mp: 165–169°. Sol in aq alkaline solns; sltly sol in EtOH, CCl_4, $CHCl_3$, and C_6H_6; insol in H_2O.

SYNS: CARBAZONE, DIPHENYLTHIO- □ DITHIZON □ DITHIZONE □ 3-FORMAZANTHIOL, 1,5-DIPHENYL- □ (PHENYLAZO)THIOFORMIC ACID, 2-PHENYL-HYDRAZIDE □ SEMICARBAZIDE, 1-PHENYL-4-(PHENYLIMINO)-3-THIO- □ THIOFORMIC ACID, PHENYLAZO-, PHENYLHYDRAZIDE □ USAF EK-3092

CONSENSUS REPORTS: Reported in EPA TSCA Inventory.

SAFETY PROFILE: Poison by intravenous and intraperitoneal routes. Can cause eye injury and glycosuria. When heated to decomposition it emits highly toxic fumes of NO_x and SO_x.

DWN800   CAS: 102-08-9   HR: 2
DIPHENYLTHIOUREA
mf: $C_{13}H_{12}N_2S$ mw: 228.33

PROP: White to faint gray powder or leaflets from alc. Mp: 154°, bp: decomp, d: 1.32 @ 25°.

SYNS: DFT □ N,N'-DIPHENYLTHIOCARBAMIDE □ sym-DIPHENYLTHIOCARBAMIDE □ N,N'-DIPHENYL-THIOUREA □ sym-DIPHENYLTHIOUREA □ 1,3-DIPHENYLTHIOUREA □ 1,3-DIPHENYL-2-THIOUREA □ 2-FENYLOTIOMOCZNIK (POLISH) □ RHENOCURE CA □ STABILISATOR C □ SULFOCARBANILIDE □ THIOCARBANILIDE □ USAF EK-245 □ VALKACIT CA

CONSENSUS REPORTS: Reported in EPA TSCA Inventory.

SAFETY PROFILE: Moderately toxic by ingestion and intraperitoneal routes. Experimental reproductive effects. When heated to decomposition it emits highly toxic fumes of SO_x and NO_x.

DWO800   CAS: 136-35-6   HR: 2
1,3-DIPHENYLTRIAZENE
mf: $C_{12}H_{11}N_3$ mw: 197.26

PROP: Golden-yellow crystals from pet ether. Mp: 98–99°, bp: explodes, vap d: 6.8.

SYNS: CELLOFOR (CZECH) □ DAAB □ DIAZO-AMINOBENZEN (CZECH) □ DIAZOAMINOBENZENE □ p-DIAZOAMINOBENZENE □ DIAZOAMINOBENZ-OL (GERMAN) □ N-(PHENYLAZO)ANILINE

CONSENSUS REPORTS: EPA Genetic Toxicology Program. Reported in EPA TSCA Inventory.

SAFETY PROFILE: Questionable carcinogen with experimental tumorigenic data. Mutation data reported. Strongly explosive when shocked or heated to 98°C. Mixture with acetic anhydride explodes when warmed. When heated to decomposition it emits toxic fumes of NO_x.

DWQ000   CAS: 7727-21-1   HR: 3
DIPOTASSIUM PERSULFATE
DOT: UN 1492
mf: $H_2O_8S_2•2K$ mw: 272.34

PROP: White, odorless, colorless, triclinic crystals. Mp: decomp @ 100°, d: 2.477. Decomp on heating to $K_2S_2O_7$ with loss of O_2. Mod sol in H_2O.

SYNS: ANTHION □ PEROXYDISULFURIC ACID DIPOTASSIUM SALT □ POTASSIUM PEROXY-DISULFATE □ POTASSIUM PEROXYDISULPHATE □ POTASSIUM PERSULFATE (DOT)

CONSENSUS REPORTS: Reported in EPA TSCA Inventory.

DOT CLASSIFICATION: 5.1; Label: Oxidizer

SAFETY PROFILE: Moderately toxic by ingestion. An irritant and allergen. A powerful oxidizer. Flammable when exposed to heat or by chemical reaction. Can react with reducing materials. It liberates oxygen above 100° when dry or @ about 50° when in solution. When heated to decomposition it emits highly toxic fumes of SO_x, S_2O_8, and K_2O.

DWQ800   CAS: 3248-28-0   HR: 3
DIPROPIONYL PEROXIDE
mf: $C_6H_{10}O_4$ mw: 146.15

PROP: Crystals.

SYNS: BIS(1-OXOPROPYL)PEROXIDE □
DIPROPIONYL PEROXIDE, >28% in solution (DOT) □
PEROXIDE, BIS(1-OXOPROPYL) □ PROPIONYL
PEROXIDE (DOT)
DOT CLASSIFICATION: Forbidden
SAFETY PROFILE: Moderately toxic by
inhalation. The pure material explodes at
room temperature. When heated to
decomposition it emits acrid smoke and
fumes. See also PEROXIDES.

DWQ875   CAS: 106-19-4   HR: 2
DIPROPYL ADIPATE
mf: $C_{12}H_{22}O_4$ mw: 230.34
SYN: DI-n-PROPYL ADIPATE
CONSENSUS REPORTS: Reported in EPA
TSCA Inventory.
SAFETY PROFILE: Moderately toxic by
some routes. Experimental reproductive
effects. An experimental teratogen. When
heated to decomposition it emits acrid
smoke and fumes.

DWR000   CAS: 142-84-7   HR: 3
DIPROPYLAMINE
DOT: UN 2383
mf: $C_6H_{15}N$ mw: 101.22
PROP: Water-white liquid; amine odor. Mp:
−63°, bp: 110°, flash p: 63°F (OC), d: 0.741
@ 20°, vap d: 3.5.
SYNS: DI-n-PROPYLAMINE □ n-DIPROPYLAMINE □
N-PROPYL-1-PROPANAMINE □ RCRA WASTE NUMBER
U110
CONSENSUS REPORTS: Reported in EPA
TSCA Inventory.
DOT CLASSIFICATION: 3; Label:
Flammable Liquid
SAFETY PROFILE: Poison by ingestion.
Moderately toxic by skin contact and
inhalation. A skin irritant. A very dangerous
fire hazard, when exposed to heat or flame.
Can react with oxidizers. Explosion hazard
is unknown. Keep away from heat and open
flame. To fight fire, use foam, CO_2, dry
chemical. When heated to decomposition it
emits toxic fumes of NO_x. See also
AMINES.

DWT200   CAS: 34590-94-8   HR: 2
DIPROPYLENE GLYCOL METHYL
ETHER
mf: $C_7H_{16}O_3$ mw: 148.23
PROP: Liquid. Bp: 190°, d: 0.951, vap d:
5.11, flash p: 185°F. IDLH 600 ppm.
SYNS: ARCOSOLV □ DIPROPYLENE GLYCOL
MONOMETHYL ETHER □ DOWANOL DPM □
DOWANOL-50B □ UCAR SOLVENT 2LM
CONSENSUS REPORTS: Reported in EPA
TSCA Inventory. Glycol ether compounds
are on the Community Right-To-Know List.
OSHA PEL: TWA 100 ppm; STEL 150 ppm
(skin)
ACGIH TLV: TWA 100 ppm; STEL 150
ppm (skin)
DFG MAK: 50 ppm (310 mg/m³)
SAFETY PROFILE: Mildly toxic by
ingestion and skin contact. An experimental
skin and human eye irritant. A mild allergen.
Combustible when exposed to heat or
flame; can react with oxidizing materials. To
fight fire, use dry chemical, CO_2, mist, foam.
When heated to decomposition it emits
acrid smoke and irritating fumes. See also
GLYCOL ETHERS.

DWT600   CAS: 123-19-3   HR: 3
DIPROPYL KETONE
DOT: UN 2710
mf: $C_7H_{14}O$ mw: 114.21
PROP: Colorless, refractive liquid. Bp: 144°,
mp: −32.6°, vap press: 5.2 mm @ 20°, flash
p: 120°F (CC), d: 0.815, vap d: 3.93.
SYNS: BUTYRONE (DOT) □ GBL □ HEPTAN-4-ONE
□ 4-HEPTANONE □ PROPYL KETONE
CONSENSUS REPORTS: Reported in EPA
TSCA Inventory.
OSHA PEL: TWA 50 ppm
ACGIH TLV: TWA 50 ppm
DOT CLASSIFICATION: 3; Label:
Flammable Liquid
SAFETY PROFILE: Moderately toxic by
ingestion, inhalation, and skin contact. A
skin and eye irritant. Flammable liquid when
exposed to heat or flame; can react with
oxidizing materials. To fight fire, use CO_2,
dry chemical, alcohol foam, fog, and mist.
When heated to decomposition it emits

acrid smoke and fumes. See also
KETONES.

DWU000   CAS: 628-85-3   HR: 3
DIPROPYL MERCURY
mf: $C_6H_{14}Hg$ mw: 286.79
PROP: Colorless liquid. Immiscible in water.
D: 2.0208, bp: 190°. Sol in Et_2O; less sol in
EtOH. IDLH 10 mg/m³ (as Hg).
CONSENSUS REPORTS: Mercury and its
compounds are on the Community Right-
To-Know List.
OSHA PEL: TWA 0.01 mg(Hg)/m³; STEL
0.03 mg/m³ (skin)
ACGIH TLV: TWA 0.01 mg(Hg)/m³; BEI:
35 µg/g creatinine total inorganic mercury in
urine preshift; 15 µg/g creatinine total
inorganic mercury in blood at end of shift at
end of workweek.
DFG MAK: Confirmed Animal Carcinogen
with Unknown Relevance to Humans
NIOSH REL: (Mercury, Organo) TWA 0.01
mg/m³; STEL 0.03 mg/m³ (skin)
SAFETY PROFILE: Poison by
intraperitoneal route. Violent reaction with
iodine. When heated to decomposition it
emits toxic fumes of Hg. See also
MERCURY COMPOUNDS, ORGANIC.

DWV000   CAS: 7664-98-4   HR: 3
DIPROPYLOXOSTANNANE
mf: $C_6H_{14}OSn$ mw: 220.89
PROP: Polymeric powder.
SYNS: DIPROPYLTIN OXIDE □ KYSLICNIK DI-N-
PROPYLCINICITY (CZECH)
OSHA PEL: TWA 0.1 mg(Sn)/m³ (skin)
ACGIH TLV: TWA 0.1 mg(Sn)/m³; STEL
0.2 mg(Sn)/m³ (skin).
NIOSH REL: (Organotin Compounds) TWA
0.1 mg(Sn)/m³
SAFETY PROFILE: Poison by ingestion. An
eye and severe skin irritant. When heated to
decomposition it emits acrid smoke and
irritating fumes. See also TIN
COMPOUNDS.

DWV400   CAS: 16066-38-9   HR: 2
DI-n-PROPYL PEROXYDICARBONATE

mf: $C_8H_{14}O_6$ mw: 206.22
SYNS: PEROXYDICARBONIC ACID DIPROPYL ESTER
□ n-PROPYL PERCARBONATE
CONSENSUS REPORTS: Reported in EPA
TSCA Inventory.
SAFETY PROFILE: Moderately toxic by
ingestion and skin contact. When heated to
decomposition it emits acrid smoke and
irritating fumes.

DWV500   CAS: 131-16-8   HR: 2
DIPROPYL PHTHALATE
mf: $C_{14}H_{18}O_4$ mw: 250.32
PROP: Bp: 317.5°, d: 1.078, flash p: >230°F.
SYNS: 1,2-BENZENEDICARBOXYLIC ACID,
DIPROPYL ESTER □ DI-n-PROPYL PHTHALATE □
PHTHALIC ACID, DIPROPYL ESTER
CONSENSUS REPORTS: Reported in EPA
TSCA Inventory.
SAFETY PROFILE: Moderately toxic by
intraperitoneal route. Experimental
reproductive effects. An irritant.
Combustible when exposed to heat and
flame. When heated to decomposition it
emits acrid smoke and irritating fumes.

DWW700   CAS: 67730-10-3   HR: 3
DIPYRIDO(1,2-a:3',2'-d)IMIDAZOL-2-
AMINE
mf: $C_{10}H_8N_4$ mw: 184.22
SYNS: 2-AMINODIPYRIDO(1,2-a:3',2'-d)-IMIDAZOLE □
GLU-P-2
CONSENSUS REPORTS: IARC Cancer
Review: Group 2B IMEMDT 7,56,87;
Animal Sufficient Evidence IMEMDT
40,235,86.
SAFETY PROFILE: Confirmed carcinogen
with experimental carcinogenic data.
Experimental reproductive effects. Human
mutation data reported. When heated to
decomposition it emits toxic fumes of NO_x.

DWX200   CAS: 20738-78-7   HR: 3
DI-3-PYRIDYLMERCURY
mf: $C_{10}H_8HgN_2$ mw: 356.79
PROP: IDLH 10 mg/m³ (as Hg).
CONSENSUS REPORTS: Mercury and its
compounds are on the Community Right-
To-Know List.

OSHA PEL: CL 0.1 mg(Hg)/m³ (skin)
ACGIH TLV: TWA 0.1 mg(Hg)/m³ (skin);
BEI: 35 µg/g creatinine total inorganic
mercury in urine preshift; 15 µg/g creatinine
total inorganic mercury in blood at end of
shift at end of workweek.
DFG MAK: Confirmed Animal Carcinogen
with Unknown Relevance to Humans
NIOSH REL: (Mercury, Aryl and Inorganic)
CL 0.1 mg/m³ (skin)
SAFETY PROFILE: Poison by intravenous
route. When heated to decomposition it
emits very toxic fumes of NO_x and Hg. See
also MERCURY COMPOUNDS,
ORGANIC.

DWX800   CAS: 85-00-7   HR: 3
DIQUAT
mf: $C_{12}H_{12}N_2 \cdot 2Br$ mw: 344.08
PROP: Yellow crystals. Mp: 355°. Sol in
water.
SYNS: AQUACIDE □ DEIQUAT □ DEXTRONE □
9,10-DIHYDRO-8a,10,-DIAZONIAPHENANTHRENE
DIBROMIDE □ 9,10-DIHYDRO-8a,10a-DIAZONIA-
PHENANTHRENE(1,1'-ETHYLENE-2,2'-BIPY-
RIDYLIUM)DIBROMIDE □ 5,6-DIHYDRO-DIPYRIDO-
(1,2a;2,1c)PYRAZINIUM DIBROMIDE □ 6,7-
DIHYDROPYRIDO(1,2a;2',1'-c)PYRAZINEDIUM
DIBROMIDE □ DIQUAT DIBROMIDE □ 1,1'-
ETHYLENE-2,2'-BIPYRIDYLIUM DIBROMIDE □
ETHYLENE DIPYRIDYLIUM DIBROMIDE □ 1,1-
ETHYLENE 2,2-DIPYRIDYLIUM DIBROMIDE □ 1,1'-
ETHYLENE-2,2'-DIPYRIDYLIUM DIBROMIDE □ FB/2 □
FEGLOX □ PREEGLONE □ REGLON □ REGLONE □
WEEDTRINE-D
CONSENSUS REPORTS: EPA Genetic
Toxicology Program.
OSHA PEL: TWA 0.5 mg/m³
ACGIH TLV: TWA Total Dust 0.5 mg/m³;
Respirable Dust: 0.1 mg/m³ (skin); Not
Classifiable as a Human Carcinogen
SAFETY PROFILE: Poison by ingestion,
subcutaneous, intravenous, and
intraperitoneal routes. Experimental
teratogenic and reproductive effects. A skin
and eye irritant. Human mutation data
reported. When heated to decomposition it
emits very toxic fumes of NO_x and Br⁻. See
also PARAQUAT.

DWY800   CAS: 1464-43-3   HR: 3
3,3'-DISELENODIALANINE
mf: $C_6H_{12}N_2O_4Se_2$ mw: 334.12
SYNS: SELENIUM CYSTINE □ SELENOCYSTINE
CONSENSUS REPORTS: Selenium and its
compounds are on the Community Right-
To-Know List.
OSHA PEL: TWA 0.2 mg(Se)/m³
ACGIH TLV: TWA 0.2 mg(Se)/m³
DFG MAK: 0.1 mg(Se)/m³
SAFETY PROFILE: Poison by
intraperitoneal route. Mutation data
reported. When heated to decomposition it
emits very toxic fumes of NO_x and Se. See
also SELENIUM COMPOUNDS.

DWZ100   CAS: 70145-55-0   HR: 3
β,β'-DISELENODIPROPIONIC ACID,
**   SODIUM SALT**
mf: $C_6H_9O_4Se_2 \cdot Na$ mw: 326.06
SYN: PROPIONIC ACID, 3,3'-DISELENODI-, SODIUM
SALT
OSHA PEL: TWA 0.2 mg(Se)/m³
ACGIH TLV: TWA 0.2 mg(Se)/m³
SAFETY PROFILE: Poison by
intraperitoneal route. When heated to
decomposition it emits toxic fumes of Se.

DXA000   HR: 3
DISILANE
mf: H_6Si_2 mw: 62.22
PROP: Gas; repulsive odor. Mp: −132.5°,
bp: −14.5°, d: 0.686 @ −25°/4°.
SYN: SILICOETHANE
SAFETY PROFILE: Poison by inhalation.
Dangerous when exposed to heat or flame
or by chemical reaction; can react with
oxidizing materials. Ignites spontaneously in
air. Reacts violently with CCl_4, $CHCl_3$, O_2,
and SF_6. See also HYDRIDES.

DXC200   CAS: 7775-11-3   HR: 3
DISODIUM CHROMATE
mf: $CrO_4 \cdot 2Na$ mw: 161.98
PROP: Yellow crystals. Mp: 780°. Sol in
H_2O; fairly insol in MeOH and EtOH.
IDLH Ca [15 mg/m³ {as Cr(VI)}].
SYNS: CHROMATE of SODA □ CHROMIUM
DISODIUM OXIDE □ CHROMIUM SODIUM OXIDE □

NEUTRAL SODIUM CHROMATE □ SODIUM CHROMATE (DOT) □ SODIUM CHROMATE (VI)
CONSENSUS REPORTS: IARC Cancer Review: Group 1 IMEMDT 49,49,90; Human Inadequate Evidence IMEMDT 23,205,80; Human Sufficient Evidence IMEMDT 49,49,90; Animal Inadequate Evidence IMEMDT 23,205,80. Reported in EPA TSCA Inventory. EPA Genetic Toxicology Program. Chromium and its compounds are on the Community Right-To-Know List.
OSHA PEL: Cl 0.1 mg(CrO$_3$)/m^3
ACGIH TLV: TWA 0.05 mg(CrO$_3$)/m^3
NIOSH REL: (Chromium(VI)) TWA 25 μg(Cr(VI))/m^3; CL 50 μg/m^3/15M
SAFETY PROFILE: Poison by skin contact, intraperitoneal, intravenous, subcutaneous, and intradermal routes. Experimental reproductive effects. Mutation data reported. A powerful oxidizer. When heated to decomposition it emits toxic fumes of Na$_2$O. See also CHROMIUM COMPOUNDS.

DXC400   CAS: 144-33-2   HR: 3
DISODIUM CITRATE
mf: C$_6$H$_6$O$_7$•2Na   mw: 236.10
PROP: White crystals or granular powder; odorless. Mp: loses water @ 150°, bp: decomp @ red heat. Sol in water; insol in alc.
SYNS: DISODIUM HYDROGEN CITRATE □ NATRIUM CITRICUM (GERMAN) □ SODIUM CITRATE (FCC)
CONSENSUS REPORTS: Reported in EPA TSCA Inventory.
SAFETY PROFILE: Poison by intravenous route. Moderately toxic by intraperitoneal and subcutaneous routes. When heated to decomposition it emits toxic fumes of Na$_2$O.

DXC900   CAS: 52207-48-4   HR: 3
DISODIUM S,S′-(2-DIMETHYLAMINO-1,3-PROPANEDIYL)BIS-(THIOSULFATE)
mf: C$_5$H$_{11}$NO$_6$S$_4$•2Na   mw: 355.39
SYNS: DIMEHYPO □ DIMEHYPO JUMBO □ SHA CHONG DAN □ SHA CHONG SHUANG □

THIOSULFURIC ACID, S,S′-(2-(DIMETHYLAMINO)-1,3-PROPANEDIYL) ESTER, DISODIUM SALT
SAFETY PROFILE: A poison by ingestion and subcutaneous routes. Mutation data reported. When heated to decomposition it emits toxic vapors of NO$_x$ and SO$_x$.

DXD200   CAS: 142-59-6   HR: 3
DISODIUM ETHYLENE-1,2-BISDITHIOCARBAMATE
mf: C$_4$H$_6$N$_2$S$_4$•2Na   mw: 256.34
PROP: Crystals. Sol in water.
SYNS: CARBON D □ CHEM BAM □ DINATRIUM-AETHYLENBISDITHIOCARBAMAT (GERMAN) □ DINATRIUM-(N,N′-AETHYLEN-BIS(DITHIOCARBA-MAT)) (GERMAN) □ DINATRIUM-(N,N′-ETHYLEEN-BIS(DITHIOCARBAMAAT)) (DUTCH) □ DISODIUM ETHYLENEBIS(DITHIOCARBAMATE) □ DITHANE A-40 □ DITHANE D-14 □ DSE □ 1,2-ETHANEDIYLBIS-CARBAMODITHIOIC ACID DISODIUM SALT □ N,N′-ETHYLENE BIS(DITHIOCARBAMATE de SODIUM) (FRENCH) □ ETHYLENEBIS(DITHIOCARBAMATE) DISODIUM SALT □ ETHYLENEBIS(DITHIOCARBAMIC ACID) DISODIUM SALT □ N,N′-ETILEN-BIS(DITIO-CARBAMMATO) di SODIO (ITALIAN) □ NABAM □ NABAME (FRENCH) □ PARZATE □ SPRING-BAK
CONSENSUS REPORTS: Reported in EPA TSCA Inventory. EPA Genetic Toxicology Program.
SAFETY PROFILE: Poison by ingestion. Moderately toxic by intraperitoneal route. Experimental teratogenic and reproductive effects. Mutation data reported. When heated to decomposition it emits very toxic fumes of NO$_x$, Na$_2$O, and SO$_x$. See also CARBAMATES.

DXD400   CAS: 7414-83-7   HR: 3
DISODIUM ETIDRONATE
mf: C$_2$H$_6$O$_7$P$_2$•2Na   mw: 249.99
SYNS: DIDRONEL R □ DISODIUM DIHYDROGEN-(1-HYDROXYETHYLIDENE)DIPHOSPHONATE □ DISODIUM ETHANOL-1,1-DIPHOSPHONATE □ DISODIUM ETHYDRONATE □ EITDRONATE DISODIUM □ ETHANE-1-HYDROXY-1,1-DIPHOSPHON-IC ACID DISODIUM SALT □ (1-HYDROXYETHYL-IDENE)DIPHOSPHONIC ACID DISODIUM SALT □ SODIUM ETHIDRONATE □ SODIUM ETHYDRONATE □ SODIUM ETIDRONATE
CONSENSUS REPORTS: Reported in EPA TSCA Inventory.

SAFETY PROFILE: Poison by intravenous and subcutaneous routes. Moderately toxic by ingestion. An experimental teratogen. Other experimental reproductive effects. When heated to decomposition it emits toxic fumes of PO_x and Na_2O.

DXE000   CAS: 16893-85-9   HR: 3
DISODIUM HEXAFLUOROSILICATE
DOT: UN 2674

mf: $F_6Si•2Na$ mw: 188.07

PROP: Colorless hexagonal crystals. Fluorescent when activated by Ti(IV). Practically insol in H_2O; insol in EtOH.

SYNS: DESTRUXOL APPLEX □ (2-)-DISODIUM HEXAFLUOROSILICATE □ DISODIUM SILICO-FLUORIDE □ ENS-ZEM WEEVIL BAIT □ ENT 1,501 □ FLUOSILICATE de SODIUM □ NATRIUMSILICO-FLUORID (GERMAN) □ ORTHO EARWIG BAIT □ ORTHO WEEVIL BAIT □ PRODAN □ PSC CO-OP WEEVIL BAIT □ SAFSAN □ SALUFER □ SILICON SODIUM FLUORIDE □ SODIUM FLUOROSILICATE □ SODIUM FLUOSILICATE □ SODIUM HEXAFLUORO-SILICATE □ SODIUM HEXAFLUOSILICATE □ SODIUM SILICOFLUORIDE (DOT) □ SUPER PRODAN

CONSENSUS REPORTS: Reported in EPA TSCA Inventory.

OSHA PEL: TWA 2.5 mg(F)/m^3

ACGIH TLV: TWA 2.5 mg(F)/m^3; BEI: 3 mg/g creatinine of fluorides in urine prior to shift; 10 mg/g creatinine of fluorides in urine at end of shift.

NIOSH REL: (Inorganic Fluorides) TWA 2.5 mg(F)/m^3

DOT CLASSIFICATION: 6.1; Label: KEEP AWAY FROM FOOD

SAFETY PROFILE: Poison by ingestion and subcutaneous routes. A skin and severe eye irritant. An insecticide. When heated to decomposition it emits very toxic fumes of F^- and Na_2O.

DXE500   CAS: 4691-65-0   HR: 2
DISODIUM INOSINATE
mf: $C_{10}H_{11}N_4O_8P•2Na$ mw: 392.20

PROP: Colorless to white crystals; characteristic taste. Sol in water; sltly sol in alc; insol in ether.

SYNS: DISODIUM IMP □ DISODIUM-5'-INOSINATE □ DISODIUM INOSINE-5'-MONOPHOSPHATE □

DISODIUM INOSINE-5'-PHOSPHATE □ IMP DISODIUM SALT □ 5'-IMP DISODIUM SALT □ IMP SODIUM SALT □ INOSINE-5'-MONOPHOSPHATE DISODIUM □ INOSIN-5'-MONOPHOSPHATE DISODIUM □ SODIUM INOSINATE □ SODIUM-5'-INOSINATE

CONSENSUS REPORTS: Reported in EPA TSCA Inventory.

SAFETY PROFILE: Moderately toxic by several routes. An experimental teratogen. Mutation data reported. When heated to decomposition it emits toxic fumes of PO_x, NO_x, and Na_2O.

DXE800   CAS: 7631-95-0   HR: 3
DISODIUM MOLYBDATE
mf: $MoO_4•2Na$ mw: 205.92

PROP: White solid. Mp: 686°. Sol in H_2O. IDLH 1000 mg/m^3 (as Mo).

SYNS: MOLYBDIC ACID, DISODIUM SALT □ NATRIUMMOLYBDAT (GERMAN) □ SODIUM MOLYBDATE □ SODIUM MOLYBDATE(VI)

CONSENSUS REPORTS: Reported in EPA TSCA Inventory.

OSHA PEL: TWA 5 mg(Mo)/m^3

ACGIH TLV: TWA Soluble Compounds: TWA 0.5 mg(Mo)/m^3 Confirmed Animal Carcinogen with Unknown Relevance to Humans

SAFETY PROFILE: Poison by intraperitoneal route. Moderately toxic by subcutaneous and intravenous routes. Experimental reproductive effects. Mutation data reported. When heated to decomposition it emits toxic fumes of Na_2O. See also MOLYBDENUM COMPOUNDS.

DXF000   CAS: 15467-20-6   HR: 3
DISODIUM NITRILOTRIACETATE
mf: $C_6H_7NO_6•2Na$ mw: 235.12

SYNS: N,N-BIS(CARBOXYMETHYL)GLYCINE DISODIUM SALT □ DISODIUM HYDROGEN NITRILOTRIACETATE □ GLYCINE, N,N-BIS(CARBOXYMETHYL)-, DISODIUM SALT (9CI) □ KIRESUTO NTB □ NITRILOTRIACETIC ACID, DISODIUM SALT

CONSENSUS REPORTS: IARC Cancer Review: Group 2B IMEMDT 48,181,90; Animal Sufficient Evidence IMEMDT

48,181,90. Reported in EPA TSCA Inventory.
SAFETY PROFILE: Confirmed carcinogen. Moderately toxic by ingestion. When heated to decomposition it emits toxic fumes of NO_x and Na_2O.

DXF800   CAS: 7758-16-9   HR: 3
DISODIUM PYROPHOSPHATE
mf: $H_2O_7P_2 \cdot Na_2$ mw: 221.94
PROP: White, crystalline powder or monoclinic lattice. D: 1.862, mp: 220° (decomp). Sol in water.
SYNS: DINATRIUMPYROPHOSPHAT (GERMAN) □ DIPHOSPHORIC ACID, DISODIUM SALT □ DISODIUM DIHYDROGEN PYROPHOSPHATE □ DISODIUM DIPHOSPHATE □ SODIUM ACID PYROPHOSPHATE (FCC) □ SODIUM PYROPHOSPHATE
CONSENSUS REPORTS: Reported in EPA TSCA Inventory.
SAFETY PROFILE: Poison by intravenous route. Moderately toxic by ingestion and subcutaneous routes. An irritant to skin, eyes, and mucous membranes. When heated to decomposition it emits toxic fumes of PO_x and Na_2O.

DXG000   CAS: 13410-01-0   HR: 3
DISODIUM SELENATE
mf: $O_4Se \cdot 2Na$ mw: 188.94
PROP: Colorless, rhombic crystals. D: 3.098. Very sol in H_2O.
SYNS: NATRIUMSELENIAT (GERMAN) □ P-40 □ SEL-TOX SSO2 and SS-20 □ SODIUM SELENATE
CONSENSUS REPORTS: IARC Cancer Review: Group 3 IMEMDT 7,56,87. Selenium and its compounds are on the Community Right-To-Know List. EPA Genetic Toxicology Program. Reported in EPA TSCA Inventory. EPA Extremely Hazardous Substances List.
OSHA PEL: TWA 0.2 mg(Se)/m^3
ACGIH TLV: TWA 0.2 mg(Se)/m^3
DFG MAK: 0.1 mg(Se)/m^3
DOT CLASSIFICATION: 6.1; Label: Poison, Corrosive
SAFETY PROFILE: Poison by ingestion, intravenous, subcutaneous, and intraperitoneal routes. Questionable

carcinogen with experimental carcinogenic and teratogenic data. Human systemic effects by ingestion: EKG changes, hypermotility, diarrhea, and liver impairment. Experimental reproductive effects. Effects similar to those of arsenic. Mutation data reported. A pesticide. When heated to decomposition it emits toxic fumes of Se and Na_2O. See also SELENIUM COMPOUNDS and ARSENIC COMPOUNDS.

DXG025   CAS: 2583-80-4   HR: 1
DISODIUM 2-(4-STYRYL-3-SULFOPHEN-YL)-7-SULFO-2H-NAPHTHO(1,2-d)TRIAZOLE
mf: $C_{24}H_{15}N_3O_6S_2 \cdot 2Na$ mw: 551.52
SYNS: 2H-NAPHTHO(1,2-d)TRIAZOLE, 2-(4-STYRYL-3-SULFOPHENYL)-7-SULFO-, DISODIUM SALT □ NAPHTHO(1,2-d)TRIAZOLE-7-SULFONIC ACID,2-(4-(2-PHENYLETHENYL)-3-SULFOPHENYL)-, DISODIUM □ 2-STILBENESULFONIC ACID, 4-(7-SULFO-2H-NAPHTHO(1,2-d)TRIAZOL-2-YL)-, DISODIUM SALT
CONSENSUS REPORTS: Reported in EPA TSCA Inventory.
SAFETY PROFILE: A skin and eye irritant. When heated to decomposition it emits toxic fumes of NO_x and SO_x

DXG035   CAS: 1330-43-4   HR: 1
DISODIUM TETRABORATE
mf: $B_4Na_2O_7$ mw: 201.22
SYNS: ANHYDROUS BORAX □ BORATES, TETRA, SODIUM SALT, anhydrous (OSHA) □ BORAX GLASS □ BORIC ACID, DISODIUM SALT □ FR 28 □ FUSED BORAX □ RASORITE 65 □ SODIUM BIBORATE □ SODIUM TETRABORATE □ SODIUM TETRABORATE ($Na_2B_4O_7$)
CONSENSUS REPORTS: Reported in EPA TSCA Inventory.
OSHA PEL: TWA 10 mg/m^3
ACGIH TLV: TWA 1 mg/m^3
SAFETY PROFILE: A nuisance dust. Experimental reproductive effects. When heated to decomposition it emits toxic vapors of B.

D

DXG650   HR: D
DISTEARYL THIODIPROPIONATE
SAFETY PROFILE: When heated to
decomposition emits toxic fumes of SO_x.

DXG810   CAS: 64741-61-3   HR: 3
DISTILLATES (PETROLEUM), HEAVY CATALYTIC CRACKED
SYN: HEAVY CATALYTICALLY CRACKED
DISTILLATE
CONSENSUS REPORTS: IARC Cancer
Review: Animal Sufficient Evidence
IMEMDT 45,39,89. Reported in EPA
TSCA Inventory.
SAFETY PROFILE: Confirmed carcinogen.
When heated to decomposition it emits
acrid smoke and irritating vapors.

DXG840   CAS: 64741-59-9   HR: 3
DISTILLATES (PETROLEUM), LIGHT CATALYTIC CRACKED
SYN: LIGHT CATALYTICALLY CRACKED DISTILLATE
CONSENSUS REPORTS: IARC Cancer
Review: Animal Sufficient Evidence
IMEMDT 45,39,89. Reported in EPA
TSCA Inventory.
SAFETY PROFILE: Confirmed carcinogen.
Moderately toxic by ingestion. A severe skin
and eye irritant. When heated to
decomposition it emits acrid smoke and
irritating vapors.

DXH250   CAS: 97-77-8   HR: 3
DISULFIRAM
mf: $C_{10}H_{20}N_2S_4$ mw: 296.56
PROP: Yellow-white crystals. Mp: 70°. Sol
in CS_2, $CHCl_3$, C_6H_6, and EtOH.
SYNS: ABSTENSIL □ ABSTINYL □ ALCOPHOBIN □
ALK-AUBS □ ANTABUS □ ANTABUSE □ ANTADIX □
ANTAENYL □ ANTAETHAN □ ANTAETHYL □
ANTAETIL □ ANTALCOL □ ANTETAN □ ANTETHYL
□ ANTETIL □ ANTEYL □ ANTIAETHAN □
ANTIETANOL □ ANTI-ETHYL □ ANTIETIL □
ANTIKOL □ ANTIVITIUM □ AVERSAN □ AVERZAN □
(BIS(DIETHYLAMINO)THIOXOMETHYL) DISULPHIDE
□ BIS(DIETHYLTHIOCARBAMOYL) DISULFIDE □
BIS(N,N-DIETHYLTHIOCARBAMOYL) DISULFIDE □
BIS(N,N-DIETHYLTHIOCARBAMOYL) DISULPHIDE □
BONIBAL □ CONTRALIN □ CONTRAPOT □
CRONETAL □ DICUPRAL □ DISETIL □ DISULFAN □
DISULFURAM □ DISULPHURAM □ 1,1'-DITHIOBIS(N,N-
DIETHYLTHIOFORMAMIDE) □ EKAGOM TEDS □
EPHORRAN □ ESPENAL □ ESPERAL □ ETABUS □
ETHYLDITHIOURAME □ ETHYLDITHIURAME □
ETHYL THIRAM □ ETHYL THIUDAD □ ETHYL
THIURAD □ ETHYL TUADS □ ETHYL TUEX □
EXHORAN □ EXHORRAN □ HOCA □ KROTENAL □
NCI-C02959 □ NOCBIN □ NOXAL □ REFUSAL □ RO-
SULFIRAM □ STOPAETHYL □ STOPETHYL □
STOPETYL □ TATD □ TENURID □ TENUTEX □
TETD □ TETIDIS □ TETRADIN □ TETRADINE □
TETRAETHYLTHIOPEROXYDICARBONIC DIAMIDE □
TETRAETHYLTHIRAM DISULPHIDE □
TETRAETHYLTHIURAM □ TETRAETHYLTHIURAM
DISULFIDE □ TETRAETHYLTHIURAM DISULPHIDE □
N,N,N',N'-TETRAETHYLTHIURAM DISULPHIDE □
TETRAETIL □ TETURAM □ TETURAMIN □ THIOSAN
□ THIOSCABIN □ THIRERANIDE □ THIURAM E □
THIURANIDE □ TILLRAM □ TIURAM □ TTD □ TTS □
USAF B-33
CONSENSUS REPORTS: IARC Cancer
Review: Group 3 IMEMDT 7,56,87; Animal
Inadequate Evidence IMEMDT 12,85,76.
NCI Carcinogenesis Bioassay (feed); No
Evidence: mouse, rat NCITR* NCI-CG-
TR-16,79. Reported in EPA TSCA
Inventory.
OSHA PEL: TWA 2 mg/m³
ACGIH TLV: TWA 2 mg/m³; Not
Classifiable as a Human Carcinogen
DFG MAK: 2 mg/m³
SAFETY PROFILE: A human poison by
ingestion. An experimental poison by
intraperitoneal route. Toxic symptoms when
accompanied by ingestion of alcohol.
Human systemic effects by ingestion:
jaundice, joint changes. An experimental
teratogen. Other experimental reproductive
effects. Questionable carcinogen with
experimental neoplastigenic data. See also
THIRAM.

DXH325   CAS: 298-04-4   HR: 3
DISULFOTON
DOT: NA 2783
mf: $C_8H_{19}O_2PS_3$ mw: 274.42
SYNS: BAY 19639 □ BAYER 19639 □ O,O-DIAETHYL-S-
(2-AETHYLTHIO-AETHYL)-DITHIOPHOSPHAT
(GERMAN) □ O,O-DIAETHYL-S-(3-THIA-PENTYL)-
DITHIOPHOSPHAT (GERMAN) □ O,O-DIETHYL-S-(2-
ETHYLMERCAPTOETHYL) DITHIOPHOSPHATE □ O,O-
DIETHYL-S-(2-ETHYLTHIOETHYL) PHOSPHORO-
DITHIOATE □ O,O-DIETHYL-S-(2-ETHYLTHIO-

ETHYL)-DITHIOFOSFAAT (DUTCH) □ O,O-DIETHYL-2-ETHYLTHIOETHYL PHOSPHORODITHIOATE □ O,O-DIETHYL-S-2-(ETHYLTHIO)ETHYL PHOSPHORO-DITHIOATE □ O,O-DIETHYL-S-(2-ETHYLTHIOETHYL) THIOTHIONOPHOSPHATE □ O,O-DIETIL-S-(2-ETILTIO-ETIL)-DITIOFOSFATO (ITALIAN) □ DI-SYSTON □ DITHIODEMETON □ DITHIOPHOSPHATE de O,O-DIETHYLE et de S-(2-ETHYLTHIO-ETHYLE) □ DITHIOSYSTOX □ DUTION □ EKATIN TD □ ENT 23,437 □ O,O-ETHYL S-2(ETHYLTHIO)ETHYL PHOSPHORODITHIOATE □ ETHYL THIOMETON □ ETHYLTHIOMETON B □ S-2-(ETHYLTHIO)ETHYL O,O-DIETHYL ESTER OF PHOSPHORODITHIOIC ACID □ FRUMIN AL □ FRUMIN G □ GLEBOFOS □ M-74 □ M-74 (PESTICIDE) □ PHOSPHORODITHIONIC ACID, S-2-(ETHYLTHIO)ETHYL-O,O-DIETHYL ESTER □ RCRA WASTE NUMBER P039 □ SOLVIREX □ THIODEMETON □ VUAGT 1-4 □ VUAGT 1964

CONSENSUS REPORTS: EPA Extremely Hazardous Substances List. EPA Genetic Toxicology Program.

OSHA PEL: TWA 0.1 mg/m³ (skin)

ACGIH TLV: TWA 0.05 mg/m³ (skin); Not Classifiable as a Human Carcinogen

SAFETY PROFILE: Poison by ingestion, inhalation, skin contact, intraperitoneal, and intravenous routes. Human mutation data reported. When heated to decomposition it emits very toxic SO_x and PO_x. See also various demeton entries and ESTERS.

DXI480   CAS: 153049-45-7   HR: 3
DITHIADENOXIDE

mf: $C_{17}H_{19}NOS_2 \cdot C_4H_4O_4$ mw: 433.57

SYNS: DITHIADENOXID HYDROGEN MALEATE □ 1-PROPANAMINE, N,N-DIMETHYL-3-THIENO(2,3-C)(2)BENZOTHIEPIN-4(9H)-YLIDNEN-,S-OXIDE, (E)-, (Z)-2-BUTENEDIOATE (1:1)

SAFETY PROFILE: A poison by intravenous route. Moderately toxic by ingestion. When heated to decomposition it emits toxic vapors of NO_x and SO_x.

DXJ800   CAS: 1141-88-4   HR: 3
2,2'-DITHIOBISANILINE

mf: $C_{12}H_{12}N_2S_2$ mw: 248.38

PROP: Leaflets or needles from EtOH (aq). Mp: 93°. Sol in acids or EtOH; sltly sol in H_2O.

SYNS: BIS(o-AMINOPHENYL)DISULFIDE □ BIS(2-AMINOPHENYL)DISULFIDE □ 1,1'-BIS(2-AMINOPHENYL)DISULFIDE □ O,O'-DIAMINO DIPHENYL DISULFIDE □ O,O-DITHIO-BIS-ANILINE □ 2,2'-DITHIODIANILINE □ USAF AB-315

CONSENSUS REPORTS: Reported in EPA TSCA Inventory.

SAFETY PROFILE: Poison by intravenous and intraperitoneal routes. Moderately toxic by ingestion. A severe eye irritant. When heated to decomposition it emits very toxic fumes of NO_x and SO_x.

DXL800   CAS: 541-53-7   HR: 3
DITHIOBIURET

mf: $C_2H_5N_3S_2$ mw: 135.22

PROP: Crystals or needles from water. Mp: 181°, bp: decomp, d: 1.522 @ 30°. Sol in H_2O, EtOH, and Me_2CO.

SYNS: DTB □ RCRA WASTE NUMBER P049 □ 2-THIO-1-(THIOCARBAMOYL)UREA □ USAF B-44 □ USAF EK-P-6281

CONSENSUS REPORTS: Reported in EPA TSCA Inventory. EPA Extremely Hazardous Substances List.

SAFETY PROFILE: Poison by ingestion and intraperitoneal routes. When heated to decomposition it emits highly toxic fumes of SO_x and NO_x.

DXM600   CAS: 1892-29-1   HR: 3
DITHIODIGLYCOL

mf: $C_4H_{10}O_2S_2$ mw: 154.26

PROP: Syrupy liquid. Mp: 17°, bp: 160–162° @ 0.1 mm.

SYNS: 2,2-DITHIODIETHANOL □ USAF TH-9

CONSENSUS REPORTS: Reported in EPA TSCA Inventory.

SAFETY PROFILE: Poison by intraperitoneal route. When heated to decomposition it emits toxic fumes of SO_x.

DXN600   CAS: 333-29-9   HR: 3
DITHIOLANE IMINOPHOSPHATE

mf: $C_7H_{14}NO_2PS_3$ mw: 271.37

SYNS: AC-43064 □ AMERICAN CYANAMID AC 43,064 □ CL-43,064 □ CYALANE □ CYCLIC ETHYLENE (DIETHOXYPHOSPHINOTHIOYL)DITHIOIMIDOCARB ONATE □ CYCLIC ETHYLENE ESTER of (DIETHOXYPHOSPHINOTHIOYL)DITHIOIMIDOCARB ONIC ACID □ CYLAN □ CYOLAN □ CYOLANE INSECTICIDE □ 2-(DIETHOXYPHOSPHINYLIMINO)-1,3-DITHIOLANE □ O,O-DIETHYL 1,3-DITHIOLAN-2-

YLIDENEPHOSPHORAMIDOTHIOATE □ DIETHYL-N-1,3-DITHIOLANYL-2-IMINO PHOSPHATE □ DITHIOLANE □ 1,3-DITHIOLAN-2-YLIDENE-PHOSPHORAMIDOTHIOIC ACID DIETHYL ESTER □ 1,3-DITHIOLAN-2-YLIDENE-PHOSPHORAMIDOTHIOIC ACID-O,O-DIETHYL ESTER □ ENT 25,809 □ IMINOPHOSPHATE □ PHOSFOLAN

SAFETY PROFILE: Poison by ingestion and skin contact. An insecticide. When heated to decomposition it emits very toxic fumes of NO_x and SO_x.

DXN830   CAS: 21709-44-4   HR: 3
o-(1,3-DITHIOLAN-2-YL)PHENYL DI-METHYLCARBAMATE

mf: $C_{12}H_{15}NO_2S_2$ mw: 269.40

SYNS: C 13963 □ C-13963 □ CARBAMIC ACID, DIMETHYL-, 2-(1,3-DITHIOLAN-2-YL)PHENYL ESTER □ CARBAMIC ACID, DIMETHYL-, o-(1,3-DITHIOLAN-2-YL)PHENYL ESTER

SAFETY PROFILE: A poison by ingestion. When heated to decomposition it emits toxic vapors of NO_x and SO_x.

DXO000   CAS: 572-48-5   HR: 3
DITHION

mf: $C_{17}H_{21}O_5PS$ mw: 368.41

PROP: Crystals nearly insol in water. Mp: 88°.

SYNS: O,O-DIETHYL-7-HYDROXY-3,4-TETRAMETHY-LENE COUMARINYL PHOSPHOROTHIOATE □ O,O-DIETHYL-O-(7,8,9,10-TETRAHYDRO-6-OXOBENZO-(C)CHROMAN-3-YL)PHOSPHOROTHIOATE □ O,O-DIETHYL-O-(7,8,9,10-TETRAHYDRO-6-OXO-6H-DIBENZO(b,d)PYRAN-3-YL)PHOSPHOROTHIOATE □ O,O-DIETHYL-O-(3,4-TETRAMETHYLENE-COUMARINYL-7) THIOPHOSPHATE □ DITHIONE □ ENT 24,986 □ 7-HYDROXY-3,4-TETRAMETHYLENE-COUMARIN-O,O-DIETHYL THIOPHOSPHATE

SAFETY PROFILE: Poison by ingestion. When heated to decomposition it emits very toxic fumes of PO_x and SO_x. See also PARATHION.

DXO200   CAS: 79-40-3   HR: 3
DITHIOOXAMIDE

mf: $C_2H_4N_2S_2$ mw: 120.20

PROP: Red crystals.

SYNS: DITHIOOXALDIIMIDIC ACID □ DITHIOXAMIDE □ ETHANEDITHIOAMIDE □ HYDRORUBEANIC ACID □ RUBEANE □ RUBEANIC ACID □ RVK □ USAF EK-4394 □ USAF MK-6

CONSENSUS REPORTS: Reported in EPA TSCA Inventory.

SAFETY PROFILE: Poison by ingestion, intraperitoneal, and intravenous routes. When heated to decomposition it emits very toxic fumes of NO_x and SO_x.

DXP200   CAS: 97-39-2   HR: 3
DI-o-TOLYLGUANIDINE

mf: $C_{15}H_{17}N_3$ mw: 239.35

PROP: White crystals from alc (aq). Mp: 179°, d: 1.10 @ 20°/4°, vap d: 8.24.

SYNS: DIORTHOTOLYLGUANIDINE □ 1,3-DI-o-TOLYLGUANIDINE □ DOTG ACCELERATOR □ USAF A-6598 □ VULKACIT DOTG/C

CONSENSUS REPORTS: Reported in EPA TSCA Inventory.

SAFETY PROFILE: Poison by ingestion and intraperitoneal routes. When heated to decomposition it emits toxic fumes of NO_x.

DXP600   CAS: 137-97-3   HR: 3
DI-o-TOLYLTHIOUREA

mf: $C_{15}H_{16}N_2S$ mw: 256.39

PROP: Crystals or needles from alc. Mp: 165–166°, vap d: 8.85. Sol in dichloroethane.

SYNS: N,N'-BIS(2-METHYLPHENYL)THIOUREA □ 1,3-BIS(o-TOLYL)-2-THIOUREA □ 2,2'-DIMETHYLTHIOCARBANILIDE □ DI-o-TOLUYLTHIOUREA □ USAF EK-1651

CONSENSUS REPORTS: Reported in EPA TSCA Inventory.

SAFETY PROFILE: Poison by intraperitoneal route. Moderately toxic by ingestion. When heated to decomposition it emits very toxic fumes of NO_x and SO_x.

DXQ500   CAS: 330-54-1   HR: 2
DIURON

mf: $C_9H_{10}Cl_2N_2O$ mw: 233.11

PROP: Crystals. Mp: 153.5–155°. Sltly sol in water and hydrocarbon solvents.

SYNS: AF 101 □ CEKIURON □ CRISURON □ DAILON □ DCMU □ DIATER □ 3-(3,4-DICHLOOR-FENYL)-1,1-DIMETHYLUREUM (DUTCH) □ DICHLORFENIDIM □ 3-(3,4-DICHLOROPHENOL)-1,1-DIMETHYLUREA □ N'-(3,4-DICHLOROPHENYL)-N,N-DIMETHYLUREA □ 1-(3,4-DICHLOROPHENYL)-3,3-DIMETHYLUREE (FRENCH) □ 3-(3,4-DICHLOR-PHENYL)-1,1-DIMETHYL-HARNSTOFF (GERMAN) □ 3-(3,4-DICLORO-FENYL)-1,1-DIMETIL-UREA (ITALIAN) □ 1,1-DIMETHYL-3-(3,4-DICHLOROPHENYL)UREA □ DI-ON □ DIREX 4L □ DIUREX □ DIUROL □ DIURON 4L □ DMU □ DREXEL

□ DREXEL DIURON 4L □ DURAN □ DYNEX □ FARMCO DIURON □ HERBATOX □ HW 920 □ KARMEX □ KARMEX DIURON HERBICIDE □ KARMEX DW □ MARMER □ SUP'R FLO □ TELVAR □ TELVAR DIURON WEED KILLER □ UNIDRON □ USAF P-7 □ USAF XR-42 □ VONDURON

CONSENSUS REPORTS: Reported in EPA TSCA Inventory. EPA Genetic Toxicology Program. Chlorophenol compounds are on the Community Right-To-Know List.
OSHA PEL: TWA 10 mg/m^3
ACGIH TLV: TWA 10 mg/m^3
NIOSH REL: (Diuron) TWA 10 mg/m^3
SAFETY PROFILE: Moderately toxic by ingestion and intraperitoneal routes. Questionable carcinogen with experimental tumorigenic and teratogenic data. Mutation data reported. When heated to decomposition it emits highly toxic fumes of Cl$^-$ and NO$_x$. See also CHLOROPHENOLS.

DXQ740   CAS: 1321-74-0   HR: 1
DIVINYLBENZENE
mf: C$_{10}$H$_{10}$ mw: 130.20
SYNS: BENZENE, DIVINYL- □ VINYLSTYRENE
CONSENSUS REPORTS: Reported in EPA TSCA Inventory.
ACGIH TLV: TWA 10 ppm
NIOSH REL: (divinyl benzene) TWA 10 ppm
SAFETY PROFILE: Mildly toxic by ingestion. An eye irritant. Combustible. When heated to decomposition it emits acrid smoke and irritating fumes.

DXQ745   CAS: 108-57-6   HR: 2
DIVINYLBENZENE
mf: C$_{10}$H$_{10}$ mw: 130.20
PROP: Pale straw-colored liquid. Bp: 195–200°, mp: −66.1°, d: 0.918 grams per cubic centimeter, flash p: 165F°. Not misc in water; sol in ether and methanol.
SYNS: m-DIVINYLBENZEN □ m-DIVINYLBENZENE □ m-VINYLSTYRENE
CONSENSUS REPORTS: Reported in EPA TSCA Inventory.
OSHA PEL: 10 ppm
ACGIH TLV: 10 ppm
SAFETY PROFILE: An eye irritant. Combustible. When heated to decomposition it emits acrid smoke and irritating fumes.

DXQ750   HR: D
DIVINYLBENZENE COPOLYMER

SAFETY PROFILE: When heated to decomposition it emits acrid smoke and irritating fumes.

DXS700   HR: 2
Δ-DODECALACTONE
mf: C$_{12}$H$_{22}$O$_2$ mw: 198.31
PROP: Colorless to yellow liquid; coconut-fruity odor. Refr index: 1.458–1.461, flash p: 151°F. Very sol in alc, propylene glycol, veg oil; insol in water.
SYN: FEMA No. 2401
SAFETY PROFILE: Combustible liquid. When heated to decomposition it emits acrid smoke and irritating fumes.

DXT000   CAS: 112-54-9   HR: 1
1-DODECANAL
mf: C$_{12}$H$_{24}$O   mw: 184.36
PROP: Crystals. Mp: 44.5°, bp: 184–185° @ 100 mm. Reported in pine-needle, lime, sweet-orange, and a dozen other essential oils (FCTXAV 11,477,73). Colorless to light-yellow liquid; fatty odor. D: 0.826–0.836, refr index: 1.433–1.439, flash p: 180°F. Sol in alc, fixed oils, propylene glycol; insol in glycerin, water.
SYNS: C-12 ALDEHYDE, LAURIC □ 1-DODECYL ALDEHYDE □ DUODECYLIC ALDEHYDE □ FEMA No. 2615 □ LAURYL ALDEHYDE (FCC)
CONSENSUS REPORTS: Reported in EPA TSCA Inventory.
SAFETY PROFILE: Mildly toxic by ingestion. A human and experimental skin irritant. Combustible liquid. When heated to decomposition it emits acrid smoke and irritating fumes. See also ALDEHYDES.

DXT800   CAS: 25103-58-6   HR: 3
tert-DODECANETHIOL
mf: C$_{12}$H$_{26}$S   mw: 202.44
PROP: White to light-yellow liquid. Bp: 200–235°, flash p: 205°F (OC), d: 0.85 @ 25°/25°, vap d: 6.98.
SYNS: tert-DODECYLMERCAPTAN □ tert. DODECYLMERKAPTAN (CZECH) □ tert-DODECYLTHIOL □ 2,3,3,4,4,5-HEXAMETHYL-2-HEXANETHIOL
CONSENSUS REPORTS: Reported in EPA TSCA Inventory.
SAFETY PROFILE: Poison by ingestion. Moderately toxic by intraperitoneal route. A skin and eye irritant. Combustible when exposed to heat or flame; can react vigorously with oxidizing materials. To fight fire, use foam, CO$_2$, dry chemical. When heated to decomposition it emits

toxic fumes of SO_x. See also MERCAPTANS and SULFATES.

DXU280   CAS: 4826-62-4   HR: 2
2-DODECENAL
mf: $C_{12}H_{22}O$ mw: 182.34
SYN: β-OCTYL ACROLEIN
CONSENSUS REPORTS: Reported in EPA TSCA Inventory.
SAFETY PROFILE: A severe skin irritant. When heated to decomposition it emits acrid smoke and irritating fumes.

DXU400   CAS: 2855-19-8   HR: 2
DODECENE EPOXIDE
mf: $C_{12}H_{24}O$ mw: 184.36
SYN: 1,2-EPOXYDODECANE
CONSENSUS REPORTS: Reported in EPA TSCA Inventory.
SAFETY PROFILE: Questionable carcinogen with experimental tumorigenic data. When heated to decomposition it emits acrid smoke and irritating fumes.

DXV000   CAS: 25377-73-5   HR: 3
DODECENYLSUCCINIC ANHYDRIDE
mf: $C_{16}H_{27}O_3$ mw: 266.38
PROP: Light-yellow, clear, visc oil. Bp: 180–182° @ 5 mm, flash p: 352°F (COC), d: 1.002 @ 25°/4°.
SYNS: DDS □ DDS A □ 2,5-FURANDIONE, 3-(DODECENYL)DIHYDRO-
CONSENSUS REPORTS: Reported in EPA TSCA Inventory.
SAFETY PROFILE: Poison by inhalation and intraperitoneal routes. An irritant and sensitizer. Combustible when exposed to heat or flame; can react with oxidizing materials. To fight fire, use foam, CO_2, dry chemical. When heated to decomposition it emits acrid smoke and irritating fumes.

DXV600   CAS: 112-53-8   HR: 2
DODECYL ALCOHOL
mf: $C_{12}H_{26}O$ mw: 186.38
PROP: Crystals from EtOH (aq) or liquid above 24°; floral odor. Mp: 24°, bp: 145–148° @ 18 mm, d: 0.830–0.836, refr index: 1.440–1.444, flash p: 260°F, autoign temp: 527°F. Sol in 2 parts of 70% alc, fixed oils, propylene glycol; insol in water, glycerin.
SYNS: ALCOHOL C-12 □ ALFOL 12 □ CACHALOT L-50 □ CO 12 □ CO-1214 □ n-DODECANOL □ 1-DODECANOL □ n-DODECYL ALCOHOL □

DUODECYL ALCOHOL □ DYTOL J-68 □ EPAL 12 □ FEMA No. 2617 □ LAURIC ALCOHOL □ LAURINIC ALCOHOL □ LAURYL 24 □ LAURYL ALCOHOL (FCC) □ n-LAURYL ALCOHOL, PRIMARY □ LOROL □ MA-1214 □ SIPOL L12
CONSENSUS REPORTS: Reported in EPA TSCA Inventory.
SAFETY PROFILE: Moderately toxic by intraperitoneal route. Mildly toxic by ingestion. A severe human skin irritant. Questionable carcinogen with experimental tumorigenic data. Combustible when exposed to heat or flame; can react with oxidizing materials. To fight fire, use dry chemical, CO_2. When heated to decomposition it emits acrid smoke and irritating fumes.

DXW000   CAS: 124-22-1   HR: 3
DODECYLAMINE
mf: $C_{12}H_{27}N$ mw: 185.40
PROP: Oil, amine odor or crystals from C_6H_6. Fp: 28.3°, vap press: 64 mm @ 170°, mp: 27–28°, bp: 247–249°. Sol in EtOH, Et_2O, C_6H_6, and $CHCl_3$.
SYNS: ALAMINE 4 □ AMINE BB □ 1-AMINODODECANE □ ARMEEN 12D □ 1-DODECANAMINE (9CI) □ n-DODECYLAMINE □ 1-DODECYLAMINE □ KEMAMINE P690 □ LAURINAMINE □ LAURYLAMINE □ n-LAURYLAMINE □ MONODODECYLAMINE □ NISSAN AMINE BB
CONSENSUS REPORTS: Reported in EPA TSCA Inventory.
SAFETY PROFILE: Poison by intraperitoneal route. Moderately toxic by ingestion. A severe skin and eye irritant. When heated to decomposition it emits toxic fumes of NO_x. See also AMINES.

DXW200   CAS: 25155-30-0   HR: 3
DODECYL BENZENE SODIUM SULFONATE
mf: $C_{18}H_{29}O_3S•Na$ mw: 348.52
PROP: White to light-yellow flakes, granules, or powder.
SYNS: AA-9 □ ABESON NAM □ BIO-SOFT D-40 □ CALSOFT F-90 □ CONCO AAS-35 □ CONOCO C-50 □ DETERGENT HD-90 □ DODECYLBENZENESULFONIC ACID SODIUM SALT □ DODECYLBENZENESULPHONATE, SODIUM SALT □ DODECYLBENZENSULFONAN SODNY (CZECH) □ MERCOL 25 □ NACCANOL NR □ NECCANOL SW □ PILOT HD-90 □ PILOT SF-40 □ RICHONATE 1850 □ SANTOMERSE 3 □ SODIUM DODECYLBENZENESULFONATE (DOT) □ SODIUM DODECYLBENZENESULFONATE, dry □ SODIUM LAURYLBENZENESULFONATE □ SOLAR 40 □ SOL SODOWA KWASU LAURYLOBENZENOSULFON-OWEGO (POLISH) □ SULFAPOL □ SULFAPOLU

(POLISH) ☐ SULFRAMIN 85 ☐ SULFRAMIN 40 FLAKES ☐ SULFRAMIN 40 GRANULAR ☐ SULFRAMIN 1238 SLURRY ☐ p-1',1',4',4'-TETRAMETHYLOKTYLBENZEN-SULFONAN SODNY (CZECH) ☐ ULTRAWET K

CONSENSUS REPORTS: Reported in EPA TSCA Inventory.

SAFETY PROFILE: Poison by intravenous route. Moderately toxic by ingestion. A skin and severe eye irritant. When heated to decomposition it emits toxic fumes of Na_2O. See also SULFONATES.

DXX000   CAS: 538-71-6   HR: 3
DODECYLDIMETHYL(2-PHENOXYETH-YL)AMMONIUM BROMIDE

mf: $C_{22}H_{40}NO•Br$ mw: 414.54

PROP: A solid. Mp: 112–113°.

SYNS: PHENODODECINIUM BROMIDE ☐ β-PHENOXYETHYLDIMETHYLDODECYLAMMONIUM BROMIDE

CONSENSUS REPORTS: Reported in EPA TSCA Inventory.

SAFETY PROFILE: Poison by intraperitoneal and intravenous routes. Mutation data reported. When heated to decomposition it emits very toxic fumes of NO_x, NH_3, and Br^-. See also BROMIDES.

DXX200   CAS: 1166-52-5   HR: 3
DODECYL GALLATE

mf: $C_{19}H_{30}O_5$ mw: 338.49

PROP: Sol in EtOH and Me_2CO.

SYNS: DODECYLESTER KYSELINY GALLOVE ☐ GALLIC ACID, DODECYL ESTER ☐ GALLIC ACID, LAURYL ESTER ☐ LAURYL GALLATE ☐ NIPAGALLIN LA ☐ PROGALLIN LA

CONSENSUS REPORTS: Reported in EPA TSCA Inventory.

SAFETY PROFILE: A poison by intraperitoneal route. When heated to decomposition it emits acrid smoke and irritating fumes.

DXX400   CAS: 2439-10-3   HR: 3
N-DODECYLGUANIDINE ACETATE

mf: $C_{13}H_{29}N_3•C_2H_4O_2$ mw: 287.51

PROP: Crystals. Mp: 136°. Sol in hot water and alc; insol in nonpolar solvs.

SYNS: AC 5223 ☐ AMERICAN CYANAMID 5223 ☐ APADODINE ☐ CARPENE ☐ CURITAN ☐ CYPREX ☐ CYPREX 65W ☐ N-DODECYLGUANIDINACETAT (GERMAN) ☐ DODECYLGUANIDINE ACETATE ☐ DODGUADINE ☐ DODINE ☐ DODINE ACETATE ☐ DODINE, mixture with GLYODIN ☐ DOGQUADINE ☐ ENT 16,436 ☐ EXPERIMENTAL FUNGICIDE 5223 ☐

LAURYLGUANIDINE ACETATE ☐ MELPREX ☐ MILPREX ☐ SYLLIT ☐ TSITREX ☐ VENTUROL ☐ VONDODINE

CONSENSUS REPORTS: Reported in EPA TSCA Inventory. EPA Genetic Toxicology Program.

SAFETY PROFILE: Poison by ingestion and inhalation. Moderately toxic by skin contact. A severe eye irritant. Questionable carcinogen with experimental tumorigenic data. A pesticide. When heated to decomposition it emits very toxic fumes of NO_x.

DXY600   CAS: 27193-86-8   HR: 2
DODECYLPHENOL (mixed isomers)

mf: $C_{18}H_{30}O$ mw: 262.48

PROP: Straw-colored liquid; phenolic odor. Bp: 154–168°, flash p: 325°F (OC), d: 0.93 @ 20°/20°, vap d: 9.04.

SYN: T-DET

CONSENSUS REPORTS: Reported in EPA TSCA Inventory.

SAFETY PROFILE: Moderately toxic by ingestion. Mildly toxic by skin contact. Combustible when exposed to heat or flame; can react with oxidizing materials. To fight fire, use CO_2, dry chemical. When heated to decomposition it emits acrid smoke and irritating fumes.

DXZ000   CAS: 7631-98-3   HR: 3
N-DODECYLSARCOSINE SODIUM SALT

mf: $C_{15}H_{30}NO_2•Na$ mw: 279.45

SYN: SODIUM-N-LAURYL SARCOSINE

CONSENSUS REPORTS: Reported in EPA TSCA Inventory.

SAFETY PROFILE: Poison by intravenous route. When heated to decomposition it emits toxic fumes of NO_x and Na_2O.

DYA800   CAS: 4484-72-4   HR: 3
DODECYLTRICHLOROSILANE

DOT: UN 1771

mf: $C_{12}H_{25}Cl_3Si$ mw: 303.81

PROP: Colorless to yellow liquid. Bp: 152–153° @ 3 mm, d: 1.026 @ 25°/25°.

SYN: TRICHLORODODECYLSILANE

CONSENSUS REPORTS: Reported in EPA TSCA Inventory.

DOT CLASSIFICATION: 8; Label: Corrosive

SAFETY PROFILE: A poison. A corrosive irritant to the eyes, skin, and mucous membranes. Readily hydrolyzed by moisture with the production of hydrochloric acid. When heated to decomposition it emits toxic fumes of Cl⁻. See also CHLOROSILANES.

DYC400   CAS: 51-61-6   HR: 3
DOPAMINE
mf: $C_8H_{11}NO_2$ mw: 153.20

SYNS: 4-(2-AMINOETHYL)PYROCATECHOL □ 3-HYDROXYTYRAMINE

SAFETY PROFILE: Poison by intravenous, intracervical, and intraperitoneal routes. Experimental teratogenic and reproductive effects. Human mutation data reported. A neurotransmitter. An adrenergic agent. When heated to decomposition it emits toxic fumes of NO_x.

DYC800   CAS: 77-21-4   HR: 3
DORIDEN
mf: $C_{13}H_{15}NO_2$ mw: 217.29

PROP: dl-Form: Crystals from ether or from ethyl acetate + pet ether. Mp: 84°. Freely sol in ethyl acetate, acetone, ether, chloroform; sol in ethanol, methanol. Practically insol in water. d-Form: Crystals. Mp: 102.5–103°, refr index: (α) (20/D) +176° (methanol). l-Form: Crystals. Mp: 102–103°, refr index: (α) (20/D) −181° (methanol).

SYNS: ALFIMID □ CC 11511 □ DORIDEN-SED □ ELRODORM □ 3-ETHYL-3-PHENYL-2,6-DIKETO-PIPERIDINE □ 3-ETHYL-3-PHENYL-2,6-DIOXOPIP-ERIDINE □ α-ETHYL-α-PHENYLGLUTARIMIDE □ 2-ETHYL-2-PHENYLGLUTARIMIDE □ 3-ETHYL-3-PHENYL-2,6-PIPERIDINEDIONE □ GIMID □ GLIMID □ GLUTATHIMID □ GLUTETHIMID □ GLUTETHIMIDE □ GLUTETIMIDE □ NOXYRON □ 3-PHENYL-3-ETHYL-2,6-DIKETOPIPERIDINE □ 3-PHENYL-3-ETHYL-2,6-DIOXOPIPERIDINE □ α-PHENYL-α-ETHYLGLUTARIC ACID IMIDE □ 2-PHENYL-2-ETHYLGLUTARIC ACID IMIDE □ α-PHENYL-α-ETHYLGLUTARIMIDE □ SARODORMIN

CONSENSUS REPORTS: EPA Genetic Toxicology Program.

SAFETY PROFILE: Poison by ingestion and intraperitoneal routes. Human systemic effects by ingestion: pupillary dilation, ataxia, somnolence, coma, and blood pressure depression. An experimental teratogen. Other experimental reproductive effects. When heated to decomposition it emits toxic fumes of NO_x. Caution: May be habit forming. This is a controlled substance (depressant) listed in the U.S. Code of Federal Regulations, Title 21 Part 1308.13 (1985).

DYE600   CAS: 523-87-5   HR: 3
DRAMAMINE

mf: $C_{17}H_{21}NO \cdot C_7H_7ClN_4O_2$ mw: 470.02

PROP: A solid. Mp: 102–107°.

SYNS: AMOSYT □ ANAUTINE □ ANDRAMINE □ AVIOMARIN □ o-BENZHYDRYLDIMETHYLAMINO-ETHANOL-8-CHLOROTHEOPHYLLINATE □ 2-(BENZHYDRYLOXY)-N,N-DIMETHYLETHYLAMINE with 8-CHLOROTHEOPHYLLINE □ CHLORANAUTINE □ DIAMARIN □ DIMENHYDRINATE □ DIPHEN-HYDRINATE □ DRAMAMIN □ DRAMARIN □ DRAMYL □ DROMYL □ ELDODRAM □ ETHYLAMINE-2-(DIPHENYLMETHOXY)-N,N-DIMETHYL, compound with 8-CHLOROTHEOPHYLLINE (1:1) □ GRAVINOL □ GRAVOL □ MENHYDRINATE □ NCI-C60639 □ NEO-NAVIGAN □ NOVAMIN □ NOVAMINE □ PERMITAL □ REISE-ENGLETTEN □ SUPREMAL □ TEODRAMIN □ TRAVELIN □ TRAVELMIN □ VOMEX A □ XAMAMINA

CONSENSUS REPORTS: Reported in EPA TSCA Inventory.

SAFETY PROFILE: Poison by intraperitoneal and intravenous routes. Moderately toxic by ingestion. A drug much used for motion sickness. Human systemic effects by ingestion: arrhythmias, convulsions, distorted perceptions, hallucinations, intracranial pressure increase. Mutation data reported. When heated to decomposition it emits very toxic fumes of NO_x and Cl⁻. See also AMINES.

DYG000   HR: 3
DYNAMITE
DOT: UN 0081

PROP: Major constituent is nitroglycerin (85ESA3 8,739,68).

SYN: GELATINE DYNAMITE

DOT CLASSIFICATION: 1.1D; Label: EXPLOSIVE, BLASTING TYPE A

SAFETY PROFILE: A high explosive used industrially in construction and mining. The name generally refers to a mixture containing as its principal explosive ingredient either glyceryl trinitrate (nitroglycerin) or ammonium nitrate, suitably sensitized. It does not apply to black blasting powders, chlorate powders, and other deflagrating mixtures. While this material is a powerful explosive when detonated by shock or heat, it is only moderately hazardous. It can react vigorously with oxidizing materials. Dangerous; shock and heat will cause it to explode; when heated to decomposition it emits highly toxic fumes of NO_x and CO, etc. See also NITROGLYCERIN; EXPLOSIVES, HIGH; and NITRATES.

An ordinary blasting cap or an electric blasting cap is used for detonating a charge of dynamite. The various classes and grades of dynamite are made from mixtures composed of an explosive compound or a mixture of explosive compounds, a "dope," and an antiacid. If any of the explosive ingredients are in a liquid state they are referred to as the "explosive oil," which is usually composed of glyceryl trinitrate (nitroglycerin) and about 25–30% of ethylene glycol dinitrate. The latter compound depresses the freezing point of the nitroglycerin and renders the dynamite low-freezing. Other compounds may also be used as freezing point depressants. The explosive oil is absorbed by carbonaceous materials that have entirely replaced kieselguhr (diatomaceous earth), formerly used exclusively as the absorbent or "dope" in dynamites. This type of "dope" does not enter into the explosive reaction. Wood pulp is now most commonly used as the absorber, either alone or mixed in suitable proportions with flour, starch, etc.

The absorbents may be mixed with an oxidizer such as sodium nitrate, in which case an active "dope" is formed. For neutralizing any acid that may be present, about 1% of an antiacid (calcium carbonate or zinc oxide) is added to the mixture. The explosive oil is mixed into the "dope." The strength of a kieselguhr dynamite, when detonated, is derived only from the explosive oil, since kieselguhr is inert. A mixture of this kind is known as a straight dynamite. On the other hand, an active "dope" (an admixture of carbonaceous absorbents with an oxidizer) furnishes explosive strength in addition to that derived from the explosive ingredients.

By replacing a part of the explosive oil of a straight dynamite with ammonium nitrate, so that the latter becomes the principal explosive ingredient, a mixture known as an ammonia dynamite is obtained.

When the explosive oil is gelatinized the explosive is known as a gelatin or an ammonia gelatin dynamite.

Blasting gelatin is a gelatinized mass of an elastic nature obtained by incorporating nitrocotton with an explosive oil into which is mixed about 1% of antiacid.

Dynamites may be in bulk form (bag powder) or in cartridge form, the most common size being 1.25 inch in diameter and 8 inches long, although, for holes of small diameter, cartridges as small as 7/8 inch in diameter are also used. In large-diameter well-drill holes for quarry blasting, cartridge diameters up to 10 inches and lengths up to 30 inches may be used. These upper limits or 50 pounds in weight of each cartridge are imposed by the DOT Regulations, and the maximum length of 30 inches applies to all cartridge diameters between 4 and 10 inches.

An integral part of a stick of dynamite is the paraffined paper wrapper that not only holds the ingredients together but enters into the explosive reaction.

The wrapper also affords some measure of protection from moderate exposure to dampness. For blasting in wet operations, a gelatinized dynamite that resists the absorption of water is used.

The strength of straight dynamite is graded by its explosive oil content (% by weight), while for any other class of dynamite, the strength is determined experimentally in comparison with the various grades of the straight dynamites. For example, a 40% straight dynamite is one that contains 40% of explosive oil; a 40%-strength ammonia dynamite, as determined by tests, equals a 40% straight dynamite in strength. In other words a 40%-strength ammonia dynamite will release the same energy as an equivalent weight of a 40% straight dynamite.

DYG600   CAS: 10025-74-8   HR: 3
DYSPROSIUM CHLORIDE
mf: Cl_3Dy mw: 268.85

PROP: Shiny, yellow crystals or hygroscopic white crystals. D: 3.67 @ 0°/4°, mp: 718°, bp: 1500°. A sol salt. Sol in water and alc.

CONSENSUS REPORTS: Reported in EPA TSCA Inventory.

SAFETY PROFILE: Poison by intraperitoneal route. Mildly toxic by ingestion. When heated to decomposition it emits toxic fumes of Cl^-. See also CHLORIDES and RARE EARTHS.

E

EAG000   **CAS: 23315-05-1**   **HR: 3**
ELAIOMYCIN
mf: $C_{13}H_{26}N_2O_3$ mw: 258.41
PROP: Neutral yellow oil. Metabolite of
Streptomyces hepaticus (NATUAS 221,765,69).
SYN: d-threo-METHOXY-3-(1-OCTENYL-O,N,N-AZOXY)-2-BUTANOL
SAFETY PROFILE: Poison by intravenous
and subcutaneous routes. Questionable
carcinogen with experimental tumorigenic
data. Causes tumors of the brain. When
heated to decomposition it emits toxic
fumes of NO_x.

EAG100   **CAS: 2169-75-7**   **HR: 3**
ELARGIN
mf: $C_{23}H_{27}NO \cdot C_6H_8O_7$ mw: 525.65
SYNS: 1-α-H,5-α-H-TROPANE, 3-α-((10,11-DIHYDRO-5H-DIBENZO(AD)CYCLOHEPTEN-5-YL)OXY)-,
CITRATE (1:1) □ BRONTIN □ BRONTINA □ BRONTINE
□ BRONTISOL □ BS 6987 □ DEPTRIN □ DEPTROPINE
CITRATE □ DIBENZHEPTROPINE □ DIBENZ-
HEPTROPINE CITRATE □ 3-α-(10,11-DIHYDRO-5H-
DIBENZO(A,D)CYCLOHEPTEN-5-YLOXY)TROPANE
CITRATE □ 3-α-((10,11-DIHYDRO-5H-DIBENZO(A,D)-
CYCLOHEPTEN-5-YL)OXY)TROPAN DIHYDROGEN
CITRATE □ ELAMOL □ ELARGYL □ SU-BRONTINE □
TROPANE, 3-α-((10,11-DIHYDRO-5H-DIBENZO(A,D)-
CYCLOHEPTEN-5-YL)OXY)-, CITRATE (1:1)
SAFETY PROFILE: A poison by ingestion,
intraperitoneal, intravenous route. When
heated to decomposition it emits toxic
vapors of NO_x.

EAI000   **CAS: 549-18-8**   **HR: 3**
ELAVIL HYDROCHLORIDE
mf: $C_{20}H_{23}N \cdot ClH$ mw: 313.90
PROP: Minute crystals. Mp: 196–197°. Very
sol in H_2O.
SYNS: AMITID □ AMITRIL □ AMITRIPTYLINE
CHLORIDE □ AMITRYPTYLINE HYDROCHLORIDE □
DAMILEN HYDROCHLORIDE □ DEPREX □ 10,11-
DIHYDRO-N,N-DIMETHYL-5H-DIBENZO(a,d)-
CYCLOHEPTENE-Δ^{5,γ}-PROPYLAMINE HCL □ 3-(3-
DIMETHYLAMINOPROPYLIDENE)-1:2-4:5-

DIBENZOCYCLOHEPTA-1:4-DIENE □ 5-(3-DIMETHYL-
AMINOPROPYLIDENE)DIBENZO(a,d)(1,4)CYCLOHEPTA
DIENE HYDROCHLORIDE □ DOMICAL □ ELAVIL □
ENDEP □ LENTIZOL □ MIKETORIN □ PROHEPT-
ADIEN MONOHYDROCHLORIDE □ SAROTEN □
SAROTENE □ SK-AMITRIPTYLINE □ TRYPTIZOL □
TRYPTIZOL HYDROCHLORIDE
CONSENSUS REPORTS: Reported in EPA
TSCA Inventory.
SAFETY PROFILE: Poison by ingestion,
intraperitoneal, intravenous, and
subcutaneous routes. Human systemic
effects by ingestion: convulsions, respiratory
depression, changes in sleep, hallucinations,
muscle contractions, somnolence, blood
pressure decrease, coma, cyanosis, dyspnea,
and ataxia. An experimental teratogen.
Other experimental reproductive effects.
Mutation data reported. Used in the
treatment of depression. When heated to
decomposition it emits very toxic fumes of
HCl and NO_x. See also ELAVIL.

EAJ500   **CAS: 19526-81-9**   **HR: 2**
EMAZOL RED B
mf: $C_{18}H_{16}N_2O_{10}S_3 \cdot 2Na$ mw: 562.52
CONSENSUS REPORTS: Reported in EPA
TSCA Inventory.
SAFETY PROFILE: Questionable
carcinogen with experimental tumorigenic
data. When heated to decomposition it
emits very toxic fumes of NO_x, Na_2O, and
SO_x.

EAM000   **HR: 3**
EMETINE ANTIMONY IODIDE
PROP: Percentage composition = 34%
emetine and 14% antimony (AJTMAQ
10,249,30).
SYN: ANTIMONY EMETINE IODIDE
CONSENSUS REPORTS: Antimony and its
compounds are on the Community Right-
To-Know List.

OSHA PEL: TWA 0.5 mg(Sb)/m^3
ACGIH TLV: TWA 0.5 mg(Sb)/m^3;
Proposed: (inhalable fraction) 0.1 mg/m^3;
Not Classifiable as a Human Carcinogen)
NIOSH REL: (Antimony) TWA 0.5 mg/m^3
SAFETY PROFILE: Poison by ingestion.
Human systemic effects by ingestion:
nausea, vomiting, and other gastrointestinal
effects. When heated to decomposition it
emits very toxic fumes of I$^-$, NO$_x$, and Sb.
See also ANTIMONY COMPOUNDS,
EMETINE and IODIDES.

EAN000   CAS: 316-42-7   HR: 3
1-EMETINE DIHYDROCHLORIDE
mf: C$_{29}$H$_{40}$N$_2$O$_4$•2ClH   mw: 553.63
PROP: Needles. Mp: 255° (decomp); sinters
at 2°.
SYNS: AMEBICIDE □ EMETINE, DIHYDRO-
CHLORIDE □ (−)-EMETINE DIHYDROCHLORIDE □
EMETINE HYDROCHLORIDE □ NSC-33669
CONSENSUS REPORTS: EPA Extremely
Hazardous Substances List. Reported in
EPA TSCA Inventory.
SAFETY PROFILE: A poison by ingestion,
intraperitoneal, subcutaneous, and
intravenous routes. Human systemic effects:
diarrhea, distorted perceptions, dyspnea,
hallucinations, hypermotility, nausea or
vomiting. A human eye irritant. When
heated to decomposition it emits very toxic
fumes of Cl$^-$ and NO$_x$.

EAN525   CAS: 149950-60-7 HR: 3
EMIVIRINE
mf: C$_{17}$H$_{22}$N$_2$O$_3$ mw: 302.41
SYNS: 1-(ETHOXYMETHYL)-5-(1-METHYLETHYL)-6-
(PHENYLMETHYL)-2,4(1H,3H)-PYRIMIDINEDIONE □ 6-
BENZYL-1-(ETHOXYMETHYL)-5-ISOPROPYLURACIL □
I-EBU □ MKC-442 □ 2,4(1H,3H)-PYRIMIDINEDIONE, 1-
(ETHOXYMETHYL)-5-(1-METHYLETHYL)-6-
(PHENYLMETHYL)-
SAFETY PROFILE: Moderately toxic by
ingestion. When heated to decomposition it
emits toxic vapors of NO$_x$.

EAP000   CAS: 8015-30-3   HR: 3
ENAVID
mf: C$_{21}$H$_{26}$O$_2$•C$_{20}$H$_{26}$O$_2$ mw: 608.93

PROP: Mixture of 98.5% (17-α)-19-
norpregn-4-en-20-yn-3-one, 17-hydroxy-
and 1.5% (17-α)-19-norpregna-1,3,5(10)-
trien-20-yn-17-ol,3-methoxy- (IARC**
6,193,74).
SYNS: CONOVID □ CONOVID E □ ENIDREL □
ENOVID □ ENOVID-E □ ETHINYLESTRADIOL-3-
METHYL ETHER and NORETHYNODRED (1:50) □
MESTRANOL mixed with NORETHYNODREL □
NORETHANDROL □ NORETHYNODREL and
ETHINYLESTRADIOL-3-METHYL ETHER (50:1) □
NORETHYNODREL mixed with MESTRANOL
CONSENSUS REPORTS: IARC Cancer
Review: Animal Sufficient Evidence
IMEMDT 6,191,74. EPA Genetic
Toxicology Program.
SAFETY PROFILE: Confirmed carcinogen
producing liver tumors in women by
ingestion. Experimental carcinogenic,
neoplastigenic, and tumorigenic data.
Human reproductive effects by ingestion:
menstrual cycle changes or disorders;
abnormalities of the uterus, cervix, and
vagina; and changes in fertility. A human
teratogen that causes developmental
abnormalities of the urogenital system.
Experimental reproductive effects. Mutation
data reported. A steroid. When heated to
decomposition it emits acrid smoke and
fumes.

EAQ750   CAS: 115-29-7   HR: 3
ENDOSULFAN
mf: C$_9$H$_6$Cl$_6$O$_3$S   mw: 406.91
PROP: A mixture of 2 isomers, brown
crystals, nearly insol in water; sol in most org
solvs. Mp: (α) 106°, mp: (β) 212°, d: 1.745 @
20°/20°.
SYNS: BENZOEPIN □ BEOSIT □ BIO 5,462 □
CHLORTHIEPIN □ CRISULFAN □ CYCLODAN □
DEVISULPHAN □ ENDOCEL □ ENDOSOL □
ENDOSULPHAN □ ENSURE □ ENT 23,979 □ FMC 5462
□ 1,2,3,4,7,7-HEXACHLOROBICYCLO(2.2.1)HEPTEN-5,6-
BIOXYMETHYLENESULFITE □ α,β-1,2,3,4,7,7-HEXA-
CHLOROBICYCLO(2.2.1)-2-HEPTENE-5,6-BISOXY-
METHYLENE SULFITE □ HEXACHLOROHEXA-
HYDROMETHANO-2,4,3-BENZODIOXATHIEPIN-3-
OXIDE □ 6,7,8,9,10,10-HEXACHLORO-1,5,5a,6,9,9a-
HEXAHYDRO-6,9-METHANO-2,4,3-BENZODIOXA-
THIEPIN-3-OXIDE □ 1,4,5,6,7,7-HEXACHLORO-5-
NORBORNENE-2,3-DIMETHANOL CYCLIC SULFITE □

HILDAN □ HOE 2,671 □ INSECTOPHENE □ KOP-
THIODAN □ MALIX □ NCI-C00566 □ NIA 5462 □
NIAGARA 5,462 □ OMS 570 □ RCRA WASTE NUMBER
P050 □ SULFUROUS ACID, cyclic ester with 1,4,5,6,7,7-
HEXACHLORO-5-NORBORNENE-2,3-DIMETHANOL □
THIFOR □ THIMUL □ THIODAN □ THIOFOR □
THIOMUL □ THIONEX □ THIOSULFAN □ THIO-
SULFAN TIONEL □ TIOVEL

CONSENSUS REPORTS: EPA Extremely
Hazardous Substances List. NCI
Carcinogenesis Bioassay (feed); No
Evidence: mouse, rat NCITR* NCI-CG-
TR-62,77.
OSHA PEL: TWA 0.1 mg/m³ (skin)
ACGIH TLV: TWA 0.1 mg/m³ (skin); Not
Classifiable as a Human Carcinogen
SAFETY PROFILE: Poison by ingestion,
inhalation, skin contact, intraperitoneal, and
subcutaneous routes. Experimental
teratogenic and reproductive effects.
Questionable carcinogen with experimental
tumorigenic and neoplastigenic data. Human
systemic effects: convulsions, cyanosis.
Human mutation data reported. A central
nervous system stimulant producing
convulsions. A highly toxic organochlorine
pesticide that does not accumulate
significantly in human tissue. Absorption is
normally slow, but is increased by alcohols,
oil, and emulsifiers. When heated to
decomposition it emits toxic fumes of Cl⁻
and SO$_x$. See also CHLORIDES and
SULFITES.

EAR000   CAS: 145-73-3   HR: 3
ENDOTHAL
mf: C$_8$H$_{10}$O$_5$ mw: 186.18
PROP: Solid. Mp: 144°. Sol in water.
SYNS: AQUATHOL □ 3,6-ENDOOXOHEXAHYDRO-
PHTHALIC ACID □ ENDOTHALL □ ENDOTHAL
TECHNICAL □ 3,6-ENDOXOHEXAHYDROPHTHALIC
ACID □ 3,6-endo-EPOXY-1,2-CYCLOHEXANEDI-
CARBOXYLIC ACID □ HEXAHYDRO-3,6-endo-
OXYPHTHALIC ACID □ HYDOUT □ HYDROTHAL-47 □
7-OXABICYCLO(2.2.1)HEPTANE-2,3-DICARBOXYLIC
ACID □ RCRA WASTE NUMBER P088 □ TRI-
ENDOTHAL
SAFETY PROFILE: Poison by ingestion.
Very irritating to skin, eyes, and mucous
membranes. Causes diarrhea. When heated

to decomposition it emits acrid smoke and
fumes.

EAT500   CAS: 72-20-8   HR: 3
ENDRIN
mf: C$_{12}$H$_8$Cl$_6$O   mw: 380.90
PROP: White crystals. Mp: decomp @ 200°.
Sol in Me$_2$CO, C$_6$H$_6$, xylene; sltly sol in CCl$_4$
and hexane. IDLH 2 mg/m³.
SYNS: COMPOUND 269 □ ENDREX □ ENDRINE
(FRENCH) □ ENT 17,251 □ HEXACHLOROEPOXY-
OCTAHYDRO-endo,endo-DIMETHANONAPHT-HALENE
□ 3,4,5,6,9,9-HEXACHLORO-1a,2,2a,3,6,6a,7,7aOCTA-
HYDRO-2,7:3,6-DIMETHANONAPHTH(2,3-b)OXIRENE □
HEXADRIN □ MENDRIN □ NCI-C00157 □ NENDRIN □
RCRA WASTE NUMBER P051
CONSENSUS REPORTS: IARC Cancer
Review: Group 3 IMEMDT 7,56,87; Animal
Inadequate Evidence IMEMDT 5,157,74;
Human Inadequate Evidence IMEMDT
5,157,74. NCI Carcinogenesis Bioassay
(feed); No Evidence: mouse, rat NCITR*
NCI-CG-TR-12,79. EPA Genetic
Toxicology Program. EPA Extremely
Hazardous Substances List.
OSHA PEL: TWA 0.1 mg/m³ (skin)
ACGIH TLV: TWA 0.1 mg/m³ (skin); Not
Classifiable as a Human Carcinogen
DFG MAK: 0.1 mg/m³
SAFETY PROFILE: Poison by ingestion,
skin contact, and intravenous routes.
Experimental teratogenic and reproductive
effects. Questionable carcinogen. Mutation
data reported. A central nervous system
stimulant. Highly toxic to birds, fish, and
humans. Many cases of fatal poisoning have
been attributed to it. Does not accumulate
in human tissue. In humans, ingestion of 1
mg/kg has caused symptoms. A dangerous
fire hazard. Mixtures with parathion dissolve
very exothermically in petroleum solvents
and may cause an air-vapor explosion. See
also ALDRIN.

EAT900   CAS: 13838-16-9   HR: 2
ENFLURANE
mf: C$_3$H$_2$ClF$_5$O   mw: 184.50
PROP: Bp: 56.5°, vap p: 188.6 mm @ 22°.
SYNS: ANESTHETIC COMPOUND No. 347 □ 2-
CHLORO-1-(DIFLUOROMETHOXY)-1,1,2-

E

TRIFLUOROETHANE □ 2-CHLORO-1,1,2-TRIFLUORO-ETHYL DIFLUOROMETHYL ETHER □ COMPOUND 347 □ ETHRANE □ METHYLFLURETHER □ NSC-115944 □ OHIO 347

CONSENSUS REPORTS: Reported in EPA TSCA Inventory.

ACGIH TLV: TWA 75 ppm; Not Classifiable as a Human Carcinogen

DFG MAK: 20 ppm

NIOSH REL: (Waste Anesthetic Gases and Vapors) CL 2 ppm/1H

SAFETY PROFILE: Mildly toxic by inhalation, ingestion, and subcutaneous routes. Human systemic effects by inhalation: decreased urine volume or anuria. An experimental teratogen. Experimental reproductive effects. Human mutation data reported. An eye irritant. Questionable carcinogen with experimental carcinogenic data. An anesthetic. When heated to decomposition it emits very toxic fumes of F^- and Cl^-. See also ETHERS.

EAV500   CAS: 33419-42-0   HR: 3
EPE

mf: $C_{29}H_{32}O_{13}$ mw: 588.61

PROP: Crystals from MeOH. Mp: 236–251°.

SYNS: DEMETHYL-EPIODOPHYLLOTOXIN ETHYLIDENE GLUCOSIDE □ 4-DEMETHYLEPIO-DOPHYLLOTOXIN-β,d-ETHYLIDENEGLUCOSIDE □ 4'-DEMETHYLEPIPODOPHYLLOTOXIN-9-(4,6-O-ETHYLIDENE-β-d-GLUCOPYRANOSIDE □ 4'-DEMETHYLEPIPODOPHYLLOTOXIN ETHYLIDENE-β,d-GLUCOSIDE □ 4-DEMETHYL-EPIPODOPHYLLO-TOXIN-β,d-ETHYLIDEN-GLUCOSIDE □ 4'-O-DEMETH-YL-1-O-(4,6-O-ETHYLIDENE-β,d-GLUCOPYRANOSYL)-EPIPODOPHYLLOTOXIN □ ETOPOSIDE □ NK 171 □ NSC-141540 □ VEPESID □ VP 16213

SAFETY PROFILE: Poison by ingestion, intraperitoneal, intravenous, and subcutaneous routes. An experimental teratogen. Human systemic effects by ingestion and inhalation: agranulocytosis, aplastic anemia, and other changes in bone marrow. Experimental reproductive effects. Human mutation data reported. When heated to decomposition it emits acrid smoke and fumes.

EAW000   CAS: 299-42-3   HR: 3
EPHEDRINE

mf: $C_{10}H_{15}NO$ mw: 165.26

PROP: White granules. Mp: 79° (dl), mp: 40° (l), bp: 225° (decomp). Sol in ether and chloroform.

SYNS: BIOPHEDRIN □ ECIPHIN □ EFEDRIN □ EPHEDRAL □ EPHEDRATE □ EPHEDREMAL □ EPHEDRIN □ l-EPHEDRINE □ l(−)-EPHEDRINE □ EPHEDRITAL □ EPHEDROL □ EPHEDROSAN □ EPHEDROTAL □ EPHEDSOL □ EPHENDRONAL □ EPHOXAMIN □ FEDRIN □ α-HYDROXY-β-METHYL AMINE PROPYLBENZENE □ 1-HYDROXY-2-METHYLAMINO-1-PHENYLPROPANE □ I-SEDRIN □ ISOFEDROL □ KRATEDYN □ MANADRIN □ MANDRIN □ (−)-α-(1-METHYLAMINOETHYL)BENZYL ALCOHOL □ 1-α-(1-METHYLAMINOETHYL)BENZYL ALCOHOL □ 1-2-METHYLAMINO-1-PHENYLPROPAN-OL □ N-METHYLNOREPHEDRINE □ NASOL □ 1-PHENYL-2-METHYLAMINOPROPANOL □ SANEDRINE □ VENCIPON □ ZEPHROL

CONSENSUS REPORTS: Reported in EPA TSCA Inventory.

SAFETY PROFILE: A human poison by an unspecified route. An experimental poison by intravenous, subcutaneous, intramuscular, and intraperitoneal routes. Moderately toxic by ingestion and parenteral routes. Causes rapid pulse, rise in blood pressure, and other actions similar to epinephrine. An experimental teratogen. Used in production of drugs of abuse. Has been known to cause allergic sensitization. When heated to decomposition it emits toxic fumes of NO_x.

EAW500   CAS: 50-98-6   HR: 3
EPHEDRINE HYDROCHLORIDE

mf: $C_{10}H_{15}NO \cdot ClH$ mw: 201.72

PROP: A solid. Mp: 218°.

SYNS: BENZENEMETHANOL, α-(1-(METHYLAMINO)ETHYL)-, HYDROCHLORIDE, (R-(R*,S*))- □ EPHEDRINE HYDROCHLORIDE □ l-EPHEDRINE, HYDROCHLORIDE □ N-METHYL-β-OXY-β-PHENYLISOPROPYLAMINHYDROCHLORID

CONSENSUS REPORTS: Reported in EPA TSCA Inventory.

SAFETY PROFILE: A poison by ingestion, intraperitoneal, subcutaneous, and intravenous routes. Human systemic effects by intradermal route: skin effects. When

E

heated to decomposition it emits very toxic fumes of Cl⁻ and NO_x. See also EPHEDRINE.

EAW995   CAS: 24221-86-1   HR: 3
d-EPHEDRINE HYDROCHLORIDE

mf: $C_{10}H_{15}NO \cdot ClH$ mw: 201.72
PROP:
PROP: A solid. Mp: 217–218°.
CONSENSUS REPORTS: Reported in EPA TSCA Inventory.
SAFETY PROFILE: A poison by intraperitoneal, subcutaneous, and intravenous routes. When heated to decomposition it emits very toxic fumes of HCl and NO_x. See also EPHEDRINE.

EAX000   CAS: 50-98-6   HR: 3
l-EPHEDRINE HYDROCHLORIDE

mf: $C_{10}H_{15}NO \cdot ClH$ mw: 201.72
PROP: Crystals. Mp: 187°–188°. One gram dissolves in 4 ml water.
SYNS: EPHEDRINE HYDROCHLORIDE □ (–)-EPHEDRINE HYDROCHLORIDE □ (R-(R*,S*))-α-(1-(METHYLAMINO)ETHYL)BENZENEMETHANOL HYDROCHLORIDE
CONSENSUS REPORTS: Reported in EPA TSCA Inventory.
SAFETY PROFILE: Poison by ingestion, intraperitoneal, intravenous, subcutaneous, and intramuscular routes. Moderately toxic by parenteral route. Human systemic effects by intradermal route: local anesthetic. When heated to decomposition it emits very toxic fumes of HCl and NO_x. See also EPHEDRINE.

EAX500   CAS: 134-71-4   HR: 3
dl-EPHEDRINE HYDROCHLORIDE

mf: $C_{10}H_{15}NO \cdot ClH$ mw: 201.72
PROP: A solid. Mp: 188–189.5°.
SYNS: EPHETONIN □ EPHETONINE □ dl-α-(1-(METHYLAMINO)ETHYL) BENZYL ALCOHOL HYDROCHLORIDE □ 1-PHENYL-2-METHYL-AMINOPROPANOL-1 □ RACEPHEDRINE HYDROCHLORIDE
CONSENSUS REPORTS: Reported in EPA TSCA Inventory.
SAFETY PROFILE: Poison by subcutaneous, intravenous, and

intraperitoneal routes. Moderately toxic by ingestion. Human systemic effects: cardiac changes, nausea or vomiting, sweating. When heated to decomposition it emits very toxic fumes of HCl and NO_x. See also EPHEDRINE.

EAY500   CAS: 134-72-5   HR: 3
l-EPHEDRINE SULFATE

mf: $C_{20}H_{30}N_2O_2 \cdot H_2O_4S$ mw: 428.60
PROP: White microcrystalline powder. Mp: 245°–248° (decomposes). Sol in water: >=100 mg/mL @ 20°.
SYNS: ISOFEDROL □ 1-α-(1-(METHYLAMINO)ETHYL)-BENZYL ALCOHOL SULFATE □ NCI-C55652 □ 1-PHENYL-2-METHYLAMINE-PROPANOL-1-SULFATE
CONSENSUS REPORTS: NTP Carcinogenesis Studies (feed); No Evidence: mouse, rat NTPTR* NTP-TR-307,86. Reported in EPA TSCA Inventory.
SAFETY PROFILE: Poison by intravenous, intraperitoneal, and subcutaneous routes. Moderately toxic by ingestion. Human systemic effects by intravenous route: increased pulse rate and blood pressure. When heated to decomposition it emits very toxic fumes of NO_x and SO_x. See also EPHEDRINE.

EAZ500   CAS: 106-89-8   HR: 3
EPICHLOROHYDRIN

DOT: UN 2023
mf: C_3H_5ClO mw: 92.53
PROP: Colorless, mobile liquid; irritating chloroform-like odor. Bp: 117.9°, fp: −57.1°, flash p: 105.1°F (OC) (40°C), mp: −25.6°C, d: 1.1761 @ 20°/20°, vap press: 10 mm @ 16.6°, vap d: 3.29. IDLH 75 ppm.
SYNS: 1-CHLOOR-2,3-EPOXY-PROPAAN (DUTCH) □ 1-CHLOR-2,3-EPOXY-PROPAN (GERMAN) □ 1-CHLORO-2,3-EPOXYPROPANE □ 3-CHLORO-1,2-EPOXYPRO-PANE □ epi-CHLOROHYDRIN □ (CHLOROMETHYL)-ETHYLENE OXIDE □ CHLOROMETHYLOXIRANE □ 2-(CHLOROMETHYL)OXIRANE □ CHLOROPROPYLENE OXIDE □ γ-CHLOROPROPYLENE OXIDE □ 3-CHLORO-1,2-PROPYLENE OXIDE □ 1-CLORO-2,3-EPOSSIPROPANO (ITALIAN) □ ECH □ EPICHLOOR-HYDRINE (DUTCH) □ EPICHLORHYDRIN (GERMAN) □ EPICHLORHYDRINE (FRENCH) □ EPICHLOROP-HYDRIN □ α-EPICHLOROHYDRIN □ (dl)-α-

EPICHLOROHYDRIN □ EPICHLOROHYDRYNA (POLISH) □ EPICLORIDRINA (ITALIAN) □ 1,2-EPOXY-3-CHLOROPROPANE □ 2,3-EPOXYPROPYL CHLORIDE □ GLYCEROL EPICHLORHYDRIN □ RCRA WASTE NUMBER U041 □ SKEKhG

CONSENSUS REPORTS: NTP 10th Report on Carcinogens. IARC Cancer Review: Group 2A IMEMDT 7,202,87; Animal Sufficient Evidence IMEMDT 11,131,76. EPA Genetic Toxicology Program. Community Right-To-Know List. EPA Extremely Hazardous Substances List. Reported in EPA TSCA Inventory.

OSHA PEL: TWA 2 ppm (skin)

ACGIH TLV: TWA 0.5 ppm (skin); Animal Carcinogen

DFG MAK: DFG TRK: Animal Carcinogen, Suspected Human Carcinogen

NIOSH REL: Minimize exposure

DOT CLASSIFICATION: 6.1; Label: Poison

SAFETY PROFILE: Confirmed carcinogen with experimental carcinogenic data. Poison by ingestion, skin contact, intravenous, and intraperitoneal routes. Moderately toxic by inhalation. An experimental teratogen. Other experimental reproductive effects. Human systemic effects by inhalation: respiratory, nose, and eyes. Human mutation data reported. A skin and eye irritant. A sensitizer. Flammable liquid when exposed to heat or flame. Explosive reaction with aniline. Reaction with trichloroethylene forms the explosive dichloroacetylene. Ignition on contact with potassium tert-butoxide. Violent reaction with sulfuric acid or isopropylamine. Exothermic polymerization on contact with strong acids, caustic alkalies, aluminum, aluminum chloride, iron(III) chloride, or zinc. When heated to decomposition it emits toxic fumes of Cl⁻.

**EBA100   CAS: 516-95-0   HR: 2
EPIDEHYDROCHOLESTERIN**
mf: $C_{27}H_{48}O$ mw: 388.75
PROP: Needles from EtOH. Mp: 185–186°.
SYNS: CHOLESTAN-3-OL, (3-α-5α-)-(9CI) □ 5-α-CHOLESTAN-3-α-OL (8CI) □ α-CHOLESTANOL (7CI) □ epi-CHOLESTANOL □ EPICHOLESTANOL

CONSENSUS REPORTS: Reported in EPA TSCA Inventory.

SAFETY PROFILE: Moderately toxic by ingestion. An experimental teratogen. Experimental reproductive effects. When heated to decomposition it emits acrid smoke and irritating fumes.

**EBD700   CAS: 2104-64-5   HR: 3
EPN**
mf: $C_{14}H_{14}NO_4PS$ mw: 323.32
PROP: Liquid or pale-yellow crystals with an aromatic odor. D: 1.268 @ 25°, mp: 36°. Nearly insol in water; sol in org solvs. IDLH 5 mg/m³.
SYNS: O-AETHYL-O-n(4-NITROPHENYL)-PHENYL-MONOTHIOPHOSPHONAT (GERMAN) □ ENT 17,798 □ O-ESTER-p-NITROPHENOL with O-ETHYL PHENYL PHOSPHONOTHIOATE □ ETHOXY-4-NITROPHEN-OXYPHENYLPHOSPHINE SULFIDE □ O-ETHYL-O-((4-NITROFENYL)-FENYL)-MONOTHIOFOSFONAAT (DUTCH) □ O-ETHYL O-(4-NITROPHENYL)BENZENE-THIONOPHOSPHONATE □ ETHYL-p-NITROPHENYL BENZENETHIONOPHOSPHONATE □ ETHYL-p-NITROPHENYL BENZENETHIOPHOSPHATE □ ETHYL-p-NITROPHENYL BENZENETHIOPHOSPHON-ATE □ ETHYL-p-NITROPHENYL PHENYLPHOS-PHONOTHIOATE □ O-ETHYL-O-(4-NITROPHENYL) PHENYLPHOSPHONOTHIOATE □ O-ETHYL-O-p-NITROPHENYL PHENYLPHOSPHONOTHIOLATE □ O-ETHYL-O-p-NITROPHENYL PHENYLPHOSPHORO-THIOATE □ ETHYL-p-NITROPHENYL THIO-NOBENZENEPHOSPHATE □ ETHYL-p-NITROPHENYL THIONOBENZENEPHOSPHONATE □ O-ETHYL-PHENYL-p-NITROPHENYL THIOPHOSPHONATE □ O-ETIL-O-((4-NITRO-FENIL)-FENIL)-MONOTIOFOS-FONATO (ITALIAN) □ PHENYLTHIOPHOSPHONATE de O-ETHYLE et O-4-NITROPHENYLE (FRENCH) □ PIN □ SANTOX □ THIONOBENZENEPHOSPHONIC ACID ETHYL-p-NITROPHENYL ESTER

CONSENSUS REPORTS: EPA Farm Worker Field Reentry FEREAC 39, 16888,74. EPA Extremely Hazardous Substances List.

OSHA PEL: TWA 0.5 mg/m³ (skin)

ACGIH TLV: TWA 0.1 mg/m³ (skin); Not Classifiable as a Human Carcinogen

DFG MAK: 0.5 mg/m³

SAFETY PROFILE: Poison by ingestion, skin contact, and intraperitoneal routes. An experimental teratogen. A cholinesterase inhibitor. This material is extremely hazardous on contact with skin, inhalation, or ingestion. A highly toxic insecticide. When heated to decomposition it emits highly toxic fumes of SO_x, PO_x, NO_x, and phosphine. See also PARATHION, NITRO COMPOUNDS OF AROMATIC HYDROCARBONS, PHOSPHINE, and SULFIDES.

EBH525   HR: D
EPOXIDIZED SOYBEAN OIL
PROP: Iodine number maximum of 6, oxirane oxygen minimum of 6.0 percent.
SAFETY PROFILE: When heated to decomposition it emits acrid smoke and irritating fumes.

EBJ500   CAS: 930-22-3   HR: 3
3,4-EPOXY-1-BUTENE
mf: C_4H_6O mw: 70.10
PROP: Liquid. Mp: $-135°$, bp: $67°$, flash p: $<-58°F$ (CC), d: 0.869, autoign temp: $806°F$, vap d: 2.41.
SYNS: BUTADIENE MONOEPOXIDE □ BUTADIENE MONOXIDE □ 1,2-EPOXYBUTENE-3 □ VINYLOXIRANE
CONSENSUS REPORTS: Reported in EPA TSCA Inventory.
SAFETY PROFILE: A poison by intraperitoneal route. Questionable carcinogen with experimental tumorigenic data. Mutation data reported. A very dangerous fire hazard when exposed to heat or flame; can react with oxidizing materials. To fight fire, use CO_2, dry chemical, water spray. When heated to decomposition it emits acrid smoke and fumes.

EBQ700   CAS: 4016-11-9   HR: 3
1,2-EPOXY-3-ETHOXYPROPANE
DOT: UN 2752
mf: $C_5H_{10}O_2$ mw: 102.15
PROP: Bp: $160°-163°$ @ 14-15 mm.
SYNS: (ETHOXYMETHYL)OXIRANE □ ETHYL GLYCIDYL ETHER □ OXIRANE, (ETHOXYMETHYL)-(9CI) □ PROPANE, 1,2-EPOXY-3-ETHOXY-
CONSENSUS REPORTS: Reported in EPA TSCA Inventory.

DOT CLASSIFICATION: 3; Label: Flammable Liquid
SAFETY PROFILE: Mutation data reported. A flammable liquid. When heated to decomposition it emits acrid smoke and irritating vapors.

EBR000   CAS: 96-09-3   HR: 3
1,2-EPOXYETHYLBENZENE
mf: C_8H_8O mw: 120.16
PROP: Colorless liquid. Bp: $194.2°$, flash p: $165°F$ (OC), fp: $-36.7°$, d: 1.0469 @ $25°/4°$, vap d: 4.14.
SYNS: EPOXYETHYLBENZENE (8CI) □ EPOXY-STYRENE □ α,β-EPOXYSTYRENE □ NCI-C54977 □ PHENETHYLENE OXIDE □ 1-PHENYL-1,2-EPOXYETHANE □ PHENYLETHYLENE OXIDE □ PHENYLOXIRANE □ 1-PHENYLOXIRANE □ 2-PHENYLOXIRANE □ STYRENE EPOXIDE □ STYRENE OXIDE □ STYRENE-7,8-OXIDE □ STYRYL OXIDE
CONSENSUS REPORTS: NTP 10th Report on Carcinogens. IARC Cancer Review: Group 2A IMEMDT 7,56,87; Animal Sufficient Evidence IMEMDT 36,245,85; Human No Adequate Data IMEMDT 36,245,85. Reported in EPA TSCA Inventory. EPA Genetic Toxicology Program.
SAFETY PROFILE: Confirmed carcinogen with experimental carcinogenic, tumorigenic, and teratogenic data. Moderately toxic by ingestion, inhalation, skin contact, and intraperitoneal routes. Experimental reproductive effects. Human mutation data reported. A skin and eye irritant. Flammable when exposed to heat, flame, or oxidizers; can react with oxidizing materials. To fight fire, use foam, CO_2, dry chemical. When heated to decomposition it emits acrid smoke and fumes.

EBU100   CAS: 68071-23-8   HR: 1
EPOXYGUAIENE
mf: $C_{15}H_{24}O$ mw: 220.39
PROP: Flagrance and flavor.
SYNS: AZULENE, 1,2,3,4,5,6,7,8-OCTAHYDRO-1,4-DIMETHYL-7-(1-METHYLETHYLIDENE)-, MONOEPOXIDE □ 1,2,3,4,5,6,7,8-OCTAHYDRO-1,4-DIMETHYL-7-(1-METHYLETHYLIDENE)AZULENE-MONOEPOXIDE

CONSENSUS REPORTS: Reported in EPA TSCA Inventory.

SAFETY PROFILE: Low toxicity by ingestion and skin contact. A skin irritant. When heated to decomposition it emits acrid smoke and irritating fumes.

EBW500   CAS: 1024-57-3   HR: 3
EPOXYHEPTACHLOR
mf: $C_{10}H_5Cl_7O$ mw: 389.30
SYNS: ENT 25,584 □ HCE □ HEPTACHLOR EPOXIDE (USDA) □ 1,4,5,6,7,8,8-HEPTACHLORO-2,3-EPOXY-2,3,3a,4,7,7a-HEXAHYDRO-4,7-METHANOINDENE □ 1,4,5,6,7,8,8-HEPTACHLORO-2,3-EPOXY-3a,4,7,7a-TETRAHYDRO-4,7-METHANOINDAN □ 2,3,4,5,6,7,7-HEPTACHLORO-1a,1b,5,5a,6,6a-HEXAHYDRO-2,5-METHANO-2H-INDENO(1,2-b)OXIRENE □ VELSICOL 53-CS-17

CONSENSUS REPORTS: IARC Cancer Review: Human Inadequate Evidence IMEMDT 20,129,79; Animal Inadequate Evidence IMEMDT 5,173,74; Animal Limited Evidence IMEMDT 20,129,79. EPA Genetic Toxicology Program.
ACGIH TLV: TWA 0.05 mg/m³ (skin); Animal Carcinogen

SAFETY PROFILE: Confirmed carcinogen with experimental carcinogenic data. Poison by ingestion and intravenous routes. Human mutation data reported. When heated to decomposition it emits toxic fumes of Cl⁻. See also HEPTACHLOR.

EBX500   CAS: 7320-37-8   HR: 2
1,2-EPOXYHEXADECANE
mf: $C_{16}H_{32}O$ mw: 240.48
SYNS: HEXADECENE EPOXIDE □ NCI-C55538
CONSENSUS REPORTS: Reported in EPA TSCA Inventory.

SAFETY PROFILE: Questionable carcinogen with experimental tumorigenic data. Mutation data reported. When heated to decomposition it emits acrid smoke and fumes.

ECH500   CAS: 106-90-1   HR: 3
2,3-EPOXYPROPYL ACRYLATE
mf: $C_6H_8O_3$ mw: 128.14
PROP: Insol in water. Bp: 57.2° @ 2 mm, flash p: 141°F (OC), d: 1.1, vap d: 4.4.

SYNS: 2,3-EPOXY-1-PROPANOL ACRYLATE □ 2,3-EPOXYPROPYL ESTER ACRYLIC ACID □ GLYCIDYL ACRYLATE □ GLYCIDYL PROPENATE □ 2-PROPENOIC ACID OXIRANYLMETHYL ESTER
CONSENSUS REPORTS: Reported in EPA TSCA Inventory.

SAFETY PROFILE: A poison by ingestion, inhalation, and skin contact. Mutation data reported. A skin and severe eye irritant. Flammable liquid when exposed to heat or flame. Can react vigorously with oxidizers. To fight fire, use foam, dry chemical, CO_2. When heated to decomposition it emits acrid smoke and fumes. See also ESTERS.

ECL500   HR: D
EPOXY RESINS, CURED
SAFETY PROFILE: Most cured resins have little or no toxicity. If curing is incomplete there may be residues of highly toxic curing agents such as the organic amines: m-phenylene diamine, diethylene triamine, tetraethylene pentamine, and hexamethylene tetramine, as well as phthalic anhydride and related compounds. When heated to decomposition they emit highly toxic fumes. See also various epoxy hardeners and POLYMERS, INSOLUBLE.

ECM500   HR: 3
EPOXY RESINS, UNCURED
SYN: POLYMERS of EPICHLOROHYDRIN and 2,2-BIS(4-HYDROXYPHENYL)PIPERAZINE
SAFETY PROFILE: Animal experiments have shown disturbed blood formation. The degree of toxicity of uncured epoxy resins varies and is partly dependent on the extent of unreacted curing agents. See also other epoxy resin entries and POLYMERS, INSOLUBLE. When heated to decomposition they emit acrid smoke and fumes.

ECU750   CAS: 12126-59-9   HR: 3
EQUIGYNE
PROP: Conjugated forms of natural mixed estrogens, principally sodium estrone sulfate and sodium equilin sulfate, or synthetic estrogen piperazine estrone sulfate.

SYNS: AMNESTROGEN □ CES □ CLIMESTRONE □ CO-ESTRO □ CONEST □ CONESTRON □ CONJES □ CONJUGATED ESTROGENS □ CONJUTABS □ EQUIGYNE □ ESTRATAB □ ESTRIFOL □ ESTROATE □ ESTROCON □ ESTROMED □ ESTROPAN □ EVEX □ FEMACOID □ FEMEST □ FEM H □ FEMOGEN □ FORMATRIX □ GANEAKE □ GENISIS □ GLYESTRIN □ KESTRIN □ MENEST □ MENOGEN □ MENOTAB □ MENOTROL □ MILPREM □ MSMED □ NEO-ESTRONE □ NOVOCONESTRON □ OESTRILIN □ OESTRO-FEMINAL □ OESTROPAK MORNING □ OVEST □ PALOPAUSE □ PAR ESTRO □ PMB □ PREMARIN □ PRESOMEN □ PROMARIT □ SK-ESTROGENS □ SODESTRIN-H □ TAG-39 □ THEOGEN □ TRANSANNON □ TROCOSONE □ ZESTE

CONSENSUS REPORTS: IARC Cancer Review: Animal Limited Evidence IMEMDT 7,283,87.

SAFETY PROFILE: Suspected human carcinogen producing tumors of the vascular system and liver. Human reproductive effects: changes in female fertility. When heated to decomposition it emits toxic fumes of Na_2O. See also individual components.

ECW520   CAS: 16680-47-0   HR: 3
EQUILIN SODIUM SULFATE
mf: $C_{18}H_{20}O_5S \cdot Na$ mw: 371.43

PROP: Buff colored powder. Sol in water.

SYNS: EQUILIN, SULFATE, SODIUM SALT (6CI) □ ESTRA-1,3,5(10),7-TETRAEN-17-ONE, 3-HYDROXY-, HYDROGEN SULFATE SODIUM SALT (8CI) □ ESTRA-1,3,5(10),7-TETRAEN-17-ONE, 3-(SULFOOXY)-, SODIUM SALT □ SODIUM EQUILIN 3-MONOSULFATE □ SODIUM EQUILIN SULFATE

CONSENSUS REPORTS: NTP 10th Report on Carcinogens.

SAFETY PROFILE: Confirmed carcinogen. When heated to decomposition it emits toxic fumes of SO_x.

ECX500   CAS: 10138-41-7   HR: 3
ERBIUM CHLORIDE
mf: Cl_3Er mw: 273.61

PROP: Hygroscopic pink crystals. Mp: 776°. Sol in H_2O and EtOH.

SYN: ERBIUM TRICHLORIDE

CONSENSUS REPORTS: Reported in EPA TSCA Inventory.

SAFETY PROFILE: Poison by intraperitoneal and subcutaneous routes. Moderately toxic by ingestion. When heated to decomposition it emits toxic fumes of Cl^-. See also RARE EARTHS.

ECY500   CAS: 10168-80-6   HR: 3
ERBIUM(III) NITRATE (1:3)
mf: $N_3O_9 \cdot Er$ mw: 353.29

PROP: Reddish crystals or hygroscopic pink solid. Sol in H_2O and EtOH.

SYN: NITRIC ACID, ERBIUM (3+) SALT

CONSENSUS REPORTS: Reported in EPA TSCA Inventory.

SAFETY PROFILE: Poison by intravenous and intraperitoneal routes. When heated to decomposition it emits toxic fumes of NO_x. See also RARE EARTHS and NITRATES.

EDC500   CAS: 379-79-3   HR: 3
ERGOTAMINE TARTRATE
mf: $C_{66}H_{70}N_{10}O_{10} \cdot C_4H_6O_6$ mw: 1313.56

PROP: A solid. Mp: 203° (decomp). Analgesic specific for migraine.

SYNS: ERGAM □ ERGATE □ ERGOMAR □ ERGOSTAT □ ERGOTAMINE BITARTRATE □ ERGOTARTRATE □ ETIN □ EXMIGRA □ FEMERGIN □ GOTAMINE TARTRATE □ GYNERGEN □ LINGRAINE □ LINGRAN □ NEO-ERGOTIN □ RIGETAMIN □ SECAGYN □ SECUPAN

CONSENSUS REPORTS: Reported in EPA TSCA Inventory. EPA Genetic Toxicology Program. EPA Extremely Hazardous Substances List.

SAFETY PROFILE: Poison by ingestion, intravenous, and subcutaneous routes. An experimental teratogen. Human systemic effects by ingestion: hallucinations, distorted perceptions, convulsions, nausea or vomiting, blood pressure elevation. Experimental reproductive effects. Human mutation data reported. Used in production of drugs of abuse. When heated to decomposition it emits toxic fumes of NO_x. See also ERGOT.

EDC650   **CAS: 12510-42-8**   **HR: 3**
ERIONITE
PROP: Transparent to translucent colered cryastal in white, green, gray, orange. Distinct cleavage.
DFG MAK: Confirmed Human Carcinogen
SAFETY PROFILE: Confirmed carcinogen with experimental carcinogenic and tumorigenic data.

EDC700   **CAS: 66733-21-9**   **HR: 3**
ERIONITE (CAKNA
mf: $Al_2O_{18}Si_7 \cdot 1/2Ca \cdot 7H_2O \cdot 1/2Na$ mw: 715.68
CONSENSUS REPORTS: NTP 10th Report on Carcinogens, 2000:Known to be human carcinogen. IARC Cancer Review: Group 1 IMSUDL 7,203,1987; Human Sufficient Evidence IMEMDT 42,225,1987; Animal Sufficient Evidence IMEMDT 42,225,1987.
SAFETY PROFILE: Confirmed human carcinogen.

EDE600   **HR: D**
ERYTHORBIC ACID
PROP: White or sltly yellow crystals or powder. Mp: 164–171° (decomp). Sol in water, alc; sltly sol in glycerin.
SYN: d-ARABOASCORBIC ACID
SAFETY PROFILE: When heated to decomposition it emits acrid smoke and irritating fumes.

EDH500   **CAS: 114-07-8**   **HR: 3**
ERYTHROMYCIN
mf: $C_{37}H_{67}NO_{13}$ mw: 734.05
PROP: White or sltly yellow, crystalline powder; odorless. Crystals from Me_2CO (aq) or $CHCl_3$. Mp: 136–140°, mp: 190–193° (double mp). Freely sol in alc, chloroform, and ether; very sltly sol in water.
SYNS: DOTYCIN □ EM □ E-MYCIN □ ERYCIN □ ERYTHROCIN □ ERYTHROGRAN □ ERYTHROGUENT □ ERYTHROMYCIN A □ ILOTYCIN □ PANTOMICINA □ PROPIOCINE □ ROBIMYCIN
CONSENSUS REPORTS: EPA Genetic Toxicology Program.
SAFETY PROFILE: Poison by intravenous and intramuscular routes. Moderately toxic

by ingestion, intraperitoneal, and subcutaneous routes. An experimental teratogen. Other experimental reproductive effects. Mutation data reported. When heated to decomposition it emits toxic fumes of NO_x.

EDM000   **CAS: 20977-05-3**   **HR: 3**
ESCIN, SODIUM SALT
mf: $C_{55}H_{85}O_{24} \cdot Na$ mw: 1153.39
PROP: A mixture of saponins occurring in the seed of the horse chestnut tree (ARZNAD 12,815,62).
SYNS: A-4760 □ AESCIN SODIUM SALT □ AESCUSAN SODIUM SALT □ Na-AESCINAT □ REPARIL SODIUM SALT □ SODIUM AESCINATE
SAFETY PROFILE: Poison by ingestion, intravenous, intraperitoneal, and subcutaneous routes. Experimental teratogenic and reproductive effects. When heated to decomposition it emits toxic fumes of Na_2O. See also SAPONIN and other escin entries.

EDN100   **HR: D**
ESTERASE-LIPASE
PROP: Derived from *Mucor miehei*.
SAFETY PROFILE: When heated to decomposition it emits acrid smoke and irritating fumes.

EDN500   **HR: D**
ESTERS
PROP: A large group of organic compounds that correspond structurally to salts in inorganic chemistry. They are considered to be derived from acids by the replacement of hydrogen by an organic alkyl radical. Esters of acetic acid are called acetates and esters of carbonic acid are called carbonates. The esterification of a fatty acid RCOOH, by an alcohol R'OH, yields the fatty ester RCOOR'. The most common alcohol used is methanol, yielding the methyl ester $RCOOCH_3$. The methyl esters of fatty acids have higher vapor pressures than the corresponding acids.

E

SAFETY PROFILE: No general statement can be made as to the toxicity of esters. Many are highly volatile and hence can act as asphyxiants or narcotics. Skin contact, as well as inhalation, may be an important route of absorption for those esters that are volatile and have a high solvent action. The degree of toxicity ranges from mildly toxic to poison. Esters generally hydrolyze upon contact with moisture; hence, a rough guide to the toxicity of a given ester may be the sum of the toxicities of the products of hydrolysis. Incompatible with nitrates. When heated to decomposition they emit acrid smoke and fumes.

EDO000 CAS: 50-28-2 HR: 3
ESTRADIOL
mf: $C_{18}H_{24}O_2$ mw: 272.42
PROP: Needles out of benzene, acetone; leaflets or needles from alc. Mp: 178°, bp: decomp. Sol in dioxone, alc, and ether.
SYNS: ALTRAD □ BARDIOL □ DIHYDRO-FOLLICULAR HORMONE □ DIHYDROFOLLICULIN □ DIHYDROMENFORMON □ DIHYDROTHEELIN □ 3,17-β-DIHYDROXYESTRA-1,3,5(10)-TRIENE □ 3,17-β-DIHYDROXY-1,3,5(10)-ESTRATRIENE □ DIHYDROXY-ESTRIN □ 3,17-β-DIHYDROXYOESTRA-1,3,5-TRIENE □ 3,17-β-DIHYDROXY-1,3,5(10)-OESTRATRIENE □ DIHYDROXYOESTRIN □ DIMENFORMON □ DIMENFORMON PROLONGATUM □ DIOGYN □ DIOGYNETS □ E² □ 3,17-EPIDIHYDROXYESTRATRI-ENE □ 3,17-EPIDIHYDROXYOESTRATRIENE □ ESTRADIOL-17-β □ α-ESTRADIOL □ β-ESTRADIOL □ cis-ESTRADIOL □ d-ESTRADIOL □ 3,17-β-ESTRADIOL □ 17-β-ESTRADIOL □ 17-β-OH-ESTRADIOL □ 17-β-OH-OESTRADIOL □ d-3,17-β-ESTRADIOL □ ESTRALDINE □ ESTRA-1,3,5(10)-TRIENE-3,17-β-DIOL □ 17-β-ESTRA-1,3,5(10)-TRIENE-3,17-DIOL □ 1,3,5-ESTRATRIENE-3,17-β-DIOL □ ESTROVITE □ FEMESTRAL □ FEMOGEN □ GYNERGON □ GYNESTREL □ GYNOESTRYL □ LAMDIOL □ MACRODIOL □ MACROL □ MICRODIOL □ NORDICOL □ NSC-9895 □ OESTERGON □ OESTRADIOL □ α-OESTRADIOL □ β-OESTRADIOL □ cis-OESTRADIOL □ d-OESTRADIOL □ 3,17-β-OESTRADIOL □ d-3,17-β-OESTRADIOL □ OESTRA-1,3,5(10)-TRIENE-3,17-β-DIOL □ OESTRADIOL R □ OESTRADIOL-17-β □ 17-β-OESTRA-1,3,5(10)-TRIENE-3,17-DIOL □ OESTROGLANDOL □ OESTROGYNAL □ OVAHORMON □ OVASTEROL □ OVASTEVOL □ OVOCICLINA □ OVOCYCLIN □ OVOCYCLINE □

OVOCYLIN □ PRIMOFOL □ PROFOLIOL □ PROGYNON □ PROGYNON-DH □ SYNDIOL
CONSENSUS REPORTS: NTP 10th Report on Carcinogens. IARC Cancer Review: Human Limited Evidence IMEMDT 21,279,79; Animal Sufficient Evidence IMEMDT 21,279,79; IMEMDT 6,99,74. EPA Genetic Toxicology Program.
SAFETY PROFILE: Confirmed carcinogen with experimental carcinogenic, neoplastigenic, tumorigenic, and teratogenic data. A promoter. Human reproductive effects by ingestion: fertility effects. Experimental reproductive effects. Human mutation data reported. A steroid hormone much used in medicine. When heated to decomposition it emits acrid smoke and irritating fumes.

EDP000 CAS: 50-50-0 HR: 3
ESTRADIOL-3-BENZOATE
mf: $C_{25}H_{28}O_3$ mw: 376.53
PROP: White or sltly yellow to brownish crystalline powder; odorless. Mp: 193°. Almost insol in water; sol in alc, acetone, and dioxane; sparingly sol in vegetable oils; sltly sol in ether.
SYNS: BENOVOCYLIN □ BENZHORMOVARINE □ BENZOATE d'OESTRADIOL (FRENCH) □ BENZO-ESTROFOL □ BENZOFOLINE □ BENZO-GYNOESTRYL □ BENZOIC ACID ESTRADIOL □ DIFFOLLISTEROL □ DIFOLLICULINE □ DIHYDROESTRIN BENZOATE □ DIHYDRO-FOLLICULIN BENZOATE □ DIMENFORM-ON BENZOATE □ DIMENFORMONE □ DIOGYN B □ EBZ □ ESTON-B □ ESTRADIOL BENZOATE □ β-ESTRADIOL BENZOATE □ ESTRADIOL-17-β-BENZOATE □ ESTRADIOL-17-β-3-BENZOATE □ β-ESTRADIOL-3-BENZOATE □ 17-β-ESTRADIOL BENZOATE □ 17-β-ESTRADIOL-3-BENZOATE □ ESTRADIOL MONOBENZOATE □ 17-β-ESTRADIOL MONOBENZOATE □ ESTRA-1,3,5(10)-TRIENE-3,17-DIOL (17-β)-3-BENZOATE □ ESTRA-1,3,5(10)-TRIENE-3,17-β-DIOL, 3-BENZOATE □ 1,3,5(10)-ESTRATRIENE-3,17-β-DIOL 3-BENZOATE □ FEMESTRONE □ FOLLICORMON □ FOLLIDRIN □ GRAAFINA □ de GRAAFINA □ GYNECORMONE □ GYNFORMONE □ HIDROESTRON □ HORMOGYNON □ HYDROXYESTRIN BENZOATE □ MEE □ ODB □ OESTRADIOL BENZOATE □ β-OESTRADIOL BENZOATE □ OESTRADIOL-3-BENZOATE □ β-OESTRADIOL-3-BENZOATE □ 17-β-OESTRADIOL-3-BENZOATE □ OESTRADIOL MONOBENZOATE □

OESTRAFORM (BDH) □ 1,3,5(10)-OESTRATRIENE-3,17-β-DIOL 3-BENZOATE □ OVAHORMON BENZOATE □ OVASTEROL-B □ OVEX □ OVOCYCLIN BENZOATE □ OVOCYCLIN M □ OVOCYCLIN-MB □ PRIMOGYN B □ PRIMOGYN BOLEOSUM □ PRIMOGYN I □ PROGYNON B □ PROGYNON BENZOATE □ RECTHORMONE OESTRADIOL □ SOLESTRO □ UNISTRADIOL

CONSENSUS REPORTS: IARC Cancer Review: Animal Sufficient Evidence IMEMDT 21,279,79.

SAFETY PROFILE: Confirmed carcinogen with experimental carcinogenic, tumorigenic, and teratogenic data. Human reproductive effects by intramuscular route: menstrual cycle changes and disorders. Experimental reproductive effects. Mutation data reported. A steroid. When heated to decomposition it emits acrid smoke and irritating fumes. See also ESTRADIOL.

EDR000   CAS: 113-38-2   HR: 3
ESTRADIOL DIPROPIONATE
mf: $C_{24}H_{32}O_4$ mw: 384.56
PROP: Leaflets from MeOH (aq). Mp: 104–105°.
SYNS: AGOFOLLIN □ DIMENFORMON DIPROPIONATE □ DIOVOCYCLIN □ DIOVOCYLIN □ DIPROPIONATE d'OESTRADIOL (FRENCH) □ DIPROSTRON □ ENDOFOLLICOLINA D.P. □ β-ESTRADIOL DIPROPIONATE □ ESTRADIOL-3,17-DIPROPIONATE □ β-ESTRADIOL-3,17-DIPROPIONATE □ 3,17-β-ESTRADIOL DIPROPIONATE □ 17-β-ESTRADIOL DIPROPIONATE □ ESTRA-1,3,5(10)-TRIENE-3,17-DIOL (17-β)-DIPROPIONATE □ 1,3,5(10)-ESTRATRIENE-3,17-β-DIOL DIPROPIONATE □ ESTROICI □ ESTRONEX □ FOLLICYCLIN P □ NACYCLYL □ OESTRADIOL DIPROPIONATE □ β-OESTRADIOL DIPROPIONATE □ OESTRADIOL-3,17-DIPROPIONATE □ 3,17-β-OESTRADIOL DIPROPIONATE □ 17-β-OESTRADIOL DIPROPIONATE □ OVOCYCLIN DIPROPIONATE □ OVOCYCLIN-P □ PROGYNON-DP

CONSENSUS REPORTS: IARC Cancer Review: Animal Sufficient Evidence IMEMDT 21,279,79. EPA Genetic Toxicology Program.

SAFETY PROFILE: Confirmed carcinogen with experimental carcinogenic, tumorigenic, and teratogenic data. A poison by intravenous and parenteral routes. Experimental reproductive effects. A drug

for the treatment of menopause. When heated to decomposition it emits acrid smoke and irritating fumes. See also ESTRADIOL.

EDS000   CAS: 28014-46-2   HR: 3
ESTRADIOL POLYESTER with PHOS-PHORIC ACID
mf: $(C_{18}H_{24}O_2 \bullet H_3O_4P)_x$
SYNS: ESTRADIOL PHOSPHATE POLYMER □ ESTRADURIN □ (17-β)-ESTRA-1,3,5(10)-TRIENE-3,17-DIOL POLYMER with PHOSPHORIC ACID □ OESTRADIOL PHOSPHATE POLYMER □ OESTRADIOL POLYESTER with PHOSPHORIC ACID □ PEP □ POLY(ESTRADIOL PHOSPHATE) □ POLYOESTRADIOL PHOSPHATE

CONSENSUS REPORTS: IARC Cancer Review: Animal Sufficient Evidence IMEMDT 21,279,79.

SAFETY PROFILE: Confirmed carcinogen producing liver tumors. An experimental teratogen. A drug used in cancer treatment. When heated to decomposition it emits toxic fumes of PO_x. See also ESTRADIOL, ESTERS, POLYMERS, and PHOSPHORIC ACID.

EDS100   CAS: 979-32-8   HR: 3
ESTRADIOL-17-VALERATE
mf: $C_{24}H_{32}O_3$ mw: 368.56
PROP: A solid. Mp: 144–145°.
SYNS: ALTADIOL □ DELADIOL □ DELAHORMONE UNIMATIC □ DELESTROGEN □ DELESTROGEN 4X □ DURA-ESTRADIOL □ ESTRADIOL VALERATE □ ESTRADIOL 17-β-VALERATE □ ESTRADIOL VALERIANATE □ (17-β)-ESTRA-1,3,5(10)-TRIENE-3,17-DIOL-17-PENTANOATE (9CI) □ ESTRAVEL □ FEMOGEX □ NEOFOLLIN □ PHARLON □ PROGYNON □ PROGYNON-DEPOT □ PROGYNOVA

SAFETY PROFILE: Suspected carcinogen with carcinogenic and teratogenic data. Experimental reproductive effects. When heated to decomposition it emits acrid smoke and irritating fumes. See also ESTRADIOL.

EDU500   CAS: 50-27-1   HR: 3
ESTRIOL
mf: $C_{18}H_{24}O_3$ mw: 288.42

PROP: Small, white crystals. D: 0.965, mp: 214.6°, bp: 214.6°.

SYNS: AACIFEMINE □ COLPOVISTER □ DESTRIOL □ DEUSLON-A □ ESTRA-1,3,5(10)-TRIENE-3,16-α,17-β-TRIOL □ 1,3,5-ESTRATRIENE-3-β,16-α,17-β-TRIOL □ (16-α,17-β)-ESTRA-1,3,5(10)-TRIENE-3,16,17-TRIOL □ ESTRATRIOL □ 3,16-α,17-β-ESTRIOL □ 16-α,17-β-ESTRIOL □ ESTRIOLO (ITALIAN) □ FOLLICULAR HORMONE HYDRATE □ GYNAESAN □ HEMOSTYPTANON □ HOLIN □ HORMOMED □ HORMONIN □ 16-α-HYDROXYESTRADIOL □ 16-α-HYDROXYOESTRADIOL □ KLIMORAL □ NSC-12169 □ OE3 □ OESTRA-1,3,5(10)-TRIENE-3,16-α,17-β-TRIOL □ 1,3,5-OESTRATRIENE-3-β-3,16-α,17-β-TRIOL □ (16-α,17-β)-OESTRA-1,3,5(10)-TRIENE-3,16,17-TRIOL □ OESTRATRIOL □ OESTRIOL □ 3,16-α,17-β-OESTRIOL □ 16-α,17-β-OESTRIOL □ ORGASTYPTIN □ OVESTERIN □ OVESTIN □ OVESTINON □ OVESTRION □ STIPTANON □ SYNAPAUSE □ THEELOL □ THULOL □ TRIDESTRIN □ 3,16-α,17-β-TRIHYDROXY-Δ-1,3,5-ESTRATRIENE □ 3,16-α,17-β-TRIHYDROXYESTRA-1,3,5(10)-TRIENE □ 3,16-α,17-β-TRIHYDROXY-Δ-1,3,5-OESTRATRIENE □ TRIHYDROXYESTRIN □ 3,16-α,17-β-TRIHYDROXYOESTRA-1,3,5(10)-TRIENE □ TRIHYDROXYOESTRIN □ TRIODURIN □ TRIOVEX

CONSENSUS REPORTS: IARC Cancer Review: Animal Limited Evidence IMEMDT 21,327,79; Human Limited Evidence IMEMDT 21,327,79; Animal Inadequate Evidence IMEMDT 6,117,74.

SAFETY PROFILE: Suspected carcinogen with experimental carcinogenic, neoplastigenic, tumorigenic, and teratogenic data. Other experimental reproductive effects. Mutation data reported. A steroid drug for the treatment of menopause. When heated to decomposition it emits acrid smoke and irritating fumes.

EDV000 CAS: 53-16-7 **HR: 3**
ESTRONE
mf: $C_{18}H_{22}O_2$ mw: 270.40

PROP: White crystals from EtOH trimorphic. Mp: 254°. Insol in water; sol in alc, benzene, ether, and chloroform.

SYNS: AQUACRINE □ CRISTALLOVAR □ CRYSTOGEN □ DESTRONE □ DISYNFORMON □ ENDOFOLLICULINA □ ESTERONE □ 1,3,5-ESTRATRIEN-3-OL-17-ONE □ 1,3,5(10)-ESTRATRIEN-3-OL-17-ONE □ Δ-1,3,5-ESTRATRIEN-3-β-OL-17-ONE □ ESTRIN □ ESTROL □ ESTRON □ ESTRONA (SPANISH) □ ESTRONE-A □ ESTRUGENONE □ ESTRUSOL □

FEMESTRONE INJECTION □ FEMIDYN □ FOLIKRIN □ FOLIPEX □ FOLISAN □ FOLLESTRINE □ FOLLICULAR HORMONE □ FOLLICULIN □ FOLLICULINE BENZOATE □ FOLLICUNODIS □ FOLLIDRIN □ GLANDUBOLIN □ HIESTRONE □ HORMOFOLLIN □ HORMOVARINE □ 3-HYDROXYESTRA-1,3,5(10)-TRIEN-17-ONE □ 3-HYDROXY-17-KETO-ESTRA-1,3,5-TRIENE □ 3-HYDROXY-17-KETO-OESTRA-1,3,5-TRIENE □ 3-HYDROXY-OESTRA-1,3,5(10)-TRIEN-17-ONE □ 3-HYDROXY-1,3,5(10)-OESTRATRIEN-17-ONE □ KESTRONE □ KETODESTRIN □ KETOHYDROXY-ESTRATRIENE □ KETOHYDROXYESTRIN □ KETOHYDROXYOESTRIN □ KOLPON □ MENAGEN □ MENFORMON □ Δ-1,3,5-OESTRATRIEN-3-β-OL-17-ONE □ 1,3,5-OESTRATRIEN-3-OL-17-ONE □ 1,3,5(10)-OESTRATRIEN-3-OL-17-ONE □ OESTRIN □ OESTROFORM □ OESTRONE □ OESTROPEROS □ OVEX □ OVIFOLLIN □ PERLATAN □ SOLLICULIN □ THEELIN □ THELESTRIN □ THELYKININ □ THYNESTRON □ TOKOKIN □ UNDEN □ WNYESTRON

CONSENSUS REPORTS: NTP 10th Report on Carcinogens. IARC Cancer Review: Human Limited Evidence IMEMDT 21,343,79; Animal Sufficient Evidence IMEMDT 6,123,74; IMEMDT 21,343,79. Reported in EPA TSCA Inventory.

SAFETY PROFILE: Confirmed carcinogen with experimental carcinogenic, neoplastigenic, tumorigenic, and teratogenic data. A poison by intraperitoneal and subcutaneous routes. Human reproductive effects by implantation: spermatogenesis and impotence. Mutation data reported. A steroid drug for the treatment of menopause and ovariectomy symptoms. When heated to decomposition it emits acrid smoke and irritating fumes.

EDV500 CAS: 2393-53-5 **HR: 3**
ESTRONE BENZOATE
mf: $C_{25}H_{26}O_3$ mw: 374.51

SYNS: BENZOATE d'OESTRONE (FRENCH) □ 3-(BENZOYLOXY)ESTRA-1,3,5(10)-TRIEN-17-ONE □ 3-HYDROXYESTRA-1,3,5(10)-TRIEN-17-ONE BENZOATE □ KETOHYDROXYESTRIN BENZOATE □ OESTRON-BENZOAT (GERMAN)

CONSENSUS REPORTS: IARC Cancer Review: Animal Limited Evidence IMEMDT 21,343,79.

SAFETY PROFILE: Suspected carcinogen with experimental carcinogenic, neoplastigenic, and tumorigenic data. A steroid. When heated to decomposition it emits acrid smoke and irritating fumes.

EDV600   CAS: 438-67-5   HR: 3
ESTRONE SODIUM SULFATE
mf: $C_{18}H_{22}O_5S$•Na   mw: 373.45

SYNS: CONESTORAL □ ESTRA-1,3,5(10)-TRIEN-17-ONE, 3-(SULFOXY)-, SODIUM SALT (9CI) □ ESTRONE, HYDROGEN SULFATE, SODIUM SALT □ ESTRONE SULFATE SODIUM □ ESTRONE SULFATE SODIUM SALT □ ESTRONE-3-SULFATE SODIUM SALT □ EVEX □ MORESTIN □ OESTRONE-3-SULPHATE SODIUM SALT □ SODIUM ESTRONE SULFATE □ SODIUM ESTRONE-3-SULFATE

CONSENSUS REPORTS: NTP 10th Report on Carcinogens.

SAFETY PROFILE: Confirmed carcinogen. Experimental reproductive effects. When heated to decomposition it emits toxic fumes of SO_x.

EDV700   CAS: 15686-63-2   HR: 3
ETABENZARONE
mf: $C_{23}H_{27}NO_3$ mw: 365.51

SYNS: BENZOFURAN, 3-(p-(2-(DIETHYLAMINO)-ETHOXY)BENZOYL)-2-ETHYL- □ p-(2-(DIETHYL-AMINO)ETHOXY)PHENYL 2-ETHYL-3-BENZOFURAN-YL KETONE □ KETONE, p-(2-(DIETHYLAMINO)ETHO-XY)PHENYL 2-ETHYL-3-BENZOFURANYL □ L 2642-LABAZ □ METHANONE, (4-(2-(DIETHYLAMINO)ETHO-XY)PHENYL)(2-ETHYL-3-BENZOFURANYL)-(9CI)

DOT CLASSIFICATION: 3; Label: Flammable Liquid

SAFETY PROFILE: A poison by intraperitoneal route. A flammable liquid. When heated to decomposition it emits toxic vapors of NO_x.

EDW500   CAS: 1837-57-6   HR: 3
ETHACRIDINE LACTATE
mf: $C_{15}H_{15}N_3O$•$C_3H_6O_3$ mw: 343.42

PROP: Pale-yellow crystals from EtOH/Et$_2$O. Mp: 235°.

SYNS: ACRINOL □ ACROLACTINE □ 2-AETHOXY-6,9-DIAMINOACRIDINLACTAT (GERMAN) □ 2,5-DIAMINO-7-ETHOXYACRIDINE LACTATE □ 6,9-DIAMINO-2-ETHOXYACRIDINE LACTATE MONOHYDRATE □ ETHODIN □ 2-ETHOXY-6,9-DIAMINOACRIDINE LACTATE □ 2-ETHOXY-6,9-DIAMINOACRIDINE LACTATE HYDRATE □ 2-ETHOXY-6,9-DIAMINOACRIDINIUM LACTATE □ FLAVITROL □

METIFEX □ RIMAON □ RIVANOL □ RIVINOL □ VUCINE

CONSENSUS REPORTS: Reported in EPA TSCA Inventory.

SAFETY PROFILE: Poison by subcutaneous, intraperitoneal, and intravenous routes. Experimental reproductive effects. An antiseptic. When heated to decomposition it emits toxic fumes of NO_x.

EDZ000   CAS: 74-84-0   HR: 3
ETHANE
DOT: UN 1035/UN 1961
mf: C_2H_6 mw: 30.08

PROP: Colorless, odorless, flammable gas. Mp: −172°, bp: −88.6°, lel: 3.0%, uel: 12.5%, fp: −183.2°, d: 0.446 @ 0° (liquid), autoign temp: 959°F, vap d: 1.04, flash p: −202°F. Sol in EtOH, liquid O$_2$; sltly sol in H$_2$O.

SYNS: BIMETHYL □ DIMETHYL □ ETHANE, compressed (UN 1035) (DOT) □ ETHANE, refrigerated liquid (UN 1961) (DOT) □ ETHYL HYDRIDE □ METHYLMETHANE

CONSENSUS REPORTS: Reported in EPA TSCA Inventory.

DOT CLASSIFICATION: 2.1; Label: Flammable Gas

SAFETY PROFILE: A simple asphyxiant. See ARGON for properties of simple asphyxiants. A very dangerous fire hazard when exposed to heat or flame; can react vigorously with oxidizing materials. Moderate explosion hazard when exposed to flame. To fight fire, stop flow of gas. Incompatible with chlorine, dioxygenyl tetrafluoroborate, oxidizing materials, heat or flame. When heated to decomposition it emits acrid smoke and irritating fumes.

EEA500   CAS: 107-15-3   HR: 3
1,2-ETHANEDIAMINE
DOT: UN 1604
mf: $C_2H_8N_2$ mw: 60.12

PROP: Volatile, colorless, clear, thick, strongly alkaline, hygroscopic liquid; ammonia-like odor. Mp: 8.5°, bp: 117.2°, flash p: 110°F (CC), d: 0.8994 @ 20°/4°, vap press: 10.7 mm @ 20°, vap d: 2.07, autoign temp: 725°F. Sol in EtOH and H_2O (with hydration); insol in C_6H_6; sltly sol in Et_2O. IDLH 1000 ppm.

SYNS: AETHALDIAMIN (GERMAN) □ AETHYLENE-DIAMIN (GERMAN) □ 1,2-DIAMINOAETHAN (GERMAN) □ 1,2-DIAMINO-ETHAAN (DUTCH) □ 1,2-DIAMINOETHANE □ 1,2-DIAMINO-ETHANO (ITALIAN) □ DIMETHYLENEDIAMINE □ ETHYLE-ENDIAMINE (DUTCH) □ ETHYLENEDIAMINE (OSHA) □ 1,2-ETHYLENEDIAMINE □ ETHYLENE-DIAMINE (FRENCH) □ NCI-C60402

CONSENSUS REPORTS: Reported in EPA TSCA Inventory. EPA Extremely Hazardous Substances List.

OSHA PEL: TWA 10 ppm

ACGIH TLV: TWA 10 ppm; Not Classifiable as a Human Carcinogen

DFG MAK: 10 ppm (25 mg/m³)

DOT CLASSIFICATION: 8; Label: Corrosive, Flammable Liquid

SAFETY PROFILE: A human poison by inhalation. Experimental poison by inhalation, intraperitoneal, subcutaneous, and intravenous routes. Moderately toxic by ingestion and skin contact. Experimental reproductive effects. Corrosive. A severe skin and eye irritant. An allergen and sensitizer. Mutation data reported. Flammable liquid when exposed to heat, flame, or oxidizers. Can react violently with acetic acid, acetic anhydride, acrolein, acrylic acid, acrylonitrile, allyl chloride, CS_2, chlorosulfonic acid, epichlorohydrin, ethylene chlorohydrin, HCl, mesityl oxide, HNO_3, oleum, $AgClO_4$, H_2SO_4, β-propiolactone, or vinyl acetate. To fight fire, use CO_2, dry chemical, alcohol foam. When heated to decomposition it emits toxic fumes of NO_x and NH_3. See also AMINES.

EEA700   CAS: 81861-89-4   HR: 3
N,N'-(1,2-ETHANEDIOXYSULFINYL)-BIS(S-METHYL-N-METHYLCARB-AMOYLOXYTHIOACETIMIDATE)
mf: $C_{12}H_{22}N_4O_8S_4$ mw: 478.62

SYN: ETHANIMIDOTHIOIC ACID, N,N'-(1,2-ETHANEDIYL-BIS(OXYSULFINYL(METHYL-IMINO)CARBONYLOXY)BIS-, DIMETHYL ESTER

SAFETY PROFILE: A poison by ingestion. When heated to decomposition it emits toxic vapors of NO_x and SO_x.

EEB000   CAS: 540-63-6   HR: 3
1,2-ETHANEDITHIOL
mf: $C_2H_6S_2$ mw: 94.20
PROP: A liquid. D: 1.124, bp: 146°.

SYNS: 1,2-DIMERCAPTOETHANE □ DITHIOETHYL-ENEGLYCOL □ DITHIOGLYCOL □ ETHYLENE DIMERCAPTAN □ α-ETHYLENE DIMERCAPTAN □ ETHYLENE DITHIOGLYCOL □ ETHYLENEDITHIOL □ ETHYL HYDROPERSULFIDE

CONSENSUS REPORTS: Reported in EPA TSCA Inventory.

SAFETY PROFILE: Poison by ingestion, intraperitoneal, and intravenous routes. When heated to decomposition it emits very toxic fumes of SO_x. See also MERCAPTANS.

EEB050   CAS: 81861-94-1   HR: 3
N,N'-(1,2-ETHANEDITHIOSULFINYL)-BIS(S-METHYL-N-METHYLCARB-AMOYLOXYTHIOACETIMIDATE)
mf: $C_{12}H_{22}N_4O_6S_6$ mw: 510.74

SYN: ETHANIMIDOTHIOIC ACID, N,N'-(1,2-ETHANE-DIYLBIS(THIOSULFINYL(METHYLIMINO)CARBONYLO XY))BIS-, DIMETHYL ESTER

SAFETY PROFILE: A poison by ingestion. When heated to decomposition it emits toxic vapors of NO_x and SO_x.

EEC600   CAS: 141-43-5   HR: 3
ETHANOLAMINE
DOT: UN 2491
mf: C_2H_7NO mw: 61.10
PROP: Colorless, viscous, hygroscopic liquid with ammonia-like odor. Bp: 170.5°, fp: 10.5°, flash p: 200°F (OC), d: 1.012 @ 25°/4°, vap press: 6 mm @ 60°, vap d: 2.11. Misc in water and alc; sltly sol in benzene; sol in chloroform. IDLH 30 ppm.

SYNS: AETHANOLAMIN (GERMAN) □ 2-AMINO-AETHANOL (GERMAN) □ 2-AMINOETANOLO (ITALIAN) □ 2-AMINOETHANOL (MAK) □ β-AMINOETHYL ALCOHOL □ COLAMINE □ ETANOL-AMINA (ITALIAN) □ β-ETHANOLAMINE □ ETHANOL-AMINE, solution (DOT) □ ETHYLOLAMINE □ GLYCIN-OL □ β-HYDROXYETHYLAMINE □ 2-HYDROXYETH-YLAMINE □ MEA □ MONOAETHANOLAMIN

(GERMAN) □ MONOETHANOLAMINE □ OLAMINE □ THIOFACO M-50 □ USAF EK-1597

CONSENSUS REPORTS: Reported in EPA TSCA Inventory.

OSHA PEL: TWA 3 ppm; STEL 6 ppm

ACGIH TLV: TWA 3 ppm; STEL 6 ppm

DFG MAK: 2 ppm (5.1 mg/m^3)

DOT CLASSIFICATION: 8; Label: Corrosive

SAFETY PROFILE: Poison by intraperitoneal route. Moderately toxic by ingestion, skin contact, subcutaneous, intravenous, and intramuscular routes. A corrosive irritant to skin, eyes, and mucous membranes. Human mutation data reported. Flammable when exposed to heat or flame. A powerful base. Reacts violently with acetic acid, acetic anhydride, acrolein, acrylic acid, acrylonitrile, cellulose, chlorosulfonic acid, epichlorohydrin, HCl, HF, mesityl oxide, HNO$_3$, oleum, H$_2$SO$_4$, β-propiolactone, vinyl acetate. To fight fire, use foam, alcohol foam, dry chemical. When heated to decomposition it emits toxic fumes of NO$_x$. See also AMINES.

EEE000   CAS: 20398-06-5   HR: 3
ETHANOL THALLIUM(1+) SALT

mf: C$_2$H$_6$O•Tl   mw: 250.45

PROP: Sol in org solvs. IDLH 15 mg/m^3 (as Tl).

SYN: ETHYL ALCOHOL THALLIUM (I)

CONSENSUS REPORTS: Reported in EPA TSCA Inventory.

OSHA PEL: TWA 0.1 mg(Tl)/m^3 (skin)

ACGIH TLV: TWA 0.1 mg(Tl)/m^3 (skin)

SAFETY PROFILE: Poison by ingestion. When heated to decomposition it emits toxic fumes of Tl. See also THALLIUM COMPOUNDS.

EEE200   CAS: 38527-91-2   HR: 3
ETHAPHOS

mf: C$_{11}$H$_{15}$Cl$_2$O$_3$PS   mw: 329.19

SYNS: ETAFOS □ ETAPHOS □ PHOSPHOROTHIOIC ACID, o-(2,4-DICHLOROPHENYL) o-ETHYL S-PROPYL ESTER □ PROTHIOFOS-OXON □ TOKUTHION OXON

SAFETY PROFILE: A poison by ingestion. Low toxicity by inhalation. When heated to decomposition it emits toxic vapors of PO$_x$, SO$_x$, and Cl$^-$.

EEG000   CAS: 88-12-0   HR: 3
1-ETHENYL-2-PYRROLIDINONE

mf: C$_6$H$_9$NO   mw: 111.16

PROP: Colorless liquid, water-sol. Bp: 148° @ 100 mm, fp: 13.5°, flash p: 209°F (OC), d: 1.04 @ 25°, autoign temp: 687°F, vap d: 3.8, fire p: 213°F.

SYNS: VINYLBUTYROLACTAM □ N-VINYLPYR-ROLIDINONE □ N-VINYL-2-PYRROLIDINONE □ 1-VINYL-2-PYRROLIDINONE □ VINYLPYRROLIDONE □ N-VINYLPYRROLIDONE □ N-VINYL-2-PYRROLIDONE (ACGIH) □ 1-VINYL-2-PYRROLIDONE □ V-PYROL

CONSENSUS REPORTS: IARC Cancer Review: Group 3 IMEMDT 7,56,87. Reported in EPA TSCA Inventory.

ACGIH TLV: 0.05 ppm; Confirmed Animal Carcinogen

DFG MAK: Confirmed Animal Carcinogen, Suspected Human Carcinogen

SAFETY PROFILE: Confirmed carcinogen. Moderately toxic by ingestion, inhalation, and skin contact. A severe eye irritant. Probably irritating and narcotic in high concentrations. Combustible when exposed to heat or flame; can react vigorously with oxidizing materials. To fight fire, use alcohol foam, CO$_2$, dry chemical. When heated to decomposition it emits highly toxic fumes of NO$_x$.

EEG500   HR: 3
ETHERS

PROP: Organic compounds in which an oxygen atom is interposed between two carbon atoms in the structure of the molecule.

SAFETY PROFILE: The simpler ethers such as ethyl ether, isopropyl ether, etc., are powerful narcotics that in large doses can cause death. The danger from ethers is usually acute and seldom chronic. Aftereffects to ether intoxication are uncommon although continued exposure to small concentrations (not enough to cause an overt symptom) has been known to cause loss of appetite, excessive thirst, and fatigue.

The most common ethers, such as ethyl, methyl, and diisopropyl, are particularly dangerous fire and explosion hazards when exposed to heat, flame, or sparks. They can react violently with strong oxidizers. Many plant and laboratory fires and explosions have resulted from their high flammability and tendency to form explosive peroxides. The common ethers are easily ignited and have low flash points. The diethyl, ethyl tert-butyl, ethyl tert-pentyl, and diisopropyl ethers are very hazardous. Methyl tert-alkyl ethers are relatively safe. Besides the risk of explosion from air mixtures of ether vapors, ethers tend to form peroxides upon standing. For some ethers peroxide levels do not reach dangerous concentrations (e.g., diethyl ether, ethyl vinyl ether, tetrahydrofuran, p-dioxane, 1,1-diethoxyethane, and the dimethyl ethers of ethylene glycol). When ethers containing peroxides are heated they can detonate. It is necessary to control smoking, open flames, or even the use of hot plates in areas where low-molecular-weight ethers are apt to reach 1% concentration or more in air. Only electrical equipment of explosion-proof type (Group C classification) is permitted to be operated in ether areas. Ethers should not be stored near powerful oxidizers or in areas of high fire hazard. They should be kept cool and the containers electrically grounded to avoid sparks.

Dangerous; shock or heat can cause gaseous ethers to escape from their containers and create flammable or even explosive conditions. Incompatible with oxidizing materials, BI_3. See also ETHYL ETHER.

EEH000   CAS: 126-52-3   HR: 3
ETHINAMATE
mf: $C_9H_{13}NO_2$ mw: 167.23
PROP: Rods or needles from cyclohexane. Mp: 96–98°, bp: 118–122° @ 3 mm. Very sol in EtOH; sltly sol in hexane; very sltly sol in H_2O.

SYNS: CARBAMATE de l'ETHINYLCYCLOHEXANOL (FRENCH) □ 1-ETHINYLCYCLOHEXYL CARBAMATE □ 1-ETHINYLCYCLOHEXYL CARBONATE □ 1-ETHYNYL-CYCLOHEXANOL CARBAMATE □ 1-ETHYNYLCYCLO-HEXYL CARBAMATE □ 1-ETHYNYLCYCLOHEXYL ESTER CARBAMIC ACID □ ETINAMATE □ USAF EL-42 □ VALAMINA □ VALAMINETTEN □ VALMID □ VALMIDATE □ VOLAMIN

CONSENSUS REPORTS: Reported in EPA TSCA Inventory.

SAFETY PROFILE: A deadly human poison. Experimental poison by ingestion, intravenous, subcutaneous, and intraperitoneal routes. An experimental teratogen. When heated to decomposition it emits toxic fumes of NO_x. See also CARBAMATES.

EEH500   CAS: 57-63-6   HR: 3
ETHINYL ESTRADIOL
mf: $C_{29}H_{24}O_2$ mw: 296.44
PROP: Crystals from MeOH (aq). Mp: 145–146°.
SYNS: 3,17-β-DIHYDROXY-17-α-ETHYNYL-1,3,5(10)-ESTRATRIENE □ 3,17-β-DIHYDROXY-17-α-ETHYNYL-1,3,5(10)-OESTRATRIENE □ ESTROGEN □ 17-α-ETHIN-YL-3,17-DIHYDROXY-$\Delta^{1,3,5}$-ESTRATRIENE □ 17-α-ETHINYL-3,17-DIHYDROXY-$\Delta^{1,3,5}$-OESTRATRIENE □ 17-ETHINYLESTRADIOL □ 17-ETHINYL-3,17-ESTRADIOL □ 17-α-ETHINYLESTRADIOL □ 17-α-ETHINYL-17-β-ESTRADIOL □ 17-α-ETHINYLESTRA-1,3,5(10)-TRIENE-3,17-β-DIOL □ ETHINYLESTRIOL □ ETHINYLOESTRADIOL □ 17-ETHINYL-3,17-OESTRADI-OL □ ETHINYL-OESTRANOL □ 17-α-ETHINYL-OESTRA-1,3,5(10)-TRIENE-3,17-β-DIOL □ 17-α-ETHINYL-$\Delta^{1,3,5(10)}$OESTRATRIENE-3,17-β-DIOL □ ETHINYLO-ESTRIOL □ 17-ETHYNYL-3,17-DIHYDROXY-1,3,5-OESTRATRIENE □ ETHYNYLESTRADIOL □ 17-α-ETHYNYLESTRADIOL □ 17-α-ETHYNYLESTRADIOL-17-β □ 17-α-ETHYNYL-1,3,5(10)-ESTRATRIENE-3,17-β-DIOL □ 17-α-ETHYNYLESTRA-1,3,5(10)-TRIENE-3,17-β-DIOL □ ETHYNYLOESTRADIOL □ 17-ETHYNYL-OESTRADIOL □ 17-α-ETHYNYLOESTRADIOL □ 17-α-ETHYNYL-17-β-OESTRADIOL □ 17-α-ETHYNYLO-ESTRADIOL-17-β □ 17-ETHYNYLOESTRA-1,3,5(10)-TRIENE-3,17-β-DIOL □ 17-α-ETHYNYL-1,3,5-OESTRATRIENE-3,17-β-DIOL □ 17-α-ETHYNYL-1,3,5(10)-OESTRATRIENE-3,17-β-DIOL □ 17-α-ETHYNYLO-ESTRA-1,3,5(10)-TRIENE-3,17-β-DIOL □ 19-NOR-17-α-PREGNA-1,3,5(10)-TRIEN-2-YNE-3,17-DIOL □ (17-α)-19-NORPREGNA-1,3,5(10)-TRIEN-20-YNE-3,17,DIOL

CONSENSUS REPORTS: NTP 10th Report on Carcinogens. IARC Cancer Review:

Human Limited Evidence IMEMDT 21,233,79; Animal Sufficient Evidence IMEMDT 6,77,74; IMEMDT 21,233,79. Reported in EPA TSCA Inventory.

SAFETY PROFILE: Confirmed carcinogen with experimental carcinogenic, tumorigenic, and neoplastigenic data. Poison by intraperitoneal route. Moderately toxic by ingestion. Human systemic effects by ingestion: glandular effects. An experimental teratogen. Experimental reproductive effects. Human mutation data reported. When heated to decomposition it emits acrid smoke and irritating fumes. See also ESTRADIOL.

EEH520   CAS: 8015-12-1   HR: 3
**ETHINYL ESTRADIOL and
        NORETHINDRONE ACETATE**
mf: $C_{22}H_{28}O_3 \cdot C_{20}H_{24}O_2$ mw: 636.94
SYNS: ANOVLAR 21 □ CONTROVLAR □ ETHINYL OESTRADIOL mixed with NORETHISTERONE ACETATE □ GYN-ANOVLAR □ GYNONLAR 21 □ MINORLAR □ MINOVLAR □ NORETHINDRONE ACETATE and ETHINYLESTRADIOL □ NORETHISTERONE ACETATE mixed with ETHINYL OESTRADIOL □ NORLESTRIN □ PRIMODOS

SAFETY PROFILE: Suspected human carcinogen producing lung and liver tumors. Experimental neoplastigenic and tumorigenic data. Human and experimental teratogenic and reproductive effects. When heated to decomposition it emits acrid smoke and irritating fumes.

EEH550   CAS: 68-23-5   HR: 3
17-α-ETHINYL-5,10-ESTRENOLONE
mf: $C_{20}H_{26}O_2$ mw: 298.46
PROP: Crystals from MeOH. Mp: 180–181.5°.
SYNS: 17-ETHINYL-5(10)-ESTRAENEOLONE □ 17-α-ETHINYL-ESTRA(5,10)ENEOLONE □ 17-α-ETHYNYL-5(10)-ESTREN-17-OL-3-ONE □ 17-α-ETHYNYLESTR-5(10)-EN-17-β-OL-3-ONE □ 17-α-ETHYNYL-ESTR-5(10)-EN-3-ON-17-β-OL □ 17-α-ETHYNYL-17-HYDROXYESTR-5(10)-EN-3-ONE □ 17-α-ETHYNYL-17-HYDROXY-5(10)-ESTREN-3-ONE □ 17-α-ETHYNYL-17-β-HYDROXY-5(10)-ESTREN-3-ONE □ 17-α-ETHYNYL-17-β-HYDROXY-ESTR-5(10)-EN-3-ONE □ 17-α-ETHINYL-17-β-HYDROXY-$Δ^{5(10)}$-ESTREN-3-ONE □ 17-α-ETHYNYL-17-β-HYDRO-

XY-$Δ^{5(10)}$-ESTREN-3-ONE □ 17-α-ETHYNYL-17-β-HYDROXY-3-OXO-$Δ^{5(10)}$-ESTRENE □ 17-α-ETHYNYL-19-NOR-5(10)-ANDROSTEN-17-β-OL-3-ONE □ 17-α-ETHINYL-$Δ^{5,10-19}$-NORTESTOSTERONE □ 17-β-HYDROXY-17-α-ETHINYL-5(10)-ESTREN-3-ONE □ 17-HYDROXY-19-NOR-17-α-PREGN-5(10)-EN-20-YN-3-ONE □ (17-α)-17-HYDROXY-19-NORPREGN-5(10)-EN-20-YN-3-ONE □ 17-HYDROXY(17-α)-19-NORPREGN-5(10)-EN-20-YN-3-ONE □ LYNESTROL □ NORETHINODREL □ 19-NOR-ETHINYL-5,10-TESTOSTERONE □ NORETHINYNODREL □ NORETHYNODRAL □ NORETHYNODREL □ 19-NORETHYNODREL □ NSC-15432 □ SC-4642

CONSENSUS REPORTS: IARC Cancer Review: Animal Limited Evidence IMEMDT 21,461,79; Animal Sufficient Evidence IMEMDT 6,191,74.

SAFETY PROFILE: Suspected carcinogen with experimental tumorigenic data. Human and experimental reproductive effects. Mutation data reported. When heated to decomposition it emits acrid smoke and irritating fumes.

EEH600   CAS: 563-12-2   HR: 3
ETHION
mf: $C_9H_{22}O_4P_2S_4$ mw: 384.49
PROP: Oily liquid. Mp: −13°, d: 1.31 @ 20°/4°. Sol in Me_2CO, C_6H_6, $CHCl_3$, EtOH, and Et_2O; sltly sol in H_2O; mod sol in ligroin.
SYNS: AC 3422 □ BIS(S-(DIETHOXYPHOSPHINOTHIOYL)MERCAPTO)METHANE □ BLADAN □ DIETHION □ EMBATHION □ ENT 24,105 □ ETHANOX □ ETHIOL □ ETHODAN □ ETHYL METHYLENE PHOSPHORODITHIOATE □ FMC-1240 □ FOSFONO 50 □ HYLEMOX □ ITOPAZ □ KWIT □ METHANEDITHIOL-S,S-DIESTER with O,O-DIETHYL ESTER PHOSPHORODITHIOIC ACID □ METHYLEEN-S,S'-BIS(O,O-DIETHYL-DITHIOFOSFAAT) (DUTCH) □ S,S'-METHYLEN-BIS(O,O-DIAETHYL-DITHIOPHO-SPHAT) (GERMAN) □ METHYLENE-S,S'-BIS(O,O-DIAETHYL-DITHIOPHOSPHAT) (GERMAN) □ S,S'-METHYLENE O,O,O',O'-TETRAETHYL PHOSPHORODI-THIOATE □ NIAGARA 1240 □ NIALATE □ PHOSPHOTOX E □ RHODIACIDE □ RHODOCIDE □ RODOCID □ RP 8167 □ SOPRATHION □ O,O,O',O'-TETRAAETHYL-BIS(DITHIOPHOSPHAT) (GERMAN) □ O,O,O',O'-TETRAETHYL S,S'-METHYLENEBISPHOS-PHORDITHIOATE □ O,O,O',O'-TETRAETHYL-S,S'-METHYLENEBISPHOSPHORODITHIOATE □ TETRAETHYL S,S'-METHYLENE BIS(PHOSPHORO-THIOLOTHIONATE) □ O,O,O',O'-TETRAETHYL S,S'-METHYLENE DI(PHOSPHORODITHIOATE) □ VEGFRU FOSMITE

CONSENSUS REPORTS: EPA: Farm Worker Field Reentry FEREAC 39,16888,74. EPA Genetic Toxicology Program. EPA Extremely Hazardous Substances List.
OSHA PEL: TWA 0.4 mg/m³ (skin)
ACGIH TLV: TWA 0.05 mg/m³ (skin); Not Classifiable as a Human Carcinogen
SAFETY PROFILE: Poison by ingestion, skin contact, and intraperitoneal routes. Human systemic effects by ingestion: flaccid paralysis without anesthesia, motor activity changes, fever, and inhibition of cholinesterase. When heated to decomposition it emits highly toxic fumes of SO_x and PO_x. See also PARATHION.

EEI000   CAS: 67-21-0   HR: 3
dl-ETHIONINE
mf: $C_6H_{13}NO_2S$ mw: 163.26
PROP: Crystals from alc (aq). Mp: 267–268°, decomp @ 273°.
SYNS: AETHIONIN □ 2-AMINO-4-(ETHYLTHIO)BUTYRIC ACID □ dl-2-AMINO-4-(ETHYLTHIO)BUTYRIC ACID □ CN 8676 □ ETH □ ETHIONIN □ ETHIONINE □ (±)-ETHIONINE □ S-ETHYL-HOMOCYSTEINE □ S-ETHYL-dl-HOMOCYSTEINE □ NSC-751 □ U-1434
CONSENSUS REPORTS: EPA Genetic Toxicology Program. Reported in EPA TSCA Inventory.
SAFETY PROFILE: Mildly toxic by ingestion and intraperitoneal routes. Suspected carcinogen with experimental carcinogenic and tumorigenic data. An experimental teratogen. Experimental reproductive effects. Mutation data reported. When heated to decomposition it emits toxic fumes of SO_x and NO_x.

EEJ000   CAS: 61791-14-8   HR: 2
ETHOMEEN C/15
PROP: A polyoxyethylene (5%) cocoa amine in which alkyl bonds link C_8–C_{18} carbons, which consists of dodecyl (47%), undecyl (18%), decyl (9%), octyl (8%), hexadecyl (10%), and octadecyl (5%) (FCTXAV 8,249,70).

CONSENSUS REPORTS: Reported in EPA TSCA Inventory.
SAFETY PROFILE: Moderately toxic by ingestion. An eye irritant. When heated to decomposition it emits acrid smoke and irritating fumes.

EEK100   CAS: 59-06-3   HR: D
ETHOPABATE
mf: $C_{12}H_{15}NO_4$ mw: 237.25
PROP: Odorless white to pink crystals from MeOH (aq). Mp: 148–149°. Sol in methanol, ethanol, acetone, acetonitrile, isopropanol, p-dioxane, ethyl acetate, and methylene chloride.
SYNS: 4-ACETAMIDO-2-ETHOLXBENZOIC ACID METHYL ESTER □ 2-ETHOXY-4-ACETAMIDOBENZOID ACID METHYL ESTER □ ETHYL PABATE □ METHYL 4-ACETAMIDO-2-ETHOXYBENZOATE
SAFETY PROFILE: When heated to decomposition it emits acrid smoke and irritating fumes.

EEK500   CAS: 627-03-2   HR: D
ETHOXYACETIC ACID
mf: $C_4H_8O_3$ mw: 104.12
PROP: A liquid. D: 1.1021 @ 20°/4°, bp: 206–207°.
SYNS: ACETIC ACID, ETHOXY- □ 2-ETHOXYACETIC ACID
CONSENSUS REPORTS: Reported in EPA TSCA Inventory.
SAFETY PROFILE: Experimental reproductive effects. When heated to decomposition it emits acrid smoke and irritating fumes.

EEL100   CAS: 1321-31-9   HR: D
ETHOXYANILINE
DOT: UN 2311
mf: $C_8H_{11}NO$ mw: 137.20
PROP: A solid.
SYNS: BENZENAMINE, ar-ETHOXY- □ PHENETIDINE □ PHENETIDINES (DOT)
DOT CLASSIFICATION: 6.1; Label: KEEP AWAY FROM FOOD
SAFETY PROFILE: Mutation data reported. When heated to decomposition it emits toxic vapors of NO_x.

EER500   CAS: 103-75-3   HR: 3
2-ETHOXY DIHYDROPYRAN
mf: $C_7H_{12}O_2$ mw: 128.19
PROP: D: 1.0, bp: 143°, flash p: 111°F (OC).
SYNS: 2-ETHOXY-2,3-DIHYDRO-γ-PYRAN □ 2-ETHOXY-3,4-DIHYDRO-1,2-PYRAN □ 2-ETHOXY-3,4-DIHYDRO-2H-PYRAN
CONSENSUS REPORTS: Reported in EPA TSCA Inventory.
SAFETY PROFILE: Moderately toxic by skin contact. Mildly toxic by ingestion and inhalation. A skin and eye irritant. Flammable liquid when exposed to flame, sparks, and oxidizers. To fight fire, use dry chemical, foam, fog. When heated to decomposition it emits acrid smoke and irritating fumes.

EES100   CAS: 69929-16-4   HR: 1
8-ETHOXY-2,6-DIMETHYLOCTENE-2
mf: $C_{12}H_{24}O$ mw: 184.36
PROP: Clean, sweet rose-like odor.
SYNS: CITRONELLYL ETHYL ETHER □ 2-OCTENE, 8-ETHOXY-2,6-DIMETHYL-
CONSENSUS REPORTS: Reported in EPA TSCA Inventory.
SAFETY PROFILE: Low toxicity by ingestion and skin contact. A skin irritant. When heated to decomposition it emits acrid smoke and irritating fumes.

EES200   CAS: 14857-34-2   HR: 3
ETHOXYDIMETHYLSILANE
mf: $C_4H_{12}OSi$ mw: 104.25
PROP: Bp: 54°, d: 0.757 @ 20°. Flash pt: 23° F
SYNS: DIMETHYLETHOXYSILANE (ACGIH) □ SILANE, ETHOXYDIMETHYL-
CONSENSUS REPORTS: Reported in EPA TSCA Inventory.
ACGIH TLV: TWA 0.5 ppm; STEL: 1.5 ppm
SAFETY PROFILE: An inhalation hazard. Experimental reproductive effects. A flammable liquid. When heated to decomposition it emits acrid smoke and irritating vapors.

EES350   CAS: 110-80-5   HR: 3
2-ETHOXYETHANOL
DOT: UN 1171
mf: $C_4H_{10}O_2$ mw: 90.14
PROP: Colorless liquid; practically odorless. Bp: 135.1°, lel: 1.8%, uel: 14%, fp: −70°, flash p: 202°F (CC), d: 0.9360 @ 15°/15°, autoign temp: 455°F, vap press: 3.8 mm @ 20°, vap d: 3.10. Misc in H_2O, EtOH, Et_2O, and Me_2CO. IDLH 500 ppm.
SYNS: ATHYLENGLYKOL-MONOATHYLATHER (GERMAN) □ CELLOSOLVE (DOT) □ CELLOSOLVE SOLVENT □ DOWANOL EE □ EKTASOLVE EE □ ETHER MONOETHYLIQUE de l'ETHYLENE-GLYCOL (FRENCH) □ ETHYL CELLOSOLVE □ ETHYLENE GLYCOL ETHYL ETHER □ ETHYLENE GLYCOL MONOETHYL ETHER □ ETHYLENE GLYCOL MONOETHYL ETHER (DOT) □ ETOKSYETYLOWY ALKOHOL (POLISH) □ GLYCOL ETHER EE □ GLYCOL ETHYL ETHER □ GLYCOL MONOETHYL ETHER □ HYDROXY ETHER □ JEFFERSOL EE □ NCI-C54853 □ OXITOL □ POLY-SOLV EE
CONSENSUS REPORTS: Reported in EPA TSCA Inventory. Glycol ether compounds are on the Community Right-To-Know List.
OSHA PEL: TWA 200 ppm (skin)
ACGIH TLV: TWA 5 ppm (skin); BEI: 100 mg/g creatinine of 2-ethoxyacetic acid in urine end of shift at end of workweek
DFG MAK: 5 ppm (19 mg/m³)
NIOSH REL: (Glycol Ethers) Reduce to lowest level
DOT CLASSIFICATION: 3; Label: Flammable Liquid
SAFETY PROFILE: Moderately toxic by ingestion, skin contact, intravenous, and intraperitoneal routes. Mildly toxic by inhalation and subcutaneous routes. An experimental teratogen. Other experimental reproductive effects. A mild eye and skin irritant. Combustible when exposed to heat or flame; can react with oxidizing materials. Moderate explosion hazard in the form of vapor when exposed to heat or flame. Mixture with hydrogen peroxide + polyacrylamide gel + toluene is explosive when dry. To fight fire, use alcohol foam, dry chemical. See also GLYCOL ETHERS.

EES400   CAS: 111-15-9   HR: 3
2-ETHOXYETHYL ACETATE
DOT: UN 1172

mf: $C_6H_{12}O_3$ mw: 132.18

PROP: Colorless liquid with a mild, pleasant, ester-like odor. Mp: −61°, bp: 156.4°, flash p: 117°F (COC), lel: 1.7%, fp: −61.7°, d: 0.9748 @ 20°/20°, autoign temp: 715°F, vap press: 1.2 mm @ 20°, vap d: 4.72. IDLH 500 ppm.

SYNS: ACETATE de CELLOSOLVE (FRENCH) □ ACETATE de l'ETHER MONOETHYLIQUE de l'ETHYLENE-GLYCOL (FRENCH) □ ACETATE d'ETHYLGLYCOL (FRENCH) □ ACETATO di CELLO-SOLVE (ITALIAN) □ ACETIC ACID-2-ETHOXYETHYL ESTER □ 2-AETHOXY-AETHYLACET-AT (GERMAN) □ AETHYLENGLYKOLAETHERACETAT (GERMAN) □ CELLOSOLVE ACETATE (DOT) □ CSAC □ EKTASOLVE EE ACETATE SOLVENT □ ETHOXY ACETATE □ 2-ETHOXYETHANOL ACETATE □ 2-ETHOXYETHANOL, ESTER with ACETIC ACID □ 2-ETHOXY-ETHYL-ACETAAT (DUTCH) □ ETHOXYETHYL ACETATE □ β-ETHOXYETHYL ACETATE □ 2-ETHOXYETHYLE, ACETATE de (FRENCH) □ ETHYL CELLOSOLVE ACETAAT (DUTCH) □ ETHYLENE GLYCOL ETHYL ETHER ACETATE □ ETHYLENE GLYCOL MONOETH-YL ETHER ACETATE (MAK, DOT) □ ETHYLGLYKOL-ACETAT (GERMAN) □ 2-ETOSSIETIL-ACETATO (ITALIAN) □ GLYCOL ETHER EE ACETATE □ GLYCOL MONOETHYL ETHER ACETATE □ OCTAN ETOK-SYETYLU (POLISH) □ OXYTOL ACETATE □ POLY-SOLV EE ACETATE

CONSENSUS REPORTS: Reported in EPA TSCA Inventory. Glycol ether compounds are on the Community Right-To-Know List.

OSHA PEL: TWA 100 ppm (skin)

ACGIH TLV: TWA 5 ppm (skin); BEI: 100 mg/g creatinine of 2-ethoxyacetic acid in urine end of shift at end of workweek

DFG MAK: 5 ppm (27 mg/m³)

DOT CLASSIFICATION: 3; Label: Flammable Liquid

SAFETY PROFILE: Moderately toxic by ingestion and intraperitoneal routes. A skin and eye irritant. An experimental teratogen. Other experimental reproductive effects. Flammable liquid when exposed to heat or flame; can react with oxidizing materials. Moderate explosion hazard in the form of vapor when heated. Mild explosions have occurred at the end of distillations. To fight fire, use alcohol foam, CO_2, dry chemical. When heated to decomposition it emits acrid smoke and irritating fumes. See also GLYCOL ETHERS.

EEU100   HR: D
ETHOXYLATED MONO- and DIGLYC-ERIDES

PROP: Mix of stearate, palmitate, and lesser amounts of myristate partial esters of glycerin condensed with approx. 20 moles of ethylene oxide per mole of α-monoglyceride reaction mixtures. (FCC III) Pale, sltly yellow, oily liquid; mildly bitter taste. Sol in water, alc, xylene; sltly sol in mineral oil, vegetable oil.

SYNS: POLYGLYCERATE (60) □ POLYOXYETHYLENE (20) MONO- and DIGLYCERIDES of FATTY ACIDS

SAFETY PROFILE: When heated to decomposition it emits acrid smoke and irritating fumes.

EEV200   CAS: 87-13-8   HR: 2
ETHOXYMETHYLENEMALONIC ACID, ETHYL ESTER

mf: $C_{10}H_{16}O_5$ mw: 216.26

PROP: Colorless liquid. Bp: 279°–281°. Flash pt: 311° F. D: 1.070

SYNS: DIETHYL EMME □ DIETHYL (ETHOXY-METHYLENE)MALONATE □ MALONIC ACID, (ETHOXYMETHYLENE)-, DIETHYL ESTER □ TL 1483

CONSENSUS REPORTS: Reported in EPA TSCA Inventory.

SAFETY PROFILE: Moderately toxic by ingestion. A skin irritant. A combustible liquid. When heated to decomposition it emits acrid smoke and irritating fumes.

EFA100   CAS: 622-62-8   HR: 3
4-ETHOXYPHENOL

mf: $C_8H_{10}O_2$ mw: 138.18

PROP: Leaflets. Mp: 66–67°, bp: 246–247°, d: 1.07 @ 25°/4°.

SYNS: ETHER MONOETHYLIQUE de l'HYDRO-QUINONE □ p-ETHOXYPHENOL □ 4-ETHYLO-XYPHENOL □ HYDROQUINONE MONOETHYL ETHER □ p-HYDROXYPHENETOLE □ PHENOL, 4-ETHOXY-(9CI)

CONSENSUS REPORTS: Reported in EPA TSCA Inventory.

SAFETY PROFILE: Poison by intraperitoneal route. Experimental reproductive effects. An irritant. When heated to decomposition it emits acrid smoke and irritating fumes.

EFE000   CAS: 150-69-6   HR: 3
4-ETHOXYPHENYLUREA
mf: $C_9H_{12}N_2O_2$ mw: 180.23
PROP: Needle-like crystals or plates from water. Mp: 174°.
SYNS: p-AETHOXYPHYLHARNSTOFF (GERMAN) □ DULCINE □ N-(4-ETHOXYPHENYL)UREA □ p-ETHOXYPHENYLUREA □ NCI-C02073 □ PHENETHYL-CARBAMID (GERMAN) □ p-PHENETOLCARBAMID (GERMAN) □ p-PHENETOLCARBAMIDE □ p-PHENETOLECARBAMIDE □ p-PHENETYLUREA □ SUCROL □ SUESSTOFF □ VALZIN
CONSENSUS REPORTS: IARC Cancer Review: Group 3 IMEMDT 7,56,87; Animal Inadequate Evidence IMEMDT 12,97,76.
SAFETY PROFILE: Human poison by ingestion. Moderately toxic experimentally by ingestion. Human systemic effects by ingestion: somnolence, hallucinations, distorted perceptions, and changes in motor activity. In adults 20 to 40 g produces dizziness, nausea, methemoglobinemia, cyanosis, and hypotension. Questionable carcinogen with experimental tumorigenic data. When heated to decomposition it emits toxic fumes of NO_x.

EFL000   CAS: 112-50-5   HR: 1
ETHOXYTRIGLYCOL
mf: $C_8H_{18}O_4$ mw: 178.26
PROP: Bp: 255.4°, flash p: 275°F (OC), d: 1.0208 @ 20°/20°, vap press: 0.01 mm @ 20°.
SYNS: DOWANOL TE □ 2-(2-(2-ETHOXYETHOXY)-ETHOXY)ETHANOL □ ETHOXYTRIETHYLENE GLYCOL □ POLY-SOLV TE □ TRIETHYLENE GLYCOL ETHYL ETHER □ TRIETHYLENE GLYCOL MONOETHYL ETHER □ TRIGLYCOL MONOETHYL ETHER
CONSENSUS REPORTS: Reported in EPA TSCA Inventory. Glycol ether compounds are on the Community Right-To-Know List.
SAFETY PROFILE: Mildly toxic by ingestion and skin contact. An eye irritant. Combustible when exposed to heat or flame; can react with oxidizing materials. To

fight fire, use foam, alcohol foam, CO_2, dry chemical. When heated to decomposition it emits acrid smoke and irritating fumes. See also GLYCOL ETHERS.

EFQ500   CAS: 529-65-7   HR: 2
N-ETHYLACETANILIDE
mf: $C_{10}H_{13}NO$ mw: 163.24
PROP: White crystals, faint odor. Mp: 54°, bp: 258°, d: 0.994, vap d: 5.62.
SYNS: ACETETHYLANILIDE □ ETHYLACETANILIDE
CONSENSUS REPORTS: Reported in EPA TSCA Inventory.
SAFETY PROFILE: Moderately toxic by ingestion. Can react with oxidizing materials. To fight fire, use foam, CO_2, dry chemical. When heated to decomposition it emits toxic fumes of NO_x.

EFR000   CAS: 141-78-6   HR: 3
ETHYL ACETATE
DOT: UN 1173
mf: $C_4H_8O_2$ mw: 88.12
PROP: A volatile, flammable, colorless liquid with fragrant fruity odor. Mp: −83.6°, bp: 77.15°, ULC: 85–90, lel: 2.2%, uel: 11%, flash p: 24°F, d: 0.8946 @ 25°, autoign temp: 800°F, vap press: 100 mm @ 27.0°, vap d: 3.04. Misc with alc, ether, glycerin, volatile oils, water @ 54°, and most org solvs. IDLH 2000 ppm [10%LEL].
SYNS: ACETIC ETHER □ ACETIDIN □ ACETOXY-ETHANE □ AETHYLACETAT (GERMAN) □ ESSIG-ESTER (GERMAN) □ ETHYLACETAAT (DUTCH) □ ETHYL ACETIC ESTER □ ETHYLE (ACETATE d') (FRENCH) □ ETHYL ETHANOATE □ ETILE (ACETATO di) (ITALIAN) □ FEMA No. 2414 □ OCTAN ETYLU (POLISH) □ RCRA WASTE NUMBER U112 □ VINEGAR NAPHTHA
CONSENSUS REPORTS: Reported in EPA TSCA Inventory. EPA Genetic Toxicology Program.
OSHA PEL: TWA 400 ppm
ACGIH TLV: TWA 400 ppm; Not Classifiable as a Human Carcinogen
DFG MAK: 400 ppm (1500 mg/m³)
DOT CLASSIFICATION: 3; Label: Flammable Liquid

SAFETY PROFILE: Poison by inhalation. Moderately toxic by intraperitoneal and subcutaneous routes. Mildly toxic by ingestion. Human systemic effects by inhalation: olfactory changes, conjunctiva irritation, and pulmonary changes. Human eye irritant. Mutation data reported. Irritating to mucous surfaces, particularly the eyes, gums, and respiratory passages, and is also mildly narcotic. On repeated or prolonged exposures, it causes conjunctival irritation and corneal clouding. It can cause dermatitis. High concentrations have a narcotic effect and can cause congestion of the liver and kidneys. Chronic poisoning has been described as producing anemia, leucocytosis (transient increase in the white blood cell count), and cloudy swelling, and fatty degeneration of the viscera. A synthetic flavoring substance and adjuvant.

Highly flammable liquid. A very dangerous fire hazard when exposed to heat or flame; can react vigorously with oxidizing materials. Moderate explosion hazard when exposed to flame. Potentially explosive reaction with lithium tetrahydroaluminate. Ignites on contact with potassium tert-butoxide. Violent reaction with chlorosulfonic acid, (LiAlH$_2$ + 2-chloromethyl furan), oleum. To fight fire, use CO$_2$, dry chemical, or alcohol foam. When heated to decomposition it emits acrid smoke and irritating fumes. See also ESTERS.

EFS000   CAS: 141-97-9   HR: 2
ETHYL ACETYL ACETATE
mf: C$_6$H$_{10}$O$_3$ mw: 130.16
PROP: Colorless liquid; fruity odor. Bp: 180.8°, fp: −45°, flash p: 185°F (COC), autoign temp: 563°F, d: 1.0282 @ 20°/20°, refr index: 1.418, vap press: 1 mm @ 28.5°, vap d: 4.48. Misc in most org solvents; sol in dil alkalies precipitated with CO$_2$; sltly sol in H$_2$O.
SYNS: ACETOACETIC ACID, ETHYL ESTER □ ACETOACETIC ESTER □ ACTIVE ACETYL ACETATE □ DIACETIC ETHER □ EAA □ ETHYL ACETOACETATE

(FCC) □ ETHYL ACETYLACETONATE □ ETHYL BENZYL ACETOACETATE □ ETHYL-3-OXO-BUTANOATE □ ETHYL-3-OXOBUTYRATE □ FEMA No. 2415 □ 3-OXOBUTANOIC ACID ETHYL ESTER
CONSENSUS REPORTS: Reported in EPA TSCA Inventory.
SAFETY PROFILE: Moderately toxic by ingestion. A skin and eye irritant. Combustible liquid when exposed to heat or flame; can react with oxidizing materials. Explosive reaction when heated with Zn + tribromoneopentyl alcohol or 2,2,2-tris(bromomethyl)ethanol. To fight fire, use alcohol foam, CO$_2$, dry chemical. When heated to decomposition it emits acrid smoke and irritating fumes. See also ESTERS.

EFS500   CAS: 107-00-6   HR: 3
ETHYL ACETYLENE
mf: C$_4$H$_6$ mw: 54
PROP: A colorless, highly flammable gas. Bp: 8.3°, d: 0.669 @ 0°/0°, mp: −130°, flash p: <30°F (TOC), <7°C (Gas >8°C).
SYNS: 1-BUTYNE □ ETHYL ACETYLENE, INHIBITED □ ETHYLETHYNE
CONSENSUS REPORTS: Reported in EPA TSCA Inventory.
SAFETY PROFILE: Probably an asphyxiant. A very dangerous fire hazard when exposed to heat, open flame, or powerful oxidizers. A dangerous explosion hazard. To fight fire, stop flow of gas. See also ACETYLENE and ACETYLENE COMPOUNDS.

EFS600   CAS: 539-88-8   HR: 1
ETHYL 3-ACETYLPROPIONATE
mf: C$_7$H$_{12}$O$_3$ mw: 144.19
PROP: Liquid. D: 1.012, bp: 205–206°. Very sol in water; miscible with alc.
SYNS: ETHYL KETOVALERATE □ ETHYL 4-KETOVALERATE □ ETHYL LAEVULINATE □ ETHYL LEVULATE □ ETHYL 4-OXOPENTANOATE □ ETHYL 4-OXOVALERATE □ LEVULINIC ACID, ETHYL ESTER □ PENTANOIC ACID, 4-OXO-, ETHYL ESTER (9CI)
CONSENSUS REPORTS: Reported in EPA TSCA Inventory.

SAFETY PROFILE: A skin irritant. When heated to decomposition it emits acrid smoke and irritating fumes.

EFT000   CAS: 140-88-5   HR: 3
ETHYL ACRYLATE
DOT: UN 1917
mf: $C_5H_8O_2$ mw: 100.13
PROP: Colorless liquid; acrid, penetrating odor. Mp: −71.2°, bp: 99.8°, fp: <−72°, lel: 1.8%, flash p: 60°F (OC), d: 0.916–0.919, vap press: 29.3 mm @ 20°, vap d: 3.45. Misc with alc, ether; sltly sol in water. IDLH 300 ppm.
SYNS: ACRYLATE d'ETHYLE (FRENCH) □ ACRYLIC ACID ETHYL ESTER □ ACRYLSAEUREAETHYLESTER (GERMAN) □ AETHYLACRYLAT (GERMAN) □ ETHOXYCARBONYLETHYLENE □ ETHYLACRYLAAT (DUTCH) □ ETHYLAKRYLAT (CZECH) □ ETHYL PROPENOATE □ ETHYL-2-PROPENOATE □ ETIL ACRILATO (ITALIAN) □ ETILACRILATULUI (ROMANIAN) □ FEMA No. 2418 □ NCI-C50384 □ 2-PROPENOIC ACID, ETHYL ESTER (MAK) □ RCRA WASTE NUMBER U113
CONSENSUS REPORTS: IARC Cancer Review: Group 2B IMEMDT 7,56,87; Animal Sufficient Evidence IMEMDT 39,81,86; Animal Inadequate Evidence IMEMDT 19,47,79; Human Inadequate Evidence IMEMDT 19,47,79. NTP Carcinogenesis Studies (gavage); Clear Evidence: mouse, rat NTPTR* NTP-TR-259,86. Reported in EPA TSCA Inventory. Community Right-To-Know List.
OSHA PEL: TWA 5 ppm; STEL 25 ppm (skin)
ACGIH TLV: TWA 5 ppm; STEL 15 ppm; Suspected Human Carcinogen
DFG MAK: 5 ppm (21 mg/m³)
DOT CLASSIFICATION: 3; Label: Flammable Liquid
SAFETY PROFILE: Confirmed carcinogen with experimental carcinogenic data. Poison by ingestion and inhalation. Moderately toxic by skin contact and intraperitoneal routes. Human systemic effects by inhalation: eye, olfactory, and pulmonary changes. A skin and eye irritant. Characterized in its terminal stages by dyspnea, cyanosis, and convulsive movements. It caused severe local irritation of the gastroenteric tract; and toxic degenerative changes of cardiac, hepatic, renal, and splenic tissues were observed. It gave no evidence of cumulative effects. When applied to the intact skin of rabbits, the ethyl ester caused marked local irritation, erythema, edema, thickening, and vascular damage. Animals subjected to a fairly high concentration of these esters suffered irritation of the mucous membranes of the eyes, nose, and mouth as well as lethargy, dyspnea, and convulsive movements. A substance that migrates to food from packaging materials.

Flammable liquid. A very dangerous fire hazard when exposed to heat or flame; can react vigorously with oxidizing materials. Violent reaction with chlorosulfonic acid. To fight fire, use CO_2, dry chemical, or alcohol foam. When heated to decomposition it emits acrid smoke and irritating fumes. See also ESTERS.

EFT500   CAS: 462-95-3   HR: 3
ETHYLAL
DOT: UN 2373
mf: $C_5H_{12}O_2$ mw: 104.17
PROP: Bp: 89°, flash p: <69.8°F. Sol in H_2O.
SYN: DIETHOXYMETHANE (DOT)
CONSENSUS REPORTS: Reported in EPA TSCA Inventory.
DOT CLASSIFICATION: 3; Label: Flammable Liquid
SAFETY PROFILE: Moderately toxic by ingestion. Flammable when exposed to heat or flame; can react vigorously with oxidizers. When heated to decomposition it emits acrid smoke and irritating fumes.

EFU000   CAS: 64-17-5   HR: 3
ETHYL ALCOHOL
DOT: UN 1170/UN 1986/UN 1987
mf: C_2H_6O mw: 46.08
PROP: Clear, colorless, very mobile liquid; fragrant odor and burning taste. Bp: 78.32°,

ULC: 70, lel: 3.3%, uel: 19% @ 60°, fp: −117°, flash p: 55.6°F, d: 0.7893 @ 20°/4°, autoign temp: 793°F, vap press: 40 mm @ 19°, vap d: 1.59, refr index: 1.364. Misc in water, alc, chloroform, ether, and most org solvs. IDLH 3300 ppm [10%LEL].

SYNS: ABSOLUTE ETHANOL □ AETHANOL (GERMAN) □ AETHYLALKOHOL (GERMAN) □ ALCOHOL □ ALCOHOL, anhydrous □ ALCOHOL, dehydrated □ ALCOHOLS, n.o.s. (UN 1987) (DOT) □ ALCOHOLS, toxic, n.o.s. (UN 1986) (DOT) □ ALCOOL ETHYLIQUE (FRENCH) □ ALCOOL ETILICO (ITALIAN) □ ALGRAIN □ ALKOHOL (GERMAN) □ ALKOHOLU ETYLOWEGO (POLISH) □ ANHYDROL □ COLOGNE SPIRIT □ ETANOLO (ITALIAN) □ ETHANOL (MAK) □ ETHANOL 200 PROOF □ ETHANOL SOLUTIONS (UN 1170) (DOT) □ ETHYLALCOHOL (DUTCH) □ ETHYL ALCOHOL, anhydrous □ ETHYL ALCOHOL SOLUTIONS (UN 1170) (DOT) □ ETHYL HYDRATE □ ETHYL HYDROXIDE □ ETYLOWY ALKOHOL (POLISH) □ FERMENTATION ALCOHOL □ GRAIN ALCOHOL □ JAYSOL □ JAYSOL S □ METHYLCARBINOL □ MOLASSES ALCOHOL □ NCI-C03134 □ POTATO ALCOHOL □ SD ALCOHOL 23-HYDROGEN □ SPIRIT □ SPIRITS of WINE □ TECSOL

CONSENSUS REPORTS: IARC Cancer Review: Human Sufficient Evidence IMEMDT 44,259,88. Reported in EPA TSCA Inventory. EPA Genetic Toxicology Program.

OSHA PEL: TWA 1000 ppm
ACGIH TLV: TWA 1000 ppm; Not Classifiable as a Human Carcinogen
DFG MAK: 500 ppm (960 mg/m³)
DOT CLASSIFICATION: 3; Label: Flammable Liquid (UN 1987, UN 1170); DOT Class: 3; Label: Flammable Liquid, Poison (UN 1986)
SAFETY PROFILE: Confirmed human carcinogen for ingestion of beverage alcohol. Experimental tumorigenic and teratogenic data. Moderately toxic to humans by ingestion. Moderately toxic experimentally by intravenous and intraperitoneal routes. Mildly toxic by inhalation and skin contact. Human systemic effects by ingestion and subcutaneous routes: sleep disorders, hallucinations, distorted perceptions, convulsions, motor activity changes, ataxia, coma, antipsychotic,

headache, pulmonary changes, alteration in gastric secretion, nausea or vomiting, other gastrointestinal changes, menstrual cycle changes, and body temperature decrease. Can also cause glandular effects in humans. Human reproductive effects by ingestion, intravenous, and intrauterine routes: changes in female fertility index. Effects on newborn include: changes in Apgar score, neonatal measures or effects, and drug dependence. Experimental reproductive effects. Human mutation data reported. An eye and skin irritant.

The systemic effect of ethanol differs from that of methanol. Ethanol is rapidly oxidized in the body to carbon dioxide and water, and, in contrast to methanol, no cumulative effect occurs. Though ethanol possesses narcotic properties, concentrations sufficient to produce this effect are not reached in industry. Concentrations below 1000 ppm usually produce no signs of intoxication. Exposure to concentrations over 1000 ppm may cause headache, irritation of the eyes, nose, and throat, and, if continued for an hour, drowsiness and lassitude, loss of appetite, and inability to concentrate. There is no concrete evidence that repeated exposure to ethanol vapor results in cirrhosis of the liver. Ingestion of large doses can cause alcohol poisoning. Repeated ingestions can lead to alcoholism. It is a central nervous system depressant.

Flammable liquid when exposed to heat or flame; can react vigorously with oxidizers. To fight fire, use alcohol foam, CO_2, dry chemical. Explosive reaction with the oxidized coating around potassium metal. Ignites and then explodes on contact with acetic anhydride + sodium hydrogen sulfate. Reacts violently with acetyl bromide (evolves hydrogen bromide), dichloromethane + sulfuric acid + nitrate or nitrite, disulfuryl difluoride, tetrachlorosilane + water, and strong oxidants. Ignites on contact with disulfuric acid + nitric acid, phosphorus(III) oxide, platinum, potassium-tert-butoxide + acids. Forms explosive products in reaction with ammonia + silver nitrate (forms silver nitride and silver fulminate), magnesium perchlorate (forms ethyl perchlorate), nitric acid + silver (forms silver fulminate), silver nitrate (forms ethyl nitrate), silver(I) oxide + ammonia or hydrazine (forms silver nitride and silver fulminate), sodium (evolves hydrogen gas). Incompatible with acetyl chloride, BrF_5, $Ca(OCl)_2$, ClO_3, CrO_3, $Cr(OCl)_2$, (cyanuric acid + H_2O), H_2O_2, HNO_3, (H_2O_2 + H_2SO_4), (I + CH_3OH + HgO), [$Mn(ClO_4)_2$ + 2,2-dimethoxy propane], $Hg(NO_3)_2$, $HClO_4$, perchlorates, (H_2SO_4 + permanganates), $HMnO_4$, KO_2, $KOC(CH_3)_3$, $AgClO_4$, NaH_3N_2, $UO_2(ClO_4)_2$.

EFU400   CAS: 75-04-7   HR: 3
ETHYLAMINE
DOT: UN 1036/UN 2270
mf: C_2H_7N mw: 45.10
PROP: Colorless gas or liquid; strong ammonia-like odor. Bp: 16.6°, flammable, lel: 4.95%, uel: 20.75%, fp: −80.6°, flash p: −0.4°F, d: 0.662 @ 20°/4°, autoign temp: 725°F, vap d: 1.56, vap press: 400 mm @ 20°. Misc with water, alc, and ether; salted out by NaOH. IDLH 600 ppm.
SYNS: AETHYLAMINE (GERMAN) □ AMINOETHANE □ 1-AMINOETHANE □ ETHANAMINE □ ETHYL-AMINE (UN 1036) (DOT) □ ETHYLAMINE, aqueous solution with not <50% but not >70% ethylamine (UN 2270) (DOT) □ ETILAMINA (ITALIAN) □ ETYLOAMINA (POLISH) □ MONOETHYLAMINE (DOT) □ MONOETHYLAMINE, anhydrous (DOT)
CONSENSUS REPORTS: Reported in EPA TSCA Inventory.
OSHA PEL: TWA 10 ppm
ACGIH TLV: TWA 5 ppm; 15 ppm STEL (skin)
DFG MAK: 5 ppm (9.4 mg/m³)
DOT CLASSIFICATION: 2.1; Label: Flammable Gas (UN 1036); DOT Class: 3; Label: Flammable Liquid (UN 2270)
SAFETY PROFILE: A poison by ingestion, skin contact, and intravenous routes. Moderately toxic by inhalation. A severe eye irritant. A very dangerous fire hazard when exposed to heat or flame. Moderate explosion hazard when exposed to spark or flame. Keep away from heat and open flame, can react vigorously with oxidizing materials. To fight fire, stop flow of gas, use alcohol foam, dry chemical. Incompatible with cellulose nitrate or oxidizers. When heated to decomposition it emits toxic fumes of NO_x. See also AMINES.

EFX000   CAS: 94-09-7   HR: 3
ETHYL-4-AMINOBENZOATE
mf: $C_9H_{11}NO_2$ mw: 165.21
PROP: Crystals or needles from alc. Mp: 92°, bp: 183–184° @ 14 mm.
SYNS: AMERICAINE □ p-AMINOBENZOIC ACID ETHYL ESTER □ 4-AMINOBENZOIC ACID ETHYL ESTER □ ANESTHESIN □ ANESTHONE □ BENZOCAINE □ ETHYL AMINOBENZOATE □ ETHYL-p-AMINOBENZOATE □ KELOFORM □ NORCAIN □ ORTHESIN □ PARATHESIN □ TOPCAINE
CONSENSUS REPORTS: Reported in EPA TSCA Inventory.
SAFETY PROFILE: Poison by ingestion and intraperitoneal routes. Human systemic effects by rectal route: methemoglobinemia/carboxyhemoglobinemia in infants. A skin irritant and a mild sensitizer. Local contact may cause contact dermatitis. Used as a topical anesthetic and as a sun-screening agent. When heated to decomposition it emits highly toxic fumes of NO_x. See also ETHYL ALCOHOL and ESTERS.

EGA500   CAS: 110-73-6   HR: 3
2-ETHYLAMINOETHANOL
mf: $C_4H_{11}NO$ mw: 89.16
PROP: Oily liquid with faint odor, fumes in air. Bp: 169–170°, flash p: 160°F (OC), d: 0.92, vap d: 3.06. Sol in H_2O.
SYNS: 2-(ETHYLAMINO)ETHANOL □ 2-N-MONOETHYLAMINOETHANOL
CONSENSUS REPORTS: Reported in EPA TSCA Inventory.
SAFETY PROFILE: Poison by skin contact. Moderately toxic by ingestion and intraperitoneal routes. A skin and severe eye irritant. Flammable when exposed to heat or

flame; can react vigorously with oxidizers. To fight fire, use alcohol foam, dry chemical, CO_2. When heated to decomposition it emits toxic fumes of NO_x.

EGI000   CAS: 13275-68-8   HR: 3
2-ETHYLAMINO-1,3,4-THIADIAZOLE
mf: $C_4H_7N_3S$ mw: 129.20
SYNS: CL 19217 4090L 7-5525 □ 2-ETHYLAMINOTHIADIAZOLE □ NSC-4730
CONSENSUS REPORTS: Reported in EPA TSCA Inventory.
SAFETY PROFILE: Poison by intraperitoneal and subcutaneous routes. Experimental teratogenic effects. When heated to decomposition it emits very toxic fumes of NO_x and SO_x.

EGI750   CAS: 541-85-5   HR: 3
ETHYL AMYL KETONE
DOT: UN 2271
mf: $C_8H_{16}O$ mw: 128.24
PROP: Liquid; mild, fruity odor. Bp: 157–162°, d: 0.822 @ 20°/20°, flash p: 138°F. Sol in many org solvs. IDLH 100 ppm.
SYNS: ETHYL sec-AMYL KETONE □ 3-HEPTANONE, 5-METHYL- □ 3-METHYL-5-HEPTANONE □ 5-METHYL-3-HEPTANONE
CONSENSUS REPORTS: Reported in EPA TSCA Inventory.
OSHA PEL: TWA 25 ppm
ACGIH TLV: TWA 10 ppm
DOT CLASSIFICATION: 3; Label: Flammable Liquid
SAFETY PROFILE: Moderately irritating to skin, eyes, and mucous membranes by inhalation and ingestion. Narcotic in high concentration. Flammable liquid when exposed to heat, sparks, or flame. When heated to decomposition it emits acrid smoke. To fight fire, use foam, CO_2, dry chemical. See also KETONES.

EGK000   CAS: 103-69-5   HR: 3
N-ETHYLANILINE
DOT: UN 2272
mf: $C_8H_{11}N$ mw: 121.20

PROP: Clear, yellow-brown oily liquid. Mp: −63.5°, bp: 204°, d: 0.963 @ 20°/4°, fp: −80°, vap press: 1 mm @ 38.5°, vap d: 4.18, flash p: 185°F (OC).
SYNS: AETHYLANILIN (GERMAN) □ ANILINOETHANE □ N-ETHYLAMINOBENZENE □ ETHYLANILINE □ N-ETHYLBENZENAMINE □ N-ETHYLBENZENAMINO □ ETHYLPHENYLAMINE
CONSENSUS REPORTS: Reported in EPA TSCA Inventory.
DOT CLASSIFICATION: 6.1; Label: KEEP AWAY FROM FOOD
SAFETY PROFILE: Poison by ingestion and intraperitoneal routes. Moderately toxic by an unspecified route. Mildly toxic by skin contact. An allergen. Flammable when exposed to heat or flame; can react with oxidizing materials. To fight fire, use dry chemical, CO_2, foam. Hypergolic reaction with red fuming nitric acid. When heated to decomposition or on contact with acid or acid fumes it emits highly toxic fumes of aniline and NO_x.

EGK500   CAS: 578-54-1   HR: 3
2-ETHYLANILINE
DOT: UN 2273
mf: $C_8H_{11}N$ mw: 121.20
PROP: Yellow liquid, darkens upon standing. Mp: −63.5°, bp: 215°, flash p: 185°F (OC), d: 0.98 @ 25°/25°, vap d: 4.17.
SYNS: o-AMINOETHYLBENZENE □ ANILINE, o-ETHYL-(8CI) □ BENZENAMINE, 2-ETHYL-(9CI) □ 2-ETHYL ANILINE □ 2-ETHYLANILINE (DOT) □ 2-ETHYLBENZENAMINE
CONSENSUS REPORTS: Reported in EPA TSCA Inventory.
DOT CLASSIFICATION: 6.1; Label: KEEP AWAY FROM FOOD
SAFETY PROFILE: A poison. Moderately toxic by ingestion. Flammable when exposed to heat or flame; can react with oxidizing materials. To fight fire, use foam, CO_2, dry chemical. When heated to decomposition it emits highly toxic fumes of aniline and NO_x. See also N-ETHYLANILINE.

E

EGL000   CAS: 589-16-2   HR: 3
4-ETHYLANILINE
mf: C$_8$H$_{11}$N   mw: 121.20
PROP: D: 0.963, mp: −5°, bp: 213–214°.
Insol in water; misc in alc and ether.
SYNS: 1-AMINO-4-ETHYLBENZENE □ p-
ETHYLANILINE
CONSENSUS REPORTS: Reported in EPA
TSCA Inventory.
SAFETY PROFILE: Poison by ingestion,
intravenous, and intraperitoneal routes.
Mutation data reported. When heated to
decomposition it emits toxic fumes of NO$_x$.
See also N-ETHYLANILINE.

EGM000   CAS: 87-25-2   HR: 2
ETHYL ANTHRANILATE
mf: C$_9$H$_{11}$NO$_2$ mw: 165.21
PROP: Colorless to amber liquid; floral,
orange-blossom odor. D: 1.115–1.120, refr
index: 1.563–1.566, flash p: 151°F. Sol in
alc, fixed oils, propylene glycol.
SYNS: o-AMINOBENZOIC ACID, ETHYL ESTER □
ETHYL-o-AMINOBENZOATE □ FEMA No. 2421
CONSENSUS REPORTS: Reported in EPA
TSCA Inventory.
SAFETY PROFILE: Moderately toxic by
ingestion. A skin irritant. Combustible
liquid. When heated to decomposition it
emits toxic fumes of NO$_x$.

EGP500   CAS: 100-41-4   HR: 3
ETHYL BENZENE
DOT: UN 1175
mf: C$_8$H$_{10}$ mw: 106.18
PROP: Colorless liquid; aromatic odor. Bp:
136.2°, fp: −94.9°, flash p: 59°F, d: 0.8669
@ 20°/4°, autoign temp: 810°F, vap press:
10 mm @ 25.9°, vap d: 3.66, lel: 1.2%, uel:
6.8%. Misc in alc and ether; insol in NH$_3$;
sol in SO$_2$. IDLH 800 ppm [10%LEL].
SYNS: AETHYLBENZOL (GERMAN) □ EB □
ETHYLBENZEEN (DUTCH) □ ETHYLBENZOL □
ETILBENZENE (ITALIAN) □ ETYLOBENZEN (POLISH)
□ NCI-C56393 □ PHENYLETHANE
CONSENSUS REPORTS: Reported in EPA
TSCA Inventory. EPA Genetic Toxicology
Program. Community Right-To-Know List.
OSHA PEL: TWA 100 ppm; STEL 125 ppm

ACGIH TLV: TWA 100 ppm; STEL 125
ppm; Confirmed Animal Carcinogen with
Unknown Revelance to Humans; BEI: 1.5
g/g creatinine of manelic acid) in urine at
end of shift at end of workweek
DFG MAK: 100 ppm (440 mg/m^3)
NIOSH REL: (Ethyl Benzene) TWA 100
ppm; STEL 125 ppm
DOT CLASSIFICATION: 3; Label:
Flammable Liquid
SAFETY PROFILE: Moderately toxic by
ingestion and intraperitoneal routes. Mildly
toxic by inhalation and skin contact. An
experimental teratogen. Other experimental
reproductive effects. Human systemic
effects by inhalation: eye, sleep, and
pulmonary changes. An eye and skin irritant.
Human mutation data reported. The liquid
is an irritant to the skin and mucous
membranes. A concentration of 0.1% of the
vapor in air is an irritant to human eyes, and
a concentration of 0.2% is extremely
irritating at first, then causes dizziness,
irritation of the nose and throat, and a sense
of constriction in the chest. Exposure of
guinea pigs to 1% concentration has been
reported as causing ataxia, loss of
consciousness, tremor of the extremities,
and finally death through respiratory failure.
The pathological findings were congestion
of the brain and lungs with edema.
   A very dangerous fire and explosion
hazard when exposed to heat or flame; can
react vigorously with oxidizing materials. To
fight fire, use foam, CO$_2$, dry chemical.
Emitted from modern building materials
(CENEAR 69,22,91). When heated to
decomposition it emits acrid smoke and
irritating fumes.

EGR000   CAS: 93-89-0   HR: 2
ETHYL BENZOATE
mf: C$_9$H$_{10}$O$_2$ mw: 150.19
PROP: Colorless liquid; heavy fruity odor.
Mp: −34.6°, bp: 213.4°, flash p: >204°F, d:
1.048 @ 20°/20°, fp: −34°, refr index:
1.502–1.506, vap press: 1 mm @ 44.0°, vap
d: 5.17, autoign temp: 914°F. Sol in alc,

fixed oils, and propylene glycol; insol in glycerin, water @ 212°; misc in petroleum, chloroform, and ether.

SYNS: BENZOIC ETHER ☐ ESSENCE of NIOBE ☐ FEMA No. 2422

CONSENSUS REPORTS: Reported in EPA TSCA Inventory.

SAFETY PROFILE: Moderately toxic by ingestion. Mildly toxic by skin contact. A skin and eye irritant. Combustible liquid when exposed to heat or flame; can react with oxidizing materials. To fight fire, use foam, CO_2, dry chemical. When heated to decomposition it emits acrid smoke and irritating fumes. See also ESTERS.

EGV000   CAS: 105-36-2   HR: 3
ETHYL BROMACETATE
DOT: UN 1603
mf: $C_4H_7BrO_2$ mw: 167.02
PROP: Colorless to straw-colored liquid. Bp: 158.8°, fp: <−20°, flash p: 118°F, d: 1.514 @ 13°/4°, vap d: 5.8. Insol in water; misc in alc and ether.

SYNS: ANTOL ☐ BROMOACETIC ACID, ETHYL ESTER ☐ ETHOXYCARBONYLMETHYL BROMIDE ☐ ETHYL BROMOACETATE ☐ ETHYL-α-BROMOACETATE ☐ ETHYL MONOBROMOACETATE

CONSENSUS REPORTS: Reported in EPA TSCA Inventory.

DOT CLASSIFICATION: 6.1; Label: Poison
SAFETY PROFILE: A poison. An irritant to skin, eyes, and mucous membranes. Questionable carcinogen with experimental neoplastigenic data. Flammable liquid when exposed to heat, flame, and oxidizers. Will react with water or steam to produce toxic and corrosive fumes. To fight fire, use water as a fire blanket. When heated to decomposition or on contact with acid or acid fumes, it emits highly toxic fumes of Br^-. See also BROMIDES.

EGV400   CAS: 74-96-4   HR: 3
ETHYL BROMIDE
DOT: UN 1891
mf: C_2H_5Br mw: 108.98
PROP: Colorless, volatile liquid. Mp: −119°, bp: 38.4°, fp: −125.5°, lel: 6.7%, uel: 11.3%,

flash p: <−4°F, d: 1.451 @ 20°/4°, autoign temp: 952°F, vap press: 400 mm @ 21°, vap d: 3.76. IDLH 2000 ppm.

SYNS: BROMIC ETHER ☐ BROMOETHANE ☐ BROMURE d'ETHYLE ☐ ETYLU BROMEK (POLISH) ☐ HALON 2001 ☐ HYDROBROMIC ETHER ☐ MONOBROMOETHANE ☐ NCI-C55481

CONSENSUS REPORTS: NTP Carcinogenesis Studies (inhalation); Clear Evidence: mouse NTPTR* NTP-TR-363,89; (inhalation); Some Evidence: rat NTPTR* NTP-TR-363,89. EPA Genetic Toxicology Program. Reported in EPA TSCA Inventory.

OSHA PEL: TWA 200 ppm; STEL 250 ppm
ACGIH TLV: TWA 5 ppm (skin); Animal Carcinogen
DFG MAK: Animal Carcinogen, Suspected Human Carcinogen
DOT CLASSIFICATION: 6.1; Label: Poison
SAFETY PROFILE: Confirmed carcinogen. Moderately toxic by ingestion and intraperitoneal routes. Mildly toxic by inhalation. An eye and skin irritant. Physiologically, it is an anesthetic and narcotic. Its vapors are markedly irritating to the lungs on inhalation for even short periods. It can produce acute congestion and edema. Liver and kidney damage in humans has been reported. It is much less toxic than methyl bromide, but more toxic than ethyl chloride. It is a preparative hazard. Dangerously flammable by heat, open flame (sparks), oxidizers. Moderately explosive when exposed to flame. Reacts with water or steam to produce toxic and corrosive fumes. Vigorous reaction with oxidizing materials. To fight fire, use CO_2, dry chemical. Readily decomposes when heated to emit toxic fumes of Br^-. See also BROMIDES.

EGV500   CAS: 4824-78-6   HR: 3
ETHYL BROMOPHOS
mf: $C_{10}H_{12}BrCl_2O_3PS$ mw: 394.06
PROP: Pale-yellow liquid. D: 1.52–1.55 (tech) @ 20°, bp: 122–123° @ 0.004 mm.

SYNS: 4-BROMO-2,5-DICHLOROPHENOL-o-ESTER with O,O-DIETHYL PHOSPHOROTHIOATE ☐ O-(4-BROMO-

2,5-DICHLOROPHENYL)-O,O-DIETHYL PHOS-
PHOROTHIOATE □ O-(4-BROMO-2,5-DICHLOROPHEN-
YL)-O,O-DIETHYLPHOSPHOROTHIONATE □
BROMOFOS-ETHYL □ BROMOPHOSETHYL □ CELA S-
2225 □ O,O-DIAETHYL-O-(4-BROM-2,5-DICHLOR)-
PHENYL-MONOTHIOPHOSPHAT (GERMAN) □ O,O-
DIAETHYL-O-(2,5-DICHLOR-4-BROMPHENYL)-
THIONOPHOSPHAT (GERMAN) □ O,O-DIETHYL-O-(4-
BROOM-2,5-DICHLOOR-FENYL)-MONOTHIOFOSFAAT
(DUTCH) □ O,O-DIETHYL O-2,5-DICHLORO-4-
BROMOPHENYL-PHOSPHOROTHIOATE □ O,O-
DIETHYL O-(2,5-DICHLORO-4-BROMOPHENYL)
THIOPHOSPHATE □ O,O-DIETIL-O-(4-BROMO-2,5-
DICLORO-FENIL)-MONOTIOFOSFATO (ITALIAN) □
ENT 27,258 □ FILARIOL □ NEXAGAN □ OMS-659 □ S
2225 □ THIOPHOSPHATE de O,O-DIETHYLE et de O-(2,5-
DICHLORO-4-BROMO) PHENYLE (FRENCH)

CONSENSUS REPORTS: Chlorophenol compounds are on the Community Right-To-Know List.

SAFETY PROFILE: Poison by ingestion. Moderately toxic by skin contact and inhalation. An insecticide. When heated to decomposition it emits very toxic fumes of Br^-, PO_x, SO_x, and Cl^-. See also CHLOROPHENOLS.

EGV600   CAS: 3404-63-5   HR: 3
2-ETHYLBUTADIENE
mf: C_6H_{10} mw: 82.16
SYNS: 1,3-BUTADIENE, 2-ETHYL- □ 2-ETHYL-1,3-
BUTADIENE □ 1-PENTENE, 3-METHYLENE-
SAFETY PROFILE: A poison by intravenous route. When heated to decomposition it emits acrid smoke and irritating vapors.

EGW000   CAS: 97-95-0   HR: 3
2-ETHYLBUTANOL
DOT: UN 2275
mf: $C_6H_{14}O$ mw: 102.20
PROP: Clear liquid. Bp: 144–146°, flash p: 135°F (COC), d: 0.8328, vap press: 0.9 mm @ 20°, vap d: 3.4.
SYNS: 2-ETHYLBUTANOL-1 □ 2-ETHYL-1-BUTANOL
□ 2-ETHYLBUTYL ALCOHOL □ sec-HEXANOL (DOT) □
sec-HEXYL ALCOHOL □ 3-METHYLOLPENTANE □ sec-
PENTYLCARBINOL □ 3-PENTYLCARBINOL □
PSEUDOHEXYL ALCOHOL
CONSENSUS REPORTS: Reported in EPA TSCA Inventory.

DOT CLASSIFICATION: 3; Label: Flammable Liquid
SAFETY PROFILE: Moderately toxic by ingestion and skin contact. A skin and severe eye irritant. Flammable liquid when exposed to heat or flame; can react with oxidizing materials. To fight fire, use dry chemical, CO_2, foam, fog. When heated to decomposition it emits acrid smoke and irritating fumes. See also ALCOHOLS.

EGW500   CAS: 760-21-4   HR: 3
2-ETHYL-1-BUTENE
mf: C_6H_{12} mw: 84.18
PROP: A liquid. Flash p: $<-4°$, autoign temp: 599°F, d: 0.69, vap d: 2.9, bp: 64.7°.
CONSENSUS REPORTS: Reported in EPA TSCA Inventory.
SAFETY PROFILE: A human eye irritant. A very dangerous fire hazard when exposed to heat, flames, or oxidizers. To fight fire, use dry chemical, CO_2, foam, spray. When heated to decomposition it emits acrid smoke and irritating fumes.

EGZ000   CAS: 3953-10-4   HR: 3
2-ETHYLBUTYLACRYLATE
mf: $C_9H_{16}O_2$ mw: 156.25
PROP: Clear, colorless liquid. Bp: 82° @ 10 mm, fp: $-70°$, flash p: 125°F (OC), d: 0.8964 @ 20°/20°, vap press: 1.7 mm @ 20°.
SYNS: 2-ETHYLBUTYL ESTER, ACRYLIC ACID □ 2-
ETHYLBUTYLESTER KYSELINY AKRYLOVE □ 2-
PROPENOIC ACID-2-ETHYLBUTYL ESTER
CONSENSUS REPORTS: Reported in EPA TSCA Inventory.
SAFETY PROFILE: Mildly toxic by ingestion and skin contact. An eye and severe skin irritant. Flammable liquid when exposed to heat or flame; can react with oxidizing materials. To fight fire, use foam, CO_2, dry chemical. When heated to decomposition it emits acrid smoke and irritating fumes. See also ESTERS.

E

EHA000   CAS: 617-79-8   HR: 3
2-ETHYLBUTYLAMINE
mf: $C_6H_{15}N$ mw: 101.22
PROP: Water-white liquid. Bp: 125°, flash p: 64°F (OC), d: 0.739 @ 20°/20°, vap d: 3.5.
SYN: 2-ETHYL-1-BUTANAMINE
SAFETY PROFILE: Poison by ingestion. Moderately toxic by skin contact. A skin and severe eye irritant. A very dangerous fire hazard when exposed to heat or flame; can react vigorously with oxidizing materials. Keep away from heat and open flame. To fight fire, use dry chemical, CO_2, foam. When heated to decomposition it emits toxic fumes of NO_x. See also AMINES.

EHA500   CAS: 628-81-9   HR: 3
ETHYL BUTYL ETHER
DOT: UN 1179
mf: $C_6H_{14}O$ mw: 102.20
PROP: Colorless liquid. Bp: 92°, mp: −124°, flash p: 40°F, d: 0.7528 @ 20°/20°, vap d: 3.52. Insol in water; misc in alc and ether.
SYN: ETHER ETHYLBUTYLIQUE (FRENCH)
CONSENSUS REPORTS: Reported in EPA TSCA Inventory.
DOT CLASSIFICATION: 3; Label: Flammable Liquid
SAFETY PROFILE: Moderately toxic by ingestion. A skin and eye irritant. A very dangerous fire hazard when exposed to heat or flame; can react vigorously with oxidizing materials. Keep away from heat and open flame. To fight fire, use alcohol foam, CO_2, dry chemical. When heated to decomposition it emits acrid smoke and irritating fumes. See also ETHERS.

EHA600   CAS: 106-35-4   HR: 3
ETHYL BUTYL KETONE
mf: $C_7H_{14}O$ mw: 114.21
PROP: Clear mobile liquid; fatty odor. Mp: −36.7°, bp: 149–152°, flash p: 115°F (OC), d: 0.8198 @ 20°/20°, vap d: 3.93. Misc with alc, ether, water @ 149°. IDLH 1000 ppm.
SYNS: AETHYLBUTYLKETON (GERMAN) □ n-BUTYL ETHYL KETONE □ EPTAN-3-ONE (ITALIAN) □ ETHYLBUTYLCETONE (FRENCH) □ ETHYLBUTYL-KETON (DUTCH) □ ETILBUTILCHETONE (ITALIAN) □

FEMA No. 2545 □ HEPTAN-3-ON (DUTCH, GERMAN) □ HEPTAN-3-ONE □ 3-HEPTANONE
CONSENSUS REPORTS: Reported in EPA TSCA Inventory.
OSHA PEL: TWA 50 ppm
ACGIH TLV: TWA 50 ppm; STEL 75 ppm
DOT CLASSIFICATION: 3; Label: Flammable Liquid
SAFETY PROFILE: Moderately toxic by ingestion and inhalation. A skin and eye irritant. A flammable liquid. Can react with oxidizing materials. To fight fire, use foam, CO_2, dry chemical. See also KETONES.

EHC000   CAS: 4549-44-4   HR: 3
ETHYL-N-BUTYLNITROSAMINE
mf: $C_6H_{14}N_2O$ mw: 130.22
SYNS: AETHYL-N-BUTYL-NITROSOAMIN (GERMAN) □ N-ETHYL-N-NITROSOBUTYLAMINE □ N-NITROSO-N-BUTYLETHYLAMINE □ N-NITROSOETHYL-N-BUTYLAMINE
SAFETY PROFILE: Poison by ingestion and intravenous routes. Questionable carcinogen with experimental carcinogenic and tumorigenic data. Mutation data reported. When heated to decomposition it emits toxic fumes of NO_x. See also NITROSAMINES.

EHC900   CAS: 68037-57-0   HR: 3
2-ETHYLBUTYL SILICATE
SYNS: POLYBIS(2-ETHYLBUTYL)SILOXANE □ SILICIC ACID, 2-ETHYLBUTYL ESTER
CONSENSUS REPORTS: Reported in EPA TSCA Inventory.
SAFETY PROFILE: Poison by ingestion route. A skin irritant. When heated to decomposition it emits acrid smoke and irritating fumes.

EHE000   CAS: 105-54-4   HR: 3
ETHYL n-BUTYRATE
DOT: UN 1180
mf: $C_6H_{12}O_2$ mw: 116.18
PROP: Colorless liquid; banana-pineapple odor. D: 0.900 @ 0°/4°, refr index: 1.391, mp: −100.8°, fp: −93.3°, bp: 121.6°, flash p: 79°F. Sol in water, fixed oils, propylene

glycol; misc in alc and ether; insol in glycerin @ 121°.

SYNS: BUTANOIC ACID ETHYL ESTER □ BUTYRIC ETHER □ ETHYL BUTANOATE □ ETHYL BUTYRATE (DOT,FCC) □ FEMA No. 2427

CONSENSUS REPORTS: Reported in EPA TSCA Inventory.

DOT CLASSIFICATION: 3; Label: Flammable Liquid

SAFETY PROFILE: Mildly toxic by ingestion. A skin irritant. Flammable liquid when exposed to heat or flame; can react vigorously with oxidizing materials. When heated to decomposition it emits acrid smoke and irritating fumes. See also ESTERS.

EHE500   CAS: 110-38-3   HR: 2
ETHYL CAPRATE

mf: $C_{12}H_{24}O_2$ mw: 200.36

PROP: Colorless liquid; oily, brandy odor. Bp: 243°, d: 0.863, refr index: 1.424, vap d: 6.9, flash p: 212°F. Sol in fixed oils; insol in glycerin, propylene glycol @ 243°.

SYNS: CAPRIC ACID ETHYL ESTER □ DECANOIC ACID, ETHYL ESTER □ ETHYL CAPRINATE □ ETHYL DECANOATE (FCC) □ ETHYL DECYLATE □ FEMA No. 2432

CONSENSUS REPORTS: Reported in EPA TSCA Inventory.

SAFETY PROFILE: A skin irritant. Combustible liquid when exposed to heat or flame; can react with oxidizing materials. When heated to decomposition it emits acrid smoke and irritating fumes. See ESTERS and ETHERS.

EHF000   CAS: 123-66-0   HR: 3
ETHYL CAPROATE

mf: $C_8H_{16}O_2$ mw: 144.24

PROP: Colorless liquid; mild wine odor. Bp: 163°, flash p: 130°F (OC), d: 0.867–0.871, refr index: 1.406–1.409, vap d: 5.0. Sol in fixed oils; sltly sol in propylene glycol; insol in glycerin @ 166.

SYNS: ETHYL BUTYLACETATE (DOT) □ ETHYL HEXANOATE (FCC) □ FEMA No. 2439

CONSENSUS REPORTS: Reported in EPA TSCA Inventory.

SAFETY PROFILE: A skin irritant. Flammable liquid when exposed to heat or flame; can react with oxidizing materials. When heated to decomposition it emits acrid smoke and irritating fumes. To fight fire, use CO_2, foam, dry chemical. See also ESTERS.

EHG100   CAS: 9004-57-3   HR: 1
ETHYLCELLULOSE

PROP: Ethyl ether of cellulose. White to light tan powder. Sol in some org solvs; insol in water, glycerin, and propylene glycol.

SYNS: AMPACET E/C □ CELLULOSE ETHYL □ CELLULOSE ETHYLATE □ ETs □ ETHOCEL □ ETHOCEL 150 □ ETHOCEL 890 □ ETHOCEL E7 □ ETHOCEL E50 □ ETHOCEL MED □ ETHOCEL N7 □ ETHOCEL N10 □ ETHOCEL N200 □ ETHOCEL STD □ ETs (POLYSACCHARIDE) □ G 50 □ G 200 □ G 50 (POLYSACCHARIDE) □ N 5 □ NIXON E/C □ SPT 50 CPS □ T 100 □ T 100 (POLYSACCHARIDE)

CONSENSUS REPORTS: Reported in EPA TSCA Inventory.

SAFETY PROFILE: Low toxicity by ingestion and skin contact. A skin irritant. When heated to decomposition it emits acrid smoke and irritating fumes.

EHG500   CAS: 105-39-5   HR: 3
ETHYL CHLORACETATE

DOT: UN 1181

mf: $C_4H_7ClO_2$ mw: 122.56

PROP: Colorless liquid; fruity, pungent odor. Irritant to the eyes. Bp: 143.6°, fp: −26.6°, flash p: 100°F, d: 1.159 @ 20°/4°, vap press: 10 mm @ 37.5°, vap d: 4.3. Insol in water; misc in alc and ether.

SYNS: CHLOROACETIC ACID, ETHYL ESTER □ ETHYL CHLOROACETATE □ ETHYL-α-CHLOROACETATE □ ETHYL CHLOROETHANOATE □ ETHYL MONOCHLORACETATE □ ETHYL MONOCHLOROACETATE

CONSENSUS REPORTS: Reported in EPA TSCA Inventory.

DOT CLASSIFICATION: 6.1; Label: Poison

SAFETY PROFILE: Poison by skin contact and subcutaneous routes. A severe eye irritant. Questionable carcinogen with experimental neoplastigenic data.

Flammable liquid; a dangerous fire hazard when exposed to heat or flame; can react vigorously with oxidizing materials. Will react with water or steam to produce toxic and corrosive fumes. Vigorous reaction with sodium cyanide. To fight fire, use water, foam, CO_2, dry chemical. When heated to decomposition it emits highly toxic fumes of Cl^-.

EHH000   CAS: 75-00-3   HR: 3
ETHYL CHLORIDE
DOT: UN 1037
mf: C_2H_5Cl mw: 64.52
PROP: Colorless liquid or gas which is volatile at room temp; ether-like odor, burning taste. Bp: 12.3°, lel: 3.8%, uel: 15.4%, fp: −142.5°, flash p: −58°F (CC), d: 0.917 @ 6°/6°, autoign temp: 966°F, vap press: 1000 mm @ 20°, vap d: 2.22; misc in alc and ether. Sltly sol in water. IDLH 3800 ppm [10%LEL].
SYNS: AETHYLCHLORID (GERMAN) □ AETHYLIS □ AETHYLIS CHLORIDUM □ ANODYNON □ CHELEN □ CHLOORETHAAN (DUTCH) □ CHLORETHYL □ CHLORIDUM □ CHLOROAETHAN (GERMAN) □ CHLOROETHANE □ CHLORURE d'ETHYLE (FRENCH) □ CHLORYL □ CHLORYL ANESTHETIC □ CLOROETANO (ITALIAN) □ CLORURO DI ETILE (ITALIAN) □ ETHER CHLORATUS □ ETHER HYDROCHLORIC □ ETHER MURIATIC □ ETYLU CHLOREK (POLISH) □ HYDROCHLORIC ETHER □ KELENE □ MONOCHLORETHANE □ MURIATIC ETHER □ NARCOTILE □ NCI-C06224
CONSENSUS REPORTS: Reported in EPA TSCA Inventory. Community Right-To-Know List.
OSHA PEL: TWA 1000 ppm
ACGIH TLV: TWA 1000 ppm
DFG MAK: Confirmed Animal Carcinogen with Unknown Relevance to Humans
NIOSH REL: (Chloroethane) Handle with caution
DOT CLASSIFICATION: 2.1; Label: Flammable Gas
SAFETY PROFILE: Suspected carcinogen with experimental carcinogenic and neoplastigenic data. Mildly toxic by inhalation. An irritant to skin, eyes, and mucous membranes. The liquid is harmful to the eyes and can cause some irritation. In the case of guinea pigs, the symptoms attending exposure are similar to those caused by methyl chloride, except that the signs of lung irritation are not as pronounced. It gives some warning of its presence because it is irritating, but it is possible to tolerate exposure to it until one becomes unconscious. It is the least toxic of all the chlorinated hydrocarbons. It can cause narcosis, although the effects are usually transient.

A very dangerous fire hazard when exposed to heat or flame; can react vigorously with oxidizing materials. Severe explosion hazard when exposed to flame. Reacts with water or steam to produce toxic and corrosive fumes. Incompatible with potassium. To fight fire, use carbon dioxide. When heated to decomposition it emits toxic fumes of phosgene and Cl^-. See also CHLORINATED HYDROCARBONS, ALIPHATIC.

EHJ500   CAS: 38915-14-9   HR: 3
9-((3-ETHYL-2-CHLOROETHYL)AMINO-
PROPYLAMINO)-4-METHOXY-
ACRIDINE DIHYDROCHLORIDE
mf: $C_{21}H_{26}ClN_3O•2ClH$ mw: 444.87
SYNS: ICR 377 □ 4-METHOXY-9-(3-(ETHYL-2-CHLOROETHYL)
SAFETY PROFILE: Poison by intravenous route. Questionable carcinogen with experimental neoplastigenic data. Mutation data reported. When heated to decomposition it emits very toxic fumes of Cl^- and NO_x.

EHK500   CAS: 541-41-3   HR: 3
ETHYL CHLOROFORMATE
DOT: UN 1182
mf: $C_3H_5ClO_2$ mw: 108.53
PROP: Colorless liquid. Mp: −80.6°, bp: 94°, flash p: 35.6°F, d: 1.1442 @ 15°/4°, vap d: 3.74, autoign temp: 932°F. Decomp in water; misc in alc, benzene, ether, and

chloroform. Misc in EtOH, C_6H_6, $CHCl_3$, and Et_2O; prac insol in H_2O.

SYNS: CHLORAMEISENSAEUREAETHYLESTER (GERMAN) □ CHLOROCARBONATE D'ETHYLE (FRENCH) □ CHLOROFORMIC ACID ETHYL ESTER □ ECF □ ETHYLCHLOORFORMIAAT (DUTCH) □ ETHYL CHLOROCARBONATE (DOT) □ ETHYLE, CHLORO-FORMIAT d' (FRENCH) □ ETIL CLOROCARBONATO (ITALIAN) □ ETIL CLOROFORMIATO (ITALIAN) □ TL 423

CONSENSUS REPORTS: Reported in EPA TSCA Inventory. Community Right-To-Know List.

DOT CLASSIFICATION: 6.1; Label: Poison, Flammable Liquid, Corrosive

SAFETY PROFILE: Poison by ingestion, inhalation, and intraperitoneal routes. Moderately toxic by skin contact. Corrosive. An eye, skin, and mucous membrane irritant. A very dangerous fire hazard when exposed to heat or flame; can react vigorously with oxidizing materials. Reacts with water or steam to produce toxic and corrosive fumes. To fight fire, use CO_2, dry chemical. When heated to decomposition it emits highly toxic fumes of Cl^-.

EHN000   CAS: 103-36-6   HR: 2
ETHYL-trans-CINNAMATE

mf: $C_{11}H_{12}O_2$ mw: 176.23

PROP: Nearly colorless, oily liquid; faint cinnamon odor. D: 1.049 @ 20°/4°, refr index: 1.558–1.561, bp: 271°, mp: 9°, flash p: 212°F. Misc in alc, ether, fixed oils; insol in glycerin, water @ 272°.

SYNS: ETHYL CINNAMATE (FCC) □ ETHYL-β-PHENYLACRYLATE □ ETHYL-3-PHENYLPROPENOATE □ FEMA No. 2430

CONSENSUS REPORTS: Reported in EPA TSCA Inventory.

SAFETY PROFILE: Moderately toxic by ingestion. Combustible liquid. When heated to decomposition it emits acrid smoke and irritating fumes. See also ESTERS.

EHO200   CAS: 10544-63-5   HR: 3
ETHYL CROTONATE

DOT: UN 1862

mf: $C_6H_{10}O_2$ mw: 114.16

PROP: Bp: 142–143°, d: 0.918, flash p: 36°F.

SYNS: 2-BUTENOIC ACID, ETHYL ESTER □ CROTONIC ACID, ETHYL ESTER □ ETHYLESTER KYSELINY KROTONOVE

CONSENSUS REPORTS: Reported in EPA TSCA Inventory.

DOT CLASSIFICATION: 3; Label: Flammable Liquid

SAFETY PROFILE: Slightly toxic by ingestion. Corrosive. An eye irritant and lachrymator. A flammable liquid. When heated to decomposition it emits acrid smoke and irritating fumes.

EHP500   CAS: 105-56-6   HR: 3
ETHYL CYANOACETATE

DOT: UN 2666

mf: $C_5H_7NO_2$ mw: 113.13

PROP: Colorless to pale straw-colored liquid. Mp: −22.5°, bp: 207°, flash p: 230°F, d: 1.06 @ 25°/25°, vap press: 1 mm @ 67.8°, vap d: 3.9. Insol in H_2O; sol in NH_3 aq.

SYNS: CYANACETATE ETHYLE (GERMAN) □ CYANOACETIC ACID ETHYL ESTER □ CYANOACETIC ESTER □ ESTERE CIANOACETICO □ ETHYL CYANOACETATE □ ETHYL CYANOETHANOATE □ ETHYLESTER KYSELINY KYANOCTOVE □ MALONIC ACID ETHYL ESTER NITRILE □ USAF KF-25

CONSENSUS REPORTS: Reported in EPA TSCA Inventory. Cyanide and its compounds are on the Community Right-To-Know List.

DOT CLASSIFICATION: 6.1; Label: KEEP AWAY FROM FOOD

SAFETY PROFILE: Poison by ingestion. Moderately toxic by intraperitoneal and subcutaneous routes. Combustible when exposed to heat or flame; can react with oxidizing materials. Will react with water or steam to produce toxic and flammable vapors. To fight fire, use CO_2, dry chemical. When heated to decomposition or on contact with acid or acid fumes it emits highly toxic fumes of CN^-. See also NITRILES.

E

EHP700   CAS: 7085-85-0   HR: 2
ETHYL CYANOACRYLATE
mf: $C_6H_7NO_2$ mw: 125.14
PROP: Colorless liquid with a distinctive odor. D: 1.06, Mp: −22°, bp: 54°–56° @ 3 MM Hg.
SYNS: ACE-E 50 □ ACE-EE □ ACRYLIC ACID, 2-CYANO-, ETHYL ESTER (6CI,7CI,8CI) □ ADHESIVE 502 □ ARON ALPHA D □ ARON ALPHA 402X □ BLACK MAX □ CA 3 □ CA 3 (ADHESIVE) □ CA 8-3A □ CEMEDINE 3000RP □ CEMEDINE 3000RP TYPE-II □ CEMEDINE 3000RS □ CEMEDINE 3000RS TYPE-II □ CN 2 □ CN 4 □ CYANOBOND W100 □ CYANOBOND W300 □ CYANON 5MSP □ CYANON S □ DA 737S □ 910EM □ ETHYL α-CYANOACRYLATE □ ETHYL 2-CYANO-ACRYLATE □ ETHYL 2-CYANO-2-PROPENOATE □ N 135 □ 2-PROPENOIC ACID, 2-CYANO-, ETHYL ESTER □ PTR-E 3 □ PTR-E 40 □ SICOMET 8400 □ SUPER 3-1000 □ SUPER GLUE □ TK 200 □ TK 201
CONSENSUS REPORTS: Reported in EPA TSCA Inventory.
ACGIH TLV: TWA 0.2 ppm
SAFETY PROFILE: An inhalation hazard. When heated to decomposition it emits toxic vapors of NO_x.

EHT000   CAS: 5459-93-8   HR: 3
N-ETHYL(CYCLOHEXYL)AMINE
mf: $C_8H_{17}N$ mw: 127.26
PROP: A liquid with fishy odor. Flash p: 86°F (OC), bp: 164, d: 0.8, vap d: 4.4. Sltly sol in water
SYN: N-ETHYL-CYCLOHEXYLAMINE
CONSENSUS REPORTS: Reported in EPA TSCA Inventory.
SAFETY PROFILE: Moderately toxic by ingestion, inhalation, and skin contact. A severe skin and eye irritant. A very dangerous fire hazard when exposed to heat or flame; can react vigorously with oxidizing materials. To fight fire, use alcohol foam, mist, spray, dry chemical. See also AMINES.

EHY050   CAS: 3674-13-3   HR: 3
ETHYL 2,3-DIBROMOPROPANOATE
mf: $C_5H_8Br_2O_2$ mw: 259.95
SYNS: ETHYL α,β-DIBROMOPROPIONATE □ ETHYL 2,3-DIBROMOPROPIONATE □ PROPANOIC ACID, 2,3-DIBROMO-, ETHYL ESTER □ PROPIONIC ACID, 2,3-DIBROMO-, ETHYL ESTER

SAFETY PROFILE: A poison by ingestion and skin contact. A severe skin and eye irritant. When heated to decomposition it emits toxic vapors of Br^-.

EIB600   CAS: 73263-81-7   HR: 3
ETHYL 3-((DIETHYLPHOSPHINO-THIOYL)OXY)-2-BUTENOATE
mf: $C_{10}H_{19}O_5PS$ mw: 282.32
SYNS: 2-BUTENOIC ACID-3-(DIETHOXY PHOSPHINOTHIOYL)ETHYL ESTER □ 2-BUTENOIC ACID, 3-((DIETHYLPHOSPHINOTHIOYL)OXY)-, ETHYL ESTER □ RPR-5 □ RPR-V
SAFETY PROFILE: A poison by ingestion. When heated to decomposition it emits toxic vapors of SO_x and PO_x.

EID000   CAS: 389-08-2   HR: 3
1-ETHYL-1,4-DIHYDRO-7-METHYL-4-OXO-1,8-NAPHTHYRIDINE-3-CARBOXYLIC ACID
mf: $C_{12}H_{12}N_2O_3$ mw: 232.26
PROP: Pale buff crystals. Mp: 229–230°. Sol in $CHCl_3$; mod sol in EtOH and MeOH.
SYNS: ACIDE 1-ETIL-7-METIL-1,8-NAFTIRIDIN-4-ONE-3-CARBOSSILICO (ITALIAN) □ ACIDE NALIDIXICO (ITALIAN) □ ACIDE NALIDIXIQUE (FRENCH) □ BETAXINA □ 3-CARBOXY-1-ETHYL-7-METHYL-1,8-NAPHTHIDIN-4-ONE □ CHINOIN □ CYBIS □ 1,4-DIHYDRO-1-ETHYL-7-METHYL-4-OXO-1,8-NAPHTHYRIDINE-3-CARBOXYLIC ACID □ DIXIBEN □ 1-ETHYL-7-METHYL-1,4-DIHYDRO-1,8-NAPHTHY-RIDINE-4-ONE-3-CARBOXYLIC ACID □ 1-ETHYL-7-METHYL-1,4-DIHYDRO-1,8-NAPHTHYRIDIN-4-ONE-3-CARBOXYLIC ACID □ 1-ETHYL-7-METHYL-4-OXO-1,4-DIHYDRO-1,8-NAPHTHYRIDINE-3-CARBOXYLIC ACID □ EUCISTEN □ INNOXALON □ KUSNARIN □ NA □ NALIDIC ACID □ NALIDICRON □ NALIDIXIC ACID □ NALIDIXIN □ NALITUCSAN □ NARIGIX □ NCI-C56199 □ NEGRAM □ NEVIGRAMON □ NICELATE □ NOGRAM □ NSC-82174 □ POLEON □ SPECIFEN □ URALGIN □ URIBEN □ URODIXIN □ UROMAN □ URONEG □ WIN 18,320 □ WINTOMYLON
CONSENSUS REPORTS: EPA Genetic Toxicology Program.
SAFETY PROFILE: Poison by intravenous and intraperitoneal routes. Moderately toxic by ingestion and subcutaneous routes. An experimental teratogen. Human systemic effects: convulsions, hyperglycemia, sweating, and blood changes in children. Experimental reproductive effects.

Questionable carcinogen with experimental carcinogenic and tumorigenic data. Human mutation data reported. Used as an antibacterial agent and urinary tract antiseptic. When heated to decomposition it emits toxic fumes of NO_x.

EID200   CAS: 68-90-6   HR: 3
2-ETHYL-3-(3',5'-DIIODO-4'-HYDROXY-
BENZOYL)-CUMARONE
mf: $C_{17}H_{12}I_2O_3$ mw: 518.09
SYNS: AETHYL-2-(3',5'-DIJOD-4'-OXYBENZOYL)-3 CUMARON □ ALGOCOR □ AMPLIVIX □ BENZIODARON □ BENZIODARONE □ BENZOFURAN, 3-(3,5-DIIODO-4-HYDROXYBENZOYL)-2-ETHYL- □ CARDIVIX □ CAROFAM □ CORONAL-CRINOS □ 3,5-DIIODO-4-HYDROXYPHENYL 2-ETHYL-3-BENZO-FURANYL KETONE □ DILAFURANE □ DILA-VASAL □ KETONE, 3,5-DIIODO-4-HYDROXYPHENYL 2-ETHYL-3-BENZOFURANYL □ L 2329 □ 2329 LABAZ □ RETRANGOR
DOT CLASSIFICATION: 3; Label: Flammable Liquid
SAFETY PROFILE: A poison by intraperitoneal and intravenous routes. Moderately toxic by ingestion. A flammable liquid. When heated to decomposition it emits toxic vapors of I^-.

EID250   CAS: 4568-83-6   HR: 3
ETHYL-2-(DIIODO-3,5 HYDROXY-4
BENZOYL)5-FURANNE
mf: $C_{13}H_{10}I_2O_3$ mw: 468.03
SYNS: DB 138 □ 3,5-DIIODO-4-HYDROXYPHENYL 5-ETHYL-2-FURYL KETONE □ KETONE, 3,5-DIIODO-4-HYDROXYPHENYL 5-ETHYL-2-FURYL
DOT CLASSIFICATION: 3; Label: Flammable Liquid
SAFETY PROFILE: A poison by intraperitoneal and intravenous routes. When heated to decomposition it emits toxic vapors of NO_x and I^-.

EIF000   CAS: 77-81-6   HR: 3
ETHYL DIMETHYLAMIDOCYANO-
PHOSPHATE
mf: $C_5H_{11}N_2O_2P$ mw: 162.15
PROP: A colorless to brownish liquid. Bp: decomp @ 238°, fp: −49.4°, flash p: 172°F, d: 1.073 @ 25°, vap press: 0.07 mm @ 25°, vap d: 5.63.

SYNS: DIMETHYLAMIDOETHOXYPHOSPHORYL CYANIDE □ DIMETHYLAMINOCYANPHOSPHOR-SAEUREAETHYLESTER (GERMAN) □ DIMETHYL-PHOSPHORAMIDOCYANIDIC ACID, ETHYL ESTER □ ETHYL N,N-DIMETHYLAMINO CYANOPHOSPHATE □ ETHYL DIMETHYLPHOSPHORAMIDOCYANIDATE □ ETHYL-N,N-DIMETHYLPHOSPHORAMIDOCYANIDATE □ GA □ GELAN I □ Le-100 □ MCE □ T-2104 □ TABOON A □ TABUN □ TL 1578 □ TRILON 83
CONSENSUS REPORTS: EPA Extremely Hazardous Substances List. Cyanide and its compounds are on the Community Right-To-Know List. Reported in EPA TSCA Inventory.
SAFETY PROFILE: Human poison by inhalation, skin contact, and intravenous routes. Experimental poison by ingestion, inhalation, skin contact, subcutaneous, intravenous, intraperitoneal, and intramuscular routes. A nerve gas. Vapor does not penetrate skin; liquid does so rapidly. The primary physiological action is on the sympathetic nervous system, causing a vasoparesis (partial paralysis of the vasomotor nerves, which control the diameter of the blood vessels). Vapors when inhaled can cause nausea, vomiting, and diarrhea, which can be followed by muscular twitching and convulsions. Flammable when exposed to heat or flame; can react with oxidizing materials. When heated to decomposition it emits very toxic fumes of PO_x, CN^-, and NO_x. See also PARATHION and CYANIDE.

EIG000   CAS: 50782-69-9   HR: 3
ETHYL-S-DIMETHYLAMINOETHYL
METHYLPHOSPHONOTHIOLATE
mf: $C_{11}H_{26}NO_2PS$ mw: 267.41
PROP: Colorless to straw colored liquid. Mp: −50°, bp: 298°, d: 1.0083 g/ml. Flash pt: 70.6°C.
SYNS: S-(2-DIISOPROPYLAMINOETHYL)-O-ETHYL METHYL PHOSPHONOTHIOLATE □ ETHYL-S-DIISOPROPYLAMINOETHYL METHYLTHIOPHOS-PHONATE □ O-ETHYL-S-2-DIISOPROPYLAMINO-ETHYL METHYLPHOSPHONOTHIOTE □ METHYL-PHOSPHONOTHIOIC ACID-S-(2-(BIS(METHYLETHYL)-AMINO)ETHYL)o-ETHYL ESTER □ VX

CONSENSUS REPORTS: Reported in EPA TSCA Inventory. EPA Extremely Hazardous Substances List.

SAFETY PROFILE: Human poison by skin contact. Experimental poison by intraperitoneal and subcutaneous routes. An experimental teratogen. Other experimental reproductive effects. Human systemic effects by ingestion and intravenous routes: hallucinations and distorted perceptions, blood pressure increase, hypermotility, diarrhea, nausea and vomiting, visual field changes, sleep disturbance. A chemical warfare agent. A combustible liquid. When heated to decomposition it emits very toxic fumes of SO_x and NO_x.

EIK000   CAS: 78-78-4   HR: 3
ETHYLDIMETHYLMETHANE
DOT: UN 1265
mf: C_5H_{12} mw: 72.17

PROP: Colorless liquid with pleasant odor. Fp: $-160.5°$, bp: $30-30.2°$, flash p: $<-60°F$ (CC), d: 0.620 @ $20°/4°$, vap press: 595 mm @ 21.1°, vap d: 2.48, lel: 1.4%, uel: 7.6%.

SYNS: ISOAMYLHYDRIDE □ ISOPENTANE (DOT) □ 2-METHYLBUTANE

CONSENSUS REPORTS: Reported in EPA TSCA Inventory.

ACGIH TLV: TWA 600 ppm
DFG MAK: 1000 ppm (3000 mg/m³)
NIOSH REL: (Alkanes) TWA 350 mg/m³
DOT CLASSIFICATION: 3; Label: Flammable Liquid

SAFETY PROFILE: Mildly toxic and narcotic by inhalation. See also PENTANE. Flammable liquid. A very dangerous fire and explosion hazard when exposed to heat, flame, or oxidizers. Keep away from sparks, heat, or open flame; can react with oxidizing materials. To fight fire, use foam, CO_2, dry chemical. When heated to decomposition it emits acrid smoke and irritating fumes.

EIL100   HR: 2
2-ETHYL-3,5(6)-DIMETHYLPYRAZINE
mf: $C_8H_{12}N_2$ mw: 136.20

PROP: Colorless to sltly yellow liquid; roasted-cocoa odor. D: 0.950–0.980, refr index: 1.500, flash p: 154°F. Sol in water, org solvs.

SYN: FEMA No. 3149

SAFETY PROFILE: Combustible liquid. When heated to decomposition it emits toxic fumes of NO_x.

EIM100   CAS: 77405-29-9   HR: 3
ETHYLDIPHENYLTIN ACETATE
mf: $C_{16}H_{18}O_2Sn$ mw: 361.03

SYNS: ACETOXYDIPHENYLETHYLSTANNANE □ STANNANE, ACETOXYDIPHENYLETHYL- □ STANNANE, (ACETYLOXY)ETHYLDIPHENYL-

ACGIH TLV: TWA 0.1 mg(Sn)/m³. STEL 0.2 mg/m³ (skin). Not classifiable as a human carcinogen

SAFETY PROFILE: A poison by ingestion. When heated to decomposition it emits toxic vapors of Sn.

EIN000   CAS: 13194-48-4   HR: 3
O-ETHYL-S,S-DIPROPYLPHOSPHORO-
DITHIOATE
mf: $C_8H_{19}O_2PS_2$ mw: 242.36

PROP: A yellow liquid. D: 1.094 @ $20°/4°$, bp: $86-91°$ @ 0.2 mm. Sltly sol in H_2O; very sol in most org solvs.

SYNS: ENT 27,318 □ ETHOPROP □ ETHOPROPHOS □ O-ETHYL-S,S-DIPROPYL ESTER, PHOSPHORO-DITHIOIC ACID □ JOLT □ MOBIL V-C 9-104 □ MOCAP □ PROPHOS □ V-C CHEMICAL V-C 9-104 □ VIRGINIA CAROLINA VC 9-104

CONSENSUS REPORTS: EPA Extremely Hazardous Substances List.

SAFETY PROFILE: Poison by ingestion and skin contact. A cholinesterase inhibitor type of insecticide. When heated to decomposition it emits very toxic fumes of PO_x and SO_x. See also PARATHION.

EIN500   CAS: 759-94-4   HR: 3
S-ETHYL-N,N-DI-N-PROPYLTHIO-
CARBAMATE
mf: $C_9H_{19}NOS$ mw: 189.35

PROP: A liquid. D: 0.955 @ 30°, bp: 127° @ 20 mm. Sltly sol in H_2O.

SYNS: S-AETHYL-N,N-DIPROPYLTHIOLCARBAMAT (GERMAN) □ DIPROPYLCARBAMOTHIOIC ACID-S-

ETHYL ESTER □ N,N-DIPROPYLTHIOCARBAMIC
ACID-S-ETHYL ESTER □ EPTAM □ EPTC □ S-ETHYL-
N,N-DIPROPYLTHIOCARBAMATE □ ETHYL DI-N-
PROPYLTHIOLCARBAMATE □ ETHYL-N,N-
DIPROPYLTHIOLCARBAMATE □ ETHYL-N,N-DI-N-
PROPYLTHIOLCARBAMATE □ FDA 1541 □ GENEP
EPTC □ R-1608 □ STAUFFER R 1608 □ TORBIN

CONSENSUS REPORTS: Reported in EPA
TSCA Inventory. EPA Genetic Toxicology
Program.

SAFETY PROFILE: Poison by ingestion,
inhalation, and intravenous routes.
Moderately toxic by skin contact route.
Mutation data reported. An herbicide. When
heated to decomposition it emits very toxic
fumes of NO_x and SO_x. See also
CARBAMATES.

EIO000   CAS: 74-85-1   HR: 3
ETHYLENE
DOT: UN 1038/UN 1962
mf: C_2H_4 mw: 28.06
PROP: Colorless gas; odorless and tasteless.
Bp: $-103.9°$, mp: $-169.4°$, lel: 2.7%, uel:
36%, d: 0.610 @ 0°, autoign temp: 914°F,
vap d: 0.98, fp: $-181°$. Sltly sol in H_2O; very
sol in EtOH, Et_2O; sol in Me_2CO and C_6H_6.
SYNS: ACETENE □ ATHYLEN (GERMAN) □
BICARBURETTED HYDROGEN □ ELAYL □ ETHENE □
ETHYLENE, compressed (DOT) □ ETHYLENE, refrigerated
liquid (DOT) □ LIQUID ETHYLENE □ OLEFIANT GAS
CONSENSUS REPORTS: Reported in EPA
TSCA Inventory. Community Right-To-
Know List.
ACGIH TLV: Simple asphyxiant; Not
Classifiable as a Human Carcinogen;
(Proposed: 100 ppm; Not Classifiable as a
Human Carcinogen)
DFG MAK: Confirmed Animal Carcinogen
with Unknown Relevance to Humans
DOT CLASSIFICATION: 2.1; Label:
Flammable Gas
SAFETY PROFILE: Suspected carcinogen.
A simple asphyxiant. High concentrations
cause anesthesia. A common air
contaminant. It is phytotoxic. A very
dangerous fire hazard when exposed to heat
or flame. Moderate explosion hazard when
exposed to flame. A flammable gas. To fight

fire, stop flow of gas, use CO_2, dry chemical,
or fine water spray. Mixtures with aluminum
chloride explode in the presence of nickel
catalysts, methyl chloride, or nitromethane.
Explosive reaction with
bromotrichloromethane (at 120°C/51 bar),
carbon tetrachloride (25–100°C/30 bar).
Explosive reaction with chlorine catalyzed
by sunlight or UV light or in the presence of
mercury(I) oxide, mercury(II) oxide, or
silver oxide. Mixtures with
chlorotrifluoroethylene polymerize
explosively when exposed to 50 kV gamma
rays at 308 krad/hr. Has been involved in
industrial accidents. Violent polymerization
is catalyzed by copper above 400°C/54 bar.
Incompatible with $AlCl_3$, (CCl_4 + benzoyl
peroxide), (bromotrichloromethane +
$AlCl_3$), O_3, CCl_4, Cl_2, NO_x,
tetrafluoroethylene trifluorohypofluorite.
When heated to decomposition it emits
acrid smoke and irritating fumes.

EIP000   CAS: 2274-11-5   HR: 3
ETHYLENE ACRYLATE
mf: $C_8H_{10}O_4$ mw: 170.18
PROP: Colorless liquid witn a penetrating
acrid odor. Sol in ethanol, ether and
chloroform. Slightly sol in water.
SYNS: ACRYLIC ACID, ETHYLENE ESTER □ ACRYLIC
ACID, ETHYLENE GLYCOL DIESTER □ ETHYLDIOL
ACRILATE (RUSSIAN) □ ETHYLENE DIACRYLATE □
ETHYLENE GLYCOL DIACRYLATE □ 2-PROPENOIC
ACID-1,2-ETHANEDIYL ESTER
CONSENSUS REPORTS: Reported in EPA
TSCA Inventory.
SAFETY PROFILE: Poison by ingestion and
inhalation. Moderately toxic by skin contact.
A skin and severe eye irritant. Mutation data
reported. When heated to decomposition it
emits acrid smoke and irritating fumes. See
also ESTERS.

EIQ000   CAS: 124-05-0   HR: 2
ETHYLENE BIS(CHLOROFORMATE)
mf: $C_4H_4Cl_2O_4$ mw: 186.98
SYNS: 1,2-BIS((CHLOROCARBONYL)OXY)ETHANE □
CARBONOCHLORIDE ACID, 1,2-ETHANEDIYL ESTER
□ ETHYLENE CHLOROFORMATE □ ETHYLENE

GLYCOL DI(CHLOROFORMATE) □ ETHYLENE GLY-COL, BISCHLOROFORMATE

CONSENSUS REPORTS: Reported in EPA TSCA Inventory.

DOT CLASSIFICATION: 6.1; Label: Poison, Corrosive

SAFETY PROFILE: Moderately toxic by ingestion and skin contact. Mildly toxic by inhalation. When heated to decomposition it emits toxic fumes of Cl⁻.

EIQ200   CAS: 4431-24-7   HR: 3
ETHYLENEBIS-(DIPHENYLARSINE)
mf: $C_{26}H_{24}As_2$ mw: 486.34
PROP: Colorless crystals from MeOH or EtOH. Mp: 100–103°.
SYN: ARSINE, ETHYLENEBIS(DIPHENYL)-
OSHA PEL: TWA 0.5 mg(As)/m³
SAFETY PROFILE: Poison by intravenous route. An irritant. When heated to decomposition it emits toxic fumes of As.

EIQ500   HR: D
ETHYLENEBIS(DITHIOCARBAMATO)-
MANGANESE and ZINC ACETATE
(50:1)
SYNS: MANEB plus ZINC ACETATE (50:1) □ ZINC ACETATE plus MANEB (1:50)
CONSENSUS REPORTS: Manganese and its compounds, as well as zinc and its compounds, are on the Community Right-To-Know List.
SAFETY PROFILE: An experimental teratogen. When heated to decomposition it emits very toxic fumes of NO_x, ZnO, and SO_x. See also ZINC COMPOUNDS, MANGANESE COMPOUNDS, and CARBAMATES.

EIS000   CAS: 62207-76-5   HR: 3
N,N'-ETHYLENE BIS(3-FLUOROSALI-
CYLIDENEIMINATO)COBALT(II)
mf: $C_{16}H_{12}CoF_2N_2O_2$ mw: 361.23
SYNS: BIS(3-FLUOROSALICYLALDEHYDE)-ETHYLENEDIIMINE-COBALT □ FLUOMINE □ FLUOMINE DUST
CONSENSUS REPORTS: Cobalt and its compounds are on the Community Right-To-Know List. EPA Extremely Hazardous

Substances List. Reported in EPA TSCA Inventory.
SAFETY PROFILE: Poison by ingestion and inhalation. A skin and eye irritant. When heated to decomposition it emits very toxic fumes of F⁻ and NO_x. See also COBALT COMPOUNDS.

EIT000   CAS: 67-42-5   HR: 3
(ETHYLENEBIS(OXYETHYLENENITRIL
O))TETRAACETIC ACID
mf: $C_{14}H_{24}N_2O_{10}$ mw: 380.40
SYNS: 1,2-BIS(2-DICARBOXYMETHYLAMINOETHOXY)ETHANE □ 6,9-DIOXA-3,12-DIAZATETRADECANEDIOIC ACID, 3,12-BIS(CARBOXYMETHYL)-(9CI) □ EBONTA □ EGTA □ ETHYLENEDIOXYBIS(ETHYLENEAMINO)TETRAACET IC ACID □ ETHYLENE GLYCOL BIS(AMINOETHYL ETHER)TETRAACETATE □ ETHYLENE GLYCOL BIS(β-AMINOETHYL ETHER)TETRAACETATE □ ETHYLENE GLYCOL BIS(β-AMINOETHYL ETHER)-N,N'-TETRA-ACETIC ACID □ ETHYLENE GLYCOL BIS(2-AMINO-ETHYL ETHER)TETRAACETATE □ ETHYLENE GLYCOL BIS(2-AMINOETHYL ETHER)-N,N,N',N'-TETRAACETIC ACID □ GLYCOL-ETHERDIAMINETE-TRAACETIC ACID
CONSENSUS REPORTS: Glycol ether compounds are on the Community Right-To-Know List. Reported in EPA TSCA Inventory.
SAFETY PROFILE: Poison by intraperitoneal route. Moderately toxic by ingestion. When heated to decomposition it emits toxic fumes of NO_x. See also GLYCOL ETHERS.

EIU800   CAS: 107-07-3   HR: 3
ETHYLENE CHLOROHYDRIN
DOT: UN 1135
mf: C_2H_5ClO mw: 80.52
PROP: Colorless liquid; faint, ethereal odor. Mp: −69°, fp: −67.5°, bp: 128.8°, flash p: 140°F (OC), d: 1.197 @ 20°/4°, autoign temp: 797°F, vap press: 10 mm @ 30.3°, vap d: 2.78, lel: 4.9%, uel: 15.9%. Misc in water. IDLH 7 ppm.
SYNS: AETHYLENECHLORHYDRIN (GERMAN) □ 2-CHLOORETHANOL (DUTCH) □ 2-CHLORAETHANOL (GERMAN) □ 2-CHLORETHANOL (GERMAN) □ Δ-CHLOROETHANOL □ 2-CHLOROETHANOL (MAK) □ β-CHLOROETHYL ALCOHOL □ 2-CHLOROETHYL ALCOHOL □ CHLOROETHYLOWY ALKOHOL (POLISH)

□ 2-CLOROETANOLO (ITALIAN) □ ETHYLEEN-CHLOORHYDRINE (DUTCH) □ ETHYLENE GLYCOL, CHLOROHYDRIN □ GLICOL MONOCLORIDRINA (ITALIAN) □ GLYCOL CHLOROHYDRIN □ GLYCOL-MONOCHLOORHYDRINE (DUTCH) □ GLYCOL MONO-CHLOROHYDRIN □ GLYCOMONOCHLORHYDRIN □ MONOCHLORHYDRINE du GLYCOL (FRENCH) □ 2-MONOCHLOROETHANOL □ NCI-C50135

CONSENSUS REPORTS: NTP Carcinogenesis Studies (dermal); No Evidence: mouse, rat NTPTR* NTP-TR-275,85; Reported in EPA TSCA Inventory. EPA Genetic Toxicology Program. EPA Extremely Hazardous Substances List.

OSHA PEL: CL 1 ppm (skin)

ACGIH TLV: CL 1 ppm (skin); Not Classifiable as a Human Carcinogen

DFG MAK: 1 ppm (3.3 mg/m²)

DOT CLASSIFICATION: 6.1; Label: Poison

SAFETY PROFILE: A poison by ingestion, inhalation, skin contact, intraperitoneal, intravenous, and subcutaneous routes. Moderately toxic to humans by inhalation. It can affect the nervous system, liver, spleen, and lungs. An experimental teratogen. Mutation data reported. A severe eye and mild skin irritant. Flammable liquid when exposed to heat, flame, or oxidizers. To fight fire, use alcohol foam, CO_2, dry chemical. Violent reaction with chlorosulfonic acid, ethylene diamine, sodium hydroxide. Reacts with water or steam to produce toxic and corrosive fumes. Potentially violent reaction with oxidizing materials. When heated to decomposition it emits highly toxic fumes of Cl^- and phosgene. See also CHLORINATED HYDROCARBONS, ALIPHATIC.

EIU900   CAS: 3741-32-0   HR: 3

ETHYLENE CHLOROTHIO-ARSENATE(III)

mf: $C_2H_4AsClS_2$ mw: 202.55

SYNS: 2-CHLORO-4,5-DIHYDRO-1,3,2-DITHIARSENOLE □ 1,3,2-DITHIARSENOLE, 2-CHLORODIHYDRO- □ 1,3,2-DITHIARSOLANE, 2-CHLORO-

SAFETY PROFILE: A poison by ingestion. When heated to decomposition it emits toxic vapors of SO_x, As, and Cl^-.

EIV000   CAS: 64-02-8   HR: 3

N,N′-ETHYLENEDIAMINEDIACETIC ACID TETRASODIUM SALT

mf: $C_{10}H_{12}N_2O_8•4Na$ mw: 380.20

PROP: Amorphous powder.

SYNS: AQUAMOLLIN □ CALSOL □ CELON E □ CELON H □ CELON IS □ CHEELOX BF □ CHEELOX BR-33 □ CHELON 100 □ CHEMCOLOX 200 □ COMPLEXONE □ CONIGON BC □ DISTOL 8 □ EDATHANIL TETRASODIUM □ EDETATE SODIUM □ EDETIC ACID TETRASODIUM SALT □ EDTA, SODIUM SALT □ EDTA TETRASODIUM SALT □ ENDRATE TETRASODIUM □ N,N′-1,2-ETHANEDIYLBIS(N-(CARBOXYMETHYL))GLYCINE TETRASODIUM SALT □ ETHYLENEBIS(IMINODIACETIC ACID) TETRASODIUM SALT □ ETHYLENEDIAMINETETRAACETIC ACID, TETRASODIUM SALT □ HAMP-ENE 100 □ HAMP-ENE 215 □ HAMP-ENE 220 □ HAMP-ENE Na4 □ IRGALON □ KALEX □ KEPMPLEX 100 □ KOMPLXON □ METAQUEST C □ NERVANAID B LIQUID □ NERVANID B □ NULLAPON B □ NULLAPON BF-78 □ NULLAPON BFC CONC □ PERMA KLEER 50 CRYSTALS □ PERMA KLEER TETRA CP □ QUESTEX 4 □ SEQUESTRENE 30A □ SEQUESTRENE Na 4 □ SEQUESTRENE ST □ SODIUM EDETATE □ SODIUM EDTA □ SODIUM ETHYLENEDIAMINE-TETRA-ACETATE □ SODIUM ETHYLENEDIAMINE-TETRA-ACETIC ACID □ SODIUM SALT of ETHYLENE-DIAMINETETRAACETIC ACID □ SYNTES 12A □ SYNTRON B □ TETRACEMIN □ TETRASODIUM EDTA □ TETRASODIUM ETHYLENEDIAMINETETRA-ACETATE □ TETRASODIUM ETHYLENEDIAMINETE-TRACETATE □ TETRASODIUM (ETHYLENEDI-NITRILO)TETRAACETATE □ TETRASODIUM SALT of EDTA □ TETRASODIUM SALT of ETHYLENEDI-AMINETETRACETIC ACID □ TETRINE □ TRILON B □ TST □ TYCLAROSOL □ VERSENE 100 □ VERSENE POWDER □ WARKEELATE PS-43

CONSENSUS REPORTS: Reported in EPA TSCA Inventory.

SAFETY PROFILE: Poison by intraperitoneal route. A skin and eye irritant. When heated to decomposition it emits toxic fumes of NO_x and Na_2O. See also ETHYLENEDIAMINETETRAACETIC ACID, DISODIUM SALT.

EIW000   CAS: 333-18-6   HR: 3

ETHYLENEDIAMINE HYDROCHLORIDE

mf: $C_2H_8N_2•2ClH$ mw: 133.04

PROP: Monoclinic prisms. Mp: subl. Sol in water; insol in alc and ether.

SYNS: CHLOR-ETHAMINE □ 1,2-DIAMINOETHANE DIHYDROCHLORIDE □ 1,2-ETHANEDIAMINE,

DIHYDROCHLORIDE □ ETHYLENEDIAMMONIUM CHLORIDE

CONSENSUS REPORTS: Reported in EPA TSCA Inventory.

SAFETY PROFILE: Poison by intramuscular route. Moderately toxic by ingestion. Experimental teratogenic and reproductive effects. When heated to decomposition it emits very toxic fumes of HCl and NO$_x$.

**EIX000   CAS: 60-00-4   HR: 3
ETHYLENEDIAMINETETRAACETIC
  ACID**

mf: C$_{10}$H$_{16}$N$_2$O$_8$ mw: 292.28

PROP: Colorless crystals from water (dimorphic). Mp: 220° (decomp @ 240°). Sltly water-sol; insol in common org solvs.

SYNS: ACIDE ETHYLENEDIAMINETETRACETIQUE (FRENCH) □ 3,6-BIS(CARBOXYMETHYL)-3,5-DIAZOOCTANEDIOIC ACID □ CELON A □ CELON ATH □ CHEELOX BF ACID □ CHEMCOLOX 340 □ COMPLEXON II □ EDATHAMIL □ EDETIC ACID □ EDTA (chelating agent) □ EDTA ACID □ ENDRATE □ N,N'-1,2-ETHANEDIYLBIS(N-(CARBOXYMETH-YL))GLYCINE □ ETHYLENEDIAMINETETRAACETATE □ ETHYLENEDIAMINE-N,N,N',N'-TETRAACETIC ACID □ ETHYLENEDINITRILOTETRAACETIC ACID □ HAMP-ENE ACID □ HAVIDOTE □ METAQUEST A □ NERVANAID B ACID □ NULLAPON BF ACID □ PERMA KLEER 50 ACID □ QUESTEX 4H □ SEQ 100 □ SEQUESTRENE AA □ SEQUESTRIC ACID □ SEQUESTROL □ TETRINE ACID □ TITRIPLEX □ TRICON BW □ TRILON BW □ VERSENE ACID □ VINKEIL 100 □ WARKEELATE ACID

CONSENSUS REPORTS: Reported in EPA TSCA Inventory. EPA Genetic Toxicology Program.

SAFETY PROFILE: Poison by intraperitoneal route. Experimental teratogenic and reproductive effects. Mutation data reported. A general-purpose chelating and complexing agent. When heated to decomposition it emits toxic fumes of NO$_x$.

**EIX500   CAS: 139-33-3   HR: 3
ETHYLENEDIAMINETETRAACETIC
  ACID, DISODIUM SALT**

mf: C$_{10}$H$_{14}$N$_2$O$_8$•2Na   mw: 336.24

PROP: White crystalline powder. Sol in water.

SYNS: CHELADRATE □ CHELAPLEX III □ CHELATON III □ COMPLEXON III □ d'E.D.T.A. DISODIQUE (FRENCH) □ DISODIUM DIACID ETHYLENEDIAMINETETRAACETATE □ DISODIUM DIHYDROGEN ETHYLENEDIAMINETETRAACETATE □ DISODIUM DIHYDROGEN(ETHYLENE-DINITRILO)TETRAACETATE □ DISODIUM EDATHAMIL □ DISODIUM EDETATE □ DISODIUM EDTA (FCC) □ DISODIUM ETHYLENEDIAMINETE-TRAACETATE □ DISODIUM ETHYLENEDI-AMINETETRAACETIC ACID □ DISODIUM (ETHYLENEDINITRILO)TETRAACETATE □ DISODIUM (ETHYLENEDINITRILO)TETRAACETIC ACID □ DISODIUM SALT of EDTA □ DISODIUM SEQUESTRENE □ DISODIUM TETRACEMATE □ DISODIUM VERSENATE □ DISODIUM VERSENE □ EDATHAMIL DISODIUM □ EDETATE DISODIUM □ EDTA, DISODIUM SALT □ ENDRATE DISODIUM □ N,N'-1,2-ETHANEDIYLBIS(N-(CARBOXYMETHYL)GLY-CINE) DISODIUM SALT □ ETHYLENEBIS(IMINODI-ACETIC ACID) DISODIUM SALT □ ETHYLENEDI-AMINETETRAACETATE DISODIUM SALT □ (ETHYLENEDINITRILO)-TETRAACETIC ACID DISODIUM SALT □ F 1 (complexon) □ KIRESUTO B □ METAQUEST B □ PERMA KLEER 50 CRYSTALS DISODIUM SALT □ SELEKTON B 2 □ SEQUESTRENE SODIUM 2 □ SODIUM VERSENATE □ TETRACEMATE DISODIUM □ TITRIPLEX III □ TRILON BD □ TRIPLEX III □ VERESENE DISODIUM SALT □ VERSENE SODIUM 2

CONSENSUS REPORTS: Reported in EPA TSCA Inventory. EPA Genetic Toxicology Program.

SAFETY PROFILE: Poison by intraperitoneal and intravenous routes. Moderately toxic by ingestion. Experimental teratogenic and reproductive effects. Mutation data reported. The calcium disodium salt of EDTA is used as a chelating agent in treating lead poisoning. When heated to decomposition it emits toxic fumes of NO$_x$ and Na$_2$O.

**EIY500   CAS: 106-93-4   HR: 3
1,2-ETHYLENE DIBROMIDE**

DOT: UN 1605

mf: C$_2$H$_4$Br$_2$ mw: 187.88

PROP: Colorless, heavy liquid; sweet odor. Bp: 131.4°, fp: 9.3°, flash p: none, d: 2.178 @ 20°/4, mp: 10°, vap press: 17.4 mm @ 30°, vap d: 6.48. IDLH 100 ppm.

SYNS: AETHYLENBROMID (GERMAN) □ BROMOFUME □ BROMURO di ETILE (ITALIAN) □ CELMIDE □ DBE □ 1,2-DIBROMAETHAN (GERMAN) □ 1,2-DIBROMOETANO (ITALIAN) □ α,β-DIBROMOETHANE □ sym-DIBROMOETHANE □ 1,2-DIBROMOETHANE (MAK) □ DIBROMURE d'ETHYLENE (FRENCH) □ 1,2-DIBROOMETHAAN (DUTCH) □ DOWFUME 40 □ DOWFUME EDB □ DOWFUME W-8 □ DWUBROMOETAN (POLISH) □ EDB □ EDB-85 □ E-D-BEE □ ENT 15,349 □ ETHYLENE BROMIDE □ FUMO-GAS □ GLYCOL BROMIDE □ GLYCOL DIBROMIDE □ ISCOBROME D □ KOPFUME □ NCI-C00522 □ NEPHIS □ PESTMASTER □ PESTMASTER EDB-85 □ RCRA WASTE NUMBER U067 □ SOILBROM-40 □ SOILBROM-85 □ SOILFUME □ UNIFUME

CONSENSUS REPORTS: NTP 10th Report on Carcinogens. IARC Cancer Review: Group 2A IMEMDT 7,204,87; Animal Sufficient Evidence IMEMDT 15,195,77. NCI Carcinogenesis Bioassay (gavage); Clear Evidence: mouse, rat NCITR* NCI-CG-TR-86,78; NTP Carcinogenesis Bioassay (inhalation); Clear Evidence: mouse, rat NTPTR* NTP-TR-210,82. EPA Genetic Toxicology Program. Community Right-To-Know List. Reported in EPA TSCA Inventory.

OSHA PEL: TWA 20 ppm; CL 30 ppm; Pk 50 ppm/5M/8H

ACGIH TLV: Animal Carcinogen

DFG MAK: DFG TRK: Animal Carcinogen, Suspected Human Carcinogen

NIOSH REL: (EDB) 0.045 ppm; CL 1 mg/m³/15M

DOT CLASSIFICATION: 6.1; Label: Poison

SAFETY PROFILE: Confirmed carcinogen with experimental carcinogenic, neoplastigenic, and teratogenic data. Human poison by ingestion. Experimental poison by ingestion, skin contact, intraperitoneal, and possibly other routes. Moderately toxic by inhalation and rectal routes. Human systemic effects by ingestion: hypermotility, diarrhea, nausea or vomiting, decreased urine volume or anuria. Experimental reproductive effects. Human mutation data reported. A severe skin and eye irritant. Implicated in worker sterility. When heated to decomposition it emits toxic fumes of

Br⁻. See also ETHYLENE DICHLORIDE and BROMIDES.

EIY600   **CAS: 107-06-2**   **HR: 3**
ETHYLENE DICHLORIDE
DOT: UN 1184
mf: $C_2H_4Cl_2$ mw: 98.96
PROP: Colorless, clear liquid; pleasant odor, sweet taste. Bp: 83.5°, ULC: 60−70, lel: 6.2%, uel: 15.9%, fp: −35.7°, flash p: 56°F, d: 1.257 @ 20°/4°, autoign temp: 775°F, vap press: 100 mm @ 29.4°, vap d: 3.35, refr index: 1.445 @ 20°. Sol in alc, ether, acetone, carbon tetrachloride; sltly sol in water. IDLH 50 ppm.

SYNS: AETHYLENCHLORID (GERMAN) □ BICHLORURE d'ETHYLENE (FRENCH) □ BORER SOL □ BROCIDE □ CHLORURE d'ETHYLENE (FRENCH) □ CLORURO di ETHENE (ITALIAN) □ 1,2-DCE □ DESTRUXOL BORER-SOL □ 1,2-DICHLOORETHAAN (DUTCH) □ 1,2-DICHLOR-AETHAN (GERMAN) □ DICHLOREMULSION □ DI-CHLOR-MULSION □ α,β-DICHLOROETHANE □ sym-DICHLOROETHANE □ DICHLORO-1,2-ETHANE (FRENCH) □ 1,2-DICHLOROETHANE □ DICHLOROETHYLENE □ 1,2-DICLOROETANO (ITALIAN) □ DUTCH LIQUID □ DUTCH OIL □ EDC □ ENT 1,656 □ ETHANE DICHLORIDE □ ETHYLEENDICHLORIDE (DUTCH) □ ETHYLENE CHLORIDE □ 1,2-ETHYLENE DICHLORIDE □ GLYCOL DICHLORIDE □ NCI-C00511 □ RCRA WASTE NUMBER U077

CONSENSUS REPORTS: NTP 10th Report on Carcinogens. IARC Cancer Review: Group 2B IMEMDT 7,56,87; Human Limited Evidence IMEMDT 20,429,79; Animal Sufficient Evidence IMEMDT 20,429,79. NCI Carcinogenesis Bioassay (gavage); Clear Evidence: mouse, rat NCITR* NCI-CG-TR-55,78. EPA Genetic Toxicology Program. Reported in EPA TSCA Inventory.

OSHA PEL: TWA 1 ppm; STEL 2 ppm

ACGIH TLV: TWA 10 ppm; Not Classifiable as a Human Carcinogen

DFG MAK: Confirmed Animal Carcinogen, Suspected Human Carcinogen

NIOSH REL: (Ethylene Dichloride) TWA 1 ppm; CL 2 ppm/15M

DOT CLASSIFICATION: 3; Label: Flammable Liquid, Poison

SAFETY PROFILE: Confirmed carcinogen with experimental carcinogenic, neoplastigenic, and tumorigenic data. An experimental transplacental carcinogen. A human poison by ingestion. Poison experimentally by intravenous and subcutaneous routes. Moderately toxic by inhalation, skin contact, and intraperitoneal routes. Human systemic effects by ingestion and inhalation: flaccid paralysis without anesthesia (usually neuromuscular blockage), somnolence, cough, jaundice, nausea or vomiting, hypermotility, diarrhea, ulceration or bleeding from the stomach, fatty liver degeneration, change in cardiac rate, cyanosis, and coma. It may also cause dermatitis, edema of the lungs, toxic effects on the kidneys, and severe corneal effects. A strong narcotic. Experimental teratogenic and reproductive effects. A skin and severe eye irritant, and strong local irritant. Its smell and irritant effects warn of its presence at relatively safe concentrations. Human mutation data reported.

Flammable liquid. A dangerous fire hazard if exposed to heat, flame, or oxidizers. Moderately explosive in the form of vapor when exposed to flame. Violent reaction with Al, N_2O_4, NH_3, dimethylaminopropylamine. Can react vigorously with oxidizing materials and emit vinyl chloride and HCl. To fight fire, use water, foam, CO_2, dry chemicals. When heated to decomposition it emits highly toxic fumes of Cl^- and phosgene. See also CHLORINATED HYDROCARBONS, ALIPHATIC.

EJC035   CAS: 629-17-4   HR: 3
ETHYLENEDITHIOCYANATE
mf: $C_4H_4N_2S_2$ mw: 144.22
SYNS: 1,2-BIS(THIOCYANATO)ETHANE □ 1,2-DITHIOCYANATOETHANE □ 1,2-DITHIOCYANO-ETHANE □ 1,2-ETHANEDIYL THIOCYANATE □ ETHYLENE THIOCYANATE □ THIOCYANIC ACID, ETHYLENE ESTER □ THIOCYANIC ACID, 1,2-ETHANEDIYL ESTER
SAFETY PROFILE: A poison by ingestion, skin contact, and intraperitoneal routes.

When heated to decomposition it emits toxic vapors of NO_x and SO_x.

EJC500   CAS: 107-21-1   HR: 3
ETHYLENE GLYCOL
mf: $C_2H_6O_2$ mw: 62.08
PROP: Colorless, sweet-tasting, hygroscopic, viscid, poisonous liquid. Fp: −13°, mp: −15.6°, bp: 197.5°, lel: 3.2%, flash p: 232°F (CC), d: 1.113 @ 25°/25°, autoign temp: 752°F, vap d: 2.14, vap press: 0.05 mm @ 20°. Misc in H_2O, EtOH, MeOH, Me_2CO, AcOH, and Py. Immisc in $CHCl_3$, CCl_4, Et_2O, C_6H_6, CS_2, and ligroin.
SYNS: ATHYLENGLYKOL (GERMAN) □ 1,2-DIHYDROXYETHANE □ DOWTHERM SR 1 □ 1,2-ETHANEDIOL □ ETHYLENE ALCOHOL □ ETHYLENE DIHYDRATE □ GLYCOL □ GLYCOL ALCOHOL □ LUTROL-9 □ MACROGOL 400 BPC □ M.E.G. □ MONOETHYLENE GLYCOL □ NCI-C00920 □ NORKOOL □ TESCOL □ UCAR 17
CONSENSUS REPORTS: EPA Genetic Toxicology Program. Community Right-To-Know List. Reported in EPA TSCA Inventory.
OSHA PEL: CL 50 ppm
ACGIH TLV: CL 50 ppm (vapor)
DFG MAK: 10 ppm (26 mg/m³)
SAFETY PROFILE: Human poison by ingestion. (Lethal dose for humans reported to be 100 mL.) Moderately toxic to humans by an unspecified route. Moderately toxic experimentally by ingestion, subcutaneous, intravenous, and intramuscular routes. Human systemic effects by ingestion and inhalation: eye lachrymation, general anesthesia, headache, cough, respiratory stimulation, nausea or vomiting, pulmonary, kidney, and liver changes. If ingested it causes initial central nervous system stimulation followed by depression. Later, it causes potentially lethal kidney damage. Very toxic in particulate form upon inhalation. An experimental teratogen. Other experimental reproductive effects. Human mutation data reported. A skin, eye, and mucous membrane irritant.

Combustible when exposed to heat or flame; can react vigorously with oxidants.

Moderate explosion hazard when exposed to flame. Ignites on contact with chromium trioxide, potassium permanganate, and sodium peroxide. Mixtures with ammonium dichromate, silver chlorate, sodium chlorite, and uranyl nitrate ignite when heated to 100°C. Can react violently with chlorosulfonic acid, oleum, H_2SO_4, $HClO_4$, and P_2S_5. Aqueous solutions may ignite silvered copper wires that have an applied D.C. voltage. To fight fire, use alcohol foam, water, foam, CO_2, dry chemical. When heated to decomposition it emits acrid smoke and irritating fumes.

EJD759   CAS: 111-55-7   HR: 2
ETHYLENE GLYCOL DIACETATE
mf: $C_6H_{10}O_4$ mw: 146.16
PROP: Colorless liquid or crystals. Mp: −31°, bp: 186–187°, flash p: 205°F (OC), fp: −31°, d: 1.128 @ 0°/4°, vap press: 1 mm @ 38.3°, vap d: 5.04. Misc in EtOH, Et₂O; sltly sol in H_2O.
SYNS: 1,2-ETHANEDIOL DIACETATE □ ETHYLENE ACETATE □ ETHYLENE GLYCOL ACETATE □ GLYCOL DIACETATE
CONSENSUS REPORTS: Reported in EPA TSCA Inventory.
SAFETY PROFILE: Moderately toxic by intraperitoneal route. Mildly toxic by ingestion and skin contact. An eye irritant. Combustible when exposed to heat or flame; can react with oxidizing materials. To fight fire, use alcohol foam, CO_2, dry chemical. When heated to decomposition it emits acrid smoke and irritating fumes.

EJE500   CAS: 629-14-1   HR: 3
ETHYLENE GLYCOL DIETHYL ETHER
DOT: UN 1153
mf: $C_6H_{14}O_2$ mw: 118.20
PROP: Colorless liquid; slight ethereal odor. Mp: −74°, bp: 123.5°, flash p: 95°F (OC), d: 0.8417 @ 20°/20°, autoign temp: 406°F, vap d: 6.56, vap press: 9.4 mm.
SYNS: 1,2-DIETHOXYETHANE □ DIETHYL CELLOSOLVE (DOT) □ ETHYL GLYME

CONSENSUS REPORTS: Reported in EPA TSCA Inventory. Glycol ether compounds are on the Community Right-To-Know List.
DOT CLASSIFICATION: 3; Label: Flammable Liquid
SAFETY PROFILE: Moderately toxic by ingestion. Mildly toxic by inhalation. An experimental teratogen. Experimental reproductive effects. An eye irritant. An aprotic solvent. A very dangerous fire hazard when exposed to heat or flame; can react with oxidizing materials. To fight fire, use CO_2, dry chemical. See also GLYCOL ETHERS and various cellosolve entries.

EJF000   CAS: 629-15-2   HR: 3
ETHYLENE GLYCOL DIFORMATE
mf: $C_4H_6O_4$ mw: 118.10
PROP: Liquid. Mp: −10°, bp: 177°, flash p: 200°F (OC), d: 1.2277 @ 20°/20°, vap d: 4.07.
SYNS: ETHYLENE FORMATE □ GLYCOL DIFORMATE
CONSENSUS REPORTS: Reported in EPA TSCA Inventory.
SAFETY PROFILE: Poison by ingestion. A severe eye irritant. Flammable when exposed to heat or flame; can react with oxidizing materials. To fight fire, use CO_2, dry chemical. When heated to decomposition it emits acrid smoke and irritating fumes.

EJG000   CAS: 628-96-6   HR: 3
ETHYLENE GLYCOL DINITRATE
mf: $C_2H_4N_2O_6$ mw: 152.08
PROP: Yellow liquid. Mp: −22.3°, bp: 105.5° @ 19 mm, explodes @ 114°, d: 1.483 @ 8°, vap d: 5.25. IDLH 75 mg/m³.
SYNS: DINITROGLICOL (ITALIAN) □ DINITRO-GLYCOL □ EGDN □ ETHANEDIOL DINITRATE □ ETHYLENE DINITRATE □ ETHYLENE NITRATE □ ETHYLENGLYKOLDINITRAT (CZECH) □ GLYCOLDINITRAAT (DUTCH) □ GLYCOL DINITRATE □ GLYCOL (DINITRATE DE) (FRENCH) □ GLYKOL-DINITRAT (GERMAN) □ NITROGLYCOL □ NITRO-GLYKOL (CZECH)
CONSENSUS REPORTS: Reported in EPA TSCA Inventory.
OSHA PEL: STEL 0.1 mg/m³ (skin)
ACGIH TLV: TWA 0.05 ppm (skin)

DFG MAK: 0.05 ppm (0.32 mg/m^3)
NIOSH REL: (Nitroglycerin) CL 0.1
mg/m^3/20M
DOT CLASSIFICATION: Forbidden
SAFETY PROFILE: Poison by
subcutaneous route. Can cause lowered
blood pressure leading to headache,
dizziness, and weakness. Used as an
explosive. When heated to decomposition it
emits toxic fumes of NO$_x$. See also
NITRATES.

EJH500   CAS: 109-86-4   HR: 3
ETHYLENE GLYCOL METHYL ETHER
DOT: UN 1188
mf: C$_3$H$_8$O$_2$ mw: 76.11
PROP: Colorless liquid; mild, agreeable
odor. Misc in water, alc, ether, benzene. Bp:
124.5°, fp: −86.5°, flash p: 115°F (OC), lel:
2.5%, uel: 14%, d: 0.9660 @ 20°/4°, autoign
temp: 545°F, vap press: 6.2 mm @ 20°, vap
d: 2.62. IDLH 200 ppm.
SYNS: AETHYLENGLYKOL-MONOMETHYLAETHER
(GERMAN) □ DOWANOL EM □ EGM □ EGME □
ETHER MONOMETHYLIQUE de l'ETHYLENE-GLYCOL
(FRENCH) □ ETHYLENE GLYCOL MONOMETHYL
ETHER (MAK, DOT) □ GLYCOL ETHER EM □ GLYCOL
METHYL ETHER □ GLYCOL MONOMETHYL ETHER □
JEFFERSOL EM □ MECS □ 2-METHOXY-AETHANOL
(GERMAN) □ 2-METHOXYETHANOL (ACGIH) □
METHOXYHYDROXYETHANE □ METHYL CELLO-
SOLVE (OSHA, DOT) □ METHYL ETHOXOL □ METHYL
GLYCOL □ METHYL GLYKOL (GERMAN) □ METHYL
OXITOL □ METIL CELLOSOLVE (ITALIAN) □
METOKSYETYLOWY ALKOHOL (POLISH) □ 2-
METOSSIETANOLO (ITALIAN) □ MONOMETHYL
ETHER of ETHYLENE GLYCOL □ POLY-SOLV EM □
PRIST
CONSENSUS REPORTS: Reported in EPA
TSCA Inventory. Community Right-To-
Know List.
OSHA PEL: TWA 25 ppm (skin)
ACGIH TLV: TWA 5 ppm (skin)
DFG MAK: 5 ppm (16 mg/m^3)
NIOSH REL: TWA (Glycol Ethers) Reduce
to lowest level
DOT CLASSIFICATION: 3; Label:
Flammable Liquid
SAFETY PROFILE: Moderately toxic to
humans by ingestion. Moderately toxic

experimentally by ingestion, inhalation, skin
contact, intraperitoneal, and intravenous
routes. Human systemic effects by
inhalation: change in motor activity,
tremors, and convulsions. Experimental
teratogenic and reproductive effects. A skin
and eye irritant. Mutation data reported.
When used under conditions that do not
require the application of heat, this material
probably presents little hazard to health.
However, in the manufacture of fused
collars which require pressing with a hot
iron, cases have been reported showing
disturbance of the hemopoietic system with
or without neurological signs and
symptoms. The blood picture may resemble
that produced by exposure to benzene. Two
cases reported had severe aplastic anemia
with tremors and marked mental dullness.
The persons affected had been exposed to
vapors of methyl "Cellosolve," ethanol,
methanol, ethyl acetate, and petroleum
naphtha.

Flammable liquid when exposed to heat or
flame. A moderate explosion hazard. Can
react with oxidizing materials to form
explosive peroxides. To fight fire, use
alcohol foam, CO$_2$, dry chemical. When
heated to decomposition it emits acrid
smoke and irritating fumes. See also
GLYCOL ETHERS.

EJJ500   CAS: 110-49-6   HR: 3
ETHYLENE GLYCOL MONOMETHYL
**   ETHER ACETATE**
DOT: UN 1189
mf: C$_5$H$_{10}$O$_3$ mw: 118.15
PROP: Colorless liquid; pleasant, sweet,
ether odor. Bp: 143°, fp: −70°, flash p:
111°F (CC), d: 1.005 @ 20°/20°, vap d:
4.07, lel: 1.7%, uel: 8.2%. Sol in water.
IDLH 200 ppm.
SYNS: ACETATE de l'ETHER MONOMETHYLIQUE de
l'ETHYLENE-GLYCOL (FRENCH) □ ACETATE de
METHYLE GLYCOL (FRENCH) □ ACETATO di METIL
CELLOSOLVE (ITALIAN) □ AETHYLENGLYKOL-
METHYLAETHERACETAT (GERMAN) □ ETHYLENE
GLYCOL METHYL ETHER ACETATE □ GLYCOL
ETHER EM ACETATE □ GLYCOL MONOMETHYL

ETHER ACETATE □ MeCsAc □ 2-METHOXYAETHYL-
ACETAT (GERMAN) □ 2-METHOXYETHANOL,
ACETATE □ 2-METHOXY-ETHYL ACETAAT (DUTCH)
□ 2-METHOXYETHYL ACETATE (ACGIH) □ 2-
METHOXYETHYLE, ACETATE de (FRENCH) □ METHYL
CELLOSOLYE ACETAAT (DUTCH) □ METHYL
CELLOSOLVE ACETATE (OSHA, DOT) □ METHYL
GLYCOL ACETATE □ METHYL GLYCOL MONO-
ACETATE □ METHYLGLYKOLACETAT (GERMAN) □ 2-
METOSSIETILACETATO (ITALIAN)

CONSENSUS REPORTS: Glycol ether compounds are on the Community Right-To-Know List. Reported in EPA TSCA Inventory.
OSHA PEL: TWA 25 ppm (skin)
ACGIH TLV: TWA 5 ppm (skin)
DFG MAK: 5 ppm (25 mg/m³)
DOT CLASSIFICATION: 3; Label: Flammable Liquid
SAFETY PROFILE: Moderately toxic by ingestion, intraperitoneal, and subcutaneous routes. Mildly toxic by inhalation and skin contact. Human systemic effects by inhalation: eye lachrymation, cough, and pulmonary changes. Experimental reproductive effects. Mutation data reported. An inhalation irritant in humans. An eye irritant. Flammable liquid when exposed to heat or flame; can react with oxidizing materials. A moderate explosion hazard. To fight fire, use CO₂, dry chemical. When heated to decomposition it emits acrid smoke and irritating fumes. See also GLYCOL ETHERS.

EJM500   CAS: 111-60-4   HR: 3
ETHYLENE GLYCOL STEARATE
mf: C₂₀H₄₀O₃   mw: 328.60
SYNS: CLINDROL SEG □ EMEREST 2350 □ EMPILAN 2848 □ ETHYLENE GLYCOL, MONOSTEARATE □ GLYCOL MONOSTEARATE □ GLYCOL STEARATE □ 2-HYDROXYETHYL ESTER STEARIC ACID □ IVORIT □ LIPO EGMS □ MONTHYBASE □ MONTHYLE □ PARASTARIN □ PRODHYBASE ETHYL □ S 151 □ SEDETOL □ STEARIC ACID, MONOESTER with ETHYLENE GLYCOL □ TEGO-STEARATE □ USAF KE-11
CONSENSUS REPORTS: Reported in EPA TSCA Inventory.
SAFETY PROFILE: Poison by intraperitoneal route. A skin irritant. Used in

cosmetics. When heated to decomposition it emits acrid smoke and irritating fumes. See also ESTERS.

EJM900   CAS: 151-56-4   HR: 3
ETHYLENEIMINE
DOT: UN 1185
mf: C₂H₅N   mw: 43.08
PROP: Oily, water-white liquid. Pungent ammonia-like odor. Bp: 55–56°, fp: −71.5°, flash p: 12°F, d: 0.832 @ 20°/4°, autoign temp: 608°F, vap press: 160 mm @ 20°, vap d: 1.48, lel: 3.6%, uel: 46%. Misc in water. IDLH 100 ppm.
SYNS: AETHYLENIMIN (GERMAN) □ AMINOETHYLENE □ AZACYCLOPROPANE □ AZIRANE □ AZIRIDIN (GERMAN) □ AZIRIDINE □ DIHYDROAZIRENE □ DIHYDRO-1H-AZIRINE □ DIMETHYLENEIMINE □ DIMETHYLENIMINE □ EI □ ENT 50,324 □ ETHYLEENIMINE (DUTCH) □ ETHYL-ENE IMINE, INHIBITED (DOT) □ ETHYLENIMINE □ ETHYLIMINE □ ETILENIMINA (ITALIAN) □ RCRA WASTE NUMBER P054 □ TL 337
CONSENSUS REPORTS: IARC Cancer Review: Group 3 IMEMDT 7,56,87; Animal Sufficient Evidence IMEMDT 9,37,75. Community Right-To-Know List. EPA Extremely Hazardous Substances List. Reported in EPA TSCA Inventory. EPA Genetic Toxicology Program.
OSHA PEL: TWA 1 mg/m³ (skin); Cancer Suspect Agent
ACGIH TLV: TWA 0.5 ppm (skin); Animal Carcinogen
DFG MAK: DFG TRK: Animal Carcinogen, Suspected Human Carcinogen
NIOSH REL: (Ethyleneimine) TWA use 29 CFR 1910.1012
DOT CLASSIFICATION: 6.1; Label: Poison, Flammable Liquid
SAFETY PROFILE: Confirmed carcinogen with experimental carcinogenic, neoplastigenic, tumorigenic, and teratogenic data. Other experimental reproductive effects. Poison by ingestion, skin contact, inhalation, and intraperitoneal routes. Human mutation data reported. A skin, mucous membrane, and severe eye irritant. An allergic sensitizer of skin. Causes opaque

cornea, keratoconus, and necrosis of cornea (experimentally). Has been known to cause severe human eye injury. Drinking of carbonated beverages is recommended as an antidote to this material in stomach.

A very dangerous fire and explosion hazard when exposed to heat, flame, or oxidizers. Reacts violently with acids, aluminum chloride + substituted anilines, acetic acid, acetic anhydride, acrolein, acrylic acid, allyl chloride, CS_2, Cl_2, chlorosulfonic acid, epichlorohydrin, glyoxal, HCl, HF, HNO_3, oleum, β-propiolactone, Ag, NaOCl, H_2SO_4, vinyl acetate. Reacts with chlorinating agents (e.g., sodium hypochlorite solution) to form the explosive 1-chloroaziridine. Reacts with silver or its alloys to form explosive silver derivatives. Dangerous; heat and/or the presence of catalytically active metals or chloride ions can cause a violent exothermic reaction. To fight fire, use alcohol foam, CO_2, dry chemical. When heated to decomposition it emits acrid smoke and irritating fumes.

EJN500   CAS: 75-21-8   HR: 3
ETHYLENE OXIDE
DOT: UN 1040
mf: C_2H_4O mw: 44.06
PROP: Colorless gas at room temperature. Mp: −111.3°, bp: 10.7°, ULC: 100, lel: 3.0%, uel: 100%, flash p: −4°F, d: 0.8711 @ 20°/20°, autoign temp: 804°F, vap press: 1095 mm @ 20°, vap d: 1.52. Misc in water and alc; very sol in ether. IDLH 800 ppm.
SYNS: AETHYLENOXID (GERMAN) □ AMPROLENE □ ANPROLENE □ ANPROLINE □ DIHYDROOXIRENE □ DIMETHYLENE OXIDE □ ENT 26,263 □ E.O. □ 1,2-EPOXYAETHAN (GERMAN) □ EPOXYETHANE □ 1,2-EPOXYETHANE □ ETHENE OXIDE □ ETHYLE-ENOXIDE (DUTCH) □ ETHYLENE (OXYDE d') (FRENCH) □ ETILENE (OSSIDO di) (ITALIAN) □ ETO □ ETYLENU TLENEK (POLISH) □ FEMA No. 2433 □ MERPOL □ NCI-C50088 □ OXACYCLOPROPANE □ OXANE □ OXIDOETHANE □ α,β-OXIDOETHANE □ OXIRAAN (DUTCH) □ OXIRANE □ OXYFUME □ OXYFUME 12 □ RCRA WASTE NUMBER U115 □ STERILIZING GAS ETHYLENE OXIDE 100% □ T-GAS
CONSENSUS REPORTS: NTP 10th Report on Carcinogens. IARC Cancer Review:

Group 2A IMEMDT 7,205,87; Animal Inadequate Evidence IMEMDT 11,157,76; Human Inadequate Evidence IMEMDT 36,189,85; Animal Sufficient Evidence IMEMDT 36,189,85. Community Right-To-Know List. EPA Extremely Hazardous Substances List. Reported in EPA TSCA Inventory. EPA Genetic Toxicology Program.
OSHA PEL: TWA 1 ppm; Cancer Hazard
ACGIH TLV: TWA 1 ppm; Suspected Human Carcinogen
DFG MAK: DFG TRK: Animal Carcinogen, Suspected Human Carcinogen
NIOSH REL: (Ethylene Oxide) TWA 0.1 ppm; CL 5 ppm/10M/D
DOT CLASSIFICATION: 2.3; Label: Poison Gas, Flammable Gas
SAFETY PROFILE: Confirmed human carcinogen with experimental carcinogenic, tumorigenic, neoplastigenic, and teratogenic data. Poison by ingestion, intraperitoneal, subcutaneous, and intravenous routes. Moderately toxic by inhalation. Human systemic effects by inhalation: convulsions, nausea, vomiting, olfactory and pulmonary changes. Experimental reproductive effects. Mutation data reported. A skin and eye irritant. An irritant to mucous membranes of respiratory tract. High concentrations can cause pulmonary edema.

Highly flammable liquid or gas. Severe explosion hazard when exposed to flame. To fight fire, use alcohol foam, CO_2, dry chemical. Violent polymerization occurs on contact with ammonia, alkali hydroxides, amines, metallic potassium, acids, covalent halides (e.g., aluminum chloride, iron(III) chloride, tin(IV) chloride, aluminum oxide, iron oxide, rust). Explosive reaction with glycerol at 200°. Rapid compression of the vapor with air causes explosions. Incompatible with bases, alcohols, air, m-nitroaniline, trimethyl amine, copper, iron chlorides, iron oxides, magnesium perchlorate, mercaptans, potassium, tin chlorides, contaminants, alkane thiols, bromoethane. When heated to

decomposition it emits acrid smoke and irritating fumes.

EJO000   CAS: 8070-50-6   HR: 3
ETHYLENE OXIDE, mixed with CARBON DIOXIDE
DOT: UN 1041/UN 1952
PROP: Contains less than 10% carbon dioxide (NTIS** PB225-283).
SYNS: ANHYDRIDE CARBONIQUE et OXYDE d'ETHYLENE MELANGES (FRENCH) □ ETHYLENE OXIDE and CARBON DIOXIDE MIXTURES (DOT) □ OXYFUME 20 □ OXYFUME 30
DOT CLASSIFICATION: 2.1; Label: Flammable Gas (UN 1041)
SAFETY PROFILE: A poison. Mildly toxic by inhalation. Used for the sterilization of vacuum chambers. See also ETHYLENE OXIDE.

EJO025   HR: D
ETHYLENE OXIDE POLYMER
SAFETY PROFILE: When heated to decomposition it emits acrid smoke and irritating fumes.

EJP500   CAS: 420-12-2   HR: 3
ETHYLENE SULFIDE
mf: C_2H_4S mw: 60.12
PROP: Colorless liquid. Bp: 55-56° decomp, d: 1.0368 @ 0°/4°, vap d: 2.07.
SYNS: AETHYLENSULFID (GERMAN) □ 2,3-DIHYDROTHIIRENE □ ETHYLENE EPISULFIDE □ ETHYLENE EPISULPHIDE □ ETHYLENE SULPHIDE □ THIACYCLOPROPANE □ THIIRANE
CONSENSUS REPORTS: IARC Cancer Review: Group 3 IMEMDT 7,56,87; Animal Limited Evidence IMEMDT 11,257,76. Reported in EPA TSCA Inventory.
SAFETY PROFILE: Poison by ingestion, intraperitoneal, and subcutaneous routes. Mildly toxic by inhalation. A skin, eye, and mucous membrane irritant. Questionable carcinogen with experimental tumorigenic data. Can react with oxidizing materials. When heated to decomposition, or on contact with acid or acid fumes, it emits highly toxic fumes of SO_x. See also SULFIDES.

EJU000   CAS: 60-29-7   HR: 3
ETHYL ETHER
DOT: UN 1155
mf: $C_4H_{10}O$ mw: 74.14
PROP: A clear, volatile liquid; sweet, pungent odor. Mp: −116.2°, bp: 34.6°, ULC: 100, lel: 1.85%, uel: 36%, flash p: −49°F, d: 0.7135 @ 20°/4°, autoign temp: 320°F, vap press: 442 mm @ 20°, vap d: 2.56. Sol in H_2SO_4; sltly sol in H_2O; misc in most org solvs. IDLH 1900 ppm [10%LEL].
SYNS: AETHER □ ANAESTHETIC ETHER □ ANESTHESIA ETHER □ ANESTHETIC ETHER □ DIAETHYLAETHER (GERMAN) □ DIETHYL ETHER (DOT) □ DIETHYL OXIDE □ DWUETYLOWY ETER (POLISH) □ ETERE ETILICO (ITALIAN) □ ETHER □ ETHER ETHYLIQUE (FRENCH) □ ETHOXYETHANE □ 1,1'-OXYBISETHANE □ OXYDE d'ETHYLE (FRENCH) □ RCRA WASTE NUMBER U117 □ SOLVENT ETHER
CONSENSUS REPORTS: IARC Cancer Review: Animal No Adequate Data IMEMDT 7,93,87. Reported in EPA TSCA Inventory. EPA Genetic Toxicology Program.
OSHA PEL: TWA 400 ppm; STEL 500 ppm
ACGIH TLV: TWA 400 ppm; STEL 500 ppm
DFG MAK: 400 ppm (1200 mg/m³)
DOT CLASSIFICATION: 3; Label: Flammable Liquid
SAFETY PROFILE: Moderately toxic to humans by ingestion. Poison experimentally by subcutaneous route. Moderately toxic by intraperitoneal and intravenous routes. Mildly toxic by inhalation. Human systemic effects by inhalation: olfactory changes. Mutation data reported. A severe eye and moderate skin irritant. Ethyl ether is not corrosive or dangerously reactive. It must not be considered safe for individuals to inhale or ingest. It is a depressant of the central nervous system and is capable of producing intoxication, drowsiness, stupor, and unconsciousness. Death due to respiratory failure may result from severe and continued exposure.

A very dangerous fire and explosion hazard when exposed to heat or flame. A storage hazard. It auto-oxidizes to form explosive polymeric 1-oxy-peroxides. Explosive reaction with boron triazide, bromine trifluoride, bromine pentafluoride, perchloric acid, uranyl nitrate + light, wood pulp extracts + heat. Violent reaction or ignition on contact with halogens (e.g., bromine, chlorine), interhalogens (e.g., iodine heptafluoride), oxidants (e.g., silver perchlorate, nitrosyl perchlorate, nitryl perchlorate, chromyl chloride, fluorine nitrate, permanganic acid, nitric acid, hydrogen peroxide, peroxodisulfuric acid, iodine(VII) oxide, sodium peroxide, ozone, and liquid air), sulfur and sulfur compounds (e.g., sulfur when dried with peroxidized ether, sulfuryl chloride). Can react vigorously with acetyl peroxide, air, bromoazide, ClF_3, CrO_3, $Cr(OCl)_2$, $LiAlH_2$, $NOClO_4$, O_2, $NClO_2$, (H_2SO_4 + permanganates), K_2O_2, [$(C_2H_5)_3Al$ + air], [$(CH_3)_3Al$ + air]. To fight fire, use alcohol foam, CO_2, dry chemical. Used in production of drugs of abuse. When heated to decomposition it emits acrid smoke and irritating fumes. See also ETHERS.

EKL000   CAS: 109-94-4   HR: 3
ETHYL FORMATE
DOT: UN 1190
mf: $C_3H_6O_2$ mw: 74.09
PROP: Colorless, mobile flammable liquid; sharp, pleasant, rum-like odor. Mp: −79°, bp: 54.3°, lel: 2.7%, uel: 13.5%, flash p: −4°F (CC), d: 0.9236 @ 20°/20°, refr index: 1.359, autoign temp: 851°F, vap press: 100 mm @ 5.4°, vap d: 2.55. Misc in EtOH, Et_2O, C_6H_6; sltly sol in and gradually hydrated by H_2O. IDLH 1500 ppm.
SYNS: AETHYLFORMIAT (GERMAN) □ AREGINAL □ ETHYLE (FORMIATE d') (FRENCH) □ ETHYLFORMIAAT (DUTCH) □ ETHYL FORMIC ESTER □ ETHYL METHANOATE □ ETILE (FORMIATO di) (ITALIAN) □ FEMA No. 2434 □ FORMIC ACID, ETHYL ESTER □ FORMIC ETHER □ MROWCZAN ETYLU (POLISH)
CONSENSUS REPORTS: Reported in EPA TSCA Inventory.
OSHA PEL: TWA 100 ppm
ACGIH TLV: TWA 100 ppm
DFG MAK: 100 ppm (310 mg/m³)

DOT CLASSIFICATION: 3; Label: Flammable Liquid
SAFETY PROFILE: Moderately toxic by ingestion and subcutaneous routes. Mildly toxic by skin contact and inhalation. A powerful inhalation irritant in humans. A skin and eye irritant. Questionable carcinogen with experimental tumorigenic data. Highly flammable liquid. A very dangerous fire and explosion hazard when exposed to heat, flame, or oxidizers. To fight fire, use alcohol foam, spray, mist, dry chemical. When heated to decomposition it emits acrid smoke and irritating fumes. See also ESTERS.

EKM200   CAS: 4016-11-9   HR: 3
ETHYL GLYCIDYL ETHER
DOT: UN 2752
mf: $C_5H_{10}O_2$ mw: 102.15
SYNS: 1,2-EPOXY-3-ETHOXYPROPANE □ 1,2-EPOXY-3-ETHOXYPROPANE (DOT) □ (ETHOXYMETHYL)-OXIRANE □ OXIRANE, (ETHOXYMETHYL)-(9CI) □ PROPANE, 1,2-EPOXY-3-ETHOXY-
CONSENSUS REPORTS: Reported in EPA TSCA Inventory.
DOT CLASSIFICATION: 3; Label: Flammable Liquid
SAFETY PROFILE: Mutation data reported. A flammable liquid. When heated to decomposition it emits acrid smoke and irritating vapors.

EKN000   CAS: 2642-71-9   HR: 3
ETHYL GUTHION
mf: $C_{12}H_{16}N_3O_3PS_2$ mw: 345.40
PROP: Needles. D: 1.284 @ 20°/4°, mp: 53°, bp: 111° @ 0.001 mm.
SYNS: ATHYL-GUSATHION □ AZINFOS-ETHYL (DUTCH) □ AZINOS □ AZINPHOS-AETHYL (GERMAN) □ AZINPHOS ETHYL □ AZINPHOS-ETILE (ITALIAN) □ BAY 16225 □ BAYER 16259 □ BENZOTRIAZINE derivative of an ETHYL DITHIOPHOSPHATE □ COTNION-ETHYL □ CYRSTHION □ O,O-DIAETHYL-S-(4-OXOBENZO-TRIAZIN-3-METHYL)-DITHIOPHOSPHAT (GERMAN) □ O,O-DIAETHYL-S-((4-OXO-3H-1,2,3-BENZOTRIAZIN-3-YL)-METHYL)-DITHIOPHOSPHAT (GERMAN) □ O,O-DIETHYL-S-(4-OXO-3H-1,2,3-BENZOTRIAZINE-3-YL)-METHYL-DITHIOPHOSPHATE □ O,O-DIETHYL-S-((4-OXO-3H-1,2,3-BENZOTRIAZIN-3-YL)-METHYL)-DITHIO FOSFAAT (DUTCH) □ O,O-DIETHYL-S-(4-OXOBENZO-

TRIAZINO-3-METHYL)PHOSPHORODITHIOATE □
O,O-DIETHYL PHOSPHORODITHIOATE S-ester with 3-
(MERCAPTOMETHYL)-1,2,3-BENZOTRIAZIN-4(3H)-ONE
□ O,O-DIETIL-S-((4-OXO-3H-1,2,3-BENZOTRIAZIN-3-IL)-
METIL)-DITIOFOSFATO (ITALIAN) □ 3,4-DIHYDRO-4-
OXO-3-BENZOTRIAZINYLMETHYL O,O-DIETHYL
PHOSPHORODITHIOATE □ S-(3,4-DIHYDRO-4-OXO-
1,2,3-BENZOTRIAZIN-3-YLMETHYL) O,O-DIETHYL
PHOSPHORODITHIOATE □ ENT 22,014 □ ETHYL
GUSATHION □ GUSATHION A □ GUTHION (ETHYL)
□ R 1513 □ TRIAZOTION (RUSSIAN)
CONSENSUS REPORTS: EPA Extremely
Hazardous Substances List.
SAFETY PROFILE: Poison by ingestion,
inhalation, skin contact, and intraperitoneal
route. A cholinesterase inhibitor type of
insecticide. When heated to decomposition
it emits toxic fumes of SO_x, PO_x, and NO_x.
See also PARATHION.

EKN050   CAS: 106-30-9   HR: 3
ETHYL HEPTANOATE
mf: $C_9H_{18}O_2$ mw: 158.24
PROP: Colorless liquid; wine-brandy odor.
D: 0.867–0.872, refr index: 1.411, fp:
−66.1°, bp: 188.6°, flash p: 149°F. Misc in
alc, chloroform, fixed oils; sltly sol in
propylene glycol.
SYNS: COGNAC OIL □ ENANTHYLIC ETHER □
ETHYL ENANTHATE □ ETHYL HEPTANOATE □
ETHYL n-HEPTANOATE □ ETHYL HEPTOATE □
ETHYL HEPTYLATE □ ETHYL OENANTHATE □
ETHYL OENANTHYLATE □ FEMA No. 2437 □
OENANTHIC ETHER
CONSENSUS REPORTS: Reported in EPA
TSCA Inventory.
SAFETY PROFILE: Low toxicity by
ingestion and skin contact. Flammable liquid
when exposed to heat, sparks, or flame.
When heated to decomposition it emits
acrid smoke and irritating fumes.

EKQ000   CAS: 104-76-7   HR: 2
2-ETHYLHEXANOL
mf: $C_8H_{18}O$ mw: 130.26
PROP: Clear liquid. Bp: 184–185°, mp:
<−76°, flash p: 178°F (81°C), n: (20/D)
1.4300, d: 0.834 @ 20°/20°, vap press: 0.2
mm @ 20°, vap d: 4.49. Sol in about 720
parts water and in many org solvs.

SYNS: 1-AETHYLHEXANOL (GERMAN) □ 2-ETHYL-1-
HEXANOL □ 2-ETHYLHEXYL ALCOHOL
CONSENSUS REPORTS: Reported in EPA
TSCA Inventory.
SAFETY PROFILE: Moderately toxic by
ingestion, skin contact, intraperitoneal,
subcutaneous, and parenteral routes. An
experimental teratogen. Other experimental
reproductive effects. A severe eye and
moderate skin irritant. Mutation data
reported. A dangerous fire hazard when exposed to heat or flame; can react vigorously
with oxidizing materials. To fight fire, use
foam, CO_2, dry chemical. When heated to
decomposition it emits acrid smoke and
fumes. See also ALCOHOLS.

EKR500   CAS: 1632-16-2   HR: 3
2-ETHYL-1-HEXENE
mf: C_8H_{16} mw: 112.24
PROP: Colorless liquid. Bp: 120°, d: 0.7270
@ 20°/20°, vap d: 3.87.
SYNS: 2-ETHYL HEXENE-1 □ USAF DO-21
CONSENSUS REPORTS: Reported in EPA
TSCA Inventory.
SAFETY PROFILE: Poison by
intraperitoneal route. Mildly toxic by
inhalation. A skin and eye irritant.
Combustible when exposed to heat or
flame; can react with oxidizing materials.
When heated to decomposition it emits
acrid smoke and irritating fumes.

EKS500   CAS: 104-75-6   HR: 3
2-ETHYL HEXYLAMINE
DOT: UN 2276
mf: $C_8H_{19}N$ mw: 129.28
PROP: A clear, miscible liquid. Bp: 169.2°,
flash p: 140°F (OC), d: 0.7894 @ 20°/20°,
vap press: 1.2 mm @ 20°, vap d: 4.45.
SYN: 1-AMINO-2-ETHYLHEXAN (CZECH)
CONSENSUS REPORTS: Reported in EPA
TSCA Inventory.
DOT CLASSIFICATION: 8; Label: Corrosive
SAFETY PROFILE: Poison by
intraperitoneal route. Moderately toxic by
ingestion, inhalation, and skin contact.
Corrosive. A severe skin and eye irritant.

Flammable liquid when exposed to heat or flame; can react with oxidizing materials. To fight fire, use alcohol foam, CO_2, dry chemical. When heated to decomposition it emits toxic fumes of NO_x. See also AMINES.

EKV000   CAS: 94-96-2   HR: 2
ETHYL HEXYLENE GLYCOL
mf: $C_8H_{18}O_2$ mw: 146.26
PROP: Practically colorless, somewhat viscous, odorless liquid. Bp: 243.1°, flash p: 260°F (OC), fp: −40°, d: 0.9422 @ 20°/20°, vap press: <0.01 mm @ 20°, vap d: 5.03. Sltly sol in H_2O; misc in $CHCl_3$, EtOH, and Et_2O.
SYNS: CARBIDE 6-12 ◻ COMPOUND 6-12 INSECT REPELLENT ◻ ENT 375 ◻ ETHOHEXADIOL ◻ ETHYL HEXANEDIOL ◻ 2-ETHYLHEXANEDIOL-1,3 ◻ 2-ETHYLHEXANE-1,3-DIOL ◻ 2-ETHYL-1,3-HEXANEDI-OL ◻ 2-ETHYL-3-PROPYL-1,3-PROPANEDIOL ◻ 3-HYDROXYMETHYL-n-HEPTAN-4-OL ◻ 6-12-INSECT REPELLENT ◻ OCTYLENE GLYCOL ◻ REPELLENT 612 ◻ RUTGERS 612
CONSENSUS REPORTS: Reported in EPA TSCA Inventory.
SAFETY PROFILE: Moderately toxic by ingestion and skin contact. Experimental teratogenic and reproductive effects. A skin and severe eye irritant. Used as an insecticide, insect repellent, and in hair care preparations. Combustible when exposed to heat or flame; can react with oxidizing materials. To fight fire, use alcohol foam, foam, dry chemical. When heated to decomposition it emits acrid smoke and irritating fumes.

ELD000   CAS: 3031-74-1   HR: 3
ETHYL HYDROPEROXIDE
mf: $C_2H_6O_2$ mw: 62.07
SYNS: ETHYL HYDROGEN PEROXIDE ◻ HYDROPEROXIDE, ETHYL
DOT CLASSIFICATION: Forbidden
SAFETY PROFILE: Explodes violently when superheated. The barium salt is heat- and impact-sensitive. Explosive reaction with hydroiodic acid or finely divided silver. When heated to decomposition it emits acrid smoke and irritating fumes. See also PEROXIDES.

ELG100   CAS: 58066-96-9   HR: 3
N-ETHYL-N-(2-HYDROXYETHYL)-3-METHYL-4-NITROSOANILINE
mf: $C_{11}H_{16}N_2O_2$ mw: 208.29
SYNS: ETHANOL, 2-(ETHYL(3-METHYL-4-NITROSO-PHENYL)AMINO)- ◻ 2-(ETHYL(3-METHYL-4-NITROSOPHENYL)AMINO)ETHANOL
SAFETY PROFILE: A poison by ingestion. When heated to decomposition it emits toxic vapors of NO_x.

ELG500   CAS: 13147-25-6   HR: 3
ETHYL-2-HYDROXYETHYLNITRO-SAMINE
mf: $C_4H_{10}N_2O_2$ mw: 118.16
SYNS: AETHYL-AETHANOL-NITROSOAMIN (GERMAN) ◻ EENA ◻ EHEN ◻ N-ETHYL-N-HYDROXYETHYLNITROSAMINE ◻ 2-(ETHYL-NITROSAMINO)ETHANOL ◻ N-NITROSO-AETHYLAETHANOLAMIN (GERMAN) ◻ N-NITROSOETHYLETHANOLAMINE ◻ N-NITROSO-ETHYL-2-HYDROXYETHYLAMINE ◻ N-NITROSO-N-ETHYL-N-(2-HYDROXYETHYL)AMINE
CONSENSUS REPORTS: IARC Cancer Review: Animal Limited Evidence IMEMDT 17,83,78. EPA Genetic Toxicology Program.
SAFETY PROFILE: Suspected carcinogen with experimental carcinogenic, neoplastigenic, and tumorigenic data. Mutation data reported. Explodes when heated to 170°C. When heated to decomposition it emits toxic fumes of NO_x. See also NITROSAMINES.

ELI600   CAS: 39544-02-0   HR: 3
1-(2-ETHYL-7-(2-HYDROXY-3-((1-METHYLETHYL)AMINO)PROPOXY)-4-BENZOFURANYL) ETHANONE
mf: $C_{18}H_{25}NO_4$ mw: 319.44
SYNS: ETHANONE, 1-(2-ETHYL-7-(2-HYDROXY-3-((1-METHYLETHYL)AMINO)PROPOXY)-4-BENZO-FURANYL)- ◻ 2-ETHYL-4-ACETYL-7-(2-HYDROXY-3-ISOPROPYLAMINOPROPOXY)BENZOFURAN ◻ KETONE, 2-ETHYL-7-(2-HYDROXY-3-(ISOPROPYL-AMINO)PROPOXY)-4-BENZOFURANYL METHYL
DOT CLASSIFICATION: 3; Label: Flammable Liquid

SAFETY PROFILE: A poison by intravenous route. A flammable liquid. When heated to decomposition it emits toxic vapors of NO_x.

ELJ600   CAS: 22208-25-9   HR: 3
2-ETHYL-2-(HYDROXYMETHYL)-1,3-PROPANEDIOL TRIACETOACETATE
mf: $C_{18}H_{26}O_9$ mw: 386.44

SYNS: ACETOACETIC ACID, TRIESTER WITH 2-ETHYL-2-(HYDROXYMETHYL)-1,3-PROPANEDIOL □ ACETOACETIC ACID, 1,1,1-TRIHYDROXY-METHYL-PROPANE TRIESTER □ BUTANOIC ACID, 3-OXO-, 2-((1,3-DIOXOBUTOXY)METHYL)-2-ETHYL-1,3-PROPANE-DIYL ESTER □ 1,3-PROPANEDIOL, 2-ETHYL-2-(HYDROXYMETHYL)-, TRIACETOACETATE □ TRIMETHYLOLPROPANE TRIACETOACETATE

CONSENSUS REPORTS: Reported in EPA TSCA Inventory.

SAFETY PROFILE: A poison by ingestion and skin contact. When heated to decomposition it emits acrid smoke and irritating vapors.

ELL500   CAS: 70-70-2   HR: 3
ETHYL-p-HYDROXYPHENYL KETONE
mf: $C_9H_{10}O_2$ mw: 150.19

PROP: Needles or prisms from H_2O. Mp: 149°.

SYNS: FRENANTOL □ FRENOHYPON □ H-365 □ p-HYDROXYPHENYL-1-PROPANONE □ 1-(4-HYDROXY-PHENYL)-1-PROPANONE □ HYDROXYPROPIO-PHENONE □ p-HYDROXYPROPIOPHENONE □ 4-HYDROXYPROPIOPHENONE □ HYPOPHENON □ p-OXYPROPIOPHENONE □ PAROXON □ PAROXY-PROPIONE □ PHP □ POP □ PROFENONE □ p-PROPIONYLPHENOL □ USAF EK-3302

CONSENSUS REPORTS: Reported in EPA TSCA Inventory.

DOT CLASSIFICATION: 3; Label: Flammable Liquid

SAFETY PROFILE: Poison by intraperitoneal, subcutaneous, and parenteral routes. An experimental teratogen. Other experimental reproductive effects. A flammable liquid. When heated to decomposition it emits acrid smoke and irritating fumes. See also KETONES.

ELN500   CAS: 75-37-6   HR: 3
ETHYLIDENE DIFLUORIDE
mf: $C_2H_4F_2$ mw: 66.06

PROP: Colorless gas. Mp: −117.0°, bp: −26.5°, d: 1.004 @ 25°, vap d: 2.28.

SYNS: ALGOFRENE TYPE 67 □ DIFLUOROETHANE □ 1,1-DIFLUOROETHANE □ ETHYLENE FLUORIDE □ ETHYLIDENE FLUORIDE □ FC 152a □ FREON 152 □ GENETRON 100 □ HALOCARBON 152A

CONSENSUS REPORTS: Reported in EPA TSCA Inventory. EPA Genetic Toxicology Program.

SAFETY PROFILE: Mildly toxic by inhalation. Mutation data reported. Narcotic in high concentration. A very dangerous fire hazard when exposed to heat or flame; can react vigorously with oxidizing materials. See also FLUORIDES.

ELO500   CAS: 16219-75-3   HR: 2
ETHYLIDENE NORBORNENE
mf: C_9H_{12} mw: 120.21

PROP: Bp: 70.2–70.4° @ 58 mm.

SYNS: 5-ETHYLIDENEBICYCLO(2.2.1)HEPT-2-ENE □ 5-ETHYLIDENE-2-NORBORNENE

CONSENSUS REPORTS: Reported in EPA TSCA Inventory.

OSHA PEL: CL 5 ppm
ACGIH TLV: CL 5 ppm

SAFETY PROFILE: Moderately toxic by ingestion. Mildly toxic by inhalation and skin contact. Human systemic effects by inhalation: conjunctiva, olfactory, and taste changes. A skin irritant. When heated to decomposition it emits acrid smoke and irritating fumes.

ELS000   CAS: 97-62-1   HR: 3
ETHYL ISOBUTYRATE
DOT: UN 2385
mf: $C_6H_{12}O_2$ mw: 116.18

PROP: Colorless, volatile liquid; fruity, aromatic odor. Mp: −88°, bp: 110–111°, d: 0.869, vap press: 40 mm @ 33.8°, vap d: 4.01, refr index: 1.385, flash p: <64.4°F.

SYNS: ETHYL ISOBUTANOATE □ ETHYLISOBUTY-RATE (DOT) □ ETHYL-2-METHYLPROPANOATE □ ETHYL-2-METHYLPROPIONATE □ FEMA No. 2428 □ ISOBUTYRIC ACID, ETHYL ESTER □ 2-METHYL-PROPIONIC ACID, ETHYL ESTER

CONSENSUS REPORTS: Reported in EPA TSCA Inventory.

DOT CLASSIFICATION: 3; Label: Flammable Liquid
SAFETY PROFILE: Moderately toxic by intraperitoneal route. A skin irritant. Flammable liquid. A very dangerous fire hazard when exposed to heat or flame; can react vigorously with oxidizing materials. To fight fire, use foam, CO_2, dry chemical. When heated to decomposition it emits acrid smoke and irritating fumes. See also ESTERS.

**ELS500   CAS: 109-90-0   HR: 3
ETHYL ISOCYANATE**
DOT: UN 2481
mf: C_3H_5NO mw: 71.09
PROP: Pungent smelling liquid. Bp: 60°, d: 0.90 @ 20°/4°, vap d: 2.45.
SYNS: ETHYL ISOCYANATE (DOT) □ ISOCYANATOETHANE □ ISOCYANIC ACID, ETHYL ESTER
CONSENSUS REPORTS: Reported in EPA TSCA Inventory.
DOT CLASSIFICATION: 3; Label: Flammable Liquid, Poison; DOT Class: 6.1; Label: Poison; DOT Class: 6.1; Label: Poison, Flammable Liquid; DOT Class: 3; Label: Flammable Liquid, Poison
SAFETY PROFILE: Poison by intravenous route. Mutation data reported. A flammable liquid. When heated to decomposition it emits toxic fumes of NO_x. See also CYANATES.

**ELU000   CAS: 1570-45-2   HR: 3
ETHYL ISONICOTINATE**
mf: $C_8H_9NO_2$ mw: 151.18
PROP: Bp: 219–220°.
SYNS: ISONICOTINIC ACID, ETHYL ESTER □ 4-PYRIDINECARBOXYLIC ACID, ETHYL ESTER
CONSENSUS REPORTS: Reported in EPA TSCA Inventory.
SAFETY PROFILE: Poison by intravenous route. When heated to decomposition it emits toxic fumes of NO_x.

**ELY700   CAS: 106-33-2   HR: 2
ETHYL LAURATE**
mf: $C_{14}H_{28}O_2$ mw: 228.37

PROP: Colorless, oily liquid; fruity-floral odor. D: 0.858, refr index: 1.430, bp: 163° @ 25 mm, flash p: 212°F. Misc in alc, chloroform, ether; insol in water @ 269°.
SYNS: ETHYL DODECANOATE □ FEMA No. 2441
SAFETY PROFILE: Combustible liquid. When heated to decomposition it emits acrid smoke and irritating fumes.

**EMA500   CAS: 105-53-3   HR: 2
ETHYL MALONATE**
mf: $C_7H_{12}O_4$ mw: 160.19
PROP: Clear, colorless liquid; fruit-like odor. Bp: 198.9°, fp: −49.8°, flash p: 200°F (OC), d: 1.055 @ 20°/4°, refr index: 1.413–1.416, vap press: 1 mm @ 40.0°, vap d: 5.52. Sol in fixed oils, propylene glycol; sltly sol in alc, water; insol in glycerin, mineral oil @ 200°.
SYNS: CARBETHOXYACETIC ESTER □ DICARBETHOXYMETHANE □ DIETHYL MALONATE (FCC) □ DIETHYL PROPANEDIOATE □ FEMA No. 2375 □ MALONIC ACID, DIETHYL ESTER □ MALONIC ESTER □ METHANEDICARBOXYLIC ACID, DIETHYL ESTER □ PROPANEDIOIC ACID, DIETHYL ESTER
CONSENSUS REPORTS: Reported in EPA TSCA Inventory.
SAFETY PROFILE: Mildly toxic by ingestion. A skin irritant. Combustible liquid when exposed to heat or flame; can react with oxidizing materials. To fight fire, use water to blanket fire, foam, CO_2, dry chemical. When heated to decomposition it emits acrid smoke and irritating fumes. See also ESTERS.

**EMA600   CAS: 4940-11-8   HR: 2
ETHYL MALTOL**
mf: $C_7H_8O_3$ mw: 140.15
PROP: White crystalline powder; sweet fruity taste. Mp: 90°. Sol in water, alc, propylene glycol, chloroform.
SYNS: 2-ETHYL-3-HYDROXY-4H-PYRAN-4-ONE □ 2-ETHYL PYROMECONIC ACID □ 3-HYDROXY-2-ETHYL-4-PYRONE
CONSENSUS REPORTS: Reported in EPA TSCA Inventory.
SAFETY PROFILE: Moderately toxic by ingestion and subcutaneous routes. Mutation data reported. When heated to

decomposition it emits acrid smoke and irritating fumes.

EMB100   CAS: 75-08-1   HR: 3
ETHYL MERCAPTAN
DOT: UN 2363

mf: C_2H_6S mw: 62.14

PROP: Colorless, volatile liquid with penetrating garlic-like odor. Fp: $-148°$, mp: $-121°$, bp: $36.1°$, lel: 2.8%, uel: 18.2%, d: 0.83907 @ $20°/4°$, autoign temp: 570°F, vap d: 2.14, flash p: $<-0.4°F$. Sol in EtOH, Et_2O, alkalies; very sltly sol in H_2O. IDLH 500 ppm.

SYNS: AETHANETHIOL (GERMAN) □ AETHYL-MERCAPTAN (GERMAN) □ ETANTIOLO (ITALIAN) □ ETHAANTHIOL (DUTCH) □ ETHANETHIOL □ ETHYL HYDROSULFIDE □ ETHYLMERCAPTAAN (DUTCH) □ ETHYLMERKAPTAN (CZECH) □ ETHYL SULF-HYDRATE □ ETHYL THIOALCOHOL □ ETILMER-CAPTANO (ITALIAN) □ LPG ETHYL MERCAPTAN 1010 □ THIOETHANOL □ THIOETHYL ALCOHOL

CONSENSUS REPORTS: Reported in EPA TSCA Inventory.

OSHA PEL: TWA 0.5 ppm

ACGIH TLV: TWA 0.5 ppm

DFG MAK: 0.5 ppm (1.3 mg/m³)

NIOSH REL: (n-Alkane Mono Thiols) CL 0.5 ppm/15M

DOT CLASSIFICATION: 3; Label: Flammable Liquid; DOT Class: 6.1; Label: Poison, Flammable Liquid

SAFETY PROFILE: Moderately toxic by ingestion, inhalation, and intraperitoneal routes. A skin and eye irritant. Inhalation causes central nervous system effects in humans. A very dangerous fire hazard when exposed to heat or flame; can react vigorously with oxidizing materials. A moderate explosion hazard when exposed to spark or flame. Violent reaction with $Ca(OCl)_2$. Will react with water or steam to produce toxic and flammable vapors. To fight fire, use CO_2, dry chemical. When heated to decomposition or on contact with acid or acid fumes it emits highly toxic fumes of SO_x. See also MERCAPTANS.

EME050   CAS: 2597-93-5   HR: 3
ETHYLMERCURICHLORENDIMIDE
mf: $C_{11}H_7Cl_6HgNO_2$ mw: 598.48

PROP: IDLH 10 mg/m³ (as Hg).

SYNS: 50-CS-46 □ EMMI □ N-(ETHYLMERCURI)-1,4,5,6,7,7-HEXACHLOROBICYCLO(2.2.1)HEPT-5-ENE-2,3-DICARBOXIMIDE □ N-ETHYLMERCURI-3,4,5,6,7,7-HEXACHLORO-3,6-ENDOMETHYLENE-1,2,3,6-TETRAHYDROPHTHALIMIDE □ N-ETHYLMERCURI-1,2,3,6-TETRAHYDRO-3,6-ENDOMETHANO-3,4,5,6,7,7-HEXACHLOROPHTHALIMIDE □ 1,4,5,6,77-HEXACHL-ORO-N-(ETHYLMERCURI)-5-NORBORNENE-2,3-DICAR-BOXIMIDE

CONSENSUS REPORTS: Mercury and its compounds are on the Community Right-To-Know List.

OSHA PEL: CL 0.1 mg(Hg)/m³ (skin)

ACGIH TLV: TWA 0.1 mg(Hg)/m³ (skin); BEI: 35 µg/g creatinine total inorganic mercury in urine preshift; 15 µg/g creatinine total inorganic mercury in blood at end of shift at end of workweek.

DFG MAK: Confirmed Animal Carcinogen with Unknown Relevance to Humans

NIOSH REL: (Mercury, Aryl and Inorganic) CL 0.1 mg/m³ (skin)

SAFETY PROFILE: Poison by ingestion. When heated to decomposition it emits very toxic fumes of Cl^-, Hg, and NO_x. See also MERCURY COMPOUNDS.

EME500   CAS: 517-16-8   HR: 3
ETHYLMERCURY-p-TOLUENE
SULFONAMIDE
mf: $C_{15}H_{17}HgNO_2S$ mw: 475.98

PROP: Crystals from EtOH; pungent, garlic-like odor. Mp: 156°. Practically insol in water. IDLH 10 mg/m³ (as Hg).

SYNS: CERESAN M □ COMPOUND-1452-F □ EMTS □ N-ETHYLMERCURI-N-PHENYL-p-TOLUENE-SULFONAMIDE □ N-(ETHYLMERCURI)-p-TOLUENE-SULFONANILIDE □ N-(ETHYLMERCURI)-p-TOLUENE-SULPHONANILIDE □ ETHYLMERCURY p-TOLUENE-SULFANILIDE □ ETHYLMERCURY-p-TOLUENESULFO-NANILIDE □ ETHYL(N-PHENYL-p-TOLUENESULFON-AMIDATO)MERCURY □ ETHYL(N-PHENYL-p-TOLUENESULFONAMIDO)MERCURY □ ETHYL(p-TOLUENESULFONANILIDATO)MERCURY □ GRANOSAN M □ (N-PHENYL-p-TOLUENESUL-FONAMIDO)ETHYLMERCURY

CONSENSUS REPORTS: Mercury and its compounds are on the Community Right-

To-Know List. EPA Genetic Toxicology Program.

OSHA PEL: TWA 0.01 mg(Hg)/m³; STEL 0.03 mg/m³ (skin)

ACGIH TLV: TWA 0.01 mg(Hg)/m³; BEI: 35 μg/g creatinine total inorganic mercury in urine preshift; 15 μg/g creatinine total inorganic mercury in blood at end of shift at end of workweek.

DFG MAK: Confirmed Animal Carcinogen with Unknown Relevance to Humans

NIOSH REL: (Mercury, Organo) TWA 0.01 mg/m³; STEL 0.03 mg/m³ (skin)

SAFETY PROFILE: Poison by ingestion and intraperitoneal route. Mutation data reported. A fungicide. When heated to decomposition it emits very toxic fumes of Hg, NO$_x$, and SO$_x$. See also MERCURY COMPOUNDS.

EMF000   CAS: 97-63-2   HR: 3
ETHYL METHACRYLATE
DOT: UN 2277
mf: $C_6H_{10}O_2$ mw: 114.16
PROP: A liquid. Mp: <−75°, bp: 119°, lel: 1.8%, uel: saturation, flash p: 68°F (OC), d: 0.911 @ 25°/25°, vap d: 3.94.
SYNS: ETHYL METHACRYLATE, INHIBITED (DOT) □ ETHYL-α-METHYL ACRYLATE □ ETHYL-2-METHYL-ACRYLATE □ ETHYL-2-METHYL-2-PROPENOATE □ 2-METHYL-2-PROPENOIC ACID, ETHYL ESTER □ RCRA WASTE NUMBER U118 □ RHOPLEX AC-33
CONSENSUS REPORTS: Reported in EPA TSCA Inventory.
DOT CLASSIFICATION: 3; Label: Flammable Liquid
SAFETY PROFILE: Moderately toxic by ingestion and intraperitoneal routes. Mildly toxic by inhalation. Experimental teratogenic and reproductive effects. A skin irritant. A very dangerous fire and explosion hazard when exposed to heat, sparks, or flame; can react with oxidizing materials. To fight fire, use CO_2, dry chemical. When heated to decomposition it emits acrid smoke and irritating fumes.

EMF500   CAS: 62-50-0   HR: 3
ETHYL METHANESULFONATE

mf: $C_3H_8O_3S$ mw: 124.17
SYNS: EMS □ ENT 26,396 □ ETHYL ESTER of METHANESULFONIC ACID □ ETHYL ESTER of METHYLSULFONIC ACID □ ETHYL ESTER of METHYLSULPHONIC ACID □ ETHYL METHANE-SULPHONATE □ ETHYL METHANSULFONATE □ ETHYL METHANSULPHONATE □ HALF-MYLERAN □ METHANESULPHONIC ACID ETHYL ESTER □ METHYLSULFONIC ACID, ETHYL ESTER □ NSC-26805 □ RCRA WASTE NUMBER U119
CONSENSUS REPORTS: NTP 10th Report on Carcinogens. IARC Cancer Review: Group 2B IMEMDT 7,56,87; Animal Sufficient Evidence IMEMDT 7,245,74. Reported in EPA TSCA Inventory. EPA Genetic Toxicology Program.
SAFETY PROFILE: Confirmed carcinogen with experimental carcinogenic, neoplastigenic, tumorigenic, and teratogenic data. Poison by ingestion and intraperitoneal routes. Experimental reproductive effects. Human mutation data reported. When heated to decomposition it emits toxic fumes of SO$_x$. See also SULFONATES and ESTERS.

EMP600   CAS: 7452-79-1   HR: 2
ETHYL 2-METHYLBUTYRATE
mf: $C_7H_{14}O_2$ mw: 130.19
PROP: Colorless liquid; strong, apple-like odor. D: 0.861–0.866, refr index: 1.396, bp: 133.5°, flash p: 153°F. Sol in alc, propylene glycol; misc in fixed oils; very sltly sol in water.
SYN: FEMA No. 2443
SAFETY PROFILE: Combustible liquid. When heated to decomposition it emits acrid smoke and irritating fumes.

EMQ500   CAS: 105-40-8   HR: 2
ETHYL-N-METHYLCARBAMATE
mf: $C_4H_9NO_2$ mw: 103.14
PROP: Needles. Mp: 54°, bp: 170°.
SYNS: ETHYLESTER KYSELINY METHYLKARBAMINOVE □ ETHYL METHYLCARBAMATE □ METHYLCARBAMIC ACID, ETHYL ESTER □ N-METHYL URETHAN □ METHYLURETHANE □ N-METHYLURETHANE

CONSENSUS REPORTS: Reported in EPA TSCA Inventory.

SAFETY PROFILE: Moderately toxic by subcutaneous route. Experimental teratogenic effects. Questionable carcinogen with experimental tumorigenic data. Mutation data reported. When heated to decomposition it emits toxic fumes of NO_x. See also CARBAMATES.

EMR100   CAS: 1942-78-5   HR: 3
o-ETHYL-S-(3-METHYL-4-CHLORO-PHENYL)ETHYL PHOSPHONODI-THIOATE

mf: $C_{11}H_{16}ClOPS_2$ mw: 294.81

SYNS: S-(4-CHLORO-3-METHYLPHENYL) o-ETHYL ETHYLPHOSPHONODITHIOATE □ ENT 27,045 □ N 4446 □ PHOSPHONODITHIOIC ACID, ETHYL-, S-(4-CHLORO-m-TOLYL) o-ETHYL ESTER □ PHOSPHONODITHIOIC ACID, ETHYL-, S-(4-CHLORO-3-METHYLPHENYL)o-ETHYL ESTER □ STAUFFER N-4446

SAFETY PROFILE: A poison by ingestion and skin contact. When heated to decomposition it emits toxic vapors of PO_x, SO_x, and Cl^-.

EMT000   CAS: 540-67-0   HR: 3
ETHYL METHYL ETHER

DOT: UN 1039

mf: C_3H_8O mw: 60.11

PROP: Colorless liquid or gas at room temp. Bp: 10.8°, lel: 2.0%, uel: 10.1%, flash p: −35°F (CC), d: 0.7260 @ 0°/4°, autoign temp: 374°F, vap d: 2.07.

SYNS: ETHOXYMETHANE □ ETHYL METHYL ETHER (DOT) □ METHOXYETHANE □ METHYL ETHYL ETHER (DOT)

DOT CLASSIFICATION: 2.1; Label: Flammable Gas

SAFETY PROFILE: Has anesthetic properties. A very dangerous fire and moderate explosion hazard when exposed to heat or flame; can react vigorously with oxidizing materials (e.g., air, O_2). To fight fire, use alcohol foam, CO_2, dry chemical. See also ETHERS.

EMU500   CAS: 96-29-7   HR: 3
ETHYL METHYL KETOXIME

mf: C_4H_9NO mw: 87.14

PROP: A liquid. D: 0.9232 @ 20°/4°, mp: −29.5°, bp: 152°.

SYNS: 2-BUTANONE, OXIME □ ETHYL METHYL KETONE OXIME □ ETHYL-METHYLKETONOXIM □ MEK-OXIME □ METHYL ETHYL KETOXIME □ SKINO #2 □ TROYKYD ANTI-SKIN B □ USAF AM-3 □ USAF DO-44 □ USAF EK-906

CONSENSUS REPORTS: Reported in EPA TSCA Inventory.

DOT CLASSIFICATION: 3; Label: Flammable Liquid

SAFETY PROFILE: Poison by intraperitoneal route. Moderately toxic by subcutaneous route. May explode if heated. Reacts with sulfuric acid to form an explosive product. When heated to decomposition it emits toxic fumes of NO_x.

ENB500   CAS: 115-38-8   HR: 3
5-ETHYL-N-METHYL-5-PHENYL-BARBITURIC ACID

mf: $C_{13}H_{14}N_2O_3$ mw: 246.29

PROP: A solid. Mp: 176°.

SYNS: ENFENEMAL □ ENPHENEMAL □ N-ETHYLMETHYLPHENYLBARBITURIC ACID □ 5-ETHYL-1-METHYL-5-PHENYLBARBITURIC ACID □ 5-ETHYL-1-METHYL-5-PHENYL-2,4,6(1H,3H,5H)-PYRIMIDINETRIONE □ 5-ETHYL-5-PHENYL-N-METHYLBARBITURIC ACID □ ISONAL □ ISONAL (ROUSSEL) □ MEBARAL □ MEBEREL □ MENTA-BAL □ MEPHOBARBITAL □ MEPHOBARBITONE □ MEPHY-TAL □ METHYL-CALMINAL □ 1-METHYL-5-ETHYL-5-PHENYLBARBITURIC ACID □ METHYLPHENO-BARBITAL □ N-METHYLPHENOBARBITAL □ 1-METHYLPHENOBARBITAL □ METHYLPHENOBARBIT-ONE □ N-METHYLPHENOBARBITOL □ METHYL-PHENYLBARBITURIC ACID □ N-METHYL-5-PHENYL-5-ETHYLBARBITAL □ 1-METHYL-5-PHENYL-5-ETHYLB-ARBITURIC ACID □ METYLFENEMAL □ METYNA □ MORBUSAN □ PHEMETONE □ PHEMITON □ PHEMIT-ONE □ 5-PHENYL-5-ETHYL-3-METHYLBARBITURIC ACID □ PROMINAL

SAFETY PROFILE: Poison by ingestion and intraperitoneal routes. A human teratogen by an unspecified route with developmental abnormalities of the cardiovascular (circulatory) system. When heated to decomposition it emits toxic NO_x. See also BARBITURATES.

ENC000   CAS: 77-83-8   HR: 2

ETHYL METHYLPHENYLGLYCIDATE

mf: $C_{12}H_{14}O_3$ mw: 206.26

PROP: Colorless to yellowish liquid; strawberry-like odor. D: 1.086–1.112, refr index: 1.504–1.513, flash p: 273°F. Sol in fixed oils, propylene glycol; insol in glycerin.

SYNS: C-16 ALDEHYDE □ EMPG □ α-β-EPOXY-β-METHYLHYDROCINNAMIC ACID, ETHYL ESTER □ ETHYL α,β-EPOXY-β-METHYLHYDROCINNAMATE □ ETHYL 2,3-EPOXY-3-METHYL-3-PHENYLPROPIONATE □ ETHYL ESTER of 2,3-EPOXY-3-PHENYLBUTANOIC ACID □ FEMA No. 2444 □ FRAESEOL □ 3-METHYL-3-PHENYLGLYCIDIC ACID ETHYL ESTER □ STRAWBERRY ALDEHYDE

CONSENSUS REPORTS: Reported in EPA TSCA Inventory.

SAFETY PROFILE: Mildly toxic by ingestion. Mutation data reported. Combustible liquid. When heated to decomposition it emits acrid smoke and irritating fumes. See also ALDEHYDES.

ENF050   CAS: 86073-23-6   HR: 3
o-ETHYL S-1-METHYLPROPYL S-1,1-DIMETHYLETHYL PHOSPHORO-DITHIOATE

mf: $C_{10}H_{23}O_2PS_2$ mw: 270.42

SYNS: S-sec-BUTYL S-tert-BUTYL o-ETHYL PHOSPHORODITHIOATE □ PHOSPHORODITHIOIC ACID, S-(1,1-DIMETHYLETHYL) o-ETHYL S-(1-METHYLPROPYL) ESTER

SAFETY PROFILE: A poison by ocular route. When heated to decomposition it emits toxic vapors of PO_x, and SO_x.

ENF200   CAS: 15707-23-0   HR: 2
2-ETHYL-3-METHYLPYRAZINE

mf: $C_7H_{10}N_2$ mw: 122.17

PROP: Colorless to sltly yellow liquid; strong, raw potato odor. D: 0.980–0.999 @ 20°, refr index: 1.502. Sol in water, org solvs.

SYN: FEMA No. 3155

CONSENSUS REPORTS: Reported in EPA TSCA Inventory.

SAFETY PROFILE: Moderately toxic by ingestion. When heated to decomposition it emits acrid smoke and irritating fumes.

ENK500   CAS: 6746-59-4   HR: 3
ETHYL MORPHINE HYDROCHLORIDE DIHYDRATE

mf: $C_{19}H_{23}NO_3 \cdot ClH \cdot 2H_2O$ mw: 385.93

PROP: White, microscopic, crystalline powder. Mp: 125° (decomp), vap d: 13.3.

SYNS: 7,8-DIDEHYDRO-4,5-α-EPOXY-3-ETHOXY-17-METHYLMORPHINAN-6-α-OL HYDROCHLORIDE DIHYDRATE □ DIONIN □ ETHYLMORPHINE HYDROCHLORIDE

SAFETY PROFILE: Poison by subcutaneous route. Can be habit forming. An experimental teratogen. When heated to decomposition it emits very toxic fumes of NO_x and HCl. See also CODEINE and MORPHINE.

ENL000   CAS: 100-74-3   HR: 3
N-ETHYLMORPHOLINE

mf: $C_6H_{13}NO$ mw: 115.20

PROP: Colorless liquid. Bp: 138°, flash p: 89.6°F (OC), d: 0.916 @ 20°/20°, vap d: 4.00. IDLH 100 ppm.

SYNS: 4-ETHYLMORPHOLINE □ NEM

CONSENSUS REPORTS: Reported in EPA TSCA Inventory.

OSHA PEL: TWA 5 ppm (skin)

ACGIH TLV: TWA 5 ppm (skin)

SAFETY PROFILE: Poison by intravenous route. Moderately toxic by ingestion. Mildly toxic by inhalation. A skin and severe eye irritant. A very dangerous fire hazard when exposed to heat or flame; can react vigorously with oxidizing materials. To fight fire, use alcohol foam, foam, CO_2, dry chemical. When heated to decomposition it emits toxic fumes of NO_x.

ENL850   CAS: 124-06-1   HR: 2
ETHYL MYRISTATE

mf: $C_{16}H_{23}O_2$ mw: 256.42

PROP: Colorless to pale-yellow liquid; waxy odor or crystals from (Me_2CO). D: 0.857, mp: 12.3°, bp: 295°, refr index: 1.434, flash p: 212°F.

SYN: FEMA No. 2445

SAFETY PROFILE: Combustible liquid. When heated to decomposition it emits acrid smoke and irritating fumes.

ENM500　CAS: 625-58-1　HR: 3
ETHYL NITRATE

mf: $C_2H_5NO_3$　mw: 91.08

PROP: Colorless liquid, pleasant odor, sweet taste. Mp: −112°, bp: 87.7°, lel: 3.8%, flash p: 50°F (CC), d: 1.004 @ 20°/4°, vap d: 3.14. Sol in water.

SYNS: NITRIC ACID, ETHYL ESTER □ NITRIC ETHER

CONSENSUS REPORTS: Reported in EPA TSCA Inventory.

SAFETY PROFILE: A poison by intraperitoneal route. Mutation data reported. A very dangerous fire hazard when exposed to heat or flame; can react vigorously with oxidizing materials. A moderate explosion hazard when exposed to heat (explodes @ 185°F). To fight fire, use foam, CO_2, dry chemical, water to blanket fire. Incompatible with Lewis acids. When heated to decomposition it emits toxic fumes of NO_x. See also NITRATES and ESTERS.

ENN000　CAS: 109-95-5　HR: 3
ETHYL NITRITE

DOT: UN 1194

mf: $C_2H_5NO_2$　mw: 75.08

PROP: Colorless or yellowish liquid or gas; highly aromatic, ethereal odor. Decomp on standing. Very sltly sol in water; misc in alc and ether. Bp: 17°, lel: 3.0%, uel: 50%, explodes at 194°F, flash p: −31°F (CC), d: 0.900 @ 15.5°, autoign temp: 194°F, vap d: 2.59. Can explode >90°C.

SYNS: ETHYLESTER KYSELINY DUSITE □ ETHYL NITRITE □ ETHYL NITRITE SOLUTIONS (DOT) □ NITROSYL ETHOXIDE □ NITROUS ETHER □ NITROUS ETHYL ETHER

CONSENSUS REPORTS: Reported in EPA TSCA Inventory.

DOT CLASSIFICATION: 3; Label: Flammable Liquid, Poison

SAFETY PROFILE: Poison by inhalation and ingestion. Narcotic in high concentrations. Lowers blood pressure. Methemoglobinemia has been reported. A very dangerous fire and severe explosion hazard when exposed to heat or flame. A powerful oxidizer. May explode when heated above 90°C. Highly dangerous when heated to decomposition or on contact with acid or acid fumes. To fight fire, use foam, CO_2, dry chemical, or water spray. When heated to decomposition it emits toxic fumes of NO_x. See also NITRITES and ETHERS.

ENV000　CAS: 759-73-9　HR: 3
1-ETHYL-1-NITROSOUREA

mf: $C_3H_7N_3O_2$　mw: 117.13

PROP: Pale-yellow crystals. Mp: 103° (decomp).

SYNS: AENH (GERMAN) □ AETHYLNITROSO-HARNSTOFF (GERMAN) □ ENU □ N-ETHYL-N-NITROSOCARBAMIDE □ ETHYLNITROSOUREA □ N-ETHYL-N-NITROSO-UREA □ NEU □ NITROSOETHYL-UREA □ NSC-45403 □ RCRA WASTE NUMBER U176

CONSENSUS REPORTS: NTP 10th Report on Carcinogens. IARC Cancer Review: Group 2A IMEMDT 7,56,87; Human Limited Evidence IMEMDT 17,191,78; Animal Sufficient Evidence IMEMDT 17,191,78; IMEMDT 1,135,72. EPA Genetic Toxicology Program. Community Right-To-Know List. Reported in EPA TSCA Inventory.

SAFETY PROFILE: Confirmed carcinogen with experimental carcinogenic, neoplastigenic, tumorigenic, and teratogenic data. Poison by ingestion, subcutaneous, intraperitoneal, and intravenous routes. Human mutation data reported. When heated to decomposition it emits toxic fumes of NO_x. See also N-NITROSO COMPOUNDS.

ENV500　CAS: 139-94-6　HR: 3
1-ETHYL-3-(5-NITRO-2-THIAZOLYL) UREA

mf: $C_6H_8N_4O_3S$　mw: 216.24

PROP: A solid. Mp: 228° (decomp).

SYNS: N-ETHYL-N'-(5-NITRO-2-THIAZOLYL)UREA □ HEPZIDE □ NCI-C03792 □ NITHIAZID □ NITHIAZIDE

CONSENSUS REPORTS: IARC Cancer Review: Group 3 IMEMDT 7,56,87; Animal Limited Evidence IMEMDT 31,179,83. NCI Carcinogenesis Bioassay (feed); Clear

Evidence: mouse, rat NCITR* NCI-CG-TR-146,79.

SAFETY PROFILE: Suspected carcinogen with experimental carcinogenic data. Moderately toxic by ingestion. Mutation data reported. When heated to decomposition it emits very toxic fumes of NO_x and PO_x.

ENW000   CAS: 123-29-5   HR: 2
ETHYL NONANOATE
mf: $C_{11}H_{22}O_2$ mw: 186.33
PROP: Colorless liquid; fruity, cognac odor. D: 0.863–0867, refr index: 1.420, flash p: 185°F. Misc with alc, propylene glycol; insol in water.
SYNS: ETHYL NONYLATE □ ETHYL PELARGONATE □ FEMA No. 2447 □ NONANOIC ACID, ETHYL ESTER □ WINE ETHER
CONSENSUS REPORTS: Reported in EPA TSCA Inventory.
SAFETY PROFILE: Mildly toxic by ingestion. A skin irritant. Combustible liquid. When heated to decomposition it emits acrid smoke and irritating fumes.

ENY000   CAS: 106-32-1   HR: 2
ETHYL OCTANOATE
mf: $C_{10}H_{20}O_2$ mw: 172.30
PROP: Colorless liquid; wine-brandy fruit odor. D: 0.865–0.869, refr index: 1.417, flash p: 185°F. Sol in fixed oils; sltly sol in propylene glycol; insol in glycerin, water @ 209°.
SYNS: ETHYL CAPRYLATE □ ETHYL OCTYLATE □ FEMA No. 2449 □ OCTANOIC ACID, ETHYL ESTER
CONSENSUS REPORTS: Reported in EPA TSCA Inventory.
SAFETY PROFILE: Mildly toxic by ingestion. A skin irritant. Combustible liquid. When heated to decomposition it emits acrid smoke and irritating fumes. See also ESTERS.

ENY500   CAS: 122-51-0   HR: 3
ETHYL ORTHOFORMATE
DOT: UN 2524
mf: $C_7H_{16}O_3$ mw: 148.23
PROP: Clear liquid; pungent sweet odor. Fp: 30°, bp: 145.9°, flash p: 86°F (CC), d: 0.895

@ 20°/20°, vap press: 10 mm @ 40.5°, vap d: 5.11.
SYNS: AETHON □ ETHONE □ ETHYLESTER KYSELINY ORTHOMRAVENCI (CZECH) □ 1,1',1'-(METH-YLIDYNETRIS(OXY))TRIS(ETHANE) □ ORTHOFORMIC ACID, ETHYL ESTER □ ORTHOFORMIC ACID, TRI-ETHYL ESTER □ ORTHOMRAVENCAN ETHYLNATY (CZECH) □ TRIETHOXYMETHANE □ TRIETHYL ORTHOFORMATE
CONSENSUS REPORTS: Reported in EPA TSCA Inventory.
DOT CLASSIFICATION: 3; Label: Flammable Liquid
SAFETY PROFILE: Moderately toxic by ingestion. Mildly toxic by inhalation, skin contact, and subcutaneous routes. A skin and eye irritant. A very dangerous fire hazard when exposed to heat or flame; can react vigorously with oxidizing materials. To fight fire, use foam, CO_2, dry chemical. When heated to decomposition it emits acrid smoke and irritating fumes. See also ESTERS.

EOD000   CAS: 22750-93-2   HR: 3
ETHYL PERCHLORATE
mf: $C_2H_5ClO_4$ mw: 128.52
PROP: Oil. Bp: 74°.
SYN: PERCHLORIC ACID, ETHYL ESTER
DOT CLASSIFICATION: Forbidden
SAFETY PROFILE: Possibly the most explosive chemical known. Very sensitive to impact, friction, and heat. Upon decomposition it emits toxic fumes of Cl^-. See also PERCHLORATES.

EOE100   CAS: 123-07-9   HR: 3
4-ETHYLPHENOL
mf: $C_8H_{10}O$ mw: 122.18
SYN: PHENOL, p-ETHYL-
CONSENSUS REPORTS: Reported in EPA TSCA Inventory.
DOT CLASSIFICATION: 6.1; Label: KEEP AWAY FROM FOOD
SAFETY PROFILE: Poison by intravenous route. When heated to decomposition it emits acrid smoke and irritating vapors.

EOH000   CAS: 101-97-3   HR: 2

ETHYL PHENYLACETATE

mf: $C_{10}H_{12}O_2$ mw: 164.22

PROP: Colorless liquid; sweet, honey odor. Bp: 227°, d: 1.033 @ 20°, refr index: 1.496–1.500, vap d: 5.67, flash p: 100°C. Sol in fixed oils; insol in glycerin, propylene glycol, water.

SYNS: BENZENEACETIC ACID, ETHYL ESTER (9CI) □ ETHYL BENZENEACETATE □ ETHYL PHENACETATE □ ETHYL-2-PHENYLETHANOATE □ ETHYL-α-TOLU-ATE □ FEMA No. 2452 □ PHENYLACETIC ACID, ETHYL ESTER □ α-TOLUIC ACID, ETHYL ESTER

CONSENSUS REPORTS: Reported in EPA TSCA Inventory.

SAFETY PROFILE: Moderately toxic by ingestion. Combustible liquid. Mutation data reported. When heated to decomposition it emits acrid smoke and irritating fumes. See also ESTERS.

EOJ500   CAS: 6368-72-5   HR: 2
N-ETHYL-1-((p-(PHENYLAZO)PHENYL)-AZO)-2-NAPHTHYLAMINE

mf: $C_{24}H_{21}N_5$ mw: 379.50

SYNS: CERES RED 7B □ C.I. 26050 □ C.I. SOLVENT RED 19 □ N-ETHYL-1-((p-(PHENYLAZO)PHENYL)AZO)-2-NAPHTHALENAMINE □ N-ETHYL-1-((4-(PHENYL-AZO)PHENYL)AZO)-2-NAPHTHALENAMINE □ N-ETHYL-1-((4-(PHENYLAZO)PHENYL)AZO)-2-NAPHTH-YLAMINE □ FAT RED 7B □ HEXATYPE CARMINE B □ LACQUER RED V3B □ OIL VIOLET □ ORGANOL BORDEAUX B □ (PHENYLAZO-4-PHENYLAZO)-1-ETHYLAMINO-2-NAPHTHALENE □ 1-(4-PHENYLAZO-PHENYLAZO)-2-ETHYLAMINONAPHTHALENE □ SOLVENT RED 19 □ SPECIAL BLUE X 2137 □ SUDAN RED 7B □ SUDANROT 7B □ TYPOGEN CARMINE

CONSENSUS REPORTS: IARC Cancer Review: Group 3 IMEMDT 7,56,87; Animal Inadequate Evidence IMEMDT 8,253,75. Reported in EPA TSCA Inventory.

SAFETY PROFILE: Questionable carcinogen with experimental tumorigenic data. Mutation data reported. When heated to decomposition it emits toxic fumes of NO_x.

EOK000   CAS: 50-06-6   HR: 3
5-ETHYL-5-PHENYLBARBITURIC ACID

mf: $C_{12}H_{12}N_2O_3$ mw: 232.26

PROP: Crystals in three modifications, one stable and two unstable. Mp: 156–157°. Sol in EtOH; mod sol in Et_2O; sltly sol in $CHCl_3$; prac insol in C_6H_6.

SYNS: ACIDO-5-FENIL-5-ETILBARBITURICO (ITALIAN) □ ADONAL □ AGRYPNAL □ AMYLOFENE □ APHENYLBARBIT □ AUSTROMINAL □ BARBAPIL □ BARBENYL □ BARBILEHAE (BARBILETTAE) □ BARBI-PHENYL □ BARBITA □ BARBONAL □ BARBOPHEN □ BARTOL □ BIALMINAL □ CABRONAL □ CALMINAL □ CODIBARBITA □ CRATECIL □ DEZIBARBITUR □ DORMIRAL □ DUNERYL □ ENSODORM □ EPIDORM □ EPISEDAL □ ESKABARB □ 5-ETHYL-5-PHENYL-2,4,6-(1H,3H,5H)PYRIMIDINETRIONE □ ETILFEN □ FENBIT-AL □ FENOBARBITAL □ FENYLETTAE □ GARDEPAN-YL □ GLYSOLETTEN □ HAPLOS □ HENNOLETTEN □ HYPNOGEN □ HYPNO-TABLINETTEN □ LEFEBAR □ LEPHEBAR □ LEPINAL □ LINASEN □ LIQUITAL □ LUBERGAL □ LUMEN □ LUMESETTES □ LUMINAL □ LUMOFRIDETTEN □ LURAMIN □ NEUROBARB □ NOPTIL □ NOVA-PHENO □ PARKOTAL □ PHENAEM-AL □ PHENOBAL □ PHENOBARBITAL □ PHENO-BARBITONE □ PHENOBARBITURIC ACID □ PHENO-LURIC □ PHENOMET □ PHENONYL □ PHENOTURIC □ PHENYLETHYLBARBITURATE □ PHENYL-ETHYL-BARBITURIC ACID □ 5-PHENYL-5-ETHYLBARBITURIC ACID □ PHENYLETHYLMALONYLUREA □ PHENYL-ETTEN □ PHENYRAL □ PHOB □ POLCOMINAL □ PROMPTONAL □ SEDA-TABLINEN □ SEDICAT □ SEDIZORIN □ SEDOFEN □ SEDONAL □ SEDOPHEN □ SEVENAL □ SK-PHENOBARBITAL □ SOMBUTOL □ SOMNOLETTEN □ SOMNOSAN □ SOMONAL □ STARIFEN □ STENTAL EXTENTABS □ TALPHENO □ THENOBARBITAL □ THEOMINAL □ TRIABARB □ TRIDEZIBARBITUR □ VERSOMNAL □ ZADONAL

CONSENSUS REPORTS: EPA Genetic Toxicology Program. IARC Cancer Review: Group 2B IMEMDT 7,313,87; Animal Sufficient Evidence IMEMDT 7,313,87; Human Inadequate Evidence IMEMDT 13,157,77.

SAFETY PROFILE: Confirmed carcinogen with experimental carcinogenic, neoplastigenic, tumorigenic, and teratogenic data. A human poison by ingestion. An experimental poison by ingestion, intraperitoneal, subcutaneous, intravenous, and rectal routes. Human systemic effects by ingestion: somnolence, motor activity changes, pulmonary changes, allergic dermatitis, and fever. Human reproductive effects by ingestion: drug dependence and other postnatal measures or effects. Human teratogenic effects include developmental abnormalities of the central nervous system, body wall, musculoskeletal, respiratory, gastrointestinal, and urogenital systems. Experimental reproductive effects. Human mutation data reported. Used as a drug in the treatment of epilepsy, and as a hypnotic and sedative. When heated to decomposition it emits toxic fumes of NO_x. See also BARBITURATES.

EOK600   **CAS: 121-39-1**   **HR: 2**
ETHYL PHENYLGLYCIDATE
mf: $C_{11}H_{12}O_3$ mw: 192.23
PROP: Colorless liquid; strong strawberry odor. D: 1.120, refr index: 1.516–1.521. Sol in alc, chloroform, ether; insol in water.
SYNS: ETHYL-α,β-EPOXYHYDROCINNAMATE □ ETHYL-α,β-EPOXY-α-PHENYLPROPIONATE □ ETHYL-3-PHENYLGLYCIDATE □ FEMA No. 2454
CONSENSUS REPORTS: Reported in EPA TSCA Inventory.
SAFETY PROFILE: Moderately toxic by ingestion. Mutation data reported. When heated to decomposition it emits acrid smoke and irritating fumes. See also ESTERS.

EOL500   **CAS: 93-55-0**   **HR: 3**
ETHYL PHENYL KETONE
mf: $C_9H_{10}O$ mw: 134.19
PROP: Water-white to light amber liquid or crystals. Mp: 19–20°, bp: 218.0°, d: 1.009 @ 25°/25°, vap press: 1 mm @ 50.0°. Insol in water; misc in alc, ether.
SYNS: PHENYL ETHYL KETONE □ 1-PHENYL-1-PROPANONE □ PROPIONYLBENZENE □ PROPIOPHENONE □ USAF EK-1235
CONSENSUS REPORTS: Reported in EPA TSCA Inventory.
DOT CLASSIFICATION: 3; Label: Flammable Liquid
SAFETY PROFILE: Poison by intraperitoneal route. Moderately toxic by subcutaneous route. Mildly toxic by ingestion and skin contact. A skin and eye irritant. A flammable liquid when exposed to heat or flame; can react with oxidizing materials. To fight fire, use foam, CO_2, dry chemical. When heated to decomposition it

emits acrid smoke and irritating fumes. See also KETONES.

EOM650   **CAS: 296269-54-0 HR: 3**
α-(5-ETHYL-1-PHENYL-1H-PYRAZOL-4-YL)-1-PIPERIDINEBUTANOL
mf: $C_{20}H_{29}N_3O$ mw: 327.47
SAFETY PROFILE: A poison by ingestion. When heated to decomposition it emits toxic vapors of NO_x.

EOP600   **CAS: 993-43-1**   **HR: 2**
ETHYL PHOSPHONOTHIOIC DICHLORIDE
DOT: NA 2927
mf: $C_2H_5Cl_2PS$ mw: 163.00
SYNS: DICHLOROETHYLPHOSPHINE SULFIDE □ ETHYL PHOSPHONOTHIOIC DICHLORIDE, anhydrous (DOT) □ ETHYLPHOSPHONOTHIONIC DICHLORIDE □ ETHYL PHOSPHONOTHIOYL DICHLORIDE □ ETHYLTHIONOPHOSPHONYL DICHLORIDE □ ETHYLTHIOPHOSPHONIC DICHLORIDE □ PHOSPHONOTHIOIC DICHLORIDE, ETHYL-
CONSENSUS REPORTS: Reported in EPA TSCA Inventory.
DOT CLASSIFICATION: 6.1; Label: Poison, Corrosive
SAFETY PROFILE: A corrosive. When heated to decomposition it emits toxic vapors of SO_x, PO_x, and Cl^-.

EOQ000   **CAS: 1498-40-4**   **HR: 2**
ETHYL PHOSPHONOUS DICHLORIDE
DOT: UN 2845
mf: $C_2H_5Cl_2P$ mw: 130.94
PROP: A liquid with pungent, foul odor. D: 1.26 @ 20°/4°, bp: 111–112° @ 722 mm.
SYNS: DICHLOROETHYLPHOSPHINE □ DICHLOROMETHYL PHOSPHINE □ ETHYL PHOSPHONOUS DICHLORIDE, anhydrous (DOT) □ TL 373
DOT CLASSIFICATION: 6.1; Label: Poison, Spontaneously Combustible
SAFETY PROFILE: Moderately toxic by inhalation. Corrosive. A severe irritant to skin, eyes, and mucous membranes. When heated to decomposition it emits very toxic fumes of PO_x, Cl^-, and phosphine. See also PHOSPHINE.

EOR000   **CAS: 1498-51-7**   **HR: 2**

ETHYL PHOSPHORODICHLORIDATE
DOT: NA 2927
mf: $C_2H_5Cl_2O_2P$ mw: 162.94
PROP: Colorless pungent liquid. D: 1.38, bp: 167°.
SYNS: DICHLOROPHOSPHORIC ACID, ETHYL ESTER □ PHOSPHORODICHLORIDIC ACID, ETHYL ESTER
CONSENSUS REPORTS: Reported in EPA TSCA Inventory.
DOT CLASSIFICATION: 6.1; Label: Poison, Corrosive
SAFETY PROFILE: A corrosive material that is very toxic to tissue. A severe eye, skin, and mucous membrane irritant. When heated to decomposition it emits very toxic fumes of Cl⁻ and PO_x.

EOR525   HR: D
ETHYLPHTHALYL ETHYL GLYCOLATE
SAFETY PROFILE: When heated to decomposition it emits acrid smoke and irritating fumes.

EOS000   CAS: 104-90-5   HR: 3
5-ETHYL-α-PICOLINE
DOT: UN 2300
mf: $C_8H_{11}N$ mw: 121.20
PROP: Liquid. Bp: 174°, d: 0.9184 @ 23°/4°, flash p: 165° (OC).
SYNS: ALDEHYDECOLLIDINE □ ALDEHYDINE □ COLLIDINE, ALDEHYDECOLLIDINE □ 3-ETHYL-6-METHYLPYRIDINE □ 5-ETHYL-2-METHYLPYRIDINE □ 5-ETHYL-2-PICOLINE □ MEP □ 2-METHYL-5-ETHYLPYRIDINE □ 6-METHYL-3-ETHYLPYRIDINE □ METHYL ETHYL PYRIDINE (DOT) □ 2-METHYL-5-ETHYLPYRIDINE (DOT)
CONSENSUS REPORTS: Reported in EPA TSCA Inventory. EPA Genetic Toxicology Program.
DOT CLASSIFICATION: 6.1; Label: KEEP AWAY FROM FOOD
SAFETY PROFILE: Poison by ingestion and subcutaneous routes. Moderately toxic by skin contact. Mildly toxic by inhalation. Corrosive. A severe skin and eye irritant. Flammable when exposed to heat or flame; can react vigorously with oxidizers. Potentially explosive reaction with nitric acid at 145°C/14.5 bar. To fight fire, use alcohol

foam. When heated to decomposition it emits acrid smoke and irritating fumes. See also ALDEHYDES.

EOS500   CAS: 766-09-6   HR: 3
1-ETHYLPIPERIDINE
DOT: UN 2386
mf: $C_7H_{15}N$ mw: 113.23
PROP: A liquid. D: 0.824 @ 20°/4°, bp: 128°, flash p: 66.2°F.
SYN: N-AETHYLPIPERIDIN (GERMAN)
CONSENSUS REPORTS: Reported in EPA TSCA Inventory.
DOT CLASSIFICATION: 3; Label: Flammable Liquid
SAFETY PROFILE: Poison by intravenous and subcutaneous routes. An eye irritant. A very dangerous fire hazard when exposed to heat or flame; can react vigorously with oxidizing materials. When heated to decomposition it emits toxic fumes of NO_x.

EPB500   CAS: 105-37-3   HR: 3
ETHYL PROPIONATE
DOT: UN 1195
mf: $C_5H_{10}O_2$ mw: 102.15
PROP: Colorless liquid; fruity, rum odor. Mp: −72.6°, bp: 99°, flash p: 54°F (CC), d: 0.891 @ 20°/4°, fp: −73°, refr index: 1.383, autoign temp: 824°F, vap press: 40 mm @ 27.2°, vap d: 3.52, lel: 1.9%, uel: 11%. Misc with alc, ether, propylene glycol; sol in water and fixed oils.
SYNS: FEMA No. 2456 □ PROPIONATE d'ETHYLE (FRENCH) □ PROPIONIC ACID, ETHYL ESTER □ PROPIONIC ETHER
CONSENSUS REPORTS: Reported in EPA TSCA Inventory.
DOT CLASSIFICATION: 3; Label: Flammable Liquid
SAFETY PROFILE: Moderately toxic by ingestion and intraperitoneal routes. A skin and eye irritant. A flammable liquid. A very dangerous fire and explosion hazard when exposed to heat or flame; can react vigorously with oxidizing materials. To fight fire, use foam, CO_2, dry chemical. When heated to decomposition it emits acrid

smoke and irritating fumes. See also ETHERS.

EPC125   CAS: 628-32-0   HR: 3
ETHYL PROPYL ETHER
DOT: UN 2615

mf: $C_5H_{12}O$ mw: 88.15

PROP: A liquid. D: 0.8, bp: 63.6°, flash p: <−4°F, lel: 1.7%, uel: 9%.

SYNS: 1-ETHOXYPROPANE ☐ ETHYL n-PROPYL ETHER ☐ PROPANE, 1-ETHOXY-(9CI) ☐ PROPYL ETHYL ETHER

DOT CLASSIFICATION: 3; Label: Flammable Liquid

SAFETY PROFILE: A slight inhalation hazard. Very dangerous fire and explosion hazard when exposed to heat or open flame. To fight fire, use alcohol foam. When heated to decomposition it emits acrid smoke and fumes. See also ETHERS.

EPC500   CAS: 297-97-2   HR: 3
ETHYL PYRAZINYL PHOSPHORO-
THIOATE
mf: $C_8H_{13}N_2O_3PS$ mw: 248.26

PROP: Amber liquid or oil. Mp: −1.7°, bp: 80°, n: (25/D) 1.5131, vap press @ 30°: 0.003 mm Hg. Sltly sol in water; misc with most org solvs.

SYNS: AC 18133 ☐ AMERCIAN CYANAMID 18133 ☐ CL 18133 ☐ CYNEM ☐ O,O-DIAETHYL-O-(PYRAZIN-2YL)-MONOTHIOPHOSPHAT (GERMAN) ☐ O,O-DIAETHYL-O-(2-PYRAZINYL)-THIONOPHOSPHAT (GERMAN) ☐ O,O-DIETHYL-O,2-PYRAZINYL PHOSPHOROTHIOATE ☐ DIETHYL-O-2-PYRAZINYL PHOSPHOROTHIONATE ☐ O,O-DIETHYL-O-2-PYRAZINYL PHOSPHOTHION-ATE ☐ O,O-DIETHYL-O-PYRAZINYL THIOPHOSPHATE ☐ EN 18133 ☐ ENT 25,580 ☐ EXPERIMENTAL NEMA-TOCIDE 18,133 ☐ NEMAFOS ☐ NEMAPHOS ☐ NEMA-TOCIDE ☐ PHOSPHOROTHIOIC ACID-O,O-DIETHYL-O-2-PYRAZINYL ESTER ☐ PYRAZINOL-O-ESTER with O,O-DIETHYL PHOSPHOROTHIOATE ☐ RCRA WASTE NUMBER P404 ☐ THIONAZIN ☐ ZINOPHOS

CONSENSUS REPORTS: EPA Extremely Hazardous Substances List. Reported in EPA TSCA Inventory. Community Right-To-Know List.

SAFETY PROFILE: Poison by ingestion, skin contact, and ocular routes. A cholinesterase inhibitor type of insecticide. When heated to decomposition it emits

highly toxic fumes of NO_x, PO_x, and SO_x. See also PARATHION.

EPC700   CAS: 2687-91-4   HR: 2
1-ETHYL-2-PYRROLIDINONE
mf: $C_6H_{11}NO$ mw: 113.18

PROP: A liquid. Bp: 97° @ 20 mm, d: 0.992, flash p: 169°F.

SYNS: N-ETHYLPYRROLIDINONE ☐ N-ETHYLPYR-ROLIDONE ☐ 2-PYRROLIDINONE, 1-ETHYL-

CONSENSUS REPORTS: Reported in EPA TSCA Inventory.

SAFETY PROFILE: Moderately toxic by ingestion. An eye irritant. Combustible. When heated to decomposition it emits toxic fumes of NO_x.

EPF550   CAS: 78-10-4   HR: 3
ETHYL SILICATE
DOT: UN 1292

mf: $C_8H_{20}O_4Si$ mw: 208.37

PROP: Colorless liquid. Mp: −77°, bp: 165–166°, flash p: 125°F (52°C), d: 0.933 @ 20°/4°, n: (25/D) 1.3818. Viscosity 0.6 cps. Practically insol in water with slow decomp. Miscible with alc. IDLH 700 ppm.

SYNS: ETHYL ORTHOSILICATE ☐ ETYLU KRZEMIAN (POLISH) ☐ EXTREMA ☐ SILICATE D'ETHYLE (FRENCH) ☐ SILICIC ACID TETRAETHYL ESTER ☐ TEOS ☐ TETRAETHOXYSILANE ☐ TETRAETHYL ORTHOSILICATE ☐ TETRAETHYL ORTHOSILICATE (DOT) ☐ TETRAETHYL SILICATE ☐ TETRAETHYL SILICATE (DOT)

CONSENSUS REPORTS: Reported in EPA TSCA Inventory.

OSHA PEL: TWA 10 ppm

ACGIH TLV: TWA 10 ppm

DFG MAK: 10 ppm (86 mg/m³)

DOT CLASSIFICATION: 3; Label: Flammable Liquid

SAFETY PROFILE: Poison by intravenous route. Moderately toxic by other routes. A skin, mucous membrane, and severe eye irritant. Narcotic in high concentrations. Flammable liquid when exposed to heat or flame; can react vigorously with oxidizing materials. When heated to decomposition it emits acrid smoke and fumes. See also ESTERS.

E

EPH000   CAS: 352-93-2   HR: 3
ETHYL SULFIDE
DOT: UN 2375
mf: $C_4H_{10}S$ mw: 90.20
PROP: Liquid; garlic-like odor. Mp: $-102°$,
fp: $-102.05°$, bp: 92–93°, d: 0.837 @
20°/4°, vap d: 3.11, flash p: 14°F. Insol in
water.
SYNS: DIETHYLSULFID (CZECH) □ DIETHYL
SULFIDE (DOT) □ DIETHYLTHIOETHER □ ETHYL
MONOSULFIDE □ ETHYLTHIOETHANE □ ETHYL
THIOETHER □ SULFODOR (CZECH) □ 3-THIAPENT-
ANE □ 1,1'-THIOBISETHANE □ THIOETHYL ETHER
CONSENSUS REPORTS: Reported in EPA
TSCA Inventory.
DOT CLASSIFICATION: 3; Label:
Flammable Liquid
SAFETY PROFILE: Mildly toxic by
ingestion. A skin and eye irritant. A very
dangerous fire hazard when exposed to heat,
flame, or sparks; can react vigorously with
oxidizers. Reacts with water, steam, acids, or
acid fumes to produce toxic and flammable
vapors. To fight fire, use water spray or
mist, dry chemical, CO_2, foam. When heated
to decomposition it yields highly toxic
fumes of SO_x. See also SULFIDES.

EPI400   CAS: 67465-28-5   HR: 3
3-(3-ETHYLSULFONYL)PENTYL
**   PIPERIDINO KETONE**
mf: $C_{13}H_{25}NO_3S$ mw: 275.45
SYN: KETONE, 3-(3-ETHYLSULFONYL)PENTYL
PIPERIDINO
DOT CLASSIFICATION: 3; Label:
Flammable Liquid
SAFETY PROFILE: A poison by
intraperitoneal route. A flammable liquid.
When heated to decomposition it emits
toxic vapors of NO_x and SO_x.

EPJ000   CAS: 20941-65-5   HR: 2
ETHYL TELLURAC
mf: $C_{20}H_{40}N_4S_8$•Te   mw: 720.72
PROP: Orange-yellow powder. D: 1.44, mp:
108–118°.
SYNS: DIETHYLDITHIOCARBAMIC ACID TELLURIUM
SALT □ NCI-C02857 □ TELLURIUM DIETHYLDI-
THIOCARBAMATE □ TETRAKIS(DIETHYLCARBAMO-

DITHIOATO-S,S')TELLURIUM □ TETRAKIS(DIETHYL-
DITHIOCARBAMATO)TELLURIUM
CONSENSUS REPORTS: IARC Cancer
Review: Group 3 IMEMDT 7,56,87; Animal
Inadequate Evidence IMEMDT 12,115,76.
NCI Carcinogenesis Bioassay (feed); No
Evidence: mouse, rat NCITR* NCI-CG-
TR-152,79; Results Indefinite: mouse, rat
NCITR* NCI-CG-TR-152,79. Reported in
EPA TSCA Inventory.
OSHA PEL: TWA 0.1 mg(Te)/m³
ACGIH TLV: TWA 0.1 mg(Te)/m³
SAFETY PROFILE: Questionable
carcinogen with experimental tumorigenic
data. When heated to decomposition it
emits very toxic fumes of NO_x, SO_x, and Te.
See also TELLURIUM COMPOUNDS and
CARBAMATES.

EPP000   CAS: 542-90-5   HR: 3
ETHYL THIOCYANATE
mf: C_3H_5NS mw: 87.15
PROP: A liquid. D: 1.020 @ 16°, mp:
$-85.5°$, bp: 145°. Insol in water; misc in alc
and ether.
SYNS: AETHYLRHODANID (GERMAN) □ ETHYL
RHODANATE □ ETHYL SULFOCYANATE □
THIOCYANATOETHANE □ THIOCYANIC ACID,
ETHYL ESTER
CONSENSUS REPORTS: EPA Extremely
Hazardous Substances List. Community
Right-To-Know List. Reported in EPA
TSCA Inventory.
SAFETY PROFILE: Poison by ingestion,
subcutaneous, intraperitoneal, and
intravenous routes. When heated to
decomposition it emits very toxic fumes of
NO_x and SO_x. See also THIOCYANATES.

EPR600   CAS: 625-53-6   HR: 3
ETHYL THIOUREA
mf: $C_3H_8N_2S$ mw: 104.19
SYN: 1-ETHYLTHIOUREA
CONSENSUS REPORTS: Reported in EPA
TSCA Inventory.
SAFETY PROFILE: Poison by ingestion.
Experimental teratogenic effects. Mutation
data reported. When heated to

decomposition it emits toxic fumes of SO_x and NO_x.

EPW500   CAS: 80-40-0   HR: 2
ETHYL TOSYLATE
mf: $C_9H_{12}O_3S$ mw: 200.27
PROP: Crystals from alc. Mp: 33°, bp: 173° @ 15 mm, flash p: 316°F (CC), d: 1.17, vap d: 6.98.
SYNS: ETHYL-p-METHYL BENZENESULFONATE □ ETHYL PTS □ ETHYL-p-TOLUENESULFONATE □ ETHYL-p-TOSYLATE □ p-TOLUOLSULFONSAEURE-AETHYL ESTER (GERMAN)
CONSENSUS REPORTS: Reported in EPA TSCA Inventory. EPA Genetic Toxicology Program.
SAFETY PROFILE: Moderately toxic by subcutaneous and intraperitoneal routes. Questionable carcinogen with experimental tumorigenic data. Mutation data reported. Combustible when exposed to heat or flame; can react with oxidizing materials. To fight fire, use CO_2, dry chemical. When heated to decomposition it emits highly toxic fumes of SO_x. See also SULFONATES and ESTERS.

EPY000   CAS: 327-98-0   HR: 3
ETHYL TRICHLOROPHENYLETHYL-
**   PHOSPHONOTHIOATE**
mf: $C_{10}H_{12}Cl_3O_2PS$ mw: 333.60
PROP: Amber liquid. D: 1.365 @ 20°/4°, bp: 108° @ 0.01 mm.
SYNS: O-AETHYL-O-(2,4,5-TRICHLORPHENYL)-AETHYLTHIONOPHOSPHONAT (GERMAN) □ AGRISIL □ AGRITOX □ BAYER 5081 □ BAYER 37289 □ BAYER S 4400 □ CHEMAGRO 37289 □ ENT 25,712 □ O-ETHYL-O-2,4,5-TRICHLOROPHENYL ETHYLPHOSPHONO-THIOATE □ FENOPHOSPHON □ PHYTOSOL □ STAUFFER N-3049 □ TRICHLORONAT □ 2,4,5-TRI-CHLOROPHENOL-O-ESTER with O-ETHYL ETHYL-PHOSPHONOTHIOATE □ WIRKSTOFF 37289
CONSENSUS REPORTS: Chlorophenol compounds are on the Community Right-To-Know List. EPA Extremely Hazardous Substances List.
SAFETY PROFILE: Poison by ingestion and skin contact. Moderately toxic by inhalation. An insecticide. When heated to decomposition it emits very toxic fumes of Cl^-, PO_x, and SO_x. See also CHLOROPHENOLS.

EPY500   CAS: 115-21-9   HR: 3
ETHYL TRICHLOROSILANE
DOT: UN 1196
mf: $C_2H_5Cl_3Si$ mw: 163.51
PROP: Liquid. Mp: −105.6°, bp: 99.5°, flash p: 72°F (OC), d: 1.24 @ 25°/25°, vap d: 5.6.
SYNS: ETHYL SILICON TRICHLORIDE □ ETHYL-TRICHLOROSILANE (DOT) □ SILANE, TRICHLORO-ETHYL- □ SILICANE, TRICHLOROETHYL- □ TRICHL-OROETHYLSILANE □ TRICHLOROETHYLSILICANE
CONSENSUS REPORTS: EPA Extremely Hazardous Substances List. Reported in EPA TSCA Inventory.
DOT CLASSIFICATION: 3; Label: Flammable Liquid, Corrosive
SAFETY PROFILE: Poison by inhalation and intraperitoneal routes. Moderately toxic by ingestion. A skin and severe eye irritant. A very dangerous fire hazard when exposed to heat, flame, or oxidizers; will react with water or steam to produce heat and toxic and corrosive fumes; can react vigorously with oxidizing materials. To fight fire, use CO_2, dry chemical. When heated to decomposition it emits highly toxic fumes of Cl^- and phosgene. See also CHLOROSILANES.

EQD100   CAS: 67590-56-1   HR: 3
4-ETHYL-2,6,7-TRIOXA-1-
**   ARSABICYCLO(2.2.2)OCTANE**
mf: $C_6H_{11}AsO_3$ mw: 206.09
SYN: 2,6,7-TRIOXA-1-ARSABICYCLO(2.2.2)OCTANE, 4-ETHYL-
OSHA PEL: TWA 0.5 mg(As)/m^3
SAFETY PROFILE: Poison by intravenous route. When heated to decomposition it emits toxic fumes of As.

EQD875   CAS: 625-52-5   HR: 1
1-ETHYLUREA
mf: $C_3H_8N_2O$ mw: 88.13
SYNS: ETHYLUREA □ N-ETHYLUREA □ UREA, ETHYL- □ UREA, 1-ETHYL-

CONSENSUS REPORTS: EPA Genetic Toxicology Program. Reported in EPA TSCA Inventory.
SAFETY PROFILE: Mildly toxic by parenteral route. Experimental reproductive effects. Mutation data reported. When heated to decomposition it emits toxic fumes of NO_x.

EQE000   HR: 3
ETHYLUREA and SODIUM NITRITE (2:1)
SYNS: AETHYLHARNSTOFF und NATRIUMNITRIT (GERMAN) □ AETHYLHARNSTOFF und NITRIT (GERMAN) □ SODIUM NITRITE and ETHYLUREA (1:2)
SAFETY PROFILE: Suspected carcinogen with experimental carcinogenic, neoplastigenic, tumorigenic, and teratogenic data. Experimental reproductive effects. When heated to decomposition it emits toxic fumes of NO_x and Na_2O. See also SODIUM NITRITE.

EQF000   CAS: 121-32-4   HR: 2
ETHYL VANILLIN
mf: $C_9H_{10}O_3$ mw: 166.19
PROP: Fine, crystalline needles; vanilla odor. Mp: 76.5°, flash p: 212°F. Sol in alc, chloroform, ether, propylene glycol; sltly sol in water.
SYNS: BOURBONAL □ ETHAVAN □ ETHOVAN □ 3-ETHOXY-4-HYDROXYBENZALDEHYDE □ ETHYL-PROTAL □ FEMA No. 2464 □ 4-HYDROXY-3-ETHOXY-BENZALDEHYDE □ PROTOCATECHUIC ALDEHYDE ETHYL ETHER □ QUANTROVANIL □ VANILLAL □ VANIROM
CONSENSUS REPORTS: Reported in EPA TSCA Inventory.
SAFETY PROFILE: Moderately toxic by ingestion, intraperitoneal, subcutaneous, and intravenous routes. A human skin irritant. Mutation data reported. When heated to decomposition it emits acrid smoke and irritating fumes. See also ALDEHYDES and ETHERS.

EQF500   CAS: 109-92-2   HR: 3
ETHYL VINYL ETHER
DOT: UN 1302
mf: C_4H_8O mw: 72.12

PROP: Colorless, volatile liquid. Fp: −115°, bp: 35.6°, flash p: <−50°F, d: 0.754, autoign temp: 395°F, vap press: 428 mm @ 20°, lel: 1.7%, uel: 28%, vap d: 2.5. Sltly sol in water.
SYNS: ETHOXY ETHENE □ EVE □ VINAMAR □ VINYL ETHYL ETHER □ VINYL ETHYL ETHER, inhibited (DOT)
CONSENSUS REPORTS: Reported in EPA TSCA Inventory.
DOT CLASSIFICATION: 3; Label: Flammable Liquid
SAFETY PROFILE: Mildly toxic by ingestion. Mutation data reported. A skin irritant. A very dangerous fire and explosion hazard when exposed to heat or flame; can react vigorously with oxidizing materials. To fight fire, use alcohol foam, foam, CO_2, dry chemical. Explosive polymerization is catalyzed by methane sulfonic acid. When heated to decomposition it emits acrid smoke and irritating fumes. See also ETHERS.

EQJ500   CAS: 297-76-7   HR: 3
ETHYNODIOL ACETATE
mf: $C_{24}H_{32}O_4$ mw: 384.56
PROP: Crystals from MeOH (aq) or crystals from Me_2CO/hexane. Mp: 129–132°.
SYNS: CERVICUNDIN □ 3-β,17-β-DIACETOXY-17-α-ETHYNYL-4-OESTRENE □ 3-β,17-β-DIACETOXY-19-NOR-17-α-PREGN-4-EN-20-YNE □ ETHINODIOL DIACETATE □ ETHYNODIOL DIACETATE □ β-ETHYNODIOL DIACETATE □ 17-α-ETHYNYL-3,17-DIHYDROXY-4-ESTRENE DIACETATE □ 17-α-ETHYNYLESTR-4-ENE-3-β,17-β-DIOL ACETATE □ 17-α-ETHYNYL-4-ESTRENE-3-β,17-DIOL DIACETATE □ 17-α-ETHYNYL-4-ESTRENE-3-β,17-β-DIOL DIACETATE □ 17-α-ETHYNYL-19-NORANDROST-4-ENE-3-β,17-β-DIOL DIACETATE □ FEMULEN □ LUTO-METRODIOL □ METRODIOL □ METRODIOL DIACETATE □ (3-β,17-α)-19-NORPREGN-4-EN-20-YNE-3,17-DIOL DIACETATE □ OVULEN 50
CONSENSUS REPORTS: IARC Cancer Review: Animal Limited Evidence IMEMDT 21,387,79; Animal Sufficient Evidence IMEMDT 6,173,74.
SAFETY PROFILE: Suspected carcinogen. Human reproductive effects by ingestion: menstrual cycle changes. Experimental reproductive effects. Mutation data reported. A steroid. When heated to

decomposition it emits acrid smoke and irritating fumes.

EQP000   CAS: 29767-20-2   HR: 3
ETP

mf: $C_{32}H_{32}O_{13}S$ mw: 656.70

PROP: Crystals from EtOH. Mp: 242–246°.

SYNS: 4'-DEMETHYLEPIPODOPHYLLOTOXIN-9-(4,6-O-2-THENYLIDENE-β-d-GLUCOPYRANOSIDE □ 4'-DEMETHYL-EPIPODOPHYLLOTOXIN-β-d-THENYLID-ENE-GLUCOSIDE □ 4'-DEMETHYL 1-O-(4,6-O,O-(2-THENYLIDENE)-β-d-GLUCOPYRANOSYL)EPIPODO-PHYLLOTOXIN □ EPT □ NSC-122819 □ PTG □ TENIPOSIDE □ VEHAM-SANDOZ □ VEHEM □ VM-26 □ VUMON

SAFETY PROFILE: Poison by intraperitoneal and subcutaneous routes. An experimental teratogen. Human systemic effects by ingestion and intravenous route: anorexia, nausea or vomiting, leukopenia, agranulocytosis and aplastic anemia of the blood, bone marrow changes, and hair changes. Experimental reproductive effects. Human mutation data reported. When heated to decomposition it emits very toxic fumes of SO_x.

EQQ000   CAS: 8000-48-4   HR: 3
EUCALYPTUS OIL

PROP: From steam distillation of leaves of *Eucalyptus globulus* Labillardiere. Chief constituent is eucalyptol (FCTXAV 13,19,75). Colorless to pale-yellow liquid; spicy odor and taste. Composition: eucalyptol, aldehydes, d-pinene. Mp: −15.4° (approx), d: 0.905–0.925 @ 25°/25°.

SYNS: DINKUM OIL □ EUKALYPTUS OEL (GERMAN) □ OIL OF EUCALYPTUS

CONSENSUS REPORTS: Reported in EPA TSCA Inventory.

SAFETY PROFILE: A human poison by ingestion. Moderately toxic by skin contact. Human systemic effects by ingestion: ciliary eye spasms, nausea or vomiting, respiratory depression, somnolence, sweating. A skin irritant. When heated to decomposition it emits acrid smoke and irritating fumes. See also ALDEHYDES.

EQR500   CAS: 97-53-0   HR: 2
EUGENOL

mf: $C_{10}H_{12}O_2$ mw: 164.22

PROP: Colorless or yellowish liquid or oil; pungent, clove odor. D: 1.064–1.070, refr index: 1.540, fp: −9°, bp: 248°, flash p: 219°F. Sol in alc, chloroform, ether, volatile oils; very sltly sol in water.

SYNS: 4-ALLYLGUAIACOL □ 4-ALLYL-1-HYDROXY-2-METHOXYBENZENE □ 4-ALLYL-2-METHOXYPHENOL □ CARYOPHYLLIC ACID □ EUGENIC ACID □ Fa 100 □ FEMA No. 2467 □ 1-HYDROXY-2-METHOXY-4-ALLYLBENZENE □ 4-HYDROXY-3-METHOXYALLYLBENZENE □ 1-HYDROXY-2-METHOXY-4-PROP-2-ENYLBENZENE □ 2-METHOXY-4-ALLYLPHENOL □ 2-METHOXY-4-PROP-2-ENYLPHENOL □ 2-METHOXY-4-(2-PROPENYL)PHENOL □ 2-METOKSY-4-ALLILOFENOL (POLISH) □ NCI-C50453 □ SYNTHETIC EUGENOL

CONSENSUS REPORTS: IARC Cancer Review: Group 3 IMEMDT 7,56,87; Animal Limited Evidence IMEMDT 36,75,85. NTP Carcinogenesis Studies (feed); Equivocal Evidence: mouse NTPTR* NTP-TR-223,83; No Evidence: rat NTPTR* NTP-TR-223,83. Reported in EPA TSCA Inventory. EPA Genetic Toxicology Program.

SAFETY PROFILE: Moderately toxic by ingestion, intraperitoneal, and subcutaneous routes. Human mutation data reported. A human skin irritant. Questionable carcinogen with experimental carcinogenic and tumorigenic data. Combustible liquid. When heated to decomposition it emits acrid smoke and irritating fumes. See also ALLYL COMPOUNDS.

EQS000   CAS: 93-28-7   HR: 2
EUGENOL ACETATE

mf: $C_{12}H_{14}O_3$ mw: 206.26

PROP: Solid or pale-yellow liquid or plates from alc; mild clove odor. D: 1.87, mp: 30–31°, bp: 281.2°, flash p: 151°F. Insol in water; sol in alc and ether.

SYNS: ACETEUGENOL □ 1-ACETOXY-2-METHOXY-4-ALLYLBENZENE □ ACETYLEUGENOL □ 4-ALLYL-2-METHOXYPHENOL ACETATE □ 1,3,4-EUGENOL ACETATE □ EUGENYL ACETATE □ FEMA No. 2469

CONSENSUS REPORTS: Reported in EPA TSCA Inventory.

SAFETY PROFILE: Moderately toxic by ingestion. A skin irritant. Combustible liquid. When heated to decomposition it emits acrid smoke and irritating fumes. See also EUGENOL, ALLYL COMPOUNDS, and ESTERS.

ERA500   CAS: 10025-76-0   HR: 3
EUROPIUM CHLORIDE
mf: Cl$_3$Eu   mw: 258.31

PROP: Hygroscopic yellow needles. Mp: 623°.

SYN: EUROPIC CHLORIDE

CONSENSUS REPORTS: Reported in EPA TSCA Inventory.

SAFETY PROFILE: Poison by intraperitoneal route. Moderately toxic by ingestion. When heated to decomposition it emits very toxic fumes of Cl⁻. See also RARE EARTHS.

ERF000   HR: 3
EXPLOSIVES, HIGH
SAFETY PROFILE: High explosives (HE) are those that decompose by detonation. This is a very rapid (nearly instantaneous), and hence violent, process. An explosion may be initiated by sudden shock, high temperatures, or a combination of the two. The conditions under which many explosives will explode are well known.

An explosion may be initiated by elevated temperature alone, as in the following cases.

(1) Mercury fulminate by 15-second exposure to 200°C or 1-second exposure to 340°C will be set off.

(2) Trinitrotoluene will be set off by exposure to 500°C for 1 second.

(3) Tetryl will detonate in 1000 seconds at 160°C or in 0.1 second at 500°C.

(4) Picric acid will detonate in 9 seconds at 300°C or 1 second at 355°C.

An explosion of HE may also be initiated by severe shock. Sensitivity of explosives to shock may be measured in several ways, such as the impact pendulum method and the drop test. In the impact pendulum test, a heavy pendulum swings down over a sample of explosive in a dished, inclined container so arranged that there is very little clearance between the pendulum and the sample. Thus, the effect of contact between the sample and the pendulum bob is one of a combination of shock and rubbing. The height from which the pendulum is allowed to swing to explode the sample is a measure of the sensitivity of the sample to this test. The drop test consists of placing a sample upon an anvil and allowing a 5-pound weight to drop on it. The height from which the weight must drop to explode the sample is a measure of the sample's sensitivity to shock.

The table below shows the results of a drop test upon several samples. These results must be considered as relative and not by any means absolute. A solid explosive in a tightly fitting container is much more sensitive to shock.

(1) mercury fulminate = 2 in. at 5 lbs.
(2) nitroglycerin = 4 in. at 5 lbs.
(3) tetryl = 8 in. at 5 lbs.
(4) picric acid = 14 in. at 5 lbs.
(5) trinitrotoluene = 20 in. at 5 lbs.
(6) black powder (a low explosive) = 30 in. at 5 lbs.*

*From Explosions, Their Anatomy and Destructiveness, by C. S. Robinson (McGraw-Hill).

Another test for explosives is the speed at which a detonation travels. This speed is usually in the range of thousands of m/sec. Speed of detonation is found to be a function of the kind of explosive and state of compaction. There is an optimum state of compaction beyond which the explosive tends to become "deadpressed," in which state it is difficult to make the whole sample explode. Below the point of optimum compaction the rate of detonation is found to be directly proportional to the density of the sample. Some maximum detonation rates are listed below in m/sec for some common explosives:

(1) nitroglycerin, 8500
(2) PETN, 8100

(3) tetryl, 7700

(4) picric acid, 7400

(5) trinitrotoluene, 7400

(6) lead azide, 4900

(7) mercury fulminate, 4800

(8) ammonium nitrate, 1100

(9) low explosives, 1000

It has been found that upon detonation, an explosive can cause a nearby sample of explosive to detonate "sympathetically." The distance over which one charge can detonate another is a function of the amount of energy produced by the first explosion and the medium through which the shock wave is propagated to the second charge of explosive. For instance, the relationship for air (very approximately) would be expected to be: weight of explosive in lbs/(distance in ft)^3 = 4. Thus, to calculate the maximum distance for a possible sympathetic detonation of 40,000 lbs of explosive, the calculation is:

D^3 = (40,000)/4

D^3 = 10,000

D = 22 ft (approximately).

According to C. S. Robinson, the formula is more nearly:

weight of explosive = 4 × (distance)$^{2.25}$

The power of the shock wave is much more rapidly attenuated in water, wood, etc., than in air, which means that if a shield of water or wood is interposed between piles of explosive the distance between them may be lessened.

Liquid Oxygen: Though not itself explosive, liquid oxygen can be dangerous when blended with highly flammable or carbonaceous materials. In this combination it is used in coal mining, quarrying, strip mining, open-cut ore mining, and in rocket fuels. Its use underground or in confined places is not recommended by the U.S. Bureau of Mines because it evolves a great deal of carbon monoxide. This type of explosive has many safety advantages. For instance, it is not itself an explosive until mixed with a flammable absorbent, a process that can be done at the last moment before firing. However, once the explosive has been made up, it is very flammable and when it catches fire it will usually detonate. Liquid oxygen explosives are not stored, as they deteriorate rapidly and lose a great deal of their explosive power in a short time.

A very dangerous fire hazard when exposed to heat or by chemical reaction with powerful oxidizing or reducing agents.

A moderate to dangerous explosion hazard when severely shocked or heated, depending upon the kind of explosive, state of compaction, degree of confinement, etc. Practically all high explosives used commercially require a detonator or cap to set them off, as compared to an igniter needed to set off black blasting powder.

Detonating Devices: To develop the desired disruptive effect of an explosive, some means must be adopted to "set off," "fire," or "detonate" it without killing or maiming the persons doing the blasting. Several devices or methods are being utilized, all with a view to having this work done as safely and efficiently as possible. There are two general types of devices or methods of getting explosives into action, namely, igniters and detonators. The former merely conveys a flame to the explosive mass and ignites it, while the latter transmits (originally through ignition of a small quantity of highly explosive substance by an arc, a flame, or a spark) a sharp blow that causes the explosive to disassociate, or detonate, or burn with very great rapidity. Igniters include squibs (plain and electric), fuse, and delay igniters; detonators include blasting caps (plain and electric), delay electric blasting caps, delay electric igniters with caps, and Cordeau-Bickford detonating fuse.

The squib is a small-diameter tube of straw or paper filled with quick-burning powder and having a relatively slow-burning "match head" attached to one end; the latter is ignited or lighted by an ordinary match or other flame, and its relatively slow burning allows the person handling the ignition to retire before the fire is communicated to the quick-burning material in the tube. Squibs are by no means either safe or efficient, even though still used to a considerable extent, especially in coal mining. Electric squibs are somewhat similar to ordinary squibs, except that the ignition is accomplished by means of an electric arc; electric squibs are much more satisfactory from a safety viewpoint than ordinary squibs.

A fuse (or, as it is sometimes called, "safety fuse") consists of a fine-grained black powder core covered with cotton hemp or jute to form a ropelike material about 3/16 inch in diameter; one end of the fuse is brought in contact with the powder charge or with a detonating "cap," and the other end (usually several feet away from the explosive) is lighted by a flame from a match or open light. The fine-grained black powder burns gradually and somewhat slowly (about 30 to 40 seconds to the linear foot of fuse) until it reaches the explosive (black powder) or the detonating cap (if some form of dynamite is used), giving the blaster time to get in the clear before the main explosion takes place. Fuses are much safer than squibs, but have their own hazards and must be used with care.

Delay electric igniters usually are a combination of electric igniters and fuses, the latter being ignited by the igniters within the blasting hole, the fuse transmitting the ignition to the explosive. Delay igniters usually are much safer than fuses, particularly for coal-mine use; but they, too, have their hazards. Delay blasting is by no means a safe procedure in coal mining, though it is a standard and relatively safe practice in metal mining and tunneling, if sensible precautions are taken.

Blasting caps or detonators are metallic cylinders (usually copper) closed at one end, about 3/16 inch in diameter, and usually less than 2 inches in length, partly filled with a small amount of relatively easily fired or "detonated" compound, the resultant shock or blow when fired being sufficient, when embedded in dynamite, to fire or detonate the dynamite mass. Ordinary blasting caps usually are fired or detonated by the flame of the fuse, the end of the latter being inserted into the open end of the detonator or cap and placed in contact with the highly explosive material in the interior of the metallic capsule or cap. Caps are extremely hazardous to handle, as they are likely to be detonated by heat, friction, or a relatively moderate blow; however, they are relatively safe if handled carefully. Partial proof of this is the fact that they are manufactured and shipped by the thousands daily and accidents are decidedly rare, primarily because the caps are at all times handled with utmost care.

Electric blasting caps are somewhat similar to ordinary caps or detonators, but the cap is fired by electricity. The electric wires are so placed in the capsule or cap that when attached to an electric current an arc is formed within the cap, which detonates the sensitive explosive material in the cap. A hazard in the use of electric blasting caps is unexpected explosions due to radio or radar-induced electric currents that may activate the cap.

Delay electric blasting caps or detonators are somewhat similar to ordinary electric blasting caps, except that several time intervals in blasting are obtained by having the electric arc ignite a short piece of fuse or some slow-burning substance before it reaches the highly sensitive detonating material in the capsule or cap. Numerous time-interval delays are obtained; in general, delay electric blasting caps are relatively safe and effective even in wet holes, though they ought not be used in coal mining if explosions of gas or dust are to be avoided. Delay electric igniters with caps or detonators are a combination of electric igniter and blasting cap, usually with suitable lengths of fuse between to give the desired delay; they have some advantages but are relatively unsafe and should not be used in coal mining.

The Cordeau-Bickford denotating fuse is a combined fuse and detonator in the form of a lead tube about 1/4 inch in diameter filled throughout its length with a high explosive, trinitrotoluene (TNT). It is fired by a fuse and an ordinary detonator or cap or by an electric cap; when fired, it detonates throughout its length (which may be up to or over 100 feet) almost instantaneously, the explosion wave traveling at a rate of about 17,500 ft/sec. Although somewhat expensive, it is relatively safe to handle and is particularly effective in deep-well drill holes in quarry and similar work, as it detonates simultaneously throughout its length, adding effectiveness to the main body of explosive that it detonates. It fires black powder as well as high explosives (dynamite, etc.), and is obtainable in lengths of approximately 500 feet wound on spools.

See also EXPLOSIVES, PERMITTED; DYNAMITE, NITROGLYCERIN, AMMONIUM NITRATE, and NITRATES.

ERF500   HR: 3
EXPLOSIVES, LOW
DOT: UN 0027/UN 0028
PROP: Black powder is composed of saltpeter, charcoal and sulfur in the approximate proportions of 6:1:1. ("A" blasting powder uses KNO_3 and "B" blasting powder uses $NaNO_3$.)
SYNS: "A" BLASTING POWDER □ "B" BLASTING POWDER □ BLACK BLASTING POWDER □ BLACK POWDER, compressed (DOT) □ BLACK POWDER, granular or as a meal (UN 0027) (DOT) □ BLACK POWDER, in pellets (UN 0028) (DOT) □ BLASTING POWDER □ GUN-POWDER □ GUNPOWDER, compressed (UN 0028) (DOT) □ GUNPOWDER, granular or as a meal (UN 0027) (DOT) □ GUNPOWDER, in pellets (UN 0028) (DOT) □ RIFLE POWDER
DOT CLASSIFICATION: EXPLOSIVE 1.1D; Label: EXPLOSIVE 1.1D
SAFETY PROFILE: Low explosives are explosives that deflagrate; this differentiates them both in composition and properties from high explosives, which detonate. A deflagrating explosive is one that burns progressively over a relatively sustained period of time, in contrast with a detonating explosive, which decomposes almost instantaneously. A dangerous fire hazard when exposed to heat or flame or by chemical reaction.

Black powder is the most treacherous explosive material used today and it is regarded as one of the worst explosive hazards known. When ignited unconfined it burns with explosive violence and will explode if ignited under even slight confinement. It can be ignited easily by very small sparks, heat, and friction. It is the slowest acting of all explosives. It has a shearing and heaving action tending to blast materials into large, firm fragments. The action derives from a relatively slow development of gas pressure so that it must be carefully loaded and closely confined. It is subject to rapid deterioration in the presence of moisture, but if kept dry it retains its explosive properties for many years. It is used to ignite smokeless powder, propelling charges, airplane flares, and bursting charges of hand grenades, as a bursting charge in shrapnel, practice bombs, practice trench-mortar shells, in saluting charges, smoke-puff charges, time and percussion fuses, pellets, primers and primer detonators, and in expelling charges of pyrotechnic signals.

Although most safety experts now look upon black blasting powder with disfavor, it is one of the oldest and most generally used explosives in commercial work. It burns with extreme rapidity instead of detonating as high explosives do. It is highly sensitive to flame, sparks, or friction and gives off much flame, which is hot and of great length of duration. These properties make it extremely hazardous for use in mines (especially coal mines) and quarries. The gases given off in detonation are not only very hot but frequently contain harmful constituents. Notwithstanding its numerous deficiencies, from a safety standpoint it has action characteristics that make it valuable in both coal mining and quarrying, though it has relatively little utility in metal mining. It is difficult to use effectively in wet places and this is its main disadvantage from an efficiency standpoint.

Most black powder fires start from sparks. Ignition results in an explosion so quickly that no attempt can be made to fight the fire. Every effort should be made to prevent fires from reaching stores of black powder; but if this fails, fire-fighting forces should be withdrawn to a distance of at least 800 feet from the fire and should protect themselves against an explosion by seeking any cover available or by lying flat on the ground. If an explosion does occur, every effort should be made to prevent flames from spreading to neighboring magazines. Fire-fighting forces should be cautioned against approaching a fire that may involve black powder to avoid being trapped or injured by an explosion.

The following safety rules should be strictly enforced and obeyed. Open no containers in a magazine in which explosives or ammunitions are stored. This should be done only in a building free from all other explosives or ammunitions, or in suitable weather in the open, at least 100 feet from the nearest magazine. The quantity at or near such an operation should be limited to 100 pounds. Only safety tools should be used in opening or closing containers or in other operations involving black powder. Processes should be so laid out as to bring about frequent grounding of all operators handling this material. Safety shoes (non-insulating) should be worn in all rooms where black powder is handled and by all persons engaged in handling black powder. The wearing of all nonconductive shoes, such as rubber, is prohibited. If black powder is handled on or over a concrete floor, the floor should be covered with a tarpaulin or other suitable material. Loose black powder is extremely dangerous. Whenever it is necessary to handle loose black powder, not over 50 pounds should be permitted at or near such operations. If

black powder is spilled on benches or floors, all work should be stopped until it has been removed and the explosive hazard of any remaining dust or particles has been neutralized with water. Rooms or buildings in which black powder is handled should be inspected frequently for dust, and all such dust should be immediately removed with water. The empty powder containers should be washed out, as explosions are said to have occurred from "empty" containers.

If dry and in good condition, black powder burns rapidly, especially in small grain size, with a yellow or pinkish-blue flame and dense smoke.

Pellet powder is black blasting powder in consolidated (pellet or stick) form rather than in grains or granules, and it has few if any real advantages over black blasting powder, notwithstanding the fairly prevalent idea that it is a "safe" explosive.

Smokeless powders have a composition somewhat different from that of black blasting powder and are used chiefly for sportsmen's ammunition and, more widely, for military purposes. They are decidedly sensitive to flame and impact but ordinarily are so packaged that if reasonable judgment is used they are relatively harmless.

ERG000   HR: 3
EXPLOSIVES, PERMITTED
SAFETY PROFILE: "Permissible" explosives are essentially high explosives (dynamite) modified by the introduction of "dopes." The function of the dopes in general is to decrease flame temperature, and to a smaller extent, the length and duration of flame, when the explosive is converted from a solid into a gas, i.e., when it is fired or detonated. The designation "permissible" is given to an explosive of modified dynamite type after it has passed certain tests designated by the Federal Bureau of Mines. The permissible character of such explosives depends not only upon the ingredients in the explosive, but also on certain well-defined specifications as to

handling and use. As with the dynamites, there are several different types and grades: "permissibles," hydrated "permissibles," organic nitrate "permissibles," nitroglycerin "permissibles," ammonium perchlorate "permissibles," and gelatin "permissibles." Essentially all of those now used to any extent are in either the ammonium nitrate or the gelatin classes. See also DYNAMITE.

The ammonium nitrate "permissible" explosives contain relatively little nitroglycerin and relatively large proportions of ammonium nitrate. The latter is an explosive but one less sensitive to impact, sparks, and flames than nitroglycerin. This type of permissible explosive is now used extensively, as it has a rather wide range of strength, rate of detonation, density, size of cartridge, etc., and can be utilized not only in dry but also to some extent in fairly wet holes if charged carefully and fired promptly.

Gelatin "permissible" explosives are more suitable than ammonium nitrate "permissible" ones for wet holes, and in general are stronger and more violent than the ammonium nitrate types.

All "permissible" explosives are strong, and must be used in relatively small quantities (less than 1.5 pounds) per hole to retain their permissibility. They give off considerable quantities of toxic gases on detonation, and, while much safer than black blasting powder or dynamite, must be stored, handled, and used with care.

Classification upon Basis of Toxic Gases: All "permissible" explosives, when detonated, emit some toxic gases and a much larger volume of nontoxic gases. In order that the toxic products may not become a menace to the life or health of miners, no explosive is now or can become "permissible" if upon detonation it evolves more than 158 liters (5.5 cu ft) of toxic gases per 1.5-pound charge as determined by tests in the Bichel pressure gauge. Classification upon the basis of the volume of toxic gases produced by 580 g (1.5-pounds) of explosive is as

follows: *Class A,* not more than 53 liters; *Class B,* between 53 and 106 liters; and *Class C,* between 106 and 158 liters. (These classifications are not to be confused with the I.C.C. Classification of explosives.)

Field tests were made with a 1.5 pound charge of a "permissible" explosive that produced, in the Bichel gauge, the maximum allowable quantity of poisonous gases (158 liters per 1.5 pounds); these tests indicated that in a narrow entry, without artificial ventilation, 1800 ppm of carbon monoxide (the only poisonous gas present) was produced, as shown by analysis of an air sample taken 2 minutes after the shot. Another sample of the air taken 2 minutes later contained 800 ppm of carbon monoxide. Under no conditions should miners or shot firers return to the place until the poisonous gases have been removed by adequate ventilation.

It is provided further that, in accordance with the provisions and conditions, explosives enumerated on the "permissible" lists of the Bureau of Mines are "permissible" in use only when they satisfy the following requirements:

1. The explosive must be in all respects similar to the sample submitted by the manufacturer for test, and the diameters of the cartridges used must be those that have been approved.

2. Electric detonators (not fuse and detonators) must be used of not less efficiency than No. 6, the detonation charge of which shall consist of a 1 g mixture of 80 parts of mercury fulminate and 20 parts of potassium chlorate (or their equivalents), and the required electric firing must be done by means of a "permissible" type blasting unit.

3. The explosive must be stored in surface magazines under proper conditions, so that it will not undergo change in character, and after being taken underground it must be used in less than 36 hours.

4. The coal to be blasted must be undercut or equivalently relieved; to prevent blow-

through, all portions of the borehole must be at least 18 inches from relief in any direction; to prevent blowouts, the charge must be properly confined with not less than 2 feet of clay (if the length of the hole will not permit the charge desired and 2 feet of stemming, at least half the length of the hole shall be filled with stemming) or other incombustible stemming and not be on the solid; to prevent the hole being on the solid it shall be at least 6 inches shorter than the depth of the undercut or equivalent relief, and, when placed adjacent to the roofs, ribs, or floor, all but 12 inches at the rear of the hole must be at least 6 inches from the adjacent surface as projected into the coal to be blasted, and all parts of the hole shall be free from the adjacent surface as projected into the coal to be blasted; the shot must not be a dependent shot; and the shot hole must be cleaned before charging.

5. The quantity used for a shot (1) must not be in excess of 680 g (1.5 pounds) when fired in accordance with these requirements and (2) when used under certain additional requirements or restrictions must not be in excess of 1361 g (3 pounds). For charges of over 1.5 pounds, the following additional requirements must be observed: (a) shot holes must be 6 feet or more in length; (b) explosives must be charged in a continuous train, with no cartridges deliberately deformed or crushed, with all cartridges in contact with each other, and with the end cartridges touching the rear of the hole and the stemming, respectively; (c) examination for gas must be made in the blasting area before and after a shot is fired; (d) the "permissible" explosive must be one showing toxic gas emission that will place it either in Class A or Class B.

6. The region in which the blasting is done must be kept well protected by rock dust or otherwise be in accordance with Bureau of Mines inspection standards.

7. The shot must not be fired in the presence of a dangerous percentage of firedamp. Examination for firedamp is to be made at the blasting area before shooting in a gassy mine.

See also AMMONIUM NITRATE, AZIDES, DYNAMITE, FULMINATES, NITRATES, NITROGLYCERIN, PENTAERYTHRITOL TETRANITRATE, and PICRIC ACID

F

FAB850
FATTY ACIDS
HR: D

PROP: Consists of capric, caprylic, lauric, myristic, oleic, palmitic, and stearic acids manufactured from fats and oils derived from edible sources.

SAFETY PROFILE: When heated to decomposition it emits acrid smoke and irritating fumes.

FAG018　CAS: 3564-09-8　HR: 3
FD&C RED No. 1

mf: $C_{19}H_{16}N_2O_7S_2 \cdot 2Na$　mw: 494.47

SYNS: A.F. RED No. 1 □ CERVEN KUMIDINOVA □ C.I. 16155 □ C.I. FOOD RED 6 □ C.I. FOOD RED 6, DISODIUM SALT □ DISODIUM-3-HYDROXY-4-((2,4,5-TRIMETHYLPHENYL)AZO)-2,7-NAPHTHALENE-DISULFONATE □ DISODIUM-3-HYDROXY-4-((2,4,5-TRIMETHYLPHENYL)AZO)-2,7-NAPHTHALENE-DISULFONIC ACID □ DISODIUM-3-HYDROXY-4-((2,4,5-TRIMETHYLPHENYL)AZO)-2,7-NAPHTHALENE-DISULPHONATE □ DISODIUM-3-HYDROXY-4-((2,4,5-TRIMETHYLPHENYL)AZO)-2,7-NAPHTHALENE-DISULPHONIC ACID □ DOLKWAL PONCEAU 3R □ EXT. D&C RED No. 15 □ 3-HYDROXY-4-((2,4,5-TRIMETHYLPHENYL)AZO)-2,7-NAPHTHALENE-DISULFONIC ACID, DISODIUM SALT □ 3-HYDROXY-4-((2,4,5-TRIMETHYLPHENYL)AZO)-2,7-NAPHTHAL-ENEDISULPHONIC ACID, DISODIUM SALT □ MAPLE PONCEAU 3R □ PONCEAU 3R □ SODIUM CUMENE-AZO-β-NAPHTHOL DISULPHONATE □ USACERT RED No. 1

CONSENSUS REPORTS: IARC Cancer Review: Group 2B IMEMDT 7,56,87; Animal Sufficient Evidence IMEMDT 8,199,75.

SAFETY PROFILE: Confirmed carcinogen with experimental carcinogenic and tumorigenic data. Mutation data reported. When heated to decomposition it emits toxic fumes of NO_x and SO_x.

FAG120　CAS: 1694-09-3　HR: 3
FD&C VIOLET No. 1

mf: $C_{39}H_{41}N_3O_6S_2 \cdot Na$　mw: 734.94

PROP: Black powder. Mp: 245°–250° (decomposes).

SYNS: ACID VIOLET □ A.F. VIOLET No 1 □ AIZEN FOOD VIOLET No 1 □ BENZYL VIOLET □ BENZYL VIOLET 3B □ CALCOCID VIOLET 4BNS □ C.I. 42640 □ C.I. FOOD VIOLET 2 □ COOMASSIE VIOLET □ DISPERSED VIOLET 12197 □ FORMYL VIOLET S4BN □ PERGACID VIOLET 2B □ SOLAR VIOLET 5BN □ WOOL VIOLET

CONSENSUS REPORTS: IARC Cancer Review: Group 2B IMEMDT 7,56,87; Animal Sufficient Evidence IMEMDT 16,153,78. EPA Genetic Toxicology Program. Reported in EPA TSCA Inventory.

SAFETY PROFILE: Confirmed carcinogen with experimental carcinogenic and tumorigenic data. Mutation data reported. When heated to decomposition it emits very toxic fumes of NO_x, NH_3, Na_2O, and SO_x.

FAG130　CAS: 85-84-7　HR: 2
FD&C YELLOW No. 3

mf: $C_{16}H_{13}N_3$　mw: 247.32

PROP: Red plates from EtOH. Mp: 102–104°. Sol in EtOH and AcOH.

SYNS: A.F YELLOW No. 2 □ 1-BENZENE-AZO-β-NAPHTHYLAMINE □ 1-BENZENEAZO-2-NAPHTHYLAMINE □ CERISOL YELLOW AB □ C.I. 11380 □ C.I. FOOD YELLOW 10 □ C.I. SOLVENT YELLOW 5 □ DOLKWAL YELLOW AB □ EXT. D&C YELLOW No. 9 □ GRASAL YELLOW □ JAUNE AB □ OIL YELLOW A □ 1-(PHENYLAZO)-2-NAPHTHALENAMINE □ 1-(PHENYLAZO)-2-NAPHTHYLAMINE □ YELLOW AB □ YELLOW No. 2

CONSENSUS REPORTS: IARC Cancer Review: Group 3 IMEMDT 7,56,87; Animal No Evidence IMEMDT 8,279,75. Reported in EPA TSCA Inventory. EPA Genetic Toxicology Program.

SAFETY PROFILE: Moderately toxic by ingestion and subcutaneous routes. Questionable carcinogen with experimental tumorigenic data. Mutation data reported.

When heated to decomposition it emits toxic fumes of NO_x.

FAG135   CAS: 131-79-3   HR: 3
FD&C YELLOW No. 4
mf: $C_{17}H_{15}N_3$ mw: 261.35

SYNS: A.F. YELLOW No. 3 □ CERISOL YELLOW TB □ C.I. 11390 □ C.I. FOOD YELLOW 11 □ DOLKWAL YELLOW OB □ EXT. D&C YELLOW No. 10 □ JAUNE OB □ 1-(2-METHYLPHENYL)AZO-2-NAPHTHALENAMINE □ 1-((2-METHYLPHENYL)AZO)-2-NAPHTHALENAMINE □ 1-(2-METHYLPHENYL)AZO-2-NAPHTHYLAMINE □ OIL YELLOW OB □ o-TOLUENE-1-AZO-2-NAPHTHYL-AMINE □ 1-(o-TOLYLAZO)-2-NAPHTHYLAMINE □ YELLOW OB □ ZLUT MASELNA OB □ ZLUT ROZPO-USTEDLOVA 6

CONSENSUS REPORTS: IARC Cancer Review: Group 3 IMEMDT 7,56,87; Animal Sufficient Evidence IMEMDT 8,287,75. EPA Genetic Toxicology Program.

SAFETY PROFILE: A poison by ingestion. Moderately toxic by intraperitoneal and subcutaneous routes. Questionable carcinogen with experimental tumorigenic data. May be contaminated with the carcinogen β-naphthylamine. Mutation data reported. When heated to decomposition it emits toxic fumes of NO_x. See also AROMATIC AMINES.

FAG140   CAS: 1934-21-0   HR: 1
FD&C YELLOW No. 5
mf: $C_{16}H_9N_4O_9S_2 \cdot 3Na$ mw: 534.38

PROP: Yellow-orange powder. Sol in water, conc sulfuric acid.

SYNS: ACID LEATHER YELLOW T □ ACID YELLOW 23 □ ACID YELLOW T □ ACILAN YELLOW GG □ A.F. YELLOW NO. 4 □ AIREDALE YELLOW T □ AIZEN TARTRAZINE □ AMACID YELLOW T □ ATUL TARTR-AZINE □ BUCACID TARTRAZINE □ CALCOCID YELLOW MCG □ CALCOCID YELLOW XX □ CANA-CERT TARTRAZINE □ 3-CARBOXY-5-HYDROXY-1-p-SULFOPHENYL-4-p-SULFOPHENYLAZOPYRAZOLE TRISODIUM SALT □ CERTICOL TARTRAZOL YELLOW S □ CILEFA YELLOW T □ C.I. 640 □ C.I. 19140 □ C.I. ACID YELLOW 23 □ C.I. ACID YELLOW 23, TRISODIUM SALT □ C.I. FOOD YELLOW 4 □ CURON FAST YELLOW 5G □ D and C YELLOW NO. 5 □ DOLKWAL TARTRA-ZINE □ DYE FD and C YELLOW NO. 5 □ E 102 □ EDICOL SUPRA TARTRAZINE N □ EGG YELLOW A □ ERIO TARTRAZINE □ EUROCERT TARTRAZINE □ FD & C YELLOW NO. 5 TARTRAZINE □ FENAZO YELLOW

T □ FOOD YELLOW 4 □ FOOD YELLOW 5 □ FOOD YELLOW NO. 4 □ HD TARTRAZINE □ HD TARTRAZINE SUPRA □ HEXACERT YELLOW NO. 5 □ HEXACOL TARTRAZINE □ HIDAZID TARTRAZINE □ HISPACID FAST YELLOW T □ HYDRAZINE YELLOW □ HYDROXINE YELLOW L □ KAKO TARTRAZINE □ KAYAKU FOOD COLOUR YELLOW NO. 4 □ KAYAKU TARTRAZINE □ KCA FOODCOL TARTRAZINE PF □ KCA TARTRAZINE PF □ KITON YELLOW T □ LAKE YELLOW □ LEMON YELLOW A □ LEMON YELLOW A GEIGY □ L-GELB 2 □ MAPLE TARTRAZOL YELLOW □ MITSUI TARTRAZINE □ NAPHTOCARD YELLOW O □ NEKLACID YELLOW T □ OXANAL YELLOW T □ SAN-EI TARTRAZINE □ SCHULTZ NO. 737 □ SUGAI TARTRAZINE □ TARTAR YELLOW FS □ TARTAR YELLOW N □ TARTAR YELLOW PF □ TARTAR YELLOW S □ TARTRAN YELLOW □ TARTRAPHENINE □ TARTRAZINE □ TARTRAZINE A EXPO T □ TARTRAZINE B □ TARTRAZINE B.P.C. □ TARTRAZINE EXTRA PURE A □ TARTRAZINE FD & C YELLOW #5 □ TARTRAZINE FQ □ TARTRAZINE G □ TARTRAZINE LAKE □ TARTRAZINE LAKE YELLOW N □ TARTR-AZINE M □ TARTRAZINE MCGL □ TARTRAZINE N □ TARTRAZINE NS □ TARTRAZINE O □ TARTRAZINE T □ TARTRAZINE XX □ TARTRAZINE XXX □ TARTR-AZINE YELLOW □ TARTRAZOL BPC □ TARTRAZOL YELLOW □ TARTRINE YELLOW O □ TRISODIUM 3-CARBOXY-5-HYDROXY-1-p-SULFOPHENYL-4-p-SULFOPHENYLAZOPYRAZOLE □ TRISODIUM SALT of 3-CARBOXY-5-HYDROXY-1-SULFOPHENYLAZOPY-RAZOLE □ UNITERTRACID YELLOW TE □ USACERT YELLOW NO. 5 □ VONDACID TARTRAZINE □ WOOL YELLOW □ XYLENE FAST YELLOW GT □ Y-4 □ 1310 YELLOW □ 1409 YELLOW □ YELLOW LAKE 69 □ YELLOW NO. 5 □ YELLOW NO. 5 FDC □ ZLUT KYSELA 23 □ ZLUT PIGMENT 100 □ ZLUT POTRAVINARSKA 4

CONSENSUS REPORTS: Reported in EPA TSCA Inventory. EPA Genetic Toxicology Program.

SAFETY PROFILE: Low toxicity by ingestion. Mutation data reported. When heated to decomposition it emits very toxic fumes of NO_x, SO_x, and Na_2O.

FAG150   CAS: 2783-94-0   HR: 1
FD&C YELLOW No. 6
mf: $C_{16}H_{10}N_2O_7S_2Na_2$ mw: 452.36

PROP: Orange powder. Sol in water, conc sulfuric acid; sltly sol in abs alc.

SYNS: ACID YELLOW TRA □ A.F. YELLOW NO. 5 □ AIZEN FOOD YELLOW NO. 5 □ ALABASTER NO. 3 □ ATUL SUNSET YELLOW FCF □ CANACERT SUNSET YELLOW FCF □ CERTICOL SUNSET YELLOW CFS □ CERTOLAKE SUNSET YELLOW □ C.I. 15985 □ C.I.

FOOD YELLOW 3 □ C.I. FOOD YELLOW 3, DISODIUM SALT □ CILEFA ORANGE S □ DISODIUM SALT of 1-p-SULPHOPHENYLAZO-2-NAPHTHOL-6-SULPHONIC ACID □ DISPERSED ORANGE 11348 □ DISPERSED YELLOW 12116 □ DOLKWAL SUNSET YELLOW □ DYE FDC YELLOW LAKE 6 □ DYE FD & C YELLOW LAKE 6 □ DYE FDC YELLOW NO. 6 □ DYE FD & C YELLOW NO. 6 □ DYE SUNSET YELLOW □ E 110 □ EDICOL SUPRA YELLOW FC □ ENIACID SUNSET YELLOW □ EUROCERT ORANGE FCF □ FD & C NO. 6 □ FD and C NO. 6 □ FD and C YELLOW 6 □ FD and C YELLOW LAKE NO. 6 □ FD and C YELLOW NO. 6 □ FD & C YELLOW NO. 6 ALUMINIUM LAKE □ FDC YELLOW NO. 6 □ FOODCOL SUNSET YELLOW FCF □ FOOD YELLOW 3 □ FOOD YELLOW 6 □ GELBORANGE-S □ HD SUNSET YELLOW FCF □ HD SUNSET YELLOW FCF SUPRA □ HEXACOL SUNSET YELLOW F & F SUPRA □ HEXACOL SUNSET YELLOW FCF □ HEXACOL SUNSET YELLOW FCF SUPRA □ HEXACOL SUNSET YELLOW FCP □ 6-HYDROXY-5-((p-SULFOPHENYL)AZO)-2-NAPHTHAL-ENESULFONIC ACID, DISODIUM SALT □ 6-HYDROXY-5-((4-SULFOPHENYL)AZO)-2-NAPHTHALENESULFONIC ACID, DISODIUM SALT □ 6-HYDROXY-5-((p-SULPHO-PHENYL)AZO)-2-NAPHTHALENESULPHONIC ACID, DISODIUM SALT □ 6-HYDROXY-5-((4-SULPHOPHEN-YL)AZO)-2-NAPHTHALENESULPHONIC ACID, DISODIUM SALT □ KCA FOODCOL SUNSET YELLOW FCF □ JAUNE ORANGE S □ JAUNE SOLEIL □ L-ORANGE 2 □ L. ORANGE Z2010 □ MAPLE SUNSET YELLOW FCF □ NCI-C53907 □ ORANGE II R □ ORANGE PAL □ ORANGE RGL CONC. SPECIALLY PURE □ ORANGE YELLOW S □ ORANGE YELLOW S.AF □ ORANGE YELLOW S.FQ □ PARA ORANGE □ STANDACOL SUNSET YELLOW FCF □ 1-p-SULFO-PHENYLAZO-2-HYDROXYNAPHTHALENE-6-SULFONATE, DISODIUM SALT □ 1-p-SULFOPHENYL-AZO-2-NAPHTHOL-6-SULFONIC ACID, DISODIUM SALT □ 1-p-SULPHOPHENYLAZO-2-NAPHTHOL-6-SULPHONIC ACID, DISODIUM SALT □ SUN ORANGE A GEIGY □ SUNSET YELLOW □ SUNSET YELLOW BSS □ SUNSET YELLOW FCF □ SUNSET YELLOW FCF SUPRA □ SUNSET YELLOW FU □ SUNSET YELLOW FU SUPRA □ SUNSET YELLOW LAKE □ SUN YELLOW □ SUN YELLOW A-CE □ SUN YELLOW A-FDC □ SUN YELLOW EXTRA CONC. A EXPORT □ SUN YELLOW EXTRA PURE A □ SUN YELLOW FCF □ USACERT YELLOW NO. 6 □ USACERT FD & C YELLOW NO. 6 □ USALAKE FD & C YELLOW NO. 6 LAKE □ 1351 YELLOW □ 1899 YELLOW □ YELLOW NO. 6 □ YELLOW ORANGE S □ YELLOW ORANGE S SPECIALLY PURE □ YELLOW ORANGE SPECIALLY PURE 85 □ YELLOW SF FOR FOOD □ YELLOW SUN □ YELLOW SY FOR FOOD □ ZLUT POTRAVINARSKA 3

CONSENSUS REPORTS: IARC Cancer Review: Group 3 IMEMDT 7,56,87; Animal Inadequate Evidence IMEMDT 8,257,75. Reported in EPA TSCA Inventory.

SAFETY PROFILE: Low toxicity by ingestion and intraperitoneal routes. Mutation data reported. When heated to decomposition it emits very toxic fumes of NO_x and SO_x.

FAK000   CAS: 22224-92-6   HR: 3
FENAMIPHOS
mf: $C_{13}H_{22}NO_3PS$ mw: 303.39
PROP: A solid. Mp: 49°. Sltly sol in H_2O.
SYNS: O-AETHYL-O-(3-METHYL-4-METHYLTHIOPHENYL)-ISOPROPYLAMIDO-PHOSPHORSAEURE ESTER (GERMAN) □ BAY 68138 □ ENT 27,572 □ ETHYL-3-METHYL-4-(METHYLTHIO)PHENYL(1-METHYLETHYL)PHOSPHORAMIDATE □ ETHYL-4-(METHYLTHIO)-m-TOLYL ISOPROPYL PHOSPHORAMIDATE □ ISOPROPYLAMINO-O-ETHYL-(4-METHYLMERCAPTO-3-METHYLPHENYL)PHOSPHATE □ 1-(METHYLETHYL)-ETHYL 3-METHYL-4-(METHYLTHIO)PHENYL PHOSPHORAMIDATE □ NEMACUR □ NSC-195106 □ PHANAMIPHOS

CONSENSUS REPORTS: EPA Extremely Hazardous Substances List.
OSHA PEL: TWA 0.1 mg/m³ (skin)
ACGIH TLV: TWA 0.1 mg/m³ (skin); Not Classifiable as a Human Carcinogen
SAFETY PROFILE: Poison by ingestion, inhalation, and skin contact. When heated to decomposition it emits very toxic fumes of NO_x, PO_x, and SO_x.

FAK100   CAS: 60168-88-9   HR: 2
FENARIMOL
mf: $C_{17}H_{12}Cl_2N_2O$ mw: 331.21
PROP: White, odorless crystals. Mp: 117–119°. Practically insol in water; sol in most org solvs.
SYNS: BLOC □ (2-CHLOROPHENYL)-α-(4-CHLOROPHENYL)-5-PYRIMIDINEMETHANOL □ α-(2-CHLOROPHENYL)-α-(4-CHLOROPHENYL)-5-PYRIMIDINEMETHANOL □ EL 222 □ RIMIDIN □ RUBIGAN

CONSENSUS REPORTS: Reported in EPA TSCA Inventory.
SAFETY PROFILE: Moderately toxic by ingestion. Experimental reproductive effects. Mutation data reported. When heated to decomposition it emits toxic fumes of Cl^- and NO_x.

FAM100   CAS: 7424-00-2   HR: D
FENCHLONINE
mf: $C_9H_{10}ClNO_2$ mw: 199.65
PROP: Crystals from MeOH (aq). Mp: 238–240°.
SYNS: ALANINE, 3-(p-CHLOROPHENYL)-, dl- □ dl-p-CHLOROPHENYLALANINE □ dl-4-CHLOROPHENYLALANINE □ (±)-p-CHLORPHENYLALANINE □ CP 10,188 □ C-PAL □ FENCLONIN □ FENCLONINE
CONSENSUS REPORTS: Reported in EPA TSCA Inventory.
SAFETY PROFILE: An experimental teratogen. When heated to decomposition it emits toxic fumes of NO_x and Cl^-.

FAP000   CAS: 8006-84-6   HR: 2
FENNEL OIL
PROP: From steam distillation of *Foeniculum vulgare* Miller (Fam. *Umbelliferae*) (FCTXAV 12,807,74). Colorless to pale-yellow liquid; odor and taste of fennel.
SYNS: BITTER FENNEL OIL □ FENCHEL OEL (GERMAN) □ OIL OF FENNEL
CONSENSUS REPORTS: Reported in EPA TSCA Inventory.
SAFETY PROFILE: Moderately toxic by ingestion. Mutation data reported. A severe skin irritant. When heated to decomposition it emits acrid smoke and irritating fumes.

FAQ800   CAS: 115-90-2   HR: 3
FENSULFOTHION
mf: $C_{11}H_{17}O_4PS_2$ mw: 308.37
PROP: Yellow oil. D: 1.202 @ 20°/4°, bp: 138–141° @ 0.01 mm. Sol in most org solvs; sltly sol in H_2O.
SYNS: BAY 25141 □ BAYER S767 □ CHEMAGRO 25141 □ DASANIT □ O,O-DIAETHYL-O-4-METHYLSULFINYL-PHENYL-MONOTHIOPHOSPHAT (GERMAN) □ O,O-DIETHYL-O-(p-(METHYLSULFINYL)PHENYL) PHOS-PHOROTHIOATE □ O,O-DIETHYL-O-p-(METHYL-SULFINYL)PHENYL THIOPHOSPHATE □ DMSP □ ENT 24,945 □ S 767 □ TERRACUR P
CONSENSUS REPORTS: EPA Genetic Toxicology Program. EPA Extremely Hazardous Substances List.
OSHA PEL: TWA 0.1 mg/m³
ACGIH TLV: TWA 0.1 mg/m³; Not Classifiable as a Human Carcinogen

SAFETY PROFILE: A poison by ingestion, inhalation, and skin contact. Experimental reproductive effects. A pesticide. When heated to decomposition it emits very toxic fumes of SO_x and PO_x.

FAR100   CAS: 51630-58-1   HR: 3
FENVALERATE
mf: $C_{25}H_{22}ClNO_3$ mw: 419.93
PROP: Clear, yellow, viscous liquid at 23°. D: 1.17, n: (20/D) 1.5533, bp: 300° @ 37 mm. Solubility at 20° (g/L): acetone >450, chloroform >450, methanol >450, hexane 77. Insol in water. Decomp gradually between 150 and 300°.
SYNS: BELMARK □ α-CYANO-3-PHENOXYBENZYL-2-(4-CHLOROPHENYL)ISOVALERATE PYDRIN □ α-CYANO-3-PHENOXYBENZYL-2-(4-CHLOROPHENYL)-3-METHYLBUTYRATE □ CYANO(3-PHENOXYPHENYL)-METHYL 4-CHLORO-α-(1-METHYLETHYL)BENZENE-ACETATE □ ECTRIN □ PHENVALERATE □ PYDRIN □ S 5602 □ SANMARTON □ SD 43775 □ SUMICIDIN □ SUMIFLY □ SUMIPOWER □ WL 43775
CONSENSUS REPORTS: Cyanide and its compounds are on the Community Right-To-Know List.
SAFETY PROFILE: Poison by ingestion, intravenous, and intracerebral routes. Moderately toxic by skin contact. Experimental reproductive effects. Mutation data reported. Highly toxic to fish and bees. Corrosive, causes eye damage. A skin irritant. When heated to decomposition it emits toxic fumes of Cl^-, NO_x, and CN^-. See also CYANIDE.

FAS000   CAS: 14484-64-1   HR: 3
FERBAM
mf: $C_9H_{18}N_3S_6•Fe$ mw: 416.51
PROP: Black solid or powder. Mp: 180° (decomp). Decomposes upon prolonged storage or in contact with moisture. Sltly sol in H_2O; sol in Me_2CO, $CHCl_3$, Py, and MeCN. IDLH 800 mg/m³.
SYNS: AAFERTIS □ BERCEMA FERTAM 50 □ CARBAMATE □ DIMETHYLCARBAMODITHIOIC ACID, IRON COMPLEX □ DIMETHYLCARBAMODITHIOIC ACID, IRON(3+) SALT □ DIMETHYLDITHIOCARBAMIC ACID, IRON SALT □ DIMETHYLDITHIOCARBAMIC ACID, IRON(3+) SALT □ EISENDIMETHYLDITHIO-

CARBAMAT (GERMAN) □ EISEN(III)-TRIS(N,N-DIMETHYLDITHIOCARBAMAT) (GERMAN) □ ENT 14,689 □ FERBAM 50 □ FERBAM, IRON SALT □ FERBECK □ FERMATE FERBAM FUNGICIDE □ FERMOCIDE □ FERRADOW □ FERRIC DIMETHYL-DITHIOCARBAMATE □ FUKLASIN ULTRA □ HEXA-FERB □ HOKMATE □ IRON DIMETHYLDITHIO-CARBAMATE □ KARBAM BLACK □ KNOCKMATE □ NIACIDE □ SUP'R FLO FERBAM FLOWABLE □ TRIFUNGOL □ TRIS(DIMETHYLCARBAMODITHIO-ATO-S,S')IRON □ TRIS(DIMETHYLDITHIOCARBAM-ATO)IRON □ TRIS(N,N-DIMETHYLDITHIOCARBAM-ATO) IRON(III) □ VANCIDE FE95

CONSENSUS REPORTS: IARC Cancer Review: Group 3 IMEMDT 7,56,87; Animal Inadequate Evidence IMEMDT 12,121,76. Reported in EPA TSCA Inventory. EPA Genetic Toxicology Program.

OSHA PEL: TWA Total Dust: 10 mg/m^3; Respirable Fraction: 5 mg/m^3

ACGIH TLV: TWA 10 mg/m^3; Not Classifiable as a Human Carcinogen

DFG MAK: 15 mg/m^3

SAFETY PROFILE: Poison by intraperitoneal route. Moderately toxic by ingestion. Experimental teratogenic and reproductive effects. Questionable carcinogen with experimental carcinogenic and tumorigenic data. Mutation data reported. A fungicide. When heated to decomposition it emits very toxic fumes of NO$_x$ and SO$_x$. See also CARBAMATES.

**FAU000   CAS: 7705-08-0   HR: 3
FERRIC CHLORIDE**

DOT: UN 1773/UN 2582

mf: Cl$_3$Fe   mw: 162.20

PROP: Black-brown solid or hygroscopic dark-green or black crystals. Mp: 303°, bp: 315°, d: 2.90 @ 25°, vap press: 1 mm @ 194.0°. Aq solns are strongly acidic. Sol in H$_2$O to give hydrates; sol in MeOH and Et$_2$O.

SYNS: CHLORURE PERRIQUE □ FERRIC CHLORIDE (UN 1733) (DOT) □ FERRIC CHLORIDE, solution (UN 2582) (DOT) □ FLORES MARTIS □ IRON CHLORIDE □ IRON(III) CHLORIDE □ IRON TRICHLORIDE □ PERCHLORURE de FER

CONSENSUS REPORTS: Reported in EPA TSCA Inventory. EPA Genetic Toxicology Program.

OSHA PEL: TWA 1 mg(Fe)/m^3

ACGIH TLV: TWA 1 mg(Fe)/m^3

DOT CLASSIFICATION: 8; Label: Corrosive

SAFETY PROFILE: Poison by ingestion and intravenous routes. Experimental reproductive effects. Corrosive. Probably an eye, skin, and mucous membrane irritant. Mutation data reported. Reacts with water to produce toxic and corrosive fumes. Catalyzes potentially explosive polymerization of ethylene oxide, chlorine + monomers (e.g., styrene). Forms shock-sensitive explosive mixtures with some metals (e.g., potassium, sodium). Violent reaction with allyl chloride. When heated to decomposition it emits highly toxic fumes of HCl.

**FAW100   CAS: 2338-05-8   HR: D
FERRIC CITRATE**

mf: C$_6$H$_5$FeO$_7$ mw: 244.95

PROP: White or red crystals; odorless with slt metallic taste. Sol in water.

SYN: IRON(III) CITRATE

SAFETY PROFILE: When heated to decomposition it emits acrid smoke and irritating fumes.

**FAX000   CAS: 7783-50-8   HR: 3
FERRIC FLUORIDE**

mf: F$_3$Fe   mw: 112.85

PROP: White or green crystals. D: 3.87, mp: 1000°. Sltly sol in H$_2$O, HF, EtOH, and Et$_2$O.

SYNS: IRON FLUORIDE □ IRON TRIFLUORIDE

CONSENSUS REPORTS: Reported in EPA TSCA Inventory.

OSHA PEL: TWA 2.5 mg(F)/m^3; TWA 1 mg(Fe)/m^3

ACGIH TLV: TWA 2.5 mg(F)/m^3; BEI: 3 mg/g creatinine of fluorides in urine prior to shift; 10 mg/g creatinine of fluorides in urine at end of shift; 1 mg(Fe)/m^3

NIOSH REL: TWA (Inorganic Fluorides) 2.5 mg(F)/m^3

SAFETY PROFILE: Moderately toxic by intraperitoneal route. Mutation data reported. When heated to decomposition it emits toxic fumes of F⁻. See also FLUORIDES.

FAY200   CAS: 10421-48-4   HR: 2
FERRIC NITRATE
DOT: UN 1466
mf: $N_3O_9 \cdot Fe$ mw: 241.88
SYNS: FERRIC NITRATE (DOT) □ IRON NITRATE □ IRON (III) NITRATE, ANHYDROUS □ IRON TRINITRATE □ NITRIC ACID, IRON(3+) SALT
ACGIH TLV: TWA 1 mg/(Fe)/m³
DOT CLASSIFICATION: 5.1; Label: Oxidizer
CONSENSUS REPORTS: Reported in EPA TSCA Inventory.
SAFETY PROFILE: Mutation data reported. A reactive oxidizer.

FAZ525   CAS: 10058-44-3   HR: D
FERRIC PYROPHOSPHATE
mf: $Fe_4(P_2O_7)_3 \cdot xH_2O$ mw: 745.22
PROP: Tan to yellow powder. Sol in mineral acids; insol in water.
SYN: IRON PYROPHOSPHATE
SAFETY PROFILE: When heated to decomposition it emits very toxic fumes of PO_x.

FBA000   CAS: 10028-22-5   HR: 3
FERRIC SULFATE
mf: $Fe_2O_{12}S_3$ mw: 399.88
PROP: Yellow solid or crystals, or gray-white powder. Sltly soluble in H_2O; soly increased if trace $FeSO_4$ present.
SYNS: DIIRON TRISULFATE □ IRON PERSULFATE □ IRON SESQUISULFATE □ IRON SULFATE (2:3) □ IRON(III) SULFATE □ IRON TERSULFATE □ SULFURIC ACID, IRON (3⁺) SALT (3:2)
CONSENSUS REPORTS: EPA FIFRA 1988 pesticide subject to registration or re-registration. Reported in EPA TSCA Inventory.
ACGIH TLV: TWA 1 mg(Fe)/m³
SAFETY PROFILE: A poison by intraperitoneal route. Mutation data reported. When heated to decomposition it

emits toxic fumes of SO_x and Fe⁻. See also SULFATES and other ferric salts.

FBC000   CAS: 102-54-5   HR: 3
FERROCENE
mf: $C_{10}H_{10}Fe$ mw: 186.05
PROP: Orange crystals from alc (aq); camphor odor. Mp: 172.5–173°, bp: 249°, subl @ >100°, volatile in steam. Insol in water; sol in alcohol and ether.
SYNS: BISCYCLOPENTADIENYLIRON □ DI-2,4-CYCLOPENTADIEN-1-YL IRON □ DICYCLOPENTA-DIENYL IRON (OSHA, ACGIH) □ IRON BIS(CYCLO-PENTADIENE) □ IRON DICYCLOPENTADIENYL
CONSENSUS REPORTS: Reported in EPA TSCA Inventory.
OSHA PEL: TWA Total Dust: 10 mg/m³; Respirable Fraction: 5 mg/m³
ACGIH TLV: TWA 10 mg/m³
SAFETY PROFILE: Poison by intraperitoneal and intravenous routes. Moderately toxic by ingestion. Questionable carcinogen with experimental tumorigenic data. Mutation data reported. Flammable; reacts violently with NH_4ClO_4. When heated to decomposition it emits acrid smoke and irritating fumes.

FBC100   CAS: 1336-80-7   HR: 3
FERROCHOLINATE
mf: $C_6H_{10}FeO_{10} \cdot C_5H_{14}NO$ mw: 402.21
PROP: Greenish-brown, reddish-brown, or brown amorph solid or powder with glistening surface upon fracture. Sol in water, acids, and alkalies.
SYNS: CHELAFER □ CHEL-IRON □ FERRIC CHOLINE CITRATE □ FERROLIP □ IRON CHOLINE CITRATE COMPLEX
SAFETY PROFILE: Poison by intravenous and intraperitoneal routes. Mildly toxic by ingestion. When heated to decomposition it emits toxic fumes of NO_x. See also CHOLINE.

FBD000   CAS: 11114-46-8   HR: 3
FERROCHROME (exothermic)
SYNS: CARBON FERROCHROMIUM □ CHROME FERROALLOY □ CHROMIUM ALLOY, Cr,C,Fe,N,Si □ CHROMIUM ALLOY, BASE, Cr,C,Fe,N,Si

(FERROCHROMIUM) □ FERROCHROME □ exothermic FERROCHROME □ FERROCHROMIUM

CONSENSUS REPORTS: IARC Cancer Review: Animal Inadequate Evidence IMEMDT 23,205,80. Reported in EPA TSCA Inventory. Chromium and its compounds are on the Community Right-To-Know List.

OSHA PEL: TWA 1 mg(Cr)/m³
ACGIH TLV: TWA 0.5 mg(Cr)/m³; Not Classifiable as a Carcinogen
SAFETY PROFILE: Poison by inhalation. Questionable carcinogen. See also CHROMIUM COMPOUNDS.

FBE000   CAS: 12604-53-4   HR: 3
FERROMANGANESE (exothermic)
SYN: exothermic FERROMANGANESE (DOT)
CONSENSUS REPORTS: Reported in EPA TSCA Inventory. Manganese and its compounds are on the Community Right-To-Know List.
SAFETY PROFILE: The dust will burn violently and give off toxic fumes of MnO_2. See also MANGANESE COMPOUNDS.

FBG000   CAS: 8049-17-0   HR: 3
FERROSILICON
DOT: UN 1408
mf: FeSi   mw: 83.90
PROP: Crystalline, metallic solid. Fe + Si, d: 5.4. Containing 30% or more but not more than 70% silicon (FEREAC 41,15972,76).
SYN: FERROSILICON, containing more than 30% but less than 90% SILICON (DOT)
CONSENSUS REPORTS: Reported in EPA TSCA Inventory.
DOT CLASSIFICATION: 4.3; Label: Dangerous When Wet
SAFETY PROFILE: Moderate inhalation hazard. Low skin toxicity. Reaction with moisture releases hydrogen and acetylene gases, which then ignite; impurities in the alloy may liberate such poisonous and reactive gases as phosphine and arsine. Dry mixtures with sodium hydroxide react incandescently when water is added. Reaction with acid, acid fumes, or oxidizing materials can emit toxic fumes. Reaction

hazards increase with decreasing particle size.

FBH000   CAS: 3094-87-9   HR: 2
FERROUS ACETATE
mf: $C_4H_6O_4$•Fe   mw: 173.95
PROP: White or colorless crystals.
SYNS: ACETIC ACID, IRON(2+) SALT □ IRON(2+) ACETATE □ IRON(II) ACETATE □ IRON DIACETATE
CONSENSUS REPORTS: Reported in EPA TSCA Inventory.
OSHA PEL: TWA 1 mg(Fe)/m³
ACGIH TLV: TWA 1 mg(Fe)/m³
SAFETY PROFILE: Moderately toxic by subcutaneous route. When heated to decomposition it emits acrid smoke and irritating fumes.

FBI000   CAS: 7758-94-3   HR: 3
FERROUS CHLORIDE
DOT: UN 1759/UN 1760
mf: Cl_2Fe   mw: 126.75
PROP: White crystals when pure; hygroscopic. Green to yellow, deliquescent crystals. Mp: 676°, bp: 1012°, d: 3.16, vap press: 10 mm @ 700°. Sol in H_2O; insol in Et_2O; sltly sol in C_6H_6.
SYNS: IRON(II) CHLORIDE (1:2) □ FERROUS CHLORIDE, solution (NA 1760) (DOT) □ FERROUS CHLORIDE, solid (NA 1759) (DOT) □ IRON DICHLORIDE □ IRON PROTOCHLORIDE
CONSENSUS REPORTS: Reported in EPA TSCA Inventory. EPA Genetic Toxicology Program.
OSHA PEL: TWA 1 mg(Fe)/m³
ACGIH TLV: TWA 1 mg(Fe)/m³
DOT CLASSIFICATION: 8; Label: Corrosive
SAFETY PROFILE: Poison by ingestion and intraperitoneal routes. Mutation data reported. Corrosive. Probably an irritant to the eyes, skin, and mucous membranes. Can react violently with ethylene oxide, K, Na. When heated to decomposition it emits toxic fumes of Cl⁻. See also CHLORIDES and IRON.

FBJ100   CAS: 141-01-5   HR: 3
FERROUS FUMARATE
mf: $C_4H_2O_4$•Fe   mw: 169.91

PROP: Reddish-orange to reddish-brown granular powder or solid; odorless, almost tasteless. D: 2.435. Solubility at 25° in water: 0.14 g/100 mL; in alc <0.01 g/100 mL. Solubility in acid is limited by liberation of fumaric acid. Insol in EtOH.

SYNS: CPIRON □ ERCO-FER □ ERCOFERRO □ FEOSTAT □ FEROTON □ FERROFUME □ FERRONAT □ FERRONE □ FERROTEMP □ FERRUM □ FERSAMAL □ FIRON □ FUMAFER □ FUMAR-F □ FUMIRON □ GALFER □ HEMOTON □ IRCON □ IRON FUMARATE □ METERFER □ METERFOLIC □ ONE-IRON □ PALAFER □ TOLERON □ TOLFERAIN □ TOLIFER

CONSENSUS REPORTS: Reported in EPA TSCA Inventory.

OSHA PEL: TWA 1 mg(Fe)/m³

ACGIH TLV: TWA 1 mg(Fe)/m³

SAFETY PROFILE: Poison by intraperitoneal route. Moderately toxic by ingestion and subcutaneous routes. Human systemic effects by ingestion: dyspnea, nausea or vomiting, somnolence. When heated to decomposition it emits acrid smoke and irritating fumes. See also FUMARIC ACID.

FBK000   CAS: 299-29-6   HR: 3
FERROUS GLUCONATE
mf: $C_{12}H_{22}O_{14}$•Fe   mw: 446.19

PROP: Yellowish-gray or pale-greenish-yellow, fine powder or granules with slt odor of burned sugar. Sol in water and glycerin; insol in alc.

SYNS: FERGON □ FERGON PREPARATIONS □ FERLUCON □ FERRONICUM □ GLUCO-FERRUM □ IROMIN □ IRON GLUCONATE □ IROX (GADOR) □ NIONATE □ RAY-GLUCIRON

CONSENSUS REPORTS: Reported in EPA TSCA Inventory.

OSHA PEL: TWA 1 mg(Fe)/m³

ACGIH TLV: TWA 1 mg(Fe)/m³

SAFETY PROFILE: Poison by intraperitoneal and intravenous routes. Moderately toxic by ingestion. Human systemic effects by ingestion: hypermotility, diarrhea, nausea, and vomiting. Experimental reproductive effects. Questionable carcinogen with experimental tumorigenic data. When heated to

decomposition it emits acrid smoke and irritating fumes.

FBN100   CAS: 7720-78-7   HR: 3
FERROUS SULFATE
mf: O_4S•Fe   mw: 151.91

PROP: Grayish white to buff powder. Slowly sol in water; insol in alc.

SYNS: COPPERAS □ DURETTER □ DUROFERON □ EXSICCATED FERROUS SULFATE □ EXSICCATED FERROUS SULPHATE □ FEOSOL □ FEOSPAN □ FER-IN-SOL □ FERO-GRADUMET □ FERRALYN □ FERRO-GRADUMET □ FERROSULFAT (GERMAN) □ FERRO-SULFATE □ FERRO-THERON □ FERSOLATE □ GREEN VITRIOL □ IRON MONOSULFATE □ IRON PROTO-SULFATE □ IRON(II) SULFATE (1:1) □ IRON VITRIOL □ IROSPAN □ IROSUL □ SLOW-FE □ SULFERROUS □ SULFURIC ACID, IRON(2^+) SALT (1:1)

CONSENSUS REPORTS: Reported in EPA TSCA Inventory. EPA Genetic Toxicology Program.

OSHA PEL: TWA 1 mg(Fe)/m³

ACGIH TLV: TWA 1 mg(Fe)/m³

SAFETY PROFILE: A human poison by ingestion. Moderately toxic to humans by an unspecified route. An experimental poison by ingestion, intraduodenal, intraperitoneal, intravenous, and subcutaneous routes. Human systemic effects by ingestion: aggression, somnolence, brain recording changes, diarrhea, nausea or vomiting, bleeding from the stomach, coma. Questionable carcinogen with experimental tumorigenic data. Experimental teratogenic and reproductive effects. Mutation data reported. Potentially explosive reaction with methyl isocyanoacetate at 25°. May ignite on contact with arsenic trioxide + sodium nitrate. When heated to decomposition it emits toxic fumes of SO_x. See also IRON COMPOUNDS.

FBO000   CAS: 7782-63-0   HR: 3
FERROUS SULFATE HEPTAHYDRATE
mf: O_4S•Fe•$7H_2O$ mw: 278.05

PROP: Pale blue green monoclinic, hygroscopic crystals or granules; odorless with a salt taste. D: 2.99–3.08, mp: 64°. Sol in H_2O to give $[Fe(H_2O)_6]_2$; acidic soln insol in EtOH.

SYNS: COPPERAS □ FEOSOL □ FER-IN-SOL □ FERO-GRADUMET □ FERROUS SULFATE (FCC) □ FESOFOR □ FESOTYME □ GREEN VITROL □ HAEMOFORT □ IRONATE □ IRON(II) SULFATE (1:1), HEPTAHYDRATE □ IRON VITROL □ IROSUL □ MOL-IRON □ PRESFERSUL □ SULFERROUS

OSHA PEL: TWA 1 mg(Fe)/m³

ACGIH TLV: TWA 1 mg(Fe)/m³

SAFETY PROFILE: Poison by intravenous, intraperitoneal, and subcutaneous routes. Moderately toxic by ingestion and rectal routes. Mutation data reported. When heated to decomposition it emits toxic fumes of SO_x.

FBP000   CAS: 12604-58-9   HR: 2
FERROVANADIUM DUST

PROP: A gray to black dust. IDLH 500 mg/m³.

CONSENSUS REPORTS: Reported in EPA TSCA Inventory.

OSHA PEL: TWA 1 mg/m³; STEL 3 mg/m³

ACGIH TLV: TWA 1 mg/m³; STEL 3 mg/m³

DFG MAK: 1 mg/m³

NIOSH REL: (Vanadium) TWA 1.0 mg(V)/m³

SAFETY PROFILE: Can cause pulmonary damage. Combustible when exposed to heat or flame. See also VANADIUM and IRON.

FBQ000   HR: 2
FIBROUS GLASS

PROP: Is of a borosilicate variety, of low alkalinity, and consists of calcia-alumina-silicate (85INA8 5,270,86).

SYNS: FIBERGLASS □ FIBROUS GLASS DUST (ACGIH) □ GLASS □ GLASS FIBERS

OSHA PEL: TWA 15 mg/m³ (total dust); 5 mg/m³ (nuisance dust)

ACGIH TLV: TWA 10 mg/m³ (dust)

NIOSH REL: TWA 5 mg/m³ (total fibrous glass)

SAFETY PROFILE: Suspected carcinogen with experimental carcinogenic, neoplastigenic, and tumorigenic data by inhalation and other routes. Human mutation data reported. Used as thermal and acoustic insulation.

The possibility of lung problems due to inhalation of fine particles or flakes or fibers of fiberglass has often been raised. The extensive medical research so far reported has shown no consistent evidence of chronic health effects in workers who are exposed to man-made vitreous fibers. In some studies where massive doses of fine-diameter fibers were implanted into mice, cancer development in the pleura was noted. Also some animal studies involving injection of fibers into the trachae resulted in a minimal fibrosis.

Exposure to glass fibers sometimes causes irritation of the skin and, less frequently, irritation of the eyes, nose, or throat. This is not an allergic reaction, but simply a mechanical irritation. Skin irritation typically is experienced by individuals who are newly exposed to fibrous glass and it usually diminishes after several days of exposure. Good personal and industrial hygiene practices minimize the amount of discomfort experienced.

FBS000   CAS: 9001-33-6   HR: 3
FICIN

PROP: A proteolytic enzyme in the crude latex of the fig tree *Ficus* (JPETAB 71,20,41). White powder. Very sol in water.

SYNS: DEBRICIN □ FICUS PROTEASE □ FICUS PROTEINASE □ HIGUEROXYL DELABARRE □ TL 367

CONSENSUS REPORTS: Reported in EPA TSCA Inventory.

SAFETY PROFILE: Poison by inhalation and intravenous routes. Mildly toxic by ingestion. When heated to decomposition it emits toxic fumes.

FBU000   CAS: 59536-65-1   HR: 3
FIREMASTER BP-6

PROP: Consists mainly of penta-, hexa-, and heptabromobiphenyl, with lesser amounts of tetra- and other brominated biphenyls (ENVRAL 10,390,75).

SYNS: HEXABROMOBIPHENYL (technical grade) □ PBB □ POLYBROMINATED BIPHENYLS
CONSENSUS REPORTS: IARC Cancer Review: Animal Inadequate Evidence IMEMDT 18,107,78. Polybrominated biphenyl compounds are on the Community Right-To-Know List.
SAFETY PROFILE: Poison by ingestion. Experimental teratogenic and reproductive effects. Questionable carcinogen with experimental carcinogenic and tumorigenic data. Mutation data reported. When heated to decomposition it emits very toxic Br⁻. See also POLYBROMINATED BIPHENYLS.

FBU509   CAS: 67774-32-7   HR: 3
FIREMASTER FF-1
PROP: 2,4,5,2',4',5'-hexabromobiphenyl is the predominant isomer (LANCAO 2,602,77).
SYNS: 2,4,5,2',4',5'-HEXABROMOBIPHENYL □ PBB □ POLYBROMINATED BIPHENYL □ POLYBROMINATED BIPHENYL (FF-1)
CONSENSUS REPORTS: NTP 10th Report on Carcinogens. IARC Cancer Review: Group 2B IMEMDT 7,321,87; Human Inadequate Evidence IMEMDT 41,261,86; Animal Sufficient Evidence IMEMDT 41,261,86. NTP Carcinogenesis Studies (gavage); Clear Evidence: mouse, rat NTPTR* NTP-TR-244,83. Polybrominated biphenyl compounds are on the Community Right-To-Know List.
SAFETY PROFILE: Confirmed carcinogen with experimental carcinogenic, neoplastigenic, and teratogenic data. Experimental reproductive effects. When heated to decomposition it emits very toxic fumes of Br⁻. See also POLYBROMINATED BIPHENYLS.

FBU850   HR: D
FIR NEEDLE OIL, CANADIAN TYPE
PROP: Found in the needles and twigs of *Abies balsamea* L. Mill (Fam. *Pinaceae*) (FCTXAV 13,449,75). Colorless to faintly yellow liquid; pleasant odor. Sol in fixed oils, mineral oil; sltly sol in propylene glycol; insol in glycerin.

SYN: BALSAM FIR OIL
SAFETY PROFILE: When heated to decomposition it emits acrid smoke and irritating fumes.

FBV000   CAS: 8021-29-2   HR: 1
FIR NEEDLE OIL, SIBERIAN
PROP: Found in the needles and twigs of *Abies sibirica* Ledeb. (Fam. *Pinaceae*) (FCTXAV 13,449,75). Colorless to faintly yellow liquid. Sol in fixed oils, mineral oil; insol in glycerin, propylene glycol.
SYN: PINE NEEDLE OIL
CONSENSUS REPORTS: Reported in EPA TSCA Inventory.
SAFETY PROFILE: Mildly toxic by ingestion. A human and experimental skin irritant. When heated to decomposition it emits acrid smoke and irritating fumes.

FDD125   CAS: 16872-11-0   HR: 2
FLUOBORIC ACID
DOT: UN 1775
mf: $BF_4 \cdot H$ mw: 87.82
PROP: Colorless liquid. Bp: 130°.
SYNS: BORATE(1-), TETRAFLUORO-, HYDROGEN □ BOROFLUORIC ACID □ FLUOBORIC ACID (DOT) □ HYDROFLUOBORIC ACID □ HYDROGEN TETRAFLUOROBORATE □ TETRAFLUOROBORIC ACID
CONSENSUS REPORTS: Reported in EPA TSCA Inventory.
OSHA PEL: TWA 2.5 mg(F)/m³
ACGIH TLV: TWA 2.5 mg(F)/m³; BEI: 3 mg/g creatinine of fluorides in urine prior to shift; 10 mg/g creatinine of fluorides in urine at end of shift.
NIOSH REL: (Fluorides, inorganic) TWA 2.5 mg(F)/m³
DOT CLASSIFICATION: 8; Label: Corrosive
SAFETY PROFILE: A corrosive acid. When heated to decomposition it emits toxic vapors of B and F⁻.

FDE000   CAS: 30223-48-4   HR: D
FLUORACIZINE
mf: $C_{20}H_{21}F_3N_2OS$ mw: 394.49
SYN: 10-DIETHYLAMINOPROPIONYL-3-TRIFLUOROMETHYL PHENOTHIAZINE HYDROCHLORIDE

SAFETY PROFILE: Experimental teratogenic effects. When heated to decomposition it emits very toxic fumes of SO_x, NO_x, and F^-.

FDF000   CAS: 206-44-0   HR: 3
FLUORANTHENE
mf: $C_{16}H_{10}$ mw: 202.26
PROP: A polycyclic hydrocarbon. Colorless solid. Needles or plates from alc. Mp: 110°, bp: 250–251° @ 60 mm, vap press: 0.01 mm @ 20°.
SYNS: 1,2-BENZACENAPHTHENE □ BENZO(jk)-FLUORENE □ IDRYL □ 1,2-(1,8-NAPHTHALENEDI-YL)BENZENE □ 1,2-(1,8-NAPHTHYLENE)BENZENE □ RCRA WASTE NUMBER U120
CONSENSUS REPORTS: IARC Cancer Review: Group 3 IMEMDT 7,56,87; Animal No Evidence IMEMDT 32,355,83. Reported in EPA TSCA Inventory. EPA Genetic Toxicology Program.
SAFETY PROFILE: Poison by intravenous route. Moderately toxic by ingestion and skin contact. Questionable carcinogen with experimental tumorigenic data. Human mutation data reported. Combustible when exposed to heat or flame. When heated to decomposition it emits acrid smoke and irritating fumes.

FDI000   CAS: 153-78-6   HR: 3
FLUOREN-2-AMINE
mf: $C_{13}H_{11}N$ mw: 181.25
PROP: Crystals from EtOH (aq). Mp: 129°.
SYNS: AMINOFLUOREN (GERMAN) □ 2-AMINOFLUORENE □ 2-FLUORENAMINE □ 2-FLUORENEAMINE
CONSENSUS REPORTS: EPA Genetic Toxicology Program.
SAFETY PROFILE: Suspected carcinogen with experimental carcinogenic, neoplastigenic, and tumorigenic data. Poison by intraperitoneal route. Mutation data reported. When heated to decomposition it emits toxic fumes of NO_x. See also AMINES.

FDR000   CAS: 53-96-3   HR: 3
N-FLUOREN-2-YL ACETAMIDE

mf: $C_{15}H_{13}NO$ mw: 223.29
SYNS: AAF □ 2-AAF □ 2-ACETAMIDOFLUORENE □ 2-ACETAMINOFLUORENE □ ACETOAMINOFLUOR-ENE □ 2-ACETYLAMINO-FLUOREN (GERMAN) □ N-ACETYL-2-AMINOFLUORENE □ 2-ACETYLAMINO-FLUORENE (OSHA) □ AZETYLAMINOFLUOREN (GERMAN) □ FAA □ 2-FAA □ 2-FLUORENYL-ACETAMIDE □ N-2-FLUORENYLACETAMIDE □ RCRA WASTE NUMBER U005
CONSENSUS REPORTS: NTP 10th Report on Carcinogens. Community Right-To-Know List. EPA Genetic Toxicology Program. Reported in EPA TSCA Inventory.
OSHA PEL: Cancer Suspect Agent
NIOSH REL: (2-Acetylaminofluorene) TWA use 29 CFR 1910.1014
SAFETY PROFILE: Confirmed human carcinogen with experimental carcinogenic, neoplastigenic, tumorigenic, and teratogenic data. Moderately toxic by ingestion and intraperitoneal routes. Experimental reproductive effects. Human mutation data reported. When heated to decomposition it emits toxic fumes of NO_x.

FEI100   CAS: 28920-43-6   HR: 3
9-FLUORENYLMETHYL CHLORO-FORMATE
mf: $C_{15}H_{11}ClO_2$ mw: 258.71
SYNS: CARBONOCHLORIDIC ACID, 9H-FLUOREN-9-YLMETHYL ESTER □ FORMIC ACID, CHLORO-, FLUOREN-9-YLMETHYL ESTER
DOT CLASSIFICATION: 6.1; Label: Poison, Corrosive
SAFETY PROFILE: A poison. Mutation data reported. A corrosive. When heated to decomposition it emits toxic vapors of Cl^-.

FEV000   CAS: 2321-07-5   HR: 3
FLUORESCEIN
mf: $C_{20}H_{12}O_5$ mw: 332.32
PROP: Orange-red, crystalline powder. Mp: 314–316° with decomp.
SYNS: 9-(o-CARBOXYPHENYL)-6-HYDROXY-3-ISOXANTHENONE □ 9-(o-CARBOXYPHENYL)-6-HYDROXY-3H-XANTHEN-3-ONE □ C.I. 45330 □ C.I. 45350 (FREE ACID) □ C.I. SOLVENT YELLOW 94 □ D&C YELLOW No. 7 □ 3',6'-DIHYDROXYFLUORAN □ DIHYDROXYFLUORANE □ 3',6'-DIHYDROXYSPIRO-(ISOBENZOFURAN-1(3H),9'(9H)-XANTHEN)-3-ONE □

3,6-FLUORANDIOL □ 3',6'-FLUORANDIOL □ FLUO-RESCEINE □ HIDACID FLUORESCEIN □ RESOR-CINOLPHTHALEIN □ SOAP YELLOW F □ 11712 YELLOW

CONSENSUS REPORTS: Reported in EPA TSCA Inventory. EPA Genetic Toxicology Program.

SAFETY PROFILE: Poison by intravenous route. Moderately toxic by intraperitoneal route. Mutation data reported. When heated to decomposition it emits acrid smoke and irritating fumes. See also FLUORESCEIN SODIUM.

FEV100 CAS: 3570-80-7 HR: 3
FLUORESCEIN MERCURIC ACETATE
mf: $C_{24}H_{16}Hg_2O_9$ mw: 849.58

SYNS: FLUORESCEIN MERCURIACETATE □ FLUORESCEIN MERCURY ACETATE □ FMA □ FMA (analytical reagent) □ MERCURY, BIS(ACETATO)(mu-(3',6'-DIHYDROXY-2',7'-FLUORANDIYL))DI-

CONSENSUS REPORTS: EPA TSCA Chemical Inventory.

ACGIH TLV: TWA 0.1 mg(Hg)/m³ (skin); BEI: 35 μg/g creatinine total inorganic mercury in urine preshift; 15 μg/g creatinine total inorganic mercury in blood at end of shift at end of workweek.

DFG MAK: Confirmed Animal Carcinogen with Unknown Relevance to Humans

NIOSH REL: (Mercury, Aryl and Inorganic) CL 0.1 mg/m³ (skin)

SAFETY PROFILE: Highly toxic. Mutation data reported. When heated to decomposition it emits toxic fumes of Hg.

FEW000 CAS: 518-47-8 HR: 2
FLUORESCEIN SODIUM
mf: $C_{20}H_{10}O_5 \cdot 2Na$ mw: 376.28

PROP: Orange-red, hygroscopic powder. Sol in water; sltly sol in alc.

SYNS: AIZEN URANINE □ CALCOCID URANINE B4315 □ 9-o-CARBOXYPHENYL-6-HYDROXY-3-ISOXANTHONE, DISODIUM SALT □ CERTIQUAL FLUORESCEINE □ C.I. 766 □ C.I. ACID YELLOW 73 □ C.I. 45350 DISODIUM SALT □ D&C YELLOW No. 8 □ DISODIUM-6-HYDROXY-3-OXO-9-XANTHENE-o-BENZOATE □ FLUORESCEIN SODIUM B.P □ FLUORESCEIN, soluble □ FLUOR-I-STRIP A.T. □ FUL-GLO □ FUNDUSCEIN □ FURANIUM □ HIDACID URANINE □ NCI-C54706 □ RESORCINOL PHTHALEIN

SODIUM □ SODIUM FLUORESCEIN □ SODIUM FLUORESCEINATE □ SODIUM SALT of HYDROXY-o-CARBOXY-PHENYL-FLUORONE □ SOLUBLE FLUORESCEIN □ SPIRO(ISOBENZOFURAN)-1(3H),9'-(9H)XANTHENE-3-ONE, 3',6'-DIHYDROXY-DISODIUM SALT □ URANIN □ URANINE A EXTRA □ URANINE USP XII □ URANINE YELLOW □ 11824 YELLOW □ 12417 YELLOW

CONSENSUS REPORTS: Reported in EPA TSCA Inventory.

SAFETY PROFILE: Moderately toxic by intraperitoneal route. Mildly toxic by ingestion. Human systemic effects by intravenous route: arrhythmias, eye hemorrhage, nausea or vomiting. Experimental reproductive effects. Questionable carcinogen with experimental tumorigenic data. Mutation data reported. When heated to decomposition it emits toxic fumes of Na_2O.

FEY000 HR: 2
FLUORIDES
OSHA PEL: TWA 2.5 mg(F)/m³

ACGIH TLV: TWA 2.5 mg(F)/m³; BEI: 3 mg/g creatinine of fluorides in urine prior to shift; 10 mg/g creatinine of fluorides in urine at end of shift.

DFG MAK: 2.5 mg/m³; BAT: 7 mg/kg creatinine in urine at end of exposure; 4 mg/kg creatinine in urine about 16 hours after end of exposure

NIOSH REL: TWA 2.5 mg(F)/m³

SAFETY PROFILE: Inorganic fluorides are generally highly irritating and toxic. Acute effects resulting from exposure to fluorine compounds are due to HF. Chronic fluorine poisoning, or "fluorosis," occurs among miners of cryolite, and consists of a sclerosis of the bones, caused by fixation of the calcium by the fluorine. There may also be some calcification of the ligaments. The teeth are mottled, and there is osteosclerosis and osteomalacia. The bony and ligamentous changes are demonstrable by x-ray. The estimated human lethal dose is 2.5 to 5.9 g of F⁻. Large doses can cause very severe nausea, vomiting, diarrhea, abdominal burning, and cramp-like pains. It is not taken up by the thyroid and does not interfere with iodine uptake. Can cause or aggravate attacks of asthma and severe bone changes, making normal movements painful. Some signs of pulmonary fibrosis are noted. Some enzyme systems effects are reported. Irritants to the eyes, skin, and mucous membranes. Loss of weight, anorexia, anemia, wasting and cachexia, and dental defects are among the common findings in chronic fluorine poisoning. There may be an eosinophilia and impairment of growth in young workers. Symptoms of intoxication include gastric, intestinal, circulatory, respiratory and nervous complaints, and skin rashes. When heated to decomposition, or on contact with acid or acid fumes, they emit highly toxic fumes of F⁻.

Organic fluorides are generally less toxic than other halogenated hydrocarbons. Fluorocarbons are chemically inert to most materials but can react violently with barium, sodium, and potassium. Fluoroamides react violently with lithium tetrahydroaluminate and with sodium at very high temperatures. Some fluorinated cyclopropenyl methyl ethers react violently with water or methanol. Some fluorodinitro compounds of methane and ethane are sensitive explosives. When heated to decomposition they emit toxic fumes of F⁻. Common air contaminants.

FEZ000   CAS: 7782-41-4   HR: 3
FLUORINE
DOT: UN 1045
mf: F_2 mw: 38.00
PROP: Pale-yellow gas (turning white at −2°) which reacts with most organic and inorganic materials. Powerful oxidant. Mp: −218°, bp: −187°, d: 1.14 @ −200°, 1.108 @ −188°, vap d: 1.695. IDLH 25 ppm.
SYNS: BIFLUORIDEN (DUTCH) □ FLUOR (DUTCH, FRENCH, GERMAN, POLISH) □ FLUORINE, compressed (DOT) □ FLUORO (ITALIAN) □ FLUORURES ACIDE (FRENCH) □ FLUORURI ACIDI (ITALIAN) □ RCRA WASTE NUMBER P056 □ SAEURE FLUORIDE (GERMAN)

CONSENSUS REPORTS: EPA Extremely Hazardous Substances List. Reported in EPA TSCA Inventory.
OSHA PEL: TWA 0.1 ppm
ACGIH TLV: TWA 1 ppm; STEL 2 ppm
DFG MAK: 0.1 ppm (0.16 mg/m³)
DOT CLASSIFICATION: 2.3; Label: Poison Gas, Oxidizer
SAFETY PROFILE: A poison gas. A skin, eye, and mucous membrane irritant. A most powerful caustic irritant to tissue. Mutation data reported. A very dangerous fire and explosion hazard. A powerful oxidizer. Reacts violently with many materials.

Explosive or potentially explosive reaction with ammonia, cesium fluoride + fluorocarboxylic acids, cesium heptafluoropropoxide, 1- or 2-fluoriminoperfluoropropane, graphite, halocarbons (e.g., carbon tetrachloride, chloroform, perfluorocyclobutane, iodoform, 1,2-dichlorotetrafluoroethane), liquid hydrocarbons (e.g., anthracene, turpentine), hydrogen, hydrogen + oxygen, hydrogen fluoride + seleninyl fluoride + heat, nitric acid, silver cyanide, sulfur dioxide, carbon monoxide, sodium acetate, sodium bromate, stainless steel, water.

Reacts to form explosive products with alkanes + oxygen (forms peroxides), cyanoguanidine, perchloric acid (forms fluorine perchlorate gas), potassium chlorate (forms fluorine perchlorate gas), potassium hydroxide (forms potassium trioxide). Forms explosive mixtures with acetonitrile + chlorine fluoride, ice.

Ignition or violent reaction on contact with acetylene, ceramic materials, covalent halides (e.g., chromyl chloride, phosphorus pentachloride, phosphorus trichloride, phosphorus trifluoride, boron trichloride, silicon tetrachloride), halogens (e.g., bromine, iodine, chlorine + spark or heating to 100°C), dicyanogen, gaseous hydrocarbons (e.g., town gas, methane, benzene), hydrogen halide gases or concentrated solutions (e.g., hydrogen bromide, hydrogen chloride, hydrogen iodide, hydrogen fluoride), metal acetylides and carbides (e.g., monocesium acetylide, cesium acetylide, lithium acetylide, rubidium acetylide, tungsten carbide, ditungsten carbide, zirconium dicarbide, uranium dicarbide), metal cyano complexes [e.g., potassium hexacyanoferrate(II), lead hexacyanoferrate(III), potassium hexa-cyanoferrate(III)], metal hydrides (e.g., copper hydride, potassium hydride, sodium hydride), metal iodides (e.g., lead iodide, calcium iodide, mercury iodide, potassium iodide), metals, metal salts, metal silicides (e.g., calcium disilicide, lithium hexasilicide), nickel(IV) oxide, nonmetals (e.g., boron, yellow or red phosphorus, selenium, tellurium, silicon, carbon, charcoal, sulfur), oxygenated organic compounds (e.g., methanol, ethanol, 3-methyl butanol, acetaldehyde, trichloroacetaldehyde, acetone, lactic acid, benzoic acid, salicylic acid, ethyl acetate, methyl borate), nonmetal oxides (e.g., arsenic trioxide, nitrogen oxide, dinitrogen tetroxide), oxygen + polymers [e.g., phenol-formaldehyde resins (bakelite), polyacrylonitrile-butadiene (Buna N), polyamides (nylons), polychloropene (neoprene), polyethylene, polytrifluoropropylmethylsiloxane, polyvinylchloride-vinyl acetate (Tygon), polyvinylidene fluoride-hexafluoropropylene (Viton), polyurethane foam, polymethyl methacrylate (Perspex), polytetrafluooethylene (Teflon)], sulfides (e.g., antimony trisulfide, carbon disulfide vapor, chromium (II) sulfide, hydrogen sulfide, barium sulfide, potassium sulfide, zinc sulfide, molybdenum sulfide), xenon + catalysts (e.g., nickel fluoride, silver difluoride, nickel(III) oxide, silver (I) oxide).

Incandescent reaction with boron nitride, hexalithium disilicide + heat, metal borides, metal oxides (e.g., nickel(II) oxide, alkali metal oxides, alkaline earth oxides), nitrogenous bases (e.g., aniline, dimethylamine, pyridine), gallic acid.

Incompatible with cesium heptafluoro propoxide, cyanoguanidine, halocarbons, hexalithium disilicide, seleninyl fluoride, hydrogen sulfide, oxygen, sodium acetate, sodium bromate, sodium dicyanamides, most organic matter, H-containing molecules, oxides of S, N, P, alkali metals,and alkaline earths. It reacts violently with halogen acids, hydrazine, ClO_2, coke, cyanamide, cyanides, KNO_3, (PbO + glycerol), CCl_4, silicides, silicates, trinitromethane, alkenes, alkyl benzenes, CS_2, $Cr(OCl)_2$, Al, Tl, Sn, Sb, As, natural gas, liquid air, perfluoropropionyl fluoride, polyvinyl chloride acetate. Many reactions go on even at $<-160°$. Reacts with water or steam to produce heat and toxic and corrosive fumes. Used as one component of liquid rocket fuel and in chemical lasers. See also FLUORIDES.

FFF000   CAS: 640-19-7   HR: 3
FLUOROACETAMIDE
mf: C_2H_4FNO mw: 77.07
PROP: Needles from $CHCl_3$. Mp: 107–109°. Sol in H_2O and Me_2CO. Sltly sol in $CHCl_3$.
SYNS: AFL 1081 □ COMPOUND 1081 □ FAA □ FLUORAKIL 100 □ 2-FLUOROACETAMIDE □ FLUOROACETIC ACID AMIDE □ FUSSOL □ MEGATOX □ MONOFLUOROACETAMIDE □ NAVRON □ RCRA WASTE NUMBER P057 □ RODEX □ YANOCK
CONSENSUS REPORTS: EPA Extremely Hazardous Substances List. Reported in EPA TSCA Inventory.
SAFETY PROFILE: A human poison by an unspecified route. Poison experimentally by ingestion, skin contact, intraperitoneal, subcutaneous, and intravenous routes. Human systemic effects by unspecified route: convulsions, coma, nausea and vomiting. Experimental reproductive effects. Mutation data reported. Used as an insecticide and rodenticide. When heated to decomposition it emits very toxic fumes of F^- and NO_x. See also FLUORIDES.

FFR000   CAS: 359-06-8   HR: 3
FLUOROACETYL CHLORIDE
mf: C_2H_2ClFO mw: 96.49
PROP: Liquid. Rodenticide.

CONSENSUS REPORTS: EPA Extremely Hazardous Substances List. Reported in EPA TSCA Inventory.
SAFETY PROFILE: Poison by inhalation. When heated to decomposition it emits very toxic fumes of Cl$^-$ and F$^-$. See also FLUORIDES and CHLORIDES.

FFY000   CAS: 371-40-4   HR: 3
4-FLUOROANILINE
mf: C$_6$H$_6$FN   mw: 111.13
PROP: D: 1.1724, mp: −0.82°, bp: 184–186°.
SYNS: BENZENAMINE, 4-FLUORO-(9CI) □ 4-FLUO-RANILIN □ p-FLUOROANILINE □ 4-FLUOROBENZEN-AMINE □ p-FLUOROPHENYLAMINE
CONSENSUS REPORTS: Reported in EPA TSCA Inventory. EPA Genetic Toxicology Program.
SAFETY PROFILE: Poison by ingestion. Mutation data reported. A severe skin and eye irritant. When heated to decomposition it emits very toxic fumes of NO$_x$ and F$^-$.

FGA000   CAS: 462-06-6   HR: 3
FLUOROBENZENE
DOT: UN 2387
mf: C$_6$H$_5$F   mw: 96.11
PROP: Colorless liquid. D: 1.024, mp: −40°, bp: 85.2°, flash p: 5°F, d: 1.024, vap d: 3.31. Insol in water; misc in alc and ether.
SYN: PHENYL FLUORIDE
CONSENSUS REPORTS: Reported in EPA TSCA Inventory.
DOT CLASSIFICATION: 3; Label: Flammable Liquid
SAFETY PROFILE: Mildly toxic by ingestion and inhalation. A very dangerous fire hazard when exposed to heat, flame, or oxidizers. To fight fire, use water spray, mist, foam, dry chemical, CO$_2$. When heated to decomposition it emits toxic fumes of F$^-$.

FGA100   CAS: 5430-13-7   HR: 3
4-FLUOROBENZENEARSONIC ACID
mf: C$_6$H$_6$AsFO$_3$ mw: 220.04
PROP: A solid. Mp: 240° (decomp).
SYN: BENZENEARSONIC ACID, p-FLUORO-
OSHA PEL: TWA 0.5 mg(As)/m^3

SAFETY PROFILE: Poison by intravenous route. When heated to decomposition it emits toxic fumes of As and F$^-$.

FGA200   CAS: 200398-40-9 HR: 3
1-(2-(4-(6-FLUORO-1,2-BENZISOXAZOL-3-YL)-1-PIPERIDINYL)ETHYL)-3-PHENYL-2-IMIDAZOLIDINONE
mf: C$_{23}$H$_{25}$FN$_4$O$_2$ mw: 408.48
SYN: S18327
SAFETY PROFILE: A poison by ingestion, intravenous and subcutaneous routes. When heated to decomposition it emits toxic vapors of NO$_x$ and F$^-$.

FGI100   CAS: 59417-86-6   HR: 3
6-FLUOROBENZO(a)PYRENE
mf: C$_{20}$H$_{11}$F   mw: 270.31
SAFETY PROFILE: Suspected carcinogen with experimental carcinogenic and neoplastigenic data. When heated to decomposition it emits toxic fumes of F$^-$.

FHC200   CAS: 353-16-2   HR: 3
3-FLUOROBUTYL ISOCYANATE
DOT: UN 2206/UN 2207/UN 2478/UN 3080
mf: C$_5$H$_8$FNO   mw: 117.14
SYN: ISOCYANIC ACID, 4-FLUOROBUTYL ESTER
DOT CLASSIFICATION: 6.1; Label: KEEP AWAY FROM FOOD (UN2207); 6.1; Label: Poison (UN2206); 6.1; Label: Poison, Flammable Liquid (UN3080); 3; Label: Flammable Liquid, Poison (UN2478)
SAFETY PROFILE: A poison. A flammable liquid. When heated to decomposition it emits toxic vapors of NO$_x$ and CN$^-$.

FHG000   CAS: 1893-33-0   HR: 3
FLUOROBUTYROPHENONE
mf: C$_{21}$H$_{30}$FN$_3$O$_2$ mw: 375.54
SYNS: DIPIPERAL □ DIPIPERON □ DIPIPERONE □ FLOROPIPAMIDE □ 1'-(3-(p-FLUOROBENZOYL)PROP-YL)(1,4'-BIPIPERIDINE)-4-CARBOXAMIDE □ 1-(3-(p-FLUOROBENZOYL)PROPYL)-4-PIPERIDINOISO-NIPACOTAMIDE □ 4'-FLUORO-4-(4-N-PIPERIDINO-4-CARBAMIDOPIPERIDINO)BUTYROPHENONE □ p-FLUORO-γ-(4-PIPERIDINO-4-CARBAMOYLPIPERID-INO)BUTYROPHENONE □ FPA □ MCN-JR-3345 □

PIPAMPERONE □ PIPANEPERONE □ PIPERONYL □ PROPITAN □ R 3345

SAFETY PROFILE: Poison by ingestion, subcutaneous, and intravenous routes. An experimental teratogen. When heated to decomposition it emits very toxic fumes of F^- and NO_x.

FIB000   **CAS: 353-36-6**   **HR: 3**
FLUOROETHANE
DOT: UN 2453
mf: C_2H_5F mw: 48.06
PROP: A gas with ethereal odor. Mp: $-143.2°$, bp: $-37.7°$, d: 0.8158 @ $-37.7°$, vap d: 1.66. Sol in EtOH and Et_2O; sltly sol in H_2O.
SYNS: ETHYL FLUORIDE (DOT) □ MONOFLUORO-ETHANE □ R161
CONSENSUS REPORTS: Reported in EPA TSCA Inventory.
DOT CLASSIFICATION: 2.1; Label: Flammable Gas
SAFETY PROFILE: A poison by inhalation. A very dangerous fire hazard when exposed to heat, flames, or oxidizers. To fight fire, stop flow of gas. When heated to decomposition it emits toxic fumes of F^-.

FIC000   **CAS: 144-49-0**   **HR: 3**
FLUOROETHANOIC ACID
DOT: UN 2642
mf: $C_2H_3FO_2$ mw: 78.05
PROP: Colorless solid or crystals. Mp: 35.3°, d: 1.393 @ 36 mm, bp: 167–168°. Sol in water and alc.
SYNS: ACIDE-MONOFLUORACETIQUE (FRENCH) □ ACIDO MONOFLUOROACETIO (ITALIAN) □ CYMONIC ACID □ FAA □ FLUOROACETATE □ FLUOROACETIC ACID □ 2-FLUOROACETIC ACID □ FLUOROACETIC ACID (DOT) □ GIFBLAAR POISON □ HFA □ MFA □ MONOFLUORAZIJNZUUR (DUTCH) □ MONOFLUO-RESSIGSAEURE (GERMAN) □ MONOFLUOROACET-ATE □ MONOFLUOROACETIC ACID
CONSENSUS REPORTS: EPA Extremely Hazardous Substances List. Reported in EPA TSCA Inventory.
DOT CLASSIFICATION: 6.1; Label: Poison
SAFETY PROFILE: Poison by ingestion, subcutaneous, intraperitoneal, and

intravenous routes. Affects the human central nervous system, causing convulsions and ventricular fibrillation. When heated to decomposition it emits toxic fumes of F^- and Na_2O. See also SODIUM FLUOROACETATE.

FIE000   **CAS: 371-62-0**   **HR: 3**
2-FLUOROETHANOL
mf: C_2H_5FO mw: 64.07
PROP: A liquid. Fp: $-26.45°$, bp: 103.35° @ 757 mm. Sol in H_2O.
SYNS: β-FLUOROETHANOL □ TL 741
CONSENSUS REPORTS: EPA Extremely Hazardous Substances List. Reported in EPA TSCA Inventory.
SAFETY PROFILE: Poison by inhalation, intraperitoneal, subcutaneous, and intravenous routes. When heated to decomposition it emits very toxic fumes of F^-.

FIH100   **CAS: 462-27-1**   **HR: 3**
2-FLUOROETHYL CHLOROFORMATE
mf: $C_3H_4ClFO_2$ mw: 126.52
SYNS: CHLOROFORMIC ACID 2-FLUOROETHYL ESTER □ 2-FLUORETHYLESTER KYSELINY CHLORMRAVENCI □ FORMIC ACID, CHLORO-, 2-FLUOROETHYL ESTER □ TL 751
DOT CLASSIFICATION: 6.1; Label: Poison, Corrosive
SAFETY PROFILE: Poison by inhalation. A corrosive. When heated to decomposition it emits toxic vapors of F^- and Cl^-.

FJK000   **CAS: 593-53-3**   **HR: 3**
FLUOROMETHANE
mf: CH_3F mw: 34.03
PROP: Colorless gas; agreeable, ether-like odor. D: (liquid) 0.8774 @ $-78°$, (gas) 1.1951 (air = 1), (gas) 1.0813 (oxygen = 1), mp: $-141.8°$, bp: $-75.7°$ @ 872 mm, $-78.2°$ @ 760 mm. Mod sol in H_2O; sol in EtOH and Et_2O.
SYNS: FREON 41 □ METHYL FLUORIDE
CONSENSUS REPORTS: Reported in EPA TSCA Inventory.
OSHA PEL: TWA 2.5 mg(F)/m^3

ACGIH TLV: TWA 2.5 mg(F)/m³; BEI: 3 mg/g creatinine of fluorides in urine prior to shift; 10 mg/g creatinine of fluorides in urine at end of shift.
NIOSH REL: (Fluorides, Inorganic) TWA 2.5 mg(F)/m³
SAFETY PROFILE: Narcotic in high concentrations. Acts as a simple asphyxiant. Burns with evolution of hydrogen fluoride. The flame is about as colorless as that of alcohol. When heated to decomposition it emits toxic fumes of F⁻.

FKI000   CAS: 1622-79-3   HR: 3
4'-FLUORO-4-(4-METHYLPIPERIDINO)-BUTYROPHENONE HYDRO-CHLORIDE
mf: $C_{16}H_{22}FNO•ClH$ mw: 299.85
PROP: A solid. Mp: 209–211°.
SYNS: BURONIL □ EUNERPAN □ FG 5111 □ METHYLPERONE HYDROCHLORIDE □ γ-(4-METHYLPIPERIDINE)-p-FLUOROBUTYROPHENONE HYDROCHLORIDE
SAFETY PROFILE: Poison by ingestion, subcutaneous, and intravenous routes. Experimental teratogenic effects. A neuroleptic drug used to treat anxiety and confusion. When heated to decomposition it emits very toxic fumes of F⁻, NOₓ, and HCl.

FKK035   CAS: 326800-76-4 HR: 3
3-(6-FLUORO-2-NAPHTHALENYL)-2,5-DIHYDRO-1,2-DIMETHYL-1H-PYRR-OLE, EL-(2R,3R)-2,3-DIHYDROXY-BUTANEDIOATE (1:1)
mf: $C_{16}H_{16}FN•C_4H_6O_6$ mw: 391.40
SAFETY PROFILE: A poison by subcutaneous route. When heated to decomposition it emits toxic vapors of NOₓ and F⁻.

FKQ100   CAS: 399-24-6   HR: 3
9-FLUORONONYL PHENYL KETONE
mf: $C_{16}H_{23}FO$ mw: 250.39
SYN: DECANOPHENONE, 10-FLUORO-
DOT CLASSIFICATION: 3; Label: Flammable Liquid
SAFETY PROFILE: A poison by intraperitoneal route. A flammable liquid.

When heated to decomposition it emits toxic vapors of F⁻ and Cl⁻.

FKT050   CAS: 326-52-3   HR: 3
8-FLUOROOCTYL PHENYL KETONE
mf: $C_{15}H_{21}FO$ mw: 236.36
SYN: NONANOPHENONE, 9-FLUORO-
DOT CLASSIFICATION: 3; Label: Flammable Liquid
SAFETY PROFILE: A poison by intraperitoneal route. A flammable liquid. When heated to decomposition it emits toxic vapors of F⁻.

FLL100   CAS: 315706-75-3 HR: 3
3-(4-FLUOROPHENYL)-N-(4-PROPYL-CYCLOHEXYL)-2-PROPENAMIDE
mf: $C_{18}H_{24}FNO$ mw: 289.39
SAFETY PROFILE: A poison by ingestion. When heated to decomposition it emits toxic vapors of NOₓ and F⁻.

FLR100   CAS: 407-99-8   HR: 3
3-FLUOROPROPYL ISOCYANATE
DOT: UN 2206/UN 2207/UN 2478/UN 3080
mf: C_4H_6FNO mw: 103.11
SYN: ISOCYANIC ACID, 3-FLUOROPROPYL ESTER
DOT CLASSIFICATION: 6.1; Label: KEEP AWAY FROM FOOD (UN2207); 6.1; Label: Poison (UN2206); 6.1; Label: Poison, Flammable Liquid (UN3080); 3; Label: Flammable Liquid, Poison (UN2478)
SAFETY PROFILE: A poison by intraperitoneal route. A flammable liquid. When heated to decomposition it emits toxic vapors of NOₓ and CN⁻.

FLU000   CAS: 1649-18-9   HR: 3
4'-FLUORO-4-(4-(2-PYRIDYL)-1-PIP-ERAZINYL)BUTYROPHENONE
mf: $C_{19}H_{22}FN_3O$ mw: 327.44
PROP: A solid. Mp: 73–75°.
SYNS: AZAPERONE (USDA) □ AZEPERONE □ EUCALMYL □ FLUOPERIDOL □ 1-(3-(4-FLUORO-BENZOYL)PROPYL)-4-(2-PYRIDYL)PIPERAZINE □ 1-(4-FLUOROPHENYL)-4-(4-(2-PYRIDINYL)-1-PIPERAZINYL)-1-BUTANONE □ R 1929 □ STRESNIL □ SUICALM

SAFETY PROFILE: Poison by ingestion, intravenous, intraperitoneal, and subcutaneous routes. When heated to decomposition it emits very toxic fumes of F^- and NO_x.

FLZ000   CAS: 7789-21-1   HR: 3
FLUOROSULFURIC ACID

DOT: UN 1777
mf: FHO_3S mw: 100.07
PROP: Colorless, fumes in moist air; highly corrosive liquid. Mp: $-89°$, bp: $163°$, d: 1.726 @ $20°$. Sol in HOAc, $PhNO_2$, Et_2O; insol in CCl_4, CS_2.
SYNS: FLUOROSULFONIC ACID (DOT) □ FLUO-SULFONIC ACID (DOT)
CONSENSUS REPORTS: Reported in EPA TSCA Inventory.
OSHA PEL: TWA 2.5 mg(F)/m^3
ACGIH TLV: TWA 2.5 mg(F)/m^3; BEI: 3 mg/g creatinine of fluorides in urine prior to shift; 10 mg/g creatinine of fluorides in urine at end of shift.
NIOSH REL: (Inorganic Fluorides) TWA 2.5 mg(F)/m^3
DOT CLASSIFICATION: 8; Label: Corrosive
SAFETY PROFILE: Probably a poison by inhalation. A corrosive irritant to the skin, eyes, and mucous membranes. See also FLUORIDES, SULFURIC ACID, and FLUOROSULFONATES.

FLZ050   CAS: 17902-23-7   HR: 3
5-FLUORO-1-(TETRAHYDROFURAN-2-YL)URACIL

mf: $C_8H_9FN_2O_3$ mw: 200.19
SYNS: CARZONAL □ CITOFUR □ COPAROGIN □ EXONAL □ FENTAL □ F-5-FU □ FLUOROFUR □ 5-FLUORO-1-(TETRAHYDRO-2-FURANYL)-2,4-PYRIMI-DINEDIONE □ 5-FLUORO-1-(TETRAHYDRO-2-FURANYL)-2,4(1H,3H)-PYRIMIDINEDIONE □ 5-FLUORO-1-(TETRAHYDRO-3-FURYL)URACIL □ FRANROZE □ FTORAFUR □ FULAID □ FULFEEL □ FURAFLUOR □ FUROFUTRAN □ FUTRAFUL □ LAMAR □ LIFRIL □ MJF-12264 □ NEBERK □ NITOBANIL □ NSC-148958 □ PYRIMIDINE-DEOXYRIBOSE N1-2'-FURANIDYL-5-FLUOROURACIL □ RIOL □ SINOFLUR-OL □ SUNFRAL □ TEFSIEL C □ TEGAFUR □ 1-(TETRA-HYDROFURAN-2-YL)-5-FLUOROURACIL □ N^1-(2-TETRAHYDROFURYL)-5-FLUOROURACIL □ THFU

CONSENSUS REPORTS: Reported in EPA TSCA Inventory.
SAFETY PROFILE: Poison by ingestion. Moderately toxic to humans by intravenous route. Moderately toxic experimentally by intraperitoneal, intravenous, and subcutaneous routes. Experimental teratogenic data. Human systemic effects: nausea and vomiting. Experimental reproductive effects. Questionable human carcinogen producing gastrointestinal tumors. Human mutation data reported. Used as an anti-cancer agent. When heated to decomposition it emits very toxic fumes of F^- and NO_x.

FMC000   CAS: 352-32-9   HR: 3
p-FLUOROTOLUENE

mf: C_7H_7F mw: 110.14
PROP: Colorless liquid. D: 1.001, mp: $-56°$, bp: $116-117°$, flash p: $50°F$. Insol in water; sol in alc and ether.
CONSENSUS REPORTS: Reported in EPA TSCA Inventory.
SAFETY PROFILE: Moderately toxic by parenteral route. A very dangerous fire hazard when exposed to heat or flame; can react vigorously with oxidizing materials. When heated to decomposition it emits toxic fumes of F^-. See also FLUORIDES.

FME000   CAS: 1983-10-4   HR: 3
FLUOROTRIBUTYLSTANNANE

mf: $C_{12}H_{27}FSn$ mw: 309.08
PROP: White solid. Mp: $250°-257°$.
SYNS: TRIBUTYLSTANNANE FLUORIDE □ TRIBUTYLTIN FLUORIDE
CONSENSUS REPORTS: Reported in EPA TSCA Inventory.
OSHA PEL: TWA 0.1 mg(Sn)/m^3 (skin)
ACGIH TLV: TWA 0.1 mg(Sn)/m^3; STEL 0.2 mg(Sn)/m^3 (skin).
DFG MAK: 0.0021 ppm (0.05 mg/m^3)
NIOSH REL: (Organotin Compounds) TWA 0.1 mg(Sn)/m^3
SAFETY PROFILE: Poison by ingestion. Many tributyl tin compounds are highly toxic to marine life. Mutation data reported.

When heated to decomposition it emits toxic fumes of F⁻. See also TIN COMPOUNDS and FLUORIDES.

FMM000   CAS: 51-21-8   HR: 3
FLUOROURACIL
mf: $C_4H_3FN_2O_2$ mw: 130.09
PROP: Crystals from H_2O. Mp: 282–283° (decomp).
SYNS: ADRUCIL □ ARUMEL □ CARZONAL □ EFFLUDERM (free base) □ EFUDEX □ EFUDIX □ 5-FLUORACIL (GERMAN) □ FLUOROBLASTIN □ FLUOROPLEX □ 5-FLUORO-2,4-PYRIMIDINEDIONE □ 5-FLUORO-2,4(1H,3H)-PYRIMIDINEDIONE □ 5-FLUOROURACIL □ 5-FLUORPROPYRIMIDINE-2,4-DIONE □ 5-FLUORURACIL (GERMAN) □ FLURACIL □ FLURI □ FLURIL □ 5-FU □ NSC-19893 □ RO 2-9757 □ TIMAZIN □ U-8953 □ ULUP
CONSENSUS REPORTS: IARC Cancer Review: Group 3 IMEMDT 7,210,87; Human Inadequate Evidence IMEMDT 26,217,81; Animal Inadequate Evidence IMEMDT 26,217,81. Reported in EPA TSCA Inventory. EPA Genetic Toxicology Program. EPA Extremely Hazardous Substances List.
SAFETY PROFILE: Poison by ingestion, intraperitoneal, subcutaneous, and intravenous routes. Moderately toxic by parenteral and rectal routes. Experimental teratogenic and reproductive effects. Human systemic effects: EKG changes, bone marrow changes, cardiac, pulmonary, and gastrointestinal effects. Human mutation data reported. A human skin irritant. Questionable carcinogen. When heated to decomposition it emits very toxic fumes of F⁻ and NO_x.

FMQ000   CAS: 17617-23-1   HR: 3
FLURAZEPAM
mf: $C_{21}H_{23}ClFN_3O$ mw: 387.92
PROP: White rods from ether-pet ether. Mp: 77–82°.
SYNS: 7-CHLORO-1-(2-(DIETHYLAMINO)ETHYL)-5-(2-FLUOROPHENYL)-1H-1,4-BENZODIAZEPIN-2(3H)-ONE □ FELMANE □ NOCTOSOM □ Ro-5-6901/3 □ STAURODERM
SAFETY PROFILE: Poison by intravenous routes. Moderately toxic by ingestion and

subcutaneous routes. Experimental reproductive effects. Caution: May be habit-forming. This is a controlled substance (depressant) listed in the U.S. Code of Federal Regulations, Title 21 Part 1308.14. When heated to decomposition it emits very toxic fumes of Cl⁻, F⁻ and NO_x. See also DIAZEPAM.

FMT000   CAS: 59-30-3   HR: 3
FOLIC ACID
mf: $C_{19}H_{19}N_7O_6$ mw: 441.45
PROP: A member of the vitamin B complex. Odorless orange-yellow needles or platelets from H_2O at pH 3. Sol in dilute alkali hydroxide and carbonate solns; sltly sol in water; insol in lipid solvents, acetone, alc, chloroform, ether.
SYNS: l-N-(p-(((-2-AMINO-4-HYDROXY-6-PTERIDIN-YL)METHYL)AMINO)BENZOYL)GLUTAMIC ACID □ FOLACIN □ FOLATE □ FOLCYSTEINE □ NSC-3073 □ PTEGLU □ PTEROYLGLUTAMIC ACID □ PTEROYL-l-GLUTAMIC ACID □ PTEROYLMONOGLUTAMIC ACID □ PTEROYL-l-MONOGLUTAMIC ACID □ USAF CB-13 □ VITAMIN Bc □ VITAMIN M
CONSENSUS REPORTS: Reported in EPA TSCA Inventory.
SAFETY PROFILE: Poison by intraperitoneal and intravenous routes. Experimental teratogenic effects. Mutation data reported. When heated to decomposition it emits toxic fumes of NO_x.

FMU045   CAS: 944-22-9   HR: 3
FONOFOS
mf: $C_{10}H_{15}OPS_2$ mw: 246.34
PROP: Insecticde.
SYNS: O-AETHYL-S-PHENYL-AETHYL-DITHIO-PHOSPHONAT (GERMAN) □ DIFONATE □ DYFONATE □ DYPHONATE □ ENT 25,796 □ O-ETHYL-S-PHENYL ETHYLDITHIOPHOSPHONATE □ O-ETHYL-S-PHENYL ETHYLPHOSPHONODITHIOATE □ N 2790 □ STAUFFER N 2790
CONSENSUS REPORTS: EPA Genetic Toxicology Program. EPA Extremely Hazardous Substances List.
OSHA PEL: TWA 0.1 mg/m³ (skin)
ACGIH TLV: TWA 0.1 mg/m³ (skin); Not Classifiable as a Human Carcinogen

SAFETY PROFILE: Poison by ingestion and skin contact. An insecticide. When heated to decomposition it emits very toxic fumes of PO_x and SO_x.

FMU059   CAS: 2650-18-2   HR: 3
FOOD BLUE 1

mf: $C_{37}H_{36}N_2O_9S_3 \cdot 2H_3N$ mw: 783.01

PROP: Dark blue crystals or powder. Sol in H_2O.

SYNS: ACID BLUE 9 □ ACILAN TURQUOISE BLUE AE □ A.F. BLUE No. 1 □ AIZEN BRILLIANT BLUE FCF □ ALPHAZURINE □ AMACID BLUE FG CONC □ BLEU BRILLIANT FCF □ 11388 BLUE □ BRILLIANT BLUE □ BUCACID AZURE BLUE □ CALCOCID BLUE EG □ C.I. 671 □ C.I. 42090 □ C.I. ACID BLUE 9, DIAMMONIUM SALT □ C.I. DIRECT BROWN 78, DIAMMONIUM SALT □ C.I. FOOD BLUE 2 □ D&C BLUE No. 4 □ DISULPHINE LAKE BLUE EG □ EDICOL SUPRA BLUE E6 □ ERIOGLAUCINE □ ERIOSKY BLUE □ FENAZO BLUE XR □ HIDACID AZURE BLUE □ H.K. FORMULA No. K. 7117 □ KITON PURE BLUE L □ MAPLE BRILLIANT BLUE FCF □ NEPTUNE BLUE BRA CONCENTRATION □ PATENT BLUE AE □ PEACOCK BLUE X-1756 □ SCHULTZ No. 770 □ TRIANTINE LIGHT BROWN 3RN □ XYLENE BLUE VSG

CONSENSUS REPORTS: Community Right-To-Know List. Reported in EPA TSCA Inventory. EPA Genetic Toxicology Program.

SAFETY PROFILE: Human poison by intravenous route. Human systemic effects by intravenous route: muscle contractions or spasticity and dyspnea. Questionable carcinogen with experimental neoplastigenic and tumorigenic data. Mutation data reported. When heated to decomposition it emits very toxic fumes of NH_3, NO_x, and SO_x.

FMU070   CAS: 3761-53-3   HR: 2
FOOD RED No. 101

mf: $C_{18}H_{14}N_2O_7S_2 \cdot 2Na$ mw: 480.44

PROP: Dark red crystals or powder. Sol in H_2O; sltly sol in EtOH and Me_2CO; insol in Et_2O and C_6H_6.

SYNS: ACETACID RED J □ ACIDAL PONCEAU G □ ACID LEATHER RED KPR □ ACID PONCEAU R □ ACID RED 26 □ ACID SCARLET □ ACILAN PONCEAU RRL □ AHCOCID FAST SCARLET R □ AIZEN PONCEAU RH □ AMACID LAKE SCARLET 2R □ BRILLIANT PONCEAU G

□ CALCOCID 2RIL □ CALCOLAKE SCARLET 2R □ CERTICOL PONCEAU MXS □ CERVEN KYSELA 26 □ C.I. 79 □ C.I. 16150 □ C.I. ACID RED 26 □ C.I. ACID RED 26, DISODIUM SALT □ C.I. FOOD RED 5 □ COLACID PONCEAU SPECIAL □ D&C RED No. 5 □ 4-((2,4-DIMETHYLPHENYL)AZO)-3-HYDROXY-2,7-NAPHTHALENEDISULFONIC ACID, DISODIUM SALT □ 4-((2,4-DIMETHYLPHENYL)AZO)-3-HYDROXY-2,7-NAPHTHALENEDISULPHONIC ACID, DISODIUM SALT □ DISODIUM (2,4-DIMETHYLPHENYLAZO)-2-HYDRO-XYNAPHTHALENE-3,6-DISULFONATE □ DISODIUM (2,4-DIMETHYLPHENYLAZO)-2-HYDROXYNAPHTHAL-ENE-3,6-DISULPHONATE □ DISODIUM SALT of 1-(2,4-XYLYLAZO)-2-NAPHTHOL-3,6-DISULFONIC ACID □ DISODIUM SALT of 1-(2,4-XYLYLAZO)-2-NAPHTHOL-3,6-DISULPHONIC ACID □ EDICOL PONCEAU RS □ FENAZO SCARLET 2R □ HEXACOL PONCEAU MX □ HIDACID SCARLET 2R □ 3-HYDROXY-4-(2,4-XYLYL-AZO)-3,7-NAPHTHALENEDISULFONIC ACID, DISOD-IUM SALT □ 3-HYDROXY-4-(2,4-XYLYLAZO)-3,7-NAPHTHALENEDISULPHONIC ACID, DISODIUM SALT □ JAVA PONCEAU 2R □ KITON PONCEAU R □ LAKE PONCEAU □ NAPHTHALENE LAKE SCARLET R □ NEKLACID RED RR □ NEW PONCEAU 4R □ PAPER RED HRR □ PIGMENT PONCEAU R □ PONCEAU BNA □ PONCEAU R (BIOLOGICAL STAIN) □ PONCEAU XYLIDINE (BIOLOGICAL STAIN) □ 1695 RED □ RED R □ SCARLET R □ SCHULTZ No. 95 □ TERTRACID PONCEAU 2R □ XYLIDINE PONCEAU □ 1-XYLYLAZO-2-NAPHTHOL-3,6-DISULFONIC ACID, DISODIUM SALT □ 1-XYLYLAZO-2-NAPHTHOL-3,6-DISULPHONIC ACID, DISODIUM SALT □ 1-(2,4-XYLYLAZO)-2-NAPHTHOL-3,6-DISULPHONIC ACID, DISODIUM SALT

CONSENSUS REPORTS: IARC Cancer Review: Group 3 IMEMDT 7,56,87; Animal Sufficient Evidence IMEMDT 8,189,75. Reported in EPA TSCA Inventory.

SAFETY PROFILE: Moderately toxic by intraperitoneal route. Questionable carcinogen with experimental carcinogenic and tumorigenic data. Mutation data reported. When heated to decomposition it emits toxic fumes of NO_x and SO_x.

FMU080   CAS: 2611-82-7   HR: 2
FOOD RED No. 102

mf: $C_{20}H_{14}N_2O_{10}S_3 \cdot 3Na$ mw: 607.51

SYNS: ACIDAL BRIGHT PONCEAU 3R □ ACID BRILLIANT SCARLET 3R □ ACID PONCEAU 4R □ ACID RED 18 □ ACID SCARLET 3R □ ACILAN SCARLET V3R □ AIZEN BRILLIANT SCARLET 3RH □ ATUL ACID SCARLET 3R □ BRILLIANT PONCEAU 3R □ BRILLIANT SCARLET □ BUCACID BRILLIANT SCARLET 3R □ CALCOCID BRILLIANT SCARLET 3RN □ CERTICOL PONCEAU 4RS □ CERVEN KOSENILOVA A □ CILEFA

PONCEAU 4R □ COCCINE □ COCHENILLEROT A □ COCHINEAL RED A □ COLACID PONCEAU 4R □ C.I. 185 □ C.I. 16255 □ C.I. ACID RED 18 □ C.I. FOOD RED 7 □ CRIMSON SX □ CUROL BRIGHT RED 4R □ DAISHIKI BRILLIANT SCARLET 3R □ EDICOL SUPRA PONCEAU 4R □ EUROCERT COCHINEAL RED A □ FENAZO SCARLET 3R □ FOOD RED 6 □ FOOD RED 7 □ HD PONCEAU 4R □ HEXACOL PONCEAU 4R □ HIDACID FAST SCARLET 3R □ HISPACID BRILLIANT SCARLET 3RF □ JAVA SCARLET 3R □ KAYAKU ACID BRILLIANT SCARLET 3R □ KITON SCARLET 4R □ KOCHINEAL RED A FOR FOOD □ 1,3-NAPHTHALENEDISULFONIC ACID, 7-HYDROXY-8-((4-SULFO-1-NAPHTHYL)AZO)-, TRISODIUM SALT □ NAPHTHALENE INK SCARLET 4R □ NEKLACID RED 3R □ NEUCOCCIN □ NEW COCCIN □ PONCEAU 4R □ PONCEAU 4R ALUMINUM LAKE □ PONTACYL SCARLET RR □ PURPLE RED □ ROUGE de COCHENILLE A □ SAN-EI BRILLIANT SCARLET 3R □ STRAWBERRY RED A GEIGY □ SYMULON ACID BRILLIANT SCARLET 3R □ TAKAOKA BRILLIANT SCARLET 3R □ VICTORIA SCARLET 3R

CONSENSUS REPORTS: Reported in EPA TSCA Inventory.

SAFETY PROFILE: Moderately toxic by intraperitoneal route. Questionable carcinogen with experimental tumorigenic data. Mutation data reported. When heated to decomposition it emits toxic fumes of NO_x and SO_x.

FMU100   **HR: D**
FOOD STARCH, MODIFIED
PROP: White powders; tasteless and odorless. Insol in water, alc, ether, chloroform.
SAFETY PROFILE: When heated to decomposition it emits acrid smoke and irritating fumes.

FMV000   CAS: 50-00-0   HR: 3
FORMALDEHYDE
DOT: UN 1198/UN 2209
mf: CH_2O mw: 30.03
PROP: Clear, water-white, very sltly acid gas or liquid; pungent odor. Pure formaldehyde is not available commercially because of its tendency to polymerize. It is sold as aqueous solns containing from 37 to 50% formaldehyde by weight and varying amounts of methanol. Some alcoholic solns are used industrially, and the physical properties and hazards may be greatly influenced by the solvent. Lel: 7.0%, uel: 73.0%, autoign temp: 806°F, mp: −92°, d: 1.083, bp: −21°, flash p: (37%, methanol-free) 185°F, flash p: (15%, methanol-free) 122°F. Sol in H_2O and most org solvs except pet ether. IDLH 20 ppm.

SYNS: ALDEHYDE FORMIQUE (FRENCH) □ ALDEIDE FORMICA (ITALIAN) □ BFV □ FA □ FANNOFORM □ FORMALDEHYD (CZECH, POLISH) □ FORMALDEHYDE, solution (DOT) □ FORMALIN □ FORMALIN 40 □ FORMALIN (DOT) □ FORMALINA (ITALIAN) □ FORMALINE (GERMAN) □ FORMALIN-LOESUNGEN (GERMAN) □ FORMALITH □ FORMIC ALDEHYDE □ FORMOL □ FYDE □ HOCH □ IVALON □ KARSAN □ LYSOFORM □ METHANAL □ METHYL ALDEHYDE □ METHYLENE GLYCOL □ METHYLENE OXIDE □ MORBOCID □ NCI-C02799 □ OPLOSSINGEN (DUTCH) □ OXOMETHANE □ OXYMETHYLENE □ PARAFORM □ POLYOXYMETHYLENE GLYCOLS □ RCRA WASTE NUMBER U122 □ SUPERLYSOFORM

CONSENSUS REPORTS: NTP 10th Report on Carcinogens. IARC Cancer Review: Group 2A IMEMDT 7,211,87; Human Inadequate Evidence IMEMDT 29,345,82; Animal Sufficient Evidence IMEMDT 29,345,82. EPA Genetic Toxicology Program. Reported in EPA TSCA Inventory. EPA Extremely Hazardous Substances List.

OSHA PEL: TWA 0.75 ppm; STEL 2 ppm
ACGIH TLV: CL 0.3 ppm (sensitizer); Suspected Human Carcinogen
DFG MAK: 0.5 ppm (0.6 mg/m³); Confirmed Animal Carcinogen with Unknown Relevance to Humans
NIOSH REL: (Formaldehyde) Limit to lowest feasible level
DOT CLASSIFICATION: 9; Label: None (UN 2209); DOT Class: 3; Label: Flammable Liquid (UN 1198)
SAFETY PROFILE: Confirmed carcinogen with experimental carcinogenic, tumorigenic, and teratogenic data. Human poison by ingestion. Experimental poison by ingestion, skin contact, inhalation, intravenous, intraperitoneal, and subcutaneous routes. Human systemic effects by inhalation: lachrymation, olfactory changes, aggression, and pulmonary

changes. Experimental reproductive effects. Human mutation data reported. A human skin and eye irritant. If swallowed it causes violent vomiting and diarrhea that can lead to collapse. Frequent or prolonged exposure can cause hypersensitivity leading to contact dermatitis, possibly of an eczematoid nature. An air concentration of 20 ppm is quickly irritating to eyes. A common air contaminant.

Flammable liquid when exposed to heat or flame; can react vigorously with oxidizers. A moderate explosion hazard when exposed to heat or flame. The gas is a more dangerous fire hazard than the vapor. Should formaldehyde be involved in a fire, irritating gaseous formaldehyde may be evolved. When aqueous formaldehyde solutions are heated above their flash points, a potential for an explosion hazard exists. High formaldehyde concentration or methanol content lowers the flash point. Reacts with sodium hydroxide to yield formic acid and hydrogen. Reacts with NO_x at about 180°; the reaction becomes explosive. Also reacts violently with perchloric acid + aniline, performic acid, nitromethane, magnesium carbonate, H_2O_2. Moderately dangerous because of irritating vapor that may exist in toxic concentrations locally if storage tank is ruptured. To fight fire, stop flow of gas (for pure form); alcohol foam for 37% methanol-free form. When heated to decomposition it emits acrid smoke and fumes. See also ALDEHYDES.

FMY000　　CAS: 75-12-7　　HR: 3
FORMAMIDE
mf: CH_3NO　　mw: 45.05
PROP: Colorless, odorless, hygroscopic, sltly viscous, oily liquid. Mp: 2.5°, fp: 2.6°, vap press: 29.7 mm @ 129.4°, flash p: 310°F (COC), bp: 70.5° @ 1 mm, d: 1.134 @ 20°/40°, 1.1292 @ 25°/4°. Misc in H_2O, MeOH; very sltly sol in Et_2O, C_6H_6; insol in $CHCl_3$, hexane.
SYNS: CARBAMALDEHYDE □ METHANAMIDE

CONSENSUS REPORTS: Reported in EPA TSCA Inventory. EPA Genetic Toxicology Program.
OSHA PEL: TWA 20 ppm; STEL 30 ppm
ACGIH TLV: TWA 10 ppm (skin)
SAFETY PROFILE: Poison by skin contact and subcutaneous routes. Moderately toxic by ingestion, intraperitoneal, and intramuscular routes. An irritant to skin, eyes, and mucous membranes. Experimental teratogenic and reproductive effects. An eye irritant. Mutation data reported. Combustible when exposed to heat or flame; can react vigorously with oxidizing materials. Incompatible with I_2, pyridine, SO_3. When heated to decomposition it emits toxic fumes of NO_x. Has exploded while in storage.

FNA000　　CAS: 64-18-6　　HR: 3
FORMIC ACID
DOT: UN 1779
mf: CH_2O_2　　mw: 46.03
PROP: Colorless, fuming liquid; pungent, penetrating odor. Bp: 100.8°, fp: 8.2°, flash p: 156°F (OC), d: 1.220 @ 20°/4°, 1.220 @ 20°/4°, mp: 8.4°, autoign temp: 1114°F, vap press: 40 mm @ 24.0°, vap d: 1.59, flash p: (90% soln) 122°F, autoign temp: (90% soln) 813°F, lel: (90% soln) 18%, uel: (90% soln) 57%. Misc in H_2O, EtOH, Et_2O; mod sol in C_6H_6. IDLH 30 ppm.
SYNS: ACIDE FORMIQUE (FRENCH) □ ACIDO FORMICO (ITALIAN) □ AMEISENSAEURE (GERMAN) □ AMINIC ACID □ FORMYLIC ACID □ HYDROGEN CARBOXYLIC ACID □ KWAS METANIOWY (POLISH) □ METHANOIC ACID □ MIERENZUUR (DUTCH) □ RCRA WASTE NUMBER U123
CONSENSUS REPORTS: Reported in EPA TSCA Inventory.
OSHA PEL: TWA 5 ppm
ACGIH TLV: 5 ppm; STEL 10 ppm
DFG MAK: 5 ppm (9.5 mg/m³)
DOT CLASSIFICATION: 8; Label: Corrosive
SAFETY PROFILE: Poison by inhalation, intravenous, and intraperitoneal routes. Moderately toxic by ingestion. Mutation data reported. Corrosive. A skin and severe eye irritant. A substance migrating to food from

packaging materials. Combustible liquid when exposed to heat or flame; can react vigorously with oxidizing materials. Explosive reaction with furfuryl alcohol, H_2O_2, $Tl(NO_3)_3 \cdot 3H_2O$, nitromethane, P_2O_5. To fight fire, use CO_2, dry chemical, alcohol foam. When heated to decomposition it emits acrid smoke and irritating fumes.

FNK025   CAS: 100-50-5   HR: 3
4-FORMYLCYCLOHEXENE
DOT: UN 2498
mf: $C_7H_{10}O$ mw: 110.17
PROP: Oil. Mp: $-96.1°$, bp: $163.5-164.5°$.
SYNS: 3-CYCLOHEXENE-1-CARBOXALDEHYDE □ 1,2,3,6-TETRAHYDROBENZALDEHYDE (DOT) □ 1,2,5,6-TETRAHYDROBENZALDEHYDE
CONSENSUS REPORTS: Reported in EPA TSCA Inventory.
DOT CLASSIFICATION: 3; Label: Flammable Liquid
SAFETY PROFILE: Moderately toxic by ingestion, inhalation, and skin contact. Corrosive. An eye, skin, and mucous membrane irritant. Flammable liquid. When heated to decomposition it emits acrid smoke and irritating fumes. See also ALDEHYDES.

FNW000   CAS: 758-17-8   HR: 3
N-FORMYL-N-METHYLHYDRAZINE
mf: $C_2H_6N_2O$ mw: 74.10
SYNS: FORMIC ACID, METHYLHYDRAZIDE □ 1-FORMYL-1-METHYLHYDRAZINE □ N-METHYL-N-FORMYLHYDRAZINE □ MFH
SAFETY PROFILE: Suspected carcinogen with experimental carcinogenic data. Poison by ingestion route. Mutation data reported. When heated to decomposition it emits toxic fumes of NO_x. See also HYDRAZINE.

FOI000   CAS: 6804-07-5   HR: 2
2-FORMYLQUINOXALINE-1,4-DIOXIDE CARBOMETHOXYHYDRAZONE
mf: $C_{11}H_{10}N_4O_4$ mw: 262.25
PROP: Minute yellow crystals. Mp: $239.5-240°$. Insol in H_2O.

SYNS: CARBADOX (USDA) □ FORTIGRO □ GS 6244 □ MECADOX □ (2-QUINOXALINYLMETHYLENE)-HYDRAZINECARBOXYLIC ACID METHYL ESTER-N,N'-DIOXIDE
SAFETY PROFILE: Moderately toxic by ingestion. Questionable carcinogen with experimental carcinogenic effects. Human mutation data reported. When heated to decomposition it emits toxic fumes of NO_x.

FOO000   CAS: 76-13-1   HR: 2
FREON 113
mf: $C_2Cl_3F_3$ mw: 187.37
PROP: Colorless gas. Mp: $-36.4°$, bp: $45.8°$, d: 1.5702, autoign temp: $1256°F$. IDLH 2000 ppm.
SYNS: ARCTON 63 □ ARKLONE P □ DAIFLON S 3 □ FLUOROCARBON 113 □ FREON 113TR-T □ FRIGEN 113a □ GENETRON 113 □ HALOCARBON 113 □ ISCEON 113 □ KAISER CHEMICALS 11 □ KHLADON 113 □ R 113 □ REFRIGERANT 113 □ TRICHLOROTRIFLU-OROETHANE □ 1,1,2-TRICHLORO-1,2,2-TRIFLUORO-ETHANE (OSHA, ACGIH, MAK) □ UCON 113 □ UCON FLUOROCARBON 113 □ UCON 113/HALOCARBON 113
CONSENSUS REPORTS: Reported in EPA TSCA Inventory.
OSHA PEL: TWA 1000 ppm; STEL 1250 ppm
ACGIH TLV: TWA 1000 ppm; STEL 1250 ppm; Not Classifiable as a Human Carcinogen
DFG MAK: 500 ppm (3900 mg/m³)
SAFETY PROFILE: Mildly toxic by ingestion and inhalation. Affects the central nervous system in humans. A skin irritant. Combustible when exposed to heat or flame. Incompatible with Al, Ba, Li, Sm, NaK alloy, Ti. See also CHLORINATED HYDROCARBONS, ALIPHATIC; and FLUORIDES.

FOO509   CAS: 76-14-2   HR: 1
FREON 114
mf: $C_2Cl_2F_4$ mw: 170.92
PROP: Colorless, practically odorless, noncorrosive, nonirritating, nonflammable gas. Faint, ether-like odor in high concentrations. D: 1.5312, mp: $-94°$, bp: $4.1°$, n: (0/D) 1.3092. Insol in water; sol in alc and ether. IDLH 15,000 ppm.

SYNS: ARCTON 33 □ ARCTON 114 □ CRYOFLUORAN □ CRYOFLUORANE □ sym-DICHLOROTETRAFLUORO-ETHANE □ 1,2-DICHLORO-1,1,2,2-TETRAFLUOROETH-ANE (MAK) □ DICHLOROTETRAFLUOROETHANE (OSHA, ACGIH) □ F 114 □ FC 114 □ FLUORANE 114 □ FLUOROCARBON 114 □ FRIGEN 114 □ FRIGIDERM □ GENETRON 114 □ GENETRON 316 □ HALOCARBON 114 □ LEDON 114 □ PROPELLANT 114 □ R 114 □ 1,1,2,2-TETRAFLUORO-1,2-DICHLOROETHANE □ UCON 114

CONSENSUS REPORTS: Reported in EPA TSCA Inventory.

OSHA PEL: TWA 1000 ppm

ACGIH TLV: TWA 1000 ppm; Not Classifiable as a Human Carcinogen

DFG MAK: 1000 ppm (7100 mg/m³)

SAFETY PROFILE: An asphyxiant. See also DICHLOROTETRAFLUOROETHANE.

FOO560   CAS: 39432-81-0   HR: 1
FREON 502

DOT: UN 1973

mf: $C_2ClF_5 \cdot CHClF_2$ mw: 240.94

SYNS: CHLORODIFLUOROMETHANE and CHLOROPENTAFLUOROETHANE MIXTURE (DOT) □ ETHANE, CHLOROPENTAFLUORO-, mixt. with CHLORODIFLUOROMETHANE □ R502 (DOT) □ REFRIGERANT 502

DOT CLASSIFICATION: 2.2; Label: Nonflammable Gas

SAFETY PROFILE: A simple asphyxiant. When heated to decomposition it emits toxic vapors of NO_x and Cl^-.

FOO562   CAS: 50815-73-1   HR: 1
FREON 503

DOT: UN 2599

SYNS: CHLOROTRIFLUOROMETHANE mixed with TRIFLUOROMETHANE □ CHLOROTRIFLUORO-METHANE and TRIFLUOROMETHANE AZEOTROPIC MIXTURE (DOT) □ METHANE, CHLOROTRIFLUORO-, mixt. with TRIFLUOROMETHANE (9CI) □ METHANE, TRIFLUORO-, mixt. with CHLOROTRIFLUOROMETHANE □ R503 (DOT)

DOT CLASSIFICATION: 2.2; Label: Nonflammable Gas

SAFETY PROFILE: A simple asphyxiant. When heated to decomposition it emits toxic vapors of NO_x and Cl^-.

FOP000   HR: 3
FUEL OIL

DOT: NA 1993

PROP: A petroleum fraction consisting of a complex mixture of aromatic, paraffinic, olefinic, and naphthenic hydrocarbons. Brown, sltly viscous liquid. Flash p: 100°F, d: <1, autoign temp: 494°F.

SYNS: AUTOMOTIVE DIESEL OIL □ DIESEL FUEL (DOT) □ DIESEL OIL (PETROLEUM) □ DIESEL OILS □ DIESEL TEST FUEL □ FUELS, DIESEL □ OLEJ NAPEDOWY III

CONSENSUS REPORTS: IARC Cancer Review: Group 3 IMEMDT 45,219,89; Human Inadequate Evidence IMEMDT 45,219,89.

DOT CLASSIFICATION: 3; Label: None

SAFETY PROFILE: Mildly toxic by ingestion. A moderate skin irritant. Questionable carcinogen. Flammable when exposed to heat or flame; can react vigorously with oxidizing materials. To fight fire, use CO_2, dry chemical. When heated to decomposition it emits acrid smoke and irritating fumes. See also DIESEL EXHAUST, DIESEL EXHAUST EXTRACT, DIESEL EXHAUST PARTICLES, DIESEL FUEL MARINE.

FOP200   CAS: 68476-33-5   HR: 3
FUEL OIL, RESIDUAL

SYN: RESIDUAL(HEAVY) FUEL OIL

CONSENSUS REPORTS: IARC Cancer Review: Group 2B IMEMDT 45,239,89; Animal Sufficient Evidence IMEMDT 45,239,89. Reported in EPA TSCA Inventory.

SAFETY PROFILE: Confirmed carcinogen. When heated to decomposition it emits acrid smoke and irritating vapors.

FOQ000   CAS: 4368-28-9   HR: 3
FUGU POISON

mf: $C_{11}H_{17}N_3O_8$ mw: 319.31

PROP: Crystals.

SYNS: MACULOTOXIN □ SPHEROIDINE □ TARICHATOXIN □ TETRODONTOXIN □ TETRODOTOXIN □ TETRODOXIN □ TTX

SAFETY PROFILE: Poison by ingestion, intraperitoneal, subcutaneous, and intravenous routes. When heated to decomposition it emits toxic fumes of NO$_x$.

FOS000   HR: 3
FULMINATES

SAFETY PROFILE: Variable toxicity. A very dangerous fire hazard when exposed to heat or flame. Severe explosion hazard when shocked or exposed to heat or flame. See also various fulminates and EXPLOSIVES, HIGH.

The fulminates are a group of explosives that are very sensitive to heat, impact, and friction when dry. They should be kept moist until ready for use. If compressed beyond 25,000 psi they become what is known as "deadpressed," i.e., not capable of being exploded by flame. Fulminates are subject to deterioration when stored in hot climates. They decompose completely and violently when detonated. They can be ignited with a flame or "spit," with a fuse, or with an electrically heated wire. They are widely used as initiators or primers for detonation of high explosives or the ignition of powder. They are commonly used in combination with substances that provide a more prolonged blow and a bigger flame than fulminates alone. In the reinforced type of detonator, fulminates are made more effective by the addition of a more sensitive and powerful high explosive such as tetryl. This material is generally used in the manufacture of caps and detonators for initiating explosions for military, industrial, and sporting purposes.

All precautions required for protection of magazines apply to storage of these materials. They should not be handled when frozen. Wet fulminate of mercury or wet floor coverings containing small quantities of fulminates may be burned on windrows of flammable material. Nonexplosive products are formed by neutralizing fulminates with cold sodium thiosulfate. All floors, tables, and walls where the dry fulminates have been used should be washed with this solution. In the manufacture of mercury fulminate, the fumes given off are toxic and flammable. Care is required to prevent fulminate dust from being carried off in the exhaust system: deposits thus made have caused explosions. Careful attention should be given to cleanliness as foreign or gritty materials in the product may cause an unexpected explosion. The floors on which fulminates are used should be covered with 1/16-inch cloth-inserted rubber packing or its equal. All cracks and crevices should be covered. The walls of these rooms should be covered with glazed, waterproof material. Frequent washing with neutralizing solution is necessary. In manufacture, the fulminate is dried on muslin squares on a drying table. Drying tables may be heated with hot water or the dry house may be heated with an air blower system to between 50 and 60°. Primer caps and detonators loaded with fulminate of mercury are less sensitive than the dry bulk material but must be handled with great care. Fires involving these assemblies should be treated the same as for the bulk material. They will explode as soon as fire reaches them. Stocks in an assembly or loading room should be kept as small as possible. Examples of fulminates commonly used in the explosive industry are mercury fulminate, copper fulminate, and silver fulminate.

FOS050   CAS: 506-85-4   HR: 3
FULMINIC ACID

mf: CHNO   mw: 43.02

PROP: Oily liquid with prussic acid odor.

CONSENSUS REPORTS: Cyanide and its compounds are on the Community Right-To-Know List.

DOT CLASSIFICATION: Forbidden

SAFETY PROFILE: An unstable explosive sensitive to heat, shock, or friction. When heated to decomposition it emits toxic fumes of NO$_x$.

FOT000 CAS: 6029-87-4 HR: 3
FULVINE

mf: $C_{16}H_{23}NO_5$ mw: 309.40

PROP: Prisms from Me_2CO. Mp: 212–213°.

SYN: CRISPATINE

SAFETY PROFILE: Poison by intraperitoneal route. Experimental teratogenic and reproductive effects. Mutation data reported. When heated to decomposition it emits toxic fumes of NO_x.

FOU000 CAS: 110-17-8 HR: 3
FUMARIC ACID

mf: $C_4H_4O_4$ mw: 116.08

PROP: White, monoclinic, prismatic, crystals, needles, or leaflets; odorless. Mp: 300–302° (sealed tube), d: 1.635 @ 20°/4°, bp: 290°. Sol in EtOH; sltly sol in H_2O, Et_2O, and Me_2CO; prac insol in C_6H_6.

SYNS: ALLOMALEIC ACID □ BOLETIC ACID □ trans-BUTENEDIOIC ACID □ (E)-BUTENEDIOIC ACID □ trans-1,2-ETHYLENEDICARBOXYLIC ACID □ (E)1,2-ETHYLENEDICARBOXYLIC ACID □ KYSELINA FUMAROVA (CZECH) □ LICHENIC ACID □ NSC-2752 □ U-1149 □ USAF EK-P-583

CONSENSUS REPORTS: Reported in EPA TSCA Inventory.

SAFETY PROFILE: Poison by intraperitoneal route. Mildly toxic by ingestion and skin contact. A skin and eye irritant. Mutation data reported. Combustible when exposed to heat or flame; can react vigorously with oxidizing materials. When heated to decomposition it emits acrid smoke and irritating fumes.

FOY000 CAS: 627-63-4 HR: 2
FUMARYL CHLORIDE

DOT: UN 1780

mf: $C_4H_2Cl_2O_2$ mw: 152.96

PROP: Clear, straw-colored liquid. Mp: 158–160°, d: 1.408 @ 20°/4°.

SYNS: CHLORURE de FUMARYLE (FRENCH) □ DICHLORID KYSELINY FUMAROVE (CZECH) □ FUMAROYL CHLORIDE □ FUMARYLCHLORID (CZECH) □ TL 189

CONSENSUS REPORTS: Reported in EPA TSCA Inventory.

DOT CLASSIFICATION: 8; Label: Corrosive

SAFETY PROFILE: Moderately toxic by ingestion, inhalation, and skin contact. A skin, eye, and mucous membrane irritant. A corrosive agent. Will react with water or steam to produce toxic and corrosive fumes. When heated to decomposition it emits highly toxic fumes of phosgene and HCl.

FPB875 CAS: 35554-44-0 HR: 3
FUNGAFLOR

mf: $C_{14}H_{14}Cl_2N_2O$ mw: 297.20

PROP: Solidified oil. Sltly sol in org solvs; poorly sol in water.

SYNS: (±)-1-(β-(ALLYLOXY)-2,4-DICHLOROPHENE-THYL)IMIDAZOLE □ 1-(2-(2,4-DICHLOROPHENYL)-2-(2-PROPENYLOXY)ETHYL)-1H-IMIDAZOLE □ 1-(2-(2,4-DICHLORPHENYL)-2-(PROPENYLOXY)AETHYL)-1H-IMIDAZOLE □ ENILOCONAZOL (SP) □ IMAVEROL □ IMAZALIL □ R 23979

SAFETY PROFILE: Poison by ingestion and intraperitoneal routes. Experimental reproductive effects. A skin and eye irritant. When heated to decomposition it emits toxic fumes of Cl^- and NO_x.

FPC200 CAS: 58501-21-6 HR: 3
FUNGIZONE INTRAVENOUS

mf: $C_{47}H_{73}NO_{17} \cdot C_{24}H_{40}O_4 \cdot Na$ mw: 1339.84

SYNS: AMPHOTERICIN B DEOXYCHOLATE □ AMPHOTERICIN B SODIUM DESOXYCHOLATE □ AMPHOTERICIN B, MIXT. WITH (3-α,5-β,12-α)-3,12-DIHYDROXYCHOLAN-24-OIC ACIDMONOSODIUM SALT □ DESOXYCHOLATE AMPHOTERICIN B

SAFETY PROFILE: A poison by intravenous and intraperitoneal route. Human systemic effects. When heated to decomposition it emits toxic vapors of NO_x.

FPI150 CAS: 3031-51-4 HR: 3
l-FURALTADONE HYDROCHLORIDE

mf: $C_{13}H_{16}N_4O_6 \cdot ClH$ mw: 360.79

PROP: Yellow crystals. Decomposes @ 206°.

SYNS: FURMETHONOL □ l-5-(MORPHOLINOMETH-YL)-3-((5-NITROFURFURYLIDENE)AMINO)-2-OXAZOLIDINONEHYDROCHLORIDE □ NF-260

CONSENSUS REPORTS: IARC Cancer Review: Group 2B IMEMDT 7,56,87;

Animal Limited Evidence IMEMDT 7,161,74.
SAFETY PROFILE: Suspected carcinogen with experimental carcinogenic data. Poison by intravenous route. Moderately toxic by ingestion. Mutation data reported. When heated to decomposition it emits very toxic fumes of HCl and NO_x.

FPK000   CAS: 110-00-9   HR: 3
FURAN
DOT: UN 2389
mf: C_4H_4O mw: 68.08
PROP: Water white volatile liquid. Mp: −85.65°, bp: 32°, lel: 2.3%, uel: 14.3%, flash p: −32°F, d: 0.964 @ 0°, vap d: 2.35. Sol in EtOH, Et_2O; insol in H_2O.
SYNS: AXOLE □ DIVINYLENE OXIDE □ 1,4-EPO-XY-1,3-BUTADIENE □ FURAN (DOT) □ FURFURAN □ NCI-C56202 □ OXACYCLOPENTADIENE □ OXOLE □ RCRA WASTE NUMBER U124 □ TETROLE
CONSENSUS REPORTS: NTP 10th Report on Carcinogens. IARC Cancer Review: Animal Sufficient Evidence IMEMDT NTP-TR-402,93. EPA Extremely Hazardous Substances List. Reported in EPA TSCA Inventory.
DOT CLASSIFICATION: 3; Label: Flammable Liquid
SAFETY PROFILE: Confirmed carcinogen. Poison by inhalation and intraperitoneal routes. Moderately toxic by ingestion and skin contact. Experimental reproductive effects. A narcotic. Mutation data reported. The exposure concentration limit of 10 ppm together with its low boiling point requires that adequate ventilation be provided in areas where this chemical is handled. Contact with liquid must be avoided since this chemical can be absorbed through the skin. Washing thoroughly with soap and water followed by prolonged rinsing should be done immediately after accidental contact.
A very dangerous fire hazard when exposed to heat or flame; can react with oxidizing materials. Unstabilized, it may form unstable peroxides on exposure to air and should always be tested before distillation. Washing with an aqueous solution of ferrous sulfate slightly acidified with sodium bisulfate will remove these peroxides. Confirm by test. Contact with acids can initiate a violent exothermic reaction. Moderate explosion hazard when exposed to flame. Furan's low boiling point makes it easy to obtain explosive concentrations of the vapor in inadequately ventilated areas. To fight fire, use CO_2, dry chemical. When heated to decomposition it emits acrid smoke and irritating fumes. See also PEROXIDES.

FPM000   CAS: 98-02-2   HR: 3
2-FURANMETHANETHIOL
DOT: UN 1228/UN 3071
mf: C_5H_6OS mw: 114.17
PROP: A liquid with very disagreeable odor. Bp: 160°.
SYNS: FURFURYL MERCAPTAN □ USAF B-58
CONSENSUS REPORTS: Reported in EPA TSCA Inventory.
DOT CLASSIFICATION: 3; Label: Flammable Liquid, Poison (UN 1228); DOT Class: 6.1; Label: Poison, Flammable Liquid (UN 3071)
SAFETY PROFILE: Poison by intraperitoneal route. Experimental reproductive effects. Used as a flavoring in chocolate, fruit, nuts, and coffee. When heated to decomposition it emits toxic fumes of SO_x. See also MERCAPTANS.

FPQ000   CAS: 9000-21-9   HR: 2
FURCELLERAN GUM
PROP: Vegetable gum from *Furcellaria fastigiata* (Fam. *Rodophyceae*) available as an odorless white powder. Sol in warm water.
SYN: BURTONITE 44
CONSENSUS REPORTS: Reported in EPA TSCA Inventory.
SAFETY PROFILE: Moderately toxic by ingestion. When heated to decomposition it emits acrid smoke and fumes.

F

FPQ875   CAS: 98-01-1   HR: 3
FURFURAL
DOT: UN 1199

mf: $C_5H_4O_2$ mw: 96.09

PROP: Colorless–yellowish liquid; almond-like odor. Bp: 161.7° @ 764 mm, lel: 2.1%, uel: 19.3%, flash p: 140°F (CC), d: 1.154–1.158, refr index: 1.522–1.528, autoign temp: 600°F, vap d: 3.31. Sol in water; misc with alc. IDLH 100 ppm.

SYNS: ARTIFICIAL ANT OIL □ FEMA No. 2489 □ FURAL □ 2-FURALDEHYDE □ FURALE □ 2-FURANALDEHYDE □ 2-FURANCARBONAL □ 2-FURANCARBOXALDEHYDE □ 2-FURFURAL □ FURFURALDEHYDE □ FURFURALE (ITALIAN) □ FURFUROL □ FURFUROLE □ 2-FURIL-METANALE (ITALIAN) □ FUROLE □ α-FUROLE □ 2-FURYL-METHANAL □ NCI-C56177 □ PYROMUCIC ALDEHYDE □ RCRA WASTE NUMBER U125

CONSENSUS REPORTS: NTP Carcinogenesis Studies (gavage); Clear Evidence: mouse NCITR* NTP-TR-382,90, Some Evidence: rat NCITR* NTP-TR-382,90. EPA Genetic Toxicology Program. Reported in EPA TSCA Inventory.

OSHA PEL: TWA 2 ppm (skin)

ACGIH TLV: TWA 2 ppm (skin); Animal Carcinogen: BEI: 200 mg/g creatinine of total furoic acid in urine at end of shift

DFG MAK: Confirmed Animal Carcinogen with Unknown Relevance to Humans

DOT CLASSIFICATION: 3; Label: Flammable Liquid

SAFETY PROFILE: Confirmed carcinogen. Poison by ingestion, intraperitoneal, subcutaneous, intravenous, and intramuscular routes. Moderately toxic by inhalation and skin contact. Human mutation data reported. A skin and eye irritant. Mutation data reported. The liquid is dangerous to the eyes. The vapor is irritating to mucous membranes and is a central nervous system poison. However, its low volatility reduces its toxicity effect. Ingestion of furfural has produced cirrhosis of the liver in rats. In industry there is a tendency to minimize the danger of acute effects resulting from exposure to it. This is particularly true because of its low volatility.

Flammable liquid when exposed to heat or flame; can react with oxidizing materials. Moderate explosion hazard when exposed to heat or flame or by chemical reaction. An exothermic polymerization of almost explosive violence can occur upon contact with strong mineral acids or alkalies. Keep away from heat and open flames. Mixture with sodium hydrogen carbonate ignites spontaneously. To fight fire, use alcohol foam, CO_2, dry chemical. When heated to decomposition it emits acrid smoke and irritating fumes.

FPU000   CAS: 98-00-0   HR: 3
FURFURYL ALCOHOL
DOT: UN 2874

mf: $C_5H_6O_2$ mw: 98.11

PROP: Clear, colorless, mobile liquid. Mp: −31°, lel: 1.8%, uel: 16.3% (between 72 and 122°), bp: 171° @ 750 mm, flash p: 167°F (OC), d: 1.129 @ 20°/4°, autoign temp: 915°F, vap press: 1 mm @ 31.8°, vap d: 3.37. Misc in H_2O; very sol in EtOH and Et_2O. IDLH 75 ppm.

SYNS: 2-FURANCARBINOL □ 2-FURANMETHANOL □ FURFURAL ALCOHOL □ 2-FURFURYLALKOHOL (CZECH) □ FURYL ALCOHOL □ α-FURYLCARBINOL □ 2-FURYLCARBINOL □ (2-FURYL)METHANOL □ 2-HYDROXYMETHYLFURAN □ NCI-C56224

CONSENSUS REPORTS: Reported in EPA TSCA Inventory.

OSHA PEL: TWA 10 ppm; STEL 15 ppm (skin)

ACGIH TLV: TWA 10 ppm; STEL 15 ppm (skin)

DFG MAK: 10 ppm (41 mg/m³)

NIOSH REL: (Furfuryl Alcohol) TWA 200 mg/m³

DOT CLASSIFICATION: 6.1; Label: KEEP AWAY FROM FOOD

SAFETY PROFILE: Poison by ingestion, skin contact, and subcutaneous routes. Moderately toxic by inhalation and intraperitoneal routes. Mutation data reported. An eye irritant. Flammable when exposed to heat or flame; can react with oxidizing materials. Moderate explosion hazard when exposed to heat or flame. Reacts violently with acids (e.g., formic acid, cyanoacetic acid + heat). Ignites on contact with 85% hydrogen peroxide. To fight fire, use alcohol foam, CO_2, dry chemical. When heated to decomposition it emits acrid smoke and fumes.

FPW000 CAS: 617-89-0 HR: 3
FURFURYLAMINE
DOT: UN 2526
mf: C_5H_7NO mw: 97.13
PROP: Light straw-colored liquid or oil. Bp: 146°, flash p: 99°F (OC), fp: −70°, d: 1.0502 @ 25°, vap d: 3.35. Misc in water.
SYNS: 2-FURANMETHYLAMINE □ 1-(2-FURYL)METHYLAMINE □ USAF Q-1
CONSENSUS REPORTS: Reported in EPA TSCA Inventory.
DOT CLASSIFICATION: 3; Label: Flammable Liquid
SAFETY PROFILE: Poison by intraperitoneal route. A skin, eye, and mucous membrane irritant. A dangerous fire hazard when exposed to heat or flame; can react with oxidizing materials. To fight fire, use foam, CO_2, dry chemical. When heated to decomposition it emits toxic fumes of NO_x. See also AMINES.

FPX028 CAS: 1197-40-6 HR: 3
2-(2-FURFURYL)FURAN
mf: $C_9H_8O_2$ mw: 148.17
SYNS: DIFURYLMETHANE □ DI-α-FURYLMETHANE □ DI-2-FURYLMETHANE □ 2,2'-DIFURYLMETHANE □ FURAN, 2,2'-METHYLENEDI- □ FURAN, 2,2'-METHYLENEBIS- □ 2,2'-METHYLENEBISFURAN
SAFETY PROFILE: A poison by skin contact. Low toxicity by inhalation. A moderate eye irritant. When heated to decomposition it emits acrid smoke and irritating vapors.

FQL200 CAS: 4682-94-4 HR: 3
2-FURYL p-HYDROXYPHENYL KETONE
mf: $C_{11}H_8O_3$ mw: 188.19
SYNS: DB 133 □ HYDROXY-4 BENZOYL-2-FURANNE □ KETONE, 2-FURYL p-HYDROXYPHENYL

DOT CLASSIFICATION: 3; Label: Flammable Liquid
SAFETY PROFILE: A poison by intraperitoneal route. A flammable liquid. When heated to decomposition it emits acrid smoke and irritating vapors.

FQM100 CAS: 699-18-3 HR: 3
2-FURYL-1-NITROETHENE
mf: $C_6H_5NO_3$ mw: 139.12
SYNS: FURAN, 2-(2-NITROVINYL)- □ G-0 □ β-NITROVINYLFURAN
SAFETY PROFILE: A poison by ingestion. Moderately toxic by inhalation. Mutation data reported. When heated to decomposition it emits toxic vapors of NO_x.

FQN000 CAS: 3688-53-7 HR: 3
2-(2-FURYL)-3-(5-NITRO-2-FURYL)ACRYLAMIDE
mf: $C_{11}H_8N_2O_5$ mw: 248.21
SYNS: AF-2 (preservative) □ FF □ FURYLAMIDE □ FURYLFURAMIDE □ α-2-FURYL-5-NITRO-2-FURAN-ACYRLAMIDE □ 2-(2-FURYL)-3-(5-NITRO-2-FURYL)-ACRYLIC ACID AMIDE □ α-(FURYL)-β-(5-NITRO-2-FURYL)ACRYLIC AMIDE □ TOFURON
CONSENSUS REPORTS: IARC Cancer Review: Group 2B IMEMDT 7,56,87; Human Inadequate Evidence IMEMDT 31,47,83; Animal Sufficient Evidence IMEMDT 31,47,83. EPA Genetic Toxicology Program.
SAFETY PROFILE: Confirmed carcinogen with experimental carcinogenic, neoplastigenic, and teratogenic data. Poison by ingestion. Experimental reproductive effects. Human mutation data reported. When heated to decomposition it emits toxic fumes of NO_x.

FQO050 CAS: 6453-98-1 HR: 3
3-FURYL PHENYL KETONE
mf: $C_{11}H_8O_2$ mw: 172.19
SYNS: 3-BENZOYLFURAN □ 3-FURANYLPHENYLMETHANONE □ KETONE, 3-FURYL PHENYL □ METHANONE, 3-FURFANYLPHENYL-(9CI)
DOT CLASSIFICATION: 3; Label: Flammable Liquid

SAFETY PROFILE: A poison by intraperitoneal route. A flammable liquid. When heated to decomposition it emits acrid smoke and irritating vapors.

FQR000   CAS: 23255-69-8   HR: 3
FUSARENONE X

mf: $C_{17}H_{22}O_8$ mw: 354.39

PROP: Crystals. Mp: 181–184°. Isolated from the culture filtrate of *Fusarium nivale* (34ZHAD -,163,71).

SYNS: 4-ACETYLOXY-12,13-EPOXY-3,7,15-TRIHYDRO-XY-(3-α,4-β,7-β)-TRICHOTHEC-9-EN-8-ONE □ NIVAL-ENOL-4-O-ACETATE □ 3,7,15-TRIHYDROXY-4-ACETO-XY-8-OXO-12,13-EPOXY-Δ^9-TRICHOTHECENE □ 3,7,15-TRIHYDROXYSCIRP-4-ACETOXY-9-EN-8-ONE

CONSENSUS REPORTS: IARC Cancer Review: Group 3 IMEMDT 7,56,87; Animal Inadequate Evidence IMEMDT 11,169,76; Animal Inadequate Evidence IMEMDT 31,153,83.

SAFETY PROFILE: Poison by ingestion, subcutaneous, intravenous, and intraperitoneal routes. An experimental teratogen. Experimental reproductive effects. Questionable carcinogen with experimental tumorigenic data. Human mutation data reported. When heated to decomposition it emits acrid smoke and irritating fumes.

FQS000   CAS: 21259-20-1   HR: 3
FUSARIOTOXIN T 2

mf: $C_{24}H_{34}O_9$ mw: 466.58

PROP: Needles. Mp: 151–152°. A strain of *F. tricinctum* isolated from infected corn (AJVRAH 32,1843,71).

SYNS: 4,15-DIACETOXY-8-(3-METHYLBUTYRYLOXY)-12,13-EPOXY-Δ-9-TRICHOTHECEN-3-OL □ 4-β,15-DIACETOXY-8-α-(3-METHYLBUTYRYLOXY)-3-α-HYDROXY-12,13-EPOXYTRICHOTHEC-9-ENE □ 3-HYDROXY-4,15-DIACETOXY-8-(3-METHYLBUTYRYLO-XY)-12,13-EPOXY-Δ^9-TRICHOTHECENE □ INSARIO-TOXIN □ 8-ISOVALERATE □ 8-(3-METHYLBUTYRYLO-XY)-DIACETOXYSCIRPENOL □ NSC-138780 □ T-2 MYCOTOXIN □ TOXIN T2 □ T^2-TRICHOTHECENE

CONSENSUS REPORTS: IARC Cancer Review: Group 3 IMEMDT 7,56,87; Animal Inadequate Evidence IMEMDT 31,265,83. EPA Genetic Toxicology Program.

SAFETY PROFILE: Poison by ingestion, intramuscular, subcutaneous, intraperitoneal, intracerebral, and intravenous routes. Moderately toxic by inhalation. Experimental teratogenic and reproductive effects. A skin irritant. Questionable carcinogen with experimental neoplastigenic data. Mutation data reported. When heated to decomposition it emits acrid smoke and irritating fumes.

FQT000   CAS: 8013-75-0   HR: 3
FUSEL OIL

DOT: UN 1201

PROP: Colorless to pale-yellow liquid; odorless. D: 0.807–0.813, refr index: 1.405–1.410. Composition of grain fusel oil is methanol, ethanol, acetaldehyde, and other alcohols (ARGEAR 33,49,69).

SYNS: FEMA No. 2497 □ FUSELOEL (GERMAN) □ FUSEL OIL, REFINED (FCC) □ HUILE de FUSEL (FRENCH)

CONSENSUS REPORTS: Reported in EPA TSCA Inventory.

DOT CLASSIFICATION: 3; Label: Flammable Liquid

SAFETY PROFILE: May contain carcinogens. Experimental reproductive effects. Mutation data reported. Flammable liquid when exposed to heat or flame; can react vigorously with oxidizing materials. When heated to decomposition it emits acrid smoke and fumes. See also individual components.

FQU875   CAS: 13674-87-8   HR: 2
FYROL FR 2

mf: $C_9H_{15}Cl_6O_4P$ mw: 430.91

PROP: Viscous liquid. Bp: (5) 236–237°, n: (20/D) 1.5022. Solubility in water: 100 ppm.

SYNS: 1,3-DICHLORO-2-PROPANOL PHOSPHATE (3:1) □ EMULSION 212 □ FOSFORAN TROJ-(1,3-DWUCHL-OROIZOPROPYLOWY) (POLISH) □ PF 38 □ PHOSPHOR-IC ACID TRIS(1,3-DICHLORO-2-PROPYL)ESTER □ TCPP □ TDCPP □ TRIS(1-CHLOROMETHYL-2-CHLOROETH-YL)PHOSPHATE □ TRIS(1,3-DICHLOROISOPROPYL)-PHOSPHATE □ TRIS-(1,3-DICHLORO-2-PROPYL)-PHOSPHATE

CONSENSUS REPORTS: EPA Genetic Toxicology Program. Reported in EPA TSCA Inventory.

SAFETY PROFILE: Moderately toxic by ingestion. An experimental teratogen. Experimental reproductive effects. Questionable carcinogen with experimental carcinogenic data. Mutation data reported. When heated to decomposition it emits toxic fumes of Cl^- and PO_x.

G

GAV050   HR: D
α-GALACTOSIDASE
PROP: Derived from *Mortierella vinaceae* var. *raffinoseutilizer*.
SYN: ATCC No. 20034
SAFETY PROFILE: When heated to decomposition it emits acrid smoke and irritating fumes.

GBK000   CAS: 1303-00-0   HR: 3
GALLIUM ARSENIDE
mf: AsGa   mw: 144.64
PROP: Black solid or cubic crystals with dark-gray metallic sheen. Mp: 1238°, d: 5.31.
SYN: GALLIUM MONOARSENIDE
CONSENSUS REPORTS: Reported in EPA TSCA Inventory. Arsenic and its compounds are on the Community Right-To-Know List.
OSHA PEL: OSHA: Cancer Hazard
ACGIH TLV: TWA 0.01 mg(As)/m³; Human Carcinogen; BEI: 35 μ (As)/L inorganic arsenic and methylated metabolites in urine
NIOSH REL: (Gallium Arsenide) CL 0.002 mg(As)/m³/15M
SAFETY PROFILE: Confirmed carcinogen. Mildly toxic by intraperitoneal route. Most arsenic compounds are poisons. Can react with steam, acids, and acid fumes to evolve the deadly poisonous arsine. Molten gallium arsenide attacks quartz. When heated to decomposition it emits very toxic fumes of As. See also ARSENIC COMPOUNDS and GALLIUM COMPOUNDS.

GBW025   CAS: 64741-58-8   HR: 3
GAS OILS (petroleum), light vacuum
SYN: LIGHT GAS OIL
CONSENSUS REPORTS: IARC Cancer Review: Animal Sufficient Evidence IMEMDT 45,39,89. Reported in EPA TSCA Inventory.

SAFETY PROFILE: Confirmed carcinogen with experimental carcinogenic data. Pulmonary aspiration can cause severe pneumonitis. A flammable liquid. When heated to decomposition it emits acrid smoke and irritating fumes.

GBY000   CAS: 8006-61-9   HR: 3
GASOLINE
DOT: UN 1203/UN 1257
PROP: Clear, aromatic, volatile liquid; a mixture of aliphatic hydrocarbons. Flash p: −50°F, d: <1.0, vap d: 3.0–4.0, ULC: 95–100, lel: 1.3%, uel: 6.0%, autoign temp: 536–853°F, bp: initially 39°, after 10% distilled = 60°, after 50% = 110°, after 90% = 170°, final bp: = 204°. Insol in water; freely sol in abs alc, ether, chloroform, and benzene.
SYNS: MOTOR SPIRIT (DOT) □ NATURAL GASOLINE (DOT) □ PETROL (DOT)
CONSENSUS REPORTS: Reported in EPA TSCA Inventory.
OSHA PEL: TWA 300 ppm; STEL 500 ppm
ACGIH TLV: TWA 300 ppm; STEL 500 ppm; Animal Carcinogen
DOT CLASSIFICATION: 3; Label: Flammable Liquid
SAFETY PROFILE: Confirmed carcinogen. Mildly toxic by inhalation. Human systemic effects by inhalation: cough, conjunctiva irritation, hallucinations or distorted perceptions. Repeated or prolonged dermal exposure causes dermatitis. Can cause blistering of skin. Inhalation or ingestion can cause central nervous system depression. Pulmonary aspiration can cause severe pneumonitis. Some addiction has been reported from inhalation of fumes. Even brief inhalations of high concentrations can cause a fatal pulmonary edema. The vapors are considered to be moderately poisonous. If its con-

centration in air is sufficiently high to reduce the oxygen content below that needed to maintain life, it acts as a simple asphyxiant. A human eye irritant. Gasoline is a common air contaminant. A very dangerous fire and explosion hazard when exposed to heat or flame; can react vigorously with oxidizing materials. To fight fire, use foam, CO_2, dry chemical.

GCE100   HR: 3
GASOLINE, UNLEADED
PROP: A clear water-like liquid, strong aromatic hydrocarbon odor. Boiling range: 85°–437° F. Flash pt: −45° F.
SYNS: UNLEADED GASOLINE □ UNLEADED MOTOR GASOLINE
SAFETY PROFILE: Suspected carcinogen with experimental carcinogenic data. Moderately toxic by inhalation. Pulmonary aspiration can cause severe pneumonitis. Skin irritant. Flammable liquid. When heated to decomposition it emits acrid smoke and irritating fumes.

GCM350   CAS: 446-72-0   HR: D
GENISTEIN
mf: $C_{15}H_{10}O_5$ mw: 270.23
PROP: Prisms from EtOH (aq). Mp: 301–302° (decomp). Rectangular or six-sided rods from 60% alc. Dendritic needles from ether. Mp: 301–302° (decomp). Sol in the usual org solvs; practically insol in water; sol in dil alkalies, with yellow color.
SYNS: GENISTEOL □ GENISTERIN □ PRUNETOL □ SOPHORICOL □ 4',5,7-TRIHYDROXYISOFLAVONE □ 5,7,4'-TRIHYDROXYISOFLAVONE
CONSENSUS REPORTS: Reported in EPA TSCA Inventory.
SAFETY PROFILE: Experimental reproductive effects. Human mutation data reported. When heated to decomposition it emits acrid smoke and irritating fumes.

GCO000   CAS: 1403-66-3   HR: 3
GENTAMYCIN
PROP: Veterinary vaccine. Bp: >100° C, d: 1.0.

SYNS: GARAMYCIN □ GENTAMYCIN □ GENTAMYCIN-CREME (GERMAN) □ UROMYCINE
CONSENSUS REPORTS: EPA Genetic Toxicology Program.
SAFETY PROFILE: Poison by intravenous, intraperitoneal, intramuscular, and subcutaneous routes. Mildly toxic by ingestion. Experimental teratogenic and reproductive effects. Mutation data reported. Human systemic effects: change in motor activity, changes in vestibular functions, distorted perceptions, eye hemorrhage, hallucinations, kidney changes, motor activity changes, trigeminal nerve sensory changes, vestibular function changes, visual field changes. Affects the peripheral nervous system by intravenous route. An antibiotic. When heated to decomposition it emits acrid smoke and irritating fumes. See also other gentamycin entries.

GCS000   CAS: 1405-41-0   HR: 3
GENTAMYCIN SULFATE
PROP: White to buff colored powder or slightly yellow liquid. Bp: (decomposes). Highly sol in water.
SYNS: GARAMYCIN □ GENOPTIC □ GENOPTIC S.O.P. □ GM SULFATE □ NSC-82261 □ SCH 9724
SAFETY PROFILE: Poison by intravenous, intraperitoneal, and intramuscular routes. Moderately toxic by subcutaneous route. Human systemic effects: level changes for metals other than Na/K/Fe/Ca/P/Cl. An experimental teratogen. Other experimental reproductive effects. Mutation data reported. When heated to decomposition it emits very toxic fumes of SO_x. See also other gentamycin entries.

GCU000   CAS: 490-79-9   HR: 3
GENTISIC ACID
mf: $C_7H_6O_4$ mw: 154.13
PROP: Needles, monoclinic prisms from water. Mp: 204.5–205°. Sol in water (more so in hot water), alc, ether; practically insol in carbon disulfide, chloroform, ether.
SYNS: 2,5-DHBA □ 2,5-DIHYDROXYBENZOIC ACID □ GENTISATE □ HYDROQUINONECARBOXYLIC ACID □ 5-HYDROXYSALICYLIC ACID

CONSENSUS REPORTS: Reported in EPA TSCA Inventory.

SAFETY PROFILE: Poison by intravenous route. Moderately toxic by ingestion and intraperitoneal routes. Experimental teratogenic and reproductive effects. Mutation data reported. When heated to decomposition it emits acrid smoke and irritating fumes.

GCY000   CAS: 105-86-2   HR: 1
GERANIOL FORMATE
mf: $C_{11}H_{18}O_2$ mw: 182.29

PROP: Colorless to pale-yellow liquid or oil; rose odor. D: 0.906–0.920, refr index: 1.457–1.466, flash p: 205°F, bp: 113–114° @ 15 mm. Sol in alc, fixed oils; insol in glycerin, propylene glycol, water @ 216°.

SYNS: trans-3,7-DIMETHYL-2,6-OCTADIEN-1-OL FORMATE □ 3,7-DIMETHYL-2,6-OCTADIENYL ESTER FORMIC ACID (E) □ trans-3,7-DIMETHYL-2,6-OCTADIEN-1-YL FORMATE □ FEMA No. 2514 □ FORMIC ACID, GERANIOL ESTER □ GERANYL FORMATE (FCC)

CONSENSUS REPORTS: Reported in EPA TSCA Inventory.

SAFETY PROFILE: Low toxicity by ingestion and skin contact. A human skin irritant and an experimental eye irritant. Combustible liquid. When heated to decomposition it emits acrid smoke and irritating fumes. See also ESTERS.

GDA000   CAS: 8000-46-2   HR: 1
GERANIUM OIL ALGERIAN TYPE
PROP: From steam distillation of leaves from *Pelargonium graveolens* l'Her (Fam. *Geraniaceae*). Contains geraniol and geranyl tiglate (FCTXAV 14,659,76). Yellow liquid; odor of rose and geraniol. D: 0.886–0.898, refr index: 1.454–1.472 @ 20°. Sol in fixed oils, mineral oil; insol in glycerin.

SYNS: GERANIUM OIL □ OIL OF GERANIUM □ OIL OF PELARGONIUM □ OIL OF ROSE GERANIUM □ OIL ROSE GERANIUM ALGERIAN □ PELARGONIUM OIL □ ROSE GERANIUM OIL ALGERIAN

CONSENSUS REPORTS: Reported in EPA TSCA Inventory.

SAFETY PROFILE: A skin irritant. When heated to decomposition it emits acrid smoke and irritating fumes.

GDE400   CAS: 689-67-8   HR: 1
GERANYL ACETONE
mf: $C_{13}H_{22}O$ mw: 194.35

PROP: Clear, colorless liquid. D: 0.907 g/mL @ 15°, bp: 242°. Sol in water: <1 mg/mL @ 20°

SYNS: DIHYDROPSEUDOIONONE □ α-β-DIHYDROPSEUDOIONONE □ 6,10-DIMETHYL-UNDECA-5,9-DIEN-2-ONE □ 5,9-UNDECADIEN-2-ONE, 6,10-DIMETHYL-

CONSENSUS REPORTS: Reported in EPA TSCA Inventory.

SAFETY PROFILE: A skin irritant. When heated to decomposition it emits acrid smoke and irritating fumes.

GDE800   HR: 2
GERANYL BENZOATE
mf: $C_{17}H_{22}O_2$ mw: 258.36

PROP: Sltly yellow liquid; floral odor resembling that of ylang ylang oil. D: 0.978–0.984, refr index: 1.513–1.518, flash p: 212°F. Misc in alc, chloroform; insol in water @ 305°.

SYNS: 3,7-DIMETHYL-2,6-OCTADIEN-1-YL BENZOATE □ FEMA No. 2511

SAFETY PROFILE: Combustible liquid. When heated to decomposition it emits acrid smoke and irritating fumes.

GDG100   CAS: 40267-72-9   HR: 1
GERANYL ETHYL ETHER
mf: $C_{12}H_{22}O$ mw: 182.34

PROP: Diffusive green liquid. Used in fragrances and perfumes.

SYNS: 1-ETHOXY-3,7-DIMETHYL-2,6-OCTADIENE □ ETHYL GERANYL ETHER □ 2,6-OCTADIENE, 1-ETHOXY-3,7-DIMETHYL-

CONSENSUS REPORTS: Reported in EPA TSCA Inventory.

SAFETY PROFILE: Low toxicity by ingestion and skin contact. A skin irritant. When heated to decomposition it emits acrid smoke and irritating fumes.

G

GDK000   **CAS: 109-20-6**   **HR: 1**
GERANYL ISOVALERATE
mf: $C_{15}H_{26}O_2$ mw: 238.41
SYNS: trans-3,7-DIMETHYL-2,6-OCTADIENYL ISO-
PENTANOATE □ (E)-ISOVALERIC ACID-3,7-DIMETH-
YL-2,6-OCTADIENYL ESTER □ (E)-3-METHYLBUTYRIC
ACID-3,7-DIMETHYL-2,6-OCTADIENYL ESTER
CONSENSUS REPORTS: Reported in EPA
TSCA Inventory.
SAFETY PROFILE: A skin irritant. When
heated to decomposition it emits acrid
smoke and irritating fumes. See also
ESTERS.

GDM100   **CAS: 65405-73-4**   **HR: 1**
GERANYL OXYACETALDEHYDE
mf: $C_{12}H_{20}O_2$ mw: 196.32
SYNS: ACETALDEHYDE, ((3,7-DIMETHYL-2,6-
OCTADIENYL)OXY)-, (E)- □ GERANOXY
ACETALDEHYDE
CONSENSUS REPORTS: Reported in EPA
TSCA Inventory.
SAFETY PROFILE: A skin irritant. When
heated to decomposition it emits acrid
smoke and irritating fumes.

GDM400   **CAS: 102-22-7**   **HR: 2**
GERANYL PHENYLACETATE
mf: $C_{18}H_{24}O_2$ mw: 272.39
PROP: Yellow liquid; honey-rose odor. D:
0.971–0.978, refr index: 1.507–1.511, flash p:
212°F. Misc in alc, chloroform, ether; insol
in water.
SYNS: ACETIC ACID, PHENYL-, 3,7-DIMETHYL-2,6-
OCTADIENYL ESTER, (E)-(8CI) □ BENZENEACETIC
ACID, 3,7-DIMETHYL-2,6-OCTADIENYL ESTER, (E)- □
3,7-DIMETHYL-2,6-OCTADIEN-1-YL PHENYLACETATE
□ trans-3,7-DIMETHYL-2,6-OCTADIEN-1-YL PHENYL-
ACETATE □ GERANYL α-TOLUATE □ FEMA No. 2516
CONSENSUS REPORTS: Reported in EPA
TSCA Inventory.
SAFETY PROFILE: Low toxicity by
ingestion and skin contact. Combustible
liquid. When heated to decomposition it
emits acrid smoke and irritating fumes.

GDM450   **CAS: 105-90-8**   **HR: 2**
GERANYL PROPIONATE
mf: $C_{13}H_{22}O_2$ mw: 210.32
PROP: Colorless liquid; rosy, fruity odor. D:
0.896–0.913, refr index: 1.456–1.464, flash p:

212°F. Sol in alc, fixed oils; insol in glycerin,
propylene glycol, water @ 253°.
SYNS: (E)-3,7-DIMETHYL-2,6-OCTADIEN-1-OL PROP-
IONATE □ 3,7-DIMETHYL-2,6-OCTADADIEN-1-YL
PROPIONATE □ trans-3,7-DIMETHYL-2,6-OCTADIEN-1-
YL PROPIONATE □ 2,6-OCTADIEN-1-OL, 3,7-DIMETH-
YL-, PROPIONATE, (E)-(8CI) □ FEMA No. 2517
CONSENSUS REPORTS: Reported in EPA
TSCA Inventory.
SAFETY PROFILE: Low toxicity by
ingestion and skin contact. Combustible
liquid. When heated to decomposition it
emits acrid smoke and irritating fumes.

GDO200   **CAS: 6902-91-6**   **HR: 3**
GERMACRONE
mf: $C_{15}H_{22}O$ mw: 218.34
SYN: 3,7-CYCLODECADIEN-1-ONE, 3,7-DIMETHYL-10-
(1-METHYLETHYLIDENE)-, (3E,7E)-
SAFETY PROFILE: A poison by ingestion.
When heated to decomposition it emits
acrid smoke and irritating vapors.

GDS000   **CAS: 1310-53-8**   **HR: 3**
GERMANIC OXIDE (crystalline)
CONSENSUS REPORTS: Reported in EPA
TSCA Inventory.
SAFETY PROFILE: Poison by
intraperitoneal route. When heated to
decomposition it emits acrid smoke and
irritating fumes.

GDW000   **CAS: 13450-92-5**   **HR: 3**
GERMANIUM BROMIDE
mf: Br_4Ge mw: 392.23
PROP: Gray-white, octahedral crystals. Mp:
26.1°, bp: 186.5°, d: 3.232 @ 29°/29°.
Decomp in H_2O; sol in EtOH, Et_2O, and
C_6H_6.
SYN: GERMANIUM TETRABROMIDE
CONSENSUS REPORTS: Reported in EPA
TSCA Inventory.
SAFETY PROFILE: Poison by intravenous
route. When heated to decomposition it
emits very toxic fumes of Br^-. See also
GERMANIUM COMPOUNDS.

GDY000   CAS: 10038-98-9   HR: 3
GERMANIUM CHLORIDE
mf: Cl₄Ge   mw: 214.39
PROP: Colorless, mobile liquid. Fumes in air. Peculiar acidic odor but can be distinguished from that of concentrated HCl. Mp: −49.5°, bp: 83.1°, d: 1.879 @ 20°/20°. Volatile @ room temp. Decomp in H₂O; sol in EtOH, Et₂O. Very sol in dil HCl; insol in conc HCl, and H₂SO₄.
SYN: EXTREMA □ GERMANIUM TETRACHLORIDE
CONSENSUS REPORTS: Reported in EPA TSCA Inventory. EPA Genetic Toxicology Program.
SAFETY PROFILE: Poison by intravenous route. Mildly toxic by inhalation. A skin, severe eye, and mucous membrane irritant. Will react violently with water or steam to produce toxic and corrosive fumes. When heated to decomposition it emits toxic fumes of Cl⁻. See also GERMANIUM COMPOUNDS.

GEA000   HR: 2
GERMANIUM COMPOUNDS
SAFETY PROFILE: Germanium compounds are considered to be of a low order of toxicity, but rare instances of poisoning have been reported in the literature. Experimental LD50 values are typically about 100–1000 mg/kg for parenteral route and 500–5000 mg/kg for ingestion. The experimental animals suffer from hypothermia, diarrhea, and respiratory and cardiac failure. Inhalation of large amounts of GeCl₄ by experimental animals causes necrosis of the tracheal epithelium, bronchitis, and interstitial pneumonia. These effects were not apparent with chronic inhalation of 7 mg/m³. The tetrachloride and tetrafluoride are eye, skin, and mucous membrane irritants. Alkyl germanium compounds are much less toxic than the corresponding tin or lead compounds. Tributyl germanium and germanium tetrachloride are mutagens. Dimethyl germanium is a teratogen. Chronic ingestion of 1000 ppm or 100 ppm of germanium dioxide in water has been shown to inhibit growth in chickens. No effect was seen at 5 ppm. It has been found that the dioxide stimulates the generation of red blood cells, but it is believed to be relatively nontoxic. Buffered germanium dioxides in solution have been found to be nonirritating to the skin. Germanium hydride is a hemolytic gas and has been shown to have toxic properties at a concentration of 100 ppm. It can cause death at a concentration of 150 ppm. Otherwise, little is known about the toxicity of organic germanium compounds except that they may resemble other organometals in having higher toxicity than inorganic forms. When germanium is given in sublethal amounts, it causes a pronounced tolerance. Interest is high in this material because of its close chemical relationship to arsenic.

GEC000   CAS: 1310-53-8   HR: 3
GERMANIUM DIOXIDE
mf: GeO₂   mw: 104.59
PROP: Soluble form: Hexagonal, colorless crystals. Mp: 1115.0°, d: 4.703 @ 18°. Insoluble form: Tetragonal crystals, mp: 1086 ± 5°, d: 6.239.
SYNS: GERMANIA □ GERMANIC ACID □ GERMANIUM OXIDE □ GERMANIUM OXIDE (GeO2)
CONSENSUS REPORTS: Reported in EPA TSCA Inventory.
SAFETY PROFILE: Poison by intraperitoneal route. Moderately toxic by ingestion and subcutaneous routes. Experimental teratogenic and reproductive effects. When heated to decomposition it emits acrid smoke and irritating fumes. See also GERMANIUM COMPOUNDS.

GEI100   CAS: 7782-65-2   HR: 3
GERMANIUM TETRAHYDRIDE
DOT: UN 2192
mf: GeH₄   mw: 76.63
PROP: Colorless gas. Mp: −165°, bp: −90°, d: 1.523 @ −142°/4°. Insol in H₂O; sol in aq NH₃, aq NaOCl; sltly sol in hot HCl.

SYNS: GERMANE (DOT) □ GERMANIUM HYDRIDE □ MONOGERMANE

CONSENSUS REPORTS: Reported in EPA TSCA Inventory.

ACGIH TLV: TWA 0.2 ppm

DOT CLASSIFICATION: 2.3; Label: Poison Gas, Flammable Gas

SAFETY PROFILE: Poison by inhalation. Moderately toxic by ingestion. A hemolytic gas. Ignites spontaneously in air. Incompatible with Br$_2$. See also HYDRIDES, GERMANIUM COMPOUNDS, and GERMANIUM.

GEM000   CAS: 77-06-5   HR: 2
GIBBERELLIC ACID
mf: C$_{19}$H$_{22}$O$_6$ mw: 346.41
PROP: A plant-growth-promoting hormone. White crystals or crystalline powder from alc/pet ether. Mp: 233–235° (decomp). Sltly sol in water, ether; sol in methanol, ethanol, acetone, aq solns of sodium bicarbonate and sodium acetate; moderately sol in ethyl acetate.
SYNS: BERELEX □ BRELLIN □ CEKUGIB □ FLORA-LTONE □ GA □ GIBBERELLIN □ GIBBREL □ GIB-SOL □ GIB-TABS □ GROCEL □ NCI-C55823 □ PRO-GIBB □ 2,4a,7-TRIHYDROXY-1-METHYL-8-METHYLENEGIBB-3-ENE-1,10-CARBOXYLIC ACID 1-4-LACTONE
CONSENSUS REPORTS: EPA Genetic Toxicology Program. Reported in EPA TSCA Inventory.
SAFETY PROFILE: Mildly toxic by ingestion. Questionable carcinogen with experimental tumorigenic data. Mutation data reported. When heated to decomposition it emits acrid smoke and irritating fumes.

GEQ000   CAS: 8007-08-7   HR: 2
GINGER OIL
PROP: From steam distillation of ground rhizomes of *Zingiber officinale* Roscoe (Fam. *Zingiberaceae*) (FCTXAV 12,807,74). Yellow liquid; odor of ginger. D: 0.870–0.882, refr index: 1.488 @ 20°. Sol in fixed oils, mineral oil, alc; insol in glycerin, propylene glycol.
CONSENSUS REPORTS: Reported in EPA TSCA Inventory.

SAFETY PROFILE: Moderately toxic by ingestion and intraperitoneal routes. A skin irritant. Mutation data reported. When heated to decomposition it emits acrid smoke and irritating fumes.

GFA000   CAS: 15879-93-3   HR: 3
α-d-GLUCOCHLORALOSE
mf: C$_8$H$_{11}$Cl$_3$O$_6$ mw: 309.54
PROP: Needles from EtOH or Et$_2$O. Mp: 187°.
SYNS: AGC □ ALFAMAT □ ANHYDROGLUCO-CHLORAL □ APHOSAL □ CHLORALOSANE □ α-CHLORALOSE □ CHLOROALOSANE □ DULCIDOR □ GLUCOCHLORAL □ GLUCOCHLORALOSE □ KALMET-TUMSOMNIFERUM □ MONOTRICHLOR-AETHYLIDEN-α-GLUCOSE (GERMAN) □ MUREX □ SOMIO □ 1,2-o-(2,2,2-TRICHLOROETHYLIDENE)-α-d-GLUCOFURAN-OSE
CONSENSUS REPORTS: Reported in EPA TSCA Inventory.
SAFETY PROFILE: Poison by ingestion, subcutaneous, and intraperitoneal routes. Questionable carcinogen with experimental tumorigenic data. When heated to decomposition it emits toxic fumes of Cl⁻.

GFA200   CAS: 90-80-2   HR: D
GLUCONO-Δ-LACTONE
mf: C$_6$H$_{10}$O$_6$ mw: 178.16
PROP: White crystalline powder. Decomp @ 153°. Sol in water, sltly sol in alc.
SYNS: DELTAGLUCONOLACTONE □ d-GLUCONIC Δ-LACTONE □ GLUCONOLACTONE
CONSENSUS REPORTS: Reported in EPA TSCA Inventory.
SAFETY PROFILE: Mutation data reported. When heated to decomposition it emits acrid smoke and irritating fumes.

GFG000   CAS: 50-99-7   HR: 2
d-GLUCOSE
mf: C$_6$H$_{12}$O$_6$ mw: 180.18
PROP: Colorless crystals or white crystalline or granular powder; odorless with sweet taste. D: 1.544, mp: 146°. Sol in water; sltly sol in alc. α Form: (monohydrate) crystals from water. Mp: 83°. α Form: (anhydrous) crystals from hot ethanol or water. Mp:

146°. Very sparingly sol in abs alc, ether, acetone; sol in hot glacial acetic acid, pyridine, aniline. β Form: crystals from hot H_2O + ethanol, from dil acetic acid, or from pyridine; mp: 148–155°.

SYNS: CARTOSE □ CERELOSE □ CORN SUGAR □ DEXTROPUR □ DEXTROSE (FCC) □ DEXTROSE, anhydrous □ DEXTROSOL □ GLUCOLIN □ GLUCOSE □ d-GLUCOSE, anhydrous □ GLUCOSE LIQUID □ GRAPE SUGAR □ SIRUP

CONSENSUS REPORTS: Reported in EPA TSCA Inventory. EPA Genetic Toxicology Program.

SAFETY PROFILE: Mildly toxic by ingestion. An experimental teratogen. Experimental reproductive effects. Questionable carcinogen with experimental tumorigenic data. Mutation data reported. Potentially explosive reaction with potassium nitrate + sodium peroxide when heated in a sealed container. Mixtures with alkali release carbon monoxide when heated. When heated to decomposition it emits acrid smoke and irritating fumes.

GFO000   CAS: 56-86-0   HR: 1
l-GLUTAMIC ACID
mf: $C_5H_9NO_4$ mw: 147.15

PROP: A nonessential amino acid present in all complete proteins. White rhombic crystals from alc (aq), or crystalline powder. Mp (dl form): 194°, d (dl form): 1.4601 @ 20°/4°, mp (l form): 224–225°, d (l form): 1.538 @ 20°/4°. Sltly sol in water.

SYNS: α-AMINOGLUTARIC ACID □ l-2-AMINO-GLUTARIC ACID □ 2-AMINOPENTANEDIOIC ACID □ 1-AMINOPROPANE-1,3-DICARBOXYLIC ACID □ GLUSATE □ GLUTACID □ GLUTAMIC ACID □ α-GLUTAMIC ACID □ d-GLUTAMIENSUUR □ GLUT-AMINIC ACID □ l-GLUTAMINIC ACID □ GLUTAMINOL □ GLUTATON

CONSENSUS REPORTS: Reported in EPA TSCA Inventory.

SAFETY PROFILE: Human systemic effects by ingestion and intravenous routes: headache and nausea or vomiting. When heated to decomposition it emits toxic fumes of NO_x.

GFO050   CAS: 56-85-9   HR: 1
GLUTAMINE
mf: $C_5H_{10}N_2O_3$ mw: 146.17

PROP: l-Form (natural): Fine opaque needles from water or dil ethanol. Decomp at 185–186°. Sol in water; practically insol in methanol, ethanol, ether, benzene, acetone, ethyl acetate, chloroform. dl-Form: prisms from dil acetone. Mp: 185–186°.

SYNS: 2-AMINOGLUTARAMIC ACID □ l-2-AMINO-GLUTARAMIDIC ACID □ CEBROGEN □ GLUMIN □ GLUTAMIC ACID AMIDE □ GLUTAMIC ACID-5-AMIDE □ γ-GLUTAMINE □ l-GLUTAMINE (9CI, FCC) □ LEVO-GLUTAMID □ LEVOGLUTAMIDE □ STIMULINA

CONSENSUS REPORTS: Reported in EPA TSCA Inventory.

SAFETY PROFILE: Mildly toxic by ingestion. Human systemic effects: euphoria. When heated to decomposition it emits toxic fumes of NO_x.

GFQ000   CAS: 111-30-8   HR: 3
GLUTARALDEHYDE
mf: $C_5H_8O_2$ mw: 100.13

PROP: Oil. Bp: 71–72° @ 10 mm.

SYNS: CIDEX □ GLUTARAL □ GLUTARALDEHYD (CZECH) □ GLUTARDIALDEHYDE □ GLUTARIC DIALDEHYDE □ NCI-C55425 □ 1,5-PENTANEDIAL □ 1,5-PENTANEDIONE □ POTENTIATED ACID GLUTARALDEHYDE □ SONACIDE

CONSENSUS REPORTS: Reported in EPA TSCA Inventory.

OSHA PEL: CL 0.2 ppm

ACGIH TLV: CL 0.05 ppm (skin, sensitizer); Not Classifiable as a Human Carcinogen

DFG MAK: 0.1 ppm (0.42 mg/m³)

SAFETY PROFILE: Poison by ingestion, intravenous, and intraperitoneal routes. Moderately toxic by inhalation, skin contact, and subcutaneous routes. Experimental teratogenic and reproductive effects. A severe eye and human skin irritant. Mutation data reported. When heated to decomposition it emits acrid smoke and irritating fumes. See also ALDEHYDES.

GFW000   CAS: 70-18-8   HR: 2
GLUTATHIONE
mf: $C_{10}H_{17}N_3O_6S$ mw: 307.36

PROP: Colorless prisms out of alc. Mp: 195° decomp in hot water; insol in abs alc, ether, and acid. Freely sol in H_2O, dil alc, liquid ammonia, and dimethylformamide.

SYNS: COPREN □ DELTATHIONE □ GLUTATHIONE (reduced) □ GLUTATIOL □ GLUTATIONE □ GLUTIDE □ GLUTINAL □ GSH □ ISETHION □ NEUTHION □ TATHIONE □ TRIPTIDE

CONSENSUS REPORTS: EPA Genetic Toxicology Program. Reported in EPA TSCA Inventory.

SAFETY PROFILE: Moderately toxic by intravenous route. Experimental reproductive effects. Human mutation data reported. When heated to decomposition it emits very toxic fumes of SO_x and NO_x.

GGA000   CAS: 56-81-5   HR: 3
GLYCERIN

mf: $C_3H_8O_3$ mw: 92.11

PROP: Colorless or pale-yellow liquid; odorless, syrupy, sweet and warm taste. Mp: 17.9 (solidifies at a much lower temp), bp: 290° (part decomp), ULC: 10–20, flash p: 320°F, d: 1.260 @ 20°/4°, autoign temp: 698°F, vap press: 0.0025 mm @ 50°, vap d: 3.17. Misc in H_2O, EtOH; insol in C_6H_6, $CHCl_3$, and CCl_4.

SYNS: GLYCERIN, anhydrous □ GLYCERIN, synthetic □ GLYCERINE □ GLYCERITOL □ GLYCEROL □ GLYCYL ALCOHOL □ GROCOLENE □ MOON □ 1,2,3-PRO-PANETRIOL □ STAR □ SUPEROL □ SYNTHETIC GLYCERIN □ 90 TECHNICAL GLYCERINE □ TRIHY-DROXYPROPANE □ 1,2,3-TRIHYDROXYPROPANE

CONSENSUS REPORTS: Reported in EPA TSCA Inventory.

OSHA PEL: TWA Total Mist: 10 mg/m³; Respirable Fraction: 5 mg/m³

ACGIH TLV: TWA 10 mg/m³ (vapor)

SAFETY PROFILE: Poison by subcutaneous route. Mildly toxic by ingestion. Human systemic effects by ingestion: headache and nausea or vomiting. Experimental reproductive effects. Human mutation data reported. A skin and eye irritant. In the form of mist it is a nuisance particulate and inhalation irritant.

Combustible liquid when exposed to heat, flame, or powerful oxidizers. Mixtures with hydrogen peroxide are highly explosive. Ignites on contact with potassium permanganate, calcium hypochlorite. Mixture with nitric acid + sulfuric acid forms the explosive glyceryl nitrate. Mixture with perchloric acid + lead oxide forms explosive perchlorate esters. Confined mixture with chlorine explodes if heated to 70–80°. Can react violently with acetic anhydride, aniline + nitrobenzene, $Ca(OCl)_2$, CrO_3, Cr_2O_3, F_2 + PbO, phosphorus triiodide, ethylene oxide + heat, $KMnO_4$, K_2O_2, $AgClO_4$, Na_2O_2, NaH. Energetic reaction with sodium hydride. Mixture with nitric acid + hydrofluoric acid is a storage hazard due to gas evolution. To fight fire, use alcohol foam, CO_2, dry chemical. When heated to decomposition it emits acrid smoke and fumes.

GGA850   HR: D
GLYCEROL ESTER of PARTIALLY DIMERIZED ROSIN

PROP: Hard, pale amber-colored resin. Sol in acetone, benzene; insol in water.

SAFETY PROFILE: When heated to decomposition it emits acrid smoke and irritating fumes.

GGA865   HR: D
GLYCEROL ESTER of POLYMERIZED ROSIN

PROP: Hard, pale amber resin. Sol in acetone, benzene; insol in water, alc.

SAFETY PROFILE: When heated to decomposition it emits acrid smoke and irritating fumes.

GGA875   HR: D
GLYCEROL ESTER of WOOD ROSIN

PROP: Hard, pale amber resin. Sol in acetone, benzene; insol in water.

SAFETY PROFILE: When heated to decomposition it emits acrid smoke and irritating fumes.

GGA900   **HR: D**
GLYCEROL-LACTO PALMITATE
SAFETY PROFILE: When heated to decomposition it emits acrid smoke and irritating fumes.

GGA925   CAS: 25496-72-4   HR: D
GLYCEROL MONOOLEATE
PROP: Clear amber or pale yellow liquid. Insol in water.
SAFETY PROFILE: When heated to decomposition it emits acrid smoke and irritating fumes.

GGG000   CAS: 96-11-7   HR: 2
GLYCEROL TRIBROMOHYDRIN
mf: $C_3H_5Br_3$ mw: 280.81
PROP: A liquid. Mp: 16.5°, bp: 220°, d: 2.43 @ 23°. Insol in H_2O; sol in EtOH and Et_2O.
SYNS: GLYCERYL TRIBROMOHYDRIN □ sym-TRIBROMOPROPANE □ 1,2,3-TRIBROMOPROPANE
CONSENSUS REPORTS: EPA Genetic Toxicology Program. Reported in EPA TSCA Inventory.
SAFETY PROFILE: Moderately toxic by ingestion. Experimental reproductive effects. Mutation data reported. When heated to decomposition it emits toxic fumes of Br⁻. See also BROMIDES.

GGI000   CAS: 38571-73-2   HR: 3
GLYCEROL (TRI(CHLOROMETHYL))-ETHER
mf: $C_6H_{11}Cl_3O_3$ mw: 237.52
SYN: TRIS-1,2,3-(CHLOROMETHOXY)PROPANE
CONSENSUS REPORTS: IARC Cancer Review: Group 2A IMEMDT 7,56,87; Animal Sufficient Evidence IMEMDT 15,301,77.
SAFETY PROFILE: Confirmed carcinogen with experimental neoplastigenic and tumorigenic data. When heated to decomposition it emits toxic fumes of Cl⁻. See also ETHERS.

GGR200   CAS: 25496-72-4   HR: 1
GLYCERYL MONOOLEATE
mf: $C_{21}H_{40}O_4$ mw: 356.61

PROP: Yellow semi-solid. Mp: 86° F.
SYNS: ADCHEM GMO □ AJAX GMO □ ALDO 40 □ ALDO MO-FG □ DUR-EM 204 □ EMCOL O □ EMERY OLEIC ACID ESTER 2221 □ EMRITE 6009 □ GLYCERINE MONOOLEATE □ GLYCERIN MONOOLEATE □ GLYCEROL MONOOLEATE □ GLYCEROL OLEATE □ GLYCERYL OLEATE □ GMO 8903 □ HAROWAX L 9 □ LOXIOL G 10 □ MONOGLYCERYL OLEATE □ MONO-OLEIN □ MONOOLEOYLGLYCEROL □ 9-OCTADECE-NOIC ACID (Z)-, MONOESTER with 1,2,3-PROPANETRI-OL (9CI) □ OLEIC ACID GLYCEROL MONOESTER □ OLEIC ACID MONOGLYCERIDE □ OLEOYLGLYCER-OL □ OLEYLMONOGLYCERIDE □ OLICINE □ RIKEM-AL O 71D □ RIKEMAL OL 100 □ S 1096 □ S 1097 □ SINNOESTER OGC □ S 1096R □ SUNSOFT O 30B □ SUPEOL
CONSENSUS REPORTS: EPA TSCA Chemical Inventory, JUNE 1993. Reported in EPA TSCA Inventory.
SAFETY PROFILE: A skin and eye irritant. When heated to decomposition it emits acrid smoke and irritating fumes.

GGU400   CAS: 555-43-1   HR: D
GLYCERYL TRISTEARATE
PROP: Crystals from Et_2O or hexane. Mp: 73.5°.
SAFETY PROFILE: When heated to decomposition it emits acrid smoke and irritating fumes.

GGW000   CAS: 765-34-4   HR: 3
GLYCIDALDEHYDE
DOT: UN 2622
mf: $C_3H_4O_2$ mw: 72.07
PROP: Colorless liquid. Bp: 113°, d: 1.1403 @ 20°/4°.
SYNS: EPIHYDRINALDEHYDE □ EPIHYDRINE ALDEHYDE □ 2,3-EPOXYPROPANAL □ 2,3-EPOXY-1-PROPANAL □ 2,3-EPOXYPROPIONALDEHYDE □ GLYCIDAL □ GLYCIDYLALDEHDYE □ OXIRANE-CARBOXALDEHYDE □ RCRA WASTE NUMBER U126
CONSENSUS REPORTS: IARC Cancer Review: Group 2B IMEMDT 7,56,87; Animal Sufficient Evidence IMEMDT 11,175,76. EPA Genetic Toxicology Program.
DOT CLASSIFICATION: 3; Label: Flammable Liquid, Poison
SAFETY PROFILE: Confirmed carcinogen with experimental carcinogenic,

neoplastigenic, and tumorigenic data. Poison by ingestion, skin contact, intraperitoneal, and intravenous routes. Moderately toxic by inhalation. Human systemic effects by inhalation: changes in central nervous system electrical activity, olfactory changes, and excitement. Mutation data reported. A human eye irritant. Powerful skin sensitizer and mucous membrane irritant. Flammable when exposed to heat, flame, or oxidizing materials. When heated to decomposition it emits acrid smoke and irritating fumes. See also ALDEHYDES.

GGW500   CAS: 556-52-5   HR: 3
GLYCIDOL
mf: $C_3H_6O_2$ mw: 74.09
PROP: Colorless liquid. D: 1.165 @ 0°/4°, bp: 167° (decomp). Entirely sol in water, alc, and ether. IDLH 150 ppm.
SYNS: EPIHYDRIN ALCOHOL □ 2,3-EPOXYPROPAN-OL □ 2,3-EPOXY-1-PROPANOL □ 2,3-EPOXY-1-PROP-ANOL (OSHA) □ GLYCIDE □ GLYCIDYL ALCOHOL □ 3-HYDROXY-1,2-EPOXYPROPANE □ METHANOL, OXIRANYL- □ NCI-C55549
CONSENSUS REPORTS: NTP 10th Report on Carcinogens. Reported in EPA TSCA Inventory. EPA Genetic Toxicology Program.
OSHA PEL: TWA 25 ppm
ACGIH TLV: TWA 2 ppm; Animal Carcinogen
DFG MAK: 50 ppm (150 mg/m³)
SAFETY PROFILE: Confirmed carcinogen with carcinogenic data reported. Poison by intraperitoneal route. Moderately toxic by ingestion, inhalation, and skin contact. Experimental teratogenic and reproductive effects. A skin irritant. Human mutation data reported. Animal experiments suggest somewhat lower toxicity than for related epoxy compounds. Readily absorbed through the skin. Causes nervous excitation followed by depression. Explodes when heated or in the presence of strong acids, bases, metals (e.g., copper, zinc), and metal salts (e.g., aluminum chloride, iron(III) chloride, tin(IV) chloride). When heated to

decomposition it emits acrid smoke and fumes. See also DIGLYCIDYL ETHER.

GGY200   CAS: 3033-77-0   HR: 3
GLYCIDYL-TRIMETHYL-AMMONIUM CHLORIDE
mf: $C_6H_{14}NO \bullet Cl$ mw: 151.66
PROP: Clear liquid, colorless to light yellow. Practically odorless.
SYNS: (2,3-EPOXYPROPYL)TRIMETHYLAMMONIUM CHLORIDE □ GLYTAC □ GLYTAC A 100 □ G-MAC □ OXIRANEMETHANAMINIUM, N,N,N-TRIMETHYL-, CHLORIDE (9CI) □ TRIMETHYLGLYCIDYLAMMONIUM CHLORIDE □ N,N,N-TRIMETHYLOXIRANEMETHAN-AMINIUM CHLORIDE
CONSENSUS REPORTS: Reported in EPA TSCA Inventory.
DFG MAK: Confirmed Animal Carcinogen, Suspected Human Carcinogen
SAFETY PROFILE: Suspected carcinogen. Poison by subcutaneous route. Mutation data reported. When heated to decomposition it emits toxic fumes of Cl⁻, NH₃, and NOₓ. See also AMMONIUM CHLORIDE.

GHA000   CAS: 56-40-6   HR: 2
GLYCINE
mf: $C_2H_5NO_2$ mw: 75.08
PROP: White crystals from alc (aq); odorless, sweet taste. The simplest amino acid and the principal amino acid in sugar cane. Mp: 262° (decomp), d: 1.1607. Sol in water; insol in alc and ether.
SYNS: AMINOACETIC ACID □ GLYCOLIXIR □ HAMPSHIRE GLYCINE
CONSENSUS REPORTS: Reported in EPA TSCA Inventory.
SAFETY PROFILE: Moderately toxic by intravenous route. Mildly toxic by ingestion. Mutation data reported. When heated to decomposition it emits toxic fumes of NOₓ.

GHI100   CAS: 6011-14-9   HR: D
GLYCINONITRILE HYDROCHLORIDE
mf: $C_2H_4N_2 \bullet ClH$ mw: 92.54
PROP: A solid. Mp: 165° (decomp). Sltly sol in EtOH and Et₂O.
SYNS: ACETONITRILE, AMINO-, MONOHYDRO-CHLORIDE (9CI) □ AMINOACETONITRILE HYDRO-

CHLORIDE □ GLYCINONITRILE, MONOHYDRO-CHLORIDE

CONSENSUS REPORTS: Reported in EPA TSCA Inventory.

SAFETY PROFILE: An experimental teratogen. Experimental reproductive effects. When heated to decomposition it emits toxic fumes of NO_x and HCl.

GIA000   CAS: 36734-19-7   HR: 2
GLYCOPHEN
mf: $C_{13}H_{13}Cl_2N_3O_3$ mw: 330.19
PROP: Crystals. Mp: 136°.
SYNS: CHIPCO 26019 □ 3-(3,5-DICHLOROPHENYL)-N-(1-METHYLETHYL)-2,4-DIOXO-1-IMIDAZOLIDINE-CARBOXAMIDE □ GLYCOPHENE □ IPRODIONE □ 1-ISOPROPYL CARBAMOYL-3-(3,5-DICHLOROPHENYL)-HYDANTOIN □ LFA 2043 □ MRC 910 □ PROMIDIONE □ ROP 500 F □ ROVRAL □ RP 26019
SAFETY PROFILE: Moderately toxic by ingestion. When heated to decomposition it emits very toxic fumes of NO_x and Cl⁻.

GIC000   CAS: 596-51-0   HR: 3
GLYCOPYRRONIUM BROMIDE
mf: $C_{19}H_{28}NO_3 \cdot Br$ mw: 398.39
PROP: Crystals from butanone. Mp: 193.2–194.5°.
SYNS: ASECRYL □ 1,1-DIMETHYL-3-HYDROXYPYR-ROLIDINIUM BROMIDE-α-CYCLOPENTYLMANDEL-ATE □ GASTRODYN □ GLYCOPYRROLATE □ GLYCOPYRROLATE BROMIDE □ NODAPTON □ ROBANUL □ ROBINUL □ TARODYL □ TARODYN
SAFETY PROFILE: Poison by intravenous and intraperitoneal routes. Moderately toxic by ingestion and subcutaneous routes. Experimental reproductive effects. When heated to decomposition it emits very toxic fumes of NO_x and Br⁻. See also BROMIDES.

GIE100   CAS: 53956-04-0   HR: 2
GLYCYRRHIZIC ACID, AMMONIUM SALT
mf: $C_{42}H_{63}O_{16} \cdot xH_3N$ mw: 943.33
PROP: Needles from alc (aq).
SYNS: AMMONIATED GLYCYRRHIZIN □ AMMON-IUM GLYCYRRHIZINATE □ α-D-GLUCOPYRANO-SIDURONIC ACID, (3-β,20-β)-20-CARBOXY-11-OXO-30-NOROLEAN-12-EN-3-YL 2-O-β-D-GLUCOPYRANURO-NOSYL-, AMMONIATE □ MONOAMMONIUM GLYCYRRHIZINATE

CONSENSUS REPORTS: Reported in EPA TSCA Inventory.

SAFETY PROFILE: Moderately toxic by intravenous route. An experimental teratogen. Experimental reproductive effects. Mutation data reported. When heated to decomposition it emits toxic fumes of NH_3.

GIG000   CAS: 1405-86-3   HR: 3
GLYCYRRHIZINIC ACID
mf: $C_{42}H_{62}O_{16}$ mw: 823.04
PROP: Crystals from glacial acetic acid or hygroscopic powder. Mp: 220° (approx). Intensely sweet taste. Freely sol in hot water, alc; practically insol in ether. The active component of licorice (BMJOAE 1,488,77).
SYNS: GLYCYRON □ GLYCYRRHETINIC ACID GLYCOSIDE □ GLYCYRRHIZIC ACID □ GLYCYR-RHIZIC ACID (8CI) □ 18-β-GLYCYRRHIZIC ACID □ GLYCYRRHIZIN □ β-GLYCYRRHIZIN □ LIQUORICE
CONSENSUS REPORTS: Reported in EPA TSCA Inventory.
SAFETY PROFILE: Poison by intravenous route. Moderately toxic by ingestion and intraperitoneal routes. Human systemic effects by ingestion: somnolence and changes in the metabolism of phosphorus. When heated to decomposition it emits acrid smoke and irritating fumes.

GIK000   CAS: 107-22-2   HR: 3
GLYOXAL
mf: $C_2H_2O_2$ mw: 58.04
PROP: Yellow prisms or irregular pieces turning white on cooling. D: 1.29 @ 20°/4°. Opaque @ 10°, mp: 15°, bp: (776) 51°. The vapors are green and burn with a purple flame, n: (20.5/D) 1.3826. Sol in anhyd solvents, pH of a 40% aq soln: 2.1–2.7, d: (20/4) 1.27.
SYN: AEROTEX GLYOXAL 40
CONSENSUS REPORTS: Reported in EPA TSCA Inventory.
ACGIH TLV: TWA 0.1 mg/m³ (sensitizer); Not Classifiable as a Human Carcinogen).
SAFETY PROFILE: Low toxicity by ingestion and skin contact. A skin irritant. A

powerful reducing agent. May explode on contact with air. Polymerizes violently on contact with water. During storage it may spontaneously polymerize and ignite. Reacts violently with chlorosulfonic acid, ethylene imine, HNO_3, oleum, NaOH, can cause violent reactions. Can explode during manufacture. When heated to decomposition it emits acrid smoke and irritating fumes. See also ALDEHYDES.

GIS000   CAS: 7440-57-5   HR: 1
GOLD
af: Au   aw: 196.97
PROP: Cubic, yellow, ductile, metallic crystals. Forms red, blue, or violet colloidal suspensions. Physical properties depend on mechanical treatment. Mp: 1064.76°, bp: 2700°, d: 19.3 (liquid) 17.0 @ 1063°, vap press: 1 mm @ 1869°, hardness: (Mohs') 2.5–3.0, (Brinell's) 18.5.
SYNS: BURNISH GOLD □ C.I. 77480 □ C.I. PIGMENT METAL 3 □ COLLOIDAL GOLD □ GOLD FLAKE □ GOLD LEAF □ GOLD POWDER □ MAGNESIUM GOLD PURPLE □ SHELL GOLD
CONSENSUS REPORTS: Reported in EPA TSCA Inventory.
SAFETY PROFILE: Poison by intravenous route. Questionable carcinogen with experimental tumorigenic data by implantation. Can form explosive compounds with NH_3, NH_4OH + aqua regia, H_2O_2. Incompatible with mixtures containing chlorides, bromides, or iodides (if they can generate nascent halogens), some oxidizing materials (especially those containing halogens), alkali cyanides, thiocyanate solutions, and double cyanides. See also GOLD COMPOUNDS.

GIW176   CAS: 13453-07-1   HR: 2
GOLD CHLORIDE
mf: $AuCl_3$ mw: 303.33
PROP: Claret-red crystals; mp: 254° (decomp), bp: subl at 265°, d: 3.9.
SYNS: AURIC CHLORIDE □ GOLD(III) CHLORIDE □ GOLD TRICHLORIDE
CONSENSUS REPORTS: Reported in EPA TSCA Inventory.

SAFETY PROFILE: Experimental reproductive effects. Human mutation data reported. Reaction with ammonia or ammonium salts yields fulminating gold, a heat-, friction-, and impact-sensitive explosive similar to mercury and silver fulminates. See also GOLD COMPOUNDS and CHLORIDES. When heated to decomposition it emits toxic fumes of Cl^-.

GJC000   CAS: 12244-57-4   HR: 3
GOLD SODIUM THIOMALATE
mf: $C_4H_3AuO_4S•2Na$ mw: 390.08
PROP: Yellowish-white powder. Sol in H_2O; insol in EtOH and Et_2O.
SYNS: AuTM □ ((1,2-DICARBOXYETHYL)THIO)GOLD DISODIUM SALT □ (DIHYDROGEN MERCAPTOSUC-CINATO)GOLD DISODIUM SALT □ DISODIUM AUROTHIOMALATE □ (MERCAPTOBUTANEDIO-ATO(1-))GOLD DISODIUM SALT □ MERCAPTOSUC-CINIC ACID, GOLD SODIUM SALT □ MYOCHRYSINE □ MYOCRISIN □ SODIUM AUROTHIOMALATE □ TAURE(o)DON
SAFETY PROFILE: Poison by subcutaneous and intramuscular routes. Moderately toxic by intravenous route. Human systemic effects: aggression, agranulocytosis, aplastic anemia, cell count changes, changes in circulation, cholestatic jaundice, dermatitis, encephalitis, fasciculations, flaccid paralysis without anesthesia, hemorrhage, hepatitis (hepatocellular necrosis), increased body temperature, interstitial fibrosis, muscle weakness, proteinuria, recording from peripheral motor nerve, depressed renal function tests, somnolence, structural changes in nerve sheath, thrombocytopenia, uncharacterized allergic reaction, changes in blood, teeth, and supporting structures. Experimental teratogenic and reproductive effects. When heated to decomposition it emits very toxic Na_2O and SO_x.

GJS300   HR: D
GRAPE COLOR EXTRACT
SAFETY PROFILE: When heated to decomposition it emits acrid smoke and irritating fumes.

GJU000   CAS: 8016-20-4   HR: 2
GRAPEFRUIT OIL
PROP: From the fresh peel of *Citrus paradisi*
Macfayden (*Citrus decumana L.*). Yellow
liquid. Sol in fixed oils, mineral oil; sltly sol
in propylene glycol; insol in glycerin.
SYNS: GRAPEFRUIT OIL, coldpressed □ GRAPEFRUIT
OIL, expressed □ OIL OF GRAPEFRUIT □ OIL OF
SHADDOCK
CONSENSUS REPORTS: Reported in EPA
TSCA Inventory.
SAFETY PROFILE: A skin irritant.
Questionable carcinogen with experimental
tumorigenic data. Mutation data reported.
When heated to decomposition it emits
acrid smoke and irritating fumes.

GJU600   CAS: 68085-85-8   HR: 3
GRENADE
DOT: NA 0349/UN 0110/UN 0284/UN
0285/UN 0292/UN 0293/UN 0318/UN
0372/UN 0452
mf: $C_{23}H_{19}ClF_3NO_3$ mw: 449.88
PROP: Yellow-brown viscous liquid.
Characteristic odor. Decomp below BP at
275 C, mp: 10 C. Flash pt: 80° C. Insol in
water.
SYNS: CYCLOPROPANECARBOXYLIC ACID, 3-(2-
CHLORO-3,3,3-TRIFLUORO-1-PROPENYL)-2,2-
DIMETHYL-, CYANO(3-PHENOXYPHENYL)METHYL
ESTER □ CYHALOTHRIN □ CYHALOTHRINE □
GRENADES, empty primed (NA0349) (DOT) □ GREN-
ADES, hand or rifle, with bursting charge (UN0284, UN0285,
UN0292, UN0293) (DOT) □ GRENADES, practice, hand or
rifle (UN0452, UN0110, UN0318, UN0372) (DOT) □ ICI
146814 □ ICI-PP 563 □ PP 563
DOT CLASSIFICATION: Explosive 1.4S;
Label: None (NA 0349, UN 0110);
Explosive 1.1D; Label: Explosive 1.1D (UN
0284); Explosive 1.1F; Label: Explosive 1.1F
(UN 0292); Explosive 1.2F; Label:
Explosive 1.2F (UN 0293); Explosive 1.4G;
Label: Explosive 1.4G (UN 0452);
Explosive 1.3G; Label: Explosive 1.3G (UN
0318); Explosive 1.2G; Label: Explosive
1.2G (UN0372); Explosive 1.2D; Label:
Explosive 1.2D (UN0285)
SAFETY PROFILE: A poison by ingestion
and inhalation. An explosive. A combustible

liquid. When heated to decomposition it
emits toxic vapors of NO_x, F^-, and Cl^-.

GKE000   CAS: 126-07-8   HR: 3
GRISOFULVIN
mf: $C_{17}H_{17}ClO_6$ mw: 352.79
PROP: Octahedra crystals from C_6H_6. Mp:
225–226°.
SYNS: AMUDANE □ BIOGRISIN-FP □ 7-CHLORO-
4,6,2'-TRIMETHOXY-6'-METHYLGRIS-2'-EN-3,4'-DIONE
□ DELMOFULVINA □ FULCIN □ FULCINE □ FULVIC-
AN GRISACTIN □ FULVICIN □ FULVINA □ FULVISTA-
TIN □ FUNGIVIN □ GREOSIN □ GRESFEED □ GRICIN
□ GRIFULVIN □ GRISACTIN □ GRISCOFULVIN □
GRISEFULINE □ GRISEO □ (+)-GRISEOFULVIN □
GRISEOFULVIN-FORTE □ GRISEOFULVINUM □
GRISETIN □ GRISOVIN □ GRIS-PEG □ GRYSIO □
GUSERVIN □ LAMORYL □ LIKUDEN □ MURFULVIN □
NEO-FULCIN □ NSC-34533 □ PONCYL □ SPIROFULVIN
□ SPOROSTATIN □ USAF SC-2
CONSENSUS REPORTS: IARC Cancer
Review: Group 2B IMEMDT 7,56,87;
Animal Sufficient Evidence IMEMDT
10,153,76. EPA Genetic Toxicology
Program.
SAFETY PROFILE: Confirmed carcinogen
with experimental neoplastigenic and
teratogenic data. Poison by intravenous and
intraperitoneal routes. Moderately toxic by
subcutaneous route. Human mutation data
reported. Experimental reproductive effects.
Used as an antibiotic, pharmaceutical, and
veterinary drug. When heated to
decomposition it emits toxic fumes of Cl^-.

GKI000   CAS: 90-05-1   HR: 3
GUAIACOL
mf: $C_7H_8O_2$ mw: 124.15
PROP: Clear, pale-yellow liquid, solid,
prisms, or needles. Characteristic odor,
darkens on exposure to air and light. D
(crystals): 1.129, d (liquid): about 1.112,
(needles), mp: 32°, (prisms), bp: 202–209°,
fp: −3.2°, flash p: 180°F (OC), d: 1.097 @
25°/25°. Misc with alc, chloroform, ether,
oils, glacial acetic acid; sltly sol in pet ether;
sol in NaOH soln.
SYNS: GUAICOL □ o-HYDROXYANISOLE □ 2-HY-
DROXYANISOLE □ 1-HYDROXY-2-METHOXY-
BENZENE □ o-METHOXYPHENOL □ 2-METHOXY-

G

PHENOL □ METHYLCATECHOL □ PYROGUAIAC ACID

CONSENSUS REPORTS: Reported in EPA TSCA Inventory. EPA Genetic Toxicology Program.

SAFETY PROFILE: Human poison by ingestion. Experimental poison by intravenous route. Mildly toxic by skin contact and inhalation. Human systemic effects by ingestion: tremors and gastrointestinal changes. Human mutation data reported. An eye and severe skin irritant. Ingestion produces burning in the mouth and throat. Flammable when exposed to heat or flame; can react with oxidizing materials. To fight fire, use foam, CO_2, dry chemical. Protect from light. Used as an expectorant. When heated to decomposition it emits acrid smoke and irritating fumes. See also PHENOL.

GLA000   CAS: 506-93-4   HR: 3
GUANIDINE MONONITRATE
DOT: UN 1467
mf: $CH_5N_3 \cdot HNO_3$ mw: 122.11
PROP: White granules. Mp: 214°.
SYN: GUANIDINE NITRATE (DOT)
CONSENSUS REPORTS: Reported in EPA TSCA Inventory.
DOT CLASSIFICATION: 5.1; Label: Oxidizer
SAFETY PROFILE: Moderately toxic by ingestion. A severe skin and an eye irritant. A powerful oxidizer. Flammable when shocked or exposed to heat or flame. A stable, flashless, non-hygroscopic high explosive used as a blasting explosive in combination with charcoal and inorganic nitrates. Keep away from heat and open flame. When heated to decomposition it emits very toxic fumes of HNO_3 and NO_x. See also NITRATES, GUANIDINE MONOHYDROCHLORIDE, and EXPLOSIVES, HIGH.

GLI000   CAS: 73-40-5   HR: 2
GUANINE
mf: $C_5H_5N_5O$ mw: 151.15

PROP: Crystals. Usually amorph. Decomp: >360° with partial subl. Very sol in ammonia water, aq KOH solns, dil acids; very sltly sol in alc, ether; insol in water.
SYNS: 2-AMINOHYPOXANTHINE □ MEARLMAID
CONSENSUS REPORTS: Reported in EPA TSCA Inventory. EPA Genetic Toxicology Program.
SAFETY PROFILE: Questionable carcinogen with experimental tumorigenic data. Human mutation data reported. When heated to decomposition it emits toxic fumes of NO_x.

GLS000   CAS: 118-00-3   HR: 3
GUANOSINE
mf: $C_{10}H_{13}N_5O_5$ mw: 283.28
PROP: A component of nucleic acids. Mp: 239° (decomp).
SYNS: GR □ GUANINE, 9-β-D-RIBOFURANOSYL- □ GUANINE RIBOSIDE □ 2(3H)-IMINO-9-β-D-RIBO-FURANOSYL-9H-PURIN-6(1H)-ONE □ INOSINE, 2-AMINO- □ RIBOFURANOSIDE, GUANINE-9, β-D- □ 9-β-D-RIBOFURANOSYLGUANINE □ USAF CB-11 □ VERNINE
CONSENSUS REPORTS: Reported in EPA TSCA Inventory.
SAFETY PROFILE: Poison by intravenous route. Moderately toxic by intraperitoneal route. Experimental teratogenic effects. Human mutation data reported. When heated to decomposition it emits toxic fumes of NO_x.

GLS800   CAS: 5550-12-9   HR: 2
GUANYLIC ACID SODIUM SALT
mf: $C_{10}H_{14}N_5O_8P \cdot 2Na$ mw: 409.24
PROP: Colorless to white crystals; characteristic taste. Sol in water; sltly sol in alc; insol in ether.
SYNS: DISODIUM GMP □ DISODIUM-5'-GMP □ DISODIUM GUANYLATE (FCC) □ DISODIUM-5'-GUANYLATE □ GMP DISODIUM SALT □ 5'-GMP DISODIUM SALT □ GMP SODIUM SALT □ SODIUM GMP □ SODIUM GUANOSINE-5'-MONOPHOSPHATE □ SODIUM GUANYLATE □ SODIUM-5'-GUANYLATE
CONSENSUS REPORTS: Reported in EPA TSCA Inventory.
SAFETY PROFILE: Moderately toxic by intraperitoneal, subcutaneous, and

intravenous routes. Mildly toxic by ingestion. Mutation data reported. When heated to decomposition it emits toxic fumes of PO_x, NO_x, and Na_2O.

GLU000   CAS: 9000-30-0   HR: 1
GUAR GUM

PROP: Yellowish-white powder; odorless. Sol in water; insol in oils, grease, hydrocarbons, ketones, esters. Obtained from the ground endosperms of *Cyanopsis tetragonoloan L. Taub* (Fam. *Leguminosae*).

SYNS: 1212A □ A-20D □ BURTONITE V-7-E □ CYAMOPSIS GUM □ DEALCA TP1 □ DEALCA TP2 □ DECORPA □ GALACTASOL □ GENDRIV 162 □ GUAR □ GUARAN □ GUAR FLOUR □ GUM CYAMOPSIS □ GUM GUAR □ INDALCA AG □ INDALCA AG-BV □ INDALCA AG-HV □ JAGUAR □ JAGUAR 6000 □ JAGUAR A 20 B □ JAGUAR A 20D □ JAGUAR A 40F □ JAGUAR GUM A-20-D □ JAGUAR No. 124 □ JAGUAR PLUS □ J 2Fp □ LYCOID DR □ NCI-C50395 □ REGONOL □ REIN GUARIN □ SUPERCOL G.F. □ SUPERCOL U POWDER □ SYNGUM D 46D □ UNI-GUAR

CONSENSUS REPORTS: NTP Carcinogenesis Bioassay (feed); No Evidence: mouse, rat NTPTR* NTP-TR-229,82. Reported in EPA TSCA Inventory. EPA Genetic Toxicology Program.

SAFETY PROFILE: Mildly toxic by ingestion. Experimental reproductive effects. When heated to decomposition it emits acrid smoke and irritating fumes.

GLY000   CAS: 9000-28-6   HR: 1
GUM GHATTI

PROP: The gummy exudation from the stem of *Anogeissus latifolia*. Colorless to pale-yellow tears; almost odorless. Sltly sol in water.

SYN: INDIAN GUM

CONSENSUS REPORTS: Reported in EPA TSCA Inventory.

SAFETY PROFILE: Mildly toxic by ingestion. When heated to decomposition it emits acrid smoke and irritating fumes.

GLY100   CAS: 9000-29-7   HR: 2
GUM GUAIAC

PROP: From wood of *guajacum officinale L.* or *Guajacum sanctum L.* (Fam. *Zygophyllaceae*). Brown solid; balsamic odor, sltly acrid taste. Sol in alc, ether, chloroform, solns of alkalies; sltly sol in carbon disulfide, benzene.

SYN: GUAIAC GUM

SAFETY PROFILE: Moderately toxic by ingestion. When heated to decomposition it emits acrid smoke and irritating fumes.

G

H

HAF375   CAS: 12298-43-0   HR: 1
HALLOYSITE
PROP: White to gray or brown solid.
DFG MAK: Confirmed Animal Carcinogen
with Unknown Relevance to Humans
SAFETY PROFILE: Suspected carcinogen
with experimental tumorigenic data.

HAF600   HR: D
HALOFUGINONE HYDROBROMIDE
mf: $C_{16}H_{18}Br_2ClN_3O_3$ mw: 495.612
PROP: Crystals.
SYNS: 7-BROMO-6-CHLOROFEBRIFUGINE
HYDROBROMIDE □ 7-BROMO-6-CHLORO-3-[3-(3-
HYDROXY-2-PIPERDINYL)-2-OXOPROPYL]-4(3H)-
QUINAZOLINONE HYDROBROMIDE
SAFETY PROFILE: When heated to
decomposition it emits acrid smoke and
irritating fumes.

HAK000   CAS: 7789-20-0   HR: D
HEAVY WATER
mf: D_2O mw: 20.02
PROP: Colorless, odorless liquid. Mp: 3.81°,
triple point temp: 3.82°, bp: 101.42°. Critical
temp: 371.5°, d: 1.1044. Heat is evolved on
mixing with normal water.
SYNS: DEUTERIUM OXIDE □ DIDEUTERIUM OXIDE
□ HEAVY WATER-d2 □ WATER-d2 (9CI) □ WATER²-H2
CONSENSUS REPORTS: EPA Genetic
Toxicology Program. Reported in EPA
TSCA Inventory.
SAFETY PROFILE: Experimental
reproductive effects. See also WATER.

HAM500   CAS: 7440-59-7   HR: 1
HELIUM
DOT: UN 1046/UN 1963
af: He   aw: 4.00
PROP: Colorless, odorless, tasteless,
monatomic, non-toxic, inert gas. Forms no
normal chemical compounds. Mp: −272.2°

@ 26 atm, bp: −268.9°, d: (gas): 0.1785 g/L
@ 0°, d: (liquid): 0.147 @ −270.8°.
SYNS: HELIUM, compressed (UN 1046) (DOT) □
HELIUM, refrigerated liquid (cryogenic liquid) (UN 1963)
(DOT)
CONSENSUS REPORTS: Reported in EPA
TSCA Inventory.
DOT CLASSIFICATION: 2.2; Label:
Nonflammable Gas
SAFETY PROFILE: A simple asphyxiant.
A nonflammable gas. See ARGON for a
description of simple asphyxiants.

HAO875   CAS: 1317-60-8   HR: 2
HEMATITE
PROP: Consists mainly of Fe_2O_3 (IARC**
1,29,71).
SYNS: BLOODSTONE □ HAEMATITE □ IRON ORE □
RED IRON ORE
CONSENSUS REPORTS: IARC Cancer
Review: Group 3, Indefinite IMSUPP
4,254,82. Reported in EPA TSCA
Inventory.
SAFETY PROFILE: Questionable
carcinogen.

HAO900   CAS: 635-65-4   HR: D
HEMATOIDIN
mf: $C_{33}H_{36}N_4O_6$ mw: 584.73
PROP: Orange-brown crystals. Sltly sol org
solvs; insol in H_2O.
SYNS: BILINE-8,12-DIPROPIONIC ACID, 1,10,19,22,23-
,24-HEXAHYDRO-2,7,13,17-TETRAMETHYL-1,19-DIOXO-
3,18-DIVINYL- □ BILIRUBIN □ BILIRUBIN IX-α □
HEMETOIDIN □ PRINCIPAL BILE PIGMENT
CONSENSUS REPORTS: Reported in EPA
TSCA Inventory.
SAFETY PROFILE: An experimental
teratogen. Experimental reproductive
effects. When heated to decomposition it
emits toxic fumes of NO_x.

HAP100 CAS: 16478-59-4 HR: 3
HEMICHOLINIUM

mf: $C_{24}H_{34}N_2O_4$ mw: 414.60

SYNS: 2,2'-(BIPHENYL)-4,4'-DIYLBIS(2-HYDROXY-4,4-DIMETHYLMORPHOLINIUM) □ 2,2'-(4,4'-BIPHENYL-ENE)BIS(2-HYDROXY-4,4-DIMETHYLMORPHOLINIUM) □ MORPHOLINIUM, 2,2'-(4,4'-BIPHENYLENE)BIS(2-HYDROXY-4,4-DIMETHYL- □ MORPHOLINIUM, 2,2'-(1,1'-BIPHENYL)-4,4'-DIYLBIS(2-HYDROXY-4,4-DIMETHYL-

SAFETY PROFILE: A poison by ingestion. When heated to decomposition it emits toxic vapors of NO_x.

HAQ000 CAS: 312-45-8 HR: 3
HEMICHOLINIUM-3-DIBROMIDE

mf: $C_{24}H_{34}N_2O_4 \cdot 2Br$ mw: 574.42

SYNS: 2,2'-(1,1'-BIPHENYL)-4,4'-DIYLBIS(2-HYDROXY-4,4-DIMETHYL)-MORPHOLINIUM DIBROMIDE □ HC-3 □ HEMICHOLINE □ HEMICHOLINIUM-3 □ HEMICHOLINIUM BROMIDE □ HEMICHOLINIUM-3-BROMIDE □ HEMICHOLINIUM DIBROMIDE

CONSENSUS REPORTS: Reported in EPA TSCA Inventory.

SAFETY PROFILE: A poison by subcutaneous, intraperitoneal, and intravenous routes. See also BROMIDES. When heated to decomposition it emits very toxic fumes of NO_x and Br^-.

HAR000 CAS: 76-44-8 HR: 3
HEPTACHLOR

mf: $C_{10}H_5Cl_7$ mw: 373.30

PROP: Crystals. Mp: 96°. Nearly insol in water; sol in org solvs. IDLH 35 mg/m³.

SYNS: AGROCERES □ 3-CHLOROCHLORDENE □ DRINOX □ E 3314 □ ENT 15,152 □ EPTACLORO (ITALIAN) □ 1,4,5,6,7,8,8-EPTACLORO-3a,4,7,7a-TETRAIDRO-4,7-endo-METANO-INDENE (ITALIAN) □ GPKh □ H-34 □ HEPTACHLOOR (DUTCH) □ 1,4,5,6,7,8,8-HEPTACHLOOR-3a,4,7,7a-TETRAHYDRO-4,7-endo-METHANO-INDEEN (DUTCH) □ HEPTACHLORE (FRENCH) □ 3,4,5,6,7,8,8-HEPTACHLORODICYCLO-PENTADIENE □ 3,4,5,6,7,8,8a-HEPTACHLORODICYCLO-PENTADIENE □ 1,4,5,6,7,8,8-HEPTACHLORO-3a,4,7,7a-TETRAHYDRO-4,7-ENDOMETHANOINDENE □ 1,4,5,6,7,10,10-HEPTACHLORO-4,7,8,9,-TETRAHYDRO-4,7-ENDOMETHYLENEINDENE □ 1,4,5,6,7,8,8a-HEPTACHLORO-3a,4,7,7a-TETRAHYDRO-4,7-METHANOINDANE □ 1,4,5,6,7,8,8-HEPTACHLORO-3a,4,7,7a-TETRAHYDRO-4,7-METHANOINDENE □ 1(3a),4,5,6,7,8,8-HEPTACHLORO-3a(1),4,7,7a-TETRAHYDRO-4,7-METHANOINDENE □ 1,4,5,6,7,8,8-HEPTACHLORO-3a,4,7,7a-TETRAHYDRO-4,7-METHANOL-1H-INDENE □ 1,4,5,6,7,8,8-HEPTACHLORO-3a,4,7,7,7a-TETRAHYDRO-4,7-METHYLENE

INDENE □ 1,4,5,6,7,8,8-HEPTACHLOR-3a,4,7,7,7a-TETRAHYDRO-4,7-endo-METHANO-INDEN (GERMAN) □ HEPTAGRAN □ HEPTAMUL □ NCI-C00180 □ RCRA WASTE NUMBER P059 □ RHODIACHLOR □ VELSICOL 104

CONSENSUS REPORTS: IARC Cancer Review: Group 3 IMEMDT 7,146,87; Human Inadequate Evidence IMEMDT 20,129,79; Animal Inadequate Evidence IMEMDT 5,173,74; Animal Sufficient Evidence IMEMDT 20,129,79. NCI Carcinogenesis Bioassay (feed) Clear Evidence: mouse NCITR* NCI-CG-TR-9,77; Results negative: rat NCITR* NCI-CG-TR-9,77. EPA Genetic Toxicology Program. Community Right-To-Know List.

OSHA PEL: TWA 0.5 mg/m³ (skin)

ACGIH TLV: 0.05 mg/m³ (skin); Animal Carcinogen

DFG MAK: 0.5 mg/m³, Confirmed Animal Carcinogen with Unknown Relevance to Humans

SAFETY PROFILE: Confirmed carcinogen with experimental carcinogenic data. A poison by ingestion, skin contact, intraperitoneal, and intravenous routes. Human mutation data reported. Acute exposure and chronic doses have caused liver damage. See also closely related chlordane. In humans, a dose of 1–3 g can cause serious symptoms, especially where liver impairment is the case. Acute symptoms include tremors, convulsions, kidney damage, respiratory collapse, and death. When heated to decomposition it emits toxic fumes of Cl^-.

NOTE: The EPA has canceled registration of pesticides containing heptachlor with the exception of its use for termite control by subsurface ground insertion external to the dwelling.

HAR100 CAS: 35822-46-9 HR: 3
1,2,3,4,6,7,8-HEPTACHLORODIBENZO-p-DIOXIN

mf: $C_{12}HCl_7O_2$ mw: 425.28

SYNS: DIBENZO(B,E)(1,4)DIOXIN, 1,2,3,4,6,7,8-HEPTACHLORO- □ DIBENZO-p-DIOXIN, 1,2,3,4,6,7,8-HEPTACHLORO- □ HEPTACHLORODIBENZO-p-

DIOXIN □ 1,2,3,4,6,7,8-HEPTACHLORODIBENZO-
DIOXIN

CONSENSUS REPORTS: IARC Cancer
Review: Group 3 IMEMDT 7,56,87; Animal
Inadequate Evidence IMEMDT 15,41,77;
Human No Adequate Data IMEMDT
15,41,77.
SAFETY PROFILE: A poison by ingestion.
Questionable carcinogen. When heated to
decomposition it emits toxic vapors of Cl⁻.

HAS075   CAS: 1763-23-1   HR: 3
1,1,2,2,3,3,4,4,5,5,6,6,7,7,8,8,8-HEPTA-
DECAFLUORO-1-OCTANESULFONIC
ACID
mf: $C_8HF_{17}O_3S$ mw: 500.13
SYNS: PERFLUOROOCTANESULFONIC ACID □ PFOS
SAFETY PROFILE: A poison by ingestion.
Experimental reproductive effects. When
heated to decomposition it emits toxic
vapors of SO_x and F⁻.

HAV450   CAS: 5910-85-0   HR: 3
2,4-HEPTADIENAL
mf: $C_7H_{10}O$ mw: 110.17
PROP: Sltly yellow liquid; green odor. Refr
index: 1.478–1.480, flash p: 140°F. Sol in
alc, fixed oils, water.
SYNS: FEMA No. 3164 □ HEPTADIENAL-2,4 □
trans,trans-2,4-HEPTADIENAL □ 2,4-HEPTDAIENAL
SAFETY PROFILE: Poison by skin contact.
Moderately toxic by ingestion. A severe skin
irritant. Flammable liquid when exposed to
heat, sparks, or flame. When heated to
decomposition it emits acrid smoke and
fumes.

HAW000   CAS: 3794-64-7   HR: 2
HEPTAFLUOROBUTANOIC ACID,
SILVER SALT
mf: $C_4F_7O_2•Ag$ mw: 320.91
PROP: IDLH 10 mg/m³ (as Ag).
SYN: HEPTAFLUORMASELNAN STRIBRNY (CZECH)
CONSENSUS REPORTS: Silver and its
compounds are on the Community Right-
To-Know List. Reported in EPA TSCA
Inventory.
OSHA PEL: TWA 0.01 mg(Ag)/m³
ACGIH TLV: TWA 0.01 mg(Ag)/m³

SAFETY PROFILE: Moderately toxic by
ingestion. A skin and severe eye irritant.
When heated to decomposition it emits
toxic fumes of F⁻. See also SILVER
COMPOUNDS.

HAX500   CAS: 375-22-4   HR: 3
HEPTAFLUOROBUTYRIC ACID
mf: $C_4HF_7O_2$ mw: 214.05
PROP: Colorless liquid; sharp, butyric acid
odor. Mp: −19.9°, bp: 120.8–121°, d: 1.65
@ 20°/4°.
CONSENSUS REPORTS: Reported in EPA
TSCA Inventory.
SAFETY PROFILE: A poison by
intraperitoneal route. Probably an eye, skin,
and mucous membrane irritant. Will react
with water or steam to produce corrosive
fumes. When heated to decomposition it
emits toxic fumes of F⁻.

HAY300   CAS: 27636-85-7   HR: 1
HEPTAFLUOROIODOPROPANE
mf: C_3F_7I mw: 295.93
PROP: A liquid. D: 2.06 @ 20°/4°, bp:
41.2°.
SYNS: HEPTAFLUORJODPROPAN □ IODOHEPTA-
FLUOROPROPANE □ PROPANE, HEPTAFLUORO-
IODO-
CONSENSUS REPORTS: Reported in EPA
TSCA Inventory.
SAFETY PROFILE: An eye irritant. When
heated to decomposition it emits toxic
fumes of Cl⁻ and I⁻.

HBA550   CAS: 105-21-5   HR: 1
γ-HEPTALACTONE
mf: $C_7H_{12}O_2$ mw: 128.19
PROP: Colorless, sltly oily liquid; coconut,
sweet, malty, caramel odor. D: 0.997–1.004
@ 20°, refr index: 1.439–1.445. Misc in alc,
fixed oils; very sltly sol in water.
SYNS: FEMA No. 2539 □ HEPTANOLIDE-1,4 □
HEPTANOLIDE-4,1 □ 4-HYDROXYHEPTANOIC ACID
LACTONE □ 4-HYDROXYHEPTANOIC ACID, γ-
LACTONE □ γ-PROPIOBUTYROLACTONE
CONSENSUS REPORTS: Reported in EPA
TSCA Inventory.

H

SAFETY PROFILE: A skin irritant. When heated to decomposition it emits acrid smoke and irritating fumes.

HBB500   CAS: 111-71-7   HR: 3
HEPTANAL
DOT: UN 3056

mf: $C_7H_{14}O$ mw: 114.18

PROP: Colorless liquid; penetrating, fruity odor. D: 0.814–0.819, refr index: 1.412–1.420, mp: −43.3°, bp: 152.8°, flash p: 93°F. Sol in alc, ether, fixed oils; sltly sol in water @ 153°; misc in alc, ether.

SYNS: ENANTHAL □ ENANTHALDEHYDE □ ENAN-THOLE □ FEMA No. 2540 □ HEPTALDEHYDE □ n-HEPTALDEHYDE □ OENANTHAL □ OENANTH-ALDEHYDE □ OENANTHIC ALDEHYDE □ OENANTHOL

CONSENSUS REPORTS: Reported in EPA TSCA Inventory.

DOT CLASSIFICATION: 3; Label: Flammable Liquid

SAFETY PROFILE: Mildly toxic by ingestion. Flammable liquid. When heated to decomposition it emits acrid smoke.

HBC500   CAS: 142-82-5   HR: 3
HEPTANE
DOT: UN 1206

mf: C_7H_{16} mw: 100.23

PROP: Colorless liquid. Bp: 98.52°, lel: 1.05%, uel: 6.7%, mp: −91.61°, flash p: 25°F (CC), d: 0.684 @ 20°/4°, autoign temp: 433.4°F, vap press: 40 mm @ 22.3°, vap d: 3.45. Sltly sol in alc; misc in ether and chloroform; insol in water. IDLH 750 ppm.

SYNS: DIPROPYL METHANE □ EPTANI (ITALIAN) □ GETTYSOLVE-C □ HEPTAN (POLISH) □ n-HEPTANE □ HEPTANEN (DUTCH) □ HEPTYL HYDRIDE

CONSENSUS REPORTS: Reported in EPA TSCA Inventory.

OSHA PEL: TWA 400 ppm; STEL 500 ppm

ACGIH TLV: TWA 400 ppm; STEL 500 ppm

DFG MAK: 500 ppm (2100 mg/m³)

NIOSH REL: TWA (Alkanes) 350 mg/m³

DOT CLASSIFICATION: 3; Label: Flammable Liquid

SAFETY PROFILE: Poison by intravenous route. Mildly toxic by inhalation. Human systemic effects by inhalation: hallucinations. Narcotic in high concentrations. A volatile, flammable liquid when exposed to heat or flame. Can react vigorously with oxidizing materials. Moderately explosive when exposed to heat or flame. Violent reaction with phosphorus + chlorine. To fight fire, use foam, CO_2, dry chemical. When heated to decomposition it emits acrid smoke and fumes.

HBD500   CAS: 1639-09-4   HR: 3
1-HEPTANETHIOL
DOT: UN 1228/UN 3071

mf: $C_7H_{16}S$ mw: 132.29

PROP: A liquid with powerful odor. Bp: 173–176°.

SYNS: HEPTYL MERCAPTAN □ n-HEPTYLMERCAPT-AN □ USAF EK-2122

CONSENSUS REPORTS: Reported in EPA TSCA Inventory.

NIOSH REL: (n-Alkane Mono Thiols) CL 0.5 ppm/15M

DOT CLASSIFICATION: 3; Label: Flammable Liquid, Poison (UN 1228); DOT Class: 6.1; Label: Poison, Flammable Liquid (UN 3071)

SAFETY PROFILE: A poison. Toxic by inhalation. A flammable liquid. When heated to decomposition it emits very toxic fumes of SO_x. See also MERCAPTANS.

HBE500   CAS: 543-49-7   HR: 2
2-HEPTANOL
mf: $C_7H_{16}O$ mw: 116.23

PROP: Liquid. Bp: 160.4°, flash p: 160°F (OC), d: 0.8344 @ 0°, vap press: 1 mm @ 14.6°, vap d: 4.01. Insol in water; sol in alc, ether, and benzene.

SYNS: AMYL METHYL CARBINOL □ HEPTANOL-2 □ 2-HYDROXYHEPTANE □ METHYL AMYL CARBINOL

CONSENSUS REPORTS: Reported in EPA TSCA Inventory.

SAFETY PROFILE: Moderately toxic by ingestion and skin contact. A skin and severe eye irritant. Combustible when exposed to heat and flame; can react vigorously with oxidizers. To fight fire, use foam, CO_2, dry chemical. See also ALCOHOLS.

HBF600   CAS: 32941-30-3   HR: 3
4-HEPTANOYLPYRIDINE
mf: $C_{12}H_{17}NO$ mw: 191.30

SYNS: 1-HEPTANONE, 1-(4-PYRIDYL)- □ HEPTYL 4-PYRIDYL KETONE □ KETONE, HEPTYL 4-PYRIDYL □ PYRIDINE, 4-HEPTANOYL- □ 1-(4-PYRIDYL)-1-HEPTANONE

DOT CLASSIFICATION: 3; Label: Flammable Liquid

SAFETY PROFILE: A poison by intraperitoneal route. A flammable liquid. When heated to decomposition it emits toxic vapors of NO_x.

HBI800   HR: 3
cis-4-HEPTEN-1-AL
mf: $C_7H_{12}O$ mw: 112.17

PROP: Sltly yellow liquid; fatty odor. Refr index: 1.432–1.436, flash p: 68°F. Sol in alc, fixed oils; insol in water.

SYNS: FEMA No. 3289 □ 4-HEPTENAL □ n-PROP-YLIDENE BUTYRALDEHYDE

SAFETY PROFILE: Flammable liquid. When heated to decomposition it emits acrid smoke and irritating fumes.

HBJ000   CAS: 592-76-7   HR: 3
n-HEPTENE
DOT: UN 2278

mf: C_7H_{14} mw: 98.21

PROP: Colorless liquid, insol in water, sol in ether. D: 0.6969 @ 20°, mp: −10°, bp: 93.6°, flash p: <30.2°F, autoign temp: 707°F.

SYNS: 1-n-HEPTENE □ 1-HEPTYLENE

CONSENSUS REPORTS: Reported in EPA TSCA Inventory.

DOT CLASSIFICATION: 3; Label: Flammable Liquid

SAFETY PROFILE: A simple asphyxiant. See ARGON for a description of simple asphyxiants. Dangerous fire hazard when exposed to heat, flame, or oxidizers. Unknown explosion hazard. To fight fire,

use foam, dry chemical, CO_2. When heated to decomposition it emits acrid smoke and fumes.

HBL500   CAS: 111-70-6   HR: 2
HEPTYL ALCOHOL
mf: $C_7H_{16}O$ mw: 116.23

PROP: Colorless liquid; citrus odor. Mp: −34.6°, bp: 175.8°, d: 0.824 @ 20°/4°, refr index: 1.423–1.427, flash p: 160°F. Misc in alc, fixed oils, ether; sltly sol in water @ 175°.

SYNS: l'ALCOOL n-HEPTYLIQUE PRIMAIRE (FRENCH) □ ENANTHIC ALCOHOL □ FEMA No. 2548 □ n-HEPTANOL □ 1-HEPTANOL □ n-HEPTANOL-1 (FRENCH) □ 1-HYDROXYHEPTANE

CONSENSUS REPORTS: Reported in EPA TSCA Inventory.

SAFETY PROFILE: Moderately toxic by ingestion and skin contact. Mildly toxic by inhalation. Mutation data reported. Combustible liquid. Can react with oxidizing materials. When heated to decomposition it emits acrid smoke and fumes. See also ALCOHOLS.

HBL600   CAS: 111-68-2   HR: 3
1-HEPTYLAMINE
DOT: UN 2733/UN 2734

mf: $C_7H_{17}N$ mw: 115.25

SYNS: 1-AMINOHEPTANE □ 1-HEPTANAMINE □ HEPTYLAMINE □ n-HEPTYLAMINE

CONSENSUS REPORTS: Reported in EPA TSCA Inventory.

DOT CLASSIFICATION: 8; Label: Corrosive, Flammable Liquid (UN 2734); DOT Class: 3; Label: Flammable Liquid, Corrosive (UN 2733)

SAFETY PROFILE: Poison by intraperitoneal route. A flammable liquid. When heated to decomposition it emits toxic vapors of NO_x.

HBN600   CAS: 64049-21-4   HR: 3
HEPTYLDICHLORARSINE
mf: $C_7H_{15}AsCl_2$ mw: 245.04

SYNS: ARSINE, DICHLOROHEPTYL- □ TL 229

OSHA PEL: TWA 0.5 mg(As)/m³

SAFETY PROFILE: Poison by skin contact. Slightly toxic by inhalation. When heated to decomposition it emits toxic fumes of As and Cl⁻.

HBO500   CAS: 112-23-2   HR: 1
HEPTYL FORMATE
mf: $C_8H_{16}O_2$ mw: 144.24
SYNS: FORMIC ACID, HEPTYL ESTER □ HEPTANOL, FORMATE □ n-HEPTYL METHANOATE
CONSENSUS REPORTS: Reported in EPA TSCA Inventory.
SAFETY PROFILE: A skin irritant. When heated to decomposition it emits acrid smoke and fumes. See also ESTERS.

HBP300   HR: D
HEPTYLPARABEN
mf: $C_{14}H_{20}O_3$ mw: 236.31
PROP: Small colorless crystals or white crystalline powder; odorless, burning taste. Mp: 48–51°. Sol in alc, ether; very sltly sol in water.
SYN: n-HEPTYL p-HYDROXYBENZOATE
SAFETY PROFILE: When heated to decomposition it emits acrid smoke and irritating fumes.

HBP400   CAS: 3648-21-3   HR: D
HEPTYL PHTHALATE
mf: $C_{22}H_{34}O_4$ mw: 362.56
SYNS: 1,2-BENZENEDICARBOXYLIC ACID, DIHEPTYL ESTER (9CI) □ DIHEPTYL PHTHALATE □ DI-n-HEPTYL PHTHALATE □ PHTHALIC ACID, DIHEPTYL ESTER
CONSENSUS REPORTS: Reported in EPA TSCA Inventory.
SAFETY PROFILE: An experimental teratogen. When heated to decomposition it emits acrid smoke and irritating fumes.

HBP450   CAS: 713-95-1   HR: 1
n-HEPTYL-Δ-VALEROLACTONE
mf: $C_{12}H_{22}O_2$ mw: 198.34
SYNS: Δ-DODECALACTONE □ 5-HYDROXYDO-DECANOIC ACID LACTONE □ 5-HYDROXYDO-DECANOIC ACID Δ-LACTONE □ 2H-PYRAN-2-ONE, 6-HEPTYLTETRAHYDRO-
CONSENSUS REPORTS: Reported in EPA TSCA Inventory.

SAFETY PROFILE: A skin irritant. When heated to decomposition it emits acrid smoke and irritating fumes.

HBU400   CAS: 24292-52-2   HR: D
HESPERIDIN METHYLCHALCONE
mf: $C_{29}H_{36}O_{15}$ mw: 624.65
SYN: CHALCONE, 2',3,4'-TRIHYDROXY-4,6'-DIMETHO-XY-, 4'-(6-O-(6-DEOXY-α-l-MANNOPYRANOSYL)-β-d-GLUCOPYRANOSIDE)
CONSENSUS REPORTS: Reported in EPA TSCA Inventory.
SAFETY PROFILE: Experimental reproductive effects. When heated to decomposition it emits acrid smoke and irritating fumes.

HBU415   CAS: 40626-35-5   HR: 3
HETEROPHOS
mf: $C_{11}H_{17}O_3PS$ mw: 260.31
SYNS: HETEROFOS □ PHOSPHOROTHIOIC ACID, o-ETHYL o-PHENYL S-PROPYL ESTER □ PHOSTIL
SAFETY PROFILE: A poison by ingestion. When heated to decomposition it emits toxic vapors of SO_x and PO_x.

HCA500   CAS: 36355-01-8   HR: 1
HEXABROMOBIPHENYL
mf: $C_{12}H_4Br_6$ mw: 627.62
SYNS: HBB □ NCI-C53634 □ POLYBROMINATED BIPHENYL
CONSENSUS REPORTS: Polybrominated biphenyl compounds are on the Community Right-To-Know List. Reported in EPA TSCA Inventory.
SAFETY PROFILE: Mildly toxic by ingestion and skin contact. When heated to decomposition it emits toxic fumes of Br⁻. See also POLYBROMINATED BIPHENYLS.

HCA650   CAS: 75625-24-0   HR: 3
1,2,3,4,6,7-HEXABROMONAPHTHAL-ENE
mf: $C_{10}H_2Br_6$ mw: 601.58
SYN: NAPHTHALENE, 1,2,3,4,6,7-HEXABROMO-
SAFETY PROFILE: A poison by ingestion. When heated to decomposition it emits toxic vapors of Br⁻.

HCA700   CAS: 4808-30-4   HR: 3
1,1,1,3,3,3-HEXABUTYLDISTAN-
NTHIANE
mf: $C_{24}H_{54}SSn_2$ mw: 612.22
PROP: Colorless oil. Bp: 208° (decomp).
SYNS: BIS(TRIBUTYLTIN)SULFIDE □ DISTAN-
NATHIANE, HEXABUTYL-(9CI) □ DISTANNTHIANE,
HEXABUTYL- □ HEXABUTYLDISTANNTHIANE □
TRIBUTYLTIN SULFIDE
CONSENSUS REPORTS: Reported in EPA
TSCA Inventory.
OSHA PEL: TWA 0.1 mg(Sn)/m³
ACGIH TLV: TWA 0.1 mg(Sn)/m³; STEL
0.2 mg/m³ (skin)
NIOSH REL: (Organotin Compounds) 10H
TWA 0.1 mg(Sn)/m³
SAFETY PROFILE: Poison by
intraperitoneal route. Moderately toxic by
ingestion. When heated to decomposition it
emits toxic fumes of SO_x and Sn.

HCB000   CAS: 13007-92-6   HR: 3
HEXACARBONYLCHROMIUM
mf: C_6CrO_6 mw: 220.06
PROP: Colorless crystals from
methylcyclohexane or by sublimation. Mp:
152–155°. Sltly sol in CCl_4; insol in H_2O,
EtOH, and Et_2O. IDLH Ca [15 mg/m³ {as
Cr(VI)}].
SYNS: CHROMIUM CARBONYL (MAK) □ CHROMIUM
CARBONYL (OC-6-11) (9CI) □ CHROMIUM
HEXACARBONYL □ HEXACARBONYL CHROMIUM
CONSENSUS REPORTS: IARC Cancer
Review: Animal Inadequate Evidence
IMEMDT 23,205,80; Chromium and its
compounds are on the Community Right-
To-Know List. Reported in EPA TSCA
Inventory.
OSHA PEL: CL 0.1 mg(CrO₃)/m³
ACGIH TLV: TWA 0.05 mg(Cr)/m³;
Confirmed Human Carcinogen
DFG MAK: DFG TRK: Confirmed Animal
Carcinogen with Unknown Relevance to
Humans
NIOSH REL: (Chromium(VI)) TWA 0.001
mg(Cr(VI))/m³
SAFETY PROFILE: Confirmed carcinogen
with experimental tumorigenic data. Poison
by ingestion and intravenous routes.

Mutation data reported. Explodes at 210°C.
See also CHROMIUM COMPOUNDS and
CARBONYLS.

HCC500   CAS: 118-74-1   HR: 3
HEXACHLOROBENZENE
DOT: UN 2729
mf: C_6Cl_6 mw: 284.76
PROP: Needles from 2-propanaol. Mp:
226°, bp: 323–326°, flash p: 468°F, vap
press: 1 mm @ 114.4°, vap d: 9.8, d: 2.44.
Insol in water; sol in benzene; very sltly sol
in hot alc; sol in hot ether and chloroform.
SYNS: AMATIN □ ANTICARIE □ BUNT-CURE □
BUNT-NO-MORE □ CEKU C.B. □ CO-OP HEXA □
ESACHLOROBENZENE (ITALIAN) □ GRANOX NM □
HCB □ HEXA C.B. □ HEXACHLORBENZOL
(GERMAN) □ JULIN'S CARBON CHLORIDE □ NO
BUNT □ NO BUNT 40 □ NO BUNT 80 □ NO BUNT
LIQUID □ PENTACHLOROPHENYL CHLORIDE □
PERCHLOROBENZENE □ PHENYL PERCHLORYL □
RCRA WASTE NUMBER U127 □ SAATBEIZFUNGIZID
(GERMAN) □ SANOCID □ SANOCIDE □ SMUT-GO □
SNIECIOTOX
CONSENSUS REPORTS: NTP 10th Report
on Carcinogens. IARC Cancer Review:
Group 2B IMEMDT 7,219,87; Animal
Sufficient Evidence IMEMDT 20,155,79;
Human Limited Evidence IMEMDT
20,155,79. Community Right-To-Know List.
Reported in EPA TSCA Inventory. EPA
Genetic Toxicology Program.
ACGIH TLV: TWA 0.002 (skin); Animal
Carcinogen
DFG MAK: Not Classifiable as a Human
Carcinogen; BAT: 15 µg/dL in
plasma/serum
DOT CLASSIFICATION: 6.1; Label: KEEP
AWAY FROM FOOD
SAFETY PROFILE: Confirmed carcinogen
with experimental carcinogenic,
neoplastigenic, and teratogenic data. A
human poison by an unspecified route.
Experimental reproductive effects. Mildly
toxic experimentally by inhalation. Mutation
data reported. A fungicide. Combustible
when exposed to heat or flame. Violent
reaction with dimethylformamide. To fight
fire, use CO_2, dry chemical. When heated to
decomposition it emits highly toxic fumes of

Cl⁻. See also CHLORINATED HYDROCARBONS, AROMATIC.

HCD250   CAS: 87-68-3   HR: 3
HEXACHLOROBUTADIENE
DOT: UN 2279
mf: C_4Cl_6 mw: 260.74
PROP: A liquid. D: 1.682 @ 20°/4°, mp: −21°, bp: 211–215°, autoign temp: 1130°F, vap d: 8.99.
SYNS: DOLEN-PUR □ GP-40-66:120 □ HCBD □ HEXACHLOR-1,3-BUTADIEN (CZECH) □ HEXA-CHLORO-1,3-BUTADIENE (MAK) □ 1,1,2,3,4,4-HEXACHLORO-1,3-BUTADIENE □ PERCHLORO-BUTADIENE □ RCRA WASTE NUMBER U128
CONSENSUS REPORTS: IARC Cancer Review: Group 3 IMEMDT 7,56,87; Animal Suspected IMEMDT 20,179,79. Community Right-To-Know List. Reported in EPA TSCA Inventory.
OSHA PEL: TWA 0.02 ppm
ACGIH TLV: TWA 0.02 ppm (skin); Animal Carcinogen
DFG MAK: Confirmed Animal Carcinogen with Unknown Relevance to Humans
DOT CLASSIFICATION: 6.1; Label: KEEP AWAY FROM FOOD
SAFETY PROFILE: Suspected carcinogen with experimental carcinogenic data. Poison by ingestion and intraperitoneal routes. Moderately toxic by inhalation and skin contact. A skin and eye irritant. An experimental teratogen. Experimental reproductive effects. Mutation data reported. Combustible when exposed to heat or flame; can react vigorously with oxidizing materials. To fight fire, use dry chemical, CO_2, alcohol foam, water spray, fog, mist. Reacts with bromine perchlorate to form an explosive product. When heated to decomposition it emits very toxic fumes of Cl⁻. A solvent, heat transfer fluid, transformer, hydraulic fluid, and wash liquor.

HCE500   CAS: 77-47-4   HR: 3
HEXACHLOROCYCLOPENTADIENE
DOT: UN 2646
mf: C_5Cl_6 mw: 272.75

PROP: Greenish-yellow to amber-colored liquid with a pungent odor. Fp: −2°, mp: 9.9°, bp: 234°, flash p: none (OC), d: 1.70 @ 25°/4°, vap d: 9.42.
SYNS: C-56 □ GRAPHLOX □ HCCPD □ HEXA-CHLORCYKLOPENTADIEN (CZECH) □ HEXACHLORO-1,3-CYCLOPENTADIENE □ 1,2,3,4,5,5-HEXACHLORO-1,3-CYCLOPENTADIENE □ HEXACHLOROCYCLO-PENTADIENE (ACGIH,DOT,OSHA) □ HRS 1655 □ NCI-C55607 □ PCL □ PERCHLOROCYCLOPENTADIENE □ RCRA WASTE NUMBER U130
CONSENSUS REPORTS: EPA Extremely Hazardous Substances List. Community Right-To-Know List. Reported in EPA TSCA Inventory.
OSHA PEL: TWA 0.01 ppm
ACGIH TLV: TWA 0.01 ppm; Not Classifiable as a Human Carcinogen
DOT CLASSIFICATION: 6.1; Label: Poison
SAFETY PROFILE: A deadly poison by inhalation. Moderately toxic by ingestion and skin contact. Experimental teratogenic effects. Corrosive. A severe skin and eye irritant. May explode on contact with sodium. When heated to decomposition it emits toxic fumes of Cl⁻. See also CHLORINATED HYDROCARBONS, ALIPHATIC.

HCF500   HR: 3
1,2,3,6,7,8-HEXACHLORODIBENZO-p-DIOXIN mixed with 1,2,3,7,8,9-HEXACHLORODIBENZO-p-DIOXIN
PROP: Composed of 67% of 1,2,3,7,8,9-hexachlorodibenzo-p-dioxin and 31% of 1,2,3,6,7,8-hexachlorodibenzo-p-dioxin NCITR* NCI-CG-TR-198,80.
SYNS: 1,2,3,7,8,9-HEXACHLORODIBENZO-p-DIOXIN mixed with 1,2,3,6,7,8-HEXACHLORODIBENZO-p-DIOXIN □ NCI-C03703
CONSENSUS REPORTS: NCI Carcinogenesis Bioassay (gavage); Clear Evidence: mouse, rat NCITR* NCI-CG-TR-198,80. NCI Carcinogenesis Bioassay (dermal); No Evidence: mouse NCITR* NCI-CG-TR-202,80.
SAFETY PROFILE: Suspected carcinogen with experimental carcinogenic data. A deadly poison by ingestion. When heated to

decomposition it emits very toxic fumes of Cl⁻ and dioxin.

HCI000   CAS: 67-72-1   HR: 3
HEXACHLOROETHANE

mf: C_2Cl_6 mw: 236.72

PROP: Rhombic, triclinic, or cubic crystals from $EtOH/Et_2O$; colorless, camphor-like odor. Mp: 186.6° (subl), d: 2.091, vap press: 1 mm @ 32.7°, bp: 186.8° (triple point). Sol in alc, benzene, chloroform, ether, oils; insol in water.

SYNS: AVLOTANE □ CARBON HEXACHLORIDE □ DISTOKAL □ DISTOPAN □ DISTOPIN □ EGITOL □ ETHANE HEXACHLORIDE □ ETHYLENE HEXACHLORIDE □ FALKITOL □ FASCIOLIN □ HEXACHLOR-AETHAN (GERMAN) □ 1,1,1,2,2,2-HEXACHLOROETHANE □ HEXACHLOROETHYLENE □ MOTTENHEXE □ NCI-C04604 □ PERCHLORO-ETHANE □ PHENOHEP □ RCRA WASTE NUMBER U131

CONSENSUS REPORTS: NTP 10th Report on Carcinogens. IARC Cancer Review: Group 3 IMEMDT 7,56,87; Animal Limited Evidence IMEMDT 20,467,79. NCI Carcinogenesis Bioassay (gavage); Clear Evidence: mouse NCITR* NCI-CG-TR-68,78. NCI Carcinogenesis Bioassay (gavage); No Evidence: rat NCITR* NCI-CG-TR-68,78. Community Right-To-Know List. Reported in EPA TSCA Inventory. EPA Genetic Toxicology Program.

OSHA PEL: TWA 1 ppm (skin)
ACGIH TLV: TWA 1 ppm; Suspected Human Carcinogen
DFG MAK: 1 ppm (9.8 mg/m³)
NIOSH REL: (Hexachloroethane) Reduce to lowest level

SAFETY PROFILE: Confirmed carcinogen with experimental carcinogenic data. A poison by intravenous route. Moderately toxic by intraperitoneal route. Mildly toxic by ingestion. Experimental reproductive effects. Mutation data reported. Liver injury has resulted from exposure to this material. An insecticide. Slightly explosive by spontaneous chemical reaction. Dehalogenation of this material by reaction with alkalies, metals, etc., will produce spontaneously explosive chloroacetylenes.

When heated to decomposition it emits highly toxic fumes of Cl⁻ and phosgene. See also CHLORINATED HYDROCARBONS, ALIPHATIC.

HCK500   CAS: 1335-87-1   HR: 3
HEXACHLORONAPHTHALENE

mf: $C_{10}H_2Cl_6$ mw: 334.82

PROP: White solid. IDLH 2 mg/m³.

SYNS: HALOWAX 1014 □ HEXACHLORNAFTALEN

CONSENSUS REPORTS: Community Right-To-Know List. Reported in EPA TSCA Inventory.

OSHA PEL: TWA 0.2 mg/m³ (skin)
ACGIH TLV: TWA 0.2 mg/m³ (skin)
NIOSH REL: TWA 0.2 mg/m³ (skin)

SAFETY PROFILE: A poison by ingestion, skin contact, and inhalation. Causes severe acne-form eruptions and toxic narcosis of liver. Absorbed by skin. When heated to decomposition it emits toxic fumes of Cl⁻. See also CHLORINATED HYDROCARBONS, AROMATIC.

HCK550   CAS: 103426-96-6 HR: D
1,2,3,4,6,7-HEXACHLORONAPHTHAL-ENE

mf: $C_{10}H_2Cl_6$ mw: 334.82

SYNS: 1,2,3,4,6,7-HEXACHLORINATED NAPHTHALANE □ NAPHTHALENE, 1,2,3,4,6,7-HEXACHLORO-

SAFETY PROFILE: Experimental reproductive effects. When heated to decomposition it emits toxic vapors of Cl⁻.

HCK600   CAS: 103426-97-7 HR: 3
1,2,3,5,6,7-HEXACHLORONAPHTHAL-ENE

mf: $C_{10}H_2Br_6$ mw: 601.58

SYN: NAPHTHALENE, 1,2,3,5,6,7-HEXACHLORO-

SAFETY PROFILE: A poison by ingestion. When heated to decomposition it emits toxic vapors of Br⁻.

HCL000   CAS: 70-30-4   HR: 3
HEXACHLOROPHENE

DOT: UN 2875
mf: $C_{13}H_6Cl_6O_2$ mw: 406.89

H

PROP: Crystals, water insol. Mp: 165°. Sol in alc, acetone, ether, chloroform, propylene glycol, polyethylene glycols, olive oil, cottonseed oil, dil solns of alkalies.
SYNS: ACIGENA □ ALMEDERM □ AT 7 □ B32 □ BILEVON □ BIS(2-HYDROXY-3,5,6-TRICHLOROPHEN-YL)METHANE □ BIS-2,3,5-TRICHLOR-6-HYDROXY-FENYLMETHAN (CZECH) □ BIS(3,5,6-TRICHLORO-2-HYDROXYPHENYL)METHANE □ COMPOUND G-11 □ COTOFILM □ DERMADEX □ 2,2'-DIHYDROXY-3,3',5,5',6,6'-HEXACHLORODIPHENYLMETHANE □ 2,2'-DIHYDROXY-3,5,6,3',5',6'-HEXACHLORODIPHENYL-METHANE □ EXOFENE □ FOMAC □ FOSTRIL □ G-11 □ GAMOPHENE □ G-ELEVEN □ GERMA-MEDICA □ HCP □ HEXABALM □ 2,2',3,3',5,5'-HEXACHLORO-6,6'-DIHYDROXYDIPHENYLMETHANE □ HEXACHLORO-FEN (CZECH) □ HEXACHLOROPHANE □ HEXA-CHLOROPHEN □ HEXACHLOROPHENE (DOT) □ HEXAFEN □ HEXIDE □ HEXOPHENE □ HEXOSAN □ ISOBAC 20 □ 2,2'-METHYLENEBIS(3,4,6-TRICHLORO-PHENOL) □ NABAC □ NCI-C02653 □ NEOSEPT □ PHISODANV □ PHISOHEX □ RCRA WASTE NUMBER U132 □ RITOSEPT □ SEPTISOL □ SEPTOFEN □ STERAL □ STERASKIN □ SURGI-CEN □ SUROFENE □ TERSASEPTIC □ TRICHLOROPHENE □ TURGEX
CONSENSUS REPORTS: IARC Cancer Review: Group 3 IMEMDT 7,56,87; Human Inadequate Evidence IMEMDT 20,241,79. NCI Carcinogenesis Bioassay (feed); No Evidence: rat NCITR* NCI-CG-TR-40,78. Reported in EPA TSCA Inventory. Chlorophenols are on the Community Right-To-Know List.
DOT CLASSIFICATION: 6.1; Label: KEEP AWAY FROM FOOD
SAFETY PROFILE: A human poison by ingestion. An experimental poison by ingestion, intraperitoneal, and intravenous routes. Moderately toxic by skin contact. Human systemic effects by ingestion: cardiomyopathy (damage to the heart muscle), nausea or vomiting, diarrhea, shock. Unspecified human reproductive effects. Experimental teratogenic and reproductive effects. An eye and human skin irritant. Questionable carcinogen with experimental neoplastigenic and tumorigenic data. Strong concentrations may be irritating, but ordinary use of 1–2% solutions is not.

For many years, the toxicologic hazard of hexachlorophene was unrecognized and the compound had a wide and virtually unrestricted use. However, studies by FDA scientists demonstrated that brain lesions occur from exposure in both rats and monkeys treated at levels only slightly higher than those of persons using soaps, toothpaste, shampoos, and a variety of other household products and cosmetics containing it. The FDA has now restricted sale of hexachlorophene, and most preparations containing higher levels of the compound are available only through prescription. In the recent FDA studies, it was found that 2 weeks after onset of exposure, rats fed 500 ppm (25 mg/kg/day) of hexachlorophene in their diet showed weakness in their hindquarters that progressed to paralysis. Microscopic examination of the brain and spinal cord of these rats revealed a particular edema of the white matter resembling spongy degeneration noted in infants. When the animals were removed from the diet, they recovered gradually over a period of weeks. Similar symptoms were noted in monkeys. Following ingestion of hexachlorophene, early symptoms are primarily gastrointestinal in nature and include anorexia, nausea, vomiting, abdominal cramps, and diarrhea. Dehydration is sometimes severe and may be associated with shock.

Used as a germicidal agent. An additive permitted in the feed and drinking water of animals and/or for the treatment of food-producing animals. Also permitted in food for human consumption. When heated to decomposition it emits toxic fumes of Cl⁻. See also CHLOROPHENOLS.

HCL500   CAS: 116-16-5   HR: 3
HEXACHLORO-2-PROPANONE
DOT: UN 2661
mf: C_3Cl_6O mw: 264.73
PROP: Liquid. Bp: 203°, fp: −2°, vap d: 9.2, d: 1.74 @ 12°/12°. Sltly sol in water.

SYNS: ACETONE, HEXACHLORO- ☐ GC-1106 ☐ HCA ☐ HCA WEEDKILLER ☐ HEXACHLOROACETONE (DOT) ☐ 1,1,1,3,3,3-HEXACHLORO-2-PROPANONE ☐ 2-PROPANONE, HEXACHLORO-

CONSENSUS REPORTS: Reported in EPA TSCA Inventory.

DOT CLASSIFICATION: 6.1; Label: KEEP AWAY FROM FOOD

SAFETY PROFILE: A poison by ingestion. Moderately toxic by inhalation, and skin contact. Mutation data reported. When heated to decomposition it emits toxic fumes of Cl⁻.

HCM000   CAS: 1888-71-7   HR: 3
HEXACHLOROPROPENE

mf: C_3Cl_6 mw: 248.73

PROP: D: 1.76 @ 20°/4°, bp: 209–210°.

SYN: HEXACHLOROPROPYLENE

CONSENSUS REPORTS: Reported in EPA TSCA Inventory.

SAFETY PROFILE: A poison by inhalation and intraperitoneal routes. A powerful irritant. Mutation data reported. When heated to decomposition it emits toxic fumes of Cl⁻. See also CHLORINATED HYDROCARBONS, ALIPHATIC.

HCM500   CAS: 68-36-0   HR: 2
α,α′-HEXACHLOROXYLENE

mf: $C_8H_4Cl_6$ mw: 312.82

PROP: Crystals from hexane. Mp: 108–110°.

SYNS: 1:4-BIS-TRICHLOROMETHYL BENZENE ☐ α,α,α,α′,α′,α′-HEXACHLORO-p-XYLENE

CONSENSUS REPORTS: Reported in EPA TSCA Inventory.

SAFETY PROFILE: Moderately toxic by ingestion. Experimental reproductive effects. When heated to decomposition it emits toxic fumes of Cl⁻. See also CHLORINATED HYDROCARBONS, AROMATIC.

HCN000   CAS: 3734-48-3   HR: 2
4,5,6,7,8,8-HEXACHLOR-Δ¹,⁵-TETRAHY-DRO-4,7-METHANOINDEN

mf: $C_{10}H_6Cl_6$ mw: 338.86

SYNS: ADDUKT HEXACHLORCYKLOPENTADIENU S CYKLOPENTADIENEM (CZECH) ☐ CHLORDENE ☐ 4,5,6,7,8,8-HEXACHLORO-3a,4,7,7a-TETRAHYDRO-4,7-METHANOINDENE ☐ 4,7-METHANOINDENE, 4,5,6,7,8,8-HEXACHLORO-3a,4,7,7a-TETRAHYDRO-

CONSENSUS REPORTS: Reported in EPA TSCA Inventory.

SAFETY PROFILE: Moderately toxic by ingestion, inhalation, and skin contact. An experimental teratogen. A severe eye irritant. When heated to decomposition it emits toxic fumes of Cl⁻. See also CHLORINATED HYDROCARBONS, ALIPHATIC.

HCO500   CAS: 143-27-1   HR: 3
1-HEXADECANAMINE

mf: $C_{16}H_{35}N$ mw: 241.52

PROP: A solid. Mp: 46.2°, bp: 162–165° @ 5.2 mm.

SYNS: ALAMINE 6 ☐ ARMEEN 16D ☐ CETYLAMIN (GERMAN) ☐ CETYLAMINE ☐ N-HEXADECYLAMINE ☐ PALMITYLAMINE

CONSENSUS REPORTS: Reported in EPA TSCA Inventory.

SAFETY PROFILE: A poison by intraperitoneal route. An experimental teratogen. Other experimental reproductive effects. When heated to decomposition it emits toxic fumes of NO_x. See also AMINES.

HCP000   CAS: 36653-82-4   HR: 3
1-HEXADECANOL

mf: $C_{16}H_{34}O$ mw: 242.50

PROP: Solid or leaf-like crystals. Mp: 50°, bp: 178–182° @ 12 mm, d: 0.8176 @ 50°/4°. Insol in water; sol in alc, chloroform, ether.

SYNS: ADOL ☐ ALCOHOL C-16 ☐ ATALCO C ☐ CACHALOT C-50 ☐ CETAFFINE ☐ CETAL ☐ CETALOL CA ☐ CETYL ALCOHOL ☐ CETYLIC ALCOHOL ☐ CETYLOL ☐ CO-1670 ☐ CRODACOL-CAS ☐ CYCLAL CETYL ALCOHOL ☐ DYTOL F-11 ☐ EPAL 16NF ☐ ETHAL ☐ ETHOL ☐ HEXADECANOL ☐ n-HEXA-DECANOL ☐ HEXADECAN-1-OL ☐ HEXADECYL ALCOHOL ☐ n-HEXADECYL ALCOHOL ☐ LOROL 24 ☐ LOXANOL K ☐ PALMITYL ALCOHOL ☐ PRODUCT 308

CONSENSUS REPORTS: Reported in EPA TSCA Inventory.

SAFETY PROFILE: Moderately toxic by ingestion and intraperitoneal routes. An eye and human skin irritant. Flammable when exposed to heat or flame; can react with oxidizing materials. To fight fire, use foam, CO_2, dry chemical. When heated to decomposition it emits acrid smoke and fumes. See also ALCOHOLS.

HCP550   CAS: 59130-69-7   HR: 2
HEXADECYL 2-ETHYLHEXANOATE
mf: $C_{24}H_{48}O_2$ mw: 368.72
PROP: Water white.
SYNS: CETYL 2-ETHYLHEXANOATE □ HEXANOIC ACID, 2-ETHYL-, HEXADECYL ESTER □ PERCELINE OIL
CONSENSUS REPORTS: Reported in EPA TSCA Inventory.
SAFETY PROFILE: A severe skin irritant. When heated to decomposition it emits acrid smoke and irritating fumes.

HCQ000   CAS: 5894-60-0   HR: 2
HEXADECYLTRICHLOROSILANE
DOT: UN 1781
mf: $C_{16}H_{33}Cl_3Si$ mw: 359.93
PROP: Colorless to yellow liquid. D: 0.996 @ 25°/25°, bp: 269°, flash p: 295°F (COC).
SYN: TRICHLOROHEXADECYLSILANE
CONSENSUS REPORTS: Reported in EPA TSCA Inventory.
DOT CLASSIFICATION: 8; Label: Corrosive
SAFETY PROFILE: A corrosive irritant to skin, eyes, and mucous membranes. Combustible when exposed to heat or flame. When heated to decomposition or on contact with water it emits toxic fumes of Cl^- and HCl. See also CHLOROSILANES.

HCQ500   CAS: 57-09-0   HR: 3
HEXADECYLTRIMETHYLAMMONIUM
BROMIDE
mf: $C_{19}H_{42}N•Br$ mw: 364.53
PROP: Microcrystals. Mp: 237–243°. Sol in H_2O, EtOH; insol in C_6H_6 and Et_2O.
SYNS: ACETOQUAT CTAB □ BROMAT □ CEE DEE □ CENTIMIDE □ CETAB □ CETAROL □ CETAVLON □ CETRIMIDE □ CETRIMONIUM BROMIDE □ CETYL-AMINE □ CETYLTRIMETHYLAMMONIUM BROMIDE □ N-CETYLTRIMETHYLAMMONIUM BROMIDE □ CIRRASOL-OD □ CTAB □ CYCLOTON V □ N-HEXA-DECYLTRIMETHYLAMMONIUM BROMIDE □ N-HEXADECYL-N,N,N-TRIMETHYLAMMONIUM BROMIDE □ (1-HEXADECYL)TRIMETHYLAMMONIUM BROMIDE □ LISSOLAMINE □ MICOL □ POLLACID □ QUAMONIUM □ SUTICIDE □ TRIMETHYLCETYL-AMMONIUM BROMIDE □ N,N,N-TRIMETHYL-1-HEXADECANAMINIUM BROMIDE □ TRIMETHYL-HEXADECYLAMMONIUM BROMIDE
CONSENSUS REPORTS: Reported in EPA TSCA Inventory.
SAFETY PROFILE: A poison by ingestion, intravenous, intraperitoneal, and subcutaneous routes. Experimental teratogenic and reproductive effects. A skin and severe eye irritant. When heated to decomposition it emits very toxic fumes of NH_3, NO_x, and Br^-. See also BROMIDES.

HCQ600   CAS: 42296-74-2   HR: 3
HEXADIENE
DOT: UN 2458
mf: C_6H_{10} mw: 82.16
DOT CLASSIFICATION: 3; Label: Flammable Liquid
SAFETY PROFILE: A flammable liquid. When heated to decomposition it emits acrid smoke and irritating vapors.

HCR500   CAS: 592-42-7   HR: 3
1,5-HEXADIENE
mf: C_6H_{10} mw: 82.16
PROP: Liquid. D: 0.691, mp: −141°, bp: 59.6°, flash p: −50.80°F. Insol in water.
SYNS: BIALLYL □ DIALLYL □ HEXA-1,5-DIENE
CONSENSUS REPORTS: Reported in EPA TSCA Inventory.
SAFETY PROFILE: A very dangerous fire and explosion hazard when exposed to heat, flame, or oxidizers. When heated to decomposition it emits acrid smoke and fumes.

HCX050   CAS: 1112-63-6   HR: 3
HEXAETHYLDISTANNOXANE
mf: $C_{12}H_{30}OSn_2$ mw: 427.80
PROP: Air-sensitive liquid. D: 1.377 @ 20°, bp: 272°.
SYNS: DISTANNOXANE, HEXAETHYL- □ 1,1,1,3,3,3-HEXAETHYLDISTANNOXANE
OSHA PEL: TWA 0.1 mg(Sn)/m³

ACGIH TLV: TWA 0.1 mg(Sn)/m³; STEL 0.2 mg/m³ (skin)
NIOSH REL: (Organotin Compound): 10H TWA 0.1 mg(Sn)/m³
SAFETY PROFILE: Poison by intraperitoneal route. When heated to decomposition it emits toxic fumes of Sn.

HCX100   CAS: 994-50-3   HR: 3
HEXAETHYLDISTANNTHIANE
mf: $C_{12}H_{30}SSn_2$ mw: 443.86
PROP: A liquid. D: 1.431 @ 20°, bp: 187–188° @ 20 mm.
SYNS: DISTANNTHIANE, HEXAETHYL- □ 1,1,1,3,3,3-HEXAETHYLDISTANNTHIANE
OSHA PEL: TWA 0.1 mg(Sn)/m³
ACGIH TLV: TWA 0.1 mg(Sn)/m³; STEL 0.2 mg/m³ (skin)
NIOSH REL: (Organotin Compounds): 10H TWA 0.1 mg(Sn)/m³
SAFETY PROFILE: Poison by intraperitoneal route. When heated to decomposition it emits toxic fumes of SO_x and Sn.

HCY000   CAS: 757-58-4   HR: 3
HEXAETHYL TETRAPHOSPHATE
DOT: UN 1611
mf: $C_{12}H_{30}O_{13}P_4$ mw: 506.30
PROP: Liquid. Mp: −40°; bp: decomp above 150°.
SYNS: BLADAN □ BLADAN BASE □ ETHYL TETRAPHOSPHATE □ ETHYL TETRAPHOSPHATE, HEXA- □ HET □ HETP □ HEXAETHYLTETRAFOSFAT □ HEXAETHYL TETRAPHOSPHATE, liquid or solid (DOT) □ HTP □ RCRA WASTE NUMBER P062 □ TETRAPHOSPHATE HEXAETHYLIQUE (FRENCH)
DOT CLASSIFICATION: 6.1; Label: Poison, KEEP AWAY FROM FOOD
SAFETY PROFILE: A poison by ingestion, skin contact, intraperitoneal, subcutaneous, intravenous, and intramuscular routes. When heated to decomposition it emits toxic fumes of PO_x. See also TETRAETHYL PYROPHOSPHATE.

HCZ000   CAS: 684-16-2   HR: 3
HEXAFLUOROACETONE
DOT: UN 2420

mf: C_3F_6O mw: 166.03
PROP: A colorless, nonflammable gas. D: 1.65 @ 25°, fp: −129°, bp: −26°.
SYNS: 6FK □ NCI-C56440
CONSENSUS REPORTS: Reported in EPA TSCA Inventory.
OSHA PEL: TWA 0.1 ppm (skin)
ACGIH TLV: TWA 0.1 ppm (skin)
DOT CLASSIFICATION: 2.3; Label: Poison Gas
SAFETY PROFILE: A poison by ingestion. Moderately toxic by inhalation. A poisonous irritant to the skin, eyes, and mucous membranes. An experimental teratogen. Other experimental reproductive effects. When heated to decomposition it emits toxic fumes of F⁻. See also FLUORINE and FLUORIDES.

HDA000   CAS: 10543-95-0   HR: 3
HEXAFLUOROACETONE HYDRATE
DOT: UN 2552
mf: $C_3F_6O \cdot H_2O$ mw: 184.05
PROP: Bp: 57° @ 93 mm.
SYN: HEXAFLUORO-2-PROPANONE HYDRATE
DOT CLASSIFICATION: 6.1; Label: Poison
SAFETY PROFILE: Poison by ingestion and skin contact. When heated to decomposition it emits toxic fumes of F⁻. See also HEXAFLUOROACETONE.

HDC500   CAS: 920-66-1   HR: 3
HEXAFLUOROISOPROPANOL
mf: $C_3H_2F_6O$ mw: 168.05
PROP: A liquid. D: 1.46, bp: 57–58°.
SYNS: 1,1,1,3,3,3-HEXAFLUORO-2-PROPANOL □ HFIP
CONSENSUS REPORTS: Reported in EPA TSCA Inventory.
SAFETY PROFILE: A poison by intravenous and intraperitoneal routes. Moderately toxic by ingestion. Mildly toxic by inhalation. When heated to decomposition it emits toxic fumes of F⁻. See also FLUORIDES.

HDE000   CAS: 16940-81-1   HR: 3
HEXAFLUOROPHOSPHORIC ACID
DOT: UN 1782

mf: F$_6$HP   mw: 145.98
PROP: Clear oil. Mp: 31°, d: 1.65. Strong aq solns fume in air and gradually decomp.
SYN: HYDROGEN HEXAFLUOROPHOSPHATE
CONSENSUS REPORTS: Reported in EPA TSCA Inventory.
OSHA PEL: TWA 2.5 mg(F)/m^3
ACGIH TLV: TWA 2.5 mg(F)/m^3; BEI: 3 mg/g creatinine of fluorides in urine prior to shift; 10 mg/g creatinine of fluorides in urine at end of shift.
NIOSH REL: (Inorganic Fluorides) TWA 2.5 mg(F)/m^3
DOT CLASSIFICATION: 8; Label: Corrosive
SAFETY PROFILE: A poison by all routes. A corrosive irritant to skin, eyes, and mucous membranes. When heated to decomposition it emits highly toxic F$^-$ and PO$_x$. See also HYDROFLUORIC ACID and PHOSPHORIC ACID.

HDF000   CAS: 116-15-4   HR: 3
HEXAFLUOROPROPENE
DOT: UN 1858
mf: C$_3$F$_6$ mw: 150.03
PROP: Gas. Mp: −156°, bp: −29°, d: 1.583 @ −40°/4°.
SYNS: HEXAFLUOROPROPYLENE (DOT) □ PERFLUOROPROPENE □ PERFLUOROPROPYLENE
CONSENSUS REPORTS: Reported in EPA TSCA Inventory.
ACGIH TLV: 0.1 ppm
DOT CLASSIFICATION: 2.2; Label: Nonflammable Gas
SAFETY PROFILE: Mildly toxic by inhalation. Explosive reaction with Grignard reagents (e.g., phenylmagnesium bromide). Reacts with tetrafluorethylene + air to form explosive peroxides. When heated to decomposition it emits toxic fumes of F$^-$.

HDF050   CAS: 428-59-1   HR: D
HEXAFLUOROPROPENE EPOXIDE
DOT: NA 1956
mf: C$_3$F$_6$O   mw: 166.03
PROP: Bp: −27°.
SYNS: HEXAFLUOROEPOXYPROPANE □ HEXAFLUORO-1,2-EPOXYPROPANE □ HEXA-FLUOROPROPENE OXIDE □ HEXAFLUORO-

PROPYLENE OXIDE (DOT) □ OXIRANE, TRIFLUORO-(TRIFLUOROMETHYL)- □ PERFLUORO(METHYLO-XIRANE) □ PERFLUOROPROPYLENE OXIDE □ PROPANE, 1,2-EPOXY-1,1,2,3,3,3-HEXAFLUORO- □ PROPYLENE OXIDE HEXAFLUORIDE □ (TRIFLUORO-METHYL)TRIFLUOROOXIRANE □ TRIFLUORO-(TRIFLUOROMETHYL)OXIRANE
CONSENSUS REPORTS: Reported in EPA TSCA Inventory.
DOT CLASSIFICATION: 2.2; Label: Nonflammable Gas
SAFETY PROFILE: Nonflammable gas shipped under pressure. When heated to decomposition it emits toxic vapors of F$^-$.

HDG000   CAS: 111-49-9   HR: 3
HEXAHYDRO-1H-AZEPINE
DOT: UN 2493
mf: C$_6$H$_{13}$N   mw: 99.20
PROP: A liquid. D: 0.864 @ 22°/4°, bp: 136–137°, flash p: 71.6°F. Partially misc in water.
SYNS: AZACYCLOHEPTANE □ 1-AZACYCLO-HEPTANE □ CYCLOHEXAMETHYLENIMINE □ G 0 □ HEXAHYDROAZEPINE □ HEXAMETHYLENE IMINE (DOT) □ HEXAMETHYLENIMINE □ HOMOPIPERID-INE □ PERHYDROAZEPINE
CONSENSUS REPORTS: Reported in EPA TSCA Inventory.
DOT CLASSIFICATION: 3; Label: Flammable Liquid, Corrosive
SAFETY PROFILE: Moderately toxic by ingestion and subcutaneous routes. Mildly toxic by inhalation. A corrosive irritant to the eyes, skin, and mucous membranes. A dangerous fire hazard when exposed to heat or flame; can react vigorously with oxidizers. When heated to decomposition it emits toxic fumes of NO$_x$.

HDY000   CAS: 531-18-0   HR: 3
HEXA(HYDROXYMETHYL)MELAMINE
mf: C$_9$H$_{18}$N$_6$O$_6$ mw: 306.33
SYNS: CILAG 61 □
HEXAKIS(HYDROXYMETHYL)MELAMINE □ HEXAKIS-(HYDROXYMETHYL)-1,3,5-TRIAZINE-2,4,6-TRIAMINE □ HEXAMETHYLOLMELAMIN (CZECH) □ HEXAMETH-YLOLMELAMINE □ RESLOOM M 75 □ (1,3,5-TRIAZINE-2,4,6-TRIYLTRINITRILO)HEXAKIS METHANOL □ (s-TRIAZINE-2,4,6-TRIYLTRINITRILO)HEXAMETHANOL □ 2,4,6-TRIS(BIS(HYDROXYMETHYL)AMINO)-s-TRIAZ-

INE □ 2,4,6-TRIS(DI(HYDROXYMETHYL)AMINO)-1,3,5-TRIAZINE
CONSENSUS REPORTS: Reported in EPA TSCA Inventory.
SAFETY PROFILE: A poison by intravenous route. A skin and eye irritant. When heated to decomposition it emits toxic fumes of NO_x.

HDY100   CAS: 3750-18-3   HR: 3
HEXAISOBUTYLDITIN
mf: $C_{24}H_{54}Sn_2$ mw: 580.16
SYNS: BIS(TRIISOBUTYLSTANNANE) □ DISTAN-NANE, HEXAISOPROPYL-
OSHA PEL: TWA 0.1 mg(Sn)/m³
ACGIH TLV: TWA 0.1 mg(Sn)/m³; STEL 0.2 mg/m³ (skin)
NIOSH REL: (Organotin Compounds): 10H TWA 0.1 mg(Sn)/m³
SAFETY PROFILE: Poison by intravenous route. When heated to decomposition it emits toxic fumes of Sn.

HEC000   CAS: 87-85-4   HR: 2
HEXAMETHYLBENZENE
mf: $C_{12}H_{18}$ mw: 162.30
PROP: Plates from ethanol. Mp: 165.5°, bp: 265°. Insol in water; very sol in ether.
CONSENSUS REPORTS: Reported in EPA TSCA Inventory.
SAFETY PROFILE: Questionable carcinogen with experimental neoplastigenic data. Potentially explosive reaction with nitromethane. When heated to decomposition it emits acrid smoke and fumes.

HEI500   CAS: 100-97-0   HR: 3
HEXAMETHYLENETETRAMINE
DOT: UN 1328
mf: $C_6H_{12}N_4$ mw: 140.22
PROP: Odorless, volatile, rhombic crystals from alc. Mp: 280° (sublimes), flash p: 482°F, d: 1.33 @ −5°. Sol in water; very sltly sol in hot ether or hot water.
SYNS: ACETO HMT □ AMINOFORM □ AMMOFORM □ AMMONIOFORMALDEHYDE □ CYSTAMIN □ CYSTOGEN □ ESAMETILENTETRAMINA (ITALIAN) □ FORMAMINE □ FORMIN □ HEXAFORM □

HEXAMETHYLENAMINE □ HEXAMETHYLENEAMINE □ HEXAMETHYLENETETRAAMINE □ HEXAMETHYL-ENTETRAMIN (GERMAN) □ HEXAMINE (DOT) □ HEXILMETHYLENAMINE □ HMT □ METHAMIN □ METHENAMINE □ PREPARATION AF □ RESOTROPIN □ 1,3,5,7-TETRAAZAADAMANTANE □ URITONE □ UROTROPIN □ UROTROPINE
CONSENSUS REPORTS: EPA Genetic Toxicology Program. Reported in EPA TSCA Inventory.
DOT CLASSIFICATION: 4.1; Label: Flammable Solid
SAFETY PROFILE: A poison by subcutaneous route. Moderately toxic by ingestion and intraperitoneal routes. Questionable carcinogen with experimental tumorigenic data. An irritant to skin, eyes, and mucous membranes. Some persons suffer a skin rash if they come in contact with this material or the fumes evolved when it is heated. Human mutation data reported. Pure hexamethylenetetramine may be taken internally in small amounts and has been used in medicine as a urinary antiseptic. Its major industrial use is in the manufacture of phenolic resins.

Combustible when exposed to heat or flame. Can react with oxidizing materials. Explosive reaction with acetic acid + acetic anhydride + ammonium nitrate + nitric acid, 1-bromopenta borane(9) (above 90°C), iodoform (at 178°C), iodine (at 138°C). Reaction with nitric acid + acetic anhydride forms the military explosives RDX and HMX. Reacts violently with Na_2O_2. When heated to decomposition it emits toxic fumes of formaldehyde and NO_x. See also AMINES.

HEK000   CAS: 680-31-9   HR: 3
HEXAMETHYLPHOSPHORAMIDE
mf: $C_6H_{18}N_3OP$ mw: 179.24
PROP: Clear, colorless, mobile liquid; spicy odor. Mp: 7°, bp: 233°, d: 1.024 @ 25°/25°, vap d: 6.18. Misc in water.
SYNS: EASTMAN INHIBITOR HPT □ ENT 50,882 □ HEMPA □ HEXAMETAPOL □ HEXAMETHYLPHOS-PHORIC ACID TRIAMIDE (MAK) □ HEXAMETHYL-PHOSPHORIC TRIAMIDE □ N,N,N,N,N,N-HEXA-

METHYLPHOSPHORIC TRIAMIDE ◻ HEXA-
METHYLPHOSPHOROTRIAMIDE ◻ HEXAMETHYL-
PHOSPHOTRIAMIDE ◻ HEXMETHYLPHOSPHOR-
AMIDE ◻ HMPA ◻ HMPT ◻ HPT ◻ MEMPA ◻
PHOSPHORIC TRIS(DIMETHYLAMIDE) ◻
PHOSPHORYL HEXAMETHYLTRIAMIDE ◻
TRI(DIMETHYLAMINO)PHOSPHINE OXIDE ◻
TRIS(DIMETHYLAMINO)PHOSPHINE OXIDE ◻
TRIS(DIMETHYLAMINO)PHOSPHORUS OXIDE
CONSENSUS REPORTS: NTP 10th Report
on Carcinogens. IARC Cancer Review:
Group 2B IMEMDT 7,56,87; Animal
Sufficient Evidence IMEMDT 15,211,77.
Community Right-To-Know List. Reported
in EPA TSCA Inventory. EPA Genetic
Toxicology Program.
ACGIH TLV: Animal Carcinogen, Suspected
Human Carcinogen
DFG MAK: Animal Carcinogen, Suspected
Human Carcinogen
SAFETY PROFILE: Confirmed carcinogen
with experimental carcinogenic and
tumorigenic data. Moderately toxic by
ingestion, skin contact, intraperitoneal, and
intravenous routes. Experimental
reproductive effects. Human mutation data
reported. When heated to decomposition it
emits very toxic fumes of phosphine, PO_x,
and NO_x.

HEM000   CAS: 66-25-1   HR: 3
1-HEXANAL
DOT: UN 1207
mf: $C_6H_{12}O$ mw: 100.18
PROP: Colorless liquid; powerful fatty-green
odor. Reported in about a dozen essential
oils (FCTXAV 11,95,73). Mp: −56.3°, bp:
131°, flash p: 90°F (OC), d: 0.808–0.812,
refr index: 1.402–1.407, vap press: 8.6 mm
@ 20°, vap d: 3.45. Sol in alc, fixed oils,
propylene glycol; very sltly sol in water.
SYNS: ALDEHYDE C-6 ◻ CAPROALDEHYDE ◻
CAPROIC ALDEHYDE ◻ CAPRONALDEHYDE ◻ n-
CAPROYLALDEHYDE ◻ FEMA No. 2557 ◻
HEXALDEHYDE (DOT) ◻ HEXANAL
CONSENSUS REPORTS: Reported in EPA
TSCA Inventory.
DOT CLASSIFICATION: 3; Label:
Flammable Liquid

SAFETY PROFILE: Mildly toxic by
ingestion and inhalation. An irritant to skin
and eyes. Flammable liquid. A dangerous
fire hazard when exposed to heat or flame;
can react vigorously with oxidizing
materials. When heated to decomposition it
emits acrid smoke and fumes. See also
ALDEHYDES.

HEM500   CAS: 628-02-4   HR: 2
HEXANAMIDE
mf: $C_6H_{13}NO$ mw: 115.20
PROP: A solid. Mp: 100°. Mod sol in hot
water.
SYNS: CAPROAMIDE ◻ CAPRONAMIDE ◻ NCI-
C02142
CONSENSUS REPORTS: Reported in EPA
TSCA Inventory.
SAFETY PROFILE: Questionable
carcinogen with experimental carcinogenic
data. When heated to decomposition it
emits toxic fumes such as NO_x.

HEN000   CAS: 110-54-3   HR: 3
n-HEXANE
DOT: UN 1208
mf: C_6H_{14} mw: 86.20
PROP: Colorless clear liquid; faint odor. Fp:
−93.6°, bp: 69°, ULC: 90–95, lel: 1.2%, uel:
7.5%, flash p: −9.4°F, d: 0.655 @ 25°/4°,
autoign temp: 437°F, vap press: 100 mm @
15.8°, vap d: 2.97. Insol in water; misc in
chloroform, ether, alc. Very volatile liquid.
IDLH 1100 ppm [10%LEL].
SYNS: ESANI (ITALIAN) ◻ GETTYSOLVE-B ◻
HEKSAN (POLISH) ◻ HEXANE (DOT) ◻ HEXANEN
(DUTCH) ◻ HEXANES (FCC) ◻ NCI-C60571
CONSENSUS REPORTS: Reported in EPA
TSCA Inventory.
OSHA PEL: TWA 50 ppm
ACGIH TLV: TWA 50 ppm (skin); BEI: 5
mg(2,5-hexanedione)/g creatinine in urine at
end of shift; 40 ppm n-hexane in end-
exhaled air during shift
DFG MAK: 50 ppm (180 mg/m³)
NIOSH REL: TWA (Alkanes) 350 mg/m³
DOT CLASSIFICATION: 3; Label:
Flammable Liquid

SAFETY PROFILE: Slightly toxic by ingestion and inhalation. Human systemic effects: hallucinations, structural change in nerve or sheath. Experimental teratogenic and reproductive effects. Mutation data reported. An eye irritant. Can cause a motor neuropathy in exposed workers. May be irritating to respiratory tract and narcotic in high concentrations. Inhalation of 5000 ppm for 1/6 hour produces marked vertigo; 2500–1000 ppm for 12 hours produces drowsiness, fatigue, loss of appetite, paresthesia in distal extremities; 2500–500 ppm for 1/6 hour produces muscle weakness, cold pulsation in extremities, blurred vision, headache, anorexia, and onset of polyneuropathy; 2000 ppm for 1/6 hour produces no symptoms; 1000–500 ppm for 3–6 months produces fatigue, loss of appetite, distal paresthesia. Dangerous if abused.

Flammable liquid. A very dangerous fire and explosion hazard when exposed to heat or flame; can react vigorously with oxidizing materials. Mixtures with dinitrogen tetraoxide may explode at 28°. To fight fire, use CO_2, dry chemical. When heated to decomposition it emits acrid smoke and fumes.

HEO000   CAS: 124-09-4   HR: 3
1,6-HEXANEDIAMINE
DOT: UN 1783/UN 2280
mf: $C_6H_{16}N_2$ mw: 116.24
PROP: Colorless leaflets, long needles by sublimation; odor of piperidine. Mp: 39–42°, bp: 205°. Absorbs water and CO_2 from air; very sol in water; sltly sol in alc, benzene.
SYNS: 1,6-DIAMINOHEXANE □ HEXAMETHYLENE-DIAMINE □ 1,6-HEXAMETHYLENEDIAMINE □ HEXAMETHYLENEDIAMINE, solid (UN 2280) (DOT) □ HEXAMETHYLENEDIAMINE, solution (UN 1783) (DOT) □ HMDA □ NCI-C61405
CONSENSUS REPORTS: Reported in EPA TSCA Inventory.
ACGIH TLV: TWA 0.5 ppm
DOT CLASSIFICATION: 8; Label: Corrosive
SAFETY PROFILE: Poison by intravenous and intraperitoneal routes. Moderately toxic

by ingestion, inhalation, and skin contact. An experimental teratogen. A corrosive irritant to skin, eyes, and mucous membranes. Combustible when exposed to heat or flame; can react with oxidizing materials. See also AMINES.

HEQ200   CAS: 3848-24-6   HR: 3
2,3-HEXANEDIONE
mf: $C_6H_{10}O_2$ mw: 114.16
PROP: Bp: 128°, d: 0.934, flash p: 83°F.
SYNS: ACETYLBUTYRYL □ METHYL PROPYL DIKETONE
CONSENSUS REPORTS: Reported in EPA TSCA Inventory.
SAFETY PROFILE: A skin irritant. Flammable liquid. When heated to decomposition it emits acrid smoke and irritating fumes.

HEQ500   CAS: 110-13-4   HR: 2
2,5-HEXANEDIONE
mf: $C_6H_{10}O_2$ mw: 114.16
PROP: Colorless liquid. Gradually turns yellow. Mp: −9°, bp: 188°, flash p: 174°F (CC), d: 0.970 @ 20°/4°, autoign temp: 920°, vap d: 3.94. Misc in water and alc.
SYNS: ACETONYL ACETONE □ α,β-DIACETYLETHANE □ 1,2-DIACETYLETHANE □ 2,5-DIKETOHEXANE
CONSENSUS REPORTS: Reported in EPA TSCA Inventory.
SAFETY PROFILE: Moderately toxic by ingestion and inhalation. Mildly toxic by skin contact. An eye irritant. Experimental reproductive effects. Combustible when exposed to heat or flame; can react with oxidizing materials. To fight fire, use CO_2, dry chemical (multi-purpose dry chemical), water spray or mist, alcohol foam. When heated to decomposition it emits acrid smoke and fumes.

HER500   CAS: 628-73-9   HR: 3
HEXANENITRILE
mf: $C_6H_{11}N$ mw: 97.18
PROP: Bp: 163.6°.
SYNS: CAPRONITRILE □ NC5

CONSENSUS REPORTS: Reported in EPA TSCA Inventory. Cyanide and its compounds are on the Community Right-To-Know List.
SAFETY PROFILE: A poison by intravenous and subcutaneous routes. Moderately toxic by ingestion. When heated to decomposition it emits toxic fumes of NO_x and CN^-. See also NITRILES.

HES000   CAS: 111-31-9   HR: 2
1-HEXANETHIOL
DOT: UN 1228/UN 3071
mf: $C_6H_{14}S$ mw: 118.26
SYNS: HEXYL MERCAPTAN □ USAF EK-4628
CONSENSUS REPORTS: Reported in EPA TSCA Inventory.
NIOSH REL: (n-Alkane Mono Thiols) CL 0.5 ppm/15M
DOT CLASSIFICATION: 3; Label: Flammable Liquid, Poison (UN 1228); DOT Class: 6.1; Label: Poison, Flammable Liquid (UN 3071)
SAFETY PROFILE: Moderately toxic by inhalation and ingestion. A flammable liquid. When heated to decomposition it emits very toxic fumes of SO_x. See also MERCAPTANS.

HET500   CAS: 131-73-7   HR: 3
2,4,6,2',4',6'-HEXANITRODIPHENYL-
    AMINE
DOT: UN 0079
mf: $C_{12}H_5N_7O_{12}$ mw: 439.24
PROP: Prisms from AcOH. Mp: 245–246°. Sol in AcOH.
SYNS: AURANTIA □ BIS(2,4,6-TRINITRO-PHENYL)-AMIN (GERMAN) □ C.I. 10360 □ DIPHENYLAMINE, HEXANITRO- □ DIPICRYLAMINE □ DIPICRYLAMINE (DOT) □ DIPIKRYLAMIN □ DPA □ ESANITRO-DIFENILAMINA (ITALIAN) □ 2,2',4,4',6,6'-HEXANITRO-DIFENYLAMIN □ HEXANITRODIFENYLAMINE (DUTCH) □ HEXANITRODIPHENYLAMINE □ HEXANITRODIPHENYLAMINE (FRENCH) □ 2,2',4,4',6,6'-HEXANITRODIPHENYLAMINE □ 2,4,6,2',4',6'-HEXANITRODIPHENYLAMINE □ HEXYL (GERMAN, DUTCH)
CONSENSUS REPORTS: Reported in EPA TSCA Inventory.

DOT CLASSIFICATION: EXPLOSIVE 1.1D; Label: EXPLOSIVE 1.1D
SAFETY PROFILE: Questionable carcinogen with experimental neoplastigenic data. Mutation data reported. A powerful and violent explosive used as a booster explosive; its use is superior to TNT. It is not as good for this purpose as tetryl, but is extremely stable and much safer to handle. See also NITRO COMPOUNDS of AROMATIC HYDROCARBONS.

HET675   CAS: 918-37-6   HR: 3
HEXANITROETHANE
mf: $C_2N_6O_{12}$ mw: 300.06
PROP: A solid. Mp: 150° (decomp). Sol in CCl_4 and CH_2Cl_2.
SYN: ETHANE, HEXANITRO-
CONSENSUS REPORTS: Reported in EPA TSCA Inventory.
DOT CLASSIFICATION: Forbidden
SAFETY PROFILE: A powerful oxidant which explodes above 140°C. Explosive reaction with boron. Hypergolic reaction with dimethyl hydrazine or other strong organic bases. Forms powerfully explosive mixtures with nitrogen containing organic compounds (e.g., 2-nitroaniline). Upon decomposition it emits toxic fumes of NO_x. See also NITRO COMPOUNDS.

HEU000   CAS: 142-62-1   HR: 2
HEXANOIC ACID
DOT: UN 2829
mf: $C_6H_{12}O_2$ mw: 116.18
PROP: Oily, colorless liquid; odor of Limburger cheese. Fp: −1.5°, bp: 205.0°, flash p: 215°F (COC), d: 0.9295 @ 20°/20°, refr index: 1.415–1.418, vap press: 0.18 mm @ 20°, vap d: 4.0, autoign temp: 716°F. Sol in alkalies, EtOH, and Et_2O; sltly sol in H_2O.
SYNS: BUTYLACETIC ACID □ CAPROIC ACID □ n-CAPROIC ACID □ CAPRONIC ACID □ FEMA No. 2559 □ HEXACID 698 □ n-HEXANOIC ACID □ n-HEXOIC ACID □ PENTIFORMIC ACID □ PENTYLFORMIC ACID
CONSENSUS REPORTS: Reported in EPA TSCA Inventory.
DOT CLASSIFICATION: 8; Label: Corrosive

SAFETY PROFILE: Moderately toxic by ingestion, skin contact, intraperitoneal, and subcutaneous routes. Mutation data reported. A corrosive material. A skin and severe eye irritant. Combustible when exposed to heat or flame; can react with oxidizing materials. To fight fire, use CO_2, dry chemical, fog, mist. When heated to decomposition it emits acrid smoke and fumes.

HEV000   CAS: 591-78-6   HR: 3
2-HEXANONE
mf: $C_6H_{12}O$ mw: 100.18
PROP: Clear liquid; odor of nail-polish remover. Mp: $-56.9°$, bp: $127.2°$, lel: 1.22%, uel: 8.0%, flash p: 95°F (OC), d: 0.830 @ 0°/4°, vap press: 10 mm @ 38.8°, vap d: 3.45, autoign temp: 991°F. Sltly sol in water; sol in alc and ether. IDLH 1600 ppm.
SYNS: BUTYL METHYL KETONE □ n-BUTYL METHYL KETONE □ HEXANONE-2 □ MBK □ METHYL n-BUTYL KETONE (ACGIH) □ MNBK
CONSENSUS REPORTS: Reported in EPA TSCA Inventory.
OSHA PEL: TWA 5 ppm
ACGIH TLV: TWA 5 ppm; STEL 10 ppm
DFG MAK: 5 ppm (21 mg/m³)
NIOSH REL: (Ketones) TWA 4 mg/m³
DOT CLASSIFICATION: 3; Label: Flammable Liquid
SAFETY PROFILE: Moderately toxic by ingestion and intraperitoneal routes. Mildly toxic by inhalation and skin contact. Experimental teratogenic and reproductive effects. Human systemic effects by inhalation: unspecified eye effects, headache, nausea or vomiting. A skin and eye irritant. Dangerous fire and explosion hazard when exposed to heat or flame; can react with oxidizing materials. To fight fire, use alcohol foam, CO_2, dry chemical. See also KETONES.

HEW050   CAS: 23389-74-4   HR: 3
4-HEXANOYLPYRIDINE
mf: $C_{11}H_{15}NO$ mw: 177.27
SYNS: 1-HEXANONE, 1-(4-PYRIDYL)- □ KETONE, PENTYL 4-PYRIDYL □ PENTYL 4-PYRIDYL KETONE □

PYRIDINE, 4-HEXANOYL- □ 1-(4-PYRIDYL)-1-HEXANONE
DOT CLASSIFICATION: 3; Label: Flammable Liquid
SAFETY PROFILE: A poison by intraperitoneal route. A flammable liquid. When heated to decomposition it emits toxic vapors of NO_x.

HEW200   CAS: 7328-05-4   HR: 3
HEXAPROPYLDISTANNTHIANE
mf: $C_{18}H_{42}SSn_2$ mw: 528.04
SYNS: DISTANNTHIANE, HEXAPROPYL- □ 1,1,1,3,3,3-HEXAPROPYLDISTANNTHIANE
OSHA PEL: TWA 0.1 mg(Sn)/m³
ACGIH TLV: TWA 0.1 mg(Sn)/m³; STEL 0.2 mg/m³ (skin)
NIOSH REL: (Organotin Compounds): 10H TWA 0.1 mg(Sn)/m³
SAFETY PROFILE: Poison by intraperitoneal route. When heated to decomposition it emits toxic fumes of SO_x and Sn.

HEZ800   CAS: 14023-01-9   HR: 3
HEXAUREA CHROMIC CHLORIDE
mf: $C_6H_{24}CrN_{12}O_6 \cdot 3Cl$ mw: 518.77
PROP: Green needles. Sol in H_2O; insol in EtOH. IDLH 25 mg/m³ [as Cr(III)].
SYNS: CHROMIUM CHLORIDE, HEXAUREA □ CHROMIUM(3+), HEXAKIS(UREA-O)-, TRICHLORIDE, (OC-6-11)-(9CI) □ CHROMIUM(3+), HEXAKIS(UREA)-, TRICHLORIDE (8CI) □ CHROMIUM(III) HEXA-UREA CHLORIDE
OSHA PEL: TWA 0.5 mg(Cr)/m³
ACGIH TLV: TWA 0.5 mg(Cr)/m³; Not Classifiable as a Carcinogen.
SAFETY PROFILE: Poison by intravenous route. When heated to decomposition it emits toxic fumes of NO_x, Cr, and Cl⁻.

HFA300   CAS: 51235-04-2   HR: 2
HEXAZINONE
mf: $C_{12}H_{20}N_4O_2$ mw: 252.36
PROP: Crystals. Very sol in $CHCl_3$, MeOH; sol in Me_2CO, C_6H_6; mod sol in H_2O.
SYNS: 3-CYCLOHEXYL-6-(DIMETHYLAMINO)-1-METHYL-s-TRIAZINE-2,4(1H,3H)-DIONE □ 3-CYCLO-HEXYL-6-(DIMETHYLAMINO)-1-METHYL-1,3,5-TRIAZINE-2,4(1H,3H)-DIONE □ DPX 3674 □ VELPAR □ VELPAR WEED KILLER

CONSENSUS REPORTS: Reported in EPA TSCA Inventory.

SAFETY PROFILE: Moderately toxic by ingestion and intraperitoneal routes. Mildly toxic by skin contact. Experimental reproductive effects. An eye irritant. When heated to decomposition it emits toxic fumes of NO_x.

HFA500   CAS: 505-57-7   HR: 3
2-HEXENAL

mf: $C_6H_{10}O$ mw: 98.16

PROP: Colorless liquid. Flash pt: 104° F.

SYNS: HEX-2-ENAL □ HEX-2-EN-1-AL □ HEXYLENIC ALDEHYDE □ LEAF ALDEHYDE

CONSENSUS REPORTS: Reported in EPA TSCA Inventory.

SAFETY PROFILE: A poison by intraperitoneal route. Mutation data reported. A combustible liquid. When heated to decomposition it emits acrid smoke and fumes. See also ALDEHYDES.

HFA525   HR: 3
trans-2-HEXEN-1-AL

mf: $C_6H_{10}O$ mw: 98.15

PROP: Pale-yellow liquid; fruity, vegetable odor. D: 0.841–0.848, refr index: 1.445–1.449, flash p: 100°F. Sol in alc, propylene glycol, fixed oils; very sltly sol in water.

SYN: FEMA No. 2560

SAFETY PROFILE: Flammable liquid. When heated to decomposition it emits acrid smoke and irritating fumes.

HFB000   CAS: 592-41-6   HR: 3
1-HEXENE

mf: C_6H_{12} mw: 84.158

PROP: Colorless liquid. Bp: 64.5°, mp: −139.9°, flash p: −14.8°F, d: 0.6732 @ 20°/4°, vap press: 310 mm @ 38°, vap d: 3.0, lel: 1.2%, uel: 6.9%

SYNS: BUTYL ETHYLENE □ HEXENE □ HEXYLENE

CONSENSUS REPORTS: Reported in EPA TSCA Inventory.

ACGIH TLV: TWA 50 ppm

SAFETY PROFILE: Moderately toxic irritant to skin, eyes, and mucous membranes. A very dangerous fire and explosion hazard when exposed to heat, flame, or oxidizers. Can react vigorously with oxidizing materials. To fight fire, use dry chemical, CO_2, foam. When heated to decomposition it emits acrid smoke and fumes.

HFD500   CAS: 928-95-0   HR: 3
2-HEXEN-1-OL, (E)-

mf: $C_6H_{12}O$ mw: 100.18

PROP: Colorless liquid; fruity-green odor. Bp: 158–160°, d: 0.836–0.841, refr index: 0.437–1.442, flash p: 129°F. Sol in alc, propylene glycol, fixed oils; very sltly sol in water.

SYNS: FEMA No. 2562 □ 2-HEXENOL □ trans-2-HEXENOL □ trans-2-HEXEN-1-OL (FCC)

CONSENSUS REPORTS: Reported in EPA TSCA Inventory.

SAFETY PROFILE: Moderately toxic by ingestion. Mildly toxic by skin contact. A skin irritant. Flammable liquid when exposed to heat, sparks, or flame. When heated to decomposition it emits acrid smoke and fumes. See also ALCOHOLS.

HFE000   CAS: 928-96-1   HR: 3
cis-3-HEXENOL

mf: $C_6H_{12}O$ mw: 100.18

PROP: Colorless liquid; powerful grassy-green odor. D: 0.846–0.850, refr index: 1.43–1.441, bp: 157°, flash p: 111°F. Sol in alc, propylene glycol, fixed oils; very sltly sol in water.

SYNS: BLAETTERALKOHOL □ FEMA No. 2563 □ β-γ-HEXENOL □ cis-3-HEXEN-1-OL (FCC) □ LEAF ALCOHOL

CONSENSUS REPORTS: Reported in EPA TSCA Inventory.

SAFETY PROFILE: A poison by intraperitoneal route. Mildly toxic by ingestion. Flammable liquid. When heated to decomposition it emits acrid smoke and fumes. See also ALCOHOLS.

HFE100   CAS: 2497-18-9   HR: 1
2-HEXEN-1-OL ACETATE
mf: $C_8H_{14}O_2$ mw: 142.22
PROP: Green grassy, spicy, fruity ester like aroma.
SYNS: HEX-2-ENYL ACETATE □ 2-HEXENYL ACETATE □ 2-HEXEN-1-YL-ACETATE □ (E)-2-HEXENYL ACETATE □ trans-2-HEXENYL ACETATE
CONSENSUS REPORTS: Reported in EPA TSCA Inventory.
SAFETY PROFILE: A skin irritant. When heated to decomposition it emits acrid smoke and irritating fumes.

HFE520   CAS: 41519-23-7   HR: 1
cis-3-HEXENYL ISOBUTYRATE
mf: $C_{10}H_{18}O_2$ mw: 170.28
SYNS: ENT 33,348 □ β,γ-HEXENYL ISOBUTANOATE □ PROPANOIC ACID, 2-METHYL-, 3-HEXENYL ESTER, (Z)-
CONSENSUS REPORTS: Reported in EPA TSCA Inventory.
SAFETY PROFILE: A skin irritant. When heated to decomposition it emits acrid smoke and irritating fumes.

HFE550   HR: 2
cis-3-HEXENYL 2-METHYLBUTYRATE
mf: $C_{11}H_{20}O_2$ mw: 184.28
PROP: Colorless liquid; powerful, fruity odor like that of unripe apples. D: 0.876–0.880, refr index: 1.430, flash p: 153°F. Sol in alc, fixed oils; insol in water.
SYN: FEMA No. 3497
SAFETY PROFILE: Combustible liquid. When heated to decomposition it emits acrid smoke and irritating fumes.

HFE625   CAS: 42436-07-7   HR: 1
cis-3-HEXENYL PHENYLACETATE
mf: $C_{14}H_{18}O_2$ mw: 218.32
SYNS: BENZENEACETIC ACID, 3-HEXENYL ESTER, (Z)- □ β,γ-HEXENYL α-TOLUATE
CONSENSUS REPORTS: Reported in EPA TSCA Inventory.
SAFETY PROFILE: A skin irritant. When heated to decomposition it emits acrid smoke and irritating fumes.

HFE650   CAS: 33467-74-2   HR: 1
cis-3-HEXENYL PROPIONATE
mf: $C_9H_{16}O_2$ mw: 156.25
SYNS: 3-HEXEN-1-OL, PROPANOATE (Z)- □ β,γ-HEXENYL PROPANOATE
CONSENSUS REPORTS: Reported in EPA TSCA Inventory.
SAFETY PROFILE: A skin irritant. When heated to decomposition it emits acrid smoke and irritating fumes.

HFG500   CAS: 108-10-1   HR: 3
HEXONE
DOT: UN 1245
mf: $C_6H_{12}O$ mw: 100.18
PROP: Colorless mobile liquid; fruity, ethereal odor. Fp: −80.2°, bp: 116.8°, lel: 1.4%, uel: 7.5%, flash p: 62.6°F, d: 0.801, autoign temp: 858°F, vap press: 16 mm @ 20°. Misc with alc, ether; sol in water. IDLH 500 ppm.
SYNS: FEMA No. 2731 □ HEXON (CZECH) □ ISOBUTYL-METHYLKETON (CZECH) □ ISOBUTYL METHYL KETONE □ ISOPROPYLACETONE □ METHYL-ISOBUTYL-CETONE (FRENCH) □ METHYLISOBUTYL-KETON (DUTCH, GERMAN) □ METHYL ISOBUTYL KETONE (ACGIH, DOT) □ 4-METHYL-PENTAN-2-ON (DUTCH, GERMAN) □ 4-METHYL-2-PENTANON (CZECH) □ 2-METHYL-4-PENTANONE □ 4-METHYL-2-PENTANONE (FCC) □ METILISOBUTILCHETONE (ITALIAN) □ 4-METILPENTAN-2-ONE (ITALIAN) □ METYLOIZOBUTYLOKETON (POLISH) □ MIBK □ MIK □ RCRA WASTE NUMBER U161 □ SHELL MIBK
CONSENSUS REPORTS: Reported in EPA TSCA Inventory. Community Right-To-Know List.
OSHA PEL: TWA 50 ppm; STEL 75 ppm
ACGIH TLV: TWA 50 ppm; STEL 75 ppm; BEI: 2 mg/L of MIBK in urine at end of shift; (Proposed: 30 ppm; STEL 75 ppm; Confirmed Animal Carcinogen with Unknown Revelance to Humans)
DFG MAK: 20 ppm (83 mg/m³)
NIOSH REL: (Ketones) TWA 200 mg/m³
DOT CLASSIFICATION: 3; Label: Flammable Liquid
SAFETY PROFILE: A poison by intraperitoneal route. Moderately toxic by ingestion. Mildly toxic by inhalation. Very irritating to the skin, eyes, and mucous

H

membranes. An experimental teratogen. A human systemic irritant by inhalation. Narcotic in high concentration. Flammable liquid when exposed to heat, flame, or oxidizers. Ignites on contact with potassium-tert-butoxide. Moderately explosive in the form of vapor when exposed to heat or flame. May form explosive peroxides upon exposure to air. Can react vigorously with reducing materials. To fight fire, use alcohol foam, CO_2, dry chemical. Incompatible with air, potassium-tert-butoxide. See also KETONES.

HFG700   CAS: 17597-95-4   HR: 1
HEXOXYACETALDEHYDE DIMETHYL-
   ACETAL
mf: $C_{10}H_{22}O_3$ mw: 190.32
SYNS: ACETALDEHYDE, (HEXYLOXY)-, DIMETHYL ACETAL □ β-HEXOXYACETALDEHYDE DIMETHYL-ACETAL □ 2-HEXOXYACETALDEHYDE DIMETHYLACETAL
CONSENSUS REPORTS: Reported in EPA TSCA Inventory.
SAFETY PROFILE: A skin irritant. When heated to decomposition it emits acrid smoke and irritating fumes.

HFI500   CAS: 142-92-7   HR: 3
HEXYL ACETATE
mf: $C_8H_{16}O_2$ mw: 144.24
PROP: Colorless liquid; fruity odor. D: 0.878, mp: −60.9°, bp: 171.5°, refr index: 1.407, flash p: 109°F. Insol in water; very sol in alc and ether.
SYNS: ACETIC ACID HEXYL ESTER □ FEMA No. 2565 □ n-HEXYL ACETATE (FCC) □ 1-HEXYL ACETATE □ HEXYL ALCOHOL, ACETATE □ HEXYL ETHANOATE
CONSENSUS REPORTS: Reported in EPA TSCA Inventory.
SAFETY PROFILE: Low toxicity by ingestion and skin contact. A skin and eye irritant. A flammable liquid. When heated to decomposition it emits acrid smoke and fumes. See also ESTERS.

HFJ000   CAS: 108-84-9   HR: 3
sec-HEXYL ACETATE

DOT: UN 1233
mf: $C_8H_{16}O_2$ mw: 144.24
PROP: Clear liquid; pleasant odor. Bp: 146.3°, fp: −63.8°, flash p: 113°F (COC), d: 0.8598 @ 20°/20°, vap press: 3.8 mm @ 20°, vap d: 4.97. IDLH 500 ppm.
SYNS: ACETIC ACID-1,3-DIMETHYLBUTYL ESTER □ 1,3-DIMETHYLBUTYL ACETATE □ MAAC □ METHYL-AMYL ACETATE □ METHYL AMYL ACETATE (DOT) □ METHYLISOAMYL ACETATE □ METHYLISOBUTYL-CARBINOL ACETATE □ METHYLISOBUTYLCARBINYL ACETATE □ 4-METHYL-2-PENTANOL, ACETATE □ 4-METHYL-2-PENTYL ACETATE
CONSENSUS REPORTS: Reported in EPA TSCA Inventory.
OSHA PEL: TWA 50 ppm
ACGIH TLV: TWA 50 ppm
DFG MAK: 50 ppm (300 mg/m³)
DOT CLASSIFICATION: 3; Label: Flammable Liquid
SAFETY PROFILE: Mildly toxic by ingestion, skin contact, and inhalation. Human systemic effects by inhalation: conjunctiva irritation, unspecified changes in olfactory and respiratory systems. A skin and human eye irritant. Flammable liquid when exposed to heat or flame; can react with oxidizing materials. To fight fire, use alcohol foam, CO_2, dry chemical. See also ESTERS.

HFJ500   CAS: 111-27-3   HR: 3
n-HEXYL ALCOHOL
mf: $C_6H_{14}O$ mw: 102.20
PROP: Colorless liquid. Fp: −46.7°, bp: 157.2°, flash p: 145°F, d: 0.8204 @ 20°/20°, vap press: 1 mm @ 24.4°, vap d: 3.52. Misc in alc, ether; sltly sol in water.
SYNS: AMYLCARBINOL □ CAPROYL ALCOHOL □ EPAL 6 □ FEMA No. 2567 □ HEXANOL □ n-HEXANOL □ 1-HEXANOL □ HEXYL ALCOHOL □ 1-HYDROXY-HEXANE □ PENTYLCARBINOL
CONSENSUS REPORTS: Reported in EPA TSCA Inventory.
SAFETY PROFILE: Poison by intravenous route. Moderately toxic by ingestion and skin contact. A skin and severe eye irritant. Flammable liquid when exposed to heat, sparks, or flame. Can react with oxidizing

materials. To fight fire, use alcohol foam, CO_2, dry chemical. See also ALCOHOLS.

HFJ600   CAS: 26401-20-7   HR: 3
tert-HEXYL ALCOHOL
mf: $C_6H_{14}O$ mw: 102.20
SYN: tert-HEXANOL (9CI, DOT)
SAFETY PROFILE: Poison by ingestion and intravenous routes. When heated to decomposition it emits acrid smoke and fumes. Flammable when exposed to heat or flame; can react vigorously with oxidizing materials. A fire hazard. See also ALCOHOLS.

HFK000   CAS: 111-26-2   HR: 3
HEXYLAMINE
DOT: UN 2733/UN 2734
mf: $C_6H_{15}N$ mw: 101.22
PROP: Liquid. Mp: −22.9°, fp: −19°, bp: 131.4°, flash p: 85°F (OC), d: 0.7675 @ 20°/20°, vap d: 3.49.
SYNS: 1-AMINOHEXANE □ 1-HEXANAMINE □ N-HEXYLAMINE □ MONO-N-HEXYLAMINE
CONSENSUS REPORTS: Reported in EPA TSCA Inventory.
DOT CLASSIFICATION: 8; Label: Corrosive, Flammable Liquid (UN 2734); DOT Class: 3; Label: Flammable Liquid, Corrosive (UN 2733)
SAFETY PROFILE: A poison by intraperitoneal route. Moderately toxic by ingestion, inhalation, and skin contact. A severe skin and eye irritant. Dangerous fire hazard when exposed to heat or flame; can react with oxidizing materials. To fight fire, use alcohol foam, CO_2, dry chemical. Upon decomposition it emits toxic fumes of NO_x. See also AMINES.

HFM600   CAS: 19089-92-0   HR: 1
HEXYL-2-BUTENOATE
mf: $C_{10}H_{18}O_2$ mw: 170.28
PROP: Colorless liquid; fruity odor. D: 0.880, refr index: 1.428–1.449. Sol in alc, fixed oils; insol in water, propylene glycol.
SYNS: 2-BUTENOIC ACID, HEXYL ESTER □ FEMA No. 3354 □ n-HEXYL 2-BUTENOATE □ HEXYL CROTONATE

CONSENSUS REPORTS: Reported in EPA TSCA Inventory.
SAFETY PROFILE: Low toxicity by ingestion and skin contact. A skin irritant. When heated to decomposition it emits acrid smoke and irritating fumes.

HFM700   CAS: 2639-63-6   HR: 1
1-HEXYL BUTYRATE
mf: $C_{10}H_{20}O_2$ mw: 172.30
PROP: Flavor and frgagrance checmical.
SYNS: BUTYRIC ACID, HEXYL ESTER □ HEXYL BUTANOATE □ n-HEXYL BUTANOATE □ n-HEXYL n-BUTANOATE □ HEXYL BUTYRATE □ n-HEXYL BUTYRATE
CONSENSUS REPORTS: Reported in EPA TSCA Inventory.
SAFETY PROFILE: A skin irritant. When heated to decomposition it emits acrid smoke and irritating fumes.

HFN500   CAS: 20740-05-0   HR: 3
n-HEXYL CARBORANE
mf: $C_8H_{24}B_{10}$ mw: 228.42
PROP: A liquid. Bp: 326.7° @ 760 mm.
SYNS: HEXYLDICARBADODECABORANE(12) □ NHC
CONSENSUS REPORTS: Reported in EPA TSCA Inventory.
SAFETY PROFILE: Poison by intravenous route. Moderately toxic by intraperitoneal route. An experimental teratogen. A skin irritant. When heated to decomposition it emits toxic fumes of boron. See also BORANES and BORON COMPOUNDS.

HFO500   CAS: 101-86-0   HR: 2
HEXYL CINNAMALDEHYDE
mf: $C_{15}H_{20}O$ mw: 216.35
PROP: Pale-yellow liquid; jasmine odor. D: 0.953–0.959, mp: 4°, bp: 169° @ 20 mm, refr index: 1.548–1.552. Sol in fixed oils; insol in propylene glycol, glycerin.
SYNS: FEMA No. 2569 □ α-HEXYLCINNAMALDEHYDE (FCC) □ HEXYL CINNAMIC ALDEHYDE □ α-HEXYLCINNAMIC ALDEHYDE □ α-n-HEXYL-β-PHENYLACROLEIN □ 2-(PHENYLMETHYLENE)OCTANOL
CONSENSUS REPORTS: Reported in EPA TSCA Inventory.

SAFETY PROFILE: Moderately toxic by ingestion. A skin irritant. When heated to decomposition it emits acrid smoke and fumes. See also ALDEHYDES.

HFO700   CAS: 95-41-0   HR: 3
2-n-HEXYL-2-CYCLOPENTEN-1-ONE
mf: $C_{11}H_{18}O$ mw: 166.29
PROP: Bp: 126–129° @ 30 mm.
SYNS: 2-CYCLOPENTEN-1-ONE, 2-HEXYL- □ DIHYDRO-ISOJASMONE
CONSENSUS REPORTS: Reported in EPA TSCA Inventory.
SAFETY PROFILE: Poison by intravenous route. A skin irritant. When heated to decomposition it emits acrid smoke and irritating fumes.

HFP600   CAS: 64049-22-5   HR: 3
HEXYLDICHLORARSINE
mf: $C_6H_{13}AsCl_2$ mw: 231.01
SYNS: ARSINE, DICHLOROHEXYL- □ TL 231
OSHA PEL: TWA 0.5 mg(As)/m³
SAFETY PROFILE: Poison by skin contact. When heated to decomposition it emits toxic fumes of As and Cl⁻.

HFP875   CAS: 107-41-5   HR: 2
HEXYLENE GLYCOL
mf: $C_6H_{14}O_2$ mw: 118.20
PROP: Mild odor, colorless liquid, water-sol. Bp: 197.1°, fp: −50°, flash p: 205°F (OC), d: 0.9234 @ 20°/20°, vap press: 0.05 mm @ 20°.
SYNS: 2,4-DIHYDROXY-2-METHYLPENTANE □ DIOLANE □ 1,2-HEXANEDIOL □ ISOL □ 2-METHYL PENTANE-2,4-DIOL □ 2-METHYL-2,4-PENTANEDIOL □ PINAKON □ α,α,α'-TRIMETHYLTRIMETHYLENE GLYCOL
CONSENSUS REPORTS: Reported in EPA TSCA Inventory.
OSHA PEL: CL 25 ppm
ACGIH TLV: CL 25 ppm
DFG MAK: 10 ppm
SAFETY PROFILE: Moderately toxic by ingestion and intraperitoneal routes. Mildly toxic by skin contact. Human systemic effects by inhalation: conjunctiva and other eye, olfactory, and pulmonary changes.

Mutation data reported. Combustible when exposed to heat or flame; can react with oxidizing materials. To fight fire, use foam, CO_2, dry chemicals. When heated to decomposition it emits acrid smoke and fumes. See also GLYCOLS.

HFQ550   CAS: 2349-07-7   HR: 1
n-HEXYL ISOBUTYRATE
mf: $C_{10}H_{20}O_2$ mw: 172.30
SYNS: HEXYL ISOBUTANOATE □ n-HEXYL ISOBUTANOATE □ HEXYL ISOBUTYRATE □ 1-HEXYL ISOBUTYRATE □ ISOBUTYRIC ACID, HEXYL ESTER □ PROPANOIC ACID, 2-METHYL-, HEXYL ESTER (9CI)
CONSENSUS REPORTS: Reported in EPA TSCA Inventory.
SAFETY PROFILE: Low toxicity by ingestion and skin contact. A skin irritant. When heated to decomposition it emits acrid smoke and irritating fumes.

HFQ600   HR: D
HEXYL ISOVALERATE
mf: $C_{11}H_{22}O_2$ mw: 186.30
PROP: Colorless liquid; pungent, fruity odor. D: 0.853, refr index: 1.417. Sol in alc, fixed oils; insol in water.
SYN: FEMA No. 3500
SAFETY PROFILE: When heated to decomposition it emits acrid smoke and irritating fumes.

HFR100   CAS: 18431-36-2   HR: 3
n-HEXYLMERCURIC BROMIDE
mf: $C_6H_{13}BrHg$ mw: 365.69
SYNS: BROMOHEXYLMERCURY □ HEXYLMERCURIC BROMIDE □ HEXYL MERCURY BROMIDE □ HMB □ MERCURY, BROMOHEXYL
OSHA PEL: TWA 0.01 mg(Hg)/m³; CL 0.03 mg(Hg)/m³ (skin)
ACGIH TLV: TWA 0.01 mg(Hg)/m³; BEI: 35 µg/g creatinine total inorganic mercury in urine preshift; 15 µg/g creatinine total inorganic mercury in blood at end of shift at end of workweek.
DFG MAK: Confirmed Animal Carcinogen with Unknown Relevance to Humans
NIOSH REL: (Mercury, Organo) TWA 0.01 mg/m³; STEL 0.03 mg/m³ (skin)

SAFETY PROFILE: Poison by subcutaneous route. Mutation data reported. When heated to decomposition it emits toxic fumes of Hg and Br⁻.

HFR200   CAS: 10032-15-2   HR: 3
HEXYL 2-METHYLBUTYRATE

mf: $C_{11}H_{22}O_2$ mw: 186.33

PROP: Colorless liquid; strong, fresh-green, fruity odor. D: 0.854, refr index: 1.416–1.421, flash p: 122°F. Sol in alc, fixed oils; insol in water.

SYNS: FEMA No. 3499 □ 2-METHYLBUTANOIC ACID, n-HEXYL ESTER

CONSENSUS REPORTS: Reported in EPA TSCA Inventory.

SAFETY PROFILE: Low toxicity by ingestion and skin contact. A skin irritant. Flammable liquid when exposed to heat, sparks, or flame. When heated to decomposition it emits acrid smoke and irritating fumes.

HFT550   CAS: 25961-89-1   HR: 3
2-(2-(2-(HEXYLOXY)ETHOXY)ETHOXY)-ETHANOL

mf: $C_{12}H_{26}O_4$ mw: 234.38

SYNS: CHR 9 □ C6E3 □ ETHANOL, 2-(2-(2-(HEXYLOXY)ETHOXY)ETHOXY)- □ SRI 10163-71 □ TRIETHYLENE GLYCOL MONOHEXYL ETHER

CONSENSUS REPORTS: Reported in EPA TSCA Inventory.

SAFETY PROFILE: A poison by ingestion. When heated to decomposition it emits acrid smoke and irritating vapors.

HFV500   CAS: 136-77-6   HR: 3
HEXYLRESORCINOL

mf: $C_{12}H_{18}O_2$ mw: 194.30

PROP: Colorless liquid to pale-yellow, heavy liquid becoming solid on standing at room temp; needles from benzene or pet ether. Pungent odor, sharp astringent taste. Bp: 179°, mp: 67.5–69°. Very sol in water; sol in benzene, ether, acetone, chloroform, alc, vegetable oils; sltly sol in pet ether.

SYNS: ASCARYL □ CAPROKOL □ CRYSTOIDS □ CYSTOIDS ANTHELMINTIC □ 4-HEXYL-1,3-BENZENEDIOL □ 4-HEXYL-1,3-DIHYDROXYBENZENE □ HEXYLRESORCIN (GERMAN) □ 4-HEXYLRESORC-INE □ p-HEXYLRESORCINOL □ 4-HEXYLRESORCINOL □ 4-n-HEXYLRESORCINOL □ NCI-C55787 □ S.T. 37 □ SUCRETS □ WORM-AGEN

CONSENSUS REPORTS: Reported in EPA TSCA Inventory.

SAFETY PROFILE: A poison by ingestion and intraperitoneal routes. Moderately toxic by subcutaneous route. Experimental reproductive effects. Questionable carcinogen with experimental tumorigenic data. Mutation data reported. An eye irritant. Concentrated solutions can cause burns on the skin and mucous membranes in humans. An anthelmintic and topical antiseptic. When heated to decomposition it emits acrid smoke and fumes.

HFX500   CAS: 928-65-4   HR: 3
HEXYLTRICHLOROSILANE

DOT: UN 1784

mf: $C_6H_{13}Cl_3Si$ mw: 219.63

PROP: A liquid. D: 1.11 @ 20°/4°, bp: 191–192°.

CONSENSUS REPORTS: Reported in EPA TSCA Inventory.

DOT CLASSIFICATION: 8; Label: Corrosive

SAFETY PROFILE: A poison by ingestion and inhalation. Corrosive. A severe irritant to skin, eyes, and mucous membranes. When heated to decomposition or in reaction with water or steam it produces toxic and corrosive fumes of Cl⁻ and HCl. See also CHLOROSILANES.

HGA100   CAS: 68917-43-1   HR: 1
HIBAWOOD OIL

PROP: Yellow viscous liquid. Fresh sweet cedar woody aroma. Perfume uses.

SYN: OIL, HIBAWOOD

CONSENSUS REPORTS: Reported in EPA TSCA Inventory.

SAFETY PROFILE: A skin irritant. When heated to decomposition it emits acrid smoke and irritating fumes.

HGC000   CAS: 1936-15-8   HR: 2
HISPACID FAST ORANGE 2G

mf: $C_{16}H_{10}N_2O_7S_2 \cdot 2Na$ mw: 452.38

PROP: Orange colored hygroscopic powder. Sol in H_2O; sltly sol in EtOH.

SYNS: ACIDAL FAST ORANGE □ ACID FAST ORANGE EGG □ ACID LEATHER ORANGE PGW □ ACID LIGHT ORANGE G □ ACID ORANGE 10 □ ACILAN ORANGE GX □ APOCID ORANGE 2G □ ATUL ACID CRYSTAL ORANGE G □ BRASILAN ORANGE 2G □ BUCACID FAST ORANGE G □ CALCOCID FAST LIGHT ORANGE 2G □ CERTICOL ORANGE GS □ CETIL LIGHT ORANGE GG □ C.I. 27 □ C.I. ACID ORANGE 10 □ C.I. FOOD ORANGE 4 □ CRYSTAL ORANGE 2G □ D&C ORANGE No. 3 □ ENIACID LIGHT ORANGE G □ ERIO FAST ORANGE AS □ FAST LIGHT ORANGE GA □ HEXACOL ORANGE GG CRYSTALS □ HIDACID FAST ORANGE G □ 7-HYDROXY-8-(PHENYLAZO)-1,3-NAPHTHALENEDISULFONIC ACID, DISODIUM SALT □ 7-HYDROXY-8-(PHENYLAZO)-1,3-NAPHTHALENEDISULPHONIC ACID, DISODIUM SALT □ INK ORANGE JSN □ INTRACID FAST ORANGE G □ JAVA ORANGE 2G □ KITON FAST ORANGE G □ NAPHTHALENE FAST ORANGE 2GS □ NCI-C53838 □ NEKLACID FAST ORANGE 2G □ ORANGE #10 □ ORANGE G (biological stain) □ ORANGE G DYE □ ORANGE G (indicator) □ ORANZ G (POLISH) □ 1-PHENYLAZO-2-NAPHTHOL-6,8-DISULFONIC ACID, DISODIUM SALT □ 1-PHENYLAZO-2-NAPHTHOL-6,8-DISULPHONIC ACID, DISODIUM SALT □ SCHULTZ No. 39 □ SOLAR LIGHT ORANGE GX □ STANDACOL ORANGE G □ SULFACID LIGHT ORANGE J □ TERTRACID LIGHT ORANGE G □ UNITERTRACID LIGHT ORANGE G □ VENDACID LIGHT ORANGE 2G □ WOOL ORANGE 2G □ XYLENE FAST ORANGE G

CONSENSUS REPORTS: IARC Cancer Review: Group 3 IMEMDT 7,56,87; Animal Inadequate Evidence IMEMDT 8,181,75. Reported in EPA TSCA Inventory. EPA Genetic Toxicology Program.

SAFETY PROFILE: Experimental reproductive effects. Questionable carcinogen. Mutation data reported. Used as a drug and cosmetic colorant. When heated to decomposition it emits very toxic SO_x, Na_2O, and NO_x.

HGC500   CAS: 569-65-3   HR: 3
HISTAMETHIZINE

mf: $C_{25}H_{27}ClN_2$ mw: 390.99

PROP: Bp: 230° @ 2 mm.

SYNS: ANCOLAN □ BONADETTES □ BONADOXIN □ BONAMINE □ CALMONAL □ CHICLIDA □ 1-(p-CHLOROBENZHYDRYL)-4-(m-METHYLBENZYL)DIETH-YLENEDIAMINE □ 1-p-CHLOROBENZHYDRYL-4-m-METHYLBENZYLPIPERAZINE □ 1-(p-CHLORO-α-PHENYLBENZYL)-4-(m-METHYLBENZYL)PIPERAZINE

□ HISTAMETHINE □ HISTAMETIZINE □ HISTA-METIZYNE □ ITINEROL □ LONGIFENE □ MAREX □ MECLIZINE □ MECLOZINE □ NAVICALM □ NEO-ISTAFENE □ NEO-SUPRIMAL □ NEO-SUPRIMEL □ PARACHLORAMINE □ PEREMESIN □ POSTAFEN □ SABARI □ SEA-LEGS □ SIGURAN □ SUBARI □ SUPRIMAL □ TRAVELON □ UCB 170 □ VIBAZINE □ VOMISSELS

CONSENSUS REPORTS: Reported in EPA TSCA Inventory.

SAFETY PROFILE: A poison by intravenous route. Moderate toxicity by ingestion and intramuscular routes. Human reproductive effects by an unspecified route: reduced viability of newborn. Experimental teratogenic and reproductive effects. An antihistamine. When heated to decomposition it emits very toxic fumes of Cl^- and NO_x.

HGD500   CAS: 56-92-8   HR: 3
HISTAMINE DICHLORIDE

mf: $C_5H_9N_3 \cdot 2ClH$ mw: 184.09

PROP: Prisms of aqueous ethyl alc. Mp: 239–246° (decomp). Sol in water, methanol; sltly sol in alc; insol in ether.

SYN: HISTAMINE DIHYDROCHLORIDE

CONSENSUS REPORTS: Reported in EPA TSCA Inventory.

SAFETY PROFILE: Poison by subcutaneous, intravenous, intraperitoneal, and parenteral routes. Moderately toxic by ingestion. Experimental reproductive effects. When heated to decomposition it emits very toxic fumes of Cl^- and NO_x. See also HISTAMINE.

HGE000   CAS: 51-74-1   HR: 3
HISTAMINE DIPHOSPHATE

mf: $C_5H_9N_3 \cdot 2H_3O_4P$ mw: 307.17

PROP: Powder.

SYNS: 4-(2-AMINOETHYL)IMIDAZOLE BIS(DI-HYDROGEN PHOSPHATE) □ 4-(2-AMINOETHYL)-IMIDAZOLE DI-ACID PHOSPHATE □ HISTAMINE ACID PHOSPHATE □ HISTAMINE PHOSPHATE (1:2) □ 1H-IMIDAZOLE-4-ETHANAMINE PHOSPHATE (1:2)

CONSENSUS REPORTS: Reported in EPA TSCA Inventory.

SAFETY PROFILE: A deadly poison by intravenous and parenteral routes.

Moderately toxic by ingestion and intraperitoneal routes. When heated to decomposition it emits very toxic fumes of NO_x and PO_x.

HGE700 CAS: 71-00-1 HR: D
HISTIDINE
mf: $C_6H_9N_3O_2$ mw: 155.18
PROP: l-Histidine, the natural form. White needles, plates, or crystalline powder; sltly bitter taste. Decomp 287° (softens at 277°). Solubility in water at 25°: 41.9 g/L. Sol in water; very sltly sol in alc; insol in ether.
SYNS: l-α-AMINO-4(OR 5)-IMIDAZOLEPROPIONIC ACID □ GLYOXALINE-5-ALANINE □ l-HISTIDINE (FCC)
CONSENSUS REPORTS: Reported in EPA TSCA Inventory.
SAFETY PROFILE: Human mutation data reported. When heated to decomposition it emits toxic fumes of NO_x.

HGE800 HR: D
HISTIDINE MONOHYDROCHLORIDE
mf: $C_6H_9N_3O_2 \cdot HCl \cdot H_2O$ mw: 209.63
PROP: White needles, plates, or crystalline powder; sltly bitter taste. Decomp 250°. Sol in water; insol in alc, ether.
SYNS: l-α-AMINO-4(OR 5)-IMIDAZOLEPROPIONIC ACID MONOHYDROCHLORIDE □ GLYOXALINE-5-ALANINE MONOHYDROCHLORIDE
SAFETY PROFILE: When heated to decomposition it emits toxic fumes of NO_x.

HGG000 CAS: 10138-62-2 HR: 3
HOLMIUM CHLORIDE
mf: Cl_3Ho mw: 271.28
PROP: Hygroscopic, bright yellow, crystalline solid. Mp: 718°. Sol in water.
CONSENSUS REPORTS: Reported in EPA TSCA Inventory.
SAFETY PROFILE: A poison by intraperitoneal route. Mildly toxic by ingestion. When heated to decomposition it emits highly toxic fumes of Cl^-. See also HOLMIUM and RARE EARTHS.

HGI900 CAS: 505-66-8 HR: 3
HOMOPIPERAZINE

mf: $C_5H_{12}N_2$ mw: 100.19
PROP: Hygroscopic solid. Mp: 38–40°, bp: 169°, flash p: 148°F.
SYNS: 1,4-DIAZACYCLOHEPTANE □ 1H-1,4-DIAZEPINE, HEXAHYDRO- □ HEXAHYDRO-1,4-DIAZEPINE □ TRIMETHYLENEETHYLENEDIAMINE
CONSENSUS REPORTS: Reported in EPA TSCA Inventory.
SAFETY PROFILE: Moderately toxic by ingestion and skin contact. A corrosive and severe skin and eye irritant. Flammable liquid when exposed to heat, sparks, or flame. When heated to decomposition it emits toxic fumes of NO_x.

HGK800 HR: D
HOPS OIL
PROP: From steam distillation of cones from female *Humulus lupulus* L. or *Humulus americanus* Nutt. (Fam. *Moraceae*). Yellow liquid; aromatic odor. D: 0.825–0.926, refr index: 1.470–1.494 @ 20°. Sol in fixed oils, mineral oil; insol in glycerin, propylene glycol.
SAFETY PROFILE: When heated to decomposition it emits acrid smoke and irritating fumes.

HGL680 CAS: 89213-87-6 HR: 3
HUMAN ATRIAL NATRIURETIC PEPTIDE (99-126)
mf: $Cl_{27}H_{203}N_{45}O_{39}S_3$ mw: 2512.81
SYNS: α-ATRIOPEPTIN (HUMAN) □ ATRIOPEPTIN-33(RAT), 1-DE-l-LEUCINE-2-DE-l-ALANINE-3-DEGLYCINE-4-DE-l-PROLINE-5-DE-l-ARGININE-1 7-l-METHIONINE- □ ANTRIOPEPTIN (HUMAN α-COMPONENT) □ ATRIOPEPTIN-28 (HUMAN) □ CARPERITIDE □ HORSE ATRIAL NATRIURETIC PEPTIDE-28 □ HUMAN ATRIAL NATRIURETIC FACTOR (99-126) □ α-hmn ATRIAL NATRIURETIC HORMONE □ HUMAN ATRIAL NATRIURETIC PEPTIDE (1-28) (99-126) □ HUMAN ATRIOPEPTIN(1-28) □ HUMAN ATRIOPEPTIN(99-126) □ (99-126)-hmn PROATRIOPEPTIN □ SUN-4936 □ TRIOPEPTIN (HUMAN α-COMPONENT)
SAFETY PROFILE: A poison by intravenous route. Experimental reproductive effects. When heated to decomposition it emits toxic vapors of NO_x and SO_x.

HGL920　CAS: 94948-59-1　HR: 3
HUMAN RECOMBINANT TUMOR NEC-ROSIS FACTOR-α

SYNS: RECOMBINANT HUMAN TUMOR NECROSIS FACTOR α □ RECOMBINANT HUMAN TUMOR NECROSIS FACTOR-α □ TNF-α □ TUMOR NECROSIS FACTOR-α

SAFETY PROFILE: A poison by intravenous route. Experimental reproductive effects. Mutation data reported. When heated to decomposition it emits acrid smoke and irritating vapors.

HGO600　CAS: 461-72-3　HR: D
HYDANTOIN

mf: $C_3H_4N_2O_2$　　mw: 100.09

PROP: Needles from methanol. Mp: 220°. Sltly sol in water or ether; sol in alc and in solns of fixed alkali hydroxides.

SYNS: GLYCOLYLUREA □ 2,4-IMIDAZOLIDINED-IONE (9CI)

CONSENSUS REPORTS: Reported in EPA TSCA Inventory.

SAFETY PROFILE: An experimental teratogen. When heated to decomposition it emits toxic fumes of NO_x.

HGP000　CAS: 109-78-4　HR: 3
HYDRACRYLONITRILE

mf: C_3H_5NO　　mw: 71.09

PROP: Colorless to straw-colored liquid or oil. Fp: −46°, bp: 220° decomp, flash p: 265°F (OC), d: 1.0404 @ 25°, vap press: 0.08 mm @ 25°, vap d: 2.45. Misc with water, acetone, methyl ethyl ketone, and ethanol. Sltly sol in ether; insol in benzene, pet ether, carbon disulfide, and carbon tetrachloride.

SYNS: 2-CYANOETHANOL □ 2-CYANOETHYL ALCOHOL □ ETHYLENE CYANOHYDRIN □ GLYCOL CYANOHYDRIN □ β-HPN □ 3-HYDROXYPROPANE-NITRILE □ β-HYDROXYPROPIONITRILE □ 3-HYDROXYPROPIONITRILE □ METHANOLACETO-NITRILE □ USAF RH-7

CONSENSUS REPORTS: Reported in EPA TSCA Inventory. Cyanide compounds are on the Community Right-To-Know List.

SAFETY PROFILE: Poison by inhalation. Moderately toxic by ingestion and intraperitoneal routes. Mildly toxic by skin contact. A skin and eye irritant. Combustible when exposed to heat or flame. Reacts violently with mineral acids (e.g., chlorosulfonic acid, oleum, sulfuric acid), amines, or inorganic bases (e.g., NaOH). Reacts with water or steam to produce toxic and flammable vapors. To fight fire, use CO_2, dry chemical, alcohol foam. When heated to decomposition or on contact with acid or acid fumes it emits highly toxic fumes of CN^-. See also NITRILES.

HGP495　CAS: 86-54-4　HR: 3
HYDRALAZINE

mf: $C_8H_8N_4$　　mw: 160.20

PROP: Yellow crystal from MeOH. Mp: 171–173°.

SYNS: APPRESSIN □ APRESOLIN □ APREZOLIN □ BA5968 □ C-5068 □ C 5968 □ CIBA 5968 □ HIDRALAZIN □ HIPOFTALIN □ HYDRALLAZINE □ HYDRAZINO-PHTHALAZINE □ 1-HYDRAZINOPHTHALAZINE □ HYPOPHTHALIN □ IDRALAZINA (ITALIAN) □ 1(2H)-PHTHALAZINONE HYDRAZONE

CONSENSUS REPORTS: IARC Cancer Review: Group 3 IMEMDT 7,222,87; Human Inadequate Evidence IMEMDT 24,85,80.

SAFETY PROFILE: Poison by ingestion, intravenous, intraperitoneal, and subcutaneous routes. Human systemic effects by ingestion: allergic dermatitis, cardiomyopathy, changes in coronary arteries. Human teratogenic effects by an unspecified route: developmental abnormalities of the blood and lymphatic system. Questionable carcinogen. Mutation data reported. When heated to decomposition it emits toxic fumes of NO_x.

HGP500　CAS: 304-20-1　HR: 3
HYDRALAZINE HYDROCHLORIDE

mf: $C_8H_8N_4 \cdot ClH$　　mw: 196.66

PROP: Yellow crystals. Decomp @ 273°. Very sltly sol in ether.

SYNS: AISELAZINE □ APPRESINUM □ APRELAZINE □ APRESAZIDE □ APRESINE □ APRESOLIN □ APRESOLINE-ESIDRIX □ APRESOLINE HYDRO-CHLORIDE □ APREZOLIN □ BA 5968 □ CIBA 5968 □ DRALZINE □ HIDRALAZIN □ HIPOFTALIN □ HYDRALAZINE CHLORIDE □ HYDRALAZINE

MONOHYDROCHLORIDE □ HYDRALLAZINE HYDROCHLORIDE □ HYDRAPRESS □ 1-HYDRA-ZINOPHTHALAZINE HYDROCHLORIDE □ 1-HYDRAZINOPHTHALAZINE MONOHYDROCHLORIDE □ HYPERAZIN □ HYPOPHTHALIN □ HYPOS □ IPOLINA □ LOPRESS □ NOR-PRESS 25 □ 1(2H)-PHTHALAZINONE HYDRAZONE HYDROCHLORIDE □ 1(2H)-PHTHALAZINONE, HYDRAZONE, MONO-HYDROCHLORIDE □ PRAPARAT 5968 □ ROLAZINE □ SERPASIL APRESOLINE No. 2

CONSENSUS REPORTS: IARC Cancer Review: Animal Limited Evidence IMEMDT 24,85,80. Reported in EPA TSCA Inventory.

SAFETY PROFILE: Suspected carcinogen with experimental neoplastigenic data. A poison by ingestion, subcutaneous, intravenous, and intraperitoneal routes. Human mutation data reported. An experimental teratogen. An antihypertensive agent. When heated to decomposition it emits very toxic NO_x and HCl.

HGS000   CAS: 302-01-2   HR: 3
HYDRAZINE
DOT: UN 2029
mf: H_4N_2 mw: 32.06
PROP: Colorless, oily, fuming liquid or white crystals. Mp: 254°, bp: 113.5°, flash p: 100°F (OC), d: 1.1011 @ 15° (liquid), autoign temp: can vary from 74°F in contact with iron rust, 270°F in contact with black iron, 313°F in contact with stainless steel, 518°F in contact with glass. Vap d: 1.1, lel: 4.7%, uel: 100%. Misc in H_2O, alcohols; sltly sol in org solvs. IDLH 50 ppm.
SYNS: DIAMIDE □ DIAMINE □ HYDRAZINE, anhydrous (DOT) □ HYDRAZINE AQUEOUS SOLUTIONS with >64% hydrazine, by weight (DOT) □ HYDRAZYNA (POLISH) □ RCRA WASTE NUMBER U133

CONSENSUS REPORTS: NTP 10th Report on Carcinogens. IARC Cancer Review: Group 2B IMEMDT 7,223,87; Animal Sufficient Evidence IMEMDT 4,127,74. EPA Extremely Hazardous Substances List. Community Right-To-Know List. Genetic Toxicology Program. Reported in EPA TSCA Inventory.
OSHA PEL: TWA 0.1 ppm (skin)

ACGIH TLV: TWA 0.01 ppm (skin), Confirmed Animal Carcinogen.
DFG MAK: DFG TRK: Animal Carcinogen, Suspected Human Carcinogen
NIOSH REL: (Hydrazines) CL 0.04 mg/m^3/2H
DOT CLASSIFICATION: 3; Label: Flammable Liquid, Poison, Corrosive
SAFETY PROFILE: Confirmed carcinogen with experimental carcinogenic, neoplastigenic, and tumorigenic data. A poison by ingestion, skin contact, intraperitoneal, and intravenous routes. Moderately toxic by inhalation. An experimental teratogen. Other experimental reproductive effects. Human mutation data reported. A powerful reducing agent that is corrosive to the eyes, skin, and mucous membranes. May cause skin sensitization as well as systemic poisoning. Hydrazine and some of its derivatives may cause damage to the liver and destruction of red blood cells.

Flammable liquid. A very dangerous fire hazard when exposed to heat, flame, or oxidizing agents. Severe explosion hazard when exposed to heat or flame, or by chemical reaction. Explodes on contact with barium oxide; calcium oxide; chromate salts; chromium dioxide; dicyanofurazan; mercury oxide; trioxygen difluoride; N-haloimides; potassium; silver compounds; sodium hydroxide; titanium compounds (at 130°). Potentially explosive reactions with alkali metals; NH_3; Cl_2; chromates; CuO; Cu^{++} salts; F_2; metallic oxides; Ni; $Ni(ClO_4)_2$; O_2; liquid O_2; $K_2Cr_2O_7$; $Na_2Cr_2O_7$; tetryl; zinc diamide; $Zn(C_2H_5)_2$. Forms sensitive, explosive mixtures with 2-chloro-5-methyl-nitrobenzene; metal salts [e.g., cadmium perchlorate; copper chlorate (heat-sensitive); manganese nitrate (heat-sensitive); mercury(I) chloride; mercury(II) chloride; mercury(I) nitrate; mercury(II) nitrate; tin(II) chloride]; methanol + nitromethane; air; lithium perchlorate; sodium perchlorate; sodium. Ignites on contact with cotton waste + heavy metals; dinitrogen oxide;

rhenium + alumina; catalysts; nitric acid; hydrogen peroxide; N,2,4,6-tetranitroaniline; rust + heat. Ignites spontaneously in air when absorbed on earth, asbestos, cloth, wood. Violent reaction with 1-chloro-2,4-dinitrobenzene; oxidants (e.g., iron oxide; chlorates; peroxides); thiocarbonyl azide thiocyanate. Vigorous reaction with benzene-seleninic acid or anhydride; carbon dioxide + stainless steel; copper oxide; lead oxide; potassium peroxodisulfate; ruthenium(III) chloride. On contact with metal catalysts (e.g., platinum black; Raney nickel; copper-iron oxide; molybdenum; molybdenum oxides; iridium), it decomposes to ammonia, hydrogen and nitrogen gases which may ignite or explode. A hypergolic reaction with dinitrogen tetraoxide is the basis of a liquid rocket fuel mixture. The vapor will burn without air. It is a powerful explosive. It is very sensitive and must not be used without full and complete instructions from the manufacturer for handling, storage, and disposal. Dangerous; when heated to decomposition it emits highly toxic fumes of NO_x and NH_3.

HGU000   CAS: 57-56-7   HR: 3
HYDRAZINE CARBOXAMIDE
mf: CH_5N_3O mw: 75.09

PROP: Prisms from EtOH. Mp: 96°.

SYNS: AMINOUREA □ CARBAMIC ACID HYDRAZIDE □ CARBAMOYLHYDRAZINE □ CARBAMYLHYDRA-ZINE □ CARBAZAMIDE □ SEMICARBAZIDE

CONSENSUS REPORTS: Reported in EPA TSCA Inventory.

SAFETY PROFILE: A poison by ingestion, intraperitoneal, subcutaneous, and intravenous routes. Human systemic effects by intravenous route: convulsions. Questionable carcinogen with experimental tumorigenic data. Mutation data reported. When heated to decomposition it emits toxic fumes of NO_x. See also HYDRAZINE.

HGU025   CAS: 14931-40-9   HR: 3
HYDRAZINE, COMPD. WITH BORANE (1:1)
mf: BH_7N_2 mw: 45.90

SYNS: BORANE, COMPD. WITH N2H4 □ BORON, (HYDRAZINE-N)TRIHYDRO-, (T-4)- □ BORON, (HYDRAZINE-KAPPAN)TRIHYDRO-, (T-4)- □ HYDRAZINE BORANE (1:1) □ (T-4)-(HYDRAZINE-KAPPAN)TRIHYDROBORON

SAFETY PROFILE: A poison by ingestion, inhalation, and subcutaneous routes. May be unstable at room temperature. When heated to decomposition it emits toxic vapors of NO_x and B.

HGU100   CAS: 13537-45-6   HR: 3
HYDRAZINE DIFLUORIDE
mf: $H_4N_2 \cdot 2FH$ mw: 72.08

SYN: HYDRAZINE, DIHYDROFLUORIDE

OSHA PEL: TWA 2.5 mg(F)/m³

ACGIH TLV: TWA 2.5 mg(F)/m³; BEI: 3 mg/g creatinine of fluorides in urine prior to shift; 10 mg/g creatinine of fluorides in urine at end of shift.

NIOSH REL: (Fluorides, Inorganic): 10H TWA 2.5 mg(F)/m³

SAFETY PROFILE: Poison by intravenous route. When heated to decomposition it emits toxic fumes of NO_x and HF.

HGU500   CAS: 10217-52-4   HR: 3
HYDRAZINE HYDRATE
DOT: UN 2030

mf: $H_4N_2 \cdot H_2O$ mw: 50.08

PROP: Colorless, fuming, refractive liquid. Mp: −51.7°, bp: 118.5° @ 740 mm, d: 1.03 @ 21°. Faint characteristic odor. A strong base, very corrosive; attacks glass, rubber, and cork. Very powerful reducing agent. Misc with water and alc; insol in chloroform and ether.

SYNS: HYDRAZINE AQUEOUS SOLUTIONS, with not >64% hydrazine, by weight (DOT) □ HYDRAZINE HYDRATE, with not >64% hydrazine, by weight (DOT)

CONSENSUS REPORTS: EPA Genetic Toxicology Program.

NIOSH REL: (Hydrazines) CL 0.04 mg/m³/2H

DOT CLASSIFICATION: 8; Label: Corrosive, Poison

SAFETY PROFILE: A poison by ingestion, inhalation, skin contact, and subcutaneous routes. A corrosive irritant to the eyes, skin, and mucous membranes. Incompatible with HgO, Na, $SnCl_2$, 2,4-dinitrochlorobenzene. When heated to decomposition it emits toxic fumes of NO_x. See also HYDRAZINE.

HGV000   CAS: 2644-70-4   HR: 3
HYDRAZINE, HYDROCHLORIDE
mf: $H_4N_2•ClH$ mw: 68.52
SYNS: HYDRAZINE MONOCHLORIDE □ HYDRA- ZINIUM CHLORIDE □ HYDRAZINIUM MONOCHLORIDE
CONSENSUS REPORTS: Reported in EPA TSCA Inventory.
NIOSH REL: (Hydrazines) CL 0.04 mg/m³/2H
SAFETY PROFILE: A poison by ingestion, intravenous, and intraperitoneal routes. Mutation data reported. When heated to decomposition it emits very toxic fumes of Cl^- and NO_x. See also HYDRAZINE.

HGW500   CAS: 10034-93-2   HR: 3
HYDRAZINE SULFATE (1:1)
mf: $H_4N_2•H_2O_4S$ mw: 130.14
PROP: Colorless crystals. D: 1.378, mp: 254° (decomp). Sol in water; insol in alc; very sol in hot water.
SYNS: HS □ HYDRAZINE HYDROGEN SULFATE □ HYDRAZINE MONOSULFATE □ HYDRAZINE SULPHATE □ HYDRAZINIUM SULFATE □ HYDRAZONIUM SULFATE □ IDRAZINA SOLFATO (ITALIAN) □ NSC-150014 □ SIRAN HYDRAZINU (CZECH)
CONSENSUS REPORTS: NTP 10th Report on Carcinogens. IARC Cancer Review: Animal Sufficient Evidence IMEMDT 4,127,74. Community Right-To-Know List. Reported in EPA TSCA Inventory. EPA Genetic Toxicology Program.
NIOSH REL: (Hydrazines) CL 0.04 mg/m³/2H
SAFETY PROFILE: Confirmed carcinogen with experimental carcinogenic, neoplastigenic, and tumorigenic data. A poison by ingestion and intraperitoneal routes. Human systemic effects by ingestion: paresthesia (abnormal sensations), somnolence, nausea or vomiting. An experimental teratogen. Human mutation data reported. An eye irritant. A reducing agent. When heated to decomposition it emits very toxic fumes of SO_x and NO_x. See also HYDRAZINE and SULFATES.

HHA100   CAS: 73953-53-4   HR: 3
HYDRAZINIUM TRIFLUOROSTANNITE
mf: $H_4N_2•F_3HSn$ mw: 208.76
SYN: HYDRAZINE, TRIFLUOROSTANNITE
OSHA PEL: TWA 2 mg(Sn)/m³; TWA 2.5 mg(F)/m³
ACGIH TLV: TWA 2 mg(Sn)/m³; TWA 2.5 mg(F)/m³; BEI: 3 mg/g creatinine of fluorides in urine prior to shift; 10 mg/g creatinine of fluorides in urine at end of shift.
NIOSH REL: (Fluorides, Inorganic): 10H TWA 2.5 mg(F)/m³
SAFETY PROFILE: Poison by intravenous route. When heated to decomposition it emits toxic fumes of NO_x, Sn, and F^-.

HHC000   CAS: 109-84-2   HR: 3
2-HYDRAZINOETHANOL
mf: $C_2H_8N_2O$ mw: 76.12
PROP: Colorless, sltly viscous liquid. Mp: −70°, bp: 148–152° @ 25 mm, flash p: 224°F, vap d: 2.63, d: 1.11. Misc with water; sol in lower alcs; sltly sol in ether.
SYNS: BOH □ HYDROXYETHYL HYDRAZINE □ β- HYDROXYETHYLHYDRAZINE □ N-(2-HYDROXY- ETHYL)HYDRAZINE □ OMAFLORA
CONSENSUS REPORTS: Reported in EPA TSCA Inventory. EPA Genetic Toxicology Program.
SAFETY PROFILE: Poison by ingestion. Questionable carcinogen with experimental carcinogenic data. Mutation data reported. Combustible when exposed to heat or flame; can react with oxidizing materials. To fight fire, use foam, CO_2, dry chemical. When heated to decomposition it emits toxic fumes such as NO_x.

H

HHG000   CAS: 122-66-7   HR: 3
HYDRAZOBENZENE
mf: $C_{12}H_{12}N_2$ mw: 184.26
PROP: Light or yellow crystals from
ethanol. D: 1.58, mp: 126–127°, bp:
decomp. Very sltly sol in water; insol in
acetylene.
SYNS: N,N'-BIANILINE □ sym-DIPHENYLHYDRAZ-
INE □ 1,2-DIPHENYLHYDRAZINE □ HYDRAZOBEN-
ZEN (CZECH) □ HYDRAZODIBENZENE □ NCI-C01854
□ RCRA WASTE NUMBER U109
CONSENSUS REPORTS: NTP 10th Report
on Carcinogens. NCI Carcinogenesis
Bioassay (feed); Clear Evidence: mouse, rat
NCITR* NCI-CG-TR-92,78. Community
Right-To-Know List. Reported in EPA
TSCA Inventory.
DFG MAK: Animal Carcinogen, Suspected
Human Carcinogen
SAFETY PROFILE: Confirmed carcinogen
with experimental carcinogenic and
tumorigenic data. Poison by ingestion.
Mutation data reported. When heated to
decomposition it emits toxic fumes of NO_x.

HHG500   CAS: 7782-79-8   HR: 3
HYDRAZOIC ACID
mf: HN_3 mw: 43.04
PROP: Colorless liquid; intolerable pungent
odor. Mp: −80°, bp: 37°, d: 1.09 @ 25°/4°.
Very sol in water.
SYNS: AZOIMIDE □ DIAZOIMIDE □ HYDROGEN
AZIDE □ HYDRONITRIC ACID □ STICKSTOF-
FWASSERSTOFFSAEURE (GERMAN) □ TRIAZOIC ACID
CONSENSUS REPORTS: Reported in EPA
TSCA Inventory. EPA Genetic Toxicology
Program.
ACGIH TLV: CL 0.1 ppm (vapor)
DFG MAK: 0.1 ppm (0.18 mg/m³)
SAFETY PROFILE: Poison by
intraperitoneal route. Mildly toxic by
inhalation. A severe irritant to skin, eyes,
and mucous membranes. Continued
inhalation causes central nervous system
problems in humans (changes in EEG,
somnolence, cough, headache, change in
heart rate, fall in blood pressure, collapse,
chills, and fever). High concentrations can
cause fatal convulsions. Chronic exposure

has been reported to cause injury to kidneys
and spleen, hypotension, palpitation, ataxia,
weakness. A dangerously sensitive explosive
hazard when shocked or exposed to heat.
Reacts with heavy metals to form very
unstable heavy metal azides. Reacts violently
with Cd, Cu, Ni, HNO_3, F_2. When heated to
decomposition it emits toxic fumes of NO_x.
See also AZIDES.

HHI500   CAS: 10034-85-2   HR: 3
HYDRIODIC ACID
DOT: UN 1787/UN 2197
mf: HI   mw: 127.91
PROP: Acrid gas; colorless when freshly
made, but rapidly turns yellowish or brown
on exposure to light or air. Mp: −50.8°, bp:
−35.38° @ 5 atm, d: 5.66 g/L @ 0°. Keep
protected from light and air, preferably not
above 3°. When heated, decomp to H_2 and
I_2. Aq solns are strongly acid. Attacks
natural rubber. Misc with water and alc.
SYNS: ANHYDROUS HYDRIODIC ACID □ HYDRI-
ODIC ACID, solution (UN 1787) (DOT) □ HYDROGEN
IODIDE □ HYDROGEN IODIDE, anhydrous (UN 2197)
(DOT)
CONSENSUS REPORTS: Reported in EPA
TSCA Inventory.
DOT CLASSIFICATION: 8; Label: Corrosive
(UN 1787); DOT Class: 2.2; Label:
Nonflammable Gas, Corrosive (UN 2197)
SAFETY PROFILE: Poison by ingestion and
inhalation. A corrosive and poisonous
irritant to skin, eyes, and mucous
membranes. Explodes on contact with ethyl
hydroperoxide. Ignites on contact with
magnesium, perchloric acid, potassium +
heat, potassium chlorate + heat, oxidants
(e.g., fluorine, dinitrogen trioxide, dinitrogen
tetraoxide, fuming nitric acid). Violent
reaction with $HClO_4$ + Mg, O_3, metals.
Potentially violent reaction with
phosphorus. Reacts with water or steam to
produce toxic and corrosive fumes. When
heated to decomposition it emits highly
toxic fumes of I^-. See also IODIDES.

HHJ000   **CAS: 10035-10-6**   **HR: 3**
HYDROBROMIC ACID
DOT: UN 1048/UN 1788
mf: BrH   mw: 80.92
PROP: Colorless with an acrid odor, or pale-yellow liquid. Mp: $-87°$, bp: $-66.5°$, d: 3.50 g/L @ 0°. Misc with water, alc. Keep protected from light. IDLH 30 ppm.
SYNS: ACIDE BROMHYDRIQUE (FRENCH) □ ACIDO BROMIDRICO (ITALIAN) □ ANHYDROUS HYDRO-BROMIC ACID □ BROMOWODOR (POLISH) □ BROMWASSERSTOFF (GERMAN) □ BROOMWATER-STOF (DUTCH) □ HYDROBROMIC ACID SOLUTION, >49% hydrobromic acid (UN 1788) (DOT) □ HYDRO-BROMIC ACID SOLUTION, not >49% hydrobromic acid (UN 1788) (DOT) □ HYDROGEN BROMIDE (ACGIH,OSHA,-MAK) □ HYDROGEN BROMIDE, anhydrous (UN 1048) (DOT)
CONSENSUS REPORTS: Reported in EPA TSCA Inventory.
OSHA PEL: CL 3 ppm
ACGIH TLV: CL 3 ppm; (Proposed: CL 2 ppm)
DFG MAK: 2 ppm (6.7 mg/m^3)
DOT CLASSIFICATION: 8; Label: Corrosive (UN 1788); DOT Class: 2.3; Label: Poison Gas, Corrosive (UN 1048)
SAFETY PROFILE: A poison gas. A corrosive irritant to the eyes, skin, and mucous membranes. Reacts violently with F_2, Fe_2O_3, NH_3, O_3. When heated to decomposition or in reaction with water or steam it emits toxic and corrosive fumes of Br^- and HBr. See also BROMIDES.

HHJ500   **HR: 3**
HYDROCARBON GAS
DOT: UN 1023/UN 1964/UN 1965
PROP: Contains hydrogen, methane, carbon monoxide, lel: 5.3%, uel: 31%, autoign temp: 1200°F.
SYNS: COAL GAS (UN 1023) (DOT) □ HYDROCARBON GASES, COMPRESSED, N.O.S. (UN 1964) (DOT) □ HY-DROCARBON GASES, LIQUEFIED, N.O.S. (UN 1965) (DOT) □ HYDROCARBON GASES MIXTURES, COM-PRESSED, N.O.S. (UN 1964) (DOT) □ HYDROCARBON GASES MIXTURES, LIQUEFIED, N.O.S. (UN 1965) (DOT)
DOT CLASSIFICATION: 2.1; Label: Flammable Gas (UN 1964, UN 1965); DOT Class: 2.3; Label: Poison Gas, Flammable Gas (UN 1023)

SAFETY PROFILE: A poison by inhalation. Very dangerous fire hazard when exposed to heat or flame; can react vigorously with oxidizing materials. Moderately explosive when exposed to heat or flame. To fight fire, stop flow of gas; CO_2, dry chemical, or water spray. See also CARBON MONOXIDE, HYDROGEN, and METHANE.

HHK050   **CAS: 17692-34-1**   **HR: 3**
HYDROCHLORBENZETHYLAMINE
mf: $C_{23}H_{31}ClN_2O_3$ mw: 419.01
SYNS: 8-(4-(4-CHLOROPHENYLPHENYLMETHYL)-PIPERAZINYL)-3,6-DIOXAOCTANOL □ 3,6-DIOXA-OCTANOL, 8-(4-(4-CHLOROPHENYLPHENYL-METHYL)PIPERAZINYL)- □ ETHANOL, 2-(2-(2-(4-(p-CHLORO-α-PHENYLBENZYL)-1-PIPERAZINYL)-ETHOXY)ETHOXY)- □ ETHANOL,- 2-(2-(2-(4-((4-CHLOROPHENYL)PHENYLMETHYL)-1-PIPERAZIN-YL)ETHOXY)ETHOXY)- □ ETODROXINE □ ETODRO-XYZINE □ VESPARAX-WIRKSTOFF □ UCB 1414
SAFETY PROFILE: A poison by intravenous. Moderately toxic by ingestion. When heated to decomposition it emits toxic vapors of NO_x and Cl^-.

HHL000   **CAS: 7647-01-0**   **HR: 3**
HYDROCHLORIC ACID
DOT: UN 1050/UN 1789/UN 2186
mf: ClH   mw: 36.46
PROP: Colorless, corrosive, gas or fuming liquid; strongly corrosive with pungent odor. Dissolves in H_2O to give a strong, highly corrosive acid. Mp: $-114.3°$, bp: $-84.8°$, d: (gas) 1.639 g/L @ 0°, (liquid) 1.194 @ $-26°$, vap press: 4.0 atm @ 17.8°. Very sol in H_2O; sol in MeOH, EtOH, and Et_2O. IDLH 50 ppm.
SYNS: ACIDE CHLORHYDRIQUE (FRENCH) □ ACIDO CLORIDRICO (ITALIAN) □ ANHYDROUS HYDRO-CHLORIC ACID □ CHLOORWATERSTOF (DUTCH) □ CHLOROHYDRIC ACID □ CHLOROWODOR (POLISH) □ CHLORWASSERSTOFF (GERMAN) □ HYDRO-CHLORIC ACID, solution (UN 1789) (DOT) □ HYDRO-CHLORIDE □ HYDROGEN CHLORIDE, anhydrous (UN 1050) (DOT) □ HYDROGEN CHLORIDE, refrigerated liquid (UN 2186) (DOT) □ MURIATIC ACID □ SPIRITS of SALT
CONSENSUS REPORTS: EPA Extremely Hazardous Substances List. Community Right-To-Know List. Reported in EPA

TSCA Inventory. EPA Genetic Toxicology Program.

OSHA PEL: CL 5 ppm

ACGIH TLV: CL 2 ppm; Not Classifiable as a Human Carcinogen

DFG MAK: 5 ppm (7.6 mg/m³)

DOT CLASSIFICATION: 8; Label: Corrosive (UN 1789); DOT Class: 2.3; Label: Poison Gas, Corrosive (UN 1050, UN 2186)

SAFETY PROFILE: A human poison by an unspecified route. Mildly toxic to humans by inhalation. Moderately toxic experimentally by ingestion. A corrosive irritant to the skin, eyes, and mucous membranes. Mutation data reported. An experimental teratogen. A concentration of 35 ppm causes irritation of the throat after short exposure. In general, hydrochloric acid causes little trouble in industry other than from accidental splashes and burns. It is a common air contaminant and is heavily used in industry.

Nonflammable gas. Explosive reaction with alcohols + hydrogen cyanide, potassium permanganate, sodium, tetraselenium tetranitride. Ignition on contact with fluorine, hexalithium disilicide, metal acetylides or carbides (e.g., cesium acetylide, rubidium acetylide). Violent reactions with acetic anhydride, 2-amino ethanol, NH_4OH, Ca_3P_2, chlorosulfonic acid, 1,1-difluoroethylene, ethylene diamine, ethylene imine, oleum, $HClO_4$, β-propiolactone, propylene oxide, ($AgClO_4$ + CCl_4), NaOH, H_2SO_4, U_3P_4, vinyl acetate, CaC_2, CsC_2H, Cs_2C_2, Mg_3B_2, $HgSO_4$, RbC_2H, Rb_2C_2, Na. Vigorous reaction with aluminum, chlorine + dinitroanilines (evolves gas). Potentially dangerous reaction with sulfuric acid releases HCl gas. When heated to decomposition it emits toxic fumes of Cl⁻. See also HYDROGEN CHLORIDE.

HHM000   CAS: 8007-56-5   HR: 3
HYDROCHLORIC ACID, mixed with NITRIC ACID (3:1)
DOT: UN 1798
mf: ClH•HNO₃

PROP: Yellow, fuming, corrosive, volatile liquid; suffocating odor. Misc with water.

SYNS: AQUA REGIA □ NITROHYDROCHLORIC ACID (DOT) □ NITROHYDROCHLORIC ACID, diluted (DOT) □ NITROMURIATIC ACID (DOT)

DOT CLASSIFICATION: 8; Label: Corrosive

SAFETY PROFILE: A corrosive irritant to the eyes, skin, and mucous membranes. When heated to decomposition it emits very toxic HCl, HNO_3, Cl⁻, and NO_x. See also HYDROCHLORIC ACID, NITRIC ACID, and NITROSYL CHLORIDE.

HHP000   CAS: 104-53-0   HR: 3
HYDROCINNAMALDEHYDE
mf: $C_9H_{10}O$ mw: 134.19

PROP: Colorless to sltly yellow liquid; strong floral, hyacinth odor. Bp: 221–224°, d: 1.010–1.020, refr index: 1.520–1.532, flash p: 203°F. Misc with alc, ether; insol in water.

SYNS: BENZENEPROPANAL □ BENZYLACETALDEHYDE □ DIHYDROCINNAMALDEHYDE □ FEMA No. 2887 □ HYDROCINNAMIC ALDEHYDE □ 3-PHENYLPROPANAL □ 3-PHENYL-1-PROPANAL □ 3-PHENYLPROPIONALDEHYDE (FCC) □ β-PHENYLPROPIONALDEHYDE □ 3-PHENYLPROPYL ALDEHYDE

CONSENSUS REPORTS: Reported in EPA TSCA Inventory.

SAFETY PROFILE: A poison by intravenous route. A human skin irritant. Combustible liquid. When heated to decomposition it emits acrid smoke and fumes. See also ALDEHYDES.

HHP050   CAS: 122-97-4   HR: 2
HYDROCINNAMIC ALCOHOL
mf: $C_9H_{12}O$ mw: 136.21

PROP: Colorless sltly viscous liquid; sweet, hyacinth-mignonette odor. D: 0.998–1.002, bp: 235°, refr index: 1.524–1.528, flash p: 228°F. Sol in fixed oils, propylene glycol; insol in glycerin.

SYNS: 3-BENZENEPROPANOL □ FEMA No. 2885 □ HYDROCINNAMYL ALCOHOL □ (3-HYDROXYPROPYL)BENZENE □ γ-PHENYLPROPANOL □ 3-PHENYLPROPANOL □ 3-PHENYL-1-PROPANOL (FCC) □ PHENYLPROPYL ALCOHOL □ γ-PHENYLPROPYL ALCOHOL □ 3-PHENYLPROPYL ALCOHOL

CONSENSUS REPORTS: Reported in EPA TSCA Inventory.

SAFETY PROFILE: Moderately toxic by ingestion. Mildly toxic by skin contact. A skin irritant. Combustible liquid. When heated to decomposition it emits toxic fumes. See also ALCOHOLS.

HHP500   CAS: 122-72-5   HR: 2
HYDROCINNAMYL ACETATE
mf: $C_{11}H_{14}O_2$ mw: 178.25
PROP: Colorless liquid; spicy, floral odor. D: 1.012, refr index: 1.494, flash p: 212°F. Sol in alc; insol in water.
SYNS: FEMA No. 2890 □ 3-PHENYL-1-PROPANOL ACETATE □ PHENYLPROPYL ACETATE □ 3-PHENYLPROPYL ACETATE (FCC) □ 3-PHENYL-1-PROPYL ACETATE
CONSENSUS REPORTS: Reported in EPA TSCA Inventory.
SAFETY PROFILE: Mildly toxic by ingestion. Combustible liquid. When heated to decomposition it emits acrid smoke and fumes.

HHQ550   CAS: 122-74-7   HR: 1
HYDROCINNAMYL PROPIONATE
mf: $C_{12}H_{16}O_2$ mw: 192.28
SYNS: BENZENEPROPANOL, PROPANOATE (9CI) □ PHENYLPROPYL PROPIONATE □ β-PHENYLPROPYL PROPIONATE □ 3-PHENYLPROPYL PROPIONATE □ 1-PROPANOL, 3-PHENYL-, PROPIONATE
CONSENSUS REPORTS: Reported in EPA TSCA Inventory.
SAFETY PROFILE: Low toxicity by ingestion and skin contact. A skin irritant. When heated to decomposition it emits acrid smoke and irritating fumes.

HHR000   CAS: 125-04-2   HR: 2
HYDROCORTISONE SODIUM SUCCINATE
mf: $C_{25}H_{35}O_9 \cdot Na$ mw: 502.59
PROP: White, odorless, hygroscopic, amorph solid or powder. Mp: 169–171°. Very sol in water and alc; insol in chloroform; very sltly sol in acetone.
SYNS: A-HYDROCORT □ BUCCALSONE □ CORLAN □ CORTISOL HEMISUCCINATE SODIUM SALT □ CORTISOL SODIUM HEMISUCCINATE □ CORTISOL

SODIUM SUCCINATE □ CORTISOL-21-SODIUM SUCCINATE □ CORTISOL SUCCINATE, SODIUM SALT □ EL-CORTELAN SOLUBLE □ EMI-CORLIN □ FLEBOCORTID □ HYCORACE □ HYDROCORTISONE-21-SODIUM SUCCINATE □ 21-(HYDROGEN SUCCINATE)-CORTISOL, MONOSODIUM SALT □ INTRACORT □ NORDICORT □ ORALSONE □ SODIUM HYDRO-CORTISONE SUCCINATE □ SODIUM HYDRO-CORTISONE-21-SUCCINATE □ SOLU-CORTEF □ SOLU-GLYC □ U 4905
SAFETY PROFILE: Moderately toxic by intraperitoneal route. Experimental teratogenic and reproductive effects. When heated to decomposition it emits toxic fumes of Na_2O. See also other hydrocortisone entries.

HHR500   CAS: 119-84-6   HR: 3
HYDROCOUMARIN
mf: $C_9H_8O_2$ mw: 148.17
PROP: Colorless to pale-yellow liquid or crystals; coconut odor. Mp: 25°, bp: 272°, d: 1.186, refr index: 1.555, flash p: 266°F.
SYNS: 1,2-BENZODIHYDROPYRONE (FCC) □ 2-CHROMANONE □ DIHYDROCOUMARIN □ 3,4-DIHYDROCOUMARIN □ o-HYDROXY-HYDROCIN-NAMIC ACID-Δ-LACTONE □ FEMA No. 2381 □ MELILO-TIN □ MELILOTOL □ NCI-C55890 □ 2-OXOCHROMAN □ USAF DO-12
CONSENSUS REPORTS: Reported in EPA TSCA Inventory.
SAFETY PROFILE: A poison by intraperitoneal route. Moderately toxic by ingestion. A skin irritant. Combustible liquid. When heated to decomposition it emits acrid smoke and fumes.

HHS000   CAS: 74-90-8   HR: 3
HYDROCYANIC ACID
DOT: NA 1051/UN 1613/UN 1614
mf: CHN   mw: 27.03
PROP: Very volatile liquid or colorless gas smelling of bitter almonds. Mp: −13°, bp: 25.7°, lel: 5.6%, uel: 40%, flash p: 0°F (CC), d: 0.715 @ 0°, autoign temp: 1000°F, vap press: 400 mm @ 9.8°, vap d: 0.932. Misc in water, alc, and ether. IDLH 50 ppm.
SYNS: ACIDE CYANHYDRIQUE (FRENCH) □ ACIDO CIANIDRICO (ITALIAN) □ AERO liquid HCN □ BLAUSA-EURE (GERMAN) □ BLAUWZUUR (DUTCH) □ CARBON HYDRIDE NITRIDE (CHN) □ CYAANWATERSTOF

(DUTCH) □ CYANWASSERSTOFF (GERMAN) □ CY-CLON □ CYCLONE B □ CYJANOWODOR (POLISH) □ EVERCYN □ FORMIC ANAMMONIDE □ FORMO-NITRILE □ HYDROCYANIC ACID, aqueous solutions <5% HCN (NA 1613) (DOT) □ HYDROCYANIC ACID, aqueous solutions not >20% hydrocyanic acid (UN 1613) (DOT) □ HYDROCYANIC ACID (PRUSSIC), unstabilized (DOT) □ HYDROGEN CYANIDE □ HYDROGEN CYANIDE (ACGIH,OSHA) □ HYDROGEN CYANIDE, anhydrous, stabilized (UN 1051) (DOT) □ HYDROGEN CYANIDE, anhydrous, stabilized, absorbed in a porous inert material (UN 1614) (DOT) □ PRUSSIC ACID □ PRUSSIC ACID, UNSTABILIZED □ RCRA WASTE NUMBER P063 □ ZACLONDISCOIDS

CONSENSUS REPORTS: EPA Extremely Hazardous Substances List. Community Right-To-Know List. Reported in EPA TSCA Inventory.

OSHA PEL: STEL 4.7 ppm (skin)
ACGIH TLV: CL 4.7 ppm (skin)
DFG MAK: 10 ppm (11 mg/m³)
NIOSH REL: (Cyanide) CL 5 mg(CN)/m³/10M

DOT CLASSIFICATION: 6.1; Label: Poison (NA 1613, UN 1613, UN 1614); DOT Class: Forbidden (unstabilized); DOT Class: 6.1; Label: Poison, Flammable Liquid (UN 1051)

SAFETY PROFILE: A deadly human and experimental poison by all routes. Hydrocyanic acid and the cyanides are true protoplasmic poisons, combining in the tissues with the enzymes associated with cellular oxidation. They thereby render the oxygen unavailable to the tissues and cause death through asphyxia. The suspension of tissue oxidation lasts only while the cyanide is present; upon its removal, normal function is restored, provided death has not already occurred. HCN does not combine easily with hemoglobin, but it does combine readily with methemoglobin to form cyanmethemoglobin. This property is utilized in the treatment of cyanide poisoning when an attempt is made to induce methemoglobin formation. The presence of cherry-red venous blood in cases of cyanide poisoning is due to the inability of the tissues to remove the oxygen from the blood. Exposure to concentrations of 100–200 ppm for periods of 30–60 minutes can cause death. In cases of acute cyanide poisoning death is extremely rapid, although sometimes breathing may continue for a few minutes. In less acute cases, there is cyanosis, headache, dizziness, unsteadiness of gait, a feeling of suffocation, and nausea. Where the patient recovers, there is rarely any disability.

Very dangerous fire hazard when exposed to heat, flame, or oxidizers. Can polymerize explosively at 50–60°C or in the presence of traces of alkali. Severe explosion hazard when exposed to heat or flame or by chemical reaction with oxidizers. The anhydrous liquid is stabilized at or below room temperature by the addition of acid. The gas forms explosive mixtures with air. Reacts violently with acetaldehyde. To fight fire, use CO_2, non-alkaline dry chemical, foam. When heated to decomposition or in reaction with water, steam, acid, or acid fumes it produces highly toxic fumes of CN^-. An insecticide. See also CYANIDE.

HHS600   CAS: 16872-11-0   HR: 2
HYDROFLUOBORIC ACID

DOT: UN 1775
mf: $BF_4 \cdot H$ mw: 87.82
PROP: Colorless liquid. Bp: 130° (decomposes). Miscible in water.
SYNS: BORATE(1-), TETRAFLUORO-, HYDROGEN □ BOROFLUORIC ACID □ FLUOBORIC ACID □ FLUO-BORIC ACID (DOT) □ HYDROGEN TETRA-FLUOROBORATE □ TETRAFLUOROBORIC ACID

CONSENSUS REPORTS: Reported in EPA TSCA Inventory.

OSHA PEL: TWA 2.5 mg(F)/m³
NIOSH REL: (Fluorides, inorganic) TWA 2.5 mg(F)/m³

DOT CLASSIFICATION: 8; Label: Corrosive
SAFETY PROFILE: A severe corrosive. A severe skin and eye irritant. When heated to decomposition it emits toxic vapors of boron and F^-.

HHU500   CAS: 7664-39-3   **HR: 3**
HYDROFLUORIC ACID
DOT: UN 1052/UN 1790
mf: FH   mw: 20.01
PROP: Clear, colorless, nonflammable, fuming, corrosive liquid or gas. One of the most acidic substances known, but aq solns are only weakly acid. Dissolves silica to give H_2SiF_6. Mp: $-83.1°$, bp: $19.54°$, d: 0.901 g/L (gas), 0.699 @ $22°$ (liquid), vap press: 400 mm @ $2.5°$. Very sol in H_2O, EtOH; sltly sol in Et_2O. IDLH 30 ppm.
SYNS: ACIDE FLUORHYDRIQUE (FRENCH) □ ACIDO FLUORIDRICO (ITALIAN) □ FLUOROWODOR (POLISH) □ FLUORWASSERSTOFF (GERMAN) □ FLUORWATER-STOF (DUTCH) □ HYDROFLUORIC ACID, solution, >60% strength (UN 1790) (DOT) □ HYDROFLUORIC ACID, solution, not >60% strength (UN 1790) (DOT) □ HYDROFLUORIDE □ HYDROGEN FLUORIDE, anhydrous (UN 1052) (DOT) □ RCRA WASTE NUMBER U134 □ RUBIGINE
CONSENSUS REPORTS: EPA Extremely Hazardous Substances List. Community Right-To-Know List. EPA Genetic Toxicology Program. Reported in EPA TSCA Inventory.
OSHA PEL: TWA 3 ppm; STEL 6 ppm (F)
ACGIH TLV: CL 3 ppm (F); BEI: 3 mg/g creatinine of fluorides in urine prior to shift; 10 mg/g creatinine of fluorides in urine at end of shift.
DFG MAK: 3 ppm (2.5 mg/m³); BAT 7.0 mg/g creatinine in urine at end of shift
NIOSH REL: (HF) TWA 2.5 mg(F)/m³; CL 5.0 mg(F)/m³/15M
DOT CLASSIFICATION: 8; Label: Corrosive, Poison
SAFETY PROFILE: A human poison by inhalation. A poison experimentally by inhalation, subcutaneous, and intraperitoneal routes. A corrosive irritant to skin, eyes (@ 0.05 mg/L), and mucous membranes. Experimental teratogenic effects. Experimental reproductive effects. Mutation data reported. Inhalation of the vapor may cause ulcers of the upper respiratory tract. Concentrations of 50–250 ppm are dangerous, even for brief exposures. Hydrofluoric acid produces severe skin burns that are slow in healing. The subcutaneous tissues may be affected, becoming blanched and bloodless. Gangrene of the affected areas may follow. It is a common air contaminant.

Explosive reaction with cyanogen fluoride; glycerol + nitric acid; sodium (with aqueous acid); methanesulfonic acid (evolves oxygen difluoride that explodes). Violent reaction with As_2O_3; P_2O_5; acetic anhydride; 2-amino ethanol; NH_4OH; $HBiO_3$; bismuthic acid (evolves oxygen); CaO; chlorosulfonic acid; ethylene diamine; ethylene imine; F_2; mercury(II) oxide + organic materials (above 0°C); n-phenylazopiperidine; potassium permanganate; potassium tetrafluorosilicate(2-)(evolves silicon tetrafluoride gas); (HNO_3 + lactic acid); oleum; β-propiolactone; propylene oxide; Na; NaOH; H_2SO_4; vinyl acetate; HgO; sodium tetrafluoro silicate; n-phenyl azo piperidine. Incandescent reaction of liquid HF with oxides (e.g., arsenic trioxide, calcium oxide). Dangerous storage hazard with nitric acid + lactic acid; nitric acid + propylene glycol (mixtures evolve gas which may burst a sealed container). Reacts with water or steam to produce toxic and corrosive fumes. When heated to decomposition it emits highly corrosive fumes of F^-. See also FLUORIDES.

HHV000   **HR: 3**
HYDROFLUORIC ACID mixed with SULFURIC ACID
DOT: UN 1786
SYNS: HYDROFLUORIC and SULFURIC ACIDS, MIXTURE (DOT) □ SULFURIC AND HYDROFLUORIC ACIDS, MIXTURE (DOT)
DOT CLASSIFICATION: 8; Label: Corrosive, Poison
SAFETY PROFILE: Poison by ingestion, inhalation, and skin contact. A corrosive irritant to the eyes, skin and mucous membranes. When heated to decomposition it emits very toxic fumes of HF and SO_x. See also HYDROFLUORIC ACID and SULFURIC ACID.

HHW500   CAS: 1333-74-0   HR: 3
HYDROGEN

DOT: UN 1049/UN 1966

mf: H_2 mw: 2.02

PROP: Stable, colorless, odorless, tasteless gas. Forms compounds with almost every other element. Mp: $-259.18°$, bp: $-252.8°$, lel: 4.1%, uel: 74.2%, d: 0.0899 g/L, autoign temp: 752°F, vap d: 0.069. Very low solubility in most liquids.

SYNS: HYDROGEN (DOT) □ HYDROGEN, compressed (DOT) □ HYDROGEN, refrigerated liquid (DOT)

CONSENSUS REPORTS: Reported in EPA TSCA Inventory.

DOT CLASSIFICATION: 2.1; Label: Flammable Gas

SAFETY PROFILE: Practically no toxicity except that it may asphyxiate. Highly dangerous fire and severe explosion hazard when exposed to heat, flame, or oxidizers. Flammable or explosive when mixed with air, O_2, chlorine. To fight fire, stop flow of gas.

Explodes on contact with bromine trifluoride; chlorine trifluoride; fluorine; hydrogen peroxide + catalysts; acetylene + ethylene. Explodes when heated with calcium carbonate + magnesium; 3,4-dichloronitrobenzene + catalysts; vegetable oils + catalysts; ethylene + nickel catalysts; difluorodiazene (above 90°C); 2-nitroanisole (above 250°C/34 bar + 12% catalyst); copper(II) oxide; nitryl fluoride (above 200°C); polycarbon monofluoride (above 500°C).

Forms sensitive explosive mixtures with bromine; chlorine; iodine heptafluoride (heat- or spark-sensitive); chlorine dioxide; dichlorine oxide; iodine heptafluoride (heat- or spark-sensitive); dinitrogen oxide; dinitrogen tetraoxide; oxygen (gas); 1,1,1-trisazidomethylethane + palladium catalyst. Mixtures with liquid nitrogen react with heat to form an explosive product.

Violent reaction or ignition with air + catalysts (platinum and similar metals containing adsorbed oxygen or hydrogen); bromine; iodine; dioxane + nickel; lithium; nitrogen trifluoride; oxygen difluoride; palladium + isopropyl alcohol; 3-methyl-2-penten-4-yn-1-ol; lead trifluoride; bromine fluoride (ignition on contact); nickel + oxygen; fluorine perchlorate (ignition on contact); xenon hexafluoride (violent reaction); nitrogen oxide + oxygen (ignition above 360°C); palladium powder + 2-propanol + air (spontaneous ignition); platinum catalyst; polycarbon monofluoride (ignition above 400°C).

Vigorous exothermic reaction with benzene + Raney nickel catalyst; metals (e.g., lithium; calcium; barium; strontium; sodium; potassium; above 300°C); palladium(II) oxide; palladium trifluoride; 1,1,1-tris(hydroxymethyl)nitromethane + nickel catalyst.

HHW560   CAS: 8016-14-6   HR: D
HYDROGENATED FISH OIL

PROP: Oil. Mp: $>32°$.

SAFETY PROFILE: When heated to decomposition it emits acrid smoke and irritating fumes.

HHW800   CAS: 61788-32-7   HR: 3
HYDROGENATED TERPHENYLS

PROP: Complex mixtures of o-, m-, and p-terphenyls in various stages of hydrogenation. Five such stages exist for each of the three above isomers.

CONSENSUS REPORTS: Reported in EPA TSCA Inventory.

OSHA PEL: TWA 0.5 ppm

ACGIH TLV: TWA 0.5 ppm

NIOSH REL: (Hydrogenated Terphenyls) TWA 0.5 ppm

SAFETY PROFILE: Contact with hot coolant can cause severe damage to lungs, skin, and eyes from burns. May cause chronic damage to liver, kidney, and blood-forming organs; metabolic disorders. Inhalation has caused bronchopneumonia. When heated to decomposition they emit acrid smoke and fumes.

HHX000   CAS: 7647-01-0   HR: 3
HYDROGEN CHLORIDE
mf: ClH   mw: 36.46
PROP: Colorless, corrosive, nonflammable gas. Pungent odor, fumes in air. D: 1.639 @ −137.77°, bp: −154.37° @ 1.0 mm.
CONSENSUS REPORTS: EPA Extremely Hazardous Substances List. EPA Genetic Toxicology Program. Reported in EPA TSCA Inventory.
OSHA PEL: CL 5 ppm
ACGIH TLV: CL 5 ppm
DFG MAK: 5 ppm (7 mg/m³)
SAFETY PROFILE: A highly corrosive irritant to the eyes, skin, and mucous membranes. Mildly toxic by inhalation. Explosive reaction with alcohols + hydrogen cyanide, potassium permanganate, sodium (with aqueous HCl), tetraselenium tetranitride. Ignition on contact with aluminum-titanium alloys (with HCl vapor), fluorine, hexalithium disilicide, metal acetylides or carbides (e.g., cesium acetylide, rubidium acetylide). Violent reaction with 1,1-difluoroethylene. Vigorous reaction with aluminum, chlorine + dinitroanilines (evolves gas). Potentially dangerous reaction with sulfuric acid releases HCl gas. Adsorption of the acid onto silicon dioxide is exothermic. See also HYDROGEN CHLORIDE (AEROSOL) and HYDROCHLORIC ACID.

HIB005   HR: 2
HYDROGEN PEROXIDE, 8% to 20%
DOT: UN 2984
mf: H₂O₂   mw: 34.02
SYN: HYDROGEN PEROXIDE, solution, 8% to 20% (DOT)
DOT CLASSIFICATION: 5.1; Label: Oxidizer (UN 2984)
SAFETY PROFILE: Moderately toxic by ingestion. Experimental reproductive effects. A moderate oxidizer.

HIB010   CAS: 7722-84-1   HR: 2
HYDROGEN PEROXIDE, 30%
DOT: UN 2014/UN 2984

mf: H₂O₂   mw: 34.02
PROP: IDLH 75 ppm.
SYNS: ALBONE 35 □ ALBONE 50 □ ALBONE 70 □ ALBONE 35CG □ ALBONE 50CG □ ALBONE 70CG □ HYDROGEN PEROXIDE, solution, 30% □ HYDROGEN PEROXIDE, aqueous solutions with not <8% but <20% hydrogen peroxide (UN 2984) (DOT) □ HYDROGEN PEROXIDE, aqueous solutions with >40%, not >60% hydrogen peroxide (UN 2014) □ INTEROX □ KASTONE □ PERONE 30 □ PERONE 35 □ PERONE 50
CONSENSUS REPORTS: IARC Cancer Review: Group 3 IMEMDT 7,56,87, Animal Limited Evidence IMEMDT 36,285,85. Reported in EPA TSCA Inventory.
DOT CLASSIFICATION: 5.1; Label: Oxidizer, Corrosive (UN 2014); DOT Class: 5.1; Label: Oxidizer (UN 2984)
SAFETY PROFILE: Moderately toxic by ingestion. Questionable carcinogen with experimental carcinogenic data. Mutation data reported. See HYDROGEN PEROXIDE.

HIB050   CAS: 7722-84-1   HR: 3
HYDROGEN PEROXIDE, 90%
DOT: UN 2014
mf: H₂O₂   mw: 34.02
PROP: Colorless, unstable, heavy liquid or, at low temp, a crystalline solid; bitter taste. D: 1.71 @ −20°, 1.46 @ 0°, vap press: 1 mm @ 15.3°, unstable, mp: −0.43°, bp: 152°. Misc with water; sol in ether; insol in pet ether. Decomposed by many org solvs.
SYNS: ALBONE □ DIHYDROGEN DIOXIDE □ HIOXYL □ HYDROGEN DIOXIDE □ HYDROGEN PEROXIDE, stabilized with >60% hydrogen peroxide (DOT) □ HYDROPEROXIDE □ INHIBINE □ OXYDOL □ PERHYDROL □ PERONE □ PEROSSIDO di IDROGENO (ITALIAN) □ PEROXAN □ PEROXIDE □ PEROXYDE d'HYDROGENE (FRENCH) □ SUPEROXOL □ T-STUFF □ WASSERSTOFFPEROXID (GERMAN) □ WATERSTOFPEROXYDE (DUTCH)
CONSENSUS REPORTS: IARC Cancer Review: Group 3 IMSUDL 7,56,87; Human No Adequate Data IMEMDT 36,285,85; Animal Limited Evidence IMEMDT 36,285,85. EPA Extremely Hazardous Substances List. Reported in EPA TSCA Inventory. EPA Genetic Toxicology Program. EPA FIFRA 1988 pesticide subject to registration or re-registration.

H

OSHA PEL: TWA 1 ppm
ACGIH TLV: TWA 1 ppm; Animal
Carcinogen
DFG MAK: 1 ppm (1.4 mg/m³)
NIOSH REL: (Hydrogen peroxide) TWA 1.0
ppm
DOT CLASSIFICATION: 5.1; Label:
Oxidizer, Corrosive
SAFETY PROFILE: Confirmed carcinogen
with experimental tumorigenic data.
Moderately toxic by inhalation, ingestion,
and skin contact. A corrosive irritant to skin,
eyes, and mucous membranes. Human
mutation data reported. A very powerful
oxidizer.

Pure H_2O_2, its solutions, vapors, and mists
are very irritating to body tissue. This
irritation can vary from mild to severe
depending upon the concentration of H_2O_2.
For instance, solutions of H_2O_2 of 35 wt%
and over can easily cause blistering of the
skin. Irritation caused by H_2O_2 that does not
subside upon flushing the affected part with
water should be treated by a physician. The
eyes are particularly sensitive to this
material. It is a common air contaminant.

A dangerous fire hazard by chemical
reaction with flammable materials. H_2O_2 is a
powerful oxidizer, particularly in the
concentrated state. It is important to keep
containers covered because the contents of
uncovered containers are much more prone
to react with flammable vapors, gases, etc.;
and, if uncovered, the water from an H_2O_2
solution can evaporate, concentrating the
material and thus increasing the fire hazard.
For instance, solutions of H_2O_2 in
concentration in excess of 65 wt% heat up
spontaneously when decomposed to H_2O +
$1/2 O_2$. Thus, 90 wt% solutions, when
caused to decompose rapidly due to the
introduction of a catalytic decomposition
agent, can get quite hot and perhaps start
fires.

A severe explosion hazard when highly
concentrated or when pure H_2O_2 is exposed
to: heat, mechanical impact, or detonation
of a blasting cap, or is caused to decompose
catalytically by metals (in order of decreasing
effectiveness: osmium; palladium; platinum;
iridium; gold; silver; manganese; cobalt;
copper; lead). Explodes on contact with
alcohols + H_2SO_4; acetal + acetic acid +
heat; acetic acid + n-heterocycles (above
50°); 2-amino-4-methyloxazole + iron(II)
catalyst; aromatic hydrocarbons +
trifluoroacetic acid; azeliac acid + sulfuric
acid (above 45°); benzenesulfonic anhydride;
tert-butanol + sulfuric acid; carboxylic acids;
3,5-dimethyl-3-hexanol + sulfuric acid;
diphenyl diselenide (above 53°); 2-
ethoyxethanol + polyacrylamide gel +
toluene + heat; gadolinium hydroxide
(above 80°); gallium + hydrochloric acid;
hydrogen + palladium catalysts (has caused
major industrial explosions); iron(II) sulfate
+ 2-methylpyridine + sulfuric acid; iron(II)
sulfate + nitric acid + sodium
carboxymethylcellulose (when evaporated);
nitric acid + ketones (e.g., 2-butanone, 3-
pentanone, cyclopentanone, cyclohexanone,
3-methylcyclohexanone), trioxane (sensitive
to heat, shock, or on contact with lead),
methanol + tert-amines + platinum
catalysts; nitric acid + soils; nitrogenous
bases (e.g., ammonia, hydrazine hydrate, 1,1-
dimethylhydrazine); organic compounds
(e.g., glycerol, acetic acid, ethanol, aniline,
quinoline, 2-phenyl-1,1-dimethylethanol,
cellulose, charcoal); organic materials +
sulfuric acid (especially if confined); water +
oxygenated compounds (e.g., acetaldehyde,
acetic acid, acetone, ethanol, formaldehyde,
formic acid, methanol, 2-propanol,
propionaldehyde); sulfuric acid (during
evaporation); tetrahydrothiophene; vinyl
acetate; alcohols + tin chloride; P_2O_5; P;
H_2O; HNO_3; Sb_2S_3; As_2S_3; Cl_2 + KOH +
chlorosulfonic acid; CuS; FeS; formic acid +
organic matter; H_2Se; hydrazine; PbO_2;
PbO; PbS; MnO_2; HgO; Hg_2O; MoS_2;
organic matter, (2-methyl-1-phenyl-2-
propanol + sulfuric acid); $KMnO_4$; $NaIO_3$;
thiodiglycol; uns-dimethyl hydrazine; $FeSO_4$
+ 2-methylpyridine + H_2SO_4; HgO +
HNO_3.

Forms unstable explosive products in reaction with acetaldehyde + desiccants (forms polyethylidine peroxide); acetic acid (forms peracetic acid); acetic + 3-thietanol; acetic anhydride; acetone (forms explosive peroxides); alcohols (products are shock- and heat-sensitive); carboxylic acids (e.g., formic acid, acetic acid, tartaric acid), diethyl ether, ethyl acetate, formic acid + metaboric acid, ketene (forms diacetyl peroxide); mercury(II) oxide + nitric acid (forms mercury(II) peroxide); thiourea + nitric acid; polyacetoxyacrylic acid lactone + poly(2-hydroxyacrylic acid) + sodium hydroxide.

Ignition on contact with furfuryl alcohol; powdered metals (e.g., magnesium; iron); wood. Violent reaction with aluminum isopropoxide + heavy metal salts; charcoal; coal; dimethylphenylphosphine; hydrogen selenide; lithium tetrahydroaluminate; metals (e.g., potassium, sodium, lithium); metal oxides (e.g., cobalt oxide, iron oxide, lead oxide, lead hydroxide, manganese oxide, mercury oxide, nickel oxide); metal salts (e.g., calcium permanganate); methanol + phosphoric acid; 4-methyl-2,4,6-triazatricyclo [5.2.2.0^{2,6}] undeca-8-ene-3,5-dione + potassium hydroxide; α-phenylselenoketones; phosphorus; phosphorus(V) oxide; tin(II) chloride; unsaturated organic compounds.

BEWARE: Although many mixtures of H_2O_2 and organic materials do not explode upon contact, the resultant combination is detonatable either upon catching fire or by impact. The detonation velocity of aqueous solutions of H_2O_2 has been found to be about 6500 m/s for solutions of between 96 and 100 wt% H_2O_2. Another source of H_2O_2 explosions is the sealing of the material in strong containers. Under such conditions, even gradual decomposition of H_2O_2 to H_2O + 1/2 O_2 can cause large pressures to build up in the containers, which may then burst explosively. Highly dangerous; when heated, shocked, or contaminated, the concentrated material can explode or start fires.

HIB500   CAS: 124-43-6   HR: 1
HYDROGEN PEROXIDE with UREA (1:1)
DOT: UN 1511
mf: $CH_4N_2O \cdot H_2O_2$ mw: 94.09
PROP: White crystals. Mp: 75–85° (decomp).
SYNS: CARBAMIDE PEROXIDE □ GLY-OXIDE □ HYDROGEN PEROXIDE CARBAMIDE □ HYDROPERIT □ HYPEROL □ ORTIZON □ PERCARBAMIDE □ PERHYDRIT □ PERHYDROL-UREA □ THENARDOL □ UREA DIOXIDE □ UREA HYDROGEN PEROXIDE (DOT) □ UREA HYDROGEN PEROXIDE SALT □ UREA HYDROPEROXIDE □ UREA PEROXIDE (DOT)
CONSENSUS REPORTS: Reported in EPA TSCA Inventory.
DOT CLASSIFICATION: 5.1; Label: Oxidizer
SAFETY PROFILE: An irritant to skin, eyes, and mucous membranes. An FDA over-the-counter drug. When heated to decomposition it emits toxic fumes of NO_x. See also individual components and PEROXIDES, ORGANIC.

HIC000   CAS: 7783-07-5   HR: 3
HYDROGEN SELENIDE
DOT: UN 2202
mf: H_2Se mw: 80.98
PROP: Colorless gas. Mp: −64°, bp: −41.4°, d: 3.614 g/L (gas), 2.12 @ −42° (liquid), vap press: 10 atm @ 23.4°. Flammable. Disagreeable odor. Sol in carbonyl chloride and carbon disulfide. IDLH 1 ppm.
SYNS: ELECTRONIC E-2 □ HYDROGEN SELENIDE, anhydrous (DOT) □ SELENIUM HYDRIDE
CONSENSUS REPORTS: Selenium and its compounds are on the Community Right-To-Know List. EPA Extremely Hazardous Substances List. Reported in EPA TSCA Inventory.
OSHA PEL: TWA 0.05 ppm (Se)
ACGIH TLV: TWA 0.05 ppm (Se)
DFG MAK: 0.05 ppm (0.17 mg/m³)
DOT CLASSIFICATION: 2.3; Label: Poison Gas, Flammable Gas
SAFETY PROFILE: A deadly poison by inhalation. Very poisonous irritant to skin, eyes, and mucous membranes. Causes central nervous system effects in humans.

An allergen. Can cause damage to the lungs and liver as well as conjunctivitis. It has been found that repeated exposures to concentrations of 0.3 ppm prove fatal to experimental animals by causing a pneumonitis, as well as injury to the liver and spleen. Causes garlic odor of breath, dizziness, nausea. Concentrations of 0.3 ppm are readily detected by odor, but there is no noticeable irritant effect at that level. Concentrations of 1.5 ppm or higher are strongly irritating to the eyes and nasal passages.

As in the case of hydrogen sulfide, the odor of hydrogen selenide in concentrations below 1 ppm disappears rapidly because of olfactory fatigue. The odor and irritating effects do not offer a dependable warning to workers who may be exposed to gradually increasing amounts and therefore become used to it. Due to its extreme toxicity and irritating effects, it seldom is allowed to reach a concentration in which it is flammable in air. Very little data are available on possible chronic effects of this material, but it is logical to assume that when the concentration of this gas is low enough to avoid the irritant effects, only the systemic effects will be noticeable.

Dangerous fire hazard when exposed to heat or flame; will react vigorously with powerful oxidizing agents, such as H_2O_2, HNO_3. Dangerous; forms explosive mixtures with air; keep away from heat and open flame. See also SELENIUM COMPOUNDS and HYDRIDES.

HIC500   CAS: 7783-06-4   HR: 3
HYDROGEN SULFIDE
DOT: UN 1053
mf: H_2S mw: 34.08
PROP: Colorless, flammable, poisonous gas; offensive odor. Mp: $-85.5°$, bp: $-60.4°$, d: -60, (gas) 0.993, lel: 4%, uel: 46%, autoign temp: 500°F, d: 1.539 g/L @ 0°, vap press: 20 atm @ 25.5°, vap d: 1.189. IDLH 100 ppm.

SYNS: ACIDE SULFHYDRIQUE (FRENCH) □ HYDROGENE SULFURE (FRENCH) □ HYDROGEN SULFURIC ACID □ IDROGENO SOLFORATO (ITALIAN) □ RCRA WASTE NUMBER U135 □ SCHWEFELWAS-SERSTOFF (GERMAN) □ SIARKOWODOR (POLISH) □ STINK DAMP □ SULFURETED HYDROGEN □ SULFUR HYDRIDE □ ZWAVELWATERSTOF (DUTCH)
CONSENSUS REPORTS: EPA Extremely Hazardous Substances List. Reported in EPA TSCA Inventory.
OSHA PEL: TWA 10 ppm; STEL 15 ppm
ACGIH TLV: (Proposed: TWA 1 ppm; STEL 5 ppm)
DFG MAK: 10 ppm (14 mg/m³)
NIOSH REL: (Hydrogen Sulfide) CL 15 mg/m³/10M
DOT CLASSIFICATION: 2.3; Label: Poison Gas, Flammable Gas
SAFETY PROFILE: A human poison by inhalation. A severe irritant to eyes and mucous membranes. Experimental reproductive effects. An asphyxiant. Human systemic effects by inhalation: coma, chronic pulmonary edema. Low concentrations of 20–150 ppm cause irritation of the eyes; slightly higher concentrations may cause irritation of the upper respiratory tract, and, if exposure is prolonged, pulmonary edema may result. The irritant action has been explained on the basis that H_2S combines with the alkali present in moist surface tissues to form sodium sulfide, a caustic. With higher concentration the action of the gas on the nervous system becomes more prominent. A 30-minute exposure to 500 ppm results in headache, dizziness, excitement, staggering gait, diarrhea, and dysuria, followed sometimes by bronchitis or bronchopneumonia.

The action of small amounts on the nervous system is one of depression; in larger amounts, it stimulates, and with very high amounts the respiratory center is paralyzed. Exposures of 800–1000 ppm may be fatal in 30 minutes, and high concentrations are instantly fatal. Fatal hydrogen sulfide poisoning may occur even more rapidly than that following exposure to a similar concentration of HCN. H_2S does not combine with the hemoglobin of the blood; its asphyxiant action is due to paralysis of the respiratory center. With repeated exposures to low concentrations, conjunctivitis, photophobia, corneal bullae, tearing, pain, and blurred vision are the commonest findings. High concentrations may cause rhinitis, bronchitis, and occasionally pulmonary edema. Exposure to very high concentrations results in immediate death. Chronic poisoning results in headache, inflammation of the conjunctivae and eyelids, digestive disturbances, weight loss, and general debility. It is a common air contaminant.

It is an insidious poison since sense of smell may be fatigued. The odor and irritating effects do not offer a dependable warning to workers who may be exposed to gradually increasing amounts and therefore become used to it.

Very dangerous fire hazard when exposed to heat, flame, or oxidizers. Moderately explosive when exposed to heat or flame. Explodes on contact with oxygen difluoride; nitrogen trichloride; bromine pentafluoride; chlorine trifluoride; dichlorine oxide; silver fulminate. Potentially explosive reaction with copper + oxygen. Explosive reaction when heated with perchloryl fluoride (above 100°C), oxygen (above 280°C). Reacts with 4-bromobenzenediazonium chloride to form an explosive product.

Ignites on contact with metal oxides (e.g., barium peroxide, chromium trioxide, copper oxide, lead dioxide, manganese dioxide, nickel oxide, silver(I) oxide, silver(II) oxide, sodium peroxide, thallium(III) oxide, mercury oxide, calcium oxide, nickel oxide), oxidants (e.g., silver bromate, heptasilver nitrate octaoxide, dibismuth dichromium nonaoxide, mercury(I) bromate, lead(II) hypochlorite, copper chromate, fluorine, nitric acid, sodium peroxide, lead(IV) oxide), rust, soda-lime + air. Reacts violently with NI_3, NF_3, p-bromobenzenediazonium chloride, OF_2, F_2, Cu, ClO, BrF_5,

acetaldehyde, $(BaO + Hg_2O + air)$, $(BaO + NiO + air)$, hydrated iron oxide, phenyl diazonium chloride, $(NaOH + CaO + air)$. Incandescent reaction with chromium trioxide. Vigorous reaction with metal powders (e.g., copper, tungsten). When heated to decomposition it emits highly toxic fumes of SO_x. To fight fire, stop flow of gas.

HIC600   CAS: 15181-46-1   HR: 2
HYDROGEN SULFITE
mf: HO_3S mw: 81.07
SYNS: BISULFITE □ BISULPHITE □ HYDROSULFITE ANION □ SULFITE LYE
DOT CLASSIFICATION: 8; Label: Corrosive
SAFETY PROFILE: Mutation data reported. A corrosive. When heated to decomposition it emits toxic fumes of SO_x. See also SULFITES.

HIH000   CAS: 123-31-9   HR: 3
HYDROQUINONE
DOT: UN 2662
mf: $C_6H_6O_2$ mw: 110.12
PROP: Colorless, hexagonal prisms; needles from water. Mp: 172°, bp: 285° @ 730 mm, flash p: 329°F (CC), d: 1.358 @ 20°/4°, autoign temp: 960°F (CC), vap press: 1 mm @ 132.4°, vap d: 3.81. Very sol in alc and ether; sltly sol in benzene. Keep well closed and protected from light. IDLH 50 mg/m³.
SYNS: ARCTUVIN □ BENZENE, p-DIHYDROXY- □ p-BENZENEDIOL □ 1,4-BENZENEDIOL □ BENZOHY-DROQUINONE □ BENZOQUINOL □ BLACK AND WHITE BLEACHING CREAM □ 1,4-DIHYDROXY-BENZEEN (DUTCH) □ 1,4-DIHYDROXYBENZEN (CZECH) □ DIHYDROXYBENZENE □ p-DIHYDROXY-BENZENE □ 1,4-DIHYDROXYBENZENE □ DIHYDRO-XYBENZENE (OSHA) □ 1,4-DIHYDROXY-BENZOL (GERMAN) □ 1,4-DIIDROBENZENE (ITALIAN) □ p-DIOXOBENZENE □ p-DIOXYBENZENE □ ELDOPA-QUE □ ELDOQUIN □ HYDROCHINON (CZECH, POLISH) □ HYDROQUINOL □ HYDROQUINOLE □ α-HYDROQUINONE □ p-HYDROQUINONE □ HYDROQUINONE, liquid or solid (DOT) □ p-HYDROXY-PHENOL □ IDROCHINONE (ITALIAN) □ NCI-C55834 □ QUINOL □ β-QUINOL □ TECQUINOL □ TENOX HQ □ TEQUINOL □ USAF EK-356

CONSENSUS REPORTS: IARC Cancer Review: Group 3 IMEMDT 7,56,87; Animal Inadequate Evidence IMEMDT 15,155,77. Community Right-To-Know List. EPA Extremely Hazardous Substances List. EPA Genetic Toxicology Program. Reported in EPA TSCA Inventory.

OSHA PEL: TWA 2 mg/m³
ACGIH TLV: TWA 2 mg/m³; Animal Carcinogen; (Proposed: 1 mg/m³ (sensitizer); Confirmed Animal Carcinogen with Unknown Revelance to Humans)
DFG MAK: Animal Carcinogen, Suspected Human Carcinogen
NIOSH REL: (Hydroquinone) CL 2.0 mg/m³/15M
DOT CLASSIFICATION: 6.1; Label: KEEP AWAY FROM FOOD
SAFETY PROFILE: Confirmed carcinogen. A human poison by ingestion. An experimental poison by ingestion, intraperitoneal, intravenous, and subcutaneous routes. Human systemic effects by ingestion: pulse rate increase without fall in blood pressure, cyanosis, coma. An active allergen. Human mutation data reported. A severe human skin irritant. Experimental reproductive data.

Absorption of this material by tissues can cause symptoms of illness that resemble those induced by its o- or m-isomers. For instance, the ingestion of 1 g by an adult or a smaller quantity by a child may induce tinnitis, nausea, dizziness, a sensation of suffocation, an increased rate of respiration, vomiting, pallor, muscular twitching, headache, dyspnea, cyanosis, delirium, and collapse. The literature contains reports of fatal cases that have been caused by the ingestion of 5–12 g. Cases of dermatitis have resulted from skin contact, and have also followed the application of an antiseptic oil that apparently contained traces of hydroquinone added as an antioxidant. The report also describes cases of keratitis and discoloration of the conjunctiva among personnel exposed to this material in concentrations ranging from 10 to 30 mg of the vapor or dust per cubic meter of air. It is considered to be more toxic than phenol. The inhalation of vapors, particularly when liberated at high temperatures, must be avoided. If this material accidentally comes into contact with the skin, it should be removed at once and the affected area washed with plenty of soap and water.

Combustible when exposed to heat or flame; can react with oxidizing materials. Potentially explosive reaction with oxygen at 90°C/100 bar. Violent reaction with NaOH. Slight explosion hazard when exposed to heat. To fight fire, use water, CO_2, dry chemical.

HII600   CAS: 90-87-9   HR: 2
HYDROTROPIC ALDEHYDE DIMETHYL ACETAL
mf: $C_{11}H_{16}O_2$ mw: 180.27
PROP: Odor of ylang.
SYNS: 1,1-DIMETHOXY-2-PHENYLPROPANE □ HYDROTROPALDEHYDE DIMETHYL ACETAL □ 2-PHENYLPROPIONALDEHYDE DIMETHYL ACETAL
CONSENSUS REPORTS: Reported in EPA TSCA Inventory.
SAFETY PROFILE: Moderately toxic by ingestion. A skin irritant. When heated to decomposition it emits acrid smoke and irritating fumes.

HIM000   CAS: 103-90-2   HR: 3
4'-HYDROXYACETANILIDE
mf: $C_8H_9NO_2$ mw: 151.18
PROP: Prisms from EtOH. Mp: 169–170.5°. Sol in hot H_2O.
SYNS: ABENSANIL □ ACAMOL □ ACETAGESIC □ ACETALGIN □ p-ACETAMIDOPHENOL □ 4-ACETAMIDOPHENOL □ ACETAMINOPHEN □ p-ACETAMINOPHENOL □ N-ACETYL-p-AMINOPHENOL □ p-ACETYLAMINOPHENOL □ ALGOTROPYL □ ALPINYL □ ALVEDON □ AMADIL □ ANAFLON □ ANELIX □ ANHIBA □ APADON □ APAMIDE □ APAP □ BEN-U-RON □ BICKIE-MOL □ CALPOL □ CETADOL □ CLIXODYNE □ DATRIL □ DIAL-A-GESIC □ DIROX □ DOLIPRANE □ DYMADON □ ENELFA □ ENERIL □ EXDOL □ FEBRILIX □ FEBRO-GESIC □ FEBROLIN □ FENDON □ FINIMAL □ G 1 □ GELOCATIL □ HEDEX □ HOMOOLAN □ p-HYDROXYACETANILIDE □ 4-HYDROXYACETANILIDE □ N-(4-HYDROXYPHENYL)-ACETAMIDE □ JANUPAP □ KORUM □ LESTEMP □

LIQUAGESIC □ LONARID □ LYTECA SYRUP □ MOMENTUM □ MULTIN □ NAPA □ NAPRINOL □ NCI-C55801 □ NOBEDON □ PACEMO □ PANADOL □ PANETS □ PANEX □ PANOFEN □ PARACETAMOLE □ PARACETAMOLO (ITALIAN) □ PARACETANOL □ PARAPAN □ PARASPEN □ PARMOL □ PEDRIC □ PYRINAZINE □ TABALGIN □ TAPAR □ TEMLO □ TEMPANAL □ TEMPRA □ TRALGON □ TUSSAPAP □ TYLENOL □ VALADOL □ VALGESIC

CONSENSUS REPORTS: Reported in EPA TSCA Inventory. EPA Genetic Toxicology Program.

SAFETY PROFILE: Suspected carcinogen with experimental carcinogenic and tumorigenic data. A human poison by ingestion. An experimental poison by intraperitoneal and subcutaneous routes. Moderately toxic by intravenous route. Human systemic effects by ingestion: changes in exocrine pancreas, diarrhea, nausea, irritability, somnolence, general anesthesia, fever, hepatitis, kidney tubule damage. Experimental teratogenic and reproductive effects. Human mutation data reported. Used as an analgesic and antipyretic. When heated to decomposition it emits toxic fumes of NO_x.

HIM500   CAS: 107-16-4   HR: 3
HYDROXYACETONITRILE
mf: C_2H_3NO mw: 57.06
PROP: Bp: 183° (slt decomp).
SYNS: CYANOMETHANOL □ FORMALDEHYDE CYANOHYDRIN □ GLYCOLIC NITRILE □ GLYCOLONITRILE □ GLYCONITRILE □ 2-HYDROXYACETONITRILE □ HYDROXY-METHYLINITRILE □ USAF A-8565

CONSENSUS REPORTS: EPA Extremely Hazardous Substances List. Reported in EPA TSCA Inventory. Cyanide and its compounds are on the Community Right-To-Know List.

NIOSH REL: (Nitriles) CL 5 mg/m³/15M

SAFETY PROFILE: A poison by ingestion, skin contact, inhalation, intraperitoneal, ocular, and subcutaneous routes. An eye irritant. May undergo spontaneous and violent decomposition. Traces of alkali promote violent polymerization. When heated to decomposition it emits toxic fumes of NO_x and CN^-. See also NITRILES.

HIN500   CAS: 118-93-4   HR: 3
o-HYDROXYACETOPHENONE
mf: $C_8H_8O_2$ mw: 136.16
PROP: Greenish-yellow liquid or oil, highly refractive, minty odor. Mp: 95°, vap d: 4.69, bp: 213° @ 717 mm.
SYNS: ACETOPHENONE, 2'-HYDROXY-(8CI) □ o-ACETYLPHENOL □ 2-ACETYLPHENOL □ ETHANONE, 1-(2-HYDROXYPHENYL)-(9CI) □ 2'-HYDROXY-ACETOPHENONE □ o-HYDROXYPHENYL METHYL KETONE □ USAF KE-20

CONSENSUS REPORTS: Reported in EPA TSCA Inventory.

DOT CLASSIFICATION: 3; Label: Flammable Liquid

SAFETY PROFILE: Poison by intraperitoneal route. When heated to decomposition it emits acrid smoke and fumes. See also KETONES.

HIP000   CAS: 53-95-2   HR: 3
N-HYDROXY-N-ACETYL-2-AMINOFLUORENE
mf: $C_{15}H_{13}NO_2$ mw: 239.29
SYNS: FLUORENYL-2-ACETHYDROXAMIC ACID □ N-FLUOREN-2-YL ACETOHYDROXAMIC ACID □ N-2-FLUORENYL ACETOHYDROXAMIC ACID □ N-HYDROXY-AAF □ N-HYDROXY-2-ACETAMIDO-FLUORENE □ 2-(N-HYDROXYACETAMIDO)-FLUORENE □ N-HYDROXY-2-ACETYLAMINO-FLUORENE □ N-HYDROXY-2-FAA □ N-HYDROXY-N-(2-FLUORENYL)ACETAMIDE □ NOHFAA

CONSENSUS REPORTS: EPA Genetic Toxicology Program.

SAFETY PROFILE: Suspected carcinogen with experimental carcinogenic, neoplastigenic, and tumorigenic data. A poison by intraperitoneal route. Experimental teratogenic and other reproductive effects. Human mutation data reported. When heated to decomposition it emits toxic fumes of NO_x.

HIS100   CAS: 52098-13-2   HR: 3
2-(HYDROXYACETYL)INDOLE
mf: $C_{10}H_9NO_2$ mw: 175.20

H

SYNS: KETONE, HYDROXYMETHYL 2-INDOLYL- □ HYDROXYMETHYL 2-INDOLYL KETONE
DOT CLASSIFICATION: 3; Label: Flammable Liquid
SAFETY PROFILE: A poison by intraperitoneal route. A flammable liquid. When heated to decomposition it emits toxic vapors of NO_x.

HIS120   CAS: 52098-14-3   HR: 3
2-(HYDROXYACETYL)-1-METHYLIN-DOLE
mf: $C_{11}H_{11}NO_2$ mw: 189.23
SYNS: 2-(2-HYDROXYACETYL)-1-METHYLINDOLE □ HYDROXYMETHYL 1-METHYL-2-INDOLYL KETONE □ KETONE, HYDROXYMETHYL 1-METHYL-2-INDOLYL-
DOT CLASSIFICATION: 3; Label: Flammable Liquid
SAFETY PROFILE: A poison by intraperitoneal route. A flammable liquid. When heated to decomposition it emits toxic vapors of NO_x.

HIS130   CAS: 52098-15-4   HR: 3
2-(HYDROXYACETYL)-3-METHYLIN-DOLE
mf: $C_{11}H_{11}NO_2$ mw: 189.23
SYNS: HYDROXYMETHYL 3-METHYL-2-INDOLYL KETONE □ KETONE, HYDROXYMETHYL 3-METHYL-2-INDOLYL-
DOT CLASSIFICATION: 3; Label: Flammable Liquid
SAFETY PROFILE: A poison by intraperitoneal route. A flammable liquid. When heated to decomposition it emits toxic vapors of NO_x.

HIS140   CAS: 23518-13-0   HR: 3
3-(HYDROXYACETYL)-1-METHYLIN-DOLE
mf: $C_{11}H_{11}NO_2$ mw: 189.23
SYNS: INDOLE, 3-(HYDROXYACETYL)-1-METHYL □ KETONE, HYDROXYMETHYL 1-METHYL-3-INDOLYL
DOT CLASSIFICATION: 3; Label: Flammable Liquid
SAFETY PROFILE: A poison by intraperitoneal route. A flammable liquid. When heated to decomposition it emits toxic vapors of NO_x.

HIS150   CAS: 27463-04-3   HR: 3
3-(2-HYDROXYACETYL)-2-METHYLIN-DOLE
mf: $C_{11}H_{11}NO_2$ mw: 189.23
SYNS: INDOLE, 3-(HYDROXYACETYL)-2-METHYL □ KETONE, HYDROXYMETHYL 2-METHYL-3-INDOLYL
DOT CLASSIFICATION: 3; Label: Flammable Liquid
SAFETY PROFILE: A poison by intraperitoneal route. A flammable liquid. When heated to decomposition it emits toxic vapors of NO_x.

HJB225   CAS: 1239-31-2   HR: 1
3-β-HYDROXYANDROSTEN-17-ONE ACETATE
mf: $C_{21}H_{32}O_3$ mw: 332.53
SYNS: (3-β,5-α)-3-(ACETYLOXY)ANDROSTAN-17-ONE □ ANDROSTAN-17-ONE, 3-(ACETYLOXY)-, (3-β,5-α)- □ DEHYDROEPIANDROSTERONE ACETATE □ EPIANDROSTERONE, DEHYDRO-, ACETATE □ ANDROSTEN-17-ONE, 3-β-HYDROXY-, ACETATE
SAFETY PROFILE: Low toxicity by ingestion. Questionable carcinogen with experimental data reported. Experimental reproductive effects. When heated to decomposition it emits acrid smoke and irritating vapors.

HJE400   CAS: 6318-57-6   HR: 3
2-HYDROXY-p-ARSANILIC ACID
mf: $C_6H_8AsNO_4$ mw: 233.07
SYNS: 4-AMINO-2-HYDROXYBENZENEARSONIC ACID □ BENZENEARSONIC ACID, 4-AMINO-2-HYDROXY-
OSHA PEL: TWA 0.5 mg(As)/m³
SAFETY PROFILE: Poison by intravenous route. When heated to decomposition it emits toxic fumes of NO_x and As.

HJF000   CAS: 1689-82-3   HR: 3
4-HYDROXYAZOBENZENE
mf: $C_{12}H_{10}N_2O$ mw: 198.24
PROP: Orange, rhombic crystals from ethanol. Mp: 155–156°, bp: 220–230°. Very sol in ether.
SYNS: p-BENZENEAZOPHENOL □ C.I. SOLVENT YELLOW 7 □ p-HYDROXYAZOBENZENE □ p-PHENYLAZOPHENOL □ 4-PHENYLAZOPHENOL

CONSENSUS REPORTS: IARC Cancer Review: Group 3 IMEMDT 7,56,87; Animal Inadequate Evidence IMEMDT 8,157,75. Reported in EPA TSCA Inventory.

SAFETY PROFILE: A poison by intraperitoneal route. Moderately toxic by ingestion. Questionable carcinogen. When heated to decomposition it emits toxic fumes of NO_x.

HJF500   CAS: 3567-69-9   HR: 2
4-HYDROXY-3,4'-AZODI-1-NAPHTH-
ALENESULFONIC ACID, DISODIUM
SALT

mf: $C_{20}H_{12}N_2O_7S_2 \cdot 2Na$ mw: 502.44

PROP: Dark red crystals or powder. Sol in H_2O; mod sol in EtOH; insol in Me_2CO.

SYNS: ACETACID RED B □ ACID BRILLIANT RUBINE 2G □ ACID CHROME BLUE BA □ ACID FAST RED FB □ ACID RUBINE □ AIREDALE CARMOISINE □ AMACID CHROME BLUE R □ ATUL CRYSTAL RED F □ AZORUBIN □ BRASILAN AZO RUBINE 2NS □ BRILLIANT CRIMSON RED □ CARMOISIN (GERMAN) □ CARMOISINE ALUMINUM LAKE □ CARMOISINE SUPRA □ CERTICOL CARMOISINE S □ CHROME FAST BLUE 2R □ C.I. 14720 □ C.I. ACID RED 14, DISODIUM SALT □ C.I. FOOD RED 3 □ CRIMSON EMBL □ DIADEM CHROME BLUE R □ DISODIUM SALT of 2-(4-SULPHO-1-NAPHTHYLAZO)-1-NAPHTHOL-4-SULPHONIC ACID □ DISODIUM-2-(4-SULFO-1-NAPHTHYLAZO)-1-NAPHTHOL-4-SULFONATE □ DISODIUM-2-(4-SULPHO-1-NAPHTHYLAZO)-1-NAPHTHOL-4-SULPHONATE □ EDICOL SUPRA CARMOISINE WS □ ENIACID BRILLIANT RUBINE 3B □ EUROCERT AZORUBINE □ EXTRACT D&C RED No. 10 □ FENAZO RED C □ FOOD RED 5 □ FRUIT RED A EXTRA YELLOWISH GEIGY □ HEXACOL CARMOISINE □ HIDACID AZO RUBINE □ 4-HYDROXY-3,4'-AZODI-1-NAPHTHALENESULPHONIC ACID, DISODIUM SALT □ 4-HYDROXY-3-((4-SULFO-1-NAPHTHALENYL)AZO)-1-NAPHTHALENESULFONIC ACID, DISODIUM SALT □ JAVA RUBINE N □ KARMESIN □ KENACHROME BLUE 2R □ KITON CRIMSON 2R □ LIGHTHOUSE CHROME BLUE 2R □ NACARAT A EXPORT □ NCI-C53849 □ NEKLACID RUBINE W □ NYLOMINE ACID RED P4B □ OMEGA CHROME BLUE FB □ POLOXAL RED 2B □ PONTACYL RUBINE R □ RED #14 □ 11959 RED □ SCHULTZ Nr. 208 (GERMAN) □ SOLAR RUBINE □ SOLOCHROME BLUE FB □ STANDACOL CARMOISINE □ 2-(4-SULFO-1-NAPHTHYLAZO)-1-NAPHTHOL-4-SULFONIC ACID, DISODIUM SALT □ TERTRACID RED CA □ TERTROCHROME BLUE FB

CONSENSUS REPORTS: IARC Cancer Review: Group 3 IMEMDT 7,56,87; Animal Inadequate Evidence IMEMDT 8,83,75. NTP Carcinogenesis Bioassay (feed); No Evidence: mouse, rat NTPTR* NTP-TR-220,82. Reported in EPA TSCA Inventory. EPA Genetic Toxicology Program.

SAFETY PROFILE: Moderately toxic by intraperitoneal and intravenous routes. Questionable carcinogen. Mutation data reported. When heated to decomposition it emits very toxic fumes of SO_x, Na_2O, and NO_x.

HJH500   CAS: 1333-39-7   HR: 3
HYDROXYBENZENESULFONIC ACID

DOT: UN 1803

mf: $C_6H_6O_4S$ mw: 174.18

SYNS: PHENOLSULFONIC ACID □ PHENOLSULFONIC ACID, liquid (DOT) □ SULFOCARBOLIC ACID

CONSENSUS REPORTS: Reported in EPA TSCA Inventory.

DOT CLASSIFICATION: 8; Label: Corrosive

SAFETY PROFILE: Poison by subcutaneous and intraperitoneal routes. Moderately toxic by ingestion. When heated to decomposition it emits toxic vapors of SO_x.

HJI100   CAS: 99-06-9   HR: 2
3-HYDROXYBENZOIC ACID

mf: $C_7H_6O_3$ mw: 138.13

PROP: Needles from H_2O. Mp: 202°.

SYNS: ACIDO-m-IDROSSIBENZOICO (ITALIAN) □ 3-CARBOXYPHENOL □ m-HBA □ m-HYDROXYBENZOIC ACID □ m-SALICYLIC ACID

CONSENSUS REPORTS: Reported in EPA TSCA Inventory.

SAFETY PROFILE: Moderately toxic by ingestion and intraperitoneal routes. An experimental teratogen. When heated to decomposition it emits acrid smoke and fumes.

HJL000   CAS: 120-47-8   HR: 2
p-HYDROXYBENZOIC ACID ETHYL
ESTER

mf: $C_9H_{10}O_3$ mw: 166.19

PROP: Crystals. Mp: 116°, bp: 297–298° (decomp).

SYNS: ASEPTOFORM E □ BONOMOLD OE □ p-CAR-BETHOXYPHENOL □ EASEPTOL □ ETHYL-p-HYDRO-XYBENZOATE □ ETHYL PARABEN □ ETHYL PARA-SEPT □ p-HYDROXYBENZOIC ETHYL ESTER □ NIP-AGIN A □ NIPAZIN A □ p-OXYBENZOESAEUREAETH-YLESTER (GERMAN) □ SOLBROL A □ TEGOSEPT E

CONSENSUS REPORTS: Reported in EPA TSCA Inventory.

SAFETY PROFILE: Moderately toxic by ingestion and intraperitoneal routes. Experimental teratogenic effects. Mutation data reported. When heated to decomposition it emits acrid smoke and fumes. See also ESTERS.

HJL500   CAS: 99-76-3   HR: 2
p-HYDROXYBENZOIC ACID METHYL ESTER

mf: $C_8H_8O_3$ mw: 152.16

PROP: Colorless crystals or white crystalline powder; faint odor and burning taste. Mp: 131°, bp: 270–280° (decomp). Sol in alc, ether, and propylene glycol; sltly sol in water, glycerin, fixed oils, benzene, and carbon tetrachloride.

SYNS: ABIOL □ ASEPTOFORM □ MASEPTOL □ METHYLBEN □ METHYL CHEMOSEPT □ METHYL ESTER of p-HYDROXYBENZOIC ACID □ METHYL p-HYDROXYBENZOATE □ METHYL p-OXYBENZOATE □ METHYLPARABEN (FCC) □ METHYL PARAHYDRO-XYBENZOATE □ METHYL PARASEPT □ METOXYDE □ MOLDEX □ NIPAGIN □ p-OXYBENZOESAEURE-METHYLESTER (GERMAN) □ PARABEN □ PARASEPT □ PARIDOL □ PRESERVAL M □ SEPTOS □ SOLBROL M □ TEGOSEPT M

CONSENSUS REPORTS: Reported in EPA TSCA Inventory.

SAFETY PROFILE: Moderately toxic by ingestion, subcutaneous, and intraperitoneal routes. Mutation data reported. When heated to decomposition it emits acrid smoke and fumes. See also ESTERS.

HJQ350   CAS: 3817-11-6   HR: 3
4-HYDROXYBUTYLBUTYLNITROS-AMINE

mf: $C_8H_{18}N_2O_2$ mw: 174.28

SYNS: BBN □ BBNOH □ BHBN □ BUTANOL (4)-BUTYL-NITROSAMINE □ BUTYL-BUTANOL(4)-NITROSAMIN □ BUTYL-BUTANOL-NITROSAMINE □ N-BUTYL-N-(4-HYDROXYBUTYL)NITROSAMINE □ n-BUTYL-(4-HYDROXYBUTYL)NITROSAMINE □ 4-(BUTYLNITROSAMINO)-1-BUTANOL □ 4-(n-BUTYLNITROSAMINO)-1-BUTANOL □ DIBUTYL-AMINE, 4-HYDROXY-N-NITROSO- □ HBBN □ NBHA □ N-NITROSO-n-BUTYL-(4-HYDROXYBUTYL)AMINE □ OH-BBN

CONSENSUS REPORTS: NTP 10th Report on Carcinogens. IARC Cancer Review: Animal Sufficient Evidence IMEMDT 17,51,78.

SAFETY PROFILE: Confirmed carcinogen with experimental carcinogenic and neoplastigenic data. Moderately toxic by ingestion. Mutation data reported. When heated to decomposition it emits toxic fumes of NO_x.

HJS500   CAS: 502-85-2   HR: 2
4-HYDROXYBUTYRIC ACID SODIUM SALT

mf: $C_4H_7O_3 \cdot Na$ mw: 126.10

PROP: Crystals from EtOH. Mp: 145–146°.

SYNS: GAMMA OH □ γ-HYDROXYBUTYRATE SODIUM SALT □ SODIUM-γ-HYDROXYBUTYRATE □ SODIUM-4-HYDROXYBUTYRATE □ SODIUM OXYBATE □ SOMSANIT □ WY-3478

CONSENSUS REPORTS: Reported in EPA TSCA Inventory.

SAFETY PROFILE: Moderately toxic by intraperitoneal route. Mildly toxic by ingestion. Experimental reproductive effects. Human systemic effects: coma, distorted perceptions, hallucinations, nausea or vomiting. When heated to decomposition it emits toxic fumes of Na_2O.

HJS850   CAS: 1083-57-4   HR: 2
3-HYDROXY-p-BUTYROPHENETIDIDE

mf: $C_{12}H_{17}NO_3$ mw: 223.30

SYNS: BETADID □ BUCETIN □ BUTANAMIDE, N-(4-ETHOXYPHENYL)-3-HYDROXY- □ BUTYRANILIDE, 4'-ETHOXY-3-HYDROXY- □ 4'-ETHOXY-3-HYDROXY-BUTYRANILIDE □ β-HYDROXYBUTYRIC ACID-p-PHENETIDIDE □ β-OXYBUTTERSAEURE-p-PHENETIDID

CONSENSUS REPORTS: Reported in EPA TSCA Inventory.

SAFETY PROFILE: Moderately toxic by ingestion and intraperitoneal routes. Questionable carcinogen with experimental carcinogenic and neoplastigenic data. Mutation data reported. When heated to decomposition it emits toxic fumes of NO_x.

HJV700   HR: 2
HYDROXYCITRONELLAL DIMETHYL ACETAL

mf: $C_{12}H_{26}O_3$ mw: 218.34

PROP: Colorless liquid; floral odor. D: 0.925, refr index: 1.441, flash p: 212°F. Sol in fixed oils, propylene glycol; insol in glycerin.

SYNS: FEMA No. 2585 □ 7-HYDROXY-3,7-DIMETHYL OCTANAL:ACETAL

SAFETY PROFILE: Combustible liquid. When heated to decomposition it emits acrid smoke and irritating fumes.

HKA123   CAS: 193551-21-2 HR: 3
trans-1-(4-HYDROXYCYCLOHEXYL)-4-(4-FLUOROPHENYL)-5-(2-METHOXY-PYRIMIDIN-4-YL)IMIDAZOLE

mf: $C_{20}H_{21}FN_4O_2$ mw: 368.41

SYN: CYCLOHEXANOL, 4-(4-(4-FLUOROPHENYL)-5-(2-METHOXY-4-PYRIMIDINYL)-1H-IMIDAZOL-1-YL)-, trans-

SAFETY PROFILE: A poison by ingestion. When heated to decomposition it emits toxic vapors of NO_x and F^-.

HKA300   CAS: 25316-40-9   HR: 3
HYDROXYDAUNORUBICIN HYDRO-CHLORIDE

mf: $C_{27}H_{29}NO_{11} \cdot ClH$ mw: 580.03

PROP: Orange-red needles. Mp: 204–205° (decomp).

SYNS: ADM HYDROCHLORIDE □ ADR □ ADRIACIN □ ADRIAMYCIN □ ADRIAMYCIN, HYDROCHLORIDE □ ADRIBLASTIN □ ADRIBLASTINE □ DOX HYDRO-CHLORIDE □ DOXORUBICIN □ DOXORUBICIN HY-DROCHLORIDE □ FI 106 □ FI 6804

CONSENSUS REPORTS: NTP 10th Report on Carcinogens. EPA Genetic Toxicology Program.

SAFETY PROFILE: Poison by subcutaneous, intramuscular, intravenous, and intraperitoneal routes. Moderately toxic by ingestion. Experimental reproductive

effects. Human systemic effects: changes in kidney tubules, cardiomyopathy, acute pulmonary edema. Questionable carcinogen with experimental tumorigenic data. Mutation data reported. An antineoplastic and immunosuppressive agent. When heated to decomposition it emits toxic fumes of NO_x and HCl.

HKC000   CAS: 75-60-5   HR: 3
HYDROXYDIMETHYLARSINE OXIDE

DOT: UN 1572

mf: $C_2H_7AsO_2$ mw: 138.01

PROP: Colorless, odorless crystals from MeOH, 2-propanol, and EtOH. Mp: 192°. Very sol in H_2O; sol in EtOH; insol in Et_2O.

SYNS: ACIDE CACODYLIQUE (FRENCH) □ ACIDE DIMETHYLARSINIQUE (FRENCH) □ AGENT BLUE □ ANSAR □ ARSAN □ ARSINIC ACID, DIMETHYL-(9CI) □ BOLLS-EYE □ CACODYLIC ACID (DOT) □ CHEXMATE □ COTTON AIDE HC □ DILIC □ DIMETHYLARSENIC ACID □ DIMETHYLARSINIC ACID □ DMAA □ ERASE □ KYSELINA KAKODYLOVA □ MONCIDE □ MONTAR □ PHYTAR □ PHYTAR 138 □ PHYTAR 560 □ PHYTAR 600 □ RAD-E-CATE 25 □ RCRA WASTE NUMBER U136 □ SALVO □ SILVISAR 510

CONSENSUS REPORTS: IARC Cancer Review: Animal Inadequate Evidence IMEMDT 23,39,80. Arsenic and its compounds are on the Community Right-To-Know List. Reported in EPA TSCA Inventory. EPA Genetic Toxicology Program.

OSHA PEL: TWA 0.5 mg(As)/m³

ACGIH TLV: BEI: 35 μ (As)/L inorganic arsenic and methylated metabolites in urine

DOT CLASSIFICATION: 6.1; Label: Poison

SAFETY PROFILE: Poison by an unspecified route. Moderately toxic by ingestion and intraperitoneal routes. Experimental teratogenic and reproductive effects. A skin and eye irritant. Questionable carcinogen with experimental tumorigenic data. Mutation data reported. Used as an herbicide, defoliant, and silvicide. Hazardous when water solution is in contact with active metals, e.g., Fe, Al, Zn. When heated to decomposition it emits toxic fumes of As.

See also ARSINE and ARSENIC COMPOUNDS.

HKC500   CAS: 124-65-2   HR: 3
HYDROXYDIMETHYLARSINE OXIDE, SODIUM SALT
DOT: UN 1688
mf: $C_2H_6AsO_2•Na$ mw: 159.99
SYNS: ALKARSODYL □ ANSAR 160 □ ANSAR 560 □ ARSECODILE □ ARSINIC ACID, DIMETHYL-, SODIUM SALT (9CI) □ ARSYCODILE □ BOLLS-EYE □ CACODYLATE de SODIUM (FRENCH) □ CACODYLIC ACID SODIUM SALT □ CHEMAID □ DIMETHYLARSINAT SODNY □ ((DIMETHYLARSINO)OXY)SODIUM-As-OXIDE □ DUTCH-TREAT □ HYDROXYDIMETHYL-ARSINE OXIDE, SODIUM SALT □ KAKODYLAN DODNY □ PHYTAR 560 □ RAD-E-CATE □ RAD-E-CATE 16 □ RAD-E-CATE 25 □ RAD-E-CATE 35 □ SILVISAR □ SODIUM CACODYLATE (DOT) □ SODIUM DIMETHYL-ARSINATE □ SODIUM DIMETHYLARSONATE □ SODIUM SALT of CACODYLIC ACID
CONSENSUS REPORTS: Arsenic and its compounds are on the Community Right-To-Know List. EPA Extremely Hazardous Substances List. Reported in EPA TSCA Inventory.
OSHA PEL: TWA 0.5 mg(As)/m³
ACGIH TLV: BEI: 35 μ (As)/L inorganic arsenic and methylated metabolites in urine
DOT CLASSIFICATION: 6.1; Label: Poison
SAFETY PROFILE: Poison by ingestion. Experimental teratogenic and other reproductive effects. When heated to decomposition it emits toxic fumes of As and Na_2O. See also ARSINE and ARSENIC COMPOUNDS.

HKC600   HR: D
6-HYDROXY-3,7-DIMETHYLOCTANOIC ACID LACTONE
mf: $C_{10}H_{18}O_2$ mw: 170.24
PROP: Colorless solid; maple syrup odor. D: 0.966, refr index: 1.457–1.461. Sol in alc; very sltly sol in water.
SYN: FEMA No. 3355
SAFETY PROFILE: When heated to decomposition it emits acrid smoke and irritating fumes.

HKE600   CAS: 609-99-4   HR: 3
2-HYDROXY-3,5-DINITROBENZOIC ACID
mf: $C_7H_4N_2O_7$ mw: 228.13
SYNS: BENZOIC ACID, 2-HYDROXY-3,5-DINITRO-(9CI) □ 3,5-DINITROSALICYLIC ACID □ SALICYLIC ACID, 3,5-DINITRO-
CONSENSUS REPORTS: Reported in EPA TSCA Inventory.
DOT CLASSIFICATION: Forbidden
SAFETY PROFILE: Poison by ingestion. When heated to decomposition it emits toxic vapors of NO_x.

HKJ000   CAS: 106-11-6   HR: 3
2-(2-HYDROXYETHOXY)ETHYL ESTER STEARIC ACID
mf: $C_{22}H_{44}O_4$ mw: 372.66
SYNS: AQUA CERA □ ATLAS G 2146 □ CERASYNT □ CLINDROL SDG □ DIETHYLENE GLYCOL, MONO-ESTER with STEARIC ACID □ DIETHYLENE GLYCOL MONOSTEARATE □ DIETHYLENE GLYCOL STEAR-ATE □ DIGLYCOL MONOSTEARATE □ DIGLYCOL STEARATE □ EMCOL DS-50 CAD □ GLYCO STEARIN □ NONEX 411 □ PROMUL 5080 □ USAF KE-8
CONSENSUS REPORTS: Reported in EPA TSCA Inventory.
SAFETY PROFILE: A poison by intraperitoneal route. When heated to decomposition it emits acrid smoke and fumes. See also ESTERS.

HKS780   CAS: 2809-21-4   HR: 3
1-HYDROXYETHYLIDENE-1,1-DIPHOS-PHONIC ACID
mf: $C_2H_8O_7P_2$ mw: 206.03
PROP: Syrup or crystals from AcOH aq. Mp: 105°. Sol in H_2O, EtOH, and MeOH.
SYNS: DEQUEST 2010 □ DEQUEST 2015 □ DEQUEST Z 010 □ EHDP □ ETHANE-1-HYDROXY-1,1-DIPHO-SPHONATE □ 1,1,1-ETHANETRIOL DIPHOSPHONATE □ ETIDRONIC ACID □ FERROFOS 510 □ HEDP □ 1-HYDROXY-1,1-DIPHOSPHONOETHANE □ HYDROXY-ETHANEDIPHOSPHONIC ACID □ 1-HYDROXY-ETHANEDIPHOSPHONIC ACID □ OXYETHYL-IDENEDIPHOSPHONIC ACID □ PHOSPHONIC ACID, 1-HYDROXY-1,1-ETHANEDIYL ESTER □ PHOSPHONIC ACID, (1-HYDROXYETHYLIDENE)BIS- □ 1000SL □ TURPINAL SL
CONSENSUS REPORTS: Reported in EPA TSCA Inventory.

SAFETY PROFILE: Moderately toxic by ingestion. When heated above 200° it decomposes violently to produce toxic fumes of phosphine, phosphoric acid, and PO_x.

HKW500 CAS: 13743-07-2 HR: 3
1-(2-HYDROXYETHYL)-1-NITROSO-UREA

mf: $C_3H_7N_3O_3$ mw: 133.13

SYNS: HENU □ HNU □ N-NITROSOHYDROXY-ETHYLUREA □ NITROSO-2-HYDROXYETHYLUREA □ 1-NITROSO-1-(2-HYDROXYETHYL)UREA

CONSENSUS REPORTS: EPA Genetic Toxicology Program.

SAFETY PROFILE: Suspected carcinogen with experimental carcinogenic and tumorigenic data. A poison by intraperitoneal route. Experimental reproductive effects. Mutation data reported. When heated to decomposition it emits toxic fumes of NO_x. See also N-NITROSO COMPOUNDS.

HKY650 CAS: 69521-64-8 HR: 3
(2-HYDROXYETHYL)-2-PROPENE-NITRILE

mf: C_5H_7NO mw: 97.13

SYNS: 2-ACRYLONITRILE, 1-HYDROXYETHYL- □ 2-(1-HYDROXYETHYL)ACRYLONITRILE □ 2-PROPENE-NITRILE, (2-HYDROXYETHYL)-

SAFETY PROFILE: A poison by ingestion and skin contact. A severe eye irritant. When heated to decomposition it emits toxic vapors of NO_x.

HLB400 CAS: 9005-27-0 HR: 1
HYDROXYETHYL STARCH

SYNS: ESSEX 1360 □ ESSEX GUM 1360 □ ETHYLEX GUM 2020 □ HAS (GERMAN) □ HES □ HESPANDER □ HESPANDER INJECTION □ HYDROXYATHYLSTARKE (GERMAN) □ o-(HYDROXYETHYL)STARCH □ 2-HYDROXYETHYL STARCH □ o-(2-HYDROXYETHYL)-STARCH □ 2-HYDROXYETHYL STARCH ETHER □ PENFORD 260 □ PENFORD 280 □ PENFORD 290 □ PENFORD P 208 □ PLASMASTERIL □ STARCH HYDROXYETHYL ETHER □ TAPIOCA STARCH HYDROXYETHYL ETHER

CONSENSUS REPORTS: Reported in EPA TSCA Inventory.

SAFETY PROFILE: Mildly toxic by intravenous route. Experimental teratogenic and reproductive effects. When heated to decomposition it emits acrid smoke and fumes.

HLB500 CAS: 27375-52-6 HR: 1
4'-(2-HYDROXYETHYLSULFONYL)-ACETANILIDE

mf: $C_{10}H_{13}NO_4S$ mw: 243.30

SYNS: p-ACETAMINOFENYL-β-HYDROXYETHYL-SULFON □ p-ACETAMINOFENYL-2-HYDROXY-ETHYLSULFON □ ACETANILIDE, 4'-(2-HYDROXY-ETHYLSULFONYL)-

CONSENSUS REPORTS: Reported in EPA TSCA Inventory.

SAFETY PROFILE: An eye irritant. When heated to decomposition it emits toxic fumes of NO_x and SO_x.

HLK600 CAS: 39552-01-7 HR: 3
7-(2-HYDROXY-3-(ISOPROPYLAMINO)-PROPOXY)-2-BENZOFURANYL METHYL KETONE

mf: $C_{16}H_{21}NO_4$ mw: 291.38

SYNS: 2-ACETYL-7-(2-HYROXY-3-ISOPROPYL-AMINOPROPOXY)BENZOFURAN □ BEFUNOLOL □ ETHANONE, 1-(7-(2-HYDROXY-3-((1-METHYLETHYL)-AMINO)PROPOXY)-2-BENZOFURANYL)-(9CI) □ 1-(7-(2-HYDROXY-3-((1-METHYLETHYL)AMINO)PROPOXY)-2-BENZOFURANYL)ETHANONE □ KETONE, 7-(2-HYDROXY-3-(ISOPROPYLAMINO)PROPOXY)-2-BENZO-FURANYL METHYL

DOT CLASSIFICATION: 3; Label: Flammable Liquid

SAFETY PROFILE: A poison by intravenous route. A flammable liquid. When heated to decomposition it emits toxic vapors of NO_x.

HLM500 CAS: 7803-49-8 HR: 3
HYDROXYLAMINE

mf: H_3NO mw: 33.04

PROP: Colorless or white, thermally unstable, hygroscopic, liquid needles. Decomp rapidly at room temp. Mp: 34.0°, bp: 110.0°, flash p: explodes at 265°F, d: 1.227, vap press: 10 mm @ 47.2°. Decomp in hot water. Very sol in liquid ammonia, water, and methanol; Sol in acids; sltly sol in ether, benzene, carbon disulfide, and chloroform.

SYN: OXAMMONIUM

CONSENSUS REPORTS: Reported in EPA TSCA Inventory. EPA Genetic Toxicology Program.

SAFETY PROFILE: A poison by intraperitoneal and subcutaneous routes. A corrosive irritant to the eyes, skin, and mucous membranes. Locally it is irritating, and systemically it can cause methemoglobinemia. Human mutation data reported. Dangerous fire hazard when exposed to heat, flame, and oxidizers. May ignite spontaneously in air if a large surface area is exposed (e.g., precipitate on paper). Explodes in air when heated above 70°C. Explosive reaction with potassium dichromate, chromium trioxide, powdered zinc + heat. Forms the heat-sensitive explosive bis(hydroxylamide) in reaction with zinc or calcium. Ignites on contact with copper(II) sulfate, metals (e.g., sodium), oxidants (e.g., barium peroxide, barium oxide, lead dioxide, potassium permanganate, chlorine), phosphorus chlorides (e.g., phosphorus trichloride, phosphorus pentachloride). Incompatible with carbonyls, pyridine. Vigorous reaction with hypochlorites. When heated to decomposition it emits toxic fumes of NO_x. See also AMINES.

HLN700   HR: D
HYDROXYLATED LECITHIN

PROP: Light yellow liquid to paste; characteristic odor. Moderatelly sol in water.

SAFETY PROFILE: When heated to decomposition it emits acrid smoke and irritating fumes.

HLO400   CAS: 61792-05-0   HR: 3
HYDROXYMERCURI-o-NITROPHENOL

mf: $C_6H_5HgNO_4$ mw: 355.71

PROP: IDLH 10 mg/m³ (as Hg).

SYNS: HYDROXY(4-HYDROXY-3-NITROPHENYL)-MERCURY □ MERCURY, HYDROXY(4-HYDROXY-3-NITROPHENYL)-

ACGIH TLV: TWA 0.1 mg(Hg)/m³ (skin); BEI: 35 µg/g creatinine total inorganic mercury in urine preshift; 15 µg/g creatinine total inorganic mercury in blood at end of shift at end of workweek.

DFG MAK: Confirmed Animal Carcinogen with Unknown Relevance to Humans

NIOSH REL: (Mercury, Aryl and Inorganic) CL 0.1 mg/m³ (skin)

SAFETY PROFILE: Poison by intraperitoneal route. When heated to decomposition it emits toxic fumes of NO_x and Hg.

HLS500   CAS: 4756-45-0   HR: 3
o-(2-HYDROXY-4-METHOXYBENZOYL)-
BENZOIC ACID

mf: $C_{15}H_{12}O_5$ mw: 272.27

SYN: 2'-CARBOXY-2-HYDROXY-4-METHOXY-BENZOPHENONE(o-(2-HYDROXY-p-ANISOYL)BENZ-OIC ACID)

CONSENSUS REPORTS: Reported in EPA TSCA Inventory.

SAFETY PROFILE: A poison by intravenous route. When heated to decomposition it emits acrid smoke and fumes.

HLU500   CAS: 924-42-5   HR: 2
N-(HYDROXYMETHYL)ACRYLAMIDE

mf: $C_4H_7NO_2$ mw: 101.12

PROP: Crystals. Mp: 74–75°.

SYNS: N-(HYDROXYMETHYL)-2-PROPENAMIDE □ N-METHANOLACRYLAMIDE □ N-METHYLOLACRYL-AMIDE □ MONOMETHYLOLACRYLAMIDE □ NCI-C60333 □ URAMINE T 80

CONSENSUS REPORTS: Reported in EPA TSCA Inventory.

SAFETY PROFILE: Moderately toxic by ingestion and intraperitoneal routes. Experimental reproductive effects. May undergo spontaneous combustion in storage. Mutation data reported. When heated to decomposition it emits toxic fumes of NO_x.

HLX925   CAS: 78246-54-5   HR: 3
4-(HYDROXYMETHYL)BENZENEDI-
AZONIUM TETRAFLUOROBORATE
mf: $C_7H_7N_2O \cdot BF_4$ mw: 221.97
SYNS: BENZENEDIAZONIUM, 4-(HYDROXYMETH-
YL)-, TETRAFLUOROBORATE(1-) □ HMBD
SAFETY PROFILE: Suspected carcinogen
with experimental carcinogenic data.
Mutation data reported. When heated to
decomposition it emits toxic fumes of NO_x,
B, and F^-.

HMB500   CAS: 80-71-7   HR: 2
2-HYDROXY-3-METHYL-2-CYCLOPENT-
EN-1-ONE
mf: $C_6H_8O_2$ mw: 112.14
PROP: White crystalline powder; nutty odor,
maple odor in dilute solutions. Flash p:
212°F. Sol in alc, propylene glycol; sltly sol
in fixed oils, water.
SYNS: CORYLON □ CORYLONE □ CYCLOTEN □
FEMA No. 2700 □ MAPLE LACTONE □ 3-METHYL-
CYCLOPENTANE-1,2-DIONE □ METHYL CYCLO-
PENTENOLONE (FCC)
CONSENSUS REPORTS: Reported in EPA
TSCA Inventory.
SAFETY PROFILE: Moderately toxic by
ingestion and intraperitoneal routes. Human
mutation data reported. Combustible liquid.
When heated to decomposition it emits
acrid smoke and fumes.

HMF000   CAS: 568-75-2   HR: 3
7-HYDROXYMETHYL-12-METHYL-
BENZ(a)ANTHRACENE
mf: $C_{20}H_{16}O$ mw: 272.36
SYNS: 7-HM-12-MBA □ 12-
METHYBENZ(a)ANTHRACENE-7-METHANOL □ 7-
OHM-MBA □ 7-OHM-12-MBA
CONSENSUS REPORTS: EPA Genetic
Toxicology Program.
SAFETY PROFILE: Poison by intravenous
route. Experimental teratogenic and
reproductive effects. Questionable
carcinogen with experimental carcinogenic,
neoplastigenic, and tumorigenic data.
Mutation data reported. When heated to
decomposition it emits acrid smoke and
fumes.

HMK100   CAS: 90-01-7   HR: D
2-HYDROXYMETHYLPHENOL
mf: $C_7H_8O_2$ mw: 124.15
PROP: Needles or plates from H_2O or
Et_2O. D: 1.16 @ 25°, mp: 87°.
SYNS: BENZENEMETHANOL, 2-HYDROXY- (9CI) □
BENZYL ALCOHOL, o-HYDROXY- □ DIATHESIN □ α-
2-DIHYDROXYTOLUENE □ 2-HYDROXYBENZENE-
METHANOL □ o-HYDROXYBENZYL ALCOHOL □ 2-
HYDROXYBENZYL ALCOHOL □ o-(HYDROXYMETH-
YL)PHENOL □ o-METHYLOLPHENOL □ 2-METHYL-
OLPHENOL □ SAL □ SALICYL ALCOHOL □
SALIGENIN □ SALIGENOL
CONSENSUS REPORTS: Reported in EPA
TSCA Inventory.
SAFETY PROFILE: An experimental
teratogen. Experimental reproductive
effects. When heated to decomposition it
emits acrid smoke and irritating fumes.

HMK200   CAS: 13911-65-4   HR: 3
HYDROXYMETHYLPHENYLARSINE
OXIDE
mf: $C_7H_9AsO_2$ mw: 200.08
PROP: White needles from $CHCl_3$. Mp:
176°, bp: 151–152° @ 0.1 mm.
SYNS: ARSENIC ACID, METHYLPHENYL-(9CI) □
ARSINE OXIDE, HYDROXYMETHYLPHENYL- □
METHYLPHENYLARSINIC ACID □ METHYL-
PHENYLARSONIC ACID
OSHA PEL: TWA 0.5 mg(As)/m³
SAFETY PROFILE: Poison by intravenous
route. When heated to decomposition it
emits toxic fumes of As.

HMP100   CAS: 118-29-6   HR: D
N-(HYDROXYMETHYL)PHTHALIMIDE
mf: $C_9H_7NO_3$ mw: 177.17
PROP: Crystals. Mp: 142–145°.
SYNS: HYDROXYMETHYLPHTHALIMIDE □ 1H-
ISOINDOLE-1,3(2H)-DIONE, 2-(HYDROXYMETHYL)- □
N-METHYLOLPHTHALIMIDE □ OXYMETHYLPHTH-
ALIMIDE □ PHTHALIMIDE, N-(HYDROXYMETHYL)- □
PHTHALIMIDOMETHYL ALCOHOL
CONSENSUS REPORTS: Reported in EPA
TSCA Inventory.
SAFETY PROFILE: An experimental
teratogen. When heated to decomposition it
emits toxic fumes of NO_x.

HMR550   CAS: 17773-41-0   HR: 3
2-HYDROXY-4-(METHYLTHIO)BUTANE-
NITRILE
mf: C_5H_9NOS mw: 131.21
SYNS: CP 4517 □ BUTANENITRILE, 2-HYDROXY-4-
(METHYLTHIO)- □ METHYLTHIOPROPIONIC
CYANOHYDRIN □ MHA NITRILE
SAFETY PROFILE: A poison by ingestion
and skin contact. Low toxicity by ingestion.
inhalation. A moderate skin irritant. When
heated to decomposition it emits toxic
vapors of NO_x and SO_x.

HMY000   CAS: 121-19-7   HR: 3
4-HYDROXY-3-NITROBENZENE-
ARSONIC ACID
mf: $C_6H_6AsNO_6$ mw: 263.05
PROP: Pale-yellow needles or plates from
H_2O. Sltly sol in hot H_2O; insol in Et_2O and
EtOAc; very sol in MeOH, EtOH, and
Me_2CO.
SYNS: AKLOMIX-3 □ 4-HYDROXY-3-NITROPHENYL-
ARSONIC ACID □ NCI-C56508 □ 3N4HPA □ NITRO
ACID 100 percent □ 2-NITRO-1-HYDROXYBENZENE-4-
ARSONIC ACID □ 3-NITRO-4-HYDROXYBENZENE-
ARSONIC ACID □ 3-NITRO-4-HYDROXYPHENYL-
ARSONIC ACID □ NITROPHENOLARSONIC ACID □
NSC-2101 □ REN O-SAL □ RISTAT □ ROXARSONE
(USDA)
CONSENSUS REPORTS: NTP
Carcinogenesis Studies (Feed); Equivocal
Evidence rat NTPTR* NTP-TR-345,89;
(Feed); No Evidence mouse NTPTR* NTP-
TR-345,89. Arsenic and its compounds are
on the Community Right-To-Know List.
Reported in EPA TSCA Inventory.
OSHA PEL: TWA 0.5 mg(As)/m³
ACGIH TLV: BEI: 35 μ (As)/L inorganic
arsenic and methylated metabolites in urine
SAFETY PROFILE: Poison by ingestion and
intraperitoneal routes. Mutation data
reported. Questionable carcinogen. When
heated to decomposition it emits very toxic
fumes of NO_x and As. See also ARSENIC
COMPOUNDS.

HNK575   CAS: 99071-30-4   HR: 3
(4-HYDROXY-m-PHENYLENE)BIS-
(ACETATOMERCURY)
mf: $C_{10}H_{10}Hg_2O_5$ mw: 611.38
SYNS: DIACETOXYMERCURIPHENOL □ MERCURY,
(4-HYDROXY-m-PHENYLENE)BIS(ACETATO)- □
PHENOL, 2,4-BIS(ACETOXYMERCURI)-
ACGIH TLV: TWA 0.1 mg(Hg)/m³ (skin);
BEI: 35 μg/g creatinine total inorganic
mercury in urine preshift; 15 μg/g creatinine
total inorganic mercury in blood at end of
shift at end of workweek.
DFG MAK: Confirmed Animal Carcinogen
with Unknown Relevance to Humans
SAFETY PROFILE: Poison by
intraperitoneal route. When heated to
decomposition it emits toxic fumes of Hg.

HNK950   CAS: 64058-44-2   HR: 3
4-(m-HYDROXYPHENYL)-1-METHYL-
ISONIPECOTINOYL METHYL
KETONE
mf: $C_{14}H_{19}NO_2$ mw: 233.34
SYN: METHYL (4-(m-HYDROXYPHENYL)-1-METHYL)-
4-PIPERIDYL KETONE
DOT CLASSIFICATION: 3; Label:
Flammable Liquid
SAFETY PROFILE: A poison by
intravenous route. A flammable liquid.
When heated to decomposition it emits
toxic vapors of NO_x.

HNT600   CAS: 999-61-1   HR: 3
2-HYDROXYPROPYL ACRYLATE
mf: $C_6H_{10}O_3$ mw: 130.16
SYNS: ACRYLIC ACID-2-HYDROXYPROPYL ESTER □
β-HYDROXYPROPYL ACRYLATE □ 1,2-PROPANEDIOL-
1-ACRYLATE □ 2-PROPENOIC ACID-2-HYDROXY-
PROPYL ESTER □ PROPYLENE GLYCOL
MONOACRYLATE
OSHA PEL: TWA 0.5 ppm (skin)
ACGIH TLV: TWA 0.5 ppm (sensitizer)
SAFETY PROFILE: Poison by ingestion and
subcutaneous routes. See also ESTERS.
When heated to decomposition it emits
acrid smoke and fumes.

HNU500   CAS: 94-13-3   HR: 3
p-HYDROXYPROPYL BENZOATE
mf: $C_{10}H_{12}O_3$ mw: 180.22
PROP: Colorless crystals or white powder.
Mp: 96–97°. Sltly sol in water; sol in alc,
ether.

SYNS: ASEPTOFORM P □ BETACIDE P □ BONOMOLD OP □ 4-HYDROXYBENZOIC ACID PROPYL ESTER □ p-HYDROXYBENZOIC ACID PROPYL ESTER □ NIPASOL □ p-OXYBENZOESAEUREPRO-PYLESTER (GERMAN) □ PARABEN □ PARASEPT □ PASEPTOL □ PRESERVAL P □ PROPYL p-HYDROXY-BENZOATE □ n-PROPYL p-HYDROXYBENZOATE □ PROPYLPARABEN (FCC) □ PROPYLPARASEPT □ PROTABEN P □ TEGOSEPT P

CONSENSUS REPORTS: Reported in EPA TSCA Inventory. EPA Genetic Toxicology Program.

SAFETY PROFILE: Poison by intraperitoneal route. Moderately toxic by subcutaneous route. Mildly toxic by ingestion. An allergen. When heated to decomposition it emits acrid smoke and fumes.

HNV000   CAS: 9004-64-2   HR: 3
HYDROXYPROPYL CELLULOSE

PROP: White powder. Sol in water and org solvs.

SYNS: HYDROXYPROPYL ETHER of CELLULOSE □ KLUCEL

CONSENSUS REPORTS: Reported in EPA TSCA Inventory.

SAFETY PROFILE: A poison by intravenous route. Slightly toxic by ingestion. When heated to decomposition it emits acrid smoke and fumes.

HNX000   CAS: 9004-65-3   HR: 1
HYDROXYPROPYL METHYLCELLUL-OSE

PROP: White fibrous or granular powder. Sol in water, org solvs; insol in anhyd alc, ether, and chloroform.

SYN: METHOCEL HG

CONSENSUS REPORTS: Reported in EPA TSCA Inventory.

SAFETY PROFILE: Mildly toxic by intraperitoneal route. When heated to decomposition it emits acrid smoke and fumes.

HNX500   CAS: 61499-28-3   HR: 3
1-(((2-HYDROXYPROPYL)NITROSO)-AMINO)ACETONE

mf: $C_6H_{12}N_2O_3$ mw: 160.20

SYNS: HPOP □ 1-((2-HYDROXYPROPYL)NITROSOAMINO)-2-PROPANONE □ N-NITROSO(2-HYDROXYPROPYL)(2-OXOPROPYL)-AMINE

SAFETY PROFILE: Suspected carcinogen with experimental carcinogenic, neoplastigenic, and tumorigenic data. A poison by subcutaneous route. Mutation data reported. When heated to decomposition it emits toxic fumes of NO_x. See also NITROSAMINES.

HNX600   CAS: 21905-32-8   HR: 3
2-HYDROXYPROPYL PHENYL ARSINIC ACID

mf: $C_9H_{13}AsO_3$ mw: 244.14

SYNS: ARSINE OXIDE, HYDROXY(2-HYDROXYPROPYL)PHENYL- □ HYDROXY(2-HYDROXYPROPYL)PHENYLARSINE OXIDE

OSHA PEL: TWA 0.5 mg(As)/m³

SAFETY PROFILE: Poison by intravenous route. When heated to decomposition it emits toxic fumes of As.

HOA575   CAS: 114-03-4   HR: 2
dl-HYDROXYTRYPTOPHAN

mf: $C_{11}H_{12}N_2O_3$ mw: 220.25

PROP: Rods or needles from EtOH (aq). Decomposes @ 298–300°. Mod sol in water; sol in 50% boiling alc.

SYNS: 5-HYDROXYTRYPTOPHAN □ (±)-5-HYDROXY-TRYPTOPHAN □ dl-5-HYDROXYTRYPTOPHAN □ PRETONINE

CONSENSUS REPORTS: Reported in EPA TSCA Inventory.

SAFETY PROFILE: Moderately toxic by intraperitoneal route. An experimental teratogen. When heated to decomposition it emits toxic fumes of NO_x.

HOA600   CAS: 4350-09-8   HR: 3
5-HYDROXY-l-TRYPTOPHAN

mf: $C_{11}H_{12}N_2O_3$ mw: 220.25

PROP: Pale-pink needles. Mp: 273° (decomp).

SYNS: l-5-HTP □ l-5-HYDROXYTRYPTOPHAN □ l-TRYPTOPHAN, 5-HYDROXY-, (9CI)

CONSENSUS REPORTS: Reported in EPA TSCA Inventory.

SAFETY PROFILE: Poison by ingestion, intraperitoneal, subcutaneous, and intravenous routes. An experimental teratogen. Other experimental reproductive effects. When heated to decomposition it emits toxic fumes of NO_x.

HOC000   CAS: 15922-78-8   HR: 3
1-HYDROXY-2-(1H)-PYRIDINETHIONE SODIUM SALT

mf: $C_5H_5NOS \cdot Na$ mw: 150.16

SYNS: OMACIDE 24 □ SODIUM OMADINE □ SODIUM PYRIDINETHIONE □ SODIUM PYRITHIONE SQ 3277 □ SQ 3277

DFG MAK: 1 mg/m³

SAFETY PROFILE: Poison by intraperitoneal and intravenous routes. Moderately toxic by ingestion and subcutaneous routes. When heated to decomposition it emits very toxic fumes of Na_2O, NO_x, and SO_x.

HOE600   CAS: 89-86-1   HR: D
4-HYDROXYSALICYLIC ACID

mf: $C_7H_6O_4$ mw: 154.13

PROP: Crystals from H_2O. Mp: 218–219°.

SYNS: BENZOIC ACID, 2,4-DIHYDROXY- (9CI) □ 4-CARBOXYRESORCINOL □ 2,4-DHBA □ 2,4-DIHY-DROXYBENZOIC ACID □ p-HYDROXYSALICYLIC ACID □ β-RESORCINOLIC ACID □ β-RESORCYLIC ACID

CONSENSUS REPORTS: Reported in EPA TSCA Inventory.

SAFETY PROFILE: An experimental teratogen. When heated to decomposition it emits acrid smoke and irritating fumes.

HOF000   CAS: 26782-43-4   HR: 3
HYDROXYSENKIRKINE

mf: $C_{19}H_{27}NO_7$ mw: 381.47

PROP: Isolated from the plant *Crotalaria laburnifolia*.

SYN: 8,12,18-TRIHYDROXY-4-METHYL-11,16-DIOXOSENECIONANIUM

CONSENSUS REPORTS: IARC Cancer Review: Group 3 IMEMDT 7,56,87; Animal Inadequate Evidence IMEMDT 10,265,76.

SAFETY PROFILE: Poison by ingestion. Questionable carcinogen with experimental tumorigenic data. When heated to decomposition it emits toxic fumes of NO_x.

HOH500   CAS: 79-57-2   HR: 3
5-HYDROXYTETRACYCLINE

mf: $C_{22}H_{24}N_2O_9$ mw: 460.48

PROP: Light-yellow crystals or needles from MeOH (aq). Mp: 181–182° (decomp) (hydrate).

SYNS: ADAMYCIN □ ANTIBIOTIC TM 25 □ BERKMYCEN □ BIOSTAT □ BIOSTAT PA □ DABICYCLINE □ FANTERRIN □ GEOMYCIN □ LENOCYCLINE □ LIQUAMYCIN LA 200 □ MACOCYN □ MYCOSHIELD TMQTHC 20 □ NCI-C56473 □ OKSISYKLIN □ OTC □ OXITETRACYCLIN □ OXYMYCIN □ OXYMYKOIN □ OXYTERRACIN □ OXYTERRACINE □ OXYTERRACYNE □ OXYTETRACYCLINE □ OXYTETRACYCLINE AMPHOTERIC □ RIOMITSIN □ RYOMYCIN □ TAOMYCIN □ TAOMYXIN □ TERRAFUNGINE □ TERRAMITSIN □ TERRAMYCIN □ TETRACYCLINE, 5-HYDROXY- □ TETRAN

CONSENSUS REPORTS: Reported in EPA TSCA Inventory.

SAFETY PROFILE: A poison by intravenous route. Moderately toxic by ingestion, subcutaneous, and intraperitoneal routes. Human systemic effects by ingestion: hemorrhage, dermatitis, and unspecified effects on teeth and supporting structures. Human reproductive effects by an unspecified route: abnormal postnatal measures or effects. Experimental teratogenic and reproductive effects. Mutation data reported. When heated to decomposition it emits toxic fumes of NO_x. See also TETRACYCLINE and various tetracycline derivatives.

HOI000   CAS: 2058-46-0   HR: 3
5-HYDROXYTETRACYCLINE HYDRO-CHLORIDE

mf: $C_{22}H_{24}N_2O_9 \cdot ClH$ mw: 496.94

PROP: Needles from MeOH; yellow platelets from H_2O.

SYNS: BISOLVOMYCIN □ HYDROCYCLIN □ LIQUAMYCIN INJECTABLE □ NSC-9169 □ OTETRYN □ OXLOPAR □ OXYJECT 100 □ OXYTETRACYCLINE HYDROCHLORIDE □ TERAMYCIN HYDROCHLORIDE □ TETRAMINE □ TETRAN HYDROCHLORIDE

CONSENSUS REPORTS: NTP Carcinogenesis Studies (feed); Equivocal Evidence: rat NTPTR* NTP-TR-315,87. NTP Carcinogenesis Studies (feed); No Evidence: mouse NTPTR* NTP-TR-315,87. Reported in EPA TSCA Inventory.

SAFETY PROFILE: Poison by intravenous route. Moderately toxic by subcutaneous route. Mildly toxic by ingestion. Experimental teratogenic and reproductive effects. Questionable carcinogen with experimental tumorigenic data. Mutation data reported. When heated to decomposition it emits very toxic fumes of HCl and NO$_x$. See also TETRACYCLINE and various tetracycline derivatives.

HOI245   CAS: 90035-14-6   HR: 3
4-HYDROXY-3-(1,2,3,4-TETRAHYDRO-3-(4-(4-(TRIFLUOROMETHYL)PHEN-OXY) PHENYL)-1-NAPHTHALEN-YL)2H-1-BENZOPYRAN-2-ONE

mf: C$_{32}$H$_{23}$F$_3$O$_4$ mw: 528.55

SAFETY PROFILE: A poison by ingestion. When heated to decomposition it emits toxic vapors of F$^-$.

HOJ150   CAS: 3569-58-2   HR: 3
HYDROXYTHIOSPASMIN

mf: C$_{18}$H$_{27}$O$_3$S•I   mw: 450.41

SYNS: OXYSONIUM IODIDE □ SULFONIUM, (2-((CYCLOHEXYLHYDROXYPHENYLACETYL)OXY)ETHYL)DIMETHYL-, IODIDE □ SULFONIUM, (2-HYDROXY-ETHYL)DIMETHYL-, IODIDE, α-PHENYLCYCLO-HEXANEGLYCOLATE

SAFETY PROFILE: A poison by ingestion and intravenous routes. When heated to decomposition it emits toxic vapors of SO$_x$ and I$^-$.

HON000   CAS: 76-87-9   HR: 3
HYDROXYTRIPHENYLSTANNANE

mf: C$_{18}$H$_{16}$OSn   mw: 367.03

PROP: White powder. Mp: 116–120° (decomp). Mod sol in most org solvs; very sltly sol in H$_2$O.

SYNS: DOWCO 186 □ DU-TER □ ENT 28,009 □ FENOLOVO □ FENTIN HYDROXIDE □ FINTINE HYDROXYDE (FRENCH) □ FINTIN HYDROXID (GERMAN) □ FINTIN HYDROXYDE (DUTCH) □

FINTIN IDROSSIDO (ITALIAN) □ HAITIN □ HYDROXYDE de TRIPHENYL-ETAIN (FRENCH) □ HYDROXYTRIPHENYLTIN □ IDROSSIDO DI STAGNO TRIFENILE (ITALIAN) □ NCI-C00260 □ SUZU H □ TPTH □ TRIFENYL-TINHYDROXYDE (DUTCH) □ TRIPHENYLTIN HYDROXIDE (USDA) □ TRIPHENYLTIN OXIDE □ TRIPHENYL-ZINNHYDROXID (GERMAN) □ TUBOTIN □ VANCIDE KS

CONSENSUS REPORTS: NCI Carcinogenesis Bioassay (feed); No Evidence: mouse, rat NCITR* NCI-CG-TR-139,78. Reported in EPA TSCA Inventory.

OSHA PEL: TWA 0.1 mg(Sn)/m^3 (skin)

ACGIH TLV: TWA 0.1 mg(Sn)/m^3; STEL 0.2 mg(Sn)/m^3 (skin).

NIOSH REL: (Organotin Compounds) TWA 0.1 mg(Sn)/m^3

SAFETY PROFILE: A poison by ingestion and intraperitoneal routes. Moderately toxic by an unspecified route. A severe eye irritant. Experimental teratogenic and reproductive effects. Mutation data reported. When heated to decomposition it emits acrid smoke and fumes. See also TIN COMPOUNDS.

HON800   CAS: 114-03-4   HR: 3
dl-HYDROXYTRYPTOPHAN

mf: C$_{11}$H$_{12}$N$_2$O$_3$ mw: 220.25

SYNS: 5-HYDROXYTRYPTOPHAN □ (±)-5-HYDROXY-TRYPTOPHAN

CONSENSUS REPORTS: Reported in EPA TSCA Inventory.

SAFETY PROFILE: Poison by intraperitoneal route. An experimental teratogen. When heated to decomposition it emits toxic fumes of NO$_x$.

HOO000   CAS: 4350-09-8   HR: 3
5-HYDROXY-I-TRYPTOPHAN

mf: C$_{11}$H$_{12}$N$_2$O$_3$ mw: 220.25

SYNS: l-5-HTP □ l-5-HYDROXYTRYPTOPHAN

CONSENSUS REPORTS: Reported in EPA TSCA Inventory.

SAFETY PROFILE: Poison by ingestion, subcutaneous, intravenous, and intraperitoneal routes. Experimental teratogenic and reproductive effects. When heated to decomposition it emits toxic fumes of NO_x.

HOO100   CAS: 56-69-9   HR: 3
5-HYDROXYTRYPTOPHANE
mf: $C_{11}H_{12}N_2O_3$ mw: 220.25

PROP: dl form: minute rods or needles, decomp @ 298–300°; l form: crystals; d form: crystals.

SYNS: 5-HTP □ HYDROXYTRYPTOPHAN □ 5-HYDROXYTRYPTOPHAN □ NCI-C56644 □ USAF CB-96

CONSENSUS REPORTS: Reported in EPA TSCA Inventory.

SAFETY PROFILE: Poison by intraperitoneal route. Experimental reproductive effects. When heated to decomposition it emits toxic fumes of NO_x.

HOP259   CAS: 16870-90-9   HR: 3
7-HYDROXYXANTHINE
mf: $C_5H_4N_4O_3$ mw: 168.13

SYN: XANTHINE-7-N-OXIDE

SAFETY PROFILE: Poison by intraperitoneal route. Questionable carcinogen with experimental carcinogenic and neoplastigenic data. When heated to decomposition it emits toxic fumes of NO_x. See also other hydroxyxanthine entries.

HOT500   CAS: 114-49-8   HR: 3
HYOSCINE HYDROBROMIDE
mf: $C_{17}H_{21}NO_4$•BrH   mw: 384.31

PROP: A solid. Mp: 193–194° (anhyd).

SYNS: BELDAVRIN □ EUSCOPOL □ HYDROSCINE HYDROBROMIDE □ HYOSCINE BROMIDE □ HYO-SCINE F HYDROBROMIDE □ (−)-HYOSCINE HYDRO-BROMIDE □ 1-HYOSCINE HYDROBROMIDE □ HYOSCYINE HYDROBROMIDE □ HYSCO □ ISOSCOPIL □ KWELLS □ SCOPAMIN □ SCOPOLAMINE BROMIDE □ (−)-SCOPOLAMINE BROMIDE □ SCOPOLAMINE HYDROBROMIDE □ (−)-SCOPOLAMINE HYDROBROM-IDE □ SCOPOLAMINIUM BROMIDE □ SCOPOLAM-MONIUM BROMIDE □ SCOPOS □ SEREEN □ TRIPTONE

CONSENSUS REPORTS: Reported in EPA TSCA Inventory. EPA Genetic Toxicology Program.

SAFETY PROFILE: A poison by intravenous and intramuscular routes. Moderately toxic by ingestion,

subcutaneous, intraduodenal, and intraperitoneal routes. Experimental teratogenic and reproductive effects. Human mutation data reported. When heated to decomposition it emits very toxic fumes of NO_x and HBr. See also SCOPOLAMINE.

HOU000   CAS: 101-31-5   HR: 3
(−)-HYOSCYAMINE
mf: $C_{17}H_{23}NO_3$ mw: 289.41

PROP: White, crystalline alkaloid. Mp: 108–111°. Very sol in alc, dil acids.

SYNS: (−)-ATROPINE □ DATURINE □ HYOSCYAMINE □ 1-HYOSCYAMINE □ (−)-TROPIC ACID ESTER with TROPINE

CONSENSUS REPORTS: Reported in EPA TSCA Inventory.

SAFETY PROFILE: A deadly human poison by an unspecified route. An experimental poison by intravenous route. This is one of the atropine alkaloids and is very toxic, acting very much like atropine. It has the same effect on the central nervous system but twice the effect on the peripheral nerves. The symptoms of poisoning are dryness of the throat and mouth, marked difficulty in swallowing, and a sensation of burning and thirst. The vision becomes impaired through dilation and loss of accommodation, and the eyes present a rather prominent, brilliant, staring appearance. The voice is husky and the tongue is red. When heated to decomposition it emits highly toxic fumes of NO_x.

HOU500   HR: 3
HYPOCHLORITES
PROP: Salts of hypochlorous acid.

SAFETY PROFILE: Toxic by ingestion and inhalation. Powerful irritants to the skin, eyes, and mucous membranes. Flammable by chemical reaction with reducing agents. These are powerful oxidizers particularly at higher temperatures, when chlorine and then oxygen are evolved, or in the presence of moisture or carbon dioxide. With urea, they form the highly explosive NCl_3. Dangerous; when heated or on contact with acid or acid fumes, they emit highly toxic fumes of Cl⁻. React with water or steam to produce toxic and corrosive fumes of Cl⁻ and HCl. See also HYPOCHLOROUS ACID for more reactivity information.

HOV000   CAS: 7790-92-3   HR: 3
HYPOCHLOROUS ACID
mf: ClHO   mw: 52.46
PROP: Greenish-yellow liquid (aq soln). Decomp to Cl_2, O_2, and $HClO_4$; very weak acid. Strong oxidizing agent. Protect from light. HOCl solns decomp to Cl_2 + O_2 + some ClO_3 slowly. Forms a dihydrate. Can be stored only in aq soln. Sol in H_2O.
CONSENSUS REPORTS: Reported in EPA TSCA Inventory.
SAFETY PROFILE: An experimental teratogen. Explodes on contact with ammonia. Ignites on contact with arsenic. Mixture with acetic anhydride is a sensitive explosive. Incompatible with alcohols. When heated to decomposition it emits toxic fumes of Cl⁻. See also HYPOCHLORITES.

HOV500   CAS: 7778-54-3   HR: 3
HYPOCHLOROUS ACID, CALCIUM SALT
DOT: UN 1748
mf: $Cl_2O_2 \cdot Ca$ mw: 142.98
PROP: White powder. Compound contains 39% or less available chlorine (FEREAC 41,15972,76). Disproportionates in aq soln forming $CaCl_2$ and $Ca(ClO_3)_2$. Decomp on heating to $CaCl_2$ + O_2. At high temps the reaction becomes explosive. Mp: 100°. Very sol in H_2O; insol in EtOH.
SYNS: B-K POWDER □ BLEACHING POWDER □ BLEACHING POWDER, containing 39% or less chlorine (DOT) □ CALCIUM CHLOROHYDROCHLORITE □ CALCIUM HYPOCHLORIDE □ CALCIUM HYPOCHLORITE □ CALCIUM OXYCHLORIDE □ CAPORIT □ CCH □ CHLORIDE of LIME (DOT) □ CHLORINATED LIME (DOT) □ HTH □ HY-CHLOR □ LIME CHLORIDE □ LO-BAX □ LOSANTIN □ PERCHLORON □ PITTCHLOR □ PITTCIDE □ PITTCLOR □ SENTRY

CONSENSUS REPORTS: Reported in EPA TSCA Inventory.
DOT CLASSIFICATION: 5.1; Label: Oxidizer
SAFETY PROFILE: Moderately toxic by ingestion. Can cause severe irritation of skin and mucous membranes and emit fumes capable of causing pulmonary edema. Mutation data reported. A powerful oxidizer.

The bulk material may ignite or explode in storage. Traces of water may initiate the reaction. A rapid exothermic decomposition above 175°C releases oxygen and chlorine. Moderately explosive in its solid form when heated. Explosive reaction with acetic acid + potassium cyanide, amines, ammonium chloride, carbon or charcoal + heat, carbon tetrachloride + heat, N,N-dichloromethylamine + heat, ethanol, methanol, iron oxide, rust, 1-propanethiol, isobutanethiol, turpentine. Potentially explosive reaction with sodium hydrogen sulfate + starch + sodium carbonate. Reaction with acetylene or nitrogenous bases forms explosive products.

Ignites on contact with algicide, hydroxy compounds (e.g., glycerol, diethylene glycol monomethyl ether, phenol), organic sulfur compounds. Violent reaction with organic matter (above 100°C), sulfur. Vigorous reaction with nitromethane, reducing materials. Flammable by chemical reaction with combustible materials, e.g., anthracene, grease, oil, mercaptans, methyl carbitol, nitromethane, organic matter, propylmercaptan.

Deflagration occurs in contact with combustible substances. Dangerous; when heated to decomposition or on contact with acid or acid fumes, it emits highly toxic fumes of HCl and explodes. Reacts with water or steam to produce toxic and corrosive fumes of Cl⁻ and HCl.

HOW500   CAS: 14448-38-5   HR: 3
HYPONITROUS ACID
mf: $H_2N_2O_2$ mw: 62.03

PROP: White deliquescent plates. Very sol in EtOH; mod sol in Et_2O, $CHCl_3$, and C_6H_6.

SYN: N-NITROSOHYDROXYLAMINE

DOT CLASSIFICATION: Forbidden

SAFETY PROFILE: Many N-nitroso compounds are carcinogens. Incompatible with potassium hydroxide. When heated to decomposition it emits toxic fumes of NO_x. See also N-NITROSO COMPOUNDS.

I

IAN000 **CAS: 5034-77-5** **HR: 3**
IMIDAZOLE MUSTARD
mf: $C_8H_{12}Cl_2N_6O$ mw: 279.16
SYNS: BIC □ 5-(3,3-BIS(2-CHLOROETHYL)-1-
TRIAZENO)IMIDAZOLE-4-CARBOXAMIDE □ NCI-
C01616 □ NSC-82196 □ SRI 2489 □ TIC MUSTARD
CONSENSUS REPORTS: NCI
Carcinogenesis Studies (ipr); Clear Evidence:
mouse, rat CANCAR 40,1935,77.
SAFETY PROFILE: Suspected carcinogen
with experimental neoplastigenic and
tumorigenic data. Poison by ingestion and
intraperitoneal routes. Experimental
teratogenic effects. Human systemic effects
by intravenous route: nausea. When heated
to decomposition it emits very toxic fumes
of Cl^- and NO_x.

IAQ000 **CAS: 96-45-7** **HR: 3**
2-IMIDAZOLIDINETHIONE
mf: $C_3H_6N_2S$ mw: 102.17
PROP: White crystals, needles, or prisms
from EtOh or pentanol. Mp: 197–200°.
Water solubility: 9 g/100 mL @ 30°. Often
occurs as a main degradation product of the
metal salts of ethylene bis-dithiocarbamic
acid. Sol in H_2O; insol in Et_2O, $CHCl_3$,
C_6H_6, and Me_2CO.
SYNS: AKROCHEM ETU-22 □ 4,5-DIHYDROIMIDAZ-
OLE-2(3H)-THIONE □ ETHYLENE THIOUREA □ N,N'-
ETHYLENETHIOUREA □ 1,3-ETHYLENE-2-THIOUREA
□ l'ETHYLENE THIOUREE (FRENCH) □ ETU □ 2-
MERCAPTOIMIDAZOLINE □ 2-MERKAPTOIMID-
AZOLIN (CZECH) □ NA-22 □ NCI-C03372 □ PENNAC
CRA □ RCRA WASTE NUMBER U116 □ RODANIN S-62
(CZECH) □ SODIUM-22 NEOPRENE ACCELERATOR □
2-THIOL-DIHYDROGLYOXALINE □ USAF EL-62 □
VULKACIT NPV/C2 □ WARECURE C
CONSENSUS REPORTS: NTP 10th Report
on Carcinogens. IARC Cancer Review:
Group 2B IMEMDT 7,207,87; Animal
Sufficient Evidence IMEMDT 7,45,74.
Community Right-To-Know List. EPA

Genetic Toxicology Program. Reported in
EPA TSCA Inventory.
DFG MAK: Confirmed Animal Carcinogen
with Unknown Relevance to Humans
NIOSH REL: (ETU) Use encapsulated form;
minimize exposure
SAFETY PROFILE: Confirmed carcinogen
with experimental carcinogenic data. Poison
by ingestion and intraperitoneal routes.
Experimental teratogenic and reproductive
effects. Mutation data reported. An eye
irritant. When heated to decomposition it
emits very toxic fumes of NO_x and SO_x.

IAR000 **HR: 3**
2-IMIDAZOLIDINETHIONE mixed with
    SODIUM NITRITE
SYNS: ETHYLENETHIOUREA mixed with SODIUM
NITRITE □ SODIUM NITRITE mixed with ETHYLENE-
THIOUREA
SAFETY PROFILE: Suspected carcinogen.
2-Imidazolidinethione and sodium nitrite are
experimental carcinogens. Experimental
teratogenic and reproductive data. Sodium
nitrite is a poison. Mutation data reported.
When heated to decomposition it emits very
toxic fumes of SO_x, Na_2O, and NO_x. See
also SODIUM NITRITE and 2-
IMIDAZOLIDINETHIONE.

IAT275 **CAS: 159081-23-9** **HR: 3**
2-(1H-IMIDAZOL-4-YLMETHYL)-8H-
    INDENO(1,2-D)THIAZOLE MONO-
    FUMARATE
mf: $C_{14}H_{11}N_3S•C_4H_4O_4$ mw: 369.40
SYNS: 8H-INDENO(1,2-D)THIAZOLE, 2-(1H-
IMIDAZOL-4-YLMETHYL)-, (E)-2-BUTENEDIOATE (1:1)
□ YM-31636
SAFETY PROFILE: A poison by ingestion
and subcutaneous routes. When heated to
decomposition it emits toxic vapors of NO_x.

IBA000   CAS: 2465-27-2   HR: 3
4,4'-(IMIDOCARBONYL)BIS(N,N-DIMETHYLAMINE) MONOHYDRO-CHLORIDE
mf: $C_{17}H_{21}N_3 \cdot ClH \cdot H_2O$ mw: 321.89
PROP: Golden-yellow plates from H_2O.
Mp: 267°. Sltly sol in cold H_2O.
SYNS: ADC AURAMINE O □ AIZEN AURAMINE □ AURAMINE (MAK) □ AURAMINE HYDROCHLORIDE □ AURAMINE O (BIOLOGICAL STAIN) □ AURAMINE YELLOW □ 4,4'-BIS(DIMETHYLAMINO)BENZHYDRYLIDENIMINE HYDROCHLORIDE □ 4,4'-BIS(DIMETHYLAMINO)BENZOPHENONE-IMINE HYDROCHLORIDE □ 1,1-BIS(p-DIMETHYLAMINO-PHENYL)METHYLENIMINEHYDROCHLORIDE □ CALCOZINE YELLOW OX □ 4,4'-CARBONIMIDOYL-BIS(N,N-DIMETHYLBENZENAMINE)MONOHYDRO-CHLORIDE □ C.I. 41000 □ C.I. BASIC YELLOW 2 □ C.I. BASIC YELLOW 2, MONOHYDROCHLORIDE □ MITSUI AURAMINE O
CONSENSUS REPORTS: Reported in EPA TSCA Inventory. EPA Genetic Toxicology Program.
DFG MAK: Animal Carcinogen, Suspected Human Carcinogen
SAFETY PROFILE: Confirmed carcinogen with experimental neoplastigenic and tumorigenic data. Poison by skin contact, ingestion, and intraperitoneal routes. Human mutation data reported. A chelating agent that might disturb trace element metabolism if taken into the body. Used as a biological stain. When heated to decomposition it emits very toxic fumes of NO_x and HCl.

IBB000   CAS: 492-80-8   HR: 3
4,4'-(IMIDOCARBONYL)BIS(N,N-DIMETHYLANILINE)
mf: $C_{17}H_{21}N_3$ mw: 267.41
PROP: Yellow needles or yellow plates from EtOH. Mp: 136°. Insol in water; sol in EtOH, Et_2O, and acids; sltly sol in H_2O.
SYNS: APYONINE AURAMINE BASE □ AURAMINE (MAK) □ AURAMINE BASE □ BIS(p-DIMETHYL-AMINOPHENYL)METHYLENEIMINE □ BRILLIANT OIL YELLOW □ 4,4'-CARBONIMIDOYLBIS(N,N-DIMETHYLBENZENAMINE) □ C.I. 41000B □ C.I. BASIC YELLOW 2, FREE BASE □ C.I. SOLVENT YELLOW 34 □ 4,4'-DIMETHYLAMINOBENZOPHENONIMIDE □ GLAURAMINE □ RCRA WASTE NUMBER U014 □ TETRAMETHYLDIAMINODIPHENYLACETIMINE □ WAXOLINE YELLOW O □ YELLOW PYOCTANINE

CONSENSUS REPORTS: IARC Cancer Review: Group 2B IMEMDT 7,118,87; Human Sufficient Evidence IMEMDT 1,69,72; Animal Sufficient Evidence IMEMDT 1,69,72. Community Right-To-Know List. Reported in EPA TSCA Inventory.
DFG MAK: Animal Carcinogen, Suspected Human Carcinogen
SAFETY PROFILE: Confirmed human carcinogen with experimental carcinogenic, neoplastigenic, and tumorigenic data. Poison by intraperitoneal route. Human mutation data reported. Used as an antiseptic. When heated to decomposition it emits toxic fumes of NO_x.

IBS000   CAS: 606-23-5   HR: 3
1,3-INDANDIONE
mf: $C_9H_6O_2$ mw: 146.15
PROP: Crystals or liquid; needles from ligroin. Mp: 129–131° decomp. Very sltly sol in cold water; sol in hot alc, benzene.
SYNS: 1,3-DIKETOHYDRINDENE □ 1H-INDENE-1,3(2H)-DIONE
CONSENSUS REPORTS: Reported in EPA TSCA Inventory.
SAFETY PROFILE: A poison by intraperitoneal route. Experimental teratogenic and reproductive effects. When heated to decomposition it emits acrid smoke and fumes.

IBX000   CAS: 95-13-6   HR: 2
INDENE
mf: C_9H_8 mw: 116.17
PROP: Liquid from coal tars. D: 0.9968 @ 20°/4°, mp: −1.8°, bp: 181.6°. Water-insol, but misc in org solvs.
SYN: INDONAPHTHENE
CONSENSUS REPORTS: Reported in EPA TSCA Inventory.
OSHA PEL: TWA 10 ppm
ACGIH TLV: TWA 10 ppm; (Proposed: 5 ppm)
NIOSH REL: (Indene) TWA 10 ppm
SAFETY PROFILE: Low toxicity by ingestion, inhalation, and possibly other

routes. Irritating to skin, eyes, and mucous membranes. It has exploded during nitration with (H_2SO_4 + HNO_3). When heated to decomposition it emits acrid smoke and fumes.

IBZ000   CAS: 193-39-5   HR: 3
INDENO(1,2,3-cd)PYRENE
mf: $C_{22}H_{12}$ mw: 276.34

PROP: Yellow crystals from cyclohexane; bright-yellow plates from pet ether/C_6H_6. Mp: 161–163.5°.

SYNS: 1,10-(o-PHENYLENE)PYRENE □ 1,10-(1,2-PHENYLENE)PYRENE □ 2,3-PHENYLENEPYRENE □ 2,3-o-PHENYLENEPYRENE □ RCRA WASTE NUMBER U137

CONSENSUS REPORTS: NTP 10th Report on Carcinogens. IARC Cancer Review: Group 2B IMEMDT 7,56,87; Animal Sufficient Evidence IMEMDT 32,373,83; IMEMDT 3,229,73. Reported in EPA TSCA Inventory.

SAFETY PROFILE: Confirmed carcinogen with experimental carcinogenic and tumorigenic data. Mutation data reported. When heated to decomposition it emits acrid smoke and fumes.

ICF000   CAS: 7440-74-6   HR: 3
INDIUM
af: In   aw: 114.82

PROP: Soft, silvery-white, malleable and ductile metal. Liquid wets glass; stable in dry air. Slowly oxidized in moist air. Reacts with halogens, S, Se, Te, As, P on heating. Dissolves in Hg. Not affected by alkalies. Has plastic properties at cryogenic temps. Mp: 156.61°, bp: 2080°, d: 7.31 @ 20°. Insol in H_2O in bulk form; sol in most acids.

CONSENSUS REPORTS: Reported in EPA TSCA Inventory.

OSHA PEL: TWA 0.1 mg(In)/m^3
ACGIH TLV: TWA 0.1 mg(In)/m^3

SAFETY PROFILE: A poison by subcutaneous route. It affects the liver, heart, kidneys, and the blood. Teratogenic effects. Inhalation of indium compounds may cause damage to the respiratory system. Hydrated indium oxide is a poison by

intravenous route. Flammable in the form of dust when exposed to heat or flame. Incandesces. Explosive reaction with dinitrogen tetraoxide + acetonitrile. Violent reaction with mercury(II) bromide at 350°C. Mixtures with sulfur ignite when heated.

ICI000   CAS: 13770-61-1   HR: 2
INDIUM NITRATE
mf: InN_3O_9 mw: 300.85

PROP: Crystal white powder.

CONSENSUS REPORTS: Reported in EPA TSCA Inventory.

SAFETY PROFILE: Experimental teratogenic and reproductive effects. A severe skin irritant. When heated to decomposition it emits toxic fumes of NO_x. See also INDIUM and NITRATES.

ICI300   CAS: 22398-80-7   HR: 3
INDIUM PHOSPHIDE
mf: InP   mw: 145.79

SYN: INDIUM MONOPHOSPHIDE

CONSENSUS REPORTS: Reported in EPA TSCA Inventory.

SAFETY PROFILE: A poison by intratracheal route. Low toxicity by ingestion and intraperitoneal route. Experimental reproductive effects. When heated to decomposition it emits toxic vapors of PO_x and In.

ICJ000   CAS: 13464-82-9   HR: 3
INDIUM SULFATE
mf: $O_{12}S_3 \cdot In_2$ mw: 517.82

PROP: Grayish-white, hygroscopic powder. D: 3.44. Sol in water. Keep well-closed.

SYNS: INDISULFAT (GERMAN) □ SULFURIC ACID, INDIUM SALT

CONSENSUS REPORTS: Reported in EPA TSCA Inventory.

SAFETY PROFILE: A poison by intravenous and subcutaneous routes. Moderately toxic by ingestion. When heated to decomposition it emits toxic fumes of SO_x. See also INDIUM and SULFATES.

ICK000   **CAS: 10025-82-8**   **HR: 3**
INDIUM TRICHLORIDE
mf: Cl₃In   mw: 221.17
PROP: Yellowish, deliquescent crystals. D:
4.0, mp: 586°, sublimes @ 500°, bp: volatile
@ 600°. Very sol in water. Keep tightly
closed.
SYN: INDIUM CHLORIDE
CONSENSUS REPORTS: Reported in EPA
TSCA Inventory.
SAFETY PROFILE: A poison by
subcutaneous, intraperitoneal, and
intravenous routes. Mutation data reported.
When heated to decomposition it emits
toxic fumes of Cl⁻. See also INDIUM.

ICM000   **CAS: 120-72-9**   **HR: 3**
INDOLE
mf: C₈H₇N   mw: 117.16
PROP: Colorless to yellowish scales or
crystals from water. Intense fecal odor. Mp:
52°, bp: 253°; volatile with steam. Sol in hot
water, alc, ether, petroleum ether; insol in
mineral oil, glycerin.
SYNS: 1-AZAINDENE □ 1-BENZAZOLE □
BENZOPYRROLE □ 2,3-BENZOPYRROLE □ FEMA No.
2593 □ INDOL (GERMAN) □ KETOLE
CONSENSUS REPORTS: Reported in EPA
TSCA Inventory.
SAFETY PROFILE: A poison by
intraperitoneal and subcutaneous routes.
Moderately toxic by ingestion and skin
contact. A severe eye irritant. Questionable
carcinogen with experimental carcinogenic
and tumorigenic data. When heated to
decomposition it emits toxic fumes of NOₓ.

ICN000   **CAS: 87-51-4**   **HR: 3**
1H-INDOLE-3-ACETIC ACID
mf: C₁₀H₉NO₂   mw: 175.20
PROP: Colorless leaves from benzene.
Crystals from CHCl₃. Mp: 165–168°. Very
sltly sol in cold water; sol in alc, ether, and
acetic acid; insol in chloroform.
SYNS: HETEROAUXIN □ IAA □ β-INDOLEACETIC
ACID □ β-INDOLE-3-ACETIC ACID □ 3-INDOLE-
ACETIC ACID □ INDOLYLACETIC ACID □ α-INDOL-
3-YL-ACETIC ACID □ β-INDOLYLACETIC ACID □

INDOLYL-3-ACETIC ACID □ 3-INDOLYLACETIC ACID
□ RHIZOPIN □ ω-SKATOLE CARBOXYLIC ACID
CONSENSUS REPORTS: Reported in EPA
TSCA Inventory.
SAFETY PROFILE: A poison by
intraperitoneal route. Questionable
carcinogen with experimental tumorigenic
and teratogenic data. Mutation data
reported. When heated to decomposition it
emits toxic fumes of NOₓ.

ICP000   **CAS: 133-32-4**   **HR: 3**
1H-INDOLE-3-BUTANOIC ACID
mf: C₁₂H₁₃NO₂   mw: 203.26
PROP: White crystals from C₆H₆/pet ether
or powder. Mp: 124°. Sol in acetone and
ether; insol in water and chloroform.
SYNS: HORMEX ROOTING POWDER □ HORMODIN
□ IBA □ INDOLE BUTYRIC □ INDOLE BUTYRIC
ACID □ β-INDOLEBUTYRIC ACID □ γ-(INDOLE-3)-
BUTYRIC ACID □ 3-INDOLEBUTYRIC ACID □ γ-
(INDOL-3-YL)BUTYRIC ACID □ INDOLYL-3-BUTYRIC
ACID □ 3-INDOLYL-γ-BUTYRIC ACID □ γ-(3-
INDOLYL)BUTYRIC ACID □ 4-(INDOLYL)BUTYRIC
ACID □ 4-(INDOL-3-YL)BUTYRIC ACID □ 4-(3-
INDOLYL)BUTYRIC ACID □ JIFFY GROW □ ROO-
TONE
CONSENSUS REPORTS: Reported in EPA
TSCA Inventory.
SAFETY PROFILE: A poison by ingestion
and intraperitoneal routes. Mutation data
reported. Used for promoting and
accelerating root formation of plant
clippings. When heated to decomposition it
emits toxic fumes of NOₓ.

ICR000   **CAS: 91-56-5**   **HR: 3**
INDOLE-2,3-DIONE
mf: C₈H₅NO₂   mw: 147.14
PROP: Orange crystals. Mp: 203.5°
(decomp). Sltly sol in H₂O.
SYNS: o-AMINOBENZOYLFORMIC ANHYDRIDE □
2,3-DIKETOINDOLINE □ 2,3-DIOXOINDOLINE □ 2,3-
INDOLINEDIONE □ ISATIC ACID LACTAM □ ISATIN
□ ISATINIC ACID ANHYDRIDE □ 2,3-KETOINDO-
LINE
CONSENSUS REPORTS: Reported in EPA
TSCA Inventory.
SAFETY PROFILE: Poison by ingestion.
Moderately toxic by intraperitoneal route.
Experimental reproductive effects. When

heated to decomposition it emits toxic fumes of NO_x. See also ANHYDRIDES.

ICS100   CAS: 68527-79-7   HR: 1
INDOLENE
mf: $C_{18}H_{25}NO$ mw: 271.44
SYNS: HYDROXYCITRONELLAL-INDOLE (SCHIFF BASE) □ HYDROXYCITRONELLYLIDENE-INDOLE □ 7-OCTEN-2-OL, 2,6-DIMETHYL-8-(1H-INDOL-1-YL)-
CONSENSUS REPORTS: Reported in EPA TSCA Inventory.
SAFETY PROFILE: A skin irritant. When heated to decomposition it emits toxic fumes of NO_x.

ICW000   CAS: 771-51-7   HR: 3
3-INDOLYLACETONITRILE
mf: $C_{10}H_8N_2$ mw: 156.20
PROP: A solid. Mp: 36.5–37°, bp: 157° @ 0.2 mm.
SYNS: 3-(CYANOMETHYL)INDOLE □ 3-INDOL-ACETONITRILE □ INDOLEACETONITRILE □ INDOLE-3-ACETONITRILE □ 1H-INDOLE-3-ACETONITRILE □ INDOLYLACETONITRILE □ USAF CB-29
CONSENSUS REPORTS: Reported in EPA TSCA Inventory. Cyanide and its compounds are on the Community Right-To-Know List.
SAFETY PROFILE: A poison by intraperitoneal and subcutaneous routes. When heated to decomposition it emits toxic NO_x and CN^-. See also NITRILES.

ICW100   CAS: 73747-53-2   HR: 3
INDOL-1-YL ETHYL KETONE
mf: $C_{11}H_{11}NO$ mw: 173.23
SYNS: INDOLE, 1-PROPIONYL- □ N-PROPIONYLINDOLE
DOT CLASSIFICATION: 3; Label: Flammable Liquid
SAFETY PROFILE: A poison by intravenous route. A flammable liquid. When heated to decomposition it emits toxic vapors of NO_x.

ICW200   CAS: 34559-71-2   HR: 3
2-INDOLYL METHOXYMETHYL KETONE
mf: $C_{11}H_{12}NO_2$ mw: 190.24

SYNS: KETONE, 2-INDOLYL METHOXYMETHYL- □ 2-(METHOXYACETYL)INDOLE
DOT CLASSIFICATION: 3; Label: Flammable Liquid
SAFETY PROFILE: A poison by intraperitoneal route. A flammable liquid. When heated to decomposition it emits toxic vapors of NO_x.

ICY100   CAS: 703-80-0   HR: 3
INDOL-3-YL METHYL KETONE
mf: $C_{10}H_9NO$ mw: 159.20
SYNS: ACETYL-3-INDOLE □ 3-ACETYLINDOLE
DOT CLASSIFICATION: 3; Label: Flammable Liquid
SAFETY PROFILE: A poison by intraperitoneal route. A flammable liquid. When heated to decomposition it emits toxic vapors of NO_x.

ICZ150   CAS: 30256-74-7   HR: 3
INDOLYL-3-MORPHOLINOMETHYL KETONE
mf: $C_{14}H_{16}N_2O_2$ mw: 244.32
SYN: KETONE, 3-INDOLYL MORPHOLINOMETHYL
DOT CLASSIFICATION: 3; Label: Flammable Liquid
SAFETY PROFILE: A poison by intravenous route. A flammable liquid. When heated to decomposition it emits toxic vapors of NO_x.

ICZ200   CAS: 30256-73-6   HR: 3
INDOLYL-3-PIPERIDINOMETHYL KETONE
mf: $C_{15}H_{18}N_2O$ mw: 242.35
SYN: KETONE, 3-INDOLYL PIPERIDINOMETHYL
DOT CLASSIFICATION: 3; Label: Flammable Liquid
SAFETY PROFILE: A poison by intravenous route. A flammable liquid. When heated to decomposition it emits toxic vapors of NO_x.

IDA000   CAS: 53-86-1   HR: 3
INDOMETHACIN
mf: $C_{19}H_{16}ClNO_4$ mw: 357.81
PROP: Crystals from tert-butyl alcohol. One form: mp: 155°; another form: mp: 162°. Sol

in ethanol, ether, acetone, and castor oil; insol in water.

SYNS: AMUNO □ ARTRACIN □ ARTRINOVO □ ARTRIVIA □ N-p-CHLORBENZOYL-5-METHOXY-2-METHYLINDOLE-3-ACETIC ACID □ 1-(p-CHLORO-BENZOYL)-5-METHOXY-2-METHYLINDOLE-3-ACETIC ACID □ 1-(p-CHLOROBENZOYL)-2-METHYL-5-METHOXYINDOLE-3-ACETIC ACID □ 1-(p-CHLORO-BENZOYL)-2-METHYL-5-METHOXY-3-INDOLE-ACETIC ACID □ α-(1-(p-CHLOROBENZOYL)-2-METHYL-5-METHOXY-3-INDOLYL)ACETIC ACID □ 1-p-CLORO-BENZOIL-5-METOXI-2-METILINDOL-3-ACIDO ACETICO (SPANISH) □ CONFORTID □ DOLOVIN □ IDOMETHINE □ IMBRILON □ INACID □ INDOCID □ INDOMECOL □ INDOMED □ INDOMETHAZINE □ INDOMETICINA (SPANISH) □ INDOPTIC □ INDO-RECTOLMIN □ INDO-TABLINEN □ INFLAZON □ INTEBAN SP □ LAUSIT □ METACEN □ METARTRIL □ METHAZINE □ METINDOL □ MEZOLIN □ MIKAMETAN □ MOBILAN □ NCI-C56144 □ REUMACIDE □ SADOREUM □ TANNEX

CONSENSUS REPORTS: Reported in EPA TSCA Inventory.

SAFETY PROFILE: A poison by ingestion, intravenous, intraperitoneal, and subcutaneous routes. Human systemic effects by ingestion: aplastic anemia, changes in kidney tubules, decreased urine volume, diarrhea, fibrous hepatitis, hemorrhage, hypermotility, liver changes, necrotic stomach changes, retinal changes. Human teratogenic effects by ingestion and intravenous routes: developmental abnormalities of the respiratory system and urogenital system, homeostasis, other neonatal effects. Experimental teratogenic and reproductive effects. Mutation data reported. When heated to decomposition it emits very toxic Cl⁻ and NO$_x$.

IDA100   CAS: 74252-25-8   HR: 3
INDOMETHACIN SODIUM TRIHYDRATE

mf: $C_{19}H_{15}ClNO_4 \cdot 3H_2O \cdot Na$ mw: 401.85

SYNS: INDOCIN I.V. □ 1H-INDOLE-3-ACETIC ACID, 1-(4-CHLOROBENZOYL)-5-METHOXY-2-METHYL-, SODIUM SALT, TRIHYDRATE □ SODIUM 1-(p-CHLOROBENZOYL)-5-METHOXY-2-METHYLINDOLE-3-ACETATE TRIHYDRATE

SAFETY PROFILE: A poison by ingestion, subcutaneous, and intravenous routes.

When heated to decomposition it emits toxic vapors of NO$_x$ and Cl⁻.

IDE000   CAS: 58-63-9   HR: 2
INOSINE

mf: $C_{10}H_{12}N_4O_5$ mw: 268.26

PROP: A solid. Mp: 215° (decomp).

SYNS: ATOREL □ HXR □ HYPOXANTHINE NUCLEOSIDE □ HYPOXANTHINE RIBONUCLEOSIDE □ HYPOXANTHINE RIBOSIDE □ HYPOXANTHINE-d-RIBOSIDE □ HYPOXANTHOSINE □ INO □ INOSIE □ β-INOSINE □ OXIAMIN □ PANTHOLIC-L □ RIBONOSINE □ SELFER □ TROPHICARDYL

CONSENSUS REPORTS: Reported in EPA TSCA Inventory.

SAFETY PROFILE: Moderately toxic by intraperitoneal route. Mutation data reported. When heated to decomposition it emits toxic fumes of NO$_x$.

IDE300   CAS: 87-89-8   HR: D
INOSITOL

mf: $C_6H_{12}O_6$ mw: 180.16

PROP: White crystals or crystalline powder; odorless with a sweet taste. Mp: 225°. Sol in water; insol in ether, chloroform.

SYNS: cis-1,2,3,5-trans-4,6-CYCLOHEXANEHEXOL □ i-INOSITOL □ meso-INOSITOL

SAFETY PROFILE: When heated to decomposition it emits acrid smoke and irritating fumes.

IDJ700   HR: 1
IODATES

SAFETY PROFILE: Salts of iodic acid. Variable toxicity. Generally eye, skin, and mucous membrane irritants. Powerful oxidizers. Similar to bromates and chlorates. Contamination of iodates with organic matter may produce explosive mixtures. Iodates are used in bread as an improving agent for the dough. When heated to decomposition they emit toxic fumes of I⁻. See also specific compounds.

IDL000   **HR: 2**
IODIDES
ACGIH TLV: Proposed: (inhalable fraction)
0.1 mg/m³; Not Classifiable as a Human
Carcinogen)
SAFETY PROFILE: Similar in toxicity to
bromides. Prolonged absorption of iodides
may produce "iodism," which is manifested
by skin rash, running nose, headache, and
irritation of mucous membranes. In severe
cases, the skin may show pimples, boils,
redness, black-and-blue spots, hives, and
blisters. Weakness, anemia, loss of weight,
and general depression may occur. Generally
very soluble in water and easily absorbed
into the body. The iodides of copper(I),
lead(II), silver(I), and mercury(II) are poorly
soluble in water. When heated to
decomposition they can emit highly toxic
fumes of I⁻ and iodine compounds. See also
IODINE.

IDL100   **HR: D**
IODINATED CASEIN
SAFETY PROFILE: When heated to
decomposition it emits acrid smoke and
irritating fumes.

IDM000   CAS: 7553-56-2   **HR: 3**
IODINE
mf: I_2 mw: 253.80
PROP: Rhombic, violet-black crystals with
metallic luster; flakes with characteristic
odor, sharp acrid taste. Sublimes slowly at
room temp. Mp: 113.5°, bp: 185.24°, d: 4.93
(solid @ 25°), vap press: 1 mm @ 38.7°, vap
press: (solid): 0.030 mm @ 0°. Sltly sol in
H_2O. Sol in many org solvs. IDLH 2 ppm.
SYNS: IODE (FRENCH) □ IODINE CRYSTALS □
IODINE SUBLIMED □ IODIO (ITALIAN) □ JOD
(GERMAN, POLISH) □ JOOD (DUTCH)
CONSENSUS REPORTS: Reported in EPA
TSCA Inventory.
OSHA PEL: CL 0.1 ppm
ACGIH TLV: CL 0.1 ppm; (Proposed: 0.01
ppm; STEL 0.1 ppm; Not Classifiable as a
Human Carcinogen)
DFG MAK: 0.1 ppm (1.1 mg/m³)

SAFETY PROFILE: A human poison by
ingestion and possibly other routes. An
experimental poison by intravenous and
subcutaneous routes. Moderately toxic by
inhalation. Human systemic effects by
ingestion: diarrhea, evidence of thyroid
hyperfunction. Experimental reproductive
effects. Mutation data reported. The effect
of iodine vapor upon the body is similar to
that of chlorine and bromine, but it is more
irritating to the lungs. Serious exposures are
seldom encountered in industry due to the
low volatility of the solid at ordinary room
temperatures. Signs and symptoms are
irritation and burning of the eyes,
lachrymation, coughing, and irritation of the
nose and throat. Ingestion of large
quantities causes abdominal pain, nausea,
vomiting, diarrhea. In severe cases,
purging, excessive thirst, and circulatory
failure may develop. Doses of 2–3 g have
been fatal. Chronic ingestion of large
amounts (200 mg/day) results in thyroid
disease.

Explosive reaction with acetylene,
antimony powder, hafnium powder + heat,
tetraamine copper(II) sulfate + ethanol,
trioxygen difluoride (possibly ignition),
polyacetylene (at 113°C). Forms sensitive,
explosive mixtures with potassium (impact-
and heat-sensitive), sodium (shock-
sensitive), oxygen difluoride (heat-sensitive).
Reacts to form explosive products with
ammonia, ammonia + lithium 1-heptynide,
ammonia + potassium, butadiene + ethanol
+ mercuric oxide, silver azide.

Ignition on contact with bromine
pentafluoride (or violent reaction), chlorine
trifluoride, fluorine, metals (powdered) +
water, aluminum-titanium alloys + heat,
metal acetylides (e.g., cesium acetylide,
copper(I) acetylide, lithium acetylide,
rubidium acetylide), nonmetals (e.g., boron
ignites at 700°C), phosphorus, sodium
phosphinate. Violent reaction with
acetaldehyde, aluminum + diethyl ether,
dipropylmercury, titanium (above 113°C).
Incandescent reaction with cesium oxide

(above 150°C), bromine trifluoride, metal acetylides or carbides [e.g., barium acetylide (above 122°C), calcium acetylide (above 305°C), strontium acetylide (above 182°C), zirconium acetylide (above 400°C)].

Incompatible with ethanol, ethanol + butadiene, ethanol + phosphorus, ethanol + methanol + HgO, formamide + pyridine + sulfur trioxide, formamide, halogens or interhalogens (e.g., chlorine), mercuric oxide, metals (e.g., aluminum, lithium, magnesium), metal carbides (e.g., lithium carbide, zirconium carbide), oxygen, pyridine, sodium hydride, sulfides.

When heated to decomposition it emits toxic fumes of I^- and various iodine compounds. Reacts vigorously with reducing materials. See also IODIDES.

IDN000   CAS: 14696-82-3   HR: 3
IODINE AZIDE
mf: IN_3 mw: 168.93

PROP: Bright yellow crystals; stable in the dark for several days. Sol in Et_2O, CH_2Cl_2, and MeCN.

SYNS: IODINE AZIDE (dry) (DOT) □ IODINE(I) AZIDE □ IODOAZIDE □ NITROGEN IODIDE

ACGIH TLV: Proposed: (inhalable fraction) 0.1 mg/m³; Not Classifiable as a Human Carcinogen)

DOT CLASSIFICATION: Forbidden

SAFETY PROFILE: A very shock- and friction-sensitive explosive. Incompatible with sulfur-containing alkenes. When heated to decomposition it emits very toxic fumes of I^- and NO_x. See also IODINE.

IDS000   CAS: 7790-99-0   HR: 3
IODINE MONOCHLORIDE
DOT: UN 1792
mf: ClI   mw: 162.38

PROP: Black crystals or reddish-brown liquid. Exists in α, β forms; crystals α form (stable) black needles; sol in water, alc, ether, CS_2, acetic acid. Red-brown crystals or oily liquid. Mp: (α) 27°, (β) 14°, bp: 97.4° (decomp @ 100°), d (α) 3.1822 @ 0°, (β) 3.24 @ 34°.

SYNS: IODINE CHLORIDE □ PROTOCHLORURE d'IODE (FRENCH) □ WIJS' CHLORIDE

CONSENSUS REPORTS: Reported in EPA TSCA Inventory.

DOT CLASSIFICATION: 8; Label: Corrosive

SAFETY PROFILE: A poison by ingestion. Moderately toxic by skin contact. A corrosive irritant to skin, eyes, and mucous membranes. Moderately explosive when exposed to heat. Reacts with water or steam to produce toxic and corrosive fumes. Dangerous reactions with metals e.g., sodium (mixture explodes on impact). potassium (explodes on contact). aluminum (ignition after a delay period). Reacts violently with Al foil. CdS. PbS. organic matter. P. PCl_3. rubber. Ag_2S. ZnS. When heated to decomposition it emits highly toxic fumes of Cl^- and I^- and may explode. See also IODINE and CHLORIDES.

IDT000   CAS: 7783-66-6   HR: 3
IODINE PENTAFLUORIDE
DOT: UN 2495
mf: F_5I mw: 221.90

PROP: Yellow liquid. Mp: 9.43°, bp: 100.5°, d: 3.19 @ 25°. Fumes in air, attacks glass, especially when hot.

SYN: PENTAFLUOROIODINE

CONSENSUS REPORTS: Reported in EPA TSCA Inventory.

OSHA PEL: TWA 2.5 mg(F)/m³

ACGIH TLV: TWA 2.5 mg(F)/m³; BEI: 3 mg/g creatinine of fluorides in urine prior to shift; 10 mg/g creatinine of fluorides in urine at end of shift.

DOT CLASSIFICATION: 5.1; Label: Oxidizer, Poison

NIOSH REL: (Fluorides, Inorganic) TWA 2.5 mg(F)/m³

SAFETY PROFILE: A poison. Probably an irritant to the eyes, skin, and mucous membranes. A powerful oxidizer. Explosive reaction with benzene (above 50°C), diethyl-aminotrimethyl silane, dimethyl sulfoxide, limonene + tetrafluoroethylene (polymerization), potassium, molten sodium, tetraio-doethylene. Reaction with organic compounds results in charring and then ignition. Violent reaction with water, potassium hydroxide. Incandescent reaction with calcium carbide, potassium hydride, metals, and nonmetals (e.g., boron, silicon, red phosphorus, sulfur, arsenic, antimony, bismuth, molybdenum, tungsten). When heated to decomposition it emits very toxic fumes of F^- and I^-. See also IODINE and FLUORIDES.

IDW000   CAS: 144-48-9   HR: 3
IODOACETAMIDE
mf: C_2H_4INO mw: 184.97
PROP: Flaky crystals from H_2O. Mp: 95°.
SYNS: α-IODOACETAMIDE □ 2-IODOACETAMIDE □ MONOIODOACETAMIDE □ SURAUTO □ USAF D-1
CONSENSUS REPORTS: EPA Genetic Toxicology Program. Reported in EPA TSCA Inventory.
SAFETY PROFILE: A poison by ingestion, intraperitoneal, and intravenous routes. Questionable carcinogen with experimental tumorigenic data. Human mutation data reported. When heated to decomposition it emits very toxic fumes of I^- and NO_x.

IDZ000   CAS: 64-69-7   HR: 3
IODOACETIC ACID
mf: $C_2H_3IO_2$ mw: 185.95
PROP: Colorless or white crystals; plates from ligroin. Mp: 82–83°. Sol in water and alc; very sltly sol in ether.
SYNS: IA □ IODOACETATE □ MIA □ MONO-IODOACETATE □ MONOIODOACETIC ACID
CONSENSUS REPORTS: Reported in EPA TSCA Inventory.
SAFETY PROFILE: A poison by ingestion, subcutaneous, and intravenous routes. Experimental teratogenic effects. Questionable carcinogen with experimental neoplastigenic and tumorigenic data. Human mutation data reported. When heated to decomposition it emits toxic fumes of I^-. See also IODINE.

IDZ100   CAS: 18312-12-4   HR: 3
N-(IODOACETYL)-3-AZABICYCLO-

(3.2.2)NONANE
mf: $C_{10}H_{16}INO$ mw: 293.17
SYNS: 3-AZABICYCLO(3.2.2)NONANE, 3-(IODOACETYL)- □ KETONE, 3-AZABICYCLO(3.2.2)NONYL IODOMETHYL
DOT CLASSIFICATION: 3; Label: Flammable Liquid
SAFETY PROFILE: A poison by intravenous route. A flammable liquid. When heated to decomposition it emits toxic vapors of NO_x and I^-.

IDZ200   CAS: 143-86-2   HR: 3
1-IODOACETYL-α-α-DIPHENYL-4-PIPERIDINEMETHANOL
mf: $C_{20}H_{22}INO_2$ mw: 435.33
SYNS: KETONE, 4-(DIPHENYLHYDROXYMETHYL)PIPERIDINO IODOMETHYL □ 4-PIPERIDINEMETHANOL, 1-(IODOACETYL)-α-α-DIPHENYL-
DOT CLASSIFICATION: 3; Label: Flammable Liquid
SAFETY PROFILE: A poison by intraperitoneal route. A flammable liquid. When heated to decomposition it emits toxic vapors of NO_x and I^-.

IEH000   CAS: 513-48-4   HR: 3
2-IODOBUTANE
DOT: UN 2390
mf: C_4H_9I mw: 184.03
PROP: Flash p: 14°F.
SYNS: sec-BUTYL IODIDE □ 2-JODBUTAN
CONSENSUS REPORTS: Reported in EPA TSCA Inventory. EPA Genetic Toxicology Program.
ACGIH TLV: Proposed: (inhalable fraction) 0.1 mg/m³; Not Classifiable as a Human Carcinogen)
DOT CLASSIFICATION: 3; Label: Flammable Liquid
SAFETY PROFILE: Questionable carcinogen with experimental neoplastigenic data. Mutation data reported. A very dangerous fire hazard when exposed to heat or flame; can react vigorously with oxidizing materials. When heated to decomposition it emits toxic fumes of I^-. See also IODIDES.

IEI700   CAS: 302542-42-3 HR: 3
4-IODO-3,5-DIMETHYL-N-(2-METHYL-
PHENYL)-1H-PYRAZOLE-1-ACET-
AMIDE
mf: $C_{14}H_{16}IN_3O$ mw: 369.21
SAFETY PROFILE: A poison by
subcutaneous and ocular routes. Moderately
toxic by ingestion. When heated to
decomposition it emits toxic vapors of NO_x
and I^-.

IEI740   CAS: 302542-63-8 HR: 3
4-IODO-3,5-DIMETHYL-N-(4-METHYL-
PHENYL)-1H-PYRAZOLE-1-
ACETAMIDE
mf: $C_{14}H_{16}IN_3O$ mw: 369.21
SAFETY PROFILE: A poison by
subcutaneous ocular routes. Moderately
toxic by ingestion. When heated to
decomposition it emits toxic vapors of NO_x
and I^-.

IEL800   CAS: 5634-39-9   HR: 3
2-(1-IODOETHYL)-1,3-DIOXOLANE-4-
METHANOL
mf: $C_6H_{11}IO_3$ mw: 258.07
SYNS: 1,3-DIOXOLANE-4-METHANOL, 2-(1-
IODOETHYL)- □ IODINATED GLYCEROL □
IODOPROPYLIDENE GLYCEROL □ NCI-C55469 □
ORGANIDIN
CONSENSUS REPORTS: NTP
Carcinogenesis Studies (gavage) Some
Evidence: mouse, rat NTPTR* NTP-TR-
340,90.
SAFETY PROFILE: Suspected carcinogen
with carcinogenic and neoplastigenic data.
Mutation data reported. Experimental
reproductive effects. When heated to
decomposition it emits toxic vapors of I^-.

IEP000   CAS: 75-47-8   HR: 3
IODOFORM
mf: CHI_3 mw: 393.72
PROP: Yellow powder or crystals; plates
from Me_2CO; disagreeable odor. D: 4.1, mp:
120° (approx), bp: subl. Decomp at high
temp, evolving iodine. Volatile with steam.
Very sol in water, benzene, acetone; sltly sol
in pet ether.

SYNS: NCI-C04568 □ TRIIODOMETHANE
CONSENSUS REPORTS: NCI
Carcinogenesis Bioassay (gavage); No
Evidence: mouse, rat NCITR* NCI-CG-
TR-110,78. Reported in EPA TSCA
Inventory.
OSHA PEL: TWA 0.6 ppm (skin)
ACGIH TLV: TWA 0.6 ppm
SAFETY PROFILE: A poison by ingestion.
Moderately toxic by inhalation, skin contact,
and subcutaneous routes. Mutation data
reported. Used as an antiseptic, disinfectant
on superficial wounds, and in female
reproductive tract. 1:1 mixtures with
hexamethylenetetramine explode at 178°C.
Incompatible with mercuric oxide, calomel,
silver nitrate, tannin, balsam Peru directly
mixed, Li, acetone. When heated to
decomposition it emits toxic fumes of I^-.
See also IODIDES.

IEY000   CAS: 141-76-4   HR: 2
3-IODOPROPIONIC ACID
mf: $C_3H_5IO_2$ mw: 199.98
PROP: (a) Needles from water. D: 1.857,
mp: 93–94°, bp: decomp. Sltly sol in water;
very sol in alc and ether. (b) Needles. Mp:
44.5–45.5°, bp: 105°. Very sol in EtOH,
Et_2O; sol in hot H_2O; very sltly sol in cold
H_2O.
CONSENSUS REPORTS: Reported in EPA
TSCA Inventory.
SAFETY PROFILE: Moderately toxic by
skin contact. Questionable carcinogen with
experimental tumorigenic data. Mutation
data reported. When heated to
decomposition it emits toxic fumes of I^-.

IFW000   CAS: 6901-97-9   HR: 1
α-IONONE
mf: $C_{13}H_{20}O$ mw: 192.33
PROP: Colorless oil; woody, violet odor. D:
0.930, refr index: 1.497–1.502, bp: 136.1. Sol
in alc, fixed oils, propylene glycol; sltly sol in
water; misc in ether; insol in glycerin.
SYNS: α-CYCLOCITRYLIDENEACETONE □ FEMA
No. 2594 □ 4-(2,6,6-TRIMETHYL-2-CYCLOHEXEN-1-YL)-
3-BUTEN-2-ONE

CONSENSUS REPORTS: Reported in EPA TSCA Inventory.

SAFETY PROFILE: Mildly toxic by ingestion. When heated to decomposition it emits acrid smoke and fumes. See also IONONE.

IFX000   CAS: 14901-07-6   HR: 2
β-IONONE

mf: $C_{13}H_{20}O$ mw: 192.33

PROP: Colorless oil; woody odor. D: 0.944, refr index: 1.517–1.522, bp: 150°, flash p: 234°F. Sol in alc, fixed oils, propylene glycol; sltly sol in water; misc in ether; insol in glycerin.

SYNS: β-CYCLOCITRYLIDENEACETONE □ FEMA No. 2595 □ 4-(2,6,6-TRIMETHYL-1-CYCLOHEXEN-1-YL)-3-BUTEN-2-ONE

CONSENSUS REPORTS: Reported in EPA TSCA Inventory.

SAFETY PROFILE: Moderately toxic by intraperitoneal route. Mildly toxic by ingestion. Combustible liquid. When heated to decomposition it emits acrid smoke and fumes. See also IONONE.

IGH000   CAS: 14885-29-1   HR: 2
IPROPRAN

mf: $C_7H_{11}N_3O_2$ mw: 169.21

PROP: Plates. Mp: 62–63°.

SYNS: IPRONIDAZOLE (USDA) □ 2-ISOPROPYL-1-METHYL-5-NITROIMIDAZOLE □ 1-METHYL-2-(1-METHYLETHYL)-5-NITRO-1H-IMIDAZOLE □ RO 7-1554

SAFETY PROFILE: Moderately toxic by ingestion. Mutation data reported. When heated to decomposition it emits toxic fumes of NO_x.

IGJ550   CAS: 136572-09-3 HR: 3
IRINOTECTAN HYDROCHLORIDE HYDRATE

mf: $C_{33}H_{38}N_4O_6 \cdot ClH \cdot 3H_2O$ mw: 677.27

SYN: (1,4'-BIPIPERIDINE)-1'-CARBOXYLIC ACID, 4,11-DIETHYL-3,4,12,14-TETRAHYDRO-4-HYDROXY-3,14-DIOXO-1H-PYRANO(3',4':6,7)INDOLIZINO(1,2-B)QUINOLIN-9-YL ESTER, MONOHYDRO-CHLORIDE,TRIHYDRATE, (S)-

SAFETY PROFILE: A poison by intravenous route. Moderately toxic by

ingestion. When heated to decomposition it emits toxic vapors of NO_x, HCl, and Cl⁻.

IGK800   CAS: 7439-89-6   HR: 3
IRON

af: Fe   aw: 55.85

PROP: Silvery-white metal, relatively soft when pure. Traces of impurities have profound effect on physical props (steels). Rapidly oxidized, especially in damp air (rust). Attacked by dil acids. Passivated by HNO_3. From decomposition of iron pentacarbonyl: dark-gray powder. From electrodeposition: lusterless, gray-black powder. From chemical reduction: gray-black powder. Mp: 1535°, bp: 27° @ 3000 mm.

SYNS: ANCOR EN 80/150 □ ARMCO IRON □ CARBONYL IRON □ EFV 250/400 □ EO 5A □ FERROVAC E □ GS 6 □ IRON, CARBONYL (FCC) □ IRON, ELECTROLYTIC □ IRON, ELEMENTAL □ IRON, REDUCED (FCC) □ LOHA □ NC 100 □ PZh2M □ PZhO □ REMKO □ SUY-B 2 □ 3ZhP

CONSENSUS REPORTS: Reported in EPA TSCA Inventory.

SAFETY PROFILE: Poison by intraperitoneal route. Questionable carcinogen with experimental tumorigenic data. Human systemic effects: irritability, nausea or vomiting, normocytic anemia. Iron is potentially toxic in all forms and by all routes of exposure. The inhalation of large amounts of iron dust results in iron pneumoconiosis (arc welder's lung). Chronic exposure to excess levels of iron (>50–100 mg Fe/day) can result in pathological deposition of iron in the body tissues, the symptoms of which are fibrosis of the pancreas, diabetes mellitus, and liver cirrhosis.

As with other metals, it becomes more reactive as it is more finely divided. Ultrafine iron powder is pyrophoric and potentially explosive. Explosive or violent reaction with ammonium nitrate + heat, ammonium peroxodisulfate, chloric acid, chlorine trifluoride, chloroformamidinium nitrate, bromine pentafluoride + heat (with iron powder), air + oil (with iron dust), sodium acetylide. Ignites on contact with chlorine, dinitrogen tetraoxide, liquid fluorine, hydrogen peroxide (with iron powder), nitryl fluoride + heat, peroxyformic acid, potassium perchlorate, potassium dichromate, sodium peroxide (at 240°), polystyrene + friction or spark (iron powder). Mixtures of iron dust with air + water may ignite on drying. Reduced iron reacts with water to produce explosive hydrogen gas. Catalyzes the exothermic polymerization of acetaldehyde. See also IRON COMPOUNDS, IRON DUST, and FERROUS ION.

IGM000   CAS: 10102-50-8   HR: 3
IRON(II) ARSENATE (3:2)
DOT: UN 1608
mf: $As_2O_8 \cdot 3Fe$ mw: 445.39
SYNS: ARSENATE of IRON, FERROUS ☐ FERROUS ARSENATE (DOT) ☐ FERROUS ARSENATE, solid (DOT) ☐ IRON ARSENATE (DOT)
CONSENSUS REPORTS: Arsenic and its compounds are on the Community Right-To-Know List.
OSHA PEL: OSHA: Cancer Hazard
ACGIH TLV: TWA 0.01 mg/m³; Confirmed Human Carcinogen; BEI: 35 μ (As)/L inorganic arsenic and methylated metabolites in urine
NIOSH REL: (Inorganic Arsenic) CL 0.002 mg(As)/m³/15M
DOT CLASSIFICATION: 6.1; Label: Poison
SAFETY PROFILE: Confirmed human carcinogen. A deadly poison by various routes. A pesticide. When heated to decomposition it emits toxic fumes of As. See also ARSENIC COMPOUNDS and IRON COMPOUNDS.

IGN000   CAS: 10102-49-5   HR: 3
IRON(III) ARSENATE (1:1)
DOT: UN 1606
mf: $AsO_4 \cdot Fe$ mw: 194.77
SYNS: ARSENATE of IRON, FERRIC ☐ FERRIC ARSENATE, solid (DOT)
CONSENSUS REPORTS: Arsenic and its compounds are on the Community Right-To-Know List.

OSHA PEL: OSHA: Cancer Hazard
ACGIH TLV: TWA 0.01 mg/m³; Confirmed Human Carcinogen; BEI: 35 μ (As)/L inorganic arsenic and methylated metabolites in urine
NIOSH REL: (Inorganic Arsenic) CL 0.002 mg(As)/m³/15M
DOT CLASSIFICATION: 6.1; Label: Poison
SAFETY PROFILE: Confirmed human carcinogen. A deadly poison. A pesticide. When heated to decomposition it emits toxic fumes of As. See also ARSENIC COMPOUNDS and IRON COMPOUNDS.

IGO000   CAS: 63989-69-5   HR: 3
IRON(III)-o-ARSENITE PENTAHYDRATE
DOT: UN 1607
mf: $As_2Fe_2O_6 \cdot Fe_2O_3 \cdot 5H_2O$ mw: 607.34
PROP: Brown-yellow powder.
SYNS: FERRIC ARSENITE, BASIC ☐ FERRIC ARSENITE, solid (DOT)
CONSENSUS REPORTS: Arsenic and its compounds are on the Community Right-To-Know List.
OSHA PEL: TWA 0.01 mg(As)/m³; Cancer Hazard
ACGIH TLV: TWA 0.01 mg/m³; Confirmed Human Carcinogen; BEI: 35 μ (As)/L inorganic arsenic and methylated metabolites in urine
NIOSH REL: (Inorganic Arsenic) CL 0.002 mg(As)/m³/15M
DOT CLASSIFICATION: 6.1; Label: Poison
SAFETY PROFILE: Confirmed human carcinogen. A deadly poison. When heated to decomposition it emits toxic fumes of As. See also ARSENIC COMPOUNDS and IRON COMPOUNDS.

IGQ050   HR: D
IRON CAPRYLATE
SAFETY PROFILE: When heated to decomposition it emits acrid smoke and irritating fumes.

IGS000   CAS: 9004-66-4   HR: 3
IRON-DEXTRAN COMPLEX
PROP: For human use, it is a sterile dark-brown colloidal solvent, water-soluble. Approximate molecular weight is 180,000 (IARC** 2,161,72).
SYNS: A 100 (pharmaceutical) □ B 75 □ CHINOFER □ DEXTRAN ION COMPLEX □ DEXTROFER 75 □ EISENDEXTRAN (GERMAN) □ Fe-DEXTRAN □ FENATE □ FERDEX 100 □ FERRIC DEXTRAN □ FERRIDEXTRAN □ FERRODEXTRAN □ FERRO-FLUKIN 75 □ FERROGLUCIN □ FERROGLUKIN 75 □ IMFERON □ IMPOSIL □ IRO-JEX □ IRON DEXTRAN □ IRON DEXTRAN INJECTION □ IRON HYDROGEN-ATED DEXTRAN □ IRONORM INJECTION □ MY-OFER 100 □ POLYFER □ PROLONGAL □ RCRA WASTE NUMBER U139 □ URSOFERRAN
CONSENSUS REPORTS: NTP 10th Report on Carcinogens. IARC Cancer Review: Group 2B IMEMDT 7,226,87; Human Inadequate Evidence IMEMDT 2,161,73; Animal Sufficient Evidence IMEMDT 2,161,73.
SAFETY PROFILE: Confirmed carcinogen producing tumors at site of application. Experimental carcinogenic, neoplastigenic, tumorigenic, and teratogenic data. Moderately toxic by ingestion and several other routes. Other experimental reproductive effects. See also IRON COMPOUNDS.

IGV000   CAS: 12068-85-8   HR: 3
IRON DISULFIDE
mf: FeS_2 mw: 119.97
SYNS: IRON PYRITES □ IRON SULFIDE
CONSENSUS REPORTS: Reported in EPA TSCA Inventory.
SAFETY PROFILE: A poison by inhalation and ingestion. The powdered sulfide ignites spontaneously in air and some air-powder mixtures may be explosive. Trace carbon lowers the ignition temperature in air to 228°C and increases the sensitivity of air-dust mixtures. Heats up spontaneously and ignites with combustibles. Incompatible with water. When heated to decomposition or in reaction with acid or acid fumes it emits very toxic fumes of SO_x. See also

IRON COMPOUNDS, SULFIDES, and HYDROGEN SULFIDE.

IGW000   HR: 3
IRON DUST
PROP: Silvery-white, tenacious, lustrous, ductile metal. Mp: 1535°, bp: 3000°, d: 7.86, vap press: 1 mm @ 1787°. Iron dust from open hearth furnace contained 52% iron (85AGAF -,480,76).
SAFETY PROFILE: Iron dust can cause conjunctivitis, choroiditis, retinitis, and siderosis of tissues if iron contacts and remains in these tissues. Iron ore dust can cause palpebral conjunctivitis, massive pulmonary fibrosis, and an increased incidence of lung cancer. Questionable carcinogen with experimental neoplastigenic data.
  Flammable in the form of dust when exposed to heat or flame. Reacts violently with Cl_2, ClF_3, F_2, H_2O_2, NO_2, P, Na_2C_2, H_2SO_4, air, water, polystyrene. Moderately explosive in the form of dust when exposed to heat or flame. To fight fire, use special mixtures of dry chemical. See also IRON.

IGW500   CAS: 79-69-6   HR: 2
α-IRONE
mf: $C_{14}H_{22}O$ mw: 206.36
SYNS: 3-BUTEN-2-ONE, 4-(2,5,6,6-TETRAMETHYL-2-CYCLOHEXEN-1-YL)-(9CI) □ 4-(2,5,6,6-TETRAMETHYL-2-CYCLO-HEXEN-1-YL)-3-BUTEN-2-ONE
CONSENSUS REPORTS: Reported in EPA TSCA Inventory.
SAFETY PROFILE: Questionable carcinogen with experimental tumorigenic data. When heated to decomposition it emits acrid smoke and irritating fumes.

IHB700   HR: D
IRON NAPHTHENATE
SAFETY PROFILE: When heated to decomposition it emits acrid smoke and irritating fumes.

IHB800   CAS: 12645-50-0   HR: 3
IRON NICKEL ZINC OXIDE
SYNS: NICKEL ZINC FERRATE □ NICKEL ZINC FERRITE □ 1000 NN FERRITE
CONSENSUS REPORTS: NTP 10th Report on Carcinogens. Reported in EPA TSCA Inventory.
ACGIH TLV: TWA 1 mg(Ni)/m^3
SAFETY PROFILE: Confirmed human carcinogen. An experimental teratogen. Experimental reproductive effects.

IHB900   CAS: 10421-48-4   HR: 2
IRON (III) NITRATE, ANHYDROUS
DOT: UN 1466
mf: N$_3$O$_9$•Fe   mw: 241.88
SYNS: FERRIC NITRATE □ FERRIC NITRATE (DOT) □ IRON NITRATE □ IRON TRINITRATE □ NITRIC ACID, IRON(3+) SALT
CONSENSUS REPORTS: Reported in EPA TSCA Inventory.
ACGIH TLV: TWA 1 mg(Fe)/m^3
DOT CLASSIFICATION: 5.1; Label: Oxidizer
SAFETY PROFILE: Mutation data reported. A reactive oxidizer. When heated to decomposition it emits toxic vapors of NO$_x$.

IHC450   CAS: 1309-37-1   HR: 3
IRON OXIDE
mf: Fe$_2$O$_3$ mw: 159.70
PROP: Dark-red powder. Insol in H$_2$O. IDLH 2500 mg/m^3 (as Fe).
SYNS: ANCHRED STANDARD □ ANHYDROUS IRON OXIDE □ ANHYDROUS OXIDE of IRON □ ARMENIAN BOLE □ BAUXITE RESIDUE □ BLACK OXIDE of IRON □ BLENDED RED OXIDES of IRON □ BURNTISLAND RED □ BURNT SIENNA □ BURNT UMBER □ CALCOTONE RED □ CAPUT MORTUUM □ C.I. 77491 □ C.I. PIGMENT RED 101 □ COLCOTHAR □ COLLOIDAL FERRIC OXIDE □ CROCUS MARTIS ADSTRINGENS □ DEANOX □ EISENOXYD □ ENGLISH RED □ FERRIC OXIDE □ FERRUGO □ INDIAN RED □ IRON(III) OXIDE □ IRON OXIDE RED □ IRON SESQUIOXIDE □ JEWELER'S ROUGE □ LEVANOX RED 130A □ LIGHT RED □ MANUFACTURED IRON OXIDES □ MARS BROWN □ MARS RED □ NATURAL IRON OXIDES □ NATURAL RED OXIDE □ OCHRE □ PRUSSIAN BROWN □ RADDLE □ 11554 RED □ RED IRON OXIDE □ RED OCHRE □ ROUGE □ RUBIGO □ SIENNA □ SPECULAR IRON □ STONE RED □ SUPRA □ SYNTHETIC IRON OXIDE □ VENETIAN RED □ VITRIOL RED □ VOGEL'S IRON RED □ YELLOW FERRIC OXIDE □ YELLOW OXIDE of IRON
CONSENSUS REPORTS: IARC Cancer Review: Group 3 IMEMDT 7,216,87; Human Limited Evidence IMEMDT 1,29,72; Animal No Evidence IMEMDT 1,29,72. Reported in EPA TSCA Inventory.
OSHA PEL: Dust and Fume: TWA 10 mg(Fe)/m^3; Rouge: TWA Total Dust: 10 mg/m^3; Respirable Fraction: 5 mg/m^3
ACGIH TLV: TWA 5 mg(Fe)/m^3 (vapor, dust); Not Classifiable as a Human Carcinogen; Rouge: 10 mg/m^3; Not Classifiable as a Human Carcinogen
DFG MAK: 1.5 mg/m^3 calculated as fine dust
NIOSH REL: (Iron Oxide, Dust and Fume) TWA 5 mg/m^3
SAFETY PROFILE: A poison by subcutaneous route. Questionable carcinogen with experimental tumorigenic data. Catalyzes the potentially explosive polymerization of ethylene oxide. Explosive reaction when heated with guanidinium perchlorate. Reaction with carbon monoxide may form an explosive product. Potentially violent reaction with hydrogen peroxide. The wet oxide reacts explosively with molten aluminum-magnesium alloys. Violent reaction when heated with powdered aluminum, calcium disilicide, magnesium, metal acetylides (e.g., calcium acetylide + iron(III) chloride (on ignition), cesium acetylide (incandescent reaction when warmed), rubidium acetylide). Reacts violently with Al, Ca(OCl)$_2$, N$_2$H$_4$, ethylene oxide. See also IRON and IRON COMPOUNDS.

IHE000   HR: 3
IRON OXIDE, CHROMIUM OXIDE, and NICKEL OXIDE FUME
PROP: Fume composed of iron(+3) oxide:chromium(+3) oxide:nickel(+2)oxide, 6:1:1 (BJIMAG 29,169,72).
SYNS: CHROMIUM OXIDE, NICKEL OXIDE, and IRON OXIDE FUME □ NICKEL OXIDE, IRON OXIDE, and CHROMIUM OXIDE FUME
CONSENSUS REPORTS: NTP 10th Report on Carcinogens. Nickel and its compounds,

as well as chromium and its compounds, are on the Community Right-To-Know List.
OSHA PEL: TWA 1 mg(Ni)/m³; CL 0.1 mg (CrO₃)/m³
ACGIH TLV: TWA 1 mg(Ni)/m³; TWA 0.05 mg(Ni)/m³; Confirmed Human Carcinogen
NIOSH REL: (Chromium (VI)) TWA 0.025 mg(Cr(VI))/m³; CL 0.05/15M; (Inorganic Nickel) TWA 0.015 mg(Ni)/m³
SAFETY PROFILE: Confirmed human carcinogen. See also individual components; NICKEL COMPOUNDS, IRON COMPOUNDS, and CHROMIUM COMPOUNDS.

IHG000   CAS: 8047-67-4   HR: 3
IRON OXIDE, SACCHARATED
PROP: Saccharated oxide of iron (JNCIAM 24,109,60).
SYNS: COLLIRON I.V. □ FEOJECTIN □ FERRIC OXIDE, SACCHARATED □ FERRIC SACCHARATE IRON OXIDE (MIX.) □ FERRIVENIN □ IRON SACCHARATE □ IRON SUGAR □ IVIRON □ NEO-FERRUM □ PROFERRIN □ SACCHARATED FERRIC OXIDE □ SACCHARATED IRON □ SUCROFER
CONSENSUS REPORTS: IARC Cancer Review: Animal Sufficient Evidence IMEMDT 2,161,73.
SAFETY PROFILE: Confirmed carcinogen with experimental neoplastigenic and tumorigenic data. A poison by intravenous route. See also IRON COMPOUNDS.

IHG100   CAS: 1332-37-2   HR: 3
IRON OXIDE, spent
DOT: UN 1376
SYNS: FERROUS FERRITE □ IRON OXIDE □ IRON OXIDE RED 130B □ IRON SPONGE, spent obtained from coal gas purification (DOT) □ MIO 40GN □ SIFERRIT
CONSENSUS REPORTS: Reported in EPA TSCA Inventory.
DOT CLASSIFICATION: 4.2; Label: Spontaneously Combustible
SAFETY PROFILE: Flammable solid. Keep away from sparks and flames.

IHG500   CAS: 13463-40-6   HR: 3
IRON PENTACARBONYL
DOT: UN 1994

mf: C₅FeO₅   mw: 195.90
PROP: Yellow to dark-red viscous liquid. Mp: −25°, Fp: −20° (to −19°), bp: 103.0°, flash p: 5°F, d: 1.453 @ 25°/4°, vap press: 40 mm @ 30.3°. Sol in hexane.
SYNS: FER PENTACARBONYLE (FRENCH) □ IRON CARBONYL □ PENTACARBONYLIRON
CONSENSUS REPORTS: EPA Extremely Hazardous Substances List. Reported in EPA TSCA Inventory.
OSHA PEL: TWA 0.1 ppm (Fe); STEL 0.2 ppm
ACGIH TLV: TWA 0.1 ppm (Fe); STEL 0.2 ppm
DFG MAK: 0.1 ppm (0.81 mg/m³)
DOT CLASSIFICATION: 6.1; Label: Poison, Flammable Liquid
SAFETY PROFILE: A poison by inhalation, skin contact, ingestion, subcutaneous, and intravenous routes. Inhalation causes dizziness, nausea, and vomiting. If continued, unconsciousness follows. Often there is a delayed reaction of chest pain, cough, and difficult breathing. There may be cyanosis and circulatory collapse. In fatal cases, death occurs from the fourth to eleventh day with pneumonitis and injury to kidneys, liver, and brain. Iron carbonyl is less toxic than nickel carbonyl.

A very dangerous fire and moderate explosion hazard when exposed to heat or flame; can react vigorously with oxidizing materials. Warning: pyrophoric in air. Mixtures with nitrogen oxide explode above 50°C. Violent reaction with zinc + transition metal halides (e.g., cobalt halides, rhodium halides, ruthenium halides). Mixtures with acetic acid + water produce a pyrophoric powder. To fight fire, use water, foam, CO₂, dry chemical. See also CARBONYLS and IRON COMPOUNDS.

IHN200   CAS: 118-48-9   HR: D
ISATOIC ACID ANHYDRIDE
mf: C₈H₅NO₃   mw: 163.14
PROP: Crystals from EtOH or dioxan. Mp: 243° (decomp).
SYNS: 2H-3,1-BENZOXAZINE-2,4(1H)-DIONE □ IA □ ISATOIC ANHYDRIDE

CONSENSUS REPORTS: Reported in EPA TSCA Inventory.

SAFETY PROFILE: Experimental reproductive effects. When heated to decomposition it emits toxic fumes of NO_x.

IHN300   CAS: 68000-78-2   HR: 3
ISEPAMICIN DISULFATE

mf: $C_{22}H_{43}N_5O_{12} \cdot 2H_2O_4S$ mw: 765.86

SYN: d-STREPTAMINE, o-6-AMINO-6-DEOXY-α-d-GLUCOPYRANOSYL-(1-4)-o-(3-DEOXY-4-C-METHYL-3-(METHYLAMINO)-β-l-ARABINOPYRANOSYL-(1-6))-N^1)-(3-AMINO-2-HYDROXY-1-OXOPROPYL)-2-DEOXY-, (S)-, SULFATE (1:2) (SALT)

SAFETY PROFILE: A poison by intravenous route. Moderately toxic by intraperitoneal, subcutaneous, and intramuscular routes. Low toxicity by ingestion. When heated to decomposition it emits toxic vapors of NO_x and SO_x.

IHO850   CAS: 123-92-2   HR: 3
ISOAMYL ACETATE

mf: $C_7H_{14}O_2$ mw: 130.21

PROP: Colorless liquid; banana-like odor. Bp: 142.0°, ULC: 55–60, lel: 1% @ 212°F, uel: 7.5%, flash p: 77°F, d: 0.876, refr index: 1.400, autoign temp: 680°F, vap d: 4.49. Misc in alc, ether, ethyl acetate, fixed oils; sltly sol in water; insol in glycerin, propylene glycol. IDLH 1000 ppm.

SYNS: ACETIC ACID, ISOPENTYL ESTER □ BANANA OIL □ FEMA No. 2055 □ ISOAMYL ETHANOATE □ ISOPENTYL ACETATE □ ISOPENTYL ALCOHOL ACETATE □ 3-METHYLBUTYL ACETATE □ 3-METHYL-1-BUTYL ACETATE □ 3-METHYLBUTYL ETHANOATE □ PEAR OIL

CONSENSUS REPORTS: Reported in EPA TSCA Inventory.

OSHA PEL: TWA 100 ppm

ACGIH TLV: TWA 150 ppm

DFG MAK: 50 ppm

SAFETY PROFILE: Mildly toxic by ingestion, inhalation, and subcutaneous routes. Exposure to concentrations of about 1000 ppm for 1 hour can cause headache, fatigue, pulmonary irritation, and serious toxicity effects. Highly flammable liquid when exposed to heat or flame; can react vigorously with reducing materials.

Moderately explosive in the form of vapor when exposed to heat or flame. To fight fire, use alcohol foam, CO_2, dry chemical. When heated to decomposition it emits acrid smoke and fumes.

IHP000   CAS: 123-51-3   HR: 3
ISOAMYL ALCOHOL

mf: $C_5H_{12}O$ mw: 88.17

PROP: Clear liquid; pungent, repulsive taste. Bp: 132°, ULC: 35–40, lel: 1.2%, uel: 9.0% @ 212°F, flash p: 109°F (CC), d: 0.813, autoign temp: 662°F, vap d: 3.04, mp: −117.2°. Sol in water @ 14°; misc in alc and ether. IDLH 500 ppm.

SYNS: ALCOOL AMILICO (ITALIAN) □ ALCOOL ISOAMYLIQUE (FRENCH) □ AMYLOWY ALKOHOL (POLISH) □ FERMENTATION AMYL ALCOHOL □ ISOAMYL ALKOHOL (CZECH) □ ISO-AMYLALKOHOL (GERMAN) □ ISOAMYLOL □ ISOBUTYLCARBINOL □ ISOPENTANOL □ ISOPENTYL ALCOHOL □ 2-METHYL-4-BUTANOL □ 3-METHYL BUTANOL □ 3-METHYLBUTAN-1-OL □ 3-METHYL-1-BUTANOL (CZECH) □ 3-METIL-BUTANOLO (ITALIAN)

CONSENSUS REPORTS: Reported in EPA TSCA Inventory.

OSHA PEL: TWA 100 ppm; STEL 125 ppm

ACGIH TLV: TWA 100 ppm; STEL 125 ppm

DFG MAK: 100 ppm (370 mg/m³)

SAFETY PROFILE: A poison by intraperitoneal and intravenous routes. Moderately toxic by ingestion and skin contact. A skin and human eye irritant. Human systemic effects by inhalation: olfactory effects, conjunctiva irritation, respiratory changes. Questionable carcinogen with experimental carcinogenic data. Flammable liquid when exposed to heat or flame; can react vigorously with reducing materials. Slight explosion hazard when exposed to flame. Explosive reaction with hydrogen trisulfide. To fight fire, use alcohol foam, CO_2, dry chemical. When heated to decomposition it emits acrid smoke and fumes. Used as a flotation agent, a solvent, and in organic synthesis.

IHP010   CAS: 584-02-1   HR: 3
ISOAMYL ALCOHOL
mf: $C_5H_{12}O$ mw: 88.17
PROP: Liquid; acetone-like odor. Bp:
115.6°, d: 0.815 @ 25°/4°, flash p: 66°F, lel:
1.2%, uel: 9%. Sol in alc, ether; sltly sol in
water.
SYNS: DIETHYL CARBINOL □ DIETHYLCARBINOL
□ 3-PENTANOL □ PENTANOL-3 □ PENTAN-3-OL
CONSENSUS REPORTS: Reported in EPA
TSCA Inventory.
OSHA PEL: TWA 100 ppm; STEL 125 ppm
ACGIH TLV: TWA 100 ppm; STEL 125
ppm
DFG MAK: 100 ppm (360 mg/m³)
SAFETY PROFILE: Moderately toxic by
ingestion, skin contact, and intraperitoneal
routes. A severe eye and mild skin irritant.
Dangerous fire and explosion hazard when
exposed to heat, flame, or oxidizing
materials. Used as a flotation agent, a
solvent, and in organic synthesis. When
heated to decomposition it emits acrid
smoke and irritating fumes. See also
ALCOHOLS.

IHP100   CAS: 94-46-2   HR: 2
ISOAMYL BENZOATE
mf: $C_{12}H_{16}O$ mw: 176.28
PROP: Colorless to pale-yellow liquid;
pungent, fruity odor. D: 0.986–0.992, refr
index: 1.492, flash p: 212°F.
SYNS: AMYL BENZOATE □ BENZOIC ACID, 1-(3-
METHYL)BUTYL ESTER □ FEMA No. 2058 □
ISOPENTYL BENZOATE □ 1-(3-METHYL)BUTYL
BENZOATE
CONSENSUS REPORTS: Reported in EPA
TSCA Inventory.
SAFETY PROFILE: Mildly toxic by
ingestion. A skin irritant. Combustible
liquid. When heated to decomposition it
emits acrid smoke and irritating fumes. See
also ESTERS.

IHP400   CAS: 106-27-4   HR: 3
ISOAMYL BUTYRATE
mf: $C_9H_{18}O_2$ mw: 158.24
PROP: Colorless liquid; fruity odor. D:
0.860, refr index: 1.409–1.414, flash p:

149°F. Sol in alc, fixed oils; insol in glycerin,
propylene glycol, water @ 179°.
SYNS: BUTANOIC ACID, 3-METHYLBUTYL ESTER
(9CI) □ FEMA No. 2060 □ ISOAMYL BUTANOATE □
ISOAMYL BUTYLATE □ ISOAMYL-n-BUTYRATE □
ISOPENTYL BUTANOATE □ ISOPENTYL BUTYRATE
□ 3-METHYLBUTYL BUTYRATE
CONSENSUS REPORTS: Reported in EPA
TSCA Inventory.
SAFETY PROFILE: A skin irritant.
Flammable liquid when exposed to heat,
sparks, or flame. When heated to
decomposition it emits acrid smoke and
irritating fumes.

IHP500   CAS: 2035-99-6   HR: 2
ISOAMYL CAPRYLATE
mf: $C_{13}H_{26}O_2$ mw: 214.39
PROP: Colorless to pale yellow translucent
liquid with fresh fruity, apple-pineapple,
green sweet aroma. D: 0.838–0.878 @ 25°,
bp: 222°. Flash pt: 85° (closed cup). Sol in
water.
SYNS: ISOAMYL OCTANOATE □ ISOPENTYL
OCTANOATE □ OCTANOIC ACID, ISOPENTYL ESTER
CONSENSUS REPORTS: Reported in EPA
TSCA Inventory.
SAFETY PROFILE: A skin irritant. A
combustible liquid. When heated to
decomposition it emits acrid smoke and
irritating fumes.

IHR220   CAS: 26760-64-5   HR: 3
ISOAMYLENE
DOT: UN 2371
mf: C_5H_{10} mw: 70.15
PROP: Liquid; disagreeable odor. Mp:
−124°, bp: 30.1°, lel: 1.6%, uel: 8.7%, flash
p: 0°F (OC), d: 0.643, vap d: 2.42, autoign
temp: 527°F.
SYNS: tert-AMYLENE □ ISOPENTENE □ ISOPENT-
ENES (DOT) □ METHYLBUTENE □ 2-METHYL-
BUTENE
CONSENSUS REPORTS: Reported in EPA
TSCA Inventory.
DOT CLASSIFICATION: 3; Label:
Flammable Liquid
SAFETY PROFILE: Moderately toxic by
ingestion and inhalation. Narcotic in high

concentration. A simple asphyxiant. Extremely flammable. Moderately explosive when exposed to heat, flame, or powerful oxidizers. To fight fire, use alcohol foam, spray, mist, dry chemical. When heated to decomposition it emits acrid smoke and irritating fumes.

IHS000   CAS: 110-45-2   HR: 3
ISOAMYL FORMATE
mf: $C_6H_{12}O_2$ mw: 116.18
PROP: Clear liquid; fruity odor. Bp: 123.3°, d: 0.877 @ 20°, refr index: 1.396, vap press: 10 mm @ 17.1°, flash p: 127°F. Misc with alc, ether, propylene glycol; very sltly sol in water; insol in glycerin.
SYNS: FEMA No. 2069 □ FORMIC ACID, ISOPENTYL ESTER □ ISOAMYL METHANOATE □ ISOPENTYL ALCOHOL, FORMATE □ ISOPENTYL FORMATE □ 3-METHYLBUTYL FORMATE
CONSENSUS REPORTS: Reported in EPA TSCA Inventory.
SAFETY PROFILE: Moderately toxic by ingestion. A skin irritant. This material is very irritating and can cause narcosis. The symptoms are usually transient in nature, but it is possible upon severe or prolonged exposure to have serious consequences. Flammable liquid when exposed to heat, sparks, or flame. Can react with oxidizing materials. When heated to decomposition it emits acrid smoke and fumes.

IHS100   CAS: 68133-73-3   HR: 1
ISOAMYL GERANATE
mf: $C_{15}H_{26}O_2$ mw: 238.41
SYN: 2,6-OCTADIENOIC ACID, 3,7-DIMETHYL-, ISOPENTYL ESTER, (E)-
CONSENSUS REPORTS: Reported in EPA TSCA Inventory.
SAFETY PROFILE: A skin irritant. When heated to decomposition it emits acrid smoke and irritating fumes.

IHU100   CAS: 2198-61-0   HR: 2
ISOAMYL HEXANOATE
mf: $C_{11}H_{22}O_2$ mw: 186.33
PROP: Colorless liquid; fruity odor. D: 0.858–0.863, refr index: 1.418–1.422, flash p:

190°F. Sol in alc, fixed oils; insol in glycerin, propylene glycol, water @ 222°.
SYNS: AMYL HEXANOATE □ FEMA No. 2075 □ ISOAMYL CAPROATE □ ISOAMYL HEXANOATE □ ISOPENTYL HEXANOATE □ ISOPENTYL-n-HEXANOATE
CONSENSUS REPORTS: Reported in EPA TSCA Inventory.
SAFETY PROFILE: A mild skin irritant. Combustible liquid. When heated to decomposition it emits acrid smoke and irritating fumes.

IHV050   CAS: 56011-02-0   HR: 1
ISOAMYL PHENYLETHYL ETHER
mf: $C_{13}H_{20}O$ mw: 192.33
PROP: Odor of chamomile, or pineapple.
SYNS: ANTHER □ BENZENE, (2-(3-METHYLBUTO-XY)ETHYL)- □ (2-(3-METHYLBUTOXY)ETHYL)-BENZENE □ PHENYLETHYL ISOAMYL ETHER
CONSENSUS REPORTS: Reported in EPA TSCA Inventory.
SAFETY PROFILE: Low toxicity by ingestion and skin contact. A skin irritant. When heated to decomposition it emits acrid smoke and irritating fumes.

IHX600   CAS: 125-12-2   HR: 2
ISOBORNYL ACETATE
mf: $C_{12}H_{20}O_2$ mw: 196.29
PROP: Colorless liquid; camphoraceous, piney, balsamic odor. D: 0.980, refr index: 1.462, flash p: 212°F. Sol in alc, fixed oils; sltly sol in propylene glycol; insol in water @ 227°.
SYN: BICYCLO(2.2.1)HEPTAN-2-OL, 1,7,7-TRIMETHYL-, ACETATE, EXO- □ FEMA No. 2160 □ ISOBORNYL ACETATE □ PICHTOSIN □ PICHTOSINE
CONSENSUS REPORTS: Reported in EPA TSCA Inventory.
SAFETY PROFILE: Low toxicity by ingestion. Combustible liquid. When heated to decomposition it emits acrid smoke and irritating fumes.

IHZ000   CAS: 115-31-1   HR: 3
ISOBORNYL THIOCYANATOACETATE
mf: $C_{13}H_{19}NO_2S$ mw: 253.39

PROP: Yellow, oily liquid; terpene-like odor. D: 1.1465 @ 25°/4°, bp: 95° @ 0.06 mm, flash p: 82°C (180°F). Very sol in alc, benzene, chloroform, and ether; insol in water.

SYNS: BORNATE □ CIDALON □ ENT 92 □ ISOBORNEOL THIOCYANATOACETATE □ ISOBORNYL THIOCYANOACETATE □ TERPINYL THIOCYANOACETATE □ THANISOL □ THANITE □ THIOCYANATOACETIC ACID ISOBORNYL ESTER

CONSENSUS REPORTS: Reported in EPA TSCA Inventory.

SAFETY PROFILE: A poison by intraperitoneal route. Moderately toxic by ingestion. Sltly toxic by skin contact. Very irritating to eyes, mucous membranes, and skin. Flammable when exposed to heat or flame; can react vigorously with oxidizing materials. When heated to decomposition it emits very toxic fumes of NO_x and SO_x. See also THIOCYANATES and ESTERS. Used as an FDA over-the-counter drug; an insecticide and fly spray.

IIC000   CAS: 115-11-7   HR: 3
ISOBUTENE
DOT: UN 1055
mf: C_4H_8 mw: 56.12
PROP: Volatile liquid or easily liquefied gas. Bp: −6.9°, fp: −140.3°, flash p: <14°F, d: 0.600, autoign temp: 869°F, lel: 1.8%, uel: 9.6%. Insol in water; very sol in alc, ether, sulfuric acid.

SYNS: γ-BUTYLENE □ ISOBUTYLENE (DOT) □ LIQUEFIED PETROLEUM GAS (DOT) □ 2-METHYLPROPENE

CONSENSUS REPORTS: Reported in EPA TSCA Inventory.

ACGIH TLV: (Proposed: TWA 250 ppm; Not Classifiable as a Human Carcinogen.)

DOT CLASSIFICATION: 2.1; Label: Flammable Gas

SAFETY PROFILE: A simple asphyxiant; may have narcotizing action. A very dangerous fire and explosion hazard when exposed to heat or flame. Can react vigorously with oxidizing materials. To fight fire, stop flow of gas. When heated to decomposition it emits acrid smoke and fumes.

IIJ000   CAS: 110-19-0   HR: 3
ISOBUTYL ACETATE
DOT: UN 1213
mf: $C_6H_{12}O_2$ mw: 116.18
PROP: Colorless, neutral liquid; fruit-like odor. Mp: −98.9°, bp: 118°, flash p: 64°F (CC) (18°), d: 0.8685 @ 15°, refr index: 1.389, vap press: 10 mm @ 12.8°, autoign temp: 793°F, vap d: 4.0, lel: 2.4%, uel: 10.5%. Very sol in alc, fixed oils, propylene glycol; sltly sol in water. IDLH 1300 ppm [10%LEL].

SYNS: ACETATE d'ISOBUTYLE (FRENCH) □ ACETIC ACID, ISOBUTYL ESTER □ ACETIC ACID-2-METHYL-PROPYL ESTER □ FEMA No. 2175 □ ISOBUTYLESTER KYSELINY OCTOVE □ 2-METHYLPROPYL ACETATE □ 2-METHYL-1-PROPYL ACETATE □ β-METHYLPROPYL ETHANOATE

CONSENSUS REPORTS: Reported in EPA TSCA Inventory.

OSHA PEL: TWA 150 ppm
ACGIH TLV: TWA 150 ppm
DFG MAK: 100 ppm (480 mg/m³)
DOT CLASSIFICATION: 3; Label: Flammable Liquid

SAFETY PROFILE: Mildly toxic by ingestion and inhalation. A skin and eye irritant. Upon absorption by the body it can hydrolyze to acetic acid and isobutanol. Highly flammable liquid. A very dangerous fire and moderate explosion hazard when exposed to heat, flame, or oxidizers. To fight fire, use alcohol foam, CO_2, dry chemical. When heated to decomposition it emits acrid smoke and fumes. See also ESTERS and n-BUTYL ACETATE.

IIK000   CAS: 106-63-8   HR: 3
ISOBUTYL ACRYLATE
DOT: UN 2527
mf: $C_7H_{12}O_2$ mw: 128.19
PROP: Clear colorless liquid with ester like odor. Bp: 139°, mp: −61°.

SYNS: ACRYLIC ACID ISOBUTYL ESTER □ ISOBUTYL ACRYLATE, inhibited (DOT) □ ISOBUTYL PROPENOATE □ ISOBUTYL-2-PROPENOATE □ Z-

METHYLPROPYL ACRYLATE □ 2-PROPENOIC ACID-2-METHYLPROPYL ESTER

CONSENSUS REPORTS: Reported in EPA TSCA Inventory.

DOT CLASSIFICATION: 3; Label: Flammable Liquid

SAFETY PROFILE: Moderately toxic by skin contact and intraperitoneal routes. Mildly toxic by inhalation and ingestion. A skin irritant. Flammable when exposed to heat or flame; can react vigorously with oxidizing materials. When heated to decomposition it emits acrid smoke and toxic fumes. See also ESTERS.

IIL000   CAS: 78-83-1   HR: 3
ISOBUTYL ALCOHOL

DOT: UN 1212

mf: $C_4H_{10}O$ mw: 74.14

PROP: Clear, colorless, refractive, mobile liquid; sweet odor. Flammable. Bp: 107.90°, flash p: 82°F, ULC: 40–45, lel: 1.2%, uel: 10.9% @ 212°F, fp: −108°, d: 0.800, autoign temp: 800°F, vap press: 10 mm @ 21.7°, vap d: 2.55. Sltly sol in water; misc with alc and ether. IDLH 1600 ppm.

SYNS: ALCOOL ISOBUTYLIQUE (FRENCH) □ FEMA No. 2179 □ FERMENTATION BUTYL ALCOHOL □ 1-HYDROXYMETHYLPROPANE □ ISOBUTANOL (DOT) □ ISOBUTYLALKOHOL (CZECH) □ ISOPROPYL-CARBINOL □ 2-METHYL PROPANOL □ 2-METHYL-PROPAN-1-OL □ 2-METHYL-1-PROPANOL □ 2-METHYLPROPYL ALCOHOL □ RCRA WASTE NUMBER U140

CONSENSUS REPORTS: Reported in EPA TSCA Inventory.

OSHA PEL: TWA 50 ppm

ACGIH TLV: TWA 50 ppm

DFG MAK: 100 ppm (310 mg/m³)

DOT CLASSIFICATION: 3; Label: Flammable Liquid

SAFETY PROFILE: Poison by intravenous and intraperitoneal routes. Moderately toxic by ingestion and skin contact. Mildly toxic by inhalation. A severe skin and eye irritant. Questionable carcinogen with experimental carcinogenic and tumorigenic data. Mutation data reported. Flammable liquid. Dangerous fire hazard when exposed to heat or flame. Moderately explosive in the form of vapor when exposed to heat, flame, or oxidizers. Ignites on contact with chromium trioxide. Reacts with aluminum at 100° to form explosive hydrogen gas. Keep away from heat and open flame. To fight fire, use alcohol foam, CO_2, dry chemical. When heated to decomposition it emits acrid smoke and fumes. See also ALCOHOLS.

IIM000   CAS: 78-81-9   HR: 3
ISOBUTYLAMINE

DOT: UN 1214

mf: $C_4H_{11}N$ mw: 73.16

PROP: Colorless liquid. Mp: −85.5°, bp: 68.6°, flash p: 15°F, d: 0.731 @ 20°/20°, vap press: 100 mm @ 18.8°, autoign temp: 712°F, vap d: 2.5. Misc with water, alc, and ether.

SYNS: 1-AMINO-2-METHYLPROPANE □ MONOISO-BUTYLAMINE □ NSC-8028 □ 1-PROPANAMINE, 2-METHYL- □ VALAMINE

CONSENSUS REPORTS: Reported in EPA TSCA Inventory.

DFG MAK: 5 ppm (15 mg/m³)

DOT CLASSIFICATION: 3; Label: Flammable Liquid

SAFETY PROFILE: A poison by ingestion. A powerful irritant to skin, eyes, and mucous membranes. Skin contact can cause blistering. Inhalation can cause headache and dryness of nose and throat. A very dangerous fire hazard when exposed to heat or flame. Can react vigorously with oxidizing materials. To fight fire, use dry chemical, foam, CO_2, alcohol foam. When heated to decomposition it emits toxic fumes of NO_x. See also AMINES.

IIN000   CAS: 538-93-2   HR: 3
ISOBUTYLBENZENE

mf: $C_{10}H_{14}$ mw: 134.24

PROP: Liquid. Insol in water; sol in alc and ether. Mp: −51.5°, bp: 170.5°, flash p: 131°F (CC), d: 0.867 @ 20°/4°, autoign temp: 806°F, vap press: 1 mm @ 14.1°, vap d: 4.62, lel: 0.8%, uel: 6.0%.

SYN: 2-METHYL-1-PHENYLPROPANE

CONSENSUS REPORTS: Reported in EPA TSCA Inventory.

SAFETY PROFILE: Mildly toxic by ingestion. An irritant and possibly narcotic. Flammable liquid when exposed to heat, sparks, or flame. Can react with oxidizing materials. Moderate explosion hazard when exposed to heat or flame. To fight fire, use foam, CO_2, dry chemical. When heated to decomposition it emits acrid smoke and fumes.

IIN300   HR: D
ISOBUTYL-2-BUTENOATE
mf: $C_8H_{14}O_2$ mw: 142.19

PROP: Colorless liquid; powerful fruity odor. D: 0.880, refr index: 1.426–1.430. Sol in alc, propylene glycol, fixed oils; sltly sol in water.

SYN: FEMA No. 3432

SAFETY PROFILE: When heated to decomposition it emits acrid smoke and irritating fumes.

IIP000   CAS: 4439-24-1   HR: 3
ISOBUTYL CELLOSOLVE
mf: $C_6H_{14}O_2$ mw: 118.20

PROP: Colorless liquid. D: 0.903 @ 20°/4°, bp: 157–158°; misc in water, alc, ether.

SYNS: EKTASOLVE EIB □ ETHYLENE GLYCOL MONOISOBUTYL ETHER □ 2-ISOBUTOXYETHANOL

CONSENSUS REPORTS: Glycol ether compounds are on the Community Right-To-Know List. Reported in EPA TSCA Inventory.

SAFETY PROFILE: Poison by ingestion. Moderately toxic by skin contact. Mildly toxic by inhalation. A skin irritant. When heated to decomposition it emits acrid smoke and fumes. See also GLYCOL ETHERS.

IIQ000   CAS: 122-67-8   HR: 2
ISOBUTYL CINNAMATE
mf: $C_{13}H_{16}O_2$ mw: 204.29

PROP: Colorless liquid; sweet, fruity odor. D: 1.001, refr index: 1.539–1.541, flash p:

212°F. Misc with alc, chloroform, ether, fixed oils; insol in water.

SYNS: CINNAMIC ACID, ISOBUTYL ESTER □ FEMA No. 2193 □ LABDANOL □ 3-PHENYL-2-PROPENOIC ACID, 2-METHYLPROPYL ESTER

CONSENSUS REPORTS: Reported in EPA TSCA Inventory.

SAFETY PROFILE: A skin irritant. Combustible liquid. When heated to decomposition it emits acrid smoke and fumes. See also ESTERS.

IIR000   CAS: 542-55-2   HR: 3
ISOBUTYL FORMATE
DOT: UN 2393

mf: $C_5H_{10}O_2$ mw: 102.15

PROP: Liquid. D: 0.885 @ 20°/4°, mp: −95.3°, bp: 98.2°, flash p: <70°F, autoign temp: 608°F, lel: 2.0%, uel: 8%. Sol in water @ 22°; misc in alc and ether.

SYNS: ISOBUTYLESTER KYSELINY MRAVENCI □ ISO-BUTYL FORMATE □ FORMIC ACID, ISOBUTYL ESTER □ TETRYL FORMATE

CONSENSUS REPORTS: Reported in EPA TSCA Inventory.

DOT CLASSIFICATION: 3; Label: Flammable Liquid

SAFETY PROFILE: Moderately toxic by ingestion. A very dangerous fire hazard when exposed to heat, open flame, or oxidizers. A moderate explosion hazard when exposed to heat or flame. To fight fire, use water spray, foam, CO_2, dry chemical. When heated to decomposition it emits acrid smoke and fumes. See also ESTERS.

IIU000   CAS: 15687-27-1   HR: 3
p-ISOBUTYLHYDRATROPIC ACID
mf: $C_{13}H_{18}O_2$ mw: 206.31

PROP: White to off white powder. Odorless. Relatively insol in water.

SYNS: ACIDE (ISOBUTYL-4 PHENYL)-2 PROPIONI-QUE (FRENCH) □ ADRAN □ ANFLAGEN □ ARTRIL 300 □ BLUTON □ BRUFANIC □ BRUFEN □ BUBUR-ONE □ BUTYLENIN □ DOLGIN □ EMODIN □ EPOBRON □ IBUFEN □ IBUPROCIN □ IBUPROFEN □ IP-82 □ 4-ISOBUTYLHYDRATROPIC ACID □ 2-(4-ISOBUTYLPHENYL)PROPANOIC ACID □ α-p-ISOBUT-YLPHENYLPROPIONIC ACID □ α-(4-ISOBUTYLPHEN-

YL)PROPIONIC ACID □   2-(p-ISOBUTYLPHENYL)PRO-
PIONIC ACID □   LAMIDON □   LIPTAN □   α-METHYL-
4-(2-METHYLPROPYL)BENZENEACETIC ACID □
MOTRIN □   MYNOSEDIN □   NAPACETIN □   NO-
BFELON □   NOBFEN □   NOBGEN □   R.D. 13621 □
REBUGEN □   ROIDENIN

CONSENSUS REPORTS: Reported in EPA
TSCA Inventory.

SAFETY PROFILE: A human poison by
ingestion. Poison experimentally by
subcutaneous and intraperitoneal routes.
Moderately toxic by ingestion and rectal
routes. Human systemic effects: eye effects,
dermatitis, increased body temperature,
hepatitis, allergic reaction with multiple
organ involvement, diplopia. Human
reproductive effects by ingestion: menstrual
cycle changes or disorders. Experimental
teratogenic and reproductive effects. An
FDA over-the-counter drug used as an
analgesic and anti-inflammatory agent.
When heated to decomposition it emits
acrid smoke and fumes.

IIV000   CAS: 6104-30-9   HR: 2
ISOBUTYLIDENEDIUREA
mf: $C_6H_{14}N_4O_2$ mw: 174.24
PROP: White crystals. Mp: 195°–205°, d:
0.55. Sol in water.
SYNS: 1,1-DIUREIDISOBUTANE □   DIUREIDO-
ISOBUTANE □   IBDU □   ISOBUTYLDIUREA □
ISOBUTYLENEDIUREA □   1,1'-ISOBUTYLIDENE-
BISUREA □   ISODUR □   N,N"-(2-METHYLPROPYL-
IDENE)BISUREA (9CI)

CONSENSUS REPORTS: Reported in EPA
TSCA Inventory.

SAFETY PROFILE: Questionable
carcinogen with experimental tumorigenic
data. When heated to decomposition it
emits toxic fumes of NO_x.

IIW000   CAS: 97-85-8   HR: 3
ISOBUTYL ISOBUTYRATE
DOT: UN 2528
mf: $C_8H_{16}O_2$ mw: 144.24
PROP: Liquid with fruity odor. Mp: −81°,
bp: 147.5°, d: 0.850–0.860 @ 20°/20°, vap
press: 10 mm @ 39.9°. Insol in water; misc
with alc.

SYNS: ISOBUTYLISOBUTYRATE (DOT) □   ISOBUTYR-
IC ACID, ISOBUTYL ESTER □   2-METHYLPROPYL ISO-
BUTYRATE □   2-METHYLPROPYLPROPANOIC ACID-2-
METHYLPROPYL ESTER (9CI)

CONSENSUS REPORTS: Reported in EPA
TSCA Inventory.

DOT CLASSIFICATION: 3; Label:
Flammable Liquid

SAFETY PROFILE: Mildly toxic by
ingestion and inhalation. An insect repellent.
Flammable when exposed to heat or flame.
Can react with oxidizing materials. When
heated to decomposition it emits acrid
smoke and fumes. See also ISOBUTYL
ALCOHOL.

IIY000   CAS: 97-86-9   HR: 3
ISOBUTYL METHACRYLATE
DOT: UN 2283
mf: $C_8H_{14}O_2$ mw: 142.22
PROP: Clear liquid. Mp: −33°, bp: 155°, d:
0.885–0.889. Insol in water.
SYNS: ISOBUTYLESTER KYSELINY METHAKRYLOVE
□   ISOBUTYL-α-METHACRYLATE □   METHACRYLIC
ACID, ISOBUTYL ESTER □   2-METHYL-2-PROPENOIC
ACID-2-METHYLPROPYL ESTER □   2-METHYLPROPYL
METHACRYLATE

CONSENSUS REPORTS: Reported in EPA
TSCA Inventory.

DOT CLASSIFICATION: 3; Label:
Flammable Liquid

SAFETY PROFILE: Moderately toxic by
intraperitoneal route. Mildly toxic by
ingestion. Experimental teratogenic and
reproductive effects. Flammable when
exposed to heat or flame. When heated to
decomposition it emits acrid smoke and
fumes. See also ESTERS.

IJD000   CAS: 542-56-3   HR: 3
ISOBUTYL NITRITE
mf: $C_4H_9NO_2$ mw: 103.14
PROP: Liquid. D: 0.870 @ 22°/4°, bp:
67–68°. Sltly sol in and decomp in water;
misc in alc.
SYNS: IBN □   NCI-C61052 □   NITROUS ACID,
ISOBUTYL ESTER □   NITROUS ACID, 2-METHYL-
PROPYL ESTER

CONSENSUS REPORTS: Reported in EPA
TSCA Inventory.

ACGIH TLV: CL 1 200 ppm; Confirmed Animal Carcinogen

SAFETY PROFILE: A poison by ingestion and intraperitoneal routes. Mildly toxic by inhalation. Human systemic effects by ingestion: carboxyhemoglobinemia, blood pressure lowering, change in heart rate. Mutation data reported. When heated to decomposition it emits toxic fumes of NO_x. See also ESTERS and NITRITES.

IJF400   **HR: 2**
ISOBUTYL PHENYLACETATE
mf: $C_{12}H_{16}O_2$ mw: 192.23
PROP: Colorless liquid; rose, honey-like odor. D: 0.984–0.988, refr index: 1.486, flash p: 241°F. Sol in alc, fixed oils; insol in glycerin, propylene glycol, water.
SYN: FEMA No. 2210
SAFETY PROFILE: Combustible liquid. When heated to decomposition it emits acrid smoke and irritating fumes.

IJN000   **CAS: 87-19-4**   **HR: 2**
ISOBUTYL SALICYLATE
mf: $C_{11}H_{14}O_3$ mw: 194.25
PROP: Colorless liquid; orchid odor. D: 1.062–1.066, refr index: 1.507, flash p: 250°F. Sol in fixed oils; insol in glycerin, propylene glycol.
SYNS: FEMA No. 2213 □ ISOBUTYL-o-HYDROXYBENZOATE □ SALICYLIC ACID, ISOBUTYL ESTER
CONSENSUS REPORTS: Reported in EPA TSCA Inventory.
SAFETY PROFILE: Moderately toxic by ingestion. Combustible liquid. When heated to decomposition it emits acrid smoke and fumes. See also ESTERS.

IJQ000   **CAS: 109-53-5**   **HR: 3**
ISOBUTYL VINYL ETHER
DOT: UN 1304
mf: $C_6H_{12}O$ mw: 100.18
PROP: Liquid. Mp: −112°, bp: 82.9–83.2°, flash p: 16°F, d: 0.76 @ 25°/4°, vap d: 3.45.
SYNS: IVE □ VINOFLEX MO 400* □ VINYL ISOBUTYL ETHER (DOT) □ VINYL ISOBUTYL ETHER, inhibited (DOT)

CONSENSUS REPORTS: Reported in EPA TSCA Inventory.
DOT CLASSIFICATION: 3; Label: Flammable Liquid
SAFETY PROFILE: Very mildly toxic by ingestion, inhalation, and skin contact. A very dangerous fire hazard when exposed to heat, flame, oxidizers. Severe explosion hazard when exposed to sparks or open flame. Can react vigorously with oxidizing materials. When heated to decomposition it emits acrid smoke and fumes. To fight fire, use alcohol foam, CO_2, dry chemical. See also ETHERS.

IJS000   **CAS: 78-84-2**   **HR: 3**
ISOBUTYRALDEHYDE
DOT: UN 2045
mf: C_4H_8O mw: 72.12
PROP: Transparent, colorless, highly refractive liquid; pungent odor. Mp: −65°, bp: 64°, flash p: −40°F (CC), d: 0.783–0.788, autoign temp: 434°F, lel: 1.6%, uel: 10.6%, vap d: 2.5. Sol in water; misc in alc, ether, benzene, carbon disulfide, acetone, toluene, chloroform.
SYNS: FEMA No. 2220 □ ISOBUTANAL □ ISOBUTYL-ALDEHYDE □ ISOBUTYL ALDEHYDE (DOT) □ ISOBUTYRALDEHYD (CZECH) □ ISOBUTYRIC ALDEHYDE □ 2-METHYLPROPANAL □ 2-METHYL-1-PROPANAL □ 2-METHYLPROPIONALDEHYDE □ NCI-C60968 □ VALINE ALDEHDYE
CONSENSUS REPORTS: Community Right-To-Know List. Reported in EPA TSCA Inventory.
DOT CLASSIFICATION: 3; Label: Flammable Liquid
SAFETY PROFILE: Moderately toxic by ingestion. Mildly toxic by skin contact and inhalation. A severe skin and eye irritant. Highly flammable liquid. A very dangerous fire hazard when exposed to heat, flame, or oxidizers. Moderately explosive in the form of vapor when exposed to heat or flame. Can react vigorously with reducing materials. When heated to decomposition it emits acrid smoke and fumes. To fight fire, use dry chemical, CO_2, mist, or foam. See also ALDEHYDES.

IJU000   CAS: 79-31-2   HR: 3
ISOBUTYRIC ACID
DOT: UN 2529
mf: $C_4H_8O_2$ mw: 88.12
PROP: Colorless liquid; pungent odor of rancid butter. Mp: $-47°$, bp: $118–119°$ @ 35 mm, flash p: $132°F$ (TOC), d: 0.949 @ $20°/4°$, refr index: 1.392, vap press: 1 mm @ $14.7°$, vap d: 3.04, autoign temp: $935°F$. Misc with alc, chloroform, ether, fixed oils, glycerin, propylene glycol; insol in water.
SYNS: ACETIC ACID, DIMETHYL- □ DIMETHYL-ACETIC ACID □ FEMA No. 2222 □ ISOBUTYRIC ACID (DOT) □ ISOPROPYLFORMIC ACID □ KYSELINA ISOMASELNA □ α-METHYLPROPIONIC ACID □ 2-METHYLPROPIONIC ACID □ PROPIONIC ACID, 2-METHYL-
CONSENSUS REPORTS: Reported in EPA TSCA Inventory.
DOT CLASSIFICATION: 3; Label: Flammable Liquid
SAFETY PROFILE: A poison by ingestion. Moderately toxic by skin contact. A corrosive irritant to the eyes, skin, and mucous membranes. Flammable liquid when exposed to heat or flame; can react with oxidizing materials. To fight fire, use alcohol foam, CO_2, dry chemical. When heated to decomposition it emits acrid smoke and fumes.

IJV000   CAS: 103-28-6   HR: 2
ISOBUTYRIC ACID, BENZYL ESTER
mf: $C_{11}H_{14}O_2$ mw: 178.25
PROP: Colorless liquid; floral, jasmine odor. D: 1.001–1.005, refr index: 1.489, bp: $229°$, flash p: $212°F$. Sol in alc, fixed oils; sltly sol in propylene glycol; insol in glycerin @ $229°$.
SYNS: BENZYL ISOBUTYRATE (FCC) □ BENZYL-2-METHYL PROPIONATE □ FEMA No. 2141
CONSENSUS REPORTS: Reported in EPA TSCA Inventory.
SAFETY PROFILE: Moderately toxic by ingestion. Combustible liquid. When heated to decomposition it emits acrid smoke and fumes. See also ESTERS.

IJW000   CAS: 97-72-3   HR: 3
ISOBUTYRIC ANHYDRIDE

DOT: UN 2530
mf: $C_8H_{14}O_3$ mw: 158.22
PROP: Liquid, decomp in water. Bp: $73–75°$ @ 18 mm, d: 0.951–0.956 @ $20°/20°$, vap d: 5.5, flash p: $139°F$, autoign temp: $665°F$.
CONSENSUS REPORTS: Reported in EPA TSCA Inventory.
DOT CLASSIFICATION: 3; Label: Flammable Liquid
SAFETY PROFILE: A corrosive irritant to skin, eyes, and mucous membranes. Flammable liquid when exposed to heat, flame, or oxidizers. To fight fire, use alcohol foam, fog, dry chemical, CO_2. When heated to decomposition it emits acrid smoke and fumes. See also ANHYDRIDES.

IJX000   CAS: 78-82-0   HR: 3
ISOBUTYRONITRILE
DOT: UN 2284
mf: C_4H_7N mw: 69.12
PROP: Colorless liquid, sltly sol in water, very sol in alc and ether. D: 0.773 @ $20°/20°$, bp: $107°$, mp: $-75°$, flash p: $46.4°F$.
SYNS: 2-CYANOPROPANE □ DIMETHYLACETO-NITRILE □ ISOPROPYL CYANIDE □ ISOPROPYLK-YANID □ ISOPROPYL NITRILE □ α-METHYLPROP-ANENITRILE □ 2-METHYLPROPANENITRILE □ 2-METHYLPROPIONITRILE
CONSENSUS REPORTS: Cyanide and its compounds are on the Community Right-To-Know List. EPA Extremely Hazardous Substances List. Reported in EPA TSCA Inventory.
NIOSH REL: (Nitriles) TWA 22 mg/m^3
DOT CLASSIFICATION: 3; Label: Flammable Liquid, Poison
SAFETY PROFILE: A poison by ingestion, skin contact, and subcutaneous routes. Mildly toxic by inhalation. A skin irritant. A very dangerous fire hazard when exposed to heat or flame. When heated to decomposition it emits toxic fumes of NO_x and CN^-. See also NITRILES.

IKE000   CAS: 513-37-1   HR: 3
ISOCROTYL CHLORIDE
mf: C_4H_7Cl mw: 90.56

PROP: Liquid. D: 0.919 @ 20°/4°, bp: 68°.
SYNS: α-CHLOROISOBUTYLENE □ 1-CHLORO-2-METHYLPROPENE □ 1-CHLORO-2-METHYL-1-PROPENE □ β,β-DIMETHYLVINYL CHLORIDE □ NCI-C54819
CONSENSUS REPORTS: NTP 10th Report on Carcinogens. NTP Carcinogenesis Studies (gavage); Clear Evidence: mouse, rat NTPTR* NTP-TR-316,86.
SAFETY PROFILE: Confirmed carcinogen with experimental carcinogenic and neoplastigenic data. Moderately toxic by ingestion. Mildly toxic by inhalation. A local irritant and narcotic in high concentration. When heated to decomposition it emits toxic fumes of Cl⁻.

IKG000   CAS: 103-65-1   HR: 3
ISOCUMENE
DOT: UN 2364
mf: C_9H_{12} mw: 120.21
PROP: Clear liquid. Insol in water; misc in alc and ether. Mp: −92.2°, bp: 159.2°, flash p: 86°F (CC), d: 0.862, vap press: 10 mm @ 43.4°, vap d: 4.14, autoign temp: 842°F, lel: 0.8%, uel: 6%.
SYNS: 1-PHENYLPROPANE □ n-PROPYLBENZENE □ PROPYL BENZENE (DOT)
CONSENSUS REPORTS: Reported in EPA TSCA Inventory.
DOT CLASSIFICATION: 3; Label: Flammable Liquid
SAFETY PROFILE: Mildly toxic by ingestion and inhalation. A very dangerous fire hazard when exposed to heat, flame, or oxidizers; can react with oxidizing materials. A moderate explosion hazard in the form of vapor when exposed to heat or flame. To fight fire, use foam, CO_2, dry chemical. Emitted from modern building materials (CENEAR 6922,91). When heated to decomposition it emits acrid smoke and fumes.

IKG349   HR: D
ISOCYANATES
PROP: Liquids with sharp fruity odor in high levels.

SAFETY PROFILE: Compounds containing the isocyanate radical –NCO. Derivatives of isocyanic acid (cyanic acid). Usually the term refers to a diisocyanate. Inorganic isocyanates are only slightly toxic. Organic isocyanates (diisocyanates) can cause local irritation and allergic reactions. When heated to decomposition they emit toxic fumes of NO_x.

IKG700   CAS: 30674-80-7   HR: 3
2-ISOCYANATOETHYL METHACRYL-ATE
mf: $C_7H_9NO_3$ mw: 155.17
PROP: Bp: 87–89° @ 10 mm.
SYNS: β-ISOCYANATOETHYL METHACRYLATE □ METHACRYLOYLOXYETHYL ISOCYANATE
CONSENSUS REPORTS: EPA Extremely Hazardous Substances List. Reported in EPA TSCA Inventory.
SAFETY PROFILE: Poison by inhalation. Moderately toxic by ingestion. Experimental reproductive effects. When heated to decomposition it emits toxic fumes of NO_x. See also ESTERS and ISOCYANATES.

IKG725   CAS: 614-68-6   HR: 3
1-ISOCYANATO-2-METHYLBENZENE
mf: C_8H_7NO mw: 133.16
SYNS: BENZENE, 1-ISOCYANATO-2-METHYL- □ ISOCYANIC ACID, o-TOLYL ESTER □ o-METHYLPHENYL ISOCYANATE □ 2-METHYLPHENYL ISOCYANATE □ o-TOLUENE ISOCYANATE □ o-TOLYL ISOCYANATE □ 2-TOLYL ISOCYANATE
SAFETY PROFILE: A poison by inhalation and intraperitoneal routes. Moderately toxic by ingestion. When heated to decomposition it emits toxic vapors of NO_x.

IKG800   CAS: 2094-99-7   HR: 3
1-(1-ISOCYANATO-1-METHYLETHYL)-3-(1-METHYLETHENYL)BENZENE
DOT: UN 2207/UN 2478/UN 3080
mf: $C_{13}H_{15}NO$ mw: 201.29
SYNS: BENZENE, 1-(1-ISOCYANATO-1-METHYLETHYL)-3-(1-METHYLETHENYL)- □ α-α-DIMETHYL-m-ISOPROPENYL BENZYL ISOCYANATE □ ISOCYANIC ACID, m-ISOPROPENYL-α-α-DIMETHYL BENZYL ESTER □ m-TMI

DOT CLASSIFICATION: 6.1; Label: KEEP AWAY FROM FOOD (UN 2207); DOT Class: 6.1; Label: Poison; DOT Class: 6.1; Label: Poison, Flammable Liquid (UN 3080); DOT Class: 3; Label: Flammable Liquid, Poison (UN 2478)
SAFETY PROFILE: Poison by inhalation. Moderately toxic by ingestion. When heated to decomposition it emits toxic vapors of NO_x.

IKG900   CAS: 24801-88-5   HR: 3
3-ISOCYANATOPROPYLTRIETHO-XYSILANE
mf: $C_{10}H_{21}NO_4Si$ mw: 247.41
SYNS: I 7840 □ γ-ISOCYANATOPROPYLTRIETHOXY-SILANE □ ISOCYANIC ACID, 3-(TRIETHOXYSILYL)-PROPYL ESTER □ SILANE, TRIETHOXY(3-ISOCY-ANATOPROPYL)- □ TRIETHOXY(3-ISOCYANATO-PROPYL)SILANE □ Y 9030 □ YH 9030
CONSENSUS REPORTS: Reported in EPA TSCA Inventory.
SAFETY PROFILE: A poison by is and skin contact. Moderately toxic by inhalation. A mild skin and severe eye irritant. When heated to decomposition it emits toxic vapors of NO_x.

IKH000   CAS: 1943-83-5   HR: 2
ISOCYANIC ACID-2-CHLOROETHYL ESTER
DOT: UN 2206/UN 2207/UN 2478/UN 3080
mf: C_3H_4ClNO mw: 105.53
PROP: Crystals. D: 1.237 @ 20°, fp: 56°, bp: 135°.
SYNS: 2-CHLORETHYLISOKYANAT □ 2-CHLORO-ETHYL ISOCYANATE □ CIC □ NSC-87418
CONSENSUS REPORTS: EPA Genetic Toxicology Program. Reported in EPA TSCA Inventory.
DOT CLASSIFICATION: 6.1; Label: KEEP AWAY FROM FOOD (UN 2207); DOT Class: 6.1; Label: Poison (UN 2206); DOT Class: 6.1; Label: Poison, Flammable Liquid (UN 3080); DOT Class: 3; Label: Flammable Liquid, Poison (UN 2478)
SAFETY PROFILE: Moderately toxic by inhalation. Human mutation data reported.

A flammable liquid. When heated to decomposition it emits very toxic fumes of Cl^- and NO_x. See also ISOCYANATES and ESTERS.

IKH099   CAS: 102-36-3   HR: 3
ISOCYANIC ACID-3,4-DICHLOROPHEN-YL ESTER
DOT: UN 2206/UN 2207/UN 2478/UN 3080
mf: $C_7H_3Cl_2NO$ mw: 188.01
PROP: Crystals. Mp: 42–43°, bp: 112° @ 12 mm.
SYNS: 3,4-DICHLORFENYLISOKYANAT □ 3,4-DICHLOROPHENYL ISOCYANATE
CONSENSUS REPORTS: EPA Extremely Hazardous Substances List.
DOT CLASSIFICATION: 6.1; Label: KEEP AWAY FROM FOOD (UN 2207); DOT Class: 6.1; Label: Poison (UN 2206); DOT Class: 6.1; Label: Poison, Flammable Liquid (UN 3080); DOT Class: 3; Label: Flammable Liquid, Poison (UN 2478)
SAFETY PROFILE: Poison by inhalation. A flammable liquid. When heated to decomposition it emits toxic fumes of Cl^- and NO_x. See also ISOCYANATES and ESTERS.

IKH780   CAS: 1984-23-2   HR: 3
1-ISOCYANO-4-NITROBENZENE
mf: $C_7H_4N_2O_2$ mw: 148.13
SYNS: BENZENE, 1-ISOCYANO-4-NITRO- □ p-NITROPHENYL ISOCYANIDE □ 4-NITROPHENYL ISOCYANIDE □ p-NITROPHENYL ISONITRILE □ PHENYL ISOCYANIDE, p-NITRO-
SAFETY PROFILE: A poison by subcutaneous route. Low toxicity by ingestion. When heated to decomposition it emits toxic vapors of NO_x.

IKM000   CAS: 29964-84-9   HR: 2
ISODECYL METHACRYLATE
mf: $C_{14}H_{26}O_2$ mw: 226.40
PROP: Clear, colorless to yellowish liquid with slightly ester-like odor. Bp: > 250°. Flash pt:~ 70° C. Insol in water.
SYNS: AGELFLEX FM-10 □ METHACRYLIC ACID, ISODECYL ESTER

CONSENSUS REPORTS: Reported in EPA TSCA Inventory.

SAFETY PROFILE: Moderately toxic by intraperitoneal route. Experimental teratogenic and reproductive effects. A combustible liquid. When heated to decomposition it emits acrid smoke and fumes. See also ESTERS.

IKM100 CAS: 35158-25-9 HR: 1
ISODIHYDROLAVANDULYL ALDEHYDE

mf: $C_{10}H_{18}O$ mw: 154.28

PROP: Flavor and fragrance chemical.

SYNS: 2-HEXEN-1-AL, 2-ISOPROPYL-5-METHYL- □ 2-HEXEN-1-AL, 5-METHYL-2-(1-METHYLETHYL)- □ 2-ISOPROPYL-5-METHYL-2-HEXEN-1-AL

CONSENSUS REPORTS: Reported in EPA TSCA Inventory.

SAFETY PROFILE: A skin irritant. When heated to decomposition it emits acrid smoke and irritating fumes.

IKO000 CAS: 465-73-6 HR: 3
ISODRIN

mf: $C_{12}H_8Cl_6$ mw: 364.90

PROP: Crystals. Mp: 240–242°.

SYNS: COMPOUND 711 □ ENT 19,244 □ EXPERIMENTAL INSECTICIDE 711 □ 1,2,3,4,10,10-HEXACHLORO-1,4,4a,5,8,8a-HEXAHYDRO-1,4-endo,endo-5,8-DIMETHANONAPHTHALENE □ 1,2,3,4,10,10-HEXACHLORO-1,4,4a,5,8,8a-HEXAHYDRO-1,4,5,8-endo,endo-DIMETHANONAPHTHALENE □ RCRA WASTE NUMBER P060

CONSENSUS REPORTS: EPA Extremely Hazardous Substances List. Reported in EPA TSCA Inventory. Chlorophenol compounds are on the Community Right-To-Know List.

SAFETY PROFILE: A poison by ingestion and skin contact. Causes liver injury, acne, and skin rashes. When heated to decomposition it emits toxic fumes of Cl⁻. See also CHLOROPHENOLS and ALDRIN.

IKQ000 CAS: 97-54-1 HR: 2
ISOEUGENOL

mf: $C_{10}H_{12}O_2$ mw: 164.22

PROP: Pale-yellow oil; carnation odor. D: 1.079–1.085, refr index: 1.572–1.577, mp:

−10°, bp: 266°. cis Form: liquid, bp: 133° @ 11 mm, d: 1.088 @ 20°/4°. trans Form: crystals, mp: 33°, bp: 140° @ 12 mm, d: 1.087 @ 20°/4°, flash p: 212°F. Sol in fixed oils, propylene glycol; very sltly sol in water; misc in alc and ether; insol in glycerin.

SYNS: FEMA No. 2468 □ 1-HYDROXY-2-METHOXY-4-PROPENYLBENZENE □ 4-HYDROXY-3-METHOXY-1-PROPENYLBENZENE □ 2-METHOXY-4-PROPENYL-PHENOL □ NCI-C60979 □ 4-PROPENYLGUAIACOL

CONSENSUS REPORTS: Reported in EPA TSCA Inventory.

SAFETY PROFILE: Moderately toxic by ingestion. A moderate human skin irritant. Human mutation data reported. Combustible liquid. When heated to decomposition it emits acrid smoke and fumes. See also EUGENOL.

IKR000 CAS: 93-16-3 HR: 3
1,3,4-ISOEUGENOL METHYL ETHER

mf: $C_{11}H_{14}O_2$ mw: 178.25

PROP: Colorless to pale-yellow liquid; clove-carnation odor. D: 1.047, refr index: 1.566, flash p: 212°F. Sol in fixed oils; insol in glycerin, propylene glycol.

SYNS: 1,2-DIMETHOXY-4-PROPENYLBENZENE □ FEMA No. 2476 □ ISOEUGENYL METHYL ETHER □ ISOHOMOGENOL □ METHYL ISOEUGENOL (FCC) □ 4-PROPENYL VERATROLE

CONSENSUS REPORTS: Reported in EPA TSCA Inventory.

SAFETY PROFILE: Poison by intravenous route. Moderately toxic by intraperitoneal route. A skin irritant. Combustible liquid. When heated to decomposition it emits acrid smoke and fumes. See also EUGENOL and ETHERS.

IKS600 CAS: 107-83-5 HR: 3
ISOHEXANE

mf: C_6H_{14} mw: 86.20

PROP: Liquid or oil. Fp: −154°, bp: 60.3°, lel: 1.0%, uel: 7.0%, flash p: 20°F (CC), d: 0.669, vap d: 3.00, autoign temp: 583°F.

SYNS: 1,2-DIMETHYLBUTANE □ 2-METHYL-PENTANE

CONSENSUS REPORTS: Reported in EPA TSCA Inventory.

OSHA PEL: TWA 500 ppm; STEL 1000 ppm

ACGIH TLV: TWA 500 ppm; STEL 1000 ppm (hexane isomer)

DFG MAK: 200 ppm (720 mg/m^3)

NIOSH REL: (Alkanes) TWA 350 mg/m^3

SAFETY PROFILE: A human eye irritant. A very dangerous fire hazard when exposed to heat, flame, or oxidizers. Explosive in the form of vapor when exposed to heat or flame. Keep away from sparks, heat, or open flame; can react vigorously with oxidizing materials. To fight fire, use foam, CO_2, dry chemical. When heated to decomposition it emits acrid smoke and irritating fumes. See also HEXANE.

IKX000   **CAS: 73-32-5**   **HR: 1**
ISOLEUCINE

mf: $C_6H_{13}NO_2$ mw: 131.17

PROP: Crystals from EtOH (aq). An essential amino acid; many isomeric forms. White crystalline powder; bitter taste. Mp: (dl) 292° (decomp), (l) 283–284° (decomp). Sltly sol in water; nearly insol in alc; insol in ether.

SYNS: ACETIC ACID, AMINO-sec-BUTYL- □ 2-AMINO-3-METHYLPENTANOIC ACID □ α-AMINO-β-METHYLVALERIC ACID □ ISOLEUCINE □ l-ISOLEUCINE (FCC) □ NORVALINE, 3-METHYL- □ VALERIC ACID, 2-AMINO-3-METHYL-

CONSENSUS REPORTS: Reported in EPA TSCA Inventory.

SAFETY PROFILE: Mildly toxic by intraperitoneal route. When heated to decomposition it emits toxic fumes of NO_x.

IKX010   **CAS: 443-79-8**   **HR: D**
dl-ISOLEUCINE

mf: $C_6H_{13}NO_2$ mw: 131.17

PROP: White crystalline powder from EtOH; sltly bitter taste. Mp: 292° (decomp). Sol in water; insol in alc, ether.

SYN: dl-2-AMINO-3-METHYLVALERIC ACID

SAFETY PROFILE: When heated to decomposition it emits toxic fumes of NO_x.

ILG200   **CAS: 5275-02-5**   **HR: 3**
3-ISONIPECOTYLINDOLE

mf: $C_{14}H_{16}N_2O$ mw: 228.32

SYN: KETONE, 3-INDOLYL 4-PIPERIDYL

DOT CLASSIFICATION: 3; Label: Flammable Liquid

SAFETY PROFILE: A poison by intravenous route. A flammable liquid. When heated to decomposition it emits toxic vapors of NO_x.

ILK000   **CAS: 503-01-5**   **HR: 3**
ISONYL

mf: $C_9H_{19}N$ mw: 141.29

PROP: Colorless, oily liquid; characteristic amine odor, water-insol. Mp: 176–178°.

SYNS: ISOMETHEPTENE □ 6-METHYLAMINO-2-METHYLHEPTENE □ METHYLISOOCTENYLAMINE □ 2-METHYL-6-METHYLAMINO-2-HEPTENE □ METHYL-OCTENYLAMINE □ OCTANIL □ OCTIN □ OCTINUM □ OCTON □ N-1,5-TRIMETHYL-4-HEXENYLAMINE

CONSENSUS REPORTS: Reported in EPA TSCA Inventory.

SAFETY PROFILE: A poison by ingestion, intravenous, intraperitoneal, and subcutaneous routes. Can cause headache, nausea, and dizziness in humans. When heated to decomposition it emits toxic fumes of NO_x.

ILL000   **CAS: 26952-21-6**   **HR: 2**
ISOOCTYL ALCOHOL

mf: $C_8H_{18}O$ mw: 130.26

SYN: ISOOCTANOL

CONSENSUS REPORTS: Reported in EPA TSCA Inventory.

OSHA PEL: TWA 50 ppm (skin)

ACGIH TLV: TWA 50 ppm (skin)

NIOSH REL: (Isooctyl Alcohol) TWA 50 ppm (skin)

SAFETY PROFILE: Moderately toxic by ingestion and skin contact. A skin and severe eye irritant. When heated to decomposition it emits acrid smoke and fumes. See also ALCOHOLS.

ILM000   **CAS: 543-82-8**   **HR: 3**
2-ISOOCTYLAMINE

mf: $C_8H_{19}N$ mw: 129.28

PROP: dl-Form: Viscous liquid; fishy odor. Bp: 154–156°, n: (24/D) 1.4200.

SYNS: AMIDRINE □ 2-AMINO-6-METHYLHEPTANE □ 6-AMINO-2-METHYLHEPTANE □ α,ε-DIMETHYL-HEXYLAMINE □ 1,5-DIMETHYLHEXYLAMINE □ 2-METHYL-6-AMINOHEPTANE □ 2-METHYL-2-HEPTYL-AMINE □ 6-METHYL-2-HEPTYLAMINE □ 2-METIL-6-AMINO-EPTANO (ITALIAN) □ OCTODRINE □ SKF 51 □ VAPORPAC

CONSENSUS REPORTS: Reported in EPA TSCA Inventory.

SAFETY PROFILE: A poison by intramuscular route. Moderately toxic by ingestion and subcutaneous routes. When heated to decomposition it emits toxic fumes of NO_x. See also AMINES.

ILR100   CAS: 27554-26-3   HR: 2
ISOOCTYL PHTHALATE
mf: $C_{24}H_{38}O_4$ mw: 390.62
PROP: Colorless viscous liquid. Bp: 370°. Flash pt: 227° C (CC). Insol in water.
SYNS: 1,2-BENZENEDICARBOXYLIC ACID, DIISOOCTYL ESTER □ BIS(6-METHYLHEPTYL)ESTER of PHTHALIC ACID □ CORFLEX 880 □ DIISOOCTYL PHTHALATE □ FLEXOL PLASTICIZER DIP □ HEXAPLAS M/O

CONSENSUS REPORTS: Reported in EPA TSCA Inventory.

SAFETY PROFILE: Moderately toxic by ingestion. Mildly toxic by skin contact. A skin irritant. A combustible liquid. When heated to decomposition it emits acrid smoke and irritating fumes.

IMB000   CAS: 110-46-3   HR: 3
ISOPENTYL NITRITE
mf: $C_5H_{11}NO_2$ mw: 117.17
PROP: Transparent, flammable liquid; penetrating, fragrant odor. Unstable; decomp on exposure to air and light. D: 0.872 @ 20°/4°, bp: 97–99°, autoign temp: 408°F, vap d: 4.0, flash p: <73.4°F.
SYNS: ISOAMYL NITRITE □ ISOPENTYL ALCOHOL NITRITE □ 3-METHYLBUTANOL NITRITE □ 3-METHYLBUTYL NITRITE □ NITROUS ACID-3-METHYL BUTYL ESTER □ VAPOROLE

CONSENSUS REPORTS: Reported in EPA TSCA Inventory.

SAFETY PROFILE: Poison by intravenous and intraperitoneal routes. Moderately toxic by ingestion. Mildly toxic by inhalation.

Mutation data reported. A recreational drug said to enhance sexual enjoyment in humans by inhalation. Dangerous fire hazard when exposed to spark, heat, oxidizers, or flame. Forms an explosive mixture in air or O_2. Vapors will explode when heated. When heated to decomposition it emits toxic fumes of NO_x. See also NITRITES.

IME000   CAS: 87-20-7   HR: 2
ISOPENTYL SALICYLATE
mf: $C_{12}H_{16}O_3$ mw: 208.28
PROP: Colorless liquid; pleasant odor. D: 1.047, refr index: 1.503–1.509, flash p: 271°F. Misc with alc, chloroform, ether, fixed oils; insol in glycerin, propylene glycol, water.
SYNS: FEMA No. 2084 □ ISOAMYL o-HYDROXY-BENZOATE □ ISOAMYL SALICYLATE (FCC) □ ISOPENTYL-2-HYDROXYPHENYL METHANOATE □ 3-METHYLBUTYL 2-HYDROXYBENZOATE □ SALICYLIC ACID, ISOPENTYL ESTER

CONSENSUS REPORTS: Reported in EPA TSCA Inventory.

SAFETY PROFILE: Experimental reproductive effects. Combustible liquid. When heated to decomposition it emits acrid smoke and fumes.

IMF400   CAS: 78-59-1   HR: 3
ISOPHORONE
mf: $C_9H_{14}O$ mw: 138.23
PROP: Practically water-white liquid. Bp: 215.2°, flash p: 184°F (OC), d: 0.9229, autoign temp: 864°F, vap press: 1 mm @ 38.0°, vap d: 4.77, lel: 0.8%, uel: 3.8%. IDLH 200 ppm.
SYNS: ISOACETOPHORONE □ ISOFORON □ ISOFORONE (ITALIAN) □ IZOFORON (POLISH) □ NCI-C55618 □ 1,1,3-TRIMETHYL-3-CYCLOHEXENE-5-ONE □ 3,5,5-TRIMETHYL-2-CYCLOHEXENE-1-ONE □ 3,5,5-TRIMETHYL-2-CYCLOHEXEN-1-ON (GERMAN, DUTCH) □ 3,5,5-TRIMETIL-2-CICLOESEN-1-ONE (ITALIAN)

CONSENSUS REPORTS: NTP Carcinogenesis Studies (gavage); Some Evidence: rat NTPTR* NTP-TR-291,86; (gavage); Equivocal Evidence: mouse NTPTR* NTP-TR-291,86. Reported in EPA TSCA Inventory.

OSHA PEL: TWA 4 ppm
ACGIH TLV: CL 5 ppm
DFG MAK: 2 ppm (11 mg/m³)
NIOSH REL: TWA (Ketones) 23 mg/m³
SAFETY PROFILE: Moderately toxic by ingestion. Mildly toxic by inhalation. Human systemic effects by inhalation: olfactory changes, conjunctiva irritation, and respiratory changes. Human systemic irritant by inhalation. A skin and severe eye irritant. Questionable carcinogen with experimental carcinogenic data. Mutation data reported. Considered to be more toxic than mesityl oxide. However, due to its low volatility, it is not a dangerous industrial hazard. The response of guinea pigs and rats to repeated inhalation of the vapors indicates that it is one of the most toxic of the ketones. It is chiefly a kidney poison. It can cause irritation, lachrymation, possible opacity of the cornea, and necrosis of the cornea (experimental). It is irritating at the level of 25 ppm to humans. In animal experiments death during exposure was usually due to narcosis, but occasionally due to irritation of the lungs.

Flammable and explosive when exposed to heat or flame; can react with oxidizing materials. To fight fire, use foam, CO₂, dry chemical. See also KETONES.

IMG000   CAS: 4098-71-9   HR: 3
ISOPHORONE DIISOCYANATE
DOT: UN 2290/UN 2906
mf: C₁₂H₁₈N₂O₂   mw: 222.32
PROP: D: 1.062 @ 20°/4°, bp: 217° @ 100 mm.
SYNS: CYCLOHEXANE, 5-ISOCYANATO-1-(ISOCYANATOMETHYL)-1,3,3-TRIMETHYL-(9CI) □ IPDI □ 3-ISOCYANATOMETHYL-3,5,5-TRIMETHYLCYCLO-HEXYLISOCYANATE □ ISOPHORONE DIAMINE DIISOCYANATE □ ISOPHORONEDIISOCYANATE, solution, 70%, by weight (DOT) □ TRIISOCYANATO-ISOCYANURATE, solution, 70%, by weight (DOT)
CONSENSUS REPORTS: EPA Extremely Hazardous Substances List. Reported in EPA TSCA Inventory.
OSHA PEL: TWA 0.005 ppm (skin)
ACGIH TLV: TWA 0.005 ppm (skin)

DFG MAK: 0.01 ppm (0.092 mg/m³)
NIOSH REL: (Diisocyanates) 10H TWA 0.005 ppm; CL 0.02 ppm/10M
DOT CLASSIFICATION: 3; Label: Flammable Liquid; DOT Class: 6.1; Label: KEEP AWAY FROM FOOD (UN 2290)
SAFETY PROFILE: Poison by inhalation. Moderately toxic by skin contact. A flammable liquid. When heated to decomposition it emits toxic fumes of NOₓ and CN⁻. See also ISOCYANATES.

IMG500   CAS: 7027-11-4   HR: 3
ISOPHORONENITRILE
mf: C₁₀H₁₅NO   mw: 165.26
SYNS: CYCLOHEXANECARBONITRILE, 5-OXO-1,3,3-TRIMETHYL- □ 5-OXO-1,3,3-TRIMETHYLCYCLOHEX-ANECARBONITRILE □ 3,5,5-TRIMETHYL-5-CYANO-CYCLOHEXANONE
SAFETY PROFILE: A poison by ingestion. Low toxicity by skin contact. A mild eye irritant. When heated to decomposition it emits toxic vapors of NOₓ.

IMH000   CAS: 3778-73-2   HR: 3
ISOPHOSPHAMIDE
mf: C₇H₁₅Cl₂N₂O₂P   mw: 261.11
PROP: White crystals. Mp: 48°–51°. Sol in water.
SYNS: A 4942 □ ASTA Z 4942 □ N,N-BIS(β-CHLORO-ETHYL)-AMINO-N'-O-PROPYLENE-PHOSPHORIC ACID ESTER DIAMIDE □ 2,3-(N,N(1)-BIS(2-CHLOROETHYL)-DIAMIDO)-1,3,2-OXAZAPHOSPHORIDINOXY □ N,3-BIS(2-CHLOROETHYL)TETRAHYDRO-2H-1,3,2-OXA-ZAPHOSPHORIN-2-AMINE 2-OXIDE □ N-(2-CHLORA-ETHYL)-N'-(2 CHLOROETHYL)-N'-o-PROPYLEN-PHOSPHORSAEUREESTER-DIAMID (GERMAN) □ 3-(2-CHLOROETHYL)-2-((2-CHLOROETHYL)AMINO)-PERHYDRO-2H-1,3,2-OXAZAPHOSPHORINE OXIDE □ 3-(2-CHLOROETHYL)-2-((2-CHLOROETHYL)AMINO)-TETRAHYDRO-2H-1,3,2-OXAZAPHOSPHORINE-2-OXIDE □ N-(2-CHLOROETHYL)-N'-(2-CHLOROETH-YL)-N',O-PROPYLENEPHOSPHORIC ACID DIAMIDE □ N-(2-CHLOROETHYL)-N'-(2-CHLOROETHYL)-N',O-PROPYLENEPHOSPHORIC ACID ESTER DIAMIDE □ CYFOS □ HOLOXAN □ IFOSFAMID □ IFOSFAMIDE □ IPHOSPHAMIDE □ ISOENDOXAN □ ISOFOSF-AMIDE □ MITOXANA □ MJF 9325 □ NAXAMIDE □ NCI-C01638 □ NSC-109724 □ Z 4942
CONSENSUS REPORTS: IARC Cancer Review: Group 3 IMEMDT 7,56,87; Animal Limited Evidence IMEMDT 26,237,81.

NCI Carcinogenesis Bioassay (ipr); Clear Evidence: mouse, rat NCITR* NCI-CG-TR-32,77. EPA Genetic Toxicology Program.

SAFETY PROFILE: Suspected carcinogen with experimental carcinogenic and neoplastigenic data. A poison by ingestion, intraperitoneal, subcutaneous, and intravenous routes. Human systemic effects by ingestion and intravenous routes: nausea or vomiting; proteinuria, hematuria, inflammation, necrosis or scarring of the bladder, and other kidney, ureter, or bladder changes; changes in hair covering the skin; leukopenia (decreased white blood cell count), thrombocytopenia (decrease in the number of blood platelets); hallucinations, distorted perceptions; tumorigenic effects (active as an anti-cancer agent). Experimental teratogenic and reproductive effects. Human mutation data reported. When heated to decomposition it emits very toxic Cl^-, NO_x, and PO_x.

IMS000   CAS: 78-79-5   HR: 3
ISOPRENE
DOT: UN 1218
mf: C_5H_8 mw: 68.13
PROP: Colorless, volatile liquid. Mp: −146.7°, bp: 34°, flash p: −65°F, d: 0.6806 @ 20°/4°, autoign temp: 428°F, vap press: 400 mm @ 15.4°, vap d: 2.35; fp: −145.95°. Insol in water; misc in alc and ether.
SYNS: ISOPRENE, INHIBITED (DOT) □ β-METHYLBIVINYL □ 2-METHYLBUTADIENE □ 2-METHYL-1,3-BUTADIENE (DOT)
CONSENSUS REPORTS: NTP 10th Report on Carcinogens. Reported in EPA TSCA Inventory.
DOT CLASSIFICATION: 3; Label: Flammable Liquid
SAFETY PROFILE: Confirmed carcinogen. Mildly toxic by inhalation. Irritating to skin, eyes, and mucous membranes. A concentration of 2% in air is not narcotic to mice but produces bronchial irritation. Highly dangerous fire hazard when exposed to heat, flame, or oxidizers. Reacts with air

to form dangerously unstable peroxides that can explode after concentration by evaporation. Ignites on contact with oxygen + ozone. Reacts with ozone to form explosive peroxides. Explosive reaction with vinylamine. Violent reaction with chlorosulfonic acid, HNO_3, oleum, H_2SO_4. Can react vigorously with reducing materials. To fight fire, use CO_2, dry chemical. When heated to decomposition it emits acrid smoke and fumes. See also PEROXIDES.

IMU000   CAS: 75-33-2   HR: 3
ISOPROPANETHIOL
DOT: UN 1228/UN 3071
mf: C_3H_8S mw: 76.17
PROP: Liquid, extremely powerful unpleasant odor. Mp: −130.7°, bp: 58–60°, d: 0.814 @ 60°/60°F, boiling range: 51–55°, flash p: −30°F. Sltly sol in water; misc in alc and ether.
SYNS: ISOPROPYL MERCAPTAN (DOT) □ ISOPROP-YLTHIOL □ 2-MERCAPTOPROPANE □ 1-METHYL-ETHANETHIOL □ 2-PROPANETHIOL □ 2-PROPYL MERCAPTAN
CONSENSUS REPORTS: Reported in EPA TSCA Inventory.
DOT CLASSIFICATION: 3; Label: Flammable Liquid, Poison (UN 1228); DOT Class: 6.1; Label: Poison, Flammable Liquid (UN 3071)
SAFETY PROFILE: Probably moderately toxic by inhalation. A flammable liquid and very dangerous fire hazard when exposed to heat, flame, or oxidizers. When heated to decomposition it emits highly toxic fumes of SO_x. See also MERCAPTANS.

IMW000   CAS: 513-42-8   HR: 3
ISOPROPENYL CARBINOL
DOT: UN 2614
mf: C_4H_8O mw: 72.12
PROP: Liquid. D: 0.852 @ 20°/4°C, bp: 114.5°. Sol in water @ 25°.
SYNS: METHALLYL ALCOHOL (DOT) □ 2-METHYL-2-PROPEN-1-OL
CONSENSUS REPORTS: Reported in EPA TSCA Inventory.

DOT CLASSIFICATION: 3; Label: Flammable Liquid
SAFETY PROFILE: Moderately toxic by ingestion and skin contact. Mildly toxic by inhalation. A skin irritant. Flammable when exposed to heat or flame. When heated to decomposition it emits acrid smoke and fumes. See also ALCOHOLS and ALLYL COMPOUNDS.

INA500   CAS: 109-59-1   HR: 2
2-ISOPROPOXYETHANOL
mf: $C_5H_{12}O_2$ mw: 104.17
PROP: Bp: 144° @ 743 mm.
SYNS: DOWANOL EIPAT □ ETHYLENE GLYCOL ISOPROPYL ETHER □ ETHYLENE GLYCOL, MONOISOPROPYL ETHER □ β-HYDROXYETHYL ISOPROPYL ETHER □ ISOPROPYL CELLOSOLVE □ ISOPROPYL GLYCOL □ MONOISOPROPYL ETHER of ETHYLENE GLYCOL
CONSENSUS REPORTS: Glycol ether compounds are on the Community Right-To-Know List. Reported in EPA TSCA Inventory.
OSHA PEL: TWA 25 ppm
ACGIH TLV: TWA 25 ppm
DFG MAK: 5 ppm (22 mg/m³)
SAFETY PROFILE: Moderately toxic by skin contact and intraperitoneal routes. Mildly toxic by inhalation and ingestion. A skin and eye irritant. When heated to decomposition it emits acrid smoke and fumes. See also GLYCOL ETHERS.

INE062   CAS: 77276-08-5   HR: 3
2-ISOPROPOXYPHENYL (METHYL)(T-BUTOXYSULFINYL)CARBAMATE
mf: $C_{15}H_{23}NO_5S$ mw: 329.45
SYNS: CARBAMIC ACID, ((((1,1-DIMETHYLETHOXY)-SULFINYL)METHYL)-, 2-(1-METHYLETHOXY)PHENYL ESTER □ 2-(1-METHYLETHOXY)PHENYL ((((1,1-DIMETHYLETHOXY)SULFINYL)METHYL)CARBAMATE
SAFETY PROFILE: A poison by ingestion. When heated to decomposition it emits toxic vapors of NO_x and SO_x.

INE100   CAS: 108-21-4   HR: 3
ISOPROPYL ACETATE
DOT: UN 1220
mf: $C_5H_{10}O_2$ mw: 102.15

PROP: Colorless, aromatic liquid. Mp: −73°, bp: 88.4°, lel: 1.8%, uel: 7.8%, fp: −69.3°, flash p: 40°F, d: 0.874 @ 20°/20°, autoign temp: 860°F, vap press: 40 mm @ 17.0°. Sltly sol in water; misc in alc, ether, fixed oils. IDLH 1800 ppm.
SYNS: ACETATE d'ISOPROPYLE (FRENCH) □ ACETIC ACID ISOPROPYL ESTER □ ACETIC ACID-1-METHYLETHYL ESTER (9CI) □ 2-ACETOXYPROPANE □ FEMA No. 2926 □ ISOPROPILE (ACETATO di) (ITALIAN) □ ISOPROPYLACETAAT (DUTCH) □ ISOPROPYLACETAT (GERMAN) □ ISOPROPYL (ACETATE d') (FRENCH) □ ISOPROPYLESTER KYSELINY OCTOVE □ 2-PROPYL ACETATE
CONSENSUS REPORTS: Reported in EPA TSCA Inventory.
OSHA PEL: TWA 250 ppm; STEL 310 ppm
ACGIH TLV: TWA 100 ppm; STEL 200 ppm
DFG MAK: 200 ppm (850 mg/m³)
DOT CLASSIFICATION: 3; Label: Flammable Liquid
SAFETY PROFILE: Moderately toxic by ingestion. Mildly toxic by inhalation. See also ESTERS. Human systemic irritant effects by inhalation and systemic eye effects by an unspecified route. Narcotic in high concentration. Chronic exposure can cause liver damage. Highly flammable liquid. Dangerous fire hazard when exposed to heat, flame, or oxidizers. Moderately explosive when exposed to heat or flame. Dangerous; keep away from heat and open flame; can react vigorously with oxidizing materials. To fight fire, use foam, CO_2, dry chemical. See also ESTERS.

INH000   CAS: 2210-25-5   HR: 3
N-ISOPROPYLACRYLAMIDE
mf: $C_6H_{11}NO$ mw: 113.18
PROP: Cream colored powder. Mp: 64°. Sol in water.
SYNS: ISOPROPYL ACRYLAMIDE □ ISOPROPYL-AMID KYSELINY AKRYLOVE □ N-(1-METHYLETHYL)-2-PROPENAMIDE □ NIPAM □ 2-PROPENAMIDE, N-(1-METHYLETHYL)-(9CI)
CONSENSUS REPORTS: Reported in EPA TSCA Inventory.
SAFETY PROFILE: Poison by ingestion. Moderately toxic by intraperitoneal route.

Experimental reproductive effects. Mutation data reported. When heated to decomposition it emits toxic fumes of NO_x. See also AMIDES.

INJ000   CAS: 67-63-0   HR: 3
ISOPROPYL ALCOHOL
DOT: UN 1219
mf: C_3H_8O mw: 60.11
PROP: Clear, colorless liquid; slt odor, sltly bitter taste. Mp: −88.5 to −89.5°, bp: 82.5°, lel: 2.5%, uel: 12%, flash p: 53°F (CC), d: 0.7854 @ 20°/4°, refr index: 1.377 @ 20°, vap d: 2.07, ULC: 70, fp: −89.5°, autoign temp: 852°F. Misc with water, alc, ether, chloroform; insol in salt solns. IDLH 2000 ppm [10%LEL].
SYNS: ALCOOL ISOPROPILICO (ITALIAN) □ ALCOOL ISOPROPYLIQUE (FRENCH) □ DIMETHYL-CARBINOL □ ISOHOL □ ISOPROPANOL (DOT) □ ISO-PROPYLALKOHOL (GERMAN) □ LUTOSOL □ PETROHOL □ i-PROPANOL (GERMAN) □ PROPAN-2-OL □ 2-PROPANOL □ sec-PROPYL ALCOHOL (DOT) □ i-PROPYLALKOHOL (GERMAN) □ SPECTRAR
CONSENSUS REPORTS: IARC Cancer Review: Group 3 IMEMDT 7,229,87. The isopropyl alcohol strong acid manufacturing process is on the Community Right-To-Know List. EPA Genetic Toxicology Program. Reported in EPA TSCA Inventory.
OSHA PEL: TWA 400 ppm; STEL 500 ppm
ACGIH TLV: TWA 200 ppm; STEL 400 ppm; Not Classifiable as a Human Carcinogen
DFG MAK: 200 ppm (500 mg/m³)
NIOSH REL: (Isopropyl Alcohol) TWA 400 ppm; CL 800 ppm/15M
DOT CLASSIFICATION: 3; Label: Flammable Liquid
SAFETY PROFILE: Moderately toxic to humans by an unspecified route. Moderately toxic experimentally by intravenous and intraperitoneal routes. Mildly toxic by skin contact. Human systemic effects by ingestion or inhalation: flushing, pulse rate decrease, blood pressure lowering, anesthesia, narcosis, headache, dizziness, mental depression, hallucinations, distorted perceptions, dyspnea, respiratory depression, nausea or vomiting, coma. Experimental teratogenic and reproductive effects. Mutation data reported. An eye and skin irritant. Questionable carcinogen.

The single lethal dose for a human adult is about 250 mL, although as little as 100 mL can be fatal. It can cause corneal burns and eye damage. Acts as a local respiratory irritant and in high concentration as a narcotic. It has good warning properties because it causes a mild irritation of the eyes, nose, and throat at a concentration level of 400 ppm. It may induce a mild narcosis, the effects of which are usually transient, and it is somewhat less toxic than the normal isomer, but twice as volatile.

There is some evidence that humans can acquire a slight tolerance to this material. It is absorbed by the skin, but single or repeated applications on the skin of rats, rabbits, dogs, or human beings induced no untoward effects. It acts very much like ethanol in regard to absorption, metabolism, and elimination but with a stronger narcotic action. Chronic injuries have been detected in animals. Workers producing isopropanol show an excess of sinus and laryngeal cancers. This may be caused, completely or in part, by the by-product, isopropyl oil. Humans have ingested up to 20 mL diluted with water and noticed only a sensation of heat and slight lowering of the blood pressure. There are, however, reports of serious illness from as little as 10 mL taken internally. A common air contaminant.

Flammable liquid. A very dangerous fire hazard when exposed to heat, flame, or oxidizers. Moderately explosive when exposed to heat or flame. Reacts with air to form dangerous peroxides. The presence of 2-butanone increases the reaction rate for peroxide formation. Hydrogen peroxide sharply reduces the autoignition temperature. Violent explosive reaction when heated with aluminum isopropoxide + crotonaldehyde + heat. Forms explosive mixtures with trinitromethane, hydrogen peroxide (similar in power and sensitivity to glyceryl nitrate). Reacts with barium perchlorate to form the highly explosive propyl perchlorate. Ignites on contact with dioxygenyl tetrafluoroborate, chromium trioxide, potassium tert-butoxide (after a delay). Reacts with oxygen to form dangerously unstable peroxides. Vigorous reaction with sodium dichromate + sulfuric acid, aluminum (after a delay period). Reacts violently with H_2 + Pd, nitroform, oleum, $COCl_2$, Al triisopropoxide, oxidants. Can react vigorously with oxidizing materials. To fight fire, use CO_2, dry chemical, alcohol foam. When heated to decomposition it emits acrid smoke and fumes. See also ALCOHOLS.

INK000   CAS: 75-31-0   HR: 3
ISOPROPYLAMINE
DOT: UN 1221
mf: C_3H_9N mw: 59.13
PROP: Colorless liquid; amino odor. Mp: $-101.2°$, flash p: $-35°F$ (OC), d: 0.694 @ $15°/4°$, autoign temp: $756°F$, d: 2.03, bp: $33–34°$, lel: 2.3%, uel: 10.4%. Misc with water, alc, and ether.
SYNS: 2-AMINO-PROPAAN (DUTCH) □ 2-AMINO-PROPAN (GERMAN) □ 2-AMINOPROPANE □ 2-AMINO-PROPANO (ITALIAN) □ ISOPROPILAMINA (ITALIAN) □ 1-METHYLETHYLAMINE □ MONOISOPROPYLAMINE □ 2-PROPANAMINE □ sec-PROPYLAMINE □ 2-PROPYLAMINE
CONSENSUS REPORTS: Reported in EPA TSCA Inventory.
OSHA PEL: TWA 5 ppm; STEL 10 ppm
ACGIH TLV: TWA 5 ppm; STEL 10 ppm
DFG MAK: 5 ppm (12 mg/m³)
DOT CLASSIFICATION: 3; Label: Flammable Liquid; DOT Class: 3; Label: Flammable Liquid, Corrosive
SAFETY PROFILE: Poison by skin contact. Moderately toxic by ingestion. Mildly toxic by inhalation. A severe skin and eye irritant. Occasionally contact causes sensitization. Narcotic in high concentration. Very dangerous fire hazard and moderate explosion hazard when exposed to sparks, heat, flame, or oxidizers. Can react

vigorously with oxidizing materials. Reacts with perchloryl fluoride to form an explosive liquid. Incompatible with 1-chloro-1,3-epoxypropane. To fight fire, use alcohol foam, CO_2, dry chemical. When heated to decomposition it emits toxic fumes of NO_x. See also AMINES.

INN400   CAS: 109-56-8   HR: 2
N-ISOPROPYLAMINOETHANOL
mf: $C_5H_{13}NO$ mw: 103.19
PROP: Oil. Bp: 150–152° @ 13 mm. Sol in H_2O.
SYNS: ETHANOLISOPROPYLAMINE □ ETHANOL, 2-(ISOPROPYLAMINO)- □ ETHANOL, 2-((1-METHYLETHYL)AMINO)- (9CI) □ (N-HYDROXYETHYL)ISOPROPYLAMINE □ ISOPROPYLAMINOETHANOL □ 2-ISOPROPYLAMINOETHANOL □ N-ISOPROPYLETHANOLAMINE □ MONOISOPROPYLAMINOETHANOL
CONSENSUS REPORTS: Reported in EPA TSCA Inventory.
SAFETY PROFILE: Moderately toxic by ingestion. Experimental reproductive effects. When heated to decomposition it emits toxic fumes of NO_x.

INX000   CAS: 768-52-5   HR: 2
N-ISOPROPYLANILINE
mf: $C_9H_{13}N$ mw: 135.23
PROP: Oil. Bp: 206–208°.
CONSENSUS REPORTS: Reported in EPA TSCA Inventory.
OSHA PEL: TWA 2 ppm (10 mg/m³)(skin)
ACGIH TLV: TWA 2 ppm (skin)
NIOSH REL: (N-Isopropylaniline) TWA 2 ppm (skin)
SAFETY PROFILE: Moderately toxic by ingestion and inhalation. When heated to decomposition it emits toxic fumes of NO_x.

INY000   CAS: 479-92-5   HR: 3
4-ISOPROPYLANTIPYRINE
mf: $C_{14}H_{18}N_2O$ mw: 230.34
PROP: A solid. Mp: 103°.
SYNS: 1,2-DIHYDRO-1,5-DIMETHYL-4-((1-METHYLETHYL)AMINO)-2-PHENYL-3H-PYRAZOL-3-ONE □ ISOPROPYLANTIPYRIN □ ISOPROPYLANTIPYRINE □ 4-ISOPROPYL-2,3-DIMETHYL-1-PHENYL-3-PYRAZOLIN-5-ONE □ ISOPROPYLPHENAZONE □ ISOPYRINE □ LARODON □ 1-PHENYL-2,3-DIMETHYL-4-ISOPROPYL-

3-PYRAZOLIN-5-ONE □ 1-PHENYL-2,3-DIMETHYL-4-ISOPROPYLPYRAZOL-5-ONE □ PROPYPHENAZONE

CONSENSUS REPORTS: Reported in EPA TSCA Inventory.

SAFETY PROFILE: A poison by ingestion and intraperitoneal routes. When heated to decomposition it emits toxic fumes of NO_x.

IOB000   CAS: 80-15-9   HR: 3
ISOPROPYLBENZENE
**   HYDROPEROXIDE**

mf: $C_9H_{12}O_2$ mw: 152.21

PROP: A liquid. Bp: 153°, flash p: 175°F, d: 1.05. The hydroperoxide of cumene.

SYNS: CUMEENHYDROPEROXYDE (DUTCH) □ CUMENE HYDROPEROXIDE (DOT) □ CUMENE HYDROPEROXIDE, TECHNICALLY PURE (DOT) □ CUMENT HYDROPEROXIDE □ CUMENYL HYDROPEROXIDE □ CUMOLHYDROPEROXID (GERMAN) □ CUMYL HYDROPEROXIDE □ α-CUMYL HYDROPEROXIDE □ CUMYL HYDROPEROXIDE, TECHNICAL PURE (DOT) □ α,α-DIMETHYLBENZYL HYDROPEROXIDE (MAK) □ HYDROPEROXYDE de CUMENE (FRENCH) □ HYDROPEROXYDE de CUMYLE (FRENCH) □ IDROPEROSSIDO di CUMENE (ITALIAN) □ IDROPEROSSIDO di CUMOLO (ITALIAN) □ RCRA WASTE NUMBER U096

CONSENSUS REPORTS: Community Right-To-Know List. Reported in EPA TSCA Inventory. EPA Genetic Toxicology Program.

DFG MAK: Moderate Skin Effects

SAFETY PROFILE: A poison by ingestion and intraperitoneal routes. Moderately toxic by skin contact, inhalation and, subcutaneous routes. Mutation data reported. A skin and eye irritant. Questionable carcinogen with experimental tumorigenic data. A strong oxidizing agent. Flammable when exposed to heat or flame; can react with reducing materials. Its use in industry has resulted in many explosions. Storage above 109°C may cause explosive decomposition. Potentially explosive reactions with acids or reductants. Violent or explosive reaction when heated with solutions of 1,2-dibromo-1,2-diisocyanatoethane polymers in benzene. Violent decomposition on contact with cobalt, copper, copper alloys, lead alloys,

mineral acids. Vigorous exothermic reaction on contact with charcoal. When heated to decomposition it emits acrid smoke and fumes. To fight fire, use foam, CO_2, dry chemical. See also PEROXIDES.

IOD050   CAS: 13816-33-6   HR: 2
p-ISOPROPYLBENZONITRILE

mf: $C_{10}H_{11}N$ mw: 145.22

PROP: A liquid. Bp: 85° @ 0.7 mm.

SYNS: BENZONITRILE, p-ISOPROPYL- □ BENZONITRILE, 4-(1-METHYLETHYL)- □ CUMINYL NITRILE □ p-CYANOCUMENE □ 4-(1-METHYLETHYL)-BENZONITRILE

CONSENSUS REPORTS: Reported in EPA TSCA Inventory.

SAFETY PROFILE: Moderately toxic by ingestion. A skin irritant. When heated to decomposition it emits toxic fumes of NO_x.

IOI000   CAS: 5419-55-6   HR: 3
ISOPROPYL BORATE

DOT: UN 2616

mf: $C_9H_{21}BO_3$ mw: 188.11

PROP: Colorless moisture-sensitive liquid. Mp: −59°, bp: 141.0–142.4°, flash p: 82°F (TCC), d: 0.8138 @ 25°. Sol in non-hydroxylic solvs.

SYNS: BORIC ACID, TRIISOPROPYL ESTER □ TRIISOPROPYL BORATE (DOT)

CONSENSUS REPORTS: Reported in EPA TSCA Inventory.

DOT CLASSIFICATION: 3; Label: Flammable Liquid

SAFETY PROFILE: A poison by intravenous route. Moderately toxic by ingestion. An eye irritant. A flammable liquid and dangerous fire hazard when exposed to heat, flame, or oxidizers. Can react vigorously with oxidizing materials. To fight fire, use foam, CO_2, dry chemical. See also ESTERS and BORON COMPOUNDS.

IOJ000   CAS: 1746-77-6   HR: 2
ISOPROPYL CARBAMATE

mf: $C_4H_9NO_2$ mw: 103.14

PROP: Prisms. Mp: 60–61°, bp: 200°C. Very sol in water, alc, and ether.
SYNS: CARBAMIC ACID, ISOPROPYL ESTER □ CARBAMIC ACID-1-METHYLETHYL ESTER □ ISOPROPYLESTER KYSELINY KARBAMINOVE
CONSENSUS REPORTS: Reported in EPA TSCA Inventory.
SAFETY PROFILE: Moderately toxic by subcutaneous route. Mutation data reported. Questionable carcinogen with experimental neoplastigenic data. When heated to decomposition it emits toxic fumes of NO_x. See also CARBAMATES.

IOJ500   CAS: 5412-01-1   HR: 3
ISOPROPYL CARBITOL
mf: $C_7H_{16}O_3$ mw: 148.23
SYNS: DIETHYLENE GLYCOL ISOPROPYL ETHER □ ETHANOL, 2-(2-ISOPROPOXYETHOXY)- □ ETHANOL, 2-(2-(1-METHYLETHOXY)ETHOXYL)- □ GLYCOSOLVE DIP □ 2-(2-(1-METHYLETHOXY)ETHOXYL)ETHANOL
SAFETY PROFILE: A poison by ingestion and skin contact. A moderate eye irritant. When heated to decomposition it emits acrid smoke and irritating vapors.

IOL000   CAS: 108-23-6   HR: 3
ISOPROPYL CHLOROCARBONATE
DOT: UN 2407
mf: $C_4H_7ClO_2$ mw: 122.56
PROP: A clear, colorless, volatile liquid with a pungent, irritating odor. D: 1.078 @ 20°/4°, bp: 105°, flash p: 28°C (TOC), 20°C (TCC), fp: −80°, fire point: 40°C (TOC), autoign temp: >500°, vap d: 4.2 @ 20°, refr index: 1.3974 @ 20°, lel: 4%, uel: 15%, vap press: 72 mm @ 70°F, bulk d: 9.0 lbs/gal, percent volatile: 100%. Sol in aromatic or aliphatic hydrocarbon solvents, ethyl ether, acetone, and chloroform. Insol in water and alc. Decomp slowly in cold water, faster in hot water. Misc in ether and benzene. A phosgene derivative.
SYNS: CARBONOCHLORIDE ACID-1-METHYL ESTER □ CHLOROFORMIC ACID ISOPROPYL ESTER □ ISOPROPYL CHLOROFORMATE □ ISOPROPYL CHLOROMETHANOATE
CONSENSUS REPORTS: EPA Extremely Hazardous Substances List. Reported in EPA TSCA Inventory.

DOT CLASSIFICATION: 3; Label: Flammable Liquid, Corrosive, Poison
SAFETY PROFILE: A poison by skin contact and ingestion. Moderately toxic by inhalation. Ingestion of even small amounts can be fatal. A skin and severe eye irritant. Inhalation of a small amount can cause immediate lachrymation, coughing, choking, and respiratory distress. Death may result from pulmonary edema which may not appear for several hours after exposure. A dangerous fire and moderate explosion hazard when exposed to heat, spark, or flame. Self-reactive. Iron salts may catalyze a potentially explosive thermal decomposition. Incompatible with water, iron, metal salts, acids, alkalies, amines, alcohols. Stable under refrigeration below 20°, but one reference (1973) reports that it has exploded while stored in a refrigerator. Present-day formulations appear to be more stable. Temperatures above 20° can cause decomposition. When heated to decomposition it emits acrid smoke and fumes.

IOO222   HR: D
ISOPROPYL CITRATE
PROP: Viscous colorless syrup. Crystallizes upon standing.
SAFETY PROFILE: When heated to decomposition it emits acrid smoke and irritating fumes.

IOO300   CAS: 4621-04-9   HR: 2
4-ISOPROPYLCYCLOHEXANOL
mf: $C_9H_{18}O$ mw: 142.27
PROP: Fragrance chemical.
SYNS: CYCLOHEXANOL, p-ISOPROPYL- □ p-ISOPROPYLCYCLOHEXANOL
CONSENSUS REPORTS: Reported in EPA TSCA Inventory.
SAFETY PROFILE: Moderately toxic by ingestion. A skin irritant. When heated to decomposition it emits acrid smoke and irritating fumes.

IOY000   CAS: 94-11-1   HR: 3
ISOPROPYL-2,4-D ESTER
mf: $C_{11}H_{12}Cl_2O_3$ mw: 263.13
PROP: Bp: 139–140° @ 1 mm.
SYNS: (2,4-DICHLOROPHENOXY)ACETIC ACID, ISOPROPYL ESTER □ (2-4-DICHLOROPHENOXY)ACETIC ACID-1-METHYLETHYL ESTER (9CI) □ 2,4-D ISOPROPYL ESTER □ ESTERON 44 □ WEEDONE 128
CONSENSUS REPORTS: IARC Cancer Review: Animal Inadequate Evidence IMEMDT 15,111,77.
SAFETY PROFILE: Poison by ingestion. Experimental teratogenic and reproductive effects. Questionable carcinogen with experimental tumorigenic data. Used as a pesticide. When heated to decomposition it emits toxic fumes of Cl⁻. See also ESTERS.

IOZ750   CAS: 108-20-3   HR: 3
ISOPROPYL ETHER
DOT: UN 1159
mf: $C_6H_{14}O$ mw: 102.20
PROP: Colorless liquid; ethereal odor. Mp: −60°, bp: 68.5°, lel: 1.4%, uel: 7.9%, flash p: −18°F (CC), d: 0.719 @ 25°, autoign temp: 830°F, vap press: 150 mm @ 25°, vap d: 3.52. Misc in water. IDLH 1400 ppm [10%LEL].
SYNS: DIISOPROPYL ETHER □ DIISOPROPYL OXIDE □ ETHER ISOPROPYLIQUE (FRENCH) □ 2-ISOPROPOXYPROPANE □ IZOPROPYLOWY ETER (POLISH)
OSHA PEL: TWA 500 ppm
ACGIH TLV: TWA 250 ppm; STEL 310 ppm
DFG MAK: 500 ppm (2100 mg/m³)
DOT CLASSIFICATION: 3; Label: Flammable Liquid
SAFETY PROFILE: Moderately toxic by intraperitoneal route. Mildly toxic by ingestion, inhalation, and skin contact. A skin irritant. A very dangerous fire hazard and severe explosion hazard when exposed to heat, flame, sparks, or oxidizers. Under some conditions shock will explode it. Dangerous; on exposure to air it rapidly forms very sensitive, explosive peroxides that precipitate as crystals. Violent reaction with chlorosulfonic acid, HNO_3. Potentially dangerous reaction with propionyl chloride can burst a sealed container. Reacts vigorously with oxidizing materials. To fight fire, use alcohol foam, CO_2, foam, dry chemical. When heated to decomposition it emits acrid smoke and fumes. See also ETHERS.

IPC000   CAS: 625-55-8   HR: 3
ISOPROPYL FORMATE
mf: $C_4H_8O_2$ mw: 88.12
PROP: Clear liquid. Bp: 68.3°, flash p: 22°F (CC), d: 0.873, autoign temp: 905°F, vap press: 100 mm @ 17.8°, vap d: 3.03.
SYN: FORMIC ACID, ISOPROPYL ESTER
CONSENSUS REPORTS: Reported in EPA TSCA Inventory.
SAFETY PROFILE: A poison by ingestion. A toxic fumigant. A very dangerous fire hazard when exposed to heat, spark, or flame. Can react vigorously with oxidizing materials. When heated to decomposition it emits acrid smoke and fumes. To fight fire, use alcohol foam, foam, CO_2, dry chemical. See also ESTERS.

IPD000   CAS: 4016-14-2   HR: 2
ISOPROPYL GLYCIDYL ETHER
mf: $C_6H_{12}O_2$ mw: 116.18
PROP: A liquid. IDLH 400 ppm.
SYNS: 1,2-EPOXY-3-ISOPROPOXYPROPANE □ 2,3-EPOXYPROPYL ISOPROPYL ETHER □ GLYCIDYL ISOPROPYL ETHER □ IGE □ IGE (OSHA) □ (ISOPROPOXYMETHYL)OXIRANE □ ISOPROPYL EPOXYPROPYL ETHER □ 3-ISOPROPYLOXYPROPYL-ENE OXIDE □ ((1-METHYLETHOXY)METHYL)-OXIRANE □ NCI-C56439 □ OXIRANE, ((1-METHYL-ETHOXY)METHYL)-(9CI)
CONSENSUS REPORTS: Glycol ether compounds are on the Community Right-To-Know List. Reported in EPA TSCA Inventory.
OSHA PEL: TWA 50 ppm; STEL 75 ppm
ACGIH TLV: TWA 50 ppm; STEL 75 ppm
DFG MAK: Confirmed Animal Carcinogen with Unknown Relevance to Humans
NIOSH REL: (Glycidyl Ethers) CL 240 mg/m³/15M

SAFETY PROFILE: Suspected carcinogen. Moderately toxic by ingestion. Mildly toxic by inhalation and skin contact. A skin and eye irritant. Mutation data reported. When heated to decomposition it emits acrid smoke and fumes. See also GLYCOL ETHERS.

IPO000   CAS: 25068-38-6   HR: 2
4,4'-ISOPROPYLIDENEDIPHENOL, POLYMER with 1-CHLORO-2,3-EPOXYPROPANE
mf: $(C_{15}H_{16}O_2 \cdot C_3H_5ClO)_x$
SYNS: EPIDIAN 5 □ EPON 828
CONSENSUS REPORTS: EPA Genetic Toxicology Program. Reported in EPA TSCA Inventory.
SAFETY PROFILE: Moderately toxic by intraperitoneal route. Very slightly toxic by ingestion. Experimental teratogenic effects. Other experimental reproductive effects. A skin and eye irritant. When heated to decomposition it emits toxic fumes of Cl$^-$. See also other 4,4'-isopropylidenediphenol entries.

IPS100   CAS: 73791-43-2   HR: 3
ISOPROPYL ISOBUTYL ARSINIC ACID
mf: $C_7H_{17}AsO_2$ mw: 208.16
SYNS: ARSINE OXIDE, HYDROXYISOBUTYLISOPROPYL- □ HYDROXYISOBUTYLISOPROPYLARSINE OXIDE
OSHA PEL: TWA 0.5 mg(As)/m^3
SAFETY PROFILE: Poison by intravenous route. When heated to decomposition it emits toxic fumes of As.

IPX000   CAS: 107-44-8   HR: 3
ISOPROPYL METHANEFLUOROPHOS-PHONATE
mf: $C_4H_{10}FO_2P$ mw: 140.11
PROP: Bp: 147°, fp: −58°, d: 1.100 @ 20°, vap press: 1.57 mm @ 20°, vap d: 4.86.
SYNS: FLUOROISOPROPOXYMETHYLPHOSPHINE OXIDE □ GB □ IMPF □ ISOPROPHYL METHYL-PHOSPHONOFLUORIDATE □ ISOPROPOXYMETHYL-PHORYL, FLUORIDE □ ISOPROPYL METHYLFLUORO-PHOSPHATE □ ISOPROPYL METHYLPHOSPHONO-FLUORIDATE □ O-ISOPROPYL METHYLPHOSPHONO-FLUORIDATE □ ISOPROPYL-METHYL-PHOSPHORYL

FLUORIDE □ METHYLFLUOROPHOSPHORIC ACID, ISOPROPYL ESTER □ METHYLFLUORPHOSPHOR-SAEUREISOPROPYLESTER (GERMAN) □ METHYL-PHOSPHONOFLUORIDIC ACID ISOPROPYL ESTER □ METHYLPHOSPHONOFLUORIDIC ACID-1-METHYLETHYL ESTER □ MFI □ SARIN □ SARIN II □ T-144 □ T-2106 □ TL 1618 □ TRILONE 46
CONSENSUS REPORTS: EPA Extremely Hazardous Substances List. Reported in EPA TSCA Inventory.
SAFETY PROFILE: A deadly human poison by skin contact and inhalation. (A small drop on the skin can kill a man.) A deadly experimental poison by ingestion, inhalation, skin contact, subcutaneous, intravenous, intramuscular, and intraperitoneal routes. Human systemic effects: muscle weakness, bronchiolar constriction, nausea or vomiting, flaccid paralysis without anesthesia, miosis (pupillary constriction), cholinesterase inhibition. A "nerve gas" used as a chemical warfare agent. To fight fire, use foam, CO$_2$, dry chemical. When heated to decomposition or reacted with steam, it emits very toxic fumes of F$^-$ and PO$_x$. See also PARATHION.

IPY000   CAS: 926-06-7   HR: 3
ISOPROPYLMETHANESULFONATE
mf: $C_4H_{10}O_3S$ mw: 138.20
SYNS: IMS □ ISOPROPYL MESYLATE □ ISOPROPYL METHANE SULPHONATE □ METHANESULFONIC ACID-1-METHYLETHYL ESTER
CONSENSUS REPORTS: EPA Genetic Toxicology Program. Reported in EPA TSCA Inventory.
SAFETY PROFILE: Poison by ingestion. An experimental teratogen. Experimental reproductive effects. Questionable carcinogen with experimental carcinogenic and tumorigenic data. Human mutation data reported. When heated to decomposition it emits toxic fumes of SO$_x$. See also SULFONATES.

IQN000   CAS: 110-27-0   HR: 1
ISOPROPYL MYRISTATE
mf: $C_{17}H_{34}O_2$ mw: 270.44
PROP: Liquid of low viscosity, odorless. Bp: 192.6° @ 20 mm, decomp @ 208°, d:

0.8532 @ 20°. Sol in castor oil, cottonseed oil, acetone, chloroform, ethyl acetate, ethanol, toluene, and mineral oil. Insol in water, glycerol, and propylene glycerol. Dissolves many waxes.
SYNS: BISOMEL □ CRODAMOL IPM □ DELTYLEXTRA □ EMCOL-IM □ EMEREST 2314 □ ISOMYST □ ISOPROPYL TETRADECANOATE □ JA-FA IPM □ KESSCO ISOPROPYL MYRISTATE □ KESSCO-MIR □ MYRISTIC ACID, ISOPROPYL ESTER □ PLYMOUTM IPM □ PROMYR □ STARFOL IPM □ STEPAN D-50 □ TEGESTER □ TETRADECANOIC ACID, ISOPROPYL □ TETRADECANOIC ACID, 1-METHYLETHYL ESTER □ 1-TRIDECANECARBOXYLIC ACID, ISOPROPYL ESTER □ UNIMATE IPM □ WICKENOL 101
CONSENSUS REPORTS: Reported in EPA TSCA Inventory.
SAFETY PROFILE: Low toxicity by ingestion and skin contact. A human skin irritant. When heated to decomposition it emits toxic smoke and fumes.

IQP000   CAS: 1712-64-7   HR: 3
ISOPROPYL NITRATE
DOT: UN 1222
mf: $C_3H_7NO_3$ mw: 105.11
PROP: A liquid. Bp: 102°, d: 1.036 @ 19°, flash p: 51.8°F, uel: 100%.
SYN: NITRIC ACID, ISOPROPYL ESTER
CONSENSUS REPORTS: Reported in EPA TSCA Inventory.
DOT CLASSIFICATION: 3; Label: Flammable Liquid
SAFETY PROFILE: Mildly toxic by inhalation. Flammable liquid and dangerous fire hazard when exposed to heat, spark, or flames. Ignites spontaneously when compressed. The pure vapor ignites spontaneously at very low temperatures and pressures. An explosive of low sensitivity. It can be used as a rocket monopropellant. When heated to decomposition it emits toxic fumes of NO_x. Incompatible with Lewis acids. See also NITRATES.

IQQ000   CAS: 541-42-4   HR: 3
ISOPROPYL NITRITE
mf: $C_3H_7NO_2$ mw: 89.10

PROP: Bp: 39–40° @ 745 mm, flash p: <50°F.
SYNS: ISOPROPYLESTER KYSELINY DUSITE □ NITROUS ACID, 1-METHYLETHYL ESTER (9CI) □ 2-PROPANOL NITRITE
CONSENSUS REPORTS: Reported in EPA TSCA Inventory.
SAFETY PROFILE: Moderately toxic by inhalation. Can cause vasodilation with fall in blood pressure, tachycardia, headache. Large doses can cause methemoglobinuria and cyanosis. Severe poisoning results in shock that can be fatal. A very dangerous fire hazard when exposed to heat, spark, or flame. When heated to decomposition it emits toxic fumes of NO_x. See also NITRITES.

IQU000   HR: 3
ISOPROPYL OILS
PROP: A by-product of isopropyl alcohol manufacture composed of trimeric and tetrameric polypropylene + small amounts of benzene, toluene, alkyl benzenes, polyaromatic ring compounds, hexane, heptane, acetone, ethanol, isopropyl ether, and isopropyl alcohol (IARC** 15,225,77).
CONSENSUS REPORTS: IARC Cancer Review: Animal Inadequate Evidence IMEMDT 15,223,77; Human Limited Evidence IMEMDT 15,223,77.
DFG MAK: Suspected Carcinogen
SAFETY PROFILE: Suspected carcinogen with experimental neoplastigenic data. When heated to decomposition they emit acrid smoke and fumes.

IQX100   CAS: 88-69-7   HR: 3
o-ISOPROPYLPHENOL
mf: $C_9H_{12}O$ mw: 136.21
PROP: Colorless or yellow liquid. Mp: 16°, bp: 215°, d: 1.102. Flash pt: 88°.
SYNS: 2-ISOPROPYLPHENOL □ 2-(1-METHYLETHYL)PHENOL □ PHENOL, o-ISOPROPYL- □ PHENOL, 2-(1-METHYLETHYL)-(9CI) □ PRODOX 131
CONSENSUS REPORTS: Reported in EPA TSCA Inventory.
DOT CLASSIFICATION: 6.1; Label: KEEP AWAY FROM FOOD

SAFETY PROFILE: Poison by intravenous route. A combustible liquid. When heated to decomposition it emits acrid smoke and irritating vapors.

IQZ000   CAS: 99-89-8   HR: 3
p-ISOPROPYLPHENOL
mf: $C_9H_{12}O$ mw: 136.21
PROP: Needles from pet ether. D: 0.990 @ 20°, mp: 61°, bp: 223–225°. Very sltly sol in water; sol in alc @ 25°, sol in ether @ 25°.
SYNS: AUSTRALOL □ p-CUMENOL □ 4-ISOPROPYLPHENOL □ 4-(1-METHYLETHYL)PHENOL □ PRODOX 133
CONSENSUS REPORTS: Reported in EPA TSCA Inventory.
DOT CLASSIFICATION: 6.1; Label: KEEP AWAY FROM FOOD
SAFETY PROFILE: A poison by intraperitoneal and intravenous routes. Moderately toxic by ingestion. When heated to decomposition it emits acrid smoke and fumes. See also PHENOL.

IRF000   CAS: 55-91-4   HR: 3
ISOPROPYL PHOSPHOROFLUORIDATE
mf: $C_6H_{14}FO_3P$ mw: 184.17
PROP: Oily liquid. Mp: −82°, bp: 46° @ 5 mm, d: 1.07 (approx), vap d: 5.24. Slightly sol in water.
SYNS: DFP □ DIFLUPYL □ DIFLUROPHATE □ DIISOPROPOXYPHOSPHORYL FLUORIDE □ DIISOPROPYL FLUOROPHOSPHATE □ O,O-DIISOPROPYL FLUOROPHOSPHATE □ DIISOPROPYL FLUOROPHOSPHONATE □ DIISOPROPYLFLUORO-PHOSPHORIC ACID ESTER □ DIISOPROPYLFLUOR-PHOSPHORSAEUREESTER (GERMAN) □ DIISOPROP-YL PHOSPHOFLUORIDATE □ DIISOPROPYL PHOS-PHOROFLUORIDATE □ O,O'-DIISOPROPYL PHOSP-HORYL FLUORIDE □ DYFLOS □ FLOROPRYL □ FLUOPHOSPHORIC ACID, DIISOPROPYL ESTER □ FLUORODIISOPROPYL PHOSPHATE □ FLUOROPRYL □ FLUOSTIGMINE □ ISOFLUOROPHATE □ ISOFLUROPHATE □ ISOPROPYL FLUOPHOSPHATE □ NEOGLAUCIT □ PF-3 □ PHOSPHOROFLUORIDIC ACID, DIISOPROPYL ESTER □ RCRA WASTE NUMBER P043 □ T-1703 □ TL 466
CONSENSUS REPORTS: EPA Extremely Hazardous Substances List. Reported in EPA TSCA Inventory.

SAFETY PROFILE: A poison by ingestion, inhalation, skin contact, intraperitoneal, subcutaneous, intramuscular, ocular, and intravenous routes. Moderately toxic by parenteral route. Human systemic effects by inhalation: miosis (pupillary constriction) and headache. Experimental reproductive effects. Used as a basis for "nerve gases." An insecticide. Ingestion can cause damage to eyes, nausea, vomiting, diarrhea, and central nervous system disturbances. An FDA proprietary drug. Used as a miotic agent. When heated to decomposition it emits toxic fumes of F^- and PO_x. See also PARATHION.

IRG100   CAS: 3772-26-7   HR: 2
1-ISOPROPYL-2-PYRROLIDINONE
mf: $C_7H_{13}NO$ mw: 127.21
PROP: Liquid. Mp: 18°, bp: 216°, d: 0.971, flash p: 212°F.
SYNS: N-ISOPROPYLPYRROLIDINONE □ 2-PYRROLIDINONE, 1-ISOPROPYL- □ 2-PYRRO-LIDINONE, 1-(1-METHYLETHYL)-(9CI)
CONSENSUS REPORTS: Reported in EPA TSCA Inventory.
SAFETY PROFILE: Moderately toxic by ingestion. A skin and eye irritant. Combustible liquid. When heated to decomposition it emits toxic fumes of NO_x.

IRN100   CAS: 1733-25-1   HR: 1
ISOPROPYL TIGLATE
mf: $C_8H_{14}O_2$ mw: 142.22
SYNS: 2-BUTENOIC ACID, 2-METHYL, 1-ISOPROPYL ESTER (E)- □ ISOPROPYL 2-METHYL-2-BUTENOATE □ ISOPROPYL α-METHYL CROTONATE
CONSENSUS REPORTS: Reported in EPA TSCA Inventory.
SAFETY PROFILE: A skin irritant. When heated to decomposition it emits acrid smoke and irritating fumes.

IRQ100   CAS: 67590-57-2   HR: 3
4-ISOPROPYL-2,6,7-TRIOXA-1-ARSABI-
CYCLO(2.2.2)OCTANE
mf: $C_7H_{13}AsO_3$ mw: 220.12
SYN: 2,6,7-TRIOXA-1-ARSABICYCLO(2.2.2)OCTANE, 4-ISOPROPYL-
OSHA PEL: TWA 0.5 mg(As)/m³

SAFETY PROFILE: Poison by intravenous route. When heated to decomposition it emits toxic fumes of As.

IRR050   CAS: 120373-24-2   HR: 3
ISOPROPYL UNOPROSTONE
mf: $C_{25}H_{44}O_5$ mw: 424.69

SYNS: 13,14-DIHYDRO-15-KETO-20-ETHYL-PGF2 □ 5-HEPTENOIC ACID, 7-(3,5-DIHYDROXY-2-(3-OXODEC-YL)CYCLOPENTYL)-, 1-METHYLETHYL ESTER,(1R-(1α(Z),2β,3α,5α))- □ (+)-ISOPROPYL, Z-7-((1R,2R,3R,5S)-3,5-DIHYDROXY-2-(3-OXODECYL)CYCLOPENTYL)-HEPT-5-ENOATE □ RESCULA □ UF 021 □ UNO-PROSTONE ISOPROPYL ESTER

SAFETY PROFILE: A poison by intravenous route. Moderately toxic by subcutaneous route. When heated to decomposition it emits acrid smoke and irritating vapors.

IRX000   CAS: 119-65-3   HR: 3
ISOQUINOLINE
mf: C_9H_7N mw: 129.17

PROP: Crystals, oil or liquid; pungent odor. Hygroscopic platelets when solid. D: 1.10 @ 20°/4°, mp: 26.48°, bp: 243°. Almost insol in water; misc with many org solvs, acids; sol in EtOH, Et₂O, Me₂CO, and C₆H₆.

SYNS: 2-AZANAPHTHALENE □ 2-BENZAZINE □ BENZO(c)PYRIDINE □ LEUCOLINE

CONSENSUS REPORTS: Reported in EPA TSCA Inventory. EPA Genetic Toxicology Program.

SAFETY PROFILE: A poison by ingestion. Moderately toxic by skin contact. A severe skin and eye irritant. When heated to decomposition it emits toxic fumes of NO_x.

IRY000   CAS: 94-86-0   HR: 2
ISOSAFROEUGENOL
mf: $C_{11}H_{14}O_2$ mw: 178.25

PROP: White crystalline powder; vanilla odor. Flash p: 212°F. Sol in fixed oils; insol in water.

SYNS: 6-ETHOXY-m-ANOL □ 1-ETHOXY-2-HYDRO-XY-4-PROPENYLBENZENE □ FEMA No. 2922 □ HYDROXY METHYL ANETHOL □ PROPENYL-GUAETHOL (FCC) □ VANITROPE

CONSENSUS REPORTS: Reported in EPA TSCA Inventory.

SAFETY PROFILE: Moderately toxic by ingestion. Combustible liquid. When heated to decomposition it emits acrid smoke and fumes.

IRZ000   CAS: 120-58-1   HR: 3
ISOSAFROLE
mf: $C_{10}H_{10}O_2$ mw: 162.20

PROP: Liquid; odor of anise. Bp: 253°, mp: 8.2°.

SYNS: 1,2-METHYLENEDIOXY-4-PROPENYLBENZ-ENE □ 3,4-METHYLENEDIOXY-1-PROPENYL BENZENE □ 5-(1-PROPENYL)-1,3-BENZODIOXOLE □ 4-PROPENYLCATECHOL METHYLENE ETHER □ 4-PROPENYL-1,2-METHYLENEDIOXYBENZENE □ RCRA WASTE NUMBER U141

CONSENSUS REPORTS: IARC Cancer Review: Group 3 IMEMDT 7,56,87; Animal Sufficient Evidence IMEMDT 1,169,72. Reported in EPA TSCA Inventory. EPA Genetic Toxicology Program.

SAFETY PROFILE: Poison by intraperitoneal and intravenous routes. Moderately toxic by ingestion and subcutaneous routes. A skin irritant. Questionable carcinogen with experimental carcinogenic and tumorigenic data. Used as a pesticide. When heated to decomposition it emits acrid smoke and fumes.

ISC550   CAS: 58958-60-4   HR: 1
ISOSTEARYL NEOPENTANOATE
mf: $C_{23}H_{46}O_2$ mw: 354.69

SYNS: CERAPHYL 375 □ CYCLOCHEM INEO □ 2,2-DIMETHYLPROPANOIC ACID ISOOCTADECYL ESTER □ PROPANOIC ACID, 2,2-DIMETHYL-, ISOOCTADECYL ESTER □ SCHERCEMOL 85

CONSENSUS REPORTS: Reported in EPA TSCA Inventory.

SAFETY PROFILE: An eye irritant. When heated to decomposition it emits acrid smoke and irritating fumes.

ISD066   CAS: 55965-84-9   HR: 3
ISOTHIAZOLINONE CHLORIDE
mf: $C_4H_5NOS•C_4H_4ClNOS$ mw: 264.76

SYNS: BIO-PERGE □ 3(2H)-ISOTHIAZOLONE, 5-CHLORO-2-METHYL-, MIXT. WITH 2-METHYL-3(2H)-

ISOTHIAZOLONE □ KATHON 886 □ KATHON CG □ KATHON LX □ KATHON 886MW □ KATHON 886 W □ KATHON WT □ KKM 43 □ KB □ KATHON BIOCIDE □ KATHON RH 886

SAFETY PROFILE: A poison by ingestion. Mutation data reported. When heated to decomposition it emits toxic vapors of NO_x, SO_x, and Cl^-.

**ISE000   CAS: 556-61-6   HR: 3
ISOTHIOCYANATOMETHANE**
DOT: UN 2477
mf: C_2H_3NS mw: 73.12
PROP: Crystalline. Mp: 36°, bp: 119°, d: 1.069. Very sltly sol in water; misc in alc and ether.
SYNS: DI-TRAPEX □ EP-161E □ ISOTHIOCYANATE de METHYLE (FRENCH) □ ISOTHIOCYANIC ACID, METHYL ESTER □ ISOTIOCIANATO di METILE (ITALIAN) □ METHYLISOTHIOCYANAAT (DUTCH) □ METHYL-ISOTHIOCYANAT (GERMAN) □ METHYL ISOTHIOCYANATE (DOT) □ METHYL MUSTARD OIL □ METHYLSENFOEL (GERMAN) □ MIC □ MIT □ MITC □ MORTON WP-161E □ TRAPEX □ TRAPEX-40 □ TRAPEXIDE □ VORLEX □ VORTEX □ WN 12
CONSENSUS REPORTS: EPA Extremely Hazardous Substances List. Reported in EPA TSCA Inventory.
DOT CLASSIFICATION: 3; Label: Flammable Liquid, Poison
SAFETY PROFILE: A poison by ingestion, skin contact, and subcutaneous routes. Very irritating to skin, eyes, and mucous membranes. Human systemic effects by ingestion: convulsions, change in motor activity, coma. An agricultural chemical and pesticide. Flammable when exposed to heat or flame; can react vigorously with oxidizing materials. When heated to decomposition it emits very toxic fumes of NO_x and SO_x. See also THIOCYANATES.

**ISQ000   CAS: 103-72-0   HR: 3
ISOTHIOCYANIC ACID, PHENYL ESTER**
mf: C_7H_5NS mw: 135.19
PROP: Pale-yellow liquid. Mp: −21°, bp: 221°, d: 1.1282. Insol in water; sol in alc and ether.
SYNS: BENZENE-1-ISOTHIOCYANATE □ ISOTHIO-CYANATOBENZENE □ PHENYL ISOTHIOCYANATE □

PHENYL MUSTARD OIL □ PHENYLSENFOEL (GERMAN) □ PITC □ THIOCARBANIL □ USAF M-4
CONSENSUS REPORTS: Reported in EPA TSCA Inventory.
SAFETY PROFILE: A poison by ingestion, intraperitoneal, and subcutaneous routes. An experimental teratogen. Mutation data reported. When heated to decomposition, or on contact with acid or acid fumes, it emits highly toxic fumes of cyanides and SO_x. See also THIOCYANATES.

**ISR000   CAS: 62-56-6   HR: 3
ISOTHIOUREA**
DOT: UN 2877
mf: CH_4N_2S mw: 76.13
PROP: White powder or crystals; rhombohedra or needles from EtOH. Mp: 177°, bp: decomp, d: 1.405. Sol in H_2O, EtOH; sltly sol in Et_2O.
SYNS: PSEUDOTHIOUREA □ RCRA WASTE NUMBER U219 □ SULOUREA □ THIOCARBAMATE □ THIOCARBAMIDE □ β-THIOPSEUDOUREA □ THIOUREA (DOT) □ 2-THIOUREA □ THU □ TSIZP 34 □ USAF EK-497
CONSENSUS REPORTS: NTP 10th Report on Carcinogens. IARC Cancer Review: Group 2B IMEMDT 7,56,87; Animal Sufficient Evidence IMEMDT 7,95,74. EPA Genetic Toxicology Program. Reported in EPA TSCA Inventory.
DFG MAK: Confirmed Animal Carcinogen with Unknown Relevance to Humans
SAFETY PROFILE: Confirmed carcinogen with experimental carcinogenic, neoplastigenic, and tumorigenic data. A human poison by an unspecified route. An experimental poison by ingestion. An eye irritant. Human mutation data reported. Human systemic effects by ingestion: hemorrhage, granulocytopenia (reduction in number of granulocytes), and changes in cell count (unspecified). May cause depression of bone marrow with anemia, leukopenia, and thrombocytopenia. May also cause allergic skin eruptions. Causes hepatic tumors upon chronic administration. Experimental teratogenic and reproductive effects. May react violently with acrolein.

Incompatible with acrylaldehyde, H_2O_2, HNO_3. When heated to decomposition it emits very toxic fumes of NO_x and SO_x.

ISU000   CAS: 503-74-2   HR: 3
ISOVALERIC ACID

mf: $C_5H_{10}O_2$ mw: 102.15

PROP: Colorless liquid or oil; acid taste, disagreeable rancid-cheese odor. Solidifies @ $-37°$, d: 0.931 @ $20°/4°$, refr index: 1.403, mp: $-37.6°$, bp: $175–177°$. Sol in water @ $16°$; misc in alc, chloroform, ether.

SYNS: DELPHINIC ACID □ FEMA No. 3102 □ ISOPENTANOIC ACID (DOT) □ ISOPROPYLACETIC ACID □ ISOVALERIANIC AICD □ 3-METHYL-BUTANOIC ACID □ β-METHYLBUTYRIC ACID □ 3-METHYLBUTYRIC ACID

CONSENSUS REPORTS: Reported in EPA TSCA Inventory.

SAFETY PROFILE: A poison by skin contact. Moderately toxic by ingestion and intravenous routes. A corrosive skin and eye irritant. When heated to decomposition it emits acrid smoke and fumes.

ISV000   CAS: 2835-39-4   HR: 3
ISOVALERIC ACID, ALLYL ESTER

mf: $C_8H_{14}O_2$ mw: 142.22

SYNS: ALLYL ISOVALERATE □ ALLYL ISOVALERI-ANATE □ ALLYL 3-METHYLBUTYRATE □ FEMA No. 2045 □ 3-METHYLBUTANOIC ACID, 2-PROPENYL ESTER □ 3-METHYLBUTYRIC ACID, ALLYL ESTER □ NCI-C54717 □ 2-PROPENYL ISOVALERATE □ 2-PROPENYL 3-METHYLBUTANOATE

CONSENSUS REPORTS: IARC Cancer Review: Group 3 IMEMDT 7,56,87; Animal Limited Evidence IMEMDT 36,69,85. NTP Carcinogenesis Studies (gavage); Clear Evidence: mouse, rat NTPTR* NTP-TR-253,83. Reported in EPA TSCA Inventory.

SAFETY PROFILE: Suspected carcinogen with experimental carcinogenic and tumorigenic data. A poison by ingestion. Moderately toxic by skin contact. A skin irritant. When heated to decomposition it emits acrid smoke and fumes. See also ALLYL COMPOUNDS and ESTERS.

ISW000   CAS: 103-38-8   HR: 2
ISOVALERIC ACID, BENZYL ESTER

mf: $C_{12}H_{16}O_2$ mw: 192.28

PROP: Colorless liquid; fruity apple odor. D: 0.985–0.9911, refr index: 1.486, flash p: 212°F. Sol in alc, fixed oils; sltly sol in propylene glycol; insol in glycerin, water @ 246°.

SYNS: BENZYL ISOVALERATE (FCC) □ BENZYL-3-METHYLBUTANOATE □ BENZYL-3-METHYL BUTYRATE □ FEMA No. 2152 □ ISOPENTANOIC ACID, PHENYLMETHYL ESTER □ ISOPROPYL ACETIC ACID, BENZYL ESTER □ 3-METHYLBUTANOIC ACID, PHENYLETHYL ESTER

CONSENSUS REPORTS: Reported in EPA TSCA Inventory.

SAFETY PROFILE: Low toxicity by ingestion and skin contact. A skin irritant. Combustible liquid. When heated to decomposition it emits acrid smoke and fumes. See also ESTERS.

ISX000   CAS: 109-19-3   HR: 3
ISOVALERIC ACID, BUTYL ESTER

mf: $C_9H_{18}O_2$ mw: 158.27

PROP: Colorless to pale-yellow liquid; fruity odor. Vap d: 5.45, bp: 150°, d: 0.851–0.857, refr index: 1.407. Misc with alc, fixed oils; sltly sol in propylene glycol; insol in water.

SYNS: n-BUTYL ISOPENTANOATE □ n-BUTYL ISOVALERATE □ 1-BUTYL ISOVALERATE □ BUTYL ISOVALERIANATE □ BUTYL 3-METHYLBUTYRATE □ FEMA No. 2218 □ 3-METHYLBUTANOIC ACID, BUTYL ESTER

CONSENSUS REPORTS: Reported in EPA TSCA Inventory.

SAFETY PROFILE: Low toxicity by ingestion and skin contact. A skin irritant. Flammable when exposed to heat, flame, sparks, and oxidizers. To fight fire, use alcohol foam, dry chemical, spray, mist, fog. When heated to decomposition it emits acrid smoke and fumes. See also ESTERS.

ISY000   CAS: 108-64-5   HR: 3
ISOVALERIC ACID, ETHYL ESTER

mf: $C_7H_{14}O_2$ mw: 130.21

PROP: Colorless, oily liquid; apple odor. Flash p: 77°F, d: 0.868 @ $20°/20°$, refr index: 1.395–1.399, bp: 135°, mp: $-99°$.

Misc with alc, fixed oils, benzene, ether; sol in propylene glycol; sltly sol in water @ 135°.

SYNS: ETHYL ISOVALERATE (FCC) ☐ FEMA No. 2463 ☐ 3-METHYLBUTANOIC ACID, ETHYL ESTER ☐ 3-METHYLBUTYRIC ACID, ETHYL ESTER

CONSENSUS REPORTS: Reported in EPA TSCA Inventory.

SAFETY PROFILE: Moderately toxic by intraperitoneal route. Mildly toxic by ingestion. A skin irritant. Flammable liquid when exposed to heat, flame, or sparks. When heated to decomposition it emits acrid smoke and fumes. See also ESTERS.

ISZ000   CAS: 35154-45-1   HR: 1
(Z)-ISOVALERIC ACID-3-HEXENYL

mf: $C_{11}H_{20}O_2$ mw: 184.31

PROP: Colorless liquid; sweet, apple odor. D: 0.869–0.874, refr index: 1.439–1.435. Sol in alc, propylene glycol, fixed oils; insol in water.

SYNS: AI3-35966 ☐ FEMA No. 3498 ☐ cis-3-HEXENYL ISOVALERATE (FCC)

CONSENSUS REPORTS: Reported in EPA TSCA Inventory.

SAFETY PROFILE: A skin irritant. When heated to decomposition it emits acrid smoke and fumes.

ITB000   CAS: 659-70-1   HR: 2
ISOVALERIC ACID, ISOPENTYL ESTER

mf: $C_{10}H_{20}O_2$ mw: 172.30

PROP: Colorless liquid; apple odor. D: 0.851–0.857, refr index: 1.411, bp: 190.5°, flash p: 162°F. Misc in alc, fixed oils; sltly sol in propylene glycol; insol in water.

SYNS: FEMA No. 2085 ☐ ISOAMYL ISOVALERATE (FCC) ☐ ISOPENTYL ISOVALERATE

CONSENSUS REPORTS: Reported in EPA TSCA Inventory.

SAFETY PROFILE: Mildly toxic by ingestion. A skin irritant. Combustible liquid. When heated to decomposition it emits acrid smoke and fumes. See also ESTERS.

ITC000   CAS: 556-24-1   HR: 3
ISOVALERIC ACID, METHYL ESTER

DOT: UN 2400

mf: $C_6H_{12}O_2$ mw: 116.18

SYNS: 3-METHYLBUTANOIC ACID, METHYL ESTER ☐ METHYL ISOPENTANOATE ☐ METHYL ISO-VALERATE ☐ METHYLISOVALERATE (DOT) ☐ METHYL-3-METHYLBUTANOATE ☐ METHYL-3-METHYLBUTYRATE

CONSENSUS REPORTS: Reported in EPA TSCA Inventory.

DOT CLASSIFICATION: 3; Label: Flammable Liquid

SAFETY PROFILE: Mildly toxic by ingestion and very slightly toxic by inhalation. Flammable when exposed to heat or flame; can react vigorously with oxidizing materials. When heated to decomposition it emits acrid smoke and fumes. See also ESTERS.

ITD875   CAS: 70288-86-7   HR: 3
IVERMECTIN

SYNS: 22,23-DIHYDROAVERMECTIN B1 ☐ HYVERMECTIN ☐ MK 933

SAFETY PROFILE: Poison by subcutaneous route. When heated to decomposition it emits toxic fumes of NO_x.

J

JAT000   CAS: 128-58-5   HR: 1
JADE GREEN BASE
mf: $C_{36}H_{20}O_4$ mw: 516.56

SYNS:C.I. 59825 □ DIMETHOXYVIOLANTHRONE □ 16,17-DIMETHOXYVIOLANTHRONE □ ZELEN OSTANTHRENOVA BRILANTNI FFB (CZECH)

CONSENSUS REPORTS: Reported in EPA TSCA Inventory.

SAFETY PROFILE: An eye irritant. When heated to decomposition it emits acrid smoke and fumes.

JCS000   CAS: 469-59-0   HR: 3
JERVINE
mf: $C_{27}H_{39}NO_3$ mw: 425.67

PROP: Needles from (methanol + water). Mp: 243.5–244.5°. An alkamine isolated from *Veratrum album*.

SAFETY PROFILE: Poison by ingestion, intravenous, and subcutaneous routes. An experimental teratogen. Experimental reproductive effects. A natural toxin found in some plants. When heated to decomposition it emits toxic fumes of NO_x.

JEA000   CAS: 8002-68-4   HR: 2
JUNIPER BERRY OIL
PROP: A volatile oil. Principal constituents include d-pinene, camphene, 1-terpineol-4, and other oxygenated constituents. From steam distillation of the fruit of *Juniperus communis L.* (Fam. *Cupressaceae*) (FCTXAV 14,307,76). Colorless to faint green-yellow liquid; aromatic bitter taste. Sol in fixed oils, mineral oil; insol in glycerin, propylene glycol.

SYNS:JUNIPER OIL □ OIL OF JUNIPER BERRY □ OILS, JUNIPER □ WACHOLDERBEER OEL (GERMAN)

CONSENSUS REPORTS: Reported in EPA TSCA Inventory.

SAFETY PROFILE: Mildly toxic by ingestion. A human skin irritant. An allergen. A systemic irritant. If taken internally, a severe kidney irritation similar to that caused by turpentine may result. When heated to decomposition it emits acrid smoke and fumes. See also individual components.

K

KAL000   CAS: 59-01-8   HR: 3
KANAMYCIN
mf: $C_{18}H_{36}N_4O_{11}$ mw: 484.58
PROP: Crystals from EtOH.
SYNS: CANTREX □ 4,6-DIAMINO-2-HYDROXY-1,3-
CYCLOHEXANE-3,6'-DIAMINO-3,6'-DIDEOXYDI-α-d-
GLUCOSIDE □ 4,6-DIAMINO-2-HYDROXY-1,3-
CYCLOHEXYLENE 3,6'-DIAMINO-3,6'-DIDEOXYDI-d-
GLUCOPYRANOSIDE □ KANAMICINA (ITALIAN) □
KANAMYCIN A □ KANAMYTREX □ KANTREX □ KM
□ KM (the antibiotic)
CONSENSUS REPORTS: Reported in EPA
TSCA Inventory. EPA Genetic Toxicology
Program.
SAFETY PROFILE: Poison by intravenous
and intramuscular routes. Moderately toxic
by ingestion, intraperitoneal, and
subcutaneous routes. An experimental
teratogen. Experimental reproductive
effects. Mutation data reported. When
heated to decomposition it emits toxic
fumes of NO_x.

KBB600   CAS: 1332-58-7   HR: 1
KAOLIN
PROP: Fine white to light-yellow powder;
earth taste. Insol in ether, alc, dil acids, and
alkali solutions.
SYNS: ALTOWHITES □ BENTONE □ CONTINENTAL
□ DIXIE □ EMATHLITE □ FITROL □ FITROL DESIC-
CATE 25 □ GLOMAX □ HYDRITE □ KAOPAOUS □
KAOPHILLS-2 □ LANGFORD □ MCNAMEE □
PARCLAY □ PEERLESS □ SNOW TEX
OSHA PEL: TWA Total Dust: 10 mg/m³;
Respirable Fraction: 5 mg/m³
ACGIH TLV: TWA 2 mg/m³; Respirable
Fraction; Not Classifiable as a Human
Carcinogen
SAFETY PROFILE: A nuisance dust.

KBK000   CAS: 9000-36-6   HR: 1
KARAYA GUM

PROP: Dried exudate of the tree *Sterculia
ureus* Roxburgh (Fam. *Sterculiaceae*). Fine,
white powder; slt odor of acetic acid. Insol
in alc; swells in water to a gel.
SYNS: GUM STERCULIA □ STERCULIA GUM
CONSENSUS REPORTS: Reported in EPA
TSCA Inventory.
SAFETY PROFILE: Very mildly toxic by
ingestion. A mild allergen.

KBU000   CAS: 39472-31-6   HR: 3
KARMINOMYCIN
SYNS: CARMINOMYCIN □ o-
DEMETHYLDAUNOMYCIN
SAFETY PROFILE: Poison by ingestion,
intravenous, intraperitoneal, and
subcutaneous routes. Human systemic
effects by intravenous route: anorexia,
hallucinations and distorted perceptions,
thrombosis, nausea or vomiting, fatty liver
degeneration, impaired liver function,
endocrine changes, and leukopenia (reduced
white blood cell count). An experimental
teratogen. Experimental reproductive
effects. Mutation data reported. When
heated to decomposition it emits acrid
smoke and fumes.

KDK700   HR: D
KELP
PROP: Dehydrated seaweed, dark green to
brown; salty, characteristic taste. From
*Macrocystis pyrifera, Laminaria digitata,
Laminaria saccharina, and Laminaria cloustoni.*
SAFETY PROFILE: When heated to
decomposition it emits acrid smoke and
irritating fumes.

KEA000   CAS: 143-50-0   HR: 3
KEPONE
mf: $C_{10}Cl_{10}O$ mw: 490.60

PROP: A chlorinated polycyclic ketone. A crystalline material. Mp: decomp @ 350°. Sltly water-sol; sol in alc, ketones, and acetic acid. Readily hydrates on exposure to room temperature and humidity; normally used as a mono- to trihydrate (NCIBR*).
SYNS: CHLORDECONE □ CIBA 8514 □ COMPOUND 1189 □ 1,2,3,5,6,7,8,9,10,10-DECACHLORO(5.2.1.0^{2,6}.-0^{3,9}.0^{5,8})DECANO-4-ONE □ DECACHLOROKETONE □ DECACHLORO-1,3,4-METHENO-2H-CYCLOBUTA-(cd)PENTALEN-2-ONE □ DECACHLOROOCTAHYDRO-KEPONE-2-ONE □ DECACHLOROOCTAHYDRO-1,3,4-METHENO-2H-CYCLOBUTA(cd)PENTALEN-2-ONE □ 1,1a,3,3a,4,5,5,5a,5b,6-DECACHLOROOCTAHYDRO-1,3,4-METHENO-2H-CYCLOBUTA(cd)PENTALEN-2-ONE □ DECACHLOROPENTACYCLO(5.2.1.0^{2,6}.0^{3,9}.0^{5,8})DECAN-4-ONE □ DECACHLOROPENTACYCLO(5.3.0.0^{2,6}.-0^{4,10}.0^{5,9})DECAN-3-ONE □ DECACHLOROTETRA-CYCLODECANONE □ DECACHLOROTETRAHYDRO-4,7-METHANOINDENEONE □ ENT 16,391 □ GENERAL CHEMICALS 1189 □ MEREX □ NCI-C00191 □ RCRA WASTE NUMBER U142
CONSENSUS REPORTS: NTP 10th Report on Carcinogens. IARC Cancer Review: Group 2B IMEMDT 7,56,87; Human Limited Evidence IMEMDT 20,67,79; Animal Sufficient Evidence IMEMDT 20,67,79. EPA Genetic Toxicology Program.
DFG MAK: Confirmed Animal Carcinogen with Unknown Relevance to Humans
NIOSH REL: (Kepone) CL 0.001 mg/m³/15M
SAFETY PROFILE: Confirmed carcinogen with experimental carcinogenic data. Poison by ingestion, skin contact. Experimental teratogenic and reproductive effects. Mutation data reported. Inhalation, absorption, or ingestion by humans can lead to central nervous system, liver, and kidney damage, including bizarre symptoms caused by damage to the nervous system. Usually, the symptoms are tremors, ataxia, skin changes, hyperexcitability, hyperactivity, muscle spasms, testicular atrophy, low sperm count, estrogenic effects, sterility, breast enlargement, liver lesions, and cancer. An insecticide and fungicide. Registration suspended by the USEPA.

KEK000   CAS: 8008-20-6   HR: 3
KEROSENE
DOT: UN 1223
PROP: A pale-yellow to water-white, oily liquid. Bp: 175–325°, ULC: 40, flash p: 150–185°F, d: 0.80 to <1.0, lel: 0.7%, uel: 5.0%, autoign temp: 410°F, vap d: 4.5. Insol in water; misc with other pet solvents. A mixture of petroleum hydrocarbons, chiefly of the methane series having from 10–16 carbon atoms per molecule.
SYNS: COAL OIL □ DEOBASE □ KEROSINE □ KEROSINE (petroleum) □ STRAIGHT-RUN KEROSENE
CONSENSUS REPORTS: IARC Cancer Review: Group 2A IMEMDT 45,39,89; Animal Limited Evidence IMEMDT 45,39,89. Reported in EPA TSCA Inventory.
ACGIH TLV: 200 mg/m³ (skin); Confirmed Animal Carcinogen
NIOSH REL: (Kerosene) TWA 100 mg/m³
DOT CLASSIFICATION: 3; Label: Flammable Liquid
SAFETY PROFILE: Suspected carcinogen. Poison by intravenous and intratracheal routes. Moderately toxic to animals by ingestion. A severe skin irritant. Mutation data reported. Human systemic effects by ingestion and intravenous routes: somnolence, hallucinations and distorted perceptions, coughing, nausea or vomiting, and fever. Aspiration of vomitus can cause serious pneumonitis, particularly in young children. Combustible when exposed to heat or flame; can react with oxidizing materials. Moderately explosive in the form of vapor when exposed to heat or flame. When heated to decomposition it emits acrid smoke and fumes. To fight fire, use foam, CO_2, dry chemical.

KEK100   CAS: 64742-47-8   HR: 3
KEROSENE (PETROLEUM), hydro-treated
SYN: HYDROTREATED KEROSENE
CONSENSUS REPORTS: IARC Cancer Review: Animal Limited Evidence

IMEMDT 45,39,89. Reported in EPA TSCA Inventory.

SAFETY PROFILE: Suspected carcinogen. A combustible liquid. When heated to decomposition it emits acrid smoke and irritating vapors.

KEU000   **CAS: 463-51-4**   **HR: 3**
KETENE
mf: C_2H_2O mw: 42.04
PROP: Colorless gas with disagreeable taste and pungent odor. Decomp in water. Mp: $-150°$, bp: $-56°$, vap d: 1.45. Decomp in alc. Fairly sol in Me_2CO; sol in H_2O, ether, and acetone. IDLH 5 ppm.
SYNS: CARBOMETHENE □ ETHENONE □ KETO-ETHYLENE
CONSENSUS REPORTS: Reported in EPA TSCA Inventory. EPA Genetic Toxicology Program.
OSHA PEL: TWA 0.5 ppm; STEL 1.5 ppm
ACGIH TLV: TWA 0.5 ppm; STEL 1.5 ppm
DFG MAK: 0.5 ppm (0.87 mg/m³)
SAFETY PROFILE: Poison by inhalation. Moderately toxic by ingestion. Can cause pulmonary edema. Reacts with hydrogen peroxide to form the explosive diacetyl peroxide. When heated to decomposition it emits acrid smoke and fumes.

KFA000   **CAS: 674-82-8**   **HR: 3**
KETENE DIMER
DOT: UN 2521
mf: $C_4H_4O_2$ mw: 84.08
PROP: Colorless, nonhygroscopic liquid; pungent odor. Mp: $-6.5°$, bp: $127.4°$, d: 1.0897, vap d: 2.9, flash p: 93°F (TOC). Decomp in water. Insol in water.
SYNS: 3-BUTENO-β-LACTONE □ DIKETENE □ DIKETENE, inhibited (DOT) □ 4-METHYLENE-2-OXETANONE
CONSENSUS REPORTS: Reported in EPA TSCA Inventory.
DOT CLASSIFICATION: 3; Label: Flammable Liquid, Poison
SAFETY PROFILE: Moderately toxic by ingestion and skin contact. A skin and severe eye irritant. Flammable when exposed to heat or flame; can react vigorously with

oxidizing materials. A violent polymerization reaction is catalyzed by acids, bases, or sodium acetate. A storage hazard. Self-initiated exothermic dimerization is explosive. To fight fire, use alcohol foam. When heated to decomposition it emits acrid smoke and fumes.

KFK000   **CAS: 469-79-4**   **HR: 3**
KETOBEMIDONE
mf: $C_{15}H_{21}NO_2$ mw: 247.37
PROP: Crystals. Mp: $201–202°$. Sol in H_2O; sltly sol in EtOH.
SYNS: A 21 LUNDBECK □ CETOBEMIDON □ CETO-BEMIDONE □ CIBA 7115 □ CLIRADON □ CLIRADONE □ CYMIDON □ ETHYL (4-(m-HYDROXYPHENYL)-1-METHYL)-4-PIPERIDYL KETONE □ HOECHST 10720 □ K 4710 □ KETONE, ETHYL 4-(m-HYDROXYPHENYL)-1-METHYLPIPERIDYL □ 1-PROPANONE, 1-(4-(3-HYDRO-XYPHENYL)-1-METHYL-4-PIPERIDINYL)-(9CI) □ 1-PROPANONE, 1-(4-(m-HYDROXYPHENYL)-1-METHYL-4-PIPERIDYL)- □ WIN 1539
DOT CLASSIFICATION: 3; Label: Flammable Liquid
SAFETY PROFILE: Poison by intravenous route. When heated to decomposition it emits toxic fumes of NO_x. See also KETONES.

KFK200   **CAS: 53494-70-5**   **HR: 3**
Δ-KETOENDRIN
mf: $C_{12}H_8Cl_6O$ mw: 380.90
PROP: Crystalline solid. Bp:$>280°$ C.
SYNS: ENDRIN KETONE □ Δ-KETO 153 □ 2,5,7-METHENO-3H-CYCLOPENTA(A)PENTALEN-3-ONE, 3B,4,5,6,6,6A-HEXACHLORODECAHYDRO-, (2-α-3A-β,3B-β,4-β,5-β,6A-β,7-α-7A-β,8R*)- □ SD 2614
DOT CLASSIFICATION: 3; Label: Flammable Liquid
SAFETY PROFILE: A poison by ingestion. A flammable liquid. When heated to decomposition it emits toxic vapors of Cl⁻.

KGA000   **HR: D**
KETONES
PROP: Liquid organic compounds containing the carbonyl group C=O attached to two alkyl groups. Derived from secondary alcohols by oxidation. Acetone, which is

K

dimethyl ketone, is the most familiar of this group of compounds.

SAFETY PROFILE: No general statement can be made as to the toxicity of ketones. Some are highly volatile and hence may have narcotic or anesthetic effects. Skin absorption, as well as inhalation, may be an important route of entry into the body. None of the ketones has been shown to have a high degree of chronic toxicity. Some are dangerous fire hazards. They react violently with aldehydes, HNO_3, HNO_3 + H_2O_2, $HClO_4$. A variety of peroxides can be formed from the reactions of ketones and hydrogen peroxide. Many of these peroxides are explosives sensitive to heat and shock. Common air contaminants. See also ACETONE, DIETHYL KETONE, and METHYL ETHYL KETONE.

KGK120   CAS: 58013-13-1   HR: 3
1-(2-KETO-2-(3'-PYRIDYL)ETHYL)-4-(2'-CHLOROPHENYL)PIPERAZINE

mf: $C_{17}H_{18}ClN_3O$ mw: 315.83

SYNS: KETONE, 4-(o-CHLOROPHENYL)-1-PIPER-AZINYLMETHYL 3-PYRIDYL- □ PIPERAZINE, 1-(o-CHLOROPHENYL)-4-NICOTINOYLMETHYL-

DOT CLASSIFICATION: 3; Label: Flammable Liquid

SAFETY PROFILE: A poison by intraperitoneal route. A flammable liquid. When heated to decomposition it emits toxic vapors of NO_x and Cl^-.

KGK130   CAS: 58013-14-2   HR: 3
1-(2-KETO-2-(3'-PYRIDYL)ETHYL)-4-(2'-METHOXYPHENYL)PIPERAZINE

mf: $C_{18}H_{21}N_3O_2$ mw: 311.42

SYNS: KETONE, 4-(o-METHOXYPHENYL)-1-PIPERAZINYLMETHYL 3-PYRIDYL- □ PIPERAZINE, 1-(o-METHOXYPHENYL)-4-NICOTINOYLMETHYL-

DOT CLASSIFICATION: 3; Label: Flammable Liquid

SAFETY PROFILE: A poison by intraperitoneal route. A flammable liquid. When heated to decomposition it emits toxic vapors of NO_x.

KHU000   CAS: 74278-22-1   HR: 3
KROMAD

PROP: Contains 5% cadmium sebacate, 5% potassium chromate, 1% malachite green, and 16% thiram (FMCHA2 -,D176,80).

CONSENSUS REPORTS: Cadmium and its compounds, as well as chromium and its compounds, are on the Community Right-To-Know List.

OSHA PEL: TWA 5 μg(Cd)/m³

ACGIH TLV: TWA 0.002 mg(Cd)/m³ (respirable dust), Suspected Human Carcinogen); BEI: 5 μg/g creatinine in urine; 5 μg/L in blood

DFG MAK: DFG BAT: Blood 1.5 μg/dL; Urine 15 μg/dL; Suspected Carcinogen

NIOSH REL: (Cadmium) Reduce to lowest feasible level

SAFETY PROFILE: Confirmed human carcinogen. Poison by ingestion. Moderately toxic by skin contact. When heated to decomposition it emits toxic fumes of K_2O, Cd, and Cr. See also CADMIUM COMPOUNDS, POTASSIUM CHROMATE, and THIRAM.

L

L

LAE350   **HR: D**
LACTASE ENZYME PREPARATIONS
from KLUYVEROMYCES LACTIS
PROP: Derived from *Kluyveromyces lactis*.
SAFETY PROFILE: When heated to
decomposition it emits acrid smoke and
irritating fumes.

LAG000   CAS: 50-21-5   HR: 2
LACTIC ACID
mf: $C_3H_6O_3$ mw: 90.09
PROP: Yellow to colorless crystals or syrupy
50% liquid. Mp: 16.8°, bp: 122° @ 15 mm,
d: 1.249 @ 15°. Volatile with superheated
steam. Sol in alc and furfurol; sltly sol in
ether; insol in chloroform, pet ether, carbon
disulfide. Misc in water, (alc + ether).
SYNS: ACETONIC ACID □ ETHYLIDENELACTIC
ACID □ 1-HYDROXYETHANECARBOXYLIC ACID □ 2-
HYDROXYPROPANOIC ACID □ 2-HYDROXYPROPION-
IC ACID □ α-HYDROXYPROPIONIC ACID □ KYSELINA
2-HYDROXYPROPANOVA □ KYSELINA MLECNA
(CZECH) □ dl-LACTIC ACID □ MILCHSAEURE
(GERMAN) □ MILK ACID □ ORDINARY LACTIC ACID
□ PROPANOIC ACID, 2-HYDROXY- □ PROPEL □
PROPIONIC ACID, 2-HYDROXY- □ RACEMIC LACTIC
ACID □ SY-83
CONSENSUS REPORTS: Reported in EPA
TSCA Inventory.
SAFETY PROFILE: Moderately toxic by
ingestion and rectal routes. Mutation data
reported. A severe skin and eye irritant.
Mixtures with nitric acid + hydrofluoric acid
may react vigorously and are storage
hazards. When heated to decomposition it
emits acrid smoke and irritating fumes.

LAH000   CAS: 64059-26-3   HR: 3
LACTIC ACID, BERYLLIUM SALT
SYN: BERYLLIUM LACTATE
CONSENSUS REPORTS: IARC Cancer
Review: Group 1 IMEMDT 58,41,93;
Human Sufficient Evidence IMEMDT
58,41,93; Animal Sufficient Evidence
IMEMDT 1,17,72; Animal Sufficient
Evidence IMEMDT 23,143,80; Animal
Sufficient Evidence IMEMDT 58,41,93.
Beryllium and its compounds are on the
Community Right-To-Know List.
OSHA PEL: TWA 0.002 mg(Be)/m³; STEL
0.005 mg(Be)/m³/30M; CL 0.025
mg(Be)/m³
ACGIH TLV: TWA 0.002 mg(Be)/m³;
Confirmed Human Carcinogen; (Proposed:
TWA 0.00005 mg(Be)/m³; STEL 0.0002
mg(Be)/m³ (sensitizer) (skin); Confirmed
Human Carcinogen)
NIOSH REL: CL (Beryllium) not to exceed
0.0005 mg(Be)/m³
SAFETY PROFILE: Confirmed
carcinogen. Poison by intravenous route.
When heated to decomposition it emits
very toxic fumes of Be. See also
BERYLLIUM COMPOUNDS.

LAJ000   CAS: 97-64-3   HR: 3
LACTIC ACID, ETHYL ESTER
DOT: UN 1192
mf: $C_5H_{10}O_3$ mw: 118.15
PROP: Colorless liquid; mild odor. Bp: 154°,
ULC: 30–35, lel: 1.55% @ 212°F, flash p:
115°F (CC), flash p (technical): 131°F, d:
1.029–1.032, refr index: 1.410–1.420,
autoign temp: 752°F, vap d: 4.07. Very sol in
alc, ether, chloroform, water.
SYNS: ACTYLOL □ ACYTOL □ ETHYLESTER KYSEL-
INY MLECNE □ ETHYL α-HYDROXYPROPIONATE □
ETHYL 2-HYDROXYPROPIONATE □ ETHYL LACTATE
(DOT,FCC) □ FEMA No. 2440 □ LACTATE d'ETHYLE
(FRENCH) □ SOLACTOL
CONSENSUS REPORTS: Reported in EPA
TSCA Inventory.
DOT CLASSIFICATION: 3; Label:
Flammable Liquid

SAFETY PROFILE: Moderately toxic by intraperitoneal, subcutaneous, and intravenous routes. Low oral and skin contact toxicity. A flammable liquid when exposed to heat or flame; can react with oxidizing materials. Slight explosion hazard in the form of vapor when exposed to flame. To fight fire, use foam, CO_2, dry chemical. When heated to decomposition it emits acrid smoke and irritating fumes. See also IRON COMPOUNDS.

LAL000   CAS: 5905-52-2   HR: 3
LACTIC ACID, IRON(2+) SALT (2:1)
mf: $C_6H_{10}O_6 \bullet Fe$ mw: 234.01
PROP: Greenish-white crystals; slight peculiar odor. Moderately sol in water; sltly sol in alc.
SYNS: FERROUS LACTATE □ IRON(2+) LACTATE
OSHA PEL: TWA 1 mg(Fe)/m³
ACGIH TLV: TWA 1 mg(Fe)/m³
SAFETY PROFILE: Poison by ingestion. Questionable carcinogen with experimental tumorigenic data. When heated to decomposition it emits acrid smoke and irritating fumes. See also IRON COMPOUNDS.

LAL100   CAS: 18917-93-6 HR: 3
LACTIC ACID, MAGNESIUM SALT
mf: $C_6H_{10}O_6 \bullet Mg$ mw: 202.47
PROP: White, crystalline powder with no odor. Mp: >200°. Water sol: 5.2 g in 100g @ 25°.
SYNS: BIS(LACTATO)MAGNESIUM □ MAGNESIUM, BIS(2-HYDROXYPROPANOATO-O(1),O(2))-, (T-4)-(9CI) □ MAGNESIUM, BIS(LACTATO)-(8CI) □ MAGNESIUM LACTATE
DOT CLASSIFICATION: 4.2; Label: Spontaneously Combustible
CONSENSUS REPORTS: Reported in EPA TSCA Inventory.
SAFETY PROFILE: A poison by intravenous route. When heated to decomposition it emits toxic vapors of magnesium.

LAM000   CAS: 72-17-3   HR: 2
LACTIC ACID, MONOSODIUM SALT
mf: $C_3H_5O_3 \bullet Na$ mw: 112.07
PROP: Hygroscopic solid; slt salt taste.
SYNS: 2-HYDROXYPROPANOIC ACID MONOSODIUM SALT □ LACOLIN □ LACTIC ACID SODIUM SALT □ PER-GLYCERIN □ SODIUM LACTATE
CONSENSUS REPORTS: Reported in EPA TSCA Inventory.
SAFETY PROFILE: Moderately toxic by intraperitoneal route. An eye irritant. When heated to decomposition it emits toxic fumes of Na_2O.

LAQ000   CAS: 78-97-7   HR: 3
LACTONITRILE
mf: C_3H_5NO mw: 71.09
PROP: Straw-colored liquid. Mp: −40°, bp: 103° @ 50 mm, fp: −34°, flash p: 170°F (TCC), d: 0.9834 @ 25°, vap d: 2.45.
SYNS: 2-HYDROXYPROPANNITRIL □ 2-HYDROXY-PROPIONITRILE □ NSC-7764 □ PROPIONITRILE, 2-HYDROXY-
CONSENSUS REPORTS: EPA Extremely Hazardous Substances List. Reported in EPA TSCA Inventory. Cyanide and its compounds are on the Community Right-To-Know List.
SAFETY PROFILE: Poison by ingestion, skin contact, and subcutaneous routes. Moderately toxic by inhalation. In the presence of alkali, it evolves HCN. Combustible when exposed to heat or flame; can react vigorously with oxidizing materials. To fight fire, use foam, CO_2, dry chemical. When heated to decomposition it emits toxic fumes of CN^- and NO_x. See also NITRILES.

LAR000   CAS: 63-42-3   HR: 2
LACTOSE
mf: $C_{12}H_{22}O_{11}$ mw: 342.34
PROP: Colorless, rhombic crystals; faintly sweet taste. D: 1.525 @ 20°, mp: 202° (anhyd), bp: decomp. Sol in water; insol in alc and ether.
SYNS: 4-(β-d-GALACTOSIDO)-d-GLUCOSE □ LACTIN □ LACTOBIOSE □ d-LACTOSE □ MILK SUGAR □ SACCHARUM LACTIN
CONSENSUS REPORTS: Reported in EPA TSCA Inventory.

SAFETY PROFILE: Moderately toxic by intravenous route. Questionable carcinogen with experimental tumorigenic and teratogenic data. Mixtures with oxidants (e.g., potassium chlorate, potassium nitrate, or potassium perchlorate) may be explosion hazards. When heated to decomposition it emits acrid smoke and irritating fumes.

LAR400   HR: D
LACTYLATED FATTY ACID ESTERS of GLYCEROL and PROPYLENE GLYCOL

PROP: Soft to hard waxy solid. Dispersible in hot water; moderately sol in hot isopropanol, benzene, chloroform, soybean oil.
SYN: PROPYLENE GLYCOL LACTOSTEARATE
SAFETY PROFILE: When heated to decomposition it emits acrid smoke and irritating fumes.

LAS200   CAS: 91465-08-6   HR: 3
LAMBDA-CYHALOTHRIN

mf: $C_{23}H_{19}ClF_3NO_3$ mw: 449.88
SYNS: CYCLOPROPANECARBOXYLIC ACID, 3-(2-CHLORO-3,3,3-TRIFLUORO-1-PROPENYL)-2,2-DIMETHYL-,CYANO(3-PHENOXYPHENYL)METHYL ESTER, (1-α(S*),3-α(Z))-(+−)- □ CYHALOTHRIN K □ ICON □ KARATE □ LAMBDA-CYHALOTHRIN TECHNICAL □ PP 321 □ LAMBDA-CYHALOTHRIN
SAFETY PROFILE: A poison by ingestion, intravenous. Moderately toxic by skin contact and inhalation. Experimental reproductive effects. Mutation data reported. When heated to decomposition it emits toxic vapors of NO_x, Cl^-, and Cl^-.

LAU550   HR: D
LANOLIN, anhydrous

PROP: Yellow-white semisolid. Insol in water; sol in chloroform, ether.
SYN: WOOL FAT
SAFETY PROFILE: When heated to decomposition it emits acrid smoke and irritating fumes.

LAV000   CAS: 7439-91-0   HR: 3
LANTHANUM

af: La   aw: 138.91

PROP: Silvery-white, malleable and ductile metal element soft enough to cut with a knife. Very reactive rare earth metal. Mp: 920°, bp: 3464°, d: 6.166 @ 25°.
SAFETY PROFILE: Poison by intravenous route. Lanthanum and other lanthanoids can cause delayed blood clotting leading to hemorrhages. Has caused liver injury in experimental animals. The dust is a dangerous fire hazard when exposed to flame; can react vigorously with oxidizing materials. Violent reaction with nitric acid, phosphorus (above 400°C), air, halogens. Moderately explosive in the form of dust when exposed to flame or by chemical reaction. Incompatible with H_2O, C, N, B, Se, Si, S. See also RARE EARTHS and POWDERED METALS.

LAW000   CAS: 917-70-4   HR: 2
LANTHANUM ACETATE

mf: $C_2H_4O_2 \cdot xLa$ mw: 1032.43
PROP: White powder. Sol in water.
SYNS: LANTHANACETAT (GERMAN) □ LANTHANUM TRIACETATE
CONSENSUS REPORTS: Reported in EPA TSCA Inventory.
SAFETY PROFILE: Moderately toxic by intraperitoneal and subcutaneous routes. Mildly toxic by ingestion. Mutation data reported. When heated to decomposition it emits acrid smoke and irritating fumes. See also LANTHANUM and RARE EARTHS.

LAX000   CAS: 10099-58-8   HR: 3
LANTHANUM CHLORIDE

mf: Cl_3La mw: 245.26
PROP: Deliquescent, heptahydrate: triclinic white crystals. Mp: 860°, bp: 18° @ 1730 mm. Sol in H_2O, EtOH, Py, DMSO, and tributyl phosphate; sltly sol in THF and Me_2CO.
CONSENSUS REPORTS: Reported in EPA TSCA Inventory. EPA Genetic Toxicology Program.
SAFETY PROFILE: Poison by intraperitoneal and intravenous routes. Moderately toxic by subcutaneous route.

Mildly toxic by ingestion. Experimental reproductive effects. Mutation data reported. When heated to decomposition it emits toxic fumes of Cl$^-$. See also RARE EARTHS, LANTHANUM, and CHLORIDES.

LBA000   CAS: 10099-59-9   HR: 3
LANTHANUM NITRATE
mf: $N_3O_9 \cdot La$ mw: 324.94
PROP: Hexahydrate; white, deliquescent crystals. Mp: approx 40°, bp: 126°. Very sol in water, alc. Keep well-stoppered.
CONSENSUS REPORTS: Reported in EPA TSCA Inventory.
SAFETY PROFILE: Poison by intraperitoneal route. Mildly toxic by ingestion. Experimental reproductive effects. Mutation data reported. When heated to decomposition it emits toxic fumes of NO_x. See also NITRATES, LANTHANUM, and RARE EARTHS.

LBF500   CAS: 11054-70-9   HR: 3
LASALOCID
mf: $C_{35}H_{54}O_8$ mw: 602.89
PROP: Veterinary antibiotic.
SYN: ANTIBIOTIC X 537
SAFETY PROFILE: Poison by ingestion and intraperitoneal routes. An eye and skin irritant. When heated to decomposition it emits acrid smoke and irritating fumes.

LBG000   CAS: 303-34-4   HR: 3
LASIOCARPINE
mf: $C_{21}H_{33}NO_7$ mw: 411.55
PROP: Plates from pet ether. Mp: 95.5–97°. An alkaloid isolated from *H. Lasiocarpum*.
SYNS: HELIOTRIDINE ESTER with LASIOCARPUM and ANGELIC ACID □ NCI-C01478 □ RCRA WASTE NUMBER U143
CONSENSUS REPORTS: IARC Cancer Review: Group 2B IMEMDT 7,56,87; Animal Limited Evidence IMEMDT 10,281,76. NCI Carcinogenesis Bioassay (feed); No Evidence: mouse, rat NCITR* NCI-CG-TR-39,78. EPA Genetic Toxicology Program.

SAFETY PROFILE: Suspected carcinogen with experimental carcinogenic data. Poison by ingestion, intravenous, intraperitoneal, and parenteral routes. Human mutation data reported. When heated to decomposition it emits toxic fumes of NO_x. See also ESTERS.

LBK000   CAS: 8006-78-8   HR: 2
LAUREL LEAF OIL
PROP: Main constituent is cineole. From steam distillation of the leaves of *Laurus nobilis L.* (Fam. *Lauraceae*). Yellow liquid; aromatic and spicy odor. D: 0.905–0.929, refr index: 1.465 at 20°. Sol in fixed oils, mineral oil, propylene glycol; insol in glycerin.
SYNS: BAY LEAF OIL □ BAY OIL □ BOIS D'INDE □ MYRCIA OIL □ OIL OF BAY □ OIL OF MYRCIA □ PIMENTA RACEMOSA OIL □ WEST INDIAN BAY OIL
CONSENSUS REPORTS: Reported in EPA TSCA Inventory.
SAFETY PROFILE: Moderately toxic by ingestion. A skin irritant. When heated to decomposition it emits acrid smoke and irritating fumes. See also CAJEPUTOL.

LBL000   CAS: 143-07-7   HR: 3
LAURIC ACID
mf: $C_{12}H_{24}O_2$ mw: 200.36
PROP: Colorless, needle-like crystals from EtOH; slt odor of bay oil. Mp: 44°, bp: 225° @ 100 mm, d: 0.883, vap press: 1 mm @ 121.0°. Insol in water; sol in chloroform, benzene, alc, ether, and petroleum ether.
SYNS: DODECANOIC ACID □ DODECOIC ACID □ DUODECYLIC ACID □ HYDROFOL ACID 1255 □ HYSTRENE 9512 □ LAUROSTEARIC ACID □ NEO-FAT 12 □ NINOL AA-62 EXTRA □ 1-UNDECANECARBOXYLIC ACID □ WECOLINE 1295
CONSENSUS REPORTS: Reported in EPA TSCA Inventory.
SAFETY PROFILE: Poison by intravenous route. Mildly toxic by ingestion. Questionable carcinogen with experimental neoplastigenic data. A skin and eye irritant. Mutation data reported. Combustible when exposed to heat or flame; can react with oxidizing materials. When heated to

decomposition it emits acrid smoke and irritating fumes.

LBR000   CAS: 105-74-8   HR: 3
LAUROYL PEROXIDE
mf: $C_{24}H_{46}O_4$ mw: 398.70
PROP: White, tasteless, coarse powder; faint odor. Mp: 53–55°.
SYNS: ALPEROX C □ BIS(1-OXODODECYL)PEROX-IDE □ DILAUROYL PEROXIDE □ DILAUROYL PEROX-IDE, TECHNICAL PURE (DOT) □ DODECANOYL PEROXIDE □ DYP-97 F □ LAUROX □ LAUROYL PEROXIDE, TECHNICALLY PURE (DOT) □ LAURYDOL □ LYP 97 □ PEROXYDE de LAUROYLE (FRENCH)
CONSENSUS REPORTS: IARC Cancer Review: Group 3 IMEMDT 7,56,87; Animal Inadequate Evidence IMEMDT 36,315,85. Reported in EPA TSCA Inventory.
DFG MAK: Mild skin effects
SAFETY PROFILE: Questionable carcinogen with experimental tumorigenic data. A powerful oxidizing agent. It is a corrosive irritant to the eyes and mucous membranes and can cause burns. A dangerous fire hazard. When heated to decomposition it emits acrid smoke and fumes. See also PEROXIDES, ORGANIC.

LBU100   CAS: 27176-87-0   HR: 2
LAURYLBENZENESULFONIC ACID
mf: $C_{18}H_{30}O_3S$ mw: 326.54
PROP: Opaque pale green thick liquid with a pine odor. Bp: >212° F, d: 1.214. Sol in water.
SYNS: BENZENESULFONIC ACID, DODECYL- □ BIO-SOFT S 100 □ CALSOFT LAS 99 □ DOBANIC ACID 83 □ DOBANIC ACID JN □ n-DODECYLBENZENESULFONIC ACID □ DODECYLBENZENESULFONIC ACID (DOT) □ DODECYLBENZENESULPHONIC ACID □ E 7256 □ ELFAN WA SULPHONIC ACID □ MARLON AS 3 □ NACCONOL 98SA □ NANSA 1042P □ NANSA SSA □ RICHONIC ACID B □ SULFRAMIN ACID 1298 □ WITCO 1298 SULFONIC ACID
DOT CLASSIFICATION: 8; Label: Corrosive
CONSENSUS REPORTS: Reported in EPA TSCA Inventory.
SAFETY PROFILE: Moderately toxic by ingestion. A corrosive. When heated to decomposition it emits toxic vapors of SO_x.

LBW000   CAS: 93-23-2   HR: 3
LAURYLISOQUINOLINIUM BROMIDE
mf: $C_{21}H_{32}N•Br$ mw: 378.45
PROP: Deep amber, water-sol liquid; pleasant, characteristic odor.
SYNS: 2-DODECYLISOQUINOLINIUM BROMIDE □ INTEXSAN LQ75 □ ISOTHAN
CONSENSUS REPORTS: Reported in EPA TSCA Inventory.
SAFETY PROFILE: Poison by ingestion. A severe eye irritant. Combustible when exposed to heat or flame. Incompatible with oxidizing materials. An FDA over-the-counter drug. When heated to decomposition it emits toxic fumes of Br^- and NO_x. See also BROMIDES.

LBX000   CAS: 112-55-0   HR: 3
LAURYL MERCAPTAN
DOT: UN 1228/UN 3071
mf: $C_{12}H_{26}S$ mw: 202.44
PROP: Water-white to pale-yellow liquid. Mp: −7°, bp: 115–177°, flash p: 262°F (OC), d: 0.849 @ 15.5°/15.5°.
SYNS: 1-DODECANETHIOL □ DODECYL MER-CAPTAN (ACGIH) □ m-DODECYL MERCAPTAN □ 1-DODECYL MERCAPTAN □ m-LAURYL MERCAPTAN □ 1-MERCAPTODODECANE □ NCI-C60935 □ PENNFLOAT M □ PENNFLOAT S
CONSENSUS REPORTS: Reported in EPA TSCA Inventory.
ACGIH TLV: (Proposed: 0.1 ppm (sensitizer))
NIOSH REL: (n-Alkane Mono Thiols) CL 0.5 ppm/15M
DOT CLASSIFICATION: 3; Label: Flammable Liquid, Poison (UN 1228); DOT Class: 6.1; Label: Poison, Flammable Liquid (UN 3071)
SAFETY PROFILE: Inhalation hazard. Mutation data reported. Combustible when exposed to heat or flame. To fight fire, use alcohol foam. When heated to decomposition it emits toxic fumes of SO_x. See also MERCAPTANS.

L

LCA000   CAS: 8022-15-9   HR: 1
LAVANDIN OIL
PROP: Main constituent is Linalool.
Prepared by steam distillation of the
flowering stalks of the plants *Lavandula
hybrida reverchon, Lavandula abrialis* (Fam.
Labiatae), *Lavandula officinalis*, or *Lavandula
latifolia*. Yellow liquid; camphoraceous odor
of lavender. D: 0.885, refr index: 1.460 @
20°. Sol in fixed oils, propylene glycol,
mineral oil; insol in glycerin.
SYNS: ABRIAL LAVANDIN OIL □ LAVANDIN
ABSOLUTE □ LAVANDIN BENZOL ABSOLUTE □ OIL
OF LAVANDIN □ OIL OF LAVANDIN, ABRIAL TYPE
CONSENSUS REPORTS: Reported in EPA
TSCA Inventory.
SAFETY PROFILE: A skin irritant. When
heated to decomposition it emits acrid
smoke and irritating fumes.

LCA100   CAS: 20777-39-3   HR: 1
LAVANDULYL ACETATE
mf: $C_{12}H_{20}O_2$ mw: 196.32
PROP: Mobile, clear, light yellow liquid with
lavendar fragrance.
SYNS: 4-HEXEN-1-OL, 5-METHYL-2-(1-METHYL-
ETHENYL)-, ACETATE □ 5-METHYL-2-(1-METHYL-
ETHENYL)-4-HEXEN-1-OL ACETATE
CONSENSUS REPORTS: Reported in EPA
TSCA Inventory.
SAFETY PROFILE: A skin irritant. When
heated to decomposition it emits acrid
smoke and irritating fumes.

LCD000   CAS: 8000-28-0   HR: 1
LAVENDER OIL
PROP: Found in the flowers of *Lavandula
officinalis* Chaix et Villars, *Lavandula vera* De
Candolle (Fam. *Labiatae*). The main
constituent is linalyl acetate. A colorless to
yellow liquid; characteristic odor and taste of
lavender flowers. D: 0.875, refr index:
1.459–1.470 @ 20°.
SYNS: LAVENDEL OEL (GERMAN) □ OIL OF
LAVENDER
CONSENSUS REPORTS: Reported in EPA
TSCA Inventory.
SAFETY PROFILE: Mildly toxic by
ingestion. A skin irritant. When heated to

decomposition it emits acrid smoke and
irritating fumes.

LCF000   CAS: 7439-92-1   HR: 3
LEAD
af: Pb   aw: 207.19
PROP: Bluish-gray, soft, weak, ductile metal
which tarnishes in moist air. Otherwise
stable to O_2 and H_2O at ordinary temp. Mp:
327.43°, bp: 1740°, d: 11.34 @ 20°/4°, vap
press: 1 mm @ 973°. Dissolves in dil
HNO_3, acetic acid, HCl (slowly). Sol in alkali
solns. Attacked at room temp by F_2 and Cl_2.
IDLH 100 mg/m³ (as Pb).
SYNS: C.I. 77575 □ C.I. PIGMENT METAL 4 □ GLOVER
□ LEAD FLAKE □ LEAD S2 □ OLOW (POLISH) □
OMAHA □ OMAHA & GRANT □ SI □ SO
CONSENSUS REPORTS: IARC Cancer
Review: Group 2B IMEMDT 7,230,87;
Animal Inadequate Evidence IMEMDT
23,325,80. Lead and its compounds are on
the Community Right-To-Know List.
Reported in EPA TSCA Inventory. EPA
Genetic Toxicology Program.
OSHA PEL: TWA 0.05 mg(Pb)/m³
ACGIH TLV: TWA 0.15 mg(Pb)/m³; BEI: 50
µg(lead)/L in blood; 150 µg(lead)/g
creatinine in urine
DFG MAK: 0.1 mg/m³; BAT: 70 µg(lead)/L
in blood; 30 µg(lead)/L in blood of women
less than 45 years old
NIOSH REL: TWA (Inorganic Lead) 0.10
mg(Pb)/m³
SAFETY PROFILE: Poison by ingestion.
Moderately toxic by intraperitoneal route.
Questionable carcinogen. Human systemic
effects by ingestion and inhalation: loss of
appetite, anemia, malaise, insomnia,
headache, irritability, muscle and joint pains,
tremors, flaccid paralysis without anesthesia,
hallucinations and distorted perceptions,
muscle weakness, gastritis, and liver changes.
The major organ systems affected are the
nervous system, blood system, and kidneys.
Lead encephalopathy is accompanied by
severe cerebral edema, increase in cerebral
spinal fluid pressure, proliferation and
swelling of endothelial cells in capillaries and

arterioles, proliferation of glial cells, neuronal degeneration, and areas of focal cortical necrosis in fatal cases. Experimental evidence now suggests that blood levels of lead below 10 µg/dL can have the effect of diminishing the IQ scores of children. Low levels of lead impair neurotransmission and immune system function and may increase systolic blood pressure. Reversible kidney damage can occur from acute exposure. Chronic exposure can lead to irreversible vascular sclerosis, tubular cell atrophy, interstitial fibrosis, and glomerular sclerosis. Severe toxicity can cause sterility, abortion, and neonatal mortality and morbidity. An experimental teratogen. Experimental reproductive effects. Human mutation data reported. Very heavy intoxication can sometimes be detected by formation of a dark line on the gum margins, the so-called lead line.

When lead is ingested, much of it passes through the body unabsorbed, and is eliminated in the feces. The greater portion of the lead that is absorbed is caught by the liver and excreted, in part, in the bile. For this reason, larger amounts of lead are necessary to cause toxic effects by this route, and a longer period of exposure is usually necessary to produce symptoms. On the other hand, upon inhalation, absorption takes place easily from the respiratory tract and symptoms tend to develop more quickly. For industry, inhalation is much more important than is ingestion. For the general population, exposure to lead occurs from inhaled air, dust of various types, and food and water, with an approximate 50/50 division between inhalation and ingestion routes. Lead occurs in water in either dissolved or particulate form. At low pH, lead is more easily dissolved. Chemical treatment to soften water increases the solubility of lead. Adults absorb about 5–15% of ingested lead and retain less than 5%. Children absorb about 50% and retain about 30%.

Lead produces a brittleness of the red blood cells so that they hemolyze with but slight trauma; the hemoglobin is not affected. Due to their increased fragility, the red cells are destroyed more rapidly in the body than is normal, producing an anemia that is rarely severe. The loss of circulating red cells stimulates the production of new young cells, which, on entering the bloodstream, are acted upon by the circulating lead, with resultant coagulation of their basophilic material. These cells, after suitable staining, are recognized as "stippled cells." There is no uniformity of opinion regarding the effect of lead on the white blood cells.

In addition to its effect on the red blood cells, lead produces a damaging effect on the organs or tissues with which it comes in contact. No specific or characteristic lesion is produced. Autopsies in deaths attributed to lead poisoning and experimental work on animals have shown pathological lesions of the kidneys, liver, male gonads, nervous system, blood vessels, and other tissues. None of these changes, however, has been found consistently. In cases of severe lead poisoning, the amount of lead found in the blood is frequently in excess of 0.07 mg per 100 cc of whole blood. The urinary lead excretion generally exceeds 0.1 mg per liter of urine.

Flammable in the form of dust when exposed to heat or flame. Moderately explosive in the form of dust when exposed to heat or flame. Mixtures of hydrogen peroxide + trioxane explode on contact with lead. Rubber gloves containing lead may ignite in nitric acid. Violent reaction on ignition with chlorine trifluoride, concentrated hydrogen peroxide, ammonium nitrate (below 200° with powdered lead), sodium acetylide (with powdered lead). Incompatible with NaN_3, Zr, disodium acetylide, oxidants. Can react vigorously with oxidizing materials. A common air contaminant. When heated to

L

decomposition it emits highly toxic fumes of Pb. See also LEAD COMPOUNDS.

LCH000   CAS: 1335-32-6   HR: 2
LEAD ACETATE, BASIC

mf: $C_4H_{10}O_8Pb_3$ mw: 807.71

PROP: White powder or white monoclinic or gelatinous solid. Very sol in cold water.

SYNS: BASIC LEAD ACETATE □ BIS(ACETATO)-TETRAHYDROXYTRILEAD □ BIS(ACETO)DIHYDRO-XYTRILEAD □ BLA □ LEAD MONOSUBACETATE □ LEAD SUBACETATE □ MONOBASIC LEAD ACETATE □ RCRA WASTE NUMBER U146 □ SUBACETATE LEAD

CONSENSUS REPORTS: IARC Cancer Review: Group 3 IMEMDT 7,230,87; Animal Sufficient Evidence IMEMDT 23,325,80; IMEMDT 1,40,72; Human Limited Evidence IMEMDT 23,325,80. Lead and its compounds are on the Community Right-To-Know List. Reported in EPA TSCA Inventory. EPA Genetic Toxicology Program.

SAFETY PROFILE: Experimental reproductive effects. Questionable carcinogen with experimental carcinogenic, neoplastigenic, and tumorigenic data. Mutation data reported. When heated to decomposition it emits toxic fumes of Pb. See also LEAD and LEAD COMPOUNDS.

LCJ000   CAS: 6080-56-4   HR: 3
LEAD ACETATE(II), TRIHYDRATE

mf: $C_4H_6O_4 \bullet Pb \bullet 3H_2O$ mw: 379.35

PROP: White crystals, odorless. Mp: 75°, d: 2.55. Sol in water: 1 g in 1.6 ml water at ambient temperature.

SYNS: ACETIC ACID, LEAD(2+) SALT TRIHYDRATE □ BIS(ACETATO)TRIHYDROXYTRILEAD □ BLEIAZETAT (GERMAN) □ LEAD ACETATE TRIHYDRATE □ LEAD DIACETATE TRIHYDRATE □ PLUMBOUS ACETATE

CONSENSUS REPORTS: IARC Cancer Review: Animal Sufficient Evidence IMEMDT 1,40,72. EPA Genetic Toxicology Program. Lead and its compounds are on the Community Right-To-Know List.

OSHA PEL: TWA 0.05 mg(Pb)/m³

NIOSH REL: (Inorganic Lead) TWA 0.10 mg(Pb)/m³

SAFETY PROFILE: Confirmed carcinogen with experimental carcinogenic and teratogenic data. Poison by intraperitoneal route. Moderately toxic by subcutaneous route. Experimental reproductive effects. Mutation data reported. When heated to decomposition it emits toxic fumes of Pb. See also LEAD COMPOUNDS.

LCK000   CAS: 7784-40-9   HR: 3
LEAD ACID ARSENATE

DOT: UN 1617

mf: $AsHO_4 \bullet Pb$ mw: 347.12

PROP: White crystals or white, transparent, monoclinic leaflets or fluffy powder. Insol in cold H_2O; sltly sol in hot H_2O; sol in HNO_3.

SYNS: ACID LEAD ARSENATE □ ACID LEAD ORTHOARSENATE □ ARSENATE of LEAD □ ARSINET-TE □ DIBASIC LEAD ARSENATE □ GYPSINE □ LEAD ARSENATE □ LEAD ARSENATE, solid (DOT) □ LEAD ARSENATE (standard) □ ORTHO L10 DUST □ ORTHO L40 DUST □ SCHULTENITE □ SECURITY □ SOPRABEL □ STANDARD LEAD ARSENATE □ TALBOT

CONSENSUS REPORTS: IARC Cancer Review: Human Sufficient Evidence IMEMDT 23,39,80; Animal Inadequate Evidence IMEMDT 1,40,72; IMEMDT 1,40,72. Arsenic and its compounds, as well as lead and its compounds, are on the Community Right-To-Know List. Reported in EPA TSCA Inventory.

OSHA PEL: TWA 0.05 mg(Pb)/m³; 0.01 mg(As)/m³; Cancer Hazard

ACGIH TLV: TWA 0.01 mg/m³; Confirmed Human Carcinogen; BEI: 35 μ (As)/L inorganic arsenic and methylated metabolites in urine

NIOSH REL: (Inorganic Lead) TWA 0.10 mg(Pb)/m³; (Inorganic Arsenic) CL 0.002 mg(As)/m³/15M

DOT CLASSIFICATION: 6.1; Label: Poison

SAFETY PROFILE: Confirmed human carcinogen. A poison by ingestion. Used as an insecticide and herbicide. When heated to decomposition it emits very toxic fumes of As and Pb. See also ARSENIC

COMPOUNDS and LEAD COMPOUNDS.

LCK100   CAS: 3687-31-8   HR: 3
LEAD ARSENATE
DOT: UN 1617

mf: $As_2O_8 \cdot 3Pb$ mw: 899.41

PROP: Odorless, white, heavy powder. Mp: 280°. Insol in water.

SYNS: ARSENIC ACID, LEAD(2+) SALT (2:3) □ ARSINETTE □ GYPSINE □ NU REXFORM □ ORTHO L10 DUST □ SOPRABEL □ TALBOT

OSHA PEL: OSHA: Cancer Hazard

DOT CLASSIFICATION: 6.1; Label: Poison

SAFETY PROFILE: Confirmed carcinogen. A poison by ingestion. When heated to decomposition it emits toxic vapors of lead and arsenic.

LCL000   CAS: 10031-13-7   HR: 3
LEAD(II) ARSENITE
DOT: UN 1618

mf: $As_2O_4 \cdot Pb$ mw: 421.03

PROP: White powder or white crystals. D: 5.85. Insol in water; sol in dil HNO_3.

SYN: LEAD ARSENITES (DOT)

CONSENSUS REPORTS: Arsenic and its compounds, as well as lead and its compounds, are on the Community Right-To-Know List.

OSHA PEL: TWA 0.05 mg(Pb)/m³; 0.01 mg(As)/m³; Cancer Hazard

ACGIH TLV: TWA 0.01 mg/m³; Confirmed Human Carcinogen; BEI: 35 μ (As)/L inorganic arsenic and methylated metabolites in urine

NIOSH REL: (Inorganic Lead) TWA 0.10 mg(Pb)/m³; (Inorganic Arsenic) CL 0.002 mg(As)/m³/15M

DOT CLASSIFICATION: 6.1; Label: Poison

SAFETY PROFILE: Confirmed human carcinogen. A poison. When heated to decomposition it emits very toxic fumes of Pb and As. See also LEAD COMPOUNDS and ARSENIC COMPOUNDS.

LCM000   CAS: 13424-46-9   HR: 3
LEAD(II) AZIDE
DOT: UN 0129

mf: N_6Pb mw: 291.25

PROP: Colorless needles or white powder. Explodes @ 350° or when shocked. Sltly sol in cold water; very sol in acetic acid; insol in NH_4OH.

SYNS: LEAD AZIDE (dry) (DOT) □ LEAD AZIDE, wetted with not <20% water or mixture of alcohol and water, by weight (DOT)

CONSENSUS REPORTS: Lead and its compounds are on the Community Right-To-Know List. Reported in EPA TSCA Inventory.

OSHA PEL: TWA 0.05 mg(Pb)/m³

ACGIH TLV: TWA 0.15 mg(Pb)/m³

NIOSH REL: (Inorganic Lead) TWA 0.10 mg(Pb)/m³

DOT CLASSIFICATION: Forbidden (dry)

SAFETY PROFILE: A deadly poison. An explosive sensitive to shock or heating to 250°C. Will explode spontaneously during crystallization. Mixtures with calcium stearate may explode spontaneously. May explode spontaneously after prolonged contact with copper, zinc, or their alloys (e.g., brass). Incompatible with CS_2. Used in commercial blasting caps and military ammunition. When heated it emits highly toxic fumes of Pb and NO_x. See also LEAD COMPOUNDS, AZIDES, and EXPLOSIVES, HIGH.

LCP000   CAS: 598-63-0   HR: 2
LEAD CARBONATE
mf: $CO_3 \cdot Pb$ mw: 267.20

PROP: White, heavy powder or crystals. D: 6.61, mp: 315°, decomp @ 400° leaving residue of PbO. Insol in water, alc; sol in acetic acid, dil HNO_3 (effervescence).

SYNS: CARBONIC ACID, LEAD(2+) SALT (1:1) □ CERUSSETE □ DIBASIC LEAD CARBONATE □ LEAD(2+) CARBONATE □ WHITE LEAD

CONSENSUS REPORTS: IARC Cancer Review: Animal Inadequate Evidence IMEMDT 23,325,80; IMEMDT 1,40,72. Lead and its compounds are on the Community Right-To-Know List. Reported in EPA TSCA Inventory.

OSHA PEL: TWA 0.05 mg(Pb)/m³

ACGIH TLV: TWA 0.15 mg(Pb)/m³
NIOSH REL: (Inorganic Lead) TWA 0.10 mg(Pb)/m³
SAFETY PROFILE: Moderately toxic by ingestion. Human systemic effects by ingestion: gastrointestinal contractions, jaundice, brain degenerative changes, convulsions, nausea or vomiting. Experimental reproductive effects. Questionable carcinogen. Ignites spontaneously and burns fiercely in fluorine. When heated to decomposition it emits toxic fumes of Pb. See also LEAD COMPOUNDS.

LCQ000   CAS: 7758-95-4   HR: 2
LEAD CHLORIDE
mf: Cl_2Pb mw: 278.09
PROP: White crystals from water. Mp: 501°, bp: 950°, d: 5.85, vap press: 1 mm @ 547°. Somewhat sol in cold water, more sol in hot water; very sol in ammonium chloride, NH_4NO_3, alkali hydroxides, and alc.
SYNS: LEAD(2+) CHLORIDE □ LEAD(II) CHLORIDE □ LEAD DICHLORIDE □ PLUMBOUS CHLORIDE
CONSENSUS REPORTS: IARC Cancer Review: Animal Inadequate Evidence IMEMDT 23,325,80. Lead and its compounds are on the Community Right-To-Know List. Reported in EPA TSCA Inventory. EPA Genetic Toxicology Program.
OSHA PEL: TWA 0.05 mg(Pb)/m³
ACGIH TLV: TWA 0.15 mg(Pb)/m³
NIOSH REL: (Inorganic Lead) TWA 0.10 mg(Pb)/m³
SAFETY PROFILE: Moderately toxic by ingestion. An experimental teratogen. Experimental reproductive effects. Questionable carcinogen. Human mutation data reported. Explosive reaction with calcium when heated slightly. When heated to decomposition it emits very toxic fumes of Pb and Cl⁻. See also LEAD and LEAD COMPOUNDS.

LCR000   CAS: 7758-97-6   HR: 3
LEAD CHROMATE
mf: $CrO_4 \cdot Pb$ mw: 323.19
PROP: Yellow or orange-yellow powder. Stable orange-yellow monoclinic cryst; unstable yellow orthorhombic form, and orange-red tetragonal form, stable above 7°. Mp: 844°, bp: decomp, d: 6.3. One of the most insol salts. Insol in acetic acid; sol in solns of fixed alkali hydroxides, dil HNO_3. IDLH Ca [15 mg/m³ {as Cr(VI)}].
SYNS: CANARY CHROME YELLOW 40-2250 □ CHROMATE de PLOMB (FRENCH) □ CHROME GREEN □ CHROME LEMON □ CHROME YELLOW □ CHROMIC ACID, LEAD(2+) SALT (1:1) □ CHROMIUM YELLOW □ C.I. 77600 □ C.I. PIGMENT YELLOW 34 □ COLOGNE YELLOW □ C.P. CHROME YELLOW LIGHT □ CROCOITE □ DAINICHI CHROME YELLOW G □ GIALLO CROMO (ITALIAN) □ KING'S YELLOW □ LEAD CHROMATE(VI) □ LEIPZIG YELLOW □ LEMON YELLOW □ PARIS YELLOW □ PIGMENT GREEN 15 □ PLUMBOUS CHROMATE □ PURE LEMON CHROME L3GS
CONSENSUS REPORTS: NTP 10th Report on Carcinogens. IARC Cancer Review: Group 1 IMEMDT 7,165,87; Animal Inadequate Evidence IMEMDT 2,100,73; Animal Sufficient Evidence IMEMDT 23,205,80; Human Sufficient Evidence IMEMDT 23,205,80. Lead and its compounds, as well as chromium and its compounds, are on the Community Right-To-Know List. Reported in EPA TSCA Inventory. EPA Genetic Toxicology Program.
OSHA PEL: TWA 0.05 mg(Pb)/m³; CL 0.1 mg(CrO₃)/m³
ACGIH TLV: 0.05 mg(Cr)/m³; Human Carcinogen
DFG MAK: Confirmed Animal Carcinogen with Unknown Relevance to Humans
NIOSH REL: (Chromium(VI)) TWA 0.001 mg(Cr(VI))/m³; (Inorganic Lead) TWA 0.10 mg(Pb)/m³
SAFETY PROFILE: Confirmed carcinogen with experimental neoplastigenic and tumorigenic data. Poison by intraperitoneal route. Mildly toxic by ingestion. Human mutation data reported. Potentially explosive reactions with azodyestuffs (e.g., dinitro-aniline orange, chlorinated para red). Violent reaction with aluminum + dinitronaphthal-

ene + heat. Forms pyrophoric mixtures with sulfur, tantalum, and iron(III) hexacyanoferrate(4−) (e.g., brunswick green pigment, prussian blue pigment). When heated to decomposition it emits toxic fumes of Pb. See also LEAD COMPOUNDS and CHROMIUM COMPOUNDS.

LCS000   CAS: 18454-12-1   HR: 3
LEAD CHROMATE, BASIC
mf: CrO$_4$Pb•OPb   mw: 546.38
PROP: Red, amorphous or crystalline solid. Mp: 920°. Insol in H$_2$O; sol in acid, alkali. IDLH Ca [15 mg/m^3 {as Cr(VI)}].
SYNS: ARANCIO CROMO (ITALIAN) □ AUSTRIAN CINNABAR □ BASIC LEAD CHROMATE □ CHINESE RED □ CHROME ORANGE □ CHROMIUM LEAD OXIDE □ C.I. 77601 □ C.I. PIGMENT ORANGE 21 □ C.I. PIGMENT RED □ C.P. CHROME LIGHT 2010 □ C.P. CHROME ORANGE DARK 2030 □ C.P. CHROME ORANGE MEDIUM 2020 □ DAINICHI CHROME ORANGE R □ GENUINE ACETATE CHROME ORANGE □ GENUINE ORANGE CHROME □ INDIAN RED □ INTERNATIONAL ORANGE 2221 □ IRGACHROME ORANGE OS □ LEAD CHROMATE OXIDE (MAK) □ LEAD CHROMATE, RED □ LIGHT ORANGE CHROME □ No. 156 ORANGE CHROME □ ORANGE CHROME □ ORANGE NITRATE CHROME □ PALE ORANGE CHROME □ PERSIAN RED □ PURE ORANGE CHROME M □ RED LEAD CHROMATE □ VYNAMON ORANGE CR
CONSENSUS REPORTS: IARC Cancer Review: Human Sufficient Evidence IMEMDT 23,205,80; Animal Limited Evidence IMEMDT 23,205,80. Lead and its compounds, as well as chromium and its compounds, are on the Community Right-To-Know List. Reported in EPA TSCA Inventory.
OSHA PEL: TWA 0.05 mg(Pb)/m^3; CL 0.1 mg(CrO$_3$)/m^3
ACGIH TLV: TWA 0.05 mg(Cr)/m^3; TWA 0.15 mg(Pb)/m^3
DFG MAK: Suspected Carcinogen
NIOSH REL: (Chromium(VI)) TWA 0.001 mg(Cr(VI))/m^3; (Inorganic Lead) TWA 0.10 mg(Pb)/m^3
SAFETY PROFILE: Suspected human carcinogen with experimental carcinogenic, neoplastigenic, and tumorigenic data.

Human mutation data reported. When heated to decomposition it emits very toxic fumes of Pb. See also LEAD COMPOUNDS and CHROMIUM COMPOUNDS.

LCT000   HR: 3
LEAD COMPOUNDS
CONSENSUS REPORTS: Lead and its compounds are on the Community Right-To-Know List.
SAFETY PROFILE: Some are experimental neoplastigens and tumorigens. Lead poisoning is one of the commonest of occupational diseases. The presence of lead-bearing materials or lead compounds in an industrial plant does not necessarily result in exposure on the part of the worker. The lead must be in such form, and so distributed, as to gain entrance into the body or tissues of the worker in measurable quantity; otherwise no exposure can be said to exist. Some lead compounds are carcinogens of the lungs and kidneys.

Mode of entry into body: 1. By inhalation of the dust, fumes, mists, or vapors. (Common air contaminants.) 2. By ingestion of lead compounds trapped in the upper respiratory tract or introduced into the mouth on food, tobacco, fingers, or other objects. 3. Through the skin; this route is of special importance in the case of organic compounds of lead, such as lead tetraethyl. In the case of the inorganic forms of lead, this route is of no practical importance. Significant quantities of lead can be ingested from water that has been sitting in pipes with lead solder. Some water coolers may also have this type of solder.

Lead is a cumulative poison. Increasing amounts build up in the body and eventually reach a point at which symptoms and disability occur. See LEAD for symptoms of overexposure.

The toxicity of the various lead compounds appears to depend upon several factors: (1) the solubility of the compound in the body fluids; (2) the fineness of the particles of the compound (solubility is greater in proportion to the fineness of the particles); (3) conditions under which the compound is being used. Where a lead compound is used as a powder, contamination of the atmosphere will be much less if the powder is kept damp. Of the various lead compounds, the carbonate, the monoxide, and the sulfate are considered to be more toxic than metallic lead or other lead compounds. Lead arsenate is very toxic due to the presence of the arsenic radical. Organolead compounds are rapidly absorbed by the respiratory and gastrointestinal systems and through the skin. Tetraethyl lead is converted in the body to triethyl lead which is a more severe neurotoxin than inorganic lead. Diagnostic mobilization of lead with calcium EDTA may be useful in questionable cases. When heated to decomposition they emit toxic fumes of Pb. See also LEAD and specific compounds.

LCU000   CAS: 592-05-2   HR: 3
LEAD(II) CYANIDE
DOT: UN 1620
mf: C_2N_2Pb mw: 259.23
PROP: Yellowish-white powder. Sol in cold H_2O; sol in hot H_2O; sol in (aq) KCN.
SYNS: C.I. 77610 □ C.I. PIGMENT YELLOW 48 □ CYANURE de PLOMB (FRENCH) □ LEAD CYANIDE (DOT)
CONSENSUS REPORTS: Lead and its compounds, as well as cyanide and its compounds, are on the Community Right-To-Know List.
OSHA PEL: TWA 0.05 mg(Pb)/m³
ACGIH TLV: TWA 0.15 mg(Pb)/m³
NIOSH REL: (Inorganic Lead) TWA 0.10 mg(Pb)/m³
DOT CLASSIFICATION: 6.1; Label: Poison
SAFETY PROFILE: Poison by intraperitoneal route. Violent reaction with Mg. A fire hazard and a powerful oxidizer. When heated to decomposition it emits very toxic fumes of Pb, CN⁻, and NO$_x$. See also LEAD COMPOUNDS and CYANIDE.

LCV000   CAS: 301-04-2   HR: 3
LEAD DIACETATE
DOT: UN 1616
mf: $C_4H_6O_4•Pb$ mw: 325.29
PROP: Trihydrate: colorless crystals or white granules or powder. Sltly acetic odor, slowly effloresces. D: 2.55, mp: 75° (when rapidly heated), decomp above 200°. Very sol in glycerin and water; sltly sol in EtOH.
SYNS: ACETATE de PLOMB (FRENCH) □ ACETIC ACID LEAD(2+) SALT □ BLEIACETAT (GERMAN) □ DIBASIC LEAD ACETATE □ LEAD ACETATE □ LEAD(2+) ACETATE □ LEAD(II) ACETATE □ LEAD DIBASIC ACETATE □ NORMAL LEAD ACETATE □ PLUMBOUS ACETATE □ RCRA WASTE NUMBER U144 □ SALT of SATURN □ SUGAR of LEAD
CONSENSUS REPORTS: NTP 10th Report on Carcinogens. IARC Cancer Review: Group 3 IMEMDT 7,230,87; Animal Sufficient Evidence IMEMDT 23,325,80; IMEMDT 1,40,72; Human Limited Evidence IMEMDT 23,325,80. Lead and its compounds are on the Community Right-To-Know List. Reported in EPA TSCA Inventory. EPA Genetic Toxicology Program.
OSHA PEL: TWA 0.05 mg(Pb)/m³
ACGIH TLV: TWA 0.15 mg(Pb)/m³
NIOSH REL: (Inorganic Lead) TWA 0.10 mg(Pb)/m³
DOT CLASSIFICATION: 6.1; Label: KEEP AWAY FROM FOOD
SAFETY PROFILE: Confirmed carcinogen with experimental neoplastigenic, tumorigenic, and teratogenic data. Poison by ingestion, intraperitoneal, subcutaneous, and intravenous routes. Experimental reproductive effects. Human systemic effects by ingestion: brain degenerative changes, convulsions, nausea or vomiting, hepatitis normocytic anemia. Human mutation data reported. Used as a color additive in hair dyes, an insecticide, an astringent, and a sedative. Incompatible with $KBrO_3$, acids, soluble sulfates, citrates, tartrates, chlorides, carbonates, alkalies, tannin phosphates, resorcinol, salicylic acid, phenol, chloral hydrate, sulfites, vegetable infusions, tinctures. When heated to

decomposition it emits toxic fumes of Pb. See also LEAD COMPOUNDS.

LCV100   CAS: 1344-40-7   HR: 3
LEAD DIBASIC PHOSPHITE
DOT: UN 2989
mf: $HO_5PPb_3•1/2H_2O$ mw: 742.56
SYNS: C.I. 77620 ☐ DIBASIC LEAD METAPHOSPHATE ☐ DIBASIC LEAD PHOSPHITE ☐ LEAD OXIDE PHOSPHONATE, HEMIHYDRATE ☐ LEAD PHOSPHITE, dibasic (DOT)
ACGIH TLV: TWA 0.15 mg(Pb)/m³
DOT CLASSIFICATION: 4.1; Label: Flammable Solid
SAFETY PROFILE: A poison by ingestion. A flammable solid. When heated to decomposition it emits toxic vapors of lead and PO_x.

LCW000   CAS: 19010-66-3   HR: 2
LEAD DIMETHYLDITHIOCARBAMATE
mf: $C_6H_{12}N_2S_4•Pb$ mw: 447.63
PROP: Solid or pale-yellow needles from Me_2CO. Mp: 258°, d: 2.5.
SYNS: BIS(DIMETHYLCARBAMODITHIOATO-S,S')LEAD ☐ BIS(DIMETHYLDITHIOCARBAMIA-TO)LEAD ☐ DIMETHYLDITHIOCARBAMIC ACID, LEAD SALT ☐ METHYL LEADATE ☐ NCI-C02891
CONSENSUS REPORTS: IARC Cancer Review: Group 3 IMEMDT 7,230,87; Animal Inadequate Evidence IMEMDT 12,131,76. NCI Carcinogenesis Bioassay (feed); No Evidence: mouse, rat NCITR* NCI-CG-TR-151,79. Lead and its compounds are on the Community Right-To-Know List. Reported in EPA TSCA Inventory.
NIOSH REL: (Inorganic Lead) TWA 0.10 mg(Pb)/m³
SAFETY PROFILE: Questionable carcinogen with experimental tumorigenic data. Mutation data reported. Combustible when exposed to heat or flame. When heated to decomposition it emits very toxic fumes of Pb, NO_x, and SO_x. See also LEAD COMPOUNDS and CARBAMATES.

LCX000   CAS: 1309-60-0   HR: 3
LEAD DIOXIDE
DOT: UN 1872
mf: O_2Pb mw: 239.19
PROP: Brown, hexagonal crystals or dark-brown powder. Mp: decomp @ 290°, d: 9.375. Liberates O_2 when heated. Insol in water; sol in HCl evolving chlorine; sol in alkali iodide solns liberating iodine; sol in hot caustic alkali soln.
SYNS: BIOXYDE de PLOMB (FRENCH) ☐ C.I. 77580 ☐ LEAD BROWN ☐ LEAD(IV) OXIDE ☐ LEAD OXIDE BROWN ☐ LEAD PEROXIDE (DOT) ☐ LEAD SUPEROXIDE ☐ PEROXYDE de PLOMB (FRENCH)
CONSENSUS REPORTS: Reported in EPA TSCA Inventory. Lead and its compounds are on the Community Right-To-Know List.
OSHA PEL: TWA 0.05 mg(Pb)/m³
ACGIH TLV: TWA 0.15 mg(Pb)/m³
NIOSH REL: (Inorganic Lead) TWA 0.10 mg(Pb)/m³
DOT CLASSIFICATION: 5.1; Label: Oxidizer
SAFETY PROFILE: Poison by intraperitoneal route. A powerful oxidizer. Probably a severe eye, skin, and mucous membrane irritant. Explosive reaction with warm potassium or sodium, cesium acetylide at 350°C, boron (when ground), yellow phosphorus (when ground), sulfinyl dichloride. Mixtures with silicon (2:1 silicon/lead dioxide) are used as initiators and heat to 1100°C when exposed to flame. Mixtures with zirconium can deflagrate (burn explosively) and are sensitive to friction, ignition, and static electricity. Violent reaction or ignition with chlorine trifluoride, hydrogen sulfide, nitrogen compounds (e.g., hydroxylamine), red phosphorus, sulfur (when ground), sulfur + sulfuric acid, peroxyformic acid. Violent reactions with powdered aluminum, Al_4C_3, metal acetylides or carbides, H_2O_2, magnesium, nonmetal halides, performic acid, phenyl hydrazine, $S(OCl)_2$. Vigorous reaction with seleninyl chloride, metal sulfides + heat (e.g., calcium sulfide, strontium sulfide, or barium sulfide). Incandescent reaction with powdered molybdenum or tungsten when heated, warm phosphorus trichloride, sulfur dioxide.

L

Metal oxides increase the explosive sensitivity of nitroalkanes (e.g., nitromethane, nitroethane). Can react vigorously with reducing materials. When heated to decomposition it emits toxic fumes of Pb. See also LEAD COMPOUNDS and PEROXIDES.

LDB000   CAS: 22904-40-1   HR: 2
LEAD DISODIUM ETHYLENEDINITRILO-
TETRACETATE

mf: $C_{10}H_{12}N_2O_8 \cdot 2Na \cdot Pb$ mw: 541.41
SYNS: LEAD DISODIUM EDTA □ LEAD(2-), ((ETHYLENEDINITRILO)TETRAACETATO)-, DISODIUM
CONSENSUS REPORTS: Reported in EPA TSCA Inventory. Lead and its compounds are on the Community Right-To-Know List.
NIOSH REL: (Lead, Inorganic) 10H TWA 0.10 mg(Pb)/m^3
SAFETY PROFILE: Moderately toxic by intraperitoneal and intravenous routes. When heated to decomposition it emits very toxic fumes of NO_x, Na_2O, and Pb. See also LEAD COMPOUNDS.

LDC000   CAS: 69029-52-3   HR: 2
LEAD DROSS

SYN: LEAD SCRAP
CONSENSUS REPORTS: Reported in EPA TSCA Inventory. Lead and its compounds are on the Community Right-To-Know List.
OSHA PEL: TWA 0.05 mg(Pb)/m^3
ACGIH TLV: TWA 0.15 mg(Pb)/m^3
NIOSH REL: (Inorganic Lead) TWA 0.10 mg(Pb)/m^3
SAFETY PROFILE: A corrosive irritant to the eyes, skin, and mucous membranes. When heated to decomposition it emits toxic fumes of lead. See also LEAD.

LDE000   CAS: 13814-96-5   HR: 3
LEAD FLUOBORATE

mf: $B_2F_8 \cdot Pb$ mw: 380.81
PROP: Clear liquid. Fully miscible in water.
SYN: TETRAFLUORO BORATE(1-) LEAD(2+)
CONSENSUS REPORTS: Reported in EPA TSCA Inventory. Lead and its compounds are on the Community Right-To-Know List.

OSHA PEL: TWA 0.05 mg(Pb)/m^3; TWA 2.5 mg(Pb)/m^3
ACGIH TLV: TWA 0.15 mg(Pb)/m^3
NIOSH REL: (Lead, Inorganic) 10H TWA 0.10 mg(Pb)/m^3
SAFETY PROFILE: Poison by ingestion. When heated to decomposition it emits very toxic fumes of Pb, F$^-$, and BO_x. See also LEAD COMPOUNDS and FLUORIDES.

LDF000   CAS: 7783-46-2   HR: 3
LEAD(II) FLUORIDE

mf: F_2Pb mw: 245.19
PROP: Colorless solid. D: (orthorhombic) 8.445, d: (cubic) 7.750, mp: 855°, bp: 1290°, vap press: 10 mm @ 904°. Sltly sol in H_2O; sol in HNO_3; insol in NH_3 and Me_2CO.
SYNS: LEAD DIFLUORIDE □ LEAD FLUORIDE (DOT) □ PLOMB FLUORURE (FRENCH) □ PLUMBOUS FLUORIDE
CONSENSUS REPORTS: Lead and its compounds are on the Community Right-To-Know List. Reported in EPA TSCA Inventory.
OSHA PEL: TWA 0.05 mg(Pb)/m^3; TWA 2.5 mg(F)/m^3
ACGIH TLV: TWA 0.15 mg(Pb)/m^3; TWA 2.5 mg(F)/m^3; BEI: 3 mg/g creatinine of fluorides in urine prior to shift; 10 mg/g creatinine of fluorides in urine at end of shift.
NIOSH REL: (Inorganic Lead) TWA 0.10 mg(Pb)/m^3
SAFETY PROFILE: Moderately toxic by ingestion and subcutaneous routes. Vigorous reaction with fluorine. Incompatible with CaC_2. When heated to decomposition it emits very toxic fumes of Pb and F$^-$. See also LEAD COMPOUNDS and FLUORIDES.

LDG000   CAS: 25808-74-6   HR: 3
LEAD(II) FLUOROSILICATE

mf: $F_6Si \cdot Pb \cdot 2H_2O$ mw: 385.32
PROP: Monoclinic, colorless powder. Mp: decomp. Sol in cold H_2O; very sol in hot H_2O.
SYN: HEXAFLUOROSILICATE (2-1) LEAD(II) SALT DIHYDRATE

CONSENSUS REPORTS: Lead and its compounds are on the Community Right-To-Know List. Reported in EPA TSCA Inventory.

OSHA PEL: TWA 0.05 mg(Pb)/m³

ACGIH TLV: TWA 0.15 mg(Pb)/m³

NIOSH REL: (Lead, Inorganic):10H TWA 0.10 mg(Pb)/m³

DOT CLASSIFICATION: 6.1; Label: KEEP AWAY FROM FOOD

SAFETY PROFILE: Poison by ingestion. When heated to decomposition it emits very toxic fumes of F⁻ and Pb. See also LEAD COMPOUNDS.

LDM000   CAS: 12709-98-7   HR: 2
LEAD-MOLYBDENUM CHROMATE

PROP: IDLH 1000 mg/m³ (as Mo). IDLH Ca [15 mg/m³ {as Cr(VI)}].

SYNS: CHROMIC ACID, LEAD and MOLYBDENUM SALT □ CHROMIC ACID LEAD SALT with LEAD MOLYBDATE □ C.I. PIGMENT RED 104 □ LEAD CHROMATE, SULPHATE and MOLYBDATE □ MOLYBDENUM-LEAD CHROMATE □ MOLYBDENUM ORANGE

CONSENSUS REPORTS: Lead and its compounds, as well as chromium and its compounds, are on the Community Right-To-Know List.

OSHA PEL: TWA CL 0.1 mg(CrO₃)/m³; TWA 0.05 mg(Pb)/m³; TWA 5 mg(Mo)/m³

ACGIH TLV: TWA 0.05 mg(Cr)/m³; TWA Soluble Compounds: TWA 0.5 mg(Mo)/m³ Confirmed Animal Carcinogen with Unknown Relevance to Humans; TWA 0.15 mg(Pb)/m³

NIOSH REL: (Chromium(VI)) TWA 0.001 mg(Cr(VI))/m³; (Inorganic Lead) TWA 0.10 mg(Pb)/m³

SAFETY PROFILE: Questionable carcinogen with experimental neoplastigenic and tumorigenic data. Human mutation data reported. A powerful oxidizer. Probably a severe eye, skin, and mucous membrane irritant. When heated to decomposition it emits toxic fumes of Pb, chromium trioxide, and Mo. See also LEAD COMPOUNDS, MOLYBDENUM COMPOUNDS, and CHROMIUM COMPOUNDS.

LDN000   CAS: 1317-36-8   HR: 2
LEAD MONOXIDE

mf: OPb   mw: 223.19

PROP: Exists in 2 forms: (1) red to reddish-yellow, tetragonal crystals; stable at ordinary temps. (2) Yellow, orthorhombic crystals; stable >489°. D: 9.53, mp: 897°. Insol in water, alc; sol in acetic acid, dil HNO₃, warm solns of fixed alkali hydroxides.

SYNS: C.I. 77577 □ C.I. PIGMENT YELLOW 46 □ LEAD OXIDE □ LEAD(II) OXIDE □ LEAD OXIDE YELLOW □ LEAD PROTOXIDE □ LITHARGE □ LITHARGE YELLOW L-28 □ MASSICOT □ MASSICOTITE □ PLUMBOUS OXIDE □ YELLOW LEAD OCHER

CONSENSUS REPORTS: IARC Cancer Review: Animal Inadequate Evidence IMEMDT 23,325,80. Reported in EPA TSCA Inventory. EPA Genetic Toxicology Program. Lead and its compounds are on the Community Right-To-Know List.

OSHA PEL: TWA 0.05 mg(Pb)/m³

ACGIH TLV: TWA 0.15 mg(Pb)/m³

NIOSH REL: (Inorganic Lead) TWA 0.10 mg(Pb)/m³

SAFETY PROFILE: Moderately toxic by ingestion and intraperitoneal routes. Mutation data reported. A skin irritant. Questionable carcinogen. Avoid breathing dust. Wash thoroughly after contact with the material and before eating or smoking. Explosive reaction with rubidium acetylide at 200°C, zirconium + heat, silicon + aluminum + heat, chlorine + ethylene (at 100°C), perchloric acid + glycerin. Violent or explosive thermite reaction when heated with aluminum powder. Violent or explosive reaction with chlorinated rubber (above 200°C), fluoroelastomers (at 200°C), peroxyformic acid. Violent reaction or ignition with hydrogen trisulfide. May ignite spontaneously with linseed oil, dichloromethylsilane, fluorine + glycerin. Vigorous reaction with silicon + heat. Incandescent reaction with warm aluminum carbide, lithium acetylide, boron, seleninyl chloride. Incompatible with chlorine, perchloric acid, metal acetylides, metals, nonmetals. Mixtures of lead oxide with glycerin have been used as a jointing

L

compound and may explode when exposed to powerful oxidizers. When heated to decomposition it emits toxic fumes of Pb. Used in manufacturing of storage batteries, ceramic products, paints, and rubber. See also LEAD COMPOUNDS.

LDO000   CAS: 10099-74-8   HR: 3
LEAD(II) NITRATE (1:2)
DOT: UN 1469
mf: $N_2O_6 \cdot Pb$ mw: 331.21
PROP: White crystals. Mp: decomp @ 470°, d: 4.53 @ 20°. Very sol in H_2O; mod sol in EtOH.
SYNS: LEAD DINITRATE □ LEAD NITRATE □ LEAD(2+) NITRATE □ LEAD(II) NITRATE □ NITRATE de PLOMB (FRENCH) □ NITRIC ACID, LEAD(2+) SALT
CONSENSUS REPORTS: IARC Cancer Review: Animal Inadequate Evidence IMEMDT 23,325,80. Reported in EPA TSCA Inventory. Lead and its compounds are on the Community Right-To-Know List.
OSHA PEL: TWA 0.05 mg(Pb)/m^3
ACGIH TLV: TWA 0.15 mg(Pb)/m^3
NIOSH REL: (Inorganic Lead) TWA 0.10 mg(Pb)/m^3
DOT CLASSIFICATION: 5.1; Label: Oxidizer, Poison
SAFETY PROFILE: Poison by intravenous and intraperitoneal routes. Moderately toxic by ingestion. Experimental teratogenic and reproductive effects. Questionable carcinogen. Probably a severe eye, skin, and mucous membrane irritant. Mutation data reported. A powerful oxidizer. Explodes on contact with red-hot carbon, cyclopentadienylsodium (at 100–130°C), potassium acetate + heat. Reacts violently with ammonium thiocyanate, carbon, lead hypophosphite. When heated to decomposition it emits very toxic fumes of Pb and NO_x. Used as a mordant, a chemical reagent, and in production of matches and pyrotechnics. See also LEAD COMPOUNDS and NITRATES.

LDP000   CAS: 51317-24-9   HR: 3
LEAD NITRORESORCINATE
DOT: NA 0473
mf: $C_6H_5NO_4 \cdot xPb$ mw: 1605.45
SYNS: INITIATING EXPLOSIVE LEAD MONONITRORESORCINATE (DOT) □ LEAD MONONITRORESORCINATE (DRY) (DOT)
CONSENSUS REPORTS: Lead and its compounds are on the Community Right-To-Know List.
DOT CLASSIFICATION: EXPLOSIVE 1.1A; Label: EXPLOSIVE 1.1A; DOT Class: Forbidden (dry)
SAFETY PROFILE: Poison by ingestion and inhalation. An explosive. When heated to decomposition it emits very toxic fumes of NO_x and Pb. See also LEAD TRINITRORESORCINATE, LEAD COMPOUNDS, NITRO COMPOUNDS of AROMATIC HYDROCARBONS, and EXPLOSIVES, HIGH.

LDQ000   CAS: 1120-46-3   HR: 2
LEAD(II) OLEATE (1:2)
mf: $C_{36}H_{66}O_4 \cdot Pb$ mw: 770.21
PROP: White, ointment-like granules or mass. Insol in water; sol in alc, benzene, ether, oil, turpentine.
SYNS: OLEIC ACID LEAD SALT □ OLEIC ACID, LEAD(2+) SALT (2:1)
CONSENSUS REPORTS: Lead and its compounds are on the Community Right-To-Know List. Reported in EPA TSCA Inventory.
OSHA PEL: TWA 0.05 mg(Pb)/m^3
ACGIH TLV: TWA 0.15 mg(Pb)/m^3
NIOSH REL: (Inorganic Lead) TWA 0.10 mg(Pb)/m^3
SAFETY PROFILE: Moderately toxic by ingestion. Used as a grease it may explode in hot-running bearings. When heated to decomposition it emits toxic fumes of Pb. Used in varnishes and high-pressure lubricants. See also LEAD COMPOUNDS.

LDS000   CAS: 1314-41-6   HR: 3
LEAD OXIDE RED
mf: O_4Pb_3 mw: 685.57
PROP: Bright red powder or crystals. Evolves O_2 on heating. Mp: 830° (decomp),

bp: 1472°, d: 8.32–9.16, vap press: 1 mm @ 943°. Insol in H_2O, EtOH; sol in AcOH.

SYNS: C.I. 77578 □ C.I. PIGMENT RED 105 □ DILEAD(II) LEAD(IV) OXIDE □ GOLD SATINOBRE □ LEAD ORTHOPLUMBATE □ LEAD TETRAOXIDE □ MINERAL ORANGE □ MINERAL RED □ MINIUM □ MINIUM NON-SETTING RL-95 □ ORANGE LEAD □ PARIS RED □ PLUMBOPLUMBIC OXIDE □ RED LEAD □ RED LEAD OXIDE □ SANDIX □ SATURN RED □ TRILEAD TETROXIDE

CONSENSUS REPORTS: Lead and its compounds are on the Community Right-To-Know List. Reported in EPA TSCA Inventory.

OSHA PEL: TWA 0.05 mg(Pb)/m^3
ACGIH TLV: TWA 0.15 mg(Pb)/m^3
NIOSH REL: (Inorganic Lead) TWA 0.10 mg(Pb)/m^3

SAFETY PROFILE: Poison by intraperitoneal route. Moderately toxic by ingestion. Combustible by chemical reaction with reducing agents. An oxidizing agent. Explodes on contact with peroxyformic acid. Ignites on contact with dichloromethylsilane. Incandescent reaction with seleninyl chloride. One-percent fresh red lead decreases the explosion temperature of 2,4,6-trinitrotoluene to 192°C. Incompatible with Al, $CsHC_2$, (F_2 + glycerin), H_2S_3, (glycerin + $HClO_4$), $RbHC_2$, (Si + Al), Na, SO_3, Ti, Zr. Mixtures of lead oxide with glycerin have been used as a jointing compound and may explode when exposed to powerful oxidizers. When heated to decomposition it emits toxic fumes of Pb. See also LEAD COMPOUNDS.

LDS499   CAS: 13637-76-8   HR: 3
LEAD(II) PERCHLORATE
DOT: UN 1470
mf: Cl_2O_8Pb mw: 406.10
PROP: White, deliquescent solid. Stable at 2° but begins to decomp above 2°. Very sol in H_2O.

SYNS: LEAD DIPERCHLORATE □ LEAD PERCHLOR-ATE □ LEAD PERCHLORATE, solid or solution (DOT) □ LEAD(2+) PERCHLORATE

CONSENSUS REPORTS: Lead and its compounds are on the Community Right-

To-Know List. Reported in EPA TSCA Inventory.

ACGIH TLV: TWA 0.15 mg(Pb)/m^3
DOT CLASSIFICATION: 5.1; Label: Oxidizer, Poison

SAFETY PROFILE: Solutions in methanol are sensitive explosives when no moisture is present. When heated to decomposition it emits toxic fumes of Cl⁻ and Pb. Used as a corrosion-inhibiting pigment in primers and paints, and in making storage batteries. See also LEAD COMPOUNDS and PERCHLORATES.

LDT000   CAS: 63916-96-1   HR: 3
LEAD(II) PERCHLORATE,
   HEXAHYDRATE (1:2:6)
mf: Cl_2O_8•Pb•$6H_2O$ mw: 514.21
SYN: PERCHLORATE ACID, LEAD SALT, HEXAHYDRATE

OSHA PEL: TWA 0.05 mg(Pb)/m^3
ACGIH TLV: TWA 0.15 mg(Pb)/m^3
NIOSH REL: (Inorganic Lead) TWA 0.10 mg(Pb)/m^3

CONSENSUS REPORTS: Lead and its compounds are on the Community Right-To-Know List.

DOT CLASSIFICATION: 5.1; Label: Oxidizer

SAFETY PROFILE: Poison by intraperitoneal route. An oxidizer. When heated to decomposition it emits very toxic fumes of Pb and Cl⁻. See also LEAD COMPOUNDS and PERCHLORATES.

LDU000   CAS: 7446-27-7   HR: 3
LEAD(II) PHOSPHATE (3:2)
mf: O_8P_2•3Pb   mw: 811.51
PROP: Hexagonal, colorless crystals or white powder. Mp: 1014°, d: 6.9–7.3. Insol in water, alc; sol in HNO_3, fixed alkali hydroxides.

SYNS: BLEIPHOSPHAT (GERMAN) □ C.I. 77622 □ LEAD ORTHOPHOSPHATE □ LEAD PHOSPHATE □ LEAD PHOSPHATE (3:2) □ LEAD(2+) PHOSPHATE □ NORMAL LEAD ORTHOPHOSPHATE □ PHOSPHORIC ACID, LEAD(2+) SALT (2:3) □ PLUMBOUS PHOSPHATE □ TRILEAD PHOSPHATE

CONSENSUS REPORTS: NTP 10th Report on Carcinogens. IARC Cancer Review:

Group 2B IMEMDT 7,230,87; Animal Sufficient Evidence IMEMDT 23,325,80; IMEMDT 1,40,72; Human Limited Evidence IMEMDT 23,325,80. Lead and its compounds are on the Community Right-To-Know List. Reported in EPA TSCA Inventory.
OSHA PEL: TWA 0.05 mg(Pb)/m^3
ACGIH TLV: TWA 0.15 mg(Pb)/m^3
NIOSH REL: TWA 0.10 mg(Pb)/m^3
SAFETY PROFILE: Confirmed carcinogen with experimental carcinogenic and tumorigenic data. When heated to decomposition it emits very toxic fumes of Pb and PO$_x$. See also LEAD COMPOUNDS.

LDW000   CAS: 10099-76-0   HR: 2
LEAD SILICATE
mf: O$_3$Si•Pb   mw: 283.28
PROP: White crystals. One stable and two metastable modifications are known. Mp: 766°, d: 6.49. Insol in water.
CONSENSUS REPORTS: Lead and its compounds are on the Community Right-To-Know List. Reported in EPA TSCA Inventory.
OSHA PEL: TWA 0.05 mg(Pb)/m^3
ACGIH TLV: TWA 0.15 mg(Pb)/m^3
NIOSH REL: (Inorganic Lead) TWA 0.10 mg(Pb)/m^3
SAFETY PROFILE: When heated to decomposition it emits toxic fumes of Pb. Used in paints, electrode position process in the automotive industry, as a heating stabilizer. See also LEAD COMPOUNDS.

LDX000   CAS: 7428-48-0   HR: 1
LEAD STEARATE
mf: C$_{18}$H$_{36}$O$_2$•xPb   mw: 1734.87
PROP: White powder. Insol in water; sol in hot alc. Mp: 115.7°.
SYNS: BLEISTEARAT (GERMAN) □ STEARIC ACID, LEAD SALT
CONSENSUS REPORTS: Lead and its compounds are on the Community Right-To-Know List. Reported in EPA TSCA Inventory.

ACGIH TLV: TWA 0.15 mg(Pb)/m^3
NIOSH REL: (Lead, inorganic) TWA 0.10 mg(Pb)/m^3
SAFETY PROFILE: Mildly toxic by ingestion. When heated to decomposition it emits toxic fumes of Pb. See also LEAD COMPOUNDS.

LDY000   CAS: 7446-14-2   HR: 3
LEAD(II) SULFATE (1:1)
DOT: UN 1794
mf: O$_4$S•Pb   mw: 303.25
PROP: White to yellow-green crystals. Strong oxidant. Hydrolyzes to form PbO$_2$. Mp: 1170° (decomp @ 1000°), d: 6.2. Insol in alc; sol in NaOH, ammonium acetate, or tartrate soln + conc HI. Practically insol in water; somewhat more sol in dil HCl or HNO$_3$.
SYNS: ANGLISLITE □ BLEISULFAT (GERMAN) □ C.I. 77630 □ C.I. PIGMENT WHITE 3 □ FAST WHITE □ FREEMANS WHITE LEAD □ LEAD BOTTOMS □ LEAD DROSS (DOT) □ LEAD SULFATE, solid, containing more than 3% free acid (DOT) □ MILK WHITE □ MULHOUSE WHITE □ SULFATE de PLOMB (FRENCH) □ SULFURIC ACID, LEAD(2+) SALT (1:1)
CONSENSUS REPORTS: Lead and its compounds are on the Community Right-To-Know List. Reported in EPA TSCA Inventory.
OSHA PEL: TWA 0.05 mg(Pb)/m^3
ACGIH TLV: TWA 0.15 mg(Pb)/m^3
NIOSH REL: (Inorganic Lead) TWA 0.10 mg(Pb)/m^3
DOT CLASSIFICATION: 8; Label: Corrosive
SAFETY PROFILE: Poison by intraperitoneal route. Moderately toxic by ingestion. Human mutation data reported. A corrosive irritant to skin, eyes, and mucous membranes. Violent or explosive reaction with potassium. When heated to decomposition it emits very toxic fumes of Pb and SO$_x$. Used in batteries, lithography, rapid-drying oil varnishes, weighting fabrics. See also LEAD COMPOUNDS and SULFATES.

LDZ000   CAS: 1314-87-0   HR: 2
LEAD SULFIDE
mf: PbS   mw: 239.25
PROP: Silvery, metallic crystals or black powder; *p*-type semiconductor when S rich, and an *n*-type when Pb rich. Mp: 1114°, bp: 1281° (subl), d: 7.5, vap press: 1 mm @ 852°. Insol in water; sol in HNO_3, hot dil HCl.
SYNS: C.I. 77640 □ GALENA □ NATURAL LEAD SULFIDE □ PLUMBOUS SULFIDE
CONSENSUS REPORTS: Lead and its compounds are on the Community Right-To-Know List. Reported in EPA TSCA Inventory.
OSHA PEL: TWA 0.05 mg(Pb)/m³
ACGIH TLV: TWA 0.15 mg(Pb)/m³
NIOSH REL: TWA 0.10 mg(Pb)/m³
SAFETY PROFILE: Moderately toxic by intraperitoneal route. Mildly toxic by ingestion. Violent reaction with ICl, H_2O_2. When heated to decomposition it emits very toxic fumes of Pb and SO_x. Used in glazing earthenware, as a friction additive in clutch facings and disc brakes. See also SULFIDES and LEAD COMPOUNDS.

LEA000   CAS: 815-84-9   HR: 2
LEAD(II) TARTRATE (1:1)
mf: $C_4H_4O_6$•Pb   mw: 355.27
PROP: White, crystalline powder or solid. D: 2.54 @ 19°.
CONSENSUS REPORTS: Lead and its compounds are on the Community Right-To-Know List. Reported in EPA TSCA Inventory.
ACGIH TLV: TWA 0.15 mg(Pb)/m³
NIOSH REL: (Inorganic Lead) TWA 0.10 mg(Pb)/m³
SAFETY PROFILE: Moderately toxic by intraperitoneal route. When heated to decomposition it emits toxic fumes of Pb. See also LEAD COMPOUNDS.

LED000   CAS: 12060-00-3   HR: 2
LEAD TITANATE
mf: O_3Ti•Pb   mw: 303.09

PROP: Ferroelectric pale-yellow solid. Forms tetragonal crystals below 4° and cubic crystals above 4°. D: 7.52. Insol in water.
SYN: TITANIC ACID, LEAD SALT
CONSENSUS REPORTS: Lead and its compounds are on the Community Right-To-Know List. Reported in EPA TSCA Inventory.
OSHA PEL: TWA 0.05 mg(Pb)/m³
ACGIH TLV: TWA 0.15 mg(Pb)/m³
NIOSH REL: TWA 0.10 mg(Pb)/m³
SAFETY PROFILE: Moderately toxic by intraperitoneal route. When heated to decomposition it emits toxic fumes of Pb. See also LEAD COMPOUNDS and TITANIUM COMPOUNDS.

LEE000   CAS: 63918-97-8   HR: 3
LEAD TRINITRORESORCINATE
DOT: UN 0130
mf: $C_6HN_3O_8Pb$ mw: 450.29
PROP: Orange-yellow, monoclinic crystals. Mp: explodes @ 311°, d: 2.9–3.1.
SYNS: LEAD STYPHNATE □ LEAD STYPHNATE (dry) (DOT) □ LEAD STYPHNATE, wetted or lead trinitroresorcinate, wetted with not <20% water or mixt. (DOT)
CONSENSUS REPORTS: Lead and its compounds are on the Community Right-To-Know List. Reported in EPA TSCA Inventory.
DOT CLASSIFICATION: Forbidden
SAFETY PROFILE: A poisonous material. A very shock-, heat-, and friction-sensitive priming explosive. It has detonated spontaneously when dry. Explodes when heated to 311°. Upon decomposition it emits very toxic fumes of NO_x and Pb. See also LEAD COMPOUNDS, NITRATES, and EXPLOSIVES, HIGH.

LEF180   HR: D
LECITHIN
PROP: A complex mixture from soybeans and other plants. Light-yellow to brown semisolid; slt nutlike odor, bland taste.
SAFETY PROFILE: When heated to decomposition it emits acrid smoke and irritating fumes.

LEH000   CAS: 8007-02-1   HR: 1
LEMONGRASS OIL WEST INDIAN
PROP: Main constituent is citral. From steam distillation of freshly cut and partially dried grasses of *Cymbopogon citratus* (STAPF) and *Andropogon nardus var. ceriferus* (Hack). Light-yellow to brown liquid; light lemon odor. D: 0.869–0.894, refr index: 1.483. Sol in mineral oil, propylene glycol; insol in water,
SYNS: GUATEMALA LEMONGRASS OIL □ MADAGASCAR LEMONGRASS OIL □ OIL OF LEMONGRASS, WEST INDIAN □ WEST INDIAN LEMONGRASS OIL
CONSENSUS REPORTS: Reported in EPA TSCA Inventory.
SAFETY PROFILE: Low toxicity by ingestion and skin contact. A skin irritant. When heated to decomposition it emits acrid smoke and irritating fumes. See also 3,7-DIMETHYL-2,6-OCTADIENAL.

LEI000   CAS: 8008-56-8   HR: 2
LEMON OIL
PROP: Expressed from the peel of the fruit of *Citrus limon L.* Burmann filius (Fam. *Rutaceae*). Pale-yellow liquid; taste and odor of lemon peel. D: 0.849, refr index: 1.473 @ 20°. Misc with dehydrated alc, glacial acetic acid.
SYNS: CEDRO OIL □ LEMON OIL, COLDPRESSED (FCC) □ LEMON OIL, EXPRESSED □ OIL OF LEMON □ ZITRONEN OEL (GERMAN)
CONSENSUS REPORTS: Reported in EPA TSCA Inventory.
SAFETY PROFILE: Moderately toxic by ingestion. A skin irritant. Questionable carcinogen with experimental tumorigenic data. When heated to decomposition it emits acrid smoke and irritating fumes.

LEI025   HR: 1
LEMON OIL, desert type, coldpressed
PROP: Expressed without heat from the peel of the fruit of *Citrus limon L. Burmann filius* (Fam. *Rutaceae*). Pale-yellow liquid; taste and odor of lemon peel. D: 0.846, refr index: 1.473 @ 20°. Misc with dehydrated alc, glacial acetic acid.

SYN: OIL OF LEMON, desert type, coldpressed
SAFETY PROFILE: A skin irritant. When heated to decomposition it emits acrid smoke and irritating fumes.

LEN000   CAS: 21609-90-5   HR: 3
LEPTOPHOS
mf: $C_{13}H_{10}BrCl_2O_2PS$ mw: 412.07
PROP: Solid. Mp: 70.2°
SYNS: ABAR □ O-(4-BROMO-2,5-DICHLOROPHENYL)-O-METHYL PHENYLPHOSPHONOTHIOATE □ O-(2,5-DICHLORO-4-BROMOPHENYL)-O-METHYL PHENYL-THIOPHOSPHONATE □ FOSVEL □ K62-105 □ MBCP □ O-METHYL-O-(4-BROMO-2,5-DICHLOROPHEN-YL)PHENYL THIOPHOSPHONATE □ O-METHYL-O-2,5-DICHLORO-4-BROMOPHENYL PHENYLTHIO-PHOSPHONATE □ NK 711 □ PHENYLPHOSPHONO-THIOIC ACID O-(4-BROMO-2,5-BROMO-2,5-DICHLORO-PHENYL) O-METHYL ESTER □ PHOSVEL □ PSL □ VELSICOL 506 □ VELSICOL VCS 506
SAFETY PROFILE: Poison by ingestion, skin contact, intraperitoneal, and subcutaneous routes. An experimental teratogen. Mutation data reported. Used in insecticides. When heated to decomposition it emits very toxic fumes of SO_x, PO_x, Br^-, and Cl^-.

LER000   CAS: 328-38-1   HR: 1
dl-LEUCINE
mf: $C_6H_{13}NO_2$ mw: 131.17
PROP: dl Form (synthetic form): leaflets from water; odorless with sweet taste. Mp: 290° (decomp). Plates from alc. Mp: 293° (sealed tube). Sol in water; sltly sol in alc; insol in ether.
SYN: dl-2-AMINO-4-METHYLVALERIC ACID
CONSENSUS REPORTS: Reported in EPA TSCA Inventory.
SAFETY PROFILE: Mildly toxic by intraperitoneal route. When heated to decomposition it emits toxic fumes of NO_x.

LES000   CAS: 61-90-5   HR: 2
l-LEUCINE
mf: $C_6H_{13}NO_2$ mw: 131.20
PROP: An essential amino acid; occurs in isomeric forms. White crystals. Mp (dl): 332° with decomp, mp (l): 295°, d: 1.239 @ 18°/4°. l Form (natural): glistening,

hexagonal plates from aq alc. D: 1.291 @ 18°, subl @ 145–148°, decomp @ 293–295°. Sol in water; sltly sol in alc; insol in ether.

SYNS: α-AMINOISOCAPROIC ACID □ 2-AMINO-4-METHYLPENTANOIC ACID □ α-AMINO-γ-METHYL-VALERIC ACID □ 2-AMINO-4-METHYLVALERIC ACID □ 1,2-AMINO-4-METHYLVALERIC ACID □ LEUCIN (GERMAN) □ LEUCINE □ NORVALINE, 4-METHYL- □ VALERIC ACID, 2-AMINO-4-METHYL-

CONSENSUS REPORTS: Reported in EPA TSCA Inventory.

SAFETY PROFILE: Moderately toxic by subcutaneous route. An experimental teratogen. Experimental reproductive effects. When heated to decomposition it emits toxic fumes of NO$_x$.

LEY000   CAS: 57-22-7   HR: 3
LEUROCRISTINE
mf: C$_{46}$H$_{56}$N$_4$O$_{10}$ mw: 825.06
SYNS: LCR □ NCI-C04864 □ NSC-67574 □ ONCOVIN □ 22-OXOVINCALEUKOBLASTINE □ VCR □ VINCRISTINE □ VINCRYSTINE □ VINKRISTIN

CONSENSUS REPORTS: NCI Carcinogenesis Studies (ipr); No Evidence: mouse, rat CANCAR 40,1935,77. EPA Genetic Toxicology Program.

SAFETY PROFILE: Poison by parenteral, intraperitoneal, and intravenous routes. Human systemic effects: sensory change involving peripheral nerves, flaccid paralysis without anesthesia, somnolence, anorexia, convulsions or effect on seizure threshold, nausea or vomiting, changes in blood cell count and bone marrow, pulmonary and gastrointestinal changes. Experimental reproductive effects. Questionable carcinogen with experimental tumorigenic and teratogenic data. Human mutation data reported. A skin irritant. When heated to decomposition it emits toxic fumes of NO$_x$.

LEZ000   CAS: 2068-78-2   HR: 3
LEUROCRISTINE SULFATE (1:1)
mf: C$_{46}$H$_{56}$N$_4$O$_{10}$•H$_2$O$_4$S   mw: 923.14
PROP: Crystals from alc.
SYNS: KYOCRISTINE □ LILLY 37231 □ NSC-67574 □ ONCOVIN □ VCR SULFATE □ VINCRISTINE SULFATE

ONCORIN □ VINCRISTINSULFAT (GERMAN) □ VINCRISUL

CONSENSUS REPORTS: IARC Cancer Review: Group 3 IMEMDT 7,372,87; Human Inadequate Evidence IMEMDT 26,365,81; Animal Inadequate Evidence IMEMDT 26,365,81.

SAFETY PROFILE: Poison by intraperitoneal and intravenous routes. An experimental teratogen. Experimental reproductive effects. Questionable carcinogen. Human mutation data reported. When heated to decomposition it emits very toxic fumes of NO$_x$ and SO$_x$.

LFA000   CAS: 6649-23-6   HR: 3
LEVAMISOLE
mf: C$_{11}$H$_{12}$N$_2$S   mw: 204.31
PROP: Cancer drug.
SYNS: 6-PHENYL-2,3,5,6-TETRAHYDROIMIDAZO(2,1-b)THIAZOLE □ 2,3,5,6-TETRAHYDRO-6-PHENYLIMI-DAZO(2,1-b)THIAZOLE

SAFETY PROFILE: Poison by ingestion, intravenous, intraperitoneal, and subcutaneous routes. Human systemic effects by ingestion: coma, skin dermatitis and irritation, and fever. When heated to decomposition it emits very toxic fumes of NO$_x$ and SO$_x$.

LFA020   CAS: 16595-80-5   HR: 3
LEVAMISOLE HYDROCHLORIDE
mf: C$_{11}$H$_{12}$N$_2$S•ClH   mw: 240.77
SYNS: CITARIN L □ DECARIS □ IMIDAZO(2,1-β)THIAZOLE MONOHYDROCHLORIDE □ KW-2-LE-T □ LEVAMISOLE □ LEV HYDROCHLORIDE □ LEVOMYSOL HYDROCHLORIDE □ NEMICIDE □ NIRATIC HYDROCHLORIDE □ NIRATIC-PURON HYDROCHLORIDE □ NSC-177023 □ R-12,564 □ RIPERCOL-L □ SOLASKIL □ STIMAMIZOL HYDRO-CHLORIDE □ (−)-2,3,5,6-TETRAHYDRO-6-PHENYL-IMIDAZO(2,1-b)THIAZOLE HYDROCHLORIDE □ l-(−)-2,3,5,6-TETRAHYDRO-6-PHENYL-IMIDAZO(2,1-B)THIAZOLE HYDROCHLORIDE □ 1-TETRAMISOLE HYDROCHLORIDE □ TRAMISOL □ TRAMISOLE □ WORM-CHEK

SAFETY PROFILE: Poison by ingestion, intraperitoneal, subcutaneous, intravenous, and intramuscular routes. An experimental teratogen. Human systemic effects by ingestion: thrombocytopenia. Experimental reproductive effects. Mutation data reported. When heated to decomposition it emits very toxic fumes of NO_x, SO_x, and HCl.

LFF000 **CAS: 1403-17-4** **HR: 3**
LEVORIN
mf: $C_{63}H_{85}N_{21}O_{19}$ mw: 1440.69
SYNS: CANDEPTIN □ CANDIMON □ VANOBID
SAFETY PROFILE: Poison by intraperitoneal route. Moderately toxic by ingestion and subcutaneous routes. Experimental teratogenic data reported. Other experimental reproductive effects. When heated to decomposition it emits toxic fumes of NO_x.

LFG000 **CAS: 77-07-6** **HR: 3**
LEVORPHANOL
mf: $C_{17}H_{23}NO$ mw: 257.41
SYNS: levo-DROMORAN □ (−)-3-HYDROXY-N-METHYLMORPHINAN □ LEVORPHAN
SAFETY PROFILE: Poison by ingestion, subcutaneous, and intravenous routes. Experimental reproductive effects reported. When heated to decomposition it emits toxic fumes of NO_x.

LFG050 **CAS: 55-03-8** **HR: 3**
LEVOTHYROXINE SODIUM
mf: $C_{15}H_{11}I_4NO_4$•Na mw: 799.86
PROP: Pentahydrate, triclinic crystals or cream-colored powder; odorless and tasteless. Somewhat hygroscopic. D: 2.381. Solubility @ 25° in water: about 15 mg/100 mL. Sol in mineral acids and in solns of alkali hydroxides and carbonates. More sol in alc; very sltly sol in chloroform and ether.
SYNS: DATHROID □ EFEROX □ ELTROXIN □ EUTHYROX □ LAEVOXIN □ LETTER □ LEVAXIN □ LEVOROXINE □ LEVOTHROID □ LEVOTHYROX □ LEVOTHYROXINE SODIUM □ MONOSODIUM THYROXINE □ OROXINE □ SODIUM LEVOTHY-ROXINE □ SODIUM THYROXIN □ SODIUM THYROXINATE □ SODIUM THYROXINE □ SODIUM l-THYROXINE □ SYNTHROID □ SYNTHROID SODIUM □ 3,3',5,5'-TETRAIODO-l-THYRONINE, SODIUM SALT □ THYROXEVAN □ l-THYROXINE MONOSODIUM SALT □ THYROXINE SODIUM SALT □ l-THYROXINE SODIUM SALT

CONSENSUS REPORTS: Reported in EPA TSCA Inventory.
SAFETY PROFILE: Poison by ingestion, subcutaneous, and intraperitoneal routes. An experimental teratogen. Experimental reproductive effects. Human systemic effects: blood pressure elevation, change in heart rate, coma, convulsions, diarrhea, hypermotility, pulse rate increase, thyroid hypofunction. When heated to decomposition it emits toxic fumes of I^-, NO_x, and Na_2O.

LFI000 **CAS: 7660-25-5** **HR: 2**
LEVULOSE
mf: $C_6H_{12}O_6$ mw: 180.18
PROP: White, hygroscopic crystals or crystalline powder; odorless with sweet taste. D: 1.6. Sol in methanol, ethanol, water.
SYNS: FRUCTOSE (FCC) □ FRUIT SUGAR □ FRUTABS □ LAEVORAL □ LAEVOSAN □ LEVUGEN
SAFETY PROFILE: Questionable carcinogen with experimental tumorigenic data. When heated to decomposition it emits acrid smoke and fumes.

LFK000 **CAS: 58-25-3** **HR: 3**
LIBRIUM
mf: $C_{16}H_{14}ClN_3O$ mw: 299.78
SYNS: CD 2 □ CDP □ CHLORDIAZEPOXIDE □ CHLORIDIAZEPIDE □ CHLORIDIAZEPOXIDE □ CHLORODIAZEPOXIDE □ 7-CHLORO-2-METHYL-AMINO-5-PHENYL-3H-1,4-BENZODIAZEPINE 4-OXIDE □ 7-CHLORO-2-METHYLAMINO-5-PHENYL-3H-1,4-BENZODIAZEPIN 4-OXIDE □ 7-CHLORO-N-METHYL-5-PHENYL-3H-1,4-BENZODIAZEPIN-2-AMINE-4-OXIDE □ CLOPOXIDE □ CLORDIAZEPOSSIDO (ITALIAN) □ 7-CLORO-2-METILAMINO-5-FENIL-3H-1,4-BENZOIDI-AZEPINA 4-OSSIDO (ITALIAN) □ DECACIL □ EDEN □ ELENIUM □ IFIBRIUM □ KALMOCAPS □ LIBRAX □ LIBRININ □ LIBRITABS □ MESURAL □ METHAMINO-DIAZEPOXIDE □ MILDMEN □ NAPOTON □ PSICO-SAN □ RADEPUR □ VIOPSICOL
CONSENSUS REPORTS: EPA Genetic Toxicology Program.
SAFETY PROFILE: Poison by ingestion, intraperitoneal, intravenous, subcutaneous, and intramuscular routes. Human male reproductive effects by ingestion: impotence. Human systemic effects by ingestion: sleep, euphoria, somnolence,

ataxia, and antianxiety. An experimental teratogen. Experimental reproductive effects. Mutation data reported. Has been implicated in development of aplastic anemia. Used as a pharmaceutical and veterinary drug. When heated to decomposition it emits very toxic fumes of NO_x.

LFT100   CAS: 91-51-0   HR: 1
LILIAL-METHYLANTHRANILATE,
   Schiff's base
mf: $C_{22}H_{27}NO_2$ mw: 337.50
SYNS: ANTHRANILIC ACID, N-(3-(p-tert-BUTYLPHEN-YL)-2-METHYLPROPYLIDENE)-, METHYL ESTER □ BENZOIC ACID,2-((3-(4-(1,1-DIMETHYLETHYL)PHEN-YL)-2-METHYLPROPYLIDENE)AMINO)-, METHYL-ESTER □ METHYL-N-(p-tert-BUTYL-α-METHYLHYDRO-CINNAMYLIDENE) ANTHRANILATE □ VERDANTIOL
CONSENSUS REPORTS: Reported in EPA TSCA Inventory.
SAFETY PROFILE: Low toxicity by ingestion and skin contact. A skin irritant. When heated to decomposition it emits toxic fumes of NO_x.

LFU000   CAS: 5989-27-5   HR: 3
d-LIMONENE
mf: $C_{10}H_{16}$ mw: 136.26
PROP: Colorless liquid or oil; citrus odor. Bp: 175.5–176°, d: 0.8402 @ 25°/4°, refr index: 1.471. Misc with alc, fixed oils; sltly sol in glycerin; insol in propylene glycol, water.
SYNS: FEMA No. 2633 □ (+)-4-ISOPROPENYL-1-METHYLCYCLOHEXENE □ d-(+)-LIMONENE □ (+)-R-LIMONENE □ d-p-MENTHA-1,8-DIENE □ p-MENTHA-1,8-DIENE □ (R)-1-METHYL-4-(1-METHYLETHENYL)-CYCLOHEXENE □ NCI-C55572
CONSENSUS REPORTS: Reported in EPA TSCA Inventory.
SAFETY PROFILE: Poison by intravenous route. Moderately toxic by intraperitoneal and intraduodenal routes. Mildly toxic by ingestion. Experimental reproductive effects. Questionable carcinogen with experimental tumorigenic and teratogenic data. Reacts explosively with iodine pentafluoride + tetrafluoroethylene (the pentafluoride reacts exothermically with the

inhibitor and initiates explosive polymerization of the TFE). When heated to decomposition it emits acrid smoke and irritating fumes. Used as a food additive, flavor agent, packaging material, as an inhibitor of tetrafluoroethylene polymerization, and as a gallstone solubilizer.

LFV000   CAS: 96-08-2   HR: 2
LIMONENE DIOXIDE
mf: $C_{10}H_{16}O_2$ mw: 168.26
SYNS: 1,2,8,9-DIEPOXYLIMONENE □ 1,2:8,9-DIEPO-XYMENTHANE □ 1,2:8,9-DIEPOXY-p-MENTHANE □ DIPENTENE DIOXIDE □ EPOXIDE 269 □ 4-(1,2-EPO-XY-1-METHYLETHYL)-1-METHYL-7-OXABICYCLO-(4.1.0)HEPTANE □ UNOXAT EPOXIDE 269
CONSENSUS REPORTS: Reported in EPA TSCA Inventory.
SAFETY PROFILE: Moderately toxic by skin contact and intramuscular routes. Mildly toxic by ingestion. A skin irritant. Questionable carcinogen with experimental tumorigenic data. When heated to decomposition it emits acrid smoke and irritating fumes.

LFX000   CAS: 78-70-6   HR: 2
LINALOOL
mf: $C_{10}H_{18}O$ mw: 154.28
PROP: Colorless liquid; odor similar to that of bergamot oil and French lavender. D: 0.858–0.868 @ 25°, refr index: 1.461, bp: 195–199°, flash p: 172°F. Sol in alc, ether, fixed oils, propylene glycol; insol in glycerin.
SYNS: ALLO-OCIMENOL □ 2,6-DIMETHYL-2,7-OCTADIENE-6-OL □ 2,6-DIMETHYLOCTA-2,7-DIEN-6-OL □ 3,7-DIMETHYLOCTA-1,6-DIEN-3-OL □ 3,7-DIMETHYL-1,6-OCTADIEN-3-OL □ FEMA No. 2635 □ LINALOL □ LINALYL ALCOHOL
CONSENSUS REPORTS: Reported in EPA TSCA Inventory.
SAFETY PROFILE: Moderately toxic by ingestion. Mildly toxic by skin contact. A skin irritant. A synthetic flavoring substance and adjuvant. When heated to decomposition it emits acrid smoke and irritating fumes.

L

LFY333   CAS: 29171-20-8   HR: 3
LINALOOL, DEHYDRO-
mf: $C_{10}H_{16}O$ mw: 152.26
SYNS: 3,7-DIMETHYL-6-OCTEN-1-YN-3-OL □ DEHY-
DROLINALOOL □ DEHYDRO-β-LINALOOL □ 6-OCT-
EN-1-YN-3-OL, 3,7-DIMETHYL-
SAFETY PROFILE: A poison by ingestion
and intraperitoneal routes. A mild skin and
moderate eye irritant. When heated to
decomposition it emits acrid smoke and
irritating vapors.

LFY600   CAS: 115-95-7   HR: 2
LINALYL ACETATE
mf: $C_{12}H_{20}O_2$ mw: 196.32
PROP: Clear, colorless, oily liquid; odor of
bergamot. Bp: 108–110°, d: 0.898–0.914,
flash p: 185°F. Sol in alc, ether, diethyl
phthalate, benzyl benzoate, mineral oil, fixed
oils; sltly sol in propylene glycol; insol in
water, glycerin.
SYNS: ACETIC ACID LINALOOL ESTER □
BERGAMIOL □ 3,7-DIMETHYL-1,6-OCTADIEN-3-OL
ACETATE □ 3,7-DIMETHYL-1,6-OCTADIEN-3-YL
ACETATE □ FEMA No. 2636 □ LICAREOL ACETATE □
LINALOL ACETATE □ LINALOOL ACETATE
CONSENSUS REPORTS: Reported in EPA
TSCA Inventory.
SAFETY PROFILE: Mildly toxic by
ingestion. A severe skin irritant.
Combustible liquid. When heated to
decomposition it emits acrid smoke and
irritating fumes. See also ESTERS.

LFZ000   CAS: 126-64-7   HR: 2
LINALYL BENZOATE
mf: $C_{17}H_{22}O_2$ mw: 258.39
PROP: Found in the essential oils of ylang-
ylang and tuberose (FCTXAV 14,443,76).
Yellow to brown-yellow liquid; tuberose
odor. D: 0.980–0.999, refr index:
1.505–1.520, flash p: 208°F. Sol in
chloroform, alc, ether; insol in water.
SYNS: 3,7-DIMETHYL-1,6-OCTADIEN-3-OL BENZO-
ATE □ 3,7-DIMETHYL-1,6-OCTADIEN-3-YL BENZOATE
□ 1,5-DIMETHYL-1-VINYL-4-HEXEN-1-OL BENZOATE
□ 1,5-DIMETHYL-1-VINYL-4-HEXEN-1-YL BENZOATE
□ FEMA No. 2638
CONSENSUS REPORTS: Reported in EPA
TSCA Inventory.

SAFETY PROFILE: Low toxicity by
ingestion and skin contact. A skin irritant.
Combustible liquid. When heated to
decomposition it emits acrid smoke and
irritating fumes.

LGA050   HR: 2
LINALYL FORMATE
mf: $C_{11}H_{18}O_2$ mw: 182.26
PROP: Colorless liquid; citrus, herbaceous
odor. D: 0.910–0.918, refr index:
1.453–1.458, flash p: 189°F. Sol in alc, fixed
oils; sltly sol in propylene glycol, water; insol
in glycerin @ 202°.
SYNS: 3,7-DIMETHYL-1,6-OCTADIEN-3-YL FORMATE
□ FEMA No. 2642
SAFETY PROFILE: Combustible liquid.
When heated to decomposition it emits
acrid smoke and irritating fumes.

LGB000   CAS: 78-35-3   HR: 2
LINALYL ISOBUTYRATE
mf: $C_{14}H_{24}O_2$ mw: 224.38
PROP: Colorless liquid; fresh, rosy odor. D:
0.882–0.888, refr index: 1.446–1.451, flash p:
212°F. Misc with alc, chloroform, ether;
insol in water @ 20°.
SYNS: 3,7-DIMETHYL-1,6-OCTADIEN-3-OL ISOBU-
TYRATE □ 3,7-DIMETHYL-1,6-OCTADIEN-3-YL
ISOBUTYRATE □ 1,5-DIMETHYL-1-VINYL-4-HEXENYL
ESTER, ISOBUTYRIC ACID □ FEMA No. 2640 □
LINALOOL ISOBUTYRATE
CONSENSUS REPORTS: Reported in EPA
TSCA Inventory.
SAFETY PROFILE: Combustible liquid.
Mildly toxic by ingestion. When heated to
decomposition it emits acrid smoke and
irritating fumes.

LGD000   CAS: 154-21-2   HR: 3
LINCOMYCIN
mf: $C_{18}H_{34}N_2O_6S$ mw: 406.60
PROP: Amorphous. Sol in methanol,
ethanol, butanol, isopropanol, ethyl acetate,
n-butyl acetate, amyl acetate, etc. Moderately
sol in water.
SYNS: ALBIOTIC □ LINCOCIN □ LINCOLCINA □
LINCOLNENSIN □ LINCOMYCINE (FRENCH) □ NSC-
70731 □ U-10149

CONSENSUS REPORTS: EPA Genetic Toxicology Program.
SAFETY PROFILE: Poison by intramuscular route. Moderately toxic by ingestion and intraperitoneal routes. When heated to decomposition it emits very toxic fumes of SO_x and NO_x.

LGF825   CAS: 68411-30-3   HR: 3
LINEAR ALKYLBENZENE SULFONATE, SODIUM SALT
PROP: Anionic surfactant.
SYNS: LAS-Na □ LAS, SODIUM SALT □ STRAIGHT-CHAIN ALKYL BENZENE SULFONATE
CONSENSUS REPORTS: Reported in EPA TSCA Inventory.
SAFETY PROFILE: Poison by intravenous route. Moderately toxic by ingestion and subcutaneous routes. An experimental teratogen. Human systemic effects: lachrymation, somnolence, hypermotility, diarrhea. When heated to decomposition it emits toxic fumes of SO_x and Na_2O. See also SULFONATES.

LGF900   CAS: 3999-01-7   HR: D
LINOLEAMIDE
PROP: An antistatic agent, opacifier, viscosity controlling agent.
SYN: LINOLEIC ACID AMIDE
SAFETY PROFILE: When heated to decomposition it emits acrid smoke and irritating fumes.

LGG000   CAS: 60-33-3   HR: 3
LINOLEIC ACID
mf: $C_{18}H_{32}O_2$ mw: 280.50
PROP: Colorless oil, easily oxidized by air. D: 0.9038 @ 18°/4°, mp: −12°, bp: 230° @ 16 mm. Sol in ether and ethanol; misc with dimethyl formamide, fat solvents, oils.
SYNS: LEINOLEIC ACID □ 9,12-LINOLEIC ACID □ cis,cis-9,12-OCTADECADIENOIC ACID □ cis-9,cis-12-OCTADECADIENOIC ACID □ 9,12-OCTADECADIENOIC ACID
CONSENSUS REPORTS: Reported in EPA TSCA Inventory.
SAFETY PROFILE: A poison by intraperitoneal route. A human skin irritant.

Ingestion can cause nausea and vomiting. When heated to decomposition it emits acrid smoke and irritating fumes.

LGK000   CAS: 8001-26-1   HR: 2
LINSEED OIL
PROP: Yellowish liquid, peculiar odor, bland taste. Sltly sol in alc; misc with chloroform, ether, pet ether, carbon disulfide, oil, turpentine. Bp: 343°, mp: −19°, d: 0.93, flash p: (raw oil) 432°F (CC), flash p: (boiled) 403°F (CC), autoign temp: 650°F. From seed of *Linum usitatissimum*.
SYNS: GROCO □ L-310
CONSENSUS REPORTS: Reported in EPA TSCA Inventory.
SAFETY PROFILE: An allergen and skin irritant to humans. Combustible liquid when exposed to heat or flame; can react with oxidizing materials. Subject to spontaneous heating. Violent reaction with Cl_2. To fight fire, use CO_2, dry chemical.

LGL000   HR: 2
LIQUEFIED CARBON DIOXIDE
mf: CO_2 mw: 44.0
PROP: Heavy gas or liquid under pressure. Mp: −56.6° @ 3952 mm, bp: −78.5° (subl), d: 1.977 g/L @ 0°, (liquid) 1.101 @ −37°.
SYN: LIQUID CARBONIC GAS
SAFETY PROFILE: Contact with skin or living tissue can cause frostbite-like burns. This material is stable when very cold. Solid CO_2 goes directly (sublimes) to gaseous CO_2, which is mainly an asphyxiant. See also CARBON DIOXIDE.

LGM000   CAS: 68476-85-7   HR: 3
LIQUEFIED PETROLEUM GAS
DOT: UN 1075
PROP: IDLH 2000 ppm [10%LEL].
SYNS: LPG □ L.P.G. (OSHA, ACGIH) □ PETROLEUM GAS, LIQUEFIED □ PETROLEUM GASES, liquefied or liquefied petroleum gas (DOT)
CONSENSUS REPORTS: Reported in EPA TSCA Inventory.
OSHA PEL: TWA 1000 ppm
NIOSH REL: TWA 350 mg/m³; CL 1800 mg/m³/15M

ACGIH TLV: TWA 1000 ppm
DOT CLASSIFICATION: 2.1; Label:
Flammable Gas
SAFETY PROFILE: Olefinic impurities may
lend a narcotic effect or it may act as a
simple asphyxiant. A very dangerous fire
hazard when exposed to heat or flame. Can
react with oxidizing materials. To fight fire,
use CO_2, dry chemical, water spray. Used as
a fuel refrigerant, propellant, and raw
material in chemical synthesis.

LGO000   CAS: 7439-93-2   HR: 3
LITHIUM
DOT: UN 1415
af: Li   aw: 6.94
PROP: Silver-colored, light, malleable,
lustrous metal which tarnishes in air, turning
black owing to formation of Li_3N, Li_2O,
LiOH, and Li_2CO_3. Reacts vigorously with
H_2O but not quite as violently as the heavier
alkali metals; mixture of isotopes Li^6 and Li^7.
Mp: 180.5°, bp: 1340°, d: 0.534 @ 25°, vap
press: 1 mm @ 723°. Keep under mineral
oil or other liquid free from O_2 or water. Sol
in NH_3 (l) or blue-black soln.
SYNS: LITHIUM METAL (DOT) □ LITHIUM METAL,
IN CARTRIDGES (DOT)
CONSENSUS REPORTS: Reported in EPA
TSCA Inventory.
DOT CLASSIFICATION: 4.3; Label:
Dangerous When Wet
SAFETY PROFILE: See LITHIUM
COMPOUNDS for a discussion of the
toxicity of the lithium ion. See SODIUM for
a discussion of the toxicity of metallic
lithium.

A very dangerous fire hazard when
exposed to heat or flame. The powder may
ignite spontaneously in air. The solid metal
ignites above 180°C. It will burn in oxygen,
nitrogen, or carbon dioxide, and will
continue to burn in sand or sodium
carbonate. The use of most types of fire
extinguishers (e.g., water, foam, carbon
dioxide, halocarbons, sodium carbonate,
sodium chloride, and other dry powders)
may cause an explosion. Molten lithium is
extremely reactive and attacks such
otherwise inert materials as sand, concrete,
and ceramics.

Explosive reaction with bromobenzene,
carbon + lithium tetrachloroaluminate +
sulfinyl chloride, diazomethane. Forms very
friction- and impact-sensitive explosive
mixtures with halogens (e.g., bromine,
iodine (above 200°C)), halocarbons (e.g.,
bromoform, carbon tetrabromide, carbon
tetrachloride, carbon tetraiodide,
chloroform, dichloromethane,
diiodomethane, fluorotrichloromethane,
tetrachloroethylene, trichloroethylene, 1,1,2-
trichloro-trifluoroethane).

Violent reaction with acetonitrile, sulfur,
mercury (potentially explosive), metal oxides
(e.g., chromium(III) oxide (at 185°C),
molybdenum trioxide (at 180°C), niobium
pentoxide (at 320°C), titanium dioxide (at
200–400°C), tungsten trioxide (at 200°C),
vanadium pentoxide (at 394°C)), iron(II)
sulfide (at 260°C), manganese telluride (at
230°C), hot water, bromine pentafluoride
(may ignite with lithium powder), platinum
(at about 540°C), trifluoromethyl
hypofluorite (at about 170°C), arsenic,
beryllium, maleic anhydride, carbides,
carbon dioxide, carbon monoxide + water,
chlorine, chromium, chromium trichloride,
cobalt alloys, iron sulfide, diborane,
manganese alloys, nickel alloys, nitric acid,
nitrogen, organic matter, oxygen,
phosphorus, rubber, silicates, $NaNO_2$,
Ta_2O_5, Fe alloys, V, $ZrCl_4$, CHI_3,
trifluoromethylhypofluorite.

Ignition on contact with carbon + sulfinyl
chloride (when ground), nitric acid
(becomes violent), viton poly(1,1-
difluorethylene-hexafluoropropylene),
chlorine tri- and penta-fluorides (hypergolic
reaction), diborane (forms a complex that is
pyrophoric), hydrogen (above 300°C).

Incandescent reaction with ethylene +
heat, nitrogen + metal chlorides (e.g.,
chromium trichloride, zirconium
tetrachloride, nitryl fluoride (at 200°C)).
Incompatible with atmospheric gases,

bromine pentafluoride, diazomethane, metal chlorides, metal oxides, nonmetal oxides.

When burned it emits toxic fumes of LiO_2 and hydroxide. Reacts vigorously with water or steam to produce heat and hydrogen. Can react vigorously with oxidizing materials. To fight fire, use special mixtures of dry chemical, soda ash, graphite. NOTE: Water, sand, carbon tetrachloride, and carbon dioxide are ineffective.

LGO100   CAS: 546-89-4   HR: 2
LITHIUM ACETATE
mf: $C_2H_3O_2 \cdot Li$ mw: 67.00
PROP: White crystalline powder. D: 1.25 g/cm³, mp: 280°–285°.
SYNS: ACETIC ACID, LITHIUM SALT □ QUILONE
CONSENSUS REPORTS: Reported in EPA TSCA Inventory.
DOT CLASSIFICATION: 4.2; Label: Spontaneously Combustible
SAFETY PROFILE: Moderately toxic by ingestion and subcutaneous routes. Mutation data reported. Spontaneously combustible; take special storage and handling precautions. When heated to decomposition it emits toxic fumes of Li.

LGS000   CAS: 17476-04-9   HR: 3
LITHIUM ALUMINUM TRI-tert-BUTOXYHYDRIDE
mf: $C_{12}H_{18}AlLiO_3$ mw: 244.22
PROP: White air- and moisture-sensitive tabular crystals from diglyme. Sol in Et_2O, THF, and diglyme.
CONSENSUS REPORTS: Reported in EPA TSCA Inventory.
ACGIH TLV: TWA 2 mg(Al)/m³
SAFETY PROFILE: Poison by intravenous route. When heated to decomposition it emits acrid smoke and irritating fumes. See also ALUMINUM and LITHIUM COMPOUNDS.

LGT000   CAS: 7782-89-0   HR: 3
LITHIUM AMIDE
mf: H_2LiN mw: 22.97

PROP: White or colorless crystalline solid or powder. Moisture-sensitive. Mp: 380–400°, d: 1.178 @ 17.50°. Reacts with H_2O to form LiOH and NH_3. Reacts with N_2O to form LiN_3 with evolution of H_2O. Subl in NH_3 current. Insol in anhydrous ether, benzene, and toluene. Sltly sol in EtOH; insol in Et_2O and C_6H_6.
SYNS: LITHAMIDE □ LITHIUM AMIDE, POWDERED
CONSENSUS REPORTS: Reported in EPA TSCA Inventory.
SAFETY PROFILE: A powerful irritant to skin, eyes, and mucous membranes. Flammable when exposed to heat or flame. Ammonia is liberated and lithium hydroxide is formed when this compound is exposed to moisture. Reacts violently with water or steam to produce toxic and flammable vapors. Vigorous reaction with oxidizing materials. Exothermic reaction with acid or acid fumes. When heated to decomposition it emits very toxic fumes of LiO, NH_3, and NO_x. Used in synthesis of drugs, vitamins, steroids, and other organics. See also LITHIUM COMPOUNDS, AMIDES, AMMONIA, and LITHIUM HYDROXIDE.

LGU000   CAS: 305-97-5   HR: 3
LITHIUM ANTIMONY THIOMALATE
mf: $C_{12}H_9O_{12}S_3Sb \cdot 6Li$ mw: 604.78
PROP: Very sol in H_2O.
SYNS: ANTHIOLIMINE □ ANTHIOMALINE □ LITHIUM ANTIMONIOTHIOMALATE □ MERCAPTO-SUCCINIC ACID ANTIMONATE(III) HEXALITHIUM SALT □ MERCAPTOSUCCINIC ACID-S-ANTIMONY DERIVATIVE LITHIUM SALT □ MERCAPTOSUCCINIC ACID, THIOANTIMONATE(III), DILITHIUM SALT □ 2,2',2''-(STIBILIDYNETRIS(THIO))TRIS-BUTANEDIOIC ACID HEXALITHIUM SALT
CONSENSUS REPORTS: Antimony and its compounds are on the Community Right-To-Know List.
OSHA PEL: TWA 0.5 mg(Sb)/m³
ACGIH TLV: TWA 0.5 mg(Sb)/m³
NIOSH REL: (Antimony) TWA 0.5 mg(Sb)/m³
SAFETY PROFILE: Poison by intraperitoneal and intravenous routes. Human systemic effects by ingestion and

intravenous routes: hallucinations, distorted perceptions, nausea or vomiting, skin dermatitis, and fever. An anthelmintic agent. When heated to decomposition it emits very toxic fumes of SO_x, Sb, and Li_2O. See also ANTIMONY COMPOUNDS and LITHIUM COMPOUNDS.

LGZ000   CAS: 554-13-2   HR: 3
LITHIUM CARBONATE (2:1)
mf: $CO_3 \cdot 2Li$ mw: 73.89

PROP: White, slightly alkaline, crystalline powder. Decomp on heating by CO_2 loss. D: 2.11 @ 17.5°, mp: 720°. Insol in alc @ 17.5°. Sltly sol in H_2O; insol in EtOH and Me_2CO.

SYNS: CAMCOLIT □ CANDAMIDE □ CARBOLITH □ CARBONIC ACID, DILITHIUM SALT □ CARBONIC ACID LITHIUM SALT □ CEGLUTION □ CP-15467-61 □ DILITHIUM CARBONATE □ ESKALITH □ HYPNOREX □ LIMAS □ LISKONUM □ LITHANE □ LITHICARB □ LITHINATE □ LITHIUM CARBONATE □ LITHOBID □ LITHONATE □ LITHOTABS □ NSC-16895 □ PLENUR □ PRIADEL □ QUILONUM RETARD

CONSENSUS REPORTS: Reported in EPA TSCA Inventory.

SAFETY PROFILE: Human carcinogenic data. Poison by intraperitoneal and intravenous routes. Moderately toxic by ingestion and subcutaneous routes. Human systemic effects by ingestion: toxic psychosis, tremors, changes in fluid intake, muscle weakness, increased urine volume, nausea or vomiting, allergic dermatitis. Human reproductive effects by ingestion: effects on newborn, including Apgar score changes and other neonatal measures or effects. Human teratogenic effects by ingestion: developmental abnormalities of the cardiovascular system, central nervous system, musculoskeletal and gastrointestinal systems. An experimental teratogen. Experimental reproductive effects. Experimental carcinogen producing leukemia and thyroid tumors. Human mutation data reported. Used in the treatment of manic-depressive psychoses. Incompatible with fluorine. See also LITHIUM COMPOUNDS.

LHB000   CAS: 7447-41-8   HR: 3
LITHIUM CHLORIDE
mf: ClLi   mw: 42.39

PROP: Cubic, white, deliquescent, extremely hygroscopic crystals. Mp: 610°, bp: 1350°, d: 2.068 @ 25°, vap press: 1 mm @ 547°. Very sol in H_2O, MeOH; sol in Me_2CO.

SYNS: CHLORKU LITU (POLISH) □ CHLORURE de LITHIUM (FRENCH)

CONSENSUS REPORTS: Reported in EPA TSCA Inventory. EPA Genetic Toxicology Program.

SAFETY PROFILE: Human poison by ingestion. Experimental poison by intravenous and intracerebral routes. Moderately toxic by subcutaneous and intraperitoneal routes. Experimental teratogenic and reproductive effects. Human systemic effects by ingestion: somnolence, tremors, nausea or vomiting. An eye and severe skin irritant. Human mutation data reported. Questionable carcinogen with experimental neoplastigenic data. This material has been recommended and used as a substitute for sodium chloride in "salt-free" diets, but cases have been reported in which the ingestion of lithium chloride has produced dizziness, ringing in the ears, visual disturbances, tremors, and mental confusion. In most cases, the symptoms disappeared when use was discontinued. Prolonged absorption may cause disturbed electrolyte balance, impaired renal function. Reaction is violent with BrF_3. When heated to decomposition it emits toxic fumes of Cl^-. Used for dehumidification in the air conditioning industry. Also used to obtain lithium metal. See also LITHIUM COMPOUNDS.

LHF000   CAS: 7789-24-4   HR: 3
LITHIUM FLUORIDE
mf: FLi   mw: 25.94

PROP: Fine, white powder or cubic crystals. Mp: 848°, bp: 1676°, d: 2.635 @ 20°, vap press: 1 mm @ 1047°. Sol in acids. Sltly sol in H_2O; insol in EtOH.

SYNS: LITHIUM FLUORURE (FRENCH) □ TLD 100

CONSENSUS REPORTS: Reported in EPA TSCA Inventory.

OSHA PEL: TWA 2.5 mg(F)/m³

ACGIH TLV: TWA 2.5 mg(F)/m³; BEI: 3 mg/g creatinine of fluorides in urine prior to shift; 10 mg/g creatinine of fluorides in urine at end of shift.

NIOSH REL: (Inorganic Fluorides) TWA 2.5 mg(F)/m³

SAFETY PROFILE: Poison by ingestion and subcutaneous routes. When heated to decomposition it emits toxic fumes of F⁻. Used as a flux in enamels, glasses, glazes, and welding. See also FLUORIDES and LITHIUM COMPOUNDS.

LHH000   CAS: 7580-67-8   HR: 3
LITHIUM HYDRIDE
DOT: UN 1414/UN 2805
mf: HLi   mw: 7.95
PROP: White, translucent, moisture sensitive crystals. Mp: 688.7 (decomp), d: 0.76–0.77. Darkens rapidly on exposure to light. Reacts with H_2O to form LiOH and H_2, dissoc above mp to form Li metal and H_2. IDLH 0.5 mg/m³.
SYNS: HYDRURE de LITHIUM (FRENCH) □ LITHIUM HYDRIDE (UN 1414) (DOT) □ LITHIUM HYDRIDE, fused solid (UN 2805) (DOT)
CONSENSUS REPORTS: Reported in EPA TSCA Inventory. EPA Extremely Hazardous Substances List.
OSHA PEL: TWA 0.025 mg/m³
ACGIH TLV: TWA 0.025 mg/m³
DFG MAK: 0.025 mg/m³
DOT CLASSIFICATION: 4.3; Label: Dangerous When Wet
SAFETY PROFILE: Poison by inhalation. A severe eye, skin, and mucous membrane irritant. Upon contact with moisture, lithium hydroxide is formed. The LiOH formed is very caustic and therefore highly toxic, particularly to lungs and respiratory tract, skin, and mucous membranes. The powder ignites spontaneously in air. The solid can ignite spontaneously in moist air. Mixtures of the powder with liquid oxygen are explosive. Ignites on contact with dinitrogen

oxide, oxygen + moisture. To fight fire, use special mixtures of dry chemical. See also LITHIUM COMPOUNDS and HYDRIDES.

LHI100   CAS: 1310-65-2   HR: 3
LITHIUM HYDROXIDE
DOT: UN 2679/UN 2680
mf: HLiO   mw: 23.95
PROP: Solid.
SYNS: LITHIUM HYDROXIDE (Li(OH)) (9CI) □ LITHIUM HYDROXIDE, monohydrate or lithium hydroxide, solid (DOT) □ LITHIUM HYDROXIDE, solution (DOT)
CONSENSUS REPORTS: Reported in EPA TSCA Inventory.
DOT CLASSIFICATION: 8; Label: Corrosive
SAFETY PROFILE: Poison by ingestion and subcutaneous routes. Mildly toxic by inhalation. A corrosive. When heated to decomposition it emits toxic fumes of Li.

LHJ000   CAS: 13840-33-0   HR: 1
LITHIUM HYPOCHLORITE
DOT: UN 1471
mf: ClO•Li   mw: 58.39
PROP: White powder. Very sol in H_2O.
SYN: LITHIUM HYPOCHLORITE COMPOUND, dry, containing more than 39% available chlorine (DOT)
CONSENSUS REPORTS: Reported in EPA TSCA Inventory.
DOT CLASSIFICATION: 5.1; Label: Oxidizer
SAFETY PROFILE: A powerful oxidizer. An eye, skin, and mucous membrane irritant. When heated to decomposition it emits very toxic fumes of Li_2O and Cl⁻. Used for swimming pool chlorination, and as a laundry bleach. See also LITHIUM COMPOUNDS and HYPOCHLORITES.

LHK000   CAS: 64082-35-5   HR: 3
LITHIUM IRON SILICON
DOT: UN 2830
mf: FeLiSi   mw: 90.88
PROP: Dark, crystalline, brittle, metallic lumps or powder; evolves a flammable gas when in contact with moisture.
SYN: LITHIUM FERRO SILICON
DOT CLASSIFICATION: 4.3; Label: Dangerous When Wet

L

SAFETY PROFILE: Flammable solid which evolves a flammable gas when exposed to moisture, steam, or acid fumes. Flammable when exposed to heat or flame. See also LITHIUM COMPOUNDS.

LHL000   CAS: 867-55-0   HR: 2
LITHIUM LACTATE
mf: $C_3H_5O_3 \cdot Li$ mw: 96.02
SYN: LACTIC ACID LITHIUM SALT
CONSENSUS REPORTS: Reported in EPA TSCA Inventory.
DOT CLASSIFICATION: 4.2; Label: Spontaneously Combustible
SAFETY PROFILE: Moderately toxic by ingestion and subcutaneous routes. When heated to decomposition it emits acrid smoke and irritating fumes. See also LITHIUM COMPOUNDS.

LHM000   CAS: 26134-62-3   HR: 3
LITHIUM NITRIDE
DOT: UN 2806
mf: Li_3N mw: 34.82
PROP: Brownish-red or purplish hexagonal crystals; sensitive to atmosphere, slowly decomp on contact with moisture. D: 1.3, mp: 813°. Reacts with H_2O with formation of LiOH and evolution of NH_3.
CONSENSUS REPORTS: Reported in EPA TSCA Inventory.
DOT CLASSIFICATION: 4.3; Label: Dangerous When Wet
SAFETY PROFILE: A powerful reducing agent. Upon contact with moisture, it decomposes into lithium hydroxide, lithium compounds, and ammonia. The powder may ignite spontaneously in moist air. Flammable at elevated temperatures; ignites and burns intensely in air. Violent reaction with silicon tetrafluoride, copper(I) chloride + heat. To fight fire, use dry chemical, sand, graphite; avoid use of water or carbon tetrachloride. When heated to decomposition it emits very toxic fumes of Li_2O and NO_x. Used as a strong reducing agent in organic synthesis and a solid electrolyte in lithium batteries. See also LITHIUM COMPOUNDS and NITRIDES.

LHM800   CAS: 63255-29-8   HR: 3
LITHIUM OXYBUTYRATE
mf: $C_4H_7O_3 \cdot Li$ mw: 110.05
SYNS: BUTANOIC ACID, 4-HYDROXY-, MONOLITHIUM SALT □ 4-HYDROXYBUTANOIC ACID MONOLITHIUM SALT □ LITHIUM HYDROXYBUTYRATE □ LITHIUM γ-HYDROXYBUTYRATE
DOT CLASSIFICATION: 4.2; Label: Spontaneously Combustible
SAFETY PROFILE: Moderately toxic by intravenous and intraperitoneal routes. Warning: This substance is spontaneously combustible. When heated to decomposition it emits toxic vapors of lithium.

LHO000   CAS: 12031-80-0   HR: 3
LITHIUM PEROXIDE
DOT: UN 1472
mf: Li_2O_2 mw: 45.88
PROP: Fine, white powder or sandy-yellow, granular material. Moisture sensitive colorless hexagonal crystals. Mp: decomp, d: 2.14 @ 20°. Reacts with H_2O with formation of H_2O_2. Decomp on heating with O_2 evolution. Reacts with CO_2 to form Li_2CO_3 and O_2.
CONSENSUS REPORTS: Reported in EPA TSCA Inventory.
DOT CLASSIFICATION: 5.1; Label: Oxidizer
SAFETY PROFILE: A powerful oxidizer and irritant to skin, eyes, and mucous membranes. A very dangerous fire hazard because it is an extremely powerful oxidizing agent. Will react with water or steam to produce heat; on contact with reducing materials, can react vigorously. See also LITHIUM COMPOUNDS, PEROXIDES, and PEROXIDES, INORGANIC.

LHP000   CAS: 68848-64-6   HR: 3
LITHIUM SILICON
DOT: UN 1417
PROP: Solid. Composition: Li + Si.
DOT CLASSIFICATION: 4.3; Label: Dangerous When Wet

SAFETY PROFILE: A very dangerous fire hazard in the form of dust when exposed to heat or flame or by chemical reaction with moisture or acids. In contact with water, silane and hydrogen are evolved. Slightly explosive in the form of dust when exposed to flame. Will react with water or steam to produce flammable vapors; on contact with oxidizing materials, can react vigorously; on contact with acid or acid fumes, can emit toxic and flammable fumes. To fight fire, use CO_2, dry chemical. See also LITHIUM, SILICON, and POWDERED METALS.

LHQ100   CAS: 4485-12-5   HR: 3
LITHIUM STEARATE

mf: $C_{18}H_{35}O_2 \cdot Li$ mw: 290.47

PROP: Opacifiers, viscosity controlling agent.

SYNS: LITHALURE □ LITHIUM OCTADECANOATE □ LITHOLITE □ OCTADECANOIC ACID, LITHIUM SALT □ STAVINOR □ STEARIC ACID, LITHIUM SALT

CONSENSUS REPORTS: Reported in EPA TSCA Inventory.

ACGIH TLV: TWA 10 mg/m³, total dust

DOT CLASSIFICATION: 4.2; Label: Spontaneously Combustible

SAFETY PROFILE: Low toxicity by ingestion. Warning: This substance is spontaneously combustible. When heated to decomposition it emits toxic vapors of lithium.

LHS000   CAS: 16853-85-3   HR: 3
LITHIUM TETRAHYDROALUMINATE

DOT: UN 1410/UN 1411

mf: $AlH_4 \cdot Li$ mw: 37.96

PROP: White, microcrystalline lumps; moisture sensitive crystals. Reacts violently with H_2O. Decomp on heating to form LiH, Al, and H_2. Sol in Et_2O; sltly sol in hydrocarbons.

SYNS: ALUMINUM LITHIUM HYDRIDE □ LITHIUM ALANATE □ LITHIUM ALUMINOHYDRIDE □ LITHIUM ALUMINUM HYDRIDE (DOT) □ LITHIUM ALUMINUM HYDRIDE, ETHEREAL (DOT) □ LITHIUM ALUMINUM TETRAHYDRIDE

CONSENSUS REPORTS: Reported in EPA TSCA Inventory.

ACGIH TLV: TWA 2 mg(Al)/m³

DOT CLASSIFICATION: 4.3; Label: Dangerous When Wet (UN 1410); DOT Class: 4.3; Label: Dangerous When Wet, Flammable Liquid (UN 1411)

SAFETY PROFILE: Stable in dry air at room temperature. It decomposes above 125° forming Al, H_2, and lithium hydride. Very powerful reducer. Can ignite if pulverized even in a dry box. Reacts violently with air, acids, alcohols, benzoyl peroxide, boron trifluoride etherate, (2-chloromethyl furan + ethyl acetate), diethylene glycol dimethyl ether, diethyl ether, 1,2-dimethoxyethane, dimethyl ether, methyl ethyl ether, (nitriles + H_2O), perfluorosuccinamide, (perfluorosuccinamide + H_2O), tetrahydrofuran, water. To fight fire, use dry chemical, including special formulations of dry chemicals as recommended by the supplier of the lithium aluminum hydride. Do not use water, fog, spray, or mist. Incompatible with bis(2-methoxy-ethyl)ether, CO_2, BF_3, diethyl etherate, dibenzoyl peroxide, 3,5-dibromocyclopentene, 1,2-dimethoxy ethane, ethyl acetate, fluoro amides, pyridine, tetrahydrofuran. Used as a reducing agent in the preparation of pharmaceuticals. See also ALUMINUM, LITHIUM COMPOUNDS, and HYDRIDES.

LHT000   CAS: 16949-15-8   HR: 3
LITHIUM TETRAHYDROBORATE

DOT: UN 1413

mf: $BH_4 \cdot Li$ mw: 21.79

PROP: Colorless orthorhombic crystals. Moisture sensitive. Decomp on heating with formation of constituents in elemental form. Mp: 284°. Sol in Et_2O and THF.

SYN: LITHIUM BOROHYDRIDE (DOT)

CONSENSUS REPORTS: Reported in EPA TSCA Inventory.

DOT CLASSIFICATION: 4.3; Label: Dangerous When Wet

SAFETY PROFILE: Poison by ingestion, inhalation, and skin contact. Flammable; can liberate H_2. Incompatible with H_2O as moisture on fibers of cellulose or as liquid. See also LITHIUM, BORON COMPOUNDS, and HYDRIDES.

LIA000   CAS: 9000-40-2   HR: 1
LOCUST BEAN GUM
PROP: From the ground endosperms of *Ceratonia ailiqua (L.) Taub.* (Fam. *Leguminosae*). White powder; odorless and tasteless but acquires a leguminous taste when boiled in water. A galactomannan polysaccharide. Mw: 310,000 (approx). Insol in most org solvs.
SYNS: ALGAROBA □ CAROB BEAN GUM □ CAROB FLOUR □ NCI-C50419 □ ST. JOHN'S BREAD □ SUPERCOL
CONSENSUS REPORTS: NTP Carcinogenesis Bioassay (feed); No Evidence: mouse, rat NTPTR* NTP-TR-221,82. Reported in EPA TSCA Inventory.
SAFETY PROFILE: Mildly toxic by ingestion. When heated to decomposition it emits acrid smoke and irritating fumes.

LIC000   CAS: 8012-74-6   HR: 3
LONDON PURPLE
DOT: UN 1621
DOT CLASSIFICATION: 6.1; Label: Poison
SAFETY PROFILE: A poison. When heated to decomposition it emits very toxic fumes of As and NO_x. See also ARSENIC and ANILINE.

LII000   CAS: 8016-31-7   HR: 2
LOVAGE OIL
PROP: The constituents include d-α-terpineol, butyl dihydrophthalides, butyl tetrahydrophthalides, coumarin, aldehydes, acetic acid, and isovaleric acid. From steam distillation of fresh root of *Levisticum officinale L. Koch* syn. *Angelica levisticum, Baillon* (Fam. *Umbelliferae*). Yellow to green to brown liquid; strong odor and taste. D: 1.034–1.057, refr index: 1.536–1.554 @ 20°.

Sol in fixed oils; sltly sol in mineral oil; insol in glycerin, propylene glycol.
CONSENSUS REPORTS: Reported in EPA TSCA Inventory.
SAFETY PROFILE: Moderately toxic by ingestion. A skin irritant. When heated to decomposition it emits acrid smoke and irritating fumes. See also constituents as listed.

LII100   CAS: 107097-80-3 HR: 3
LOXIGLUMIDE
mf: $C_{21}H_{30}Cl_2N_2O_5$ mw: 461.43
SYNS: CR 1505 □ d,l-4-(3,4-DICHLOROBENZOYL-AMINO)-5-(N-3-METHOXYPROPYLPENTYLAMINO)-5-OXO-PENTANOIC ACID □ (+−)-4-((3,4-DICHLORO-BENZYL)AMINO)-5-((3-METHOXYPROPYL)PENTYL-AMINO)-5-OXOPENTANOIC ACID □ PENTANOIC ACID, 4-((3,4-DICHLOROBENZOYL)AMINO)-5-((3-METHOXYPROPYL)PENTYLAMINO)-5-OXO-, (+−)-
SAFETY PROFILE: A poison by intravenous. Moderately toxic by ingestion, intraperitoneal, and subcutaneous routes. When heated to decomposition it emits toxic vapors of NO_x and Cl⁻.

LIM000   CAS: 3105-97-3   HR: 3
LUCANTHONE METABOLITE
mf: $C_{20}H_{24}N_2O_2S$ mw: 356.52
PROP: Crystals from isopropyl acetate. Mp: 100.6–102.8°.
SYNS: 1-((2-(DIETHYLAMINO)ETHYL)AMINO)-4-(HYDROXYMETHYL)THIOXANTHEN-9-ONE □ 1-((2-(DIETHYLAMINO)ETHYL)AMINO)-4-(HYDROXYMETH-YL)9H-THIOXANTHEN-9-ONE □ HYCANTHON □ HYCANTHONE □ NSC-134434 □ WIN 24933
CONSENSUS REPORTS: EPA Genetic Toxicology Program.
SAFETY PROFILE: Poison by subcutaneous, intravenous, and intramuscular routes. Moderately toxic by ingestion. Experimental teratogenic effects. Human mutation data reported. Questionable carcinogen with experimental carcinogenic data. When heated to decomposition it emits very toxic fumes of NO_x and SO_x.

LIU305   CAS: 34973-08-5   HR: 3
LUTEINIZING HORMONE-RELEASING
**   FACTOR (PIG), ACETATE (SALT)**
mf: $C_{55}H_{75}H_{17}O_{13} \cdot xH_2H_4O_2$ mw:
1227.89
SYNS: CYSTORELIN □ GONADORELIN ACETATE □
LUPROLITE ACETATE □ LUTEINIZING HORMONE-
RELEASING FACTOR (SWINE), ACETATE (SALT) □
LUTREPULSE
SAFETY PROFILE: A poison by
intravenous route. Moderately toxic by
ingestion and subcutaneous routes.
Experimental reproductive effects. When
heated to decomposition it emits acrid
smoke and irritating vapors.

LIU370   CAS: 9034-40-6   HR: 3
LUTEINIZING HORMONE-RELEASING
**   HORMONE**
mf: $C_{55}H_{75}N_{17}O_{13}$ mw: 1182.33
PROP: Neurohumoral hormone produced
in the hypothalamus that stimulates the
secretion of the pituitary hormones LH
(luteinizing hormone) and FSH (follicle-
stimulating hormone), which in turn
produce changes resulting in the induction
of ovulation.
SYNS: AY 24034 □ CYSTORELIN □ FERTIRAL □ Gn-
RH □ GONADORELIN □ GONADOTROPIN-
RELEASING FACTOR □ GONADOTROPIN RELEASING
HORMONE □ KRYPTOCUR □ LH RELEASING FACTOR
□ LH-RELEASING HORMONE □ LH-RF □ LHRH □ LH-
RH □ LH-RH/FSH-RH □ LRF □ LRH □ LULIBERIN □
LUTEINIZING HORMONE-RELEASING FACTOR □
LUTEOSTIMULIN □ LUTRELEF □ OVARELIN □
RELEFACT LH-RH □ SYNTHETIC LH-RH
CONSENSUS REPORTS: Reported in EPA
TSCA Inventory. EPA Genetic Toxicology
Program.
SAFETY PROFILE: An experimental
teratogen. Human reproductive effects in
women by subcutaneous route: menstrual
cycle changes and other unspecified effects.
Experimental reproductive effects. Used in
the treatment of oligospermia and male
infertility. See also LUTEINIZING
HORMONE and other luteinizing
hormone-releasing hormone entries.

LIV000   CAS: 21884-44-6   HR: 3
LUTEOSKYRIN
mf: $C_{30}H_{22}O_{12}$ mw: 574.52
PROP: Yellow rectangular crystals or
needles from EtOH or Me₂CO. Mp: 278°
(decomp). Anthraquinoid hepatotoxin of
Penicillium islandicum sopp (JJEMAG
41,177,71).
SYNS: 5H,6H-6,5A,13A,14-(1,2,3,4)BUTANETETRA-
CYCLOOCTA(1,2-B:5,6-B')DINAPHTHALENE □ 8,8'-
DIHYDROXY-RUGULOSIN □ FLAVOMYCELIN □ (−)-
LUTEOSKYRIN
CONSENSUS REPORTS: IARC Cancer
Review: Group 3 IMEMDT 7,56,87; Animal
Limited Evidence IMEMDT 10,163,76.
SAFETY PROFILE: Poison by ingestion,
intraperitoneal, subcutaneous, and
intravenous routes. Questionable carcinogen
with experimental carcinogenic and
tumorigenic data. Human mutation data
reported. When heated to decomposition it
emits acrid smoke and irritating fumes.

LIW000   CAS: 10099-66-8   HR: 3
LUTETIUM CHLORIDE
mf: Cl_3Lu mw: 281.32
PROP: Colorless crystals or hygroscopic
white solid. Subl above 750°, mp: 905°. Sol
in water and EtOH.
CONSENSUS REPORTS: Reported in EPA
TSCA Inventory.
SAFETY PROFILE: Poison by
intraperitoneal route. Mildly toxic by
ingestion. When heated to decomposition it
emits toxic fumes of Cl⁻. See also RARE
EARTHS.

LIY000   CAS: 10099-67-9   HR: 3
LUTETIUM(III) NITRATE (1:3)
mf: $N_3O_9 \cdot Lu$ mw: 361.00
SYN: NITRIC ACID, LUTETIUM(3+) SALT
CONSENSUS REPORTS: Reported in EPA
TSCA Inventory.
SAFETY PROFILE: Poison by
intraperitoneal route. When heated to
decomposition it emits toxic fumes of NOₓ.
See also NITRATES and RARE EARTHS.

L

LJB000   CAS: 583-58-4   HR: 3
3,4-LUTIDINE
mf: C_7H_9N mw: 107.17
PROP: A liquid. Bp: 163.5–164.5°, d: 0.928
@ 25°/4°. Sol in EtOH and Et_2O.
SYN: 3,4-DIMETHYLPYRIDINE
CONSENSUS REPORTS: Reported in EPA
TSCA Inventory.
SAFETY PROFILE: Poison by skin contact.
Moderately toxic by ingestion and
inhalation. When heated to decomposition it
emits toxic fumes of NO_x.

LJE000   CAS: 8015-14-3   HR: 3
LYNDIOL
mf: $C_{21}H_{26}O_2 \cdot C_{20}H_{28}O$ mw: 594.95
PROP: Pharmaceutical.
SYNS: LYNESTRENOL mixed with MESTRANOL □
LYNESTROL mixed with MESTRANOL □ LYNOESTREN-
OL mixed with MESTRANOL □ MESTRANOL mixed with
LYNESTRENOL □ MESTRANOL mixed with LYNESTROL
□ NORACYCLINE □ OVANON □ OVARIOSTAT
(FRENCH) □ RESTOVAR □ SISTOMETRENOL
CONSENSUS REPORTS: EPA Genetic
Toxicology Program.
SAFETY PROFILE: Suspected human
carcinogen producing liver tumors. An
experimental teratogen. Human systemic
effects by ingestion: dyspnea, nausea or
vomiting, and fever. Experimental
reproductive effects. Used as an oral
contraceptive. When heated to
decomposition it emits acrid smoke and
irritating fumes.

LJG000   HR: 3
**d-LYSERGIC ACID DIETHYLAMIDE
    TARTRATE**
mf: $C_{40}H_{50}N_6O_2 \cdot C_4H_6O_6 \cdot 2CH_4O$ mw:
861.16
SYNS: 9,10-DIDEHYDRO-N,N-DIETHYL-6-METHYL-
ERGOLINE-8-β-CARBOXAMIDE-d-TARTRATE with

METHANOL (1:2) □ LSD TARTRATE □ LYSERGIC ACID
DIETHYLAMIDE TARTRATE
SAFETY PROFILE: Poison by
intraperitoneal route. An experimental
teratogen. Experimental reproductive
effects. Human mutation data reported.
When heated to decomposition it emits
toxic fumes of NO_x.

LJM700   CAS: 56-87-1   HR: D
LYSINE
mf: $C_6H_{14}N_2O_2$ mw: 146.22
PROP: l-Lysine: Needles from water,
hexagonal plates from dil alc. Darkens at
210°; decomp 224.5°. Very freely sol in
water; very sltly sol in alc; practically insol in
ether.
SYNS: AMINUTRIN □ α,ε-DIAMINOCAPROIC ACID □
2,6-DIAMINOHEXANOIC ACID □ l-LYSINE (9CI) □ l-(+)-
LYSINE □ LYSINE ACID
CONSENSUS REPORTS: EPA Genetic
Toxicology Program. Reported in EPA
TSCA Inventory.
SAFETY PROFILE: An experimental
teratogen. Experimental reproductive
effects. When heated to decomposition it
emits toxic fumes of NO_x.

LJO000   CAS: 657-27-2   HR: 1
l-LYSINE MONOHYDROCHLORIDE
mf: $C_6H_{14}N_2O_2 \cdot ClH$ mw: 182.68
PROP: White powder. Mp: 235–236°. Sol in
water; insol in alc and ether. Crystals from
dil ethanol. Mp: 263–264° (decomp) when
anhyd.
SYNS: 2,6-DIAMINOHEXANOIC ACID HYDRO-
CHLORIDE □ l-LYSINE HYDROCHLORIDE □ LYSINE
MONOHYDROCHLORIDE
CONSENSUS REPORTS: Reported in EPA
TSCA Inventory.
SAFETY PROFILE: Mildly toxic by
ingestion. When heated to decomposition it
emits very toxic fumes of HCl and NO_x.

M

MAC750 CAS: 7439-95-4 HR: 3
MAGNESIUM
DOT: UN 1418/UN 1869/UN 2950
af: Mg aw: 24.31
PROP: Hexagonal, light, silvery-white crystals. The bulk metal tarnishes in air. Mp: 651°, bp: 1100°, d: 1.74 @ 5°, d: 1.738 @ 20°, vap press: 1 mm @ 621°.
SYNS: MAGNESIO (ITALIAN) □ MAGNESIUM (UN 1869) (DOT) □ MAGNESIUM ALLOYS, powder (UN 1418) (DOT) □ MAGNESIUM ALLOYS with >50% magnesium in pellets, turnings or ribbons (UN 1869) (DOT) □ MAGNESIUM CLIPPINGS □ MAGNESIUM GRANULES, coated particle size not <149 microns (UN 2950) (DOT) □ MAGNESIUM PELLETS □ MAGNESIUM POWDERED □ MAGNESIUM, powder (UN 1418) (DOT) □ MAGNESIUM RIBBONS □ MAGNESIUM TURNINGS (DOT) □ RMC
CONSENSUS REPORTS: Reported in EPA TSCA Inventory.
DOT CLASSIFICATION: 4.3; Label: Dangerous When Wet (UN 2950); DOT Class: 4.1; Label: Flammable Solid (UN 1869); DOT Class: 4.3; Label: Danger When Wet, Spontaneously Combustible
SAFETY PROFILE: Inhalation of dust and fumes can cause metal fume fever. The powdered metal ignites readily on the skin causing burns. Particles embedded in the skin can produce gaseous blebs that heal slowly.
  A dangerous fire hazard in the form of dust or flakes when exposed to flame or oxidizing agents. In solid form, magnesium is difficult to ignite because heat is conducted rapidly away from the source of ignition; it must be heated above its melting point before it will burn. However, in finely divided form, it may be ignited by a spark or the flame of a match. Magnesium fires do not flare up violently unless there is moisture present. Therefore, it must be kept away from water, moisture, etc. It may

ignited spontaneously when the material is finely divided and damp, particularly with water-oil emulsion. Moderately explosive in the form of dust when exposed to flame. Also, magnesium reacts with moisture, acids, etc., to evolve hydrogen, a highly dangerous fire and explosion hazard.
  Explosive reaction or ignition with calcium carbonate + hydrogen + heat, gold cyanide + heat, mercury cyanide + heat, silver oxide + heat, fused nitrates, phosphates, or sulfates (e.g., ammonium nitrate, metal nitrates), chloroformamidinium nitrate + water (when ignited with powder). The powder may explode on contact with halocarbons (e.g., chloromethane, chloroform, or carbon tetrachloride), and explodes when sparked in dichlorodifluoromethane. Hypergolic reaction with nitric acid + 2-nitroaniline. Mixtures of powdered magnesium and methanol are more powerful than some military explosives. Mixtures of magnesium powder + water can be detonated. Reacts with acetylenic compounds including traces of acetylene found in ethylene gas to form explosive magnesium acetylide.
  Violent reactions with ammonium salts, chlorate salts, beryllium fluoride, boron diiodophosphide, carbon tetrachloride + methanol, 1,1,1-trichloroethane, 1,2-dibromoethane, halogens or interhalogens (e.g., fluorine, chlorine, bromine, iodine vapor, chlorine trifluoride, iodine heptafluoride), hydrogen iodide, metal oxides + heat (e.g., beryllium oxide, cadmium oxide, copper oxide, mercury oxide, molybdenum oxide, tin oxide, zinc oxide), nitrogen (when ignited), silicon dioxide powder + heat, polytetrafluoroethylene powder + heat,

sulfur + heat, tellurium + heat, barium peroxide, nitric acid vapor, hydrogen peroxide, ammonium nitrate, sodium iodate + heat, sodium nitrate + heat, dinitrogen tetraoxide (when ignited), lead dioxide. Ignites in carbon dioxide at 780°C, molten barium carbonate + water, fluorocarbon polymers + heat, carbon tetrachloride or trichloroethylene (on impact), dichlorodifluoromethane + heat.

Incompatible with ethylene oxide, metal oxosalts, oxidants, potassium carbonate, Al + $KClO_4$, [$Ba(NO_3)_2$ + BaO_2 + Zn], bromobenzyl trifluoride, CaC, carbonates, $CHCl_3$, [$CuSO_4$ (anhydrous) + NH_4NO_3 + $KClO_3$ + H_2O], $CuSO_4$, (H_2 + $CaCO_3$), CH_3Cl, NO_2, liquid oxygen, metal cyanides (e.g., cadmium cyanide, cobalt cyanide, copper cyanide, lead cyanide, nickel cyanide, zinc cyanide), performic acid, phosphates, $KClO_3$, $KClO_4$, $AgNO_3$, $NaClO_4$, (Na_2O_2 + CO_2), sulfates, trichloroethylene, Na_2O_2.

To fight fire, operators and firefighters can approach a magnesium fire to within a few feet if no moisture is present. Water and ordinary extinguishers, such as CO_2, carbon tetrachloride, etc., should not be used on magnesium fires. G-1 powder or powdered talc should be used on open fires. Dangerous when heated; burns violently in air and emits fumes; will react with water or steam to produce hydrogen. See also MAGNESIUM COMPOUNDS.

MAD025   CAS: 14644-70-3   HR: 3
MAGNESIUM AMMONIUM ARSENATE DIHYDRATE
mf: $AsO_4 \cdot H_3N \cdot Mg \cdot 2H_2O$ mw: 216.31
SYNS: AMMONIUM MAGNESIUM ARSENATE □ AMMONIUM MAGNESIUM ARSENATE DIHYDRATE □ ARSENIC ACID, AMMONIUM MAGNESIUM SALT, HYDRATE (1:1:1:2) □ ARSENIC ACID (H3-AS-O4), AMMONIUM MAGNESIUM SALT, (1:1:1) □ MAGNESIUM AMMONIUM ARSENATE
SAFETY PROFILE: A poison by ingestion, skin contact, and intraperitoneal routes. When heated to decomposition it emits toxic vapors of NH_4^-, Mn, and As.

MAE750   HR: D
MAGNESIUM COMPOUNDS
SAFETY PROFILE: Variable toxicity. The inhalation of fumes of freshly sublimed magnesium oxide may cause metal fume fever. There is no evidence that magnesium produces true systemic poisoning. Particles of metallic magnesium or magnesium alloy that perforate the skin or gain entry through cuts and scratches may produce a severe local lesion characterized by the evolution of gas and acute inflammatory reaction, frequently with necrosis. The condition has been called a "chemical gas gangrene." Gaseous blebs may develop within 24 hours of the injury. The inflammatory response is marked at the site of injury and there may be signs of lymphangitis. The lesion is very slow to heal. The most serious hazard presented by magnesium is the danger from burns. Protection necessary for personnel handling and processing magnesium is usually no different from that necessary for other metals. The toxicity of magnesium compounds is usually that of the anion. When heated to decomposition it emits toxic fumes of MgO. See also MAGNESIUM and specific compounds.

MAF000   CAS: 69011-63-8   HR: 3
MAGNESIUM DROSS (HOT)
SYN: MAGNESIUM DROSS, wet or hot (DOT)
CONSENSUS REPORTS: Reported in EPA TSCA Inventory.
DOT CLASSIFICATION: Forbidden
SAFETY PROFILE: See MAGNESIUM COMPOUNDS.

MAF500   CAS: 7783-40-6   HR: 2
MAGNESIUM FLUORIDE
mf: F_2Mg mw: 62.31
PROP: Luminous substance; faint violet, luminous tetragonal crystals. Dissolves in HNO_3 soln. Mp: 1263°, bp: 2239°, d: 2.9–3.2. Practically insol in water; sltly sol in dil acids.
SYNS: AFLUON □ IRTRAN 1 □ MAGNESIUM FLUORURE (FRENCH) □ SELLAITE

CONSENSUS REPORTS: Reported in EPA TSCA Inventory.
OSHA PEL: TWA 2.5 mg(F)/m³
ACGIH TLV: TWA 2.5 mg(F)/m³; BEI: 3 mg/g creatinine of fluorides in urine prior to shift; 10 mg/g creatinine of fluorides in urine at end of shift.
NIOSH REL: TWA 2.5 mg(F)/m³
SAFETY PROFILE: Moderately toxic by ingestion. When heated to decomposition it emits toxic fumes of F⁻. See also MAGNESIUM and FLUORIDES.

**MAF600   CAS: 16949-65-8   HR: 2
MAGNESIUM FLUOSILICATE**
DOT: UN 2853
mf: F₆Si•Mg   mw: 166.40
PROP: Crystalline white, lightly odored granules. D: 1.788 @ 20°, Mp: 100°. Sol in water.
SYNS: HEXAFLUOROSILICATE(2-) MAGNESIUM (1:1) □ MAGNESIUM FLUOROSILICATE (DOT) □ SILICATE(2-), HEXAFLUORO-, MAGNESIUM (1:1)
DOT CLASSIFICATION: 6.1; Label: KEEP AWAY FROM FOOD
CONSENSUS REPORTS: Reported in EPA TSCA Inventory.
SAFETY PROFILE: Moderately toxic by ingestion. When heated to decomposition it emits toxic vapors of magnesium and F⁻.

**MAG100   HR: D
MAGNESIUM GLYCEROPHOSPHATE**
SAFETY PROFILE: When heated to decomposition it emits acrid smoke and irritating fumes.

**MAG250   CAS: 18972-56-0   HR: 3
MAGNESIUM HEXAFLUOROSILICATE**
mf: F₆Si•Mg•6H₂O   mw: 274.52
PROP: White, efflorescent, trigonal, crystalline powder. Decomp on heating with simultaneous loss of H₂O and SiF₄. Very sol in H₂O; insol in EtOH.
SYNS: EULAVA SM □ FLUOSILICATE de MAGNESIUM (FRENCH) □ MAGNESIUM FLUOSILICATE □ MAGNESIUM SILICOFLUORIDE
OSHA PEL: TWA 2.5 mg(F)/m³

ACGIH TLV: TWA 2.5 mg(F)/m³; BEI: 3 mg/g creatinine of fluorides in urine prior to shift; 10 mg/g creatinine of fluorides in urine at end of shift.
NIOSH REL: TWA 2.5 mg(F)/m³
SAFETY PROFILE: Poison by ingestion. Moderately toxic by subcutaneous route. When heated to decomposition it emits toxic fumes of F⁻. See also FLUORIDES and MAGNESIUM COMPOUNDS.

**MAG750   CAS: 1309-42-8   HR: 2
MAGNESIUM HYDROXIDE**
mf: H₂MgO₂   mw: 58.33
PROP: Amorphous, white powder, odorless or colorless hexagonal crystals. Readily dissolves in aq acids forming corresponding salts. Decomp on heating at 3° releasing MgO + H₂O. D: 2.36, mp: decomp @ 350°. Sol in solns of ammonium salts and dilute acids; almost insol in water and alc.
SYNS: BASCHEM 12 □ HYDRO-MAG MA □ MAGNESIA MAGMA □ MAGNESIUM HYDRATE □ MAGOX □ MARINCO H □ MILK OF MAGNESIA □ MINT-O-MAG □ NEMALITE
CONSENSUS REPORTS: Reported in EPA TSCA Inventory.
SAFETY PROFILE: Moderately toxic by intraperitoneal route. Human systemic effects: chlorine level changes, coma, somnolence. Incompatible with maleic anhydride, phosphorus. See also MAGNESIUM COMPOUNDS.

**MAH000   CAS: 10377-60-3   HR: 3
MAGNESIUM(II) NITRATE (1:2)**
DOT: UN 1474
mf: N₂O₆•Mg   mw: 148.33
PROP: Hygroscopic colorless cubic crystals. Decomp on heating forming MgO and nitrogen oxides. The dihydrate, [Mg(NO₃)₂•2H₂O]: mw: 184.37, forms white crystals (prisms). D: 2.0256 @ 25°, mp: 129.0°. The hexahydrate, [Mg(NO₃)₂•6H₂O] mw: 256.43, forms monoclinic, colorless, deliq crystals. D: 1.464, mp: 95°, bp: −5H₂O @ 330°.
SYNS: MAGNESIUM NITRATE (DOT) □ NITRIC ACID, MAGNESIUM SALT (2:1)

CONSENSUS REPORTS: Reported in EPA TSCA Inventory.

DOT CLASSIFICATION: 5.1; Label: Oxidizer

SAFETY PROFILE: Probably a severe irritant to the eyes, skin, and mucous membranes. A powerful oxidizer. Violent decomposition on contact with dimethylformamide. When heated to decomposition it emits toxic fumes of NO_x. See also NITRATES and MAGNESIUM COMPOUNDS.

MAH500   CAS: 1309-48-4   HR: 2
MAGNESIUM OXIDE

mf: MgO   mw: 40.31

PROP: White, bulky, very fine, odorless powder; or colorless cubic crystals, moisture sensitive. Mp: 2832°, bp: 3600°, d: 3.65–3.75. Very sltly sol in water; sol in dil acids; insol in alc. IDLH 750 mg/m³.

SYNS: AKRO-MAG □ ANIMAG □ CALCINED BRUCITE □ CALCINED MAGNESIA □ CALCINED MAGNESITE □ GRANMAG □ MAGCAL □ MAGCHEM 100 □ MAGLITE □ MAGNESIA □ MAGNESIA USTA □ MAGNESIUM OXIDE FUME (ACGIH) □ MAGNEZU TLENEK (POLISH) □ MAGOX □ MAGOX 85 □ MAGOX 90 □ MAGOX 95 □ MAGOX 98 □ MAGOX OP □ MARMAG □ OXYMAG □ PERICLASE □ SEAWATER MAGNESIA

CONSENSUS REPORTS: Reported in EPA TSCA Inventory.

OSHA PEL: Fume: Total Dust: TWA 10 mg/m³; Respirable Fraction: 5 mg/m³

ACGIH TLV: TWA 10 mg/m³ (fume); Not Classifiable as a Human Carcinogen

DFG MAK: 1.5 mg/m³ (fume)

SAFETY PROFILE: Inhalation of the fumes can produce a febrile reaction and leucocytosis in humans. Questionable carcinogen with experimental tumorigenic data. Violent reaction or ignition in contact with interhalogens (e.g., bromine pentafluoride, chlorine trifluoride). Incandescent reaction with phosphorus pentachloride. See also MAGNESIUM COMPOUNDS.

MAH750   CAS: 14452-57-4   HR: 3
MAGNESIUM PEROXIDE

DOT: UN 1476

mf: MgO_2 mw: 56.31

PROP: White powder; tasteless and odorless. Insol in water and slowly decomp evolving O_2; sol in dil acids. Keep container closed.

SYNS: IXPER 25M □ MAGNESIUM PEROXIDE, solid (DOT)

CONSENSUS REPORTS: Reported in EPA TSCA Inventory.

DOT CLASSIFICATION: 5.1; Label: Oxidizer

SAFETY PROFILE: A powerful oxidizer. Probably a severe irritant to the eyes, skin, and mucous membranes. Flammable by chemical reaction with acidic materials and moisture; an oxidizing agent. Dangerous; reacts vigorously with reducing agents; will decompose violently in or near a fire. See also MAGNESIUM COMPOUNDS and PEROXIDES, INORGANIC.

MAI000   CAS: 12057-74-8   HR: 3
MAGNESIUM PHOSPHIDE

DOT: UN 2011

mf: Mg_3P_2 mw: 134.87

PROP: Moisture sensitive bright yellow cubic crystals.

SYNS: FOSFURI di MAGNESIO (ITALIAN) □ MAGNESIUMFOSFIDE (DUTCH) □ PHOSPHURE de MAGNESIUM (FRENCH)

CONSENSUS REPORTS: Reported in EPA TSCA Inventory.

DOT CLASSIFICATION: 4.3; Label: Dangerous When Wet, Poison

SAFETY PROFILE: A poison. Flammable when exposed to heat, flame, or oxidizing materials. Ignites when heated in chlorine, bromine, or iodine vapors. Incandescent reaction with nitric acid. Reacts with water to evolve flammable phosphine gas. When heated to decomposition it emits toxic fumes of PO_x and phosphine. See also MAGNESIUM and PHOSPHIDES.

MAJ000   CAS: 1343-90-4   HR: 1
MAGNESIUM SILICATE HYDRATE

mf: $Mg_2O_8Si_3 \cdot H_2O$ mw: 278.91

PROP: Fine white powder; odorless and tasteless. Insol in water and alc.

SAFETY PROFILE: A human skin irritant. See also MAGNESIUM and SILICATES.

MAJ030 CAS: 557-04-0 HR: 1
MAGNESIUM STEARATE
mf: $C_{36}H_{70}O_4 \cdot Mg$ mw: 591.37
SYNS: MAGNESIUM STEARATE □ MAGNESIUM STEARATE (ACGIH) □ OCTADECANOIC ACID, MAGNESIUM SALT
PROP: Fine white bulky powder; faint, characteristic odor. Insol in water, alc, ether.
CONSENSUS REPORTS: Reported in EPA TSCA Inventory.
ACGIH TLV: TWA 10 mg/m³
DOT CLASSIFICATION: 4.2; Label: Spontaneously Combustible
SAFETY PROFILE: Slightly toxic by ingestion. Spontaneously combustible. When heated to decomposition it emits acrid smoke and toxic fumes.

MAJ250 CAS: 7487-88-9 HR: 3
MAGNESIUM SULFATE (1:1)
mf: $O_4S \cdot Mg$ mw: 120.37
PROP: Opaque, hygroscopic, colorless, orthorhombic needles or granular crystalline powder; odorless with cooling, bitter, salt taste. Mp: 1127°. Very sol in H_2O; sol in EtOH, Et_2O; insol in Me_2CO.
SYNS: EPSOM SALTS □ MAGNESIUM SULPHATE
CONSENSUS REPORTS: Reported in EPA TSCA Inventory.
SAFETY PROFILE: A poison by intravenous route. Moderately toxic by ingestion, intraperitoneal, and subcutaneous routes. Human systemic effects: heart changes, cyanosis, flaccid paralysis with appropriate anesthesia. An experimental teratogen. Mutation data reported. Potentially explosive reaction when heated with ethoxyethynyl alcohols (e.g., 1-ethoxy-3-methyl-1-butyn-3-ol). When heated to decomposition it emits toxic fumes of SO_x. See also SULFATES.

MAK275 CAS: 62959-43-7 HR: 3
MAGNESIUM VALPROATE
mf: $C_{16}H_{30}O_4 \cdot Mg$ mw: 310.77

SYNS: MAGNESIUM DIPROPYLACETATE □ MV □ PENTANOIC ACID, 2-PROPYL-, MAGNESIUM SALT
DOT CLASSIFICATION: 4.2; Label: Spontaneously Combustible
SAFETY PROFILE: Moderately toxic by ingestion. Danger: A spontaneously combustible solid. When heated to decomposition it emits toxic vapors of magnesium.

MAK700 CAS: 121-75-5 HR: 3
MALATHION
mf: $C_{10}H_{19}O_6PS_2$ mw: 330.38
PROP: Brown to yellow liquid; characteristic odor. D: 1.23 @ 25°/4°, mp: 2.9°, bp: 156° @ 0.7 mm. Misc in org solvs; sltly water-sol. IDLH 250 mg/m³.
SYNS: AMERICAN CYANAMID 4,049 □ S-(1,2-BIS-(AETHOXY-CARBONYL)-AETHYL)-O,O-DIMETHYL-DITHIOPHOSPHAT (GERMAN) □ S-(1,2-BIS(CARBETHO-XY)ETHYL)-O,O-DIMETHYL DITHIOPHOSPHATE □ S-(1,2-BIS(ETHOXY-CARBONYL)-ETHYL)-O,O-DIMETHYL-DITHIOFOSFAAT (DUTCH) □ S-(1,2-BIS(ETHOXY-CARBONYL)ETHYL)-O,O-DIMETHYL PHOSPHORODI-THIOATE □ S-1,2-BIS(ETHOXYCARBONYL)ETHYL-O,O-DIMETHYL THIOPHOSPHATE □ S-(1,2-BIS(ETOSSI-CARBONIL)-ETIL)-O,O-DIMETIL-DITIOFOSFATO (ITALIAN) □ CALMATHION □ CARBETHOXY MALA-THION □ CARBETOVUR □ CARBETOX □ CARBOFOS □ CARBOPHOS □ CELTHIGN □ CHEMATHION □ CIMEXAN □ COMPOUND 4049 □ CYTHION □ DET-MOL MA □ DETMOL MA 96% □ S-(1,2-DICARBETHO-XYETHYL)-O,O-DIMETHYLDITHIOPHOSPHATE □ DICARBOETHOXYETHYL-O,O-DIMETHYL PHOS-PHORODITHIOATE □ 1,2-DI(ETHOXYCARBONYL)-ETHYL-O,O-DIMETHYL PHOSPHORODITHIOATE □ S-(1,2-DI(ETHOXYCARBONYL)ETHYL) DIMETHYL PHOS-PHOROTHIOLOTHIONATE □ DIETHYL (DIMETHOXY-PHOSPHINOTHIOYLTHIO) BUTANEDIOATE □ DIETHYL (DIMETHOXYPHOSPHINOTHIOYLTHIO)-SUCCINATE □ DIETHYL MERCAPTOSUCCINATE-O,O-DIMETHYL DITHIOPHOSPHATE, S-ESTER □ DIETHYL MERCAPTOSUCCINATE-O,O-DIMETHYL PHOSPHORO-DITHIOATE □ DIETHYL MERCAPTOSUCCINATE-O,O-DIMETHYL THIOPHOSPHATE □ DIETHYL MERCAP-TOSUCCINATE-S-ESTER with O,O-DIMETHYLPHOSPH-ORODITHIOATE □ DIETHYL MERCAPTOSUCCINIC ACID-O,O-DIMETHYL PHOSPHORODITHIOATE □ ((DIMETHOXYPHOSPHINOTHIOYL)THIO)BUTANEDIO-IC ACID DIETHYL ESTER □ O,O-DIMETHYL-S-(1,2-BIS(ETHOXYCARBONYL)ETHYL)DITHIOPHOSPHATE □ O,O-DIMETHYL-S-1,2-(DICARBAETHOXYAETHYL)-DITHIOPHOSPHAT (GERMAN) □ O,O-DIMETHYL-S-(1,2-DICARBETHOXYETHYL) DITHIOPHOSPHATE □ O,O-DIMETHYL-S-(1,2-DICARBETHOXYETHYL)-PHOSPHORODITHIOATE □ O,O-DIMETHYL-S-(1,2-

M

DICARBETHOXYETHYL) THIOTHIONOPHOSPHATE □ O,O-DIMETHYL-S-1,2-DI(ETHOXYCARBAMYL)ETHYL PHOSPHORODITHIOATE □ O,O-DIMETHYL-S-1,2-DIKARBETOXYLETHYLDITIOFOSFAT (CZECH) □ O,O-DIMETHYLDITHIOPHOSPHATE DIETHYLMERCAPTO-SUCCINATE □ DITHIOPHOSPHATE de O,O-DIMETHY-LE et de S-(1,2-DICARBOETHOXYETHYLE) (FRENCH) □ EL 4049 □ EMMATOS □ EMMATOS EXTRA □ ENT 17,034 □ S-ESTER with O,O-DIMETHYL PHOSPHORO-THIOATE □ ETHIOLACAR □ ETIOL □ EXPERIMENT-AL INSECTICIDE 4049 □ EXTERMATHION □ FORMAL □ FORTHION □ FOSFOTHION □ FOSFOTION □ FOUR THOUSAND FORTY-NINE □ FYFANON □ HILTHION □ HILTHION 25WDP □ INSECTICIDE No. 4049 □ KARBOFOS □ KOP-THION □ KYPFOS □ MALACIDE □ MALAFOR □ MALAGRAN □ MALAKILL □ MALAMAR □ MALAMAR 50 □ MALAPHELE □ MALAPHOS □ MALASOL □ MALASPRAY □ MALATHION ULV CONCENTRATE □ MALATHIOZOO □ MALATHON □ MALATHYL LV CONCENTRATE & ULV CONCENTRATE □ MALATION (POLISH) □ MALATOL □ MALATOX □ MALDISON □ MALMED □ MALPHOS □ MALTOX □ MALTOX MLT □ MERCAPTOSUCCINIC ACID DIETHYL ESTER □ MERCAPTOTHION □ MERCAPTOTION (SPANISH) □ MLT □ MOSCARDA □ NCI-C00215 □ OLEOPHOSPHOTHION □ ORTHO MALATHION □ PHOSPHORODITHIOIC ACID-O,O-DIMETHYL ESTER-S-ESTER with DIETHYL MERCAPTOSUCCINATE □ PHOSPHOTHION □ PRIODERM □ SADOFOS □ SADOPHOS □ SF 60 □ SIPTOX I □ SUMITOX □ TAK □ TM-4049 □ VEGFRU MALATOX □ VETIOL □ ZITHIOL

CONSENSUS REPORTS: IARC Cancer Review: Group 3 IMEMDT 7,56,87; Animal No Evidence IMEMDT 30,103,83; NCI Carcinogenesis Bioassay (feed); No Evidence: mouse, rat NCITR* NCI-CG-TR-24,78; No Evidence: rat NCITR* NCI-CG-TR-192,79. EPA Genetic Toxicology Program.

OSHA PEL: TWA Total Dust: 10 mg/m³; Respirable Fraction: 5 mg/m³ (skin)

ACGIH TLV: TWA 1 mg/m³ (skin); Not Classifiable as a Human Carcinogen

DFG MAK: 15 mg/m³

NIOSH REL: (Malathion) TWA 15 mg/m³

SAFETY PROFILE: A human poison by ingestion and skin contact. Can penetrate intact skin. An experimental poison by ingestion, inhalation, intraperitoneal, intravenous, intraarterial, and subcutaneous routes. Human systemic effects by ingestion: coma, blood pressure depression, and difficulty in breathing. Questionable carcinogen. An experimental teratogen. Other experimental reproductive effects. Human mutation data reported. Has caused allergic sensitization of the skin. An organic phosphate cholinesterase inhibitor. When heated to decomposition it emits toxic fumes of PO_x and SO_x. See also PHOSPHATES and PARATHION.

MAK900   CAS: 110-16-7   HR: 2
MALEIC ACID
DOT: NA 2215
mf: $C_4H_4O_4$ mw: 116.08
PROP: White crystals; faint acidulous odor. Prisms from H_2O with repulsive astringent taste. Mp: 138–139°, bp: 135° (decomp), d: 1.590 @ 20°/4°, vap d: 4.0. Sol in H_2O and EtOH.
SYNS: cis-BUTENEDIOIC ACID □ (Z)-BUTENEDIOIC ACID □ cis-1,2-ETHYLENEDICARBOXYLIC ACID □ MALEINIC ACID □ MALENIC ACID □ TOXILIC ACID
CONSENSUS REPORTS: Reported in EPA TSCA Inventory.
DOT CLASSIFICATION: 8; Label: Corrosive
SAFETY PROFILE: Moderately toxic by ingestion and skin contact. Passes through intact skin. A skin and severe eye irritant and a corrosive. Believed to be more toxic than its isomer, fumeric acid. Combustible when exposed to heat or flame. When heated to decomposition it emits acrid smoke and irritating fumes.

MAL250   CAS: 128-53-0   HR: 3
MALEIC ACID-N-ETHYLIMIDE
mf: $C_6H_7NO_2$ mw: 125.14
PROP: Crystals; irritating odor. Mp: 45°, bp: 210°.
SYNS: N-ETHYLMALEIMIDE □ MALEIC ACID N-ETHYLIMIDE □ MALEIMIDE, N-ETHYL- □ NEM □ USAF B-121
CONSENSUS REPORTS: Reported in EPA TSCA Inventory.
SAFETY PROFILE: Poison by intraperitoneal route. Human mutation data reported. Vapors are highly irritating. When heated to decomposition it emits toxic fumes of NO_x.

MAM000   CAS: 108-31-6   HR: 3
MALEIC ANHYDRIDE
DOT: UN 2215
mf: $C_4H_2O_3$ mw: 98.06
PROP: Fused black or white crystals. Orthorhombic needles from $CHCl_3$ or by subl. Mp: 52.8°, bp: 202°, flash p: 215°F (CC), d: 1.48 @ 20°/4°, autoign temp: 890°F, vap press: 1 mm @ 44.0°, vap d: 3.4, lel: 1.4%, uel: 7.1%. Sol in dioxane, water @ 30° forming maleic acid; very sltly sol in alc and ligroin. IDLH 10 mg/m³.
SYNS: cis-BUTENEDIOIC ANHYDRIDE □ 2,5-DIHYDROFURAN-2,5-DIONE □ 2,5-FURANDIONE □ MALEIC ACID ANHYDRIDE (MAK) □ RCRA WASTE NUMBER U147 □ TOXILIC ANHYDRIDE
CONSENSUS REPORTS: Community Right-To-Know List. Reported in EPA TSCA Inventory.
OSHA PEL: TWA 0.25 ppm
ACGIH TLV: TWA 0.1 ppm (skin, sensitizer); Not Classifiable as a Human Carcinogen
DFG MAK: 0.1 ppm (0.41 mg/m³)
DOT CLASSIFICATION: 8; Label: Corrosive
SAFETY PROFILE: Poison by ingestion and intraperitoneal routes. Moderately toxic by skin contact. A corrosive irritant to eyes, skin, and mucous membranes. Can cause pulmonary edema. Questionable carcinogen with experimental tumorigenic data. Mutation data reported. A pesticide. Combustible when exposed to heat or flame; can react vigorously on contact with oxidizing materials. Explosive in the form of vapor when exposed to heat or flame. Reacts with water or steam to produce heat. Violent reaction with bases (e.g., sodium hydroxide, potassium hydroxide, calcium hydroxide), alkali metals (e.g., sodium, potassium), amines (e.g., dimethylamine, triethylamine), lithium, pyridine. To fight fire, use alcohol foam. Incompatible with cations. When heated to decomposition (above 150°C) it emits acrid smoke and irritating fumes. See also ANHYDRIDES.

MAM750   CAS: 541-59-3   HR: 3
MALEIMIDE
mf: $C_4H_3NO_2$ mw: 97.08
PROP: Plates. Mp: 93°.
SYNS: MALEINIMIDE □ PYRROLE-2,5-DIONE □ 3-PYRROLINE-2,5-DIONE
CONSENSUS REPORTS: Reported in EPA TSCA Inventory.
SAFETY PROFILE: Poison by intraperitoneal and intravenous routes. An experimental teratogen. Experimental reproductive effects. When heated to decomposition it emits toxic fumes of NO_x.

MAN000   CAS: 6915-15-7   HR: 3
MALIC ACID
mf: $C_4H_6O_5$ mw: 134.10
PROP: White or colorless crystals; acid taste. Exhibits isomeric forms (dl, l, and d). D (dl): 1.601, d (d or l): 1.595 @ 20°/40, mp (dl): 128°, mp (d or l): 100°, bp (dl): 150°, bp (d or l): 140° (decomp). Very sol in water and alc; sltly sol in ether.
SYNS: BUTANEDIOIC ACID, HYDROXY-(9CI) □ DEOXYTETRARIC ACID □ HYDROXYBUTANEDIOIC ACID □ HYDROXYSUCCINIC ACID □ α-HYDROXY-SUCCINIC ACID □ KYSELINA HYDROXYBUTANDIOVA (CZECH) □ KYSELINA JABLECNA (CZECH) □ POMALUS ACID □ SUCCINIC ACID, HYDROXY-
CONSENSUS REPORTS: Reported in EPA TSCA Inventory.
SAFETY PROFILE: A poison by intraperitoneal route. Moderately toxic by ingestion. A skin and severe eye irritant. When heated to decomposition it emits acrid smoke and irritating fumes.

MAO250   CAS: 109-77-3   HR: 3
MALONONITRILE
DOT: UN 2647
mf: $C_3H_2N_2$ mw: 66.07
PROP: White powder or crystals. D: 1.049 @ 34°/4°, mp: 30.5°, bp: 220°, flash p: 266°F (TOC). Sol in H_2O, EtOH, Et_2O, and C_6H_6.
SYNS: CYANOACETONITRILE □ DICYANOMETH-ANE □ DWUMETYLOSULFOTLENKU (POLISH) □ MALONIC DINITRILE □ METHYLENE CYANIDE □ NITRIL KYSELINY MALONOVE (CZECH) □ PROPANE-

M

DINITRILE □ RCRA WASTE NUMBER U149 □ USAF A-4600

CONSENSUS REPORTS: Cyanide and its compounds are on the Community Right-To-Know List. EPA Extremely Hazardous Substances List. Reported in EPA TSCA Inventory.

NIOSH REL: (Nitriles) TWA 8 mg/m^3

DOT CLASSIFICATION: 6.1; Label: Poison

SAFETY PROFILE: Poison by ingestion, skin contact, subcutaneous, intravenous, and intraperitoneal routes. A severe eye irritant. Combustible when exposed to heat or flame. Polymerizes violently when heated to 130°C or on contact with strong base. May spontaneously explode when stored at 70–80°C. To fight fire, use water, fog, spray, foam. When heated to decomposition it emits toxic fumes of NO_x and CN^-. See also NITRILES.

MAO300   CAS: 9050-36-6   HR: D
MALTODEXTRIN

mf: $(C_6H_{10}O_5)_n$

PROP: White powder or solution from partial hydrolysis of corn starch.

SAFETY PROFILE: When heated to decomposition it emits acrid smoke and irritating fumes.

MAO350   CAS: 118-71-8   HR: 2
MALTOL

mf: $C_6H_6O_3$ mw: 126.12

PROP: White crystalline powder, needles, or prisms from toluene or $CHCl_3$ with odor of caramel/butterscotch. Mp: 161–162°.1.3 Sol in water, alc, glycerin, and propylene glycol. Mod sol in water; sol in alc.

SYNS: CORPS PRALINE □ 3-HYDROXY-2-METHYL-4H-PYRAN-4-ONE □ 3-HYDROXY-2-METHYL-γ-PYRONE □ 3-HYDROXY-2-METHYL-4-PYRONE □ LARIXIC ACID □ LARIXINIC ACID □ 2-METHYL-3-HYDROXY-4-PYRONE □ 2-METHYL-3-OXY-γ-PYRONE □ 2-METHYL PYROMECONIC ACID □ PALATONE □ TALMON □ VETOL

CONSENSUS REPORTS: Reported in EPA TSCA Inventory.

SAFETY PROFILE: Moderately toxic by ingestion, intraperitoneal, and subcutaneous routes. A skin irritant. Human mutation data reported. When heated to decomposition it emits acrid smoke and irritating fumes.

MAO500   CAS: 69-79-4   HR: 2
MALTOSE

mf: $C_{12}H_{22}O_{11}$ mw: 342.31

PROP: Colorless needles or crystals. D: 1.540 @ 17°, mp: 102–103° (decomp). Very sol in water; very sltly sol in cold alc; insol in ether.

SYNS: 4-(α-d-GLUCOPYRANOSIDO)-α-GLUCOPYRAN-OSE □ 4-(α-d-GLUCOSIDO)-d-GLUCOSE □ MALTOBI-OSE □ d-MALTOSE □ MALT SUGAR □ α-MALT SUGAR

CONSENSUS REPORTS: Reported in EPA TSCA Inventory.

SAFETY PROFILE: Experimental teratogenic and reproductive effects. Questionable carcinogen with experimental tumorigenic data. When heated to decomposition it emits acrid smoke and irritating fumes.

MAO900   HR: D
MANDARIN OIL, COLDPRESSED

PROP: From expression of peel of *Citrus reticulata* Blanco var. *Mandarin*. Clear orange to brown-orange liquid; orange odor. D: 0.846. Sol in fixed oils, mineral oil; slt sol in propylene glycol; insol in glycerin.

SAFETY PROFILE: When heated to decomposition it emits acrid smoke and irritating fumes.

MAP000   CAS: 90-64-2   HR: 3
MANDELIC ACID

mf: $C_8H_8O_3$ mw: 152.16

PROP: Large, white crystals or powder; faint odor. Bp: decomp, d: 1.30, mp: 117–119°. Sol in water, alc, and ether. Darkens and decomp on prolonged exposure to light.

SYNS: ACIDO MANDELICO □ AMYGDALIC ACID □ AMYGDALINIC ACID □ GLYCOLIC ACID, PHENYL- □ α-HYDROXYPHENYLACETIC ACID □ α-HYDROXY-α-TOLUIC ACID □ KYSELINA 2-FENYL-2-HYDROXYETH-ANOVA □ KYSELINA MANDLOVA □ PARAMANDELIC ACID □ PHENYLGLYCOLIC ACID □ PHENYLHYDRO-XYACETIC ACID □ RACEMIC MANDELIC ACID □ α-TOLUIC ACID, α-HYDROXY- □ UROMALINE

CONSENSUS REPORTS: Reported in EPA TSCA Inventory.

SAFETY PROFILE: Poison by intramuscular route. Moderately toxic by ingestion. Continued absorption can cause kidney irritation. When heated to decomposition it emits acrid smoke and irritating fumes.

MAP250   CAS: 532-28-5   HR: 3
MANDELIC ACID NITRILE
mf: C_8H_7NO mw: 133.16
PROP: Needles or yellow, viscous liquid. Mp: 28.5–29.5°, bp: 170° decomp, d: 1.124.
SYNS: AMYGDALONITRILE □ BENZALDEHYDE CYANOHYDRIN □ BENZALDEHYDKYANHYDRIN (CZECH) □ HYDROXYPHENYLACETONITRILE □ NITRIL KYSELINY MANDLOVE (CZECH) □ PHENYL-GLYCOLONITRILE

CONSENSUS REPORTS: Cyanide and its compounds are on the Community Right-To-Know List. Reported in EPA TSCA Inventory.

SAFETY PROFILE: Poison by intravenous and subcutaneous routes. Mutation data reported. A severe eye irritant. When heated to decomposition it emits toxic fumes of NO_x and CN^-. See also NITRILES.

MAP750   CAS: 7439-96-5   HR: 3
MANGANESE
af: Mn   aw: 54.94
PROP: Reddish-gray or silvery, brittle, metallic element. Reacts with H_2O or steam to give H_2. Oxidizes superficially in air. Mp: 1244°, bp: 2060°, d: 7.20, vap press: 1 mm @ 1292°. IDLH 500 mg/m³ (as Mn).
SYNS: COLLOIDAL MANGANESE □ MANGACAT □ MANGAN (POLISH) □ MANGAN NITRIDOVANY (CZECH) □ TRONAMANG

CONSENSUS REPORTS: Manganese and its compounds are on the Community Right-To-Know List. Reported in EPA TSCA Inventory.

OSHA PEL: Fume: TWA 1 mg/m³; STEL 3 mg/m³; Compounds: CL 5 mg/m³
ACGIH TLV: Fume: 1 mg/m³; STEL 3 mg/m³; Dust and Compounds: TWA 5 mg/m³; (Proposed: TWA 0.2 mg/m³)

DFG MAK: 0.5 mg/m³
SAFETY PROFILE: Human systemic effects by inhalation: degenerative brain changes, change in motor activity, muscle weakness. A skin and eye irritant. Questionable carcinogen with experimental tumorigenic data. Flammable and moderately explosive in the form of dust or powder when exposed to flame. The dust may be pyrophoric in air and may explode when heated in carbon dioxide. Mixtures of aluminum dust and manganese dust may explode in air. Mixtures with ammonium nitrate may explode when heated. The powdered metal ignites on contact with fluorine, chlorine + heat, hydrogen peroxide, bromine pentafluoride, sulfur dioxide + heat. Violent reaction with NO_2 and oxidants. Incandescent reaction with phosphorus, nitryl fluoride, nitric acid. Will react with water or steam to produce hydrogen; can react with oxidizing materials. To fight fire, use special dry chemical. See also MANGANESE COMPOUNDS.

MAQ000   CAS: 638-38-0   HR: 2
MANGANESE ACETATE
mf: $C_4H_6O_4 \cdot Mn$ mw: 173.04
PROP: Pale-red crystals. Sol in H_2O, MeOH, EtOH, AcOH; insol in Me_2CO.
SYNS: ACETIC ACID MANGANESE(II) SALT (2:1) □ DIACETYLMANGANESE □ MANGANESE(2+) ACETATE □ MANGANESE(II) ACETATE □ MANGANESE DIACETATE □ MANGANOUS ACETATE □ OCTAN MANGANATY (CZECH)

CONSENSUS REPORTS: Manganese and its compounds are on the Community Right-To-Know List. Reported in EPA TSCA Inventory.

OSHA PEL: CL 5 mg(Mn)/m³
ACGIH TLV: TWA 5 mg(Mn)/m³
SAFETY PROFILE: Moderately toxic by ingestion. Mutation data reported. Used in food packaging. When heated to decomposition it emits acrid smoke and irritating fumes. See also MANGANESE COMPOUNDS.

M

MAQ500   CAS: 14024-58-9   HR: 2
MANGANESE ACETYLACETONATE
mf: $C_{10}H_{14}O_4Mn$ mw: 253.18
PROP: Colorless crystals. Sltly sol in H_2O,
EtOH, and Me_2CO.
SYN: MANGANOUS ACETYLACETONATE
CONSENSUS REPORTS: Manganese and
its compounds are on the Community
Right-To-Know List. Reported in EPA
TSCA Inventory.
OSHA PEL: CL 5 mg(Mn)/m³
ACGIH TLV: TWA 5 mg(Mn)/m³
SAFETY PROFILE: Questionable
carcinogen with experimental neoplastigenic
and tumorigenic data. When heated to
decomposition it emits acrid smoke and
irritating fumes. See also MANGANESE
COMPOUNDS.

MAR000   CAS: 7773-01-5   HR: 3
MANGANESE(II) CHLORIDE (1:2)
mf: Cl_2Mn mw: 125.84
PROP: Cubic, deliquescent, pink crystals.
Mp: 654°, bp: 1225°, d: 2.977 @ 25°. Sol in
water.
SYNS: MANGANESE DICHLORIDE □ MANGANOUS
CHLORIDE
CONSENSUS REPORTS: Manganese and
its compounds are on the Community
Right-To-Know List. Reported in EPA
TSCA Inventory. EPA Genetic Toxicology
Program.
OSHA PEL: CL 5 mg(Mn)/m³
ACGIH TLV: TWA 0.03 mg(Mn)/m³
SAFETY PROFILE: Poison by
intraperitoneal, subcutaneous, intramuscular,
intravenous, and parenteral routes.
Moderately toxic by ingestion. Experimental
teratogenic and reproductive effects.
Questionable carcinogen with experimental
carcinogenic data. Mutation data reported.
Explosive reaction when heated with zinc
foil. Reacts violently with potassium or
sodium. When heated to decomposition it
emits toxic fumes of Cl⁻. See also
MANGANESE COMPOUNDS and
CHLORIDES.

MAR260   CAS: 10024-66-5   HR: D
MANGANESE CITRATE
mf: $Mn_3(C_6H_5O_7)_2$ mw: 543.02
PROP: Pale orange or pinkish-white
powder.
SAFETY PROFILE: When heated to
decomposition it emits acrid smoke and
irritating fumes.

MAR500   HR: 3
MANGANESE COMPOUNDS
CONSENSUS REPORTS: Manganese and
its compounds are on the Community
Right-To-Know List.
SAFETY PROFILE: Some are experimental
tumorigens. Can cause central nervous
system and pulmonary system damage by
inhalation of fumes and dust. Very few
poisonings have occurred from ingestion.
Chronic manganese poisoning is a clearly
characterized disease that results from
inhalation of fumes or dusts of manganese.
Exposure to heavy concentrations of dusts
or fumes for as little as three months may
produce the condition, but usually cases
develop after 1–3 years of exposure. The
central nervous system is the chief site of
damage. If cases are removed from exposure
shortly after appearance of symptoms, some
improvement in the patient's condition
frequently occurs, though there may be
some residual disturbances in gait and
speech. When well established, however, the
disease results in permanent disability.
Exposure to dusts and fumes can possibly
increase the incidence of upper respiratory
infections and pneumonia. Chronic
manganese poisoning usually begins with
complaints of languor and sleepiness. This is
followed by weakness in the legs and the
development of stolid, mask-like faces. The
patient speaks with a slow monotonous
voice. Then muscular twitching appears,
varying from a fine tremor of the hands to
coarse, rhythmical movements of the arms,
legs, and trunk. Nocturnal cramps of the
legs appear about the same time. There is a
slight increase in tendon reflexes, ankle and

patellar clonus, and a typical Parkinsonian slapping gait. The handwriting may be quite minute. The symptoms may simulate progressive bulbar paralysis, postencephalitic Parkinsonism, multiple sclerosis, amyotrophic lateral sclerosis, and progressive lenticular degeneration (Wilson's Disease). Often a history of exposure is the only aid in establishing the diagnosis. Manganese compounds are common air contaminants.

MAS000   CAS: 1313-13-9   HR: 3
MANGANESE DIOXIDE
mf: MnO_2 mw: 86.94

PROP: Tetragonal crystals. Inert to most acids except when heated whence it functions as an oxidizing agent. With concentrated HCl, Cl_2 is evolved. With H_2SO_4 at 1° O_2 is evolved and an Mn(III) acid sulfate is formed. Mp: loses O_2 @ 535°, d: 5.0. Insol in water, nitric or cold sulfuric acid.

SYNS: BLACK MANGANESE OXIDE □ BOG MANG-ANESE □ BRAUNSTEIN (GERMAN) □ BRUINSTEEN (DUTCH) □ CEMENT BLACK □ C.I. 77728 □ C.I. PIGMENT BLACK 14 □ C.I. PIGMENT BROWN 8 □ MANGAANBIOXYDE (DUTCH) □ MANGAANDIOXYDE (DUTCH) □ MANGANDIOXID (GERMAN) □ MANGAN-ESE BINOXIDE □ MANGANESE (BIOSSIDO di) (ITALIAN) □ MANGANESE (BIOXYDE de) (FRENCH) □ MANGANESE BLACK □ MANGANESE (DIOSSIDO di) (ITALIAN) □ MANGANESE (DIOXYDE de) (FRENCH) □ MANGANESE OXIDE □ MANGANESE(IV) OXIDE □ MANGANESE PEROXIDE □ MANGANESE SUPEROXIDE □ PYROLUSITE BROWN

CONSENSUS REPORTS: Manganese and its compounds are on the Community Right-To-Know List. Reported in EPA TSCA Inventory.

OSHA PEL: CL 5 mg(Mn)/m³

ACGIH TLV: TWA 0.03 mg(Mn)/m³

SAFETY PROFILE: Poison by intravenous and intratracheal routes. Moderately toxic by subcutaneous route. Experimental reproductive effects. A powerful oxidizer. Flammable by chemical reaction. It must not be heated or rubbed in contact with easily oxidizable matter. Violent thermite reaction when heated with aluminum. Potentially

explosive reaction with hydrogen peroxide, peroxomonosulfuric acid, chlorates + heat, anilinium perchlorate. Ignition on contact with hydrogen sulfide. Violent reaction with oxidizers, potassium azide (when warmed), diboron tetrafluoride, Incandescent reaction with calcium hydride, chlorine trifluoride, rubidium acetylide (at 350°C). Vigorous reaction with hydroxylaminium chloride. Incompatible with H_2O_2, H_2SO_5, Na_2O_2. Keep away from heat and flammable materials. See also MANGANESE COMPOUNDS.

MAS500   CAS: 12427-38-2   HR: 2
MANGANESE(II)
ETHYLENEBIS(DITHIOCARBAMATE)
DOT: UN 2210/UN 2968

mf: $C_4H_6N_2S_4•Mn$ mw: 265.30

PROP: Yellow powder or crystals. Slowly decomp on prolonged exp to air, rapid decomp in acid. May be stabilized by presence of formaldehyde or various other compds. Mp: 131° (decomp). Sol in water and common org solvs.

SYNS: AAMANGAN □ AKZO CHEMIE MANEB □ BASF-MANEB SPRITZPULVER □ CARBAMIC ACID, ETHYLENEBIS(DITHIO)-, MANGANESE SALT □ CHEM NEB □ CHLOROBLE M □ CR 3029 □ CURZATE M □ DELSENE M □ DITHANE M 22 □ DITHANE M 22 SPECIAL □ ENT 14,875 □ 1,2-ETHANEDIYLBIS(CARB-AMODITHIOATO)(2−)-MANGANESE □ 1,2-ETHANE-DIYLBISCARBAMODITHIOIC ACID MANGANESE COMPLEX □ 1,2-ETHANEDIYLBISCARBAMODITHIOIC ACID, MANGANESE(2+) SALT (1:1) □ 1,2-ETHANEDIYL-BISMANEB, MANGANESE(2+) SALT (1:1) □ ETHYLENE-BISDITHIOCARBAMATE MANGANESE □ N,N'-ETHYLENE BIS(DITHIOCARBAMATE MANGANEUX) (FRENCH) □ ETHYLENEBIS(DITHIOCARBAMATO) MANGANESE □ ETHYLENEBIS(DITHIOCARBAMIC ACID) MANGANESE SALT □ ETHYLENEBIS(DITHIO-CARBAMIC ACID) MANGANOUS SALT □ 1,2-ETHYL-ENEDIYLBIS(CARBAMODITHIOATO)MANGANESE □ N,N'-ETILEN-BIS(DITIOCARBAMMATO) di MANGAN-ESE (ITALIAN) □ F 10 (pesticide) □ GRIFFIN MANEX □ KYPMAN 80 □ LONOCOL M □ MANAM □ MANEB □ MANEB 80 □ MANEB (UN 2210) (DOT) □ MANEBA □ MANEBE (FRENCH) □ MANEBE 80 □ MANEBGAN □ MANEB PREPARATIONS with not <60% maneb (UN 2210) (DOT) □ MANEB PREPARATIONS, stabilized against self-heating (UN 2968) (DOT) □ MANEB, stabilized (UN 2968) (DOT) □ MANEB ZL4 □ MANESAN □ MANEX □ MANGAAN(II)-(N,N'-ETHYLEEN-BIS(DITHIOCARBAMA-

AT)) (DUTCH) □ MANGAN(II)-(N,N'-AETHYLEN-BIS(DITHIOCARBAMAT)) (GERMAN) □ MANGANESE ETHYLENE-1,2-BISDITHIOCARBAMATE □ MANGANESE(II) ETHYLENE DI(DITHIOCARBAMATE) □ MANZATE □ MANZATE 200 □ MANZATE D □ MANZATE MANEB FUNGICIDE □ MANZEB □ MANZIN □ M-DIPHAR □ MEB □ MnEBD □ NEREB □ NESPOR □ PLANTIFOG 160M □ POLYRAM M □ REMASAN CHLOROBLE M □ RHODIANEBE □ SOPRANEBE □ SUPERMAN MANEB F □ SUP'R FLO □ TERSAN-LSR □ TRIMANGOL □ TRIMANGOL 80 □ TUBOTHANE □ UNICROP MANEB □ VANCIDE MANEB 80

CONSENSUS REPORTS: IARC Cancer Review: Group 3 IMEMDT 7,56,87; Animal Inadequate Evidence IMEMDT 12,137,76. Community Right-To-Know List. EPA Genetic Toxicology Program.

OSHA PEL: CL 5 mg(Mn)/m³

ACGIH TLV: TWA 5 mg(Mn)/m³

DOT CLASSIFICATION: 4.2; Label: Spontaneously Combustible, Dangerous When Wet (UN 2210); DOT Class: 4.3; Label: Dangerous When Wet (UN 2968)

SAFETY PROFILE: Moderately toxic by ingestion. Experimental teratogenic and reproductive effects. Questionable carcinogen with experimental carcinogenic and tumorigenic data. Mutation data reported. A fungicide. May ignite spontaneously in air. When heated to decomposition it emits highly toxic fumes of NO_x and SO_x. See also MANGANESE COMPOUNDS and CARBAMATES.

MAS750   CAS: 7782-64-1   HR: 3
MANGANESE(II) FLUORIDE
mf: F_2Mn mw: 92.94

PROP: Pink, tetragonal crystals or reddish powder. Mp: 856°, d: 3.98. Insol in alc; sol in dilute hydrofluoric acid, concentrated hydrochloric or nitric acid. Sltly sol in water.

SYNS: MANGANESE FLUORIDE □ MANGANESE FLUORURE (FRENCH)

CONSENSUS REPORTS: Manganese and its compounds are on the Community Right-To-Know List. Reported in EPA TSCA Inventory.

OSHA PEL: TWA 2.5 mg(F)/m³; Cl 5 mg(Mn)/m³

ACGIH TLV: TWA 2.5 mg(F)/m³; BEI: 3 mg/g creatinine of fluorides in urine prior to shift; 10 mg/g creatinine of fluorides in urine at end of shift; TWA 0.03 mg(Mn)/m³

NIOSH REL: TWA 2.5 mg(F)/m³

SAFETY PROFILE: Poison by subcutaneous route. When heated to decomposition it emits toxic fumes of F^-. See also MANGANESE COMPOUNDS and FLUORIDES.

MAS810   HR: D
MANGANESE GLYCEROPHOSPHATE
mf: $C_3H_7MnO_6P \cdot xH_2O$ mw: 225.00

PROP: White or pink powder; odorless and tasteless. Sol in citric acid solution. Sltly sol in water; insol in alc.

SAFETY PROFILE: When heated to decomposition emits toxic fumes of manganese.

MAS818   HR: D
MANGANESE LINOLEATE
SAFETY PROFILE: When heated to decomposition it emits acrid smoke and irritating fumes.

MAS900   CAS: 10377-66-9   HR: 2
MANGANESE(II) NITRATE
DOT: UN 2724

mf: $N_2O_6 \cdot Mn$ mw: 178.96

SYNS: MANGANESE DINITRATE □ MANGANESE NITRATE □ MANGANESE NITRATE (DOT) □ MANGANESE(2+) NITRATE □ MANGANESE (II) NITRATE, ANHYDROUS □ MANGANOUS DINITRATE □ MANGANOUS NITRATE □ NITRIC ACID, MANGANESE(2+) SALT

OSHA PEL: CL 5 mg(Mn)/m³

ACGIH TLV: TWA 0.03 mg(Mn)/m³

DOT CLASSIFICATION: 5.1; Label: Oxidizer

CONSENSUS REPORTS: Reported in EPA TSCA Inventory.

SAFETY PROFILE: Mutation data reported. When heated to decomposition it emits toxic vapors of manganese.

MAT250   CAS: 1344-43-0   HR: 2
MANGANESE(II) OXIDE
mf: MnO   mw: 70.94

PROP: Grass-green powder. D: 5.45, mp: 1850°, converted to Mn_3O_4 if heated in air. Thermal decomp of $MnCO_3$ at 420–422° produces pyrophoric MnO which rapidly becomes brown. Basic oxide: insol in H_2O. Relatively easily reduced to Mn by metals but not by H_2 or CO. Sol in acids. Insol in water.

SYNS: CASSEL GREEN □ C.I. 77726 □ MANGANESE GREEN □ MANGANESE MONOXIDE □ MANGANOUS OXIDE □ NU-MANESE □ ROSENSTHIEL

CONSENSUS REPORTS: Manganese and its compounds are on the Community Right-To-Know List. Reported in EPA TSCA Inventory.

OSHA PEL: CL 5 mg(Mn)/m³

ACGIH TLV: TWA 0.03 mg(Mn)/m³

SAFETY PROFILE: Moderately toxic by intratracheal and subcutaneous routes. Violent reaction with hydrogen peroxide, $Ca(OCl)_2$, F_2, H_2O_2. See also MANGANESE COMPOUNDS.

MAT500   CAS: 1317-34-6   HR: 2
MANGANESE(III) OXIDE

mf: Mn_2O_3 mw: 157.88

PROP: Fine, black powder. D: 4.50, mp: 871–887° (decomp). Insol in water; sol in HCl, evolving chlorine.

SYNS: CASSEL BROWN □ C.I. 77727 □ C.I. NATURAL BROWN 8 □ COLOGNE EARTH □ COLOGNE UMBER □ CULLEN EARTH □ DIMANGANESE TRIOXIDE □ MANGANESE MANGANATE □ MANGANESE SESQUIOXIDE □ MANGANESE TRIOXIDE □ MANGANIC OXIDE □ RUBENS BROWN □ SOLUBLE VAN DYKE BROWN □ VAN DYKE BROWN □ WALNUT STAIN

CONSENSUS REPORTS: Manganese and its compounds are on the Community Right-To-Know List. Reported in EPA TSCA Inventory.

OSHA PEL: CL 5 mg(Mn)/m³

ACGIH TLV: TWA 0.03 mg(Mn)/m³

SAFETY PROFILE: Moderately toxic by subcutaneous and intratracheal routes. See also MANGANESE COMPOUNDS.

MAU250   CAS: 7785-87-7   HR: 3
MANGANESE(II) SULFATE (1:1)

mf: $O_4S•Mn$ mw: 151.00

PROP: Pink granular powder or pale pink crystals; odorless. Thermally very stable. Mp: 700°, bp: decomp @ 850°, d: 3.25. Very sol in water, more so in boiling water; sltly sol in MeOH and EtOH; insol in alc.

SYNS: MANGANOUS SULFATE □ MAN-GRO □ NCI-C61143 □ SORBA-SPRAY Mn □ SULFURIC ACID, MANGANESE(2+) SALT

CONSENSUS REPORTS: Manganese and its compounds are on the Community Right-To-Know List. Reported in EPA TSCA Inventory. EPA Genetic Toxicology Program.

OSHA PEL: CL 5 mg(Mn)/m³

ACGIH TLV: TWA 0.03 mg(Mn)/m³

SAFETY PROFILE: Poison by intraperitoneal route. Questionable carcinogen with experimental neoplastigenic data. An experimental teratogen. Experimental reproductive effects. Mutation data reported. When heated to decomposition it emits toxic fumes of SO_2, SO_3, and Mn oxides. See also MANGANESE COMPOUNDS and SULFATES.

MAV100   HR: D
MANGANESE TALLATE

SAFETY PROFILE: When heated to decomposition it emits acrid smoke and irritating fumes.

MAV550   CAS: 1317-35-7   HR: 2
MANGANESE TETROXIDE

mf: Mn_3O_4 mw: 228.82

PROP: Brownish-black powder or black crystals or dark red powder when finely divided. Reacts with F_2 to give MnF_3 and some MnF_2. D: 4.7. Insol in water; sol in HCl, liberating chlorine.

SYNS: MANGANESE OXIDE □ MANGANOMANGANIC OXIDE □ TRIMANGANESE TETRAOXIDE □ TRIMANGANESE TETROXIDE

CONSENSUS REPORTS: Manganese and its compounds are on the Community Right-To-Know List. Reported in EPA TSCA Inventory.

OSHA PEL: TWA 1 mg(Mn)/m³

ACGIH TLV: TWA 1 mg(Mn)/m³

M

DFG MAK: 1 mg/m³
SAFETY PROFILE: Experimental reproductive effects. Reacts violently @ <100°. See also MANGANESE COMPOUNDS.

MAV750　CAS: 12108-13-3　HR: 3
MANGANESE TRICARBONYL METHYLCYCLOPENTADIENYL
mf: $C_9H_7MnO_3$　　mw: 218.10
PROP: Yellow liquid. D: 1.388 @ 20°/4°, mp: 1.5°, bp: 233°. Almost insol in H_2O; misc in nonpolar solvs.
SYNS: AK-33X □ ANTIKNOCK-33 □ CI-2 □ COMBUSTION IMPROVER-2 □ MANGANESE, (METHYLCYCLOPENTADIENYL)TRICARBONYL- □ METHYLCYCLOPENTADIENYL MANGANESE TRICARBONYL □ METHYLCYCLOPENTADIENYL MANGANESE TRICARBONYL (OSHA) □ 2-METHYLCYCLOPENTADIENYL MANGANESETRICARBONYL □ 2-METHYLCYCLOPENTADIENYL MANGANESE TRICARBONYL (ACGIH) □ METHYLCYKLOPENTADIENTRIKARBONYLMANGANIUM □ MMT □ TRICARBONYL(METHYLCYCLOPENTADIENYL)MANGANESE
CONSENSUS REPORTS: EPA Extremely Hazardous Substances List. Manganese and its compounds are on the Community Right-To-Know List. Reported in EPA TSCA Inventory.
OSHA PEL: TWA 0.2 mg(Mn)/m³ (skin)
ACGIH TLV: TWA 5 mg(Mn)/m³
SAFETY PROFILE: Poison by ingestion, inhalation, skin contact, intravenous, and intraperitoneal routes. A skin irritant. When heated to decomposition it emits toxic fumes of CO. See also MANGANESE COMPOUNDS and CARBONYLS.

MAW100　CAS: 87-78-5　HR: 3
MANNITOL
mf: $C_6H_{14}O_6$　　mw: 182.20
SAFETY PROFILE: A poison by intravenous route. Human systemic effects. When heated to decomposition it emits acrid smoke and irritating vapors.

MAW250　CAS: 15825-70-4　HR: 3
MANNITOL HEXANITRATE
DOT: UN 0133
mf: $C_6H_8N_6O_{18}$　　mw: 452.17

PROP: Colorless crystals or needles from EtOH. Bp: explodes @ 120°, d: 1.603 @ 0°. Mp: 106–108°. Long needles in regular clusters from alc. Sol in alc and ether; insol in water.
SYNS: DILANGIL □ HEXANITROL □ HYPERTENAIN □ MANEXIN □ MANHEXIN □ MANICOLE □ MANITE □ MANNEX □ MANNITOL HEXANITRATE (dry) (DOT) □ MANNITOL HEXANITRATE, wetted with not <40% water, by weight or mixture (NA 0133) (DOT) □ d-MANNITOL HEXANITRATE □ MANNITRIN □ MAXITATE □ MEDEMANOL □ NITRANITOL □ NITRO MANNITE □ NITROMANNITE (dry) (DOT) □ NITROMANNITE, wetted with not <40% water, by weight or mixture (NA 0133) (DOT) □ NITROMANNITOL □ SDM No. 5
CONSENSUS REPORTS: Reported in EPA TSCA Inventory.
DOT CLASSIFICATION: Forbidden (dry); DOT Class: EXPLOSIVE 1.1A; Label: EXPLOSIVE 1.1A
SAFETY PROFILE: Moderately toxic by ingestion and inhalation causing a fall in blood pressure that may result in weakness, headache, and dizziness. Chronic exposure may produce methemoglobinemia with cyanosis. A powerful explosive sensitive to shock or heat. Upon decomposition it emits toxic fumes of NO_x. See also NITRATES and EXPLOSIVES, HIGH.

MAW750　CAS: 551-74-6　HR: 3
MANNOMUSTINE DIHYDROCHLORIDE
mf: $C_{10}H_{23}Cl_2N_2O_4$•2ClH　　mw: 378.13
PROP: Crystals from 80% ethanol. Mp: 239–241° (decomp). Sol in water; sltly sol in ethanol.
SYNS: 1,6-BIS-(CHLOROETHYLAMINO)-1,6-DESOXY-d-MANNITOLDIHYDROCHLORIDE □ 1,6-BIS-(CHLOROETHYLAMINO)-1,6-DIDEOXY-d-MANNITOLDIHYDROCHLORIDE □ 1,6-DIDEOXY-1,6-DI(2-CHLOROETHYLAMINO)-d-MANNITOLDIHYDROCHLORIDE □ MANNITOL MUSTARD DIHYDROCHLORIDE □ NSC-9698
CONSENSUS REPORTS: IARC Cancer Review: Group 3 IMEMDT 7,56,87; Animal Sufficient Evidence IMEMDT 9,157,75.
SAFETY PROFILE: Poison by intravenous, subcutaneous, and parenteral routes. Experimental reproductive effects. Questionable carcinogen with experimental carcinogenic and neoplastigenic data.

Human mutation data reported. A drug used for the treatment of malignant neoplasms. When heated to decomposition it emits very toxic fumes of HCl and NO$_x$. See also MANNOMUSTINE.

**MBU500   CAS: 8015-01-8   HR: 1
MARJORAM OIL, SPANISH**
PROP: Main constituent is cineole. From steam distillation of the flowering plant material from the shrub *Thymus mastichina L.* (Fam. *Labiatae*) (FCTXAV 14,443,76). Faintly yellow liquid. D: 0.904–0.920, refr index: 1.463 @ 20°. Sol in fixed oils. Insol in glycerin, propylene glycol, mineral oil.
SYNS: OIL OF MARJORAM, SPANISH □ SPANISH MARJORAM OIL
CONSENSUS REPORTS: Reported in EPA TSCA Inventory.
SAFETY PROFILE: Low toxicity by ingestion and skin contact. A skin irritant. When heated to decomposition it emits acrid smoke and irritating fumes.

**MBX250   CAS: 1104-22-9   HR: 2
MECLIZINE DIHYDROCHLORIDE**
mf: C$_{25}$H$_{27}$ClN$_2$•2ClH   mw: 463.91
PROP: A solid. Mp: 224° (decomp). Very sol in chloroform and pyridine; sltly sol in dilute acids, alc; insol in H$_2$O, ether.
SYNS: ANCOLAN DIHYDROCHLORIDE □ 1-p-CHLORBENZHYDRYL-m-METHYLBENZYLPIPERAZINE DIHYDROCHLORIDE
CONSENSUS REPORTS: Reported in EPA TSCA Inventory.
SAFETY PROFILE: Moderately toxic by ingestion and intraperitoneal routes. An experimental teratogen. Experimental reproductive effects. When heated to decomposition it emits very toxic fumes of Cl⁻ and NO$_x$.

**MCA000   CAS: 71-58-9   HR: 3
MEDROXYPROGESTERONE ACETATE**
mf: C$_{24}$H$_{34}$O$_4$ mw: 386.58
PROP: Crystals from Me$_2$CO (aq). White to off-white, odorless, crystalline powder. Melting range 207–209°. Insol in water; freely sol in chloroform; sparingly sol in alc.

SYNS: 17-α-ACETOXY-6-α-METHYLPREGN-4-ENE-3,20-DIONE □ 17-ACETOXY-6-α-METHYLPROGESTER-ONE □ (6-α)-17-(ACETYLOXY)-6-METHYLPREG-4-ENE-3,20-DIONE □ DEPO-PROVERA □ FARLUTIN □ 17-HYDROXY-6-α-METHYLPREGN-4-ENE-3,20-DIONE ACETATE □ 17-α-HYDROXY-6-α-METHYLPREGN-4-ENE-3,20-DIONE ACETATE □ 17-α-HYDROXY-6-α-METHYLPROGESTERONE ACETATE □ 6-α-METHYL-17-α-ACETOXYPREGN-4-ENE-3,20-DIONE □ 6-α-METHYL-17-α-ACETOXYPROGESTERONE □ 6-α-METHYL-17-α-HYDROXYPROGESTERONE ACETATE □ 6-α-METHYL-4-PREGNENE-3,20-DION-17-α-OL ACETATE □ METIPREGNONE □ NOGEST □ ORAGEST □ PERLUTEX □ REPROMIX
CONSENSUS REPORTS: IARC Cancer Review: Group 2B IMEMDT 7,289,87; Animal Limited Evidence IMEMDT 21,417,79; IMEMDT 6,157,74; Human Inadequate Evidence IMEMDT 21,417,79. Reported in EPA TSCA Inventory. EPA Genetic Toxicology Program.
SAFETY PROFILE: Suspected carcinogen with experimental carcinogenic, neoplast-igenic, tumorigenic, and teratogenic data. Human systemic effects by intravenous route: increased intraocular pressure. Human teratogenic effects by an unspecified route: developmental abnormalities of the urogenital system. Human reproductive effects by multiple routes: spermatogenesis, menstrual cycle changes or disorders, postpartum effects, female fertility effects, abortion, newborn behavioral effects. Human mutation data reported. Experimental reproductive effects. A drug for the treatment of secondary amenorrhoea and dysfunctional uterine bleeding. When heated to decomposition it emits acrid smoke and irritating fumes.

**MCB000   CAS: 108-78-1   HR: 2
MELAMINE**
mf: C$_3$H$_6$N$_6$ mw: 126.15
PROP: Monoclinic, colorless prisms or crystals. Mp: 347°, bp: sublimes, d: 1.573 @ 250°, vap press: 50 mm @ 315°, vap d: 4.34. Sltly sol in water. Very sltly sol in hot alc; insol in ether.
SYNS: AERO □ AMMELIDE □ CYANURAMIDE □ CYANURIC TRIAMIDE □ CYANUROTRIAMIDE □

CYANUROTRIAMINE □ CYMEL □ HICOPHOR PR □ ISOMELAMINE □ NCI-C50715 □ PLURAGARD □ PLURAGARD C 133 □ TEOHARN □ THEOHARN □ 2,4,6-TRIAMINO-s-TRIAZINE □ 2,4,6-TRIAMINO-1,3,5-TRIAZINE □ 1,3,5-TRIAZINE-2,4,6-TRIAMINE □ s-TRIAZINE, 2,4,6-TRIAMINO- □ VIRSET 656-4

CONSENSUS REPORTS: IARC Cancer Review: Group 3 IMEMDT 7,56,87; Animal Inadequate Evidence IMEMDT 39,333,86. NTP Carcinogenesis Bioassay (feed); No Evidence: mouse NTPTR* NTP-TR-245,83 (feed); Clear Evidence: rat NTPTR* NTP-TR-245,83. Community Right-To-Know List. Reported in EPA TSCA Inventory.

SAFETY PROFILE: Moderately toxic by ingestion and intraperitoneal routes. An eye, skin, and mucous membrane irritant. Causes dermatitis in humans. Questionable carcinogen with experimental carcinogenic and tumorigenic data. Experimental reproductive effects. Mutation data reported. When heated to decomposition it emits toxic fumes of NO_x and CN^-.

MCB380   CAS: 2919-66-6   HR: D
MELENGESTROL ACETATE
mf: $C_{25}H_{32}O_4$ mw: 396.57
PROP: A solid. Mp: 224–226°.
SYNS: 17-(ACETYLOXY)-6-METHYL-16-METHYLENE-PREGNA-4,6-DIENE-3,20-DIONE (9CI) □ 6-DEHYDRO-16-METHYLENE-6-METHYL-17-ACETOXYPROGESTER-ONE □ 17-HYDROXY-6-METHYL-16-METHYLENE-PREGNA-4,6-DIENE-3,20-DIONE, ACETATE □ MGA □ MGA 100 (STEROID)
SAFETY PROFILE: An experimental teratogen. Experimental reproductive effects. When heated to decomposition it yields acrid smoke and irritating fumes.

MCB500   CAS: 3771-19-5   HR: 3
MELIPAN
mf: $C_{20}H_{22}O_3$ mw: 310.42
SYNS: 2-METHYL-2-(4-(1,2,3,4-TETRAHYDRO-1-NAPHTHALENYL)PHENOXY)PROPANOIC ACID □ 2-METHYL-2-(4-(1,2,3,4-TETRAHYDRO-1-NAPHTHYL)-PHENOXY)PROPANOIC ACID □ α-METHYL-α-(p-1,2,3,4-TETRAHYDRONAPHTH-1-YLPHENOXY)PROPIONIC ACID □ 2-METHYL-2-(p-(1,2,3,4-TETRAHYDRO-1-NAPHTHYL)PHENOXY)PROPIONIC ACID □ NAFENO-IC ACID □ NAFENOPIN □ SU-13437 □ TPIA

CONSENSUS REPORTS: IARC Cancer Review: Group 2B IMEMDT 7,56,87; Human Limited Evidence IMEMDT 24,125,80; Animal Sufficient Evidence IMEMDT 24,125,80.

SAFETY PROFILE: Confirmed carcinogen with experimental carcinogenic and tumorigenic data. Moderately toxic by ingestion. Mutation data reported. A drug for the treatment of hypercholesterolemia or hypertriglyceridemia. When heated to decomposition it emits acrid smoke and irritating fumes.

MCB625   HR: D
MENTHA ARVENSIS, OIL
PROP: From *Mentha arvensis var. piperascens Holmes (forma piperascens Malinvaud)* (Fam. *Cabiatae*) (CCPTAY 24,559,81). Colorless to yellow liquid, minty odor. D: 0.888–0.908, refr index: 1.458 @ 20°. Sol in fixed oils, mineral oil, propylene glycol; insol in glycerin.
SYNS: CORNMINT OIL, PARTIALLY DEMENT-HOLIZED □ MENTHA ARVENSIS OIL, PARTIALLY DEMENTHOLIZED (FCC)
SAFETY PROFILE: An experimental teratogen. Experimental reproductive effects. When heated to decomposition it emits acrid smoke and irritating fumes.

MCB750   CAS: 99-85-4   HR: 2
p-MENTHA-1,4-DIENE
mf: $C_{10}H_{16}$ mw: 136.26
PROP: Colorless liquid or oil; citrus odor. D: 0.841, refr index: 1.473–1.477, bp: 183°. Sol in alc, fixed oils; insol in water.
SYNS: FEMA No. 3559 □ 1-METHYL-4-ISOPROPYL-CYCLOHEXADIENE-1,4 □ γ-TERPINENE (FCC)
CONSENSUS REPORTS: Reported in EPA TSCA Inventory.
SAFETY PROFILE: Moderately toxic by ingestion. A skin irritant. When heated to decomposition it emits acrid smoke and irritating fumes.

MCC000   CAS: 99-83-2   HR: 3
p-MENTHA-1,5-DIENE
mf: $C_{10}H_{16}$ mw: 136.26

PROP: Colorless to sltly yellow liquid; mint odor. D: 0.835–0.865, refr index: 1.471–1.477, flash p: 120°F. Sol in alc; insol in water.
SYNS: α-FELLANDRENE □ FEMA No. 2856 □ 4-ISOPROPYL-1-METHYL-1,5-CYCLOHEXADIENE □ 5-ISOPROPYL-2-METHYL-1,3-CYCLOHEXADIENE □ 2-METHYL-5-ISOPROPYL-1,3-CYCLOHEXADIENE □ α-PHELLANDRENE (FCC)
CONSENSUS REPORTS: Reported in EPA TSCA Inventory.
SAFETY PROFILE: Mildly toxic by ingestion. A severe human skin irritant. Incompatible with air. Flammable liquid when exposed to heat, sparks, or flame. When heated to decomposition it emits acrid smoke and irritating fumes.

MCC250   CAS: 138-86-3   HR: 3
p-MENTHA-1,8-DIENE
DOT: UN 2052
mf: $C_{10}H_{16}$ mw: 136.26
PROP: Liquid. D: 0.842 @ 20°/4°, mp: −96.9°, bp: 177°. Insol in water; misc in alc and ether.
SYNS: ACINTENE DP □ ACINTENE DP DIPENTENE □ CAJEPUTENE □ CINENE □ DIPANOL □ DIPENTENE □ INACTIVE LIMONENE □ KAUTSCHIN □ LIMONENE □ dl-LIMONENE □ 1,8(9)-p-MENTHADIENE □ 1-METHYL-4-ISOPROPENYL-1-CYCLOHEXENE □ NESOL □ Δ-1,8-TERPODIENE □ UNITENE
CONSENSUS REPORTS: Reported in EPA TSCA Inventory.
DOT CLASSIFICATION: 3; Label: Flammable Liquid
SAFETY PROFILE: A skin irritant. Flammable when exposed to heat or flame; can react vigorously with oxidizing materials. When heated to decomposition it emits acrid smoke and irritating fumes.

MCC500   CAS: 5989-54-8   HR: 1
(S)-(−)-p-MENTHA-1,8-DIENE
mf: $C_{10}H_{16}$ mw: 136.26
PROP: Colorless liquid or oil; light odor. D: 0.837–0.841, refr index: 1.469–1.473, bp: 177.6–177.8°. Misc in alc, fixed oils; insol in water.
SYNS: 1-LIMONENE □ (−)-LIMONENE (FCC) □ 1-METHYL-4-(1-METHYLETHENYL)-(S)-CYCLOHEXENE

CONSENSUS REPORTS: Reported in EPA TSCA Inventory.
SAFETY PROFILE: A skin irritant. When heated to decomposition it emits acrid smoke and irritating fumes.

MCE250   CAS: 1074-95-9   HR: 2
p-MENTHAN-3-ONE racemic
mf: $C_{10}H_{18}O$ mw: 154.28
PROP: Several stereoisomers found in nature; 1-menthone found in essential oils of Russian and American peppermint, geranium, *Andropogon fragrans, Mentha timija, Mentha arvensis,* and others; d-menthone found in essential oils of *Barosma pulchellum, Nepeta japonica maxim,* and others; d-isomenthone isolated from *Micromeriabissinica benth., Pelargonium tometosum jacquin,* and others; 1-isomenthone identified in *Reunion geranium, Pelargonium capitatum,* and others (FCTXAV 14,443,76). Flash p: 156°F.
SYNS: FEMA No. 2667 □ 2-ISOPROPYL-5-METHYL-CYCLOHEXAN-1-ONE, racemic □ MENTHONE, racemic
CONSENSUS REPORTS: Reported in EPA TSCA Inventory.
SAFETY PROFILE: Moderately toxic by ingestion. A skin irritant. Combustible liquid. When heated to decomposition it emits acrid smoke and irritating fumes.

MCE750   CAS: 7786-67-6   HR: 2
p-MENTH-8-EN-3-OL
mf: $C_{10}H_{18}O$ mw: 154.28
PROP: Colorless liquid; mint odor. D: 0.904–0.913, refr index: 1.470–1.475. Misc in alc, ether, fixed oils; sltly sol in water.
SYNS: FEMA No. 2962 □ ISOPULEGOL (FCC) □ 8(9)-p-MENTHEN-3-OL □ 1-METHYL-4-ISOPROPENYL-CYCLOHEXAN-3-OL
CONSENSUS REPORTS: Reported in EPA TSCA Inventory.
SAFETY PROFILE: Moderately toxic by ingestion and skin contact. When heated to decomposition it emits acrid smoke and irritating fumes.

MCF525   CAS: 31375-17-4   HR: 3
MENTHENYL KETONE
mf: $C_{13}H_{22}O$ mw: 194.35

M

SYNS: 1-(p-MENTHEN-6-YL)-1-PROPANONE ☐
NERONE ☐ 1-PROPANONE, 1-p-MENTH-6-EN-2-YL-
DOT CLASSIFICATION: 3; Label:
Flammable Liquid
CONSENSUS REPORTS: Reported in EPA
TSCA Inventory.
SAFETY PROFILE: A skin irritant. A
flammable liquid. When heated to
decomposition it emits acrid smoke and
irritating fumes.

MCF750   CAS: 89-78-1   HR: 3
MENTHOL
mf: $C_{10}H_{20}O$ mw: 156.26
PROP: Hexagonal crystals or granules;
peppermint taste and odor. D: 0.890 @
15°/15°, vap press: 1 mm @ 56.0°, vap d:
5.38, mp: 41–43°, bp: 212°, flash p: 199°F.
Very sol in alc, chloroform, ether, pet ether,
glacial acetic acid, liquid petrolatum; sltly sol
in water.
SYNS: CYCLOHEXANOL, 2-ISOPROPYL-5-METHYL- ☐
FEMA No. 2665 ☐ HEXAHYDROTHYMOL ☐ 2-ISOPROP-
YL-5-METHYLCYCLOHEXANOL ☐ p-MENTHAN-3-OL
☐ l-MENTHOL ☐ 5-METHYL-2-(1-METHYLETHYL)CY-
CLOHEXANOL ☐ PEPPERMINT CAMPHOR ☐ TRA-
KILL TRACHEAL MITE KILLER
CONSENSUS REPORTS: Reported in EPA
TSCA Inventory.
SAFETY PROFILE: Poison by intravenous
route. Moderately toxic by ingestion and
intraperitoneal routes. A severe eye irritant.
Incompatible with phenol, β-naphthol,
resorcinol or thymol in trituration,
potassium permanganate, chromium
trioxide, pyrogallol. Combustible liquid.
When heated to decomposition it emits
acrid smoke and irritating fumes.

MCG000   CAS: 15356-70-4   HR: 2
dl-MENTHOL
mf: $C_{10}H_{20}O$ mw: 156.30
PROP: White crystals. D: 0.904 @ 15°/15°,
mp: 38°, bp: 216°. Sol in water: < 1mg/mL
@ 21°.
SYNS: FEMA No. 2665 ☐ 4-ISOPROPYL-1-METHYL-
CYCLOHEXAN-3-OL ☐ dl-3-p-MENTHANOL ☐ 3-p-
MENTHOL ☐ MENTHOL racemic ☐ MENTHOL racemique
(FRENCH) ☐ 5-METHYL-2-(1-METHYLETHYL)-
CYCLOHEXANOL (1-α,2-β,5-α)

CONSENSUS REPORTS: NCI
Carcinogenesis Bioassay (feed); No
Evidence: mouse, rat NCITR* NCI-GC-
TR-98,79. Reported in EPA TSCA
Inventory.
SAFETY PROFILE: Moderately toxic by
ingestion, intraperitoneal, and subcutaneous
routes. An eye and skin irritant. When
heated to decomposition it emits acrid
smoke and irritating fumes. See also
MENTHOL and l-MENTHOL.

MCG250   CAS: 2216-51-5   HR: 3
l-MENTHOL
mf: $C_{10}H_{20}O$ mw: 156.30
PROP: Crystals from MeOH with strong
peppermint odor. Mp: 42.5–43°, bp: 216°.
Sltly sol in H_2O; very sol in org solvs. Found
in high concentrations in oils of Peppermint
(*Mentha Piperita*), and Japanese Mint Oil
(*Mentha Arvensis*), and in lower
concentrations in Reunion Geranium Oil,
and in a large number of essential oils;
prepared by isolation from *Mentha arvensis*
oils (FCTXAV 14,443,76).
SYNS: FEMA No. 2665 ☐ (−)-MENTHYL ALCOHOL ☐
(1R-(1-α,2-β,5-α))-5-METHYL-2-(1-METHYLETHYL)-
CYCLOHEXANOL ☐ U.S.P. MENTHOL
CONSENSUS REPORTS: Reported in EPA
TSCA Inventory.
SAFETY PROFILE: Poison by intravenous
route. Moderately toxic by ingestion,
intraperitoneal, and subcutaneous routes. An
eye irritant. Mutation data reported. When
heated to decomposition it emits acrid
smoke and irritating fumes.

MCG275   CAS: 89-80-5   HR: 2
MENTHONE
mf: $C_{10}H_{18}O$ mw: 154.28
PROP: Colorless liquid; mint odor. D:
0.888–0.895, refr index: 1.448–1.453. Sol in
alc, fixed oils; very sltly sol in water.
SYNS: FEMA No. 2667 ☐ l-p-MENTHAN-3-ONE ☐ l-
MENTHONE (FCC) ☐ p-MENTHONE ☐ trans-MENTH-
ONE ☐ trans-5-METHYL-2-(1-METHYLETHYL)-
CYCLOHEXANONE
CONSENSUS REPORTS: Reported in EPA
TSCA Inventory.

SAFETY PROFILE: Moderately toxic by ingestion, intravenous, and subcutaneous routes. Mutation data reported. When heated to decomposition it emits acrid smoke and irritating fumes. See also KETONES.

MCG500   CAS: 16409-45-3   HR: 2
dl-MENTHYL ACETATE
mf: $C_{12}H_{22}O_2$ mw: 198.34
PROP: Colorless liquid; characteristic minty odor. D: 0.919 @ 20°/4°, refr index: 1.443–1.450, bp: 227°, flash p: 197°F. Sltly sol in water, glycerin; misc with alc, ether, propylene glycol, fixed oils.
SYNS: FEMA No. 2668 □ MENTHOL, ACETATE (8CI) □ MENTHYL ACETATE □ MENTHYL ACETATE racemic □ p-MENTH-3-YL ESTER-dl-ACETIC ACID
SAFETY PROFILE: Mildly toxic by ingestion. A skin irritant. Combustible liquid. When heated to decomposition it emits acrid smoke and irritating fumes.

MCG750   CAS: 2623-23-6   HR: 1
1-p-MENTH-3-YL ACETATE
mf: $C_{12}H_{22}O$ mw: 182.34
PROP: Colorless liquid; minty odor. D: 0.919–0.924, refr index: 1.443–1.447. Sol in alc, propylene glycol, fixed oils; sltly sol in water, glycerin.
SYNS: FEMA No. 2668 □ 1-2-ISOPROPYL-5-METHYL-CYCLOHEXAN-1-OL ACETATE □ (−)-MENTHYL ACETATE □ l-MENTHYL ACETATE (FCC) □ 1-p-MENTH-3-YL ACETATE □ (R-(1α,2β,5α))-5-METHYL-2-(1-METHYLETHYL)-CYCLOHEXANOL ACETATE (9CI)
CONSENSUS REPORTS: Reported in EPA TSCA Inventory.
SAFETY PROFILE: Low toxicity by ingestion and skin contact. A skin irritant. When heated to decomposition it emits acrid smoke and irritating fumes.

MCI750   CAS: 33396-37-1   HR: 3
MEPROSCILLARIN
mf: $C_{31}H_{44}O_8$ mw: 544.75
PROP: A solid. Mp: 213–217°.
SYN: CLIFT
SAFETY PROFILE: Poison by ingestion and intravenous routes. Experimental

reproductive effects. When heated to decomposition it emits acrid smoke and irritating fumes.

MCJ500   HR: 2
MERCAPTANS
PROP: Compounds containing the -SH group bound to carbon. Also called thiols.
SAFETY PROFILE: Generally they have a very offensive odor that may cause nausea and headache. High concentrations of vapor can produce unconsciousness with cyanosis, cold extremities, and rapid pulse. A common air contaminant. Dangerous; when heated to decomposition they almost always emit highly toxic fumes of SO_x. They may react with water, steam, or acids to produce toxic and flammable vapors. Aliphatic mercaptans are flammable. They can react violently with powerful oxidizers such as $Ca(OCl)_2$.

MCK000   CAS: 4822-44-0   HR: 3
α-MERCAPTOACETANILIDE
mf: C_8H_8NOS mw: 166.23
PROP: Needles from EtOH or H_2O. Mp: 110.5–111°.
SYNS: 2-MERCAPTOACETANILIDE □ THIOGLYCOLANILIDE □ THIOGLYCOLIC ACID ANILIDE □ USAF EK-6583
CONSENSUS REPORTS: Reported in EPA TSCA Inventory.
SAFETY PROFILE: Poison by intraperitoneal route. When heated to decomposition it emits very toxic fumes of NO_x and SO_x. See also MERCAPTANS.

MCM750   CAS: 123-93-3   HR: 3
MERCAPTODIACETIC ACID
mf: $C_4H_6O_4S$ mw: 150.16
PROP: A white powder or crystals from EtOAc/C_6H_6. Mp: 128°. Very sol in water.
SYNS: (CARBOXYMETHYLTHIO)ACETIC ACID □ DIMETHYLSULFIDE-α,α'-DICARBOXYLIC ACID □ THIODIGLYCOLIC ACID □ β,β'-THIODIGLYCOLIC ACID □ 2,2'-THIODIGLYCOLIC ACID □ THIODIGLYCOLLIC ACID □ USAF CB-36 □ USAF E-2
CONSENSUS REPORTS: Reported in EPA TSCA Inventory.

M

SAFETY PROFILE: Poison by intraperitoneal route. When heated to decomposition or on contact with acid or acid fumes it emits toxic fumes of SO_x. See also MERCAPTOACETIC ACID.

MCN250 CAS: 60-24-2 HR: 3
2-MERCAPTOETHANOL
DOT: UN 2966
mf: C_2H_6OS mw: 78.14
PROP: Water-white, mobile liquid with faint characteristic odor. Bp: 157–158° (decomp) @ 742 mm, flash p: 165°F (COC), d: 1.1168 @ 20°/20°, vap press: 1.0 mm @ 20°, vap d: 2.69. Pure liquid is misc with water, alc, ether, and benzene.
SYNS: EMERY 5791 □ 1-ETHANOL-2-THIOL □ 2-HYDROXY-1-ETHANETHIOL □ 2-HYDROXYETHYL MERCAPTAN □ 2-ME □ MERCAPTOETHANOL □ β-MERCAPTOETHANOL □ MONOTHIOETHYLENE-GLYCOL □ 2-THIOETHANOL □ THIOGLYCOL (DOT) □ THIOMONOGLYCOL □ USAF EK-4196
CONSENSUS REPORTS: Reported in EPA TSCA Inventory.
DOT CLASSIFICATION: 6.1; Label: Poison
SAFETY PROFILE: Poison by ingestion, skin contact, and intraperitoneal routes. Moderately toxic by intravenous route. A skin and severe eye irritant. Human mutation data reported. A combustible liquid when exposed to heat, flame, or oxidizers. To fight fire, use alcohol foam, CO_2, dry chemical. When heated to decomposition it emits highly toxic fumes of SO_x. See also MERCAPTANS.

MCO500 CAS: 60-56-0 HR: 3
2-MERCAPTO-1-METHYLIMIDAZOLE
mf: $C_4H_6N_2S$ mw: 114.18
PROP: Leaflets from EtOH. Mp: 146–148°, bp: 280° (decomp).
SYNS: BASOLAN □ DANANTIZOL □ FAVISTAN □ FRENTIROX □ MERCAPTAZOLE □ MERCAZOLYL □ METAZOLO □ METHIAMAZOLE □ 1-METHYLIMIDAZ-OLE-2-THIOL □ 1-METHYL-2-MERCAPTOIMIDAZOLE □ METIZOL □ METOTHYRINE □ 1-METYLO-2-MER-KAPTOIMIDAZOLEM (POLISH) □ STRUMAZOLE □ TAPAZOLE □ THACAPZOL □ THIAMAZOLE □ THYCAPSOL □ USAF EL-30

CONSENSUS REPORTS: Reported in EPA TSCA Inventory.
SAFETY PROFILE: Poison by subcutaneous route. Moderately toxic by ingestion and intraperitoneal routes. Human teratogenic effects. An experimental teratogen. Experimental reproductive effects. Questionable carcinogen with experimental neoplastigenic data. Human mutation data reported. An antithyroid drug. When heated to decomposition it emits very toxic fumes of NO_x and SO_x. See also MERCAPTANS.

MCQ500 CAS: 4988-64-1 HR: 3
MERCAPTOPURINE RIBONUCLEOSIDE
mf: $C_{10}H_{12}N_4O_4S$ mw: 284.32
SYNS: 6-MERCAPTOPURINE RIBOSIDE □ NSC-4911 □ RIBOFURANOSIDE, 9H-PURINE-6-THIOL-9 □ RIBOSYL-6-THIOPURINE □ THIONOSINE □ 6-THIOPURINE RIBONUCLEOSIDE □ 6-THIOPURINE RIBOSIDE □ TIOINOSINE
SAFETY PROFILE: Poison by an unspecified route. Moderately toxic by ingestion, subcutaneous, and intraperitoneal routes. An experimental teratogen. Human mutation data reported. When heated to decomposition it emits very toxic fumes of SO_x and NO_x. See also MERCAPTANS.

MCQ750 CAS: 3811-73-2 HR: 3
2-MERCAPTOPYRIDINE-N-OXIDE SODIUM SALT
mf: $C_5H_5NOS•Na$ mw: 150.16
SYNS: (1-HYDROXY-2-PYRIDINETHIONE), SODIUM SALT, TECH □ 2-PYRIDINETHIOL-1-OXIDE SODIUM SALT □ SODIUM PYRITHIONE
CONSENSUS REPORTS: Reported in EPA TSCA Inventory.
DFG MAK: 1 mg/m³
SAFETY PROFILE: Poison by intraperitoneal and intravenous routes. Moderately toxic by ingestion, subcutaneous and parenteral routes. Used in preservation of cosmetics. When heated to decomposition it emits very toxic fumes of Na_2O, NO_x, and SO_x. See also MERCAPTANS.

MCR750   CAS: 52-67-5   HR: 3
d,3-MERCAPTOVALINE
mf: $C_5H_{11}NO_2S$ mw: 149.23
PROP: A solid. Mp: 202–206° (rapid
heating).
SYNS: CUPRENIL □ CUPRIMINE □ DEPEN □ DI-
METHYLCYSTEINE □ β,β-DIMETHYLCYSTEINE □ d-
MERCAPTOVALINE □ METALCAPTASE □ PCA □ d-
PENAMINE □ PENICILLAMIN □ (S)-PENICILLAMIN □
PENICILLAMINE □ d-PENICILLAMINE □ REDUCED-d-
PENICILLAMINE □ TROLOVOL
SAFETY PROFILE: Poison by
intraperitoneal route. Moderately toxic by
subcutaneous and intravenous routes. Mildly
toxic by ingestion. An experimental
teratogen. Human systemic effects by
ingestion: agranulocytosis, dermatitis, fever,
hemorrhage, increased body temperature,
dermatitis, leukopenia, proteinuria,
thrombocytopenia. Human teratogenic
effects by an unspecified route:
developmental abnormalities of the
craniofacial areas, skin, and skin appendages,
and body wall. Experimental reproductive
effects. Questionable human carcinogen
producing leukemia. Mutation data reported.
Used in the treatment of rheumatoid
arthritis, metal poisonings, and cystinuria.
When heated to decomposition it emits very
toxic fumes of NO_x and SO_x. See also
MERCAPTANS.

MCS600   CAS: 66499-61-4   HR: 3
MERCURIBIS-o-NITROPHENOL
mf: $C_{12}H_8HgN_2O_6$ mw: 476.81
PROP: IDLH 10 mg/m³ (as Hg).
SYNS: BIS(4-HYDROXY-3-NITROPHENYL)MERCURY
□ MERCURY, BIS(4-HYDROXY-3-NITROPHENYL)-
ACGIH TLV: TWA 0.1 mg(Hg)/m³ (skin);
BEI: 35 μg/g creatinine total inorganic
mercury in urine preshift; 15 μg/g creatinine
total inorganic mercury in blood at end of
shift at end of workweek.
DFG MAK: Confirmed Animal Carcinogen
with Unknown Relevance to Humans
NIOSH REL: (Mercury, Aryl and Inorganic)
CL 0.1 mg/m³ (skin)
SAFETY PROFILE: Poison by
intraperitoneal route. When heated to

decomposition it emits toxic fumes of NO_x
and Hg.

MCS750   CAS: 1600-27-7   HR: 3
MERCURIC ACETATE
DOT: UN 1629
mf: $C_4H_6O_4•Hg$ mw: 318.69
PROP: White crystals or powder;
photosensitive; slt acetic odor. D: 3.280, mp:
178–180° (overheating causes decomp). Sol
in H_2O and AcOH. IDLH 10 mg/m³ (as
Hg).
SYNS: ACETIC ACID, MERCURY(2+) SALT □ BIS-
(ACETYLOXY)MERCURY □ DIACETOXYMERCURY □
MERCURIACETATE □ MERCURIC DIACETATE □
MERCURY ACETATE □ MERCURY(2+) ACETATE □
MERCURY(II) ACETATE □ MERCURY DIACETATE □
MERCURYL ACETATE
CONSENSUS REPORTS: EPA Extremely
Hazardous Substances List. Mercury and its
compounds are on the Community Right-
To-Know List. EPA Genetic Toxicology
Program. Reported in EPA TSCA
Inventory.
OSHA PEL: CL 0.1 mg(Hg)/m³ (skin)
ACGIH TLV: TWA 0.1 mg(Hg)/m³ (skin);
BEI: 35 μg/g creatinine total inorganic
mercury in urine preshift; 15 μg/g
creatinine total inorganic mercury in blood
at end of shift at end of workweek.
DFG MAK: Confirmed Animal Carcinogen
with Unknown Relevance to Humans
NIOSH REL: (Mercury, Aryl and Inorganic)
CL 0.1 mg/m³ (skin)
DOT CLASSIFICATION: 6.1; Label: Poison
SAFETY PROFILE: Poison by ingestion,
intravenous, intraperitoneal, and
subcutaneous routes. Moderately toxic by
skin contact. An experimental teratogen.
Experimental reproductive effects. Mutation
data reported. When heated to
decomposition it emits toxic fumes of Hg.
See also MERCURY COMPOUNDS.

MCT500   CAS: 21908-53-2   HR: 3
MERCURIC OXIDE
DOT: UN 1641
mf: HgO   mw: 216.59

M

PROP: Heavy, bright orange-red or orange-yellow powder. Mp: decomp @ 500°, d: 11.14. Practically insol in water; sol in dil HCl or HNO₃. Protect from light. IDLH 10 mg/m³ (as Hg).

SYNS: C.I. 77760 □ MERCURIC OXIDE, RED □ MERCURIC OXIDE, solid (DOT) □ MERCURIC OXIDE, YELLOW □ MERCURY(II) OXIDE □ OXYDE de MERCURE (FRENCH) □ QUECKSILBEROXID (GERMAN) □ RED OXIDE of MERCURY □ RED PRECIPITATE □ SANTAR □ YELLOW MERCURIC OXIDE □ YELLOW OXIDE of MERCURY □ YELLOW PRECIPITATE

CONSENSUS REPORTS: EPA Extremely Hazardous Substances List. Mercury and its compounds are on the Community Right-To-Know List. Reported in EPA TSCA Inventory.

OSHA PEL: CL 0.1 mg(Hg)/m³ (skin)

ACGIH TLV: TWA 0.1 mg(Hg)/m³ (skin); BEI: 35 μg/g creatinine total inorganic mercury in urine preshift; 15 μg/g creatinine total inorganic mercury in blood at end of shift at end of workweek.

NIOSH REL: (Mercury, Aryl and Inorganic) CL 0.1 mg/m³ (skin)

DOT CLASSIFICATION: 6.1; Label: Poison

SAFETY PROFILE: Poison by ingestion, skin contact, intraperitoneal, and intramuscular routes. An experimental teratogen. Experimental reproductive effects. An FDA over-the-counter drug. Used for treating fruit trees. Flammable by chemical reactions. A powerful oxidizer. Explosive reaction with acetyl nitrate, butadiene + ethanol + iodine (at 35°C), chlorine + hydrocarbons (e.g., methane, ethylene), diboron tetrafluoride, hydrogen peroxide + traces of nitric acid, reducing agents (e.g., hydrazine hydrate, phosphinic acid). Forms heat- or impact-sensitive explosive mixtures with nonmetals (e.g., phosphorus, sulfur), metals (e.g., magnesium, potassium, sodium-potassium alloy). Reacts violently with hydrogen trisulfide (on ignition), hydrazine hydrate, hydrogen peroxide, hypophosphorous acid, iodine + methanol or ethanol, phospham, acetyl nitrate, S₂Cl₂, reductants. Incandescent reaction with phospham.

When heated to decomposition it emits highly toxic fumes of Hg. See also MERCURY COMPOUNDS, INORGANIC.

MCU000   CAS: 5970-32-1   HR: 3
MERCURIC SALICYLATE
DOT: UN 1644
mf: C₇H₄HgO₃   mw: 336.70

PROP: White-yellow or pinkish, odorless powder. Insol in water or alc; sol in warm solns of alkali halides. IDLH 10 mg/m³ (as Hg).

SYNS: MERCURIC SALICYLATE, solid (DOT) □ MERCURISALICYLIC ACID □ MERCURY SALICYLATE □ MERCURY SUBSALICYLATE

CONSENSUS REPORTS: Mercury and its compounds are on the Community Right-To-Know List.

OSHA PEL: CL 0.1 mg(Hg)/m³ (skin)

ACGIH TLV: TWA 0.1 mg(Hg)/m³ (skin); BEI: 35 μg/g creatinine total inorganic mercury in urine preshift; 15 μg/g creatinine total inorganic mercury in blood at end of shift at end of workweek.

DFG MAK: Confirmed Animal Carcinogen with Unknown Relevance to Humans

NIOSH REL: (Mercury, Aryl and Inorganic) CL 0.1 mg/m³ (skin)

DOT CLASSIFICATION: 6.1; Label: Poison

SAFETY PROFILE: Poison by subcutaneous and intramuscular routes. An FDA over-the-counter drug. Incompatible with alkali iodides. When heated to decomposition it emits toxic fumes of Hg. See also MERCURY COMPOUNDS.

MCU250   CAS: 592-85-8   HR: 3
MERCURIC SULFOCYANATE
DOT: UN 1646
mf: C₂HgN₂S₂   mw: 316.79

PROP: White, odorless powder; sltly sol in cold water; more sol in boiling water (decomp); sol in dil HCl. Protect from light. IDLH 10 mg/m³ (as Hg).

SYNS: BIS(THIOCYANATO)-MERCURY □ MERCURIC SULFOCYANIDE □ MERCURIC SULFOCYANATE, solid (DOT) □ MERCURIC THIOCYANATE □ MERCURIC THIOCYANATE, solid (DOT) □ MERCURY

DITHIOCYANATE □ MERCURY THIOCYANATE (DOT) □ MERCURY(II) THIOCYANATE □ THIOCYANIC ACID, MERCURY(2+) SALT

CONSENSUS REPORTS: Mercury and its compounds are on the Community Right-To-Know List. Reported in EPA TSCA Inventory.

OSHA PEL: CL 0.1 mg(Hg)/m³ (skin)

ACGIH TLV: TWA 0.1 mg(Hg)/m³ (skin); BEI: 35 µg/g creatinine total inorganic mercury in urine preshift; 15 µg/g creatinine total inorganic mercury in blood at end of shift at end of workweek.

DFG MAK: Confirmed Animal Carcinogen with Unknown Relevance to Humans

NIOSH REL: (Mercury, Aryl and Inorganic) CL 0.1 mg/m³ (skin)

DOT CLASSIFICATION: 6.1; Label: Poison

SAFETY PROFILE: A poison by ingestion and intraperitoneal routes. Moderately toxic by skin contact. Thermally unstable and decomposition may be vigorous. When heated to decomposition it emits very toxic fumes of Hg, NOₓ, SOₓ, and CN⁻. See also MERCURY COMPOUNDS and CYANATES.

MCU750   CAS: 55-68-5   HR: 3
MERCURIPHENYL NITRATE
DOT: UN 1895

mf: C₆H₅HgNO₃   mw: 339.71

PROP: Crystals. Mp: 176–186°. Insol in cold water. IDLH 10 mg/m³ (as Hg).

SYNS: FENYLMERKURINITRAT □ MERPHENYL NITRATE □ MERSOLITE 7 □ NITRIC ACID, PHENYL-MERCURY SALT □ PHE-MER-NITE □ PHENALCO □ PHENITOL □ PHENMERZYL NITRATE □ PHENYL-MERCURIC NITRATE □ PHENYLMERCURY NITRATE □ PHERMERNITE

CONSENSUS REPORTS: Mercury and its compounds are on the Community Right-To-Know List. Reported in EPA TSCA Inventory. EPA Genetic Toxicology Program.

OSHA PEL: CL 0.1 mg(Hg)/m³ (skin)

ACGIH TLV: TWA 0.1 mg(Hg)/m³ (skin); BEI: 35 µg/g creatinine total inorganic mercury in urine preshift; 15 µg/g creatinine total inorganic mercury in blood at end of shift at end of workweek.

DFG MAK: Confirmed Animal Carcinogen with Unknown Relevance to Humans

NIOSH REL: (Mercury, Aryl and Inorganic) CL 0.1 mg/m³ (skin)

DOT CLASSIFICATION: 6.1; Label: Poison

SAFETY PROFILE: Poison by intravenous route. FDA over-the-counter drug. When heated to decomposition it emits very toxic fumes of Hg and NOₓ. See also MERCURY COMPOUNDS and NITRATES.

MCV000   CAS: 129-16-8   HR: 3
MERCUROCHROME

mf: C₂₀H₁₀Br₂HgO₆•2Na   mw: 752.69

PROP: Iridescent green scales. Sol in H₂O (soln carmine-red); insol in EtOH, Me₂CO, CHCl₃, and Et₂O. IDLH 10 mg/m³ (as Hg).

SYNS: ASCEPTICHROME □ ASEPTICHROME □ CHROMARGYRE □ 2,7-DIBROMO-4-HYDROXY-MERCURIFLUORESCEINE DISODIUM SALT □ DISOD-IUM-2,7-DIBROM-4-HYDROXY-MERCURI-FLUORESC-EIN □ DISODIUM-2',7'-DIBROMO-4'-(HYDROXYMER-CURY)FLUORESCEIN □ DOMF □ FLAVUROL □ FLUO-ROCHROME □ GALLOCHROME □ GYNOCHROME □ MERBROMIN □ MERCURANINE □ MERCUROCHRO-ME-220 SOLUBLE □ MERCUROCOL □ MERCUROME □ MERCUROPHAGE □ PLANOCHROME

CONSENSUS REPORTS: Mercury and its compounds are on the Community Right-To-Know List.

NIOSH REL: (Mercury, Aryl and Inorganic) CL 0.1 mg/m³ (skin)

SAFETY PROFILE: Poison by intravenous and subcutaneous routes. Mutation data reported. Relatively nonirritating and nontoxic to damaged skin or tissue. A topical antiseptic. An FDA over-the-counter drug. When heated to decomposition it emits very toxic fumes including fumes of Na₂O, Br⁻, and Hg. See also MERCURY COMPOUNDS, ORGANIC.

MCV250   CAS: 12002-19-6   HR: 3
MERCUROL
DOT: UN 1639

PROP: Colorless to brownish powder. Contains 20% mercury. IDLH 10 mg/m³ (as Hg).

SYN: MERCURY NUCLEATE, solid (DOT)

CONSENSUS REPORTS: Mercury and its compounds are on the Community Right-To-Know List.

NIOSH REL: (Mercury, Aryl and Inorganic) CL 0.1 mg/m³ (skin)

DOT CLASSIFICATION: 6.1; Label: Poison

SAFETY PROFILE: A poison. When heated to decomposition it emits toxic fumes of Hg. See also MERCURY COMPOUNDS.

MCW000   CAS: 7546-30-7   HR: 3
MERCUROUS CHLORIDE
mf: ClHg   mw: 236.04

PROP: White, odorless, tasteless, heavy powder or crystals. Subl @ 400°, d: 7.150. Insol in water, alc, and ether. Protect from light. Sunlight causes it to decomp into mercuric chloride and metallic Hg. IDLH 10 mg/m³ (as Hg).

SYNS: CALOGREEN □ CALOMEL □ CALOMELANO (ITALIAN) □ CALOSAN □ CHLORURE MERCUREUX (FRENCH) □ C.I. 77764 □ CLORURO MERCUROSO (ITALIAN) □ CYCLOSAN □ KALOMEL (GERMAN) □ MERCUROCHLORID (DUTCH) □ MERCURY(I) CHLOR-IDE □ MERCURY MONOCHLORIDE □ MERCURY PROTOCHLORIDE □ MILD MERCURY CHLORIDE □ PRECIPITE BLANC □ QUECKSILBER(I)-CHLORID (GERMAN) □ QUECKSILBER CHLORUER (GERMAN) □ SUBCHLORIDE of MERCURY

CONSENSUS REPORTS: Mercury and its compounds are on the Community Right-To-Know List. EPA Genetic Toxicology Program. Reported in EPA TSCA Inventory.

OSHA PEL: CL 0.1 mg(Hg)/m³ (skin)

ACGIH TLV: TWA 0.1 mg(Hg)/m³ (skin); BEI: 35 µg/g creatinine total inorganic mercury in urine preshift; 15 µg/g creatinine total inorganic mercury in blood at end of shift at end of workweek; (Proposed: (inhalable fraction) 0.1 mg/m³; Not Classifiable as a Human Carcinogen)

NIOSH REL: (Mercury, Aryl and Inorganic) CL 0.1 mg/m³ (skin)

SAFETY PROFILE: Poison by ingestion and intraperitoneal routes. Moderately toxic by

skin contact. Mutation data reported. A fungicide. An FDA over-the-counter drug. Incompatible with bromides, iodides, alkali chlorides, sulfates, sulfites, carbonates, hydroxides, lime water, ammonia, golden antimony sulfide, cyanides, copper salts, hydrogen peroxide, iodine, iodoform, lead salts, silver salts, sulfides. When heated to decomposition it emits very toxic fumes of Cl⁻ and Hg. See also MERCURY COMPOUNDS.

MCW250   CAS: 7439-97-6   HR: 3
MERCURY
DOT: NA 2809
af: Hg   aw: 200.59

PROP: Silvery, heavy, mobile liquid at room temp, freezing to a white solid. Solid: tin-white, ductile, malleable mass that can be cut with a knife. A liquid metallic element. Colorless vapor. When heated, reacts with O_2 (historically important reaction), S, halogens. Reacts with conc HNO_3, but not with dil non-oxidizing acids Mp: −38.89°, bp: 356.9°, d: 13.534 @ 25°, vap press: 2 × 10^{-3} mm @ 25°. IDLH 10 mg/m³ (as Hg).

SYNS: COLLOIDAL MERCURY □ KWIK (DUTCH) □ MERCURE (FRENCH) □ MERCURIO (ITALIAN) □ MERCURY, METALLIC (DOT) □ NCI-C60399 □ QUECKSILBER (GERMAN) □ QUICK SILVER □ RCRA WASTE NUMBER U151 □ RTEC (POLISH)

CONSENSUS REPORTS: Mercury and its compounds are on the Community Right-To-Know List.

OSHA PEL: Vapor: TWA 0.05 mg/m³ (skin)

ACGIH TLV: TWA 0.025 mg(Hg)/m³ (skin); Not Classifiable as a Carcinogen; BEI 35 µg/g creatinine total inorganic mercury in urine, preshift

NIOSH REL: (Mercury, Aryl and Inorganic) CL 0.1 mg/m³ (skin)

DOT CLASSIFICATION: 8; Label: Corrosive

SAFETY PROFILE: Poison by inhalation. Human systemic effects by inhalation: wakefulness, muscle weakness, anorexia, headache, tinnitus, hypermotility, diarrhea, liver changes, dermatitis, fever. An experimental teratogen. Experimental reproductive effects. Questionable carcinogen with experimental tumorigenic data. Human mutation data reported. Used in dental applications, electronics, and chemical synthesis.

May explode on contact with 3-bromopropyne, alkynes + silver perchlorate, ethylene oxide, lithium, methylsilane + oxygen (explodes when shaken), peroxyformic acid, chlorine dioxide, tetracarbonylnickel + oxygen. May react with ammonia to form an explosive product. Mixtures with methyl azide are shock- and spark-sensitive explosives. The vapor ignites on contact with boron diiodophosphide. Reacts violently with acetylenic compounds (e.g., acetylene, sodium acetylide, 2-butyne-1,4-diol + acid), metals (e.g., aluminum, calcium, potassium, sodium, rubidium, exothermic formation of amalgams), Cl_2, ClO_2, CH_3N_3, Na_2C_2, nitromethane. Incompatible with methyl azide, oxidants. When heated to decomposition it emits toxic fumes of Hg. See also MERCURY COMPOUNDS.

MCW349   CAS: 68833-55-6   HR: 3
MERCURY ACETYLIDE (DOT)
mf: C_2HHg mw: 225.62
PROP: IDLH 10 mg/m³ (as Hg).
SYN: MERCURY ACETYLIDE
ACGIH TLV: TWA 0.01 mg(Hg)/m³; BEI: 35 µg/g creatinine total inorganic mercury in urine preshift; 15 µg/g creatinine total inorganic mercury in blood at end of shift at end of workweek.
DFG MAK: Confirmed Animal Carcinogen with Unknown Relevance to Humans
NIOSH REL: (Mercury, Organo) TWA 0.01 mg/m³; STEL 0.03 mg/m³ (skin)
DOT CLASSIFICATION: Forbidden
SAFETY PROFILE: Extremely reactive. When heated to decomposition it emits toxic fumes of Hg.

MCW500   CAS: 10124-48-8   HR: 3
MERCURY AMIDE CHLORIDE
DOT: UN 1630
mf: ClH_2HgN mw: 252.071

PROP: White, pulverized lumps or powder. IDLH 10 mg/m³ (as Hg).
SYNS: AMINOMERCURIC CHLORIDE □ AMMONIATED MERCURY □ MERCURIC AMMONIUM CHLORIDE, solid □ MERCURIC CHLORIDE, AMMONIATED □ MERCURY AMINE CHLORIDE □ MERCURY AMMONIATED □ WHITE MERCURY PRECIPITATED □ WHITE PRECIPITATE
CONSENSUS REPORTS: Mercury and its compounds are on the Community Right-To-Know List. Reported in EPA TSCA Inventory.
OSHA PEL: CL 0.1 mg(Hg)/m³ (skin)
ACGIH TLV: TWA 0.1 mg(Hg)/m³ (skin); BEI: 35 µg/g creatinine total inorganic mercury in urine preshift; 15 µg/g creatinine total inorganic mercury in blood at end of shift at end of workweek.
DFG MAK: Confirmed Animal Carcinogen with Unknown Relevance to Humans
NIOSH REL: (Mercury, Organo) TWA 0.01 mg/m³; STEL 0.03 mg/m³ (skin)
DOT CLASSIFICATION: 6.1; Label: Poison
SAFETY PROFILE: A poison by ingestion. Moderately toxic by skin contact. Explosive reaction with halogens or amine metal salts. When heated to decomposition it emits very toxic fumes of Cl⁻, NO_x, and Hg. See also MERCURY COMPOUNDS.

MCX000   CAS: 38232-63-2   HR: 3
MERCURY(I) AZIDE
mf: Hg_2N_6 mw: 485.22
PROP: White, light-sensitive solid. Bp: 220°. Sltly sol in H_2O. IDLH 10 mg/m³ (as Hg).
SYNS: MERCUROUS AZIDE (DOT) □ MERCURY AZIDE
CONSENSUS REPORTS: Mercury and its compounds are on the Community Right-To-Know List.
OSHA PEL: CL 0.1 mg(Hg)/m³ (skin)
ACGIH TLV: TWA 0.1 mg(Hg)/m³ (skin); BEI: 35 µg/g creatinine total inorganic mercury in urine preshift; 15 µg/g creatinine total inorganic mercury in blood at end of shift at end of workweek.
DOT CLASSIFICATION: Forbidden
SAFETY PROFILE: Poison. Explodes on heating in air. When heated to

M

decomposition it emits very toxic fumes of NO_x and Hg. See also AZIDES and MERCURY COMPOUNDS.

MCX500   CAS: 583-15-3   HR: 3
MERCURY(II) BENZOATE
DOT: UN 1631
mf: $C_{14}H_{10}O_4 \cdot Hg$ mw: 442.83
PROP: White, crystalline, odorless powder or solid. Mp: 120–129°. Very sol in NaCl soln; insol in alc. IDLH 10 mg/m³ (as Hg).
SYNS: MERCURIC BENZOATE □ MERCURIC BENZOATE, solid (DOT)
CONSENSUS REPORTS: Mercury and its compounds are on the Community Right-To-Know List.
OSHA PEL: CL 0.1 mg(Hg)/m³ (skin)
ACGIH TLV: TWA 0.1 mg(Hg)/m³ (skin); BEI: 35 µg/g creatinine total inorganic mercury in urine preshift; 15 µg/g creatinine total inorganic mercury in blood at end of shift at end of workweek.
DFG MAK: Confirmed Animal Carcinogen with Unknown Relevance to Humans
NIOSH REL: (Mercury, Aryl and Inorganic) CL 0.1 mg/m³ (skin)
DOT CLASSIFICATION: 6.1; Label: Poison
SAFETY PROFILE: A poison. When heated to decomposition it emits toxic fumes of Hg. See also MERCURY COMPOUNDS.

MCX750   CAS: 10031-18-2   HR: 3
MERCURY(I) BROMIDE (1:1)
mf: BrHg   mw: 280.50
PROP: White-yellow, odorless, tetragonal crystals or powder. Darkens on exposure to light. D: 7.307, vap d: 19.3. Sublimes @ approx 390° (decomp). Insol in water, alc, and ether; decomp by hot HCl or alkali bromides. Protect from light. IDLH 10 mg/m³ (as Hg).
SYN: MERCUROUS BROMIDE, solid (DOT)
CONSENSUS REPORTS: Mercury and its compounds are on the Community Right-To-Know List.
OSHA PEL: CL 0.1 mg(Hg)/m³ (skin)
ACGIH TLV: TWA 0.1 mg(Hg)/m³ (skin); BEI: 35 µg/g creatinine total inorganic

mercury in urine preshift; 15 µg/g creatinine total inorganic mercury in blood at end of shift at end of workweek.
NIOSH REL: (Mercury, Aryl and Inorganic) CL 0.1 mg/m³ (skin)
SAFETY PROFILE: A poison. When heated to decomposition it emits very toxic fumes of Br^- and Hg. See also MERCURY COMPOUNDS and BROMIDES.

MCY000   CAS: 7789-47-1   HR: 3
MERCURY(II) BROMIDE (1:2)
mf: Br_2Hg mw: 360.41
PROP: White crystals or sublimable, colorless, crystalline powder or yellow liquid. Sensitive to light. Mp: 238°, bp: 318° (subl), d: 6.109 @ 25°, vap press: 1 mm @ 136.5°. Very sol in hot alc, methanol, HCl, HBr, alkali bromide solns; sltly sol in chloroform. IDLH 10 mg/m³ (as Hg).
SYNS: MERCURIC BROMIDE □ MERCURIC BROMIDE, solid
CONSENSUS REPORTS: Mercury and its compounds are on the Community Right-To-Know List. Reported in EPA TSCA Inventory.
OSHA PEL: CL 0.1 mg(Hg)/m³ (skin)
ACGIH TLV: TWA 0.1 mg(Hg)/m³ (skin); BEI: 35 µg/g creatinine total inorganic mercury in urine preshift; 15 µg/g creatinine total inorganic mercury in blood at end of shift at end of workweek.
NIOSH REL: (Mercury, Aryl and Inorganic) CL 0.1 mg/m³ (skin)
SAFETY PROFILE: A poison by ingestion, skin contact, and intraperitoneal routes. Vigorous reaction with indium at 350°C. Incompatible with sodium and potassium. When heated to decomposition it emits very toxic fumes of Br^- and Hg. See also MERCURY COMPOUNDS and BROMIDES.

MCY475   CAS: 7487-94-7   HR: 3
MERCURY(II) CHLORIDE
DOT: UN 1624
mf: Cl_2Hg mw: 271.50

PROP: White colorless crystals or powder. Mp: 280°, bp: 302°, d: 5.440 @ 25°, vap press: 1 mm @ 136.2°. IDLH 10 mg/m³ (as Hg).
SYNS: BICHLORIDE of MERCURY □ BICHLORURE de MERCURE (FRENCH) □ CALOCHLOR □ CHLORID RTUTNATY (CZECH) □ CHLORURE MERCURIQUE (FRENCH) □ CLORURO di MERCURIO (ITALIAN) □ CORROSIVE MERCURY CHLORIDE □ CORROSIVE SUBLIMATE □ MERCURIC CHLORIDE (DOT) □ MERCURY BICHLORIDE □ MERCURY PERCHLORIDE □ NCI-C60173 □ PERCHLORIDE of MERCURY □ QUECKSILBER CHLORID (GERMAN) □ SUBLIMAT (CZECH) □ SULEMA (RUSSIAN) □ TL 898
CONSENSUS REPORTS: IARC Cancer Review: Group 3 IMEMDT 58,239,93; Animal Limited Evidence IMEMDT 58,239,93; Human Inadequate Evidence IMEMDT 58,239,93. Mercury and its compounds are on the Community Right-To-Know List. EPA Genetic Toxicology Program. Reported in EPA TSCA Inventory. EPA Extremely Hazardous Substances List.
OSHA PEL: CL 0.1 mg(Hg)/m³ (skin)
ACGIH TLV: TWA 0.1 mg(Hg)/m³ (skin); BEI: 35 µg/g creatinine total inorganic mercury in urine preshift; 15 µg/g creatinine total inorganic mercury in blood at end of shift at end of workweek.
NIOSH REL: (Mercury, Aryl and Inorganic) CL 0.1 mg/m³ (skin)
DOT CLASSIFICATION: 6.1; Label: Poison
SAFETY PROFILE: A human poison by ingestion. Poison experimentally by ingestion, skin contact, and subcutaneous routes. Human systemic effects by ingestion: respiratory obstruction, nausea or vomiting, urine volume decrease or anuria. Human reproductive effects by ingestion: terminates pregnancy. Experimental teratogenic and reproductive effects. Human mutation data reported. Questionable carcinogen. A severe eye and skin irritant. Reaction with sodium aci-nitromethanide + acids forms the explosive mercury fulminate. Reacts violently with K, Na. When heated to decomposition it emits toxic fumes of Hg.

See also MERCURY COMPOUNDS and CHLORIDES.

MCZ000   HR: 3
MERCURY COMPOUNDS, INORGANIC
PROP: IDLH 10 mg/m³ (as Hg).
CONSENSUS REPORTS: Mercury and its compounds are on the Community Right-To-Know List.
SAFETY PROFILE: Mercury is a general protoplasmic poison; after absorption it circulates in the blood and is stored in the liver, kidneys, spleen, and bone. In industrial poisoning, the principal effect is upon the central nervous system, the mouth, and gums. The cardinal symptoms of industrial mercury poisoning are stomatitis, tremors, and psychic disturbances. Usually the first complaints are of excessive salivation and painful chewing. In severe cases there may be gingivitis with loosening of the teeth, and a dark line on the gum margins resembling the "lead line." The psychic disturbance (so called "erethism") includes loss of memory, insomnia, lack of confidence, irritability, vague fears, and depression. The dermatitis produced by fulminate of mercury takes the form of small, discrete ulcers on the exposed parts, and is usually accompanied by conjunctivitis and inflammation of the mucous membranes of the nose and throat. In humans, it is readily absorbed by the respiratory tract (elemental mercury vapor, dusts of mercury compounds), intact skin, and the gastrointestinal tract. Occasional incidental swallowing of metallic mercury may be without harm. Spilled and heated elemental mercury is particularly hazardous. A number of mercury compounds, in addition to the fulminate, can cause skin irritation and be absorbed through the skin. They are strong allergens and common air contaminants. Acute toxicity: Soluble salts have violent corrosive effects on skin and mucous membranes, cause severe nausea, vomiting, abdominal pain, bloody diarrhea, kidney damage, and death usually within 10 days. Many mercury compounds are

M

explosively unstable or undergo hazardous reactions. When heated to decomposition they emit toxic fumes of Hg.

MDA000   HR: 3
MERCURY COMPOUNDS, ORGANIC
PROP: IDLH 10 mg/m³ (as Hg).
CONSENSUS REPORTS: Mercury and its compounds are on the Community Right-To-Know List.
DFG MAK: 0.01 mg/m³
SAFETY PROFILE: The customary grouping of all organic mercurials in a single category is not fully justified by the toxicity of the compounds. Alkyl mercurials have very high toxicity; aryl compounds, particularly the phenyls, are much less toxic, and the organomercurials used in therapeutics are less toxic. The alkyls and aryls commonly cause skin burns and other forms of irritation, and both can be absorbed through the skin. Fatal poisoning has occurred due to exposure to alkyl mercurials and permanent damage to the brain has been reported. Phenyl mercurials appear to be no more toxic than metallic mercury. Organic mercury compounds, like organic lead compounds, seem to have an affinity for lipid-containing organs, resulting in central nervous system disturbances such as from tetraethyl lead. These are common air contaminants. Many mercury compounds are explosively unstable or undergo hazardous reactions. When heated to decomposition they emit highly toxic fumes of Hg.

MDA250   CAS: 592-04-1   HR: 3
MERCURY(II) CYANIDE
DOT: UN 1636
mf: C_2HgN_2 mw: 252.63
PROP: Colorless, odorless, transparent prisms; darkened by light. Two forms: grayish crystals (low P-form); dark brown solid (high P-form). Decomp @ 320°, d: 3.996. Sol in H_2O; sol in EtOH and ether. IDLH 10 mg/m³ (as Hg).

SYNS: CYANURE de MERCURE (FRENCH) □ MERCURIC CYANIDE, solid (DOT)
CONSENSUS REPORTS: Reported in EPA TSCA Inventory. Mercury and its compounds, as well as cyanide and its compounds, are on the Community Right-To-Know List.
OSHA PEL: CL 0.1 mg(Hg)/m³ (skin)
ACGIH TLV: TWA 0.1 mg(Hg)/m³ (skin); BEI: 35 μg/g creatinine total inorganic mercury in urine preshift; 15 μg/g creatinine total inorganic mercury in blood at end of shift at end of workweek.
NIOSH REL: (Mercury, Aryl and Inorganic) CL 0.1 mg/m³ (skin)
DOT CLASSIFICATION: 6.1; Label: Poison
SAFETY PROFILE: Poison by ingestion, subcutaneous, intravenous, and intraperitoneal routes. Human systemic effects by ingestion: nausea or vomiting, hypermotility, diarrhea, kidney changes, somnolence. Hydrolyzes to toxic fumes. A friction- and impact-sensitive explosive. It may initiate detonation of liquid hydrogen cyanide. Incompatible with fluorine, magnesium, sodium nitrite. When heated to decomposition it emits very toxic fumes of Hg, NO_x, and CN^-. See also CYANIDE and MERCURY COMPOUNDS.

MDA500   CAS: 1335-31-5   HR: 3
MERCURY CYANIDE OXIDE
DOT: UN 1642
mf: $C_2Hg_2N_2O$ mw: 469.22
PROP: White, orthorhombic crystals or crystalline powder from water. D: 4.44. Sol in water. IDLH 10 mg/m³ (as Hg).
SYNS: MERCURIC OXYCYANIDE □ MERCURIC OXYCYANIDE, solid (desensitized) (DOT) □ MERCURY OXYCYANIDE
CONSENSUS REPORTS: Mercury and its compounds, as well as cyanide and its compounds, are on the Community Right-To-Know List.
OSHA PEL: CL 0.1 mg(Hg)/m³ (skin)
ACGIH TLV: TWA 0.1 mg(Hg)/m³ (skin); BEI: 35 μg/g creatinine total inorganic mercury in urine preshift; 15 μg/g creatinine

total inorganic mercury in blood at end of shift at end of workweek.

NIOSH REL: (Mercury, Aryl and Inorganic) CL 0.1 mg/m³ (skin)

DOT CLASSIFICATION: Forbidden; DOT Class: 6.1; Label: Poison (desensitized)

SAFETY PROFILE: Poison by intravenous route. An explosive sensitive to friction, impact, or heat. The commercial product is stabilized by excess mercury(II) cyanide. When heated to decomposition it emits very toxic fumes of Hg, CN⁻, and NO$_x$. See also MERCURY COMPOUNDS and CYANIDE.

MDC000   CAS: 628-86-4   HR: 3
MERCURY(II) FULMINATE

DOT: UN 0135

mf: $C_2HgN_2O_2$ mw: 284.63

PROP: White solid. Mp: explodes, d: 4.42. IDLH 10 mg/m³ (as Hg).

SYNS: FULMINATE of MERCURY □ FULMINATE of MERCURY (dry) (DOT) □ FULMINATING MERCURY (DOT) □ MERCURY FULMINATE, wetted with not <20% water, or mixture (UN 0135) (DOT) □ RCRA WASTE NUMBER P065

CONSENSUS REPORTS: Mercury and its compounds are on the Community Right-To-Know List. Reported in EPA TSCA Inventory.

OSHA PEL: CL 0.1 mg(Hg)/m³ (skin)

ACGIH TLV: TWA 0.1 mg(Hg)/m³ (skin); BEI: 35 µg/g creatinine total inorganic mercury in urine preshift; 15 µg/g creatinine total inorganic mercury in blood at end of shift at end of workweek.

NIOSH REL: (Mercury, Organo) TWA 0.01 mg/m³; STEL 0.03 mg/m³ (skin)

DOT CLASSIFICATION: Forbidden; DOT Class: EXPLOSIVE 1.1A; Label: EXPLOSIVE 1.1A (UN 0135)

SAFETY PROFILE: An explosive sensitive to flame, heat, impact, friction, intense radiation, or contact with sulfuric acid. Self-explodes. Dangerously flammable; should be kept moist until used. Incompatible with sulfuric acid. When heated to decomposition it emits very toxic fumes of Hg and NO$_x$.

See also MERCURY COMPOUNDS and FULMINATES.

MDC500   CAS: 63937-14-4   HR: 3
MERCURY(I) GLUCONATE

DOT: UN 1637

mf: $C_6H_{11}O_7 \bullet Hg$ mw: 395.76

PROP: White solid. IDLH 10 mg/m³ (as Hg).

SYNS: MERCUROUS GLUCONATE □ MERCUROUS GLUCONATE, solid (DOT)

CONSENSUS REPORTS: Mercury and its compounds are on the Community Right-To-Know List.

OSHA PEL: CL 0.1 mg(Hg)/m³ (skin)

ACGIH TLV: TWA 0.1 mg(Hg)/m³ (skin); BEI: 35 µg/g creatinine total inorganic mercury in urine preshift; 15 µg/g creatinine total inorganic mercury in blood at end of shift at end of workweek.

DFG MAK: Confirmed Animal Carcinogen with Unknown Relevance to Humans

DOT CLASSIFICATION: 6.1; Label: Poison

SAFETY PROFILE: A poison. When heated to decomposition it emits toxic fumes of Hg. See also MERCURY COMPOUNDS, ORGANIC.

MDC750   CAS: 7783-30-4   HR: 3
MERCURY(I) IODIDE

DOT: UN 1638

mf: HgI   mw: 327.49

PROP: Heavy, odorless, yellow, tetragonal crystals or amorphous powder. D: 7.70, mp: 290° when rapidly heated (partial decomp). Insol in water, alc, and ether; sol in solns of mercurous or mercuric nitrates. Protect from light. IDLH 10 mg/m³ (as Hg).

SYNS: IODURE de MERCURE (FRENCH) □ MERCUROUS IODIDE □ MERCURY IODIDE (DOT) □ MERCURY IODIDE, solution (DOT) □ MERCURY PROTOIODIDE □ YELLOW MERCURY IODIDE

CONSENSUS REPORTS: Mercury and its compounds are on the Community Right-To-Know List.

ACGIH TLV: Proposed: (inhalable fraction) 0.1 mg/m³; Not Classifiable as a Human Carcinogen)

M

NIOSH REL: (Mercury, Aryl and Inorganic) CL 0.1 mg/m³ (skin)
DOT CLASSIFICATION: 6.1; Label: Poison
SAFETY PROFILE: Poison by ingestion and intraperitoneal routes. When heated to decomposition it emits very toxic fumes of Hg and I⁻. See also MERCURY and IODIDES.

MDD000   CAS: 7774-29-0   HR: 3
MERCURY(II) IODIDE
mf: HgI₂   mw: 454.39
PROP: Scarlet, heavy, odorless, almost tasteless powder. Sensitive to light. D: 6.28, mp: 259°, bp: approx 350° (subl). Very sol in alkali iodides, HgCl₂, Na₂S₂O₃; very sltly sol in water. IDLH 10 mg/m³ (as Hg).
SYNS: HYDRARGYRUM BIJODATUM (GERMAN) □ MERCURIC IODIDE □ MERCURIC IODIDE, solid □ MERCURIC IODIDE, solution □ MERCURIC IODIDE, RED □ MERCURY BINIODIDE □ RED MERCURIC IODIDE
CONSENSUS REPORTS: Mercury and its compounds are on the Community Right-To-Know List. Reported in EPA TSCA Inventory.
OSHA PEL: CL 0.1 mg(Hg)/m³ (skin)
ACGIH TLV: TWA 0.1 mg(Hg)/m³ (skin); BEI: 35 μg/g creatinine total inorganic mercury in urine preshift; 15 μg/g creatinine total inorganic mercury in blood at end of shift at end of workweek; (Proposed: (inhalable fraction) 0.1 mg/m³; Not Classifiable as a Human Carcinogen)
NIOSH REL: (Mercury, Aryl and Inorganic) CL 0.1 mg/m³ (skin)
SAFETY PROFILE: A human poison by ingestion. Poison experimentally by ingestion, skin contact, and intraperitoneal routes. An experimental teratogen. Violent reaction with chlorine trifluoride. When heated to decomposition it emits very toxic fumes of Hg and I⁻. See also MERCURY COMPOUNDS and IODIDES.

MDD250   CAS: 7774-29-0   HR: 3
MERCURY(II) IODIDE (solution)
mf: HgI₂   mw: 454.39

PROP: Scarlet red powder. Odorless. D: 6.36 @ 25°, mp: ~259°, bp: ~350°. Sol in water: 6mg/100 g @ 25°. IDLH 10 mg/m³ (as Hg).
CONSENSUS REPORTS: Reported in EPA TSCA Inventory. Mercury and its compounds are on the Community Right-To-Know List.
OSHA PEL: CL 0.1 mg(Hg)/m³ (skin)
ACGIH TLV: TWA 0.1 mg(Hg)/m³ (skin); BEI: 35 μg/g creatinine total inorganic mercury in urine preshift; 15 μg/g creatinine total inorganic mercury in blood at end of shift at end of workweek; (Proposed: (inhalable fraction) 0.1 mg/m³; Not Classifiable as a Human Carcinogen)
NIOSH REL: (Inorganic Mercury) TWA 0.05 mg(Hg)/m³
SAFETY PROFILE: A poison. When heated to decomposition it emits very toxic fumes of Hg and I⁻. See also MERCURY(II) IODIDE.

MDD750   CAS: 115-09-3   HR: 3
MERCURY METHYLCHLORIDE
mf: CH₃ClHg   mw: 251.08
PROP: White crystals with characteristic odor or plates from EtOH. D: 4.063, mp: 170°. IDLH 10 mg/m³ (as Hg).
SYNS: CASPAN □ CHLOROMETHYLMERCURY □ METHYLMERCURIC CHLORIDE □ METHYLMERCURY CHLORIDE □ MMC □ MONOMETHYL MERCURY CHLORIDE
CONSENSUS REPORTS: Mercury and its compounds are on the Community Right-To-Know List. EPA Genetic Toxicology Program.
OSHA PEL: TWA 0.01 mg(Hg)/m³; STEL 0.03 mg/m³ (skin)
ACGIH TLV: TWA 0.01 mg(Hg)/m³; BEI: 35 μg/g creatinine total inorganic mercury in urine preshift; 15 μg/g creatinine total inorganic mercury in blood at end of shift at end of workweek.
DFG MAK: Confirmed Animal Carcinogen with Unknown Relevance to Humans
NIOSH REL: TWA 0.05 mg(Hg)/m³
SAFETY PROFILE: Poison by ingestion, intramuscular, intravenous, and

intraperitoneal routes. Questionable carcinogen with experimental carcinogenic and teratogenic data. Human mutation data reported. Experimental reproductive effects. When heated to decomposition it emits very toxic fumes of Cl⁻ and Hg. See also MERCURY COMPOUNDS.

MDE250   CAS: 631-60-7   HR: 3
MERCURY MONOACETATE
DOT: UN 1629
mf: $C_2H_3O_2 \cdot Hg$ mw: 259.64
PROP: Light sensitive, colorless scales or plates from aq AcOH. Mp: decomp. Sol in dil acetic acid; insol in alc, ether. IDLH 10 mg/m³ (as Hg).
SYNS: MERCUROUS ACETATE □ MERCUROUS ACETATE, solid (DOT) □ MERCURY ACETATE
CONSENSUS REPORTS: Mercury and its compounds are on the Community Right-To-Know List.
OSHA PEL: CL 0.1 mg(Hg)/m³ (skin)
ACGIH TLV: TWA 0.1 mg(Hg)/m³ (skin); BEI: 35 µg/g creatinine total inorganic mercury in urine preshift; 15 µg/g creatinine total inorganic mercury in blood at end of shift at end of workweek.
DFG MAK: Confirmed Animal Carcinogen with Unknown Relevance to Humans
NIOSH REL: (Mercury, Aryl and Inorganic) CL 0.1 mg/m³ (skin)
DOT CLASSIFICATION: 6.1; Label: Poison
SAFETY PROFILE: A poison by ingestion and intraperitoneal routes. Moderately toxic by skin contact. When heated to decomposition it emits toxic fumes of Hg. See also MERCURY COMPOUNDS.

MDE750   CAS: 10415-75-5   HR: 3
MERCURY(I) NITRATE (1:1)
DOT: UN 1627
mf: $NO_3 \cdot Hg$ mw: 262.60
PROP: Crystals. IDLH 10 mg/m³ (as Hg).
SYNS: MERCUROUS NITRATE, solid (DOT) □ NITRATE MERCUREUX (FRENCH) □ NITRIC ACID, MERCURY(I) SALT
CONSENSUS REPORTS: Mercury and its compounds are on the Community Right-

To-Know List. Reported in EPA TSCA Inventory.
OSHA PEL: CL 0.1 mg(Hg)/m³ (skin)
ACGIH TLV: TWA 0.1 mg(Hg)/m³ (skin); BEI: 35 µg/g creatinine total inorganic mercury in urine preshift; 15 µg/g creatinine total inorganic mercury in blood at end of shift at end of workweek.
NIOSH REL: (Mercury, Aryl and Inorganic) CL 0.1 mg/m³ (skin)
DOT CLASSIFICATION: 6.1; Label: Poison
SAFETY PROFILE: Poison by ingestion and intraperitoneal routes. Moderately toxic by skin contact. A powerful oxidizer. Explodes on contact with red-hot carbon. Mixtures with phosphorus are impact-sensitive explosives. When heated to decomposition it emits very toxic fumes of Hg and NOₓ. See also MERCURY COMPOUNDS.

MDF000   CAS: 10045-94-0   HR: 3
MERCURY(II) NITRATE (1:2)
DOT: UN 1625
mf: $N_2O_6 \cdot Hg$ mw: 324.61
PROP: White-yellowish, deliq powder. Mp: 79°, bp: decomp, d: 4.39. IDLH 10 mg/m³ (as Hg).
SYNS: MERCURIC NITRATE □ MERCURY NITRATE □ MERCURY PERNITRATE □ NITRATE MERCURIQUE (FRENCH) □ NITRIC ACID, MERCURY(II) SALT
CONSENSUS REPORTS: Mercury and its compounds are on the Community Right-To-Know List. Reported in EPA TSCA Inventory.
OSHA PEL: CL 0.1 mg(Hg)/m³ (skin)
ACGIH TLV: TWA 0.1 mg(Hg)/m³ (skin); BEI: 35 µg/g creatinine total inorganic mercury in urine preshift; 15 µg/g creatinine total inorganic mercury in blood at end of shift at end of workweek.
NIOSH REL: (Mercury, Aryl and Inorganic) CL 0.1 mg/m³ (skin)
DOT CLASSIFICATION: 6.1; Label: Poison
SAFETY PROFILE: Poison by ingestion, skin contact, intraperitoneal, and subcutaneous routes. A powerful oxidizer. Probably an eye, skin, and mucous membrane irritant. Reacts with acetylene to

M

form the explosive mercury acetylide which is sensitive to heat, friction, or contact with sulfuric acid. Reaction with ethanol forms the explosive mercury fulminate. Reaction with isobutene forms an unstable explosive product. Forms explosive mixtures with phosphine (heat- and impact-sensitive), potassium cyanide (heat-sensitive), and sulfur. Violent reaction with phosphinic acid, hypophosphoric acid, unsaturated hydrocarbons, aromatics. Vigorous reaction with petroleum hydrocarbons. When heated to decomposition it emits very toxic fumes of Hg and NO_x. See also MERCURY COMPOUNDS, INORGANIC; and NITRATES.

MDF050   CAS: 73128-65-1   HR: 3
MERCURY, NITRILOTRIACETATE
mf: $C_{12}H_{12}HgN_2O_{12}$ mw: 576.85
PROP: IDLH 10 mg/m³ (as Hg).
SYNS: ACETIC ACID, NITRILOTRI-, MERCURY(II) COMPLEX □ MERCURATE(4-), BIS(N,N-BIS(CARBOXY-METHYL)GLYCINATO(3-)-N,O,O',O'')-, TETRAHYDROGEN
ACGIH TLV: TWA 0.01. STEL 0.03 mg/m³ (skin)
NIOSH REL: (MERCURY, ORGANO) TWA 0.01 mg/m³. STEL 0.03 mg/m³ (Sk)
SAFETY PROFILE: A poison by ingestion and intraperitoneal routes. When heated to decomposition it emits toxic vapors of NO_x and Hg.

MDF250   CAS: 1191-80-6   HR: 3
MERCURY OLEATE
DOT: UN 1640
mf: $C_{36}H_{66}O_4$•Hg   mw: 763.61
PROP: Yellowish-brown, somewhat transparent, ointment-like mass; odor of oleic acid. Practically insol in water; sltly sol in alc and ether; very sol in oils. Protect from light. IDLH 10 mg/m³ (as Hg).
SYNS: MERCURIC OLEATE, solid (DOT) □ OLEATE of MERCURY
CONSENSUS REPORTS: Mercury and its compounds are on the Community Right-To-Know List. Reported in EPA TSCA Inventory.

OSHA PEL: CL 0.1 mg(Hg)/m³ (skin)
ACGIH TLV: TWA 0.1 mg(Hg)/m³ (skin); BEI: 35 µg/g creatinine total inorganic mercury in urine preshift; 15 µg/g creatinine total inorganic mercury in blood at end of shift at end of workweek.
DFG MAK: Confirmed Animal Carcinogen with Unknown Relevance to Humans
NIOSH REL: (Mercury, Aryl and Inorganic) CL 0.1 mg/m³ (skin)
DOT CLASSIFICATION: 6.1; Label: Poison
SAFETY PROFILE: A poison. An FDA over-the-counter drug. When heated to decomposition it emits toxic fumes of Hg. See also MERCURY COMPOUNDS.

MDF350   CAS: 7784-37-4   HR: 3
MERCURY(II) ORTHOARSENATE
DOT: UN 1623
mf: $AsHO_4$•Hg   mw: 340.52
PROP: Yellow powder. Mp: decomp. Insol in water; sol in HCl or HNO_3. IDLH 10 mg/m³ (as Hg).
SYN: MERCURIC ARSENATE
CONSENSUS REPORTS: Mercury and its compounds, as well as arsenic and its compounds, are on the Community Right-To-Know List.
OSHA PEL: 0.01 mg(As)/m³; Cancer Hazard; CL 0.1 mg(Hg)/m³ (skin)
ACGIH TLV: TWA 0.01 mg/m³; Confirmed Human Carcinogen; BEI: 35 µ (As)/L inorganic arsenic and methylated metabolites in urine; BEI: 35 µg/g creatinine total inorganic mercury in urine preshift; 15 µg/g creatinine total inorganic mercury in blood at end of shift at end of workweek.
NIOSH REL: (Arsenic, Inorganic) CL 0.002 mg(As)/m³/15M; (Mercury, Aryl and Inorganic) CL 0.1 mg/m³ (skin)
DOT CLASSIFICATION: 6.1; Label: Poison
SAFETY PROFILE: Confirmed human carcinogen. A poison. When heated to decomposition it emits very toxic fumes of Hg and As. See also MERCURY and ARSENIC COMPOUNDS.

MDF750   CAS: 15829-53-5   HR: 3
MERCURY(I) OXIDE

mf: Hg$_2$O   mw: 417.22

PROP: Black to grayish-black powder. Mp: decomp @ 100°, d: 9.8. Insol in water; sol in HNO$_3$. Protect from light. IDLH 10 mg/m^3 (as Hg).

SYNS: MERCUROUS OXIDE, BLACK, solid (DOT) □ QUECKSILBEROXID (GERMAN)

CONSENSUS REPORTS: Mercury and its compounds are on the Community Right-To-Know List. Reported in EPA TSCA Inventory.

OSHA PEL: CL 0.1 mg(Hg)/m^3 (skin)
ACGIH TLV: TWA 0.1 mg(Hg)/m^3 (skin); BEI: 35 μg/g creatinine total inorganic mercury in urine preshift; 15 μg/g creatinine total inorganic mercury in blood at end of shift at end of workweek.
NIOSH REL: (Mercury, Aryl and Inorganic) CL 0.1 mg/m^3 (skin)

SAFETY PROFILE: A poison. Flammable by chemical reaction; an oxidizer. Explosive reaction with hydrogen peroxide, chlorine + ethylene. Reacts violently with molten potassium, molten sodium, S, (H$_2$S + BaO + air). Forms explosive mixtures with nonmetals [e.g., phosphorus (impact-sensitive), sulfur (friction-sensitive)]. Incompatible with alkali metals, reducing materials. Dangerous; when heated to decomposition it emits highly toxic fumes of Hg. See also MERCURY COMPOUNDS, INORGANIC.

MDG500   CAS: 7783-35-9   HR: 3
MERCURY(II) SULFATE (1:1)

mf: O$_4$S•Hg   mw: 296.65

PROP: White, light-sensitive crystalline powder; odorless. Mp: decomp, d: 6.47. Sol in HCl, hot dilute H$_2$SO$_4$, concentrated solns of NaCl. Protect from light. IDLH 10 mg/m^3 (as Hg).

SYNS: MERCURIC SULFATE, solid □ MERCURY BISULFATE □ MERCURY PERSULFATE □ SULFATE MERCURIQUE (FRENCH) □ SULFURIC ACID, MERCURY(2+) SALT (1:1)

CONSENSUS REPORTS: Mercury and its compounds are on the Community Right-To-Know List. Reported in EPA TSCA Inventory.

OSHA PEL: CL 0.1 mg(Hg)/m^3 (skin)
ACGIH TLV: TWA 0.1 mg(Hg)/m^3 (skin); BEI: 35 μg/g creatinine total inorganic mercury in urine preshift; 15 μg/g creatinine total inorganic mercury in blood at end of shift at end of workweek.
NIOSH REL: (Mercury, Aryl and Inorganic) CL 0.1 mg/m^3 (skin)

SAFETY PROFILE: Poison by ingestion and intraperitoneal routes. Moderately toxic by skin contact. When heated to decomposition it emits very toxic fumes of Hg and SO$_x$. See also MERCURY COMPOUNDS.

MDI000   CAS: 54-64-8   HR: 3
MERTHIOLATE SODIUM

mf: C$_9$H$_9$HgO$_2$S•Na   mw: 404.82

PROP: Air-stable cream crystals or powder, unstable in light. Sol in H$_2$O, EtOH; insol in Et$_2$O and C$_6$H$_6$. IDLH 10 mg/m^3 (as Hg).

SYNS: ((o-CARBOXYPHENYL)THIO)ETHYLMERCURY SODIUM SALT □ ELCIDE 75 □ ELICIDE □ o-(ETHYL-MERCURITHIO)BENZOIC ACID SODIUM SALT □ ETHYLMERCURITHIOSALICYLIC ACID SODIUM SALT □ MERCUROTHIOLATE □ MERFAMIN □ MERTHIOL-ATE □ MERTHIOLATE SALT □ MERTORGAN □ MERZONIN SODIUM □ SET □ SODIUM ETHYLMER-CURIC THIOSALICYLATE □ SODIUM-o-(ETHYLMER-CURITHIO)BENZOATE □ SODIUM ETHYLMERCURI-THIOSALICYLATE □ SODIUM MERTHIOLATE □ THIMEROSALATE □ THIMEROSOL □ THIOMER-SALATE

CONSENSUS REPORTS: Mercury and its compounds are on the Community Right-To-Know List. EPA Genetic Toxicology Program.

OSHA PEL: CL 0.1 mg(Hg)/m^3 (skin)
ACGIH TLV: TWA 0.1 mg(Hg)/m^3 (skin); BEI: 35 μg/g creatinine total inorganic mercury in urine preshift; 15 μg/g creatinine total inorganic mercury in blood at end of shift at end of workweek.
NIOSH REL: (Mercury, Aryl and Inorganic) CL 0.1 mg/m^3 (skin)

SAFETY PROFILE: Poison by ingestion, subcutaneous, and intravenous routes. Experimental teratogenic and reproductive effects. An eye irritant. Questionable

M

carcinogen with experimental neoplastigenic data. Mutation data reported. An ophthalmic preservative, a topical anti-infective, topical veterinary antibacterial and antifungal agent. An FDA over-the-counter drug. When heated to decomposition it emits very toxic fumes of Hg, Na_2O, and SO_x. See also MERCURY COMPOUNDS.

MDI500   CAS: 54-04-6   HR: 3
MESCALINE
mf: $C_{11}H_{17}NO_3$ mw: 211.29
PROP: Crystals. Mp: 35–36°, bp: 180° @ 11 mm. Mod sol in water; sol in alc, chloroform, and benzene; practically insol in ether and pet ether.
SYNS: MEZCALINE □ MEZCLINE □ 3,4,5-TRIMETHO-XYBENZENEETHANAMINE □ 3,4,5-TRIMETHOXY-PHENETHYLAMINE
SAFETY PROFILE: Poison by intraperitoneal and intravenous routes. Moderately toxic by ingestion route. An experimental teratogen. Other experimental reproductive effects. Human systemic effects by intramuscular route: euphoria and hallucinations, distorted perceptions. A psychotoimetic agent (a drug of abuse). When heated to decomposition it emits toxic fumes of NO_x.

MDJ750   CAS: 141-79-7   HR: 3
MESITYL OXIDE
DOT: UN 1229
mf: $C_6H_{10}O$ mw: 98.16
PROP: Oily, colorless liquid; strong odor. Mp: −59°, bp: 130.0°, flash p: 87°F (CC), d: 0.8539 @ 20°/4°, autoign temp: 652°F, vap press: 10 mm @ 26.0°, vap d: 3.38. Solidifies @ 41.5°; somewhat sol in water @ 20°. Misc in alc and ether and with most organic liquids.
SYNS: ISOBUTENYL METHYL KETONE □ ISOPROP-YLIDENEACETONE □ MESITYLOXID (GERMAN) □ MESITYLOXYDE (DUTCH) □ METHYL ISOBUTENYL KETONE □ 4-METHYL-3-PENTENE-2-ONE □ 4-METHYL-3-PENTEN-2-ON (DUTCH, GERMAN) □ 2-METHYL-2-PENTEN-4-ONE □ 4-METHYL-3-PENTEN-2-ONE □ 4-METIL-3-PENTEN-2-ONE (ITALIAN) □ OSSIDO di MESITILE (ITALIAN) □ OXYDE de MESITYLE (FRENCH)

CONSENSUS REPORTS: Reported in EPA TSCA Inventory.
OSHA PEL: TWA 15 ppm; STEL 25 ppm
ACGIH TLV: TWA 15 ppm; STEL 25 ppm
DFG MAK: 25 ppm (100 mg/m³)
NIOSH REL: (Ketones) TWA 40 mg/m³
DOT CLASSIFICATION: 3; Label: Flammable Liquid
SAFETY PROFILE: Poison by intraperitoneal route. Moderately toxic by ingestion. Mildly toxic by inhalation and skin contact. Human systemic effects by inhalation: conjunctiva irritation. This compound is highly irritating to all tissues on contact; its vapors also are irritating. High concentrations are narcotic. It is readily absorbed through intact skin. Single exposures tend to indicate that this ketone has greater acute and narcotic action than isophorone. It can have harmful effects upon the kidneys and liver, and may damage the eyes and lungs to a serious degree. Prolonged exposure can injure liver, kidneys, and lungs. It can cause opaque cornea, keratoconus, and extensive necrosis of cornea. Dangerous fire hazard when exposed to heat, sparks, or flame; can react with oxidizing materials. Reacts violently with 2-amino ethanol, chlorosulfonic acid, ethylene diamine, HNO_3, oleum, H_2SO_4. An insect repellent. To fight fire, use alcohol foam, CO_2, dry chemical. When heated to decomposition it emits acrid smoke and irritating fumes. See also KETONES.

MDL500   CAS: 2244-11-3   HR: 3
MESOXALYLUREA MONOHYDRATE
mf: $C_4H_2N_2O_4 \cdot H_2O$ mw: 160.10
PROP: White crystals, become pink on exposure to air. Colorless, aqueous solution imparts pink color to skin. Mp: 170° (decomp); sol in water and alc.
SYNS: ALLOXAN MONOHYDRATE □ MESOXALYL-CARBAMIDE MONOHYDRATE □ 2,4,5,6(1H,3H)-PYRIMI-DINETETRONE HYDRATE □ 2,4,5,6-TETRAOXOHEXA-HYDROPYRIMIDINE HYDRATE
SAFETY PROFILE: Poison by intravenous route. An experimental teratogen. Experimental reproductive effects. Mutation

data reported. When heated to decomposition it emits toxic fumes of NO_x.

MDM100 CAS: 57837-19-1 HR: 2
METALAXYL
mf: $C_{15}H_{21}NO_4$ mw: 279.35
PROP: White crystals.
SYNS: DL-ALANINE, N-(2,6-DIMETHYLPHENYL)-N-(METHOXYACETYL)-, METHYL ESTER (9CI) □ APRON □ APRON 2E □ APRON FL □ CG 117 □ CGA 48988 □ D,L-N-(2,6-DIMETHYLPHENYL)-N-(2'-METHOXYACET-YL)ALANINATE de METHYLE □ N-(2,6-DIMETHYL-PHENYL)-N-(METHOXYACETYL)-ALANINE METHYL ESTER □ N-(2,6-DIMETHYLPHENYL)-N-(METHOXY-ACETYL)-DL-ALANINE METHYL ESTER □ METALAXIL □ METAXANIN □ RIDOMIL □ RIDOMIL 2E □ SUBDUE □ SUBDUE 2E □ SUBDUE 5SP
SAFETY PROFILE: Moderately toxic by ingestion. When heated to decomposition it emits toxic fumes of NO_x.

MDM775 CAS: 587-98-4 HR: 2
METANIL YELLOW
mf: $C_{18}H_{15}N_3O_3S•Na$ mw: 376.41
PROP: Brownish-yellow powder or crystals. Sol in water, alc; moderately sol in benzene and ether; sltly sol in acetone.
SYNS: ACIDIC METANIL YELLOW □ ACID LEATHER YELLOW PRW □ ACID LEATHER YELLOW R □ ACID METANIL YELLOW □ ACID YELLOW 36 □ AIZEN METANIL YELLOW □ AMACID YELLOW M □ BRASILAN METANIL YELLOW □ BUCACID METANIL YELLOW □ CALCOCID YELLOW MXXX □ C.I. 13065 □ C.I. ACID YELLOW 36 □ C.I. ACID YELLOW 36 MONOSODIUM SALT □ DIACID METANIL YELLOW □ ENIACID METANIL YELLOW GN □ EXT D&C YELLOW No. 1 □ FENAZO YELLOW M □ HIDACID METANIL YELLOW □ HISPACID YELLOW MG □ JAVA METANIL YELLOW G □ KITON ORANGE MNO □ KITON YELLOW MS □ METANILE YELLOW O □ METANIL YELLOW 1955 □ METANIL YELLOW C □ METANIL YELLOW E □ METANIL YELLOW EXTRA □ METANIL YELLOW F □ METANIL YELLOW G □ METANIL YELLOW GRIESBACH □ METANIL YELLOW K □ METANIL YELLOW KRSU □ METANIL YELLOW M3X □ METANIL YELLOW O □ METANIL YELLOW PL □ METANIL YELLOW S □ METANIL YELLOW SUPRA P □ METANIL YELLOW VS □ METANIL YELLOW WS □ METANIL YELLOW Y □ METANIL YELLOW YK □ MITSUI METANIL YELLOW □ MONOAZO □ REMADERM YELLOW HPR □ SHIKISO METANIL YELLOW □ SYMULON METANIL YELLOW □ TAKAOKA METANIL YELLOW □ TERTRACID YELLOW M □ TROPAEOLIN G □ VONDACID METANIL YELLOW G □ 11363 YELLOW □ YODOCHROME METANIL YELLOW
CONSENSUS REPORTS: Reported in EPA TSCA Inventory.
SAFETY PROFILE: Moderately toxic by intraperitoneal route. Experimental reproductive effects. Human mutation data reported. When heated to decomposition it emits toxic fumes of SO_x, NO_x, and Na_2O. See also AROMATIC AMINES and SULFONATES.

MDN250 CAS: 79-41-4 HR: 3
METHACRYLIC ACID
DOT: UN 2531
mf: $C_4H_6O_2$ mw: 86.10
PROP: Corrosive liquid or colorless crystals; repulsive odor. Mp: 16°, bp: 163°, flash p: 171°F (COC), d: 1.014 @ 25° (glacial), vap press: 1 mm @ 25.5°. Sol in warm water; misc with alc, ether.
SYNS: ACRYLIC ACID, 2-METHYL- □ KYSELINA METHAKRYLOVA □ METHACRYLIC ACID (ACGIH, OSHA) □ METHACRYLIC ACID, inhibited (DOT) □ α-METHYLACRYLIC ACID □ 2-METHYLPROPENOIC ACID □ 2-PROPENOIC ACID, 2-METHYL-(9CI) □ PROPIONIC ACID, 2-METHYLENE-
CONSENSUS REPORTS: Reported in EPA TSCA Inventory.
OSHA PEL: TWA 20 ppm (skin)
ACGIH TLV: TWA 20 ppm
DOT CLASSIFICATION: 8; Label: Corrosive
SAFETY PROFILE: Poison by intraperitoneal route. Moderately toxic by ingestion and skin contact. Corrosive to skin, eyes, and mucous membranes. Mutation data reported. Flammable when exposed to heat, flame, or oxidizers. A storage hazard; exothermic polymerization may occur spontaneously. To fight fire, use alcohol foam, spray, mist, dry chemical. When heated to decomposition it emits acrid smoke and irritating fumes.

MDN500 CAS: 79-39-0 HR: 3
METHACRYLIC ACID AMIDE
mf: C_4H_7NO mw: 85.12
PROP: A solid. Mp: 105–107°.

SYNS: METHACRYLIC AMIDE □ 2-
METHYLACRYLAMIDE □ α-METHYL ACRYLIC AMIDE
□ 2-METHYLPROPENAMIDE □ USAF RH-1
CONSENSUS REPORTS: Reported in EPA
TSCA Inventory.
SAFETY PROFILE: Poison by
intraperitoneal route. Moderately toxic by
ingestion. Human systemic effects by
inhalation: degenerative brain changes and
liver and kidney changes. When heated to
decomposition it emits toxic fumes of NO_x.

MDN525   HR: D
METHACRYLIC ACID-DIVINYLBENZENE
COPOLYMER
SAFETY PROFILE: When heated to
decomposition it emits acrid smoke and
irritating fumes.

MDO750   CAS: 76-99-3   HR: 3
METHADONE
mf: $C_{21}H_{27}NO$ mw: 309.49
PROP: Narcotic analgesic.
SYNS: ADANON □ AMIDONE □ DIAMINON □
DOLOPHINE □ HEPTADONE □ HEPTANON □
KETALGIN □ MECODIN □ PHENADONE □
PHYSEPTONE □ POLAMIDONE
CONSENSUS REPORTS: EPA Genetic
Toxicology Program.
SAFETY PROFILE: Poison by ingestion,
intraperitoneal, intravenous, subcutaneous,
and intraduodenal routes. Human systemic
effects: coma, nausea or vomiting,
respiratory changes, respiratory depression,
somnolence. An experimental teratogen.
Experimental reproductive effects. Caution:
Abuse leads to habituation or addiction.
When heated to decomposition it emits
toxic fumes of NO_x. See also
METHADONE HYDROCHLORIDE.

MDP240   CAS: 15284-15-8   HR: 3
d-METHADONE HYDROCHLORIDE
mf: $C_{21}H_{27}NO•ClH$ mw: 345.95
PROP: Crystals from 2-propanol. Mp:
243–244°.
SYN: d-DOLOPHINE HYDROCHLORIDE
SAFETY PROFILE: Poison by intravenous
and intraperitoneal routes. Human systemic

effects by ingestion: euphoria, somnolence
(general depressed activity), and analgesia.
Caution: Abuse leads to habituation or
addiction. When heated to decomposition it
emits toxic fumes of Cl⁻ and NO_x. See also
METHADONE HYDROCHLORIDE.

MDP770   CAS: 5967-73-7   HR: 3
l-METHADONE HYDROCHLORIDE
mf: $C_{21}H_{27}NO•ClH$ mw: 345.95
PROP: dl Form: Crystals. Mp: 245–246°.
SYNS: 1-6-DIMETHYLAMINO-4,4-DIPHENYL-3-
HEPTANONE HYDROCHLORIDE □ LEVADONE □
LEVOTHYL □ POLAMIDON
SAFETY PROFILE: Poison by intravenous,
subcutaneous, and intraperitoneal routes.
Human systemic effects by ingestion:
hallucinations, distorted perceptions, and
analgesia. Caution: Abuse leads to
habituation or addiction. When heated to
decomposition it emits very toxic fumes of
Cl⁻ and NO_x. See also METHADONE
HYDROCHLORIDE.

MDQ250   CAS: 438-41-5   HR: 3
METHAMINODIAZEPOXIDE HYDRO-
CHLORIDE
mf: $C_{16}H_{14}ClN_3O•ClH$ mw: 336.24
PROP: Crystals from MeOH. Mp: 213°.
SYNS: ANSIACAL □ A-POXIDE □ BENT □ BENZO-
DIAPIN □ CALMODEN □ CEBRUM □ CHLORDI-
AZACHEL □ CHLORDIAZEPOXIDE HYDROCHLORIDE
□ CHLORDIAZEPOXIDE MONOHYDROCHLORIDE □
CHLORIDEAZEPOXIDE HYDROCHLORIDE □ 7-CHL-
ORO-2-METHYLAMINO-5-PHENYL-3H-1,4-BENZODI-
AZEPIN, 4-OXIDE, HYDROCHLORIDE □ 7-CHLORO-N-
METHYL-5-PHENYL-EH-1,4-BENZODIAZEPIN-2-
AMINE-4-OXIDE, MONOHYDROCHLORIDE □ CORAX
□ DIAZACHEL (OBS.) □ DROXOL □ ELENIUM □
EQUIBRAL □ J-LIBERTY □ KALMOCAPS □ LABICAN □
LENTOTRAN □ LIBRIUM □ LIBRIUM HYDRO-
CHLORIDE □ METHAMINODIAZEPINE HYDROCHL-
ORIDE □ MILDMEN □ MURCIL □ NAPOTON □
NOVOSED □ PSICHIAL □ PSICOSAN □ RELIBERAN □
RO 5-0690 □ SEREN VITA □ SK-LYGEN □ SOPHIAMIN
□ TENSINYL □ TIMOSIN □ TRAKIPEAL □ VIANSIN □
VIOPSICOL
CONSENSUS REPORTS: EPA Genetic
Toxicology Program.
SAFETY PROFILE: Poison by
intraperitoneal and intravenous routes.

Moderately toxic by ingestion and subcutaneous routes. An experimental teratogen. Experimental reproductive effects. Human systemic effects: ataxia, distorted perceptions, hallucinations, somnolence, and surface EEG changes. Mutation data reported. A minor tranquilizer. When heated to decomposition it emits very toxic fumes of HCl and NO_x.

MDQ500   CAS: 826-10-8   HR: 3
l-METHAMPHETAMINE HYDRO-
** CHLORIDE**
mf: $C_{10}H_{15}N \cdot ClH$ mw: 185.72
PROP: Crystals; bitter taste. Mp: 170–175°. Sol in water, alc, and chloroform; almost insol in ether.
SYNS: ADIPEX □ l-DESOXYEPHEDRINE HYDRO-CHLORIDE □ (−)-N-α-DIMETHYLPHENETHYLAMINE HYDROCHLORIDE □ "METH" □ l-N-METHYL-β-PHENYLISOPROPYLAMINE HYDROCHLORIDE □ "SPEED" □ SYNDROX
SAFETY PROFILE: Poison by ingestion, intravenous, intraperitoneal, and subcutaneous routes. An experimental teratogen. Experimental reproductive effects. A powerful central nervous system stimulant. *Caution:* Excessive use may lead to tolerance and habituation. When heated to decomposition it emits very toxic fumes of HCl and NO_x. See also BENZEDRINE.

MDQ750   CAS: 74-82-8   HR: 3
METHANE
DOT: UN 1971/UN 1972
mf: CH_4 mw: 16.05
PROP: Colorless, odorless, tasteless, flammable gas; needles when solid. Mp: −182.6°, bp: −161.5°, lel: 5.3%, uel: 15%, fp: −183.2°, d: 0.554 @ 0°/4° (air = 1) or 0.7168 g/L, autoign temp: 650°, vap d: 0.6, flash p: −368.6°F. Sol in water, alc, and ether.
SYNS: FIRE DAMP □ MARSH GAS □ METHANE, compressed (UN 1971) (DOT) □ METHANE, refrigerated liquid (cryogenic liquid) (UN 1972) (DOT) □ METHYL HYDRIDE □ NATURAL GAS, compressed (with high methane content) (UN 1971) (DOT) □ NATURAL GAS, refrigerated liquid (cryogenic liquid) (with high methane content) (UN 1972) (DOT)

CONSENSUS REPORTS: Reported in EPA TSCA Inventory.
DOT CLASSIFICATION: 2.1; Label: Flammable Gas
SAFETY PROFILE: A simple asphyxiant. Very dangerous fire and explosion hazard when exposed to heat or flame. Reacts violently with powerful oxidizers (e.g., bromine pentafluoride, chlorine trifluoride, chlorine, fluorine, iodine heptafluoride, dioxygenyl tetrafluoroborate, dioxygen difluoride, trioxygen difluoride, liquid oxygen, ClO_2, NF_3, OF_2). Incompatible with halogens or interhalogens in air (forms explosive mixtures). Explosive in the form of vapor when exposed to heat or flame. To fight fire, stop flow of gas. See also ARGON for a description of asphyxiants.

MDQ825   CAS: 7526-26-3   HR: 3
METHANEPHOSPHONIC ACID, DIPHEN-
** YL ESTER**
mf: $C_{13}H_{13}O_3P$ mw: 248.23
SYNS: DIPHENYL METHANEPHOSPHONATE □ DIPHENYL METHYLPHOSPHONATE □ PHOSPHONIC ACID, METHYL-, DIPHENYL ESTER
SAFETY PROFILE: A poison by ingestion. When heated to decomposition it emits toxic vapors of PO_x.

MDR250   CAS: 75-75-2   HR: 3
METHANESULFONIC ACID
mf: CH_4O_3S mw: 96.11
PROP: Solid or liquid. D: 1.4812 @ 18°/4°, mp: 20°, bp: 167° @ 10 mm. Sol in water, alc, and ether. Corrosive to iron, steel, brass, copper, and lead.
SYN: KYSELINA METHANSULFONOVA (CZECH)
CONSENSUS REPORTS: Reported in EPA TSCA Inventory.
SAFETY PROFILE: Poison by ingestion and intraperitoneal routes. May be corrosive to skin, eyes, and mucous membranes. Explosive reaction with ethyl vinyl ether. Incompatible with hydrogen fluoride. When heated to decomposition it emits toxic fumes of SO_x. See also SULFONATES.

M

MDT250   CAS: 340-56-7   HR: 3
METHAQUALONE HYDROCHLORIDE
mf: $C_{16}H_{14}N_2O \cdot ClH$ mw: 286.78
PROP: Crystals. Mp: 255–265°. Sol in ether, ethanol; almost insol in water.
SYNS: MELSEDIN □ METHYLQUINAZOLONE HYDROCHLORIDE □ 2-METHYL-3-TOLYLCHIN-AZOLON-4 HYDROCHLORIDE (GERMAN) □ 2-METHYL-3-o-TOLYL-4(3H)-QUINAZOLINONE HYDROCHLORIDE □ 2-METHYL-3-(o-TOLYL)-4-QUINAZOLONE HYDROCHLORIDE □ MTQ HYDROCHLORIDE □ OPTIMIL □ PAREST □ SOMNAFAC □ TUAZOLE
SAFETY PROFILE: Poison by ingestion, intraperitoneal, and intravenous routes. An experimental teratogen. Experimental reproductive effects. Human systemic effects: convulsions, dyspnea, pulse rate increase. When heated to decomposition it emits very toxic fumes of NO_x and HCl.

MDT600   CAS: 51-57-0   HR: 3
METHEDRINE
mf: $C_{10}H_{15}N \cdot ClH$ mw: 185.72
PROP: Crystals from EtOH. Mp: 172°.
SYNS: ADIPEX □ DEOFED □ d-DEOXYEPHEDRINE HYDROCHLORIDE □ DESOXO-5 □ d-DESOXYE-PHEDRINE HYDROCHLORIDE □ DESOXYFED □ DESOXYN □ DESOXYNE □ DESTIM □ DESYPHED □ DEXOVAL □ DEXTIM □ DOXYFED □ DRINALFA □ EFROXINE □ EUFODRIANL □ GERVOT □ ISOPHEN □ METAMPHETAMINE HYDROCHLORIDE □ (+)-METH-AMPHETAMINE CHLORIDE □ METHAMPHETAMINE HYDROCHLORIDE □ (+)-METHAMPHETAMINE HYDROCHLORIDE □ d-METHAMPHETAMINE HYDROCHLORIDE □ METHAMPHETAMINIUM CHLORIDE □ METHEDRINE HYDROCHLORIDE □ METHYLAMPHETAMINE HYDROCHLORIDE □ d-METHYLAMPHETAMINE HYDROCHLORIDE □ N-METHYLAMPHETAMINE HYDROCHLORIDE □ METHYLISOMYN □ NORODIN HYDROCHLORIDE □ PERVITIN □ PHILOPON □ SOXYSYMPAMINE □ SYNDROX □ TONEDRON
CONSENSUS REPORTS: EPA Genetic Toxicology Program. Reported in EPA TSCA Inventory.
SAFETY PROFILE: Poison by ingestion, subcutaneous, intravenous, and intraperitoneal routes. An experimental teratogen. Experimental reproductive effects. Questionable carcinogen with experimental neoplastigenic data. When heated to decomposition it emits toxic fumes of NO_x and HCl. See also BENZEDRINE and various amphetamines.

MDT740   CAS: 59-51-8   HR: 2
dl-METHIONINE
mf: $C_5H_{11}NO_2S$ mw: 149.23
PROP: White crystalline platelets; characteristic odor. Mp: 281° (decomp). Sol in water, dil acids, and alkalies; very sltly sol in alc; insol in ether.
SYNS: ACIMETION □ BANTHIONINE □ CYNARON □ DYPRIN □ LOBAMINE □ MEONINE □ MERTIONIN □ METHILANIN □ (±)-METHIONINE □ METIONE □ NESTON
CONSENSUS REPORTS: EPA Genetic Toxicology Program. Reported in EPA TSCA Inventory.
SAFETY PROFILE: Moderately toxic by ingestion and other routes. An experimental teratogen. Experimental reproductive effects. When heated to decomposition it emits toxic fumes of SO_x and NO_x. See also l-METHIONINE.

MDT750   CAS: 63-68-3   HR: 1
l-METHIONINE
mf: $C_5H_{11}NO_2S$ mw: 149.23
PROP: White, crystalline powder or platelets; faint odor. Mp: 281° (decomp), d: 1.340. Sol in water, dil acids, and alkalies; insol in abs alc, alc, benzene, acetone, ether.
SYNS: l-α-AMINO-γ-METHYLMERCAPTOBUTYRIC ACID □ l(−)-AMINO-γ-METHYLTHIOBUTYRIC ACID □ 2-AMINO-4-(METHYLTHIO)BUTYRIC ACID □ CYMETHION □ LIQUIMETH □ METHIONINE □ l-(−)-METHIONINE □ l-γ-METHYLTHIO-α-AMINOBUTYRIC ACID
CONSENSUS REPORTS: Reported in EPA TSCA Inventory. EPA Genetic Toxicology Program.
SAFETY PROFILE: Mildly toxic by ingestion and intraperitoneal routes. Human mutation data reported. An experimental teratogen. Experimental reproductive effects. An essential sulfur-containing amino acid. When heated to decomposition it emits very toxic fumes of NO_x and SO_x.

MDU500 **CAS: 309-36-4** **HR: 3**
METHOHEXITAL SODIUM
mf: $C_{14}H_{17}N_2O_3 \cdot Na$ mw: 284.32
PROP: Minute crystals. Mp: 60–64°. Sol in water.
SYNS: 5-ALLYL-1-METHYL-5-(1-METHYL-2-PENTYN-YL)BARBITURIC ACID SODIUM SALT □ BREVIMYTAL □ BREVITAL SODIUM □ BRIETAL SODIUM □ ENALLYNYMAL SODIUM □ LILLY 22451 □ METHO-HEXITONE SODIUM □ 1-METHYL-5-ALLYL-5-(1-METHYL-2-PENTYNYL)BARBITURIC ACID SODIUM SALT □ SODIUM-dl-5-ALLYL-1-METHYL-5-(1-METHYL-2-PENTYNYL)BARBITURATE □ SODIUM METHO-HEXITAL □ SODIUM METHOHEXITONE □ SODIUM dl-1-METHYL-5-ALLYL-5-(1-METHYL-2-PENTYNYL)-BARBITURATE
SAFETY PROFILE: Poison by intravenous and implant routes. Human systemic effects by intravenous route: blood pressure lowering, gastrointestinal effects, and allergic dermatitis. An FDA proprietary drug. *Caution:* Excessive use may lead to addiction or habituation. Allergenic effects by intravenous route. When heated to decomposition it emits toxic fumes of Na_2O and NO_x. See also BARBITURATES.

MDU600 **CAS: 16752-77-5** **HR: 3**
METHOMYL
mf: $C_5H_{10}N_2O_2S$ mw: 162.23
PROP: White, crystalline solid; sulfurous odor. Mp: 78–79°. Moderately sol in water; very sol in Me_2CO, EtOH, MeOH, and toluene.
SYNS: DU PONT INSECTICIDE 1179 □ ENT 27,341 □ INSECTICIDE 1,179 □ LANNATE □ MESOMILE □ ((METHYL N-((METHYLAMINO)CARBONYL)OXY)ETH-ANIMIDO)THIOATE □ METHYL-N-((METHYLCAR-BAMOYL)OXY)THIOACETIMIDATE □ S-METHYL N-[METHYLCARBAMOYLOXY]THIOACETIMIDATE □ 2-METHYLTHIO-ACETALDEHYD-O-(METHYLCARB-AMOYL)-OXIM (GERMAN) □ 2-METHYLTHIO-PROP-IONALDEHYD-O-(METHYLCARBAMOYL)-OXIM (GERMAN) □ METOMIL (ITALIAN) □ NU-BAIT II □ NUDRIN □ RCRA WASTE NUMBER P066 □ 3-THI-ABUTAN-2-ONE, O-(METHYLCARBAMOYL)OXIME □ WL 18236
CONSENSUS REPORTS: EPA Genetic Toxicology Program. EPA Extremely Hazardous Substances List.
OSHA PEL: TWA 2.5 mg/m³
ACGIH TLV: TWA 2.5 mg/m³; Not Classifiable as a Human Carcinogen

SAFETY PROFILE: Poison by ingestion, inhalation, and subcutaneous routes. Mildly toxic by skin contact. When heated to decomposition it emits very toxic fumes of NO_x and SO_x.

MDU750 **CAS: 522-23-6** **HR: 3**
METHOPHENAZINE DIFUMARATE
mf: $C_{31}H_{36}ClN_3O_5S \cdot C_8H_4O_8$ mw: 826.33
PROP: A solid. Mp: 190–194°.
SYNS: FRENOLON DIFUMARATE □ METHOPHEN-AZATE ACID FUMARATE □ PHRENOLAN □ T-82 DIFUMARATE □ 3,4,5-TRIMETHOXY-BENZOIC ACID 2-(4-(3-(2-CHLOROPHENOTHIAZIN-10-YL)PROPYL)-1-PIPERAZINYL)ETHYL ESTER, DIFUMARATE
SAFETY PROFILE: Poison by intraperitoneal, subcutaneous, and intravenous routes. Moderately toxic by ingestion. An experimental teratogen. When heated to decomposition it emits very toxic fumes of Cl^-, SO_x, and NO_x. See also ESTERS.

MDV500 **CAS: 59-05-2** **HR: 3**
METHOTREXATE
mf: $C_{20}H_{22}N_8O_5$ mw: 454.50
SYNS: AMETHOPTERIN □ 4-AMINO-4-DEOXY-N[10]-METHYLPTEROYLGLUTAMATE □ 4-AMINO-4-DEOXY-N[10]-METHYLPTEROYLGLUTAMIC ACID □ 4-AMINO-10-METHYLFOLIC ACID □ 4-AMINO-N[10]-METHYLPTEROYLGLUTAMIC ACID □ ANTIFOLAN □ N-BISMETHYLPTEROYLGLUTAMIC ACID □ CL-14377 □ l-(+)-N-(p-(((2,4-DIAMINO-6-PTERIDINYL)METHYL)-METHYLAMINO)BENZOYL)GLUTAMIC ACID □ EMT 25,299 □ EMTEXATE □ HDMTX □ METHOPTERIN □ METHOTEXTRATE □ METHYLAMINOPTERIN □ MTX □ NCI-C04671 □ NSC-740 □ R 9985
CONSENSUS REPORTS: IARC Cancer Review: Group 3 IMEMDT 7,241,87; Animal Inadequate Evidence IMEMDT 26,267,81; Human Inadequate Evidence IMEMDT 26,267,81. NCI Carcinogenesis Studies (ipr); No Evidence: mouse, rat CANCAR 40,1935,77. Reported in EPA TSCA Inventory.
SAFETY PROFILE: A human poison by intraspinal route. Poison experimentally by ingestion, intravenous, subcutaneous, and intraperitoneal routes. Human teratogenic

M

effects by ingestion: developmental abnormalities of the craniofacial area and the musculoskeletal system. Human systemic effects by multiple routes: thrombocytopenia (decrease in the number of blood platelets), bone marrow changes, other blood changes, cerebral spinal fluid effects, eye effects, blood pressure lowering, cough, dyspnea, fibrosis (pneumoconiosis), cyanosis, gastrointestinal effects, fatty liver degeneration, hepatitis, impairment of liver function tests, other liver changes, fever, effects on inflammation or mediation of inflammation, leukopenia. Human mutation data reported. Experimental reproductive effects. A human eye irritant. Questionable human carcinogen producing leukemia, Hodgkin's disease, and skin tumors. An FDA proprietary drug. A chemotherapeutic agent. When heated to decomposition it emits toxic fumes including NO_x.

MDW000 CAS: 61-16-5 HR: 3
METHOXAMINE HYDROCHLORIDE
mf: $C_{11}H_{17}NO_3 \cdot ClH$ mw: 247.71
PROP: Crystals. Mp: 212–216°. Very sol in water; practically insol in ether, benzene, and chloroform.
SYNS: 2-AMINO-1-(2,5-DIMETHOXYPHENYL)-1-PROPANOL HYDROCHLORIDE □ α-(1-AMINOETHYL)-2,5-DIMETHOXYBENZYL ALCOHOL HYDROCHLORIDE □ β-(2,5-DIMETHOXYPHENYL)-β-HYDROXYISOPROPYLAMINE HYDROCHLORIDE □ β-HYDROXY-β-(2,5-DIMETHOXYPHENYL)-ISOPROPYLAMINE HYDROCHLORIDE □ PRESSOMIN HYDROCHLORIDE □ VASOXINE □ VASOXINE HYDROCHLORIDE □ VASOXYL HYDROCHLORIDE
SAFETY PROFILE: Poison by ingestion, intravenous, and intraperitoneal routes. Experimental reproductive effects. An FDA proprietary drug. When heated to decomposition it emits very toxic fumes of HCl and NO_x.

MDW275 CAS: 625-45-6 HR: 2
METHOXYACETIC ACID
mf: $C_3H_6O_3$ mw: 90.09
PROP: A liquid. D: 1.177 @ 20°/4°, bp: 203–204°.
SYN: 2-METHOXYACETIC ACID

CONSENSUS REPORTS: Reported in EPA TSCA Inventory.
DFG MAK: 5 ppm
SAFETY PROFILE: Moderately toxic by ingestion. An experimental teratogen. Experimental reproductive effects. Mutation data reported. When heated to decomposition it emits acrid smoke and irritating fumes.

MDW750 CAS: 100-06-1 HR: 3
4'-METHOXYACETOPHENONE
mf: $C_9H_{10}O_2$ mw: 150.19
PROP: Colorless to pale-yellow fused solid; hawthorn odor. Plates from alc. Mp: 38–39°, bp: 250°. Sol in fixed oils, propylene glycol; misc in glycerin.
SYNS: ACETANISOLE (FCC) □ p-ACETYLANISOLE □ 4-ACETYLANISOLE □ BANANOTE □ ETHANONE, 1-(4-METHOXYPHENYL)-(9CI) □ FEMA No. 2005 □ LINARODIN □ 4-METHOXYACETOFENON □ 4-METHOXYACETOPHENONE □ p-METHOXYACETOPHENONE □ 4-METHOXYPHENYL METHYL KETONE □ p-METHOXYPHENYL METHYL KETONE □ NOVATONE □ VANANOTE
CONSENSUS REPORTS: Reported in EPA TSCA Inventory.
DOT CLASSIFICATION: 3; Label: Flammable Liquid
SAFETY PROFILE: Moderately toxic by ingestion. Human systemic effects by inhalation: pulse rate increase without fall in blood pressure and blood pressure elevation. A skin irritant. Flammable liquid. When heated to decomposition it emits acrid smoke and irritating fumes. See also KETONES.

MDX300 CAS: 52098-17-6 HR: 3
2-(METHOXYACETYL)-1-METHYLINDOLE
mf: $C_{12}H_{13}NO_2$ mw: 203.26
SYNS: KETONE, METHOXYMETHYL 1-METHYL-2-INDOLYL □ METHOXYMETHYL 1-METHYL-2-INDOLYL KETONE
DOT CLASSIFICATION: 3; Label: Flammable Liquid
SAFETY PROFILE: A poison by intraperitoneal route. A flammable liquid.

When heated to decomposition it emits toxic vapors of NO$_x$.

MDX310 CAS: 52098-18-7 HR: 3
2-(METHOXYACETYL)-3-METHYLIN-DOLE

mf: $C_{12}H_{13}NO_2$ mw: 203.26

SYNS: KETONE, METHOXYMETHYL 3-METHYL-2-INDOLYL- □ METHOXYMETHYL 3-METHYL-2-INDOLYL KETONE

DOT CLASSIFICATION: 3; Label: Flammable Liquid

SAFETY PROFILE: A poison by intraperitoneal route. A flammable liquid. When heated to decomposition it emits toxic vapors of NO$_x$.

MED500 CAS: 105-13-5 HR: 2
p-METHOXYBENZYL ALCOHOL

mf: $C_8H_{10}O_2$ mw: 138.18

PROP: Needles or colorless liquid; floral odor. D: 1.113 @ 15°/15°, refr index: 1.543, mp: 25°, bp: 135–136° @ 12 mm, flash p: 210°F. Insol in water; sol in alc, ether, and fixed oils; sltly sol in glycerin.

SYNS: ANISE ALCOHOL □ ANISIC ALCOHOL □ p-ANISOL ALCOHOL □ ANISYL ALCOHOL (FCC) □ FEMA No. 2099 □ 4-METHOXYBENZENEMETHANOL □ 4-METHOXYBENZYL ALCOHOL

CONSENSUS REPORTS: Reported in EPA TSCA Inventory.

SAFETY PROFILE: Moderately toxic by ingestion. A skin irritant. Combustible liquid. When heated to decomposition it emits acrid smoke and irritating fumes. See also ALCOHOLS.

MEI450 CAS: 72-43-5 HR: 3
METHOXYCHLOR

mf: $C_{16}H_{15}Cl_3O_2$ mw: 345.66

PROP: Dimorphic crystals from MeOH. Mp: 78°, vap d: 12. IDLH 5000 mg/m³.

SYNS: 2,2-BIS(p-ANISYL)-1,1,1-TRICHLOROETHANE □ 1,1-BIS(p-METHOXYPHENYL)-2,2,2-TRICHLOROETH-ANE □ 2,2-BIS(p-METHOXYPHENYL)-1,1,1-TRICHLORO-ETHANE □ CHEMFORM □ DIANISYLTRICHLOROETH-ANE □ 2,2-DI-p-ANISYL-1,1,1-TRICHLOROETHANE □ DIMETHOXY-DDT □ p,p'-DIMETHOXYDIPHENYL-TRICHLOROETHANE □ DIMETHOXY-DT □ 2,2-DI-(p-METHOXYPHENYL)-1,1,1-TRICHLOROETHANE □ DI(p-METHOXYPHENYL)-TRICHLOROMETHYL METHANE

□ DMDT □ p,p'-DMDT □ ENT 1,716 □ MARALATE □ MARLATE □ METHOXCIDE □ METHOXO □ p,p'-METHOXYCHLOR □ METHOXY-DDT □ METO-KSYCHLOR (POLISH) □ METOX □ MOXIE □ NCI-C00497 □ RCRA WASTE NUMBER U247 □ 1,1,1-TRICHLOR-2,2-BIS(4-METHOXY-PHENYL)-AETHAN (GERMAN) □ 1,1,1-TRICHLORO-2,2-BIS(p-ANISYL)-ETHANE □ 1,1,1-TRICHLORO-2,2-BIS(p-METHOXY-PHENOL)ETHANOL □ 1,1,1-TRICHLORO-2,2-BIS(p-METHOXYPHENYL)ETHANE □ 1,1,1-TRICHLORO-2,2-DI(4-METHOXYPHENYL)ETHANE □ 1,1'-(2,2,2-TRICHL-OROETHYLIDENE)BIS(4-METHOXYBENZENE)

CONSENSUS REPORTS: IARC Cancer Review: Group 3 IMEMDT 7,56,87; Animal No Evidence IMEMDT 20,259,79; Animal Inadequate Evidence IMEMDT 5,193,74. NCI Carcinogenesis Bioassay (feed); No Evidence: mouse, rat NCITR* NCI-CG-TR-35,78. EPA Genetic Toxicology Program. Community Right-To-Know List.

OSHA PEL: TWA Total Dust: 10 mg/m³; 5 mg/m³

ACGIH TLV: TWA 10 mg/m³; Not Classifiable as a Human Carcinogen

DFG MAK: 15 mg/m³

SAFETY PROFILE: Suspected carcinogen with experimental carcinogenic, tumorigenic, and teratogenic data. Moderately toxic by intraperitoneal and skin contact routes. Human systemic effects by skin contact: somnolence. Experimental reproductive effects. Mutation data reported. When heated to decomposition it emits highly toxic fumes of Cl$^-$. See also DDT and CHLOROPHENOLS.

MEL100 CAS: 107746-52-1 HR: 3
N-METHOXY-3-(3,5-DI-tert-BUTYL-4-HYDROXYBENZYLIDENE)-2-PYRROLIDONE

mf: $C_{20}H_{29}NO_3$ mw: 331.50

SYNS: E 5110 □ 3-((3,5-BIS(1,1-DIMETHYLETHYL)-4-HYDROXYPHENYL)METHYLENE)-1-METHOXY-2-PYRROLIDINONE □ 2-PYRROLIDINONE, 3-((3,5-BIS(1,1-DIMETHYLETHYL)-4-HYDROXYPHENYL)METHYL-ENE)-1-METHOXY-

SAFETY PROFILE: A poison by ingestion, intraperitoneal, and parenteral routes. When heated to decomposition it emits toxic vapors of NO$_x$.

M

MEL500   CAS: 1918-00-9   HR: 2
2-METHOXY-3,6-DICHLOROBENZOIC ACID

mf: $C_8H_6Cl_2O_3$ mw: 221.04

PROP: Colorless crystals. Decomposes below boiling point at 200C. Mp: 114°–116°. Sol in water.

SYNS: ACIDO (3,6-DICLORO-2-METOSSI)-BENZOICO (ITALIAN) □ BANEX □ BANLEN □ BANVEL □ BANVEL HERBICIDE □ BRUSH BUSTER □ COMPOUND B DICAMBA □ DIANAT (RUSSIAN) □ DIANATE □ DICAMBA (DOT) □ 3,6-DICHLOOR-2-METHOXY-BENZOEIZUUR (DUTCH) □ 3,6-DICHLOR-3-METHOXY-BENZOESAEURE (GERMAN) □ 3,6-DICHLORO-o-ANISIC ACID □ 2,5-DICHLORO-6-METHOXYBENZOIC ACID □ 3,6-DICHLORO-2-METHOXYBENZOIC ACID □ MDBA □ MEDIBEN □ VELSICOL COMPOUND "R" □ VELSICOL 58-CS-11

CONSENSUS REPORTS: EPA Genetic Toxicology Program.

SAFETY PROFILE: Moderately toxic by ingestion. Mutation data reported. When heated to decomposition it emits toxic fumes of Cl⁻.

MEO750   CAS: 151-38-2   HR: 3
METHOXYETHYL MERCURIC ACETATE

mf: $C_5H_{10}HgO_3$ mw: 318.74

PROP: Crystals or needles from pet ether. Mp: 42°. Water-sol. IDLH 10 mg/m³ (as Hg).

SYNS: ACETATO(2-METHOXYETHYL)MERCURY □ CEKUSIL UNIVERSAL A □ LANDISAN □ MEMA □ MERCURAN □ MERCURY, ACETOXY(2-METHOXY-ETHYL)- □ METHOXYETHYL MERCURIC ACETATE □ METHOXYETHYLMERCURY ACETATE □ 2-METHOXY-ETHYLMERKURIACETAT □ PANOGEN □ PANOGEN M □ PANOGEN METOX □ RADOSAN

CONSENSUS REPORTS: EPA Extremely Hazardous Substances List. Mercury and its compounds are on the Community Right-To-Know List.

OSHA PEL: CL 0.1 mg(Hg)/m³ (skin)

ACGIH TLV: TWA 0.1 mg(Hg)/m³ (skin); BEI: 35 µg/g creatinine total inorganic mercury in urine preshift; 15 µg/g creatinine total inorganic mercury in blood at end of shift at end of workweek.; STEL 0.03 mg(Hg)/m³

DFG MAK: Confirmed Animal Carcinogen with Unknown Relevance to Humans

NIOSH REL: (Mercury, Organo) TWA 0.01 mg/m³; STEL 0.03 mg/m³ (skin)

SAFETY PROFILE: Poison by ingestion. Mutation data reported. A fungicide. When heated to decomposition it emits toxic fumes of Hg. See also MERCURY COMPOUNDS.

MEP250   CAS: 123-88-6   HR: 3
2-METHOXYETHYLMERCURY CHLORIDE

mf: C_3H_7ClHgO mw: 295.14

PROP: Crystals or powder from ligroin. Mp: 68–68.5°. Sol in Me_2CO, EtOH; mod sol in H_2O. IDLH 10 mg/m³ (as Hg).

SYNS: AGALLO FORTE □ AGALLOL □ AGALLOLAT □ AGALOL □ ARATAN □ ARETAN □ ARETAN 6 □ ATIRAN □ BAGALOL □ BAYTAN □ CEKUSIL UNIVERSAL C □ CELMER □ CERESAN-UNIVERSAL NASSBEIZE □ CERESAN UNIVERSAL NAZBEIZE □ CHLORO(2-METHOXYETHYL)MERCURY □ CURESAN □ CURETAN □ EMISAN 6 □ FALISAN □ GRAMISAN □ HIGOSAN □ MEMC □ MERCHLORATE □ METHOXY-AETHYLQUECKSILBERCHLORID □ (β-METHOXYETH-YL)MERCURIC CHLORIDE □ METHOXYETHYL MER-CURIC CHLORIDE □ 2-METHOXYETHYLMERCURIC CHLORIDE □ METHOXYETHYLMERCURY CHLORIDE □ β-METHOXYETHYLMERCURY CHLORIDE □ 2-METHOXYETHYLMERKURICHLORID □ SEDRESAN □ TAFASAN □ TAFASAN 6W □ TAYSSATO □ TRIADIMENOL

CONSENSUS REPORTS: Mercury and its compounds are on the Community Right-To-Know List. EPA Genetic Toxicology Program.

OSHA PEL: TWA 0.01 mg(Hg)/m³; STEL 0.03 mg/m³ (skin)

ACGIH TLV: TWA 0.01 mg(Hg)/m³; BEI: 35 µg/g creatinine total inorganic mercury in urine preshift; 15 µg/g creatinine total inorganic mercury in blood at end of shift at end of workweek.

DFG MAK: Confirmed Animal Carcinogen with Unknown Relevance to Humans

NIOSH REL: (Mercury, Organo) TWA 0.01 mg/m³; STEL 0.03 mg/m³ (skin)

SAFETY PROFILE: Poison by ingestion and subcutaneous routes. Experimental reproductive effects. Human mutation data reported. Used to control pineapple disease of sugarcane. When heated to decomposition it emits very toxic fumes of Hg and Cl⁻. See also MERCURY COMPOUNDS and CHLORIDES.

MEP500   CAS: 61738-05-4   HR: 3
1-METHOXY ETHYL METHYLNITRO-SAMINE

mf: $C_4H_{10}N_2O_2$ mw: 118.16

SYN: 1-METHOXY-AETHYL-METHYLNITROSAMIN (GERMAN)

SAFETY PROFILE: Poison by ingestion. Questionable carcinogen with experimental carcinogenic data. When heated to decomposition it emits toxic fumes of NO_x. See also NITROSAMINES.

MES000   CAS: 131-57-7   HR: 3
4-METHOXY-2-HYDROXYBENZO-PHENONE

mf: $C_{14}H_{12}O_3$ mw: 228.26

PROP: Crystals from 2-propanol. Mp: 66°.

SYNS: BENZOPHENONE-3 □ CYASORB UV 9 □ 2-HYDROXY-4-METHOXYBENZOPHENONE □ (2-HYDROXY-4-METHOXYPHENYL)PHENYLMETHAN-ONE □ MOB □ NCI-C60957 □ NSC-7778 □ OXYBENZ-ONE □ SPECTRA-SORB UV 9 □ SYNTASE 62 □ USAF CY-9 □ UVINUL M 40

CONSENSUS REPORTS: Reported in EPA TSCA Inventory.

SAFETY PROFILE: Poison by intraperitoneal route. Mildly toxic by ingestion. Experimental reproductive effects. Mutation data reported. When heated to decomposition it emits acrid smoke and irritating fumes.

MES550   CAS: 90293-50-8   HR: 3
9-(1-(METHOXYIMINO)ETHYL)-6-OXO-N,N,2,2,5-PENTAMETHYL-7-OXA-3,4-DITHIA-5,8-DIAZADEC-8-EN-10-AMIDE

mf: $C_{13}H_{24}N_4O_4S_2$ mw: 364.53

SAFETY PROFILE: A poison by ingestion. When heated to decomposition it emits toxic vapors of NO_x and SO_x.

MES900   CAS: 63845-31-8   HR: 3
5-METHOXY-1H-INDOL-3-YL 4-PIPER-IDYLMETHYL KETONE MONO-

HYDROCHLORIDE

mf: $C_{16}H_{20}N_2O_2 \cdot ClH$ mw: 308.84

SYNS: ETHANONE, 1-(5-METHOXY-1H-INDOL-3-YL)-2-(4-PIPERIDINYL)-, MONOHYDROCHLORIDE (9CI) □ (5-METHOXY-3-INDOLYL)-(4-PIPERIDYL-METHYL)-KETON HYDROCHLORID

DOT CLASSIFICATION: 3; Label: Flammable Liquid

SAFETY PROFILE: A poison by intravenous route. A flammable liquid. When heated to decomposition it emits toxic vapors of NO_x and HCl.

MEX250   CAS: 107-70-0   HR: 3
4-METHOXY-4-METHYL-2-PENTANONE

DOT: UN 2293

mf: $C_7H_{14}O_2$ mw: 130.21

PROP: Colorless liquid. Bp: 147°–163°. Sol in water. Flash pt: 60° F.

SYN: 4-METHOXY-4-METHYLPENTAN-2-ONE (DOT)

CONSENSUS REPORTS: Reported in EPA TSCA Inventory.

DOT CLASSIFICATION: 3; Label: Flammable Liquid

SAFETY PROFILE: Moderately toxic by ingestion and skin contact. Mildly toxic by inhalation. A skin irritant. Flammable when exposed to heat or flame, can react vigorously with oxidizing materials. When heated to decomposition it emits acrid smoke and irritating fumes.

MEX350   HR: 3
2-METHOXY-3(5)-METHYLPYRAZINE

mf: $C_6H_8N_2O$ mw: 124.14

PROP: Colorless liquid; roasted-hazelnut odor. D: 1.000–1.090 @ 20°, refr index: 1.506, flash p: 131°F. Sol in water, org solvs.

SYN: FEMA No. 3183

SAFETY PROFILE: Flammable liquid when exposed to heat, sparks, or flame. When heated to decomposition it emits toxic fumes of NO_x.

MEY200   CAS: 33918-12-6   HR: 3
2-(METHOXY(METHYLTHIO)PHOSPHIN-YLIMINO)-3-METHYL-1,3-THIAZO-LINE

mf: $C_6H_{13}N_2O_2PS_2$ mw: 240.30

M

SYNS: 2-(O,S-DIMETHYLTHIOPHOSPHORYLIMINO)-3-METHYLTHIAZOLIDINE □ PHOSPHORAMIDOTHIO-IC ACID, (3-METHYL-2-THIAZOLIDINYLIDENE)-, O,S-DIMETHYL ESTER □ 1,3-THIAZOLIDINE, 2-(METHO-XY(METHYLTHIO)PHOSPHINYLIMINO)-3-METHYL-

SAFETY PROFILE: A poison by ingestion and skin contact. When heated to decomposition it emits toxic vapors of NO_x, SO_x, and PO_x.

MFB400　CAS: 75965-74-1　HR: 3
7-METHOXY-2-NITRONAPHTHO(2,1-b)FURAN

mf: $C_{13}H_9NO_4$　　mw: 243.23
PROP: Crystalline powder. Mp: 108° C.
SYNS: 2-NITRO-7-METHOXYNAPHTHO(2,1-b)FURAN □ R7000

SAFETY PROFILE: Suspected carcinogen with experimental carcinogenic and tumorigenic data. Mutation data reported. When heated to decomposition it emits toxic fumes of NO_x.

MFC700　CAS: 150-76-5　HR: 3
4-METHOXYPHENOL

mf: $C_7H_8O_2$　　mw: 124.15
PROP: White, waxy solid, or leaflets from water. Mp: 52.5°, bp: 246°, d: 1.55 @ 20°/20°.
SYNS: HYDROQUINONE MONOMETHYL ETHER □ MEQUINOL □ p-METHOXYPHENOL □ MME □ MONO-METHYL ETHER HYDROQUINONE □ USAF AN-7

CONSENSUS REPORTS: Reported in EPA TSCA Inventory. EPA Genetic Toxicology Program.
OSHA PEL: TWA 5 mg/m³
ACGIH TLV: TWA 5 mg/m³
SAFETY PROFILE: Poison by intraperitoneal route. A skin irritant. When heated to decomposition it emits acrid smoke and fumes. See also ETHERS.

MFE250　CAS: 104-01-8　HR: 2
p-METHOXYPHENYLACETIC ACID

mf: $C_9H_{10}O_3$　　mw: 166.19
PROP: A solid. Mp: 85–87°.
SYNS: ANISYL FORMATE □ 2-(p-ANISYL)ACETIC ACID □ HOMOANISIC ACID □ 4-METHOXY-BENZENEACETIC ACID □ p-METHOXYBENZYL FORMATE □ 4-METHOXYPHENYLACETIC ACID □ MOPA

CONSENSUS REPORTS: Reported in EPA TSCA Inventory.
SAFETY PROFILE: Moderately toxic by ingestion and intraperitoneal routes. Questionable carcinogen with experimental neoplastigenic data. When heated to decomposition it emits acrid smoke and irritating fumes.

MFF250　CAS: 3647-17-4　HR: 3
N-(p-METHOXYPHENYL)-1-AZIRIDINE-CARBOXAMIDE

mf: $C_{10}H_{12}N_2O_2$　　mw: 192.24
SYNS: 1-(1-AZIRIDINYL)-N-(p-METHOXY-PHENYL)FORMAMIDE □ p-METHOXYPHENYL-N-CARBAMOYLAZIRIDINE

SAFETY PROFILE: Poison by intravenous route. Questionable carcinogen with experimental neoplastigenic data. When heated to decomposition it emits toxic fumes of NO_x.

MFF580　CAS: 104-20-1　HR: 3
4-p-METHOXYPHENYL-2-BUTANONE

mf: $C_{11}H_{14}O_2$　　mw: 178.25
PROP: Colorless to pale-yellow liquid; sweet, floral odor. D: 1.042–1.048, refr index: 1.517–1.521, flash p: 212°F.
SYNS: ANISYLACETONE □ 2-BUTANONE, 4-(p-METHOXYPHENYL)-(6CI,7CI,8CI) □ ENT 20,279 □ FEMA No. 2672 □ 4-METHOXYBENZYLACETONE □ p-METHOXYPHENYLBUTANONE □ 4-(p-METHOXY-PHENYL)-2-BUTANONE □ RASPBERRY KETONE METHYL ETHER

DOT CLASSIFICATION: 3; Label: Flammable Liquid
SAFETY PROFILE: Low oral and skin toxicity. A flammable liquid. When heated to decomposition it emits acrid smoke and irritating fumes.

MFH760　CAS: 17766-68-6　HR: 3
4-(o-METHOXYPHENYL)PIPERAZINYL 3,4,5-TRIMETHOXYPHENYL KETONE

mf: $C_{21}H_{26}N_2O_5$　　mw: 386.49
SYNS: KETONE, 4-(o-METHOXYPHENYL)PIPERAZINYL 3,4,5-TRIMETHOX-YPHENYL □ 1-(o-METHOXYPHENYL)-4-(3,4,5-TRI-METHOXYBENZOYL)PIPERAZINE □ PIPERAZINE, 1-(o-METHOXYPHENYL)-4-(3,4,5-TRIMETHOXY-BENZOYL)-

DOT CLASSIFICATION: 3; Label: Flammable Liquid
SAFETY PROFILE: A poison by intraperitoneal route. A flammable liquid. When heated to decomposition it emits toxic vapors of NO_x.

MFH770   CAS: 17766-70-0   HR: 3
4-(p-METHOXYPHENYL)PIPERAZINYL 3,4,5-TRIMETHOXYPHENYL KETONE
mf: $C_{21}H_{26}N_2O_5$ mw: 386.49
SYNS: KETONE, 4-(p-METHOXYPHENYL)PIPERAZIN-YL 3,4,5-TRIMETHOXYPHENYL □ 1-(p-METHOXY-PHENYL)-4-(3,4,5-TRIMETHOXYBENZOYL)PIPERAZINE □ PIPERAZINE, 1-(p-METHOXYPHENYL)-4-(3,4,5-TRIM-ETHOXYBENZOYL)-
DOT CLASSIFICATION: 3; Label: Flammable Liquid
SAFETY PROFILE: A poison by intraperitoneal route. A flammable liquid. When heated to decomposition it emits toxic vapors of NO_x.

MFH930   CAS: 14548-47-1   HR: 3
p-METHOXYPHENYL 4-PYRIDYL KETONE
mf: $C_{13}H_{11}NO_2$ mw: 213.25
SYNS: p-ANISYL 4-PYRIDYL KETONE □ KETONE, p-ANISYL 4-PYRIDYL □ KETONE, (p-METHOXYPHENYL) 4-PYRIDYL □ 4-(4-METHOXYBENZOYL)PYRIDINE □ PYRIDINE, 4-(4-METHOXYBENZOYL)-
DOT CLASSIFICATION: 3; Label: Flammable Liquid
SAFETY PROFILE: A poison by intraperitoneal route. A flammable liquid. When heated to decomposition it emits toxic vapors of NO_x.

MFM000   CAS: 5332-73-0   HR: 3
3-METHOXYPROPYLAMINE
mf: $C_4H_{11}NO$ mw: 89.16
PROP: Colorless liquid. Mp: $-75.7°$, bp: $116°$, flash p: $90°F$ (TOC), d: $0.8615 @ 30°$, vap press: 20 mm @ $30°$, vap d: 3.07.
SYN: 3-MPA
CONSENSUS REPORTS: Reported in EPA TSCA Inventory.
SAFETY PROFILE: Poison by intravenous route. Irritating to skin, eyes, and mucous membranes. Dangerous fire hazard when exposed to heat or flame; can react with oxidizing materials. To fight fire, use CO_2, dry chemical. When heated to decomposition it emits toxic fumes of NO_x. See also AMINES.

MFN285   CAS: 3149-28-8   HR: 1
2-METHOXYPYRAZINE
mf: $C_5H_6N_2O$ mw: 110.12
PROP: Colorless to yellow liquid; nutty, cocoalike odor. D: 1.110–1.140 @ $20°$, bp: $60–61°$ @ 29 mm, refr index: 1.508. Sol in alc; insol in water @ $61°$.
SYN: FEMA No. 3302
SAFETY PROFILE: Skin and eye irritant. When heated to decomposition it emits toxic fumes of NO_x.

MFT500   CAS: 127-25-3   HR: 2
METHYL ABIETATE
mf: $C_{21}H_{32}O_2$ mw: 316.47
PROP: Colorless to yellow thick liquid; almost odorless. Flash p: $356°F$ (OC), vap d: 10.9, d: $1.040 @ 20°/20°$, bp: $360–365°$ with decomp. Insol in water, misc in alc and ether, the usual org solvs, and with aliphatic hydrocarbons. From the esterification of the resinous residue of turpentine (FCTXAV 12,807,74).
SYNS: ABIETIC ACID, METHYL ESTER □ METHYL ESTER of WOOD ROSIN □ METHYL ESTER of WOOD ROSIN, partially hydrogenated (FCC)
CONSENSUS REPORTS: Reported in EPA TSCA Inventory.
SAFETY PROFILE: Low toxicity by ingestion and skin contact. A skin irritant. Probably slightly toxic. Combustible liquid when exposed to heat or flame; can react with oxidizing materials. To fight fire, use CO_2, dry chemical. When heated to decomposition it emits acrid smoke and irritating fumes.

MFT750   CAS: 79-16-3   HR: 2
METHYLACETAMIDE
mf: C_3H_7NO mw: 73.11
PROP: Needles. Mp: $30.55°$, bp: $206°$. Sol in H_2O, EtOH, and C_6H_6; insol in ligroin.

M

SYNS: N-METHYLACETAMIDE □ MONOMETHYL-
ACETAMIDE

CONSENSUS REPORTS: Reported in EPA
TSCA Inventory.

SAFETY PROFILE: Moderately toxic by
intraperitoneal and subcutaneous routes.
Mildly toxic by ingestion and intravenous
routes. An experimental teratogen.
Experimental reproductive effects. Mutation
data reported. When heated to
decomposition it emits toxic fumes of NO_x.

MFW100   CAS: 79-20-9   HR: 3
METHYL ACETATE

DOT: UN 1231

mf: $C_3H_6O_2$ mw: 74.09

PROP: Pleasant smelling, colorless, volatile
liquid. Mp: −98.7°, lel: 3.1%, uel: 16%, bp:
57.8°, ULC: 85–90, flash p: 14°F, d:
0.92438, autoign temp: 935°F, vap press:
100 mm @ 9.4°, vap d: 2.55. Moderately sol
in water; misc in alc, ether. IDLH 3100 ppm
[10%LEL].

SYNS: ACETATE de METHYLE (FRENCH) □ ACETIC
ACID METHYL ESTER □ DEVOTON □ ETHYL ESTER
of MONOACETIC ACID □ METHYLACETAAT (DUTCH)
□ METHYLACETAT (GERMAN) □ METHYLE (ACETATE
de) (FRENCH) □ METHYLESTER KISELINY OCTOVE
(CZECH) □ METHYL ETHANOATE □ METILE
(ACETATO di) (ITALIAN) □ OCTAN METYLU (POLISH)
□ TERETON

CONSENSUS REPORTS: Reported in EPA
TSCA Inventory.

OSHA PEL: TWA 200 ppm; STEL 250 ppm

ACGIH TLV: TWA 200 ppm; STEL 250
ppm

DFG MAK: 200 ppm (610 mg/m³)

DOT CLASSIFICATION: 3; Label:
Flammable Liquid

SAFETY PROFILE: Moderately toxic by
several routes. A human systemic irritant by
inhalation. A moderate skin and eye irritant.
Mutation data reported. Dangerous fire
hazard when exposed to heat, flame, or
oxidizers. Moderate explosion hazard when
exposed to heat or flame. When heated to
decomposition it emits acrid smoke and
fumes. See also ESTERS.

MFW250   CAS: 122-00-9   HR: 3
4′-METHYL ACETOPHENONE

mf: $C_9H_{10}O$ mw: 134.19

PROP: Colorless liquid or needles; fruity,
actophenone odor. D: 0.996–1.004, refr
index: 1.530–1.535, mp: 28°, bp: 225° @
736 mm, flash p: 198°F. Sol in fixed oils,
propylene glycol; insol in glycerin.

SYNS: p-ACETYLTOLUENE □ ETHANONE, 1-(4-
METHYLPHENYL)-(9CI) □ FEMA No. 2677 □
MELILOTAL □ p-METHYL ACETOPHENONE □ 1-
METHYL-4-ACETYLBENZENE □ METHYL-p-TOLYL
KETONE

CONSENSUS REPORTS: Reported in EPA
TSCA Inventory.

DOT CLASSIFICATION: 3; Label:
Flammable Liquid

SAFETY PROFILE: Moderately toxic by
ingestion. A human skin irritant. A
flammable liquid. When heated to
decomposition it emits acrid smoke and
irritating fumes. See also KETONES.

MFW500   CAS: 520-45-6   HR: 3
METHYLACETOPYRONONE

mf: $C_8H_8O_4$ mw: 168.16

PROP: White crystals or crystalline powder.
Mp: 109°, bp: 269.0°, vap press: 1 mm @
91.7°, vap d: 5.8. Moderately sol in water
and org solvs.

SYNS: 2-ACETYL-5-HYDROXY-3-OXO-4-HEXENOIC
ACID Δ-LACTONE □ 3-ACETYL-6-METHYL-2,4-
PYRANDIONE □ 3-ACETYL-6-METHYLPYRANDIONE-
2,4 □ 3-ACETYL-6-METHYL-2H-PYRAN-2,4(3H)-DIONE
□ DEHYDRACETIC ACID □ DEHYDROACETIC ACID
(FCC) □ DHA □ DHS

CONSENSUS REPORTS: Reported in EPA
TSCA Inventory.

SAFETY PROFILE: Poison by ingestion.
Moderately toxic by intraperitoneal route.
Questionable carcinogen with experimental
tumorigenic data. Combustible when
exposed to heat or flame. When heated to
decomposition it emits acrid smoke and
irritating fumes.

MFX250   CAS: 105-45-3   HR: 2
METHYL ACETYLACETATE

mf: $C_5H_8O_3$ mw: 116.13

PROP: Colorless liquid. Mp: 27.5°, bp: 170°, flash p: 170°F, autoign temp: 536°F, d: 1.077, vap d: 4.00. Misc in water.
SYNS: ACETOACETIC METHYL ESTER □ METHYL-ACETOACETATE □ METHYL ACETYLACETONATE □ METHYL-3-OXOBUTYRATE □ 3-OXOBUTANOIC ACID METHYL ESTER
CONSENSUS REPORTS: Reported in EPA TSCA Inventory.
SAFETY PROFILE: Moderately toxic by ingestion. A skin and severe eye irritant. Combustible when exposed to heat or flame. To fight fire, use foam, CO_2, dry chemical. When heated to decomposition it emits acrid smoke and irritating fumes. See also ESTERS.

MFX590   CAS: 74-99-7   HR: 3
METHYL ACETYLENE
mf: C_3H_4 mw: 40.07
PROP: Gas. Mp: −104°, lel: 1.7%, bp: −23.3°, vap press: 3876 mm @ 20°, d: 1.787 g/L @ 0°, vap d: 1.38. Mod sol in H_2O; sol in EtOH and Et_2O. IDLH 1700 ppm [10%LEL].
SYNS: ACETYLENE, METHYL- □ ALLYLENE □ PROPINE □ PROPYNE (OSHA)
CONSENSUS REPORTS: Reported in EPA TSCA Inventory.
OSHA PEL: TWA 1000 ppm
ACGIH TLV: TWA 1000 ppm
DFG MAK: 1000 ppm (1700 mg/m³)
SAFETY PROFILE: This compound is a simple anesthetic and in high concentration is an asphyxiant. Mutation data reported. Dangerous fire hazard when exposed to heat or flame; can react vigorously with oxidizing materials. Explosive in the form of vapor when exposed to heat or flame. Localized heating of liquid-containing cylinders to 95°C may cause an explosion. Product of reaction with silver nitrate ignites at 150°C. A commercial mixture containing 30% propyne in MAPP gas is similar to ethylene in potential hazards and handling requirements. To fight fire, stop flow of gas. When heated to decomposition it emits acrid smoke and irritating fumes. See also ACETYLENE COMPOUNDS.

MFX600   CAS: 59355-75-8   HR: 3
METHYL ACETYLENE-PROPADIENE MIXTURE
DOT: UN 1060
PROP: IDLH 3400 ppm [10%LEL].
SYNS: MAPP (OSHA) □ METHYL ACETYLENE and PROPADIENE MIXTURES, stabilized (DOT)
OSHA PEL: TWA 1000 ppm; STEL 1250 ppm
ACGIH TLV: TWA 1000 ppm; STEL 1250 ppm
DFG MAK: 1000 ppm (1650 mg/m³)
DOT CLASSIFICATION: 2.1; Label: Flammable Gas
SAFETY PROFILE: A flammable gas mixture. To fight fire, stop flow of gas. When heated to decomposition it emits acrid smoke and irritating fumes.

MGA250   CAS: 78-85-3   HR: 3
METHYLACRYLALDEHYDE
DOT: UN 2396
mf: C_4H_6O mw: 70.10
PROP: Colorless liquid or lachrymatory oil. Mp: −81°C, bp: 68–70°, flash p: 35°F (OC), d: 0.830 @ 20°/4°, vap press: 120 mm @ 20°, vap d: 2.42. Sol in water.
SYNS: ACROLEIN, 2-METHYL- □ ISOBUTENAL □ METHACRALDEHYDE (DOT) □ METHACROLEIN □ METHACRYLIC ALDEHYDE □ METHAKRYLALDEHYD □ α-METHYLACROLEIN □ 2-METHYLACROLEIN □ METHYLACRYLALDEHYDE □ 2-METHYLPROPENAL (CZECH) □ NSC-8260 □ 2-PROPENAL, 2-METHYL-
CONSENSUS REPORTS: Reported in EPA TSCA Inventory.
DOT CLASSIFICATION: 3; Label: Flammable Liquid, Poison
SAFETY PROFILE: Poison by ingestion and skin contact. Moderately toxic by inhalation. Severe eye and skin irritant. Mutation data reported. Dangerously flammable liquid when exposed to heat, flame, or oxidizers. Can react vigorously with oxidizing materials. To fight fire, use CO_2, alcohol foam, foam, dry chemical. When heated to decomposition it emits acrid smoke and irritating fumes. See also ALDEHYDES.

M

MGA300   CAS: 1187-59-3   HR: 2
N-METHYLACRYLAMIDE
mf: C_4H_7NO mw: 85.12

SYNS: ACRYLAMIDE, N-METHYL- □ 2-
PROPENAMIDE, N-METHYL- (9CI)

CONSENSUS REPORTS: Reported in EPA
TSCA Inventory.

SAFETY PROFILE: Moderately toxic by
ingestion. Experimental reproductive
effects. Mutation data reported. When
heated to decomposition it emits toxic
fumes of NO_x.

MGA500   CAS: 96-33-3   HR: 3
METHYL ACRYLATE
DOT: UN 1919

mf: $C_4H_6O_2$ mw: 86.10

PROP: Colorless liquid; acrid odor. D:
0.9561 @ 20°/4°, mp: −76.5°, bp: 85° @
608 mm, lel: 2.8%, uel: 25%, fp: −75°, flash
p: 27°F (OC), vap press: 100 mm @ 28°,
vap d: 2.97. Sol in alc and ether. IDLH 250
ppm.

SYNS: ACRYLATE de METHYLE (FRENCH) □
ACRYLIC ACID METHYL ESTER (MAK) □ ACRYL-
SAEUREMETHYLESTER (GERMAN) □ CURITHANE 103
□ METHOXYCARBONYLETHYLENE □ METHYL-
ACRYLAAT (DUTCH) □ METHYL-ACRYLAT (GERMAN)
□ METHYL ACRYLATE, INHIBITED (DOT) □ METHYL
PROPENATE □ METHYL PROPENOATE □ METHYL-2-
PROPENOATE □ METILACRILATO (ITALIAN) □
PROPENOIC ACID METHYL ESTER □ 2-PROPENOIC
ACID METHYL ESTER

CONSENSUS REPORTS: IARC Cancer
Review: Group 3 IMEMDT 7,56,87; Animal
Inadequate Evidence IMEMDT 39,99,86;
Human Inadequate Evidence IMEMDT
19,47,79. Community Right-To-Know List.
Reported in EPA TSCA Inventory.

OSHA PEL: TWA 10 ppm (skin)

ACGIH TLV: TWA 2 ppm (skin, sensitizer);
Not Classifiable as a Human Carcinogen

DFG MAK: 5 ppm (18 mg/m³)

DOT CLASSIFICATION: 3; Label:
Flammable Liquid

SAFETY PROFILE: Poison by ingestion and
intraperitoneal routes. Moderately toxic by
skin contact. Mildly toxic by inhalation.
Human systemic effects by inhalation:
olfaction effects, eye effects, and respiratory

effects. A skin and eye irritant. Mutation
data reported. Chronic exposure has
produced injury to lungs, liver, and kidneys
in experimental animals. Questionable
carcinogen. Dangerously flammable when
exposed to heat, flame, or oxidizers.
Dangerous explosion hazard in the form of
vapor when exposed to heat, sparks, or
flame. Can react vigorously with oxidizing
materials. A storage hazard; it forms
peroxides, which may initiate exothermic
polymerization. To fight fire, use foam,
CO_2, dry chemical. When heated to
decomposition it emits acrid smoke and
irritating fumes. See also ESTERS.

MGA750   CAS: 126-98-7   HR: 3
METHYLACRYLONITRILE
mf: C_4H_5N mw: 67.10

PROP: Colorless liquid. Mp: −36°, bp:
90.3°, d: 0.805, vap press: 40 mm @ 12.8°,
flash p: 55°F. Insol in water.

SYNS: 2-CYANOPROPENE-1 □ ISOPROPENE CYAN-
IDE □ ISOPROPENYLNITRILE □ METHACRYLONI-
TRILE, inhibited □ α-METHYLACRYLONITRILE □ 2-
METHYLPROPENENITRILE □ RCRA WASTE NUMBER
U152 □ USAF ST-40

CONSENSUS REPORTS: EPA Extremely
Hazardous Substances List. Cyanide and its
compounds are on the Community Right-
To-Know List. Reported in EPA TSCA
Inventory.

OSHA PEL: TWA 1 ppm (skin)

ACGIH TLV: TWA 1 ppm (skin)

DOT CLASSIFICATION: 3; Label:
Flammable Liquid, Poison

SAFETY PROFILE: Poison by ingestion,
inhalation, skin contact, and intraperitoneal
routes. An eye irritant. A dangerous fire
hazard when exposed to heat, flame, or
sparks. When heated to decomposition it
emits toxic fumes of NO_x and CN^-. See also
NITRILES.

MGA850   CAS: 109-87-5   HR: 3
METHYLAL
DOT: UN 1234

mf: $C_3H_8O_2$ mw: 76.11

PROP: Colorless volatile liquid; pungent odor. Mp: −104.8°, bp: 42.3°, d: 0.864 @ 20°/4°, vap press: 330 mm @ 20°, vap d: 2.63, autoign temp: 459°F, flash p: −0.4°F. IDLH 2200 ppm [10%LEL].
SYNS: ANESTHENYL □ DIMETHOXYMETHANE □ DIMETHYL FORMAL □ FORMAL □ FORMALDEHYDE DIMETHYLACETAL □ METHYLENE DIMETHYL ETHER □ METYLAL (POLISH)
OSHA PEL: TWA 1000 ppm
ACGIH TLV: TWA 1000 ppm
DFG MAK: 1000 ppm (3200 mg/m³)
DOT CLASSIFICATION: 3; Label: Flammable Liquid
SAFETY PROFILE: Moderately toxic by subcutaneous route. Mildly toxic by ingestion and inhalation. Can cause injury to lungs, liver, kidneys, and the heart. A narcotic and anesthetic in high concentrations. A very dangerous fire hazard when exposed to heat, flame, or oxidizers. Moderately explosive when exposed to heat or flame. May ignite or explode when heated with oxygen. To fight fire, use foam, CO₂, dry chemical. When heated to decomposition it emits acrid smoke and irritating fumes. See also ETHERS.

MGB150   CAS: 67-56-1   HR: 3
METHYL ALCOHOL
DOT: UN 1230
mf: CH₄O   mw: 32.05
PROP: Clear, colorless, very mobile liquid; slt alcoholic odor when pure; crude material may have a repulsive pungent odor. Bp: 64.8°, lel: 6.0%, uel: 36.5%, ULC: 70, fp: −97.8°, d: 0.7915 @ 20°/4°, flash p: 54°F, autoign temp: 878°F, vap press: 100 mm @ 21.2°, vap d: 1.11. Misc in water, ethanol, ether, benzene, ketones, and most other org solvs. Part misc in pet ether. IDLH 6000 ppm.
SYNS: ALCOOL METHYLIQUE (FRENCH) □ ALCOOL METILICO (ITALIAN) □ CARBINOL □ COLONIAL SPIRIT □ COLUMBIAN SPIRITS (DOT) □ METANOLO (ITALIAN) □ METHANOL □ METHYLALKOHOL (GERMAN) □ METHYL HYDROXIDE □ METHYLOL □ METYLOWY ALKOHOL (POLISH) □ MONOHYDROXY-

METHANE □ PYROXYLIC SPIRIT □ RCRA WASTE NUMBER U154 □ WOOD ALCOHOL (DOT) □ WOOD NAPHTHA □ WOOD SPIRIT
CONSENSUS REPORTS: Community Right-To-Know List. Reported in EPA TSCA Inventory. EPA Genetic Toxicology Program.
OSHA PEL: TWA 200 ppm; STEL 250 ppm (skin)
ACGIH TLV: TWA 200 ppm; STEL 250 ppm (skin); BEI: 15 mg/L of methanol in urine at end of shift
DFG MAK: 200 ppm (270 mg/m³); BAT: 30 mg/L in urine at end of shift
NIOSH REL: TWA 200 ppm; CL 800 ppm/15M
DOT CLASSIFICATION: 3; Label: Flammable Liquid, Poison
SAFETY PROFILE: A human poison by ingestion. Poison experimentally by skin contact. Moderately toxic experimentally by intravenous and intraperitoneal routes. Mildly toxic by inhalation. Human systemic effects: changes in circulation, cough, dyspnea, headache, lachrymation, nausea or vomiting, optic nerve neuropathy, respiratory effects, visual field changes. An experimental teratogen. Experimental reproductive effects. An eye and skin irritant. Human mutation data reported. A narcotic.

Its main toxic effect is exerted upon the nervous system, particularly the optic nerves and possibly the retinae. The condition can progress to permanent blindness. Once absorbed, methanol is only very slowly eliminated. Coma resulting from massive exposures may last as long as 2–4 days. In the body, the products formed by its oxidation are formaldehyde and formic acid, both of which are toxic. Because of the slow elimination, methanol should be regarded as a cumulative poison. Though single exposures to fumes may cause no harmful effect, daily exposure may result in the accumulation of sufficient methanol in the body to cause illness. Death from ingestion

M

of less than 30 mL has been reported. A common air contaminant.

Flammable liquid. Dangerous fire hazard when exposed to heat, flame, or oxidizers. Explosive in the form of vapor when exposed to heat or flame. Explosive reaction with chloroform + sodium methoxide, diethyl zinc. Violent reaction with alkyl aluminum salts, acetyl bromide, chloroform + sodium hydroxide, CrO_3, cyanuric chloride, (I + ethanol + HgO), $Pb(ClO_4)_2$, $HClO_4$, P_2O_3, $(KOH + CHCl_3)$, nitric acid. Incompatible with beryllium dihydride, metals (e.g., potassium, magnesium), oxidants (e.g., barium perchlorate, bromine, sodium hypochlorite, chlorine, hydrogen peroxide), potassium tert-butoxide, carbon tetrachloride + metals (e.g., aluminum, magnesium, zinc), dichloromethane. Dangerous; can react vigorously with oxidizing materials. To fight fire, use alcohol foam. When heated to decomposition it emits acrid smoke and irritating fumes.

MGC225   CAS: 12263-85-3   HR: 3
METHYL ALUMINUM SESQUIBROMIDE
mf: $C_3H_9Al_2Br_3$ mw: 338.81
PROP: Colorless liquid. Bp: 100–120° @ 50 mm.
SYNS: ALUMINUM, TRIBROMOTRIMETHYLDI- □ METHYL ALUMINIUM SESQUIBROMIDE □ METHYL ALUMINUM SESQUIBROMIDE □ TRIBROMO-TRIMETHYLDIALUMINUM
CONSENSUS REPORTS: Reported in EPA TSCA Inventory.
ACGIH TLV: TWA 2 mg(Al)/m³
SAFETY PROFILE: A flammable solid. Danger from spontaneous combustion. When heated to decomposition it emits toxic fumes of Br⁻.

MGC230   CAS: 12542-85-7   HR: 3
METHYL ALUMINUM SESQUI-CHLORIDE
mf: $C_3H_9Al_2Cl_3$ mw: 205.43
PROP: Clear liquid. D: 0.877, flash p: 1°F, bp: 120–140°.

SYNS: ALUMINUM, TRICHLOROTRIMETHYLDI- □ METHYL ALUMINIUM SESQUICHLORIDE □ TRICHLOROTRIMETHYLDIALUMINUM
CONSENSUS REPORTS: Reported in EPA TSCA Inventory.
ACGIH TLV: TWA 2 mg(Al)/m³
SAFETY PROFILE: A flammable liquid. Danger from spontaneous combustion. When heated to decomposition it emits toxic fumes of Cl⁻.

MGC250   CAS: 74-89-5   HR: 3
METHYLAMINE
DOT: UN 1061/UN 1235
mf: CH_5N mw: 31.07
PROP: Colorless, flammable gas or liquid; powerful ammonia-like odor. Frequently encountered as strong aq soln. Bp: 6.3°, lel: 4.95%, uel: 20.75%, mp: −93.5°, flash p: 32°F (CC), d: 0.662 @ 20°/4°, autoign temp: 806°F, vap d: 1.07. Fuming liquid when liquefied: d: 0.699 @ −10.8°/4°. Sol in alc; misc with ether. IDLH 100 ppm.
SYNS: AMINOMETHANE □ CARBINAMINE □ MERCURIALIN □ METHANAMINE (9CI) □ METHYL-AMINE (ACGIH,OSHA) □ METHYLAMINE, anhydrous (UN 1061) (DOT) □ METHYLAMINE, aqueous solution (UN 1235) (DOT) □ METHYLAMINEN (DUTCH) □ METILAMINE (ITALIAN) □ METYLOAMINA (POLISH) □ MONOMETHYLAMINE
CONSENSUS REPORTS: Reported in EPA TSCA Inventory.
OSHA PEL: TWA 10 ppm
ACGIH TLV: TWA 5 ppm; STEL 15 ppm
DFG MAK: 10 ppm (13 mg/m³)
DOT CLASSIFICATION: 2.3; Label: Poison Gas, Flammable Gas (UN 1061); DOT Class: 3; Label: Flammable Liquid, Corrosive (UN 1235)
SAFETY PROFILE: Poison by subcutaneous route. Moderately toxic by inhalation. A severe skin irritant. Mutation data reported. A strong base. Flammable gas at ordinary temperature and pressure. Very dangerous fire hazard when exposed to heat, flame, or sparks. Explosive when exposed to heat or flame. To fight fire, stop flow of gas. Forms an explosive mixture with nitromethane. When heated to

decomposition it emits toxic fumes of NO_x. See also AMINES.

MGG000   CAS: 109-83-1   HR: 3
2-METHYLAMINOETHANOL

mf: C_3H_9NO mw: 75.11

PROP: Viscous liquid; fishy odor. Corrosive to skin, cork, and metals. A strong base. D: 0.9, vap d: 2.9, bp: 156°, flash p: 165°F (OC). Misc with water, alc, and ether.

SYNS: β-(METHYLAMINO)ETHANOL □ N-METHYL-AMINOETHANOL □ N-METHYLETHANOLAMINE □ METHYLETHYLOLAMINE □ METHYL(β-HYDROXY-ETHYL)AMINE □ MONOMETHYL-AMINOAETHANOL (GERMAN) □ MONOMETHYLAMINOETHANOL □ N-MONOMETHYLAMINOETHANOL □ USAF DO-50

CONSENSUS REPORTS: Reported in EPA TSCA Inventory.

SAFETY PROFILE: Poison by intraperitoneal route. Moderately toxic by ingestion and subcutaneous routes. A corrosive irritant to skin, eyes, and mucous membranes. Flammable when exposed to heat, flame, or oxidizers. To fight fire, use alcohol foam. When heated to decomposition it emits toxic fumes such as NO_x. See also AMINES and ALCOHOLS.

MGL600   CAS: 7698-91-1   HR: 3
N-METHYL-4-AMINO-1,2,5-SELENA-DIAZOLE-3-CARBOXAMIDE

mf: $C_4H_6N_4OSe$ mw: 205.10

SYNS: 4-AMINO-N-METHYL-1,2,5-SELENADIAZOLE-3-CARBOXAMIDE □ NSC-93169 □ 1,2,5-SELEN-ADIAZOLE-3-CARBOXAMIDE, 4-AMINO-N-METHYL-

OSHA PEL: TWA 0.2 mg(Se)/m³
ACGIH TLV: TWA 0.2 mg(Se)/m³
SAFETY PROFILE: Poison by intraperitoneal route. When heated to decomposition it emits toxic fumes of NO_x and Se.

MGN500   CAS: 110-43-0   HR: 3
METHYL n-AMYL KETONE

DOT: UN 1110

mf: $C_7H_{14}O$ mw: 114.21

PROP: Colorless, mobile liquid; penetrating, fruity odor; or light yellow oil. Bp: 151.5°, flash p: 120°F (OC), autoign temp: 991°F,

vap d: 3.94, d: 0.8197 @ 15°/4°. Very sltly sol in water; sol in alc and ether. IDLH 800 ppm.

SYNS: AMYL-METHYL-CETONE (FRENCH) □ n-AMYL METHYL KETONE □ AMYL METHYL KETONE (DOT) □ FEMA No. 2544 □ 2-HEPTANONE □ METHYL-AMYL-CETONE (FRENCH) □ METHYL AMYL KETONE (DOT) □ METHYL PENTYL KETONE

CONSENSUS REPORTS: Reported in EPA TSCA Inventory.

OSHA PEL: TWA 100 ppm
ACGIH TLV: TWA 50 ppm
NIOSH REL: (Ketones) TWA 465 mg/m³
DOT CLASSIFICATION: 3; Label: Flammable Liquid

SAFETY PROFILE: Moderately toxic by ingestion. Mildly toxic by inhalation and skin contact. A skin irritant. A flammable liquid when exposed to heat or flame; can react with oxidizing materials. To fight fire, use foam, CO_2, dry chemical. When heated to decomposition it emits acrid smoke and fumes. See also KETONES.

MGN750   CAS: 100-61-8   HR: 3
METHYLANILINE

DOT: UN 2294

mf: C_7H_9N mw: 107.17

PROP: Colorless or sltly yellow liquid; becomes brown on exposure to air. Mp: −57°, d: 0.989 @ 20°/4°, bp: 194–197°. Sol in alc, ether; sltly sol in water. IDLH 100 ppm.

SYNS: ANILINOMETHANE □ BENZENAMINE, N-METHYL-(9CI) □ (METHYLAMINO)BENZENE □ N-METHYLAMINOBENZENE □ N-METHYLANILINE □ N-METHYLANILINE (ACGIH,DOT) □ N-METHYL-BENZENAMINE □ METHYLPHENYLAMINE □ N-METHYLPHENYLAMINE □ MONOMETHYLANILINE □ N-MONOMETHYLANILINE □ MONOMETHYL ANILINE (OSHA) □ N-PHENYLMETHYLAMINE

CONSENSUS REPORTS: Reported in EPA TSCA Inventory.

OSHA PEL: TWA 0.5 ppm (skin)
ACGIH TLV: TWA 0.5 ppm (skin)
DFG MAK: 0.5 ppm (2.2 mg/m³)
DOT CLASSIFICATION: 6.1; Label: KEEP AWAY FROM FOOD

SAFETY PROFILE: Poison by ingestion and intravenous routes. Moderately toxic by

M

subcutaneous route. When heated to decomposition it emits toxic fumes of NO_x.

MGO500   CAS: 102-50-1   HR: 3
2-METHYL-p-ANISIDINE
mf: $C_8H_{11}NO$ mw: 137.20
PROP: Crystals from ligroin. Mp: 29–30°, bp: 146–147° @ 23 mm.
SYNS: m-CRESIDINE □ 4-METHOXY-2-METHYL-ANILINE □ 4-METHOXY-2-METHYLBENZENAMINE □ 2-METHYL-4-METHOXYANILINE □ NCI-C02993
CONSENSUS REPORTS: IARC Cancer Review: Group 3 IMEMDT 7,56,87; Animal Inadequate Evidence IMEMDT 27,91,82. NCI Carcinogenesis Bioassay (gavage); Clear Evidence: rat NCITR* NCI-CG-TR-105,78; (gavage); Inadequate Studies: mouse NCITR* NCI-CG-TR-105,78. Reported in EPA TSCA Inventory.
SAFETY PROFILE: Suspected carcinogen with experimental carcinogenic and tumorigenic data. Mutation data reported. When heated to decomposition it emits toxic fumes of NO_x.

MGO750   CAS: 120-71-8   HR: 3
5-METHYL-o-ANISIDINE
mf: $C_8H_{11}NO$ mw: 137.20
PROP: Needles from pet ether. Mp: 51–52°, bp: 235°.
SYNS: m-AMINO-p-CRESOL, METHYL ESTER □ 3-AMINO-p-CRESOL METHYL ESTER □ 1-AMINO-2-METHOXY-5-METHYLBENZENE □ 3-AMINO-4-METHOXYTOLUENE □ 2-AMINO-4-METHYLANISOLE □ AZOIC RED 36 □ C.I. AZOIC RED 83 □ CRESIDINE □ p-CRESIDINE □ KRESIDIN □ KREZIDINE □ 2-METHOXY-5-METHYLANILINE □ 2-METHOXY-5-METHYL-BENZENAMINE (9CI) □ 4-METHOXY-m-TOLUIDINE □ 4-METHYL-2-AMINOANISOLE □ NCI-C02982
CONSENSUS REPORTS: NTP 10th Report on Carcinogens. IARC Cancer Review: Group 2B IMEMDT 7,56,87; Human Limited Evidence IMEMDT 27,91,82; Animal Sufficient Evidence IMEMDT 27,91,82. NCI Carcinogenesis Bioassay (feed); Clear Evidence: mouse, rat NCITR* NCI-CG-TR-142,79. Reported in EPA TSCA Inventory. Community Right-To-Know List.

DFG MAK: Confirmed Animal Carcinogen, Suspected Human Carcinogen
SAFETY PROFILE: Confirmed carcinogen with experimental carcinogenic and neoplastigenic data. Moderately toxic by ingestion. A skin and eye irritant. Mutation data reported. When heated to decomposition it emits toxic fumes of NO_x. See also ESTERS.

MGP000   CAS: 104-93-8   HR: 3
p-METHYL ANISOLE
mf: $C_8H_{10}O$ mw: 122.18
PROP: Liquid, found in oil of ylang-ylang, Cananga, and others (FCTXAV 12,385,74). Colorless liquid; ylang-ylang odor. D: 0.996–0.970, refr index: 1.510–1.513, bp: 175–176°, flash p: 144°F. Sol in fixed oils; insol in glycerin, propylene glycol.
SYNS: p-CRESOL METHYL ETHER □ p-CRESYL METHYL ETHER □ FEMA No. 2681 □ p-METHOXY-TOLUENE □ 4-METHOXYTOLUENE □ 4-METHYL-1-METHOXYBENZENE □ 4-METHYLPHENOL METHYL ETHER □ METHYL-p-TOLYL ETHER □ p-TOLYL METHYL ETHER
CONSENSUS REPORTS: Reported in EPA TSCA Inventory.
SAFETY PROFILE: Moderately toxic by ingestion. A skin irritant. Flammable liquid when exposed to heat, sparks, or flame. When heated to decomposition it emits acrid smoke and irritating fumes.

MGP750   CAS: 779-02-2   HR: 2
9-METHYLANTHRACENE
mf: $C_{15}H_{12}$ mw: 192.27
PROP: Needles from EtOH. Mp: 81.5°, bp: 196–197° @ 12 mm.
CONSENSUS REPORTS: Reported in EPA TSCA Inventory.
SAFETY PROFILE: Questionable carcinogen with experimental tumorigenic data. Human mutation data reported. When heated to decomposition it emits acrid smoke and irritating fumes.

MGQ250   CAS: 85-91-6   HR: 3
N-METHYLANTHRANILIC ACID,
**   METHYL ESTER**
mf: $C_9H_{11}NO_2$ mw: 165.21
PROP: Pale-yellow liquid; grape-like odor;
or crystals from pet ether. D: 1.126–1.132,
mp: 19°, bp: 256°, refr index: 1.578–1.581,
flash p: 196°F. Sol in fixed oils; sltly sol in
propylene glycol; insol in water, glycerin.
SYNS: DIMETHYL ANTHRANILATE (FCC) □ FEMA
No. 2718 □ 2-METHYLAMINO METHYL BENZOATE □
METHYL METHYLAMINOBENZOATE □ METHYL-N-
METHYL ANTHRANILATE □ MMA
CONSENSUS REPORTS: Reported in EPA
TSCA Inventory.
SAFETY PROFILE: Poison by intravenous
route. Moderately toxic by ingestion.
Combustible liquid. When heated to
decomposition it emits toxic fumes of NO_x.
See also ESTERS.

MGQ775   CAS: 7207-97-8   HR: 3
METHYLARSINE DIIODIDE
mf: CH_3AsI_2 mw: 343.76
PROP: Crystals or needles from EtOH. Mp:
28°, bp: 125° @ 16 mm. Sltly sol in water.
SYNS: ARSINE, DIIODOMETHYL- □ ARSONOUS
DIIODIDE, METHYL-(9CI) □ DIIODOMETHYLARSINE
□ METHYLDIIODOARSINE
OSHA PEL: TWA 0.5 mg(As)/m³
ACGIH TLV: Proposed: (inhalable fraction)
0.1 mg/m³; Not Classifiable as a Human
Carcinogen)
SAFETY PROFILE: Poison by intravenous
route. When heated to decomposition it
emits toxic fumes of As and I⁻.

MGR500   CAS: 52-88-0   HR: 3
8-METHYLATROPINIUM NITRATE
mf: $C_{18}H_{26}NO_3$•NO_3 mw: 366.46
PROP: A solid. Mp: 166–168°.
SYNS: ATROPINE METHONITRATE □ ATROPINE
METHYL NITRATE □ EKOMINE □ EUMIDRINA □
EUMYDRIN □ EUROPEN □ HARVATRATE □ 3-α-
HYDROXY-8-METHYL-1-α-H,5-α-H-TROPANIUM
NITRATE (±)-TROPATE (ESTER) □ dl-HYOSCYAMINE
METHYLNITRATE □ dl-HYOSYAMINE METHYL-
NITRATE □ METANITE □ METHYL ATROPINE
NITRATE □ N-METHYLATROPINE NITRATE □ N-
METHYLATROPINIUM NITRATE □ METROPINE □
PYLOSTROPIN

CONSENSUS REPORTS: Reported in EPA
TSCA Inventory.
SAFETY PROFILE: Poison by
intraperitoneal, intravenous, and
subcutaneous routes. Moderately toxic by
ingestion. Human systemic effects:
mydriasis. Experimental reproductive
effects. When heated to decomposition it
emits toxic fumes of NO_x. See also
ATROPINE and NITRATES.

MGS750   CAS: 592-62-1   HR: 3
METHYLAZOXYMETHYL ACETATE
mf: $C_4H_8N_2O_3$ mw: 132.14
PROP: Bp: 191°.
SYNS: ACETIC ACID, (METHYL-ONN-AZOXY)-
METHYL ESTER □ CYCASIN ACETATE □ MAM AC □
MAM ACETATE □ METHYLAZOXYMETHANOL
ACETATE □ METHYLAZOXYMETHYLESTER
KYSELINY OCTOVE (CZECH) □ (METHYL-ONN-
AZOXY)METHANOL, ACETATE (ester)
CONSENSUS REPORTS: IARC Cancer
Review: Group 2B IMEMDT 7,56,87;
Animal Sufficient Evidence IMEMDT
10,121,76. EPA Genetic Toxicology
Program.
SAFETY PROFILE: Confirmed carcinogen
with experimental carcinogenic,
neoplastigenic, tumorigenic, and
teratogenic data. Poison by ingestion,
intraperitoneal, and intravenous routes. An
experimental teratogen. Experimental
reproductive effects. Human mutation data
reported. When heated to decomposition it
emits toxic fumes of NO_x. See also
ESTERS.

MGZ000   CAS: 1155-38-0   HR: 2
7-METHYLBENZ(a)ANTHRACENE-5,6-
**   OXIDE**
mf: $C_{19}H_{14}O$ mw: 258.33
SYN: 5,6-EPOXY-5,6-DIHYDRO-7-METHYLBENZ(A)
ANTHRACENE
CONSENSUS REPORTS: EPA Genetic
Toxicology Program.
SAFETY PROFILE: Questionable
carcinogen with experimental neoplastigenic
and tumorigenic data. Mutation data

reported. When heated to decomposition it emits acrid smoke and irritating fumes.

MHA500   CAS: 101-41-7   HR: 2
METHYL BENZENEACETATE
mf: $C_9H_{10}O_2$ mw: 150.19

PROP: Colorless liquid; honey, jasmine odor. D: 1.062, refr index: 1.503–1.509, bp: 215°, vap d: 5.18, flash p: 192°F. Sol in alc, fixed oils; insol in glycerin, propylene glycol, water @ 215°.

SYNS: BENZENEACETIC ACID, METHYL ESTER □ FEMA No. 2733 □ METHYL PHENYLACETATE (FCC) □ METHYL-α-TOLUATE □ PHENYLACETIC ACID, METHYL ESTER

CONSENSUS REPORTS: Reported in EPA TSCA Inventory.

SAFETY PROFILE: Moderately toxic by ingestion and skin contact. A skin irritant. Combustible liquid. When heated to decomposition it emits acrid smoke and irritating fumes. See also ESTERS.

MHA750   CAS: 93-58-3   HR: 2
METHYL BENZENECARBOXYLATE
DOT: UN 2938

mf: $C_8H_8O_2$ mw: 136.16

PROP: Colorless liquid; fragrant odor. Mp: −12.5°, bp: 199.6°, flash p: 181°F, d: 1.082–1.088, refr index: 1.515, vap press: 1 mm @ 39.0°, vap d: 4.69. Sol in alc, fixed oils, propylene glycol, water @ 30°; misc in ether; insol in glycerin.

SYNS: ESSENCE OF NIOBE □ FEMA No. 2683 □ METHYL BENZOATE (FCC, DOT) □ METHYLESTER KYSELINY BENZOOVE □ NIOBE OIL □ OIL OF NIOBE □ OXIDATE LE

CONSENSUS REPORTS: Reported in EPA TSCA Inventory.

DOT CLASSIFICATION: 6.1; Label: KEEP AWAY FROM FOOD

SAFETY PROFILE: Moderately toxic by ingestion. Mildly toxic by skin contact. A skin and eye irritant. Combustible liquid when exposed to heat or flame; can react with oxidizing materials. To fight fire, use foam, CO_2, dry chemical, water to blanket fire. When heated to decomposition it emits acrid smoke and irritating fumes.

MHC250   CAS: 615-15-6   HR: 3
METHYL-2-BENZIMIDAZOLE
mf: $C_8H_8N_2$ mw: 132.18

PROP: Needles from water. Mp: 175–176°. Sol in hot water, NaOH; sltly sol in alc and ether.

SYN: 2-METHYLBENZIMIDAZOLE

CONSENSUS REPORTS: Reported in EPA TSCA Inventory.

SAFETY PROFILE: Poison by intravenous route. Moderately toxic by ingestion. Experimental reproductive effects. Mutation data reported. When heated to decomposition it emits toxic fumes of NO_x.

MHC750   CAS: 10605-21-7   HR: 2
METHYL BENZIMIDAZOLE-2-YL
CARBAMATE
mf: $C_9H_9N_3O_2$ mw: 191.21

PROP: Light-gray powder. Mp: 302–307° (decomp). Sol in DMF; sltly sol in most solvs.

SYNS: BAS-3460 □ BAS 67054 □ BAVISTIN □ BCM □ BENZIMIDAZOLE-2-CARBAMIC ACID, METHYL ESTER □ N-2-(BENZIMIDAZOLYL) CARBAMATE □ 1H-BENZIMIDAZOL-2-YLCARBAMIC ACID METHYL ESTER □ BMC □ CARBENDAZIM □ CARBENDAZIME □ CARBENDAZOL □ CARBENDAZOLE □ CARBENDAZYM □ CEKUDAZIM □ CTR 6669 □ CUSTOS □ DELSENE □ DEROSAL □ EQUITDAZIN □ HOE 17411 □ KEMDAZIN □ MBC □ 2-(METHOXY-CARBONYL-AMINO)-BENZIMIDAZOL □ 2-(METHOXYCARBONYL-AMINO)-BENZIMIDAZOLE □ METHYL 1H-BENZ-EMEDAZOL-2-YLCARBAMATE □ METHYL 2-BEN-ZIMIDAZOLECARBAMATE □ PILLARSTIN □ STEMPOR □ TRITICOL □ U-32.104

CONSENSUS REPORTS: Reported in EPA TSCA Inventory. EPA Genetic Toxicology Program.

SAFETY PROFILE: Moderately toxic by skin contact. Mildly toxic by ingestion. An experimental teratogen. Experimental reproductive effects. Human mutation data reported. An agricultural chemical and pesticide. When heated to decomposition it emits toxic fumes of NO_x. See also CARBAMATES.

MHF750   CAS: 134-84-9   HR: 3
4-METHYL BENZOPHENONE
mf: $C_{14}H_{12}O$ mw: 196.26
PROP: Crystals from pet ether in two forms.
Mp: 59–60° (stable form), bp: 326°.
SYNS: p-BENZOPHENONE, METHYL- □ PHENYL p-
TOLYL KETONE □ USAF DO-54
CONSENSUS REPORTS: Reported in EPA
TSCA Inventory.
DOT CLASSIFICATION: 3; Label:
Flammable Liquid
SAFETY PROFILE: Poison by
intraperitoneal route. A flammable liquid.
When heated to decomposition it emits
acrid smoke and irritating fumes.

MHI250   CAS: 553-97-9   HR: 3
2-METHYL-p-BENZOQUINONE
mf: $C_7H_6O_2$ mw: 122.13
PROP: Yellow plates or needles. Mp: 69°.
SYNS: METHYL-p-BENZOQUINONE □ METHYL-1,4-
BENZOQUINONE □ 2-METHYLBENZOQUINONE-1,4 □
2-METHYL-1,4-QUINONE □ p-TOLUQUINONE □ 1,4-
TOLUQUINONE
CONSENSUS REPORTS: Reported in EPA
TSCA Inventory.
SAFETY PROFILE: Poison by ingestion.
When heated to decomposition it emits
acrid smoke and irritating fumes.

MHI300   CAS: 1123-91-7   HR: 3
5-METHYL-2,1,3-BENZOSELENADIA-
ZOLE
mf: $C_7H_6N_2Se$ mw: 197.11
SYN: 2,1,3-BENZOSELENADIAZOLE, 5-METHYL-
OSHA PEL: TWA 0.2 mg(Se)/m³
ACGIH TLV: TWA 0.2 mg(Se)/m³
SAFETY PROFILE: Poison by intravenous
route. When heated to decomposition it
emits toxic fumes of NO_x and Se.

MHI500   CAS: 2818-88-4   HR: 3
2-METHYLBENZOSELENAZOLE
mf: C_8H_7NSe mw: 196.12
SYN: 2-METHYLBENZSELENAZOL (CZECH)
CONSENSUS REPORTS: Selenium and its
compounds are on the Community Right-
To-Know List. Reported in EPA TSCA
Inventory.

OSHA PEL: TWA 0.2 mg(Se)/m³
ACGIH TLV: TWA 0.2 mg(Se)/m³
SAFETY PROFILE: Poison by intravenous
route. Moderately toxic by ingestion. An eye
and skin irritant. When heated to
decomposition it emits very toxic fumes of
NO_x and Se. See also SELENIUM
COMPOUNDS.

MHL000   CAS: 31431-39-7   HR: 2
METHYL-5-BENZOYL BENZIMIDAZOLE-
2-CARBAMATE
mf: $C_{16}H_{13}N_3O_3$ mw: 295.32
PROP: Crystals from AcOH/MeOH. Mp:
288.5°.
SYNS: BANTENOL □ 2-BENZIMIDAZOLECARBAMIC
ACID, 5-BENZOYL-, METHYL ESTER □ N-2 (5-BENZO-
YL-BENZIMIDAZOLE) CARBAMATE de METHYLE □ 5-
BENZOYL-2-BENZIMIDAZOLECARBAMIC ACID
METHYL ESTER □ N-(BENZOYL-5, BENZIMIDAZOL-
YL)-2, CARBAMATE de METHYLE □ (5-BENZOYL-1H-
BENZIMIDAZOL-2-YL)-CARBAMIC ACID METHYL
ESTER □ BESANTIN □ LOMPER □ MBDZ □ MEB-
ENDAZOLE (USDA) □ MEBENVET □ METHYL 5-
BENZOYL BENZIMIDAZOLE-2-CARBAMATE □ METH-
YL 5-BENZOYL-2-BENZIMIDAZOLECARBAMATE □
NOVERME □ OVITELMIN □ PANTELMIN □ R 17635 □
TELMIN □ VERMICIDIN □ VERMIRAX □ VERMOX □
VERPANYL
CONSENSUS REPORTS: EPA Genetic
Toxicology Program.
SAFETY PROFILE: Moderately toxic by
ingestion and intraperitoneal routes. Human
mutation data reported. An experimental
teratogen. Experimental reproductive
effects. When heated to decomposition it
emits toxic fumes of NO_x. See also
CARBAMATES.

MHM100   HR: D
METHYLBENZYL ACETATE
mf: $C_{10}H_{12}O_2$ mw: 164.20
PROP: Colorless liquid; sweet, nutty odor.
D: 1.030–1.035, refr index: 1.501. Sol in
fixed oils; sltly sol in propylene glycol; insol
in glycerin.
SYN: TOLYL ACETATE
SAFETY PROFILE: When heated to
decomposition it emits acrid smoke and
irritating fumes.

M

MHN750   CAS: 10309-79-2   HR: 3
1-METHYL-2-BENZYLHYDRAZINE
mf: $C_8H_{12}N_2$ mw: 136.22
PROP: Bp: 117° @ 20 mm.
SYN: 1-BENZYL-2-METHYLHYDRAZINE
SAFETY PROFILE: Poison by
subcutaneous route. Questionable
carcinogen with experimental carcinogenic,
tumorigenic, and teratogenic data. Mutation
data reported. When heated to
decomposition it emits toxic fumes of NO_x.

MHP250   CAS: 937-40-6   HR: 3
N-METHYL-N-BENZYLNITROSAMINE
mf: $C_8H_{10}N_2O$ mw: 150.20
SYNS: METHYL-BENZYL-NITROSOAMIN (GERMAN)
□ N-METHYL-N-NITROSOBENZYLAMINE □ N-
NITROSOBENZYLMETHYLAMINE □ N-NITROSO-
METHYLBENZYLAMINE
CONSENSUS REPORTS: EPA Genetic
Toxicology Program.
SAFETY PROFILE: Poison by ingestion.
Questionable carcinogen with experimental
tumorigenic data. Mutation data reported.
When heated to decomposition it emits
toxic fumes of NO_x. See also
NITROSAMINES.

MHR200   CAS: 74-83-9   HR: 3
METHYL BROMIDE
mf: CH_3Br mw: 94.95
PROP: Colorless, transparent, volatile liquid
or gas; burning taste, chloroform-like odor.
Bp: 3.56°, lel: 13.5%, uel: 14.5%, fp: −93°,
flash p: none, d: 1.732 @ 0°/0°, autoign
temp: 998°F, vap d: 3.27, vap press: 1824
mm @ 25°. Sltly sol in water. IDLH 250
ppm.
SYNS: BROM-METHAN (GERMAN) □ BROMOMET-
ANO (ITALIAN) □ BROMOMETHANE □ BROMO-O-
GAS □ BROMURE de METHYLE (FRENCH) □ BRO-
MURO di METILE (ITALIAN) □ BROOMMETHAAN
(DUTCH) □ DAWSON 100 □ DOWFUME □ DOWFUME
MC-2 SOIL FUMIGANT □ EDCO □ EMBAFUME □
FUMIGANT-1 (OBS.) □ HALON 1001 □ ISCOBROME □
KAYAFUME □ MB □ MBX □ MEBR □ METAFUME □
METHOGAS □ METHYLBROMID (GERMAN) □
METYLU BROMEK (POLISH) □ MONOBROMO-
METHANE □ PESTMASTER (OBS.) □ PROFUME (OBS.)
□ R 40B1 □ RCRA WASTE NUMBER U029 □ ROTOX □
TERABOL □ TERR-O-GAS 100 □ ZYTOX

CONSENSUS REPORTS: IARC Cancer
Review: Group 3 IMEMDT 7,245,87;
Human Inadequate Evidence IMEMDT
41,187,86; Animal Limited Evidence
IMEMDT 41,187,86. Reported in EPA
TSCA Inventory. Community Right-To-
Know List. EPA Extremely Hazardous
Substances List.
OSHA PEL: TWA 5 ppm (skin)
ACGIH TLV: TWA 1 ppm (skin); Not
Classifiable as a Human Carcinogen
DFG MAK: Confirmed Animal Carcinogen
with Unknown Relevance to Humans
NIOSH REL: (Monohalomethanes) Reduce
to lowest level
SAFETY PROFILE: Suspected carcinogen
with experimental carcinogenic data. A
human poison by inhalation. Human
systemic effects by inhalation: anorexia,
nausea or vomiting. Corrosive to skin; can
produce severe burns. Human mutation data
reported. A powerful fumigant gas that is
one of the most toxic of the common
organic halides. It is hemotoxic and narcotic
with delayed action. The effects are
cumulative and damaging to nervous system,
kidneys, and lung. Central nervous system
effects include blurred vision, mental
confusion, numbness, tremors, and speech
defects.
    Methyl bromide is reported to be eight
times more toxic on inhalation than ethyl
bromide. Moreover, because of its greater
volatility, it is a much more frequent cause
of poisoning. Death following acute
poisoning is usually caused by its irritant
effect on the lungs. In chronic poisoning,
death is due to injury to the central nervous
system. Fatal poisoning has always resulted
from exposure to relatively high
concentrations of methyl bromide vapors
(from 8600 to 60,000 ppm). Nonfatal
poisoning has resulted from exposure to
concentrations as low as 100–500 ppm. In
addition to injury to the lung and central
nervous system, the kidneys may be
damaged, with development of albuminuria
and, in fatal cases, cloudy swelling and/or

tubular degeneration. The liver may be enlarged. There are no characteristic blood changes.

Mixtures of 10–15 percent with air may be ignited with difficulty. Moderately explosive when exposed to sparks or flame. Forms explosive mixtures with air within narrow limits at atmospheric pressure, with wider limits at higher pressure. The explosive sensitivity of mixtures with air may be increased by the presence of aluminum, magnesium, zinc, or their alloys. Incompatible with metals, dimethyl sulfoxide, ethylene oxide. To fight fire, use foam, water, CO_2, dry chemical. When heated to decomposition it emits toxic fumes of Br^-. See also BROMIDES.

MHR250   CAS: 96-32-2   HR: 3
METHYL BROMOACETATE
DOT: UN 2643
mf: $C_3H_5BrO_2$ mw: 152.99
PROP: Liquid. Bp: 51–52° @ 15 mm.
SYNS: BROMOACETIC ACID METHYL ESTER ☐ METHYL α-BROMOACETATE ☐ METHYLESTER KYSELINY BROMOCTOVE ☐ METHYL MONOBROMO-ACETATE
CONSENSUS REPORTS: Reported in EPA TSCA Inventory.
DOT CLASSIFICATION: 6.1; Label: Poison
SAFETY PROFILE: Poison by intravenous route. When heated to decomposition it emits toxic fumes of Br^-. See also ESTERS.

MHS550   CAS: 1679-09-0   HR: 3
2-METHYL-2-BUTANETHIOL
mf: $C_5H_{12}S$ mw: 104.23
SYNS: tert-AMYLMERCAPTAN ☐ tert-AMYLTHIOL ☐ 2-BUTANETHIOL, 2-METHYL- ☐ tert-PENTYL MER-CAPTAN
SAFETY PROFILE: A poison by ingestion, inhalation, and skin contact. A mild eye irritant. When heated to decomposition it emits toxic vapors of SO_x.

MHS750   CAS: 137-32-6   HR: 3
2-METHYL BUTANOL-1
mf: $C_5H_{12}O$ mw: 88.15

PROP: Colorless liquid. D: 0.81–0.82 @ 20°, fp: <−70°, bp: 128°, flash p: 122°F (OC), vap d: 3.0, lel: 1.4%, uel: 9.0%. Sltly sol in water; misc with alc and ether.
SYNS: dl-sec-BUTYLCARBINOL ☐ 2-METHYL-BUTANOL
CONSENSUS REPORTS: Reported in EPA TSCA Inventory.
SAFETY PROFILE: Moderately toxic by skin contact and intraperitoneal routes. Mildly toxic by ingestion. An eye, skin, and mucous membrane irritant. Can cause deafness, delirium, headache, nausea, and vomiting. Flammable liquid when exposed to heat, flame, or oxidizers. Explosive in the form of vapor when exposed to heat or flame. Incompatible with H_2S_3. To fight fire, use alcohol foam, spray, mist, dry chemical. When heated to decomposition it emits acrid smoke and irritating fumes. See also ALCOHOLS.

MHT000   CAS: 563-46-2   HR: 3
2-METHYL-1-BUTENE
mf: C_5H_{10} mw: 70.14
PROP: Colorless, extremely volatile liquid or gas. D: 0.7, vap d: 2.4, bp: 31.05°, flash p: −4°F. Insol in water.
CONSENSUS REPORTS: Reported in EPA TSCA Inventory.
SAFETY PROFILE: A simple asphyxiant. Very dangerous fire hazard when exposed to heat, flame, or oxidizers. To fight fire, use dry chemical, CO_2, foam. When heated to decomposition it emits acrid smoke and irritating fumes.

MHT250   CAS: 563-45-1   HR: 3
3-METHYL-1-BUTENE
mf: C_5H_{10} mw: 70.14
PROP: Colorless, very volatile liquid or gas; disagreeable odor. Bp: 20.1°, d: 0.65 @ 20°/20°, fp: −137.5°, flash p: 19.4°F, vap d: 2.4, lel: 1.5%, uel: 9.1%. Insol in water; sol in alc.
CONSENSUS REPORTS: Reported in EPA TSCA Inventory.

M

SAFETY PROFILE: Very dangerous fire hazard when exposed to heat, flame, or oxidizers. Explosive in the form of vapor when exposed to heat or flame. To fight fire, use alcohol foam, mist, spray, dry chemical, CO_2. When heated to decomposition it emits acrid smoke and irritating fumes. See also 2-METHYL-1-BUTENE.

MHU150   CAS: 5205-11-8   HR: 2
3-METHYL-2-BUTENYL BENZOATE
mf: $C_{12}H_{14}O_2$ mw: 190.26
SYNS: BENZOIC ACID, 3-METHYL-2-BUTENYL ESTER □ 2-BUTEN-1-OL, 3-METHYL-, BENZOATE □ PRENYL BENZOATE
CONSENSUS REPORTS: Reported in EPA TSCA Inventory.
SAFETY PROFILE: Moderately toxic by ingestion. A skin irritant. When heated to decomposition it emits acrid smoke and irritating fumes.

MHU750   CAS: 97-88-1   HR: 3
2-METHYL BUTYLACRYLATE
DOT: UN 2227
mf: $C_8H_{14}O_2$ mw: 142.22
PROP: Colorless liquid; ester odor. Bp: 163°, flash p: 126°F (TOC), lel: 2%, uel: 8%, autoign temp: 562°F, vap press: 4.9 mm @ 20°, d: 0.895 @ 20°/4°, vap d: 4.8.
SYNS: BUTIL METACRILATO (ITALIAN) □ BUTYL-METHACRYLAAT (DUTCH) □ N-BUTYL METH-ACRYLATE □ BUTYL-2-METHACRYLATE □ BUTYL-2-METHYL-2-PROPENOATE □ METHACRYLATE de BUTYLE (FRENCH) □ METHACRYLSAEUREBUTYL-ESTER (GERMAN) □ 2-METHYL-BUTYLACRYLAAT (DUTCH) □ 2-METHYL-BUTYLACRYLAT (GERMAN)
CONSENSUS REPORTS: Reported in EPA TSCA Inventory.
DOT CLASSIFICATION: 3; Label: Flammable Liquid
SAFETY PROFILE: Moderately toxic by intraperitoneal route. Mildly toxic by ingestion, inhalation, and skin contact. An experimental teratogen. Experimental reproductive effects. A skin irritant. Flammable liquid when exposed to heat or flame. Explosive in the form of vapor when exposed to heat or flame. Violent polymerization can be caused by heat, moisture, oxidizers. To fight fire, use foam, dry chemical, CO_2. When heated to decomposition it emits acrid smoke and irritating fumes.

MHV000   CAS: 110-68-9   HR: 3
N-METHYL-n-BUTYLAMINE
DOT: UN 2945
mf: $C_5H_{13}N$ mw: 87.19
PROP: Liquid. D: 0.7335, bp: 91.1°, vap d: 3.0, flash p: 35.6°F. Sol in water.
SYNS: METHYLBUTYLAMINE □ N-(METHYL) BUTYL AMINE
CONSENSUS REPORTS: Reported in EPA TSCA Inventory.
DOT CLASSIFICATION: 3; Label: Flammable Liquid
SAFETY PROFILE: Poison by intravenous route. Moderately toxic by ingestion, skin contact, and intraperitoneal routes. Mildly toxic by inhalation. A skin and severe eye irritant. Flammable liquid when exposed to heat, sparks, or flame. To fight fire, use alcohol foam. When heated to decomposition it emits toxic fumes of NO_x. See also AMINES.

MHV750   CAS: 4435-53-4   HR: 3
METHYL-1,3-BUTYLENE GLYCOL ACETATE
DOT: UN 2708
mf: $C_7H_{14}O_3$ mw: 146.21
PROP: Liquid; bitter taste and acrid odor. Bp: 135°, flash p: 170°F, d: 0.952–0.958 @ 20°/20°, vap d: 5.05.
SYNS: ACETIC ACID-3-METHOXYBUTYL ESTER □ BUTOXYL □ 3-METHOXYBUTYL ACETATE □ 3-METHOXYBUTYLESTER KYSELINY OCTOVE
CONSENSUS REPORTS: Reported in EPA TSCA Inventory.
DOT CLASSIFICATION: 3; Label: Flammable Liquid
SAFETY PROFILE: Mildly toxic by ingestion. A skin and eye irritant. A flammable liquid. Flammable when exposed to heat or flame; can react with oxidizing materials. To fight fire, use alcohol foam,

CO₂, dry chemical. When heated to decomposition it emits acrid smoke and irritating fumes.

MHV859   CAS: 1634-04-4   HR: 3
METHYL tert-BUTYL ETHER
DOT: UN 2398
mf: C₅H₁₂O   mw: 88.17
PROP: Bp: 54°, d: 0.741 @ 20°/4°. Sltly sol in water.
SYNS: 2-METHOXY-2-METHYLPROPANE □ METHYL 1,1-DIMETHYLETHYL ETHER □ METHYL tert-BUTYL ETHER (DOT) □ MTBE □ PROPANE, 2-METHOXY-2-METHYL- (9CI)
CONSENSUS REPORTS: Community Right-To-Know List. Reported in EPA TSCA Inventory.
ACGIH TLV: TWA 50 ppm; Confirmed Animal Carcinogen with Unknown Revelance to Humans
DOT CLASSIFICATION: 3; Label: Flammable Liquid
SAFETY PROFILE: Poison by intravenous route. Slightly toxic by ingestion and inhalation. Flammable when exposed to heat or flame. When heated to decomposition it emits acrid smoke and irritating fumes. See also ETHERS.

MHW350   CAS: 71016-15-4   HR: 3
N-3-METHYLBUTYL-N-1-METHYL ACETONYLNITROSAMINE
mf: C₉H₁₈N₂O₂   mw: 186.29
PROP: Pale-yellow oil.
SYNS: 3-((ISOPENTYL)NITROSOAMINO)-2-BUTAN-ONE □ MAMBNA
SAFETY PROFILE: Suspected carcinogen with experimental carcinogenic, neoplastigenic, and tumorigenic data. Mutation data reported. When heated to decomposition it emits toxic fumes of NOₓ. See also NITROSAMINES.

MHW500   CAS: 7068-83-9   HR: 3
METHYLBUTYLNITROSAMINE
mf: C₅H₁₂N₂O   mw: 116.19
PROP: A liquid. Bp: 107.1–107.7° @ 40 mm.

SYNS: MBNA □ METHYL-BUTYL-NITROSAMIN (GERMAN) □ METHYL-N-BUTYLNITROSAMINE □ N-METHYL-N-NITROSOBUTYLAMINE □ N-NITROSO-N-BUTYLMETHYLAMINE □ N-NITROSOMETHYL-N-BUTYLAMINE □ NMBA
CONSENSUS REPORTS: EPA Genetic Toxicology Program.
SAFETY PROFILE: Poison by ingestion, inhalation, intraperitoneal, and subcutaneous routes. Questionable carcinogen with experimental carcinogenic and tumorigenic data. Mutation data reported. When heated to decomposition it emits toxic fumes of NOₓ. See also NITROSAMINES.

MHY000   CAS: 623-42-7   HR: 3
METHYL-n-BUTYRATE
DOT: UN 1237
mf: C₅H₁₀O₂   mw: 102.13
PROP: Colorless liquid. Mp: <−97°, bp: 102.3°, flash p: 57°F (CC), d: 0.919 @ 0°/4°, vap press: 40 mm @ 29.6°, vap d: 3.53. Sltly sol in water; misc in alc and ether.
SYNS: METHYL n-BUTANOATE □ METHYL BUTYRATE
CONSENSUS REPORTS: Reported in EPA TSCA Inventory.
DOT CLASSIFICATION: 3; Label: Flammable Liquid
SAFETY PROFILE: Moderately toxic by ingestion and skin contact. A skin irritant. A very dangerous fire hazard when exposed to heat, flame, or oxidizers. Can react vigorously with oxidizing materials. To fight fire, use alcohol foam, CO₂, dry chemical. When heated to decomposition it emits acrid smoke and irritating fumes. See also ESTERS.

MHY550   CAS: 7568-37-8   HR: 3
METHYL CADMIUM AZIDE
mf: CH₃CdN₃   mw: 97.13
PROP: Hygroscopic crystals; stable to 3°. Insol in nonpolar solvs.
CONSENSUS REPORTS: Cadmium and its compounds are on the Community Right-To-Know List.
OSHA PEL: TWA 5 μg(Cd)/m³

M

ACGIH TLV: TWA 0.002 mg(Cd)/m³ (respirable dust), Suspected Human Carcinogen); BEI: 5 μg/g creatinine in urine; 5 μg/L in blood

DFG MAK: DFG BAT: Blood 1.5 μg/dL; Urine 15 μg/dL, Suspected Carcinogen

NIOSH REL: (Cadmium) Reduce to lowest feasible level

SAFETY PROFILE: Confirmed human carcinogen. Hydrolysis reaction in the presence of moisture forms the explosive hydrogen azide gas. When heated to decomposition it emits toxic fumes of Cd and NOₓ. See also CADMIUM COMPOUNDS and AZIDES.

MHZ000   CAS: 598-55-0   HR: 3
METHYL CARBAMATE

mf: $C_2H_5NO_2$ mw: 75.07

PROP: Needles. Bp: 177°, mp: 52–54°. Very sol in water, alc.

SYNS: BENDIOCARB □ METHYLURETHAN □ METHYLURETHANE □ NCI-C55594 □ URETHYLANE

CONSENSUS REPORTS: IARC Cancer Review: Group 3 IMEMDT 7,56,87; Animal Inadequate Evidence IMEMDT 12,151,76. EPA Genetic Toxicology Program. Reported in EPA TSCA Inventory.

SAFETY PROFILE: Poison by ingestion and intraperitoneal routes. Questionable carcinogen with experimental carcinogenic and tumorigenic data. Mutation data reported. When heated to decomposition it emits toxic fumes of NOₓ. See also CARBAMATES.

MIA250   CAS: 2631-40-5   HR: 3
METHYLCARBAMIC ACID-o-CUMENYL
**   ESTER**

mf: $C_{11}H_{15}NO_2$ mw: 193.27

PROP: Crystals. Mp: 88–93°.

SYNS: BAY 39731 □ BAY 105807 □ BAYER 39731 □ CARBAMIC ACID, METHYL-, o-ISOPROPYLPHENYL ESTER □ CARBAMIC ACID, METHYL-, 2-(1-METHYL-ETHYL)PHENYL ESTER □ o-CUMENYL METHYL-CARBAMATE □ ENT 25,670 □ ETROFOLAN □ HYTOX □ ISOPROCARB □ ISOPROCARBE □ o-ISOPROPYL-PHENOL METHYLCARBAMATE □ o-ISOPROPYL-PHENYL N-METHYLCARBAMATE □ 2-ISOPROPYL-PHENYL N-METHYLCARBAMATE □ KHE 0145 □ 2-(1-

METHYLETHYL)PHENYL METHYLCARBAMATE □ MIPC □ MIPCIN □ MIPCINE □ MIPSIN □ OMS-32 □ PHENOL, O-ISOPROPYL-, METHYLCARBAMATE □ PHENOL, 2-(1-METHYLETHYL)-, METHYLCARBAMATE (9CI) □ PPC 3 □ RO 7-5050

CONSENSUS REPORTS: Reported in EPA TSCA Inventory. EPA Genetic Toxicology Program.

SAFETY PROFILE: Poison by ingestion, intravenous, and intraperitoneal routes. Moderately toxic by skin contact. When heated to decomposition it emits toxic fumes of NOₓ. See also CARBAMATES. Used for controlling leafhoppers, planthoppers, and bugs in rice and cacao.

MIB750   CAS: 1129-41-5   HR: 3
METHYLCARBAMIC ACID m-TOLYL
**   ESTER**

mf: $C_9H_{11}NO_2$ mw: 165.21

SYNS: CARBAMIC ACID, METHYL-, 3-METHYLPHEN-YL ESTER (9CI) □ CARBAMIC ACID, METHYL-, 3-TOLYL ESTER □ m-CRESYL METHYLCARBAMATE □ DICRES-YL □ DICRESYL N-METHYLCARBAMATE □ DRC 3341 □ KUMIAI □ METACRATE □ METHOLCARB □ m-METHYLPHENYL METHYLCARBAMATE □ 3-METHYL-PHENYL N-METHYLCARBAMATE □ METOLCARB □ MTMC □ S 1065 □ m-TOLYLESTER KYSELINY METHYL-KARBAMINOVE □ m-TOLYL N-METHYLCARBAMATE □ 3-TOLYL-N-METHYLCARBAMATE □ TSUMACIDE □ TSUMAUNKA

CONSENSUS REPORTS: EPA Extremely Hazardous Substances List. Reported in EPA TSCA Inventory. EPA Genetic Toxicology Program.

SAFETY PROFILE: Poison by ingestion and skin contact. Moderately toxic by inhalation. Mutation data reported. When heated to decomposition it emits toxic fumes of NOₓ. See also CARBAMATES.

MID860   CAS: 66637-25-0   HR: 3
2-(o-(METHYLCARBAMOYL)OXIMINO)-
**   3,3-DIMETHYLTETRAHYDRO-1,4-**
**   THIAZIN-5-ONE**

mf: $C_8H_{13}N_3O_3S$ mw: 231.30

SYNS: 3,3-DIMETHYL-2,5-THIOMORPHOLINEDIONE 2-(o-((METHYLAMINO)CARBONYL)OXIME) □ 2,5-THIOMORPHOLINEDIONE, 3,3-DIMETHYL-, 2-(o-((METHYLAMINO)CARBONYL)OXIME)

SAFETY PROFILE: A poison by ingestion. When heated to decomposition it emits toxic vapors of NO_x and SO_x.

MID900   CAS: 64049-00-9   HR: 3
(3-(N-METHYLCARBAMOYLOXY)PHEN-
YL)TRIMETHYL-ARSONIUM IODIDE
mf: $C_{13}H_{21}AsNO_2 \cdot I$ mw: 425.17

SYNS: ARSONIUM, (3-HYDROXYPHENYL)DIETHYLMETHYL-, IODIDE, METHYLCARBAMATE □ CARBAMIC ACID, N-METHYL-, 3-DIETHYLARSINOPHENYL ESTER, METHIODIDE □ TL-1504

OSHA PEL: TWA 0.5 mg(As)/m³
ACGIH TLV: Proposed: (inhalable fraction) 0.1 mg/m³; Not Classifiable as a Human Carcinogen)
SAFETY PROFILE: Poison by subcutaneous route. When heated to decomposition it emits toxic fumes of NO_x, As, and I^-.

MIF000   CAS: 616-38-6   HR: 3
METHYL CARBONATE
DOT: UN 1161
mf: $C_3H_6O_3$ mw: 90.09
PROP: Colorless liquid; pleasant odor. Mp: 0.5°, d: 1.065 @ 17°/4°, flash p: 66°F (OC), bp: 90.91°. Misc with acids and alkalies; sol in most org solvs; insol in water.
SYN: DIMETHYL CARBONATE
CONSENSUS REPORTS: Reported in EPA TSCA Inventory.
DOT CLASSIFICATION: 3; Label: Flammable Liquid
SAFETY PROFILE: Moderately toxic by intraperitoneal route. Mildly toxic by ingestion. An irritant. Violent reaction or ignition on contact with potassium-tert-butoxide. A very dangerous fire hazard when exposed to heat, open flames (sparks), or oxidizers. To fight fire, use alcohol foam. When heated to decomposition it emits acrid smoke and irritating fumes.

MIF750   CAS: 3121-61-7   HR: 3
METHYL CELLOSOLVE ACRYLATE
mf: $C_6H_{10}O_3$ mw: 130.16

PROP: Liquid. Bp: 61° @ 17 mm, flash p: 180°F (OC), d: 1.0134 @ 20°, vap d: 4.49.
SYNS: ACRYLIC ACID-2-METHOXYETHYL ESTER □ ETHYLENE GLYCOL MONOMETHYL ETHER ACRYLATE □ GLYCOL MONOMETHYL ETHER ACRYLATE □ 2-METHOXYETHANOL, ACRYLATE
CONSENSUS REPORTS: Glycol ether compounds are on the Community Right-To-Know List. Reported in EPA TSCA Inventory.
SAFETY PROFILE: Poison by skin contact. Moderately toxic by ingestion and inhalation. Experimental reproductive effects. A skin irritant. Flammable when exposed to heat or flame; can react with oxidizing materials. To fight fire, use foam, CO_2, dry chemical. When heated to decomposition it emits acrid smoke and irritating fumes. See also GLYCOL ETHERS.

MIF760   CAS: 9004-67-5   HR: 3
METHYL CELLULOSE
PROP: White, fibrous powders. Sol in water, some org solvs.
SYNS: ADULSIN □ BAGOLAX □ BUFAPTO METHALOSE □ BULKALOID □ CELACOL M □ CELACOL M20 □ CELACOL M450 □ CELACOL MM □ CELACOL MM 10P □ CELACOL M 20P □ CELLAPRET □ CELLOGRAN □ CELLOTHYL □ CELLULOSE METHYL □ CELLULOSE METHYLATE □ CELLUMETH □ CETHYLOSE □ CETHYTIN □ CULMINAL K 42 □ EDISOL M □ HYDROLOSE □ MAPOLOSE M25 □ MAPOLOSE 60SH50 □ MCO 8000 □ MC 4000 cP □ MC 20000S □ MELLOSE □ METHOCEL 10 □ METHOCEL 15 □ METHOCEL 181 □ METHOCEL 400 □ METHOCEL 4000 □ METHOCEL A □ METHOCEL CHG □ METHO-CEL 400CPS □ METHOCEL 4000CPS □ METHOCEL MC □ METHOCEL MC 25 □ METHOCEL MC4000 □ METHO-CEL MC 8000 □ METHOCEL SM 100 □ METHULOSE □ METHYL CELLULOSE-A □ METHYL CELLULOSE ETHER □ METOLOSE MC 8000 □ METOLOSE 60SH □ METOLOSE 60SH400 □ METOLOSE SM 15 □ METOL-OSE SM 100 □ METOLOSE SM 4000 □ MMTs-BTR □ MTs □ NAPOLONE □ NICEL □ RHOMELLOSE □ SYNCEL-OSE □ TYLOSE 444 □ TYLOSE A4S □ TYLOSE MF □ TYLOSE MH □ TYLOSE MH20 □ TYLOSE MH50 □ TYLOSE MH300 □ TYLOSE MH1000 □ TYLOSE MH2000 □ TYLOSE MH300P □ TYLOSE MH4000 □ TYLOSE SAP □ TYLOSE SL □ TYLOSE SL 100 □ TYLOSE SL 400 □ TYLOSE SL 600 □ TYLOSE TWA □ USP METHYLCEL-

LULOSE □ VISCOL □ VISCONTRAN L52 □ VISCOSOL □ WALSRODER MC 20000S

CONSENSUS REPORTS: Reported in EPA TSCA Inventory.

SAFETY PROFILE: A poison by intraperitoneal route. When heated to decomposition it emits acrid smoke and irritating fumes.

MIF762 CAS: 4224-87-7 HR: 3
4-METHYLCHALCONE

mf: $C_{16}H_{14}O$ mw: 222.30

SYNS: CHALCONE, 4-METHYL-(6CI,7CI,8CI) □ (4-METHYLBENZYLIDENE)ACETOPHENONE □ p-METHYLCHALCONE □ 3-(4-METHYLPHENYL)-1-PHENYL-2-PROPEN-1-ONE □ PHENYL p-METHYLSTYRYL KETONE □ 2-PROPEN-1-ONE, 3-(4-METHYLPHENYL)-1-PHENYL-

DOT CLASSIFICATION: 3; Label: Flammable Liquid

SAFETY PROFILE: A poison by intraperitoneal route. Moderately toxic by ingestion. A flammable liquid. When heated to decomposition it emits acrid smoke and irritating vapors.

MIF765 CAS: 74-87-3 HR: 3
METHYL CHLORIDE

DOT: UN 1063

mf: CH_3Cl mw: 50.49

PROP: Colorless gas; ethereal odor and sweet taste. D: 0.918 @ 20°/4°, mp: −97°, bp: −23.7°, flash p: <32°F, lel: 8.1%, uel: 17%, autoign temp: 1170°F, vap d: 1.78. Sltly sol in water; misc with chloroform, ether, glacial acetic acid; sol in alc. IDLH 2000 ppm.

SYNS: ARTIC □ CHLOOR-METHAAN (DUTCH) □ CHLOR-METHAN (GERMAN) □ CHLOROMETHANE □ CHLORURE de METHYLE (FRENCH) □ CLOROMET-ANO (ITALIAN) □ CLORURO di METILE (ITALIAN) □ METHYLCHLORID (GERMAN) □ METYLU CHLOREK (POLISH) □ MONOCHLOROMETHANE □ R 40 □ RCRA WASTE NUMBER U045

CONSENSUS REPORTS: IARC Cancer Review: Group 3 IMEMDT 7,246,87; Human Inadequate Evidence IMEMDT 41,161,86; Animal Inadequate Evidence IMEMDT 41,161,86. Reported in EPA TSCA Inventory. EPA Genetic Toxicology Program.

OSHA PEL: TWA 50 ppm; STEL 100 ppm

ACGIH TLV: TWA 50 ppm; STEL 100 ppm; Not Classifiable as a Human Carcinogen

DFG MAK: 50 ppm (100 mg/m³); Suspected Carcinogen

NIOSH REL: (Monohalomethanes) TWA Reduce to lowest level

DOT CLASSIFICATION: 2.1; Label: Flammable Gas

SAFETY PROFILE: Suspected carcinogen. Very mildly toxic by inhalation. An experimental teratogen. Other experimental reproductive effects. Human mutation data reported. Human systemic effects by inhalation: convulsions, nausea or vomiting, and unspecified effects on the eye.

Methyl chloride has slight irritant properties and may be inhaled without noticeable discomfort. It has some narcotic action, but this effect is weaker than that of chloroform. Acute poisoning, characterized by the narcotic effect, is rare in industry. In exposures to high concentrations, dizziness, drowsiness, incoordination, confusion, nausea and vomiting, abdominal pains, hiccoughs, diplopia, and dimness of vision are followed by delirium, convulsions, and coma. Death may be immediate; however, if the exposure is not fatal, recovery is usually slow. Degenerative changes in the central nervous system are not uncommon. The liver, kidneys, and bone marrow may be affected, with resulting acute nephritis and anemia. Death resulting from degenerative changes in the heart, liver, and especially the kidneys may occur several days after exposure. Repeated exposure to low concentrations causes damage to the central nervous system and, less frequently, to the liver, kidneys, bone marrow, and cardiovascular system. Hemorrhages into the lungs, intestinal tract, and dura have been reported. Sprayed on the skin, chloromethane produces anesthesia through freezing of the tissues as it evaporates.

Flammable gas. Very dangerous fire hazard when exposed to heat, flame, or powerful oxidizers. Moderate explosion hazard when exposed to flame and sparks. Explodes on contact with interhalogens (e.g., bromine trifluoride, bromine pentafluoride), magnesium and alloys, potassium and alloys, sodium and alloys, zinc. Potentially explosive reaction with aluminum when heated to 152° in a sealed container. Mixtures with aluminum chloride + ethylene react exothermically and then explode when pressurized to above 30 bar. May ignite on contact with aluminum chloride or powdered aluminum. To fight fire, stop flow of gas and use CO_2, dry chemical, or water spray. When heated to decomposition it emits highly toxic fumes of Cl⁻. See also CHLORINATED HYDROCARBONS, ALIPHATIC.

MIF775   CAS: 96-34-4   HR: 3
METHYL CHLOROACETATE
DOT: UN 2295
mf: $C_3H_5ClO_2$ mw: 108.53
PROP: Colorless liquid. D: 1.238, mp: −33°, bp: 130–132°. Insol in water; misc with alc, ether.
SYNS: CHLOROACETIC ACID METHYL ESTER □ METHYL CHLOROACETATE (DOT) □ METHYLESTER KYSELINY CHLOROCTOVE □ METHYL MONOCHL-ORACETATE □ METHYL MONOCHLOROACETATE □ MONOCHLOROACETIC ACID METHYL ESTER
CONSENSUS REPORTS: Reported in EPA TSCA Inventory.
DFG MAK: 1 ppm
DOT CLASSIFICATION: 6.1; Label: Poison
SAFETY PROFILE: Poison by ingestion. Moderately toxic by inhalation and subcutaneous routes. Flammable when exposed to heat or flame; can react vigorously with oxidizing materials. When heated to decomposition it emits toxic fumes of Cl⁻.

MIG000   CAS: 79-22-1   HR: 3
METHYL CHLOROCARBONATE
DOT: UN 1238
mf: $C_2H_3ClO_2$ mw: 94.50

PROP: Colorless liquid. Bp: 71.4°, d: 1.223 @ 20°/4°, vap d: 3.26, flash p: 54°F, autoign temp: 940°F. Sltly sol in water with gradual decomp; misc with alc, benzene, chloroform, and ether.
SYNS: CHLORAMEISENSAEURE METHYLESTER (GERMAN) □ CHLOROCARBONATE de METHYLE (FRENCH) □ CHLOROCARBONIC ACID METHYL ESTER □ CHLOROFORMIC ACID METHYL ESTER □ MCF □ METHOXYCARBONYL CHLORIDE □ METHYLCHLOORFORMIAT (DUTCH) □ METHYL CHLOROFORMATE (DOT) □ METILCLOROFORMIATO (ITALIAN) □ RCRA WASTE NUMBER U156
CONSENSUS REPORTS: EPA Extremely Hazardous Substances List. Reported in EPA TSCA Inventory.
DOT CLASSIFICATION: 6.1; Label: Poison, Flammable Liquid, Corrosive
SAFETY PROFILE: Poison by ingestion, inhalation, and intraperitoneal routes. Moderately toxic by skin contact. Human systemic effects by inhalation: conjunctiva irritation and respiratory effects. Corrosive to skin, eyes, and mucous membranes. Very dangerous fire hazard when exposed to heat sources, sparks, flame, or oxidizers. Reacts with water or steam to produce toxic and corrosive fumes. When heated to decomposition it emits toxic fumes of Cl⁻, methyl chloroformate, and phosgene.

MIH275   CAS: 71-55-6   HR: 3
METHYL CHLOROFORM
DOT: UN 2831
mf: $C_2H_3Cl_3$ mw: 133.40
PROP: Colorless, nonflammable liquid. Bp: 74.1°, fp: −32.5°, flash p: none, d: 1.3376 @ 20°/4°, vap press: 100 mm @ 20.0°. Insol in water; sol in acetone, benzene, carbon tetrachloride, methanol, ether. IDLH 700 ppm.
SYNS: AEROTHENE TT □ CHLOROETENE □ CHLOROETHENE □ CHLOROTHANE NU □ CHLOROTHENE □ CHLOROTHENE (inhibited) □ CHLOROTHENE NU □ CHLOROTHENE VG □ CHLORTEN □ INHIBISOL □ METHYLTRICHLORO-METHANE □ NCI-C04626 □ RCRA WASTE NUMBER U226 □ SOLVENT 111 □ STROBANE □ α-T □ 1,1,1-TCE □ 1,1,1-TRICHLOORETHAAN (DUTCH) □ 1,1,1-TRICHLORAETHAN (GERMAN) □ TRICHLORO-1,1,1-

ETHANE (FRENCH) □ 1,1,1-TRICHLOROETHANE □ α-TRICHLOROETHANE □ 1,1,1-TRICLOROETANO (ITALIAN) □ TRI-ETHANE

CONSENSUS REPORTS: IARC Cancer Review: Group 3 IMEMDT 7,56,87; Animal Inadequate Evidence IMEMDT 20,515,79. NCI Carcinogenesis Bioassay (gavage); Inadequate Studies: mouse, rat NCITR* NCI-CG-TR-3,77. Community Right-To-Know List. Reported in EPA TSCA Inventory. EPA Genetic Toxicology Program.

OSHA PEL: TWA 350 ppm; STEL 450 ppm
ACGIH TLV: TWA 350 ppm; STEL 450 ppm; BEI: 10 mg/L trichloroacetic acid in urine at end of workweek; Not Classifiable as a Human Carcinogen
DFG MAK: 200 ppm (1100 mg/m³); BAT: 55 μg/dL in blood after several shifts
NIOSH REL: (1,1,1-Trichloroethane) CL 350 ppm/15M
DOT CLASSIFICATION: 6.1; Label: KEEP AWAY FROM FOOD
SAFETY PROFILE: Poison by intravenous route. Moderately toxic by ingestion, inhalation, skin contact, subcutaneous, and intraperitoneal routes. An experimental teratogen. Human systemic effects by ingestion and inhalation: conjunctiva irritation, hallucinations or distorted perceptions, motor activity changes, irritability, aggression, hypermotility, diarrhea, nausea or vomiting and other gastrointestinal changes. Experimental reproductive effects. Questionable carcinogen. Mutation data reported. A human skin irritant. An experimental skin and severe eye irritant. Narcotic in high concentrations. Causes a proarrhythmic activity that sensitizes the heart to epinephrine-induced arrhythmias. This sometimes will cause cardiac arrest, particularly when this material is massively inhaled as in drug abuse for euphoria.

Under the proper conditions it can undergo hazardous reactions with aluminum oxide + heavy metals, dinitrogen tetraoxide, inhibitors, metals (e.g., magnesium, aluminum, potassium, potassium-sodium alloy), sodium hydroxide, N₂O₄, oxygen. When heated to decomposition it emits toxic fumes of Cl⁻. Used as a cleaning solvent, as a chemical intermediate to produce vinylidene chloride, and as a propellant in aerosol cans. See also CHLORINATED HYDROCARBONS, ALIPHATIC.

MIJ500   CAS: 127-33-3   HR: 3
METHYLCHLORTETRACYCLINE
mf: $C_{21}H_{21}ClN_2O_8$ mw: 464.89
PROP: Sesquihydrate. Mp: 174–178° (decomp).
SYNS: 7-CHLORO-6-DEMETHYLTETRACYCLINE □ CHLORTETRACYCLINE, 6-DEMETHYL- □ DECLOMYCIN □ DEMECLOCYCLINE □ DEMETHYL-CHLOROTETRACYCLINE □ 6-DEMETHYLCHLORO-TETRACYCLINE □ 6-DEMETHYL-7-CHLOROTETRA-CYCLINE □ DEMETHYLCHLORTETRACYCLIN □ DEMETHYLCHLORTETRACYCLINE □ 6-DEMETHYL-CHLORTETRACYCLINE □ 6-DEMETHYL-7-CHLOR-TETRACYCLINE □ DEMETHYLCHLORTETRACYCLINE BASE □ 6-DEMETIL-7-CLOROTETRACICLINA □ DMCT □ LEDERMYCIN □ MEXOCINE □ RP 10192 □ TETRACYCLINE, 7-CHLORO-6-DEMETHYL-
SAFETY PROFILE: Poison by intravenous and intraperitoneal routes. Human systemic effects by ingestion: diabetes insipidus, urine volume increase, other changes in urine composition, dermatitis, changes in the nails, allergic rhinitis, serum sickness, effects on cyclic nucleotides. Human reproductive effects by an unspecified route: postnatal measures or effects on newborn. An experimental teratogen. Experimental reproductive effects. Human mutation data reported. When heated to decomposition it emits very toxic fumes of Cl⁻ and NOₓ.

MIJ750   CAS: 56-49-5   HR: 3
3-METHYLCHOLANTHRENE
mf: $C_{21}H_{16}$ mw: 268.37
PROP: Pale-yellow needles from benzene. Mp: 179–180°, bp: 280° @ 80 mm, d: 1.28 @ 20°. Sol in benzene, xylene, toluene; sltly sol in amyl alc; insol in water.
SYNS: 1,2-DIHYDRO-3-METHYL-BENZ(j)ACEAN-THRYLENE □ 3-MCA □ METHYLCHOLANTHRENE □

20-METHYLCHOLANTHRENE ☐ RCRA WASTE NUMBER U157

CONSENSUS REPORTS: Reported in EPA TSCA Inventory. EPA Genetic Toxicology Program.

SAFETY PROFILE: Suspected carcinogen with experimental carcinogenic, neoplastigenic, and tumorigenic data. Poison by intravenous and intraperitoneal routes. Experimental teratogenic and reproductive effects. Human mutation data reported. When heated to decomposition it emits acrid smoke and irritating fumes.

MIN500   CAS: 3697-24-3   HR: 3
5-METHYLCHRYSENE
mf: $C_{19}H_{14}$ mw: 242.33
PROP: Needles from C_6H_6/EtOH. Mp: 118–119°.
CONSENSUS REPORTS: NTP 10th Report on Carcinogens. IARC Cancer Review: Group 3 IMEMDT 7,56,87; Animal Sufficient Evidence IMEMDT 32,379,83. EPA Genetic Toxicology Program.
SAFETY PROFILE: Confirmed carcinogen with experimental carcinogenic, neoplastigenic, and tumorigenic data. Mutation data reported. When heated to decomposition it emits acrid smoke and irritating fumes. See also other methylchrysene entries.

MIO000   CAS: 101-39-3   HR: 2
α-METHYLCINNAMALDEHYDE
mf: $C_{10}H_{10}O$ mw: 146.20
PROP: Yellow liquid; cinnamon odor. D: 1.035–1.039, refr index: 1.602–1.607, flash p: 174°F. Sol in fixed oils, propylene glycol; insol in glycerin.
SYNS: FEMA No. 2697 ☐ METHYL CINNAMIC ALDEHYDE ☐ α-METHYLCINNAMIC ALDEHYDE ☐ α-METHYLCINNIMAL ☐ 2-METHYL-3-PHENYL-2-PROPENAL
CONSENSUS REPORTS: Reported in EPA TSCA Inventory.
SAFETY PROFILE: Moderately toxic by ingestion. A skin irritant. Combustible liquid. When heated to decomposition it

emits acrid smoke and irritating fumes. See also ALDEHYDES.

MIO500   CAS: 103-26-4   HR: 2
METHYL CINNAMATE
mf: $C_{10}H_{10}O_2$ mw: 162.20
PROP: White to sltly yellow crystals; fruity odor. D: 1.042 @ 36°/0°, mp: 33.4°, bp: 263°, flash p: 212°F. Very sol in alc, ether; sol in fixed oils, glycerin, propylene glycol; insol in water.
SYNS: FEMA No. 2698 ☐ METHYL CINNAMYLATE ☐ METHYL-3-PHENYLPROPENOATE ☐ 3-PHENYL-2-PROPENOIC ACID METHYL ESTER (9CI)
CONSENSUS REPORTS: Reported in EPA TSCA Inventory.
SAFETY PROFILE: Moderately toxic by ingestion. Combustible liquid. When heated to decomposition it emits acrid smoke and irritating fumes.

MIP750   CAS: 92-48-8   HR: 3
6-METHYLCOUMARIN
mf: $C_{10}H_8O_2$ mw: 160.18
PROP: White needles from benzene; coconut odor. Needles from alc. Mp: 73–76°, flash p: 153°F, bp: 303° @ 725 mm. Very sol in EtOH, Et_2O, and C_6H_6; sltly sol in pet ether.
SYNS: FEMA No. 2690 ☐ 6-MC ☐ 6-METHYL-2H-1-BENZOPYRAN-2-ONE ☐ 6-METHYLBENZOPYRONE ☐ 6-METHYL-1,2-BENZOPYRONE ☐ 6-METHYL-COUMARINIC ANHYDRIDE ☐ NCI-C55812 ☐ TONCARINE
CONSENSUS REPORTS: EPA Genetic Toxicology Program. Reported in EPA TSCA Inventory.
SAFETY PROFILE: Poison by subcutaneous route. Moderately toxic by ingestion. A skin irritant. Mutation data reported. Combustible liquid. When heated to decomposition it emits acrid smoke and irritating fumes.

MIQ075   CAS: 137-05-3   HR: 2
METHYL 2-CYANOACRYLATE
mf: $C_9H_{13}NO_2$ mw: 111.11
PROP: Thick, clear colorless liquid; sharp odor. Bp: 47–48° @ 2 mm.

SYNS: ADHERE □ COAPT □ α-CYANOACRYLIC ACID METHYL ESTER □ 2-CYANOACRYLIC ACID, METHYL ESTER □ CYANOLIT □ EASTMAN 910 □ EASTMAN 910 ADHESIVE □ EASTMAN 910 MONOMER □ MECRILAT □ MECRYLATE □ METHYL CYANOACRYLATE □ METHYL α-CYANOACRYLATE □ SUPER GLUE

CONSENSUS REPORTS: Reported in EPA TSCA Inventory.

OSHA PEL: TWA 2 ppm; STEL 4 ppm

ACGIH TLV: TWA 0.2 ppm

DFG MAK: 2 ppm (9.2 mg/m^3)

SAFETY PROFILE: Moderately toxic by ingestion and inhalation routes. Experimental reproductive effects. A human eye irritant. Can bond the eyelids or skin surfaces instantly. Mutation data reported. When heated to decomposition it emits toxic fumes of NO_x and CN^-.

MIQ740   CAS: 108-87-2   HR: 3
METHYLCYCLOHEXANE

DOT: UN 2296

mf: C_7H_{14} mw: 98.21

PROP: Colorless liquid. Mp: −126.4°, lel: 1.2%, uel: 6.7%, bp: 100.3°, flash p: 25°F (CC), d: 0.7864 @ 0°/4°, 0.769 @ 20°/4°, vap press: 40 mm @ 22.0°, vap d: 3.39, autoign temp: 482°F. IDLH 1200 ppm [LEL].

SYNS: CYCLOHEXYLMETHANE □ HEXAHYDRO-TOLUENE □ METYLOCYKLOHEKSAN (POLISH) □ SEXTONE B □ TOLUENE HEXAHYDRIDE

CONSENSUS REPORTS: Reported in EPA TSCA Inventory.

OSHA PEL: TWA 400 ppm

ACGIH TLV: TWA 400 ppm

DFG MAK: 500 ppm (2000 mg/m^3)

DOT CLASSIFICATION: 3; Label: Flammable Liquid

SAFETY PROFILE: Moderately toxic by ingestion. Mildly toxic by inhalation and skin contact. This material does not cause irritation to the eyes and nose, and, even at the level of 500 ppm, exhibits only a very faint odor. Therefore, it cannot be said to have any warning properties. It is believed to be about three times as toxic as hexane, and has caused death by tetanic spasm in animals. In sublethal concentrations, it causes narcosis and anesthesia. Dangerous fire hazard and moderate explosion hazard when exposed to heat, flame, or oxidizers. To fight fire, use foam, CO_2, dry chemical. When heated to decomposition it emits acrid smoke and fumes.

MIQ745   CAS: 25639-42-3   HR: 3
METHYLCYCLOHEXANOL

DOT: UN 2617

mf: $C_7H_{14}O$ mw: 114.21

PROP: Colorless, viscous liquid; aromatic, menthol-like odor. Bp: 155–180°, flash p: 154°F (CC), autoign temp: 565°F, d: 0.924 @ 15.5°/15.5°, vap d: 3.93. IDLH 500 ppm.

SYNS: HEXAHYDROCRESOL □ HEXAHYDROMETH-YLPHENOL □ METHYLCYCLOHEXANOL (ACGIH, DOT, OSHA) □ METHYL CYCLOHEXANOLS, Fp not >60.5 degrees C (DOT) □ METYLOCYKLOHEKSANOL (POLISH)

OSHA PEL: TWA 50 ppm

ACGIH TLV: TWA 50 ppm

DFG MAK: 50 ppm (235 mg/m^3)

DOT CLASSIFICATION: 3; Label: Flammable Liquid

SAFETY PROFILE: Moderately toxic by ingestion and subcutaneous routes. Mildly toxic by skin contact. Human system effects by inhalation: antipsychotic, unspecified liver and kidney effects. Combustible when exposed to heat, flame, or oxidizers. On heating it emits acrid fumes; can react with oxidizing materials. To fight fire, use alcohol foam, CO_2, dry chemical.

MIR250   CAS: 1331-22-2   HR: 3
METHYLCYCLOHEXANONE

DOT: UN 2297

mf: $C_7H_{12}O$ mw: 112.19

PROP: Water-white to pale-yellow liquid; acetone-like odor. Mp: −14°C, bp: 160–170°, flash p: 118°F (CC), d: 0.925 @ 15°/5°, vap d: 3.86. Insol in water; sol in ether and alc.

SYN: METYLOCYKLOHEKSANON (POLISH)

CONSENSUS REPORTS: Reported in EPA TSCA Inventory.

DOT CLASSIFICATION: 3; Label: Flammable Liquid

SAFETY PROFILE: Moderately toxic by ingestion. Mildly toxic by skin contact. A toxic compound that can damage the kidneys and the liver. It is similar to cyclohexanol in its toxic action, although it is somewhat less active. Harmful exposure in industry is rare. Experimental animals can withstand prolonged exposures of 0.02–0.05% by volume in air. Flammable liquid when exposed to heat, sparks, or flame. Can react violently with HNO_3 and other oxidizers. To fight fire, use foam, CO_2, dry chemical. When heated to decomposition it emits acrid smoke and irritating fumes. See also KETONES.

MIR500   CAS: 583-60-8   HR: 3
2-METHYLCYCLOHEXANONE
mf: $C_7H_{12}O$ mw: 112.19

PROP: Liquid. D: 0.925 @ 20°/4°, mp: −14°, bp: 165.1°. Insol in water; sol in alc and ether. IDLH 600 ppm.

SYNS: 2-METHYL-CYCLOHEXANON (GERMAN, DUTCH) □ o-METHYLCYCLOHEXANONE □ 1-METH-YLCYCLOHEXAN-2-ONE □ 2-METILCICLOESANONE (ITALIAN)

CONSENSUS REPORTS: Reported in EPA TSCA Inventory.

OSHA PEL: TWA 50 ppm; STEL 75 ppm (skin)

ACGIH TLV: TWA 50 ppm (skin)

DFG MAK: 50 ppm (230 mg/m³)

SAFETY PROFILE: Poison by intraperitoneal route. Moderately toxic by ingestion and skin contact. When heated to decomposition it emits acrid smoke and irritating fumes. See also KETONES and other methylcyclohexanone entries.

MIT625   CAS: 541-91-3   HR: 3
3-METHYL-1-CYCLOPENTADECANONE
mf: $C_{16}H_{30}O$ mw: 238.46

SYNS: CYCLOPENTADECANONE, 3-METHYL- □ 3-METHYLCYCLOPENTADECANONE □ MOSCHUS KETONE □ MUSCONE □ MUSKONE

CONSENSUS REPORTS: Reported in EPA TSCA Inventory.

DOT CLASSIFICATION: 3; Label: Flammable Liquid

SAFETY PROFILE: Low oral and skin contact toxicity. A skin irritant. A flammable liquid. When heated to decomposition it emits acrid smoke and irritating fumes.

MIU500   CAS: 96-37-7   HR: 3
METHYLCYCLOPENTANE
DOT: UN 2298

mf: C_6H_{12} mw: 84.18

PROP: Colorless liquid or solid. Mp: −142.5°, bp: 71.8°, flash p: <20°F, d: 0.750 @ 20°/4°, vap press: 100 mm @ 17.9°, vap d: 2.9. Insol in water; sol in ether.

SYN: METHYL CYCLOPENTANE (DOT)

CONSENSUS REPORTS: Reported in EPA TSCA Inventory.

DOT CLASSIFICATION: 3; Label: Flammable Liquid

SAFETY PROFILE: Mildly toxic by inhalation. Probably irritating and narcotic in high concentration. Very dangerous fire hazard when exposed to heat, flame, or oxidizers. Can react vigorously with oxidizing materials. To fight fire, use foam, CO_2, dry chemical. When heated to decomposition it emits acrid smoke and irritating fumes.

MIV300   CAS: 63884-38-8   HR: 3
METHYLCYCLOPROPANECARBONYLH
YDRAZINE
mf: $C_5H_{10}N_2O$ mw: 114.17

SYNS: CYCLOPROPANE, 1-(N-AMINO)CARBAMOYL-2-METHYL- □ HYDRAZINE, N-((2-METHYLCYCLOPROP-YL)CARBONYL)- □ KETONE, 2-METHYLCYCLOPROP-YL HYDRAZINO □ USAF A-14980

DOT CLASSIFICATION: 3; Label: Flammable Liquid

SAFETY PROFILE: A poison by intraperitoneal route. A flammable liquid. When heated to decomposition it emits toxic vapors of NO_x.

MJR775   CAS: 234-17-3   HR: 3
7,8-METHYLENEDIOXYISOQUINOLINE
mf: $C_{10}H_7NO_2$ mw: 173.17

SYN: 1,3-DIOXOLO(4,5-H)ISOQUINOLINE

SAFETY PROFILE: A poison by ingestion. When heated to decomposition it emits toxic vapors of NO_x.

MIW050   CAS: 7011-83-8   HR: 1
4-METHYLDECANOLIDE
mf: $C_{11}H_{20}O_2$ mw: 184.31
SYNS: DECANOIC ACID, 4-HYDROXY-4-METHYL-, γ-LACTONE □ α-METHYL DECALACTONE
CONSENSUS REPORTS: Reported in EPA TSCA Inventory.
SAFETY PROFILE: A skin irritant. When heated to decomposition it emits acrid smoke and irritating fumes.

MIW075   CAS: 7289-52-3   HR: 1
METHYL n-DECYL ETHER
mf: $C_{11}H_{24}O$ mw: 172.35
SYNS: DECANE, 1-METHOXY- □ DECYL METHYL ETHER □ ETHER, DECYL METHYL □ 1-METHOXYDECANE □ METHYL DECYL ETHER
CONSENSUS REPORTS: Reported in EPA TSCA Inventory.
SAFETY PROFILE: Low toxicity by ingestion and skin contact. A skin irritant. When heated to decomposition it emits acrid smoke and irritating fumes.

MIW250   CAS: 2587-90-8   HR: 3
METHYL DEMETON METHYL
mf: $C_5H_{13}O_3PS_2$ mw: 216.27
PROP: Pale-yellow oil or liquid. Bp: 89° @ 0.15 mm, d: 1.207 @ 20°/4°. Sol in water at room temp, sol in org solvs.
SYNS: CEBETOX □ CYMETOX □ DEMEPHION □ ISONITOX □ 2-(METHYLTHIO)-ETHANETHIOL-O,O-DIMETHYL PHOSPHOROTHIOATE □ 2-(METHYLTHIO)-ETHANETHIOL-S-ESTER with O,O-DIMETHYL PHOSPHOROTHIOATE □ TINOX
CONSENSUS REPORTS: EPA Extremely Hazardous Substances List.
SAFETY PROFILE: Poison by ingestion and skin contact. Mutation data reported. *Caution:* It is a cholinesterase inhibitor. When heated to decomposition it emits very toxic fumes of PO_x and SO_x. See also PARATHION and various demeton entries.

MIW500   CAS: 477-30-5   HR: 3
N-METHYL-N-DESACETYLCOLCHICINE
mf: $C_{21}H_{25}NO_5$ mw: 371.47
PROP: A solid. Mp: 186°.
SYNS: ALKALOID H 3, from COLCHICUM ANTUMNALE □ BENZO(a)HEPTALEN-9(5H)-ONE, 6,7-DIHYDRO-1,2,3,10-TETRAMETHOXY-7-(METHYL-AMINO)-, (S)- □ C-12669 □ CIBA 12669A □ COLCEMID □ COLCEMIDE □ COLCHAMIN □ COLCHAMINE □ COLCHICINE, 7-DEACETAMIDO-7-(METHYLAMINO)- □ COLCHICINE, DEACETYL-N-METHYL- □ COLEMID □ DEACETYLMETHYLCOLCHICINE □ DEACETYL-N-METHYLCOLCHICINE □ N-DEACETYL-N-METHYL-COLCHICINE □ DEMECOLCIN □ DEMECOLCINE □ DESACETYLMETHYLCOLCHICINE □ N-DESACETYL-METHYLCOLCHICINE □ N-DESACETYL-N-METHYL-COLCHICINE □ DESMECOLCHINE □ DESMECOLCINE □ 6,7-DIHYDRO-1,2,3,10-TETRAMETHOXY-7-(METHYL-AMINO)-BENZO(α)HEPTALEN-9 (5H)-ONE □ KOL-CHAMIN □ KOLCHICIN □ KOLKAMIN □ METHYL-COLCHICINE □ N-METHYL-N-DEACETYLCOLCHICINE □ N-METHYLDEMECOLCINE □ N-METHYL-N-DESACETYLCOLCHICINE □ NSC-3096 □ OMAIN □ OMAINE □ REICHSTEIN'S F □ SANTAVY'S SUBSTANCE F □ SUBSTANCE F
CONSENSUS REPORTS: EPA Genetic Toxicology Program.
SAFETY PROFILE: Poison by ingestion, intraperitoneal, parenteral, intravenous, and intramuscular routes. Human systemic effects by ingestion: (skin and appendages) hair effects. Human mutation data reported. An experimental teratogen. Experimental reproductive effects. When heated to decomposition it emits toxic fumes of NO_x.

MJE500   CAS: 892-17-1   HR: 3
11-METHYL-15,16-DIHYDRO-17-OXO-CYCLOPENTA(a)PHENANTHRENE
mf: $C_{18}H_{14}O$ mw: 246.32
SYNS: 15,16-DIHYDRO-11-METHYLCYCLOPENTA(a)-PHENANTHREN-17-ONE □ 15,16-DIHYDRO-11-METHYL-17H-CYCLOPENTA(a)PHENANTHREN-17-ONE □ 11-METHYL-15,16-DIHYDRO-17H-CYCLOPENTA(a)PHENANTHREN-17-ONE
CONSENSUS REPORTS: EPA Genetic Toxicology Program.
SAFETY PROFILE: Suspected carcinogen with experimental carcinogenic, neoplastigenic, and tumorigenic data. Mutation data reported. When heated to

decomposition it emits acrid smoke and irritating fumes.

MJE900   CAS: 16881-77-9   HR: 3
METHYLDIMETHOXYSILANE

mf: $C_3H_{10}O_2Si$ mw: 106.22

SYNS: DIMETHOXYMETHYLSILANE □ SILANE, DIMETHOXYMETHYL-

SAFETY PROFILE: A poison by ingestion, subcutaneous, and intravenous routes. A moderate eye irritant. When heated to decomposition it emits acrid smoke and irritating vapors.

MJL500   CAS: 110-26-9   HR: 3
N,N'-METHYLENEBIS(ACRYLAMIDE)

mf: $C_7H_{10}N_2O_2$ mw: 154.19

PROP: Colorless, crystalline, stable, white powder. Mp: 185° (with decomp), d: 1.235 @ 30°, vap d: 5.31.

SYNS: METHYLENEBISACRYLAMIDE □ N,N'-METHYLENEDIACRYLAMIDE □ N,N'-METHYLIDEN-EBISACRYLAMIDE

CONSENSUS REPORTS: Reported in EPA TSCA Inventory.

SAFETY PROFILE: Poison by ingestion. Experimental reproductive effects. Mutation data reported. When heated to decomposition it emits toxic fumes of NO_x. See also AMIDES.

MJM200   CAS: 101-14-4   HR: 3
4,4'-METHYLENE BIS(2-CHLORO-
ANILINE)

mf: $C_{13}H_{12}Cl_2N_2$ mw: 267.17

PROP: Tan solid.

SYNS: BIS AMINE □ CURALIN M □ CURENE 442 □ CYANASET □ DI-(4-AMINO-3-CHLOROPHENYL)METH-ANE □ DI-(4-AMINO-3-CLOROFENIL)METANO (ITALIAN) □ 4,4'-DIAMINO-3,3'-DICHLORODIPHENYL-METHANE □ 3,3'-DICHLOR-4,4'-DIAMINODIPHENYL-METHAN (GERMAN) □ 3,3'-DICHLORO-4,4'-DIAMINODIPHENYLMETHANE □ 3,3'-DICLORO-4,4'-DIAMINODIFENILMETANO (ITALIAN) □ MBOCA □ METHYLENE-4,4'-BIS(o-CHLOROANILINE) □ p,p'-METHYLENEBIS(α-CHLOROANILINE) □ p,p'-METHYL-ENEBIS(o-CHLOROANILINE) □ 4,4'-METHYLENE(BIS)-CHLOROANILINE □ 4,4'-METHYLENEBIS(o-CHLORO-ANILINE) □ 4,4'-METHYLENEBIS-2-CHLOROBENZ-ENAMINE □ METHYLENE-BIS-ORTHOCHLOROANIL-INE □ 4,4-METILENE-BIS-o-CLOROANILINA (ITALIAN) □ MOCA □ RCRA WASTE NUMBER U158

CONSENSUS REPORTS: NTP 10th Report on Carcinogens. IARC Cancer Review: Group 2A IMEMDT 7,246,87; Animal Sufficient Evidence IMEMDT 4,65,74. EPA Genetic Toxicology Program. Community Right-To-Know List. Reported in EPA TSCA Inventory.

OSHA PEL: TWA 0.02 ppm (skin)

ACGIH TLV: TWA 0.01 ppm; Suspected Human Carcinogen

DFG MAK: Animal Carcinogen, Suspected Human Carcinogen

NIOSH REL: (MOCA): TWA 0.003 mg/m^3 (Skin)

SAFETY PROFILE: Confirmed carcinogen with experimental carcinogenic and tumorigenic data. Poison by ingestion and intraperitoneal routes. Mutation data reported. Flammable liquid. Reactive with active metals such as sodium, potassium, magnesium, or zinc. When heated to decomposition it emits very toxic fumes of Cl^- and NO_x. See also AMINES.

MJM500   CAS: 97-23-4   HR: 3
2,2'-METHYLENEBIS(4-CHLOROPHEN-
OL)

mf: $C_{13}H_{10}Cl_2O_2$ mw: 269.13

PROP: Crystals, nearly insol in water. Mp: 178°, vap press: 10^{-4} mm @ 100°. Sltly sol in toluene.

SYNS: ANTHIPHEN □ ANTIPHEN □ BIS(5-CHLOR-2-HYDROXYPHENYL)-METHAN □ BIS(5-CHLORO-2-HYDROXYPHENYL)METHANE □ BIS-2-HYDROXY-5-CHLORFENYLMETHAN □ BIS(2-HYDROXY-5-CHLOROPHENYL)METHANE □ DDDM □ DDM □ DICESTAL □ DICHLOORFEEN □ 5,5'-DICHLORO-2,2'-DIHYDROXYDIPHENYLMETHANE □ DICHLOROFEN □ DI-(5-CHLORO-2-HYDROXYPHENYL)METHANE □ 4,4'-DICHLORO-2,2'-METHYLENEDIPHENOL □ DICHLOROPHEN □ DICHLOROPHEN B □ DICHLOR-OPHENE □ DICHLORPHEN □ DIDROXAN □ DIDROX-ANE □ 2,2'-DIHYDROXY-5,5'-DICHLORODIPHENYL-METHANE □ DIPHENTHANE 70 □ FUNGICIDE FX □ G 4 □ GH □ HYOSAN □ KORIUM □ O,O-METHYLEEN-BIS(4-CHLOORFENOL) □ 2,2'-METHYLENEBIS(4-CHLOROPHENOL) □ O,O-METILEN-BIS(4-CLORO-FENOLO) □ PANACIDE □ PARABIS □ PLATH-LYSE □ PREVENTAL □ PREVENTOL □ PREVENTOL GD □

M

PREVENTOL GDC □ SUPER MOSSTOX □ TAENIATOL □ TENIATHANE □ TENIATOL □ WESPURIL

CONSENSUS REPORTS: Chlorophenol compounds are on the Community Right-To-Know List. Reported in EPA TSCA Inventory.

SAFETY PROFILE: Poison by intravenous route. Moderately toxic by ingestion. A skin and severe eye irritant. Mutation data reported. Can cause cramps and diarrhea. Possibly similar to DDT. An FDA over-the-counter drug. An anthelmintic. When heated to decomposition it emits toxic fumes of Cl^-. See also DDT and CHLOROPHENOLS.

MJM600   CAS: 5124-30-1   HR: 3
METHYLENE BIS(4-CYCLOHEXYLISO-CYANATE)

mf: $C_{15}H_{22}NO_2$ mw: 262.39
PROP: Colorless liquid.
SYNS: BIS(4-ISOCYANATOCYCLOHEXYL)METHANE □ DICYCLOHEXYLMETHANE-4,4'-DIISOCYANATE □ HYDROGENATED MDI □ METHYLENE BIS(4-CYCLOHEXYLISOCYANATE) (ACGIH,OSHA) □ NACCONATE H 12
CONSENSUS REPORTS: Reported in EPA TSCA Inventory.
OSHA PEL: CL 0.01
ACGIH TLV: TWA 0.005 ppm
NIOSH REL: (Dicyclohexylmethane 4,4'-diisocyanate) TWA CL 0.01 ppm
SAFETY PROFILE: Poison by inhalation. Mildly toxic by ingestion. When heated to decomposition it emits very toxic fumes of NO_x and CN^-.

MJN000   CAS: 101-61-1   HR: 3
4,4'-METHYLENE BIS(N,N'-DIMETHYL-ANILINE)

mf: $C_{17}H_{22}N_2$ mw: 254.41
PROP: Crystals, leaflets, or plates from EtOH or ligroin. Mp: 91°, bp: 390°. Sol in Me_2CO.
SYNS: p,p'-BIS(DIMETHYLAMINO)DIPHENYLMETHANE □ 4,4'-BIS(DIMETHYLAMINO)DIPHENYLMETHANE □ BIS(p-DIMETHYLAMINOPHENYL)METHANE □ BIS(p-(N,N-DIMETHYLAMINO)PHENYL)METHANE □ p,p'-BIS(N,N-DIMETHYLAMINOPHENYL)METHANE □ p,p'-

DIMETHYLAMINODIPHENYLMETHANE □ METHANE BASE □ 4,4'-METHYLENEBIS(N,N-DIMETHYL)BENZEN-AMINE □ MICHLER'S BASE □ MICHLER'S HYDRIDE □ MICHLER'S METHANE □ NCI-C01990 □ TETRA-BASE □ TETRAMETHYLDIAMINODIPHENYLMETHANE □ p,p'-TETRAMETHYLDIAMINODIPHENYLMETHANE □ 4,4'-TETRAMETHYLDIAMINODIPHENYLMETHANE

CONSENSUS REPORTS: NTP 10th Report on Carcinogens. IARC Cancer Review: Group 3 IMEMDT 7,56,87; Animal Limited Evidence IMEMDT 27,119,82. NCI Carcinogenesis Bioassay (feed); Clear Evidence: mouse, rat NCITR* NCI-CG-TR-186,79. EPA Genetic Toxicology Program. Reported in EPA TSCA Inventory. Community Right-To-Know List.

DFG MAK: Confirmed Animal Carcinogen, Suspected Human Carcinogen

DOT CLASSIFICATION: 3; Label: Flammable Liquid

SAFETY PROFILE: Confirmed carcinogen with experimental carcinogenic, neoplastigenic, and tumorigenic data. Moderately toxic by ingestion. Mutation data reported. A flammable liquid. When heated to decomposition it emits toxic fumes of NO_x.

MJN250   CAS: 88-24-4   HR: 3
2,2'-METHYLENEBIS(4-ETHYL-6-tert-BUTYLPHENOL)

mf: $C_{25}H_{36}O_2$ mw: 368.61
SYNS: AGIDOL 7 □ ANTAGE W 500 □ ANTIOXIDANT 425 □ AO 425 □ BIS(2-HYDROXY-3-tert-BUTYL-5-ETHYLPHENYL)METHANE □ CHEMANOX 22 □ CYANOX 425 □ 2,2'-METHYLENEBIS(6-tert-BUTYL-4-ETHYLPHENOL) □ NOCRAC NS 5 □ PHENOL, 2,2'-METHYLENEBIS(6-(1,1-DIMETHYLETHYL))-4-ETHYL-(9CI) □ PLASTANOX 425 ANTIOXIDANT □ USAF CY-6 □ YOSHINOX 425
CONSENSUS REPORTS: Reported in EPA TSCA Inventory.
SAFETY PROFILE: Poison by intraperitoneal route. Experimental reproductive effects. When heated to decomposition it emits acrid smoke and irritating fumes.

MJN750 **CAS: 139-25-3** **HR: 3**
5,5'-METHYLENEBIS(2-ISOCYANATO)-TOLUENE

mf: $C_{17}H_{14}N_2O_2$ mw: 278.33

SYNS: 3,3'-DIMETHYLDIPHENYLMETHANE-4,4'-DIISOCYANATE □ ISOCYANIC ACID, ESTER with DI-o-TOLUENEMETHANE

CONSENSUS REPORTS: Reported in EPA TSCA Inventory.

NIOSH REL: (Diisocyanates) TWA 0.005 ppm; CL 0.02 ppm/10M

SAFETY PROFILE: Poison by intravenous route. A sensitizer. When heated to decomposition it emits toxic fumes of NO_x and CN^-.

MJO250 **CAS: 838-88-0** **HR: 3**
4,4'-METHYLENEBIS(2-METHYLANIL-INE)

mf: $C_{15}H_{18}N_2$ mw: 226.35

PROP: Pale-amber crystals from EtOH. Mp: 158–159°.

SYNS: BIS-4-AMINO-3-METHYLFENYLMETHAN (CZECH) □ 3,3'-DIMETHYL-4,4'-DIAMINODIPHENYL-METHANE □ MBOT □ ME-MDA □ 4,4'-METHYLENE-BIS(2-METHYLBENZENAMINE) □ 4,4'-METHYLENE DI-o-TOLUIDINE

CONSENSUS REPORTS: IARC Cancer Review: Group 2B IMEMDT 7,248,87; Animal Limited Evidence IMEMDT 4,73,74. Reported in EPA TSCA Inventory.

DFG MAK: Confirmed Animal Carcinogen, Suspected Human Carcinogen

SAFETY PROFILE: Confirmed carcinogen with experimental carcinogenic data. Moderately toxic by ingestion. An eye irritant. Mutation data reported. When heated to decomposition it emits toxic fumes of NO_x.

MJP400 **CAS: 101-68-8** **HR: 3**
METHYLENE BISPHENYL ISOCYANATE

DOT: UN 2489

mf: $C_{15}H_{10}N_2O_2$ mw: 250.27

PROP: Crystals or yellow fused solid. Mp: 37.2°, bp: 184° @ 3 mm, d: 1.19 @ 50°, vap press: 0.001 mm @ 40°. IDLH 75 mg/m³.

SYNS: BIS(p-ISOCYANATOPHENYL)METHANE □ BIS(1,4-ISOCYANATOPHENYL)METHANE □ BIS(4-ISOCYANATOPHENYL)METHANE □ CARADATE 30 □ DESMODUR 44 □ DIFENIL-METAN-DIISOCIANATO (ITALIAN) □ DIFENYLMETHAAN-DIISSOCYANAAT (DUTCH) □ 4-4'-DIISOCYANATE de DIPHENYLMETH-ANE (FRENCH) □ 4,4'-DIISOCYANATODIPHENYL-METHANE □ DIPHENYLMETHAN-4,4'-DIISOCYANAT (GERMAN) □ DIPHENYL METHANE DIISOCYANATE □ p,p'-DIPHENYLMETHANE DIISOCYANATE □ 4,4'-DIPHENYLMETHANE DIISOCYANATE □ DIPHENYL-METHANE 4,4'-DIISOCYANATE (DOT) □ HYLENE M50 □ ISONATE □ MDI □ METHYLENEBIS(4-ISOCYANATOBENZENE) □ 1,1-METHYLENEBIS(4-ISOCYANATOBENZENE) □ METHYLENEBIS(p-PHENYLENE ISOCYANATE) □ METHYLENEBIS(4-PHENYLENE ISOCYANATE) □ p,p'-METHYLENEBIS-(PHENYL ISOCYANATE) □ METHYLENEBIS(p-PHENYL ISOCYANATE) □ METHYLENEBIS(4-PHENYL ISOCYANATE) □ 4,4'-METHYLENEBIS(PHENYL ISOCYANATE) □ 4,4'-METHYLENEDIPHENYL DIISOCYANATE □ METHYLENEDI-p-PHENYLENE DIISOCYANATE □ METHYLENEDI-p-PHENYLENE ISOCYANATE □ 4,4'-METHYLENEDIPHENYLENE ISOCYANATE □ METHYLENE DI(PHENYLENE ISOCYANATE) (DOT) □ 4,4'-METHYLENEDIPHENYL ISOCYANATE □ NACCONATE 300 □ NCI-C50668 □ RUBINATE 44

CONSENSUS REPORTS: IARC Cancer Review: Group 3 IMEMDT 7,56,87. Reported in EPA TSCA Inventory. Community Right-To-Know List.

OSHA PEL: CL 0.02 ppm

ACGIH TLV: 0.005 ppm

DFG MAK: 0.05 mg/m³; Confirmed Animal Carcinogen with Unknown Relevance to Humans

NIOSH REL: (Diisocyanates) TWA 0.005 ppm; CL 0.02 ppm/10M

DOT CLASSIFICATION: 6.1; Label: KEEP AWAY FROM FOOD; DOT Class: 6.1; Label: Poison; DOT Class: 6.1; Label: Poison, Flammable Liquid; DOT Class: 3; Label: Flammable Liquid, Poison

SAFETY PROFILE: Poison by inhalation. Mildly toxic by ingestion. Human systemic effects by inhalation: increased immune response and body temperature. A skin and eye irritant. An allergic sensitizer. Questionable carcinogen. Mutation data reported. A flammable liquid. When heated to decomposition it emits toxic fumes of NO_x and SO_x. See also CYANATES.

M

MJP450   **CAS: 75-09-2**   **HR: 3**
METHYLENE CHLORIDE
DOT: UN 1593
mf: CH_2Cl_2 mw: 84.93
PROP: Colorless, volatile liquid; odor of chloroform. Bp: 39.8°, lel: 15.5% in O_2, uel: 66.4% in O_2, fp: −96.7°, d: 1.326 @ 20°/4°, autoign temp: 1139°F, vap press: 380 mm @ 22°, vap d: 2.93, refr index: 1.424 @ 20 L. Sol in water; misc with alc, acetone, chloroform, ether, and carbon tetrachloride.
SYNS: AEROTHENE MM □ CHLORURE de METHYL-ENE (FRENCH) □ DCM □ DICHLOROMETHANE (MAK, DOT) □ FREON 30 □ METHANE DICHLORIDE □ METHYLENE BICHLORIDE □ METHYLENE DICHLORIDE □ METYLENU CHLOREK (POLISH) □ NCI-C50102 □ R 30 □ RCRA WASTE NUMBER U080 □ SOLAESTHIN □ SOLMETHINE
CONSENSUS REPORTS: NTP 10th Report on Carcinogens. IARC Cancer Review: Group 2B IMEMDT 7,194,87; Human Inadequate Evidence IMEMDT 41,43,86; Animal Sufficient Evidence IMEMDT 41,43,86; Animal Inadequate Evidence IMEMDT 20,449,79. NTP Carcinogenesis Studies (inhalation); Clear Evidence: mouse, rat NTPTR* NTP-TR-306,86. Reported in EPA TSCA Inventory. EPA Genetic Toxicology Program. Community Right-To-Know List.
OSHA PEL: 25 ppm
ACGIH TLV: TWA 50 ppm; Animal Carcinogen
DFG MAK: Confirmed Animal Carcinogen with Unknown Relevance to Humans; 100 ppm (350 mg/m³); BAT: 5% CO-Hb in blood at end of shift;
NIOSH REL: (Methylene Chloride) Reduce to lowest feasible level
DOT CLASSIFICATION: 6.1; Label: KEEP AWAY FROM FOOD
SAFETY PROFILE: Confirmed carcinogen with experimental carcinogenic and tumorigenic data. Poison by intravenous route. Moderately toxic by ingestion, subcutaneous, and intraperitoneal routes. Mildly toxic by inhalation. Human systemic effects by ingestion and inhalation: paresthesia, somnolence, altered sleep time, convulsions, euphoria, and change in cardiac rate. An experimental teratogen. Experimental reproductive effects. An eye and severe skin irritant. Human mutation data reported. It is flammable in the range of 12–19% in air but ignition is difficult. It will not form explosive mixtures with air at ordinary temperatures. Mixtures in air with methanol vapor are flammable. It will form explosive mixtures with an atmosphere having a high oxygen content, in liquid O_2, N_2O_4, K, Na, NaK. Explosive in the form of vapor when exposed to heat or flame. Reacts violently with Li, NaK, potassium-tert-butoxide, (KOH + N-methyl-N-nitrosourea). It can be decomposed by contact with hot surfaces and open flame, and then yield toxic fumes that are irritating and give warning of their presence. When heated to decomposition it emits highly toxic fumes of phosgene and Cl⁻. See also CHLORINATED HYDROCARBONS, ALIPHATIC.

MJP750   **CAS: 1208-52-2**   **HR: 3**
2,4'-METHYLENEDIANILINE
mf: $C_{13}H_{14}N_2$ mw: 198.29
PROP: Leaflets from C_6H_6. Mp: 88–89°, bp: 222° @ 9 mm.
SYNS: 2',4-BIS(AMINOPHENYL)METHANE □ 2,4'-DIAMINODIPHENYLMETHAN (GERMAN) □ o,p'-DIAMINODIPHENYLMETHANE □ 2,4'-DIAMINODI-PHENYLMETHANE □ 2,4'-DIPHENYLMETHANEDI-AMINE □ 2,4'-METHYLENEBIS(ANILINE)
CONSENSUS REPORTS: Reported in EPA TSCA Inventory.
DFG MAK: Animal Carcinogen, Suspected Human Carcinogen
SAFETY PROFILE: Suspected carcinogen. Moderately toxic by subcutaneous route. When heated to decomposition it emits toxic fumes of NO_x.

MJQ000   **CAS: 101-77-9**   **HR: 3**
4,4'-METHYLENEDIANILINE
DOT: UN 2651
mf: $C_{13}H_{14}N_2$ mw: 198.29

PROP: Tan flakes, lumps, or pearly leaflets from benzene; faint amine-like odor. Mp: 93°, flash p: 440°F, bp: 232° @ 9 mm.

SYNS: 4-(4-AMINOBENZYL)ANILINE □ ANCAMINE TL □ ARALDITE HARDENER 972 □ BENZENAMINE, 4,4'-METHYLENEBIS- □ BIS-p-AMINOFENYLMETHAN □ BIS(p-AMINOPHENYL)METHANE □ BIS(4-AMINO-PHENYL)METHANE □ CURITHANE □ DADPM □ DAPM □ DDM □ p,p'-DIAMINODIFENYLMETHAN □ 4,4'-DIAMINODIPHENYLMETHAN □ DIAMINODI-PHENYLMETHANE □ p,p'-DIAMINODIPHENYL-METHANE □ 4,4'-DIAMINODIPHENYLMETHANE □ 4,4'-DIAMINODIPHENYLMETHANE (DOT) □ DI-(4-AMINOPHENYL)METHANE □ DIANILINOMETHANE □ 4,4'-DIPHENYLMETHANEDIAMINE □ EPICURE DDM □ EPIKURE DDM □ HT 972 □ JEFFAMINE AP-20 □ MDA □ METHYLENEBIS(ANILINE) □ 4,4'-METHYL-ENEBISANILINE □ 4,4'-METHYLENEBIS(BENZENE-AMINE) □ METHYLENEDIANILINE □ p,p'-METHYL-ENEDIANILINE □ 4,4-METHYLENEDIANILINE (ACGIH) □ SUMICURE M □ TONOX

CONSENSUS REPORTS: NTP 10th Report on Carcinogens. IARC Cancer Review: Group 2B IMEMDT 7,56,87; Animal Sufficient Evidence IMEMDT 39,347,86; Animal Inadequate Evidence IMEMDT 4,79,74. Community Right-To-Know List. Reported in EPA TSCA Inventory.

ACGIH TLV: TWA 0.1 ppm (skin); Animal Carcinogen

DFG MAK: Animal Carcinogen, Suspected Human Carcinogen

DOT CLASSIFICATION: 6.1; Label: KEEP AWAY FROM FOOD

SAFETY PROFILE: Confirmed carcinogen with experimental tumorigenic data. Human poison by ingestion. Poison by subcutaneous and intraperitoneal routes. Human systemic effects by ingestion: rigidity, jaundice, other liver changes. An eye irritant. Mutation data reported. It is not rapidly absorbed through the skin. Combustible when exposed to heat or flame. When heated to decomposition it emits highly toxic fumes of aniline and NO_x.

MJQ100   CAS: 13552-44-8   HR: 3
4,4'-METHYLENEDIANILINE DIHYDRO-
**   CHLORIDE**
mf: $C_{13}H_{14}N_2 \cdot 2ClH$ mw: 271.21

SYNS: BENZENAMINE, 4,4'-METHYLENEBIS-, DIHYDROCHLORIDE □ p,p'-METHYLENEDIANILINE DIHYDROCHLORIDE □ NCI-C54604

CONSENSUS REPORTS: NTP 10th Report on Carcinogens. IARC Cancer Review: Animal Sufficient Evidence IMEMDT 39,347,86. NTP Carcinogenesis Studies (oral); Clear Evidence: mouse, rat NTPTR* NTP-TR-248,83. Reported in EPA TSCA Inventory.

NIOSH REL: (4,4'-Methylenedianiline) TWA reduce to lowest level

SAFETY PROFILE: Confirmed carcinogen with experimental carcinogenic and neoplastigenic data. Mutation data reported. When heated to decomposition it emits toxic fumes of NO_x and HCl. See also ANILINE DYES.

MJQ500   CAS: 156-72-9   HR: 3
METHYLENE DIMETHANESULFONATE
mf: $C_3H_8O_6S_2$ mw: 204.23

SYNS: ENT 51,799 □ METHANESULFONIC ACID, METHYLENE ESTER □ METHYLENE BIS(METHANE-SULFONATE)

CONSENSUS REPORTS: EPA Genetic Toxicology Program.

SAFETY PROFILE: Poison by intraperitoneal route. An experimental teratogen. Experimental reproductive effects. Mutation data reported. When heated to decomposition it emits toxic fumes of SO_x. See also SULFONATES.

MJT500   CAS: 6317-18-6   HR: 3
METHYLENE DITHIOCYANATE
mf: $C_3H_2N_2S_2$ mw: 130.19

PROP: A solid. Mp: 102°.

SYN: METHYLENDIRHODANID (CZECH, GERMAN)

CONSENSUS REPORTS: Reported in EPA TSCA Inventory.

SAFETY PROFILE: Poison by ingestion, intravenous, and subcutaneous routes. When heated to decomposition it emits very toxic fumes of NO_x and SO_x. See also THIOCYANATES.

M

MJW000   CAS: 112-61-8   HR: 2
METHYL ESTER STEARIC ACID
mf: $C_{19}H_{38}O_2$ mw: 298.57
PROP: Liquid to semi-solid. Mp: 38°, bp: 215° @ 15 mm, flash p: 307°F (CC), d: 0.860. Sol in water and ether.
SYNS: EMERY 2218 □ METHOLENE 2218 □ METHYL OCTADECANOATE □ METHYL STEARATE □ OCTADECANOIC ACID, METHYL ESTER
CONSENSUS REPORTS: Reported in EPA TSCA Inventory.
SAFETY PROFILE: Questionable carcinogen with experimental tumorigenic data. Combustible when exposed to heat or flame; can react with oxidizing materials. To fight fire, use CO_2, dry chemical. When heated to decomposition it emits acrid smoke and irritating fumes. See also ESTERS.

MJW250   CAS: 1912-28-3   HR: D
METHYL ETHANE SULPHONATE
mf: $C_3H_8O_3S$ mw: 124.17
PROP: Oil. Bp: 197.5–200.5°.
SYNS: MES □ METHYL ETHANE SULFONATE
CONSENSUS REPORTS: Reported in EPA TSCA Inventory.
SAFETY PROFILE: Experimental reproductive effects. Mutation data reported. When heated to decomposition it emits toxic fumes of SO_x. See also SULFONATES.

MJW500   CAS: 115-10-6   HR: 3
METHYL ETHER
DOT: UN 1033
mf: C_2H_6O mw: 46.08
PROP: Colorless gas; ether odor. Mp: −138.5°, bp: −23.7°, lel: 3.4%, uel: 27%, flash p: −42°F (CC), autoign temp: 662°F, vap d: 1.617, d: 0.661 (air = 1). Sol in alc, water, ether.
SYNS: DIMETHYL ETHER (DOT) □ OXYBISMETHANE □ WOOD ETHER
CONSENSUS REPORTS: Reported in EPA TSCA Inventory.
DFG MAK: 1000 ppm (1900 mg/m³)
DOT CLASSIFICATION: 2.1; Label: Flammable Gas

SAFETY PROFILE: Slightly toxic by inhalation. Very dangerous fire hazard when exposed to heat, flame, or oxidizers. Dangerous explosion hazard when exposed to flame, sparks, etc. Violent reaction with AlH_3 and $LiAlH_2$. Keep in closed container away from heat and open flame. To fight fire, stop flow of gas. When heated to decomposition it emits acrid smoke and irritating fumes. See also ETHERS and ETHYL ETHER.

MJY500   CAS: 25057-89-0   HR: 2
3-(1-METHYLETHYL)-1H-2,1,3-BENZO-THIAZAIN-4(3H)-ONE-2,2-DIOXIDE
mf: $C_{10}H_{12}N_2O_3S$ mw: 240.30
PROP: Crystals or powder. Sltly sol in H_2O, C_6H_6; sol in Me_2CO, $CHCl_3$, and EtOH.
SYNS: BAS 351-H □ BASAGRAN □ BENDIOXIDE □ BENTAZON □ 3-ISOPROPYL-2,1,3-BENZOTHIADIAZIN-ON-(4)-2,2-DIOXID (GERMAN) □ 3-ISOPROPYL-1H-2,1,3-BENZOTHIADIAZIN-4(3H)-ONE-2,2-DIOXIDE
CONSENSUS REPORTS: Reported in EPA TSCA Inventory. EPA Genetic Toxicology Program.
SAFETY PROFILE: Moderately toxic by ingestion and skin contact. An experimental teratogen. Other experimental reproductive effects. When heated to decomposition it emits very toxic fumes of SO_x and NO_x.

MJY550   HR: D
METHYL ETHYL CELLULOSE
PROP: White fibrous solid or powder. Disperses in water.
SAFETY PROFILE: When heated to decomposition it emits acrid smoke and irritating fumes.

MKA000   CAS: 31218-83-4   HR: 3
(E)-1-METHYLETHYL-3-((ETHYL-AMINO)METHOXYPHOSPHINOTHIO-YL)OXY-2-BUTENOATE
mf: $C_{10}H_{20}NO_4PS$ mw: 281.34
PROP: D: 1.13 @ 20°/4°, bp: 87–89° @ 0.005 mm. Spar sol in H_2O.
SYNS: BLOTIC □ ENT 27,989 □ (3)-O-2-ISOPROPOXY-CARBONYL-1-METHYLVINYL-O-METHYL ETHYLPHOSPHORAMIDOTHIOATE □

PROPETAMPHOS □ SAFROTIN □ SAN 52 139 I □ SANDOZ 52139 □ VEL 4283

SAFETY PROFILE: Poison by ingestion. Moderately toxic by skin contact. When heated to decomposition it emits very toxic fumes of PO_x, SO_x, and NO_x.

MKA250   CAS: 64-65-3   HR: 3
3-METHYL-3-ETHYLGLUTARIMIDE

mf: $C_8H_{13}NO_2$ mw: 155.22

PROP: Platelets from H_2O or Me_2CO. Mp: 123.5–124°. Sol in H_2O.

SYNS: AHYPNON □ BEMEGRIDE □ 2,6-DIOXO-4-METHYL-4-ETHYLPIPERIDINE □ 4-ETHYL-4-METHYL-2,6-DIOXOPIPERIDINE □ β-ETHYL-β-METHYL-GLUTARIMIDE □ 3-ETHYL-3-METHYLGLUTARIMIDE □ 4-ETHYL-4-METHYL-2,6-PIPERIDINEDIONE □ EUKRATON □ MALYSOL □ MEGIMIDE □ 4-METHYL-4-ETHYL-2,6-DIOXOPIPERIDINE □ β-METHYL-β-ETHYLGLUTARIMIDE □ MIKEDIMIDE

CONSENSUS REPORTS: Reported in EPA TSCA Inventory.

SAFETY PROFILE: Poison by ingestion, intravenous, intraperitoneal, intramuscular, subcutaneous, and parenteral routes. Human systemic effects by ingestion: wakefulness, hallucinations, distorted perceptions, toxic psychosis. An analeptic, central nervous system stimulant; used to counteract barbiturate poisoning. When heated to decomposition it emits toxic fumes of NO_x.

MKA300   CAS: 28846-43-7   HR: 3
9-(3'-METHYL-4'-ETHYLIDENE-THIO-SEMICARBAZIDO)ACRIDINE

mf: $C_{17}H_{16}N_4S$ mw: 308.43

SYN: ACETALDEHYDE, 4-(9-ACRIDINYL)-2-METHYL-3-THIOSEMICARBAZONE

SAFETY PROFILE: A poison by ingestion. When heated to decomposition it emits toxic vapors of NO_x and SO_x.

MKA400   CAS: 78-93-3   HR: 3
METHYL ETHYL KETONE

DOT: UN 1193

mf: C_4H_8O mw: 72.12

PROP: Colorless liquid; acetone-like odor. Fp: −85.9°, bp: 79.57°, lel: 1.8%, uel: 11.5%, flash p: 22°F (TOC), d: 0.80615 @ 20°/20°, vap press: 71.2 mm @ 20°, autoign temp: 960°F, vap d: 2.42, ULC: 85–90. Misc with alc, ether, fixed oils, and water. IDLH 3000 ppm.

SYNS: AETHYLMETHYLKETON (GERMAN) □ BUTANONE 2 (FRENCH) □ 2-BUTANONE (OSHA) □ ETHYL METHYL CETONE (FRENCH) □ ETHYL-METHYLKETON (DUTCH) □ ETHYL METHYL KETONE (DOT) □ FEMA No. 2170 □ MEK □ METHYL ACETONE (DOT) □ METILETILCHETONE (ITALIAN) □ METYLOETYLOKETON (POLISH) □ RCRA WASTE NUMBER U159

CONSENSUS REPORTS: Community Right-To-Know List. EPA Genetic Toxicology Program. Reported in EPA TSCA Inventory.

OSHA PEL: TWA 200 ppm; STEL 300 ppm

ACGIH TLV: TWA 200 ppm; STEL 300 ppm; BEI: 2 mg(MEK)/L in urine at end of shift

DFG MAK: 200 ppm (600 mg/m³)

NIOSH REL: (Ketones) TWA 590 mg/m³

DOT CLASSIFICATION: 3; Label: Flammable Liquid

SAFETY PROFILE: Moderately toxic by ingestion, skin contact, and intraperitoneal routes. Human systemic effects by inhalation: conjunctiva irritation and unspecified effects on the nose and respiratory system. An experimental teratogen. A strong irritant. Human eye irritation @ 350 ppm. Affects peripheral nervous system and central nervous system. See also KETONES. Highly flammable liquid. Reaction with hydrogen peroxide + nitric acid forms a heat- and shock-sensitive explosive product. Ignition on contact with potassium tert-butoxide. Mixture with 2-propanol will produce explosive peroxides during storage. Vigorous reaction with chloroform + alkali. Incompatible with chlorosulfonic acid, oleum. To fight fire, use alcohol foam, CO_2, dry chemical. Used in production of drugs of abuse. When heated to decomposition it emits acrid smoke and fumes.

MKA500   CAS: 1338-23-4   HR: 3
METHYL ETHYL KETONE PEROXIDE

mf: $C_8H_{16}O_4$ mw: 176.24

PROP: Colorless liquid.

SYNS: BUTANOX LPT □ BUTANOX M 50 □ BUTANOX M 105 □ CADOX □ CHALOXYD MEKP-HA 1 □ CHALOXYD MEKP-LA 1 □ ESPERFOAM FR □ ETHYL METHYL KETONE PEROXIDE □ FR 222 □ HI-POINT 90 □ HI-POINT 180 □ HI-POINT PD-1 □ KETONOX □ LUCIDOL DELTAX □ LUPERSOL □ LUPERSOL DDA 30 □ LUPERSOL DDM □ LUPERSOL Δ-X □ LUPERSOL DNF □ LUPERSOL DSW □ MEK PEROXIDE □ MEKP (OSHA) □ METHYL ETHYL KETONE HYDROPEROX-IDE □ METHYL ETHYL KETONE PEROXIDE, in solution with >9% by weight active oxygen (DOT) □ METHYLETH-YLKETONHYDROPEROXIDE □ NCI-C55447 □ PERMEK N □ QUICKSET EXTRA □ QUICKSET SUPER □ RCRA WASTE NUMBER U160 □ SPRAYSET MEKP □ THERM-ACURE

CONSENSUS REPORTS: Reported in EPA TSCA Inventory.

OSHA PEL: CL 0.7 ppm

ACGIH TLV: CL 0.2 ppm

DFG MAK: Organic Peroxide, moderate skin irritant

DOT CLASSIFICATION: Forbidden

SAFETY PROFILE: Poison by intraperitoneal route. Moderately toxic by ingestion and inhalation. Human systemic effects by ingestion: changes in structure or function of esophagus, nausea or vomiting, other gastrointestinal effects. A moderate skin and eye irritant. Questionable carcinogen with experimental tumorigenic data. A shock-sensitive explosive. When heated to decomposition it emits acrid smoke and irritating fumes. See also KETONES and PEROXIDES.

MKA750   CAS: 624-46-4   HR: 3
METHYL ETHYL KETONE SEMI-
**   CARBAZONE**

mf: $C_5H_{11}N_3O$ mw: 129.19

SYN: 2-BUTANONE, SEMICARBAZONE

DOT CLASSIFICATION: 3; Label: Flammable Liquid

SAFETY PROFILE: Poison by intravenous and intraperitoneal routes. When heated to decomposition it emits toxic fumes of NO_x. See also KETONES.

MKB000   CAS: 10595-95-6   HR: 3
N,N-METHYLETHYLNITROSAMINE

mf: $C_3H_8N_2O$ mw: 88.13

PROP: Yellow liquid. Bp: 67° @ 40 mm.

SYNS: ETHYLMETHYLNITROSAMINE □ METHYL-AETHYLNITROSAMIN (GERMAN) □ METHYLETHYL-NITROSAMINE □ N-METHYL-N-NITROSO-ETHAMINE □ N-METHYL-N-NITROSOETHYLAMINE □ NEMA □ N-NITROSOETHYLMETHYLAMINE □ N-NITROSOMETHYLETHYLAMINE (MAK) □ NMEA

CONSENSUS REPORTS: IARC Cancer Review: Group 2B IMEMDT 7,56,87; Animal Limited Evidence IMEMDT 17,221,78. EPA Genetic Toxicology Program.

DFG MAK: Animal Carcinogen, Suspected Human Carcinogen

SAFETY PROFILE: Confirmed carcinogen with experimental carcinogenic and tumorigenic data. Poison by ingestion. Experimental reproductive effects. Mutation data reported. When heated to decomposition it emits toxic fumes of NO_x. See also NITROSAMINES.

MKB250   CAS: 50-12-4   HR: 3
3-METHYL-5-ETHYL-5-PHENYLHYD-
**   ANTOIN**

mf: $C_{12}H_{14}N_2O_2$ mw: 218.28

SYNS: EPILAN □ 5-ETHYL-3-METHYL-5-PHENYL-HYDANTOIN □ 5-ETHYL-3-METHYL-5-PHENYL-2,4(3H,5H)-IMIDAZOLEDIONE □ 5-ETHYL-3-METHYL-5-PHENYLIMIDAZOLIDIN-2,4-DIONE □ 3-ETHYL-NIRVANOL □ GEROT-EPILAN □ INSULTON □ MEPHENYTOIN □ MESANTOIN □ METHOIN □ METHYL HYDANTOIN □ 3-METHYL-5,5-PHENYL-ETHYLHYDANTOIN □ NSC-34652 □ PHENANTOIN □ PHENYLETHYLMETHYLHYDANTOIN □ SACERNO □ SEDANTOINAL □ TRIANTOIN

SAFETY PROFILE: Poison by ingestion and intraperitoneal routes. Human systemic effects by ingestion: somnolence, hemorrhage, changes in teeth and supporting structures. Human mutation data reported. An experimental teratogen. An FDA proprietary drug used as an anticonvulsant. When heated to decomposition it emits toxic fumes of NO_x.

MKB320   CAS: 37902-85-5   HR: 3
METHYL 3-((ETHYL(PROPYLAMINO)-
PHOSPHINOTHIOYL)OXY)-2-
BUTENOATE
mf: $C_{10}H_{20}NO_3PS$ mw: 265.34
SYNS: 2-BUTENOIC ACID, 3-((ETHYL(PROPYL-
AMINO)PHOSPHINOTHIOYL)OXY)-, METHYL ESTER □
2-CARBOMETHOXY-1-METHYLVINYL-N-PROPYL
ETHYLPHOSPHONOAMIDOTHIOATE
SAFETY PROFILE: A poison by ingestion
and skin contact. When heated to
decomposition it emits toxic vapors of NO_x,
SO_x, and PO_x.

MKB750   CAS: 72-33-3   HR: 3
3-METHYLETHYNYLESTRADIOL
mf: $C_{21}H_{26}O_2$ mw: 310.47
PROP: Crystals from Me_2CO. Mp:
150–151°.
SYNS: COMPOUND 33355 □ DELTA-MVE □ ETHIN-
YLESTRADIOL-3-METHYL ETHER □ 17-α-ETHINYL
ESTRADIOL 3-METHYL ETHER □ ETHINYL-
OESTRADIOL-3-METHYL ETHER □ 17-α-ETHINYL
OESTRADIOL-3-METHYL ETHER □ ETHYNYL-
ESTRADIOL-3-METHYL ETHER □ 17-ETHYNYLESTRA-
DIOL-3-METHYL ETHER □ 17-α-ETHYNYLESTRADIOL-
3-METHYL ETHER □ (+)-17-α-ETHYNYL-17-β-HYDRO-
XY-3-METHOXY-1,3,5(10)-ESTRATRIENE □ (+)-17-α-
ETHYNYL-17-β-HYDROXY-3-METHOXY-1,3,5(10)-
OESTRATRIENE □ 17-ETHYNYL-3-METHOXY-1,3,5(10)-
ESTRATRIEN-17-β-OL □ 17-α-ETHYNYL-3-METHOXY-
1,3,5(10)-ESTRATRIEN-17-β-OL □ 17-α-ETHYNYL-3-
METHOXY-17-β-HYDROXY-Δ-1,3,5(10)-ESTRATRIENE □
17-α-ETHYNYL-3-METHOXY-17-β-HYDROXY-Δ-1,3,5(10)-
OESTRATRIENE □ 17-ETHYNYL-3-METHOXY-1,3,5(10)-
OESTRATIEN-17-β-OL □ ETHYNYLOESTRADIOL
METHYL ETHER □ 17-ETHYNYLOESTRADIOL-3-
METHYL ETHER □ 17-α-ETHYNYLOESTRADIOL
METHYL ETHER □ 17-α-ETHYNYLOESTRADIOL-3-
METHYL ETHER □ MESTRANOL □ MESTRENOL □ 3-
METHOXY-17-α-ETHINYLESTRADIOL □ 3-METHOXY-
17-α-ETHINYLOESTRADIOL □ 3-METHOXY-17-α-
ETHYNOESTRADIOL □ 3-METHOXYETHYNYL-
ESTRADIOL □ 3-METHOXY-17-α-ETHYNYLESTRADIOL
□ 3-METHOXY-17-α-ETHYNYL-1,3,5(10)-ESTRATRIEN-
17-β-OL □ 3-METHOXYETHYNYLOESTRADIOL □ 3-
METHOXY-17-ETHYNYLOESTRADIOL-17-β □ 3-
METHOXY-17-α-ETHYNYL-1,3,5(10)-OESTRATRIEN-17-
β-OL □ 3-METHOXY-17-α-19-NORPREGNA-K,3,5(10)-
TRIEN-20-YN-17-OL □ 3-METHOXY-19-NOR-17-α-
PREGNA-1,3,5(10)-TRIEN-10-YN-17-OL □ (17-α)-3-
METHOXY-19-NORPREGNA-1,3,5(10)-TRIEN-20-YN-17-
OL □ 3-METHYLETHYNYLOESTRADIOL

CONSENSUS REPORTS: NTP 10th Report
on Carcinogens. IARC Cancer Review:
Human Limited Evidence IMEMDT
21,257,79; Animal Sufficient Evidence
IMEMDT 6,87,74; IMEMDT 21,257,79.
SAFETY PROFILE: Confirmed carcinogen
with experimental neoplastigenic,
tumorigenic, and teratogenic data.
Moderately toxic by subcutaneous route.
Human reproductive effects by ingestion:
changes in ovaries and fallopian tubes,
fertility effects. Experimental reproductive
effects. Mutation data reported. An FDA
proprietary drug. A steroid used in oral
contraceptives. When heated to
decomposition it emits acrid smoke and
irritating fumes.

MKG500   CAS: 123-39-7   HR: 2
N-METHYLFORMAMIDE
mf: C_2H_5NO mw: 59.08
PROP: Bp: 180–185°, flash p: <71.6°F, d:
1.01 @ 19°. Sol in H_2O and EtOH; insol in
Et_2O.
SYNS: METHYLFORMAMIDE □
MONOMETHYLFORMAMIDE □ NSC-3051
CONSENSUS REPORTS: Reported in
EPA TSCA Inventory. EPA Genetic
Toxicology Program.
SAFETY PROFILE: Moderately toxic by
ingestion, intraperitoneal, intravenous,
intramuscular, and subcutaneous routes. An
experimental teratogen. Experimental
reproductive effects. An eye irritant. A very
dangerous fire hazard when exposed to heat
or flame. Violent reaction with benzene
sulfonyl chloride. When heated to
decomposition it emits toxic fumes of NO_x.

MKG750   CAS: 107-31-3   HR: 3
METHYL FORMATE
DOT: UN 1243
mf: $C_2H_4O_2$ mw: 60.06
PROP: Colorless liquid; agreeable odor. Mp:
−99.8°, bp: 31.5°, lel: 5.9%, uel: 20%, flash
p: −2.2°F, d: 0.98149 @ 15°/4°, 0.975 @
20°/4°, autoign temp: 869°F, vap press: 400
mm @ 16°/0°, vap d: 2.07. Solidifies at

M

about 100°. Moderately sol in water, methyl alcohol; misc in alc. IDLH 4500 ppm.

SYNS: FORMIATE de METHYLE (FRENCH) □ METHYLE (FORMIATE de) (FRENCH) □ METHYL-FORMIAAT (DUTCH) □ METHYLFORMIAT (GERMAN) □ METHYL METHANOATE □ METIL (FORMIATO di) (ITALIAN)

CONSENSUS REPORTS: Reported in EPA TSCA Inventory.

OSHA PEL: TWA 100 ppm; STEL 150 ppm

ACGIH TLV: TWA 100 ppm; STEL 150 ppm

DFG MAK: 50 ppm (120 mg/m³)

DOT CLASSIFICATION: 3; Label: Flammable Liquid

SAFETY PROFILE: Moderately toxic by ingestion. Inhalation of vapor can cause irritation to nasal passages and conjunctiva, optic neuritis, narcosis, retching, and death from pulmonary irritation. Industrial fatalities have occurred only with exposure to high concentrations. Flammable liquid. Very dangerous fire hazard when exposed to heat or flame; can react vigorously with oxidizing materials. Explosive in the form of vapor when exposed to heat or flame. Reacts with methanol + sodium methoxide to form an explosive product. To fight fire, use alcohol foam, CO_2, dry chemical. When heated to decomposition it emits acrid smoke and irritating fumes.

MKH000   CAS: 534-22-5   HR: 3
2-METHYLFURAN

DOT: UN 2301

mf: C_5H_6O mw: 82.11

PROP: Colorless, mobile liquid; ether-like odor. Bp: 63.7°, fp: −88.7°, flash p: −22°F, d: 0.827 @ 20°/4°, vap press: 139 mm @ 20°, vap d: 2.8. Sltly sol in water.

SYNS: METHYLFURAN □ SILVAN (CZECH)

CONSENSUS REPORTS: Reported in EPA TSCA Inventory.

DOT CLASSIFICATION: 3; Label: Flammable Liquid

SAFETY PROFILE: Poison by ingestion. Moderately toxic by inhalation. An eye irritant. Mutation data reported. Very dangerous fire hazard when exposed to heat

or flame; can react vigorously with oxidizing materials. To fight fire, use CO_2, dry chemical. When heated to decomposition it emits acrid smoke and irritating fumes. See also ETHERS for explosion hazard.

MKH600   CAS: 611-13-2   HR: 3
METHYL FUROATE

mf: $C_6H_6O_3$ mw: 126.12

PROP: Mp: 164°, bp: 181°, d: 1.179.

SYNS: FURAN-α-CARBOXYLIC ACID METHYL ESTER □ 2-FUROIC ACID, METHYL ESTER □ METHYL 2-FURANCARBOXYLATE □ METHYL 2-FUROATE □ METHYL PYROMUCATE □ PYROMUCIC ACID METHYL ESTER

CONSENSUS REPORTS: Reported in EPA TSCA Inventory.

SAFETY PROFILE: Poison by intraperitoneal route. A skin irritant and lachrymator. When heated to decomposition it emits acrid smoke and irritating fumes.

MKJ250   CAS: 141-59-3   HR: 3
METHYL HEPTANETHIOL

DOT: UN 3023

mf: $C_8H_{18}S$ mw: 146.32

PROP: Liquid. Bp: 159–166°, d: 0.848 @ 60°/60°F, vap d: 5.0, flash p: 115°F (OC). Insol in water.

SYNS: tert-OCTANETHIOL □ T-OCTYL MERCAPTAN □ tert-OCTYLMERCAPTAN □ terc.OKTANTHIOL □ 2-PENTANETHIOL, 2,4,4-TRIMETHYL- □ 2,4,4-TRIMETH-YL-2-PENTANETHIOL

CONSENSUS REPORTS: Reported in EPA TSCA Inventory.

DOT CLASSIFICATION: 6.1; Label: Poison, Flammable Liquid

SAFETY PROFILE: Poison by ingestion. Irritating to eyes. A flammable liquid when exposed to heat, flame, or oxidizers. To fight fire, use foam, alcohol foam. When heated to decomposition it emits very toxic fumes of SO_x. See also MERCAPTANS.

MKK000   CAS: 110-93-0   HR: 3
6-METHYL-5-HEPTEN-2-ONE

mf: $C_8H_{14}O$ mw: 126.22

PROP: Sltly yellow liquid; citrus-lemongrass odor. D: 0.846–0.851, refr index:

1.438–1.442, mp: −67.1°, bp: 173–174°, flash p: 122°F. Insol in water; misc in alc, ether, and chloroform.

SYNS: FEMA No. 2707 □ 5-HEPTEN-2-ONE, 6-METHYL- □ METHYL HEPTENONE

CONSENSUS REPORTS: Reported in EPA TSCA Inventory.

SAFETY PROFILE: Moderately toxic by ingestion. A skin irritant. Flammable liquid when exposed to heat, sparks, or flame. When heated to decomposition it emits acrid smoke and irritating fumes. See also KETONES.

MKL300   CAS: 13706-86-0   HR: 1
5-METHYL-2,3-HEXANEDIONE
mf: $C_7H_{12}O_2$ mw: 128.19
PROP: D: 0.908 @ 22°/4°, bp: 138°.
SYNS: ACETYLISOPENTANOYL □ ACETYL ISOVALERYL □ 2,3-HEXANEDIONE, 5-METHYL-

CONSENSUS REPORTS: Reported in EPA TSCA Inventory.

SAFETY PROFILE: A skin irritant. When heated to decomposition it emits acrid smoke and irritating fumes.

MKM800   CAS: 77267-50-6   HR: 3
METHYL-N-(((((HEXYLOXY)SULFINYL)-
**   METHYLAMINO)CARBONYL)OXY)ET**
**   HANIMIDOTHIOATE**
mf: $C_{11}H_{22}N_2O_4S_2$ mw: 310.47
SYNS: ETHANIMIDOTHIOIC ACID, N-(((((HEXYLO-XY)SULFINYL)METHYLAMINO)CARBONYL)OXY)-, METHYL ESTER □ S-METHYL N-(N'-METHYL-N'-HEXOXYSULFINYLCARBAMOYLOXY)THIOACETIMID ATE

SAFETY PROFILE: A poison by ingestion. When heated to decomposition it emits toxic vapors of NO_x and SO_x.

MKN000   CAS: 60-34-4   HR: 3
METHYL HYDRAZINE
DOT: UN 1244
mf: CH_6N_2 mw: 46.09
PROP: Colorless, hydroscopic liquid; ammonia-like odor. D: 0.874 @ 25°, mp: −20.9°, bp: 87.8°, vap d: 1.6, flash p: 73.4°F, fp: −52.4°, autoign temp: 196°, lel: 2.5%, uel: 97 ± 2%. Sltly sol in water; sol in alc, hydrocarbons, and ether; misc with

hydrazine. Strong reducing agent. IDLH 20 ppm.

SYNS: HYDRAZOMETHANE □ 1-METHYL HYDRAZINE □ METHYLHYDRAZINE (DOT) □ METYLOHYDRAZYNA (POLISH) □ MMH □ MONOMETHYL HYDRAZINE □ RCRA WASTE NUMBER P068

CONSENSUS REPORTS: Community Right-To-Know List. EPA Extremely Hazardous Substances List. Reported in EPA TSCA Inventory. EPA Genetic Toxicology Program.

OSHA PEL: CL 0.2 ppm (skin)
ACGIH TLV: TWA 0.01 ppm (skin); Suspected Human Carcinogen
NIOSH REL: CL 0.08 mg/m^3/2H
DOT CLASSIFICATION: 6.1; Label: Poison, Flammable Liquid, Corrosive

SAFETY PROFILE: Suspected carcinogen with experimental carcinogenic, neoplastigenic, tumorigenic, and teratogenic data. Poison by inhalation, ingestion, skin contact, intraperitoneal, subcutaneous, and intravenous routes. Experimental reproductive effects. Human mutation data reported. Corrosive to skin, eyes, and mucous membranes. May self-ignite in air. Very dangerous fire hazard when exposed to heat or flame. To fight fire, use alcohol foam, CO_2, dry chemical. Explosive in the form of vapor when exposed to heat or flame. A powerful reducing agent. It is hypergolic with many oxidants (e.g., dinitrogen tetraoxide and hydrogen peroxide). When heated to decomposition it emits toxic fumes of NO_x. See also HYDRAZINE.

MKN250   CAS: 7339-53-9   HR: 3
METHYLHYDRAZINE HYDROCHLORIDE
mf: $CH_6N_2 \cdot ClH$ mw: 82.55
NIOSH REL: (Hydrazines) CL 0.08 mg/m^3/2H

SAFETY PROFILE: Poison by ingestion, intraperitoneal, and intravenous routes. When heated to decomposition it emits very toxic fumes of Cl^- and NO_x. See also METHYLHYDRAZINE.

M

MKP500   CAS: 90-33-5   HR: 2
4-METHYL-7-HYDROXYCOUMARIN
mf: $C_{10}H_8O_3$ mw: 176.18
PROP: Needles from EtOH. Mp: 194–195°.
SYNS: BILCOLIC □ BILICANTE □ CANTABILINE □ COUMARIN 4 □ EUROGALE □ 7-HYDROXY-4-METHYLCOUMARIN □ 7-HYDROXY-4-METHYL-2-OXO-2H-1-BENZOPYRAN □ HYMECROMONE □ MEDILLA □ MENDIAXON □ 4-METHYLUMBEL-LIFERON (CZECH) □ β-METHYLUMBELLIFERONE □ 4-METHYLUMBELLIFERONE □ OMEGA 127 □ PILOT 447
CONSENSUS REPORTS: Reported in EPA TSCA Inventory. EPA Genetic Toxicology Program.
SAFETY PROFILE: Moderately toxic by ingestion and intraperitoneal routes. An experimental teratogen. Experimental reproductive effects. When heated to decomposition it emits acrid smoke and irritating fumes.

MKQ600   CAS: 174175-11-2 HR: 3
(+−)-(E)-METHYL-2-((E)-HYDRO-XYIMINO)-5-NITRO-6-METHOXY-3-HEXENEAMIDE
mf: $C_8H_{13}N_3O_5$ mw: 231.21
SYNS: 3-HEXENAMIDE, 2-(HYDROXYIMINO)-6-METHOXY-4-METHYL-5-NITRO-, (2E,3E)- □ NOR-1
SAFETY PROFILE: Suspected carcinogen. When heated to decomposition it emits toxic vapors of NO_x.

MKV500   CAS: 876-83-5   HR: 3
2-METHYL-1,3-INDANDIONE
mf: $C_{10}H_8O_2$ mw: 160.18
PROP: A solid. Mp: 83–84°.
SAFETY PROFILE: Poison by intraperitoneal route. An experimental teratogen. Experimental reproductive effects. When heated to decomposition it emits acrid smoke and irritating fumes.

MKV750   CAS: 83-34-1   HR: 3
β-METHYLINDOLE
mf: C_9H_9N mw: 131.19
PROP: Leaves from ligroin. Mp: 95°, bp: 265° @ 755 mm. Sol in cold water, alc, chloroform, ether, and benzene.
SYNS: 3-METHYLINDOLE □ 3-METHYL-1H-INDOLE □ 3-MI □ SCATOLE □ SKATOL □ SKATOLE

CONSENSUS REPORTS: Reported in EPA TSCA Inventory.
SAFETY PROFILE: Poison by ingestion, intravenous, and intraperitoneal routes. Moderately toxic by subcutaneous route. When heated to decomposition it emits toxic fumes of NO_x.

MKW200   CAS: 74-88-4   HR: 3
METHYL IODIDE
DOT: UN 2644
mf: CH_3I mw: 141.94
PROP: Colorless liquid with pleasant odor, turns brown on exposure to light. Mp: −66.4°, bp: 42.5°, d: 2.279 @ 20°/4°, vap press: 400 mm @ 25.3°, vap d: 4.89. Sol in water @ 15°, misc in alc and ether. IDLH 100 ppm.
SYNS: HALON 10001 □ IODOMETANO (ITALIAN) □ IODOMETHANE □ IODURE de METHYLE (FRENCH) □ JOD-METHAN (GERMAN) □ JOODMETHAAN (DUTCH) □ METHYLJODID (GERMAN) □ METHYLJODIDE (DUTCH) □ METYLU JODEK (POLISH) □ MONOIODURO di METILE (ITALIAN) □ RCRA WASTE NUMBER U138
CONSENSUS REPORTS: IARC Cancer Review: Group 3 IMEMDT 7,56,87; Animal Limited Evidence IMEMDT 41,213,86. Community Right-To-Know List. EPA Genetic Toxicology Program. Reported in EPA TSCA Inventory.
OSHA PEL: TWA 2 ppm (skin)
ACGIH TLV: TWA 0.01 ppm; Animal Carcinogen; (Proposed: (inhalable fraction) 0.1 mg/m³; Not Classifiable as a Human Carcinogen)
DFG MAK: Animal Carcinogen, Suspected Human Carcinogen
NIOSH REL: (Monohalomethanes) Reduce to lowest level
DOT CLASSIFICATION: 6.1; Label: Poison
SAFETY PROFILE: Confirmed carcinogen with experimental neoplastigenic and tumorigenic data. A poison by ingestion, intraperitoneal, and subcutaneous routes. Moderately toxic by inhalation and skin contact. A human skin irritant. Human mutation data reported. A strong narcotic and anesthetic. Explosive reaction with

trialkylphosphines, silver chlorite. Violent reaction with oxygen (at 300°C), sodium. When heated to decomposition it emits toxic fumes of I⁻. See also IODIDES.

MKW450   CAS: 110-12-3   HR: 3
METHYL ISOAMYL KETONE
DOT: UN 2302
mf: $C_7H_{14}O$ mw: 114.21
PROP: Colorless, stable liquid; pleasant odor. Bp: 144°, d: 0.8132 @ 20°/20°, fp: −73.9°, flash p: 110°F (OC). Sltly sol in water; misc with most org solvs.
SYNS: ISOAMYL METHYL KETONE □ ISOPENTYL METHYL KETONE □ KETONE, METHYL ISOAMYL □ 2-METHYL-5-HEXANONE □ 5-METHYL-2-HEXANONE □ 5-METHYLHEXAN-2-ONE (DOT) □ MIAK
CONSENSUS REPORTS: Reported in EPA TSCA Inventory.
OSHA PEL: TWA 50 ppm
ACGIH TLV: TWA 50 ppm
NIOSH REL: Ketones (Methyl Isoamyl Ketone) TWA 230 mg/m³
DOT CLASSIFICATION: 3; Label: Flammable Liquid
SAFETY PROFILE: Moderately toxic by ingestion and intraperitoneal routes. Mildly toxic by inhalation and skin contact. A flammable liquid when exposed to heat, flame, or oxidizers. To fight fire, use dry chemical, CO_2, foam, fog. When heated to decomposition it emits acrid smoke and irritating fumes. See also KETONES.

MKW600   CAS: 108-11-2   HR: 3
METHYL ISOBUTYL CARBINOL
DOT: UN 2053
mf: $C_6H_{14}O$ mw: 102.20
PROP: Clear liquid. Bp: 131.8°, fp: <−90° (sets to a glass), flash p: 106°F, d: 0.8079 @ 20°/20°, vap press: 2.8 mm @ 20°, vap d: 3.53, lel: 1.0%, uel: 5.5%. IDLH 400 ppm.
SYNS: ALCOOL METHYL AMYLIQUE (FRENCH) □ ISOBUTYL METHYL CARBINOL □ ISOBUTYLMETHYL-METHANOL □ MAOH □ METHYL AMYL ALCOHOL □ METHYLISOBUTYL CARBINOL □ 2-METHYL-4-PENTANOL □ 4-METHYLPENTANOL-2 □ 4-METHYL-2-PENTANOL (MAK) □ METILAMIL ALCOHOL (ITALIAN) □ 4-METILPENTAN-2-OLO (ITALIAN) □ MIBC □ MIC □ 3-MIC

CONSENSUS REPORTS: Reported in EPA TSCA Inventory.
OSHA PEL: TWA 25 ppm; STEL 40 ppm (skin)
ACGIH TLV: TWA 25 ppm; STEL 40 ppm (skin)
DFG MAK: 25 ppm (110 mg/m³)
DOT CLASSIFICATION: 3; Label: Flammable Liquid
SAFETY PROFILE: Moderately toxic by ingestion, skin contact, and intraperitoneal routes. Mildly toxic by inhalation. A skin and severe eye irritant. Inhalation of high concentrations can cause anesthesia. Flammable liquid when exposed to heat or flame; can react with oxidizing materials. A moderate explosion hazard when exposed to heat or flame. To fight fire, use alcohol foam. When heated to decomposition it emits acrid smoke and fumes. See also ALCOHOLS.

MKX250   CAS: 624-83-9   HR: 3
METHYL ISOCYANATE
DOT: UN 2480
mf: C_2H_3NO mw: 57.06
PROP: Liquid; sharp, unpleasant odor. D: 0.9599 @ 20°/20°, bp: 43–45°, flash p: <5°F. IDLH 3 ppm.
SYNS: ISOCYANATE de METHYLE (FRENCH) □ ISO-CYANATOMETHANE □ ISOCYANIC ACID, METHYL ESTER □ METHYLISOCYANAAT (DUTCH) □ METHYL ISOCYANAT (GERMAN) □ METHYL ISOCYANATE, solutions (DOT) □ METIL ISOCIANATO (ITALIAN) □ MIC □ RCRA WASTE NUMBER P064 □ TL 1450
CONSENSUS REPORTS: Reported in EPA TSCA Inventory.
OSHA PEL: TWA 0.02 ppm (skin)
ACGIH TLV: TWA 0.02 ppm (skin)
DFG MAK: 0.01 ppm (0.025 mg/m³)
DOT CLASSIFICATION: 6.1; Label: Poison, Flammable Liquid; DOT Class: 3; Label: Flammable Liquid, Poison
SAFETY PROFILE: Poison by inhalation, ingestion, and skin contact. Human systemic effects by inhalation: conjunctiva irritation, olfactory and pulmonary changes. An experimental teratogen. Other experimental reproductive effects. Mutation data

M

reported. A severe eye, skin, and mucous membrane irritant and a sensitizer. It can be absorbed through the skin. Exposure to high concentrations of the vapor can cause blindness; lung damage, including edema, permanent fibrosis, emphysema, and bronchitis; and gynecological effects. Most deaths are a result of lung tissue damage. This was the predominant cause of death in the release of MIC in 1984 at Bhopal, India. Effects of cyanide poisoning have been noted but this may be due to impurities. A flammable liquid and a very dangerous fire hazard when exposed to heat, flame, or oxidizers. To fight fire, use spray, foam, CO_2, dry chemical. Exothermic reaction with water. When heated to decomposition it emits toxic fumes of NO_x and CN^-. See also ISOCYANATES.

MKY500   CAS: 814-78-8   HR: 3
METHYL ISOPROPENYL KETONE
DOT: UN 1246
mf: C_5H_8O mw: 84.119
PROP: Oil. Flash p: 69.8°F, d: 0.855 @ 20°/20°, lel: 1.8%, uel: 9.0%, vap d: 2.9, bp: 98°.
SYNS: 3-METHYL-3-BUTEN-2-ON (GERMAN) □ 2-METHYL-1-BUTEN-3-ONE □ METHYL ISOPROPENYL KETONE INHIBITED (DOT)
CONSENSUS REPORTS: Reported in EPA TSCA Inventory.
DOT CLASSIFICATION: 3; Label: Flammable Liquid
SAFETY PROFILE: Poison by ingestion and skin contact. Moderately toxic by inhalation and intraperitoneal routes. A skin and severe eye irritant. A dangerous fire hazard when exposed to heat or flame. Explosive in the form of vapor when exposed to heat or flame. When heated to decomposition it emits acrid smoke and irritating fumes. See also KETONES.

MLA250   CAS: 99-86-5   HR: 2
1-METHYL-4-ISOPROPYLCYCLO-HEXADIENE-1,3
mf: $C_{10}H_{16}$ mw: 136.26

PROP: Colorless liquid or oil; lemon odor. D: 0.834 @ 20°/4°, refr index: 1.475–1.480, bp: 173.5–174.8° @ 755 mm. Insol in water; misc in alc, ether, fixed oils.
SYNS: FEMA No. 3558 □ p-MENTHA-1,3-DIENE □ 1-METHYL-4-ISOPROPYL-1,3-CYCLOHEXADIENE □ α-TERPINENE (FCC)
CONSENSUS REPORTS: Reported in EPA TSCA Inventory.
SAFETY PROFILE: Moderately toxic by ingestion. When heated to decomposition it emits acrid smoke and irritating fumes.

MLA300   HR: D
5-METHYL-2-ISOPROPYL-2-HEXENAL
mf: $C_{10}H_{18}O$ mw: 154.25
PROP: Sltly yellow liquid; herbaceous, woody, fruity, chocolate odor. D: 0.845–0.860, refr index: 1.448. Sol in alc, fixed oils; insol in water, propylene glycol.
SYN: FEMA No. 3406
SAFETY PROFILE: When heated to decomposition it emits acrid smoke and irritating fumes.

MLA750   CAS: 563-80-4   HR: 3
METHYL ISOPROPYL KETONE
DOT: UN 2397
mf: $C_5H_{10}O$ mw: 86.15
PROP: Colorless liquid; acetone-like odor. D: 0.805 @ 16°/4°, bp: 93–94°.
SYNS: ISOPROPYL METHYL KETONE □ 3-METHYL-2-BUTANONE □ 3-METHYL BUTAN-2-ONE (DOT) □ MIPK
CONSENSUS REPORTS: Reported in EPA TSCA Inventory.
OSHA PEL: TWA 200 ppm
ACGIH TLV: TWA 200 ppm
DOT CLASSIFICATION: 3; Label: Flammable Liquid
SAFETY PROFILE: Poison by ingestion. Mildly toxic by inhalation and skin contact. Mutation data reported. A skin and eye irritant. Flammable when exposed to heat or flame; can react vigorously with oxidizing materials. When heated to decomposition it emits acrid smoke and irritating fumes. See also KETONES.

MLC750　　CAS: 75-86-5　　HR: 3
2-METHYLLACTONITRILE
DOT: UN 1541
mf: C_4H_7NO　　mw: 85.12
PROP: Mp: $-20°$, bp: $82°$ @ 23 mm, d:
0.932 @ $19°$, autoign temp: $1270°F$, flash p:
$165°F$, vap d: 2.93. Very sol in H_2O; spar sol
in pet ether.
SYNS: ACETONCIANHIDRINEI (ROUMANIAN) □
ACETONCIANIDRINA (ITALIAN) □ ACETONCYA-
ANHYDRINE (DUTCH) □ ACETONCYANHYDRIN
(GERMAN) □ ACETONECYANHYDRINE (FRENCH) □
ACETONE CYANOHYDRIN (ACGIH,DOT) □
ACETONKYANHYDRIN (CZECH) □ CYANHYDRINE
d'ACETONE (FRENCH) □ α-HYDROXYISOBUTYR-
ONITRILE □ 2-HYDROXY-2-METHYLPROPIONITRILE
□ RCRA WASTE NUMBER P069 □ USAF RH-8
CONSENSUS REPORTS: Cyanide and its
compounds are on the Community Right-
To-Know List. Reported in EPA TSCA
Inventory. EPA Extremely Hazardous
Substances List.
ACGIH TLV: CL 4.7 ppm (skin)
NIOSH REL: (Nitriles) CL 4 mg/m³/15M
DOT CLASSIFICATION: 6.1; Label: Poison
SAFETY PROFILE: Poison by ingestion,
skin contact, inhalation, intraperitoneal, and
subcutaneous routes. Readily decomposes to
HCN and acetone. Keep cool and do not
store for long periods. Combustible when
exposed to heat or flame. To fight fire, use
CO_2, dry chemical, alcohol foam. Vigorous
reaction with H_2SO_4. When heated to
decomposition it emits toxic fumes of CN^-.
See also NITRILES, HYDROCYANIC
ACID, and ACETONE.

MLE000　　CAS: 75-16-1　　HR: 3
METHYLMAGNESIUM BROMIDE (ethyl
**　　ether solution)**
DOT: UN 1928
mf: CH_3BrMg　　mw: 119.26
PROP: Sol in Et_2O and THF; insol in
hydrocarbons. Concentration of ethyl ether
is not over 40%.
SYN: METHYL MAGNESIUM BROMIDE in ETHYL
ETHER (DOT)
CONSENSUS REPORTS: Reported in EPA
TSCA Inventory.

DOT CLASSIFICATION: 4.3; Label:
Dangerous When Wet, Flammable Liquid
SAFETY PROFILE: May ignite
spontaneously in air. A very dangerous fire
hazard when exposed to heat or flame; can
react vigorously with oxidizing materials.
When heated to decomposition it emits
acrid smoke and irritating fumes. See also
ETHERS, MAGNESIUM COMPOUNDS,
and BROMIDES.

MLE650　　CAS: 74-93-1　　HR: 3
METHYL MERCAPTAN
DOT: UN 1064
mf: CH_4S　　mw: 48.11
PROP: Gas; odor of rotten cabbage. Mp:
$-123.1°$, vap d: 1.66, lel: 3.9%, uel: 21.8%,
bp: $5.95°$, d: 0.8665 @ $20°/4°$, solidifies @
$-123°$, flash p: $-0.4°F$. Sol in water. IDLH
150 ppm. IDLH 10 mg/m³ (as Hg).
SYNS: MERCAPTAN METHYLIQUE (FRENCH) □
METANTIOLO (ITALIAN) □ METHAANTHIOL (DUTCH)
□ METHANETHIOL □ METHANTHIOL (GERMAN) □
METHYLMERCAPTAAN (DUTCH) □ METILMERCAP-
TANO (ITALIAN) □ RCRA WASTE NUMBER U153 □
THIOMETHANOL
CONSENSUS REPORTS: EPA Extremely
Hazardous Substances List. Reported in
EPA TSCA Inventory.
OSHA PEL: TWA 0.5 ppm
DFG MAK: 0.5 ppm (1 mg/m³)
NIOSH REL: (n-Alkane Monothiols) CL 0.5
ppm/15M
DOT CLASSIFICATION: 2.3; Label: Poison
Gas, Flammable Gas
SAFETY PROFILE: Poison by inhalation.
Mutation data reported. A common air
contaminant. Very dangerous fire hazard
when exposed to heat or flame; can react
vigorously with oxidizing materials.
Explosive in the form of vapor when
exposed to heat or flame. Reacts with water,
steam, or acids to produce toxic and
flammable vapors. Violent reaction with
mercury(II) oxide. To fight fire, use alcohol
foam, CO_2, dry chemical. Upon
decomposition it emits highly toxic fumes of
SO_x. See also MERCAPTANS.

M

MLF250   CAS: 502-39-6   HR: 3
METHYLMERCURIC DICYANDIAMIDE
mf: $C_3H_6HgN_4$ mw: 298.72
PROP: IDLH 10 mg/m³ (as Hg).
SYNS: AGROSOL □
CYANO(METHYLMERCURI)GUANIDINE □ GUANID-
INE, CYANO-, METHYLMERCURY deriv. □ MEMA □
METHYLMERCURIC CYANOGUANIDINE □ METHYL-
MERCURIC DICYANDIAMIDE □ METHYLMERCURY
DICYANDIAMIDE □ METHYLMERKURIDIKYANDI-
AMID □ MMD □ MORSODREN □ MORTON EP-227 □
MORTON SOIL DRENCH □ MORTON SOIL-DRENCH-C
□ PANDRINOX □ PANO-DRENCH 4 □ PANODRIN A-13
□ PANOGEN □ PANOGEN 15 □ PANOGEN 43 □
PANOGEN PX □ PANOGEN TURF FUNGICIDE □
PANOGEN TURF SPRAY □ PANOSPRAY 30 □ R 8 □ R 8
(fungicide) □ ZAPRAWA NASIENNA PLYNNA
CONSENSUS REPORTS: Mercury and its
compounds are on the Community Right-
To-Know List. EPA Extremely Hazardous
Substances List.
OSHA PEL: TWA 0.01 mg(Hg)/m³; STEL
0.03 mg/m³ (skin)
ACGIH TLV: TWA 0.01 mg(Hg)/m³ (skin);
STEL 0.03 mg(Hg)/m³; BEI: 35 µg/g
creatinine total inorganic mercury in urine
preshift; 15 µg/g creatinine total inorganic
mercury in blood at end of shift at end of
workweek.
DFG MAK: Confirmed Animal Carcinogen
with Unknown Relevance to Humans
NIOSH REL: (Mercury, Organo) TWA 0.01
mg/m³; STEL 0.03 mg/m³ (skin)
SAFETY PROFILE: Poison by ingestion and
intraperitoneal routes. An experimental
teratogen. Experimental reproductive
effects. Mutation data reported. When
heated to decomposition it emits very toxic
fumes of Hg and NO_x. See also MERCURY
COMPOUNDS.

MLF500   CAS: 5902-79-4   HR: 3
METHYLMERCURICHLORENDIMIDE
mf: $C_{10}H_5Cl_6HgNO_2$ mw: 584.45
PROP: IDLH 10 mg/m³ (as Hg).
SYNS: MEMMI □ N-(METHYLMERCURI)-1,4,5,6,7,7-
HEXACHLOROBICYCLO(2.2.1)HEPT-5-ENE-2,3-
DICARBOXIMIDE □ N-METHYLMERCURI-1,2,3,6-
TETRAHYDRO-3,6-ENDOMETHANO-3,4,5,6,7,7-
HEXACHLOROPHTHALIMIDE □ N-METHYLMERCURI-

1,2,3,6-TETRAHYDRO-3,6-METHANO-3,4,5,6,7,7-
HEXACHLOROPHTHALIMIDE
CONSENSUS REPORTS: Mercury and its
compounds are on the Community Right-
To-Know List.
OSHA PEL: TWA 0.01 mg(Hg)/m³; STEL
0.03 mg/m³ (skin)
ACGIH TLV: TWA 0.01 mg(Hg)/m³; BEI:
35 µg/g creatinine total inorganic mercury in
urine preshift; 15 µg/g creatinine total
inorganic mercury in blood at end of shift at
end of workweek.
DFG MAK: Confirmed Animal Carcinogen
with Unknown Relevance to Humans
NIOSH REL: (Mercury, Aryl and Inorganic)
CL 0.1 mg/m³ (skin)
SAFETY PROFILE: Poison by ingestion.
When heated to decomposition it emits very
toxic fumes of Cl^-, Hg, and NO_x. See also
MERCURY COMPOUNDS and
CHLORIDES.

MLF550   CAS: 22967-92-6   HR: 3
METHYLMERCURY
mf: CH_3Hg mw: 215.63
SYNS: METHYL-MERCURY(1+) (9CI) □ METHYL-
MERCURY(I) CATION □ METHYLMERCURY ION □
METHYLMERCURY ION(1+)
CONSENSUS REPORTS: Mercury and its
compounds are on the Community Right-
To-Know List.
OSHA PEL: TWA 0.01 mg(Hg)/m³; STEL
0.03 mg/m³ (skin)
ACGIH TLV: TWA 0.01 mg(Hg)/m³; BEI:
35 µg/g creatinine total inorganic mercury in
urine preshift; 15 µg/g creatinine total
inorganic mercury in blood at end of shift at
end of workweek.
DFG MAK: Confirmed Animal Carcinogen
with Unknown Relevance to Humans
NIOSH REL: (Mercury, Organo) TWA 0.01
mg/m³; STEL 0.03 mg/m³ (skin)
SAFETY PROFILE: A poison. An
experimental teratogen. Experimental
reproductive effects. Mutation data
reported. Used as a fungicide. When heated
to decomposition it emits toxic fumes of
Hg. See also MERCURY COMPOUNDS,
ORGANIC.

MLG000   CAS: 1184-57-2   HR: 3
METHYLMERCURY HYDROXIDE
mf: CH_4HgO mw: 232.64
PROP: Flakes with unpleasant odor. Mp: 137°. IDLH 10 mg/m³ (as Hg).
SYNS: HYDROXYMETHYLMERCURY □ METHYL-MERCURIC HYDROXIDE
CONSENSUS REPORTS: Mercury and its compounds are on the Community Right-To-Know List. EPA Genetic Toxicology Program.
OSHA PEL: TWA 0.01 mg(Hg)/m³; STEL 0.03 mg/m³ (skin)
ACGIH TLV: TWA 0.01 mg(Hg)/m³; BEI: 35 µg/g creatinine total inorganic mercury in urine preshift; 15 µg/g creatinine total inorganic mercury in blood at end of shift at end of workweek.
DFG MAK: Confirmed Animal Carcinogen with Unknown Relevance to Humans
NIOSH REL: (Mercury, Organo) TWA 0.01 mg/m³; STEL 0.03 mg/m³ (skin)
SAFETY PROFILE: Poison by intraperitoneal route. An experimental teratogen. Experimental reproductive effects. Human mutation data reported. When heated to decomposition it emits toxic fumes of Hg. See also MERCURY COMPOUNDS.

MLH000   CAS: 86-85-1   HR: 3
METHYLMERCURY QUINOLINOLATE
mf: $C_{10}H_9HgNO$ mw: 359.79
PROP: Pale-yellow crystals from hexane. Mp: 90°. IDLH 10 mg/m³ (as Hg).
SYNS: ARTHO LM □ LIQUI-SAN □ LM SEED PROTECTANT □ METASOL □ METAZOL □ 8-(METHYLMERCURIOXY)QUINOLINE □ METHYL-MERCURY β-HYDROXYQUINOLATE □ METHYL-MERCURY 8-HYDROXYQUINOLINATE □ METHYL-MERCURY OXINATE □ METHYLMERCURY OXYQUINOLINATE □ ORTHO-LM APPLE SPRAY □ ORTHO LM CONCENTRATE □ ORTHO LM SEED PROTECTANT □ 8-(QUINOLINOLATO)METHYL MERCURY □ 8-QUINOLINOL, MERCURY COMPLEX
CONSENSUS REPORTS: Mercury and its compounds are on the Community Right-To-Know List.
OSHA PEL: TWA 0.01 mg(Hg)/m³; STEL 0.03 mg/m³ (skin)
ACGIH TLV: TWA 0.01 mg(Hg)/m³; BEI: 35 µg/g creatinine total inorganic mercury in urine preshift; 15 µg/g creatinine total inorganic mercury in blood at end of shift at end of workweek.
DFG MAK: Confirmed Animal Carcinogen with Unknown Relevance to Humans
NIOSH REL: (Mercury, Aryl and Inorganic) CL 0.1 mg/m³ (skin)
SAFETY PROFILE: Poison by ingestion. A pesticide. When heated to decomposition it emits very toxic fumes of Hg and NO$_x$. See also MERCURY COMPOUNDS.

MLH100   CAS: 102280-93-3 HR: 3
METHYL-MERCURY TOLUENE-
   SULPHAMIDE
mf: $C_9H_{14}Hg_2N_2O_2S$ mw: 615.49
PROP: IDLH 10 mg/m³ (as Hg).
SYNS: N,N-BIS(METHYLQUECKSILBER)-p-TOLUOL-SULFAMID □ MERCURY, (((p-TOLYL)SULFAMOYL)-IMINO)BIS(METHYL)- □ METHYL-QUECKSILBER-TOLUOLSULFAMID □ (((p-TOLYL)SULFAMOYL)IMINO)-BIS(METHYLMERCURY)
ACGIH TLV: TWA 0.01 mg(Hg)/m³; BEI: 35 µg/g creatinine total inorganic mercury in urine preshift; 15 µg/g creatinine total inorganic mercury in blood at end of shift at end of workweek.
DFG MAK: Confirmed Animal Carcinogen with Unknown Relevance to Humans
SAFETY PROFILE: Poison by ingestion. When heated to decomposition it emits toxic fumes of NO$_x$, SO$_x$, and Hg.

MLH500   CAS: 66-27-3   HR: 3
METHYL MESYLATE
mf: $C_2H_6O_3S$ mw: 110.14
PROP: A liquid. D: 1.29 @ 20°/4°, bp: 203° @ 753 mm. Decomp in water. Sol in alc and ether.
SYNS: as-DIMETHYL SULPHATE □ METHANE-SULPHONIC ACID METHYL ESTER □ METHYL ESTER of METHANESULFONIC ACID □ METHYL ESTER of METHANESULPHONIC ACID □ METHYL METHANE-SULFONATE □ METHYL METHANESULPHONATE □ METHYL METHANSULFONAT (GERMAN) □ METHYL METHANSULFONATE □ METHYL METHANSULPHON-ATE □ MMS □ NSC-50256

CONSENSUS REPORTS: NTP 10th Report on Carcinogens. IARC Cancer Review: Group 3 IMEMDT 7,56,87; Animal Sufficient Evidence IMEMDT 7,253,74. Reported in EPA TSCA Inventory. EPA Genetic Toxicology Program.
SAFETY PROFILE: Confirmed carcinogen with carcinogenic and neoplastigenic data. Poison by ingestion, intraperitoneal, intravenous, and subcutaneous routes. Human mutation data reported. Experimental teratogenic and reproductive effects. When heated to decomposition it emits toxic fumes of SO_x. See also SULFONATES and ESTERS.

MLH750   CAS: 80-62-6   HR: 3
METHYL METHACRYLATE
DOT: NA 1247
mf: $C_5H_8O_2$ mw: 100.13
PROP: Colorless liquid; sharp, fruity odor. Mp: $-50°$, bp: $101.0°$, flash p: $50°F$ (OC), d: 0.936 @ $20°/4°$, vap press: 40 mm @ $25.5°$, vap d: 3.45, lel: 2.1%, uel: 12.5%. Very sltly sol in water. Sol in Me_2CO. IDLH 1000 ppm.
SYNS: ACRYLIC ACID, 2-METHYL-, METHYL ESTER □ DIAKON □ METAKRYLAN METYLU (POLISH) □ METHACRYLATE de METHYLE (FRENCH) □ METH-ACRYLIC ACID, METHYL ESTER (MAK) □ METHA-CRYLSAEUREMETHYL ESTER (GERMAN) □ METHYLESTER KYSELINY METHAKRYLOVE □ METHYLMETHACRYLAAT (DUTCH) □ METHYL-METHACRYLAT (GERMAN) □ METHYL METHACRYL-ATE MONOMER, INHIBITED (DOT) □ METHYL-α-METHYLACRYLATE □ METHYL-2-METHYL-2-PROPENOATE □ 2-METHYL-2-PROPENOIC ACID METHYL ESTER □ METIL METACRILATO (ITALIAN) □ MME □ "MONOCITE" METHACRYLATE MONOMER □ NCI-C50680 □ 2-PROPENOIC ACID, 2-METHYL-, METHYL ESTER □ RCRA WASTE NUMBER U162
CONSENSUS REPORTS: IARC Cancer Review: Group 3 IMEMDT 7,56,87; Human Inadequate Evidence IMEMDT 19,187,79; Animal Inadequate Evidence IMEMDT 19,187,79. NTP Carcinogenesis Studies (inhalation); No Evidence: mouse, rat NTPTR* NTP-TR-314,86. Reported in EPA TSCA Inventory. Community Right-To-Know List.

OSHA PEL: TWA 100 ppm
ACGIH TLV: TWA 50 ppm; STEL 100 ppm (sensitizer); Not Classifiable as a Human Carcinogen
DFG MAK: 50 ppm (210 mg/m³)
SAFETY PROFILE: Moderately toxic by inhalation and intraperitoneal routes. Mildly toxic by ingestion. Human systemic effects by inhalation: sleep effects, excitement, anorexia, and blood pressure decrease. Experimental teratogenic and reproductive effects. Mutation data reported. A skin and eye irritant. Questionable carcinogen with experimental tumorigenic data. A common air contaminant.

A very dangerous fire hazard when exposed to heat or flame; can react with oxidizing materials. Explosive in the form of vapor when exposed to heat or flame. The monomer may undergo spontaneous, explosive polymerization. Reacts in air to form a heat-sensitive explosive product (explodes on evaporation at 60°C). May ignite on contact with benzoyl peroxide. Potentially violent reaction with the polymerization initiators azoisobutyronitrile, dibenzoyl peroxide, di-tert-butyl peroxide, propionaldehyde. To fight fire, use foam, CO_2, dry chemical. When heated to decomposition it emits acrid smoke and irritating fumes. See also ESTERS.

MLI900   CAS: 143390-89-0 HR: 2
METHYL (E)-2-METHOXYIMINO-(2-(o-TOLYLOXYMETHYL)PHENYL)ACETATE
mf: $C_{18}H_{19}NO_4$ mw: 313.35
SYNS: BENZENEACETIC ACID, α-(METHOXYIMINO)-2-((2-METHYLPHENOXY)METHYL)-, METHYL ESTER, (α-E)- □ KRESOXIM-METHYL □ KRESOXIM-METHYL TECHNICAL
SAFETY PROFILE: Moderately toxic by skin contact. Low toxicity by ingestion and inhalation. Questionable carcinogen with experimental data reported. When heated to decomposition it emits toxic vapors of NO_x.

MLJ500   CAS: 926-93-2   HR: 3
1-METHYL-6-(1-METHYLALLYL)-2,5-
DITHIOBIUREA
mf: $C_7H_{14}N_4S_2$ mw: 218.37
PROP: A solid. Mp: 198–200° (decomp).
SYNS: AIMAX □ AY-61122 □ COMPOUND 33,828 □
DITHIOCARBAMOYLHYDRAZINE □ I.C.I. 33,828 □
MATCH □ METALLIBURE □ METHALLIBURE □ N-((1-
METHYLALLYL)THIOCARBAMOYL)-N'-(METHYLTHIO-
CARBAMOYL)HYDRAZINE □ 1-α-METHYLALLYLTHIO-
CARBAMOYL-2-METHYLTHIOCARBAMOYLHY-
DRAZINE □ 1-METHYL-6-(1-METHYLALLYL)DITHIOBI-
UREA □ NSC-69536 □ SUISYNCHRON □ TURISYN-
CHRON
SAFETY PROFILE: Poison by intravenous
route. Slightly toxic by ingestion. An
experimental teratogen. Human
reproductive effects by unspecified route:
menstrual cycle changes or disorders.
Experimental reproductive effects. Mutation
data reported. When heated to
decomposition it emits very toxic fumes of
SO_x and NO_x.

MLL100   CAS: 64029-07-8   HR: 3
2-(o-(N-METHYL-N-(N'-METHYL-N'-
ETHOXYCARBONYLAMINOSULFEN
YL)CARBAMOYL)OXIMINO)-1,4-
DITHIANE
mf: $C_{10}H_{17}N_3O_4S_3$ mw: 339.48
SYN: CARBAMIC ACID, (((((1,4-DITHIAN-2-
YLIDENEAMINO)OXY)CARBONYL)METHYLAMINO)TH
IO)METHYL-,ETHYL ESTER
SAFETY PROFILE: A poison by ingestion.
When heated to decomposition it emits
toxic vapors of NO_x and SO_x.

MLL250   CAS: 80-48-8   HR: 3
METHYL-p-METHYLBENZENE-
SULFONATE
mf: $C_8H_{10}O_3S$ mw: 186.24
PROP: Light-brown crystals; crystals of
ethyl ligroin. D: 1.230–1.238 @ 25°/25°,
vap d: 6.45, mp: 28°. Insol in water; sol in
benzene; very sol in alc and ether.
SYNS: METHYLESTER KYSELINY p-TOLUEN-
SULFONOVE (CZECH) □ METHYL-4-METHYLBENZ-
ENESULFONATE □ METHYL-p-TOLUENESULFONATE
□ METHYL TOLUENE-4-SULFONATE □ METHYL
TOSYLATE □ METHYL-p-TOSYLATE □ p-TOLUOL-
SULFONSAEURE METHYL ESTER (GERMAN)

CONSENSUS REPORTS: Reported in EPA
TSCA Inventory.
SAFETY PROFILE: Poison by ingestion and
subcutaneous routes. An eye and severe skin
irritant. A vesicant and skin sensitizer.
Questionable carcinogen with experimental
tumorigenic data. When heated to
decomposition it emits toxic fumes of SO_x.
See also ESTERS and SULFONATES.

MLL600   CAS: 53955-81-0   HR: 3
METHYL 2-METHYLBUTYRATE
mf: $C_6H_{12}O_2$ mw: 116.16
PROP: Colorless liquid; sweet, fruity, apple-
like odor. D: 0.879, refr index: 1.393–1.397,
flash p: 91°F. Sol in alc, fixed oils; insol in
water.
SYNS: FEMA No. 2719 □ METHYL 2-
METHYLBUTANOATE
SAFETY PROFILE: Flammable liquid.
When heated to decomposition it emits
acrid smoke and irritating fumes.

MLO900   CAS: 64029-08-9   HR: 3
METHYL((METHYL(((((5-METHYL-1,3-
OXATHIOLAN-4-YLIDENE)AMINO)-
OXY)CARBONYL)AMINO)THIO)
CARBAMIC ACID, ETHYL ESTER
mf: $C_{10}H_{17}N_3O_5S_2$ mw: 323.42
SAFETY PROFILE: A poison by ingestion.
When heated to decomposition it emits
toxic vapors of NO_x and SO_x.

MLP250   CAS: 36304-84-4   HR: 3
d-3-METHYL-N-METHYLMORPHINAN
PHOSPHATE
mf: $C_{18}H_{25}N•H_3O_4P$ mw: 353.44
PROP: Bitter-tasting crystals. Sltly sol in
H_2O and MeOH; insol in Me_2CO, EtOH,
$CHCl_3$, and Et_2O.
SYNS: ASTOMIN □ AT-17 PHOSPHATE □ DIMEMOR-
FAN PHOSPHATE □ (9-α,13-α,14-α)-3,17-DIMETHYL-
MORPHINAN PHOSPHATE □ 3,17-DIMETHYL-9-α,13-
α,14-α-MORPHINAN PHOSPHATE
SAFETY PROFILE: Poison by intravenous,
intraperitoneal, and subcutaneous routes.
Moderately toxic by ingestion. An
experimental teratogen. Experimental
reproductive effects. Used as an antitussive

M

agent. When heated to decomposition it emits very toxic fumes of NO_x and PO_x. See also PHOSPHATES.

MLW600   CAS: 21621-75-0   HR: 3
METHYL 5-METHYL-1-(2-QUINOLYL)-4-PYRAZOLYL KETONE
mf: $C_{15}H_{13}N_3O$ mw: 251.31
SYN: KETONE, METHYL 5-METHYL-1-(2-QUINOLYL)-4-PYRAZOLYL
DOT CLASSIFICATION: 3; Label: Flammable Liquid
SAFETY PROFILE: A poison by intravenous route. A flammable liquid. When heated to decomposition it emits toxic vapors of NO_x.

MLW630   CAS: 21621-73-8   HR: 3
METHYL 5-METHYL-1-(2-QUINOXALIN-YL)-4-PYRAZOLYL KETONE
mf: $C_{14}H_{12}N_4O$ mw: 252.30
SYN: KETONE, METHYL 5-METHYL-1-(2-QUINOXALINYL)-4-PYRAZOLYL
DOT CLASSIFICATION: 3; Label: Flammable Liquid
SAFETY PROFILE: A poison by intravenous route. A flammable liquid. When heated to decomposition it emits toxic vapors of NO_x.

MLX820   CAS: 62382-23-4   HR: 3
2-METHYL-2-(METHYLTHIO)PROPAN-OL-o-((N-METHYL-N-MORPHOLINO-SULFENYL)CARBAMOYL)OXIME
mf: $C_{11}H_{21}N_3O_3S_2$ mw: 307.47
SYNS: 2-METHYL-2-(METHYLTHIO)PROPANOL-o-((METHYL(4-MORPHOLINYLTHIO)AMINO)CARBON-YL)OXIME □ 2-METHYL-2-(METHYLTHIO)PROPION-ALDEHYDE-o-(N-METHYL-N-(4-MORPHOLINOSUL-FENYL)CARBAMOYL)OXIME □ PROPANOL, 2-METHYL-2-(METHYLTHIO)-, o-((METHYL(4-MORPHOLINYLTHIO)AMINO)CARBONYL)OXIME
SAFETY PROFILE: A poison by ingestion. When heated to decomposition it emits toxic vapors of NO_x and SO_x.

MMA250   CAS: 109-02-4   HR: 3
N-METHYL MORPHOLINE
DOT: UN 2535
mf: $C_5H_{11}NO$ mw: 101.17

PROP: Liquid. Flash p: 75.2°F, d: 0.9, vap d: 3.5, bp: 115°.
SYNS: METHYLMORPHOLINE (DOT) □ 4-METHYLMORPHOLINE □ MORPHOLINE, N-METHYL-
CONSENSUS REPORTS: Reported in EPA TSCA Inventory.
DOT CLASSIFICATION: 3; Label: Flammable Liquid, Corrosive
SAFETY PROFILE: Moderately toxic by ingestion and skin contact. Mildly toxic by inhalation. An irritant to skin, eyes, and mucous membranes. Flammable when exposed to heat or flame, can react vigorously with oxidizing materials. When heated to decomposition it emits toxic fumes of NO_x.

MMA600   CAS: 110147-48-3   HR: 3
METHYL 10-(3-MORPHOLINOPROPYL)-PHENOTHIAZIN-2-YL KETONE
mf: $C_{21}H_{24}N_2O_2S$ mw: 368.53
SYNS: 3-ACETYL-10-(3'-MORPHOLINOPROPYL)-PHENOTHIAZIN □ KETONE, METHYL 10-(3-MORPHOLINOPROPYL)PHENOTHIAZIN-2-YL
DOT CLASSIFICATION: 3; Label: Flammable Liquid
SAFETY PROFILE: A poison by intravenous route. Moderately toxic by ingestion. A flammable liquid. When heated to decomposition it emits toxic vapors of NO_x and SO_x.

MMD500   CAS: 58-27-5   HR: 3
2-METHYL-1,4-NAPHTHOQUINONE
mf: $C_{11}H_8O_2$ mw: 172.19
PROP: Bright-yellow crystals from ligroin, AcOH (aq), or EtOH. Insol in H_2O; spar sol in EtOH.
SYNS: AQUAKAY □ AQUINONE □ HEMODAL □ KAERGONA □ KANONE □ KAPPAXAN □ KARCON □ KAREON □ KATIV-G □ KAYKLOT □ KAYQUINONE □ KIPCA □ KLOTTONE □ KOAXIN □ KOLKLOT □ K-THROMBYL □ K-VITAN □ MENADION □ MENADI-ONE □ MENAPHTHON □ MENAPHTONE □ 2-METHYL-1,4-NAPHTHALENDIONE □ 2-METHYL-1,4-NAPHTHALENEDIONE □ 2-METHYL-1,4-NAPHTH-OCHINON (GERMAN) □ 3-METHYL-1,4-NAPHTHO-QUINONE □ MITENON □ MNQ □ NSC-4170 □ PANOSINE □ PROKAYVIT □ SYNKAY □ THYLO-QUINONE □ USAF EK-5185 □ VITAMIN K2(O) □ VITAMIN K3

CONSENSUS REPORTS: Reported in EPA TSCA Inventory.

SAFETY PROFILE: Poison by ingestion, intraperitoneal, and subcutaneous routes. Experimental teratogenic effects. Questionable carcinogen with experimental tumorigenic data. Human mutation data reported. When heated to decomposition it emits acrid smoke and irritating fumes.

MME500   CAS: 10546-24-4   HR: 3
3-METHYL-2-NAPHTHYLAMINE
mf: $C_{11}H_{11}N$ mw: 157.23
PROP: Crystals from pet ether. Mp: 135–135.5°.
SAFETY PROFILE: Suspected carcinogen with experimental carcinogenic and tumorigenic data. When heated to decomposition it emits toxic fumes of NO_x. See also AMINES.

MME809   CAS: 5903-13-9   HR: 3
N-METHYL-N-(1-NAPHTHYL)FLUORO-
ACETAMIDE
mf: $C_{13}H_{12}FNO$ mw: 217.26
SYNS: 1-(N-ACETAMIDOFLUOROMETHYL)-NAPHTH-ALENE □ DP X 1410 □ FAM □ 2-FLUORO-N-METHYL-N-1-NAPHTHALENYLACETAMIDE □ 2-FLUORO-N-METHYL-N-1-NAPHTHYLACETAMIDE □ N-METHYL-N-(1-NAPHTHYL)MONOFLUOROACETAMIDE □ MFNA □ MNFA □ NISSOL EC
SAFETY PROFILE: Poison by ingestion, skin contact, subcutaneous, intravenous, and intraperitoneal routes. Experimental reproductive effects. When heated to decomposition it emits very toxic fumes of F^- and NO_x.

MMF500   CAS: 598-58-3   HR: 3
METHYL NITRATE
mf: CH_3NO_3 mw: 77.05
PROP: Colorless liquid. Bp: 65° (explodes), d: 1.208 @ 20°/4°, vap d: 2.66, mp: −83°. Sltly sol in water; sol in alc, ether.
SYN: NITRIC ACID METHYL ESTER
DOT CLASSIFICATION: Forbidden
SAFETY PROFILE: Poison by ingestion. Moderately toxic by inhalation. A dangerous fire and explosion hazard by spontaneous chemical reaction. A very shock- and heat-sensitive explosive. Explodes when heated to 65°C. It does not require external O_2 for combustion. A rocket fuel. When heated to decomposition it emits toxic fumes of NO_x. See also NITRATES.

MMF750   CAS: 624-91-9   HR: 3
METHYL NITRITE
mf: CH_3NO_2 mw: 61.05
PROP: Gas above 10.4°F. Mp: −17°, bp: −12°, d: 0.991 @ 15°. Sol in alc, ether.
SYN: NITROUS ACID, METHYL ESTER
CONSENSUS REPORTS: Reported in EPA TSCA Inventory.
DOT CLASSIFICATION: Forbidden
SAFETY PROFILE: Moderately toxic by inhalation. Mutation data reported. Narcotic in high concentration. A very dangerous fire and explosion hazard when exposed to heat or flame. A heat-sensitive explosive more powerful than ethyl nitrite. When heated to decomposition it emits toxic fumes of NO_x. See also NITRITES and n-AMYL NITRITE.

M

MMG000   CAS: 129-15-7   HR: 3
2-METHYL-1-NITROANTHRAQUINONE
mf: $C_{15}H_9NO_4$ mw: 267.25
SYNS: 2-METHYL-1-NITRO-9,10-ANTHRACENE-DIONE □ NCI-C01923 □ 1-NITRO-2-METHYL-ANTHRAQUINONE □ 1-N-2-MA (RUSSIAN)
CONSENSUS REPORTS: IARC Cancer Review: Group 2B IMEMDT 7,56,87; Animal Sufficient Evidence IMEMDT 27,205,82. NCI Carcinogenesis Bioassay (feed); Clear Evidence: rat NCITR* NCI-CG-TR-29,78; (feed); Clear Evidence: mouse IJCNAW 19,117,77.
SAFETY PROFILE: Confirmed carcinogen with experimental carcinogenic and neoplastigenic data. Moderately toxic by intraperitoneal route. Mutation data reported. When heated to decomposition it emits toxic fumes of NO_x. See also NITRO COMPOUNDS of AROMATIC HYDROCARBONS.

MMN250   CAS: 443-48-1   HR: 3
2-METHYL-5-NITROIMIDAZOLE-1-
ETHANOL

mf: $C_6H_9N_3O_3$ mw: 171.18

PROP: Cream-colored crystals from EtOAc.
Mp: 158–160°.

SYNS: ACROMONA □ ANAGIARDIL □ ATRIVYL □
BAYER 5360 □ BEXON □ CLONT □ CONT □ DANIZOL
□ DEFLAMON-WIRKSTOFF □ EFLORAN □ ELYZOL □
ENTIZOL □ 1-(β-ETHYLOL)-2-METHYL-5-NITRO-3-
AZAPYRROLE □ EUMIN □ FLAGEMONA □ FLAGESOL
□ FLAGIL □ FLAGYL □ GIATRICOL □ GINEFLAVIR □
1-HYDROXYETHYL-2-METHYL-5-NITROIMIDAZOLE □
1-(β-HYDROXYETHYL)-2-METHYL-5-NITROIMIDAZ-
OLE □ 1-(2-HYDROXYETHYL)-2-METHYL-5-NITRO-
IMIDAZOLE □ 1-(2-HYDROXY-1-ETHYL)-2-METHYL-5-
NITROIMIDAZOLE □ KLION □ MERONIDAL □ 2-
METHYL-1-(2-HYDROXYETHYL)-5-NITROIMIDAZOLE
□ 2-METHYL-3-(2-HYDROXYETHYL)-4-NITROIMIDAZ-
OLE □ METRONIDAZ □ METRONIDAZOL □
METRONIDAZOLO □ MONAGYL □ NALOX □ NEO-
TRIC □ NIDA □ NOVONIDAZOL □ NSC-50364 □
ORVAGIL □ 1-(β-OXYETHYL)-2-METHYL-5-NITRO-
IMIDAZOLE □ RP 8823 □ SANATRICHOM □ SC 10295 □
TRICHAZOL □ TRICHOCIDE □ TRICHOMOL □
TRICHOMONACID "PHARMACHIM" □ TRICHOPOL □
TRICOM □ TRICOWAS B □ TRIKOJOL □ TRIMEKS □
TRIVAZOL □ VAGILEN □ VAGIMID □ VERTISAL

CONSENSUS REPORTS: NTP 10th Report
on Carcinogens. IARC Cancer Review:
Group 2B IMEMDT 7,250,87; Animal
Sufficient Evidence IMEMDT 13,113,77.
EPA Genetic Toxicology Program.

SAFETY PROFILE: Confirmed carcinogen
with experimental carcinogenic,
neoplastigenic, tumorigenic, and teratogenic
data. Moderately toxic by ingestion,
intraperitoneal, and subcutaneous routes.
Human systemic effects by ingestion:
paresthesia, nerve or sheath structural
changes, eye changes, tremors, fever,
jaundice and other liver changes, hearing-
acuity changes, somnolence, and ataxia.
Experimental reproductive effects. Human
mutation data reported. When heated to
decomposition it emits toxic fumes of NO_x.

MMP000   CAS: 70-25-7   HR: 3
N-METHYL-N'-NITRO-N-NITROSO-
GUANIDINE

DOT: NA 0473

mf: $C_2H_5N_5O_3$ mw: 147.12

PROP: Crystals from MeOH. Mp: 118°
(decomp), bp: 89–97° @ 225°.

SYNS: GUANIDINE, N-METHYL-N'-NITRO-N-
NITROSO-(9CI) □ METHYLNITRONITROSOGUANID-
INE □ N-METHYL-N'-NITRO-N-NITROSOGUANIDINE
□ 1-METHYL-3-NITRO-1-NITROSOGUANIDINE □ N-
METHYL-N-NITROSONITROGUANIDIN □ N-METHYL-
N-NITROSO-N'-NITROGUANIDINE □ 1-METHYL-1-
NITROSO-3-NITROGUANIDINE □ N-METYLO-N'-
NITRO-N-NITROZOGUANIDYNY □ MNG □ MNNG □
N'-NITRO-N-NITROSO-N-METHYLGUANIDINE □ N-
NITROSO-N-METHYLNITROGUANIDINE □
NITROSOGUANIDINE (DOT) □ NSC-9369 □ RCRA
WASTE NUMBER U163

CONSENSUS REPORTS: NTP 10th Report
on Carcinogens. IARC Cancer Review:
Group 2A IMEMDT 7,248,87; Animal
Sufficient Evidence IMEMDT 4,183,74.
Reported in EPA TSCA Inventory. EPA
Genetic Toxicology Program.

DOT CLASSIFICATION: EXPLOSIVE
1.1A; Label: EXPLOSIVE 1.1A

SAFETY PROFILE: Confirmed carcinogen
with experimental carcinogenic,
tumorigenic, and teratogenic data. Poison by
ingestion, intraperitoneal, and intravenous
routes. Moderately toxic by subcutaneous
route. Experimental reproductive effects.
Human mutation data reported. An
explosive sensitive to heat or impact.
Flammable when exposed to heat or flame;
can react vigorously with oxidizing
materials. When heated to decomposition it
emits very toxic fumes of NO_x. See also N-
NITROSO COMPOUNDS.

MMS200   CAS: 60153-49-3   HR: 3
3-METHYLNITROSAMINOPROPIO-
NITRILE

mf: $C_4H_7N_3O$ mw: 113.14

SYNS: MNPN □ PROPANENITRILE, 3-(METHYL-
NITROSOAMINO)-

CONSENSUS REPORTS: IARC Cancer
Review: Group 2B IMEMDT 7,56,87,
Animal Sufficient Evidence IMEMDT
37,263,85.

SAFETY PROFILE: Confirmed carcinogen
with experimental carcinogenic and

tumorigenic data. When heated to decomposition it emits toxic fumes of NO_x.

MMS500   CAS: 64091-91-4   HR: 3
4-(N-METHYL-N-NITROSAMINO)-1-(3-PYRIDYL)-1-BUTANONE
mf: $C_{10}H_{13}N_3O_2$ mw: 207.26

SYNS: 4-(N-METHYL-N-NITROSOAMINO)-4-(3-PYRIDYL)-1-BUTANONE □ N-METHYL-N-NITROSO-4-OXO-4-(3-PYRIDYL)BUTYL AMINE □ 4-(NITROSO-AMINO-N-METHYL)-1-(3-PYRIDYL)-1-BUTANONE □ 4-(N-NITROSO-N-METHYLAMINO)-1-(3-PYRIDYL)-1-BUTANONE □ NNK

CONSENSUS REPORTS: NTP 10th Report on Carcinogens. IARC Cancer Review: Group 2B IMEMDT 7,56,87; Animal Sufficient Evidence IMEMDT 37,209,85. EPA Genetic Toxicology Program.

DOT CLASSIFICATION: 3; Label: Flammable Liquid

SAFETY PROFILE: Confirmed carcinogen with experimental carcinogenic and neoplastigenic data. An experimental teratogen. Mutation data reported. A flammable liquid. When heated to decomposition it emits toxic fumes of NO_x. See also N-NITROSO COMPOUNDS.

MMT000   CAS: 7417-67-6   HR: 3
METHYLNITROSOACETAMIDE
mf: $C_3H_6N_2O_2$ mw: 102.11

SYNS: METHYLNITROSOACETAMID (GERMAN) □ N-METHYL-N-NITROSOACETAMIDE □ N-NITROSO-N-METHYLACETAMIDE

CONSENSUS REPORTS: EPA Genetic Toxicology Program.

SAFETY PROFILE: Poison by ingestion. Questionable carcinogen with experimental tumorigenic data. Mutation data reported. When heated to decomposition it emits toxic fumes of NO_x. See also N-NITROSO COMPOUNDS.

MMT500   CAS: 4549-43-3   HR: 3
N-METHYL-N-NITROSOALLYLAMINE
mf: $C_4H_8N_2O$ mw: 100.14

SYNS: METHYLALLYLNITROSAMIN (GERMAN) □ METHYLALLYLNITROSAMINE □ N-METHYL-N-NITROSO-2-PROPEN-1-AMINE □ N-NITROSOALLYL-METHYLAMINE □ NITROSOMETHYLALLYLAMINE □ N-NITROSOMETHYLALLYLAMINE

SAFETY PROFILE: Poison by ingestion and intravenous routes. Questionable carcinogen with experimental tumorigenic data. Mutation data reported. When heated to decomposition it emits toxic fumes of NO_x. See also N-NITROSO COMPOUNDS.

MMT750   CAS: 3684-97-7   HR: 3
2-(N-METHYL-N-NITROSO)AMINO-ACETONITRILE
mf: $C_3H_5N_3O$ mw: 99.11

SYNS: N-NITROSOMETHYLAMINACETONITRIL (GERMAN) □ N-NITROSOMETHYLAMINO-ACETONITRILE

CONSENSUS REPORTS: Cyanide and its compounds are on the Community Right-To-Know List.

SAFETY PROFILE: Poison by ingestion. Questionable carcinogen with experimental tumorigenic data. When heated to decomposition it emits toxic fumes of NO_x and CN^-. See also N-NITROSO COMPOUNDS and NITRILES.

MMU250   CAS: 614-00-6   HR: 3
N-METHYL-N-NITROSOANILINE
mf: $C_7H_8N_2O$ mw: 136.17

PROP: A liquid. D: 1.124 @ 20°/4°, mp: 13°, bp: 128–128.4° @ 19 mm.

SYNS: N-METHYL-N-NITROSOBENZENAMINE □ METHYLPHENYLNITROSAMINE □ MNA □ NITROSOMETHYLANILINE □ N-NITROSO-N-METHYLANILINE □ N-NITROSOMETHYLPHENYL-AMINE (MAK) □ NMA □ PHENYLMETHYLNITRO-SAMINE

CONSENSUS REPORTS: Reported in EPA TSCA Inventory. EPA Genetic Toxicology Program.

DFG MAK: Animal Carcinogen, Suspected Human Carcinogen

SAFETY PROFILE: Confirmed carcinogen with experimental carcinogenic, neoplastigenic, tumorigenic, and teratogenic data. Poison by ingestion and intraperitoneal routes. Mutation data reported. When heated to decomposition it emits toxic fumes of NO_x. See also N-NITROSO COMPOUNDS.

M

MMW775   CAS: 25355-61-7   HR: 3
1-METHYL-1-NITROSO-3-(p-CHLORO-PHENYL)UREA

mf: $C_8H_8ClN_3O_2$ mw: 213.64

SYNS: 1-(p-CHLOROPHENYL)-3-METHYL-3-NITROSOUREA □ 3-(p-CHLOROPHENYL)-1-METHYL-1-NITROSOUREA □ N-METHYL-N'-(p-CHLOROPHENYL)-N-NITROSOUREA

SAFETY PROFILE: Suspected carcinogen with experimental carcinogenic data. Mutation data reported. When heated to decomposition it emits toxic fumes of Cl⁻ and NO_x. See also N-NITROSO COMPOUNDS.

MMX000   CAS: 33868-17-6   HR: 3
METHYLNITROSOCYANAMIDE

mf: $C_2H_3N_3O$ mw: 85.08

SYN: MNC

SAFETY PROFILE: Poison by ingestion. Questionable carcinogen with experimental tumorigenic data. Mutation data reported. When heated to decomposition it emits toxic fumes of NO_x. See also N-NITROSO COMPOUNDS.

MMX250   CAS: 615-53-2   HR: 3
N-METHYL-N-NITROSOETHYL-CARBAMATE

mf: $C_4H_8N_2O_3$ mw: 132.14

SYNS: ETHYL ESTER of METHYLNITROSO-CARBAMIC ACID □ N-METHYL-N-NITROSOCARBAMIC ACID, ETHYL ESTER □ METHYLNITROSOURETHAN (GERMAN) □ METHYLNITROSOURETHANE □ N-METHYL-N-NITROSO-URETHANE □ MNU □ NITROSOMETHYLURETHAN (GERMAN) □ NITROSO-METHYLURETHANE □ N-NITROSO-N-METHYL-URETHANE □ NMUM □ NMUT □ RCRA WASTE NUMBER U178

CONSENSUS REPORTS: IARC Cancer Review: Group 2B IMEMDT 7,56,87; Animal Sufficient Evidence IMEMDT 4,211,74. Reported in EPA TSCA Inventory. EPA Genetic Toxicology Program.

SAFETY PROFILE: Confirmed carcinogen with experimental carcinogenic, tumorigenic, and teratogenic data. Poison by ingestion, intraperitoneal, subcutaneous, and intravenous routes. Moderately toxic by inhalation. Experimental reproductive effects. Mutation data reported. Has been implicated as a transplacental brain carcinogen. Combustible when exposed to heat, sparks, open flame, and powerful oxidizers. Explodes when heated. A storage hazard. When heated to decomposition it emits toxic fumes of NO_x. See also NITROSAMINES and CARBAMATES.

MMY500   CAS: 21561-99-9   HR: 3
1-METHYL-1-NITROSO-3-PHENYLUREA

mf: $C_8H_9N_3O_2$ mw: 179.20

SYNS: N-METHYL-N-NITROSO-N'-PHENYLUREA □ METHYLPHENYLNITROSOUREA □ N-METHYL-N'-PHENYL-N-NITROSOUREA □ MPNU □ NITROSO-METHYLPHENYLUREA □ 3-PHENYL-1-METHYL-1-NITROSOHARNSTOFF (GERMAN)

SAFETY PROFILE: Suspected carcinogen with experimental carcinogenic and tumorigenic data. Mutation data reported. When heated to decomposition it emits toxic fumes of NO_x. See also N-NITROSO COMPOUNDS.

MNA000   CAS: 924-46-9   HR: 3
N-METHYL-N-NITROSO-1-PROPAN-AMINE

mf: $C_4H_{10}N_2O$ mw: 102.16

PROP: Yellow liquid. Bp: 90.4–91.2° @ 40 mm.

SYNS: METHYL-N-PROPYLNITROSAMINE □ METHYLPROPYLNITROSOAMINE □ MPN □ NITROSOMETHYLPROPYLAMINE □ NITROSOMETH-YL-N-PROPYLAMINE

CONSENSUS REPORTS: EPA Genetic Toxicology Program.

SAFETY PROFILE: Suspected carcinogen with experimental carcinogenic, neoplastigenic, and tumorigenic data. Poison by subcutaneous route. An experimental teratogen. Mutation data reported. When heated to decomposition it emits toxic fumes of NO_x. See also N-NITROSO COMPOUNDS.

MNA750   CAS: 684-93-5   HR: 3
N-METHYL-N-NITROSOUREA

mf: $C_2H_5N_3O_2$ mw: 103.10

PROP: Yellow solid or plates from Et_2O. Mp: 124° (decomp).
SYNS: METHYLNITROSO-HARNSTOFF (GERMAN) □ N-METHYL-N-NITROSO-HARNSTOFF (GERMAN) □ METHYLNITROSOUREA □ 1-METHYL-1-NITROSO-UREA □ METHYLNITROSOUREE (FRENCH) □ MNU □ N-NITROSO-N-METHYLCARBAMIDE □ N-NITROSO-N-METHYL-HARNSTOFF (GERMAN) □ NITROSO-METHYLUREA □ N-NITROSO-N-METHYLUREA □ 1-NITROSO-1-METHYLUREA □ NMH □ NMU □ NSC-23909 □ RCRA WASTE NUMBER U177 □ SKI 24464 □ SRI 859
CONSENSUS REPORTS: NTP 10th Report on Carcinogens. IARC Cancer Review: Group 2A IMEMDT 7,56,87; Human Limited Evidence IMEMDT 17,227,78; Animal Sufficient Evidence IMEMDT 17,227,78, IMEMDT 1,125,72. EPA Genetic Toxicology Program. Community Right-To-Know List. Reported in EPA TSCA Inventory.
SAFETY PROFILE: Confirmed carcinogen with experimental carcinogenic, neoplastigenic, tumorigenic, and teratogenic data. Poison by ingestion and intravenous routes. Experimental reproductive effects. Human mutation data reported. Explodes at room temperature. Can detonate with $(KOH + CH_2Cl_2)$. When heated to decomposition it emits toxic fumes of NO_x. See also N-NITROSO COMPOUNDS.

MNB500   CAS: 53153-66-5   HR: 2
3-METHYL-2(3)-NONENENITRILE
mf: $C_{10}H_{17}N$ mw: 151.28
SYNS: CITGRENILE □ 2-NONENENITRILE, 3-METHYL-
CONSENSUS REPORTS: Reported in EPA TSCA Inventory.
SAFETY PROFILE: Moderately toxic by ingestion. A skin irritant. Low toxicity by skin contact. When heated to decomposition it emits toxic fumes of NO_x and CN^-.

MNC175   CAS: 7786-29-0   HR: 1
2-METHYLOCTANAL
mf: $C_9H_{18}O$ mw: 142.27
SYNS: METHYLHEXYLACETALDEHYDE □ α-METHYLOCTANAL □ OCTANAL, 2-METHYL-

CONSENSUS REPORTS: Reported in EPA TSCA Inventory.
SAFETY PROFILE: Low toxicity by ingestion and skin contact. A skin irritant. When heated to decomposition it emits acrid smoke and irritating fumes.

MND050   CAS: 5340-36-3   HR: 2
3-METHYL-3-OCTANOL
mf: $C_9H_{20}O$ mw: 144.29
SYNS: AMYLETHYLMETHYLCARBINOL □ APROL 161 □ 3-METHYLOCTAN-3-OL □ 3-OCTANOL, 3-METHYL-
CONSENSUS REPORTS: Reported in EPA TSCA Inventory.
SAFETY PROFILE: Moderately toxic by ingestion. Low toxicity by skin contact. A skin irritant. When heated to decomposition it emits acrid smoke and irritating fumes.

MND100   CAS: 24089-00-7   HR: 2
3-METHYL-1-OCTEN-OL
mf: $C_9H_{18}O$ mw: 142.27
SYNS: APROL 160 □ 1-OCTEN-3-OL, 3-METHYL-
CONSENSUS REPORTS: Reported in EPA TSCA Inventory.
SAFETY PROFILE: Moderately toxic by ingestion. Low toxicity by skin contact. A severe skin irritant. When heated to decomposition it emits acrid smoke and irritating fumes.

MND275   CAS: 111-12-6   HR: 2
METHYL 2-OCTYNOATE
mf: $C_9H_{14}O_2$ mw: 154.23
PROP: Colorless to sltly yellow liquid; powerful, unpleasant odor; violet odor when diluted. Bp: 94° @ 10 mm, d: 0.919, refr index: 1.446, flash p: 212°F. Sol in fixed oils; sltly sol in propylene glycol; insol in glycerin.
SYNS: FEMA No. 2729 □ FOLIONE □ METHYL HEPTINE CARBONATE □ METHYL 2-OCTINATE
CONSENSUS REPORTS: Reported in EPA TSCA Inventory.
SAFETY PROFILE: Moderately toxic by ingestion and skin contact. A moderate skin and eye irritant. A combustible liquid. When heated to decomposition it emits acrid smoke and irritating fumes.

MND700   CAS: 27970-32-7   HR: 3
3-METHYL-1,3-OXAZOLIDINE
mf: C_4H_9NO mw: 87.14

SYNS: N-METHYLOXAZOLIDINE □ 3-METHYLOXAZOLIDINE □ OXAZOLIDINE, 3-METHYL-

SAFETY PROFILE: A poison by ingestion and skin contact. A severe eye irritant. When heated to decomposition it emits toxic vapors of NO_x.

MNH000   CAS: 298-00-0   HR: 3
METHYL PARATHION
DOT: NA 2783/NA 3018

mf: $C_8H_{10}NO_5PS$ mw: 263.22

PROP: Crystals or solid; garlic-like odor. Vap d: 9.1, mp: 35–36°, d: 1.358 @ 20°/4°, bp: 158° @ 2 mm. Sol in most org solvs.

SYNS: A-GRO □ AZOFOS □ AZOPHOS □ BAY 11405 □ BAY E-601 □ BLADAN-M □ CEKUMETHION □ DALF □ DEVITHION □ O,O-DIMETHYL-O-p-NITROFENYL-ESTER KYSELINY THIOFOSFORECNE (CZECH) □ O,O-DIMETHYL-O-(4-NITROFENYL)-MONOTHIOFOSFAAT (DUTCH) □ O,O-DIMETHYL-O-(4-NITRO-PHENYL)-MONOTHIOPHOSPHAT (GERMAN) □ DIMETHYL p-NITROPHENYL MONOTHIOPHOSPHATE □ O,O-DIMETHYL-O-(p-NITROPHENYL) PHOSPHOROTHIO-ATE □ O,O-DIMETHYL-O-(4-NITROPHENYL) PHOSPHOROTHIOATE □ DIMETHYL 4-NITROPHENYL PHOSPHOROTHIONATE □ O,O-DIMETHYL-O-(p-NITROPHENYL)-THIONOPHOSPHAT (GERMAN) □ O,O-DIMETHYL-O-(4-NITROPHENYL)-THIONO-PHOSPHAT (GERMAN) □ DIMETHYL-p-NITROPHENYL THIONPHOSPHATE □ DIMETHYL p-NITROPHENYL THIOPHOSPHATE □ O,O-DIMETHYL-O-p-NITROPHENYL THIOPHOSPHATE □ DIMETHYL PARATHION □ O,O-DIMETIL-O-(4-NITRO-FENIL)-MONOTIOFOSFATO (ITALIAN) □ DREXEL METHYL PARATHION 4E □ ENT 17,292 □ FOLIDOL M □ GEARPHOS □ ME-PARATHION □ MEPATON □ MEPTOX □ METACIDE □ METAFOS □ METAPHOR □ METAPHOS □ METHYL-E 605 □ METHYL FOSFERNO □ METHYL NIRAN □ METHYLTHIOPHOS □ METILPARATION (HUNGARIAN) □ METRON □ METYLOPARATION (POLISH) □ METYLPARATION (CZECH) □ NCI-C02971 □ p-NITROPHENYLDIMETHYL-THIONOPHOSPHATE □ NITROX □ OLEOVOFOTOX □ PARAPEST M-50 □ PARATAF □ M-PARATHION □ PARATHION METHYL □ PARATHION-METILE (ITALIAN) □ PARATOX □ PENNCAP-M □ RCRA WASTE NUMBER P071 □ SINAFID M-48 □ SIXTY-THREE SPECIAL E.C. INSECTICIDE □ TEKWAISA □ THIO-PHENIT □ THIOPHOSPHATE de O,O-DIMETHYLE et de O-(4-NITROPHENYLE) (FRENCH) □ THYLFAR M-50 □ TOLL □ VERTAC METHYL PARATHION TECHNISCH

80% □ VOFATOX □ WOFATOS □ WOFATOX □ WOFOTOX

CONSENSUS REPORTS: IARC Cancer Review: Group 3 IMEMDT 7,56,87; Animal No Evidence IMEMDT 30,131,83. NCI Carcinogenesis Bioassay (feed); No Evidence: mouse, rat NCITR* NCI-CG-TR-157,79. EPA Genetic Toxicology Program. EPA Extremely Hazardous Substances List.

OSHA PEL: TWA 0.2 mg/m³ (skin)

ACGIH TLV: TWA 0.2 mg/m³ (skin); Not Classifiable as a Human Carcinogen

NIOSH REL: (Methyl Parathion) TWA 0.2 mg/m³

DOT CLASSIFICATION: 6.1; Label: Poison

SAFETY PROFILE: Poison by inhalation, ingestion, skin contact, subcutaneous, intravenous, and intraperitoneal routes. Fatal poisoning can result from skin or eye contact after very brief exposure to concentrated solution. Experimental teratogenic and reproductive effects. Questionable carcinogen. Human mutation data reported. A cholinesterase inhibitor type of insecticide. When heated to decomposition it emits very toxic fumes of NO_x, PO_x, and SO_x. See also PARATHION.

MNI500   CAS: 96-14-0   HR: 3
3-METHYLPENTANE
mf: C_6H_{14} mw: 86.18

PROP: Flash p: 19.4°F, lel: 1.2%, uel: 7.0%, bp: 63.3°, fp: −118° (sets to a glass), d: 0.669 @ 15°/15°, vap press: 100 mm @ 10.5°, vap d: 2.97.

SYN: DIETHYLMETHYL METHANE

CONSENSUS REPORTS: Reported in EPA TSCA Inventory.

DFG MAK: 200 ppm (720 mg/m³)

SAFETY PROFILE: May have narcotic or anesthetic properties. A very dangerous fire hazard when exposed to heat or flame; can react vigorously with oxidizing materials. Explosive in the form of vapor when exposed to heat or flame. To fight fire, use foam, CO_2, dry chemical. When heated to

decomposition it emits acrid smoke and irritating fumes.

MNP450   CAS: 16018-21-6   HR: 3
1-(α-METHYLPHENETHYL)-4-PHENYL-PIPERAZINE DIHYDROCHLORIDE

mf: $C_{19}H_{24}N_2 \cdot 2ClH$ mw: 353.37

SYN: PIPERAZINE, 1-(α-METHYLPHENETHYL)-4-PHENYL-, DIHYDROCHLORIDE

SAFETY PROFILE: A poison by intravenous route. When heated to decomposition it emits toxic vapors of NO_x, HCl, and Cl^-.

MNQ000   CAS: 113-45-1   HR: 3
METHYL PHENIDYL ACETATE

mf: $C_{14}H_{19}NO_2$ mw: 233.34

PROP: Crystals from EtOH (aq). Mp: 74–75°.

SYNS: CALOCAIN □ CENTEDEIN □ CENTREDIN □ MERIDIL □ METHYLPHENIDAN □ METHYL PHENIDATE □ METHYL α-PHENYL-α-(2-PIPERIDYL)-ACETATE □ NCI-C56280 □ PHENIDYLATE □ α-PHENYL-2-PIPERIDINEACETIC ACID METHYL ESTER □ PLIMASINE □ RITALIN □ RITALINE □ RITCHER WORKS

SAFETY PROFILE: Poison experimentally by ingestion, intraperitoneal, intravenous, and subcutaneous routes. Moderately toxic to humans by intravenous route. Human systemic effects by intravenous route: dyspnea. An experimental teratogen. When heated to decomposition it emits toxic fumes of NO_x.

MNR250   CAS: 140-39-6   HR: 2
4-METHYLPHENYL ACETATE

mf: $C_9H_{10}O_2$ mw: 150.19

PROP: Colorless liquid; strong floral odor. D: 1.044 @ 16°, refr index: 1.499–1.502, bp: 212–213° (decomp @ 360°), mp: 220°, vap d: 5.18, flash p: 203°F. Sol in fixed oils, propylene glycol; misc in alc and ether; insol in water, glycerin.

SYNS: ACETIC ACID-4-METHYLPHENYL ESTER □ p-ACETOXYTOLUENE □ 4-ACETOXYTOLUENE □ p-CRESOL ACETATE □ p-CRESYL ACETATE (FCC) □ FEMA No. 3073 □ 4-METHYLBENZOIC ACID METHYL ESTER □ p-METHYLPHENYL ACETATE □ NARCEOL □ PARACRESYL ACETATE □ p-TOLYL ACETATE □ p-TOLYL ETHANOATE

CONSENSUS REPORTS: Reported in EPA TSCA Inventory.

SAFETY PROFILE: Moderately toxic by ingestion and skin contact. Combustible liquid. When heated to decomposition it emits toxic smoke and irritating fumes. See also ESTERS.

MNT075   HR: 2
METHYL PHENYLCARBINYL ACETATE

mf: $C_{10}H_{12}O_2$ mw: 164.20

PROP: Colorless liquid; gardenia odor. D: 1.023, refr index: 1.493–1.497, flash p: 176°F. Sol in fixed oils, glycerin; insol in water.

SYNS: FEMA No. 2684 □ α-PHENYL ETHYL ACETATE

SAFETY PROFILE: Combustible liquid. When heated to decomposition it emits acrid smoke and irritating fumes.

MNT500   CAS: 20240-98-6   HR: 3
1-(2-METHYLPHENYL)-3,3-DIMETHYL-TRIAZENE

mf: $C_9H_{13}N_3$ mw: 163.25

SYNS: 3,3-DIMETHYL-1-(o-METHYLPHENYL)TRIAZENE □ 3,3-DIMETHYL-1-(o-TOLYL)TRIAZENE □ 1-(o-METHYLPHENYL)-3,3-DIMETHYL-TRIAZEN (GERMAN) □ 1-(o-METHYLPHENYL)-3,3-DIMETHYL-TRIAZENE

SAFETY PROFILE: Poison by ingestion. Moderately toxic by subcutaneous route. Experimental reproductive effects. Questionable carcinogen with experimental carcinogenic data. When heated to decomposition it emits toxic fumes of NO_x.

MNU050   CAS: 73791-44-3   HR: 3
METHYL(2-PHENYLETHYL)ARSINIC ACID

mf: $C_9H_{13}AsO_2$ mw: 228.14

SYN: ARSINE OXIDE, HYDROXYMETHYL-PHENETHYL-

OSHA PEL: TWA 0.5 mg(As)/m^3

SAFETY PROFILE: Poison by intravenous route. When heated to decomposition it emits toxic fumes of As.

M

MNU250   CAS: 13256-11-6   HR: 3
METHYL-PHENYLETHYL-NITROS-
  AMINE
mf: $C_9H_{12}N_2O$ mw: 164.23
SYNS: N-METHYL-N-NITROSOPHENETHYLAMINE □
METHYL(2-PHENYLAETHYL)NITROSAMIN (GERMAN)
□ N-NITROSO-N-METHYL-2-PHENYLETHYLAMINE
SAFETY PROFILE: Poison by ingestion.
Questionable carcinogen with experimental
carcinogenic and tumorigenic data. Mutation
data reported. When heated to
decomposition it emits toxic fumes of NO_x.
See also NITROSAMINES.

MNV750   CAS: 1707-14-8   HR: 3
3-METHYL-2-PHENYLMORPHOLINE
  HYDROCHLORIDE
mf: $C_{11}H_{15}NO•ClH$ mw: 213.73
PROP: Crystals from EtOH/Et$_2$O. Mp:
180–181°.
SYNS: A 66 HYDROCHLORIDE □ MARSIN □ 3-
METHYL-2-PHENYLTETRAHYDRO-2H-1,4-OXAZINE
HYDROCHLORIDE □ NEO-ZINE □ PHENMETRAZINE
HYDROCHLORIDE □ 2-PHENYL-3-METHYLTETRA-
HYDRO-1,4-OXAZINE HYDROCHLORIDE □ PRELUDIN
HYDROCHLORIDE □ PROBESE-P HYDROCHLORIDE
□ PSYCHAMINE A 66 HYDROCHLORIDE □ USAF GE-1
SAFETY PROFILE: Poison by ingestion,
intravenous, intraperitoneal, and
subcutaneous routes. Human reproductive
effects. Human teratogenic effects by
ingestion: developmental abnormalities of
the respiratory and gastrointestinal systems,
and effects on newborn including neonatal
measures or effects. When heated to
decomposition it emits very toxic fumes of
NO_x and HCl.

MNW790   CAS: 315706-79-7 HR: 3
3-(4-METHYLPHENYL)-N-(4-PROPYLCY-
  CLOHEXYL)-2-PROPENAMIDE
mf: $C_{19}H_{27}NO$ mw: 285.43
SAFETY PROFILE: A poison by ingestion.
When heated to decomposition it emits
toxic vapors of NO_x.

MNX310   CAS: 302542-49-0 HR: 3
N-(3-METHYLPHENYL)-1H-PYRAZOLE-
  1-ACETAMIDE
mf: $C_{12}H_{13}N_3O$ mw: 215.25

SAFETY PROFILE: A poison by
subcutaneous and ocular routes. Moderately
toxic by ingestion. When heated to
decomposition it emits toxic vapors of NO_x.

MNX420   CAS: 296269-53-9 HR: 3
α-(5-METHYL-1-PHENYL-1H-PYRAZOL-
  4-YL)-1-PIPERIDINEBUTANOL
mf: $C_{19}H_{27}N_3O$ mw: 313.44
SAFETY PROFILE: A poison by ingestion.
When heated to decomposition it emits
toxic vapors of NO_x.

MOB399   CAS: 676-97-1   HR: 3
METHYL PHOSPHONIC DICHLORIDE
DOT: NA 9206
mf: CH_3Cl_2OP mw: 132.91
PROP: Low melting solid with pungent
odor. Easily hydrolyzed. Mp: 32°, bp: 162°.
CONSENSUS REPORTS: EPA Extremely
Hazardous Substances List. Reported in
EPA TSCA Inventory.
DOT CLASSIFICATION: 6.1; Label: Poison,
Corrosive
SAFETY PROFILE: Poison by inhalation.
A corrosive irritant to the eyes, skin,
and mucous membranes. When heated
to decomposition it emits toxic fumes
of Cl^- and PO_x. See also CHLORIDES.

MOC000   CAS: 676-98-2   HR: 2
METHYL PHOSPHONOTHIOIC DICHLO-
  RIDE, pyrophoric liquid (DOT)
DOT: NA 2845
mf: CH_3Cl_2PS mw: 148.97
PROP: Liquid with pungent odor. D:
1.35–1.43 @ 20°/4°, bp: 177–178°.
DOT CLASSIFICATION: 6.1; Label: Poison,
Spontaneously Combustible
SAFETY PROFILE: A corrosive irritant to
skin, eyes, and mucous membranes. When
heated to decomposition it emits very toxic
fumes of Cl^-, PO_x, and SO_x. See also
PHOSPHATES, CHLORIDES, and
SULFIDES.

MOC250   CAS: 676-83-5   HR: 3
METHYLPHOSPHONOUS DICHLORIDE
PROP: A liquid with pungent, foul odor. D: 1.30 @ 20°/4°, bp: 80–81° @ 729 mm.
CONSENSUS REPORTS: Reported in EPA TSCA Inventory.
SAFETY PROFILE: A poison. A corrosive irritant to the skin, eyes, and mucous membranes. Flammable when exposed to heat or flame; can react vigorously with oxidizing materials. When heated to decomposition it emits very toxic fumes of Cl⁻ and PO_x. See also HYDROCHLORIC ACID.

MOD250   CAS: 109-01-3   HR: 3
N-METHYLPIPERAZINE
mf: $C_5H_{12}N_2$ mw: 100.19
PROP: A hygroscopic solid; typical amine-like odor. D: 0.9031 at 20°/20°, mp: 65.5°, bp: 134–136°, flash p: 108°F (OC), vap d: 3.5.
SYN: 1-METHYLPIPERAZINE
CONSENSUS REPORTS: Reported in EPA TSCA Inventory.
SAFETY PROFILE: Poison by intraperitoneal route. Moderately toxic by inhalation, ingestion, and skin contact. A severe skin and eye irritant. Flammable liquid when exposed to heat or flame; can react with oxidizing materials. To fight fire, use alcohol foam, CO_2, dry chemical. When heated to decomposition it emits toxic fumes of NO_x.

MOG500   CAS: 626-67-5   HR: 3
N-METHYLPIPERIDINE
DOT: UN 2399
mf: $C_6H_{13}N$ mw: 99.20
PROP: A liquid. D: 0.820 @ 20°/4°, bp: 107°, flash p: <73.4°F.
SYN: 1-METHYLPIPERIDINE (DOT)
CONSENSUS REPORTS: Reported in EPA TSCA Inventory.
DOT CLASSIFICATION: 3; Label: Flammable Liquid
SAFETY PROFILE: Poison by intraperitoneal and subcutaneous routes. A

very dangerous fire hazard when exposed to heat or flame. When heated to decomposition it emits toxic fumes of NO_x.

MOH290   CAS: 101831-65-6 HR: 3
2-METHYLPIPERIDINE β-NAPHTHO-
AMIDE
mf: $C_{17}H_{18}NO$ mw: 252.36
SYN: KETONE, 2-METHYLPIPERIDINO 2-NAPHTHYL
DOT CLASSIFICATION: 3; Label: Flammable Liquid
SAFETY PROFILE: A poison by intravenous route. A flammable liquid. When heated to decomposition it emits toxic vapors of NO_x.

MOL300   CAS: 98271-51-3   HR: 3
METHYL 10-(3-PIPERIDINOPROPYL)-
PHENOTHIAZIN-2-YL KETONE
mf: $C_{22}H_{26}N_2OS$ mw: 366.56
SYN: KETONE, METHYL 10-(3-PIPERIDINO-PROPYL)PHENOTHIAZIN-2-YL
DOT CLASSIFICATION: 3; Label: Flammable Liquid
SAFETY PROFILE: A poison by subcutaneous route. A flammable liquid. When heated to decomposition it emits toxic vapors of NO_x and SO_x.

MOR500   CAS: 83-43-2   HR: 2
METHYLPREDNISOLONE
mf: $C_{22}H_{30}O_5$ mw: 374.52
PROP: Crystals. Mp: 228–237°.
SYNS: MEDROL □ MEDROL DOSEPAK □ MEDRONE □ Δ^1-6-α-METHYLHYDROCORTISONE □ 6-α-METHYLPREDNISOLONE □ METRISONE □ NSC-19987 □ 11-β,17,21-TRIHYDROXY-6-α-METHYLPREGNA-1,4-DIENE-3,20-DIONE □ 11-β,17-α,21-TRIHYDROXY-6-α-METHYL-1,4-PREGNADIENE-3,20-DIONE □ URBASON □ URBASONE □ WYACORT
CONSENSUS REPORTS: Reported in EPA TSCA Inventory.
SAFETY PROFILE: Moderately toxic by intraperitoneal route. A steroid hormone. Human systemic effects include arrhythmias, blood pressure lowering, heart rate changes, increased body temperature, pulse rate increase, respiratory depression. When

M

heated to decomposition it emits acrid smoke and irritating fumes.

MOR750   CAS: 75-28-5   HR: 3
2-METHYLPROPANE
DOT: UN 1969

mf: C_4H_{10} mw: 58.14

PROP: Colorless gas. Fp: $-145°$, bp: $-10.2°$, lel: 1.9%, uel: 8.5%, d: 0.5572 @ 20°, autoign temp: 864°F, vap d: 2.01. Sol in EtOH, Et_2O, and $CHCl_3$; spar sol in H_2O.

SYNS: ISOBUTANE □ ISOBUTANE (DOT) □ ISOBUTANE MIXTURES (DOT)

CONSENSUS REPORTS: Reported in EPA TSCA Inventory.

DFG MAK: 1000 ppm (2400 mg/m³)

DOT CLASSIFICATION: 2.1; Label: Flammable Gas

SAFETY PROFILE: An asphyxiant. A common air contaminant. A very dangerous fire and explosion hazard when exposed to heat, flame, or oxidizers. To fight fire, stop flow of gas. When heated to decomposition it emits acrid smoke and irritating fumes.

MOS000   CAS: 75-66-1   HR: 3
2-METHYL-2-PROPANETHIOL
mf: $C_4H_{10}S$ mw: 90.20

PROP: Mobile liquid; heavy skunk odor. Mp: $-0.5°$, bp: $63.7-64.2°$, d: 0.79–0.82 @ 15.5°/15.5°, flash p: $<-20°F$, vap d: 3.1, n: (25/D) 1.41984. Sltly sol in water; very sol in alc, ether, and liquid H_2S.

SYNS: tert-BUTANETHIOL □ tert-BUTYL MERCAPTAN

CONSENSUS REPORTS: Reported in EPA TSCA Inventory.

SAFETY PROFILE: Moderately toxic by intraperitoneal route. Mildly toxic by ingestion. An eye irritant. A very dangerous fire hazard when exposed to heat or flame. Can react vigorously with oxidizing materials. To fight fire, use alcohol foam, dry chemical, mist, fog. When heated to decomposition or on contact with acid or acid fumes it emits highly toxic fumes of SO_x. See also MERCAPTANS.

MOT000   CAS: 554-12-1   HR: 3
METHYL PROPIONATE
DOT: UN 1248

mf: $C_4H_8O_2$ mw: 88.12

PROP: Colorless liquid. Mp: $-87.0°$, bp: 79.8°, flash p: 28°F (CC) (−2°C), d: 0.937 @ 4°, autoign temp: 876°F, vap press: 40 mm @ 11.0°, vap d: 3.03, lel: 2.50%, uel: 13%, d: 0.915 @ 20°/4°. Sol in water @ 20°; misc in alc and ether.

SYNS: METHYL PROPANOATE □ METHYL PROPYLATE □ PROPANOIC ACID, METHYL ESTER □ PROPIONATE de METHYLE (FRENCH)

CONSENSUS REPORTS: Reported in EPA TSCA Inventory.

DOT CLASSIFICATION: 3; Label: Flammable Liquid

SAFETY PROFILE: Moderately toxic by ingestion. Mildly toxic by inhalation. A skin irritant. A very dangerous fire hazard when exposed to heat, flame, or oxidizers. Explosive in the form of vapor when exposed to heat or flame. To fight fire, use foam, CO_2, dry chemical. When heated to decomposition it emits acrid smoke and irritating fumes.

MOT800   CAS: 73791-45-4   HR: 3
METHYLPROPYLARSINIC ACID
mf: $C_4H_{11}AsO_2$ mw: 166.07

SYN: ARSINE OXIDE, HYDROXYMETHYLPROPYL-

OSHA PEL: TWA 0.5 mg(As)/m³

SAFETY PROFILE: Poison by intravenous route. When heated to decomposition it emits toxic fumes of As.

MOU820   CAS: 315706-71-9   HR: 3
4-METHYL-N-(4-PROPYLCYCLOHEX-
YL)BENZAMIDE
mf: $C_{17}H_{25}NO$ mw: 259.39

SAFETY PROFILE: A poison by ingestion. When heated to decomposition it emits toxic vapors of NO_x.

MOU830   CAS: 557-17-5   HR: 3
METHYL PROPYL ETHER
DOT: UN 2612

mf: $C_4H_{10}O$ mw: 74.14

SYNS: ETHER, METHYL PROPYL □ α-METHOXY PROPANE □ 1-METHOXYPROPANE □ METHYL n-PROPYL ETHER □ METOPRYL □ NEOTHYL □ PROPANE, 1-METHOXY-(9CI)

DOT CLASSIFICATION: 3; Label: Flammable Liquid

SAFETY PROFILE: A poison by inhalation. A flammable liquid. When heated to decomposition it emits acrid smoke and irritating vapors.

MOV000   CAS: 3766-81-2   HR: 3
2-(1-METHYLPROPYL)PHENYL METH-YLCARBAMATE

mf: $C_{12}H_{17}NO_2$ mw: 207.30

PROP: A low-melting solid. Mp: 32°.

SYNS: BARIZON □ BASSA □ BAY 41637 □ BAYCARB □ BAYER 41367C □ BAYER 41637 □ BPMC □ 2-sek.BUTYLFENYLESTER KYSELINY METHYLKAR-BAMINOVE (CZECH) □ o-sec-BUTYLPHENYL METHYLCARBAMATE □ 2-sec-BUTYLPHENYL N-METHYLCARBAMATE □ CARVIL □ FENOBCARB □ FENOBUCARB □ GEOCARB 50EC □ HOPCIN □ METHYLCARBAMIC ACID o-sec-BUTYLPHENYL ESTER □ OSBAC □ PHENOL, 2-(1-METHYLPROPYL)-, METHYLCARBAMATE

CONSENSUS REPORTS: Reported in EPA TSCA Inventory. EPA Genetic Toxicology Program.

SAFETY PROFILE: Poison by ingestion, skin contact, intravenous, and intraperitoneal routes. Used as an insecticide. When heated to decomposition it emits toxic fumes of NO_x. See also CARBAMATES.

MOW750   CAS: 109-08-0   HR: 3
2-METHYLPYRAZINE

mf: $C_5H_6N_2$ mw: 94.13

PROP: A liquid; nutty, cocoa odor. Mp: −29°, bp: 136–137° @ 737 mm, flash p: 122°F (COC), d: 1.029 @ 20°/4°, refr index: 1.504, vap d: 3.2. Misc with water, alc, acetone, fixed oils.

SYN: FEMA No. 3309

CONSENSUS REPORTS: Reported in EPA TSCA Inventory.

SAFETY PROFILE: Moderately toxic by ingestion and intraperitoneal routes. Mutation data reported. Flammable liquid when exposed to heat, sparks, or flame. Can

react with oxidizing materials. To fight fire, use water spray, foam, dry chemical, CO_2. When heated to decomposition it emits highly toxic fumes of NO_x.

MOY000   CAS: 109-06-8   HR: 3
2-METHYLPYRIDINE

mf: C_6H_7N mw: 93.14

PROP: Colorless liquid or oil; strong unpleasant odor. Mp: −70°, bp: 129°, flash p: 102°F (OC), d: 0.95 @ 15°/4°, autoign temp: 1000°F, vap press: 10 mm @ 24.4°, vap d: 3.2. Very sol in water; misc in alc and ether.

SYNS: α-METHYLPYRIDINE □ 2-PICOLINE □ α-PICOLINE □ o-PICOLINE □ RCRA WASTE NUMBER U191

CONSENSUS REPORTS: Reported in EPA TSCA Inventory.

SAFETY PROFILE: Poison by intraperitoneal route. Moderately toxic by ingestion and skin contact. Mildly toxic by inhalation. A skin and severe eye irritant. Mutation data reported. Flammable liquid when exposed to heat or flame. To fight fire, use CO_2, dry chemical. Mixtures with hydrogen peroxide + iron(II) sulfate + sulfuric acid may ignite and then explode. When heated to decomposition it emits toxic fumes of NO_x. See also 4-METHYLPYRIDINE.

MOY250   CAS: 108-89-4   HR: 3
4-METHYLPYRIDINE

mf: C_6H_7N mw: 93.14

PROP: Colorless liquid; disagreeable odor. Bp: 145°, fp: 3.7°, d: 0.9571 @ 15°/4°, vap d: 3.21, flash p: 134°F (OC). Sol in H_2O, EtOH, and Et_2O.

SYNS: γ-PICOLINE □ 4-PICOLINE □ p-PICOLINE

CONSENSUS REPORTS: Reported in EPA TSCA Inventory.

SAFETY PROFILE: Poison by ingestion and intraperitoneal routes. Moderately toxic by skin contact. Mildly toxic by inhalation. A severe skin and eye irritant. Flammable liquid when exposed to heat, flames, oxidizers. To fight fire, use alcohol foam. When heated to decomposition it emits

M

toxic fumes of NO_x. See also 2-METHYLPYRIDINE.

MPB250   CAS: 120-94-5   HR: 3
1-METHYLPYRROLIDINE
mf: $C_5H_{11}N$ mw: 85.15
PROP: Colorless to yellow liquid; penetrating amine-like odor. Fp: $-90°$, bp: 80.5°, d: 0.8054 @ 20°/20°, flash p: 37.4°F, vap d: 2.9. Misc in H_2O.
SYN: N-METHYLTETRAHYDROPYRROLE
CONSENSUS REPORTS: Reported in EPA TSCA Inventory.
SAFETY PROFILE: Poison by intraperitoneal and intravenous routes. This material is strongly alkaline. Liquid and vapors are corrosive to the skin, eyes, or mucous membranes. A very dangerous fire hazard; keep away from sparks, heat sources, and powerful oxidizers. Keep in closed containers. To fight fire, use alcohol foam. When heated to decomposition it emits highly toxic fumes of NO_x. See also AMMONIA.

MPF200   CAS: 872-50-4   HR: 3
N-METHYLPYRROLIDONE
mf: C_5H_9NO mw: 99.15
PROP: Colorless liquid; mild odor. Fp: $-24°$, mp: $-17°$, bp: 202°, flash p: 204°F (OC), d: 1.027 @ 25°/4°, vap d: 3.4.
SYNS: N-METHYLPYRROLIDINONE □ N-METHYL-2-PYRROLIDINONE □ 1-METHYL-2-PYRROLIDINONE □ 1-METHYL-5-PYRROLIDINONE □ METHYLPYR-ROLIDONE □ 1-METHYL-2-PYRROLIDONE □ N-METHYL-2-PYRROLIDONE □ M-PYROL □ NMP
CONSENSUS REPORTS: Reported in EPA TSCA Inventory.
DFG MAK: 19 ppm (80 mg/m³)
SAFETY PROFILE: Poison by intravenous route. Moderately toxic by ingestion and intraperitoneal routes. Mildly toxic by skin contact. An experimental teratogen. Experimental reproductive effects. Mutation data reported. Combustible when exposed to heat, open flame, or powerful oxidizers. To fight fire, use foam, CO_2, dry chemical. When heated to decomposition it emits toxic fumes of NO_x.

MPF300   CAS: 126268-14-2   HR: 3
3-((1-METHYLPYRROL-2-YL)METHY-LENEAMINO)-4-(PIPERIDINOMETH-YL)-2-OXAZOLIDONE
mf: $C_{15}H_{21}N_4O_2$ mw: 289.40
SYN: 2-OXAZOLIDONE, 3-((1-METHYLPYRROL-2-YL)METHYLENEAMINO)-4-(PIPERIDINOMETHYL)-
SAFETY PROFILE: Moderately toxic by ingestion. When heated to decomposition it emits toxic vapors of NO_x.

MPH500   CAS: 504-15-4   HR: 3
5-METHYLRESORCINOL
mf: $C_7H_8O_2$ mw: 124.15
PROP: Crystals from water or leaflets from $CHCl_3$. Mp: 107.5°, bp: 287–290°.
SYNS: 1,3-DIHYDROXY-5-METHYLBENZENE □ 3,5-DIHYDROXYTOLUENE □ 5-METHYL-1,3-BENZENDI-OL □ 5-METHYLRESORCINOL ORCINOL □ ORCIN □ ORCINOL
CONSENSUS REPORTS: EPA Genetic Toxicology Program. Reported in EPA TSCA Inventory.
SAFETY PROFILE: Poison by subcutaneous and intravenous routes. Moderately toxic by ingestion and intraperitoneal routes. Mildly toxic by skin contact. When heated to decomposition it emits acrid smoke and irritating fumes.

MPI000   CAS: 119-36-8   HR: 3
METHYL SALICYLATE
mf: $C_8H_8O_3$ mw: 152.16
PROP: Colorless, yellowish, or reddish oily liquid; odor and taste of wintergreen. Mp: $-8.6°$, bp: 223.3°, ULC: 20–25, flash p: 214°F (CC), fp: $-1.2°$, d: 1.1840 @ 25°/25°, refr index: 1.535, autoign temp: 850°F, vap press: 1 mm @ 54.0°, vap d: 5.24. Sltly sol in water @ 222° (decomp); sol in chloroform, ether, alc, glacial acetic acid. From steam distillation of leaves from *Gaultheria procumbens L.* (Fam. *Ericacaae*) or from the bark of *Betula lenta L.* (Fam. *Betulaceae*).
SYNS: ACIDE ANISIQUE (FRENCH) □ ACIDE METHYL-o-BENZOIQUE (FRENCH) □ o-ANISIC ACID □ BETULA OIL □ FEMA No. 2745 □ GAULTHERIA OIL, ARTIFICIAL □ o-HYDROXYBENZOIC ACID, METHYL ESTER □ 2-HYDROXYBENZOIC ACID METHYL ESTER □ o-METHOXYBENZOIC ACID □ 2-METHOXYBENZ-

OIC ACID □ METHYL-o-HYDROXYBENZOATE □ METYLESTER KYSELINY SALICYLOVE (CZECH) □ NATURAL WINTERGREEN OIL □ OIL OF WINTER-GREEN □ SALICYLIC ACID, METHYL ESTER □ SWEET BIRCH OIL □ SYNTHETIC WINTERGREEN OIL □ TEABERRY OIL □ WINTERGREEN OIL (FCC) □ WINTERGREEN OIL, SYNTHETIC

CONSENSUS REPORTS: Reported in EPA TSCA Inventory.

SAFETY PROFILE: Human poison by ingestion. Moderately toxic to humans by an unspecified route. Moderately toxic experimentally by intraperitoneal, intravenous, and subcutaneous routes. An experimental teratogen. Human systemic effects by ingestion: flaccid paralysis without anesthesia, general anesthesia, dyspnea, nausea, vomiting, and respiratory stimulation. Experimental reproductive effects. A severe skin and eye irritant. Ingestion of relatively small amounts has caused severe poisoning and death. Combustible liquid when exposed to heat or flame; can react with oxidizing materials. To fight fire, use CO_2, dry chemical. When heated to decomposition it emits acrid smoke and irritating fumes. See also ESTERS.

MPI750   CAS: 681-84-5   HR: 3
METHYL SILICATE
DOT: UN 2606
mf: $C_4H_{12}O_4Si$ mw: 152.25
PROP: Clear liquid. Vap d: 5.25, d: 1.03 @ 22°/4°, mp: 4–5°, bp: 120–121°.
SYNS: METHYL ORTHOSILICATE □ METHYL ORTHOSILICATE (DOT) □ SILICIC ACID, METHYL ESTER of ortho- □ TETRAMETHOXYSILANE □ TETRAMETHYL SILICATE □ TETRAMETHYLSILIKAT □ TL 190

CONSENSUS REPORTS: Reported in EPA TSCA Inventory.

OSHA PEL: TWA 1 ppm
ACGIH TLV: TWA 1 ppm
DOT CLASSIFICATION: 3; Label: Flammable Liquid, Poison
SAFETY PROFILE: Poison by intraperitoneal route. Moderately toxic by inhalation. Mildly toxic by skin contact. A severe eye irritant. This material can cause extensive necrosis (experimentally), keratoconus, and opaque cornea. It also causes severe human eye injuries, as well as necrosis of corneal cells, which progresses long after exposure has ceased. It is destructive and its effects resist treatment. Permanent blindness is possible from exposure to it. The kidney seems to be most subject to injury regardless of the mode of exposure. Pulmonary edema has also occurred. This material is more toxic than either ethyl silicate or silicic acid, although it has been thought that the injury caused is largely due to the action of the silicic acid. Flammable when exposed to heat or flame; can react vigorously with oxidizing materials. Potentially violent reaction with metal hexafluorides (e.g., rhenium, molybdenum, tungsten). When heated to decomposition it emits acrid smoke and irritating fumes. See also SILICATES, and SILICA, AMORPHOUS HYDRATED.

MPK250   CAS: 98-83-9   HR: 3
α-METHYL STYRENE
DOT: UN 2303
mf: C_9H_{10} mw: 118.19
PROP: Colorless liquid. D: 0.913 @ 17°/4°, mp: −24.0°, bp: 167–170°. Insol in water; misc in alc and ether. IDLH 700 ppm.
SYNS: ISOPROPENIL-BENZOLO (ITALIAN) □ ISOPROPENYL-BENZEEN (DUTCH) □ ISO-PROPENYLBENZENE □ ISOPROPENYL-BENZOL (GERMAN) □ as-METHYLPHENYLETHYLENE □ α-METHYLSTYREEN (DUTCH) □ α-METHYL-STYROL (GERMAN) □ α-METIL-STIROLO (ITALIAN) □ β-PHENYLPROPENE □ 2-PHENYLPROPENE □ β-PHENYLPROPYLENE □ 2-PHENYLPROPYLENE

CONSENSUS REPORTS: Reported in EPA TSCA Inventory.

OSHA PEL: TWA 50 ppm; STEL 100 ppm
ACGIH TLV: TWA 50 ppm; STEL 100 ppm; (Proposed: 20 ppm; Confirmed Animal Carcinogen with Unknown Revelance to Humans)
DFG MAK: 100 ppm (490 mg/m³)
DOT CLASSIFICATION: 3; Label: Flammable Liquid

SAFETY PROFILE: Mildly toxic by inhalation. Human systemic effects by inhalation: irritant effects. A skin and eye irritant. Flammable when exposed to heat or flame; can react vigorously with oxidizing materials. When heated to decomposition it emits acrid smoke and irritating fumes.

MPN275   CAS: 22262-19-7   HR: 3
o-(METHYLTELLURO)BENZOIC ACID
mf: $C_8H_7O_2Te•Na$ mw: 285.74
SYN: BENZOIC ACID, o-(METHYLTELLURO)-, SODIUM SALT
ACGIH TLV: TWA 0.1 mg(Te)/m^3
SAFETY PROFILE: Poison by intraperitoneal route. When heated to decomposition it emits toxic fumes of Te.

MPN500   CAS: 58-18-4   HR: 3
17-METHYLTESTOSTERONE
mf: $C_{20}H_{30}O_2$ mw: 302.50
PROP: Crystals from EtOAc/isooctane. Mp: 164–166°.
SYNS: ANDROMETH □ ANDROSAN □ ANDROSAN (tablets) □ ANDROSTEN □ 4-ANDROSTENE-17-α-METHYL-17-β-OL-3-ONE □ ANERTAN □ ANERTAN (tablets) □ DELATESTRYL □ DIANABOL □ DUMOGRAN □ GLOSSO STERANDRYL □ HOMANDREN □ HORMALE □ 17-β-HYDROXY-17-METHYLANDROST-4-EN-3-ONE □ MALESTRONE □ MALOGEN □ MASENONE □ MASTESTONA □ MESTERONE □ METANDREN □ 17-METHYLTESTOSTERON □ METHYLTESTOSTERONE □ 17-α-METHYLTESTOSTER-ONE □ METRONE □ M.T. MUCORETTES □ NABOLIN □ NEO-HOMBREOL-M □ NSC-9701 □ NU MAN □ ORAVIRON □ ORETON-M □ ORETON METHYL □ STENOLON □ STERONYL □ SYNANDRETS □ SYNANDROTABS □ TESTHORMONE □ TESTORA □ TESTOVIRON □ TESTRED
CONSENSUS REPORTS: Reported in EPA TSCA Inventory. EPA Genetic Toxicology Program.
SAFETY PROFILE: Poison by intraperitoneal route. Moderately toxic by ingestion. Human teratogenic effects by ingestion: developmental abnormalities of the urogenital system. Experimental teratogenic and reproductive effects. Human systemic effects: cholestatic jaundice, weight loss or decreased weight gain. Questionable human carcinogen producing liver tumors. A synthetic androgenic steroid. When heated to decomposition it emits acrid smoke and irritating fumes. See also TESTOSTERONE.

MPO800   CAS: 22056-53-7   HR: 3
METHYL 1,4,5,6-TETRAHYDRO-2-METHYLCYCLOPENTA(b)PYRROL-3-YL KETONE
mf: $C_{10}H_{13}NO$ mw: 163.24
SYN: KETONE, METHYL 1,4,5,6-TETRAHYDRO-2-METHYLCYCLOPENTA(b)PYRROL-3-YL
DOT CLASSIFICATION: 3; Label: Flammable Liquid
SAFETY PROFILE: A poison by intraperitoneal route. A flammable liquid. When heated to decomposition it emits toxic vapors of NO_x.

MPQ250   CAS: 13183-79-4   HR: 3
1-METHYL-1H-TETRAZOLE-5-THIOL
mf: $C_2H_4N_4S$ mw: 116.16
PROP: IDLH 2500 mg/m^3.
SYN: 1-METHYL-5-MERCAPTO-1,2,3,4-TETRAZOLE
CONSENSUS REPORTS: Reported in EPA TSCA Inventory.
SAFETY PROFILE: Poison by intraperitoneal route. Experimental reproductive effects. When heated to decomposition it emits very toxic fumes of NO_x and SO_x. See also SULFIDES.

MPQ900   CAS: 1128-05-8   HR: 3
METHYL 3-THIANAPHTHENYL KETONE
mf: $C_{10}H_8OS$ mw: 176.24
DOT CLASSIFICATION: 3; Label: Flammable Liquid
SAFETY PROFILE: A poison by parenteral route. A flammable liquid. When heated to decomposition it emits toxic vapors of SO_x.

MPR300   CAS: 13679-74-8   HR: 3
METHYLTHIENYLCETONE
mf: C_7H_8OS mw: 140.21
SYNS: ETHANONE, 1-(5-METHYL-2-THIENYL)- □ KETONE, METHYL 5-METHYL-2-THIENYL
DOT CLASSIFICATION: 3; Label: Flammable Liquid

SAFETY PROFILE: A poison by intraperitoneal route. A flammable liquid. When heated to decomposition it emits toxic vapors of SO$_x$.

MPS600   CAS: 23611-64-5   HR: 3
6-METHYL-2-THIO-2H-1,3-BENZOXA-
ZINE-2,4(3H)-DIONE
mf: C$_9$H$_7$NO$_2$S   mw: 193.23
SYN: 2H-1,3-BENZOXAZINE-2,4(3H)-DIONE, 6-METHYL-2-THIO-
SAFETY PROFILE: A poison by ingestion. When heated to decomposition it emits toxic vapors of NO$_x$ and SO$_x$.

MPT000   CAS: 556-64-9   HR: 3
METHYL THIOCYANATE
mf: C$_2$H$_3$NS   mw: 73.12
PROP: Liquid. D: 1.068 @ 20°, mp: −51°, bp: 130–133°. Very sltly sol in water; misc in alc and ether.
SYNS: METHYLRHODANID (GERMAN) □ METHYL SULFOCYANATE □ METHYLTHIOKYANAT
CONSENSUS REPORTS: Reported in EPA TSCA Inventory. EPA Extremely Hazardous Substances List.
SAFETY PROFILE: Poison by ingestion, intravenous, and subcutaneous routes. When heated to decomposition it emits very toxic fumes of NO$_x$ and SO$_x$. See also THIOCYANATES.

MPU000   CAS: 342-69-8   HR: 3
METHYLTHIOINOSINE
mf: C$_{11}$H$_{14}$N$_4$O$_4$S   mw: 298.35
PROP: Crystals from EtOH. Mp: 165–167°.
SYNS: 6-METHYLMERCAPTOPURINE RIBONUCLEO-SIDE □ 6-METHYLMERCAPTOPURINE RIBOSIDE □ 6-METHYL-9-RIBOFURANOSYLPURINE-6-THIOL □ 6-METHYL-MP-RIBOSIDE □ 6-METHYLTHIOINOSINE □ 6-(METHYLTHIO)PURINE RIBONUCLEOSIDE □ 6-METHYLTHIOPURINE RIBOSIDE □ NCI-C04784 □ NSC-40774 □ β-d-RIBOSYL-6-METHYLTHIOPURINE □ SQ 21977
CONSENSUS REPORTS: NCI Carcinogenesis Studies (ipr); Equivocal Evidence: mouse CANCAR 40,1935,77; (ipr); No Evidence: rat CANCAR 40,1935,77.

SAFETY PROFILE: Poison by intraperitoneal route. Experimental teratogenic effects. Questionable carcinogen with experimental tumorigenic data. Mutation data reported. When heated to decomposition it emits very toxic fumes of SO$_x$ and NO$_x$.

MPU600   CAS: 66637-32-9   HR: 3
3-METHYL-2,5-THIOMORPHOLINEDI-
ONE 2-(o-((METHYLAMINO)-
CARBONYL)OXIME)
mf: C$_7$H$_{11}$N$_3$O$_3$S   mw: 217.27
SYNS: 2-(o-(METHYLCARBAMOYL)OXIMINO)-3-METHYLTETRAHYDRO-1,4-THIAZIN-5-ONE □ 2,5-THIOMORPHOLINEDIONE, 3-METHYL-, 2-(o-((METHYLAMINO)CARBONYL)OXIME)
SAFETY PROFILE: A poison by ingestion. When heated to decomposition it emits toxic vapors of NO$_x$ and SO$_x$.

MPV400   CAS: 3268-49-3   HR: 2
3-(METHYLTHIO)PROPIONALDEHYDE
DOT: UN 2785
mf: C$_4$H$_8$OS   mw: 104.18
SYNS: METHIONAL □ β-(METHYLMERCAPTO)PROPIONALDEHYDE □ 3-(METHYLMERCAPTO)PROPIONALDEHYDE □ METHYLMERCAPTOPROPIONIC ALDEHYDE □ 3-(METHYLTHIO)PROPANAL □ β-(METHYLTHIO)-PROPIONALDEHYDE □ PROPANAL, 3-(METHYLTHIO)-(9CI) □ PROPIONALDEHYDE, 3-(METHYLTHIO)- □ THIA-4-PENTANAL (DOT)
CONSENSUS REPORTS: Reported in EPA TSCA Inventory.
DOT CLASSIFICATION: 6.1; Label: KEEP AWAY FROM FOOD
SAFETY PROFILE: Moderately toxic by ingestion. When heated to decomposition it emits toxic vapors of SO$_x$.

MPW500   CAS: 56-04-2   HR: 3
6-METHYLTHIOURACIL
mf: C$_5$H$_6$N$_2$OS   mw: 142.19
PROP: Bitter crystals or colorless liquid; odor of onions. Prisms from water. Mp: 326–331° (decomp), sublimes readily. Very sltly sol in ether, cold water, alkaline hydroxides, NH$_3$; sltly sol in alc, acetone; almost insol in benzene, chloroform.

M

SYNS: ALKIRON □ ANTIBASON □ BASECIL □ BASETHYRIN □ 2,3-DIHYDRO-6-METHYL-2-THIOXO-4(1H)-PYRIMIDINONE □ 2-MERCAPTO-4-HYDROXY-6-METHYLPYRIMIDINE □ 2-MERCAPTO-6-METHYL-PYRIMID-4-ONE □ 2-MERCAPTO-6-METHYL-4-PYRIMIDONE □ METACIL □ METHIACIL □ METHIOCIL □ 6-METHYL-2-THIO-2,4-(1H3H)PYRIMIDINEDIONE □ METHYLTHIOURACIL □ 4-METHYL-2-THIOURACIL □ 6-METHYL-2-THIOURACIL □ 4-METHYLURACIL □ 6-METIL-TIOURACILE (ITALIAN) □ MTU □ MURACIL □ ORCANON □ PROSTRUMYL □ RCRA WASTE NUMBER U164 □ STRUMACIL □ THIMECIL □ THIOMECIL □ 2-THIO-6-METHYL-1,3-PYRIMIDIN-4-ONE □ 6-THIO-4-METHYLURACIL □ THIOMIDIL □ 2-THIO-4-OXO-6-METHYL-1,3-PYRIMIDINE □ THIORYL □ THIOTHYMIN □ THIOTHYRON □ THIURYL □ THYREONORM □ THYREOSTAT □ THYRIL □ TIOMERACIL □ TIORALEM □ TIOTIRON □ USAF EK-6454

CONSENSUS REPORTS: IARC Cancer Review: Group 2B IMEMDT 7,56,87; Animal Sufficient Evidence IMEMDT 7,53,74. Reported in EPA TSCA Inventory.

SAFETY PROFILE: Confirmed carcinogen with experimental carcinogenic, neoplastigenic, tumorigenic, and teratogenic data. Poison by intraperitoneal route. Moderately toxic by ingestion. Human teratogenic and reproductive effects by an unspecified route: developmental abnormalities of the endocrine system and effects on newborn including neonatal measures or effects. Experimental reproductive effects. Used to treat hyperthyroidism. When heated to decomposition it emits very toxic fumes of NO_x and SO_x.

MQC320   CAS: 55391-24-7   HR: 3
4-((N-METHYL-N-TRICHLOROMETH-ANESULFENYL)CARBAMOYL-OXIMINO)-1,3-DITHIOLANE

mf: $C_6H_7Cl_3N_2O_2S_3$ mw: 341.68

SYN: 1,3-DITHIOLAN-4-ONE, o-((METHYL((TRI-CHLOROMETHYL)THIO)AMINO)CARBONYL)OXIME

SAFETY PROFILE: A poison by ingestion. When heated to decomposition it emits toxic vapors of NO_x, SO_x, and Cl^-.

MQC500   CAS: 75-79-6   HR: 3
METHYLTRICHLOROSILANE

DOT: UN 1250

mf: CH_3Cl_3Si mw: 149.48

PROP: A liquid. D: 1.27 @ 20°/4°, bp: 66°.

SYNS: METHYL-TRICHLORSILAN (CZECH) □ SILANE, TRICHLOROMETHYL- □ TRICHLOR-METHYLSILAN

CONSENSUS REPORTS: Reported in EPA TSCA Inventory. EPA Extremely Hazardous Substances List.

DOT CLASSIFICATION: 3; Label: Flammable Liquid, Corrosive

SAFETY PROFILE: Poison by inhalation and intraperitoneal routes. Moderately toxic by ingestion. A severe irritant to skin, eyes, and mucous membranes. Flammable when exposed to heat or flame; can react vigorously with oxidizing materials. When heated to decomposition it emits toxic fumes of Cl^-. See also CHLOROSILANES.

MQC750   CAS: 993-16-8   HR: 2
METHYLTRICHLOROSTANNANE

mf: CH_3Cl_3Sn mw: 240.08

PROP: Crystals from pet ether or C_6H_6 or by sublimation. Mp: 53°.

SYNS: METHYLTIN TRICHLORIDE □ METHYL-TRICHLOROTIN □ MONOMETHYLTIN TRICHLORIDE □ TRICHLOROMETHYLSTANNANE □ TRICHLORO-METHYLTIN

CONSENSUS REPORTS: Reported in EPA TSCA Inventory.

OSHA PEL: TWA 0.1 mg(Sn)/m³ (skin)

ACGIH TLV: TWA 0.1 mg(Sn)/m³; STEL 0.2 mg(Sn)/m³ (skin).

NIOSH REL: (Organotin Compounds) TWA 0.1 mg(Sn)/m³

SAFETY PROFILE: Moderately toxic by an unspecified route. Experimental reproductive effects. When heated to decomposition it emits toxic fumes of Cl^-. See also TIN COMPOUNDS and CHLORIDES.

MQE100   CAS: 73747-54-3   HR: 3
(α-METHYL-m-TRIFLUOROMETHYL-PHENETHYLAMINOMETHYL) PIPERIDINO KETONE

mf: $C_{17}H_{23}F_3N_2O$ mw: 328.42

SYN: KETONE, (α-METHYL-m-TRIFLUOROMETHYL-PHENETHYLAMINOMETHYL) PIPERIDINO

DOT CLASSIFICATION: 3; Label: Flammable Liquid
SAFETY PROFILE: A poison by intraperitoneal route. A flammable liquid. When heated to decomposition it emits toxic vapors of NO_x and F^-.

MQH000 CAS: 5137-55-3 HR: 3
METHYLTRIOCTYLAMMONIUM CHLORIDE
mf: $C_{25}H_{54}N•Cl$ mw: 404.25
SYNS: ALIQUAT 336 ☐ ALIQUAT 336N ☐ ALIQUAT 336-PTC ☐ N-METHYL-N,N-DIOCTYL-1-OCTAN-AMINIUM CHLORIDE ☐ METHYLTRICAPRYLYL-AMMONIUM CHLORIDE ☐ 1-OCTANAMINIUM, N-METHYL-N,N-DIOCTYL-, CHLORIDE (9CI) ☐ TRICAPRYLMETHYLAMMONIUM CHLORIDE ☐ TRICAPRYLYLMETHYLAMMONIUM CHLORIDE ☐ TRIOCTYLMETHYLAMMONIUM CHLORIDE
CONSENSUS REPORTS: Reported in EPA TSCA Inventory.
SAFETY PROFILE: Poison by ingestion and intraperitoneal routes. When heated to decomposition it emits very toxic fumes of NO_x, NH_3, and Cl^-. See also CHLORIDES.

MQH100 CAS: 22223-55-8 HR: 3
4-METHYL-2,6,7-TRIOXA-1-ARSABI-CYCLO(2.2.2)OCTANE
mf: $C_5H_9AsO_3$ mw: 192.06
PROP: Prisms. Mp: 42–43°, bp: 45° @ 0.1 mm.
SYN: 2,6,7-TRIOXA-1-ARSABICYCLO(2.2.2)OCTANE, 4-METHYL-
OSHA PEL: TWA 0.5 mg(As)/m³
SAFETY PROFILE: Poison by intravenous route. When heated to decomposition it emits toxic fumes of As.

MQH500 CAS: 57583-34-3 HR: 2
METHYLTRIS(2-ETHYLHEXYLOXY-CARBONYLMETHYLTHIO)-STANNANE
mf: $C_{31}H_{60}O_6S_3Sn$ mw: 743.78
CONSENSUS REPORTS: Reported in EPA TSCA Inventory.
OSHA PEL: TWA 0.1 mg(Sn)/m³ (skin)
ACGIH TLV: TWA 0.1 mg(Sn)/m³; STEL 0.2 mg(Sn)/m³ (skin).

NIOSH REL: (Organotin Compounds) TWA 0.1 mg(Sn)/m³
SAFETY PROFILE: Moderately toxic by ingestion. When heated to decomposition it emits toxic fumes of SO_x. See also TIN COMPOUNDS and CARBONYLS.

MQH750 CAS: 953-17-3 HR: 3
METHYL TRITHION
mf: $C_9H_{12}ClO_2PS_3$ mw: 314.81
PROP: Light-yellow to amber liquid. D: 1.34–1.35 @ 20°/20°, mp: −18°. Very spar sol in H_2O; misc in most org solvs.
SYNS: ((p-CHLOROPHENYL)THIO)METHANETHIOL-S-ESTER with O,O-DIMETHYL PHOSPHORODITHIOATE ☐ S-(((p-CHLOROPHENYL)THIO)METHYL) O,O-DI-METHYL PHOSPHORODITHIOATE ☐ S-(((4-CHLORO-PHENYL)THIO)METHYL) O,O-DIMETHYLPHOSPHORO-DITHIOATE ☐ DIMETHYL-p-CHLOROPHENYLTHIO-METHYL DITHIOPHOSPHATE ☐ O,O-DIMETHYL-S-(p-CHLOROPHENYLTHIOMETHYL)PHOSPHORODITHIOA TE ☐ O,O-DIMETHYLTHIOPHOSPHORIC ACID, p-CHLOROPHENYL ESTER ☐ ENT 25,599 ☐ G-29288 ☐ GEIGY G-29288 ☐ METHYLCARBOPHENOTHION ☐ R-1492 ☐ STAUFFER R-1492 ☐ TRI-ME
SAFETY PROFILE: Poison by ingestion and skin contact routes. A cholinesterase inhibitor type of insecticide. When heated to decomposition it emits very toxic fumes of Cl^-, PO_x, and SO_x. See also PARATHION.

MQI550 CAS: 110-41-8 HR: 1
2-METHYLUNDECANAL
mf: $C_{12}H_{24}O$ mw: 184.32
PROP: Colorless to sltly yellow liquid; fatty odor. D: 0.822–0.830, refr index: 1.431. Sol in alc, fixed oils, propylene glycol; insol in glycerin.
SYNS: ALDEHYDE C-12, MNA ☐ ALDEHYDE M.N.A. ☐ FEMA No. 2749 ☐ METHYL n-NONYL ACETAL-DEHYDE ☐ METHYL NONYL ACETIC ALDEHYDE ☐ METHYLNONYLACETALDEHYDE
CONSENSUS REPORTS: Reported in EPA TSCA Inventory.
SAFETY PROFILE: Low toxicity by ingestion and skin contact. When heated to decomposition it emits acrid smoke and irritating fumes.

MQI750   CAS: 626-48-2   HR: 2
6-METHYLURACIL
mf: $C_5H_6N_2O_2$ mw: 126.13

PROP: Crystals from glacial acetic acid or water. Mp: 311–312°. Decomp above 300°.

SYNS: 6-METHYL-2,4(1H,3H)-PYRIMIDINEDIONE □ 4-METHYLURACIL □ PSEUDOTHYMINE

CONSENSUS REPORTS: Reported in EPA TSCA Inventory.

SAFETY PROFILE: Moderately toxic by ingestion. Questionable carcinogen with experimental neoplastigenic and teratogenic data. Experimental reproductive effects. When heated to decomposition it emits toxic fumes of NO_x.

MQK750   CAS: 108-22-5   HR: 3
METHYLVINYL ACETATE
DOT: UN 2403

mf: $C_5H_8O_2$ mw: 100.13

PROP: Water-white liquid. Mp: −92.9°, bp: 96.6° @ 746 mm, flash p: 60°F (OC), d: 0.9226 @ 20°/20°, vap d: 3.45.

SYNS: ACETIC ACID, ISOPROPENYL ESTER □ ISOPROPENYL ACETATE □ ISOPROPENYL ACETATE (DOT) □ ISOPROPENYLESTER KYSELINY OCTOVE □ 1-PROPEN-2-YL ACETATE

CONSENSUS REPORTS: Reported in EPA TSCA Inventory.

DOT CLASSIFICATION: 3; Label: Flammable Liquid

SAFETY PROFILE: Moderately toxic by ingestion. A skin, eye, and mucous membrane irritant. A very dangerous fire hazard when exposed to heat or flame; can react vigorously with oxidizing materials. To fight fire, use alcohol foam, CO_2, dry chemical. When heated to decomposition it emits acrid smoke and irritating fumes.

MQL750   CAS: 107-25-5   HR: 3
METHYL VINYL ETHER
DOT: UN 1087

mf: C_3H_6O mw: 58.09

PROP: Colorless, easily liquefied gas or colorless liquid. Bp: 8.0°, d: 0.7500, vap d: 2.0, fp: −121.6°, vap press: 1052 mm @ 20°, flash p: −68.8°F, lel: 2.6%, uel: 39.0%.

SYNS: METHOXYETHENE □ VINYL METHYL ETHER (DOT)

CONSENSUS REPORTS: Reported in EPA TSCA Inventory.

DOT CLASSIFICATION: 2.1; Label: Flammable Gas

SAFETY PROFILE: Mildly toxic by ingestion. A very dangerous fire hazard when exposed to heat, flame, or oxidizers. Explosive in the form of vapor when exposed to heat or flame. Can react vigorously with oxidizing materials. To fight fire, stop flow of gas. Potentially explosive reaction with halogens (e.g., bromine, chlorine) or hydrogen halides (e.g., hydrogen bromide, hydrogen chloride). Reaction with acids forms acetaldehyde. Weak acids catalyze the exothermic polymerization of the ether. The unstabilized ether can form dangerous peroxides. When heated to decomposition it emits acrid smoke and irritating fumes. See also ETHERS and PEROXIDES.

MQM150   CAS: 73160-32-4   HR: 3
METHYL VINYL OXIMINO SILANE
mf: $C_{11}H_{22}N_2O_2Si$ mw: 242.44

SYNS: 2-BUTANONE, O,O'-(ETHENYLMETHYLSILYLENE)DIOXIME, (E,Z)- □ 2-BUTANONE, O,O'-(ETHENYLMETHYLSILYLENE)-DIOXIME, (2E,2'Z)-

SAFETY PROFILE: A poison by ingestion and skin contact. Experimental reproductive effects. When heated to decomposition it emits toxic vapors of NO_x. Mutation data reported.

MQP500   CAS: 129-49-7   HR: 3
METHYSERGIDE DIMALEATE
mf: $C_{21}H_{27}N_3O_2 \cdot C_4H_4O_4$ mw: 469.59

PROP: Mp: 187–188° (decomp). Sol in methanol; less sol in water; insol in abs ethanol.

SYNS: 1-(HYDROXYMETHYL)PROPYLAMIDE of 1-METHYL-(+)-LYSERGIC ACID HYDROGEN MALEATE □ METHYLSERGIDE BIMALEATE □ SANSERT

SAFETY PROFILE: Poison by ingestion and intravenous routes. Experimental reproductive effects. Human mutation effects reported. When heated to decomposition it emits toxic fumes of NO$_x$.

MQQ000   CAS: 5800-19-1   HR: 3
METIAPINE
mf: C$_{19}$H$_{21}$N$_3$S   mw: 323.49
PROP: A solid. Mp: 99–107°.
SYN: 2-METHYL-11-(4-METHYL-1-PIPERAZINYL)-DIBENZO(b,f)(1,4)THIAZEPINE
SAFETY PROFILE: Poison by intraperitoneal route. Moderately toxic by ingestion. An experimental teratogen. Experimental reproductive effects. When heated to decomposition it emits very toxic fumes of NO$_x$ and SO$_x$.

MQR200   CAS: 1178-29-6   HR: 3
METOSERPATE HYDROCHLORIDE
mf: C$_{24}$H$_{32}$N$_2$O$_5$•ClH   mw: 465.04
PROP: A solid. Mp: 240–242°. Sol in water.
SYNS: METHYL-18-EPIRESERPATE METHYL ETHER HYDROCHLORIDE □ PACITRAN □ SU-9064 □ SU 8842 HYDROCHLORIDE
SAFETY PROFILE: Poison by ingestion and intravenous routes. When heated to decomposition it emits toxic fumes of NO$_x$ and HCl.

MQR275   CAS: 21087-64-9   HR: 3
METRIBUZIN
mf: C$_8$H$_{14}$N$_4$OS   mw: 214.32
PROP: White crystalline solid; mild odor. Mp: 125°. Sltly sol in water.
SYNS: 4-AMINO-6-tert-BUTYL-3-(METHYLTHIO)-1,2,4-TRIAZIN-5-ONE □ 4-AMINO-6-tert-BUTYL-3-METHYL-THIO-as-TRIAZIN-5-ONE □ 4-AMINO-6-(1,1-DIMETHYL-ETHYL)-3-(METHYLTHIO)-1,2,4-TRIAZIN-5(4H)-ONE □ BAY 61597 □ BAY DIC 1468 □ BAYER 6159H □ BAYER 6443H □ BAYER 94337 □ DIC 1468 □ LEXONE □ SENCOR □ SENCORAL □ SENCORER □ SENCOREX
CONSENSUS REPORTS: EPA Genetic Toxicology Program.
OSHA PEL: TWA 5 mg/m^3
ACGIH TLV: TWA 5 mg/m^3; Not Classifiable as a Human Carcinogen
NIOSH REL: (Metribuzin) TWA 5 mg/m^3
SAFETY PROFILE: Poison by ingestion and intraperitoneal routes. Low toxicity by skin contact. When heated to decomposition it emits very toxic fumes of NO$_x$ and SO$_x$.

MQR750   CAS: 7786-34-7   HR: 3
MEVINPHOS
mf: C$_7$H$_{13}$O$_6$P   mw: 224.17
PROP: Misc in H$_2$O and org solvs except pet ether. IDLH 4 ppm.
SYNS: APAVINPHOS □ α-2-CARBOMETHOXY-1-METHYLVINYL DIMETHYL PHOSPHATE □ 2-CARBOMETHOXY-1-PROPEN-2-YL DIMETHYL PHOSPHATE □ CMDP □ COMPOUND 2046 □ 3-((DIMETHOXYPHOSPHINYL)OXY)-2-BUTENOIC ACID METHYL ESTER □ O,O-DIMETHYL-O-(2-CARBO-METHOXY-1-METHYLVINYL) PHOSPHATE □ DIMETHYL-1-CARBOMETHOXY-1-PROPEN-2-YL PHOSPHATE □ DIMETHYL ESTER PHOSPHORIC ACID ESTER with METHYL 3-HYDROXYCROTONATE □ O,O-DIMETHYL-O-2-METHOXYCARBONYL-1-METHYL-VINYL-PHOSPHAT (GERMAN) □ DIMETHYL 2-METHOXYCARBONYL-1-METHYLVINYL PHOSPHATE □ DIMETHYL METHOXYCARBONYLPROPENYL PHOSPHATE □ DIMETHYL (1-METHOXYCARBOXY-PROPEN-2-YL)PHOSPHATE □ O,O-DIMETHYL O-(1-METHYL-2-CARBOXYVINYL) PHOSPHATE □ DURAPHOS □ ENT 22,374 □ FOSDRIN □ GESFID □ GESTID □ 3-HYDROXYCROTONIC ACID METHYL ESTER DIMETHYL PHOSPHATE □ MENIPHOS □ MENITE □ (2-METHOXYCARBONYL-1-METHYL-VINYL)-DIMETHYL-FOSFAAT (DUTCH) □ (2-METHOXYCARBONYL-1-METHYL-VINYL)-DIMETHYL-PHOSPHAT (GERMAN) □ 2-METHOXYCARBONYL-1-METHYLVINYL DIMETHYLPHOSPHATE □ 1-METHOXYCARBONYL-1-PROPEN-2-YL DIMETHYL PHOSPHATE □ (1-METHOXYCARBOXYPROPEN-2-YL)PHOSPHORIC ACID, DIMETHYL ESTER □ METHYL-3-(DIMETHOXYPHOSPHINYLOXY)CROTONATE □ (2-METOSSICARBONIL-1-METIL-VINIL)-DIMETIL-FOSFATO (ITALIAN) □ MEVINFOS (DUTCH) □ OS 2046 □ PHOSDRIN (OSHA) □ PHOSFENE □ PHOSPHATE de DIMETHYLE et de 2-METHOXYCARBONYL-1 METHYLVINYLE (FRENCH) □ PHOSPHENE (FRENCH)
CONSENSUS REPORTS: EPA Genetic Toxicology Program. EPA Extremely Hazardous Substances List.
OSHA PEL: TWA 0.01 ppm; STEL 0.03 ppm (skin)
ACGIH TLV: TWA 0.01 mg/m^3 (skin); Not Classifiable as a Human Carcinogen
DFG MAK: 0.01 ppm (0.093 mg/m^3)
SAFETY PROFILE: Poison by ingestion, inhalation, skin contact, subcutaneous, intravenous, and intraperitoneal routes. Human systemic effects by ingestion: peripheral motor nerve recording changes. An insecticide. When heated to decomposition it emits toxic fumes of PO$_x$.

M

MQS225   CAS: 3704-09-4   HR: 2
MIBOLERONE
mf: $C_{20}H_{30}O_2$ mw: 302.50
PROP: Crystalline solid from
Me_2CO/hexane. Solubility in deionized
water: 0.0454 mg/mL @ 37°.
SYNS: CHEQUE □ (7-α,17-β)-17-HYDROXY-7,17-
DIMETHYL-ESTR-4-EN-3-ONE (9CI) □ 17-β-HYDROXY-
7-α,17-DIMETHYLESTR-4-EN-3-ONE □ MATENON □
MIBOLERON □ U 10997
SAFETY PROFILE: Experimental
teratogenic and reproductive effects.
Questionable carcinogen with experimental
neoplastigenic data. When heated to
decomposition it emits acrid smoke and
irritating fumes.

MQS250   CAS: 12001-26-2   HR: 2
MICA
PROP: Containing less than 1% crystalline
silica (FEREAC 39,23540,74). IDLH 1500
mg/m³.
SYNS: MICA SILICATE □ SUZORITE MICA
OSHA PEL: TWA Respirable Fraction: 3
mg/m³
ACGIH TLV: TWA Respirable Fraction: 3
mg/m³
NIOSH REL: (Silicates <1% Crystalline
Silica) TWA 3 mg/m³
SAFETY PROFILE: The dust is injurious to
lungs. See SILICATES.

MQS500   CAS: 90-94-8   HR: 3
MICHLER'S KETONE
mf: $C_{17}H_{20}N_2O$ mw: 268.39
PROP: Leaves from ethanol. Mp: 179°, bp:
>360° decomp. Insol in water; very sol in
benzene; sol in alc; very sltly sol in ether.
SYNS: p,p'-BIS(N,N-DIMETHYLAMINO)BENZO-
PHENONE □ 4,4'-BIS(DIMETHYLAMINO)BENZO-
PHENONE □ BIS(p-(N,N-DIMETHYLAMINO)PHENYL)-
KETONE □ BIS(4-(DIMETHYLAMINO)PHENYL)-
METHANONE □ p,p'-MICHLER'S KETONE □ NCI-
C02006 □ TETRAMETHYLDIAMINOBENZOPHENONE
CONSENSUS REPORTS: NTP 10th Report
on Carcinogens. NCI Carcinogenesis
Bioassay (feed); Clear Evidence: mouse, rat
NCITR* NCI-CG-TR-181,79. Reported in
EPA TSCA Inventory. EPA Genetic
Toxicology Program.
DFG MAK: Confirmed Animal Carcinogen
with Unknown Relevance to Humans
DOT CLASSIFICATION: 3; Label:
Flammable Liquid
SAFETY PROFILE: Confirmed human
carcinogen with experimental carcinogenic
and neoplastigenic data. A poison by
ingestion. Mutation data reported. A
flammable liquid. When heated to
decomposition it emits toxic fumes of NO_x.
See also KETONES.

MQU075   HR: D
MILK-CLOTTING ENZYME from BACIL-
**   LUS CEREUS**
PROP: Derived from *Bacillus cereus* (Fam.
Bacillaceae).
SAFETY PROFILE: When heated to
decomposition it emits acrid smoke and
irritating fumes.

MQU120   HR: D
MILK-CLOTTING ENZYME from MUCOR
**   MIEHEI**
PROP: Derived from *Mucor miehei Cooney et
Emerson* (Fam. *Mucoraceae*).
SAFETY PROFILE: When heated to
decomposition it emits acrid smoke and
irritating fumes.

MQU750   CAS: 57-53-4   HR: 3
MILTOWN
mf: $C_9H_{18}N_2O_4$ mw: 218.29
PROP: Crystals with bitter taste. Mp:
104–106°.
SYNS: AMEPROMAT □ AMOSENE □ ANASTRESS □
ANATHYLMON □ ANDAKSIN □ ANDAXIN □
ANEURAL □ ANEUXRAL □ ANSIATAN □ ANSIL □
ANSIOWAS □ ANURAL □ ANXIETIL □ APASCIL □
ARCOBAN □ ARTOLON □ ATRAXINE □ AYERMATE □
BAMD 400 □ BIOBAMAT □ BROBAMATE □ CALMADIN
□ CALMAX □ CALMIREN □ CANQUIL-400 □ CAP-O-
TRAN □ CIRPONYL □ CRESTANIL □ CYPRON □
DAPAZ □ DICANDIOL □ 2,2-DI(CARBAMOYLOXY-
METHYL)PENTANE □ DIVERON □ DORMABROL □
ECUANIL □ EDENAL □ ENORDEN □ EPICUR □
EQUANIL SUSPENSION □ EQUILIUM □ EQUINIL □
ERINA □ ESTASIL □ FAS-CILE □ GADEXYL □

HARMONIN □ HARTOL □ HOLBAMATE □ IPSOTIAN □ KESSOBAMATE □ KLORT □ LARTEN □ LEPENIL □ LEPETOWN □ LETYL □ LIBIOLAN □ MADIOL □ MARBATE □ MARGONIL □ MENDEL □ MEPAMTIN □ MEPAVLON □ MEPIOSINE □ MEPOSED □ MEPRANIL □ MEPROBAM □ MEPROBAMAT (GERMAN) □ MEPROBAMATE □ MEPROBAMATO (ITALIAN) □ MEPROCOMPREN □ MEPROCON CMC □ MEPRODIL □ MEPROLEAF □ MEPROSAN □ MEPROTABS □ MEPROZINE □ MEPTRAN □ 2-METHYL-2-N-PROPYL-1,3-PROPANEDIOL DICARBAMATE □ 2-METHYL-2-PROPYLTRIMETHYLENE CARBAMATE □ METRACTYL □ MILPREM □ MILTANN □ NEO-TRAN □ NEPHENTINE □ OROLEVOL □ PANCALMA □ PAN-TRANQUIL □ PEREQUIL □ PLACIDON □ PROCALMIDOL □ PROQUANIL □ QUIETIDON □ RESTENIL □ ROBAMATE □ SEDABAMATE □ SERIL □ SPANTRAN □ TRANQUILAN □ TRELMAR □ URBIL □ VISTABAMATE □ WARDAMATE □ WYSEALS □ ZIRPON

CONSENSUS REPORTS: Reported in EPA TSCA Inventory. EPA Genetic Toxicology Program.

SAFETY PROFILE: Human poison by unspecified routes. Moderately toxic to humans and experimentally by ingestion. Experimental poison by intravenous, intraperitoneal, and subcutaneous routes. An experimental teratogen. Human systemic effects by ingestion: coma, blood pressure decrease, regional or general arteriolar constriction, dyspnea, cyanosis, respiratory depression, nausea or vomiting, and allergic skin dermatitis. Experimental reproductive effects. Mutation data reported. Implicated in aplastic anemia. Used as a tranquilizer. When heated to decomposition it emits toxic fumes of NO_x. See also CARBAMATES.

MQV250   CAS: 1401-55-4   HR: 3
MIMOSA TANNIN
SYNS: ACACIA MOLLISSIMA TANNIN □ TANNIN from MIMOSA

SAFETY PROFILE: Poison by intravenous and intraperitoneal routes. Questionable carcinogen with experimental tumorigenic data. When heated to decomposition it emits acrid smoke and irritating fumes. See also TANNIN.

MQV500   HR: D
MINERAL DUSTS
SAFETY PROFILE: Variable toxicity. From the economic and toxicity standpoints, the most important are those containing free silica, which can cause silicosis upon inhalation of sufficient quantity. These include sand, sandstone, quartz, and flint. They consist mainly of silica in the form of quartz; diatomaceous earth, which is essentially amorphous silica; and granite, which contains 20–40% quartz. Minerals that contain combined silica in the form of silicates but no free silica are generally less capable of causing silicosis. Asbestos, however, can cause a fibrotic lung condition of its own, known as asbestosis, and lung cancer. (See also various asbestos entries.) Mica and talc dust are also considered somewhat hazardous. Non-siliceous minerals, like limestone, marble, dolomite, etc., that do not contain toxic elements, do not ordinarily present any significant dust hazard. Minerals containing toxic elements, such as cryolite, which contains fluorine, and pyrolusite, which contains manganese, may cause systemic poisoning upon inhalation or ingestion of sufficient quantity. In any event, the minerals are usually less reactive than synthetic compounds of the same elements and, in fact, may be relatively inert by comparison. These are common air contaminants. See also specific materials.

MQV750   CAS: 8012-95-1   HR: 2
MINERAL OIL
PROP: Colorless, oily liquid; practically tasteless and odorless. D: 0.83–0.86 (light), 0.875–0.905 (heavy), flash p: 444°F (OC), ULC: 10–20. Insol in water and alc; sol in benzene, chloroform, and ether. A mixture of liquid hydrocarbons from petroleum. IDLH 2500 mg/m³.

SYNS: ADEPSINE OIL □ ALBOLINE □ BAYOL F □ BLANDLUBE □ CRYSTOSOL □ DRAKEOL □ FONOLINE □ GLYMOL □ KAYDOL □ KONDREMUL □ MINERAL OIL, WHITE (FCC) □ MOLOL □ NEO-CULTOL □ NUJOL □ OIL MIST, MINERAL (OSHA,

ACGIH) □ PARAFFIN OIL □ PAROL □ PAROLEINE □ PENETECK □ PENRECO □ PERFECTA □ PETROGAL- AR □ PETROLATUM, liquid □ PRIMOL 335 □ PROTOPET □ SAXOL □ TECH PET F □ WHITE MINERAL OIL

CONSENSUS REPORTS: Reported in EPA TSCA Inventory.

OSHA PEL: Oil Mist: TWA 5 mg/m³

ACGIH TLV: (Proposed: TWA 0.2 mg/m³ (metal working poorly refined); TWA 5 mg/m³ (metal working pure, highly refined))

SAFETY PROFILE: A human teratogen by inhalation that causes testicular tumors in the fetus. Inhalation of vapor or particulates can cause aspiration pneumonia. A skin and eye irritant. Highly purified food grades are of low toxicity. Questionable human carcinogen producing gastrointestinal tumors. Slightly combustible liquid when exposed to heat or flame. To fight fire, use dry chemical, CO_2, foam. When heated to decomposition it emits acrid smoke and fumes.

MQV755   CAS: 64741-49-7   HR: 3
MINERAL OIL, PETROLEUM CONDEN- SATES, VACUUM TOWER

SYNS: CONDENSATES (PETROLEUM), VACUUM TOWER (9CI) □ VACUUM RESIDUUM

CONSENSUS REPORTS: IARC Cancer Review: Group 1 IMEMDT 7,252,87; Animal Sufficient Evidence IMEMDT 33,87,84. Reported in EPA TSCA Inventory.

SAFETY PROFILE: Confirmed carcinogen. When heated to decomposition it emits acrid smoke and irritating fumes.

MQV760   CAS: 64742-18-3   HR: 3
MINERAL OIL, PETROLEUM DISTIL- LATES, ACID-TREATED HEAVY NAPHTHENIC (mild or no solvent- refining or hydrotreatment)

SYNS: ACID-TREATED HEAVY NAPHTHENIC DISTILLATE □ DISTILLATES (PETROLEUM), ACID- TREATED HEAVY NAPHTHENIC (9CI)

CONSENSUS REPORTS: IARC Cancer Review: Group 1 IMEMDT 7,252,87; Animal Sufficient Evidence IMEMDT 33,87,84. Reported in EPA TSCA Inventory.

SAFETY PROFILE: Confirmed carcinogen. When heated to decomposition it emits acrid smoke and irritating fumes.

MQV765   CAS: 64742-20-7   HR: 2
MINERAL OIL, PETROLEUM DISTIL- LATES, ACID-TREATED HEAVY PARAFFINIC (severe solvent-refining and/or hydrotreatment)

CONSENSUS REPORTS: IARC Cancer Review: Group 3 IMEMDT 7,252,87; Animal Inadequate Evidence IMEMDT 33,87,84. Reported in EPA TSCA Inventory.

SAFETY PROFILE: Questionable carcinogen. When heated to decomposition it emits acrid smoke and irritating fumes.

MQV770   CAS: 64742-19-4   HR: 3
MINERAL OIL, PETROLEUM DISTIL- LATES, ACID-TREATED LIGHT NAPHTHENIC (mild or no solvent- refining or hydrotreatment)

SYNS: ACID-TREATED LIGHT NAPHTHENIC DISTILLATE □ DISTILLATES (PETROLEUM), ACID- TREATED LIGHT NAPHTHENIC (9CI)

CONSENSUS REPORTS: IARC Cancer Review: Group 1 IMEMDT 7,252,87; Animal Sufficient Evidence IMEMDT 33,87,84. Reported in EPA TSCA Inventory.

SAFETY PROFILE: Confirmed carcinogen. When heated to decomposition it emits acrid smoke and irritating fumes.

MQV775   CAS: 64742-21-8   HR: 3
MINERAL OIL, PETROLEUM DISTIL- LATES, ACID-TREATED LIGHT PARAFFINIC (mild or no solvent- refining or hydrotreatment)

SYNS: ACID-TREATED LIGHT PARAFFINIC DISTILLATE □ DISTILLATES (PETROLEUM), ACID- TREATED LIGHT PARAFFINIC (9CI)

CONSENSUS REPORTS: IARC Cancer Review: Group 1 IMEMDT 7,252,87; Animal Sufficient Evidence IMEMDT 33,87,84. Reported in EPA TSCA Inventory.

SAFETY PROFILE: Confirmed carcinogen. When heated to decomposition it emits acrid smoke and irritating fumes.

MQV776   CAS: 64742-68-3   HR: 3
MINERAL OIL, PETROLEUM DISTILLATES CATALYTIC DEWAXED HEAVY NAPHTHENIC (mild or no solvent-refining or hydrotreatment)

SYNS: CATALYTIC-DEWAXED HEAVY NAPHTHENIC DISTILLATE □ NAPHTHENIC OILS (PETROLEUM), CATALYTIC DEWAXED HEAVY (9CI)

CONSENSUS REPORTS: IARC Cancer Review: Group 1 IMEMDT 7,252,87; Animal Sufficient Evidence IMEMDT 33,87,84. Reported in EPA TSCA Inventory.

SAFETY PROFILE: Confirmed carcinogen. When heated to decomposition it emits acrid smoke and irritating fumes.

MQV777   CAS: 64742-69-4   HR: 3
MINERAL OIL, PETROLEUM DISTILLATES CATALYTIC DEWAXED LIGHT NAPHTHENIC (mild or no solvent-refining or hydrotreatment)

SYNS: CATALYTIC-DEWAXED LIGHT NAPHTHENIC DISTILLATE □ NAPHTHENIC OILS (PETROLEUM), CATALYTIC DEWAXED LIGHT (9CI)

CONSENSUS REPORTS: IARC Cancer Review: Group 1 IMEMDT 7,252,87; Animal Sufficient Evidence IMEMDT 33,87,84. Reported in EPA TSCA Inventory.

SAFETY PROFILE: Confirmed carcinogen. When heated to decomposition it emits acrid smoke and irritating fumes.

MQV778   CAS: 64742-70-7   HR: 3
MINERAL OIL, PETROLEUM DISTILLATES CATALYTIC DEWAXED HEAVY PARAFFINIC (mild or no solvent-refining or hydrotreatment)

SYNS: CATALYTIC-DEWAXED HEAVY PARAFFINIC DISTILLATE □ PARAFFIN OILS (PETROLEUM), CATALYTIC DEWAXED HEAVY (9CI)

CONSENSUS REPORTS: IARC Cancer Review: Group 1 IMEMDT 7,252,87; Animal Sufficient Evidence IMEMDT

33,87,84. Reported in EPA TSCA Inventory.

SAFETY PROFILE: Confirmed carcinogen. When heated to decomposition it emits acrid smoke and irritating fumes.

MQV779   CAS: 64742-71-8   HR: 3
MINERAL OIL, PETROLEUM DISTILLATES CATALYTIC DEWAXED LIGHT PARAFFINIC (mild or no solvent-refining or hydrotreatment)

SYNS: CATALYTIC-DEWAXED LIGHT PARAFFINIC DISTILLATE □ PARAFFIN OILS (PETROLEUM), CATALYTIC DEWAXED LIGHT (9CI)

CONSENSUS REPORTS: IARC Cancer Review: Group 1 IMEMDT 7,252,87; Animal Sufficient Evidence IMEMDT 33,87,84. Reported in EPA TSCA Inventory.

SAFETY PROFILE: Confirmed carcinogen. When heated to decomposition it emits acrid smoke and irritating fumes.

MQV780   CAS: 64741-53-3   HR: 3
MINERAL OIL, PETROLEUM DISTILLATES, HEAVY NAPHTHENIC

SYNS: DISTILLATES (PETROLEUM), HEAVY NAPHTHENIC (9CI) □ HEAVY NAPHTHENIC DISTILLATE □ HEAVY NAPHTHENIC DISTILLATES (PETROLEUM)

CONSENSUS REPORTS: IARC Cancer Review: Group 1 IMEMDT 7,252,87; Animal Sufficient Evidence IMEMDT 33,87,84. Reported in EPA TSCA Inventory.

SAFETY PROFILE: Confirmed carcinogen with experimental neoplastigenic data. Mutation data reported. When heated to decomposition it emits acrid smoke and irritating fumes.

MQV785   CAS: 64741-51-1   HR: 3
MINERAL OIL, PETROLEUM DISTILLATES, HEAVY PARAFFINIC

SYNS: DISTILLATES (PETROLEUM), HEAVY PARAFFINIC (9CI) □ HEAVY PARAFFINIC DISTILLATE

CONSENSUS REPORTS: IARC Cancer Review: Group 1 IMEMDT 7,252,87; Animal Sufficient Evidence IMEMDT

M

33,87,84. Reported in EPA TSCA Inventory.
SAFETY PROFILE: Confirmed carcinogen. Mutation data reported.

MQV790   CAS: 64742-52-5   HR: 3
MINERAL OIL, PETROLEUM DISTIL-LATES, HYDROTREATED (mild) HEAVY NAPHTHENIC

SYNS: DISTILLATES (PETROLEUM), HYDRO-TREATED (mild) HEAVY NAPHTHENIC (9CI) □ HYDROTREATED (mild) HEAVY NAPHTHENIC DISTILLATE □ HYDROTREATED (mild) HEAVY NAPHTHENIC DISTILLATES (PETROLEUM) □ PETROLEUM DISTILLATES, HYDROTREATED (mild) HEAVY NAPHTHENIC

CONSENSUS REPORTS: IARC Cancer Review: Group 1 IMEMDT 7,252,87; Animal Inadequate Evidence IMEMDT 33,87,84. Reported in EPA TSCA Inventory.
SAFETY PROFILE: Confirmed carcinogen with experimental tumorigenic data. Low toxicity by ingestion and skin contact. A severe skin irritant. Mutation data reported. When heated to decomposition it emits acrid smoke and irritating fumes.

MQV795   CAS: 64742-54-7   HR: 3
MINERAL OIL, PETROLEUM DISTIL-LATES, HYDROTREATED (mild) HEAVY PARAFFINIC

SYNS: DISTILLATES (PETROLEUM), HYDROTREATED (mild) HEAVY PARAFFINIC (9CI) □ HYDROTREATED (mild) HEAVY PARAFFINIC DISTILLATE

CONSENSUS REPORTS: IARC Cancer Review: Group 1 IMEMDT 7,252,87; Animal Sufficient Evidence IARC 33,87,84. Reported in EPA TSCA Inventory.
SAFETY PROFILE: Confirmed carcinogen. Low toxicity by ingestion and skin contact. When heated to decomposition it emits acrid smoke and irritating fumes.

MQV800   CAS: 64742-53-6   HR: 3
MINERAL OIL, PETROLEUM DISTIL-LATES, HYDROTREATED (mild) LIGHT NAPHTHENIC

SYNS: DISTILLATES (PETROLEUM), HYDRO-TREATED (mild) LIGHT NAPHTHENIC (9CI) □

HYDROTREATED (mild) LIGHT NAPHTHENIC DISTILLATE □ HYDROTREATED (mild) LIGHT NAPHTHENIC DISTILLATES (PETROLEUM)

CONSENSUS REPORTS: IARC Cancer Review: Group 1 IMEMDT 7,252,87; Animal Sufficient Evidence IMEMDT 33,87,84. Reported in EPA TSCA Inventory.
SAFETY PROFILE: Confirmed carcinogen with experimental neoplastigenic data. Low toxicity by ingestion and skin contact. A severe skin irritant. When heated to decomposition it emits acrid smoke and irritating fumes.

MQV805   CAS: 64742-55-8   HR: 3
MINERAL OIL, PETROLEUM DISTIL-LATES, HYDROTREATED (mild) LIGHT PARAFFINIC

SYNS: DISTILLATES (PETROLEUM), HYDROTREATED (mild) LIGHT PARAFFINIC (9CI) □ HYDROTREATED (mild) LIGHT PARAFFINIC DISTILLATE

CONSENSUS REPORTS: IARC Cancer Review: Group 1 IMEMDT 7,252,87; Animal Sufficient Evidence IMEMDT 33,87,84. Reported in EPA TSCA Inventory.
SAFETY PROFILE: Confirmed carcinogen. When heated to decomposition it emits acrid smoke and irritating fumes.

MQV810   CAS: 64741-52-2   HR: 3
MINERAL OIL, PETROLEUM DISTIL-LATES, LIGHT NAPHTHENIC

SYNS: DISTILLATES (PETROLEUM), LIGHT NAPHTHENIC (9CI) □ LIGHT NAPHTHENIC DISTILLATE □ LIGHT NAPHTHENIC DISTILLATES (PETROLEUM)

CONSENSUS REPORTS: IARC Cancer Review: Group 1 IMEMDT 7,252,87; Animal Sufficient Evidence IMEMDT 33,87,84. Reported in EPA TSCA Inventory.
SAFETY PROFILE: Confirmed carcinogen with experimental neoplastigenic data. When heated to decomposition it emits acrid smoke and irritating fumes.

MQV815   **CAS: 64741-50-0**   **HR: 3**
MINERAL OIL, PETROLEUM DISTIL-
LATES, LIGHT PARAFFINIC
SYNS: DISTILLATES (PETROLEUM), LIGHT
PARAFFINIC (9CI) □ LIGHT PARAFFINIC DISTILLATE
CONSENSUS REPORTS: IARC Cancer
Review: Group 1 IMEMDT 7,252,87;
Animal Sufficient Evidence IMEMDT
33,87,84. Reported in EPA TSCA
Inventory.
SAFETY PROFILE: Confirmed carcinogen.
Low toxicity by ingestion and skin contact.
A skin irritant. Mutation data reported.
When heated to decomposition it emits
acrid smoke and irritating fumes.

MQV820   **CAS: 64742-63-8**   **HR: 3**
MINERAL OIL, PETROLEUM DISTIL-
LATES, SOLVENT-DEWAXED
HEAVY NAPHTHENIC (mild or no
solvent-refining or hydrotreatment)
SYNS: DISTILLATES (PETROLEUM), SOLVENT-
DEWAXED HEAVY NAPHTHENIC (9CI) □ SOLVENT-
DEWAXED HEAVY NAPHTHENIC DISTILLATE
CONSENSUS REPORTS: IARC Cancer
Review: Group 1 IMEMDT 7,252,87;
Animal Sufficient Evidence IMEMDT
33,87,84. Reported in EPA TSCA
Inventory.
SAFETY PROFILE: Confirmed carcinogen.
When heated to decomposition it emits
acrid smoke and irritating fumes.

MQV825   **CAS: 64742-65-0**   **HR: 3**
MINERAL OIL, PETROLEUM DISTIL-
LATES, SOLVENT-DEWAXED
HEAVY PARAFFINIC (mild or no
solvent-refining or hydrotreatment)
SYNS: DISTILLATES (PETROLEUM), SOLVENT-
DEWAXED HEAVY PARAFFINIC (9CI) □ PETROLEUM
DISTILLATES, SOLVENT-DEWAXED HEAVY PARAF-
FINIC □ SOLVENT-DEWAXED HEAVY PARAFFINIC
DISTILLATE
CONSENSUS REPORTS: IARC Cancer
Review: Group 1 IMEMDT 7,252,87;
Animal Sufficient Evidence IMEMDT
33,87,84. Reported in EPA TSCA
Inventory.
SAFETY PROFILE: Confirmed carcinogen
with experimental tumorigenic data. Low

toxicity by ingestion and skin contact. When
heated to decomposition it emits acrid
smoke and irritating fumes.

MQV835   **CAS: 64742-64-9**   **HR: 3**
MINERAL OIL, PETROLEUM DISTIL-
LATES, SOLVENT-DEWAXED LIGHT
NAPHTHENIC (mild or no solvent-
refining or hydrotreatment)
SYNS: DISTILLATES (PETROLEUM), SOLVENT-
DEWAXED LIGHT NAPHTHENIC (9CI) □ SOLVENT-
DEWAXED LIGHT NAPHTHENIC DISTILLATE
CONSENSUS REPORTS: IARC Cancer
Review: Group 1 IMEMDT 7,252,87;
Animal Sufficient Evidence IMEMDT
33,87,84. Reported in EPA TSCA
Inventory.
SAFETY PROFILE: Confirmed carcinogen.
When heated to decomposition it emits
acrid smoke and irritating fumes.

MQV840   **CAS: 64742-56-9**   **HR: 3**
MINERAL OIL, PETROLEUM DISTIL-
LATES, SOLVENT-DEWAXED LIGHT
PARAFFINIC (mild or no solvent-
refining or hydrotreatment)
SYNS: DISTILLATES (PETROLEUM), SOLVENT-
DEWAXED LIGHT PARAFFINIC (9CI) □ SOLVENT-
DEWAXED LIGHT PARAFFINIC DISTILLATE
CONSENSUS REPORTS: IARC Cancer
Review: Group 1 IMEMDT 7,252,87;
Animal Sufficient Evidence IMEMDT
33,87,84. Reported in EPA TSCA
Inventory.
SAFETY PROFILE: Confirmed carcinogen.
When heated to decomposition it emits
acrid smoke and irritating fumes.

MQV845   **CAS: 64741-96-4**   **HR: 3**
MINERAL OIL, PETROLEUM DISTIL-
LATES, SOLVENT-REFINED (mild)
HEAVY NAPHTHENIC
SYNS: DISTILLATES (PETROLEUM), SOLVENT-
REFINED (mild) HEAVY NAPHTHENIC (9CI) □
NAPHTHENIC BASE LUBE STOCK □ SOLVENT-
REFINED (mild) HEAVY NAPHTHENIC DISTILLATE
CONSENSUS REPORTS: IARC Cancer
Review: Group 1 IMEMDT 7,252,87;
Animal Sufficient Evidence IMEMDT

M

33,87,84. Reported in EPA TSCA Inventory.

SAFETY PROFILE: Confirmed carcinogen. Low toxicity by ingestion and skin contact. When heated to decomposition it emits acrid smoke and irritating fumes.

MQV850   CAS: 64741-88-4   HR: 2
MINERAL OIL, PETROLEUM DISTIL-
LATES, SOLVENT-REFINED (mild)
HEAVY PARAFFINIC
SYNS: DISTILLATES (PETROLEUM), SOLVENT-REFINED (mild) HEAVY PARAFFINIC (9CI) □ SOLVENT-REFINED (mild) HEAVY PARAFFINIC DISTILLATE
CONSENSUS REPORTS: IARC Cancer Review: Group 1 IMEMDT 7,252,87; Animal Sufficient Evidence IMEMDT 33,87,84. Reported in EPA TSCA Inventory.
SAFETY PROFILE: Questionable carcinogen. When heated to decomposition it emits acrid smoke and irritating fumes.

MQV852   CAS: 64741-97-5   HR: 3
MINERAL OIL, PETROLEUM DISTIL-
LATES, SOLVENT-REFINED (mild)
LIGHT NAPHTHENIC
SYNS: DISTILLATES (PETROLEUM), SOLVENT-REFINED (mild) LIGHT NAPHTHENIC (9CI) □ SOLVENT-REFINED (mild) LIGHT NAPHTHENIC DISTILLATE
CONSENSUS REPORTS: IARC Cancer Review: Group 1 IMEMDT 7,252,87; Animal Sufficient Evidence IMEMDT 33,87,84. Reported in EPA TSCA Inventory.
SAFETY PROFILE: Confirmed carcinogen. When heated to decomposition it emits acrid smoke and irritating fumes.

MQV855   CAS: 64741-89-5   HR: 3
MINERAL OIL, PETROLEUM DISTIL-
LATES, SOLVENT-REFINED (mild)
LIGHT PARAFFINIC
SYNS: DISTILLATES (PETROLEUM), SOLVENT-REFINED (mild) LIGHT PARAFFINIC (9CI) □ EMULSIFIABLE OIL □ HORTICULTURAL SPRAY OIL □ SOLVENT-REFINED (mild) LIGHT PARAFFINIC DISTILLATE □ SUPERIOR OIL
CONSENSUS REPORTS: IARC Cancer Review: Group 1 IMEMDT 7,252,87;

Animal Sufficient Evidence IMEMDT 33,87,84. Reported in EPA TSCA Inventory.

SAFETY PROFILE: Confirmed carcinogen. Low toxicity by ingestion and skin contact. When heated to decomposition it emits acrid smoke and irritating fumes.

MQV857   CAS: 64742-11-6   HR: 3
MINERAL OIL, PETROLEUM EX-
TRACTS, HEAVY NAPHTHENIC
DISTILLATE SOLVENT
SYNS: EXTRACTS (PETROLEUM), HEAVY NAPHTHENIC DISTILLATE SOLVENT (9CI) □ HEAVY NAPHTHENIC DISTILLATE SOLVENT EXTRACT
CONSENSUS REPORTS: IARC Cancer Review: Group 1 IMEMDT 7,252,87; Animal Sufficient Evidence IMEMDT 33,87,84. Reported in EPA TSCA Inventory.
SAFETY PROFILE: Confirmed carcinogen. An experimental teratogen. Experimental reproductive effects. When heated to decomposition it emits acrid smoke and irritating fumes.

MQV859   CAS: 64742-04-7   HR: 3
MINERAL OIL, PETROLEUM EXTRAC-
TS, HEAVY PARAFFINIC DISTIL-
LATE SOLVENT
SYNS: EXTRACTS (PETROLEUM), HEAVY PARAFFINIC DISTILLATE SOLVENT (9CI) □ HEAVY PARAFFINIC DISTILLATE, SOLVENT EXTRACT
CONSENSUS REPORTS: IARC Cancer Review: Group 1 IMEMDT 7,252,87; Animal Sufficient Evidence IMEMDT 33,87,84. Reported in EPA TSCA Inventory.
SAFETY PROFILE: Confirmed carcinogen. When heated to decomposition it emits acrid smoke and irritating fumes.

MQV860   CAS: 64742-03-6   HR: 3
MINERAL OIL, PETROLEUM EXTRAC-
TS, LIGHT NAPHTHENIC DISTIL-
LATE SOLVENT
SYNS: EXTRACTS (PETROLEUM), LIGHT NAPHTHENIC DISTILLATE SOLVENT (9CI) □ LIGHT NAPHTHENIC DISTILLATE, SOLVENT EXTRACT

CONSENSUS REPORTS: IARC Cancer Review: Group 1 IMEMDT 7,252,87; Animal Sufficient Evidence 33,87,84. Reported in EPA TSCA Inventory.
SAFETY PROFILE: Confirmed carcinogen. When heated to decomposition it emits acrid smoke and irritating fumes.

MQV862   CAS: 64742-05-8   HR: 3
MINERAL OIL, PETROLEUM EXTRACTS, LIGHT PARAFFINIC DISTILLATE SOLVENT

SYNS: EXTRACTS (PETROLEUM), LIGHT PARAFFINIC DISTILLATE SOLVENT (9CI) □ LIGHT PARAFFINIC DISTILLATE, SOLVENT EXTRACT
CONSENSUS REPORTS: IARC Cancer Review: Group 1 IMEMDT 7,252,87; Animal Sufficient Evidence IMEMDT 33,87,84. Reported in EPA TSCA Inventory.
SAFETY PROFILE: Confirmed carcinogen. Low toxicity by ingestion and skin contact. A severe skin irritant. When heated to decomposition it emits acrid smoke and irritating fumes.

MQV863   CAS: 64742-10-5   HR: 3
MINERAL OIL, PETROLEUM EXTRACTS, RESIDUAL OIL SOLVENT

SYNS: EXTRACTS (PETROLEUM), RESIDUAL OIL SOLVENT (9CI) □ RESIDUAL OIL SOLVENT EXTRACT
CONSENSUS REPORTS: IARC Cancer Review: Group 1 IMEMDT 7,252,87; Animal Sufficient Evidence IMEMDT 33,87,84. Reported in EPA TSCA Inventory.
SAFETY PROFILE: Confirmed carcinogen. When heated to decomposition it emits acrid smoke and irritating fumes.

MQV872   CAS: 64742-17-2   HR: 3
MINERAL OIL, PETROLEUM RESIDUAL OILS, ACID-TREATED

SYNS: ACID-TREATED RESIDUAL OIL □ RESIDUAL OILS (PETROLEUM), ACID-TREATED (9CI)
CONSENSUS REPORTS: IARC Cancer Review: Group 1 IMEMDT 7,252,87; Animal Sufficient Evidence IMEMDT 33,87,84.

SAFETY PROFILE: Confirmed carcinogen. When heated to decomposition it emits acrid smoke and irritating fumes.

MQV875   CAS: 8042-47-5   HR: 2
MINERAL OIL, WHITE

SYNS: DRAKEOL □ KAYDOL □ PAROL □ PENETECK □ SLAB OIL (OBS.) □ WHITE MINERAL OIL
CONSENSUS REPORTS: IARC Cancer Review: Group 3 IMEMDT 7,252,87; Animal Inadequate Evidence IMEMDT 33,87,84. Reported in EPA TSCA Inventory.
SAFETY PROFILE: Highly purified food grades are of low toxicity. Questionable carcinogen. When heated to decomposition it emits acrid smoke and irritating fumes.

MQW500   CAS: 2385-85-5   HR: 3
MIREX

mf: $C_{10}Cl_{12}$ mw: 545.50
PROP: Very white, odorless crystals or solid. Decomp @ 485°. Water-insol; sol in dioxane and benzene.
SYNS: BICHLORENDO □ CG-1283 □ DECHLORANE 4070 □ DODECACHLOROOCTAHYDRO-1,3,4-METHENO-2H-CYCLOBUTA(c,d)PENTALENE □ 1,1a,2,2,3,3a,4,5,5, 5a,5b,6-DODECACHLOROOCTAHYDRO-1,3,4-METHENO-1H-CYCLOBUTA(c,d)PENTALENE □ DODECA-CHLOROPENTACYCLODECANE □ DODECACHLORO-PENTACYCLO(3,2,2,0^{2,6},0^{3,9},0^{5,10})DECANE □ ENT 25,719 □ FERRIAMICIDE □ HEXACHLOROCYCLOPENTADI-ENEDIMER □ 1,2,3,4,5,5-HEXACHLORO-1,3-CYCLO-PENTADIENE DIMER □ HRS 1276 □ NCI-C06428 □ PERCHLORODIHOMOCUBANE □ PERCHLOROPENTA-CYCLODECANE □ PERCHLOROPENTACYCLO-(5.2.1.0^{2,6}.0^{3,9}.0^{5,8})DECANE
CONSENSUS REPORTS: NTP 10th Report on Carcinogens. IARC Cancer Review: Group 2B IMEMDT 7,56,87; Human Limited Evidence IMEMDT 20,283,79; Animal Sufficient Evidence IMEMDT 20,283,79; IMEMDT 5,203,74. EPA Genetic Toxicology Program.
SAFETY PROFILE: Confirmed carcinogen with experimental carcinogenic, tumorigenic, and teratogenic data. Poison by ingestion. Moderately toxic by inhalation and skin contact. An experimental teratogen.

M

Experimental reproductive effects. Mutation data reported. A persistent insecticide that is toxic to non-target species. It can bioaccumulate. See also CHLORINATED HYDROCARBONS, ALIPHATIC.

MQY325   CAS: 63642-17-1   HR: 3
MNCO

mf: $C_7H_{14}N_4O_4$ mw: 218.25

SYNS: N$^\Delta$-(N-METHYL-N-NITROSOCARBAMOYL)-l-ORNITHINE ☐ N^5-(METHYLNITROSOCARBAMOYL)-l-ORNITHINE ☐ N^5-(N-METHYL-N-NITROSOCARBAMO-YL)-l-ORNITHINE

SAFETY PROFILE: Suspected carcinogen with experimental carcinogenic and tumorigenic data. Mutation data reported. When heated to decomposition it emits toxic fumes of NO$_x$. See also N-NITROSO COMPOUNDS.

MRA075   HR: D
MODIFIED POLYACRYLAMIDE RESINS

PROP: Produced by copolymerization of acrylamide with not more than 5-mole percent of β-methacrylyloxyethyl trimethylammonium methyl sulfate.

SYN: POLYACRYLAMIDE RESINS, MODIFIED

SAFETY PROFILE: When heated to decomposition it emits acrid smoke and irritating fumes.

MRA250   CAS: 11015-37-5   HR: 3
MOENOMYCIN

PROP: Produced by *Streptomyces roseoflavus* (85ERAY 1,740,78).

SYNS: BAMBERMYCIN ☐ FLAVOMYCIN ☐ FLAVO-PHOSPHOLIPOL ☐ MENOMYCIN ☐ MOENOMYCIN A

SAFETY PROFILE: Poison by intravenous route. Moderately toxic by subcutaneous route.

MRC000   CAS: 12656-85-8   HR: 3
MOLYBDATE ORANGE

PROP: IDLH 1000 mg/m³ (as Mo).

SYNS: CHROME VERMILION ☐ C.I. 77605 ☐ C.I. PIGMENT RED 104 ☐ KROLOR ORANGE RKO 786D ☐ LEAD CHROMATE MOLYBDATE SULFATE RED ☐ MINERAL FIRE RED 5DDS ☐ MINERAL FIRE RED 5GS ☐ MOLYBDATE ORANGE Y 786D ☐ MOLYBDATE ORANGE YE 421D ☐ MOLYBDATE ORANGE YE 698D ☐ MOLYBDATE RED ☐ MOLYBDATE RED AA3 ☐ MOLYBDEN RED ☐ MOLYBDENUM RED ☐ NCI-C54626 ☐ RENOL MOLYBDATE RED RGS ☐ VYNAMON SCARLET BY

CONSENSUS REPORTS: IARC Cancer Review: Group 1 IMEMDT 49,49,90; Human Sufficient Evidence IMEMDT 49,49,90. Chromium and its compounds are on the Community Right-To-Know List. Reported in EPA TSCA Inventory.

OSHA PEL: TWA 5 mg(Mo)/m³

ACGIH TLV: TWA Soluble Compounds: TWA 0.5 mg(Mo)/m³ Confirmed Animal Carcinogen with Unknown Relevance to Humans

SAFETY PROFILE: Confirmed carcinogen. Dusts are poison by inhalation. See MOLYBDENUM COMPOUNDS and CHROMIUM COMPOUNDS.

MRC250   CAS: 7439-98-7   HR: 3
MOLYBDENUM

af: Mo   aw: 95.94

PROP: Lustrous, cubic, silver-white metallic crystals or gray-black powder. Fairly soft when pure. Less reactive than Cr to acids. Combines with O_2 on heating to give MoO_3. Mp: 2626°, bp: 5560°, d: 10.2, vap press: 1 mm @ 3102°. IDLH 5000 mg/m³ (as Mo).

SYN: MOLYBDATE

CONSENSUS REPORTS: Reported in EPA TSCA Inventory.

OSHA PEL: Soluble Compounds: TWA 5 mg(Mo)/m³; Insoluble Compounds: TWA Total Dust: 10 mg/m³; Respirable Fraction: 5 mg/m³

ACGIH TLV: Insoluble Compounds: TWA 10 mg(Mo)/m³; Soluble Compounds: TWA 0.5 mg(Mo)/m³ Confirmed Animal Carcinogen with Unknown Relevance to Humans

DFG MAK: (Insoluble Compounds) 4 mg/m³; (Soluble Compounds) 5 mg/m³

SAFETY PROFILE: Poison by intratracheal route. Mutation data reported. An experimental teratogen. Experimental reproductive effects. Flammable or explosive in the form of dust when exposed to heat or flame. Violent reaction with oxidants (e.g., bromine trifluoride, bromine pentafluoride. chlorine trifluoride, potassium perchlorate, nitryl fluoride, fluorine, iodine pentafluoride, sodium peroxide, lead dioxide). When heated to decomposition it emits toxic fumes of Mo. See also POWDERED METALS and MOLYBDENUM COMPOUNDS.

MRC650   CAS: 12007-97-5   HR: 3
MOLYBDENUM BORIDE
mf: B_5Mo_2 mw: 245.93
PROP: A solid. Mp: 2335°, d: 7.48. IDLH 1000 mg/m³ (as Mo).
CONSENSUS REPORTS: Reported in EPA TSCA Inventory.
OSHA PEL: TWA 15 mg(Mo)/m³
ACGIH TLV: Insoluble Compounds: inhalable fraction, 10 mg(Mo)/m³, 3 mg(Mo)/m³, respirable fraction.
SAFETY PROFILE: Poison by intratracheal route. Moderately toxic by ingestion and intraperitoneal routes. When heated to decomposition it emits toxic fumes of boron and Mo.

MRC750   HR: 3
MOLYBDENUM COMPOUNDS
PROP: IDLH 1000 mg/m³ (as Mo).
SAFETY PROFILE: Poison by subcutaneous and intraperitoneal routes. Molybdenum and its compounds are highly toxic based upon animal experiments. Symptoms of acute poisoning include severe gastrointestinal irritation with diarrhea, coma, and deaths from heart failure. Experimental animals exposed to high levels (57 mg Mo/m³) of molybdenum dust for 120 days (4 hours/day) accumulated Mo in the lungs, spleen, and heart, and showed a decrease of DNA and RNA in the liver, kidneys, and spleen. Workers exposed to Mo or MoO_3 (concentrations of 1–19 mg Mo/m³) over a period of 3–7 years have suffered from pneumoconiosis. Inhalation of molybdenum dust from alloys or carbides can cause "hard-metal lung disease." It is suggested that suitable precautions should

be taken against human inhalation of significant amounts of the more soluble molybdenum compounds. MoO_3 and Na_2MoO_4 are soluble. $CaMoO_4$, MoO, and MoS_2 are insoluble. Hexavalent molybdenum compounds are readily absorbed through the gastrointestinal tract. Coal-fired electrical power plants can be significant sources of molybdenum. Application of some fertilizers may raise molybdenum concentrations in ground water. Molybdenum is rapidly excreted by the body. Molybdenum is an important trace element in the normal growth and development of plants. It is found also in animal tissue, although its precise function is unknown. It is a common air contaminant. See also specific compounds.

MRD250   CAS: 18868-43-4   HR: 3
MOLYBDENUM OXIDE
mf: MoO_2 mw: 127.94
PROP: Brown-black powder or dark violet blue crystals. Insol in H_2O, acids, and alkalies. IDLH 1000 mg/m³ (as Mo).
SYNS: MOLYBDENUM DIOXIDE □ MOLYBDENUM(IV) OXIDE
CONSENSUS REPORTS: Reported in EPA TSCA Inventory.
OSHA PEL: TWA 10 mg(Mo)/m³, total dust; TWA 5 mg(Mo)/m³, respirable fraction
ACGIH TLV: Insoluble Compounds: inhalable fraction, 10 mg(Mo)/m³, 3 mg(Mo)/m³, respirable fraction.
SAFETY PROFILE: Poison by subcutaneous route. Incandescent reaction with air. When heated to decomposition it emits toxic fumes of Mo. See also MOLYBDENUM COMPOUNDS.

MRD500   CAS: 10241-05-1   HR: 3
MOLYBDENUM PENTACHLORIDE
DOT: UN 2508
mf: Cl_5Mo mw: 273.19
PROP: Green-black solid, dark-red as liquid or vapor. Hygroscopic, reacting with water and air. Mp: 194°, bp: 268°, d: 2.9. Sol in dry

M

ether, dry alc, and other anhydrous org solvs. IDLH 1000 mg/m³ (as Mo).
CONSENSUS REPORTS: Reported in EPA TSCA Inventory. EPA Genetic Toxicology Program.
OSHA PEL: TWA 5 mg(Mo)/m³
ACGIH TLV: TWA Soluble Compounds: TWA 0.5 mg(Mo)/m³ Confirmed Animal Carcinogen with Unknown Relevance to Humans
DOT CLASSIFICATION: 8; Label: Corrosive
SAFETY PROFILE: A poison. A corrosive irritant to skin, eyes, and mucous membranes. Reacts with moisture to form hydrochloric acid. When heated to decomposition it emits toxic fumes of Mo and Cl⁻. See also MOLYBDENUM COMPOUNDS and HYDROCHLORIC ACID.

MRE000   CAS: 1313-27-5   HR: 3
MOLYBDENUM TRIOXIDE
mf: MoO_3 mw: 143.94
PROP: White or yellow to sltly bluish powder, granules, or solid. Photosensitive. An acidic oxide. Mp: 795°; bp: 1155°; d: 4.696 @ 26°/4°. Sol in 1000 parts water, in concentrated mineral acids, solutions of alkali hydroxides. Sol in ammonia or potassium bitartrate, solidifying to a yellowish-white mass. IDLH 1000 mg/m³ (as Mo).
SYNS: MOLYBDENUM(VI) OXIDE □ MOLYBDIC ANHYDRIDE □ MOLYBDIC TRIOXIDE
CONSENSUS REPORTS: Reported in EPA TSCA Inventory.
OSHA PEL: TWA 5 mg(Mo)/m³
ACGIH TLV: TWA Soluble Compounds: TWA 0.5 mg(Mo)/m³ Confirmed Animal Carcinogen with Unknown Relevance to Humans
SAFETY PROFILE: Poison by ingestion, subcutaneous, and intraperitoneal routes. Human systemic effects by inhalation: pulmonary fibrosis and cough. Questionable carcinogen with experimental neoplastigenic data. A powerful irritant. Explodes on contact with molten magnesium. Violent

reaction with interhalogens (e.g., bromine pentafluoride, chlorine trifluoride). Incandescent reaction with hot sodium, potassium, or lithium. When heated to decomposition it emits toxic fumes of Mo. See also MOLYBDENUM COMPOUNDS.

MRE225   CAS: 17090-79-8   HR: 3
MONENSIC ACID
mf: $C_{36}H_{62}O_{11}$ mw: 670.98
PROP: Crystals. Mp: 103–105° (monohydrate). Very stable under alkaline conditions. Sltly sol in water; more sol in hydrocarbons; very sol in other org solvs.
SYNS: A 3823A □ ELANCOBAN □ MONELAN □ MONENSIN (USDA) □ MONENSIN A
SAFETY PROFILE: Poison by ingestion and intraperitoneal routes. An eye and skin irritant. When heated to decomposition it emits acrid smoke and irritating fumes.

MRF000   CAS: 7558-63-6   HR: 2
MONOAMMONIUM GLUTAMATE
mf: $C_5H_9NO_4 \cdot H_3N$ mw: 164.19
PROP: White crystalline powder; odorless. Sol in water; insol in common org solvs.
SYNS: AMMONIUMGLUTAMINAT (GERMAN) □ MAG □ MONOAMMONIUM l-GLUTAMATE
CONSENSUS REPORTS: Reported in EPA TSCA Inventory.
SAFETY PROFILE: Moderately toxic by intraperitoneal route. When heated to decomposition it emits toxic fumes including NO_x and NH_3. See also MONOSODIUM GLUTAMATE.

MRF525   CAS: 131-70-4   HR: D
MONOBUTYL PHTHALATE
mf: $C_{12}H_{14}O_4$ mw: 222.26
SYNS: BUTYL HYDROGEN PHTHALATE □ MBP □ MONO-n-BUTYL PHTHALATE
CONSENSUS REPORTS: Reported in EPA TSCA Inventory.
SAFETY PROFILE: Experimental reproductive effects. When heated to decomposition it emits acrid smoke and irritating fumes.

MRH000 CAS: 315-22-0 HR: 3
MONOCROTALINE
mf: $C_{16}H_{23}NO_6$ mw: 325.40
PROP: Prisms from abs ethanol. Decomp @ 202–203°.
SYNS: CROTALINE □ 14,19-DIHYDRO-12,13-DIHYDROXY(13-α,14-α)-20-NORCROTALANAN-11,15-DIONE □ MONOCRATILIN □ NCI-C56462 □ NSC-28693
CONSENSUS REPORTS: IARC Cancer Review: Group 2B IMEMDT 7,56,87; Animal Limited Evidence IMEMDT 10,291,76. EPA Genetic Toxicology Program.
SAFETY PROFILE: Suspected carcinogen with experimental carcinogenic data. Poison by ingestion, intravenous, intraperitoneal, and subcutaneous routes. Human mutation data reported. When heated to decomposition it emits toxic fumes of NO_x.

MRH209 CAS: 6923-22-4 HR: 3
MONOCROTOPHOS
mf: $C_7H_{14}NO_5P$ mw: 223.19
PROP: Crystals or a reddish-brown solid; mild ester odor. Bp: 125°, mp: 54–55°. Sol in H_2O, Me_2CO; insol in hexane.
SYNS: APADRIN □ AZODRIN (OSHA) □ AZODRIN-71 □ AZODRIN PESTICIDE □ BILOBRAN □ BILOBORN □ C 1414 □ CIBA 1414 □ COROPHOS □ CRISODIN □ CRISODRIN □ 3-(DIMETHOXYPHOSPHINYLOXY)N-METHYL-cis-CROTONAMIDE □ O,O-DIMETHYL-O-(2-N-METHYLCARBAMOYL-1-METHYL-VINYL)-FOSFAAT (DUTCH) □ O,O-DIMETHYL-O-(2-N-METHYLCARB-AMOYL-1-METHYL)-VINYL-PHOSPHAT (GERMAN) □ O,O-DIMETHYL-O-(2-N-METHYLCARBAMOYL-1-METHYL-VINYL) PHOSPHATE □ (E)-DIMETHYL 1-METHYL-3-(METHYLAMINO)-3-OXO-1-PROPENYL □ DIMETHYL-1-METHYL-2-(METHYLCARBAMOYL)-VINYLPHOSPHATE, cis PHOSPHATE □ DIMETHYL PHOSPHATE ESTER of 3-HYDROXY-N-METHYL-cis-CROTONAMIDE □ DIMETHYL PHOSPHATE of 3-HYDROXY-N-METHYL-cis-CROTONAMINE □ O,O-DIMETIL-O-(2-N-METILCARBAMOIL-1-METIL-VINIL)-FOSFATO (ITALIAN) □ ENT 27,129 □ HAZODRIN □ 3-HYDROXY-N-METHYL-cis-CROTONAMIDE DIMETHYL PHOSPHATE □ cis-1-METHYL-2-METHYL CARBAMOYL VINYL PHOSPHATE □ MONOCIL □ MONOCIL 40 □ MONOCRON □ MONOCROTOPHOS □ MONOCROT-OPHOS (ACGIH,OSHA) □ MONODRIN □ NUVACRON □ NUVACRON 20 □ OMS 834 □ PHOSPHATE de DIMETHYLE et de 2-METHYLCARBAMOYL 1-METHYL VINYLE □ PHOSPHORIC ACID, DIMETHYL ESTER, ESTER with cis-3-HYDROXY-N-METHYLCROTONAMIDE □ PHOSPHORIC ACID, DIMETHYL 1-METHYL-3-(METHYLAMINO)-3-OXO-1-PROPENYL ESTER, (E)- □ PILLARDRIN □ PLANTDRIN □ SD 9129 □ SHELL SD 9129 □ SUSVIN □ ULVAIR
CONSENSUS REPORTS: EPA Genetic Toxicology Program. EPA Extremely Hazardous Substances List.
OSHA PEL: TWA 0.25 mg/m^3
ACGIH TLV: TWA 0.05 mg/m^3; (skin); Not Classifiable as a Human Carcinogen
SAFETY PROFILE: Poison by ingestion, inhalation, skin contact, intraperitoneal, subcutaneous, and intravenous routes. Mutation data reported. Use may be restricted. When heated to decomposition it emits very toxic NO_x and PO_x.

MRH215 HR: D
MONO- and DIGLYCERIDES
PROP: Yellow liquids to ivory-colored plastics to hard solids; bland odor and taste. Sol in alc, ethyl acetate, chloroform, other chlorinated hydrocarbons; insol in water.
SAFETY PROFILE: When heated to decomposition it emits acrid smoke and irritating fumes.

MRH225 HR: D
MONO-, DI-, and TRIPOTASSIUM CITRATE
SAFETY PROFILE: When heated to decomposition it emits acrid smoke and irritating fumes.

MRH235 HR: D
MONO-, DI-, and TRISTEARYL CITRATE
SAFETY PROFILE: When heated to decomposition it emits acrid smoke and irritating fumes.

MRI750 CAS: 39801-14-4 HR: 3
8-MONOHYDRO MIREX
mf: $C_{10}HCl_{11}$ mw: 511.06
SYNS: HYDROMIREX □ PHOTOMIREX □ 1,2,3,4,5,5,6,7,9,10,10-UNDECACHLOROPENTA-CYCLO(5.3.0.0^{2,6}.0^{3,9}.0^{4,8})DECANE
SAFETY PROFILE: Poison by ingestion. Experimental teratogenic effects. Questionable carcinogen with experimental tumorigenic data. Mutation data reported.

M

When heated to decomposition it emits toxic fumes of Cl⁻. See also MIREX.

MRI785   HR: D
MONOISOPROPYL CITRATE
SAFETY PROFILE: When heated to decomposition it emits acrid smoke and irritating fumes.

MRK500   CAS: 19473-49-5   HR: 1
MONOPOTASSIUM GLUTAMATE
mf: $C_5H_8NO_4 \cdot K$ mw: 185.24
PROP: White, free-flowing, hygroscopic crystalline powder; practically odorless. Freely sol in water; sltly sol in alc.
SYNS: l-GLUTAMIC ACID, MONOPOTASSIUM SALT □ MONOPOTASSIUM l-GLUTAMATE (FCC) □ MPG □ POTASSIUM GLUTAMATE □ POTASSIUM GLUT-AMINATE
CONSENSUS REPORTS: Reported in EPA TSCA Inventory. EPA Genetic Toxicology Program.
SAFETY PROFILE: Mildly toxic by ingestion. Human systemic effects by ingestion: headache. When heated to decomposition it emits toxic fumes of K_2O and NO_x.

MRL500   CAS: 142-47-2   HR: 2
MONOSODIUM GLUTAMATE
mf: $C_5H_8NO_4 \cdot Na$ mw: 169.13
PROP: White or almost white crystals or powder; slt peptone-like odor, meal-like taste. Very sol in water; sltly sol in alc.
SYNS: ACCENT □ AJINOMOTO □ CHINESE SEASONING □ GLUTACYL □ GLUTAMIC ACID, SODIUM SALT □ GLUTAMMATO MONOSODICO (ITALIAN) □ GLUTAVENE □ MONOSODIO-GLUTAMMATO (ITALIAN) □ MONOSODIUM-l-GLUTAMATE (FCC) □ α-MONOSODIUM GLUTAMATE □ MSG □ NATRIUMGLUTAMINAT (GERMAN) □ RL-50 □ SODIUM GLUTAMATE □ SODIUM l-GLUTAMATE □ l(+) SODIUM GLUTAMATE □ VETSIN □ ZEST
CONSENSUS REPORTS: Reported in EPA TSCA Inventory. EPA Genetic Toxicology Program.
SAFETY PROFILE: Moderately toxic by intravenous route. Mildly toxic by ingestion and other routes. An experimental teratogen. Other experimental reproductive effects. Human systemic effects by ingestion and intravenous routes: somnolence, hallucinations and distorted perceptions, headache, dyspnea, nausea or vomiting, dermatitis. The cause of "Chinese restaurant syndrome." When heated to decomposition it emits toxic fumes of NO_x and Na_2O.

MRL750   CAS: 2163-80-6   HR: 3
MONOSODIUM METHYLARSONATE
mf: $CH_4AsO_3 \cdot Na$ mw: 161.96
PROP: Crystals from H_2O. Mp: 113–116°. Very sol in H_2O; sol in MeOH; insol in most org solvs.
SYNS: ANSAR 170 □ ARSONATE liquid □ ASAZOL □ BUENO □ DACONATE 6 □ DAL-E-RAD □ HERB-ALL □ HERBAN M □ MERGE □ MESAMATE □ MESAMATE CONCENTRATE □ METHYLARSENIC ACID, SODIUM SALT □ MONATE □ MONOSODIUM ACID METHANE-ARSONATE □ MONOSODIUM ACID METHARSONATE □ MONOSODIUM METHANEARSONATE □ MONO-SODIUM METHANEARSONIC ACID □ MSMA □ NCI-C60071 □ PHYBAN □ SILVISAR 550 □ SODIUM ACID METHANEARSONATE □ SODIUM METHANEARSON-ATE □ TARGET MSMA □ TRANS-VERT □ WEED 108 □ WEED-E-RAD □ WEED-HOE
CONSENSUS REPORTS: Arsenic and its compounds are on the Community Right-To-Know List. EPA Genetic Toxicology Program.
OSHA PEL: TWA 0.5 mg(As)/m³
ACGIH TLV: BEI: 35 μ (As)/L inorganic arsenic and methylated metabolites in urine
SAFETY PROFILE: Poison by unspecified route. Moderately toxic by ingestion. A skin and eye irritant. When heated to decomposition it emits toxic fumes of As and Na_2O. See also ARSENIC COMPOUNDS.

MRM750   CAS: 96-27-5   HR: 3
MONOTHIOGLYCEROL
mf: $C_3H_8O_2S$ mw: 108.17
PROP: A liquid. Bp: 118° @ 5 mm, d: 1.248 @ 20°/20°. Sol in water.
SYNS: 1-MERCAPTOGLYCEROL □ 1-MERCAPTO-2,3-PROPANEDIOL □ 3-MERCAPTO-1,2-PROPANEDIOL □ α-MONOTHIOGLYCEROL □ THIOGLYCERIN □ α-THIOGLYCEROL □ 1-THIOGLYCEROL □ THIOVANOL □ USAF B-40 □ USAF CB-37

CONSENSUS REPORTS: Reported in EPA TSCA Inventory.

SAFETY PROFILE: Poison by intraperitoneal and intravenous routes. Experimental reproductive effects. Mutation data reported. Flammable when exposed to heat or flame; can react with oxidizing materials. When heated to decomposition it emits highly toxic fumes of SO_x. See also MERCAPTANS.

MRN500   CAS: 480-16-0   HR: 2
MORIN

mf: $C_{15}H_{10}O_7$ mw: 302.25

PROP: Anhydrous needles from abs alc. Pale-yellow needles from AcOH (aq). Decomp @ 285–290°, mp: 303–304°. Crystallized with 1 or 2 moles of water. Sltly sol in water, ether, and acetic acid; very sol in alc.

SYNS: A1-MORIN □ AURANTICA □ BOIS D'ARC (FRENCH) □ CALICO YELLOW □ C.I. 75660 □ C.I. NATURAL YELLOW 8 □ C.I. NATURAL YELLOW 11 □ 2'-HYDROXYPELARGIDENOLON 1522 □ MORIN □ OSAGE ORANGE □ OSAGE ORANGE CRYSTALS □ OSAGE ORANGE EXTRACT □ 2',3,4',5,7-PENTAHYDRO-XYFLAVONE □ 2',4',3,5,7-PENTAHYDROXYFLAVONE □ 3,5,7,2',4'-PENTAHYDROXYFLAVONE □ 3,5,7,2',4'-PENTAHYDROXYFLAVONOL □ TOXYLON POMIFERUM □ ZLUT PRIRODNI 8 □ ZLUT PRIRODNI 11

CONSENSUS REPORTS: Reported in EPA TSCA Inventory.

SAFETY PROFILE: Moderately toxic by intraperitoneal route. Mutation data reported. When heated to decomposition it emits acrid smoke and irritating fumes.

MRO500   CAS: 57-27-2   HR: 3
(−)-MORPHINE

mf: $C_{17}H_{19}NO_3$ mw: 285.37

PROP: White, crystalline alkaloid. Short, orthorhombic, columnar prisms from anisole. Mp: 254–256° (decomp) (anhyd), subl @ 190–200°. Very sol in EtOAc, Me_2CO; sol in EtOH; spar sol in H_2O, Et_2O, and $CHCl_3$.

SYNS: MORFINA (ITALIAN) □ MORPHIA □ MORP-HINA □ MORPHINE □ MORPHINISM □ MORPHINUM □ MORPHIUM □ 4a,5,7a,8-TETRAHYDRO-12-METHYL-9H-9,9c-IMINOETHANOPHENANTHRO(4,5-bcd)FURAN-3,5-DIOL

SAFETY PROFILE: Poison experimentally by ingestion, intracerebral, intraperitoneal, subcutaneous, and intravenous routes. Human reproductive effects by an unspecified route: effects on newborn, including drug dependence. Experimental reproductive effects. Mutation data reported.

Morphine is the constituent of opium most responsible for its toxic effects. When taken orally, the effects of morphine poisoning begin to appear in 20–40 minutes; if taken hypodermically, the symptoms appear much earlier and narcotism is more likely to follow the early symptoms. Abuse leads to habituation or addiction. Individual susceptibility varies greatly and children are more susceptible than adults. When heated to decomposition it emits toxic fumes of NO_x.

MRO750   CAS: 52-26-6   HR: 3
MORPHINE HYDROCHLORIDE

mf: $C_{17}H_{19}NO_3 \cdot ClH$ mw: 321.83

PROP: Trihydrate: White flakes or crystalline powder; bitter taste. Mp: approx 200° (decomp). Dissolves slowly in glycerin; insol in chloroform, ether.

SYNS: 7,8-DIDEHYDRO-4,5-α-EPOXY-17-METHYL-MORPHINAN-3,6-α-DIOL HYDROCHLORIDE □ 7,8-DIDEHYDRO-4,5-α-EPOXY-17-METHYLMORPHINE HYDROCHLORIDE □ MORPHINE CHLORHYDRATE □ MORPHINE CHLORIDE

CONSENSUS REPORTS: EPA Genetic Toxicology Program.

SAFETY PROFILE: Poison by ingestion, intraperitoneal, intravenous, parenteral, and subcutaneous routes. An experimental teratogen. Experimental reproductive effects. Abuse leads to habituation or addiction. When heated to decomposition it emits very toxic fumes of NO_x and HCl. See also MORPHINE.

MRP250   CAS: 64-31-3   HR: 3
MORPHINE SULFATE

mf: $C_{34}H_{38}N_2O_6 \cdot H_2O_4S$ mw: 668.82

SYN: MORPHINE SULPHATE

SAFETY PROFILE: Poison by subcutaneous, intravenous, intraperitoneal, and intramuscular routes. Moderately toxic by ingestion and parenteral routes. An experimental teratogen. Experimental reproductive effects. Mutation data reported. Used as a narcotic. Abuse leads to habituation or addiction. When heated to decomposition it emits very toxic fumes of NO_x and SO_x. See also MORPHINE and SULFATES.

MRP750   CAS: 110-91-8   HR: 3
MORPHOLINE
DOT: UN 2054
mf: C_4H_9NO mw: 87.14
PROP: Colorless, hygroscopic oil; amine odor. Fp: $-7.5°$, bp: $128.9°$, flash p: $100°F$ (OC), autoign temp: $590°F$, vap press: 10 mm @ $23°$, vap d: 3.00, mp: $-4.9°$, d: 1.007 @ $20°/4°$. Volatile with steam; misc with water evolving some heat; misc with acetone, benzene, ether, castor oil, methanol, ethanol, ethylene, glycol, linseed oil, turpentine, pine oil. Immiscible with concentrated NaOH solns. IDLH 1400 ppm [10%LEL].
SYNS: DIETHYLENE IMIDE OXIDE □ DIETHYLENE IMIDOXIDE □ DIETHYLENE OXIMIDE □ DIETHYL-ENIMIDE OXIDE □ MORPHOLINE, AQUEOUS MIXTURE (DOT) □ 1-OXA-4-AZACYCLOHEXANE □ TETRAHYDRO-p-ISOXAZINE □ TETRAHYDRO-1,4-ISOXAZINE □ TETRAHYDRO-1,4-OXAZINE □ TETRAHYDRO-2H-1,4-OXAZINE
CONSENSUS REPORTS: Reported in EPA TSCA Inventory. EPA Genetic Toxicology Program.
OSHA PEL: TWA 20 ppm (skin); STEL 30 ppm (skin)
ACGIH TLV: TWA 20 ppm (skin); Not Classifiable as a Human Carcinogen
DFG MAK: 10 ppm (36 mg/m³)
DOT CLASSIFICATION: 3; Label: Flammable Liquid
SAFETY PROFILE: Moderately toxic by ingestion, inhalation, skin contact, and intraperitoneal routes. Mutation data reported. A corrosive irritant to skin, eyes, and mucous membranes. Can cause kidney damage. Questionable carcinogen with experimental neoplastigenic data. Flammable liquid. A very dangerous fire hazard when exposed to flame, heat, or oxidizers; can react with oxidizing materials. To fight fire, use alcohol foam, CO_2, dry chemical. Mixtures with nitromethane are explosive. May ignite spontaneously in contact with cellulose nitrate of high surface area. When heated to decomposition it emits highly toxic fumes of NO_x.

MRR115   CAS: 28846-42-6   HR: 3
9-(MORPHOLINOAMINO)ACRIDINE MONO(METHYL SULFATE)
mf: $C_{17}H_{17}N_3O•CH_4O_4S$ mw: 391.48
SYN: ACRIDINE, 9-(MORPHOLINOAMINO)-, MONO(METHYL SULFATE)
SAFETY PROFILE: A poison by ingestion. When heated to decomposition it emits toxic vapors of NO_x, SO_x, and Cl^-.

MRR760   CAS: 7157-29-1   HR: 3
4-MORPHOLINOCARBONYL-2,3-TETRAMETHYLENEQUINOLINE
mf: $C_{18}H_{20}N_2O_2$ mw: 296.40
SYNS: ACRIDINE, 9-(MORPHOLINOCARBONYL)-1,2,3,4-TETRAHYDRO- □ ACRIDINE, 1,2,3,4-TETRAHYDRO-9-(MORPHOLINOCARBONYL)- □ KETONE, MORPHOLINO(1,2,3,4-TETRAHYDRO-9-ACRIDINYL)
DOT CLASSIFICATION: 3; Label: Flammable Liquid
SAFETY PROFILE: A poison by intraperitoneal route. A flammable liquid. When heated to decomposition it emits toxic vapors of NO_x.

MRU077   CAS: 7101-65-7   HR: 3
MORPHOLINO(7,8,9,10-TETRAHYDRO-11-(6H-CYCLOHEPTA(b)QUINOLIN-YL)) KETONE
mf: $C_{19}H_{22}N_2O_2$ mw: 310.43
SYN: 6H-CYCLOHEPTA(b)QUINOLINE, 7,8,9,10-TETRAHYDRO-11-(MORPHOLINOCARBONYL)-
DOT CLASSIFICATION: 3; Label: Flammable Liquid
SAFETY PROFILE: A poison by intraperitoneal route. A flammable liquid.

When heated to decomposition it emits toxic vapors of NO_x.

MRU253   CAS: 112885-41-3 HR: 3
MOSAPRIDE
mf: $C_{21}H_{25}ClFN_3O_3$ mw: 421.90
SYN: BENZAMIDE, 4-AMINO-5-CHLORO-2-ETHOXY-N-((4-((4-FLUOROPHENYL)METHYL)-2-MORPHOLIN-YL)MET HYL)-
SAFETY PROFILE: A poison by ingestion. When heated to decomposition it emits toxic vapors of NO_x, F^-, and Cl^-.

MRU900   CAS: 87-56-9   HR: 3
MUCOCHLORIC ACID
mf: $C_4H_2Cl_2O_3$ mw: 168.96
PROP: Plates from H_2O. Mp: 127°.
SYNS: ALDEHYDODICHLOROMALEIC ACID □ 2-BUTENOIC ACID, 2,3-DICHLOR-4-OXO-, (Z)-(9CI) □ α-β-DICHLORO-β-FORMYL ACRYLIC ACID □ 3,4-DICHL-ORO-2-HYDROXYCROTONOLACTONE □ 3,4-DICHL-ORO-2-HYDROXYCROTONOLACTONIC ACID □ DI-CHLOROMALEALDEHYDIC ACID □ 2,3-DICHLORO-MALEIC ALDEHYDE ACID □ 2,3-DICHLORO-4-OXO-2-BUTENOIC ACID □ KYSELINA MUKOCHLOROVA □ MALEALDEHYDIC ACID, DICHLORO-4-OXO-, (Z)-(9CI)
CONSENSUS REPORTS: Reported in EPA TSCA Inventory.
SAFETY PROFILE: Poison by ingestion. Moderate skin and severe eye irritant. Questionable carcinogen with experimental tumorigenic data. Mutation data reported. When heated to decomposition it emits toxic fumes of Cl^-.

MRW775   CAS: 88671-89-0 HR: D
MYCLOBUTANIL
mf: $C_{15}H_{17}ClN_4$ mw: 288.81
SYNS: 2-P-CHLOROPHENYL-2-(1H-1,2,4-TRIAZOL-1-YLMETHYL)HEXANENITRILE □ NU-FLOW M □ NOVA □ NOVA W □ RALLY □ RH 3866 □ RH-53,866 □ SYSTHANE □ SYSTHANE 6 FLO □ α-BUTYL-α(4-CHLOROPHENYL)-1H-1,2,4-THIAZOLE-1-PROPAN-ENITRILE
SAFETY PROFILE: Moterately toxic by ingestion, inhalation, and skin contact. Experimental reproductive effects. When heated to decomposition emits toxic fumes of NO_x, SO_x, Cl^-.

MRZ150   CAS: 123-35-3   HR: 3
MYRCENE
mf: $C_{10}H_{16}$ mw: 136.26
PROP: Colorless to pale-yellow liquid or oil; sweet, balsamic odor. D: 0.789, refr index: 1.466–1.471, bp: 116°, flash p: 99°F. Sol in alc, fixed oils; insol in water.
SYNS: FEMA No. 2762 □ 3-METHYLENE-7-METHYL-1,6-OCTADIENE □ 7-METHYL-3-METHYLENE-1,6-OCTADIENE
CONSENSUS REPORTS: Reported in EPA TSCA Inventory.
SAFETY PROFILE: Low toxicity by ingestion and skin contact. Experimental reproductive effects. A moderate skin and eye irritant. A flammable liquid. When heated to decomposition it emits acrid smoke and irritating fumes.

MSA250   CAS: 544-63-8   HR: 3
MYRISTIC ACID
mf: $C_{14}H_{28}O_2$ mw: 228.36
PROP: White or faintly yellow crystals from methanol. Mp: 54°, bp: 250.5° @ 100 mm, d: 0.8622 @ 54°/4°. Sol in abs alc, methanol, ether, pet ether, benzene, chloroform; insol in water.
SYNS: CRODACID □ EMERY 655 □ HYDROFOL ACID 1495 □ HYSTRENE 9014 □ 1-TRIDECANECARBOXYLIC ACID □ TETRADECANOIC ACID □ n-TETRADECOIC ACID □ UNIVOL U 316S
CONSENSUS REPORTS: Reported in EPA TSCA Inventory.
SAFETY PROFILE: Poison by intravenous route. Mutation data reported. An eye and human skin irritant. When heated to decomposition it emits acrid smoke and irritating fumes.

MSB500   CAS: 2748-88-1   HR: 3
MYRISTYL-γ-PICOLINIUM CHLORIDE
mf: $C_{20}H_{36}N \cdot Cl$ mw: 326.02
PROP: A solid. Mp: 73–74°.
SYNS: QUATRESIN □ WET-TONE B
CONSENSUS REPORTS: Reported in EPA TSCA Inventory.
SAFETY PROFILE: Poison by ingestion, intraperitoneal, intravenous, and subcutaneous routes. When heated to

M

decomposition it emits very toxic fumes of NO_x and Cl^-.

N

NAH600   CAS: 8030-30-6   HR: 3
NAPHTHA
DOT: UN 1255/UN 1256/UN 1270/UN 2553
PROP: Dark straw-colored to colorless liquid. Bp: 149–216°, flash p: 107°F (CC), d: 0.862–0.892, autoign temp: 531°F. Sol in benzene, toluene, xylene, etc. Made from American coal oil and consists chiefly of pentane, hexane, and heptane (XPHPAW 255,43,40). IDLH 1000 ppm [10%LEL].
SYNS: AMSCO H-J □ AMSCO H-SB □ BENZIN B70 □ HI-FLASH NAPHTHA □ HYDROTREATED NAPHTHA □ NAPHTHA □ NAPHTHA COAL TAR (OSHA) □ NAPH-THA (UN2553) (DOT) □ NAPHTHA, hydrotreated □ NAPHTHA, petroleum (UN1255) (DOT) □ NAPHTHA, solvent (UN1256) (DOT) □ PETROLEUM BENZIN □ PETROLEUM-DERIVED NAPHTHA □ PETROLEUM DISTILLATES (NAPHTHA) □ PETROLEUM OIL (UN1270) (DOT) □ SUPER VMP
CONSENSUS REPORTS: Reported in EPA TSCA Inventory.
OSHA PEL: TWA 100 ppm
ACGIH TLV: TWA 300 ppm
NIOSH REL: (Refined Petroleum Solvents) 10H TWA 350 mg/m^3; CL 1800 mg/m^3/15M
DOT CLASSIFICATION: 3; Label: Flammable Liquid
SAFETY PROFILE: A human poison via intravenous route. Experimental carcinogenic effects reported by skin contact. Human systemic effects by intravenous route: dyspnea, respiratory stimulation, and other unspecified respiratory effects. Mildly toxic by inhalation. Can cause unconsciousness, which may be followed by coma, stentorious breathing, and bluish tint to the skin. Recovery follows removal from exposure. In mild form, intoxication resembles drunkenness. On a chronic basis, no true poisoning; sometimes headache, lack of appetite, dizziness, sleeplessness, indigestion, and nausea. A common air contaminant. Flammable liquid when exposed to heat or flame; can react with oxidizing materials. Keep containers tightly closed. Slight explosion hazard. To fight fire, use foam, CO_2, dry chemical. See also NAPHTHA, PENTANE, HEXANE, and HEPTANE.

NAJ500   CAS: 91-20-3   HR: 3
NAPHTHALENE
DOT: UN 1334/UN 2304
mf: $C_{10}H_8$ mw: 128.18
PROP: Aromatic odor; white, crystalline, volatile flakes. Plates from EtOH with characteristic odor. Mp: 80.1°, bp: 217.9°, flash p: 174°F (OC), d: 1.162, lel: 0.9%, uel: 5.9%, vap press: 1 mm @ 52.6°, vap d: 4.42, autoign temp: 1053°F (567°C). Sol in alc, benzene; insol in water; very sol in ether, CCl_4, CS_2, hydronaphthalenes, and in fixed and volatile oils. IDLH 250 ppm.
SYNS: CAMPHOR TAR □ MIGHTY 150 □ MOTH BALLS (DOT) □ MOTH FLAKES □ NAFTALEN (POLISH) □ NAPHTHALENE, crude or refined (DOT) □ NAPHT-HALENE, molten (DOT) □ NAPHTHALIN (DOT) □ NAPHTHALINE □ NAPHTHENE □ NCI-C52904 □ RCRA WASTE NUMBER U165 □ TAR CAMPHOR □ WHITE TAR
CONSENSUS REPORTS: Reported in EPA TSCA Inventory. EPA Genetic Toxicology Program. Community Right-To-Know List.
OSHA PEL: TWA 10 ppm; STEL 15 ppm
ACGIH TLV: TWA 10 ppm; STEL 15 ppm; Not Classifiable as a Human Carcinogen
DFG MAK: Confirmed Animal Carcinogen with Unknown Relevance to Humans
DOT CLASSIFICATION: 4.1; Label: Flammable Solid
SAFETY PROFILE: Human poison by ingestion. Experimental poison by ingestion,

intravenous, and intraperitoneal routes. Moderately toxic by subcutaneous route. An experimental teratogen. Experimental reproductive effects. An eye and skin irritant. Can cause nausea, headache, diaphoresis, hematuria, fever, anemia, liver damage, vomiting, convulsions, and coma. Poisoning may occur by ingestion of large doses, inhalation, or skin absorption. Questionable carcinogen with experimental tumorigenic data. Flammable when exposed to heat or flame; reacts with oxidizing materials. Explosive reaction with dinitrogen pentaoxide. Reacts violently with CrO_3, aluminum chloride + benzoyl chloride. Fires in the benzene scrubbers of coke oven gas plants have been attributed to oxidation of naphthalene. Explosive in the form of vapor or dust when exposed to heat or flame. To fight fire, use water, CO_2, dry chemical. When heated to decomposition it emits acrid smoke and irritating fumes.

NAK500   CAS: 86-87-3   HR: 3
1-NAPHTHALENEACETIC ACID
mf: $C_{12}H_{10}O_2$ mw: 186.22
PROP: Needles from water; white, odorless crystals. Mp: 134.5–135.5°. Only sltly water-sol; sol in approx 30 parts alc; very sol in acetone, ether, chloroform.
SYNS: AGRONAA □ ALPHASPRA □ ANA □ APPL-SET □ CELMONE □ FRUITONE □ KLINGTITE □ LIQUI-STIK □ NAA 800 □ NAFUSAKU □ α-NAPHTHALENE-ACETIC ACID □ NAPHTHALENE-1-ACETIC ACID □ α-NAPHTHYLACETIC □ NAPHTHYLACETIC ACID □ α-NAPHTHYLACETIC ACID □ 1-NAPHTHYLACETIC ACID □ α-NAPHTHYLENEACETIC ACID □ α-NAPHTH-YLESSIGSAEURE (GERMAN) □ NAPHYL-1-ESSIG-SAEURE (GERMAN) □ NIAGARA-STIK □ NU-TONE □ PARMONE □ PHYMONE □ PIMACOL-SOL □ PLANO-FIX □ PLUCKER □ PRIMACOL □ ROOTONE □ STAFAST □ STIK □ STOP-DROP □ TEKKAM □ TIP-OFF □ TRANSPLANTONE □ TRE-HOLD □ VARDHAK
CONSENSUS REPORTS: Reported in EPA TSCA Inventory.
SAFETY PROFILE: Poison by intraperitoneal route. Moderately toxic by ingestion. Mutation data reported. A skin, mucous membrane, and severe eye irritant. Can cause depression. A pesticide. When

heated to decomposition it emits acrid smoke and irritating fumes.

NAM000   CAS: 2243-62-1   HR: 3
1,5-NAPHTHALENEDIAMINE
mf: $C_{10}H_{10}N_2$ mw: 158.22
PROP: Prisms from H_2O, EtOH, or Et_2O. Mp: 190° (subl). Sol in hot H_2O.
SYNS: 1,5-DIAMINONAPHTHALENE □ 1,5-NAPHTHYLENEDIAMINE □ NCI-C03021
CONSENSUS REPORTS: IARC Cancer Review: Group 3 IMEMDT 7,56,87; Animal Limited Evidence IMEMDT 27,127,82. NCI Carcinogenesis Bioassay (feed); Clear Evidence: mouse, rat NCITR* NCI-CG-TR-143,78. EPA Genetic Toxicology Program.
SAFETY PROFILE: Suspected carcinogen with experimental carcinogenic, neoplastigenic, tumorigenic data. Experimental reproductive effects. Mutation data reported. When heated to decomposition it emits toxic fumes of NO_x. See also AMINES.

NAM500   CAS: 3173-72-6   HR: 2
1,5-NAPHTHALENE DIISOCYANATE
mf: $C_{12}H_6N_2O_2$ mw: 210.20
PROP: White to light-yellow crystals.
SYNS: 1,5-DIISOCYANATONAPHTHALENE □ ISOCYANIC ACID-1,5-NAPHTHYLENE ESTER
CONSENSUS REPORTS: IARC Cancer Review: Group 3 IMEMDT 7,56,87. Reported in EPA TSCA Inventory.
DFG MAK: 0.01 ppm (0.087 mg/m³)
SAFETY PROFILE: A powerful allergen. An irritant. Questionable carcinogen. When heated to decomposition it emits toxic fumes of NO_x. See also ISOCYANATES.

NAP500   CAS: 91-60-1   HR: 3
2-NAPHTHALENETHIOL
mf: $C_{10}H_8S$ mw: 160.24
PROP: Crystals from ethanol; disagreeable odor. Mp: 81°, bp: 286°. Very sol in ethanol, ether, pet ether; sltly sol in water; sltly volatile with steam.
SYNS: β-MERCAPTONAPHTHALENE □ 2-MERCAPT-ONAPHTHALENE □ β-NAPHTHALENETHIOL □

NAPHTHALENE-2-THIOL □ β-NAPHTHYL MERCAPT-
AN □ 2-NAPHTHYL MERCAPTAN □ 2-NAPHTHYL
THIOL □ RENACIT 1 □ RPA 2 □ RPA NO. 2 □ THIO-
NAPHTHOL □ THIO-β-NAPHTHOL □ β-THIONAPHTH-
OL □ 2-THIONAPHTHOL □ USAF CY-4 □ VULCAMEL
TBN
DOT CLASSIFICATION: 3; Label:
Flammable Liquid, Poison
CONSENSUS REPORTS: Reported in EPA
TSCA Inventory.
SAFETY PROFILE: Poison by ingestion and
intraperitoneal routes. A mosquito larvicide.
When heated to decomposition it emits
highly toxic fumes of SO_x. See also
MERCAPTANS.

NAQ540   CAS: 64741-55-5   HR: 3
NAPHTHA (PETROLEUM), LIGHT
CATALYTIC CRACKED
SYN: LIGHT CATALYTICALLY CRACKED NAPHTHA
CONSENSUS REPORTS: IARC Cancer
Review: Animal Limited Evidence
IMEMDT 45,39,89. Reported in EPA
TSCA Inventory.
SAFETY PROFILE: Low toxicity by
ingestion, inhalation, and skin contact. A
suspected carcinogen. A skin irritant. When
heated to decomposition it emits acrid
smoke and irritating vapors.

NAQ560   CAS: 64741-46-4   HR: 3
NAPHTHA (PETROLEUM), LIGHT
STRAIGHT-RUN
SYN: LIGHT STRAIGHT-RUN NAPHTHA
CONSENSUS REPORTS: IARC Cancer
Review: Animal Limited Evidence
IMEMDT 45,39,89. Reported in EPA
TSCA Inventory.
SAFETY PROFILE: A suspected
carcinogen. When heated to decomposition
it emits acrid smoke and irritating vapors.

NAR000   CAS: 1338-24-5   HR: 2
NAPHTHENIC ACID
PROP: Odorless crystals. D: 1.034, mp: 31°,
bp: 233°. Sltly water-sol.
SYNS: AGENAP □ NAPHID □ SUNAPTIC ACID B □
SUNAPTIC ACID C
CONSENSUS REPORTS: Reported in EPA
TSCA Inventory.

SAFETY PROFILE: Moderately toxic by
ingestion and intraperitoneal routes. When
heated to decomposition it emits acrid
smoke and irritating fumes.

NAR500   CAS: 61789-51-3   HR: 3
NAPHTHENIC ACID, COBALT SALT
DOT: UN 2001
PROP: Brown, amorph powder or bluish-
red solid. Flash p: 120°F, d: 0.9, autoign
temp: 529°F. Water-insol; sol in oil, alc,
ether. Contains 6% cobalt (AMIHAB
12,477,55).
SYNS: COBALT NAPHTHENATE, POWDER (DOT) □
NAPHTHENATE de COBALT (FRENCH)
CONSENSUS REPORTS: Cobalt and its
compounds are on the Community Right-
To-Know List. Reported in EPA TSCA
Inventory.
DOT CLASSIFICATION: 4.1; Label:
Flammable Solid
SAFETY PROFILE: Moderately toxic by
ingestion. Flammable when exposed to heat
or flame. When heated to decomposition it
emits acrid smoke and irritating fumes. See
also COBALT COMPOUNDS.

NAS000   CAS: 1338-02-9   HR: 3
NAPHTHENIC ACID, COPPER SALT
PROP: A solid. Flash p: 100°F, d: 1.055.
Contains 8% copper (AMIHAB 12,477,55).
SYNS: CHAPCO Cu-NAP □ CNC □ COPPER
NAPHTHENATE □ COPPER UVERSOL □ CUNAPSOL □
CUPRINOL □ TROYSAN COPPER 8% □ WILTZ-65 □
WITTOX C
CONSENSUS REPORTS: Copper and its
compounds are on the Community Right-
To-Know List. Reported in EPA TSCA
Inventory.
SAFETY PROFILE: A poison by ingestion.
A pesticide. A dangerous fire hazard when
exposed to heat or flame; can react with
oxidizing materials. To fight fire, use foam,
CO_2, dry chemical. See also COPPER
COMPOUNDS.

NAS500   CAS: 61790-14-5   HR: 3
NAPHTHENIC ACID, LEAD SALT
mf: $C_7H_{12}O_2 \cdot xPb$ mw: 1578.52

N

PROP: Contains 24% lead (AMIHAB 12,477,55).
SYNS: CYCLOHEXANECARBOXYLIC ACID, LEAD SALT □ LEAD NAPHTHENATE
CONSENSUS REPORTS: IARC Cancer Review: Animal Inadequate Evidence IMEMDT 23,325,80. Lead and its compounds are on the Community Right-To-Know List. Reported in EPA TSCA Inventory.
SAFETY PROFILE: A poison. Moderately toxic by intraperitoneal route. Mildly toxic by ingestion. Questionable carcinogen with experimental tumorigenic data. When heated to decomposition it emits toxic fumes of lead. See also LEAD COMPOUNDS.

NAT500   CAS: 192-65-4   HR: 3
NAPHTHO(1,2,3,4-def)CHRYSENE
mf: $C_{24}H_{14}$ mw: 302.38
PROP: Pale-yellow crystals from xylene. Mp: 225° (*in vacuo*).
SYNS: DB(a,e)P □ DIBENZO(a,e)PYRENE □ 1,2,4,5-DIBENZOPYRENE
CONSENSUS REPORTS: NTP 10th Report on Carcinogens. IARC Cancer Review: Group 2B IMEMDT 7,56,87; Animal Sufficient Evidence IMEMDT 32,327,83; IMEMDT 3,201,73.
SAFETY PROFILE: Confirmed carcinogen with experimental neoplastigenic and tumorigenic data. When heated to decomposition it emits acrid smoke and irritating fumes.

NAW500   CAS: 90-15-3   HR: 3
1-NAPHTHOL
mf: $C_{10}H_8O$ mw: 144.18
PROP: Colorless crystals or prisms; odor of phenol, disagreeable taste. Mp: 96°, bp: 282.5°, d: 1.0954 @ 98.7°/4°, vap press: 1 mm @ 94.0°. Very sltly sol in water; sol in alc and ether.
SYNS: BASF URSOL ERN □ C.I. 76605 □ C.I. OXIDATION BASE 33 □ DURAFUR DEVELOPER D □ FOURAMINE ERN □ FOURRINE 99 □ FOURRINE ERN □ FURRO ER □ α-HYDROXYNAPHTHALENE □ 1-HYDROXYNAPHTHALENE □ NAKO TRB □ 1-NAPHTHALENOL □ α-NAPHTHOL □ TERTRAL ERN □ URSOL ERN □ ZOBA ERN

CONSENSUS REPORTS: Reported in EPA TSCA Inventory. EPA Genetic Toxicology Program.
SAFETY PROFILE: Poison by ingestion and intraperitoneal routes. Moderately toxic by skin contact. An experimental teratogen. Experimental reproductive effects. A severe eye and skin irritant. Mutation data reported. Ingestion of large amounts can cause nephritis, vomiting, diarrhea, circulatory collapse, anemia, convulsions, and death. Can cause kidney irritation and injury to cornea and lens of the eye. Combustible when exposed to heat or flame. When heated to decomposition it emits acrid smoke and irritating fumes.

NAX000   CAS: 135-19-3   HR: 3
2-NAPHTHOL
mf: $C_{10}H_8O$ mw: 144.18
PROP: White to yellowish-white crystals; slt phenolic odor. Plates by subl. Mp: 121–123°, bp: 285–286°, d: 1.22, flash p: 307°F, vap press: 10 mm @ 145.5°, vap d: 4.97. Darkens with age or exposure to light. Subl when heated; distills in vacuum. Sltly sol in water; more sol in boiling water, glycerol, olive oil, solns of alkali hydroxides. Very sol in alc, ether; sol in chloroform.
SYNS: AZOGEN DEVELOPER A □ C.I. 37500 □ C.I. AZOIC COUPLING COMPONENT 1 □ C.I. DEVELOPER 5 □ DEVELOPER A □ DEVELOPER AMS □ DEVELOPER BN □ DEVELOPER SODIUM □ β-HYDROXYNAPHTHALENE □ 2-HYDROXYNAPHTHALENE □ ISONAPHTHOL □ β-MONOXYNAPHTHALENE □ β-NAFTOL (DUTCH) □ 2-NAFTOL (DUTCH) □ β-NAFTOLO (ITALIAN) □ 2-NAFTOLO (ITALIAN) □ 2-NAPHTHALENOL □ NAPHTHOL B □ β-NAPHTHOL □ β-NAPHTHYL ALCOHOL □ β-NAPHTHYL HYDROXIDE □ β-NAPHTOL (GERMAN) □ 2-NAPHTOL (FRENCH)
CONSENSUS REPORTS: Reported in EPA TSCA Inventory. EPA Genetic Toxicology Program.
SAFETY PROFILE: Poison by ingestion, inhalation, and subcutaneous routes. Mutation data reported. A skin and eye irritant. Combustible when exposed to heat or flame. To fight fire, use CO_2, dry chemical. Incompatible with antipyrine,

camphor, phenol, ferric salts, menthol, potassium permanganate and other oxidizing materials, urethane.

NBA500　　CAS: 130-15-4　　HR: 3
1,4-NAPHTHOQUINONE

mf: $C_{10}H_6O_2$　　mw: 158.16

PROP: Yellow triclinic; odor of benzoquinone; needles from EtOH or pet ether. Mp: 125–126°, d: 1.422. Very sltly sol in cold water; very sol in hot alc; sol in ether, benzene, chloroform, carbon bisulfide, acetic acid, alkali hydroxide solns. Volatile with steam.

SYNS: 1,4-DIHYDRO-1,4-DIKETONAPHTHALENE □ 1,4-NAPHTHALENEDIONE □ α-NAPHTHOQUINONE □ RCRA WASTE NUMBER U166 □ USAF CY-10

CONSENSUS REPORTS: Reported in EPA TSCA Inventory.

SAFETY PROFILE: Poison by ingestion, intravenous, subcutaneous, and intraperitoneal routes. Experimental reproductive effects. Questionable carcinogen with experimental tumorigenic data. When heated to decomposition it emits acrid smoke and irritating fumes.

NBE500　　CAS: 91-59-8　　HR: 3
β-NAPHTHYLAMINE

DOT: UN 1650

mf: $C_{10}H_9N$　　mw: 143.20

PROP: White to faint pink, lustrous leaflets from water; faint aromatic odor. Mp: 113°, d: 1.061 @ 98°/4°, vap press: 1 mm @ 108.0°, bp: 294°. Sol in hot water, alc, and ether.

SYNS: 2-AMINONAFTALEN (CZECH) □ 2-AMINONAPHTHALENE □ BETA-NAFTYLOAMINA (POLISH) □ C.I. 37270 □ FAST SCARLET BASE B □ NA □ β-NAFTILAMINA (ITALIAN) □ β-NAFTYLAMIN (CZECH) □ 2-NAFTYLAMINE (DUTCH) □ β-NAFTYLOAMINA (POLISH) □ 2-NAPHTHALAMINE □ 2-NAPHTHALENAMINE □ β-NAPHTHYLAMIN (GERMAN) □ 2-NAPHTHYLAMIN (GERMAN) □ 2-NAPHTHYLAMINE □ 6-NAPHTHYLAMINE □ 2-NAPHTHYLAMINE MUSTARD □ RCRA WASTE NUMBER U168 □ USAF CB-22

CONSENSUS REPORTS: NTP 10th Report on Carcinogens. IARC Cancer Review: Group 1 IMEMDT 7,261,87; Animal

Sufficient Evidence IMEMDT 4,97,74; Human Sufficient Evidence IMEMDT 4,97,74. Community Right-To-Know List. EPA Genetic Toxicology Program. Reported in EPA TSCA Inventory.

OSHA PEL: Cancer Suspect Agent

ACGIH TLV: Confirmed Human Carcinogen

DFG MAK: Human Carcinogen

NIOSH REL: (β-Naphthylamine) TWA use 29 CFR 1910.1009

DOT CLASSIFICATION: 6.1; Label: Poison

SAFETY PROFILE: Confirmed human carcinogen with experimental neoplastigenic and tumorigenic data. Long and continued exposure to even small amounts may produce tumors and cancers of the bladder. Poison by intraperitoneal route. Moderately toxic by ingestion. Experimental reproductive effects. Human mutation data reported. A very toxic chemical in any of its physical forms, such as flake, lump, dust, liquid, or vapor. It can be absorbed into the body through the lungs, the gastrointestinal tract, or the skin. Combustible when exposed to heat or flame. At elevated temperatures it evolves a vapor that is flammable and explosive. Incompatible with nitrous acid. When heated to decomposition it emits toxic fumes of NO_x. See also AROMATIC AMINES and 1-NAPHTHYLAMINE.

NBE700　　CAS: 134-32-7　　HR: 3
1-NAPHTHYLAMINE

DOT: UN 2077

mf: $C_{10}H_9N$　　mw: 143.20

PROP: White crystals, reddening on exposure to air; unpleasant odor. Needles from EtOH (aq) or Et_2O. Mp: 50°, bp: 300.8°, flash p: 315°F, d: 1.131, vap press: 1 mm @ 104.3°, vap d: 4.93. Sublimes, volatile with steam. Sol in 590 parts water; very sol in alc, ether. Keep well closed and away from light.

SYNS: ALFANAFTILAMINA (ITALIAN) □ ALFA-NAFTYLOAMINA (POLISH) □ 1-AMINONAFTALEN (CZECH) □ 1-AMINONAPHTHALENE □ C.I. AZOIC DIAZO COMPONENT 114 □ 1-NAFTILAMINA (SPANISH) □ α-NAFTYLAMIN (CZECH) □ 1-NAFTYL-

AMINE (DUTCH) □ NAPHTHALIDINE □ 1-NAPHTHYLAMIN (GERMAN) □ α-NAPHTHYLAMINE □ RCRA WASTE NUMBER U167

CONSENSUS REPORTS: IARC Cancer Review: Group 3 IMEMDT 7,260,87; Animal Inadequate Evidence IMEMDT 4,87,74; Human Limited Evidence IMEMDT 4,87,74. EPA Genetic Toxicology Program. Community Right-To-Know List. Reported in EPA TSCA Inventory.

OSHA PEL: Cancer Suspect Agent

NIOSH REL: (α-Naphthylamine) TWA use 29 CFR 1910.1004

DOT CLASSIFICATION: 6.1; Label: KEEP AWAY FROM FOOD

SAFETY PROFILE: Confirmed carcinogen with experimental tumorigenic data. Along with β-naphthylamine and benzidine, it has been incriminated as a cause of urinary bladder cancer. Poison by subcutaneous and intraperitoneal routes. Moderately toxic by ingestion. Mutation data reported. Combustible when exposed to heat or flame. Incompatible with nitrous acid. To fight fire, use dry chemical, CO_2, mist, spray. When heated to decomposition it emits toxic fumes of NO_x. See also 2-NAPHTHYLAMINE and AROMATIC AMINES.

NBE850   CAS: 86-65-7   HR: 2
2-NAPHTHYLAMINE-6,8-DISULFONIC ACID

mf: $C_{10}H_9NO_6S_2$ mw: 303.32

PROP: Needles from H_2O. Sol in H_2O.

SYNS: AMIDO-G-ACID □ 7-AMINO-1,3-NAPHTH-ALENEDISULFONIC ACID

CONSENSUS REPORTS: Reported in EPA TSCA Inventory.

SAFETY PROFILE: Questionable carcinogen with experimental neoplastigenic data. When heated to decomposition it emits toxic fumes of NO_x and SO_x.

NBH200   CAS: 3759-61-3   HR: 3
1-NAPHTHYL CHLOROFORMATE

mf: $C_{11}H_7ClO_2$ mw: 206.63

SYNS: CARBANOCHLORIDIC ACID, NAPHTHYL ESTER □ CARBONOCHLORIDIC ACID, 1-NAPHTH-ALENYL ESTER □ CHLOROFORMIC ACID, ESTER with 1-NAPHTHOL □ CHLOROFORMIC ACID 1-NAPHTHYL ESTER □ CHLOROMROWCZAN 1-NAFTYLU (CZECH) □ 1-NAPHTHYL CHLOROCARBONATE □ α-NAPHTHYL CHLOROFORMATE

CONSENSUS REPORTS: Reported in EPA TSCA Inventory.

DOT CLASSIFICATION: 6.1; Label: Poison, Corrosive

SAFETY PROFILE: Poison by intravenous and intraperitoneal routes. Moderately toxic by ingestion. A severe eye and skin irritant. When heated to decomposition it emits toxic fumes of Cl⁻. See also ESTERS.

NBJ500   CAS: 7090-25-7   HR: 3
1-NAPHTHYL METHYLNITROSO-CARBAMATE

mf: $C_{12}H_{10}N_2O_3$ mw: 230.24

SYNS: DENAPON, NITROSATED (JAPANESE) □ METHYL-NITROSOCARBAMIC ACID-1-NAPHTHYL ESTER □ 1-NAPHTHYL-N-METHYL-N-NITROSO-CARBAMATE □ N-NITROSOCARBARYL □ NITROSO-NAC

CONSENSUS REPORTS: IARC Cancer Review: Animal Sufficient Evidence IMEMDT 12,37,76. EPA Genetic Toxicology Program.

SAFETY PROFILE: Confirmed carcinogen with experimental carcinogenic, neoplastigenic, and tumorigenic data. Human mutation data reported. When heated to decomposition it emits toxic fumes of NO_x. See also N-NITROSO COMPOUNDS and CARBAMATES.

NBL000   CAS: 93-46-9   HR: 2
2-NAPHTHYL-p-PHENYLENEDIAMINE

mf: $C_{26}H_{20}N_2$ mw: 360.48

SYNS: ACETO DIPP □ AGERITE WHITE □ DI-β-NAPHTHYL-p-PHENYLDIAMINE □ DI-β-NAPHTHYL-p-PHENYLENEDIAMINE □ N,N'-DI-β-NAPHTHYL-p-PHENYLENEDIAMINE □ sym-DI-β-NAPHTHYL-p-PHENYLENEDIAMINE □ DNPD □ DWU-β-NAFTYLO-p-FENYLODWUAMINA (POLISH) □ NONOX CL □ TISPERSE MB-2X

CONSENSUS REPORTS: Reported in EPA TSCA Inventory.

SAFETY PROFILE: A human skin irritant. An experimental skin and eye irritant. Mutation data reported. Questionable carcinogen with experimental tumorigenic data. When heated to decomposition it emits toxic fumes of NO_x. See also AMINES.

NBR000 **CAS: 500-38-9** **HR: 2**
NDGA
mf: $C_{18}H_{22}O_4$ mw: 302.40
PROP: Crystals from acetic acid. Mp: 184–185°. Sol in methanol, ethanol, and ether; sltly sol in hot water and chloroform; nearly insol in benzene and petroleum ether.
SYNS: 1,4-BIS(3,4-DIHYDROXYPHENYL)-2,3-DIMETH-YLBUTANE □ DIHYDRONORGUAIARETIC ACID □ β,γ-DIMETHYL-α,Δ-BIS(3,4-DIHYDROXYPHENYL)BUTANE □ 4,4'-(2,3-DIMETHYLTETRAMETHYLENE)DIPYRO-CATECHOL □ NORDIHYDROGUAIARETIC ACID □ NORDIHYDROGUAIRARETIC ACID
SAFETY PROFILE: Moderately toxic by intraperitoneal route. An antioxidant food additive. When heated to decomposition it emits acrid smoke and irritating fumes.

NBU000 **CAS: 57-33-0** **HR: 3**
NEMBUTAL SODIUM
mf: $C_{11}H_{18}N_2O_3$•Na   mw: 249.30
PROP: White, crystalline powder. Sol in water and alc; insol in ether.
SYNS: AUROPAN □ BARPENTAL □ BIOSEDAN □ BUTYLONE □ CARBRITAL □ CONTINAL □ DIABUTAL □ EMBUTAL □ ETAMINAL SODIUM □ ETHAMINAL SODIUM □ 5-ETHYL-5-(1-METHYLBUTYL)BARBITURIC ACID SODIUM SALT □ 5-ETHYL-5-(1-METHYLBUTYL)-2,4,6(1H,3H,5H)-PYRIMIDINETRIONE MONOSODIUM SALT (9CI) □ EUTHATAL □ IPRAL SODIUM □ ISOBARB □ MEBUBARBITAL SODIUM □ MEBUMAL NATRIUM □ MEBUMAL SODIUM □ MINTAL □ NAPENTAL □ PACIFAN □ PALAPENT □ PENBAR □ PENTABARBIT-AL SODIUM □ PENTAL □ PENTOBARBITONE SODIUM □ PENTONAL □ PENTYL □ PROPYLMETHYLCARBIN-YLETHYL BARBITURIC ACID SODIUM SALT □ RIVADORN □ SAGATAL □ SODITAL □ SODIUM ETHAMINAL □ SODIUM-5-ETHYL-5-(1-METHYL-BUTYL)BARBITURATE □ SODIUM NEMBUTAL □ SODIUM-PENT □ SODIUM PENTABARBITAL □ SODIUM PENTABARBITONE □ SODIUM PENTO-BARBITAL □ SODIUM PENTOBARBITONE □ SODIUM PENTOBARBITURATE □ SOLUBLE PENTOBARBITAL □

SOMNOPENTYL □ SONISTAN □ SONTOBARBITAL NABITONE □ SOPENTAL □ SOTYL □ VETBUTAL
SAFETY PROFILE: Poison by ingestion, intraperitoneal, subcutaneous, intravenous, intraduodenal, intramuscular, intracerebral, parenteral, and rectal routes. An experimental teratogen. Other experimental reproductive effects. Human systemic effects by ingestion: wakefulness, change in motor activity, ataxia, and antipsychotic effects. Mutation data reported. When heated to decomposition it emits toxic fumes of NO_x and Na_2O. See also BARBITURATES.

NBW100 **CAS: 108944-67-8** **HR: 3**
NEOCURDIONE
mf: $C_{15}H_{24}O_2$ mw: 236.35
SYN: 6-CYCLODECENE-1,4-DIONE, 6,10-DIMETHYL-3-(1-METHYLETHYL)-, (3R,6E,10S)-
SAFETY PROFILE: A poison by ingestion. When heated to decomposition it emits acrid smoke and irritating vapors.

NBX000 **CAS: 7440-00-8** **HR: 3**
NEODYMIUM
af: Nd   aw: 144.24
PROP: It is a bright, silvery, lustrous, very reactive rare-earth metal that tarnishes quickly in air. Bp: 3074°, d: 7.003, mp: approx 1024°. Dissolves in dil acid. Reacts slowly with cold water, faster with hot water.
CONSENSUS REPORTS: Reported in EPA TSCA Inventory.
SAFETY PROFILE: Human systemic effects by intracerebral route: blood changes. It may be an anticoagulant lanthanoid. Care in handling is advised. Flammable in the form of dust when exposed to heat or flame. Slight explosion hazard in the form of dust when exposed to flame. Can react violently with air, halogens, N_2. Violent reaction with phosphorus above 400°C. Many of its compounds are poisons. See also RARE EARTHS and various neodymium compounds.

N

NBY000   CAS: 10024-93-8   HR: 3
NEODYMIUM CHLORIDE
mf: Cl_3Nd mw: 250.59
PROP: A gas. Mp: 7°, bp: 16°. Sol in water, alc.
CONSENSUS REPORTS: Reported in EPA TSCA Inventory.
SAFETY PROFILE: Poison by intravenous, intraperitoneal, and subcutaneous routes. Moderately toxic by ingestion. A skin and eye irritant. When heated to decomposition it emits very toxic fumes of Cl^-. See also NEODYMIUM, CHLORIDES, and RARE EARTHS.

NCB000   CAS: 10045-95-1   HR: 3
NEODYMIUM(III) NITRATE (1:3)
mf: $N_3O_9 \cdot Nd$ mw: 330.27
PROP: Triclinic hygroscopic violet crystals. Sol in H_2O and EtOH.
SYN: NITRIC ACID, NEODYMIUM SALT
CONSENSUS REPORTS: Reported in EPA TSCA Inventory.
SAFETY PROFILE: Poison by intraperitoneal and intravenous routes. Moderately toxic by ingestion. When heated to decomposition it emits very toxic fumes of NO_x. See also NITRATES and NEODYMIUM.

NCC000   CAS: 1313-97-9   HR: 1
NEODYMIUM OXIDE
mf: Nd_2O_3 mw: 336.48
PROP: Light blue solid (red fluorescence) or powder with hexagonal structure. Mp: 2233°, bp: 3760°. Very stable; sol in dil acids; insol in water.
SYNS: DINEODYMIUM TRIOXIDE □ NEODYMIA □ NEODYMIUM(III) OXIDE □ NEODYMIUM(3+) OXIDE □ NEODYMIUM SESQUIOXIDE □ NEODYMIUM TRIOXIDE
CONSENSUS REPORTS: Reported in EPA TSCA Inventory.
SAFETY PROFILE: Low toxicity by ingestion. See also NEODYMIUM and RARE EARTHS.

NCE000   CAS: 1404-04-2   HR: 3
NEOMYCIN
PROP: An antibiotic.
SYNS: MYACYNE □ MYCIFRADIN □ NEOMCIN □ NIVEMYCIN □ VONAMYCIN POWDER V
SAFETY PROFILE: Poison by intraperitoneal, intravenous, and subcutaneous routes. Moderately toxic by ingestion. Human systemic effects: changes in hearing acuity, liver tubule changes, and decreased urine volume or anuria. Mutation data reported. When heated to decomposition it emits acrid smoke and irritating fumes.

NCG000   CAS: 1405-10-3   HR: 3
NEOMYCIN SULFATE
PROP: White odorless powder. Bp: Degrades before boiling. Sol in water
SYNS: BIOSOL VETERINARY □ FRADIOMYCIN SULFATE □ LIDAMYCIN CREME □ MYCAIFRADIN SULFATE □ MYCIFRADIN-N □ MYCIGIENT □ NEOBIOTIC □ NEO-MANTLE CREME □ NEOMIX □ NEOMYCINE SULFATE □ NEOMYCIN SULPHATE □ OTOBIOTIC □ QUINTESS-N □ USAF CB-19
CONSENSUS REPORTS: Reported in EPA TSCA Inventory. EPA Genetic Toxicology Program.
SAFETY PROFILE: Poison by intramuscular, intravenous, and subcutaneous routes. Human systemic effects by ingestion: somnolence, hallucinations and distorted perceptions, and anorexia. A human skin irritant. When heated to decomposition it emits very toxic fumes of SO_x. See also NEOMYCIN.

NCG500   CAS: 7440-01-9   HR: 1
NEON
DOT: UN 1065/UN 1913
af: Ne   aw: 20.18
PROP: Colorless, monatomic, non-toxic, gaseous element; tasteless and odorless. Inert except towards F_2. Mp: −248.67°, bp: −245.9°, d: (liquid) 1.204 @ −245.9°; (gas) 0.89994 g/L @ 0°. Sol in water.
SYNS: NEON, compressed (UN 1065) (DOT) □ NEON, refrigerated liquid (cryogenic liquid) (UN 1913) (DOT)

CONSENSUS REPORTS: Reported in EPA TSCA Inventory.
ACGIH TLV: Simple Asphyxiant
DOT CLASSIFICATION: 2.2; Label: Nonflammable Gas
SAFETY PROFILE: An inert asphyxiant gas. See also ARGON.

NCH000   CAS: 463-82-1   HR: 3
NEOPENTANE
DOT: UN 2044
mf: C_5H_{12} mw: 72.17
PROP: Liquid or gas at room temp. Solidifies @ $-19.8°$, bp: $9.5°$, d: $0.613°$ @ $0°/0°$ (liquid), flash p: $<19.4°F$, lel: 1.4%, uel: 7.5%. Insol in water.
SYNS: 2,2-DIMETHYLPROPANE ◻ 2,2-DIMETHYL-PROPANE, other than pentane and isopentane (DOT) ◻ tert-PENTANE
CONSENSUS REPORTS: Reported in EPA TSCA Inventory.
ACGIH TLV: TWA 600 ppm
DFG MAK: 1000 ppm (3000 mg/m³)
NIOSH REL: TWA 120 ppm; CL 610 ppm/15M
DOT CLASSIFICATION: 2.1; Label: Flammable Gas
SAFETY PROFILE: Poison by intraperitoneal route. An inhalation hazard. Both the gas and the liquid are flammable when exposed to heat or flame; can react vigorously with oxidizing materials. When heated to decomposition it emits acrid smoke and irritating fumes.

NCI500   CAS: 126-99-8   HR: 3
NEOPRENE
DOT: UN 1991
mf: C_4H_5Cl mw: 88.54
PROP: Colorless liquid. An oil-resistant synthetic rubber made by the polymerization of chloroprene. D: 0.958 @ $20°/20°$, bp: $59.4°$, flash p: $-4°F$, lel: 4.0%, uel: 20%, vap d: 3.0, brittle point: $-35°$, softens @ approx $80°$. Sltly sol in water; misc in alc and ether. IDLH 300 ppm.
SYNS: 2-CHLOOR-1,3-BUTADIEEN (DUTCH) ◻ 2-CHLOR-1,3-BUTADIEN (GERMAN) ◻ CHLOROBUTADI-ENE ◻ 2-CHLOROBUTA-1,3-DIENE ◻ 2-CHLORO-1,3-BUTADIENE ◻ CHLOROPREEN (DUTCH) ◻ CHLOR-OPREN (GERMAN, POLISH) ◻ CHLOROPRENE ◻ β-CHLOROPRENE (OSHA, MAK) ◻ CHLOROPRENE, inhibited (DOT) ◻ CHLOROPRENE, uninhibited (DOT) ◻ 2-CLORO-1,3-BUTADIENE (ITALIAN) ◻ CLOROPRENE (ITALIAN)
CONSENSUS REPORTS: NTP 10th Report on Carcinogens. IARC Cancer Review: Group 3 IMEMDT 7,160,87; Animal Inadequate Evidence IMEMDT 19,131,79; Human Inadequate Evidence IMEMDT 19,131,79. Reported in EPA TSCA Inventory. Community Right-To-Know List.
OSHA PEL: TWA 10 ppm (skin)
ACGIH TLV: TWA 10 ppm (skin)
DFG MAK: Animal Carcinogen, Suspected Human Carcinogen
NIOSH REL: CL (Chloroprene) 1 ppm/15M
DOT CLASSIFICATION: 3; Label: Flammable Liquid, Poison (UN 1991); DOT Class: Forbidden
SAFETY PROFILE: Confirmed carcinogen. Poison by ingestion, intravenous, and subcutaneous routes. Moderately toxic by inhalation. An experimental teratogen. Experimental reproductive effects. Human mutation data reported. Human exposure has caused dermatitis, conjunctivitis, corneal necrosis, anemia, temporary loss of hair, nervousness, and irritability. Exposure to the vapor can cause respiratory tract irritation leading to asphyxia. Other effects are central nervous system depression, drop in blood pressure, severe degenerative changes in the liver, kidneys, lungs, and other vital organs. A very dangerous fire hazard when exposed to heat or flame. Explosive in the form of vapor when exposed to heat or flame. To fight fire, use alcohol foam. Auto-oxidizes in air to form an unstable peroxide that catalyzes exothermic polymerization of the monomer. Incompatible with liquid or gaseous fluorine. When heated to decomposition it emits toxic fumes of Cl^-. See also CHLORINATED HYDROCARBONS, ALIPHATIC.

NCJ500   CAS: 457-60-3   HR: 3
NEOSALVARSAN

mf: $C_{13}H_{13}As_2N_2H_2O_4S \cdot Na$ mw: 466.17

PROP: Yellow powder.

SYNS: ((5-(3-AMINO-4-HYDROXYPHENYL)ARSENO)-2-HYDROXYANILINO)METHANOL SULFOXYLATE SODIUM □ ARSEVAN □ ARSPHENAMINE METHYLE-NESULFOXYLIC ACID SODIUM SALT □ COLLUNOVAR □ COLLUNOVER □ 3,3'-DIAMINO-4,4'-DIHYDROXY ARSENOBENZENE METHYLENESULFOXYLATE SODIUM □ MIARSENOL □ NEOARSOLUIN □ NEOARSPHENAMINE □ NOVARSAN □ NOVARSENO-BENZOL □ NOVARSENOBILLON □ VETARSENO-BILLON

CONSENSUS REPORTS: Arsenic and its compounds are on the Community Right-To-Know List.

OSHA PEL: TWA 0.5 mg(As)m³

ACGIH TLV: BEI: 35 μ (As)/L inorganic arsenic and methylated metabolites in urine

SAFETY PROFILE: Poison by intravenous and subcutaneous routes. When heated to decomposition it emits very toxic fumes of As, NO_x, Na_2O, and SO_x. See also ARSENIC COMPOUNDS.

NCL500   CAS: 59-42-7   HR: 3
NEOSYNEPHRINE

mf: $C_9H_{13}NO_2$ mw: 167.23

PROP: A solid. Mp: 177°.

SYNS: 1-α-HYDROXY-β-METHYLAMINO-3-HYDRO-XY-1-ETHYLBENZENE □ (R)-3-HYDROXY-α-((METHYL-AMINO)METHYL)BENZENEMETHANOL □ 1-m-HYDROXY-α-((METHYLAMINO)METHYL)-BENZYL ALCOHOL □ (−)-m-HYDROXY-α-(METHYLAMINO-METHYL)BENZYL ALCOHOL □ 1-1-(m-HYDROXY-PHENYL)-2-METHYLAMINOETHANOL □ 1-(3-HYDROXYPHENYL)-N-METHYLETHANOLAMINE □ ISOPHRIN □ MESATON □ METAOXEDRIN □ METASYMPATOL □ METASYNEPHRINE □ m-METHYLAMINOETHANOLPHENOL □ MEZATON □ R(−)-MEZATON □ m-OXEDRINE □ (−)-m-OXEDRINE □ PHENYLEPHRINE □ (−)-PHENYLEPHRINE □ R(−)-PHENYLEPHRINE □ m-SYMPATHOL □ m-SYMPATOL □ m-SYNEPHRINE □ VISADRON

SAFETY PROFILE: Poison by ingestion, subcutaneous, intravenous, intraperitoneal, and intraduodenal routes. Human systemic effects by ocular route: blood pressure increase. An experimental teratogen. Other experimental reproductive effects. A nasal decongestant. When heated to decomposition it emits toxic fumes of NO_x.

NCM800   CAS: 26552-50-1   HR: 3
NEPTAL

mf: $C_{12}H_{15}HgNO_6$ mw: 469.87

SYNS: ACETIC ACID, (o-((2-HYDROXY-3-HYDROXY-MERCURI)PROPYL)CARBAMOYL)PHENOXY- □ (o-((2-HYDROXY-3-(HYDROXYMERCURY)PROPYL)CAR-BAMOYL)PHENOXY)ACETIC ACID □ HYDROXY-MERCURIPROPANOLAMIDE of o-CARBOXYPHENOXY-ACETIC ACID □ MERCURY, (3-(α-CARBOXY-o-ANIS-AMIDO)-2-HYDROXYPROPYL)HYDROXY-

ACGIH TLV: TWA 0.1 mg(Hg)/m³ (skin); BEI: 35 μg/g creatinine total inorganic mercury in urine preshift; 15 μg/g creatinine total inorganic mercury in blood at end of shift at end of workweek.

DFG MAK: Confirmed Animal Carcinogen with Unknown Relevance to Humans

NIOSH REL: (Mercury, Aryl and Inorganic) CL 0.1 mg/m³ (skin)

SAFETY PROFILE: Poison by intravenous route. When heated to decomposition it emits toxic fumes of NO_x and Hg.

NCN600   CAS: 13997-19-8   HR: D
NEQUINATE

mf: $C_{22}H_{23}NO_4$ mw: 365.43

PROP: Crystals. Mp: 287–288°.

SYNS: 7-(BENZYLOXY)-6-N-BUTYL-1,4-DIHYDRO-4-OXO-3-QUINOLINECARBOXYLIC ACID METHYL ESTER □ 7-(BENZYLOXY)-6-N-BUTYL-4-HYDROXY-3-QUINOLINECARBOXYLIC ACID METHYL □ 3-METHOXYCARBONYL-6-N-BUTYL-7-BENZYLOXY-4-OXOQUINOLINE □ STATYL

SAFETY PROFILE: When heated to decomposition it emits acrid smoke and irritating fumes.

NCN800   CAS: 56001-43-5   HR: 1
NEROLIDYL ACETATE

mf: $C_{17}H_{28}O_2$ mw: 264.45

SYNS: 1,6,10-DODECATRIEN-3-OL, 3,7,11-TRIMETHYL, ACETATE, (S-(Z))- □ 3,7,11-TRIMETHYL-1,6,10-DODECA-TRIEN-3-YL ACETATE

CONSENSUS REPORTS: Reported in EPA TSCA Inventory.

SAFETY PROFILE: A skin irritant. When heated to decomposition it emits acrid smoke and irritating fumes.

NCP500   CAS: 7783-33-7   HR: 3
NESSLER REAGENT
DOT: UN 1643
mf: HgI$_4$2K   mw: 786.39
PROP: Light yellow liquid. IDLH 10 mg/m^3 (as Hg).
SYNS: CHANNING'S SOLUTION □ MERCURIC POTASSIUM IODIDE □ MERCURIC POTASSIUM IODIDE, solid (DOT) □ MERCURY(II) POTASSIUM IODIDE □ POTASSIUM IODOHYDRARGYRATE □ POTASSIUM MERCURIC IODIDE □ POTASSIUM TETRAIODOMERCURATE(II) □ TETRAIODOMERCUR-ATE(2+), DIPOTASSIUM
CONSENSUS REPORTS: Mercury and its compounds are on the Community Right-To-Know List. Reported in EPA TSCA Inventory.
OSHA PEL: CL 0.1 mg(Hg)/m^3 (skin)
ACGIH TLV: TWA 0.1 mg(Hg)/m^3 (skin); BEI: 35 μg/g creatinine total inorganic mercury in urine preshift; 15 μg/g creatinine total inorganic mercury in blood at end of shift at end of workweek; (Proposed: (inhalable fraction) 0.1 mg/m^3; Not Classifiable as a Human Carcinogen)
NIOSH REL: (Mercury, Aryl and Inorganic) CL 0.1 mg/m^3 (skin)
DOT CLASSIFICATION: 6.1; Label: Poison
SAFETY PROFILE: A poison. Moderately toxic by skin contact and intraperitoneal routes. When heated to decomposition it emits very toxic fumes of Hg, K$_2$O, and I$^-$. See also MERCURY.

NCQ820   CAS: 464-45-9   HR: 1
NGAI CAMPHOR
mf: C$_{10}$H$_{18}$O   mw: 154.28
PROP: Crystals from pet ether. Mp: 208–209°, bp: 212°.
SYNS: 1-2-BORNANOL □ (−)-BORNEOL □ (1S,2R,4S)-(−)-1-BORNEOL □ 1-BORNYL ALCOHOL □ 1-2-CAMPHANOL □ LINDEROL
CONSENSUS REPORTS: Reported in EPA TSCA Inventory. EPA Genetic Toxicology Program.
SAFETY PROFILE: Mildly toxic by ingestion. A skin irritant. When heated to decomposition it emits acrid smoke and irritating fumes. See also ALCOHOLS.

NCQ900   CAS: 59-67-6   HR: 3
NIACIN
mf: C$_6$H$_5$NO$_2$ mw: 123.12
PROP: The anti-pellagra vitamin. Colorless needles or white crystalline powder from water or alc; slt odor. Mp: 236°, subl above mp, d: 1.473. Sol in water and boiling alc; insol in most lipid solvents. Nonhygroscopic and stable in air.
SYNS: ACIDE NICOTINIQUE (FRENCH) □ ACIDUM NICOTINICUM □ AKOTIN □ ANTI-PELLAGRA VITAMIN □ APELAGRIN □ BIONIC □ 3-CARBOXYPYRIDINE □ DASKIL □ DAVITAMON PP □ DIREKTAN □ EFACIN □ NAH □ NAOTIN □ NICACID □ NICAMIN □ NICANGIN □ NICO □ NICO-400 □ NICOBID □ NICOCAP □ NICOCIDIN □ NICOCRISINA □ NICODAN □ NICODELMINE □ NICOLAR □ NICONACID □ NICONAT □ NICONAZID □ NICOROL □ NICOSIDE □ NICO-SPAN □ NICOSYL □ NICOTAM-IN □ NICOTENE □ NICOTIL □ NICOTINE ACID □ NICOTINIC ACID □ NICOTINIPCA □ NICOTINOYL HYDRAZINE □ NICOTINSAEURE (GERMAN) □ NICOVASAN □ NICOVASEN □ NICOVEL □ NICYL □ NIPELLEN □ PELLAGRAMIN □ PELLAGRA PREVENT-IVE FACTOR □ PELLAGRIN □ PELONIN □ PEVITON □ PP FACTOR □ P.P. FACTOR-PELLAGRA PREVENTIVE FACTOR □ PYRIDINE-3-CARBONIC ACID □ PYRIDINE-β-CARBOXYLIC ACID □ PYRIDINE-3-CARBOXYLIC ACID □ 3-PYRIDINECARBOXYLIC ACID □ PYRIDINE-CARBOXYLIQUE-3 (FRENCH) □ S115 □ SK-NIACIN □ TINIC □ VITAPLEX N □ WAMPOCAP
CONSENSUS REPORTS: Reported in EPA TSCA Inventory.
SAFETY PROFILE: Poison by intraperitoneal route. Moderately toxic by ingestion, intravenous, and subcutaneous routes. Human systemic effects: change in clotting factors, changes in platelet count. Questionable carcinogen with experimental carcinogenic data. When heated to decomposition it emits toxic fumes of NO$_x$.

NCR000   CAS: 98-92-0   HR: 2
NIACINAMIDE
mf: C$_6$H$_6$N$_2$O   mw: 122.14
PROP: Colorless needles or white crystalline powder; odorless with a bitter taste. Mp: 129°, d: 1.40, bp: 150–160° @ 0.0005. Very sol in water, ether, glycerin.
SYNS: ACID AMIDE □ AMIDE PP □ AMINICOTIN □ AMIXICOTYN □ AMNICOTIN □ AUSTROVIT PP □

BENICOT □ DELONIN AMIDE □ DIPEGYL □ DIPIGYL □ ENDOBION □ FACTOR PP □ HANSAMID □ INOVITAN PP □ NAM □ NANDERVIT-N □ NIACEVIT □ NIAMIDE □ NICAMIDE □ NICAMINA □ NICAMINDON □ NICASIR □ NICOBION □ NICOFORT □ NICOGEN □ NICOMIDOL □ NICOSAN 2 □ NICOTA □ NICOTAMIDE □ NICOTILAMIDE □ NICOTILIL-AMIDO □ NICOTINE ACID AMIDE □ NICOTINIC ACID AMIDE □ NICOTINIC AMIDE □ NICOTINSAEUREAM-ID (GERMAN) □ NICOTOL □ NICOTYLAMIDE □ NICOVEL □ NICOVIT □ NICOVITOL □ NICOZYMIN □ NIKO-TAMIN □ NIKOTINSAEUREAMID (GERMAN) □ NIOCINAMIDE □ NIOZYMIN □ PELMIN □ PELMINE □ PELONIN AMIDE □ PP-FACTOR □ PYRIDINE-3-CARBOXYLIC ACID AMIDE □ 3-PYRIDINECARBOXYL-IC ACID AMIDE □ VI-NICOTYL □ VI-NICTYL □ VITAMIN B3 □ VITAMIN PP □ WITAMINA PP

CONSENSUS REPORTS: Reported in EPA TSCA Inventory.

SAFETY PROFILE: Moderately toxic by ingestion, intravenous, intraperitoneal, and subcutaneous routes. Mutation data reported. When heated to decomposition it emits toxic fumes of NO_x.

NCR025   HR: D
NIACINAMIDE ASCORBATE

PROP: Lemon yellow-colored powder; very slt odor. Mp: 141–145°. Sol in water, alc; sltly sol in chloroform, ether, glycerin; insol in benzene.

SYN: NIACINAMIDE ASCORBIC ACID COMPLEX

SAFETY PROFILE: When heated to decomposition it emits acrid smoke and irritating fumes.

NCW500   CAS: 7440-02-0   HR: 3
NICKEL

af: Ni   aw: 58.71

PROP: A silvery-white, hard, malleable, and ductile metal. Crystallizes as metallic cubes. D: 8.90 @ 25°, vap press: 1 mm @ 1810°, mp: 1455°, bp: 2920°. Stable in air at room temp. IDLH 10 mg/m³ (as Ni).

SYNS: C.I. 77775 □ Ni 270 □ NICHEL (ITALIAN) □ NICKEL 270 □ NICKEL (DUST) □ NICKEL PARTICLES □ NICKEL SPONGE □ Ni 0901-S □ Ni 4303T □ NP 2 □ RANEY ALLOY □ RANEY NICKEL

CONSENSUS REPORTS: NTP 10th Report on Carcinogens. IARC Cancer Review: Group 1 IMEMDT 7,264,87; Animal

Sufficient Evidence IMEMDT 11,75,76; Animal Inadequate Evidence IMEMDT 2,126,73. Community Right-To-Know List. Reported in EPA TSCA Inventory.

OSHA PEL: TWA Soluble Compounds: 0.1 mg(Ni)/m³; Insoluble Compounds: 1 mg(Ni)/m³

ACGIH TLV: TWA 0.1 mg(Ni)/m³; Not Suspected as a Human Carcinogen)

DFG MAK: DFG TRK: Human Carcinogen

NIOSH REL: (Inorganic Nickel) TWA 0.015 mg(Ni)/m³

SAFETY PROFILE: Confirmed carcinogen with experimental carcinogenic, neoplastigenic, and tumorigenic data. Poison by ingestion, intratracheal, intraperitoneal, subcutaneous, and intravenous routes. An experimental teratogen. Ingestion of soluble salts causes nausea, vomiting, and diarrhea. Mutation data reported. Hypersensitivity to nickel is common and can cause allergic contact dermatitis, pulmonary asthma, conjunctivitis, and inflammatory reactions around nickel-containing medical implants and prostheses. Powders may ignite spontaneously in air. Reacts violently with F_2, NH_4NO_3, hydrazine, NH_3, (H_2 + dioxane), performic acid, P, Se, S, (Ti + $KClO_3$). Incompatible with oxidants (e.g., bromine pentafluoride, peroxyformic acid, potassium perchlorate, chlorine, nitryl fluoride, ammonium nitrate), Raney-nickel catalysts may initiate hazardous reactions with ethylene + aluminum chloride, p-dioxane, hydrogen, hydrogen + oxygen, magnesium silicate, methanol, organic solvents + heat, sulfur compounds. Nickel catalysts have caused many industrial accidents.

NCX000   CAS: 373-02-4   HR: 3
NICKEL(II) ACETATE (1:2)

mf: $C_4H_6O_4 \cdot Ni$ mw: 176.81

PROP: Green prisms or hygroscopic green powder. Mp: decomp, d: 1.798. IDLH 10 mg/m³ (as Ni).

SYNS: ACETIC ACID, NICKEL(2+) SALT □ NICKELOUS ACETATE

CONSENSUS REPORTS: NTP 10th Report on Carcinogens. IARC Cancer Review: Group 1 IMEMDT 7,264,87. Nickel and its compounds are on the Community Right-To-Know List. Reported in EPA TSCA Inventory.
OSHA PEL: TWA 0.1 mg (Ni)/m^3
ACGIH TLV: TWA 0.2 mg(Ni)/m^3; Human Carcinogen)
NIOSH REL: (Inorganic Nickel) TWA 0.015 mg(Ni)/m^3
SAFETY PROFILE: Confirmed carcinogen with experimental neoplastigenic and tumorigenic data. Poison by ingestion, intraperitoneal, and subcutaneous routes. Experimental reproductive effects. Mutation data reported. When heated to decomposition it emits irritating fumes. See also NICKEL COMPOUNDS.

NCX500   CAS: 6018-89-9   HR: 3
NICKEL ACETATE TETRAHYDRATE
mf: $C_4H_6O_4 \cdot Ni \cdot 4H_2O$ mw: 248.89
PROP: Green crystals. Sol in H_2O and alcohols. IDLH 10 mg/m^3 (as Ni).
SYNS: ACETIC ACID, NICKEL(+2) SALT, TETRAHYDRATE □ NICKEL(II) ACETATE TETRAHYDRATE □ NICKEL DIACETATE TETRAHYDRATE □ NICKELOUS ACETATE TETRAHYDRATE
CONSENSUS REPORTS: NTP 10th Report on Carcinogens. Nickel and its compounds are on the Community Right-To-Know List.
OSHA PEL: TWA 0.1 mg (Ni)/m^3
ACGIH TLV: TWA 0.2 mg(Ni)/m^3; Human Carcinogen)
NIOSH REL: (Inorganic Nickel) TWA 0.015 mg(Ni)/m^3
SAFETY PROFILE: Confirmed human carcinogen. Poison by ingestion and intraperitoneal routes. Mutation data reported. When heated to decomposition it emits acrid smoke and irritating fumes. See also NICKEL(II) ACETATE and NICKEL COMPOUNDS.

NCY050   CAS: 15699-18-0   HR: 3
NICKEL AMMONIUM SULFATE
mf: $O_8S_2 \cdot Ni \cdot 2H_4N$ mw: 286.93

PROP: Crystals. IDLH 10 mg/m^3 (as Ni).
SYNS: AMMONIUM DISULFATONICKELATE(II) □ AMMONIUM NICKEL SULFATE □ SULFURIC ACID, AMMONIUM NICKEL(2+) SALT (2:2:1)
CONSENSUS REPORTS: NTP 10th Report on Carcinogens. IARC Cancer Review: Animal Limited Evidence IMEMDT 49,257,90. Reported in EPA TSCA Inventory.
OSHA PEL: TWA 1 mg(Ni)/m^3
ACGIH TLV: TWA 1 mg(Ni)/m^3; Not Classifiable as a Human Carcinogen
SAFETY PROFILE: Confirmed human carcinogen. Poison by ingestion. When heated to decomposition it emits toxic fumes of NO_x, SO_x, and Ni.

NCY100   CAS: 12035-52-8   HR: 3
NICKEL ANTIMONIDE
mf: NiSb   mw: 180.46
PROP: Light copper-red solid. Mp: 1158°. IDLH 10 mg/m^3 (as Ni).
SYNS: ANTIMONY, compounded with NICKEL (1:1) □ NICKEL MONOANTIMONIDE
CONSENSUS REPORTS: NTP 10th Report on Carcinogens. IARC Cancer Review: Animal Limited Evidence IMEMDT 49,257,90. Reported in EPA TSCA Inventory.
ACGIH TLV: TWA 1 mg(Ni)/m^3
SAFETY PROFILE: Confirmed human carcinogen with experimental carcinogenic data. When heated to decomposition it emits toxic fumes of Ni and Sb.

NCY110   CAS: 27016-75-7   HR: 3
NICKEL ARSENIDE
mf: AsNi   mw: 133.63
PROP: Copper-red to pink metallic opaque crystals. D: 7.8. IDLH 10 mg/m^3 (as Ni).
SYN: NICKEL MONOARSENIDE
CONSENSUS REPORTS: NTP 10th Report on Carcinogens. IARC Cancer Review: Animal Limited Evidence IMEMDT 49,257,90. Reported in EPA TSCA Inventory.
ACGIH TLV: TWA 0.01 mg/m^3; Confirmed Human Carcinogen; BEI: 35 μ (As)/L

N

inorganic arsenic and methylated metabolites in urine; TWA 1 mg(Ni)/m³
SAFETY PROFILE: Confirmed human carcinogen. When heated to decomposition it emits toxic vapors of nickel and arsenic.

NCY125   CAS: 12255-10-6   HR: 3
NICKEL ARSENIDE SULFIDE
mf: AsNiS   mw: 165.69
PROP: IDLH 10 mg/m³ (as Ni).
SYN: NICKEL SULFARSENIDE
CONSENSUS REPORTS: NTP 10th Report on Carcinogens.
OSHA PEL: OSHA: Cancer Hazard
ACGIH TLV: TWA 0.01 mg/m³; Confirmed Human Carcinogen; BEI: 35 μ (As)/L inorganic arsenic and methylated metabolites in urine; TLV: TWA 1 mg(Ni)/m³
SAFETY PROFILE: Confirmed human carcinogen with experimental carcinogenic and neoplastigenic data. When heated to decomposition it emits toxic fumes of Ni, As, and SO_x.

NCY500   CAS: 3333-67-3   HR: 3
NICKEL(II) CARBONATE (1:1)
mf: CNiO₃   mw: 118.72
PROP: Rhombic, light-green crystals or solid. Mp: decomp. IDLH 10 mg/m³ (as Ni).
SYNS: BASIC NICKEL CARBONATE □ CARBONIC ACID, NICKEL SALT (1:1) □ C.I. 77779 □ NICKELOUS CARBONATE
CONSENSUS REPORTS: NTP 10th Report on Carcinogens. IARC Cancer Review: Group 1 IMEMDT 7,264,87; Animal Sufficient Evidence IMEMDT 11,75,76. Nickel and its compounds are on the Community Right-To-Know List. Reported in EPA TSCA Inventory.
OSHA PEL: TWA 1 mg(Ni)/m³
ACGIH TLV: TWA 1 mg(Ni)/m³; Not Classifiable as a Human Carcinogen
DFG MAK: DFG TRK: 0.5 mg/m³; Human Carcinogen
NIOSH REL: (Inorganic Nickel) TWA 0.015 mg(Ni)/m³

SAFETY PROFILE: Confirmed carcinogen with experimental carcinogenic and tumorigenic data. Poison by subcutaneous route. Mutation data reported. See also NICKEL COMPOUNDS.

NCY600   CAS: 12607-70-4   HR: 3
NICKEL CARBONATE HYDROXIDE
mf: CH₄Ni₃O₇   mw: 304.18
PROP: IDLH 10 mg/m³ (as Ni).
SYNS: BASIC NICKEL(II) CARBONATE □ CARBONIC ACID, NICKEL SALT, BASIC □ NICKEL, (CARBON-ATO(2-))TETRAHYDROXYTRI-
CONSENSUS REPORTS: NTP 10th Report on Carcinogens. IARC Cancer Review: Animal Limited Evidence IMEMDT 49,257,90. Reported in EPA TSCA Inventory.
SAFETY PROFILE: Confirmed human carcinogen. When heated to decomposition it emits toxic vapors of nickel.

NCZ000   CAS: 13463-39-3   HR: 3
NICKEL CARBONYL
DOT: UN 1259
mf: C₄NiO₄   mw: 170.75
PROP: Colorless, volatile liquid or needles; oxidizes in air. Mp: −19.3°, bp: 43°, lel: 2% @ 20°, d: 1.3185 @ 17°, vap press: 400 mm @ 25.8°, flash p: <−4°. Oxidizes in air. Sol in alc, benzene, chloroform, acetone, and carbon tetrachloride. IDLH 10 mg/m³ (as Ni).
SYNS: NICHEL TETRACARBONILE (ITALIAN) □ NICKEL CARBONYLE (FRENCH) □ NICKEL TETRACARBONYL □ NICKEL TETRACARBONYLE (FRENCH) □ NIKKELTETRACARBONYL (DUTCH) □ RCRA WASTE NUMBER P073 □ TETRACARBONYL NICKEL
CONSENSUS REPORTS: NTP 10th Report on Carcinogens. IARC Cancer Review: Animal Sufficient Evidence IMEMDT 7,264,87; Animal Limited Evidence IMEMDT 49,257,90. EPA Extremely Hazardous Substances List. Nickel and its compounds are on the Community Right-To-Know List. Reported in EPA TSCA Inventory.
OSHA PEL: TWA 0.001 ppm (Ni)

ACGIH TLV: TWA 0.05 mg(Ni)/m³
DFG MAK: DFG TRK: Animal Carcinogen, Suspected Human Carcinogen
NIOSH REL: (Nickel Carbonyl) TWA 0.001 ppm
DOT CLASSIFICATION: 6.1; Label: Poison, Flammable Liquid
SAFETY PROFILE: Confirmed carcinogen with experimental carcinogenic, tumorigenic, and teratogenic data. A human poison by inhalation. Poison experimentally by inhalation, intravenous, subcutaneous, and intraperitoneal routes. An experimental teratogen. Other experimental reproductive effects. Human systemic effects by inhalation: somnolence, fever, and other pulmonary changes. Vapors may cause coughing, dyspnea (difficult breathing), irritation, congestion and edema of the lungs, tachycardia (rapid pulse), cyanosis, headache, dizziness, and weakness. Toxicity by inhalation is believed to be caused by both the nickel and carbon monoxide liberated in the lungs. Chronic exposure may cause cancer of lungs, nasal sinuses. Sensitization dermatitis is fairly common. Probably the most hazardous compound of nickel in the workplace. A common air contaminant. It is lipid soluble and can cross biological membranes (e.g., lung alveolus, blood-brain barrier, placental barrier). A very dangerous fire hazard when exposed to heat, flame, or oxidizers. Moderate explosion hazard when exposed to heat or flame. Explodes when heated to about 60°. Explosive reaction with liquid bromine, mercury + oxygen, oxygen + butane. Violent reaction with dinitrogen tetraoxide, air, oxygen. Reacts with tetrachloropropadiene to form the extremely sensitive explosive dicarbonyl trichloropropenyl dinickel chloride dimer. Can react with oxidizing materials. To fight fire, use water, foam, CO₂, dry chemical. When heated to decomposition or on contact with acid or acid fumes, it emits highly toxic fumes of carbon monoxide. See also NICKEL COMPOUNDS and CARBONYLS.

NDA000   CAS: 7791-20-0   HR: 3
NICKEL(II) CHLORIDE HEXAHYDRATE (1:2:6)
mf: $Cl_2Ni \cdot 6H_2O$ mw: 237.73
PROP: A: Yellow, deliq scales; b: monoclinic, green crystals. A: $NiCl_2$; b: $NiCl_2 \cdot 6H_2O$; mw (a): 129.60, mw (b): 237.70, mp (a): subl, bp (a): 987°, d (a): 3.55, vap press: 1 mm @ 671°. Sol in water. IDLH 10 mg/m³ (as Ni).
CONSENSUS REPORTS: NTP 10th Report on Carcinogens. Nickel and its compounds are on the Community Right-To-Know List.
OSHA PEL: TWA 0.1 mg (Ni)/m³
ACGIH TLV: TWA 0.2 mg(Ni)/m³; Human Carcinogen)
NIOSH REL: (Inorganic Nickel) TWA 0.015 mg(Ni)/m³
SAFETY PROFILE: Confirmed human carcinogen. Poison by intraperitoneal and intravenous routes. Experimental reproductive effects. Mutation data reported. Violent reaction with potassium. When heated to decomposition it emits very toxic fumes of Cl⁻. See also NICKEL COMPOUNDS and CHLORIDES.

NDA100   CAS: 12018-18-7   HR: 3
NICKEL CHROMATE
mf: Cr_2NiO_4 mw: 226.71
PROP: Blue white to steel gray color solid. Mp 1903°. bp: 2642°. Insol in water. IDLH 10 mg/m³ (as Ni).
SYNS: CHROMIUM NICKEL OXIDE □ NICKEL CHROMITE □ NICKEL CHROMIUM OXIDE
CONSENSUS REPORTS: NTP 10th Report on Carcinogens. IARC Cancer Review: Group 3 IMEMDT 49,49,90; Animal Inadequate Evidence IMEMDT 49,49,90; Human Inadequate Evidence IMEMDT 49,49,90. Reported in EPA TSCA Inventory.
ACGIH TLV: TWA 0.1 mg(Ni)/m³; Not Classifiable as a Human Carcinogen

N

SAFETY PROFILE: Confirmed human carcinogen. When heated to decomposition it emits toxic vapors of nickel and Cr⁻.

NDA500   CAS: 1271-28-9   HR: 3
NICKEL, COMPOUND with pi-CYCLO-PENTADIENYL (1:2)

mf: $C_{10}H_{10} \cdot Ni$ mw: 188.91

PROP: Dark-green needles from pet ether. Mp: 171–173°. IDLH 10 mg/m³ (as Ni).

SYNS: pi-CYCLOPENTADIENYL COMPOUND with NICKEL □ DI-pi-CYCLOPENTADIENYLNICKEL □ NICKEL BISCYCLOPENTADIENE □ NICKELOCENE

CONSENSUS REPORTS: NTP 10th Report on Carcinogens. IARC Cancer Review: Group 1 IMEMDT 7,264,87; Animal Sufficient Evidence IMEMDT 11,75,76; Animal Inadequate Evidence IMEMDT 2,126,73. Nickel and its compounds are on the Community Right-To-Know List. Reported in EPA TSCA Inventory.

SAFETY PROFILE: Confirmed carcinogen with experimental carcinogenic, neoplastigenic, and tumorigenic data. Poison by intraperitoneal and intramuscular routes. Moderately toxic by ingestion. When heated to decomposition it emits acrid smoke and irritating fumes. See also NICKEL COMPOUNDS.

NDB000   HR: 3
NICKEL COMPOUNDS

PROP: IDLH 10 mg/m³ (as Ni).

CONSENSUS REPORTS: NTP 10th Report on Carcinogens. Nickel and its compounds are on the Community Right-To-Know List.

OSHA PEL: TWA Soluble: Compounds: 0.1 mg(Ni)/m³; Insoluble Compounds: 1 mg(Ni)/m³

ACGIH TLV: (insoluble) TWA 0.05 mg(Ni)/m³; Human Carcinogen; (soluble) TWA 0.1 mg/m³;; Not Classifiable as a Carcinogen

DFG MAK: Human Carcinogen

NIOSH REL: (Inorganic Nickel) TWA 0.015 mg(Ni)/m³

SAFETY PROFILE: Many are human carcinogens by inhalation. Nickel and many of its compounds are poisons and carcinogens. All airborne nickel contaminating dusts are regarded as carcinogenic by inhalation. Nickel carbonyl is probably the most hazardous compound of nickel in the workplace. It is carcinogenic and highly irritating to the lungs and can produce asphyxia by decomposing to form carbon monoxide. Nickel chloride ($NiCl_2$), sulfate ($NiSO_4 \cdot 6H_2O$), nitrate [$Ni(NO_3)_2 \cdot 6H_2O$], carbonate ($NiCO_3$), hydroxide [$Ni(OH)_2$], and acetate [$Ni(COOCH_3)_2$] are the salts of greatest commercial importance.

Ingestion of large doses of nickel compounds (1–3 mg/kg) has been shown to cause intestinal disorders, convulsions, and asphyxia. Hypersensitivity to nickel is common and can cause allergic contact dermatitis, pulmonary asthma, conjunctivitis, and inflammatory reactions around nickel-containing medical implants and prostheses. The most common effect resulting from exposure to nickel compounds is the development of "nickel itch." It occurs primarily in persons doing nickel-plating and is most frequent under conditions of high temperature and humidity, when the skin is moist, and mainly affects the hands and arms. There is marked variation in individual susceptibility to the dermatitis.

Nickel refinery workers experience increased mortality rates from cancer of the lungs and nasal cavities attributed to inhalation of airborne nickel compounds. Cancer develops in rodents after administration of Ni_3S_2, NiO, and $Ni(CO)_4$. Nickel chloride, sulfate, carbonate, and carbonyl are experimental teratogens.

Pulmonary damage develops in rodents chronically exposed to aerosols of nickel dust, $NiCl_2$, or NiO. Divalent nickel salts cause hyperglycemia, immune system effects, kidney damage, liver damage, and heart effects in experimental animals by parenteral administration. These compounds are common air contaminants. See also NICKEL and specific compounds.

NDB500   CAS: 557-19-7   HR: 3
NICKEL CYANIDE (solid)
DOT: UN 1653
mf: C_2N_2Ni mw: 110.75
PROP: Apple-green plates or powder. Sol in MeOAc. IDLH 10 mg/m³ (as Ni).
SYNS: NICKEL CYANIDE (DOT) □ RCRA WASTE NUMBER P074
CONSENSUS REPORTS: NTP 10th Report on Carcinogens. Cyanide and its compounds, as well as nickel and its compounds, are on the Community Right-To-Know List. Reported in EPA TSCA Inventory.
OSHA PEL: TWA 0.1 mg (Ni)/m³
ACGIH TLV: TWA 0.2 mg(Ni)/m³; Human Carcinogen)
DOT CLASSIFICATION: 6.1; Label: Poison
SAFETY PROFILE: Confirmed human carcinogen. A poison. Incandescent reaction when heated with magnesium. When heated to decomposition it emits very toxic fumes of CN⁻. See also CYANIDE and NICKEL COMPOUNDS.

NDC000   CAS: 14708-14-6   HR: 3
NICKEL(II) FLUOBORATE
mf: $B_2F_8 \cdot Ni$ mw: 232.33
PROP: Green crystals from 6H₂O. Sol in H₂O and EtOH. IDLH 10 mg/m³ (as Ni).
SYNS: NICKEL BOROFLUORIDE □ NICKEL FLUOROBORATE □ NICKELOUS TETRAFLUORO-BORATE □ NICKEL(II) TETRAFLUOROBORATE □ TL 1091
CONSENSUS REPORTS: NTP 10th Report on Carcinogens. Nickel and its compounds are on the Community Right-To-Know List. Reported in EPA TSCA Inventory.
OSHA PEL: TWA 0.1 mg (Ni)/m³; TWA 2.5 mg(F)/m³; BEI: 3 mg/g creatinine of fluorides in urine prior to shift; 10 mg/g creatinine of fluorides in urine at end of shift.
ACGIH TLV: TWA 0.2 mg(Ni)/m³; Human Carcinogen)

NIOSH REL: (Inorganic Nickel) TWA 0.015 mg(Ni)/m³
SAFETY PROFILE: Confirmed carcinogen. Moderately toxic by ingestion and inhalation. When heated to decomposition it emits very toxic fumes of F⁻. See also FLUORIDES, BORON, and NICKEL COMPOUNDS.

NDC500   CAS: 10028-18-9   HR: 3
NICKEL(II) FLUORIDE (1:2)
mf: F_2Ni mw: 96.71
PROP: Green tetragonal crystals or light brown-green prisms or light yellow powder; inert to conc acids. D: 4.63. Sltly water-sol; decomp by boiling water; insol in alc, ether. IDLH 10 mg/m³ (as Ni).
SYNS: NICKEL DIFLUORIDE □ NICKELOUS FLUORIDE
CONSENSUS REPORTS: NTP 10th Report on Carcinogens. Nickel and its compounds are on the Community Right-To-Know List. Reported in EPA TSCA Inventory.
OSHA PEL: TWA 0.1 mg (Ni)/m³; TWA 2.5 mg(F)/m³; BEI: 3 mg/g creatinine of fluorides in urine prior to shift; 10 mg/g creatinine of fluorides in urine at end of shift.
ACGIH TLV: TWA 0.2 mg(Ni)/m³; Human Carcinogen)
NIOSH REL: (Inorganic Nickel) TWA 0.015 mg(Ni)/m³
SAFETY PROFILE: NTP 10th Report on Carcinogens. Reacts violently with potassium. Chronic exposure may cause mottling of teeth, changes in bones. Mutation data reported. When heated to decomposition it emits toxic fumes of F⁻. See also FLUORIDES and NICKEL COMPOUNDS.

NDD000   CAS: 26043-11-8   HR: 3
NICKEL(II) FLUOSILICATE (1:1)
mf: $F_6Si \cdot Ni$ mw: 200.80
PROP: IDLH 10 mg/m³ (as Ni).
SYN: HEXAFLUOROSILICATE (2−), NICKEL
CONSENSUS REPORTS: NTP 10th Report on Carcinogens. Nickel and its compounds

N

are on the Community Right-To-Know List. Reported in EPA TSCA Inventory.
OSHA PEL: TWA 0.1 mg (Ni)/m³
ACGIH TLV: TWA 1 mg(Ni)/m³; Not Classifiable as a Human Carcinogen
NIOSH REL: (Inorganic Nickel) TWA 0.015 mg(Ni)/m³
SAFETY PROFILE: Confirmed human carcinogen. Poison by ingestion. When heated to decomposition it emits toxic fumes of F⁻. See also NICKEL COMPOUNDS.

NDD500   CAS: 56668-59-8   HR: 3
NICKEL-GALLIUM ALLOY
PROP: Alloy of nickel (60%) and gallium (40%) (JDREAF 29,1023,60).
PROP: IDLH 10 mg/m³ (as Ni).
SYN: GALLIUM-NICKEL ALLOY
CONSENSUS REPORTS: Nickel and its compounds are on the Community Right-To-Know List.
OSHA PEL: TWA 1 mg(Ni)/m³
ACGIH TLV: TWA 1 mg(Ni)/m³; Not Classifiable as a Human Carcinogen
NIOSH REL: (Inorganic Nickel) TWA 0.015 mg(Ni)/m³
SAFETY PROFILE: Suspected carcinogen with experimental tumorigenic data. See also NICKEL COMPOUNDS and GALLIUM COMPOUNDS.

NDE000   CAS: 12054-48-7   HR: 3
NICKEL(II) HYDROXIDE
mf: H_2NiO_2 mw: 92.73
PROP: Light-green crystals or amorphous. IDLH 10 mg/m³ (as Ni).
SYNS: NICKEL DIHYDROXIDE □ NICKELOUS HYDROXIDE
CONSENSUS REPORTS: NTP 10th Report on Carcinogens. IARC Cancer Review: Group 1 IMEMDT 7,264,87; Animal Sufficient Evidence IMEMDT 11,75,76. Nickel and its compounds are on the Community Right-To-Know List. Reported in EPA TSCA Inventory.
OSHA PEL: TWA 0.1 mg (Ni)/m³

ACGIH TLV: TWA 0.2 mg(Ni)/m³; Human Carcinogen)
NIOSH REL: (Inorganic Nickel) TWA 0.015 mg(Ni)/m³
SAFETY PROFILE: Confirmed carcinogen with experimental carcinogenic and tumorigenic data. Poison by subcutaneous route. See also NICKEL COMPOUNDS.

NDE010   CAS: 12125-56-3   HR: 3
NICKEL(III) HYDROXIDE
mf: H_3NiO_3 mw: 109.74
PROP: IDLH 10 mg/m³ (as Ni).
SYNS: NICKEL BLACK □ NICKELIC HYDROXIDE
CONSENSUS REPORTS: NTP 10th Report on Carcinogens. IARC Cancer Review: Animal Sufficient Evidence IMEMDT 11,75,76.
OSHA PEL: TWA 1 mg(Ni)/m³
ACGIH TLV: TWA 0.05 mg(Ni)/m³; Confirmed Carcinogen
NIOSH REL: (NICKEL, INORGANIC) 10H TWA 0.015 mg(Ni)/m³
SAFETY PROFILE: Confirmed carcinogen. Poison by intravenous route. When heated to decomposition it emits toxic fumes of Ni.

NDE500   CAS: 59978-65-3   HR: 3
NICKEL IRON SULFIDE
mf: $FeNi_4S_4$ mw: 418.93
PROP: Closely related to pyrite but contains up to 20% nickel. IDLH 10 mg/m³ (as Ni).
SYNS: IRON NICKEL SULFIDE □ NICKEL-IRON SULFIDE MATTE
CONSENSUS REPORTS: NTP 10th Report on Carcinogens. Nickel and its compounds are on the Community Right-To-Know List.
OSHA PEL: TWA 1 mg(Ni)/m³
ACGIH TLV: TWA 1 mg(Ni)/m³; Not Classifiable as a Human Carcinogen
NIOSH REL: (Inorganic Nickel) TWA 0.015 mg(Ni)/m³
SAFETY PROFILE: Confirmed human carcinogen with experimental carcinogenic and neoplastigenic data. When heated to decomposition it emits toxic fumes of SO_x. See also NICKEL COMPOUNDS, SULFIDES, and IRON COMPOUNDS.

NDF500   CAS: 1313-99-1   HR: 3
NICKEL MONOXIDE
mf: NiO   mw: 74.71
PROP: Cubic, green-black crystals; yellow
when hot. Mp: 1984°, d: 7.45. Insol in water;
sol in acids, NH_3 (aq). IDLH 10 mg/m³ (as
Ni).
SYNS: BUNSENITE □ C.I. 77777 □ GREEN NICKEL
OXIDE □ NICKELOUS OXIDE □ NICKEL OXIDE
(MAK) □ NICKEL(II) OXIDE (1:1) □ NICKEL
PROTOXIDE
CONSENSUS REPORTS: NTP 10th Report
on Carcinogens. IARC Cancer Review:
Group 1 IMEMDT 7,264,87; Animal
Inadequate Evidence IMEMDT 2,126,73;
Animal Sufficient Evidence IMEMDT
11,75,76. Nickel and its compounds are on
the Community Right-To-Know List.
Reported in EPA TSCA Inventory.
OSHA PEL: TWA 1 mg(Ni)/m³
ACGIH TLV: TWA 1 mg(Ni)/m³; Not
Classifiable as a Human Carcinogen
DFG MAK: DFG TRK: 0.5 mg/m³; Human
Carcinogen
NIOSH REL: (Inorganic Nickel) TWA 0.015
mg(Ni)/m³
SAFETY PROFILE: Confirmed carcinogen
with experimental carcinogenic and
tumorigenic data. Poison by intratracheal,
intravenous, and subcutaneous routes.
Mutation data reported. Can react violently
with fluorine, hydrogen peroxide, hydrogen
sulfide, iodine, barium oxide + air. See also
NICKEL COMPOUNDS.

NDG000   CAS: 13138-45-9   HR: 3
NICKEL(II) NITRATE (1:2)
DOT: UN 2725
mf: $N_2O_6 \cdot Ni$ mw: 182.73
PROP: Green, deliquescent crystals. Mp:
56.7°, bp: 136.7°, d: 2.05. IDLH 10 mg/m³
(as Ni).
SYNS: NICKEL NITRATE □ NITRIC ACID, NICKEL(II)
SALT
CONSENSUS REPORTS: NTP 10th Report
on Carcinogens. Nickel and its compounds
are on the Community Right-To-Know List.
Reported in EPA TSCA Inventory.
OSHA PEL: TWA 0.1 mg (Ni)/m³

ACGIH TLV: TWA 0.2 mg(Ni)/m³; Human
Carcinogen)
NIOSH REL: (Inorganic Nickel) TWA 0.015
mg(Ni)/m³
DOT CLASSIFICATION: 5.1; Label: Oxidizer
SAFETY PROFILE: Confirmed human
carcinogen. Poison by intravenous route.
Experimental reproductive effects. Mutation
data reported. A powerful oxidizer. When
heated to decomposition it emits very toxic
fumes of NO_x. See also NICKEL
COMPOUNDS and NITRATES.

NDG500   CAS: 13478-00-7   HR: 3
NICKEL(II) NITRATE, HEXAHYDRATE
   (1:2:6)
mf: $N_2O_6 \cdot Ni \cdot 6H_2O$ mw: 290.85
PROP: Green, deliquescent crystals. Mp:
56.7°, bp: 136.7°, d: 2.05. Sol in 0.4 parts
water or alc. Keep well closed. IDLH 10
mg/m³ (as Ni).
SYNS: NICKEL(2⁺) NITRATE, HEXAHYDRATE □
NITRIC ACID, NICKEL(2+) SALT, HEXAHYDRATE
CONSENSUS REPORTS: NTP 10th Report
on Carcinogens. Nickel and its compounds
are on the Community Right-To-Know List.
OSHA PEL: TWA 0.1 mg (Ni)/m³
ACGIH TLV: TWA 0.2 mg(Ni)/m³; Human
Carcinogen
NIOSH REL: (Inorganic Nickel) TWA
0.015 mg(Ni)/m³
SAFETY PROFILE: Confirmed carcinogen.
Moderately toxic by ingestion. When heated
to decomposition it emits toxic fumes of
NO_x. See also NICKEL(II) NITRATE.

NDG550   CAS: 17861-62-0   HR: 3
NICKEL NITRITE
DOT: UN 2726
mf: $N_2O_4 \cdot Ni$ mw: 150.73
PROP: IDLH 10 mg/m³ (as Ni).
SYNS: NICKEL DINITRITE □ NICKEL NITRITE □
NITROUS ACID, NICKEL(2+) SALT
CONSENSUS REPORTS: NTP 10th Report
on Carcinogens.
OSHA PEL: TWA 1 mg(Ni)/m³
ACGIH TLV: TWA 0.05 mg(Ni)/m³;
Confirmed Carcinogen
DOT CLASSIFICATION: 5.1; Label: Oxidizer

N

SAFETY PROFILE: Confirmed carcinogen. When heated to decomposition it emits toxic fumes of NO_x and Ni.

NDH000 CAS: 7718-54-9 HR: 3
NICKELOUS CHLORIDE
mf: Cl_2Ni mw: 129.61
PROP: Sparkling golden yellow scales. Mp: 1001° (sealed tube). Sublimes @ 9°. Sol in H_2O and polar org solvs. IDLH 10 mg/m³ (as Ni).
SYNS: NICKEL CHLORIDE □ NICKEL(II) CHLORIDE (1:2)
CONSENSUS REPORTS: NTP 10th Report on Carcinogens. IARC Cancer Review: Animal Limited Evidence IMEMDT 49,257,90. Nickel and its compounds are on the Community Right-To-Know List. Reported in EPA TSCA Inventory. EPA Genetic Toxicology Program.
OSHA PEL: TWA 0.1 mg (Ni)/m³
ACGIH TLV: TWA 0.2 mg(Ni)/m³; Human Carcinogen)
NIOSH REL: (Inorganic Nickel) TWA 0.015 mg(Ni)/m³
SAFETY PROFILE: Confirmed human carcinogen. Poison by ingestion, intravenous, intramuscular, and intraperitoneal routes. An experimental teratogen. Experimental reproductive effects. Mutation data reported. When heated to decomposition it emits very toxic fumes of Cl⁻. See also NICKEL COMPOUNDS.

NDH500 CAS: 1314-06-3 HR: 3
NICKEL PEROXIDE
mf: Ni_2O_3 mw: 165.42
PROP: Gray-black powder or solid. Mp: loses O_2 @ 600°, d: 4.83. Decomp about 600° into NiO and O_2. Insol in water; very sltly sol in cold acid; dissolved by hot HCl, evolving Cl_2; dissolved by hot H_2SO_4 or HNO_3, evolving O_2. IDLH 10 mg/m³ (as Ni).
SYNS: DINICKEL TRIOXIDE □ NICKELIC OXIDE □ NICKEL OXIDE □ NICKEL OXIDE PEROXIDE □ NICKEL SESQUIOXIDE □ NICKEL TRIOXIDE

CONSENSUS REPORTS: NTP 10th Report on Carcinogens. IARC Cancer Review: Animal Inadequate Evidence IMEMDT 49,257,90. Nickel and its compounds are on the Community Right-To-Know List. Reported in EPA TSCA Inventory.
OSHA PEL: TWA 0.1 mg (Ni)/m³
ACGIH TLV: TWA 1 mg(Ni)/m³; Not Classifiable as a Human Carcinogen
NIOSH REL: (Inorganic Nickel) TWA 0.015 mg(Ni)/m³
SAFETY PROFILE: Confirmed human carcinogen. Poison by subcutaneous route. Mutation data reported. Hazardous reaction with hydrogen peroxide. Presence of the oxide increases the sensitivity of nitroalkanes (e.g., nitromethane, nitroethane, 1-nitropropane) to heat. See also NICKEL COMPOUNDS and PEROXIDES.

NDI000 CAS: 14220-17-8 HR: 3
NICKEL POTASSIUM CYANIDE
mf: $C_4N_4Ni•2K$ mw: 240.99
PROP: Red-orange crystals from H_2O. Very sol in H_2O. IDLH 10 mg/m³ (as Ni).
SYNS: DIPOTASSIUM NICKEL TETRACYANIDE □ DIPOTASSIUM TETRACYANONICKELATE □ POTASSIUM TETRACYANONICKELATE □ POTASSIUM TETRACYANONICKELATE(II)
CONSENSUS REPORTS: NTP 10th Report on Carcinogens. Nickel and its compounds, as well as cyanide and its compounds, are on the Community Right-To-Know List. Reported in EPA TSCA Inventory.
OSHA PEL: TWA 0.1 mg (Ni)/m³
ACGIH TLV: TWA 0.2 mg(Ni)/m³; Human Carcinogen)
NIOSH REL: (Inorganic Nickel) TWA 0.015 mg(Ni)/m³
DOT CLASSIFICATION: 6.1; Label: Poison, KEEP AWAY FROM FOOD
SAFETY PROFILE: Confirmed human carcinogen. Poison by ingestion. When heated to decomposition it emits very toxic fumes of NO_x, K_2O, and CN⁻. See also NICKEL COMPOUNDS and CYANIDE.

NDI500 HR: 3
NICKEL REFINERY DUST
PROP: *Analysis:* Cupric oxide (3.4%), nickel sulfate (20.0%), nickel sulfide (57.0%), nickel oxide (6.3%), cobalt oxide (1.0%), ferric oxide (1.8%), silicon dioxide (1.2%), misc (2.0%), water (7.3%) (CNREA8 22,158,62). IDLH 10 mg/m³ (as Ni).
CONSENSUS REPORTS: NTP 10th Report on Carcinogens. IARC Cancer Review: Human Sufficient Evidence IMEMDT 2,126,73. Nickel and its compounds are on the Community Right-To-Know List.
OSHA PEL: TWA 1 mg(Ni)/m³
ACGIH TLV: TWA 0.2 mg(Ni)/m³; Human Carcinogen
NIOSH REL: (Inorganic Nickel) TWA 0.015 mg(Ni)/m³
SAFETY PROFILE: Confirmed carcinogen with experimental carcinogenic and neoplastigenic data. A human carcinogen. Moderately toxic by intramuscular route. When heated to decomposition it emits toxic fumes of SO_x. See also NICKEL, NICKEL COMPOUNDS, and individual components.

NDJ000 CAS: 13520-61-1 HR: 3
NICKEL(2+) SALT PERCHLORIC ACID HEXAHYDRATE
mf: $Cl_2O_8 \cdot Ni \cdot 6H_2O$ mw: 365.73
PROP: Hygroscopic blue-green crystals. Mp: 209° (sealed tube). Sol in H_2O, EtOH, Me_2CO, and DMF. IDLH 10 mg/m³ (as Ni).
SYN: NICKEL(2+) PERCHLORATE, HEXAHYDRATE
CONSENSUS REPORTS: NTP 10th Report on Carcinogens. Nickel and its compounds are on the Community Right-To-Know List.
OSHA PEL: TWA 0.1 mg (Ni)/m³
ACGIH TLV: TWA 0.2 mg(Ni)/m³; Human Carcinogen)
NIOSH REL: TWA 15 μg(Ni)/m³
DOT CLASSIFICATION: 5.1; Label: Oxidizer
SAFETY PROFILE: Confirmed human carcinogen. Poison by intraperitoneal route. A powerful oxidizer. Mixtures with 2,2-dimethoxypropane explode when heated above 65°C. When heated to decomposition it emits toxic fumes of Cl⁻. See also NICKEL COMPOUNDS and PERCHLORATES.

NDJ399 CAS: 12255-80-0 HR: 3
NICKEL SUBARSENIDE
mf: As_2Ni_5 mw: 443.39
PROP: IDLH 10 mg/m³ (as Ni).
SYN: NICKEL ARSENIDE (As2-Ni5)
CONSENSUS REPORTS: NTP 10th Report on Carcinogens. IARC Cancer Review: Animal Limited Evidence IMEMDT 49,257,90.
OSHA PEL: OSHA: Cancer Hazard
ACGIH TLV: TWA 0.01 mg/m³; Confirmed Human Carcinogen; BEI: 35 μ (As)/L inorganic arsenic and methylated metabolites in urine; TWA 0.2 mg(Ni)/m³
SAFETY PROFILE: Confirmed human carcinogen with experimental carcinogenic data. Moderately toxic by ingestion. When heated to decomposition it emits toxic fumes of As.

NDJ400 CAS: 12256-33-6 HR: 3
NICKEL SUBARSENIDE
mf: As_8Ni_{11} mw: 1245.17
PROP: IDLH 10 mg/m³ (as Ni).
SYN: NICKEL ARSENIDE (As8-Ni11)
CONSENSUS REPORTS: NTP 10th Report on Carcinogens. IARC Cancer Review: Animal Limited Evidence IMEMDT 49,257,90.
OSHA PEL: OSHA: Cancer Hazard
ACGIH TLV: TWA 0.01 mg/m³; Confirmed Human Carcinogen; BEI: 35 μ (As)/L inorganic arsenic and methylated metabolites in urine; TLV: TWA 0.05 mg(Ni)/m³
SAFETY PROFILE: Confirmed human carcinogen with experimental carcinogenic data. When heated to decomposition it emits toxic fumes of As.

NDJ475 CAS: 12137-13-2 HR: 3
NICKEL SUBSELENIDE
mf: Ni_3Se_2 mw: 334.05

N

PROP: IDLH 10 mg/m³ (as Ni).
SYNS: NICKEL SELENIDE □ NICKEL SELENIDE (3:2) CRYSTALLINE
CONSENSUS REPORTS: Confirmed human carcinogen.
OSHA PEL: TWA Soluble: Compounds: 0.1 mg(Ni)/m³; Insoluble Compounds: 1 mg(Ni)/m³; TWA 0.2 mg(Se)/m³
ACGIH TLV: TWA 0.2 mg(Ni)/m³; Human Carcinogen; TWA 0.2 mg(Se)/m³
DFG MAK: DFG TRK: 0.5 mg/m³; Human Carcinogen
NIOSH REL: (Inorganic Nickel) TWA 0.015 mg(Ni)/m³
SAFETY PROFILE: Confirmed carcinogen with experimental carcinogenic data. Mutation data reported.

NDJ500   CAS: 12035-72-2   HR: 3
NICKEL SUBSULFIDE
mf: Ni₃S₂   mw: 240.25
PROP: Golden yellow solid. Metallic conductor. Mp: 2140°. IDLH 10 mg/m³ (as Ni).
SYNS: HEAZLEWOODITE □ NICKEL SUBSULPHIDE □ NICKEL SULFIDE □ α-NICKEL SULFIDE (3:2) CRYSTALLINE □ NICKEL SULPHIDE □ NICKEL TRITADISULPHIDE
CONSENSUS REPORTS: NTP 10th Report on Carcinogens. IARC Cancer Review: Group 1 IMEMDT 7,264,87; Animal Sufficient Evidence IMEMDT 2,126,73, IMEMDT 11,75,76. Nickel and its compounds are on the Community Right-To-Know List. Reported in EPA TSCA Inventory.
OSHA PEL: TWA 1 mg(Ni)/m³
ACGIH TLV: TWA 0.1 mg(Ni)/m³; Human Carcinogen
NIOSH REL: TWA 0.015 mg(Ni)/m³
SAFETY PROFILE: Confirmed carcinogen with experimental carcinogenic, neoplastigenic, tumorigenic, and teratogenic data. Poison by intraperitoneal and intratracheal routes. An experimental teratogen. Other experimental reproductive effects. Human mutation data reported. When heated to decomposition it emits

toxic fumes of SOₓ. See also NICKEL COMPOUNDS and SULFIDES.

NDK000   CAS: 13770-89-3   HR: 3
NICKEL (II) SULFAMATE
mf: H₄N₂NiO₆S₂   mw: 250.89
PROP: IDLH 10 mg/m³ (as Ni).
CONSENSUS REPORTS: NTP 10th Report on Carcinogens. Nickel and its compounds are on the Community Right-To-Know List. Reported in EPA TSCA Inventory.
OSHA PEL: TWA 0.1 mg (Ni)/m³
ACGIH TLV: TWA 0.2 mg(Ni)/m³; Human Carcinogen)
NIOSH REL: (Inorganic Nickel) TWA 0.015 mg(Ni)/m³
SAFETY PROFILE: Confirmed human carcinogen. Poison by intraperitoneal route. When heated to decomposition it emits very toxic fumes of SOₓ and NOₓ. See also NICKEL COMPOUNDS.

NDK500   CAS: 7786-81-4   HR: 3
NICKEL SULFATE
mf: O₄S•Ni   mw: 154.77
PROP: Cubic yellow crystals. Mp: 840°, d: 3.68. IDLH 10 mg/m³ (as Ni).
SYNS: NCI-C60344 □ NICKELOUS SULFATE □ NICKEL SULFATE (1:1) □ NICKEL(II) SULFATE □ NICKEL(II) SULFATE (1:1) □ NICKEL(2⁺)SULFATE (1:1) □ SULFURIC ACID, NICKEL(2⁺)SALT □ SULFURIC ACID, NICKEL(2⁺) SALT (1:1)
CONSENSUS REPORTS: NTP 10th Report on Carcinogens. Nickel and its compounds are on the Community Right-To-Know List. Reported in EPA TSCA Inventory. EPA Genetic Toxicology Program.
OSHA PEL: TWA 0.1 mg (Ni)/m³
ACGIH TLV: TWA 0.2 mg(Ni)/m³; Human Carcinogen)
NIOSH REL: (Inorganic Nickel) TWA 0.015 mg(Ni)/m³
SAFETY PROFILE: Confirmed human carcinogen with experimental tumorigenic data. Poison by intravenous, intraperitoneal, and subcutaneous routes. Human mutation data reported. A human skin irritant. When heated to decomposition it emits very toxic

fumes of SO_x. See also NICKEL COMPOUNDS and SULFATES.

NDL000   CAS: 10101-97-0   HR: 3
NICKEL(II) SULFATE HEXAHYDRATE (1:1:6)

mf: $NiO_4S \bullet 6H_2O$ mw: 262.89

PROP: IDLH 10 mg/m³ (as Ni).

SYNS: NICKEL MONOSULFATE HEXAHYDRATE □ NICKEL SULFATE HEXAHYDRATE □ NICKEL (II) SULFATE HEXAHYDRATE □ NICKEL SULPHATE HEXAHYDRATE □ SULFURIC ACID, NICKEL(2+) SALT, HEXAHYDRATE

CONSENSUS REPORTS: NTP 10th Report on Carcinogens. Nickel and its compounds are on the Community Right-To-Know List. EPA Genetic Toxicology Program.

OSHA PEL: TWA 0.1 mg (Ni)/m³

ACGIH TLV: TWA 0.2 mg(Ni)/m³; Human Carcinogen)

NIOSH REL: (Inorganic Nickel) TWA 0.015 mg(Ni)/m³

SAFETY PROFILE: Confirmed human carcinogen. Poison by ingestion, intravenous, and subcutaneous routes. Experimental reproductive effects. Human mutation data reported. When heated to decomposition it emits toxic fumes of SO_x. See also NICKEL SULFATE.

NDL100   CAS: 16812-54-7   HR: 3
NICKEL SULFIDE

mf: NiS   mw: 90.77

PROP: Yellow or black crystalline solid. Mp: 976°, bp: decomposes @ 2047°. IDLH 10 mg/m³ (as Ni).

SYNS: MONONICKEL MONOSULFIDE □ NICKEL MONOSULFIDE □ NICKELOUS SULFIDE □ NICKEL(II) SULFIDE □ NICKEL(2+) SULFIDE

CONSENSUS REPORTS: NTP 10th Report on Carcinogens. IARC Cancer Review: Animal Sufficient Evidence IMEMDT 49,257,90. Nickel and its compounds are on the Community Right-To-Know List.

OSHA PEL: TWA 1 mg(Ni)/m³

ACGIH TLV: TWA 0.2 mg(Ni)/m³; Human Carcinogen

DFG MAK: DFG TRK: 0.5 mg/m³; Human Carcinogen

NIOSH REL: TWA 0.015 mg(Ni)/m³

SAFETY PROFILE: Confirmed carcinogen with experimental carcinogenic and neoplastigenic data. Mutation data reported. When heated to decomposition it emits toxic fumes of SO_x. See also NICKEL COMPOUNDS and SULFIDES.

NDL425   CAS: 12142-88-0   HR: 3
NICKEL TELLURIDE

mf: NiTe   mw: 186.31

PROP: IDLH 10 mg/m³ (as Ni).

CONSENSUS REPORTS: NTP 10th Report on Carcinogens. Reported in EPA TSCA Inventory.

OSHA PEL: TWA 1 mg(Ni)/m³; Confirmed Carcinogen; 1 mg(Te)/m³

ACGIH TLV: TWA 0.1 mg(Te)/m³

NIOSH REL: (Nickel, Inorganic) TWA 0.015 mg(Ni)/m³

SAFETY PROFILE: Confirmed carcinogen with experimental carcinogenic and tumorigenic data.

NDL500   CAS: 12035-39-1   HR: 3
NICKEL TITANIUM OXIDE

mf: NiO_3Ti mw: 154

PROP: IDLH 10 mg/m³ (as Ni).

SYNS: NICKEL-TITANATE □ TITANIUM NICKEL OXIDE

CONSENSUS REPORTS: NTP 10th Report on Carcinogens. Nickel and its compounds are on the Community Right-To-Know List. Reported in EPA TSCA Inventory.

OSHA PEL: TWA 1 mg(Ni)/m³

ACGIH TLV: TWA 0.2 mg(Ni)/m³; Human Carcinogen)

NIOSH REL: (Inorganic Nickel) TWA 0.015 mg(Ni)/m³

SAFETY PROFILE: Confirmed human carcinogen with experimental carcinogenic and tumorigenic data. See also NICKEL COMPOUNDS and TITANIUM COMPOUNDS.

NDM000   CAS: 27848-84-6   HR: 3
NICOTERGOLINE

mf: $C_{24}H_{26}BrN_3O_3$ mw: 484.44

PROP: A solid. Mp: 136–138°.
SYNS: 8-β-((5-BROMONICOTINOYLOXY)METHYL)-
1,6-DIMETHYL-10-α-METHOXYERGOLINE ☐ FI 6714 ☐
10-METHOXY-1,6-DIMETHYLERGOLINE-8-
METHANOL-5-BROMO-3-PYRIDINECARBOXYLATE
(ester) ☐ 10-METHOXY-1,6-DIMETHYL-ERGOLIN-8-β-
METHANOL-(5-BROMNICOTINAT) (GERMAN) ☐ 1-
METHYL-LUMILYSERGOL-8-(5-BROMONICOTINATE)-
10-METHYL ETHER ☐ MNE ☐ NARGOLINE ☐
NICERGOLIN (GERMAN) ☐ NICERGOLINE ☐ NIMER-
GOLINE ☐ SERMION
SAFETY PROFILE: Poison by intravenous
route. Moderately toxic by ingestion and
subcutaneous routes. An experimental
teratogen. Other experimental reproductive
effects. A vasodilator. When heated to
decomposition it emits very toxic fumes of
Br^- and NO_x.

NDN000   CAS: 54-11-5   HR: 3
NICOTINE
DOT: UN 1654
mf: $C_{10}H_{14}N_2$ mw: 162.26
PROP: An alkaloid from tobacco. In its pure
state, a colorless and almost odorless oil;
sharp burning taste. Mp: $<-80°$, bp: 247.3°
(partial decomp), lel: 0.75%, uel: 4.0%, d:
1.0092 @ 20°, autoign temp: 471°F, vap
press: 1 mm @ 61.8°, vap d: 5.61. Volatile
with steam; misc with water below 60°; very
sol in alc, chloroform ether, pet ether, and
kerosene oils. IDLH 5 mg/m³.
SYNS: BLACK LEAF ☐ DESTRUXOL ORCHID SPRAY
☐ EMO-NIK ☐ ENT 3,424 ☐ FLUX MAAG ☐ FUMETO-
BAC ☐ MACH-NIC ☐ 1-METHYL-2-(3-PYRIDYL)PYR-
ROLIDINE ☐ 3-(N-METHYLPYRROLIDINO)PYRIDINE
☐ 3-(1-METHYL-2-PYRROLIDINYL)PYRIDINE ☐ (S)-3-(1-
METHYL-2-PYRROLIDINYL)PYRIDINE (9CI) ☐ l-3-(1-
METHYL-2-PYRROLIDYL)PYRIDINE ☐ (−)-3-(1-
METHYL-2-PYRROLIDYL)PYRIDINE ☐ NIAGARA P.A.
DUST ☐ NICOCIDE ☐ NICO-DUST ☐ NICO-FUME ☐
NICOTINA (ITALIAN) ☐ (−)-NICOTINE ☐ l-NICOTINE
☐ NICOTINE, liquid (DOT) ☐ NICOTINE, solid (DOT) ☐
NICOTINE ALKALOID ☐ NIKOTIN (GERMAN) ☐
NIKOTYNA (POLISH) ☐ ORTHO N-4 DUST ☐ ORTHO
N-5 DUST ☐ PYRIDINE, 3-(TETRAHYDRO-1-METHYL-
PYRROL-2-YL) ☐ β-PYRIDYL-α-N-METHYLPYRROLID-
INE ☐ RCRA WASTE NUMBER P075 ☐ TENDUST ☐ dl-
TETRAHYDRONICOTYRINE ☐ XL ALL INSECTICIDE
CONSENSUS REPORTS: EPA Extremely
Hazardous Substances List. Reported in
EPA TSCA Inventory. EPA Genetic
Toxicology Program.
OSHA PEL: TWA 0.5 mg/m³ (skin)
ACGIH TLV: TWA 0.5 mg/m³ (skin)
DFG MAK: 0.07 ppm; 0.47 mg/m³
DOT CLASSIFICATION: 6.1; Label: Poison
SAFETY PROFILE: A deadly human poison
by unspecified route. Experimental poison
by ingestion, skin contact, intraperitoneal,
subcutaneous, intravenous, intramuscular,
parenteral, intratracheal, and intraduodenal
routes. Human systemic effects by rectal
route: hallucinations, distorted perceptions,
nausea or vomiting. Human teratogenic
effects by ingestion: developmental
abnormalities of the cardiovascular system.
Human blood pressure effects. Can be
absorbed by intact skin. An experimental
teratogen. Experimental reproductive
effects. "Nicotinism," poisoning by nicotine,
is characterized by stimulation and
subsequent depression of the central and
autonomic nervous systems. Death can
result from respiratory paralysis. Mutation
data reported. Used as a pesticide and in
veterinary medicine as an external
parasiticide. Combustible when exposed to
heat or flame. Moderately explosive in the
form of vapor when exposed to heat or
flame. Can react with oxidizing materials. To
fight fire, use alcohol foam, dry chemical,
CO_2. When heated to decomposition it
emits NO_x, CO, and other highly toxic
fumes. See also SMOKELESS TOBACCO.

NDP400   CAS: 2820-51-1   HR: 3
NICOTINE HYDROCHLORIDE
DOT: UN 1656
mf: $C_{10}H_{14}N_2 \cdot xClH$ mw: 417.48
SYNS: CHLORHYDRATE de NICOTINE (FRENCH) ☐
NICOTINE HYDROCHLORIDE (d,l) ☐ NICOTINE
HYDROCHLORIDE, solution (DOT)
DOT CLASSIFICATION: 6.1; Label: Poison
SAFETY PROFILE: Poison by intravenous,
subcutaneous, and intraperitoneal routes.
When heated to decomposition it emits very
toxic fumes of Cl^-, NO_x, and CO. See also
NICOTINE.

NDR000   CAS: 29790-52-1   HR: 3
NICOTINE MONOSALICYLATE
DOT: UN 1657
mf: $C_{10}H_{14}N_2 \cdot C_7H_6O_3$ mw: 300.39
PROP: Six-sided plates. Mp: 118°. Very sol in water or alc.
SYN: NICOTINE SALICYLATE (DOT)
DOT CLASSIFICATION: 6.1; Label: Poison
SAFETY PROFILE: A poison. Symptoms of exposure: extreme nausea, vomiting, evacuation of bowel and bladder, mental confusion, twitching, convulsions. Base is readily absorbed through mucous membranes and intact skin. Institute treatment immediately. When heated to decomposition it emits toxic fumes of NO_x and CO. See also NICOTINE.

NDR500   CAS: 65-30-5   HR: 3
NICOTINE SULFATE
DOT: UN 1658
mf: $C_{20}H_{26}N_4 \cdot O_4S$ mw: 418.56
SYNS: ENT 2,435 □ 1-1-METHYL-2-(3-PYRIDYL)-PYRROLIDINE SULFATE □ (S)-3-(1-METHYL-2-PYRROLIDINYL)PYRIDINE SULFATE (2:1) □ 1-3-(1-METHYL-2-PYRROLIDYL)PYRIDINE SULFATE □ NICOTINE SULFATE □ NICOTINE SULFATE, solid or solution (DOT) □ NIKOTINSULFAT □ PYRIDINE, 3-(1-METHYL-2-PYRROLIDINYL)-, (S)-, SULFATE (2:1) □ PYRROLIDINE, 1-METHYL-2-(3-PYRIDYL)-, SULFATE □ SULFATE de NICOTINE (FRENCH)
CONSENSUS REPORTS: EPA Extremely Hazardous Substances List. Reported in EPA TSCA Inventory.
DOT CLASSIFICATION: 6.1; Label: Poison
SAFETY PROFILE: Poison by ingestion and skin contact. When heated to decomposition it emits very toxic fumes of SO_x and organic fumes. See also NICOTINE and SULFATES.

NDS500   CAS: 65-31-6   HR: 3
NICOTINE TARTRATE (1:2)
DOT: UN 1659
mf: $C_{10}H_{14}N_2 \cdot 2C_4H_6O_6$ mw: 462.46
SYNS: (S)-3-(1-METHYL-2-PYRROLIDINYL-PYRIDINE (R)-(R,R))-2,3-DIHYDROXYBUTANEDIOATE (1:2) □ NICOTINE ACID TARTRATE □ NICOTINE BITARTR-ATE □ NICOTINE HYDROGEN TARTRATE □ (−)-NICOTINE HYDROGEN TARTRATE □ NICOTINE

TARTRATE □ NICOTINE TARTRATE (DOT) □ TARTRATE de NICOTINE (FRENCH)
DOT CLASSIFICATION: 6.1; Label: Poison
SAFETY PROFILE: Poison by ingestion, intravenous, intraperitoneal, and subcutaneous routes. An experimental teratogen. Experimental reproductive effects. When heated to decomposition it emits toxic fumes of NO_x and CO. See also NICOTINE.

NDW520   CAS: 58013-12-0   HR: 3
1-NICOTINOYLMETHYL-4-PHENYL-PIPERAZINE
mf: $C_{17}H_{19}N_3O$ mw: 281.39
SYNS: KETONE, 4-PHENYL-1-PIPERAZINYLMETHYL-3-PYRIDYL- □ 1-(2-KETO-2-(3'-PYRIDYL)ETHYL)-4-(PHENYL)PIPERAZINE
DOT CLASSIFICATION: 3; Label: Flammable Liquid
SAFETY PROFILE: A poison by intraperitoneal route. A flammable liquid. When heated to decomposition it emits toxic vapors of NO_x.

NDY000   CAS: 555-84-0   HR: 3
NIFURADENE
mf: $C_8H_8N_4O_4$ mw: 224.20
PROP: Lemon-yellow solid from nitromethane. Mp: 261.5–263° (decomp).
SYNS: NF 246 □ 1-(((5-NITRO-2-FURANYL)METHYL-ENE)AMINO)-2-IMIDAZOLIDINONE □ N-(5-NITRO-2-FURFURYLIDENE)-1-AMINO-2-IMIDAZOLIDONE □ N-(5-NITRO-2-FURFURYLIDENEAMINO)-2-IMIDAZO-LIDINONE □ 1-((5-NITROFURFURYLIDENE)AMINO)-2-IMIDAZOLIDINONE □ NSC-6470 □ OXAFURADENE □ OXYFURADENE □ RENAFUR
CONSENSUS REPORTS: IARC Cancer Review: Group 2B IMEMDT 7,56,87; Animal Limited Evidence IMEMDT 7,181,74.
SAFETY PROFILE: Suspected carcinogen with experimental carcinogenic data. Moderately toxic by ingestion and intraperitoneal routes. When heated to decomposition it emits toxic fumes of NO_x.

NDY500   CAS: 3570-75-0   HR: 3
NIFURTHIAZOLE
mf: $C_8H_6N_4O_4S$ mw: 254.24

PROP: Bright-yellow plates. Mp: 215.5° (decomp).
SYNS: AS-17665 □ FNT □ FORMIC 2-(4-(5-NITRO-FURYL)-2-THIAZOLYL)HYDRAZIDE □ 2-(2-FORMYL-HYDRAZINO)-4-(5-NITRO-2-FURYL)THIAZOLE □ NEFURTHIAZOLE □ 2-(4-(5-NITRO-2-FURANYL)-2-THIAZOLYL)-HYDRAZINECARBOXALDEHYDE □ NSC-525334
CONSENSUS REPORTS: IARC Cancer Review: Group 2B IMEMDT 7,56,87; Animal Sufficient Evidence IMEMDT 7,151,74. EPA Genetic Toxicology Program.
SAFETY PROFILE: Confirmed carcinogen with experimental carcinogenic and neoplastigenic data. Mutation data reported. When heated to decomposition it emits very toxic fumes of NO_x and SO_x.

NDY550   CAS: 189624-85-9 HR: 3
(–)-NIGALDIPINE HYDROCHLORIDE
mf: $C_{41}H_{42}N_4O_6$•2ClH   mw: 759.79
SYNS: (S)-(+)-AE 0047 □ (S)-AE0047 □ B 859-35 □ B 8509-035 □ BY 935 □ DEXNIGULDIPINE HYDRO-CHLORIDE □ (R)-NIGALDIPINE HYDROCHLORIDE □ 3,5-PYRIDINEDICARBOXYLIC ACID, 1,4-DIHYDRO-2,6-DIMETHYL-4-(3-NITROPHENYL)-, 2-(4-(4-(DIPHENYL-METHYL)-1-PIPERAZINYL)PHENYL)ETHYL METHYL ESTER, DIHYDROCHLORIDE, (S)-(+)- □ (S)-(+)-WATANI-DIPINE HYDROCHLORIDE □ WATANIDIPINE HY-DROCHLORIDE (S)-ENANTIOMER
SAFETY PROFILE: A poison by ingestion and intravenous routes. When heated to decomposition it emits toxic vapors of NO_x, HCl, and Cl⁻.

NDZ000   CAS: 7440-03-1   HR: 2
NIOBIUM
mf: Nb   mw: 92.91
PROP: Steel-gray, cubic crystals. Mp: 2468 ± 10°, bp: 4742°, d: 8.57. Sol in aqua regia, fused alkali. An element that occurs throughout nature.
SYNS: COLUMBIUM □ NIOBIUM-93 □ NIOBIUM ELEMENT □ VN 1
CONSENSUS REPORTS: Reported in EPA TSCA Inventory.
SAFETY PROFILE: An eye and severe skin irritant. Very low toxicity by ingestion and intraperitoneal routes. No reports of human intoxication. Can cause kidney damage. Experimentally, there is a moderate fibrogenic effect on the lungs after intratracheal administration. Some niobium is found in all parts of the body. Flammable in the form of dust when exposed to flame or by chemical reaction. Moderately explosive in the form of dust when exposed to flame. Ignites in fluorine, chlorine (at 205°C). Incandescent reaction with bromine trifluoride.

NEA000   CAS: 10026-12-7   HR: 3
NIOBIUM CHLORIDE
mf: Cl_5Nb mw: 270.16
PROP: Yellow, very deliq, monoclinic crystals. Decomp in moist air evolving HCl. D: 2.75, mp: 204.7–209.5°, bp: approx 250°, subl @ 125°. Sol in inert org solvs, HCl, carbon tetrachloride.
SYNS: COLUMBIUM PENTACHLORIDE □ NIOBIUM PENTACHLORIDE
CONSENSUS REPORTS: Reported in EPA TSCA Inventory. EPA Genetic Toxicology Program.
SAFETY PROFILE: Poison by intraperitoneal route. Moderately toxic by ingestion. May cause kidney injury. When heated to decomposition it emits very toxic fumes of Cl⁻. See also NIOBIUM and CHLORIDES.

NED000   HR: 3
NITRATES
PROP: Organic nitrates are usually termed nitro compounds. These compounds are a combination of the nitro (—NO_2) group and an organic radical. However, this term is often used to denote nitric acid esters of an organic material. Inorganic nitrates are compounds of metals with the monovalent NO_3 radical.
SAFETY PROFILE: Large amounts taken by mouth may have serious or even fatal effects. The symptoms are dizziness, abdominal cramps, vomiting, bloody diarrhea, weakness, convulsions, and collapse. Small, repeated doses may lead to weakness, general depression, headache, and mental impairment. Also, there is some

implication of increased cancer incidence among those exposed.

Flammable by spontaneous chemical reaction; practically all nitrates are powerful oxidizing agents. Some nitrates may explode when shocked, exposed to heat or flame, or by spontaneous chemical reaction (see also EXPLOSIVES, HIGH). All the inorganic nitrates act as oxygen carriers; under proper conditions, these can give up their oxygen to other materials which may in turn detonate. Ammonium nitrate has all the properties of the other nitrates, but it is also able to detonate by itself under certain conditions. It is therefore a high explosive, although very insensitive to impact and difficult to detonate. In the pure state, it requires a combination of an initiator and a high explosive. It is a relatively safe high explosive, which, however, must be stored in a cool, ventilated place away from acute fire hazards and easily oxidized materials. Ammonium nitrate must not be confined because, if a fire should start, confinement can cause detonation with extremely violent results.

Violent reaction with Al, BP, cyanide, esters, PN_2H, P, NaCN, $SnCl_2$, sodium hypophosphite, thiocyanates. Dangerous disaster hazard due to fire and explosion hazard. When heated to decomposition it emits toxic fumes of NO_x. They are powerful oxidizing agents that may cause violent reaction with reducing materials. Nitrates should be protected carefully in storage.

NED500 **CAS: 7697-37-2** **HR: 3**
NITRIC ACID
DOT: UN 2031
mf: HNO_3 mw: 63.02
PROP: Transparent, colorless or yellowish, fuming, suffocating, caustic and corrosive liquid. Pure acid; decomp especially in light. Mp: −42°, bp: 83°, d: 1.50269 @ 25°/4°. IDLH 25 ppm.
SYNS: ACIDE NITRIQUE (FRENCH) □ ACIDO NITRICO (ITALIAN) □ AQUA FORTIS □ AZOTIC ACID

□ AZOTOWY KWAS (POLISH) □ HYDROGEN NITRATE □ KYSELINA DUSICNE □ NITRIC ACID, over 40% (DOT) □ NITRIC ACID other than red fuming with >70% nitric acid (DOT) □ NITRIC ACID other than red fuming with not >70% nitric acid (DOT) □ SALPETERSAEURE (GERMAN) □ SALPETERZUUROPLOSSINGEN (DUTCH)
CONSENSUS REPORTS: EPA Extremely Hazardous Substances List. Reported in EPA TSCA Inventory. EPA Genetic Toxicology Program. Community Right-To-Know List.
OSHA PEL: TWA 2 ppm; STEL 4 ppm
ACGIH TLV: TWA 2 ppm; STEL 4 ppm
DFG MAK: 2 ppm (5.2 mg/m³)
NIOSH REL: (Nitric Acid) TWA 2 ppm
DOT CLASSIFICATION: 8; Label: Corrosive, Oxidizer, Poison
SAFETY PROFILE: Human poison by ingestion. An experimental teratogen. Experimental reproductive effects. Corrosive to eyes, skin, mucous membranes, and teeth. Causes upper respiratory irritation that may seem to clear up, only to return in a few hours and more severely. Depending on environmental factors the vapor will consist of a mixture of the various oxides of nitrogen and nitric acid. Flammable by chemical reaction with reducing agents. It is a powerful oxidizing agent.

Explosive reaction with acetic anhydride, acetone + acetic acid (in storage), acetone + hydrogen peroxide, acetone + sulfuric acid (if confined), alcohols, alkane thiols, 2-aminothiazole, 2-aminothiazole + sulfuric acid, dinitrogen tetraoxide or sulfuric acid + aromatic amines (e.g., aniline, n-ethylamine, o-toluidine, xylidine, p-phenylenediamine), benzidine (hypergolic), benzonitrile + sulfuric acid, 5-acetylamino-3-bromoben-zo(b)thiophene, 1,4-bis(methoxymethyl)-2,3,5,5-tetramethylbenzene (at 80°C), 1,3-bis(trifluoromethyl)benzene + sulfuric acid (vapors are initiated by spark), tert-butyl-m-xylene, cadmium phosphide, chlorobenzene, cotton + rubber + sulfuric acid + water (mixture has caused industrial explosions), crotonaldehyde (hypergolic), cyclohexyl-amine, 1,2-diaminoethanebis(trimethylgold), diethyl ether, diethyl ether + sulfuric acid,

1,1-dimethyl hydrazine (hypergolic), dimethyl sulfide, dimethyl sulfide + 1,4-dioxane, dimethyl sulfoxide + water, dinitrogen tetraoxide + nitrogenous fuels (hypergolic with triethylamine, dimethylhydrazine, mixo-xylidine), dioxane + perchloric acid, diphenyldistibene, divinyl ether (hypergolic), ethane sulfonamide, 5-ethyl-2-methylpyridine, fat + sulfuric acid (when confined), 2-formylamino-1-phenyl-1,3-propanediol, fluorine, furfurylidene ketones (hypergolic), glycerin + sulfuric acid, hexalithium disilicide, 2,2,4,4,6,6-hexamethyl-trithiane, hydrazine (hypergolic), hydrocarbons, hydrocarbons + 1,1-dimethylhydrazine, hydrogen peroxide + soils, ion exchange resins, magnesium + 2-nitroaniline (hypergolic), metal acetylides (e.g., cesium acetylide, rubidium acetylide), metal cyanides, metal hexacyanoferrates (3-) or (4-), metal thiocyanates, 4-methylcyclohexanone (above 75°C), methylthiophene, nitrobenzene + sulfuric acid, 1-nitronaphthalene + sulfuric acid, nonmetal hydrides (e.g., arsine, phosphine, tetraborane(10), stibene), phosphorus, organic materials + oxidizers (e.g., perchloric acid, potassium chlorate, sulfuric acid), phenylacetylene + 1,1-dimethylhydrazine, phosphine derivatives (e.g., phosphine, phosphonium iodide, ethyl phosphine, tris(iodomercury) phosphine), tetraphosphorus diiodo triselenide, phosphorus trichloride, polyurethane foam, propiophenone + sulfuric acid, pyrocatechol (hypergolic), potassium phosphinate + heat, resorcinol, rubber, silicone oil, silver buten-3-ymide, sulfur dioxide, 1,3,5-triacetylhexahydro-1,3,5-triazine + trifluoroacetic anhydride, triazine + trifluoroacetic anhydride, zinc ethoxide. Explosive or hypergolic reaction with various hydrocarbons (e.g., acetylene derivatives, benzene, 3-carene, cashew nut shell oil, cyclopentadienes, dicyclopentadiene, dienes, hexamethylbenzene, mesitylene, burning petroleum products, toluene, turpentine, p-xylene).

Ignition on contact with acetone, alcohols + disulfuric acid, alcohols + potassium permanganate, aliphatic amines, O-alkyl ethylene dithiophosphate, ammonia, anilinium nitrate + inorganic materials (e.g., copper(I) chloride, potassium permanganate, sodium pentacyanonitrosylferrate, vanadium(V) oxide, ammonium metavanadate, sodium metavanadate, aromatic amines + metal compounds (e.g., ammonium metavanadate), copper(I) oxide, copper(II) oxide, iron(III) chloride, iron(III) oxide, potassium chromate, potassium dichromate, potassium hexacyanoferrate(I), potassium hexacyanoferrate(III), sodium metavanadate, sodium pentacyanonitrosylferrate(II), vanadium(V) oxide), dichromates + organic fuels (e.g., ammonium dichromate, potassium dichromate, potassium chromate, cyclohexanol, 2-cresol, 3-cresol, furfural), diphenyl tin, lead-containing rubber, metals (e.g., lithium, sodium, magnesium), nonmetal hydrides (e.g., hydrogen iodide, hydrogen selenide, hydrogen sulfide, hydrogen telluride), phosphorus vapor, nickel tetraphosphide, tetraphosphorus iodide, polysilylene, turpentine + catalysts (e.g., concentrated sulfuric acid, iron(III) chloride, ammonium metavanadate, copper(II) chloride), wood.

Forms explosive mixtures with acetic acid, acetic acid + sodium hexahydroxyplatinate(IV), acetic anhydride + hexamethylenetetramine acetate, acetoxyethylene glycol, ammonium nitrate, anilinium nitrate, 1,2-dichloroethane, dichloroethylene, dichloromethane, diethylaminoethanol, 3,6-dihydro-1,2,2H-oxazine, dimethyl ether, 1,3-dinitrobenzene, disodium phenyl orthophosphate, 2-hexenal (heat sensitive), hydrofluoric acid + lactic acid, hydrofluoric acid + propylene glycol + silver nitrate, hydrogen peroxide + ketones (e.g., 2-butanone, 3-pentanone, cyclopentanone, cyclohexanone, 3-

methylcyclohexanone), hydrogen peroxide + mercury(II) oxide, metals (e.g., titanium, uranium, tin), metal salicylates, nitroaromatics (e.g., mono- and di-nitrobenzenes, di- and tri-nitrotoluenes), nitrobenzene + water, nitromethane, salicylic acid.

Incompatible with 4-acetoxy-3-methoxybenzaldehyde, acetylene, acrolein, acrylonitrile, acrylonitrile + methacrylate copolymer, allyl alcohol, allyl chloride, 2-amino ethanol, ammonium hydroxide, aniline, anion exchange resins, antimony, SbH_3, arsenic hydride, arsine + boron tribromide, benzo[b]thiophene derivatives, N-benzyl-N-ethylaniline + sulfuric acid, bismuth, boron, boron decahydride, B_4H_{10}, boron phosphide, bromine pentafluoride, butanethiol, 2,6-di-tert-butyl phenol, calcium hypophosphite, carbon, $C_2H_5PH_2$, cellulose, Cs_2C_2, chlorine trifluoride, 4-chloro-2-nitroaniline, chlorosulfonic acid, coal, CuN_3, Cu_3N_2, copper(I) nitride, cresol, cumene, cyanides, cyclic ketones, cyclohexanol + cyclohexanone, diborane, di-2-6-butoxyethylether (butex), dichromate + anion exchange resins, diisopropylether, dimethylaminomethylferrocene + water, uns-dimethyl hydrazine, diphenylmercury, epichlorohydrin, ethanol, m-ethylaniline, ethylene diamine, ethylene imine, 5-ethyl-2-picoline, formaldehyde + impurities, formic acid + heat, formic acid + urea, furfural, furfuryl alcohol, germanium, glycerol + hydrofluoric acid, glyoxal, hydrogen iodide, HN_3, hydrogen peroxide, hydrogen selenide, hydrogen sulfide, H_2Te, indane + sulfuric acid, FeO, iron(II) oxide powder, isoprene, ketones + hydrogen peroxide, lactic acid + HF, Li_6Si_2, magnesium silicide, magnesium phosphide, magnesium-titanium alloy, manganese, mesityl oxide, mesitylene, metals (e.g., bismuth powder, germanium powder, uranium powder, molten zinc), 2-methyl-5-ethyl pyridine, NdP, nonmetal powders (e.g., boron, silicon, arsenic, carbon), n-butyr-aldehyde, oleum, phosphorus halides, phthalic acid, phthalic anhydride, polyalk-enes (e.g., polyethylene, polypropylene), polydibromosilane, polyethylene oxide derivatives, KH_2PO_2, β-propiolactone, propylene oxide, pyridine, Rb_2C_2, reductants, selenium, selenium iodophosphide, silver + ethanol, sodium, sodium hydroxide, NaN_3, sucrose, sulfamic acid, sulfuric acid, sulfuric acid + $C_6H_5CH_3$, sulfuric acid + glycerides, sulfur halides (e.g., sulfur dichloride, sulfur dibromide, disulfur dibromide), sulfuric acid + terephthalic acid, terpenes, thioaldehydes, thiocyanates, thioketones, thiophene, titanium, titanium alloy, titanium-magnesium alloy, toluidine, triazine, tricadmium diphosphide, triethylgallium monoetherate, trimagnesium diphosphide, 2,4,6-trimethyltrioxane, uranium, uranium disulfide, uranium-neodymium alloy, uranium-neodymium-zirconium alloy, vinylacetate, vinylidene chloride, zinc, zirconium-uranium alloys.

Will react with water or steam to produce heat and toxic and corrosive fumes. To fight fire, use water. When heated to decomposition emits highly toxic fumes of NO_x and hydrogen nitrate.

NEE500   CAS: 7697-37-2   HR: 3
NITRIC ACID (RED FUMING)
DOT: UN 2032
mf: HNO_3 mw: 63.02
PROP: Colorless to yellow to red corrosive liquid. D: >1.480. Contains from 8 to 17% NO_2 (AMIHBC 10,418,54).
SYNS: NITRIC ACID, FUMING (DOT) □ NITRIC ACID, RED FUMING (DOT) □ NITROUS FUMES □ RED FUMING NITRIC ACID □ RFNA
CONSENSUS REPORTS: EPA Genetic Toxicology Program.
DOT CLASSIFICATION: 8; Label: Corrosive, Oxidizer, Poison
SAFETY PROFILE: Poison by inhalation. A corrosive irritant to skin, eyes, and mucous membranes. A very dangerous fire hazard and very powerful oxidizing agent. Can react explosively with many reducing agents. Will react with water or steam to produce heat and toxic, corrosive, and flammable vapors.

When heated to decomposition it emits highly toxic fumes of NO$_x$. See also NITRIC ACID.

NEG100 CAS: 10102-43-9 HR: 3
NITRIC OXIDE
DOT: UN 1660
mf: NO   mw: 30.01
PROP: Colorless non-flammable gas or blue liquid and solid. With O$_2$ gives brown NO$_2$. Mp: −161°, bp: −151.18°, d: 1.3402 g/L; liquid, 1.269 @ −150°; gas, 1.04 g/L. Sltly sol in water. IDLH 100 ppm.
SYNS: BIOXYDE d'AZOTE (FRENCH) □ NITROGEN MONOXIDE □ OXYDE NITRIQUE (FRENCH) □ RCRA WASTE NUMBER P076 □ STICKMONOXYD (GERMAN)
CONSENSUS REPORTS: Reported in EPA TSCA Inventory. EPA Extremely Hazardous Substances List.
OSHA PEL: TWA 25 ppm
ACGIH TLV: TWA 25 ppm
NIOSH REL: (Oxides of Nitrogen) TWA 25 ppm
DOT CLASSIFICATION: 2.3; Label: Poison Gas
SAFETY PROFILE: A poison gas. A severe eye, skin, and mucous membrane irritant. A systemic irritant by inhalation. Mutation data reported. Exposure may occur whenever nitric acid acts upon organic material, such as wood, sawdust, and refuse; it occurs when nitric acid is heated, and when organic nitro compounds are burned, for example, celluloid, cellulose nitrate (guncotton), and dynamite. The action of nitric acid upon metals, as in metal etching and pickling, also liberates the fumes. In high-temperature welding, as with the oxyacetylene or electric torch, the nitrogen and oxygen of the air unite to form oxides of nitrogen. Automobile exhaust and power plant emissions are also sources of NO$_x$. Exposure occurs in many manufacturing processes when nitric acid is made or used. Oxides of nitrogen have been implicated as a cause of acid rain.

The oxides of nitrogen are somewhat soluble in water, reacting with it to form nitric and nitrous acids. This is the action that takes place deep in the respiratory system. The acids formed are irritating and can cause congestion in the throat and bronchi and edema of the lungs. The acids are neutralized by the alkalies present in the tissues, with the formation of nitrates and nitrites. The latter may cause some arterial dilation, fall in blood pressure, headache, and dizziness, and there may be some formation of methemoglobin. However, the nitrite effect is of secondary importance.

Because of their relatively low solubility in water, the nitrogen oxides are initially only slightly irritating to the mucous membranes of the upper respiratory tract. Their warning power is therefore low, and dangerous amounts of the fumes may be breathed before the worker notices any real discomfort. Higher concentrations (60–150 ppm) cause immediate irritation of the nose and throat, with coughing and burning in the throat and chest. These symptoms often clear upon breathing fresh air, and the worker may feel well for several hours. Some 6–24 hours after exposure, a sensation of tightness and burning in the chest develops, followed by shortness of breath, sleeplessness, and restlessness. Dyspnea and air hunger may increase rapidly with development of cyanosis and loss of consciousness followed by death. In cases that recover from the pulmonary edema, there is usually no permanent disability, but pneumonia may develop later. Concentrations of 100–150 ppm are dangerous for short exposures of 30–60 minutes. Concentrations of 200–700 ppm may be fatal after even very short exposures.

Continued exposure to low concentrations of the fumes, insufficient to cause pulmonary edema, is said to result in chronic irritation of the respiratory tract, with cough, headache, loss of appetite, dyspepsia, corrosion of the teeth, and gradual loss of strength.

Exposure to NO$_x$ is always potentially serious, and persons so exposed should be kept under close observation for at least 48 hours.

An oxidizer. The liquid is a sensitive explosive. Explosive reaction with carbon disulfide (when ignited), methanol (when ignited), pentacarbonyl iron (at 50°C), phosphine + oxygen, sodium diphenylketyl, dichlorine oxide, fluorine, nitrogen trichloride, ozone, perchloryl fluoride (at 100–300°C), vinyl chloride. Reacts to form explosive products with dienes (e.g., 1,3-butadiene, cyclopentadiene, propadiene).

Can react violently with acetic anhydride, Al, amorphous boron, BaO, BCl_3, $CsHC_2$, calcium, carbon + potassium hydrogen tartrate, charcoal, ClO, pyrophoric chromium, 1,2-dichloroethane, dichloroethylene, ethylene, fuels, hydrocarbons, hydrogen + oxygen, Na_2O, uns-dimethyl hydrazine, NH_3, $CHCl_3$, Fe, Mg, Mn, CH_2Cl_2, olefins, phosphorus, PNH_2, PH_3, potassium, potassium sulfide, propylene, rubidium acetylide, Na, S, tungsten carbide, trichloroethylene, 1,1,1-trichloroethane, uns-tetrachloroethane, uranium, uranium dicarbide. Will react with water or steam to produce heat and corrosive fumes; can react vigorously with reducing materials.

NEH500
NITRILES
HR: 3

PROP: Nitriles are organic compounds containing the (—C≡N) grouping, e.g., acrylonitrile (CH_2:CHC≡N).
CONSENSUS REPORTS: Cyanide and its compounds are on the Community Right-To-Know List.
SAFETY PROFILE: Nitriles are organic cyanides; acrylonitrile, propionitrile, and some others resemble cyanides in toxicity. Other nitriles, such as cyanamides and cyanates, have no cyanide effect. Can react violently with ($LiAlH_4$ + H_2O). The nitriles may be used as insecticides. Many are flammable. When heated to decomposition they emit highly toxic fumes of CN^-. See also specific compounds and CYANIDE.

NEI000   CAS: 18662-53-8   HR: 3
NITRILOTRIACETIC ACID TRISODIUM SALT MONOHYDRATE
mf: $C_6H_6NO_6$•3Na•H_2O mw: 311.16
PROP: White odorless crystalline powder. Mp: 340°, d: 1.782. Sol in water.
SYNS: N,N-BIS(CARBOXYMETHYL)GLYCINE TRISODIUM SALT MONOHYDRATE □ NCI-C01445 □ NITRILOACETIC ACID TRISODIUM SALT MONOHYDRATE □ NTA SODIUM HYDRATE □ TRISODIUM NITRILOTRIACETATE MONOHYDRATE
CONSENSUS REPORTS: NCI Carcinogenesis Bioassay (feed); Clear Evidence: mouse, rat NCITR* NCI-CG-TR-6,77. Cyanide and its compounds are on the Community Right-To-Know List.
SAFETY PROFILE: Suspected carcinogen with experimental carcinogenic, tumorigenic, and teratogenic data. Moderately toxic by intraperitoneal route. Human mutation data reported. When heated to decomposition it emits toxic fumes of NO_x, CN^-, and Na_2O.

NEI800   CAS: 4097-89-6   HR: 3
2,2',2"-NITRILOTRIS(ETHYLAMINE)
mf: $C_6H_{18}N_4$ mw: 146.28
SYNS: 4-(2-AMINOETHYL)DIETHYLENETRIAMINE □ DIETHYLENETRIAMINE, 4-(2-AMINOETHYL)- □ N,N-BIS(2-AMINOETHYL)-1,2-ETHANEDIAMINE □ 1,2-ETHANEDIAMINE, N,N-BIS(2-AMINOETHYL)- □ NITRILOTRIS(ETHYLAMINE) □ TREN □ TREN HP □ TRI(2-AMINOETHYL)AMINE □ 2,2',2"-TRIAMINO-TRIETHYLAMINE □ β,β',β"-TRIAMINOTRIETHYL-AMINE □ 2,2',2"-TRIAMINOTRIS(ETHYLAMINE) □ TRIS(AMINOETHYL)AMINE □ TRIS(β-AMINOETHYL)AMINE
SAFETY PROFILE: A poison by ingestion and skin contact. When heated to decomposition it emits toxic vapors of NO_x.

NEJ000
NITRITES
HR: 3

PROP: Salts of nitrous acid.
SAFETY PROFILE: Large amounts taken by mouth may produce nausea, vomiting, cyanosis (due to methemoglobin formation), collapse, and coma. Repeated small doses cause a fall in blood pressure, rapid pulse, headache, and visual disturbances. They have been implicated in an increased

incidence of cancer. They may react with organic amines in the body to form carcinogenic nitrosamines. Organic nitrites are used to treat angina pectoris. Fire hazards are variable. They are generally powerful oxidizers. On contact with readily oxidized materials, a violent reaction such as a fire or explosion may ensue. Explosion hazards are also variable. Organic nitrites may decompose violently in contact with NH_4, salts, cyanide, KCN. Dangerous; shock may explode them; can react vigorously with reducing materials. When heated to decomposition they emit highly toxic fumes of NO_x. See also SODIUM NITRITE and specific compounds.

NEJ500   CAS: 602-87-9   HR: 3
5-NITROACENAPHTHENE
mf: $C_{12}H_9NO_2$ mw: 199.22
PROP: A solid. Mp: 101.5–102.5°. Sol in Et_2O, EtOH, ligroin, and hot H_2O.
SYNS: 1,2-DIHYDRO-5-NITRO-ACENAPHTHYLENE □ 5-NAN □ NCI-C01967 □ 5-NITROACENAPHTHYLENE □ 5-NITROACENAPTHENE □ 5-NITRONAPHTHALENE ETHYLENE
CONSENSUS REPORTS: IARC Cancer Review: Group 2B IMEMDT 7,56,87; Animal Sufficient Evidence IMEMDT 16,319,78. NCI Carcinogenesis Bioassay (feed); Clear Evidence: mouse, rat NCITR* NCI-CG-TR-118,78. Reported in EPA TSCA Inventory. EPA Genetic Toxicology Program.
DFG MAK: Animal Carcinogen, Suspected Human Carcinogen
SAFETY PROFILE: Confirmed carcinogen with experimental carcinogenic and neoplastigenic data. Mutation data reported. When heated to decomposition it emits toxic fumes of NO_x. See also NITRO COMPOUNDS OF AROMATIC HYDROCARBONS.

NEL000   CAS: 1777-84-0   HR: 2
3-NITRO-p-ACETOPHENETIDIDE
mf: $C_{10}H_{12}N_2O_4$ mw: 224.24

PROP: Yellow needles in water. Mp: 103–104°. Sol in abs alc, ether, and chloroform.
SYNS: 4-ACETAMINO-2-NITROPHENETOLE □ N-(4-ETHOXY-3-NITRO)PHENYLACETAMIDE □ N-(4-ETHOXYPHENYL)-3'-NITROACETAMIDE □ NCI-C01978 □ 2-NITRO-4-ACETAMINOFENETOL (CZECH) □ 3-NITRO-p-ACETOPHENETIDE □ 5-NITRO-p-ACETOPHENETIDIDE □ 3'-NITRO-p-ACETO-PHENETIDIN
CONSENSUS REPORTS: NCI Carcinogenesis Bioassay (feed); Clear Evidence: mouse NCITR* NCI-CG-TR-133,79; (feed); No Evidence: rat NCITR* NCI-CG-TR-133,79. Reported in EPA TSCA Inventory.
SAFETY PROFILE: An eye irritant. Mutation data reported. Questionable carcinogen with experimental carcinogenic, neoplastigenic, and tumorigenic data. When heated to decomposition it emits toxic fumes of NO_x. See also NITRO COMPOUNDS OF AROMATIC HYDROCARBONS.

NEM480   CAS: 119-34-6   HR: 3
2-NITRO-4-AMINOPHENOL
mf: $C_6H_6N_2O_3$ mw: 154.14
PROP: Dark-red plates or needles. Mp: 131°.
SYNS: 4-AMINO-2-NITROPHENOL □ C.I. 76555 □ FOURRINE 57 □ FOURRINE BROWN PR □ FOURRINE BROWN PROPYL □ 4-HYDROXY-3-NITROANILINE □ NCI-C03963 □ o-NITRO-p-AMINOPHENOL □ OXIDATION BASE 25
CONSENSUS REPORTS: IARC Cancer Review: Group 3 IMEMDT 7,56,87; Animal Inadequate Evidence IMEMDT 16,43,78. NCI Carcinogenesis Bioassay (feed); No Evidence: mouse NCITR* NCI-CG-TR-94,78; Clear Evidence: rat NCITR* NCI-CG-TR-94,78. Reported in EPA TSCA Inventory. EPA Genetic Toxicology Program.
DFG MAK: Confirmed Animal Carcinogen with Unknown Relevance to Humans
SAFETY PROFILE: Suspected carcinogen with experimental carcinogenic data. Very poisonous by intraperitoneal route. Moderately toxic by ingestion. A severe eye

irritant. Mutation data reported. When heated to decomposition it emits toxic fumes of NO$_x$.

NEN500   CAS: 99-09-2   HR: 3
m-NITROANILINE

DOT: UN 1661

mf: C$_6$H$_6$N$_2$O$_2$ mw: 138.14

PROP: Yellow, rhombic crystals or needles from water. D: 0.9011 @ 25°/4°, mp: 114°, bp: 306.4°. Sol in water, alc, and ether.

SYNS: m-AMINONITROBENZENE □ 1-AMINO-3-NITROBENZENE □ AZOBASE MNA □ C.I. 37030 □ C.I. AZOIC DIAZO COMPONENT 7 □ DAITO ORANGE BASE R □ DEVOL ORANGE R □ DIAZO FAST ORANGE R □ FAST ORANGE R SALT □ HILTONIL FAST ORANGE R BASE □ MNA □ NAPHTOELAN ORANGE R BASE □ m-NITRANILINE □ m-NITROAMINOBENZENE □ 3-NITROANILINE □ 3-NITROBENZENAMINE □ m-NITROPHENYLAMINE □ ORANGE BASE IRGA I

CONSENSUS REPORTS: Reported in EPA TSCA Inventory. EPA Genetic Toxicology Program.

DOT CLASSIFICATION: 3; Label: Flammable Liquid

SAFETY PROFILE: Poison by ingestion and intraperitoneal routes. Mutation data reported. Absorbed through the skin and by inhalation of the dust. Acute exposure may cause methemoglobinemia cyanosis. Chronic exposure may cause liver damage. Flammable liquid. Decomposes exothermically at 247°C. Possibly explosive reaction with ethylene oxide at 130°C. When heated to decomposition it emits toxic fumes of NO$_x$. See also o-NITROANILINE, p-NITROANILINE, and ANILINE DYES.

NEO000   CAS: 88-74-4   HR: 3
o-NITROANILINE

DOT: UN 1661

mf: C$_6$H$_6$N$_2$O$_2$ mw: 138.14

PROP: Orange-yellow crystals from water. Mp: 72°, bp: 284.5°, vap press: 1 mm @ 104°, d: 0.9015 @ 25°/4°. Sltly sol in cold water; sol in hot water, alc, and ether.

SYNS: 1-AMINO-2-NITROBENZENE □ AZOENE FAST ORANGE GR BASE □ AZOGENE FAST ORANGE GR □

AZOIC DIAZO COMPONENT 6 □ BRENTAMINE FAST ORANGE GR BASE □ C.I. 37025 □ C.I. AZOIC DIAZO COMPONENT 6 □ DEVOL ORANGE B □ DIAZO FAST ORANGE GR □ FAST ORANGE BASE GR □ FAST ORANGE BASE JR □ FAST ORANGE GR BASE □ FAST ORANGE O BASE □ HILTONIL FAST ORANGE GR BASE □ o-NITROANILINE □ 2-NITROANILINE □ ORANGE BASE CIBA II □ ORANGE BASE IRGA II □ ORANGE SALT CIBA II □ ORANGE SALT IRGA II

CONSENSUS REPORTS: Reported in EPA TSCA Inventory.

DOT CLASSIFICATION: 6.1; Label: Poison

SAFETY PROFILE: A poison. Moderately toxic by ingestion. Mildly toxic by skin contact. Mutation data reported. Mixtures with magnesium are hypergolic on contact with nitric acid. Forms extremely explosive addition compounds with hexanitroethane. Vigorous reaction with sulfuric acid above 200°C. When heated to decomposition it emits toxic fumes of NO$_x$. See also m-NITROANILINE, p-NITROANILINE, and ANILINE DYES.

NEO500   CAS: 100-01-6   HR: 3
p-NITROANILINE

DOT: UN 1661

mf: C$_6$H$_6$N$_2$O$_2$ mw: 138.14

PROP: Bright-yellow powder or pale-yellow needles from water. Mp: 148.5°, bp: 332°, flash p: 390°F (CC), d: 1.424, vap press: 1 mm @ 142.4°. Sol in water, alc, ether, benzene, methanol. IDLH 300 mg/m³.

SYNS: p-AMINONITROBENZENE □ 1-AMINO-4-NITROBENZENE □ ANILINE, 4-NITRO- □ AZOAMINE RED ZH □ AZOIC DIAZO COMPONENT 37 □ BENZEN-AMINE, 4-NITRO-(9CI) □ C.I. 37035 □ C.I. AZOIC DIAZO COMPONENT 37 □ C.I. DEVELOPER 17 □ DEVELOPER P □ DEVOL RED GG □ DIAZO FAST RED GG □ FAST RED BASE GG □ FAST RED BASE 2J □ FAST RED 2G BASE □ FAST RED GG BASE □ FAST RED MP BASE □ FAST RED P BASE □ NAPHTOELAN RED GG BASE □ NCI-C60786 □ p-NITRANILINE □ 4-NITRANILINE □ NITRAZOL CF EXTRA □ p-NITROANILINA □ p-NITROANILINE □ 4-NITROBENZENAMINE □ p-NITROPHENYLAMINE □ PNA □ RCRA WASTE NUMBER P077 □ RED 2G BASE □ SHINNIPPON FAST RED GG BASE

CONSENSUS REPORTS: Reported in EPA TSCA Inventory. EPA Genetic Toxicology Program.

OSHA PEL: TWA 3 mg/m³ (skin)

ACGIH TLV: TWA 3 mg/m³ (skin); Not Classifiable as a Human Carcinogen

DFG MAK: 1 ppm (5.7 mg/m³)

DOT CLASSIFICATION: 6.1; Label: Poison

SAFETY PROFILE: Poison by ingestion, intravenous, and intraperitoneal routes. Moderately toxic by intramuscular route. Mutation data reported. Acute symptoms of exposure are headache, nausea, vomiting, weakness and stupor, cyanosis and methemoglobinemia. Chronic exposure can cause liver damage. Experimental reproductive effects. Combustible when exposed to heat or flame. See NITRATES for explosion and disaster hazards. To fight fire, use water spray or mist, foam, dry chemical, CO_2. Vigorous reaction with sulfuric acid above 200°C. Reaction with sodium hydroxide at 130°C under pressure may produce the explosive sodium-4-nitrophenoxide. When heated to decomposition it emits toxic fumes of NO_x. See also m-NITROANILINE, o-NITROANILINE, NITRO COMPOUNDS OF AROMATIC HYDROCARBONS, and ANILINE DYES.

NEQ500   CAS: 99-59-2   HR: 3
5-NITRO-o-ANISIDINE

mf: $C_7H_8N_2O_3$ mw: 168.17

PROP: Red needles from alc. D: 1.207 @ 156°, mp: 118°. Sol in hot benzene, alc, and acetic acid; very sltly sol in ligroin.

SYNS: 2-AMINO-1-METHOXY-4-NITROBENZENE □ 3-AMINO-4-METHOXYNITROBENZENE □ 2-AMINO-4-NITROANISOLE □ o-ANISIDINE NITRATE □ AZOAMINE SCARLET □ AZOGENE ECARLATE R □ AZOIC DIAZO COMPONENT 13 BASE □ C.I. AZOIC DIAZO COMPONENT 13 □ C.I. 37130 □ FAST SCARLET R □ 2-METHOXY-5-NITROANILINE □ 2-METHOXY-5-NITROBENZENAMINE □ NCI-C01934 □ 3-NITRO-6-METHOXYANILINE □ 5-NITRO-2-METHOXYANILINE

CONSENSUS REPORTS: IARC Cancer Review: Group 3 IMEMDT 7,56,87; Animal Limited Evidence IMEMDT 27,133,82.

NCI Carcinogenesis Bioassay (feed); Clear Evidence: mouse, rat NCITR* NCI-CG-TR-127,78. Reported in EPA TSCA Inventory. EPA Genetic Toxicology Program. Community Right-To-Know List.

SAFETY PROFILE: Suspected carcinogen with experimental carcinogenic and tumorigenic data. Moderately toxic by ingestion. Mutation data reported. When heated to decomposition it emits toxic fumes of NO_x. See also NITRO COMPOUNDS OF AROMATIC HYDROCARBONS.

NER000   CAS: 91-23-6   HR: 3
o-NITROANISOLE

mf: $C_7H_7NO_3$ mw: 153.15

PROP: Colorless crystals. D: 1.254 @ 20°/4°, mp: 9.5–10.5°, bp: 277°. Insol in water; sol in alc, ether.

SYNS: 1-METHOXY-2-NITROBENZENE □ 2-METHOXYNITROBENZENE □ NCI-C60388 □ 2-NITROANISOLE □ o-NITROPHENYL METHYL ETHER

CONSENSUS REPORTS: NTP 10th Report on Carcinogens. NCI Carcinogenesis Studies (feed): Clear Evidence: mouse, rat NTPTR* NTP-TR-416,93. Reported in EPA TSCA Inventory. EPA Genetic Toxicology Program.

DFG MAK: Animal Carcinogen, Suspected Human Carcinogen

SAFETY PROFILE: Confirmed carcinogen. Moderately toxic by ingestion. Mutation data reported. Explosive reaction with sodium hydroxide + zinc. Vigorous reaction with hydrogen + catalyst (at 250°C/34bar). When heated to decomposition it emits toxic fumes of NO_x. See also p-NITROANISOLE.

NER500   CAS: 100-17-4   HR: 3
p-NITROANISOLE

mf: $C_7H_7NO_3$ mw: 153.15

PROP: Prisms from alc. D: 1.233 @ 20°, mp: 54°, bp: 274°. Sltly sol in cold pet ether; sol in water; very sol in alc, boiling pet ether, and ether.

SYNS: p-METHOXYNITROBENZENE □ 1-METHOXY-4-NITROBENZENE □ 4-METHOXYNITROBENZENE □ p-NITROANISOL □ 4-NITROANISOLE

CONSENSUS REPORTS: Reported in EPA TSCA Inventory. EPA Genetic Toxicology Program.

SAFETY PROFILE: A poison by ingestion. Mutation data reported. Can explode in presence of Ni. When heated to decomposition it emits toxic fumes of NO_x. See also o-NITROANISOLE and NITRO COMPOUNDS OF AROMATIC HYDROCARBONS.

NEX000   CAS: 98-95-3   HR: 3
NITROBENZENE
DOT: UN 1662
mf: $C_6H_5NO_2$ mw: 123.12

PROP: Bright-yellow crystals or pale-yellow to colorless, oily liquid; odor of volatile almond oil. Mp: 6°, bp: 210–211°, ULC: 20–30%, lel: 1.8% @ 200°F, flash p: 190°F (CC), d: 1.205 @ 15°/4°, autoign temp: 900°F, vap press: 1 mm @ 44.4°, vap d: 4.25. Volatile with steam; sol in about 500 parts water; very sol in alc, benzene, ether, oils. IDLH 200 ppm.

SYNS: ESSENCE of MIRBANE □ ESSENCE of MYRBANE □ MIRBANE OIL □ NCI-C60082 □ NITROBENZEEN (DUTCH) □ NITROBENZEN (POLISH) □ NITROBENZENE, liquid (DOT) □ NITROBENZOL (DOT) □ NITROBENZOL, liquid (DOT) □ OIL of MIRBANE (DOT) □ OIL of MYRBANE □ RCRA WASTE NUMBER U169

CONSENSUS REPORTS: Community Right-To-Know List. EPA Extremely Hazardous Substances List. Reported in EPA TSCA Inventory.

OSHA PEL: TWA 1 ppm (skin)

ACGIH TLV: TWA 1 ppm (skin); Animal Carcinogen; BEI: 5 mg/g creatinine of total p-nitrophenol in urine at end of shift at end of workweek

DFG MAK: Confirmed Animal Carcinogen with Unknown Relevance to Humans; BAT: 100 µg/L of aniline in blood after several shifts

DOT CLASSIFICATION: 6.1; Label: Poison

SAFETY PROFILE: Confirmed carcinogen. Human poison by an unspecified route. Poison experimentally by subcutaneous and intravenous routes. Moderately toxic by ingestion, skin contact, and intraperitoneal routes. Human systemic effects by ingestion: general anesthetic, respiratory stimulation, and vascular changes. An experimental teratogen. Experimental reproductive effects. Mutation data reported. An eye and skin irritant. Can cause cyanosis due to formation of methemoglobin. It is absorbed rapidly through the skin. The vapors are hazardous.

An oxidant. Combustible when exposed to heat and flame. Moderate explosion hazard when exposed to heat or flame. Explosive reaction with solid or concentrated alkali + heat (e.g., sodium hydroxide or potassium hydroxide), aluminum chloride + phenol (at 120°C), aniline + glycerol + sulfuric acid, nitric + sulfuric acid + heat. Forms explosive mixtures with aluminum chloride, oxidants (e.g., fluorodinitromethane, uranium perchlorate, tetranitromethane, sodium chlorate, nitric acid, nitric acid + water, peroxodisulfuric acid, dinitrogen tetraoxide), phosphorus pentachloride, potassium, sulfuric acid. Reacts violently with aniline + glycerin, N_2O, $AgClO_4$. To fight fire, use water, foam, CO_2, dry chemical. Incompatible with potassium hydroxide. When heated to decomposition it emits toxic fumes of NO_x. See also NITRO COMPOUNDS OF AROMATIC HYDROCARBONS.

NEX500   CAS: 5410-29-7   HR: 3
o-NITROBENZENEARSONIC ACID
mf: $C_6H_6AsNO_5$ mw: 247.05

PROP: Light yellow needles from H_2O. Mp: 194–196°. Spar sol in EtOH, $CHCl_3$, and cold Me_2CO. Probably dehydrated in AcOH.

CONSENSUS REPORTS: Arsenic and its compounds are on the Community Right-

To-Know List. Reported in EPA TSCA Inventory.

OSHA PEL: TWA 0.5 mg(As)/m³

SAFETY PROFILE: Poison by ingestion. When heated to decomposition it emits very toxic fumes of NO_x and As. See also ARSENIC COMPOUNDS and NITRO COMPOUNDS OF AROMATIC HYDROCARBONS.

NFA500   CAS: 22751-24-2   HR: 3
3-NITROBENZENEDIAZONIUM PER-
** CHLORATE**
mf: $C_6H_4ClN_3O_6$ mw: 249.58

SYN: m-NITROBENZENE DIAZONIUM PERCHLORATE (DOT)

DOT CLASSIFICATION: Forbidden

SAFETY PROFILE: An extremely shock- and heat-sensitive explosive. Upon decomposition it emits toxic fumes of Cl^- and NO_x. See also NITRO COMPOUNDS OF AROMATIC HYDROCARBONS.

NFB500   CAS: 98-47-5   HR: 3
3-NITROBENZENESULFONIC ACID
mf: $C_6H_5NO_5S$ mw: 203.18

PROP: Hygroscopic leaflets or deliquescent plates. Mp: 70°, bp: decomp. Very sol in water; sol in alc and alkali; insol in ether.

SYNS: KYSELINA NITROBENZEN-m-SULFONOVA (CZECH) □ m-NITROBENZENESULFONIC ACID

CONSENSUS REPORTS: Reported in EPA TSCA Inventory.

SAFETY PROFILE: A skin and severe eye irritant. Decomposes violently at about 200°. Mixture with sulfuric acid + sulfur trioxide may explode above 150°C. When heated to decomposition it emits toxic fumes of SO_x and NO_x. See also NITRO COMPOUNDS OF AROMATIC HYDROCARBONS and SULFONATES.

NFJ000   CAS: 2338-12-7   HR: 3
5-NITROBENZOTRIAZOLE
DOT: UN 0385
mf: $C_6H_4N_4O_2$ mw: 164.14

PROP: Long, pale-yellow needles from H_2O. Mp: 212°.

SYNS: 1H-BENZOTRIAZOLE, 6-NITRO- □ 5-NITROBENZOTRIAZOL (DOT) □ 5-NITRO-1H-BENZOTRIAZOLE □ 6-NITRO-1H-BENZOTRIAZOLE

CONSENSUS REPORTS: Reported in EPA TSCA Inventory.

DOT CLASSIFICATION: EXPLOSIVE 1.1D; Label: EXPLOSIVE 1.1D

SAFETY PROFILE: Poison by intraperitoneal and intravenous routes. An explosive. When heated to decomposition it emits toxic fumes of NO_x. See also NITRO COMPOUNDS OF AROMATIC HYDROCARBONS.

NFJ500   CAS: 98-46-4   HR: 3
3-NITROBENZOTRIFLUORIDE
mf: $C_7H_4F_3NO_2$ mw: 191.12

PROP: Thin, pale, straw-colored, oily liquid with aromatic odor. Mp: −5°, bp: 202.8°, flash p: 217°F (OC), d: 1.437 @ 15.5°/15.5°, vap press: 0.3 mm @ 25°.

SYNS: m-NITROBENZOTRIFLUORIDE (DOT) □ m-NITROTRIFLUOROTOLUENE □ m-NITROTRIFLUOR-TOLUOL (GERMAN) □ m-(TRIFLUOROMETHYL)-NITROBENZENE □ 3-TRIFLUOROMETHYLNITRO-BENZENE □ α,α,α-TRIFLUORO-m-NITROTOLUENE □ USAF MA-5

CONSENSUS REPORTS: Reported in EPA TSCA Inventory.

SAFETY PROFILE: Poison by intraperitoneal route. Moderately toxic by ingestion, inhalation, and subcutaneous routes. Combustible when exposed to heat or flame. When heated to decomposition it emits very toxic fumes of F^- and NO_x. See also FLUORIDES and NITRO COMPOUNDS OF AROMATIC HYDROCARBONS.

NFK100   CAS: 122-04-3   HR: 2
4-NITROBENZOYL CHLORIDE
mf: $C_7H_4ClNO_3$ mw: 185.57

PROP: Bright-yellow needles from pet ether; pungent odor. Mp: 75°, bp: 205°. Decomposed by water and alc; sol in ether. Sol in Py and toluene. Keep well closed.

SYNS: p-NITROBENZOIC ACID CHLORIDE □ 4-NITROBENZOIC ACID CHLORIDE □ p-NITROBENZO-YL CHLORIDE

CONSENSUS REPORTS: Reported in EPA TSCA Inventory.

SAFETY PROFILE: Moderately toxic by ingestion. Experimental reproductive effects. Mutation data reported. When heated to decomposition it emits toxic fumes of NO_x and Cl^-. See also NITRO COMPOUNDS OF AROMATIC HYDROCARBONS and CHLORIDES.

NFQ000   CAS: 92-93-3   HR: 3
p-NITROBIPHENYL
mf: $C_{12}H_9NO_2$ mw: 199.22
PROP: Yellow needles from alc. Mp: 113–114°, bp: 223–224°. Insol in water; sltly sol in cold alc; very sol in ether.
SYNS: 4-NITRODIPHENYL □ p-PHENYL-NITRO-BENZENE □ 4-PHENYL-NITROBENZENE □ PNB
CONSENSUS REPORTS: IARC Cancer Review: Group 3 IMEMDT 7,56,87; Animal Limited Evidence IMEMDT 4,113,74. Reported in EPA TSCA Inventory.
OSHA PEL: OSHA: Cancer Suspect Agent
ACGIH TLV: Confirmed Human Carcinogen
DFG MAK: Confirmed Animal Carcinogen, Suspected Human Carcinogen
NIOSH REL: (4-Nitrobiphenyl) TWA use 29 CFR 1910.1003
SAFETY PROFILE: Confirmed carcinogen with experimental carcinogenic, neoplastigenic, and tumorigenic data. Poison by intraperitoneal route. Moderately toxic by ingestion. Mutation data reported. When heated to decomposition it emits toxic fumes of NO_x. See also NITRO COMPOUNDS OF AROMATIC HYDROCARBONS.

NFS525   CAS: 100-00-5   HR: 3
p-NITROCHLOROBENZENE
DOT: UN 1578
mf: $C_6H_4ClNO_2$ mw: 157.56
PROP: D: 1.520, mp: 83°, bp: 242°, flash p: 110°. Insol in water; sltly sol in alc; very sol in CS_2 and ether. IDLH 100 mg/m³.
SYNS: 1-CHLOOR-4-NITROBENZEEN (DUTCH) □ 1-CHLOR-4-NITROBENZOL (GERMAN) □ p-CHLORO-NITROBENZENE □ 4-CHLORONITROBENZENE □ 1-CHLORO-4-NITROBENZENE □ 4-CHLORO-1-NITROBENZENE □ 1-CLORO-4-NITROBENZENE (ITALIAN) □ p-NITROCHLOORBENZEEN (DUTCH) □ p-NITROCHLOROBENZENE solid (DOT) □ p-NITROCHLOROBENZOL (GERMAN) □ p-NITROCLOROBENZENE (ITALIAN) □ PNCB

CONSENSUS REPORTS: IARC Cancer Review: Group 3 IMEMDT 65,263,96; Human Inadequate Evidence IMEMDT 65,263,96; Animal Inadequate Evidence IMEMDT 65,263,96. Reported in EPA TSCA Inventory.
OSHA PEL: TWA 1 mg/m³ (skin)
ACGIH TLV: TWA 0.1 ppm (skin); Animal Carcinogen
DFG MAK: Confirmed Animal Carcinogen with Unknown Relevance to Humans
NIOSH REL: (p-Nitrochlorobenzene) lowest feasible conc. (skin)
DOT CLASSIFICATION: 6.1; Label: Poison
SAFETY PROFILE: Confirmed carcinogen with experimental carcinogenic data. A poison by ingestion. Experimental reproductive effects. Mutation data reported. Flammable liquid when exposed to heat, sparks, or flame. May explode on heating. Potentially violent reaction with sodium methoxide. When heated to decomposition it emits very toxic fumes of NO_x and Cl^-. See also other chloronitrobenzene entries and NITRO COMPOUNDS OF AROMATIC HYDROCARBONS.

NFS700   CAS: 121-17-5   HR: 3
3-NITRO-4-CHLOROBENZOTRIFLUORIDE
DOT: UN 2307
mf: $C_7H_3ClF_3NO_2$ mw: 225.56
PROP: Bp: 113–114° @ 26 mm.
SYNS: 4-CHLORO-3-NITRO-α,α,α-TRIFLUORO-TOLUENE □ 3-NITRO-4-CHLORO-α,α,α-TRIFLUORO-TOLUENE
DOT CLASSIFICATION: 6.1; Label: Poison
SAFETY PROFILE: Poison by ingestion and intravenous routes. Human mutation data reported. When heated to decomposition it emits toxic fumes of F^-, Cl^-, and NO_x. See also CHLORINATED HYDROCARBONS, AROMATIC; and

NITRO COMPOUNDS OF AROMATIC HYDROCARBONS.

NFT400   CAS: 7496-02-8   HR: 3
6-NITROCHRYSENE
mf: $C_{18}H_{11}NO_2$ mw: 273.30
PROP: Crystalline solid. Mp: 210°.
CONSENSUS REPORTS: NTP 10th Report on Carcinogens. IARC Cancer Review: Group 2B IMEMDT 46,267,89; Animal Sufficient Evidence IMEMDT 46,267,89; Human No Adequate Data IMEMDT 46,267,89.
SAFETY PROFILE: Confirmed carcinogen with experimental carcinogenic data. Mutation data reported. When heated to decomposition it emits toxic fumes of NO_x. See also CHRYSENE and NITRO COMPOUNDS OF AROMATIC HYDROCARBONS.

NFT459   HR: 3
NITRO COMPOUNDS
PROP: Compounds of the form $C-NO_2$ or $N-NO_2$.
SAFETY PROFILE: The presence of a C— or N— linked nitro group in an organic compound can significantly decrease its reactivity and stability. Nitrate esters, with the nitro group linked to O, are very unstable. Nitro alkanes are mild oxidants but high temperatures and pressures may cause them to react violently. Polynitroalkanes may be explosive. Alkalies react with nitro alkanes to form explosive metal salts. The presence of metal oxides increases the thermal sensitivity of the lower nitro alkanes (e.g., nitromethane, nitroethane, and 1-nitropropane). Many nitro alkenes are highly reactive and may be explosive. Compounds with more than one nitro group (polynitroalkyls, such as trinitromethane and dinitroacetonitrile) are generally explosive. See also specific compounds and NITRO COMPOUNDS OF AROMATIC HYDROCARBONS.

NFT500   HR: 3
NITRO COMPOUNDS of AROMATIC HYDROCARBONS
SAFETY PROFILE: The mono-, di-, and trinitrobenzenes are absorbed chiefly through the skin and through inhalation of the dust or vapor when these materials are heated. The dinitrobenzenes are believed to be somewhat more toxic than the mononitrobenzenes and more toxic than aniline. The effect of di- and trinitrobenzene on the body is similar to that of aniline and mononitrobenzene, with reduction of the oxygen-carrying power of the blood and depression of the nervous system being responsible for most of the symptoms following acute exposure. Poisoning with the solid nitro compounds is usually slower and less severe than is the case with the liquid nitro and amino benzenes since absorption is less rapid. Thus, chronic poisoning occurs more frequently than acute, the picture observed in the chronic form being one of anemia, moderate cyanosis, fatigue, slight dizziness, headache, insomnia, and loss of weight. Prolonged chronic exposure may result in damage to the liver and kidneys, with production of acute yellow atrophy, toxic hepatitis, and fatty degeneration of the kidneys. The introduction of one or more Cl atoms into the nitrobenzene ring results in the forma-tion of chloronitrobenzene compounds or nitrochlors. The chloro-mono-nitrobenz-enes have essentially the same toxic effect as nitrobenzene. The Cl derivatives of dinitrobenzene, on the other hand, while resembling dinitrobenzene in their systemic effects, are much more irritating to the skin. They act as direct irritants and, in addition, may cause sensitization.

Dangerous; many of these compounds are highly flammable and some are explosive, especially those with more than one nitro group on the ring (polynitroaryls, such as trinitrobenzene, trinitrotoluene, tetranitro-N-methylaniline, trinitrophenol). The presence of alkali increases the thermal sensitivity of the explosive materials. Industrial explosions have occurred in this manner. When heated to decomposition they evolve highly toxic fumes of NO_x. See specific nitro compounds.

NFW210   CAS: 23611-69-0   HR: 3
8-NITRO-DIHYDRO-1,3-BENZOXAZINE-2-THIONE-4-ONE

mf: $C_8H_4N_2O_4S$ mw: 224.20

SYNS: 2H-1,3-BENZOXAZINE-2,4(3H)-DIONE, 8-NITRO-2-THIO- ◻ 8-NITRO-2-THIO-2H-1,3-BENZOXAZINE-2,4(3H)-DIONE

SAFETY PROFILE: A poison by ingestion. When heated to decomposition it emits toxic vapors of NO_x and SO_x.

NFY500   CAS: 79-24-3   HR: 3
NITROETHANE

DOT: UN 2842

mf: $C_2H_5NO_2$ mw: 75.08

PROP: Oily, colorless liquid with pleasant but slightly irritating odor. Mp: −90°, bp: 114.0°, fp: −50°, d: 1.046 @ 25°/25°, autoign temp: 778°F, flash p: 106°F, decomp @ 335–382°, lel: 4.0%, vap press: 15.6 mm @ 20°, vap d: 2.58. Misc in MeOH, EtOH, Et_2O, $CHCl_3$; sltly sol in hot water; insol in cold H_2O. IDLH 1000 ppm.

SYN: NITROETAN (POLISH)

CONSENSUS REPORTS: Reported in EPA TSCA Inventory. EPA Genetic Toxicology Program.

OSHA PEL: TWA 100 ppm

ACGIH TLV: TWA 100 ppm

DFG MAK: 100 ppm (310 mg/m³)

DOT CLASSIFICATION: 3; Label: Flammable Liquid

SAFETY PROFILE: Poison by intraperitoneal route. Moderately toxic by ingestion. Mildly toxic by inhalation. Causes injury to liver and kidneys. An eye and mucous membrane irritant. Flammable liquid when exposed to heat, sparks, flame, or oxidizers. To fight fire, use alcohol foam, CO_2, dry chemical; water can blanket fire. Incompatible with $Ca(OH)_2$, hydrocarbons, hydroxides, inorganic bases, KOH, NaOH, metal oxides, Explodes when heated. When

heated to decomposition it emits toxic fumes of NO_x. See also NITRO COMPOUNDS.

NGB000   CAS: 607-57-8   HR: 3
2-NITROFLUORENE

mf: $C_{13}H_9NO_2$ mw: 211.23

PROP: Needles from 50% acetic acid. Mp: 157–158°.

CONSENSUS REPORTS: IARC Cancer Review: Group 2B IMEMDT 46,291,89; Animal Sufficient Evidence IMEMDT 46,277,89; Human No Adequate Data IMEMDT 46,277,89. Reported in EPA TSCA Inventory. EPA Genetic Toxicology Program.

SAFETY PROFILE: Confirmed carcinogen with experimental carcinogenic and tumorigenic data. Moderately toxic by intraperitoneal route. Human mutation data reported. When heated to decomposition it emits toxic fumes of NO_x. See also NITRO COMPOUNDS OF AROMATIC HYDROCARBONS.

NGE000   CAS: 67-20-9   HR: 3
NITROFURANTOIN

mf: $C_8H_6N_4O_5$ mw: 238.18

PROP: Orange-yellow needles from Me_2CO (aq). Mp: 258–262°.

SYNS: BENKFURAN ◻ BERKFURIN ◻ CHEMI-OFURAN ◻ CYANTIN ◻ DANTAFUR ◻ FURADANTIN ◻ FURADONIN ◻ FURANTOIN ◻ FUROBACTINA ◻ ITURAN ◻ MACRODANTIN ◻ NCI-C55196 ◻ N-(5-NITRO-2-FURFURYLIDENE)-1-AMINOHYDANTOIN ◻ N-(5-NITROFURFURYLIDENE)-1-AMINOHYDANTOIN ◻ 1-((5-NITROFURFURYLIDENE)AMINO)HYDANTOIN ◻ NSC-2107 ◻ N-TOIN ◻ ORAFURAN ◻ PARFURAN ◻ URIZEPT ◻ USAF EA-2 ◻ WELFURIN ◻ ZOOFURIN

CONSENSUS REPORTS: IARC Cancer Review: Group 3 IMEMDT 50,211,90; Animal Limited Evidence IMEMDT 50,211,90; Human Inadequate Evidence IMEMDT 50,211,90. Reported in EPA TSCA Inventory. EPA Genetic Toxicology Program.

SAFETY PROFILE: Poison by ingestion and intraperitoneal routes. Human systemic effects: peripheral motor nerve recording

N

changes, ataxia, changes in urine composition, and hemolysis with or without anemia. Human reproductive effects by ingestion: spermatogenesis. An experimental teratogen. Other experimental reproductive effects. Questionable carcinogen with experimental neoplastigenic data. Human mutation data reported. When heated to decomposition it emits toxic fumes of NO_x.

NGE500   CAS: 59-87-0   HR: 3
NITROFURAZONE

mf: $C_6H_6N_4O_4$ mw: 198.16

PROP: Odorless, lemon-yellow crystals; bitter aftertaste. Darkens upon prolonged exposure to light. Mp: 236–240° (decomp). Sol in water; sltly sol in alc, propylene glycol; sol in alkaline solns; insol in ether.

SYNS: ALDOMYCIN □ ALFUCIN □ AMIFUR □ BABROCID □ BIOFUREA □ CHEMOFURAN □ COCAFURIN □ COXISTAT □ DERMOFURAL □ DYNAZONE □ ELDEZOL □ FEDACIN □ FLAVAZONE □ FRACINE □ FURACILLIN □ FURACINETTEN □ FURACOCCID □ FURACORT □ FURACYCLINE □ FURALDON □ FURAN-OFTENO □ FURAPLAST □ FURASEPTYL □ FURAZONE □ FURESOL □ FURFURIN □ FUVACILLIN □ HEMOFURAN □ IBIOFURAL □ MAMMEX □ MONOFURACIN □ NCI-C56064 □ NEFCO □ NF □ NIFUZON □ 5-NITROFURALDEHYDE SEMICARBAZIDE □ 6-NITROFURALDEHYDE SEMICARBAZIDE □ 5-NITRO-2-FURALDEHYDE SEMICARBAZONE □ 5-NITROFURAN-2-ALDEHYDE SEMICARBAZONE □ 5-NITRO-2-FURANCARBOX-ALDEHYDE SEMICARBAZONE □ 2((5-NITRO-2-FURANYL)METHYLENE)HYDRAZINECARBOXAMIDE □ 5-NITROFURFURAL SEMICARBAZONE □ (5-NITRO-2-FURFURYLIDENEAMINO)UREA □ NITROZONE □ NSC-2100 □ OTOFURAN □ SANFURAN □ SPRAY-DERMIS □ SPRAY-FORAL □ U-6421 □ USAF EA-4 □ VABROCID □ VADROCID □ VETERINARY NITROFURAZONE □ YATROCIN

CONSENSUS REPORTS: IARC Cancer Review: Group 3 IMEMDT 50,195,90; Animal Limited Evidence IMEMDT 50,195,90; Animal Inadequate Evidence IMEMDT 7,171,74; Human No Adequate Data IMEMDT 7,171,74; Human Inadequate Evidence IMEMDT 50,195,90. Reported in NTP Carcinogenesis Studies (feed); Clear Evidence: rat, mouse NTPTR* NTP-TR-337,88. Reported in EPA TSCA Inventory. EPA Genetic Toxicology Program.

SAFETY PROFILE: Poison by ingestion and intraperitoneal routes. Moderately toxic by subcutaneous route. Questionable carcinogen with experimental carcinogenic, neoplastigenic, tumorigenic, and teratogenic data. Experimental reproductive effects. A human sensitizer. Human mutation data reported. When heated to decomposition it emits toxic fumes of NO_x.

NGG500   CAS: 67-45-8   HR: 3
3-((5-NITROFURFURYLIDENE)AMINO)-2-OXAZOLIDONE

mf: $C_8H_7N_3O_5$ mw: 225.18

PROP: Yellow crystals. Mp: 275° (decomp).

SYNS: BIFURON □ CORIZIUM □ DIAFURON □ ENTEROTOXON □ FURAXONE □ FURAZOL □ FURAZOLIDON □ FURAZOLIDONE (USDA) □ FURAZON □ FURIDON □ FUROVAG □ FUROX □ FUROXAL □ FUROXANE □ FUROXONE SWINE MIX □ FUROZOLIDINE □ GIARDIL □ GIARLAM □ MEDARON □ NEFTIN □ NG-180 □ NICOLEN □ NIFULIDONE □ NIFURAN □ 3-(((5-NITRO-2-FURANYL)METHYLENE)AMINO)-2-OXAZOLIDINONE □ NITROFURAZOLIDONE □ NITROFURAZOLIDON-UM □ 3-(5'-NITROFURFURALAMINO)-2-OXAZOLIDONE □ N-(5-NITRO-2-FURFURYLIDENE)-3-AMINOOXAZ-OLIDINE-2-ONE □ N-(5-NITRO-2-FURFURYLIDENE)-3-AMINO-2-OXAZOLIDONE □ NITROFUROXON □ 3-((5-NITROFURYLIDENE)AMINO)-2-OXAZOLIDONE □ 5-NITRO-N-(2-OXO-3-OXAZOLIDINYL)-2-FURANMETH-ANIMINE □ PURADIN □ ROPTAZOL □ SCLAVENTER-OL □ TIKOFURAN □ TOPAZONE □ TRICHOFURON □ TRICOFURON □ USAF EA-1 □ VIOFURAGYN

CONSENSUS REPORTS: IARC Cancer Review: Group 3 IMEMDT 7,56,87; Animal Inadequate Evidence IMEMDT 31,141,83. Reported in EPA TSCA Inventory. EPA Genetic Toxicology Program.

SAFETY PROFILE: Poison by ingestion and intraperitoneal routes. Human systemic effects by ingestion: dyspnea, respiratory depression, and eosinophilia. Experimental reproductive effects. Human mutation data reported. Questionable carcinogen. When heated to decomposition it emits toxic fumes of NO_x.

NGI500   CAS: 712-68-5   HR: 3
2-(5-NITRO-2-FURYL)-5-AMINO-1,3,4-
   THIADIAZOLE
mf: $C_6H_4N_4O_3S$ mw: 212.20
SYNS: 2-AMINO-5-(5-NITRO-2-FURYL)-1,3,4-
THIADIAZOLE □ 5-AMINO-2-(5-NITRO-2-FURYL)-1,3,4-
THIADIAZOLE □ FURIDIAZINE □ 5-(5-NITRO-2-
FURANYL)-1,3,4-THIADIAZOL-2-AMINE □ 5-(5-NITRO-2-
FURYL)-2-AMINO-1,3,4-THIADIAZOLE
CONSENSUS REPORTS: IARC Cancer
Review: Group 2B IMEMDT 7,56,87;
Animal Limited Evidence IMEMDT
7,143,74.
SAFETY PROFILE: Suspected carcinogen
with experimental carcinogenic and
neoplastigenic data. Experimental
reproductive effects. When heated to
decomposition it emits very toxic fumes of
NO_x and SO_x. See also NITRO
COMPOUNDS OF AROMATIC
HYDROCARBONS.

NGI800   CAS: 75198-31-1   HR: 3
3-(5-NITRO-2-FURYL)-IMIDAZO(1,2-
   a)PYRIDINE
mf: $C_{11}H_7N_3O_3$ mw: 229.21
SYN: NFIP
SAFETY PROFILE: Suspected carcinogen
with experimental carcinogenic data. When
heated to decomposition it emits toxic
fumes of NO_x.

NGM500   CAS: 24554-26-5   HR: 3
N-(4-(5-NITRO-2-FURYL)-2-THIAZOL-
   YL)FORMAMIDE
mf: $C_8H_5N_3O_4S$ mw: 239.22
SYNS: FANFT □ 2-FORMYLAMINO-4-(5-NITRO-2-
FURYL)THIAZOLE □ N-(4-(5-NITRO-2-FURYL)-2-
THIAZOLYL)FORMAMID (GERMAN)
CONSENSUS REPORTS: EPA Genetic
Toxicology Program.
SAFETY PROFILE: Suspected carcinogen
with experimental carcinogenic, tumorigenic
data. Mutation data reported. Experimental
reproductive effects. When heated to
decomposition it emits very toxic fumes of
SO_x and NO_x.

NGN500   CAS: 42011-48-3   HR: 3
N-(4-(5-NITRO-2-FURYL)-2-THIAZOLYL)-

2,2,2-TRIFLUOROACETAMIDE
mf: $C_9H_4F_3N_3O_4S$ mw: 307.22
SYNS: 2-(2,2,2-TRIFLUOROACETAMIDO)-4-(5-NITRO-
2-FURYL)THIAZOLE □ 2,2,2-TRIFLUORO-N-(4-(5-
NITRO-2-FURYL)-2-THIAZOLYL)ACETAMIDE
SAFETY PROFILE: Suspected carcinogen
with experimental carcinogenic data.
Mutation data reported. When heated to
decomposition it emits very toxic fumes of
F^-, NO_x, and SO_x.

NGP500   CAS: 7727-37-9   HR: 1
NITROGEN
DOT: UN 1066/UN 1977
mf: N_2 mw: 28.02
PROP: Colorless, odorless, very stable,
nonflammable gas; generally unreactive;
colorless liquid or cubic crystals at low temp.
Mp: $-210.0°$, d: 1.2506 g/L @ 0°, d (liquid):
0.808 g/cm^3 @ $-195.8°$. Condenses to a
liquid. Sltly sol in water; sol in liquid
ammonia, alc.
SYNS: NITROGEN, compressed (UN 1066) (DOT) □
NITROGEN, refrigerated liquid (cryogenic liquid) (UN 1977)
(DOT) □ NITROGEN GAS
CONSENSUS REPORTS: Reported in EPA
TSCA Inventory.
DOT CLASSIFICATION: 2.2; Label:
Nonflammable Gas
SAFETY PROFILE: Low toxicity. In high
concentrations it is a simple asphyxiant.
The release of nitrogen from solution in the
blood, with formation of small bubbles, is
the cause of most of the symptoms and
changes found in compressed air illness
(caisson disease). It is a narcotic at high
concentration and high pressure. Both the
narcotic effects and the bends are hazards of
compressed air atmospheres such as found
in underwater diving. Nonflammable gas.
Can react violently with lithium,
neodymium, titanium under the proper
conditions. See also ARGON.

NGQ500   CAS: 10025-85-1   HR: 3
NITROGEN CHLORIDE
mf: Cl_3N mw: 120.37
PROP: Very unstable, volatile, yellowish oil
or rhombic crystals; pungent odor. Mp:

<−40°, explodes above 60°, bp: <71°, d: 1.653, vap press: 150 mm @ 20°. Sol in CCl_4, $CHCl_3$.

SYNS: AGENE □ CHLORINE NITRIDE □ NITROGEN TRICHLORIDE □ NITROGEN TRICHLORIDE (DOT) □ TRICHLORAMINE □ TRICHLORINE NITRIDE

DOT CLASSIFICATION: Forbidden

SAFETY PROFILE: Moderately toxic by inhalation. An irritant to the eyes, skin, mucous membranes, and a systemic central nervous system irritant. An explosive sensitive to impact, light, and ultrasound. The solid explodes on melting. The liquid explodes above 60°C. Concentrated solutions are also explosive. Explosive decomposition is initiated by contact with: concentrated ammonia, arsenic, dinitrogen tetraoxide, hydrogen sulfide, hydrogen trisulfide, nitrogen oxide, organic matter, ozone, phosphine, phosphorus, potassium cyanide, potassium hydroxide solutions, selenium, hydrogen chloride, hydrogen fluoride, hydrogen bromide, hydrogen iodide. Mixtures with chlorine + hydrogen are potentially explosive. Upon decomposition it emits toxic fumes of Cl^- and NO_x. See also CHLORIDES.

NGR500   CAS: 10102-44-0   HR: 3
NITROGEN DIOXIDE

mf: NO_2 mw: 46.01

PROP: Brown gas or colorless solid to yellow liquid; irritating odor. Reacts with H_2O giving HNO_3 + NO. Mp: −9.3° (yellow liquid), bp: 21° (red-brown gas with decomp), d: 1.491 @ 0°, vap press: 400 mm @ 80°. Liquid below 21.15°. Sol in concentrated sulfuric acid, nitric acid. Corrosive to steel when wet. IDLH 20 ppm.

SYNS: AZOTE (FRENCH) □ AZOTO (ITALIAN) □ NITRITO □ NITROGEN PEROXIDE □ RCRA WASTE NUMBER P078 □ STICKSTOFFDIOXID (GERMAN) □ STIKSTOFDIOXYDE (DUTCH)

CONSENSUS REPORTS: EPA Extremely Hazardous Substances List. Reported in EPA TSCA Inventory. EPA Genetic Toxicology Program.

OSHA PEL: STEL 1 ppm

ACGIH TLV: TWA 3 ppm; STEL 5 ppm; Not Classifiable as a Human Carcinogen

DFG MAK: 5 ppm (9 mg/m³)

NIOSH REL: CL (Oxides of Nitrogen) 1 ppm/15M

SAFETY PROFILE: Experimental poison by inhalation. Moderately toxic to humans by inhalation. An experimental teratogen. Other experimental reproductive effects. Human systemic effects by inhalation: pulmonary vascular resistance changes, cough, dyspnea, and other pulmonary changes. Mutation data reported. Violent reaction with cyclohexane, F_2, formaldehyde, alcohols, nitrobenzene, petroleum, toluene. When heated to decomposition it emits toxic fumes of NO_x. See also NITRIC OXIDE.

NGS500   CAS: 13847-65-9   HR: 3
NITROGEN FLUORIDE OXIDE

mf: F_3NO mw: 87.01

PROP: Colorless gas. Strong oxidizing agent. Stable in glass. Resistant to hydrolysis. Mp: −160°, bp: −87°.

SYNS: AMOX □ TRIFLUOROAMINE OXIDE

OSHA PEL: TWA 2.5 mg(F)/m³

ACGIH TLV: TWA 2.5 mg(F)/m³; BEI: 3 mg/g creatinine of fluorides in urine prior to shift; 10 mg/g creatinine of fluorides in urine at end of shift.

NIOSH REL: (Inorganic Fluorides) TWA 2.5 mg(F)/m³

SAFETY PROFILE: Poison by inhalation and intraperitoneal routes. A skin, eye, and mucous membrane irritant. When heated to decomposition it emits very toxic fumes of F^- and NO_x. See also FLUORIDES.

NGT500   CAS: 63907-41-5   HR: 3
NITROGEN MONOXIDE, mixed with
   NITROGEN TETROXIDE

DOT: UN 1975

PROP: Containing up to 33.2% by weight nitric oxide (FEREAC 41,15972,76).

SYNS: AZOTU TLENKI (POLISH) □ NITRIC OXIDE and NITROGEN TETROXIDE MIXTURES □ NITROGEN TETROXIDE-NITRIC OXIDE MIXTURE

DOT CLASSIFICATION: 2.3; Label: Poison Gas, Oxidizer

SAFETY PROFILE: A poison. Moderately toxic by inhalation. When heated to decomposition it emits toxic fumes of NO_x. See also components as listed.

NGU000   CAS: 10024-97-2   HR: 2
NITROGEN OXIDE

DOT: UN 1070/UN 2201

mf: N_2O mw: 44.02

PROP: Colorless nonflammable gas, liquid, or cubic crystals; slt sweet odor. Mp: $-90.8°$, bp: $-88.49°$, d: 1.977 g/L (liquid 1.226 @ $-89°$). Sltly sol in H_2O; sol in Et_2O; freely sol in $CHCl_3$ and EtOH.

SYNS: DINITROGEN MONOXIDE □ FACTITIOUS AIR □ HYPONITROUS ACID ANHYDRIDE □ LAUGHING GAS □ NITROUS OXIDE (DOT) □ NITROUS OXIDE, compressed (UN 1070) (DOT) □ NITROUS OXIDE, refrigerated liquid (UN 2201) (DOT)

CONSENSUS REPORTS: Reported in EPA TSCA Inventory. EPA Genetic Toxicology Program.

OSHA PEL: OSHA PEL (Shipyard): Simple asphyxiant-inert gas and vapor

ACGIH TLV: 50 ppm; Not Classifiable as a Human Carcinogen

DFG MAK: 100 ppm (180 mg/m³)

NIOSH REL: (Waste Anesthetic Gases and Vapors) TWA 25 ppm

DOT CLASSIFICATION: 2.2; Label: Nonflammable Gas

SAFETY PROFILE: Moderately toxic by inhalation. Human systemic effects by inhalation: general anesthetic, decreased pulse rate without blood pressure fall, and body temperature decrease. An experimental teratogen. Experimental reproductive effects. Mutation data reported. An asphyxiant. Does not burn but is flammable by chemical reaction and supports combustion. Moderate explosion hazard; it can form an explosive mixture with air. Violent reaction with Al, B, hydrazine, LiH, LiC_6H_5, PH_3, Na, tungsten carbide. Also self-explodes at high temperatures.

NGU500   CAS: 10544-72-6   HR: 3
NITROGEN TETROXIDE

DOT: UN 1067

mf: N_2O_4 mw: 92.02

PROP: Brown paramagnetic gas or liquid; white diamagnetic solid. Stable in glass vessels. D: 1.493 @ $20°$, mp: $-11.2°$, bp: $21.15°$. Nitrogen tetroxide is a dimer of nitrogen dioxide (AIHAAP 23,457,62).

SYNS: DINITROGEN TETROXIDE □ DINITROGEN TETROXIDE, liquefied (DOT) □ NITROGEN DIOXIDE, DI-

CONSENSUS REPORTS: Reported in EPA TSCA Inventory.

DOT CLASSIFICATION: 2.3; Label: Poison Gas, Oxidizer

SAFETY PROFILE: A poison. Moderately toxic by inhalation. When heated to decomposition it emits toxic fumes of NO_x. See also NITROGEN MONOXIDE.

NGW000   CAS: 7783-54-2   HR: 3
NITROGEN TRIFLUORIDE

DOT: UN 2451

mf: F_3N mw: 71.01

PROP: Colorless, odorous gas; odor of mold. Mp: $-208.5°$, bp: $-129°$, d (liquid): 1.537 @ $-129°$; d: (liquid @ bp) 1.885. Sltly sol in H_2O. IDLH 1000 ppm.

SYN: NITROGEN FLUORIDE

CONSENSUS REPORTS: Reported in EPA TSCA Inventory.

OSHA PEL: TWA 10 ppm

ACGIH TLV: TWA 10 ppm; BEI: 3 mg/g creatinine of fluorides in urine prior to shift; 10 mg/g creatinine of fluorides in urine at end of shift.

NIOSH REL: (Inorganic Fluorides) TWA 2.5 mg(F)/m³

DOT CLASSIFICATION: 2.2; Label: Nonflammable Gas, Oxidizer

SAFETY PROFILE: A poison. Mildly toxic by inhalation. Prolonged absorption may cause mottling of teeth, skeletal changes. Severe explosion hazard by chemical reaction with reducing agents, particularly when under pressure. A very dangerous fire hazard; a very powerful oxidizer; otherwise inert at normal temperatures and pressures.

N

Reacts violently when ignited with H_2. When pure (dry) it does not attack glass or mercury at normal temperatures. Can react violently with NH_3, CO, diborane, H_2, H_2S, CH_4, tetrafluorohydrazine. Can react vigorously with reducing materials. Particularly hazardous under pressure. Incompatible with charcoal, hydrogen-containing compounds, tetrafluorohydrazine. When heated to decomposition it emits highly toxic fumes of F^-. See also FLUORIDES.

NGW500   CAS: 13444-85-4   HR: 3
NITROGEN TRIIODIDE

mf: NI_3 mw: 394.7

PROP: Black crystals or brown-black powder. Mp: explodes, bp: subl in vacuum. Insol in H_2O, EtOH and Et_2O.

SYN: NITROGEN IODIDE

ACGIH TLV: Proposed: (inhalable fraction) 0.1 mg/m³; Not Classifiable as a Human Carcinogen)

DOT CLASSIFICATION: Forbidden

SAFETY PROFILE: A severe explosion hazard when shocked, exposed to heat or flame, or by spontaneous chemical reaction. It has no known uses as an explosive because it is far too sensitive in the dry state to store or handle safely. If this material must be worked with, it should be kept wet. A convenient way of keeping it wet is with ether; when it is needed in the dry state, it simply has to be taken out into the open and the ether will evaporate, leaving it perfectly dry. When dry, it will explode when given the slightest touch, vibration, or rise in temperature. Even a puff of air directed into it can cause it to detonate. It is a high explosive and is very violent. Incompatible with O_3. H_2S, Cl_2, Br_2, acids. See also IODIDES.

NGY000   CAS: 55-63-0   HR: 3
NITROGLYCERIN

DOT: UN 0143/UN 0144/UN 1204/UN 3064

mf: $C_3H_5N_3O_9$ mw: 227.11

PROP: Colorless to yellow liquid; sweet burning taste; sensitive to shock. Mp: 13°, bp: explodes @ 218°, d: 1.599 @ 15°/15°, vap press: 1 mm @ 127°, vap d: 7.84, autoign temp: 518°F, decomp @ 50–60°, fp: 13°. Volatile @ 100°. Misc with ether, acetone, glacial acetic acid, ethyl acetate, benzene, nitrobenzene, pyridine, chloroform, ethylene bromide, dichloroethylene; sltly sol in pet ether, glycerin. Misc in most org solvs; prac insol in H_2O. IDLH 75 mg/m³.

SYNS: ANGIBID □ ANGININE □ ANGIOLINGUAL □ ANGORIN □ BLASTING GELATIN (DOT) □ BLASTING OIL □ CARDAMIST □ GILUCOR NITRO □ GLONOIN □ GLYCERINTRINITRATE (CZECH) □ GLYCEROL, NITRIC ACID TRIESTER □ GLYCEROLTRINITRAAT (DUTCH) □ GLYCEROL TRINITRATE □ GLYCEROL (TRINITRATE de) (FRENCH) □ GLYCERYL NITRATE □ GLYCERYL TRINITRATE □ GTN □ KLAVI KORDAL □ LENITRAL □ MYOCON □ MYOGLYCERIN □ NG □ NIGLYCON □ NIONG □ NITORA □ NITRIC ACID TRIESTER OF GLYCEROL □ NITRIN □ NITRINE □ NITRINE-TDC □ NITRO-DUR □ NITROGLICERINA (ITALIAN) □ NITROGLICERYNA (POLISH) □ NITROGLYCERIN, desensitized, not <40% non-volatile water insoluble phlegmatizer (UN 0143) (DOT) □ NITROGLYCER-IN, liquid, not desensitized (DOT) □ NITROGLYCERIN, solution in alcohol, with >1% but not >5% nitroglycerin (UN 3064) (DOT) □ NITROGLYCERIN, solution in alcohol, with >1% but not >10% nitroglycerin (UN 0144) (DOT) □ NITROGLYCERIN, solution in alcohol, with not >1% nitroglycerin (UN 1204) (DOT) □ NITROGLYCERINE □ NITROGLYCERIN, SPIRITS OF □ NITROGLYCEROL □ NITROGLYN □ NITROL □ NITROLAN □ NITRO-LENT □ NITROLETTEN □ NITROLINGUAL □ NITROLOWE □ NITROMEL □ NITRONET □ NITRONG □ NITRORECTAL □ NITROSTABILIN □ NITROSTAT □ NITROZELL RETARD □ NK-843 □ NTG □ NYSCONITRINE □ PERGLOTTAL □ PROPANETRIOL TRINITRATE □ 1,2,3-PROPANETRIOL, TRINITRATE □ 1,2,3-PROPANETRIYL NITRATE □ RCRA WASTE NUMBER P081 □ SK-106N □ S.N.G □ SOLUTION GLYCERYL TRINITRATE □ SOUP □ SPIRIT OF GLONOIN □ SPIRIT OF GLYCERYL TRINITRATE □ SPIRIT OF TRINITROGLYCERIN □ TEMPONITRIN □ TNG □ TRINALGON □ TRINITRIN □ TRINITRO-GLYCERIN □ TRINITROGLYCEROL □ TRINITROL □ VASOGLYN

CONSENSUS REPORTS: Reported in EPA TSCA Inventory. Community Right-To-Know List.

OSHA PEL: STEL 0.1 mg/m³ (skin)

ACGIH TLV: TWA 0.05 ppm (skin)

DFG MAK: 0.05 ppm (0.47 mg/m³) (skin)
NIOSH REL: CL (Nitroglycerin or EGDN)
0.1 mg/m³/20M
DOT CLASSIFICATION: EXPLOSIVE
1.1D; Label: EXPLOSIVE 1.1D, Poison
(UN 0143); DOT Class: Forbidden (not
desensitized); DOT Class: 3; Label:
Flammable Liquid (UN 3064, UN 1204);
DOT Class: EXPLOSIVE 1.1D; Label:
EXPLOSIVE 1.1D (UN 0144)
SAFETY PROFILE: Human poison by an
unspecified route. Poison experimentally by
ingestion, intraperitoneal, subcutaneous, and
intravenous routes. An experimental
teratogen. Other experimental reproductive
effects. A skin irritant. Questionable
carcinogen with experimental tumorigenic
data. Mutation data reported. It can cause
respiratory difficulties and death due to
respiratory paralysis by ingestion. The acute
symptoms of nitroglycerin poisoning are
headaches, nausea, vomiting, abdominal
cramps, convulsions, methemoglobinemia,
circulatory collapse and reduced blood
pressure, excitement, vertigo, fainting,
respiratory rales, and cyanosis. Toxic effects
may occur by ingestion, inhalation of dust,
or absorption through intact skin. Human
systemic effects by intravenous route:
encephalitis, miosis, corneal damage. Used
as a vasodilator and as an explosive.

A very dangerous fire hazard when
exposed to heat, flame, or by spontaneous
chemical reaction. A severe explosion
hazard when shocked or exposed to O_3,
heat, or flame. Nitroglycerin is a powerful
explosive, very sensitive to mechanical
shock, heat, or UV radiation. Small
quantities of it can readily be detonated by a
hammer blow on a hard surface, particularly
when it has been absorbed in filter paper. It
explodes when heated to 215°C. Frozen
nitroglycerin is somewhat less sensitive than
the liquid. However, a half-thawed or
partially thawed mixture is more sensitive
than either one. When heated to
decomposition it emits toxic fumes of NO_x.

See also EXPLOSIVES, HIGH; and
DYNAMITE.

**NHA500   CAS: 556-88-7   HR: 3
α-NITROGUANIDINE**
DOT: UN 0282/UN 1336
mf: $CH_4N_4O_2$ mw: 104.09
PROP: Two forms: (more stable, þ&☐ -
form) long, thin, flat needles or prisms from
H_2O; fan-like clusters of small, thin,
elongated plates from H_2O. Mp: 246°,
decomp: 225–250°. Sltly sol in alc,
concentrated acids, cold solns of alkalies; sol
in water; very sltly sol in ether.
SYNS: GUANIDINE, NITRO- ☐ NITROGUANIDINE ☐
2-NITROGUANIDINE ☐ NITROGUANIDINE, dry or
wetted with <20% water, by weight (UN 0282) (DOT) ☐
NITROGUANIDINE, wetted (UN 1336) (DOT) ☐ PICRITE
(the explosive) ☐ PICRITE, dry or wetted with <20% water, by
weight (UN 0282) (DOT) ☐ PICRITE, wetted with not <20%
water, by weight (UN 1336) (DOT)
CONSENSUS REPORTS: Reported in EPA
TSCA Inventory.
DOT CLASSIFICATION: EXPLOSIVE
1.1D; Label: EXPLOSIVE 1.1D (UN 0282);
DOT Class: 4.1; Label: Flammable Solid
(UN 1336)
SAFETY PROFILE: Poison by
intraperitoneal route. Moderately toxic by
ingestion. Mutation data reported. A very
dangerous fire hazard when exposed to
heat, flame, or by chemical reaction with
oxidizers. A severe explosion hazard when
shocked or exposed to heat or flame. It is
about as powerful as TNT. It is normally
mixed with colloided nitrocellulose or
ammonium nitrate and paraffin wax. Can
react vigorously with oxidizing materials and
the derivatives can be explosive. The
mercury and silver salts and other
derivatives are much more impact-sensitive.
When heated to decomposition it emits
highly toxic fumes of NO_x. See also NITRO
COMPOUNDS.

**NHE000   CAS: 4812-22-0   HR: 3
3-NITRO-3-HEXENE**
mf: $C_6H_{11}NO_2$ mw: 129.18

SAFETY PROFILE: Poison by intraperitoneal route. Moderately toxic by ingestion and skin contact. A severe eye and skin irritant. Questionable carcinogen with experimental neoplastigenic data. When heated to decomposition it emits toxic fumes of NO_x. See also NITRO COMPOUNDS.

NHE600   CAS: 102107-61-9 HR: 3
3-NITRO-4-HYDROXYPHENYLARSEN-
    OUS ACID
mf: $C_6H_4AsNO_4$ mw: 229.03
SYNS: 4-ARSENOSO-2-NITROPHENOL □ PHENOL, 4-ARSENOSO-2-NITRO-
OSHA PEL: TWA 0.5 mg(As)/m³
ACGIH TLV: BEI: 35 μ (As)/L inorganic arsenic and methylated metabolites in urine
SAFETY PROFILE: Poison by intravenous route. When heated to decomposition it emits toxic fumes of NO_x and As.

NHI500   CAS: 7046-61-9   HR: 3
NITROIMINODIETHYLENEDIISOCYANIC
    ACID
mf: $C_6H_8N_4O_4$ mw: 200.18
SYN: 3-NITRO-3-AZAPENTANE-1,5-DIISOCYANATE
CONSENSUS REPORTS: Reported in EPA TSCA Inventory.
NIOSH REL: (Diisocyanates) TWA 0.005 ppm; CL 0.02 ppm/10M
SAFETY PROFILE: Poison by intravenous route. A sensitizer. When heated to decomposition it emits toxic fumes of NO_x.

NHK800   CAS: 24458-48-8   HR: 2
NITROL
mf: $C_{10}H_{13}N_3O_3$ mw: 223.26
SYNS: BENZENAMINE, N-(2-METHYL-2-NITRO-PROPYL)-p-NITROSO-(9CI) □ CP 25017 □ N-(2-METHYL-2-NITROPROPYL)-p-NITROSOANILINE □ N-(2-METHYL-2-NITROPROPYL)-4-NITROSOBENZAMINE □ NITROL (PROMOTER) □ N-(p-NITROSOANILINO-METHYL)-2-NITROPROPANE
CONSENSUS REPORTS: Reported in EPA TSCA Inventory.
SAFETY PROFILE: Moderately toxic by ingestion. Questionable carcinogen with experimental carcinogenic data. When

heated to decomposition it emits toxic fumes of NO_x.

NHK900   CAS: 133-58-4   HR: 3
NITROMERSOL
mf: $C_7H_5HgNO_3$ mw: 351.72
PROP: Yellow powder or granules, odorless and tasteless. Sol in boiling glacial acetic acid, insol in water, ether, acetone, or alc.
SYNS: MERCURY, (2-METHYL-5-NITROPHENOLATO(2-)-C^6,O^1)-(9CI) □ METAPHEN □ 5-METHYL-2-NITRO-7-OXA-8-MERCURABICYCLO(4.2.0)-OCTA-1,3,5-TRIENE □ 4-NITRO-5-HYDROXYMERCURI-ORTHOCRESOL □ NITROMERSOL SOLUTION □ 7-OXA-8-MERCURABICYCLO(4.2.0)OCTA-1,3,5-TRIENE, 5-METHYL-2-NITRO-
ACGIH TLV: TWA 0.1 mg(Hg)/m³ (skin); BEI: 35 μg/g creatinine total inorganic mercury in urine preshift; 15 μg/g creatinine total inorganic mercury in blood at end of shift at end of workweek.
DFG MAK: Confirmed Animal Carcinogen with Unknown Relevance to Humans
NIOSH REL: (Mercury, Organo) TWA 0.01 mg/m³; STEL 0.03 mg/m³ (skin)
SAFETY PROFILE: Poison by intraperitoneal and intravenous routes. When heated to decomposition it emits toxic fumes of NO_x and Hg.

NHM500   CAS: 75-52-5   HR: 3
NITROMETHANE
DOT: UN 1261
mf: CH_3NO_2 mw: 61.05
PROP: A poisonous oily liquid; moderate to strong disagreeable odor. Bp: 101°, lel: 7.3%, fp: −29°, flash p: 95°F (CC), d: 1.1322 @ 25°/4°, autoign temp: 785°F, vap press: 27.8 mm @ 20°, vap d: 2.11. Sol in EtOH, Et_2O, DMF; sltly sol in H_2O. IDLH 750 ppm.
SYNS: NITROCARBOL □ NITROMETAN (POLISH)
CONSENSUS REPORTS: Reported in EPA TSCA Inventory.
OSHA PEL: TWA 100 ppm
ACGIH TLV: TWA 20 ppm; Confirmed Animal Carcinogen with Unknown Relevance to Humans
DFG MAK: 100 ppm (250 mg/m³)

DOT CLASSIFICATION: 3; Label: Flammable Liquid

SAFETY PROFILE: Poison by ingestion and intraperitoneal routes. Moderately toxic by intravenous route. Mildly toxic by inhalation. In humans it may cause anorexia, nausea, vomiting, diarrhea, kidney injury, and liver damage.

A very dangerous fire hazard when exposed to heat, oxidizers, or flame. May explode by detonation, heat, or shock. Its sensitivity is increased when mixed with acids, bases, acetone, aluminum powder, ammonium salts + organic solvents, bis(2-aminoethyl)amine, 1,2-diaminoethane + N,2,4,6-tetranitro-N-methyl aniline, haloforms (e.g., chloroform, bromoform), hydrazine + methanol. Ignites when mixed with alkyl metal halides (e.g., diethylaluminum bromide, dimethylaluminum bromide, ethylaluminum bromide iodide, methyl zinc iodide, methylaluminum diiodide). Can react violently with $AlCl_3$ + organic matter, $Ca(OH)_2$, m-methyl aniline, $Ca(OCl)_2$, hexamethylbenzene, hydrocarbons, inorganic bases, hydroxides, organic amines, KOH, formaldehyde, nitric acid, metal oxides, 1,2-diaminomethane, lithium perchlorate, sodium hydride. Reacts with aqueous silver nitrate to form the explosive silver fulminate. When heated to decomposition it emits toxic fumes of NO_x. See also NITROALKANES.

**NHN500   CAS: 598-57-2   HR: 2
N-NITROMETHYLAMINE**

mf: $CH_4N_2O_2$ mw: 76.07

PROP: Plates from Et_2O. Very sol in H_2O, EtOH, C_6H_6, and $CHCl_3$; less sol in Et_2O; spar sol in pet ether.

SYNS: METHANAMINE, N-NITRO-(9CI) □ METHYL-NITRAMINE □ METHYL NITRAMINE (dry) (DOT) □ NITRAMINE, METHYL- □ N-NITROMETHANAMINE

DOT CLASSIFICATION: Forbidden

SAFETY PROFILE: Moderately toxic by intraperitoneal route. Questionable carcinogen with carcinogenic data. Mutation data reported. Explodes on contact with concentrated sulfuric acid. When heated to decomposition it emits toxic fumes of NO_x. See also NITRO COMPOUNDS and AMINES.

**NHP990   CAS: 27254-36-0   HR: 3
NITRONAPHTHALENE**

DOT: UN 2538

mf: $C_{10}H_7NO_2$ mw: 173.18

PROP: Yellow crystals. Mp: 59°–61°, bp: (sublimes). Insol in water.

SYNS: MONONITRONAPHTHALENE □ NAPHTHALENE, MONONITRO- □ NAPHTHALENE, NITRO- □ NITRONAPHTHALENE □ NITRONAPHTHALENE (DOT)

DOT CLASSIFICATION: 4.1; Label: Flammable Solid

SAFETY PROFILE: Poison by ingestion. Moderately toxic by intraperitoneal route. A skin, eye, and mucous membrane irritant. A flammable solid when exposed to heat or flame. To fight fire, use CO_2, dry chemical, or water spray. Explosive reaction with nitric acid + sulfuric acid above 60°C. Forms a sensitive explosive mixture with tetranitromethane. When heated to decomposition it emits toxic fumes of NO_x. See also 1-NITRONAPHTHALENE, 2-NITRONAPHTHALENE, and NITRO COMPOUNDS OF AROMATIC HYDROCARBONS.

**NHQ000   CAS: 86-57-7   HR: 3
1-NITRONAPHTHALENE**

mf: $C_{10}H_7NO_2$ mw: 173.18

PROP: Yellow crystals. Bp: 304°, flash p: 327°F (CC), d: 1.331 @ 4°/4°, vap d: 5.96, mp: 59–61°. Insol in water; sol in CS_2, alc, chloroform, and ether.

SYNS: NCI-C01956 □ α-NITRONAPHTHALENE

CONSENSUS REPORTS: IARC Cancer Review: Group 3 IMEMDT 46,291,89; Animal Inadequate Evidence IMEMDT 46,291,89; Human No Adequate Data IMEMDT 46,291,89. NCI Carcinogenesis Bioassay (feed); No Evidence: mouse, rat NCITR* NCI-CG-TR-64,78. Reported in EPA TSCA Inventory.

N

DFG MAK: Confirmed Animal Carcinogen with Unknown Relevance to Humans
SAFETY PROFILE: Poison by intraperitoneal route. Mutation data reported. A skin, eye, and mucous membrane irritant. Flammable solid and combustible liquid when exposed to heat or flame. To fight fire, use CO_2, dry chemical, or water spray. Explosive reaction with nitric acid + sulfuric acid above 60°C. Forms a sensitive explosive mixture with tetranitromethane. When heated to decomposition it emits toxic fumes of NO_x. See also 2-NITRONAPHTHALENE and NITRO COMPOUNDS OF AROMATIC HYDROCARBONS.

NHQ500   CAS: 581-89-5   HR: 3
2-NITRONAPHTHALENE
mf: $C_{10}H_7NO_2$ mw: 173.18
PROP: Colorless solution when dissolved in ethanol. Mp: 79°, bp: 165° @ 15 mm. Insol in water; very sol in alc and ether.
SYN: β-NITRONAPHTHALENE
CONSENSUS REPORTS: IARC Cancer Review: Group 3 IMEMDT 46,303,89; Animal Inadequate Evidence IMEMDT 46,303,89; Human No Adequate Data IMEMDT 46,303,89. EPA Genetic Toxicology Program.
DFG MAK: DFG TRK: Animal Carcinogen, Suspected Human Carcinogen
NIOSH REL: (2-Nitronaphthalene) lowest feasible conc
SAFETY PROFILE: Confirmed carcinogen with experimental tumorigenic data. Moderately toxic by ingestion and intraperitoneal routes. Mutation data reported. A skin and lung irritant. For explosion and disaster hazards, see NITRATES. Combustible when exposed to heat or flame. When heated to decomposition it emits toxic fumes of NO_x. See also 1-NITRONAPHTHALENE.

NHS500   CAS: 13826-86-3   HR: 3
NITRONIUM TETRAFLUORO-
    BORATE(1−)

mf: BF_4NO_2 mw: 132.82
PROP: Colorless crystals; easily hydrolyized. Spar sol in MeCN and $MeNO_2$.
CONSENSUS REPORTS: Reported in EPA TSCA Inventory.
OSHA PEL: TWA 2.5 mg(F)/m^3
ACGIH TLV: TWA 2.5 mg(F)/m^3; BEI: 3 mg/g creatinine of fluorides in urine prior to shift; 10 mg/g creatinine of fluorides in urine at end of shift.
NIOSH REL: (Inorganic Fluorides) TWA 2.5 mg(F)/m^3
SAFETY PROFILE: Poison by intravenous route. A powerful nitrating agent when dissolved in sulfolane. Incompatible with tetrahydrothiophene-1,1-dioxide, organics, α-propanol, methane, ethane @ −78° or higher. When heated to decomposition it emits very toxic fumes of F^- and NO_x. See also BORON COMPOUNDS.

NIE600   CAS: 554-84-7   HR: 3
3-NITROPHENOL
DOT: UN 2648
mf: $C_6H_5NO_3$ mw: 139.12
PROP: Monoclinic crystals from HCl. Mp: 97°, bp: 194° @ 70 mm, d: 1.485 @ 20°/4°. Decomposes when distilled at ordinary pressure. Sol in hot water and dil acids, caustic solns; insol in pet ether.
SYNS: m-HYDROXYNITROBENZENE □ 3-HYDROXY-NITROBENZENE □ m-NITROFENOL □ m-NITRO-PHENOL (DOT)
CONSENSUS REPORTS: EPA Genetic Toxicology Program. Reported in EPA TSCA Inventory.
DOT CLASSIFICATION: 6.1; Label: KEEP AWAY FROM FOOD
SAFETY PROFILE: Poison by ingestion, subcutaneous, and intraperitoneal routes. Moderately toxic by skin contact. A skin and severe eye irritant. When heated to decomposition it emits toxic fumes of NO_x. See also other nitrophenol entries.

NIF000   CAS: 100-02-7   HR: 3
4-NITROPHENOL
DOT: UN 1663

mf: $C_6H_5NO_3$ mw: 139.12
PROP: Colorless to sltly yellow; odorless crystals, sweet then burning taste. D: 1.270 @ 120°/4°, mp: 113–114° (subl). Sltly sol in cold water; very sol in alc, chloroform, ether; sol in alkali solns, hydroxides, and carbonates.
SYNS: 4-HYDROXYNITROBENZENE □ NCI-C55992 □ 4-NITROFENOL (DUTCH) □ p-NITROPHENOL (DOT) □ PARANITROFENOL (DUTCH) □ PARANITROFENOLO (ITALIAN) □ PARANITROPHENOL (FRENCH, GERMAN) □ RCRA WASTE NUMBER U170
CONSENSUS REPORTS: EPA Genetic Toxicology Program. Community Right-To-Know List. Reported in EPA TSCA Inventory.
DOT CLASSIFICATION: 6.1; Label: KEEP AWAY FROM FOOD
SAFETY PROFILE: Poison by ingestion, subcutaneous, intraperitoneal, intravenous, and intramuscular routes. Moderately toxic by skin contact. Human mutation data reported. Its exothermic decomposition causes a dangerously fast pressure increase. Mixtures with diethyl phosphite may explode when heated. When heated to decomposition it emits toxic fumes of NO_x. See also other nitrophenol entries and NITRO COMPOUNDS OF AROMATIC HYDROCARBONS.

NIF010   CAS: 88-75-5   HR: 3
o-NITROPHENOL
DOT: UN 1663
mf: $C_6H_5NO_3$ mw: 139.12
PROP: Light-yellow crystals, needles, or prisms from EtOH or Et_2O; aromatic odor. Mp: 45°, bp: 214.5°, d: 1.495 @ 20°, vap press: 1 mm @ 49.3°. Sol in water; very sol in alc, ether, benzene, CS; volatile with steam.
SYNS: 2-HYDROXYNITROBENZENE □ o-NITROFEN-OL □ 2-NITROPHENOL
CONSENSUS REPORTS: EPA Genetic Toxicology Program. Reported in EPA TSCA Inventory. Community Right-To-Know List.
DOT CLASSIFICATION: 6.1; Label: KEEP AWAY FROM FOOD

SAFETY PROFILE: Poison by ingestion, subcutaneous, intravenous, and intraperitoneal routes. Moderately toxic by intramuscular route. Can cause liver and kidney damage. The liquid phenol reacts violently with KOH. Product of reaction with chlorosulfuric acid decomposes violently at 24°C. When heated to decomposition it emits toxic fumes of NO_x. See also NITRO COMPOUNDS OF AROMATIC HYDROCARBONS.

NII200   CAS: 3644-32-4   HR: 3
p-NITROPHENOXYTRIBUTYLTIN
mf: $C_{18}H_{31}NO_3Sn$ mw: 428.19
SYN: STANNANE, (p-NITROPHENOXY)TRIBUTYL-
OSHA PEL: TWA 0.1 mg(Sn)/m^3 (skin)
ACGIH TLV: TWA 0.1 mg(Sn)/m^3; STEL 0.2 mg/m^3 (skin)
NIOSH REL: (Organotin Compound) 10H TWA 0.1 mg(Sn)/m^3
SAFETY PROFILE: Poison by intravenous route. When heated to decomposition it emits toxic fumes of NO_x and Sn.

NIJ500   CAS: 98-72-6   HR: 3
4-NITROPHENYLARSONIC ACID
mf: $C_6H_6AsNO_5$ mw: 247.05
PROP: Pale-yellow leaflets from H_2O. Mp: 298–300°. Spar sol in cold MeOH, H_2O, and EtOH.
SYNS: NITARSONE □ 4-NITROBENZENEARSONIC ACID □ p-NITROPHENYLARSONIC ACID □ RAS-26
CONSENSUS REPORTS: Arsenic and its compounds are on the Community Right-To-Know List.
OSHA PEL: TWA 0.5 mg(As)/m^3
ACGIH TLV: BEI: 35 μ (As)/L inorganic arsenic and methylated metabolites in urine
SAFETY PROFILE: Poison by ingestion and intravenous routes. When heated to decomposition it emits very toxic fumes of NO_x and As. See also ARSENIC COMPOUNDS and NITRO COMPOUNDS OF AROMATIC HYDROCARBONS.

N

NIX500   CAS: 108-03-2   HR: 3
1-NITROPROPANE

mf: $C_3H_7NO_2$ mw: 89.11

PROP: Colorless oily liquid; irritant to mucous membranes. Fp: $-108°$, bp: 132°, flash p: 93°F (TCC), d: 1.003 @ 20°/20°, autoign temp: 789°F, vap press: 7.5 mm @ 20°, vap d: 3.06, lel: 2.2%. Sltly sol in water; misc with alc, ether, and many org solvs. IDLH 1000 ppm.

SYN: 1-NP

CONSENSUS REPORTS: Reported in EPA TSCA Inventory.

OSHA PEL: TWA 25 ppm

ACGIH TLV: TWA 25 ppm; Not Classifiable as a Human Carcinogen

DFG MAK: 25 ppm (92 mg/m³)

SAFETY PROFILE: Poison by ingestion and intraperitoneal routes. Mildly toxic by inhalation. A human eye irritant. Human systemic effects by inhalation: conjunctiva irritation. Mutation data reported. Very dangerous fire hazard when exposed to heat, open flame, or oxidizers. Reacts violently with $Ca(OH)_2$, hydrocarbons, hydroxides, inorganic bases. May explode on heating. Metal oxides increase its sensitivity to thermal ignition. To fight fire, use alcohol foam, CO_2, dry chemical, water spray. When heated to decomposition it emits toxic fumes of NO_x. See also 2-NITROPROPANE, NITROALKANES, and NITRO COMPOUNDS.

NIY000   CAS: 79-46-9   HR: 3
2-NITROPROPANE

DOT: UN 2608

mf: $C_3H_7NO_2$ mw: 89.11

PROP: Colorless liquid. Bp: 120°, fp: $-93°$, flash p: 82°F (TCC), d: 0.992 @ 20°/20°, autoign temp: 802°F, vap press: 10 mm @ 15.8°, vap d: 3.06, lel: 2.6%. Misc with org solvs; sol in water, alc, and ether. IDLH 100 ppm.

SYNS: DIMETHYLNITROMETHANE □ ISONITRO-PROPANE □ NIPAR S-20 □ NIPAR S-20 SOLVENT □ NIPAR S-30 SOLVENT □ NITROISOPROPANE □ NITROPROPANE □ NITROPROPANE (DOT) □ β-

NITROPROPANE □ 2-NP □ RCRA WASTE NUMBER U171

CONSENSUS REPORTS: NTP 10th Report on Carcinogens. IARC Cancer Review: Group 2B IMEMDT 7,56,87; Human Inadequate Evidence IMEMDT 29,331,82; Animal Sufficient Evidence IMEMDT 29,331,82. Community Right-To-Know List. EPA Genetic Toxicology Program. Reported in EPA TSCA Inventory.

OSHA PEL: TWA 10 ppm

ACGIH TLV: TWA 10 ppm; Animal Carcinogen

DFG MAK: DFG TRK: Animal Carcinogen, Suspected Human Carcinogen

NIOSH REL: (2-Nitropropane) TWA reduce to lowest feasible level

DOT CLASSIFICATION: 3; Label: Flammable Liquid

SAFETY PROFILE: Confirmed carcinogen with experimental carcinogenic, tumorigenic, and teratogenic data. Poison by intraperitoneal route. Moderately toxic by ingestion and inhalation. Human systemic effects by inhalation: anorexia, hypermotility, diarrhea, nausea or vomiting. An experimental teratogen. Other experimental reproductive effects. Mutation data reported. Can cause liver and kidney injury, methemoglobinemia, and cyanosis. Very dangerous fire hazard when exposed to heat, open flame, or oxidizers. May explode on heating. Violent reactions with chlorosulfonic acid, oleum. May react with amines + heavy metal oxides (e.g., mercury oxide or silver oxide) to form explosive salts. May ignite on contact with mixtures of carbon + hopcalite, which are used in some respirators. Hopcalite is a catalyst consisting of coprecipitated copper(II) oxide and manganese(IV) oxide. To fight fire, use alcohol foam, CO_2, dry chemical, water spray. When heated to decomposition it emits toxic fumes of NO_x.

NIY600   CAS: 315706-72-0 HR: 3
4-NITRO-N-(4-PROPYLCYCLOHEXYL)-
**   BENZAMIDE**
mf: $C_{16}H_{22}N_2O_3$ mw: 290.36
SAFETY PROFILE: A poison by ingestion.
When heated to decomposition it emits
toxic vapors of NO_x.

NJA000   CAS: 5522-43-0   HR: 3
3-NITROPYRENE
mf: $C_{16}H_9NO_2$ mw: 247.26
PROP: Yellow needles from MeCN. Mp:
151–152°.
SYN: 1-NITROPYRENE
CONSENSUS REPORTS: NTP 10th Report
on Carcinogens. IARC Cancer Review:
Group 2B IMEMDT 46,321,89; Animal
Sufficient Evidence IMEMDT 46,321,89;
Human No Adequate Data IMEMDT
46,321,89. Reported in EPA TSCA
Inventory.
DFG MAK: Confirmed Animal Carcinogen
with Unknown Relevance to Humans
SAFETY PROFILE: Confirmed carcinogen
with experimental carcinogenic, neoplastigenic, and tumorigenic data. Human
mutation data reported. When heated to
decomposition it emits toxic fumes of NO_x.

NJA100   CAS: 57835-92-4   HR: 3
4-NITROPYRENE
mf: $C_{16}H_9NO_2$ mw: 247.26
PROP: Orange needles from
Me_2CO/MeOH. Mp: 196–197.5°.
CONSENSUS REPORTS: NTP 10th Report
on Carcinogens. IARC Cancer Review:
Group 2B IMEMDT 46,367,89; Animal
Sufficient Evidence IMEMDT 46,367,89;
Human No Adequate Data IMEMDT
46,367,89.
DFG MAK: Confirmed Animal Carcinogen
with Unknown Relevance to Humans
SAFETY PROFILE: Confirmed carcinogen
with experimental carcinogenic data.
Mutation data reported. When heated to
decomposition it emits toxic fumes of NO_x.

NJA500   CAS: 1124-33-0   HR: 3

4-NITROPYRIDINE-N-OXIDE
mf: $C_5H_4N_2O_3$ mw: 140.11
PROP: Pale-yellow rhombic crystals from
EtOH. Mp: 159–160°
SYN: 4-NITROPYRIDINE-1-OXIDE
CONSENSUS REPORTS: EPA Extremely
Hazardous Substances List. EPA Genetic
Toxicology Program. Reported in EPA
TSCA Inventory.
SAFETY PROFILE: Poison by ingestion and
skin contact. Questionable carcinogen with
experimental tumorigenic data. Mutation
data reported. Mixtures with diethyl-1,4-
dihydro-2,6-dimethylpyridine-3,5-
dicarboxylate explode when heated above
130°C. When heated to decomposition it
emits toxic fumes of NO_x.

NJD500   CAS: 607-35-2   HR: 3
8-NITROQUINOLINE
mf: $C_9H_6N_2O_2$ mw: 174.17
PROP: Monoclinic crystals from alcohol.
Mp: 91–92°. Sol in hot water, alc, ether, and
benzene.
CONSENSUS REPORTS: EPA Genetic
Toxicology Program. Reported in EPA
TSCA Inventory.
SAFETY PROFILE: Poison by
intraperitoneal route. Experimental
reproductive effects. Questionable
carcinogen with experimental carcinogenic
data. Mutation data reported. When heated
to decomposition it emits toxic fumes of
NO_x. See also other nitroquinoline entries.

NJF000   CAS: 56-57-5   HR: 3
4-NITROQUINOLINE-N-OXIDE
mf: $C_9H_6N_2O_3$ mw: 190.17
PROP: Yellow brown crystals or powder.
Mp: 154°. Sltly sol in water.
SYNS: 4-NITROCHINOLIN N-OXID (SWEDISH) □ 4-
NITROQUINOLINE-1-OXIDE □ 4-NQO
CONSENSUS REPORTS: EPA Genetic
Toxicology Program.
SAFETY PROFILE: Suspected carcinogen
with experimental carcinogenic, neoplasti-
genic, and tumorigenic data. Poison by
intraperitoneal and subcutaneous routes. An

N

experimental teratogen. Other experimental reproductive effects. Human mutation data reported. When heated to decomposition it emits toxic fumes of NO_x.

NJH000   HR: 3
NITROSAMINES
PROP: Compounds that have the chemical group $=N-N=O$ attached to an alkyl or aryl group. They are formed by reaction between an amine and NO_x or nitrites.
SAFETY PROFILE: Confirmed carcinogen of the lung, nasal sinus, brain, esophagus, stomach, liver, bladder, and kidney. They are often produced in food as by-products from processing and preparation. They are found in whiskey, herbicides, and cosmetics, as well as in tanneries, rubber factories, and iron foundries. They can be formed within the body by reaction of amine-containing foods or drugs with the nitrites resulting from bacterial conversion of nitrates. See also N-NITROSO COMPOUNDS.

NJN000   CAS: 60599-38-4   HR: 3
N-NITROSOBIS(2-OXOPROPYL)AMINE
mf: $C_6H_{10}N_2O_3$ mw: 158.18
SYNS: BIS-(2-OXOPROPYL)-N-NITROSAMINE □ BOP □ DI-OXO-DI-N-PROPYLNITROSAMINE □ 2,2'-DIOXO-DI-N-PROPYLNITROSAMINE □ 2,2'-DIOXO-N-NITRO-SODIPROPYLAMINE □ N,N-DI(2-OXOPROPYL)NITROS-AMINE □ 2,2'-DIOXOPROPYL-N-PROPYLNITROSAM-INE □ DOPN □ N-NITROSO-N,N-DI(2-OXYPROPYL)-AMINE □ (NITROSOIMINO)DIACETONE
SAFETY PROFILE: Suspected carcinogen with experimental carcinogenic, neoplastigenic, tumorigenic data. Poison by ingestion and subcutaneous routes. Human mutation data reported. When heated to decomposition it emits toxic fumes of NO_x. See also N-NITROSO COMPOUNDS.

NJT500   HR: 3
NITROSO COMPOUNDS
PROP: Compounds of the form $C-N=O$ or $N-N=O$. Organic nitrogen compounds.
SAFETY PROFILE: Usually highly toxic carcinogens, teratogens, and mutagens by almost all routes of exposure. Some of these

compounds may have hazardous instabilities under the appropriate conditions. When heated to decomposition they emit very toxic fumes of NO_x. See also specific compounds.

NJT550   HR: 3
N-NITROSO COMPOUNDS
PROP: A class of organic compounds of the form $R_2-N-N=O$ or $R=N-N=O$.
SAFETY PROFILE: Many members of this class are toxins, carcinogens, teratogens, and mutagens. Sources of exposure to N-nitroso compounds are: formation in the environment and absorption from food, water, air, or industrial and consumer products; formation in the body from precursors in food, water, or air; from tobacco; and from naturally occurring compounds. Some are used in the production of rubber and they may be formed as by-products in industrial processes. Nitrosamines have been found in food and cosmetics. N-nitroso compounds can be formed from the reaction of nitrates with nitrites under acidic conditions. These conditions can occur in the environment, mouth, and stomach. Nitrites are formed in the mouth by the action of bacteria on nitrates. Nitrosatable substances in the environment include secondary and tertiary amines, quaternary ammonium compounds, ureas, carbamates, and guanidines. Many of the resulting N-nitroso compounds are experimental carcinogens and mutagens. See also individual compounds, NITROSAMINES, CARBAMATES, and SMOKELESS TOBACCO.

NJW500   CAS: 55-18-5   HR: 3
N-NITROSODIETHYLAMINE
mf: $C_4H_{10}N_2O$ mw: 102.16
PROP: Yellow oil. D: 0.9422 @ 20°/4°, bp: 176.9°. Sol in water, alc, and ether.
SYNS: DANA □ DEN □ DENA □ DIAETHYL-NITROSAMIN (GERMAN) □ DIETHYLNITROSAMINE □ N,N-DIETHYLNITROSAMINE □ DIETHYLNITROSO-AMINE □ N-ETHYL-N-NITROSO-ETHANAMINE □ NDEA □ N-NITROSODIAETHYLAMIN (GERMAN) □

NITROSODIETHYLAMINE □ RCRA WASTE NUMBER U174
CONSENSUS REPORTS: NTP 10th Report on Carcinogens. IARC Cancer Review: Group 2A IMEMDT 7,56,87; Animal Sufficient Evidence IMEMDT 1,107,72, IMEMDT 17,83,78, IMEMDT 28,151,82; Human Limited Evidence IMEMDT 17,83,78. NCI Carcinogenesis Studies (ipr); Clear Evidence: mouse, rat RRCRBU 52,1,75. Reported in EPA TSCA Inventory. Community Right-To-Know List.
DFG MAK: Animal Carcinogen, Suspected Human Carcinogen
SAFETY PROFILE: Confirmed carcinogen with experimental carcinogenic, neoplastigenic, and tumorigenic data. Poison by ingestion, intravenous, intraperitoneal, and subcutaneous routes. An experimental teratogen. Other experimental reproductive effects. Human mutation data reported. A transplacental carcinogen. When heated to decomposition it emits toxic fumes of NO_x. See also N-NITROSO COMPOUNDS and AMINES.

NKA000   CAS: 601-77-4   HR: 3
N-NITROSODIISOPROPYLAMINE
mf: $C_6H_{14}N_2O$ mw: 130.22
SYNS: DIISOPROPYLNITROSAMIN (GERMAN) □ N-NITROSODI-i-PROPYLAMINE (MAK)
DFG MAK: Animal Carcinogen, Suspected Human Carcinogen
SAFETY PROFILE: Confirmed carcinogen with experimental carcinogenic and tumorigenic data. Moderately toxic by ingestion. When heated to decomposition it emits toxic fumes of NO_x. See also N-NITROSO COMPOUNDS and AMINES.

NKA600   CAS: 62-75-9   HR: 3
N-NITROSODIMETHYLAMINE
mf: $C_2H_6N_2O$ mw: 74.10
PROP: Yellow liquid; sol in water, alc, and ether. Bp: 152°, d: 1.005 @ 20°/4°.
SYNS: DIMETHYLNITROSAMIN (GERMAN) □ DIMETHYLNITROSAMINE □ N,N-DIMETHYLNITROSAMINE □ DIMETHYLNITROSOAMINE □ DMN □ DMNA □ N-METHYL-N-NITROSOMETHANAMINE □

NDMA □ NITROSODIMETHYLAMINE □ RCRA WASTE NUMBER P082
CONSENSUS REPORTS: NTP 10th Report on Carcinogens. IARC Cancer Review: Group 2A IMEMDT 7,56,87; Animal Sufficient Evidence IMEMDT 17,125,78; IMEMDT 1,95,72; Human Limited Evidence IMEMDT 17,125,78; Human Inadequate Evidence IMEMDT 1,95,72. Reported in EPA TSCA Inventory. EPA Genetic Toxicology Program. Community Right-To-Know List. EPA Extremely Hazardous Substances List.
OSHA PEL: Cancer Suspect Agent
ACGIH TLV: Animal Carcinogen
DFG MAK: Animal Carcinogen, Suspected Human Carcinogen
SAFETY PROFILE: Confirmed carcinogen with experimental carcinogenic, neoplastigenic, tumorigenic, and teratogenic data. A transplacental carcinogen. Human poison by ingestion. Experimental poison by ingestion, inhalation, intraperitoneal, subcutaneous, and intravenous routes. Human systemic effects by ingestion: ulceration or bleeding from small intestine, nausea or vomiting, and fever. Experimental reproductive effects. Human mutation data reported. Has caused fatal liver disease in humans. When heated to decomposition it emits toxic fumes of NO_x. See also NITROSAMINES.

NKB500   CAS: 156-10-5   HR: 3
p-NITROSODIPHENYLAMINE
mf: $C_{12}H_{10}N_2O$ mw: 198.24
PROP: Green plates with bluish luster (from benzene) or steel-blue prisms or plates (from ether + H_2O). Mp: 144–145°. Sltly sol in water or pet ether; very sol in alc, ether, benzene, chloroform.
SYNS: NAUGARD TKB □ NCI-C02244 □ p-NITROSODIFENYLAMIN (CZECH) □ 4-NITROSODIPHENYLAMINE □ p-NITROSO-N-PHENYLANILINE □ 4-NITROSO-N-PHENYLANILINE □ 4-NITROSO-N-PHENYLBENZENAMINE □ N-PHENYL-p-NITROSOANILINE □ TKB
CONSENSUS REPORTS: IARC Cancer Review: Group 3 IMEMDT 7,56,87; Animal

Inadequate Evidence IMEMDT 27,227,82. NCI Carcinogenesis Bioassay (feed); Clear Evidence: mouse, rat NCITR* NCI-CG-TR-190,79. Community Right-To-Know List. Reported in EPA TSCA Inventory. SAFETY PROFILE: Poison by intravenous route. Suspected carcinogen with experimental carcinogenic and neoplastigenic data. Moderately toxic by ingestion. Mutation data reported. An eye irritant. When heated to decomposition it emits toxic fumes of NO_x. See also N-NITROSO COMPOUNDS and AMINES.

NKB700   CAS: 621-64-7   HR: 3
N-NITROSODI-N-PROPYLAMINE
mf: $C_6H_{14}N_2O$ mw: 130.22
SYNS: DI-n-PROPYLNITROSAMINE □ DIPROPYL-NITROSOAMINE □ DPN □ DPNA □ NDPA □ N-NITROSODIPROPYLAMINE □ N-NITROSO-N-DIPROPYLAMINE □ N-NITROSO-N-PROPYLPROPAN-AMINE □ N-NITROSO-N-PROPYL-1-PROPANAMINE □ RCRA WASTE NUMBER U111
CONSENSUS REPORTS: NTP 10th Report on Carcinogens. IARC Cancer Review: Group 2B IMEMDT 7,56,87; Animal Sufficient Evidence IMEMDT 17,177,78; Human Limited Evidence IMEMDT 17,177,78. EPA Genetic Toxicology Program. Community Right-To-Know List. Reported in EPA TSCA Inventory.
DFG MAK: Animal Carcinogen, Suspected Human Carcinogen
SAFETY PROFILE: Confirmed carcinogen with experimental carcinogenic, neoplastigenic, tumorigenic data. Moderately toxic by ingestion and subcutaneous routes. An experimental teratogen. Human mutation data reported. When heated to decomposition it emits toxic fumes of NO_x. See also NITROSAMINES.

NKC000   CAS: 17608-59-2   HR: 3
N-NITROSOEPHEDRINE
mf: $C_{10}H_{14}N_2O_2$ mw: 194.26
SYNS: α-(1-(N-METHYL-N-NITROSOAMINO)ETHYL)-BENZYL ALCOHOL □ 2-(N-METHYL-N-NITROSO-AMINO)-1-PHENYL-1-PROPANOL

SAFETY PROFILE: Suspected carcinogen with experimental carcinogenic data. Poison by intraperitoneal route. Mutation data reported. When heated to decomposition it emits toxic fumes of NO_x. See also N-NITROSO COMPOUNDS.

NKD000   CAS: 612-64-6   HR: 3
N-NITROSO-N-ETHYL ANILINE
mf: $C_8H_{10}N_2O$ mw: 150.20
PROP: Yellow oil. D: 1.087 @ 20°/4°, bp: 119–120° @ 15 mm. Insol in water.
SYNS: ETHYLNITROSOANILINE □ N-ETHYL-N-NITROSOBENZENAMINE □ NEA □ NITROSO-ETHYLANILINE □ N-NITROSOETHYLPHENYLAMINE (MAK)
CONSENSUS REPORTS: EPA Genetic Toxicology Program.
DFG MAK: Animal Carcinogen, Suspected Human Carcinogen
SAFETY PROFILE: Confirmed carcinogen. Poison by ingestion and intraperitoneal routes. An experimental teratogen. Mutation data reported. Many N-nitroso compounds are carcinogens. When heated to decomposition it emits toxic fumes of NO_x. See also N-NITROSO COMPOUNDS.

NKE500   CAS: 614-95-9   HR: 3
N-NITROSO-N-ETHYLURETHAN
mf: $C_5H_{10}N_2O_3$ mw: 146.17
SYNS: AETHYLNITROSOURETHAN (GERMAN) □ ENU □ ETHYLESTER KYSELINY N-ETHYL-N-NITRO-SOKARBAMINOVE □ ETHYL N-ETHYLNITROSO-CARBAMATE □ ETHYLNITROSOCARBAMIC ACID, ETHYL ESTER □ N-ETHYL-N-NITROSOCARBAMIC ACID ETHYL ESTER □ ETHYL NITROSOURETHAN □ N-ETHYL-N-NITROSOURETHAN □ ETHYL NITROSO-URETHANE □ N-ETHYL-N-NITROSOURETHANE □ NEU □ NITROSOETHYLURETHAN □ N-NITROSO-N-ETHYLURETHAN □ NITROSO-N-ETHYLURETHANE □ NSC-24890
CONSENSUS REPORTS: EPA Genetic Toxicology Program.
SAFETY PROFILE: Poison by intravenous route. Experimental teratogenic effects. Mutation data reported. Questionable carcinogen with experimental carcinogenic data. When heated to decomposition it emits toxic fumes of NO_x. See also N-

NITROSO COMPOUNDS and CARBAMATES.

NKF000   CAS: 13256-13-8   HR: 3
N-NITROSO-N-ETHYLVINYLAMINE
mf: $C_4H_8N_2O$ mw: 100.14
SYNS: AETHYL-VINYL-NITROSOAMIN (GERMAN) □ N-ETHYL-N-NITROSOETHENAMINE □ N-ETHYL-N-NITROSOETHENYLAMINE □ N-ETHYL-N-NITROSO-VINYLAMINE □ ETHYLVINYLNITROSAMINE □ N-NITROSOETHYLVINYLAMINE □ VINYLETHYL-NITROSAMIN (GERMAN) □ VINYLETHYLNITROS-AMINE
CONSENSUS REPORTS: EPA Genetic Toxicology Program.
SAFETY PROFILE: Poison by ingestion, intravenous, and subcutaneous routes. Experimental teratogenic effects. Questionable carcinogen with experimental carcinogenic and tumorigenic data. When heated to decomposition it emits toxic fumes of NO_x. See also N-NITROSO COMPOUNDS and AMINES.

NKH000   CAS: 674-81-7   HR: 3
NITROSOGUANIDINE
mf: CH_4N_4O mw: 88.09
PROP: A solid.
SYNS: INITIATING EXPLOSIVE NITROSO-GUANIDINE □ NITROSOGUANIDIN (GERMAN) □ N-NITROSOGUANIDINE
SAFETY PROFILE: Poison by intraperitoneal route. Many N-nitroso compounds are carcinogens. An explosive. Dangerous when stored in sealed containers as it decomposes to release nitrogen. When heated to decomposition it emits toxic fumes of NO_x. See also N-NITROSO COMPOUNDS and EXPLOSIVES, HIGH.

NKI000   CAS: 932-83-2   HR: 3
N-NITROSOHEXAHYDROAZEPINE
mf: $C_6H_{12}N_2O$ mw: 128.20
SYNS: HEXAHYDRO-1-NITROSO-1H-AZEPINE □ N-6-MI □ N-NITROSOAZACYCLOHEPTANE □ N-NITROSOHEXAMETHYLENEIMINE □ NITROSOHEX-AMETHYLENIMINE □ N-NITROSOPERHYDRO-AZEPINE

SAFETY PROFILE: Suspected carcinogen with experimental carcinogenic, tumorigenic, and teratogenic data. Poison by ingestion, intraperitoneal, and subcutaneous routes. Experimental reproductive effects. Mutation data reported. When heated to decomposition it emits toxic fumes of NO_x. See also N-NITROSO COMPOUNDS.

NKK500   CAS: 3715-92-2   HR: 3
N-NITROSOIMIDAZOLIDINETHIONE
mf: $C_3H_5ON_3S$ mw: 131.17
SYNS: N-NITROSOETHYLENETHIOUREA □ NO-ETU
CONSENSUS REPORTS: EPA Genetic Toxicology Program.
SAFETY PROFILE: Poison by ingestion. Experimental teratogenic and reproductive effects. Questionable carcinogen with experimental neoplastigenic data. Mutation data reported. Many N-nitroso compounds are carcinogens. When heated to decomposition it emits very toxic fumes of NO_x and SO_x. See also N-NITROSO COMPOUNDS.

NKL000   CAS: 3844-63-1   HR: 3
1-NITROSOIMIDAZOLIDINONE
mf: $C_3H_5N_3O_2$ mw: 115.11
PROP: A solid. Mp: 101.5–101.8°.
SYNS: ETHYLENENITROSOUREA □ N-NITRO-2-IMIDAZOLIDONE □ 1-NITRO-2-IMIDAZOLIDONE □ 1-NITROSO-2-IMIDAZOLIDINONE □ N-NITROSO-IMIDAZOLIDON (GERMAN) □ N-NITROSOIMID-AZOLIDONE □ NSC-73438 □ SRI 1869
CONSENSUS REPORTS: EPA Genetic Toxicology Program.
SAFETY PROFILE: Poison by intraperitoneal and subcutaneous routes. Mutation data reported. Questionable carcinogen with experimental tumorigenic data. Many N-nitroso compounds are carcinogens. When heated to decomposition it emits toxic fumes of NO_x. See also N-NITROSO COMPOUNDS.

NKM000   CAS: 1116-54-7   HR: 3
NITROSOIMINO DIETHANOL
mf: $C_4H_{10}N_2O_3$ mw: 134.16

N

SYNS: BIS(β-HYDROXYAETHYL)NITROSAMIN (GERMAN) □ BIS(β-HYDROXYETHYL)NITROSAMINE □ DIAETHANOLNITROSAMIN (GERMAN) □ DIETHANO-LNITROSOAMINE □ 2,2'-DIHYDROXY-N-NITROSO-DIETHYLAMINE □ 2,2'-IMINODI-N-NITROSOETHAN-OL □ NCI-C55583 □ NDELA □ N-NITROSOAMINO-DIETHANOL □ N-NITROSOBIS(2-HYDROXYETHYL)-AMINE □ N-NITROSODIETHANOLAMIN (GERMAN) □ N-NITROSODIETHANOLAMINE (MAK) □ 2,2'-(NITROSOIMINO)BISETHANOL □ RCRA WASTE NUMBER U173

CONSENSUS REPORTS: NTP 10th Report on Carcinogens. IARC Cancer Review: Group 2B IMEMDT 7,56,87; Animal Sufficient Evidence IMEMDT 17,77,78; Human Limited Evidence IMEMDT 17,77,78. Reported in EPA TSCA Inventory.

DFG MAK: Animal Carcinogen, Suspected Human Carcinogen

SAFETY PROFILE: Confirmed carcinogen with experimental carcinogenic, neoplastigenic, and tumorigenic data. Mildly toxic by ingestion. Mutation data reported. When heated to decomposition it emits toxic fumes of NO_x. See also N-NITROSO COMPOUNDS and ALCOHOLS.

NKO400   CAS: 71752-69-7   HR: 3
NITROSOISOPROPANOLUREA
mf: $C_4H_9N_3O_3$ mw: 147.16
SYNS: 1-(2-HYDROXYPROPYL)-1-NITROSOUREA □ NITROSO-2-HYDROXY-N-PROPYLUREA □ N-NITRO-SO-2-HYDROXY-N-PROPYLUREA

SAFETY PROFILE: Suspected carcinogen with experimental carcinogenic and tumorigenic data. Mutation data reported. When heated to decomposition it emits toxic fumes of NO_x. See also N-NITROSO COMPOUNDS.

NKQ100   CAS: 16219-99-1   HR: 3
N-NITROSO-N-METHYL-4-AMINOPY-
RIDINE
mf: $C_6H_7N_3O$ mw: 137.16
SYNS: 4-NITROSOMETHYLAMINOPYRIDINE □ 4-PYRIDINAMINE, N-METHYL-N-NITROSO- □ PYRID-INE, 4-NITROSOMETHYLAMINO- □ PYRIDINE, 4-(METHYLNITROSAMINO)- □ N-METHYL-N-NITROSO-4-AMINOPYRIDINE

SAFETY PROFILE: A poison by ingestion. Mutation data reported. When heated to decomposition it emits toxic vapors of NO_x.

NKT500   CAS: 5432-28-0   HR: 3
N-NITROSO-N-METHYLCYCLOHEXYL-
AMINE
mf: $C_7H_{14}N_2O$ mw: 142.23
SYNS: METHYLCYCLOHEXYLNITROSAMIN (GERMAN) □ METHYLCYCLOHEXYLNITROSAMINE □ N-METHYL-N-NITROSOCYCLOHEXYLAMINE □ N-NITROSOMETHYLCYCLOHEXYLAMINE

SAFETY PROFILE: Poison by ingestion, intravenous, and intraperitoneal routes. Questionable carcinogen with experimental tumorigenic data. Mutation data reported. Many N-nitroso compounds are carcinogens. When heated to decomposition it emits toxic fumes of NO_x. See also N-NITROSO COMPOUNDS and AMINES.

NKU000   CAS: 55090-44-3   HR: 3
NITROSOMETHYL-n-DODECYLAMINE
mf: $C_{13}H_{28}N_2O$ mw: 228.43
SYNS: N-METHYL-N-NITROSOLAURYLAMINE □ N-NITROSO-N-METHYL-N-DODECYLAMIN (GERMAN) □ N-NITROSO-N-METHYL-N-DODECYLAMINE □ NMDDA

CONSENSUS REPORTS: EPA Genetic Toxicology Program.

SAFETY PROFILE: Suspected carcinogen with experimental carcinogenic and tumorigenic data. Moderately toxic by ingestion. Mutation data reported. When heated to decomposition it emits toxic fumes of NO_x. See also N-NITROSO COMPOUNDS and AMINES.

NKU875   CAS: 35631-27-7   HR: 3
NITROSO-5-METHYLOXAZOLIDONE
mf: $C_4H_8N_2O_2$ mw: 116.14
SYNS: 5-METHYL-3-NITROSO-1,3-OXAZOLIDINE □ NITROSO-5-METHYL-1,3-OXAZOLIDINE □ N-NITROSO-5-METHYL-1,3-OXAZOLIDINE

SAFETY PROFILE: Suspected carcinogen with experimental carcinogenic and tumorigenic data. Mutation data reported. When heated to decomposition it emits toxic fumes of NO_x. See also N-NITROSO COMPOUNDS.

NKV000 CAS: 55984-51-5 HR: 3
N-NITROSOMETHYL-2-OXOPROPYL-
AMINE
mf: $C_4H_8N_2O_2$ mw: 116.14
SYNS: 1-(METHYLNITROSOAMINO)2-PROPANONE □
MOP □ NMOP
SAFETY PROFILE: Suspected carcinogen
with experimental carcinogenic and
tumorigenic data. Poison by subcutaneous
route. Mutation data reported. When heated
to decomposition it emits toxic fumes of
NO_x. See also N-NITROSO
COMPOUNDS.

NKW500 CAS: 16339-07-4 HR: 3
1-NITROSO-4-METHYLPIPERAZINE
mf: $C_5H_{11}N_3O$ mw: 129.19
SYNS: N'-METHYL-N-NITROSOPIPERAZINE □ 1-
METHYL-4-NITROSOPIPERAZINE □ N-NITROSO-N'-
METHYLPIPERAZIN (GERMAN) □ N-NITROSO-N'-
METHYLPIPERAZINE
SAFETY PROFILE: Poison by ingestion.
Questionable carcinogen with experimental
tumorigenic data. Mutation data reported.
Many N-nitroso compounds are
carcinogens. When heated to decomposition
it emits toxic fumes of NO_x. See also N-
NITROSO COMPOUNDS.

NKY000 CAS: 4549-40-0 HR: 3
N-NITROSOMETHYLVINYLAMINE
mf: $C_3H_6N_2O$ mw: 86.11
SYNS: N-METHYL-N-NITROSO-ETHENYLAMINE □
N-METHYL-N-NITROSOVINYLAMINE □ METHYL-
VINYLNITROSAMIN (GERMAN) □ METHYLVINYL-
NITROSAMINE □ MVNA □ NMVA □ RCRA WASTE
NUMBER P084
CONSENSUS REPORTS: NTP 10th Report
on Carcinogens. IARC Cancer Review:
Group 2B IMEMDT 7,56,87; Animal
Sufficient Evidence IMEMDT 17,257,78;
Human Limited Evidence IMEMDT
17,257,78. Community Right-To-Know List.
EPA Genetic Toxicology Program.
SAFETY PROFILE: Confirmed carcinogen
with experimental tumorigenic data. Poison
by ingestion and inhalation. When heated to
decomposition it emits toxic fumes of NO_x.
See also N-NITROSO COMPOUNDS and
AMINES.

NKZ000 CAS: 59-89-2 HR: 3
4-NITROSOMORPHOLINE
mf: $C_4H_8N_2O_2$ mw: 116.14
SYNS: N-NITROSOMORPHOLIN (GERMAN) □
NITROSOMORPHOLINE □ N-NITROSOMORPHOLINE
(MAK) □ NMOR
CONSENSUS REPORTS: NTP 10th Report
on Carcinogens. IARC Cancer Review:
Group 2B IMEMDT 7,56,87; Animal
Sufficient Evidence IMEMDT 17,263,78;
Human Limited Evidence IMEMDT
17,263,78. Community Right-To-Know List.
EPA Genetic Toxicology Program.
DFG MAK: Animal Carcinogen, Suspected
Human Carcinogen
SAFETY PROFILE: Confirmed carcinogen
with experimental carcinogenic,
neoplastigenic, and tumorigenic data. Poison
by ingestion, intraperitoneal, subcutaneous,
and intravenous routes. Moderately toxic by
inhalation. Human mutation data reported.
Experimental reproductive effects. When
heated to decomposition it emits toxic
fumes of NO_x. See also N-NITROSO
COMPOUNDS.

NLB500 CAS: 132-53-6 HR: 2
2-NITROSO-1-NAPHTHOL
mf: $C_{10}H_7NO_2$ mw: 173.18
PROP: Yellow needles from H_2O or C_6H_6.
Sol in AcOH, EtOH and alkalies.
CONSENSUS REPORTS: Reported in EPA
TSCA Inventory.
SAFETY PROFILE: Moderately toxic by
ingestion. Many nitroso compounds are
carcinogens. Questionable carcinogen with
experimental tumorigenic data. When heated
to decomposition it emits toxic fumes of
NO_x. See also NITROSO COMPOUNDS.

NLD500 CAS: 16543-55-8 HR: 3
N'-NITROSONORNICOTINE
mf: $C_9H_{11}N_3O$ mw: 177.23
SYNS: NICOTINE, 1'-NITROSO-1'-DEMETHYL- □ 1'-
NITROSO-1'-DEMETHYLNICOTINE □ 1-NITROSO-2-(3-
PYRIDYL)PYRROLIDINE □ 3-(1-NITROSO-2-PYRRO-
LIDINYL)PYRIDINE □ PYRIDINE, 3-(1-NITROSO-2-
PYRROLIDINYL)-, (S)-

N

CONSENSUS REPORTS: NTP 10th Report on Carcinogens. IARC Cancer Review: Group 2B IMEMDT 7,56,87; Animal Sufficient Evidence IMEMDT 17,281,78; Human Limited Evidence IMEMDT 17,281,78. Community Right-To-Know List. EPA Genetic Toxicology Program.
SAFETY PROFILE: Confirmed carcinogen with experimental carcinogenic, neoplastigenic, and tumorigenic data. Low toxicity by intraperitoneal route. Mutation data reported. When heated to decomposition it emits toxic fumes of NO_x. See also N-NITROSO COMPOUNDS.

NLE000   CAS: 39884-52-1   HR: 3
N-NITROSOOXAZOLIDINE
mf: $C_3H_6N_2O_2$ mw: 102.11
SYNS: N-NITROSOOXAZOLIDIN (GERMAN) □ N-NITROSO-1,3-OXAZOLIDINE □ 3-NITROSO-OXAZOLIDINE □ NITROSOOXAZOLIDONE
SAFETY PROFILE: Suspected carcinogen with experimental carcinogenic and tumorigenic data. Moderately toxic by ingestion. Mutation data reported. When heated to decomposition it emits toxic fumes of NO_x. See also N-NITROSO COMPOUNDS.

NLF200   CAS: 104-91-6   HR: 3
NITROSOPHENOL
mf: $C_6H_5NO_2$ mw: 123.12
PROP: Pale-yellow, orthorhombic needles. Mp: 144° (decomp). Sltly sol in water; sol in dilute alkalies, alc, ether, and acetone.
SYNS: p-NITROSOPHENOL □ 4-NITROSOPHENOL □ QUINONE MONOXIME □ QUINONE OXIME
CONSENSUS REPORTS: Reported in EPA TSCA Inventory.
SAFETY PROFILE: Poison by parenteral and intraperitoneal routes. Mutation data reported. An irritant and sensitizer. Many nitroso compounds are carcinogens. A very dangerous fire and explosion hazard. When exposed to heat or flame, it burns explosively. Contamination by acid or alkali may cause ignition. Can heat spontaneously and cause fire. When heated to decomposition it emits toxic fumes of NO_x. See also NITROSO COMPOUNDS.

NLH000   CAS: 13256-23-0   HR: 3
N-NITROSO-4-PICOLYLETHYLAMINE
mf: $C_8H_{11}N_3O$ mw: 165.22
SYNS: AETHYL-4-PICOLYLNITROSAMIN (GERMAN) □ 4-((ETHYLNITROSAMINO)METHYL)PYRIDINE
SAFETY PROFILE: Poison by ingestion and intravenous routes. Questionable carcinogen with experimental tumorigenic data. Many N-nitroso compounds are carcinogens. When heated to decomposition it emits toxic fumes of NO_x. See also N-NITROSO COMPOUNDS and AMINES.

NLJ500   CAS: 100-75-4   HR: 3
N-NITROSOPIPERIDINE
mf: $C_5H_{10}N_2O$ mw: 114.17
PROP: Light-yellow oil. D: 1.063 @ 18.5°/4°, bp: 217–218°. Sol in water; very sol in acid solns.
SYNS: NITROSOPIPERIDIN □ N-NITROSO-PIPERIDIN □ 1-NITROSOPIPERIDINE □ N-N-PIP □ NO-Pip □ NPIP □ PYRIDINE, HEXAHYDRO-N-NITROSO- □ RCRA WASTE NUMBER U179 □ TL 266
CONSENSUS REPORTS: NTP 10th Report on Carcinogens. IARC Cancer Review: Group 2B IMEMDT 7,56,87; Human Limited Evidence IMEMDT 17,287,78; Animal Sufficient Evidence IMEMDT 17,287,78; IMEMDT 28,151,82. Community Right-To-Know List. EPA Genetic Toxicology Program. Reported in EPA TSCA Inventory.
DFG MAK: Animal Carcinogen, Suspected Human Carcinogen
SAFETY PROFILE: Confirmed carcinogen with experimental carcinogenic, neoplastigenic, and tumorigenic data. Poison by ingestion, intravenous, and subcutaneous routes. An experimental teratogen. Human mutation data reported. When heated to decomposition it emits toxic fumes of NO_x. See also N-NITROSO COMPOUNDS.

NLM500   CAS: 39603-53-7   HR: 3
1-(NITROSOPROPYLAMINO)-2-PROPANOL

mf: $C_6H_{14}N_2O_2$ mw: 146.22

SYNS: 2-HPPN □ β-HPPN □ β-HYDROXYPROPYL-PROPYLNITROSAMINE □ (2-HYDROXYPROPYL)-PROPYLNITROSOAMINE □ N-NITROSO-2-HYDROXY-n-PROPYL-n-PROPYLAMINE

CONSENSUS REPORTS: IARC Cancer Review: Animal Sufficient Evidence IMEMDT 17,177,78.

SAFETY PROFILE: Confirmed carcinogen with experimental carcinogenic and tumorigenic data. Moderately toxic by subcutaneous route. An experimental teratogen. Mutation data reported. When heated to decomposition it emits toxic fumes of NO_x. See also N-NITROSO COMPOUNDS.

NLO500   CAS: 816-57-9   HR: 3
N-NITROSO-N-PROPYLUREA
mf: $C_4H_9N_3O_2$ mw: 131.16

SYNS: NITROSOPROPYLUREA □ NITROSO-N-PROPYLUREA □ NPU □ PNU □ N-PROPYLNITRO-SOHARNSTOFF (GERMAN) □ N-PROPYLNITROSO-UREA □ 1-PROPYL-1-NITROSOUREA

CONSENSUS REPORTS: EPA Genetic Toxicology Program.

SAFETY PROFILE: Suspected carcinogen with experimental carcinogenic, neoplastigenic, tumorigenic data. An experimental teratogen. Mutation data reported. When heated to decomposition it emits toxic fumes of NO_x. See also N-NITROSO COMPOUNDS.

NLP500   CAS: 930-55-2   HR: 3
N-NITROSOPYRROLIDINE
mf: $C_4H_8N_2O$ mw: 100.14

SYNS: N-NITROSOPYRROLIDIN (GERMAN) □ 1-NITROSOPYRROLIDINE □ N-N-PYR □ NO-PYR □ NPYR □ RCRA WASTE NUMBER U180 □ TETRAHYDRO-N-NITROSOPYRROLE

CONSENSUS REPORTS: NTP 10th Report on Carcinogens. IARC Cancer Review: Group 2B IMEMDT 7,56,87; Animal Sufficient Evidence IMEMDT 17,313,78; Human Limited Evidence IMEMDT 17,313,78. EPA Genetic Toxicology Program. Reported in EPA TSCA Inventory.

DFG MAK: Animal Carcinogen, Suspected Human Carcinogen

SAFETY PROFILE: Confirmed carcinogen with experimental carcinogenic, neoplastigenic, and tumorigenic data. Poison by ingestion and subcutaneous routes. Human mutation data reported. When heated to decomposition it emits toxic fumes of NO_x. See also N-NITROSO COMPOUNDS.

NLR500   CAS: 13256-22-9   HR: 3
N-NITROSOSARCOSINE
mf: $C_3H_6N_2O_3$ mw: 118.11

SYNS: GLYCINE, N-METHYL-N-NITROSO- □ N-METHYL-N-NITROSOGLYCINE □ N-NITROSO-METHYLGLYCINE □ NITROSO SARKOSIN □ NSAR

CONSENSUS REPORTS: NTP 10th Report on Carcinogens. IARC Cancer Review: Group 2B IMEMDT 7,56,87; Animal Sufficient Evidence IMEMDT 17,327,78; Human Limited Evidence IMEMDT 17,327,78.

SAFETY PROFILE: Confirmed carcinogen with experimental tumorigenic data. Mildly toxic by ingestion. When heated to decomposition it emits toxic fumes of NO_x. See also N-NITROSO COMPOUNDS.

NLY750   CAS: 88208-15-5   HR: 3
NITROSO-3,4,5-TRIMETHYLPIPERAZ-INE
mf: $C_7H_{15}N_3O$ mw: 157.25

SYNS: N-NITROSO-3,4,5-TRIMETHYLPIPERAZINE □ 1-NITROSO-3,4,5-TRIMETHYLPIPERAZINE

SAFETY PROFILE: Suspected carcinogen with experimental carcinogenic and tumorigenic data. Mutation data reported. When heated to decomposition it emits toxic fumes of NO_x. See also N-NITROSO COMPOUNDS.

NMB000   CAS: 9056-38-6   HR: 3
NITROSTARCH
DOT: UN 0146/UN 1337

PROP: Solid.

SYNS: NITROSTARCH, dry or wetted with <20% water, by weight (UN 0146) (DOT) □ NITROSTARCH, wetted with not <20% water, by weight (UN 1337) (DOT)

CONSENSUS REPORTS: Reported in EPA TSCA Inventory.

DOT CLASSIFICATION: EXPLOSIVE 1.1D; Label: EXPLOSIVE 1.1D (UN 0146); DOT Class: 4.1; Label: Flammable Solid (UN 1337)

SAFETY PROFILE: A very dangerous fire and explosion hazard when exposed to heat, flame, shock, or oxidizers. It is a powerful high explosive. Nitrostarch is not a definite compound, but a mixture of various nitric acid esters of starch with different degrees of nitration. When heated to decomposition it emits toxic fumes of NO_x. See also NITRO COMPOUNDS.

NMH000   CAS: 2696-92-6   HR: 3
NITROSYL CHLORIDE
DOT: UN 1069
mf: ClNO   mw: 65.46
PROP: Orange-yellow gas condensing to a deep red liquid; irritating odor. The dry gas slowly attacks glass. Mp: $-61.5°$, bp: $-6.4°$, d (liquid): 1.250 @ 30°, vap d: 2.3, vap press: 76 mm @ 50°. Non-explosive, very corrosive. Liquid @ $-5.5°$, solid @ $-61.5°$. Sol in fuming H_2SO_4.
SYN: NITROGEN OXYCHLORIDE
CONSENSUS REPORTS: Reported in EPA TSCA Inventory.
DOT CLASSIFICATION: 2.3; Label: Poison Gas, Corrosive
SAFETY PROFILE: Poison by inhalation and ingestion. A corrosive irritant to skin, eyes, and mucous membranes. Inhalation may cause pulmonary edema and hemorrhage. Potentially explosive reaction with acetone + platinum. Mixtures with hydrogen + oxygen ignite spontaneously. When heated to decomposition it emits very toxic fumes of Cl^- and NO_x.

NMJ000   CAS: 7782-78-7   HR: 3
NITROSYLSULFURIC ACID
mf: HNO_5S mw: 127.07
PROP: White crystals or prisms, stable in dry air. Mp: 73.5°, decomp @ 73.5°. Sol in sulfuric acid, decomp in water.

SYNS: CHAMBER CRYSTALS □ NITRO ACID SULFITE □ NITROSONIUM BISULFITE □ NITROSYL HYDROGEN SULFATE □ NITROSYL SULFATE □ SULFURIC ACID, MONOANHYDRIDE with NITROUS ACID

CONSENSUS REPORTS: Reported in EPA TSCA Inventory.

SAFETY PROFILE: A poison. A corrosive irritant to skin, eyes, and mucous membranes. Explosive reaction above 50°C with 2-chloro-4,6-dinitroaniline and 4-chloro-2,6-dinitroaniline. Potentially explosive reaction with dinitroaniline. When heated to decomposition it emits toxic fumes of SO_x and NO_x. See also SULFATES and NITRATES.

NMO500   CAS: 99-08-1   HR: 3
m-NITROTOLUENE
DOT: UN 1664
mf: $C_7H_7NO_2$ mw: 137.15
PROP: Crystals or liquid. Mp: 16°, flash p: 233°F (CC), d: 1.1630 @ 15°/4°, vap press: 1 mm @ 50.2°, vap d: 4.72, bp: 231.9°. Misc with alc, ether; sol in benzene; sol in water @ 30°. IDLH 200 ppm.
SYNS: m-METHYLNITROBENZENE □ 3-METHYLNITROBENZENE □ MNT □ 3-NITROTOLUENE □ 3-NITROTOLUOL
CONSENSUS REPORTS: Reported in EPA TSCA Inventory.
OSHA PEL: TWA 2 ppm (skin)
ACGIH TLV: TWA 2 ppm (skin)
DFG MAK: 5 ppm (28 mg/m³)
DOT CLASSIFICATION: 6.1; Label: Poison
SAFETY PROFILE: Poison by ingestion. Moderately toxic by inhalation. Combustible when exposed to heat, flame, or oxidizers. To fight fire, use water, CO_2, dry chemical. Probably an explosive. When heated to decomposition it emits toxic fumes of NO_x. See also other methylnitrobenzene entries and NITRO COMPOUNDS OF AROMATIC HYDROCARBONS.

NMO525   CAS: 88-72-2   HR: 3
o-NITROTOLUENE
DOT: UN 1664
mf: $C_7H_7NO_2$ mw: 137.15

PROP: Yellowish oily liquid. Crystals in two forms; transparent needles or snow white crystals. Mp: −10°, bp: 222.3°, flash p: 223°F (CC), d: 1.1622 @ 19°/15°, vap press: 1 mm @ 50°, vap d: 4.72. Insol in water; sol in SO_2 and pet ether; misc in alc, benzene, and ether. Sltly sol in NH_3. IDLH 200 ppm.

SYNS: 2-METHYLNITROBENZENE □ o-METHYL-NITROBENZENE □ 2-NITROTOLUENE □ ONT

CONSENSUS REPORTS: Reported in EPA TSCA Inventory.

OSHA PEL: TWA 2 ppm (skin)

ACGIH TLV: TWA 2 ppm (skin)

DFG MAK: Animal Carcinogen, Suspected Human Carcinogen

DOT CLASSIFICATION: 6.1; Label: Poison

SAFETY PROFILE: Confirmed carcinogen. A poison. Moderately toxic by ingestion. Mucous membrane effects by inhalation. Mutation data reported. Combustible when exposed to heat or open flame. To fight fire, use water spray, fog, foam, CO_2. Potentially explosive reaction with alkali (e.g., sodium hydroxide). When heated to decomposition it emits toxic fumes of NO_x. See also other methylnitrobenzene entries and NITRO COMPOUNDS OF AROMATIC HYDROCARBONS.

NMO550   CAS: 99-99-0   HR: 3
p-NITROTOLUENE
DOT: UN 1664
mf: $C_7H_7NO_2$ mw: 137.15
PROP: Yellowish crystals from alc. Bp: 238.3°, flash p: 223°F (CC), d: 1.286, vap press: 1 mm @ 53.7°, vap d: 4.72. Mp: 53–54°. Insol in water; sol in alc, benzene, ether, chloroform, and acetone. IDLH 200 ppm.

SYNS: 4-METHYLNITROBENZENE □ p-METHYL NITROBENZENE □ NCI-C60537 □ 4-NITROTOLUENE □ 4-NITROTOLUOL □ PNT

CONSENSUS REPORTS: Reported in EPA TSCA Inventory.

OSHA PEL: TWA 2 ppm (skin)

ACGIH TLV: TWA 2 ppm (skin)

DFG MAK: 5 ppm (28 mg/m³)

DOT CLASSIFICATION: 6.1; Label: Poison

SAFETY PROFILE: A poison. Moderately toxic by ingestion, inhalation, and intraperitoneal routes. Mildly toxic by skin contact. Mutation data reported. Combustible when exposed to heat or flame. To fight fire, use CO_2, dry chemical, foam. The residue from vacuum distillation may explode spontaneously. Reacts with sodium to form an ignitable product. Violent reaction with concentrated sulfuric acid (above 160°C), sulfuric acid + sulfur trioxide (above 52°C). Mixtures with tetranitromethane are sensitive high explosives. May explode on standing. It has been involved in plant scale explosions. When heated to decomposition it emits toxic fumes of NO_x. See also other methylnitrobenzene entries and NITRO COMPOUNDS OF AROMATIC HYDROCARBONS.

NMP500   CAS: 99-55-8   HR: 3
5-NITRO-o-TOLUIDINE
mf: $C_7H_8N_2O_2$ mw: 152.17
PROP: Yellow prisms from EtOH. Mp: 107°.

SYNS: AMARTHOL FAST SCARLET G BASE □ AMARTHOL FAST SCARLET G SALT □ 1-AMINO-2-METHYL-5-NITROBENZENE □ 2-AMINO-4-NITROTOLUENE □ AZOENE FAST SCARLET GC BASE □ AZOENE FAST SCARLET GC SALT □ AZOFIX SCARLET G SALT □ AZOGENE FAST SCARLET G □ AZOIC DIAZO COMPONENT 12 □ BENZENAMINE, 2-METHYL-5-NITRO- □ C.I. 37105 □ C.I. AZOIC DIAZO COMPONENT 12 □ DAINICHI FAST SCARLET G BASE □ DAITO SCARLET BASE G □ DEVOL SCARLET B □ DEVOL SCARLET G SALT □ DIABASE SCARLET G □ DIAZO FAST SCARLET G □ FAST RED SG BASE □ FAST SCARLET BASE G □ FAST SCARLET BASE J □ FAST SCARLET G □ FAST SCARLET G BASE □ FAST SCARLET GC BASE □ FAST SCARLET M4NT BASE □ FAST SCARLET T BASE □ FAST SCARLET G SALT □ FAST SCARLET J SALT □ HILTONIL FAST SCARLET G BASE □ HILTONIL FAST SCARLET GC BASE □ HILTONIL FAST SCARLET G SALT □ KAYAKU SCARLET G BASE □ LAKE SCARLET G BASE □ LITHOSOL ORANGE R BASE □ 2-METHYL-5-NITROANILINE □ 6-METHYL-3-NITROANILINE □ 2-METHYL-5-NITRO-BENZENEAMINE □ MITSUI SCARLET G BASE □ NAPHTHANIL SCARLET G BASE □ NAPHTOELAN

FAST SCARLET G BASE □ NAPHTOELAN FAST SCARLET G SALT □ NCI-C01843 □ 4-NITRO-2-AMINOTOLUENE □ 3-NITRO-6-METHYLANILINE □ 5-NITRO-2-METHYLANILINE □ 5-NITRO-2-TOLUIDINE □ PNOT □ RCRA WASTE NUMBER U181 □ SCARLET BASE CIBA II □ SCARLET BASE IRGA II □ SCARLET BASE NSP □ SCARLET G BASE □ SUGAI FAST SCARLET G BASE □ SYMULON SCARLET G BASE
CONSENSUS REPORTS: IARC Cancer Review: Group 3 IMEMDT 48,169,90; Animal Limited Evidence IMEMDT 48,169,90. NCI Carcinogenesis Bioassay (feed); Clear Evidence: mouse NCITR* NCI-CG-TR-107,78. Reported in EPA TSCA Inventory.
ACGIH TLV: TWA (inhalable fraction) 1 mg/m³; Confirmed Animal Carcinogen with Unknown Revelance to Humans)
DFG MAK: Animal Carcinogen, Suspected Human Carcinogen
SAFETY PROFILE: Confirmed carcinogen with experimental carcinogenic data. Moderately toxic by ingestion. Mutation data reported. Decomposes exothermically when heated to 150°C. When heated to decomposition it emits toxic fumes of NO_x. See also NITRO COMPOUNDS OF AROMATIC HYDROCARBONS.

NMP620   CAS: 932-64-9   HR: 3
NITROTRIAZOLONE
DOT: UN 0490
SYNS: 3-NITRO-1,2,4-TRIAZOL-5-ONE □ NTO □ NTO (DOT) □ 1,2,4-TRIAZOL-5-ONE, 3-NITRO-
DOT CLASSIFICATION: Explosive 1.1D; Label: Explosive 1.1D
SAFETY PROFILE: Low toxicity by ingestion and skin contact. A skin and eye irritant. An unstable explosive. When heated to decomposition it emits toxic vapors of NO_x.

NMQ000   CAS: 464-10-8   HR: 3
NITROTRIBROMOMETHANE
mf: CBr_3NO_2 mw: 297.75
PROP: Crystals or liquid. Mp: 10°, bp: 127°, d: 2.79 @ 20°.
SYN: TRIBROMONITROMETHANE

CONSENSUS REPORTS: Reported in EPA TSCA Inventory.
SAFETY PROFILE: Poison by intraperitoneal route. In vapor form it is highly toxic by ingestion and inhalation and on contact with skin, eyes, and mucous membranes. An explosive. When heated to decomposition it emits very toxic fumes of Br^- and NO_x. See also NITRO COMPOUNDS and BROMIDES.

NMQ500   CAS: 556-89-8   HR: 3
NITROUREA
DOT: UN 0147
mf: $CH_3N_3O_3$ mw: 105.06
PROP: Crystals or leaflets or prisms from EtOH; or platelets from EtOH/ligroin. Mp: 159°. Sol in hot H_2O (aq solns, unstable). Very sol in Me_2CO, AcOH; sol in EtOH; spar sol in $CHCl_3$, C_6H_6, and ligroin.
SYNS: N-NITROCARBAMIDE □ N-NITROUREA □ 1-NITROUREA
CONSENSUS REPORTS: Reported in EPA TSCA Inventory.
DOT CLASSIFICATION: EXPLOSIVE 1.1D; Label: EXPLOSIVE 1.1D
SAFETY PROFILE: A very dangerous fire hazard when exposed to heat or flame. A severe explosion hazard when shocked or exposed to heat. Can react vigorously with oxidizing materials. It is a high explosive. Incompatible with mercuric and silver salts. When heated to decomposition it emits highly toxic fumes of NO_x. See also EXPLOSIVES, HIGH; and NITRATES.

NMR000   CAS: 7782-77-6   HR: 3
NITROUS ACID
mf: HNO_2 mw: 47.02
PROP: Known only as pale blue aq soln. Stable in the forms of its many salts. Dehydration yields N_2O_3.
SYNS: KYSELINA DUSITE □ NITROSYL HYDROXIDE
CONSENSUS REPORTS: EPA Genetic Toxicology Program. Reported in EPA TSCA Inventory.
SAFETY PROFILE: Mutation data reported. Flammable by chemical reaction; a powerful

oxidizer. Explodes on contact with phosphorus trichloride. Reacts violently with PH_3 and PCl_3. Reactions with 1-amino-5-nitrophenol, ammonium decahydroborate(2−), hydrazine (product is hydrogen azide) may give explosive products. Incompatible with anilines (e.g., 4-bromoaniline, 2-chloroaniline, 3-chloroaniline, 2-nitroaniline, 3-nitroaniline, 4-nitroaniline, aniline), semicarbazone, silver nitrate. When heated to decomposition it emits highly toxic fumes of NO_x. See also NITRIC OXIDE.

NMS000   CAS: 25168-04-1   HR: 3
NITROXYLENE
DOT: NA 1665
mf: $C_8H_9NO_2$ mw: 151.18
PROP: Light-yellow liquid. Mp: 2°, bp: 244°, d: 1.135 g/cm³ @ 15°/4°.
SYNS: NITRODIMETHYLBENZENE □ NITROXYLENE □ NITROXYLENES (o-, m-, p-) (DOT)
DOT CLASSIFICATION: 6.1; Label: Poison
SAFETY PROFILE: A poison. Moderately toxic by ingestion. When heated to decomposition it emits toxic fumes of NO_x. See also NITRO COMPOUNDS OF AROMATIC HYDROCARBONS.

NMT500   CAS: 10022-50-1   HR: 3
NITRYL FLUORIDE
mf: FNO_2 mw: 65.01
PROP: Colorless gas; pungent odor; stable in glass vessels. Hydrol to HNO_3 and HF; highly reactive. Mp: −166.0°, bp: −72.4°, d: (liquid) 1.796 @ bp, d: (solid) 1.924.
SAFETY PROFILE: Poison by inhalation. A severe irritant to skin, eyes, and mucous membranes. A powerful oxidizing agent. This gas is intensely reactive. Explosive reaction with hydrogen at 200–300°C. Ignites on contact with antimony, arsenic, boron, iodine, phosphorus, selenium. Ignites when warmed with bismuth, carbon, chromium, lead, sulfur. Incandescent reaction with aluminum, cadmium, cobalt, iron, molybdenum, nickel, potassium, sodium, thorium, titanium, tungsten, uran-ium, vanadium, zinc, zirconium, lithium (at 200–300°C), manganese (at 200–300°C). Incompatible with metals, nonmetals. When heated to decomposition it emits toxic fumes of F^- and NO_x. See also FLUORIDES.

NMV735   CAS: 27753-52-2   HR: 2
NONABROMOBIPHENYL
mf: $C_{12}HBr_9$ mw: 864.32
SYNS: 1,1-BIPHENYL, NONABROMO- □ BROMKAL 80-9D
CONSENSUS REPORTS: Reported in EPA TSCA Inventory.
SAFETY PROFILE: Questionable carcinogen with experimental carcinogenic data. When heated to decomposition it emits toxic fumes of Br^-.

NMV760   CAS: 557-48-2   HR: 1
trans,cis-2,6-NONADIENAL
mf: $C_9H_{14}O$ mw: 138.23
PROP: Sltly yellow liquid; powerful, violet-cucumber odor. D: 0.850–0.870, refr index: 1.470, bp: 94–98° @ 11 mm. Sol in alc, fixed oils; insol in water.
SYNS: CUCUMBER ALDEHYDE □ FEMA No. 3317 □ 2,6-NONADIENAL □ trans,cis-2,6-NONADIENAL □ trans-2,cis-6-NONADIENAL □ VIOLET LEAF ALDEHYDE
CONSENSUS REPORTS: Reported in EPA TSCA Inventory.
SAFETY PROFILE: Low toxicity by ingestion and skin contact. A moderate skin irritant. When heated to decomposition it emits acrid smoke and irritating fumes.

NMV775   CAS: 5910-87-2   HR: D
trans,trans-2,4-NONADIENAL
mf: $C_9H_{14}O$ mw: 138.21
PROP: Slightly yellow liquid; strong, fatty, floral odor. D: 0.850–0.870, bp: 97–98° @ 10 mm, refr index: 1.522. Sol in alc, fixed oils; insol in water.
SYN: FEMA No. 3212
SAFETY PROFILE: When heated to decomposition it emits acrid smoke and irritating fumes.

NMV780   CAS: 28069-72-9   HR: 1

trans,cis-2,6-NONADIENOL

mf: C₉H₁₆O   mw: 140.25

PROP: White to yellow liquid; powerful, vegetable odor. D: 0.860–0.880, bp: 98–100° @ 11 mm, refr index: 1.464. Insol in water.

SYNS: CUCUMBER ALCOHOL □ FEMA No. 2780 □ NONADIENOL □ 2-trans-6-cis-NONADIEN-1-OL □ VIOLET LEAF ALCOHOL

CONSENSUS REPORTS: Reported in EPA TSCA Inventory.

SAFETY PROFILE: Low toxicity by ingestion and skin contact. A skin irritant. When heated to decomposition it emits acrid smoke and irritating fumes.

NMW500   CAS: 124-19-6   HR: 2
1-NONANAL

mf: C₉H₁₈O   mw: 142.27

PROP: Colorless to light-yellow liquid; citrus-rose odor. D: 0.820–0.830, refr index: 1.422–1.429, flash p: 162°F, bp: 190–192°. Found in at least 20 essential oils, including rose and citrus oils and several species of pine oil (FCTXAV 11, 95,73). Sol in alc, fixed oils, propylene glycol; insol in glycerin.

SYNS: ALDEHYDE C-9 □ C-9 ALDEHYDE □ FEMA No. 2782 □ NCI-C61018 □ 1-NONALDEHYDE □ 1-NONYL ALDEHYDE □ PELARGONIC ALDEHYDE

CONSENSUS REPORTS: Reported in EPA TSCA Inventory.

SAFETY PROFILE: A severe skin irritant. Combustible liquid. Mutation data reported. When heated to decomposition it emits acrid smoke and irritating fumes. See also ALDEHYDES.

NMX000   CAS: 111-84-2   HR: 3
NONANE

mf: C₉H₂₀   mw: 128.29

PROP: Colorless liquid. Mp: −53.7°, fp: −51°, bp: 150.7°, lel: 0.8%, uel: 2.9%, flash p: 88°F (CC), d: 0.718 @ 20°/4°, autoign temp: 374°F, vap press: 10 mm @ 38.0°, vap d: 4.41. Insol in water; sol in abs alc and ether.

SYN: SHELLSOL 140

CONSENSUS REPORTS: Reported in EPA TSCA Inventory.

OSHA PEL: TWA 200 ppm

ACGIH TLV: TWA 200 ppm

DOT CLASSIFICATION: 3; Label: Flammable Liquid

SAFETY PROFILE: Poison by intravenous route. Mildly toxic by inhalation. Irritating to respiratory tract. Narcotic in high concentrations. A very dangerous fire hazard when exposed to heat or flame; can react with oxidizing materials. Explosive in the form of vapor when exposed to heat or flame. Emitted from modern building materials (CENEAR 69,22,91). To fight fire, use CO_2, dry chemical. When heated to decomposition it emits acrid smoke and irritating fumes.

NMY000   CAS: 112-05-0   HR: 3
NONANOIC ACID

mf: C₉H₁₈O₂   mw: 158.27

PROP: Oily, irritant, colorless liquid, with characteristic odor. Fp: 12.24°, mp: 12°, bp: 254°, d: 0.9055 @ 20°/4°. Very sltly sol in water.

SYNS: CIRRASOL 185A □ EMFAC 1202 □ HEXACID C-9 □ n-NONOIC ACID □ n-NONYLIC ACID □ 1-OCTANECARBOXYLIC ACID □ PELARGIC ACID □ PELARGON (RUSSIAN) □ PELARGONIC ACID

CONSENSUS REPORTS: Reported in EPA TSCA Inventory.

SAFETY PROFILE: Poison by intravenous route. Moderately toxic by ingestion. A severe skin and eye irritant. When heated to decomposition it emits acrid smoke and irritating fumes.

NNA100   CAS: 762-13-0   HR: 1
NONANOYL PEROXIDE

DOT: UN 2130

mf: C₁₈H₃₄O₄   mw: 314.52

SYNS: DIPELARGONYL PEROXIDE □ PELARGONOYL PEROXIDE □ PELARGONYL PEROXIDE □ PELARGONYL PEROXIDE, technically pure (DOT) □ PEROXIDE, BIS(1-OXONONYL)

CONSENSUS REPORTS: Reported in EPA TSCA Inventory.

DOT CLASSIFICATION: 3; Label: Flammable Liquid

SAFETY PROFILE: A peroxide. Handle with care. When heated to decomposition it emits acrid smoke and irritating vapors.

NNA300 CAS: 2463-53-8 HR: 2
2-NONENAL
mf: $C_9H_{16}O$ mw: 140.25
PROP: White to sltly yellow liquid; fatty, violet odor. D: 0.850–0.870, refr index: 1.457. Sol in alc, fixed oils; insol in water.
SYNS: FEMA No. 3213 □ β-HEXYLACROLEIN □ NON-2-ENAL □ 2-NONEN-1-AL □ α-NONENYL ALDEHYDE □ trans-2-NONENAL (FCC)
CONSENSUS REPORTS: Reported in EPA TSCA Inventory.
SAFETY PROFILE: Moderately toxic by skin contact. Mildly toxic by ingestion. A severe skin irritant. Mutation data reported. When heated to decomposition it emits acrid smoke and irritating fumes. See also ALDEHYDES.

NNA325 CAS: 2277-19-2 HR: 3
6-NONENAL, (Z)-
mf: $C_9H_{16}O$ mw: 140.25
PROP: A liquid. Flash p: 130°F.
SYNS: cis-6-NONENAL □ cis-6-NONEN-1-AL □ (Z)-6-NONENAL
CONSENSUS REPORTS: Reported in EPA TSCA Inventory.
SAFETY PROFILE: Low toxicity by ingestion and skin contact. A skin irritant. Flammable liquid when exposed to heat, sparks, or flame. When heated to decomposition it emits acrid smoke and irritating fumes.

NNA530 HR: D
cis-6-NONEN-1-OL
mf: $C_9H_{18}O$ mw: 142.23
PROP: White to slightly yellow liquid; powerful, melonlike odor. D: 0.850–0.870, refr index: 1.448. Insol in water.
SYN: FEMA No. 3465
SAFETY PROFILE: When heated to decomposition it emits acrid smoke and irritating fumes.

NNB300 CAS: 26027-38-3 HR: 3
NONOXYNOL-9
mf: $(C_2H_4O)_n•C_{15}H_{24}O$
PROP: Very stable. Compounds with $n<15$ are yellow to almost colorless liquids; $n>20$ are pale-yellow to off-white pastes or waxes. Compounds with $n<6$ are sol in oil; higher n's are sol in water. Almost colorless liquid. D: 1.06, fp: 26°F, pour point: 37°F, flash p: 535–555°F, visc (25°): 175–250 cp. Sol in water, ethanol, ethylene glycol, ethylene dichloride, xylene, corn oil. Insol in Stoddard solvent, deodorized kerosene, low-viscosity white mineral oil.
SYNS: N-9 □ NONYLPHENOXYPOLY(ETHYLENEOXY)ETHANOL □ NP-9
CONSENSUS REPORTS: Reported in EPA TSCA Inventory.
SAFETY PROFILE: Poison by intraperitoneal route. Experimental reproductive effects. Mutation data reported. An active ingredient in contraceptive jellies, foams, and creams. Combustible when exposed to heat or flames. When heated to decomposition it emits acrid smoke and irritating fumes. See also ALCOHOLS.

NNB400 CAS: 143-13-5 HR: 2
NONYL ACETATE
mf: $C_{11}H_{22}O_2$ mw: 186.29
PROP: Colorless liquid; fruity odor. D: 0.864, refr index: 1.422, flash p: 153°F. Sol in alc, ether; insol in water.
SYN: FEMA No. 2788
SAFETY PROFILE: Combustible liquid. When heated to decomposition it emits acrid smoke and irritating fumes.

NNB500 CAS: 143-08-8 HR: 2
n-NONYL ALCOHOL
mf: $C_9H_{20}O$ mw: 144.29
PROP: Colorless to yellow liquid; rose-citrus odor. D: 0.827 @ 20°/4°, refr index: 1.43–1.435, mp: −5°, bp: 213.5°, flash p: 169°F. Insol in water; misc in alc, ether, chloroform.

SYNS: ALCOHOL C-9 □ FEMA No. 2789 □ NONALOL □ NONAN-1-OL □ 1-NONANOL □ NONYL ALCOHOL □ OCTYL CARBINOL □ PELARGONIC ALCOHOL

CONSENSUS REPORTS: Reported in EPA TSCA Inventory.

SAFETY PROFILE: Mildly toxic by ingestion, skin contact, and inhalation. Experimental reproductive effects. Combustible liquid. When heated to decomposition it emits acrid smoke and irritating fumes. See also ALCOHOLS.

NNE000   CAS: 5283-67-0   HR: 2
NONYLTRICHLOROSILANE

DOT: UN 1799

mf: $C_9H_{19}Cl_3Si$ mw: 261.72

CONSENSUS REPORTS: Reported in EPA TSCA Inventory.

DOT CLASSIFICATION: 8; Label: Corrosive

SAFETY PROFILE: A corrosive irritant to skin, eyes, and mucous membranes. When heated to decomposition it emits toxic fumes of Cl^-. See also CHLOROSILANES.

NNE100   CAS: 13257-44-8   HR: 1
2-NONYNAL DIMETHYLACETAL

mf: $C_{11}H_{20}O_2$ mw: 184.31

SYNS: 2-NONYN-1-AL DIMETHYLACETAL □ 2-NONYNAL, DIMETHYL ACETAL □ PARMAVERT

CONSENSUS REPORTS: Reported in EPA TSCA Inventory.

SAFETY PROFILE: Low toxicity by ingestion and skin contact. A skin irritant. When heated to decomposition it emits acrid smoke and irritating fumes.

NNG000   CAS: 121-46-0   HR: 3
NORBORNADIENE

DOT: UN 2251

mf: C_7H_8 mw: 92.15

PROP: A liquid. Fp: −19.1°, bp: 90.3°.

SYNS: BICYCLO(2.2.1)HEPTADIENE □ 2,5-NORBORNADIENE

CONSENSUS REPORTS: EPA Extremely Hazardous Substances List. Reported in EPA TSCA Inventory.

DOT CLASSIFICATION: 3; Label: Flammable Liquid

SAFETY PROFILE: Poison by intravenous route. Moderately toxic by ingestion and intraperitoneal routes. Mildly toxic by inhalation. A flammable liquid. When heated to decomposition it emits acrid smoke and irritating fumes.

NNM000   CAS: 492-41-1   HR: 3
(−)-NOREPHEDRINE

mf: $C_9H_{13}NO$ mw: 151.23

PROP: A solid. Mp: 51°.

SYNS: α-(1-AMINOETHYL)-BENZYL ALCOHOL □ 2-AMINO-1-PHENYL-1-PROPANOL □ FENILPROPAN-OLAMINA (ITALIAN) □ MYDRIATIN □ PHENYL-PROPANOLAMINE □ PPA □ PROPADRINE □ USAF CS-6

CONSENSUS REPORTS: Reported in EPA TSCA Inventory.

SAFETY PROFILE: A human poison by ingestion. Poison experimentally by intravenous, subcutaneous, and intraperitoneal routes. Moderately toxic by an unspecified route. Human systemic effects by ingestion: sleep, increased pulse rate without blood pressure decrease, and chronic pulmonary edema or congestion, convulsions, headache, and blood pressure elevation. Used in production of drugs of abuse. When heated to decomposition it emits toxic fumes of NO_x.

NNO500   CAS: 51-41-2   HR: 3
l-NOREPINEPHRINE

mf: $C_8H_{11}NO_3$ mw: 169.20

PROP: Microcrystals. Mp: 216.5–218° (decomp). Sol in water.

SYNS: ADRENOR □ AKTAMIN □ l-2-AMINO-1-(3,4-DIHYDROXYPHENYL)ETHANOL □ (R)-4-(2-AMINO-1-HYDROXYETHYL)-1,2-BENZENEDIOL □ l-α-(AMINOMETHYL)-3,4-DIHYDROXYBENZYL ALCOHOL □ (−)-α-(AMINOMETHYL)PROTOCATECHUYL ALCOHOL □ ARTERENOL □ l-ARTERENOL □ l-1-(3,4-DIHYDROXYPHENYL)-2-AMINOETHANOL □ l-3,4-DIHYDROXYPHENYLETHANOLAMINE □ LEVART-ERENOL □ LEVOARTERENOL □ LEVONOR-ADRENALINE □ LEVONOREPINEPHRINE □ LEVO-PHED □ (−)-NORADREC □ NORADRENALIN □ NORADRENALINA (ITALIAN) □ NORADRENALINE □ (−)-NORADRENALINE □ d-(−)-NORADRENALINE □ l-NORADRENALINE □ NORADRENLINE □ NOR-

ARTRINAL □ NOREPINEPHRINE □ (−)-NOR-EPINEPHRINE □ NOREPIRENAMINE □ SYMPATHIN E

SAFETY PROFILE: Poison by ingestion, intraperitoneal, subcutaneous, and intravenous routes. An experimental teratogen. Experimental reproductive effects. Human mutation data reported. A sympathomimetic vasopressor. When heated to decomposition it emits toxic fumes of NO_x.

NNP500   **CAS: 68-22-4**   **HR: 3**
19-NORETHISTERONE
mf: $C_{20}H_{26}O_2$ mw: 298.46
PROP: Crystals from EtOAc. Mp: 203–204°.
SYNS: 17-α-ETHINYLESTRA-4-EN-17-β-OL-3-ONE □ 17-α-ETHINYL-17-β-HYDROXY-Δ:4-ESTREN-3-ONE □ 17-α-ETHINYL-19-NORTESTOSTERONE □ 17-α-ETHYNYL-4-ESTREN-17-OL-3-ONE □ 17-α-ETHYNYL-17-HYDROXY-4-ESTREN-3-ONE □ 17-α-ETHYNYL-17-β-HYDROXY-19-NORANDROST-4-EN-3-ONE □ 17-α-ETHYNYL-19-NORANDROST-4-EN-17-β-OL-3-ONE □ 17-α-ETHYNYL-19-NOR-4-ANDROSTEN-17-β-OL-3-ONE □ 17-α-ETHYNYL-19-NORTESTOSTERONE □ (17-α)-17-HYDROXY-19-NORPREGN-4-EN-20-YN-3-ONE □ 17-β-HYDROXY-19-NORPREGN-4-EN-20-YN-3-ONE □ 17-HYDROXY-19-NOR-17-α-PREGN-4-EN-20-YN-3-ONE □ 19-NOR-ETHINYL-4,5-TESTOSTERONE □ 19-NOR-17-α-ETHYNYLANDROSTEN-17-β-OL-3-ONE □ 19-NOR-17-α-ETHYNYL-17-β-HYDROXY-4-ANDROSTEN-3-ONE □ 19-NOR-17-α-ETHYNYLTESTOSTERONE □ NORLUTIN

CONSENSUS REPORTS: NTP 10th Report on Carcinogens. IARC Cancer Review: Animal Sufficient Evidence IMEMDT 7,294,87. EPA Genetic Toxicology Program.
SAFETY PROFILE: Confirmed carcinogen with experimental carcinogenic, tumorigenic, and teratogenic data. Mildly toxic by ingestion. Human systemic effects by ingestion: dermatitis and androgenic effects. Human teratogenic effects: developmental abnormalities of the musculoskeletal system and urogenital system; and behavioral effects in the newborn. Human reproductive effects: spermatogenesis; testes, epididymis, sperm duct changes; impotence; male breast development; other male effects; ovaries, fallopian tube changes; menstrual cycle changes or disorders; uterus, cervix, vagina effects; postpartum effects; changes in female fertility. Experimental reproductive effects. Human mutation data reported. When heated to decomposition it emits acrid smoke and irritating fumes.

NNQ100   **HR: D**
NORFLURAZON
SYN: 4-CHLORO-5-(METHYLAMINO)-2-(α,α,α-TRIFLUORO-m-TOLYL)-3(2H)-PYRIDAZINONE
SAFETY PROFILE: When heated to decomposition emits toxic fumes of NO_x, F^-, and Cl^-.

NOA600   **CAS: 10540-29-1**   **HR: 3**
NOVADEX
mf: $C_{26}H_{29}NO$ mw: 371.56
PROP: Crystals from pet ether. Mp: 96–98°. cis-Form base: mp: 72–74° from methanol.
SYNS: cis-1-(p-(2-(N,N-DIMETHYLAMINO)ETHOXY)PHENYL)-1,2-DIPHENYL-BUT-1-ENE □ cis-N,N-DIMETHYL-2-(p-(1,2-DIPHENYL-1-BUTENYL)PHENOXY)ETHYLAMINE □ (Z)-2-(p-(1,2-DIPHENYL-1-BUTENYL)-PHENOXY)-N,N-DIMETHYL-ETHYLAMINE □ ICI 46,474 □ TAMOXIFEN

CONSENSUS REPORTS: NTP 10th Report on Carcinogens. IARC Cancer Review: Group 1 IMEMDT 66,253,96; Human Sufficient Evidence IMEMDT 66,253,96 (Benefits outweigh risk for breast cancer patients. 22 Feb, 1997). EPA Genetic Toxicology Program.
SAFETY PROFILE: Confirmed human carcinogen. Moderately toxic by ingestion and intraperitoneal routes. Human systemic effects by an unspecified route: nausea or vomiting, leukopenia, thrombocytopenia, and skin changes. An experimental teratogen. Other experimental reproductive effects. Human mutation data reported. When heated to decomposition it emits toxic fumes of NO_x.

NOB000   **CAS: 1476-53-5**   **HR: 3**
NOVOBIOCIN, MONOSODIUM SALT
mf: $C_{31}H_{35}N_2O_{11} \cdot Na$ mw: 634.67
PROP: A solid. Mp: 210–215° (decomp).
SYNS: ALBAMYCIN □ ALBAMYCIN SODIUM □ CATHOMYCIN SODIUM □ CATHOMYCIN SODIUM

LYOVAC □ INAMYCIN □ MONOSODIUM NOVOBIOCIN □ NOVOBIOCIN MONOSODIUM □ NOVOBIOCIN, SODIUM derivative □ SODIUM ALBAMYCIN □ SODIUM NOVOBIOCIN □ U-6591
SAFETY PROFILE: Poison by intraperitoneal and subcutaneous routes. Moderately toxic by ingestion. Mutation data reported. An antibiotic with serious side effects which include liver and blood disease. It may also promote the development of resistant strains of staphylococcus. When heated to decomposition it emits toxic fumes of NO_x and Na_2O.

NOB800   CAS: 105-14-6   HR: 3
NOVOL KETONE
mf: $C_9H_{19}NO$ mw: 157.29
SYNS: DF 493 □ 1-(DIETHYLAMINO)-4-PENTANONE □ 5-DIETHYLAMINO-2-PENTANONE □ 2-PENTAN-ONE, 5-(DIETHYLAMINO)-
DOT CLASSIFICATION: 3; Label: Flammable Liquid
SAFETY PROFILE: A poison by intravenous route. A flammable liquid. When heated to decomposition it emits toxic vapors of NO_x.

NOF000   HR: 1
NUISANCE DUSTS and AEROSOLS
SAFETY PROFILE: Variable toxicity depending upon composition. Cause local irritation of eyes, nose, throat, and lungs. Some may lead to chronic bronchitis, emphysema, and bronchial asthma. Dermatitis may result from short contact. Asthma, angioneurotic edema, hives, etc., may result from short periods of inhalation. A topic eczema, angioneurotic edema, hives, etc., may also result from prolonged contact. A common air contaminant. Nuisance aerosols do evoke some tissue response in the lung upon inhalation of sufficient amounts. However, this reaction is potentially reversible and leaves no scar tissue.

NOG500   CAS: 8008-45-5   HR: 2
NUTMEG OIL, EAST INDIAN
PROP: Major components are α- and β-pinene, camphene, myristicin, dipentene, and sabanene. Found in fruit of *Myristica fragrans Houttuyn* (Fam. *Myristicaceae*). Prepared by steam distillation of dried nutmeg (FCTXAV 14,601,76). Colorless to pale-yellow liquid; odor and taste of nutmeg. East Indian: d: 0.880–0.910, refr index: 1.474–1.488; West Indian: d: 0.854–0.880, refr index: 1.469–1.476 @ 20°. Sol in fixed oils, mineral oil; sltly sol in cold alc; very sol in hot alc, chloroform, ether; insol in glycerin, propylene glycol.
SYNS: MYRISTICA OIL □ NUTMEG OIL □ OIL of MYRISTICA □ OIL of NUTMEG
CONSENSUS REPORTS: Reported in EPA TSCA Inventory.
SAFETY PROFILE: Moderately toxic by ingestion. Low toxicity by skin contact. An experimental teratogen. Experimental reproductive effects. Mutation data reported. A skin irritant. When heated to decomposition it emits acrid smoke and irritating fumes.

NOH500   CAS: 1400-61-9   HR: 3
NYSTATIN
mf: $C_{46}H_{83}NO_{18}$ mw: 938.30
PROP: Yellow to light-tan powder; odor suggestive of cereals. Mp: decomp >160°. Sparingly sol in methanol and ethanol; very sltly sol in water; insol in chloroform, ether, and benzene.
SYNS: BIOFANAL □ CANDEX □ CANDIO-HERMAL □ DIASTATIN □ MORONAL □ MYCOSTATIN □ MYCOSTATIN 20 □ NILSTAT □ NYSTAN □ NYSTATINE □ NYSTAVESCENT □ O-V STATIN
CONSENSUS REPORTS: EPA Genetic Toxicology Program.
SAFETY PROFILE: Poison by intraperitoneal and intravenous routes. Moderately toxic by subcutaneous route. Mildly toxic by ingestion. An experimental teratogen. Mutation data reported. An antibiotic. When heated to decomposition it emits toxic fumes of NO_x.

O

OAJ000   **CAS: 3268-87-9**   **HR: 3**
OCTACHLORODIBENZODIOXIN
mf: $C_{12}Cl_8O_2$ mw: 459.72
PROP: Colorless crystals or needles from trichlorobenzene. Mp: 330–332°.
SYNS: NCI-C03678 □ OCDD □ OCTACHLORO-DIBENZO(b,e)(1,4)DIOXIN □ OCTACHLORODIBENZO-p-DIOXIN □ 1,2,3,4,6,7,8,9-OCTACHLORODIBENZO-DIOXIN
CONSENSUS REPORTS: IARC Cancer Review: Animal Inadequate Evidence IMEMDT 15,41,77.
SAFETY PROFILE: Poison by ingestion. An experimental teratogen. An eye irritant. Questionable carcinogen with experimental tumorigenic data. When heated to decomposition it emits toxic fumes of Cl⁻.

OAP000   **CAS: 2234-13-1**   **HR: 3**
OCTACHLORONAPHTHALENE
mf: $C_{10}Cl_8$ mw: 403.70
PROP: Crystals from cyclohexane. Mp: 197.5–198°, bp: 246–250° @ 0.5 mm. Sol in C_6H_6.
CONSENSUS REPORTS: EPA Extremely Hazardous Substances List. Community Right-To-Know List. Reported in EPA TSCA Inventory.
OSHA PEL: TWA 0.1 mg/m³ (skin); STEL 0.3 mg/m³
ACGIH TLV: TWA 0.1 mg/m³ (skin); STEL 0.3 mg/m³
NIOSH REL: TWA 0.1 mg/m³; STEL 0.3 mg/m³ (skin)
SAFETY PROFILE: Poison by inhalation, ingestion, and skin contact. When heated to decomposition it emits highly toxic fumes of Cl⁻. See also CHLORINATED HYDROCARBONS, AROMATIC.

OAT000   **CAS: 2223-93-0**   **HR: 3**
OCTADECANOIC ACID, CADMIUM
SALT
mf: $C_{36}H_{72}O_4 \cdot Cd$ mw: 681.48
SYNS: ALAIXOL II □ CADMIUM(II) STEARATE □ CADMIUM OCTADECANOATE □ CADMIUM STEARATE □ KADMIUMSTEARAT (GERMAN) □ STABILISATOR SCD □ STABILIZER SCD □ STEARIC ACID, CADMIUM SALT
CONSENSUS REPORTS: EPA Extremely Hazardous Substances List. Cadmium and its compounds are on the Community Right-To-Know List. Reported in EPA TSCA Inventory.
OSHA PEL: TWA 5 µg(Cd)/m³
ACGIH TLV: TWA 0.002 mg(Cd)/m³ (respirable dust), Suspected Human Carcinogen); BEI: 5 µg/g creatinine in urine; 5 µg/L in blood
DFG MAK: DFG BAT: Blood 1.5 µg/dL; Urine 15 µg/dL, Suspected Carcinogen
NIOSH REL: (Cadmium) Reduce to lowest feasible level.
SAFETY PROFILE: Confirmed human carcinogen. Poison by inhalation. Moderately toxic by ingestion. Human systemic effects by inhalation: hallucinations or distorted perceptions; nausea or vomiting, other gastrointestinal effects; weight loss or decreased weight gain; cardiac effects. When heated to decomposition it emits toxic fumes of Cd. See also CADMIUM COMPOUNDS.

OAV000   **CAS: 31566-31-1**   **HR: 3**
OCTADECANOIC ACID, MONOESTER
with 1,2,3-PROPANETRIOL
mf: $C_{21}H_{42}O_4$ mw: 358.63
PROP: Pure-white or cream-colored, wax-like solid; faint odor. Mp: 58–59°, d: 0.97. Sol in (hot) alc, oils, and hydrocarbons.
SYNS: ABRACOL S.L.G □ ADMUL □ ADVAWAX 140 □ ALDO-28 □ ALDO-72 □ ALDO HMS □ ALDO MS □ ALDO MSA □ ALDO MSLG □ ARLACEL 161 □ ARLACEL 169 □ ARMOSTAT 801 □ ATMOS 150 □ ATMUL 67 □

ATMUL 84 □ ATMUL 124 □ CEFATIN □ CELINHOL -A □ CERASYNT 1000-D □ CERASYNT S □ CERASYNT SD □ CERASYNT SE □ CERASYNT WM □ CITOMULGAN M □ CYCLOCHEM GMS □ DERMAGINE □ DISTEARIN □ DREWMULSE TP □ DREWMULSE V □ DRUMULSE AA □ EMCOL CA □ EMCOL MSK □ EMEREST 2400 □ EMEREST 2401 □ EMUL P.7 □ ESTOL 603 □ GLYCERIN MONOSTEARATE □ GLYCEROL MONOSTEARATE □ GLYCERYL MONOSTEARATE □ GROCOR 5500 □ GROCOR 6000 □ HODAG GMS □ IMWITOR 191 □ IMWITOR 900K □ KESSCO 40 □ LIPO GMS 410 □ LIPO GMS 450 □ LIPO GMS 600 □ MONELGIN □ MONO-STEARIN □ OGEEN 515 □ OGEEN GRB □ OGEEN M □ ORBON □ PROTACHEM GMS □ SEDETINE □ STARFOL GMS 450 □ STARFOL GMS 600 □ STARFOL GMS 900 □ STEARIC ACID, MONOESTER with GLYCEROL □ STEARIC MONOGLYCERIDE □ TEGIN □ TEGIN 503 □ TEGIN 515 □ UNIMATE GMS □ USAF KE-7 □ WITCONOL MS □ WITCONOL MST

CONSENSUS REPORTS: Reported in EPA TSCA Inventory.

SAFETY PROFILE: Poison by intraperitoneal route. When heated to decomposition it emits acrid smoke and irritating fumes. See also ESTERS.

OAX000   CAS: 112-92-5   HR: 3
1-OCTADECANOL

mf: $C_{18}H_{38}O$ mw: 270.56

PROP: Colorless solid or flakes. Fp: 57.95°, mp: 58°, bp: 202° @ 10 mm, d: 0.8124 @ 59°/4°.

SYNS: ADOL □ ADOL 68 □ ATALCO S □ CO-1895 □ CO-1897 □ CRODACOL-S □ DECYL OCTYL ALCOHOL □ DYTOL E-46 □ LOROL 28 □ OCTADECANOL □ n-OCTADECANOL □ OCTA DECYL ALCOHOL □ n-OCTADECYL ALCOHOL □ POLAAX □ SIPOL S □ SIPONOL S □ STEAROL □ STEARYL ALCOHOL □ STERAFFINE □ USP XIII STEARYL ALCOHOL

CONSENSUS REPORTS: Reported in EPA TSCA Inventory.

SAFETY PROFILE: Mildly toxic by ingestion. Questionable carcinogen with experimental neoplastigenic data. A skin and eye irritant. Flammable when exposed to heat or flame; can react with oxidizing materials. To fight fire, use foam, CO_2, dry chemical. When heated to decomposition it emits acrid smoke and irritating fumes. See also ALCOHOLS.

OBC000   CAS: 124-30-1   HR: 3
OCTADECYLAMINE

mf: $C_{18}H_{39}N$ mw: 269.58

PROP: Crystals. Mp: 49–52°, bp: 183.0–183.1° @ 5 mm.

SYNS: ADOGENEN 142 □ ALAMINE 7 □ ALAMINE 7D □ AMINE AB □ 1-AMINOOCTADECANE □ ARMEEN 18 □ ARMEEN 18D □ ARMOFILM □ CRODAMINE 1.18D □ FARMIN 80 □ KEMAMINE P990 □ MONOOCTADECYLAMINE □ NISSAN AMINE AB □ n-OCTADECYLAMINE □ 1-OCTADECYLAMINE □ OKTADECYLAMIN (CZECH) □ STEARAMINE □ STEARYLAMINE □ n-STEARYLAMINE

CONSENSUS REPORTS: Reported in EPA TSCA Inventory.

SAFETY PROFILE: Poison by intraperitoneal route. Moderately toxic by ingestion. A skin irritant. When heated to decomposition it emits toxic fumes of NO_x. See also AMINES.

OBG000   CAS: 112-96-9   HR: 3
OCTADECYL ISOCYANATE

DOT: UN 2207/UN 2478/UN 3080

mf: $C_{19}H_{37}NO$ mw: 295.57

PROP: Mp: 15–18°, bp: 150–180° @ 0.75 mm, d: 0.86 @ 25°.

SYNS: ISOCYANIC ACID, OCTADECYL ESTER □ TONCO-70

CONSENSUS REPORTS: Reported in EPA TSCA Inventory.

DOT CLASSIFICATION: 6.1; Label: KEEP AWAY FROM FOOD (UN 2207); DOT Class: 6.1; Label: Poison (UN 2206); DOT Class: 6.1; Label: Poison, Flammable Liquid (UN 3080); DOT Class: 3; Label: Flammable Liquid, Poison (UN 2478)

SAFETY PROFILE: Poison by intravenous route. A flammable liquid. When heated to decomposition it emits toxic fumes of NO_x. See also ISOCYANATES.

OBI000   CAS: 112-04-9   HR: 2
OCTADECYLTRICHLOROSILANE

DOT: UN 1800

mf: $C_{18}H_{37}Cl_3Si$ mw: 387.99

PROP: A liquid. D: 0.95 @ 4°, bp: 160° @ 13 mm.

SYN: TRICHLOROOCTADECYLSILANE

CONSENSUS REPORTS: Reported in EPA TSCA Inventory.

DOT CLASSIFICATION: 8; Label: Corrosive

SAFETY PROFILE: A corrosive irritant to skin, eyes, and mucous membranes. Reacts with water or steam to produce toxic and corrosive fumes. When heated to decomposition it emits toxic fumes of Cl⁻. See also CHLOROSILANES.

OBO000 CAS: 360-89-4 HR: 1
1,1,1,2,3,4,4,4-OCTAFLUORO-2-BUTENE

DOT: UN 2422

mf: C₄F₈   mw: 200.04

PROP: A liquid. Mp: −139°, fp: −135°, bp: 1.2°.

SYNS: FC-1318 □ OCTAFLUOROBUTENE-2 □ OCTAFLUOROBUT-2-ENE (DOT) □ PERFLUOROBUT-2-ENE □ PERFLUORO-2-BUTENE (DOT)

CONSENSUS REPORTS: Reported in EPA TSCA Inventory. EPA Genetic Toxicology Program.

DOT CLASSIFICATION: 2.2; Label: Nonflammable Gas

SAFETY PROFILE: Mildly toxic by inhalation. Mutation data reported. When heated to decomposition it emits toxic fumes of F⁻. See also FLUORIDES.

OBU100 CAS: 1317-70-0 HR: 2
OCTAHEDRITE (MINERAL)

mf: O₂Ti   mw: 79.90

SYNS: ANATASE □ TIOXIDE A-HR

CONSENSUS REPORTS: IARC Cancer Review: Group 3 IMEMDT 47,307,89; Animal Limited Evidence IMEMDT 47,307,89; Human Inadequate Evidence IMEMDT 47,307,89. Reported in EPA TSCA Inventory.

SAFETY PROFILE: A questionable carcinogen.

OBY000 CAS: 20917-49-1 HR: 3
OCTAHYDRO-1-NITROSOAZOCINE

mf: C₇H₁₄N₂O   mw: 142.23

SYNS: NHMI □ N-NITROSOAZACYCLOOCTANE □ NITROSOHEPTAMETHYLENEIMINE □ N-NITROSOHEPTAMETHYLENEIMINE □ NITROSO-HEPTAMETHYLENIMIN (GERMAN)

SAFETY PROFILE: Poison by ingestion and subcutaneous routes. Questionable carcinogen with experimental carcinogenic and tumorigenic data. Mutation data reported. When heated to decomposition it emits toxic fumes of NOₓ. See also N-NITROSO COMPOUNDS.

OCE000 CAS: 104-50-7 HR: 1
γ-OCTALACTONE

mf: C₈H₁₄O₂   mw: 142.22

PROP: Colorless to pale-yellow liquid; coconut odor. D: 0.970–0.980, bp: 132–133° @ 5.5 mm, refr index: 1.443–1.447. Sol in alc; sltly sol in water.

SYNS: γ-n-BUTYL-γ-BUTYROLACTONE □ FEMA No. 2798 □ 5-HYDROXYOCTANOIC ACID LACTONE □ OCTANOLIDE-1,4 □ TETRAHYDRO-6-PROPYL-2H-PYRAN-2-ONE

CONSENSUS REPORTS: Reported in EPA TSCA Inventory.

SAFETY PROFILE: Mildly toxic by ingestion. A skin irritant. When heated to decomposition it emits acrid smoke and irritating fumes.

OCM000 CAS: 152-16-9 HR: 3
OCTAMETHYLPYROPHOSPHORAMIDE

mf: C₈H₂₄N₄O₃P₂   mw: 286.30

PROP: Viscous liquid. Mp: 20–21°, bp: 154° @ 2.0 mm, d: 1.09 @ 25°/4°. Misc with water; sol in most org solvs; almost insol in higher aliphatic hydrocarbons.

SYNS: BIS(BISDIMETHYLAMINOPHOSPHONOUS)ANHYDRIDE □ BIS(DIMETHYLAMINO)PHOSPHONOUS ANHYDRIDE □ BIS(DIMETHYLAMINO)PHOSPHORIC ANHYDRIDE □ BIS-N,N,N',N'-TETRAMETHYLPHOSPHORODIAMIDIC ANHYDRIDE □ ENT 17,291 □ LETHALAIRE G-59 □ OCTAMETHYL-DIFOSFORZUURTETRAMIDE (DUTCH) □ OCTAMETHYLDIPHOSPHORAMIDE □ OCTAMETHYL-DIPHOSPHORSAEURE-TETRAMID (GERMAN) □ OCTAMETHYL PYROPHOSPHORTETRAMIDE □ OCTAMETHYL TETRAMIDO PYROPHOSPHATE □ OMPA □ OMPACIDE □ OMPATOX □ OMPAX □ OTTOMETIL-PIROFOSFORAMMIDE (ITALIAN) □ PESTOX □ PESTOX 3 □ PESTOX III □ PYROPHOSPHORIC ACID OCTAMETHYLTETRAAMIDE □ PYROPHOSPHORYLTETRAKISDIMETHYLAMIDE □ RCRA WASTE NUMBER P085 □ SCHRADAN □

SCHRADANE (FRENCH) □ SYSTAM □ SYSTOPHOS □ SYTAM □ TETRAKISDIMETHYLAMINOPHOS-PHONOUS ANHYDRIDE
CONSENSUS REPORTS: EPA Extremely Hazardous Substances List. Reported in EPA TSCA Inventory.
SAFETY PROFILE: Poison by ingestion, inhalation, skin contact, intraperitoneal, intravenous, subcutaneous, and ocular routes. Human systemic effects by ingestion: a cholinesterase inhibitor. Has been found to inhibit peripheral cholinesterase without pronounced effects on the central nervous system. An insecticide. When heated to decomposition it emits toxic fumes of NO_x and PO_x. See also PARATHION and ANHYDRIDES.

OCM100   CAS: 1624-01-7   HR: 3
OCTAMETHYLSILANETETRAMINE
mf: $C_8H_{24}N_4Si$ mw: 204.45
SYNS: SILANE 48-12 TETRAKIS □ SILANETETRAM-INE, OCTAMETHYL- □ TETRAKIS(DIMETHYL-AMINO)SILANE
SAFETY PROFILE: A poison by ingestion and skin contact. A severe eye irritant. When heated to decomposition it emits toxic vapors of NO_x.

OCO000   CAS: 124-13-0   HR: 3
1-OCTANAL
mf: $C_8H_{16}O$ mw: 128.24
PROP: Found in about 20 essential oils, including a number of citrus oils (FCTXAV 11,95,73). Colorless to light-yellow liquid; fatty-orange odor. Bp: 163.4°, flash p: 125°F (CC), d: 0.821 @ 20°/4°, refr index: 1.417–1.425, vap d: 4.41. Sol in alc, fixed oils, propylene glycol; insol in glycerin.
SYNS: ALDEHYDE C-8 □ C-8 ALDEHYDE □ FEMA No. 2797 □ OCTANALDEHYDE □ n-OCTYL ALDEHYDE
CONSENSUS REPORTS: Reported in EPA TSCA Inventory.
SAFETY PROFILE: Mildly toxic by ingestion and skin contact. A skin and eye irritant. Flammable liquid when exposed to heat, sparks, or flame. Can react with oxidizing materials. To fight fire, use foam, CO_2, dry chemical. See also ALDEHYDES.

OCU000   CAS: 111-65-9   HR: 3
OCTANE
DOT: UN 1262
mf: C_8H_{18} mw: 114.26
PROP: Clear liquid. Bp: 125.8°, lel: 1.0%, uel: 4.7%, fp: −56.5°, flash p: 56°F, d: 0.7036 @ 20°/4°, autoign temp: 428°F, vap press: 10 mm @ 19.2°, vap d: 3.86. Insol in water; sltly sol in alc, ether; misc with benzene. IDLH 1000 ppm [10%LEL].
SYNS: n-OCTANE □ OKTAN (POLISH) □ OKTANEN (DUTCH) □ OTTANE (ITALIAN)
CONSENSUS REPORTS: Reported in EPA TSCA Inventory.
OSHA PEL: TWA 300 ppm; STEL 375 ppm
ACGIH TLV: TWA 300 ppm
DFG MAK: 500 ppm (2400 mg/m³)
NIOSH REL: (Alkanes) TWA 350 mg/m³
DOT CLASSIFICATION: 3; Label: Flammable Liquid
SAFETY PROFILE: Poison by intravenous route. May act as a simple asphyxiant. See also ARGON for a description of simple asphyxiants. A narcotic in high concentration. Human dermal exposure to undiluted octane for five hours resulted in blister formation but no anesthesia; exposure for one hour caused diffuse burning sensation. A very dangerous fire hazard and severe explosion hazard when exposed to heat, flame, or oxidizers. When heated to decomposition it emits acrid smoke and irritating fumes. See also ALKANES.

OCY000   CAS: 124-07-2   HR: 2
OCTANOIC ACID
mf: $C_8H_{16}O_2$ mw: 144.24
PROP: Colorless, oily liquid or crystals; unpleasant odor, burning rancid taste. D: 0.91 @ 20°, bp: 240°, mp: 17°. Sol in alkalies, EtOH, $CHCl_3$; spar sol in hot H_2O.
SYNS: C-8 ACID □ CAPRYLIC ACID □ n-CAPRYLIC ACID □ 1-HEPTANECARBOXYLIC ACID □ HEXACID 898 □ NEO-FAT 8 □ OCTIC ACID □ n-OCTOIC ACID □ n-OCTYLIC ACID
CONSENSUS REPORTS: Reported in EPA TSCA Inventory.

SAFETY PROFILE: Moderately toxic by intravenous route. Mildly toxic by ingestion. Mutation data reported. A skin irritant. Yields irritating vapors that can cause coughing. When heated to decomposition it emits acrid smoke and irritating fumes.

OCY100   CAS: 589-98-0   **HR: 2**
3-OCTANOL
mf: $C_8H_{18}O$ mw: 130.28
PROP: Colorless liquid; strong, nutty odor. D: 0.816–0.821, bp: 176–177.5°, refr index: 1.425. Sol in alc, fixed oils; insol in water.
SYNS: AMYLETHYLCARBINOL □ ETHYLAMYLCARB-INOL □ ETHYL-n-AMYLCARBINOL □ FEMA No. 3581 □ OCTANOL-3 □ D-n-OCTANOL
SAFETY PROFILE: A moderate skin and eye irritant. When heated to decomposition it emits acrid smoke and irritating fumes.

ODG000   CAS: 111-13-7   **HR: 3**
2-OCTANONE
mf: $C_8H_{16}O$ mw: 128.24
PROP: Colorless liquid; pleasant apple odor. D: 0.813–0.818, refr index: 1.414–1.418, fp: −16°, mp: −20.9°, bp: 173.5°, vap d: 4.4, flash p: 160°F. Sltly sol in water; sol in alc, hydrocarbons, ether, esters.
SYNS: FEMA No. 2802 □ METHYL HEXYL KETONE (FCC)
CONSENSUS REPORTS: Reported in EPA TSCA Inventory.
DOT CLASSIFICATION: 3; Label: Flammable Liquid
SAFETY PROFILE: Poison by ingestion. Moderately toxic by intraperitoneal route. A skin irritant. Flammable liquid when exposed to heat, flame, or oxidizers. To fight fire, use foam, alcohol foam. When heated to decomposition it emits acrid smoke and irritating fumes. See also ETHER and KETONES.

ODI000   CAS: 106-68-3   **HR: 3**
3-OCTANONE
DOT: UN 2271
mf: $C_8H_{16}O$ mw: 128.24
PROP: Liquid; fruity odor. Bp: 157–162°, d: 0.822 @ 20°/20°, flash p: 138°F.

SYNS: AMYL ETHYL KETONE □ EAK □ ETHYL AMYL KETONE □ 5-METHYL-3-HEPTANONE (OSHA)
CONSENSUS REPORTS: Reported in EPA TSCA Inventory.
OSHA PEL: TWA 25 ppm
DOT CLASSIFICATION: 3; Label: Flammable Liquid
SAFETY PROFILE: Poison by intraperitoneal route. Moderately irritating to skin, eyes, and mucous membranes by inhalation. Narcotic in high concentration. Flammable liquid when exposed to heat, sparks, flame, or oxidizers. To fight fire, use foam, CO_2, dry chemical. When heated to decomposition it emits acrid smoke. See also KETONES.

ODQ800   **HR: D**
trans-2-OCTEN-1-AL
mf: $C_8H_{14}O$ mw: 126.20
PROP: Slightly yellow liquid; green odor. D: 0.830–0.850, refr index: 1.421–1.424. Sol in alc, fixed oils; sltly sol in water.
SYN: FEMA No. 3215
SAFETY PROFILE: When heated to decomposition it emits acrid smoke and irritating fumes.

ODW000   CAS: 3391-86-4   **HR: 3**
1-OCTEN-3-OL
mf: $C_8H_{16}O$ mw: 128.24
PROP: Oil.
SYNS: AMYLVINYLCARBINOL □ MATSUTAKE ALCOHOL (JAPANESE)
CONSENSUS REPORTS: Reported in EPA TSCA Inventory.
SAFETY PROFILE: Poison by ingestion and intravenous routes. Moderately toxic by skin contact. When heated to decomposition it emits acrid smoke and irritating fumes.

ODW030   **HR: D**
1-OCTEN-3-YL ACETATE
mf: $C_{10}H_{18}O_2$ mw: 170.24
PROP: Colorless liquid; metallic, mushroom odor. D: 0.865–0.886, refr index: 1.414–1.434 @ 25°. Sol in fixed oils; insol in water, propylene glycol.
SYNS: FEMA No. 3587 □ PINOCARVEOL

SAFETY PROFILE: When heated to decomposition it emits acrid smoke and irritating fumes.

OEG000   CAS: 112-14-1   HR: 2
1-OCTYL ACETATE
mf: $C_{10}H_{20}O_2$ mw: 172.30

PROP: Colorless liquid; orange-jasmine odor. D: 0.865, refr index: 1.418–1.421, mp: −38.5°, bp: 210°, flash p: 190°F. Insol in water; misc with alc, ether, fixed oils.

SYNS: ACETATE C-8 □ ACETIC ACID, OCTYL ESTER □ CAPRYLYL ACETATE □ FEMA No. 2806 □ 1-OCTANOL ACETATE □ n-OCTANYL ACETATE □ OCTYL ACETATE □ n-OCTYL ACETATE □ OCTYL ALCOHOL ACETATE

CONSENSUS REPORTS: Reported in EPA TSCA Inventory.

SAFETY PROFILE: Moderately toxic by ingestion. A skin irritant. Combustible liquid. When heated to decomposition it emits acrid smoke and irritating fumes. See also ESTERS.

OEG100   HR: 2
3-OCTYL ACETATE
mf: $C_{10}H_{20}O_2$ mw: 172.27

PROP: Colorless liquid; rosy, minty odor. D: 0.856–0.860, refr index: 1.414, fp: 190°. Sol in alc, propylene glycol, fixed oils; sltly sol in water.

SYN: FEMA No. 3583

SAFETY PROFILE: Combustible liquid. When heated to decomposition it emits acrid smoke and irritating fumes.

OEI000   CAS: 111-87-5   HR: 3
OCTYL ALCOHOL
mf: $C_8H_{18}O$ mw: 130.28

PROP: Colorless liquid with penetrating aromatic odor. D: 0.827 @ 20°, mp: −16.7°, bp: 194.5°, flash p: 178°F. Sol in water; misc in alc, ether, and chloroform. Found in several citrus oils and at least 10 other natural sources (FCTXAV 11,95,73).

SYNS: ALCOHOL C-8 □ ALFOL 8 □ CAPRYL ALCOHOL □ CAPRYLIC ALCOHOL □ DYTOL M-83 □ EPAL 8 □ FEMA No. 2800 □ HEPTYL CARBINOL □ 1-HYDROXYOCTANE □ LOROL 20 □ OCTANOL □ n-OCTANOL □ 1-OCTANOL (FCC) □ OCTILIN □ OCTYL ALCOHOL, NORMAL-PRIMARY □ PRIMARY OCTYL ALCOHOL □ SIPOL L8

CONSENSUS REPORTS: Reported in EPA TSCA Inventory.

SAFETY PROFILE: Poison by intravenous route. Moderately toxic by ingestion. Mutation data reported. A skin irritant. Combustible liquid when exposed to heat or flame; can react with oxidizing materials. To fight fire, use water foam, fog, alcohol foam, dry chemical, CO_2. See also ALCOHOLS.

OEK010   CAS: 693-16-3   HR: 3
2-OCTYLAMINE
DOT: UN 2733

mf: $C_8H_{19}N$ mw: 129.28

SYNS: 2-AMINOOCTANE □ CAPRYLAMINE □ HEPTYLAMINE, 1-METHYL- □ 1-METHYLHEPTYL-AMINE □ 2-OCTANAMINE

CONSENSUS REPORTS: Reported in EPA TSCA Inventory.

DOT CLASSIFICATION: 8; Label: Corrosive, Flammable Liquid (UN 2734); DOT Class: 3; Label: Flammable Liquid, Corrosive (UN 2733)

SAFETY PROFILE: Poison by intravenous route. A flammable liquid. When heated to decomposition it emits toxic vapors of NO_x.

OES000   CAS: 113-48-4   HR: 2
N-OCTYL BICYCLOHEPTENE DICARB-OXIMIDE
mf: $C_{17}H_{25}NO_2$ mw: 275.43

SYNS: BICYCLO(2.2.1)HEPTENE-2-DICARBOXYLIC ACID, 2-ETHYLHEXYLIMIDE □ ENDOMETHYLENE-TETRAHYDROPHTHALIC ACID, N-2-ETHYLHEXYL IMIDE □ ENT 8,184 □ N-(2-ETHYLHEXYL)BICYCLO-(2,2,1)-HEPT-5-ENE-2,3-DICARBOXIMIDE □ N-2-ETHYLHEXYLIMIDEENDOMETHYLENETETRAHYDR OPHTHALIC ACID □ N-(2-ETHYLHEXYL)-5-NORBORNENE-2,3-DICARBOXIMIDE □ 2-(2-ETHYLHEXYL)-3a,4,7,7a-TETRAHYDRO-4,7-METHANO-1H-ISOINDOLE-1,3(2H)-DIONE □ MGK-264 □ OCTACIDE 264 □ N-OCTYLBICYCLO-(2.2.1)-5-HEPTENE-2,3-DICARBOXIMIDE □ PYRODONE □ SYNERGIST 264 □ VAN DYK 264

SAFETY PROFILE: Moderately toxic by ingestion, skin contact, and intraperitoneal routes. Experimental reproductive effects. Large doses can cause central nervous system stimulation followed by depression.

When heated to decomposition it emits toxic fumes of NO_x.

OEY100   CAS: 112-32-3   HR: 1
OCTYL FORMATE
mf: $C_9H_{18}O_2$ mw: 158.27
PROP: Colorless liquid; fruity odor. D: 0.869, refr index: 1.418. Sol in fixed oils, propylene glycol; insol in glycerin.
SYNS: FEMA No. 2809 □ FORMIC ACID, OCTYL ESTER □ n-OCTYL FORMATE
CONSENSUS REPORTS: Reported in EPA TSCA Inventory.
SAFETY PROFILE: A skin irritant. When heated to decomposition it emits acrid smoke and irritating fumes.

OFA000   CAS: 1034-01-1   HR: 3
OCTYL GALLATE
mf: $C_{15}H_{22}O_5$ mw: 282.37
SYNS: n-OCTYL ESTER of 3,4,5-TRIHYDROXYBENZOIC ACID □ OKTYLESTER KYSELINY GALLOVE
CONSENSUS REPORTS: Reported in EPA TSCA Inventory.
SAFETY PROFILE: A poison by intraperitoneal route. Moderately toxic by ingestion. When heated to decomposition it emits acrid smoke and irritating fumes.

OFE000   CAS: 26530-20-1   HR: 2
2-OCTYL-4-ISOTHIAZOLIN-3-ONE
mf: $C_{11}H_{19}NOS$ mw: 213.37
SYNS: KATHON LP PRESERVATIVE □ KATHON SP 70 □ MICRO-CHEK 11 □ MICRO-CHEK SKANE □ OCTHILINONE □ 2-OCTYL-3(2H)-ISOTHIAZOLONE □ PANCIL □ RH 893 □ SKANE M8
CONSENSUS REPORTS: Reported in EPA TSCA Inventory.
DFG MAK: 0.05 mg/m³
SAFETY PROFILE: Moderately toxic by ingestion and skin contact. A skin and severe eye irritant. A mildewcide. When heated to decomposition it emits very toxic fumes of SO_x and NO_x. See also KETONES.

OFE030   CAS: 141-59-3   HR: 3
tert-OCTYLMERCAPTAN (DOT)

DOT: UN 3023
mf: $C_8H_{18}S$ mw: 146.32
SYNS: tert-OCTANETHIOL □ T-OCTYL MERCAPTAN □ terc.OKTANTHIOL □ 2-PENTANETHIOL, 2,4,4-TRIMETHYL- □ 2,4,4-TRIMETHYL-2-PENTANETHIOL
CONSENSUS REPORTS: Reported in EPA TSCA Inventory.
DOT CLASSIFICATION: 6.1; Label: Poison, Flammable Liquid
SAFETY PROFILE: A poison. A flammable liquid. When heated to decomposition it emits toxic vapors of SO_x.

OFG100   CAS: 29806-73-3   HR: 1
OCTYL PALMITATE
mf: $C_{24}H_{48}O_2$ mw: 368.72
SYNS: CERAPHYL 368 □ 2-ETHYLHEXYL PALMITATE □ HEXADECANOIC ACID, 2-ETHYLHEXYL ESTER (9CI) □ PALMITIC ACID, 2-ETHYLHEXYL ESTER □ WICKENOL 155
CONSENSUS REPORTS: Reported in EPA TSCA Inventory.
SAFETY PROFILE: Low toxicity by ingestion and skin contact. A skin irritant. When heated to decomposition it emits acrid smoke and irritating fumes.

OFK000   CAS: 27193-28-8   HR: 3
OCTYL PHENOL
mf: $C_{14}H_{22}O$ mw: 206.36
PROP: White to light pink flakes. Bp: 280–283°, fp: 72–74°, d: 0.941 @ 24°/24°.
SYN: USAF RH-6
CONSENSUS REPORTS: Reported in EPA TSCA Inventory.
DOT CLASSIFICATION: 6.1; Label: KEEP AWAY FROM FOOD
SAFETY PROFILE: Poison by intraperitoneal route. Combustible when exposed to heat or flame; can react vigorously with oxidizing materials. When heated to decomposition it emits acrid smoke and irritating fumes. See also PHENOL.

OGE000   CAS: 5283-66-9   HR: 2
OCTYLTRICHLOROSILANE
DOT: UN 1801
mf: $C_8H_{17}Cl_3Si$ mw: 247.69

O

PROP: Fuming liquid. D: 1.07 @ 20°/4°, bp: 224–226°.

CONSENSUS REPORTS: Reported in EPA TSCA Inventory.

DOT CLASSIFICATION: 8; Label: Corrosive

SAFETY PROFILE: A corrosive irritant to skin, eyes, and mucous membranes. Will react with water or steam to produce toxic and corrosive fumes. When heated to decomposition it emits toxic fumes of Cl⁻. See also CHLOROSILANES.

OGG000   CAS: 3091-25-6   HR: 2
OCTYLTRICHLOROSTANNANE

mf: $C_8H_{17}Cl_3Sn$ mw: 338.29

SYNS: MONO-n-OCTYLTIN TRICHLORIDE □ MONO-N-OCTYL-ZINN-TRICHLORID (GERMAN) □ STAN-NANE, TRICHLOROOCTYL- □ TIN, OCTYL-, TRICHLORIDE

CONSENSUS REPORTS: Reported in EPA TSCA Inventory.

OSHA PEL: TWA 0.1 mg(Sn)/m³ (skin)

ACGIH TLV: TWA 0.1 mg(Sn)/m³; STEL 0.2 mg(Sn)/m³ (skin).

DFG MAK: 0.1 mg(Sn)/m³ calculated as total dust

NIOSH REL: (Organotin Compounds) TWA 0.1 mg(Sn)/m³

SAFETY PROFILE: Moderately toxic by an unspecified route. Mildly toxic by ingestion. When heated to decomposition it emits toxic fumes of Cl⁻. See also TIN COMPOUNDS.

OGI000   CAS: 27107-89-7   HR: 2
OCTYLTRIS(2-ETHYLHEXYLOXY-
CARBONYLMETHYLTHIO)STAN-
NANE

mf: $C_{38}H_{74}O_6S_3Sn$ mw: 841.99

SYNS: ACETIC ACID, ((OCTYLSTANNYLIDYNE)-TRITHIO)TRI-, TRIS(2-ETHYLHEXYL) ESTER □ MONO-n-OCTYL-TIN-TRIS-(2-ETHYLHEXYLMERCAPTO-ACETATE) □ TIN, OCTYL-, TRIS(ISOOCTYLTHIO GLYCOLLATE)

CONSENSUS REPORTS: Reported in EPA TSCA Inventory.

OSHA PEL: TWA 0.1 mg(Sn)/m³ (skin)

ACGIH TLV: TWA 0.1 mg(Sn)/m³; STEL 0.2 mg(Sn)/m³ (skin).

DFG MAK: 0.1 mg(Sn)/m³ calculated as total dust

NIOSH REL: (Organotin Compounds) TWA 0.1 mg(Sn)/m³

SAFETY PROFILE: Moderately toxic by ingestion. When heated to decomposition it emits toxic fumes of SO_x. See also TIN COMPOUNDS.

OGI200   HR: D
ODORLESS LIGHT PETROLEUM
HYDROCARBONS

PROP: Liquid; faint odor. Bp: 300–650°.

SAFETY PROFILE: When heated to decomposition it emits acrid smoke and irritating fumes.

OGK000   CAS: 8015-79-0   HR: 3
OIL OF CALAMUS, GERMAN

PROP: Extract of *Acorus calamus L.*, (Fam. *Araceae*). Containing: asarone, eugenol; esters of acetic and heptylic acids. Volatile oil. Yellow to yellowish-brown liquid (viscid); aromatic odor, bitter taste. D: 0.960–0.9707 @ 20°/20°. Very sltly sol in water; misc with alc. Keep well closed, cool, and protected from light.

SYNS: CALAMUS OIL □ KALMUS OEL (GERMAN) □ OIL OF SWEET FLAG

CONSENSUS REPORTS: Reported in EPA TSCA Inventory.

SAFETY PROFILE: Poison by intraperitoneal route. Moderately toxic by ingestion. Questionable carcinogen with experimental tumorigenic data. When heated to decomposition it emits acrid smoke and irritating fumes. See also individual components.

OGM800   HR: D
OIL OF LIME OIL, COLDPRESSED

PROP: Expressed from the peel of *Citrus aurantofolia* Swingle (Mexican type) or *Citrus latifolia* (Tahitian type). Yellow to brown-green liquid. Sol in fixed oils, mineral oil; insol glycerin, propylene glycol.

SAFETY PROFILE: When heated to decomposition it emits acrid smoke and irritating fumes.

OGM850   CAS: 8008-26-2   HR: 2
OIL OF LIME, distilled

PROP: From distillation of juice or crushed fruit of *Citrus aurantofolia Swingle*. Colorless to green-yellow liquid. Sol in fixed oils, mineral oil; insol in glycerin, propylene glycol.

SYNS: DISTILLED LIME OIL □ LIME OIL □ LIME OIL, distilled (FCC) □ OILS, LIME

CONSENSUS REPORTS: Reported in EPA TSCA Inventory.

SAFETY PROFILE: A skin irritant. Questionable carcinogen with experimental tumorigenic data. Mutation data reported. When heated to decomposition it emits acrid smoke and irritating fumes.

OGQ100   CAS: 8007-12-3   HR: 2
OIL OF MACE

PROP: From steam distillation of dried arillode of the ripe seed of *Myristica fragrans Houtt.* (Fam. *Myristicaceae*). Colorless to pale-yellow liquid; odor and taste of nutmeg. East Indian: d: 0.880–0.930, refr index: 1.474–1.488; West Indian: d: 0.854–0.880, refr index: 1.469–1.480 @ 20°. Sol in fixed oils, mineral oil; sltly sol in cold alc; very sol in hot alc, chloroform, ether; insol in glycerin, propylene glycol.

SYNS: NCI-C56484 □ MACE OIL □ OIL OF NUTMEG, expressed

CONSENSUS REPORTS: Reported in EPA TSCA Inventory. EPA Genetic Toxicology Program.

SAFETY PROFILE: Moderately toxic by ingestion. A skin irritant. Human ingestion causes symptoms similar to volatile oil of nutmeg. Human systemic effects: arrhythmias, distorted perceptions, hallucinations, toxic psychosis. When heated to decomposition it emits acrid smoke and irritating fumes.

OGY000   CAS: 8008-57-9   HR: 2
OIL OF ORANGE

PROP: Yellow to deep-orange liquid; characteristic orange taste and odor. D: 0.842–0.846 @ 25°/25°, refr index: 1.472 @ 20°. Sol in 2 vols 90% alc, in 1 vol glacial acetic acid; sltly sol in water; misc with abs alc, carbon disulfide. Keep well closed, cool, and protected from light. Oil expressed from the peel of *Citrus sinensis L. Osbeck* (Fam. *Rutaceae*) (BJCAAI 13,92,59).

SYNS: NEAT OIL OF SWEET ORANGE □ OIL OF SWEET ORANGE □ ORANGE OIL □ ORANGE OIL, coldpressed (FCC) □ SWEET ORANGE OIL

CONSENSUS REPORTS: Reported in EPA TSCA Inventory.

SAFETY PROFILE: A skin irritant. Questionable carcinogen with experimental neoplastigenic data. When heated to decomposition it emits acrid smoke and irritating fumes.

OGY220   CAS: 8006-87-9   HR: 1
OIL OF SANDALWOOD, EAST INDIAN

mf: $C_{15}H_{24}O$ mw: 220.39

PROP: From steam distillation of the ground dried wood of *Santalus album L.* (FCTXAV 12,807,74). Colorless to sltly yellow viscous oily liquid; sandalwood odor. D: 0.965–0.973, refr index: 1.505, bp: 166–167° @ 14 mm. Very sol in alc, fixed oils, propylene glycol; insol in water, glycerin.

SYNS: ARHEOL □ EAST INDIAN SANDALWOOD OIL □ FEMA No. 3006 □ OILS, SANDALWOOD □ OIL OF SANTAL □ SANTAL OIL □ α-SANTALOL (FCC)

CONSENSUS REPORTS: Reported in EPA TSCA Inventory.

SAFETY PROFILE: Low toxicity by ingestion and skin contact. A skin irritant. When heated to decomposition it emits acrid smoke and irritating fumes.

OHI000   CAS: 8006-80-2   HR: 3
OIL OF SASSAFRAS

PROP: Yellow to reddish-yellow liquid; characteristic odor and taste of sassafras. D: 1.065–1.077 @ 25°/25°. Very sltly sol in water; sol in 2 vols 90% alc. Keep well closed, cool, and protected from light. 80% safrol (27ZTAP 3,106,69).

SYN: SASSAFRAS OIL

CONSENSUS REPORTS: Reported in EPA TSCA Inventory.

SAFETY PROFILE: Human poison by ingestion. A skin irritant. When heated to decomposition it emits acrid smoke and irritating fumes.

OHI200   CAS: 85-86-9   HR: 3
OIL RED

mf: $C_{22}H_{16}N_4O$ mw: 352.42

PROP: Dark red crystalline powder. Mp: 199° (decomp). Sol in alkalies, EtOH, and C_6H_6; insol in H_2O.

SYNS: ATUL OIL RED G □ BENZENEAZOBENZENE-AZO-β-NAPHTHOL □ BRASILAZINA OIL SCARLET □ CERASIN RED □ CERASINROT □ CEROTINSCHARL-ACH R □ CERTIQUAL OIL RED □ CERVEN ROZPOUSTEDLOVA 23 □ C.I. 23 □ C.I. 26100 □ C.I. SOLVENT RED 23 □ D & C RED NO. 17 □ FAST OIL SCARLET III □ FAST RED R □ FAT RED (BLUISH) □ FAT RED G □ FAT RED HRR □ FAT RED R □ FAT RED RS □ FAT SCARLET LB □ FAT SOLUBLE RED ZH □ GRASAL BRILLIANT RED G □ FETTPONCEAU G □ FETTROT □ FETTSCHARLACH □ FETTSCHARLACH LB □ MOTIROT 2R □ OIL RED 6566 □ OIL RED AS □ OIL RED B □ OIL RED 3B □ OIL RED G □ OIL RED 3G □ OIL RED O □ OIL SCARLET □ OIL SCARLET AS □ OIL SCARLET G □ ORGANOL RED BS □ ORGANOL SCARLET □ 1-((4-(PHENYLAZO)PHENYL)AZO)-2-NAPHTHALENOL □ 1-((p-PHENYLAZO)PHENYL)AZO-2-NAPHTHOL □ PONCEAU INSOLUBLE OLG □ PYRONALROT B □ 111440 RED □ RED ZH □ ROT C □ ROT G □ ROUGE CERASINE □ SCARLET B FAT SOLUBLE □ SCHULTZ NO. 31 □ SILOTRAS SCARLET TB □ SOMALIA RED III □ SOUDAN III □ STEARIX SCARLET □ SUDAN III □ SUDAN G □ SUDAN G III □ SUDAN III (G) □ SUDAN P III □ SUDAN RED III □ TETRAZOBENZENE-β-NAPHTHOL □ TONEY RED □ TONY RED

CONSENSUS REPORTS: IARC Cancer Review: Group 3 IMEMDT 7,56,87; Animal Inadequate Evidence IMEMDT 8,241,75. Reported in EPA TSCA Inventory.

SAFETY PROFILE: Poison by intraperitoneal route. Moderately toxic by subcutaneous and intrapleural routes. Questionable carcinogen. Mutation data reported. When heated to decomposition it emits toxic fumes of NO_x.

OHM700   CAS: 112-90-3   HR: 2
OLEAMINE

mf: $C_{18}H_{37}N$ mw: 267.56

PROP: Hazy clear liquid with slt amine/lemon odor. Bp: 100°, d: 1.03. Sol in water.

SYNS: ALAMINE 11 □ ARMEEN O □ KEMAMINE P 989 □ NORAM O □ cis-9-OCTADECENYLAMINE □ OLEINAMINE □ OLEYLAMIN (GERMAN) □ OLEYL AMINE

CONSENSUS REPORTS: Reported in EPA TSCA Inventory.

SAFETY PROFILE: Moderately toxic by intraperitoneal route. An experimental teratogen. Experimental reproductive effects. When heated to decomposition it emits toxic fumes of NO_x. See also AMINES.

OHM900   CAS: 3922-90-5   HR: 3
OLEANDOMYCIN

mf: $C_{35}H_{61}NO_{12}$ mw: 687.97

SYNS: AMIMYCIN □ ANTIBIOTIC PA-105 □ OLEANDOMYCINE □ PA 775 □ ROMICIL

SAFETY PROFILE: A poison by intravenous route. Low toxicity by ingestion. When heated to decomposition it emits toxic vapors of NO_x.

OHO000   CAS: 6696-47-5   HR: 3
OLEANDOMYCIN HYDROCHLORIDE

mf: $C_{35}H_{61}NO_{12} \cdot ClH$ mw: 724.43

PROP: Long needles from ethyl acetate. Mp: 134–135°. Very sol in water.

SYN: OLEANDOMYCIN MONOHYDROCHLORIDE

SAFETY PROFILE: Poison by intravenous route. Moderately toxic by ingestion and subcutaneous routes. When heated to decomposition it emits very toxic fumes of NO_x and HCl.

OHO200   CAS: 7060-74-4   HR: 3
OLEANDOMYCIN PHOSPHATE

mf: $C_{35}H_{61}NO_{12} \cdot H_3O_4P$ mw: 785.97

PROP: Sol in DMSO.

SYN: MATROMYCIN

SAFETY PROFILE: Poison by intravenous route. Moderately toxic by ingestion and subcutaneous routes. When heated to decomposition it emits toxic fumes of PO_x and NO_x. See also PHOSPHATES.

OHS000   HR: 3
OLEFINS
PROP: Unsaturated aliphatic hydrocarbons having one or more double bonds.
SAFETY PROFILE: Unsaturated aliphatic hydrocarbons do not differ greatly from paraffins, particularly insofar as their toxic effect on working personnel is concerned. Ethylene and some of its homologs occur in manufactured and natural gases. Ethylene can be used as an anesthetic, and on inhalation in sufficient quantity it can be an asphyxiant. However, the greatest hazard from its use is the danger of fire and explosion. Prolonged or repeated exposures to high concentrations of various olefins have caused certain toxic effects in animals, such as liver damage and hyperplasia of the bone marrow (due to butene-2), but no corresponding effects due to industrial exposures have been discovered in human beings. The diolefins butadiene and isoprene are more irritating than paraffins or mono-olefins of the same volatility. The α-olefins (e.g., 1-octene, 1-octadecene) are particularly reactive because the double bond is on the first carbon. In general the olefins have comparatively low toxicity, but are fire and explosion hazards.

OHU000   CAS: 112-80-1   HR: 3
OLEIC ACID
mf: $C_{18}H_{34}O_2$ mw: 282.52
PROP: Colorless liquid; odorless when pure. Mp: 12° (labile form), mp: 16° (stable form), bp: 203–205° @ 5 mm, bp: 286° @ 100 mm, flash p: 372°F (CC), d: 0.895 @ 25°/25°, autoign temp: 685°F, vap press: 1 mm @ 176.5°. Insol in water; misc in alc and ether.
SYNS: CENTURY CD FATTY ACID □ EMERSOL 210 □ EMERSOL 213 □ EMERSOL 6321 □ EMERSOL 233LL □ EMERSOL 221 LOW TITER WHITE OLEIC ACID □ EMERSOL 220 WHITE OLEIC ACID □ GLYCON RO □ GLYCON WO □ GROCO 2 □ GROCO 4 □ GROCO 5L □ HY-PHI 1055 □ HY-PHI 1088 □ HY-PHI 2066 □ HY-PHI

2088 □ HY-PHI 2102 □ INDUSTRENE 105 □ INDUSTRENE 205 □ INDUSTRENE 206 □ K 52 □ l'ACIDE OLEIQUE (FRENCH) □ METAUPON □ NEO-FAT 90-04 □ NEO-FAT 92-04 □ cis-Δ^9-OCTADECENOIC ACID □ cis-OCTADEC-9-ENOIC ACID □ cis-9-OCTA-DECENOIC ACID □ 9,10-OCTADECENOIC ACID □ PAMOLYN □ RED OIL □ TEGO-OLEIC 130 □ VOPCOLENE 27 □ WECOLINE OO □ WOCHEM No. 320
CONSENSUS REPORTS: Reported in EPA TSCA Inventory.
DFG MAK: Not Classifiable as a Human Carcinogen
SAFETY PROFILE: Poison by intravenous route. Mildly toxic by ingestion. Mutation data reported. A human skin and eye irritant. Questionable carcinogen with experimental tumorigenic data. Combustible when exposed to heat or flame. To fight fire, use CO_2, dry chemical. The peroxidized acid explodes on contact with aluminum. Potentially dangerous reaction with perchloric acid + heat. When heated to decomposition it emits acrid smoke and irritating fumes.

OHW000   CAS: 112-62-9   HR: 2
cis-OLEIC ACID, METHYL ESTER
mf: $C_{19}H_{36}O_2$ mw: 296.55
PROP: Oil. D: 0.874 @ 20°/4°, bp: 212–213° @ 15 mm. Insol in water; misc in alc and ether.
SYNS: EMEREST 2301 □ EMEREST 2801 □ EMERY 2219 □ EMERY 2310 □ EMERY OLEIC ACID ESTER 2301 □ KEMESTER 105 □ KEMESTER 115 □ KEMESTER 205 □ KEMESTER 213 □ METHYL-9-OCTADECENOATE □ METHYL cis-9-OCTADECENOATE □ METHYL (Z)-9-OCTADECENOATE □ METHYL OLEATE □ (Z)-9-OCTADECENOIC ACID METHYL ESTER
CONSENSUS REPORTS: Reported in EPA TSCA Inventory.
SAFETY PROFILE: Questionable carcinogen with experimental tumorigenic data by skin contact. When heated to decomposition it emits acrid smoke and irritating fumes.

OHY000   CAS: 143-18-0   HR: 1
OLEIC ACID, POTASSIUM SALT
mf: $C_{18}H_{34}O_2 \cdot K$ mw: 321.62
PROP: Pale white powder. Sol in water.

SYNS: 9-OCTADECENOIC ACID (Z)-, POTASSIUM SALT □ POTASSIUM cis-9-OCTADECENOIC ACID □ POTASSIUM OLEATE □ TRENAMINE D-200 □ TRENAMINE D-201

CONSENSUS REPORTS: Reported in EPA TSCA Inventory.

SAFETY PROFILE: An eye irritant. When heated to decomposition it emits toxic fumes of K_2O.

OIA000   CAS: 143-19-1   HR: 3
OLEIC ACID, SODIUM SALT
mf: $C_{18}H_{33}O_2 \cdot Na$ mw: 304.50

PROP: White powder; slt tallow odor. Mp: 232–235°.

SYNS: EUNATROL □ OLATE FLAKES □ SODIUM OLEATE

CONSENSUS REPORTS: Reported in EPA TSCA Inventory.

SAFETY PROFILE: Poison by intravenous route. Migrates to food from packaging materials. Combustible when exposed to heat or flame. When heated to decomposition it emits toxic fumes of Na_2O.

OIM000   CAS: 8050-07-5   HR: 1
OLIBANUM GUM
PROP: Contains 3–8% volatile oil (pinene, dipentene, etc.), 60% resins, 20% gum (polysaccharide fraction), and 6–8% bassorin (FCTXAV 16,637,78). A gum from the trees *Boswellia carterii* Birdw. and other *Boswellia* species (Fam. *Burseraceae*).

SYN: FRANKINCENSE GUM

CONSENSUS REPORTS: Reported in EPA TSCA Inventory.

SAFETY PROFILE: A skin irritant. When heated to decomposition it emits acrid smoke and irritating fumes.

OIQ000   CAS: 8001-25-0   HR: 2
OLIVE OIL
PROP: Yellow oil; pleasing, delicate flavor. Mp: −6°, flash p: 437°F (CC), autoign temp: 650°F, d: 0.909–0.915 @ 25°/25°. Becomes rancid on exposure to air. Sltly sol in alc; misc with ether, chloroform, carbon

disulfide. From fruit of *Olea europaea* (85DIA2 2,196,77).

CONSENSUS REPORTS: Reported in EPA TSCA Inventory. EPA Genetic Toxicology Program.

SAFETY PROFILE: Moderately toxic by intraperitoneal route. A human skin irritant. Combustible when exposed to heat or flame; can react with oxidizing materials. Some spontaneous heating. To fight fire, use CO_2, dry chemical. When heated to decomposition it emits acrid smoke and irritating fumes.

OIU850   CAS: 43143-11-9   HR: D
OMADINE MDS
mf: $C_{10}H_8N_2O_2S_2 \cdot O_4S \cdot Mg$ mw: 372.69

SYNS: MAGNESIUM SULFATE adduct of 2,2-DITHIO-BIS-PYRIDINE 1-OXIDE □ SULFURIC ACID, MAGNES-IUM SALT (1:1), compounded with 2,2'-DITHIOBIS(PYRIDINE) 1,1'-OXIDE

CONSENSUS REPORTS: Reported in EPA TSCA Inventory.

SAFETY PROFILE: Experimental reproductive effects. When heated to decomposition it emits toxic fumes of SO_x and NO_x.

OIY000   CAS: 26354-18-7   HR: 3
OMP-2
mf: $(C_{16}H_{32}O_2Sn \cdot C_5H_8O_2)_x$

PROP: Trialkyltin methacrylate polymer (NTIS** AD-A062–138).

SYNS: 2-METHYL-2-PROPENOIC ACID METHYL ESTER, POLYMER with TRIBUTYL((2-METHYL-1-OXO-2-PROPENYL)OXY)STANNANE □ TRIBUTYL(METH-ACRYLOYLOXY)-STANNANE POLYMER with METHYL METHACRYLATE (8CI)

CONSENSUS REPORTS: Reported in EPA TSCA Inventory.

SAFETY PROFILE: Poison by ingestion and inhalation. A skin and eye irritant. When heated to decomposition it emits acrid smoke and irritating fumes. See also TIN COMPOUNDS and ORGANOMETALS.

OJD200   HR: 1
ONION OIL
PROP: From steam distillation of bulbs of *Allium ceoa* L. (Fam. *Lillaceae*). Clear amber

liquid; strong pungent odor and taste of onion. Sol in fixed oils, mineral oil, alc; insol in glycerin, propylene glycol.

SYN: OIL OF ONION

SAFETY PROFILE: Skin irritant. When heated to decomposition it emits acrid smoke and irritating fumes.

OJG000   HR: 3
OPIUM

PROP: Air-dried, milky exudation from incised, unripe capsules of *Papaver somniferum L.* or *P. album Mill.* Morphine is the most important alkaloid and occurs to the extent of 10–16%.

SYN: GUM OPIUM

SAFETY PROFILE: Poison by ingestion. Mutation data reported. Use may lead to habituation and addiction. A narcotic, sedative, analgesic, and hypnotic. Source of morphine, codeine, papaverine, thebaine, etc. Can cause nausea, vomiting, constipation, and respiratory problems. Combustible when exposed to heat or flame. See also MORPHINE.

OJK340   CAS: 8050-89-3   HR: 1
OREGON BALSAM

PROP: Pale yellow liquid eith fresh pine odor.

SYNS: BALSAM FIR, OREGON □ BALSAM, OREGON □ FIR BALSAM OREGON □ DOUGLAS FIR OIL □ OILS, DOUGLAS FIR

SAFETY PROFILE: Low toxicity by ingestion and skin contact. When heated to decomposition it emits acrid smoke and irritating vapors.

OJK325   CAS: 15139-76-1   HR: D
ORANGE B

mf: $C_{22}H_{16}N_4Na_2O_9S_2$ mw: 590.50

PROP: Dull orange crystals.

SYN: 1-(4-SULFOPHENYL)-3-ETHYLCARBOXY-4-(4-SULFONAPHTHYLAZO)-5-HYDROXYPYRAZOLE

SAFETY PROFILE: When heated to decomposition emits toxic fumes of SO_x.

OJM000   HR: 3
ORGANOMETALS

PROP: Compounds containing carbon and a metal. Ordinarily, metallic carbonates (calcium carbonate, etc.) are excluded and also metallic salts of common organic acids. Examples of organic metal compounds are Grignard compounds, such as methyl magnesium iodide (CH_3MgI), and metallic alkyls, such as butyllithium (C_4H_9Li), tetraethyllead, triethyl aluminum, tetrabutyl titanate, sodium methylate, copper phthalocyanine, and metallocenes. Also, there are many organotin compounds, such as monoalkyltins, monoaryltins, dialkyltins, diaryltins, trialkyltins, triaryltins, tetraalkyltins, and tetraaryltins.

SAFETY PROFILE: Many are highly toxic or flammable. As an example, organotin compounds are poisons by ingestion and intravenous routes. Irritating to skin, eyes, and mucous membranes. Can damage lung tissue and the liver. Trialkyltins are most toxic as a group. Next are the dialkyltins and the monoalkyltins. In each major organotin group the ethyltin derivative is the most toxic, followed by the methyltins. This group of compounds is constantly growing in importance, but there is relatively little toxicity information on most of them. Alkyl compounds of lead, tin, mercury, and aluminum are known to be highly toxic. Less is known about other organometals, but for the most part they are highly reactive chemically and therefore dangerous, if only on direct contact. It is prudent to exercise great caution in handling organometals, particularly the alkyl forms. Many organolithium compounds are explosive. See also individual compounds.

OJO000   CAS: 8007-11-2   HR: 3
ORIGANUM OIL

PROP: Main constituent is carvacrol. From steam distillation of the herb *Thymus capitatus* Hoffm. et Link (FCTXAV 12,807,74). Yellow to dark red-brown liquid; pungent spicy odor of thyme oil. D: 0.935–0.960, refr index: 1.502 @ 20°. Sol in fixed oil, propylene glycol, mineral oil; insol in glycerin.

SYN: OIL OF ORIGANUM

CONSENSUS REPORTS: Reported in EPA TSCA Inventory.

SAFETY PROFILE: Poison by skin contact. Moderately toxic by ingestion. A severe skin irritant. When heated to decomposition it emits acrid smoke and irritating fumes. See also CARVACROL.

OJW000   CAS: 341-69-5   HR: 3
ORPHENADRINE HYDROCHLORIDE

mf: $C_{18}H_{23}NO \cdot ClH$ mw: 305.88

PROP: Crystals. Mp: 156–157°. Sol in water, alc, chloroform; sltly sol in acetone, benzene; almost insol in ether.

SYNS: BF 5930 □ BG 5930 □ BROCADISIPAL □ BROCASIPAL □ BS 5930 □ 2-DIMETHYLAMINOETHYL-2-METHYLBENZHYDRYL ETHERHYDROCHLORIDE □ N,N-DIMETHYL-2-(o-METHYL-α-PHENYLBENZYLO-XY)ETHYLAMINE HYDROCHLORIDE □ DISIPAL HYDROCHLORIDE □ MEPHENAMINE HYDROCHLORIDE □ MEPHENAMIN HYDROCHLORIDE

SAFETY PROFILE: Poison by ingestion, intravenous, intraperitoneal, and subcutaneous routes. Human systemic effects: mydriasis (pupillary dilation), hallucinations, distorted perceptions, pulse rate increase, intracranial pressure increase. An experimental teratogen. Experimental reproductive effects. When heated to decomposition it emits toxic fumes of NO_x and HCl.

OJW100   HR: D
ORRIS ROOT OIL

PROP: From steam distillation of peeled, dried, aged rhizomes of *Iris pallida L.* (Fam. *Iridaceae*). Light yellow to brown solid at room temp. Mp: 38–50°. Sol in fixed oils, mineral oil, propylene glycol; insol in glycerin.

SAFETY PROFILE: When heated to decomposition it emits acrid smoke and irritating fumes.

OKE000   CAS: 7440-04-2   HR: 3
OSMIUM

af: Os   aw: 190.20

PROP: A lustrous, bluish-white, extremely hard and dense, brittle metal. Unaffected by air, H_2O, acids. Dissolves in molten alkalies. When heated oxidizes to OsO_4 (toxic). Metal smells because of vol OsO_4. D: 22.57.

SYN: METALLIC OSMIUM

CONSENSUS REPORTS: Reported in EPA TSCA Inventory.

SAFETY PROFILE: Poison by intravenous route. An irritant to eyes and mucous membranes. The principal effects of exposure are ocular disturbances and an asthmatic condition caused by inhalation. Furthermore, it causes dermatitis and ulceration of the skin upon contact. When osmium is heated, it gives off a pungent, poisonous fume of osmium tetroxide. One case of osmium poisoning reported in the literature resulted from the inhalation of osmium, which gave rise to a capillary bronchitis and dermatitis. The tetroxide vapor has a pronounced and nauseating odor that should be taken as a warning of a possibly toxic concentration in the atmosphere, and personnel should immediately move to an area of fresh air. The metal itself is not highly toxic. Flammable in the form of dust when exposed to heat or flame. Slight explosion hazard in the form of dust when exposed to heat or flame. Violent reaction or ignition with chlorine trichloride or oxygen difluoride. Ignites when heated to 100°C with fluorine. Incandescent reaction in phosphorus vapor. When heated to decomposition it emits toxic fumes of OsO_4. See also OSMIUM TETROXIDE.

OKK000   CAS: 20816-12-0   HR: 3
OSMIUM TETROXIDE

DOT: UN 2471

mf: O_4Os mw: 254.20

PROP: (A) Yellow, monoclinic, colorless crystals; (B) yellow mass; pungent, chlorine-like odor. Mp (A): 39.5°, mp: (B): 41°, bp: 130° (subl), d: 4.906 @ 22°, vap press (A): 10 mm @ 26.0°, vap press (B): 10 mm @

31.3°. Sol in CCl_4, C_6H_6, EtOH, Et_2O; spar sol in dil H_2SO_4 and H_2O. IDLH 1 mg/m³.

SYNS: OSMIC ACID □ OSMIUM(VIII) OXIDE □ RCRA WASTE NUMBER P087

CONSENSUS REPORTS: Community Right-To-Know List. EPA Genetic Toxicology Program. Reported in EPA TSCA Inventory.

OSHA PEL: TWA 0.0002 ppm; STEL 0.0006 ppm (Os)

ACGIH TLV: TWA 0.0002 ppm; STEL 0.0006 ppm (Os)

DFG MAK: 0.0002 ppm (0.002 mg/m³)

DOT CLASSIFICATION: 6.1; Label: Poison

SAFETY PROFILE: Poison by ingestion, inhalation, and intraperitoneal routes. Human systemic effects by inhalation: lachrymation and other eye effects and structural or functional changes in trachea or bronchi. Experimental reproductive effects. Mutation data reported. Explodes on contact with 1-methylimidazole. Catalytic decomposition of hydrogen peroxide can be hazardous. See also OSMIUM.

OKS000   CAS: 630-60-4   HR: 3
OUABAIN

mf: $C_{29}H_{44}O_{12}$ mw: 584.73

PROP: Crystals from H_2O. Mp: 190°. A natural plant product (JPETAB 49,561,33).

SYNS: ACOCANTHERIN □ ASTROBAIN □ GRATIBAIN □ GRATUS STROPHANTHIN □ G-STROPHANTHIN □ OUABAGENIN-l-RHAMNOSID (GERMAN) □ OUABAGENIN-l-RHAMNOSIDE □ OUABAINE □ OUBAIN □ PUROSTROPHAN □ STROPHANTHIN G □ STROPHOPERM

CONSENSUS REPORTS: EPA Extremely Hazardous Substances List. Reported in EPA TSCA Inventory.

SAFETY PROFILE: Poison by ingestion, intramuscular, intraperitoneal, intravenous, subcutaneous, and parenteral routes. Moderately toxic by intraduodenal route. A cardiac stimulant. Mutation data reported. When heated to decomposition it emits acrid smoke and irritating fumes.

OKW100   CAS: 3391-83-1   HR: 1
11-OXAHEXADECANOLIDE

mf: $C_{15}H_{28}O_3$ mw: 256.43

PROP: Musk fragrance.

SYNS: 1,7-DIOXACYCLOHEPTADECAN-17-ONE □ 16-HYDROXY-11-OXAHEXADECANOIC ACID, ω-LACTONE □ MUSK R 1

CONSENSUS REPORTS: Reported in EPA TSCA Inventory.

SAFETY PROFILE: Low toxicity by ingestion and skin contact. A skin irritant. When heated to decomposition it emits acrid smoke and irritating fumes.

OKW110   CAS: 6707-60-4   HR: 2
12-OXAHEXADECANOLIDE

mf: $C_{15}H_{28}O_3$ mw: 256.43

PROP: Liquid with characterisitc odor. Flash pt: 151° C. D: 0.981–0.987, bp: 170°.

SYNS: CERVOLIDE □ 1,6-DIOXACYCLOHEPTA-DECAN-17-ONE □ HIBISCOLIDE □ 16-HYDROXY-12-OXAHEXADECANOIC ACID, ω-LACTONE

CONSENSUS REPORTS: Reported in EPA TSCA Inventory.

SAFETY PROFILE: A skin irritant. A combustible liquid. When heated to decomposition it emits acrid smoke and irritating fumes.

OKY000   HR: 3
OXALATES

PROP: Salts of oxalic acid.

SAFETY PROFILE: Poisons by ingestion and inhalation. Powerful irritants. Oxalates are corrosive to tissue and produce local irritation. When ingested they have a caustic effect on the mouth, esophagus, and stomach. The soluble oxalates are readily absorbed from the gastrointestinal tract and can cause severe damage to the kidneys. Oxalates are common components of poisonous plants. When heated to decomposition they emit toxic and irritating fumes. See also OXALIC ACID.

OLA000   CAS: 144-62-7   HR: 3
OXALIC ACID

mf: $C_2H_2O_4$ mw: 90.04

PROP: Orthorhombic colorless crystals from water. Mp: 101.5° (anhyd) 189°, d: 1.65 @ 18.5°/4°. Very sol in H_2O; mod sol

O

in EtOH; spar sol in Et_2O. IDLH 500 mg/m³.

SYNS: ACIDE OXALIQUE (FRENCH) □ ACIDO OSSALICO (ITALIAN) □ ETHANEDIOIC ACID □ ETHANEDIONIC ACID □ KYSELINA STAVELOVA (CZECH) □ NCI-C55209 □ OXAALZUUR (DUTCH) □ OXALSAEURE (GERMAN)

CONSENSUS REPORTS: Reported in EPA TSCA Inventory.

OSHA PEL: TWA 1 mg/m³; STEL 2 mg/m³
ACGIH TLV: TWA 1 mg/m³; STEL 2 mg/m³

SAFETY PROFILE: Poison by subcutaneous route. Moderately toxic by ingestion. A skin and severe eye irritant. Acute oxalic poisoning results from ingestion of a solution of the acid. There is marked corrosion of the mouth, esophagus, and stomach, with symptoms of vomiting, burning abdominal pain, collapse, and sometimes convulsions. Death may follow quickly. The systemic effects are attributed to the removal by the oxalic acid of the calcium in the blood. The renal tubules become obstructed by the insoluble calcium oxalate, and there is profound kidney disturbance. The chief effects of inhalation of the dusts or vapor are severe irritation of the eyes and upper respiratory tract, gastrointestinal disturbances, albuminuria, gradual loss of weight, increasing weakness and nervous system complaints, ulceration of the mucous membranes of the nose and throat, epistaxis, headache, irritation, and nervousness. Oxalic acid has a caustic action on the skin and may cause dermatitis; a case of early gangrene of the fingers resembling that caused by phenol has been described. More severe cases may show albuminuria, chronic cough, vomiting, pain in the back, and gradual emaciation and weakness. The skin lesions are characterized by cracking and fissuring of the skin and the development of slow-healing ulcers. The skin may be bluish in color, and the nails brittle and yellow. Violent reaction with furfuryl alcohol, Ag, $NaClO_3$, NaOCl. When heated to decomposition it emits acrid smoke and irritating fumes. See also OXALATES.

OLO000   CAS: 471-46-5   HR: 3
OXAMIDE
mf: $C_2H_4N_2O_2$ mw: 88.08
PROP: Triclinic needles. Decomp @ 350°, d: 1.667 @ 20°/4°. Sltly sol in hot water, alc.
SYNS: AMID KYSELINY STAVELOVE (CZECH) □ 1-CARBAMOYLFORMIMIDIC ACID □ ETHANEDIAMIDE □ OXALAMIDE □ OXALIC ACID DIAMIDE □ OXAMID (CZECH) □ OXAMIMIDIC ACID
CONSENSUS REPORTS: Reported in EPA TSCA Inventory.
SAFETY PROFILE: Poison by ingestion and intraperitoneal routes. An eye irritant. When heated to decomposition it emits toxic fumes of NO_x. See also AMIDES.

OLS000   CAS: 10039-54-0   HR: 3
OXAMMONIUM SULFATE
DOT: UN 2865
mf: $H_6N_2O_2 \cdot H_2O_4S$ mw: 164.16
PROP: A crystalline material. Mp: 177°. Sol in water.
SYNS: BIS(HYDROXYLAMINE) SULFATE □ HYDRO-XYLAMINE NEUTRAL SULFATE □ HYDROXYLAMINE SULFATE □ HYDROXYLAMINE SULFATE (2:1) □ HYDROXYLAMMONIUM SULFATE
CONSENSUS REPORTS: Reported in EPA TSCA Inventory.
DOT CLASSIFICATION: 8; Label: Corrosive
SAFETY PROFILE: Poison by skin contact and intraperitoneal routes. Mutation data reported. A corrosive irritant to skin, eyes, and mucous membranes. Moderately explosive when exposed to heat or by chemical reaction. In the presence of alkalies at elevated temperatures, free hydroxylamine is liberated and may decompose explosively. When heated to decomposition it emits toxic fumes of SO_x and NO_x. See also AMINES and SULFATES.

OMG000   CAS: 60607-34-3   HR: 3
OXATIMIDE
mf: $C_{27}H_{30}N_4O$ mw: 426.61
SYNS: 1-(3-(4-(DIPHENYLMETHYL)-1-PIPERAZINYL)-PROPYL)-2-BENZIMIDAZOLINONE □ 1-(3-(4-(DIPHENYLMETHYL)-1-PIPERAZINYL)PROPYL)-1,3-

DIHYDRO-2H-BENZIMIDAZOL-2-ONE □ KW-4354 □ OXATOMIDA □ OXATOMIDE □ R 35443 □ TINSET

SAFETY PROFILE: Poison by ingestion, intraperitoneal, and intravenous routes. An experimental teratogen. Experimental reproductive effects. Used to treat allergies and asthma. When heated to decomposition it emits toxic fumes of NO_x.

OMM300   CAS: 6542-37-6   HR: 3
1H,3H,5H-OXAZOLO(3,4-C)OXAZOLE-7A(7H)-METHANOL

mf: $C_6H_{11}NO_3$ mw: 145.18

SYNS: OXAZOLIDINE T □ BONDING AGENT M 3 □ GDUE □ 5-(HYDROXYMETHYL)-1-AZA-3,7-DIOXA-BICYCLO(3.3.0)OCTANE □ M 3 (CURING AGENT)

CONSENSUS REPORTS: EPA FIFRA 1988 pesticide subject to registration or re-registration.

CONSENSUS REPORTS: Reported in EPA TSCA Inventory.

SAFETY PROFILE: A poison by ingestion. When heated to decomposition it emits toxic vapors of NO_x.

OMW000   CAS: 503-30-0   HR: 2
OXETANE

mf: C_3H_6O mw: 58.09

PROP: Oily liquid with agreeable odor. D: 0.8930 @ 25°/4°, bp: 480° @ 750 mm. Sol in water.

SYNS: CYCLOOXABUTANE □ 1,3-EPOXYPROPANE □ OXACYCLOBUTANE □ OXETAN □ α-γ-PROPANE OXIDE □ 1,3-PROPYLENE OXIDE □ TRIMETHYLENE OXIDE □ TRIMETHYLENOXID (GERMAN)

CONSENSUS REPORTS: Reported in EPA TSCA Inventory.

SAFETY PROFILE: Moderately toxic by subcutaneous route. May be narcotic in high concentrations. Questionable carcinogen with experimental tumorigenic data. Mutation data reported. When heated to decomposition it emits acrid smoke and irritating fumes.

ONY000   CAS: 1707-95-5   HR: 3
2-(3-OXO-1-INDANYLIDENE)-1,3-INDANDIONE

mf: $C_{18}H_{10}O_3$ mw: 274.28

PROP: Yellow plates. Mp: 208–210°.

SYN: BINDON

SAFETY PROFILE: Poison by intraperitoneal route. An experimental teratogen. Experimental reproductive effects. When heated to decomposition it emits acrid smoke and irritating fumes.

OOE000   CAS: 1949-20-8   HR: 3
OXOLAMINE CITRATE

mf: $C_{14}H_{19}N_3O•C_6H_8O_7$ mw: 437.50

PROP: Crystals. Sltly sol in water and alc.

SYNS: 5-β-DIETHYLAMINOETHYL-3-PHENYL-1,2,4-OXADIAZOLE CITRATE □ 3-PHENYL-5-(β-(DIETHYLAMINO)ETHYL)-1,2,4-OXADIAZOLE CITRATE

SAFETY PROFILE: Poison by intraperitoneal route. Moderately toxic by ingestion. Experimental teratogenic and reproductive effects. Questionable carcinogen with experimental carcinogenic data. When heated to decomposition it emits toxic fumes of NO_x. See also AMINES and other oxolamine entries.

OOK100   CAS: 306-12-7   HR: 3
OXOPHENARSINE

mf: $C_6H_6AsNO_2$ mw: 199.05

PROP: Bright-yellow deliquescent powder or white crystals. Mp: 133° (decomp).

SYNS: 2-AMINO-4-ARSENOSOPHENOL □ (3-AMINO-4-HYDROXYPHENYL)ARSENOUS ACID □ PHENARSEN □ PHENOL, 2-AMINO-4-ARSENOSO-

OSHA PEL: TWA 0.5 mg(As)/m³

SAFETY PROFILE: Poison by ingestion and intravenous routes. When heated to decomposition it emits toxic fumes of NO_x and As.

OOK175   CAS: 63981-21-5   HR: 3
2-OXO-2-(PHENYLAMINO)ETHYL SELENOCYANATE

mf: $C_9H_8N_2OSe$ mw: 239.15

SYNS: SELENOCYANIC ACID, 2-OXO-2-(PHENYLAMINO)ETHYL ESTER □ SELENOCYANIC ACID, ESTER WITH α-HYDROXYACETANILIDE

ACGIH TLV: TWA 0.2 mg(Se)/m³

SAFETY PROFILE: A poison by ingestion and intraperitoneal routes. When heated to

O

decomposition it emits toxic vapors of NO_x and Se.

OOK200   CAS: 5415-07-6   HR: 3
β-OXO-α-PHENYLBENZENEPROPANE-NITRILE

mf: $C_{15}H_{11}NO$ mw: 221.27

SYNS: ACETONITRILE, BENZOYLPHENYL- □ BENZENEPROPANENITRILE, β-OXO-α-PHENYL-(9CI) □ α-BENZOYLBENZYL CYANIDE □ BENZOYLPHENYL-ACETONITRILE □ α-BENZOYLPHENYLACETONITRILE □ α-CYANOBENZYL PHENYL KETONE □ α-CYANODEOXYBENZOIN

DOT CLASSIFICATION: 3; Label: Flammable Liquid

SAFETY PROFILE: A poison by intraperitoneal route. A flammable liquid. When heated to decomposition it emits toxic vapors of NO_x.

OOO100   CAS: 40942-73-2   HR: 1
3-(2-OXOPROPYL)-2-PENTYLCYCLO-PENTANONE

mf: $C_{13}H_{22}O_2$ mw: 210.35

SYNS: CYCLOPENTANONE, 3-(2-OXOPROPYL)-2-PENTYL- □ MAGNOLIONE □ PENTYLCYCLOPENT-ANONEPROPANONE

CONSENSUS REPORTS: Reported in EPA TSCA Inventory.

SAFETY PROFILE: Low toxicity by ingestion and skin contact. A skin irritant. When heated to decomposition it emits acrid smoke and irritating fumes.

OPE000   CAS: 80-51-3   HR: 3
OXYBIS(BENZENESULFONYL HYDRAZIDE)

DOT: UN 2951

mf: $C_{12}H_{14}N_4O_5S_2$ mw: 358.42

PROP: White powder. Decomp temp. 150°–160°.

SYNS: BENZENESULFONIC ACID, OXYBIS-, DIHYDRAZIDE (9CI) □ CELLMIC S □ CELOGEN OT □ CENITRON OB □ DIPHENYLOXIDE-4,4'-DISULFOHYDRAZIDE (DOT) □ NITROPORE OBSH □ OBSH □ p,p'-OXYBISBENZENE DISULFONYLHYDRAZIDE □ OXYBISBENZENESULFONIC ACID DIHYDRAZIDE □ p,p'-OXYBIS(BENZENESULFONYL) HYDRAZINE

CONSENSUS REPORTS: Reported in EPA TSCA Inventory.

ACGIH TLV: ACGIH TWA: TLV 0.1 mg/m³

DOT CLASSIFICATION: 4.1; Label: Flammable Solid

SAFETY PROFILE: Mutation data reported. A flammable solid. When heated to decomposition it emits very toxic fumes of NO_x and SO_x.

OPG100   CAS: 53061-10-2   HR: 3
1,1'-(OXYBIS(METHYLENESULFON-YL))BIS(2-CHLOROETHANE)

mf: $C_6H_{12}Cl_2O_5S_2$ mw: 299.20

SYNS: BIS(2-CHLOROETHYLSULFONYLMETHYL)-ETHER □ ETHANE, 1,1'-(OXYBIS(METHYLENESULF-ONYL))BIS(2-CHLORO-

SAFETY PROFILE: A poison by ingestion and intraperitoneal routes. Moderately toxic by ingestion. When heated to decomposition it emits toxic vapors of SO_x and Cl^-.

OPM000   CAS: 101-80-4   HR: 3
4,4'-OXYDIANILINE

mf: $C_{12}H_{12}N_2O$ mw: 200.26

PROP: Colorless crystals. Mp: 187°, bp: >300°.

SYNS: p-AMINOPHENYL ETHER □ 4-AMINOPHENYL ETHER □ BIS(p-AMINOPHENYL)ETHER □ BIS(4-AMINOPHENYL)ETHER □ DADPE □ 4,4'-DIAMINO-BIPHENYLOXIDE □ DIAMINODIPHENYL ETHER □ p,p'-DIAMINODIPHENYL ETHER □ 4,4-DIAMINODI-PHENYL ETHER □ 4,4'-DIAMINODIPHENYL OXIDE □ 4,4'-DIAMINOPHENYL ETHER □ NCI-C50146 □ OXYBIS(4-AMINOBENZENE) □ p,p'-OXYBIS(ANILINE) □ 4,4'-OXYBISANILINE □ 4,4'-OXYBISBENZENAMINE □ OXYDIANILINE □ p,p'-OXYDIANILINE □ 4,4'-OXYDIPHENYLAMINE □ OXYDI-p-PHENYLENEDIAMINE

CONSENSUS REPORTS: NTP 10th Report on Carcinogens. IARC Cancer Review: Group 2B IMEMDT 7,56,87; Animal Sufficient Evidence IMEMDT 29,203,82; Animal Inadequate Evidence IMEMDT 16,301,78. NCI Carcinogenesis Bioassay (feed); Clear Evidence: mouse, rat NCITR* NCI-CG-TR-205,80. Reported in EPA TSCA Inventory.

DFG MAK: Animal Carcinogen, Suspected Human Carcinogen

SAFETY PROFILE: Confirmed carcinogen with experimental carcinogenic, neoplastigenic, and tumorigenic data. Poison by intraperitoneal route. Moderately toxic by ingestion. Mutation data reported. When heated to decomposition it emits toxic fumes of NO_x.

OPO000 CAS: 106-75-2 HR: 2
OXYDIETHYLENE BIS(CHLORO- FORMATE)
mf: $C_6H_8Cl_2O_5$ mw: 231.04
SYNS: CARBONOCHLORIDIC ACID, OXYDI-2,1-ETHANEDIYL ESTER □ DIETHYLENE GLYCOL, BISCHLOROFORMATE □ FORMIC ACID, CHLORO-, OXYDIETHYLENE ESTER □ OXYDIETHYLENE CHLOROFORMATE
CONSENSUS REPORTS: Reported in EPA TSCA Inventory.
DOT CLASSIFICATION: 6.1; Label: Poison, Corrosive
SAFETY PROFILE: Moderately toxic by ingestion, inhalation, and skin contact. When heated to decomposition it emits toxic fumes of Cl^-.

OQU100 CAS: 42874-03-3 HR: D
OXYFLUORFEN
mf: $C_{15}H_{11}ClF_3NO4$ mw: 361.72
PROP: Orange crystal solid. Sol in water and most solids.
SYNS: 2-CHLORO-1-(3-ETHOXY-4-NITROPHENOXY)-4-(TRIFLUOROMETHYL)BENZENE □ 2-CHLORO-α-α-α-TRIFLUORO-p-TOLYL-3-ETHOXY-4-NITROPHENYL ETHER □ GOAL □ KOLTAR □ OXYFLUORFENE □ RH-2915
SAFETY PROFILE: When heated to decomposition emits toxic fumes of Cl^-, F^-, and NO_2.

OQW000 CAS: 7782-44-7 HR: 3
OXYGEN
DOT: UN 1072/UN 1073
mf: O_2 mw: 32.00
PROP: Colorless, odorless, tasteless gas, liquid, or hexagonal crystals. Condenses to paramagnetic blue liquid and blue solid. Reacts with all elements except He, Ne, Ar. Supports combustion. D: (liquid) 1.14 @ $-183.0°$, d: (solid) 1.426 @ $-252.5°$, vap d: 1.429 @ 0°. D: (gas) 1.429 g/L @ 0°, mp: $-218.4°$, bp: $-182.96°$. Very sltly sol in H_2O; more sol in org solvs.
SYNS: OXYGEN, compressed (UN 1072) (DOT) □ OXYGEN, refrigerated liquid (cryogenic liquid) (UN 1073) (DOT)
CONSENSUS REPORTS: Reported in EPA TSCA Inventory. EPA Genetic Toxicology Program.
DOT CLASSIFICATION: 2.2; Label: Nonflammable Gas, Oxidizer
SAFETY PROFILE: Human systemic effects by inhalation: cough and other pulmonary changes. Human teratogenic effects by inhalation: developmental abnormalities of the fetal cardiovascular system. Mutation data reported. Not toxic as gas. In liquid form it can cause severe "burns" and tissue damage on contact with the skin due to extreme cold.

An oxidant. Though itself nonflammable, it is essential to combustion. Even a slight increase in the oxygen content of the air above the normal 21% greatly increases the oxidation or burning rate (and the hazard) of many materials. Exclusion of O_2 from the neighborhood of a fire is one of the principal methods of extinguishment. Avoid smoking, flames, electric sparks. Liquid O_2 can explode on contact with readily oxidizable materials, especially at high temperatures. Under the proper conditions of temperature, pressure, and reagent concentration it can react violently with acetaldehyde, acetylene, acetone, secondary alcohols (e.g., 2-propanol, 2-butanol) aluminum, $Al(BH_4)_3$, AlH_3, aluminum-titanium alloys, alkali metals (lithium, cesium, potassium, rubidium, sodium, potassium), ammonia, ammonia + platinum, asphalt, CCl_4, chlorinated hydrocarbons, cyanogen, barium, benzene, 1,4-benzenediol + 1-propanol, benzoic acid, $Be(BH_4)_2$, biological materials + ether, BAs_2Br_3, B_2H_{10}, B_2H_6, boron tribromide, boron trichloride, bromine + chlorotrifluoroethylene, butane + $Ni(CO)_4$, carbon disulfide, carbon disulfide + mercury + anthracene, carbon monoxide, CsH, calcium, calcium phosphide, copper + hydrogen sulfide, $C_{10}H_{14}$, cyclohexane-1,2-dione bis(phenyl-hydrazone), cyclooctatetraene, diborane, diboron tetrafluoride, dimethoxymethane, dimethylketene, dimethyl sulfide, diphenyl ethylene, disilane, ethers (e.g., diethyl ether, diisopropyl ether, tetrahydrofuran, dioxane, ethyl ether), fibrous fabrics, fluorine + hydrogen, fuels, germanium, glycerol, halocarbons (e.g., 1,1,1-trichloroethane, trichloroethylene, chlorotrifluoroethylene, bromotrifluoroethylene), hydrazine, hydrocarbons (e.g., 1,1-diphenylethylene, gasoline, cyclohexane, ethylene, cumene, p-xylene, but-3-yne), hydrocarbons + promo-ters (e.g., methyl nitrate, nitromethane, ethyl nitrate, tetrafluorohydrazine), hydrogen, hydrogen sulfide, lithiated dialkylnitroso-amines, magnesium, metals, metal hydrides (e.g., sodium hydride, uranium hydride, lithium hydride, potassium hydride, rubidium hydride, cesium hydride, magnesium hydride), methane, methoxy-cyclooctatetraene, 4-methoxytoluene, $Ni(CO)_4$ + butane, nonmetal hydrides (e.g., diborane, tetraborane(10), phosphine, pentaborane(11), pentaborane(9), decaborane(14), aluminum tetrahydrobor-ate), oil films, organic matter, (OF_2 + H_2O), phosphorus, phosphorus tribromide, phosphorus trifluoride, phosphorus(III) oxide, polymers [e.g., foam rubber, neoprene, polytetrafluoroethylene (teflon)], polytetrafluoroethylene + stainless steel, polyurethane, polyvinyl chloride, propylene oxide, K_2O_2, rhenium, trirhenium nona-chloride, rubber + ozone, rubberized fabric, selenium, NaH, sodium hydroxide + tetra-methyldisiloxane, strontium, tetracarbony-lnickel, tetracarbonylnickel + mercury, tetrafluoroethylene, tetrafluorohydrazine, tetrasilane, titanium and alloys, trisilane, CH_2Cl_2, oil, paraformaldehyde, wood, charcoal. Compressed O_2 is shipped in steel cylinders under high pressure. If these containers are broken due to shock or exposed to high temperature, an explosion and fire may result.

ORA000   CAS: 7783-41-7   HR: 3
OXYGEN DIFLUORIDE
DOT: UN 2190
mf: F_2O mw: 54.00
PROP: Colorless gas or yellowish-brown liquid. Stable to 2°. Stable in dry glass vessels. Reacts with Hg, but is less reactive than Cl_2O. D: (liquid) 1.90 @ −224°, mp: −223.8°, bp: −144.8°. Sltly sol in water. IDLH 0.5 ppm.
SYNS: FLUORINE MONOXIDE □ FLUORINE OXIDE □ OXYGEN FLUORIDE
OSHA PEL: CL 0.05 ppm
ACGIH TLV: CL 0.05 ppm
DOT CLASSIFICATION: 2.3; Label: Poison Gas, Oxidizer
SAFETY PROFILE: Poison by inhalation. Human systemic effects by inhalation: chronic pulmonary edema or congestion. A corrosive skin, eye, and mucous membrane irritant. Attacks lungs with delayed appearance of symptoms. A very powerful oxidizer. Must be kept away from contact with reducing agents. Explosive reaction with adsorbents (e.g., silica gel, alumina, molecular sieve), diborane, halogens + heat, metal halides, aluminum chloride, antimony pentachloride (at 150°C), tungsten + heat, hydrogen sulfide, liquid nitrogen oxide, nitrosyl fluoride, charcoal, sulfur tetrafluor-ide. Forms spark-sensitive explosive mixtures with water or combustible gases (e.g., carbon monoxide, hydrogen, methane). Ignites on contact with diborane tetrafluor-ide, nonmetals (e.g., red phosphorus, boron powder, silicon), phosphorus(V) oxide, nitrogen oxide gas. Incandescent reaction with metals (e.g., aluminum, barium, cadmium, magnesium, strontium, zinc, zirconium, lithium (above 400°C)), potas-sium (above 400°C), sodium. Incompatible with NH_3, As_2O_3, Cl_2 + Cu, CrO_3, Ir, O_3, O_2 + H_2O, Pd, Pt, Rh, Ru, SiO_2. When heated to decomposition it emits highly toxic fumes of F^-. See also FLUORIDES.

ORI400   CAS: 17297-82-4   HR: 3
OXYPENDYL HYDROCHLORIDE
mf: $C_{20}H_{26}N_4OS \cdot 2ClH$ mw: 443.48
SYNS: D 704 □ 10-(3-(4-OXYAETHYL-PIPERAZINO)-
PROPYL-(1))-4-AZAPHENTHIAZIN DIHYDROCHLORID
□ 1-PIPERAZINEETHANOL, 4-(3-(10H-PYRIDO(3,2-
B)(1,4)BENZOTHIAZIN-10-YL)PROPYL)-,DIHYDRO-
CHLORIDE □ 10-(3'-(1"-β-HYDROXYETHYL-4"-PIPERAZ-
INYL)-PROPYL)-THIOPHENYLPYRIDYAMINE HYDRO-
CHLORIDE □ PERVETRAL
SAFETY PROFILE: A poison by
intravenous, intraperitoneal and
subcutaneous routes. Moderately toxic by
ingestion. When heated to decomposition it
emits toxic vapors of NO_x, SO_x, and Cl^-.

ORS000   CAS: 39603-54-8   HR: 3
β-OXYPROPYLPROPYLNITROSAMINE
mf: $C_6H_{12}N_2O_2$ mw: 144.20
SYNS: N-NITROSO-2-OXO-N-PROPYL-N-PROPYL-
AMINE □ 1-(NITROSOPROPYLAMINO)-2-PROPANONE
□ 2-OXI-PROPYL-PROPYLNITROSAMIN (GERMAN) □
2-OXO-PROPYL-PROPYLNITROSAMINE □ (2-
OXOPROPYL)PROPYLNITROSOAMINE
SAFETY PROFILE: Suspected carcinogen
with experimental carcinogenic,
neoplastigenic, and tumorigenic data.
Moderately toxic by subcutaneous route. An
experimental teratogen. Mutation data
reported. When heated to decomposition it
emits toxic fumes of NO_x. See also
NITROSAMINES.

ORW000   CAS: 10028-15-6   HR: 3
OZONE
mf: O_3 mw: 48.00
PROP: Blue or violet-black solid or unstable
colorless gas or dark-blue liquid;
characteristic odor at low concentration.
Mp: −193°, bp: −111.9°, d: (gas) 2.144 g/L,
1.71 @ −183°, d: (liquid) 1.614 g/mL @
−195.4°. Sltly sol in water. IDLH 5 ppm.
SYNS: OZON (POLISH) □ TRIATOMIC OXYGEN
CONSENSUS REPORTS: EPA Extremely
Hazardous Substances List. Reported in
EPA TSCA Inventory. EPA Genetic
Toxicology Program.
OSHA PEL: TWA 0.1 ppm; STEL 0.3 ppm
ACGIH TLV: TWA 0.05 ppm (heavy work),
0.08 ppm (moderate work), 0.10 (light

work); all workloads < 2 hours 20 ppm; Not
Classifiable as a Human Carcinogen
DFG MAK: Confirmed Animal Carcinogen
with Unknown Relevance to Humans
SAFETY PROFILE: A human poison by
inhalation. Human systemic effects by
inhalation: visual field changes,
lachrymation, headache, decreased pulse rate
with fall in blood pressure, dermatitis,
cough, dyspnea, respiratory stimulation and
other pulmonary changes. Experimental
teratogenic and reproductive effects. Human
mutation data reported. A skin, eye, upper
respiratory system, and mucous membrane
irritant. Questionable carcinogen with
experimental neoplastigenic and tumorigenic
data. Can be a safe water disinfectant in low
concentration. Concentration of 0.015 ppm
of ozone in air produces a barely detectable
odor. Concentrations of 1 ppm produce a
disagreeable sulfur like odor and may cause
headache and irritation of eyes and the
upper respiratory tract; symptoms disappear
after leaving the exposure.

A powerful oxidizing agent. Dangerous
chemical reaction with acetylene, alkenes,
alkylmetals (e.g., dimethylzinc, diethylzinc),
antimony, aromatic compounds (e.g.,
benzene, aniline), benzene + oxygen +
rubber, bromine, charcoal + potassium
iodide, citronellic acid, combustible gases
(e.g., carbon monoxide, ethylene, nitrogen
oxide, ammonia, phosphine), (diallyl methyl
carbinol + acetic acid), trans-2,3-dichloro-2-
butene, dicyanogen, dienes + oxygen,
diethyl ether, 1,1-difluoroethylene, N_2O_5,
ethylene + formyl fluoride, fluoroethylene,
liquid hydrogen, hydrogen + oxygen
difluoride, hydrogen bromide, hydrogen
iodide, 4-hydroxy-4-methyl-1,6-heptadiene,
2,3-hydroxy-2,2,4-trimethyl-3-pentenoic acid
lactone, isopropylidene compounds,
nitrogen, NO_2, NO, nitrogen trichloride,
nitrogen triiodide, nitroglycerin, organic
liquids, organic matter, oxygen + rubber
powder, oxygen fluorides (e.g., dioxygen
difluoride, dioxygen trifluoride), silica gel,
stibine, tetrafluorohydrazine,

tetramethylammonium hydroxide, trifluoroethylene, unsaturated acetals. A severe explosion hazard in liquid form when shocked, exposed to heat or flame, or in concentrated form by chemical reaction with powerful reducing agents. Incompatible with rubber; dinitrogen tetraoxide. See also OZONIDES and PEROXIDES.

P

PAD500 CAS: 7647-10-1 HR: 3
PALLADIUM(2+) CHLORIDE
mf: Cl_2Pd mw: 177.30
PROP: Dark brown, deliq crystals. D: 4.0 @ 18°, mp: 678–680° (decomp). Sol in water, alc, acetone, and hydrochloric acid.
SYNS: NCI-C60184 □ PALLADIUM CHLORIDE □ PALLADOUS CHLORIDE
CONSENSUS REPORTS: Reported in EPA TSCA Inventory. EPA Genetic Toxicology Program.
SAFETY PROFILE: Poison by intraperitoneal, intravenous, and intratracheal routes. Moderately toxic by ingestion. Experimental reproductive effects. A skin irritant. Questionable carcinogen with experimental carcinogenic data. Human mutation data reported. When heated to decomposition it emits highly toxic fumes of Cl⁻. See also PALLADIUM.

PAE000 CAS: 8014-19-5 HR: 1
PALMAROSA OIL
PROP: From steam distillation of the grass *Cymbopogon Martini* Stapf. Var. Motia, mainly *Geraniol* (FCTXAV 12,807,74). Yellow oily liquid. D: 0.879–0.892, refr index: 1.473 @ 20°. Sol in fixed oils, propylene glycol, mineral oil; insol in glycerin.
SYNS: GERANIUM OIL, EAST INDIAN TYPE □ GERANIUM OIL, TURKISH TYPE □ OIL of PALMAROSA
CONSENSUS REPORTS: Reported in EPA TSCA Inventory.
SAFETY PROFILE: Low toxicity by ingestion and skin contact. A skin irritant. When heated to decomposition it emits acrid smoke and irritating fumes.

PAE240 HR: D
PALMITAMIDE
PROP: Semisynthetic compound derived from palm oil.

SYN: PALMITIC ACID AMIDE
SAFETY PROFILE: When heated to decomposition it emits acrid smoke and irritating fumes.

PAE250 CAS: 57-10-3 HR: 3
PALMITIC ACID
mf: $C_{17}H_{32}O_2$ mw: 256.48
PROP: Colorless plates or white crystalline powder; slt characteristic odor and taste. D: 0.849 @ 70°/4°, mp: 63–64°, bp: 271.5° @ 100 mm. Insol in water; very sltly sol in pet ether; sol in abs ether, chloroform.
SYNS: CETYLIC ACID □ EMERSOL 140 □ EMERSOL 143 □ HEXADECANOIC ACID □ n-HEXADECOIC ACID □ HEXADECYLIC ACID □ HYDROFOL □ HYSTRENE 8016 □ INDUSTRENE 4516 □ 1-PENTADECANE-CARBOXYLIC ACID
CONSENSUS REPORTS: Reported in EPA TSCA Inventory.
SAFETY PROFILE: A poison by intravenous route. A human skin irritant. Questionable carcinogen with experimental neoplastigenic data. When heated to decomposition it emits acrid smoke and irritating fumes.

PAE500 CAS: 8002-75-3 HR: D
PALM OIL
PROP: Liquid.
SYNS: OILS, PALM □ PALM BUTTER
CONSENSUS REPORTS: Reported in EPA TSCA Inventory.
SAFETY PROFILE: An experimental teratogen. Experimental reproductive effects. When heated to decomposition it emits acrid smoke and irritating fumes.

PAE300 HR: D
PALM OIL (UNHYDROGENATED)
PROP: From the pulp of the fruit of the oil palm *Elaeis guineensis*. A deep orange-red fatty

semisolid @ 21–27°; characteristic sweet nutty flavor.

SAFETY PROFILE: When heated to decomposition it emits acrid smoke and irritating fumes.

PAE750   CAS: 12174-11-7   HR: 3
PALYGORSCITE
PROP: White or gray monoclinic or orthorhombic crystals.
SYNS: ACTIVATED ATTAPULGITE □ ATTACLAY □ ATTACLAY X 250 □ ATTACOTE □ ATTAGEL □ ATTAGEL 40 □ ATTAGEL 50 □ ATTAGEL 150 □ ATTAPULGITE □ ATTASORB □ DILUEX □ MIN-U-GEL 200 □ MIN-U-GEL 400 □ MIN-U-GEL FG □ PALYGORS-KIT (GERMAN) □ PERMAGEL □ PHARMASORB-COLLOIDAL □ RVM-FG □ 200U/P-RVM □ X 250 □ ZEOGEL
CONSENSUS REPORTS: IARC Cancer Review: Group 3 IMEMDT 7,117,87; Animal Limited Evidence IMEMDT 42,159,87; Human Inadequate Evidence IMEMDT 42,159,87.
DFG MAK: Animal Carcinogen, Suspected Human Carcinogen
SAFETY PROFILE: Suspected carcinogen with experimental neoplastigenic and tumorigenic data. When heated to decomposition it emits acrid smoke and irritating fumes.

PAG200   CAS: 81-13-0   HR: 2
d-PANTHENOL
mf: $C_9H_{19}NO_4$ mw: 205.29
PROP: Viscous oil, somewhat hygroscopic liquid; sltly bitter taste. D: (20°/20°) 1.2, bp: 118–120°, easily decomp on distillation. Freely sol in water, alc, methanol, ether; sltly sol in glycerin. Natural pH about 9.5.
SYNS: ALCOPAN-250 □ BEPANTHEN □ BEPANTH-ENE □ BEPANTOL □ COZYME □ DEXPANTHENOL (FCC) □ d-(+)-2,4-DIHYDROXY-N-(3-HYDROXYPROP-YL)-3,3-DIMETHYLBUTYRAMIDE □ D-P-A INJECTION □ ILOPAN □ MOTILYN □ PANADON □ PANTHENOL □ d(+)-PANTHENOL (FCC) □ PANTHODERM □ PANTOL □ PANTOTHENOL □ d-PANTOTHENOL □ PANTOTHENYL ALCOHOL □ d-PANTOTHENYL ALCOHOL □ d(+)-PANTOTHENYL ALCOHOL □ THENALTON □ ZENTINIC

CONSENSUS REPORTS: Reported in EPA TSCA Inventory.
SAFETY PROFILE: Moderately toxic by intravenous route. A skin and eye irritant. When heated to decomposition it emits toxic fumes of NO_x. See also AMIDES.

PAG500   CAS: 9001-73-4   HR: 3
PAPAIN
PROP: White to gray, sltly hygroscopic powder. Sol in water and glycerin; insol in other common org solvs. The most thermostatic enzyme known, digests protein. Isolated from the latex of the green fruit and leaves of *Carcia papaya L.* (IJMRAQ 67,499,78).
SYNS: ARBUZ □ CAROID □ NEMATOLYT □ PAPAYOTIN □ SUMMETRIN □ TROMASIN □ VEGETABLE PEPSIN □ VELARDON □ VERMIZYM
CONSENSUS REPORTS: Reported in EPA TSCA Inventory.
SAFETY PROFILE: Poison by intraperitoneal route. Human systemic effects by ingestion: changes in structure or function of esophagus. Experimental teratogenic and reproductive effects. An allergen. When heated to decomposition it emits toxic fumes of NO_x.

PAH000   CAS: 58-74-2   HR: 3
PAPAVERINE
mf: $C_{20}H_{21}NO_4$ mw: 339.42
PROP: Colorless, rhombic needles or prisms from EtOH/Et₂O. Mp: 147°, bp: decomp, d: 1.337 @ 20°/4°. Insol in water; sol in hot benzene, glacial acetic acid, acetone; sltly sol in chloroform, carbon tetrachloride, pet ether.
SYNS: 1-((3,4-DIMETHOXYPHENYL)METHYL)-6,7-DIMETHOXYISOQUINOLINE □ 6,7-DIMETHOXY-1-VERATRYLISOQUINOLINE □ PAPANERINE □ PAPAVERINA (ITALIAN)
SAFETY PROFILE: Poison by ingestion, intramuscular, subcutaneous, intradermal, intraperitoneal, and intravenous routes. Human systemic effects: coma, somnolence. Its central nervous system action is about midway between those of morphine and codeine, and large doses do not produce the

amount of excitement caused by codeine or the soporific action of morphine. Mutation data reported. A cerebral vasodilator and smooth muscle relaxant. Combustible when exposed to heat or flame. When heated to decomposition it emits toxic fumes of NO_x. See also MORPHINE.

PAH250   CAS: 61-25-6   HR: 3
PAPAVERINE CHLOROHYDRATE
mf: $C_{20}H_{21}NO_4•ClH$ mw: 375.88
PROP: Rods from H_2O. Mp: 221–222° (decomp).
SYNS: ARTEGODAN □ CARDOVERINA □ CEPA-VERIN □ CERESPAN □ CHLORHYDRATE de PAPAVER-INE (FRENCH) □ 6,7-DIMETHOXY-1-VERATRYLISO-QUINOLINE HYDROCHLORIDE □ DIPAV □ DISPAMIL □ DYNOVAS □ ISOQUINOLINE, 6,7-DIMETHOXY-1-VERATRYL-, HYDROCHLORIDE □ LAPAV □ NCI-C56359 □ OPTENYL □ PAMEION □ PAPAVARINE CHLORHYDRATE □ PAPAVERINE CHLOROHYDRATE □ PAPAVERINE HYDROCHLORIDE □ PAPAVERINE MONOHYDROCHLORIDE □ PAPAVERIN-HCL (GERMAN) □ PAP H □ PAVABID □ PAVAGRANT □ PAVAKEY □ PAVASED □ PAVATEST □ SPASMO-NIT □ 6,7,3',4'-TETRAMETHOXY-1-BENZYLISOQUINOLINE HYDROCHLORIDE □ THERAPAV □ VASAL □ VASOSPAN
CONSENSUS REPORTS: Reported in EPA TSCA Inventory.
SAFETY PROFILE: Poison by ingestion, intraperitoneal, intraduodenal, intravenous, and subcutaneous routes. Human systemic effects: metabolic acidosis, pulse rate increase. An experimental teratogen. Mutation data reported. When heated to decomposition it emits very toxic fumes of NO_x and HCl. See also PAPAVERINE.

PAH280   HR: D
PAPRIKA OLEORESIN
PROP: Derived from organic solvent extraction of ground dried pod of mild capsicum *Capsicum annuum* L.
SAFETY PROFILE: When heated to decomposition it emits acrid smoke and irritating fumes.

PAH750   CAS: 8002-74-2   HR: 2
PARAFFIN

PROP: Colorless or white, translucent wax; odorless. D: approx 0.90, mp: 50–57°. Insol in water, alc; sol in benzene, chloroform, ether, carbon disulfide, oils; misc with fats.
SYNS: PARAFFIN WAX □ PARAFFIN WAX FUME (ACGIH)
CONSENSUS REPORTS: Reported in EPA TSCA Inventory.
OSHA PEL: Fume: TWA 2 mg/m³ (fume)
ACGIH TLV: Fume: TWA 2 mg/m³ (fume)
NIOSH REL: (Paraffin Wax Fume) TWA 2 mg/m³
SAFETY PROFILE: A skin and eye irritant. Questionable carcinogen with experimental tumorigenic data by implant route. Many paraffin waxes contain carcinogens. Fumes cause lung damage. See also PARAFFIN HYDROCARBONS.

PAH770   HR: 2
PARAFFIN HYDROCARBONS
PROP: Gas, liquid or solid depending on temp and pressure.
SAFETY PROFILE: The effects of the paraffin hydrocarbons vary with the volatility. The gaseous hydrocarbons, such as methane, ethane, etc., have but slight anesthetic effects and are hazardous only when present in sufficient concentration to dilute the oxygen to a point below that necessary to sustain life. With the volatile liquid hydrocarbons, or with the next-higher fraction, the anesthetic action predominates, and with the higher molecular weights or with the less volatile compounds, the anesthetic increases, but at the same time an irritant action becomes more pronounced. For information concerning toxic and hazardous properties of these materials, see the individual compounds. Paraffins are common air contaminants. Can be a dangerous fire hazard depending on volatility.

PAH780   CAS: 63449-39-8   HR: 3
PARAFFIN WAXES and HYDROCAR-BON WAXES, CHLORINATED
PROP: Liquid and solid. Low solin water.

P

SYNS: CERECHLOR 54 □ CERECLOR □ CERECLOR 30 □ CERECLOR 42 □ CERECLOR 48 □ CERECLOR 52 □ CERECLOR 54 □ CERECLOR 70 □ CERECLOR 51L □ CERECLOR 56L □ CERECLOR 63L □ CERECLOR 65L □ CERECLOR 70L □ CERECLOR 50LV □ CERECLOR S 42 □ CERECLOR S52 □ CERECLOR S70 □ CHLORCOSANE □ CHLOREZ 700 □ CHLOREZ 700HMP □ CHLORIN-ATED PARAFFINS □ CHLOROPARAFFINE 40G □ CHLOROWAX □ CHLOROWAX 40 □ CHLOROWAX 50 □ CHLOROWAX 70 □ CHLOROWAX 500C □ CHLORO-WAX 70S □ CHLOROWAX S 70 □ CLORAFIN □ CRECHLOR S 45 □ FLEXCHLOR □ NCI-C53587 □ PAROIL CHLOREZ □ UNICHLOR □ UNICHLOR 50

CONSENSUS REPORTS: Reported in EPA TSCA Inventory.

DFG MAK: Confirmed Animal Carcinogen with Unknown Relevance to Humans

SAFETY PROFILE: Suspected carcinogen. A skin and eye irritant. When heated to decomposition it emits toxic vapors of Cl⁻.

PAH800   CAS: 108171-26-2 HR: 3
PARAFFIN WAXES and HYDROCAR-BON WAXES, CHLORINATED (C_{12}, 60% CHLORINE)

PROP: Clear to light amber liquid. D: 1.35 @ 25°/25°, bp: Decomposes. Sol in water: <1 mg/mL @ 24°.

SYN: CHLORINATED PARAFFINS (C_{12}, 60% CHLOR-INE)

CONSENSUS REPORTS: NTP 10th Report on Carcinogens. IARC Cancer Review: Group 2B IMEMDT 48,55,90; Animal Sufficient Evidence IMEMDT 48,55,90. NTP Carcinogenesis Studies (gavage); Clear Evidence: mouse, rat NTPTR* NTP-TR-308,86. Reported in EPA TSCA Inventory.

SAFETY PROFILE: Confirmed carcinogen with carcinogenic and neoplastigenic data. When heated to decomposition it emits acrid smoke and irritating fumes.

PAH810   CAS: 108171-27-3 HR: 3
PARAFFIN WAXES and HYDROCAR-BON WAXES, CHLORINATED (C_{23}, 43% CHLORINE)

PROP: Clear to light amber liquid. D: 1.17 @ 25°/25°, bp: decomposes. Sol in water: <1 mg/mL @ 24°. Flash pt: >232° C.

SYN: CHLORINATED PARAFFINS (C_{23}, 43% CHLORINE)

CONSENSUS REPORTS: IARC Cancer Review: Animal Limited Evidence IMEMDT 48,55,90. NTP Carcinogenesis Studies (gavage): Clear Evidence: mouse NTPTR* NTP-TR-305,86; Equivocal Evidence: rat NTPTR* NTP-TR-305,86. Reported in EPA TSCA Inventory.

SAFETY PROFILE: Suspected carcinogen with experimental carcinogenic and neoplastigenic data. A combustible liquid. When heated to decomposition it emits acrid smoke and irritating fumes.

PAI000   CAS: 30525-89-4   HR: 3
PARAFORMALDEHYDE

DOT: UN 2213

mf: $(CH_2O)_n$

PROP: White crystals; odor of formaldehyde. Flash p: 158°F, autoign temp: 572°F, mp: 163–165° (decomp). Sltly sol in cold water; moderately sol in hot water, yielding formaldehyde; sol in strong alkalies; insol in EtOH, Et₂O.

SYNS: FLO-MOR □ FORMAGENE □ PARAFORSN □ TRIFORMOL □ TRIOXYMETHYLENE

CONSENSUS REPORTS: Reported in EPA TSCA Inventory.

DOT CLASSIFICATION: 4.1; Label: None

SAFETY PROFILE: Moderately toxic by ingestion. A severe eye and skin irritant. Mutation data reported. Flammable when exposed to heat or flame; can react with oxidizing materials. To fight fire, use alcohol foam, CO_2, dry chemical. Incompatible with liquid oxygen. Dangerous; when heated to decomposition it emits toxic formaldehyde gas. See also FORMALDEHYDE.

PAI250   CAS: 123-63-7   HR: 3
PARALDEHYDE

DOT: UN 1264

mf: $C_6H_{12}O_3$ mw: 132.18

PROP: Colorless liquid; disagreeable taste, aromatic odor. Mp: 12.6°, lel: 1.3%, bp: 124.4° @ 752 mm, flash p: 62.6°F, d: 0.9943 @ 20°/4°, autoign temp: 460°F, vap d: 4.55.

Sol in water; misc with alc, ether, oils, chloroform.

SYNS: ACETALDEHYDE, TRIMER □ ELALDEHYDE □ PARACETALDEHYDE □ PARAL □ PARALDEHYD (GERMAN) □ PARALDEIDE (ITALIAN) □ PCHO □ RCRA WASTE NUMBER U182 □ TRIACETALDEHYDE (FRENCH) □ 2,4,6-TRIMETHYL-1,3,5-TRIOXAAN (DUTCH) □ 2,4,6-TRIMETHYL-s-TRIOXANE □ 2,4,6-TRIMETHYL-1,3,5-TRIOXANE □ s-TRIMETHYLTRIOXY-METHYLENE □ 2,4,6-TRIMETIL-1,3,5-TRIOSSANO (ITALIAN)

CONSENSUS REPORTS: Reported in EPA TSCA Inventory.

DOT CLASSIFICATION: 3; Label: Flammable Liquid

SAFETY PROFILE: A human poison by rectal route. Moderately toxic to humans by intramuscular route. Moderately toxic experimentally by inhalation, ingestion, intraperitoneal, and subcutaneous routes. Human systemic effects by rectal route: necrotic changes. A skin and severe eye irritant. Low doses produce hypnotic and analgesic effects. Larger doses depress the nervous system with loss of reflexes, coma, and respiratory depression leading to respiratory paralysis and death. Chronic effects include weight loss, muscular weakness, and mental fatigue. However, poisoning is rare. A hypnotic agent. Dangerous fire hazard when exposed to heat, flame, or oxidizers. Slight explosion hazard when exposed to heat or flame. Dangerous; keep away from heat and open flame. To fight fire, use alcohol foam, CO_2, dry chemical. Potentially violent reaction with nitric acid. Incompatible with alkalies, hydrocyanic acid, iodides, oxidizers. When heated to decomposition it emits acrid smoke and irritating fumes. See also ALDEHYDES.

PAI990   CAS: 4685-14-7   HR: 3
PARAQUAT
mf: $C_{12}H_{14}N_2$ mw: 186.28
PROP: Off-white Powder. Bp: 175°–180°. Sol in water.
SYNS: 1,1'-DIMETHYL-4,4'-BIPYRIDINIUM □ DIMETH-YL VIOLOGEN □ GRAMOXONE S □ METHYL

VIOLOGEN (2+) □ PARAQUAT DICATION □ PARAQUAT ION □ PRIGLONE

CONSENSUS REPORTS: EPA Genetic Toxicology Program.

OSHA PEL: Respirable Dust: TWA 0.1 mg/m³ (skin)

ACGIH TLV: TWA 0.1 mg/m³

SAFETY PROFILE: Poison by ingestion and intraperitoneal routes. Mutation data reported. Human systemic effects: changes in structure or function of esophagus, diarrhea, edema, fibrosis of lung, fibrosis, focal (pneumoconiosis), hemorrhage, jaundice, renal damage, renal function tests depressed, respiratory depression, ulceration or bleeding from stomach, vomiting. Death from anoxia may result. When heated to decomposition it emits toxic fumes of NO_x. See also PARAQUAT DICHLORIDE.

PAJ500   CAS: 10048-32-5   HR: 3
PARASCORBIC ACID
mf: $C_6H_8O_2$ mw: 112.14
PROP: Oily liquid; sweet, aromatic odor. Bp: 104–105° @ 14 mm, 119–123° @ 22 mm, d: 1.079 @ 18°/4°. Sol in water; very sol in alc, ether.
SYNS: (S)-(+)-5,6-DIHYDRO-6-METHYL-2H-PYRAN-2-ONE □ γ-HEXENOLACTONE □ 2-HEXEN-5,1-OLIDE □ D''-HEXENOLLACTONE □ 5-HYDROXY-2-HEXENOIC ACID LACTONE □ PARASORBIC ACID □ (+)-PARASORBINSAEURE (GERMAN) □ SORBIC OIL

CONSENSUS REPORTS: IARC Cancer Review: Group 3 IMEMDT 7,56,87; Animal Limited Evidence IMEMDT 10,199,76.

SAFETY PROFILE: Poison by intraperitoneal and intravenous routes. Mildly toxic by skin contact. Questionable carcinogen with experimental neoplastigenic data. When heated to decomposition it emits acrid smoke and irritating fumes.

PAK000   CAS: 56-38-2   HR: 3
PARATHION
DOT: NA 2783
mf: $C_{10}H_{14}NO_5PS$ mw: 291.28
PROP: Pale-yellow liquid. Mp: 6°, bp: 375°, d: 1.26 @ 25°/4°. Very sol in alcs, esters,

ethers, ketones, aromatic hydrocarbons; insol in water, pet ether, kerosene. IDLH 10 mg/m³.

SYNS: AAT □ AATP □ ALLERON □ APHAMITE □ ARALO □ BAY E-605 □ BAYER E-605 □ BLADAN □ COMPOUND 3422 □ COROTHION □ CORTHION □ CORTHIONE □ DANTHION □ O,O-DIAETHYL-O-(4-NITROPHENYL)-MONOTHIOPHOSPHAT (GERMAN) □ O,O-DIETHYL-O-(4-NITRO-FENIL)-MONOTHIOFOSFA-AT (DUTCH) □ O,O-DIETHYL-O-p-NITROFENYLESTER KYSELINYTHIOFOSFORECNE (CZECH) □ O,O-DIETHYL-O-p-NITROFENYLTIOFOSFAT (CZECH) □ O,O-DIETHYL-O-(p-NITROPHENYL) PHOSPHORO-THIOATE □ O,O-DIETHYL-O-(4-NITROPHENYL) PHOSPHOROTHIOATE □ O,O-DIETHYL-O-4-NITROPHENYLPHOSPHOROTHIOATE □ DIETHYL-4-NITROPHENYL PHOSPHOROTHIONATE □ DIETHYL-p-NITROPHENYLTHIONOPHOSPHATE □ O,O-DIETHYL-O-(p-NITROPHENYL)THIONOPHOSPHATE □ DIETHYL-p-NITROPHENYLTHIOPHOSPHATE □ O,O-DIETHYL-O-p-NITROPHENYL THIOPHOSPHATE □ O,O-DIETHYL-O-4-NITROPHENYL THIOPHOSPHATE □ DIETHYLPARATHION □ O,O-DIETIL-O-(4-NITRO-FENIL)-MONOTIOFOSFATO (ITALIAN) □ DNTP □ DPP □ DREXEL PARATHION 8E □ ECATOX □ EKATOX □ ENT 15,108 □ ETHLON □ ETHYL PARATHION □ FOLIDOL □ FOSFERMO □ FOSFEX □ FOSFIVE □ FOSOVA □ FOSTERN □ FOSTOX □ GEARPHOS □ GENITHION □ KOLPHOS □ KYPTHION □ LETHALAIRE G-54 □ LIROTHION □ MURFOS □ NCI-C00226 □ NIRAN □ p-NITROPHENOL, O-ESTER with O,O-DIETHYLPHOSPHOROTHIOATE □ NITROSTIGM-IN (GERMAN) □ NITROSTIGMINE □ NOURITHION □ OLEOFOS 20 □ OLEOPARAPHENE □ OLEOPARA-THION □ ORTHOPHOS □ PANTHION □ PARADUST □ PARAMAR □ PARAPHOS □ PARATHENE □ PARATH-ION, liquid (DOT) □ PARATHION-ETHYL □ PARAWET □ PESTOX PLUS □ PETHION □ PHOSKIL □ PHOSPHEMOL □ PHOSPHENOL □ PHOSPHORO-THIOIC ACID, O,O-DIETHYL-O-(4-NITROPHENYL) ESTER □ PHOSPHOSTIGMINE □ RCRA WASTE NUMBER P089 □ RHODIASOL □ RHODIATOX □ RHODIATROX □ SELEPHOS □ SIXTY-THREE SPECIAL E.C. INSECTICIDE □ SNP □ SOPRATHION □ STATHION □ STRATHION □ SULPHOS □ SUPER RODIATOX □ THIOPHOS □ THIOPHOSPHATE de O,O-DIETHYLE et de O-(4-NITROPHENYLE) (FRENCH) □ TIOFOS □ TOX 47 □ VAPOPHOS □ VITREX

CONSENSUS REPORTS: IARC Cancer Review: Group 3 IMEMDT 7,56,87; Human Inadequate Evidence IMEMDT 30,153,83; Animal Inadequate Evidence IMEMDT 30,153,83. NCI Carcinogenesis Bioassay (feed); Clear Evidence: rat NCITR* NCI-CG-TR-70,79; (feed); No Evidence: mouse NCITR* NCI-CG-TR-70,79. EPA Farm Worker Field Reentry FEREAC 39,16888,74. EPA Extremely Hazardous Substances List. Community Right-To-Know List. EPA Genetic Toxicology Program.

OSHA PEL: TWA 0.1 mg/m³ (skin)
ACGIH TLV: TWA 0.05 mg/m³ (skin); Not Classifiable as a Human Carcinogen)); BEI: 0.5 mg/g creatinine in urine at end of shift; Not Classifiable as a Human Carcinogen
DFG MAK: 0.1 mg/m³; BAT: 500 µg/L p-nitrophenol in urine after several shifts
NIOSH REL: (Parathion) TWA 0.05 mg/m³
DOT CLASSIFICATION: 6.1; Label: Poison
SAFETY PROFILE: A deadly poison by all routes. Human systemic effects by ingestion: general anesthetic; pulmonary effects; and kidney, ureter, bladder effects, true cholinesterase changes. Experimental teratogenic and reproductive effects. Questionable carcinogen with experimental carcinogenic and tumorigenic data. Human mutation data reported. A cholinesterase inhibitor. Parathion, like the other organic phosphorus poisons, acts as an irreversible inhibitor of the enzyme cholinesterase and thus allows the accumulation of large amounts of acetylcholine. When a critical level of cholinesterase depletion is reached, grave symptoms appear. Whether death is actually caused entirely by cholinesterase depletion or by the disturbance of a number of enzymes is not yet known. Recovery is apparently complete if a poisoned animal or human has time to re-form a critical amount of cholinesterase. The organism exposed remains susceptible to relatively low dosages of parathion until the cholinesterase has regenerated. Small doses at frequent intervals are, therefore, more or less additive. There is no indication that, when recovery from a given exposure is entirely complete, the exposed organism is prejudiced in any way. Combustible when exposed to heat or flame. Violent reaction with endrin. Highly dangerous; shock can shatter the container, releasing the contents.

A broad spectrum insecticide in agricultural applications. When heated to decomposition it emits highly toxic fumes of NO_x, PO_x, and SO_x.

PAK230   HR: 3
PARATHION and compressed gas mixture (DOT)

DOT: NA 1967

SYN: PHOSPHOROTHIOIC ACID-O,O-DIETHYL, O-(p-NITROPHENYL)ESTER, mixed with compressed gas

DOT CLASSIFICATION: 2.3; Label: Poison Gas

SAFETY PROFILE: A poison gas. Questionable carcinogen. When heated to decomposition it emits very toxic fumes of PO_x, SO_x, and NO_x. See also PARATHION.

PAL750   CAS: 8000-68-8   HR: 2
PARSLEY OIL

PROP: From steam distillation of above-ground parts (herb oil) or ripe seed (seed oil) of *Petroselinum sativum Hoffm.* (Fam. *Umbelligerae*). Yellow to light-brown liquid; odor of parsley. D (herb oil): 0.908–0.940, (seed oil): 1.040; refr index (herb oil): 1.503–1.530 @ 20°, (seed oil): 1.513–1.522 @ 20°. Sol in fixed oils, mineral oil; sltly sol in propylene glycol; insol in glycerin.

SYNS: OIL of PARSLEY □ PARSLEY HERB OIL (FCC) □ PARSLEY SEED OIL (FCC) □ PETERSILIENSAMEN OEL (GERMAN)

CONSENSUS REPORTS: Reported in EPA TSCA Inventory.

SAFETY PROFILE: Moderately toxic by ingestion. A human skin irritant. When heated to decomposition it emits acrid smoke and irritating fumes.

PAN100   CAS: 434-07-1   HR: 3
PAVISOID

mf: $C_{21}H_{32}O_3$ mw: 332.53

PROP: Crystals from ethyl acetate. Mp: 185–190°.

SYNS: ADROIDIN □ ADROYD □ ANADROL □ ANADROYD □ ANAPOLON □ ANASTERON □ ANASTERONAL □ ANASTERONE □ BECOREL □ CI-406 □ 4,5-DIHYDRO-2-HYDROXYMETHYLENE-17-α-METHYLTESTOSTERONE □ DYNASTEN □ HMD □ 17-

β-HYDROXY-2-HYDROXYMETHYLENE-17-α-METHYL-3-ANDROSTANONE □ 17-β-HYDROXY-2-(HYDROXY-METHYLENE)-17-METHYL-5-α-ANDROSTAN-3-ONE □ 17-β-HYDROXY-2-(HYDROXYMETHYLENE)-17-α-METHYL-5-α-ANDROSTAN-3-ONE □ 17-HYDROXY-2-(HYDROXYMETHYLENE)-17-METHYL-5-α-17-β-ANDROSTAN-3-ONE □ 2-HYDROXYMETHYLENE-17-α-METHYL-5-α-ANDROSTAN-17-β-OL-3-ONE □ 2-HYDROXYMETHYLENE-17-α-METHYL-DIHYDRO-TESTOSTERONE □ 2-(HYDROXYMETHYLENE)-17-α-METHYLDIHYDROTESTOSTERONE □ 2-HYDROXY-METHYLENE-17-α-METHYL-17-β-HYDROXY-3-ANDROSTANONE □ METHABOL □ 17-α-METHYL-2-HYDROXYMETHYLENE-17-HYDROXY-5-α-ANDROS-TAN-3-ONE □ NASTENON □ NSC-26198 □ OXIME-THOLONUM □ OXIMETOLONA □ OXITOSONA-50 □ OXYMETHALONE □ OXYMETHENOLONE □ OXYMETHOLONE □ PARDROYD □ PLENASTRIL □ PROTANABOL □ ROBORAL □ SYNASTERON □ ZENALOSYN

CONSENSUS REPORTS: NTP 10th Report on Carcinogens. IARC Cancer Review: Animal No Adequate Data IMEMDT 7,96,87.

SAFETY PROFILE: Confirmed human carcinogen producing liver tumors. Human systemic effects by ingestion: impaired liver function. An experimental teratogen. Experimental reproductive effects. When heated to decomposition it emits acrid smoke and irritating fumes. See also TESTOSTERONE.

PAO000   CAS: 8002-03-7   HR: 2
PEANUT OIL

PROP: Straw-yellow to greenish-yellow or nearly colorless oil; nutty odor and bland taste. Mp: 2.7°, flash p: 540°F, d: 0.92, autoign temp: 833°F. Misc with ether, pet ether, chloroform, carbon disulfide; sol in benzene, carbon tetrachloride, oils; very sltly sol in alc. From seed of *Arachis hypogaea* (85DIA2 2,201,77).

SYNS: ARACHIS OIL □ EARTHNUT OIL □ GROUNDNUT OIL □ INDIGENOUS PEANUT OIL □ KATCHUNG OIL □ PECAN SHELL POWDER

CONSENSUS REPORTS: Reported in EPA TSCA Inventory.

SAFETY PROFILE: A human skin irritant and mild allergen. Questionable carcinogen with experimental tumorigenic data. Mutation data reported. Combustible when exposed to heat or flame; can react with oxidizing materials. Slight spontaneous heating. To fight fire, use CO_2, dry chemical. When heated to decomposition it emits acrid smoke and irritating fumes.

PAP250   CAS: 26864-56-2   HR: 3
PENFLURIDOL
mf: $C_{28}H_{27}ClF_5NO$ mw: 524.01
PROP: White microcrystals. Mp: 105–107°. Sltly sol in water.
SYNS: 4-(4-CHLORO-α,α,α-TRIFLUORO-m-TOLYL)-1-(4,4-BIS(p-FLUOROPHENYL)BUTYL)-4-PIPERIDINOL □ McN-JR-16,341 □ R 16341 □ SEMAP □ TLP-607
SAFETY PROFILE: Poison by ingestion and intravenous routes. Experimental teratogenic and reproductive effects. A neuroleptic agent. When heated to decomposition it emits very toxic fumes of Cl^-, F^-, and NO_x.

PAP750   CAS: 90-65-3   HR: 3
PENICILLIC ACID
mf: $C_8H_{10}O_4$ mw: 170.18
PROP: Needles from pet ether or rhombic or hexagonal plates. Mp: 83–84°. Sltly sol in cold water, hot pet ether; very sol in hot water, alc, ether, benzene chloroform; insol in pentane-hexane.
SYNS: γ-KETO-β-METHOXY-Δ-METHYLENE-Δ^α-HvEXENOIC ACID □ 3-METHOXY-5-METHYL-4-OXO-2,5-HEXADIENOIC ACID □ PA □ PENCILLIC ACID
CONSENSUS REPORTS: IARC Cancer Review: Group 3 IMEMDT 7,56,87; Animal Sufficient Evidence IMEMDT 10,211,76. EPA Genetic Toxicology Program.
SAFETY PROFILE: Poison by intravenous, subcutaneous, and intraperitoneal routes. Moderately toxic by ingestion. An experimental teratogen. Experimental reproductive effects. Questionable carcinogen with experimental neoplastigenic data. Human mutation data reported. When heated to decomposition it emits acrid smoke and irritating fumes.

PAQ000   CAS: 1406-05-9   HR: 3
PENICILLIN
mf: $(CH_3)_2C_5H_3NSO(COOH)NHCOOR$ (bicyclic)
PROP: A group of isomeric and closely related antibiotic compounds with outstanding bacterial activity. An extract from *Penicillium notatum* (JPETAB 77,40,43). Different varieties of penicillin are produced by adding the proper precursors to the nutrient solution.
SYN: PENIZILLIN (GERMAN)
CONSENSUS REPORTS: EPA Genetic Toxicology Program.
SAFETY PROFILE: Poison by intraperitoneal and subcutaneous routes. Moderately toxic by intravenous route. Human reproductive effects by ingestion: abortion. Human systemic effects by intramuscular route: dermatitis. Experimental reproductive effects. Has been implicated in aplastic anemia. When heated to decomposition it emits very toxic fumes of NO_x and SO_x.

PAR500   HR: 3
PENNYROYAL OIL
PROP: Chief constituent is d-pulegone. From steam distillation of *Mentha pulegium L.* (Fam. *Labiatae*) (FCTXAV 12,807,74). Light-yellow liquid; mint odor. Sol in fixed oils, propylene glycol, mineral oil; insol in glycerin.
SYN: AMERICAN PENNYROYAL OIL
SAFETY PROFILE: Experimental poison by ingestion. A skin irritant. When heated to decomposition it emits acrid smoke and irritating fumes.

PAT750   CAS: 19624-22-7   HR: 3
PENTABORANE(9)
DOT: UN 1380
mf: B_5H_9 mw: 63.14
PROP: Colorless gas or liquid; bad odor. Mp: −46.6°, bp: 60°, d: 0.61 @ 0°, vap d: 2.2, vap press: 66 mm @ 0°, lel: 0.42%. Sol in THF, diglyme, Et_2O, and hexane. IDLH 1 ppm.
SYN: PENTABORANE (ACGIH,DOT,OSHA)

CONSENSUS REPORTS: EPA Extremely Hazardous Substances List. Reported in EPA TSCA Inventory.

OSHA PEL: TWA 0.005 ppm; STEL 0.015 ppm

ACGIH TLV: TWA 0.005 ppm; STEL 0.013 ppm

DFG MAK: 0.005 ppm (0.013 mg/m^3)

DOT CLASSIFICATION: 4.2; Label: Spontaneously Combustible, Poison

SAFETY PROFILE: Poison by inhalation and intraperitoneal routes. Dangerous fire hazard by chemical reaction; spontaneously flammable in air. Dangerous explosion hazard. To fight fire, use special fire-fighting materials; water is not effective; reacts violently with halogenated extinguishing agents. Get instructions from supplier. Explosive reaction with oxygen. Forms shock-sensitive solutions in solvents containing carbonyl, ether, or ester functions; or halogens. Incompatible with dimethyl sulfoxide. Upon decomposition it emits toxic fumes of B. See also BORANES and BORON COMPOUNDS.

PAU250   CAS: 608-71-9   HR: 3
PENTABROMOPHENOL

mf: C$_6$HBr$_5$O   mw: 488.62

PROP: Needles from EtOH. Mp: 229.5°, bp: subl. Insol in water; sltly sol in alc and ether.

SYN: PENTABROMFENOL

CONSENSUS REPORTS: Reported in EPA TSCA Inventory.

SAFETY PROFILE: Poison by intraperitoneal route. When heated to decomposition it emits toxic fumes of Br⁻. See also BROMIDES.

PAU500   CAS: 1163-19-5   HR: 2
PENTABROMOPHENYL ETHER

mf: C$_{12}$Br$_{10}$O   mw: 959.22

SYNS: BERKFLAM B 10E □ BR 55N □ BROMKAL 83-10DE □ BROMKAL 82-ODE □ DBDPO □ DECABROMO-BIPHENYL ETHER □ DECABROMOBIPHENYL OXIDE □ DECABROMODIPHENYL OXIDE □ DECABROMO-PHENYL ETHER □ DE 83R □ FR 300 □ FRP 53 □ NCI-

C55287 □ 1,1'-OXYBIS(2,3,4,5,6-PENTABROMOBENZENE) (9CI) □ SAYTEX 102 □ SAYTEX 102E □ TARDEX 100

CONSENSUS REPORTS: IARC Cancer Review: Group 3 IMEMDT 48,73,90; Animal Limited Evidence IMEMDT 48,73,90. NTP Carcinogenesis Studies (feed); Some Evidence: rat NTPTR* NTP-TR-309,86; (feed); Equivocal Evidence: mouse NTPTR* NTP-TR-309,86. Polybrominated biphenyl compounds are on the Community Right-To-Know List. Reported in EPA TSCA Inventory.

SAFETY PROFILE: Questionable carcinogen with experimental neoplastigenic data. Experimental reproductive effects. Used as a flame retardant for thermoplastics. When heated to decomposition it emits toxic fumes of Br⁻. See also ETHERS and BROMIDES.

PAV250   CAS: 25201-35-8   HR: 3
PENTACHLOROACETOPHENONE

mf: C$_8$Cl$_5$H$_3$O   mw: 292.36

PROP: Crystals from EtOH. Mp: 90°.

SYN: 2',3',4',5',6'-PENTACHLOROACETOPHENONE

SAFETY PROFILE: Poison by intraperitoneal route. Moderately toxic by ingestion and skin contact. Experimental reproductive effects. When heated to decomposition it emits toxic fumes of Cl⁻. See also CHLORINATED HYDROCARBONS, ALIPHATIC.

PAV500   CAS: 608-93-5   HR: 2
PENTACHLOROBENZENE

mf: C$_6$HCl$_5$ mw: 250.32

PROP: Needles from alc. D: 1.834 @ 17°, mp: 85–86°, bp: 275–277°; insol in water; sol in hot alc; very sol in ether.

SYNS: QCB □ RCRA WASTE NUMBER U183

CONSENSUS REPORTS: Reported in EPA TSCA Inventory.

SAFETY PROFILE: Moderately toxic by ingestion. An experimental teratogen. When heated to decomposition it emits toxic fumes of Cl⁻. See also CHLORINATED HYDROCARBONS, AROMATIC.

P

PAW250   CAS: 42279-29-8   HR: 3
PENTACHLORO DIPHENYL OXIDE
mf: $C_{12}H_5Cl_5O$ mw: 342.42
SYNS: ETHER, PENTACHLOROPHENYL □ PHENYL
ETHER PENTACHLORO
OSHA PEL: TWA 0.5 mg/m³
SAFETY PROFILE: Poison by ingestion.
When heated to decomposition it emits
toxic fumes of Cl⁻.

PAW500   CAS: 76-01-7   HR: 3
PENTACHLOROETHANE
DOT: UN 1669
mf: C_2HCl_5 mw: 202.28
PROP: Colorless liquid; chloroform-like
odor. Mp: −29°, bp: 161–162°, d: 1.6728 @
25°/4°. Insol in water; misc in alc and ether.
SYNS: ETHANE PENTACHLORIDE □ NCI-C53894 □
PENTACHLOORETHAAN (DUTCH) □ PENTACHLO-
RAETHAN (GERMAN) □ PENTACHLORETHANE
(FRENCH) □ PENTACLOROETANO (ITALIAN) □
PENTALIN □ RCRA WASTE NUMBER U184
CONSENSUS REPORTS: IARC Cancer
Review: Group 3 IMEMDT 7,56,87; Animal
Limited Evidence IMEMDT 41,99,86. NTP
Carcinogenesis Bioassay (gavage); Clear
Evidence: mouse NTPTR* NTP-TR-232,82;
(gavage); No Evidence: rat NTPTR* NTP-
TR-232,82. Reported in EPA TSCA
Inventory.
DFG MAK: 5 ppm (42 mg/m³)
DOT CLASSIFICATION: 6.1; Label: Poison
SAFETY PROFILE: Poison by inhalation
and intravenous routes. Moderately toxic by
ingestion and subcutaneous routes. An
irritant. Questionable carcinogen with
experimental carcinogenic data. Flammable
when exposed to heat or flame. Moderately
explosive by spontaneous chemical reaction.
To fight fire, use water, CO_2, dry chemical.
Dehalogenation by reaction with alkalies,
metals, etc., will produce spontaneously
explosive chloroacetylenes. Violent reaction
with NaK alloy + bromoform. Mixtures
with potassium are very shock-sensitive
explosives. When heated to decomposition
it emits highly toxic fumes of Cl⁻. See also
CHLORINATED HYDROCARBONS,
ALIPHATIC.

PAW750   CAS: 1321-64-8   HR: 3
PENTACHLORONAPHTHALENE
mf: $C_{10}H_3Cl_5$ mw: 300.38
PROP: White solid.
CONSENSUS REPORTS: Reported in EPA
TSCA Inventory.
OSHA PEL: TWA 0.5 mg/m³ (skin)
ACGIH TLV: TWA 0.5 mg/m³
DFG MAK: 0.5 mg/m³
NIOSH REL: (Pentachloronaphthalen) TWA
0.5 mg/m³ (skin)
SAFETY PROFILE: Poison by ingestion,
inhalation, and skin contact. An irritant.
Action similar to that of chlorinated
naphthalenes and chlorinated diphenyls.
Dangerous; when heated to decomposition
it emits highly toxic fumes of Cl⁻. See also
CHLORINATED HYDROCARBONS,
AROMATIC.

PAX000   CAS: 82-68-8   HR: 3
PENTACHLORONITROBENZENE
mf: $C_6Cl_5NO_2$ mw: 295.32
PROP: Colorless crystals from alc. Mp:
146°, bp: 328°, vap press: 0.013 mm @ 25°.
SYNS: AVICOL □ BATRILEX □ BRASSICOL □
EARTHCIDE □ FARTOX □ FOLOSAN □ FOMAC 2 □
FUNGICLOR □ GC 3944-3-4 □ KOBU □ KOBUTOL □ KP
2 □ NCI-C00419 □ OLPISAN □ PCNB □ PENTACHLO-
RNITROBENZOL (GERMAN) □ PENTAGEN □ PKhNB
□ QUINTOCENE □ QUINTOZEN □ QUINTOZENE □
RCRA WASTE NUMBER U185 □ SANICLOR 30 □
TERRACHLOR □ TERRAFUN □ TILCAREX □ TRI-PCNB
□ TRITISAN
CONSENSUS REPORTS: IARC Cancer
Review: Group 3 IMEMDT 7,56,87; Animal
Sufficient Evidence IMEMDT 5,211,74.
NCI Carcinogenesis Bioassay (feed); No
Evidence: mouse, rat NCITR* NCI-CG-
TR-61,78. Reported in EPA TSCA
Inventory. EPA Genetic Toxicology
Program.
ACGIH TLV: TWA 0.5 mg/m³; Not
Classifiable as a Human Carcinogen
SAFETY PROFILE: Moderately toxic by
ingestion. An experimental teratogen.
Questionable carcinogen with experimental
carcinogenic data. Mutation data reported.
Used as a fungicide. Dangerous; when

heated to decomposition it emits highly toxic fumes of NO$_x$ and Cl⁻. See also NITRO COMPOUNDS OF AROMATIC HYDROCARBONS.

PAX250 **CAS: 87-86-5** **HR: 3**
PENTACHLOROPHENOL
mf: C$_6$HCl$_5$O mw: 266.32
PROP: Dark-colored flakes, monoclinic prisms, and sublimed needle crystals from benzene; characteristic odor. Mp: 174°, mp: 191° (anhydrous), bp: 310° (decomp), d: 1.978, vap press: 40 mm @ 211.2°. Sol in ether, benzene; very sol in alc; insol in water; sltly sol in cold pet ether. IDLH 2.5 mg/m³.
SYNS: CHEM-TOL □ CHLOROPHEN □ CRYPTOGIL OL □ DOWCIDE 7 □ DOWCIDE 7 □ DOWCIDE EC-7 □ DOWCIDE G □ DOW PENTACHLOROPHENOL DP-2 ANTIMICROBIAL □ DUROTOX □ EP 30 □ FUNGIFEN □ GLAZD PENTA □ GRUNDIER ARBEZOL □ LAUXTOL □ LAUXTOL A □ LIROPREM □ NCI-C54933 □ NCI-C55378 □ NCI-C56655 □ PCP □ PENCHLOROL □ PENTA □ PENTACHLOORFENOL (DUTCH) □ PENTACHLOROFENOL □ PENTACHLOROPHENATE □ PENTACHLOROPHENOL (GERMAN) □ 2,3,4,5,6-PENTACHLOROPHENOL □ PENTACHLOROPHENOL, DOWCIDE EC-7 □ PENTACHLOROPHENOL, DP-2 □ PENTACHLOROPHENOL, TECHNICAL □ PENTACLOROFENOLO (ITALIAN) □ PENTACON □ PENTA-KIL □ PENTASOL □ PENWAR □ PERATOX □ PERMACIDE □ PERMAGARD □ PERMASAN □ PERMATOX DP-2 □ PERMATOX PENTA □ PERMITE □ PRILTOX □ RCRA WASTE NUMBER U242 □ SANTOBRITE □ SANTOPHEN □ SANTOPHEN 20 □ SINITUHO □ TERM-I-TROL □ THOMPSON'S WOOD FIX □ WEEDONE
CONSENSUS REPORTS: IARC Cancer Review: Group 2B IMEMDT 53,371,91; Human Limited Evidence IMEMDT 41,319,86; Animal Sufficient Evidence IMEMDT 53,371,91; IARC Cancer Review: Animal Inadequate Evidence IMEMDT 20,303,79; Human Inadequate Evidence IMEMDT 53,371,91. Chlorophenol compounds are on the Community Right-To-Know List. Reported in EPA TSCA Inventory. EPA Genetic Toxicology Program.
OSHA PEL: TWA 0.5 mg/m³ (skin)

ACGIH TLV: TWA 0.5 mg/m³ (skin); BEI: 2 mg/g creatinine in urine prior to last shift of workweek; Animal Carcinogen
DFG MAK: Confirmed Animal Carcinogen, Suspected Human Carcinogen; BAT: 1000 μg/L in plasma/serum
SAFETY PROFILE: Confirmed human carcinogen with experimental tumorigenic data. Human poison by ingestion. Poison experimentally by ingestion, skin contact, intraperitoneal, and subcutaneous routes. An experimental teratogen. Other experimental reproductive effects. A skin irritant. Mutation data reported. Acute poisoning is marked by weakness with changes in respiration, blood pressure, and urinary output. Also causes dermatitis, convulsions, and collapse. Chronic exposure can cause liver and kidney injury. Dangerous; when heated to decomposition it emits highly toxic fumes of Cl⁻. See also CHLOROPHENOLS.

PBB750 **CAS: 115-77-5** **HR: 3**
PENTAERYTHRITOL
mf: C$_5$H$_{12}$O$_4$ mw: 136.17
PROP: Ditetragolan crystals from HCl (aq). Mp: 262°, d: 1.38 @ 25°/4°.
SYNS: AUXINUTRIL □ 2,2-BIS(HYDROXYMETHYL)-1,3-PROPANEDIOL □ HERCULES P6 □ METHANE TETRAMETHYLOL □ MONOPENTEK □ PE □ PENTAERYTHRITE □ PENTEK □ TETRAHYDROXY-METHYLMETHANE □ TETRAKIS(HYDROXYMETHYL)METHANE □ TETRAMETHYLOLMETHANE
CONSENSUS REPORTS: Reported in EPA TSCA Inventory.
OSHA PEL: TWA Total Dust: 10 mg/m³; Respirable Fraction: 5 mg/m³
ACGIH TLV: TWA (nuisance particulate) 10 mg/m³ of total dust (when toxic impurities are not present, e.g., quartz <1%)
SAFETY PROFILE: Mildly toxic by ingestion. A nuisance dust. Flammable from heat or flame or oxidizers. Mixtures with thiophosphoryl chloride react when heated to form a product that ignites and then explodes on contact with air. Used in coatings, stabilizers, explosives, P.E.T.N resins, drugs, insecticides, and lubricants.

P

When heated to decomposition it emits acrid smoke and irritating fumes.

PBB800   HR: D
PENTAERYTHRITOL ESTER of PARTI-ALLY HYDROGENATED WOOD ROSIN
PROP: Hard, amber-colored solid. Sol in acetone, benzene; insol in water.
SAFETY PROFILE: When heated to decomposition it emits acrid smoke and irritating fumes.

PBC250   CAS: 78-11-5   HR: 3
PENTAERYTHRITOL TETRANITRATE
DOT: UN 0411
mf: $C_5H_8N_4O_{12}$ mw: 316.17
PROP: Crystals or prisms from $Me_2CO/EtOH$. Mp: 138–140°, bp: explodes @ 205–215°, d: 1.773 @ 20°/4°. Sol in acetone; insol in water; sltly sol in alc, ether.
SYNS: ANGICAP □ ANGITET □ ANTORA □ ARC-OTRATE □ BARITRATE □ 2,2-BISDIHYDROXYMETH-YL-1,3-PROPANEDIOL TETRANITRATE □ 2,2-BIS-(HYDROXYMETHYL)-1,3-PROPANEDIOL TETRANI-TRATE □ CHOT □ DELTRATE-20 □ 1,3-DINITRATO-2,2-BIS(NITRATOMETHYL)PROPANE □ DUOTRATE □ EL P.E.T.N. □ ERINIT □ HASETHROL □ INITIATING EXPLOSIVE PENTAERYTHRITE TETRANITRATE (DOT) □ KAYTRATE □ LOWETRATE □ MARTRATE-45 □ METRANIL □ MYCARDOL □ MYOTRATE "10" □ NCI-C55743 □ NEO-COROVAS □ NEOPENTANETETRAYL NITRATE □ NIPERYT □ NIPERYTH □ NITROPENTA □ NITROPENTAERYTHRITE □ NITROPENTAERYTHRIT-OL □ PENCARD □ PENTAERYTHRITE TETRANITR-ATE □ PENTAERYTHRITE TETRANITRATE (DOT) □ PENTAERYTHRITE TETRANITRATE, desensitized, wet (DOT) □ PENTAERYTHRITE TETRANITRATE, dry (DOT) □ PENTAERYTHRITE TETRANITRATE, with not less than 7% wax (DOT) □ PENTAERYTHRITOL TETRA-NITRATE, diluted □ PENTAFIN □ PENTESTAN-80 □ PENTETRATE UNICELLES □ PENTRATE □ PENTRIOL □ PENTRYATE 80 □ PERGITRAL □ PERIDEX-LA □ PERITRATE □ PERITYL □ PREVANGOR □ QUINTR-ATE □ RYTHRITOL □ SDM No. 23 □ SUBICARD □ TENTRATE-20 □ TETRANITROPENTAERYTHRITE □ TETRASULE □ TRANITE D-LAY □ VASITOL □ VASODIATOL □ VASO-80 UNICELLES
DOT CLASSIFICATION: Forbidden (dry); DOT Class: Explosive 1.1A; Label: EXPLOSIVE 1.1A (NA 0150); DOT Class:

Explosive 1.1D; Label: EXPLOSIVE 1.1D (UN 0411)
SAFETY PROFILE: Human systemic effects by ingestion: dermatitis. Effects are similar to those of nitroglycerin, i.e., headache, weakness, and fall in blood pressure. Very low oral toxicity. Severe explosion hazard when shocked or exposed to heat. It explodes at 215°C. On decomposition it emits highly toxic fumes of NO_x; can react vigorously with oxidizing materials. Used in detonators and explosive specialities. See also NITRATES and EXPLOSIVES, HIGH.

PBF250   CAS: 509-09-1   HR: 2
PENTAFLUOROPROPIONIC ACID SILVER SALT
mf: $C_3F_5O_2 \cdot Ag$ mw: 270.90
PROP: IDLH 10 mg/m³ (as Ag).
SYN: PENTAFLUORPROPIONAN STRIBRNY (CZECH)
CONSENSUS REPORTS: Silver and its compounds are on the Community Right-To-Know List. Reported in EPA TSCA Inventory.
OSHA PEL: TWA 0.01 mg(Ag)/m³
ACGIH TLV: TWA 0.01 mg(Ag)/m³
SAFETY PROFILE: Moderately toxic by ingestion. Skin and severe eye irritant. When heated to decomposition it emits toxic fumes of F^-. See also FLUORIDES and SILVER COMPOUNDS.

PBI500   CAS: 54-95-5   HR: 3
1,5-PENTAMETHYLENETETRAZOLE
mf: $C_6H_{10}N_4$ mw: 138.20
PROP: White, crystalline powder. Mp: 57–58°. Sol in water, alc, and ether.
SYNS: α,β-CYCLOPENTAMETHYLENETETRAZOLE □ PENTAMETHYLENETETRAZOL □ PENTAMETHYL-ENE-1,5-TETRAZOLE □ PENTYLENETETRAZOL □ 6,7,8,9-TETRAHYDRO-5-AZEPOTETRAZOLE □ 6,7,8,9-TETRAHYDRO-5H-TETRAZOLOAZEPINE □ 7,8,9,10-TETRAZABICYCLO(5.3.0)-8,10-DECADIENE □ 1,2,3,3A-TETRAZACYCLOHEPTA-8A,2-CYCLOPENTADIENE
CONSENSUS REPORTS: Reported in EPA TSCA Inventory.
SAFETY PROFILE: A human poison by ingestion and intravenous routes. Poison

experimentally by ingestion, intravenous, intraperitoneal, subcutaneous, rectal, and parenteral routes. When heated to decomposition it emits toxic fumes of NO_x.

PBK250   CAS: 109-66-0   HR: 3
n-PENTANE
DOT: UN 1265
mf: C_5H_{12} mw: 72.17
PROP: Colorless liquid. Bp: 36.1°, flash p: $<-40°F$, fp: $-129.8°$, d: 0.626 @ 20°/4°, autoign temp: 588°F, vap press: 400 mm @ 18.5°, vap d: 2.48, lel: 1.5%, uel: 7.8%. Sol in water; misc in alc, ether, org solv. IDLH 1500 ppm [10%LEL].
SYNS: AMYL HYDRIDE (DOT) □ PENTAN (POLISH) □ PENTANE (ITALIAN) □ PENTANEN (DUTCH)
CONSENSUS REPORTS: Reported in EPA TSCA Inventory.
OSHA PEL: TWA 600 ppm; STEL 750 ppm
ACGIH TLV: TWA 600 ppm
DFG MAK: 1000 ppm (3000 mg/m³)
NIOSH REL: (Alkanes) TWA 350 mg/m³
DOT CLASSIFICATION: 3; Label: Flammable Liquid
SAFETY PROFILE: Moderately toxic by inhalation and intravenous routes. Narcotic in high concentration. The liquid can cause blisters on contact. Flammable liquid. Highly dangerous fire hazard when exposed to heat, flame, or oxidizers. Severe explosion hazard when exposed to heat or flame. Shock can shatter metal containers and release contents. To fight fire, use foam, CO_2, dry chemical. When heated to decomposition it emits acrid smoke and irritating fumes.

PBL350   CAS: 600-14-6   HR: 3
2,3-PENTANEDIONE
mf: $C_5H_8O_2$ mw: 100.13
PROP: Yellow liquid. Mp: $-52°$, bp: 110–112°, d: 0.957, flash p: 66°F. Sol in water.
SYN: ACETYLPROPIONYL
CONSENSUS REPORTS: Reported in EPA TSCA Inventory.

SAFETY PROFILE: Moderately toxic by ingestion. A skin irritant. Flammable liquid. When heated to decomposition it emits acrid smoke and irritating fumes.

PBL500   CAS: 3264-82-2   HR: 3
2,4-PENTANEDIONE, NICKEL(II) DERI-
VATIVE
mf: $C_{10}H_{14}NiO_4$ mw: 256.95
CONSENSUS REPORTS: NTP 10th Report on Carcinogens. Nickel and its compounds are on the Community Right-To-Know List. Reported in EPA TSCA Inventory.
NIOSH REL: (Nickel, Inorganic) TWA 0.015 mg(Ni)/m³
SAFETY PROFILE: Confirmed human carcinogen. Poison by intraperitoneal route. When heated to decomposition it emits acrid smoke and irritating fumes. See also NICKEL COMPOUNDS.

PBL750   CAS: 17501-44-9   HR: 3
2,4-PENTANEDIONE, ZIRCONIUM
COMPLEX
mf: $C_{20}H_{28}O_8Zr$ mw: 487.70
PROP: Colorless solid. Sol in EtOH, C_6H_6, and toluene. IDLH 50 mg/m³ (as Zr).
SYN: ACETYLZIRCONIUM, ACETONATE
CONSENSUS REPORTS: Reported in EPA TSCA Inventory.
OSHA PEL: TWA 5 mg(Zr)/m³; STEL 10 mg(Zr)/m³
ACGIH TLV: TWA 5 mg(Zr)/m³; STEL 10 mg(Zr)/m³; Not Classifiable as a Human Carcinogen
DFG MAK: 1 mg(Zr)/m³
SAFETY PROFILE: Poison by intraperitoneal route. When heated to decomposition it emits acrid smoke and irritating fumes. Used as a cross-linking agent for oxygen-containing polymers. See also ZIRCONIUM COMPOUNDS.

PBM000   CAS: 110-66-7   HR: 3
1-PENTANETHIOL
DOT: UN 1228/UN 3071
mf: $C_5H_{12}S$ mw: 104.23

PROP: Water-white to yellow liquid. D: 0.857 @ 20°, bp: 123.64°, flash p: 65°F, vap press: 13.8 mm @ 25°, vap d: 3.59. Insol in water; misc in alc and ether.
SYNS: AMYL HYDROSULFIDE □ n-AMYL MERCAPT-AN □ AMYL MERCAPTAN (DOT) □ AMYL SULFHYDR-ATE □ AMYL THIOALCOHOL □ MERCAPTAN AMYLIQUE (FRENCH) □ PENTYL MERCAPTAN
CONSENSUS REPORTS: Reported in EPA TSCA Inventory.
NIOSH REL: (Thiols (n-Alkane Mono)) CL 0.5 ppm/15M
DOT CLASSIFICATION: 3; Label: Flammable Liquid, Poison (UN 1228); DOT Class: 6.1; Label: Poison, Flammable Liquid (UN 3071)
SAFETY PROFILE: Moderately toxic by inhalation. A weak sensitizer and allergen. Local contact may cause contact dermatitis. A flammable liquid and dangerous fire hazard when exposed to heat or flame; can react vigorously with oxidizing materials. Hypergolic reaction with concentrated nitric acid. To fight fire, use foam, CO_2, dry chemical. See also MERCAPTANS.

PBM750   CAS: 6032-29-7   HR: 3
2-PENTANOL
mf: $C_5H_{12}O$ mw: 88.17
PROP: Colorless liquid. Bp: 119.3°, flash p: 105°F (OC), ULC: 40-45, uel: 9.0%, lel: 1.2%, fp: −50°, d: 0.8169 @ 20°/20°, autoign temp: 650-725°F, vap d: 3.04. Sltly sol in water; misc in alc and ether. IDLH 500 ppm.
SYNS: sec-AMYL ALCOHOL □ METHYL PROPYL CARBINOL □ PENTANOL-2 □ sec-PENTYL ALCOHOL
CONSENSUS REPORTS: Reported in EPA TSCA Inventory.
OSHA PEL: TWA 100 ppm (360 mg/m³); STEL 125 ppm (450 mg/m³)
NIOSH REL: (Isoamyl Alcohol, sec-): TWA 100 ppm; STEL 125 ppm
SAFETY PROFILE: Moderately toxic by ingestion and intraperitoneal routes. A narcotic. A skin and severe eye irritant. Flammable liquid when exposed to heat or flame; can react with oxidizing materials. A severe explosion hazard when exposed to

heat or flame. To fight fire, use alcohol foam, dry chemical. When heated to decomposition it emits acrid smoke and irritating fumes. See also ALCOHOLS.

PBN250   CAS: 107-87-9   HR: 3
2-PENTANONE
DOT: UN 1249
mf: $C_5H_{10}O$ mw: 86.15
PROP: Water-white liquid; fruity, ethereal odor. D: 0.801-0.806, vap d: 3.0, bp: 102°, flash p: 45°F, autoign temp: 941°F, lel: 1.5%, uel: 8.2%. Sltly sol in water; misc with alc, ether. IDLH 1500 ppm.
SYNS: ETHYL ACETONE □ FEMA No. 2842 □ METHYL-PROPYL-CETONE (FRENCH) □ METHYL-n-PROPYL KETONE □ METHYL PROPYL KETONE (ACGIH, DOT) □ METYLOPROPYLOKETON (POLISH) □ MPK
OSHA PEL: TWA 200 ppm; STEL 250 ppm
ACGIH TLV: STEL 150 ppm
DFG MAK: 200 ppm (710 mg/m³)
NIOSH REL: TWA 530 mg/m³
DOT CLASSIFICATION: 3; Label: Flammable Liquid
SAFETY PROFILE: Moderately toxic by ingestion and intraperitoneal routes. Mildly toxic by skin contact and inhalation. Human systemic effects by inhalation: headache, nausea, irritation of the respiratory passages, eyes, and skin. A skin irritant. Mutation data reported. A highly flammable liquid. A very dangerous fire hazard when exposed to heat or flame; can react vigorously with oxidizing materials. An explosion hazard in the form of vapor when exposed to heat or flame. To fight fire, use alcohol foam. Mixtures with bromine trifluoride may explode during evaporation. When heated to decomposition it emits acrid smoke and irritating fumes. See also KETONES.

PBO500   CAS: 143-24-8   HR: 1
2,5,8,11,14-PENTAOXAPENTADECANE
mf: $C_{10}H_{22}O_5$ mw: 222.32
PROP: A liquid. D: 1.009, bp: 275-276°. Sol in H_2O.
SYNS: ANSUL ETHER 181AT □ BIS(2-(2-METH-OXYETHOXY)ETHYL) ETHER □ BIS(2-METHOXY-

ETHYL)ETHER □ DIMETHOXYTETRAETHYLENE GLYCOL □ DIMETHOXYTETRAGLYCOL □ ETHER, BIS(2-(2-METHOXYETHOXY)ETHYL) □ TETRAETHYL-ENE GLYCOL DIMETHYL ETHER □ TETRAGLYME

CONSENSUS REPORTS: Glycol ether compounds are on the Community Right-To-Know List. Reported in EPA TSCA Inventory.

SAFETY PROFILE: Mildly toxic by ingestion. Experimental reproductive effects. An eye irritant. Many glycol ethers are suspected of having dangerous human reproductive effects. When heated to decomposition it emits acrid smoke and irritating fumes. See also GLYCOL ETHERS.

PBR250   CAS: 1629-58-9   HR: 3
1-PENTEN-3-ONE
mf: C_5H_8O mw: 84.13
PROP: Bp: 102° @ 740 mm.
SYNS: ETHYL VINYL KETONE □ KETONE, ETHYL VINYL
CONSENSUS REPORTS: Reported in EPA TSCA Inventory.
DOT CLASSIFICATION: 3; Label: Flammable Liquid
SAFETY PROFILE: Poison by intravenous route. When heated to decomposition it emits acrid smoke and irritating fumes. See also KETONES.

PBS000   CAS: 60-44-6   HR: 3
PENTHIENATE BROMIDE
mf: $C_{18}H_{30}NO_3S•Br$ mw: 420.46
PROP: Pale-yellow crystals. Mp: 124.6°.
SYNS: α-CYCLOPENTYL-2-THIOPHENEGLYCOLATE DIETHYL(2-HYDROXYETHYL)METHYLAMMONIUM BROMIDE □ 2-DIETHYLAMINOETHYL-2-CYCLOPENT-YL-2-(2-THIENYL)HYDROXYACETATE METHOBROM-IDE □ 2-DIETHYLAMINOETHYL-α-CYCLOPENTYL-2-THIOPHENEGLYCOLATE METHOBROMIDE □ MONODRAL □ MONODRAL BROMIDE □ WIN 4369
SAFETY PROFILE: Poison by intravenous and subcutaneous routes. Moderately toxic by ingestion. When heated to decomposition it emits very toxic fumes of NH_3, NO_x, SO_x, and Br^-.

PBS250   CAS: 115-58-2   HR: 3
PENTOBARBITAL
mf: $C_{11}H_{18}N_2O_3$ mw: 226.31
PROP: Available either as the free acid or sodium salt. Free acid is a fine white crystalline powder that is only slightly sol in water and ethanol. Salt is a white crystalline powder or granules and is very sol in water and ethanol.
SYNS: 5-AETHYL-5-PENTYL-(2')-BARBITURSAEURE (GERMAN) □ 5-ETHYL-5-PENTYLBARBITURIC ACID
SAFETY PROFILE: Poison by intraperitoneal route. When heated to decomposition it emits toxic fumes of NO_x. See also BARBITURATES.

PBT050   CAS: 8066-33-9   HR: 3
PENTOLITE
DOT: UN 0151
mf: $C_7H_5N_3O_6•C_5H_8N_4O_{12}$ mw: 543.32
PROP: The explosive charge is a pale yellow to orange solid with a faint odor. D: 1.7 @ 20°. mp: 76°.
SYNS: PENTOLITE, dry or wetted with <15% water, by weight (DOT) □ 1,3-PROPANEDIOL, 2,2-BIS((NITROO-XY)METHYL)-, DINITRATE (ester), mixt. with 2-METHYL-1,3,5-TRINITROBENZENE
DOT CLASSIFICATION: EXPLOSIVE 1.1D; Label: EXPLOSIVE 1.1D
SAFETY PROFILE: An unstable explosive. When heated to decomposition it emits toxic vapors of NO_x.

PBT500   CAS: 71-73-8   HR: 3
PENTOTHAL SODIUM
mf: $C_{11}H_{18}N_2O_2S•Na$ mw: 265.36
PROP: Yellowish hygroscopic powder.
SYNS: 5-ETHYLDIHYDRO-5-(1-METHYLBUTYL)-2-THIOXO-4,6(1H,5H)-PYRIMIDINEDIONE, MONO-SODIUM SALT □ 5-ETHYL-5-(1-METHYLBUTYL)-2-THIOBARBITURIC ACID MONOSODIUM □ FARMOTAL □ HYPNOSTAN □ INTRAVAL SODIUM □ LEOPENTAL □ MONOSODIUM-5-ETHYL-5-(1-METHYLBUTYL) THIOBARBITURATE □ NESDONAL SODIUM □ PENTHIOBARBITAL SODIUM □ RAVONAL □ SODIUM-5-ETHYL-5-(1-METHYLBUTYL)-2-THIOBARBITURATE □ SODIUM PENTHIOBARBITAL □ SODIUM PENTOTHAL □ SODIUM PENTOTHIOBARBITAL □ SODIUM THIOPENTAL □ SODIUM THIOPENTOBARBITAL □ SODIUM THIOPENTONE □ SOLUBLE THIOPENTONE □ THIOMEBUMAL SODIUM □ THIONEMBUTAL □

P

THIOPENTAL SODIUM □ THIOPENTAL SODIUM SALT □ THIOPENTONE SODIUM □ THIOTHAL SODIUM □ THIPENTAL SODIUM □ TIOPENTAL SODIUM □ TRAPANAL □ TRAPANAL SODIUM

SAFETY PROFILE: Poison by ingestion, intraperitoneal, rectal, subcutaneous, and intravenous routes. Human systemic effects by intraarterial route: acute arterial occlusion; by rectal route: respiratory depression, body temperature decrease, general anesthetic. An experimental teratogen. Experimental reproductive effects. An intravenous anesthetic. When heated to decomposition it emits toxic fumes of NO_x and Na_2O. See also PENTOTHAL and BARBITURATES.

PBV000   CAS: 75-85-4   HR: 3
tert-PENTYL ALCOHOL

mf: $C_5H_{12}O$ mw: 88.17
PROP: Colorless liquid. Mp: −11.9°, bp: 101.8°, flash p: 105°F (CC), d: 0.809, autoign temp: 819°F, vap press: 10 mm @ 17.2°, lel: 1.2%, uel: 9%, vap d: 3.03. Sltly sol in water; sol in alc and ether.
SYNS: tert-AMYL ALCOHOL (DOT) □ AMYLENE HYDRATE □ DIMETHYLETHYLCARBINOL □ 2-METHYL BUTANOL-2 □ 2-METHYL-2-BUTANOL □ 3-METHYLBUTAN-3-OL □ tert-PENTANOL
CONSENSUS REPORTS: Reported in EPA TSCA Inventory.
SAFETY PROFILE: Moderately toxic to humans by an unspecified route. Moderately toxic experimentally by ingestion, intraperitoneal, subcutaneous, and rectal routes. Narcotic in high concentration. Flammable liquid when exposed to heat, flame, or oxidizing materials. Moderately explosive in the form of vapor when exposed to heat or flame. A hypnotic agent. When heated to decomposition it emits acrid smoke and irritating fumes.

PBV500   HR: 3
PENTYLAMINE (mixed isomers)

DOT: UN 1106
mf: $C_5H_{13}N$ mw: 87.17

PROP: Water-white liquid. Mp: −55°, bp: 104°, flash p: 45°F (OC), d: 0.7614 @ 20°/4°, vap d: 3.01, lel: 2.2%, uel: 22%.
SYNS: AMYLAMINE (mixed isomers) □ AMYLAMINES (DOT)
DOT CLASSIFICATION: 3; Label: Flammable Liquid; DOT Class: 3; Label: Flammable Liquid, Corrosive
SAFETY PROFILE: Moderately toxic by ingestion and skin contact. A severe skin irritant. Dangerous fire hazard when exposed to heat or flame; can react with oxidizing materials. Moderately explosive in the form of vapor when exposed to heat or flame. To fight fire, use alcohol foam, dry chemical. When heated to decomposition it emits toxic fumes of NO_x. See also AMINES.

PBV505   CAS: 110-58-7   HR: 3
1-PENTYLAMINE

DOT: UN 2733/UN 2734
mf: $C_5H_{13}N$ mw: 87.19
PROP: Colorless liquid. D: 0.7547, mp: −55°–−50°, bo: 104.4°. Very sol in water. Flash pt: 45° F.
SYNS: 1-AMINOPENTANE □ AMYLAMINE □ AMYL-AMINE (DOT) □ n-AMYLAMINE □ MONOAMYLAMINE □ NORLEUCAMINE □ 1-PENTANAMINE □ PENTYL-AMINE □ n-PENTYLAMINE
CONSENSUS REPORTS: Reported in EPA TSCA Inventory.
DOT CLASSIFICATION: 8; Label: Corrosive, Flammable Liquid (UN 2734); DOT Class: 3; Label: Flammable Liquid, Corrosive (UN 2733)
SAFETY PROFILE: Poison by intraperitoneal route. A corrosive. A flammable liquid. When heated to decomposition it emits toxic vapors of NO_x.

PBW500   CAS: 543-59-9   HR: 3
PENTYL CHLORIDE

mf: $C_5H_{11}Cl$ mw: 106.61
PROP: Water-white liquid; sweet odor. Mp: −99°, bp: 108.2°, flash p: 54°F (OC), d: 0.883 @ 20°/4°, autoign temp: 500°F, vap d: 3.67, lel: 1.4%, uel: 8.6%.

SYNS: n-AMYL CHLORIDE □ AMYL CHLORIDE (DOT) □ 1-CHLOROPENTANE

CONSENSUS REPORTS: Reported in EPA TSCA Inventory.

SAFETY PROFILE: Dangerous fire hazard when exposed to heat or flame; can react with oxidizing materials. Moderately explosive in the form of vapor when exposed to heat or flame. To fight fire, use foam, CO_2, dry chemical. Dangerous; when heated to decomposition it emits highly toxic fumes of phosgene and Cl^-. See also CHLORINATED HYDROCARBONS, ALIPHATIC.

PBW750   CAS: 12789-46-7   HR: 2
PENTYL ESTER PHOSPHORIC ACID
DOT: UN 2819
mf: $C_5H_{13}O_4P$ mw: 168.15
SYN: AMYL ACID PHOSPHATE (DOT)

DOT CLASSIFICATION: 8; Label: Corrosive

SAFETY PROFILE: Corrosive and irritating to the skin, eyes, and mucous membranes. When heated to decomposition it emits toxic fumes of PO_x. See also ESTERS.

PBX000   CAS: 693-65-2   HR: 3
PENTYL ETHER
mf: $C_{10}H_{22}O$ mw: 158.32

PROP: Liquid. Mp: $-69.3°$, bp: $187°$, flash p: $135°F$ (OC), d: 0.783 @ $20°/4°$, vap d: 5.46, autoign temp: $340°F$.

SYNS: AMYL ETHER □ n-AMYL ETHER □ DIAMYL ETHER □ DI-n-AMYL ETHER □ DIPENTYL ETHER □ ETHER, DI-n-PENTYL- □ 1,1'-OXYBISPENTANE □ PENTANE, 1,1'-OXYBIS-(9CI)

CONSENSUS REPORTS: Reported in EPA TSCA Inventory.

SAFETY PROFILE: Poison by intravenous route. Flammable liquid when exposed to heat or flame; reacts with oxidizing materials. To fight fire, use alcohol foam, dry chemical. When heated to decomposition it emits acrid smoke and irritating fumes. See also ETHERS.

PBX500   CAS: 10589-74-9   HR: 3
n-PENTYLNITROSOUREA
mf: $C_6H_{13}N_3O_2$ mw: 159.22

SYNS: n-AMYLNITROSOUREA □ 1-AMYL-1-NITROSOUREA □ ANU □ 1-NITROSO-1-PENTYLUREA

SAFETY PROFILE: Suspected carcinogen with experimental carcinogenic and tumorigenic data. Moderately toxic by ingestion. Mutation data reported. When heated to decomposition it emits toxic fumes of NO_x. See also N-NITROSO COMPOUNDS.

PBX800   CAS: 23389-74-4   HR: 3
PENTYL 4-PYRIDYL KETONE
mf: $C_{11}H_{15}NO$ mw: 177.27

SYNS: 1-HEXANONE, 1-(4-PYRIDYL)- □ 4-HEXANOYLPYRIDINE □ KETONE, PENTYL 4-PYRIDYL □ PYRIDINE, 4-HEXANOYL- □ 1-(4-PYRIDYL)-1-HEXANONE

DOT CLASSIFICATION: 3; Label: Flammable Liquid

SAFETY PROFILE: A poison by intraperitoneal route. A flammable liquid. When heated to decomposition it emits toxic vapors of NO_x.

PBY750   CAS: 107-72-2   HR: 2
PENTYLTRICHLOROSILANE
DOT: UN 1728
mf: $C_5H_{11}Cl_3Si$ mw: 205.60

PROP: A liquid. D: 1.13 @ $25°$, bp: $171–172°$.

SYN: AMYL TRICHLOROSILANE

CONSENSUS REPORTS: Reported in EPA TSCA Inventory.

DOT CLASSIFICATION: 8; Label: Corrosive

SAFETY PROFILE: Moderately toxic by ingestion and skin contact. Mildly toxic by inhalation. A corrosive irritant to the eyes, skin, and mucous membranes. When heated to decomposition it emits toxic fumes of Cl^-. See also CHLOROSILANES.

PCB250   CAS: 8006-90-4   HR: 2
PEPPERMINT OIL
PROP: From steam distillation of *Mentha piperita L.* (Fam. *Labiatae*). Colorless to pale yellow liquid; strong odor and taste of peppermint. D: 0.896–0.908 @ $25°/25°$, refr index: 1.459 @ $20°$.

P

SYN: PFEFFERMINZ OEL (GERMAN)
CONSENSUS REPORTS: Reported in EPA TSCA Inventory.
SAFETY PROFILE: Moderately toxic by ingestion and intraperitoneal routes. An allergen. Mutation data reported. When heated to decomposition it emits acrid smoke and irritating fumes.

PCC480   CAS: 13654-09-6   HR: 3
PERBROMOBIPHENYL

mf: $C_{12}Br_{10}$ mw: 943.22
SYNS: ADINE 0102 □ BIPHENYL, DECABROMO- □ 1,1'-BIPHENYL, 2,2',3,3',4,4',5,5',6,6'-DECABROMO- □ BERKFLAM B 10 □ DECABROMOBIPHENYL □ 2,2',3,3',4,4',5,5',6,6'-DECABROMOBIPHENYL □ DECABROMODIPHENYL □ FLAMMEX B 10
CONSENSUS REPORTS: NTP 10th Report on Carcinogens, 2000: Reasonably anticipated to be human carcinogen
SAFETY PROFILE: Confirmed HUMAN carcinogen. When heated to decomposition it emits toxic vapors of Br^-.

PCD000   HR: 3
PERCHLORATES

PROP: Composition: combinations with the monovalent $-ClO_4$ radical.
SAFETY PROFILE: Perchlorates are unstable materials, and are an irritant to the body wherever they come in contact with it. Avoid skin contact. Flammable by chemical reaction; powerful oxidizers. All perchlorates are potentially hazardous when in contact with reducing materials. Moderate explosion hazard when shocked or exposed to heat or by chemical reaction. Perchlorates, when mixed with carbonaceous material, form explosive mixtures. Many perchlorates of nitrogenous bases (e.g., hydroxylamine, urea, methylamine, ethylamine, isopropylamine, 4-ethylpyridine, diaminoethane) and organic perchlorates are explosives. Diazonium perchlorates are very dangerous. All perchlorates are considered to be fire and explosive hazards when associated with carbonaceous materials or finely divided metals. This is also true in the presence of calcium hydride, sulfur,

powdered magnesium, aluminum, zinc. Perchlorates react violently with benzene, CaH_2, charcoal, olefins, ethanol, SrH_2, S, H_2SO_4, and reducing materials. To fight fire, use water or foam. When heated to decomposition they emit toxic fumes of Cl^-. See also EXPLOSIVES, HIGH.

PCD250   CAS: 7601-90-3   HR: 3
PERCHLORIC ACID

DOT: UN 1802/UN 1873
mf: $ClHO_4$ mw: 100.46
PROP: Colorless, fuming, unstable, mobile liquid. Mp: $-112°$, bp: $19°$ @ 11 mm, d: 1.768 @ $22°$.
SYNS: PERCHLORIC ACID, >72% acid by weight (DOT) □ PERCHLORIC ACID, >50% but not >72% acid, by weight (UN 1873) (DOT) □ PERCHLORIC ACID, not >50% acid, by weight (UN 1802) (DOT)
CONSENSUS REPORTS: Reported in EPA TSCA Inventory.
DOT CLASSIFICATION: 5.1; Label: Oxidizer, Corrosive (UN 1873); DOT Class: 8; Label: Corrosive, Oxidizer (UN 1802)
SAFETY PROFILE: Poison by ingestion and subcutaneous routes. A severe irritant to the eyes, skin, and mucous membranes. A powerful oxidizer. A severe explosion hazard; the anhydrous form can explode spontaneously.

Potentially explosive reaction with acetic anhydride + acetic acid + organic materials, acetic anhydride + organic materials + transition metals (e.g., chromium, iron, nickel), acetonitrile, alcohols, azo dyes + orthoperiodic acid, bis(2-hydroxyethyl)-terephthalate + ethanol + ethylene glycol, bismuth (above 110°C), antimony (above 110°C), carbon, charcoal + chromium trioxide + heat, cellulose and derivatives + heat, combustible materials, dehydrating agents, dichloromethane + dimethylsulfoxide, diethyl ether, dimethyl ether, dioxane + nitric acid + heat, fecal material + nitric acid, graphitic carbon + nitric acid, hydrofluoric acid + structural materials, iron(II) sulfate, nitric acid + organic matter + heat, nitric acid + pyridine + sulfuric acid, nitrogenous epoxides, organic materials +

sodium hydrogen carbonate (above 200°C), phenyl acetylene (at −78°C), sodium phosphinate + heat, sulfuric acid + organic materials, sulfur trioxide. Reacts to form explosive products with aniline + form-aldehyde, ethylbenzene + thallium triacetate (at 65°C), fluorine (forms fluorine perchlor-ate), glycerol + lead oxide, hydrogen + heat, hydrogen halides, phosphine, pyridine, sulfoxides. Violent reaction or ignition with acetic acid, acetic acid + acetic anhydride, acetic anhydride, acetic anhydride + carbon tetrachloride + 2-methyl cyclohexanone, antimony compounds, azo pigments, bis-1,2-diaminopropane-cis-dichlorochrom-ium(III) perchlorate, carbon, 1,3-bis(di-n-cyclopentadienyl iron)-2-propen-1-one, CH_3OH, CCl_4, copper dichromium tetraoxide (at 120°C), DNA, dibutyl sulfoxide, dimethyl sulfoxide, ethylbenzene, glycol ethers, glycols, HNO_3, HCl, H_2SO_4, hypophosphites, iron sulfate, iodides, ketones, PbO + glycerin, methanol + triglycerides, 2-methylpropene + metal oxides, 2-methyl cyclohexanone, NI_3, nitrogenous epoxides, nitrosophenol, o-periodic acid, oleic acid, organophosphorus compounds, paper, P_2O_5 + $CHCl_3$, P_2O_5, P_2Zn_3, sodium iodide + hydroiodic acid, sodium phosphinate, steel, sulfinyl chloride, SO_3, trichloroethylene, vegetable matter, wood, zinc phosphide. When heated to decomposition it emits toxic fumes of Cl^-. See also PERCHLORATES.

PCD500   CAS: 7790-98-9   HR: 3
PERCHLORIC ACID, AMMONIUM SALT
DOT: UN 0402/UN 1442
mf: $ClO_4 \cdot H_4N$ mw: 117.50
PROP: White crystals. Mp: decomp, d: 1.95. Very sol in H_2O, liquid NH_3.
SYN: AMMONIUM PERCHLORATE (DOT)
CONSENSUS REPORTS: Reported in EPA TSCA Inventory.
DOT CLASSIFICATION: 5.1; Label: Oxidizer; DOT Class: EXPLOSIVE 1.1D; Label: EXPLOSIVE 1.1D

SAFETY PROFILE: Moderately toxic by ingestion and parenteral routes. Flammable when exposed to heat or flame or by spontaneous chemical reaction with reducing materials. A very powerful oxidizer that has caused explosions in industry. Ignites violently with combustibles. Severe explosion hazard; decomposes at 130° and explodes at 380°. When contaminated by powdered carbon, ferrocene, S, organic matter, powdered metals, nitryl perchlorate, potassium periodate, potassium permanganate, it becomes impact sensitive. Potentially explosive reactions with carbon (above 240°C), dichromium trioxide (at 270°C), cadmium oxide (at 260°C), zinc oxide (at 200°C), copper chromite, copper oxide, iron oxide, potassium permanganate, potassium dichromate, mono-, di-, tri-, or tetra-methylammonium perchlorates, metal perchlorates (e.g., lithium perchlorate, zinc perchlorate), nitrophenol-formaldehyde polymer. Mixtures with aluminum or copper burn violently when ignited. Mixtures with ethylene dinitrate ignite when stored at 60°C. When heated to decomposition it emits toxic fumes of NH_3 and Cl^-. See also PERCHLORATES and EXPLOSIVES, HIGH.

PCD750   CAS: 13465-95-7   HR: 3
PERCHLORIC ACID, BARIUM SALT·3H$_2$O
DOT: UN 1447
mf: $Cl_2O_8 \cdot Ba \cdot 3H_2O$ mw: 336.24
PROP: Colorless, hexagonal crystals. Undergoes polymorphic transitions at 2° and 3°. Decomp on heating forming $BaCl_2$ + O_2. Mp: 505°, d: 2.74. Very sol in H_2O; very sol in EtOH.
SYN: BARIUM PERCHLORATE (DOT)
CONSENSUS REPORTS: Barium and its compounds are on the Community Right-To-Know List. Reported in EPA TSCA Inventory.
OSHA PEL: TWA 0.5 mg(Ba)/m³
ACGIH TLV: TWA 0.5 mg(Ba)/m³; Not Classifiable as a Human Carcinogen

DFG MAK: 0.5 mg(Ba)/m^3
DOT CLASSIFICATION: 5.1; Label:
Oxidizer, Poison
SAFETY PROFILE: A poison. An unstable material. An oxidizer. When refluxed with an alcohol highly explosive alkyl perchlorates are formed. When heated to decomposition it emits toxic fumes of Cl$^-$. A desiccant. See also PERCHLORATES and BARIUM COMPOUNDS.

PCE000   CAS: 10034-81-8   HR: 2
PERCHLORIC ACID, MAGNESIUM SALT
DOT: UN 1475
mf: Cl$_2$O$_8$•Mg   mw: 223.21
PROP: White, colorless, deliquescent solid or hygroscopic crystals. Decomp on heating with formation of MgO. Mp: decomp @ 251°, d: 2.60 @ 25°. Very sol in H$_2$O, EtOH.
SYNS: ANHYDRONE □ DEHYDRITE □ MAGNESIUM PERCHLORATE □ PERCHLORATE de MAGNESIUM (FRENCH)
CONSENSUS REPORTS: Reported in EPA TSCA Inventory.
DOT CLASSIFICATION: 5.1; Label: Oxidizer
SAFETY PROFILE: Moderately toxic by intraperitoneal route. Severe skin and eye irritant. A powerful oxidizer which has caused many explosions in industry. Potentially explosive reactions with alkenes (above 220°C), ammonia, aryl hydrazine + ether, dimethyl sulfoxide + heat, ethylene oxide, fluorobutane + water, organic materials, phosphorus, trimethyl phosphate. Reacts to form explosive products with ethanol (forms ethyl perchlorate), cellulose + dinitrogen tetraoxide + oxygen (forms cellulose nitrate). Avoid contact with mineral acids, butyl fluorides, hydrocarbons. A drying agent. When heated to decomposition it emits toxic fumes of MgO and Cl$^-$. See also MAGNESIUM COMPOUNDS and PERCHLORATES.

PCE750   CAS: 7601-89-0   HR: 3
PERCHLORIC ACID, SODIUM SALT
DOT: UN 1502

mf: ClO$_4$•Na   mw: 122.44
PROP: White or colorless, orthorhombic, deliquescent crystals. Mp: 482° (decomp). Mod sol in H$_2$O, EtOH.
SYNS: NATRIUMPERCHLORAAT (DUTCH) □ NATRIUMPERCHLORAT (GERMAN) □ PERCHLORATE de SODIUM (FRENCH) □ SODIO (PERCLORATO DI) (ITALIAN) □ SODIUM PERCHLORATE □ SODIUM PERCHLORATE (DOT)
CONSENSUS REPORTS: Reported in EPA TSCA Inventory.
DOT CLASSIFICATION: 5.1; Label: Oxidizer
SAFETY PROFILE: Moderately toxic by ingestion and intraperitoneal routes. A powerful oxidizer. Forms explosive mixture with acetone, 1,3-butylene glycol, 2,3-butylene glycol, CaH$_2$, charcoal, diaminoethane, dimethyl formamide, ethanolamine, ethylene glycol, formamide, galactose, glycerin, hydrazine, water, NH$_4$NO$_3$, Mg, reducing agents, SrH$_2$, urea. When heated to decomposition it emits toxic fumes of Cl$^-$ and Na$_2$O. See also PERCHLORATES.

PCF275   CAS: 127-18-4   HR: 3
PERCHLOROETHYLENE
DOT: UN 1897
mf: C$_2$Cl$_4$ mw: 165.82
PROP: Colorless liquid; chloroform-like odor. Mp: −23.35°, fp: −22.35°, bp: 121.20°, d: 1.6311 @ 15°/4°, vap press: 15.8 mm @ 22°, vap d: 5.83. IDLH 150 ppm.
SYNS: ANKILOSTIN □ ANTISOL 1 □ CARBON BICHLORIDE □ CARBON DICHLORIDE □ CZTERO-CHLOROETYLEN (POLISH) □ DIDAKENE □ DOW-PER □ ENT 1,860 □ ETHYLENE TETRACHLORIDE □ FEDAL-UN □ NCI-C04580 □ NEMA □ PERAWIN □ PERCHLOORETHYLEEN, PER (DUTCH) □ PERCHLOR □ PERCHLORAETHYLEN, PER (GERMAN) □ PER-CHLORETHYLENE □ PERCHLORETHYLENE, PER (FRENCH) □ PERCLENE □ PERCLOROETILENE (ITALIAN) □ PERCOSOLVE □ PERK □ PERKLONE □ PERSEC □ RCRA WASTE NUMBER U210 □ TETLEN □ TETRACAP □ TETRACHLOORETHEEN (DUTCH) □ TETRACHLORAETHEN (GERMAN) □ TETRACHLORO-ETHENE □ TETRACHLOROETHYLENE (DOT, ACGIH) □ 1,1,2,2-TETRACHLOROETHYLENE □ TETRACLORO-ETENE (ITALIAN) □ TETRALENO □ TETRALEX □ TETRAVEC □ TETROGUER □ TETROPIL

CONSENSUS REPORTS: NTP 10th Report on Carcinogens. IARC Cancer Review: Group 2B IMEMDT 7,355,87; Animal Limited Evidence IMEMDT 20,491,79. NCI Carcinogenesis Bioassay (gavage); Clear Evidence: mouse NCITR* NCI-CG-TR-13,77 (inhalation); Clear Evidence: mouse, rat NTPTR* NTP-TR-311,86 (gavage); Inadequate Studies: rat NCITR* NCI-CG-TR-13,77. Reported in EPA TSCA Inventory. EPA Genetic Toxicology Program. Community Right-To-Know List.
OSHA PEL: TWA 25 ppm
ACGIH TLV: TWA 25 ppm; Animal Carcinogen; BEI: 3.5 mg/L trichloroacetic acid in urine at end of shift at end of workweek
DFG MAK: Confirmed Animal Carcinogen with Unknown Relevance to Humans; BAT: blood 100 μg/dL
NIOSH REL: (Tetrachloroethylene) Minimize workplace exposure
DOT CLASSIFICATION: 6.1; Label: KEEP AWAY FROM FOOD
SAFETY PROFILE: Confirmed carcinogen with experimental carcinogenic, and neoplastigenic data. Experimental poison by intravenous route. Moderately toxic to humans by inhalation, with the following effects: local anesthetic, conjunctiva irritation, general anesthesia, hallucinations, distorted perceptions, coma, and pulmonary changes. Moderately experimentally toxic by ingestion, inhalation, intraperitoneal, and subcutaneous routes. An experimental teratogen. Experimental reproductive effects. Human mutation data reported. An eye and severe skin irritant. The liquid can cause injuries to the eyes; however, with proper precautions it can be handled safely. The symptoms of acute intoxication from this material are the result of its effects upon the nervous system. Can cause dermatitis, particularly after repeated or prolonged contact with the skin. Irritates the gastrointestinal tract upon ingestion. It may be handled in the presence or absence of air, water, and light with any of the common construction materials at temperatures up to 140°. This material is extremely stable and resists hydrolysis. A common air contaminant. Reacts violently under the proper conditions with Ba, Be, Li, N_2O_4, metals, NaOH. When heated to decomposition it emits highly toxic fumes of Cl^-. See also CHLORINATED HYDROCARBONS, ALIPHATIC.

PCF300   CAS: 594-42-3   HR: 3
PERCHLOROMETHYL MERCAPTAN
DOT: UN 1670
mf: CCl_4S mw: 185.87
PROP: Yellow, oily liquid. Bp: slt decomp @ 149°, d: 1.700 @ 20°, vap d: 6.414. IDLH 10 ppm.
SYNS: CLAIRSIT □ MERCAPTAN METHYLIQUE PERCHLORE (FRENCH) □ PCM □ PERCHLORMETHYL-MERKAPTAN (CZECH) □ RCRA WASTE NUMBER P118 □ TRICHLOROMETHANE SULFENYL CHLORIDE □ TRICHLOROMETHYLSULFENYL CHLORIDE □ TRICHLOROMETHYLSULPHENYL CHLORIDE
CONSENSUS REPORTS: EPA Extremely Hazardous Substances List. Reported in EPA TSCA Inventory.
OSHA PEL: TWA 0.1 ppm
ACGIH TLV: TWA 0.1 ppm
DOT CLASSIFICATION: 6.1; Label: Poison
SAFETY PROFILE: Poison by ingestion, inhalation, and intravenous routes. A severe skin, eye, and mucous membrane irritant. When heated to decomposition it emits very toxic fumes of Cl^- and SO_x. See also MERCAPTANS.

PCF750   CAS: 7616-94-6   HR: 2
PERCHLORYL FLUORIDE
DOT: UN 3083
mf: $ClFO_3$ mw: 102.45
PROP: Colorless, noncorrosive gas stable to 5°; stable to water; characteristic sweet odor. Mp: −146°, bp: −46.8°, d: (liquid) 1.434. d: (gas) 0.637. IDLH 100 ppm.
SYNS: CHLORINE FLUORIDE OXIDE □ CHLORINE OXYFLUORIDE
CONSENSUS REPORTS: Reported in EPA TSCA Inventory.
OSHA PEL: TWA 3 ppm; STEL 6 ppm

ACGIH TLV: TWA 3 ppm; STEL 6 ppm
DOT CLASSIFICATION: 2.3; Label: Poison Gas, Oxidizer
SAFETY PROFILE: A poison gas which forms methemoglobin in the body and destroys red cells causing anemia, anorexia, and cyanosis. Recovery is said to be rapid, leaving no permanent physiological damage. Can be absorbed through the skin. Its odor can be detected as low as 10 ppm although this cannot be relied upon as an indication of toxic concentration in air. While nonflammable, it supports combustion. It is a powerful oxidizer. Moderately explosive. Potentially explosive reactions with combustible gases or vapors. benzene + aluminum trichloride, benzocyclobutene + butyllithium + potassium tert-butoxide, calcium acetylide, potassium cyanide, potassium thiocyanate, sodium iodide, charcoal, ethyl-4-fluorobenzoylacetate, hydrocarbons, hydrogen sulfide, nitrogen oxide, sulfur dichloride, vinylidene chloride, 3α-hydroxy-5β-androstane-11,17-dione-17-hydrazone, lithiated compounds, 2-lithio(dimethylaminomethyl)ferroxene, methyl-2-bromo-5,5-ethylene dioxy(2,2,1)-bicycloheptane-7-carboxylate, aliphatic heterocyclic amines, sodium methoxide + methanol, vinylidene chloride. Reacts to form explosive products with nitrogenous bases (e.g., isopropylamine, isobutylamine, aniline, phenyl hydrazine, 1,2-diphenyl hydrazine), sawdust, lampblack. Violent reaction with finely divided organic materials. A fluorinating agent in chemical synthesis, and an oxidant in rocket fuel. When heated to decomposition it emits toxic fumes of F$^-$ and Cl$^-$. See also FLUORINE and PERCHLORATES.

PCG500   CAS: 76-42-6   HR: 3
PERCODAN
mf: $C_{18}H_{21}NO_4$ mw: 315.40
PROP: Long rods from EtOH. Mp: 218–220°. Sol in EtOH, CHCl$_3$, insol in H$_2$O, and Et$_2$O.

SYNS: DIHYDROHYDROXYCODEINONE □ DIHYDRO-14-HYDROXYCODEINONE □ 14-HYDROXYDIHYDROCODEINONE □ OXYCODEIN-ONE □ PERCOBARB
SAFETY PROFILE: Poison by intraperitoneal route. Moderately toxic by subcutaneous route. When heated to decomposition it emits toxic fumes of NO$_x$.

PCG725   CAS: 335-76-2   HR: 3
PERFLUORODECANOIC ACID
mf: $C_{10}HF_{19}O_2$ mw: 514.11
PROP: A liquid. Bp: 130° @ 30 mm.
SYNS: NDFDA □ NONADECAFLUORODECANOIC ACID □ NONADECAFLUORO-n-DECANOIC ACID □ PERFLUORO-N-DECANOIC ACID □ PFDA
CONSENSUS REPORTS: Reported in EPA TSCA Inventory.
SAFETY PROFILE: Poison by ingestion and intraperitoneal routes. An experimental teratogen. Experimental reproductive effects. Mutation data reported. When heated to decomposition it emits toxic fumes of F$^-$.

PCI550   CAS: 536-59-4   HR: 2
PERILLA ALCOHOL
mf: $C_{10}H_{16}O$ mw: 152.26
PROP: Bp: 119–121°/11 mm Hg, d: 0.960, flash p: >230°F.
SYNS: CYCLOHEX-1-ENE-1-METHANOL, 4-(1-METH-YLETHENYL)- □ DIHYDROCUMINYL ALCOHOL □ 4-ISOPROPENYL-CYCLOHEX-1-ENE-1-METHANOL □ p-MENTHA-1,8-DIEN-7-OL □ PERILLOL □ PERILLYL ALCOHOL
CONSENSUS REPORTS: Reported in EPA TSCA Inventory.
SAFETY PROFILE: Moderately toxic by ingestion. A severe skin irritant. Combustible liquid. When heated to decomposition it emits acrid smoke and irritating fumes.

PCI750   CAS: 553-84-4   HR: 3
PERILLA KETONE
mf: $C_{10}H_{14}O_2$ mw: 166.24
PROP: Light-yellow liquid. Bp: 130° @ 30 mm. A potent lung toxin from the mint plant *Perilla frutescens* (SCIEAS 197,573,77).

SYNS: 1-(3-FURANYL)-4-METHYL-1-PENTANONE □
β-FURYL ISOAMYL KETONE □ 1-(3-FURYL)-4-METHYL-
1-PENTANONE □ PURPLE MINT PLANT EXTRACT
DOT CLASSIFICATION: 3; Label:
Flammable Liquid
SAFETY PROFILE: Poison by intravenous
and intraperitoneal routes. A potent
pulmonary edemagenic agent (experimental).
May also be hazardous to humans. A
flammable liquid. When heated to
decomposition it emits acrid smoke and
irritating fumes.

PCJ230   CAS: 107133-36-8 HR: 3
PERINDOPRIL tert-BUTYLAMINE
mf: $C_{19}H_{32}N_2O_5 \cdot C_4H_{11}N$ mw: 441.69
SYNS: 1H-INDOLE-2-CARBOXYLIC ACID, 1-(2-((1-
(ETHOXYCARBONYL)BUTYL)AMINO)-1-OXO-
PROPYL)OCTAHYDRO-,(2S-(1(R*(R*)),2α,3Aβ,7Aβ))-,
COMPD. WITH 2-METHYL-2-PROPANAMINE (1:1) □
MCN-A 2833-109 □ PERINDOPRIL ERBUMINE □ S 9490-3
SAFETY PROFILE: A poison by
intravenous route. Moderately toxic by
ingestion. When heated to decomposition it
emits toxic vapors of NO_x.

PCJ400   CAS: 93763-70-3   HR: 1
PERLITE
PROP: Average density of 0.13. Expands
when finely ground and heated. Natural
glass, amorphous mineral consisting of
fused sodium potassium aluminum silicate,
containing <1% quartz.
OSHA PEL: TWA Total Dust: 15 mg/m³;
Respirable Fraction: 5 mg/m³
ACGIH TLV: TWA (nuisance particulate) 10
mg/m³ of total dust; Not Classifiable as a
Human Carcinogen (when toxic impurities
are not present, e.g., quartz <1%)
NIOSH REL: (Perlite, Respirable Fraction)
TWA 5 mg/m³
SAFETY PROFILE: Slightly toxic by
ingestion. A nuisance dust.

PCJ500   HR: 3
PERMANGANATES
PROP: Compounds containing an MnO_4^-
radical.

CONSENSUS REPORTS: Manganese and
its compounds are on the Community
Right-To-Know List.
SAFETY PROFILE: Poisons. Many are
strong oxidizing agents, hence irritating.
Flammable by chemical reaction with
reducing agents. Moderately explosive when
shocked or exposed to heat. Silver
permanganate and other metallic
permanganates may detonate when exposed
to high temperatures or when they are
involved in fires or severely shocked. Store
in a cool, ventilated area, away from acute
fire hazards and easily oxidized materials.
They may be disposed of by dissolving in
water since practically all permanganates are
soluble in water. They can react vigorously
on contact with reducing materials.
Incompatible with acetic acid, acetic
anhydride, $H_2SO_4 + C_6H_6$. See also
MANGANESE COMPOUNDS.

PCK000   CAS: 7787-36-2   HR: 2
PERMANGANIC ACID, BARIUM SALT
DOT: UN 1448
mf: $Mn_2O_8 \cdot Ba$ mw: 375.22
PROP: Dark purple crystals. Sparingly sol in
H_2O.
SYN: BARIUM PERMANGANATE
CONSENSUS REPORTS: Barium and its
compounds and manganese and its
compounds are on the Community Right-
To-Know List.
OSHA PEL: TWA 0.5 mg(Ba)/m³; CL 5
mg(Mn)/m³
ACGIH TLV: TWA 0.5 mg(Ba)/m³; Not
Classifiable as a Human Carcinogen
DFG MAK: 0.5 mg(Ba)/m³
DOT CLASSIFICATION: 5.1; Label:
Oxidizer, Poison
SAFETY PROFILE: Probably an irritant. A
powerful oxidizer. When heated to
decomposition it emits acrid smoke and
irritating fumes. See also BARIUM
COMPOUNDS.

P

PCL000   HR: 3
PEROXIDES, INORGANIC
SAFETY PROFILE: Variable toxicity. They
may cause injury on contact with skin or
mucous membranes. Moderate to dangerous
fire hazard by chemical reaction with
reducing agents and contaminants; strong
oxidizing agents; contact with moisture may
produce much heat. Moderate explosion
hazard; heat, shock, or catalysts can cause
violent decomposition. Contact with
reducing agents may give rise to explosively
violent reactions. See also HYDROGEN
PEROXIDE and SODIUM PEROXIDE.

PCL250   HR: 3
PEROXIDES, ORGANIC
PROP: Organic compounds containing the
—OO— group.
SAFETY PROFILE: Often highly toxic and
irritating to the skin, eyes, and mucous
membranes. Dangerous fire hazard by
chemical reaction with reducing agents or
exposure to heat. They readily release
oxygen and thus are powerful oxidizers.
Severe explosion hazard when shocked,
exposed to heat, or by spontaneous
chemical reaction. Many peroxides are very
unstable. Upon contact with reducing
materials, such as organic matter or
thiocyanates, an explosion can occur.

Many solvents form dangerous levels of
peroxides during storage: e.g., dipropyl
ether, divinylacetylene, vinylidene chloride,
potassium amide, sodium amide. Other
compounds form peroxides in storage but
concentration is required to reach dangerous
levels: e.g., diethyl ether, ethyl vinyl ether,
tetrahydrofuran, p-dioxane, 1,1-diethoxyeth-
ane, ethylene glycol dimethyl ether, propyne,
butadiene, dicyclopentadiene, cyclohexene,
tetrahydronaphthalenes, deca-hydrona-
phthalenes. Some monomeric materials can
form peroxides that catalyze hazardous
polymerization reactions: e.g., acrylic acid,
acrylonitrile, butadiene, 2-chlorobutadiene,
chlorotrifluoroethylene, methyl
methacrylate, styrene, tetrafluoroethylene,
vinyl acetate, vinylacetylene, vinyl chloride,
vinylidine chloride, vinylpyridine.
Compounds that contain one or two ether
functions are especially susceptible to
peroxide formation.

Peroxyacids (RCO•OOH) are some of the
most powerful oxidants of the organic
peroxides. Some of the simple peroxyacids
are peroxyformic acid, peroxyacetic acid,
peroxypivalic acid, peroxytrifluoroacetic
acid. Traces of transition metals (e.g., cobalt,
iron, manganese, nickel, vanadium) can
catalyze explosive decomposition of these
acids.

Handle all peroxides or peroxide-
containing materials with great care. See also
specific compounds.

PCL500   CAS: 79-21-0   HR: 3
PEROXYACETIC ACID
DOT: UN 3149
mf: $C_2H_4O_3$ mw: 76.06
PROP: Not over 40% peracetic acid and not
over 6% hydrogen peroxide (FEREAC
41,15972,76). Colorless liquid; strong odor.
Fp: 0.1°, bp: 105°, explodes @ 110°, flash p:
105°F (OC), d: 1.15 @ 20°. Sol in H_2O,
EtOH, Et_2O.
SYNS: ACETIC PEROXIDE □ ACETYL HYDRO-
PEROXIDE □ ACIDE PERACETIQUE (FRENCH) □
DESOXON 1 □ ESTOSTERIL □ ETHANEPEROXOIC
ACID □ HYDROGEN PEROXIDE & PEROXYACETIC
ACID MIXT., with acids, H_2O not >5% peroxyacetic acid
(DOT) □ HYDROPEROXIDE, ACETYL □ KYSELINA
PEROXYOCTOVA □ MONOPERACETIC ACID □
OSBON AC □ PERACETIC ACID □ PEROXOACETIC
ACID □ PEROXYACETIC ACID, >43% and with >6%
hydrogen peroxide (DOT) □ PROXITANE 4002
CONSENSUS REPORTS: EPA Extremely
Hazardous Substances List. Community
Right-To-Know List. Reported in EPA
TSCA Inventory.
DFG MAK: Very strong skin effects;
Confirmed Animal Carcinogen with
Unknown Relevance to Humans
DOT CLASSIFICATION: 5.1; Label:
Oxidizer, Corrosive (UN3149)
SAFETY PROFILE: Poison by ingestion.
Moderately toxic by inhalation and skin

contact. A corrosive eye, skin, and mucous membrane irritant. Questionable carcinogen with experimental tumorigenic data by skin contact. Flammable liquid. Severe explosion hazard when exposed to heat or by spontaneous chemical reaction. Explodes violently at 110°C. A powerful oxidizing agent. Explosive reaction with acetic anhydride, 5-p-chlorophenyl-2,2-dimethyl-3-hexanone. Violent reaction with ether solvents (e.g., tetrahydrofuran, diethyl ether), metal chloride solutions (e.g., calcium chloride, potassium chloride, sodium chloride), olefins, organic matter. Dangerous; keep away from combustible materials. When heated to decomposition it emits acrid smoke and irritating fumes. To fight fire, use water, foam, CO_2. Used as a polymerization initiator, curing agent, and cross-linking agent. See also PEROXIDES, ORGANIC.

PCR000   CAS: 91845-41-9   HR: 3
PETASITES JAPONICUS MAXIM
PROP: Dried flower stalk of *Petasites Japonicus Maxim* (GANNA2 64,527,73).
SYNS: COLTS FOOT □ FUKI-NO-TOH (JAPANESE)
SAFETY PROFILE: Suspected carcinogen with experimental carcinogenic and tumorigenic data. When heated to decomposition it emits acrid smoke and irritating fumes.

PCR100   HR: D
PETITGRAIN OIL, PARAGUAY TYPE
PROP: From steam distillation of the leaves of *Citrus aurantium* L. subspecies *amara*. Yellow to brown liquid; harsh bitter odor. D: 0.878–0.889, refr index: 1.455 @ 20°. Sol in fixed oils, mineral oil, propylene glycol; insol in glycerin.
SAFETY PROFILE: When heated to decomposition it emits acrid smoke and irritating fumes.

PCR250   CAS: 8002-05-9   HR: 3
PETROLEUM
DOT: UN 1267

PROP: A thick, flammable, dark-yellow to brown or green-black liquid. D: 0.780–0.970, flash p: 20–90°F. Insol in water; sol in benzene, chloroform, ether. Consists of a mixture of hydrocarbons from C_2H_6 and up, chiefly of the paraffins, cycloparaffins, or cyclic aromatic hydrocarbons, with small amounts of benzene hydrocarbons, sulfur, and oxygenated compounds. IDLH 1100 ppm [10%LEL].
SYNS: BASE OIL □ COAL LIQUID □ COAL OIL □ CRUDE OIL □ CRUDE PETROLEUM □ PETROL □ PETROLEUM CRUDE □ PETROLEUM CRUDE OIL (DOT) □ ROCK OIL □ SENECA OIL
CONSENSUS REPORTS: IARC Cancer Review: Group 3 IMEMDT 45,119,89; Animal Limited Evidence IMEMDT 45,119,89; Human Inadequate Evidence IMEMDT 45,119,89. Reported in EPA TSCA Inventory.
DOT CLASSIFICATION: 3; Label: Flammable Liquid
SAFETY PROFILE: Questionable carcinogen with experimental carcinogenic, neoplastigenic, and tumorigenic data by skin contact. A dangerous fire hazard when exposed to heat, flame, or powerful oxidizers. To fight fire, use foam, CO_2, dry chemical. When heated to decomposition it emits acrid smoke and irritating fumes. See also MINERAL OIL.

PCR500   CAS: 8052-42-4   HR: 2
PETROLEUM ASPHALT
PROP: Steam refined asphalt (IMSUAI 34,255,65).
SYNS: ASPHALT, PETROLEUM □ PETROLEUM ROOFING TAR □ ROAD ASPHALT
CONSENSUS REPORTS: Reported in EPA TSCA Inventory.
SAFETY PROFILE: Questionable carcinogen with experimental neoplastigenic data on skin contact. When heated to decomposition it emits acrid smoke and irritating fumes. See also ASPHALT.

PCS250   CAS: 8002-05-9   HR: 3
PETROLEUM DISTILLATE
DOT: UN 1268
CONSENSUS REPORTS: Reported in EPA
TSCA Inventory.
OSHA PEL: TWA 400 ppm
DOT CLASSIFICATION: 3; Label:
Flammable Liquid
SAFETY PROFILE: Human systemic effects
by parenteral route: cough, dyspnea, nausea
or vomiting. Mildly toxic by inhalation and
ingestion. Moderate skin and eye irritation.
A flammable liquid when exposed to heat or
flame. When heated to decomposition it
emits acrid smoke and irritating fumes. Used
as a vehicle for pesticides. See also
PETROLEUM and ASPHALT.

PCS260   CAS: 64742-44-5   HR: 2
PETROLEUM DISTILLATES, CLAY-
** TREATED HEAVY NAPHTHENIC**
CONSENSUS REPORTS: Reported in EPA
TSCA Inventory.
SAFETY PROFILE: Questionable
carcinogen with experimental tumorigenic
data. When heated to decomposition it
emits acrid smoke and irritating fumes.

PCS270   CAS: 64742-45-6   HR: 2
PETROLEUM DISTILLATES, CLAY-
** TREATED LIGHT NAPHTHENIC**
CONSENSUS REPORTS: Reported in EPA
TSCA Inventory.
SAFETY PROFILE: Questionable
carcinogen with experimental tumorigenic
data. When heated to decomposition it
emits acrid smoke and irritating fumes.

PCT250   CAS: 8032-32-4   HR: 3
PETROLEUM SPIRITS
DOT: UN 1271
PROP: Volatile, clear, colorless and non-
fluorescent liquid. Mp: $<-73°$, bp: 40–80°,
ULC: 95–100, lel: 1.1%, uel: 5.9%, flash p:
$<0°F$, d: 0.635–0.660, autoign temp: 550°F,
vap d: 2.50.
SYNS: BENZINE (LIGHT PETROLEUM DISTILLATE)
□ BENZOLINE □ CANADOL □ LIGROIN □ PAINTER'S
NAPHTHA □ PETROLEUM ETHER □ PETROLEUM

SPIRIT (DOT) □ REFINED SOLVENT NAPHTHA □
SKELLYSOLVE F □ SKELLYSOLVE G □ VARNISH
MARKER'S NAPHTHA □ VM and P NAPHTHA □ VM & P
NAPHTHA □ VM&P NAPHTHA □ VM & P NAPHTHA
(ACGIH,OSHA)
CONSENSUS REPORTS: Reported in EPA
TSCA Inventory.
OSHA PEL: TWA 300 ppm; STEL 400 PPM
ACGIH TLV: TWA 300 ppm; Animal
Carcinogen
NIOSH REL: (VM & P Naphtha) TWA 350
mg/m³; CL 1800 mg/m³/15M
DOT CLASSIFICATION: 3; Label:
Flammable Liquid
SAFETY PROFILE: Confirmed carcinogen.
A poison by intravenous route. Mildly toxic
by inhalation. Ingestion can cause a burning
sensation, vomiting, diarrhea, drowsiness,
and, in severe cases, pulmonary edema.
Inhalation of concentrated vapors can cause
intoxication resembling that from alcohol,
headache, nausea, coma, and hemorrhage to
various vital organs. An eye irritant. A
flammable liquid and highly dangerous fire
hazard when exposed to heat, flame, sparks,
or oxidizing materials. Explosive in the form
of vapor when exposed to heat or flame.
Highly dangerous; keep away from heat or
flame. To fight fire, use foam, CO_2, dry
chemical. When heated to decomposition it
emits acrid smoke and irritating fumes.

PCT600   HR: D
PETROLEUM WAX
PROP: Translucent; tasteless and odorless
wax. Mp: 48–93°. Insol in water; very sltly
sol in org solvs.
SYNS: MICROCRYSTALLINE WAX □ PETROLEUM
WAX, SYNTHETIC (FCC) □ REFINED PETROLEUM
WAX
SAFETY PROFILE: When heated to
decomposition it emits acrid smoke and
irritating fumes.

PCU360   CAS: 9000-70-8   HR: D
PHARMAGEL A
PROP: Sltly yellow sheets, flakes, or coarse
powder. Sol in hot H_2O, glycerol, AcOH;
insol in org solvs.

SYNS: ABSORBABLE GELATIN SPONGE □ GELATIN □ GELATINE □ GELATIN FOAM □ GELATINS □ GELFOAM □ GT □ PHARMAGEL AdB □ PHARMAGEL B □ PURAGEL □ SPONGIOFORT □ VEE GEE GELATIN
CONSENSUS REPORTS: Reported in EPA TSCA Inventory.
SAFETY PROFILE: An experimental teratogen. Experimental reproductive effects. When heated to decomposition it emits acrid smoke and irritating fumes.

PCW250   CAS: 85-01-8   HR: 3
PHENANTHRENE
mf: $C_{14}H_{10}$ mw: 178.24
PROP: Solid or monoclinic crystals; plates from EtOH. Mp: 100°, bp: 339°, d: 1.179 @ 25°, vap press: 1 mm @ 118.3°, vap d: 6.14. Insol in water; sol in CS_2, benzene, and hot alc; very sol in ether.
SYNS: COAL TAR PITCH VOLATILES: PHENANTHRENE □ PHENANTHREN (GERMAN) □ PHENANTRIN
CONSENSUS REPORTS: IARC Cancer Review: Group 3 IMEMDT 7,56,87; Animal Inadequate Evidence IMEMDT 32,419,83. Reported in EPA TSCA Inventory. EPA Genetic Toxicology Program.
OSHA PEL: TWA 0.2 mg/m³
SAFETY PROFILE: Poison by intravenous route. Moderately toxic by ingestion. Mutation data reported. A human skin photosensitizer. Questionable carcinogen with experimental neoplastigenic and tumorigenic data by skin contact. Combustible when exposed to heat or flame; can react vigorously with oxidizing materials. To fight fire, use water, foam, CO_2, dry chemical. When heated to decomposition it emits acrid smoke and irritating fumes.

PCY250   CAS: 66-71-7   HR: 3
1,10-PHENANTHROLINE
mf: $C_{12}H_8N_2$ mw: 180.22
PROP: Crystals from benzene. Mp: 93–94°, anhydr 117°. Sol in water, alc, ether.
SYNS: 4,5-DIAZAPHENANTHRENE □ ORTHOPHEN-ANTHROLINE □ β-PHENANTHROLINE □ o-PHEN-ANTHROLINE □ 1,10-o-PHENANTHROLINE

CONSENSUS REPORTS: Reported in EPA TSCA Inventory.
SAFETY PROFILE: Poison by ingestion and intraperitoneal routes. An experimental teratogen. Mutation data reported. When heated to decomposition it emits toxic fumes of NO_x.

PDB000   CAS: 578-94-9   HR: 3
PHENARSAZINE CHLORIDE
DOT: UN 1698
mf: $C_{12}H_9AsClN$ mw: 277.59
PROP: Light yellow to green granules from xylene or CCl_4, or by subl; irritating odor. Mp: 195° (dimorph), bp: 410° (decomp), d: 1.65, vap press: very low @ 20°, vap d: 9.6. Very sltly sol in water; sltly sol in benzene, brass. Corrodes iron, bronze.
SYNS: ADAMSITE □ 5-AZA-10-ARSENAANTHRA-CENE CHLORIDE □ 10-CHLORO-5,10-DIHYDROAR-SACRIDINE □ 10-CHLORO-5,10-DIHYDROPHEN-ARSAZINE □ DIPHENYLAMINECHLORARSINE □ DIPHENYLAMINECHLOROARSINE (DOT) □ DM
CONSENSUS REPORTS: Arsenic and its compounds are on the Community Right-To-Know List.
DOT CLASSIFICATION: 6.1; Label: Poison
OSHA PEL: TWA 0.5 mg(As)/m³
SAFETY PROFILE: Human poison by inhalation. Poison experimentally by intravenous route. Human systemic effects by inhalation: changes in function or structure of salivary glands, nausea or vomiting, cough. May be irritating to skin, eyes, and mucous membranes. A vomiting type of poison gas (non-persistent). When heated to decomposition it emits very toxic fumes of As and Cl⁻. See also ARSENIC COMPOUNDS.

PDB300   CAS: 73973-02-1   HR: 3
S-(10-PHENARSAZINYL)-O,O-DIISO-OCTYLPHOSPHORODITHIOATE
mf: $C_{28}H_{43}AsNO_2PS_2$ mw: 595.73
SYNS: 10-PHENARSAZINETHIOL, S-ESTER with O,O-DIISOOCTYLPHOSPHORODITHIOATE □ PHOSPHORODITHIOIC ACID, O,O-DIISOOCTYL S-(10-PHENARSAZINYL) ESTER
OSHA PEL: TWA 0.5 mg(As)/m³

SAFETY PROFILE: Poison by intravenous route. When heated to decomposition it emits toxic fumes of NO_x, SO_x, PO_x, and As.

PDB500   CAS: 92-82-0   HR: 3
PHENAZINE
mf: $C_{12}H_8N_2$ mw: 180.22
PROP: Pale-yellow-orange crystals from AcOH. Mp: 176–177°, bp: >360° (subl). Very sltly sol in water; sol in cold and hot alc, ether; mod sol in Et_2O, C_6H_6.
SYNS: AZOPHENYLENE □ DIBENZOPARADIAZINE □ DIBENZOPYRAZINE
CONSENSUS REPORTS: Reported in EPA TSCA Inventory.
SAFETY PROFILE: Poison by intraperitoneal and intravenous routes. Questionable carcinogen with experimental tumorigenic data. When heated to decomposition it emits toxic fumes of NO_x.

PDC250   CAS: 136-40-3   HR: 3
PHENAZOPYRIDINIUM CHLORIDE
mf: $C_{11}H_{11}N_5$•ClH   mw: 249.73
PROP: Red crystals; sltly bitter taste. Mp: 233–238°. Sltly sol in cold water, alc; sol in acetic acid; insol in acetone, benzene, chloroform, ether.
SYNS: AZODINE □ AZODIUM □ AZODYNE □ AZO GANTRISIN □ AZO GASTANOL □ AZO-MANDEL-AMINE □ AZOMINE □ AZO-STANDARD □ AZO-STAT □ AZOTREX □ BARIDIUM □ BISTERIL □ CYSTAMINE "MCCLUNG" □ CYSTOPYRIN □ CYSTURAL □ 2,6-DIAMINO-3-PHENYLAZOPYRIDINE HYDROCHLOR-IDE □ 2,6-DIAMINO-3-(PHENYLAZO)PYRIDINE MONOHYDROCHLORIDE □ DI-AZO □ DIRIDONE □ DOLONIL □ EUCISTIN □ GIRACID □ MALLOFEEN □ MALLOPHENE □ NC 150 □ NCI-C01672 □ NEFRECIL □ PAP □ PDP □ PHENAZO □ PHENAZODINE □ PHENAZOPYRIDINE HYDROCHLORIDE □ PHENYL-AZODIAMINOPYRIDINE HYDROCHLORIDE □ β-PHENYLAZO-α,α'-DIAMINOPYRIDINE HYDRO-CHLORIDE □ 3-PHENYLAZO-2,6-DIAMINOPYRIDINE HYDROCHLORIDE □ PHENYLAZO-α,α'-DIAMINO-PYRIDINE MONOHYDROCHLORIDE □ 3-(PHENYL-AZO)-2,6-PYRIDINEDIAMINE, HYDROCHLORIDE □ PHENYLAZOPYRIDINE HYDROCHLORIDE □ PHENYLAZO TABLETS □ PHENYL-IDIUM □ PHENYL-IDIUM 200 □ PIRID □ PIRIDACIL □ PYRAZODINE □ PYRAZOFEN □ PYREDAL □ PYRIDACIL □ PYRIDENAL □ PYRIDENE □ PYRIDIATE □ PYRIDIUM □ PYRIDIVITE □ PYRIPYRIDIUM □ PYRIZIN □ SEDURAL □ SULADYNE □ SULODYNE □ THIO-SULFIL-A FORTE □ URAZIUM □ URIDINAL □ URIPLEX □ UROBIOTIC-250 □ URODINE □ UROFEEN □ UROMIDE □ UROPHENYL □ UROPYRIDIN □ UROPYRINE □ UTOSTAN □ VESTIN □ W 1655
CONSENSUS REPORTS: NTP 10th Report on Carcinogens. IARC Cancer Review: Group 2B IMEMDT 7,312,87; Animal Sufficient Evidence IMEMDT 24,163,80; Human Limited Evidence IMEMDT 24,163,80; Animal Inadequate Evidence IMEMDT 8,117,75. NCI Carcinogenesis Bioassay (feed); Clear Evidence: mouse, rat NCITR* NCI-CG-TR-99,78.
SAFETY PROFILE: Confirmed carcinogen with experimental carcinogenic and tumorigenic data. A poison by intraperitoneal and intravenous routes. Moderately toxic by ingestion. Human systemic effects by ingestion: somnolence, cyanosis, diarrhea, nausea or vomiting, anuria or decreased urine volume, normocytic anemia, methemoglobinemia-carboxyhemoglobinemia, dehydration, changes in blood sodium levels. Mutation data reported. When heated to decomposition it emits very toxic fumes of NO_x and HCl.

PDC750   CAS: 2275-14-1   HR: 3
PHENCAPTON
mf: $C_{11}H_{15}Cl_2O_2PS_3$ mw: 377.31
SYNS: O,O-DIAETHYL-S((2,5-DICHLOR-PHENYL-THIO)-METHYL)-DITHIOPHOSPHAT (GERMAN) □ S-(2,5-DICHLORODPHENYLTHIOMETHYL) O,O-DIETHYL PHOSPHORODITHIOATE □ 2,5-DICHLOROPHENYL-THIOMETHYL O,O-DIETHYL PHOSPHORODITHIOATE □ S-(2,5-DICHLOROPHENYLTHIOMETHYL) DIETHYL PHOSPHOROTHIOLOTHIONATE □ O,O-DIETHYL-S-(2,5-DICHLOROPHENYLTHIOMETHYL) DITHIOPHOS-PHATE □ O,O-DIETHYL-S-(2,5-DICHLOROPHENYL-THIOMETHYL) DITHIOPHOSPHORAN □ O,O-DIETH-YL-S-(2,5-DICHLOROPHENYLTHIOMETHYL) PHOS-PHORODITHIOATE □ O,O-DIETHYL-S-(2,5-DICHLORO-PHENYLTHIOMETHYL) PHOSPHOROTHIOLO-THIONATE □ DITHIOPHOSPHATE de-O,O-DIETHYLE et de S(2,5-DICHLOROPHENYL) THIOMETHYLE (FRENCH) □ EENKAPTON (DUTCH) □ ENT 25,585 □ GEIGY G-28029 □ PRZEDZIORKOFOS (POLISH)
SAFETY PROFILE: Poison by ingestion. Moderately toxic by skin contact. A

cholinesterase inhibitor. When heated to decomposition it emits very toxic fumes of Cl⁻, PO$_x$, and SO$_x$. See also PARATHION.

PDD750   CAS: 60-12-8   HR: 2
PHENETHYL ALCOHOL
mf: C$_8$H$_{10}$O   mw: 122.18
PROP: Colorless liquid; floral odor of roses. Mp: −27°, bp: 220°, flash p: 216°F, d: 1.020 @ 20°/4°, vap d: 4.21. Misc with alc, ether; sol in fixed oils, glycerin, propylene glycol.
SYNS: BENZYL CARBINOL □ ETHANOL, 2-PHENYL- □ FEMA No. 2858 □ β-FENETHYLALKOHOL □ β-FENYLETHANOL □ β-HYDROXYETHYLBENZENE □ METHANOL, BENZYL- □ ORANGE OIL □ PEA □ β-PEA □ PHENETHANOL □ β-PHENETHYL ALCOHOL □ 2-PHENETHYL ALCOHOL □ β-PHENYLETHANOL □ 2-PHENYLETHANOL □ PHENYLETHYL ALCOHOL □ β-PHENYLETHYL ALCOHOL □ 2-PHENYLETHYL ALCOHOL □ ROSE OIL
CONSENSUS REPORTS: Reported in EPA TSCA Inventory.
SAFETY PROFILE: Moderately toxic by ingestion and skin contact. A skin and eye irritant. Experimental teratogenic effects. Other experimental reproductive effects. Causes severe central nervous system injury to experimental animals. Mutation data reported. Combustible when exposed to heat or flame; can react with oxidizing materials. To fight fire, use CO$_2$, dry chemical. When heated to decomposition it emits acrid smoke and irritating fumes.

PDE000   CAS: 98-85-1   HR: 3
α-PHENETHYL ALCOHOL
DOT: UN 2937
mf: C$_8$H$_{10}$O   mw: 122.18
PROP: Colorless liquid; hyacinth odor. Bp: 204°, fp: 21.4°, d: 1.015 @ 20°/20°, refr index: 1.525, vap press: 0.1 mm @ 20°, vap d: 4.21, flash p: 205°F (OC). Sol in fixed oils, propylene glycol; very sol in glycerin.
SYNS: BENZENEMETHANOL, α-METHYL- □ ETHANOL, 1-PHENYL- □ FEMA No. 2685 □ 1-FENYLETHANOL □ FENYL-METHYLKARBINOL □ α-METHYLBENZYL ALCOHOL (FCC) □ METHYLPHEN-YLCARBINOL □ METHYPHENYLMETHANOL □ NCI-C55685 □ 1-PHENYLETHANOL □ PHENYLMETHYL-CARBINOL □ STYRALLYL ALCOHOL □ STYRALYL ALCOHOL
CONSENSUS REPORTS: NTP Carcinogenesis Studies (gavage); Some Evidence: rat NTPTR* NTP-TR-369,90; (gavage); No Evidence: mouse NTPTR* NTP-TR-369,90. Reported in EPA TSCA Inventory. EPA Genetic Toxicology Program.
DOT CLASSIFICATION: 6.1; Label: KEEP AWAY FROM FOOD
SAFETY PROFILE: Poison by ingestion and subcutaneous routes. Moderately toxic by skin contact. A skin and severe eye irritant. Questionable carcinogen. Combustible when exposed to heat or flame; can react with oxidizing materials. To fight fire, use alcohol foam, foam, CO$_2$, dry chemical.

PDE250   CAS: 64-04-0   HR: 3
β-PHENETHYLAMINE
mf: C$_8$H$_{11}$N   mw: 121.20
PROP: Colorless to sltly yellow liquid; fishy odor. Bp: 197–198°, d: 0.96 @ 15.5°/15.5°, vap d: 4.18. Sol in water; very sol in alc, ether.
SYNS: β-AMINOETHYLBENZENE □ 1-AMINO-2-PHENYLETHANE □ β-PHENYLAETHYLAMIN (GERMAN) □ 1-PHENYL-2-AMINO-AETHAN (GERMAN) □ 1-PHENYL-2-AMINOETHANE □ PHENYLETHYL-AMINE □ ω-PHENYLETHYLAMINE □ 2-PHENYLETHY-LAMINE
CONSENSUS REPORTS: Reported in EPA TSCA Inventory.
SAFETY PROFILE: Poison by intraperitoneal, subcutaneous, intracervical, and intravenous routes. Moderately toxic by ingestion. A strong base. A skin irritant and possible sensitizer. When heated to decomposition it emits toxic fumes of NO$_x$. See also AMINES.

PDF750   CAS: 103-48-0   HR: 2
PHENETHYL ISOBUTYRATE
mf: C$_{12}$H$_{16}$O$_2$ mw: 192.28
PROP: Colorless to light-yellow liquid; fruity, rosy odor. D: 0.987, refr index: 1.486–1.490, flash p: 212°F. Sol in alc, fixed oils; insol in water @ 230°.

SYNS: BENZYLCARBINOL ISOBUTYRATE ☐ BENZ-YLCARBINYL ISOBUTYRATE ☐ FEMA No. 2862 ☐ PHENYLETHYL ISOBUTYRATE ☐ β-PHENYLETHYL ISOBUTYRATE ☐ 2-PHENYLETHYL ISOBUTYRATE ☐ 2-PHENYLETHYL-2-METHYLPROPIONATE

CONSENSUS REPORTS: Reported in EPA TSCA Inventory.

SAFETY PROFILE: Mildly toxic by ingestion. Combustible liquid. When heated to decomposition it emits acrid smoke and irritating fumes.

PDF775   CAS: 140-26-1   HR: 2
PHENETHYL ISOVALERATE
mf: $C_{13}H_{18}O_2$ mw: 206.31

PROP: Colorless to sltly yellow liquid; fruity, rosy odor. D: 0.973, refr index: 1.484, flash p: 212°F. Sol in alc, fixed oils; insol in water @ 263°.

SYNS: FEMA No. 2871 ☐ 3-METHYL-BUTANOIC ACID 2-PHENYLETHYL ESTER ☐ PHENETHYL ESTER ISOVALERIC ACID ☐ PHENYLETHYL ISOVALERATE ☐ β-PHENYLETHYL ISOVALERATE ☐ 2-PHENYLETHYL-3-METHYLBUTYRATE

CONSENSUS REPORTS: Reported in EPA TSCA Inventory.

SAFETY PROFILE: Mildly toxic by ingestion. Combustible liquid. When heated to decomposition it emits acrid smoke and irritating fumes. See also ESTERS.

PDF790   HR: 2
2-PHENETHYL 2-METHYLBUTYRATE
mf: $C_{13}H_{18}O_2$ mw: 206.28

PROP: Colorless liquid; floral, fruity odor. D: 0.973, refr index: 1.484, flash p: 212°F. Sol in alc, fixed oils; insol in water.

SYN: FEMA No. 3632

SAFETY PROFILE: Combustible liquid. When heated to decomposition it emits acrid smoke and irritating fumes.

PDI000   CAS: 102-20-5   HR: 2
PHENETHYL PHENYLACETATE
mf: $C_{16}H_{16}O_2$ mw: 240.32

PROP: Colorless to sltly yellow liquid above 26°; rosy, hyacinth odor. D: 1.079–1.082, flash p: 212°F. Sol in alc; insol in water.

SYNS: BENZENEACETIC ACID, 2-PHENYLETHYL ESTER ☐ BENZYLCARBINYL-α-TOLUATE ☐ FEMA No.

2866 ☐ PHENYLACETIC ACID, PHENETHYL ESTER ☐ β-PHENYLETHYL PHENYLACETATE ☐ 2-PHENYLETH-YL PHENYLACETATE ☐ 2-PHENYLETHYL-α-TOLUATE

CONSENSUS REPORTS: Reported in EPA TSCA Inventory.

SAFETY PROFILE: Moderately toxic by ingestion. Combustible liquid. When heated to decomposition it emits acrid smoke and irritating fumes. See also ESTERS.

PDI550   CAS: 40576-25-8   HR: 3
1-PHENETHYL-3-(PIPERIDINOCARBON-YL)PIPERIDINE
mf: $C_{19}H_{28}N_2O$ mw: 300.49

SYNS: KETONE, 1-PHENETHYL-3-PIPERIDYL PIPERIDINO ☐ 1-PHENETHYL-3-PIPERIDYL PIPERIDINO KETONE ☐ PIPERIDINE, 1-((1-(2-PHENYLETHYL)-3-PIPERIDINYL)CARBONYL)-

DOT CLASSIFICATION: 3; Label: Flammable Liquid

SAFETY PROFILE: A poison by intraperitoneal route. A flammable liquid. When heated to decomposition it emits toxic vapors of NO_x.

PDK200   HR: 2
PHENETHYL SALICYLATE
mf: $C_{15}H_{14}O_3$ mw: 242.27

PROP: White crystals; balsamic odor. Solidification point: 41°, flash p: 212°F. Sol in alc; insol in water.

SYN: FEMA No. 2868

SAFETY PROFILE: Combustible liquid. When heated to decomposition it emits acrid smoke and irritating fumes.

PDK790   CAS: 156-43-4   HR: 3
PHENETIDINE
mf: $C_8H_{11}NO$ mw: 137.20

PROP: A liquid. Mp: 4°, bp: 253°.

SYNS: 4-AMINOETHOXYBENZENE ☐ p-AMINO-FENETOL ☐ p-AMINOPHENETOLE ☐ 4-AMINO-PHENETOLE ☐ ANILINE, p-ETHOXY- ☐ BENZENAM-INE, 4-ETHOXY-(9CI) ☐ p-ETHOXYANILINE ☐ 4-ETHOXYANILINE ☐ 4-ETHOXYBENZENAMINE ☐ p-FENETIDIN ☐ p-PHENETIDIN

CONSENSUS REPORTS: Reported in EPA TSCA Inventory.

SAFETY PROFILE: Poison by inhalation. Moderately toxic by ingestion and intraperitoneal routes. Caution: It can be absorbed through the skin. A skin and eye irritant. Mutation data reported. When heated to decomposition it emits toxic fumes of NO_x.

PDM250   CAS: 404-82-0   HR: 3
PHENFLUORAMINE HYDROCHLORIDE
mf: $C_{12}H_{16}F_3N \cdot ClH$ mw: 267.75

PROP: A solid. Mp: 166°.

SYNS: N-ETHYL-α-METHYL-m-TRIFLUOROMETHYL-PHENETHYLAMINE □ N-ETHYL-α-METHYL-m-(TRIFLUOROMETHYL)PHENETHYLAMINE HYDROCHLORIDE □ FENFLUORAMINE HYDROCHLORIDE □ PONDERAL □ PONDERAX □ PONDIMIN □ 1-(3-TRIFLUOROMETHYLPHENYL)-2-ETHYLAMINOPROPANE HYDROCHLORIDE

SAFETY PROFILE: Poison by ingestion, intravenous, and intraperitoneal routes. Human systemic effects by ingestion: mydriasis, change in motor activity, nausea. An experimental teratogen. Other experimental reproductive effects. When heated to decomposition it emits very toxic fumes of F^-, NO_x, and HCl.

PDN750   CAS: 108-95-2   HR: 3
PHENOL
DOT: UN 1671/UN 2312/NA 2821

mf: C_6H_6O mw: 94.12

PROP: Deliquescent needles or white, crystalline mass that turns pink or red if not perfectly pure; burning taste, distinctive odor. Mp: 43°, fp: 41°, bp: 90.2° @ 25 mm, flash p: 175°F (CC), d: 1.072, autoign temp: 1319°F, vap press: 1 mm @ 40.1°, vap d: 3.24. Sol in water; misc in alc and ether. IDLH 250 ppm

SYNS: ACIDE CARBOLIQUE (FRENCH) □ BAKER'S P AND S LIQUID and OINTMENT □ BENZENOL □ CARBOLIC ACID □ CARBOLSAEURE (GERMAN) □ FENOL (DUTCH, POLISH) □ FENOLO (ITALIAN) □ HYDROXYBENZENE □ MONOHYDROXYBENZENE □ MONOPHENOL □ NCI-C50124 □ OXYBENZENE □ PHENIC ACID □ PHENOL, molten (DOT) □ PHENOL ALCOHOL □ PHENOLE (GERMAN) □ PHENYL HYDRATE □ PHENYL HYDROXIDE □ PHENYLIC ACID □ PHENYLIC ALCOHOL □ RCRA WASTE NUMBER U188

CONSENSUS REPORTS: NCI Carcinogenesis Bioassay (oral); No

Evidence: mouse, rat NCITR* NCI-CG-TR-203,80. EPA Extremely Hazardous Substances List. Community Right-To-Know List. Reported in EPA TSCA Inventory. EPA Genetic Toxicology Program.

OSHA PEL: TWA 5 ppm (skin)

ACGIH TLV: TWA 5 ppm (skin); BEI: 250 mg(total phenol)/g creatinine in urine at end of shift; Not Classifiable as a Human Carcinogen

DFG MAK: Confirmed Animal Carcinogen with Unknown Relevance to Humans; BAT: 300 mg/L at end of shift

NIOSH REL: (Phenol) TWA 20 mg/m³; CL 60 mg/m³/15M

DOT CLASSIFICATION: 6.1; Label: Poison

SAFETY PROFILE: Human poison by ingestion. An experimental poison by ingestion, subcutaneous, intravenous, parenteral, and intraperitoneal routes. Moderately toxic by skin contact. A severe eye and skin irritant. Questionable carcinogen with experimental carcinogenic and neoplastigenic data. Mutation data reported. An experimental teratogen. Absorption of phenolic solutions through the skin may be very rapid, and can cause death within 30 minutes to several hours by exposure of as little as 64 square inches of skin. Lesser exposures can cause damage to the kidneys, liver, pancreas, and spleen, and edema of the lungs. Ingestion can cause corrosion of the lips, mouth, throat, esophagus, and stomach, and gangrene. Ingestion of 1.5 g has killed. Chronic exposures can cause death from liver and kidney damage. Dermatitis resulting from contact with phenol or phenol-containing products is fairly common in industry. A common air contaminant.

Combustible when exposed to heat, flame, or oxidizers. Potentially explosive reaction with aluminum chloride + nitromethane (at 110°C/100 bar), formaldehyde, peroxydisulfuric acid, peroxymonosulfuric acid, sodium nitrite + heat. Violent reaction with aluminum chloride + nitrobenzene (at

120°C), sodium nitrate + trifluoroacetic acid, butadiene. Can react with oxidizing materials. To fight fire, use alcohol foam, CO_2, dry chemical. When heated to decomposition it emits acrid smoke and irritating fumes.

PDO750   CAS: 77-09-8   HR: 3
PHENOLPHTHALEIN
mf: $C_{20}H_{14}O_4$ mw: 318.34
PROP: Small crystals or needles. Mp: 258–262°, d: 1.299. Very sol in chloroform; sltly sol in water.
SYNS: AGORAL □ 3,3-BIS(p-HYDROXYPHENYL)-PHTHALIDE □ α-DI(p-HYDROXYPHENYL)PHTHALIDE □ DIHYDROXYPHTHALOPHENONE □ EUCHESSINA □ EVAC-Q-KIT □ EVAC-Q-KWIK □ EVAC-U-GEN □ FEEN-A-MINT GUM □ FENOLFTALEIN □ 1(3H)-ISOBENZOFURANONE, 3,3-BIS(4-HYDROXYPHENYL)- □ KOPROL □ LAXOGEN □ LILO □ NCI-C55798 □ PHTHALIDE 3,3,-BIS(p-HYDROXYPHENYL)- □ PHTHALIMETTEN □ PRULET □ PURGA □ PURGEN □ PURGOPHEN □ SPULMAKO-LAX □ TRILAX
CONSENSUS REPORTS: NTP 10th Report on Carcinogens. Reported in EPA TSCA Inventory.
SAFETY PROFILE: Confirmed carcinogen. US Food and Drug Administration recommends removal from laxative formulations. Moderately toxic by intraperitoneal route. Human systemic effects: changes in urine composition, gastritis, nausea or vomiting. Used in medicine as a laxative; in chemistry as an indicator. When heated to decomposition it emits acrid smoke and irritating fumes.

PDP250   CAS: 92-84-2   HR: 3
PHENOTHIAZINE
mf: $C_{12}H_9NS$ mw: 199.28
PROP: Yellow plates, rhombic leaflets, or diamond-shaped plates from toluene or butanol. Mp: 182°, sublimes at 130° at 1 mm, bp: 371°. Freely sol in benzene; sol in ether, hot acetic acid; sltly sol in alc and in mineral oils; practically insol in pet ether, chloroform, water.
SYNS: AFI-TIAZIN □ AGRAZINE □ ANTIVERM □ BIVERM □ CONTAVERM □ DIBENZOPARATHIAZINE □ DIBENZOTHIAZINE □ DIBENZO-1,4-THIAZINE □ ENT 38 □ FEENO □ FENOTHIAZINE (DUTCH) □ FENOTIAZINA (ITALIAN) □ FENOVERM □ FENTIAZIN □ HELMETINA □ LETHELMIN □ NEMAZENE □ NEMAZINE □ ORIMON □ PADOPHENE □ PENTHAZINE □ PHENEGIC □ PHENOSAN □ PHENOVERM □ PHENOVIS □ PHENOXUR □ PHENTHIAZINE □ RECONOX □ SOUFRAMINE □ THIODIFENYLAMINE (DUTCH) □ THIODIPHENYLAMIN (GERMAN) □ THIODIPHENYLAMINE □ TIODIFENILAMINA (ITALIAN) □ VERMITIN □ WURM-THIONAL □ XL-50
CONSENSUS REPORTS: EPA Genetic Toxicology Program. Reported in EPA TSCA Inventory.
OSHA PEL: TWA 5 mg/m³ (skin)
ACGIH TLV: TWA 5 mg/m³ (skin)
NIOSH REL: (Phenothiazine) TWA 5 mg/m³ (skin)
SAFETY PROFILE: Poison by intravenous route. Moderately toxic to humans by ingestion. Experimental reproductive effects. An insecticide. Large doses, i.e., heavy exposure, may cause hemolytic anemia and toxic degeneration of the liver. Can cause skin irritation and photosensitization. Dangerous; when heated to decomposition or on contact with acid or acid fumes it emits highly toxic fumes of SO_x and NO_x.

PDS900   CAS: 103-60-6   HR: 2
PHENOXYETHYL ISOBUTYRATE
mf: $C_{12}H_{16}O_3$ mw: 208.28
PROP: Colorless liquid; honey, roselike odor. D: 1.044, refr index: 1.492, flash p: 212°F. Misc in alc, chloroform, ether; insol in water.
SYNS: FEMA No. 2873 □ ISOBUTYRIC ACID, 2-PHENOXYETHYL ESTER (6CI,8CI) □ PROPANOIC ACID, 2-METHYL-, 2-PHENOXYETHYL ESTER
CONSENSUS REPORTS: Reported in EPA TSCA Inventory.
SAFETY PROFILE: Low toxicity by ingestion and skin contact. Combustible liquid. When heated to decomposition it emits acrid smoke and irritating fumes.

PDT250   CAS: 59-96-1   HR: 3
N-PHENOXYISOPROPYL-N-BENZYL-β-CHLOROETHYLAMINE
mf: $C_{18}H_{22}ClNO$ mw: 303.86

PROP: Crystals from pet ether. Mp: 38–49°. Sol in C_6H_6.

SYNS: A 688 □ BENSYLYTE □ 2-(N-BENZYL-2-CHLOROETHYLAMINO)-1-PHENOXYPROPANE □ BENZYL(2-CHLOROETHYL)-(1-METHYL-2-PHENOXYETHYL)AMINE □ BENZYLT □ N-(2-CHLOROETHYL)-N-(1-METHYL-2-PHENOXYETHYL)-BENZENEMETHANAMINE □ N-(2-CHLOROETHYL)-N-(1-METHYL-2-PHENOXYETHYL)BENZYLAMINE □ DIBENYLIN □ DIBENYLINE □ DIBENZYLINE □ NSC-37448 □ PHENOXYBENZAMINE

CONSENSUS REPORTS: IARC Cancer Review: Animal Sufficient Evidence IMEMDT 24,185,80; Animal Limited Evidence IMEMDT 9,223,75.

SAFETY PROFILE: Confirmed carcinogen with experimental carcinogenic and neoplastigenic data. Poison by intravenous and intracerebral routes. Moderately toxic by ingestion. Human reproductive effects by ingestion: spermatogenesis. Experimental reproductive effects. When heated to decomposition it emits very toxic fumes of Cl^- and NO_x.

PDW750   CAS: 990-73-8   HR: 3
PHENTANYL CITRATE
mf: $C_{22}H_{28}N_2O•C_6H_8O_7$ mw: 528.66
PROP: Crystals. Mp: 149–151°. Slightly sol in H_2O.
SYNS: FENTANEST □ FENTANYL CITRATE □ LEPTANAL □ McN-JR 4263 □ MCN-JR-4263-49 □ PENTANYL □ N-(1-PHENETHYL-4-PIPERIDIN-YL)PROPIONANILIDE DIHYDROGEN CITRATE □ N-(1-PHENETHYL-4-PIPERIDYL)PROPIONANILIDE CITRATE □ N-(1-PHENETHYL-4-PIPERID-YL)PROPIONANILIDE DIHYDROGEN CITRATE □ R 4263 □ R 5240 □ SUBLIMAZE □ SUBLIMAZE CITRATE

SAFETY PROFILE: Poison by ingestion, subcutaneous, and intravenous routes. When heated to decomposition it emits toxic fumes of NO_x. See also PHENTANYL.

PDX000   CAS: 101-48-4   HR: 2
PHENYLACETALDEHYDE DIMETHYL ACETAL
mf: $C_{10}H_{14}O_2$ mw: 166.24
PROP: Colorless liquid; strong odor. D: 1.000–1.006, refr index: 1.493, flash p:

194°F. Sol in fixed oils, propylene glycol; insol in glycerin.

SYNS: (2,2-DIMETHOXYETHYL)-BENZENE (9CI) □ 1,1-DIMETHOXY-2-PHENYLETHANE □ FEMA No. 2876 □ HYSCYLENE P □ PHENACETALDEHYDE DIMETHYL ACETAL □ α-TOLYL ALDEHYDE DIMETH-YL ACETAL □ VIRIDINE

CONSENSUS REPORTS: Reported in EPA TSCA Inventory.

SAFETY PROFILE: Moderately toxic by ingestion. Combustible liquid. When heated to decomposition it emits acrid smoke and irritating fumes. See also ALDEHYDES.

PDY500   CAS: 4075-79-0   HR: 2
4'-PHENYLACETANILIDE
mf: $C_{14}H_{13}NO$ mw: 211.28
PROP: Crystals from MeOH (aq). Mp: 171°.
SYNS: 4-ACETAMIDOBIPHENYL □ 4-ACETYLAMINOBIPHENYL □ 4-BIPHENYLACETAMIDE □ N-4-BIPHENYLACETAMIDE □ N-(4-BIPHENYLYL)ACETAMIDE □ p-PHENYLACETANILIDE

CONSENSUS REPORTS: Reported in EPA TSCA Inventory.

SAFETY PROFILE: Questionable carcinogen with experimental carcinogenic and tumorigenic data. Mutation data reported. When heated to decomposition it emits toxic fumes of NO_x. Used in the manufacture of plastics, resins, rubber, synthetics, dyes, and pigments.

PDY850   CAS: 103-82-2   HR: 2
PHENYLACETIC ACID
mf: $C_8H_8O_2$ mw: 136.16
PROP: Leaflets on distillation in vac; plates, tablets from pet ether; disagreeable odor of geranium. Mp: 76.5°, bp: 265.5°, d (77°/4°) 1.091, flash p: 212°F. Sltly sol in cold water; freely sol in hot water; sol in alc and ether. Solubility @ 25° in chloroform (moles/L): 4.422; in carbon tetrachloride: 1.842; in acetylene tetrachloride: 4.513; in trichlorethylene: 3.299; in tetrachlorethylene: 1.558; in pentachloroethane: 3.252.
SYNS: BENZENACETIC ACID □ BENZENEACETIC ACID □ FEMA No. 2878 □ ω-PHENYLACETIC ACID □ α-TOLUIC ACID

CONSENSUS REPORTS: Reported in EPA TSCA Inventory.

SAFETY PROFILE: Moderately toxic by ingestion, subcutaneous, and intraperitoneal routes. An experimental teratogen. Combustible liquid. Used in production of drugs of abuse. When heated to decomposition it emits acrid smoke and irritating fumes.

PEA750   CAS: 140-29-4   HR: 3
PHENYLACETONITRILE
DOT: UN 2470
mf: C_8H_7N mw: 117.16
PROP: Oily liquid; aromatic odor. Mp: −23.8°, bp: 233.5°, d: 1.0214 @ 15°/15°, vap press: 1 mm @ 60.0°. Insol in water; misc in alc, ether.
SYNS: BENZENEACETONITRILE □ BENZYL CYANIDE □ BENZYLKYANID □ BENZYL NITRILE □ (CYANOMETHYL)BENZENE □ α-CYANOTOLUENE □ ω-CYANOTOLUENE □ 2-PHENYLACETONITRILE □ PHENYLACETONITRILE, liquid (DOT) □ PHENYL ACETYL NITRILE □ α-TOLUNITRILE □ USAF KF-21
CONSENSUS REPORTS: EPA Extremely Hazardous Substances List. Community Right-To-Know List. Reported in EPA TSCA Inventory.
DOT CLASSIFICATION: 6.1; Label: KEEP AWAY FROM FOOD
SAFETY PROFILE: Poison by ingestion, inhalation, skin contact, subcutaneous, and intraperitoneal routes. A skin irritant. Explosive reaction with sodium hypochlorite. Used in production of drugs of abuse. When heated to decomposition it emits very toxic fumes of CN^- and NO_x. See also NITRILES.

PEB000   CAS: 451-40-1   HR: 3
2-PHENYLACETOPHENONE
mf: $C_{14}H_{12}O$ mw: 196.26
PROP: Plates. Mp: 60°, bp: 320°.
SYNS: BENZYL PHENYL KETONE □ DEOXYBENZO-IN □ DESOXYBENZOIN □ 1,2-DIPHENYLETHANONE □ 2-PHENYLACETOPHENONE □ PHENYL BENZYL KETONE
CONSENSUS REPORTS: Reported in EPA TSCA Inventory.
DOT CLASSIFICATION: 3; Label: Flammable Liquid

SAFETY PROFILE: Poison by intravenous route. Flammable liquid. When heated to decomposition it emits acrid smoke and irritating fumes. See also KETONES.

PEC500   CAS: 673-06-3   HR: 1
d-PHENYLALANINE
mf: $C_9H_{11}NO_2$ mw: 165.21
PROP: Needles from alc, white crystalline platelets or leaflets from water. Mp: 196° (decomp); sol in hot water; very sltly sol in alc; sltly sol in pet ether.
SYNS: dl-α-AMINO-β-PHENYLPROPIONIC ACID □ NCI-C60195 □ d-β-PHENYLALANINE □ dl-PHENYLALANINE (FCC)
CONSENSUS REPORTS: Reported in EPA TSCA Inventory.
SAFETY PROFILE: Mildly toxic by intraperitoneal route. Human systemic effects by ingestion: nausea, hypermotility, diarrhea. When heated to decomposition it emits toxic fumes of NO_x.

PEC750   CAS: 63-91-2   HR: 1
l-PHENYLALANINE
mf: $C_9H_{11}NO_2$ mw: 165.21
PROP: White crystals or crystalline powder or needles from water; slt odor and bitter taste. Mp: decomp @ 283–284° (rapid heat). Sol in water; very sltly sol in alc and ether.
SYNS: ALANINE, PHENYL- □ ALANINE, 3-PHENYL- □ l-ALANINE, PHENYL- □ (S)-α-AMINOBENZENE-PROPANOIC ACID □ α-AMINOHYDROCINNAMIC ACID □ α-AMINO-β-PHENYLPROPIONIC ACID □ ANTIBIOTIC FN 1636 □ HYDROCINNAMIC ACID, α-AMINO- □ PAL □ PHENYLALANINE □ PHENYL-α-ALANINE □ β-PHENYLALANINE □ β-PHENYL-α-ALANINE, l- □ l-β-PHENYLALANINE □ (S)-PHENYLALANINE □ 3-PHENYLALANINE
CONSENSUS REPORTS: Reported in EPA TSCA Inventory.
SAFETY PROFILE: Mildly toxic by intraperitoneal route. An experimental teratogen. Experimental reproductive effects. Human mutation data reported. When heated to decomposition it emits toxic fumes of NO_x.

PED750   CAS: 148-82-3   HR: 3
I-PHENYLALANINE MUSTARD
mf: $C_{13}H_{18}Cl_2N_2O_2$ mw: 305.23
PROP: Needles from MeOH. Mp: 182–183°
(decomp).
SYNS: ALANINE NITROGEN MUSTARD □ ALKERAN
□ AT-290 □ p-N-BIS(2-CHLOROETHYL)AMINO-l-
PHENYLALANINE □ l-3-(p-(BIS(2-CHLOROETHYL)-
AMINO)PHENYL)ALANINE □ 3-(p-(p-(BIS(2-
CHLOROETHYL)AMINO)PHENYL))-l-ALANINE □ 4-
(BIS(2-CHLOROETHYL)AMINO)-l-PHENYLALANINE □
CB 3025 □ p-N-DI(CHLOROETHYL)AMINOPHENYL-
ALANINE □ p-DI-(2-CHLOROETHYL)AMINO-l-
PHENYLALANINE □ 3-p-(DI(2-CHLOROETHYL)-
AMINO)-PHENYL-l-ALANINE □ MELPHALAN □ NCI-
C04853 □ NSC-8806 □ l-PAM □ PHENYLALANINE
NITROGEN MUSTARD □ RCRA WASTE NUMBER U150
□ l-SARCOLYSIN □ p-l-SARCOLYSIN □ SK-15673
CONSENSUS REPORTS: NTP 10th Report
on Carcinogens. IARC Cancer Review:
Group 1 IMEMDT 7,239,87; Animal
Sufficient Evidence; Human Limited
Evidence IMEMDT 9,167,75. NCI
Carcinogenesis Studies (ipr); Clear Evidence:
mouse, rat RRCRBU 52,1,75. EPA Genetic
Toxicology Program.
SAFETY PROFILE: Confirmed human
carcinogen producing leukemia and
Hodgkin's disease. Poison by ingestion,
intravenous, and intracerebral routes.
Human systemic effects by ingestion:
nausea, hypermotility, diarrhea,
agranulocytosis, thrombocytopenia. Human
reproductive effects by ingestion: menstrual
changes. Mutation data reported. A skin
irritant. Used as a poison gas. When heated
to decomposition it emits toxic fumes of Cl⁻
and NO$_x$.

PEE750   CAS: 1698-60-8   HR: 2
1-PHENYL-4-AMINO-5-CHLORPYRIDAZ-
6-ONE
mf: $C_{10}H_8ClN_3O$ mw: 221.66
PROP: Pale-yellowish solid. Mp: 205–206°.
Spar sol in H_2O, C_6H_6; sltly sol in Me_2CO,
MeOH.
SYNS: 5-AMINO-4-CHLORO-2,3-DIHYDRO-3-OXO-2-
PHENYLPYRIDAZINE □ 5-AMINO-4-CHLORO-2-
PHENYL-3(2H)-PYRIDAZINONE □ BUREX (CZECH) □
CHLORIDAZON □ 1-FENYL-4-AMINO-5-CHLOR-6-
PYRIDAZINON (CZECH) □ HS-119-1 □ PCA □

PHENOSANE □ 1-PHENYL-4-AMINO-5-CHLORO-
PYRIDAZON-(6) (GERMAN) □ 1-PHENYL-4-AMINO-5-
CHLOROPYRIDAZONE-6 □ 1-PHENYL-4-AMINO-5-
CHLORO-6-PYRIDAZONE □ PYRAMINE □ PYRAMIN
RB □ PYRAZON □ PYRAZONE □ PYRAZONL
CONSENSUS REPORTS: Reported in EPA
TSCA Inventory. EPA Genetic Toxicology
Program.
SAFETY PROFILE: Moderately toxic by
ingestion and intraperitoneal routes. A
severe eye irritant. Experimental
reproductive effects. Used as a
preemergence and early post-emergence
herbicide. When heated to decomposition it
emits very toxic fumes of Cl⁻ and NO$_x$.

PEG250   CAS: 613-37-6   HR: 2
p-PHENYLANISOLE
mf: $C_{13}H_{12}O$ mw: 184.25
PROP: Leaves or plates from alc. Mp:
91–92°, bp: 174° @ 18 mm. Sol in hot alc.
SYNS: p-METHOXYBIPHENYL □ 4-METHOXYBI-
PHENYL
CONSENSUS REPORTS: Reported in EPA
TSCA Inventory.
SAFETY PROFILE: Questionable
carcinogen with experimental tumorigenic
data. When heated to decomposition it
emits acrid smoke and irritating fumes.

PEG500   CAS: 91-40-7   HR: 3
N-PHENYLANTHRANILIC ACID
mf: $C_{13}H_{11}NO_2$ mw: 213.25
PROP: Needles or prisms from alc. Mp:
184°, decomp @ 183–184°. Very sltly sol
in hot water; sol in hot alc; very sltly sol in
ether.
SYNS: o-ANILINOBENZOIC ACID □ 2-ANILINO-
BENZOIC ACID □ 2-CARBOXYDIPHENYLAMINE □
DIPHENYLAMINE-2-CARBOXYLIC ACID □ FENAMIC
ACID □ PA □ 2-(PHENYLAMINO)BENZOIC ACID □
PHENYLANTHRANILIC ACID
CONSENSUS REPORTS: Reported in EPA
TSCA Inventory.
SAFETY PROFILE: Poison by intravenous
and intraperitoneal routes. When heated to
decomposition it emits toxic fumes of NO$_x$.

P

PEG750   CAS: 637-03-6   HR: 3
PHENYL ARSINE OXIDE
mf: C_6H_5AsO mw: 168.03
PROP: Colorless crystals from EtOH. Mp: 119–120°.
SYNS: ARSENOSOBENZENE □ ARZENE □ PHENYLARSENOXIDE
CONSENSUS REPORTS: Arsenic and its compounds are on the Community Right-To-Know List. Reported in EPA TSCA Inventory.
OSHA PEL: TWA 0.5 mg(As)/m³
SAFETY PROFILE: Poison by intravenous and intraperitoneal routes. When heated to decomposition it emits toxic fumes of As. See also ARSENIC COMPOUNDS.

PEI000   CAS: 60-09-3   HR: 3
p-(PHENYLAZO)ANILINE
mf: $C_{12}H_{11}N_3$ mw: 197.26
PROP: Yellow crystals or orange needles with blue case from EtOH. Mp: 126°, bp: 360°. Sltly sol in hot water; sol in hot alc and ether.
SYNS: AAB □ AMINOAZOBENZENE □ p-AMINOAZOBENZENE □ 4-AMINOAZOBENZENE □ 4-AMINO-1,1'-AZOBENZENE □ p-AMINOAZOBENZOL □ 4-AMINOAZOBENZOL □ p-AMINODIPHENYLIMIDE □ ANILINE YELLOW □ 4-BENZENEAZOANILINE □ BRASILAZINA OIL YELLOW G □ CERES YELLOW R □ C.I. 11000 □ C.I. SOLVENT BLUE 7 □ C.I. SOLVENT YELLOW 1 □ FAST SPIRIT YELLOW AAB □ OIL SOLUBLE ANILINE YELLOW □ OIL YELLOW AAB □ ORGANOL YELLOW □ PARAPHENOLAZO ANILINE □ 4-(PHENYLAZO)ANILINE □ 4-(PHENYLAZO)BENZEN-AMINE □ p-PHENYLAZOPHENYLAMINE □ SOLVENT YELLOW 1 □ SUDAN YELLOW R □ USAF EK-1375
CONSENSUS REPORTS: IARC Cancer Review: Group 2B IMEMDT 7,56,87; Animal Sufficient Evidence IMEMDT 8,53,75. Community Right-To-Know List. Reported in EPA TSCA Inventory.
SAFETY PROFILE: Confirmed carcinogen with experimental neoplastigenic and tumorigenic data. Poison by intraperitoneal route. An experimental teratogen. Mutation data reported. Used as a dye for lacquer, varnish, wax products, oil stains, and styrene resins. When heated to decomposition it emits toxic fumes of NO_x. See also AMINES.

PEJ250   CAS: 22670-79-7   HR: 3
N-PHENYLAZO-N-METHYLTAURINE
  SODIUM SALT
mf: $C_9H_{12}N_3O_3S \cdot Na$ mw: 265.29
SYNS: 3-METHYL-1-PHENYL-3-(2-SULFOETHYL)-TRIAZENE SODIUM SALT □ 1-PHENYL-3-METHYL-3-(2-SULFOAETHYL) NATRIUM SALZ (GERMAN) □ 1-PHENYL-3-METHYL-3-(2-SULFOETHYL)TRIAZENE, SODIUM SALT
SAFETY PROFILE: Poison by subcutaneous route. Experimental reproductive effects. Questionable carcinogen with experimental neoplastigenic data. When heated to decomposition it emits very toxic fumes of NO_x, Na_2O, and SO_x.

PEJ500   CAS: 842-07-9   HR: 2
1-(PHENYLAZO)-2-NAPHTHOL
mf: $C_{16}H_{12}N_2O$ mw: 248.30
PROP: Dark-reddish-yellow leaflets. Mp: 134°. Sol in Et_2O, CS_2, C_6H_6, pet ether, conc HCl.
SYNS: ATUL ORANGE R □ BENZENEAZO-β-NAPHTHOL □ BENZENE-1-AZO-2-NAPHTHOL □ 1-BENZOAZO-2-NAPHTHOL □ BRILLIANT OIL ORANGE R □ CALCOGAS ORANGE NC □ CALCO OIL ORANGE 7078 □ CAMPBELLINE OIL ORANGE □ CARMINAPH □ CERES ORANGE R □ CEROTINORANGE G □ C.I. 12055 □ C.I. SOLVENT YELLOW 14 □ DISPERSOL YELLOW PP □ DUNKELGELB □ ENIAL ORANGE I □ FAST OIL ORANGE □ FAST ORANGE □ FETTORANGE R □ GRASAN ORANGE R □ HIDACO OIL ORANGE □ LACQUER ORANGE VG □ MOTIORANGE R □ NCI-C53929 □ OIL ORANGE □ OLEAL ORANGE R □ ORANGE A l'HUILE □ ORANGE INSOLUBLE OLG □ ORANGE PEL □ ORANGE RESENOLE No. 3 □ ORANGE SOLUBLE A l'HUILE □ ORGANOL ORANGE □ ORIENT OIL ORANGE PS □ PETROL ORANGE Y □ 1-(PHENYLAZO)-2-NAPHTHALENOL □ 1-PHENYLAZO-β-NAPHTHOL □ PLASTORESIN ORANGE F4A □ PYRONALORANGE □ RESINOL ORANGE R □ RESOFORM ORANGE G □ SANSEL ORANGE G □ SCHARLACH B □ SILOTRAS ORANGE TR □ SOLVENT YELLOW 14 □ SOMALIA ORANGE I □ SOUDAN I □ SPIRIT ORANGE □ SPIRIT YELLOW I □ STEARIX ORANGE □ SUDAN ORANGE R □ TERTROGRAS ORANGE SV □ TOYO OIL ORANGE □ WAXAKOL ORANGE GL □ WAXOLINE YELLOW I

CONSENSUS REPORTS: IARC Cancer Review: Group 3 IMEMDT 7,56,87; Animal Sufficient Evidence IMEMDT 8,225,75. NTP Carcinogenesis Bioassay (feed); Clear Evidence: rat NTPTR* NTP-TR-226,82. Community Right-To-Know List. Reported in EPA TSCA Inventory. EPA Genetic Toxicology Program.

SAFETY PROFILE: Experimental reproductive effects. Questionable carcinogen with experimental carcinogenic, neoplastigenic, and tumorigenic data. Mutation data reported. When heated to decomposition it emits toxic fumes of NO_x. Used for coloring hydrocarbon solvents, oils, fats, waxes, shoe and floor polishes, and gasoline.

PEK000   CAS: 532-82-1   HR: 2
4-PHENYLAZO-m-PHENYLENEDIAMINE
mf: $C_{12}H_{12}N_4 \bullet ClH$ mw: 248.74
PROP: A solid. Mp: 235°.
SYNS: ASTRA CHRYSOIDINE R □ BRASILAZINA ORANGE Y □ BRILLIANT OIL ORANGE Y BASE □ CALCOZINE CHRYSOIDINE Y □ CALCOZINE ORANGE YS □ CHRYSOIDIN □ CHRYSOIDINE □ CHRYSOIDINE A □ CHRYSOIDINE B □ CHRYSOIDINE C CRYSTALS □ CHRYSOIDINE G □ CHRYSOIDINE GN □ CHRYSOID-INE HR □ CHRYSOIDINE(II) □ CHRYSOIDINE J □ CHRYSOIDINE M □ CHRYSOIDINE ORANGE □ CHRYSOIDINE PRL □ CHRYSOIDINE PRR □ CHRY-SOIDINE SL □ CHRYSOIDINE SPECIAL (biological stain and indicator) □ CHRYSOIDINE SS □ CHRYSOIDINE Y □ CHRYSOIDINE Y BASE NEW □ CHRYSOIDINE Y CRYSTALS □ CHRYSOIDINE Y EX □ CHRYSOIDINE YGH □ CHRYSOIDINE YL □ CHRYSOIDINE YN □ CHRYSOIDINE Y SPECIAL □ CHRYSOIDIN FB □ CHRYSOIDIN Y □ CHRYSOIDIN YN □ CHRYZOIDYNA F.B. (POLISH) □ C.I. 11270 □ C.I. BASIC ORANGE 2 □ C.I. BASIC ORANGE 3 □ C.I. BASIC ORANGE 2, MONOHYDROCHLORIDE □ C.I. SOLVENT ORANGE 3 □ 2,4-DIAMINOAZOBENZENE HYDROCHLORIDE □ DIAZOCARD CHRYSOIDINE G □ ELCOZINE CHRYSOIDINE Y □ LEATHER ORANGE HR □ 4-(PHENYLAZO)-1,3-BENZENEDIAMINE MONOHYDROCHLORIDE □ 4-(PHENYLAZO)-m-PHENYLENEDIAMINE MONOHYDROCHLORIDE □ PURE CHRYSOIDINE YBH □ PURE CHRYSOIDINE YD □ PYRACRYL ORANGE Y □ SUGAI CHRYSOIDINE □ TERTROPHENE BROWN CG
CONSENSUS REPORTS: IARC Cancer Review: Group 3 IMEMDT 7,169,87;

Animal Sufficient Evidence IMEMDT 8,91,75. Reported in EPA TSCA Inventory. EPA Genetic Toxicology Program.

SAFETY PROFILE: Moderately toxic by ingestion and subcutaneous routes. Questionable carcinogen with experimental tumorigenic data. Mutation data reported. When heated to decomposition it emits very toxic fumes of NO_x and HCl. Used as a colorant in textiles, paper, leather, inks, wood, and biological stains.

PEL750   CAS: 363-03-1   HR: 3
PHENYL-p-BENZOQUINONE
mf: $C_{12}H_8O_2$ mw: 184.20
PROP: Yellow leaflets from pet ether. Mp: 113.5–114°.
SYNS: p-BENZOQUINONE, 2-PHENYL- □ 2,5-CYCLOHEXADIENE-1,4-DIONE, 2-PHENYL-(9CI) □ PHENYLBENZOQUINONE □ o-PHENYLBENZO-QUINONE □ PHENYL-1,4-BENZOQUINONE □ 2-PHENYLBENZOQUINONE □ 2-PHENYL-2,5-CYCLOHEXADIENE-1,4-DIONE □ PHENYLQUINONE
CONSENSUS REPORTS: Reported in EPA TSCA Inventory.
SAFETY PROFILE: Poison by intraperitoneal route. Experimental reproductive effects. Mutation data reported. When heated to decomposition it emits acrid smoke and irritating fumes.

PEM750   CAS: 511-55-7   HR: 3
8-(p-PHENYLBENZYL)ATROPINIUM
BROMIDE
mf: $C_{30}H_{34}NO_3 \bullet Br$ mw: 536.56
PROP: Crystals. Mp: 230°. Decomp @ 220–222°.
SYNS: N-(p-BIPHENYLMETHYL)-ATROPINIUM BROMIDE □ N,4-BIPHENYL-METHYL-dl-TROPEYL-α-TROPINIUMBROMID (GERMAN) □ p-BIPHENYL-METHYL-(dl-TROPYL-α-TROPINIUM)BROMIDE □ DENDREPAR □ 4-DIPHENYLMETHYLTROPYL-TROPINIUM BROMIDE □ 4-DIPHENYLMETHYL-dl-TROPYLTROPINIUM BROMIDE □ GASTRIPON □ 3-α-HYDROXY-8-(p-PHENYLBENZYL)-1-α-H,5-α-H-TROPANIUM BROMIDE, (±)-TROPATE □ N-399 □ XENYTROPIUM BROMIDE
SAFETY PROFILE: Poison by intravenous, subcutaneous, and intraperitoneal routes. Moderately toxic by ingestion. Experimental

reproductive effects. When heated to decomposition it emits very toxic fumes of NO_x and Br⁻. See also BROMIDES.

PEO500   CAS: 108-86-1   HR: 3
PHENYL BROMIDE
DOT: UN 2514
mf: C_6H_5Br mw: 157.02
PROP: Colorless, clear, mobile liquid. Mp: −30.7°, bp: 156.2°, flash p: 124°F, d: 1.497, vap press: 10 mm @ 40°, vap d: 5.41, autoign temp: 1051°F.
SYNS: BROMOBENZENE (DOT) □ MONOBROMOBENZENE □ NCI-C55492
CONSENSUS REPORTS: Reported in EPA TSCA Inventory. EPA Genetic Toxicology Program.
DOT CLASSIFICATION: 3; Label: Flammable Liquid
SAFETY PROFILE: Moderately toxic by ingestion, subcutaneous, and intraperitoneal routes. Mildly toxic by inhalation. An eye and mucous membrane irritant. Mutation data reported. Flammable liquid when exposed to heat, sparks, or flame. Can react with oxidizing materials. To fight fire, use water to blanket fire, foam, CO_2, water spray or mist, dry chemical. Violent reaction with bromobutane + sodium when heated above 30°C. When heated to decomposition it emits toxic fumes of Br⁻. See also BROMIDES.

PEO600   CAS: 25687-87-0   HR: 3
4-PHENYL-2-BUTANONE 3-THIOSEMI-
   CARBAZONE
SYN: 2-BUTANONE, 4-PHENYL-, 3-THIOSEMICARBAZ-ONE
SAFETY PROFILE: A poison by ingestion. When heated to decomposition it emits toxic vapors of NO_x and SO_x.

PEQ750   CAS: 104-68-7   HR: 2
PHENYL CARBITOL
mf: $C_{10}H_{14}O_3$ mw: 182.24
PROP: Liquid. Bp: 207° @ 55 mm, fp: −50°, d: 1.1158 @ 20°/20°, vap press: <0.01 mm @ 20°, vap d: 6.28.

SYNS: DIETHYLENE GLYCOL MONOPHENYL ETHER □ DIETHYLENE GLYCOL PHENYL ETHER □ 2-(2-PHENOXYETHOXY)ETHANOL
CONSENSUS REPORTS: Glycol ether compounds are on the Community Right-To-Know List. Reported in EPA TSCA Inventory.
SAFETY PROFILE: Moderately toxic by ingestion and skin contact. A skin and severe eye irritant. Some glycol ethers have dangerous human reproductive effects. When heated to decomposition it emits acrid smoke and irritating fumes. See also GLYCOL ETHERS.

PER000   CAS: 122-99-6   HR: 2
PHENYL CELLOSOLVE
mf: $C_8H_{10}O_2$ mw: 138.18
PROP: Yellow, brown or clear liquid. Mp: 14°, bp: 242°, flash p: 250°F, d: 1.11 @ 20°/20°, fp: 11–13°.
SYNS: AROSOL □ DOWANOL EP □ DOWANOL EPH □ EMERESSENCE 1160 □ EMERY 6705 □ ETHYLENE GLYCOL MONOPHENYL ETHER □ ETHYLENE GLYCOL PHENYL ETHER □ 2-FENOXYETHANOL (CZECH) □ FENYL-CELLOSOLVE (CZECH) □ GLYCOL MONOPHENYL ETHER □ β-HYDROXYETHYL PHENYL ETHER □ 1-HYDROXY-2-PHENOXYETHANE □ PHENOXETHOL □ PHENOXETOL □ PHENOXY-ETHANOL □ 2-PHENOXYETHANOL □ PHENOXY-ETHYL ALCOHOL □ PHENOXYTOL □ PHENYLMONO-GLYCOL ETHER □ ROSE ETHER
CONSENSUS REPORTS: Glycol ether compounds are on the Community Right-To-Know List. Reported in EPA TSCA Inventory.
DFG MAK: 20 ppm
SAFETY PROFILE: Moderately toxic by ingestion and skin contact. A skin and severe eye irritant. Mutation data reported. Some glycol ethers have dangerous human reproductive effects. Combustible when exposed to heat or flame; can react vigorously with oxidizing materials. When heated to decomposition it emits acrid smoke and irritating fumes. To fight fire, use CO_2, dry chemical. Used as a solvent for ester-type resins. See also GLYCOL ETHERS.

PER250   CAS: 48145-04-6   HR: 2
PHENYL CELLOSOLVE ACRYLATE
mf: $C_{11}H_{12}O_3$ mw: 192.23
PROP: Colorless to yellowish lquid with
pungent odor. Mp: −36°, bp: 11°, d: 1.0998
g/cm³. Flash pt: 143° C. Sol in water.~
SYNS: CHEMLINK 160 □ EBECRYL 110 □ LIGHT
ESTER PO-A □ 2-PHENOXYETHANOL ACRYLATE □
PHENOXYETHYL ACRYLATE □ POA □ 2-PROPENOIC
ACID, 2-PHENOXYETHYL ESTER □ R 561 □ SR 339
CONSENSUS REPORTS: Reported in EPA
TSCA Inventory.
SAFETY PROFILE: Moderately toxic by
skin contact. Mildly toxic by ingestion. A
skin irritant. A combustible liquid. When
heated to decomposition it emits acrid
smoke and irritating fumes.

PER600   CAS: 64049-07-6   HR: 3
PHENYL(β-CHLOROVINYL)CHLORO-
   ARSINE
mf: $C_8H_7AsCl_2$ mw: 248.97
SYNS: ARSINE, CHLORO(2-CHLOROVINYL)PHENYL-
□ PHENYL(β-CHLOROVINYL)CHLORARSINE □ TL 59
OSHA PEL: TWA 0.5 mg(As)/m³
SAFETY PROFILE: Poison by inhalation.
When heated to decomposition it emits
toxic fumes of As and Cl⁻.

PES750   CAS: 712-50-5   HR: 3
PHENYL CYCLOHEXYL KETONE
mf: $C_{13}H_{16}O$ mw: 188.29
PROP: Needles from pet ether. Mp: 59–60°,
bp: 164–165° @ 18 mm.
SYN: USAF KF-3
CONSENSUS REPORTS: Reported in EPA
TSCA Inventory.
DOT CLASSIFICATION: 3; Label:
Flammable Liquid
SAFETY PROFILE: Poison by
intraperitoneal route. When heated to
decomposition it emits acrid smoke and
irritating fumes. See also KETONES.

PET750   CAS: 3721-28-6   HR: 3
trans-2-PHENYLCYCLOPROPYLAMINE
mf: $C_9H_{11}N$ mw: 133.21
SYNS: PARNATE □ trans-2-PHENYL-1-AMINO-
CYCLOPROPANE □ SKF 385 □ TRANILCYPROMINE □

TRANSAMINE □ TRANYLCYPRAMINE □
TRANYLCYPROMINE
SAFETY PROFILE: Poison by ingestion,
intraperitoneal, and subcutaneous routes.
Experimental reproductive effects. Mutation
data reported. When heated to
decomposition it emits toxic fumes of NO_x.
See also AMINES.

PEV500   CAS: 1754-58-1   HR: 3
O-PHENYL-N,N′-DIMETHYL PHOS-
   PHORODIAMIDATE
mf: $C_8H_{13}N_2O_2P$ mw: 200.20
PROP: Crystals from $CHCl_3/Et_2O$ or CCl_4.
Mp: 103–104°. Sltly sol in EtOAc, C_6H_6.
SYNS: DIAMIDAFOS □ DIAMIDFOS □ DOWCO 169 □
NELLITE
SAFETY PROFILE: Poison by ingestion and
skin contact. When heated to decomposition
it emits very toxic fumes of PO_x and NO_x.
A pesticide used on tobacco to control
rootknot nematodes.

PEY000   CAS: 108-45-2   HR: 3
m-PHENYLENEDIAMINE
DOT: UN 1673
mf: $C_6H_8N_2$ mw: 108.16
PROP: White crystals from C_6H_6/hexane.
Mp: 65°, bp: 287°, d: 1.139, vap press: 1 mm
@ 99.8°. Sol in water, methanol, ethanol,
chloroform, acetone; sltly sol in ether,
carbon tetrachloride; very sltly sol in
benzene, toluene.
SYNS: m-AMINOANILINE □ 3-AMINOANILINE □
APCO 2330 □ m-BENZENEDIAMINE □ 1,3-BENZENE-
DIAMINE □ C.I. 76025 □ DEVELOPER 11 □ m-
DIAMINOBENZENE □ 1,3-DIAMINOBENZENE □
DIRECT BROWN BR □ m-FENYLENDIAMIN (CZECH) □
METAPHENYLENEDIAMINE □ 1,3-PHENYLENEDI-
AMINE □ m-PHENYLENEDIAMINE (DOT) □ PHENYL-
ENEDIAMINE, META, solid (DOT)
CONSENSUS REPORTS: IARC Cancer
Review: Group 3 IMEMDT 7,56,87; Animal
Inadequate Evidence IMEMDT 16,111,78.
EPA Genetic Toxicology Program.
Reported in EPA TSCA Inventory.
ACGIH TLV: TWA 0.1 mg/m³; Not
Classifiable as a Human Carcinogen
DFG MAK: Confirmed Animal Carcinogen
with Unknown Relevance to Humans

P

DOT CLASSIFICATION: 6.1; Label: KEEP AWAY FROM FOOD

SAFETY PROFILE: Suspected carcinogen with experimental tumorigenic and teratogenic data. Poison by ingestion, intravenous, subcutaneous, and intraperitoneal routes. Mildly toxic by skin contact. Mutation data reported. Combustible when exposed to heat or flame. A hair dye ingredient. When heated to decomposition it emits toxic fumes of NO_x. See also other phenylenediamine entries and AMINES.

PEY250   CAS: 95-54-5   HR: 3
o-PHENYLENEDIAMINE

DOT: UN 1673

mf: $C_6H_8N_2$ mw: 108.16

PROP: Tan crystals or leaflets from water. Mp: 104°, bp: 257°. Sltly sol in water; very sol in alc, chloroform, ether.

SYNS: 2-AMINOANILINE □ o-BENZENEDIAMINE □ 1,2-BENZENEDIAMINE □ C.I. 76010 □ C.I. OXIDATION BASE 16 □ o-DIAMINOBENZENE □ 1,2-DIAMINO-BENZENE □ EK 1700 □ NSC-5354 □ ORTHAMINE □ 1,2-PHENYLENEDIAMINE (DOT)

CONSENSUS REPORTS: Reported in EPA TSCA Inventory. EPA Genetic Toxicology Program.

ACGIH TLV: TWA 0.1 mg/m³; Animal Carcinogen

DFG MAK: Confirmed Animal Carcinogen with Unknown Relevance to Humans

DOT CLASSIFICATION: 6.1; Label: KEEP AWAY FROM FOOD

SAFETY PROFILE: Confirmed carcinogen. Poison by ingestion and intraperitoneal routes. Moderately toxic by subcutaneous route. Mildly toxic by skin contact. Mutation data reported. A pesticide and pharmaceutical. When heated to decomposition it emits toxic fumes of NO_x. See also other phenylenediamine entries and AMINES.

PEY500   CAS: 106-50-3   HR: 3
p-PHENYLENEDIAMINE

DOT: UN 1673

mf: $C_6H_8N_2$ mw: 108.16

PROP: White to sltly red crystals or leaflets from Et₂O. Mp: 146°, flash p: 312°F, vap d: 3.72, bp: 267°. Sol in alc, chloroform, ether. IDLH 25 mg/m³.

SYNS: p-AMINOANILINE □ 4-AMINOANILINE □ BASF URSOL D □ p-BENZENEDIAMINE □ 1,4-BENZENEDIAMINE □ BENZOFUR D □ C.I. 76060 □ C.I. DEVELOPER 13 □ C.I. OXIDATION BASE 10 □ DEVELOPER 13 □ DEVELOPER PF □ p-DIAMINO-BENZENE □ 1,4-DIAMINOBENZENE □ DURAFUR BLACK R □ FENYLENODWUAMINA (POLISH) □ FOURAMINE D □ FOURRINE D □ FOURRINE 1 □ FUR BLACK 41867 □ FUR BROWN 41866 □ FURRO D □ FUR YELLOW □ FUTRAMINE D □ NAKO H □ ORSIN □ PARA □ PARAPHENYLEN-DIAMINE □ PELAGOL D □ PELAGOL DR □ PELAGOL GREY D □ PELTOL D □ 1,4-PHENYLENEDIAMINE □ PHENYLENEDIAMINE, PARA, solid (DOT) □ PPD □ RENAL PF □ SANTOFLEX IC □ TERTRAL D □ URSOL D □ USAF EK-394 □ VULKANOX 4020 □ ZOBA BLACK D

CONSENSUS REPORTS: IARC Cancer Review: Group 3 IMEMDT 7,56,87; Animal Inadequate Evidence IMEMDT 16,125,78. Community Right-To-Know List. Reported in EPA TSCA Inventory. EPA Genetic Toxicology Program.

OSHA PEL: TWA 0.1 mg/m³ (skin)

ACGIH TLV: TWA 0.1 mg/m³ (skin); Not Classifiable as a Human Carcinogen

DFG MAK: 0.1 mg/m³ as total dust; Confirmed Animal Carcinogen with Unknown Relevance to Humans

DOT CLASSIFICATION: 6.1; Label: KEEP AWAY FROM FOOD

SAFETY PROFILE: Suspected carcinogen with experimental tumorigenic data. Poison by ingestion, subcutaneous, intravenous, and intraperitoneal routes. Mildly toxic by skin contact. A human skin irritant. Mutation data reported. Implicated in aplastic anemia. Can cause fatal liver damage. The p-form is more toxic and a stronger irritant than the o- and m- isomers. When used as a hair dye it caused vertigo, anemia, gastritis, exfoliative dermatitis, and death. Has caused asthma and other respiratory symptoms in the fur-dyeing industry. Combustible when exposed to heat or flame; can react vigorously with oxidizing materials. To fight fire, use water, CO₂, dry chemical. When heated to decomposition it emits acrid smoke and irritating fumes. See also other phenylenediamine entries and AMINES.

PEY750 CAS: 541-69-5 HR: 3
m-PHENYLENEDIAMINE HYDRO-
  CHLORIDE
mf: $C_6H_8N_2 \cdot 2ClH$ mw: 181.08
PROP: Colorless needles. Very sol in water; sltly sol in alc, ether.
SYNS: m-AMINOANILINE DIHYDROCHLORIDE □ 3-AMINOANILINE DIHYDROCHLORIDE □ m-BENZENE-DIAMINE DIHYDROCHLORIDE □ 1,3-BENZENE-DIAMINE HYDROCHLORIDE □ m-DIAMINOBENZENE DIHYDROCHLORIDE □ 1,3-DIAMINOBENZENE DIHYDROCHLORIDE □ 1,3-PHENYLENEDIAMINE DIHYDROCHLORIDE □ USAF EK-206
CONSENSUS REPORTS: IARC Cancer Review: Animal Inadequate Evidence IMEMDT 16,111,78. Reported in EPA TSCA Inventory.
SAFETY PROFILE: Poison by intraperitoneal route. Questionable carcinogen with experimental tumorigenic data. Mutation data reported. When heated to decomposition it emits very toxic fumes of HCl and NO_x. See also other phenylenediamine entries and AMINES.

PFA500 CAS: 4044-65-9 HR: 3
1,4-PHENYLENEDIISOTHIOCYANIC
  ACID
mf: $C_8H_4N_2S_2$ mw: 192.26
PROP: Needles from Me_2CO; tasteless, odorless, colorless crystals. Mp: 132°.
SYNS: BISCOMATE □ BITOSCANATE □ 1,4-DIISOTHIOCYANATOBENZENE □ ISOTHIOCYANIC ACID-p-PHENYLENE ESTER □ JONIT □ PHENYLENE-1,4-DIISOTHIOCYANATE □ PHENYLENE THIOCYANATE
CONSENSUS REPORTS: Cyanide and its compounds are on the Community Right-To-Know List. EPA Extremely Hazardous Substances List. Reported in EPA TSCA Inventory.
SAFETY PROFILE: Poison by ingestion and intraperitoneal routes. Human systemic effects by ingestion: hallucinations, nausea. When heated to decomposition it emits very

toxic fumes of NO_x, CN^-, and SO_x. See also THIOCYANATES.

PFA850 CAS: 101-84-8 HR: 2
PHENYL ETHER
mf: $C_{12}H_{10}O$ mw: 170.22
PROP: Colorless low-melting crystals; geranium odor. Mp: 37–39°, bp: 259°, flash p: 239°F, d: 1.0728 @ 20°, vap d: 5.86, autoign temp: 1148°F, lel: 0.8%, uel: 1.5%. IDLH 100 ppm.
SYNS: BIPHENYL OXIDE □ DIPHENYL ETHER □ DIPHENYL OXIDE □ GERANIUM CRYSTALS □ PHENOXYBENZENE
CONSENSUS REPORTS: Reported in EPA TSCA Inventory.
OSHA PEL: Vapor: TWA 1 ppm
ACGIH TLV: TWA 1 ppm; STEL 2 ppm (vapor)
DFG MAK: 1 ppm (7.1 mg/m^3)
NIOSH REL: (Phenyl Ether, vapor) TWA 1 ppm
SAFETY PROFILE: Moderately toxic by ingestion. Prolonged exposure damages liver, spleen, kidneys, and thyroid, and upsets gastrointestinal tract. A skin and eye irritant. Combustible when exposed to heat or flame; can react with oxidizing materials. For explosion hazard, see ETHERS. To fight fire, use water, foam, CO_2, dry chemical. When heated to decomposition it emits acrid smoke and irritating fumes.

PFA860 CAS: 8004-13-5 HR: 3
PHENYL ETHER-BIPHENYL MIXTURE
mf: $C_{12}H_{10} \cdot C_{12}H_{10}O$ mw: 324.44
PROP: Eutectic mixture 73.5% phenylether and 26.5% biphenyl by weight (MELAAD 48,247,57). IDLH 10 ppm.
SYNS: BIPHENYL, mixed with BIPHENYL OXIDE (3:7) □ BIPHENYL-DIPHENYL ETHER mixture □ 1,1'-BIPHENYL, mixed with 1,1'-OXYBIS(BENZENE) □ DINIL □ DINYL □ DIPHENYL mixed with DIPHENYL OXIDE □ DIPHYL □ DOWTHERM □ DOWTHERM A
OSHA PEL: Vapor: TWA 1 ppm
NIOSH REL: (Phenyl Ether-Biphenyl Mixture, vapor) TWA 1 ppm
SAFETY PROFILE: Poison by inhalation. Moderately toxic by ingestion. Human

P

systemic effects by inhalation: unspecified effects on the sense of smell, conjunctiva irritation, and unspecified respiratory effects. A mild skin and eye irritant. When heated to decomposition it emits acrid smoke and irritating fumes.

PFB250   CAS: 103-45-7   HR: 2
2-PHENYLETHYL ACETATE
mf: $C_{10}H_{12}O_2$ mw: 164.22
PROP: Colorless liquid; sweet, rosy, honey odor. Mp: 164.2°, bp: 223.6°, fp: <−20°, flash p: 230°F, d: 1.032 @ 25°/25°, refr index: 1.497–1.501. Sol in alc, fixed oils, propylene glycol; insol in glycerin, water @ 232°.
SYNS: ACETIC ACID-2-PHENYLETHYL ESTER □ BENZYLCARBINYL ACETATE □ FEMA No. 2857 □ β-PHENETHYL ACETATE □ 2-PHENETHYL ACETATE □ β-PHENYLETHYL ACETATE
CONSENSUS REPORTS: Reported in EPA TSCA Inventory.
SAFETY PROFILE: Moderately toxic by ingestion. Mildly toxic by skin contact. A skin irritant. Combustible when exposed to heat or flame; can react vigorously with oxidizing materials. To fight fire, use alcohol foam, CO_2, and dry chemical. When heated to decomposition it emits acrid smoke and irritating fumes. See also ESTERS.

PFC500   CAS: 51-71-8   HR: 3
2-PHENYLETHYLHYDRAZINE
mf: $C_8H_{12}N_2$ mw: 136.22
PROP: Oil. Bp: 137–139° @ 12–13°.
SYNS: 1-HYDRAZINO-2-PHENYLETHANE □ NARDIL □ PHENELZINE □ PHENETHYLHYDRAZINE □ β-PHENYLETHYLHYDRAZINE □ STINERVAL □ W 1544
SAFETY PROFILE: Poison by ingestion, intraperitoneal, and subcutaneous routes. Human systemic effects by ingestion: ataxia, somnolence. An experimental teratogen. Experimental reproductive effects. Mutation data reported. Used as an antidepressant. When heated to decomposition it emits toxic fumes of NO_x.

PFC750   CAS: 156-51-4   HR: 3
β-PHENYLETHYLHYDRAZINE SULFATE

mf: $C_8H_{12}N_2 \cdot H_2O_4S$ mw: 234.30
SYNS: ALACINE □ ALAZIN □ ALAZINE □ EP-411 □ ESTINERVAL □ FELAZINE □ FENELZIN □ 1-HYDRA-ZINO-2-PHENYLETHANE HYDROGEN SULPHATE □ KALGAN □ MAO-REM □ MONOPHEN □ MONOTEN □ N-1544A □ NARDELZINE □ NARDIL □ P 1531 □ PHENALZINE □ PHENALZINE DIHYDROGEN SULFATE □ PHENALZINE HYDROGEN SULPHATE □ PHENELZIN □ PHENELZINE ACID SULFATE □ PHENELZINE BISULPHATE □ PHENELZINE SULFATE □ PHENETHYLHYDRAZINE SULFATE (1:1) □ PHENLINE □ PHENODYNE □ PHENYLAETHYL-HYDRAZIN □ β-PHENYLETHYLHYDRAZINE DIHYDROGEN SULFATE □ 2-PHENYLETHYLHYDRA-ZINE DIHYDROGEN SULPHATE □ β-PHENYLETHYL-HYDRAZINE HYDROGEN SULPHATE □ PHENYL-ETHYLHYDRAZINE SULPHATE □ S 1544 □ STINERVAL
CONSENSUS REPORTS: IARC Cancer Review: Group 3 IMEMDT 7,312,87; Human Inadequate Evidence IMEMDT 24,175,80; Animal Limited Evidence IMEMDT 24,175,80. EPA Genetic Toxicology Program.
SAFETY PROFILE: Poison by ingestion, intraperitoneal, intravenous, and subcutaneous routes. Human systemic effects by ingestion: wakefulness, blood pressure lowering, constipation, hepatitis, fibrous hepatitis. Experimental reproductive effects. Questionable carcinogen with experimental neoplastigenic data. Mutation data reported. Used as a drug for the treatment of depression. When heated to decomposition it emits very toxic fumes of SO_x and NO_x.

PFF360   CAS: 122-60-1   HR: 3
PHENYL GLYCIDYL ETHER
mf: $C_9H_{10}O_2$ mw: 150.19
PROP: IDLH 100 ppm.
SYNS: 1,2-EPOXY-3-PHENOXYPROPANE □ 2,3-EPOXYPROPYLPHENYL ETHER □ FENYL-GLYCIDYLETHER (CZECH) □ GLYCIDYL PHENYL ETHER □ PGE □ PHENOL-GLYCIDAETHER (GERMAN) □ PHENOL GLYCIDYL ETHER (MAK) □ 3-PHENOXY-1,2-EPOXYPROPANE □ PHENOXYPROPENE OXIDE □ PHENOXYPROPYLENE OXIDE □ PHENYL-2,3-EPOXYPROPYL ETHER
CONSENSUS REPORTS: IARC Cancer Review: Group 2B IMEMDT 47,237,89; Animal Sufficient Evidence IMEMDT

47,237,89. Reported in EPA TSCA Inventory. EPA Genetic Toxicology Program.
OSHA PEL: TWA 1 ppm
ACGIH TLV: TWA 0.1 ppm (skin, sensitizer) Confirmed Animal Carcinogen with Unknown Relevance to Humans
DFG MAK: Confirmed Animal Carcinogen, Suspected Human Carcinogen
NIOSH REL: (Phenyl Glycidyl Ether) CL 1 ppm/15M
SAFETY PROFILE: Confirmed carcinogen with experimental carcinogenic data. Moderately toxic by ingestion, skin contact, and subcutaneous routes. A severe eye and skin irritant. Experimental reproductive effects. Mutation data reported. When heated to decomposition it emits acrid smoke and irritating fumes. Used as a chemical intermediate. See also ETHERS.

PFI000   CAS: 100-63-0   HR: 3
PHENYLHYDRAZINE
DOT: UN 2572
mf: $C_6H_8N_2$ mw: 108.16
PROP: Yellow, monoclinic crystals, oil, or plates (usually an oil with aniline-like odor). Mp: 19.6°, bp: 243.5° (decomp), flash p: 192°F (CC), d: 1.0978 @ 20°/4°, vap press: 1 mm @ 71.8°, vap d: 3.7. Sltly sol in hot water; misc in alc, chloroform, ether, benzene. IDLH 15 ppm.
SYNS: FENILIDRAZINA (ITALIAN) □ FENYLHYDRAZINE (DUTCH) □ HYDRAZINE-BENZENE □ HYDRAZINOBENEZENE □ PHENYLHYDRAZIN (GERMAN)
CONSENSUS REPORTS: Reported in EPA TSCA Inventory.
OSHA PEL: TWA 5 ppm (skin); STEL 10 ppm
ACGIH TLV: TWA 0.1 ppm (skin); Animal Carcinogen
DFG MAK: Confirmed Animal Carcinogen with Unknown Relevance to Humans
NIOSH REL: CL 0.6 mg/m³/2H
DOT CLASSIFICATION: 6.1; Label: Poison
SAFETY PROFILE: Confirmed carcinogen with experimental carcinogenic data. Poison by ingestion, subcutaneous, and intravenous routes. Experimental reproductive effects. Mutation data reported. Ingestion or subcutaneous injection can cause hemolysis of red blood cells. Other effects are damage to the spleen, liver, kidneys, and bone marrow. The most common effect of occupational exposure is the development of dermatitis, which, in sensitized persons, may be quite severe. Systemic effects include anemia and general weakness, gastrointestinal disturbances, and injury to the kidneys. Flammable when exposed to heat, flame, or oxidizers. To fight fire, use alcohol foam. Violent reaction with 2-phenylamino-3-phenyloxazirane. Reacts with perchloryl fluoride to form an explosive product. Vigorous reaction with lead(IV) oxide. Used as a chemical reagent, in organic synthesis, and in the manufacture of dyes and drugs. Dangerous; when heated to decomposition it emits highly toxic fumes of NO_x; can react with oxidizing materials.

PFI250   CAS: 59-88-1   HR: 3
PHENYLHYDRAZINE HYDROCHLORIDE
mf: $C_6H_8N_2 \cdot ClH$ mw: 144.62
PROP: Leaflets from water; crystals from alc. Mp: 250–254° (decomp). Very sol in water; sol in alc; insol in ether.
SYNS: PHENYLHYDRAZINE MONOHYDROCHLORIDE □ PHENYLHYDRAZIN HYDROCHLORID (GERMAN) □ PHENYLHYDRAZINIUM CHLORIDE
CONSENSUS REPORTS: Reported in EPA TSCA Inventory. EPA Extremely Hazardous Substances List.
NIOSH REL: (Hydrazines) CL 0.6 mg/m³/2H
SAFETY PROFILE: Poison by ingestion, intraperitoneal, and subcutaneous routes. Experimental reproductive effects. Questionable carcinogen with experimental neoplastigenic and tumorigenic data. Mutation data reported. When heated to decomposition it emits very toxic fumes of NO_x and HCl.

PFI600   CAS: 100482-34-6   HR: 3
o-(PHENYLHYDROXYARSINO)BENZOIC
ACID
mf: $C_{13}H_{11}AsO_3$ mw: 290.16
SYNS: BENZOIC ACID, o-(PHENYLHYDROXY-
ARSINO)- □ 2-CARBOXYDIPHENYLARSINOUS ACID
OSHA PEL: TWA 0.5 mg(As)/m³
SAFETY PROFILE: Poison by intravenous
route. When heated to decomposition it
emits toxic fumes of As.

PFJ250   CAS: 100-65-2   HR: 3
β-PHENYLHYDROXYLAMINE
mf: C_6H_7NO mw: 109.14
PROP: Colorless needles from water. Mp:
81–82°. Sol in hot and cold water; very sol
in alc and ether; very sltly sol in ligroin.
SYNS: NCI-C60093 □ N-PHENYLHYDROXYLAMINE
CONSENSUS REPORTS: Reported in EPA
TSCA Inventory.
SAFETY PROFILE: Poison by ingestion and
subcutaneous routes. Human systemic
effects by skin contact: primary irritation.
Preparative hazard. Mutation data reported.
When heated to decomposition it emits
toxic fumes of NO_x.

PFJ400   CAS: 622-44-6   HR: 3
PHENYLIMIDOCARBONYL CHLORIDE
DOT: UN 1672
mf: $C_7H_5Cl_2N$ mw: 174.03
SYNS: CARBONIMIDIC DICHLORIDE, PHENYL-(9CI)
□ N-(DICHLOROMETHYLENE)ANILINE □ PHENYL-
CARBONIMIDIC DICHLORIDE □ N-PHENYLCARBO-
NIMIDIC DICHLORIDE □ PHENYL CARBYLAMINE
CHLORIDE □ PHENYLCARBYLAMINE CHLORIDE
(DOT) □ PHENYLIMIDOCARBONYL CHLORIDE □ N-
PHENYLIMIDOPHOSGENE □ PHENYLIMINO-
CARBONYL DICHLORIDE □ N-PHENYLIMINO-
CARBONYL DICHLORIDE □ PHENYLISONITRILE
DICHLORIDE
DOT CLASSIFICATION: 6.1; Label: Poison
SAFETY PROFILE: A poison by inhalation.
When heated to decomposition it emits
toxic vapors of NO_x and Cl⁻.

PFK250   CAS: 103-71-9   HR: 3
PHENYL ISOCYANATE
DOT: UN 2487
mf: C_7H_5NO mw: 119.13

PROP: Liquid, acrid odor. Mp: −30°
approx, bp: 158–168°, d: 1.1 @ 20°, vap
press: 1 mm @ 10.6°, flash p: 132°.
Decomp in water, alc; very sol in ether.
SYNS: CARBANIL □ FENYLISOKYANAT □
ISOCYANIC ACID, PHENYL ESTER □ KARBANIL □
MONDUR P □ PHENYLCARBIMIDE □ PHENYL
CARBONIMIDE
CONSENSUS REPORTS: Reported in EPA
TSCA Inventory.
DOT CLASSIFICATION: 6.1; Label: Poison;
DOT Class: 6.1; Label: Poison, Flammable
Liquid; DOT Class: 3; Label: Flammable
Liquid, Poison
SAFETY PROFILE: A poison. An irritant.
Mutation data reported. Flammable liquid
when exposed to heat or flame; can react
vigorously with oxidizing materials. Has
exploded when stirred with (cobalt
pentammine triazoperchlorate + nitrosyl
perchlorate). When heated to decomposition
it emits toxic fumes of CN⁻ and NO_x. See
also CYANATES.

PFL850   CAS: 108-98-5   HR: 3
PHENYL MERCAPTAN
DOT: UN 2337
mf: C_6H_6S mw: 110.18
PROP: A liquid; repulsive odor. Bp: 169.5°,
d: 1.973 @ 25°/4°.
SYNS: PHENOL, THIO- □ RCRA WASTE NUMBER
P014 □ THIOFENOL □ THIOPHENOL (DOT) □ USAF
XR-19
CONSENSUS REPORTS: Reported in EPA
TSCA Inventory. EPA Extremely
Hazardous Substances List.
OSHA PEL: TWA 0.5 ppm
ACGIH TLV: TWA 0.5 ppm; (Proposed: 0.1
ppm (skin))
NIOSH REL: CL 0.5 mg/m³/15M
DOT CLASSIFICATION: 6.1; Label: Poison,
Flammable Liquid
SAFETY PROFILE: Poison by ingestion,
inhalation, skin contact, and intraperitoneal
routes. A severe eye irritant. Can cause
severe dermatitis. Exposure may cause
headache and dizziness. When heated to
decomposition or on contact with acids it

emits toxic fumes of SO_x. See also MERCAPTANS.

PFM500   CAS: 100-56-1   HR: 3
PHENYLMERCURIC CHLORIDE
mf: C_6H_5ClHg mw: 313.15
PROP: Colorless leaves from benzene. Mp: 251°, bp: sublimes. Insol in water; sltly sol in hot alc; sol in pyridine, ether, benzene. IDLH 10 mg/m³ (as Hg).
SYNS: CHLORID FENYLRTUTNATY (CZECH) □ (CHLOROMERCURI)BENZENE □ FENYLMERCURI-CHLORID (CZECH) □ MERCURIPHENYL CHLORIDE □ MERFAZIN □ MERSOLITE 2 □ PHENYL CHLORO-MERCURY □ PHENYLMERCURY CHLORIDE □ PHENYLQUECKSILBERCHLORID (GERMAN) □ PMC □ STOPSPOT
CONSENSUS REPORTS: Mercury and its compounds are on the Community Right-To-Know List.
OSHA PEL: CL 0.1 mg(Hg)/m³ (skin)
ACGIH TLV: TWA 0.1 mg(Hg)/m³ (skin); BEI: 35 µg/g creatinine total inorganic mercury in urine preshift; 15 µg/g creatinine total inorganic mercury in blood at end of shift at end of workweek.
DFG MAK: Confirmed Animal Carcinogen with Unknown Relevance to Humans
NIOSH REL: (Mercury, Organo) TWA 0.01 mg/m³; STEL 0.03 mg/m³ (skin)
SAFETY PROFILE: Poison by ingestion, intraperitoneal, and subcutaneous routes. Human mutation data reported. When heated to decomposition it emits very toxic fumes of Cl⁻ and Hg. See also MERCURY COMPOUNDS and CHLORIDES.

PFN000   CAS: 14235-86-0   HR: 3
PHENYLMERCURIC DINAPHTHYL-METHANEDISULFONATE
mf: $C_{33}H_{24}Hg_2O_6S_2$ mw: 981.87
PROP: Amorphous powder. IDLH 10 mg/m³ (as Hg).
SYNS: BIS(PHENYLMERCURI)METHYLENEDINAPHTHALENE SULFONATE □ CONOTRANE □ FIBROTAN □ HYDRAPHEN □ HYDRARGAPHEN □ METHYLENE-DINAPHTHALENESULFONIC ACID BISPHENYL-MERCURI SALT □ PENOTRANE □ PHENYL MERCURIC FIXTAN □ PHENYLMERCURIC 3,3'-METHYLENEBIS(2-NAPHTHALENESULFONATE) □ PHENYLMERCURY

METHYLENEDINAPHTHALENESULFONATE □ P.M.F. □ SEPTOTAN □ VERSOTRANE
CONSENSUS REPORTS: Mercury and its compounds are on the Community Right-To-Know List.
OSHA PEL: CL 0.1 mg(Hg)/m³ (skin)
ACGIH TLV: TWA 0.1 mg(Hg)/m³ (skin); BEI: 35 µg/g creatinine total inorganic mercury in urine preshift; 15 µg/g creatinine total inorganic mercury in blood at end of shift at end of workweek.
DFG MAK: Confirmed Animal Carcinogen with Unknown Relevance to Humans
NIOSH REL: (Mercury, Aryl and Inorganic) CL 0.1 mg/m³ (skin)
SAFETY PROFILE: Poison by intraperitoneal and ingestion routes. A severe eye irritant. When heated to decomposition it emits very toxic fumes of Hg and SO_x. See also MERCURY COMPOUNDS and SULFONATES.

PFN100   CAS: 100-57-2   HR: 3
PHENYLMERCURIC HYDROXIDE
DOT: UN 1894
mf: C_6H_6HgO mw: 294.71
PROP: Powder or prisms from water. Mp: 234–237°. Sol in hot water. IDLH 10 mg/m³ (as Hg).
SYNS: HYDROXYPHENYLMERCURY □ MERCURY, HYDROXYPHENYL- □ MERSOLITE 1 □ PHENYL HYDROXYMERCURY □ PHENYLMERCURIC HYDROXIDE (DOT) □ PHENYLMERCURY HYDROXIDE
CONSENSUS REPORTS: Reported in EPA TSCA Inventory.
ACGIH TLV: TWA 0.1 mg(Hg)/m³ (skin); BEI: 35 µg/g creatinine total inorganic mercury in urine preshift; 15 µg/g creatinine total inorganic mercury in blood at end of shift at end of workweek.
DFG MAK: Confirmed Animal Carcinogen with Unknown Relevance to Humans
NIOSH REL: (Mercury, Aryl and Inorganic) CL 0.1 mg/m³ (skin)
DOT CLASSIFICATION: 6.1; Label: Poison
SAFETY PROFILE: Poison by intravenous route. Mutation data reported. When heated

P

to decomposition it emits toxic fumes of Hg.

PFO000   CAS: 103-27-5   HR: 3
PHENYLMERCURI PROPIONATE
mf: $C_9H_{10}HgO_2$ mw: 350.78
PROP: IDLH 10 mg/m³ (as Hg).
SYNS: METASOL P-6 □ PHENYLMERCURY PROP-IONATE □ PHENYL(PROPIONYLOXY)MERCURY □ PROPIONIC ACID, PHENYLMERCURY SALT
CONSENSUS REPORTS: Mercury and its compounds are on the Community Right-To-Know List. Reported in EPA TSCA Inventory.
OSHA PEL: CL 0.1 mg(Hg)/m³ (skin)
ACGIH TLV: TWA 0.1 mg(Hg)/m³ (skin); BEI: 35 µg/g creatinine total inorganic mercury in urine preshift; 15 µg/g creatinine total inorganic mercury in blood at end of shift at end of workweek.
DFG MAK: Confirmed Animal Carcinogen with Unknown Relevance to Humans
NIOSH REL: (Mercury, Aryl and Inorganic) CL 0.1 mg/m³ (skin)
SAFETY PROFILE: Probably a poison. When heated to decomposition it emits toxic fumes of Hg. See also MERCURY COMPOUNDS.

PFO500   HR: 3
PHENYLMERCURY ACETATE 95% plus
ETHYLMERCURY CHLORIDE 5%
PROP: IDLH 10 mg/m³ (as Hg).
SYN: (ACETATO)PHENYL-MERCURY mixed with CHLOROETHYL-MERCURY (19:1)
CONSENSUS REPORTS: Mercury and its compounds are on the Community Right-To-Know List. Reported in EPA TSCA Inventory.
OSHA PEL: CL 0.1 mg(Hg)/m³ (skin)
ACGIH TLV: TWA 0.1 mg(Hg)/m³ (skin); BEI: 35 µg/g creatinine total inorganic mercury in urine preshift; 15 µg/g creatinine total inorganic mercury in blood at end of shift at end of workweek.
DFG MAK: Confirmed Animal Carcinogen with Unknown Relevance to Humans
NIOSH REL: (Mercury, Inorganic) TWA 0.05 mg(Hg)/m³

SAFETY PROFILE: Poison by ingestion. When heated to decomposition it emits very toxic fumes of Hg and Cl⁻. See also MERCURY COMPOUNDS.

PFP500   CAS: 2279-64-3   HR: 3
PHENYLMERCURY UREA
mf: $C_7H_8HgN_2O$ mw: 336.76
PROP: IDLH 10 mg/m³ (as Hg).
SYNS: ABAVIT □ LEYTOSAN □ PHENYLMERCURIC UREA □ PHENYLMERCURIUREA
CONSENSUS REPORTS: Mercury and its compounds are on the Community Right-To-Know List.
OSHA PEL: CL 0.1 mg(Hg)/m³ (skin)
ACGIH TLV: TWA 0.1 mg(Hg)/m³ (skin); BEI: 35 µg/g creatinine total inorganic mercury in urine preshift; 15 µg/g creatinine total inorganic mercury in blood at end of shift at end of workweek.
DFG MAK: Confirmed Animal Carcinogen with Unknown Relevance to Humans
NIOSH REL: (Mercury, Aryl and Inorganic) CL 0.1 mg/m³ (skin)
SAFETY PROFILE: Poison by an unspecified route. When heated to decomposition it emits very toxic fumes of Hg and NOₓ. See also MERCURY COMPOUNDS.

PFR200   CAS: 10415-87-9   HR: 2
1-PHENYL-3-METHYL-3-PENTANOL
mf: $C_{12}H_{18}O$ mw: 178.30
SYNS: 3-PENTANOL, 3-METHYL-1-PHENYL- □ PHENETHYLMETHYLETHYLCARBINOL □ PHENYLETHYL METHYL ETHYL CARBINOL
CONSENSUS REPORTS: Reported in EPA TSCA Inventory.
SAFETY PROFILE: Moderately toxic by ingestion. A skin irritant. When heated to decomposition it emits acrid smoke and irritating fumes.

PFS500   CAS: 16033-21-9   HR: 3
1-PHENYL-3-MONOMETHYLTRIAZENE
mf: $C_7H_9N_3$ mw: 135.19
SYN: PMT
SAFETY PROFILE: Poison by subcutaneous route. Questionable

carcinogen with experimental neoplastigenic data. Mutation data reported. When heated to decomposition it emits toxic fumes of NO$_x$.

PFS750   CAS: 92-53-5   HR: 3
PHENYL MORPHOLINE
mf: C$_{10}$H$_{13}$NO   mw: 163.24
PROP: Crystals from ethanol, ether. D: 1.058 @ 270°, mp: 57°, bp: 259.9°. Sol in water, alc, ether.
SYNS: MORPHOLINOBENZENE □ N-PHENYLMORPHOLINE
CONSENSUS REPORTS: Reported in EPA TSCA Inventory.
SAFETY PROFILE: Poison by skin contact. Moderately toxic by ingestion. An eye irritant. When heated to decomposition it emits toxic fumes of NO$_x$.

PFT500   CAS: 135-88-6   HR: 3
N-PHENYL-β-NAPHTHYLAMINE
mf: C$_{16}$H$_{13}$N   mw: 219.30
PROP: Rhombic crystals and needles from MeOH. Mp: 107–108°, bp: 395.5°. Insol in water; sol in hot benzene; very sol in hot alc, ether.
SYNS: ACETO PBN □ AGERITE POWDER □ ANILINONAPHTHALENE □ 2-ANILINONAPHTHAL-ENE □ ANTIOXIDANT 116 □ ANTIOXIDANT PBN □ N-(2-NAPHTHYL)ANILINE □ β-NAPHTHYLPHENYL-AMINE □ 2-NAPHTHYLPHENYLAMINE □ NCI-C02915 □ NEOZONE D □ NILOX PBNA □ NONOX D □ PBNA □ PHENYL-β-NAPHTHYLAMINE □ PHENYL-2-NAPHTHYLAMINE □ N-PHENYL-2-NAPHTHYLAMINE □ STABILIZATOR AR
CONSENSUS REPORTS: IARC Cancer Review: Group 3 IMEMDT 7,318,87; Human Inadequate Evidence IMEMDT 16,325,78; Animal Limited Evidence IMEMDT 16,325,78. Reported in EPA TSCA Inventory.
ACGIH TLV: Not Classifiable as a Human Carcinogen
DFG MAK: Confirmed Animal Carcinogen with Unknown Relevance to Humans
SAFETY PROFILE: Suspected carcinogen with experimental carcinogenic, neoplastigenic, and tumorigenic data. Moderately toxic by ingestion. Human

mutation data reported. When heated to decomposition it emits toxic fumes of NO$_x$.

PFU300   CAS: 120-46-7   HR: 2
PHENYL PHENACYL KETONE
mf: C$_{15}$H$_{12}$O$_2$ mw: 224.27
SYNS: OMEGA-BENZOYLACETOPHENONE □ 2-BENZOYLACETOPHENONE □ DIBENZOYLMETHANE □ 1,3-DIPHENYL-1,3-PROPANEDIONE □ KARENZU DK2 □ 1,3-PROPANEDIONE, 1,3-DIPHENYL- □ RHODIASTAB 83
CONSENSUS REPORTS: Reported in EPA TSCA Inventory.
SAFETY PROFILE: Moderately toxic by ingestion. When heated to decomposition it emits acrid smoke and irritating vapors.

PFV250   CAS: 638-21-1   HR: 3
PHENYLPHOSPHINE
mf: C$_6$H$_7$P   mw: 110.10
PROP: Foul smelling liquid. Bp: 160–161°, d: 1.001 @ 15 mm. Insol in water; sol in alkali; very sol in alc and ether.
SYN: FENYLFOSFIN
OSHA PEL: CL 0.05 ppm
ACGIH TLV: CL 0.05 ppm
NIOSH REL: (Phenylphosphine) CL 0.05 ppm
SAFETY PROFILE: Poison by inhalation. Ignites spontaneously in air. When heated to decomposition it emits toxic fumes of PO$_x$. See also PHOSPHINE.

PFW200   CAS: 3497-00-5   HR: 2
PHENYL PHOSPHORUS THIODICHLORIDE
DOT: UN 2799
mf: C$_6$H$_5$Cl$_2$PS   mw: 211.04
SYNS: DICHLOROPHENYLPHOSPHINE SULFIDE □ PHENYLTHIONOPHOSPHONIC DICHLORIDE □ PHOSPHONOTHIOIC DICHLORIDE, PHENYL-
CONSENSUS REPORTS: Reported in EPA TSCA Inventory.
DOT CLASSIFICATION: 8; Label: Corrosive
SAFETY PROFILE: A corrosive. When heated to decomposition it emits acrid smoke and irritating vapors.

PFW210   CAS: 3497-00-5   HR: 2
PHENYL PHOSPHORUS THIODICHLOR-
IDE
DOT: UN 2799
mf: $C_6H_5Cl_2PS$ mw: 211.04
SYNS: DICHLOROPHENYLPHOSPHINE SULFIDE □
PHENYLTHIONOPHOSPHONIC DICHLORIDE □
PHOSPHONOTHIOIC DICHLORIDE, PHENYL-
CONSENSUS REPORTS: Reported in EPA
TSCA Inventory.
DOT CLASSIFICATION: 8; Label: Corrosive
SAFETY PROFILE: A corrosive. When
heated to decomposition it emits toxic
vapors of SO_x and PO_x.

PFX000   CAS: 92-54-6   HR: 3
1-PHENYLPIPERAZINE
mf: $C_{10}H_{14}N_2$ mw: 162.26
PROP: Pale-yellow oil. D: 1.0621 @ 20°/4°,
mp: 18.8°, bp: 162–164° @ 22 mm, flash p:
285°F. Insol in water; sol in alc, ether.
SYNS: 1-FENYLPIPERAZIN □ N-
PHENYLPIPERAZINE
CONSENSUS REPORTS: Reported in EPA
TSCA Inventory.
SAFETY PROFILE: Poison by ingestion and
skin contact. A skin and severe eye irritant.
Mutation data reported. Combustible when
exposed to heat or flame. It supports
combustion and decomposes to yield toxic
fumes of NO_x. To fight fire, use water,
foam, dry chemical. See also PIPERAZINE.

PFX600   CAS: 17766-63-1   HR: 3
4-PHENYLPIPERAZINYL 3,4,5-TRI-
METHOXYPHENYL KETONE
mf: $C_{20}H_{24}N_2O_4$ mw: 356.46
SYNS: KETONE, 4-PHENYLPIPERAZINYL 3,4,5-
TRIMETHOXYPHENYL □ 1-PHENYL-4-(3,4,5-
TRIMETHOXYBENZOYL)PIPERAZINE □ PIPERAZINE,
1-PHENYL-4-(3,4,5-TRIMETHOXYBENZOYL)-
DOT CLASSIFICATION: 3; Label:
Flammable Liquid
SAFETY PROFILE: A poison by
intraperitoneal route. A flammable liquid.
When heated to decomposition it emits
toxic vapors of NO_x.

PFY200   CAS: 13441-36-6   HR: 3
PHENYL (1-PIPERIDINOCYCLOHEXYL)

KETONE
mf: $C_{18}H_{25}NO$ mw: 271.44
SYNS: KETONE, PHENYL (1-
PIPERIDINOCYCLOHEXYL) □ α-(1-PIPERIDINO)-
CYCLOHEXYL PHENYL KETONE
DOT CLASSIFICATION: 3; Label:
Flammable Liquid
SAFETY PROFILE: A poison by
intravenous route. A flammable liquid.
When heated to decomposition it emits
acrid smoke and irritating vapors.

PGA750   CAS: 673-31-4   HR: 3
3-PHENYL-1-PROPANOL CARBAMATE
mf: $C_{10}H_{13}NO_2$ mw: 179.24
PROP: Leaflets from EtOH (aq). Mp:
101–104°.
SYNS: ACTOZINE □ ANSEPRON □ BENZENEPRO-
PANOL CARBAMATE □ CARBAMIC ACID-3-
PHENYLPROPYL ESTER □ 1-CARBAMOYLOXY-3-
PHENYLPROPANE □ EIRENAL □ EXTACOL □
FENPROBAMATO □ GAMAQUIL □ Hg 532 □ MH-532 □
PALMITA □ γ-PHENYLPROPYLCARBAMAT (GERMAN)
□ γ-PHENYLPROPYL CARBAMATE □ QUAMAQUIL □
SPANTOL □ TRANQUIL
SAFETY PROFILE: Poison by intravenous
and intraperitoneal routes. Moderately toxic
by ingestion. Used as a tranquilizer and
muscle relaxant. When heated to
decomposition it emits toxic fumes of NO_x.
See also CARBAMATES and ESTERS.

PGA800   HR: D
2-PHENYLPROPIONALDEHYDE DIM-
ETHYL ACETAL
mf: $C_{11}H_{16}O_2$ mw: 180.25
PROP: Colorless to sltly yellow liquid;
mushroom odor. D: 0.989–0.994, refr index:
1.492–1.497. Sol in alc, ether; insol in water.
SYNS: FEMA No. 2888 □ HYDRATROPIC ALDEHYDE
DIMETHYL ACETAL
SAFETY PROFILE: When heated to
decomposition it emits acrid smoke and
irritating fumes.

PGB850   CAS: 315706-78-6 HR: 3
3-PHENYL-N-(4-PROPYLCYCLO-
HEXYL)-2-PROPENAMIDE
mf: $C_{18}H_{25}NO$ mw: 271.40

SAFETY PROFILE: A poison by ingestion. When heated to decomposition it emits toxic vapors of NO_x.

PGE000   CAS: 3567-38-2   HR: 3
1-PHENYL-2-PROPYNYL CARBAMATE
mf: $C_{10}H_9NO_2$ mw: 175.20

PROP: A solid. Mp: 86–88°.

SYNS: BENZENEMETHANOL, α-ETHYNYL-, CARBAMATE (9CI) □ CARBAMIC ACID, 1-PHENYL-2-PROPYNYL ESTER □ CARBAMMATO di FENILETINIL-CARBINOLO □ CARFIMAT □ CARFIMATE □ CFC □ EQUILIUM □ α-ETHYNYLBENZENEMETHANOL CARBAMATE □ α-ETHYNYLBENZYL CARBAMATE □ NIRVOTIN □ PHENYLETHYNYLCARBINOL CARBAMATE

SAFETY PROFILE: Poison by ingestion and intraperitoneal routes. When heated to decomposition it emits toxic fumes of NO_x. See also CARBAMATES.

PGE550   CAS: 296269-52-8 HR: 3
α-(1-PHENYL-1H-PYRAZOL-4-YL)-1-PIPERIDINEBUTANOL
mf: $C_{18}H_{25}N_3O$ mw: 299.42

SAFETY PROFILE: A poison by ingestion. When heated to decomposition it emits toxic vapors of NO_x.

PGG750   CAS: 118-55-8   HR: 2
PHENYL SALICYLATE
mf: $C_{13}H_{10}O_3$ mw: 214.23

PROP: White, small crystals or plates from MeOH; pleasant odor and taste. D: 1.250 @ 20°/4°, mp: 43°, bp: 172–173° @ 12 mm. Sol in water, ether, benzene; very sol in hot alc.

SYNS: FENYLESTER KYSELINY SALICYLOVE □ SALOL

CONSENSUS REPORTS: Reported in EPA TSCA Inventory.

SAFETY PROFILE: Moderately toxic by ingestion. Experimental teratogenic and reproductive effects. When heated to decomposition it emits acrid smoke and irritating fumes. See also ESTERS.

PGH000   CAS: 304-06-3   HR: 3
3-PHENYLSALICYLIC ACID
mf: $C_{13}H_{10}O_3$ mw: 214.23

PROP: Rhombic crystals from alc or C_6H_6. D: 1.250 @ 20°/4°, mp: 41.4°, bp: 172–173° @ 12 mm. Sol in water, ether, benzene; very sol in hot alc.

SYN: USAF DO-59

CONSENSUS REPORTS: Reported in EPA TSCA Inventory.

SAFETY PROFILE: Poison by intraperitoneal route. When heated to decomposition it emits acrid smoke and irritating fumes.

PGH250   CAS: 1132-39-4   HR: 3
PHENYL SELENIDE
mf: $C_{12}H_{10}Se$ mw: 233.18

PROP: Liquid or needles on cooling. D: 1.323 @ 25°, mp: 2.5°, bp: 301–302°.

SYNS: BIPHENYL SELENIUM □ DIPHENYL SELENIDE

CONSENSUS REPORTS: Selenium and its compounds are on the Community Right-To-Know List. Reported in EPA TSCA Inventory.

OSHA PEL: TWA 0.2 mg(Se)/m³
ACGIH TLV: TWA 0.2 mg(Se)/m³
DFG MAK: 0.1 mg(Se)/m³

SAFETY PROFILE: Poison by ingestion. When heated to decomposition it emits toxic fumes of Se. See also SELENIUM COMPOUNDS.

PGM750   CAS: 645-48-7   HR: 3
1-PHENYLTHIOSEMICARBAZIDE
mf: $C_7H_9N_3S$ mw: 167.25

PROP: Prisms or crystals from alc. Mp: 200–201° (decomp). Sltly sol in water, ether; sol in hot alc.

SYNS: HYDRAZINECARBOTHIOAMIDE, 2-PHENYL-(9CI) □ 2-PHENYLHYDRAZINECARBOTHIOAMIDE

CONSENSUS REPORTS: Reported in EPA TSCA Inventory.

SAFETY PROFILE: Poison by ingestion route. Human mutation data reported. When heated to decomposition it emits very toxic fumes of NO_x and SO_x.

PGN100   CAS: 13738-70-0   HR: 3
α-(PHENYLTHIO)-p-TOLUIDINE
DOT: UN 1600

mf: $C_{13}H_{13}NS$ mw: 215.33

SYNS: DINITROTOLUENES, molten (DOT) □ p-TOLUIDINE, α-(PHENYLTHIO)-

DOT CLASSIFICATION: 6.1; Label: Poison

SAFETY PROFILE: A poison by ingestion and intravenous routes. When heated to decomposition it emits toxic vapors of NO_x and SO_x.

PGN250   CAS: 103-85-5   HR: 3
1-PHENYL-2-THIOUREA

mf: $C_7H_8N_2S$ mw: 152.23

PROP: Needle-like crystals from water; bitter taste. Mp: 154°, d: 1.3. Sol in water, alc, aq ether.

SYNS: NCI-C02017 □ PHENYLTHIOCARBAMIDE □ N-PHENYLTHIOUREA □ α-PHENYLTHIOUREA □ 1-PHENYLTHIOUREA □ PTC □ PTU □ RCRA WASTE NUMBER P093 □ U 6324 □ USAF EK-1569

CONSENSUS REPORTS: NCI Carcinogenesis Bioassay (feed); No Evidence: mouse, rat NCITR* NCI-CG-TR-148,78. EPA Extremely Hazardous Substances List. Reported in EPA TSCA Inventory.

SAFETY PROFILE: Poison by ingestion and intraperitoneal routes. Experimental teratogenic effects. When heated to decomposition or on contact with acid or acid fumes it emits highly toxic fumes of SO_x and NO_x. Used in medical genetics and production of rodenticide.

PGR000   CAS: 108-73-6   HR: 2
PHLOROGLUCINOL

mf: $C_6H_6O_3$ mw: 126.12

PROP: White crystals, leaflets, or plates from water; sweet taste. Mp: 218°, (anhyd) (rapid heat) subl with decomp. Sol in ether; sltly water sol.

SYNS: BENZENE-s-TRIOL □ BENZENE-1,3,5-TRIOL □ 1,3,5-BENZENETRIOL □ 3,5-DIHYDROXYPHENOL □ DILOSPAN S □ 5-OXYRESORCINOL □ PHLORO-GLUCIN □ s-TRIHYDROXYBENZENE □ sym-TRI-HYDROXYBENZENE □ 1,3,5-TRIHYDROXYBENZENE □ 1,3,5-TRIHYDROXYCYCLOHEXATRIENE

CONSENSUS REPORTS: Reported in EPA TSCA Inventory. EPA Genetic Toxicology Program.

SAFETY PROFILE: Moderately toxic by subcutaneous and intraperitoneal routes. Mildly toxic by ingestion. Experimental reproductive effects. Mutation data reported. When heated to decomposition it emits acrid smoke and irritating fumes. Used in diazo-type printing and textile dyeing, in microscopy as a bone specimen decalcifier.

PGR250   CAS: 90-00-6   HR: 3
PHLOROL

mf: $C_8H_{10}O$ mw: 122.18

PROP: Colorless liquid; phenol odor. Mp: −28°, d: 1.037 @ 12°, bp: 204.52°, turns solid <18°. Insol in water; misc in alc, benzene, glacial acetic acid, ether.

SYNS: o-ETHYLPHENOL □ 2-ETHYLPHENOL □ FLOROL □ PHENOL, o-ETHYL-

DOT CLASSIFICATION: 6.1; Label: KEEP AWAY FROM FOOD

CONSENSUS REPORTS: Reported in EPA TSCA Inventory.

SAFETY PROFILE: Poison by intraperitoneal route. Moderately toxic by ingestion. Human toxic action similar to, but less severe than, that of phenol. Questionable carcinogen with experimental neoplastigenic data. When heated to decomposition it emits acrid smoke and irritating fumes. See also PHENOL.

PGS000   CAS: 298-02-2   HR: 3
PHORATE

mf: $C_7H_{17}O_2PS_3$ mw: 260.39

PROP: Mobile liquid. Bp: 118–120° @ 0.8 mm, d: 1.156, @ 25°/4°. Insol in water; misc with carbon tetrachloride, dioxane, xylene.

SYNS: O,O-DIAETHYL-S-(AETHYLTHIO-METHYL)-DITHIOPHOSPHAT (GERMAN) □ O,O-DIETHYL-S-ETHYLMERCAPTOMETHYL DITHIOPHOSPHONATE □ O,O-DIETHYL-S-ETHYLTHIOMETHYL DITHIO-PHOSPHONATE □ O,O-DIETHYL-ETHYLTHIOMETH-YL PHOSPHORODITHIOATE □ O,O-DIETHYL-S-ETHYLTHIOMETHYL THIOTHIONOPHOSPHATE □ O,O-DIETIL-S-(ETILTIO-METIL)-DITIOFOSFATO (ITALIAN) □ DITHIOPHOSPHATE de O,O-DIETHYLE et d'ETHYLTHIOMETHYLE (FRENCH) □ ENT 24,042 □ FORAAT (DUTCH) □ GRANUTOX □ PHORAT (GERMAN) □ PHORATE-10G □ RAMPART □ RCRA

WASTE NUMBER P094 □ THIMET □ TIMET □ VEGFRU □ VERGFRU FORATOX

CONSENSUS REPORTS: EPA Extremely Hazardous Substances List. EPA Genetic Toxicology Program.

OSHA PEL: TWA 0.05 mg/m³ (skin)

ACGIH TLV: TWA 0.05 mg/m³ (skin)

SAFETY PROFILE: Poison by ingestion and skin contact routes. Experimental reproductive effects. Mutation data reported. A cholinesterase inhibitor. When heated to decomposition it emits toxic fumes of PO_x and SO_x. See also PARATHION.

PGW250   CAS: 504-20-1   HR: 2
PHORONE
mf: $C_9H_{14}O$ mw: 138.23
PROP: Yellowish-green prisms, solid, or greenish liquid. Mp: 28°, flash p: 185°F (OC), d: 0.879, vap press: 1 mm @ 42.0°, vap d: 4.8, bp: 198–199°. Sol in water, alc, and ether.
SYNS: DIISOPROPYLIDENE ACETONE □ sym-DIISOPROPYLIDENE ACETONE □ 2,6-DIMETHYL-2,5-HEPTADIEN-4-ONE □ PHORON (GERMAN)
CONSENSUS REPORTS: Reported in EPA TSCA Inventory.
SAFETY PROFILE: Moderately toxic by subcutaneous route. Combustible when exposed to heat or flame; can react with oxidizing materials. To fight fire, use foam, CO_2, dry chemical. When heated to decomposition it emits acrid smoke and irritating fumes. See also ISOPHORON.

PGW750   CAS: 947-02-4   HR: 3
PHOSFOLAN
mf: $C_7H_{14}NO_3PS_2$ mw: 255.31
PROP: Colorless to yellow solid. Mp: 39.5°, bp: 100–105° @ 0.001 mm. Sol in H_2O, Me_2CO, C_6H_6, EtOH.
SYNS: AC 47031 □ AMERICAN CYANAMID 47031 □ C.I. 47031 □ CYCLIC ETHYLENE(DIETHOXYPHOS-PHINOTHIOYL)DITHIOIMIDOCARBONATE □ CYCLIC ETHYLENE P,P-DIETHYL PHOSPHONODITHIOIMIDO-CARBONATE □ CYLAN □ CYOLANE □ CYOLANE INSECTICIDE □ (DIETHOXYPHOSPHINYL)DI-THIOIMIDOCARBONIC ACID CYCLIC ETHYLENE ESTER □ 2-(DIETHOXYPHOSPHINYLIMINO)-1,3-

DITHIOLANE □ P,P-DIETHYL CYCLIC ETHYLENE ESTER OF PHOSPHONODITHIOIMIDOCARBONIC ACID □ EI 47031 □ ENT 25,830 □ 1,2-ETHANEDITHIOL, CYCLIC ESTER with P,P-DIETHYL PHOSPHONODITHI-OIMIDOCARBONATE □ 1,2-ETHANEDITHIOL, CYCLIC S,S-ESTER with PHOSPHONODITHIOIMIDOCARBONIC ACID P,P-DIETHYL ESTER
CONSENSUS REPORTS: EPA Extremely Hazardous Substances List. Reported in EPA TSCA Inventory.
SAFETY PROFILE: Poison by ingestion and skin contact. An insecticide used against leaf-feeding larvae of cotton insect pests. When heated to decomposition it emits very toxic fumes of PO_x, SO_x, and NO_x. See also ESTERS.

PGX000   CAS: 75-44-5   HR: 3
PHOSGENE
DOT: UN 1076
mf: CCl_2O mw: 98.91
PROP: Colorless, poison gas or volatile liquid; odor of new-mown hay or green corn. Mp: −128°, bp: 7.6°, d: 1.419 @ 0°/4°, vap press: 1180 mm @ 20°, vap d: 3.4. Very sltly sol in water; very sol in benzene and acetic acid; decomp sltly in water. IDLH 2 ppm.
SYNS: CARBONE (OXYCHLORURE de) (FRENCH) □ CARBONIO (OSSICLORURO di) (ITALIAN) □ CARBON OXYCHLORIDE □ CARBONYLCHLORID (GERMAN) □ CARBONYL CHLORIDE □ CHLOROFORMYL CHLORIDE □ DIPHOSGENE □ FOSGEEN (DUTCH) □ FOSGEN (POLISH) □ FOSGENE (ITALIAN) □ KOOLSTOFOXYCHLORIDE (DUTCH) □ NCI-C60219 □ PHOSGEN (GERMAN) □ RCRA WASTE NUMBER P095
CONSENSUS REPORTS: EPA Extremely Hazardous Substances List. Community Right-To-Know List. Reported in EPA TSCA Inventory.
OSHA PEL: TWA 0.1 ppm
ACGIH TLV: TWA 0.1 ppm
DFG MAK: 0.02 ppm (0.082 mg/m³)
NIOSH REL: (Phosgene) TWA 0.1 ppm; CL 0.2 ppm/15M
DOT CLASSIFICATION: 2.3; Label: Poison Gas, Corrosive
SAFETY PROFILE: A human poison by inhalation. A severe eye, skin, and mucous membrane irritant. In the presence of moisture, phosgene decomposes to form

P

hydrochloric acid and carbon monoxide. This occurs in the bronchioles and alveoli of the lungs, resulting in pulmonary edema followed by bronchopneumonia and occasionally lung abscess. There is little immediate irritating effect upon the respiratory tract, and the warning properties of the gas are therefore very slight. There may be no immediate warning that dangerous concentrations are being inhaled. After a latent period of 2 to 24 hours, the patient complains of burning in the throat and chest, shortness of breath, and increasing dyspnea. Where the exposure has been severe, the development of pulmonary edema may be so rapid that the patient dies within 36 hours after exposure. In cases where the exposure has been less, pneumonia may develop several days after the occurrence of the accident. In patients who recover, no permanent residual disability is thought to occur. A common air contaminant.

Under the appropriate conditions it undergoes hazardous reactions with Al, tert-butyl azido formate, 2,4-hexadiyn-1,6-diol, isopropyl alcohol, K, Na, sodium azide, hexafluoroisopropylideneamino lithium, lithium. When heated to decomposition or on contact with water or steam it will react to produce toxic and corrosive fumes of CO and Cl⁻. *Caution:* Arrangements should be made for monitoring its use.

PGX275   CAS: 23783-98-4   HR: 3
cis-PHOSPHAMIDON
mf: $C_{10}H_{19}ClNO_5P$ mw: 299.72
SYNS: (Z)-2-CHLORO-N,N-DIETHYL-3-HYDROXY-CROTONAMIDE DIMETHYL PHOSPHORATE □ (Z)-PHOSPHAMIDON □ PHOSPHORIC ACID, 2-CHLORO-3-(DIETHYLAMINO)-1-METHYL-3-OXO-1-PROPENYL-,DIMETHYL ESTER, (Z)- □ PHOSPHORIC ACID, DIMETHYL ESTER, ESTER WITH 2-CHLORO-N,N-DIETHYL-3-HYDROXYCROTANAMIDE, (Z)-
CONSENSUS REPORTS: EPA FIFRA 1988 pesticide subject to registration or re-registration.
SAFETY PROFILE: A poison by ingestion and skin contact. When heated to

decomposition it emits toxic vapors of NO_x, SO_x, and Cl⁻.

PGX500   HR: 2
PHOSPHATES
SAFETY PROFILE: Alkali metal phosphates are strong caustics and therefore powerful irritants. Superphosphate is $Ca(H_2PO_4)_2/CaSO_4$. Triple superphosphate contains P_2O_5. Both are used as fertilizers. Organophosphates are often highly toxic pesticides. For an example of organic phosphates, see PARATHION. See also individual phosphates.

PGX750   HR: 3
PHOSPHIDES
PROP: A combination of a cation + elemental phosphorus.
SAFETY PROFILE: Phosphides are particularly dangerous because they tend to decompose to the very toxic phosphine upon contact with moisture or acids. Dangerous fire hazard by chemical reaction, particularly with moisture. Moderate explosion hazard. They react with water, steam, acid, or acid fumes to produce toxic and flammable phosphine gas. Can react vigorously with oxidizing materials. Dangerous; when heated to decomposition they may emit highly toxic fumes of PO_x. See also PHOSPHINE.

PGY000   CAS: 7803-51-2   HR: 3
PHOSPHINE
DOT: UN 2199
mf: H_3P mw: 34.00
PROP: Colorless gas; foul odor of decaying fish. Extremely weak base. Mp: −132.5°, bp: −87.5°, d: 1.529 g/L @ 0°, autoign temp: 212°F, lel: 1%. Sltly sol in water (giving neutral soln). IDLH 50 ppm.
SYNS: CELPHOS □ DELICIA □ DETIA GAS EX-B □ FOSFOROWODOR (POLISH) □ HYDROGEN PHOSPHIDE □ PHOSPHORUS TRIHYDRIDE □ PHOSPHORWASSERSTOFF (GERMAN) □ RCRA WASTE NUMBER P096

CONSENSUS REPORTS: EPA Extremely Hazardous Substances List. Reported in EPA TSCA Inventory.

OSHA PEL: TWA 0.3 ppm; STEL 1 ppm
ACGIH TLV: TWA 0.3 ppm; STEL 1 ppm
DFG MAK: 0.1 ppm (0.14 mg/m³)
DOT CLASSIFICATION: 2.3; Label: Poison Gas, Flammable Gas

SAFETY PROFILE: A poison by inhalation. A very toxic gas whose effects are not completely understood. The chief effects are central nervous system depression and lung irritation. There may be pulmonary edema, dilation of the heart, and hyperemia of the visceral organs. Inhalation can cause coma and convulsions leading to death within 48 hours. However, most cases recover without after-effects. Chronic poisoning, characterized by anemia, bronchitis, gastrointestinal disturbances, and visual, speech, and motor disturbances, may result from continued exposure to very low concentrations.

Very dangerous fire hazard by spontaneous chemical reaction. Moderately explosive when exposed to flame. Explosive reaction with dichlorine oxide, silver nitrate, concentrated nitric acid, nitrogen trichloride, oxygen. Reacts with mercury(II) nitrate to form an explosive product. Ignition or violent reaction with air, boron trichloride, Br_2, Cl_2, aqueous halogen solutions, iodine, metal nitrates, NO, NCl_3, NO_3, N_2O, HNO_2, K + NH_3, oxidants. The organic derivatives of phosphine (phosphines) react vigorously with halogens. To fight fire, use CO_2, dry chemical, or water spray. Dangerous; when heated to decomposition it emits highly toxic fumes of PO_x. Used as a fumigant, doping agent for electronic components, and in chemical synthesis.

PGZ899   CAS: 13598-36-2   HR: 2
PHOSPHONIC ACID
mf: H_3O_3P mw: 82.00
PROP: White very hygroscopic crystals; mass from EtOH (aq). Mp: 73.8°. Very sol in H_2O.

SYNS: ORTHOPHOSPHORUS ACID □ PHOSPHORUS TRIHYDROXIDE □ TRIHYDROXYPHOSPHINE

CONSENSUS REPORTS: Reported in EPA TSCA Inventory.

SAFETY PROFILE: Moderately toxic by ingestion. When heated to decomposition at 200°C it emits toxic fumes of PO_x and phosphine which may ignite. See also PHOSPHINE.

PHA500   CAS: 1071-83-6   HR: 3
N-(PHOSPHONOMETHYL)GLYCINE
mf: $C_3H_8NO_5P$ mw: 169.09
PROP: A solid. Sltly sol in H_2O; insol in most org solvs.

SYNS: GLYPHOSATE □ MON 0573 □ MON 2139

SAFETY PROFILE: Poison by intraperitoneal route. Moderately toxic by ingestion. Human systemic effects: arrhythmias, blood pressure lowering, body temperature increase, change in heart rate, convulsions, diarrhea, fibrosing alveolitis, fibrosis, hypermotility, respiratory depression, respiratory stimulation. Used as an herbicide. When heated to decomposition it emits very toxic fumes of NO_x and PO_x.

PHB250   CAS: 7664-38-2   HR: 3
PHOSPHORIC ACID
DOT: UN 1805
mf: H_3O_4P mw: 98.00
PROP: Colorless syrupy liquid or rhombic crystals. Mp: 42.35°, loses 1/2H_2O @ 213°, fp: 42.4°, d: 1.864 @ 25°, vap press: 0.0285 mm @ 20°. Misc with water and many org solvs. IDLH 1000 mg/m³.

SYNS: ACIDE PHOSPHORIQUE (FRENCH) □ ACIDO FOSFORICO (ITALIAN) □ FOSFORZUUROPLOSSINGEN (DUTCH) □ ORTHOPHOSPHORIC ACID □ PHOSPHOR-SAEURELOESUNGEN (GERMAN)

CONSENSUS REPORTS: Community Right-To-Know List. Reported in EPA TSCA Inventory. EPA Genetic Toxicology Program.

OSHA PEL: TWA 1 mg/m³; STEL 3 mg/m³
ACGIH TLV: TWA 1 mg/m³; STEL 3 mg/m³
DOT CLASSIFICATION: 8; Label: Corrosive

SAFETY PROFILE: Human poison by ingestion. Moderately toxic by skin contact. A corrosive irritant to eyes, skin, and mucous membranes, and a systemic irritant by inhalation. A common air contaminant. A strong acid. Mixtures with nitromethane are explosive. Reacts with chlorides + stainless steel to form explosive hydrogen gas. Potentially violent reaction with sodium tetrahydroborate. Dangerous; when heated to decomposition it emits toxic fumes of PO_x.

PHB500   CAS: 7784-30-7   HR: 2
PHOSPHORIC ACID, ALUMINUM SALT (1:1), (solution)
mf: $O_4P \cdot Al$ mw: 121.95
SYNS: ALUMINOPHOSPHORIC ACID □ ALUMINUM ACID PHOSPHATE □ ALUMINUM MONOPHOSPHATE □ ALUMINUM PHOSPHATE □ ALUMINUM PHOSPHATE (1:1) □ ALUMINUM PHOSPHATE, solution (DOT) □ ALUPHOS □ FFB 32 □ MONOALUMINUM PHOSPHATE □ PHOSPHALUGEL
CONSENSUS REPORTS: Reported in EPA TSCA Inventory.
ACGIH TLV: TWA 2 mg(Al)/m³
SAFETY PROFILE: Corrosive to the eyes, skin, and mucous membranes. When heated to decomposition it emits toxic fumes of PO_x. Used as an antacid and as a cement component, flux for ceramics, dental cement, glass, and gels. See also ALUMINUM COMPOUNDS and PHOSPHATES.

PHD250   CAS: 3254-63-5   HR: 3
PHOSPHORIC ACID DIMETHYL-p-(METHYLTHIO)PHENYL ESTER
mf: $C_9H_{13}O_4PS$ mw: 248.25
PROP: A liquid. D: 1.273 @ 21°/4°. Spar sol in H_2O; sol in Me_2CO, dioxan, CCl_4, EtOH, xylene.
SYNS: O,O-DIMETHYL O-(4-METHYLMERCAPTO-PHENYL)PHOSPHATE □ DIMETHYL-p-(METHYLTHIO)-PHENYL PHOSPHATE □ ENT 25,734 □ 4-METHYLTHIO-PHENYLDIMETHYL PHOSPHATE
CONSENSUS REPORTS: EPA Extremely Hazardous Substances List.
SAFETY PROFILE: Poison by ingestion, skin contact, and subcutaneous routes.

When heated to decomposition it emits very toxic fumes of SO_x and PO_x. See also ESTERS.

PHE500   CAS: 1623-24-1   HR: 2
PHOSPHORIC ACID, ISOPROPYL ESTER
DOT: UN 1793
mf: $C_3H_9O_4P$ mw: 140.09
SYNS: ISOPROPYL ACID PHOSPHATE solid □ ISOPROPYL PHOSPHORIC ACID
CONSENSUS REPORTS: Reported in EPA TSCA Inventory.
DOT CLASSIFICATION: 8; Label: Corrosive
SAFETY PROFILE: A highly corrosive material. Very irritating to the skin, eyes, and mucous membranes. When heated to decomposition it emits toxic fumes of PO_x. See also ESTERS.

PHF250   CAS: 13779-41-4   HR: 3
PHOSPHORODIFLUORIDIC ACID
DOT: UN 1768
mf: F_2HO_2P mw: 101.98
PROP: Clear mobile liquid which decomp when heated. Fumes in moist air. Mp: −93°, bp: 116° (decomp), d: 1.583 @ 25°/4°, vap d: 3.52.
CONSENSUS REPORTS: Reported in EPA TSCA Inventory.
OSHA PEL: TWA 2.5 mg(F)/m³
ACGIH TLV: TWA 2.5 mg(F)/m³; BEI: 3 mg/g creatinine of fluorides in urine prior to shift; 10 mg/g creatinine of fluorides in urine at end of shift.
NIOSH REL: TWA 2.5 mg(F)/m³
DOT CLASSIFICATION: 8; Label: Corrosive
SAFETY PROFILE: Toxic and corrosive material. Very irritating to the skin, eyes, and mucous membranes. When heated to decomposition it emits very toxic fumes of PO_x and F^-. Used as a catalyst. See also PHOSPHORODIFLUORIDIC ACID, anhydrous.

PHF750　CAS: 371-86-8　HR: 3
PHOSPHORODI(ISOPROPYLAMIDIC) FLUORIDE

mf: $C_6H_{15}FN_2OP$　mw: 182.21

PROP: Crystals from pet ether. Mp: 65°, bp: 125°. Spar sol in H_2O.

SYNS: BIS(ISOPROPYLAMIDO) FLUOROPHOSPHATE □ BIS(MONOISOPROPYLAMINO)FLUOROPHOSPHATE □ BIS(MONOISOPROPYLAMINO)FLUOROPHOSPHINE OXIDE □ N,N'-DIISOPROPIL-FOSFORODIAMMIDO-FLUORURO (ITALIAN) □ DI(ISOPROPYLAMIDO)PHOS-PHORYLFLUORIDE □ N,N'-DIISOPROPYL-DIAMIDO-FOSFORZUUR-FLUORIDE (DUTCH) □ N,N'-DIISOPROP-YL-DIAMIDO-PHOSPHORSAEURE-FLUORID (GERMAN) □ N,N'-DIISOPROPYLDIAMIDOPHOSPHORYL FLUORIDE □ N,N'-DIISOPROPYLPHOSPHORO-DIAMIDIC FLUORIDE □ FLUOROBISISOPROPYL-AMINO-PHOSPHINE OXIDE □ FLUORURE de N,N'-DIISOPROPYLE PHOSPHORODIAMIDE (FRENCH) □ ISOPESTOX □ MIPAFOX □ PESTON XV □ PESTOX 15 □ PESTOX XV

SAFETY PROFILE: Poison by ingestion, subcutaneous, and intraperitoneal routes. When heated to decomposition it emits very toxic fumes of F^-, NO_x, and PO_x. See also PARATHION and FLUORIDES.

PHI500　CAS: 640-15-3　HR: 3
PHOSPHORODITHIOIC ACID, O,O-DI-METHYL-S-(2-ETHYLTHIO)ETHYL ESTER

mf: $C_6H_{15}O_2PS_3$　mw: 246.36

PROP: Colorless liquid or oil. D: 1.209 @ 20°/4°, bp: 97° @ 0.2 mm. Very spar sol in water and most org solvs; sol in acetone, dioxane, acetonitrile.

SYNS: BAY 23129 □ COMPOUND M-81 □ O,O-DIMETHYL-S-(2-AETHYLTHIO-AETHYL)-DITHIO PHOSPHAT (GERMAN) □ O,O-DIMETHYL-S-(CARBONYLMETHYLMORPHOLINO) PHOSPHORO-DITHIOATE □ O,O-DIMETHYL-S-(2-ETHYLMERCAPTO-ETHYL) DITHIOPHOSPHATE □ O,O-DIMETHYL-S-2-ETHYLMERKAPTOETHYLESTER KYSELINY DITHIOFOSFORECNE (CZECH) □ O,O-DIMETHYL-S-(2-ETHYLTHIO-ETHYL)-DITHIOFOSFAAT (DUTCH) □ O,O-DIMETHYL S-(2-(ETHYLTHIO)ETHYL) PHOS-PHORODITHIOATE □ O,O-DIMETIL-S-(ETILTIO-ETIL)-DITIOFOSFATO (ITALIAN) □ DITHIOMETON (FRENCH) □ DITHIOPHOSPHATE de O,O-DIMETHYLE et de S-2-ETHYLTHIO-ETHYLE) (FRENCH) □ EKATIN □ EKATIN AEROSOL □ EKATINE-25 □ EKATIN ULV □ 2-ETHYLTHIOETHYL O,O-DIMETHYL PHOSPHORODI-THIOATE □ S-(2-(ETHYLTHIO)ETHYL) O,O-DIMETH-YLPHOSPHORODITHIONATE □ S-(2-(ETHYLTHIO)-

ETHYL)DIMETHYL PHOSPHOROTHIOLOTHIONATE □ INTRATHION □ INTRATION □ LUXISTELM □ M 81 □ SAN 230 □ THIAMETON □ THIOMETON

SAFETY PROFILE: Poison by ingestion, skin contact, inhalation, and intravenous routes. Mutation data reported. A skin and severe eye irritant. A cholinesterase inhibitor. When heated to decomposition it emits very toxic fumes of PO_x and SO_x. See also ESTERS and PARATHION.

PHJ250　CAS: 13537-32-1　HR: 2
PHOSPHOROFLUORIDIC ACID

DOT: UN 1776

mf: FH_2O_3P　mw: 99.99

PROP: Colorless oily acid of moderate strength. Slowly hydrol in neutral or weakly alkaline soln.

SYNS: FLUOROPHOSPHORIC ACID, anhydrous □ MONOFLUOROPHOSPHORIC ACID, anhydrous

OSHA PEL: TWA 2.5 mg(F)/m³

ACGIH TLV: TWA 2.5 mg(F)/m³; BEI: 3 mg/g creatinine of fluorides in urine prior to shift; 10 mg/g creatinine of fluorides in urine at end of shift.

NIOSH REL: TWA 2.5 mg(F)/m³

DOT CLASSIFICATION: 8; Label: Corrosive

SAFETY PROFILE: A corrosive and irritating material to skin, eyes, and mucous membranes. When heated to decomposition it emits very toxic fumes of F^- and PO_x. See also FLUORIDES.

PHK250　CAS: 3734-95-0　HR: 3
PHOSPHOROTHIOIC ACID-S-(((1-CYANO-1-METHYL-ETHYL)CAR-BAMOYL)METHYL)-O,O-DIETHYL ESTER

mf: $C_{10}H_{19}N_2O_4PS$　mw: 294.34

PROP: Oil. D: 1.19 @ 19°/4°. Sol in H_2O.

SYNS: α-CYANOISOPROPYLAMIDE OF THE O,O-DIETHYLTHIOPHOSPHORYL ACETIC ACID □ S-(((1-CYANO-1-METHYL-ETHYL)CARBAMOYL)METHYL) O,O-DIETHYL PHOSPHOROTHIOATE □ S-N-(1-CYANO-1-METHYLETHYL)CARBAMOYLMETHYL DIETHYL PHOSPHOROTHIOLATE □ O,O-DIAETHYL-S-(1-METHYLAETHYL)-CARBAMOYL-METHYL-MONOTHIOPHOSPHAT (GERMAN) □ O,O-DIETHYL-S-((2-CYAAN-2-METHYL-ETHYL)-CARBAMOYL)-METHYL-MONOTHIOFOSFAAT (DUTCH) □ O,O-DIETHYL-S-N-(A-CYANOISOPROPYL)CARBOMOYLMETHYL PHOS-

PHOROTHIOATE □ O,O-DIETIL-S-((2-CIAN-2-METIL-ETIL)-CARBAMOIL)-METIL-MONOTIOFOSFATO (ITALIAN) □ TARTAN □ THIOPHOSPHATE de S-N-(1-CYANO-1-METHYLETHYL)CARBAMOYLMETHYLE et de O,O-DIETHYLE (FRENCH)

CONSENSUS REPORTS: Cyanide and its compounds are on the Community Right-To-Know List.

SAFETY PROFILE: Poison by ingestion and skin contact. When heated to decomposition it emits very toxic fumes of CN^-, PO_x, SO_x, and NO_x. See also ESTERS.

**PHO500　CAS: 7723-14-0　HR: 3
PHOSPHORUS (red)**
DOT: UN 1338
af: P　　aw: 30.97
PROP: Reddish-brown powder. Bp: 280° (with ignition), mp: 590° @ 43 atm, d: 2.34, autoign temp: 500°F in air, vap d: 4.77. IDLH 5 mg/m³.
SYN: PHOSPHORUS, amorphous (DOT)
CONSENSUS REPORTS: EPA Extremely Hazardous Substances List.
DFG MAK: 0.1 mg/m³
DOT CLASSIFICATION: 4.1; Label: Flammable Solid
SAFETY PROFILE: A human poison by an unspecified route. May have white phosphorus as an impurity. Generally less reactive than white phosphorus. Dangerous fire hazard when exposed to heat or by chemical reaction with oxidizers. Can also react with reducing materials. Moderate explosion hazard by chemical reaction or on contact with organic materials. May explode on impact. To fight fire, use water. Explosive reaction with chlorosulfuric acid, hydroiodic acid, magnesium perchlorate, chromyl chloride. Forms sensitive explosive mixtures with metal halogenates (e.g., chlorates, bromates, or iodates of barium, calcium, magnesium, potassium, sodium, zinc), ammonium nitrate, mercury(I) nitrate, silver nitrate, sodium nitrate, potassium permanganate. Violent reaction or ignition with alkalies + heat, fluorine, chlorine, liquid bromine, antimony pentachloride. Reacts with hot alkalies or hydroiodic acid to form

phosphine gas, which then ignites. Incompatible with cyanogen iodide, halogen azides, halogen oxides (e.g., chlorine dioxide, dichlorine oxide, oxygen difluoride, trioxygen difluoride), interhalogens (e.g., bromine trifluoride, bromine pentafluoride, chlorine trifluoride, iodine trichloride, iodine pentafluoride), hexalithium disilicide, hydrogen peroxide, metal acetylides (e.g., rubidium acetylide, cesium acetylide, lithium acetylide, sodium acetylide, potassium acetylide), antimony pentachloride, metal oxides (e.g., copper oxide, manganese dioxide, lead oxide, mercury oxide, silver oxide, chromium trioxide), metal peroxides (e.g., lead peroxide, potassium peroxide, sodium peroxide), metals (e.g., beryllium, copper, manganese, thorium, zirconium, cerium, lanthanum, neodymium, praseodymium, osmium, platinum), metal sulfates (e.g., barium sulfate, calcium sulfate), nitric acid, nitrogen halides, nitrosyl fluoride, nitryl fluoride, nonmetal halides (e.g., boron triiodide, seleninyl chloride, sulfuryl chloride, disulfuryl chloride, disulfur dibromide), nonmetal oxides (e.g., nitrogen oxide, dinitogen tetraoxide, dinitrogen pentaoxide, sulfur trioxide, oxygen, peroxyformic acid, potassium nitride, selenium, sodium chlorite, sulfur, sulfuric acid, peroxides, oxidizing materials). When heated to decomposition it emits toxic fumes of PO_x. See also PHOSPHORUS (yellow).

**PHP010　CAS: 7723-14-0　HR: 3
PHOSPHORUS (yellow)**
DOT: UN 1381/UN 2447
mf: P_4　　mw: 123.88
PROP: Cubic, colorless crystals from ligroin or conc HCl; yellow leaflets from water; colorless to yellow, wax-like solid. Mp: 44.1°, bp: 280°, flash p: spontaneously flammable in air, d: 1.82, autoign temp: 86°F, vap press: 1 mm @ 76.6°, vap d: 4.42. Mod sol in water.
SYNS: BONIDE BLUE DEATH RAT KILLER □ COMMON SENSE COCKROACH and RAT

PREPARATIONS □ FOSFORO BIANCO (ITALIAN) □ GELBER PHOSPHOR (GERMAN) □ PHOSPHORE BLANC (FRENCH) □ PHOSPHOROUS (WHITE) □ PHOSPHORUS WHITE, molten (UN 2447) (DOT) □ PHOSPHOROUS YELLOW □ PHOSPHORUS, white or yellow, dry or under water or in solution (UN 1381) (DOT) □ PHOSPHORUS, YELLOW (ACGIH,OSHA) □ RAT-NIP □ TETRAFOSFOR (DUTCH) □ TETRAPHOSPHOR (GERMAN) □ WEISS PHOSPHOR (GERMAN) □ WHITE PHOSPHORUS □ YELLOW PHOSPHORUS

CONSENSUS REPORTS: EPA Extremely Hazardous Substances List. Reported in EPA TSCA Inventory.

OSHA PEL: TWA 0.1 mg/m^3

ACGIH TLV: TWA 0.1 mg/m^3

DOT CLASSIFICATION: 4.2; Label: Spontaneously Combustible, Poison

SAFETY PROFILE: Human poison by ingestion. Experimental poison by ingestion and subcutaneous routes. Experimental reproductive effects. Human systemic effects by ingestion: cardiomyopathy, cyanosis, nausea or vomiting, sweating. Toxic quantities have an acute effect on the liver and can cause severe eye damage. Inhalation can cause photophobia with myosis, dilation of the pupils, retinal hemorrhage, congestion of the blood vessels, and, rarely, an optic neuritis. Chronic exposure by inhalation or ingestion can cause anemia, gastrointestinal effects, and brittleness of the long bones, leading to spontaneous fractures. The most common symptom, however, of chronic phosphorus poisoning is necrosis of the jaw (phossy-jaw).

More reactive than red phosphorus. Dangerous fire hazard when exposed to heat, flame, or by chemical reaction with oxidizers. Ignites spontaneously in air. Very reactive. If combustion occurs in a confined space, it will remove the oxygen and cause asphyxiation. Dangerous explosion hazard by chemical reaction with: alkaline hydroxides, NH_4NO_3, SbF_5, $Ba(BrO_3)_2$, Be, BI_3, $Ca(BrO_3)_2$, $Mg(BrO_3)_2$, $K(BrO_3)$, $NaBrO_3$, $Zn(BrO_3)_2$, Br_2, halogens, BrF_3, BrN_3, (chlorates of Ba, Ca, Mg, K, Na, Zn), (iodates of Ba, Ca, Mg, K, Na, Zn), Ce, Cs, $CsHC_2$, Cs_3N, (charcoal + air), ClO_2, $(Cl_2 + heptane)$, ClO, ClF_3, ClO_3, chlorosulfonic acid, CrO_3, $Cr(OCl)_2$, Cu, NCl, IBr, ICl, IF_5, Fe, La, PbO_2, Li, Li_2C_2, Li_6CS, $Mg(ClO_4)_2$, Mn, HgO, $HgNO_3$, Nd, Ni, nitrates, NBr, NO_2, NBr_3, NCl_3, NOF, FNO_2, O_2, performic acid, Pt, K, KOH, K_3N, $KMnO_4$, K_2O_2, Rb, $RbHC_2$, Se_2Cl_2, $SeOCl_2$, $SeOF_2$, SeF_4, $AgNO_3$, Ag_2O, Na, Na_2C_2, $NaClO_2$, NaOH, Na_2O_2, S, SO_3, H_2SO_4, Th, $VOCl_2$, Zr, peroxyformic acid, chloro sulfuric acid, halogen azides, hexalithium disilicide. Can react vigorously with oxidizing materials. To fight fire, use water. Used in fertilizers, tracer bullets, incendiaries manufacturing, rat poison, and gas analysis. When heated to decomposition it emits highly toxic fumes of PO_x. See also PHOSPHORUS (red).

PHQ000   HR: D

PHOSPHORUS COMPOUNDS, INORGANIC

SAFETY PROFILE: Variable toxicity. Most inorganic phosphates (except phosphine) have low toxicity, but in large doses they may cause serious disturbances, particularly in calcium metabolism. Red phosphorus and phosphates are relatively harmless. White (yellow) phosphorus is highly toxic by several routes. The phosphorus halides decompose violently with water to form the halide acid and are thus severe irritants. Phosphorus sulfides behave similarly. Metaphosphates may be highly toxic, causing irritation and hemorrhages in the stomach, as well as liver and kidney damage. Phosphorus trichloride is the most used of the phosphorus halide compounds. Phosphoryl chloride is used to synthesize phosphate esters. Common air contaminants. When heated to decomposition it emits highly toxic fumes of PO_x. See also specific compounds.

PHQ500   **CAS: 7783-55-3**   HR: 3

PHOSPHORUS FLUORIDE

mf: F_3P mw: 87.97

PROP: Colorless gas. Slowly hydrol in moist air. Mp: −152°, bp: −102°, d: 3.907 g/L.
SYNS: PHOSPHOROUS TRIFLUORIDE □ PHOSPHORUS TRIFLUORIDE □ TL 75 □ TRIFLUOROPHOSPHINE
CONSENSUS REPORTS: Reported in EPA TSCA Inventory.
OSHA PEL: TWA 2.5 mg(F)/m³
ACGIH TLV: TWA 2.5 mg(F)/m³; BEI: 3 mg/g creatinine of fluorides in urine prior to shift; 10 mg/g creatinine of fluorides in urine at end of shift.
NIOSH REL: (Fluorides, Inorganic) TWA 2.5 mg(F)/m³
SAFETY PROFILE: Moderately toxic by inhalation. A severe eye, skin, and mucous membrane irritant. Explodes on contact with dioxygen difluoride. Violent reaction or ignition with borane, diborane, F_2. hexafluoroisopropylideneamino lithium, O_2. Will react with water or steam to produce toxic and corrosive fumes. Dangerous; when heated to decomposition it emits highly toxic fumes of F⁻ and PO_x. See also HYDROFLUORIC ACID, FLUORIDES, and PHOSPHORUS PENTAFLUORIDE.

PHQ750   CAS: 12037-82-0   HR: 3
PHOSPHORUS HEPTASULFIDE
DOT: UN 1339
mf: P_4S_7 mw: 348.30
PROP: Light-yellow crystals; light gray powder or fused solid. Mp: 310°, bp: 523°, d: 2.19 @ 17°.
SYN: PHOSPHORUS HEPTASULFIDE, free from yellow or white phosphorus (DOT)
DOT CLASSIFICATION: 4.1; Label: Flammable Solid
SAFETY PROFILE: A poison by ingestion. Flammable when exposed to heat or flame; can react vigorously with oxidizing materials. When heated to decomposition it emits very toxic fumes of PO_x and SO_x. See also SULFIDES and PHOSPHORUS.

PHQ800   CAS: 10025-87-3   HR: 3
PHOSPHORUS OXYCHLORIDE
DOT: UN 1810
mf: Cl_3OP mw: 153.32

PROP: Colorless to sltly yellow, mobile, fuming liquid. Very reactive to nucleophiles. Hydrolyzed rapidly by H_2O at room temp. Mp: 1.2°, bp: 105.1°, d: 1.685 @ 15.5°, vap press: 40 mm @ 27.3°, vap d: 5.3.
SYNS: FOSFOROXYCHLORID □ OXYCHLORID FOSFORECNY □ PHOSPHORUS OXYTRICHLORIDE □ PHOSPHORYL CHLORIDE
CONSENSUS REPORTS: EPA Extremely Hazardous Substances List. Reported in EPA TSCA Inventory.
OSHA PEL: TWA 0.1 ppm
ACGIH TLV: TWA 0.1 ppm
DFG MAK: 0.2 ppm (1 mg/m³)
DOT CLASSIFICATION: 8; Label: Corrosive, Poison
SAFETY PROFILE: Poison by inhalation and ingestion. A corrosive eye, skin, and mucous membrane irritant. Potentially explosive reaction with water evolves hydrogen chloride and phosphine, which then ignites. Explosive reaction with 2,6-dimethylpyridine N-oxide, dimethyl sulfoxide, ferrocene-1,1'-dicarboxylic acid, pyridine N-oxide (above 60°C), sodium + heat. Violent reaction or ignition with BI_3, carbon disulfide, 2,5-dimethyl pyrrole + dimethyl formamide, organic matter, zinc powder. Reacts with water or steam to produce heat and toxic and corrosive fumes. Incompatible with carbon disulfide, N,N-dimethyl-formamide, 2,5-dimethylpyrrole, 2,6-dimethylpyridine N-oxide, dimethylsulfoxide, ferrocene-1,l-dicarboxylic acid, water, zinc. When heated to decomposition it emits highly toxic fumes of Cl⁻ and PO_x.

PHR250   CAS: 7789-69-7   HR: 3
PHOSPHORUS PENTABROMIDE
mf: PBr_5 mw: 430.56
PROP: Reddish-yellow, rhombohedral, crystalline mass from $PhNO_2$ at room temp. Mp: decomp, bp: decomp @ 106°. Sol in carbon disulfide.
SYNS: PENTABROMO PHOSPHORANE □ PENTABROMO PHOSPHORUS □ PHOSPHORIC BROMIDE

CONSENSUS REPORTS: Reported in EPA TSCA Inventory.

SAFETY PROFILE: A poison. Corrosive to the eyes, skin and mucous membranes. Flammable by chemical reaction. Contact with moisture can cause a violent reaction and evolution of heat. Incompatible with water or steam to produce heat and toxic and corrosive fumes. When heated to decomposition it emits highly toxic fumes of $PBr_3 + Br_2$. See also BROMIDES.

PHR500   CAS: 10026-13-8   HR: 3
PHOSPHORUS PENTACHLORIDE
DOT: UN 1806

mf: Cl_5P mw: 208.22

PROP: Yellowish-white, fuming, crystalline mass; pungent odor. Mp: (under press) 148° decomp, bp: subl @ 160°, d: 4.65 g/L @ 296°, vap press: 1 mm @ 55.5°. IDLH 70 mg/m³.

SYNS: FOSFORO(PENTACHLORURO di) (ITALIAN) □ FOSFORPENTACHLORIDE (DUTCH) □ PHOSPHORE-(PENTACHLORURE de) (FRENCH) □ PHOSPHORIC CHLORIDE □ PHOSPHORPENTACHLORID (GERMAN) □ PHOSPHORUS PERCHLORIDE □ PIECIOCHLOREK FOSFORU (POLISH)

CONSENSUS REPORTS: EPA Extremely Hazardous Substances List. Reported in EPA TSCA Inventory.

OSHA PEL: TWA 1 mg/m³

ACGIH TLV: TWA 0.85 mg/m³

DFG MAK: 1 mg/m³

DOT CLASSIFICATION: 8; Label: Corrosive

SAFETY PROFILE: Poison by inhalation. Moderately toxic by ingestion. A severe eye, skin, and mucous membrane irritant. Corrosive to body tissues. Flammable by chemical reaction. Explosive reaction with chlorine dioxide + chlorine, sodium, urea + heat. Reacts to form explosive products with carbamates, 3'-methyl-2-nitrobenzanilide (product explodes on contact with air). Ignites on contact with fluorine. Reacts violently with moisture, ClO_3, hydroxyl-amine, magnesium oxide, nitrobenzene, phosphorus(III) oxide, K. To fight fire, use CO_2, dry chemical. Incompatible with aluminum, chlorine dioxide, chlorine,

diphosphorus trioxide, fluorine, hydroxylamine, magnesium oxide, 3'-methyl-2-nitrobenzanilide, nitrobenzene, sodium, urea, water. Will react with water or steam to produce heat and toxic and corrosive fumes. Used as a catalyst, chlorinating and dehydrating agent. When heated to decomposition it emits highly toxic fumes of Cl^- and PO_x.

PHR750   CAS: 7647-19-0   HR: 3
PHOSPHORUS PENTAFLUORIDE
DOT: UN 2198

mf: F_5P mw: 125.97

PROP: Colorless gas, fumes strongly in air, thermally stable. Non-electrolyte in HF(l). Interacts weakly with arenes, e.g. naphthalene. Very strong Lewis acid and complexes with NH_3, and amines; those from Et_3N, i-$PrNH_2$ are unstable. Mp: −83°, bp: −75°, d: (gas) 5.805 g/L.

CONSENSUS REPORTS: Reported in EPA TSCA Inventory.

OSHA PEL: TWA 2.5 mg(F)/m³

ACGIH TLV: TWA 2.5 mg(F)/m³; BEI: 3 mg/g creatinine of fluorides in urine prior to shift; 10 mg/g creatinine of fluorides in urine at end of shift.

DOT CLASSIFICATION: 2.3; Label: Poison Gas

NIOSH REL: (Fluorides, Inorganic) TWA 2.5 mg(F)/m³

SAFETY PROFILE: A poisonous gas. Violently irritating to skin, eyes, and mucous membranes. Inhalation may cause pulmonary edema. Reacts with water or steam to produce toxic and corrosive fumes. When heated to decomposition it emits highly toxic fumes of F^- and PO_x. See also FLUORIDES.

PHS000   CAS: 1314-80-3   HR: 3
PHOSPHORUS PENTASULFIDE
DOT: UN 1340

mf: P_2S_5 mw: 222.24

PROP: Gray to yellow-green, crystalline, deliquescent mass. Bp: 514°, d: 2.09, autoign

P

temp: 287°F, mp: 286–290°. IDLH 250 mg/m³.

SYNS: PENTASULFURE de PHOSPHORE (FRENCH) □ PHOSPHORIC SULFIDE □ PHOSPHORUS PENT-ASULFIDE, free from yellow or white phosphorus (DOT) □ PHOSPHORUS PERSULFIDE □ RCRA WASTE NUMBER U189 □ SIRNIK FOSFORECNY (CZECH) □ SULFUR PHOSPHIDE □ THIOPHOSPHORIC ANHYDRIDE

CONSENSUS REPORTS: Reported in EPA TSCA Inventory.

OSHA PEL: TWA 1 mg/m³; STEL 3 mg/m³

ACGIH TLV: TWA 1 mg/m³; STEL 3 mg/m³

DFG MAK: 1 mg/m³

DOT CLASSIFICATION: 4.3; Label: Dangerous When Wet

SAFETY PROFILE: A poison by ingestion. A severe eye and skin irritant. Readily liberates toxic hydrogen sulfide and phosphorus pentoxide and evolves heat on contact with moisture. Dangerous fire hazard in the form of dust when exposed to heat or flame. Spontaneous heating in the presence of moisture. Moderate explosion hazard in solid form by spontaneous chemical reaction. Reacts with water, steam, or acids to produce toxic and flammable vapors; can react vigorously with oxidizing materials. Incompatible with air, alcohols, water. To fight fire, use CO_2 snow, dry chemical, or sand. Used as an intermediate in manufacturing lubricant additives, insecticides, and fertilizer agents. When heated to decomposition it emits highly toxic fumes of SO_x and PO_x. See also HYDROGEN SULFIDE.

PHS250   CAS: 1314-56-3   HR: 3
PHOSPHORUS PENTOXIDE
DOT: NA 1807
mf: O_5P_2 mw: 141.94
PROP: Deliq crystals. D: 2.30, mp: 340°, subl @ 360°.
SYNS: DIPHOSPHORUS PENTOXIDE □ PHOSPHORUS(V) OXIDE □ PHOSPHORUS PENTAOXIDE

CONSENSUS REPORTS: Reported in EPA TSCA Inventory.

DFG MAK: 1 mg/m³

DOT CLASSIFICATION: 8; Label: Corrosive

SAFETY PROFILE: Poison by inhalation. A corrosive irritant to the eyes, skin, and mucous membranes. With the appropriate conditions it undergoes hazardous reactions with formic acid, hydrogen fluoride, inorganic bases, iodides, metals, methyl hydroperoxide, oxidants (e.g., bromine, pentafluoride, chlorine trifluoride, perchloric acid, oxygen difluoride, hydrogen peroxide), 3-propynol, water. When heated to decomposition it emits toxic fumes of PO_x.

PHS500   CAS: 1314-85-8   HR: 3
PHOSPHORUS SESQUISULFIDE
DOT: UN 1341
mf: P_4S_3 mw: 220.06
PROP: Lemon yellow, crystalline mass or solid. Mp: 172.5°, bp: 407°, d: 2.03, autoign temp: 212°F. Insol in H_2O and org solvs; sol in alkaline solns.
SYNS: PHOSPHORUS SESQUISULFIDE, free from yellow or white phosphorus (DOT) □ PHOSPHORUS (III) SULFIDE (IV) □ SESQUISULFURE de PHOSPHORE (FRENCH) □ TETRAPHOSPHORUS TRISULFIDE □ TRISULFURATED PHOSPHORUS

DOT CLASSIFICATION: 4.1; Label: Flammable Solid

SAFETY PROFILE: Poison by ingestion. Flammable by spontaneous ignition. When heated to decomposition it emits very toxic fumes of PO_x and SO_x. See also SULFIDES and PHOSPHORUS.

PHT250   CAS: 7789-60-8   HR: 3
PHOSPHORUS TRIBROMIDE
DOT: UN 1808
mf: Br_3P mw: 270.70
PROP: Colorless fuming liquid. Mp: −40°, bp: 175.3°, d: 2.852 @ 15°, vap press: 10 mm @ 47.8°, d: 2.87 @ 20°/4°.
SYNS: PHOSPHORUS BROMIDE (DOT) □ TRIBROMOPHOSPHINE

CONSENSUS REPORTS: Reported in EPA TSCA Inventory.

DOT CLASSIFICATION: 8; Label: Corrosive

SAFETY PROFILE: Probably highly toxic. A corrosive irritant to the eyes, skin, and mucous membranes. Will react with water, steam, or acids to produce heat, toxic and corrosive fumes. Violent reaction or ignition with calcium hydroxide + sodium carbonate, phenylpropanol, sulfuric acid, oleum, fluorosulfuric acid, chlorosulfuric acid, 1,1,1-tris(hydroxymethyl)methane, water, potassium, sodium, RuO_4. When heated to decomposition it emits very toxic fumes of Br^- and PO_x. See also PHOSPHIDES and BROMIDES.

PHT275   **CAS: 7719-12-2**   **HR: 3**
PHOSPHORUS TRICHLORIDE
DOT: UN 1809
mf: Cl_3P mw: 137.32
PROP: Clear, colorless, fuming liquid. Mp: $-111.8°$, fp: $-93.6°$, bp: $76°$, d: 1.574 @ $21°$, vap press: 100 mm @ $21°$, vap d: 4.75. Decomp by water and alc; sol in benzene, chloroform, and ether. IDLH 25 ppm.
SYNS: CHLORIDE of PHOSPHORUS □ FOSFORO-(TRICLORURO di) (ITALIAN) □ FOSFORTRICHLORIDE (DUTCH) □ PHOSPHORE(TRICHLORURE de) (FRENCH) □ PHOSPHORTRICHLORID (GERMAN) □ PHOSPHO-RUS CHLORIDE □ TROJCHLOREK FOSFORU (POLISH)
CONSENSUS REPORTS: EPA Extremely Hazardous Substances List. Reported in EPA TSCA Inventory.
OSHA PEL: TWA 0.2 ppm; STEL 0.5 ppm
ACGIH TLV: TWA 0.2 ppm; STEL 0.5 ppm
DFG MAK: 0.5 ppm (2.8 mg/m³)
DOT CLASSIFICATION: 8; Label: Corrosive, Poison
SAFETY PROFILE: Poison by ingestion and inhalation. A corrosive irritant to skin, eyes (at 2 ppm), and mucous membranes. Potentially explosive reaction with chlorobenzene + sodium, dimethyl sulfoxide, molten sodium, chromyl chloride, nitric acid, sodium peroxide, oxygen (above 100°C), tetravinyl lead. Reacts with carboxylic acids (e.g., acetic acid) to form violently unstable products. Violent reaction or ignition with Al, chromium pentafluoride, diallyl phosphite + allyl alcohol, F_2, hexafluoroisopropylideneaminolithium, hydroxylamine, iodine chloride, PbO_2, HNO_2, organic matter, potassium, selenium dioxide, sulfur acids (e.g., sulfuric acid, fluorosulfuric acid, oleum). Violent reaction with water evolves hydrogen chloride and diphosphane gas, that then ignite. Incompatible with metals or oxidants. Will react with water, steam, or acids to produce heat and toxic and corrosive fumes; can react with oxidizing materials. To fight fire, use CO_2, dry chemical. Used as a chlorinating agent, catalyst, and chemical intermediate. Dangerous; when heated to decomposition it emits highly toxic fumes of Cl^- and PO_x.

PHT750   **CAS: 12165-69-4**   **HR: 3**
PHOSPHORUS TRISULFIDE
DOT: UN 1343
mf: P_2S_3 mw: 158.12
PROP: Gray-yellow crystals. Mp: $290°$, bp: $490°$.
SYN: PHOSPHORUS TRISULFIDE, free from yellow or white phosphorus (DOT)
DOT CLASSIFICATION: 4.1; Label: Flammable Solid
SAFETY PROFILE: Flammable solid which can react with oxidizers, water, or steam to emit toxic fumes of H_2S. When heated to decomposition it emits very toxic fumes of PO_x and SO_x. See also SULFIDES and PHOSPHORUS COMPOUNDS.

PHU000   **CAS: 7789-59-5**   **HR: 3**
PHOSPHORYL BROMIDE
DOT: UN 1939/UN 2576
mf: Br_3OP mw: 286.70
PROP: Colorless plates or large flaky crystals. Yellows and decomp at high temps. Slowly decomp by H_2O or H_3PO_3 + HBr. Mp: $56°$, bp: $190°$, d: 2.882. Sol in org solvs.
SYNS: PHOSPHOROUS OXYBROMIDE □ PHOSPHOR-YL TRIBROMIDE
CONSENSUS REPORTS: Reported in EPA TSCA Inventory.
DOT CLASSIFICATION: 8; Label: Corrosive
SAFETY PROFILE: Poison by ingestion, inhalation, and skin contact. A corrosive irritant to skin, eyes, and mucous membranes. A corrosive material. Reacts with steam, water to produce much heat with toxic fumes. When heated to

decomposition it emits very toxic fumes of Br^- and PO_x. See also BROMIDES.

PHU500   CAS: 520-52-5   HR: 3
O-PHOSPHORYL-4-HYDROXY-N,N-
**   DIMETHYLTRYPTAMINE**
mf: $C_{12}H_{17}N_2O_4P$ mw: 284.28
PROP: Crystals from MeOH or H_2O. Mp: 220–228°.
SYNS: CY-39 □ 3-(2-(DIMETHYLAMINO)ETHYL)-1H-INDOL-4-OL DIHYDROGEN PHOSPHATE ESTER □ 3-2'-DIMETHYLAMINOETHYLINDOL-4-PHOSPHATE □ 3-(2-DIMETHYLAMINOETHYL)INDOL-4-YL DIHYDROGEN PHOSPHATE □ INDOCYBIN □ PSILOCIN PHOSPHATE ESTER □ PSILOCIPIN □ PSILOTSIBIN □ TEON-ANACATL
CONSENSUS REPORTS: EPA Genetic Toxicology Program.
SAFETY PROFILE: Poison by intravenous route. Moderately toxic by intraperitoneal route. Human systemic effects by ingestion and intraperitoneal routes: euphoria, hallucinations, toxic psychosis, muscle weakness, nausea or vomiting, visual field changes. Mutation data reported. When heated to decomposition it emits very toxic fumes of NO_x and PO_x.

PHU750   CAS: 12067-99-1   HR: 2
PHOSPHOTUNGSTIC ACID
PROP: White to sltly yellow crystals or powder. Sol in water, alc, ether.
SYN: TUNGSTOPHOSPHORIC ACID (8CI)
CONSENSUS REPORTS: Reported in EPA TSCA Inventory.
NIOSH REL: (Tungsten) TWA 1 mg(W)/m^3
SAFETY PROFILE: Moderately toxic by ingestion. When heated to decomposition it emits toxic fumes of PO_x.

PHW250   CAS: 88-99-3   HR: 2
PHTHALIC ACID
mf: $C_8H_6O_4$ mw: 166.14
PROP: Crystals or plates from water. Mp: 210° (decomp), d: 1.59, bp: 155° (decomp). Mod sol in H_2O, EtOH; sol in hot H_2O, alkalies; spar sol in Et_2O.
SYNS: ACIDE PHTHALIQUE (FRENCH) □ o-BEN-ZENEDICARBOXYLIC ACID □ BENZENE-1,2-DI-CARBOXYLIC ACID □ 1,2-BENZENEDICARBOXYLIC ACID □ o-DICARBOXYBENZENE
CONSENSUS REPORTS: Reported in EPA TSCA Inventory.
SAFETY PROFILE: Moderately toxic by ingestion and intraperitoneal routes. A skin and mucous membrane irritant. Combustible when heated. In the form of dust (anhydride) it can explode. Mixtures with sodium nitrite explode when heated. Violent reaction with HNO_3. When heated to decomposition it emits acrid smoke and irritating fumes. Used in synthesis of dyes and dyestuffs, in medicines and perfumes.

PHW750   CAS: 85-44-9   HR: 3
PHTHALIC ANHYDRIDE
DOT: UN 2214
mf: $C_8H_4O_3$ mw: 148.12
PROP: White, crystalline needles from alc. Mp: 131.2°, lel: 1.7%, uel: 10.4%, bp: 284°, flash p: 305°F (CC), d: 1.527 @ 4°, autoign temp: 1058°F, vap press: 1 mm @ 96.5°, vap d: 5.10. Very sltly sol in water; sol in alc; sltly sol in ether. IDLH 60 mg/m^3.
SYNS: ANHYDRIDE PHTHALIQUE (FRENCH) □ ANIDRIDE FTALICA (ITALIAN) □ 1,2-BENZENEDI-CARBOXYLIC ACID ANHYDRIDE □ 1,3-DIOXOPHTH-ALAN □ ESEN □ FTAALZUURANHYDRIDE (DUTCH) □ FTALOWY BEZWODNIK (POLISH) □ 1,3-ISOBENZO-FURANDIONE □ NCI-C03601 □ 1,3-PHTHALANDIONE □ PHTHALIC ACID ANHYDRIDE □ PHTHAL-SAEUREANHYDRID (GERMAN) □ RCRA WASTE NUMBER U190 □ RETARDER AK □ RETARDER ESEN □ RETARDER PD
CONSENSUS REPORTS: NCI Carcinogenesis Bioassay (feed); No Evidence: mouse, rat NCITR* NCI-CG-TR-159,79. Community Right-To-Know List. Reported in EPA TSCA Inventory.
OSHA PEL: TWA 1 ppm
ACGIH TLV: TWA 1 ppm (skin, sensitizer) Not Classifiable as a Human Carcinogen
DFG MAK: 1 mg/m^3 as total dust
DOT CLASSIFICATION: 8; Label: Corrosive
SAFETY PROFILE: Poison by ingestion. Experimental teratogenic effects. A corrosive eye, skin, and mucous membrane irritant. A common air contaminant. Combustible when exposed to heat or

flame; can react with oxidizing materials. Moderate explosion hazard in the form of dust when exposed to flame. The production of this material has caused many industrial explosions. Mixtures with copper oxide or sodium nitrite explode when heated. Violent reaction with nitric acid + sulfuric acid above 80°C. To fight fire, use CO₂, dry chemical. Used in plasticizers, polyester resins, and alkyd resins, dyes, and drugs. See also ANHYDRIDES.

PHX000   CAS: 85-41-6   HR: 2
PHTHALIMIDE
mf: $C_8H_5NO_2$ mw: 147.14
PROP: White to light-tan powder; needles from H_2O; prisms from AcOH; leaflets by sublimation. Mp: 238°, bp: subl. Sol in water, alc, alkali, hot ether; insol in benzene, pet ether.
SYNS: ISOINDOLE-1,3-DIONE □ 1,3-ISOINDOLED-IONE □ 1,3-ISOINDOLINEDIONE □ o-PHTHALIC IMIDE
CONSENSUS REPORTS: Reported in EPA TSCA Inventory.
SAFETY PROFILE: Moderately toxic by intraperitoneal route. Mildly toxic by ingestion. An experimental teratogen. Other experimental reproductive effects. When heated to decomposition it emits toxic fumes of NOₓ.

PHX100   CAS: 3343-28-0   HR: D
2-PHTHALIMIDOGLUTARIC ACID ANHYDRIDE
mf: $C_{13}H_9NO_5$ mw: 259.23
SYN: 2H-PYRAN-2,6(3H)-DIONE, DIHYDRO-2-PHTHALIMIDO-
CONSENSUS REPORTS: Reported in EPA TSCA Inventory.
SAFETY PROFILE: Experimental reproductive effects. When heated to decomposition it emits toxic fumes of NOₓ.

PHX250   CAS: 732-11-6   HR: 3
PHTHALIMIDOMETHYL-O,O-DIMETHYL PHOSPHORODITHIOATE
mf: $C_{11}H_{12}NO_4PS_2$ mw: 317.33
PROP: A solid. Mp: 72–72.5°.

SYNS: APPA □ DECEMTHION P-6 □ (O,O-DI-METHYL-PHTHALIMIDOMETHYL-DITHIOPHOS-PHATE) □ O,O-DIMETHYL S-(N-PHTHALIMIDOMETH-YL) DITHIOPHOSPHATE □ O,O-DIMETHYL S-PHTHALIMIDOMETHYL PHOSPHORODI-THIOATE □ ENT 25,705 □ FTALOPHOS □ IMIDAN □ KEMOLATE □ N-(MERCAPTOMETHYL)PHTHAL-IMIDE S-(O,O-DIMETHYL PHOSPHORODITHIOATE) □ PERCOLATE □ PHOSMET □ PHTHALIMIDO-O,O-DIMETHYL PHOSPHORODITHIOATE □ PHTHALOPHOS □ PMP □ PROLATE □ R 1504 □ SMIDAN □ STAUFFER R 1504
CONSENSUS REPORTS: EPA Extremely Hazardous Substances List.
SAFETY PROFILE: A human poison by ingestion. Poison experimentally by inhalation and ingestion routes. Moderately toxic by skin contact. Human systemic effects by inhalation: lachrymation, somnolence, and olfaction effects. Experimental teratogenic and reproductive effects. Mutation data reported. When heated to decomposition it emits very toxic fumes of NOₓ, POₓ, and SOₓ. See also ESTERS.

PHX550   CAS: 626-17-5   HR: 3
m-PHTHALODINITRILE
mf: $C_8H_4N_2$ mw: 128.14
PROP: Colorless crystals. Vap d: 4.42, mp: 161.5–162°, bp: sublimes. Insol in water; sol in benzene, acetone.
SYNS: 1,3-BENZENEDICARBONITRILE □ m-DICYA-NOBENZENE □ 1,3-DICYANOBENZENE □ DINITRILE of ISOPHTHALIC ACID □ IPN □ ISOFTALODINITRIL (CZECH) □ ISOPHTHALODINITRILE □ ISOPHTHAL-ONITRILE □ NITRIL KYSELINY ISOFTALOVE (CZECH) □ m-PDN
CONSENSUS REPORTS: Reported in EPA TSCA Inventory. Cyanide and its compounds are on the Community Right-To-Know List.
OSHA PEL: TWA 5 mg/m³
ACGIH TLV: TWA 5 mg/m³
SAFETY PROFILE: Poison by ingestion. An eye irritant. When heated to decomposition it emits toxic fumes of NOₓ and CN⁻. See also NITRILES.

P

PHY000 **CAS: 91-15-6** **HR: 3**
PHTHALONITRILE
mf: $C_8H_4N_2$ mw: 128.14
PROP: Needles in H_2O. Mp: 141°.
SYNS: o-DICYANOBENZENE □ 1,2-DICYANO-
BENZENE □ PHTHALIC ACID DINITRILE □
PHTHALODINITRILE □ o-PHTHALODINITRILE □
USAF ND-09
CONSENSUS REPORTS: Cyanide and its
compounds are on the Community Right-
To-Know List. Reported in EPA TSCA
Inventory.
SAFETY PROFILE: Poison by ingestion,
subcutaneous, and intraperitoneal routes.
Questionable carcinogen with experimental
tumorigenic data. When heated to
decomposition it emits toxic fumes of CN^-
and NO_x. See also NITRILES.

PIA500 **CAS: 57-47-6** **HR: 3**
PHYSOSTIGMINE
mf: $C_{15}H_{21}N_3O_2$ mw: 275.39
PROP: A solid. Mp: 86–87°.
SYNS: ERSERINE □ ESERINE □ ESEROLEIN,
METHYLCARBAMATE (ESTER) □ METHYL-CARBAMIC
ACID, ESTER with ESEROLINE □ PHYSOSTOL
CONSENSUS REPORTS: EPA Extremely
Hazardous Substances List. Reported in
EPA TSCA Inventory.
SAFETY PROFILE: A human poison by an
unspecified route. Poison experimentally by
ingestion, subcutaneous, intramuscular,
intravenous, and intraperitoneal routes.
Human systemic effects by ingestion:
nausea, dyspnea, coma, blood pressure
elevation, flaccid paralysis without
anesthesia, muscle weakness. Normally
administered by injection. Poisoning can
occur as a result of a mistake in dosage or
due to hypersensitivity of the patient within
5 to 25 minutes after administration. Death
usually results from respiratory paralysis.
Experimental reproductive effects.
Combustible when exposed to heat or
flame. When heated to decomposition it
emits toxic fumes of NO_x. See also
CARBAMATES.

PIA750 **CAS: 57-64-7** **HR: 3**
PHYSOSTIGMINE SALICYLATE (1:1)
mf: $C_{15}H_{21}N_3O_2 \cdot C_7H_6O_3$ mw: 413.52
PROP: A solid. Mp: 184–186°.
SYNS: ESERINE SALICYLATE □ PHYSOSTOL
SALICYLATE □ SALICYLIC ACID with PHYSOSTIGMINE
(1:1) □ TL-1380
CONSENSUS REPORTS: EPA Extremely
Hazardous Substances List. Reported in
EPA TSCA Inventory.
SAFETY PROFILE: Poison by ingestion,
subcutaneous, intramuscular, intravenous,
and intraperitoneal routes. Human systemic
effects: arrhythmias, nausea or vomiting.
Experimental reproductive effects. When
heated to decomposition it emits toxic
fumes of NO_x. See also
PHYSOSTIGMINE.

PIB250 **CAS: 83-86-3** **HR: 3**
PHYTIC ACID
mf: $C_6H_{18}O_{24}P_6$ mw: 660.06
PROP: Straw-colored syrup.
SYNS: ALKOVERT □ FYTIC ACID □
HEXAKIS(DIHYDROGEN PHOSPHATE) MYO-
INOSITOL □ INOSITHEXAPHOSPHORSAEURE
(GERMAN) □ INOSITOL HEXAPHOSPHATE □ MYO-
INOSISTOL HEXAKISPHOSPHATE □ MYO-INOSITOL
HEXAPHOSPHATE □ SAEURE DES PHYTINS
(GERMAN)
CONSENSUS REPORTS: Reported in EPA
TSCA Inventory.
SAFETY PROFILE: Poison by intravenous
route. When heated to decomposition it
emits toxic fumes of PO_x.

PIB600 **CAS: 150-86-7** **HR: 1**
PHYTOL
mf: $C_{20}H_{40}O$ mw: 296.60
PROP: Oily liquid. Bp: 203–204°, d: 0.859.
Insol in water; sol in org solvs.
SYNS: 2-HEXADECEN-1-OL, 3,7,11,15-TETRAMETHYL-
, (R-(R*,R*-(E)))-(9CI) □ trans-PHYTOL □ 3,7,11,15-
TETRAMETHYL-2-HEXADECEN-1-OL
CONSENSUS REPORTS: Reported in EPA
TSCA Inventory.
SAFETY PROFILE: Low toxicity by
ingestion and skin contact. A skin irritant.
When heated to decomposition it emits
acrid smoke and irritating fumes.

PIB900　　CAS: 1918-02-1　　HR: 2
PICLORAM
mf: $C_6H_3Cl_3N_2O_2$　　mw: 241.46
PROP: Crystals or powder. Mp: 218°
(decomp). Very spar sol in H_2O; spar sol in
org solvs.
SYNS: AMDON GRAZON □ 4-AMINO-3,5,6-
TRICHLOROPICOLINIC ACID □ 4-AMINO-3,5,6-
TRICHLORO-2-PICOLINIC ACID □ 4-AMINO-3,5,6-
TRICHLORPICOLINSAEURE (GERMAN) □ ATCP □
BOROLIN □ CHLORAMP (RUSSIAN) □ K-PIN □ NCI-
C00237 □ TORDON □ TORDON 10K □ TORDON 22K □
TORDON 101 MIXTURE □ 3,5,6-TRICHLORO-4-
AMINOPICOLINIC ACID
CONSENSUS REPORTS: NCI
Carcinogenesis Bioassay (feed); No
Evidence: mouse NCITR* NCI-CG-TR-
23,78; Clear Evidence: rat NCITR* NCI-
CG-TR-23,78.
OSHA PEL: TWA Total Dust: 10 mg/m³;
Respirable Fraction: 5 mg/m³
ACGIH TLV: TWA 10 mg/m³; Not
Classifiable as a Human Carcinogen
SAFETY PROFILE: Moderately toxic by
ingestion. Questionable carcinogen with
experimental carcinogenic, neoplastigenic,
tumorigenic, and teratogenic data. An
experimental teratogen. Mutation data
reported. When heated to decomposition it
emits very toxic fumes of Cl⁻ and NOₓ.

PIB920　　CAS: 108-99-6　　HR: 3
3-PICOLINE
mf: C_6H_7N　　mw: 93.14
PROP: Colorless liquid; sweetish, not
unpleasant odor. D: (15°/4°) 0.9613, bp:
143–144°, n: (24/D) 1.5043. Misc with
water, alc, ether, most solvs.
SYNS: 3-METHYLPYRIDINE □ β-PICOLINE □ m-
PICOLINE
CONSENSUS REPORTS: Reported in EPA
TSCA Inventory.
SAFETY PROFILE: Poison by intravenous
and intraperitoneal routes. Moderately toxic
by ingestion. Flammable when exposed to
heat or flame; can react vigorously with
oxidizing materials. When heated to
decomposition it emits toxic fumes of NOₓ.

PIC250　　CAS: 466-24-0　　HR: 3
PICRACONITINE
mf: $C_{32}H_{45}NO_{10}$　　mw: 603.78
PROP: Amorphous solid. Mp: 130°.
SYNS: BENZACONINE □ BENZOYLACONINE □
ISACONITINE
SAFETY PROFILE: Poison by intravenous
and intraperitoneal routes. When heated to
decomposition it emits toxic fumes of NOₓ.

PIC500　　CAS: 831-52-7　　HR: 3
PICRAMIC ACID, SODIUM SALT
DOT: UN 0235/UN 1349
mf: $C_6H_4N_3O_5$•Na　　mw: 221.12
PROP: Yellow, water-sol crystals.
SYNS: SODIUM PICRAMATE, dry or wetted with <20%
water, by weight (UN 0235) (DOT) □ SODIUM PICRAMATE,
wetted with not <20% water, by weight (UN 1349) (DOT)
CONSENSUS REPORTS: Reported in EPA
TSCA Inventory.
DOT CLASSIFICATION: EXPLOSIVE
1.3C; Label: EXPLOSIVE 1.3C (UN 0235);
DOT Class: 4.1; Label: Flammable Solid
(UN 1349)
SAFETY PROFILE: Poison by ingestion. A
flammable solid and explosive. When heated
to decomposition it emits toxic fumes of
NOₓ and Na₂O. See also PICRAMIC
ACID.

PIC750　　CAS: 63868-82-6　　HR: 3
**PICRAMIC ACID, ZIRCONIUM SALT
　　(WET)**
DOT: UN 0236/UN 1517
mf: $C_6H_5N_3O_5$•1/4Zr　　mw: 221.95
PROP: IDLH 50 mg/m³ (as Zr).
SYNS: ZIRCONIUM PICRAMATE, dry or wetted with
<20% water, by weight (UN 0236) (DOT) □ ZIRCONIUM
PICRAMATE, wetted with not <20% water, by weight (UN
1517) (DOT)
OSHA PEL: TWA 5 mg(Zr)/m³; STEL 10
mg(Zr)/m³
ACGIH TLV: TWA 5 mg(Zr)/m³; STEL 10
mg(Zr)/m³; Not Classifiable as a Human
Carcinogen
DFG MAK: 1 mg(Zr)/m³
DOT CLASSIFICATION: EXPLOSIVE
1.3C; Label: EXPLOSIVE 1.3C (UN 0236);
DOT Class: 4.1; Label: Flammable Solid
(UN 1517)

P

SAFETY PROFILE: Flammable when exposed to heat or flame; can react vigorously with oxidizing materials. When heated to decomposition it emits toxic fumes of NO_x. See also ZIRCONIUM COMPOUNDS and PICRAMIC ACID.

PIC800   CAS: 489-98-5   HR: 3
PICRAMIDE
DOT: UN 0153
mf: $C_6H_4N_4O_6$ mw: 228.14
SYNS: ANILINE, 2,4,6-TRINITRO- □ PICRAMIDE (DOT) □ TRINITROANILINE (DOT)
DOT CLASSIFICATION: Explosive 1.1D; Label: Explosive 1.1D
SAFETY PROFILE: An explosive. When heated to decomposition it emits toxic vapors of NO_x.

PID000   CAS: 88-89-1   HR: 3
PICRIC ACID
DOT: UN 0154/UN 1344
mf: $C_6H_3N_3O_7$ mw: 229.12
PROP: Colorless crystals from ligroin or conc HCl; yellow leaflets from water; or yellow liquid; very bitter. Mp: 121.8°, bp: explodes >300°, flash p: 302°F, d: 1.763, autoign temp: 572°F, vap d: 7.90. Mod sol in water. IDLH 75 mg/m³.
SYNS: ACIDE PICRIQUE (FRENCH) □ ACIDO PICRICO (ITALIAN) □ CARBAZOTIC ACID □ C.I. 10305 □ 2-HYDROXY-1,3,5-TRINITROBENZENE □ KYSELINA PIKROVA □ MELINITE □ NITROXANTHIC ACID □ PHENOL, 2,4,6-TRINITRO- □ PICRIC ACID (ACGIH, OSHA) □ PICRIC ACID, dry or wetted with <30% water, by weight (UN 0154) (DOT) □ PICRIC ACID, wet, with not <10% water (NA 1344) (DOT) □ PICRONITRIC ACID □ PIKRINEZUUR (DUTCH) □ PIKRINSAEURE (GERMAN) □ PIKRYNOWY KWAS (POLISH) □ 2,4,6-TRINITRO-FENOL (DUTCH) □ 2,4,6-TRINITROFENOLO (ITALIAN) □ TRINITROPHENOL (UN 0154) (DOT) □ TRINITRO-PHENOL, wetted with not <30% water, by weight (UN 1344) (DOT) □ 1,3,5-TRINITROPHENOL □ 2,4,6-TRINITRO-PHENYL (OSHA)
CONSENSUS REPORTS: Community Right-To-Know List. Reported in EPA TSCA Inventory. EPA Genetic Toxicology Program.
OSHA PEL: TWA 0.1 mg/m³ (skin)
ACGIH TLV: TWA 0.1 mg/m³
DFG MAK: 0.1 mg/m³

DOT CLASSIFICATION: 4.1; Label: Flammable Solid (NA 1344, UN 1344)
SAFETY PROFILE: Poison by ingestion and subcutaneous routes. Mutation data reported. An irritant and an allergen. Skin contact can cause local and systemic allergic reactions. Flammable solid when exposed to heat or flame; can react vigorously with oxidizing materials. Very unstable. A severe explosion hazard when shocked or exposed to heat. It forms salts easily, and many of its salts, known as picrates, are more sensitive explosives than picric acid. It forms unstable salts with concrete, NH_3, bases, and metals (e.g., copper, lead, mercury, and zinc). Many of these are heat-, friction-, or impact-sensitive. Mixtures with uranium perchlorate are extremely powerful explosives. Mixtures with aluminum and water ignite after a delay period. Can react vigorously with reducing materials. Used in synthesis of dyes, as a drug, to manufacture explosives and matches, to etch copper, and to make colored glass. See also NITRO COMPOUNDS OF AROMATIC HYDROCARBONS and EXPLOSIVES, HIGH.

PID200   CAS: 146-84-9   HR: 3
PICRIC ACID, SILVER(1+) SALT
DOT: UN 1347
mf: $C_6H_2N_3O_7$•Ag   mw: 335.98
SYNS: PICRAGOL □ PICROTOL □ SILVER PICRATE (dry) (DOT) □ SILVER PICRATE, wetted with not <30% water, by weight (UN1347) (DOT)
DOT CLASSIFICATION: Forbidden; 4.1; Label: Flammable Solid (UN1347)
SAFETY PROFILE: A flammable solid and unstable substance forbidden from transport. When heated to decomposition it emits toxic vapors of NO_x.

PIE500   CAS: 124-87-8   HR: 3
PICROTOXIN
DOT: UN 1584
mf: $C_{15}H_{18}O_7$•$C_{15}H_{16}O_6$ mw: 602.64
PROP: Crystals from EtOH. Mp: 203–204°. Dried fruit of *Anamerta cocculus* (L.) containing

meni-spermine, paramenispermine, 1%
picrotoxin, picrotoxic acid, cocculine
alkaloid, and 5% fat.

SYNS: COCCULIN □ COCCULUS □ COCCULUS solid
(DOT) □ COQUES DU LEVANT (FRENCH) □ FISH
BERRY □ INDIAN BERRY □ ORIENTAL BERRY □
PICROTIN, compounded with PICROTOXININ (1:1) □
PICROTOXINE

CONSENSUS REPORTS: EPA Extremely
Hazardous Substances List.

DOT CLASSIFICATION: 6.1; Label: Poison

SAFETY PROFILE: A human poison by
ingestion. Poison experimentally by most
routes. Human systemic effects by ingestion:
somnolence, gastrointestinal effects. An
alkaloid convulsant poison. When heated to
decomposition it emits acrid smoke and
irritating fumes.

PIE800   CAS: 57176-66-6   HR: 3
PIGMENT TRANSPARENT YELLOW 2K

SYNS: PIGMENT YELLOW TRANSPARENT 2K □
TRANSPARENT YELLOW 2K □ YELLOW
TRANSPARENT 2K □ YELLOW TRANSPARENT
PIGMENT K

SAFETY PROFILE: Confirmed carcinogen
with experimental carcinogenic data. When
heated to decomposition it emits acrid
smoke and irritating vapors.

PIE830   CAS: 57176-67-7   HR: 3
PIGMENT TRANSPARENT YELLOW O

SYNS: PIGMENT YELLOW TRANSPARENT O □
TRANSPARENT PIGMENT YELLOW O □
TRANSPARENT YELLOW O □ YELLOW TRANSPARENT
O □ YELLOW TRANSPARENT PIGMENT O

SAFETY PROFILE: Confirmed carcinogen
with experimental carcinogenic data. When
heated to decomposition it emits acrid
smoke and irritating vapors.

PIF000   CAS: 92-13-7   HR: 3
PILOCARPINE

mf: $C_{11}H_{16}N_2O_2$ mw: 208.29

PROP: Colorless or yellow, hygroscopic,
needle-like crystals. Mp: 34°, bp: 260° @ 5
mm.

SYNS: ALMOCARPINE □ (3S-cis)-3-ETHYLDIHYDRO-
4-((1-METHYL-1H-IMIDAZOL-5-YL)METHYL)-2(3H)-
FURANONE □ α-ETHYL-β-(HYDROXYMETHYL)-1-

METHYL-IMIDAZOLE-5-BUTYRIC ACID, γ-LACTONE □
PILOCARPOL

CONSENSUS REPORTS: Reported in EPA
TSCA Inventory.

SAFETY PROFILE: A human poison by
subcutaneous route. Poison experimentally
by ingestion, intravenous, intraperitoneal,
and subcutaneous routes. A very poisonous
alkaloid that is used to remove excess fluid
accumulations from the body. Its action on
the sweat glands makes it a powerful
sudorific. It very rarely causes death, but,
when it does, it is by paralysis of the heart or
edema of the lungs. Dangerous; on heating
to decomposition it emits toxic fumes of
NO_x.

PIF250   CAS: 54-71-7   HR: 3
PILOCARPINE MONOHYDROCHLORIDE

mf: $C_{11}H_{16}N_2O_2 \cdot ClH$ mw: 244.75

PROP: A solid. Mp: 204–205°.

SYNS: ALMOCARPINE □ AMI-PILO □ AMISTURA P □
ISOPTO-CARPINE □ MI-PILO OPHTH SOL □
PILOCARPINE HYDROCHLORIDE □ PILOCARPINE
MURIATE □ PILOCEL □ PILOMIOTIN □ PILOVISC

CONSENSUS REPORTS: Reported in EPA
TSCA Inventory.

SAFETY PROFILE: Poison by ingestion,
intraperitoneal, and intravenous routes.
Experimental teratogenic and reproductive
effects. Human systemic effects: cardiac
changes. When heated to decomposition it
emits very toxic fumes of HCl and NO_x.
See also PILOCARPINE.

PIF750   CAS: 7681-93-8   HR: 3
PIMARICIN

mf: $C_{33}H_{47}NO_{13}$ mw: 665.81

PROP: Crystals from MeOH (aq). Mp: 200°.
An antibiotic produced by a strain of
Streptomyces chattanoogensis (85ERAY 2,956,78).

SYNS: ANTIBIOTIC A-5283 □ CL 12,625 □ MYCOPHYT
□ MYPROZINE □ NATACYN □ NATAMYCIN □
PIMAFUCIN □ TENNECETIN

SAFETY PROFILE: Poison by intravenous,
intramuscular, subcutaneous, and
intraperitoneal routes. Moderately toxic by
ingestion. When heated to decomposition it

P

emits toxic fumes of NO_x. Used as an antibacterial agent.

PIG730   CAS: 8016-45-3   HR: 2
PIMENTA LEAF OIL
PROP: Main constituent is eugenol. From steam distillation of the shrub *Pimenta officinalis Lindl.* (Fam. *Myrtaceae*) (FCTXAV 12,807,74). Pale yellow to brown liquid; spicy odor. D: 1.037–1.050, refr index: 1.531 @ 20°. Sol in propylene glycol, fixed oils; insol in glycerin, mineral oil.
SYN: OIL of PIMENTA LEAF
CONSENSUS REPORTS: Reported in EPA TSCA Inventory.
SAFETY PROFILE: Moderately toxic by ingestion. A severe skin irritant. When heated to decomposition it emits acrid smoke and irritating fumes. See also EUGENOL.

PIG740   CAS: 8006-77-7   HR: 2
PIMENTA OIL
PROP: Contains eugenol. Distilled from the fruit of *Pimenta officinalis Lindley* (Fam. *Myrtaceae*). Yellow to red-yellow liquid; odor and taste of allspice. D: 1.018–1.048, refr index: 1.527–1.540 @ 20°.
SYNS: OIL OF ALLSPICE □ OIL OF PIMENTA □ OIL PIMENTA BERRIES □ OIL OF PIMENTO □ OILS, ALLSPICE □ OILS, PIMENTA □ PIMENTA BERRY OIL □ PIMENTA LEAF OIL □ PIMENTO OIL
CONSENSUS REPORTS: Reported in EPA TSCA Inventory.
SAFETY PROFILE: Moderately toxic by ingestion and skin contact. A severe skin and eye irritant. Mutation data reported. A weak sensitizer that may cause dermatitis on local contact. Eugenol is moderately toxic. Combustible. See also EUGENOL.

PIH175   CAS: 83-26-1   HR: 3
PINDONE
mf: $C_{14}H_{14}O_3$ mw: 230.28
PROP: Yellow crystals or solid. Mp: 108°. Very spar sol in H_2O; sol in most org solvs, alkalies. IDLH 100 mg/m³.
SYNS: CHEMRAT □ 2-(2,2-DIMETHYL-1-OXOPROP-YL)-1H-INDENE-1,3(2H)-DIONE □ PINDON (DUTCH) □

PIVACIN □ PIVAL □ PIVALDION (ITALIAN) □ PIVALDIONE (FRENCH) □ 2-PIVALOYL-INDAAN-1,3-DION (DUTCH) □ 2-PIVALOYL-INDAN-1,3-DION (GERMAN) □ 2-PIVALOYL-1,3-INDANDIONE □ 2-PIVALOYLINDANE-1,3-DIONE □ 2-PIVALYL-1,3-INDANDIONE □ PIVALYL VALONE □ PIVALYN □ TRI-BAN □ 2-(TRIMETIL-ACETIL)-INDAN-1,3-DIONE (ITALIAN)
CONSENSUS REPORTS: Reported in EPA TSCA Inventory.
OSHA PEL: TWA 0.1 mg/m³
ACGIH TLV: TWA 0.1 mg/m³
NIOSH REL: (Pindone) TWA 0.1 mg/m³
SAFETY PROFILE: Poison by ingestion, intravenous, and parenteral routes. Causes reduced blood clotting, which leads to hemorrhaging. Used as an anticoagulant and rodenticide. When heated to decomposition it emits acrid smoke and irritating fumes. See also WARFARIN.

PIH250   CAS: 80-56-8   HR: 3
2-PINENE
DOT: UN 2368
mf: $C_{10}H_{16}$ mw: 136.26
PROP: Liquid; odor of turpentine. Mp: −55°, bp: 155°, flash p: 91°F, d: 0.8592 @ 20°/4°, refr index: 1.464–1.468, vap press: 10 mm @ 37.3°, vap d: 4.7, autoign temp: 491°F. Insol in water; sol in alc, chloroform, ether, glacial acetic acid, fixed oils.
SYNS: ACINTENE A □ FEMA No. 2902 □ α-PINENE (DOT) □ 2,6,6-TRIMETHYLBICYCLO(3.1.1)-2-HEPT-2-ENE □ 4,6,6-TRIMETHYLBICYKLO(3,1,1)HEPT-3-EN
CONSENSUS REPORTS: Reported in EPA TSCA Inventory.
DOT CLASSIFICATION: 3; Label: Flammable Liquid
ACGIH TLV: TWA 20 ppm (sensitizer); Not Classifiable as a Human Carcinogen
SAFETY PROFILE: A deadly poison by inhalation. Moderately toxic by ingestion. An eye, mucous membrane, and severe human skin irritant. Flammable liquid. A dangerous fire hazard when exposed to heat, flame, or oxidizing materials. To fight fire, use foam, CO_2, dry chemical. Explodes on contact with nitrosyl perchlorate.

PIH400   CAS: 8000-26-8   HR: 1
PINE NEEDLE OIL, DWARF
PROP: From steam distillation of needles of *Pinus mugo* turra var. *pumilio (Haenke) Zenari* (Fam. *Pinaceae*) (FCTXAV 14,659,76). Colorless to yellow liquid; pleasant odor and a bitter, pungent taste. D: 0.853–0.871, refr index: 1.475 @ 20°.
SYNS: DWARF PINE NEEDLE OIL □ KNEE PINE OIL □ LATSCHENKIEFEROEL □ OIL of MOUNTAIN PINE □ PINUS MONTANA OIL □ PINUS PUMILIO OIL
CONSENSUS REPORTS: Reported in EPA TSCA Inventory.
SAFETY PROFILE: Mildly toxic by ingestion. A human skin irritant. When heated to decomposition it emits acrid smoke and irritating fumes.

PIH500   CAS: 8000-26-8   HR: 3
PINE NEEDLE OIL, SCOTCH
PROP: Volatile oil from steam distillation of *Pinus sylvestris L.* (Fam. *Pinaceae*) constituted of dipentene, pinene, sylvestrene, cadinene, and bornyl acetate. Yellow liquid; penetrating odor. Bp: 200–220°, flash p: 172°F (CC), d: 0.86, refr index: 1.473 @ 20°. Sol in fixed oils, mineral oil; sltly sol in propylene glycol; insol in glycerin.
SYNS: KIEFERNADEL OEL (GERMAN) □ SCOTCH PINE NEEDLE OIL
CONSENSUS REPORTS: Reported in EPA TSCA Inventory.
SAFETY PROFILE: Mildly toxic by ingestion. A weak allergen and a mild irritant. Flammable when exposed to heat or flame; can react vigorously with oxidizing materials. To fight fire, use foam, CO_2, dry chemical. When heated to decomposition it emits acrid smoke and irritating fumes. See also individual components.

PIH750   CAS: 8002-09-3   HR: 3
PINE OIL
DOT: UN 1272
PROP: Pale-yellow liquid; penetrating odor. Bp: 200–220°, flash p: 172°F (CC), d: 0.86, flash p: (steam distilled) 138°F. Insol in water; sol in org solvs.

SYNS: ARIZOLE □ OIL of PINE □ OILS, PINE □ OLEUM ABIETIS □ TERPENTINOEL (GERMAN) □ UNIPINE □ YARMOR □ YARMOR PINE OIL
CONSENSUS REPORTS: Reported in EPA TSCA Inventory.
DOT CLASSIFICATION: 3; Label: Flammable Liquid
SAFETY PROFILE: Moderately toxic by ingestion. Mildly toxic by skin contact. A weak allergen and a severe irritant to skin and mucous membranes. Human systemic effects by ingestion: excitement, ataxia, headache. A flammable liquid when exposed to heat or flame; can react with oxidizing materials. Moderate spontaneous heating. To fight fire, use foam, CO_2, dry chemical. Used as an odorant, disinfectant, solvent, wetting agent, and frothing agent.

PII500   CAS: 3819-00-9   HR: 3
PIPERACETAZINE
mf: $C_{24}H_{30}N_2O_2S$ mw: 410.62
PROP: A solid. Mp: 98–100°.
SYNS: 2-ACETYL-10-(3-(4-(β-HYDROXYETHYL)PIP-ERIDINO)PROPYL)PHENOTHIAZINE □ ETHAN □ 10-(3-(4-(2-HYDROXYETHYL)PIPERIDINO)PROPYL)-PHENOTHIAZIN-2-YL METHYL KETONE □ PC-1421 □ PSYMOD □ QUIDE □ SC 9794
DOT CLASSIFICATION: 3; Label: Flammable Liquid
SAFETY PROFILE: Poison by ingestion, subcutaneous, and intraperitoneal routes. A flammable liquid. When heated to decomposition it emits very toxic fumes of SO_x and NO_x.

PII750   CAS: 71-78-3   HR: 3
PIPERADROL HYDROCHLORIDE
mf: $C_{18}H_{21}NO•ClH$ mw: 303.86
PROP: A solid. Mp: 312–334°.
SYNS: α,α-DIPHENYL-2-PIPERIDINEMETHANOL HYDROCHLORIDE □ α-(2-PIPERIDYL)BENZHYDROL HYDROCHLORIDE □ PIPRADOL HYDROCHLORIDE □ PIPRADROL HYDROCHLORIDE □ PIRIDROL HYDROCHLORIDE □ PYRIDROL
SAFETY PROFILE: Poison by ingestion, subcutaneous, intravenous, and intraperitoneal routes. When heated to decomposition it emits very toxic fumes of HCl and NO_x.

PIJ000   CAS: 110-85-0   HR: 2
PIPERAZINE
DOT: UN 2579
mf: $C_4H_{10}N_2$ mw: 86.16
PROP: Colorless, rhombic crystals or
hygroscopic plates from EtOH with salty
taste. Mp: 106°, bp: 146°, flash p: 190°F
(OC), d: 1.1, vap d: 3.0. Very sol in water,
glycerin, glycols; insol in ether.
SYNS: ANTIREN □ 1,4-DIETHYLENEDIAMINE □
N,N-DIETHYLENE DIAMINE (DOT) □ DISPERMINE □
HEXAHYDRO-1,4-DIAZINE □ HEXAHYDROPYRAZINE
□ LUMBRICAL □ PIPERAZIDINE □ PIPERAZIN
(GERMAN) □ PIPERAZINE, anhydrous □ PYRAZINE
HEXAHYDRIDE
CONSENSUS REPORTS: Reported in EPA
TSCA Inventory.
DOT CLASSIFICATION: 8; Label: Corrosive
SAFETY PROFILE: Moderately toxic by
ingestion, skin contact, intravenous, and
subcutaneous routes. Mildly toxic by
inhalation. A skin and severe eye irritant.
Excessive absorption can cause urticaria,
vomiting, diarrhea, blurred vision, and
weakness. Combustible when exposed to
heat or flame; can react vigorously with
oxidizing materials. Explodes on contact
with dicyanofurazan. To fight fire, use
alcohol foam, mist, dry chemical, water
spray. When heated to decomposition it
emits highly toxic fumes of NO_x.

PIK000   CAS: 142-64-3   HR: 2
PIPERAZINE DIHYDROCHLORIDE
mf: $C_4H_{10}N_2$•2ClH   mw: 159.08
SYNS: DIHYDROCHLORIDE SALT OF DIETHYL-
ENEDIAMINE □ DOWZENE DHC □ PIPERAZINE
HYDROCHLORIDE
CONSENSUS REPORTS: Reported in EPA
TSCA Inventory. EPA Genetic Toxicology
Program.
OSHA PEL: TWA 5 mg/m³
ACGIH TLV: TWA 5 mg/m³
NIOSH REL: (Piperazine) TWA 5.0 mg/m³
SAFETY PROFILE: Moderately toxic by
intraperitoneal route. Mildly toxic by
ingestion. When heated to decomposition it
emits very toxic fumes of NO_x and HCl.
Used in making fiber, pharmaceuticals, and
insecticides. See also PIPERAZINE.

PIK250   CAS: 21416-87-5   HR: 3
**2,6-PIPERAZINEDIONE-4,4'-PROPYL-
ENE DIOXOPIPERAZINE**
mf: $C_{11}H_{16}N_4O_4$ mw: 268.31
PROP: A solid. Mp: 233–234°.
SYNS: (±)-1,2-BIS(3,5-DIOXOPIPERAZINE-1-
YL)PROPANE □ (±)-1,2-BIS(3,5-DIOXOPIPERAZIN-
YL)PROPANE □ ICRF-159 □ 4,4'-(1-METHYL-1,2-
ETHANEDIYL)BIS-2,6-PIPERAZINEDIONE □ NCI-
C01627 □ NSC-129943 □ RAZOXIN □ (±)-(3,5,3',5'-
TETRAOXO)-1,2-DIPIPERAZINOPROPANE
CONSENSUS REPORTS: NCI
Carcinogenesis Bioassay (ipr); Clear
Evidence: mouse, rat NCITR* NCI-CG-
TR-78,78. EPA Genetic Toxicology
Program.
SAFETY PROFILE: Suspected carcinogen
with experimental carcinogenic and
tumorigenic data. Moderately toxic by
intraperitoneal route. Experimental
teratogenic and reproductive effects. Human
systemic effects by ingestion: nausea,
thrombocytopenia, leukopenia. Mutation
data reported. When heated to
decomposition it emits toxic fumes of NO_x.

PIK450   CAS: 7280-37-7   HR: 3
PIPERAZINE ESTRONE SULFATE
mf: $C_{18}H_{22}O_5S$•$C_4H_{10}N_2$ mw: 436.62
SYNS: ESTRA-1,3,5(10)-TRIEN-17-ONE, 3-(SULFOOXY)-,
compounded with PIPERAZINE (1:1) □ ESTRONE,
HYDROGEN SULFATE, compounded with PIPERAZINE
(1:1) (8CI) □ ESTROPIPATE □ HARMOGEN □ OGEN □
PIPERAZINE, compounded with ESTRONE HYDROGEN
SULFATE (1:1) (8CI) □ PIPERAZINE, compounded with 3-
(SULFOOXY)ESTRA-1,3,5(10)-TRIEN-17-ONE (1:1) (9CI) □
SULESTREX
CONSENSUS REPORTS: Reported in NTP
10th Report on Carcinogens.
SAFETY PROFILE: Confirmed carcinogen.
When heated to decomposition it emits
toxic vapors of SO_x and NO_x.

PIL500   CAS: 110-89-4   HR: 3
PIPERIDINE
DOT: UN 2401
mf: $C_5H_{11}N$ mw: 85.17
PROP: Clear, colorless liquid; amine-like
odor. Mp: −9°, bp: 106°, flash p: 37.4°F, d:
0.8622 @ 20°/4°, vap press: 40 mm @

29.2°, vap d: 3.0. Misc with water; sol in alc, benzene, chloroform.

SYNS: AZACYCLOHEXANE □ CYCLOPENTIMINE □ CYPENTIL □ HEXAHYDROPYRIDINE □ HEXAZANE □ PENTAMETHYLENEIMINE □ PIPERIDIN (GERMAN)

CONSENSUS REPORTS: EPA Extremely Hazardous Substances List. Reported in EPA TSCA Inventory. EPA Genetic Toxicology Program.

DOT CLASSIFICATION: 3; Label: Flammable Liquid

SAFETY PROFILE: Poison by ingestion, skin contact, and intraperitoneal routes. Moderately toxic by subcutaneous route. Mildly toxic by inhalation. An experimental teratogen. Experimental reproductive effects by inhalation. A skin irritant. Mutation data reported. A very dangerous fire hazard when exposed to heat, flame, or oxidizers. Can react vigorously with oxidizing materials. To fight fire, use alcohol foam, CO_2, dry chemical. Explodes on contact with 1-perchloryl-piperidine, dicyanofurazan, N-nitrosoacetanilide. When heated to decomposition it emits highly toxic fumes of NO_x. Used in agriculture and pharmaceuticals, and as an intermediate for rubber accelerators. Used in production of drugs of abuse.

PIN200   CAS: 28846-40-4   HR: 3
9-(PIPERIDINOAMINO)ACRIDINE
mf: $C_{18}H_{19}N_3$ mw: 277.40
SYN: ACRIDINE, 9-(PIPERIDINOAMINO)-
SAFETY PROFILE: A poison by ingestion. When heated to decomposition it emits toxic vapors of NO_x.

PIU100   CAS: 7101-58-8   HR: 3
4-(1-PIPERIDYL)CARBONYL-2,3-
**   TETRAMETHYLENEQUINOLINE**
mf: $C_{19}H_{22}N_2O$ mw: 294.43
SYNS: ACRIDINE, 1,2,3,4-TETRAHYDRO-9-(PIP-ERIDINOCARBONYL)- □ KETONE, PIPERIDINO(1,2,3,4-TETRAHYDRO-9-ACRIDINYL)
DOT CLASSIFICATION: 3; Label: Flammable Liquid
SAFETY PROFILE: A poison by intraperitoneal route. A flammable liquid.

When heated to decomposition it emits toxic vapors of NO_x.

PIV600   CAS: 94-62-2   HR: 3
PIPERIN
mf: $C_{17}H_{19}NO_3$ mw: 285.37
PROP: Monoclinic prisms from alcohol; tasteless at first, but burning aftertaste. Mp: 130°. Insol in water (40 mg/liter at 18°) and in pet ether. One gram dissolves in 15 mL alc, 1.7 mL chloroform, 36 mL ether. Sol in benzene, acetic acid.
SYNS: 1-(5-(1,3-BENZODIOXOL-5-YL)-1-OXO-2,4-PENTADIENYL)PIPERIDINE (E,E)- (9CI) □ 1,3-BENZODIOXOL-5-YL-OXO-2,4-PENTADIENYL-PIPERINE □ PIPERINE □ 1-PIPEROYLPIPERIDINE
CONSENSUS REPORTS: Reported in EPA TSCA Inventory.
SAFETY PROFILE: Poison by ingestion and intraperitoneal routes. An experimental teratogen. Experimental reproductive effects. When heated to decomposition it emits toxic fumes of NO_x.

PIV750   CAS: 32248-37-6   HR: 3
PIPEROCAINE
mf: $C_{16}H_{23}NO_2$ mw: 261.40
SYNS: 3-BENZOXY-1-(2-METHYLPIPERIDINO)-PROPANE □ BENZOYL-γ-(2-METHYLPIPERIDINO)-PROPANOL □ ISOCAINE BASE □ 2-METHYL-1-PIPERIDINOPROPANOL, BENZOATE □ (2-METHYLPIPERIDINO)PROPYL BENZOATE □ γ-(2-METHYLPIPERIDYL)PROPYL BENZOATE □ METYCAINE □ NEOTHESIN
SAFETY PROFILE: Poison by intravenous and intraperitoneal routes. Moderately toxic by subcutaneous route. When heated to decomposition it emits toxic fumes of NO_x.

PIW000   CAS: 2622-26-6   HR: 3
PIPEROCYANOMAZINE
mf: $C_{21}H_{23}N_3OS$ mw: 365.53
PROP: Yellow crystals or powder. Mp: 116–117°. Sol in acids.
SYNS: 2-CYANO-10-(3-(4-HYDROXYPIPERIDINO)-PROPYL)PHENOTHIAZINE □ 2-CYANO-10-(3-(4-HYDROXY-1-PIPERIDYL)PROPYL)PHENOTHIAZINE □ CYANO-3-((HYDROXY-4-PIPERIDYL-1)-3 PROPYL)-10-PHENOTHIAZINE (FRENCH) □ F.I. 6145 □ 10-(3-(4-HYDROXYPIPERIDINO)PROPYL)PHENOTHIAZINE-2-

CARBONITRILE □ IC 6002 □ NEMACTIL □ NEULACTIL □ NEULEPTIL □ PERICIAZINE □ PERICYAZINE □ PROPERICIAZINE □ 6909 RP □ RP 8908 □ SKF 20,716 □ WH 7508

CONSENSUS REPORTS: Cyanide and its compounds are on the Community Right-To-Know List.

SAFETY PROFILE: Poison by ingestion, intraperitoneal, intravenous, and subcutaneous routes. Used as an antipsychotic agent. When heated to decomposition it emits very toxic fumes of CN^-, NO_x, and SO_x. See also NITRILES.

**PIW250   CAS: 120-57-0   HR: 2
PIPERONAL**
mf: $C_8H_6O_3$ mw: 150.14
PROP: Colorless, lustrous crystals from water; floral odor. Mp: 37°, bp: 263°, vap press: 1 mm @ 87.0°. Very sol in alc, ether; sol in propylene glycol, fixed oils; insol in water, glycerin.
SYNS: 3,4-BENZODIOXOLE-5-CARBOXALDEHYDE □ 3,4-DIHYDROXYBENZALDEHYDE METHYLENE KETAL □ DIOXYMETHYLENE-PROTOCATECHUIC ALDEHYDE □ FEMA No. 2911 □ HELIOTROPIN □ 3,4-METHYLENE-DIHYDROXYBENZALDEHYDE □ 3,4-METHYLENEDIOXYBENZALDEHYDE □ PIPERON-ALDEHYDE □ PIPERONYL ALDEHYDE □ PROTO-CATECHUIC ALDEHYDE METHYLENE ETHER
CONSENSUS REPORTS: Reported in EPA TSCA Inventory.
SAFETY PROFILE: Moderately toxic by ingestion and intraperitoneal routes. Can cause central nervous system depression. A human skin irritant. Mutation data reported. Combustible when exposed to heat or flame; can react with oxidizing materials. See also ALDEHYDES.

**PIX000   CAS: 326-61-4   HR: 2
PIPERONYL ACETATE**
mf: $C_{10}H_{10}O_4$ mw: 194.20
PROP: Crystals from CCl_4 or pet ether; heliotrope odor. Mp: 51°, bp: 153–154° @ 14 mm.
SYNS: HELIOTROPYL ACETATE □ 3,4-METHYLENE-DIOXYBENZYL ACETATE
CONSENSUS REPORTS: Reported in EPA TSCA Inventory.

SAFETY PROFILE: Moderately toxic by ingestion. A skin irritant. When heated to decomposition it emits acrid smoke and irritating fumes.

**PIX250   CAS: 51-03-6   HR: 3
PIPERONYL BUTOXIDE**
mf: $C_{19}H_{30}O_5$ mw: 338.49
PROP: Light-brown liquid; mild odor. Bp: 180° @ 1 mm, flash p: 340°F, d: 1.04–1.07 @ 20°/20°. Misc with methanol, ethanol, benzene.
SYNS: BUTACIDE □ BUTOCIDE □ BUTOXIDE □ α-(2-(2-BUTOXYETHOXY)ETHOXY)-4,5-METHYLENE-DIOXY-2-PROPYLTOLUENE □ α-(2-(2-n-BUTOXYETHO-XY)-ETHOXY)-4,5-METHYLENEDIOXY-2-PROPYLTOLUENE □ 5-((2-(2-BUTOXYETHOXY)ETHO-XY)METHYL)-6-PROPYL-1,3-BENZODIOXOLE □ BUTYL CARBITOL 6-PROPYLPIPERONYL ETHER □ BUTYL-CARBITYL (6-PROPYLPIPERONYL) ETHER □ ENT 14,250 □ FAC 5273 □ FMC 5273 □ 3,4-METHYLENDIO-XY-6-PROPYLBENZYL-n-BUTYL-DIAETHYLEN-GLYKOLAETHER (GERMAN) □ (3,4-METHYLENEDIO-XY-6-PROPYLBENZYL)(BUTYL)DIETHYLENE GLYCOL ETHER □ 3,4-METHYLENEDIOXY-6-PROPYLBENZYL-n-BUTYL DIETHYLENEGLYCOL ETHER □ NCI-C02813 □ NIA 5273 □ NUSYN-NOXFISH □ PB □ PRENTOX □ 6-(PROPYLPIPERONYL)-BUTYL CARBITYL ETHER □ 6-PROPYLPIPERONYL BUTYL DIETHYLENE GLYCOL ETHER □ 5-PROPYL-4-(2,5,8-TRIOXA-DODECYL)-1,3-BENZODIOXOL (GERMAN) □ PYBUTHRIN □ PYREN-ONE 606 □ SYNPREN-FISH
CONSENSUS REPORTS: IARC Cancer Review: Group 3 IMEMDT 7,56,87; Animal No Evidence IMEMDT 30,183,83. NCI Carcinogenesis Bioassay (feed); No Evidence: mouse, rat NCITR* NCI-CG-TR-120,79. Glycol ether compounds are on the Community Right-To-Know List. Reported in EPA TSCA Inventory.
SAFETY PROFILE: Poison by skin contact. Moderately toxic by ingestion and intraperitoneal routes. An experimental teratogen. Experimental reproductive effects. Many glycol ether compounds have dangerous human reproductive effects. Questionable carcinogen with experimental tumorigenic data. Mutation data reported. Combustible when exposed to heat or flame; can react with oxidizing materials. To fight fire, use foam, CO_2, dry chemical.

When heated to decomposition it emits acrid smoke and irritating fumes. See also GLYCOL ETHERS.

PIY500   CAS: 98-77-1   HR: 3
PIP-PIP
mf: $C_{11}H_{22}N_2S_2$ mw: 246.47
SYNS: PENTAMETHYLENEDITHIOCARBAMATE □ 1-PIPERIDINECARBODITHIOIC ACID, compounded with PIPERIDINE □ PIPERIDINIUM □ "522" RUBBER ACCELERATOR
CONSENSUS REPORTS: Reported in EPA TSCA Inventory.
SAFETY PROFILE: Poison by intraperitoneal route. A human skin irritant. An allergen. When heated to decomposition it emits very toxic fumes of NO_x and SO_x. See also CARBAMATES.

PJA500   CAS: 75-98-9   HR: 2
PIVALIC ACID
mf: $C_5H_{10}O_2$ mw: 102.15
PROP: Crystals. Mp: 35.5°, bp: 164°, d: 0.91. Very sol in alc, ether; somewhat sol in water.
SYNS: 2,2-DIMETHYLPROPANOIC ACID □ α,α-DIMETHYLPROPIONIC ACID □ 2,2-DIMETHYL-PROPIONIC ACID □ NEOPENTANOIC ACID □ tert-PENTANOIC ACID □ PROPANOIC ACID □ TRIMETHYLACETIC ACID
CONSENSUS REPORTS: Reported in EPA TSCA Inventory.
SAFETY PROFILE: Moderately toxic by ingestion and skin contact. Questionable carcinogen with experimental tumorigenic data. When heated to decomposition it emits acrid smoke and irritating fumes.

PJD000   CAS: 15663-27-1   HR: 3
cis-PLATINOUS DIAMMINE DICHLORIDE
mf: $Cl_2H_6N_2Pt$ mw: 300.07
PROP: Yellow solid. Mp: 270° (decomp). Sol in H_2O, DMF, DMSO. IDLH 4 mg/m³ (as Pt).
SYNS: CACP □ CDDP □ CISPLATINO (SPANISH) □ CISPLATYL □ CPDC □ CPDD □ DDP □ cis-DDP □ cis-DIAMMINEDICHLOROPLATINUM □ cis-DICHLORO-DIAMMINE PLATINUM(II) □ NCI-C55776 □ NEOPLA-TIN □ NSC-119875 □ PEYRONE'S CHLORIDE □ PLATIBLASTIN □ cis-PLATIN □ PLATINEX □

PLATINOL □ cis-PLATINUM(II) DIAMMINE-DICHLORIDE
CONSENSUS REPORTS: NTP 10th Report on Carcinogens. IARC Cancer Review: Group 2A IMEMDT 7,170,87; Animal Limited Evidence IMEMDT 26,151,81. Reported in EPA TSCA Inventory. EPA Genetic Toxicology Program.
OSHA PEL: TWA 0.002 mg(Pt)/m³
ACGIH TLV: TWA 0.002 mg(Pt)/m³
SAFETY PROFILE: Confirmed carcinogen with experimental carcinogenic and tumorigenic data. Poison by ingestion, intramuscular, subcutaneous, intravenous, and intraperitoneal routes. Human systemic effects: change in auditory acuity, change in kidney tubules, changes in bone marrow, corrosive to skin, depressed renal function tests, hallucinations, nausea or vomiting. Experimental teratogenic and reproductive effects. Human mutation data reported. When heated to decomposition it emits very toxic fumes of Cl^- and NO_x. See also PLATINUM COMPOUNDS.

PJD250   CAS: 10025-99-7   HR: 3
PLATINOUS POTASSIUM CHLORIDE
mf: $Cl_4Pt•K_2$ mw: 415.09
PROP: Ruby red crystals. Mp: decomp @ 250°, d: 3.499 @ 24°. Sol in water. IDLH 4 mg/m³ (as Pt).
SYNS: POTASSIUM CHLOROPLATINITE □ POTASSIUM PLATINOCHLORIDE □ POTASSIUM TETRACHLOROPLATINATE(II)
CONSENSUS REPORTS: Reported in EPA TSCA Inventory.
OSHA PEL: TWA 0.002 mg(Pt)/m³
ACGIH TLV: TWA 0.002 mg(Pt)/m³
SAFETY PROFILE: Human poison by ingestion. Poison experimentally by intraperitoneal route. Corrosive to human skin by intradermal route. Mutation data reported. Human systemic effects: eosinophilia, gastritis, renal function tests depressed. When heated to decomposition it emits toxic fumes of Cl^- and K_2O. Used as a catalyst for hydroformulations, photocatalysts, and dissociation of water. See also PLATINUM COMPOUNDS.

PJD500   CAS: 7440-06-4   HR: 2
PLATINUM
af: Pt   aw: 195.09

PROP: Silvery-white, malleable, ductile metal. Unaffected by air or H_2O. Platinum-black is velvety-black, finely divided, and this is attacked by O_2 at 5°. At 1° HBr, HI, Br_2, $FeCl_3$, NaCN (+ O_2) are sltly corrosive. No reaction with SO_2. Does not form an amalgam with Hg. Mp: 1772°, bp: 3827°, d: 21.45 @ 20°. Sol in aq regia, HCl in air, fused alkali. IDLH 4 mg/m³ (as Pt).

SYNS: C.I. 77795 □ LIQUID BRIGHT PLATINUM □ PLATIN (GERMAN) □ PLATINUM BLACK □ PLATINUM SPONGE

CONSENSUS REPORTS: Reported in EPA TSCA Inventory.

OSHA PEL: TWA (metal) 1 mg/m³; (soluble salts as Pt) 0.002 mg/m³

ACGIH TLV: TWA (metal) 1 mg/m³; (soluble salts as Pt) 0.002 mg/m³

DFG MAK: 0.002 mg/m³

NIOSH REL: (Platinum (as Pt), metal) TWA 1 mg/m³; (Platinum (as Pt), soluble salts): TWA 0.002 mg/m³

SAFETY PROFILE: Questionable carcinogen with experimental tumorigenic data by implant route. Finely divided platinum is a powerful catalyst and can be dangerous to handle. Used catalysts are especially dangerous and may be explosive. May undergo hazardous reactions with aluminum, acetone, arsenic, carbon + methanol, nitrosyl chloride, dioxygen difluoride, ethanol, hydrazine, hydrogen + air, hydrogen peroxide, lithium, methyl hydroperoxide, ozonides, peroxymonosulfuric acid, phosphorus, selenium, tellurium, vanadium dichloride + water. See also PLATINUM COMPOUNDS.

PJE000   CAS: 10025-65-7   HR: 2
PLATINUM CHLORIDE
mf: Cl_2Pt mw: 265.99

PROP: Grayish-green powder. D: 5.87. Insol in water, alc, ether, benzene, chloroform. IDLH 4 mg/m³ (as Pt).

SYNS: MURIATE of PLATINUM □ PLATINOUS CHLORIDE

CONSENSUS REPORTS: Reported in EPA TSCA Inventory.

OSHA PEL: TWA 0.002 mg(Pt)/m³

ACGIH TLV: TWA 0.002 mg(Pt)/m³

SAFETY PROFILE: Moderately toxic by ingestion. A skin irritant. Human mutation data reported. When heated to decomposition it emits toxic fumes of Cl⁻. See also PLATINUM COMPOUNDS.

PJE250   CAS: 13454-96-1   HR: 3
PLATINUM(IV) CHLORIDE
mf: Cl_4Pt mw: 336.89

PROP: Hygroscopic red-brown crystals. Sol in H_2O and Me_2CO. IDLH 4 mg/m³ (as Pt).

SYN: PLATINUM TETRACHLORIDE

CONSENSUS REPORTS: Reported in EPA TSCA Inventory. EPA Genetic Toxicology Program.

OSHA PEL: TWA 0.002 mg(Pt)/m³

ACGIH TLV: TWA 0.002 mg(Pt)/m³

SAFETY PROFILE: Poison by ingestion and intravenous routes. Experimental reproductive effects. Mutation data reported. A severe skin irritant. When heated to decomposition it emits toxic fumes of Cl⁻. See also PLATINUM COMPOUNDS.

PJE500   HR: 2
PLATINUM COMPOUNDS
PROP: IDLH 4 mg/m³ (as Pt).

SAFETY PROFILE: cis-[Pt(NH₃)₂Cl₂] is an experimental carcinogen. Exposure to complex platinum salts has been shown to cause symptoms of intoxication such as wheezing, coughing, running of the nose, chest tightness, shortness of breath, and cyanosis. Furthermore, many people working with platinum salts are troubled with dermatitis. They may become sensitized after years of exposure. Symptoms of platinum allergy include rhinitis, conjunctivitis, asthma, urticaria, and contact dermatitis. Mainly the ionic platinum chloro compounds [e.g., (NHN₄)₂(PtCl₆),

(NH$_4$)$_2$(PtCl$_4$), H$_2$(PtCl$_6$)] are responsible for this sensitivity. The bromide and iodide compounds are less effective. These platinum compounds form a platinum-protein conjugate that is the true allergen. Tetrachloroplatinates are mutagens. This seems to be true only of complex platinum salts. It does not apply to the complex salts of the other precious metals. Platinum amine nitrates and perchlorates either detonate when heated or are impact-sensitive.

PJI000   HR: 3
PLUTONIUM COMPOUNDS
SAFETY PROFILE: The toxicity of plutonium compounds is based first upon the very high radiotoxicity of the plutonium atom and secondly upon whatever atoms or combinations of atoms they might contain. Very dangerous!! Any disaster which causes quantities of plutonium or plutonium compounds to be scattered about the environment will cause great ecological stress and render areas of the land unfit for public occupancy. Long-term storage in plastic containers is not recommended, as the alpha particles can cause stress cracks and there is a potential for leakage. See also PLUTONIUM.

PJJ000   CAS: 9000-55-9   HR: 3
PODOPHYLLIN
PROP: Light-yellow powder or small yellow fragile lumps; bitter, acrid taste.
SYNS: PODOPHYLLUM □ PODOPHYLLUM RESIN
SAFETY PROFILE: Poison by ingestion, subcutaneous, and intraperitoneal routes. An irritant to skin, eyes, and mucous membranes. Questionable carcinogen with experimental neoplastigenic data. An experimental teratogen. Other experimental reproductive effects. Combustible when exposed to heat or flames. When heated to decomposition it emits acrid smoke and irritating fumes.

PJJ750   HR: 3
POLONIUM
af: Po   aw: 210
PROP: A low-melting, volatile, radioactive, naturally occurring metallic element. Mp: 254°, bp: 962°, d: 9.4.
SYN: RADIUM F
SAFETY PROFILE: Suspected carcinogen. Severe radiotoxicity. Very dangerous to handle. Radiation Hazard: Natural isotope ^{210}Po (radium-F, uranium series), $T_{0.5}$ = 138 days. Decays to stable ^{206}Pb by alphas of 5.3 MeV. When heated to decomposition it emits toxic and radioactive fumes of Po. See also PLUTONIUM.

PJK000   HR: 3
POLONIUM CARBONYL
mf: PoCO   mw: 237.01
SAFETY PROFILE: Suspected carcinogen. Poison by ingestion, inhalation, intravenous, and subcutaneous routes. When heated to decomposition it emits toxic and radioactive fumes of Po. See also CARBONYLS and POLONIUM.

PJK150   CAS: 9003-11-6   HR: D
POLOXALENE
mf:
HO(CH$_2$CH$_2$O)$_n$[CH(CH$_3$)CH$_2$O]$_n$(CH$_2$CH$_2$O)$_n$H
PROP: Liquid nonionic surfactant polymer.
SYNS: BIS[HYDROXYETHYLPOLY(ETHYLENEOXY)ETHYLPROPYLENEGLYCOL □ BLOAT GUARD □ DIPOLYOXY-ETHYLATEDPOLYPROPYLENEGLYCOL ETHER □ POLY(OXYETHYLENE)-POLY(OXYPROPYLENE)-POLY(OXYETHYLENE) POLYMER □ THERABLOAT
SAFETY PROFILE: When heated to decomposition it emits acrid smoke and irritating fumes.

PJK200   HR: D
POLOXAMER 331
PROP: Average molecular weight 3800. Colorless liquid. D: 1.02, refr index: 1.452. Very sltly sol in water; sol in alc; insol in propylene glycol, ethylene glycol.

P

SYNS: ETHYLENE OXIDE and PROPYLENE OXIDE BLOCK POLYMER □ PROPYLENE OXIDE and ETHYLENE OXIDE BLOCK POLYMER □ α-HYDRO-ω-HYDROXY-POLY(OXYRTHYLENE)-POLY(OXYPROPYL-ENE)(51-57 MOLES)POLY(OXYETHYLENE) BLOCK POLYMER

SAFETY PROFILE: When heated to decomposition it emits acrid smoke and irritating fumes.

PJL750   CAS: 1336-36-3   HR: 3
POLYCHLORINATED BIPHENYLS
DOT: UN 2315
PROP: Bp: 340–375°, flash p: 383°F (COC), d: 1.44 @ 30°. A series of technical mixtures consisting of many isomers and compounds that vary from mobile oily liquids to white crystalline solids and hard noncrystalline resins. Technical products vary in composition, in the degree of chlorination, and possibly according to batch (IARC** 7,262,74).
SYNS: AROCLOR □ AROCLOR 1016 □ AROCLOR 1221 □ AROCLOR 1232 □ AROCLOR 1242 □ AROCLOR 1248 □ AROCLOR 1254 □ AROCLOR 1260 □ AROCLOR 1262 □ AROCLOR 1268 □ AROCLOR 2565 □ AROCLOR 4465 □ AROCLOR 5442 □ BIPHENYL, POLYCHLORO- □ CHLOPHEN □ CHLOREXTOL □ CHLORINATED BIPHENYL □ CHLORINATED DIPHENYL □ CHLORINATED DIPHENYLENE □ CHLORO BIPHENYL □ CHLORO 1,1-BIPHENYL □ CLOPHEN □ DYKANOL □ FENCLOR □ FENCLOR 42 □ INERTEEN □ KANECHLOR □ KANECHLOR 300 □ KANECHLOR 400 □ MONTAR □ NOFLAMOL □ PCB □ PCBs □ PHENOCHLOR □ PHENOCLOR □ POLYCHLOROBIPHENYL □ PYRALENE □ PYRANOL □ SANTOTHERM □ SANTOTHERM FR □ SOVOL □ THERMINOL FR-1
CONSENSUS REPORTS: NTP 10th Report on Carcinogens. IARC Cancer Review: Group 2A IMEMDT 7,322,87; Human Limited Evidence IMEMDT 18,43,78. Reported in EPA TSCA Inventory.
DFG MAK: Suspected Carcinogen
NIOSH REL: TWA (Polychlorinated Biphenyls) 0.001 mg/m³
DOT CLASSIFICATION: 9; Label: CLASS 9
SAFETY PROFILE: Confirmed carcinogen with carcinogenic and tumorigenic data. Moderately toxic by ingestion. Some are poisons by other routes. Experimental reproductive effects.

Like the chlorinated naphthalenes, the chlorinated diphenyls have two distinct actions on the body, namely, a skin effect and a toxic action on the liver. This hepatotoxic action of the chlorinated diphenyls appears to be increased if there is exposure to carbon tetrachloride at the same time. The higher the chlorine content of the diphenyl compound, the more toxic it is liable to be. Oxides of chlorinated diphenyls are more toxic than the unoxidized materials. In persons who have suffered systemic intoxication, the usual signs and symptoms are nausea, vomiting, loss of weight, jaundice, edema, and abdominal pain. If the liver damage has been severe the patient may pass into a coma and die.

Combustible when exposed to heat or flame. When heated to decomposition they emit highly toxic fumes of Cl⁻. See also specific compounds.

PJM000   CAS: 11104-28-2   HR: 2
POLYCHLORINATED BIPHENYL
(AROCLOR 1221)
SYNS: AROCHLOR 1221 □ CHLORODIPHENYL (21% Cl) □ PCB
CONSENSUS REPORTS: IARC Cancer Review: Human Limited Evidence IMEMDT 18,43,78.
NIOSH REL: TWA (Polychlorinated Biphenyls) 0.001 mg/m³
SAFETY PROFILE: Suspected human carcinogen. Moderately toxic by ingestion and skin contact. Experimental reproductive effects. When heated to decomposition it emits toxic fumes of Cl⁻. Used in heat transfer, hydraulic fluids, lubricants, and insecticides. See also POLYCHLORINATED BIPHENYLS.

PJM250   CAS: 11141-16-5   HR: 2
POLYCHLORINATED BIPHENYL
(AROCLOR 1232)
SYNS: AROCLOR 1232 □ CHLORODIPHENYL (32% Cl) □ PCB
CONSENSUS REPORTS: IARC Cancer Review: Human Limited Evidence IMEMDT 18,43,78.

NIOSH REL: TWA (Polychlorinated Biphenyls) 0.001 mg/m³
SAFETY PROFILE: Suspected human carcinogen. Moderately toxic by skin contact. Mildly toxic by ingestion. When heated to decomposition it emits toxic fumes of Cl⁻. Used in heat transfer, hydraulic fluids, lubricants, and insecticides. See also POLYCHLORINATED BIPHENYLS.

PJM500   CAS: 53469-21-9   HR: 3
POLYCHLORINATED BIPHENYL (AROCLOR 1242)

PROP: IDLH 5 mg/m³.
SYNS: AROCHLOR 1242 □ AROCLOR 1242 □ CHLORIERTE BIPHENYLE, CHLORGEHALT 42% (GERMAN) □ CHLORODIPHENYL (42% Cl) (OSHA) □ CLORODIFENILI, CLORO 42% (ITALIAN) □ DIPHENYLE CHLORE, 42% de CHLORE (FRENCH) □ GECHLOREERDEDIFENYL (DUTCH) □ PCB
CONSENSUS REPORTS: IARC Cancer Review: Human Limited Evidence IMEMDT 18,43,78. EPA Genetic Toxicology Program.
OSHA PEL: TWA 1 mg/m³ (skin)
ACGIH TLV: TWA 1 mg/m³ (skin)
DFG MAK: 0.1 ppm (1.1 mg/m³)
NIOSH REL: TWA (Polychlorinated Biphenyls) 0.001 mg/m³
SAFETY PROFILE: Suspected human carcinogen. Poison by subcutaneous route. Mildly toxic by ingestion. Human systemic effects by inhalation: pulmonary and liver effects. Experimental reproductive effects. Mutation data reported. When heated to decomposition it emits toxic fumes of Cl⁻. Used in heat transfer, hydraulic fluids, lubricants, and insecticides. See also POLYCHLORINATED BIPHENYLS.

PJM750   CAS: 12672-29-6   HR: 3
POLYCHLORINATED BIPHENYL (AROCLOR 1248)

SYNS: AROCLOR 1248 □ CHLORODIPHENYL (48% Cl) □ PCB
CONSENSUS REPORTS: IARC Cancer Review: Human Limited Evidence IMEMDT 18,43,78.

NIOSH REL: TWA (Polychlorinated Biphenyls) 0.001 mg/m³
SAFETY PROFILE: Suspected human carcinogen. Moderately toxic by skin contact. Experimental teratogenic and reproductive effects. When heated to decomposition it emits toxic fumes of Cl⁻. Used in heat transfer, hydraulic fluids, lubricants, and insecticides. See also POLYCHLORINATED BIPHENYLS.

PJN000   CAS: 11097-69-1   HR: 3
POLYCHLORINATED BIPHENYL (AROCLOR 1254)

PROP: Composed of 11% tetra-, 49% penta-, 34% hexa-, and 6% heptachlorobiphenyls (FCTXAV 12,63,74). IDLH 5 mg/m³.
SYNS: AROCLOR 1254 □ AROCLOR 1254 □ CHLORIERTE BIPHENYLE, CHLORGEHALT 54% (GERMAN) □ CHLORODIPHENYL (54% Cl) (OSHA) □ CLORODIFENILI, CLORO 54% (ITALIAN) □ DIPHENYLE CHLORE, 54% de CHLORE (FRENCH) □ NCI-C02664 □ PCB
CONSENSUS REPORTS: NTP 10th Report on Carcinogens. IARC Cancer Review: Group 2A IMEMDT 7,322,87; Animal Sufficient Evidence IMEMDT 7,261,74; Animal Limited Evidence IMEMDT 18,43,78; Human Limited Evidence IMEMDT 18,43,78. NCI Carcinogenesis Bioassay (feed); Some Evidence: rat NCITR* NCI-CG-TR-38,78. EPA Genetic Toxicology Program.
OSHA PEL: TWA 0.5 mg/m³ (skin)
ACGIH TLV: TWA 0.5 mg/m³ (skin); Animal Carcinogen
DFG MAK: 0.05 ppm (0.70 mg/m³); Suspected Carcinogen
NIOSH REL: TWA (Polychlorinated Biphenyls) 0.001 mg/m³
SAFETY PROFILE: Confirmed carcinogen with experimental carcinogenic and neoplastigenic data. Poison by intravenous route. Moderately toxic by ingestion and intraperitoneal routes. Experimental teratogenic and reproductive effects. Mutation data reported. When heated to decomposition it emits toxic fumes of Cl⁻.

P

Used in heat transfer, hydraulic fluids, lubricants, and insecticides. See also POLYCHLORINATED BIPHENYLS.

PJN250   CAS: 11096-82-5   HR: 3
POLYCHLORINATED BIPHENYL
   (AROCLOR 1260)
PROP: Composed of 12% penta-, 38% hexa-, 41% hepta-, 8% octa-, and 1% nonachlorobiphenyls (FCTXAV 12,63,74).
SYNS: AROCHLOR 1260 □ AROCLOR 1260 □ CHLORODIPHENYL (60% Cl) □ CLOPHEN A60 □ PCB □ PHENOCLOR DP6
CONSENSUS REPORTS: NTP 10th Report on Carcinogens. IARC Cancer Review: Animal Limited Evidence IMEMDT 18,43,78; Human Limited Evidence IMEMDT 18,43,78.
NIOSH REL: TWA (Polychlorinated Biphenyls) 0.001 mg/m^3
SAFETY PROFILE: Confirmed carcinogen with carcinogenic and neoplastigenic data. Moderately toxic by ingestion and skin contact. Experimental reproductive effects. Mutation data reported. When heated to decomposition it emits highly toxic fumes of Cl⁻. Used in heat transfer, hydraulic fluids, lubricants, and insecticides. See also POLYCHLORINATED BIPHENYLS.

PJN500   CAS: 37324-23-5   HR: 3
POLYCHLORINATED BIPHENYL
   (AROCLOR 1262)
SYNS: AROCLOR 1262 □ CHLORODIPHENYL (62% Cl) □ PCB
CONSENSUS REPORTS: IARC Cancer Review: Human Limited Evidence IMEMDT 18,43,78.
DFG MAK: 0.1 ppm (1 mg/m^3)
NIOSH REL: (Polychlorinated Biphenyls) TWA 0.001 mg/m^3
SAFETY PROFILE: Suspected human carcinogen. Moderately toxic by skin contact. When heated to decomposition it emits toxic fumes of Cl⁻. Used in heat transfer, hydraulic fluids, lubricants, and insecticides. See also POLYCHLORINATED BIPHENYLS.

PJN750   CAS: 11100-14-4   HR: 3
POLYCHLORINATED BIPHENYL
   (AROCLOR 1268)
SYNS: AROCLOR 1268 □ CHLORODIPHENYL (68% Cl) □ PCB
CONSENSUS REPORTS: IARC Cancer Review: Human Limited Evidence IMEMDT 18,43,78.
NIOSH REL: (Polychlorinated Biphenyls) TWA 0.001 mg/m^3
SAFETY PROFILE: Suspected human carcinogen. Moderately toxic by skin contact. Used in heat transfer, hydraulic fluids, lubricants, and insecticides. When heated to decomposition it emits toxic fumes of Cl⁻. See also POLYCHLORINATED BIPHENYLS.

PJO000   CAS: 37324-24-6   HR: 3
POLYCHLORINATED BIPHENYL
   (AROCLOR 2565)
SYNS: AROCLOR 2565 □ PCB
CONSENSUS REPORTS: IARC Cancer Review: Human Limited Evidence IMEMDT 18,43,78.
NIOSH REL: (Polychlorinated Biphenyls) TWA 0.001 mg/m^3
SAFETY PROFILE: Suspected human carcinogen. Moderately toxic by skin contact. Mildly toxic by ingestion. When heated to decomposition it emits toxic fumes of Cl⁻. Used in heat transfer, hydraulic fluids, lubricants, and insecticides. See also POLYCHLORINATED BIPHENYLS.

PJO250   CAS: 11120-29-9   HR: 3
POLYCHLORINATED BIPHENYL
   (AROCLOR 4465)
SYNS: AROCLOR 4465 □ PCB
CONSENSUS REPORTS: IARC Cancer Review: Human Limited Evidence IMEMDT 18,43,78.
NIOSH REL: TWA (Polychlorinated Biphenyls) 0.001 mg/m^3
SAFETY PROFILE: Suspected human carcinogen. Moderately toxic by skin contact. Mildly toxic by ingestion. When heated to decomposition it emits toxic

fumes of Cl⁻. Used in heat transfer, hydraulic fluids, lubricants, and insecticides. See also POLYCHLORINATED BIPHENYLS.

PJO500   CAS: 37353-63-2   HR: 3
POLYCHLORINATED BIPHENYL
(KANECHLOR 300)
PROP: Average content: 60% trichlorobiphenyl, 23% tetrachlorobiphenyl, 17% dichlorobiphenyl, 1% pentachlorobiphenyl (IARC** 7,262,74).
SYNS: KANECHLOR 300 □ PCB
CONSENSUS REPORTS: IARC Cancer Review: Animal Limited Evidence IMEMDT 7,261,74; IMEMDT 18,43,78; Human Limited Evidence IMEMDT 18,43,78.
NIOSH REL: TWA (Polychlorinated Biphenyls) 0.001 mg/m³
SAFETY PROFILE: Moderately toxic by ingestion. Suspected human carcinogen. An experimental teratogen. Used in heat transfer, hydraulic fluids, lubricants, and insecticides. When heated to decomposition it emits toxic fumes of Cl⁻. See also POLYCHLORINATED BIPHENYLS.

PJO750   CAS: 12737-87-0   HR: 3
POLYCHLORINATED BIPHENYL
(KANECHLOR 400)
PROP: Average content: 44% tetrachlorobiphenyl, 33% trichlorobiphenyl, 16% pentachlorobiphenyl, 5% hexachlorobiphenyl, 3% dichlorobiphenyl (IARC** 7,262,74).
SYNS: KANECHLOR 400 □ KC-400 □ PCB
CONSENSUS REPORTS: IARC Cancer Review: Animal Limited Evidence IMEMDT 7,261,74; IMEMDT 18,43,78; Human Limited Evidence IMEMDT 18,43,78.
NIOSH REL: TWA (Polychlorinated Biphenyls) 0.001 mg/m³
SAFETY PROFILE: Suspected carcinogen with experimental neoplastigenic data. Moderately toxic by ingestion. Experimental teratogenic and reproductive effects. Human

systemic effects by ingestion: dermatitis, sweating. When heated to decomposition it emits toxic fumes of Cl⁻. See also POLYCHLORINATED BIPHENYLS.

PJP000   CAS: 37317-41-2   HR: 3
POLYCHLORINATED BIPHENYL
(KANECHLOR 500)
PROP: Average content, 55% pentachlorobiphenyl, 26.5% tetrachlorobiphenyl, 12.8% hexachloro biphenyl, and 5% trichlorobiphenyl (JNCIAM 51,1637,73).
SYNS: KANECHLOR 500 □ KC-500 □ PCB
CONSENSUS REPORTS: NTP 10th Report on Carcinogens. IARC Cancer Review: Human Limited Evidence IMEMDT 18,43,78; Animal Limited Evidence IMEMDT 18,43,78; Animal Sufficient Evidence IMEMDT 7,261,74.
NIOSH REL: TWA (Polychlorinated Biphenyls) 0.001 mg/m³
SAFETY PROFILE: Confirmed carcinogen with experimental carcinogenic and tumorigenic data. Experimental teratogenic and reproductive effects. When heated to decomposition it emits toxic fumes of Cl⁻. Used in heat transfer, hydraulic fluids, lubricants, and insecticides. See also POLYCHLORINATED BIPHENYLS.

PJQ430   HR: D
POLYDEXTROSE SOLUTION
PROP: Clear, straw-colored liquid.
SAFETY PROFILE: When heated to decomposition it emits acrid smoke and irritating fumes.

PJS750   CAS: 9002-88-4   HR: 2
POLYETHYLENE
mf: (C₂H₄)ₙ
PROP: Odorless. The high–molecular-weight compounds are tough, white leathery, resinous. D: 0.92 @ 20°/4°, mp: 85–110°. Sol in hot benzene; insol in water.
SYNS: AC 8 □ AC 394 □ AC 680 □ AC 1220 □ AC GA □ ACP 6 □ AC 8 (POLYMER) □ ACROART □ AGILENE □ ALATHON □ ALATHON 14 □ ALATHON 15 □ ALATHON 1560 □ ALATHON 6600 □ ALATHON 7026 □

P

ALATHON 7040 □ ALATHON 7050 □ ALATHON 7140 □ ALATHON 7511 □ ALATHON 5B □ ALATHON 71XHN □ ALCOWAX 6 □ ALDYL A □ ALITHON 7050 □ ALKATHENE □ ALKATHENE 17/04/00 □ ALKATHENE 22 300 □ ALKATHENE 200 □ ALKATHENE ARN 60 □ ALKATHENE WJG 11 □ ALKATHENE WNG 14 □ ALKATHENE XDG 33 □ ALKATHENE XJK 25 □ ALLIED PE 617 □ ALPHEX FIT 221 □ AMBYTHENE □ AMOCO 610A4 □ A 60-20R □ A 60-70R □ BAKELITE DFD 330 □ BAKELITE DHDA 4080 □ BAKELITE DYNH □ BARECO POLYWAX 2000 □ BARECO WAX C 7500 □ BICOLENE C □ BPE-I □ BRALEN KB 2-11 □ BRALEN RB 03-23 □ BULEN A □ BULEN A 30 □ CARLONA 58-030 □ CARLONA 900 □ CARLONA 18020 FA □ CARLONA PXB □ CHEMCOR □ CHEMPLEX 3006 □ CIPE □ COATHYLENE HA 1671 □ COURLENE-X3 □ CPE □ CPE 16 □ CPE 25 □ CRYOPOLYTHENE □ CRY-O-VAC L □ DAISOLAC □ DAPLEN □ DAPLEN 1810 H □ DFD 0173 □ DFD 0188 □ DFD 2005 □ DFD 6005 □ DFD 6032 □ DFD 6040 □ DFDJ 5505 □ DGNB 3825 □ DIOTHENE □ DIXOPAK □ DMDJ 4309 □ DMDJ 5140 □ DMDJ 7008 □ DOWLEX FILM □ DQDA 1868 □ DQWA 0355 □ DXM 100 □ DYALL □ DYLAN □ DYLAN SUPER □ DYLAN WPD 205 □ DYNH □ DYNK 2 □ ELTEX □ ELTEX 6037 □ ELTEX A 1050 □ EPOLENE C □ EPOLENE C 10 □ EPOLENE C 11 □ EPOLENE E □ EPOLENE E 10 □ EPOLENE E 12 □ EPOLENE N □ ETHENE POLYMER □ ETHERIN □ ETHEROL E □ ETHYLENE HOMOPOLYMER □ ETHYLENE POLYMER □ ETHYLENE POLYMERS (8CI) □ 23F203 □ FABRITONE PE □ FB 217 □ FERTENE □ FLAMOLIN MF 15711 □ FLOTHENE □ FM 510 □ FORTIFLEX 6015 □ FORTIFLEX A 60/500 □ FP 4 □ 2100 GP □ G-RESINS □ GREX □ GREX PP 60-002 □ GRISOLEN □ HFDB 4201 □ HI-FAX □ HI-FAX 1900 □ HI-FAX 4401 □ HI-FAX 4601 □ HIZEX □ HIZEX 5000 □ HIZEX 5100 □ HIZEX 3000B □ HIZEX 3300F □ HIZEX 7000F □ HIZEX 7300F □ HIZEX 1091J □ HIZEX 1291J □ HIZEX 1300J □ HIZEX 2100J □ HIZEX 2200J □ HIZEX 2100LP □ HIZEX 5100LP □ HIZEX 6100P □ HIZEX 3000S □ HIZEX 3300S □ HIZEX 5000S □ HOECHST PA 190 □ HOECHST WAX PA 520 □ HOSTALEN □ HOSTALEN GD 620 □ HOSTALEN GD 6250 □ HOSTALEN GF 4760 □ HOSTALEN GF 5750 □ HOSTALEN GM 5010 □ HOSTALEN GUR □ HOSTALEN HDPE □ INTERFLO □ IRAX □ IRRATHENE R □ LACQTEN 1020 □ LD 400 □ LD 600 □ LDPE 4 □ LUPOLEN 4261A □ LUPOLEN 6042D □ LUPOLEN 1010H □ LUPOLEN 1800H □ LUPOLEN 1810H □ LUPOLEN 6011H □ LUPOLEN KR 1032 □ LUPOLEN KR 1051 □ LUPOLEN KR 1257 □ LUPOLEN 6011L □ LUPOLEN L 6041D □ LUPOLEN N □ LUPOLEN 1800S □ MANOLENE 6050 □ MARLEX 9 □ MARLEX 50 □ MARLEX 60 □ MARLEX 960 □ MARLEX 6003 □ MARLEX 6009 □ MARLEX 6015 □ MARLEX 6050 □ MARLEX 6060 □ MARLEX EHM 6001 □ MARLEX M 309 □ MARLEX TR 704 □ MARLEX TR 880 □ MARLEX TR 885 □ MARLEX TR 906 □ MICROTHENE □ MICROTHENE 510 □ MICROTHENE 704 □ MICROTHENE 710 □ MICROTHENE F □ MICROTHENE FN 500 □ MICROTHENE FN 510 □ MICROTHENE MN 754-18 □ MIKROLOUR □ MIRASON 9 □ MIRASON 16 □ MIRASON M 15 □ MIRASON M 50 □ MIRASON M 68 □ MIRASON NEO 23H □ MIRATHEN □ MIRATHEN 1313 □ MIRATHEN 1350 □ MOPLEN RO-QG 6015 □ NEOPOLEN □ NEOPOLEN 30N □ NEOZEX 45150 □ NEOZEX 4010B □ NOPOL (POLYMER) □ NOVATEC JUO 80 □ NOVATEC JVO 80 □ NVC 9025 □ OKITEN G 23 □ ORIZON □ ORIZON 805 □ 6020P □ PA 130 □ PA 190 □ PA 520 □ PA 560 □ PAD 522 □ P 2010B □ PE 512 □ PE 617 □ PEN 100 □ PEP 211 □ PES 100 □ PES 200 □ PETROTHENE □ PETROTHENE LB 861 □ PETROTHENE LC 731 □ PETROTHENE LC 941 □ PETROTHENE NA 219 □ PETROTHENE NA 227 □ PETROTHENE XL 6301 □ P 4007EU □ P 4070L □ PLANIUM □ PLASKON PP 60-002 □ PLASTAZOTE X 1016 □ PLASTRONGA □ PLASTYLENE MA 2003 □ PLASTYLENE MA 7007 □ POLITEN □ POLITEN I 020 □ POLYAETHYLEN □ POLY-EM 12 □ POLY-EM 40 □ POLY-EM 41 □ POLYETHYLENE AS □ POLYETHYLENE RESINS □ POLYMIST A12 □ POLYMUL CS 81 □ POLYSION N 22 □ POLYTHENE □ POLYWAX 1000 □ POROLEN □ P 2070P □ PPE 2 □ PROCENE UF 1.5 □ PROFAX A 60-008 □ P 2020T □ P 2050T □ P 4007T □ PTS 2 □ PVP 8T □ PY 100 □ RCH 1000 □ REPOC □ RIGIDEX □ RIGIDEX 35 □ RIGIDEX 50 □ RIGIDEX TYPE 2 □ ROPOL □ ROPOTHENE OB.03-110 □ SANWAX 161P □ SCLAIR 59 □ SCLAIR 2911 □ SCLAIR 19A □ SCLAIR 96A □ SCLAIR 59C □ SCLAIR 79D □ SCLAIR 11K □ SCLAIR 19X6 □ SDP 640 □ SHOLEX 5003 □ SHOLEX 5100 □ SHOLEX 6000 □ SHOLEX 6002 □ SHOLEX F 171 □ SHOLEX F 6050C □ SHOLEX F 6080C □ SHOLEX 4250HM □ SHOLEX L 131 □ SHOLEX S 6008 □ SHOLEX SUPER □ SHOLEX XMO 314 □ SOCAREX □ SRM 1475 □ SRM 1476 □ STAFLEN E 650 □ STAMYLAN 900 □ STAMYLAN 1000 □ STAMYLAN 1700 □ STAMYLAN 8200 □ STAMYLAN 8400 □ SUMIKATHENE □ SUMIKATHENE F 101-1 □ SUMIKATHENE F 210-3 □ SUMIKATHENE F 702 □ SUMIKATHENE G 201 □ SUMIKATHENE G 202 □ SUMIKATHENE G 701 □ SUMIKATHENE G 801 □ SUMIKATHENE G 806 □ SUMIKATHENE HARD 2052 □ SUNWAX 151 □ SUPER DYLAN □ SUPRATHEN □ SUPRATHEN C 100 □ TAKATHENE □ TAKATHENE P 3 □ TAKATHENE P 12 □ TELCOTENE □ TELECOTHENE □ TENAPLAS □ TENITE 800 □ TENITE 1811 □ TENITE 2910 □ TENITE 2918 □ TENITE 3300 □ TENITE 3340 □ TROVIDUR PE □ TYRIN □ TYVEK □ UNIFOS DYOB S □ UNIFOS EFD 0118 □ VALERON □ VALSPEX 155-53 □ VELUSTRAL KPA □ VESTOLEN □ VESTOLEN A 616 □ VESTOLEN A 6016 □ WAX LE □ WJG 11 □ WNF 15 □ WVG 23 □ XL

335-1 □ XL 1246 □ XNM 68 □ XO 440 □ YUKALON EH 30 □ YUKALON HE 60 □ YUKALON K 3212 □ YUKALON LK 30 □ YUKALON MS 30 □ YUKALON PS 30 □ YUKALON YK 30 □ ZF 36 □ ZINPOL

CONSENSUS REPORTS: IARC Cancer Review: Group 3 IMEMDT 7,56,87; Animal Sufficient Evidence IMEMDT 19,157,79; Human Inadequate Evidence IMEMDT 19,157,79. Reported in EPA TSCA Inventory.

SAFETY PROFILE: Questionable carcinogen with experimental tumorigenic data by implant. Reacts violently with F_2. When heated to decomposition it emits acrid smoke and irritating fumes.

PJT000   CAS: 25322-68-3   HR: 2
POLYETHYLENE GLYCOL

mf: $(C_2H_4O)_n \cdot H_2O$

PROP: Clear, viscous liquid or white solid. D: 1.110–1.140 @ 20°, mp: 4–10°, flash p: 471°F. Sol in water, org solvs, aromatic hydrocarbons.

SYNS: ALKAPOL PEG-200 □ ALKAPOL PEG-300 □ ALKAPOL PEG-600 □ ALKAPOL PEG-6000 □ ALKAPOL PEG-8000 □ CARBOWAX □ α-HYDRO-omega-HYDROXYPOLY(OXY-1,2-ETHANEDIYL) □ JEFFOX □ JORCHEM 400 ML □ LUTROL □ PLURACOL E-200 □ PLURACOL E-300 □ PLURACOL E-400 □ PLURACOL E-600 □ PLURACOL E-1500 □ PLURACOL E-4000 □ PLURACOL E-6000 □ PLURACOL P-410 □ PLURACOL P-710 □ PLURACOL P-1010 □ PLURACOL P-2010 □ PLURACOL P-3010 □ PLURACOL P-4010 □ POLY(ETHYLENE OXIDE) □ POLY-G SERIES □ POLYOX □ POLY(OXY-1,2-ETHANEDIYL), α-HYDRO-omega-HYDROXY-

CONSENSUS REPORTS: Reported in EPA TSCA Inventory. EPA Genetic Toxicology Program.

SAFETY PROFILE: Moderately toxic by intravenous route. A skin and eye irritant. Combustible liquid when exposed to heat or flame. To fight fire, use water, foam, dry chemical. When heated to decomposition it emits acrid smoke and irritating fumes. See also other polyethylene glycol entries.

PJT200   CAS: 25322-68-3   HR: 1
POLYETHYLENE GLYCOL 200

mf: $H(OC_2H_4)_nOH$

PROP: Viscous, hydroscopic liquid with *n* about 4; slt characteristic odor. D (25°/25°) 1.127.

SYNS: CARBOWAX □ JEFFOX □ NYCOLINE □ PEG 200 □ PLURACOL E □ POLYAETHYLENGLYCOLE 200 (GERMAN) □ POLY-G □ POLYGLYCOL E □ SOLBASE

CONSENSUS REPORTS: EPA Genetic Toxicology Program. Reported in EPA TSCA Inventory.

DFG MAK: 1000 mg/m³

SAFETY PROFILE: Mildly toxic by ingestion. An eye irritant. Caution: Solvent action on some plastics. When heated to decomposition it emits acrid smoke and irritating fumes. See also other polyethylene glycol entries.

PJT225   CAS: 25322-68-3   HR: 1
POLYETHYLENE GLYCOL 300

mf: $(C_6H_{11}NO)_n$

SYNS: PEG 300 □ POLYAETHYLENGLYKOLE 300 (GERMAN)

CONSENSUS REPORTS: EPA Genetic Toxicology Program. Reported in EPA TSCA Inventory.

DFG MAK: 1000 mg/m³

SAFETY PROFILE: Mildly toxic by ingestion. When heated to decomposition it emits acrid smoke and irritating fumes. See also other polyethylene glycol entries.

PJT230   CAS: 25322-68-3   HR: 1
POLYETHYLENE GLYCOL 400

mf: $H(OC_2H_4)_nOH$

PROP: Liquid with *n* about 8.2 to 9.1. Mw: 380–420, d: 1.128, mp: 4–8°.

SYNS: PEG 400 □ POLYAETHYLENGLYKOLE 400 (GERMAN) □ POLY G 400

CONSENSUS REPORTS: EPA Genetic Toxicology Program. Reported in EPA TSCA Inventory.

DFG MAK: 1000 mg/m³

SAFETY PROFILE: Low toxicity by ingestion, intravenous, and intraperitoneal routes. When heated to decomposition it emits acrid smoke and irritating fumes. See also other polyethylene glycol entries.

PJT240   CAS: 25322-68-3   HR: 1
POLYETHYLENE GLYCOL 600
mf: $H(OC_2H_4)_nOH$
PROP: Liquid with n about 12.5 to 13.9.
Mw: 570–630, d: 1.128, mp: 20–25°.
SYNS: PEG 600 □ POLYAETHYLENGLYKOLE 600
(GERMAN)
CONSENSUS REPORTS: EPA Genetic
Toxicology Program. Reported in EPA
TSCA Inventory.
DFG MAK: 1000 mg/m³
SAFETY PROFILE: Low toxicity by
ingestion. An eye irritant. When heated to
decomposition it emits acrid smoke and
irritating fumes. See also other polyethylene
glycol entries.

PJT250   CAS: 25322-68-3   HR: 2
POLYETHYLENE GLYCOL 1000
mf: $(C_2H_4O)_n \cdot H_2O$
SYNS: CARBOWAX 1000 □ MACROGOL 1000 □ PEG
1000 □ POLYAETHYLENGLYKOLE #1000 (GERMAN) □
POLYGLYCOL 1000 □ POLYGLYCOL E1000
CONSENSUS REPORTS: Reported in EPA
TSCA Inventory. EPA Genetic Toxicology
Program.
SAFETY PROFILE: Moderately toxic by
intraperitoneal and intravenous routes.
Mildly toxic by ingestion. An eye irritant.
Questionable carcinogen with experimental
tumorigenic data. When heated to
decomposition it emits acrid smoke and
irritating fumes. See also other polyethylene
glycol entries.

PJT500   CAS: 25322-68-3   HR: 1
POLYETHYLENE GLYCOL 1500
mf: $H(OC_2H_4)_nOH$
PROP: White, free-flowing powder. D:
1.15–1.21 @ 25°/25°, fp: 44–48°.
SYNS: CARBOWAX 1500 □ α-HYDRO-ω-HYDROXY-
POLY(OXY-1,2-ETHANEDIYL) □ PEG 1500 □
POLYAETHYLENGLYKOLE 1500 (GERMAN) □
POLYOXYETHYLENE 1500
CONSENSUS REPORTS: Reported in EPA
TSCA Inventory. EPA Genetic Toxicology
Program.
SAFETY PROFILE: Mildly toxic by
ingestion. A human skin irritant. When
heated to decomposition it emits acrid

smoke and irritating fumes. See also other
polyethylene glycol entries.

PJT750   CAS: 25322-68-3   HR: 1
POLYETHYLENE GLYCOL 4000
mf: $H(OC_2H_4)_nOH$
PROP: White, free-flowing powder or white
flakes. D: 1.20–1.21 @ 25°/25°, fp: 54–58°.
SYNS: CARBOWAX 4000 □ CARSONON PEG-4000 □
MACROGOL 4000 □ PEG 4000 □
POLYAETHYLENGLYKOLE 4000 (GERMAN) □
POLYGLYCOL 4000 □ POLYGLYCOL E-4000 □
POLYGLYCOL E-4000 USP □ POLYOXYETHYLENE (75)
CONSENSUS REPORTS: Reported in EPA
TSCA Inventory. EPA Genetic Toxicology
Program.
SAFETY PROFILE: Mildly toxic by
ingestion. A skin irritant. When heated to
decomposition it emits acrid smoke and
irritating fumes. See also other polyethylene
glycol entries.

PJU000   CAS: 25322-68-3   HR: 1
POLYETHYLENE GLYCOL 6000
mf: $H(OC_2H_4)_nOH$
PROP: White, waxy solid. Mp: 58–62°, flash
p: >887°F. Water-sol.
SYNS: CARBOWAX 6000 □ PEG 6000 □
POLYAETHYLENGLYKOLE 6000 (GERMAN)
CONSENSUS REPORTS: Reported in EPA
TSCA Inventory. EPA Genetic Toxicology
Program.
SAFETY PROFILE: Mildly toxic by
ingestion. Mutation data reported. A skin
irritant. Combustible when exposed to heat
or flame. When heated to decomposition it
emits acrid smoke and irritating fumes. See
also other polyethylene glycol entries.

PJU500   CAS: 9005-08-7   HR: 3
POLYETHYLENE GLYCOL
   DISTEARATE
PROP: Polyethylene glycol distearate, low
molecular weight (JAPMA8 38,428,49).
SYNS: POLYETHYLENE GLYCOL 300 DISTEARATE □
POLYETHYLENE GLYCOL 400 (DI) STEARATE □
POLYETHYLENE GLYCOL 600 (DI) STEARATE □
POLYGLYCOL DISTEARATE
CONSENSUS REPORTS: Reported in EPA
TSCA Inventory.

SAFETY PROFILE: Poison by intravenous route. When heated to decomposition it emits acrid smoke and irritating fumes. See also POLYETHYLENE GLYCOL.

PJV250   CAS: 9004-99-3   HR: 2
POLYETHYLENE GLYCOL MONO-
**   STEARATE**
mf: $(C_2H_4O)_n•C_{18}H_{36}O_2$
SYNS: POLYOXYETHYLENE-8-MONOSTEARATE □ POLYOXYETHYLENE(8)STEARATE □ TRYDET SA SERIES
CONSENSUS REPORTS: Reported in EPA TSCA Inventory.
SAFETY PROFILE: Very slightly toxic by ingestion. Questionable carcinogen with experimental tumorigenic data. Experimental reproductive effects. When heated to decomposition it emits acrid smoke and irritating fumes. See also other polyethylene glycol monostearate entries and POLYETHYLENE GLYCOL.

PJX875   HR: D
POLYGLYCEROL ESTERS of FATTY
**   ACIDS**
PROP: Yellow to amber oily viscous liquids; light tan to brown soft solids; tan to brown waxy solids. Dispersible in water; sol in org solvs and oils.
SAFETY PROFILE: When heated to decomposition it emits acrid smoke and irritating fumes.

PJY500   CAS: 25038-54-4   HR: 2
POLY(IMINOCARBONYLPENTA-
**   METHYLENE)**
mf: $(C_6H_{11}NO)_n$
SYNS: AKULON □ ALKAMID □ AMILAN CM 1001 □ 6-AMINOHEXANOIC ACID HOMOPOLYMER □ BONAMID □ CAPRAN 80 □ CAPROAMIDE POLYMER □ CAPROLACTAM OLIGOMER □ ε-CAPROLACTAM POLYMERE (GERMAN) □ CAPRON □ CHEMLON □ DANAMID □ DULL 704 □ DURETHAN BK □ ERTALON 6SA □ GRILON □ HEXAHYDRO-2H-AZEPIN-2-ONE HOMOPOLYMER □ ITAMID □ KAPROLIT □ KAPROLON □ KAPROMIN □ KAPRON □ MARANYL F 114 □ METAMID □ MIRAMID WM 55 □ NYLON-6 □ ORGAMIDE □ PA 6 (polymer) □ PLASKON 201 □ POLICAPRAN □ POLYAMIDE 6 □ POLY(ε-

AMINOCAPROIC ACID) □ POLYCAPROAMIDE □ POLY(ε-CAPROAMIDE) □ POLYCAPROLACTAM □ POLY(ε-CAPROLACTAM) □ POLY(IMINO(1-OXO-1,6-HEXANEDIYL)) □ RELON P □ SPENCER 401 □ STILON □ TARLON XB □ TARNAMID T □ ULTRAMID BMK □ VIDLON □ WIDLON □ ZYTEL 211
CONSENSUS REPORTS: IARC Cancer Review: Group 3 IMEMDT 7,56,87; Animal Inadequate Evidence IMEMDT 19,115,75. Reported in EPA TSCA Inventory.
SAFETY PROFILE: Moderately toxic by ingestion. Mildly toxic by inhalation. Questionable carcinogen with experimental neoplastigenic data by implant route. When heated to decomposition it emits toxic fumes of NO_x.

PJY850   CAS: 26099-09-2   HR: D
POLYMALEIC ACID
SAFETY PROFILE: When heated to decomposition it emits acrid smoke and irritating fumes.

PKA850   HR: 2
POLYMERS, WATER-INSOLUBLE
SAFETY PROFILE: Many produce local tumors of the soft tissues surrounding the site of implantation. See also specific compounds.

PKA860   HR: 2
POLYMERS, WATER-SOLUBLE
SAFETY PROFILE: Many produce local tumors of the soft tissues surrounding the site of implantation and in the lungs, mucosal contact areas, organs, and tissues of retention and deposition. See also specific compounds.

PKB100   CAS: 9016-87-9   HR: 2
POLYMETHYLENEPOLYPHENYL
**   ISOCYANATE**
SYNS: CORONATE MR 200 □ CR 200 □ DESMODUR PU 1520A20 □ DESMODUR 44V20 □ E 534 □ ISOBIND 100 □ ISOCYANATE 580 □ ISONATE 390P □ ISOSET CX 11 □ KAISER NCO 20 □ LUPRANATE M 10 □ LUPRANATE M 70 □ LUPRANATE M 20S □ LUPRINATE M 20 □ MDI-CR □ MDI-CR 100 □ MDI-CR 200 □ MDI-CR 300 □ MILLIONATE 300 □ MILLIONATE MR □ MILLIONATE MR 100 □ MILLIONATE MR 200 □

MILLIONATE MR 300 □ MILLIONATE MR 340 □
MILLIONATE MR 400 □ MILLIONATE MR 500 □
MOBAY MRS □ MONDUR E 429 □ MONDUR E 441 □
MONDUR E 541 □ MONDUR MR □ MONDUR MR 200 □
MONDUR MRS □ MONDUR MRS 10 □ MR 200 □ MR 2000
□ NCO 20 □ NIAX AFPI □ PAPI □ PAPI 20 □ PAPI 27 □
PAPI 135 □ PAPI 580 □ PAPI 901 □ RUBINATE M □
RUBINATE MF 178 □ RUBINATE MF 182 □ SUMIDUR
44V10 □ SUMIDUR 44V20 □ SUMIDUR 44VM □
SUPRASEC 1042 □ SUPRASEC DC □ SYSTANATE MR □
SYSTANAT MR □ TAKENATE 300C □ TEDIMON 31 □
THANATE P 210 □ THANATE P 220 □ THANATE P 270
CONSENSUS REPORTS: IARC Cancer
Review: Group 3 IMEMDT 7,56,87;
Human No Adequate Data IMEMDT
19,303,79; Animal No Adequate Data
IMEMDT 19,303,79. Reported in EPA
TSCA Inventory.
DFG MAK: Confirmed Animal Carcinogen
with Unknown Relevance to Humans
SAFETY PROFILE: Low toxicity by
ingestion and skin contact. A questionable
carcinogen. When heated to decomposition
it emits toxic vapors of HCN.

**PKB500   CAS: 9011-14-7   HR: 2
POLYMETHYLMETHACRYLATE**
mf: $(C_5H_8O_2)_n$
SYNS: ACRYLITE □ ACRYPET □ ALUTOR M 70 □
CMW BONE CEMENT □ CRINOTHENE □ DEGALAN S
85 □ DELPET 50M □ DIAKON □ DISPASOL M □ DV 400
□ ELVACITE □ KALLOCRYL K □ KALLODENT CLEAR
□ KORAD □ LPT □ LUCITE □ METAPLEX NO □
METHACRYLIC ACID METHYL ESTER POLYMERS □
METHYL METHACRYLATE HOMOPOLYMER □
METHYL METHACRYLATE POLYMER □ METHYL
METHACRYLATE RESIN □ 2-METHYL-2-PROPENOIC
ACID METHYL ESTER HOMOPOLYMER □ ORGANIC
GLASS E 2 □ OSTEOBOND SURGICAL BONE CEMENT
□ PALACOS □ PARAGLAS □ PARAPLEX P 543 □
PERSPEX □ PLEXIGLAS □ PLEXIGUM M 920 □ PMMA
□ PONTALITE □ REPAIRSIN □ RESARIT 4000 □
RHOPLEX B 85 □ ROMACRYL □ SHINKOLITE □ SOL □
STELLON PINK □ SUMIPLEX LG □ SUPERACRYL AE □
SURGICAL SIMPLEX □ TENSOL 7 □ VEDRIL
CONSENSUS REPORTS: IARC Cancer
Review: Group 3 IMEMDT 7,56,87;
Human Inadequate Evidence IMEMDT
19,187,79; Animal Sufficient Evidence
IMEMDT 19,187,79. Reported in EPA
TSCA Inventory.

SAFETY PROFILE: Questionable
carcinogen with experimental tumorigenic
data by implant route. When heated to
decomposition it emits acrid smoke and
irritating fumes. Used as the main
constituent of acrylic sheet, molding, and
extrusion powders.

**PKC000   CAS: 1406-11-7   HR: 3
POLYMYXIN**
PROP: A series of antibiotic substances,
polypeptide (basic), sol in water. Colorless
powder. Decomp @ 228–230°.
SYN: B-71
SAFETY PROFILE: Poison by
intraperitoneal, subcutaneous, and
intravenous routes. An additive permitted in
food for human consumption.

**PKD300   CAS: 1264-72-8   HR: 3
POLYMYXIN E SULFATE**
mf: $C_{45}H_{85}N_{13}O_{10} \cdot H_2O_4S$ mw: 1066.51
SYNS: BELCOMYCIN □ COLIMYCIN SULFATE □
COLISTIN SULFAT □ COLISTIN SULFATE □ COLISTIN,
SULFATE (SALT) □ COLOMYCIN SYRUP □ POLYMYXIN
E SULFATE (SALT)
SAFETY PROFILE: A poison by ingestion,
intraperitoneal, subcutaneous, and
intravenous routes. When heated to
decomposition it emits toxic vapors of NO_x
and SO_x.

**PKF000   CAS: 9016-45-9   HR: 2
POLYOXYETHYLENE (9) NONYL
    PHENYL ETHER**
SYNS: ARKOPAL N-090 □ CARSONON N-9 □ CONCO
NI-90 □ IGEPAL CO-630 □ NEUTRONYX 600 □ PEG-9
NONYL PHENYL ETHER □ POLYETHYLENE GLYCOL
450 NONYL PHENYL ETHER □ PROTACHEM 630 □
REWOPOL HV-9 □ TERGITOL TP-9 (NONIONIC)
CONSENSUS REPORTS: Reported in EPA
TSCA Inventory. Glycol ethers are on the
Community Right-To-Know List.
SAFETY PROFILE: Moderately toxic by
ingestion and skin contact. A severe eye and
mild skin irritant in humans. Many glycol
ethers cause dangerous human reproductive
effects. When heated to decomposition it

emits acrid smoke and irritating fumes. See also GLYCOL ETHERS.

PKF500   CAS: 9002-93-1   HR: 2
POLY(OXYETHYLENE)-p-tert-OCTYLPHENYL ETHER

mf: $(C_2H_4O)_n \cdot C_{14}H_{22}O$

PROP: Mixture in which *n* varies from 5 to 15. Pale-yellow, viscous liquid. D: 1.0595. Miscible with water, alc, acetone; sol in benzene, toluene; insol in pet ether.

SYNS: ALFENOL 3 □ ALFENOL 9 □ ANTAROX A-200 □ CONCO NIX-100 □ HYDROL SW □ HYONIC PE-250 □ IGEPAL CA-63 □ MARLOPHEN 820 □ NEUTRONYX 605 □ OCTOXINOL □ OCTOXYNOL □ OCTOXYNOL 3 □ OCTOXYNOL 9 □ OCTYL PHENOL CONDENSED with 12–13 MOLES ETHYLENE OXIDE □ p-tert-OCTYL-PHENOXYPOLYETHOXYETHANOL □ OPE 30 □ PEG-9 OCTYL PHENYL ETHER □ POLYETHYLENE GLYCOL MONOETHER with p-tert-OCTYLPHENYL □ POLYE-THYLENE GLYCOL MONO(4-OCTYLPHENYL) ETHER □ POLYETHYLENE GLYCOL MONO(p-tert-OCTYL-PHENYL) ETHER □ POLYETHYLENE GLYCOL MONO-(4-tert-OCTYLPHENYL) ETHER □ POLYETHYLENE GLYCOL MONO(p-(1,1,3,3-TETRAMETHYLBUTYL)-PHENYL) ETHER □ POLYETHYLENE GLYCOL OCTYLPHENOL ETHER □ POLYETHYLENE GLYCOL 450 OCTYL PHENYL ETHER □ POLYETHYLENE GLYCOL p-OCTYLPHENYL ETHER □ POLYETHYLENE GLYCOL p-tert-OCTYLPHENYL ETHER □ POLYETHYL-ENE GLYCOL p-1,1,3,3,-TETRAMETHYLBUTYLPHENYL ETHER □ POLYOXYETHYLENE MONO(OCTYLPHEN-YL) ETHER □ POLYOXYETHYLENE (9) OCTYLPHENYL ETHER □ POLYOXYETHYLENE (13) OCTYLPHENYL ETHER □ PRECEPTIN □ TRITON X 35 □ TRITON X 45 □ TRITON X 100 □ TRITON X 102 □ TRITON X 165 □ TRITON X 305 □ TRITON X 405 □ TRITON X 705 □ TX 100

CONSENSUS REPORTS: Glycol ether compounds are on the Community Right-To-Know List. Reported in EPA TSCA Inventory.

SAFETY PROFILE: Moderately toxic by ingestion and intravenous routes. Experimental reproductive effects. Human mutation data reported. An eye and human skin irritant. Many glycol ethers cause dangerous human reproductive effects. When heated to decomposition it emits toxic fumes of NO_x. A surfactant. See also GLYCOL ETHERS.

PKG000   CAS: 9005-64-5   HR: 2
POLYOXYETHYLENE (20) SORBITAN MONOLAURATE

mf: $C_{58}H_{114}O_{26}$ mw: 1227.72

SYNS: ARMOTAN PML-20 □ CAPMUL □ DXEWMULSE POE-SML □ EMSORB 6915 □ GLYCOSPERSE L-20 □ GLYCOSPERSE L-20X □ HODAG PSML-20 □ LIPOSORB L-20 □ POE 20 SORBITAN MONOLAURATE □ POLYOXYETHYLENE (20) SORBITAN MONOLAURATE □ PROTASORB L-20 □ PSML □ SORBIMACROGOL LAURATE 300 □ SORBITAN, MONODODECANOTE, POLY(OXY-1,2-ETHANEDIYL) DERIVATIVES □ TWEEN 20

CONSENSUS REPORTS: Reported in EPA TSCA Inventory.

SAFETY PROFILE: Moderately toxic by intraperitoneal and intravenous routes. Experimental teratogenic and reproductive effects. A human skin irritant. When heated to decomposition it emits acrid smoke and irritating fumes. Used as a non-ionic surfactant.

PKI500   CAS: 25322-69-4   HR: 2
POLYPROPYLENE GLYCOL

mf: $(C_3H_8O_2)_n$

PROP: Clear, colorless liquid. Mw: 400–2000, mp: does not crystallize, flash p: 390°F, d: 1.002–1.007. Sol in water, aliphatic ketones, and alcs; insol in ether, aliphatic hydrocarbons.

SYNS: ALKAPOL PPG-1200 □ JEFFOX □ POLYPROP-YLENGLYKOL (CZECH)

CONSENSUS REPORTS: Reported in EPA TSCA Inventory.

SAFETY PROFILE: An eye irritant. Combustible liquid when exposed to heat or flame; can react with oxidizing materials. To fight fire, use foam, CO_2, dry chemical. When heated to decomposition it emits acrid smoke and irritating fumes. See also GLYCOLS.

PKI750   CAS: 25322-69-4   HR: 3
POLYPROPYLENE GLYCOL 750

SYN: P.P.G. 750

CONSENSUS REPORTS: Reported in EPA TSCA Inventory.

SAFETY PROFILE: Poison by intraperitoneal and intravenous routes.

P

When heated to decomposition it emits acrid smoke and irritating fumes.

PKL000   CAS: 9005-64-5   HR: 2
POLYSORBATE 20
PROP: Lemon- to amber-colored liquid; characteristic odor, bitter taste. Sol in water, alc, ethyl acetate, methanol, dioxane; insol in mineral oil, mineral spirits.
SYNS: GLYCOSPERSE L-20X □ POLYOXYETHYLENE (20) SORBITAN MONOLAURATE
CONSENSUS REPORTS: Reported in EPA TSCA Inventory.
SAFETY PROFILE: Moderately toxic by intravenous route. Mildly toxic by ingestion. A human skin irritant. When heated to decomposition it emits acrid smoke and irritating fumes.

PKL030   CAS: 9005-67-8   HR: 2
POLYSORBATE 60
mf: $C_{64}H_{126}O_{26}$ mw: 1311.90
PROP: Lemon- to orange-colored oily liquid; faint odor and bitter taste. Sol in water, aniline, ethyl acetate, toluene; insol in mineral oil, vegetable oil.
SYNS: CAPMUL □ LGYCOSPERSE S-20 □ LIPOSORB S-20 □ POLYOXYETHYLENE SORBITAN MONO-STEARATE □ POLYOXYETHYLENE 20 SORBITAN MONOSTEARATE □ SORBITAN, MONOOCTADECANO-ATE, POLY(OXY-1,2-ETHANEDIYL) DERIVATIVES □ TWEEN 60
CONSENSUS REPORTS: Reported in EPA TSCA Inventory.
SAFETY PROFILE: Moderately toxic by intravenous route. Experimental reproductive effects. Questionable carcinogen with experimental tumorigenic data. When heated to decomposition it emits acrid smoke and irritating fumes.

PKL050   HR: D
POLYSORBATE 65
PROP: Tan, waxy solid; faint odor, bitter taste. Sol in mineral oil, vegetable oil, mineral spirits, acetone, ether, dioxane, alc, methanol; dispersible in water, carbon tetrachloride.

SAFETY PROFILE: When heated to decomposition it emits acrid smoke and irritating fumes.

PKL100   CAS: 9005-65-6   HR: 2
POLYSORBATE 80
PROP: Yellow to orange oily liquid; faint odor, bitter taste. Sol in water, alc, fixed oils, ethyl acetate, toluene; insol in mineral oil.
SYNS: ARMOTAN PMO-20 □ ATLOX 1087 □ CAPMUL POE-O □ CRILL 10 □ DREWMULSE POE-SMO □ DURFAX 80 □ EMSORB 6900 □ ETHOXYLATED SORBITAN MONOOLEATE □ GLYCOSPERSE O-20 □ HODAG SVO 9 □ LIPOSORB O-20 □ MONITAN □ MONTANOX 80 □ NCI-C60286 □ NIKKOL TO □ OLOTHORB □ POLYOXYETHYLENE SORBITAN MONOOLEATE □ POLYOXYETHYLENE SORBITAN OLEATE □ POLYSORBAN 80 □ POLYSORBATE 80, U.S.P. □ PROTASORB O-20 □ ROMULGIN O □ SORBIMACROGOL OLEATE □ SORBITAL O 20 □ SORETHYTAN (20) MONOOLEATE □ SORLATE □ SVO 9 □ TWEEN 80
CONSENSUS REPORTS: Reported in NTP Carcinogenesis Studies (feed); Equivocal Evidence: rat NTPTR* NTP-TR-415,92; (feed); No Evidence: mouse NTPTR* NTP-TR-415,92. Reported in EPA TSCA Inventory.
SAFETY PROFILE: Moderately toxic by intravenous route. Mildly toxic by ingestion. Experimental reproductive effects. Questionable carcinogen with experimental tumorigenic data. Human mutation data reported. An eye irritant. When heated to decomposition it emits acrid smoke and irritating fumes.

PKL500   CAS: 9009-54-5   HR: 2
POLYURETHANE FOAM
SYNS: ANDUR □ CURENE □ ETHERON □ ETHERON SPONGE □ ISOURETHANE □ NCI-C56451 □ PLIOGRIP □ POLYFOAM PLASTIC SPONGE □ POLYFOAM SPONGE □ POLYURETHANE A □ POLYURETHANE ESTER FOAM □ POLYURETHANE ETHER FOAM □ POLYURETHANE SPONGE □ SPENKEL □ SPENLITE □ URETHANE POLYMERS
CONSENSUS REPORTS: IARC Cancer Review: Group 3 IMEMDT 7,56,87; Animal Sufficient Evidence IMEMDT 19,303,79.
SAFETY PROFILE: Questionable carcinogen with experimental tumorigenic

data. When heated to decomposition it emits acrid toxic fumes of CN⁻ and NO$_x$.

PKL750   CAS: 25931-01-5   HR: 2
POLYURETHANE Y-195
mf: (C$_{15}$H$_{10}$N$_2$O$_2$•C$_6$H$_{10}$O$_4$•C$_2$H$_6$O$_2$)$_x$
SYNS: ADIPIC ACID, POLYMER with ETHYLENE GLYCOL and METHYLENEDI-p-PHENYLENE ISOCYANATE □ AMCHEM R 14 □ HEXANEDIOIC ACID, POLYMER with 1,3-ETHANEDIOL and 1,1'-METHYLENEBIS(4-ISOCYANATOBENZENE) □ MUL F 66 □ R 14 □ Y 195
CONSENSUS REPORTS: Reported in EPA TSCA Inventory.
SAFETY PROFILE: Questionable carcinogen with experimental tumorigenic data. When heated to decomposition it emits toxic fumes of NO$_x$.

PKM000   HR: 3
POLYURETHANE Y-217
CONSENSUS REPORTS: IARC Cancer Review: Animal Sufficient Evidence IMEMDT 19,303,79.
SAFETY PROFILE: Confirmed carcinogen with experimental tumorigenic data. When heated to decomposition it emits very toxic fumes of NO$_x$ and CN⁻.

PKM250   CAS: 26375-23-5   HR: 2
POLYURETHANE Y-218
mf: (C$_{15}$H$_{10}$N$_2$O$_2$•C$_6$H$_{10}$O$_4$•C$_4$H$_{10}$O$_2$)$_x$
SYNS: ADIPIC ACID, POLYMER with 1,4-BUTANEDIOL and METHYLENEDI-p-PHENYLENE ISOCYANATE □ HEXANEDIOIC ACID, POLYMER with 1,4-BUTANEDIOL and 1,1'-METHYLENEBIS(4-ISOCYANATOBENZENE) □ PANDEX □ TEXIN 445D □ TPU 10M □ Y 218
CONSENSUS REPORTS: Reported in EPA TSCA Inventory.
SAFETY PROFILE: Questionable carcinogen with experimental tumorigenic data. When heated to decomposition it emits very toxic fumes of CN⁻ and NO$_x$.

PKM500   CAS: 32238-28-1   HR: 2
POLYURETHANE Y-221
mf:
(C$_{15}$H$_{10}$N$_2$O$_2$•C$_{10}$H$_{14}$O$_4$•C$_6$H$_{10}$O$_4$•C$_4$H$_{10}$O$_2$)$_x$
SYNS: ADIPIC ACID, POLYMER with 1,4-BUTANEDIOL, METHYLENEDI-p-PHENYLENE

ISOCYANATE and 2,2'-(p-PHENYLENEDIOXY)DIETHANOL □ Y 221
CONSENSUS REPORTS: Reported in EPA TSCA Inventory.
SAFETY PROFILE: Questionable carcinogen with experimental tumorigenic data. When heated to decomposition it emits very toxic fumes of CN⁻ and NO$_x$.

PKM750   HR: 3
POLYURETHANE Y-222
CONSENSUS REPORTS: IARC Cancer Review: Animal Sufficient Evidence IMEMDT 19,303,79.
SAFETY PROFILE: Confirmed carcinogen with experimental tumorigenic data. When heated to decomposition it emits very toxic fumes of CN⁻ and NO$_x$.

PKN000   CAS: 52292-20-3   HR: 2
POLYURETHANE Y-223
SYNS: TECOFLEX HR □ Y-223
CONSENSUS REPORTS: Reported in EPA TSCA Inventory.
SAFETY PROFILE: Questionable carcinogen with experimental tumorigenic data. When heated to decomposition it emits very toxic fumes of CN⁻ and NO$_x$.

PKN250   HR: 3
POLYURETHANE Y-224
CONSENSUS REPORTS: IARC Cancer Review: Animal Sufficient Evidence IMEMDT 19,303,79.
SAFETY PROFILE: Confirmed carcinogen with experimental tumorigenic data. When heated to decomposition it emits very toxic fumes of CN⁻ and NO$_x$.

PKN500   CAS: 56779-19-2   HR: 3
POLYURETHANE Y-225
SYN: 1,4-BUTANEDIAMINE, 2-METHYL-, POLYMER with α-HYDRO-ω-HYDROXYPOLY(OXY-1,4-BUTANEDIYL) and 1,1'-METHYLENEBIS(4-ISOCYANATOCYCLOHEXANE)
CONSENSUS REPORTS: IARC Cancer Review: Animal Sufficient Evidence IMEMDT 19,303,79.

P

SAFETY PROFILE: Confirmed carcinogen with experimental tumorigenic data. When heated to decomposition it emits very toxic fumes of CN⁻ and NO$_x$.

PKN750   CAS: 56386-98-2   HR: 3
POLYURETHANE Y-226
PROP: Faintly yellow solid.
CONSENSUS REPORTS: IARC Cancer Review: Animal Sufficient Evidence IMEMDT 19,303,79.
SAFETY PROFILE: Confirmed carcinogen with experimental tumorigenic data. When heated to decomposition it emits very toxic fumes of CN⁻ and NO$_x$.

PKO000   CAS: 56631-46-0   HR: 3
POLYURETHANE Y-227
CONSENSUS REPORTS: IARC Cancer Review: Animal Sufficient Evidence IMEMDT 19,303,79.
SAFETY PROFILE: Confirmed carcinogen with experimental tumorigenic data. When heated to decomposition it emits very toxic fumes of CN⁻ and NO$_x$.

PKO500   CAS: 27083-55-2   HR: 2
POLYURETHANE Y-290
mf:
$(C_{15}H_{10}N_2O_2 \cdot C_6H_{10}O_4 \cdot C_4H_{10}O_2 \cdot C_2H_6O_2)_x$
SYNS: E6 □ PPE201 □ P07 □ TEXIN 192A □ TPU 2T
CONSENSUS REPORTS: Reported in EPA TSCA Inventory.
SAFETY PROFILE: Questionable carcinogen with experimental tumorigenic data. When heated to decomposition it emits very toxic fumes of CN⁻ and NO$_x$.

PKP000   CAS: 25805-16-7   HR: 3
POLYURETHANE Y-302
mf: $(C_{15}H_{10}N_2O_2 \cdot C_4H_{10}O_2)_x$
SYNS: 1,4-BUTANEDIOL POLYMER with 1,1'-METHYLENEBIS(4-ISOCYANATOBENZENE) □ ISOCYANIC ACID, METHYLENEDI-p-PHENYLENE ESTER, POLYMER with 1,4-BUTANEDIOL □ SANPRENE LQX 31 □ Y 302
CONSENSUS REPORTS: IARC Cancer Review: Animal Sufficient Evidence IMEMDT 19,303,79.

SAFETY PROFILE: Confirmed carcinogen with experimental tumorigenic data by implant route. When heated to decomposition it emits very toxic fumes of CN⁻ and NO$_x$.

PKP250   CAS: 25036-33-3   HR: 3
POLYURETHANE Y-304
CONSENSUS REPORTS: IARC Cancer Review: Animal Sufficient Evidence IMEMDT 19,303,79.
SAFETY PROFILE: Confirmed carcinogen with experimental tumorigenic data. When heated to decomposition it emits very toxic fumes of CN⁻ and NO$_x$.

PKP750   CAS: 9002-89-5   HR: 3
POLYVINYL ALCOHOL
PROP: Colorless, white or cream, amorphous powder. Mp: decomp over 200°, flash p: 175°F (OC), d: 1.329. Sol in water. Polymer of average molecular weight 120,000 (AMPLAO 67,589,59).
SYNS: ALCOTEX 88/05 □ ALCOTEX 88/10 □ ALKOTEX □ ALVYL □ ARACET APV □ CIPOVIOL W 72 □ COVOL □ COVOL 971 □ ELVANOL □ ELVANOL 50-42 □ ELVANOL 52-22 □ ELVANOL 70-05 □ ELVANOL 71-30 □ ELVANOL 90-50 □ ELVANOL 522-22 □ ELVANOL 73125G □ EP 160 □ ETHENOL HOMO-POLYMER (9CI) □ GALVATOL 1-60 □ GELVATOL □ GELVATOL 1-30 □ GELVATOL 1-90 □ GELVATOL 3-91 □ GELVATOL 20-30 □ GELVATOL 2090 □ GH 20 □ GL 02 □ GL 03 □ GLO 5 □ GM 14 □ GOHSENOL □ GOHSENOL AH 22 □ GOHSENOL GH □ GOHSENOL GH 17 □ GOHSENOL GH 20 □ GOHSENOL GH 23 □ GOHSENOL GL 03 □ GOHSENOL GL 05 □ GOHSENOL GL 08 □ GOHSENOL GM 14 □ GOHSENOL GM 94 □ GOHSENOL GM 14L □ GOHSENOL KH 17 □ GOHSEN-OL NH 05 □ GOHSENOL NH 17 □ GOHSENOL NH 18 □ GOHSENOL NH 20 □ GOHSENOL NH 26 □ GOHSEN-OL NK 114 □ GOHSENOL NL 05 □ GOHSENOL NM 14 □ IVALON □ KURALON VP □ KURARE POVAL 1700 □ KURARE PVA 205 □ KURATE POVAL 120 □ LEMOL □ LEMOL 5-88 □ LEMOL 5-98 □ LEMOL 12-88 □ LEMOL 16-98 □ LEMOL 24-98 □ LEMOL 30-98 □ LEMOL 51-98 □ LEMOL 60-98 □ LEMOL 75-98 □ LEMOL GF-60 □ M 13/20 □ MOWIOL □ MOWIOL N 30-88 □ MOWIOL N 50-98 □ MOWIOL N 70-98 □ NH 18 □ NM 11 □ NM 14 □ POLYDESIS □ POLYSIZER 173 □ POLYVINOL □ POLYVIOL □ POLYVIOL M 13/140 □ POLYVIOL MO 5/140 □ POLYVIOL W 25/140 □ POLYVIOL W 40/140 □ POVAL 117 □ POVAL 120 □ POVAL 203 □ POVAL 205 □

POVAL 217 □ POVAL 1700 □ POVAL C 17 □ PVA □ PVA 008 □ PVS 4 □ RESISTOFLEX □ RHODOVIOL □ RHODOVIOL 4/125 □ RHODOVIOL 16/200 □ RHODOVIOL 4-125P □ RHODOVIOL R 16/20 □ SOLVAR □ SUMITEX H 10 □ VIBATEX S □ VINACOL MH □ VINALAK □ VINAROL □ VINAROL DT □ VINAROLE □ VINAROL ST □ VINAVILOL 2-98 □ VINNAROL □ VINOL □ VINOL 125 □ VINOL 205 □ VINOL 351 □ VINOL 523 □ VINOL UNISIZE □ VINYL ALCOHOL POLYMER □ VINYLON FILM 2000

CONSENSUS REPORTS: IARC Cancer Review: Group 3 IMEMDT 7,56,87; Animal Limited Evidence IMEMDT 19,341,79; Human Inadequate Evidence IMEMDT 19,341,79.

SAFETY PROFILE: Questionable carcinogen with experimental carcinogenic and tumorigenic data by implant route. Flammable when exposed to heat or flame; can react with oxidizing materials. Slight explosion hazard in the form of dust when exposed to flame. To fight fire, use alcohol foam, CO_2, dry chemical. When heated to decomposition it emits acrid smoke and irritating fumes.

PKQ059   CAS: 9002-86-2   HR: 2
POLYVINYL CHLORIDE
mf: $(C_2H_3Cl)_n$
PROP: Polymers with molecular weights ranging from 60,000 to 150,000 (CNREA8 15,333,55). White powder, d: 1.406.
SYNS: ARMODOUR □ ARON COMPOUND HW □ ASTRALON □ ATACTIC POLY(VINYL CHLORIDE) □ BLACAR 1716 □ BOLATRON □ BONLOID □ BREON □ CARINA □ CHLOROETHENE HOMOPOLYMER □ CHLOROETHYLENE POLYMER □ CHLOROSTOP □ COBEX (polymer) □ CONTIZELL □ CORVIC 55/9 □ DACOVIN □ DANUVIL 70 □ DARVIC 110 □ DARVIS CLEAR 025 □ DECELITH H □ DENKA VINYL SS 80 □ DIAMOND SHAMROCK 40 □ DORLYL □ DUROFOL P □ DYNADUR □ E 62 □ E 66P □ EKAVYL SD 2 □ E-PVC □ ESCAMBIA 2160 □ EUROPHAN □ EXON 605 □ FC 4648 □ FLOCOR □ GAFCOTE □ GENOTHERM □ GEON □ GEON LATEX 151 □ GUTTAGENA □ HALVIC 223 □ HISHIREX 502 □ HISPAVIC 229 □ HOSTALIT □ IGELITE F □ IMPROVED WILT PRUF □ KAYLITE □ KLEGECELL □ KOROSEAL □ LONZA G □ LUCOFLEX □ LUCOVYL PE □ LUTOFAN □ MARVINAL □ MIRREX MCFD 1025 □ MOVINYL 100 □ MYRAFORM □ NCI-C60797 □ NIKA-TEMP □ NIKAVINYL SG 700 □ NIPEON A 21 □ NIPOL 576 □ NORVINYL □ NOVON 712 □

ONGROVIL S 165 □ OPALON □ ORTUDUR □ PANTASOTE R 873 □ PARCLOID □ PATTINA V 82 □ PEVIKON D 61 □ PLIOVIC □ POLIVINIT □ POLY(CHLOROETHYLENE) □ POLYTHERM □ POLYVINYLCHLORID (GERMAN) □ PROTOTYPE III SOFT □ PVC (MAK) □ QSAH 7 □ QUIRVIL □ QYSA □ RAVINYL □ RUCON B 20 □ S 65 (polymer) □ SCON 5300 □ SICRON □ S-LON □ SOLVIC □ SP 60 (CHLOROCARBON) □ SUMILIT EXA 13 □ SUMITOMO PX 11 □ TAKILON □ TECHNOPOR □ TENNECO 1742 □ TK 1000 □ TROVIDUR □ TROVITHERN HTL □ U 1 (polymer) □ ULTRON □ UNICHEM □ VERON P 130/1 □ VESTOLIT B 7021 □ VINIKA KR 600 □ VINIKULON □ VINIPLAST □ VINIPLEN P 73 □ VINNOL E 75 □ VINOFLEX □ VINYLCHLON 4000LL □ VINYL CHLORIDE HOMOPOLYMER □ VINYL CHLORIDE POLYMER □ VYGEN 85 □ WELVIC G 2/5 □ WILT PRUF □ WINIDUR □ X-AB □ YUGOVINYL

CONSENSUS REPORTS: IARC Cancer Review: Group 3 IMEMDT 7,56,87; Human Inadequate Evidence IMEMDT 19,377,79; IARC Cancer Review: Animal Inadequate Evidence IMEMDT 19,377,79. Reported in EPA TSCA Inventory.
ACGIH TLV: (Proposed: 1 mg/m³; Not Classifiable as a Human Carcinogen))
DFG MAK: 1.5 mg/m³ (dust)
SAFETY PROFILE: Chronic inhalation of dusts can cause pulmonary damage, blood effects, abnormal liver function. "Meat wrapper's asthma" has resulted from the cutting of PVC films with a hot knife. Can cause allergic dermatitis. Questionable carcinogen with experimental tumorigenic data. Reacts violently with F_2. When heated to decomposition it emits toxic fumes of Cl^- and phosgene.

PKQ150   HR: D
POLYVINYLPOLYPYRROLIDONE
PROP: White, hygroscopic powder; faint bland odor. Insol in water.
SYNS: CROSPOVIDONE □ PVPP □ 1-VINYL-2-PYRROLIDONE CROSSLINKED INSOLUBLE POLYMER
SAFETY PROFILE: When heated to decomposition it emits acrid smoke and irritating fumes.

P

PKQ250   CAS: 9003-39-8   HR: 2
POLY(1-VINYL-2-PYRROLIDINONE)
**   HOMOPOLYMER**
mf: $(C_6H_9ON)_n$
PROP: A free-flowing, white, amorphous powder; or faintly yellow solid. D: 1.23–1.29. Sol in water, chlorinated hydrocarbons, alc, amines, nitroparaffins, and lower-molecular-weight fatty acids.
SYNS: AGENT AT 717 □ ALBIGEN A □ ALDACOL Q □ AT 717 □ BOLINAN □ 1-ETHENYL-2-PYRROLIDINONE HOMOPOLYMER □ 1-ETHENYL-2-PYRROLIDINONE POLYMERS □ GANEX P 804 □ HEMODESIS □ HEMODEZ □ K25 (polymer) □ KOLLIDON □ LUVISKOL □ MPK 90 □ NCI-C60582 □ NEOCOMPENSAN □ PERAGAL ST □ PERISTON □ PLASDONE □ POLYCLAR L □ POLY(1-(2-OXO-1-PYRROLIDINYL)ETHYLENE) □ POLYVIDONE □ POLY(n-VINYLBUTYROLACTAM) □ POLYVINYLPYR-ROLIDONE □ POVIDONE (USP XIX) □ PROTAGENT □ PVP (FCC) □ SUBTOSAN □ VINISIL □ N-VINYLBUTY-ROLACTAM POLYMER □ N-VINYLPYRROLIDONE POLYMER
CONSENSUS REPORTS: IARC Cancer Review: Group 3 IMEMDT 7,56,87. Reported in EPA TSCA Inventory.
SAFETY PROFILE: Mildly toxic by intraperitoneal and intravenous routes. Questionable carcinogen. When heated to decomposition it emits toxic fumes of NO_x.

PKS250   CAS: 26837-42-3   HR: 3
POLYVINYL SULFATE POTASSIUM
**   SALT**
mf: $(C_2H_4O_4S)_x.xK$
SYNS: POTASSIUM POLY(VINYL SULFATE) □ PVSK □ SULFURIC ACID, MONOETHENYL ESTER, HOMO-POLYMER, POTASSIUM SALT
CONSENSUS REPORTS: Reported in EPA TSCA Inventory.
SAFETY PROFILE: Poison by intraperitoneal and subcutaneous routes. When heated to decomposition it emits toxic fumes of SO_x and K_2O.

PKS750   CAS: 65997-15-1   HR: 1
PORTLAND CEMENT
PROP: Fine gray powder composed of compounds of lime, aluminum, silica, and iron oxide as $(4CaO \cdot Al_2O_3 \cdot Fe_2)_3$, $(3CaO Al_2O_3)$, $(3CaO \cdot SiO_2)$, and $(2CaOSiO_2)$. Small amounts of magnesia, sodium, potassium, chromium, and sulfur are also present in combined form. Containing less than 1% crystalline silica (FEREAC 39,23540,74). IDLH 5000 mg/m^3.
SYNS: CEMENT, PORTLAND □ PORTLAND CEMENT SILICATE
CONSENSUS REPORTS: Reported in EPA TSCA Inventory.
OSHA PEL: TWA Total Dust: 10 mg/m^3; Respirable Fraction: 5 mg/m^3
ACGIH TLV: TWA (nuisance particulate) 10 mg/m^3 of total dust (when toxic impurities are not present, e.g., quartz <1%); (Proposed: 10 mg/m^3; Not Classifiable as a Human Carcinogen)
DFG MAK: 5 mg/m^3
NIOSH REL: (Portland Cement, respirable fraction) TWA 5 mg/m^3; (Portland Cement, total dust): TWA 10 mg/m^3
SAFETY PROFILE: A nuisance dust. A skin irritant. See also NUISANCE DUSTS and AEROSOLS.

PKT250   CAS: 7440-09-7   HR: 3
POTASSIUM
DOT: UN 1420/UN 2257
af: K   aw: 39.10
PROP: Soft ductile, silvery-white, very reactive metal. Tarnishes in air forming oxides, carbonates, and the hydroxide. Mp: 63.65°, bp: 774°, d: 0.862 @ 20°. Sol in NH_3 (l) or blue-black soln.
SYN: POTASSIUM, metal alloys (UN 1420) (DOT)
CONSENSUS REPORTS: Reported in EPA TSCA Inventory.
DOT CLASSIFICATION: 4.3; Label: Dangerous When Wet
SAFETY PROFILE: The toxicity of potassium compounds is almost always that of the anion, not of potassium. A dangerous fire hazard. Metallic potassium reacts with moisture to form potassium hydroxide and hydrogen. The reaction evolves much heat, causing the potassium to melt and spatter. The reaction also ignites the hydrogen, which burns, or if there is any confinement,

may explode. It can ignite spontaneously in moist air. Store under mineral oil. Potassium metal will form the peroxide (K_2O_2) and the superoxide (KO_3 or K_2O_4) at room temperature even when stored under mineral oil. These oxides can explode on contact with organic materials. Metal that has oxidized on storage under oil may explode violently when handled or cut. Oxide-coated potassium should be destroyed by burning.

Danger: burning potassium is difficult to extinguish; dry powdered soda ash or graphite or special mixtures of dry chemical are recommended.

A violent explosion hazard with the following materials under required conditions of temperature, pressure, and state of division: acetylene, air, moist air, alcohols (e.g., n-propanol through n-octanol, benzyl alcohol, cyclohexanol), $AlBr_3$, ammonium nitrate + ammonium sulfate, ammonium chlorocuprate, NH_4Br, NH_4I, antimony halides, arsenic halides, AsH_3 + NH_3, Bi_2O_3, boric acid, BBr_3, carbon disulfide (impact-sensitive), solid carbon dioxide, carbon monoxide, chlorinated hydrocarbons (e.g., chloroethane, dichloroethane, dichloromethane, trichloroethane, chloroform, pentachloro-ethane, carbon tetrachloride, tetrachloro-ethane), halocarbons (e.g., bromoform, dibromomethane, diiodomethane), iodine (impact-sensitive), interhalogens (e.g., chlorine trifluoride, iodine bromide, iodine chloride, iodine pentafluoride, iodine trichloride), ClO, CrO_3, Cu_2OCl_2, CuO, ethylene oxide, fluorine, graphite, graphite + air, graphite + K_2O_2, hydrogen iodide, H_2O_2, hydrogen chloride, hydrazine, Pb_2OCl_2, PbO_2, $PbSO_4$, maleic anhydride, metal halides (e.g., calcium bromide, iron(III) bromide, iron(III) chloride, iron(II) chloride, iron(II) bromide, iron(II) iodide, cobalt(II) chloride, chromium tetrachloride, silver fluoride, mercury(II) bromide, mercury(II) chloride, mercury(II) fluoride, mercury(II) iodide, copper(I) chloride, copper(I) iodide, copper(II) bromide, copper(II) chloride, ammonium tetrachlorocuprate, zinc chlorides, bromides, or iodides, cadmium chlorides, bromides or iodides, aluminum fluorides, chlorides, or bromides, thallium(I) bromide, tin chlorides, tin iodide, arsenic trichloride, arsenic triiodide, antimony tribromides, trichlorides or triiodides, bismuth tribromides, trichlorides, or triiodides, vanadium(V) chloride, manganese(II) chloride, nickel bromide, chloride, or iodide), metal oxides (e.g., lead peroxide, mercury(I) oxide, MoO_3), nitric acid, nitrogen-containing explosives (e.g., ammonium nitrate, picric acid, nitrobenzene), nonmetal halides (e.g., diselenium dichloride, seleninyl chloride, seleninyl bromide, sulfur dichloride, sulfur dibromide, phosphorus tribromide, phosphorus trichloride, phosgene, disulfur dichloride), nonmetal oxides (e.g., dichlorine oxide, dinitrogen tetraoxide, dinitrogen pentaoxide, NO_2, P_2O_5), oxalyl dibromide, oxalyl dichloride, P_2NF, peroxides, $COCl_2$, PH_3 + NH_3, phosphorus, PCl_5, PBr_3, potassium chlorocuprate, potassium oxides (e.g., KO_3, K_2O_2, KO_2), selenium, $SeOCl_2$, $SiCl_4$, $AgIO_3$, $NaIO_3$, NH_3 + $NaNO_2$, Na_2O_2, SnI_4 + S, SnO_2, S, sulfuric acid, tellurium, thiophosphoryl fluoride, $VOCl_2$, water.

Other hazardous reactions may occur with carbon (e.g., soot, graphite, activated charcoal), dimethyl sulfoxide, ethylene oxide, chlorine, bromine vapor, hydrogen bromide, potassium iodide + magnesium bromide, chloride or iodide, maleic anhydride, mercury, copper(II) oxide, mercury(II) oxide, tin(IV) oxide, molybdenum(III) oxide, bismuth trioxide, phosphorus trichloride, sulfur dioxide, chromium trioxide.

When heated to decomposition it emits toxic fumes of K_2O.

PKT500   **CAS: 7440-09-7**   **HR: 3**
POTASSIUM (liquid alloy)
SYN: POTASSIUM, metal liquid alloy (DOT)

CONSENSUS REPORTS: Reported in EPA TSCA Inventory.
SAFETY PROFILE: A very dangerous fire hazard. When heated to decomposition in air it emits toxic fumes of K_2O. See also POTASSIUM.

PKU250   CAS: 7789-29-9   HR: 3
POTASSIUM ACID FLUORIDE
DOT: NA 1811
mf: FK•FH   mw: 78.11
PROP: Colorless, deliquescent, cubic crystals. Undergoes tetragonal to cubic transition at 1°. Mp: 238.8° (decomp). Very sol in H_2O; insol in EtOH.
SYNS: BIFLUORURE de POTASSIUM (FRENCH) □ HYDROGEN POTASSIUM FLUORIDE □ POTASSIUM BIFLUORIDE □ POTASSIUM BIFLUORIDE, solid or solution (DOT) □ POTASSIUM FLUORIDE □ POTASSIUM HYDROGEN DIFLUORIDE □ POTASSIUM HYDROGEN FLUORIDE □ POTASSIUM MONOHYDROGEN DIFLUORIDE
CONSENSUS REPORTS: Reported in EPA TSCA Inventory.
OSHA PEL: TWA 2.5 mg(F)/m^3
ACGIH TLV: TWA 2.5 mg(F)/m^3; BEI: 3 mg/g creatinine of fluorides in urine prior to shift; 10 mg/g creatinine of fluorides in urine at end of shift.
DOT CLASSIFICATION: 8; Label: Corrosive, Poison
NIOSH REL: TWA 2.5 mg(F)/m^3
SAFETY PROFILE: A poison by all routes. Corrosive to the eyes, skin, and mucous membranes. A very reactive, dangerous material. When heated to decomposition it emits toxic fumes of F^- and K_2O. See also FLUORIDES.

PKU500   CAS: 7789-29-9   HR: 3
POTASSIUM ACID FLUORIDE (solution)
SYNS: POTASSIUM BIFLUORIDE, solution (DOT) □ POTASSIUM HYDROGEN FLUORIDE, solution (DOT)
CONSENSUS REPORTS: Reported in EPA TSCA Inventory.
OSHA PEL: TWA 2.5 mg(F)/m^3
ACGIH TLV: TWA 2.5 mg(F)/m^3; BEI: 3 mg/g creatinine of fluorides in urine prior to

shift; 10 mg/g creatinine of fluorides in urine at end of shift.
DOT CLASSIFICATION: 8; Label: Corrosive, Poison
NIOSH REL: TWA 2.5 mg(F)/m^3
SAFETY PROFILE: A poison. Very corrosive and reactive. A corrosive irritant to the eyes, skin, and mucous membranes. When heated to decomposition it emits toxic fumes of F^- and K_2O. See also FLUORIDES and HYDROFLUORIC ACID.

PKV500   CAS: 10124-50-2   HR: 3
POTASSIUM ARSENITE
DOT: UN 1678
mf: AsH_3O_3•xK   mw: 399.65
PROP: White, hygroscopic powder. Sol in water.
SYNS: ARSENENOUS ACID, POTASSIUM SALT □ ARSENITE de POTASSIUM (FRENCH) □ ARSONIC ACID, POTASSIUM SALT □ KALIUMARSENIT (GERMAN) □ NSC-3060 □ POTASSIUM METAARSENITE
CONSENSUS REPORTS: NTP 10th Report on Carcinogens. IARC Cancer Review: Human Sufficient Evidence IMEMDT 23,39,80; Animal Inadequate Evidence IMEMDT 23,39,80; IMEMDT 2,48,73. Arsenic and its compounds are on the Community Right-To-Know List. EPA Extremely Hazardous Substances List.
OSHA PEL: TWA 0.01 mg(As)/m^3; Cancer Hazard
ACGIH TLV: BEI: 35 μ (As)/L inorganic arsenic and methylated metabolites in urine
NIOSH REL: CL (Inorganic Arsenic) 0.002 mg(As)/m^3/15M
DOT CLASSIFICATION: 6.1; Label: Poison
SAFETY PROFILE: Confirmed human carcinogen producing skin and liver tumors. Poison by ingestion, skin contact, subcutaneous, and intravenous routes. Human mutation data reported. Human systemic effects: dermatitis, liver changes. When heated to decomposition it emits toxic fumes of As and K_2O. Used in veterinary medicine and for chronic

dermatitis in humans. See also ARSENIC COMPOUNDS.

PKW760   CAS: 582-25-2   HR: 2
POTASSIUM BENZOATE
mf: $C_7H_5O_2 \cdot K$ mw: 160.22
SAFETY PROFILE: Combustible when exposed to heat or flame. When heated to decomposition it emits acrid smoke and irritating fumes.

PKX250   CAS: 7778-50-9   HR: 3
POTASSIUM BICHROMATE
mf: $Cr_2K_2O_7$ mw: 294.20
PROP: Bright, yellowish-red, transparent crystals; bitter, metallic taste. Mp: 398°, bp: decomp @ 500°, d: 2.69. Sol in H_2O, C_6H_6, DMSO. IDLH Ca [15 mg/m³ {as Cr(VI)}].
SYNS: BICHROMATE OF POTASH □ CHROMIC ACID, DIPOTASSIUM SALT □ DIPOTASSIUM DICHROMATE □ IOPEZITE □ KALIUMDICHROMAT (GERMAN) □ POTASSIUM DICHROMATE(VI)
CONSENSUS REPORTS: IARC Cancer Review: Human Inadequate Evidence IMEMDT 23,205,80; Animal Inadequate Evidence IMEMDT 23,205,80. Chromium and its compounds are on the Community Right-To-Know List. Reported in EPA TSCA Inventory. EPA Genetic Toxicology Program.
OSHA PEL: CL 0.1 mg(CrO₃)/m³
ACGIH TLV: TWA 0.05 mg(CrO₃)/m³
NIOSH REL: TWA (Chromium(VI)) 0.025 mg(Cr(VI))/m³; CL 0.05/15M
SAFETY PROFILE: Human poison by ingestion. An experimental poison by ingestion, intraperitoneal, intravenous, and subcutaneous routes. Human mutation data reported. An experimental teratogen. Other experimental reproductive effects. Flammable by chemical reaction. A powerful oxidizer. Explosive reaction with hydrazine. Reacts violently or ignites with H_2SO_4 + acetone, hydroxylamine, ethylene glycol (above 100°C). Forms pyrotechnic mixtures with boron + silicon, iron (ignites at 1090°C), tungsten (ignites at 1700°C). Reacts with sulfuric acid to form the strong oxidant chromic acid. Used in

photomechanical processing, chrome pigment production, and wool preservation methods. When heated to decomposition it emits toxic fumes of K_2O. See also CHROMIUM COMPOUNDS.

PKX500   CAS: 23746-34-1   HR: 2
POTASSIUM BIS(2-HYDROXYETH-YL)DITHIOCARBAMATE
mf: $C_5H_{10}NO_2S_2 \cdot K$ mw: 219.38
PROP: Crystals from MeOH. Sol in H_2O, MeOH.
SYNS: BIS(2-HYDROXYETHYL)CARBAMODITHIOIC ACID, MONOPOTASSIUM SALT □ BIS(2-HYDROXY-ETHYL)DITHIOCARBAMIC ACID, MONOPOTASSIUM SALT □ BIS(2-HYDROXYETHYL)DITHIOCARBAMIC ACID, POTASSIUM SALT
CONSENSUS REPORTS: IARC Cancer Review: Group 3 IMEMDT 7,56,87; Animal Sufficient Evidence IMEMDT 12,183,76. Reported in EPA TSCA Inventory.
SAFETY PROFILE: Questionable carcinogen with experimental carcinogenic and tumorigenic data. When heated to decomposition it emits very toxic fumes of K_2O, SO_x, and NO_x. Used as an analytical reagent for quantitative determination of mercury, gold, and copper. See also CARBAMATES.

PKX750   CAS: 7646-93-7   HR: 2
POTASSIUM BISULFATE
DOT: UN 2509
mf: $HO_4S \cdot K$ mw: 136.17
PROP: White or colorless, deliquescent, orthorhombic crystals. D: 2.24, mp: 214°. Sol in water; insol in EtOH, Me_2CO.
SYNS: ACID POTASSIUM SULFATE □ MONOPOTASSIUM SULFATE □ POTASSIUM ACID SULFATE □ POTASSIUM BISULPHATE □ POTASSIUM HYDROGEN SULFATE, solid (DOT) □ SAL ENIXUM □ SULFURIC ACID, MONOPOTASSIUM SALT
CONSENSUS REPORTS: Reported in EPA TSCA Inventory.
DOT CLASSIFICATION: 8; Label: Corrosive
SAFETY PROFILE: Moderately toxic by ingestion. A corrosive irritant to the skin, eyes, and mucous membranes. When heated to decomposition it emits toxic fumes of

SO_x and K_2O. Can form an explosive mixture. See also SULFATES.

PKY000   CAS: 14075-53-7   HR: 3
POTASSIUM BOROFLUORIDE
mf: $BF_4 \cdot K$ mw: 125.91
PROP: Rhombic or cubic, colorless, orthorhombic crystals. Mp: 570°, d: 2.498. Sltly sol in H_2O; sparingly sol in Et_2O, EtOH.
SYNS: AVOGODRITE □ POTASSIUM FLUOBORATE □ POTASSIUM FLUOROBORATE □ TETRAFLUORO-BORATE-(1−) POTASSIUM
CONSENSUS REPORTS: Reported in EPA TSCA Inventory.
OSHA PEL: TWA 2.5 mg(F)/m^3
ACGIH TLV: TWA 2.5 mg(F)/m^3; BEI: 3 mg/g creatinine of fluorides in urine prior to shift; 10 mg/g creatinine of fluorides in urine at end of shift.
NIOSH REL: TWA (Inorganic Fluorides) 2.5 mg(F)/m^3
SAFETY PROFILE: Poison by intraperitoneal route. When heated to decomposition it emits very toxic fumes of F^-, K_2O, and BO_x. Used in sand casting of aluminum and magnesium, grinding, and in resinoid grinding wheels. See also FLUORIDES and BORON COMPOUNDS.

PKY250   CAS: 13762-51-1   HR: 3
POTASSIUM BOROHYDRATE
DOT: UN 1870
mf: $BH_3 \cdot K$ mw: 52.94
PROP: White, moisture-sensitive, cubic crystals. Decomp on heating into elements. D: 1.177, mp: >400° (decomp). Very sol in H_2O; sparingly sol in EtOH, MeOH, THF; insol in Et_2O.
SYNS: BOROHYDRURE de POTASSIUM (FRENCH) □ POTASSIUM BOROHYDRIDE (DOT) □ TETR-AHYDROBORATE(1−) POTASSIUM
CONSENSUS REPORTS: Reported in EPA TSCA Inventory.
DOT CLASSIFICATION: 4.3; Label: Dangerous When Wet
SAFETY PROFILE: Poison by ingestion. Burns quietly in air. When heated to

decomposition it emits toxic fumes of K_2O. See also BORON COMPOUNDS and HYDRIDES.

PKY300   CAS: 7758-01-2   HR: 3
POTASSIUM BROMATE
DOT: UN 1484
mf: $BrO_3 \cdot K$ mw: 167.01
PROP: White or colorless, hexagonal crystals, or powder. Mp: 434° (decomp @ 370°), d: 3.27 @ 17.5°. Sol in water; sparingly sol in EtOH; insol in Me_2CO.
SYNS: BROMIC ACID, POTASSIUM SALT □ EEC No. E924
CONSENSUS REPORTS: IARC Cancer Review: Group 2B IMEMDT 7,56,87; Animal Sufficient Evidence IMEMDT 40,207,86. Reported in EPA TSCA Inventory.
DOT CLASSIFICATION: 5.1; Label: Oxidizer
SAFETY PROFILE: Confirmed carcinogen with experimental carcinogenic data. A poison by ingestion. A powerful oxidizer. An irritant to skin, eyes, and mucous membranes. Mutation data reported. Mixtures with sulfur may ignite. Violent reaction with Al, Al + dinitrotoluene @ 290°, As, C, Cu, $Pb(C_2H_3O_2)_2$, metal sulfides, organic matter, P, S. Aqueous solutions react violently with selenium. When heated to decomposition it emits very toxic fumes of Br^- and K_2O. See also BROMIDES.

PKY500   CAS: 7758-02-3   HR: 2
POTASSIUM BROMIDE
mf: BrK   mw: 119.01
PROP: Colorless, cubic, sltly hygroscopic crystals. Mp: 730°, bp: 1435°, d: 2.75 @ 25°, vap press: 1 mm @ 795°. Very sol in H_2O; sparingly sol in EtOH, Et_2O.
SYN: BROMIDE SALT OF POTASSIUM
CONSENSUS REPORTS: Reported in EPA TSCA Inventory.
SAFETY PROFILE: Moderately toxic by ingestion and intraperitoneal routes. Large doses can cause central nervous system depression. Prolonged inhalation may cause

skin eruptions. Mutation data reported. Violent reaction with BrF_3. When heated to decomposition it emits toxic fumes of K_2O and Br^-. See also BROMIDES.

PLA000   CAS: 584-08-7   HR: 3
POTASSIUM CARBONATE (2:1)
mf: $CO_3 \cdot 2K$ mw: 138.21
PROP: White, deliquescent, granular, translucent powder; odorless with alkaline taste; or hygroscopic colorless monoclinic crystals. Undergoes polymorphic change at 2°. Decomp on heating with CO_2 loss. D: 2.428 @ 19°, mp: 901°, bp: decomposes. Sol in water; insol in alc and acetone.
SYNS: CARBONIC ACID, DIPOTASSIUM SALT □ KALIUMCARBONAT (GERMAN) □ K-GRAN □ PEARL ASH □ POTASH
CONSENSUS REPORTS: Reported in EPA TSCA Inventory.
SAFETY PROFILE: Poison by ingestion. A strong caustic. Incompatible with KCO, chlorine trifluoride, magnesium. Mutation data reported. When heated to decomposition it emits toxic fumes of K_2O.

PLA250   CAS: 3811-04-9   HR: 3
POTASSIUM CHLORATE
DOT: UN 1485/UN 2427
mf: $ClO_3 \cdot K$ mw: 122.55
PROP: Transparent, monoclinic, colorless crystals, or white powder; cooling, saline taste. Becomes orthorhombic at 2°. Mp: 356°, bp: decomp @ 400°, d: 2.32. Sol in H_2O; insol in Me_2CO.
SYNS: BERTHOLLET SALT □ CHLORATE de POTASSIUM (FRENCH) □ CHLORATE of POTASH (DOT) □ FEKABIT □ KALIUMCHLORAAT (DUTCH) □ KALIUMCHLORAT (GERMAN) □ OXYMURIATE OF POTASH □ PEARL ASH □ POTASH CHLORATE (DOT) □ POTASSIO (CHLORATO di) (ITALIAN) □ POTASSIUM CHLORATE (DOT) □ POTASSIUM (CHLORATE de) (FRENCH) □ POTASSIUM CHLORATE (UN 1485) (DOT) □ POTASSIUM CHLORATE, solution (UN 2427) (DOT) □ POTASSIUM OXYMURIATE □ POTCRATE □ SALT OF TARTAR
CONSENSUS REPORTS: Reported in EPA TSCA Inventory.
DOT CLASSIFICATION: 5.1; Label: Oxidizer

SAFETY PROFILE: Moderately toxic to humans by an unspecified route. Moderately toxic experimentally by ingestion and intraperitoneal routes. A gastrointestinal tract and kidney irritant. Can cause hemolysis of red blood cells and methemoglobinemia. Toxic dose to a human is about 5 g.
A powerful oxidizer and very reactive material. It has been the cause of many industrial explosions. May explode on heating. Explosive reactions with ammonium chloride, aqua regia + ruthenium, sulfur dioxide solutions in ether or ethanol. Reacts with fluorine to form the explosive gas fluorine perchlorate.
Forms sensitive explosive mixtures with agricultural materials (e.g., peat, powdered sulfur, sawdust, thiuram), aluminum + antimony trisulfide powders, arsenic trisulfide, carbon, charcoal + potassium nitrate + sulfur, charcoal + sulfur, cyanides, cyanoguanidine, hydrocarbons, manganese dioxide + traces of organic matter, manganese dioxide + potassium hydroxide, metal + wood, metal phosphides (e.g., tricopper diphosphide, trimercury tetraphosphide), metal phosphinates (e.g., barium phosphinate), finely divided metals (e.g., aluminum, copper, magnesium, zinc, germanium, titanium, zirconium, steel, chromium), metal phosphides (e.g., tricopper diphosphide, trimercury tetraphosphide), metal sulfides (e.g., antimony trisulfide, silver sulfide), metal thiocyanates (e.g., ammonium thiocyanate, barium thiocyanate), nitric acid + organic materials, powdered nonmetals (e.g., arsenic, carbon, phosphorus, sulfur, boron), reducing agents (e.g., calcium hydride, strontium hydride, sodium phosphinate, calcium phosphinate, barium phosphinate), sugars (e.g., glucose), sulfur, sulfur + metal derivatives (e.g., cobalt, cobalt oxide, copper nitride, copper sulfate, copper chlorate), sulfuric acid, sodium amide, tannic acid.

Violent reaction or ignition with NH_3, NH_4Cl, NH_{4+} salts, ammonium sulfate, Sb_2S_3, As, barium hypophosphite, BaS, calcium hypophosphite, CaS, charcoal, Cu_3P_2, fabrics, gallic acid, hydrogen iodide, lactose, (Mg + $CuSO_4$ (anhydrous) + NH_4NO_3 + H_2O)), MnO_2, dinickel trioxide, dibasic organic acids, organic matter, $NaNH_2$, sugar + sulfuric acid, sucrose, SO_2, H_2SO_4, thiocyanates, thorium dicarbide, sodium amide, fabrics, KOH, metal hypophosphites.

When heated to decomposition it emits very toxic fumes of Cl^- and K_2O. Used in the manufacture of soap, glass, and pottery. See also CHLORATES.

PLA500   CAS: 7447-40-7   HR: 3
POTASSIUM CHLORIDE
mf: ClK   mw: 74.55

PROP: Colorless or white crystals or powder; odorless with salty taste. D: 1.987, mp: 771° (subl @ 1500°). Very sol in H_2O; sol in Et_2O; sparingly sol in EtOH.

SYNS: CHLORID DRASELNY (CZECH) □ CHLOROPOTASSURIL □ DIPOTASSIUM DICHLORIDE □ EMPLETS POTASSIUM CHLORIDE □ ENSEAL □ KALITABS □ KAOCHLOR □ KAON-Cl □ KAY CIEL □ K-LOR □ KLOTRIX □ K-PRENDE-DOME □ PFIKLOR □ POTASSIUM MONOCHLORIDE □ POTAVESCENT □ REKAWAN □ SLOW-K □ TRIPOTASSIUM TRICHLORIDE

CONSENSUS REPORTS: Reported in EPA TSCA Inventory

SAFETY PROFILE: A human poison by ingestion. Poison experimentally by ingestion, intravenous, and intraperitoneal routes. Human systemic effects by ingestion: nausea, blood clotting changes, cardiac arrhythmias. An eye irritant. Mutation data reported. Explosive reaction with BrF_3; sulfuric acid + potassium permanganate. When heated to decomposition it emits toxic fumes of K_2O and Cl^-.

PLB250   CAS: 7789-00-6   HR: 3
POTASSIUM CHROMATE(VI)
mf: CrO_4•2K   mw: 194.20

PROP: Rhombic, yellow crystals. Mp: 975°, d: 2.73 @ 18°. Sol in water; insol in alc, Me_2CO, and PhCN.

SYNS: BIPOTASSIUM CHROMATE □ CHROMATE OF POTASSIUM □ DIPOTASSIUM CHROMATE □ DIPOTASSIUM MONOCHROMATE □ NEUTRAL POTASSIUM CHROMATE □ TARAPACAITE

CONSENSUS REPORTS: IARC Cancer Review: Group 1 IMEMDT 49,49,90; Human Sufficient Evidence IMEMDT 49,49,90; Human Inadequate Evidence IMEMDT 23,205,80; Animal Inadequate Evidence IMEMDT 23,205,80. Reported in EPA TSCA Inventory. EPA Genetic Toxicology Program. Chromium and its compounds are on the Community Right-To-Know List.

OSHA PEL: CL 0.1 mg(CrO_3)/m³
ACGIH TLV: TWA 0.05 mg(Cr)/m³; Confirmed Human Carcinogen

SAFETY PROFILE: Confirmed carcinogen with experimental tumorigenic data. Poison by ingestion, intravenous, subcutaneous, and intramuscular routes. An experimental teratogen. Other experimental reproductive effects. Human mutation data reported. A powerful oxidizer. When heated to decomposition it emits toxic fumes of K_2O. Used as a mordant for wool, in the oxidizing and treatment of dyes on materials. See also CHROMIUM COMPOUNDS.

PLB500   CAS: 10141-00-1   HR: D
POTASSIUM CHROMIC SULFATE
mf: Cr•$2H_2O_4S$•K   mw: 287.26

SYNS: CHROME ALUM □ CHROME POTASH ALUM □ CHROMIC POTASSIUM SULFATE □ CHROMIC POTASSIUM SULPHATE □ CHROMIUM POTASSIUM SULFATE (1:1:2) □ CHROMIUM POTASSIUM SULPHATE □ CRYSTAL CHROME ALUM □ POTASSIUM CHROMIC SULPHATE □ POTASSIUM CHROMIUM ALUM □ POTASSIUM DISULPHATOCHROMATE(III) □ SULFURIC ACID, CHROMIUM (3+) POTASSIUM SALT (2:1:1)

CONSENSUS REPORTS: IARC Cancer Review: Group 3 IMEMDT 49,49,90; Animal Inadequate Evidence IMEMDT 49,49,90; Human Inadequate Evidence IMEMDT 49,49,90. Chromium and its compounds are on the Community Right-To-Know List. Reported in EPA TSCA

Inventory. EPA Genetic Toxicology Program.
OSHA PEL: TWA 0.5 mg(Cr)/m³
ACGIH TLV: TWA 0.5 mg(Cr)/m³; Not Classifiable as a Carcinogen
SAFETY PROFILE: Mutation data reported. Questionable carcinogen. When heated to decomposition it emits toxic fumes of K₂O. See also CHROMIUM COMPOUNDS and SULFATES.

PLB750   CAS: 866-84-2   HR: 3
POTASSIUM CITRATE
mf: C₆H₅O₇•3K   mw: 306.41
PROP: Colorless transparent crystals or white powder; odorless with salty taste. D: 1.98, decomp when heated to 230°. Deliq; sol in water and glycerol; almost insol in alc.
SYNS: CITRIC ACID, TRIPOTASSIUM SALT □ TRIPOTASSIUM CITRATE MONOHYDRATE
CONSENSUS REPORTS: Reported in EPA TSCA Inventory.
SAFETY PROFILE: Poison by intravenous route. When heated to decomposition it emits toxic fumes of K₂O.

PLC175   CAS: 13682-73-0   HR: 3
POTASSIUM CUPROCYANIDE
DOT: UN 1679
mf: C₂CuN₂•K   mw: 154.68
PROP: Monoclinic crystals. D: 2.38. Insol in water; sol in DMSO.
SYNS: COPPER(I) POTASSIUM CYANIDE □ CUPRATE(1-), DICYANO-, POTASSIUM □ CUPROUS POTASSIUM CYANIDE □ POTASSIUM COPPER(I) CYANIDE □ POTASSIUM DICYANOCUPRATE(1-)
CONSENSUS REPORTS: Reported in EPA TSCA Inventory.
ACGIH TLV: TWA 1 mg(Cu)/m³
DOT CLASSIFICATION: 6.1; Label: Poison
SAFETY PROFILE: A poison. When heated to decomposition it emits toxic fumes of CN⁻.

PLC250   CAS: 590-28-3   HR: 3
POTASSIUM CYANATE
mf: CNO•K   mw: 81.12

PROP: Colorless, tetragonal crystals. Mp: 700–900° (decomp), d: 2.056 @ 20°. Sol in water; very sltly sol in alc.
SYNS: AERO CYANATE □ ALICYANATE □ BONIDE KRAB CRABGRASS KILLER □ BULPUR □ CYANIC ACID, POTASSIUM SALT □ DED-WEED CRABGRASS KILLER □ D & P DOUBLE O CRABGRASS KILLER □ DU PONT PC CRABGRASS KILLER □ GREEN CROSS CRABGRASS KILLER □ KALIUMCYANAT (GERMAN) □ MILLER P.C. WEEDKILLER □ P.C. 80 CRABGRASS KILLER □ POTASSIUM ISOCYANATE □ WEEDANOL CYANOL □ WEEDONE CRAB GRASS KILLER
CONSENSUS REPORTS: Reported in EPA TSCA Inventory. EPA Genetic Toxicology Program. Cyanide and its compounds are on the Community Right-To-Know List.
SAFETY PROFILE: Poison by intraperitoneal route. Moderately toxic by ingestion. Causes irritation of the gastrointestinal tract. An herbicide. It is said to be slowly metabolized in the body to cyanide but does not have high toxicity of cyanides. When heated to decomposition it emits very toxic fumes of CN⁻ and K₂O.

PLC500   CAS: 151-50-8   HR: 3
POTASSIUM CYANIDE
DOT: UN 1680
mf: CN•K   mw: 65.12
PROP: Colorless water soln. Deliquescent colorless cubic crystals. Undergoes cubic to orthorhombic transition. Slt odor of bitter almonds. Mp: 622°. Very sol in H₂O; sparingly sol in EtOH. IDLH 25 mg/m³ (as CN).
SYNS: CYANIDE of POTASSIUM □ CYANIDES (OSHA) □ CYANURE de POTASSIUM (FRENCH) □ HYDROCYANIC ACID, POTASSIUM SALT □ KALIUM-CYANID (GERMAN) □ POTASSIUM CYANIDE, solution (DOT) □ RCRA WASTE NUMBER P098
CONSENSUS REPORTS: EPA Extremely Hazardous Substances List. Cyanide and its compounds are on the Community Right-To-Know List. Reported in EPA TSCA Inventory. EPA Genetic Toxicology Program.
OSHA PEL: TWA 5 mg(CN)/m³
ACGIH TLV: CL 5 mg(CN)/m³ (skin)
DFG MAK: 5 mg(CN)/m³

P

NIOSH REL: CL (Cyanide) 5 mg(CN)/m³/10M
DOT CLASSIFICATION: 6.1; Label: Poison
SAFETY PROFILE: A deadly human poison by ingestion. A experimental poison by ocular, subcutaneous, intravenous, intramuscular, and intraperitoneal routes. Experimental teratogenic and reproductive effects. Human systemic effects by ingestion: convulsions, pulse rate increase. Mutation data reported. Reacts with acids or acid fumes to liberate deadly HCN. When heated to decomposition it emits very toxic fumes of K_2O, CN^-, and NO_x. See also CYANIDE.

PLD000   CAS: 2244-21-5   HR: 2
POTASSIUM DICHLOROISO-
CYANURATE
mf: $C_3HCl_2N_3O_3•K$ mw: 237.07
PROP: White, sltly hygroscopic, crystalline powder or granules; chlorine odor. Mp: 250° (decomp).
SYNS: ACL-59 □ DICHLOROISOCYANURIC ACID POTASSIUM SALT □ DICHLORO-s-TRIAZINE-2,4,6(1H,3H,5H)-TRIONE POTASSIUM DERIV □ 1,3-DICHLORO-s-TRIAZINE-2,4,6(1H,3H,5H)TRIONE POTASSIUM SALT □ DICHLOR-s-TRIAZIN-2,4,6(1H,3H,5H)TRIONE POTASSIUM □ ISOCYANURIC ACID, DICHLORO-, POTASSIUM SALT □ POTASSIUM DICHLORO-s-TRIAZINETRIONE □ POTASSIUM TROCLOSENE □ s-TRIAZINE-2,4,6(1H,3H,5H)-TRIONE, DICHLORO-, POTASSIUM DERIV □ 1,3,5-TRIAZINE-2,4,6(1H,3H,5H)-TRIONE, 1,3-DICHLORO-, POTASSIUM SALT □ TROCLOSENE POTASSIUM
CONSENSUS REPORTS: Reported in EPA TSCA Inventory.
SAFETY PROFILE: Moderately toxic by ingestion, causing ulceration or bleeding from stomach. A skin and severe eye irritant. Causes emaciation, weakness, lethargy, diarrhea, weight loss. Autopsy indicates gastrointestinal tract irritation, tissue edema, liver and kidney congestion. A powerful oxidizer. When heated to decomposition it emits very toxic fumes of K_2O, Cl^-, and NO_x.

PLE260   CAS: 12030-88-5   HR: 3
POTASSIUM DIOXIDE

DOT: UN 2466
mf: KO_2 mw: 71.10
PROP: Moisture-sensitive orange tetragonal crystals. Decomp in H_2O with formation of KOH and evolution of O_2. Mp: 509°.
SYN: POTASSIUM SUPEROXIDE (DOT)
CONSENSUS REPORTS: Reported in EPA TSCA Inventory.
DOT CLASSIFICATION: 5.1; Label: Oxidizer
SAFETY PROFILE: Explosive reaction when heated with carbon, 2-aminophenol + tetrahydrofuran (at 65°C). Forms a friction-sensitive explosive mixture with hydrocarbons. Violent reaction with diselenium dichloride, ethanol, potassium-sodium alloy. May ignite on contact with organic compounds. Incandescent reaction with metals (e.g., arsenic, antimony, copper, potassium, tin, and zinc). When heated to decomposition it emits toxic fumes of K_2O. See also PEROXIDES.

PLF250   CAS: 13746-66-2   HR: 2
POTASSIUM FERRICYANATE
mf: $C_6FeN_6•3K$ mw: 329.27
PROP: Red crystals (dimorphic) or lemon-yellow crystals. Mp: loses $3H_2O$ @ 70°, bp: decomp, d: 1.85 @ 17°.
SYNS: HEXACYANOFERRATE(3⁻) TRIPOTASSIUM □ POTASSIUM FERRICYANIDE □ POTASSIUM HEXACYANOFERRATE(III) □ TRIPOTASSIUM HEXACYANOFERRATE
CONSENSUS REPORTS: Cyanide and its compounds are on the Community Right-To-Know List. Reported in EPA TSCA Inventory. EPA Genetic Toxicology Program.
OSHA PEL: TWA 5 mg(CN)/m³
ACGIH TLV: CL 5 mg(CN)/m³ (skin)
DFG MAK: 5 mg/m³
NIOSH REL: (Cyanide) CL 5 mg(CN)/m³/10M
SAFETY PROFILE: Moderately toxic by ingestion. Not as toxic as the simple cyanides. Mutation data reported. Explosive reaction with ammonia, chromium trioxide (above 196°C), sodium nitrite + heat. Violent reaction with $Cu(NO_3)_2$. Mixtures with chromium trioxide + silver grains ignite

with friction. When heated to decomposition or on contact with acid or acid fumes it emits highly toxic fumes of K_2O and CN^-. Used as a fixative in photography, as a metal cleaner, and for glass coatings.

PLF500   **CAS: 7789-23-3**   **HR: 3**
POTASSIUM FLUORIDE
DOT: UN 1812
mf: FK   mw: 58.10
PROP: White, crystalline, deliq powder; sharp saline taste; or deliq colorless crystals. Bp: 1500°, d: 2.48, vap press: 1 mm @ 885°, mp: 859.9°. Very sol in boiling water; insol in alc.
SYNS: FLUORURE de POTASSIUM (FRENCH) □ POTASSIUM FLUORIDE, solution (DOT) □ POTASSIUM FLUORURE (FRENCH)
CONSENSUS REPORTS: Reported in EPA TSCA Inventory. EPA Genetic Toxicology Program.
OSHA PEL: TWA 2.5 mg(F)/m³
ACGIH TLV: TWA 2.5 mg(F)/m³; BEI: 3 mg/g creatinine of fluorides in urine prior to shift; 10 mg/g creatinine of fluorides in urine at end of shift.
NIOSH REL: TWA (Inorganic Fluorides) 2.5 mg(F)/m³
DOT CLASSIFICATION: 6.1; Label: KEEP AWAY FROM FOOD
SAFETY PROFILE: Poison by ingestion and intraperitoneal routes. Moderately toxic by subcutaneous route. Experimental teratogenic effects. A corrosive irritant to the eyes, skin, and mucous membranes. Mutation data reported. A very reactive material. When heated to decomposition it emits toxic fumes of K_2O and F^-. Used in etching glass, as a preservative, as an insecticide, and in organic synthesis. See also FLUORIDES and HYDROFLUORIC ACID.

PLG000   **CAS: 23745-86-0**   **HR: 3**
POTASSIUM FLUOROACETATE
DOT: UN 2628
mf: $C_2H_3FO_2 \cdot K$ mw:117.15

PROP: The potassium salt of monofluoroacetic acid was once designated as potassium cymonate.
SYN: DICHAPETULUM CYMOSUM (HOOK) ENGL
DOT CLASSIFICATION: 6.1; Label: Poison
SAFETY PROFILE: Poison by ingestion, intravenous, parenteral, and subcutaneous routes. When heated to decomposition it emits toxic fumes of F^- and K_2O. See also FLUORIDES.

PLG500   **CAS: 16923-95-8**   **HR: 3**
POTASSIUM FLUOZIRCONATE
mf: K_2ZrF_6 mw: 283.4
PROP: Monoclinic, colorless crystals or white solid. D: 3.48.
SYN: ZIRCONIUM POTASSIUM FLUORIDE
CONSENSUS REPORTS: Reported in EPA TSCA Inventory.
OSHA PEL: TWA 2.5 mg(F)/m³
ACGIH TLV: TWA 2.5 mg(F)/m³; BEI: 3 mg/g creatinine of fluorides in urine prior to shift; 10 mg/g creatinine of fluorides in urine at end of shift.
NIOSH REL: TWA 2.5 mg(F)/m³
SAFETY PROFILE: Poison by ingestion. When heated to decomposition it emits toxic fumes of F^- and K_2O. See also FLUORIDES and ZIRCONIUM COMPOUNDS.

PLG775   **HR: D**
POTASSIUM GIBBERELLATE
mf: $C_{19}H_{21}KO_6$ mw: 384.47
PROP: White crystalline powder; odorless. Deliquescent, sol in water, alc, acetone.
SAFETY PROFILE: When heated to decomposition it emits acrid smoke and irritating fumes.

PLG800   **CAS: 299-27-4**   **HR: 2**
POTASSIUM GLUCONATE
mf: $C_6H_{12}O_7 \cdot K$ mw: 235.28
PROP: Yellowish-white crystals or powder; mild, sltly salty taste. Decomp at 180°. Freely sol in water, glycerin; practically insol in abs alc, ether, benzene, chloroform.

P

SYNS: d-GLUCONIC ACID, MONOPOTASSIUM SALT (9CI) □ GLUCONIC ACID POTASSIUM SALT □ GLUCONSAN K □ KALIUM-BETA □ KAON □ KAON ELIXIR □ KATORIN □ K-IAO □ POTALIUM □ POTASORAL □ POTASSIUM d-GLUCONATE □ POTASSURIL □ SIROKAL

CONSENSUS REPORTS: Reported in EPA TSCA Inventory.

SAFETY PROFILE: Moderately toxic by intraperitoneal route. Mildly toxic by ingestion. When heated to decomposition it emits toxic fumes of K_2O.

PLH000   CAS: 16924-00-8   HR: 3
POTASSIUM
HEPTAFLUOROTANTALATE
mf: $F_7Ta•2K$ mw: 392.15

PROP: Colorless crystals. Mp: 730°.

SYNS: POTASSIUM FLUOTANTALATE □ TANTALUM POTASSIUM FLUORIDE

CONSENSUS REPORTS: Reported in EPA TSCA Inventory.

OSHA PEL: TWA 2.5 mg(F)/m³

ACGIH TLV: TWA 2.5 mg(F)/m³; BEI: 3 mg/g creatinine of fluorides in urine prior to shift; 10 mg/g creatinine of fluorides in urine at end of shift.

NIOSH REL: TWA 2.5 mg(F)/m³

SAFETY PROFILE: Poison by ingestion and intraperitoneal routes. When heated to decomposition it emits toxic fumes of F^- and K_2O.

PLH500   CAS: 17029-22-0   HR: 3
POTASSIUM HEXAFLUOROARSENATE
mf: AsF_6K mw: 228.02

PROP: Thick, colorless plates.

SYNS: HEXAFLURATE □ NOPALMATE

CONSENSUS REPORTS: Arsenic and its compounds are on the Community Right-To-Know List. Reported in EPA TSCA Inventory.

OSHA PEL: TWA 2.5 mg(F)/m³; Cancer Hazard

ACGIH TLV: TWA 0.01 mg/m³; Confirmed Human Carcinogen; BEI: 35 μ (As)/L inorganic arsenic and methylated metabolites in urine

NIOSH REL: CL 2 μg/m³/15M

SAFETY PROFILE: Confirmed human carcinogen. Poison by intravenous route. Moderately toxic by ingestion. When heated to decomposition it emits very toxic fumes of K_2O, F^-, and As. See also FLUORIDES and ARSENIC COMPOUNDS.

PLH750   CAS: 16871-90-2   HR: 3
POTASSIUM HEXAFLUOROSILICATE
DOT: UN 2655

mf: $F_6Si•2K$ mw: 220.29

PROP: White, fine powder or colorless, cubic crystals. Moisture-sensitive. D: 2.27, mp: decomp. Sltly sol in cold water; practically insol in alc.

SYNS: POTASSIUM FLUOSILICATE □ POTASSIUM SILICOFLUORIDE (DOT)

CONSENSUS REPORTS: Reported in EPA TSCA Inventory.

OSHA PEL: TWA 2.5 mg(F)/m³

ACGIH TLV: TWA 2.5 mg(F)/m³; BEI: 3 mg/g creatinine of fluorides in urine prior to shift; 10 mg/g creatinine of fluorides in urine at end of shift.

NIOSH REL: TWA (Inorganic Fluorides) 2.5 mg(F)/m³

DOT CLASSIFICATION: 6.1; Label: KEEP AWAY FROM FOOD

SAFETY PROFILE: A poison by ingestion and subcutaneous routes. Ingestion can cause vomiting and diarrhea. A strong irritant. Incompatible with hydrofluoric acid. When heated to decomposition it emits toxic fumes of SiF_4, K_3SiF_7, and KF.

PLI000   CAS: 16919-27-0   HR: 3
POTASSIUM HEXAFLUOROTITANATE
mf: $F_6Ti•2K$ mw: 240.10

PROP: White solid.

SYNS: FLUOTITANATE de POTASSIUM (FRENCH) □ TITANIUM POTASSIUM FLUORIDE

CONSENSUS REPORTS: Reported in EPA TSCA Inventory.

OSHA PEL: TWA 2.5 mg(F)/m³

ACGIH TLV: TWA 2.5 mg(F)/m³; BEI: 3 mg/g creatinine of fluorides in urine prior to shift; 10 mg/g creatinine of fluorides in urine at end of shift.

NIOSH REL: TWA 2.5 mg(F)/m³

SAFETY PROFILE: Poison by subcutaneous route. When heated to decomposition it emits toxic fumes of K_2O and F^-. See also FLUORIDES.

PLJ250   CAS: 7693-26-7   HR: 3
POTASSIUM HYDRIDE
mf: HK   mw: 40.11
PROP: White needles. Moisture-sensitive white crystals. Mp: decomp, d: 1.43–1.47.
SAFETY PROFILE: Dangerous fire hazard by chemical reaction. Ignites spontaneously in air. Moderate explosion hazard when exposed to heat or by chemical reaction. Will react with water, steam, or acids to produce H_2 which then ignites. Can react vigorously with oxidizing materials. To fight fire, use CO_2, dry chemical. Potentially explosive reactions with o-2,4-dinitrophenylhydroxylamine, fluoroalkenes. Ignites on contact with air, oxygen + moisture, fluorine. Incompatible with Cl_2, acetic acid, acrolein, acrylonitrile, (CaC + Cl_2), ClO_2, (H_2O_2 + Cl_2), ($CHFl_3$ + CH_3OH), 1,2-dichloroethylene, maleic anhydride, (n-methyl-n-nitrosourea + CH_2Cl_2), nitroethane, NCl_3, nitromethane, nitroparaffins, o-nitrophenol, nitropropane, n-nitrosomethylurea, (nitrosomethylurea + CH_2Cl_2), H_2O, trichloroethylene, tetrahydrofuran, tetrachlorethane. When heated to decomposition it emits highly toxic fumes of K_2O. See also POTASSIUM and HYDRIDES.

PLJ500   CAS: 1310-58-3   HR: 3
POTASSIUM HYDROXIDE
DOT: UN 1813/UN 1814
mf: HKO   mw: 56.11
PROP: White or colorless, orthorhombic, deliquescent pieces, lumps, or sticks having crystalline fracture. Mp: 406°, bp: 1324°, d: 2.044. Very sol in water, alc; sol in EtOH; insol in Et_2O.
SYNS: CAUSTIC POTASH □ CAUSTIC POTASH, dry, solid, flake, bead, or granular (DOT) □ CAUSTIC POTASH, liquid or solution (DOT) □ HYDROXYDE de POTASSIUM (FRENCH) □ KALIUMHYDROXID (GERMAN) □ KALIUMHYDROXYDE (DUTCH) □ LYE □ POTASSA □

POTASSE CAUSTIQUE (FRENCH) □ POTASSIO (IDROSSIDO di) (ITALIAN) □ POTASSIUM HYDRATE (DOT) □ POTASSIUM HYDROXIDE, dry, solid, flake, bead, or granular (DOT) □ POTASSIUM HYDROXIDE, liquid or solution (DOT) □ POTASSIUM (HYDROXYDE de) (FRENCH)
CONSENSUS REPORTS: Reported in EPA TSCA Inventory.
OSHA PEL: CL 2 mg/m³
ACGIH TLV: CL 2 mg/m³
DOT CLASSIFICATION: 8; Label: Corrosive
SAFETY PROFILE: Poison by ingestion. An eye irritant and severe human skin irritant. Very corrosive to the eyes, skin, and mucous membranes. Mutation data reported. Ingestion may cause violent pain in throat and epigastrium, hematemesis, collapse. Stricture of esophagus may result if substance is not immediately fatal. Above 84° it reacts with reducing sugars to form poisonous carbon monoxide gas. Violent, exothermic reaction with water. Potentially explosive reaction with bromoform + crown ethers, chlorine dioxide, nitrobenzene, nitromethane, nitrogen trichloride, peroxidized tetrahydrofuran, 2,4,6-trinitrotoluene. Reaction with ammonium hexachloroplatinate(2-) + heat forms a heat-sensitive explosive product. Violent reaction or ignition under the appropriate conditions with acids, alcohols, p-bis(1,3-dibromoethyl)benzene, cyclopentadiene, germanium, hyponitrous acid, maleic anhydride, nitroalkanes, 2-nitrophenol, potassium peroxodisulfate, sugars, 2,2,3,3-tetrafluoropropanol, thorium dicarbide. When heated to decomposition it emits toxic fumes of K_2O. See also SODIUM HYDROXIDE.

PLK000   CAS: 7778-66-7   HR: 3
POTASSIUM HYPOCHLORITE
mf: ClHO•K   mw: 91.56
SYNS: HYPOCHLOROUS ACID, POTASSIUM SALT □ POTASSIUM CHLORIDE OXIDE
CONSENSUS REPORTS: IARC Cancer Review: Group 3 IMEMDT 52,159,91; Animal Inadequate Evidence IMEMDT 52,159,91; Human No Available Data

IMEMDT 52,159,91. Reported in EPA TSCA Inventory.

SAFETY PROFILE: A poison by all routes. Powerful irritant and corrosive to skin, eyes, and mucous membranes. Questionable carcinogen. When heated to decomposition it emits toxic fumes of K_2O and Cl^-. See also HYPOCHLORITES.

PLK250   CAS: 7758-05-6   HR: 3
POTASSIUM IODATE
mf: $IO_3 \cdot K$ mw: 214.00
PROP: Colorless, triclinic crystals or white crystalline powder. Becomes monoclinic at 72°. Also undergoes 2 phase changes below room temp, but remains triclinic. Mp: 560°, d: 3.89. Very sol in H_2O; insol in alc.
SYN: IODIC ACIODIC ACID, POTASSIUM SALT
CONSENSUS REPORTS: Reported in EPA TSCA Inventory.
SAFETY PROFILE: Poison by ingestion and intraperitoneal routes. A trace mineral added to animal feeds. Potentially explosive reaction with charcoal + ozone, metals (e.g., powdered aluminum, copper), arsenic carbon, phosphorus, sulfur, alkali metal hydrides, alkaline earth metal hydrides, antimony sulfide, arsenic sulfide, copper sulfide, tin sulfide, metal cyanides, metal thiocyanates, manganese dioxide, phosphorus. Violent reaction with organic matter. When heated to decomposition it emits very toxic fumes of I^- and K_2O. See also IODATES.

PLK500   CAS: 7681-11-0   HR: 3
POTASSIUM IODIDE
mf: IK   mw: 166.00
PROP: Colorless or white granules or colorless cubic crystals. Mp: 681°, bp: 1330°, d: 3.13, vap press: 1 mm @ 745°. Sltly hygroscopic. Very sol in water; mod sol in alc and Me_2CO; sparingly sol in Et_2O.
SYNS: K1-N □ KNOLLIDE □ POTIDE
CONSENSUS REPORTS: Reported in EPA TSCA Inventory.

ACGIH TLV: Proposed: (inhalable fraction) 0.1 mg/m³; Not Classifiable as a Human Carcinogen)
SAFETY PROFILE: Poison by intravenous route. Moderately toxic by ingestion and intraperitoneal routes. Human teratogenic effects by ingestion: developmental abnormalities of the endocrine system. Experimental teratogenic and reproductive effects. Mutation data reported. Explosive reaction with charcoal + ozone, trifluoroacetyl hypofluorite, fluorine perchlorate. Violent reaction or ignition on contact with diazonium salts, diisopropyl peroxydicarbonate, bromine pentafluoride, chlorine trifluoride. Incompatible with oxidants, BrF_3, $FClO$, metallic salts, calomel. When heated to decomposition it emits very toxic fumes of K_2O and I^-. See also IODIDES.

PLK810   CAS: 13769-43-2   HR: 3
POTASSIUM METAVANADATE
DOT: UN 2864
mf: $O_3V \cdot K$ mw: 138.04
SYNS: POTASSIUM METAVANADATE □ POTASSIUM VANADIUM TRIOXIDE □ VANADIC ACID, POTASSIUM SALT
CONSENSUS REPORTS: Reported in EPA TSCA Inventory.
ACGIH TLV: TWA 0.05 mg(V_2O_5)/m³
DOT CLASSIFICATION: 6.1; Label: Poison
SAFETY PROFILE: A poison. When heated to decomposition it emits toxic fumes of V_2O_5.

PLL500   CAS: 7757-79-1   HR: 3
POTASSIUM NITRATE
DOT: UN 1486
mf: KNO_3 mw: 140.21
PROP: Transparent, colorless, or white crystalline powder or orthorhombic crystals; odorless with a cooling, pungent, salty taste. Mp: 334°, bp: decomp @ 400°, d: 2.109 @ 16°. Sol in glycerin, water; mod sol in alc.
SYNS: KALIUMNITRAT (GERMAN) □ NITER □ NITRE □ NITRIC ACID, POTASSIUM SALT □ SALTPETER □ VICKNITE

CONSENSUS REPORTS: Reported in EPA TSCA Inventory.

DOT CLASSIFICATION: 5.1; Label: Oxidizer

SAFETY PROFILE: Poison by intravenous route. Moderately toxic by ingestion. An experimental teratogen. Experimental reproductive effects. Mutation data reported. Ingestion of large quantities may cause gastroenteritis. Chronic exposure can cause anemia, nephritis, and methemoglobinemia. When heated, reaction with calcium hydroxide + polychlorinated phenols forms extremely toxic chlorinated benzodioxins.

A powerful oxidizer. Gunpowder is a mixture of potassium nitrate + sulfur + charcoal. Explosive reaction with aluminum + barium nitrate + potassium perchlorate + water (in storage), boron + laminac + trichloroethylene. Forms explosive mixtures with lactose, powdered metals (e.g., titanium, antimony, germanium), metal sulfides (e.g., antimony trisulfide, barium sulfide, calcium sulfide, germanium monosulfide, titanium disulfide, arsenic disulfide, molybdenum disulfide), nonmetals (e.g., boron, carbon, white phosphorus, arsenic), organic materials, phosphides (e.g., copper(II) phosphide, copper monophosphide), reducing agents (e.g., sodium phosphinate, sodium thiosulfate), sodium acetate. Can react violently under the appropriate conditions with 1,3-bis(trichloromethyl)benzene, boron phosphide, F_2, calcium silicide, charcoal, chromium nitride, Na hypophosphite, (Na_2O_2 + dextrose), red phosphorus, (S + As_2S_3), thorium dicarbide, trichloroethylene, zinc, zirconium. When heated to decomposition it emits very toxic fumes of NO_x and K_2O. See also NITRATES.

PLL750   **HR: 3**
POTASSIUM NITRATE mixed with CHARCOAL and SULFUR (15:3:2)
DOT: UN 0027/UN 0028
SYNS: BLACK POWDER, compressed (DOT) □ BLACK POWDER, granular or as a meal (UN 0027) (DOT) □ BLACK

POWDER, in pellets (UN 0028) (DOT) □ BLASTING POWDER □ GUNPOWDER □ GUNPOWDER, compressed (UN 0028) (DOT) □ GUNPOWDER, granular or as a meal (UN 0027) (DOT) □ GUNPOWDER, in pellets (UN 0028) (DOT) □ RIFLE POWDER

DOT CLASSIFICATION: EXPLOSIVE 1.1D; Label: EXPLOSIVE 1.1D

SAFETY PROFILE: Heat and/or shock can cause an explosion. When heated to decomposition it emits toxic fumes of SO_x, K_2O, and NO_x. See also POTASSIUM NITRATE and EXPLOSIVES, LOW.

PLM000   **HR: 3**
POTASSIUM NITRATE mixed with SODIUM NITRITE
DOT: UN 1487
SYN: POTASSIUM NITRATE mixed (fused) with SODIUM NITRITE (DOT)

DOT CLASSIFICATION: 5.1; Label: Oxidizer

SAFETY PROFILE: A powerful oxidizer. Nitrites have been implicated as possible carcinogens. When heated to decomposition it emits toxic fumes of NO_x, K_2O, and Na_2O. See also POTASSIUM NITRATE and SODIUM NITRITE.

PLM500   **CAS: 7758-09-0**   **HR: 3**
POTASSIUM NITRITE (1:1)
DOT: UN 1488
mf: $NO_2 \cdot K$ mw: 85.11

PROP: White or sltly yellowish, deliquescent prisms or sticks, or hexagonal crystals. Mp: 440°, bp: decomp, d: 1.915. Very sol in water; sltly sol in alc.

SYNS: NITROUS ACID, POTASSIUM SALT □ POTASSIUM NITRITE (DOT)

CONSENSUS REPORTS: Reported in EPA TSCA Inventory.

DOT CLASSIFICATION: 5.1; Label: Oxidizer

SAFETY PROFILE: Poison by ingestion. Human systemic effects: tinnitus, pulse rate increase, blood pressure lowering. Experimental teratogenic and reproductive effects. Nitrites have been implicated in an increased incidence of cancer. Mutation data reported. Flammable when exposed to heat or flame. A powerful oxidizing material. Slight explosion hazard when exposed to heat. It will explode at 1000°F. Explosive reaction with potassium amide + heat, potassium cyanide or other cyanide salts + heat. Violent reaction or ignition with ammonium salts (e.g., ammonium sulfate), boron. Disproportionates on heating in the absence of air forming KNO_3 and K_2O and evolving N_2. Upon decomposition it emits toxic fumes of K_2O. See also NITRITES.

PLN050   CAS: 15611-84-4   HR: 3
POTASSIUM NITROTRICHLORO-
PLATINATE
mf: $Cl_3NO_2Pt•2K$ mw: 425.65
PROP: IDLH 4 mg/m³ (as Pt).
SYNS: DIPOTASSIUM TRICHLORONITROPLATINATE □ PLATINATE(2⁻), NITROTRICHLORO-, DIPOTASSIUM □ PLATINATE(2⁻), TRICHLORO(NITRITO-N-), DIPOTASSIUM (SP-4-2)- □ PLATINUM (2⁻), NITROTRICHLORO-, DIPOTASSIUM
OSHA PEL: TWA 0.002 mg(Pt)/m³
ACGIH TLV: TWA 0.002 mg(Pt)/m³
SAFETY PROFILE: Poison by intravenous route. When heated to decomposition it emits toxic fumes of NO_x and Pt.

PLO500   CAS: 7778-74-7   HR: 3
POTASSIUM PERCHLORATE
DOT: UN 1489
mf: $ClO_4•K$ mw: 138.55
PROP: Colorless, orthorhombic crystals or white powder. Undergoes orthorhombic to cubic transition at 3°. Decomp @ 400° and with organic matter. D: 2.52, mp: 525° ± 10°. Sol in H_2O; practically insol in EtOH; insol in Et_2O.
SYNS: ASTRUMAL □ IRENAL □ IRENAT □ PERIODIN □ POTASSIUM HYPERCHLORIDE □ POTASSIUM PERCHLORATE □ POTASSIUM PERCHLORATE, solid or solution (DOT)
CONSENSUS REPORTS: Reported in EPA TSCA Inventory.
DOT CLASSIFICATION: 5.1; Label: Oxidizer
SAFETY PROFILE: An experimental teratogen. A powerful oxidizer. Severe irritant to skin, eyes, and mucous membranes. Has been implicated in aplastic anemia. Absorption can cause methemoglobinemia and kidney injury.

It has been involved in many industrial explosions. Explodes on contact with aluminum + barium nitrate + potassium nitrate + water. Forms explosive mixtures with aluminum powder + titanium dioxide, ethylene glycol (240°C), cotton lint (245°C), furfural (270°C), lactose, metal powders (e.g., aluminum, iron, magnesium, molybdenum, nickel, tantalum, titanium), sulfur, titanium hydride. Reaction with ethanol + heat forms the explosive ethyl perchlorate. Violent reaction or ignition under the proper conditions with aluminum + aluminum fluoride, barium chromate + tungsten or titanium, boron + magnesium + silicone rubber, ferrocenium diammine-tetrakis(thiocyanato-N) chromate(1−), potassium hexacyanocobaltate(3−), Al + Mg, charcoal, F_2, Ni + Ti, reducing agents. When heated to decomposition it emits very toxic fumes of K_2O and Cl⁻. See also PERCHLORATES.

PLP000   CAS: 7722-64-7   HR: 3
POTASSIUM PERMANGANATE
DOT: UN 1490
mf: $MnO_4•K$ mw: 158.04
PROP: Air-stable, dark-purple crystals with a blue metallic sheen; sweetish astringent taste. Aq solns slowly deposit MnO_2. Mp: decomp @ <240°, d: 2.703. Sol in H_2O; mod sol in MeOH, AcOH, Me_2CO, and Py.
SYNS: CAIROX □ CHAMELEON MINERAL □ C.I. 77755 □ CONDY'S CRYSTALS □ KALIUMPER-MANGANAAT (DUTCH) □ KALIUMPERMANGANAT (GERMAN) □ PERMANGANATE de POTASSIUM (FRENCH) □ PERMANGANATE of POTASH (DOT) □ POTASSIO (PERMANGANATO di) (ITALIAN) □ POTASSIUM (PERMANGANATE de) (FRENCH)
CONSENSUS REPORTS: Manganese and its compounds are on the Community Right-To-Know List. Reported in EPA TSCA Inventory. EPA Genetic Toxicology Program.
OSHA PEL: CL 5 mg(Mn)/m³
ACGIH TLV: TWA 5 mg(Mn)/m³
DOT CLASSIFICATION: 5.1; Label: Oxidizer
SAFETY PROFILE: A human poison by ingestion. Poison experimentally by

ingestion and intravenous routes.
Moderately toxic by subcutaneous route.
Human systemic effects by ingestion:
dyspnea, nausea, other gastrointestinal
effects. Experimental reproductive effects.
Mutation data reported. A strong irritant
due to its oxidizing properties. Used in
production of drugs of abuse, as a topical
antibacterial agent, and a chemical reagent.

Flammable by chemical reaction. A
powerful oxidizer. A dangerous explosion
hazard; handle with care. Explosions may
occur in contact with organic or readily
oxidizable materials, either when dry or in
solution. Dangerous; keep away from
combustible materials.

Explodes on contact with acetic acid,
acetic anhydride, ammonium nitrate,
dimethylformamide, formaldehyde,
concentrated hydrochloric acid, potassium
chloride + sulfuric acid, sulfuric acid +
water. Forms sensitive explosive mixtures
with aluminum powder + ammonium
nitrate + glyceryl nitrate + nitrocellulose,
ammonium perchlorate, arsenic,
phosphorus, sulfur, slag wool, titanium.

Ignites on contact with Al$_4$C$_3$, dimethyl
sulfoxide, ethylene glycol, H$_2$S$_3$, HCl,
H$_2$SO$_4$, (H$_2$SO$_4$ + organic matter), (H$_2$SO$_4$
+ KCl), NH$_4$ClO$_4$, NH$_3$, NH$_4$, NO$_3$,
NH$_2$OH, organic matter, wood, oxygenated
organic compounds (e.g., ethylene glycol,
propane-1,2-diol, erythritol, mannitol,
triethanolamine, 3-chloropropane-1,2-diol,
acetaldehyde, isobutyraldehyde,
benzaldehyde, acetylacetone, esters of
ethylene glycol, lactic acid, acetic acid, oxalic
acid).

Violent reaction or ignition under the
proper conditions with acetone + tert-
butylamine, alcohols + nitric acid, aluminum
carbide, ammonia + sulfuric acid, antimony,
coal + peroxomonosulfuric acid,
dichloromethylsilane, dimethyl sulfoxide,
ethanol + sulfuric acid, glycerol,
concentrated hydrofluoric acid, hydrogen
peroxide, hydrogen trisulfide,
hydroxylamine, carbon, organic nitro

compounds, polypropylene, 3,4,4'-
trimethyldiphenyl sulfone.

When heated to decomposition it emits
toxic fumes of K$_2$O. See also
PERMANGANATES.

PLP250   CAS: 17014-71-0   HR: 3
POTASSIUM PEROXIDE
DOT: UN 1491
mf: K$_2$O$_2$ mw: 110.2
PROP: Yellow, amorph mass or white
crystals or deliquescent colorless
orthorhombic crystals. Mp: 490°.
CONSENSUS REPORTS: Reported in EPA
TSCA Inventory.
DOT CLASSIFICATION: 5.1; Label: Oxidizer
SAFETY PROFILE: Dangerous fire hazard
by spontaneous chemical reaction. It is a
very powerful oxidizer. Fires of this material
should be handled like sodium peroxide
fires. Moderate explosion hazard by
spontaneous chemical reaction. Explodes on
contact with water, forming H$_2$O$_2$ and
KOH. Violent reactions with air, Sb, As, O$_2$,
K. Vigorous reaction on contact with
reducing materials. On contact with acid or
acid fumes, it can emit toxic fumes.
Incompatible with carbon, diselenium
dichloride, ethanol, hydrocarbons, metals.
When heated to decomposition it emits
toxic fumes of K$_2$O. See also PEROXIDES,
INORGANIC.

PLR000   CAS: 16921-30-5   HR: 1
POTASSIUM PLATINIC CHLORIDE
mf: Cl$_6$Pt•K$_2$ mw: 485.99
PROP: Yellow octahedral crystals. Mp:
decomp @ 250°, d: 3.499 @ 24°. IDLH 4
mg/m^3 (as Pt).
SYNS: HEXACHLOROPLATINATE(2⁻) DIPOTASSIUM
□ PLATINIC POTASSIUM CHLORIDE □ POTASSIUM
CHLOROPLATINATE □ POTASSIUM HEXACHLORO-
PLATINATE(IV)
CONSENSUS REPORTS: Reported in EPA
TSCA Inventory.
OSHA PEL: TWA 0.002 mg(Pt)/m^3
ACGIH TLV: TWA 0.002 mg(Pt)/m^3

SAFETY PROFILE: Mutation data reported. Human systemic effects by intradermal route: dermatitis. When heated to decomposition it emits toxic fumes of K_2O and Cl^-. Used as a catalyst for carbonylation of alkynes. See also PLATINUM COMPOUNDS.

PLR250   CAS: 16731-55-8   HR: 2
POTASSIUM PYROSULFITE
mf: $O_5S_2 \cdot K$ mw: 183.22
PROP: Monoclinic plates; white crystalline powder; sulfur dioxide odor. Mp: decomp, d: 2.3. Sparingly sol in H_2O, EtOH; insol in Et_2O.
SYNS: POTASSIUM METABISULFITE (DOT, FCC) □ PYROSULFUROUS ACID, DIPOTASSIUM SALT
CONSENSUS REPORTS: IARC Cancer Review: Group 3 IMEMDT 54,131,92; Human Inadequate Evidence IMEMDT 54,131,92; Animal Inadequate Evidence IMEMDT 54,131,92. Reported in EPA TSCA Inventory. EPA Genetic Toxicology Program.
SAFETY PROFILE: Experimental reproductive effects. A very irritating material. Questionable carcinogen with experimental tumorigenic data. Mutation data reported. When heated to decomposition it emits toxic fumes of SO_x and K_2SO_3. See also SULFITES.

PLR750   CAS: 7790-59-2   HR: 3
POTASSIUM SELENATE
mf: $O_4Se \cdot 2K$ mw: 221.16
PROP: Colorless, hygroscopic, orthorhombic crystals. Undergoes orthorhombic to hexagonal phase transition at 4°. D: 3.07, mp: 1000°. Very sol in water.
SYN: SELENIC ACID, DIPOTASSIUM SALT
CONSENSUS REPORTS: Selenium and its compounds are on the Community Right-To-Know List. Reported in EPA TSCA Inventory. EPA Genetic Toxicology Program.
OSHA PEL: TWA 0.2 mg(Se)/m³
ACGIH TLV: TWA 0.2 mg(Se)/m³
DFG MAK: 0.1 mg(Se)/m³

DOT CLASSIFICATION: 6.1; Label: Poison, Corrosive
SAFETY PROFILE: Poison by intravenous route. Moderately toxic by ingestion. Experimental reproductive effects. When heated to decomposition it emits toxic fumes of Se and K_2O. See also SELENIUM COMPOUNDS.

PLS000   CAS: 3425-46-5   HR: 3
POTASSIUM SELENOCYANATE
mf: $CHNSe \cdot K$ mw: 145.09
PROP: Colorless monoclinic crystals; deliquescent. Sol in H_2O and EtOH.
SYN: SELENOCYANIC ACID, POTASSIUM SALT
CONSENSUS REPORTS: Cyanide and its compounds, as well as selenium and its compounds, are on the Community Right-To-Know List. Reported in EPA TSCA Inventory.
OSHA PEL: TWA 5 mg(CN)/m³; TWA 0.2 mg(Se)/m³
ACGIH TLV: CL 5 mg(CN)/m³ (skin); TWA 0.2 mg(Se)/m³
DFG MAK: 5 mg/m³
NIOSH REL: TWA CL 5 mg/m³/10M
SAFETY PROFILE: Poison by intraperitoneal route. When heated to decomposition it emits very toxic fumes of NO_x, CN^-, K_2O, and Se. See also SELENIUM COMPOUNDS and CYANATES.

PLS250   CAS: 506-61-6   HR: 3
POTASSIUM SILVER CYANIDE
mf: $C_2AgN_2 \cdot K$ mw: 199.01
PROP: White crystals, light-sensitive. Sol in water, acids.
SYNS: KYANOSTRIBRNAN DRASELNY (CZECH) □ RCRA WASTE NUMBER P099 □ SILVER POTASSIUM CYANIDE
CONSENSUS REPORTS: EPA Extremely Hazardous Substances List. Silver and its compounds, as well as cyanide and its compounds, are on the Community Right-To-Know List. Reported in EPA TSCA Inventory.
DOT CLASSIFICATION: 6.1; Label: Poison, KEEP AWAY FROM FOOD

OSHA PEL: TWA 5 mg(CN)/m³
ACGIH TLV: CL 5 mg(CN)/m³ (skin)
DFG MAK: 5 mg/m³
NIOSH REL: (Cyanide) CL 5
mg(CN)/m³/10M
SAFETY PROFILE: Poison by ingestion. A severe skin and eye irritant. When heated to decomposition it emits very toxic fumes of CN^-, K_2O, and NO_x. See also CYANIDE and SILVER COMPOUNDS.

PLS500   CAS: 11135-81-2   HR: 3
POTASSIUM SODIUM ALLOY
DOT: UN 1422
PROP: Low-melting alloy of sodium and potassium metals.
SYN: SODIUM POTASSIUM ALLOY, liquid and solid (DOT)
DOT CLASSIFICATION: 4.3; Label: Dangerous When Wet
SAFETY PROFILE: A low-melting alloy of Na and K. Its toxicity is due to either Na or K alone. Corrosive to the eyes, skin, and mucous membranes. Upon contact with moisture it reacts violently to evolve H_2; much heat; and a highly caustic residue of NaOH or KOH. Oxidation forms Na_2O and K_2O, which are powerful caustics.

A dangerous fire and explosion hazard. Violent or explosive reaction with O_2, water, moisture, steam, halogens, oxidizers, acids or acid fumes, giving off much heat, hydrogen, toxic and corrosive fumes, often spattering either red-hot particles or actually flaming particles. A severe explosion hazard, will react explosively under the appropriate conditions with moisture, acids, acid fumes, solid CO_2, carbon disulfide, halocarbons (e.g., CH_3Cl, carbon tetrachloride, chloroform, bromoform, 1,1,1-trichloroethane, 1,1,2-trichlorotrifluoroethane, tetrachloroethane, CH_2Cl_2, CH_2I_2), ammonium sulfate + NH_4 + NO_3, HgO, metal halides (e.g., silver halides, zinc chloride, iron(III) chloride), metal oxides (e.g., silver oxide, mercury oxide), nitrogen-containing explosives (e.g., ammonium nitrate, ammonium sulfate,

picric acid, nitrobenzene), oxalyl bromide, oxalyl chloride, pentachloroethane, K oxides, KO_2, Si, $NaHCO_3$, polytetrafluoroethylene. Reacts vigorously with oxidizing materials.

To fight fire, use G-1 powder, dry sodium chloride, dry sodium carbonate, dry calcium carbonate, dry sand, resin-coated sodium chloride, or dry soda ash. Never use water, graphite, carbon dioxide, halocarbons, or foam.

Dangerous; when heated it emits highly toxic fumes of Na_2O and K_2O. Used as a liquid coolant for nuclear reactor cores. See also SODIUM and POTASSIUM.

PLS750   CAS: 590-00-1   HR: 2
POTASSIUM SORBATE
mf: $C_6H_7O_2 \cdot K$ mw: 150.23
PROP: White crystals, crystalline powder, or pellets. Mp: 270° (decomp), d: 1.363 @ 25°/20°. Sol in alc, water.
SYNS: 2,4-HEXADIENOIC ACID POTASSIUM SALT □ SORBIC ACID, POTASSIUM SALT □ SORBISTAT-K □ SORBISTAT-POTASSIUM
CONSENSUS REPORTS: Reported in EPA TSCA Inventory. EPA Genetic Toxicology Program.
SAFETY PROFILE: Moderately toxic by intraperitoneal route. Mildly toxic by ingestion. Mutation data reported. When heated to decomposition it emits toxic fumes of K_2O.

PLS760   CAS: 12125-03-0   HR: 3
POTASSIUM STANNATE TRIHYDRATE
mf: $K_2OSn \cdot 3H_2O$ mw: 266.95
OSHA PEL: TWA 2 mg(Sn)/m³
ACGIH TLV: TWA 2 mg(Sn)/m³
SAFETY PROFILE: Poison by intravenous route. When heated to decomposition it emits toxic fumes of Sn.

PLS775   CAS: 593-29-3   HR: D
POTASSIUM STEARATE
mf: $KC_{18}H_{35}O_2$ mw: 322.57
PROP: White powder usually has fatty odor.
SYN: STEARIC ACID POTASSIUM SALT.

P

SAFETY PROFILE: When heated to decomposition it emits acrid smoke and irritating fumes.

PLT000   CAS: 7778-80-5   HR: 2
POTASSIUM SULFATE (2:1)
mf: $O_4S•2K$ mw: 174.26
PROP: Colorless to white, odorless, orthorhombic crystals; bitter salty taste. D: 2.66, mp: 1067°, bp: 1689°. Sol in water; insol in alc and acetone.
SYN: SULFURIC ACID, DIPOTASSIUM SALT
CONSENSUS REPORTS: Reported in EPA TSCA Inventory.
SAFETY PROFILE: Moderately toxic to humans by ingestion. Moderately toxic experimentally by subcutaneous route. Swallowing large doses causes severe gastrointestinal tract effects. When heated to decomposition it emits toxic fumes of K_2O and SO_x. See also SULFATES.

PLT250   CAS: 1312-73-8   HR: 3
POTASSIUM SULFIDE (2:1)
DOT: UN 1382/UN 1847
mf: K_2S mw: 110.26
PROP: Red, crystalline mass; moisture-sensitive, yellow-brown cubic crystals; deliquescent. Mp: 948°, d: 1.805 @ 14°. Very sol in H_2O; sol in EtOH; insol in Et_2O.
SYNS: POTASSIUM MONOSULFIDE □ POTASSIUM SULFIDE, anhydrous (UN 1382) (DOT) □ POTASSIUM SULFIDE, hydrated with not <30% water of crystallization (UN 1847) (DOT) □ POTASSIUM SULFIDE with <30% water of crystallization (UN 1382) (DOT)
CONSENSUS REPORTS: Reported in EPA TSCA Inventory.
DOT CLASSIFICATION: 4.2; Label: Spontaneously Combustible (UN 1382); DOT Class: 8; Label: Corrosive (UN 1847)
SAFETY PROFILE: Poison by ingestion and inhalation. Emits H_2S in contact with acids; steam. A flammable solid. Unstable; may explode on percussion or rapid heating. Ignites on contact with nitrogen oxide. Reacts with H_2O to form KOH and KSH. When heated to decomposition it emits very

toxic fumes of K_2O and SO_x. See also SULFIDES.

PLT500   CAS: 10117-38-1   HR: D
POTASSIUM SULFITE
mf: $O_3S•2K$ mw: 158.26
PROP: White or colorless, hexagonal crystals, or granular powder; odorless. Sol in water; sltly sol in alc.
SYN: SULFUROUS ACID, DIPOTASSIUM SALT
CONSENSUS REPORTS: Reported in EPA TSCA Inventory.
SAFETY PROFILE: When heated to decomposition it emits toxic fumes of SO_x and K_2O. See also SULFUROUS ACID and SULFITES.

PLU000   CAS: 7790-58-1   HR: 3
POTASSIUM TELLURITE
mf: $O_3Te•2K$ mw: 253.80
PROP: Soft glutinous mass or orthorhombic crystals.
CONSENSUS REPORTS: Reported in EPA TSCA Inventory. EPA Genetic Toxicology Program.
OSHA PEL: TWA 0.1 mg(Te)/m³
ACGIH TLV: TWA 0.1 mg(Te)/m³
SAFETY PROFILE: A poison by an unspecified route. Human systemic effects by ingestion: sleep disturbance, anorexia, nausea. When heated to decomposition it emits toxic fumes of K_2O and Te. See also TELLURIUM compounds.

PLU250   CAS: 13826-94-3   HR: 3
POTASSIUM TETRABROMOPLATINATE
mf: $Br_4Pt•2K$ mw: 592.93
PROP: Brown solid. Sol in H_2O. IDLH 4 mg/m³ (as Pt).
SYN: TETRABROMO-PLATINUM(2⁻), DIPOTASSIUM
CONSENSUS REPORTS: Reported in EPA TSCA Inventory.
OSHA PEL: TWA 0.002 mg(Pt)/m³
ACGIH TLV: TWA 0.002 mg(Pt)/m³
SAFETY PROFILE: Poison by intraperitoneal route. Mutation data reported. When heated to decomposition it emits toxic fumes of K_2O and Br⁻. See also

PLATINUM COMPOUNDS and BROMIDES.

PLU500   CAS: 591-89-9   HR: 3
POTASSIUM TETRACYANOMERCUR-
ATE(II)
DOT: UN 1626
mf: $C_2HgN_2 \cdot 2CKN$ mw: 382.87
PROP: Crystals.
SYNS: MERCURIC POTASSIUM CYANIDE (DOT) □ MERCURIC POTASSIUM CYANIDE, solid (DOT)
CONSENSUS REPORTS: Cyanide and its compounds, as well as mercury and its compounds, are on the Community Right-To-Know List.
OSHA PEL: CL 0.1 mg(Hg)/m³ (skin)
ACGIH TLV: TWA 0.1 mg(Hg)/m³ (skin); BEI: 35 µg/g creatinine total inorganic mercury in urine preshift; 15 µg/g creatinine total inorganic mercury in blood at end of shift at end of workweek.
DFG MAK: Confirmed Animal Carcinogen with Unknown Relevance to Humans
NIOSH REL: (Mercury, Aryl and Inorganic) CL 0.1 mg/m³ (skin)
DOT CLASSIFICATION: 6.1; Label: Poison
SAFETY PROFILE: A poison. May explode on contact with ammonia. When heated to decomposition it emits very toxic fumes of NO_x, Hg, K_2O, and CN^-. See also MERCURY COMPOUNDS and CYANIDE.

PLV750   CAS: 333-20-0   HR: 3
POTASSIUM THIOCYANATE
mf: $CNS \cdot K$ mw: 97.18
PROP: Colorless, orthorhombic, deliquescent crystals. D: 1.89, mp: about 173°. Very sol in water and acetone; sol in alc.
SYNS: ARTEROCYN □ ATEROCYN □ KYONATE □ POTASSIUM ISOTHIOCYANATE □ POTASSIUM RHODANATE □ POTASSIUM RHODANIDE □ POTASSIUM SULFOCYANATE □ POTASSIUM THIOCYANIDE □ RODANCA □ RHODANIDE
CONSENSUS REPORTS: Reported in EPA TSCA Inventory.
SAFETY PROFILE: A human poison by ingestion. Poison experimentally by

intravenous route. An experimental teratogen. Moderately toxic by subcutaneous and ingestion routes. Large doses can cause skin eruptions, psychoses, and collapse. Incompatible with calcium chlorite and perchloryl fluoride. When heated to decomposition it emits very toxic fumes of CN^-, K_2O, SO_x, and NO_x. See also THIOCYANATES.

PLW200   CAS: 12012-50-9   HR: 3
POTASSIUM TRICHLOROETHYLENE-
PLATINATE
mf: $C_2H_4Cl_3Pt \cdot K$ mw: 368.60
SYNS: PLATINATE(1⁻), TRICHLOROETHYLENE-, DIPOTASSIUM □ Pt-93
CONSENSUS REPORTS: Reported in EPA TSCA Inventory.
OSHA PEL: TWA 0.002 mg(Pt)/m³
ACGIH TLV: TWA 0.002 mg(Pt)/m³
SAFETY PROFILE: Poison by intraperitoneal route. When heated to decomposition it emits toxic fumes of Pt and Cl^-.

PLW285   CAS: 12298-68-9   HR: D
POTASSIUM TRIIODIDE
mf: I_3K mw: 419.80
PROP: Dark-red solid.
SYNS: IODINE-POTASSIUM IODIDE □ LUGOL'S IODINE □ LUGOL'S SOLUTION □ MICROIODIDE
CONSENSUS REPORTS: Reported in EPA TSCA Inventory.
ACGIH TLV: Proposed: (inhalable fraction) 0.1 mg/m³; Not Classifiable as a Human Carcinogen)
SAFETY PROFILE: Experimental reproductive effects. When heated to decomposition it emits toxic fumes of I^-.

PLW500   CAS: 11103-86-9   HR: 3
POTASSIUM ZINC CHROMATE
HYDROXIDE
mf: $Cr_2HO_9Zn_2 \cdot K$ mw: 418.85
SYNS: BUTTERCUP YELLOW □ CHROMIC ACID, POTASSIUM ZINC SALT (2:2:1) □ CITRON YELLOW □ POTASSIUM ZINC CHROMATE □ ZINC CHROME □ ZINC YELLOW

CONSENSUS REPORTS: IARC Cancer Review: Group 1 IMEMDT 49,49,90; Animal Sufficient Evidence IMEMDT 49,49,90; Human Sufficient Evidence IMEMDT 49,49,90; Animal Inadequate Evidence IMEMDT 23,205,80. Chromium and its compounds, as well as zinc and its compounds, are on the Community Right-To-Know List. Reported in EPA TSCA Inventory.
OSHA PEL: CL 0.1 mg(CrO$_3$)/m^3
ACGIH TLV: TWA 0.01 mg(Cr)/M^3; Confirmed Human Carcinogen
DFG MAK: Human Carcinogen
NIOSH REL: (Chromium (VI)) TWA 0.001 mg(Cr(VI))/m^3
SAFETY PROFILE: Confirmed carcinogen. Mutation data reported. When heated to decomposition it emits toxic fumes of ZnO and K$_2$O. Used as a corrosion inhibiting pigment and in steel priming. See also CHROMIUM COMPOUNDS and ZINC COMPOUNDS.

PLX750   CAS: 10361-79-2   HR: 3
PRASEODYMIUM CHLORIDE
mf: Cl$_3$Pr   mw: 247.26
PROP: Hygroscopic blue-green needles. Mp: 786°, bp: 19° @ 1700 mm. Sol in H$_2$O and EtOH.
CONSENSUS REPORTS: Reported in EPA TSCA Inventory.
SAFETY PROFILE: Poison by intraperitoneal, subcutaneous, and intravenous routes. Moderately toxic by ingestion. A skin and eye irritant. When heated to decomposition it emits toxic fumes of Cl⁻. See also PRASEODYMIUM.

PLY250   CAS: 10361-80-5   HR: 3
PRASEODYMIUM(III) NITRATE (1:3)
mf: N$_3$O$_9$•Pr   mw: 326.94
SYN: NITRIC ACID, PRASEODYMIUM(3$^+$) SALT
CONSENSUS REPORTS: Reported in EPA TSCA Inventory.
SAFETY PROFILE: Poison by intraperitoneal and intravenous routes. Moderately toxic by ingestion. When heated to decomposition it emits toxic fumes of NO$_x$. See also PRASEODYMIUM and NITRATES.

PLZ000   CAS: 53-03-2   HR: 3
PREDNISONE
mf: C$_{21}$H$_{26}$O$_5$ mw: 358.47
PROP: White, odorless, crystalline powder. Mp: 235° (with some decomp). Very sltly sol in water; sltly sol in alc, chloroform, methanol, and dioxane.
SYNS: ANCORTONE □ BICORTONE □ COLISONE □ CORTAN □ CORTANCYL □ Δ-CORTELAN □ CORTIDELT □ Δ-CORTISONE □ Δ^1-CORTISONE □ Δ-CORTONE □ COTONE □ DACORTIN □ DECORTANC-YL □ DECORTIN □ DECORTISYL □ Δ-1-DEHYDRO-CORTISONE □ 1-DEHYDROCORTISONE □ DEKORTIN □ DELTACORTELAN □ DELTACORTISONE □ DELTACORTONE □ DELTA-DOME □ DELTISONE □ 17,21-DIHYDROXYPREGNA-1,4-DIENE-3,11,20-TRIONE □ ENCORTON □ HOSTACORTIN □ IN-SONE □ JUVASON □ LISACORT □ METACORTANDRACIN □ NCI-C04897 □ NSC-10023 □ ORASONE □ PARACORT □ PRECORT □ PREDNICEN-M □ PREDNILONGA □ PREDNISON □ PREDNIZON □ 1,4-PREGNADIENE-17-α,21-DIOL-3,11,20-TRIONE □ RECTODELT □ SERVISONE □ SK-PREDNISONE □ SUPERCORTIL □ U 6020 □ ULTRACORTEN □ WOJTAB □ ZENADRID (VETERINARY)
CONSENSUS REPORTS: IARC Cancer Review: Group 3 IMEMDT 7,326,87; Human Inadequate Evidence IMEMDT 26,293,81; Animal Inadequate Evidence IMEMDT 26,293,81. NCI Carcinogenesis Studies (ipr); No Evidence: mouse CANCAR 40,1935,77; (ipr); Equivocal Evidence: rat CANCAR 40,1935,77. Reported in EPA TSCA Inventory.
SAFETY PROFILE: Poison by intraperitoneal and subcutaneous routes. Moderately toxic by intramuscular route. Human systemic effects: sensory change involving peripheral nerves, dermatitis. Experimental reproductive effects. Questionable carcinogen with experimental tumorigenic data. Mutation data reported. Has been implicated in aplastic anemia.

PMA000   CAS: 50-24-8   HR: 3
PREDONIN
mf: C$_{21}$H$_{28}$O$_5$ mw: 360.49

PROP: Crystals. Mp: 240–241° (decomp).
SYNS: CODELCORTONE □ CO-HYDELTRA □ Δ¹-CORTISOL □ DECORTIN H □ Δ¹-DEHYDROCORTISOL □ Δ¹-DEHYDROHYDROCORTISONE □ 1-DEHYDROHYDROCORTISONE □ DELCORTOL □ DELTA-CORTEF □ DELTACORTENOL □ DELTACORTRIL □ DELTA F □ DELTA-STAB □ DEXA-CORTIDELT HOSTACORTIN H □ DI-ADRESON F □ DICORTOL □ DYDELTRONE □ FERNISOLONE □ HOSTACORTIN □ HYDELTRA □ HYDELTRONE □ Δ¹-HYDROCORTISONE □ HYDRODELTALONE □ HYDRODELTISONE □ HYDRORETROCORTIN □ METACORTANDRALONE □ METICORTELONE □ METI-DERM □ PARACORTOL □ PARACOTOL □ PRECORTANCYL □ PRECORTISYL □ PREDNE-DOME □ PREDNELAN □ PREDNIS □ PREDNISOLONE □ PREDONINE □ 1,4-PREGNADIENE-3,20-DIONE-11-β,17-α,21-TRIOL □ 1,4-PREGNADIENE-11-β,17-α,21-TRIOL-3,20-DIONE □ SCHERISOLON □ STERANE □ STEROLONE □ 11-β,17,21-TRIHYDROXYPREGNA-1,4-DIENE-3,20-DIONE □ 11-β,17-α,21-TRIHYDROXYPREG-NA-1,4-DIENE-3,20-DIONE □ 11-β,17-α,21-TRIHYDRO-XY-1,4-PREGNADIENE-3,20-DIONE □ ULACORT □ ULTRACORTENE-H
CONSENSUS REPORTS: Reported in EPA TSCA Inventory. EPA Genetic Toxicology Program.
SAFETY PROFILE: A poison by intravenous and subcutaneous routes. Moderately toxic by ingestion and intraperitoneal routes. Human teratogenic effects by an unspecified route: developmental abnormalities of the central nervous system; effects on embryo or fetus: fetal death, extra embryonic structures. Human reproductive effects by an unspecified route: stillbirth. An experimental teratogen. Experimental reproductive effects. Human mutation data reported. When heated to decomposition it emits acrid smoke and irritating fumes.

PMB000   CAS: 12126-59-9   HR: 2
PREMARIN
SYNS: CEE □ CONJUGATED EQUINE ESTROGEN □ ESTROGENS, CONJUGATES
CONSENSUS REPORTS: IARC Cancer Review: Human Limited Evidence IMEMDT 21,147,79; Animal Inadequate Evidence IMEMDT 21,147,79.

SAFETY PROFILE: Suspected carcinogen with experimental tumorigenic data. Poison by intraperitoneal route. Human reproductive effects by ingestion: changes in female fertility. Experimental teratogenic effects. A steroid. When heated to decomposition it emits acrid smoke and irritating fumes.

PMB600   CAS: 68555-58-8   HR: 2
PRENYL SALICYLATE
mf: $C_{12}H_{14}O_3$ mw: 206.26
SYNS: BENZOIC ACID, 2-HYDROXY-, 3-METHYL-2-BUTENYL ESTER □ 2-BUTEN-1-OL, 3-METHYL-, SALICYLATE □ 3-METHYL-2-BUTENYL SALICYLATE
CONSENSUS REPORTS: Reported in EPA TSCA Inventory.
SAFETY PROFILE: Moderately toxic by ingestion. A skin irritant. When heated to decomposition it emits acrid smoke and irritating fumes.

PMC400   CAS: 68955-54-4   HR: 3
PRIMENE JM-T
CONSENSUS REPORTS: Reported in EPA TSCA Inventory.
SAFETY PROFILE: Poison by ingestion. A severe skin irritant.

PME500   CAS: 366-70-1   HR: 3
PROCARBAZINE HYDROCHLORIDE
mf: $C_{12}H_{19}N_3O \bullet ClH$ mw: 257.80
PROP: Crystals from MeOH. Mp: 223–236°.
SYNS: IBENZMETHYZINE HYDROCHLORIDE □ IBENZMETHYZIN HYDROCHLORIDE □ IBZ □ 1-(p-ISOPROPYLCARBAMOYLBENZYL)-2-METHYLHYDRA-ZINE HYDROCHLORIDE □ 2-(p-(ISOPROPYLCAR-BAMOYL)BENZYL)-1-METHYLHYDRAZINE HYDRO-CHLORIDE □ N-ISOPROPYL-p-(2-METHYLHY-DRAZINOMETHYL)BENZAMIDEHYDROCHLORIDE □ N-ISOPROPYL-α-(2-METHYLHYDRAZINO)-p-TOLUAMIDE HYDROCHLORIDE □ MATULANE □ MBH □ N-(1-METHYLETHYL)-4-((2-METHYLHYDRAZ-INO)METHYL)BENZAMIDE MONOHYDROCHLORIDE □ p-(N'-METHYLHYDRAZINOMETHYL)-N-ISOPROPYLBENZAMIDE HYDROCHLORIDE □ 1-METHYL-2-p-(ISOPROPYLCARBAMOYL)BENZO-HYDRAZINE HYDROCHLORIDE □ 1-METHYL-2-(p-ISOPROPYLCARBAMOYLBENZYL)HYDRAZINE HYDROCHLORIDE □ MIH HYDROCHLORIDE □ NATHULANE □ NATULAN □ NATULANAR □

NATULAN HYDROCHLORIDE □ NCI-C01810 □ NSC-77213 □ PCB HYDROCHLORIDE □ PROCARBAZIN (GERMAN) □ RO 4-6467

CONSENSUS REPORTS: NTP 10th Report on Carcinogens. IARC Cancer Review: Group 2A IMEMDT 7,327,87; Human Limited Evidence IMEMDT 26,311,81; Animal Sufficient Evidence IMEMDT 26,311,81. NCI Carcinogenesis Bioassay (ipr); Clear Evidence: mouse, rat NCITR* NCI-CG-TR-19,79; (ipr); Clear Evidence: mouse, rat RRCRBU 52,1,75. EPA Genetic Toxicology Program.

SAFETY PROFILE: Confirmed carcinogen with experimental carcinogenic, neoplastigenic, tumorigenic, and teratogenic data. Poison by an unspecified route. Moderately toxic by ingestion, subcutaneous, intravenous, and intraperitoneal routes. Experimental reproductive effects. Mutation data reported. When heated to decomposition it emits very toxic fumes of NO_x and HCl. Used as a chemotherapeutic agent.

PMF500　CAS: 58-38-8　HR: 3
PROCHLORPROMAZINE

mf: $C_{20}H_{24}ClN_3S$　mw: 373.98
PROP: Viscous liquid.
SYNS: CHLORO-3 (N-METHYLPIPERAZINYL-3 PROPYL)-10 PHENOTHIAZINE (FRENCH) □ 2-CHLORO-10-(3-(1-METHYL-4-PIPERAZINYL)-PROPYL)-PHENOTHIAZINE □ 2-CHLORO-10-(3-(4-METHYL-1-PIPERAZINYL)PROPYL)PHENOTHIAZINE □ 3-CHLORO-10-(3-(1-METHYL-4-PIPERAZINYL)PROPYL)-PHENOTHIAZINE □ CHLORPERAZINE □ COMPAZINE □ N-(γ-(4'-METHYLPIPERAZINYL-1')PROPYL)-3-CHLOROPHENOTHIAZINE □ NIPODAL □ NOVAMIN □ PROCHLOROPERAZINE □ PROCHLORPEMAZINE □ PROCHLORPERAZINE □ 6140 RP □ STEMETIL □ TEMENTIL

SAFETY PROFILE: Poison by ingestion, subcutaneous, intravenous, and intraperitoneal routes. Experimental teratogenic and reproductive effects. Human systemic effects by ingestion: headache, blood pressure elevation. Implicated in aplastic anemia. When heated to decomposition it emits very toxic fumes of SO_x, NO_x, and Cl⁻.

PMH250　CAS: 952-23-8　HR: 3
PROFLAVINE MONOHYDROCHLORIDE

mf: $C_{13}H_{11}N_3 \cdot ClH$　mw: 245.73
SYNS: 3,6-ACRIDINEDIAMINE, MONOHYDRO-CHLORIDE (9CI) □ 3,6-DIAMINOACRIDINE MONOHYDROCHLORIDE □ 3,6-DIAMINOACRIDINIUM CHLORIDE □ 3,6-DIAMINOACRIDINIUM CHLORIDE HYDROCHLORIDE □ 2,8-DIAMINOACRIDINIUM CHLORIDE MONOHYDROCHLORIDE □ PROFLAVINE HYDROCHLORIDE

CONSENSUS REPORTS: IARC Cancer Review: Group 3 IMEMDT 7,56,87; Animal Inadequate Evidence IMEMDT 24,195,80; Human No Adequate Data IMEMDT 24,195,80. NTP Carcinogenesis Bioassay (feed); Clear Evidence: rat NCITR* NCI-TR-5,77; (feed); Inadequate Studies: mouse NCITR* NCI-TR-5,77.

SAFETY PROFILE: Poison by subcutaneous route. Questionable carcinogen. Mutation data reported. When heated to decomposition it emits very toxic fumes of NO_x and HCl. Used as a drug, as a disinfectant, and as a topical antiseptic.

PMH500　CAS: 57-83-0　HR: 3
PROGESTERONE

mf: $C_{31}H_{30}O_2$　mw: 314.51
PROP: A female sex hormone. White, crystalline powder; odorless. D: 1.166 @ 23°, mp: 127–131°. Practically insol in water; sol in alc, acetone, and dioxane; sparingly sol in oils.
SYNS: CORLUTIN □ CORLUVITE □ CORPORIN □ CORPUS LUTEUM HORMONE □ CYCLOGEST □ GLANDUCORPIN □ HORMOFLAVEINE □ HORMOLUTON □ LINGUSORBS □ LIPO-LUTIN □ LUCORTEUM SOL □ LUTEAL HORMONE □ LUTEOHORMONE □ LUTEOSAN □ LUTEX □ LUTOCYCLIN □ LUTROMONE □ NALUTRON □ NSC-9704 □ PERCUTACRINE □ PIAPONON □ 3,20-PREGNENE-4 □ PREGNENEDIONE □ PREGNENE-3,20-DIONE □ Δ⁴-PREGNENE-3,20-DIONE □ PREGN-4-ENE-3,20-DIONE □ 4-PREGNENE-3,20-DIONE □ PROGEKAN □ PROGESTEROL □ β-PROGESTERONE □ PROGESTERONUM □ PROGESTIN □ PROGESTONE □ PROLIDON □ SYNGESTERONE □ SYNOVEX S □ SYNTOLUTAN

CONSENSUS REPORTS: NTP 10th Report on Carcinogens. IARC Cancer Review: Animal Limited Evidence IMEMDT

21,491,79; Animal Sufficient Evidence IMEMDT 6,135,74. EPA Genetic Toxicology Program. Reported in EPA TSCA Inventory.

SAFETY PROFILE: Confirmed carcinogen with experimental carcinogenic, neoplastigenic, tumorigenic, and teratogenic data. Poison by intravenous and intraperitoneal routes. Human teratogenic effects by ingestion and parenteral routes: developmental abnormalities of the urogenital system. Human male reproductive effects by intramuscular route: changes in spermatogenesis, the prostate, seminal vesicle, Cowper's gland and accessory glands, impotence, and breast development. Human female reproductive effects by ingestion, parenteral, and intravaginal routes: fertility changes; menstrual cycle changes and disorders; uterus, cervix, and vagina changes. Experimental reproductive effects. Human mutation data reported. When heated to decomposition it emits acrid smoke and irritating fumes.

PMH900   **HR: D**
l-PROLINE
mf: $C_5H_9NO_2$ mw: 115.13
PROP: White crystals or crystalline powder; odorless with sweet taste. Very sol in water, alc; insol in ether.
SYN: l-2-PYRROLIDINECARBOXYLIC ACID
SAFETY PROFILE: When heated to decomposition emits toxic fumes of NO_x.

PMI500   **CAS: 53-60-1**   **HR: 3**
PROMAZINE HYDROCHLORIDE
mf: $C_{17}H_{20}N_2S \cdot ClH$ mw: 320.91
PROP: White to sltly yellow, practically odorless, crystalline powder. Decomp @ 181°. Sol in water, methanol, ethanol, chloroform; insol in ether, benzene.
SYNS: 10-(γ-DIMETHYLAMINO-N-PROPYL)PHENO-THIAZINE HYDROCHLORIDE □ 10-(3-(DIMETHYL-AMINO)PROPYL)PHENOTHIAZINE HYDROCHLORIDE □ SPARINE HYDROCHLORIDE
CONSENSUS REPORTS: Reported in EPA TSCA Inventory.

SAFETY PROFILE: Poison by ingestion, subcutaneous, intravenous, intraperitoneal, and intramuscular routes. Human systemic effects by ingestion: general anesthesia, tremors, antipsychotic effects. An additive permitted in food for human consumption; also permitted in the feed and drinking water of animals and/or for the treatment of food-producing animals. When heated to decomposition it emits toxic fumes of NO_x, SO_x, and HCl.

PMI750   **CAS: 58-33-3**   **HR: 3**
PROMETHAZINE HYDROCHLORIDE
mf: $C_{17}H_{20}N_2S \cdot ClH$ mw: 320.91
PROP: Crystals. Mp: 230–232° (decomp). Very sol in water; sol in alc, chloroform; insol in acetone, ether, ethyl acetate.
SYNS: 10-(3-DIMETHYLAMINOISOPROPYL)-PHENOTHIAZINE HYDROCHLORIDE □ N-(2'-DIMETHYLAMINO-2'-METHYL)ETHYLPHENOTHIAZ-INE HYDROCHLORIDE □ N-(2-DIMETHYLAMINO-PROPYL-1)PHENOTHIAZINE HYDROCHLORIDE □ 10-(2-DIMETHYLAMINOPROPYL)PHENOTHIAZINE HYDROCHLORIDE □ 10-(2-(DIMETHYLAMINO)PROP-YL)PHENOTHIAZINE MONOHYDROCHLORIDE □ DIPHERGAN □ DORME □ FARGAN □ FELLOZINE □ FENAZIL □ FENERGAN □ GANPHEN □ HL 8700 □ LERGIGAN □ PHENCEN □ PHENERGAN HYDRO-CHLORIDE □ PLLETIA □ PRIMINE □ PROMANTINE □ PROMETHAZINE N-(2'-DIMETHYLAMINO-2'-METHYLETHYL)PHENOTHIAZINE HYDROCHLORIDE □ PROMETHAZIN (GERMAN) □ PROREX □ PROTAZINE □ REMSED □ 3277 R.P. □ N,N,α-TRIMETHYL-10H-PHENOTHIAZINE-10-ETHANAMINE MONOHYDROCHLORIDE
CONSENSUS REPORTS: Reported in EPA TSCA Inventory.
SAFETY PROFILE: Poison by ingestion, subcutaneous, intraperitoneal, and intravenous routes. Human systemic effects by ingestion: excitement, sleep, convulsions, rigidity. Experimental teratogenic effects. Other experimental reproductive effects. When heated to decomposition it emits very toxic fumes of HCl, SO_x, and NO_x. Used as an antihistamine.

PMJ500   **CAS: 154-41-6**   **HR: 3**
PROPADRINE HYDROCHLORIDE
mf: $C_9H_{13}NO \cdot ClH$ mw: 187.69

PROP: A solid. Mp: 194°. Sol in H_2O.
SYNS: α-(1-AMINOETHYL)BENZENEMETHANOL HYDROCHLORIDE □ α-(1-AMINOETHYL)BENZYL ALCOHOL HYDROCHLORIDE □ (±)-2-AMINO-1-PHENYL-1-PROPANOL HYDROCHLORIDE □ α-HYDROXY-β-AMINOPROPYLBENZENE HYDRO-CHLORIDE □ MONHYDRIN □ MUCORAMA □ MYDRIATINE □ dl-NOREPHEDRINE HYDRO-CHLORIDE □ dl-1-PHENYL-2-AMINO-1-PROPANOL MONOHYDROCHLORIDE □ PHENYLPROPANOL-AMINE HYDROCHLORIDE
CONSENSUS REPORTS: Reported in EPA TSCA Inventory.
SAFETY PROFILE: Poison by ingestion, subcutaneous, intravenous, intraperitoneal, and intramuscular routes. An experimental teratogen. When heated to decomposition it emits very toxic fumes of HCl and NO_x. Used as a raw material in cold and diet tablets.

PMJ750   CAS: 74-98-6   HR: 3
PROPANE
DOT: UN 1978
mf: C_3H_8 mw: 44.11
PROP: Colorless gas. Bp: −44.5°, flash p: −156°F, lel: 2.3%, uel: 9.5%, autoign temp: 842°F, d: 0.5852 @ −44.5°/4°, vap d: 1.56. Sol in water, alc, ether. IDLH 2100 ppm [10%LEL].
SYNS: DIMETHYLMETHANE □ PROPYL HYDRIDE
CONSENSUS REPORTS: Reported in EPA TSCA Inventory.
OSHA PEL: TWA 1000 ppm
DFG MAK: 1000 ppm (1800 mg/m³)
DOT CLASSIFICATION: 2.1; Label: Flammable Gas
SAFETY PROFILE: Central nervous system effects at high concentrations. An asphyxiant. Flammable gas. Highly dangerous fire hazard when exposed to heat or flame; can react vigorously with oxidizers. Explosive in the form of vapor when exposed to heat or flame. Explosive reaction with ClO_2. Violent exothermic reaction with barium peroxide + heat. To fight fire, stop flow of gas. When heated to decomposition it emits acrid smoke and irritating fumes.

PMK250   CAS: 78-90-0   HR: 3
1,2-PROPANEDIAMINE
DOT: UN 2258
mf: $C_3H_{10}N_2$ mw: 74.15
PROP: Flash p: 92°F (OC), d: 0.9, vap d: 2.6, bp: 118.9°.
SYNS: 1,2-DIAMINOPROPANE □ PROPYLENEDIAMINE □ PROPYLENE DIAMINE (DOT)
CONSENSUS REPORTS: Reported in EPA TSCA Inventory.
DOT CLASSIFICATION: 8; Label: Corrosive
SAFETY PROFILE: Moderately toxic by ingestion, skin contact, and subcutaneous routes. A corrosive irritant to eyes, skin, and mucous membranes. Dangerous fire hazard when exposed to heat, flames, oxidizers. To fight fire, use alcohol foam. When heated to decomposition it emits toxic fumes of NO_x. Used as an intermediate in production of petroleum and polymer additives, and surfactants. See also AMINES.

PMK500   CAS: 109-76-2   HR: 3
1,3-PROPANEDIAMINE
mf: $C_3H_{10}N_2$ mw: 74.15
PROP: Water-white liquid, amine odor. D: 0.8881 @ 20°/20°, fp: −12°, bp: 135–136°, flash p: 120°F (TOC). Completely sol in water, methanol, and ether.
SYNS: 1,3-DIAMINOPROPANE □ 1,3-PROPYLENEDIAMINE □ TRIMETHYLENEDIAMINE
CONSENSUS REPORTS: Reported in EPA TSCA Inventory.
SAFETY PROFILE: Poison by ingestion and skin contact. Experimental teratogenic effects. A severe skin and eye irritant. Flammable liquid when exposed to heat, sparks, or flame. When heated to decomposition it emits toxic fumes of NO_x. See also AMINES.

PMK800   CAS: 63843-89-0   HR: 2
PROPANEDIOIC ACID, ((3,5-BIS(1,1-
DIMETHYLETHYL)-4-HYDROXY-
PHENYL)METHYL)BUTYL-,
BIS(1,2,2,6,6- PENTAMETHYL-4-
PIPERIDINYL) ESTER
mf: $C_{42}H_{72}N_2O_5$ mw: 685.16
SYN: TINUVIN 144

CONSENSUS REPORTS: Reported in EPA TSCA Inventory.

SAFETY PROFILE: Moderately toxic by ingestion. Experimental reproductive effects. An eye irritant. When heated to decomposition it emits toxic fumes of NO_x.

PML000   CAS: 57-55-6   HR: 2
1,2-PROPANEDIOL

mf: $C_3H_8O_2$ mw: 76.11

PROP: Colorless viscous liquid; practically odorless. Bp: 188.2°, flash p: 210°F (OC), lel: 2.6%, uel: 12.6%, d: 1.0362 @ 25°/25°, autoign temp: 700°F, vap press: 0.08 mm @ 20°, vap d: 2.62, fp: −59°. Hygroscopic; misc with water, acetone, chloroform; sol in essential oils; immisc with fixed oils.

SYNS: 1,2-DIHYDROXYPROPANE □ DOWFROST □ METHYLETHYLENE GLYCOL □ METHYL GLYCOL □ MONOPROPYLENE GLYCOL □ PG 12 □ PROPANE-1,2-DIOL □ PROPYLENE GLYCOL (FCC) □ PROPYLENE GLYCOL USP □ α-PROPYLENEGLYCOL □ 1,2-PROPYLENE GLYCOL □ SIRLENE □ SOLAR WINTER BAN □ TRIMETHYL GLYCOL

CONSENSUS REPORTS: Reported in EPA TSCA Inventory. EPA Genetic Toxicology Program.

SAFETY PROFILE: Slightly toxic by ingestion, skin contact, intraperitoneal, intravenous, subcutaneous, and intramuscular routes. Human systemic effects by ingestion: general anesthesia, convulsions, changes in surface EEG. Experimental teratogenic and reproductive effects. An eye and human skin irritant. Mutation data reported. Combustible liquid when exposed to heat or flame; can react with oxidizing materials. Explosive in the form of vapor when exposed to heat or flame. May react with hydrofluoric acid + nitric acid + silver nitrate to form the explosive silver fulminate. To fight fire, use alcohol foam. When heated to decomposition it emits acrid smoke and irritating fumes.

PML270   CAS: 81861-90-7   HR: 3
N,N′-(1,3-PROPANEDIOXYSULFIN-
YL)BIS(2,3-DIHYDRO-2,2-DIMETHYL-

BENZOFURANYL-7-METHYL-
CARBAMATE)

mf: $C_{27}H_{34}N_2O_{10}S_2$ mw: 610.75

SYN: 4,8-DIOXA-3,9-DITHIA-2,10-DIAZAUN-DECANEDIOIC ACID, 2,10-DIMETHYL-, BIS(2,3-DIHYDRO-2,2-DIMETHYL-7-BENZOFURANYL) ESTER, 3,9-DIOXIDE

SAFETY PROFILE: A poison by ingestion. When heated to decomposition it emits toxic vapors of NO_x and SO_x.

PML400   CAS: 1120-71-4   HR: 3
PROPANE SULTONE

mf: $C_3H_6O_3S$ mw: 122.15

PROP: Prisms. D: 1.392, mp: 31°, bp: 155–157° @ 14 mm.

SYNS: 3-HYDROXY-1-PROPANESULFONIC ACID γ-SULTONE □ 3-HYDROXY-1-PROPANESULPHONIC ACID SULTONE □ 1,2-OXATHIOLANE-2,2-DIOXIDE □ 1-PROPANESULFONIC ACID-3-HYDROXY-γ-SULTONE □ 1,3-PROPANE SULTONE (MAK) □ RCRA WASTE NUMBER U193

CONSENSUS REPORTS: NTP 10th Report on Carcinogens. IARC Cancer Review: Group 2B IMEMDT 7,56,87; Animal Sufficient Evidence IMEMDT 4,253,74. Community Right-To-Know List. Reported in EPA TSCA Inventory. EPA Genetic Toxicology Program.

ACGIH TLV: Animal Carcinogen

DFG MAK: Animal Carcinogen, Suspected Human Carcinogen

SAFETY PROFILE: Confirmed carcinogen with experimental carcinogenic, neoplastigenic, tumorigenic, and teratogenic data. Poison by subcutaneous route. Moderately toxic by skin contact and intraperitoneal routes. Human mutation data reported. Implicated as a human brain carcinogen. A skin irritant. When heated to decomposition it emits toxic fumes of SO_x.

PML500   CAS: 107-03-9   HR: 3
1-PROPANETHIOL

DOT: UN 1228/UN 3071

mf: C_3H_8S mw: 76.17

PROP: Flash p: −4°F.

SYNS: 1-MERCAPTOPROPANE □ PROPANE-1-THIOL □ PROPYL MERCAPTAN □ N-PROPYL MERCAPTAN □ PROPYLTHIOL □ N-PROPYLTHIOL

CONSENSUS REPORTS: Reported in EPA TSCA Inventory.
NIOSH REL: (Thiols (n-Alkane Mono)) CL 0.5 ppm/15M
DOT CLASSIFICATION: 3; Label: Flammable Liquid, Poison (UN 1228); DOT Class: 6.1; Label: Poison, Flammable Liquid (UN 3071)
SAFETY PROFILE: A poison. Moderately toxic by intraperitoneal route. Mildly toxic by inhalation. A severe eye irritant. A flammable liquid and very dangerous fire hazard when exposed to heat or flame. Explodes on contact with calcium hypochlorite. When heated to decomposition it emits toxic fumes of SO_x. See also MERCAPTANS.

PMM250   CAS: 156-87-6   HR: 2
3-PROPANOLAMINE
mf: C_3H_9NO mw: 75.13
PROP: Colorless liquid; fishy odor. Bp: 168° @ 500 mm, flash p: >175°F (TOC), fp: 12.4°, d: 0.9786 @ 30°, vap press: 2.1 mm @ 60°, vap d: 2.59. Very sol in water; sol in alc; insol in ether.
SYNS: β-ALANINOL □ γ-AMINOPROPANOL □ 3-AMINOPROPANOL □ 3-AMINO-1-PROPANOL □ 3-AMINOPROPYL ALCOHOL □ 3-HYDROXYPROPYL-AMINE □ PROPANOLAMINE □ 1,3-PROPANOLAMINE
CONSENSUS REPORTS: Reported in EPA TSCA Inventory.
SAFETY PROFILE: Moderately toxic by ingestion and skin contact. An experimental teratogen. A severe skin and eye irritant. Combustible when exposed to heat or flame; can react with oxidizing materials. To fight fire, use foam, CO_2, dry chemical. When heated to decomposition it emits toxic fumes of NO_x. See also AMINES.

PMM300   CAS: 67239-28-5   HR: 3
2-PROPANOL, 1-(4-CYCLOPROPYL-
   CARBONYL-PHENOXY)-3-(1,2-
   DIHYDRO-2-IMINO-4-METHYL-
   PYRIDINO)-
mf: $C_{19}H_{22}N_2O_3$ mw: 326.43

SYN: KETONE, CYCLOPROPYL 4-((2-HYDROXY-3-(1,2-DIHYDRO-2-IMINO-4-METHYLPYRIDINO)PROPO-XY)PHENYL)
DOT CLASSIFICATION: 3; Label: Flammable Liquid
SAFETY PROFILE: A poison by intraperitoneal route. A flammable liquid. When heated to decomposition it emits toxic vapors of NO_x.

PMN450   CAS: 107-19-7   HR: 3
PROPARGYL ALCOHOL
mf: C_3H_4O mw: 56.07
PROP: Moderately volatile liquid; geranium odor. D: 0.9715 @ 20°/4°, mp: −48-−52°, bp: 114–115°, flash p: 33°C (97°F) (OC), vap press: 11.6 mm @ 20°, vap d: 1.93.
SYNS: ETHYNYLCARBINOL □ ETHYNYLMETHANOL □ 1-PROPYNE-3-OL □ 2-PROPYN-1-OL □ 3-PROPYNOL □ 2-PROPYNYL ALCOHOL □ RCRA WASTE NUMBER P102
CONSENSUS REPORTS: Reported in EPA TSCA Inventory.
OSHA PEL: TWA 1 ppm (skin)
ACGIH TLV: TWA 1 ppm (skin)
DFG MAK: 2 ppm (4.7 mg/m³)
SAFETY PROFILE: Poison by ingestion, skin contact, and subcutaneous routes. Moderately toxic by inhalation. A central nervous system depressant. A skin and mucous membrane irritant. Mutation data reported. Flammable liquid and dangerous fire hazard when exposed to heat or flame; can ignite. To fight fire, use foam, CO_2, dry chemical. Potentially explosive reactions with alkalies (when dried), sulfuric acid. Ignites on contact with phosphorus pentaoxide. Violent reaction with mercury(II) sulfate + sulfuric acid + water (at 70°C). Incompatible with oxidizing materials. When heated to decomposition it emits acrid smoke and irritating fumes. Used as a corrosion inhibitor, solvent stabilizer, soil fumigant, and chemical intermediate. See also ACETYLENE COMPOUNDS.

PMN500   CAS: 106-96-7   HR: 3
PROPARGYL BROMIDE
DOT: UN 2345

mf: C_3H_3Br mw: 118.97

PROP: Lachrymatory liquid; almost colorless; sharp odor. Fp: −61.07°, bp: 88–90°, flash p: 65°F (COC), d: 1.564–1.570, vap d: 6.87.

SYNS: γ-BROMOALLYLENE □ 3-BROMOPROPYNE (DOT) □ 3-BROMO-1-PROPYNE

CONSENSUS REPORTS: EPA Extremely Hazardous Substances List. Reported in EPA TSCA Inventory.

DOT CLASSIFICATION: 3; Label: Flammable Liquid

SAFETY PROFILE: A poison by ingestion. A dangerous fire hazard when exposed to heat or flame. The aerated liquid may be ignited by pressure. A dangerous, extremely shock-sensitive explosive. It can detonate when heated to 220°C, by impact (especially when mixed with chloropicrin), or when heated while confined. May explode on contact with copper, high copper alloys, mercury, or silver. Mixtures with trichloronitromethane are shock- and heat-sensitive explosives. Can react vigorously with oxidizing materials. To fight fire, use water, foam, CO_2, dry chemical. When heated to decomposition it emits highly toxic fumes of Br^-. See also ACETYLENE COMPOUNDS and BROMIDES.

PMN850   CAS: 139-40-2   HR: 2
PROPAZINE
mf: $C_9H_{16}ClN_7O_2$ mw: 229.75
PROP: A solid. Mp: 213–214°.

SYNS: 2,4-BIS(ISOPROPYLAMINO)-6-CHLORO-s-TRIAZINE □ 2,4-BIS(PROPYLAMINO)-6-CHLOR-1,3,5-TRIAZIN (GERMAN) □ GESAMIL □ MILOGARD □ PLANTULIN □ PRIMATOL P □ PROPASIN □ PROZINEX

SAFETY PROFILE: Moderately toxic by ingestion. Moderate eye irritation. Questionable carcinogen with experimental tumorigenic data. When heated to decomposition it emits toxic fumes of NO_x and Cl^-.

PMO250   CAS: 695-53-4   HR: 2
PROPAZONE
mf: $C_5H_7NO_3$ mw: 129.13

PROP: IDLH 50 ppm.

SYNS: AC 1198 □ BAX 1400Z □ DIMETHADIONE □ DIMETHYLOXAZOLIDINEDIONE □ 5,5-DIMETHYL-OXAZOLIDINE-2,4-DIONE □ 5,5-DIMETHYL-2,4-OXAZOLIDINEDIONE □ DMO □ EUPRACTONE □ NSC-30152

CONSENSUS REPORTS: Reported in EPA TSCA Inventory.

SAFETY PROFILE: Moderately toxic by intraperitoneal and intravenous routes. Experimental teratogenic and reproductive effects. When heated to decomposition it emits toxic fumes of NO_x. Used as an anticonvulsant.

PMO500   CAS: 115-07-1   HR: 3
PROPENE
DOT: UN 1077
mf: C_3H_6 mw: 42.09
PROP: A gas. D: (gas) 1.49 (air = 1.0), d: (liquid) 0.581 @ 0°. Mp: −185°, bp: −47.7°, autoign temp: 860°F, vap press: 10 atm @ 19.8°, lel: 2.4%, uel: 10.1%, vap d: 1.5, flash p: −162°F. Mod sol in alc.

SYNS: METHYLETHENE □ METHYLETHYLENE □ NCI-C50077 □ 1-PROPENE □ PROPYLENE (DOT, ACGIH)

CONSENSUS REPORTS: IARC Cancer Review: Group 3 IMEMDT 7,56,87. NTP Carcinogenesis Studies (inhalation); No Evidence: mouse, rat NTPTR* NTP-TR-272,85. Reported in EPA TSCA Inventory.

ACGIH TLV: Asphyxiant; Not Classifiable as a Human Carcinogen; (Proposed: 500 ppm; Not Classifiable as a Human Carcinogen)

DOT CLASSIFICATION: 2.1; Label: Flammable Gas

SAFETY PROFILE: A simple asphyxiant. No irritant effects from high concentrations in gaseous form. When compressed to liquid form, can cause skin burns from freezing effects of rapid evaporation on tissue. Questionable carcinogen. Flammable gas and very dangerous fire hazard when exposed to heat, flame, or oxidizers. Explosive in the form of vapor when exposed to heat or flame. Under unusual conditions, i.e., 955 atm pressure and 327°C,

P

it has been known to explode. Explodes on contact with trifluoromethyl hypofluorite. Explosive polymerization is initiated by lithium nitrate + sulfur dioxide. Reacts with oxides of nitrogen to form an explosive product. Dangerous; can react vigorously with oxidizing materials. To fight fire, stop flow of gas. Used in production of fabricated polymers, fibers, and solvents, in production of plastic products and resins. For effects of simple asphyxiants, see ARGON.

PMP500   CAS: 9003-07-0   HR: 2
PROPENE POLYMERS
mf: $(C_3H_6)_n$
PROP: Solid material. Mp: about 165°, d: 0.90–0.92. Insol in organic materials.
SYNS: ADMER PB 02 □ AMCO □ AMERFIL □ AMOCO 1010 □ ATACTIC POLYPROPYLENE □ AVISUN □ AZDEL □ BEAMETTE □ BICOLENE P □ CARLONA P □ CELGARD 2500 □ CHISSO 507B □ CLYSAR □ COATHYLENE PF 0548 □ DAPLEN AD □ DEXON E 117 □ EASTBOND M 5 □ ELPON □ ENJAY CD 460 □ EPOLENE M 5K □ GERFIL □ HERCOFLAT 135 □ HERCULON □ HOSTALEN PP □ HULS P 6500 □ ICI 543 □ ISOTACTIC POLYPROPYLENE □ J 400 □ LAMBETH □ LUPAREEN □ MARLEX 9400 □ MAURYLENE □ MERAKLON □ MOPLEN □ MOSTEN □ NOBLEN □ NOVAMONT 2030 □ NOVOLEN □ OLETAC 100 □ PAISLEY POLYMER □ PELLON 2506 □ POLYPRO 1014 □ POLYPROPENE □ POLYPROPYLENE □ POLYTAC □ POPROLIN □ PROFAX □ PROPATHENE □ 1-PROPENE HOMOPOLYMER (9CI) □ PROPOLIN □ PROPOPHANE □ PROPYLENE POLYMER □ REXALL 413S □ REXENE □ SHELL 5520 □ SHOALLOMER □ SYNDIOTACTIC POLYPROPYLENE □ TENITE 423 □ TRESPAPHAN □ TUFF-LITE □ ULSTRON □ VISCOL 350P □ W 101 □ WEX 1242
CONSENSUS REPORTS: IARC Cancer Review: Group 3 IMEMDT 7,56,87; Animal Limited Evidence IMEMDT 19,213,79; Human Inadequate Evidence IMEMDT 19,213,79. Reported in EPA TSCA Inventory.
SAFETY PROFILE: Moderately toxic by ingestion and intraperitoneal routes. Questionable carcinogen. When heated to decomposition it emits acrid smoke and irritating fumes. Used in injection molding for auto parts, in bottle caps, and in container closures.

PMP750   CAS: 6842-15-5   HR: 2
PROPENE TETRAMER
mf: $C_{12}H_{24}$ mw: 168.36
PROP: Colorless liquid. Mp: −31.5°, bp: 213°, d: 0.76 @ 20°/4°, vap press: 1 mm @ 47.2°, vap d: 5.81, flash p: <212°F, autoign temp: 491°F.
SYNS: PROPYLENE TETRAMER □ TETRAPRO-PYLENE
CONSENSUS REPORTS: Reported in EPA TSCA Inventory.
SAFETY PROFILE: Low toxicity by ingestion, inhalation, and skin contact. Probably irritating and narcotic in high concentration. Flammable or combustible when exposed to heat or flame; can react with oxidizing materials. To fight fire, use foam, CO_2, dry chemical. When heated to decomposition it emits acrid smoke and irritating fumes.

PMQ250   CAS: 768-03-6   HR: 3
2-PROPENOPHENONE
mf: C_9H_8O mw: 132.17
PROP: Oil. Bp: 115° @ 18 mm.
SYNS: ACRYLOPHENONE □ PHENYLVINYL KETONE
DOT CLASSIFICATION: 3; Label: Flammable Liquid
SAFETY PROFILE: A poison by intraperitoneal route. Questionable carcinogen with experimental tumorigenic data. When heated to decomposition it emits acrid smoke and irritating fumes. See also KETONES.

PMQ750   CAS: 104-46-1   HR: 3
p-PROPENYLANISOLE
mf: $C_{10}H_{12}O$ mw: 148.22
PROP: Leaves from alc or light-yellow liquid above 23°; sweet taste with anise odor. D: 0.991 @ 20°/20°, refr index: 1.557–1.561, mp: 22.5°, bp: 235.3°, flash p: 198°F. Very sltly sol in water; misc in abs alc, ether, chloroform.

SYNS: ACINTENE O □ ANETHOLE (FCC) □ ANISE CAMPHOR □ ARIZOLE □ FEMA No. 2086 □ ISOESTRAGOLE □ p-METHOXY-β-METHYLSTYRENE □ 1-(p-METHOXYPHENYL)PROPENE □ 1-METHOXY-4-PROPENYLBENZENE □ 4-METHOXYPROPENYL-BENZENE □ MONASIRUP □ NAULI "GUM" □ OIL of ANISEED □ p-1-PROPENYLANISOLE □ 4-PROPENYL-ANISOLE □ p-PROPENYLPHENYL METHYL ETHER

CONSENSUS REPORTS: Reported in EPA TSCA Inventory.

SAFETY PROFILE: Poison by ingestion. Questionable carcinogen with experimental tumorigenic data. Combustible liquid. When heated to decomposition it emits acrid smoke and irritating fumes. See also ETHERS.

PMR750 CAS: 590-21-6 HR: 3
PROPENYL CHLORIDE
mf: C_3H_5Cl mw: 76.53

PROP: Liquid. Mp: −137.4°, bp: 22.65°, flash p: <21°F, d: 0.9189°, lel: 4.5%, uel: 16%. Insol in water.

SYNS: 1-CHLOROPROPENE □ 1-CHLORO-1-PROPENE

CONSENSUS REPORTS: Reported in EPA TSCA Inventory.

SAFETY PROFILE: Moderately toxic by ingestion. Very mildly toxic by skin contact and inhalation. A skin and eye irritant. Mutation data reported. Questionable carcinogen with experimental neoplastigenic data. Very dangerous fire hazard when exposed to heat, flames (sparks), or oxidizers. Explosive in the form of vapor when exposed to heat or flame. To fight fire, use alcohol foam, dry chemical, mist, spray, fog. When heated to decomposition it emits toxic fumes of Cl⁻. See also CHLORINATED HYDROCARBONS, ALIPHATIC.

PMS500 CAS: 1797-74-6 HR: 2
2-PROPENYL PHENYLACETATE
mf: $C_{11}H_{12}O_2$ mw: 176.23

PROP: Colorless to light-yellow liquid; fruity odor of banana and honey.

SYNS: ALLYL PHENYLACETATE □ BENZENE-ACETIC ACID, 2-PROPENYL ESTER □ PHENYLACETIC ACID ALLYL ESTER

CONSENSUS REPORTS: Reported in EPA TSCA Inventory.

SAFETY PROFILE: Moderately toxic by ingestion. A human skin irritant. When heated to decomposition it emits acrid smoke and irritating fumes. See also ESTERS and ALLYL COMPOUNDS.

PMS900 CAS: 25332-09-6 HR: 3
PROPHENPYRIDAMINE HYDRO-CHLORIDE
mf: $C_{16}H_{20}N_2 \bullet ClH$ mw: 276.84

SYNS: 1-PHENYL-1-(2-PYRIDYL)-3-DIMETHYL-AMINOPROPANE HYDROCHLORIDE □ TRIMETON

SAFETY PROFILE: Poison by ingestion, intravenous, intraperitoneal, and subcutaneous routes. When heated to decomposition it emits very toxic fumes of NO_x and HCl.

PMT100 CAS: 57-57-8 HR: 3
β-PROPIOLACTONE
mf: $C_3H_4O_2$ mw: 72.07

PROP: A liquid. Mp: −33.4°, bp: 162° (decomp). Sol in H_2O.

SYNS: BETAPRONE □ BPL □ HYDRACRYLIC ACID β-LACTONE □ 3-HYDROXYPROPIONIC ACID LACTONE □ PROPANOLIDE □ PROPIOLACTONE □ 1,3-PROPIOLACTONE □ 3-PROPIOLACTONE □ β-PROPIONOLACTONE □ β-PROPRIOLACTONE (OSHA) □ β-PROPROLACTONE

CONSENSUS REPORTS: NTP 10th Report on Carcinogens. IARC Cancer Review: Group 2B IMEMDT 7,56,87; Animal Sufficient Evidence IMEMDT 4,259,74. EPA Genetic Toxicology Program. Community Right-To-Know List. EPA Extremely Hazardous Substances List. Reported in EPA TSCA Inventory.

OSHA PEL: OSHA: Carcinogen

ACGIH TLV: TWA 0.5 ppm; Animal Carcinogen

DFG MAK: Animal Carcinogen, Suspected Human Carcinogen

SAFETY PROFILE: Confirmed carcinogen with experimental carcinogenic, neoplastigenic, and tumorigenic data. Poison by inhalation. Moderately toxic by intraperitoneal route. An initiator. Human

P

mutation data reported. When heated to decomposition it emits acrid smoke and irritating fumes.

PMT750   CAS: 123-38-6   HR: 3
PROPIONALDEHYDE
DOT: UN 1275
mf: C_3H_6O mw: 58.09
PROP: Colorless, mobile liquid; suffocating odor. Mp: −81°, bp: 48°, flash p: 15–19°F (OC), d: 0.807 @ 20°/4°, lel: 2.9%, uel: 17%, vap d: 2.0, autoign temp: 405°F. Misc with alc, ether, water @ 49°.
SYNS: ALDEHYDE PROPIONIQUE (FRENCH) □ FEMA No. 2923 □ METHYLACETALDEHYDE □ NCI-C61029 □ PROPALDEHYDE □ PROPANAL □ PROPIONIC ALDEHYDE □ PROPYL ALDEHYDE □ PROPYLIC ALDEHYDE
CONSENSUS REPORTS: Community Right-To-Know List. Reported in EPA TSCA Inventory.
ACGIH TLV: TWA 20 ppm
DOT CLASSIFICATION: 3; Label: Flammable Liquid
SAFETY PROFILE: Moderately toxic by skin contact, ingestion, and subcutaneous routes. Mildly toxic by inhalation. A skin and severe eye irritant. Flammable liquid. Dangerous fire hazard when exposed to heat or flame; reacts vigorously with oxidizers. Explosive in the form of vapor when exposed to heat or flame. Vigorous polymerization reaction with methyl methacrylate. To fight fire, use alcohol foam, CO_2, dry chemical. When heated to decomposition it emits acrid smoke and irritating fumes. See also ALDEHYDES.

PMU750   CAS: 79-09-4   HR: 3
PROPIONIC ACID
DOT: UN 1848
mf: $C_3H_6O_2$ mw: 74.09
PROP: Oily liquid; pungent, disagreeable, rancid odor. D: 0.998 @ 15°/4°, mp: −21.5°, bp: 141.1°, vap press: 10 mm @ 39.7°, vap d: 2.56, autoign temp: 955°F. Misc in water, alc, ether, chloroform.
SYNS: ACIDE PROPIONIQUE (FRENCH) □ CARBOXYETHANE □ ETHANECARBOXYLIC ACID □

ETHYLFORMIC ACID □ KYSELINA PROPIONOVA □ LUPROSIL □ METACETONIC ACID □ METHYL ACETIC ACID □ PROPANOIC ACID □ PROPIONIC ACID (ACGIH,DOT,OSHA) □ PROPIONIC ACID GRAIN PRESERVER □ PROZOIN □ PSEUDOACETIC ACID □ SENTRY GRAIN PRESERVER □ TENOX P GRAIN PRESERVATIVE
CONSENSUS REPORTS: Reported in EPA TSCA Inventory.
OSHA PEL: TWA 10 ppm
ACGIH TLV: TWA 10 ppm
DFG MAK: 10 ppm (31 mg/m³)
DOT CLASSIFICATION: 8; Label: Corrosive
SAFETY PROFILE: Poison by intraperitoneal route. Moderately toxic by ingestion, skin contact, and intravenous routes. A corrosive irritant to eyes, skin, and mucous membranes. Flammable liquid. Highly flammable when exposed to heat, flame, or oxidizers. To fight fire, use alcohol foam. When heated to decomposition it emits acrid smoke and irritating fumes.

PMV250   CAS: 540-42-1   HR: 3
PROPIONIC ACID, ISOBUTYL ESTER
DOT: UN 2394
mf: $C_7H_{14}O_2$ mw: 130.21
PROP: A solid. D: 0.888 @ 0°/4°, bp: 137°.
SYNS: ISOBUTYL PROPIONATE (DOT) □ 2-METHYLPROPYL PROPIONATE □ PROPANOIC ACID, 2-METHYLPROPYL ESTER
CONSENSUS REPORTS: Reported in EPA TSCA Inventory.
DOT CLASSIFICATION: 3; Label: Flammable Liquid
SAFETY PROFILE: Mildly toxic by ingestion. Flammable when exposed to heat or flame, can react vigorously with oxidizing materials. When heated to decomposition it emits acrid smoke and irritating fumes. See also ESTERS, PROPIONIC ACID, and ISOBUTYL ALCOHOL.

PMV500   CAS: 123-62-6   HR: 2
PROPIONIC ANHYDRIDE
DOT: UN 2496
mf: $C_6H_{10}O_3$ mw: 130.16
PROP: Liquid; very rancid odor. Mp: −45°, bp: 167.0°, flash p: 165°F (OC), d: 1.012, vap press: 1 mm @ 20.6°, vap d: 4.49.

Decomp in water, alc; sol in methanol, ethanol, ether, chloroform.

SYNS: METHYLACETIC ANHYDRIDE □ PROPANOIC ANHYDRIDE □ PROPIONIC ACID ANHYDRIDE □ PROPIONYL OXIDE

CONSENSUS REPORTS: Reported in EPA TSCA Inventory.

DOT CLASSIFICATION: 8; Label: Corrosive

SAFETY PROFILE: Moderately toxic by ingestion. Mildly toxic by skin contact. A corrosive irritant to skin, eyes, and mucous membranes. Combustible when exposed to heat or flame; can react with oxidizing materials. To fight fire, use CO_2, dry chemical. When heated to decomposition it emits acrid smoke and irritating fumes. Used as an esterifying agent and dehydrating agent. See also ANHYDRIDES.

PMV750   CAS: 107-12-0   HR: 3
PROPIONONITRILE

DOT: UN 2404

mf: C_3H_5N mw: 55.09

PROP: Colorless liquid; ethereal odor. Fp: −103.5°, bp: 97.1°, d: 0.783 @ 21°/4°, vap d: 1.9, flash p: 36°F, lel: 3.1%, mp: 91.8°. Misc with alc and ether.

SYNS: CYANOETHANE □ ETHER CYANATUS □ ETHYL CYANIDE □ HYDROCYANIC ETHER □ PROPANENITRILE □ PROPIONIC NITRILE □ RCRA WASTE NUMBER P101

CONSENSUS REPORTS: Cyanide and its compounds are on the Community Right-To-Know List. EPA Extremely Hazardous Substances List. Reported in EPA TSCA Inventory.

ACGIH TLV: CL 5 mg(CN)/m³ (skin)

NIOSH REL: TWA (Nitriles) 14 mg/m³

DOT CLASSIFICATION: 3; Label: Flammable Liquid, Poison

SAFETY PROFILE: Poison by ingestion, skin contact, intravenous, and intraperitoneal routes. Moderately toxic by inhalation. Experimental teratogenic effects. Other experimental reproductive effects. A skin and eye irritant. Dangerous fire hazard when exposed to heat, flame (sparks), oxidizers. Mixture with N-bromosuccinimide may explode when heated. To fight fire, use water spray, foam, mist, CO_2, dry chemical. When heated to decomposition it emits toxic fumes of NO_x and CN^-. Used as a solvent in petroleum refining, and as a raw material for drug manufacture. See also NITRILES.

PMW500   CAS: 79-03-8   HR: 3
PROPIONYL CHLORIDE

mf: C_3H_5ClO mw: 92.53

PROP: Mp: −94°, bp: 80°, flash p: 53.6°, d: 1.065, vap d: 3.2.

SYNS: PROPANOYL CHLORIDE □ PROPIONIC ACID CHLORIDE □ PROPIONIC CHLORIDE

CONSENSUS REPORTS: Reported in EPA TSCA Inventory.

SAFETY PROFILE: A corrosive irritant to skin, eyes, and mucous membranes. Dangerous fire hazard when exposed to heat or flame; can react vigorously with oxidizing materials. Reacts with water or steam to produce toxic and corrosive fumes. Exothermic reaction with diisopropyl ether produces much gas. The reaction may be dangerous if confined. To fight fire, use CO_2, dry chemical; do not use water. When heated to decomposition it emits highly toxic fumes of Cl^-. See also HYDROCHLORIC ACID.

PNA500   CAS: 1639-60-7   HR: 3
d-PROPOXYPHENE HYDROCHLORIDE

mf: $C_{22}H_{29}NO_2 \cdot ClH$ mw: 375.98

PROP: Bitter crystals. Mp: 163–168.5°. Sol in water, alc, chloroform, acetone; insol in benzene, ether.

SYNS: ALGAFAN □ ANTALVIC □ DARVON HYDROCHLORIDE □ DEPRANCOL □ DEPROMIC □ DEVELIN □ DEXTROPROPOXYPHENE HYDRO-CHLORIDE □ DEXTROPROXYPHEN HYDROCHLOR-IDE □ d-4-DIMETHYLAMINO-3-METHYL-1,2-DIPHENYL-2-BUTANOL PROPIONATE HYDRO-CHLORIDE □ s-α-(2-(DIMETHYLAMINO)-1-METHYLETHYL)-α-PHENYLBENZENEETHANOL PROPIOATE HYDROCHLORIDE □ (+)-1,2-DIPHENYL-2-PROPIONOXY-3-METHYL-4-DIMETHYLAMINO-BUTANE HYDROCHLORIDE □ DOLENE □ DOLOCAP □ DOLOXENE □ DORAPHEN □ ERANTIN □ FEMADOL □ HARMAR □ PROPOX □ PROPOXYCHEL □ PROPOXYPHENE HYDROCHLORIDE □ (+)-PROPOXYPHENE HYDROCHLORIDE □ α-PROPOXY-

PHENE HYDROCHLORIDE □ α-d-PROPOXYPHENE HYDROCHLORIDE □ d-PROPOXYPHENE MONO-HYDROCHLORIDE □ PROXAGESIC

CONSENSUS REPORTS: Reported in EPA TSCA Inventory.

SAFETY PROFILE: A human poison by ingestion. An experimental poison by ingestion, intraperitoneal, subcutaneous, intravenous, and intramuscular routes. Human systemic effects by ingestion: anorexia, ataxia, chronic pulmonary edema, cyanosis, increased body temperature, jaundice, nausea or vomiting, sleep disturbance. Experimental teratogenic and reproductive effects. When heated to decomposition it emits very toxic fumes of HCl and NO_x. See also PROPOXYPHENE.

**PNC250   CAS: 109-60-4   HR: 3
n-PROPYL ACETATE**

DOT: UN 1276

mf: $C_5H_{10}O_2$ mw: 102.15

PROP: Clear, colorless liquid; pleasant odor. Mp: −92.5°, bp: 101.6°, flash p: 58°F, lel: 2.0%, uel: 8.0%, d: 0.887, autoign temp: 842°F, vap press: 40 mm @ 28.8°, vap d: 3.52. Misc with alc, ether; sol in water. IDLH 1700 ppm.

SYNS: ACETATE de PROPYLE NORMAL (FRENCH) □ ACETIC ACID, n-PROPYL ESTER □ 1-ACETOXY-PROPANE □ OCTAN PROPYLU (POLISH) □ PROPYL ACETATE □ 1-PROPYL ACETATE □ PROPYLESTER KYSELINY OCTOVE

CONSENSUS REPORTS: Reported in EPA TSCA Inventory.

OSHA PEL: TWA 200 ppm; STEL 250 ppm

ACGIH TLV: TWA 200 ppm; STEL 250 ppm

DFG MAK: 200 ppm (850 mg/m³)

DOT CLASSIFICATION: 3; Label: Flammable Liquid

SAFETY PROFILE: Moderately toxic by intraperitoneal and subcutaneous routes. Mildly toxic by ingestion and inhalation. Human systemic effects by inhalation: lachrymation, cough. A skin irritant. A narcotic at high concentrations. Isopropyl acetate is slightly less narcotic than normal propyl acetate. A flammable liquid and dangerous fire hazard when exposed to heat, flame, or oxidizers. Explosive in the form of vapor when exposed to heat or flame. Can react vigorously with oxidizing materials. To fight fire, use alcohol foam, CO_2, dry chemical. When heated to decomposition it emits acrid smoke and irritating fumes.

**PND000   CAS: 71-23-8   HR: 3
n-PROPYL ALCOHOL**

DOT: UN 1274

mf: C_3H_8O mw: 60.11

PROP: Clear liquid; alcohol-like odor. Mp: −127°, bp: 97.19°, flash p: 59°F (CC), ULC: 55–60, d: 0.8044 @ 20°/4°, lel: 2.1%, uel: 13.5%, autoign temp: 824°F, vap press: 10 mm @ 14.7°, vap d: 2.07. Misc in water, alc, and ether. IDLH 800 ppm.

SYNS: ALCOOL PROPILICO (ITALIAN) □ ALCOOL PROPYLIQUE (FRENCH) □ ETHYL CARBINOL □ 1-HYDROXYPROPANE □ OPTAL □ OSMOSOL EXTRA □ n-PROPANOL □ PROPANOL-1 □ 1-PROPANOL (ACGIH) □ PROPANOLE (GERMAN) □ PROPANOLEN (DUTCH) □ PROPANOLI (ITALIAN) □ PROPYL ALCOHOL □ 1-PROPYL ALCOHOL □ n-PROPYL ALKOHOL (GERMAN) □ PROPYLIC ALCOHOL □ PROPYLOWY ALKOHOL (POLISH)

CONSENSUS REPORTS: Reported in EPA TSCA Inventory. EPA Genetic Toxicology Program.

OSHA PEL: TWA 200 ppm; STEL 250 ppm

ACGIH TLV: TWA 100 ppm; Not Classifiable as a Human Carcinogen

DOT CLASSIFICATION: 3; Label: Flammable Liquid

SAFETY PROFILE: Poison by subcutaneous route. Moderately toxic by inhalation, ingestion, intraperitoneal, and intravenous routes. A skin and severe eye irritant. Questionable carcinogen with experimental carcinogenic data. Mutation data reported. A flammable liquid and dangerous fire hazard when exposed to heat, flame, or oxidizers. Explosive in the form of vapor when exposed to heat or flame. Ignites on contact with potassium-tert-butoxide. Dangerous upon exposure to heat or flame; can react vigorously with oxidizing materials. To fight fire, use alcohol foam, CO_2, dry chemical. When heated to decomposition it emits acrid smoke and irritating fumes.

PND250   CAS: 107-10-8   HR: 3
PROPYLAMINE
DOT: UN 1277
mf: C_3H_9N mw: 59.13
PROP: Colorless, alkaline liquid; strong ammonia odor. D: 0.7191 @ 20°/20°, mp: −83°, bp: 48–49°, vap press: 248 mm @ 20°, flash p: −35°F, autoign temp: 604°F, lel: 2.0%, uel: 10.4%. Misc in water, alc, ether.
SYNS: 1-AMINOPROPANE □ MONO-N-PROPYLAMINE □ PROPANAMINE □ N-PROPYLAMINE □ RCRA WASTE NUMBER U194
CONSENSUS REPORTS: Reported in EPA TSCA Inventory.
DOT CLASSIFICATION: 3; Label: Flammable Liquid, Corrosive
SAFETY PROFILE: Moderately toxic by inhalation, ingestion, and skin contact routes. A skin and severe eye irritant. Possibly a skin sensitizer. Very dangerous fire hazard when exposed to heat, flame, or oxidizers. Explosive in the form of vapor when exposed to heat or flame. To fight fire, use alcohol foam. When heated to decomposition it emits toxic fumes of NO_x. Incompatible with triethynyl aluminum. See also AMINES.

PNE250   CAS: 104-45-0   HR: 1
p-n-PROPYL ANISOLE
mf: $C_{10}H_{14}O$ mw: 150.24
PROP: Colorless to pale-yellow liquid; anise odor. D: 0.940, refr index: 1.502–1.506, bp: 215–216°, flash p: 185°F. Sol in fixed oils; insol in glycerin, propylene glycol.
SYNS: DIHYDROANETHOLE □ FEMA No. 2930 □ 1-METHOXY-4-PROPYLBENZENE □ 4-PROPYLANISOLE □ 4-n-PROPYLANISOLE
CONSENSUS REPORTS: Reported in EPA TSCA Inventory.
SAFETY PROFILE: Mildly toxic by ingestion. Mutation data reported. Combustible liquid. When heated to

decomposition it emits acrid smoke and irritating fumes.

PNG250   CAS: 627-12-3   HR: 2
PROPYL CARBAMATE
mf: $C_4H_9NO_2$ mw: 103.14
PROP: Crystals. Bp: 196°, mp: 60°, vap press: 1 mm @ 52.4°. Very sol in water, alc, ether.
SYNS: CARBAMIC ACID, PROPYL ESTER □ N-PROPYL CARBAMATE □ PROPYL URETHANE
CONSENSUS REPORTS: IARC Cancer Review: Group 3 IMEMDT 7,56,87; Animal Sufficient Evidence IMEMDT 12,201,76. Reported in EPA TSCA Inventory.
SAFETY PROFILE: Moderately toxic by ingestion and subcutaneous routes. An experimental teratogen. Questionable carcinogen with experimental neoplastigenic and tumorigenic data. Mutation data reported. When heated to decomposition it emits toxic fumes of NO_x. See also CARBAMATES.

PNG750   CAS: 2807-30-9   HR: 3
PROPYL CELLOSOLVE
mf: $C_5H_{12}O_2$ mw: 104.17
PROP: A liquid. D: 0.914 @ 15°/15°, bp: 150° @ 743 mm.
SYNS: EKTASOLVE EP □ ETHYLENE GLYCOL-MONO-PROPYL ETHER □ ETHYLENE GLYCOL-MONO-n-PROPYL ETHER □ MONOPROPYL ETHER of ETHYLENE GLYCOL □ 2-PROPOXYETHANOL
CONSENSUS REPORTS: Glycol ether compounds are on the Community Right-To-Know List. Reported in EPA TSCA Inventory.
DFG MAK: 20 ppm
SAFETY PROFILE: Moderately toxic by ingestion and skin contact. Mildly toxic by inhalation. An experimental teratogen. Experimental reproductive effects. Some glycol ethers have dangerous human reproductive effects. A skin and severe eye irritant. Flammable; can react with oxidizing materials. When heated to decomposition it emits acrid smoke and irritating fumes. See also GLYCOL ETHERS.

P

PNH000   CAS: 109-61-5   HR: 3
PROPYL CHLOROCARBONATE
DOT: UN 2740
mf: $C_4H_7ClO_2$ mw: 122.56
PROP: Colorless liquid. D: 1.090 @ 20°/4°,
bp: 114–115° @ 768 mm. Insol and sltly
decomp in water, alc; misc in ether,
benzene.
SYNS: CARBONOCHLORIDIC ACID, PROPYL ESTER
□ CHLOROFORMIC ACID PROPYL ESTER □ PROPYL
CHLOROFORMATE □ n-PROPYL CHLOROFORMATE
(DOT)
CONSENSUS REPORTS: EPA Extremely
Hazardous Substances List. Reported in
EPA TSCA Inventory.
DOT CLASSIFICATION: 6.1; Label: Poison,
Flammable Liquid, Corrosive
SAFETY PROFILE: Poison by skin contact.
Moderately toxic by ingestion and
inhalation. A corrosive irritant to skin, eyes,
and mucous membranes. Flammable when
exposed to heat or flame, can react
vigorously with oxidizing materials. When
heated to decomposition it emits toxic
fumes of Cl⁻. Used as a reactive intermediate
to polymerization initiators.

PNH522   CAS: 315706-70-8 HR: 3
N-(4-PROPYLCYCLOHEXYL)-
  BENZAMIDE
mf: $C_{16}H_{23}NO$ mw: 245.36
SAFETY PROFILE: A poison by ingestion.
When heated to decomposition it emits
toxic vapors of NO_x.

PNH533   CAS: 315706-73-1 HR: 3
N-(4-PROPYLCYCLOHEXYL)-3-(3,4,5-
  TRIMETHOXYPHENYL)-2-PROPEN-
  AMIDE
mf: $C_{21}H_{31}NO_4$ mw: 361.48
SAFETY PROFILE: A poison by ingestion.
When heated to decomposition it emits
toxic vapors of NO_x.

PNI600   CAS: 7650-84-2   HR: 3
PROPYLDIPHENYLPHOSPHINE
mf: $C_{15}H_{17}P$ mw: 228.29
SYNS: DIPHENYLPROPYLPHOSPHINE □
PHOSPHINE, DIPHENYLPROPYL-

SAFETY PROFILE: A poison by ingestion
and skin contact. A mild skin irritant. When
heated to decomposition it emits toxic
vapors of PO_x.

PNI850   CAS: 7361-89-9   HR: 3
PROPYL DISELENIDE
mf: $C_6H_{14}Se_2$ mw: 244.12
SYNS: DI-n-PROPYL DISELENIDE □ DISELENIDE,
DIPROPYL-(9CI)
OSHA PEL: TWA 0.2 mg(Se)/m³
ACGIH TLV: TWA 0.2 mg(Se)/m³
SAFETY PROFILE: Poison by intravenous
route. When heated to decomposition it
emits toxic fumes of Se.

PNJ400   CAS: 78-87-5   HR: 3
PROPYLENE DICHLORIDE
DOT: UN 1279
mf: $C_3H_6Cl_2$ mw: 112.99
PROP: Colorless liquid. Bp: 96.8°, flash p:
60°F, d: 1.1593 @ 20°/20°, vap press: 40
mm @ 19.4°, vap d: 3.9, autoign temp:
1035°F, lel: 3.4%, uel: 14.5%. IDLH 400
ppm.
SYNS: BICHLORURE de PROPYLENE (FRENCH) □
α,β-DICHLOROPROPANE □ 1,2-DICHLOROPROPANE
□ DWUCHLOROPROPAN (POLISH) □ ENT 15,406 □
NCI-C55141 □ PROPYLENE CHLORIDE □ α,β-
PROPYLENE DICHLORIDE □ RCRA WASTE NUMBER
U083
CONSENSUS REPORTS: IARC Cancer
Review: Group 3 IMEMDT 7,56,87; Animal
Limited Evidence IMEMDT 41,131,86.
NTP Carcinogenesis Studies (gavage);
Equivocal Evidence: rat NTPTR* NTP-TR-
263,86; Some Evidence: mouse NTPTR*
NTP-TR-263,86. Reported in EPA TSCA
Inventory. EPA Genetic Toxicology
Program. Community Right-To-Know List.
OSHA PEL: TWA 75 ppm; STEL 110 ppm
ACGIH TLV: TWA 75 ppm; STEL 110 ppm;
Not Classifiable as a Human Carcinogen
DFG MAK: Confirmed Animal Carcinogen
with Unknown Relevance to Humans
DOT CLASSIFICATION: 3; Label:
Flammable Liquid
SAFETY PROFILE: Suspected carcinogen
with experimental carcinogenic data.

Moderately toxic by inhalation and ingestion. Mildly toxic by skin contact. An eye irritant. Mutation data reported. Can cause liver, kidney, and heart damage. Can cause dermatitis. One of the more toxic chlorinated hydrocarbons. A suggested order of increasing toxicity is dichloromethane, trichloroethylene, carbon tetrachloride, dichloropropane, dichloroethane. Animals exposed to high concentrations often showed marked visceral congestion, fatty degeneration of the liver, kidney, and, less frequently, of the heart. They also showed areas of coagulation and necrosis of the liver. There was found to be a heavy mortality among mice exposed to 400 ppm concentrations.

A flammable liquid and very dangerous fire hazard when exposed to heat or flame. Reacts with aluminum to form aluminum chloride. This reaction, when confined, can lead to explosion. Can react vigorously with oxidizing materials. To fight fire, use water, foam, CO_2, dry chemical. When heated to decomposition it emits toxic fumes of Cl⁻. See also CHLORINATED HYDROCARBONS, ALIPHATIC.

PNJ750   CAS: 9005-37-2   HR: 1
PROPYLENE GLYCOL ALGINATE
mf: $(C_9H_{14}O_7)_8$ mw: 1873.6
PROP: White fibrous or granular powder; odorless and tasteless. Sol in water and dilute organic acids.
SYNS: HYDROXY PROPYL ALGINATE □ KELCOLOID
CONSENSUS REPORTS: Reported in EPA TSCA Inventory.
SAFETY PROFILE: Mildly toxic by ingestion. When heated to decomposition it emits acrid smoke and irritating fumes.

PNL000   CAS: 6423-43-4   HR: 3
PROPYLENE GLYCOL DINITRATE
mf: $C_3H_6N_2O_6$ mw: 166.11
SYNS: PGDN □ PROPYLENE GLYCOL-1,2-DINITRATE □ 1,2-PROPYLENE GLYCOL DINITRATE
CONSENSUS REPORTS: Reported in EPA TSCA Inventory.

OSHA PEL: TWA 0.05 ppm
ACGIH TLV: TWA 0.05 ppm (skin)
DFG MAK: 0.05 ppm (0.34 mg/m³)
SAFETY PROFILE: Poison by ingestion and subcutaneous routes. Moderately toxic by intraperitoneal and intravenous routes. Human systemic effects by inhalation: conjunctiva irritation, headache. An eye irritant. When heated to decomposition it emits toxic fumes of NO_x. See also NITRATES.

PNL225   HR: D
PROPYLENE GLYCOL MONO- and DIESTERS
PROP: Clear liquid or white to yellow beads or flakes; bland odor and taste. Insol in water; sol in alc, ethyl acetate, chloroform.
SYNS: PROPYLENE GLYCOL MONO- and DIESTERS of FATTY ACIDS □ PROPYLENE GLYCOL MONOSTEARATE
SAFETY PROFILE: When heated to decomposition it emits acrid smoke and irritating fumes.

PNL250   CAS: 107-98-2   HR: 3
PROPYLENE GLYCOL MONOMETHYL ETHER
mf: $C_4H_{10}O_2$ mw: 90.14
PROP: Colorless liquid. Mp: −96.7°, bp: 126–127°, flash p: 100°F, d: 0.919 @ 25°/25°.
SYNS: DOWANOL 33B □ DOWANOL PM □ DOWANOL PM GLYCOL ETHER □ DOWTHERM 209 □ GLYCOL ETHER PM □ METHOXY ETHER of PROPYLENE GLYCOL □ 1-METHOXY-2-PROPANOL □ POLY-SOLVE MPM □ PROPASOL SOLVENT M □ PROPYLENE GLYCOL METHYL ETHER □ PROPYLENE GLYCOL MONOMETHYL ETHER □ α-PROPYLENE GLYCOL MONOMETHYL ETHER □ PROPYLENE GLYCOL MONOMETHYL ETHER (ACGIH,OSHA) □ PROPYLENGLYKOL-MONOMETHYLAETHER □ UCAR SOLVENT LM (OBS.)
CONSENSUS REPORTS: Glycol ether compounds are on the Community Right-To-Know List. Reported in EPA TSCA Inventory.
OSHA PEL: TWA 100 ppm; STEL 150 ppm
ACGIH TLV: TWA 100 ppm; STEL 150 ppm

DFG MAK: 100 ppm (370 mg/m^3)
SAFETY PROFILE: Moderately toxic by intravenous route. Mildly toxic by ingestion, inhalation, and skin contact. Human systemic effects by inhalation: general anesthesia, nausea. A skin and eye irritant. An experimental teratogen. Many glycol ethers have dangerous human reproductive effects. Very dangerous fire hazard when exposed to heat or flame; can react with oxidizing materials. To fight fire, use foam, CO_2, dry chemical. When heated to decomposition it emits acrid smoke and irritating fumes. Used as a solvent and in solvent-sealing of cellophane. See also GLYCOL ETHERS and ETHYLENE GLYCOL MONOMETHYL ETHER.

PNL265   CAS: 108-65-6   HR: 2
**PROPYLENE GLYCOL MONOMETHYL
      ETHER ACETATE**
mf: $C_6H_{12}O_3$ mw: 132.18
SYNS: ACETIC ACID, 2-METHOXY-1-METHYLETHYL ESTER □ DOWANOL (R) PMA GLYCOL ETHER ACETATE □ 1-METHOXY-2-ACETOXYPROPANE
CONSENSUS REPORTS: Reported in EPA TSCA Inventory.
DFG MAK: 50 ppm
SAFETY PROFILE: Moderately toxic by intraperitoneal route. Slightly toxic by ingestion and skin contact. When heated to decomposition it emits acrid smoke and irritating vapors.

PNL400   CAS: 75-55-8   HR: 3
PROPYLENE IMINE
DOT: UN 1921
mf: C_3H_7N mw: 57.11
PROP: Liquid. Vap d: 2.0, flash p: 14°F. IDLH 100 ppm.
SYNS: 2-METHYLAZACYCLOPROPANE □ 2-METHYLAZIRIDINE □ METHYLETHYLENIMINE □ 2-METHYLETHYLENIMINE □ 1,2-PROPYLENEIMINE □ PROPYLENE IMINE, INHIBITED (DOT) □ RCRA WASTE NUMBER P067
CONSENSUS REPORTS: NTP 10th Report on Carcinogens. IARC Cancer Review: Group 2B IMEMDT 7,56,87; Animal Limited Evidence IMEMDT 9,61,75. EPA

Genetic Toxicology Program. Reported in EPA TSCA Inventory. EPA Extremely Hazardous Substances List. Community Right-To-Know List.
OSHA PEL: TWA 2 ppm (skin)
ACGIH TLV: TWA 2 ppm (skin); Animal Carcinogen
DFG MAK: Animal Carcinogen, Suspected Human Carcinogen
DOT CLASSIFICATION: 3; Label: Flammable Liquid
SAFETY PROFILE: Confirmed carcinogen with experimental carcinogenic data. Poison by ingestion and skin contact. Moderately toxic by inhalation. Mutation data reported. Severe eye irritant. Implicated as a brain carcinogen. A flammable liquid and very dangerous fire hazard when exposed to heat or flame; can react vigorously with oxidizing materials. Polymerizes explosively on exposure to acids or acid fumes. A storage hazard. When heated to decomposition it emits toxic fumes of NO_x.

PNL600   CAS: 75-56-9   HR: 3
PROPYLENE OXIDE
DOT: UN 1280
mf: C_3H_6O mw: 58.09
PROP: Colorless liquid; ethereal odor. Bp: 33.9°, lel: 2.8%, uel: 37%, fp: −104.4°, flash p: −35°F (TOC), d: 0.8304 @ 20°/20°, vap press: 400 mm @ 17.8°, vap d: 2.0. Sol in water, alc, and ether. IDLH 400 ppm.
SYNS: EPOXYPROPANE □ 1,2-EPOXYPROPANE □ 2,3-EPOXYPROPANE □ METHYL ETHYLENE OXIDE □ METHYL OXIRANE □ NCI-C50099 □ OXYDE de PROPYLENE (FRENCH) □ PROPENE OXIDE □ PROPYLENE EPOXIDE □ 1,2-PROPYLENE OXIDE
CONSENSUS REPORTS: NTP 10th Report on Carcinogens. IARC Cancer Review: Group 2A IMEMDT 7,328,87; Human Inadequate Evidence IMEMDT 36,227,85; Animal Sufficient Evidence IMEMDT 36,227,85; Animal Limited Evidence IMEMDT 11,191,76. Carcinogenesis Studies (inhalation); Some Evidence: rat NTPTR* NTP-TR-267,85; Clear Evidence: mouse NTPTR* NTP-TR-267,85. Reported in EPA TSCA Inventory. EPA Genetic

Toxicology Program. Community Right-To-Know List. EPA Extremely Hazardous Substances List.

OSHA PEL: TWA 20 ppm
ACGIH TLV: TWA 2 ppm (sensitizer); Confirmed Animal Carcinogen with Unknown Revelance to Humans
DFG MAK: Animal Carcinogen, Suspected Human Carcinogen
DOT CLASSIFICATION: 3; Label: Flammable Liquid
SAFETY PROFILE: Confirmed carcinogen with experimental carcinogenic, neoplastigenic, and tumorigenic data. Poison by intraperitoneal route. Moderately toxic by ingestion, inhalation, and skin contact. An experimental teratogen. Experimental reproductive effects. Human mutation data reported. A severe skin and eye irritant. Flammable liquid. A very dangerous fire and explosion hazard when exposed to heat or flame. Explosive reaction with epoxy resin and sodium hydroxide. Forms explosive mixtures with oxygen. Reacts with ethylene oxide + polyhydric alcohol to form the thermally unstable polyether alcohol. Incompatible with NH_4OH, chlorosulfonic acid, HCl, HF, HNO_3, oleum, H_2SO_4. Dangerous; can react vigorously with oxidizing materials. Keep away from heat and open flame. To fight fire, use alcohol foam, CO_2, dry chemical. When heated to decomposition it emits acrid smoke and fumes.

PNM000   CAS: 111-43-3   HR: 3
PROPYL ETHER
DOT: UN 2384
mf: $C_6H_{14}O$ mw: 102.20
PROP: Colorless liquid. Mp: $-122°$, bp: $90°$, d: 0.736 @ $20°/4°$, flash p: $70°F$, autoign temp: $419°F$. Sltly sol in water; sol in alc, ether; very volatile.
SYNS: DIPROPYL ETHER □ DIPROPYL OXIDE □ ETHER, DI-n-PROPYL- □ 1,1'-OXYBISPROPANE □ PROPANE, 1,1'-OXYBIS-(9CI)
CONSENSUS REPORTS: Reported in EPA TSCA Inventory.

DOT CLASSIFICATION: 3; Label: Flammable Liquid
SAFETY PROFILE: Poison by intravenous route. Moderately toxic by inhalation. Possibly narcotic. A flammable liquid and dangerous fire hazard when exposed to heat, flame, or oxidizers. Forms explosive peroxides. Dangerous upon exposure to heat or flame; can react vigorously with oxidizing materials. When heated to decomposition it emits acrid smoke and irritating fumes. See also ETHERS.

PNM500   CAS: 110-74-7   HR: 3
n-PROPYL FORMATE
DOT: UN 1281
mf: $C_4H_8O_2$ mw: 88.12
PROP: Colorless liquid; pleasant odor. Mp: $-93°$, bp: $82°$, flash p: $27°F$ (CC), d: 0.901 @ $20°$, vap press: 100 mm @ $29.5°$, vap d: 3.03, autoign temp: $851°F$, lel: 2.3%; misc in alc, ether; sltly sol in water.
SYNS: FORMIATE de PROPYLE (FRENCH) □ PROPYLESTER KYSELINY MRAVENCI □ PROPYL FORMATE (DOT) □ PROPYL METHANOATE
CONSENSUS REPORTS: Reported in EPA TSCA Inventory.
DOT CLASSIFICATION: 3; Label: Flammable Liquid
SAFETY PROFILE: Moderately toxic by ingestion. An irritant to skin, eyes, and mucous membranes. Narcotic in high concentration. Dangerous fire hazard when exposed to heat, flame, or oxidizers. Ignites on contact with potassium-tert-butoxide. Explosive in the form of vapor when exposed to heat or flame. To fight fire, use alcohol foam. When heated to decomposition it emits acrid smoke and irritating fumes. See also ESTERS.

PNM750   CAS: 121-79-9   HR: 3
n-PROPYL GALLATE
mf: $C_{10}H_{12}O_5$ mw: 212.22
PROP: Odorless, fine, ivory powder or crystals; sltly bitter taste. Mp: $147–149°$. Sltly sol in Me_2CO and 2-butanol.
SYNS: GALLIC ACID, PROPYL ESTER □ NIPA 49 □ NIPAGALLIN P □ PROGALLIN P □ *n*-PROPYL ESTER of

P

3,4,5-TRIHYDROXYBENZOIC ACID □ PROPYL GALLATE □ n-PROPYL-3,4,5-TRIHYDROXYBENZOATE □ TENOX PG □ 3,4,5-TRIHYDROXYBENZENE-1-PROPYLCARBOXYLATE □ 3,4,5-TRIHYDROXY-BENZOIC ACID, n-PROPYL ESTER

CONSENSUS REPORTS: NTP Carcinogenesis Bioassay (feed); No Evidence: mouse, rat NTPTR* NTP-TR-240,82. Reported in EPA TSCA Inventory.

SAFETY PROFILE: Poison by ingestion and intraperitoneal routes. Experimental teratogenic and reproductive effects. Questionable carcinogen with experimental tumorigenic data. Mutation data reported. Combustible when exposed to heat or flame; can react with oxidizing materials. When heated to decomposition it emits acrid smoke and irritating fumes.

PNP000   CAS: 110-78-1   HR: 3
PROPYL ISOCYANATE
DOT: UN 2482
mf: C_4H_7NO mw: 85.12
SYNS: 1-ISOCYANATOPROPANE □ ISOCYANIC ACID, PROPYL ESTER □ m-PROPYL ISOCYANATE □ 1-PROPYL ISOCYANATE

CONSENSUS REPORTS: Reported in EPA TSCA Inventory.

DOT CLASSIFICATION: 3; Label: Flammable Liquid, Poison; DOT Class: 6.1; Label: Poison; DOT Class: 6.1; Label: Poison, Flammable Liquid; DOT Class: 3; Label: Flammable Liquid, Poison

SAFETY PROFILE: Poison by intravenous route. A flammable liquid when exposed to heat or flame, can react vigorously with oxidizing materials. When heated to decomposition it emits toxic fumes of NO_x. See also ISOCYANATES.

PNQ500   CAS: 627-13-4   HR: 3
n-PROPYL NITRATE
DOT: UN 1865
mf: $C_3H_7NO_3$ mw: 105.11
PROP: Pale-yellow liquid; sickly odor. Bp: 110.5°, d: 1.054 @ 20°/4°, flash p: 68°F, autoign temp: 347°F (in air), lel: 2%, uel: 100%. Very sltly sol in water; sol in alc, ether. IDLH 500 ppm.

SYNS: NITRATE de PROPYLE NORMAL (FRENCH) □ NITRIC ACID, PROPYL ESTER □ PROPYLESTER KYSELINY DUSICNE □ PROPYL NITRATE

CONSENSUS REPORTS: Reported in EPA TSCA Inventory.

OSHA PEL: TWA 25 ppm; STEL 40 ppm
ACGIH TLV: TWA 25 ppm; STEL 40 ppm
DFG MAK: 25 ppm (110 mg/m³)
DOT CLASSIFICATION: 3; Label: Flammable Liquid

SAFETY PROFILE: Poison by intravenous route. Inhalation can cause hypotension and methemoglobinemia. Dangerous fire hazard when exposed to heat, flame, or oxidizers. Explosive in the form of vapor when exposed to heat or flame. A shock-sensitive explosive. It can be desensitized by the addition of 1–2% propane, butane, chloroform, dimethyl ether, or diethyl ether. When heated to decomposition it emits toxic fumes of NO_x. Used as a fuel ignition promoter, chemical intermediate, and in the manufacture of rocket fuels. See also NITRATES and ESTERS.

PNS250   CAS: 644-35-9   HR: 3
o-PROPYLPHENOL
mf: $C_9H_{12}O$ mw: 136.21
PROP: A liquid or oil. D: 1.00 @ 15°/15°, bp: 221–226°. Very sltly sol in water; sol in alc, ether.

CONSENSUS REPORTS: Reported in EPA TSCA Inventory.

DOT CLASSIFICATION: 6.1; Label: KEEP AWAY FROM FOOD

SAFETY PROFILE: Poison by ingestion. When heated to decomposition it emits acrid smoke and irritating fumes. See also p-PROPYLPHENOL.

PNV755   CAS: 1701-71-9   HR: 3
PROPYL 4-PYRIDYL KETONE
mf: $C_9H_{11}NO$ mw: 149.21
SYNS: 1-BUTANONE, 1-(4-PYRIDYL)- □ 4-BUTYRYLPYRIDINE □ KETONE, PROPYL 4-PYRIDYL □ PYRIDINE, 4-BUTYRYL- □ 1-(4-PYRIDYL)-1-BUTANONE

DOT CLASSIFICATION: 3; Label: Flammable Liquid

SAFETY PROFILE: A poison by intraperitoneal route. A flammable liquid. When heated to decomposition it emits toxic vapors of NO_x.

PNV760   CAS: 67465-26-3   HR: 3
n-PROPYLSELENINIC ACID
mf: $C_3H_8O_2Se$ mw: 155.07
SYN: SELENINIC ACID, PROPYL-
OSHA PEL: TWA 0.2 mg(Se)/m³
ACGIH TLV: TWA 0.2 mg(Se)/m³
SAFETY PROFILE: Poison by intraperitoneal route. When heated to decomposition it emits toxic fumes of Se.

PNX000   CAS: 51-52-5   HR: 3
6-PROPYL-2-THIOURACIL
mf: $C_7H_{10}N_2OS$ mw: 170.25
PROP: White, bitter, crystalline powder or crystals. Mp: 219–221°. Insol in ether, chloroform, benzene; very sol in aq solns of ammonia; very sltly sol in water.
SYNS: 2,3-DIHYDRO-6-PROPYL-2-THIOXO-4(1H)-PYRIMIDINONE □ 2-MERCAPTO-4-HYDROXY-6-N-PROPYLPYRIMIDINE □ 2-MERCAPTO-6-PROPYL-4-PYRIMIDONE □ 2-MERCAPTO-6-PROPYLPYRIMID-4-ONE □ PROCASIL □ PROPACIL □ PROPILTHIOURACIL □ 6-PROPIL-TIOURACILE (ITALIAN) □ PROPYCIL □ 6-PROPYL-2-THIO-2,4(1H,3H)PYRIMIDINEDIONE □ PROPYL-THIORIST □ PROPYLTHIOURACIL □ 4-PROPYL-2-THIOURACIL □ 6-N-PROPYLTHIOURACIL □ 6-N-PROPYL-2-THIOURACIL □ PROPYL-THYRACIL □ PROPYTHIOURACIL □ PROTHIUCIL □ PROTHIURONE □ PROTHYCIL □ PROTHYRAN □ PROTIURAL □ PTU (thyreostatic) □ T 72 □ 2-THIO-4-OXO-6-PROPYL-1,3-PYRIMIDINE □ 2-THIO-6-PROPYL-1,3-PYRIMIDIN-4-ONE □ 6-THIO-4-PROPYLURACIL □ THIURAGYL □ THYREOSTAT II
CONSENSUS REPORTS: NTP 10th Report on Carcinogens. IARC Cancer Review: Group 2B IMEMDT 7,329,87; Animal Sufficient Evidence IMEMDT 7,67,74. Reported in EPA TSCA Inventory.
SAFETY PROFILE: Confirmed carcinogen with experimental carcinogenic, neoplastigenic, and tumorigenic data. Poison by intraperitoneal route. Moderately toxic by ingestion. Human systemic effects: agranulocytosis, hepatitis, jaundice. Human teratogenic effects by ingestion: developmental abnormalities of the endocrine system and changes in newborn viability. Human and experimental teratogenic and reproductive effects. Human mutation data reported. When heated to decomposition it emits very toxic fumes of SO_x and NO_x. See also MERCAPTANS.

PNX250   CAS: 141-57-1   HR: 3
n-PROPYLTRICHLOROSILANE
DOT: UN 1816
mf: $C_3H_7Cl_3Si$ mw: 177.54
PROP: A liquid. D: 1.19 @ 20°/4°, bp: 123–124°, vap d: 6.15, flash p: 100°F.
SYNS: PROPYLTRICHLOROSILANE (DOT) □ TRICHLOROPROPYLSILANE
CONSENSUS REPORTS: Reported in EPA TSCA Inventory.
DOT CLASSIFICATION: 8; Label: Corrosive
SAFETY PROFILE: A corrosive and irritating material to skin, eyes, and mucous membranes. A dangerous fire hazard when exposed to heat or flame. Will react with water or steam to produce toxic and corrosive fumes; can react with oxidizing materials. To fight fire, use foam, CO_2, dry chemical. When heated to decomposition it emits toxic fumes of Cl^-. See also CHLOROSILANES.

PNX500   CAS: 60062-60-4   HR: 3
4-PROPYL-2,6,7-TRIOXA-1-STIBABI-CYCLO(2.2.2)OCTANE
mf: $C_7H_{13}O_3Sb$ mw: 266.95
SYN: 1,3-PROPANEDIOL, 2-(HYDROXYMETHYL)-2-PROPYL-, CYCLIC ESTER with ANTIMONIC ACID
OSHA PEL: TWA 0.5 mg(Sb)/m³
ACGIH TLV: TWA 0.5 mg(Sb)/m³
NIOSH REL: (Antimony) 10H TWA 0.5 mg(Sb)/m³
SAFETY PROFILE: Poison by intraperitoneal route. When heated to decomposition it emits toxic fumes of Sb.

POC500   CAS: 551-11-1   HR: 3
PROSTAGLANDIN F2-α
mf: $C_{20}H_{34}O_5$ mw: 354.54
PROP: Oil or solid. Mp: 25–35°.

P

SYNS: AMOGLANDIN □ 7-(3,5-DIHYDROXY-2-(3-HYDROXY-1-OCTENYL)CYCLOPENTYL)-5-HEPTENOIC ACID □ DINOPROST □ ENZAPROST □ ENZAPROST F □ PANACELAN □ PGF2-α □ PROSTALMON F □ PROSTARMON F □ PROSTIN F2-α □ (5Z,9,α,11,α,13E,15S)-9,11,15-TRIHYDROXYPROSTA-5,13-DIEN-1-OIC ACID □ 9,11,15-TRIHYDROXYPROSTA-5,13-DIEN-1-OIC ACID □ U-14583

CONSENSUS REPORTS: EPA Genetic Toxicology Program.

SAFETY PROFILE: Poison by subcutaneous, intravenous, and intramuscular routes. Moderately toxic by ingestion. Human and experimental teratogenic and experimental reproductive effects. Human reproductive effects by subcutaneous, intravenous, intramuscular, intraperitoneal, intravaginal, and intraplacental routes: postpartum depression and other maternal effects, abortion, and changes in measures of fertility. Human teratogenic effects by intraplacental route: extra embryonic structures. Human systemic effects by intravenous route: hypermotility, diarrhea, nausea or vomiting. Human mutation data reported. When heated to decomposition it emits acrid smoke and fumes.

**POC750   CAS: 38562-01-5   HR: 3
PROSTAGLANDIN F2-α-THAM**
mf: $C_{20}H_{34}NO_5 \cdot C_4H_{11}NO_3$ mw: 475.70
PROP: Crystals.
SYNS: 7-(3,5-DIHYDROXY-2-(3-HYDROXY-1-OCTEN-YL)CYCLOPENTYL)-5-HEPTENOIC ACID, THAM □ 7-(3,5-DIHYDROXY-2-(3-HYDROXY-1-OCTENYL)CYCLO-PENTYL)-5-HEPTENOIC ACID, TRIMETHAMINE SALT □ DINOPROST TROMETHAMINE (USDA) □ 583E □ LUTALYSE □ PGF2-α THAM □ PGF2-α TRIS SALT □ PGF2-α TROMETHAMINE □ PROSTAGLANDIN F2-α THAM SALT □ PROSTAGLANDIN F2a TROMETHAMINE □ THAM □ TROMETHAMINE PROSTAGLANDIN F2-α □ U-14

SAFETY PROFILE: Poison by intraperitoneal, subcutaneous, intravenous, and intramuscular routes. Moderately toxic by ingestion. Human reproductive effects by intervaginal route: terminates pregnancy, effects on fertility. Experimental teratogenic and reproductive effects. When heated to decomposition it emits toxic fumes of NO_x. See also other prostaglandin entries.

**POD000   CAS: 114-80-7   HR: 3
PROSTIGMINE BROMIDE**
mf: $C_{12}H_{19}N_2O_2 \cdot Br$ mw: 303.24
PROP: Crystals from EtOH/Et₂O.
SYNS: BENZENAMINIUM, 3-(((DIMETHYLAMINO)-CARBONYL)OXY)-N,N,N-TRIMETHYL-, BROMIDE (9CI) □ CARBAMIC ACID, DIMETHYL-, ester with (m-HYDROXYPHENYL)TRIMETHYLAMMONIUM BROMIDE □ EUSTIGMIN BROMIDE □ (m-HYDROXY-PHENYL)TRIMETHYLAMMONIUM BROMIDE DIMETHYLCARBAMATE □ 3-HYDROXYPHENYL-TRIMETHYLAMMONIUM BROMIDE DIMETHYL-CARBAMIC ESTER □ KIRKSTIGMINE BROMIDE □ LEOSTIGMINE BROMIDE □ NEOESERINE BROMIDE □ NEOSERINE BROMIDE □ NEOSTIGMINE BROMIDE □ NEOSTIGMINE METHYL BROMIDE □ PHILOS-TIGMIN BROMIDE □ PROSERINE □ PROSERINE BROMIDE □ PROSTIGMIN BROMIDE □ RCRA WASTE NUMBER U053 □ STIGMANOL BROMIDE □ STIGMOSAN BROMIDE □ SYNSTIGMIN BROMIDE □ SYNTHOSTIGMINE BROMIDE □ SYNTOSTIGMIN (tablet) □ SYNTOSTIGMIN BROMIDE □ SYNTOSTIGMINE BROMIDE □ VAGOSTIGMINE BROMIDE

SAFETY PROFILE: Poison by ingestion, subcutaneous, intravenous, and intraperitoneal routes. When heated to decomposition it emits very toxic fumes of Br^-, NH_3, and NO_x. See also CARBAMATES and BROMIDES.

**POD500   HR: 3
PROTACTINIUM**
af: Pa   aw: 231.036
PROP: A bright, lustrous metal. Mp: 1600°, d: 15.37, vap press: 5×10^{-5} mm @ 1927°. Natural isotope ^{231}Pa (actinium series), $T_{0.5}$ = 3×10^4 years, decays to radioactive ^{227}Ac by alphas of 5.0 MeV. Artificial isotope ^{233}Pa (neptunium series), $T_{0.5}$ = 27 days, decays to radioactive ^{233}U by betas of 0.15 (37%), 0.26 (58%), 0.57 (5%) MeV; emits gammas of 0.02–0.42 MeV. Natural isotope ^{234}Pa (uranium series), $T_{0.5}$ = 6.7 hours, decays to radioactive ^{234}U by betas of 0.23–1.36 MeV, emits gammas of 0.04–0.8 MeV.
SAFETY PROFILE: Confirmed carcinogen. A highly radiotoxic metallic element. An alpha emitter. It is a general hazard if

absorbed systemically. The dust and fumes are hazardous if inhaled. A severe radiation hazard.

POG300   CAS: 700-13-0   HR: D
PSEUDOCUMOHYDROQUINONE
mf: $C_9H_{12}O_2$ mw: 152.21
PROP: Needles from H_2O. Mp: 173°.
SYNS: 1,4-BENZENEDIOL, 2,3,5-TRIMETHYL- (9CI) □ ψ-CUMOHYDROQUINONE □ HYDROQUINONE, TRIMETHYL- □ TRIMETHYLHYDROQUINONE □ 2,3,5-TRIMETHYLHYDROQUINONE
CONSENSUS REPORTS: Reported in EPA TSCA Inventory.
SAFETY PROFILE: Experimental reproductive effects. When heated to decomposition it emits acrid smoke and irritating fumes.

POH750   CAS: 127-91-3   HR: 3
PSEUDOPINENE
mf: $C_{10}H_{16}$ mw: 136.26
PROP: Colorless liquid; pine odor. D: 0.864, refr index: 1.477, flash p: 88°F. Sol in fixed oils; insol in water, propylene glycol, glycerin.
SYNS: 6,6-DIMETHYL-2-METHYLENEBICYCLO(3.1.1)-HEPTANE □ FEMA No. 2903 □ NOPINEN □ NOPINENE □ β-PINENE (FCC) □ 2(10)-PINENE □ PSEUDOPINEN
CONSENSUS REPORTS: Reported in EPA TSCA Inventory.
ACGIH TLV: TWA 20 ppm (sensitizer); Not Classifiable as a Human Carcinogen
SAFETY PROFILE: Mildly toxic by ingestion. A skin irritant. Flammable liquid. When heated to decomposition it emits acrid smoke and irritating fumes.

POJ000   CAS: 126-17-0   HR: 3
PURAPURIDINE
mf: $C_{27}H_{43}NO_2$ mw: 413.71
PROP: A solid. Mp: 200–201°.
SYNS: SOLANCARPIDINE □ SOLANIDINE-S □ SOLASOD-5-EN-3-β-OL □ SOLASODINE
SAFETY PROFILE: Poison by intraperitoneal route. Moderately toxic by ingestion. Experimental teratogenic and reproductive effects. When heated to decomposition it emits toxic fumes of NO_x.

POK000   CAS: 50-44-2   HR: 3
PURINE-6-THIOL
mf: $C_5H_4N_4S$ mw: 152.19
PROP: Yellow prisms. Mp: 313–314° (decomp). Loses H_2O at 1°.
SYNS: 1,7-DIHYDRO-6H-PURINE-6-THIONE □ ISMIPUR □ LEUKERAN □ LEUPURIN □ MERC-ALEUKIN □ MERCAPTOPURIN (GERMAN) □ 6-MERCAPTOPURIN □ 6-MERCAPTOPURINE □ 7-MERCAPTO-1,3,4,6-TETRAZAINDENE □ MERCAPURIN □ MERN □ MP □ NCI-C04886 □ NSC-755 □ PURIMETHOL □ 3H-PURINE-6-THIOL □ PURINETHOL □ 6-PURINETHIOL □ THIOHYPOXANTHINE □ 6-THIOXOPURINE □ U-4748
CONSENSUS REPORTS: IARC Cancer Review: Group 3 IMEMDT 7,240,87; Animal Inadequate Evidence IMEMDT 26,249,81; Human Inadequate Evidence IMEMDT 26,249,81. NCI Carcinogenesis Studies (ipr); Equivocal Evidence: rat CANCAR 40,1935,77; (ipr); Clear Evidence: mouse CANCAR 40,1935,77. EPA Genetic Toxicology Program.
SAFETY PROFILE: Poison by ingestion, intraperitoneal, subcutaneous, parenteral, and intravenous routes. Human systemic effects by ingestion: dermatitis. Experimental teratogenic and reproductive effects. Questionable human carcinogen producing Hodgkin's disease and leukemia. Human mutation data reported. When heated to decomposition it emits very toxic fumes of SO_x and NO_x. See also MERCAPTANS.

POK400   CAS: 25683-07-2   HR: 3
PYOLUTEORIN
mf: $C_{11}H_7Cl_2NO_3$ mw: 272.09
SYNS: METHANONE, (4,5-DICHLORO-1H-PYRROL-2-YL)(2,6-DIHYDROXYPHENYL)- □ KETONE, 4,5-DICHLOROPYRROL-2-YL 2,6-DIHYDROXYPHENYL
DOT CLASSIFICATION: 3; Label: Flammable Liquid
SAFETY PROFILE: A poison by ingestion and intraperitoneal routes. A flammable liquid. When heated to decomposition it emits toxic vapors of NO_x and Cl^-.

POL500   CAS: 98-96-4   HR: 2
PYRAZINECARBOXAMIDE

P

mf: $C_5H_5N_3O$ mw: 123.13

PROP: Crystals from H_2O or EtOH. Mp: 189–191°.

SYNS: ALDINAMID □ 2-CARBAMYL PYRAZINE □ D-50 □ EPRAZIN □ MK 56 □ NCI-C01785 □ PYRAZINAMIDE □ PYRAZINEAMIDE □ PYRAZINE CARBOXYLAMIDE □ PYRAZINOIC ACID AMIDE □ TEBRAZID

CONSENSUS REPORTS: NCI Carcinogenesis Bioassay (feed); No Evidence: rat NCITR* NCI-CG-TR-48,78; (feed); Inadequate Studies: mouse NCITR* NCI-CG-TR-48,78. Reported in EPA TSCA Inventory.

SAFETY PROFILE: Moderately toxic by ingestion, subcutaneous, and intraperitoneal routes. Questionable carcinogen with experimental tumorigenic data. Human mutation data reported. When heated to decomposition it emits toxic fumes of NO_x.

POM500   CAS: 288-13-1   HR: 2
PYRAZOLE

mf: $C_3H_4N_2$ mw: 68.09

PROP: Needles or prisms from pet ether. Mp: 70°, bp: 186–188°. Sol in water, alc, ether, benzene.

SYN: 1,2-DIAZOLE

CONSENSUS REPORTS: Reported in EPA TSCA Inventory.

SAFETY PROFILE: Moderately toxic by ingestion and intraperitoneal routes. Experimental teratogenic and reproductive effects. When heated to decomposition it emits toxic fumes of NO_x.

PON250   CAS: 129-00-0   HR: 3
PYRENE

mf: $C_{16}H_{10}$ mw: 202.26

PROP: Pale-yellow plates by sublimation or colorless solid. Solutions have a slight blue color. Mp: 149–150°, d: 1.271 @ 23°, bp: 404°. Insol in water; sol in Et_2O, CS_2, C_6H_6, and toluene.

SYNS: BENZO(def)PHENANTHRENE □ PYREN (GERMAN) □ β-PYRINE

CONSENSUS REPORTS: IARC Cancer Review: Group 3 IMEMDT 7,56,87; Animal No Evidence IMEMDT 32,431,83. EPA

Extremely Hazardous Substances List. Reported in EPA TSCA Inventory. EPA Genetic Toxicology Program.

OSHA PEL: TWA 0.2 mg/m³

SAFETY PROFILE: Poison by inhalation. Moderately toxic by ingestion and intraperitoneal routes. A skin irritant. Questionable carcinogen with experimental tumorigenic data. Human mutation data reported. When heated to decomposition it emits acrid smoke and irritating fumes.

POO000   CAS: 97-11-0   HR: 2
PYRETHRIN

mf: $C_{21}H_{28}O_3$ mw: 328.49

SYNS: 2-CYCLOPENTENYL-4-HYDROXY-3-METHYL-2-CYCLOPENTEN-1-ONE CHRYSANTHEMATE □ 3-(2-CYCLOPENTEN-1-YL)-2-METHYL-4-OXO-2-CYCLOPENTEN-1-YL CHRYSANTHEMUMATE □ 3-(2-CYCLOPENTENYL)-2-METHYL-4-OXO-2-CYCLOPENTENYL CHRYSANTHEMUMMONO-CARBOXYLATE □ CYCLOPENTENYLRETHONYL CHRYSANTHEMATE □ ENT 22,952

SAFETY PROFILE: Moderately toxic by ingestion. When heated to decomposition it emits acrid smoke and irritating fumes. See also other pyrethrin entries.

POO100   CAS: 121-29-9   HR: 3
PYRETHRIN II

mf: $C_{22}H_{28}O_5$ mw: 372.50

PROP: Viscous liquid or oil. Bp: 192–193° @ 0.007 mm (decomp).

SYNS: CHRYSANTHEMUMDICARBOXYLIC ACID MONOMETHYL ESTER PYRETHROLONE ESTER □ ENT 7,543 □ PYRETHROLONE CHRYSANTHEMUM DICARBOXLIC ACID METHYL ESTER □ PYRETHROLONE ESTER of CHRYSAN-THEMUMDICARBOXYLIC ACID MONOMETHYL ESTER □ (+)-PYRETHRONYL (+)-PYRETHRATE □ PYRETRIN II

CONSENSUS REPORTS: Reported in EPA TSCA Inventory.

SAFETY PROFILE: Poison experimentally by ingestion and intravenous routes. Moderately toxic to humans by unspecified route. An allergen. When heated to decomposition it emits acrid smoke and irritating fumes. An insecticide. See also other pyrethrin entries.

POP000   CAS: 119-12-0   HR: 3
PYRIDAPHENTHION
mf: $C_{14}H_{17}N_2O_4PS$ mw: 340.36
SYNS: AMERICAN CYANAMID 12,503 □ CL 12503 □
O,O-DIETHYL O-(2,3-DIHYDRO-3-OXO-2-PHENYL-6-
PYRIDAZINYL)PHOSPHOROTHIOATE □ O,O-
DIETHYLPHOSPHOROTHIOATE, O-ESTER with 6-
HYDROXY-2-PHENYL-3(2H)-PYRIDAZINONE □ O-(1,6)-
(DIHYDRO-6-OXO-1-PHENYLPYRIDAZIN-3-YL), O,O-
DIETHYL PHOSPHOROTHIOATE □ ENT 23,968 □
OFNACK □ OFUNACK □ PYRIDAFENTHION
SAFETY PROFILE: Poison by
intraperitoneal route. Moderately toxic by
ingestion and skin contact. When heated to
decomposition it emits very toxic fumes of
SO_x, PO_x, and NO_x. Used to control
chewing and sucking insects on rice, fruits,
vegetables, and cereals.

POP250   CAS: 110-86-1   HR: 3
PYRIDINE
DOT: UN 1282
mf: C_5H_5N mw: 79.11
PROP: Colorless liquid; sharp, penetrating,
empyreumatic odor; burning taste. Bp:
115.3°, lel: 1.8%, uel: 12.4%, fp: −42°, flash
p: 68°F (CC), d: 0.982, autoign temp: 900°F,
vap press: 10 mm @ 13.2°, vap d: 2.73.
Volatile with steam; misc with water, alc,
ether. IDLH 1000 ppm.
SYNS: AZABENZENE □ AZINE □ NCI-C55301 □
PIRIDINA (ITALIAN) □ PIRYDYNA (POLISH) □
PYRIDIN (GERMAN) □ RCRA WASTE NUMBER U196
CONSENSUS REPORTS: Community
Right-To-Know List. Reported in EPA
TSCA Inventory. EPA Genetic Toxicology
Program.
OSHA PEL: TWA 5 ppm
ACGIH TLV: TWA 5 ppm; (Proposed: TWA
1 ppm, Confirmed Animal Carcinogen.)
DFG MAK: 5 ppm (16 mg/m³)
DOT CLASSIFICATION: 3; Label:
Flammable Liquid, Poison
SAFETY PROFILE: Poison by
intraperitoneal route. Moderately toxic by
ingestion, skin contact, intravenous, and
subcutaneous routes. Mildly toxic by
inhalation. A skin and severe eye irritant.
Mutation data reported. Can cause central
nervous system depression, gastrointestinal

upset, and liver and kidney damage. A
flammable liquid and dangerous fire hazard
when exposed to heat, flame, or oxidizers.
Severe explosion hazard in the form of
vapor when exposed to flame or spark.
Reacts violently with chlorosulfonic acid,
chromium trioxide, dinitrogen tetraoxide,
HNO_3, oleum, perchromates, β-
propiolactone, $AgClO_4$, H_2SO_4.
Incandescent reaction with fluorine. Reacts
to form pyrophoric or explosive products
with bromine trifluoride, trifluoromethyl
hypofluorite. Mixtures with formamide +
iodine + sulfur trioxide are storage hazards,
releasing carbon dioxide and sulfuric acid.
Incompatible with oxidizing materials.
Reacts with maleic anhydride (above 150°C)
evolving carbon dioxide. To fight fire, use
alcohol foam. When heated to
decomposition it emits highly toxic fumes of
NO_x.

POR500   CAS: 5344-27-4   HR: 3
4-PYRIDINEETHANOL
mf: C_7H_9NO mw: 123.17
SYNS: 4-ETHANOLPYRIDINE □ 4-(β-HYDROXY-
ETHYL)PYRIDINE □ 4-(2-HYDROXYETHYL)PYRIDINE
□ 2-(γ-PYRIDYL)ETHANOL □ 2-(4-PYRIDYL)ETHANOL
CONSENSUS REPORTS: Reported in EPA
TSCA Inventory.
SAFETY PROFILE: Poison by intravenous
route. Moderately toxic by ingestion. When
heated to decomposition it emits toxic
fumes of NO_x.

POS750   CAS: 94-63-3   HR: 3
**PYRIDINIUM-2-ALDOXIME-N-METHYL-
    IODIDE**
mf: $C_7H_9N_2O•I$ mw: 264.08
PROP: Water-sol crystals from EtOH. Mp:
225–226°. Sltly sol in water.
SYNS: 2-FORMYL-1-METHYLPYRIDINIUM IODIDE
OXIME □ 2-FORMYL-N-METHYLPYRIDINIUM OXIME
IODIDE □ 2-HYDROXYIMINOMETHYL-1-
METHYLPYRIDINIUM IODIDE □ 1-METHYL-2-
ALDOXIMINOPYRIDINIUM IODIDE □ 1-METHYL-2-
HYDROXYIMINOMETHYLPYRIDINIUM IODIDE □ N-
METHYLPYRIDINE-2-ALDOXIME IODIDE □ N-
METHYLPYRIDINIUM-2-ALDOXIME IODIDE □ NSC-
7760 □ PAM (CZECH) □ 2-PAM IODIDE □ PRALIDOX-

IME IODIDE □ PRALIDOXIME METHIODIDE □ PROTOPAM IODIDE □ 2-PYRIDINALDOXIM METHOJODID (GERMAN) □ PYRIDIN-2-ALDOXIN (CZECH) □ 2-PYRIDINE ALDOXIME IODOMETHYLATE □ PYRIDINE-2-ALDOXIME METHIODIDE □ PYRIDINE-2-ALDOXIME METHYL IODIDE
ACGIH TLV: Proposed: (inhalable fraction) 0.1 mg/m³; Not Classifiable as a Human Carcinogen)
SAFETY PROFILE: Poison by subcutaneous, intravenous, intramuscular, and intraperitoneal routes. Moderately toxic by ingestion. Used as an antidote to the cholinesterase inhibitors of the parathion group. When heated to decomposition it emits highly toxic fumes of NO_x and I^-.

PPC100   CAS: 15598-34-2   HR: 3
PYRIDINIUM PERCHLORATE
mf: $C_5H_6ClNO_4$ mw: 179.56
PROP: Crystals from EtOH. Mp: 147° (decomp).
DOT CLASSIFICATION: Forbidden
SAFETY PROFILE: An explosive sensitive to impact or heating above 335°C. Addition of ammonium perchlorate decreases the temperature required for initiation. When heated to decomposition it emits toxic fumes of Cl^- and NO_x. See also PERCHLORATES.

PPK250   CAS: 65-23-6   HR: 2
PYRIDOXOL
mf: $C_8H_{11}NO_3$ mw: 169.20
PROP: Needles from Me_2CO. Mp: 160°.
SYNS: ADERMINE □ BEESIX □ GRAVIDOX □ HYDOXIN □ 3-HYDROXY-4,5-DIMETHYLOL-α-PICOLINE □ 5-HYDROXY-6-METHYL-3,4-PYRIDINEDIMETHANOL □ 3-HYDROXY-2-PICOLINE-4,5-DIMETHANOL □ 2-METHYL-4,5-BIS(HYDROXY-METHYL)-3-HYDROXYPYRIDINE □ 2-METHYL-3-HYDROXY-4,5-BIS(HYDROXYMETHYL)PYRIDINE □ 2-METHYL-3-HYDROXY-4,5-DIHYDROXYMETHYL-PYRIDIN (GERMAN) □ 2-METHYL-3-HYDROXY-4,5-DI(HYDROXYMETHYL)PYRIDINE □ PYRODOXIN □ PYRIDOXINE □ VITAMIN B6
CONSENSUS REPORTS: Reported in EPA TSCA Inventory.
SAFETY PROFILE: Moderately toxic by ingestion, subcutaneous, intravenous, and intraperitoneal routes. Human systemic effects: ataxia, local anesthetic, paresthesia. When heated to decomposition it emits toxic fumes of NO_x.

PPK500   CAS: 58-56-0   HR: 3
PYRIDOXOL HYDROCHLORIDE
mf: $C_8H_{11}NO_3 \cdot ClH$ mw: 205.66
PROP: Commercial form of pyridoxine (vitamin B_6). Colorless to white platelets or crystalline powder; odorless. Rhombic crystals from EtOH/Me_2CO. Mp: 204–206° (decomp). Sol in water, alc, acetone; sltly sol in other org solvs; insol in ether.
SYNS: ADERMINE HYDROCHLORIDE □ BECILAN □ BENADON □ CAMPOVITON 6 □ HEBABIONE HYDROCHLORIDE □ HEXABETALIN □ HEXAVIBEX □ HEXERMIN □ HEXOBION □ 3-HYDROXY-4,5-DIMETHYLOL-α-PICOLINE HYDROCHLORIDE □ 5-HYDROXY-6-METHYL-3,4-PYRIDINEDICARBINOL HYDROCHLORIDE □ 5-HYDROXY-6-METHYL-3,4-PYRIDINEDIMETHANOL HYDROCHLORIDE □ 2-METHYL-3-HYDROXY-4,5-BIS(HYDROXYMETHYL)-PYRIDINE HYDROCHLORIDE □ PYRIDIPCA □ PYRIDOXINE HYDROCHLORIDE (FCC) □ PYRIDOXINIUM CHLORIDE □ PYRIDOXINUM HYDROCHLORICUM (HUNGARIAN) □ VITAMIN B6-HYDROCHLORIDE
CONSENSUS REPORTS: Reported in EPA TSCA Inventory.
SAFETY PROFILE: Poison by intravenous route. Moderately toxic by ingestion, intramuscular, and subcutaneous routes. Human reproductive effects by ingestion and intramuscular routes: postpartum changes. Experimental reproductive effects. Human mutation data reported. When heated to decomposition it emits very toxic fumes of NO_x and HCl.

PPL500   CAS: 19992-69-9   HR: 3
1-(PYRIDYL-3)-3,3-DIMETHYL
    TRIAZENE
mf: $C_7H_{10}N_4$ mw: 150.21
SYNS: 1-(PYRIDYL-3)-3,3-DIMETHYL-TRIAZEN (GERMAN) □ 1-(m-PYRIDYL)-3,3-DIMETHYL-TRIAZENE
SAFETY PROFILE: Poison by ingestion and subcutaneous routes. Questionable carcinogen with experimental neoplastigenic data. An experimental teratogen. Mutation data reported. When heated to decomposition it emits toxic fumes of NO_x.

PPO000   CAS: 144-83-2   HR: 2
N-2-PYRIDYLSULFANILAMIDE
mf: $C_{11}H_{11}N_3O_2S$ mw: 249.31
PROP: Colorless prisms in water or yellowish-white powder; darkens on exposure to light. Mp: 191–192°. Sol in water and alc; sltly sol in ether; very sol in alkali and aqueous HCl.
SYNS: ADIPLON □ COCCOCLASE □ DAGENAN □ EUBASIN □ EUBASINUM □ HAPTOCIL □ M+B 695 □ N^1-2-PYRIDYLSULFANILAMIDE □ RELBAPIRIDINA □ RONIN □ SEPTIPULMON □ 2-SULFANILYL AMINOPYRIDINE □ SULFAPYRIDINE □ 2-SULFAPYRIDINE □ SULFIDINE □ THIOSEPTAL □ TRIANON
CONSENSUS REPORTS: Reported in EPA TSCA Inventory.
SAFETY PROFILE: Moderately toxic by intraperitoneal and intravenous routes. Slightly toxic by ingestion. Experimental reproductive effects. Human mutation data reported. Questionable carcinogen with experimental tumorigenic data. When heated to decomposition it emits very toxic fumes of NO_x and SO_x.

PPP500   CAS: 68-35-9   HR: 3
N^1-2-PYRIMIDINYL-SULFANILAMIDE
mf: $C_{10}H_{10}N_4O_2S$ mw: 250.30
PROP: Yellowish-white or pinkish-white powder; darkens on exposure to light. Mp: 255–256°. Spar sol in hot H_2O.
SYNS: ADIAZINE □ 4-AMINO-N-2-PYRIMIDINYLBENZENESULFONAMIDE □ COCO-DIAZINE □ CODIAZINE □ CREMODIAZINE □ DEBENAL □ DELTAZINA □ DIAZOLONE □ DIAZYL □ ESKADIAZINE □ HONEY DIAZINE □ LIPO-DIAZINE □ LIPO-LEVAZINE □ LIQUADIAZINE □ MIRCOSULFON □ NEAZINE □ PECTA-DIAZINE, suspension □ PIRIDISIR □ PYRIMAL □ RP 2616 □ SANODIAZINE □ SILVADENE □ S.N. 112 □ SPOFADRIZINE □ STERAZINE □ SULFADIAZINE □ SULFANILAMIDOPYRIMIDINE □ 2-SULFANILYL-AMINOPYRIMIDINE □ SULFAPYRIMIDIN (GERMAN) □ 2-SULFAPYRIMIDINE □ SULPHADIAZINE □ THERADIAZINE
CONSENSUS REPORTS: Reported in EPA TSCA Inventory.
SAFETY PROFILE: Poison by intravenous route. Moderately toxic by ingestion and intraperitoneal routes. Human systemic

effects by ingestion: hematuria, anuria, general anesthesia, gastrointestinal effects. Experimental teratogenic and reproductive effects. When heated to decomposition it emits very toxic fumes of NO_x and SO_x.

PPP750   CAS: 53558-25-1   HR: 3
PYRIMINYL
mf: $C_{13}H_{12}N_4O_3$ mw: 272.29
SYNS: DLP-87 □ DLP 787 □ N-(4-NITROPHENYL)-N'-(3-PYRIDINYLMETHYL)UREA □ N-3-PYRIDYLMETHYL-N'-p-NITROPHENYLUREA □ 1-(3-PYRIDYLMETHYL)-3-(4-NITROPHENYL)UREA □ PYRIMINIL □ PYRINURON □ RH-787 □ VACOR
CONSENSUS REPORTS: EPA Extremely Hazardous Substances List. Reported in EPA TSCA Inventory.
SAFETY PROFILE: Human poison by ingestion. Human systemic effects: altered sleep time, blood pressure lowering, change in motor activity, diabetes mellitus, distorted perceptions, dyspnea, hallucinations, hyperglycemia, somnolence, structural change in nerve or sheath. When heated to decomposition it emits toxic fumes of NO_x.

PPQ500   CAS: 87-66-1   HR: 3
PYROGALLOL
mf: $C_6H_6O_3$ mw: 126.12
PROP: White, lustrous crystals, plates, or needles. Bp: 309°, d: 1.453 @ 4°/4°, vap press: 10 mm @ 167.7°, mp: 131–133°. Very sol in H_2O, EtOH, Et_2O; sltly sol in $CHCl_3$ and C_6H_6.
SYNS: 1,2,3-BENZENETRIOL □ C.I. 76515 □ C.I. OXIDATION BASE 32 □ FOURAMINE BROWN AP □ FOURRINE PG □ PYROGALLIC ACID □ 1,2,3-TRIHYDROXYBENZEN (CZECH) □ 1,2,3-TRIHYDROXY-BENZENE
CONSENSUS REPORTS: Reported in EPA TSCA Inventory.
SAFETY PROFILE: Human poison by ingestion and subcutaneous routes. An experimental poison by ingestion, subcutaneous, intravenous, and intraperitoneal routes. Experimental teratogenic and reproductive effects. Questionable carcinogen with experimental tumorigenic data. Mutation data reported.

P

Readily absorbed through the skin. Human systemic effects by ingestion: convulsions, dyspnea, gastrointestinal effects. A severe skin and eye irritant. Incompatible with alkalies, NH_3, antipyrine, phenol, iron and lead salts, iodine, $KMnO_4$. When heated to decomposition it emits acrid smoke and irritating fumes. Used as a topical antibacterial agent, as an intermediate, hair dye component, and analytical reagent.

PPR500   CAS: 7791-27-7   HR: 3
PYROSULFURYL CHLORIDE
DOT: UN 1817
mf: $Cl_2O_5S_2$ mw: 215.02
PROP: Colorless, mobile, fuming liquid with pungent odor. Mp: −37°, bp: 151°, d: 1.83, (gas): 9.6 g/L. Insol in cold water.
SYNS: CHLOROSULFONIC ANHYDRIDE □ DISULFUR PENTOXYDICHLORIDE □ DISULFURYL CHLORIDE
DOT CLASSIFICATION: 8; Label: Corrosive
SAFETY PROFILE: A very poisonous material that is also corrosive to the eyes, skin, and mucous membranes. Violent reaction with water. Vigorous reaction with phosphorus. When heated to decomposition it emits very toxic fumes of Cl^- and SO_x. See also CHLOROSULFONIC ACID.

PPS250   CAS: 109-97-7   HR: 3
PYRROLE
mf: C_4H_5N mw: 67.10
PROP: Colorless liquid, darkens on standing; mild nutty odor. Fp: −24°, flash p: 102°F (TCC), d: 0.968 @ 20°/4°, refr index: 1.507, vap d: 2.31, bp: 130–131° @ 761 mm. Sltly sol in water; very sol in alc, fixed oils, benzene, ether; insol in alkali.
SYNS: 1-AZA-2,4-CYCLOPENTADIENE □ AZOLE □ DIVINYLENIMINE □ FEMA No. 3386 □ IMIDOLE □ MONOPYRROLE
CONSENSUS REPORTS: Reported in EPA TSCA Inventory.
SAFETY PROFILE: Poison by ingestion, subcutaneous, and intraperitoneal routes. Flammable liquid when exposed to heat or flame; can react with oxidizing materials. To fight fire, use foam, CO_2, dry chemical. Violent reaction with 2-nitrobenzaldehyde.

When heated to decomposition it emits highly toxic fumes of NO_x.

PPS500   CAS: 123-75-1   HR: 3
PYRROLIDINE
DOT: UN 1922
mf: C_4H_9N mw: 71.14
PROP: Colorless, mobile liquid; penetrating, amine-like odor. Fp: −63°, flash p: 37°F (TCC), d: 0.8618 @ 20°/4°, vap press: 128 mm @ 39°, vap d: 2.45, bp: 88.5–89°. Fumes in air. Misc with water; sol in alc, ether, chloroform.
SYNS: AZACYCLOPENTANE □ TETRAHYDROPYRROLE □ TETRAMETHYLENIMINE
CONSENSUS REPORTS: Reported in EPA TSCA Inventory.
DOT CLASSIFICATION: 3; Label: Flammable Liquid
SAFETY PROFILE: Poison by ingestion and intravenous routes. Moderately toxic by inhalation. Dangerous fire hazard when exposed to heat or flame; can react vigorously with oxidizing materials. To fight fire, use alcohol foam, CO_2, dry chemical. When heated to decomposition it emits highly toxic fumes of NO_x.

PPY300   CAS: 15770-21-5   HR: 3
PYRROL-2-YL KETONE
mf: $C_9H_8N_2O$ mw: 160.19
SYNS: DI-1H-PYRROL-2-YL KETONE □ KETONE, DI-1H-2-PYRROLYL
DOT CLASSIFICATION: 3; Label: Flammable Liquid
SAFETY PROFILE: A poison by intravenous route. A flammable liquid. When heated to decomposition it emits toxic vapors of NO_x.

PQC525   CAS: 38082-89-2   HR: 3
PYX EXPLOSIVE
mf: $C_{17}H_7N_{11}O_{16}$ mw: 621.35
SYNS: 2,6-BIS(PICRYLAMINO)-3,5-DINITROPYRIDINE □ 3,5-DINITRO-N,N'-BIS(2,4,6-TRINITROPHENYL)-2,6-PYRIDINEDIAMINE □ 2,6-PYRIDINEDIAMINE, 3,5-DINITRO-N,N'-BIS(2,4,6-TRINITROPHENYL)- □ PYX
CONSENSUS REPORTS: Reported in EPA TSCA Inventory.

SAFETY PROFILE: Low toxicity by ingestion and skin contact. An eye irritant. Caution: an explosive. When heated to decomposition it emits toxic fumes of NO_x.

P

Q

QCA000   **CAS: 117-39-5**   **HR: 3**
QUERCETIN
mf: $C_{15}H_{10}O_7$ mw: 302.25
PROP: Yellow crystals. Mp: 313–314°
(decomp).
SYNS: C.I. 75670 □ C.I. NATURAL RED 1 □ C.I.
NATURAL YELLOW 10 □ CYANIDELONON 1522 □ 2-
(3,4-DIHYDROXYPHENYL)-3,5,7-TRIHYDROXY-4H-1-
BENZOPYRAN-4-ONE □ MELETIN □ NCI-C60106 □
3,5,7,3',4'-PENTAHYDROXYFLAVONE □ QUERCETINE
□ QUERCETOL □ QUERCITIN □ QUERTINE □
SOPHORETIN □ 3',4',5,7-TETRAHYDROXYFLAVAN-3-
OL □ T-GELB BZW, GRUN 1 □ XANTHAURINE
CONSENSUS REPORTS: IARC Cancer
Review: Group 3 IMEMDT 7,56,87; Animal
Limited Evidence IMEMDT 31,213,83.
NTP Carcinogenesis Studies (feed); Some
Evidence: rat NTPTR* NTP-TR-409,92.
Reported in EPA TSCA Inventory. EPA
Genetic Toxicology Program.
SAFETY PROFILE: Poison by ingestion,
subcutaneous, and intravenous routes.
Experimental teratogenic and reproductive
effects. Questionable carcinogen with
experimental carcinogenic, neoplastigenic,
and tumorigenic data. Human mutation data
reported. Used as a pharmaceutical and
veterinary drug. When heated to
decomposition it emits acrid smoke and
irritating fumes.

QDS000   **CAS: 4213-45-0**   **HR: 2**
QUINACRINE MUSTARD DIHYDRO-
    CHLORIDE
mf: $C_{23}H_{28}Cl_3N_3O \cdot 2ClH$ mw: 541.81
SYNS: 9-(4-BIS(2-CHLOROETHYL)AMINO-1-METHYL-
BUTYLAMINO)-6-CHLORO-2-METHOXYACRIDINE
DIHYDROCHLORIDE □ ICR 10 □ 2-METHOXY-6-
CHLORO-9-(4-BIS(2-CHLOROETHYL)AMINO-1-
METHYLBUTYLAMINO)ACRIDINE DIHYDRO-
CHLORIDE □ 2-METHOXY-6-CHLORO-9-(3-(ETHYL-2-
CHLOROETHYL)AMINOPROPYLAMINO)ACRIDINE
DIHYDROCHLORIDE □ QUINACRINE MUSTARD

CONSENSUS REPORTS: EPA Genetic
Toxicology Program. Reported in EPA
TSCA Inventory.
SAFETY PROFILE: Questionable
carcinogen with experimental neoplastigenic
data. Human mutation data reported.
Corrosive. When heated to decomposition it
emits very toxic fumes of Cl⁻ and NO_x. See
also QUINACRINE MUSTARD.

QFS000   **CAS: 56-54-2**   **HR: 3**
QUINIDINE
mf: $C_{20}H_{24}N_2O_2$ mw: 324.46
PROP: A solid. Mp: 174–175° (anhyd).
SYNS: CHINIDIN (GERMAN) □ CIN-QUIN □
CONCHININ □ CONQUININE □ 6'-METHOXYCIN-
CHONAN-9-OL □ α-(6-METHOXY-4-QUINOLYL)-5-
VINYL-2-QUINUCLIDINEMETHANOL □ 6-METHOXY-
α-(5-VINYL-2-QUINUCLIDINYL)-4-QUINOLINEMETH-
ANOL □ NCI-C56246 □ PITAYINE □ QUINICARDINE
□ QUINIDEX □ (+)-QUINIDINE □ β-QUININE
CONSENSUS REPORTS: Reported in EPA
TSCA Inventory.
SAFETY PROFILE: Poison by ingestion,
subcutaneous, intravenous, intramuscular,
and intraperitoneal routes. A skin irritant.
Implicated in aplastic anemia. When heated
to decomposition it emits toxic fumes of
NO_x.

QHJ000   **CAS: 130-95-0**   **HR: 3**
QUININE
mf: $C_{20}H_{24}N_2O_2$ mw: 324.46
PROP: Bulky, white, amorphous powder or
crystals; bitter taste. Mp: 174.9°. Sol in H_2O,
EtOH, C_6H_6, and $CHCl_3$.
SYNS: CHININ (GERMAN) □ (8-α,9R)-6'-METHOXY-
CINCHONAN-9-OL □ 6-METHOXYCINCHONINE □ α-
(6-METHOXY-4-QUINOYL)-5-VINYL-2-QUINCLIDINE-
METHANOL □ (−)-QUININE
CONSENSUS REPORTS: Reported in EPA
TSCA Inventory.

SAFETY PROFILE: Human poison by unspecified route. Experimental poison by subcutaneous, intravenous, intramuscular, and intraperitoneal routes. Moderately toxic experimentally by ingestion. An experimental teratogen. Human systemic effects by ingestion: visual field changes, tinnitus, and nausea or vomiting. Human teratogenic effects by ingestion: developmental abnormalities of the central nervous system, body wall, and musculoskeletal, cardiovascular, and hepatobiliary systems. Experimental reproductive effects. Mutation data reported. Can cause temporary loss of vision. Quinine dermatitis is an occupational hazard to barbers particularly, and generally to people who work with quinine tonics, medicaments, or cosmetics. An irritant to mucous membranes. Combustible when exposed to heat or flame. Decomposes on exposure to light. When heated to decomposition it emits toxic fumes of NO_x. Used to treat malaria.

QIJ000   CAS: 60-93-5   HR: 3
QUININE DIHYDROCHLORIDE
mf: $C_{20}H_{24}N_2O_2 \cdot 2ClH$ mw: 397.38
PROP: White needles or crystalline powder; odorless with very bitter taste. Mp: 180–185°. Sol in water, alc, glycerin; sltly sol in chloroform; very sltly sol in ether.
SYNS: ACID QUININE HYDROCHLORIDE □ CHIN-INDIHYDROCHLORID (GERMAN) □ 6'-METHOXY-CINCHONAN-9-OL DIHYDROCHLORIDE □ QUININE BIMURIATE □ (−)-QUININE DIHYDROCHLORIDE
CONSENSUS REPORTS: Reported in EPA TSCA Inventory.
SAFETY PROFILE: Poison by intravenous and subcutaneous routes. Moderately toxic by ingestion. Mutation data reported. When heated to decomposition it emits very toxic fumes of NO_x and HCl. See also QUININE.

QJS000   CAS: 130-89-2   HR: 3
QUININE HYDROCHLORIDE
mf: $C_{20}H_{24}N_2O_2 \cdot ClH$ mw: 360.92
PROP: Needles. Mp: 158–160° (anhyd).

SYNS: QUININE CHLORIDE □ QUININE MONOHYDROCHLORIDE □ QUININE MURIATE
CONSENSUS REPORTS: Reported in EPA TSCA Inventory.
SAFETY PROFILE: Poison by ingestion, subcutaneous, intravenous, intramuscular, and intraperitoneal routes. Human systemic effects by intravenous route: convulsions or effect on seizure threshold, muscle contraction or spasticity, and nausea or vomiting. Mutation data reported. Used as a local anesthetic. When heated to decomposition it emits very toxic fumes of NO_x and HCl. See also QUININE.

QMA000   CAS: 804-63-7   HR: 3
QUININE SULFATE
mf: $C_{20}H_{24}N_2O_2 \cdot O_4S$ mw: 420.52
PROP: White needlelike fine crystals; odorless with a very bitter taste. Mp: 205°. Sol in water, alc; sltly sol in chloroform.
SYNS: QUININE BISULFATE □ QUININE HYDROGEN SULFATE
CONSENSUS REPORTS: Reported in EPA TSCA Inventory.
SAFETY PROFILE: Human poison by ingestion. Human systemic effects by ingestion: acuity changes, blood agranulocytosis, fibrous hepatitis, flaccid paralysis without anesthesia, motor activity changes, mydriasis (pupillary dilation), nausea or vomiting, tinnitus, visual field changes. Experimental reproductive effects. Mutation data reported. When heated to decomposition it emits very toxic fumes of SO_x and NO_x. See also QUININE.

QMJ000   CAS: 91-22-5   HR: 3
QUINOLINE
DOT: UN 2656
mf: C_9H_7N mw: 129.17
PROP: Refractive, colorless liquid; peculiar odor. Mp: −14.5°, fp: −15.6°, bp: 237.7°, d: 1.0900 @ 25°/4°, autoign temp: 896°F, vap press: 1 mm @ 59.7°, vap d: 4.45. Sol in water, CS_2; misc in alc, ether.
SYNS: 1-AZANAPHTHALENE □ B-500 □ 1-BENZAZINE □ 1-BENZINE □ BENZO(b)PYRIDINE □

CHINOLEINE □ CHINOLIN □ CHINOLINE □ LEUCOL □ LEUCOLINE □ LEUKOL □ USAF EK-218
CONSENSUS REPORTS: Reported in EPA TSCA Inventory. EPA Genetic Toxicology Program. Community Right-To-Know List.
DOT CLASSIFICATION: 6.1; Label: KEEP AWAY FROM FOOD
SAFETY PROFILE: Poison by ingestion, subcutaneous, and intraperitoneal routes. Moderately toxic by skin contact. A skin and severe eye irritant. Mutation data reported. Questionable carcinogen with experimental neoplastigenic and tumorigenic data. It can cause retinitis similar to that caused by naphthalene but without causing opacity of the lens. Combustible when exposed to heat or flame. Its preparation has caused many industrial explosions. Potentially explosive reaction with hydrogen peroxide. Violent reaction with dinitrogen tetraoxide, perchromates. Incompatible with linseed oil + thionyl chloride, maleic anhydride. Unpredictably violent. When heated to decomposition it emits toxic fumes of NO_x.

QOJ200   CAS: 89-00-9   HR: 3
QUINOLINIC ACID
mf: $C_7H_5NO_4$ mw: 167.13
SYNS: CL 9140 □ 2,3-PYRIDINEDICARBOXYLIC ACID
CONSENSUS REPORTS: Reported in EPA TSCA Inventory.
SAFETY PROFILE: A poison by skin contact. Moderately toxic by ingestion. Experimental reproductive effects. A mild skinn irritant. When heated to decomposition it emits toxic vapors of NO_x.

QPA000   CAS: 148-24-3   HR: 3
8-QUINOLINOL
mf: C_9H_7NO mw: 145.17
PROP: White crystals or powder. Needles from EtOH (aq). Mp: 76°, bp: 267°. Very sltly sol in cold water; sltly sol in ether; sol in alc, dilute alkali.
SYNS: BIOQUIN □ FENNOSAN □ HYDROXY-BENZOPYRIDINE □ 8-HYDROXY-CHINOLIN (GERMAN) □ 8-HYDROXYQUINOLINE □ NCI-C55298 □ 8-OQ □ OXINE □ OXYBENZOPYRIDINE □ OXYCHINOLIN □ o-OXYCHINOLIN (GERMAN) □

OXYQUINOLINE □ 8-OXYQUINOLINE □ PHENOPYRIDINE □ 8-QUINOL □ QUINOPHENOL □ TUMEX □ USAF EK-794
CONSENSUS REPORTS: IARC Cancer Review: Group 3 IMEMDT 7,56,87; Animal Inadequate Evidence IMEMDT 13,101,77. NTP Carcinogenesis Studies (feed); No Evidence: mouse, rat NTPTR* NTP-TR-276,85. Reported in EPA TSCA Inventory. EPA Genetic Toxicology Program.
SAFETY PROFILE: Poison by intraperitoneal and subcutaneous routes. Moderately toxic by ingestion. Questionable carcinogen with experimental carcinogenic, neoplastigenic, and tumorigenic data. Experimental reproductive effects. A central nervous system stimulant. Human mutation data reported. Combustible when exposed to heat or flame. When heated to decomposition it emits highly toxic fumes of NO_x.

QQS200   CAS: 106-51-4   HR: 3
QUINONE
DOT: UN 2587
mf: $C_6H_4O_2$ mw: 108.10
PROP: Yellow crystals; characteristic chlorine-like odor from pet ether or water. Mp: 115.7°, bp: sublimes, d: 1.318 @ 20°/4°. Sol in EtOH and Et_2O; very spar sol in H_2O. IDLH 100 mg/m³.
SYNS: BENZO-CHINON (GERMAN) □ 1,4-BENZO-QUINE □ BENZOQUINONE (DOT) □ p-BENZO-QUINONE □ 1,4-BENZOQUINONE □ CHINON (DUTCH, GERMAN) □ p-CHINON (GERMAN) □ CHINONE □ CYCLOHEXADIENEDIONE □ 1,4-CYC-LOHEXADIENEDIONE □ 2,5-CYCLOHEXADIENE-1,4-DIONE □ 1,4-CYCLOHEXADIENE DIOXIDE □ 1,4-DIOSSIBENZENE (ITALIAN) □ 1,4-DIOXYBENZENE □ 1,4-DIOXY-BENZOL (GERMAN) □ NCI-C55845 □ p-QUINONE □ RCRA WASTE NUMBER U197 □ USAF P-220
CONSENSUS REPORTS: IARC Cancer Review: Group 3 IMEMDT 7,56,87; Animal Inadequate Evidence IMEMDT 15,255,77. Reported in EPA TSCA Inventory. Community Right-To-Know List. EPA Genetic Toxicology Program.
OSHA PEL: TWA 0.1 ppm
ACGIH TLV: TWA 0.1 ppm

DFG MAK: 0.1 ppm (0.45 mg/m³)
DOT CLASSIFICATION: 6.1; Label: Poison
SAFETY PROFILE: Poison by ingestion, subcutaneous, intraperitoneal, and intravenous routes. Questionable carcinogen with experimental tumorigenic data by skin contact. Human mutation data reported. Quinone has a characteristic, irritating odor. Causes severe damage to the skin and mucous membranes by contact with it in the solid state, in solution, or in the form of condensed vapors. Locally, it causes discoloration, severe irritation, erythema, swelling, and the formation of papules and vesicles, whereas prolonged contact may lead to necrosis. When the eyes become involved, it causes dangerous disturbances of vision. The moist material self-heats and decomposes exothermically above 60°C. When heated to decomposition it emits acrid smoke and fumes.

QSJ800   CAS: 73927-90-9   HR: 3
(2,3-QUINOXALINYLDITHIO)DIMETHYL-
   ### TIN
mf: $C_{10}H_{10}N_2S_2Sn$ mw: 341.03

SYN: STANNANE, DIMETHYL(2,3-QUINOX-ALINYLDITHIO)-
OSHA PEL: TWA 0.1 mg(Sn)/m³ (skin)
ACGIH TLV: TWA 0.1 mg(Sn)/m³; STEL 0.2 mg/m³ (skin)
NIOSH REL: (Organotin Compound) 10H TWA 0.1 mg(Sn)/m³
SAFETY PROFILE: Poison by intravenous route. When heated to decomposition it emits toxic fumes of NO_x, SO_x, and Sn.

QTS000   CAS: 59-40-5   HR: 2
N-(2-QUINOXALINYL)SULFANILAMIDE
mf: $C_{14}H_{12}N_4O_2S$ mw: 300.36
PROP: Sol in alkalies. Mp: 247–248°.
SYNS: 2-p-AMINOBENZENESULFONAMIDO-QUINOXALINE □ 2-p-AMINOBENZENESULPHO-NAMIDOQUINOXALINE □ N^1-2-QUINOXALINYL-SULFANILAMIDE □ N'-2-QUINOXALYLSULFANIL-AMIDE □ SULFABENZPYRAZINE □ 2-SULFANIL-AMIDOQUINOXALINE □ SULFAQUINOXALINE
CONSENSUS REPORTS: Reported in EPA TSCA Inventory.
SAFETY PROFILE: Moderately toxic by ingestion. When heated to decomposition it emits very toxic fumes of NO_x and SO_x.

R

RAV000 **HR: 3**
RADIUM
af: Ra   aw: 226.025
PROP: A radioactive alkaline earth metal. Brilliant white, tarnishes in air. Decomp in water. Mp: 700°, bp: 1737°, d: 5.5.
SAFETY PROFILE: A highly radiotoxic element. 1 g produces 3.7×10^{10} disintegrations per second. Inhalation, ingestion, or bodily exposure can lead to lung cancer, bone cancer, osteitis, skin damage, and blood dyscrasias. A common air contaminant. Radium replaces calcium in the bone structure and is a source of irradiation to the blood-forming organs. The ingestion of luminous dial paint prepared from radium caused death in many of the early dial painters before the hazard was fully understood. The data on these workers have been the source of many of the radiation precautions and the maximum permissible levels for internal emitters that are now accepted. ^{226}Ra is the parent of radon and the precautions described under ^{222}Rn should be followed. ^{228}Ra is a member of the thorium series. It was a common constituent of luminous paints, and, while its low beta energy was not a hazard, its daughters in the series may have been a causative agent in the deaths of the radium dial painters following World War I. It is metabolized in the same way as any other radium isotope and it is a source of thorium. The precautions recommended under ^{220}Rn should be followed. Highly dangerous; must be kept heavily shielded and stored away from possible dissemination by explosion, flood, etc.

Radiation Hazard: Natural isotope ^{223}Ra (Actinium-X, Actinium Series), T1/2 = 11.4 days, decays to radioactive ^{219}Rn by alphas of 5.5–5.7 MeV. Natural isotope ^{224}Ra (Thorium-X, Thorium Series), T1/2 = 3.6 days, decays to radioactive ^{220}Rn by alphas of 5.7 MeV. Natural isotope ^{226}Ra (Uranium Series), T1/2 = 1600 years, decays to radioactive ^{222}Rn by alphas of 4.8 MeV. Natural isotope ^{228}Ra (Mesothorium = 1, Thorium Series), T1/2 = 6.7 years, decays to radioactive ^{228}Ac by betas of 0.05 MeV.

RBA000 **CAS: 10043-92-2** **HR: 3**
RADON
af: Rn   aw: 222
PROP: Colorless, odorless, inert gas; very dense. Bp: −62°, d (gas @ 1 atm and 0°): 9.73 g/L, (liquid @ bp): 4.4.
CONSENSUS REPORTS: NTP 10th Report on Carcinogens.
SAFETY PROFILE: A carcinogen. A common air contaminant. Radon is a noble gas and thus is relatively unreactive. Radiation Hazard: Natural isotope ^{220}Rn (Thoron, Thorium Series), T1/2 = 55 seconds, decays to radioactive ^{216}Po by alphas of 6.3 MeV. Natural isotope ^{222}Rn (Uranium Series), T1/2 = 3.8 days, decays to radioactive ^{218}Po by alphas of 5.5 MeV. The permissible levels are given for ^{222}Rn in equilibrium with its daughters. The chief hazard from this isotope is inhalation of the gaseous element and its solid daughters, which are collected on the normal dust of the air. This material is deposited in the lungs and has been considered to be a major causative agent in the high incidence of lung cancer found in uranium miners. Radon and its daughters build up to an equilibrium value in about a month from radium compounds, while the build-up from uranium compounds is negligible. Good ventilation of areas where

radium is handled or stored is recommended to prevent accumulation of hazardous concentration of Rn and its daughters. Accumulation of radon in homes has been implicated in increased incidence of lung cancers. This accumulation is found in well-insulated buildings located over land that has concentrations of uranium.

RBF100   CAS: 26538-44-3   HR: D
RALGRO
mf: $C_{18}H_{26}O_5$ mw: 322.44

PROP: Crystals from 2-propanol (aq). Mp: 182–184°.

SYNS: 6-(6,10-DIHYDROXYUNDECYL)-β-RESORCYLIC ACID-μ-LACTONE □ FRIDERON □ MK-188 □ P1496 □ RALABOL □ RALONE □ ZEARALANOL □ ZEARANOL □ ZERANOL (USDA)

CONSENSUS REPORTS: Reported in EPA TSCA Inventory.

SAFETY PROFILE: An experimental teratogen. Experimental reproductive effects. Mutation data reported. When heated to decomposition it emits acrid smoke and irritating fumes.

RBK200   HR: D
RAPESEED OIL
PROP: Pale yellow liquid. Sol in chloroform and ether.

SYNS: COLZA OIL □ FULLY HYDROGENATED RAPESEED OIL □ LOW ERUCIC ACID RAPESEED OIL □ RAPE SEED OIL □ SUPERGLYCERINATED FULLY HYDROGENATED RAPESEED OIL

SAFETY PROFILE: When heated to decomposition it emits acrid smoke and irritating fumes.

RBU000   CAS: 5471-51-2   HR: 3
RASPBERRY KETONE
mf: $C_{10}H_{12}O_2$ mw: 164.22

PROP: Needles from water or white solid; raspberry odor. Mp: 81–86°.

SYNS: 2-BUTANONE, 4-(4-HYDROXYPHENYL)- □ FEMA No. 2588 □ FRAMBINONE □ p-HYDROXY-BENZYL ACETONE □ 4-(4-HYDROXYPHENYL)-2-BUTANONE □ 1-(p-HYDROXYPHENYL)-3-BUTANONE □ 4-(p-HYDROXYPHENYL)-2-BUTANONE (FCC) □ OXYPHENALON □ RHEOSMIN

CONSENSUS REPORTS: Reported in EPA TSCA Inventory.

DOT CLASSIFICATION: 3; Label: Flammable Liquid

SAFETY PROFILE: Poison by intraperitoneal route. Moderately toxic by ingestion. Flammable liquid. When heated to decomposition it emits acrid smoke and irritating fumes. See also KETONES.

RCA375   CAS: 21416-67-1   HR: 3
RAZOXANE
mf: $C_{11}H_{16}N_4O_4$ mw: 268.31

SYNS: ICI 59118 □ ICRF 159 □ 4,4'-PROPYLENEDI-2,6-PIPERAZINEDIONE □ RAZOXIN

SAFETY PROFILE: Suspected human carcinogen producing leukemia and skin tumors. Moderately toxic by intraperitoneal route. Human effects: normocytic anemia and thrombocytopenia. Human mutation data reported. When heated to decomposition it emits toxic fumes of NO_x.

RCF000   CAS: 39283-42-6   HR: 3
RED SQUILL
PROP: Sea onion bulbs contain a potent concentration of scilliroside, a glycoside bearing a close chemical resemblance to the Scillarens.

SYNS: BONIDE TOPZOL RAT BAITS and KILLING SYRUP □ RAT-O-CIDE RAT BAIT □ RAT'S END □ RODINE □ ROUGH & READY RAT BAIT & RAT PASTE □ SCILLIROSIDE GLYCOSIDE □ SILMURIN □ SQUILL □ TOPZOL □ URGENEA MARITIMA

SAFETY PROFILE: Poison by ingestion and intraperitoneal routes. Human systemic effects by ingestion: nausea or vomiting, decreased pulse rate, and fall in blood pressure. When heated to decomposition it emits acrid smoke and irritating fumes.

RCK725   HR: 3
REFRACTORY CERAMIC FIBERS
PROP: A mixture of ALUMINA and SILICA (1:1).

SYN: FIBERS, REFRACTORY CERAMIC

CONSENSUS REPORTS: NTP 10th Report on Carcinogens.

SAFETY PROFILE: Confirmed carcinogen with experimental tumorigenic data.

RDK000   CAS: 50-55-5   HR: 3
RESERPINE

mf: $C_{33}H_{40}N_2O_9$ mw: 608.75

PROP: White or pale-buff to sltly yellow powder; odorless. Long prisms from Me_2CO (aq). Mp: 262–266° (decomp). Insol in water; very sltly sol in alc; sol in chloroform and acetic acid.

SYNS: ABESTA □ ABICOL □ ADELFAN □ ADELPHANE □ ADELPHIN □ ADELPHIN-ESIDREX-K □ ALKARAU □ ALKASERP □ ALSERIN □ ANQUIL □ APOPLON □ APSICAL □ ARCUM R-S □ ASCOSERP □ ASCOSERPINA □ AUSTRAPINE □ BANASIL □ BANISIL □ BENAZYL □ BENDIGON □ BIOSERPINE □ BRINDERDIN □ BRISERINE □ BROSERPINE □ BUTISERPAZIDE-25 □ BUTISERPAZIDE-50 □ BUTISERPINE □ CARDIOSERPIN □ CARDITIVO □ CARRSERP □ CRYSTOSERPINE □ DAREBON □ DESERPINE □ DIUPRES □ DIUTENSEN-R □ DRENUSIL-R □ DYPERTANE COMPOUND □ EBERPINE □ EBERSPINE □ EBSERPINE □ ELFANEX □ ELSERPINE □ ENIPRESSER □ ENT 50,146 □ ESCASPERE □ ESERPINE □ ESKASERP □ GAMASERPIN □ GAMMASERPINE □ GILUCARD □ H 520 □ HELFOSERPIN □ HEXAPLIN □ HIPOSERPIL □ HISERPIA □ HYDROMOX R □ HYDROPRES □ HYDROPRES KA □ HYGROTON-RESERPINE □ HYPERCAL B □ HYPERTANE FORTE □ HYPERTENSAN □ IDOSERP □ IDSOSERP □ INTERPINA □ KEY-SERPINE □ KITINE □ KLIMANOSID □ "L," CARPSERP □ LEMISERP □ LOWESERP □ MARNITENSION SIMPLE □ MAVISERPIN □ MAYSERPINE □ MEPHASERPIN □ METATENSIN □ METHYL RESERPATE 3,4,5-TRIMETHOXYBENZOIC ACID □ METHYL RESERPATE 3,4,5-TRIMETHOXYBENZOIC ACID ESTER □ MIO-PRESSIN □ MODENOL □ NAQUIVAL □ NCI-C50157 □ NEMBU-SERPIN □ NEO-ANTITENSOL □ NEO-ANTITERSOL □ NEOSERFIN □ NEO-SERP □ NEOSLOWTEN □ ORTHOSERPINA □ PERSKLERAN □ PRESSIMEDIN □ PURSERPINE □ QUIESCIN □ RAUCAP □ RAUDIFORD □ RAUDIXIN □ RAUDIXOID □ RAUGAL □ RAULEN □ RAULOYCIN □ RAULOYDIN □ RAUMORINE □ RAUNERVIL □ RAUNORINE □ RAUNORMIN "ORZAN" □ RAUNOVA □ RAUPASIL □ RAUPOID □ RAURINE □ RAUSAN □ RAU-SED □ RAUSEDAN □ RAUSEDIL □ RAUSEDYL □ RAUSERPEN-ALK □ RAUSERPIN □ RAUSERPIN-ALK □ RAUSERPINE □ RAUSERPOL □ RAUSINGLE □ RAUTRIN □ RAUVILID □ RAUVLID □ RAUWASEDIN □ RAUWILID □ RAUWILOID □ RAUWILOID+ □ RAUWIPUR □ RAUWOLEAF □ RAUWOPUR "BYK" □ RAWILID □ RCRA WASTE NUMBER U200 □ RECIPIN □ REGROTON □ RENESE R □ R-E-S □ RESALTEX □ RESEDIN □ RESEDREX □ RESEDRIL □ RESE-LAR □ RESER-AR □ RESERBAL □ RESERCAPS □ RESERCEN □ RESERCRINE □ RESERFIA □ RESERJEN □ RESERLOR □ RESERP □ RESERPAL □ RESERPAMED □ RESERPANCA □ RESERPENE □ RESERPEX □ RESERPIDEFE □ RESERPIL □ RESERPIN □ RESERPINA □ RESERPINUM □ RESERPKA □ RESERPOID □ RESERPUR □ RESERP "WANDER" □ RESERSANA □ RESERUTIN □ RESIATRIC □ RESIDINE □ RESINE □ RESOCALM □ RESOMINE □ RESPERIN □ RESPERINE □ RESPITAL □ RESTRAN □ REZERPIN □ RISERPA □ RIVASED □ RIVASIN □ ROLSERP □ ROXEL □ ROXINOID □ ROXYNOID □ RYSER □ SALUPRES □ SALUTENSIN □ SANDRIL □ SANDRON □ SARPAGAN □ SARPAGEN □ SEDARAUPIN □ SEDARAUPINA □ SEDA-RECIPIN □ SEDA-SALUREPIN □ SEDERAUPIN □ SEDSERP □ SEOMINAL □ SERFIN □ SERFOLIA □ SEROLFIA □ SERP □ SERP-AFD □ SERPALAN □ SERPALOID □ SERPANEURONA □ SERPANRAY □ SERPASIL □ SERPASIL APRESOLINE □ SERPASIL-ESIDREX □ SERPASIL-ESIDREX NO. 1 □ SERPASIL-ESIDREX NO. 2 □ SERPASIL-ESIDREX K □ SERPASIL PREMIX □ SERPASOL □ SERPATE □ SERPATONE □ SERPAX □ SERPAZIL □ SERPAZOL □ SERPEDIN □ SERPEN □ SERPENA □ SERPENTIL □ SERPENTIN □ SERPENTINA □ SERPENTINE "PHARBIL" □ SERPICON □ SERPIL □ SERPILOID □ SERPILUM □ SERPINE □ SERPINE (Pharmaceutical) □ SERPIPUR □ SERPIVITE □ SERPLEX K □ SERPOGEN □ SERPOID □ SERPONE □ SERPRESAN □ SERPYRIT □ SERTABS □ SERTENS □ SERTENSIN □ SERTINA □ SINESALIN COMPOSITION □ SK-RESERPINE □ SOLFO SERPINE □ SUPERGAN □ TEFASERPINA □ TEMPO-RESERPINA □ TEMPOSERPINE □ TENDOSCEN-COMPR. □ TENSANYL □ TENSERLIX □ TENSERPINE "ASSIA" □ TENSERPINIE □ TENSIONAL □ TENSIONORME □ TEPSERPINE □ TERBOLAN □ TRANSERPIN □ 3,4,5-TRIMETHOXYBENZOYL METHYL RESERPATE □ TRISERPIN □ T-SERP □ TYLANDRIL □ UNILORD □ UNITENSEN □ USAF CB-27 □ VERILOID □ VIO-SERPINE □ V-SERP □ YOHIMBAN-16-CARBOXYLIC ACID derivative of BENZ(g)INDOLO(2,3-a)QUINOLIZINE

CONSENSUS REPORTS: NTP 10th Report on Carcinogens. IARC Cancer Review: Group 3 IMEMDT 7,330,87; Animal Inadequate Evidence IMEMDT 10,217,76; Human Limited Evidence IMEMDT 24,211,80; Animal Limited Evidence IMEMDT 24,211,80. NCI Carcinogenesis

Bioassay (feed); Clear Evidence: mouse, rat NCITR* NCI-CG-TR-193,80. Reported in EPA TSCA Inventory.
SAFETY PROFILE: Confirmed human carcinogen producing tumors of the skin and brain. Poison by ingestion, intravenous, subcutaneous, and intraperitoneal routes. Mutation data reported. An experimental teratogen. Human and experimental reproductive effects by ingestion: stillbirth, reduced viability, and other neonatal measures or effects. In humans, 0.014 mg/kg causes psychotropic effects. A medicine with side effects. Used as an additive permitted in the feed and drinking water of animals and/or for the treatment of food-producing animals. Also permitted in food for human consumption. A sedative. When heated to decomposition it emits toxic fumes of NO_x.

RDK100   CAS: 64741-80-6   HR: 3
RESIDUES (PETROLEUM), THERMAL CRACKED
SYN: THERMALLY CRACKED RESIDUE
CONSENSUS REPORTS: IARC Cancer Review: Group 2A IMEMDT 45,39,89; Animal Sufficient Evidence IMEMDT 45,39,89. Reported in EPA TSCA Inventory.
SAFETY PROFILE: Confirmed carcinogen. When heated to decomposition it emits acrid smoke and irritating vapors.

RDP000   HR: 3
RESIN (solution)
DOT: UN 1866
SYNS: RESIN SOLUTION, flammable (DOT) □ SOLUBOND 0-869 □ SOLUBOND 3520
DOT CLASSIFICATION: 3; Label: Flammable Liquid
SAFETY PROFILE: Flammable when exposed to heat or flame, can react vigorously with oxidizing materials. When heated to decomposition it emits acrid smoke and irritating fumes.

RDZ900   CAS: 102-29-4   HR: 3
RESORCIN MONOACETATE
mf: $C_8H_8O_3$ mw: 152.16
PROP: Oil. Bp: 283°, d: 1.226, flash p: >230°F.
SYNS: 3-ACETOXYPHENOL □ ACETYLRESORCINOL □ 1,3-BENZENEDIOL, MONOACETATE □ EURESOL □ m-HYDROXYPHENYL ACETATE □ REMONOL □ RESORCIN ACETATE □ RESORCINOL, MONO-ACETATE □ RESORCITATE
CONSENSUS REPORTS: Reported in EPA TSCA Inventory.
SAFETY PROFILE: Poison by intraperitoneal route. A severe eye irritant. Combustible liquid. When heated to decomposition it emits acrid smoke and irritating fumes.

REA000   CAS: 108-46-3   HR: 3
RESORCINOL
DOT: UN 2876
mf: $C_6H_6O_2$ mw: 110.12
PROP: Platelets from EtOH. Very white crystals, become pink on exposure to light when not perfectly pure; unpleasant sweet taste. Mp: 110°, bp: 280.5°, flash p: 261°F (CC), d: 1.285 @ 15°, autoign temp: 1126°F, vap press: 1 mm @ 108.4°, vap d: 3.79. Very sol in alc, ether, glycerin; sltly sol in chloroform; sol in water.
SYNS: m-BENZENEDIOL □ 1,3-BENZENEDIOL □ C.I. 76505 □ C.I. DEVELOPER 4 □ C.I. OXIDATION BASE 31 □ DEVELOPER R □ m-DIHYDROXYBEN-ZENE □ 1,3-DIHYDROXYBENZENE □ m-DIOXY-BENZENE □ DURAFUR DEVELOPER G □ FOUR-AMINE RS □ FOURRINE 79 □ m-HYDROQUINONE □ 3-HYDROXYCYCLOHEXADIEN-1-ONE □ m-HYDRO-XYPHENOL □ 3-HYDROXYPHENOL □ NAKO TGG □ NCI-C05970 □ PELAGOL GREY RS □ RCRA WASTE NUMBER U201 □ RESORCIN □ RESORCINE
CONSENSUS REPORTS: IARC Cancer Review: Group 3 IMEMDT 7,56,87; Animal Inadequate Evidence IMEMDT 15,155,77. Reported in EPA TSCA Inventory. EPA Genetic Toxicology Program.
OSHA PEL: TWA 10 ppm; STEL 20 ppm
ACGIH TLV: TWA 10 ppm; STEL 20 ppm; Not Classifiable as a Human Carcinogen
DOT CLASSIFICATION: 6.1; Label: KEEP AWAY FROM FOOD

SAFETY PROFILE: Human poison by ingestion. Experimental poison by ingestion, intraperitoneal, parenteral, and subcutaneous routes. Moderately toxic experimentally by skin contact and intravenous routes. Questionable carcinogen with experimental tumorigenic data. Human mutation data reported. A skin and severe eye irritant. It can cause systemic poisoning by acting as both a blood and nerve poison. In a suitable solvent, this material can readily be absorbed through human skin and can cause local hyperemia, itching, dermatitis, edema, and corrosion associated with enlargement of regional lymph glands as well as serious systemic disorders such as restlessness, methemoglobinemia, cyanosis, convulsions, tachycardia, dyspnea, and death. These same symptoms can be induced by ingestion of the material. For poisoning, treat symptomatically. Get medical advice. Used as a topical antiseptic and keratolytic agent.

Combustible when exposed to heat or flame; can react with oxidizing materials. To fight fire, use water, CO_2, dry chemical. Potentially explosive reaction with concentrated nitric acid. Incompatible with acetanilide, alkalies, ferric salts, spirit nitrous ether, urethan. When heated to decomposition it emits acrid smoke and irritating fumes.

REA100   CAS: 108-58-7   HR: 2
RESORCINOL, DIACETATE
mf: $C_{10}H_{10}O_4$ mw: 194.20
PROP: A liquid. Bp: 278, d: 1.178.
SYNS: 1,3-BENZENEDIOL, DIACETATE □ 1,3-DIACETOXYBENZENE □ 1,3-DIHYDROXYBENZENE DIACETATE □ m-PHENYLENEDIACETATE
CONSENSUS REPORTS: Reported in EPA TSCA Inventory.
SAFETY PROFILE: Moderately toxic by intraperitoneal route. A severe eye irritant. When heated to decomposition it emits acrid smoke and irritating fumes.

REF000   CAS: 101-90-6   HR: 3
RESORCINOL DIGLYCIDYL ETHER
mf: $C_{12}H_{14}O_4$ mw: 222.26
SYNS: ARALDITE ERE 1359 □ m-BIS(2,3-EPOXYPROPOXY)BENZENE □ 1,3-BIS(2,3-EPOXYPROPOXY)BENZENE □ m-BIS(GLYCIDYL-OXY)BENZENE □ 1,3-DIGLYCIDYLOXYBENZENE □ DIGLYCIDYL RESORCINOL ETHER □ ERE 1359 □ NCI-C54966 □ 2,2'-(1,3-PHENYLENEBIS(OXYMETH-YLENE))BISOXIRANE □ RDGE □ RESORCINOL BIS(2,3-EPOXYPROPYL)ETHER □ RESORCINYL DIGLYCIDYL ETHER
CONSENSUS REPORTS: NTP 10th Report on Carcinogens. IARC Cancer Review: Group 2B IMEMDT 7,56,87; Animal Sufficient Evidence IMEMDT 36,181,85; Animal Inadequate Evidence IMEMDT 11,125,76. NTP Carcinogenesis Studies (gavage); Clear Evidence: mouse, rat NTPTR* NTP-TR-257,86. Reported in EPA TSCA Inventory.
DFG MAK: Animal Carcinogen, Suspected Human Carcinogen
SAFETY PROFILE: Confirmed carcinogen with experimental carcinogenic and tumorigenic data. Poison by intraperitoneal route. Moderately toxic by ingestion. Mutation data reported. A skin irritant. When heated to decomposition it emits acrid smoke and irritating fumes. See also ETHERS.

REF050   CAS: 150-19-6   HR: 3
RESORCINOL MONOMETHYL ETHER
mf: $C_7H_8O_2$ mw: 124.15
PROP: Oil. Bp: 244°.
SYNS: m-GUAIACOL □ m-HYDROXYANISOLE □ 3-HYDROXYANISOLE □ m-METHOXYPHENOL □ 3-METHOXYPHENOL □ RESORCINOL METHYL ETHER
CONSENSUS REPORTS: Reported in EPA TSCA Inventory.
SAFETY PROFILE: Poison by ingestion and intraperitoneal routes. Moderately toxic by skin contact. Human systemic effects by inhalation: muscle weakness, headache, and irritability. Human female reproductive effects by inhalation: menstrual cycle changes and disorders. A severe eye irritant. When heated to decomposition it emits acrid smoke and irritating fumes. See also ETHERS.

R

RFP000   CAS: 480-54-6   **HR: 3**
RETRORSINE
mf: $C_{18}H_{25}NO_6$ mw: 351.44
PROP: Crystals from ethyl acetate. Mp: 217°. Readily sol in alc and chloroform; sltly sol in water, acetone, and ethyl acetate; practically insol in ether.
SYNS: 12,18-DIHYDROXY-SENECIONAN-11,16-DIONE □ β-LONGILOBINE □ cis-RETRONECIC ACID ESTER of RETRONECINE
CONSENSUS REPORTS: IARC Cancer Review: Group 3 IMEMDT 7,56,87; Animal Limited Evidence IMEMDT 10,303,76.
SAFETY PROFILE: Poison by ingestion, intraperitoneal, and intravenous routes. Questionable carcinogen with experimental neoplastigenic and tumorigenic data. Mutation data reported. When heated to decomposition it emits toxic fumes of NO_x.

RFU000   CAS: 15503-86-3   **HR: 3**
RETRORSINE-N-OXIDE
mf: $C_{18}H_{25}NO_7$ mw: 367.44
PROP: Crystals from ethanol. Mp: 145°.
SYNS: ISATIDINE □ cis-RETRONECIC ACID ESTER of RETRONECINE-N-OXIDE
CONSENSUS REPORTS: IARC Cancer Review: Group 3 IMEMDT 7,56,87; Animal Sufficient Evidence IMEMDT 10,269,76.
SAFETY PROFILE: Poison by ingestion and intraperitoneal routes. Moderately toxic by intravenous route. Questionable carcinogen with experimental neoplastigenic and tumorigenic data. Mutation data reported. When heated to decomposition it emits toxic fumes of NO_x. See also ESTERS.

RGP000   CAS: 13569-63-6   **HR: 3**
RHENIUM TRICHLORIDE
mf: Cl_3Re mw: 292.55
PROP: Sol in Me_2CO, MeOH, EtOH, HCl (aq).
CONSENSUS REPORTS: Reported in EPA TSCA Inventory.
SAFETY PROFILE: Poison by intraperitoneal route. When heated to decomposition it emits toxic fumes of Cl^-. See also RHENIUM and CHLORIDES.

RGW000   CAS: 989-38-8   **HR: 3**
RHODAMINE 6G EXTRA BASE
mf: $C_{28}H_{30}N_2O_3 \cdot ClH$ mw: 479.06
PROP: Bluish-pink crystals or powder. Sol in H_2O and EtOH.
SYNS: AIZEN RHODAMINE 6GCP □ BASIC RED 1 □ BASIC RHODAMINE YELLOW □ BASIC RHODAMINIC YELLOW □ CALCOZINE RED 6G □ CALCOZINE RHODAMINE 6GX □ CERVEN ZASADITA 1 □ C.I. 45160 □ C.I. BASIC RED 1 □ C.I. BASIC RED 1, MONOHYDROCHLORIDE □ ELCOZINE RHODAMINE 6GDN □ ELJON PINK TONER □ FANAL PINK B □ FANAL PINK GFK □ FANAL RED 25532 □ FLEXO RED 482 □ HELIOSTABLE BRILLIANT PINK B EXTRA □ MITSUI RHODAMINE □ MITSUI RHODAMINE 6GCP □ NCI-C56122 □ NYCO Liquid RED GF □ Rh 6G □ RHODAMINE 590 CHLORIDE □ RHODAMINE 69DN EXTRA □ RHODAMINE F4G □ RHODAMINE F5G □ RHODAMINE F5G CHLORIDE □ RHODAMINE F 5GL □ RHODAMINE 6GB □ RHODAMINE 6G (BIOLOGICAL STAIN) □ RHODAMINE 6GBN □ RHODAMINE 6G CHLORIDE □ RHODAMINE 6GCP □ RHODAMINE 4GD □ RHODAMINE 6GD □ RHODAMINE GDN □ RHODAMINE 5GDN □ RHODAMINE 6 GDN □ RHODAMINE 6 GDN EXTRA □ RHODAMINE 6GEX ETHYL ESTER □ RHODAMINE 6G EXTRA □ RHODAMINE 6G EXTRA BASE □ RHODAMINE 4GH □ RHODAMINE 6GH □ RHODAMINE 5GL □ RHODAMINE 6G LAKE □ RHODAMINE 6GO □ RHODAMINE 6GX □ RHODAMINE J □ RHODAMINE 6JH □ RHODAMINE 7JH □ RHODAMINE LAKE RED 6G □ RHODAMINE Y 20-7425 □ RHODAMINE ZH □ RHODAMINE 6ZH □ RHODAMINE 6Zh-DN □ RHODAMIN 6G □ SILOSUPER PINK B □ VALI FAST RED 1308 □ XANTHYLIUM, 9-(2-(ETHOXYCARBONYL)-PHENYL)-3,6-BIS(ETHYLAMINO)-2,7-DIMETHYL-, CHLORIDE
CONSENSUS REPORTS: IARC Cancer Review: Group 3 IMEMDT 7,56,87; Animal Limited Evidence IMEMDT 16,233,78; Human No Adequate Data IMEMDT 16,233,78. Reported in EPA TSCA Inventory. Community Right-To-Know List.
SAFETY PROFILE: Poison by ingestion and intraperitoneal routes. An experimental teratogen. Mutation data reported. Questionable carcinogen with experimental tumorigenic data. When heated to decomposition it emits very toxic fumes of Cl^- and NO_x.

RHA000   **CAS: 141-11-7**   **HR: 1**
RHODINYL ACETATE
mf: $C_{12}H_{22}O_2$ mw: 198.34
PROP: Mixture of acetates of geraniol and l-citronellol, found in geranium oil (FCTXAV 12,807,74). Colorless to sltly yellow liquid; fresh rose odor. D: 0.895–0.908, refr index: 1.450–1.458. Sol in alc and fixed oils; insol in glycerin, propylene glycol, and water @ 237°.
SYNS: α-CITRONELLYL ACETATE □ 3,7-DIMETHYL-7-OCTEN-1-OL ACETATE □ FEMA No. 2981 □ RHODINOL ACETATE
CONSENSUS REPORTS: Reported in EPA TSCA Inventory.
SAFETY PROFILE: Low toxicity by ingestion and skin contact. A skin irritant. When heated to decomposition it emits acrid smoke and irritating fumes.

RHA100   **HR: D**
RHODINYL FORMATE
mf: $C_{11}H_{20}O_2$ mw: 184.28
PROP: Colorless to slightly yellow liquid; leafy, rose-like odor. D: 0.901–0.908, refr index: 1.453–1.458. Sol in alc, fixed oils; insol in glycerin, propylene glycol, water @ 200°.
SYN: FEMA No. 2984
SAFETY PROFILE: When heated to decomposition it emits acrid smoke and irritating fumes.

RHF000   **CAS: 7440-16-6**   **HR: 3**
RHODIUM
af: Rh   aw: 102.91
PROP: A hard, lustrous, silvery-white, metallic element. Mp: 1966°, bp: 3727°, d: 12.41 @ 20°. Oxidizes slowly at 6°. IDLH 100 mg/m³ (as Rh).
SYNS: RH □ RHODIUM METAL (OSHA)
CONSENSUS REPORTS: Reported in EPA TSCA Inventory.
OSHA PEL: TWA Metal, Fume, Insol Compounds: 0.1 mg(Rh)/m³; Sol Compounds: 0.001 mg(Rh)/m³
ACGIH TLV: TWA (Metal) 1 mg/m³, (insoluble compounds as Rh) 1 mg/m³, (soluble compounds as Rh) 0.01 mg/m³

NIOSH REL: (Rhodium (as Rh)): TWA 0.1 mg/m³
SAFETY PROFILE: Handle carefully. It may be a sensitizer but not to the same extent as platinum. Most rhodium compounds have only moderate toxicity by ingestion. Flammable when exposed to heat or flame. Violent reaction with chlorine, bromine pentafluoride, bromine trifluoride, and OF_2. A catalytic metal.

RHK000   **CAS: 10049-07-7**   **HR: 3**
RHODIUM(III) CHLORIDE (1:3)
mf: Cl_3Rh mw: 209.26
PROP: Hygroscopic brown-red solid. Insol in org solvs and H_2O; sol in cyanide soln, alkali hydroxide soln. IDLH 100 mg/m³ (as Rh).
SYNS: RHODIUM CHLORIDE □ RHODIUM TRICHLORIDE
CONSENSUS REPORTS: Reported in EPA TSCA Inventory. EPA Genetic Toxicology Program.
OSHA PEL: TWA 0.001 mg(Rh)/m³
ACGIH TLV: TWA 1 mg(Rh)/m³
SAFETY PROFILE: Poison by ingestion, intraperitoneal, and intravenous routes. Experimental reproductive effects. Questionable carcinogen with experimental carcinogenic data. Mutation data reported. Incompatible with pentacarbonyl iron + zinc. When heated to decomposition it emits toxic fumes of Cl⁻. See also RHODIUM and CHLORIDES.

RHZ700   **CAS: 478-43-3**   **HR: 3**
RHUBARB YELLOW
mf: $C_{15}H_8O_6$ mw: 284.23
SYNS: 2-ANTHRACENECARBOXYLIC ACID, 9,10-DIHYDRO-4,5-DIHYDROXY-9,10-DIOXO- □ 2-ANTHROIC ACID, 9,10-DIHYDRO-4,5-DIHYDROXY-9,10-DIOXO- □ 9,10-DIHYDRO-4,5-DIHYDROXY-9,10-DIOXO-2-ANTHRACENECARBOXYLIC ACID □ CASSIC ACID □ CHRYSAZIN-3-CARBOXYLIC ACID □ MONORHEIN □ RHEIC ACID □ RHEIN
SAFETY PROFILE: A poison by intravenous route. Low toxicity by ingestion. When heated to decomposition it emits acrid smoke and irritating vapors.

R

RIK000   CAS: 83-88-5   HR: 3
RIBOFLAVINE
mf: $C_{17}H_{20}N_4O_6$ mw: 376.37
PROP: Orange to yellow crystals or needles; slt odor. Mp: 282° (decomp). Sltly sol in water, alc; insol in ether, chloroform.
SYNS: BEFLAVINE □ 6,7-DIMETHYL-9-d-RIBITYLISOALLOXAZINE □ 7,8-DIMETHYL-10-d-RIBITYLISOALLOXAZINE □ 7,8-DIMETHYL-10-(d-RIBO-2,3,4,5-TETRAHYDROXYPENTYL)ISOALLOXAZ-INE □ FLAVAXIN □ HYFLAVIN □ HYRE □ LACTO-FLAVIN □ LACTOFLAVINE □ RIBIPCA □ RIBODERM □ RIBOFLAVIN □ RIBOFLAVINEQUINONE □ VITAMIN B2 □ VITAMIN G
CONSENSUS REPORTS: Reported in EPA TSCA Inventory.
SAFETY PROFILE: Poison by intravenous route. Moderately toxic by intraperitoneal and subcutaneous routes. Mutation data reported. When heated to decomposition it emits toxic fumes of NO_x.

RJF800   HR: D
RICE BRAN WAX
PROP: Tan to brown hard wax. Mp: 75°. Sol in chloroform, benzene; insol in water.
SAFETY PROFILE: When heated to decomposition it emits acrid smoke and irritating fumes.

RJK100   CAS: 524-40-3   HR: 3
RICININ
mf: $C_8H_8N_2O_2$ mw: 164.18
SYNS: 1,2-DIHYDRO-4-METHOXY-1-METHYL-2-OXONICOTINONITRILE □ NICOTINONITRILE, 1,2-DIHYDRO-4-METHOXY-1-METHYL-2-OXO- □ 3-PYRIDINECARBONITRILE, 1,2-DIHYDRO-4-METHOXY-1-METHYL-2-OXO- □ RECININE □ RICININE
SAFETY PROFILE: A poison by ingestion and subcutaneous route. When heated to decomposition it emits toxic vapors of NO_x.

RKP000   CAS: 13292-46-1   HR: 3
RIFAMYCIN AMP
mf: $C_{43}H_{58}N_4O_{12}$ mw: 823.05
PROP: Red-orange platelets from Me_2CO. Mp: 183–188° (decomp).
SYNS: ARCHIDYN □ ARFICIN □ DIONE 21-ACETATE □ L-5103 □ 3-(4-METHYLPIPERAZINYL-IMINOMETHYL)-RIFAMYCIN SV □ 8-(4-METHYLPIP-ERAZINYLIMINOMETHYL) RIFAMYCIN SV □ 8-(((4-

METHYL-1-PIPERAZINYL)IMINO)METHYL)RIFAMYCIN SV □ NSC-113926 □ R/AMP □ RIFA □ RIFADINE □ RIFAGEN □ RIFALDAZINE □ RIFALDIN □ RIFAMATE □ RIFAMPICIN □ RIFAMPICINE (FRENCH) □ RIFAMPICINUM □ RIFAMPIN □ RIFAPRODIN □ RIFINAH □ RIFOBAC □ RIFOLDIN □ RIFORAL □ RIMACTAN □ RIMACTAZID □ TUBOCIN
CONSENSUS REPORTS: IARC Cancer Review: Group 3 IMEMDT 7,56,87; Animal Limited Evidence IMEMDT 24,243,80; Human No Adequate Data IMEMDT 24,243,80.
SAFETY PROFILE: Suspected carcinogen with experimental neoplastigenic and teratogenic data. Poison by intraperitoneal and intravenous routes. Moderately toxic to humans by ingestion. Moderately experimentally toxic by ingestion and subcutaneous routes. Human systemic effects by ingestion: conjunctiva irritation, iritis (inflammation of the iris), other eye effects, dermatitis. Experimental reproductive effects. Human mutation data reported. When heated to decomposition it emits toxic fumes of NO_x.

RLK890   CAS: 25875-51-8   HR: 2
ROBENIDINE
mf: $C_{15}H_{13}Cl_2N_5$ mw: 334.23
PROP: Crystals from ethanol. Mp: 289–290°.
SYNS: 1,3-BIS((p-CHLOROBENZYLIDENE)AMINO)GUANIDINE □ CARBONIMIDIC DIHYDRAZIDE, BIS((4-CHLOROPHEN-YL)METHYLENE)- □ CHEMCOCCIDE □ CHEMOC-CIDE □ CHIMCOCCIDE □ KHIMCOCCID □ KHIM-COECID □ KHIMKOKTSID □ KHIMKOKTSIDE
SAFETY PROFILE: Moderately toxic by ingestion. When heated to decomposition it emits toxic fumes of Cl^- and NO_x.

RMA000   CAS: 50471-44-8   HR: 1
RONILAN
mf: $C_{12}H_9Cl_2NO_3$ mw: 286.12
PROP: Crystals. Sol in Me_2CO, C_6H_6, $CHCl_3$, and EtOAc; spar sol in H_2O.
SYNS: BAS 352 F □ 3-(3,5-DICHLOROPHENYL)-5-ETHENYL-5-METHYL-2,4-OXAZOLIDINEDIONE □ 3-(3,5-DICHLOROPHENYL)-5-METHYL-5-VINYL-2,4-OXAZOLIDINEDIONE □ VINCLOZOLIN (GERMAN)

SAFETY PROFILE: Low toxicity by ingestion and inhalation. Mutation data reported. When heated to decomposition it emits very toxic fumes of Cl⁻ and NO_x.

RMA500   CAS: 299-84-3   HR: 3
RONNEL
mf: $C_8H_8Cl_3O_3PS$ mw: 321.54

PROP: White or crystalline powder. Mp: 41°, vap press: 8×10^{-4} mm, bp: 97° @ 0.01 mm. IDLH 300 mg/m³.

SYNS: DERMAFOSU (POLISH) □ DERMAPHOS □ O,O-DIMETHYL-O-2,4,5-TRICHLOROPHENYL PHOSPHOROTHIOATE □ DIMETHYL TRICHLOROPHENYL THIOPHOSPHATE □ O,O-DIMETHYL-O-(2,4,5-TRICHLOROPHENYL)THIO-PHOSPHATE □ O,O-DIMETHYL-O-(2,4,5-TRICHLO-RPHENYL)-THIONOPHOSPHAT (GERMAN) □ DOW ET 14 □ DOW ET 57 □ ECTORAL □ ENT 23,284 □ ET 14 □ ET 57 □ ETROLENE □ FENCHLOORFOS (DUTCH) □ FENCHLORFOS □ FENCHLORFOSU (POLISH) □ FENCHLOROPHOS □ FENCHLORPHOS □ KARLAN □ KORLAN □ KORLANE □ NANCHOR □ NANKER □ NANKOR □ THIOPHOSPHATE de O,O-DIMETHYLE et de O-(2,4,5-TRICHLOROPHENYLE) (FRENCH) □ O-(2,4,5-TRICHLOOR-FENYL)-O,O-DIMETHYL-MONOTHIOFOSFAAT (DUTCH) □ TRICHLORO-METAFOS □ 2,4,5-TRICHLOROPHENOL, O-ESTER with O,O-DIMETHYL PHOSPHOROTHIOATE □ O-(2,4,5-TRICHLOR-PHENYL)-O,O-DIMETHYL-MONOTHIO-PHOSPHAT (GERMAN) □ O-(2,4,5-TRICLORO-FENIL)-O,O-DIMETIL-MONOTIOFOSFATO (ITALIAN) □ TROLEN □ TROLENE □ VIOZENE

CONSENSUS REPORTS: Chlorophenol compounds are on the Community Right-To-Know List.

OSHA PEL: TWA 10 mg/m³

ACGIH TLV: TWA 10 mg/m³; Not Classifiable as a Human Carcinogen

SAFETY PROFILE: Poison by ingestion and intraperitoneal routes. Moderately toxic by skin contact. A cholinesterase inhibitor. An experimental teratogen. Experimental reproductive effects. Human mutation data reported. When heated to decomposition it emits very toxic fumes of Cl⁻, PO_x, and SO_x. See also PARATHION and CHLOROPHENOLS.

RMK020   CAS: 569-61-9   HR: 3
p-ROSANILINE HYDROCHLORIDE
mf: $C_{19}H_{17}N_3 \cdot ClH$ mw: 323.85

PROP: Colorless to red crystals. Mp: 268°–270°. Sol in water: <0.1 mg/ml @ 20°.

SYNS: 4-((4-AMINOPHENYL)(4-IMINO-2,5-CYCLO-HEXADIEN-1-YLIDENE)METHYL), MONOCHLORIDE □ BASIC PARAFUCHSINE □ CALCOZINE MAGENTA N □ C.I. 42500 □ C.I. BASIC RED 9, MONOHYDROCHLOR-IDE □ p-FUCHSIN □ FUCHSINE DR-001 □ FUCHSINE SPC □ 4,4'-((4-IMINO-2,5-CYCLOHEXADIEN-1-YLID-ENE)METHYLENE)DIANILINE MONOHYDROCHLOR-IDE-o-TOLUIDINE □ NCI-C54739 □ PARAFUCHSIN (GERMAN) □ PARA-MAGENTA □ PARAROSANILINE □ PARAROSANILINE CHLORIDE □ PARAROSANIL-INE HYDROCHLORIDE □ p-ROSANILINE HCL □ SCHULTZ-TAB No. 779 (GERMAN) □ 4,4'4"-TRIAMINO-TRIPHENYLMETHAN-HYDROCHLORID (GERMAN)

CONSENSUS REPORTS: NTP 10th Report on Carcinogens. IARC Cancer Review: Animal Sufficient Evidence IMEMDT 57,215,93; Animal Inadequate Evidence IMEMDT 4,57,74; Human Inadequate Evidence IMEMDT 4,57,74; Human Inadequate Evidence IMEMDT 57,215,93; Group 2B IMEMDT 57,215,93. EPA Genetic Toxicology Program. Reported in EPA TSCA Inventory.

SAFETY PROFILE: Confirmed carcinogen with experimental carcinogenic and tumorigenic data. Mildly toxic by ingestion. Mutation data reported. When heated to decomposition it emits very toxic fumes of HCl and NO_x.

RMP175   CAS: 632-69-9   HR: 2
ROSE BENGAL SODIUM
mf: $C_{20}H_2Cl_4I_4O_5 \cdot 2Na$ mw: 1017.60

PROP: Food red color.

SYNS: FOOD RED COLOR No. 105, SODIUM SALT □ FOOD RED No. 105, SODIUM SALT □ R105 SODIUM □ 9-(3',4',5',6'-TETRACHLORO-o-CARBOXYPHENYL)-6-HYDROXY-2,4,5,7-TETRAIODO-3-ISOXANTHONE•2Na □ 4,5,6,7-TETRACHLORO-2',4',5',7'-TETRAIODO-FLUORESCEIN DISODIUM SALT

CONSENSUS REPORTS: Reported in EPA TSCA Inventory.

SAFETY PROFILE: An experimental teratogen. Experimental reproductive effects. Questionable carcinogen with experimental neoplastigenic data. When heated to decomposition it emits toxic fumes of Cl⁻, I⁻, and Na_2O.

R

RMU000   CAS: 8000-25-7   HR: 1
ROSEMARY OIL
PROP: Constituents are α-pinene, camphene, and cineole. From steam distillation of flowering tops of *Rosmarinus officinalis* L. (Fam. *Labiatae*) (FCTXAV 12,807,74). Colorless to pale-yellow liquid; odor of rosemary. D: 0.894–0.912, refr index: 1.464 @ 20°.
SYNS: ROSEMARIE OIL □ ROSMARIN OIL (GERMAN)
CONSENSUS REPORTS: Reported in EPA TSCA Inventory.
SAFETY PROFILE: Low toxicity by ingestion and skin contact. A skin irritant. When heated to decomposition it emits acrid smoke and irritating fumes. See also individual components.

RNA000   CAS: 8007-01-0   HR: 2
ROSE OIL
PROP: Volatile oil from steam distillation of fresh flowers of *Rosa gallica* L. and *Rosa Damascena* Mill. and varieties of these species (Fam. *Rosaceae*). Colorless to yellow liquid; odor and taste of rose. D: 0.848–0.863 @ 30°/15°, refr index: 1.457 @ 30°.
SYNS: ATTAR ROSE □ ATTAR of ROSE □ ESSENCE of ROSE □ OIL OF ROSE □ OIL OF ROSE BLOSSOM □ OIL OF ROSE BULGARIAN □ OTTO ROSE □ OTTO of ROSE □ ROSE de GRASSE □ ROSE de MAI □ ROSEN OEL (GERMAN) □ ROSENOL □ ROSE OIL BULGARIAN □ ROSE OTTO
CONSENSUS REPORTS: Reported in EPA TSCA Inventory.
SAFETY PROFILE: Moderately toxic by skin contact. Mildly toxic by ingestion. When heated to decomposition it emits acrid smoke and irritating fumes.

RNU100   CAS: 8050-09-7   HR: 3
ROSIN CORE SOLDER PYROLYSIS PRODUCTS
SYNS: BALS 3A □ BANDIS G100 □ RONDIS R □ EM 3 □ HIGHROSIN □ HONGKONG ROSIN WW □ KE 709 □ ROSIN □ ROSIN WW □ SHIRAGIKU ROSIN □ YELLOW RESIN
CONSENSUS REPORTS: Reported in EPA TSCA Inventory.
ACGIH TLV: Reduce to as low as possible

SAFETY PROFILE: A poison by ingestion and inhalation route. When heated to decomposition it emits acrid smoke and irritating vapors.

ROF300   CAS: 135-51-3   HR: D
R SALT
mf: $C_{10}H_6O_7S_2$•2Na   mw: 348.26
SYNS: FERRICON □ 2-NAPHTHOL-3,6-DISULFONIC ACID SODIUM SALT
CONSENSUS REPORTS: Reported in EPA TSCA Inventory.
SAFETY PROFILE: An experimental teratogen. Experimental reproductive effects. When heated to decomposition it emits toxic fumes of SO_x and Na_2O. See also SULFONATES.

RPA000   CAS: 7440-17-7   HR: 3
RUBIDIUM
DOT: UN 1423
af: Rb   aw: 85.47
PROP: Soft, silvery-white metal. Tarnishes in air forming rubidium oxides, carbonates, and hydroxide. Mp: 39.5°, bp: 688°, d (solid): 1.532 @ 20°, d (liquid): 1.475 @ 39°. Sol in NH_3.
CONSENSUS REPORTS: Reported in EPA TSCA Inventory.
DOT CLASSIFICATION: 4.3; Label: Dangerous When Wet
SAFETY PROFILE: Moderately toxic by intraperitoneal route. A very reactive alkali metal (more reactive than potassium or cesium). In the body, rubidium substitutes for potassium as an intracellular ion. The ratio of Rb/K intake is important in the toxicology of rubidium. A ratio above 40% is dangerous. In rats, a failure to gain weight is the first symptom, followed by ataxia and hyperirritability. Symptoms include: skin ulcers, poor hair coat, sensitivity, and extreme nervousness leading to convulsions and death.
  A very dangerous fire and explosion hazard when exposed to heat or flame or by chemical reaction with oxidizers. Ignites on contact with air, oxygen, and halogens.

Ignites spontaneously on contact with water. Reaction with water, moisture, or steam forms explosive hydrogen gas, which then ignites. Explodes in contact with liquid bromine. Can react explosively with air, halogens, mercury, nonmetals, vanadium chloride oxide, moisture, acids, oxidizers. Violent reaction with vanadium trichloride oxide (at 60°C), Cl_2O_2, P. Molten rubidium ignites in sulfur vapor and reacts vigorously with carbon. RbOH is more basic than KOH. Storage and handling: Keep under benzene, petroleum, or other liquids not containing gaseous O_2. When heated to decomposition it emits toxic fumes of Rb_2O. See also SODIUM and SODIUM POTASSIUM ALLOY.

RPF000   CAS: 7791-11-9   HR: 2
RUBIDIUM CHLORIDE
mf: ClRb   mw: 120.92
PROP: White or colorless crystalline powder or cubic crystals. Mp: 718°, bp: 1390°, d: 2.76. Very sol in H_2O; sol in MeOH; spar sol in EtOH.
CONSENSUS REPORTS: Reported in EPA TSCA Inventory.
SAFETY PROFILE: Moderately toxic by ingestion and intraperitoneal routes. Experimental reproductive effects. Mutation data reported. Reacts violently with BrF_3. When heated to decomposition it emits toxic fumes of Cl^-, RbCl, and Rb_2O. See also RUBIDIUM and CHLORIDES.

RPK000   CAS: 13446-73-6   HR: 3
RUBIDIUM DICHROMATE
mf: $Cr_2O_7Rb_2$ mw: 386.94
PROP: Crystals. D: 3.02–3.13. IDLH Ca [15 mg/m³ {as Cr(VI)}].
CONSENSUS REPORTS: Chromium and its compounds are on the Community Right-To-Know List.
OSHA PEL: CL 0.1 mg(CrO₃)/m³
ACGIH TLV: TWA 0.05 mg(Cr)/m³; Confirmed Human Carcinogen
NIOSH REL: (Chromium(VI)) TWA 0.001 mg(Cr)/m³

SAFETY PROFILE: A confirmed carcinogen. A poison. A powerful oxidizer. When heated to decomposition it emits toxic fumes of Rb_2O. See also RUBIDIUM.

RPP000   CAS: 13446-74-7   HR: 3
RUBIDIUM FLUORIDE
mf: FRb   mw: 104.47
PROP: Hygroscopic, colorless, cubic crystals. Mp: 795°, bp: 1410°, d: 3.557, vap press: 1 mm @ 921°. Very sol in H_2O; insol in EtOH and Et_2O.
CONSENSUS REPORTS: Reported in EPA TSCA Inventory.
OSHA PEL: TWA 2.5 mg(F)/m³
ACGIH TLV: TWA 2.5 mg(F)/m³; BEI: 3 mg/g creatinine of fluorides in urine prior to shift; 10 mg/g creatinine of fluorides in urine at end of shift.
NIOSH REL: (Fluorides, Inorganic) TWA 2.5 mg(F)/m³
SAFETY PROFILE: Poison as a soluble fluoride. When heated to decomposition it emits toxic fumes of Rb_2O and F^-. See also RUBIDIUM and FLUORIDES.

RPZ000   CAS: 1310-82-3   HR: 2
RUBIDIUM HYDROXIDE
DOT: UN 2677/UN 2678
mf: HORb   mw: 102.48
PROP: Grayish-white, deliquescent, monoclinic crystals or mass; strong base. Mp: 300°, d: 3.203 @ 11°. Very sol in H_2O.
SYNS: RUBIDIUM HYDROXIDE (UN 2678) (DOT) □ RUBIDIUM HYDROXIDE SOLUTION (UN 2677) (DOT)
CONSENSUS REPORTS: Reported in EPA TSCA Inventory.
DOT CLASSIFICATION: 8; Label: Corrosive
SAFETY PROFILE: Moderately toxic by ingestion. A powerful, corrosive irritant to skin, eyes, and mucous membranes. When heated to decomposition it emits toxic fumes of Rb_2O. See also POTASSIUM HYDROXIDE and RUBIDIUM.

R

RRP000   HR: 2
RUSSIAN COMFREY ROOTS
PROP: Fresh roots dried, milled, and mixed with diet (JNCIAM 61,86578).
SYNS: COMFREY, RUSSIAN □ SYMPHYTUM OFFICINALE L
CONSENSUS REPORTS: IARC Cancer Review: Animal Limited Evidence IMEMDT 31,239,83.
SAFETY PROFILE: Questionable carcinogen with experimental carcinogenic data. When heated to decomposition it emits acrid smoke and irritating fumes.

RRU000   CAS: 7440-18-8   HR: 3
RUTHENIUM
af: Ru   aw: 101.07
PROP: Lustrous, hard, silvery metal; hexagonal crystals. Highly resistant to corrosive agents at mod temps; unaffected by air, H_2O, acids, but dissolved by molten alkalies. Unaffected by aqua regia, H_2SO_4, HCl, HF, or H_3PO_4. D: 12.45 @ 20°/4°, mp: approx 2310°, bp: approx 4150°. Stable in air.
SAFETY PROFILE: Most ruthenium compounds are poisons. Ruthenium is retained in the bones for a long time. Flammable in the form of dust when exposed to heat or flame. Violent reaction with ruthenium oxide. Explosive reaction with aqua regia + potassium chlorate. When heated to decomposition it emits very toxic fumes of RuO_x and Ru, which are highly injurious to the eyes and lung and can produce nasal ulcerations. See also RUTHENIUM COMPOUNDS.

RRZ000   CAS: 10049-08-8   HR: 3
RUTHENIUM CHLORIDE
mf: Cl_3Ru mw: 207.42
PROP: α Form: Black lustrous crystals. Insol in alc, water. β Form: Dark-brown, fluffy, hexagonal crystals. Sol in alc.
SYN: RUTHENIUM TRICHLORIDE
CONSENSUS REPORTS: EPA Genetic Toxicology Program. Reported in EPA TSCA Inventory.
SAFETY PROFILE: Poison by intraperitoneal route. Incompatible with iron pentacarbonyl and zinc. When heated to decomposition it emits toxic fumes of RuO_x and Cl^-. See also RUTHENIUM COMPOUNDS.

RSF000   HR: 3
RUTHENIUM COMPOUNDS
SAFETY PROFILE: Most ruthenium compounds are poisons or moderately toxic. Ruthenium red is an antagonist of Ca^{2+}, inhibits Ca^{2+} transport and binding in mitochondrial membranes, and inhibits Ca^{2+}-ATPase activity. They resemble osmium compounds in that when heated in air, they evolve fumes that are injurious to the eyes and lungs and can produce nasal ulcerations. When heated to decomposition they emit toxic fumes of RuO_x and Ru. See also RUTHENIUM and specific compounds.

S

SAC100 **HR: D**
SAFFRON
PROP: From the dried stigma of *Ceocus saffron* L.
SAFETY PROFILE: When heated to decomposition it emits acrid smoke and irritating fumes.

SAD000 **CAS: 94-59-7** **HR: 3**
SAFROL
mf: $C_{10}H_{10}O_2$ mw: 162.20
PROP: Colorless liquid, prisms, or crystals; sassafras odor. Mp: 11°, fp: 11.2°, bp: 231.5–232°, d: 1.0960 @ 20°, vap press: 1 mm @ 63.8°. Insol in water; very sol in alc; misc with chloroform, ether.
SYNS: 5-ALLYL-1,3-BENZODIOXOLE □ ALLYL-CATECHOL METHYLENE ETHER □ ALLYLDIOXY-BENZENE METHYLENE ETHER □ 1-ALLYL-3,4-METHYLENEDIOXYBENZENE □ 4-ALLYL-1,2-METHYLENEDIOXYBENZENE □ m-ALLYLPYRO-CATECHIN METHYLENE ETHER □ 4-ALLYLPYRO-CATECHOL FORMALDEHYDE ACETAL □ ALLYL-PYROCATECHOL METHYLENE ETHER □ 1,2-METHYL-ENEDIOXY-4-ALLYLBENZENE □ 3,4-METHYL-ENEDIOXY-ALLYLBENZENE □ 5-(2-PROPENYL)-1,3-BENZODIOXOLE □ RCA WASTE NUMBER U203 □ RHYUNO OIL □ SAFROLE □ SAFROLE MF □ SHIKIMOLE □ SHIKOMOL
CONSENSUS REPORTS: NTP 10th Report on Carcinogens. IARC Cancer Review: Group 2B IMEMDT 7,56,87; Animal Sufficient Evidence IMEMDT 10,231,76; Human No Adequate Data IMEMDT 10,231,76. Community Right-To-Know List. EPA Genetic Toxicology Program. Reported in EPA TSCA Inventory.
SAFETY PROFILE: Confirmed carcinogen with experimental carcinogenic and neoplastigenic data. Poison by intravenous route. Moderately toxic by ingestion and subcutaneous routes. Experimental reproductive effects. Human mutation data reported. A skin irritant. Combustible when exposed to heat or flame. When heated to decomposition it emits acrid smoke and irritating fumes. See also ALDEHYDES and ALLYL COMPOUNDS.

SAG100 **CAS: 6884-59-9** **HR: 3**
SALICYLALDEHYDE DIMETHYL ACETAL CARBAMATE
mf: $C_{11}H_{15}NO_4$ mw: 225.27
SYN: CARBAMIC ACID, ESTER WITH SALICYL-ALDEHYDE DIMETHYL ACETAL
SAFETY PROFILE: A poison by ingestion. When heated to decomposition it emits toxic vapors of NO_x.

SAI000 **CAS: 69-72-7** **HR: 3**
SALICYLIC ACID
mf: $C_7H_6O_3$ mw: 138.13
PROP: Powder or needles from water. D: 1.443 @ 20°/4°, mp: 158.3°, bp: 211° @ 20 mm. Sol in water, alc, ether.
SYNS: ACIDO SALICILICO (ITALIAN) □ o-HYDROXY-BENZOIC ACID □ 2-HYDROXYBENZOIC ACID □ KERALYT □ ORTHOHYDROXYBENZOIC ACID □ RETARDER W □ SA □ SAX
CONSENSUS REPORTS: Reported in EPA TSCA Inventory. EPA Genetic Toxicology Program.
SAFETY PROFILE: Poison by ingestion, intravenous, and intraperitoneal routes. Moderately toxic by subcutaneous route. An experimental teratogen. Human systemic effects by skin contact: ear tinnitus. Mutation data reported. A skin and severe eye irritant. Experimental reproductive effects. Incompatible with iron salts, spirit nitrous ether, lead acetate, iodine. Used in the manufacture of aspirin. When heated to decomposition it emits acrid smoke and irritating fumes.

S

SAO550   HR: D
SALTS of FATTY ACIDS
PROP: Consists of aluminum, calcium, magnesium, potassium, and sodium salts of capric, caprylic, lauric, myristic, oleic, palmitic, and stearic acids manufactured from fats and oils derived from edible sources.
SAFETY PROFILE: When heated to decomposition it emits acrid smoke and irritating fumes.

SAR000   CAS: 10465-27-7   HR: 3
SAMARIUM ACETATE
mf: $C_6H_9O_6 \cdot Sm$ mw: 327.50
PROP: Yellow solid. Sol in water.
SYNS: ACETIC ACID, SAMARIUM SALT □ SAMAR-IUMACETAT (GERMAN)
CONSENSUS REPORTS: Reported in EPA TSCA Inventory.
SAFETY PROFILE: Poison by intravenous and subcutaneous routes. When heated to decomposition it emits acrid smoke and irritating fumes. See also SAMARIUM and RARE EARTHS.

SAR500   CAS: 10361-82-7   HR: 3
SAMARIUM(III) CHLORIDE
mf: Cl_3Sm mw: 256.70
PROP: Very hygroscopic pale-yellow solid or yellowish-white powder. D: 4.465, mp: 686°. Sol in H_2O and EtOH; sltly sol in THF, Me_2CO, and Py.
CONSENSUS REPORTS: Reported in EPA TSCA Inventory.
SAFETY PROFILE: Poison by intraperitoneal and subcutaneous routes. Moderately toxic by ingestion. A skin and eye irritant. When heated to decomposition it emits toxic fumes of Cl^-. See also SAMARIUM and RARE EARTHS.

SAU400   HR: 2
SANTALYL ACETATE
PROP: Mixture of α- and β-isomers from acetylation of santalol. Colorless to sltly yellow liquid; sandalwood odor. D: 0.980, refr index: 1.488–1.491, flash p: 212°F. Sol in alc; insol in water.
SYN: FEMA No. 3007

SAFETY PROFILE: Combustible liquid. When heated to decomposition it emits acrid smoke and irritating fumes.

SAV000   CAS: 91-53-2   HR: 3
SANTOQUINE
mf: $C_{14}H_{19}NO$ mw: 217.34
PROP: Clear, light-yellow liquid. Mp: <0°, bp: 125° @ 2 mm, vap d: 7.48, d: 1.030 @ 25°, refr index: 1.57.
SYNS: 1,2-DIHYDRO-6-ETHOXY-2,2,4-TRIMETHYL-QUINOLINE □ 1,2-DIHYDRO-2,2,4-TRIMETHYL-6-ETHOXYQUINOLINE □ EMQ □ EQ □ 6-ETHOXY-1,2-DIHYDRO-2,2,4-TRIMETHYLQUINOLINE □ ETHOXY-QUIN (FCC) □ ETHOXYQUINE □ 6-ETHOXY-2,2,4-TRIMETHYL-1,2-DIHYDROQUINOLINE □ NIFLEX □ NIX-SCALD □ SANTOFLEX A □ SANTOFLEX AW □ SANTOQUIN □ STOP-SCALD □ 2,2,4-TRIMETHYL-6-ETHOXY-1,2-DIHYDROQUINOLINE □ USAF B-24
CONSENSUS REPORTS: EPA Genetic Toxicology Program. Reported in EPA TSCA Inventory.
SAFETY PROFILE: Poison by intraperitoneal route. Moderately toxic by ingestion. Mutation data reported. Combustible when exposed to heat or flame; can react with oxidizing materials. When heated to decomposition it emits toxic fumes of NO_x.

SAY900   HR: 2
SASSAFRAS
PROP: A yellowish-reddish, volatile oil; pungent, aromatic odor and taste. D: 1.065–1.077 @ 25°/25°. Sol in alc, ether, chloroform, glacial acetic acid, CS_2. Safrole-free ethanol extract of *Sassafras albidum* root bark (JNCIAM 60,683,78).
SYN: SASSAFRAS ALBIDUM
SAFETY PROFILE: A skin irritant. Questionable carcinogen with experimental neoplastigenic data. When heated to decomposition it emits acrid smoke and irritating fumes.

SBA000   CAS: 8016-68-0   HR: 3
SAVORY OIL (summer variety)
PROP: From steam distillation of *Saturiea hortensis L.* (Fam. *Labiatae*) (FCTXAV 14,659,76). Light-yellow to dark-brown

liquid; spicy odor. D: 0.875–0.954, refr index: 1.486–1.505 @ 20°. Sol in fixed oils, mineral oil; insol in glycerin, propylene glycol.

CONSENSUS REPORTS: Reported in EPA TSCA Inventory.

SAFETY PROFILE: Poison by skin contact. Moderately toxic by ingestion. A severe skin irritant. When heated to decomposition it emits acrid smoke and irritating fumes.

SBB000   CAS: 96-88-8   HR: 3
SCANDICAINE
mf: $C_{15}H_{22}N_2O$ mw: 246.39

SYNS: CARBOCAINE □ MEPIVACAINE □ dl-MEPI-VACAINE □ N-METHYLHEXAHYDRO-2-PICOLINIC ACID, 2,6-DIMETHYLANILIDE □ N-METHYL-2-PIPECOLIC ACID, 2,6-DIMETHYLANILIDE □ N-METHYL-2-PIPECOLIC ACID, 2,6-XYLIDIDE □ (±)-1-METHYL-2',6'-PIPECOLOXYLIDIDE □ SCANDICAIN □ SCANDICANE

CONSENSUS REPORTS: Reported in EPA TSCA Inventory.

SAFETY PROFILE: Poison by intravenous route. Moderately toxic by subcutaneous route. Experimental reproductive effects. When heated to decomposition it emits toxic fumes of NO_x.

SBC000   CAS: 10361-84-9   HR: 3
SCANDIUM CHLORIDE
mf: Cl_3Sc mw: 151.31

PROP: White crystals, deliquescent solid. Mp: 968°. Subl @ 13° (*in vacuo*). Sol in water; insol in alc.

SYN: SCANDIUM(3+) CHLORIDE

CONSENSUS REPORTS: Reported in EPA TSCA Inventory.

SAFETY PROFILE: Poison by intraperitoneal route. Moderately toxic by ingestion. When heated to decomposition it emits toxic fumes of Cl⁻. See also SCANDIUM.

SBC500   CAS: 85-83-6   HR: 2
SCARLET RED
mf: $C_{24}H_{20}N_4O$ mw: 380.48

SYNS: BIEBRICH SCARLET BPC □ BIEBRICH SCARLET RED □ BIEBRICH SCARLET R MEDICINAL □ BRASILAZINA OIL RED B □ CALCO OIL RED D □ CANDLE SCARLET B □ CANDLE SCARLET 2B □ CANDLE SCARLET G □ CERES RED BB □ CEROTINE PONCEAU 3B □ CERVEN ROZPOUSTEDLOVA 24 □ C.I. 258 □ C.I. 26105 □ C.I. SOLVENT RED 24 □ 2',3-DIMETHYL-4-(2-HYDROXYNAPHTHYLAZO)AZO-BENZENE □ DISPERSOL RED PP □ ENIAL RED IV □ FAST OIL RED B □ FAST RED BB □ FAT PONCEAU R □ FAT RED B □ FAT RED 2B □ FAT RED BB □ FAT RED BS □ FAT RED TS □ GRASAL BRILLIANT RED B □ GRASAN BRILLIANT RED B □ HIDACO OIL RED □ LACQUER RED V □ LACQUER RED VS □ LIPID CRIMSON □ 1-((2-METHYL-4-((2-METHYLPHENYL)-AZO)PHENYL)AZO)-2-NAPHTHALENOL □ OIL RED IV □ OIL RED 3 □ OIL RED 7 □ OIL RED 47 □ OIL RED 282 □ OIL RED A □ OIL RED APT □ OIL RED 2B □ OIL RED 3B □ OIL RED 4B □ OIL RED BB □ OIL RED BS □ OIL RED D □ OIL RED ED □ OIL RED F □ OIL RED GO □ OIL RED PEL □ OIL RED RC □ OIL RED RR □ OIL RED S □ OIL RED TAX □ OIL RED ZD □ OIL SCARLET □ OIL SCARLET 48 □ OLEAL RED BB □ ORGANOL RED B □ ORIENT OIL RED RR □ PHENOPLASTE ORGANOL RED B □ PLASTORESIN RED F □ RED 3R SOLUBLE IN GREASE □ RESINOL RED 2B □ RESOFORM RED G □ RUBRUM SCARLAT-INUM □ SARLACH R □ SCARLET RED, BIEBRICH □ SCARLET R (MICHAELIS) □ SCHARLACHROT □ SCHULTZ NO. 541 □ SILOTRAS RED T3B □ SOMALIA RED IV □ STEARIX RED 4B □ STEARIX RED 4S □ SUDAN IV □ SUDAN P □ SUDAN RED IV □ SUDAN RED 4BA □ SUDAN RED BB □ SUDAN RED BBA □ TERTROGRAS RED N □ o-TOLUENEAZO-o-TOLUENE-AZO-β-NAPHTHOL □ o-TOLUENEAZO-o-TOLUENE-β-NAPHTHOL □ o-TOLYLAZO-o-TOLYLAZO-β-NAPHTHOL □ o-TOLYLAZO-o-TOLYLAZO-β-NAPHTHOL □ 1-(4-o-TOLYLAZO-o-TOLYLAZO)-2-NAPHTHOL □ TOYO OIL RED BB □ WAXAKOL RED BL □ WAXOLINE RED O □ WAXOLINE RED OM □ WAXOLINE RED OS

CONSENSUS REPORTS: IARC Cancer Review: Group 3 IMEMDT 7,56,87; Animal Inadequate Evidence IMEMDT 8,217,75; Human No Adequate Data IMEMDT 8,217,75. Reported in EPA TSCA Inventory.

SAFETY PROFILE: Questionable carcinogen with experimental tumorigenic data. Mutation data reported. When heated to decomposition it emits toxic fumes of NO_x.

SBG000   CAS: 51-34-3   HR: 3
SCOPOLAMINE
mf: $C_{17}H_{21}NO_4$ mw: 303.39

PROP: Noncrystalline solid. Mp: 55°. Very sol in hot water, alc, ether, chloroform,

S

acetone; sltly sol in benzene, pet ether. Decomp on standing.

SYNS: ATROCHIN □ ATROQUIN □ 6-β,7-β-EPOXY-3-α-TROPANYL S-(−)-TROPATE □ EPOXYTROPINE TROPATE □ HYOSCINE □ (−)-HYOSCINE □ HYOSOL □ ISOPTO HYOSCINE □ 9-METHYL-3-OXA-9-AZATRI-CYCLO(3.3.1.0^{2,4})NONAN-7-OL TROPATE (ester) □ OSCINE □ SCOPINE TROPATE □ (−)-SCOPOLAMINE □ TROPIC ACID, ESTER with SCOPINE □ TROPIC ACID, 9-METHYL-3-OXA-9-AZATRICYCLO(3.3.1.0^{2,4})NON-7-YL ESTER

CONSENSUS REPORTS: EPA Genetic Toxicology Program.

SAFETY PROFILE: Poison by intravenous, intraperitoneal, and subcutaneous routes. Moderately toxic by ingestion. Human systemic effects from very small amounts by subcutaneous and intramuscular routes: changes in surface EEG, distorted perceptions, excitement, hallucinations, and mydriasis. It can cause the individual who is affected to lose a certain amount of his normal inhibitory control. It is for that reason that it has been called "truth serum." An experimental teratogen. Experimental reproductive effects. Human mutation data reported. In many cases of poisoning from this material, and even to a certain extent following its medical application, there is retention of the urine caused by paralysis of the bladder, and catheterization is necessary. The fatal dose is variable. Death has occurred from as little as 0.6 mg, while recovery has occurred from doses of 7–15 mg. An anticholinergic drug. When heated to decomposition it emits highly toxic fumes of NO$_x$. See also ESTERS.

SBL600   CAS: 57333-96-7   HR: 3
(1-α,3-β,5Z,7E,24R)-9,10-SECOCHOL-ESTA-5,7,10(19)-TRIENE-1,3,24-TRIOL

mf: C$_{27}$H$_{44}$O$_3$ mw: 416.71

SYNS: 1-α,24(R)-DIHYDROXYVITAMIN D3 □ TACALCITOL □ 9,10-SECOCHOLESTA-5,7,10(19)-TRIENE-1,3,24-TRIOL, (1-α,3-β,5Z,7E,24R)- □ TV 02

SAFETY PROFILE: A poison by ingestion, subcutaneous, and intravenous routes. When heated to decomposition it emits acrid smoke and irritating vapors.

SBN500   CAS: 7783-08-6   HR: 3
SELENIC ACID

DOT: UN 1905

mf: H$_2$O$_4$Se   mw: 144.98

PROP: Colorless liquid, hexagonal prisms, or deliquescent crystals or solid. Mp: 58°, bp: 260° (decomp), d (solid): 2.951 @ 15°, (liquid): 2.609 @ 15°. Very sol in water; sol in sulfuric acid; insol in ammonia; decomp in alc.

SYN: SELENIC ACID, liquid (DOT)

CONSENSUS REPORTS: Selenium and its compounds are on the Community Right-To-Know List. EPA Genetic Toxicology Program. Reported in EPA TSCA Inventory.

OSHA PEL: TWA 0.2 mg(Se)/m^3
ACGIH TLV: TWA 0.2 mg(Se)/m^3
DFG MAK: 0.1 mg(Se)/m^3
DOT CLASSIFICATION: 8; Label: Corrosive

SAFETY PROFILE: Selenium compounds are poisons. A corrosive irritant to skin, eyes, and mucous membranes. When heated to decomposition it emits toxic fumes of Se. See also SELENIUM COMPOUNDS.

SBN505   CAS: 10102-23-5   HR: 3
SELENIC ACID, DISODIUM SALT, DECAHYDRATE

mf: O$_4$Se•2Na•10H$_2$O   mw: 369.14

PROP: Colorless monoclinic crystals; decomp on heating by H$_2$O loss. Very sol in H$_2$O.

OSHA PEL: TWA 0.2 mg(Se)/m^3
ACGIH TLV: TWA 0.2 mg(Se)/m^3

SAFETY PROFILE: Poison by intraperitoneal route. When heated to decomposition it emits toxic fumes of Se.

SBN510   CAS: 55509-78-9   HR: 3
β-SELENINOPROPIONIC ACID

mf: C$_3$H$_6$O$_4$Se   mw: 185.05

SYN: PROPIONIC ACID, 3-SELENINO-

OSHA PEL: TWA 0.2 mg(Se)/m^3
ACGIH TLV: TWA 0.2 mg(Se)/m^3

SAFETY PROFILE: Poison by intravenous route. When heated to decomposition it emits toxic fumes of Se.

SBO000   CAS: 7783-00-8   HR: 3
SELENIOUS ACID
mf: H_2O_3Se mw: 128.98
PROP: Transparent, colorless, hexagonal crystals; may be dehydrated to SeO_2. Mp: decomp, d: 3.004 @ 15°/4°, vap press: 2 mm @ 15°. Very sol in alc; sol in water; insol in ammonia.
SYNS: RCRA WASTE NUMBER U204 □ SELENIUM DIOXIDE
CONSENSUS REPORTS: EPA Extremely Hazardous Substances List. Selenium and its compounds are on the Community Right-To-Know List. Reported in EPA TSCA Inventory.
OSHA PEL: TWA 0.2 mg(Se)/m³
ACGIH TLV: TWA 0.2 mg(Se)/m³
DFG MAK: 0.1 mg(Se)/m³
SAFETY PROFILE: Poison by ingestion, intraperitoneal, and intravenous routes. Human mutation data reported. Used as an oxidizing agent. When heated to decomposition it emits toxic fumes of Se. See also SELENIUM COMPOUNDS.

SBO100   CAS: 10431-47-7   HR: 3
SELENIOUS ACID, DIPOTASSIUM SALT (9CI)
mf: O_3Se•2K   mw: 205.16
SYNS: DIPOTASSIUM SELENITE □ POTASSIUM SELENITE (7CI)
CONSENSUS REPORTS: Reported in EPA TSCA Inventory.
DOT CLASSIFICATION: 6.1; Label: Poison, Corrosive
SAFETY PROFILE: A poison and corrosive. When heated to decomposition it emits toxic vapors of selinium.

SBO500   CAS: 7782-49-2   HR: 3
SELENIUM
DOT: UN 2658
af: Se   aw: 78.96
PROP: Steel-gray, metalloid element. Viscosity of liquid decreases with temperature. Mp: 170–217°, bp: 690°, d: 4.26–4.81, vap press: 1 mm @ 356°. Insol in water and alc; very sltly sol in ether. IDLH 1 mg/m³ (as Se).
SYNS: C.I. 77805 □ COLLOIDAL SELENIUM □ ELEMENTAL SELENIUM □ SELEN (POLISH) □ SELENIUM ALLOY □ SELENIUM BASE □ SELENIUM DUST □ SELENIUM ELEMENTAL □ SELENIUM HOMOPOLYMER □ SELENIUM METAL POWDER, NON-PYROPHORIC (DOT) □ VANDEX
CONSENSUS REPORTS: IARC Cancer Review: Group 3 IMEMDT 7,56,87. Selenium and its compounds are on the Community Right-To-Know List. Reported in EPA TSCA Inventory.
OSHA PEL: TWA 0.2 mg(Se)/m³
ACGIH TLV: TWA 0.2 mg(Se)/m³
DFG MAK: 0.1 mg(Se)/m³
DOT CLASSIFICATION: 6.1; Label: KEEP AWAY FROM FOOD
SAFETY PROFILE: Poison by inhalation and intravenous routes. Questionable carcinogen with experimental tumorigenic and teratogenic data. Occupational exposure has caused pallor, nervousness, depression, garlic odor of breath and sweat, gastrointestinal disturbances, and dermatitis. Liver damage in experimental animals. Chronic ingestion of 5 mg of selenium per day resulted in 49% morbidity in 5 Chinese villages. The main symptoms were brittle hair with intact follicles, new hair with no pigment, brittle nails with spots and streaks, skin lesions, peripheral anesthesia, acroparesthesia, pain, and hyperreflexia. Similar effects have been seen in populations with selenium blood levels of 800 µg/L. In cattle, "alkali disease" is associated with consumption of grain or plants containing 5–25 mg/kg of selenium. The symptoms are lack of vitality, loss of appetite, emaciation, deformation and shedding of hoofs, loss of hair, and erosion of joints. Consumption of plants grown in seleniferous areas can cause effects in humans and animals. Selenosis in humans has occurred from ingestion of 3.2 mg selenium per day. Selenium is an essential trace element for many species.
Reacts to form explosive products with metal amides. Can react violently with barium carbide, bromine pentafluoride, calcium carbide, chlorates, chlorine trifluoride, chromic oxide (CrO_3), fluorine,

lithium carbide, lithium silicon (Li₆Si₂), metals, nickel, nitric acid, sodium, nitrogen trichloride, oxygen, potassium, potassium bromate, rubidium carbide, zinc, silver bromate, strontium carbide, thorium carbide, uranium. When heated to decomposition it emits toxic fumes of Se. See also SELENIUM COMPOUNDS.

SBP000   CAS: 7782-49-2   HR: 3
SELENIUM (colloidal)
af: Se   aw: 78.96
CONSENSUS REPORTS: Selenium and its compounds are on the Community Right-To-Know List. Reported in EPA TSCA Inventory.
OSHA PEL: TWA 0.2 mg(Se)/m³
ACGIH TLV: TWA 0.2 mg(Se)/m³
DFG MAK: 0.1 mg(Se)/m³
SAFETY PROFILE: Poison by intravenous route. When heated to decomposition it emits toxic fumes of Se. See also SELENIUM and SELENIUM COMPOUNDS.

SBP500   HR: 3
SELENIUM COMPOUNDS
CONSENSUS REPORTS: Selenium and its compounds are on the Community Right-To-Know List.
OSHA PEL: TWA 0.2 mg(Se)/m³
ACGIH TLV: TWA 0.2 mg(Se)/m³
DFG MAK: 0.1 mg(Se)/m³
SAFETY PROFILE: Poison by inhalation and intravenous routes. Some selenium compounds are experimental carcinogens. Selenium in small amounts is essential for normal growth of some animals. Deficiency or excess is associated with serious disease in livestock. Long-term exposure may be a cause of amyotrophic lateral sclerosis in humans, just as it may cause "blind staggers" in cattle. Elemental selenium has low acute systemic toxicity, but dust or fumes can cause serious irritation of the respiratory tract. Hydrogen selenide resembles other hydrides in being highly toxic, and selenium oxychloride is a vesicant. Some organoselenium compounds have the high

toxicity of other organometals. Inorganic selenium compounds can cause dermatitis. Garlic odor of breath is a common symptom. Pallor, nervousness, depression, digestive disturbances, and death have been reported in cases of chronic exposure. Selenium compounds are common air contaminants. When heated to decomposition they emit toxic fumes of Se. See also SELENIUM and specific compounds.

SBQ000   CAS: 144-34-3   HR: 3
SELENIUM DIMETHYL-
**   DITHIOCARBAMATE**
mf: C₁₂H₂₄N₄S₈•Se   mw: 559.84
PROP: Yellow powder, crystals. D: 1.58, melting range: 140–172°.
SYNS: METHYL SELENAC □ TETRAKIS(DIMETHYL-CARBAMODITHIOATO-S,S')SELENIUM
CONSENSUS REPORTS: IARC Cancer Review: Group 2B IMEMDT 7,56,87; Animal Inadequate Evidence IMEMDT 12,161,76. Selenium and its compounds are on the Community Right-To-Know List. Reported in EPA TSCA Inventory.
OSHA PEL: TWA 0.2 mg(Se)/m³
ACGIH TLV: TWA 0.2 mg(Se)/m³
DFG MAK: 0.1 mg(Se)/m³
SAFETY PROFILE: Suspected carcinogen. Selenium compounds are poisons. When heated to decomposition it emits very toxic fumes of Se, SOₓ, and NOₓ. See also SELENIUM COMPOUNDS and CARBAMATES.

SBQ500   CAS: 7446-08-4   HR: 3
SELENIUM(IV) DIOXIDE (1:2)
mf: O₂Se   mw: 110.96
PROP: White to sltly reddish, lustrous, crystalline powder or needles; yellow liquid; green vapor. Stable to light and heat. Mp: 340–350° (subl), d: 3.95 @ 15°/15°, vap press: 1 mm @ 157.0°. Very sol in H₂O; sol in C₆H₆; sltly sol in most org solvs.
SYNS: RCRA WASTE NUMBER U204 □ SELENIOUS ANHYDRIDE □ SELENIUM DIOXIDE □ SELENIUM OXIDE
CONSENSUS REPORTS: Selenium and its compounds are on the Community Right-

To-Know List. Reported in EPA TSCA Inventory. EPA Genetic Toxicology Program.

OSHA PEL: TWA 0.2 mg(Se)/m^3
ACGIH TLV: TWA 0.2 mg(Se)/m^3
DFG MAK: 0.1 mg(Se)/m^3
SAFETY PROFILE: Poison by subcutaneous route. Mildly toxic by inhalation. Mutation data reported. Experimental reproductive effects. Incompatible with PCl$_3$. Used as an oxidizing agent. When heated to decomposition it emits toxic fumes of Se. See also SELENIUM COMPOUNDS.

SBR000   CAS: 7488-56-4   HR: 3
SELENIUM(IV) DISULFIDE (1:2)
DOT: UN 2657
mf: S$_2$Se   mw: 143.08
PROP: Red-yellow crystals. Mp: <100°, bp: decomp.
SYNS: EXSEL □ RCA WASTE NUMBER U205 □ SELENIUM DISULPHIDE (DOT) □ SELENIUM SULFIDE □ SELSUN BLUE
CONSENSUS REPORTS: Selenium and its compounds are on the Community Right-To-Know List. Reported in EPA TSCA Inventory.
OSHA PEL: TWA 0.2 mg(Se)/m^3
ACGIH TLV: TWA 0.2 mg(Se)/m^3
DFG MAK: 0.1 mg(Se)/m^3
DOT CLASSIFICATION: 6.1; Label: Poison
SAFETY PROFILE: Poison by ingestion. Used in shampoos. When heated to decomposition it emits very toxic fumes of SO$_x$ and Se. See also SELENIUM COMPOUNDS and SULFIDES.

SBR500   HR: 2
SELENIUM(IV) DISULFIDE SHAMPOO (2.5%)
SYN: SELENIUM DISULFIDE (2.5%) SHAMPOO
CONSENSUS REPORTS: Selenium and its compounds are on the Community Right-To-Know List.
OSHA PEL: TWA 0.2 mg(Se)/m^3
ACGIH TLV: TWA 0.2 mg(Se)/m^3
DFG MAK: 0.1 mg(Se)/m^3
SAFETY PROFILE: A severe eye irritant. When heated to decomposition it emits very

toxic fumes of Se and SO$_x$. See also SELENIUM(IV) DISULFIDE (1:2), SELENIUM COMPOUNDS, and SULFIDES.

SBS000   CAS: 7783-79-1   HR: 3
SELENIUM HEXAFLUORIDE
DOT: UN 2194
mf: F$_6$Se   mw: 192.96
PROP: Thermally, very stable, colorless gas. White solid at low temp. Sublimes before melting. Inert to water. Mp: −46.6°, d: 3.27 @ −46.6°. Sublimes @ −34°. IDLH 2 ppm.
SYN: SELENIUM FLUORIDE
CONSENSUS REPORTS: Selenium and its compounds are on the Community Right-To-Know List.
OSHA PEL: TWA 0.05 ppm (Se)
ACGIH TLV: TWA 0.05 ppm (Se)
DFG MAK: 0.1 mg(Se)/m^3
DOT CLASSIFICATION: 2.3; Label: Poison Gas
SAFETY PROFILE: Poison by inhalation. When heated to decomposition it emits very toxic fumes of F$^-$ and Se. See also SELENIUM COMPOUNDS and FLUORIDES.

SBT000   CAS: 7446-34-6   HR: 3
SELENIUM MONOSULFIDE
mf: SSe   mw: 111.02
PROP: Orange-yellow tablets or powder. Known only in vapor state. Mp: 111.03°, bp: decomp @ 118–119°, d: 3.056 @ 0°.
SYNS: NCI-C50033 □ SELENIUM SULFIDE □ SELENIUM SULPHIDE □ SELENSULFID (GERMAN) □ SULFUR SELENIDE
CONSENSUS REPORTS: NTP 10th Report on Carcinogens. NCI Carcinogenesis Bioassay (dermal); Inadequate Studies: mouse NCITR* NCI-CG-TR-197,80; (gavage); Clear Evidence: mouse, rat NCITR* NCI-CG-TR-194,80. Selenium and its compounds are on the Community Right-To-Know List.
OSHA PEL: TWA 0.2 mg(Se)/m^3
ACGIH TLV: TWA 0.2 mg(Se)/m^3
DFG MAK: 0.1 mg(Se)/m^3
SAFETY PROFILE: Confirmed carcinogen with experimental carcinogenic data. Poison

S

by ingestion. Mutation data reported. When heated to decomposition it emits very toxic fumes of SO_x and Se. See also SELENIUM COMPOUNDS and SULFIDES.

SBT200   CAS: 12640-89-0   HR: 3
SELENIUM OXIDE
DOT: NA 2811

mf: OSe   mw: 94.96

SYN: SELENIUM OXIDE (DOT)

CONSENSUS REPORTS: Reported in EPA TSCA Inventory.

DOT CLASSIFICATION: 6.1; Label: Poison

SAFETY PROFILE: A poison. When heated to decomposition it emits toxic vapors of selenium.

SBT500   CAS: 7791-23-3   HR: 3
SELENIUM OXYCHLORIDE
DOT: UN 2879

mf: Cl_2OSe mw: 165.86

PROP: Colorless to yellowish liquid with high dielectric construction. Mp: 10.9°, bp: 176.4°, d: 2.42 @ 22°, vap press: 1 mm @ 34.8°. Sol in org solvs.

SYNS: SELENINYL CHLORIDE □ SELENIUM CHLORIDE OXIDE

CONSENSUS REPORTS: Selenium and its compounds are on the Community Right-To-Know List. EPA Extremely Hazardous Substances List. Reported in EPA TSCA Inventory.

OSHA PEL: TWA 0.2 mg(Se)/m³

ACGIH TLV: TWA 0.2 mg(Se)/m³

DFG MAK: 0.1 mg(Se)/m³

DOT CLASSIFICATION: 8; Label: Corrosive, Poison

SAFETY PROFILE: Poison by skin contact and subcutaneous routes. Human systemic effects by skin contact with very small amounts: primary irritant, corrosive. Explodes on contact with potassium, white phosphorus. Ignites on contact with antimony. Vigorous reaction with metal oxides (e.g., silver oxide, lead(II) oxide, lead(IV) oxide, lead(II)(IV) oxide). When heated to decomposition it emits very toxic fumes of Cl⁻ and Se. See also SELENIUM COMPOUNDS and CHLORIDES.

SBU000   CAS: 10026-03-6   HR: 3
SELENIUM TETRACHLORIDE
mf: Cl_4Se mw: 220.76

PROP: Cubic white-yellow, deliq crystals. Completely dissoc in vapor phase to $SeCl_2$ + Cl_2. Hydrolyzes in moist air to H_2SeO_3 + HCl. Mp: 300° (subl @ 170–196°), bp: decomp @ 288°, vap press: 1 mm @ 74°, d: 2.6. Decomp in water and moist air. Insol in liquid bromine; decomp by dry ammonia. Mod sol in nonpolar org solvs.

SYN: SELENIUM(IV) CHLORIDE (1:4)

CONSENSUS REPORTS: Selenium and its compounds are on the Community Right-To-Know List. Reported in EPA TSCA Inventory.

OSHA PEL: TWA 0.2 mg(Se)/m³

ACGIH TLV: TWA 0.2 mg(Se)/m³

DFG MAK: 0.1 mg(Se)/m³

SAFETY PROFILE: Poison by subcutaneous route. When heated to decomposition it emits very toxic fumes of Se and Cl⁻. See also SELENIUM COMPOUNDS.

SBU150   CAS: 6512-83-0   HR: 3
2,2'-SELENOBIS(BENZOIC ACID)
mf: $C_{14}H_{10}O_4Se_2$ mw: 400.16

PROP: Crystals from AcOH/dioxan. Mp: 297° (decomp).

SYNS: ACIDE DISELINO SALICYLIQUE □ BENZOIC ACID, 2,2'-DISELENOBIS- □ DISELENO SALICYLIC ACID □ o,o'-SELENOBIS(BENZOIC ACID)

OSHA PEL: TWA 0.2 mg(Se)/m³

ACGIH TLV: TWA 0.2 mg(Se)/m³

SAFETY PROFILE: Poison by intraperitoneal and intravenous routes. When heated to decomposition it emits toxic fumes of Se.

SBU950   CAS: 53184-19-3   HR: 3
SELENO-TOLUIDINE BLUE
mf: $C_{15}H_{16}N_3Se•Cl$ mw: 352.75

SYNS: 3-AMINO-7-(DIMETHYLAMINO)-2-METHYL-PHENOSELENAZIN-5-IUM CHLORIDE □ PHENO-SELENAZIN-5-IUM, 3-AMINO-7-(DIMETHYLAMINO)-2-METHYL-, CHLORIDE

OSHA PEL: TWA 0.2 mg(Se)/m³

ACGIH TLV: TWA 0.2 mg(Se)/m³

SAFETY PROFILE: Poison by unreported route. When heated to decomposition it emits toxic fumes of NO_x and Se.

SBV000   CAS: 630-10-4   HR: 3
SELENOUREA
mf: CH_4N_2Se mw: 123.03
PROP: Prisms or needles in water. Mp: decomp @ 200°+. Sol in water, alc, ether.
SYN: RCRA WASTE NUMBER P103
CONSENSUS REPORTS: Selenium and its compounds are on the Community Right-To-Know List. Reported in EPA TSCA Inventory.
OSHA PEL: TWA 0.2 mg(Se)/m³
ACGIH TLV: TWA 0.2 mg(Se)/m³
DFG MAK: 0.1 mg(Se)/m³
SAFETY PROFILE: Poison by ingestion and intravenous routes. When heated to decomposition it emits very toxic fumes of NO_x and Se. See also SELENIUM COMPOUNDS.

SBW000   HR: 2
SELSUN
PROP: Consists of a 2.4% w/v selenium sulfide in aq suspension, also contains bentonite, sodium alkyl aryl sulfonate, sodium phosphate, glycerol monoricinoleate, citric acid, captan, and perfume (FEREAC 41,7218,76).
SYN: NCI-C54546
CONSENSUS REPORTS: NCI Carcinogenesis Bioassay (dermal); No Evidence: mouse NCITR* NCI-CG-TR-199,80. Selenium and its compounds are on the Community Right-To-Know List.
OSHA PEL: TWA 0.2 mg(Se)/m³
ACGIH TLV: TWA 0.2 mg(Se)/m³
DFG MAK: 0.1 mg(Se)/m³
SAFETY PROFILE: A severe eye irritant. Used as a pharmaceutical and veterinary drug. When heated to decomposition it emits very toxic fumes of Se, SO_x, PO_x, Na_2O, and NO_x. See also SELENIUM COMPOUNDS, BENTONITE, CITRIC ACID, and CAPTAN.

SBW500   CAS: 563-41-7   HR: 3
SEMICARBAZIDE HYDROCHLORIDE
mf: $CH_5N_3O \cdot ClH$ mw: 111.55
PROP: Prisms from dilute alc. Decomp @ 175–185°, mp: 176° (decomp). Very sol in water; very sltly sol in hot alc; insol in anhyd ether.
SYNS: AMIDOUREA HYDROCHLORIDE □ AMINO-UREA HYDROCHLORIDE □ CARBAMYLHYDRAZINE HYDROCHLORIDE □ CH □ HYDRAZINECARBOX-AMIDE MONOHYDROCHLORIDE
CONSENSUS REPORTS: IARC Cancer Review: Group 3 IMEMDT 7,56,87; Animal Sufficient Evidence IMEMDT 12,209,76. Reported in EPA TSCA Inventory. EPA Extremely Hazardous Substances List.
SAFETY PROFILE: Poison by ingestion and intraperitoneal routes. Experimental reproductive effects. Questionable carcinogen with experimental neoplastigenic and teratogenic data. Mutation data reported. When heated to decomposition it emits very toxic fumes of NO_x and HCl.

SCA350   CAS: 302-84-1   HR: D
dl-SERINE
mf: $C_3H_7NO_3$ mw: 105.09
PROP: White crystals from powder, or crystalline powder. Mp: 246° (decomp). Sol in water; insol in alc, ether.
SAFETY PROFILE: When heated to decomposition emits toxic fumes of NO_x.

SCC700   HR: D
SHELLAC, BLEACHED
PROP: From the resinous secretion, called lac, of the insect *Llaccifer (Tachardia) lacca Kerr* (Fam. *Coccidae*). Off white, amorphous, granular solid. Sol in alc; insol in water; sltly sol in acetone, ether.
SYNS: WHITE SHELLAC □ REGULAR BLEACHED SHELLAC
SAFETY PROFILE: When heated to decomposition it emits acrid smoke and irritating fumes.

SCF600   CAS: 171599-83-0 HR: 3
SILDENAFIL CITRATE
mf: $C_{22}H_{30}N_6O_4S \cdot C_6H_8O_7$ mw: 666.78
SYNS: PIPERAZINE, 1-((3-(4,7-DIHYDRO-1-METHYL-7-OXO-3-PROPYL-1H-PYRAZOLO(4,3-D)PYRIMIDIN-5-Y

L)-4-ETHOXYPHENYL)SULFONYL)-4-METHYL-, 2-HYDROXY-1,2,3-PROPANETRICARBOXYLATE (1:1) □ UK 92480-10 □ VIAGRA

SAFETY PROFILE: A poison by ingestion. Human systemic effects. When heated to decomposition it emits toxic vapors of NO_x and SO_x.

SCH001   CAS: 69012-64-2   HR: 1
SILICA, AMORPHOUS FUME

SYNS: AMORPHOUS SILICA FUME □ SILICA, AMORPHOS-FUME (ACGIH)

CONSENSUS REPORTS: Reported in EPA TSCA Inventory.

ACGIH TLV: TWA 2 mg/m³, respirable dust

SAFETY PROFILE: An inhalation hazard. Much less toxic than crystalline forms. Does not cause silicosis. See also other silica entries.

SCH002   CAS: 112945-52-5 HR: 2
SILICA, AMORPHOUS FUMED

mf: O_2Si mw: 60.09

PROP: A finely powdered microcellular silica foam with minimum SiO_2 content of 89.5%. Insol in water; sol in hydrofluoric acid.

SYNS: ACTICEL □ AEROSIL □ AMORPHOUS SILICA DUST □ AQUAFIL □ CAB-O-GRIP II □ CAB-O-SIL □ CAB-O-SPERSE □ CATALOID □ COLLOIDAL SILICA □ COLLOIDAL SILICON DIOXIDE □ DAVISON SG-67 □ DICALITE □ DRI-DIE PESTICIDE 67 □ ENT 25,550 □ FOSSIL FLOUR □ FUMED SILICA □ FUMED SILICON DIOXIDE □ LUDOX □ NALCOAG □ NYACOL □ NYACOL 830 □ NYACOL 1430 □ SANTOCEL □ SG-67 □ SILICA, AMORPHOUS □ SILICIC ANHYDRIDE □ SILICON DIOXIDE (FCC) □ SILIKILL □ VULKASIL

CONSENSUS REPORTS: IARC Cancer Review: Group 3 IMEMDT 7,341,87; Animal Inadequate Evidence IMEMDT 42,209,88; Human Inadequate Evidence IMEMDT 42,209,88. Reported in EPA TSCA Inventory.

SAFETY PROFILE: Poison by intraperitoneal, intravenous, and intratracheal routes. Moderately toxic by ingestion. An inhalation hazard. Much less toxic than crystalline forms. Questionable carcinogen with experimental carcinogenic data. Mutation data reported. Does not cause silicosis. See also other silica entries.

SCI000   CAS: 7631-86-9   HR: 1
SILICA, AMORPHOUS HYDRATED

mf: O_2Si mw: 60.09

PROP: Transparent, tasteless crystals or amorphous powder. Melts to a glass at ordinary temps. Chemically resistant to most reagents. Mp: 1716–1736°, bp: 2230°. Insol in H_2O; sol in HF (giving fluorosilicate ions). IDLH 3000 mg/m³.

SYNS: SILICA AEROGEL □ SILICA GEL □ SILICA XEROGEL □ SILICIC ACID

CONSENSUS REPORTS: IARC Cancer Review: Animal Inadequate Evidence IMEMDT 42,209,88; Human Inadequate Evidence IMEMDT 42,209,88.

OSHA PEL: TWA 6 mg/m³

ACGIH TLV: TWA (nuisance particulate) 10 mg/m³ of total dust (when toxic impurities are not present, e.g., quartz <1%)

DFG MAK: 4 mg/m³ as total dust

SAFETY PROFILE: The pure unaltered form is considered a nuisance dust. Some deposits contain small amounts of crystalline quartz and are therefore fibrogenic. When diatomaceous earth is calcined (with or without fluxing agents) some silica is converted to cristobalite and is therefore fibrogenic. Tridymite has never been detected in calcined diatomaceous earth. See also other silica entries.

SCI500   HR: 3
SILICA, CRYSTALLINE

mf: O_2Si mw: 60.09

PROP: Transparent, tasteless crystals or amorph powder. Mp: 1710°, bp: 2230°, d: (amorph) 2.2, d: (crystalline) 2.6, vap press: 10 mm @ 1732°. Practically insol in water or acids. Dissolves readily in HF, forming silicon tetrafluoride.

SYNS: AGATE □ AMETHYST □ CHALCEDONY □ CHERTS □ CRISTOBALITE □ FLINT □ ONYX □ PURE QUARTZ □ ROSE QUARTZ □ SAND □ SILICA FLOUR □ SILICON DIOXIDE □ TRIDYMITE □ TRIPOLI

CONSENSUS REPORTS: IARC Cancer Review: Group 2A IMEMDT 7,341,87; Animal Sufficient Evidence IMEMDT 42,209,88; Human Limited Evidence IMEMDT 42,209,88.

OSHA PEL: Total Dust: TWA 30 mg/m^3/2(%SiO$_2$+2) Respirable Fraction: TWA 0.05 mg/m^3
ACGIH TLV: TWA Respirable Fraction: 0.05 mg/m^3
DFG MAK: 0.15 mg/m^3
NIOSH REL: (Silica, Crystalline) TWA 50 μg/m^3
SAFETY PROFILE: Moderately toxic as an acute irritating dust. From the point of view of numbers of workers exposed and cases of disability produced, silica is the chief cause of pulmonary dust disease. The prolonged inhalation of dusts containing free silica may result in the development of a disabling pulmonary fibrosis known as silicosis. The Committee on Pneumoconiosis of the American Public Health Association defines silicosis as "a disease due to the breathing of air containing silica (SiO$_2$) characterized by generalized fibrotic changes and the development of miliary nodules in both lungs, and clinically by shortness of breath, decreased chest expansion, lessened capacity for work, absence of fever, increased susceptibility to tuberculosis (some or all of the symptoms may be present), and characteristic x-ray findings."

Silica occurs in the pure state in nature as highly fibrogenic quartz. It is the main constituent of relatively much less toxic sand, sandstone, tripoli, and diatomaceous earth. It is present in crystalline form in high amounts (up to 35%) in granite. Exposure to silica occurs in hard rock mining, in foundries, in manufacture of porcelain and pottery, in the spraying of vitreous enamels, in sandblasting, in granite-cutting and tombstone-making, in the manufacture of silica firebrick and other refractories, in grinding and polishing operations where natural abrasive wheels are used, and other occupations.

The duration of exposure which is associated with the development of silicosis varies widely for different occupations. Thus, the average duration of exposure required for the development of silicosis in sand-blasters is 2–10 years, in molders and granite cutters, about 30 years, and in hard rock miners, 10–15 years. There is also much variation in individual susceptibility; certain workers show radiological evidence of the disease years before their fellow workmen who are similarly exposed. Such susceptible individuals are, fortunately, rather rare.

The action of crystalline silica on the lungs results in the production of a diffuse, nodular fibrosis in which the parenchyma and the lymphatic systems are involved. This fibrosis is, to a certain extent, progressive, and may continue to increase for several years after exposure is terminated. Where the pulmonary reserve is sufficiently reduced, the worker complains of shortness of breath on exertion. This is the first and most common symptom in cases of uncomplicated silicosis. If severe, it may incapacitate the worker for heavy, or even light, physical exertion, and in extreme cases there may be shortness of breath even while at rest. The most common physical sign of silicosis is a limitation of expansion of the chest. There may be a dry cough, sometimes very troublesome. The characteristic radiographic appearance is one of diffuse, discrete nodulation, scattered throughout both lung fields. Where the disease advances, the shortness of breath becomes worse, and the cough more productive and troublesome. There is no fever or other evidence of systemic reaction. Further progress of the disease results in marked fatigue, extreme dyspnea and cyanosis, loss of appetite, pleuritic pain, and total incapacity to work. If tuberculosis does not supervene, the condition may eventually cause death either from cardiac failure or from destruction of lung tissue, with resultant anoxemia. In the later stages, the x-ray may show large conglomerate shadows, due to the coalescence of the silicotic nodules, with areas of emphysema between them.

Silica in some forms is used as an additive permitted in the feed and drinking water of animals and/or for the treatment of food-

S

producing animals. It is also permitted in food for human consumption. It is a common air contaminant. Reacts violently with ClF_3, MnF_3, OF_2. See also other silica entries.

SCJ000   CAS: 14464-46-1   HR: 3
SILICA, CRYSTALLINE–CRISTOBALITE

mf: O_2Si mw: 60.09

PROP: White, cubic-system crystals formed from quartz at temperatures above 1000°C (NTIS** PB246–697).

SYNS: CALCINED DIATOMITE □ CRISTOBALITE

CONSENSUS REPORTS: NTP 10th Report on Carcinogens. IARC Cancer Review: Group 2A IMEMDT 7,341,87; Animal Sufficient Evidence IMEMDT 42,209,88; Human Limited Evidence IMEMDT 42,209,88. Reported in EPA TSCA Inventory.

OSHA PEL: Total Dust: TWA 30 $mg/m^3/2(\%SiO_2+2)$ TWA Respirable Fraction: 0.05 mg/m^3

ACGIH TLV: TWA Respirable Fraction: 0.05 mg/m^3

DFG MAK: 0.15 mg/m^3

NIOSH REL: (Silica, Crystalline) TWA 50 $\mu g/m^3$

SAFETY PROFILE: Confirmed carcinogen with experimental carcinogenic and tumorigenic data. Poison by intratracheal route. An inhalation hazard. Human systemic effects by inhalation: cough, dyspnea, fibrosis. About twice as toxic as silica in causing silicosis. See also other silica entries.

SCJ500   CAS: 14808-60-7   HR: 3
SILICA, CRYSTALLINE–QUARTZ

mf: O_2Si mw: 60.09

PROP: White to reddish crystals. Stable below 8°. Low (α- quartz, stable at room temp; transforms to high (β-) quartz at 5°; the two forms are related by small rotations of the SiO_4 tetrahedron. Piezoelectric and pyroelectric. Mp: 1710°, bp: 2230°, d: 2.6. IDLH 25 mg/m^3 (cristobalite, tridymite); 50 mg/m^3 (quartz, tripoli).

SYNS: AGATE □ AMETHYST □ CHALCEDONY □ CHERTS □ FLINT □ ONYX □ PURE QUARTZ □ QUARTZ □ QUAZO PURO (ITALIAN) □ ROSE QUARTZ □ SAND □ SILICA FLOUR (powdered crystalline silica) □ SILICIC ANHYDRIDE

CONSENSUS REPORTS: NTP 10th Report on Carcinogens. IARC Cancer Review: Group 2A IMEMDT 7,341,87; Animal Sufficient Evidence IMEMDT 42,209,88; Human Limited Evidence IMEMDT 42,209,88. Reported in EPA TSCA Inventory.

OSHA PEL: Total Dust: TWA 30 $mg/m^3/2(\%SiO_2+2)$ TWA Respirable Fraction: 0.1 mg/m^3

ACGIH TLV: TWA 0.05 mg/m^3; Suspected Human Carcinogen

DFG MAK: 0.15 mg/m^3 as fine dust

NIOSH REL: TWA 50 $\mu g/m^3$; 3,000,000 fibers/m^3

SAFETY PROFILE: Confirmed carcinogen with experimental carcinogenic, tumorigenic, and neoplastigenic data. Experimental poison by intratracheal and intravenous routes. An inhalation hazard. Human systemic effects by inhalation: cough, dyspnea, liver effects. Incompatible with OF_2, vinyl acetate. See also other silica entries.

SCK000   CAS: 15468-32-3   HR: 3
SILICA, CRYSTALLINE–TRIDYMITE

mf: O_2Si mw: 60.09

PROP: White or colorless platelets or orthorhombic crystals formed from quartz @ temperatures >870° (NTIS** PB246–697). Stable from 870–914° at atmospheric pressure, but persists as a metastable phase below 8° forming low tridymite below 1° and middle tridymite from 117–1°. IDLH 25 mg/m^3.

SYNS: CHRISTENSENITE □ SILICA, CRYSTALLINE-TRIDYMITE (ACGIH, OSHA) □ TRIDIMITE (FRENCH) □ TRIDYMITE □ α-TRIDYMITE □ TRIDYMITE 118

CONSENSUS REPORTS: NTP 10th Report on Carcinogens. IARC Cancer Review: Group 2A IMEMDT 7,341,87; Animal Sufficient Evidence IMEMDT 42,209,88; Human Limited Evidence IMEMDT 42,209,88.

OSHA PEL: Total Dust: TWA 30
$mg/m^3/2(\%SiO_2+2)$ TWA 0.05 mg/m^3
ACGIH TLV: (Silica, Crystalline) TWA
Respirable Fraction: 0.05 mg/m^3
DFG MAK: 0.15 mg/m^3
NIOSH REL: TWA 50 μg/m^3
SAFETY PROFILE: Confirmed carcinogen
with experimental tumorigenic data. Poison
by intratracheal route. Human systemic
effects by inhalation: cough, dyspnea. About
twice as toxic as silica in causing silicosis.
Mutation data reported. See also other silica
entries.

SCK500   HR: 3
SILICA FLOUR
PROP: A finely ground *crystalline* silica
sometimes marketed as "Amorphous." It is
not amorphous.
SAFETY PROFILE: Toxic by inhalation. It
has been shown to produce a very high
incidence of silicosis among "silica flour"
workers. See also other silica entries.

SCK600   CAS: 60676-86-0   HR: 3
SILICA, FUSED
mf: O_2Si mw: 60.09
PROP: Made up of spherical
submicroscopic particles under 0.1 micron
in size (AMIHBC 9,389,54).
SYNS: ACCUSAND □ AMORPHOUS QUARTZ □
AMORPHOUS SILICA □ BORSIL P □ CRYPTOCRYS-
TALLINE QUARTZ □ DENKA F 90 □ DENKA FB 44 □
EF 10 □ F 44 □ F 125 □ FS 74 □ FUSED QUARTZ □
FUSED SILICA □ FUSELEX □ FUSELEX RD 120 □
FUSELEX RD 40-60 □ FUSELEX ZA 30 □ GP 7I □ GP 11I
□ MICROCRYSTALLINE QUARTZ □ MR 84 □ NALCAST
□ OPTOCIL □ OPTOCIL (QUARTZ) □ QG 100 □
QUARTZ GLASS □ QUARTZ SAND □ RANCOSIL □ RD
8 □ RD 120 □ S-COL □ SGA □ SILICA, AMORPHOUS-
FUSED (ACGIH) □ SILICA, FUSED □ SILICA, FUSED
(OSHA) □ SILICA, VITREOUS (9CI) □ SILICON DIOXIDE
□ SILICONE DIOXIDE □ SILTEX □ SPECTROSIL □
SUPRASIL □ SUPRASIL W □ VITREOSIL IR □
VITREOUS QUARTZ □ VITREOUS SILICA □ VITRIFIED
SILICA □ Y 40
CONSENSUS REPORTS: IARC Cancer
Review: Group 3 IMEMDT 7,341,87;
Animal Inadequate Evidence IMEMDT
42,39,87; Human Inadequate Evidence

IMEMDT 42,39,87. Reported in EPA
TSCA Inventory.
OSHA PEL: Total Dust: TWA 30
$mg/m^3/2(\%SiO_2+2)$ TWA 0.1 mg/m^3
ACGIH TLV: TWA 0.1 mg/m^3 (Respirable
Fraction)
DFG MAK: 0.3 mg/m^3 as fine dust
NIOSH REL: (Silica, Crystalline) TWA 0.05
mg/m^3
SAFETY PROFILE: An inhalation hazard.
Questionable carcinogen with experimental
tumorigenic data. Poison by intraperitoneal,
intravenous, and intratracheal routes. See
also other silica entries.

SCL000   CAS: 7699-41-4   HR: 3
SILICA, GEL and
AMORPHOUS–PRECIPITATED
mf: H_2O_3Si mw: 78.11
PROP: White amorphous powder. Insol in
H_2O.
SYNS: KIESELSAEURE (GERMAN) □ METASILICIC
ACID □ PRECIPITATED SILICA □ SILICA GEL □
SILICIC ACID
CONSENSUS REPORTS: IARC Cancer
Review: Group 3 IMEMDT 7,341,87;
Animal Inadequate Evidence IMEMDT
42,39,87; Human Inadequate Evidence
IMEMDT 42,39,87. Reported in EPA
TSCA Inventory.
OSHA PEL: TWA 6 mg/m^3
ACGIH TLV: TWA (nuisance particulate) 10
mg/m^3 of total dust (when toxic impurities
are not present, e.g., quartz <1%)
DFG MAK: 0.3 mg/m^3 as fine dust
SAFETY PROFILE: An inhalation hazard.
Poison by intravenous route. An eye irritant
and nuisance dust. Questionable carcinogen.
Mutation data reported. See also other
silica entries and SILICATES.

SCM500   HR: D
SILICATES
PROP: Widely occurring compounds
containing silicon, oxygen, and one or more
metals with or without hydrogen.
SAFETY PROFILE: Soluble alkaline silicates
act locally like mild alkalies. The dusts of
certain silicates, such as asbestos (hydrated
magnesium silicate) and talc, can produce

S

fibrotic changes in the lungs and are implicated as experimental carcinogens. React violently with Li. See also mica, soapstone, talc.

SCN000   HR: 2
SILICATE SOAPSTONE
mf: $3MgO•4SiO_2•H_2O$ mw: 379.31
PROP: Containing less than 1% crystalline silica (FEREAC 39,23540,74). IDLH 3000 mg/m³.
SYN: SOAPSTONE
OSHA PEL: TWA Total Dust: 6 mg/m³; Respirable Fraction: 3 mg/m³
ACGIH TLV: (Proposed: 1 mg/m³; Not Classifiable as a Human Carcinogen)
SAFETY PROFILE: Less toxic than quartz. See SILICATES.

SCN500   CAS: 15191-85-2   HR: 3
SILICIC ACID, BERYLLIUM SALT
mf: $O_4Si•2Be$ mw: 110.11
PROP: Colorless crystals. Decomp at >156°. D: 3.0.
SYNS: BERYLLIUM ORTHOSILICATE □ BERYLLIUM SILICATE □ BERYLLIUM SILICIC ACID
CONSENSUS REPORTS: IARC Cancer Review: Group 1 IMEMDT 58,41,93; Human Sufficient Evidence IMEMDT 58,41,93; Animal Sufficient Evidence IMEMDT 1,17,72; Animal Sufficient Evidence IMEMDT 23,143,80; Animal Sufficient Evidence IMEMDT 58,41,93. Beryllium and its compounds are on the Community Right-To-Know List.
OSHA PEL: TWA 0.002 mg(Be)/m³; STEL 0.005 mg(Be)/m³/30M; CL 0.025 mg(Be)/m³
ACGIH TLV: TWA 0.002 mg(Be)/m³; Confirmed Human Carcinogen; (Proposed: TWA 0.00005 mg(Be)/m³; STEL 0.0002 mg(Be)/m³ (sensitizer) (skin); Confirmed Human Carcinogen)
NIOSH REL: CL not to exceed 0.0005 mg/(Be)/m³
SAFETY PROFILE: Confirmed carcinogen with experimental carcinogenic and tumorigenic data. When heated to decomposition it emits toxic fumes of BeO.

See also BERYLLIUM COMPOUNDS and SILICATES.

SCO500   CAS: 16961-83-4   HR: 3
SILICOFLUORIC ACID
DOT: NA 1778
mf: $F_6Si•2H$ mw: 144.11
PROP: Transparent, colorless, fuming liquid. Bp: decomp.
SYNS: ACIDE FLUOROSILICIQUE (FRENCH) □ ACIDE FLUOSILICIQUE (FRENCH) □ ACIDO FLUOSILICICO (ITALIAN) □ FLUOROSILICIC ACID □ FLUOSILICIC ACID □ HEXAFLUOROKIESELSAEURE (GERMAN) □ HEXAFLUOROKIEZELZUUR (DUTCH) □ HEXAFLUOROSILICATE(2-) DIHYDROGEN □ HEXAFLUOSILICIC ACID □ HYDROFLUOSILICIC ACID □ HYDROGEN HEXAFLUOROSILICATE □ HYDRO-SILICOFLUORIC ACID □ KIEZELFLUORWATERSTOF-ZUUR (DUTCH) □ SAND ACID
CONSENSUS REPORTS: Reported in EPA TSCA Inventory.
OSHA PEL: TWA 2.5 mg(F)/m³
ACGIH TLV: TWA 2.5 mg(F)/m³; BEI: 3 mg/g creatinine of fluorides in urine prior to shift; 10 mg/g creatinine of fluorides in urine at end of shift.
DOT CLASSIFICATION: 8; Label: Corrosive
NIOSH REL: TWA 2.5 mg(F)/m³
SAFETY PROFILE: Poison by subcutaneous route. A corrosive irritant to skin, eyes, and mucous membranes. Will react with water or steam to produce toxic and corrosive fumes. When heated to decomposition it emits toxic fumes of F⁻. See also FLUORIDES.

SCP000   CAS: 7440-21-3   HR: 3
SILICON
DOT: UN 1346
af: Si   aw: 28.09
PROP: Cubic, steel-gray crystals or black-brown amorphous powder. Bulk Si is unreactive to O_2, H_2O, H halides (except HF) but dissolves in hot aq alkalies. Reactive to halogens, e.g., F_2 at room temp, Cl_2 at 3°. Acted on by N_2 at 14°. S reacts at 6°, and P at 10°. Mp: 1410°, bp: 2355°, d: 2.42 or 2.3 @ 20°, vap press: 1 mm @ 1724°. Almost insol in water; sol in molten alkali oxides.
SYNS: DEFOAMER S-10 □ SILICON POWDER, amorphous (DOT)

CONSENSUS REPORTS: Reported in EPA TSCA Inventory.

OSHA PEL: TWA Total Dust: 10 mg/m^3 of total; Respirable Fraction: 5 mg/m^3

ACGIH TLV: TWA (nuisance particulate) 10 mg/m^3 of total dust (when toxic impurities are not present, e.g., quartz <1%)

DOT CLASSIFICATION: 4.1; Label: Flammable Solid

SAFETY PROFILE: A nuisance dust. Moderately toxic by ingestion. An eye irritant. Does not occur freely in nature, but is found as silicon dioxide (silica) and as various silicates. Elemental Si is flammable when exposed to flame or by chemical reaction with oxidizers. Violent reactions with alkali carbonates, oxidants, (Al + PbO), Ca, Cs$_2$C$_2$, Cl$_2$, CoF$_2$, F$_2$, IF$_5$, MnF$_3$, Rb$_2$C$_2$, FNO, AgF, NaK alloy. When heated it will react with water or steam to produce H$_2$; can react with oxidizing materials. See also various silica entries, SILICATES, and POWDERED METALS.

SCQ000   CAS: 409-21-2   HR: 3
SILICON CARBIDE
mf: CSi   mw: 40.10

PROP: Bluish-black, iridescent, hexagonal or cubic crystals; colorless when pure. Mp: 2600°, bp: subl >2000°, decomp @ 2210°, d: 3.17. Insol in H$_2$O and acids; sol in fused KOH.

SYNS: ANNANOX CK □ BETARUNDUM □ BETA-RUNDUM ST-S □ BETARUNDUM UF □ BETARUNDUM ULTRAFINE □ CARBOFRAX M □ CARBOLON □ CARBON SILICIDE □ CARBORUNDEUM □ CARBO-RUNDUM □ CRYSTAR □ CRYSTOLON 37 □ CRYSTOL-ON 39 □ DENSIC C 500 □ DU-A 1 □ DU-A 2 □ DU-A 3 □ DU-A 4 □ DU-A 3C □ GC 10000 □ GREEN DENSIC □ GREEN DENSIC GC 800 □ HITACERAM SC 101 □ KZ 3M □ KZ 5M □ KZ 7M □ SC 9 □ SC 9 (CARBIDE) □ SC 201 □ SCW 1 □ SD-GP 6000 □ SD-GP 8000 □ SILICON CARB-IDE (ACGIH, OSHA) □ SILICON MONOCARBIDE □ SILUNDUM □ TOKAWHISKER □ UA 1 □ UA 2 □ UA 3 □ UA 4 □ UF 15 □ YE 5626

CONSENSUS REPORTS: Reported in EPA TSCA Inventory.

OSHA PEL: TWA Total Dust: 10 mg/m^3; Respirable Fraction: 5 mg/m^3

ACGIH TLV: TWA (inhalable nonfibrous) 10 mg/m^3 of total dust; (respirable fraction) 3 mg/m^3; (fibrous forms) 0.1 fibers/cc; Suspected Human Carcinogen

DFG MAK: (with fibers) Animal Carcinogen, Suspected Human Carcinogen; (without fibers) 1.5 mg/m^3

NIOSH REL: (Silicon Carbide, Total Dust) TWA 10 mg/m^3

SAFETY PROFILE: Suspected carcinogen with experimental neoplastigenic data. A nuisance dust.

SCQ500   CAS: 10026-04-7   HR: 2
SILICON CHLORIDE
DOT: UN 1818

mf: Cl$_4$Si   mw: 169.89

PROP: Colorless, fuming liquid; suffocating odor. Hydrolyzes by H$_2$O and alkalies. Mp: −70°, bp: 57.57°, d: 1.482. Misc with benzene, ether, chloroform, pet ether.

SYNS: CHLORID KREMICITY (CZECH) □ EXTREMA □ SILICIO(TETRACLORURO di) □ SILICIUMTETRA-CHLORID (GERMAN) □ SILICIUMTETRACHLORIDE (DUTCH) □ SILICIUM(TETRACHLORURE de) (FRENCH) □ SILICON TETRACHLORIDE (DOT) □ TETRACHLO-ROSILANE □ TETRACHLORURE de SILICIUM (FRENCH)

CONSENSUS REPORTS: Reported in EPA TSCA Inventory.

DOT CLASSIFICATION: 8; Label: Corrosive

SAFETY PROFILE: Mildly toxic by inhalation. A corrosive irritant to eyes, skin, and mucous membranes. Reacts with water to form HCl. Violent reaction with Na, K. When heated to decomposition it emits toxic fumes of Cl$^-$. See also CHLOROSILANES.

SCR400   CAS: 63148-62-9   HR: D
SILICONE 360
SYNS: ANTIFOAM FD 62 □ DC 360 □ KO 08 □ PMS 1.5 □ PMS 300 □ PMS 154A □ PMS 200A □ PNS 25 □ S DC 200 □ SILAK M 10 □ SILICONE DC 200 □ SILICONE DC 360 □ SILICONE DC 360 FLUID □ SILICONE RELEASE L 45 □ SILIKON ANTIFOAM FD 62 □ SILOXANES and SILICONES, DI Me □ UC LIQUID G □ UNION CARBIDE LIQUID G □ XF-13-563

CONSENSUS REPORTS: EPA Genetic Toxicology Program. Reported in EPA TSCA Inventory.

SAFETY PROFILE: An experimental teratogen. Experimental reproductive effects. See also SILICONES.

S

SDC000   HR: D
SILICONES
SYN: SILOXANES

CONSENSUS REPORTS: Organosilicon oxide polymers such as $-R_2Si-O$, where R is a monovalent organic radical.

SAFETY PROFILE: Most of the silicones that have been studied are only slightly toxic and mildly irritating; however, some may be severe irritants. May be spontaneously flammable in air. There can be toxicity due to contamination of silicones by components of manufacture.

SDF650   CAS: 7783-61-1   HR: 3
SILICON FLUORIDE
DOT: UN 1859

mf: F_4Si mw: 104.09

PROP: Colorless, fuming gas; very pungent odor. Rapidly hydrol by H_2O. Subl $-95°$, mp: $-77°$, bp: $-65°$ @ 181 mm, d: 4.67.

SYNS: SILICON TETRAFLUORIDE (DOT) □ TETRAFLUOROSILANE

CONSENSUS REPORTS: Reported in EPA TSCA Inventory.

OSHA PEL: TWA 2.5 mg(F)/m^3

ACGIH TLV: TWA 2.5 mg(F)/m^3; BEI: 3 mg/g creatinine of fluorides in urine prior to shift; 10 mg/g creatinine of fluorides in urine at end of shift.

NIOSH REL: (Inorganic Fluorides) TWA 2.5 mg(F)/m^3

DOT CLASSIFICATION: 2.3; Label: Poison Gas, Corrosive

SAFETY PROFILE: A poison. Moderately toxic by inhalation. A corrosive irritant to skin, eyes, and mucous membranes. When heated to decomposition it emits toxic fumes of F^-. See also FLUORIDES and HYDROFLUORIC ACID.

SDH575   CAS: 7803-62-5   HR: 3
SILICON TETRAHYDRIDE
DOT: UN 2203

mf: H_4Si mw: 32.13

PROP: Colorless gas with repulsive odor; slowly decomp by water. D: 0.68 @ $-185°$, mp: $-185°$, bp: $-112°$, fp: $-200°$. Insol in Et_2O, C_6H_6, $CHCl_3$, and EtOH. Insol in H_2O; decomp in aq KOH.

SYNS: MONOSILANE □ SILANE □ SILICANE

CONSENSUS REPORTS: Reported in EPA TSCA Inventory.

OSHA PEL: TWA 5 ppm

ACGIH TLV: TWA 5 ppm

DOT CLASSIFICATION: 2.1; Label: Flammable Gas

SAFETY PROFILE: Mildly toxic by inhalation. Silanes are irritating to skin, eyes, and mucous membranes. Easily ignited in air. Explosive reaction or ignition on contact with halogens or covalent halides (e.g., bromine, chlorine, carbonyl chloride, antimony pentachloride, tin(IV) chloride). Ignites in oxygen. Can react with oxidizers. It may self-explode. When heated to decomposition it burns or explodes.

SDI500   CAS: 7440-22-4   HR: 2
SILVER
af: Ag   aw: 107.868

PROP: Soft, ductile, malleable, lustrous white metal. Tarnishes in air with formation of black sulfide. Physical properties dependent on mechanical treatment. Attacked by Cl_2, S, H_2S, metal cyanides (in air), chromic, nitric, and sulfuric acids. Mp: 961°, bp: 2163°, d: 10.50 @ 20°. IDLH 10 mg/m^3 (as Ag).

SYNS: ARGENTUM □ C.I. 77820 □ SHELL SILVER □ SILBER (GERMAN) □ SILVER ATOM

CONSENSUS REPORTS: Silver and its compounds are on the Community Right-To-Know List. Reported in EPA TSCA Inventory.

OSHA PEL: Metal, Dust, and Fume: TWA 0.01 mg/m^3

ACGIH TLV: TWA (metal) 0.1 mg/m^3, (soluble compounds as Ag) 0.01 mg/m^3

DFG MAK: 0.1 mg/m^3; (salts) 0.01 mg/m^3

NIOSH REL: (Silver, metal and soluble compounds) TWA 0.01 mg/m^3

SAFETY PROFILE: Human systemic effects by inhalation: skin effects. Inhalation of dusts can cause argyrosis. Questionable carcinogen with experimental tumorigenic data. Flammable in the form of dust when exposed to flame or by chemical reaction with C_2H_2, NH_3, bromoazide, ClF_3, ethyleneimine, H_2O_2, oxalic acid, H_2SO_4, tartaric acid. Incompatible with acetylene, acetylene compounds, aziridine, bromine azide, 3-bromopropyne, carboxylic acids, copper + ethylene glycol, electrolytes + zinc, ethanol + nitric acid, ethylene oxide, ethyl hydroperoxide, ethyleneimine, iodoform, nitric acid, ozonides, peroxomonosulfuric acid, peroxyformic acid. See also POWDERED METALS and SILVER COMPOUNDS.

SDI750   CAS: 9007-35-6   HR: 3
SILVER (colloidal)
af: Ag   aw: 107.868
PROP: IDLH 10 mg/m³ (as Ag).
SYNS: ARGENTIUM CREDE □ COLLARGOL
CONSENSUS REPORTS: Silver and its compounds are on the Community Right-To-Know List. Reported in EPA TSCA Inventory.
OSHA PEL: Metal, Dust, and Fume: TWA 0.01 mg/m³
ACGIH TLV: TWA (metal) 0.1 mg/m³; (soluble compounds as Ag) 0.01 mg/m³
DFG MAK: 0.01 mg/m³
SAFETY PROFILE: Poison by ingestion and intravenous routes. See also SILVER and SILVER COMPOUNDS.

SDJ000   CAS: 13092-75-6   HR: 3
SILVER ACETYLIDE
mf: C_2HAg mw: 132.90
PROP: IDLH 10 mg/m³ (as Ag).
SYN: SILVER ACETYLIDE (dry) (DOT)
CONSENSUS REPORTS: Silver and its compounds are on the Community Right-To-Know List.
DOT CLASSIFICATION: Forbidden
SAFETY PROFILE: Severe explosion hazard. A more powerful detonator than copper acetylide. Explodes when heated to 120–140°C. Formed when silver-containing solutions contact acetylene. Upon decomposition it emits acrid smoke and irritating fumes. See also SILVER COMPOUNDS and ACETYLIDES.

SDL000   CAS: 102492-24-0   HR: 2
SILVER AMMONIUM LACTATE
mf: $C_6H_{14}AgNO_6$ mw: 304.08
PROP: IDLH 10 mg/m³ (as Ag).
CONSENSUS REPORTS: Silver and its compounds are on the Community Right-To-Know List. Reported in EPA TSCA Inventory.
OSHA PEL: TWA 0.01 mg(Ag)/m³
ACGIH TLV: TWA 0.01 mg(Ag)/m³
SAFETY PROFILE: A severe eye irritant. When heated to decomposition it emits toxic fumes of NH_3 and NO_x. See also SILVER AMMONIUM COMPOUNDS and SILVER COMPOUNDS.

SDL500   CAS: 23606-32-8   HR: 2
SILVER AMMONIUM NITRATE
mf: $AgH_4N_3O_6$ mw: 249.94
PROP: White crystals. IDLH 10 mg/m³ (as Ag).
CONSENSUS REPORTS: Silver and its compounds are on the Community Right-To-Know List. Reported in EPA TSCA Inventory.
OSHA PEL: TWA 0.01 mg(Ag)/m³
ACGIH TLV: TWA 0.01 mg(Ag)/m³
SAFETY PROFILE: A severe eye irritant. When heated to decomposition it emits very toxic fumes of NH_3 and NO_x. See also SILVER AMMONIUM COMPOUNDS, SILVER COMPOUNDS, and NITRATES.

SDM100   CAS: 7784-08-9   HR: 3
SILVER ARSENITE
DOT: UN 1683
mf: $AsO_3 \cdot 3Ag$ mw: 446.53
PROP: IDLH 10 mg/m³ (as Ag).
SYNS: ARSENIOUS ACID, TRISILVER(1+) SALT □ ARSENOUS ACID, TRISILVER(1+) SALT (9CI)
OSHA PEL: OSHA: Cancer Hazard
DOT CLASSIFICATION: 6.1; Label: Poison
SAFETY PROFILE: Confirmed carcinogen. When heated to decomposition it emits toxic fumes of As.

SDM500   CAS: 13863-88-2   HR: 3
SILVER AZIDE
mf: AgN_3 mw: 149.87

PROP: Pure white solid. IDLH 10 mg/m^3 (as Ag).

SYN: SILVER AZIDE (dry) (DOT)

CONSENSUS REPORTS: Silver and its compounds are on the Community Right-To-Know List.

OSHA PEL: TWA 0.01 mg(Ag)/m^3

ACGIH TLV: TWA 0.01 mg(Ag)/m^3

DOT CLASSIFICATION: Forbidden

SAFETY PROFILE: Explodes when heated above 270°C or on impact. Pure silver azide explodes at 340°. An electric field or irradiation by electron pulses can explode the crystals. Shock-sensitive when dry and has detonated @ 250°C. Solutions in aqueous ammonia explode above 100°C. Reacts to form more explosive products with iodine (forms iodine azide); bromine and other halogens. The presence of metal oxides or metal sulfides increases the azide's sensitivity to explosion. Mixtures with sulfur dioxide are explosive. When heated to decomposition it emits toxic fumes of NO$_x$. See also AZIDES and SILVER COMPOUNDS.

SDN000   CAS: 14104-20-2   HR: 3
SILVER BOROFLUORIDE

mf: AgBF$_4$ mw: 194.68

PROP: Deliquescent white crystals. Very sol in H$_2$O, Et$_2$O, toluene, and MeNO$_2$; mod sol in C$_6$H$_6$ and cyclohexene; insol in cyclohexane. IDLH 10 mg/m^3 (as Ag).

SYNS: SILVER FLUOROBORATE □ SILVER TETRAFLUOROBORATE

CONSENSUS REPORTS: Silver and its compounds are on the Community Right-To-Know List. Reported in EPA TSCA Inventory.

OSHA PEL: TWA 0.01 mg(Ag)/m^3; 2.5 mg(F)/m^3

ACGIH TLV: TWA 0.01 mg(Ag)/m^3; TWA 2.5 mg(F)/m^3; BEI: 3 mg/g creatinine of fluorides in urine prior to shift; 10 mg/g creatinine of fluorides in urine at end of shift.

NIOSH REL: (Fluorides, Inorganic) TWA 2.5 mg(F)/m^3

SAFETY PROFILE: Poison by intravenous route. When heated to decomposition it emits toxic fumes of F$^-$. Used for plating baths. See also SILVER COMPOUNDS, BORON COMPOUNDS, and FLUORIDES.

SDO500   HR: 3
SILVER COMPOUNDS

PROP: IDLH 10 mg/m^3 (as Ag).

CONSENSUS REPORTS: Silver and its compounds are on the Community Right-To-Know List.

SAFETY PROFILE: The water-soluble silver compounds are irritating to the skin and mucous membranes and may cause death if ingested. 50 mg of silver collargol is lethal after intravenous injection. Autopsy shows pulmonary edema, hemorrhage, and necrosis of the bone marrow, liver, and kidney. The absorption of silver compounds into the circulation and the subsequent deposition of the reduced silver in various tissues of the body may result in the production of a generalized grayish pigmentation of the skin and mucous membranes, a condition known as argyria. Ingestion of 1–30 g of soluble silver salts or long-term inhalation of a total 1–8 g of silver can cause argyrosis. The introduction of fine particles of silver through breaks in the skin produces a local pigmentation at the site of the injury. 1 mg/m^3 of silver dust causes skin effects. The condition develops slowly, usually after 2–25 years of exposure. Pigmentation is noticeable first in conjunctiva, and later in the mucous membranes of the mouth and gums and in the skin. There are no constitutional symptoms or physical disability. Persons exhibiting the condition, and who subsequently died from unrelated disease, showed, on autopsy, a deposition of silver in the blood vessel walls, kidneys, testes, pituitary, choroid plexus, and mucous membranes of the nose, maxillary antra, trachea, and bronchi. Once deposited, there is no known method by which the silver can be eliminated; the pigmentation is permanent. See also SILVER.

SDP000   CAS: 506-64-9   HR: 3
SILVER CYANIDE
DOT: UN 1684
mf: CAgN   mw: 133.89
PROP: White, odorless, tasteless powder that darkens upon exposure to light. Stable in dry air. Mp: 320° (decomp), d: 3.95. Insol in H_2O, alcohols, and dil acids. Sol in aq alkali cyanides, boiling conc HNO_3. IDLH 10 mg/m³ (as Ag).
SYNS: CYANURE d'ARGENT (FRENCH) □ KYANID STRIBRNY (CZECH) □ RCRA WASTE NUMBER P104
CONSENSUS REPORTS: Silver and its compounds, as well as cyanide and its compounds, are on the Community Right-To-Know List. Reported in EPA TSCA Inventory.
DOT CLASSIFICATION: 6.1; Label: Poison
OSHA PEL: TWA 5 mg(CN)/m³
ACGIH TLV: CL 5 mg(CN)/m³ (skin)
DFG MAK: 5 mg/m³
NIOSH REL: (Cyanide) CL 5 mg(CN)/m³/10M
SAFETY PROFILE: Deadly poison by ingestion. A skin and severe eye irritant. When heated to decomposition it emits very toxic fumes of CN^- and NO_x. Incompatible with phosphorus tricyanide, fluorine. Used in silver plating. See also SILVER COMPOUNDS and CYANIDE.

SDQ500   CAS: 7783-95-1   HR: 3
SILVER(II) FLUORIDE
mf: AgF_2 mw: 145.87
PROP: Very hygroscopic; white when pure; usually a gray-black or brownish amorphous solid. Light sensitive. D: 4.7, mp: 690°. IDLH 10 mg/m³ (as Ag).
SYNS: ARGENT FLUORURE (FRENCH) □ ARGENTIC FLUORIDE □ SILVER DIFLUORIDE
CONSENSUS REPORTS: Silver and its compounds are on the Community Right-To-Know List. Reported in EPA TSCA Inventory.
OSHA PEL: TWA 0.01 mg(Ag)/m³; 2.5 mg(F)/m³
ACGIH TLV: TWA 0.01 mg(Ag)/m³; TWA 2.5 mg(F)/m³; BEI: 3 mg/g creatinine of fluorides in urine prior to shift; 10 mg/g

creatinine of fluorides in urine at end of shift.
NIOSH REL: (Inorganic Fluorides) TWA 2.5 mg(F)/m³
SAFETY PROFILE: Poison by subcutaneous route. Powerful oxidizing agent. Mixtures with boron + water are explosive. When heated to decomposition it emits toxic fumes of F^-. See also FLUORIDES and SILVER COMPOUNDS.

SDR500   CAS: 2386-52-9   HR: 3
SILVER METHYLSULFONATE
mf: $CH_3O_3S\bullet Ag$ mw: 202.97
PROP: IDLH 10 mg/m³ (as Ag).
SYNS: METHANESULFONIC ACID, SILVER SALT □ METHANESULFONIC ACID, SILVER(1+) SALT □ SILVER METHANESULFONATE
CONSENSUS REPORTS: Silver and its compounds are on the Community Right-To-Know List. Reported in EPA TSCA Inventory.
OSHA PEL: TWA 0.01 mg(Ag)/m³
ACGIH TLV: TWA 0.01 mg(Ag)/m³
SAFETY PROFILE: Poison by intravenous route. When heated to decomposition it emits toxic fumes of SO_x. See also SULFONATES and SILVER COMPOUNDS.

SDS000   CAS: 7761-88-8   HR: 3
SILVER(I) NITRATE (1:1)
DOT: UN 1493
mf: $NO_3\bullet Ag$ mw: 169.88
PROP: Colorless, odorless, transparent, large or small white crystals. Not photosensitive when pure. Mp: 212°, bp: 444° (decomp), d: 4.352 @ 19°. Very sol in ammonia, water; sltly sol in ether. IDLH 10 mg/m³ (as Ag).
SYNS: LUNAR CAUSTIC □ NITRATE d'ARGENT (FRENCH) □ NITRIC ACID, SILVER(1+) SALT □ SILBERNITRAT □ SILVER(1+) NITRATE □ SILVER NITRATE (DOT)
CONSENSUS REPORTS: Silver and its compounds are on the Community Right-To-Know List. EPA Genetic Toxicology Program. Reported in EPA TSCA Inventory.

S

OSHA PEL: TWA 0.01 mg(Ag)/m^3
ACGIH TLV: TWA 0.01 mg(Ag)/m^3
DOT CLASSIFICATION: 5.1; Label: Oxidizer
SAFETY PROFILE: A human poison. Experimental poison by ingestion, intravenous, subcutaneous, and intraperitoneal routes. Experimental reproductive effects. Human mutation data reported. A severe eye irritant. A powerful caustic and irritant to skin, eyes, and mucous membranes. Swallowing can cause severe gastroenteritis that may be fatal. Questionable carcinogen with experimental tumorigenic data. A powerful oxidizer. Incompatible with acetylene, acetylides, alkalies, aluminum, antimony salts, arsenic, arsenites, bromides, carbon, carbonates, chlorides, ClF_3, chlorosulfuric acid, copper, creosote, ethanol, ferrous salts, hypophosphites, iodides, Mg powder with H_2O, morphine salts, NH_3 with KOH to yield black Ag_3N, oils, PH_3, phosphates, phosphonium iodide, phosphorus, plastics, sulfur, tannic acid, tartrates, thiocyanates, vegetable decoctions and extracts, zinc with NH_3 with KOH. When heated to decomposition it emits toxic fumes of NO_x. See also SILVER COMPOUNDS and NITRATES.

SDU500   CAS: 20667-12-3   HR: 3
SILVER(1+) OXIDE
mf: Ag_2O mw: 231.74
PROP: Brownish-black, heavy, odorless powder. Light sensitive. D: 7.22 @ 25°/4°. Decomp at approx 200°. Very sol in dilute nitric acid, ammonia; less sol in NaOH solns; insol in alc. IDLH 10 mg/m^3 (as Ag).
SYNS: ARGENTOUS OXIDE □ DISILVER OXIDE
CONSENSUS REPORTS: Silver and its compounds are on the Community Right-To-Know List. Reported in EPA TSCA Inventory.
OSHA PEL: TWA 0.01 mg(Ag)/m^3
ACGIH TLV: TWA 0.01 mg(Ag)/m^3
SAFETY PROFILE: A poison by intraperitoneal route. Moderately toxic by ingestion. Flammable by chemical reaction; an oxidizing agent. Explodes in contact with ammonia. Incompatible with CuO, (NH_3 +

ethanol), (hydrazine + ethanol), CO, H_2S, Mg, auric sulfide, Sb sulfide, Hg sulfide, nitroalkanes, Se, S, P, K, Na, NaK, seleninyl chloride. See also SILVER COMPOUNDS.

SDW000   HR: 3
SILVER PEROXYCHROMATE
mf: $AgCrO_5$ mw: 239.87
PROP: IDLH 10 mg/m^3 (as Ag).
CONSENSUS REPORTS: Silver and its compounds, as well as chromium and its compounds, are on the Community Right-To-Know List.
SAFETY PROFILE: Confirmed carcinogen. An oxidant. When mixed with H_2SO_4 @ −80° it explodes on slow warming to −30°. See also CHROMIUM and SILVER COMPOUNDS.

SEB000   HR: 2
SMOG
PROP: An atmospheric combination of smoke, fog, and industrial gases. Composition: Contents vary, but sulfur dioxide, oxides of nitrogen, and ozone are common components; others are sulfides, fluorides, chlorides, carbon particles, and various hydrocarbons.
SAFETY PROFILE: Moderately irritating to eyes and mucous membranes. Numerous chronic effects have been reported in susceptible populations. A common air contaminant. Possibly carcinogenic.

SEC000   HR: 3
SMOKE CONDENSATE, cigarette
SYNS: CIGARETTE SMOKE CONDENSATE □ CSC □ TOBACCO SMOKE CONDENSATE □ TOBACCO TAR
CONSENSUS REPORTS: IARC Cancer Review: Group 1 IMEMDT 7,359,87; Human Sufficient Evidence IMEMDT 38,309,86, Animal Sufficient Evidence IMEMDT 38,309,86.
SAFETY PROFILE: Confirmed carcinogen with experimental carcinogenic, neoplastigenic, and tumorigenic data. An experimental teratogen. Human mutation data reported. See also NICOTINE,

SMOKELESS TOBACCO, and various tobacco entries.

SED400

HR: 3

SMOKELESS TOBACCO

PROP: A variety of habituating substances containing tobacco as the major ingredient and used without burning. Tobacco is a product of the leaves and stems of two species of Nicotiana, *N. tabacum* (grown in North America and Western Europe) and *N. rustica* (grown in the former USSR and India). There is considerable evidence that many if not all of the forms of smokeless tobacco are human carcinogens.

The smokeless tobaccos are introduced into the body through the mouth (chewing tobacco, snuff, misshri, gudakhu, shammah, khaini, nass, naswar, or in combination with betel quid) or nose (snuff).

The various smokeless tobacco products are described below:

Chewing tobacco is placed between the cheek and gum and chewed slowly. There are three main types: plug, twist/roll, and loose-leaf.

Fine-cut tobacco was formerly classified in the United States as chewing tobacco and is now placed in the category of moist fine-cut snuff.

Gudakhu is a paste of powdered tobacco, molasses, and other ingredients used in parts of India to clean teeth.

Khaini is a mixture of tobacco and lime formed into a ball and placed in the mouth.

Kiwam is made from processed tobacco leaves. After the stalks and stems are removed, the leaves are soaked and boiled in water with flavorings and spices, crushed, then strained, leaving a paste, which is chewed.

Loose-leaf tobacco is prepared from fermented cigar leaves, sweetened with sugars, syrups, liquors, and other flavoring materials. It is packaged as batches of loose pieces or cut strips.

Mainpuri tobacco is a chewed mixture of tobacco with slaked lime, areca nut, camphor, and cloves. It is used in India.

Mishri is prepared from roasted or half-burnt tobacco that has been baked on a hot metal plate until black and then powdered. It is used primarily to clean teeth but is also used as chewing tobacco. Synonyms are masheri and misheri.

Nass is a mixture of tobacco, lime, wood-ash, and cottonseed oil chewed in Iran and the central Asian region of the former USSR.

Naswar is a mixture of powdered tobacco, slaked lime, and indigo placed on the bottom of the mouth or behind the lower lip. It is used in Afghanistan and Pakistan.

Pattiwala tobacco is a sun-cured tobacco leaf chewed with or without lime. It is used in India.

Pill is dried and pelleted Kiwam paste.

Plug tobacco is made from enriched tobacco leaves or leaf fragments wrapped in fine tobacco and pressed into flat bars or rolls. It is chewed.

Shammah is a mixture of powdered tobacco leaves with calcium or sodium carbonate and other materials, including ash, placed in the cheek or behind the lower lip. It is used in southern Saudi Arabia.

Snuff is taken through the mouth or the nose. Moist snuff is finely cut tobacco plus flavorings, with a moisture content of up to 50%. It is placed in the cheek. Dry snuff has a moisture content of less than 10% and may have flavorings. It may be sniffed through the nose, placed behind the lower lip, or in the cheek. Oriental snuff is about 50% heated calcium carbonate and calcium phosphate with some powdered cuttle-fish bone. In southern Africa, snuff is made from powdered tobacco leaves, plant ash, and sometimes oils, lemon juice, and herbs. In the United States, "dipping" refers to the ingestion use of snuff.

Twist/roll tobacco is stripped tobacco leaves rolled or twisted like a length of rope.

Zarda is tobacco leaf broken into small pieces and boiled in water with lime and spices to dryness and then colored with vegetable dyes. It is usually chewed mixed with areca nut and spices.

S

SYNS: CHEWING TOBACCO □ GUDAKHU (INDIA) □ KHAINI (INDIA) □ KIWAM (INDIA) □ MASHERI (INDIA) □ MISHERI (INDIA) □ MISHRI (INDIA) □ NASS (IRAN) □ NASWAR (PAKISTAN and AFGHANISTAN) □ PILLS (INDIA) □ SHAMMAH (SAUDI ARABIA) □ SNUFF □ ZARDA (INDIA)

SAFETY PROFILE: Tobaccos contain 0.5–5% alkaloids, predominantly as l-nicotine (>85%). Nicotine is strongly addictive and is the chief cause of tobacco dependence. It is a mild stimulant. It readily forms salts with most acids. These salts are poorly absorbed through the mucous membranes, whereas the base is easily absorbed. This explains the practice of combining lime or other alkali in conjunction with ingestion tobacco use. Nicotine and some of the other tobacco alkaloids are experimental teratogens and mutagens.

There are several known classes of carcinogens present in the smokeless tobaccos: N-nitrosamines, polynuclear aromatic hydrocarbons (PAH's), heavy metals (arsenic trioxide, lead, cadmium, and nickel compounds), and radionuclides (^{226}Ra, ^{210}Pb, and ^{210}Po). Of these, nitrosamines are present in the highest concentration (in the range of mg/kg). The concentrations of the nitrosamines are 100 times higher in tobacco than in other consumer products. Nitrosamine concentrations are higher in chewing tobacco than in cigarette smoke. The major nitrosamines in tobacco are N'-nitrosonornicotine (NNN), 4-(methylnitrosamine)-1-(3-pyridyl)-1-butanone (NNK), and N'-nitrosoanatabine (NAT). They are probably generated during curing, fermentation, and aging of the tobacco leaf from the tobacco alkaloids: nicotine, nornicotine, anatabine, anabasine, continine, myosmine, 2,3'-dipyridyl, and N'-formyl-nornicotine. They may also form in the mouth.

There is sufficient evidence that the ingestion use of snuff, chewing tobacco, and tobacco mixed with lime is carcinogenic to humans. Evidence suggests that the ingestion use of other smokeless tobacco preparations and the nasal use of snuff are carcinogenic to humans. Oral precancerous lesions are commonly observed in smokeless-tobacco users.

See also NICOTINE, ARECA NUT, BETEL QUID, N-NITROSO COMPOUNDS, NITROSAMINES, and individual compounds.

SEE000   CAS: 8006-28-8   HR: 2
SODA LIME
DOT: UN 1907
PROP: White to gray granules. Rapidly deteriorates on exposure to air.
SYN: SODA LIME with >4% sodium hydroxide (DOT)
DOT CLASSIFICATION: 8; Label: Corrosive
SAFETY PROFILE: A corrosive irritant to skin, eyes, and mucous membranes. See also SODIUM HYDROXIDE and CALCIUM OXIDE.

SEE500   CAS: 7440-23-5   HR: 3
SODIUM
DOT: UN 1428
af: Na   aw: 22.9898
PROP: Light, soft, ductile, malleable, silver-white metal. Tarnishes in air forming sodium oxides, the carbonates and the hydroxide. Mp: 97.81°, bp: 881.4°, d: 0.9710 @ 20°, autoign temp: >115° in dry air, vap press: 1.2 mm @ 400°. Sol in liq NH_3.
SYNS: NATRIUM □ SODIUM METAL (DOT)
CONSENSUS REPORTS: Reported in EPA TSCA Inventory.
DOT CLASSIFICATION: 4.3; Label: Dangerous When Wet
SAFETY PROFILE: Metallic sodium reacts exothermally with the moisture of body or tissue surfaces, causing thermal and chemical burns. Sodium in elemental form is highly reactive. Sodium reacts violently with water to form sodium hydroxide. A very dangerous fire hazard when exposed to heat and moisture. Under the appropriate conditions, it can react violently with moisture, air, $AlBr_3$, $AlCl_3$, AlF_3, NH_4 chlorocuprate, NH_4NO_3, $SbBr_3$, $SbCl_3$, SbI_3, $AsCl_3$, AsI_3, $BiBr_3$, $BiCl_3$, BiI_3, Bi_2O_3, BBr_3, bromoazide, CO_2, CO + NH_3, CCl_4, Cl_2, ClF_3, $CrCl_4$, CrO_3, CoBr, CoCl, $CuCl_2$, CuO,

FeBr$_3$, FeCl$_3$, FeBr$_2$, FeCl$_2$, FeI$_2$, hydrazine hydrate, H$_2$O$_2$, H$_2$S, HCl, HF, F$_2$, 1,2-dichloroethylene, dichloromethane, Br$_2$, hydroxylamine, iodine, iodine monochloride, iodine pentafluoride, lead oxide, maleic anhydride, manganous chloride, mercuric bromide, mercuric chloride, mercuric fluoride, mercuric iodide, mercurous chloride, mercurous oxide, methyl chloride, molybdenum trioxide, monoammonium phosphate, nitric acid, nitrogen peroxide, nitrosyl fluoride, nitrous oxide, phosgene, phosphorus, phosphorus pentafluoride, phosphorus pentoxide, phosphorus tribromide, phosphorus trichloride, phosphoryl chloride, potassium oxides, potassium ozonide, potassium superoxide, selenium, silicon tetrachloride, silver bromide, silver chloride, silver fluoride, silver iodide, sodium peroxide, stannic chloride, stannic iodide with sulfur, stannic oxide, stannous chloride, sulfur, sulfur dibromide, sulfur dichloride, sulfur dioxide, sulfuric acid, tellurium, tetrachloro-ethane, thallous bromide, thiophosphoryl bromide, trichlorethylene, vanadium pentachloride, vanadyl chloride, zinc bromide, any oxidizing material. Decomposes moisture to evolve hydrogen and heat. Reacts exothermally with halogens, acids, and halogenated hydrocarbons.

Heated sodium is spontaneously flammable in air. Can be safely stored under liquid hydrocarbons. Dangerous explosion hazard when exposed to moisture in any form!! Keep away from water at all times!! When heated in air it emits toxic fumes of sodium oxide. Reacts with water or steam to produce heat, hydrogen, and flammable vapors. Can react vigorously to explosively with oxidizing materials. To fight fire, use soda ash, dry sodium chloride, or graphite, in order of preference. When heated to decomposition it emits toxic fumes of Na$_2$O. See also SODIUM HYDROXIDE and HYDROGEN.

SEF500   CAS: 7440-23-5   HR: 3
SODIUM (dispersions)

PROP: Finely divided metallic sodium suspended in toluene, xylene, naphtha, kerosene, etc.
SYN: SODIUM, METAL DISPERSION IN ORGANIC SOLVENT
CONSENSUS REPORTS: Reported in EPA TSCA Inventory.
SAFETY PROFILE: A very dangerous fire hazard when exposed to heat or flame or by chemical reaction. These are very reactive forms of sodium that, if carelessly handled, may catch fire. After sodium has been extinguished, the burning organic vapor can be dealt with by very cautious use of a carbon dioxide extinguisher. To extinguish, see SODIUM. Do not use carbon tetrachloride. Moderate explosion hazard by chemical reaction; will react with water or steam to produce heat and hydrogen; on contact with oxidizing materials it can react vigorously, and on contact with acid or acid fumes it can emit toxic fumes. When heated it loses the solvent and emits highly toxic fumes of Na$_2$O. See also SODIUM and individual dispersant.

SEG500   CAS: 127-09-3   HR: 3
SODIUM ACETATE
mf: C$_2$H$_3$O$_2$•Na   mw: 82.04
PROP: White granular powder. Mp: 324°. Autoign temp: 1125°F, d: 1.45, mp: 58°. Decomp @ higher temp. Sol in water, alc.
SYNS: ACETIC ACID, SODIUM SALT □ NATRIUMACETAT (GERMAN) □ SODIUM ACETATE, anhydrous (FCC)
CONSENSUS REPORTS: Reported in EPA TSCA Inventory. EPA Genetic Toxicology Program.
SAFETY PROFILE: Poison by intravenous route. Moderately toxic by ingestion. A skin and eye irritant. Migrates to food from packaging materials. Violent reaction with F$_2$, KNO$_3$, diketene. When heated to decomposition it emits toxic fumes of Na$_2$O.

SEG800   CAS: 7681-38-1   HR: 2
SODIUM ACID SULFATE (solid)
DOT: UN 1821/UN 2837
mf: HO$_4$S•Na   mw: 120.06

S

PROP: White or colorless, triclinic, crystals or granules. Decomp on heating with H_2O loss to form $Na_2S_2O_7$ which on further heating decomp to form Na_2SO_4. Mp: >315° (decomp), d: 2.435 @ 13°. Sol in water; spar sol in alc.

SYNS: GBS □ NITRE CAKE □ SODIUM ACID SULFATE □ SODIUM BISULFATE, FUSED □ SODIUM HYDROGEN SULFATE, solid (UN 1821) (DOT) □ SODIUM HYDROGEN SULFATE, solution (UN 2837) (DOT) □ SODIUM PYROSULFATE □ SULFURIC ACID, MONOSODIUM SALT

CONSENSUS REPORTS: Reported in EPA TSCA Inventory.

DOT CLASSIFICATION: 8; Label: Corrosive

SAFETY PROFILE: A corrosive irritant to skin, eyes, and mucous membranes. Mutation data reported. Reacts with moisture to form sulfuric acid. Mixtures with calcium hypochlorite + starch + sodium carbonate explode when compressed. Violent reaction with acetic anhydride + ethanol may lead to ignition and a vapor explosion. Incompatible with calcium hypochlorite. When heated to decomposition it emits toxic fumes of SO_x and Na_2O. See also SULFATES.

SEH000   CAS: 9005-38-3   HR: 3
SODIUM ALGINATE
mf: $(C_6H_7O_6Na)_n$ mw: 198.11

PROP: Colorless to slightly yellow filamentous or granular solid or powder; odorless and tasteless. In water it forms a viscous colloidal soln; insol in ether, alc, chloroform.

SYNS: ALGIN □ ALGIN (polysaccharide) □ ALGINATE KMF □ ALGIPON L-1168 □ AMNUCOL □ ANTI-MIGRANT C 45 □ CECALGINE TBV □ COHASAL-1H □ DARID QH □ DARILOID QH □ DUCKALGIN □ HALLTEX □ K'-ALGILINE □ KELCO GEL LV □ KELCOSOL □ KELGIN □ KELGUM □ KELSET □ KELSIZE □ KELTEX □ KELTONE □ LAMITEX □ MANUCOL □ MANUCOL DM □ MANUTEX □ MEYPRALGIN R/LV □ MINUS □ MOSANON □ NOURALGINE □ OG 1 □ PECTALGINE □ PROCTIN □ PROTACELL 8 □ PROTANAL □ PROTATEK □ SNOW ALGIN H □ SODIUM POLYMANNURONATE □ STIPINE □ TAGAT □ TRAGAYA

CONSENSUS REPORTS: Reported in EPA TSCA Inventory.

SAFETY PROFILE: Poison by intravenous and intraperitoneal routes. When heated to decomposition it emits toxic fumes of Na_2O.

SEH500   HR: D
SODIUM n-ALKYLBENZENE SULFONATE
SAFETY PROFILE: When heated to decomposition it emits toxic fumes of SO_x and Na_2O.

SEM000   CAS: 1344-00-9   HR: 1
SODIUM ALUMINOSILICATE
PROP: Fine, white, amorphous powder or beads; odorless and tasteless. Insol in water, alc, and other org solvs.

SYNS: NCI-C55505 □ SODIUM SILICOALUMINATE

CONSENSUS REPORTS: Reported in EPA TSCA Inventory.

SAFETY PROFILE: An irritant to skin, eyes, and mucous membranes. When heated to decomposition it emits toxic fumes of Na_2O. See also SILICATES and ALUMINUM COMPOUNDS.

SEM500   CAS: 13770-96-2   HR: 3
SODIUM ALUMINUM TETRAHYDRIDE
DOT: UN 2835

mf: $AlH_4 \cdot Na$ mw: 54.01

PROP: White crystalline material; stable in dry air but sensitive to moisture. Mp: 183°; d: 1.24. Insol in Et_2O; sol in THF and diglyme.

SYNS: ALUMINATE(1−), TETRAHYDRO-, SODIUM, (T-4)-(9CI) □ ALUMINUM SODIUM HYDRIDE □ SAH 22 □ SODIUM ALUMINUM HYDRIDE (DOT) □ SODIUM TETRAHYDROALUMINATE(1−)

ACGIH TLV: TWA 2 mg(Al)/m³

DOT CLASSIFICATION: 4.3; Label: Dangerous When Wet

SAFETY PROFILE: Flammable when exposed to heat or flame. May ignite and explode on contact with water. Reacts violently with tetrahydrofuran when heated. When heated to decomposition it emits toxic fumes of Na_2O. See also ALUMINUM COMPOUNDS and HYDRIDES.

SEN000   CAS: 7782-92-5   HR: 3
SODIUM AMIDE
mf: H_2NNa mw: 39.02
PROP: White, crystalline powder or colorless orthorhombic crystals. Moisture sensitive. Mp: 210°, bp: 400°. Spar sol in liq NH_3.
SYN: SODAMIDE
CONSENSUS REPORTS: Reported in EPA TSCA Inventory.
SAFETY PROFILE: An intense irritant to tissue, skin, and eyes. Flammable by chemical reaction. Ignites or explodes with heat or grinding. Explosive reaction with moisture, chromium trioxide, potassium chlorate, halocarbons (e.g., 1,1-diethoxy-2-chloroethane), oxidants, sodium nitrite, air. Can become explosive in storage. Violent reaction with dinitrogen tetraoxide. Will react with water or steam to produce heat and toxic and corrosive fumes of sodium hydroxide and ammonia. When heated to decomposition it emits highly toxic fumes of NH_3 and Na_2O. See also AMIDES.

SEP000   CAS: 133-10-8   HR: 2
SODIUM p-AMINOSALICYLATE
mf: $C_7H_6NO_3 \cdot Na$ mw: 175.13
SYNS: p-AMINOSALICYLATE SODIUM □ p-AMINO-SALICYLIC ACID SODIUM SALT □ BACTYLAN □ LEPASEN □ NATRI-PAS □ NIPPAS □ PAMISYL SODIUM □ PASADE □ PASALON-RAKEET □ PASNAL □ PASSODICO □ SALVIS □ SANIPIROL □ SODIOPAS □ SODIUM AMINOSALICYLATE □ SODIUM p-AMINOSALICYLIC ACID □ TUBERSAN
CONSENSUS REPORTS: Reported in EPA TSCA Inventory.
SAFETY PROFILE: Moderately toxic by intraperitoneal, subcutaneous, and intravenous routes. Mildly toxic by ingestion. An experimental teratogen. Experimental reproductive effects. Human mutation data reported. When heated to decomposition it emits toxic fumes of NO_x and Na_2O.

SEP500   CAS: 10042-84-9   HR: 3
SODIUM AMINOTRIACETATE
mf: $C_6H_9NO_6 \cdot xNa$ mw: 352.09
SYNS: N,N-BIS(CARBOXYMETHYL)GLYCINE SODIUM SALT □ GLYCINE, N,N-BIS(CARBOXYMETHYL)-, SOD-IUM SALT (9CI) □ NITRILOTRIACETIC ACID SODIUM SALT □ SODIUM NITRILOACETATE □ SODIUM NITRILOTRIACETATE □ SODIUM NTA
CONSENSUS REPORTS: IARC Cancer Review: Group 2B IMEMDT 48,181,90; Animal Sufficient Evidence IMEMDT 48,181,90; Human No Adequate Data IMEMDT 48,181,90. Reported in EPA TSCA Inventory.
SAFETY PROFILE: Confirmed carcinogen by intraperitoneal route. Mutation data reported. When heated to decomposition it emits toxic fumes of NO_x and Na_2O.

SEQ000   CAS: 69-52-3   HR: 2
SODIUM AMPICILLIN
mf: $C_{16}H_{18}N_3O_4S \cdot Na$ mw: 371.42
PROP: White powder. Bp: 219°–220, mp: 215°. Sol in water.
SYNS: ALPEN-N □ AMCILL-S □ d-(−)-α-AMINOBENZYLPENICILLIN SODIUM SALT □ AMPICILLIN SODIUM □ AMPICILLIN SODIUM SALT □ BINOTAL SODIUM □ CITTERAL □ DOMICILLIN □ OMNIPEN-N □ PEN A/N □ PENBRITIN-S □ PENIALMEN □ POLYCILLIN-N □ PRINCIPEN/N □ SODIUM d-(−)-α-AMINOBENZYLPENICILLIN □ SODIUM BINOTAL □ SODIUM P-50
CONSENSUS REPORTS: IARC Cancer Review: Group 3 IMEMDT 50,153,90; Animal Limited Evidence IMEMDT 50,153,90; Human Inadequate Evidence IMEMDT 50,153,90.
SAFETY PROFILE: Moderately toxic by intraperitoneal route. Human systemic effects by ingestion: hypermotility, diarrhea, and other gastrointestinal changes. Questionable carcinogen. When heated to decomposition it emits very toxic fumes of NO_x, Na_2O, and SO_x.

SEY050   CAS: 10190-99-5   HR: 2
SODIUM ARISTOLOCHATE I
mf: $C_{17}H_{10}NO_7 \cdot Na$ mw: 363.27
SYNS: ARISTOLOCHIC ACID, SODIUM SALT □ ARISTOLOCHIC ACID I SODIUM SALT □ 8-METHOXY-6-NITROPHENANTHRO(3,4-D)-1,3-DIOXOLE-5-CARBOXYLIC ACID SODIUM SALT □ PHEN-ANTHRO(3,4-D)-1,3-DIOXOLE-5-CARBOXYLIC ACID, 8-METHOXY-6-NITRO-, SODIUM SALT
SAFETY PROFILE: Suspected carcinogen. Mutation data reported. When heated to decomposition it emits toxic vapors of NO_x.

SEY100   CAS: 13464-42-1   HR: 3
SODIUM ARSENATE
DOT: UN 1685
mf: $As_2O_7•4Na$ mw: 353.80
PROP: Pesticide.
SYNS: DIARSENIC ACID, TETRASODIUM SALT □
SODIUM ARSENATE (DOT) □ SODIUM DIARSENATE □
SODIUM PYROARSENATE □ TETRASODIUM
DIARSENATE
DOT CLASSIFICATION: 6.1; Label: Poison
SAFETY PROFILE: A poison. When heated
to decomposition it emits toxic vapors of
arsenic and Na_2O.

SEY200   CAS: 13464-37-4   HR: 3
SODIUM ARSENITE
DOT: UN 1686/UN 2027
mf: $AsO_3•3Na$ mw: 191.89
PROP: White crystal granules. Mp: 615°, d:
1.87.
SYNS: ARSENIOUS ACID (H_3AsO_3), TRISODIUM SALT
(8CI) □ ARSENOUS ACID, TRISODIUM SALT □ SODIUM
ARSENITE, aqueous solutions (UN 1686) (DOT) □ SODIUM
ARSENITE, solid (UN 2027) (DOT) □ TRISODIUM
ARSENITE
DOT CLASSIFICATION: 6.1; Label: Poison,
KEEP AWAY FROM FOOD (UN 1686);
6.1; Label: Poison (UN 2027)
SAFETY PROFILE: A poison by ingestion
and subcutaneous routes. When heated to
decomposition it emits toxic vapors of
arsenic and Na_2O.

SEY500   CAS: 7784-46-5   HR: 3
SODIUM ARSENITE
DOT: UN 1686/UN 2027
mf: $AsO_2•Na$ mw: 129.91
PROP: White or grayish-white powder or
thin flakes from melt. Commercially:
95–98% pure. Very sol in water; sltly sol in
alc.
SYNS: ARSENIOUS ACID, SODIUM SALT □ ARSENITE
de SODIUM (FRENCH) □ ARSENOUS ACID, SODIUM
SALT (9CI) □ ATLAS "A" □ CHEM PELS C □ CHEM-SEN
56 □ KILL-ALL □ PENITE □ PRODALUMNOL □
PRODALUMNOL DOUBLE □ SODANIT □ SODIUM
ARSENITE, liquid (solution) (DOT) □ SODIUM ARSENITE,
solid (DOT) □ SODIUM METAARSENITE
CONSENSUS REPORTS: IARC Cancer
Review: Group 1 IMEMDT 7,100,87;
Animal Inadequate Evidence IMEMDT
23,39,80; Human Sufficient Evidence
IMEMDT 23,39,80; Animal No Evidence
IMEMDT 2,48,73. Arsenic and its
compounds are on the Community Right-
To-Know List. Reported in EPA TSCA
Inventory. EPA Genetic Toxicology
Program. EPA Extremely Hazardous
Substances List.
OSHA PEL: TWA 0.01 mg(As)/m^3; Cancer
Hazard
ACGIH TLV: TWA 0.01 mg/m^3; Confirmed
Human Carcinogen; BEI: 35 μ (As)/L
inorganic arsenic and methylated
metabolites in urine
NIOSH REL: (Inorganic Arsenic) CL 0.002
mg(As)/m^3/15M
DOT CLASSIFICATION: 6.1; Label: Poison,
KEEP AWAY FROM FOOD (UN 1686)
SAFETY PROFILE: Confirmed human
carcinogen. Human poison by ingestion.
Experimental poison by ingestion, skin
contact, intravenous, intramuscular, and
intraperitoneal routes. An experimental
teratogen. Experimental reproductive
effects. Human mutation data reported.
Used as an herbicide and pesticide. When
heated to decomposition it emits toxic
fumes of arsenic and Na_2O. See also
ARSENIC COMPOUNDS.

SFA000   CAS: 26628-22-8   HR: 3
SODIUM AZIDE
DOT: UN 1687
mf: N_3Na mw: 65.02
PROP: Colorless, hexagonal crystals. Mp:
decomp, d: 1.846. Insol in ether; sol in liquid
ammonia.
SYNS: AZIDE □ AZIUM □ AZOTURE de SODIUM
(FRENCH) □ KAZOE □ NATRIUMAZID (GERMAN) □
NATRIUMAZIDE (DUTCH) □ NCI-C06462 □ NSC-3072 □
RCA WASTE NUMBER P105 □ SODIUM, AZOTURE de
(FRENCH) □ SODIUM, AZOTURO di (ITALIAN) □ U-3886
CONSENSUS REPORTS: Reported in EPA
TSCA Inventory. EPA Genetic Toxicology
Program. EPA Extremely Hazardous
Substances List.
OSHA PEL: As NH_3: CL 0.1 ppm; As NaN_3:
Cl 0.3 mg/m^3 (skin)
ACGIH TLV: CL 0.29 mg/m^3; Not
Classifiable as a Human Carcinogen; CL

0.11 ppm (as hydrazoic acid vapor); Not Classifiable as a Human Carcinogen
DFG MAK: 0.2 mg/m^3
DOT CLASSIFICATION: 6.1; Label: Poison
SAFETY PROFILE: Poison by ingestion, skin contact, intraperitoneal, intravenous, and subcutaneous routes. Human systemic effects by ingestion: general anesthesia, somnolence, and kidney changes. Questionable carcinogen with experimental tumorigenic data. Human mutation data reported.

Violent reaction with benzoyl chloride combined with KOH, Br$_2$, barium carbonate, CS$_2$, Cr(OCl)$_2$, Cu, Pb, HNO$_3$, BaCO$_3$, H$_2$SO$_4$, hot water, (CH$_3$)$_2$SO$_4$, dibromomalononitrile, sulfuric acid. Incompatible with acids, ammonium chloride + trichloroacetonitrile, phosgene, cyanuric chloride, 2,5-dinitro-3-methylbenzoic acid + oleum, trifluroroacryloyl chloride. Reacts with heavy metals (e.g., brass, copper, lead) to form dangerously explosive heavy metal azides, a particular problem in laboratory equipment and drain traps. When heated to decomposition it emits very toxic fumes of NO$_x$ and Na$_2$O. See also AZIDES.

SFB000   CAS: 532-32-1   HR: 3
SODIUM BENZOATE
mf: C$_7$H$_5$O$_2$•Na   mw: 144.11
PROP: White crystalline solid; odorless. Sol in water, alc.
SYNS: ANTIMOL □ BENZOATE of SODA □ BENZOATE SODIUM □ BENZOESAEURE (NA-SALZ) (GERMAN) □ BENZOIC ACID, SODIUM SALT □ SOBENATE □ SODIUM BENZOIC ACID
CONSENSUS REPORTS: Reported in EPA TSCA Inventory. EPA Genetic Toxicology Program.
SAFETY PROFILE: Poison by subcutaneous and intravenous routes. Moderately toxic by ingestion, intramuscular, and intraperitoneal routes. An experimental teratogen. Experimental reproductive effects. Mutation data reported. Larger doses of 8–10 g by mouth may cause nausea and vomiting. Small doses

have little or no effect. Combustible when exposed to heat or flame. When heated to decomposition it emits toxic fumes of Na$_2$O. See also BENZOIC ACID.

SFB200   CAS: 3184-65-4   HR: 3
SODIUM o-BENZYL-p-CHLOROPHENATE
mf: C$_{13}$H$_{11}$ClO•Na   mw: 241.68
SYNS: 2-BENZYL-4-CHLOROPHENOL, SODIUM SALT □ 4-CHLORO-2-(PHENYLMETHYL)PHENOL SODIUM SALT □ PHENOL, 4-CHLORO-2-(PHENYLMETHYL)-, SODIUM SALT □ SODIUM o-BENZYL-p-CHLOROPHENOLATE
DOT CLASSIFICATION: 6.1; Label: Poison, Corrosive
SAFETY PROFILE: A poison and corrosive. When heated to decomposition it emits toxic vapors of Cl$^-$.

SFB500   CAS: 63915-76-4   HR: 3
SODIUM BERYLLIUM MALEATE
mf: C$_8$H$_6$Be$_4$Na$_2$O$_{12}$•7H$_2$O   mw: 502.30
CONSENSUS REPORTS: IARC Cancer Review: Group 1 IMEMDT 58,41,93; Human Sufficient Evidence IMEMDT 58,41,93; Animal Sufficient Evidence IMEMDT 1,17,72; Animal Sufficient Evidence IMEMDT 23,143,80; Animal Sufficient Evidence IMEMDT 58,41,93. Beryllium and its compounds are on the Community Right-To-Know List.
OSHA PEL: TWA 0.002 mg(Be)/m^3; STEL 0.005 mg(Be)/m^3/30M; CL 0.025 mg(Be)/m^3
ACGIH TLV: TWA 0.002 mg(Be)/m^3; Confirmed Human Carcinogen; (Proposed: TWA 0.00005 mg(Be)/m^3; STEL 0.0002 mg(Be)/m^3 (sensitizer) (skin); Confirmed Human Carcinogen)
NIOSH REL: CL (Beryllium) not to exceed 0.0005 mg(Be)/m^3
SAFETY PROFILE: Confirmed carcinogen. Poison by intravenous route. When heated to decomposition it emits toxic fumes of BeO and Na$_2$O. See also BERYLLIUM COMPOUNDS.

S

SFC000   CAS: 63915-77-5   HR: 3
SODIUM BERYLLIUM TARTRATE
mf: $C_8H_4Be_4Na_2O_{13} \cdot 10H_2O$ mw: 570.34
CONSENSUS REPORTS: IARC Cancer Review: Group 1 IMEMDT 58,41,93; Human Sufficient Evidence IMEMDT 58,41,93; Animal Sufficient Evidence IMEMDT 1,17,72; Animal Sufficient Evidence IMEMDT 23,143,80; Animal Sufficient Evidence IMEMDT 58,41,93. Beryllium and its compounds are on the Community Right-To-Know List.
OSHA PEL: TWA 0.002 mg(Be)/m³; STEL 0.005 mg(Be)/m³/30M; CL 0.025 mg(Be)/m³
ACGIH TLV: TWA 0.002 mg(Be)/m³; Confirmed Human Carcinogen; (Proposed: TWA 0.00005 mg(Be)/m³; STEL 0.0002 mg(Be)/m³ (sensitizer) (skin); Confirmed Human Carcinogen)
NIOSH REL: CL (Beryllium) not to exceed 0.0005 mg(Be)/m³
SAFETY PROFILE: Confirmed carcinogen. Poison by subcutaneous and intravenous routes. When heated to decomposition it emits toxic fumes of BeO and Na_2O. See also BERYLLIUM COMPOUNDS.

SFC500   CAS: 144-55-8   HR: 1
SODIUM BICARBONATE
mf: $NaHCO_3$ mw: 84.01
PROP: White, monoclinic, crystalline powder. Decomp on heating releasing CO_2, H_2O, and Na_2CO_3. Sol in water; insol in alc.
SYNS: BAKING SODA □ BICARBONATE of SODA □ CARBONIC ACID MONOSODIUM SALT □ COL-EVAC □ JUSONIN □ MONOSODIUM CARBONATE □ NEUT □ SODA MINT □ SODIUM ACID CARBONATE □ SODIUM HYDROGEN CARBONATE
CONSENSUS REPORTS: Reported in EPA TSCA Inventory.
SAFETY PROFILE: Low toxicity by ingestion. An experimental teratogen. A nuisance dust. Human systemic effects: changes in potassium levels, increased urine volume, metabolic acidosis, nausea or vomiting, respiratory changes, sodium level changes. Mutation data reported.

SFE000   CAS: 7631-90-5   HR: 3
SODIUM BISULFITE
mf: $HO_3S \cdot Na$ mw: 104.06
PROP: White, crystalline powder; odor of sulfur dioxide, disagreeable taste. Yellow in soln. D: 1.48, mp: decomp. Very sol in hot or cold water; sltly sol in alc.
SYNS: BISULFITE de SODIUM (FRENCH) □ HYDRO-GEN SULFITE SODIUM □ SODIUM ACID SULFITE □ SODIUM BISULFITE □ SODIUM BISULFITE (1:1) □ SODIUM BISULFITE, solid (DOT) □ SODIUM BISULFITE, solution (DOT) □ SODIUM HYDROGEN SULFITE □ SODIUM HYDROGEN SULFITE, solid (DOT) □ SODIUM HYDROGEN SULFITE, solution (DOT) □ SODIUM SULFHYDRATE □ SULFUROUS ACID, MONOSODIUM SALT
CONSENSUS REPORTS: Reported in EPA TSCA Inventory. EPA Genetic Toxicology Program.
OSHA PEL: TWA 5 mg/m³
ACGIH TLV: TWA 5 mg/m³; Not Classifiable as a Human Carcinogen
DOT CLASSIFICATION: 8; Label: Corrosive
SAFETY PROFILE: Poison by intravenous and intraperitoneal routes. Moderately toxic by ingestion. A corrosive irritant to skin, eyes, and mucous membranes. Mutation data reported. An allergen. When heated to decomposition it emits toxic fumes of SO_x and Na_2O. See also SULFUROUS ACID and SULFITES.

SFF000   CAS: 1303-96-4   HR: 3
SODIUM BORATE DECAHYDRATE
mf: $B_4O_7 \cdot 2Na \cdot 10H_2O$ mw: 381.42
PROP: Hard, odorless crystals, granules, or crystalline powder. D: 1.73, mp: 75° (when rapidly heated).
SYNS: ANTIPYONIN □ BORACSU □ BORATES, TETRA, SODIUM SALT, anhydrous (OSHA, ACGIH) □ BORAX (8CI) □ BORAX DECAHYDRATE □ BORICIN □ GERTLEY BORATE □ JAIKIN □ NEOBOR □ POLYBOR □ SODIUM BIBORATE □ SODIUM BIBORATE DECAHYDRATE □ SODIUM PYROBORATE □ SODIUM PYROBORATE DECAHYDRATE □ SODIUM TETRABORATE □ SODIUM TETRABORATE DECAHYDRATE
CONSENSUS REPORTS: Reported in EPA TSCA Inventory.
OSHA PEL: TWA 10 mg/m³
ACGIH TLV: TWA 5 mg/m³

SAFETY PROFILE: Experimental poison by subcutaneous route. Moderately toxic to humans by ingestion. Moderately toxic experimentally by ingestion, intravenous, and intraperitoneal routes. Experimental reproductive effects. Mutation data reported. Ingestion of 5–10 g of borax by children can cause severe vomiting, diarrhea, shock, death. Incompatible with acids, metallic salts. When heated to decomposition it emits toxic fumes of Na_2O, boron. See also BORON COMPOUNDS. Used in ant poisons, for fly control around refuse and manure piles, as a larvicide, in manufacture of glazes, enamels, cleaning compounds, and in soldering metals.

SFF500   CAS: 16940-66-2   HR: 3
SODIUM BOROHYDRIDE
DOT: UN 1426
mf: $BH_4 \cdot Na$ mw: 37.84
PROP: White to gray-white, microcrystalline powder or lumps. Decomp on heating into constituent elements. Hygroscopic. Mp: 497° (vacuum) (decomp), d: 1.07. Reacts with hot water. Sol in liquid ammonia and "Cellosolve" ether, H_2O; mod sol in EtOH, MeOH; spar sol in Py; insol in Et_2O.
SYNS: BOROHYDRURE de SODIUM (FRENCH) □ SODIUM TETRAHYDROBORATE(1-)
CONSENSUS REPORTS: Reported in EPA TSCA Inventory.
DOT CLASSIFICATION: 4.3; Label: Dangerous When Wet
SAFETY PROFILE: Poison by ingestion and intraperitoneal routes. A strong alkali. A severe eye, skin, and mucous membrane irritant.
   Ignites in air above 288°C when exposed to spark. Potentially explosive reaction with aluminum chloride + bis(2-methoxyethyl) ether. Reacts with ruthenium salts to form a solid product which explodes when touched or on contact with water. Reacts to form dangerously explosive hydrogen gas on contact with alkali, water and other protic solvents (e.g., methanol, ethanol, ethylene glycol, phenol), aluminum chloride + bis(2-methoxyethyl)ether. Reacts violently with anhydrous acids (e.g., sulfuric, phosphoric, fluorophosphoric) to form diborane. Violent exothermic reaction with dimethyl formamide has caused industrial explosions. Mixtures with sulfuric acid may ignite. Incompatible with palladium, diborane + bis(2-methoxyethyl) ether, polyglycols, dimethylacetamide, oxidizers, metal salts, finely divided metallic precipitates of cobalt, nickel, copper, iron, and possibly other metals. Emits flammable vapors on contact with acid fumes. Materials sensitive to polymerization under alkaline conditions, such as acrylonitrile, may polymerize upon contact with sodium borohydride. Avoid storage in glass containers. When heated to decomposition it emits toxic fumes of Na_2O. See also HYDRIDES, BORON COMPOUNDS, and SODIUM COMPOUNDS.

SFG000   CAS: 7789-38-0   HR: 3
SODIUM BROMATE
DOT: UN 1494
mf: $BrO_3 \cdot Na$ mw: 150.90
PROP: White or colorless cubic crystals or crystalline powder. Decomp on heating with O_2 evolution and formation of NaBr. Odorless. Mp: 381°, d: 3.339 @ 17.5°. Very sol in H_2O; insol in EtOH.
SYNS: BROMATE de SODIUM (FRENCH) □ BROMIC ACID, SODIUM SALT □ DYETONE
CONSENSUS REPORTS: Reported in EPA TSCA Inventory.
DOT CLASSIFICATION: 5.1; Label: Oxidizer
SAFETY PROFILE: Poison by ingestion, intravenous, subcutaneous, and intraperitoneal routes. A powerful oxidizer. Violent reactions with Al, As, C, Cu, oil, F_2, metal sulfides, organic matter, P, S. Mixtures with grease are shock-sensitive explosives at 120°C. When heated to decomposition it emits toxic fumes of Na_2O and Br^-. See also BROMATES.

SFG500   CAS: 7647-15-6   HR: 2
SODIUM BROMIDE
mf: BrNa   mw: 102.90

PROP: White or colorless, hygroscopic, cubic crystals, granules, or powder; saline bitter taste. D: 3.21, mp: 747°, bp: 1390°, volatilizes at higher temp. Very sol in H_2O.
SYNS: BROMIDE SALT of SODIUM □ BROMNATRIUM (GERMAN) □ SEDONEURAL
CONSENSUS REPORTS: Reported in EPA TSCA Inventory.
SAFETY PROFILE: Moderately toxic by ingestion. Experimental reproductive effects. Incompatible with acids, alkaloidal and heavy-metal salts. When heated to decomposition it emits toxic fumes of Br^- and Na_2O. See also BROMIDES.

SFO000   CAS: 497-19-8   HR: 3
SODIUM CARBONATE (2:1)
mf: $CO_3 \cdot 2Na$ mw: 105.99
PROP: White, odorless, small crystals or monoclinic powder; alkali taste. Decomp on heating by CO_2 loss. Undergoes monoclinic (β) to monoclinic (α) transition at 3° and monoclinic (β) to hexagonal (α)) transition at 4°. Mp: 851°, bp: decomp, d: 2.509 @ 0°. Hygroscopic. Sol in water; spar sol in EtOH; insol in Me_2CO.
SYNS: CARBONIC ACID, DISODIUM SALT □ CRYSTOL CARBONATE □ DISODIUM CARBONATE □ SODA ASH □ TRONA
CONSENSUS REPORTS: Reported in EPA TSCA Inventory. EPA Genetic Toxicology Program.
SAFETY PROFILE: Poison by intraperitoneal route. Moderately toxic by inhalation and subcutaneous routes. Mildly toxic by ingestion. Experimental reproductive effects. A skin and eye irritant. It migrates to food from packaging materials. Can react violently with Al, P_2O_5, H_2SO_4, F_2, Li, 2,4,6-trinitrotoluene. When heated to decomposition it emits toxic fumes of Na_2O.

SFO500   CAS: 9004-32-4   HR: 2
SODIUM CARBOXYMETHYL
    CELLULOSE
PROP: A synthetic cellulose gum (the sodium salt of carboxy methyl cellulose not less than 99.5% on a dry weight basis, with maximum substitution of 0.95

carboxymethyl group per anhydroglucose unit, and with a minimum viscosity of 25 centipoises for 2% weight aq solutions at 25°). Colorless, odorless, hygroscopic powder or granules. Insol in most org solvs.
SYNS: AC-DI-SOL NF □ AQUAPLAST □ B10 □ BLANOSE BWM □ B 10 (polysaccharide) □ CARBOXY-METHYL CELLULOSE □ CARBOXYMETHYL CELLULOSE, SODIUM □ CARBOXYMETHYL CELLULOSE, SODIUM SALT □ CARMETHOSE □ CELLOFAS □ CELLOGEL C □ CELLPRO □ CELLUFIX FF 100 □ CELLUGEL □ CELLULOSE GLYCOLIC ACID, SODIUM SALT □ CELLULOSE GUM □ CELLULOSE SODIUM GLYCOLATE □ CMC □ CM-CELLULOSE Na SALT □ CMC 7H □ CMC SODIUM SALT □ COLLOWELL □ COPAGEL PB 25 □ COURLOSE A 590 □ DAICEL 1150 □ FINE GUM HES □ GLIKOCEL TA □ KMTS 212 □ LOVOSA □ LUCEL (polysaccharide) □ MAJOL PLX □ MODOCOLL 1200 □ NACM-CELLULOSE SALT □ NYMCEL S □ POLYFIBRON 120 □ SANLOSE SN 20A □ SARCELL TEL □ S 75M □ SODIUM CELLULOSE GLYCOLATE □ SODIUM CMC □ SODIUM CM-CELLULOSE □ SODIUM SALT of CARBOXYMETHYL-CELLULOSE □ TYLOSE 666 □ UNISOL RH
CONSENSUS REPORTS: Reported in EPA TSCA Inventory.
SAFETY PROFILE: Mildly toxic by ingestion. Experimental reproductive effects. Questionable carcinogen with experimental neoplastigenic data. It migrates to food from packaging materials. When heated to decomposition it emits toxic fumes of Na_2O. See also POLYMERS, SOLUBLE.

SFQ000   CAS: 9005-46-3   HR: 3
SODIUM CASEINATE
PROP: Coarse, white powder; odorless. Insol in water, alc.
SYNS: CASEIN and CASEINATE SALTS (FCC) □ CASEIN-SODIUM □ CASEIN, SODIUM COMPLEX □ CASEINS, SODIUM COMPLEXES □ NUTROSE
CONSENSUS REPORTS: Reported in EPA TSCA Inventory.
SAFETY PROFILE: Questionable carcinogen with experimental tumorigenic data. When heated to decomposition it emits toxic fumes of Na_2O.

SFS000   CAS: 7775-09-9   HR: 3
SODIUM CHLORATE
DOT: UN 1495/UN 2428

mf: ClO₃•Na   mw: 106.44

mf: $ClO_3 \cdot Na$ mw: 106.44

PROP: Colorless, odorless cubic crystals; cooling, saline taste. Decomp on heating with evolution of O_2. Mp: 264°, bp: decomp, d: 2.490 @ 15°. Very sol in H_2O; sol in EtOH.

SYNS: ASEX □ ATLACIDE □ ATRATOL □ B-HERBAT-OX □ CHLORATE of SODA (DOT) □ CHLORATE SALT of SODIUM □ CHLORAX □ CHLORSAEURE (GERMAN) □ DE-FOL-ATE □ DESOLET □ DREXEL DEFOL □ DROP LEAF □ EVAU-SUPER □ FALL □ GRAIN SORGHUM HARVEST-AID □ GRANEX O □ HARVEST-AID □ KLOREX □ KUSA-TOHRU □ KUSATOL □ NATRIUMCHLORAAT (DUTCH) □ NATRIUMCHLORAT (GERMAN) □ ORTHO C-1 DEFOLIANT & WEED KILLER □ OXYCIL □ RASIKAL □ SHED-A-LEAF □ SHED-A-LEAF "L" □ SODA CHLORATE (DOT) □ SODIO (CLORATO di) (ITALIAN) □ SODIUM (CHLORATE de) (FRENCH) □ SODIUM CHLORATE, aqueous solution (DOT) □ TRAVEX □ TUMBLEAF □ UNITED CHEMICAL DEFOLIANT No. 1 □ VAL-DROP

CONSENSUS REPORTS: Reported in EPA TSCA Inventory.

DOT CLASSIFICATION: 5.1; Label: Oxidizer

SAFETY PROFILE: Human poison by unspecified routes. Moderately toxic experimentally by ingestion and intraperitoneal routes. Human systemic effects by ingestion: blood hemolysis with or without anemia, methemoglobinemia-carboxyhemoglobinemia and pulmonary changes. Mutation data reported. A skin, mucous membrane, and eye irritant. Damages the red blood cells of humans when ingested.

A powerful oxidizer. It can explode on contact with flame or sparks (static discharge) and has caused many industrial explosions. May react explosively with agricultural materials (e.g., peat, powdered sulfur, sawdust, urotropine, thiuram), alkenes + potassium osmate, aluminum + rubber, ammonium salts, grease, leather, powdered metals, nonmetals, sulfides, cyanides, cyanoborane oligomer, nitrobenzene, organic matter, paint + polyethylene, phosphorus, sodium phosphinate. Violent reaction or ignition with aluminum, ammonium sulfate, Sb_2S_3, arsenic, arsenic trioxide, 1,3-bis(trichloromethylbenzene) + heat, carbon,

charcoal, MnO_2, phosphorus, potassium cyanide, osmium + heat, paper, sulfuric acid, thiocyanates, triethylene glycol + wood, wood, zinc. Can also react violently with nitrobenzene, paper, metal sulfides, dibasic organic acids, organic matter. When heated to decomposition it emits toxic fumes of Cl⁻ and Na_2O. See also CHLORATES.

SFT000   CAS: 7647-14-5   HR: 2
SODIUM CHLORIDE

mf: ClNa   mw: 58.44

PROP: Colorless, transparent cubic crystals; or white, crystalline powder. Mp: 801°, bp: 1413°, d: 2.165, vap press: 1 mm @ 865°. Sol in water, glycerin; spar sol in EtOH.

SYNS: COMMON SALT □ DENDRITIS □ EXTRA FINE 200 SALT □ EXTRA FINE 325 SALT □ HALITE □ H.G. BLENDING □ NATRIUMCHLORID (GERMAN) □ PUREX □ ROCK SALT □ SALINE □ SALT □ SEA SALT □ STERLING □ TABLE SALT □ TOP FLAKE □ USP SODIUM CHLORIDE □ WHITE CRYSTAL

CONSENSUS REPORTS: Reported in EPA TSCA Inventory. EPA Genetic Toxicology Program.

SAFETY PROFILE: Poison by intraperitoneal and intracervical routes. Moderately toxic by ingestion, intravenous, and subcutaneous routes. An experimental teratogen. Human systemic effects by ingestion: blood pressure increase. Human reproductive effects by intraplacental route: terminates pregnancy. Experimental reproductive effects. Human mutation data reported. A skin and eye irritant. When bulk sodium chloride is heated to high temperature, a vapor is emitted that is irritating, particularly to the eyes. Ingestion of large amounts of sodium chloride can cause irritation of the stomach. Improper use of salt tablets may produce this effect. Potentially explosive reaction with dichloromaleic anhydride + urea. Electrolysis of mixtures with nitrogen compounds may form the explosive nitrogen trichloride. Reaction with burning lithium forms the dangerously reactive sodium. The molten salt at 1100° reacts explosively with water. Violent reaction with

S

BrF$_3$. When heated to decomposition it emits toxic fumes of Cl$^-$ and Na$_2$O.

SFT500   CAS: 7758-19-2   HR: 3
SODIUM CHLORITE
DOT: UN 1496/UN 1908
mf: ClNaO$_2$ mw: 90.44
PROP: White monoclinic crystals or crystalline, hygroscopic powder. Decomp on heating with evolution of O$_2$. Bp: decomp @ 180–200°, mp: 257°. Very sol in H$_2$O.
SYNS: ALCIDE LD □ CHLOROUS ACID, SODIUM SALT (8CI,9CI) □ NEO SILOX D □ SODIUM CHLORITE (UN 1496) (DOT) □ SODIUM CHLORITE, solution with >5% available chlorine (UN 1908) (DOT) □ TEXTILE □ TEXTONE
CONSENSUS REPORTS: Reported in EPA TSCA Inventory.
DOT CLASSIFICATION: 5.1; Label: Oxidizer (UN 1496); DOT Class: 8; Label: Corrosive (UN 1908)
SAFETY PROFILE: Poison by ingestion. An experimental teratogen. Experimental reproductive effects. Questionable carcinogen with experimental carcinogenic data. Mutation data reported. May act as an irritant due to its oxidizing power. A powerful oxidizing agent; ignited by friction, heat, or shock. An explosive sensitive to impact or heating to 200°. Potentially explosive reaction with acids, oils, organic matter, oxalic acid + water, zinc. Violent reaction or ignition with carbon (above 60°), ethylene glycol (at 100°), phosphorus (above 50°), sodium dithionate, sulfur-containing materials. Can react vigorously on contact with reducing materials. When heated to decomposition it emits highly toxic fumes of Cl$^-$ and Na$_2$O. Used as a bleaching agent. See also CHLORITES.

SFU500   CAS: 3926-62-3   HR: 3
SODIUM CHLOROACETATE
DOT: UN 2659
mf: C$_2$H$_2$ClO$_2$•Na   mw: 116.48
PROP: White, free-flowing, odorless powder. Mp: decomp @ 200°.
SYNS: CHLOROACETIC ACID SODIUM SALT □ CHLOROCTAN SODNY (CZECH) □ DOW DEFOLIANT □ MONOXONE □ SMA □ SMCA □ SODIUM MONOCHLORACETATE

CONSENSUS REPORTS: Reported in EPA TSCA Inventory.
DOT CLASSIFICATION: 6.1; Label: KEEP AWAY FROM FOOD
SAFETY PROFILE: Poison by ingestion and intraperitoneal routes. When heated to decomposition it emits toxic fumes of Cl$^-$ and Na$_2$O. Used as an herbicide.

SFW000   CAS: 361-09-1   HR: 3
SODIUM CHOLATE
mf: C$_{24}$H$_{39}$O$_5$•Na   mw: 430.62
PROP: White lyophilized powder. Sol: 1% in ethanol.
SYNS: CHOLIC ACID, MONOSODIUM SALT □ DS-Na □ SODIUM CHOLIC ACID □ TRIHYDROXY 3-7-12 CHOLANATE de Na
CONSENSUS REPORTS: Reported in EPA TSCA Inventory.
SAFETY PROFILE: Poison by intraperitoneal and intravenous routes. Moderately toxic by ingestion. When heated to decomposition it emits toxic fumes of Na$_2$O. See also CHOLIC ACID.

SFW500   CAS: 13517-17-4   HR: 3
SODIUM CHROMATE DECAHYDRATE
mf: CrO$_4$2Na•10H$_2$O   mw: 342.18
PROP: Deliquescent yellow monoclinic crystals from H$_2$O. IDLH Ca [15 mg/m^3 {as Cr(VI)}].
SYN: CHROMIC ACID, DISODIUM SALT, DECAHYDRATE
CONSENSUS REPORTS: Chromium and its compounds are on the Community Right-To-Know List.
OSHA PEL: CL 0.1 mg(CrO$_3$)/m^3
ACGIH TLV: TWA 0.05 mg(Cr)/m^3; Confirmed Human Carcinogen
NIOSH REL: (Chromium(VI)) TWA 0.001 mg(Cr)/m^3
SAFETY PROFILE: Confirmed human carcinogen. When heated to decomposition it emits toxic fumes of Na$_2$O. See also CHROMIUM COMPOUNDS and SODIUM CHROMATE.

SFX730   CAS: 22560-50-5   HR: 3
SODIUM CLODRONATE
mf: CH$_2$Cl$_2$O$_6$P$_2$•2Na   mw: 288.85

SYNS: BM 06011 □ CHLODRONATE SODIUM □ CLODRONATE SODIUM □ DICHLOROMETHYL-ENEDIPHOSPHONIC ACID DISODIUM SALT □ DICHLOROMETHANE DIPHOSPHONATE □ DISODIUM CLODRONATE □ DISODIUM (DICHLOROMETHYLENE)BISPHOSPHONATE □ LODRONATE □ PHOSPHONIC ACID, (DICHLORO-METHYLENE)BIS-, DISODIUM SALT □ PHOSPHONIC ACID, (DICHLOROMETHYLENE)DI-, DISODIUM SALT

SAFETY PROFILE: A poison by intravenous route. Moderately toxic by ingestion, intraperitoneal routes. When heated to decomposition it emits toxic vapors of PO_x and Cl^-.

SFY500   CAS: 8068-28-8   HR: 3
SODIUM COLISTINE-METHANESULFONATE

mf: $C_{49}H_{93}N_{13}O_{22}S_4 \cdot 4Na$ mw: 1436.75
PROP: Antimicrobial.
SYNS: CLM □ COLIMYCIN M □ COLISTIMETHATE SODIUM □ COLISTIN SODIUM METHANESULFONATE □ COLISTIN SULFOMETHAT □ COLISTIN SULFO-METHATE SODIUM □ COLISTRIMETHATE SODIUM □ COLY-MYCIN INJECTABLE □ COLYSTINMETHAN-SULFONAT (GERMAN) □ PENTASODIUM COLISTIN-METHANESULFONATE □ SODIUM COLISTIMETHATE □ SODIUM COLISTIN METHANESULFONATE □ W 1929

CONSENSUS REPORTS: Reported in EPA TSCA Inventory.
SAFETY PROFILE: Human poison by intramuscular route. Experimental poison by intramuscular, intraperitoneal, subcutaneous, and intravenous routes. Mildly toxic by ingestion. An experimental teratogen. Human systemic effects by intramuscular route: convulsions or effect on seizure threshold, change in motor activity, change in kidney tubules, and urine volume decrease or anuria. Experimental reproductive effects. Used as an antibiotic. When heated to decomposition it emits very toxic fumes of NO_x, SO_x, and Na_2O.

SFZ000   HR: D
SODIUM COMPOUNDS
SAFETY PROFILE: Variable toxicity. Sodium ion as such is practically nontoxic. The toxicity of sodium compounds is frequently, though not always, due to the anion involved. The hydroxide is very corrosive, being strongly basic. Even here it

is the concentration of hydroxyl ion that is responsible for the caustic action of this material. When heated to decomposition it emits toxic fumes of Na_2O.

SFZ100   CAS: 14264-31-4   HR: 3
SODIUM CUPROCYANIDE (DOT)
DOT: UN 2316/UN 2317
mf: $C_3CuN_3 \cdot 2Na$ mw: 187.58
PROP: White powder.
SYNS: COPPER SODIUM CYANIDE □ CUPRATE(2-), TRIS(CYANO-C)-, DISODIUM □ SODIUM CUPROCYANIDE, solid (UN 2316) (DOT) □ SODIUM CUPROCYANIDE, solution (UN 2317) (DOT)

CONSENSUS REPORTS: Reported in EPA TSCA Inventory.
ACGIH TLV: TWA (fume) 0.2 mg/m³; (dust, mist) 1 mg(Cu)/m³;
DOT CLASSIFICATION: 6.1; Label: Poison
SAFETY PROFILE: A poison. When heated to decomposition it emits toxic fumes of CN^-.

SGA500   CAS: 143-33-9   HR: 3
SODIUM CYANIDE
DOT: UN 1689
mf: CNNa   mw: 49.01
PROP: White or colorless, deliquescent, cubic crystals or powder. Undergoes cubic to hexagonal transition on cooling below 10°. Mp: 563.7°, bp: 1496°, vap press: 1 mm @ 817°. Very sol in H_2O; spar sol in EtOH. IDLH 25 mg/m³ (as CN).
SYNS: CIANURO di SODIO (ITALIAN) □ CYANURE de SODIUM (FRENCH) □ CYANIDE of SODIUM □ CYANOBRIK □ CYANOGRAN □ CYMAG □ HYDROCYANIC ACID, SODIUM SALT □ KYANID SODNY (CZECH) □ RCRA WASTE NUMBER P106

CONSENSUS REPORTS: EPA Extremely Hazardous Substances List. Cyanide and its compounds are on the Community Right-To-Know List. Reported in EPA TSCA Inventory.
OSHA PEL: TWA 5 mg(CN)/m³ (skin)
ACGIH TLV: CL 5 mg(CN)/m³ (skin)
DFG MAK: 5 mg(CN)/m³
NIOSH REL: CL 5 mg(CN)/m³/10M
DOT CLASSIFICATION: 6.1; Label: Poison
SAFETY PROFILE: A deadly human poison by ingestion. A deadly experimental poison

by ingestion, intraperitoneal, subcutaneous, intravenous, parenteral, intramuscular, and ocular routes. An experimental teratogen. Human systemic effects by ingestion: hallucinations, distorted perceptions, muscle weakness, and gastritis. Experimental reproductive effects.

The volatile cyanides resemble hydrocyanic acid physiologically, inhibiting tissue oxidation and causing death through asphyxia. Cyanogen is probably as toxic as hydrocyanic acid; the nitriles are generally considered somewhat less toxic, probably because of their lower volatility. The nonvolatile cyanide salts appear to be relatively nonhazardous systemically, so long as they are not ingested and care is taken to prevent the formation of hydrocyanic acid. Workers, such as electroplaters and picklers, who are daily exposed to cyanide solutions may develop a "cyanide" rash, characterized by itching and by macular, papular, and vesicular eruptions. Frequently there is secondary infection. Exposure to small amounts of cyanide compounds over long periods of time is reported to cause loss of appetite, headache, weakness, nausea, dizziness, and symptoms of irritation of the upper respiratory tract and eyes.

Flammable by chemical reaction with heat, moisture, acid. Many cyanides evolve hydrocyanic acid rather easily. This is a flammable gas and is highly toxic. Carbon dioxide from the air is sufficiently acidic to liberate hydrocyanic acid from cyanide solutions. Explodes if melted with nitrite or chlorate @ about 450°. Violent reaction with F_2, Mg, nitrates, HNO_3, nitrites. Upon contact with acid, acid fumes, water, or steam, it will produce toxic and flammable vapors of CN^- and Na_2O. Used in the extraction of gold and silver ores, in electroplating, and in insecticides. See also CYANIDE and HYDROCYANIC ACID.

SGC000   CAS: 139-05-9   HR: 2
SODIUM CYCLAMATE
mf: $C_6H_{12}NO_3S•Na$ mw: 201.24

PROP: White, crystalline powder; practically odorless. Sol in water; almost insol in alc, benzene, chloroform, and ether.
SYNS: ASSUGRIN □ ASSUGRIN FEINUSS □ ASSUGRIN VOLLSUSS □ ASUGRYN □ CYCLAMATE □ CYCLAMATE SODIUM □ CYCLAMIC ACID SODIUM SALT □ CYCLOHEXANESULFAMIC ACID, MONOSODIUM SALT □ CYCLOHEXANESULPHAMIC ACID, MONOSODIUM SALT □ CYCLOHEXYL SULPHAMATE SODIUM □ DULZOR-ETAS □ HACHI-SUGAR □ IBIOSUC □ NATREEN □ NATRIUMZY-KLAMATE (GERMAN) □ SODIUM CYCLOHEXANE-SULFAMATE □ SODIUM CYCLOHEXANESULPHAMATE □ SODIUM CYCLOHEXYL AMIDOSULPHATE □ SODIUM CYCLOHEXYL SULFAMATE □ SODIUM CYCLOHEXYL SULFAMIDATE □ SODIUM CYCLOHEXYL SULPHAMATE □ SODIUM SUCARYL □ SUCARYL SODIUM □ SUCCARIL □ SUCROSA □ SUESSETTE □ SUESTAMIN □ SUGARIN □ SUGARON
CONSENSUS REPORTS: IARC Cancer Review: Group 3 IMEMDT 7,178,87; Animal Limited Evidence IMEMDT 22,55,80. Reported in EPA TSCA Inventory. EPA Genetic Toxicology Program.
SAFETY PROFILE: Moderately toxic by intravenous and intraperitoneal routes. Mildly toxic by ingestion. Experimental reproductive effects. Questionable carcinogen with experimental neoplastigenic, tumorigenic, and teratogenic data. Human mutation data reported. When heated to decomposition it emits very toxic fumes of Na_2O, SO_x, and NO_x.

SGD000   CAS: 4418-26-2   HR: 3
SODIUM DEHYDROACETIC ACID
mf: $C_8H_7O_4•Na$ mw: 190.14
PROP: White powder; odorless with slt characteristic taste. Mp: 109–111°. Sol in water, propylene glycol, glycerin.
SYNS: DEHYDROACETIC ACID, SODIUM SALT □ DHA-SODIUM □ HARVEN □ 4-HEXENOIC ACID, 2-ACETYL-5-HYDROXY-3-OXO, Δ-LACTONE, SODIUM derivative □ 3-(1-HYDROXYETHYLIDENE)-6-METHYL-2H-PYRAN-2,4(3H)-DIONE, SODIUM SALT □ SODIUM DEHYDROACETATE (FCC)
CONSENSUS REPORTS: Reported in EPA TSCA Inventory. EPA Genetic Toxicology Program.
SAFETY PROFILE: Poison by intravenous route. Moderately toxic by ingestion. An

experimental teratogen. Experimental reproductive effects. An eye irritant. Mutation data reported. When heated to decomposition it emits toxic fumes of Na_2O.

SGE000   CAS: 302-95-4   HR: 3
SODIUM DESOXYCHOLATE
mf: $C_{24}H_{39}O_4 \bullet Na$ mw: 414.62
SYNS: DEOXYCHOLATE SODIUM □ DEOXYCHOLIC ACID SODIUM SALT □ DIHYDROXY 3-12 CHOLANATE de Na (FRENCH) □ (3-α,5-β,12-α)-3,12-DIHYDROXY-CHOLAN-24-OIC ACID MONOSODIUM SALT □ 3-α,12-α-DIHYDROXY-5-β-CHOLAN-24-OIC ACID SODIUM SALT □ NA-DESOXYCHOLAT (GERMAN) □ NATRIUM-3-α,12-α-DIHYDROXYCHOLANAT (GERMAN) □ SODIUM DEOXYCHOLATE □ SODIUM DEOXYCHOLIC ACID
CONSENSUS REPORTS: Reported in EPA TSCA Inventory.
SAFETY PROFILE: Poison by intraperitoneal and intravenous routes. Moderately toxic by ingestion and subcutaneous routes. Experimental reproductive effects. Mutation data reported. When heated to decomposition it emits toxic fumes of Na_2O.

SGE400   CAS: 126-96-5   HR: D
SODIUM DIACETATE
mf: $C_4H_7NaO_4 \bullet xH_2O$
PROP: white crystalline powder; odor of acetic acid. Sol in water.
SYN: SODIUM HYDROGEN DIACETATE
CONSENSUS REPORTS: Reported in EPA TSCA Inventory.
SAFETY PROFILE: When heated to decomposition it emits acrid smoke and irritating fumes.

SGF500   CAS: 136-30-1   HR: 3
SODIUM DIBUTYLDITHIOCARBAMATE
mf: $C_9H_{18}NS_2 \bullet Na$ mw: 227.39
SYNS: BUTYL NAMATE □ DIBUTYL-DITHIOCARBAMIC ACID SODIUM SALT □ PENNAC □ SODIUM DBDT □ TEPIDONE □ TEPIDONE RUBBER ACCELERATOR □ USAF B-35 □ VULCACURE
CONSENSUS REPORTS: Reported in EPA TSCA Inventory.
SAFETY PROFILE: Poison by intraperitoneal route. When heated to decomposition it emits very toxic fumes of

NO_x, SO_x, and Na_2O. See also CARBAMATES.

SGG500   CAS: 2893-78-9   HR: 2
SODIUM DICHLOROCYANURATE
mf: $C_3HCl_2N_3O_3 \bullet Na$ mw: 220.96
PROP: White crystals; chlorine odor. Mp: 230–250°, water-sol.
SYNS: ACL 60 □ CDB 63 □ DICHLOROISOCYANURIC ACID SODIUM SALT (DOT) □ DIKONIT □ DIMANIN C □ FI CLOR 60S □ OCI 56 □ SDIC □ SIMPLA □ SODIUM DICHLORISOCYANURATE □ SODIUM DICHLOROISO-CYANURATE □ SODIUM-1,3-DICHLORO-1,3,5-TRIAZINE-2,4-DIONE-6-OXIDE □ 1-SODIUM-3,5-DICHLORO-s-TRIAZINE-2,4,6-TRIONE □ 1-SODIUM-3,5-DICHLORO-1,3,5-TRIAZINE-2,4,6-TRIONE □ SODIUM DICHLORO-s-TRIAZINETRIONE, dry, containing more than 39% available chlorine (DOT) □ SODIUM SALT of DICHLORO-s-TRIAZINETRIONE
CONSENSUS REPORTS: Reported in EPA TSCA Inventory.
SAFETY PROFILE: Moderately toxic to humans and animals by ingestion. An experimental teratogen. Experimental reproductive effects. A severe skin and eye irritant. Human systemic effects by ingestion: ulceration or bleeding from stomach. The other main toxic effects were gastrointestinal irritation, salivation, lachrymation, dyspnea, weakness, emaciation, lethargy, diarrhea, coma, and (following very high dosage) death after 1–8 days, with autopsy showing irritation of stomach and gastrointestinal tract, liver dysfunction, and lung congestion. The concentrated material may be a little more toxic, due to greater gastrointestinal irritation. In the dry form, it is not appreciably irritating to dry skin. However, when moist, the concentrated material is irritating to skin, and also may cause severe eye irritation.

A powerful oxidizer. Incompatible with combustible materials, ammonium salts, nitrogenous materials. Used to chlorinate swimming pools and in cleaning, bleaching, disinfecting, sanitizing. When heated to decomposition it emits very toxic fumes of Cl^-, NO_x, and Na_2O.

SGH500   CAS: 2702-72-9   HR: 3
SODIUM-2,4-DICHLOROPHENOXY-
   ACETATE
mf: $C_8H_5Cl_2O_3 \cdot Na$ mw: 243.02
SYNS: AGRION □ 2,4-DICHLOROPHENOXYACETIC
ACID, SODIUM SALT □ DICONIRT D □ 2,4-D SODIUM
SALT □ FERNOXENE □ HORMIT □ PIELIK E □
SODIUM-2,4-D □ SPRAY-HORMITE □ SPRITZ-HORMIT
CONSENSUS REPORTS: Reported in EPA
TSCA Inventory. EPA Genetic Toxicology
Program.
SAFETY PROFILE: Poison by ingestion,
intraperitoneal, subcutaneous, and
intravenous routes. Moderately toxic by skin
contact. An experimental teratogen. Human
systemic effects by inhalation: anorexia,
gastrointestinal, and liver changes. Mutation
data reported. When heated to
decomposition it emits toxic fumes of Cl^-
and Na_2O.

SGI000   CAS: 10588-01-9   HR: 3
SODIUM DICHROMATE
mf: $Cr_2O_7 \cdot 2Na$ mw: 261.98
PROP: Orange crystals. Deliquescent in
moist air. Anhydrous. Mp: 356.7°, decomp
@ about 400°, d: 2.35 @ 13°. Very sol in
water. IDLH Ca [15 mg/m³ {as Cr(VI)}].
SYNS: BICHROMATE de SODIUM (FRENCH) □
BICHROMATE of SODA □ CHROMIC ACID, DISODIUM
SALT □ CHROMIUM SODIUM OXIDE □ DISODIUM
DICHROMATE □ NATRIUMBICHROMAAT (DUTCH) □
NATRIUMDICHROMAAT (DUTCH) □ NATRIUMDI-
CHROMAT (GERMAN) □ SODIO (DICROMATO di)
(ITALIAN) □ SODIUM BICHROMATE □ SODIUM
CHROMATE □ SODIUM DICHROMATE(VI) □ SODIUM
DICHROMATE de (FRENCH)
CONSENSUS REPORTS: IARC Cancer
Review: Group 1 IMEMDT 7,165,87;
Animal Inadequate Evidence IMEMDT
2,100,73; IMEMDT 23,205,80; Human
Inadequate Evidence IMEMDT 23,205,80.
Chromium and its compounds are on the
Community Right-To-Know List. Reported
in EPA TSCA Inventory. EPA Genetic
Toxicology Program. EPA FIFRA 1988
pesticide subject to registration or re-
registration.
OSHA PEL: CL 0.1 mg/(CrO₃)/m³
ACGIH TLV: TWA 0.05 mg(Cr)/m³;
Confirmed Human Carcinogen

NIOSH REL: (chromium(VI)): TWA 0.001
mg(Cr)/m³
SAFETY PROFILE: Confirmed carcinogen
with experimental tumorigenic data. Poison
by ingestion, skin contact, intravenous,
intraperitoneal, and subcutaneous routes.
Human systemic effects by ingestion: cough,
nausea or vomiting, and sweating. Human
mutation data reported. A caustic and
irritant. A powerful oxidizer. Potentially
explosive reaction with acetic anhydride,
ethanol + sulfuric acid + heat, hydrazine.
Violent reaction or ignition with boron +
silicon (pyrotechnic), organic residues +
sulfuric acid, 2-propanol + sulfuric acid,
sulfuric acid + trinitrotoluene. Incompatible
with hydroxylamine. When heated to
decomposition it emits toxic fumes of
Na_2O. See also CHROMIUM
COMPOUNDS.

SGJ000   CAS: 148-18-5   HR: 3
SODIUM DIETHYLDITHIOCARBAMATE
mf: $C_5H_{10}NS_2 \cdot Na$ mw: 171.27
PROP: Crystals from EtOH. Mp: 95°, d: 1.1
@ 20°/20°, vap d: 5.9. Sol in water and alc.
SYNS: CUPRAL □ DDC □ DEDC □ DEDK □
DIETHYLCARBAMODITHIOIC ACID, SODIUM SALT □
DIETHYLDITHIOCARBAMATE SODIUM □ DIETHYL-
DITHIOCARBAMIC ACID SODIUM □ DIETHYL-
DITHIOCARBAMIC ACID, SODIUM SALT □ DIETHYL
SODIUM DITHIOCARBAMATE □ DITHIOCARB □
DITHIOCARBAMATE □ NCI-C02835 □ SODIUM DEDT
□ SODIUM N,N-DIETHYLDITHIOCARBAMATE □
SODIUM SALT of N,N-DIETHYLDITHIOCARBAMIC
ACID □ THIOCARB □ USAF EK-2596
CONSENSUS REPORTS: IARC Cancer
Review: Group 3 IMEMDT 7,56,87; Animal
Inadequate Evidence IMEMDT 12,217,76.
NCI Carcinogenesis Bioassay (feed); No
Evidence: mouse, rat NCITR* NCI-CG-
TR-172,79. Reported in EPA TSCA
Inventory.
DFG MAK: 2 mg/m³
SAFETY PROFILE: Moderately toxic by
ingestion, intraperitoneal, and subcutaneous
routes. Experimental reproductive effects.
Questionable carcinogen with experimental
neoplastigenic and teratogenic data. Human
mutation data reported. When heated to

decomposition it emits very toxic fumes of NO_x, SO_x, and Na_2O. Used as a pesticide. See also CARBAMATES.

SGM500   CAS: 128-04-1   HR: 2
**SODIUM N,N-DIMETHYLDITHIO-
     CARBAMATE**
mf: $C_3H_6NS_2 \cdot Na$ mw: 143.21
PROP: Crystals.
SYNS: ACETO SDD 40 □ ALCOBAM NM □ BROGDEX 555 □ CARBON S □ DIBAM □ DIMETHYLDITHIO-CARBAMIC ACID, SODIUM SALT □ DMDK □ METHYL NAMATE □ SDDC □ SHARSTOP 204 □ STA-FRESH 615 □ STERISEAL LIQUID #40 □ THIOSTOP N □ VINSTOP □ VULNOPOL NM □ WING STOP B
CONSENSUS REPORTS: Reported in EPA TSCA Inventory.
SAFETY PROFILE: Moderately toxic by ingestion and intraperitoneal routes. Mutation data reported. When heated to decomposition it emits very toxic fumes of NO_x, SO_x, and Na_2O. See also CARBAMATES.

SGP550   CAS: 25641-53-6   HR: 3
SODIUM DINITRO-o-CRESYLATE
DOT: UN 0234/UN 1348
mf: $C_7H_6N_2O_5 \cdot Na$ mw: 221.14
SYNS: o-CRESOL, DINITRO-, SODIUM SALT □ DINITRO-o-CRESOL SODIUM SALT □ DNOC-SODIUM □ 2-METHYLDINITROPHENOL SODIUM SALT □ PHENOL, 2-METHYLDINITRO-, SODIUM SALT (9CI) □ SODIUM DINITRO-o-CRESOLATE, dry or wetted with <15% water, by weight (UN 0234) (DOT) □ SODIUM DINITRO-o-CRESOLATE, wetted with not <15% water, by weight (UN 1348) (DOT)
DOT CLASSIFICATION: Explosive 1.3C; Label: Explosive 1.3C (UN 0234); 4.1; Label: Flammable Solid, Poison (UN 1348)
SAFETY PROFILE: A poison by ingestion. Dangerous: an explosive and flammable solid. When heated to decomposition it emits toxic vapors of NO_x.

SGQ500   CAS: 573-58-0   HR: 3
**SODIUM DIPHENYLDIAZO-BIS(α-
     NAPHTHYLAMINESULFONATE)**
mf: $C_{32}H_{24}N_6O_6S_2 \cdot 2Na$ mw: 698.72
PROP: Brown-red powder. Sol in H_2O.
SYNS: ATLANTIC CONGO RED □ ATUL CONGO RED □ AZOCARD RED CONGO □ BENZO CONGO RED □ BRASILAMINA CONGO 4B □ C.I. 22120 □ C.I. DIRECT

RED 28 □ C.I. DIRECT RED 28, DISODIUM SALT □ CONGO RED □ COTTON RED L □ DIACOTTON CONGO RED □ DIRECT RED 28 □ ERIE CONGO 4B □ HISPAMIN CONGO 4B □ KAYAKU CONGO RED □ MITSUI CONGO RED □ PEERAMINE CONGO RED □ SUGAI CONGO RED □ TERTRODIRECT RED C □ TRISULFON CONGO RED □ VONDACEL RED CL
CONSENSUS REPORTS: Reported in EPA TSCA Inventory.
SAFETY PROFILE: Human poison by ingestion with cardiovascular effects. Experimental poison by intravenous route. An experimental teratogen. Experimental reproductive effects. An eye irritant. Mutation data reported. When heated to decomposition it emits very toxic fumes of NO_x, SO_x, and Na_2O.

SHE350   CAS: 13601-19-9   HR: 2
SODIUM FERROCYANIDE
mf: $Na_4Fe(CN)_6 \cdot 10H_2O$ mw: 484.06
PROP: Yellow crystals or crystalline powder. Sol in water; insol in most org solvs.
SYN: YELLOW PRUSSIATE of SODA
OSHA PEL: TWA 5 mg(CN)/m^3
ACGIH TLV: TWA 5 mg(CN)/m^3 (skin)
DFG MAK: 5 mg/m^3
NIOSH REL: (Cyanide) CL 5 mg(CN)/m^3/10M
SAFETY PROFILE: When heated to decomposition emits toxic fumes of CN^-.

SHF000   CAS: 15096-52-3   HR: 3
SODIUM FLUOALUMINATE
mf: $AlF_6 \cdot 3Na$ mw: 209.95
PROP: Very white or discolored, brittle, solid or vitreous mass. Mp: 1000°, d: 2.95. Sol in concentrated H_2SO_4. IDLH 250 mg/m^3 (as F).
SYNS: ALUMINUM SODIUM FLUORIDE □ CRYOLITE □ ENT 24,984 □ KRYOLITH (GERMAN) □ NATRIUMALUMINUMFLUORID (GERMAN) □ NATRIUMHEXAFLUOROALUMINATE (GERMAN) □ SODIUM ALUMINOFLUORIDE □ SODIUM ALUMINUM FLUORIDE □ SODIUM HEXAFLUOROALUMINATE □ VILLIAUMITE
CONSENSUS REPORTS: Reported in EPA TSCA Inventory.
OSHA PEL: TWA 2.5 mg(F)/m^3
ACGIH TLV: TWA 2 mg(Al)/m^3; TWA 2.5 mg(F)/m^3; BEI: 3 mg/g creatinine of

S

fluorides in urine prior to shift; 10 mg/g creatinine of fluorides in urine at end of shift.

NIOSH REL: (Fluorides, Inorganic) TWA 2.5 mg(F)/m^3

SAFETY PROFILE: Poison by ingestion. Used as a pesticide. Mutation data reported.When heated to decomposition it emits toxic fumes of F$^-$ and Na$_2$O. See also FLUORIDES.

SHF500   CAS: 7681-49-4   HR: 3
SODIUM FLUORIDE
DOT: UN 1690

mf: FNa   mw: 41.99

PROP: Clear, lustrous, cubic crystals, or white powder or balls. Mp: 996°, bp: 1695°, d: 2 @ 41°, vap press: 1 mm @ 1077°. Mod sol in H$_2$O; spar sol in EtOH. IDLH 250 mg/m^3 (as F).

SYNS: ALCOA SODIUM FLUORIDE □ ANTIBULIT □ CAVI-TROL □ CHEMIFLUOR □ CREDO □ DISODIUM DIFLUORIDE □ FDA 0101 □ F1-TABS □ FLORIDINE □ FLOROCID □ FLOZENGES □ FLUORAL □ FLUORIDENT □ FLUORID SODNY (CZECH) □ FLUORIGARD □ FLUORINEED □ FLUORINSE □ FLUORITAB □ FLUOR-O-KOTE □ FLUORURE de SODIUM (FRENCH) □ FLURA-GEL □ FLURCARE □ FUNGOL B □ GEL II □ GELUTION □ GLEEM □ IRADICAV □ KARIDIUM □ KARIGEL □ KARI-RINSE □ LEA-COV □ LEMOFLUR □ LURIDE □ NAFEEN □ NaFPAK □ Na FRINSE □ NATRIUM FLUORIDE □ NCI-C55221 □ NUFLUOR □ OSSALIN □ OSSIN □ PEDIAFLOR □ PEDIDENT □ PENNWHITE □ PERGANTENE □ PHOS-FLUR □ POINT TWO □ PREDENT □ RAFLUOR □ RESCUE SQUAD □ ROACH SALT □ SODIUM FLUORIDE, solid and solution (DOT) □ SODIUM FLUORURE (FRENCH) □ SODIUM HYDROFLUORIDE □ SODIUM MONOFLUORIDE □ SO-FLO □ STAY-FLO □ STUDAFLUOR □ SUPER-DENT □ T-FLUORIDE □ THERA-FLUR-N □ TRISODIUM TRIFLUORIDE □ VILLIAUMITE

CONSENSUS REPORTS: Reported in EPA TSCA Inventory. EPA Genetic Toxicology Program.

OSHA PEL: TWA 2.5 mg(F)/m^3

ACGIH TLV: TWA 2.5 mg(F)/m^3; BEI: 3 mg/g creatinine of fluorides in urine prior to shift; 10 mg/g creatinine of fluorides in urine at end of shift.

NIOSH REL: TWA (Inorganic Fluorides) 2.5 mg(F)/m^3

DOT CLASSIFICATION: 6.1; Label: KEEP AWAY FROM FOOD

SAFETY PROFILE: Human poison by ingestion. Experimental poison by ingestion, skin contact, intravenous, intraperitoneal, subcutaneous, and intramuscular routes. Human systemic effects: changes in teeth and supporting structures, cyanosis, diarrhea, EKG changes, fluid intake, headache, hypermotility, increased immune response, muscle weakness, musculo-skeletal changes, nausea or vomiting, paresthesia, ptosis (drooping of the eyelid from sympathetic innervation), respiratory depression, salivary gland changes, tremors. Experimental teratogenic and reproductive effects. Human mutation data reported. A corrosive irritant to skin, eyes, and mucous membranes. Questionable carcinogen with experimental tumorigenic data. It is very phytotoxic. When heated to decomposition it emits toxic fumes of F$^-$ and Na$_2$O. Used in chemical cleaning, for fluoridation of drinking water, as a fungicide and insecticide. See also FLUORIDES.

SHG500   CAS: 62-74-8   HR: 3
SODIUM FLUOROACETATE
DOT: UN 2629

mf: C$_2$H$_2$FO$_2$•Na   mw: 100.03

PROP: Fine, white powder, or monoclinic crystals. Mp: 200°. Sol in water; spar sol in EtOH, MeOH; prac insol in Me$_2$CO, CCl$_4$. IDLH 2.5 mg/m^3.

SYNS: 1080 □ COMPOUND No. 1080 □ FLUORACET-ATO di SODIO (ITALIAN) □ FLUORESSIGSAEURE (GERMAN) □ FLUOROACETIC ACID, SODIUM SALT □ FRATOL □ FURATOL □ MONOFLUORESSIGSAEURE, NATRIUM (GERMAN) □ NATRIUMFLUORACETAAT (DUTCH) □ NATRIUMFLUORACETAT (GERMAN) □ RATBANE 1080 □ RCRA WASTE NUMBER P058 □ SODIO, FLUORACETATO di (ITALIAN) □ SODIUM FLUOACETATE □ SODIUM FLUOACETIC ACID □ SODIUM FLUORACETATE de (FRENCH) □ SODIUM MONOFLUOROACETATE □ TL 869 □ YASOKNOCK

CONSENSUS REPORTS: Reported in EPA TSCA Inventory. EPA Extremely Hazardous Substances List.

OSHA PEL: TWA 0.05 mg/m^3 (skin); STEL 0.15 mg/m^3 (skin)

ACGIH TLV: TWA 0.05 mg/m³ (skin)
DFG MAK: 0.05 mg/m³
DOT CLASSIFICATION: 6.1; Label: Poison
SAFETY PROFILE: A deadly human poison by ingestion. Experimental poison by ingestion, skin contact, intraperitoneal, subcutaneous, and intravenous routes. A very highly toxic water-soluble salt used mainly as an immediate-action rodenticide. It is absorbed rapidly by the gastrointestinal tract but slowly by the skin unless the skin is abraded or cut. It operates by blocking the Krebs cycle by formation of fluorocitric acid, which inhibits aconitase. It has an effect on either the cardiovascular or nervous system, or both, in all species and, in some species, the skeletal muscles. Humans have mixed responses, with the cardiac feature predominating. By a direct action on the heart, contractile power is lost, which leads to declining blood pressure. Ventricular premature contractions and arrhythmias are seen in all species, including humans. The central nervous system is directly attacked by sodium fluoroacetate. In humans, the action on the central nervous system produces epileptiform convulsive seizures followed by severe depression. The dangerous dose for humans is 0.5–2 mg/kg. Other species vary considerably in their response to this material, with primates and birds being the most resistant and carnivora and rodents being the most susceptible. Most domestic animals show a susceptibility falling between the two extremes indicated above. Experimental reproductive effects. When heated to decomposition it emits highly toxic fumes of Na_2O and F^-.

SHJ000   CAS: 141-53-7   HR: 2
SODIUM FORMATE
mf: $CHO_2 \cdot Na$ mw: 68.01
PROP: White, colorless, deliquescent, monoclinic crystals. Mp: 259–261°, d: 1.92 @ 20°. Very sol in H_2O; spar sol in EtOH; insol in Et_2O.
SYNS: MRAVENCAN SODNY □ SALACHLOR
CONSENSUS REPORTS: Reported in EPA TSCA Inventory.

SAFETY PROFILE: Moderately toxic by ingestion, intravenous, and subcutaneous routes. Combustible when exposed to heat or flame. When heated to decomposition it emits toxic fumes of Na_2O. See also FORMIC ACID.

SHK800   CAS: 527-07-1   HR: 1
SODIUM GLUCONATE
mf: $C_6H_{12}O_7 \cdot Na$ mw: 219.17
PROP: White to tan granular or crystalline powder. Very sol in water; sltly sol in alc; insol in ether.
SYNS: GLONSEN □ GLUCONATO di SODIO (ITALIAN) □ GLUCONIC ACID SODIUM SALT □ MONOSODIUM GLUCONATE □ PASEXON 100T □ PMP SODIUM GLUCONATE □ SODIUM d-GLUCONATE
CONSENSUS REPORTS: Reported in EPA TSCA Inventory.
SAFETY PROFILE: Low toxicity by intravenous route. When heated to decomposition it emits acrid smoke and irritating fumes.

SHM000   CAS: 12005-86-6   HR: 3
SODIUM HEXAFLUOROARSENATE
mf: $AsF_6 \cdot Na$ mw: 211.91
PROP: Colorless, rhombohedral crystals.
CONSENSUS REPORTS: Arsenic and its compounds are on the Community Right-To-Know List.
OSHA PEL: OSHA: Cancer Hazard
ACGIH TLV: TWA 0.01 mg/m³; Confirmed Human Carcinogen; BEI: 35 μ (As)/L inorganic arsenic and methylated metabolites in urine
NIOSH REL: (Arsenic, Inorganic) CL 2 μg/m³/15M
SAFETY PROFILE: Confirmed human carcinogen. Poison by intravenous route. When heated to decomposition it emits very toxic fumes of As, F^-, and Na_2O. See also FLUORIDES and ARSENIC COMPOUNDS.

SHM500   CAS: 10124-56-8   HR: 3
SODIUM HEXAMETAPHOSPHATE
mf: $O_{18}P_6 \cdot 6Na$ mw: 611.76
PROP: White powder or flakes. Sol in water.

S

SYNS: CALGON ☐ CHEMI-CHARL ☐
HEXAMETAPHOSPHATE, SODIUM SALT ☐ HMP ☐
MEDI-CALGON ☐ PHOSPHATE, SODIUM HEXAMETA
☐ POLYPHOS ☐ SHMP

CONSENSUS REPORTS: Reported in EPA
TSCA Inventory.

SAFETY PROFILE: Poison by intravenous
route. Moderately toxic by intraperitoneal
and subcutaneous routes. Mildly toxic by
ingestion. When heated to decomposition it
emits toxic fumes of PO_x and Na_2O. See
also PHOSPHATES.

SHO500   CAS: 7646-69-7   HR: 3
SODIUM HYDRIDE

DOT: UN 1427

mf: HNa   mw: 24.00

PROP: Microcrystalline, white to brownish-
gray powder. Decomp on heating into Na
and H_2. Reacts violently with H_2O, NaOH
+ H_2. Mp: 800° (decomp), d: 0.9.

SYN: NAH 80

CONSENSUS REPORTS: Reported in EPA
TSCA Inventory.

DOT CLASSIFICATION: 4.3; Label:
Dangerous When Wet

SAFETY PROFILE: The powder ignites
spontaneously in air. Flammable when
exposed to heat or flame. Potentially
explosive reaction with water, diethyl
succinate + ethyltrifluoroacetate (above
60°C), dimethyl sulfoxide + heat, sulfur
dioxide. Ignition or violent reaction with
dimethylformamide (above 50°C), ethyl
2,2,3-trifluoropropionate, oxygen (at 230°C).
Incompatible with acetylene + moisture,
glycerin, halogens, sulphur. Normal fire
extinguishers are unsuitable, use sand, ashes,
sodium chloride. The commercial material
may contain traces of sodium. When heated
to decomposition it emits toxic fumes of
Na_2O. See also HYDRIDES.

SHQ500   CAS: 1333-83-1   HR: 3
SODIUM HYDROGEN FLUORIDE

DOT: UN 2439

mf: F_2HNa mw: 62.00

PROP: White powder or colorless
rhombohedral crystals. D: 2.08. Mod sol in
water to 42,000 ppm @ 20°C.

SYNS: SODIUM HYDROGEN DIFLUORIDE ☐
SODIUM HYDROGEN FLUORIDE ☐ SODIUM
HYDROGEN FLUORIDE, solution (DOT)

CONSENSUS REPORTS: Reported in EPA
TSCA Inventory.

OSHA PEL: TWA 2.5 mg(F)/m³

ACGIH TLV: TWA 2.5 mg(F)/m³; BEI: 3
mg/g creatinine of fluorides in urine prior to
shift; 10 mg/g creatinine of fluorides in
urine at end of shift.

NIOSH REL: TWA 2.5 mg(F)/m³

DOT CLASSIFICATION: 8; Label: Corrosive

SAFETY PROFILE: This material is very
toxic to humans by ingestion; between 1
teaspoonful and 1 ounce may be fatal.
Inhalation of dust may cause irritation to
respiratory tract. Skin contact may result in
irritation and ulceration; eye contact may
cause burns. To fight fire, use water, foam,
CO_2, dry chemicals. When heated to
decomposition it emits toxic fumes of F^-
and Na_2O. See also FLUORIDES and
HYDROFLUORIC ACID.

SHR000   CAS: 16721-80-5   HR: 3
SODIUM HYDROSULFIDE

DOT: UN 2318/NA 2922/NA 2949

mf: HNaS   mw: 56.06

PROP: Colorless hexagonal crystals.
Undergoes hexagonal to cubic transition.
Mp: 350°. Very sol in H_2O; sol in EtOH.

SYNS: SODIUM BISULFIDE ☐ SODIUM HYDROGEN
SULFIDE ☐ SODIUM HYDROSULFIDE, solution (NA 2922)
(DOT) ☐ SODIUM HYDROSULFIDE, with <25% water of
crystallization (UN 2318) (DOT) ☐ SODIUM HYDRO-
SULFIDE, with not <25% water of crystallization (NA 2949)
(DOT) ☐ SODIUM MERCAPTAN ☐ SODIUM
MERCAPTIDE ☐ SODIUM SULFHYDRATE

CONSENSUS REPORTS: Reported in EPA
TSCA Inventory.

DOT CLASSIFICATION: 8; Label: Corrosive,
Poison (NA 2922); DOT Class: 4.2; Label:
Spontaneously Combustible (UN 2318);
DOT Class: 8; Label: Corrosive (UN 2949)

SAFETY PROFILE: Poison by
intraperitoneal and subcutaneous routes.
Mutation data reported. A corrosive irritant
to skin, eyes, and mucous membranes.
Flammable when exposed to heat or flame.
Spontaneous combustion. Reacts violently
with diazonium salts. Readily yields H_2S.

When heated to decomposition it emits toxic fumes of SO_x and Na_2O. See also SULFIDES and MERCAPTANS.

SHR500   CAS: 7775-14-6   HR: 3
SODIUM HYDROSULPHITE
DOT: UN 1384
mf: $O_4S_2 \cdot 2Na$ mw: 174.10
PROP: White, colorless, or yellow-white crystals. Mp: >300°. Decomp in water (hot); sltly sol in cold water; insol in acids.
SYNS: D-OX □ HYDROLIN □ K-BRITE □ REDUCTONE □ SODIUM DITHIONITE (DOT) □ SODIUM HYDROSULFITE (DOT) □ SODIUM SULFOXYLATE □ VATROLITE □ V-BRITE □ VIRCHEM □ VIRTEX CC □ VIRTEX D □ VIRTEX L □ VIRTEX RD
CONSENSUS REPORTS: Reported in EPA TSCA Inventory.
DOT CLASSIFICATION: 4.2; Label: Spontaneously Combustible
SAFETY PROFILE: Toxic and an irritant. An allergen. Flammable when exposed to heat or flame. Ignites on contact with water or sodium chlorite. To extinguish fires, flood the reacting mass with water. Decomposes violently when heated to 190°C and emits toxic fumes of SO_x and Na_2O.

SHS000   CAS: 1310-73-2   HR: 3
SODIUM HYDROXIDE
DOT: UN 1823/UN 1824
mf: HNaO   mw: 40.00
PROP: White, pieces, lumps, sticks or deliquescent, orthorhombic powder. Undergoes polymorphic transition at 2°. Readily reacts with atm CO_2 forming Na_2CO_3. Mp: 323°, bp: 1390°, d: 2.120 @ 20°/4°, vap press: 1 mm @ 739°. Very sol in water and alc; insol in Et_2O, Me_2CO. IDLH 10 mg/m³.
SYNS: CAUSTIC SODA □ CAUSTIC SODA, bead (DOT) □ CAUSTIC SODA, dry (DOT) □ CAUSTIC SODA, flake (DOT) □ CAUSTIC SODA, granular (DOT) □ CAUSTIC SODA, liquid (DOT) □ CAUSTIC SODA, solid (DOT) □ CAUSTIC SODA, solution (DOT) □ HYDROXYDE de SODIUM (FRENCH) □ LEWIS-RED DEVIL LYE □ LYE (DOT) □ NATRIUMHYDROXID (GERMAN) □ NATRIUMHYDROXYDE (DUTCH) □ SODA LYE □ SODIO(IDROSSIDO di) (ITALIAN) □ SODIUM HYDRATE (DOT) □ SODIUM HYDROXIDE, bead (DOT) □ SODIUM HYDROXIDE, dry (DOT) □ SODIUM HYDROXIDE, flake (DOT) □ SODIUM HYDROXIDE, granular (DOT) □ SODIUM HYDROXIDE, solid (DOT) □ SODIUM (HYDROXYDE de) (FRENCH) □ WHITE CAUSTIC
CONSENSUS REPORTS: Reported in EPA TSCA Inventory. EPA Genetic Toxicology Program.
OSHA PEL: CL 2 mg/m³
ACGIH TLV: CL 2 mg/m³
DFG MAK: 2 mg/m³
NIOSH REL: (Sodium Hydroxide) CL 2 mg/m³/15M
DOT CLASSIFICATION: 8; Label: Corrosive
SAFETY PROFILE: Poison by intraperitoneal route. Moderately toxic by ingestion. Mutation data reported. A corrosive irritant to skin, eyes, and mucous membranes. This material, both solid and in solution, has a markedly corrosive action upon all body tissue, causing burns and frequently deep ulceration, with ultimate scarring. Mists, vapors, and dusts of this compound cause small burns, and contact with the eyes rapidly causes severe damage to the delicate tissue. Ingestion causes very serious damage to the mucous membranes or other tissues with which contact is made. It can cause perforation and scarring. Inhalation of the dust or concentrated mist can cause damage to the upper respiratory tract and to lung tissue, depending upon the severity of the exposure. Thus, effects of inhalation may vary from mild irritation of the mucous membranes to a severe pneumonitis.

A strong base. Vigorous reaction with 1,2,4,5-tetrachlorobenzene has caused many industrial explosions and forms the extremely toxic 2,3,7,8-tetrachlorodibenzodioxin. Mixtures with aluminum + arsenic compounds form the poisonous gas arsine. Potentially explosive reaction with bromine, 4-chlorobutyronitrile, 4-chloro-2-methylphenol (in storage), nitrobenzene + heat, sodium tetrahydroborate, 2,2,2-trichloroethanol, zirconium + heat. Reacts to form explosive products with ammonia + silver nitrate (forms silver nitride), N,N'-bis(trinitroethyl)urea (in storage), cyanogen

S

azide, glycols above 230° (e.g., ethylene glycol, diethylene glycol), 3-methyl-2-penten-4-yn-1-ol, trichloroethylene (forms dichloroacetylene). Caution: Under the proper conditions of temperature, pressure, and state of division, it can ignite or react violently with acetic acid, acetaldehyde, acetic anhydride, acrolein, acrylonitrile, allyl alcohol, allyl chloride, Al, benzene-1,4-diol, chlorine trifluoride, chloroform + methanol, chlorohydrin, chloronitro-toluenes, chlorosulfonic acid, 1,2-dichloroethylene, ethylene cyanhydrin, glyoxal, HCl, HF, hydroquinone, maleic anhydride, HNO_3, nitroethane, nitromethane, nitroparaffins, nitropropane, pentol, oleum, P, P_2O_5, β-propiolactone, H_2SO_4, (CH_3OH + tetrachloro-benzene), tetrahydrofuran, water, cinnamaldehyde, diborane + octanol oxime, 2,2-dichloro-3,3-dimethylbutane, 4-methyl-2-nitrophenol, 1,1,1-trichloroethanol, trichloronitromethane, zinc. Reacts with formaldehyde hydroxide to yield formic acid and hydrogen.

Dangerous material to handle. When heated to decomposition it emits toxic fumes of Na_2O.

SHS500   CAS: 1310-73-2   HR: 3
SODIUM HYDROXIDE (liquid)
DOT: UN 1823/UN 1824
mf: HNaO   mw: 40.00
PROP: Clear to slightly turbid, colorless liquid.
SYNS: CAUSTIC SODA, solution □ LYE, solution □ SODA LYE □ SODIUM HYDRATE, solution □ SODIUM HYDROXIDE, solution (FCC) □ WHITE CAUSTIC, solution
CONSENSUS REPORTS: Reported in EPA TSCA Inventory. Community Right-To-Know List.
DOT CLASSIFICATION: 8; Label: Corrosive
SAFETY PROFILE: Poison by intraperitoneal route. Moderately toxic by ingestion. Mutation data reported. A corrosive irritant to skin, eyes, and mucous membranes. When heated to decomposition it emits toxic fumes of Na_2O.

SHU000   CAS: 138-85-2   HR: 3
SODIUM p-HYDROXYMERCURI-BENZOATE
mf: $C_7H_5HgO_3$•Na   mw: 360.70
PROP: IDLH 10 mg/m³ (as Hg).
CONSENSUS REPORTS: Mercury and its compounds are on the Community Right-To-Know List. Reported in EPA TSCA Inventory.
OSHA PEL: CL 0.1 mg(Hg)/m³ (skin)
ACGIH TLV: TWA 0.1 mg(Hg)/m³ (skin); BEI: 35 µg/g creatinine total inorganic mercury in urine preshift; 15 µg/g creatinine total inorganic mercury in blood at end of shift at end of workweek.
DFG MAK: Confirmed Animal Carcinogen with Unknown Relevance to Humans
NIOSH REL: (Mercury, Aryl and Inorganic) CL 0.1 mg/m³ (skin)
SAFETY PROFILE: Poison by intraperitoneal route. Mutation data reported. When heated to decomposition it emits toxic fumes of Hg and Na_2O. See also MERCURY COMPOUNDS.

SHU500   CAS: 7681-52-9   HR: 3
SODIUM HYPOCHLORITE
DOT: UN 1791
mf: ClHO•Na   mw: 75.45
PROP: Mp: decomp. Aq solns form $NaClO_3$ slowly in air.
SYNS: ANTIFORMIN □ B-K LIQUID □ CARREL-DAKIN SOLUTION □ CHLOROS □ CHLOROX □ CLOROX □ DAKINS SOLUTION □ DEOSAN □ HYCLORITE □ HYPOCHLORITE SOLUTIONS containing >7% available chlorine by wt. □ HYPOCHLORITE SOLUTIONS with >5% but <16% available chlorine by wt. (DOT) □ HYPOCHLORITE SOLUTIONS with 16% or more available chlorine by wt. (UN 1791) □ JAVEX □ KLOROCIN □ MILTON □ NEO-CLEANER □ NEOSEPTAL CL □ PAROZONE □ PURIN B □ SODIUM CHLORIDE OXIDE □ SODIUM OXYCHLORIDE □ SURCHLOR
CONSENSUS REPORTS: Reported in EPA TSCA Inventory. EPA Genetic Toxicology Program.
DOT CLASSIFICATION: 8; Label: Corrosive
SAFETY PROFILE: Mildly toxic by ingestion. Human systemic effects by ingestion: somnolence, blood pressure lowering, corrosive to skin, nausea or vomiting. Human mutation data reported.

An eye irritant. Corrosive and irritating by ingestion and inhalation. The anhydrous salt is highly explosive and sensitive to heat or friction. Explosive reaction with formic acid (at 55°), phenylacetonitrile. Reacts to form explosive products with amines, ammonium salts (e.g., ammonium acetate, $(NH_4)_2CO_3$, ammonium nitrate, ammonium oxalate, $(NH_4)_3PO_4$), aziridine, methanol. Violent reaction with phenyl acetonitrile, cellulose, ethyleneimine. Solutions in water are storage hazards due to oxygen evolution. When heated to decomposition it emits toxic fumes of Na_2O and Cl^-. Used as a bleach.

SHV000   CAS: 7681-53-0   HR: 3
SODIUM HYPOPHOSPHITE
mf: $H_2O_2P•Na$ mw: 87.98
PROP: Colorless, pearly, crystalline plates or white granular powder; bittersweet, saline taste. Deliq; sol in water; sltly sol in alc.
SYNS: NATRIUMHYPOPHOSPHIT (GERMAN) □ SODIUM PHOSPHINATE
CONSENSUS REPORTS: Reported in EPA TSCA Inventory.
SAFETY PROFILE: Poison by subcutaneous route. Moderately toxic by intraperitoneal route. Flammable when exposed to heat or flame. Aqueous solutions may explode on evaporation. Potentially explosive reaction with oxidants (e.g., chlorates, nitrates). Heat causes it to evolve phosphine. It can explode. When heated to decomposition it emits toxic fumes of PO_x and Na_2O. See also PHOSPHINE.

SHW000   CAS: 7681-82-5   HR: 2
SODIUM IODIDE
mf: INa   mw: 149.89
PROP: Cubic, colorless crystals. Mp: 660°, bp: 1304°, d: 3.667, vap press: 1 mm @ 767°. Very sol in H_2O, EtOH, and Me_2CO.
SYNS: ANAYODIN □ IODURIL □ JODID SODNY □ NATRIUMJODID (GERMAN) □ SODIUM IODINE □ SODIUM MONOIODIDE
CONSENSUS REPORTS: Reported in EPA TSCA Inventory.
ACGIH TLV: Proposed: (inhalable fraction) 0.1 mg/m³; Not Classifiable as a Human Carcinogen)

SAFETY PROFILE: Moderately toxic by ingestion, intravenous, and intraperitoneal routes. Human teratogenic effects by ingestion: developmental abnormalities of the endocrine system. Human reproductive effects by ingestion: effects on newborn, including postnatal measurements. A skin and eye irritant. Reacts violently with BrF_3, $HClO_4$, oxidants. When heated to decomposition it emits toxic fumes of I^- and Na_2O. See also IODIDES.

SIA500   CAS: 540-72-7   HR: 3
SODIUM ISOTHIOCYANATE
mf: CNS•Na   mw: 81.07
PROP: Colorless, deliq, orthorhombic crystals or white powder. Mp: 287°. Very sol in H_2O, EtOH, and Me_2CO.
SYNS: HAIMASED □ NATRIUMRHODANID (GERMAN) □ SCYAN □ SODIUM RHODANATE □ SODIUM RHODANIDE □ SODIUM SULFOCYANATE □ SODIUM SULFOCYANIDE □ SODIUM THIOCYANATE □ SODIUM THIOCYANIDE □ THIOCYANATE SODIUM □ USAF EK-T-434
CONSENSUS REPORTS: Reported in EPA TSCA Inventory.
SAFETY PROFILE: Poison by ingestion, intravenous, and subcutaneous routes. Moderately toxic by intraperitoneal route. Large doses taken internally cause vomiting, convulsions. Chronic poisoning is manifested by weakness, confusion, diarrhea, and skin rashes. When heated to decomposition it emits very toxic fumes of NO_x, SO_x, and Na_2O. See also THIOCYANATES.

SIB600   CAS: 151-21-3   HR: 3
SODIUM LAURYL SULFATE
mf: $C_{12}H_{26}O_4S•Na$ mw: 289.43
PROP: White to cream-colored crystals, flakes, or powder; slt odor. Sol in water.
SYNS: AQUAREX METHYL □ AVIROL 118 CONC □ CARSONOL SLS □ CONCO SULFATE WA □ CYCLORYL 21 □ DETERGENT 66 □ DODECYL ALCOHOL, HYDROGEN SULFATE, SODIUM SALT □ DODECYL SODIUM SULFATE □ DODECYL SULFATE, SODIUM SALT □ DREFT □ DUPONOL □ EMERSAL 6400 □ EMULSIFIER No. 104 □ HEXAMOL SLS □ IRIUM □ LANETTE WAX-S □ LAURYL SODIUM SULFATE □ LAURYL SULFATE, SODIUM SALT □ MAPROFIX 563 □

S

MAPROFIX WAC-LA □ NCI-C50191 □ NEUTRAZYME □ ORVUS WA PASTE □ PRODUCT No. 161 □ QUOLAC EX-UB □ REWOPOL NLS 30 □ RICHONOL C □ SIPEX OP □ SIPON WD □ SLS □ SODIUM DODECYL SULFATE □ SODIUM MONODODECYL SULFATE □ SOLSOL NEEDLES □ STANDAPOL 112 CONC □ STEPANOL WAQ □ STERLING WAQ-COSMETIC □ SULFOPON WA 1 □ SULFOTEX WALA □ SULFURIC ACID, MONO-DODECYL ESTER, SODIUM SALT □ TARAPON K 12 □ TEXAPON ZHC □ TREPENOL WA □ ULTRA SULFATE SL-1

CONSENSUS REPORTS: Reported in EPA TSCA Inventory.

SAFETY PROFILE: Poison by intravenous and intraperitoneal routes. Moderately toxic by ingestion. An experimental teratogen. A human skin irritant. An experimental eye and severe skin irritant. A mild allergen. Mutation data reported. When heated to decomposition it emits toxic fumes of SO_x and Na_2O. See also ESTERS and SULFATES.

SID000   CAS: 57-30-7   HR: 3
SODIUM LUMINAL
mf: $C_{12}H_{12}N_2O_3 \cdot Na$ mw: 255.25
PROP: White crystals.
SYNS: 5-ETHYL-5-PHENYLBARBITURIC ACID SODIUM □ 5-ETHYL-5-PHENYLBARBITURIC ACID SODIUM SALT □ 5-ETHYL-5-PHENYL-2,4,6-(1H,3H,5H)PYRIMIDINETRIONE MONOSODIUM SALT □ GARDENAL SODIUM □ LUMINAL SODIUM □ PBS □ PHENEMALUM □ PHENOBAL SODIUM □ PHENOBARBITAL ELIXIR □ PHENOBARBITAL Na □ PHENOBARBITAL SODIUM □ PHENOBARBITAL SODIUM SALT □ PHENOBARBITONE SODIUM □ PHENOBARBITONE SODIUM SALT □ PHENYLETHYL-BARBITURIC ACID, SODIUM SALT □ SODIUM-5-ETHYL-5-PHENYLBARBITURATE □ SODIUM PHENOBARBITAL □ SODIUM PHENOBARBITONE □ SODIUM PHENYLETHYLBARBITURATE □ SODIUM PHENYLETHYLMALONYLUREA □ SOL PHENOBARBITAL □ SOL PHENOBARBITONE □ SOLUBLE PHENOBARBITAL □ SOLUBLE PHENOBARBITONE

CONSENSUS REPORTS: IARC Cancer Review: Animal Sufficient Evidence IMEMDT 13,157,77. EPA Genetic Toxicology Program.

SAFETY PROFILE: Confirmed carcinogen with experimental carcinogenic, neoplastigenic, tumorigenic, and teratogenic data. Poison by ingestion, intravenous,

intraperitoneal, intraduodenal, and subcutaneous routes. Human systemic effects by ingestion: nausea or vomiting and coma. Experimental reproductive effects. Mutation data reported. Used to treat epilepsy, as a hypnotic and sedative. When heated to decomposition it emits toxic fumes of NO_x and Na_2O. See also BARBITURATES.

SIG500   CAS: 2492-26-4   HR: 2
SODIUM-2-MERCAPTOBENZO-THIAZOLE
mf: $C_7H_4NS_2 \cdot Na$ mw: 189.23
SYNS: 2-MERCAPTOBENZOTHIAZOLE SODIUM DERIVATIVE □ 2-MERCAPTOBENZOTHIAZOLE SODIUM SALT

CONSENSUS REPORTS: Reported in EPA TSCA Inventory.

SAFETY PROFILE: Moderately toxic by ingestion. When heated to decomposition it emits very toxic fumes of NO_x, SO_x, and Na_2O. See also MERCAPTANS.

SIH500   CAS: 492-18-2   HR: 3
SODIUM MERSALYL
mf: $C_{13}H_{16}HgNO_6 \cdot Na$ mw: 505.88
PROP: Bitter-tasting deliquescent solid.
IDLH 10 mg/m³ (as Hg).
SYNS: 3-(α-CARBOXY-o-ANISAMIDO)-2-METHOXY-PROPYL HYDROXYMERCURY, MONOSODIUM SALT □ o-((3-HYDROXYMERCURI-2-METHOXYPROPYL)-CARBAMOYL)PHENOXYACETIC ACID MONOSODIUM SALT □ N-(γ-HYDROXYMERCURI-β-METHOXYPROP-YL)SALICYLAMIDE-o-ACETIC ACID SODIUM SALT □ IGROSIN □ MERCURAMIDE □ MERCURITAL □ MERCUSAL □ MERSALIN □ MERSALYL □ SALURIN □ SALYRGAN □ SODIUM o-((3-(HYDROXYMERCURI)-2-METHOXYPROPYL)CARBAMOYL)PHENOXY ACETATE □ SODIUM SALICYL-(γ-HYDROXYMERCURI-β-METHOXYPROPYL)AMIDE-o-ACETATE □ URAGAN

CONSENSUS REPORTS: Mercury and its compounds are on the Community Right-To-Know List.

OSHA PEL: CL 0.1 mg(Hg)/m³ (skin)
ACGIH TLV: TWA 0.1 mg(Hg)/m³ (skin); BEI: 35 µg/g creatinine total inorganic mercury in urine preshift; 15 µg/g creatinine total inorganic mercury in blood at end of shift at end of workweek.
DFG MAK: Confirmed Animal Carcinogen with Unknown Relevance to Humans

SAFETY PROFILE: Poison by intravenous, intramuscular, and intraperitoneal routes. Used as a diuretic agent. When heated to decomposition it emits very toxic fumes of Hg, NO$_x$, and Na$_2$O. See also MERCURY COMPOUNDS.

SII000   CAS: 7681-57-4   HR: 3
SODIUM METABISULFITE
mf: O$_5$S$_2$•2Na   mw: 190.10
PROP: Colorless crystals or white to yellowish powder; odor of sulfur dioxide. Decomp on heating with ultimate formation of Na$_2$SO$_4$. Very sol in H$_2$O, NaHSO$_3$; mod sol in EtOH.
SYNS: DISODIUM DISULFITE □ DISODIUM PYROSULFITE □ DISULFUROUS ACID, DISODIUM SALT □ SODIUM DISULFITE □ SODIUM METABISULFITE □ SODIUM METABISULPHITE □ SODIUM PYROSULFITE
CONSENSUS REPORTS: Reported in EPA TSCA Inventory. EPA Genetic Toxicology Program.
OSHA PEL: TWA 5 mg/m^3
ACGIH TLV: TWA 5 mg/m^3; Not Classifiable as a Human Carcinogen
SAFETY PROFILE: An inhalation hazard. Poison by intravenous route. Moderately toxic by parenteral route. Experimental reproductive effects. Mutation data reported. When heated to decomposition it emits toxic fumes of SO$_x$ and Na$_2$O.

SII500   CAS: 10361-03-2   HR: 2
SODIUM METAPHOSPHATE
mf: O$_3$P•Na   mw: 101.96
PROP: Sodium metaphosphate exists as a number of different molecular species, some of which exhibit various crystalline forms. The vitreous sodium phosphates having a Na$_2$O/P$_2$O$_3$ mole ratio near unity are classified as sodium metaphosphates. The term also extends to short-chain vitreous compositions, the compounds of which exhibit the polyphosphate formula Na$_{n+2}$P$_n$O$_{3n+1}$ with n as low as 4–5. In such as (NaPO$_3$), n may be a small integer <3 (cyclic molecules) or a large number (polymers). Amorphous white solids. Very sol in water.

SYNS: GRAHAM'S SALT □ METAFOS □ SODIUM HEXAMETAPHOSPHATE □ SODIUM POLYPHO-SPHATES, GLASSY □ SODIUM TETRAPOLY-PHOSPHATE
CONSENSUS REPORTS: Reported in EPA TSCA Inventory.
SAFETY PROFILE: Moderately toxic by intraperitoneal route. When heated to decomposition it emits toxic fumes of Na$_2$O and PO$_x$. See also PHOSPHATES.

SIK450   CAS: 124-41-4   HR: 3
SODIUM METHYLATE
DOT: UN 1289/UN 1431
mf: CH$_3$O•Na   mw: 54.03
PROP: White, amorphous, free-flowing powder or tetragonal crystals; moisture sensitive. Mp: >300°. Decomp in air above 127°; decomp by water. Sol in methyl and ethyl alc, fats, esters.
SYNS: METHANOL, SODIUM SALT □ SODIUM METHOXIDE □ SODIUM METHYLATE (UN 1431) (DOT) □ SODIUM METHYLATE SOLUTIONS in alcohol (UN 1289) (DOT)
CONSENSUS REPORTS: Reported in EPA TSCA Inventory.
DOT CLASSIFICATION: 4.2; Label: Spontaneously Combustible, Corrosive (UN 1431); DOT Class: 3; Label: Flammable Liquid (UN 1289)
SAFETY PROFILE: A corrosive and irritating material. It hydrolyzes into methanol and sodium hydroxide. May ignite spontaneously in moist air. Flammable when exposed to heat or flame. Ignites on contact with water. Violent reaction with (CHCl$_3$ + CH$_3$OH), (methyl azide + dimethylmalonate), FClO$_3$. When heated to decomposition it emits toxic fumes of Na$_2$O.

SIL500   CAS: 3653-48-3   HR: 3
SODIUM (2-METHYL-4-CHLOROPHENO-XY)ACETATE
mf: C$_9$H$_8$ClO$_3$•Na   mw: 222.61
PROP: Powerful selective weed killer.
SYNS: AGROXONE 3 □ 4-CHLORO-2-METHYL-PHENOXYACETIC ACID SODIUM SALT □ (p-CHLORO-o-TOLYLOXY)ACETIC ACID SODIUM SALT □ CHWASTOKS □ CHWASTOX □ DIAMET □ DICOTEX 80 □ DIKOTEKS □ DIKOTEX 30 □ MCPA SODIUM SALT □

METAXONE □ METHOXONE □ (2-METHYL-4-CHLOROPHENOXY)ACETIC ACID, SODIUM SALT □ 2M-4KH SODIUM SALT □ 2M-4X □ Na MCPA □ PHENOXYLENE □ SODIUM (4-CHLORO-2-METHYL-PHENOXY)ACETATE □ SODIUM MCPA □ SYS 67ME
CONSENSUS REPORTS: Reported in EPA TSCA Inventory.
SAFETY PROFILE: Poison by intraperitoneal route. Moderately toxic by ingestion and subcutaneous routes. Experimental reproductive effects. Mutation data reported. When heated to decomposition it emits toxic fumes of Cl⁻ and Na_2O.

SIM400   HR: D
SODIUM MONO- and DIMETHYL NAPHTHALENE SULFONATE
SAFETY PROFILE: When heated to decomposition it emits toxic fumes of SO_x and Na_2O.

SIN000   CAS: 305-53-3   HR: 3
SODIUM MONOIODOACETATE
mf: $C_2H_2IO_2•Na$ mw: 207.93
PROP: White powder. Mp: 208° (decomp). Sol in H_2O.
SYNS: IODOACETATE SODIUM SALT □ IODOACETIC ACID SODIUM SALT □ USAF EK-6279
CONSENSUS REPORTS: Reported in EPA TSCA Inventory.
SAFETY PROFILE: Poison by ingestion and intraperitoneal routes. Experimental reproductive effects. Human mutation data reported. When heated to decomposition it emits toxic fumes of I⁻ and Na_2O.

SIN500   CAS: 12401-86-4   HR: 3
SODIUM MONOXIDE
DOT: UN 1825
mf: Na_2O mw: 61.98
PROP: White-gray, deliq crystals. Bp: 1275° (subl), d. 2.27.
SYNS: CALCINED SODA □ DISODIUM MONOXIDE □ DISODIUM OXIDE □ SODIUM MONOXIDE, solid (DOT) □ SODIUM OXIDE
DOT CLASSIFICATION: 8; Label: Corrosive
SAFETY PROFILE: Very corrosive and irritating to skin, eyes, and mucous membranes. Can react violently with water, nitric oxide (above 100°C). Ignites when mixed with 2,4-dinitrotoluene. Mixtures with phosphorus(V) oxide react violently when warmed or on contact with moisture. When heated to decomposition it emits toxic fumes of Na_2O. See also SODIUM HYDROXIDE.

SIO000   CAS: 1191-50-0   HR: 3
SODIUM MYRISTYL SULFATE
mf: $C_{14}H_{29}O_4S•Na$ mw: 316.48
PROP: White crystalline powder.
SYNS: 7-ETHYL-2-METHYL-4-HEXADECANOL SULFATE SODIUM SALT □ MYRISTYL SULFATE, SODIUM SALT □ NIAPROOF 4 □ SODIUM SOTRADECOL □ SODIUM TETRADECYL SULFATE □ STS □ SULFURIC ACID, MONOTETRADECYL ESTER, SODIUM SALT □ SULFURIC ACID, MYRISTYL ESTER, SODIUM SALT □ TERGITOL 4 □ TETRADECYL SODIUM SULFATE □ TETRADECYL SULFATE, SODIUM SALT □ TROMBAVAR □ TROMBOVAR
CONSENSUS REPORTS: Reported in EPA TSCA Inventory.
SAFETY PROFILE: Poison by intraperitoneal and intravenous routes. Experimental reproductive effects. When heated to decomposition it emits toxic fumes of SO_x and Na_2O. See also SULFATES.

SIO900   CAS: 7631-99-4   HR: 3
SODIUM NITRATE (1:1)
DOT: UN 1498
mf: $NO_3•Na$ mw: 85.00
PROP: Colorless, transparent, trigonal (rhombohedral), odorless crystals; saline, sltly bitter taste. Decomp on heating to form $NaNO_2$. Mp: 306.8°, bp: decomp @ 380°, d: 2.261. Deliq in moist air. Very sol in water; sol in EtOH, MeOH; practically insol in Me_2CO.
SYNS: CHILE SALTPETER □ CUBIC NITER □ NITRATE de SODIUM (FRENCH) □ NITRATINE □ NITRIC ACID, SODIUM SALT □ SODA NITER □ SODIUM NITRATE (DOT)
CONSENSUS REPORTS: Reported in EPA TSCA Inventory. EPA Genetic Toxicology Program.
DOT CLASSIFICATION: 5.1; Label: Oxidizer
SAFETY PROFILE: Human poison by ingestion. Poison by intravenous route.

Questionable carcinogen with experimental tumorigenic data. Human mutation data reported. A powerful oxidizer. It will ignite with heat or friction. Explodes when heated to over 1000°F, or when mixed with cyanides, sodium hypophosphite, boron phosphide. Forms explosive mixtures with aluminum powder, antimony powder, barium thiocyanate, metal amidosulfates, sodium, sodium phosphinate, sodium thiosulfate, sulfur + charcoal (gunpowder). Potentially violent reaction or ignition when mixed with bitumen, organic matter, calcium-silicon alloy, jute + magnesium chloride, magnesium, metal cyanides, nonmetals, peroxyformic acid, phenol + trifluoroacetic acid. Incompatible with acetic anhydride, barium thiocyanate, wood. A dangerous disaster hazard. Experimental reproductive effects. When heated to decomposition it emits toxic fumes of NO_x and Na_2O. See also NITRATES.

SIP500   CAS: 5064-31-3   HR: 3
SODIUM NITRILOTRIACETATE
mf: $C_6H_6NO_6 \cdot 3Na$ mw: 257.10
PROP: Cleaning product ingredient.
SYNS: HAMPSHIRE NTA □ NITRILOTRIACETIC ACID, TRISODIUM SALT □ NTA □ TRISODIUM NITRILOTRIACETATE □ TRISODIUM NITRILO-TRIACETIC ACID
CONSENSUS REPORTS: Reported in EPA TSCA Inventory.
SAFETY PROFILE: Poison by intraperitoneal route. Moderately toxic by ingestion. Experimental reproductive effects. Questionable carcinogen with experimental neoplastigenic data. Mutation data reported. When heated to decomposition it emits toxic fumes of NO_x and Na_2O.

SIQ500   CAS: 7632-00-0   HR: 3
SODIUM NITRITE
DOT: UN 1500
mf: $NO_2 \cdot Na$ mw: 69.00
PROP: Sltly yellowish or white orthorhombic, hygroscopic crystals, sticks, or powder; sltly salty taste. Mp: 271°, bp: decomp @ 320°, d: 2.168. Deliq in air. Sol in water; mod sol in MeOH; spar sol in Et_2O.
SYNS: ANTI-RUST □ DIAZOTIZING SALTS □ DUSITAN SODNY (CZECH) □ ERINITRIT □ FILMERINE □ NATRIUM NITRIT (GERMAN) □ NCI-C02084 □ NITRITE de SODIUM (FRENCH) □ NITROUS ACID, SODIUM SALT
CONSENSUS REPORTS: Reported in EPA TSCA Inventory. EPA Genetic Toxicology Program.
DOT CLASSIFICATION: 5.1; Label: Oxidizer
SAFETY PROFILE: Human poison by ingestion. Experimental poison by ingestion, inhalation, subcutaneous, intravenous, and intraperitoneal routes. Human systemic effects by ingestion: motor activity changes, coma, decreased blood pressure with possible pulse rate increase without fall in blood pressure, arteriolar or venous dilation, nausea or vomiting, and blood methemoglobinemia-carboxyhemoglobinemia. Experimental teratogenic and reproductive effects. An eye irritant. Questionable carcinogen with experimental neoplastigenic and tumorigenic data. Human mutation data reported. It may react with organic amines in the body to form carcinogenic nitrosamines.

Flammable; a strong oxidizing agent. In contact with organic matter, will ignite by friction. May explode when heated to over 1000°F or on contact with cyanides, NH_4^+ salts, cellulose, Li, $(K + NH_3)$, $Na_2S_2O_3$. Incompatible with aminoguanidine salts, butadiene, phthalic acid, phthalic anhydride, reductants, sodium amide, sodium disulfite, sodium thiocyanate, urea wood. When heated to decomposition it emits toxic fumes of NO_x and Na_2O. See also NITRITES.

SIT750   HR: 3
SODIUM NITRITE, mixed with SODIUM NITRATE and POTASSIUM NITRATE
DOT: NA 1499
DOT CLASSIFICATION: 5.1; Label: Oxidizer
SAFETY PROFILE: Both components are poisons. A powerful oxidizer. When heated to decomposition it emits toxic fumes of NO_x, K_2O, and Na_2O. See also SODIUM

S

NITRITE, POTASSIUM NITRITE, and NITRITES.

SIU500   CAS: 14402-89-2   HR: 3
SODIUM NITROFERRICYANIDE
mf: $C_5FeN_6O \cdot 2Na$ mw: 261.94

SYNS: DISODIUM NITROSYLPENTACYANOFERRATE □ NIPRIDE □ NITROPRUSSIDNATRIUM (GERMAN) □ SODIUM NITROPRUSSATE □ SODIUM NITROPRUSSIDE □ SODIUM NITROSYLPENTACYANO-FERRATE □ SODIUM NITROSYLPENTACYANOFER-RATE(III)

CONSENSUS REPORTS: Cyanide and its compounds are on the Community Right-To-Know List. Reported in EPA TSCA Inventory.

SAFETY PROFILE: Human poison by inhalation and intravenous routes. Experimental poison by ingestion, intraperitoneal, and intravenous routes. Human systemic effects: increased intracranial pressure, general anesthesia, change in heart rate, and metabolic acidosis. An experimental teratogen. Used as a vasodilator for short-term treatment of severe hypertension. Mixtures with sodium nitrite explode when heated. When heated to decomposition it emits toxic fumes of NO_x, CN^-, and Na_2O.

SIY250   CAS: 13721-39-6   HR: 3
SODIUM ORTHOVANADATE
mf: $O_4V \cdot 3Na$ mw: 183.91

PROP: Colorless, hexagonal prisms. Mp: 850–866°.

SYNS: SODIUM VANADATE □ SODIUM VANADIUM OXIDE □ TRISODIUM ORTHOVANADATE □ VANADIC(II) ACID, TRISODIUM SALT

CONSENSUS REPORTS: Reported in EPA TSCA Inventory.

ACGIH TLV: TWA 0.05 mg(V_2O_5)/m³

NIOSH REL: (Vanadium Compounds) CL 0.05 mg(V)/m³/15M

SAFETY PROFILE: Poison by ingestion, intraperitoneal, intravenous, and subcutaneous routes. Experimental reproductive effects. Mutation data reported. When heated to decomposition it emits toxic fumes of VO_x and Na_2O. See also VANADIUM COMPOUNDS.

SIZ025   CAS: 408-35-5   HR: 3
SODIUM PALMITATE
mf: $C_{16}H_{31}NaO_2$ mw: 278.47

PROP: White to yellow powder.

SYNS: SODIUM HEXADECANOATE □ SODIUM PENTADECANECARBOXYLATE □ SODIUM SALT of HEXADECANOIC ACID

SAFETY PROFILE: A poison by intravenous route. Mutation data reported. When heated to decomposition it emits acrid smoke and irritating fumes.

SJA000   CAS: 131-52-2   HR: 3
SODIUM PENTACHLOROPHENATE
DOT: UN 2567

mf: $C_6Cl_5O \cdot Na$ mw: 288.30

PROP: Tan powder.

SYNS: DOW DORMANT FUNGICIDE □ DOWICIDE G-ST □ NAPCLOR-G □ PENTACHLOROPHENATE SODIUM □ PENTACHLOROPHENOL, SODIUM SALT □ PENTACHLOROPHENOXY SODIUM □ PENTAPHEN-ATE □ SANTOBRITE □ SODIUM PCP □ SODIUM PENTACHLOROPHENATE (DOT) □ SODIUM PENTA-CHLOROPHENOL □ SODIUM PENTACHLORO-PHENOLATE □ SODIUM PENTACHLOROPHENOXIDE □ WEEDBEADS

CONSENSUS REPORTS: Chlorophenol compounds are on the Community Right-To-Know List. Reported in EPA TSCA Inventory. EPA Genetic Toxicology Program.

DOT CLASSIFICATION: 6.1; Label: Poison

SAFETY PROFILE: Poison by ingestion, inhalation, skin contact, intravenous, intraperitoneal, subcutaneous, and intratracheal routes. An experimental teratogen. Experimental reproductive effects. Mutation data reported. When heated to decomposition it emits toxic fumes of Cl^- and Na_2O. See also CHLOROPHENOLS.

SJB400   CAS: 15630-89-4   HR: 2
SODIUM PERCARBONATE
mf: $C_2H_2O_6 \cdot 2Na$ mw: 168.02

SYNS: FB SODIUM PERCARBONATE □ OXYPER □ PERDOX □ PEROXY SODIUM CARBONATE □ SODIUM CARBONATE PEROXIDE

CONSENSUS REPORTS: Reported in EPA TSCA Inventory.

SAFETY PROFILE: Moderately toxic by ingestion. When heated to decomposition it emits acrid smoke and irritating vapors.

SJC000   CAS: 10101-50-5   HR: 3
SODIUM PERMANGANATE
DOT: UN 1503
mf: MnO$_4$•Na   mw: 141.93
PROP: Purple to red-black crystals. Mp: decomp.
SYNS: PERMANGANATE de SODIUM (FRENCH) □ PERMANGANIC ACID, SODIUM SALT
CONSENSUS REPORTS: Manganese and its compounds are on the Community Right-To-Know List. Reported in EPA TSCA Inventory.
OSHA PEL: CL 5 mg(Mn)/m^3
ACGIH TLV: TWA 5 mg(Mn)/m^3
DOT CLASSIFICATION: 5.1; Label: Oxidizer
SAFETY PROFILE: Probably a severe irritant to the skin, eyes, and mucous membranes. A powerful oxidizer and fire hazard. Explosive reaction with acetic acid, acetic anhydride. Reacts vigorously with combustibles. When heated to decomposition it emits toxic fumes of Na$_2$O. See also MANGANESE COMPOUNDS, PERMANGANATES, and POTASSIUM PERMANGANATE.

SJC500   CAS: 1313-60-6   HR: 3
SODIUM PEROXIDE
DOT: UN 1504
mf: Na$_2$O$_2$ mw: 77.98
PROP: White powder turning yellow when heated. Moisture sensitive. Mp: 675° (decomp @ 460°), bp: decomp, d: 2.805. Decomp on heating with loss of O$_2$. Undergoes transition from hexagonal room temp phase to a phase of unknown symmetry at 5°.
SYNS: DISODIUM DIOXIDE □ DISODIUM PEROXIDE □ FLOCOOL 180 □ SODIUM DIOXIDE □ SODIUM OXIDE (Na2O2) □ SOLOZONE
CONSENSUS REPORTS: Reported in EPA TSCA Inventory.
DOT CLASSIFICATION: 5.1; Label: Oxidizer
SAFETY PROFILE: A severe irritant to skin, eyes, and mucous membranes. Dangerous fire hazard by chemical reaction; a powerful oxidizing agent. Reacts explosively or violently under the appropriate conditions with water, acids, powdered metals, acetic acid, acetic anhydride, Al, (Al + CO$_2$), aluminum + aluminum chloride, almond oil, (NH$_4$)$_2$S$_2$O$_8$, aniline, Sb, As, benzene, boron nitride, calcium acetylide, charcoal, Cu, cotton wool, (KNO$_3$ + dextrose), diethyl ether, fibrous materials + water, glucose + potassium nitrate, hexamethylene-tetramine, hydrogen sulfide, hydroxy compounds (e.g., ethanol, ethylene glycol, glycerol, sugar), magnesium, (Mg + CO$_2$), MnO$_2$, metals, metals + carbon dioxide + water, nonmetals (e.g., carbon, phosphorus, antimony, arsenic, boron, sulfur, selenium), nonmetal halides (e.g., diselenium dichloride, disulfur dichloride, phosphorus trichloride), organic matter, paraffin, K, silver chloride + charcoal, soap, Na, sodium dioxide, SCl, Sn, Zn, wood, peroxyformic acid, reducing materials. Will react with water or steam to produce heat and toxic fumes. To fight fire, use carbon dioxide or dry chemical. Combustible materials ignited by contact with sodium peroxide should be smothered with soda ash, salt or dolomite mixtures. Chemical fire extinguishers should not be used. If the fire cannot be smothered, it should be flooded with large quantities of water from a hose. When heated to decomposition it emits toxic fumes of Na$_2$O. See also SODIUM HYDROXIDE and PEROXIDES, INORGANIC.

SJE000   CAS: 7775-27-1   HR: 3
SODIUM PERSULFATE
DOT: UN 1505
mf: O$_8$S$_2$•2Na   mw: 238.10
PROP: White, crystalline powder. Sol in water; decomp by alc.
SYNS: PERSULFATE de SODIUM (FRENCH) □ SODIUM PEROXYDISULFATE
CONSENSUS REPORTS: Reported in EPA TSCA Inventory.
ACGIH TLV: TWA 0.1 mg/m^3
DOT CLASSIFICATION: 5.1; Label: Oxidizer
SAFETY PROFILE: Poison by intraperitoneal and intravenous routes. A

S

powerful oxidizer; can cause fires. When heated to decomposition it emits toxic fumes of SO_x and Na_2O. See also SULFATES.

SJF000   CAS: 139-02-6   HR: 3
SODIUM PHENOXIDE
DOT: UN 2497
mf: $C_6H_5O•Na$ mw: 116.10
PROP: White, deliq crystals.
SYNS: PHENOL SODIUM SALT □ SODIUM CARBOLATE □ SODIUM PHENATE □ SODIUM PHENOLATE, solid (DOT)
CONSENSUS REPORTS: Reported in EPA TSCA Inventory.
DOT CLASSIFICATION: 8; Label: Corrosive
SAFETY PROFILE: Poison by subcutaneous route. A corrosive irritant to skin, eyes, and mucous membranes. When heated to decomposition it emits toxic fumes of Na_2O. See also PHENOL and SODIUM HYDROXIDE.

SJH090   CAS: 7558-79-4   HR: 3
SODIUM PHOSPHATE, DIBASIC
mf: $HO_4P•2Na$ mw: 141.96
PROP: Colorless, translucent crystals or white hygroscopic powder. Sol in water; very sltly sol in alc.
SYNS: DIBASIC SODIUM PHOSPHATE □ DISODIUM HYDROGEN PHOSPHATE □ DISODIUM MONO-HYDROGEN PHOSPHATE □ DISODIUM ORTHOPHOS-PHATE □ DISODIUM PHOSPHATE □ DISODIUM PHOSPHORIC ACID □ DSP □ EXSICCATED SODIUM PHOSPHATE □ NATRIUMPHOSPHAT (GERMAN) □ PHOSPHORIC ACID, DISODIUM SALT □ SODA PHOSPHATE □ SODIUM HYDROGEN PHOSPHATE □ SODIUM MONOHYDROGEN PHOSPHATE (2:1:1)
CONSENSUS REPORTS: Reported in EPA TSCA Inventory.
SAFETY PROFILE: Poison by intravenous route. Moderately toxic by intraperitoneal, subcutaneous, and intramuscular routes. Mildly toxic by ingestion. A skin and eye irritant. When heated to decomposition it emits toxic fumes of PO_x and Na_2O. See also PHOSPHATES.

SJH100   CAS: 7558-80-7   HR: 3
SODIUM PHOSPHATE, MONOBASIC
mf: $H_2O_4P•Na$ mw: 119.98

PROP: White or colorless monoclinic crystalline powder or granules; odorless. Hygroscopic; sol in water; insol in alc.
SYNS: MONOSODIUM DIHYDROGEN PHOSPHATE □ MONOSODIUM PHOSPHATE □ MONOSORB XP-4 □ PRIMARY SODIUM PHOSPHATE □ SODIUM ACID PHOSPHATE □ SODIUM BIPHOSPHATE □ SODIUM BIPHOSPHATE anhydrous □ SODIUM DIHYDROGEN PHOSPHATE (1:2:1)
CONSENSUS REPORTS: Reported in EPA TSCA Inventory.
SAFETY PROFILE: Poison by intramuscular route. Mildly toxic by ingestion. A human and experimental eye irritant. When heated to decomposition it emits toxic fumes of PO_x and Na_2O. See also PHOSPHATES.

SJH200   CAS: 7601-54-9   HR: 2
SODIUM PHOSPHATE, TRIBASIC
mf: $O_4P•3Na$ mw: 163.94
PROP: White or colorless crystals, or crystalline powder; odorless. Sol in water; insol in alc.
SYNS: DRI-TRI □ EMULSIPHOS 440/660 □ NUTRIFOS STP □ PHOSPHORIC ACID, TRISODIUM SALT □ SODIUM PHOSPHATE □ SODIUM PHOSPHATE, anhydrous □ TRIBASIC SODIUM PHOSPHATE □ TRINATRIUMPHOSPHAT (GERMAN) □ TRISODIUM ORTHOPHOSPHATE □ TRISODIUM PHOSPHATE □ TROMETE □ TSP
CONSENSUS REPORTS: Reported in EPA TSCA Inventory.
SAFETY PROFILE: Moderately toxic by intravenous route. Mutation data reported. A strong, caustic material. When heated to decomposition it emits toxic fumes of Na_2O and PO_x. See also PHOSPHATES.

SJI500   CAS: 12058-85-4   HR: 3
SODIUM PHOSPHIDE
DOT: UN 1432
mf: PNa_3 mw: 99.94
PROP: Red crystals. Mp: decomp.
SYN: PHOSPHURE de SODIUM (FRENCH)
CONSENSUS REPORTS: Reported in EPA TSCA Inventory.
DOT CLASSIFICATION: 4.3; Label: Dangerous When Wet, Poison
SAFETY PROFILE: Flammable when exposed to heat or flame. Reacts violently

with water to yield phosphine. When heated to decomposition it emits toxic fumes of PO_x and Na_2O. See also PHOSPHIDES.

SJJ000   CAS: 51312-42-6   HR: 2
SODIUM PHOSPHOTUNGSTATE
mf: $Na_4O_2 \cdot O_5P_2 \cdot O_{36}W_{12} \cdot 18H_2O$ mw: 3372.46
SYNS: SODIUM TUNGSTOPHOSPHATE □ TUNGSTOPHOSPHORIC ACID, SODIUM SALT
CONSENSUS REPORTS: Reported in EPA TSCA Inventory.
ACGIH TLV: TWA 1 mg(W)/m³; STEL 3 mg(W)/m³
NIOSH REL: TWA 1 mg(W)/m³
SAFETY PROFILE: Moderately toxic by ingestion. When heated to decomposition it emits toxic fumes of PO_x and Na_2O. See also TUNGSTEN COMPOUNDS.

SJK000   CAS: 9003-04-7   HR: 1
SODIUM POLYACRYLATE
mf: $(C_3H_4O_2)_x \cdot xNa$
SYNS: POLYCO □ RHOTEX GS
CONSENSUS REPORTS: Reported in EPA TSCA Inventory.
SAFETY PROFILE: An eye irritant. When heated to decomposition it emits toxic fumes of Na_2O.

SJK375   CAS: 25704-18-1   HR: 1
SODIUM POLYSTYRENE SULFONATE
mf: $(C_8H_8O_3S \cdot Na)_n$
SYNS: 4-ETHENYL-BENZENESULFONIC ACID SODIUM SALT, HOMOPOLYMER (9CI) □ KAYEXALATE □ POLY(SODIUM p-STYRENESULFONATE)
CONSENSUS REPORTS: Reported in EPA TSCA Inventory.
SAFETY PROFILE: Mildly toxic by ingestion. Experimental reproductive effects. When heated to decomposition it emits toxic fumes of SO_x and Na_2O. See also SULFONATES.

SJL500   CAS: 137-40-6   HR: 2
SODIUM PROPIONATE
mf: $C_3H_5O_2 \cdot Na$ mw: 96.07
PROP: Transparent crystals or granules; nearly odorless. Very sol in water; sltly sol in alc.

SYNS: IMPEDEX □ MYCOBAN □ NAPROPION □ NATRIUMPROPIONAT □ OCUSEPTINE □ PROPANOIC ACID, SODIUM SALT □ PROPIONAN SODNY
CONSENSUS REPORTS: Reported in EPA TSCA Inventory.
SAFETY PROFILE: Moderately toxic by skin contact and subcutaneous routes. Mildly toxic by unspecified routes. An allergen. When heated to decomposition it emits toxic fumes of Na_2O.

SJM500   CAS: 13517-26-5   HR: 3
SODIUM PYROVANADATE
mf: $O_7V_2 \cdot 4Na$ mw: 305.84
PROP: Colorless, hygroscopic, hexagonal plates. Mp: 632–654°.
CONSENSUS REPORTS: Reported in EPA TSCA Inventory.
ACGIH TLV: TWA 0.05 mg(V_2O_5)/m³
NIOSH REL: CL 0.05 mg(V)/m³/15M
SAFETY PROFILE: Poison by subcutaneous and intravenous routes. When heated to decomposition it emits toxic fumes of Na_2O and VO_x. See also VANADIUM COMPOUNDS.

SJN700   CAS: 128-44-9   HR: 3
SODIUM SACCHARIN
mf: $C_7H_4NO_3S \cdot Na$ mw: 205.17
PROP: White crystals or crystalline powder; odorless, very sweet taste. Sol in water, alc.
SYNS: ARTIFICIAL SWEETENING SUBSTANZ GENDORF 450 □ CRISTALLOSE □ CRYSTALLOSE □ DAGUTAN □ KRISTALLOSE □ MADHURIN □ ODA □ SACCHARIN □ SACCHARINE SOLUBLE □ SAC-CHARINNATRIUM □ SACCHARIN, SODIUM □ SAC-CHARIN, SODIUM SALT □ SACCHARIN SOLUBLE □ SACCHAROIDUM NATRICUM □ SAXIN □ SODIUM-1,2 BENZISOTHIAZOLIN-3-ONE-1,1-DIOXIDE □ SODIUM o-BENZOSULFIMIDE □ SODIUM BENZOSULPHIMIDE □ SODIUM-o-BENZOSULPHIMIDE □ SODIUM-2-BENZOSULPHIMIDE □ SODIUM SACCHARIDE □ SODIUM SACCHARINATE □ SODIUM SACCHARINE □ SOLUBLE GLUSIDE □ SOLUBLE SACCHARIN □ SUCCARIL □ SUCRA □ o-SULFONBENZOIC ACID IMIDE SODIUM SALT □ SULPHOBENZOIC IMIDE, SODIUM SALT □ SWEETA □ SYKOSE □ WILLOSETTEN
CONSENSUS REPORTS: IARC Cancer Review: Group 2B IMEMDT 7,334,87; Animal Sufficient Evidence IMEMDT 22,111,80. EPA Genetic Toxicology

S

Program. Reported in EPA TSCA Inventory.

SAFETY PROFILE: Confirmed carcinogen with experimental carcinogenic, neoplastigenic, tumorigenic, and teratogenic data. Moderately toxic by ingestion and intraperitoneal routes. A promoter. Experimental reproductive effects. Human mutation data reported. When heated to decomposition it emits very toxic fumes of SO_x, Na_2O, and NO_x.

**SJO000   CAS: 54-21-7   HR: 3
SODIUM SALICYLATE**
mf: $C_7H_5O_3 \cdot Na$ mw: 160.11
PROP: White, odorless crystals, scales, or powder.
SYNS: ALYSINE □ ARDALL □ AROALL □ CLIN □ DIURETIN □ ENTEROSALICYL □ ENTEROSALIL □ 2-HYDROXYBENZOIC ACID MONOSODIUM SALT □ o-HYDROXYBENZOIC SODIUM SALT □ IDOCYL NOVUM □ KERASALICYL □ KEROSAL □ MAGSALYL □ NADISAL □ NEO-SALICYL □ PARBOCYL-REV □ SALICYLIC ACID, SODIUM SALT □ SALISOD □ SALSONIN □ SODIUM-o-HYDROXYBENZOATE □ SODIUM SALICYLIC ACID
CONSENSUS REPORTS: Reported in EPA TSCA Inventory.
SAFETY PROFILE: Experimental poison by subcutaneous route. Moderately toxic to humans by ingestion. Moderately toxic experimentally by ingestion, intraperitoneal, and intravenous routes. An experimental teratogen. Human systemic effects by multiple and unspecified routes: toxic psychosis, excitement, respiratory stimulation, nausea or vomiting, and sweating. Experimental reproductive effects. Mutation data reported. A powerful irritant which affects the central nervous system. Incompatible with ferric salts, mineral acids, iodine, lead acetate, silver nitrate, sodium phosphate powder. When heated to decomposition it emits toxic fumes of Na_2O.

**SJT000   CAS: 1313-85-5   HR: 3
SODIUM SELENIDE**
mf: Na_2Se mw: 124.94

PROP: Deliquescent, white to red, cubic crystals. Mp: >875°, d: 2.625 @ 10°.
SYN: DISODIUM MONOSELENIDE
CONSENSUS REPORTS: Selenium and its compounds are on the Community Right-To-Know List. Reported in EPA TSCA Inventory.
OSHA PEL: TWA 0.2 mg(Se)/m³
ACGIH TLV: TWA 0.2 mg(Se)/m³
DFG MAK: 0.1 mg(Se)/m³
SAFETY PROFILE: Poison by intraperitoneal route. Mutation data reported. When heated to decomposition it emits toxic fumes of Se and Na_2O. See also SELENIUM COMPOUNDS.

**SJT500   CAS: 10102-18-8   HR: 3
SODIUM SELENITE**
DOT: UN 2630
mf: $O_3Se \cdot 2Na$ mw: 172.94
PROP: White crystals or powder. Undergoes monoclinic to hexagonal transformation at 6°. Decomp on heating to form Na_2O and SeO_2 which decomp further to selenium and O_2. Mp: 710°. Very sol in H_2O.
SYNS: DISODIUM SELENITE □ NATRIUMSELENIT (GERMAN) □ SELENIOUS ACID, DISODIUM SALT
CONSENSUS REPORTS: IARC Cancer Review: Group 3 IMEMDT 7,56,87; Animal Inadequate Evidence IMEMDT 9,245,75. Reported in EPA TSCA Inventory. EPA Genetic Toxicology Program. EPA Extremely Hazardous Substances List. Selenium and its compounds are on the Community Right-To-Know List.
OSHA PEL: TWA 0.2 mg(Se)/m³
ACGIH TLV: TWA 0.2 mg(Se)/m³
DFG MAK: 0.1 mg(Se)/m³
DOT CLASSIFICATION: 6.1; Label: Poison
SAFETY PROFILE: Poison by ingestion, intraperitoneal, intravenous, subcutaneous, intracervical, parenteral, and intramuscular routes. Experimental teratogenic and reproductive effects. Questionable carcinogen. Human mutation data reported. When heated to decomposition it emits toxic fumes of Se and Na_2O. See also SELENIUM COMPOUNDS.

SJT600   CAS: 26970-82-1   HR: 3
SODIUM SELENITE PENTAHYDRATE
mf: $O_3Se \cdot 2Na \cdot 5H_2O$ mw: 332.01
PROP: White tetragonal crystals; decomp on heating by H_2O loss. Sol in H_2O; insol in EtOH.
SYN: SELENIOUS ACID DISODIUM SALT PENTAHYDRATE
CONSENSUS REPORTS: Selenium and its compounds are on the Community Right-To-Know List.
OSHA PEL: TWA 0.2 mg(Se)/m^3
ACGIH TLV: TWA 0.2 mg(Se)/m^3
DFG MAK: 0.1 mg(Se)/m^3
DOT CLASSIFICATION: 6.1; Label: Poison, Corrosive
SAFETY PROFILE: Poison by intramuscular and intraperitoneal routes. An experimental teratogen. Experimental reproductive effects. When heated to decomposition it emits toxic fumes of Se and Na_2O. See also SELENIUM COMPOUNDS.

SJU000   CAS: 6834-92-0   HR: 3
SODIUM SILICATE
mf: $O_3Si \cdot 2Na$ mw: 122.07
PROP: Usually a glass, also crystals. Mp: 1089°. Sol in H_2O; insol in EtOH.
SYNS: B-W □ CRYSTAMET □ DISODIUM METASILICATE □ DISODIUM MONOSILICATE □ METSO 20 □ METSO BEADS 2048 □ METSO BEADS, DRYMET □ METSO PENTABEAD 20 □ ORTHOSIL □ SODIUM METASILICATE □ SODIUM METASILICATE, anhydrous □ WATER GLASS
CONSENSUS REPORTS: Reported in EPA TSCA Inventory.
SAFETY PROFILE: Poison by ingestion and intraperitoneal routes. A caustic material which is a severe eye, skin, and mucous membrane irritant. Experimental reproductive effects. Ingestion causes gastrointestinal tract upset. Violent reaction with F_2. When heated to decomposition it emits toxic fumes of Na_2O. Used in cosmetics. See also SILICATES.

SJV000   CAS: 7757-81-5   HR: 2
SODIUM SORBATE
mf: $C_6H_7O_2 \cdot Na$ mw: 134.12
SYN: SORBIC ACID, SODIUM SALT

SAFETY PROFILE: Moderately toxic by intraperitoneal route. Mildly toxic by ingestion. Mutation data reported. Migrates to food from packaging material. When heated to decomposition it emits toxic fumes of Na_2O.

SJV500   CAS: 822-16-2   HR: 3
SODIUM STEARATE
mf: $C_{18}H_{36}O_2 \cdot Na$ mw: 306.52
SYNS: OCTADECANOIC ACID, SODIUM SALT □ SODIUM OCTADECANOATE □ STEARIC ACID, SODIUM SALT
CONSENSUS REPORTS: Reported in EPA TSCA Inventory.
SAFETY PROFILE: Poison by intravenous route. When heated to decomposition it emits toxic fumes of Na_2O.

SJV700   CAS: 25383-99-7   HR: D
SODIUM STEAROYL LACTYLATE
PROP: Cream-colored powder; caramel-like odor. Sol in hot oil or fat, dispersible in warm water.
SAFETY PROFILE: When heated to decomposition it emits acrid smoke and irritating fumes.

SJW200   HR: D
SODIUM SULFACHLOROPYRAZINE
**   MONOHYDRATE**
SAFETY PROFILE: When heated to decomposition it emits acrid smoke and irritating fumes.

SJW475   CAS: 127-58-2   HR: 2
SODIUM SULFAMERAZINE
mf: $C_{11}H_{12}N_4O_2S \cdot Na$ mw: 287.32
PROP: Crystals; bitter, caustic taste. Hygroscopic. On prolonged exposure to humid air, it absorbs CO_2 with the liberation of sulfamerazine and becomes incompletely sol in water. Its solns are alkaline to phenolphthalein (pH 10 or more). One gram dissolves in 3.6 mL water. Sltly sol in alc; insol in ether, chloroform.
SYNS: 4-AMINO-N-(4-METHYL-2-PYRIMIDINYL)-BENZENESULFONAMIDE MONOSODIUM SALT □ N^1-(4-METHYL-2-PYRIMIDINYL)SULFANILAMIDE SODIUM SALT □ SODIUM SULPHAMERAZINE □ SOLUBLE

S

SULFAMERAZINE □ SOLUMEDINE □ SULFAMERA-
ZINE SODIUM
CONSENSUS REPORTS: Reported in EPA
TSCA Inventory.
SAFETY PROFILE: Moderately toxic by
ingestion, subcutaneous, intraperitoneal, and
intravenous routes. When heated to
decomposition it emits toxic fumes of SO_x,
NO_x, and Na_2O.

SJY000   CAS: 7757-82-6   HR: 2
SODIUM SULFATE (2:1)
mf: $O_4S \cdot 2Na$ mw: 142.04
PROP: White or colorless, orthorhombic
crystals or powder; odorless. Mp: 888°, d:
2.671. Sol in water, glycerin; insol in alc.
Mod sol in H_2O.
SYNS: DISODIUM SULFATE □ NATRIUMSULFAT
(GERMAN) □ SALT CAKE □ SODIUM SULFATEn
anhydrous □ SODIUM SULPHATE □ SULFURIC ACID,
DISODIUM SALT □ THENARDITE □ TRONA
CONSENSUS REPORTS: Reported in EPA
TSCA Inventory. EPA Genetic Toxicology
Program.
SAFETY PROFILE: Moderately toxic by
intravenous route. Mildly toxic by ingestion.
An experimental teratogen. Experimental
reproductive effects. Questionable
carcinogen with experimental tumorigenic
effects. Violent reaction with Al. When
heated to decomposition it emits toxic
fumes of SO_x and Na_2O. See also
SULFATES.

SJY500   CAS: 1313-82-2   HR: 3
SODIUM SULFIDE
DOT: UN 1385
mf: Na_2S mw: 78.04
PROP: Amorphous, deliquescent, yellow-
pink or white, cubic crystals. Mp: 1172°, d:
1.856 @ 14°. Very sol in H_2O; spar sol in
EtOH; insol in Et_2O.
SYNS: SODIUM MONOSULFIDE □ SODIUM SULFIDE,
anhydrous (DOT) □ SODIUM SULFIDE with <30% water of
crystallization (DOT) □ SODIUM SULPHIDE
CONSENSUS REPORTS: Reported in EPA
TSCA Inventory.
DOT CLASSIFICATION: 4.2; Label:
Spontaneously Combustible

SAFETY PROFILE: A poison by ingestion
and intraperitoneal routes. Flammable when
exposed to heat or flame. Unstable and can
explode on rapid heating or percussion.
Reacts violently with carbon, diazonium
salts, n,n-dichloromethylamine, o-
nitroaniline diazonium salt, water. When
heated to decomposition it emits toxic
fumes of SO_x and Na_2O. See also
SULFIDES.

SJZ000   CAS: 7757-83-7   HR: 3
SODIUM SULFITE (2:1)
mf: $O_3S \cdot 2Na$ mw: 126.04
PROP: Hexagonal prisms or white powder;
odorless with salty, sulfurous taste. Bp:
decomp, d: 2.633 @ 15.4°. Sol in water; sltly
sol in alc.
SYNS: DISODIUM SULFITE □ EXSICATED SODIUM
SULFITE □ NATRIUMSULFID (GERMAN) □ SODIUM
SULFITE, anhydrous □ SODIUM SULPHITE □ SULFTECH
□ SULFUROUS ACID, SODIUM SALT (1:2)
CONSENSUS REPORTS: Reported in EPA
TSCA Inventory. EPA Genetic Toxicology
Program.
SAFETY PROFILE: Poison by intravenous
and subcutaneous routes. Moderately toxic
by ingestion and intraperitoneal routes.
Human mutation data reported. When
heated to decomposition it emits very toxic
fumes of Na_2O and SO_x. A reducing agent.
See also SULFITES.

SJZ050   HR: D
SODIUM SULFOACETATE derivatives
of MONO and DIGLYCERIDES
SYN: MONO and DIGLYCERIDES, SODIUM
SULFOACETATE DERIVATIVES
SAFETY PROFILE: When heated to
decomposition it emits acrid smoke and
irritating fumes.

SJZ100   CAS: 12034-12-7   HR: 2
SODIUM SUPEROXIDE
DOT: UN 2547
mf: NaO_2 mw: 54.99
SYN: SODIUM SUPEROXIDE (DOT)
DOT CLASSIFICATION: 5.1; Label: Oxidizer

SAFETY PROFILE: An oxidizer. When heated to decomposition it emits toxic vapors of Na_2O.

SKC000   CAS: 10101-83-4   HR: 3
SODIUM TELLURATE
mf: $O_4Te•2Na$ mw: 237.58
PROP: White powder or colorless monoclinic crystals.
SYN: SODIUM TELLURATE VI
CONSENSUS REPORTS: Reported in EPA TSCA Inventory.
OSHA PEL: TWA 0.1 mg(Te)/m³
ACGIH TLV: TWA 0.1 mg(Te)/m³
SAFETY PROFILE: Poison by ingestion, intraperitoneal, and intravenous routes. When heated to decomposition it emits toxic fumes of Te and Na_2O. See also TELLURIUM COMPOUNDS.

SKC100   CAS: 26006-71-3   HR: 3
SODIUM TELLURATE, DIHYDRATE
mf: $O_4Te•2Na•2H_2O$ mw: 273.62
PROP: Crystals. Mp: decomp.
OSHA PEL: TWA 0.1 mg(Te)/m³
ACGIH TLV: TWA 0.1 mg(Te)/m³
SAFETY PROFILE: Poison by intraperitoneal route. When heated to decomposition it emits toxic fumes of Te.

SKC500   CAS: 10102-20-2   HR: 3
SODIUM TELLURITE
mf: $O_3Te•2Na$ mw: 221.58
PROP: White powder. Sol in H_2O.
SYNS: SODIUM TELLURATE(IV) □ TELLURIC ACID, DISODIUM SALT □ TELLUROUS ACID, DISODIUM SALT
CONSENSUS REPORTS: EPA Extremely Hazardous Substances List. Reported in EPA TSCA Inventory.
OSHA PEL: TWA 0.1 mg(Te)/m³
ACGIH TLV: TWA 0.1 mg(Te)/m³
SAFETY PROFILE: Human poison by ingestion and parenteral routes. Experimental poison by ingestion, intravenous, and intraperitoneal routes. Human systemic effects by ingestion: coma; dyspnea; nausea or vomiting. Human mutation data reported. When heated to

decomposition it emits toxic fumes of Te and Na_2O. See also TELLURIUM.

SKE000   CAS: 13755-29-8   HR: 2
SODIUM TETRAFLUOROBORATE
mf: $BF_4•Na$ mw: 109.80
PROP: White orthorhombic crystals. Decomp on heating at >2° to form NaF and BF_3. Mp: 322°. Very sol in H_2O; spar sol in EtOH.
SYNS: STB □ TETRAFLUOROBORATE(1-) SODIUM
CONSENSUS REPORTS: Reported in EPA TSCA Inventory.
OSHA PEL: TWA 2.5 mg(F)/m³
ACGIH TLV: TWA 2.5 mg(F)/m³; BEI: 3 mg/g creatinine of fluorides in urine prior to shift; 10 mg/g creatinine of fluorides in urine at end of shift.
NIOSH REL: (Inorganic Fluorides) TWA 2.5 mg(F)/m³
SAFETY PROFILE: Moderately toxic by subcutaneous route. When heated to decomposition it emits toxic fumes of F⁻ and Na_2O.

SKF000   CAS: 12206-14-3   HR: 3
SODIUM TETRAPEROXYCHROMATE
mf: $CrNa_3O_8$ mw: 248.97
CONSENSUS REPORTS: Chromium and its compounds are on the Community Right-To-Know List.
OSHA PEL: CL 0.1 mg(CrO₃)/m³
ACGIH TLV: TWA 0.05 mg(Cr)/m³; Confirmed Human Carcinogen
NIOSH REL: TWA 0.025 mg(Cr(VI))/m³; CL 0.05 mg/m³/15M
SAFETY PROFILE: Confirmed human carcinogen. Explodes when heated to 115°C. When heated to decomposition it emits toxic fumes of Na_2O. See also CHROMIUM COMPOUNDS and PEROXIDES.

SKH500   CAS: 367-51-1   HR: 3
SODIUM THIOGLYCOLATE
mf: $C_2H_3O_2S•Na$ mw: 114.10
PROP: Hygroscopic crystals.
SYNS: MERCAPTOACETIC ACID SODIUM SALT □ SODIUM MERCAPTOACETATE □ SODIUM

S

THIOGLYCOLLATE □ THIOGLYCOLATESODIUM □ THIOGLYCOLIC ACID, SODIUM SALT □ USAF EK-5199
CONSENSUS REPORTS: Reported in EPA TSCA Inventory.
SAFETY PROFILE: Poison by intravenous and intraperitoneal routes. Moderately toxic by ingestion. A human skin irritant. This material yields hydrogen sulfide on decomposition. A death has been attributed to the absorption of toxic decomposition products from the use of this material in a hair permanent-waving solution. When heated to decomposition it emits toxic fumes of SO_x and Na_2O. See also SULFIDES and MERCAPTANS.

SKI000   CAS: 7772-98-7   HR: 2
SODIUM THIOSULFATE
mf: $O_3S_2•2Na$ mw: 158.10
PROP: Colorless, deliquescent, monoclinic crystals or crystalline powder. Decomp on heating with ultimate formation of Na_2SO_4. Sol in water; insol in alc.
SYNS: HYPO □ SODIUM HYPOSULFITE □ SODIUM THIOSULFATE, anhydrous
CONSENSUS REPORTS: Reported in EPA TSCA Inventory.
SAFETY PROFILE: Moderately toxic by subcutaneous route. Incompatible with metal nitrates, sodium nitrite. When heated to decomposition it emits very toxic fumes of Na_2O and SO_x. See also SODIUM THIOSULFATE and SODIUM THIOSULFATE, PENTAHYDRATE.

SKM500   CAS: 7785-84-4   HR: 3
SODIUM TRIMETAPHOSPHATE
mf: $O_9P_3•3Na$ mw: 305.88
PROP: White crystals or white crystalline powder. Sol in water.
SYN: TRIMETAPHOSPHATE SODIUM
CONSENSUS REPORTS: Reported in EPA TSCA Inventory.
SAFETY PROFILE: Poison by intravenous route. Moderately toxic by intraperitoneal route. When heated to decomposition it emits toxic fumes of PO_x and Na_2O. See also PHOSPHATES.

SKN000   CAS: 13573-18-7   HR: 3
SODIUM TRIPOLYPHOSPHATE
mf: $O_{10}P_3•5Na$ mw: 367.86
PROP: White granules or powder. Sltly hygroscopic; sol in water.
SYNS: ARMOFOS □ NATRIUMTRIPOLYPHOSPHAT (GERMAN) □ PENTASODIUM TRIPHOSPHATE □ POLY □ POLYGON □ SODIUM TRIPHOSPHATE □ STPP □ TRIPHOSPHORIC ACID, SODIUM SALT □ TRIPOLY □ TRIPOLYPHOSPHATE
SAFETY PROFILE: Poison by intravenous route. Moderately toxic by ingestion, subcutaneous, and intraperitoneal routes. Ingestion of large doses of sodium phosphates causes catharsis. Sodium meta- and pyrophosphates can cause hemorrhages from the intestine if taken internally in large doses. When heated to decomposition it emits toxic fumes of PO_x and Na_2O.

SKN500   CAS: 13472-45-2   HR: 3
SODIUM TUNGSTATE
mf: $O_4W•2Na$ mw: 293.83
PROP: White, rhombic crystals. Mp: 695°, d: 4.179. Sol in water.
SYNS: DISODIUM TETRAOXATUNGSTATE (2-) □ DISODIUM TETRAOXOTUNGSTATE (2-) □ DISODIUM TUNGSTATE □ SODIUM WOLFRAMATE
CONSENSUS REPORTS: Reported in EPA TSCA Inventory.
ACGIH TLV: TWA 5 mg(W)/m³
NIOSH REL: (Tungsten) TWA 1 mg(W)/m³
SAFETY PROFILE: Poison by ingestion, intravenous, intramuscular, and subcutaneous routes. Experimental reproductive effects. Mutation data reported. When heated to decomposition it emits toxic fumes of Na_2O. See also TUNGSTEN COMPOUNDS.

SKO575   CAS: 1198-77-2   HR: D
SODIUM URATE
mf: $C_5H_3N_4O_3•Na$ mw: 190.11
SYNS: MONOSODIUM URATE □ 1H-PURINE-2,6,8(3H)-TRIONE, 7,9-DIHYDRO-, MONOSODIUM SALT (9CI) □ URIC ACID, MONOSODIUM SALT
CONSENSUS REPORTS: Reported in EPA TSCA Inventory.
SAFETY PROFILE: Experimental reproductive effects. When heated to decomposition it emits toxic fumes of NO_x.

SKP000   CAS: 13718-26-8   HR: 3
SODIUM VANADATE
mf: $O_3V \cdot Na$ mw: 121.93
SYNS: SODIUM METAVANADATE □ VANADIC ACID, MONOSODIUM SALT
CONSENSUS REPORTS: Reported in EPA TSCA Inventory.
ACGIH TLV: TWA 0.05 mg(V_2O_5)/m^3
NIOSH REL: (Vanadium Compounds) CL 0.05 mg(V)/m^3/15M
SAFETY PROFILE: Poison by ingestion, intraperitoneal, subcutaneous, and intravenous routes. Experimental reproductive effects. Mutation data reported. When heated to decomposition it emits toxic fumes of Na_2O and VO_x. See also VANADIUM COMPOUNDS.

SKS150   CAS: 4356-33-6   HR: 3
SOLANOSIDE
mf: $C_{30}H_{46}O_8$ mw: 534.76
SYNS: BRN 4729620 □ CARD-20(22)-ENOLIDE, 3-((6-DEOXY-3-o-METHYL-α-l-MANNOPYRANOSYL)OXY)-14-HYDROXY-, (3-β, 5-β)-
SAFETY PROFILE: A poison by intravenous route. When heated to decomposition it emits acrid smoke and irritating vapors.

SKS750   HR: 3
SOOT
PROP: Soot is defined as a brown-to-black substance incidentally produced during the incomplete and uncontrolled combustion of any carbonaceous material. It is a mixture of colloidal carbon, organic tars, and refractory inorganics whose composition depends on combustion conditions. It is not unusual for the tarry component to account for more than 50 weight percent of the soot, particularly, when produced by inefficient combustion of coal or wood. Can be distinguished from carbon black on the basis of differences in physical and chemical properties.
CONSENSUS REPORTS: NTP 10th Report on Carcinogens.
SAFETY PROFILE: Confirmed human carcinogen producing skin, scrotum, or lung tumors. The tarry component and, to a lesser extent, trace inorganic impurities are believed responsible for the known health hazards attributed to soot, e.g., cancers from possible chronic contact or long-term inhalation. See also CARBON BLACK.

SKU000   CAS: 110-44-1   HR: 2
SORBIC ACID
mf: $C_6H_8O_2$ mw: 112.14
PROP: Colorless needles or white powder from alc (aq); characteristic odor. Bp: 228° (decomp), mp: 134.5°, flash p: 260°F (COC), vap press: 0.01 mm @ 20°, vap d: 3.87. Sol in hot water; very sol in alc, ether.
SYNS: (2-BUTENYLIDENE)ACETIC ACID □ CROTYLIDENE ACETIC ACID □ HEXADIENIC ACID □ HEXADIENOIC ACID □ 2,4-HEXADIENOIC ACID □ trans-trans-2,4-HEXADIENOIC ACID □ 1,3-PENTADIENE-1-CARBOXYLIC ACID □ 2-PROPENYLACRYLIC ACID □ SORBISTAT
CONSENSUS REPORTS: Reported in EPA TSCA Inventory.
SAFETY PROFILE: Moderately toxic by intraperitoneal and subcutaneous routes. Mildly toxic by ingestion. Experimental reproductive effects. A severe human and experimental skin irritant. Questionable carcinogen with experimental tumorigenic data. Mutation data reported. Combustible when exposed to heat or flame; can react with oxidizing materials. To fight fire, use water. When heated to decomposition it emits acrid smoke and irritating fumes.

SKV000   CAS: 1338-39-2   HR: 1
SORBITAN MONOLAURATE
mf: $C_{18}H_{34}O_6$ mw: 346.52
SYNS: EMSORB 2515 □ RADIASURF 7125 □ SORBITAN MONODODECANOATE □ SPAN 20
CONSENSUS REPORTS: Reported in EPA TSCA Inventory.
SAFETY PROFILE: Slightly toxic by ingestion. Questionable carcinogen with experimental neoplastigenic data. When heated to decomposition it emits acrid smoke and irritating fumes.

S

SKV100   CAS: 1338-43-8   HR: 1
SORBITAN MONOOLEATE
mf: $C_{24}H_{44}O_6$ mw: 428.68

SYNS: ARLACEL 80 □ ARMOTAN MO □ EMSORB 2500 □ GLYCOMUL O □ IONET S-80 □ LIPOSORB O □ LIPOSORB O-20 □ ML 33F □ ML 55F □ MONO-DEHYDROSORBITOL MONOOLEATE □ MONTAN 80 □ NIKKOL SO 10 □ NIKKOL SO-15 □ NIKKOL SO-30 □ NONION OP80R □ O 250 □ RADIASURF 7155 □ SORBESTER P 17 □ SORBITAN MONOOLEIC ACID ESTER □ SORBITAN O □ SORBITAN OLEATE □ SORGEN 40 □ SPAN 80

CONSENSUS REPORTS: Reported in EPA TSCA Inventory.

SAFETY PROFILE: A skin irritant. Human mutation data reported. When heated to decomposition it emits acrid smoke and irritating fumes.

SKV150   CAS: 1338-41-6   HR: 1
SORBITAN MONOSTEARATE
mf: $C_{24}H_{46}O_6$ mw: 430.70

PROP: Cream to tan-colored waxy solid; bland odor and taste. Insol in cold water, mineral spirits, acetone; dispersible in warm water; sol above 50° in mineral oil, ethyl acetate.

SYNS: ANHYDRO-d-GLUCITOL MONOOCTA-DECANOATE □ ANHYDROSORBITOL STEARATE □ ARLACEL 60 □ ARMOTAN MS □ CRILL 3 □ CRILL K 3 □ DREWSORB 60 □ DURTAN 60 □ EMSORB 2505 □ GLYCOMUL S □ HODAG SMS □ IONET S 60 □ LIPOSORB S □ LIPOSORB S-20 □ MONTANE 60 □ MS 33 □ MS 33F □ NEWCOL 60 □ NIKKOL SS 30 □ NISSAN NONION SP 60 □ NONION SP 60 □ NONION SP 60R □ RIKEMAL S 250 □ SORBITAN C □ SORBITAN MONOOCTADECANOATE □ SORBITAN STEARATE □ SORBON S 60 □ SORGEN 50 □ SPAN 55 □ SPAN 60

CONSENSUS REPORTS: EPA Genetic Toxicology Program.

SAFETY PROFILE: Very mildly toxic by ingestion. Experimental reproductive effects. A skin irritant. When heated to decomposition it emits acrid smoke and irritating fumes.

SKV195   CAS: 9005-71-4   HR: D
SORBITAN, TRISTEARATE, POLYOXY-ETHYLENE derivatives
SYNS: EMSORB 6907 □ GLYCOSPERSE TS 20 □ POLYSORBATE 65 □ SORBIMACROGOL TRISTEARATE 300 □ SORBITAN, TRIOCTADECANOATE, POLY(OXY-1,2-ETHANEDIYL) derivs. (9CI) □ TWEEN 65

CONSENSUS REPORTS: Reported in EPA TSCA Inventory.

SAFETY PROFILE: Experimental reproductive effects. When heated to decomposition it emits acrid smoke and irritating fumes.

SKV200   CAS: 50-70-4   HR: 1
SORBITOL
mf: $C_6H_{14}O_6$ mw: 182.20

PROP: White crystalline powder or needles; odorless with sweet taste (60% of sucrose). D: 1.47 @ −5°, mp: 93° (metastable form), 97.5°, (stable form), bp: 105°. Sol in water; sltly sol in methanol, ethanol, acetic acid, phenol, and acetamide; almost insol in other org solvs.

SYNS: CHOLAXINE □ DIAKARMON □ GLUCITOL □ d-GLUCITOL □ GULITOL □ l-GULITOL □ KARION □ NIVITIN □ SIONIT □ SIONON □ SORBICOLAN □ SORBITE □ d-SORBITOL □ SORBO □ SORBOL □ SORBOSTYL □ SORVILANDE

CONSENSUS REPORTS: Reported in EPA TSCA Inventory. EPA Genetic Toxicology Program.

SAFETY PROFILE: Mildly toxic by ingestion. Human systemic effects by ingestion: hypermotility and diarrhea. Mutation data reported. When heated to decomposition it emits acrid smoke and irritating fumes.

SKW825   HR: D
SOYBEAN OIL (UNHYDROGENATED)
PROP: From the seed of the legume *Glycine max*. Amber-colored oil.

SAFETY PROFILE: When heated to decomposition it emits acrid smoke and irritating fumes.

SKY000   CAS: 8008-79-5   HR: 1
SPEARMINT OIL
PROP: From steam distillation of the plant *Mentha spicata* L. (common spearmint), or of *Mentha cardiaca Gerard ex Baker* (scotch spearmint) (Fam. *Labiatae*). Contains principally carvone, phellandrene, limonene, and either dihydrocarveol acetate or dihydrocuminic acetate (FCTXAV 16,637,78). Colorless or greenish-yellow liquid; odor and taste of spearmint.

SYN: OIL OF SPEARMINT

CONSENSUS REPORTS: Reported in EPA TSCA Inventory.

SAFETY PROFILE: Mildly toxic by ingestion. Mutation data reported. A skin irritant and an allergen. When heated to decomposition it emits acrid smoke and irritating fumes. Used as a flavoring agent.

SLB500   CAS: 84837-04-7   HR: 2
SPIKE LAVENDER OIL

PROP: From steam distillation of the plant *Lavandula latifolia* Vill. (*Lavandula spica*, D.C.) (Fam. *Labiatae*). The main constituents are linalool and cineole (FCTXAV 14,443,76). Yellow liquid; lavender odor. D: 0.893–0.909, refr index: 1.463 @ 20°. Sol in fixed oils, propylene glycol; sltly sol in glycerin, mineral oil.

SYNS: LAVENDER OIL, SPIKE □ OIL OF SPIKE LAVENDER

SAFETY PROFILE: Moderately toxic by ingestion. A skin irritant. When heated to decomposition it emits acrid smoke and irritating fumes. See also LINALOOL and CAJEPUTOL.

SLD550   CAS: 22791-18-0   HR: 3
SPIRO(1,3-BENZODIOXOLE-2,1'-CYCLOHEXAN)-4-OL, METHYL-CARBAMATE

mf: $C_{14}H_{17}NO_4$ mw: 263.32

SYN: CARBAMIC ACID, METHYL-, SPIRO(1,3-BENZODIOXOLE-2,1'-CYCLOHEXAN)-4-YL ESTER

SAFETY PROFILE: A poison by ingestion. When heated to decomposition it emits toxic vapors of NO_x.

SLI325   CAS: 21736-83-4   HR: 2
STANILO

mf: $C_{14}H_{24}N_2O_7 \cdot 2ClH$ mw: 405.32

PROP: White-buff crystals or powder. Sol in H_2O and MeOH.

SYNS: DECAHYDRO-4a,7,9-TRIHYDROXY-2-METHYL-6,8-BIS(METHYLAMINO)-4H-PYRANO(2,3-b)(1,4)BENZO-DIOXIN-4-ONE DIHYDROCHLORIDE, (2R-(2-α,4a-β,5a-β,6-β,7-β,8-β,9-α,9a-α,10a-β))- □ SPECTINOMYCIN DIHYDROCHLORIDE □ SPECTINOMYCIN HYDRO-CHLORIDE

SAFETY PROFILE: Moderately toxic by intraperitoneal route. When heated to

decomposition it emits toxic fumes of NO_x and HCl.

SLI350   HR: D
STANNOUS STEARATE

SYN: TIN STEARATE

SAFETY PROFILE: When heated to decomposition it emits acrid smoke and irritating fumes.

SLJ000   CAS: 7081-44-9   HR: 2
STAPHYBIOTIC

mf: $C_{19}H_{17}ClN_3O_5S \cdot Na \cdot H_2O$ mw: 475.91

PROP: Monohydrate. Mp: 170° (decomp). Sol in H_2O.

SYNS: BACTOPEN □ BRL-1621 □ 6-(3-(o-CHLORO-PHENYL)-5-METHYL-4-ISOXAZOLECARBOXAMIDEO)-3,3-DIMETHYL-7-OXO-4-THIA-1-AZABICYCLO(3.2.0)-HEPTANE-2-CARBOXYLIC ACID, SODIUM SALT, MONOHYDRATE □ CLOXACILLIN SODIUM MONO-HYDRATE □ CLOXAPEN □ CLOXYPEN □ EKVACILLIN □ GELSTAPH □ METHOCILLIN-S □ ORBENIN SODIUM HYDRATE □ P-25 □ PROSTAPHL-IN-A □ SODIUM CLOXACILLIN MONOHYDRATE □ STAPHOBRISTOL-250 □ TEGOPEN □ TEPOGEN

SAFETY PROFILE: Moderately toxic by intraperitoneal, intramuscular, subcutaneous, and intravenous routes. Mildly toxic by ingestion. When heated to decomposition it emits very toxic fumes of Cl^-, NO_x, Na_2O, and SO_x.

SLJ500   CAS: 9005-25-8   HR: 3
STARCH DUST

SYNS: AMAIZO W 13 □ AMYLOMAIZE VII □ AMYL-UM □ AQUAPEL (POLYSACCHARIDE) □ ARGO BRAND CORN STARCH □ ARROWROOT STARCH □ CLARO 5591 □ CLEAREL □ CLEARJEL □ CORN PRODUCTS □ CPC 3005 □ CPC 6448 □ FARINEX 100 □ GALACTASOL A □ GENVIS □ HRW 13 □ KEESTAR □ MAIZENA □ MARANTA □ MELOGEL □ MELUNA □ OK PRE-GEL □ PENFORD GUM 380 □ REMYLINE Ac □ RICE STARCH □ SORGHUM GUM □ STARAMIC 747 □ STARCH □ α-STARCH □ STARCH, CORN □ STARCH (OSHA) □ STA-RX 1500 □ TAPIOCA STARCH □ TAPON □ TROGUM □ W-GUM □ W-13 STABILIZER

CONSENSUS REPORTS: Reported in EPA TSCA Inventory.

OSHA PEL: Total Dust: 15 mg/m³; Respirable Fraction: 5 mg/m³

ACGIH TLV: TWA (nuisance particulate) 10 mg/m³ of total dust (when toxic impurities are not present, e.g., quartz <1%); Not Classifiable as a Human Carcinogen

NIOSH REL: (Starch, Respirable Fraction) 5 mg/m³; (total dust) 10 mg/m³

SAFETY PROFILE: A nuisance dust. Mildly toxic by intraperitoneal route. A skin irritant. An allergen. Flammable when exposed to flame; can react with oxidizing materials. Moderately explosive when exposed to flame.

SLK000   CAS: 57-11-4   HR: 3
STEARIC ACID

mf: $C_{18}H_{36}O_2$ mw: 284.54

PROP: White, amorph solid or leaflets; slt odor and taste of tallow. Mp: 69.3°, bp: 383°, flash p: 385°F (CC), d: 0.847, autoign temp: 743°F, vap press: 1 mm @ 173.7°, vap d: 9.80. Sol in alc, ether, acetone, chloroform; insol in water.

SYNS: CENTURY 1240 □ DAR-CHEM 14 □ EMERSOL 120 □ GLYCON DP □ GLYCON S-70 □ GLYCON TP □ GROCO 54 □ 1-HEPTADECANECARBOXYLIC ACID □ HYDROFOL ACID 1655 □ HY-PHI 1199 □ HYSTRENE 80 □ INDUSTRENE 5016 □ KAM 1000 □ KAM 2000 □ KAM 3000 □ NEO-FAT 18-61 □ NEO-FAT 18-S □ OCTADECANOIC ACID □ PEARL STEARIC □ STEAREX BEADS □ STEAROPHANIC ACID □ TEGOSTEARIC 254

CONSENSUS REPORTS: Reported in EPA TSCA Inventory. EPA Genetic Toxicology Program.

SAFETY PROFILE: Poison by intravenous route. A human skin irritant. Questionable carcinogen with experimental tumorigenic data by implantation route. Combustible when exposed to heat or flame. Heats spontaneously. To fight fire, use CO_2, dry chemical. When heated to decomposition it emits acrid smoke and irritating fumes.

SLN100   HR: D
STEARYL CITRATE

SAFETY PROFILE: When heated to decomposition it emits acrid smoke and irritating fumes.

SLN200   HR: D
STEARYL MONOGLYCERIDYL CITRATE

PROP: Soft tan to white, waxy solid; tasteless. Insol in water; sol in chloroform, ethylene glycol.

SAFETY PROFILE: When heated to decomposition it emits acrid smoke and irritating fumes.

SLO100   CAS: 93334-51-1   HR: 3
STEPHANIA HERNANDIFOLIA Walp., extract

SYN: AKNADI EXTRACT

DOT CLASSIFICATION: 4.2; Label: Spontaneously Combustible

SAFETY PROFILE: A poison by intraperitoneal route. A storage and handling hazard: spontaneously combustible. When heated to decomposition it emits acrid smoke and irritating vapors.

SLP000   CAS: 10048-13-2   HR: 3
STERIGMATOCYSTIN

mf: $C_{18}H_{12}O_6$ mw: 328.34

PROP: Pale-yellow needles. Mp: 246° (decomp). A metabolite of *Aspergillus versicolor* (BJCAAI 20,134,66).

SYN: 3a,12c-DIHYDRO-8-HYDROXY-6-METHOXY-7H-FURO(3',2':4,5)FURO(2,3-C)XANTHEN-7-ONE

CONSENSUS REPORTS: IARC Cancer Review: Group 2B IMEMDT 7,56,87; Animal Sufficient Evidence IMEMDT 10,245,76; Animal Limited Evidence IMEMDT 1,175,72. EPA Genetic Toxicology Program.

SAFETY PROFILE: Confirmed carcinogen with experimental carcinogenic and tumorigenic data. Poison by ingestion and intraperitoneal routes. Human mutation data reported. When heated to decomposition it emits acrid smoke and irritating fumes.

SLQ000   CAS: 7803-52-3   HR: 3
STIBINE

DOT: UN 2676

mf: H_3Sb mw: 124.78

PROP: Thermally unstable gas, forming colorless liquid and crystals at liquid air temp. Disagreeable odor. Mp: −88°, bp: −18.4°, d: 2.26 @ −25°. Gas is sltly sol in water; very sol in alc, carbon disulfide, and org solvs. IDLH 5 ppm.

SYNS: ANTIMONWASSERSTOFFES (GERMAN) □ ANTIMONY HYDRIDE □ ANTIMONY TRIHYDRIDE □ ANTYMONOWODOR (POLISH) □ HYDROGEN ANTIMONIDE

CONSENSUS REPORTS: Antimony and its compounds are on the Community Right-To-Know List.

OSHA PEL: TWA 0.1 ppm

ACGIH TLV: TWA 0.1 ppm

DFG MAK: 0.1 ppm (0.52 mg/m³)

DOT CLASSIFICATION: 2.3; Label: Poison Gas, Flammable Gas

SAFETY PROFILE: Poison by inhalation. Potentially explosive decomposition at 200°C. Flammable when exposed to heat or flame. Explosive reaction with ammonia + heat, chlorine, concentrated nitric acid, ozone. Incompatible with oxidants. The decomposition products are hydrogen and metallic antimony. When heated to decomposition it emits toxic fumes of Sb. Used as a fumigating agent. See also ANTIMONY COMPOUNDS and HYDRIDES.

SLR000   CAS: 588-59-0   HR: 3
STILBENE

mf: $C_{14}H_{12}$ mw: 180.26

PROP: Colorless or sltly yellow crystals. Mp: 124–125°, bp: 306–307°, d: 0.9707. Insol in water; sol in 90 parts cold alc and 13 parts boiling alc; freely sol in benzene and ether.

SYNS: BENZENE, 1,1'-(1,2-ETHENEDIYL)BIS-(9CI) □ BIBENZAL □ BIBENZYLIDENE □ BIBENZYLIDINE □ α-β-DIPHENYLETHYLENE □ 1,2-DIPHENYLETHYL-ENE □ 1,1'-(1,2-ETHENEDIYL)DIBENZENE □ STILBEN (GERMAN)

CONSENSUS REPORTS: Reported in EPA TSCA Inventory.

SAFETY PROFILE: Poison by intravenous route. Moderately toxic by intraperitoneal route. Violent reaction with O_2. When heated to decomposition it emits acrid smoke and irritating fumes.

SLU500   CAS: 8052-41-3   HR: 3
STODDARD SOLVENT

PROP: Clear, colorless liquid. Composed of 85% nonane and 15% trimethyl benzene. Bp: 220–300°, flash p: 100–110°F, lel: 1.1%, uel: 6%, autoign temp: 450°F, d: 1.0. Insol in water; misc with abs alc, benzene, ether, chloroform, carbon tetrachloride, carbon disulfide, and some oils (not castor oil). IDLH 20,000 mg/m³.

SYNS: NAPHTHA SAFETY SOLVENT □ VARNOLINE □ WHITE SPIRITS

CONSENSUS REPORTS: Reported in EPA TSCA Inventory.

OSHA PEL: TWA 100 ppm

ACGIH TLV: TWA 100 ppm

NIOSH REL: (Refined Petroleum Solvents) TWA 350 mg/m³; CL 1800 mg/m³/15M

SAFETY PROFILE: Mildly toxic by inhalation. A human eye irritant. Flammable liquid when exposed to heat, sparks, or flame. Explosive in the form of vapor when exposed to heat or flame. When heated to decomposition it emits acrid fumes and may explode; can react with oxidizing materials. To fight fire, use foam, CO_2, dry chemical. See also n-NONANE.

SLW500   CAS: 57-92-1   HR: 3
STREPTOMYCIN

mf: $C_{21}H_{39}N_7O_{12}$ mw: 581.67

PROP: An antibiotic. It is a base and readily forms salts with anions.

SYNS: AGRIMYCIN 17 □ CHEMFORM □ GEROX □ HOKKO-MYCIN □ NSC-14083 □ STREPCEN □ STREPTOMICINA (ITALIAN) □ STREPTOMYCIN A □ STREPTOMYCINE □ STREPTOMYCINUM □ STREPTO-MYZIN (GERMAN)

CONSENSUS REPORTS: EPA Genetic Toxicology Program.

SAFETY PROFILE: Poison by intravenous and subcutaneous routes. Moderately toxic by ingestion and intraperitoneal routes. An experimental teratogen. Human systemic effects by ingestion and intraperitoneal routes: change in vestibular functions, blood pressure decrease, eosinophilia, respiratory depression, and pulmonary changes. Human reproductive and teratogenic effects by unspecified routes: developmental abnormalities of the eye and ear and effects on newborn including postnatal measures or effects. Toxic to kidneys and central nervous system. Has been implicated in aplastic anemia. Experimental reproductive effects. Human mutation data reported. When heated to decomposition it emits toxic fumes of NO_x.

SLY500   CAS: 3810-74-0   HR: 3
STREPTOMYCIN SESQUISULFATE

mf: $C_{42}H_{78}N_{14}O_{24} \cdot H_6O_{12}S_3$ mw: 1457.58

PROP: Powder.

SYNS: AGRI-MYCIN □ AGRISTREP □ AS-15 □ PHYTOMYCIN □ PLANTOMYCIN □ STREPCIN □ STREP-GRAN □ STREPSULFAT □ STREPTOMYCIN SULFATE □ STREPTOMYCIN SULPHATE B.P. □ STREPTOREX □ STREPVET □ VETSTREP

CONSENSUS REPORTS: Reported in EPA TSCA Inventory. EPA Genetic Toxicology Program.

SAFETY PROFILE: Poison by ingestion and intravenous routes. Moderately toxic by subcutaneous route. An experimental teratogen. Human systemic effects by intraperitoneal route: flaccid paralysis without anesthesia, motor activity changes and pulmonary changes. Experimental reproductive effects. Mutation data reported. When heated to decomposition it emits very toxic fumes of NO_x and SO_x. See also STREPTOMYCIN and SULFATES.

SMD000   CAS: 18883-66-4   HR: 3
STREPTOZOTICIN

mf: $C_8H_{15}N_3O_7$ mw: 265.26

PROP: Platelets or prisms from alc. Mp: 115° (decomp).

SYNS: 2-DEOXY-2-(((METHYL-NITROSOAMINO)CARBONYL)AMINO)-d-GLUCOPYRANOSE □ 2-DEOXY-2-(3-METHYL-3-NITROSOUREIDO)-d-GLUCOPYRANOSE □ 2-DEOXY-2-(3-METHYL-3-NITROSOUREIDO)-α(and β)-d-GLUCOPY-RANOSE □ N-d-GLUCOSYL(2)-N'-NITROSOMETHYL-HARNSTOFF (GERMAN) □ N-d-GLUCOSYL-(2)-N'-NITROSOMETHYLUREA □ NCI-C03167 □ NSC-85598 □ NSC-85998 □ RCRA WASTE NUMBER U206 □ STR □ STREPTOZOCIN □ STRZ □ STZ □ U-9889 □ ZANOSAR

CONSENSUS REPORTS: NTP 10th Report on Carcinogens. IARC Cancer Review: Group 2B IMEMDT 7,56,87; Human Limited Evidence IMEMDT 17,337,78; Animal Sufficient Evidence IMEMDT 17,337,78. NCI Carcinogenesis Studies (ipr); Clear Evidence: mouse, rat RRCRBU 52,1,75. EPA Genetic Toxicology Program.

SAFETY PROFILE: Confirmed carcinogen with experimental carcinogenic, neoplastigenic, tumorigenic, and teratogenic data. Experimental poison by intravenous, parenteral, subcutaneous, and intraperitoneal routes. Moderately toxic to humans by intravenous route. Human systemic effects: nausea or vomiting, impaired liver function, kidney changes. Human mutation data reported. Experimental reproductive effects. When heated to decomposition it emits toxic fumes of NO_x. See also NITROSAMINES.

SME500   CAS: 91724-16-2   HR: 3
STRONTIUM ARSENITE

DOT: UN 1691

mf: As_2O_4Sr mw: 301.46

PROP: White powder.

SYNS: ARSENIOUS ACID, STRONTIUM SALT □ STRONTIUM ARSENITE, solid (DOT)

CONSENSUS REPORTS: Arsenic and its compounds are on the Community Right-To-Know List.

OSHA PEL: OSHA: Cancer Hazard

ACGIH TLV: BEI: 35 μ (As)/L inorganic arsenic and methylated metabolites in urine

NIOSH REL: (Arsenic, Inorganic) CL 0.002 mg(As)/m³/15M

DOT CLASSIFICATION: 6.1; Label: Poison

SAFETY PROFILE: Confirmed human carcinogen. A deadly poison. When heated to decomposition it emits toxic fumes of As. See also ARSENIC COMPOUNDS and STRONTIUM COMPOUNDS.

SMF500   CAS: 7791-10-8   HR: 2
STRONTIUM CHLORATE

DOT: UN 1506

mf: $C_{12}O_6 \cdot Sr$ mw: 327.74

PROP: White or colorless, crystalline, orthorhombic powder. Mp: (decomp) 120°, d: 3.152. Very sol in H_2O; insol in EtOH.

SYNS: CHLORIC ACID, STRONTIUM SALT □ STRONTIUM CHLORATE, solid or solution (DOT)

CONSENSUS REPORTS: Reported in EPA TSCA Inventory.

DOT CLASSIFICATION: 5.1; Label: Oxidizer

SAFETY PROFILE: A powerful oxidizer. When heated to decomposition it emits toxic fumes of Cl⁻. See also CHLORATES.

SMH000   CAS: 7789-06-2   HR: 3
STRONTIUM CHROMATE (1:1)
mf: CrO₄•Sr   mw: 203.62
PROP: Monoclinic, yellow crystals. D: 3.895 @ 15°. Sol in HCl, HNO₃, and AcOH.
SYNS: CHROMIC ACID, STRONTIUM SALT (1:1) □ C.I. PIGMENT YELLOW 32 □ DEEP LEMON YELLOW □ STRONTIUM CHROMATE (VI) □ STRONTIUM CHROMATE 12170 □ STRONTIUM YELLOW
CONSENSUS REPORTS: NTP 10th Report on Carcinogens. IARC Cancer Review: Group 1 IMEMDT 7,165,87; Animal Sufficient Evidence IMEMDT 2,100,73; IMEMDT 23,205,80; Human Sufficient Evidence IMEMDT 23,205,80. Chromium and its compounds are on the Community Right-To-Know List. Reported in EPA TSCA Inventory.
OSHA PEL: CL 0.1 mg(CrO₃)/m³
ACGIH TLV: TWA 0.0005 ppm; Suspected Human Carcinogen
DFG MAK: DFG TRK: 0.1 mg/m³; Animal Carcinogen, Suspected Human Carcinogen
NIOSH REL: TWA 0.0001 mg(Cr(VI))/m³
SAFETY PROFILE: Confirmed human carcinogen with experimental carcinogenic and tumorigenic data. Moderately toxic by ingestion. Mutation data reported. See also CHROMIUM COMPOUNDS and STRONTIUM COMPOUNDS.

SMI000   CAS: 13814-98-7   HR: 2
STRONTIUM FLUOBORATE
mf: BF₄•1/2Sr   mw: 130.62
SYNS: TETRAFLUOROBORATE(1-) STRONTIUM (2:1) □ TL 1139
CONSENSUS REPORTS: Reported in EPA TSCA Inventory.
OSHA PEL: TWA 2.5 mg(F)/m³
ACGIH TLV: TWA 2.5 mg(F)/m³; BEI: 3 mg/g creatinine of fluorides in urine prior to shift; 10 mg/g creatinine of fluorides in urine at end of shift.
NIOSH REL: (Inorganic Fluorides) TWA 2.5 mg(F)/m³

SAFETY PROFILE: Moderately toxic by ingestion and inhalation. When heated to decomposition it emits toxic fumes of F⁻. See also STRONTIUM COMPOUNDS and FLUORIDES.

SMI500   CAS: 7783-48-4   HR: 2
STRONTIUM FLUORIDE
mf: F₂Sr   mw: 125.62
PROP: Cubic, odorless, colorless, luminescent crystals or white powder. Readily forms color centers activated by either radiation or by doping with alkali metals or rare earth metals. Forms solid solns with a wide range of fluorides. Mp: 1477°, d: 4.24, bp: 2460. Decomp by strong acids. Practically insol in H₂O; insol in EtOH and Me₂CO.
CONSENSUS REPORTS: Reported in EPA TSCA Inventory.
OSHA PEL: TWA 2.5 mg(F)/m³
ACGIH TLV: TWA 2.5 mg(F)/m³; BEI: 3 mg/g creatinine of fluorides in urine prior to shift; 10 mg/g creatinine of fluorides in urine at end of shift.
NIOSH REL: TWA 2.5 mg(F)/m³
SAFETY PROFILE: Moderately toxic by intravenous route. Mildly toxic by ingestion. When heated to decomposition it emits toxic fumes of F⁻. See also FLUORIDES and STRONTIUM COMPOUNDS.

SMK000   CAS: 10042-76-9   HR: 2
STRONTIUM(II) NITRATE (1:2)
DOT: UN 1507
mf: N₂O₆•Sr   mw: 211.64
PROP: Colorless cubic crystals. White powder. Mp: 570–645°, d: 2.986. Very sol in H₂O; spar sol in EtOH and Me₂CO.
SYNS: NITRATE de STRONTIUM (FRENCH) □ NITRIC ACID, STRONTIUM SALT □ STRONTIUM NITRATE (DOT)
CONSENSUS REPORTS: Reported in EPA TSCA Inventory.
DOT CLASSIFICATION: 5.1; Label: Oxidizer
SAFETY PROFILE: Moderately toxic by ingestion and intraperitoneal routes. A powerful oxidizer. When heated to decomposition it emits toxic fumes of NOₓ.

S

See also NITRATES and STRONTIUM COMPOUNDS.

SMK500   CAS: 1314-18-7   HR: 2
STRONTIUM PEROXIDE
DOT: UN 1509
mf: O_2Sr mw: 119.62
PROP: White powder. Mp: decomp, d: 4.56.
CONSENSUS REPORTS: Reported in EPA TSCA Inventory.
DOT CLASSIFICATION: 5.1; Label: Oxidizer
SAFETY PROFILE: A powerful oxidizer. A skin, eye, and mucous membrane irritant. Mixtures with organic materials readily ignite with friction or on contact with moisture. See also PEROXIDES and STRONTIUM COMPOUNDS.

SMM000   CAS: 1314-96-1   HR: 3
STRONTIUM SULFIDE
mf: SSr   mw: 119.68
PROP: Colorless, cubic, light-gray crystals. Dissolves in H_2O with formation of $SrOH_2$ and H_2S. Luminescent. D: 3.70 @ 15°. Forms solid solns with a wide range of metal sulfides; the magnetic properties of SrSbEuS solid solns are of particular interest.
SYNS: C.I. 77847 □ STRONTIUM MONOSULFIDE □ STRONTIUM SULPHIDE
CONSENSUS REPORTS: Reported in EPA TSCA Inventory.
SAFETY PROFILE: Poison by inhalation and ingestion. Readily decomposes to yield H_2S. Incompatible with lead(IV) oxide. When heated to decomposition it emits toxic fumes of SO_x. See also SULFIDES and STRONTIUM COMPOUNDS.

SMN500   CAS: 57-24-9   HR: 3
STRYCHNINE
DOT: UN 1692
mf: $C_{21}H_{22}N_2O_2$ mw: 334.45
PROP: Hard, white, crystalline alkaloid; very bitter taste. Mp: 268°, bp: 270°, d: 1.359 @ 18°.
SYNS: BOOMER-RID □ CERTOX □ DOLCO MOUSE CEREAL □ GOPHER BAIT □ GOPHER-GITTER □ HARE-RID □ KWIK-KIL □ MOLE DEATH □ MOUSE-NOTS □ MOUSE-RID □ MOUSE-TOX □ NUX VOMICA □ PIED PIPER MOUSE SEED □ RCRA WASTE NUMBER

P108 □ RO-DEX □ SANASEED □ STRICNINA (ITALIAN) □ STRYCHNIDIN-10-ONE □ STRYCHNIN (GERMAN) □ STRYCHNINE, solid and liquid (DOT) □ STRYCHNOS
CONSENSUS REPORTS: Reported in EPA TSCA Inventory. EPA Extremely Hazardous Substances List.
OSHA PEL: TWA 0.15 mg/m³
ACGIH TLV: TWA 0.15 mg/m³
DFG MAK: 0.15 mg/m³
DOT CLASSIFICATION: 6.1; Label: Poison
SAFETY PROFILE: Human poison by ingestion. Experimental poison by ingestion, intravenous, subcutaneous, and intraperitoneal routes. Experimental reproductive effects. An allergen. Lethal dose to man: 30–60 mg/kg. If ingested, the time of action depends upon the condition of the stomach, whether empty or full, and the nature of the food present. If taken by subcutaneous injection, the place of administration of the injection will affect the time of action. The first symptoms are a feeling of uneasiness with a heightened reflex of irritability, followed by muscular twitching in some parts of the body. With larger doses, this is followed by a sense of impending suffocation. Convulsive movements begin that have the effect of mechanically causing the patient to cry out or to shriek; then follow the characteristic spasms, which set in with violence. These are at first clonic and then tonic. There are successive attacks of spasms. With each successive attack, the symptoms become more violent, eventually resulting in death. A rodenticide. When heated to decomposition it emits toxic fumes of NO_x.

SMO500   HR: 3
STRYCHNINE SALT (solid)
DOT: UN 1692
SYN: STRYCHNINE SALTS (DOT)
DOT CLASSIFICATION: 6.1; Label: Poison
SAFETY PROFILE: A deadly poison. See also STRYCHNINE.

SMP000   CAS: 60-41-3   HR: 3
STRYCHNINE SULFATE (2:1)
mf: $C_{21}H_{22}N_2O_2 \cdot 1/2H_2O_4S$ mw: 383.49

PROP: Crystals. Mp: 200° (anhyd).
SYNS: STRYCHNINE SULFATE □ STRYCHNIDIN-10-ONE, SULFATE (2:1)
CONSENSUS REPORTS: EPA Extremely Hazardous Substances List.
SAFETY PROFILE: Poison by ingestion, intraperitoneal, intravenous, and subcutaneous routes. When heated to decomposition it emits very toxic fumes of SO_x and NO_x. See also STRYCHNINE and SULFATES.

SMP500   CAS: 82-71-3   HR: 3
STYPHNIC ACID
DOT: UN 0219/UN 0394
mf: $C_6H_3N_3O_8$ mw: 245.12
PROP: Hexagonal, yellow crystals from EtOAc; astringent taste. Mp: (dry) 179–180°. Very sol in alc, ether.
SYNS: 1,3-DIHYDROXY-2,4,6-TRINITROBENZENE □ 2,4-DIHYDROXY-1,3,5-TRINITROBENZENE □ 3-HYDROXY-2,4,6-TRINITROPHENOL □ 2,4,6-TRINITRO-BENZENE-1,3-DIOL □ 2,4,6-TRINITRO-1,3-BENZENEDI-OL □ 2,4,6-TRINITRORESORCINOL □ TRINITRO-RESORCINOL (DOT) □ TRINITRORESORCINOL, DRY (DOT) □ TRINITRORESORCINOL, wetted with less than 20% water (DOT)
CONSENSUS REPORTS: Reported in EPA TSCA Inventory.
DOT CLASSIFICATION: EXPLOSIVE 1.1D; Label: EXPLOSIVE 1.1D
SAFETY PROFILE: Very explosive. Upon decomposition it emits toxic fumes of NO_x. See also NITRO COMPOUNDS of AROMATIC HYDROCARBONS and EXPLOSIVES, HIGH.

SMQ000   CAS: 100-42-5   HR: 3
STYRENE
DOT: UN 2055
mf: C_8H_8 mw: 104.16
PROP: Colorless, refractive, oily liquid with penetrating odor. Mp: −33°, bp: 146°, lel: 1.1%, uel: 6.1%, flash p: 88°F, d: 0.9074 @ 20°/4°, autoign temp: 914°F, vap d: 3.6, fp: −33°, ULC: 40–50. Very sltly sol in water; misc in alc and ether. IDLH 700 ppm.
SYNS: CINNAMENE □ CINNAMENOL □ DIAREX HF 77 □ ETHENYLBENZENE □ NCI-C02200 □ PHENETHYLENE □ PHENYLETHENE □ PHENYL-ETHYLENE □ STIROLO (ITALIAN) □ STYREEN

(DUTCH) □ STYREN (CZECH) □ STYRENE MONOMER (ACGIH) □ STYRENE MONOMER, inhibited (DOT) □ STYROL (GERMAN) □ STYROLE □ STYROLENE □ STYRON □ STYROPOR □ VINYLBENZEN (CZECH) □ VINYLBENZENE □ VINYLBENZOL
CONSENSUS REPORTS: IARC Cancer Review: Group 2B IMEMDT 7,345,87; Animal Sufficient Evidence IMEMDT 19,231,79; Human Inadequate Evidence IMEMDT 19,231,79. NCI Carcinogenesis Bioassay (gavage); Inadequate Studies: mouse, rat NCITR* NCI-CG-TR-170,79 (gavage). Reported in EPA TSCA Inventory. EPA Genetic Toxicology Program. Community Right-To-Know List.
OSHA PEL: TWA 50 ppm; STEL 100 ppm
ACGIH TLV: TWA 20 ppm; STEL 40 ppm; Not Classifiable as a Human Carcinogen; BEI: 800 mg(mandelic acid)/L in urine at end of shift; 0.55 mg/L styrene in blood at end of shift; 0.02 mg/L styrene in blood prior to next shift
DFG MAK: 20 ppm (86 mg/m³); BAT: 2 g/L of mandelic acid in urine at end of shift
NIOSH REL: (Styrene) TWA 50 ppm; CL 100 ppm
DOT CLASSIFICATION: 3; Label: Flammable Liquid
SAFETY PROFILE: Confirmed carcinogen. Experimental poison by ingestion, inhalation, and intravenous routes. Moderately toxic experimentally by intraperitoneal route. Mildly toxic to humans by inhalation. An experimental teratogen. Human systemic effects by inhalation: eye and olfactory changes. It can cause irritation and violent itching of the eyes @ 200 ppm, lachrymation, and severe human eye injuries. Its toxic effects are usually transient and result in irritation and possible narcosis. Experimental reproductive effects. Human mutation data reported. A human skin irritant. An experimental skin and eye irritant.

The monomer has been involved in several industrial explosions. It is a storage hazard above 32°C. A very dangerous fire hazard when exposed to flame, heat, or oxidants. Explosive in the form of vapor when exposed to heat or flame. Reacts with

S

oxygen above 40°C to form a heat-sensitive explosive peroxide. Violent or explosive polymerization may be initiated by alkali-metal–graphite composites, butyllithium, dibenzoyl peroxide, other initiators (e.g., azoisobutyronitrile, di-tert-butyl peroxide). Reacts violently with chlorosulfonic acid, oleum, sulfuric acid, chlorine + iron(III) chloride (above 50°C). May ignite when heated with air + polymerizing polystyrene. Can react vigorously with oxidizing materials. To fight fire, use foam, CO_2, dry chemical. When heated to decomposition it emits acrid smoke and irritating fumes.

SMQ500　CAS: 9003-53-6　HR: 2
STYRENE POLYMER

mf: $(C_8H_8)_n$

SYNS: 3A □ A 3-80 □ AFCOLENE □ AFCOLENE 666 □ AFCOLENE S 100 □ ATACTIC POLYSTYRENE □ BACTOLATEX □ BAKELITE SMD 3500 □ BASF III □ BDH 29-790 □ BENZENE, ETHENYL-, HOMOPOLYMER (9CI) □ BEXTRENE XL 750 □ BICOLASTIC A 75 □ BICOLENE H □ BIO-BEADS S-S 2 □ BP-KLP □ BSB-S 40 □ BSB-S-E □ BUSTREN □ BUSTREN K 500 □ BUSTREN K 525-19 □ BUSTREN U 825 □ BUSTREN U 825E11 □ BUSTREN Y 825 □ BUSTREN Y 3532 □ CADCO 0115 □ CARINEX GP □ CARINEX HR □ CARINEX HRM □ CARINEX SB 59 □ CARINEX SB 61 □ CARINEX SL 273 □ CARINEX TGX/MF □ COPAL Z □ COSDEN 550 □ COSDEN 945E □ DENKA QP3 □ DIAREX 43G □ DIAREX HF 55 □ DIAREX HF 77 □ DIAREX HF 55-247 □ DIAREX HS 77 □ DIAREX HT 88 □ DIAREX HT 88A □ DIAREX HT 90 □ DIAREX HT 190 □ DIAREX HT 500 □ DIAREX YH 476 □ DORVON □ DORVON FR 100 □ DOW 360 □ DOW 456 □ DOW 665 □ DOW 860 □ DOW 1683 □ DOW MX 5514 □ DOW MX 5516 □ DYLARK 250 □ DYLENE □ DYLENE 8 □ DYLENE 8G □ DYLENE 9 □ DYLITE F 40 □ DYLITE F 40L □ 686E □ EDISTIR RB □ ESBRITE □ ESBRITE 2 □ ESBRITE 4 □ ESBRITE 4-62 □ ESBRITE 8 □ ESBRITE G 10 □ ESBRITE G-P 2 □ ESBRITE 500HM □ ESBRITE LBL □ ESCOREZ 7404 □ ESTYRENE 4-62 □ ESTYRENE G 15 □ ESTYRENE G 20 □ ESTYRENE G-P 4 □ ESTYRENE H 61 □ ESTYRENE 500SH □ ETHENYLBENZENE HOMOPOLYMER □ FC-MY 5450 □ FG 834 □ FOSTER GRANT 834 □ GEDEX □ 454H □ HF 10 □ HF 55 □ HF 77 □ HH 102 □ HHI 11 □ HI-STYROL □ HOSTYREN N □ HOSTYREN N 4000 □ HOSTYREN N 7001 □ HOSTYREN N 4000V □ HOSTYREN S □ HT 88 □ HT 88A □ HT 91-1 □ HT-F 76 □ IT 40 □ K 525 □ KB (POLYMER) □ KM □ KM (POLYMER) □ KOPLEN 2 □ KR 2537 □ KRASTEN 1.4 □ KRASTEN 052 □ KRASTEN SB □ LACQREN 506 □ LACQREN 550 □ LS 061A □ LS 1028E □ LUSTREX □

LUSTREX H 77 □ LUSTREX HH 101 □ LUSTREX HH 101 □ LUSTREX HP 77 □ LUSTREX HT 88 □ MX 4500 □ MX 5514 □ MX 5516 □ MX 5517-02 □ 168N15 □ NaPSt □ NBS 706 □ N 4000V □ OWISPOL GF □ PELASPAN 333 □ PELASPAN ESP 109s □ PICCOLASTIC □ PICCOLASTIC A □ PICCOLASTIC A 5 □ PICCOLASTIC A 25 □ PIC-COLASTIC A 50 □ PICCOLASTIC A 75 □ PICCOLASTIC C 125 □ PICCOLASTIC D □ PICCOLASTIC D-100 □ PIC-COLASTIC D 125 □ PICCOLASTIC D 150 □ PICCOLAS-TIC E 75 □ PICCOLASTIC E 100 □ PICCOLASTIC E 200 □ POLIGOSTYRENE □ POLYCO 220NS □ POLYFLEX □ POLYSTROL D □ POLYSTYRENE □ POLYSTYRENE BW □ POLYSTYRENE LATEX □ POLYSTYROL □ PRIN-TEL'S □ PRX 1195 □ PS 1 □ PS 2 □ PS 200 □ PS 209 □ PS-B □ PSB-C □ PSB-S □ PSB-S 40 □ PSB-S-E □ PS 454H □ PS 2 (POLYMER) □ PS 5 (POLYMER) □ PSV-L □ PSV-L 1 □ PSV-L 2 □ PSV-L 1S □ PY2763 □ R 3 □ R 3612 □ REXOLITE 1422 □ RHODOLNE □ S 173 □ SB 475K □ SD 188 □ SHELL 300 □ SMD 3500 □ SPS 600 □ SRM 705 □ SRM 706 □ ST 90 □ STERNITE 30 □ STERNITE ST 30VL □ ST 30UL □ STYRAFOIL □ STYRAGEL □ STYRENE POLYMERS □ STYREX C □ STYROCELL PM □ STYROFAN 2D □ STYROFLEX □ STYROFOAM □ STYROLUX □ STYRON □ STYRON 475 □ STYRON 492 □ STYRON 666 □ STYRON 678 □ STYRON 679 □ STYRON 683 □ STYRON 685 □ STRYON 686 □ STYRON 690 □ STYRON 69021 □ STYRON 440A □ STYRON 470A □ STYRON 475D □ STYRON GP □ STYRON 666K27 □ STYRON PS 3 □ STYRON T 679 □ STYRON 666U □ STYRON 666V □ STYROPIAN □ STYROPIAN FH 105 □ STYROPOL HT 500 □ STYROPOL IBE □ STYROPOL JQ 300 □ STYROPOL KA □ STYROPOR □ TC 3-30 □ TGD 5161 □ TMDE 6500 □ TOPOREX 500 □ TOPOREX 830 □ TOPOREX 550-02 □ TOPOREX 850-51 □ TOPOREX 855-51 □ TROLITUL □ TRYCITE 1000 □ 825TV □ 825TV-PS □ 475U □ U625 □ 666U □ UBATOL U 2001 □ UCC 6863 □ UP 1 □ UP 2 □ UP 27 □ UPM □ UPM703 □ UPM508L □ VESTOLEN P 5232G □ VESTYRON □ VESTYRON 512 □ VESTYRON 114-12 □ VESTYRON MB □ VESTYRON N □ VINAMUL N 710 □ VINAMUL N 7700 □ VINYL-BENZENE POLYMER □ VINYL PRODUCTS R 3612 □ X 600

CONSENSUS REPORTS: IARC Cancer Review: Group 3 IMEMDT 7,56,87; Animal Limited Evidence IMEMDT 19,231,79. Reported in EPA TSCA Inventory.

SAFETY PROFILE: Questionable carcinogen with experimental tumorigenic data by implant. When heated to decomposition it emits acrid smoke and irritating fumes. See also POLYMERS, INSOLUBLE.

SMR000 CAS: 9003-55-8 HR: 2
STYRENE POLYMER with 1,3-
BUTADIENE
SYNS: AFCOLAC B 101 ☐ ANDREZ ☐ BASE 661 ☐ 1,3-BUTADIENE-STYRENE COPOLYMER ☐ BUTADIENE-STYRENE POLYMER ☐ 1,3-BUTADIENE-STYRENE POLYMER ☐ BUTADIENE-STYRENE RESIN ☐ BUTADIENE-STYRENE RUBBER (FCC) ☐ BUTAKON 85-71 ☐ DIAREX 600 ☐ DIENOL S ☐ DOW 209 ☐ DOW LATEX 612 ☐ DST 50 ☐ DURANIT ☐ EDISTIR RB 268 ☐ ETHENYLBENZENE POLYMER with 1,3-BUTADIENE ☐ GOODRITE 1800X73 ☐ HISTYRENE S 6F ☐ HYCAR LX 407 ☐ K 55E ☐ KOPOLYMER BUTADIEN STYRENOVY (CZECH) ☐ KRO 1 ☐ LITEX CA ☐ LYTRON 5202 ☐ MARBON 9200 ☐ NIPOL 407 ☐ PHAROS 100.1 ☐ PLIOFLEX ☐ PLIOLITE S5 ☐ POLYBUTADIENE-POLYSTYRENE COPOLYMER ☐ POLYCO 2410 ☐ RICON 100 ☐ SBS ☐ SD 354 ☐ S6F HISTYRENE RESIN ☐ SKS 85 ☐ SOIL STABILIZER 661 ☐ SOLPRENE 300 ☐ STYRENE-BUTADIENE COPOLYMER ☐ STYRENE-1,3-BUTADIENE COPOLYMER ☐ STYRENE-BUTADIENE POLYMER ☐ SYNPOL 1500 ☐ THERMOPLASTIC 125 ☐ TR 201 ☐ UP 1E ☐ VESTYRON HI
CONSENSUS REPORTS: IARC Cancer Review: Group 3 IMEMDT 7,56,87; Human Inadequate Evidence IMEMDT 19,231,79. Reported in EPA TSCA Inventory.
SAFETY PROFILE: An eye irritant. Questionable carcinogen. When heated to decomposition it emits acrid smoke and irritating fumes. See also POLYMERS, INSOLUBLE.

SMS500 CAS: 122-57-6 HR: 3
STYRYL METHYL KETONE
mf: $C_{10}H_{10}O$ mw: 146.20
PROP: Plates. D: 1.035 @ 20°/2°, mp: 40–42°, bp: 260–262°. Sol in H_2SO_4, alc, chloroform, ether, benzene
SYNS: BENZALACETON (GERMAN) ☐ BENZAL-ACETONE ☐ BENZYLIDENE ACETONE ☐ 4-PHENYL-3-BUTEN-2-ONE
CONSENSUS REPORTS: Reported in EPA TSCA Inventory.
DOT CLASSIFICATION: 3; Label: Flammable Liquid
SAFETY PROFILE: Poison by intravenous route. Moderately toxic by intraperitoneal route. A skin irritant. A flammable liquid. When heated to decomposition it emits acrid smoke and irritating fumes. See also KETONES.

SMU100 HR: D
STYRYLPYRIDINIUM CHLORIDE,
DIETHYLCARBAMAZINE
SAFETY PROFILE: When heated to decomposition it emits acrid smoke and irritating fumes.

SMY000 CAS: 110-15-6 HR: 2
SUCCINIC ACID
mf: $C_4H_6O_4$ mw: 118.10
PROP: Colorless or white crystals or prisms; odorless with acid taste. Mp: 185°, bp: 235° (decomp), d: 1.564 @ 15°/4°. Sol in water; very sol in alc, ether, acetone, glycerin.
SYNS: AMBER ACID ☐ ASUCCIN ☐ BERNSTEIN-SAEURE (GERMAN) ☐ BUTANEDIOIC ACID ☐ 1,2-ETHANEDICARBOXYLIC ACID ☐ ETHYLENESUC-CINIC ACID ☐ KYSELINA JANTAROVA ☐ WORM-WOOD ☐ WORMWOOD ACID
CONSENSUS REPORTS: Reported in EPA TSCA Inventory.
SAFETY PROFILE: Moderately toxic by subcutaneous route. A severe eye irritant. Mutation data reported. When heated to decomposition it emits acrid smoke and irritating fumes.

SNB000 CAS: 123-25-1 HR: 2
SUCCINIC ACID, DIETHYL ESTER
mf: $C_8H_{14}O_4$ mw: 174.22
PROP: Colorless, mobile liquid; pleasant odor. Flash p: 230°F, mp: −21°, bp: 217.7°. Sol in alc, ether, fixed oils, water.
SYNS: BUTANEDIOIC ACID, DIETHYL ESTER ☐ DIETHYL SUCCINATE (FCC) ☐ ETHYL SUCCINATE ☐ FEMA No. 2377
CONSENSUS REPORTS: Reported in EPA TSCA Inventory.
SAFETY PROFILE: Mildly toxic by ingestion. A skin and eye irritant. Combustible liquid. Reaction with ethyl trifluoroacetate + sodium hydride may cause a fire or explosion. When heated to decomposition it emits acrid smoke and irritating fumes. See also ESTERS.

S

SNC000　　CAS: 108-30-5　　HR: 2
SUCCINIC ANHYDRIDE
mf: $C_4H_4O_3$　　mw: 100.08
PROP: Colorless needles or crystals from
$CHCl_3$. Mp: 119.6°, bp: 261°, d: 1.104, vap
press: 1 mm @ 92.0°. Very sltly sol in water,
pet ether; sltly sol in ether.
SYNS: BERNSTEINSAEURE-ANHYDRID (GERMAN) □
BUTANEDIOIC ANHYDRIDE □ DIHYDRO-2,5-
FURANDIONE □ 2,5-DIKETOTETRAHYDROFURAN □
NCI-C55696 □ SUCCINIC ACID ANHYDRIDE □
SUCCINYL OXIDE □ TETRAHYDRO-2,5-DIOXOFURAN
CONSENSUS REPORTS: IARC Cancer
Review: Group 3 IMEMDT 7,56,87; Animal
Inadequate Evidence IMEMDT 15,265,77.
Reported in EPA TSCA Inventory. EPA
Genetic Toxicology Program.
SAFETY PROFILE: Experimental
teratogenic effects. Moderately toxic by
ingestion. A severe eye irritant. Mutation
data reported. Questionable carcinogen with
experimental neoplastigenic data. When
heated to decomposition it emits acrid
smoke and irritating fumes. See also
ANHYDRIDES.

SNC500　　CAS: 123-23-9　　HR: 3
SUCCINIC PEROXIDE
mf: $C_8H_{10}O_8$　　mw: 234.18
PROP: Fine white powder; odorless with
tart taste. Mp: 133° (decomp). Mod sol in
water.
SYNS: BIS(3-CARBOXYPROPIONYL) PEROXIDE □
3,3'-(DIOXYDICARBONYL)DIPROPIONIC ACID □
SUCCINIC ACID PEROXIDE (DOT) □ SUCCINYL
PEROXIDE
CONSENSUS REPORTS: Reported in EPA
TSCA Inventory.
SAFETY PROFILE: An irritant. Explodes
on contact with flame. When heated to
decomposition it emits acrid smoke and
irritating fumes. See also PEROXIDES,
ORGANIC.

SNE000　　CAS: 110-61-2　　HR: 3
SUCCINONITRILE
mf: $C_4H_4N_2$　　mw: 80.10
PROP: Colorless, odorless, waxy material.
Mp: 58.1°, bp: 267°, flash p: 270°F, d: 1.022
@ 25°, vap press: 2 mm @ 100°, vap d: 2.1.
Sltly sol in ether, water, alc; sol in acetone.
SYNS: 1,4-BUTANEDINITRILE □ DEPRELIN □ s-
DICYANOETHANE □ 1,2-DICYANOETHANE □ DINILE
□ ETHYLENE CYANIDE □ ETHYLENE DICYANIDE □
SUCCINIC ACID DINITRILE □ SUCCINIC DINITRILE □
SUCCINODINITRILE □ SUXIL □ USAF A-9442
CONSENSUS REPORTS: Cyanide and its
compounds are on the Community Right-
To-Know List. Reported in EPA TSCA
Inventory.
ACGIH TLV: CL 5 mg(CN)/m³ (skin)
NIOSH REL: (Nitriles) TWA 20 mg/m³
SAFETY PROFILE: Poison by ingestion,
intraperitoneal, and subcutaneous routes. An
experimental teratogen. Combustible when
exposed to heat or flame. Decomposes
exothermically above 195°C. Can react with
oxidizing materials. To fight fire, use alcohol
foam, CO_2, dry chemical. When heated to
decomposition, or on contact with acid or
acid fumes, it emits highly toxic fumes of
NO_x and CN^-. See also NITRILES.

SNG600　　　　　　　　　　　HR: D
SUCCISTEARIN
SYN: STEAROYL PROPLENE GLYCOL HYDROGEN
SUCCINATE
SAFETY PROFILE: When heated to
decomposition it emits acrid smoke and
irritating fumes.

SNH000　　CAS: 57-50-1　　HR: 1
SUCROSE
mf: $C_{12}H_{22}O_{11}$　　mw: 342.34
PROP: White crystals from water or alc;
sweet taste. D: 1.587 @ 25°/4°, mp:
185–187° (decomp). Sol in water, alc; insol
in ether.
SYNS: BEET SUGAR □ CANE SUGAR □ CON-
FECTIONER'S SUGAR □ α-d-GLUCOPYRANOSYL β-d-
FRUCTOFURANOSIDE □ (α-d-GLUCOSIDO)-β-d-
FRUCTOFURANOSIDE □ GRANULATED SUGAR □
NCI-C56597 □ ROCK CANDY □ SACCHAROSE □
SACCHARUM □ SUGAR
CONSENSUS REPORTS: Reported in EPA
TSCA Inventory. EPA Genetic Toxicology
Program.
OSHA PEL: TWA Total Dust: 15 mg/m³;
Respirable Fraction: 5 mg/m³

ACGIH TLV: TWA (nuisance particulate) 10 mg/m³ of total dust (when toxic impurities are not present, e.g., quartz <1%); Not Classifiable as a Human Carcinogen
NIOSH REL: (Sucrose, Respirable Fraction) TWA 5 mg/m³; (Sucrose, total dust): TWA 10 mg/m³
SAFETY PROFILE: Mildly toxic by ingestion. An experimental teratogen. Mutation data reported. Vigorous reaction with nitric acid or sulfuric acid (forms carbon monoxide and carbon dioxide). When heated to decomposition it emits acrid smoke and irritating fumes.

SNH875   CAS: 116-45-0   HR: D
SULFABROMOMETHAZINE SODIUM
mf: $C_{12}H_{13}BrN_4O_2S$ mw: 357.22
PROP: Crystals. Decomp 250–252°. Sol in alkaline solns.
SYNS: 4-AMINO-N-(5-BROMO-4,6-DIMETHYL-2-PYRIMIDINYL)BENZENESILFANILAMIDE □ N¹-(5-BROMO-4,6-DIMETHYL-2-PYRIMIDINYL)BENZENE-SULFONAMIDE □ 5-BROMOSULFAMETHAZINE
SAFETY PROFILE: When heated to decomposition it emits acrid smoke and irritating fumes.

SNJ000   CAS: 57-68-1   HR: 2
SULFADIMETHYLDIAZINE
mf: $C_{12}H_{14}N_4O_2S$ mw: 278.36
PROP: Creamy-white powder or crystals from dioxan (aq); odorless. Decomp on exposure to light. Mp: 176° (also a range reported of 178–179°, 198–199°, and 205–207°). Sol in acetone, water, ether; sltly sol in alc.
SYNS: A-502 □ 2-(p-AMINOBENZENESULFON-AMIDO)-4,6-DIMETHYLPYRIMIDINE □ 6-(4'-AMINO-BENZOL-SULFONAMIDO)-2,4-DIMETHYLPYRIMIDIN (GERMAN) □ (p-AMINOBENZOLSULFONYL)-2-AMINO-4,6-DIMETHYLPYRIMIDIN (GERMAN) □ AZOLMETAZ-IN □ CREMOMETHAZINE □ DIAZYL □ N¹-(4,6-DIMETHYL-2-PYRIMIDINYL)SULFANILAMIDE □ N-(4,6-DIMETHYL-2-PYRIMIDYL)SULFANILAMIDE □ 4,6-DIMETHYL-2-SULFANILAMIDOPYRIMIDINE □ DIMEZATHINE □ MERMETH □ METAZIN □ NCI-C56600 □ NEASINA □ PIRMAZIN □ PRIMAZIN □ SA 111 □ SEAZINA □ SPANBOLET □ SULFADIMERAZINE □ SULFADIMETHYLPYRIMIDINE □ SULFADIMETINE □ SULFADIMEZINE □ SULFADIMIDINE □ SULFADINE □ SULFADSIMESINE □ SULFA-ISODIMERAZINE □

SULFAISODIMIDINE □ SULFAMETHIAZINE □ SULFAMETHIN □ SULFAMEZATHINE □ 2-SULFANILAMIDO-4,6-DIMETHYLPYRIMIDINE □ SULFISOMIDIN □ SULFISOMIDINE □ SULFODIMESIN □ SULFODIMEZINE □ SULMET □ SULPHADIMETHYL-PYRIMIDINE □ SULPHADIMIDINE □ SUPERSEPTIL □ VERTOLAN
CONSENSUS REPORTS: Reported in EPA TSCA Inventory.
SAFETY PROFILE: Moderately toxic by intravenous and intraperitoneal routes. Experimental teratogenic and reproductive effects. Questionable carcinogen with experimental tumorigenic data. When heated to decomposition it emits very toxic fumes of SO_x and NO_x.

SNJ100   CAS: 963-14-4   HR: D
SULFAETHOXYPYRIDAZINE
mf: $C_{12}H_{14}N_4O_3S$ mw: 294.34
PROP: Crystals. Mp: 183–184°. Sol in water at 37°.
SYNS: 4-AMINO-N-(6-ETHOXY-3-PYRIDAZINYL)-BENZENESULFONAMIDE □ N¹-(6-ETHOXY-3-PYRIDAZINYL)SULFANILAMIDE □ 6-ETHOXY-3-SULFANILAMIDOPYRIDAZINE
SAFETY PROFILE: When heated to decomposition it emits acrid smoke and irritating fumes.

SNK000   CAS: 723-46-6   HR: 3
SULFAMETHOXAZOL
mf: $C_{10}H_{11}N_3O_3S$ mw: 253.30
PROP: Yellowish-white powder. Mp: 169–173°. Insol in Et_2O.
SYNS: 4-AMINO-N-(5-METHYL-3-ISOXAZOLYL)-BENZENESULFONAMIDE □ 3-(p-AMINOPHENYL-SULPHONAMIDO)-5-METHYLISOXAZOLE □ AZO-GANTANOL □ BACTRIM □ CO-TRIMOXAZOLE □ EUSAPRIM □ FECTRIM □ GANTANOL □ N'-(5-METHYL-3-ISOXAZOLE)SULFANILAMIDE □ N'-(5-METHYL-3-ISOXAZOLYL)SULFANILAMIDE □ N'-(5-METHYLISOXAZOL-3-YL)SULPHANILAMIDE □ N¹-(5-METHYL-3-ISOXAZOLYL)SULPHANILAMIDE □ 5-METHYL-3-SULFANILAMIDOISOXAZOLE □ 5-METHYL-3-SULPHANIL-AMIDOISOXAZOLE □ METOXAL □ MS 53 □ RADONIL □ RO 4-2130 □ SINOMIN □ SEPTRA □ SEPTRAN □ SIM □ SULFAMETHALAZOLE □ SULFAMETHOXAZOLE □ SULFAMETHYLISOXAZOLE □ 3-SULFANILAMIDO-5-METHYLISOXAZOLE □ SULFISOMEZOLE □ SULPHAMETHALAZOLE □ SULPHAMETHOXAZOL □ SULPHAMETHOXAZOLE □ SULPHAMETHYL-

S

ISOXAZOLE □ 3-SULPHANILAMIDO-5-METHYL-ISOXAZOLE □ SULPHISOMEZOLE □ TRIB □ TRIMETOPRIM-SULFA

CONSENSUS REPORTS: IARC Cancer Review: Group 3 IMEMDT 7,348,87; Human Inadequate Evidence IMEMDT 24,285,80; Animal Limited Evidence IMEMDT 24,285,80. Reported in EPA TSCA Inventory.

SAFETY PROFILE: Moderately toxic by ingestion and intraperitoneal routes. Questionable carcinogen with experimental tumorigenic data. When heated to decomposition it emits very toxic fumes of NO_x and SO_x.

SNK500   CAS: 5329-14-6   HR: 3
SULFAMIC ACID
DOT: UN 2967

mf: H_3NO_3S mw: 97.10

PROP: White crystals; nonhygroscopic solid. Mp: 200° (decomp), bp: decomp, d: 203 @ 12°. Very sol in H_2O, liq NH_3, formamide.

SYNS: AMIDOSULFONIC ACID □ AMIDOSULFURIC ACID □ AMINOSULFONIC ACID □ KYSELINA AMIDOSULFONOVA (CZECH) □ KYSELINA SULFAMINOVA (CZECH) □ SULFAMIDIC ACID □ SULPHAMIC ACID (DOT)

CONSENSUS REPORTS: Reported in EPA TSCA Inventory.

DOT CLASSIFICATION: 8; Label: Corrosive

SAFETY PROFILE: Poison by intraperitoneal route. Moderately toxic by ingestion. A human skin irritant. A corrosive irritant to skin, eyes, and mucous membranes. A substance that migrates to food from packaging materials. Violent or explosive reactions with chlorine, metal nitrates + heat, metal nitrites + heat, fuming HNO_3. When heated to decomposition it emits very toxic fumes of SO_x and NO_x. See also SULFONATES.

SNM500   CAS: 63-74-1   HR: 3
SULFANILAMIDE
mf: $C_6H_8N_2O_2S$ mw: 172.22

PROP: Crystals or leaflets from alc (aq). Mp: 164.5–166.5°. Sol in glycerin, propylene glycol, HCl; almost insol in chloroform, ether, benzene, pet ether.

SYNS: ALBEXAN □ ALBOSAL □ AMBESIDE □ p-AMINOBENZENESULFAMIDE □ p-AMINO-BENZENESULFONAMIDE □ 4-AMINOBENZENE-SULFONAMIDE □ p-AMINOPHENYLSULFONAMIDE □ 4-AMINOPHENYLSULFONAMIDE □ p-ANILINE-SULFONAMIDE □ ANILINE-p-SULFONIC AMIDE □ ANTISTREPT □ BACTERAMID □ COLLOMIDE □ COLSULANYDE □ COPTICIDE □ DIPRON □ ESTREPTOCIDA □ F 1162 □ FOURNEAU 1162 □ GERISON □ GOMBARDOL □ LUSIL □ LYSOCOCCINE □ NEOCOCCYL □ ORGASEPTINE □ PABS □ PRONTALBIN □ PRONTOSIL I □ PROSEPTINE □ PROSEPTOL □ PYSOCOCCINE □ RUBIAZOL A □ SEPTAMIDE ALBUM □ SEPTINAL □ SEPTOPLEX □ STOPTON ALBUM □ STREPAMIDE □ STREPTAGOL □ STREPTOCLASE □ STREPTOL □ STREPTOSIL □ STREPTOZONE □ STREPTROCIDE □ p-SULFA-MIDOANILINE □ SULFAMIDYL □ SULFANA □ SULFANALONE □ SULFANIL □ SULFOCIDINE □ SULFONAMIDE □ SULFONAMIDE P □ SULPHAN-ILAMIDE □ THERAPOL □ WHITE STREPTOCIDE

CONSENSUS REPORTS: Reported in EPA TSCA Inventory. EPA Genetic Toxicology Program.

SAFETY PROFILE: Poison by intraperitoneal route. Moderately toxic by ingestion, subcutaneous, and intravenous routes. Human teratogenic effects by unspecified route: developmental abnormalities of the blood and lymphatic systems (including the spleen and bone marrow). Experimental reproductive effects. Questionable carcinogen with experimental carcinogenic data. Mutation data reported. Implicated in aplastic anemia. When heated to decomposition it emits very toxic fumes of NO_x and SO_x.

SNN300   CAS: 122-11-2   HR: 2
6-SULFANILAMIDO-2,4-DIMETHOXY-PYRIMIDINE
mf: $C_{12}H_{14}N_4O_4S$ mw: 310.36

PROP: Crystals from dil alc. Mp: 201–203°. Sol in dil HCl and in aq solns of sodium carbonate. Solubility in water at 37° (mg/100 mL): 4.6 at pH 4.10; 29.5 at pH 6.7; 58.0 at pH 7.06.

SYNS: ABCID □ AGRIBON □ ALBON □ 4-AMINO-N-(2,6-DIMETHOXY-4-PYRIMIDINYL)BENZENE-SULFONAMIDE □ ARNOSULFAN □ BACTROVET □ DEPOSUL □ DIASULFA □ DIASULFYL □ DIMETAZINA □ 2,6-DIMETHOXY-4-(p-AMINOBENZENESULFO-NAMIDO)PYRIMIDINE □ N^1-(2,6-DIMETHOXY-4-

PYRIMIDINYL)SULFANILAMIDE □ DIMETHOXY-SULFADIAZINE □ 2,4-DIMETHOXY-6-SULFANIL-AMIDO-1,3-DIAZINE □ 2,6-DIMETHOXY-4-SUL-FANILAMIDOPYRIMIDINE □ DINOSOL □ DORISUL □ FUXAL □ MADRIBON □ MADRIGID □ MADRIQID □ MADROXIN □ MADROXINE □ MAXULVET □ MEMCOZINE □ METOXIDON □ NEOSTREPAL □ OMNIBON □ PERSULFEN □ RADONIN □ REDIFAL □ ROSCOSULF □ SCANDISIL □ SDM □ SDMO □ SUDINE □ SULDIXINE □ SULFADIMETHOXIN □ SULFADI-METHOXINE □ SULFADIMETHOXYDIAZINE □ SULFADIMETOSSINA (ITALIAN) □ SULFADIMETOXIN □ SULFASOL □ SULFASTOP □ SULFOPLAN □ SULPHADIMETHOXINE □ SULXIN □ SYMBIO □ THERACANZAN

SAFETY PROFILE: Moderately toxic by intraperitoneal, intravenous, and subcutaneous routes. An experimental teratogen. Experimental reproductive effects. When heated to decomposition it emits toxic fumes of SO_x and NO_x.

SNN500 CAS: 127-69-5 HR: 2
5-SULFANILAMIDO-3,4-DIMETHYL-ISOXAZOLE

mf: $C_{11}H_{13}N_3O_3S$ mw: 267.33
PROP: Yellowish-white prisms. Mp: 195–198°. Spar sol in H_2O.
SYNS: ACCUZOLE □ AMIDOXAL □ 5-(p-AMINO-BENZENESULFONAMIDO)-3,4-DIMETHYLISOOXALE □ 5-(p-AMINOBENZENESULFONAMIDO)-3,4-DIM-ETHYLISOXAZOLE □ 5-(p-AMINOBENZENESUL-PHONAMIDE)-3,4-DIMETHYLISOXAZOLE □ 5-(p-AMINOBENZENESULPHONAMIDO)-3,4-DIMETHYL-ISOXAZOLE □ 4-AMINO-N-(3,4-DIMETHYL-5-ISOXAZOLYL)BENZENESULPHONAMIDE □ 5-(4-AMINOPHENYLSULFONAMIDO)-3,4-DIMETHYL-ISOXAZOLE □ 5-(4-AMINOPHENYLSULPHONAMIDO)-3,4-DIMETHYLISOXAZOLE □ ASTRAZOLO □ AZO GANTRISIN □ AZOSULFIZIN □ BACTESULF □ BARAZAE □ CHEMOUAG □ 3,4-DIMETHYLISOXALE-5-SULFANILAMIDE □ 3,4-DIMETHYLISOXALE-5-SULPHANILAMIDE □ N¹-(3,4-DIMETHYL-5-ISOXAZOL-YL)SULFANILAMIDE □ N'-(3,4)DIMETHYLISOXAZOL-5-YL-SULPHANILAMIDE □ N¹-(3,4-DIMETHYL-5-ISOXAZOLYL)SULPHANILAMIDE □ 3,4-DIMETHYL-5-SULFANILAMIDOISOXAZOLE □ 3,4-DIMETHYL-5-SULPHANILAMIDOISOXAZOLE □ 3,4-DIMETHYL-5-SULPHONAMIDOISOXAZOLE □ DORSULFAN □ ENTUSIL □ GANTRISINE □ ISOXAMIN □ J-SUL □ KORO-SULF □ NEOXAZOL □ NORILGAN-S □ NOVOSAXAZOLE □ PANCID □ RENOSULFAN □ RESOXOL □ ROXOSUL TABLETS □ SAXOSOZINE □ SODIZOLE □ SOXISOL □ STANSIN □ SULFADI-METHYLISOXAZOLE □ SULFAFURAZOL □ SULFAGAN

□ SULFASOXAZOLE □ SULFAZOLE □ SULFISIN □ SULFISOXAZOLE □ SULFIZOLE □ SULPHADIMETHYL-ISOXAZOLE □ SULPHAFURAZ □ 5-SULPHANILAMIDO-3,4-DIMETHYL-ISOXAZOLE □ SULPHISOXAZOL □ SULPHOFURAZOLE □ THIASIN □ UNISULF □ URISOXIN □ URITRISIN □ UROGAN □ VAGILIA

CONSENSUS REPORTS: IARC Cancer Review: Group 3 IMEMDT 7,347,87; Human Inadequate Evidence IMEMDT 24,275,80; Animal Inadequate Evidence IMEMDT 24,275,80. NCI Carcinogenesis Bioassay (gavage); No Evidence: mouse, rat NCITR* NCI-CG-TR-138,79. Reported in EPA TSCA Inventory.

SAFETY PROFILE: Mildly toxic by ingestion. An experimental teratogen. Questionable carcinogen. Mutation data reported. When heated to decomposition it emits very toxic fumes of SO_x and NO_x.

SNQ710 CAS: 144-80-9 HR: 1
N-SULFANYLACETAMIDE

mf: $C_8H_{10}N_2O_3S$ mw: 214.26
PROP: A solid. Mp: 181–184°.
SYNS: ACETOCID □ ACETOSULFAMIN □ N-ACETYL-4-AMINOBENZENESULFONAMIDE □ N¹-ACETYL-4-AMINOPHENYLSULFONAMIDE □ N'-ACETYL-SULFANILAMIDE □ N¹-ACETYLSULFANILAMIDE □ N-ACETYLSULFANILAMINE □ ALBAMINE □ ALBUCID □ ALESTEN □ p-AMINOBENZENESULFONACETAMIDE □ N-((4-AMINOPHENYL)SULFONYL)ACETAMIDE □ BLEPH-10 □ FORMOSULFACETAMIDE □ ISOPTO CETAMIDE □ OP-SULFA 30 □ OPTHEL-S □ REGION □ SEBIZON □ STERAMIDE □ SULAMYD □ SULF-10 □ SULFACET □ SULFACETAMIDE □ SULFACETIMIDE □ SULFACYL □ SULFANILACETAMIDE □ SULPHACET-AMIDE □ SULPHASIL □ UROSULFON □ UROSULFONE

CONSENSUS REPORTS: Reported in EPA TSCA Inventory. EPA Genetic Toxicology Program.

SAFETY PROFILE: Mildly toxic by ingestion and intravenous routes. An experimental teratogen. Mutation data reported. When heated to decomposition it emits very toxic fumes of NO_x and SO_x.

SNS000 HR: D
SULFATES
SAFETY PROFILE: Variable toxicity. In general the toxic properties of substances containing the sulfate radical are that of the

material (cation) with which the sulfate (anion) is combined. See specific compound. Violent reaction with Al, Mg. When heated to decomposition they emit toxic fumes of SO_x.

SNS100   CAS: 73825-85-1   HR: 3
SULFATOBIS(DIMETHYLSELENIDE)-PLATINUM(II) HYDRATE
PROP: IDLH 4 mg/m³ (as Pt).
SYNS: PLATINUM (II), BIS(METHYL SELENIDE)SULFATO-, HYDRATE □ NSC-281278
OSHA PEL: TWA 0.2 mg(Se)/m³
ACGIH TLV: TWA 0.2 mg(Se)/m³
SAFETY PROFILE: Poison by intravenous route. When heated to decomposition it emits toxic fumes of SO_x and Se.

SNT000   HR: 3
SULFIDES
SAFETY PROFILE: Variable toxicity. The alkaline sulfides (potassium, calcium, ammonium, and sodium) are similar in action to alkalies. They cause softening and irritation of the skin. If ingested, they are corrosive and irritating through the liberation of hydrogen sulfide and free alkali. Hydrogen sulfide is especially toxic. Sulfides of the heavy metals are generally insoluble and hence have little toxic action except through the liberation of hydrogen sulfide. Sulfides are used as fungicides. Flammable when exposed to flame or by spontaneous chemical reaction. Many sulfides ignite easily in air at room temperature. Others require a higher temperature or the presence of an oxidizer. Upon contact with moisture or acids, hydrogen sulfide is evolved. Many sulfides react violently and explosively on contact with powerful oxidizers. Evolved hydrogen sulfide can form explosive mixtures with air. Sulfides react with water, steam, or acids to produce toxic and flammable vapors of hydrogen sulfide. When heated to decomposition they emit highly toxic fumes of SO_x. See also HYDROGEN SULFIDE.

SNT100   CAS: 23248-53-5   HR: 3
1,1'-SULFINYLBIS(1,2-DICHLORO-ETHANE)
mf: $C_4H_2Cl_4OS$ mw: 239.92
SYNS: BIS(2,2-DICHLOROVINYL) SULFOXIDE □ CHEMAGRO 4965 □ ETHENE, 1,1'-SULFINYLBIS(2,2-DICHLORO- □ ENT 27,221 □ ETHANE, 1,1'-SULFINYL-BIS(1,2-DICHLORO- □ SULFOXIDE, BIS(1,2-DICHLOROVINYL)
SAFETY PROFILE: A poison by ingestion. When heated to decomposition it emits toxic vapors of SO_x and Cl^-.

SNT500   HR: 2
SULFITES
SAFETY PROFILE: Fairly large doses of sulfites can be tolerated since they are rapidly oxidized to sulfates, although if swallowed they may cause irritation of the stomach by liberating sulfurous acid. Experimentally, large doses of sodium sulfite have been shown to cause retarded growth, nerve irritation, atrophy of bone marrow, depression, and paralysis. They will react with water, steam, or acids to produce a toxic and corrosive material. When heated to decomposition they emit highly toxic fumes of SO_x.

SNW600   HR: D
SULFOMYXIN
SYN: N-SULFOMETHYL-POLYMYXIN B SODIUM SALT
SAFETY PROFILE: When heated to decomposition it emits acrid smoke and irritating fumes.

SNY000   HR: 1
SULFONATES
SAFETY PROFILE: Variable toxicity. See specific compounds. Usually irritating. When heated to decomposition or on contact with acid or acid fumes, they emit highly toxic fumes of SO_x.

SOA500   CAS: 80-08-0   HR: 3
4,4'-SULFONYLDIANILINE
mf: $C_{12}H_{12}N_2O_2S$ mw: 248.32
PROP: Crystals from alc (aq). Mp: 176°, vap d: 8.3. Nearly insol in water; sol in acetone and alc. Sol in dil acids.

SYNS: AVLOSULPHONE ☐ BIS(p-AMINOPHENYL) SULFONE ☐ BIS(4-AMINOPHENYL) SULFONE ☐ BIS(p-AMINOPHENYL)SULPHONE ☐ BIS(4-AMINOPHENYL)SULPHONE ☐ CROYSULFONE ☐ DADPS ☐ DAPSONE ☐ DDS ☐ DIAMINODIFENILSULFONA (SPANISH) ☐ p,p'-DIAMINODIPHENYL SULFONE ☐ DIAMINO-4,4'-DIPHENYL SULFONE ☐ 4,4'-DIAMINODIPHENYL SULFONE ☐ p,p-DIAMINODIPHENYL SULPHONE ☐ DIAMINO-4,4'-DIPHENYL SULPHONE ☐ DI(p-AMINOPHENYL) SULFONE ☐ DI(4-AMINOPHENYL)SULFONE ☐ DI(p-AMINOPHENYL)SULPHONE ☐ DI(4-AMINOPHENYL)SULPHONE ☐ DIAPHENYLSULFONE ☐ DIAPHENYLSULPHON ☐ DIAPHENYLSULPHONE ☐ DIPHONE ☐ DISULONE ☐ DSS ☐ DUBRONAX ☐ DUMITONE ☐ EPORAL ☐ 1358F ☐ F 1358 ☐ MALOPRIM ☐ METABOLITE C ☐ NCI-C01718 ☐ NOVOPHONE ☐ NSC-6091 ☐ SULFONA ☐ 1,1'-SULFONYLBIS(4-AMINOBENZENE) ☐ 4,4'-SULFONYLBISANILINE ☐ p,p-SULFONYLBISBENZAMINE ☐ 4,4'-SULFONYL-BISBENZAMINE ☐ p,p-SULFONYLBISBENZENAMINE ☐ p,p'-SULFONYLDIANILINE ☐ SULPHADIONE ☐ SULPHON-MERE ☐ 1,1'-SULPHONYLBIS(4-AMINO-BENZENE) ☐ p,p-SULPHONYLBISBENZAMINE ☐ 4,4'-SULPHONYLBISBENZAMINE ☐ p,p-SULPHONYL-BISBENZENAMINE ☐ 4,4'-SULPHONYLBISBENZE-NAMINE ☐ SULPHONYLDIANILINE ☐ p,p-SUL-PHONYLDIANILINE ☐ TARIMYL ☐ UDOLAC ☐ WR 448

CONSENSUS REPORTS: IARC Cancer Review: Group 3 IMEMDT 7,185,87; Animal Limited Evidence IMEMDT 24,59,80; Human Inadequate Evidence IMEMDT 24,59,80. NCI Carcinogenesis Bioassay (feed); No Evidence: mouse NCITR* NCI-CG-TR-20,77; (feed); Clear Evidence: rat NCITR* NCI-CG-TR-20,77. Reported in EPA TSCA Inventory.

SAFETY PROFILE: Poison by ingestion, intraperitoneal, and subcutaneous routes. Human systemic effects by ingestion: agranulocytosis, change in tubules and other kidney changes, cyanosis, effect on joints, hemolysis with or without anemia, jaundice, methemoglobinemia-carboxyhemoglobinemia, retinal changes, somnolence. Experimental reproductive effects. Can cause hepatitis, dermatitis, and neuritis. Questionable carcinogen with experimental carcinogenic and neoplastigenic data. Human mutation data reported. Used in leprosy treatment and veterinary medicine. When heated to decomposition it emits very toxic fumes of NO_x and SO_x. See also SULFONATES.

SOD100   CAS: 3689-24-5   HR: 3
SULFOTEP
DOT: UN 1704
mf: $C_8H_{20}O_5P_2S_2$ mw: 322.34
PROP: A pale-yellow, mobile liquid. D: 1.19 @ 25°/4°, bp: 110–113° @ 0.2 mm. Almost insol in water. IDLH 10 mg/m³.
SYNS: ASP 47 ☐ BAY-E-393 ☐ BAYER-E 393 ☐ BIS-O,O-DIETHYLPHOSPHOROTHIONIC ANHYDRIDE ☐ BLADAFUM ☐ BLADAFUME ☐ BLADAFUN ☐ DITHIO ☐ DITHIODIPHOSPHORIC ACID, TETRAETHYL ESTER ☐ DITHIOFOS ☐ DITHION ☐ DITHIONE ☐ DITHIOPHOS ☐ DI(THIOPHOSPHORIC) ACID, TETRAETHYL ESTER ☐ DITHIOPYROPHOSPHATE de TETRAETHYLE (FRENCH) ☐ DITHIOTEP ☐ E393 ☐ ENT 16,273 ☐ ETHYL THIOPYROPHOSPHATE ☐ LETHALAIRE G-57 ☐ PIROFOS ☐ PLANT DITHIO AEROSOL ☐ PLANTFUME 103 SMOKE GENERATOR ☐ PYROPHOSPHORODITHIOIC ACID, TETRAETHYL ESTER ☐ PYROPHOSPHORODITHIOIC ACID, O,O,O,O-TETRAETHYL ESTER ☐ RCRA WASTE NUMBER P109 ☐ SULFATEP ☐ SULFOTEPP ☐ TEDP ☐ TEDP (OSHA) ☐ TEDTP ☐ O,O,O,O-TETRAETHYL-DITHIONOPYRO-PHOSPHAT (GERMAN) ☐ TETRAETHYLDITHIODI-FOSFAT ☐ TETRAETHYL DITHIONOPYROPHOSPH-ATE ☐ TETRAETHYL DITHIOPYROPHOSPHATE ☐ O,O,O,O-TETRAETHYL DITHIOPYROPHOSPHATE ☐ TETRAETHYL DITHIOPYROPHOSPHATE and GAS MIXTURES, LC50 < or =2000 ppm (DOT) ☐ TETRAETHYL DITHIO PYROPHOSPHATE, liquid (DOT) ☐ O,O,O,O-TETRAETIL-DITIO-PIROFOSFATO (ITALIAN) ☐ THIOTEPP
OSHA PEL: TWA 0.2 mg/m³ (skin)
ACGIH TLV: TWA 0.2 mg/m³ (skin); Not Classifiable as a Human Carcinogen
DFG MAK: 0.0075 ppm (0.1 mg/m³)
DOT CLASSIFICATION: 6.1; Label: Poison
SAFETY PROFILE: Poison by ingestion, skin contact, inhalation, intramuscular, intraperitoneal, subcutaneous, and intravenous routes. A cholinesterase inhibitor type of insecticide. When heated to decomposition it emits toxic fumes of PO_x and SO_x. See also PARATHION.

SOD500   CAS: 7704-34-9   HR: 3
SULFUR
DOT: UN 1350/UN 2448
af: S   aw: 32.06
PROP: Rhombic yellow crystals or yellow powder. Mp: 119°, bp: 444.6°, flash p: 405°F (CC), d: 2.07, d: (liquid) 1.803, autoign temp: 450°F, vap press: 1 mm @

S

183.8°. Insol in water; sltly sol in alc, ether; sol in carbon disulfide, benzene, toluene.

SYNS: BENSULFOID □ BRIMSTONE □ COLLOIDAL SULFUR □ COLLOKIT □ COLSUL □ COROSUL D AND S □ COSAN □ CRYSTEX □ FLOWERS of SULPHUR (DOT) □ GROUND VOCLE SULPHUR □ HEXASUL □ KOCIDE □ KOLOFOG □ KOLOSPRAY □ KUMULUS □ MAGNETIC 70, 90, and 95 □ MICROFLOTOX □ PRECIPITATED SULFUR □ SOFRIL □ SPERLOX-S □ SPERSUL □ SPERSUL THIOVIT □ SUBLIMED SULFUR □ SULFIDAL □ SULFORON □ SULFUR FLOWER (DOT) □ SULKOL □ SUPER COSAN □ SULPHUR (DOT) □ SULPHUR, lump or powder (DOT) □ SULPHUR, molten (DOT) □ SULSOL □ TECHNETIUM TC 99M SULFUR COLLOID □ TESULOID □ THIOLUX □ THIOVIT

CONSENSUS REPORTS: EPA Extremely Hazardous Substances List. Reported in EPA TSCA Inventory.

DOT CLASSIFICATION: 4.1; Label: Flammable Solid

SAFETY PROFILE: Poison by ingestion, intravenous, and intraperitoneal routes. A human eye irritant. A fungicide. Chronic inhalation can cause irritation of mucous membranes. Combustible when exposed to heat or flame or by chemical reaction with oxidizers. Explosive in the form of dust when exposed to flame. Can react violently with halogens, carbides, halogenates, halogenites, zinc, uranium, tin, sodium, lithium, nickel, palladium, phosphorus, potassium, indium, calcium, boron, aluminum, (aluminum + niobium pentoxide), ammonia, ammonium nitrate, ammonium perchlorate, BrF_5, BrF_3, (Ca + VO + H_2O), $Ca(OCl)_2$, Ca_3P_2, Cs_3N, charcoal, (Cu + chlorates), ClO_2, ClO, ClF_3, CrO_3, $Cr(OCl)_2$, hydrocarbons, IF_5, IO_5, PbO_2, $Hg(NO_3)_2$, HgO, Hg_2O, NO_2, P_2O_3, (KNO_3 + As_2S_3), K_3N, $KMnO_4$, $AgNO_3$, Ag_2O, NaH, ($NaNO_3$ + charcoal), (Na + SnI_4), SCl_2, Tl_2O_3, F_2. Can react with oxidizing materials. To fight fire, use water or special mixtures of dry chemical. When heated it burns and emits highly toxic fumes of SO_x. See also NUISANCE DUSTS.

SOG500   CAS: 10545-99-0   HR: 3
SULFUR DICHLORIDE
mf: Cl_2S mw: 102.96

PROP: Reddish-brown liquid; pungent odor. Decomp on heating to Cl_2 + S_2Cl_2. Hydrol by H_2O. Mp: −122°, bp: 59°, d: 1.621 @ 15°/15°, vap d: 3.55. Sol in hexane.

SYNS: CHLORIDE of SULFUR □ CHLORINE SULFIDE □ DICHLOROSULFANE □ MONOSULFUR DICHLORIDE □ SULFUR CHLORIDE □ SULFUR CHLORIDE (MONO)

CONSENSUS REPORTS: Reported in EPA TSCA Inventory.

SAFETY PROFILE: Poison irritant and corrosive to skin, eyes, and mucous membranes. Flammable when exposed to heat or flame. Reacts violently with Al, NH_3, K, Na, acetone, dimethyl sulfoxide, water, oxidants, metals, hexafluoroisopropylidene amino lithium, and toluene. Reactive with water or steam. When heated to decomposition it emits very toxic fumes of SO_x and Cl^-.

SOH500   CAS: 7446-09-5   HR: 3
SULFUR DIOXIDE
DOT: UN 1079
mf: O_2S mw: 64.06

PROP: Colorless, nonflammable gas or liquid under pressure; pungent odor. Catalytically oxidized by air to SO_3. Mp: −75.5°, bp: −10.0°, d: (liquid) 1.434 @ 0°, vap d: 2.264 @ 0°, vap press: 2538 mm @ 21.1°. Sol in water, decreases with temp. IDLH 100 ppm.

SYNS: BISULFITE □ FERMENICIDE LIQUID □ FERMENICIDE POWDER □ SCHWEFELDIOXYD (GERMAN) □ SIARKI DWUTLENEK (POLISH) □ SULFUROUS ACID ANHYDRIDE □ SULFUROUS ANHYDRIDE □ SULFUROUS OXIDE □ SULFUR OXIDE □ SULPHUR DIOXIDE, LIQUEFIED (DOT)

CONSENSUS REPORTS: EPA Extremely Hazardous Substances List. Reported in EPA TSCA Inventory. EPA Genetic Toxicology Program.

OSHA PEL: TWA 2 ppm; STEL 5 ppm
ACGIH TLV: TWA 2 ppm; STEL 5 ppm; Not Classifiable as a Human Carcinogen; (Proposed: CL 0.25 ppm; Not Classifiable as a Human Carcinogen)
DFG MAK: 0.5 pm ($1.3mg/m^3$)
NIOSH REL: (Sulfur Dioxide) TWA 0.5 ppm

DOT CLASSIFICATION: 2.3; Label: Poison Gas

SAFETY PROFILE: A poison gas. Experimental reproductive effects. Human mutation data reported. Human systemic effects by inhalation: pulmonary vascular resistance, respiratory depression, and other pulmonary changes. Questionable carcinogen with experimental tumorigenic and teratogenic data. It chiefly affects the upper respiratory tract and the bronchi. It may cause edema of the lungs or glottis, and can produce respiratory paralysis. A corrosive irritant to eyes, skin, and mucous membranes. This material is so irritating that it provides its own warning of toxic concentration. Levels of 400–500 ppm are immediately dangerous to life. Its toxicity is comparable to that of hydrogen chloride. However, less than fatal concentration can be borne for fair periods of time with no apparent permanent damage. It is a common air contaminant.

A nonflammable gas. It reacts violently with acrolein, Al, $CsHC_2$, Cs_2O, chlorates, ClF_3, Cr, FeO, F_2, Mn, KHC_2, $KClO_3$, Rb_2C_2, Na, Na_2C_2, SnO, diaminolithiumacetylene carbide. Will react with water or steam to produce toxic and corrosive fumes. Incompatible with halogens or interhalogens, lithium nitrate, metal acetylides, metal oxides, metals, polymeric tubing, potassium chlorate, sodium hydride.

SOI000   CAS: 2551-62-4   HR: 1
SULFUR HEXAFLUORIDE

DOT: UN 1080

mf: F_6S mw: 146.06

PROP: Colorless, odorless gas of high chemical stability and inertness. Nonflammable. White sublimable solid at low temps. Stable to H_2O and to glass. Mp: −51° (subl @ −64°), vap d: 6.602, d (liquid): 1.67 @ −100°. Very insol in H_2O; slightly sol in EtOH.

SYNS: HEXAFLUORURE de SOUFRE (FRENCH) □ SULFUR FLUORIDE

CONSENSUS REPORTS: Reported in EPA TSCA Inventory.

OSHA PEL: TWA 1000 ppm

ACGIH TLV: TWA 1000 ppm

DFG MAK: 1000 ppm (6100 mg/m³)

DOT CLASSIFICATION: 2.2; Label: Nonflammable Gas

SAFETY PROFILE: This material is chemically inert in the pure state and is considered to be physiologically inert as well. However, as it is ordinarily obtainable, it can contain variable quantities of the low-sulfur fluorides. Some of these are toxic, very reactive chemically, and corrosive in nature. These materials can hydrolyze on contact with water to yield hydrogen fluoride, which is highly toxic and very corrosive. In high concentrations and when pure it may act as a simple asphyxiant. Incompatible with disilane. Vigorous reaction with disilane. May explode. When heated to decomposition emits highly toxic fumes of F^- and SO_x.

SOI500   CAS: 7664-93-9   HR: 3
SULFURIC ACID

DOT: UN 1830/UN 1832

mf: H_2O_4S mw: 98.08

PROP: Viscous, colorless oily liquid; odorless. Mp: 10.49°, d: 1.834, vap press: 1 mm @ 145.8°, bp: 290°, decomp @ 340°. Misc with water and alc (liberating great heat). IDLH 15 mg/m³.

SYNS: ACIDE SULFURIQUE (FRENCH) □ ACIDO SOLFORICO (ITALIAN) □ BOV □ DIPPING ACID □ HYDROOT □ MATTING ACID (DOT) □ NORDHAUSEN ACID (DOT) □ OIL of VITRIOL (DOT) □ SCHWEFEL-SAEURELOESUNGEN (GERMAN) □ SPENT SULFURIC ACID (DOT) □ SULPHURIC ACID □ VITRIOL BROWN OIL □ VITRIOL, OIL OF (DOT) □ ZWAVELZUURO-PLOSSINGEN (DUTCH)

CONSENSUS REPORTS: Reported in EPA TSCA Inventory.

OSHA PEL: TWA 1 mg/m³

ACGIH TLV: TWA 1 mg/m³; STEL 3 mg/m³; Suspected Human Carcinogen (Contained in strong inorganic mists); (Proposed: TWA 0.2 mg/m³; Suspected Human Carcinogen (Contained in strong inorganic mists))

S

DFG MAK: 1 mg/m³
NIOSH REL: (Sulfuric Acid) TWA 1 mg/m³
DOT CLASSIFICATION: 8; Label: Corrosive
SAFETY PROFILE: Suspected human carcinogen when contained in strong inorganic mists. A human poison. Experimental poison by inhalation. Moderately toxic by ingestion. A severe eye irritant. Extremely irritating, corrosive, and toxic to tissue, resulting in rapid destruction of tissue, causing severe burns. If much of the skin is involved, exposure is accompanied by shock, collapse, and symptoms similar to those seen in severe burns. Repeated contact with dilute solutions can cause a dermatitis, and repeated or prolonged inhalation of a mist of sulfuric acid can cause inflammation of the upper respiratory tract, leading to chronic bronchitis. Sensitivity to sulfuric acid or its mists or vapors varies with individuals. Normally 0.125–0.50 ppm may be mildly annoying, 1.5–2.5 ppm can be definitely unpleasant, and 10–20 ppm is unbearable. Workers exposed to low concentrations of the vapor gradually lose their sensitivity to its irritating action. Inhalation of concentrated vapor or mists from hot acid or oleum can cause rapid loss of consciousness with serious damage to lung tissue. Severe exposure may cause a chemical pneumonitis; erosion of the teeth due to exposure to strong acid fumes has been recognized in industry. An experimental teratogen.

This is a very powerful acidic oxidizer that can ignite or explode on contact with many materials, e.g., acetic acid, acetone cyanhydrin, (acetone + HNO_3), (acetone + $K_2Cr_2O_7$), acetonitrile, acrolein, acrylonitrile, (acrylonitrile + H_2O), (alcohols + H_2O_2), allyl alcohol, allyl chloride, NH_4OH, 2-amino ethanol, NH_4, triperchromate, aniline, (bromates + metals), BrF_5, n-butyraldehyde, carbides, $CoHC_2$, chlorates, (metals + chlorates), ClF_3, chlorosulfonic acid, Cu_3N, diisobutylene, (dimethyl benzylcarbinol + H_2O_2), epichlorohydrin, ethylene cyanhydrin, ethylene diamine, ethylene

glycol, ethylene imine, fulminates, HCl, H_2, IF_7, (indene + HNO_3), Fe, isoprene, Li_6Si_2, Hg_3N_2, mesityl oxide, metals, (HNO_3 + glycerides), p-nitrotoluene, perchlorates, $HClO_4$, (C_6H_6 + permanganates), pentasilver trihydroxydiamino phosphate, (1-phenyl-2-methyl propyl alcohol + H_2O_2), P, $P(OCN)_3$, picrates, potassium-tert-butoxide, $KClO_3$, $KMnO_4$, ($KMnO_4$ + KCl), ($KMnO_4$ + H_2O), β-propiolactone, $RbHC_2$, propylene oxide, pyridine, Na, Na_2CO_3, NaOH, steel, styrene monomer, water, vinyl acetate, (HNO_3 + toluene). When heated it emits highly toxic fumes; will react with water or steam to produce heat; can react with oxidizing or reducing materials. When heated to decomposition it emits toxic fumes of SO_x. See also SULFATES.

SOI520   CAS: 8014-95-7   HR: 3
SULFURIC ACID, fuming
DOT: NA 1831
mf: $H_2O_4S \cdot O_3S$ mw: 178.14
PROP: Heavy, fuming, yellow liquid. H_2SO_4 + up to 80% SO_3. A solution of sulfuric anhydride (sulfur trioxide) in anhydrous sulfuric acid (NTIS** PB233–098).
SYNS: DISULPHURIC ACID □ DITHIONIC ACID □ FUMING SULFURIC ACID □ OLEUM □ PYROSULPHURIC ACID □ SULFURIC ACID, fuming > or =30% free sulfur trioxide (DOT) □ SULFURIC ACID, fuming <30% free sulfur trioxide (DOT) □ SULFURIC ACID MIXTURE with SULFUR TRIOXIDE
CONSENSUS REPORTS: IARC Cancer Review: Group 1 IMEMDT 54,41,92; Human Sufficient Evidence IMEMDT 54,41,92.
NIOSH REL: TWA 1 mg/m³
DOT CLASSIFICATION: 8; Label: Corrosive; DOT Class: 8; Label: Corrosive, Poison
SAFETY PROFILE: Confirmed human carcinogen. A poison. Moderately toxic by inhalation. A corrosive irritant to skin, eyes, and mucous membranes. A very dangerous fire hazard by chemical reaction with reducing agents and carbohydrates. A severe explosion hazard by chemical reaction with acetic acid, acetic anhydride, acetonitrile, acrolein, acrylic acid, acrylonitrile, allyl

alcohol, allyl chloride, 2-amino ethanol, NH_4OH, aniline, cresol, n-butyraldehyde, cumene, dichloroethyl ether, diethylene glycol monomethyl ether, diisobutylene, epichlorohydrin, ethyl acetate, ethylene cyanohydrin, ethylene diamine, ethylene glycol, ethylene glycol monoethyl ether acetate, ethylene imine, glyoxal, HCl, HF, isoprene, isopropyl alcohol, mesityl oxide, methyl ethyl ketone, HNO_3, 2-nitropropane, β-propiolacetone, propylene oxide, pyridine, NaOH, styrene monomer, vinylidene chloride, sulfolane, vinyl acetate. Will react with water or steam to produce heat and toxic and corrosive fumes. Can react vigorously with reducing materials. When heated to decomposition it emits highly toxic fumes of SO_x. See also SULFUROUS ACID.

SOI530   CAS: 7664-93-9   HR: 3
SULFURIC ACID (mist)
mf: H_2O_4S mw: 98.08
PROP: The airborne form of sulfuric acid is an aerosol of droplets of varying diameter of aq sulfuric acid solution.
CONSENSUS REPORTS: EPA Extremely Hazardous Substances List. Reported in EPA TSCA Inventory.
ACGIH TLV: TWA 1 mg/m³
NIOSH REL: TWA 1 mg/m³
SAFETY PROFILE: Poison by inhalation. Human systemic effects by inhalation: mouth effects. When heated to decomposition it emits toxic fumes of SO_x. See also SULFURIC ACID.

SON510   CAS: 10025-67-9   HR: 3
SULFUR MONOCHLORIDE
mf: Cl_2S_2 mw: 135.02
PROP: Amber to yellowish-red, oily, fuming liquid; penetrating odor. Hydrolyzes to HCl + SO_2 + H_2S. Mp: −77°, bp: 138.0°, flash p: 245°F (CC), d: 1.6885 @ 15.5°/15.5°, autoign temp: 453°F, vap press: 10 mm @ 27.5°, vap d: 4.66. Decomp in water. Sol in CS_2 and org solvs. IDLH 5 ppm.
SYNS: CHLORIDE of SULFUR (DOT) □ DISULFUR DICHLORIDE □ SIARKI CHLOREK (POLISH) □ SULFUR CHLORIDE □ SULFUR CHLORIDE (DI) (DOT) □ SULFUR SUBCHLORIDE □ THIOSULFUROUS DICHLORIDE

CONSENSUS REPORTS: Reported in EPA TSCA Inventory.
OSHA PEL: CL 1 ppm
ACGIH TLV: CL 1 ppm
DFG MAK: 1 ppm (5.6 mg/m³)
NIOSH REL: (Sulfur Monochloride) CL 1 ppm
SAFETY PROFILE: Poison by ingestion and inhalation. A fuming, corrosive liquid very irritating to skin, eyes, and mucous membranes. It decomposes on contact with water to form the highly irritating hydrogen chloride, thiosulfuric acid, and sulfur. Its toxic effects are irritating to the upper respiratory tract, although the results of intoxication are usually transitory in nature. However, if hydrolysis is not complete in the upper respiratory tract, injury to the bronchioles and alveoli can result. A fire hazard when in contact with organic matter, P_2O_3, Na_2O_2, water, $Cr(OCl)_2$. Combustible when exposed to heat or flame. Will react with water or steam to produce heat and toxic and corrosive fumes. Can react with oxidizing materials. To fight fire, use CO_2, dry chemical. When heated to decomposition it emits highly toxic fumes of Cl^- and SO_x.

SOO500   CAS: 7782-99-2   HR: 3
SULFUROUS ACID
DOT: UN 1833
mf: H_2SO_3 mw: 82.08
PROP: Colorless liquid; suffocating sulfur odor (in solution only). Present in aq solns of SO_2. Oxidizes in air to sulfuric acid. D: approx 1.03.
SYNS: SCHWEFLIGESAEURE (GERMAN) □ SULFUR DIOXIDE, solution
CONSENSUS REPORTS: Reported in EPA TSCA Inventory.
DOT CLASSIFICATION: 8; Label: Corrosive
SAFETY PROFILE: A poison by ingestion and inhalation. A corrosive irritant to skin, eyes, and mucous membranes. Human systemic effects by ingestion: nausea or vomiting, hypermotility, diarrhea, and other

S

gastrointestinal effects. When heated to decomposition it emits highly toxic fumes of SO_x.

SOP000   CAS: 2312-35-8   HR: 3
SULFUROUS ACID, 2-(p-tert-BUTYL-PHENOXY)CYCLOHEXYL-2-PROPYNYL ESTER

mf: $C_{19}H_{26}O_4S$ mw: 350.51

PROP: Viscous liquid. D: 1.1 @ 5°. Very spar sol in H_2O.

SYNS: BPPS □ 2-(p-tert-BUTYLPHENOXY)CYCLOHEXYL PROPARGYL SULFITE □ 2-(p-tert-BUTYLPHENOXY)-CYCLOHEXYL 2-PROPYNYL SULFITE □ COMITE □ 2-(4-(1,1-DIMETHYLETHYL)PHENOXY)CYCLOHEXYL 2-PROPYNYL ESTER, SULFUROUS ACID □ 2-(4-(1,1-DIMETHYLETHYL)PHENOXY)CYCLOHEXYL 2-PROP-YNYL SULFITE □ DO 14 □ ENT 27,226 □ NAUGATUCK D-014 □ OMAIT □ OMITE □ PROPARGITE (DOT) □ UNIROYAL D014 □ U.S. RUBBER D-014

SAFETY PROFILE: Poison by skin contact. Moderately toxic by ingestion. When heated to decomposition it emits toxic fumes of SO_x. See also ESTERS and SULFUROUS ACID.

SOP500   CAS: 140-57-8   HR: 3
SULFUROUS ACID, 2-(p-tert-BUTYL-PHENOXY)-1-METHYLETHYL-2-CHLOROETHYL ESTER

mf: $C_{15}H_{23}ClO_4S$ mw: 334.89

PROP: Liquid. D: 1.145–1.1620, mp: −31.7°, bp: 175° @ 0.1 mm, vap press: <10 mm @ 25°. Misc with many org solvs; insol in water.

SYNS: ACARACIDE □ ARACIDE □ ARAMITE □ ARAMITEARARAMITE-15W □ ARATRON □ BUTYL-PHENOXYISOPROPYL CHLOROETHYL SULFITE □ 2-(p-BUTYLPHENOXY)ISOPROPYL 2-CHLOROETHYL SULFITE □ 2-(4-tert-BUTYLPHENOXY)ISOPROPYL-2-CHLOROETHYL SULFITE □ 2-(p-tert-BUTYL-PHENOXY)ISOPROPYL 2'-CHLOROETHYL SULPHITE □ 2-(p-tert-BUTYLPHENOXY)-1-METHYLETHYL 2-CHLOROETHYL ESTER of SULPHUROUS ACID □ 2-(p-BUTYLPHENOXY)-1-METHYLETHYL 2-CHLOROETHYL SULFITE □ 2-(p-tert-BUTYLPHENOXY)-1-METHYL-ETHYL-2-CHLOROETHYL SULFITE ESTER □ 2-(p-tert-BUTYLPHENOXY)-1-METHYLETHYL 2'-CHLOROETHYL SULPHITE □ 2-(p-tert-BUTYLPHENOXY)-1-METHYL-ETHYL SULPHITE of 2-CHLOROETHANOL □ 1-(p-tert-BUTYLPHENOXY)-2-PROPANOL-2-CHLOROETHYL SULFITE □ CES □ 2-CHLOROETHANOL-2-(p-tert-BUTYLPHENOXY)-1-METHYLETHYL SULFITE □ 2-CHLOROETHANOL ESTER with 2-(p-tert-BUTYL-PHENOXY)-1-METHYLETHYL SULFITE □ β-

CHLOROETHYL-β'-(p-tert-BUTYLPHENOXY)-α'-METHYLETHYL SULFITE □ β-CHLOROETHYL-β-(p-tert-BUTYLPHENOXY)-α-METHYLETHYL SULPHITE □ 2-CHLOROETHYL 1-METHYL-2-(p-tert-BUTYL-PHENOXY)ETHYL SULPHITE □ 2-CHLOROETHYL SULFUROUS ACID-2-(4-(1,1-DIMETHYLETHYL-)PHENOXY)-1-METHYLETHYL ESTER □ 2-CHLORO-ETHYL SULPHITE of 1-(p-tert-BUTYLPHENOXY)-2-PROPANOL □ COMPOUND 88R □ ENT 16,519 □ NIAGARAMITE □ ORTHO-MITE □ 88-R

CONSENSUS REPORTS: IARC Cancer Review: Group 2B IMEMDT 7,56,87; Animal Sufficient Evidence IMEMDT 5,39,74.

SAFETY PROFILE: Confirmed carcinogen with experimental carcinogenic, neoplastigenic, and tumorigenic data. Experimental poison by intraperitoneal route. Moderately toxic to humans by ingestion. Moderately toxic experimentally by ingestion. Experimental reproductive effects. A pesticide. When heated to decomposition it emits toxic fumes of Cl^- and SO_x. See also ESTERS and SULFUROUS ACID.

SOQ450   CAS: 5714-22-7   HR: 3
SULFUR PENTAFLUORIDE

mf: $F_{10}S_2$ mw: 254.12

PROP: Colorless liquid. Stable to H_2O in the presence of acid or base. Disproportionates at 1° to SF_6 + SF_4. Mp: −53°, bp: 26.7°. IDLH 1 ppm.

SYNS: SULFUR DECAFLUORIDE □ TL 70

OSHA PEL: CL 0.01 ppm

ACGIH TLV: CL 0.01 ppm

DFG MAK: 0.025 ppm (0.26 mg/m³)

NIOSH REL: (Sulfur Pentafluoride) CL 0.01 ppm

SAFETY PROFILE: Poison by intravenous route. Moderately toxic by inhalation. When heated to decomposition it emits very toxic fumes of F^- and SO_x. See also FLUORIDES and SULFIDES.

SOR000   CAS: 7783-60-0   HR: 3
SULFUR TETRAFLUORIDE

DOT: UN 2418

mf: F_4S mw: 108.06

PROP: Colorless gas. Readily hydrolyzes. Stable in dry Pyrex glass vessels. Bp: −38°, mp: −121°, d: 1.95° @ −78°. Very sol in C_6H_6.

SYN: TETRAFLUOROSULFURANE

CONSENSUS REPORTS: Reported in EPA TSCA Inventory. EPA Extremely Hazardous Substances List.

OSHA PEL: CL 0.1 ppm

ACGIH TLV: CL 0.1 ppm; BEI: 3 mg/g creatinine of fluorides in urine prior to shift; 10 mg/g creatinine of fluorides in urine at end of shift.

NIOSH REL: (Inorganic Fluorides) TWA 2.5 mg(F)/m³

DOT CLASSIFICATION: 2.3; Label: Poison Gas

SAFETY PROFILE: Poison by inhalation. A powerful irritant. Will react with water, steam, or acids to yield toxic and corrosive fumes. Incompatible with dioxygen difluoride. When heated to decomposition it emits very toxic fumes of F^- and SO_x. See also FLUORIDES.

SOR500   CAS: 7446-11-9   HR: 3
SULFUR TRIOXIDE

DOT: UN 1829/NA 1829

mf: O_3S mw: 80.06

PROP: Colorless liquid fumes in air. It exists in three forms; the most valuable commercially is the γ form (mp: 16.8°, bp: 44.8°), which has a strong tendency to polymerize to the straight-chain β form (mp β: 32.5°) and subsequently to the cross-linked α form (mp α: 62°). When the β or α forms are melted they tend to revert to the γ form liquid or ice-like crystals. (β) asbestos-like crystals; vap press: (β) 433 mm @ 250°, vap press (α): 344 mm, vap d: 2.76.

SYNS: SULFAN □ SULFURIC ANHYDRIDE □ SULFURIC OXIDE □ SULFUR TRIOXIDE, inhibited (UN 1829) (DOT) □ SULFUR TRIOXIDE, uninhibited (NA 1829) (DOT)

CONSENSUS REPORTS: EPA Extremely Hazardous Substances List. Reported in EPA TSCA Inventory.

DOT CLASSIFICATION: 8; Label: Corrosive, Poison

SAFETY PROFILE: Poison by inhalation. Human systemic effects by inhalation: cough and other pulmonary and olfactory changes. A corrosive irritant to skin, eyes, and mucous membranes. Violent reaction with O_2F_2, PbO, $NClO_2$, $HClO_4$, P, tetrafluorethylene, acetonitrile, sulfuric acid, dimethyl sulfoxide, dioxan, water, diphenylmercury, formamide, iodine, pyridine, metal oxides. Reacts with steam to form corrosive, toxic fumes of sulfuric acid. When heated to decomposition it emits toxic fumes of SO_x. See also SULFURIC ACID.

SOT000   CAS: 7791-25-5   HR: 3
SULFURYL CHLORIDE

DOT: UN 1834

mf: Cl_2O_2S mw: 134.96

PROP: Colorless liquid; pungent odor. Hydrolyzes to HCl and H_2SO_4. Mp: −54.1°, bp: 69.1°, d: 1.6674, vap press: 100 mm @ 17.8°, vap d: 4.65. Sol in C_6H_6, $CHCl_3$, and Et_2O.

SYNS: SULFONYL CHLORIDE □ SULFURIC OXYCHLORIDE

CONSENSUS REPORTS: Reported in EPA TSCA Inventory.

DOT CLASSIFICATION: 8; Label: Corrosive, Poison

SAFETY PROFILE: A corrosive irritant to skin, eyes, and mucous membranes. Questionable carcinogen with experimental tumorigenic data. Can explode with PbO_2. Will react with water or steam to produce heat and toxic and corrosive fumes. Incompatible with alkalies, diethyl ether, dimethyl sulfoxide, dinitrogen pentaoxide, lead dioxide, phosphorus. When heated to decomposition it emits highly toxic fumes of Cl^- and SO_x. See also SULFURIC ACID and HYDROCHLORIC ACID, which are formed upon hydrolysis.

SOT500   CAS: 13637-84-8   HR: 3
SULFURYL CHLORIDE FLUORIDE

mf: $ClFO_2S$ mw: 118.51

PROP: Colorless gas with pungent smell. Hydrolyzes by H_2O to HCl + HF + H_2SO_4. Mp: −124.7°, bp: 7.1°, d: 1.623 @ 0°.

SYNS: CHLORO FLUORO SULFONE □ CHLOROSULFONYL FLUORIDE □ FLUOROSULFONYL CHLORIDE □ SULFONYL CHLORIDE FLUORIDE □

S

SULFURYL CHLOROFLUORIDE □ SULFURYL FLUOROCHLORIDE □ TL 212

CONSENSUS REPORTS: Reported in EPA TSCA Inventory.

OSHA PEL: TWA 2.5 mg(F)/m³

ACGIH TLV: TWA 2.5 mg(F)/m³; BEI: 3 mg/g creatinine of fluorides in urine prior to shift; 10 mg/g creatinine of fluorides in urine at end of shift.

NIOSH REL: (Inorganic Fluorides) TWA 2.5 mg(F)/m³

SAFETY PROFILE: Poison by inhalation. When heated to decomposition it emits very toxic fumes of Cl⁻, F⁻, and SO_x. See also FLUORIDES, CHLORIDES, and SULFIDES.

SOU500   CAS: 2699-79-8   HR: 3
SULFURYL FLUORIDE

DOT: UN 2191

mf: F_2O_2S mw: 102.06

PROP: Odorless, colorless, fairly inert gas. Stable to 4°. Mp: −137°, bp: −55°, d: 1.7 @ −55° (approx). Sol in water. IDLH 200 ppm.

SYNS: FLUORURE de SULFURYLE (FRENCH) □ SULFURIC OXYFLUORIDE □ VIKANE □ VIKANE FUMIGANT

CONSENSUS REPORTS: Reported in EPA TSCA Inventory.

OSHA PEL: TWA 5 ppm; STEL 10 ppm

ACGIH TLV: TWA 5 ppm; STEL 10 ppm

DOT CLASSIFICATION: 2.3; Label: Poison Gas

SAFETY PROFILE: Poison by ingestion. Mildly toxic by inhalation. Accidental human exposure caused nausea, vomiting, cramps, itching. May be narcotic in high concentration. Experimental reproductive effects. Can react with water, steam. When heated to decomposition it emits very toxic fumes of F⁻ and SO_x. See also FLUORIDES.

SOU625   CAS: 35400-43-2   HR: 3
SULPROFOS

mf: $C_{12}H_{19}O_2PS_3$ mw: 322.46

SYNS: BAY-NTN-9306 □ BOLSTAR □ O-ETHYL-O-(4-(METHYLMERCAPTO)PHENYL)-S-N-PROPYL-PHOSPHOROTHIONOTHIOLATE □ O-ETHYL-O-(4-(METHYLTHIO)PHENYL)PHOSPHORODITHIOIC ACID-

S-PROPYL ESTER □ O-ETHYL-O-(4-(METHYL-THIO)PHENYL) S-PROPYL PHOSPHORODITHIOATE □ HELOTHION

OSHA PEL: TWA 1 mg/m³

ACGIH TLV: TWA 1 mg/m³; Not Classifiable as a Human Carcinogen; (Proposed: TWA (inhalable fraction) 0.1 mg/m³; (skin); Not Classifiable as a Human Carcinogen)

SAFETY PROFILE: Poison by ingestion. Moderately toxic by skin contact. When heated to decomposition it emits very toxic fumes of PO_x and SO_x.

SOU875   HR: D
SUNFLOWER OIL (UNHYDROGEN-ATED)

PROP: From the seed of *Helianthus annuus*. Amber-colored liquid.

SAFETY PROFILE: When heated to decomposition it emits acrid smoke and irritating fumes.

SOW000   CAS: 50-02-2   HR: 3
SUPERPREDNOL

mf: $C_{22}H_{29}FO_5$ mw: 392.51

SYNS: AEROSEB-DEX □ AZIUM □ CORSONE □ DEC-ADERM □ DECADRON □ DECASONE □ DECASPRAY □ DECTANCYL □ 1-DEHYDRO-16-α-METHYL-9-α-FLUOROHYDROCORTISONE □ DELTAFLUORENE □ DERGRAMIN □ DERONIL □ DESADRENE □ DESAMETASONE □ DEXA □ DEXACORT □ DEXA-CORTIDELT □ DEXADELTONE □ DEXAMETH □ DEXAMETHASONE ALCOHOL □ DEXONE □ DEXTELAN □ DEZONE □ DXMS □ Δ¹-9-α-FLUORO-16-α-METHYLCORTISOL □ 9-α-FLUORO-16-α-METHYL-PREDNISOLONE □ 9-α-FLUORO-16-α-METHYL-1,4-PREGNADIENE-11-β,17-α,21-TRIOL-3,20-DIONE □ 4-α-FLUORO-16-α-METHYL-11-β,17,21-TRIHYDROXY-PREGNA-1,4-DIENE-3,20-DIONE □ 9-FLUORO-11-β,17,21-TRIHYDROXY-16-α-METHYLPREGNA-1,4-DIENE-3,20-DIONE □ 9-α-FLUORO-11-β,17-α,21-TRIHYDROXY-16-α-METHYLPREGNA-1,4-DIENE-3,20-DIONE □ FORTECORTIN □ GAMMACORTEN □ HEXADECADROL □ HEXADROL □ MAXIDEX □ 16-α-METHYL-9-α-FLUORO-1-DEHYDROCORTISOL □ 16-α-METHYL-9-α-FLUORO-Δ¹-HYDROCORTISONE □ 16-α-METHYL-9-α-FLUOROPREDNISOLONE □ 16-α-METHYL-9-α-FLUORO-1,4-PREGNADIENE-11-β,17-α,21-TRIOL-3,20-DIONE □ 16-α-METHYL-9-α-FLUORO-11-β,17-α,21-TRIHYDROXYPREGNA-1,4-DIENE-3,20-DIONE □ MEXIDEX □ MILLICORTEN □ MK 125 □ ORADEXON □ SK-DEXAMETHASONE

CONSENSUS REPORTS: Reported in EPA TSCA Inventory.

SAFETY PROFILE: Poison by intraperitoneal and subcutaneous routes. An experimental teratogen. Experimental reproductive effects. Mutation data reported. When heated to decomposition it emits toxic fumes of F⁻.

SOX500   CAS: 337-47-3   HR: 3
SURITAL SODIUM
mf: $C_{12}H_{18}N_2O_2S \cdot Na$ mw: 277.37

SYNS: 5-ALLYL-5-(1-METHYLBUTYL)-2-THIOBARBITURATE SODIUM □ 5-ALLYL-5-(1-METHYL-BUTYL)-2-THIO-BARBITURIC ACID SODIUM SALT □ BURITAL SODIUM □ SODIUM-5-ALLYL-5-(1-METHYL-BUTYL)-2-THIOBARBITURATE □ SODIUM THIAMYLAL □ SURITAL □ SURITAL SODIUM (derivative) □ SURITAL SODIUM SALT □ THIAMYLAL SODIUM □ THIOMYLAL SODIUM

SAFETY PROFILE: Poison by ingestion, subcutaneous, intravenous, and intraperitoneal routes. An experimental teratogen. When heated to decomposition it emits very toxic fumes of NOₓ, SOₓ, and Na₂O.

SOY000   CAS: 122-10-1   HR: 3
SWAT
mf: $C_9H_{15}O_8P$ mw: 282.21

PROP: Yellow oil. Bp: 155–164° (mixture of isomers) @ 1.7 mm. Mod sol in H_2O.

SYNS: BOMYL □ DIMETHYL 1,3-BIS(CARBOMETHO-XY)-1-PROPEN-2-YL PHOSPHATE □ DIMETHYL-1,3-DI(CARBOMETHOXY)-1-PROPEN-2-YL PHOSPHATE □ DIMETHYL 3-(DIMETHOXYPHOSPHINYLOXY)GLUTA-CONATE □ DIMETHYL 3-HYDROXYGLUTACONATE DIMETHYL PHOSPHATE □ ENT 24,833 □ FLY BAIT GRITS □ GC 3707 □ GENERAL CHEMICALS 3707 □ 3-HYDROXYGLUTACONIC ACID, DIMETHYL ESTER, DIMETHYL PHOSPHATE □ 3-HYDROXY-2-PENTANE-DIOIC ACID, DIMETHYL ESTER, DIMETHYL PHOSPH-ATE □ PHOSPHORIC ACID, DIMETHYL ESTER, ESTER with DIMETHYL 3-HYDROXYGLUTACONATE

SAFETY PROFILE: Poison by ingestion and skin contact. Used as an insecticide. When heated to decomposition it emits very toxic fumes of POₓ.

SOY100   CAS: 68917-50-0   HR: 2
SWEET BIRCH OIL
SYNS: BLACK BIRCH OIL □ OILS, SWEET BIRCH

CONSENSUS REPORTS: Reported in EPA TSCA Inventory.

SAFETY PROFILE: Moderately toxic by ingestion. A skin irritant. When heated to decomposition it emits acrid smoke and irritating fumes.

SPA000   CAS: 3441-64-3   HR: 3
SYDNOPHEN HYDROCHLORIDE
mf: $C_{11}H_{13}N_3O \cdot ClH$ mw: 239.73

PROP: Crystals from 2-propanol/Me₂CO. Mp: 153–154°.

SYNS: 3-(α-METHYLPHENETHYL)SYDONE IMINE MONOHYDROCHLORIDE □ 3-(β-PHENYLISOPROPYL)-SIDNONIMINE HYDROCHLORIDE □ SIDNOFEN □ SYDNONE IMINE, 3-(1-METHYL-2-PHENYLETHYL)-, MONOHYDROCHLORIDE □ SYDNOPHENE

SAFETY PROFILE: Poison by intraperitoneal, intravenous, and subcutaneous routes. When heated to decomposition it emits very toxic fumes of NOₓ and HCl.

SPC500   CAS: 61-76-7   HR: 3
m-SYNEPHRINE HYDROCHLORIDE
mf: $C_9H_{13}NO_2 \cdot ClH$ mw: 203.6

PROP: A solid. Mp: 143°.

SYNS: ADRIANOL □ ALMEFRIN □ BIOMYDRIN □ CONSDRIN □ CONSDRIN HYDROCHLORIDE □ DERIZENE □ EFRICEL □ EMAGRIN □ FENILFAR □ FENOX □ FURACIN □ HISTABID □ (−)-α-HYDROXY-β-(METHYLAMINO)ETHYL-α-(3-HYDROXYBENZENE) HYDROCHLORIDE □ (R)-3-HYDROXY-α-((METHYL-AMINO)METHYL)BENZENEMETHANOL HYDRO-CHLORIDE □ 1-m-HYDROXY-α-(METHYLAMINO-METHYL)BENZYL ALCOHOL HYDROCHLORIDE □ l-1-(m-HYDROXYPHENYL)-2-METHYL-AMINOETHANOL HYDROCHLORIDE □ IDRIANOL □ ISOPHRINE □ ISOPHRIN HYDROCHLORIDE □ LEXATOL □ METAO-KSEDRIN □ METAOXEDRIN □ METAOXEDRINUM □ m-METHYLAMINOETHANOLPHENOL HYDRO-CHLORIDE □ METROXEDRINE □ MEZATON □ MYDFRIN □ NCI-C55641 □ NEOOXEDRINE □ NEOPHRYN □ NEO-SINEFRINA □ NEOSYMPATOL □ NEOSYNEPHRINE □ NEOSYNEPHRINE HYDROCHLORIDE □ NEOSYNESINE □ NEWPHRINE □ OCUSOL □ OFTALFRINE □ m-OXEDRINE □ PHENISTAN □ PHENYL-DRANE □ PHENYLEPHRINE HYDROCHLORIDE □ d-(−)-PHENYLEPHRINE HYDROCHLORIDE □ PREFRIN □ PYRISTAN □ RHINALL □ STANEPHRIN □ SUCRAPHEN □ m-SYMPATHOL □ m-SYMPATOL □ SYNASAL □ m-SYNEPHRINE □ SYNETHENATE □ URI □ VISADRON

S

CONSENSUS REPORTS: Reported in EPA
TSCA Inventory.
SAFETY PROFILE: Poison by ingestion,
intraperitoneal, subcutaneous, intravenous,
and intramuscular routes. Mutation data
reported. When heated to decomposition it
emits very toxic fumes of HCl and NO_x.

T

TAA900   CAS: 109581-93-3   HR: 3
TACROLIMUS HYDRATE
mf: $C_{44}H_{69}NO_{12} \cdot H_2O$ mw: 822.16
SYNS: FK 506 □ TSUKUBAENOLIDE HYDRATE
SAFETY PROFILE: A poison by ingestion and intravenous routes. When heated to decomposition it emits toxic vapors of NO_x.

TAB750   CAS: 14807-96-6   HR: 2
TALC
mf: $H_2O_3Si \cdot 3/4Mg$ mw: 96.33
PROP: White to grayish-white, fine powder; odorless and tasteless. Powdered native hydrous magnesium silicate. Insol in water, cold acids, or alkalies. Containing less than 1% crystalline silica. Insol in H_2O. IDLH 1000 mg/m³.
SYNS: AGALITE □ AGI TALC, BC 1615 □ ALPINE TALC USP, BC 127 □ ALPINE TALC USP, BC 141 □ ALPINE TALC USP, BC 662 □ ASBESTINE □ C.I. 77718 □ DESERTALC 57 □ EMTAL 596 □ FIBRENE C 400 □ LO MICRON TALC 1 □ LO MICRON TALC, BC 1621 □ LO MICRON TALC USP, BC 2755 □ METRO TALC 4604 □ METRO TALC 4608 □ METRO TALC 4609 □ MISTRON FROST P □ MISTRON RCS □ MISTRON 2SC □ MISTRON STAR □ MISTRON SUPER FROST □ MISTRON VAPOR □ MP 12-50 □ MP 25-38 □ MP 45-26 □ NCI-C06008 □ No. 907 METRO TALC □ NYTAL □ OOS □ OXO □ PURTALC USP □ SIERRA C-400 □ SNOWGOOSE □ STEAWHITE □ SUPREME DENSE □ TALCUM
CONSENSUS REPORTS: IARC Cancer Review: Group 3 IMEMDT 7,349,87; Animal Inadequate Evidence IMEMDT 42,185,87; Human Inadequate Evidence IMEMDT 42,185,87. Reported in EPA TSCA Inventory.
OSHA PEL: TWA 2 mg/m³
ACGIH TLV: TWA 2 mg/m³, respirable dust (use asbestos TLV if asbestos fibers are present); Not Classifiable as a Human Carcinogen; (Proposed: respirable dust 1 mg/m³; Not Classifiable as a Human Carcinogen)
DFG MAK: 2 mg/m³
NIOSH REL: Talc (containing no asbestos) 2 mg/m³
SAFETY PROFILE: The talc with less than 1 percent asbestos is regarded as a nuisance dust. Talc with greater percentage of asbestos may be a human carcinogen. A human skin irritant. Prolonged or repeated exposure can produce a form of pulmonary fibrosis (talc pneumoconiosis) which may be due to asbestos content. Questionable carcinogen with experimental tumorigenic data. A common air contaminant.

TAB775   CAS: 14807-96-6   HR: 3
TALC, containing asbestos fibers
mf: $H_2O_2Si \cdot 3/4Mg$ mw: 96.33
PROP: Mp: 1652°–1832° F, d: 2.70-2.80. Insol in water.
CONSENSUS REPORTS: IARC Cancer Review: Group 1 IMEMDT 7,349,87; Human Sufficient Evidence IMEMDT 42,185,87; Reported in EPA TSCA Inventory.
ACGIH TLV: Human Carcinogen; TWA >2 mg/m³, Respirable Dust
SAFETY PROFILE: Confirmed human carcinogen with experimental tumorigenic data.

TAC000   CAS: 8002-26-4   HR: 2
TALL OIL
PROP: Composition: Rosin acids, oleic and linoleic acids. Dark-brown liquid; acrid odor. D: 0.95, flash p: 360°F.
SYNS: LIQUID ROSIN □ TALLOL
SAFETY PROFILE: A mild allergen. A substance which migrates to food from packaging materials. Combustible when

exposed to heat or flame; can react with oxidizing materials. To fight fire, use dry chemical, CO_2. When heated to decomposition it emits acrid smoke and irritating fumes.

TAC200   CAS: 70084-70-7   HR: 2
N-TALLOWPYRROLIDINONE
SYN: 2-PYRROLIDINONE, 1-TALLOW ALKYL derivs.
CONSENSUS REPORTS: Reported in EPA TSCA Inventory.
SAFETY PROFILE: Mildly toxic by ingestion. A severe skin and eye irritant. When heated to decomposition it emits acrid smoke and irritating fumes.

TAD175   CAS: 54965-24-1   HR: 3
TAMOXIFEN CITRATE
mf: $C_{26}H_{29}NO•C_6H_8O_7$ mw: 563.70
PROP: Fine, white, crystalline powder. Mp: 140–142°. Sltly sol in water; sol in ethanol, methanol, and acetone. Hygroscopic at high relative humidities.
SYNS: trans-1-(p-β-DIMETHYLAMINOETHOXYPHEN-YL)-1,2-DIPHENYLBUT-1-ENE CITRATE □ ICI 46,474 □ KESSAR □ NOLTAM □ NOLVADEX □ TAMOFEN □ TAMOXASTA □ TAMOXIFEN CITRATE □ TERIMON □ ZYNOPLEX
CONSENSUS REPORTS: IARC Cancer Review: Human Sufficient Evidence - (Benefits outweigh risk for breast cancer patients. 22 Feb, 1997).
SAFETY PROFILE: Confirmed human carcinogen with experimental carcinogenic data. Poison by intraperitoneal route. Moderately toxic by ingestion. Experimental reproductive effects. Human systemic effects: visual field changes, retinal changes. An anti-estrogenic drug. Mutation data reported. When heated to decomposition it emits toxic fumes of NO_x.

TAD250   CAS: 72869-73-9   HR: 1
TANGELO OIL
PROP: Liquid.
CONSENSUS REPORTS: Reported in EPA TSCA Inventory.
SAFETY PROFILE: Low toxicity by ingestion and skin contact. A skin irritant.

When heated to decomposition it emits acrid smoke and irritating fumes.

TAD500   CAS: 8016-85-1   HR: 1
TANGERINE OIL
PROP: Expressed from the peels of Dancy and related varieties of *Citrus reticulata Blanco*. The components include d-limonene, n-octylaldehyde, n-decylaldehyde, citral, linalool, citronella, cadinene, terpenes, aldehydes, alcohols, and esters (FCTXAV 16,637,78). Red-orange to brown-orange liquid; orange-like odor. Sol in fixed oils, mineral oil; sltly sol in propylene glycol; insol in glycerin.
SYNS: TANGERINE OIL, COLDPRESSED (FCC) □ TANGERINE OIL, EXPRESSED (FCC)
CONSENSUS REPORTS: Reported in EPA TSCA Inventory.
SAFETY PROFILE: A skin irritant. When heated to decomposition it emits acrid smoke and irritating fumes. See also individual components, ALDEHYDES, ALCOHOLS, and ESTERS.

TAD750   CAS: 1401-55-4   HR: 3
TANNIC ACID
mf: $C_{76}H_{52}O_{46}$ mw: 1701.28
PROP: From the nutgalls of *Quercus infectoria Oliver* or seed pods of *Caesalpinia spinosa* or the nutgalls of various sumac species. Yellowish-white or brown, bulky powder or flakes; odorless with astringent taste. Mp: 200–218°, flash p: 390°F (OC), autoign temp: 980°F. Very sol in water, alc, acetone; almost insol in benzene, chloroform, ether, pet ether, carbon disulfide.
SYNS: d'ACIDE TANNIQUE (FRENCH) □ GALLO-TANNIC ACID □ GALLOTANNIN □ GLYCERITE □ TANNIN
CONSENSUS REPORTS: IARC Cancer Review: Group 3 IMSUDL 7,56,87; Human No Adequate Data IMEMDT 10,253,76; Animal Limited Evidence IMEMDT 10,253,76. Reported in EPA TSCA Inventory. EPA Genetic Toxicology Program.
SAFETY PROFILE: Poison by ingestion, intramuscular, intravenous, and

subcutaneous routes. Moderately toxic by parenteral route. Experimental reproductive effects. Questionable carcinogen with experimental carcinogenic and tumorigenic data. Mutation data reported. Combustible when exposed to heat or flame. To fight fire, use water. Incompatible with salts of heavy metals, oxidizing materials. When heated to decomposition it emits acrid smoke and irritating fumes.

TAE750   CAS: 7440-25-7   HR: 2
TANTALUM
af: Ta   aw: 180.948
PROP: Gray, malleable, soft ductile metal when pure. Very air and H_2O corrosion resistant (oxide film). Attacked by HF, and by fused alkalies, as well as by fuming H_2SO_4. Mp: 2996°, bp: 5429°, d: 16.69. Insol in water. IDLH 2500 mg/m³ (as Ta).
SYN: TANTALUM-181
CONSENSUS REPORTS: Reported in EPA TSCA Inventory.
OSHA PEL: TWA 5 mg/m³
ACGIH TLV: TWA 5 mg/m³
DFG MAK: 4 mg/m³
SAFETY PROFILE: An inhalation hazard. Some industrial skin injuries from tantalum have been reported. Systemic industrial poisoning, however, is apparently unknown. Questionable carcinogen with experimental tumorigenic data. The dry powder ignites spontaneously in air. Incompatible with bromine trifluoride, fluorine, lead chromate. See also specific tantalum compounds.

TAF000   CAS: 7721-01-9   HR: 3
TANTALUM CHLORIDE
mf: Cl_5Ta mw: 358.20
PROP: White or light-yellow crystals, powder, or solid; moisture-sensitive. Monoclinic; forms intercalated phases with graphite + Cl_2. D: 3.68, mp: 216–220°, bp: 239.3°. Decomposed by water; sol in alc. Volatilizes at 144°.
SYN: TANTALUM PENTACHLORIDE
CONSENSUS REPORTS: Reported in EPA TSCA Inventory.

SAFETY PROFILE: Poison by intraperitoneal route. Moderately toxic by ingestion. When heated to decomposition it emits toxic fumes of Cl⁻.

TAF250   CAS: 7783-71-3   HR: 3
TANTALUM FLUORIDE
mf: F_5Ta mw: 275.95
PROP: Deliq, refractive prisms. D: 4.74 @ 20°, mp: 96.8°, bp: 229.5°, vap press: 100 mm @ 130°. Sol in water, concentrated nitric acid; very sol in fuming nitric acid; sltly sol in hot carbon disulfide, hot carbon tetrachloride.
SYN: TANTALUM PENTAFLUORIDE
CONSENSUS REPORTS: Reported in EPA TSCA Inventory.
OSHA PEL: TWA 2.5 mg(F)/m³
ACGIH TLV: TWA 2.5 mg(F)/m³; BEI: 3 mg/g creatinine of fluorides in urine prior to shift; 10 mg/g creatinine of fluorides in urine at end of shift.
NIOSH REL: TWA 2.5 mg(F)/m³
SAFETY PROFILE: Poison by intravenous route. Corrosive. When heated to decomposition it emits toxic fumes of F⁻. See also TANTALUM and FLUORIDES.

TAF500   CAS: 1314-61-0   HR: 1
TANTALUM OXIDE
mf: O_5Ta_2 mw: 441.90
PROP: White solid or microcrystalline, infusible powder. Not amphoteric. Very inert. D: 8.200, mp: 1872°. Insol in water, alc, and mineral acids; sol in HF.
SYNS: TANTALIC ACID ANHYDRIDE □ TANTALUM-(V) OXIDE □ TANTALUM PENTAOXIDE □ TANTALUM PENTOXIDE
CONSENSUS REPORTS: Reported in EPA TSCA Inventory.
OSHA PEL: TWA 5 mg/m³
ACGIH TLV: TWA 5 mg/m³
SAFETY PROFILE: Mildly toxic by ingestion. Incompatible with ClF_3, BrF_3, Li. See also TANTALUM.

T

TAF700   CAS: 8016-88-4   HR: 2
TARRAGON OIL
PROP: From steam distillation of leaves, stems, and flowers from *Artemesia dracunculus L.* Pale-yellow to amber liquid; spicy licorice and sweet basil odor. Sol in fixed oils, mineral oil; insol in propylene glycol, glycerin.
SYN: ESTRAGON OIL
CONSENSUS REPORTS: Reported in EPA TSCA Inventory.
SAFETY PROFILE: Moderately toxic by ingestion. A skin irritant. When heated to decomposition it emits acrid smoke and irritating fumes.

TAF750   CAS: 87-69-4   HR: 2
TARTARIC ACID
mf: $C_4H_6O_6$ mw: 150.10
PROP: Colorless to translucent crystals or white powder; odorless with an acid taste. Mp: 170–172°. Sol in water and alc.
SYNS: BUTANEDIOIC ACID, 2,3-DIHYDROXY- □ 2,3-DIHYDROSUCCINIC ACID □ 2,3-DIHYDROXY-BUTANEDIOIC ACID □ KYSELINA 2,3-DIHYDROXY-BUTANDIOVA □ KYSELINA VINNA □ MALIC ACID, 3-HYDROXY- □ SUCCINIC ACID, 2,3-DIHYDROXY- □ l-(+)-TARTARIC ACID □ THREARIC ACID
CONSENSUS REPORTS: Reported in EPA TSCA Inventory.
SAFETY PROFILE: Moderately toxic by intravenous route. Mildly toxic by ingestion. Reaction with silver produces the unstable silver tartrate. When heated to decomposition it emits acrid smoke and irritating fumes.

TAG750   CAS: 107-35-7   HR: D
TAURINE
mf: $C_2H_7NO_3S$ mw: 125.16
PROP: Colorless, monoclinic prismatic rods from water. Mp: decomp @ 300°, bp: subl @ 110°. Very sol in water and alc; insol in anhydrous ether.
SYNS: 2-AMINOETHANESULFONIC ACID □ 2-AM-INOETHYLSULFONIC ACID □ ETHANESULFONIC ACID, 2-AMINO- □ NCI-C60606 □ O-DUE □ 2-SULFOETHYLAMINE □ TAUPHON

CONSENSUS REPORTS: Reported in EPA TSCA Inventory.
SAFETY PROFILE: Experimental reproductive effects. Mutation data reported. When heated to decomposition it emits very toxic fumes of SO_x and NO_x.

TAI000   CAS: 1746-01-6   HR: 3
TCDD
mf: $C_{12}H_4Cl_4O_2$ mw: 321.96
PROP: Colorless needles. Mp: 305°.
SYNS: 2,3,7,8-CZTEROCHLORODWUBENZO-p-DWUOKSYNY (POLISH) □ DIOKSYNY (POLISH) □ DI-OXIN (herbicide contaminant) □ DIOXINE □ NCI-C03714 □ TCDBD □ 2,3,7,8-TCDD □ 2,3,6,7-TETRACHLORODI-BENZO-p-DIOXIN □ 2,3,7,8-TETRACHLORODIBENZO-(b,e)(1,4)DIOXAN □ 2,3,7,8-TETRACHLORODIBENZO-p-DIOXIN □ 2,3,7,8-TETRACHLORODIBENZO-1,4-DIOXIN □ TETRADIOXIN
CONSENSUS REPORTS: NTP 10th Report on Carcinogens. IARC Cancer Review: Group 2B IMEMDT 7,350,87; Human Inadequate Evidence IMEMDT 15,41,77; Animal Inadequate Evidence IMEMDT 15,41,77. NTP Carcinogenesis Bioassay - (gavage); Clear Evidence: mouse, rat NTPTR* NTP-TR-209,82; (dermal). EPA Genetic Toxicology Program.
DFG MAK: Animal Carcinogen, Suspected Human Carcinogen
NIOSH REL: (Dioxin) Reduce to lowest feasible level
SAFETY PROFILE: Confirmed carcinogen with experimental carcinogenic, neoplastigenic, tumorigenic, and teratogenic data. One of the most toxic synthetic chemicals. A deadly experimental poison by ingestion, skin contact, and intraperitoneal routes. Human systemic effects by skin contact: allergic dermatitis. Experimental reproductive effects. Human mutation data reported. An eye irritant.
    TCDD is the most toxic member of the 75 dioxins. It causes death in rats by hepatic cell necrosis. Death can follow a lethal dose by weeks. Acute and subacute exposure result in wasting, hepatic necrosis, thymic atrophy, hemorrhage, lymphoid depletion, chloracne. A by-product of the manufacture

of polychlorinated phenols. It is found at low levels in 2,4,5-T, 2,4,5-trichlorophenol, and hexachlorophene. It is also formed during various combustion processes. Incineration of chemical wastes, including chlorophenols, chlorinated benzenes, and biphenyl ethers, may result in the presence of TCDD in flue gases, fly ash, and soot particles. It is immobile in contaminated soil and may be retained for years. TCDD has the potential for bio-accumulation in animals. An accident in Seveso, Italy, and inadvertent soil contamination in Missouri have resulted in abandonment of the contaminated areas. When heated to decomposition it emits toxic fumes of Cl⁻.

TAI250   CAS: 9002-84-0   HR: 2
TEFLON
mf: $(C_2F_4)_n$
PROP: Grayish-white, tough plastic. Chemically very inert. Mp: 342°, d: 2.1–2.3.
SYNS: AFLON □ ALGLOFLON □ ALGOFLON SV □ ALKATHENE RXDG33 □ AMIP 15m □ BALFON 7000 □ BDH 29-801 □ CHROMOSORB T □ DIXON 164 □ DLX-6000 □ DUROID 5870 □ EK 1108GY-A □ ETHICON PTFE □ FLUO-KEM □ FLUON □ FLUOROFLEX □ FLUOROLON 4 □ FLUOROPAK 80 □ FLUORPLAST 4 □ FTORLON 4 □ FTOROPLAST 4 □ GORE-TEX □ HALON TFEG 180 □ HEYDEFLON □ HOSTAFLON □ MOLYKOTE 522 □ POLIFEN □ POLITEF □ POLY(ETH-YLENE TETRAFLUORIDE) □ POLYFENE □ POLYFLON □ POLYTEF □ POLYTETRAFLUOROETHENE □ POLYTETRAFLUOROETHYLENE □ PTFE □ SOREFLON 604 □ TARFLEN □ TEFLON (various) □ TETRAFLUOROETHENE HOMOPOLYMER □ TETRAFLUOROETHENE POLYMER □ TETRAFLUORO-ETHYLENE HOMOPOLYMER □ TETRAFLUOROETH-YLENE POLYMERS □ TETRAN PTFE □ UNON P □ VALFLON □ VELFLON □ ZITEX H 662-124
CONSENSUS REPORTS: IARC Cancer Review: Group 3 IMEMDT 7,56,87; Animal Sufficient Evidence IMEMDT 19,285,79; Human Inadequate Evidence IMEMDT 19,285,79. Reported in EPA TSCA Inventory.
SAFETY PROFILE: The finished polymerized compound is inert under ordinary conditions. There have been reports of "polymer fume fever" in humans exposed to pyrolysis products, which also are irritants. Smoking should be prohibited in areas where this material is being fabricated or, in general, where there may be dust from it. Exposure to pyrolysis or decomposition products appears to be the chief health-related problem. Questionable carcinogen with experimental tumorigenic data by implant. Incompatible with fluorine, sodium potassium alloy. Under the proper conditions it undergoes hazardous reactions with boron, magnesium, or titanium. When heated to above 750°F it decomposes to yield highly toxic fumes of F⁻.

TAI500   CAS: 113-92-8   HR: 3
TELDRIN
mf: $C_{16}H_{19}ClN_2 \cdot C_4H_4O_4$ mw: 390.90
PROP: Sol in H_2O, EtOH, $CHCl_3$; spar sol in C_6H_6, Et_2O.
SYNS: ALLERCLOR □ ALLERGIN □ ALLERGISAN □ ALUNEX □ ANTAGONATE □ CARBINOXAMINE MALEATE □ CHLORMENE □ dl-2(-p-CHLORO-α-2-(DI-METHYLAMINO)ETHYLBENZYL)PYRIDINE BIMALEATE □ 1-p-CHLOROPHENYL-1-(2-PYRIDYL)-3-DIMETHYLAMINOPROPANE MALEATE □ CHLORO-PROPHENYLPYRIDAMINE MALEATE □ CHLOR-PHENIRAMINE MALEATE □ CHLOR-TRIMETON □ CHLOR-TRIMETON MALEATE □ CHLOR-TRIPOLON □ CLOROPIRIL □ C-METON □ 1-(N,N-DIMETHYLAM-INO)-3-(p-CHLOROPHENYL-3-α-PYRIDYL)PROPANE MALEATE □ HISTADUR □ HISTADUR DURA-TABS □ HISTALEN □ HISTAPAN □ IBIOTON □ LORPHEN □ M.P. CHLORCAPS T.D. □ NCI-C55265 □ NEORESTAMIN □ PIRIEX □ PIRITON □ POLARONIL (GERMAN) □ PYRIDAMAL-100 □ SYNISTAMIN
CONSENSUS REPORTS: NTP Carcinogenesis Studies (gavage); No Evidence: mouse, rat NTPTR* NTP-TR-317,86. Reported in EPA TSCA Inventory.
SAFETY PROFILE: Poison by ingestion, intravenous, and subcutaneous routes. Experimental reproductive effects. Used as an antihistamine. Mutation data reported. When heated to decomposition it emits very toxic fumes of Cl⁻ and NO_x.

TAI750   CAS: 7803-68-1   HR: 3
TELLURIC ACID
mf: H_6O_6Te mw: 229.66

PROP: White crystals or solid; stable to 1°. D: (monoclinic) 3.068, (cubic) 3.163. Mp: 136°. Sltly sol in concentrated nitric acid; very sol in water.
SYNS: ORTHOTELLURIC ACID □ TELLURATE □ TELLURIC(VI) ACID □ TELLURIUM HYDROXIDE
CONSENSUS REPORTS: Reported in EPA TSCA Inventory.
OSHA PEL: TWA 0.1 mg(Te)/m³
ACGIH TLV: TWA 0.1 mg(Te)/m³
SAFETY PROFILE: Poison by ingestion and intravenous routes. When heated to decomposition it emits toxic fumes of Te. See also TELLURIUM COMPOUNDS.

TAI800   CAS: 22451-06-5   HR: 3
TELLURIC ACID, DISODIUM SALT, PENTAHYDRATE
mf: $O_3Te•2Na•5H_2O$ mw: 380.65
OSHA PEL: TWA 0.1 mg(Te)/m³
ACGIH TLV: TWA 0.1 mg(Te)/m³
SAFETY PROFILE: Poison by intraperitoneal route. When heated to decomposition it emits toxic fumes of Te.

TAJ000   CAS: 13494-80-9   HR: 3
TELLURIUM
af: Te   aw: 127.60
PROP: Silvery-white, metallic, lustrous element; quite brittle. Semiconductor; poor conductor of heat. Forms golden-yellow vapor. Red when colloidal. Mp: 449.5°, bp: 989.8°, d: 6.24 @ 20°, vap press: 1 mm @ 520°. Insol in water, benzene, carbon disulfide, and all solvs with which it does not react. IDLH 25 mg/m³ (as Te).
SYNS: NCI-C60117 □ TELLOY □ TELLUR (POLISH)
CONSENSUS REPORTS: Reported in EPA TSCA Inventory.
OSHA PEL: TWA 0.1 mg(Te)/m³
ACGIH TLV: TWA 0.1 mg(Te)/m³
DFG MAK: 0.1 mg/m³
NIOSH REL: (Tellurium and Compounds) TWA 0.1 mg/m³
SAFETY PROFILE: Poison by ingestion and intratracheal routes. An experimental teratogen. Exposure causes nausea, vomiting, tremors, convulsions, respiratory arrest, central nervous system depression, and garlic odor to breath. Aerosols of tellurium, tellurium dioxide, and hydrogen telluride cause irritation of the respiratory system and may lead to the development of bronchitis and pneumonia. Experimental reproductive effects. Under the proper conditions it undergoes hazardous reactions with halogens (e.g., chlorine, fluorine), interhalogens (e.g., bromine pentafluoride, chlorine fluoride, chlorine trifluoride), metals (e.g., cadmium, potassium, sodium, platinum, tin, zinc), hexalithium disilicide, silver bromate, silver iodate. When heated to decomposition it emits toxic fumes of Te. See also TELLURIUM COMPOUNDS.

TAJ250   CAS: 10026-07-0   HR: D
TELLURIUM CHLORIDE
mf: Cl_4Te mw: 269.40
PROP: White crystals, solid; red liquid and red vapor. Hydrolyzed by moist air. Mp: 224°, bp: 380°, d: 3.260. Hygroscopic. Sol in non-polar org solvs.
SYNS: TELLURIC CHLORIDE □ TELLURIUM TETRACHLORIDE □ TETRACHLOROTELLURIUM
CONSENSUS REPORTS: Reported in EPA TSCA Inventory. EPA Genetic Toxicology Program.
OSHA PEL: TWA 0.1 mg(Te)/m³
ACGIH TLV: TWA 0.1 mg(Te)/m³
SAFETY PROFILE: Mutation data reported. Experimental reproductive effects. Incompatible with ammonia. Irritant. When heated to decomposition it emits very toxic fumes of Cl⁻ and Te. See also TELLURIUM COMPOUNDS.

TAJ500   HR: 2
TELLURIUM COMPOUNDS
SAFETY PROFILE: Elemental tellurium has relatively low toxicity. It is converted in the body to dimethyl telluride, which imparts a garlic-like odor to the breath and sweat. Heavy exposures may, in addition, result in headache, drowsiness, metallic taste, loss of appetite, nausea, tremors, convulsions, and respiratory arrest. Various tellurium salts

may also produce similar symptoms. Large doses can be fatal, as was the case following accidental administration of 2 g of sodium tellurite. Workers in an iron foundry exposed to less than 0.1 mg Te/m³ developed a garlic-like odor in breath, sweat, and urine, as well as anorexia, nausea, depression, somnolence, itchy skin, and metallic taste. When heated or on contact with acid or acid fumes they emit highly toxic fumes. See also TELLURIUM and specific compounds.

TAJ600   CAS: 13494-80-9   HR: 1
TELLURIUM (dust or fume)
CONSENSUS REPORTS: Reported in EPA TSCA Inventory. EPA Extremely Hazardous Substances List.
ACGIH TLV: TWA 0.1 mg/m³
DFG MAK: 0.1 mg/m³
SAFETY PROFILE: May cause irritation of the respiratory system and lead to bronchitis and pneumonia. When heated to decomposition it emits toxic fumes of Te. See also TELLURIUM and TELLURIUM COMPOUNDS.

TAJ750   CAS: 7446-07-3   HR: 3
TELLURIUM DIOXIDE
mf: O_2Te mw: 159.60
PROP: Dimorphous solid. Melts to a dark red liquid. Bp: 790–940°. Sol in aq NaOH, HCl; almost insol in H_2O.
SYN: TELLURIUM OXIDE
CONSENSUS REPORTS: Reported in EPA TSCA Inventory.
OSHA PEL: TWA 0.1 mg(Te)/m³
ACGIH TLV: TWA 0.1 mg(Te)/m³
SAFETY PROFILE: Poison by intratracheal route. An experimental teratogen. Experimental reproductive effects. When heated to decomposition it emits toxic fumes of Te. See also TELLURIUM COMPOUNDS.

TAK250   CAS: 7783-80-4   HR: 3
TELLURIUM HEXAFLUORIDE
DOT: UN 2195

mf: F_6Te mw: 241.60
PROP: Colorless gas; repulsive odor. White solid at low temp; subl before melting. Hydrol by H_2O and aq KOH. Mp: −37.6°, bp: −38.9° (subl), d: (solid) 4.006 @ −191°, (liquid) 2.499 @ −10°. IDLH 1 ppm.
CONSENSUS REPORTS: EPA Extremely Hazardous Substances List. Reported in EPA TSCA Inventory.
OSHA PEL: TWA 0.02 ppm (Te)
ACGIH TLV: TWA 0.02 ppm (Te)
DOT CLASSIFICATION: 2.3; Label: Poison Gas
SAFETY PROFILE: Poison by inhalation. Human skin (systemic) effects. When heated to decomposition it emits very toxic fumes of F⁻ and Te. See also FLUORIDES and TELLURIUM COMPOUNDS.

TAL250   CAS: 3383-96-8   HR: 3
TEMEPHOS
mf: $C_{16}H_{20}O_6P_2S_3$ mw: 466.48
PROP: White crystals or solid. Technical grade is a brown viscous liquid. Mp: 30°. Very spar sol in H_2O, hexane; sol in CCl_4, Et_2O, and toluene.
SYNS: ABATE □ ABATHION □ AC 52160 □ AMERICAN CYANAMID AC 52,160 □ BIOTHION □ BITHION □ CL 52160 □ DIFENTHOS □ O,O-DIMETHYL PHOSPHOROTHIOATE-O,O-DIESTER with 4,4'-THIODIPHENOL □ ECOPRO □ EI 52160 □ ENT 27,165 □ EXPERIMENTAL - INSECTICIDE 52160 □ NIMITEX □ NIMITOX □ SWEBATE □ TEMEFOS □ TEMOPHOS □ TETRAMETHYL-O,O'-THIODI-p-PHENYLENE PHOSPHOROTHIOATE □ O,O,O'O'-TETRAMETHYL-O,O'-THIODI-p-PHENYLENE PHOSPHOROTHIOATE □ O,O'-(THIODI-4,1-PHENYLENE)BIS(O,O-DIMETHYL PHOSPHOROTHIOATE) □ O,O'-(THIODI-p-PHENYLENE)-O,O,O',O'-TETRAMETHYL BIS(PHOSPHOROTHIOATE)
CONSENSUS REPORTS: Reported in EPA TSCA Inventory.
OSHA PEL: TWA Total Dust: 10 mg/m³; Respirable Fraction: 5 mg/m³
ACGIH TLV: TWA 10 mg/m³
SAFETY PROFILE: Poison by ingestion. Moderately toxic by intraperitoneal, subcutaneous, and skin contact routes. An experimental teratogen. A skin irritant. A cholinesterase inhibitor type of insecticide. When heated to decomposition it emits

toxic fumes of PO_x and SO_x. See also PARATHION.

TAM000   CAS: 10042-88-3   HR: 3
TERBIUM CHLORIDE
mf: Cl_3Tb mw: 265.27
PROP: Hygroscopic white solid. Mp: 588°, bp: 1550°. Sol in H_2O and EtOH.
CONSENSUS REPORTS: Reported in EPA TSCA Inventory.
SAFETY PROFILE: Poison by intraperitoneal route. Moderately toxic by ingestion. An eye and severe skin irritant. When heated to decomposition it emits very toxic fumes of Cl^-. See also TERBIUM, CHLORIDES, and RARE EARTHS.

TAN750   CAS: 100-21-0   HR: 2
TEREPHTHALIC ACID
mf: $C_8H_6O_4$ mw: 166.14
PROP: White crystals or powder. D: 1.51, sublimes @ 300°. Insol in water, chloroform, ether, acetic acid; sltly sol in alc; sol in alkalies.
SYNS: ACIDE TEREPHTHALIQUE (FRENCH) □ p-BENZENEDICARBOXYLIC ACID □ 1,4-BENZENEDI-CARBOXYLIC ACID □ KYSELINA TERFTALOVA -(CZECH) □ TA 12 □ TA-33MP
CONSENSUS REPORTS: Community Right-To-Know List. Reported in EPA TSCA Inventory. EPA Genetic Toxicology Program.
ACGIH TLV: TWA 10 mg/m³
SAFETY PROFILE: Moderately toxic by intravenous and intraperitoneal routes. Mildly toxic by ingestion. An eye irritant. Can explode during preparation. When heated to decomposition it emits acrid smoke and irritating fumes.

TAV250   CAS: 100-20-9   HR: 2
TEREPHTHALOYL DICHLORIDE
mf: $C_8H_4Cl_2O_2$ mw: 203.02
PROP: White, crystalline material from pet ether; musty odor. Bp: 266°, mp: 83–84°, flash p: 356°F (COC), fp: 81.4°.
SYNS: 1,4-BENZENEDICARBONYL CHLORIDE □ 1,4-BENZENEDICARBONYL DICHLORIDE □ p-PHENYL-ENEDICARBONYL DICHLORIDE □ p-PHTHALOYL

CHLORIDE □ p-PHTHALOYL DICHLORIDE □ TEREPHTHALIC ACID CHLORIDE □ TEREPHTHALIC ACID DICHLORIDE □ TEREPHTHALIC DICHLORIDE
CONSENSUS REPORTS: Reported in EPA TSCA Inventory.
SAFETY PROFILE: Moderately toxic by ingestion. Corrosive. Combustible when exposed to heat or flame. When heated to decomposition it emits toxic fumes of Cl^-. See also CHLORIDES.

TAV750   CAS: 126-92-1   HR: 3
TERGITOL 08
mf: $C_8H_{18}O_4S \cdot Na$ mw: 233.31
PROP: Clear, slightly viscous liquid. D: 1.114 g/cm³ @ 21.7°, bp: 96°–104°. Flash pt: >93° C. Sol in water: > 100mg/ml @ 20°.
SYNS: EMERSAL 6465 □ 2-ETHYL-1-HEXANOL HYDROGEN SULFATE, SODIUM SALT □ 2-ETHYL-1-HEXANOL SULFATE SODIUM SALT □ 2-ETHYLHEXYL SODIUM SULFATE □ MONO(2-ETHYLHEXYL)SULFATE SODIUM SALT □ NCI-C50204 □ NIA PROOF 08 □ PROPASTE 6708 □ SIPEX BOS □ SODIUM ETASULFATE □ SODIUM ETHASULFATE □ SODIUM(2-ETHYLHEX-YL)ALCOHOL SULFATE □ SODIUM-2-ETHYLHEXYL SULFATE □ SULFURIC ACID, MONO(2-ETHYLHEXYL)-ESTER, SODIUM SALT (8CI) □ TERGEMIST □ TERGIMIST □ TERGITOL ANIONIC 08
CONSENSUS REPORTS: Reported in EPA TSCA Inventory.
SAFETY PROFILE: Poison by intraperitoneal route. Moderately toxic by ingestion and skin contact. A skin and eye irritant. A combustible liquid. When heated to decomposition it emits very toxic fumes of SO_x and Na_2O.

TBB775   CAS: 60828-78-6   HR: 2
TERGITOL TMN-10
CONSENSUS REPORTS: Reported in EPA TSCA Inventory.
SAFETY PROFILE: Mildly toxic by ingestion. A skin and severe eye irritant. When heated to decomposition it emits acrid smoke and irritating fumes.

TBC500   CAS: 8001-50-1   HR: 3
TERPENE POLYCHLORINATES
PROP: Chlorinated mixed terpenes (IARC**
4,219,75).
SYNS: DICHLORICIDE MOTHPROOFER □ ENT 19,442
□ STROBANE
CONSENSUS REPORTS: IARC Cancer
Review: Group 3 IMEMDT 7,56,87; Animal
Sufficient Evidence IMEMDT 5,219,74.
SAFETY PROFILE: Poison by ingestion.
Questionable carcinogen with experimental
carcinogenic data. When heated to
decomposition it emits toxic fumes of Cl⁻.
See also CHLORINATED
HYDROCARBONS, AROMATIC.

TBC575   HR: D
TERPENE RESIN, NATURAL
PROP: Extracted from wood. Mp: 155°.
SAFETY PROFILE: When heated to
decomposition it emits acrid smoke and
irritating fumes.

TBC620   CAS: 92-06-8   HR: 2
m-TERPHENYL
mf: $C_{18}H_{14}$ mw: 230.32
PROP: Needles. Mp: 86–87°, bp: 379°.
IDLH 500 mg/m³.
SYNS: m-DIPHENYLBENZENE □ ISODIPHENYL-
BENZENE □ SANTOWAX M □ 1,3-TERPHENYL □ m-
TRIPHENYL
CONSENSUS REPORTS: Reported in EPA
TSCA Inventory.
OSHA PEL: CL 0.5 ppm
ACGIH TLV: TWA CL 0.5 ppm
SAFETY PROFILE: Moderately toxic by
ingestion. Combustible when exposed to
heat or flame. To fight fire, use water, CO_2,
dry chemical. When heated to
decomposition it emits acrid smoke and
irritating fumes.

TBC640   CAS: 84-15-1   HR: 2
o-TERPHENYL
mf: $C_{18}H_{14}$ mw: 230.32
PROP: Prisms from MeOH. Mp: 58–59°,
bp: 337°, flash p: >230°F. IDLH 500
mg/m³.
SYN: 1,2-DIPHENYLBENZENE

CONSENSUS REPORTS: Reported in EPA
TSCA Inventory.
OSHA PEL: CL 0.5 ppm
ACGIH TLV: TWA CL 0.5 ppm
SAFETY PROFILE: Moderately toxic by
ingestion. Combustible when exposed to
heat or flame. To fight fire, use water, CO_2,
dry chemical. When heated to
decomposition it emits acrid smoke and
irritating fumes.

TBC750   CAS: 92-94-4   HR: 2
p-TERPHENYL
mf: $C_{18}H_{14}$ mw: 230.32
PROP: Leaflets or needles from EtOH. D:
1.234 @ 0°/4°, mp: 212–213°, bp: 259° @
45 mm, flash p: 405°F (OC), vap d: 7.95. Sol
in hot benzene; very sol in hot alc; sltly sol
in ether.
SYNS: p-DIPHENYLBENZENE □ 1,4-DIPHENYL-
BENZENE □ 4-PHENYLBIPHENYL □ 4-PHENYLDI-
PHENYL □ SANTOWAX □ p-TRIPHENYL
CONSENSUS REPORTS: Reported in EPA
TSCA Inventory.
OSHA PEL: CL 0.5 ppm
ACGIH TLV: TWA CL 0.5 ppm
SAFETY PROFILE: Moderately toxic by
ingestion. Combustible when exposed to
heat or flame. To fight fire, use water, CO_2,
dry chemical. When heated to
decomposition it emits acrid smoke and
irritating fumes.

TBD000   CAS: 26140-60-3   HR: 2
TERPHENYLS
mf: $C_{18}H_{14}$ mw: 230.32
SYNS: DELOWAS S □ DELOWAX OM □ DIPHENYL-
BENZENE □ GILOTHERM OM 2 □ TERBENZENE □
TRIPHENYL
CONSENSUS REPORTS: Reported in EPA
TSCA Inventory.
OSHA PEL: CL 0.5 ppm
ACGIH TLV: TWA CL 0.5 ppm
NIOSH REL: (Terphenyls) CL 0.5 ppm
SAFETY PROFILE: Moderately toxic by
ingestion. Combustible when exposed to
heat or flame. To fight fire, use water, CO_2,
dry chemical. When heated to

T

decomposition it emits acrid smoke and irritating fumes.

TBD500   CAS: 8006-39-1   HR: 2
TERPINEOL

mf: $C_{10}H_{18}O$ mw: 154.28

PROP: A mixture of α, β, and γ isomers - (FCTXAV 12,807,74). Colorless, viscous liquid; lilac odor. D: 0.930–0.936, refr index: 1.482, flash p: 196°F. Sltly sol in water, glycerin.

SYNS: FEMA No. 3045 □ p-MENTH-1-EN-8-OL □ MIXTURE of p-METHENOLS □ α-TERPINEOL (FCC) □ TERPINEOLS

SAFETY PROFILE: Mildly toxic by ingestion. A skin irritant. Combustible liquid. When heated to decomposition it emits acrid smoke and irritating fumes. See also α-TERPINEOL.

TBE000   CAS: 586-62-9   HR: 3
TERPINOLENE

DOT: UN 2541

mf: $C_{10}H_{16}$ mw: 136.26

PROP: Colorless liquid or oil. Bp: 185°, d: 0.855, flash p: 100°F (CC). Insol in water; misc in alc, ether. Mixture of p-mentha-1,4-(8)-diene and p-mentha-2,4(8)-diene - (FCTXAV 14,659,76).

SYN: 1-METHYL-4-(1-METHYLETHYLIDENE)-CYCLOHEXENE

CONSENSUS REPORTS: Reported in EPA TSCA Inventory.

DOT CLASSIFICATION: 3; Label: Flammable Liquid

SAFETY PROFILE: Mildly toxic by ingestion. A very dangerous fire hazard when exposed to heat or flame. To fight fire, use foam, CO_2, dry chemical. Can react with oxidizing materials. When heated to decomposition it emits acrid smoke and irritating fumes.

TBE250   CAS: 80-26-2   HR: 2
TERPINYL ACETATE

mf: $C_{12}H_{20}O_2$ mw: 196.32

PROP: Colorless liquid; sweet, herbaceous odor. D: 0.966 @ 20/4°, refr index: 1.464, mp: <−50°, bp: 220° decomp, flash p:

212°F. Insol in water; sol in alc, fixed oils, mineral oil, propylene glycol.

SYNS: FEMA No. 3047 □ α-TERPINEOL ACETATE

CONSENSUS REPORTS: Reported in EPA TSCA Inventory.

SAFETY PROFILE: Mildly toxic by ingestion. Combustible liquid. When heated to decomposition it emits acrid smoke and irritating fumes.

TBE600   HR: 2
TERPINYL PROPIONATE

mf: $C_{13}H_{22}O_2$ mw: 210.32

PROP: Colorless to sltly yellow liquid; sweet, floral, lavender-like odor. D: 0.944, refr index: 1.461, flash p: 212°F. Sol in glycerin; misc in alc, chloroform, ether, fixed oils; sltly sol in propylene glycol; insol in water @ 240°.

SYNS: FEMA No. 3053 □ MENTHEN-1-YL-8 PROPIONATE

SAFETY PROFILE: Combustible liquid. When heated to decomposition it emits acrid smoke and irritating fumes.

TBF500   CAS: 58-22-0   HR: 3
TESTOSTERONE

mf: $C_{19}H_{28}O_2$ mw: 288.47

PROP: Crystals from Me_2CO (aq). Mp: 155°. Insol in water; sol in alc and ether.

SYNS: ANDROLIN □ ANDRONAQ □ ANDROST-4-EN-17β-OL-3-ONE □ Δ⁴-ANDROSTEN-17(β)-OL-3-ONE □ ANDRUSOL □ CRISTERONE T □ GENO-CRISTAUZ GREMY □ HOMOSTERONE □ 7-β-HYDROXYANDR-OST-4-EN-3-ONE □ 17-β-HYDROXY-Δ⁴-ANDROSTEN-3-ONE □ 17-β-HYDROXYANDROST-4-EN-3-ONE □ 17-β-HYDROXY-4-ANDROSTEN-3-ONE □ 17-HYDROXY-(17-β)-ANDROST-4-EN-3-ONE □ MALESTRONE (AMPS) □ MERTESTATE □ NEO-TESTIS □ ORETON-F □ ORQUISTERONE □ PERANDREN □ PERCUTACRINE ANDROGENIQUE □ PRIMOTEST □ PROMOTESTON □ SUSTANONE □ SYNANDROL F □ TESLEN □ TEST-ANDRONE □ TESTICULOSTERONE □ TESTOBASE □ TESTOPROPON □ TESTOSTEROID □ trans-TESTOST-ERONE □ TESTOSTERONE HYDRATE □ TESTOSTOS-TERONE □ TESTOVIRON SCHERING □ TESTOVIRON T □ TESTRONE □ TESTRYL □ VIRORMONE □ VIROSTERONE

CONSENSUS REPORTS: IARC Cancer Review: Animal Sufficient Evidence IMEMDT 6,209,74, IMEMDT 21,519,79;

Human Limited Evidence IMEMDT 21,519,79. Reported in EPA TSCA Inventory. EPA Genetic Toxicology Program.
SAFETY PROFILE: Confirmed carcinogen with experimental neoplastigenic and teratogenic data. Poison by intraperitoneal route. Human teratogenic effects by unspecified route: developmental abnormalities of the urogenital system. Experimental reproductive effects. Human mutation data reported. Workers engaged in manufacture and packaging have shown effects from this hormone, e.g., enlargement of the breasts in male workers. A promoter. When heated to decomposition it emits acrid smoke and irritating fumes. Used as a drug for the treatment of hypogonadism and metastatic breast cancer.

TBG000   CAS: 57-85-2   HR: 3
TESTOSTERONE PROPIONATE
mf: $C_{22}H_{32}O_3$ mw: 344.54
PROP: Stout prisms from alc and water. Mp: 118–122°. Insol in water; freely sol in alc, ether, pyridine, and other org solvs. Sol in vegetable oils.
SYNS: AGOVIRIN □ ANDROGEN □ ANDROSAN □ Δ⁴-ANDROSTENE-17-β-PROPIONATE-3-ONE □ ANDROTESTON □ ANDROTEST P □ ANDRUSOL-P □ ANERTAN □ AQUAVIRON □ BIO-TESTICULINA □ ENARMON □ HOMANDREN (amps) □ HORMOTESTON □ MASENATE □ NASDOL □ NEO-HOMBREOL □ NSC-9166 □ OKASA-MASCUL □ ORCHIOL □ ORCHISTIN □ ORETON □ ORETON PROPIONATE □ 17-(1-OXOPROPOXY)-(17-β)-ANDROST-4-EN-3-ONE □ PANESTIN □ PERANDREN □ PROPIOKAN □ RECTHORMONE TESTOSTERONE □ STERANDRYL □ SYNANDROL □ SYNERONE □ TELIPEX □ TESTA-FORM □ TESTEX □ TESTODET □ TESTODRIN □ TESTOGEN □ TESTONIQUE □ TESTORMOL □ TESTOSTERONE PROPIONATE □ TESTOSTERONE-17-PROPIONATE □ TESTOSTERONE-17-β-PROPIONATE □ TESTOVIRON □ TESTOXYL □ TESTREX □ TOSTRIN □ TP □ UNITESTON □ VULVAN
CONSENSUS REPORTS: IARC Cancer Review: Animal Sufficient Evidence IMEMDT 21,519,79. Reported in EPA TSCA Inventory.

SAFETY PROFILE: Confirmed carcinogen with experimental neoplastigenic, tumorigenic, and teratogenic data. Moderately toxic by ingestion and intraperitoneal routes. Human male reproductive effects by intramuscular and parenteral routes: changes in spermatogenesis, testes, epididymis, and sperm duct. Human female reproductive effects by intramuscular and parenteral routes: menstrual cycle changes or disorders and effects on fertility. Experimental reproductive effects. Mutation data reported. When heated to decomposition it emits acrid smoke and irritating fumes. See also TESTOSTERONE.

TBG700   CAS: 2475-45-8   HR: 1
1,4,5,8-TETRAAMINO-9,10-ANTHR-ACENEDIONE
mf: $C_{14}H_{12}N_4O_2$ mw: 268.30
SYNS: ACETATE BLUE G □ ACETOQUINONE BLUE L □ ACETOQUINONE BLUE R □ ACETYLON FAST BLUE G □ AMACEL BLUE GG □ AMACEL PURE BLUE B □ ARTISIL BLUE SAP □ BRASILAZET BLUE GR □ CELANTHRENE PURE BLUE BRS □ CELLITON BLUE G □ C.I. 64500 □ CIBACET SAPPHIRE BLUE G □ C.I. DISPERSE BLUE 1 □ CILLA BLUE EXTRA □ C.I. SOLVENT BLUE 18 □ DIACELLITON FAST BLUE R □ DISPERSE BLUE NO 1 □ DURANOL BRILLIANT BLUE CB □ FENACET BLUE G □ GRASOL BLUE 2GS □ KAYALON FAST BLUE BR □ MICROSETILE BLUE EB □ MIKETON FAST BLUE □ NACELAN BLUE G □ NCI-C54900 □ NEOSETILE BLUE EB □ NYLOQUINONE BLUE 2J □ ORACET SAPPHIRE BLUE G □ PERLITON BLUE B □ SERINYL BLUE 2G □ SUPRACET BRILLIANT BLUE 2GN □ 1,4,5,8-TETRAAMINOANTHRAQUINONE □ 1,4,5,8-TETRAMINOANTHRAQUINONE
CONSENSUS REPORTS: NTP 10th Report on Carcinogens. NTP Carcinogenesis Studies (feed); Equivocal Evidence: mouse NTPTR* NTP-TR-299-86; (feed); Clear Evidence: rat NTPTR* NTP-TR-299,86. Reported in EPA TSCA Inventory.
SAFETY PROFILE: Confirmed carcinogen with experimental carcinogenic data. Experimental reproductive effects. Mutation data reported. When heated to decomposition it emits toxic fumes of NO_x.

TBI700   CAS: 10563-26-5   HR: 3
1,5,8,12-TETRAAZADODECANE
mf: $C_8H_{22}N_4$ mw: 174.34
SYNS: N-(2-(3-AMINOPROPYLAMINO)ETHYL)-1,3-PROPANEDIAMINE □ N,N'-BIS(3-AMINOPROPYL)ETHYLENEDIAMINE □ N,N'-BIS(3-AMINOPROPYL)DIAMINOETHANE □ N,N'-DIAMINOPROPYLETHYLENEDIAMINE □ N4 AMINE □ 1,3-PROPANEDIAMINE, N,N''-ETHYLENEBIS- □ 1,3-PROPANEDIAMINE, N,N''-1,2-ETHANEDIYLBIS-
SAFETY PROFILE: A poison by skin contact. Moderately toxic by ingestion. When heated to decomposition it emits toxic vapors of NO_x.

TBJ600   HR: D
(1R,3S)3[(1'RS)(1',2',2',2'-TETRABROMOETHYL)]-2,2-DIMETHYL-CYCLOPROPANECARBOXYLIC ACID (S)-α-CYANO-3-PHENOXYBENZYL ESTER
SAFETY PROFILE: When heated to decomposition emits toxic fumes of Br^-, CN^-.

TBL500   CAS: 10428-19-0   HR: 3
TETRABUTYL DICHLOROSTANNOXANE
mf: $C_{16}H_{36}Cl_2OSn_2$ mw: 552.80
PROP: White crystals from Me_2CO. Mp: 112.5°. Sol in org solvs.
CONSENSUS REPORTS: Reported in EPA TSCA Inventory.
OSHA PEL: TWA 0.1 mg(Sn)/m^3 (skin)
ACGIH TLV: TWA 0.1 mg(Sn)/m^3; STEL 0.2 mg(Sn)/m^3 (skin).
NIOSH REL: TWA (Organotin Compounds) 0.1 mg(Sn)/m^3
SAFETY PROFILE: Poison by intravenous route. When heated to decomposition it emits toxic fumes of Cl^-. See also TIN COMPOUNDS.

TBM250   CAS: 1461-25-2   HR: 3
TETRABUTYLSTANNANE
mf: $C_{16}H_{36}Sn$ mw: 347.21
PROP: A liquid. D: 1.06 @ 20°/4°, bp: 145° @ 10 mm.
SYNS: TETRA-n-BUTYLCIN (CZECH) □ TETRABUTYL-TIN

CONSENSUS REPORTS: Reported in EPA TSCA Inventory.
OSHA PEL: TWA 0.1 mg(Sn)/m^3 (skin)
ACGIH TLV: TWA 0.1 mg(Sn)/m^3; STEL 0.2 mg(Sn)/m^3 (skin).
NIOSH REL: (Organotin Compounds) TWA 0.1 mg(Sn)/m^3
SAFETY PROFILE: Poison by intravenous, intraperitoneal, and parenteral routes. Moderately toxic by skin contact. An eye irritant. When heated to decomposition it emits acrid smoke and irritating fumes. See also TIN COMPOUNDS.

TBN200   CAS: 21239-57-6   HR: 3
TETRACARBONYL(TRIFLUOROMETHYLTHIO)MANGANESE dimer
mf: $C_{10}F_6Mn_2O_8S_2$ mw: 536.10
PROP: Orange or yellow crystals from CH_2Cl_2/hexane. Mp: 99°. Sublimes @ 0.01°.
SYN: MANGANESE, TETRACARBONYL(TRIFLUOROMETHYLTHIO)-, dimer
OSHA PEL: CL 5 mg(Mn)/m^3
ACGIH TLV: TWA 5 mg(Mn)/m^3
SAFETY PROFILE: Poison by intravenous route. When heated to decomposition it emits toxic fumes of SO_x, Mn, and F^-.

TBN740   CAS: 634-66-2   HR: 2
1,2,3,4-TETRACHLOROBENZENE
mf: $C_6H_2Cl_4$ mw: 215.88
PROP: Needles. Mp: 46–47°, bp: 254° @ 761 mm, flash p: >230°F.
CONSENSUS REPORTS: Reported in EPA TSCA Inventory.
SAFETY PROFILE: Moderately toxic by ingestion. An experimental teratogen. Experimental reproductive effects. Irritant. Combustible liquid. When heated to decomposition it emits toxic fumes of Cl^-. See also CHLORINATED HYDROCARBONS, AROMATIC.

TBO000   CAS: 15721-02-5   HR: 3
TETRACHLOROBENZIDINE
mf: $C_{12}H_8Cl_4N_2$ mw: 322.02
PROP: Orange crystals from EtOH. Mp: 137.5°.

SYNS: 2,2',5,5'-TETRACHLOROBENZIDINE □ 3,3',6,6'-TETRACHLOROBENZIDINE □ 2,2',5,5'-TETRACHLORO-(1,1'-BIPHENYL)-4,4'-DIAMINE, (9CI) □ 2,2',5,5'-TETRA-CHLORO-4,4'-DIAMINODIPHENYL

CONSENSUS REPORTS: IARC Cancer Review: Group 3 IMEMDT 7,56,87; Animal Inadequate Evidence IMEMDT 27,141,82. Reported in EPA TSCA Inventory.

SAFETY PROFILE: Suspected carcinogen with experimental carcinogenic and tumorigenic data. When heated to decomposition it emits very toxic fumes of NO_x and Cl^-.

TBO700   CAS: 32598-13-3   HR: 3
3,3',4,4'-TETRACHLOROBIPHENYL

mf: $C_{12}H_6Cl_4$ mw: 291.98

PROP: Needles from EtOH. Bp: 246°.

SYNS: 4-CB □ TCB □ 3,4,3',4'-TETRACHLOROBIPHEN-YL

SAFETY PROFILE: Poison by ingestion. An experimental teratogen. Experimental reproductive effects. Mutation data reported. When heated to decomposition it emits toxic fumes of Cl^-. See also POLYCHLORINATED BIPHENYLS.

TBP000   CAS: 76-11-9   HR: 1
1,1,1,2-TETRACHLORO-2,2-DI-   FLUOROETHANE

mf: $C_2Cl_4F_2$ mw: 203.82

PROP: D: 1.65 @ 20°/4°, mp: 38–40°, bp: 91°. IDLH 2000 ppm.

SYNS: CFC-112a □ 1,1-DIFLUORO-1,2,2,2-TETRA-CHLOROETHANE □ HALOCARBON 112a □ REFRIGERANT 112a □ 1,1,1,2-TETRACHLORO-2,2-DI-FLUOROETHANE

CONSENSUS REPORTS: Reported in EPA TSCA Inventory.

OSHA PEL: TWA 500 ppm

ACGIH TLV: TWA 500 ppm; (Proposed: 100 ppm)

DFG MAK: 1000 ppm (8500 mg/m³)

NIOSH REL: (1,1,1,2-Tetrachloro-2,2-difluoroethane) TWA 500 ppm

SAFETY PROFILE: Mildly toxic by inhalation. When heated to decomposition it emits very toxic fumes of Cl^- and F^-. Used as a refrigerant. See also FLUORIDES.

TBP050   CAS: 76-12-0   HR: 3
1,1,2,2-TETRACHLORO-1,2-DI-   FLUOROETHANE

mf: $C_2Cl_4F_2$ mw: 203.82

PROP: Liquid or crystals. Mp: 24.65°, bp: 92.8°, d: 1.6447 @ 25°, vap d: 7.03. IDLH 2000 ppm.

SYNS: CFC-112 □ 1,2-DIFLUORO-1,1,2,2-TETRACHLOROETHANE □ ETHANE, 1,1,2,2-TETRACHLORO-1,2-DIFLUORO- □ F-112 □ FC 112 □ FREON 112 □ FREON R 112 □ GENETRON 112 □ HALOCARBON 112 □ REFRIGERANT 112 □ UCON 112

CONSENSUS REPORTS: Reported in EPA TSCA Inventory.

OSHA PEL: TWA 500 ppm

ACGIH TLV: TWA 500 ppm; (Proposed: 50 ppm)

DFG MAK: 200 ppm (1700 mg/m³)

NIOSH REL: (1,1,2,2-Tetrachloro-1,2-difluoroethane) TWA 500 ppm

SAFETY PROFILE: A poison by inhalation. Moderately toxic by ingestion. A skin and eye irritant. When heated to decomposition it emits toxic fumes of F^- and Cl^-. See also CHLORINATED HYDROCARBONS, ALIPHATIC; and FLUORIDES.

TBP250   CAS: 31242-94-1   HR: 3
TETRACHLORODIPHENYL OXIDE

mf: $C_{12}H_6Cl_4O$ mw: 307.98

SYNS: PHENYL ETHER TETRACHLORO □ TETRA-CHLOROPHENYL ETHER

OSHA PEL: TWA 0.5 mg/m³

SAFETY PROFILE: Poison by ingestion. When heated to decomposition it emits toxic fumes of Cl^-. See also ETHERS.

TBP750   CAS: 25322-20-7   HR: 3
TETRACHLOROETHANE

DOT: UN 1702

mf: $C_2H_2Cl_4$ mw: 167.84

CONSENSUS REPORTS: Reported in EPA TSCA Inventory.

NIOSH REL: (Tetrachloroethane) Reduce to lowest feasible level

DOT CLASSIFICATION: 6.1; Label: Poison

SAFETY PROFILE: Poison by ingestion and inhalation. Moderately toxic by intraperitoneal route. Mildly toxic by skin

T

contact. Experimental reproductive effects. Mutation data reported. When heated to decomposition it emits toxic fumes of Cl⁻. See also CHLORINATED HYDROCARBONS, ALIPHATIC.

TBQ000   CAS: 630-20-6   HR: 2
1,1,1,2-TETRACHLOROETHANE
mf: $C_2H_2Cl_4$ mw: 167.84
PROP: Liquid. D: 1.542 @ 20°/4°, bp: 135.1°, fp: −68.1°. Sol in water; misc in alc, ether.
SYNS: NCI-C52459 □ RCRA WASTE NUMBER U208
CONSENSUS REPORTS: IARC Cancer Review: Group 3 IMEMDT 7,56,87; Animal Limited Evidence IMEMDT 41,87,86. NTP Carcinogenesis Bioassay (gavage); Clear Evidence: mouse NTPTR* NTP-TR-237,82; (gavage); No Evidence: rat NTPTR* NTP-TR-237,82. Reported in EPA TSCA Inventory.
NIOSH REL: (1,1,1,2-Tetrachloroethane) Handle with caution
SAFETY PROFILE: A skin and severe eye irritant. Questionable carcinogen with experimental carcinogenic data. Incompatible with dinitrogen tetraoxide, 2,4-dinitrophenyl disulfide, potassium, potassium hydroxide, nitrogen tetroxide, sodium, sodium potassium alloy. Mutation data reported. When heated to decomposition it emits very toxic fumes of Cl⁻.

TBQ100   CAS: 79-34-5   HR: 3
1,1,2,2-TETRACHLOROETHANE
mf: $C_2H_2Cl_4$ mw: 167.84
PROP: Heavy, colorless, mobile liquid; chloroform-like odor. Mp: −43.8°, bp: 146.4°, d: 1.600 @ 20°/4°, fp: −43.8°. IDLH 100 ppm.
SYNS: ACETYLENE TETRACHLORIDE □ BONO-FORM □ CELLON □ 1,1,2,2-CZTEROCHLOROETAN - (POLISH) □ 1,1-DICHLORO-2,2-DICHLOROETHANE □ NCI-C03554 □ RCRA WASTE NUMBER U209 □ TCE □ TETRACHLORETHANE □ 1,1,2,2-TETRACHLO-ORETHAAN (DUTCH) □ 1,1,2,2-TETRACHLORAETHAN - (GERMAN) □ 1,1,2,2-TETRACHLORETHANE (FRENCH) □ sym-TETRACHLOROETHANE □ TETRACHLORURE

d'ACETYLENE (FRENCH) □ 1,1,2,2-TETRACLOROETANO (ITALIAN) □ WESTRON
CONSENSUS REPORTS: IARC Cancer Review: Group 3 IMEMDT 7,354,87; Animal Limited Evidence IMEMDT 20,477,79; NCI Carcinogenesis Bioassay - (gavage); Clear Evidence: mouse NCITR* NCI-CG-TR-27,78; Some Evidence: rat NCITR* NCI-CG-TR-27,78. Reported in EPA TSCA Inventory. EPA Genetic Toxicology Program. Community Right-To-Know List.
OSHA PEL: TWA 1 ppm (skin)
ACGIH TLV: TWA 1 ppm (skin); Animal Carcinogen
DFG MAK: 1 ppm (7 mg/m³); Confirmed Animal Carcinogen with Unknown Relevance to Humans
NIOSH REL: (1,1,2,2-Tetrachlorethane) Reduce to lowest level
SAFETY PROFILE: Suspected carcinogen with experimental carcinogenic and tumorigenic data. Poison by inhalation, ingestion, and intraperitoneal routes. Moderately toxic by several other routes. Mutation data reported. Human central nervous system effects by ingestion and inhalation: general anesthesia, somnolence, hallucinations, and distorted perceptions. Considered the most toxic of the common chlorinated hydrocarbons. Considered to be a very severe industrial hazard and its use has been restricted or even forbidden in certain countries. It is not an inert solvent. Reacts violently with N_2O_4, 2,4-dinitrophenyl disulfide, and on contact with sodium or potassium. When heated in contact with solid potassium hydroxide, spontaneously flammable chloro- or dichloroacetylene gas is evolved. Any water can cause appreciable hydrolysis, even at room temperature, and both hydrolysis and oxidation become comparatively rapid above 110°. When heated to decomposition it emits toxic fumes of Cl⁻.

A strong irritant of eyes and mucous membranes. A concentration of 3 ppm produces a detectable odor, thus an initial warning effect. Its narcotic action is stronger than that of chloroform, but, because of its low volatility, narcosis is less severe and much less common in industrial poisoning than in the case of other chlorinated hydrocarbons. The toxic action of this material is chiefly on the liver, where it produces acute yellow atrophy and cirrhosis. Fatty degeneration of the kidneys and heart, hemorrhage into the lungs and serous membranes, and edema of the brain have also been found in fatal cases. Some reports indicate a toxic action on the central nervous system with changes in the brain and in the peripheral nerves. The effect on the blood is one of hemolysis with appearance of young cells in the circulation and a monocytosis. Due to its solvent action on the natural skin oils, dermatitis is not uncommon.

The initial symptoms resulting from exposure to the vapor are lachrymation, salivation, and irritation of the nose and throat. Continued exposure to high concentrations results in restlessness, dizziness, nausea, vomiting, and narcosis. The latter, however, is rare in industry. More commonly, exposure is less severe and most complaints are vague and related to the digestive and nervous systems. The patient's symptoms gradually progress to a more serious illness with development of toxic jaundice, liver tenderness, etc., and possibly albuminuria and edema. With serious liver damage, the jaundice increases and toxic symptoms appear, with somnolence, delirium, convulsions, and coma usually preceding death. See also ACETYLENE COMPOUNDS and CHLORIDES.

TBQ750   CAS: 1897-45-6   HR: 3
TETRACHLOROISOPHTHALONITRILE
mf: $C_8Cl_4N_2$ mw: 265.90
PROP: A solid. Mp: 250–251°.
SYNS: BRAVO □ BRAVO 6F □ BRAVO-W-75 □ CHLOROALONIL □ CHLOROTHALONIL □ CHLORTHALONIL (GERMAN) □ DAC 2797 □ DACONIL □ DACONIL 2787 FLOWABLE FUNGICIDE □ DACOSOIL □ 1,3-DICYANOTETRACHLOROBENZENE □ EXOTHERM □

EXOTHERM TERMIL □ FORTURF □ NCI-C00102 □ NOPCOCIDE □ SWEEP □ TCIN □ m-TCPN □ TERMIL □ 2,4,5,6-TETRACHLORO-3-CYANOBENZONITRILE □ m-TETRACHLOROPHTHALONITRILE □ TPN (pesticide)
CONSENSUS REPORTS: IARC Cancer Review: Group 3 IMEMDT 7,56,87; Animal Limited Evidence IMEMDT 30,319,83. NCI Carcinogenesis Bioassay (feed); Clear Evidence: rat NCITR* NCI-CG-TR-41,78. Cyanide and its compounds are on the Community Right-To-Know List. Reported in EPA TSCA Inventory. EPA Genetic Toxicology Program.
DFG MAK: Confirmed Animal Carcinogen with Unknown Relevance to Humans
SAFETY PROFILE: Suspected carcinogen with experimental carcinogenic data. Moderately toxic by skin contact and intraperitoneal routes. Mildly toxic by ingestion. Mutation data reported. When heated to decomposition it emits very toxic fumes of Cl^-, NO_x, and CN^-. See also NITRILES.

TBR000   CAS: 1335-88-2   HR: 3
TETRACHLORONAPHTHALENE
mf: $C_{10}H_4Cl_4$ mw: 265.94
PROP: Colorless to pale-yellow solid; aromatic odor. Mw: 265.96, specific gravity: 1.59–1.65, mp: 115°, bp: 311.5–360°, flash p: 410°F. Insol in water.
SYN: HALOWAX
CONSENSUS REPORTS: Reported in EPA TSCA Inventory.
OSHA PEL: TWA 2 mg/m³
ACGIH TLV: TWA 2 mg/m³
SAFETY PROFILE: Probably a poison. When heated to decomposition it emits highly toxic fumes of Cl^-. See also CHLORINATED HYDROCARBONS, AROMATIC; and POLYCHLORINATED BIPHENYLS.

TBT000   CAS: 58-90-2   HR: 3
2,4,5,6-TETRACHLOROPHENOL
mf: $C_6H_2Cl_4O$ mw: 231.88
PROP: Needles from ligroin. Mp: 70°, bp: 150° @ 15 mm.

SYNS: DOWICIDE 6 □ RCRA WASTE NUMBER U212 □ TCP □ 2,3,4,6-TETRACHLOROPHENOL

CONSENSUS REPORTS: IARC Cancer Review: Human Limited Evidence IMEMDT 41,319,86. Chlorophenol compounds are on the Community Right-To-Know List. Reported in EPA TSCA Inventory.

SAFETY PROFILE: Suspected carcinogen with experimental carcinogenic and teratogenic data. Poison by ingestion, skin contact, intraperitoneal, and subcutaneous routes. Mutation data reported. When heated to decomposition it emits toxic fumes of Cl⁻. Used as a disinfectant and a preservative for wood, latex, and leather. See also CHLOROPHENOLS.

TBV300   CAS: 3737-41-5   HR: 3
3,3,4,4-TETRACHLOROTETRAHYDRO-THIOPHENE-1,1-DIOXIDE
mf: $C_4H_4Cl_4O_2S$ mw: 257.94
SYNS: CP 78601 □ DAC 649 □ TETRAHYDRO-3,3,4,4-TETRACHLOROTHIOPHENE 1,1-DIOXIDE □ THIOPHENE, TETRAHYDRO-3,3,4,4-TETRACHLORO-, 1,1-DIOXIDE

SAFETY PROFILE: A poison by ingestion and skin contact. Moderately toxic by inhalation. A severe skin irritant. When heated to decomposition it emits toxic vapors of SO_x and Cl⁻.

TBV750   CAS: 6012-97-1   HR: 3
2,3,4,5-TETRACHLOROTHIOPHENE
mf: C_4Cl_4S mw: 221.90
PROP: Mp: 28–30°, bp: 75° @ 2 mm, refr index: 1.5910, d: 1.704, flash p: 230°F.
SYNS: 2,3,4,5-CHLOROTHIOPHENE □ ENT 25,764 □ IF (fumigant) □ PENN SALT TD-183 □ PENPHENE □ PERCHLOROTHIOPHENE □ TCTP □ TD-183 □ TETRACHLOROTHIOFENE □ TETRACHLORO-THIOPHENE

SAFETY PROFILE: Poison by ingestion, skin contact, intravenous, and intraperitoneal routes. Moderately toxic by inhalation. Combustible when exposed to heat or flame. When heated to decomposition it emits very toxic fumes of Cl⁻ and SO_x.

TBW100   CAS: 961-11-5   HR: 3
TETRACHLORVINPHOS
mf: $C_{10}H_9Cl_4O_4P$ mw: 365.96
PROP: Sol in $CHCl_3$; almost insol in H_2O.
SYNS: 2-CHLORO-1-(2,4,5-TRICHLOROPHENYL)VINYL DIMETHYL PHOSPHATE □ 2-CHLORO-1-(2,4,5-TRICHLOROPHENYL)VINYL PHOSPHORIC ACID DIMETHYL ESTER □ O,O-DIMETHYL-O-2-CHLOR-1-(2,4,5-TRICHLORPHENYL)-VINYL-PHOSPHAT (GERMAN) □ IPO 8 □ NCI-C00168 □ PHOSPHORIC ACID, 2-CHLORO-1-(2,4,5-TRICHLOROPHENYL)ETHENYL DIMETHYL ESTER □ 2,4,5-TRICHLORO-α-(CHLOROMETHYLENE)-BENZYL PHOSPHATE

CONSENSUS REPORTS: NCI Carcinogenesis Bioassay (feed); Results Positive: mouse, rat NCITR* NCI-CG-TR-33,78. Community Right-To-Know List.

SAFETY PROFILE: Suspected carcinogen with experimental carcinogenic, neoplastigenic, and tumorigenic data. Poison by ingestion. Moderately toxic by intraperitoneal route. Experimental reproductive effects. When heated to decomposition it emits toxic fumes of Cl⁻ and PO_x.

TBW250   CAS: 14323-41-2   HR: 3
TETRACYANONICKELATE(2−) DI-POTASSIUM, HYDRATE
mf: $C_4N_4Ni \cdot K \cdot H_2O$ mw: 219.91
SYN: POTASSIUM CYANONICKELATE HYDRATE

CONSENSUS REPORTS: NTP 10th Report on Carcinogens. Nickel and its compounds, as well as cyanide and its compounds, are on the Community Right-To-Know List.
OSHA PEL: TWA 0.1 mg (Ni)/m³
ACGIH TLV: TWA 0.2 mg(Ni)/m³; Human Carcinogen)

SAFETY PROFILE: Confirmed human carcinogen. Mutation data reported. Many nickel compounds are poisons. When heated to decomposition it emits very toxic fumes of CN⁻, K_2O, and NO_x. See also NICKEL COMPOUNDS and CYANIDE.

TBX250   CAS: 64-75-5   HR: 3
TETRACYCLINE HYDROCHLORIDE
mf: $C_{22}H_{24}N_2O_8 \cdot ClH$ mw: 480.94
PROP: Crystals from butanol/HCl. Mp: 214° (decomp). Very sol in water; sol in

methanol, ethanol; insol in ether, hydrocarbon solvents.

SYNS: ACHROMYCIN □ ACHROMYCIN HYDRO-CHLORIDE □ AMBRACYN □ ARTOMYCIN □ BRISTACYCLINE □ CEFRACYCLINE TABLETS □ CHLORHYDRATE de TETRACYCLINE (FRENCH) □ CYCLOPAR □ DIACYCINE □ DUMOCYCIN □ MEDAMYCIN □ MEPHACYCLIN □ NCI-C55561 □ PALTET □ PANMYCIN HYDROCHLORIDE □ PARTREX □ PIRACAPS □ POLYCYCLINE HYDROCHLORIDE □ QIDTET □ QUADRACYCLINE □ QUATREX □ REMICYCLIN □ RICYCLINE □ RO-CYCLINE □ SK-TETRACYCLINE □ STECLIN HYDROCHLORIDE □ SUBAMYCIN □ SUPRAMYCIN □ T-250 CAPSULES □ TC HYDROCHLORIDE □ TEFILIN □ TELINE □ TELOTREX □ TETRABAKAT □ TETRABLET □ TETRACAPS □ TETRACICLINA CLORIDRATO - (ITALIAN) □ TETRACOMPREN □ TETRACYCLINE CHLORIDE □ TETRA-D □ TETRALUTION □ TETRA-WEDEL □ TETROSOL □ TOPICYCLINE □ TOTOMYC-IN □ TRIPHACYCLIN □ U-5965 □ UNICIN □ UNIMYC-IN □ VETQUAMYCIN-324

CONSENSUS REPORTS: Reported in EPA TSCA Inventory.

SAFETY PROFILE: Poison by intraperitoneal and intravenous routes. Moderately toxic by ingestion and subcutaneous routes. Human systemic effects: change in taste function. An experimental teratogen. Experimental reproductive effects. Mutation data reported. When heated to decomposition it emits very toxic fumes of HCl and NO_x. See also TETRACYCLINE.

TBY500   CAS: 27196-00-5   HR: 2
TETRADECANOL, mixed isomers
mf: $C_{14}H_{30}O$ mw: 214.44

SYNS: MYRISTYL ALCOHOL (mixed isomers) □ TETRADECYL ALCOHOL

CONSENSUS REPORTS: Reported in EPA TSCA Inventory.

SAFETY PROFILE: Mildly toxic by ingestion and skin contact. Combustible when exposed to heat or flame; can react with oxidizing materials. To fight fire, use CO_2, dry chemical. When heated to decomposition it emits acrid smoke and irritating fumes.

TCA500   CAS: 139-08-2   HR: 1
TETRADECYL DIMETHYL BENZYL-AMMONIUM CHLORIDE
mf: $C_{23}H_{42}N \cdot Cl$ mw: 368.11

PROP: A solid. Mp: 60–61°.

SYNS: ARQUAD DM14B-90 □ N,N-DIMETHYL-N-TETRADECYLBENZENEMETHANAMINIUM, CHLORIDE (9CI) □ NISSAN CATION M2-100

CONSENSUS REPORTS: Reported in EPA TSCA Inventory.

SAFETY PROFILE: A skin and eye irritant. When heated to decomposition it emits very toxic fumes of NO_x, NH_3, and Cl^-.

TCB100   CAS: 17661-50-6   HR: 1
TETRADECYL STEARATE
mf: $C_{32}H_{64}O_2$ mw: 480.96

PROP: Flake at 25 C. Mp: 43°–47°.

SYNS: MYRISTYL STEARATE □ OCTADECANOIC ACID, TETRADECYL ESTER (9CI) □ STEARIC ACID, TETRADECYL ESTER □ TETRADECYL OCTADECANOATE

CONSENSUS REPORTS: Reported in EPA TSCA Inventory.

SAFETY PROFILE: An eye irritant. When heated to decomposition it emits acrid smoke and irritating fumes.

TCE250   CAS: 112-60-7   HR: 2
TETRAETHYLENE GLYCOL
mf: $C_8H_{18}O_5$ mw: 194.26

PROP: Colorless to pale straw-colored liquid. Bp: 327.3°, fp: −6°, flash p: 360°F - (OC), d: 1.1248 @ 20/20°, vap press: 1 mm @ 153.9°. Misc in water.

SYNS: HI-DRY □ 2,2'-(OXYBIS(ETHYLENEOXY))DI-ETHANOL

CONSENSUS REPORTS: Reported in EPA TSCA Inventory.

SAFETY PROFILE: Mildly toxic by ingestion. A skin and eye irritant. Combustible when exposed to heat or flame; can react with oxidizing materials. To fight fire, use alcohol foam, water, CO_2, dry chemical. When heated to decomposition it emits acrid smoke and irritating fumes.

T

TCE500   CAS: 112-57-2   HR: 3
TETRAETHYLENEPENTAMINE
DOT: UN 2320

mf: $C_8H_{23}N_5$ mw: 189.36

PROP: Viscous, hygroscopic liquid. Bp: 333°, mp: −40°, flash p: 325°F (OC), d: 0.9980 @ 20°/20°, vap press: <0.01 mm @ 20°.

SYNS: D.E.H. 26 □ 1,4,7,10,13-PENTAAZATRIDECANE

CONSENSUS REPORTS: Reported in EPA TSCA Inventory.

DOT CLASSIFICATION: 8; Label: Corrosive

SAFETY PROFILE: Poison by ingestion and intravenous routes. Moderately toxic by skin contact. Mutation data reported. A corrosive irritant to skin, eyes, and mucous membranes. Combustible when exposed to heat or flame. Can react with oxidizing materials. To fight fire, use CO_2, dry chemical. When heated to decomposition it emits toxic fumes of NO_x.

TCF000   CAS: 78-00-2   HR: 3
TETRAETHYL LEAD
DOT: NA 1649

mf: $C_8H_{20}Pb$ mw: 323.47

PROP: Colorless, oily liquid; pleasant characteristic odor. Mp: 125–150°, bp: 198–202° with decomp, d: 1.659 @ 18°, vap press: 1 mm @ 38.4°, flash p: 200°F. IDLH 40 mg/m³ (as Pb).

SYNS: CZTEROETHLEK OLOWIU (POLISH) □ LEAD, TETRAETHYL- □ NCI-C54988 □ NSC-22314 □ PIOMBO TETRA-ETILE □ RCRA WASTE NUMBER P110 □ TEL □ TETRAETHYL LEAD, liquid (DOT) □ TETRAETHYL-OLOVO □ TETRAETHYLPLUMBANE □ TETRAETHYL-PLUMBIUM

CONSENSUS REPORTS: IARC Cancer Review: Group 3 IMEMDT 7,230,87; Animal Inadequate Evidence IMEMDT 23,325,80; IMEMDT 2,150,73. EPA Extremely Hazardous Substances List. Reported in EPA TSCA Inventory. EPA Genetic Toxicology Program.

OSHA PEL: TWA 0.075 mg(Pb)/m³ (skin)

ACGIH TLV: TWA 0.1 mg(Pb)/m³ (skin); Not Classifiable as a Human Carcinogen

DFG MAK: 0.05 mg/m³

DOT CLASSIFICATION: 6.1; Label: Poison, Flammable Liquid

SAFETY PROFILE: Human poison by an unspecified route. Experimental poison by ingestion, intraperitoneal, intravenous, subcutaneous, and parenteral routes. Moderately toxic by inhalation and skin contact. Experimental teratogenic and reproductive effects. Questionable carcinogen with experimental carcinogenic data. Mutation data reported. Lead compounds are particularly toxic to the central nervous system. It is a solvent for fatty materials and has some solvent action on rubber as well. The fact that it is a lipoid solvent makes it an industrial hazard because it can cause intoxication not only by inhalation but also by absorption through the skin. Decomposes when exposed to sunlight or allowed to evaporate; forms triethyl lead, which is also a poisonous compound, as one of its decomposition products. May cause elemental lead intoxication by coming in contact with the skin.

A combustible liquid when exposed to heat, flame, or oxidizers. Can react vigorously with oxidizing materials. Exposure to air for several days may cause explosive decomposition. To fight fire, use dry chemical, CO_2, mist, foam. When heated to decomposition it emits toxic fumes of Pb. See also LEAD COMPOUNDS.

TCF250   CAS: 107-49-3   HR: 3
TETRAETHYL PYROPHOSPHATE
DOT: NA 2783/NA 3018

mf: $C_8H_{20}O_7P_2$ mw: 290.22

PROP: Water-white to amber hygroscopic liquid. D: 1.20, bp: 104–110°. Misc in H_2O and most org solvs. Spar sol in pet ether.

SYNS: BIS-O,O-DIETHYLPHOSPHORIC ANHYDRIDE □ BLADAN □ DIPHOSPHORIC ACID TETRAETHYL ESTER □ ENT 18,771 □ FOSVEX □ GRISOL □ HEPT □ HEXAMITE □ KILLAX □ KILMITE 40 □ LETHALAIRE G-52 □ LIROHEX □ MORTOPAL □ NIFOS T □ PYROPHOSPHATE de TETRAETHYLE (FRENCH) □ RCRA WASTE NUMBER P111 □ TEPP (ACGIH) □ O,O,O,O-TETRAETHYL-DIPHOSPHAT, BIS(O,O-DI-

AETHYLPHOSPHORSAEURE)-ANHYDRID (GERMAN) □
O,O,O,O-TETRAETHYL-DIFOSFAAT (DUTCH) □
TETRAETHYL PYROFOSFAAT □ TETRAETHYL
PYROPHOSPHATE, liquid (DOT) □ O,O,O,O-TETRAETIL-
PIROFOSFATO (ITALIAN) □ TETRASTIGMINE □
TETRON □ TETRON-100 □ VAPOTONE
CONSENSUS REPORTS: EPA Extremely
Hazardous Substances List.
OSHA PEL: TWA 0.05 mg/m³ (skin)
ACGIH TLV: TWA (inhalable fraction) 0.01
mg/m³ (skin)
DFG MAK: 0.005 ppm (0.060 mg/m³)
DOT CLASSIFICATION: 6.1; Label: Poison
SAFETY PROFILE: Human poison by
ingestion and intramuscular routes.
Experimental poison by ingestion, skin
contact, intraperitoneal, intramuscular,
subcutaneous, parenteral, and intravenous
routes. Human systemic effects by ingestion,
intramuscular, and parenteral routes:
paresthesia, wakefulness, excitement, muscle
contraction or spasticity, nausea or vomiting
and other gastrointestinal changes. The
action is similar to that of parathion: causing
an irreversible inhibition of the
cholinesterase molecules and the consequent
accumulation of large amounts of
acetylcholine. Small doses at frequent
intervals are largely additive. When heated to
decomposition it emits toxic fumes of PO_x.
See also PARATHION.

TCF260   **HR: 3**
TETRAETHYLPYROPHOSPHATE and
compressed gas mixtures
DOT: UN 1705
mf: $C_8H_{20}O_7P_2$ mw: 290.22
SYNS: TETRAETHYL PYROPHOSPHATE and
COMPRESSED GAS MIXTURES LC50 < or =200 ppm (UN
1705) (DOT) □ TETRAETHYL PYROPHOSPHATE and
COMPRESSED GAS MIXTURES LC50 >200 ppm, not >5000
ppm (UN 1705) (DOT)
DOT CLASSIFICATION: 2.3; Label: Poison
Gas
SAFETY PROFILE: A poison gas. When
heated to decomposition it emits toxic
fumes of PO_x. See also TETRAETHYL-
PYROPHOSPHATE.

TCF750   CAS: 597-64-8   HR: 3
TETRAETHYLSTANNANE
mf: $C_8H_{20}Sn$ mw: 234.97
PROP: Colorless liquid. D: 1.187 @ 23°,
mp: −112°, bp: 181°. Insol in water; sol in
org solvs.
SYN: TETRAETHYL TIN
CONSENSUS REPORTS: EPA Extremely
Hazardous Substances List. Reported in
EPA TSCA Inventory.
OSHA PEL: TWA 0.1 mg(Sn)/m³ (skin)
ACGIH TLV: TWA 0.1 mg(Sn)/m³; STEL
0.2 mg(Sn)/m³ (skin).
NIOSH REL: (Organotin Compounds) TWA
0.1 mg(Sn)/m³
SAFETY PROFILE: Poison by ingestion,
intravenous, and intraperitoneal routes.
When heated to decomposition it emits
acrid smoke and irritating fumes. See also T-
IN COMPOUNDS.

TCH500   CAS: 116-14-3   HR: 3
TETRAFLUOROETHYLENE
DOT: UN 1081
mf: C_2F_4 mw: 100.02
PROP: Colorless gas. Mp: −142.5°, bp:
−78.4°, lel: 11%, uel: 60%.
SYNS: FLUOROPLAST 4 □ PERFLUOROETHENE □
PERFLUOROETHYLENE □ TETRAFLUORETHYLENE
□ TETRAFLUOROETHENE □ TETRAFLUOROETHYL-
ENE, inhibited (DOT)
CONSENSUS REPORTS: NTP 10th Report
on Carcinogens. IARC Cancer Review:
Group 3 IMEMDT 7,56,87. Reported in
EPA TSCA Inventory.
ACGIH TLV: TWA 2 ppm; Confirmed
Animal Carcinogen with Unknown
Relevance to Humans
DOT CLASSIFICATION: 2.1; Label:
Flammable Gas
SAFETY PROFILE: Confirmed
carcinogen. Mildly toxic by inhalation. Can
act as an asphyxiant and may have other
toxic properties. The gas is flammable when
exposed to heat or flame. The inhibited
monomer will explode if ignited. Explosive
in the form of vapor when exposed to heat
or flame. Will explode at pressures above 2.7
bar if limonene inhibitor is not added.

T

Iodine pentafluoride depletes the limonene inhibitor and then causes explosive polymerization of the monomer. Mixtures with hexafluoropropene and air form an explosive peroxide. Reacts violently with SO_3; air; difluoromethylene dihypofluorite; dioxygen difluoride; iodine pentafluoride; oxygen. When heated to decomposition it emits highly toxic fumes of F^-. See also FLUORIDES.

TCI000   CAS: 10036-47-2   HR: 3
TETRAFLUORO HYDRAZINE
mf: N_2F_4 mw: 104.0
PROP: Colorless gas or liquid; white solid when pure. Almost totally dissociates to NF_2 at 3°. Slowly hydrolyzed by H_2O. Mp: −163°, bp: −73°, d (liquid): 1.5 @ −100°.
SYNS: DINITROGEN TETRAFLUORIDE □ PERFLUORO HYDRAZINE
CONSENSUS REPORTS: Reported in EPA TSCA Inventory.
OSHA PEL: TWA 2.5 mg(F)/m³
ACGIH TLV: TWA 2.5 mg(F)/m³; BEI: 3 mg/g creatinine of fluorides in urine prior to shift; 10 mg/g creatinine of fluorides in urine at end of shift.
NIOSH REL: TWA (Inorganic Fluorides) 2.5 mg(F)/m³
SAFETY PROFILE: A poison. An unstable explosive gas sensitive to light, heat, or contact with air or steel. At high pressures it can explode due to shock or blast. Flammable when exposed to heat or flame. Potentially explosive reaction with hydrocarbons; hydrogen; organic materials; reducing agents; oxygen. Forms explosive mixtures with alkenyl nitrates; nitrogen trifluoride. When heated to decomposition it emits highly toxic fumes of F^- and NO_x. See also FLUORIDES, HYDROFLUORIC ACID, and HYDRAZINE.

TCI100   CAS: 102489-70-3 HR: 3
2,2,3,3-TETRAFLUORO-4,7-METHANO-
2,3,5,6,8,9-HEXAHYDROBENZO-
SELENOPHENE
mf: $C_9H_{10}F_4Se$ mw: 273.15

SYN: 4,7-METHANOSELENOPHENE, OCTAHYDRO-2,2,3,3-TETRAFLUORO-
OSHA PEL: TWA 0.2 mg(Se)/m³
ACGIH TLV: TWA 0.2 mg(Se)/m³
SAFETY PROFILE: Poison by intravenous route. When heated to decomposition it emits toxic fumes of Se and F^-.

TCI250   CAS: 63886-77-1   HR: 3
TETRAFLUORO-m-PHENYLENE DIAM-
INE DIHYDROCHLORIDE
mf: $C_6H_4F_4N_2$•2ClH   mw: 253.04
SAFETY PROFILE: Poison by intraperitoneal route. Questionable carcinogen with experimental carcinogenic data. When heated to decomposition it emits very toxic fumes of F^-, NO_x, and HCl.

TCM250   CAS: 1972-08-3   HR: 3
1-trans-Δ^9-TETRAHYDROCANNABINOL
mf: $C_{21}H_{30}O_2$ mw: 314.51
PROP: Bp: 155–157° @ 0.05 mm.
SYNS: ABBOTT 40566 □ 3-PENTYL-6,6,9-TRIMETHYL-6a,7,8,10a-TETRAHYDRO-6H-DIBENZO(b,d)PYRAN-1-OL □ SP 104 □ (1)-Δ^1-TETRAHYDROCANNABINOL □ Δ^1-TETRAHYDROCANNABINOL □ (−)-Δ^1-3,4-trans-TETRA-HYDROCANNABINOL □ (−)-Δ^9-trans-TETRAHYDRO-CANNABINOL □ trans-Δ^9-TETRAHYDROCANNABINOL □ Δ^9-TETRAHYDROCANNABINOL □ THC □ Δ^1-THC □ Δ^9-THC □ 6,6,9-TRIMETHYL-3-PENTYL-7,8,9,10-TETRA-HYDRO-6H-DIBENZO(B,D)PYRAN-1-OL
CONSENSUS REPORTS: EPA Genetic Toxicology Program.
SAFETY PROFILE: Poison by intraperitoneal and intravenous routes. Moderately toxic by ingestion. Experimental reproductive effects. Questionable carcinogen with experimental tumorigenic and teratogenic data. Human mutation data reported. A hallucinatory drug. When heated to decomposition it emits acrid smoke and irritating fumes. See also CANNABIS.

TCO100   CAS: 7101-64-6   HR: 3
7,8,9,10-TETRAHYDRO-N,N-DIETHYL-
6H-CYCLOHEPTA(b)QUINOLINE-11-
CARBOXAMIDE,
mf: $C_{19}H_{23}N_2O$ mw: 295.44
SYNS: 6H-CYCLOHEPTA(b)QUINOLINE-11-CARBOXAMIDE, N,N-DIETHYL-7,8,9,10-TETRAHYDRO-

□ KETONE, DIETHYLAMINO(7,8,9,10-TETRAHYDRO-11-6H-CYCLOHEPTA(b)QUINOLINYL)

DOT CLASSIFICATION: 3; Label: Flammable Liquid

SAFETY PROFILE: A poison by intraperitoneal route. A flammable liquid. When heated to decomposition it emits toxic vapors of NO$_x$.

TCQ275   CAS: 25952-35-6   HR: 2
TETRAHYDRO-3,5-DIMETHYL-4H,1,3,5-OXADIAZINE-4-THIONE

mf: C$_5$H$_{10}$N$_2$OS   mw: 146.23

SYNS: 4H-1,3,5-OXADIAZINE-4-THIONE, TETRA-HYDRO-3,5-DIMETHYL- □ TDOT

CONSENSUS REPORTS: Reported in EPA TSCA Inventory.

SAFETY PROFILE: Moderately toxic by ingestion. An experimental teratogen. When heated to decomposition it emits toxic fumes of NO$_x$ and SO$_x$.

TCQ700   CAS: 33236-07-6   HR: 3
3,4,5,6-TETRAHYDRO-2-(α-ETHO-XYBENZYL)-5-ETHYL-5-METHYL-PYRIMIDINE

mf: C$_{16}$H$_{24}$N$_2$O   mw: 260.42

SYN: PYRIMIDINE, 3,4,5,6-TETRAHYDRO-2-(α-ETHO-XYBENZYL)-5-ETHYL-5-METHYL-

SAFETY PROFILE: A poison by ingestion. When heated to decomposition it emits toxic vapors of NO$_x$.

TCR750   CAS: 109-99-9   HR: 3
TETRAHYDROFURAN

DOT: UN 2056

mf: C$_4$H$_8$O   mw: 72.12

PROP: Colorless, mobile liquid; ether-like odor. Bp: 65.4°, flash p: 1.4°F (TCC), lel: 1.8%, uel: 11.8%, fp: −65°, d: 0.888 @ 21°/4°, vap press: 114 mm @ 15°, vap d: 2.5, autoign temp: 610°F. Misc with water, alc, ketones, esters, ethers, and hydrocarbons. IDLH 2000 ppm [10%LEL].

SYNS: BUTYLENE OXIDE □ CYCLOTETRAMETHYL-ENE OXIDE □ DIETHYLENE OXIDE □ 1,4-EPO-XYBUTANE □ FURANIDINE □ HYDROFURAN □ NCI-C60560 □ OXACYCLOPENTANE □ OXOLANE □ RCRA WASTE NUMBER U213 □ TETRAHYDROFURAAN -(DUTCH) □ TETRAHYDROFURANNE (FRENCH) □

TETRAIDROFURANO (ITALIAN) □ TETRAMETHYL-ENE OXIDE □ THF

CONSENSUS REPORTS: NTP Carcinogenesis Bioassay (feed); Clear Evidence: mouse, Some Evidence: rat NTPTR* NTP-TR-475,1998. Reported in EPA TSCA Inventory.

OSHA PEL: TWA 200 ppm; STEL 250 ppm
ACGIH TLV: TWA 200 ppm; STEL 250 ppm; BEI: 8 mg/L tetrahydrofuran in urine at end of shift
DFG MAK: 50 ppm (150 mg/m³)
DOT CLASSIFICATION: 3; Label: Flammable Liquid

SAFETY PROFILE: Moderately toxic by ingestion and intraperitoneal routes. Mildly toxic by inhalation. Human systemic effects by inhalation: general anesthesia. Mutation data reported. Irritant to eyes and mucous membranes. Narcotic in high concentrations. Reported as causing injury to liver and kidneys.

Flammable liquid. A very dangerous fire hazard when exposed to heat, flames, oxidizers. Explosive in the form of vapor when exposed to heat or flame. In common with ethers, unstabilized tetrahydrofuran forms thermally explosive peroxides on exposure to air. Stored THF must always be tested for peroxide prior to distillation. Peroxides can be removed by treatment with strong ferrous sulfate solution made slightly acidic with sodium bisulfate. Caustic alkalies deplete the inhibitor in THF and may subsequently cause an explosive reaction. Explosive reaction with KOH, NaAlH$_2$, NaOH, sodium tetrahydroaluminate. Reacts with 2-aminophenol + potassium dioxide to form an explosive product. Reacts with lithium tetrahydroaluminate or borane to form explosive hydrogen gas. Violent reaction with metal halides (e.g., hafnium tetrachloride, titanium tetrachloride, zirconium tetrachloride). Vigorous reaction with bromine, calcium hydride + heat. Can react with oxidizing materials. To fight fire, use foam, dry chemical, CO$_2$. When heated to decomposition it emits acrid smoke and

T

irritating fumes. See also 2-TETRAHYDROFURYL HYDROPEROXIDE.

TCS500 CAS: 4795-29-3 HR: 3
TETRAHYDROFURFURYLAMINE
DOT: UN 2943

mf: $C_5H_{11}NO$ mw: 101.17

PROP: Bp: 153–155°, d: 0.977, refr index: 1.454.

SYN: USAF Q-2

CONSENSUS REPORTS: Reported in EPA TSCA Inventory.

DOT CLASSIFICATION: 3; Label: Flammable Liquid

SAFETY PROFILE: Poison by intraperitoneal route. Flammable liquid when exposed to heat or flame. When heated to decomposition it emits toxic fumes of NO_x. See also AMINES.

TCU110 CAS: 90034-98-3 HR: 3
4'-(1,2,3,4-TETRAHYDRO-4-(4-HYDRO-XY-2-OXO-2H-1-BENZOPYRAN-3-YL-)-2-NAPHTHALENYL)(1,1'-BIPHEN-YL)-4-CARBONITRILE, cis-
mf: $C_{32}H_{23}NO_3$ mw: 469.56

SAFETY PROFILE: A poison by ingestion. When heated to decomposition it emits toxic vapors of NO_x.

TCU600 HR: 2
TETRAHYDROLINALOOL
mf: $C_{10}H_{22}O_2$ mw: 158.29

PROP: Colorless liquid; floral odor. D: 0.923, refr index: 1.431, flash p: 183°F. Sol in alc, fixed oils; insol in water.

SYNS: 3,7-DIMETHYL-3-OCTANOL □ FEMA No. 3060

SAFETY PROFILE: Combustible liquid. When heated to decomposition it emits acrid smoke and irritating fumes.

TCW750 CAS: 33401-94-4 HR: 3
(E)-4,5,6-TETRAHYDRO-1-METHYL-2-(2-(2-THIENYL)ETHENYL)PYRIMIDINE
mf: $C_{11}H_{14}N_2S \cdot C_4H_6O_6$ mw: 356.43

PROP: Crystals from EtOH. Mp: 147–148°.

SYNS: BANMINTH □ CP 10423-18 □ PYRANTEL TARTRATE □ PYREQUAN TARTRATE □ (E)-1,4,5,6-

TETRAHYDRO-1-METHYL-2-(2-(2-THIENYL)VINYL)-PYRIMIDINE TARTRATE (1:1)

SAFETY PROFILE: Poison by ingestion and intravenous routes. When heated to decomposition it emits very toxic fumes of NO_x and SO_x.

TDA500 CAS: 52-31-3 HR: 3
TETRAHYDROPHENOBARBITAL
mf: $C_{12}H_{16}N_2O_3$ mw: 236.30

PROP: Crystals with insipid bitter taste. Mp: 171–174°. Insol in H_2O.

SYNS: ADORM □ AMNOSED □ CAVONLY □ CYCLO-BARBITAL □ CYCLOBARBITOL □ CYCLOBARBITONE □ CYCLODORM □ CYCLOHEXENYL-ETHYL BAR-BITURIC ACID □ 5-(1-CYCLOHEXENYL)-5-ETHYL-BARBITURIC ACID □ 5-(1-CYCLOHEXEN-1-YL)-5-ETH-YLBARBITURIC ACID □ 5-(1-CYCLOHEXEN-1-YL)-5-ETHYL-2,4,6(1H,3H,5H)-PYRIMIDINETRIONE □ 5-ETH-YL-5-CYCLOHEXENYLBARBITURIC ACID □ ETHYL-HEXABITAL □ FANODORMO □ HEXEMAL □ IRIFAN □ NAMURON □ PALINUM □ PHANODORM □ PHANODORN □ PHILODORM □ PRALUMIN □ PRO-SONIL □ SONAFORM

SAFETY PROFILE: Poison by ingestion, subcutaneous, intravenous, and intraperitoneal routes. Human systemic effects by ingestion: pulmonary consolidation. Used as a central nervous system depressant, hypnotic, and sedative. When heated to decomposition it emits toxic fumes of NO_x. See also BARBITURATES.

TDB000 CAS: 85-43-8 HR: 2
TETRAHYDROPHTHALIC ACID ANHYDRIDE
DOT: UN 2698

mf: $C_8H_8O_3$ mw: 152.16

PROP: White powder. Mp: 101.9°, bp: 195° @ 50 mm, flash p: 315°F (OC), d: 1.375 @ 25°/20°, vap press: 0.01 mm @ 20°, vap d: 5.25.

SYNS: ANHYDRID KYSELINY TETRAHYDRO-FTALOVE (CZECH) □ MALEIC ANHYDRIDE adduct of BUTADIENE □ TETRAHYDROFTALANHYDRID -(CZECH) □ 3a,4,7,7a-TETRAHYDRO-1,3-ISOBENZO-FURANDIONE □ TETRAHYDROPHTHALIC ANHYDR-IDE □ Δ⁴-TETRAHYDROPHTHALIC ANHYDRIDE □ 1,2,3,6-TETRAHYDRO PHTHALIC ANHYDRIDE □ THPA

CONSENSUS REPORTS: Reported in EPA TSCA Inventory.

DOT CLASSIFICATION: 8; Label: Corrosive

SAFETY PROFILE: Moderately toxic by intraperitoneal route. Mildly toxic by ingestion. A corrosive irritant to skin, eyes, and mucous membranes. Combustible when exposed to heat or flame. Will react with water or steam to produce heat; can react with oxidizing materials. To fight fire, use water, foam, CO_2, dry chemical. When heated to decomposition it emits acrid smoke and irritating fumes. See also ANHYDRIDES.

TDC730   CAS: 110-01-0   HR: 3
TETRAHYDROTHIOPHENE

DOT: UN 2412

mf: C_4H_8S mw: 88.18

PROP: Mobile liquid with penetrating odor. Mp: $-96°$, bp: $119°$, flash p: $55°F$, d: 1.01, refr index: 1.5030. Misc in most solvs except H_2O.

SYNS: TETRAHYDROTHIOFEN □ TETRAMETHYL-ENESULFIDE □ THIACYCLOPENTANE □ THILANE □ THIOFAN □ THIOLANE □ THIOPHANE

DOT CLASSIFICATION: 3; Label: Flammable Liquid

SAFETY PROFILE: Mildly toxic by inhalation. Flammable liquid. Potentially explosive reaction with hydrogen peroxide. When heated to decomposition it emits toxic fumes of SO_x.

TDK000   CAS: 75-57-0   HR: 3
TETRAMETHYLAMMONIUM CHLORIDE

mf: $C_4H_{12}N•Cl$ mw: 109.62

PROP: Crystals. Mp: $>300°$. Hygroscopic.

SYN: USAF AN-8

CONSENSUS REPORTS: Reported in EPA TSCA Inventory.

SAFETY PROFILE: Poison by ingestion, intraperitoneal, and subcutaneous routes. When heated to decomposition it emits very toxic fumes of NO_x, NH_3, and Cl^-. See also CHLORIDES.

TDK500   CAS: 75-59-2   HR: 3
TETRAMETHYLAMMONIUM HYDROXIDE

DOT: UN 1835

mf: $C_4H_{12}N•HO$ mw: 91.18

PROP: Liquid. D: 1.

SYNS: AMMONIUM, TETRAMETHYL-, HYDROXIDE □ HYDROXYDE de TETRAMETHYLAMMONIUM - (FRENCH) □ TETRAMETHYLAMMONIUM HYDROX-IDE, liquid (DOT) □ TM

CONSENSUS REPORTS: Reported in EPA TSCA Inventory.

DOT CLASSIFICATION: 8; Label: Corrosive

SAFETY PROFILE: Poison by subcutaneous route. A powerful caustic. A corrosive irritant to skin, eyes, and mucous membranes. When heated to decomposition it emits toxic fumes of NO_x and NH_3.

TDQ750   CAS: 110-18-9   HR: 3
TETRAMETHYL ETHYLENE DIAMINE

DOT: UN 2372

mf: $C_6H_{16}N_2$ mw: 116.24

PROP: Mp: $-55°$, bp: $120–122°$, refr index: 1.4179, d: 0.770, flash p: $50°F$.

SYNS: 1,2-BIS-(DIMETHYLAMINO)-ETHANE (DOT) □ 1,2-DI-(DIMETHYLAMINO)ETHANE (DOT) □ PROPAM-INE D □ TEMED □ TETRAMEEN □ N,N,N',N'-TETRA-METHYL-1,2-DIAMINOETHANE □ N,N,N',N'-TETRA-METHYLETHYLENEDIAMINE □ TMEDA

CONSENSUS REPORTS: Reported in EPA TSCA Inventory.

DOT CLASSIFICATION: 3; Label: Flammable Liquid

SAFETY PROFILE: Moderately toxic by ingestion. Mildly toxic by skin contact. A skin and severe eye irritant. Flammable when exposed to heat or flame; can react with oxidizing materials. When heated to decomposition it emits toxic fumes of NO_x. See also AMINES.

TDR500   CAS: 75-74-1   HR: 3
TETRAMETHYL LEAD

mf: $C_4H_{12}Pb$ mw: 267.35

PROP: Colorless liquid. Mp: $-27.5°$, lel: 1.8%, bp: $110°$, d: 1.99, vap d: 9.2, flash p: $100°F$. IDLH 40 mg/m^3 (as Pb).

SYNS: TETRAMETHYLPLUMBANE □ TML

CONSENSUS REPORTS: EPA Extremely Hazardous Substances List. Lead and its compounds are on the Community Right-To-Know List. Reported in EPA TSCA Inventory.
OSHA PEL: TWA 0.075 mg(Pb)/m³ (skin)
ACGIH TLV: TWA 0.15 mg(Pb)/m³ (skin)
DFG MAK: 0.05 mg/m³
SAFETY PROFILE: Poison by ingestion, intraperitoneal, parenteral, and intravenous routes. Moderately toxic by skin contact. An experimental teratogen. Experimental reproductive effects. Lead and its compounds have dangerous central nervous system effects. A flammable liquid and very dangerous fire hazard when exposed to heat, flame, or oxidizers. Moderate explosion hazard in the form of vapor when exposed to flame. May explode when heated above 90°C. Explosive reaction with tetrachlorotrifluoromethyl phosphorane. Can react vigorously with oxidizing materials. To fight fire, use water, foam, CO_2, dry chemical. When heated to decomposition it emits toxic fumes of Pb. Used as an octane enhancer for gasoline. See also LEAD COMPOUNDS.

TDR750   CAS: 51-80-9   HR: 3
N,N,N'N'-TETRAMETHYLMETHANEDI-
   AMINE
mf: $C_5H_{14}N_2$ mw: 102.21
PROP: Liquid. Bp: 85°, d: 0.749, refr index: 1.4005, flash p: 35°F.
SYNS: N,N,N',N'-TETRAMETHYLDIAMINOMETHAN - (GERMAN) □ TETRAMETHYL METHYLENE DIAMINE - (DOT)
CONSENSUS REPORTS: Reported in EPA TSCA Inventory.
SAFETY PROFILE: Poison by intraperitoneal route. Flammable liquid and very dangerous fire hazard when exposed to powerful oxidizers, heat or open flame. When heated to decomposition it emits toxic fumes of NO_x. See also AMINES.

TDU800   CAS: 1124-11-4   HR: 3
TETRAMETHYLPYRAZINE
mf: $C_8H_{12}N_2$ mw: 136.22

PROP: White trihydrate crystals from water or powder; fermented-soybean odor. Mp: 86° (anhydrate), mp: 74–77° (hydrate), bp: 190°. Sol in alc, propylene glycol, and fixed oils; sltly sol in water.
SYNS: FEMA No. 3237 □ 2,3,5,6-TETRAMETHYL PYRAZINE (FCC)
CONSENSUS REPORTS: Reported in EPA TSCA Inventory.
SAFETY PROFILE: Poison by intravenous and intraperitoneal routes. Moderately toxic by ingestion. When heated to decomposition it emits toxic fumes of NO_x.

TDV750   CAS: 594-27-4   HR: 3
TETRAMETHYLSTANNANE
mf: $C_4H_{12}Sn$ mw: 178.85
PROP: Colorless liquid with ethereal odor. Bp: 78°, lel: 1.9%, d: 1.129 @ 25°/4°, mp: −55°, vap d: 6.2, flash p: <69.8°F.
SYN: TETRAMETHYL TIN
CONSENSUS REPORTS: Reported in EPA TSCA Inventory.
OSHA PEL: TWA 0.1 mg(Sn)/m³ (skin)
ACGIH TLV: TWA 0.1 mg(Sn)/m³; STEL 0.2 mg(Sn)/m³ (skin).
NIOSH REL: (Organotin Compounds) TWA 0.1 mg(Sn)/m³
SAFETY PROFILE: Moderately toxic by inhalation. A very dangerous fire and explosion hazard when exposed to heat, flame, or oxidizers. Powerful explosive reaction with dinitrogen tetroxide. To fight fire, use water, foam, CO_2, dry chemical. When heated it emits acrid smoke and irritating fumes. See also TIN COMPOUNDS and ORGANOMETALS.

TDW250   CAS: 3333-52-6   HR: 3
TETRAMETHYLSUCCINONITRILE
mf: $C_8H_{12}N_2$ mw: 136.22
PROP: Crystallizes in plates; almost no odor. Mp: 172° (sublimes). IDLH 5 ppm.
SYN: TMSN
CONSENSUS REPORTS: Cyanide and its compounds are on the Community Right-To-Know List.
OSHA PEL: TWA 0.5 ppm (skin)
ACGIH TLV: TWA 0.5 ppm (skin)

DFG MAK: 0.5 ppm (2.8 mg/m³)
NIOSH REL: (Nitriles) CL 6 mg/m³/15M
SAFETY PROFILE: Poison by ingestion, intraperitoneal, and intravenous routes. An experimental teratogen. A human skin irritant and allergen. In the preparation of sponge rubber, an azo compound is used that decomposes to form tetramethylsuccinonitrile or TMSN. Rats exposed to a concentration of 90 ppm exhibit their first convulsion after 1.5–2 hours or less. Rats exposed to concentration of 5.5 ppm exhibited their first convulsions in 27–31 hours and were dead in 31–46 hours. Absorbed by skin. The fatal dose in humans is thought to be about 25 mg/kg of body weight. TSN is slowly detoxified by the body. This nitrile is different from other nitriles in that thiosulfate is a poor antidote for intoxication. When heated to decomposition it emits toxic fumes of CN^- and NO_x. See also NITRILES and CYANIDE.

TDW500   CAS: 108-62-3   HR: 3
2,4,6,8-TETRAMETHYL-1,3,5,7-
**   TETROXOCANE**
DOT: UN 1332
mf: $C_8H_{16}O_4$ mw: 176.24
PROP: Crystals with menthol odor. Mp: 47°, bp: 65° @ 15 mm. Sol in most org solvs.
SYNS: ACETALDEHYDE, TETRAMER □ ANTIMILACE □ ARIOTOX □ CEKUMETA □ CORRY'S SLUG DEATH □ HALIZAN □ META □ METACETALDEHYDE □ METALDEHYD (GERMAN) □ METALDEHYDE (DOT) □ METALDEIDE (ITALIAN) □ METASON □ NAMEKIL □ SLUG-TOX
DOT CLASSIFICATION: 4.1; Label: Flammable Solid
SAFETY PROFILE: Human poison by ingestion. Human systemic effects by ingestion: convulsions or effect on seizure threshold. Moderately toxic by inhalation and skin contact. Experimental reproductive effects. Mutation data reported. A flammable solid. When heated to decomposition it emits acrid smoke and irritating fumes. See also ALDEHYDES.

TDX000   CAS: 2782-91-4   HR: 2
TETRAMETHYLTHIOUREA
mf: $C_5H_{12}N_2S$ mw: 132.25
PROP: Crystals from H_2O. Mp: 79–80°, bp: 245°.
SYNS: NA-101 □ 1,1,3,3-TETRAMETHYLTHIOUREA □ TMTU
CONSENSUS REPORTS: Reported in EPA TSCA Inventory.
SAFETY PROFILE: Moderately toxic by ingestion. Questionable carcinogen with experimental carcinogenic and teratogenic data. Human mutation data reported. When heated to decomposition it emits very toxic fumes of NO_x and SO_x. See also SULFIDES.

TDX250   CAS: 632-22-4   HR: 2
1,1,3,3-TETRAMETHYLUREA
mf: $C_5H_{12}N_2O$ mw: 116.19
PROP: Liquid, with faint pleasant odor. Bp: 177°, mp: −1.2°, d: 0.969, flash p: 167°F. Very sol in alc and ether; misc with water.
SYNS: TEMUR □ TETRAMETHYLUREA □ TETRA-METHYLUREE (FRENCH) □ TMU
CONSENSUS REPORTS: Reported in EPA TSCA Inventory.
SAFETY PROFILE: Moderately toxic by ingestion and intravenous routes. An experimental teratogen. Experimental reproductive effects. Human mutation data reported. Combustible when exposed to heat, flame, and oxidizers. To fight fire, use foam, mist, spray, dry chemicals. When heated to decomposition it emits toxic fumes of NO_x.

TDY075   CAS: 4591-46-2   HR: 3
N,2,4,6-TETRANITROANILINE
mf: $C_6H_3N_5O_8$ mw: 273.12
PROP: Yellow crystals from $Me_2CO/CHCl_3$.
SYNS: BENZENAMINE, N,2,4,6-TETRANITRO- □ N,2,4,6-TETRANITROANILINE □ 2,4,6-TRINITROPHEN-YL NITRAMINE (DOT)
DOT CLASSIFICATION: Forbidden
SAFETY PROFILE: The impure material deflagrates (burns explosively) when heated to 50°C. When heated to decomposition it

1324 TDY250 TETRANITROMETHANE

emits toxic fumes of NO_x. See also NITRO COMPOUNDS of AROMATIC HYDROCARBONS.

TDY250   CAS: 509-14-8   HR: 3
TETRANITROMETHANE
DOT: UN 1510

mf: CN_4O_8 mw: 196.05

PROP: Colorless or yellow liquid. Mp: 13°, bp: 125.7°, d: 1.650 @ 13°, vap press: 10 mm @ 22.7°. Insol in water; very sol in alc, ether. IDLH 4 ppm.

SYNS: NCI-C55947 □ RCRA WASTE NUMBER P112 □ TNM

CONSENSUS REPORTS: NTP 10th Report on Carcinogens. EPA Extremely Hazardous Substances List. Reported in EPA TSCA Inventory.

OSHA PEL: TWA 1 ppm

ACGIH TLV: TWA 0.005 ppm; Animal Carcinogen

DFG MAK: Confirmed Animal Carcinogen, Suspected Human Carcinogen

DOT CLASSIFICATION: 5.1; Label: Oxidizer, Poison

SAFETY PROFILE: Confirmed carcinogen with carcinogenic and neoplastigenic data. Poison by ingestion, inhalation, intravenous, and intraperitoneal routes. Irritating to the skin, eyes, mucous membranes, and respiratory passages, and does serious damage to the liver. Mutation data reported. It occurs as an impurity in crude TNT, and is thought to be mainly responsible for the irritating properties of that material. It can cause pulmonary edema, mild methemoglobinemia, and fatty degeneration of the liver and kidneys.

A powerful oxidizer. A very dangerous fire hazard. A severe explosion hazard when shocked or exposed to heat. May explode during distillation. Potentially explosive reaction with ferrocene, pyridine, sodium ethoxide. Mixtures with amines (e.g., aniline) ignite spontaneously and may explode. Mixtures with cotton or toluene may explode when ignited. Forms sensitive and powerful explosive mixtures with nitrobenzene, 1-nitrotoluene, 4-nitrotoluene, 1,3-dinitrobenzene, 1-nitronaphthalene, other oxygen-deficient explosives, hydrocarbons. Can react vigorously with oxidizing materials. Incompatible with aluminum. When heated to decomposition it emits highly toxic fumes of NO_x. Used as an oxidizer in rocket propellants and as an explosive. See also NITRATES; EXPLOSIVES, HIGH.

TDY600   CAS: 641-16-7   HR: 3
2,3,4,6-TETRANITROPHENOL
mf: $C_6H_2N_4O_9$ mw: 274.12

PROP: Pale-yellow crystals from $CHCl_3$. Mp: 140°.

DOT CLASSIFICATION: Forbidden

SAFETY PROFILE: A powerful explosive. When heated to decomposition it emits toxic fumes of NO_x. See also NITRO COMPOUNDS of AROMATIC HYDROCARBONS.

TEA300   CAS: 507-28-8   HR: 3
TETRAPHENYLARSENIUM CHLORIDE
mf: $C_{24}H_{20}As•Cl$ mw: 418.81

PROP: White crystals from $EtOAc/Et_2O$. Mp: 258–260°. Sol in water, alc, or methanol sltly sol in acetone.

SYNS: ARSONIUM, TETRAPHENYL-, CHLORIDE □ TETRAPHENYLARSONIUM CHLORIDE

OSHA PEL: TWA 0.5 mg(As)/m³

ACGIH TLV: BEI: 35 μ (As)/L inorganic arsenic and methylated metabolites in urine

SAFETY PROFILE: Poison by intravenous route. When heated to decomposition it emits toxic fumes of As and Cl⁻.

TEC500   CAS: 13943-58-3   HR: 1
TETRAPOTASSIUM
HEXACYANOFERRATE
mf: $C_6FeN_6•4K$ mw: 368.37

PROP: Pale-yellow crystals.

SYNS: FERRATE(4-), HEXACYANO-, TETRAPOTASS-IUM □ FERRATE(4-), HEXAKIS(CYANO-C)-, TETRA-POTASSIUM, (OC-6-11)- □ POTASSIUM FERROCYANATE □ POTASSIUM FERROCYANIDE □ POTASSIUM HEXA-CYANOFERRATE □ POTASSIUM HEXACYANOFER-RATE(II) □ TETRAPOTASSIUM FERROCYANIDE □

TETRAPOTASSIUM HEXACYANOFERRATE(II) ☐
TETRAPOTASSIUM HEXACYANOFERRATE(4-)
CONSENSUS REPORTS: Reported in EPA
TSCA Inventory.
SAFETY PROFILE: Mildly toxic by
ingestion. An experimental teratogen. When
heated to decomposition it emits toxic
fumes of NO$_x$.

TED600   CAS: 13274-42-5   HR: 3
TETRAPROPYLENEPENTAMINE
mf: C$_{12}$H$_{31}$N$_5$ mw: 245.48
SYNS: N-(3-AMINOPROPYL)-N'-(3-((3-AMINOPROPYL)-
AMINO)PROPYL)-1,3-PROPANEDIAMINE ☐
CALDOPENTAMINE ☐ DIPROPYLAMINE, 3,3'-BIS((3--
(AMINOPROPYL))AMINO)- ☐ 1,3-PROPANEDIAMINE,
N-(3-AMINOPROPYL)-N'-(3-((3-AMINOPROPYL)AMINO)-
PROPYL)- ☐ TETRATRIMETHYLENEPENTAMINE
SAFETY PROFILE: A poison by ingestion
and skin contact. A moderate skin and
severe eye irritant. When heated to
decomposition it emits toxic vapors of NO$_x$.

TED750   CAS: 3440-75-3   HR: 3
TETRAPROPYL LEAD
mf: C$_{12}$H$_{28}$Pb   mw: 379.59
PROP: Colorless liquid. D: 1.44, bp: 126° @
13 mm. Sol in benzene.
CONSENSUS REPORTS: Lead and its
compounds are on the Community Right-
To-Know List. Reported in EPA TSCA
Inventory.
SAFETY PROFILE: Poison by ingestion and
parenteral routes. When heated to
decomposition it emits toxic fumes of Pb.
See also LEAD COMPOUNDS.

TEE500   CAS: 7722-88-5   HR: 3
TETRASODIUM PYROPHOSPHATE
mf: O$_7$P$_2$•4Na   mw: 265.90
PROP: White orthorhombic crystalline
powder. Mp: 993°, d: 2.534. Sol in water;
insol in alc.
SYNS: NATRIUMPYROPHOSPHAT ☐ PHOSPHOTEX
☐ PYROPHOSPHATE ☐ SODIUM PYROPHOSPHATE -
(FCC) ☐ TETRANATRIUMPYROPHOSPHAT (GERMAN)
☐ TETRASODIUM DIPHOSPHATE ☐ TETRASODIUM
PYROPHOSPHATE, ANHYDROUS ☐ TSPP ☐ VICTOR
TSPP

CONSENSUS REPORTS: Reported in EPA
TSCA Inventory.
OSHA PEL: TWA 5 mg/m³
ACGIH TLV: TWA 5 mg/m³
SAFETY PROFILE: Poison by ingestion,
intraperitoneal, intravenous, and
subcutaneous routes. It is not a
cholinesterase inhibitor. When heated to
decomposition it emits toxic fumes of PO$_x$
and Na$_2$O.

TEF500   CAS: 109-27-3   HR: 3
TETRAZENE
DOT: UN 0114
mf: C$_2$H$_8$N$_{10}$O   mw: 188.20
PROP: Crystals.
SYNS: 4-AMIDINO-1-(NITROSAMINOAMIDINO)-1-
TETRAZENE ☐ 4-AMIDINO-1-(NITROSOAMINOAMID-
INO)-1-TETRAZENE ☐ GUANYL NITROSAMINO
GUANYL TETRAZENE ☐ 1-GUANYL-4-NITROSAM-
INOGUANYLTETRAZENE ☐ GUANYL NITROSAM-
INOGUANYLTETRAZENE (dry) (DOT) ☐ GUANYL
NITROSAMINOGUANYLTETRAZENE, wetted or tetrazene,
wetted with not <30% water or (DOT) ☐ TETRACENE ☐
TETRACENE EXPLOSIVE ☐ 1-TETRAZENE, 4-AMID-
INO-1-(NITRSOAMINOAMIDINO)-(8CI)
CONSENSUS REPORTS: Reported in EPA
TSCA Inventory.
DOT CLASSIFICATION: Forbidden (dry);
Explosive 1.1A; Label: Explosive 1.1A
SAFETY PROFILE: Many nitrosamines are
carcinogens. A very dangerous fire hazard.
A shock- and heat-sensitive high explosive
that evolves much flame. Highly dangerous.
Upon decomposition it emits highly toxic
fumes of NO$_x$. See also NITROSAMINES
and EXPLOSIVES, HIGH.

TEG250   CAS: 479-45-8   HR: 3
TETRYL
DOT: UN 0208
mf: C$_7$H$_5$N$_5$O$_8$ mw: 287.17
PROP: Yellow, monoclinic crystals or
prisms from EtOH. Mp: 131–132°, bp:
explodes @ 187°, d: 1.57 @ 19°. Spar sol in
EtOH. IDLH 750 mg/m³.
SYNS: BENZENAMINE, N-METHYL-N,2,4,6-TETRA-
NITRO-(9CI) ☐ CE ☐ N-METHYL-N,2,4,6-TETRANITRO-
ANILINE ☐ N-METHYL-N,2,4,6-TETRANITROBENZEN-
AMINE ☐ NITRAMINE ☐ PICRYLMETHYLNITRAMINE

□ PICRYLNITROMETHYLAMINE □ TETRALIT □ TETRALITE □ N,2,4,5-TETRANITRO-N-METHYLANIL-INE □ TETRIL □ 2,4,6-TETRYL □ TRINITROPHENYL-METHYLNITRAMINE □ 2,4,6-TRINITROPHENYLMETH-YLNITRAMINE □ 2,4,6-TRINITROPHENYL-N-METHYL-NITRAMINE

OSHA PEL: TWA 0.1 mg/m³ (skin)
ACGIH TLV: TWA 1.5 mg/m³
DFG MAK: Confirmed Animal Carcinogen with Unknown Relevance to Humans
DOT CLASSIFICATION: EXPLOSIVE 1.1D; Label: EXPLOSIVE 1.1D
SAFETY PROFILE: Mutation data reported. An irritant, sensitizer, and allergen. The chief effect from exposure is dermatitis. Conjunctivitis is followed by iridocyclitis, and keratitis can occur. Sensitization produced by exposure may play a part in these symptoms. Gastrointestinal effects and anemia have also been reported.

A powerful oxidant. A dangerous fire and explosion hazard. A high explosive sensitive to shock, friction, or heat. More sensitive to shock and friction than TNT. Explodes on contact with trioxygen difluoride. Ignites on contact with hydrazine. When heated to decomposition it emits toxic fumes of NO$_x$. See also NITRATES and EXPLOSIVES, HIGH.

TEH500   CAS: 50-35-1   HR: 3
THALIDOMIDE
mf: C$_{13}$H$_{10}$N$_2$O$_4$ mw: 258.25
PROP: Needles. Mp: 269–271°. Sltly sol in water, methanol, ethanol, acetone, ethylacrylate; very sol in dioxane; sol in ether.

SYNS: ALGOSEDIV □ ASIDON 3 □ ASMADION □ ASMAVAL □ BONBRAIN □ CALMORE □ CALMOREX □ CONTERGAN □ CORRONAROBETIN □ 2,6-DIOXO-3-PHTHALIMIDOPIPERIDINE □ 2-(2-6-DIOXO-3-PIPERID-INYL)1H-ISOINDOLE-1,3(2H)-DIONE □ N-(2,6-DIOXO-3-PIPERIDYL)PHTHALIMIDE □ DISTAVAL □ DISTAXAL □ DISTOVAL □ ECTILURAN □ ENTEROSEDIV □ GASTRINIDE □ GLUPAN □ GLUTANON □ GRIPPEX □ HIPPUZON □ IMIDA-LAB □ IMIDAN (PEYTA) □ IMIDENE □ ISOMIN □ K 17 □ KEDAVON □ KEVAD-ON □ LULAMIN □ NEAUFATIN □ NEO □ NEOSEDYN □ NEOSYDYN □ NEURODYN □ NEUROSEDIN □ NEVRODYN □ NIBROL □ NOCTOSEDIV □ NOXODYN

□ NSC-66847 □ PANGUL □ PANTOSEDIV □ α-PHTHALIMIDOGLUTARIMIDE □ α-(N-PHTHALIMIDO)-GLUTARIMIDE □ 2-PHTHALIMIDOGLUTARIMIDE □ 3-PHTHALIMIDOGLUTARIMIDE □ N-PHTHALOYL-GLUTAMIMIDE □ N-PHTHALYLGLUTAMIC ACID IMIDE □ N-PHTHALYL-GLUTAMINSAEURE-IMID-(GERMAN) □ α-N-PHTHALYLGLUTARAMIDE □ POLY-GIRON □ POLYGRIPAN □ PREDNI-SEDIV □ PRO-BAN M □ PROFARMIL □ PSYCHOLIQUID □ PSYCHO-TABLETS □ QUETIMID □ QUIETOPLEX □ SANDORM-IN □ SEDALIS SEDI-LAB □ SEDIMIDE □ SEDIN □ SEDISPERIL □ SEDOVAL □ SHIN-NAITO S □ SH-INNIBROL □ SLEEPAN □ SLIPRO □ SOFTENIL □ SOFTENON □ TALARGAN □ TALIMOL □ TELAGAN □ TELARGAN □ TELARGEAN □ TENSIVAL □ THALIN □ THALINETTE □ THEOPHILCHOLINE □ ULCERFEN □ VALGIS □ VALGRAINE □ YODOMIN

CONSENSUS REPORTS: EPA Genetic Toxicology Program.
SAFETY PROFILE: Poison by ingestion. Moderately toxic by skin contact and intraperitoneal routes. Human teratogenic effects by ingestion: developmental abnormalities of the musculoskeletal and cardiovascular systems. Experimental reproductive effects. Questionable carcinogen with experimental tumorigenic and teratogenic data. Human mutation data reported. It was commonly used as a prescription drug in Europe in the late 1950s and early 1960s. Its use was discontinued because it was discovered to cause serious congenital abnormalities in the fetus, notably amelia and phocomelia - (absence or deformity of the limbs, including hands and feet) when taken by a woman during early pregnancy. When heated to decomposition it emits toxic fumes of NO$_x$. Used as a sedative and hypnotic.

TEI000   CAS: 7440-28-0   HR: 3
THALLIUM
af: Tl   aw: 204.37
PROP: Bluish-white, soft, malleable metal. Tarnishes readily in air. Moderately strong base; reacts with steam or moist air releasing TlOH. Mp: 303.5°, bp: 1457°, d: 11.85 @ 20°, vap press: 1 mm @ 825°. Readily

dissolves in dil H_2SO_4 and HNO_3; does not dissolve in basic solns.

SYN: RAMOR

CONSENSUS REPORTS: Thallium and its compounds are on the Community Right-To-Know List. Reported in EPA TSCA Inventory.

OSHA PEL: TWA 0.1 mg(Tl)/m^3 (skin)
ACGIH TLV: TWA 0.1 mg(Tl)/m^3 (skin)
DFG MAK: 0.1 mg/m^3

SAFETY PROFILE: Human poison by unspecified route. Human systemic effects by ingestion: nerve or sheath structural changes, extra-ocular muscle changes, sweating, and other effects. Flammable in the form of dust when exposed to heat or flame. Violent reaction with F_2. When heated to decomposition it emits toxic fumes of Tl. Used as a rodenticide and fungicide, and in lenses and prisms, in high-density liquids. See also THALLIUM COMPOUNDS and POWDERED METALS.

TEI250   CAS: 563-68-8   HR: 3
THALLIUM ACETATE
mf: $C_2H_3O_2$•Tl   mw: 263.42

PROP: Silk-white crystals, or hygroscopic, colorless, light-sensitive monoclinic crystals from EtOH. Mp: 131°, d: 3.68. Sol in water, alc. IDLH 15 mg/m^3 (as Tl).

SYNS: RCRA WASTE NUMBER U214 □ THALLIUM(1+) ACETATE □ THALLIUM(I) ACETATE □ THALLIUM MONOACETATE □ THALLOUS ACETATE

CONSENSUS REPORTS: Thallium and its compounds are on the Community Right-To-Know List. EPA Genetic Toxicology Program. Reported in EPA TSCA Inventory.

OSHA PEL: TWA 0.1 mg(Tl)/m^3 (skin)
ACGIH TLV: TWA 0.1 mg(Tl)/m^3 (skin)

SAFETY PROFILE: Human poison by ingestion. Experimental poison by ingestion, intravenous, intraperitoneal, and subcutaneous routes. An experimental teratogen. Mutation data reported. When heated to decomposition it emits toxic

fumes of Tl. See also THALLIUM COMPOUNDS.

TEI750   CAS: 7789-40-4   HR: 3
THALLIUM BROMIDE
mf: BrTl   mw: 284.28

PROP: Air-stable, light-sensitive, yellowish-white powder, solids, or crystals. Mp: 460–480° (approx), bp: 815°, d: 7.557, vap press: 10 mm @ 522°. IDLH 15 mg/m^3 (as Tl).

CONSENSUS REPORTS: Thallium and its compounds are on the Community Right-To-Know List. Reported in EPA TSCA Inventory.

OSHA PEL: TWA 0.1 mg(Tl)/m^3 (skin)
ACGIH TLV: TWA 0.1 mg(Tl)/m^3 (skin)

SAFETY PROFILE: Poison by ingestion and subcutaneous routes. Reacts violently with Na, K. When heated to decomposition it emits very toxic fumes of Br^- and Tl. See also BROMIDES and THALLIUM COMPOUNDS.

TEJ000   CAS: 6533-73-9   HR: 3
THALLIUM(I) CARBONATE (2:1)
mf: CO_3•2Tl   mw: 468.75

PROP: Monoclinic, colorless crystals stable in air to 1°; loses CO_2 on further heating; dissolves acidic oxides. Mp: 273°, d: 7.11. Sol in H_2O; insol in EtOH. IDLH 15 mg/m^3 (as Tl).

SYNS: CARBONIC ACID, DITHALLIUM(1+) SALT □ DITHALLIUM CARBONATE □ RCRA WASTE NUMBER U215 □ THALLOUS CARBONATE □ THIOCHROMAN-4-ONE, OXIME

CONSENSUS REPORTS: Thallium and its compounds are on the Community Right-To-Know List. EPA Extremely Hazardous Substances List. Reported in EPA TSCA Inventory.

OSHA PEL: TWA 0.1 mg(Tl)/m^3 (skin)
ACGIH TLV: TWA 0.1 mg(Tl)/m^3 (skin)

SAFETY PROFILE: Poison by ingestion, skin contact, and subcutaneous routes. Experimental reproductive effects. Mutation data reported. When heated to decomposition it emits toxic fumes of Tl. See also THALLIUM COMPOUNDS.

T

TEJ100   CAS: 13453-30-0   HR: 3
THALLIUM CHLORATE
DOT: UN 2573
mf: $ClO_3 \cdot Tl$ mw: 287.82
PROP: IDLH 15 mg/m³ (as Tl).
SYN: CHLORIC ACID, THALLIUM(1+) SALT
ACGIH TLV: TWA 0.1 mg(Tl)/m³
DOT CLASSIFICATION: 5.1; Label:
Oxidizer, Poison
SAFETY PROFILE: A poison and oxidizer.
When heated to decomposition it emits
toxic fumes of Tl and Cl⁻.

TEJ250   CAS: 7791-12-0   HR: 3
THALLIUM CHLORIDE
mf: ClTl   mw: 239.82
PROP: Colorless or white powder, or
photosensitive crystals from water.
Insulator, wide gap (3eV); metallic at high
pressures. Mp: 430°, bp: 720°, d: 7.00, vap
press: 10 mm @ 517°. Sol in 260 parts cold
water, 70 parts boiling water; insol in alc.
IDLH 15 mg/m³ (as Tl).
SYNS: RCRA WASTE NUMBER U216 □ THALLIUM(1+)
CHLORIDE □ THALLIUM MONOCHLORIDE □
THALLOUS CHLORIDE
CONSENSUS REPORTS: EPA Extremely
Hazardous Substances List. Thallium and its
compounds are on the Community Right-
To-Know List. EPA Genetic Toxicology
Program. Reported in EPA TSCA
Inventory.
OSHA PEL: TWA 0.1 mg(Tl)/m³ (skin)
ACGIH TLV: TWA 0.1 mg(Tl)/m³ (skin)
SAFETY PROFILE: Poison by ingestion and
intraperitoneal routes. An experimental
teratogen. Mutation data reported.
Incompatible with F_2. When heated to
decomposition it emits very toxic fumes of
Cl⁻ and Tl. See also THALLIUM
COMPOUNDS and CHLORIDES.

TEK000   CAS: 7789-27-7   HR: 3
THALLIUM(I) FLUORIDE
mf: FTl   mw: 223.37
PROP: White, air stable solid. Colorless
orthorhombic crystals which deliquesce
when breathed upon and resolidify in dry

air. Mp: 322°, d: 8.36, bp: 826°. Very sol in
water. IDLH 15 mg/m³ (as Tl).
SYNS: THALLIUM MONOFLUORIDE □ THALLOUS
FLUORIDE
CONSENSUS REPORTS: Thallium and its
compounds are on the Community Right-
To-Know List. Reported in EPA TSCA
Inventory.
OSHA PEL: TWA 0.1 mg(Tl)/m³ (skin)
ACGIH TLV: TWA 0.1 mg(Tl)/m³ (skin)
SAFETY PROFILE: Poison by ingestion.
When heated to decomposition it emits very
toxic fumes of F⁻ and Tl. See also
THALLIUM COMPOUNDS and
FLUORIDES.

TEK500   CAS: 7790-30-9   HR: 3
THALLIUM IODIDE
mf: ITl   mw: 331.27
PROP: Triclinic yellow crystals; wide gap -
(3eV) insulator; metallic at high pressures.
Stable to dil HCl and H_2SO_4, but decomp by
HNO_3. Photosensitive, becomes gray on
exposure to light. D: 7.1, mp: 440°, bp:
824°. Sol in KI soln; practically insol in
water; insol in alc. IDLH 15 mg/m³ (as Tl).
SYNS: THALLIUM(I) IODIDE □ THALLIUM(1+) IODI-
DE □ THALLIUM MONOIODIDE □ THALLOUS IODI-
DE
CONSENSUS REPORTS: Thallium and its
compounds are on the Community Right-
To-Know List. Reported in EPA TSCA
Inventory.
OSHA PEL: TWA 0.1 mg(Tl)/m³ (skin)
ACGIH TLV: TWA 0.1 mg(Tl)/m³ (skin); -
(Proposed: (inhalable fraction) 0.1 mg/m³;
Not Classifiable as a Human Carcinogen)
SAFETY PROFILE: Poison by ingestion and
subcutaneous routes. Human systemic
effects: blood pressure lowering, change in
motor activity, muscle weakness. When
heated to decomposition it emits very toxic
fumes of Tl and I⁻. See also IODIDES and
THALLIUM COMPOUNDS.

TEK750   CAS: 10102-45-1   HR: 3
THALLIUM NITRATE
DOT: UN 2727
mf: $NO_3 \cdot Tl$ mw: 266.38

PROP: Cubic crystals or white solids from water. Mp: 206°, bp: 430°, d: 5.55. Decomp @ 450°; decomp at 800–809° *in vacuo* to give Tl_2O, NO, NO_2. IDLH 15 mg/m³ (as Tl). SYNS: NITRIC ACID, THALLIUM(1+) SALT □ RCRA WASTE NUMBER U217 □ THALLIUM MONONITRATE □ THALLOUS NITRATE

CONSENSUS REPORTS: Thallium and its compounds are on the Community Right-To-Know List. Reported in EPA TSCA Inventory.
OSHA PEL: TWA 0.1 mg(Tl)/m³ (skin)
ACGIH TLV: TWA 0.1 mg(Tl)/m³ (skin)
DOT CLASSIFICATION: 6.1; Label: Poison, Oxidizer
SAFETY PROFILE: Poison by ingestion, intravenous, intraperitoneal, and subcutaneous routes. Human systemic effects by ingestion: hypermotility, diarrhea, nausea or vomiting, and dehydration. Mutation data reported. When heated to decomposition it emits very toxic fumes of Tl and NO_x. See also THALLIUM COMPOUNDS and NITRATES.

TEL050   CAS: 1314-32-5   HR: 3
THALLIUM(III) OXIDE
mf: O_3Tl_2 mw: 456.74
PROP: Hexagonal brown-black crystals, amorph prisms, powder, or solid which reacts with $AlCl_3$ to give TlCl; vaporizes to give Tl_2O and O_2. Mp: 717° ± 5°, bp: 1196°, d(amorph): 9.65 @ 21°, d(hexagonal): 10.19 @ 22°. Insol in H_2O; decomp in acids. IDLH 15 mg/m³ (as Tl).
SYNS: DITHALLIUM TRIOXIDE □ RCRA WASTE NUMBER P113 □ THALLIC OXIDE □ THALLIUM OXIDE □ THALLIUM(3+) OXIDE □ THALLIUM PEROXIDE □ THALLIUM SESQUIOXIDE
CONSENSUS REPORTS: Thallium and its compounds are on the Community Right-To-Know List. Reported in EPA TSCA Inventory.
OSHA PEL: TWA 0.1 mg(Tl)/m³ (skin)
ACGIH TLV: TWA 0.1 mg(Tl)/m³ (skin)
SAFETY PROFILE: Poison by ingestion, intraperitoneal, and intravenous routes. Combustible by chemical reaction. Evolves O_2 @ 875°. Mixtures with sulfur or

antimony trisulfide explode when ground. Hydrogen sulfide ignites and may explode weakly on contact with the oxide. When heated to decomposition it emits toxic fumes of Tl. See also THALLIUM COMPOUNDS.

TEL500   CAS: 12039-52-0   HR: 3
THALLIUM SELENIDE
PROP: Dark-gray plates with a shiny appearance or black solid, metallic in plane, semi-conductor along *c*-axis (highly anisotropic), photoconducting. Mp: 330°. Insol in water and acid. IDLH 15 mg/m³ (as Tl).
SYNS: RCRA WASTE NUMBER P114 □ THALLIUM MONOSELENIDE
CONSENSUS REPORTS: Reported in EPA TSCA Inventory. Selenium and its compounds, as well as thallium and its compounds, are on the Community Right-To-Know List.
OSHA PEL: TWA 0.2 mg(Se)/m³
ACGIH TLV: TWA 0.2 mg(Se)/m³
DFG MAK: 0.1 mg(Se)/m³
SAFETY PROFILE: Thallium compounds and selenium compounds are poisons. When heated to decomposition it emits very toxic fumes of Tl and Se. See also THALLIUM COMPOUNDS and SELENIUM COMPOUNDS.

TEL750   CAS: 10031-59-1   HR: 3
THALLIUM SULFATE
DOT: NA 1707
mf: $O_4S•xTl$ mw: 1526.65
PROP: White or colorless rhomboid crystals. Mp: 632°, bp: decomp, d: 6.77. IDLH 15 mg/m³ (as Tl).
SYNS: RATOX □ SULFURIC ACID, THALLIUM SALT □ THALLIUM SULFATE, solid (DOT) □ ZELIO
CONSENSUS REPORTS: Thallium and its compounds are on the Community Right-To-Know List. EPA Extremely Hazardous Substances List. Reported in EPA TSCA Inventory.
OSHA PEL: TWA 0.1 mg(Tl)/m³ (skin)
ACGIH TLV: TWA 0.1 mg(Tl)/m³ (skin)
DOT CLASSIFICATION: 6.1; Label: Poison

T

SAFETY PROFILE: Poison to humans by ingestion. Experimental poison by ingestion, intravenous, and subcutaneous routes. Human systemic effects: cardiomyopathy, diarrhea, effects on recordings from peripheral motor nerve, hair changes, hypermotility, nausea or vomiting. Its main hazard is due to its accumulation, especially in liver, brain, and skeletal muscle; readily absorbed by gastrointestinal tract and skin. A cellular toxicant like arsenic. Fatal human dose is about 500 mg of thallium. Intake of thallium causes depilation. Many reported fatalities. An experimental teratogen. When heated to decomposition it emits very toxic fumes of SO_x and Tl. Pesticide for control of rats, moles, and house mice. See also THALLIUM COMPOUNDS and SULFATES.

TEM000   CAS: 7446-18-6   HR: 3
THALLIUM(I) SULFATE (2:1)
mf: $O_4S \cdot 2Tl$ mw: 504.80
PROP: Crystals or white rhomboid prisms from H_2O; isomorphous with β-K_2SO_4. Polymorphic. Mp: 632°, bp: (decomp), d: 6.77. IDLH 15 mg/m³ (as Tl).
SYNS: C.F.S. □ CSF-GIFTWEIZEN □ DITHALLIUM SULFATE □ DITHALLIUM(1+) SULFATE □ ECCOTHAL □ M7-GIFTKOERNER □ RATTENGIFTKONSERVE □ RCRA WASTE NUMBER P115 □ SULFURIC ACID, DI-THALLIUM(1+) SALT (8CI, 9CI) □ SULFURIC ACID, THALLIUM(1+) SALT (1:2) □ THALLOUS SULFATE
CONSENSUS REPORTS: EPA Extremely Hazardous Substances List. Thallium and its compounds are on the Community Right-To-Know List. Reported in EPA TSCA Inventory.
OSHA PEL: TWA 0.1 mg(Tl)/m³ (skin)
ACGIH TLV: TWA 0.1 mg(Tl)/m³ (skin)
SAFETY PROFILE: Human poison by ingestion. Experimental poison by ingestion and subcutaneous routes. Human systemic effects by ingestion: ataxia, change in heart rate, excitement, eye changes, irritability, nausea or vomiting, nerve or sheath structural changes, somnolence, wakefulness. Experimental reproductive effects. When heated to decomposition it

emits very toxic fumes of Tl and SO_x. Used as a rat poison, ant bait, and a reagent in analytical chemistry. See also THALLIUM COMPOUNDS and SULFATES.

TEM050   CAS: 63906-56-9   HR: 3
THALLIUM(II) SULFATE (1:1)
mf: $O_4S \cdot Tl$ mw: 300.43
PROP: IDLH 15 mg/m³ (as Tl).
SYN: SULFURIC ACID, THALLIUM(2+) SALT
CONSENSUS REPORTS: Thallium and its compounds are on the Community Right-To-Know List.
OSHA PEL: TWA 0.1 mg(Tl)m³ (skin)
ACGIH TLV: TWA 0.1 mg(Tl)/m³ (skin)
SAFETY PROFILE: Poison by ingestion. When heated to decomposition it emits very toxic fumes of Tl and SO_x. See also THALLIUM COMPOUNDS and SULFATES.

TEO500   CAS: 83-67-0   HR: 3
THEOBROMINE
mf: $C_7H_8N_4O_2$ mw: 180.19
PROP: White powder or monoclinic needles, bitter tasting alkaloid. Mp: 357°, subl @ 290–295°. Moderately sol in ammonia; almost insol in benzene, ether, chloroform, carbon tetrachloride.
SYNS: 3,7-DIHYDRO-3,7-DIMETHYL-1H-PURINE-2,6-DIONE □ 3,7-DIMETHYLXANTHINE □ DIUROBROM-INE □ SANTHEOSE □ SC 15090 □ TEOBROMIN □ THEOSALVOSE □ THEOSTENE □ THESAL □ THESODATE
CONSENSUS REPORTS: Reported in EPA TSCA Inventory. EPA Genetic Toxicology Program.
SAFETY PROFILE: Poison by ingestion. Moderately toxic by subcutaneous route. An experimental teratogen. Human systemic effects by ingestion: central nervous system and gastrointestinal changes. Experimental reproductive effects. Human mutation data reported. When heated to decomposition it emits toxic fumes of NO_x. Used as a diuretic, smooth muscle relaxant, cardiac stimulant, and vasodilator.

TEP000   CAS: 58-55-9   HR: 3
THEOPHYLLINE
mf: $C_7H_8N_4O_2$ mw: 180.19
PROP: Monoclinic, odorless needles or thin
monoclinic tablets; bitter taste. Mp: 268°.
Sol in hot water, alkali hydroxides,
ammonia, dil HCl, HNO_3; sltly sol in ether.
SYNS: ACET-THEOCIN □ AMINOPHYLLINE □ 3,7-DI-
HYDRO-1,3-DIMETHYL-1H-PURINE-2,6-DIONE □ 1,3-
DIMETHYLXANTHINE □ ELIXICON □ ELIXOPHYLL-
IN □ ELIXOPHYLLINE □ LANOPHYLLIN □ LIQUOPH-
YLLINE □ NSC-2066 □ OPTIPHYLLIN □ PARKOPHYLL-
IN □ PSEUDOTHEOPHYLLINE □ SLO-PHYLLIN □
SOLOSIN □ TEFAMIN □ TEOFYLLAMIN □ THEAL
TABL. □ THEOCIN □ THEOFOL □ THEOGRAD □
THEOLAIR □ THEOLIX □ THEOPHYL-225 □ THEOPH-
YLLIN □ THEOPHYLLINE, anhydrous
CONSENSUS REPORTS: Reported in EPA
TSCA Inventory. EPA Genetic Toxicology
Program.
SAFETY PROFILE: Human poison by
ingestion, parenteral, intravenous, and rectal
routes. Experimental poison by multiple
routes. An experimental teratogen. Human
systemic effects: coma, convulsions or effect
on seizure threshold, cyanosis, EKG
changes, fever and other metabolic effects,
heart arrhythmias, heart rate change,
hyperglycemia, metabolic acidosis, nausea or
vomiting, potassium-level changes,
respiratory stimulation, salivary gland
changes, somnolence, tremor. Experimental
reproductive effects. Human mutation data
reported. Used as a diuretic, cardiac
stimulant, smooth muscle relaxant, and to
treat asthma. When heated to
decomposition it emits toxic fumes of NO_x.

TER500   CAS: 19525-20-3   HR: 3
THIABENDAZOLE HYDROCHLORIDE
mf: $C_{10}H_7N_3S \cdot ClH$ mw: 237.72
SYN: 2-(4-THIAZOLYL)-BENZIMIDAZOLE,
HYDROCHLORIDE
SAFETY PROFILE: Poison by intravenous
route. Moderately toxic by ingestion and
intraperitoneal routes. When heated to
decomposition it emits very toxic fumes of
HCl, SO_x, and NO_x.

TES750   CAS: 59-43-8   HR: 3
THIAMINE CHLORIDE
mf: $C_{12}H_{17}N_4OS \cdot Cl$ mw: 300.84
PROP: Crystals from water. Mp: 120–122° -
(decomp) (anhyd).
SYNS: 3-((4-AMINO-2-METHYL-5-PYRIMIDINYL)-
METHYL)-5-(2-HYDROXYETHYL)-4-METHYL-
THIAZOLIUM CHLORIDE □ ANEURIN □ APATATE
DRAPE □ B-AMIN □ BEIVON □ BETABION □ BETAL-
IN S □ BETAXIN □ BETHIAMIN □ BEWON □
ORYZANIN □ ORYZANINE □ THIAMIN □ THIAMINE
MONOCHLORIDE □ VINOTHIAM □ VITAMIN B1 □
VITANEURON
CONSENSUS REPORTS: Reported in EPA
TSCA Inventory.
SAFETY PROFILE: Poison by
subcutaneous and intravenous routes. An
experimental teratogen. Experimental
reproductive effects. When heated to
decomposition it emits very toxic fumes of
NO_x, SO_x, and Cl^-.

TET300   CAS: 67-03-8   HR: 3
THIAMINE HYDROCHLORIDE
mf: $C_{12}H_{17}N_4OS \cdot ClH \cdot Cl$ mw: 337.30
PROP: Small white hygroscopic crystals,
plates, or crystalline powder from EtOH;
nut-like odor. Mp: 248° (decomp). Sol in
water, glycerin; sltly sol in alc; insol in ether,
benzene.
SYNS: THIAMINE CHLORIDE HYDROCHLORIDE □
THIAMINE DICHLORIDE □ THIAMIN
HYDROCHLORIDE □ THIAMINIUM CHLORIDE
HYDROCHLORIDE □ USAF CB-20 □ VITAMIN B^1 □
VITAMIN B HYDROCHLORIDE
CONSENSUS REPORTS: Reported in EPA
TSCA Inventory.
SAFETY PROFILE: Poison by intravenous
and intraperitoneal routes. Mildly toxic by
ingestion. The vitamin is destroyed by
alkalies and alkaline drugs such as
phenobarbital sodium and by oxidizing and
reducing agents. When heated to
decomposition it emits very toxic fumes of
HCl, Cl^-, SO_x, and NO_x.

TET500   CAS: 532-43-4   HR: 3
THIAMINE MONONITRATE
mf: $C_{12}H_{17}N_4OS \cdot NO_3$ mw: 327.40

PROP: White crystals or crystalline powder; slt characteristic odor. Mp: 196–200° - (decomp). Non-hygroscopic; sltly sol in water, alc, and chloroform.
SYNS: 3-(4-AMINO-2-METHYLPYRIMIDYL-5-METHYL-)-4-METHYL-5,β-HYDROXYETHYLTHIAZOLIUM NITRATE □ THIAMINE NITRATE □ VITAMIN B1 MONONITRATE □ VITAMIN B1 NITRATE
CONSENSUS REPORTS: Reported in EPA TSCA Inventory.
SAFETY PROFILE: Poison by intravenous and intraperitoneal routes. A powerful oxidizer. When heated to decomposition it emits very toxic fumes of NO_x and SO_x. See also NITRATES.

TET800   CAS: 55297-95-5   HR: 3
THIAMUTILIN
mf: $C_{28}H_{47}NO_4S$ mw: 493.82
PROP: Crystals from acetone. Mp: 147–148° (after stirring in ethyl acetate and drying at 60° and 80° overnight).
SYNS: 14-DEOXY-14-((2-DIETHYLAMINOETHYL-THIO)-ACETOXY)MUTILINE □ 14-DESOSSI-14-((2-DI-ETILAMINOETIL)MERCAPTO-ACETOSSI)MUTILIN IDROGENO FUMARATO (ITALIAN) □ 14-DESOXY-14-(-(DIETHYLAMINOETHYL)-MERCAPTO ACETOXYL)-MUTILIN HYDROGEN FUMARATE □ DYNALIN - INJECTABLE □ DYNAMUTILIN □ 81723 HFU □ SQ 14055 □ SQ 22947 □ TIAMULIN □ TIAMULINA - (ITALIAN)
SAFETY PROFILE: Poison by intramuscular and intravenous routes. Moderately toxic by ingestion and subcutaneous routes. When heated to decomposition it emits toxic fumes of SO_x and NO_x. See also MERCAPTANS.

TET900   CAS: 3268-49-3   HR: 2
THIA-4-PENTANAL (DOT)
DOT: UN 2785
mf: C_4H_8OS mw: 104.18
SYNS: METHIONAL □ β-(METHYLMERCAPTO)PROP-IONALDEHYDE □ 3-(METHYLMERCAPTO)PROP-IONALDEHYDE □ METHYLMERCAPTOPROPIONIC ALDEHYDE □ 3-(METHYLTHIO)PROPANAL □ β-(-METHYLTHIO)PROPIONALDEHYDE □ 3-(METHYL-THIO)PROPIONALDEHYDE □ PROPANAL, 3-(METHYL-THIO)-(9CI) □ PROPIONALDEHYDE, 3-(METHYLTHIO)-
CONSENSUS REPORTS: Reported in EPA TSCA Inventory.

DOT CLASSIFICATION: 6.1; Label: KEEP AWAY FROM FOOD
SAFETY PROFILE: Moderately toxic by ingestion. When heated to decomposition it emits toxic vapors of SO_x.

TEX000   CAS: 148-79-8   HR: 2
2-(THIAZOL-4-YL)BENZIMIDAZOLE
mf: $C_{10}H_7N_3S$ mw: 201.26
PROP: White to tan; odorless. Mp: 304°. Insol in water; sltly sol in alc, acetone; very sltly sol in ether, chloroform.
SYNS: APL-LUSTER □ ARBOTECT □ 4-(2-BENZI-MIDAZOLYL)THIAZOLE □ BOVIZOLE □ EPROFIL □ EQUIZOLE □ LOMBRISTOP □ MERTEC □ METASOL TK-100 □ MINTEZOL □ MINZOLUM □ MK 360 □ MYCOZOL □ NEMAPAN □ OMNIZOLE □ POLIVAL □ TBDZ □ TECTO □ THIABEN □ THIABENDAZOLE - (USDA) □ THIABENZOLE □ 2-(4-THIAZOLYL)-BENZIMIDAZOLE □ 2-(4'-THIAZOLYL)BENZIMID-AZOLE □ 2-(4-THIAZOLYL)-1H-BENZIMIDAZOLE □ THIBENZOLE □ TOP FORM WORMER
CONSENSUS REPORTS: EPA Genetic Toxicology Program. Reported in EPA TSCA Inventory.
SAFETY PROFILE: Moderately toxic by ingestion. An experimental teratogen. A questionable carcinogen. Experimental reproductive effects. Mutation data reported. When heated to decomposition it emits toxic fumes of SO_x and NO_x. See also SULFIDES.

TEX250   CAS: 72-14-0   HR: 3
N¹-2-THIAZOLYLSULFANILAMIDE
mf: $C_9H_9N_3O_2S_2$ mw: 255.33
PROP: Yellowish-white prismatic rods and six-sided plates and prisms. Dimorphic. Mp: 200–202° rods. Mp: *ca.* 1° (plates).
SYNS: 2-(p-AMINOBENZENESULFONAMIDO)-THIAZOLE □ 2-(p-AMINOBENZENESULPHONAMIDO)-THIAZOLE □ 4-AMINO-N-2-THIAZOLYLBENZENE-SULFONAMIDE □ AZOSEPTALE □ CERAZOL (sus-pension) □ CHEMOSEPT □ DUATOK □ ELEUDRON □ FORMOSULFATHIAZOLE □ M+B 760 □ NEOSTREPSAN □ NORSULFASOL □ NORSULFAZOLE □ PLANOMIDE □ POLISEPTIL □ RP 2990 □ STREPTOSILTHIAZOLE □ SULFAMUL □ 2-SULFANILAMIDOTHIAZOLE □ 2--(SULFANILYLAMINO)THIAZOLE □ SULFATHIAZOL □ SULFATHIAZOLE (USDA) □ 2-SULFONAMIDOTHI-AZOLE □ SULPHATHIAZOLE □ SULZOL □

THIACOCCINE □ THIAZAMIDE □ THIOZAMIDE □ USAF SN-9

CONSENSUS REPORTS: Reported in EPA TSCA Inventory. EPA Genetic Toxicology Program.

SAFETY PROFILE: Human poison by unspecified route. Experimental poison by intraperitoneal route. Moderately toxic by intravenous, subcutaneous, and parenteral routes. Mildly toxic by ingestion. Human systemic effects by unspecified route: conjunctiva irritation, tubule changes, and allergic skin dermatitis. Experimental reproductive effects. Questionable carcinogen with experimental tumorigenic data. Mutation data reported. When heated to decomposition it emits very toxic fumes of NO_x and SO_x.

TEX600   CAS: 51707-55-2   HR: 2
THIDIAZURON

mf: $C_9H_8N_4OS$ mw: 220.27

SYNS: DEFOLIT □ DROPP □ N-PHENYL-N'-1,2,3-THIADIAZOL-5-YL-UREA □ SN 49537 □ (N-1,2,3-THIADIAZOLYL-5)-N'-PHENYLUREA

SAFETY PROFILE: Moderately toxic by ingestion. Experimental reproductive effects. When heated to decomposition it emits toxic fumes of SO_x and NO_x.

TFA000   CAS: 62-55-5   HR: 3
THIOACETAMIDE

mf: C_2H_5NS mw: 75.14

PROP: Colorless leaflets or prisms from Et_2O; mercaptan odor. Mp: 113°. Very sol in alc; sol in water; mod sol in Et_2O, C_6H_6.

SYNS: ACETOTHIOAMIDE □ ETHANETHIOAMIDE □ RCRA WASTE NUMBER U218 □ TAA □ THIACETAMIDE □ USAF CB-21 □ USAF EK-1719

CONSENSUS REPORTS: NTP 10th Report on Carcinogens. IARC Cancer Review: Group 2B IMEMDT 7,56,87; Animal Sufficient Evidence IMEMDT 7,77,74. EPA Genetic Toxicology Program. Community Right-To-Know List. Reported in EPA TSCA Inventory.

SAFETY PROFILE: Confirmed carcinogen with experimental carcinogenic, neoplastigenic, tumorigenic, and teratogenic data. Poison by ingestion and intraperitoneal routes. Moderately toxic by subcutaneous route. Human mutation data reported. An experimental teratogen. Experimental reproductive effects. Exposure has caused liver damage. When heated to decomposition it emits very toxic fumes of NO_x and SO_x. See also SULFIDES and MERCAPTANS.

TFA500   CAS: 507-09-5   HR: 3
THIOACETIC ACID

DOT: UN 2436

mf: C_2H_4OS mw: 76.12

PROP: Colorless liquid or yellow fuming liquid; pungent, disagreeable odor. Mp: $<-17°$, bp: 93°, d: 1.074 @ 10°/4°, flash p: $<73.4°$. Sol in water; misc in alc, ether.

SYNS: ACETYL MERCAPTAN □ ETHANETHIOIC ACID □ ETHANETHIOLIC ACID □ KYSELINA THIOOCTOVA □ METHANECARBOTHIOLIC ACID □ THIACETIC ACID □ THIOLACETIC ACID □ THIONOACETIC ACID □ USAF EK-P-737

CONSENSUS REPORTS: Reported in EPA TSCA Inventory.

DOT CLASSIFICATION: 3; Label: Flammable Liquid

SAFETY PROFILE: Poison by intraperitoneal route. A very dangerous fire hazard when exposed to heat or flame. When heated to decomposition it emits toxic fumes of SO_x. See also SULFIDES and MERCAPTANS.

TFC570   CAS: 10021-35-9   HR: 3
2-THIO-2H-1,3-BENZOXAZINE-2,4(3H)-DIONE

mf: $C_8H_5NO_2S$ mw: 179.20

SYNS: 2H-1,3-BENZOXAZINE-2,4(3H)-DIONE, 2-THIO- □ 2,3-DIHYDRO-1,3-BENZOXAZINE-4H-2-THIONE-4-ONE

SAFETY PROFILE: A poison by ingestion. When heated to decomposition it emits toxic vapors of NO_x and SO_x.

TFC600   CAS: 96-69-5   HR: 3
4,4'-THIOBIS(6-tert-BUTYL-m-CRESOL)

mf: $C_{22}H_{30}O_2S$ mw: 358.58

T

PROP: Light-gray to tan powder. Mp: 150°, d: 1.10.

SYNS: BIS(3-tert-BUTYL-4-HYDROXY-6-METHYL-PHENYL) SULFIDE □ BIS(4-HYDROXY-5-tert-BUTYL-2--METHYLPHENYL) SULFIDE □ DISPERSE MB-61 □ SANTONOX □ SANTOWHITE CRYSTALS □ THIOALKOFEN BM 4 □ 4,4'-THIOBIS(2-tert-BUTYL-5--METHYLPHENOL) □ 4,4'-THIOBIS(6-tert-BUTYL-3--METHYLPHENOL) □ 4,4'-THIOBIS(3-METHYL-6-tert-BUTYLPHENOL) □ 1,1'-THIOBIS(2-METHYL-4-HYDRO-XY-5-tert-BUTYLBENZENE) □ USAF B-15 □ YOSHINOX S

CONSENSUS REPORTS: Reported in EPA TSCA Inventory.

OSHA PEL: TWA Total Dust: 10 mg/m³; Respirable Fraction: 5 mg/m³

ACGIH TLV: TWA 10 mg/m³; Not Classifiable as a Human Carcinogen

NIOSH REL: (4,4'-Thiobis(6-t-butyl-m-cresol), Dust): TWA 10 mg/m³; air: TWA 5 mg/m³

SAFETY PROFILE: Poison by intraperitoneal route and probably by ingestion and inhalation. Mutation data reported. See also SULFIDES. When heated to decomposition it emits highly toxic fumes of SO_x.

TFD500   CAS: 123-28-4   HR: 1
THIOBIS(DODECYL PROPIONATE)
mf: $C_{30}H_{58}O_4S$ mw: 514.94

PROP: White crystalline flakes; characteristic sweetish odor. Sol in org solvs; insol in water.

SYNS: ADVASTAB 800 □ ANTIOXIDANT AS □ ANTIOXIDANT LTDP □ BIS(DODECYLOXYCARBON-YLETHYL) SULFIDE □ CARSTAB DLTDP □ CYANOX LTDP □ DIDODECYL 3,3'-THIODIPROPIONATE □ DI-LAURYLESTER KYSELINY β',β'-THIODIPROPIONOVE □ DILAURYL THIODIPROPIONATE □ DILAURYL β-THIODIPROPIONATE □ DILAURYL β',β'-THIODI-PROPIONATE □ DILAURYL 3,3'-THIODIPROPIONATE □ DLT □ DLTDP □ DLTP □ DMPTP □ IPOGNOX 89 □ IRGANOX PS 800 □ LAURYL 3,3'-THIODIPROPIONATE □ LUSMIT □ MILBAN F □ NEGANOX DLTP □ PLASTANOX LTDP □ PLASTANOX LTDP ANTIOXID-ANT □ PROPANOIC ACID, 3,3'-THIOBIS-, DIDODECYL ESTER □ STABILIZER DLT □ TYOX B

CONSENSUS REPORTS: Reported in EPA TSCA Inventory.

SAFETY PROFILE: An eye irritant. When heated to decomposition it emits toxic fumes of SO_x.

TFD750   CAS: 91-71-4   HR: 3
THIOCARBAMIZINE
mf: $C_{21}H_{17}AsN_2O_5S_2$ mw: 516.44

PROP: White crystal powder. Sltly sol in water, alc, and alkali; insol in acids.

SYNS: 2,2'-(((4-((AMINOCARBONYL)AMINO)PHENYL)-ARSINIDENEBIS(THIO))BIS) BENZOIC ACID □ p-(BIS(o-CARBOXYPHENYLMERCAPTO)-ARSINO)-PHENYL-UREA □ p-CARBAMIDOPHENYL-BIS(2-CARBOXYPHEN-YLMERCAPTO)ARSINE □ 4-CARBAMIDOPHENYL BIS(o-CARBOXYPHENYLTHIO)ARSENITE □ p-CARBAMIDO-PHENYL-DI(1'-CARBOXYPHENYL-2')THIOARSENITE □ S,S-DIESTER with DITHIO-p-UREIDOBENZENE-ARSONOUS ACID o-MERCAPTOBENZOIC ACID □ DI-ESTER with o-MERCAPTOBENZOIC ACID DITHIO-p-UREIDOBENZENEARSONOUS ACID □ o-MERCAPTO-BENZOIC ACID, DIESTER with DITHIO-p-UREIDOBEN-ZENEARSONOUS ACID □ THIOCARBAMISIN □ (p-UREIDOBENZENEARSYLENEDITHIO)DI-o-BENZOIC ACID □ (p-UREIDOPHENYLARSYLENEDITHIO)DI-o-BENZOIC ACID

CONSENSUS REPORTS: Arsenic and its compounds are on the Community Right-To-Know List.

OSHA PEL: TWA 0.5 mg(As)/m³

ACGIH TLV: BEI: 35 μ (As)/L inorganic arsenic and methylated metabolites in urine

SAFETY PROFILE: Poison by intraperitoneal and intravenous routes. Moderately toxic by ingestion. When heated to decomposition it emits very toxic fumes of As, SO_x, and NO_x. See also MERCAPTANS and ARSENIC COMPOUNDS.

TFE500   HR: D
THIOCYANATES

SAFETY PROFILE: Variable toxicity. Thiocyanates are not normally dissociated into cyanide; they have a low acute toxicity. Prolonged absorption may produce various skin eruptions, running nose, and occasionally dizziness, cramps, nausea, vomiting, and mild or severe disturbances of the nervous system. Violent reactions have occurred when mixed with chlorates, nitrates, HNO_3, organic peroxides,

peroxides, $KClO_3$, $NaClO_3$. Metal thiocyanates are oxidized explosively by chlorates, nitrates at 400° in intimate mixture, HNO_3, or spark or flame ignition. When heated to decomposition or on contact with acid or acid fumes they emit highly toxic fumes of CN^-.

TFF600   CAS: 5285-87-0   HR: 3
THIOCYANIC ACID, PHENYL ESTER
mf: C_7H_5NS mw: 135.19

SYNS: PHENYLRHODANID □ PHENYL THIOCYANATE □ THIOCYANATOBENZENE

SAFETY PROFILE: A poison by ingestion, intraperitoneal, and intravenous routes. When heated to decomposition it emits toxic vapors of NO_x and SO_x.

TFI000   CAS: 139-65-1   HR: 3
4,4'-THIODIANILINE
mf: $C_{12}H_{12}N_2S$ mw: 216.32

PROP: Needles from water. Mp: 108°. Sol in aq HCl; insol in H_2O.

SYNS: BIS(p-AMINOPHENYL)SULFIDE □ BIS(4-AM-INOPHENYL) SULFIDE □ BIS(p-AMINOPHENYL)-SULPHIDE □ BIS(4-AMINOPHENYL) SULPHIDE □ p,p'-DIAMINODIPHENYL SULFIDE □ 4,4'-DIAMINODI-PHENYL SULFIDE □ p,p'-DIAMINODIPHENYL SULPHIDE □ DI(p-AMINOPHENYL) SULFIDE □ DI(p-AMINOPHENYL)SULPHIDE □ NCI-C01707 □ THIOANILINE □ 4,4'-THIOANILINE □ 4,4'-THIOBIS-(ANILINE) □ 4,4'-THIOBISBENZENAMINE □ p,p-THIODIANILINE □ THIODI-p-PHENYLENEDIAMINE

CONSENSUS REPORTS: IARC Cancer Review: Group 2B IMEMDT 7,56,87; Human Limited Evidence IMEMDT 27,147,82; Animal Sufficient Evidence IMEMDT 27,147,82; Animal Limited Evidence IMEMDT 16,343,78. NCI Carcinogenesis Bioassay (feed); Clear Evidence: mouse, rat NCITR* NCI-CG-TR-47,78. Reported in EPA TSCA Inventory. Community Right-To-Know List.

DFG MAK: Animal Carcinogen, Suspected Human Carcinogen

SAFETY PROFILE: Confirmed carcinogen with experimental carcinogenic and tumorigenic data. Poison by intravenous route. Moderately toxic by ingestion. Experimental reproductive effects. Mutation data reported. When heated to decomposition it emits very toxic fumes of NO_x and SO_x. See also SULFIDES.

TFJ100   CAS: 68-11-1   HR: 3
THIOGLYCOLIC ACID
DOT: UN 1940

mf: $C_2H_4O_2S$ mw: 92.12

PROP: Liquid, strong odor. Mp: −16.5°, bp: 108° @ 15 mm. Misc with water, alc, ether, chloroform, and benzene.

SYNS: ACIDE THIOGLYCOLIQUE (FRENCH) □ GLYCOLIC ACID, THIO- □ GLYCOLIC ACID, 2-THIO- □ KYSELINA MERKAPTOOCTOVA □ KYSELINA THIOGLYKOLOVA □ MERCAPTOACETATE □ MERCAPTOACETIC ACID □ α-MERCAPTOACETIC ACID □ 2-MERCAPTOACETIC ACID □ 2-THIOGLYCOL-IC ACID □ THIOGLYCOLLIC ACID □ THIOVANIC ACID □ USAF CB-35

CONSENSUS REPORTS: Reported in EPA TSCA Inventory.

OSHA PEL: TWA 1 ppm (skin)

ACGIH TLV: TWA 1 ppm (skin)

DOT CLASSIFICATION: 8; Label: Corrosive

SAFETY PROFILE: Poison by ingestion, skin contact, intraperitoneal, and intravenous routes. Moderately toxic by subcutaneous route. A corrosive irritant to skin, eyes, and mucous membranes. When heated to decomposition it emits toxic fumes of SO_x. See also MERCAPTANS and HYDROGEN SULFIDE.

TFJ250   CAS: 64039-27-6   HR: 3
β-THIOGUANINE DEOXYRIBOSIDE
mf: $C_{10}H_{13}N_5O_3S \cdot H_2O$ mw: 301.36

SYNS: 2-AMINO-9-(2-DEOXY-β-d-RIBOFURANOSYL)-9H-PURINE-6-THIOL HYDRATE □ β-DEOXYTHIO-GUANOSINE □ β-2'-DEOXY-6-THIOGUANOSINE MONOHYDRATE □ NSC-71261 □ β-TGDR

SAFETY PROFILE: Poison by intravenous and intraperitoneal routes. Moderately toxic by ingestion. Questionable carcinogen with experimental carcinogenic and neoplastigenic data. When heated to decomposition it emits very toxic fumes of NO_x and SO_x.

T

TFK250   CAS: 79-42-5   HR: 3
2-THIOLACTIC ACID
DOT: UN 2936
mf: $C_3H_6O_2S$ mw: 106.15
PROP: Oil. Mp: 10°, bp: 98–99° @ 14 mm.
Misc in water, alc, ether.
SYNS: α-MERCAPTOPROPANOIC ACID □ α-MER-
CAPTOPROPIONIC ACID □ 2-MERCAPTOPROPIONIC
ACID
CONSENSUS REPORTS: Reported in EPA
TSCA Inventory.
DOT CLASSIFICATION: 6.1; Label: Poison
SAFETY PROFILE: Poison by ingestion.
Moderately toxic by inhalation. When
heated to decomposition it emits toxic
fumes of SO_x. See also SULFIDES and
MERCAPTANS.

TFK270   CAS: 21259-76-7   HR: 3
THIOMERIN SODIUM
mf: $C_{16}H_{27}HgNO_6S \cdot 2Na$ mw: 608.07
PROP: Powder. Freely sol in H_2O;
practically insol in Et_2O, C_6H_6 and $CHCl_3$.
SYNS: N-(γ-CARBOXYMETHYLMERCAPTOMERCURI-
β-METHOXY)PROPYLCAMPHORAMIC ACID DISODIUM
SALT □ DISODIUM-N-(3-(CARBOXYMETHYLTHIO-
MERCURI)-2-METHOXYPROPYL)-α-CAMPHORAMATE
□ DIUCARDYN SODIUM □ MERCAPTOMERIN SODIUM
□ SODIUM MERCAPTOMERIN
CONSENSUS REPORTS: Mercury and its
compounds are on the Community Right-
To-Know List.
OSHA PEL: CL 0.1 mg(Hg)/m³ (skin)
ACGIH TLV: TWA 0.1 mg(Hg)/m³ (skin);
BEI: 35 µg/g creatinine total inorganic
mercury in urine preshift; 15 µg/g creatinine
total inorganic mercury in blood at end of
shift at end of workweek.
DFG MAK: Confirmed Animal Carcinogen
with Unknown Relevance to Humans
NIOSH REL: (Mercury, Organo) TWA 0.01
mg/m³; STEL 0.03 mg/m³ (skin)
SAFETY PROFILE: Poison by
subcutaneous and intravenous routes. When
heated to decomposition it emits very toxic
fumes of Hg, NO_x, Na_2O, and SO_x. See also
MERCURY COMPOUNDS and
MERCAPTANS.

TFL000   CAS: 7719-09-7   HR: 3
THIONYL CHLORIDE
DOT: UN 1836
mf: Cl_2OS mw: 118.96
PROP: Colorless to yellow to red liquid;
suffocating odor. Mp: −105°, bp: 78.8° @
746 mm, d: 1.640 @ 15.5°/15.5°, vap press:
100 mm @ 21.4°. Misc with benzene,
chloroform, carbon tetrachloride. Sol in
$CHCl_3$, C_6H_6, and CCl_4.
SYNS: SULFINYL CHLORIDE □ SULFUR CHLORIDE
OXIDE □ SULFUROUS DICHLORIDE □ SULFUROUS O-
XYCHLORIDE □ THIONYL DICHLORIDE
CONSENSUS REPORTS: Reported in EPA
TSCA Inventory.
OSHA PEL: CL 1 ppm
ACGIH TLV: CL 1 ppm
DOT CLASSIFICATION: 8; Label: Corrosive,
Poison
SAFETY PROFILE: Moderately toxic by
inhalation. The material itself is more toxic
than sulfur dioxide. Has a pungent odor
similar to that of sulfur dioxide; it fumes
upon exposure to air. Violent reaction with
water releases hydrogen chloride and sulfur
dioxide. Both these decomposition products
constitute serious toxicity hazards. A
corrosive irritant that causes burns to the
skin and eyes. A powerful chlorinating
agent. Potentially explosive reaction with
ammonia, bis(dimethylamino)sulfoxide -
(above 80°C), chloryl perchlorate, 1,2,3-
cyclohexanetrione trioxime + sulfur dioxide,
dimethyl sulfoxide, hexafluoroisopropylideneaminolithium. Violent reaction or
ignition with 2,4-hexadiyn-1-6-diol, o-
nitrobenzoyl acetic acid, o-nitrophenyl-
acetic acid, sodium (ignites at 300°C).
Incompatible with ammonia, dimethyl
formamide + trace iron or zinc, linseed oil +
quinoline, toluene + ethanol + water. When
heated to decomposition it emits toxic
fumes of SO_x and Cl⁻. See also
HYDROGEN CHLORIDE and SULFUR
DIOXIDE.

TFL250   CAS: 7783-42-8   HR: 2
THIONYL FLUORIDE
mf: F_2OS mw: 86.06
PROP: Colorless gas; suffocating odor; slowly hydrolyzes, less reactive than $SOCl_2$. Bp: −44°, mp: −130°, d (liq): 1.780 @ −100°, (solid): 2.095 @ −183°. Sol in ether, benzene.
SYNS: FLUORURE de THIONYLE (FRENCH) □ SULFUR DIFLUORIDE MONOXIDE □ SULFUR DI-FLUORIDE OXIDE □ SULFUROUS OXYFLUORIDE □ THIONYL DIFLUORIDE
CONSENSUS REPORTS: Reported in EPA TSCA Inventory.
OSHA PEL: TWA 2.5 mg(F)/m³
ACGIH TLV: TWA 2.5 mg(F)/m³; BEI: 3 mg/g creatinine of fluorides in urine prior to shift; 10 mg/g creatinine of fluorides in urine at end of shift.
NIOSH REL: (Inorganic Fluorides) TWA 2.5 mg(F)/m³
SAFETY PROFILE: Moderately toxic by inhalation. A severe irritant to skin, eyes, and mucous membranes. When heated to decomposition or on contact with water or steam it emits highly toxic and corrosive fumes of SO_x and F^-. See also FLUORIDES.

TFM250   CAS: 110-02-1   HR: 3
THIOPHENE
DOT: UN 2414
mf: C_4H_4S mw: 84.14
PROP: Clear, colorless liquid; slt aromatic odor similar to that of benzene. D: 1.0573 @ 25°/4°, mp: −38.3°, bp: 84.4°, flash p: 21.2°F, vap press: 40 mm @ 12.5°, vap d: 2.9. Insol in water; misc with most org solvs. May be heated to 850° without decomposition.
SYNS: CP 34 □ DIVINYLENE SULFIDE □ HUILE H50 □ THIACYCLOPENTADIENE □ THIAPHENE □ THIOFURAM □ THIOFURAN □ THIOFURFURAN □ THIOLE □ THIOPHEN □ THIOTETROLE □ USAF EK-1860
CONSENSUS REPORTS: Reported in EPA TSCA Inventory.
DOT CLASSIFICATION: 3; Label: Flammable Liquid

SAFETY PROFILE: Poison by ingestion and intraperitoneal routes. Mildly toxic by inhalation and subcutaneous routes. A very dangerous fire hazard when exposed to heat or flame. Explosive reaction with N-nitrosoacetanilide. Violent or explosive reaction with nitric acid. Incompatible with oxidizing materials. To fight fire, use foam, CO_2, dry chemical. When heated to decomposition it emits highly toxic fumes of SO_x.

TFN500   CAS: 463-71-8   HR: 3
THIOPHOSGENE
DOT: UN 2474
mf: CCl_2S mw: 114.97
PROP: Reddish, fuming, red liquid with acrid odor. Bp: 73.5°, d: 1.5085 @ 15°. Decomp in water and alc. Sltly sol in H_2O; sol in Et_2O. Slowly hydrolyzes by H_2O.
SYNS: CARBON CHLOROSULFIDE □ CARBONO-THIOIC DICHLORIDE (9CI) □ CARBONYL CHLORIDE, THIO- □ DICHLOROTHIOCARBONYL □ PHOSGENE, THIO- □ THIOCARBONIC DICHLORIDE □ THIO-CARBONYL CHLORIDE □ THIOCARBONYL DI-CHLORIDE □ THIOFOSGEN (CZECH) □ THIO-KARBONYLCHLORID (CZECH)
CONSENSUS REPORTS: Reported in EPA TSCA Inventory.
DOT CLASSIFICATION: 6.1; Label: Poison
SAFETY PROFILE: Poison by intravenous route. Moderately toxic by ingestion. A skin, mucous membrane, and severe eye irritant. When heated to decomposition it emits very toxic fumes of Cl^- and SO_x. See also PHOSGENE.

TFO000   CAS: 3982-91-0   HR: 3
THIOPHOSPHORYL CHLORIDE
DOT: UN 1837
mf: Cl_3PS mw: 169.38
PROP: Colorless, mobile liquid; pungent odor which fumes in air. Decomp slowly in water at room temp or below. Bp: 125°, fp: −40.8° (þ&□ -form), fp: −36.2° (β-form), flash p: none, d: 1.63 @ 25°/4°, vap press: 22 mm @ 25°, vap d: 5.86.
SYNS: FOSFORTHIOCHLORID □ PHOSPHOROTHIO-IC TRICHLORIDE □ PHOSPHOROTHIONIC TRICHLOR-

T

IDE □ PHOSPHOROUS SULFOCHLORIDE □ PHOS-
PHOROUS THIOCHLORIDE □ PHOSPHOROUS
TRICHLORIDE SULFIDE □ SULFIDOTRICHLORID
FOSFORECNY □ THIOCHLORID FOSFORECNY □
THIOPHOSPHORIC TRICHLORIDE □ THIOPHOSPHOR-
YL TRICHLORIDE □ TL 262 □ TRICHLOROPHOSPHINE
SULFIDE

CONSENSUS REPORTS: Reported in EPA
TSCA Inventory.

DOT CLASSIFICATION: 8; Label: Corrosive

SAFETY PROFILE: Poison by inhalation.
Moderately toxic by ingestion. A corrosive
irritant to skin, eyes, and mucous
membranes. Explosive reaction with
methylmagnesium iodide. Explosive
reaction with pentaerythritol + heat. Reacts
with water or steam to produce toxic and
corrosive fumes. When heated to
decomposition it emits highly toxic fumes of
PO_x, SO_x, and Cl^-.

TFP000   **CAS: 75-18-3**   **HR: 3**
2-THIOPROPANE

DOT: UN 1164

mf: C_2H_6S mw: 62.14

PROP: Colorless liquid; disagreeable odor.
Mp: −83.2°, lel: 2.2%, uel: 19.7%, flash p:
<0°F, fp: −98.27°, bp: 37.5–38°, d: 0.8458
@ 21°/4°, vap d: 2.14, autoign temp: 403°F.
Insol in water; sol in alc, ether.

SYNS: DIMETHYLSULFID (CZECH) □ DIMETHYL
SULFIDE (DOT) □ DIMETHYL SULPHIDE (ACGIH) □
DMS □ EXACT-S □ METHYL SULFIDE (DOT) □ METH-
YL SULPHIDE □ METHYLTHIOMETHANE □ SULFURE
de METHYLE (FRENCH) □ 2-THIAPROPANE

CONSENSUS REPORTS: Reported in EPA
TSCA Inventory.

ACGIH TLV: (Proposed: 10 ppm)

DOT CLASSIFICATION: 3; Label:
Flammable Liquid

SAFETY PROFILE: Poison by inhalation.
Moderately toxic by ingestion. A skin and
severe eye irritant. A very dangerous fire
hazard when exposed to heat or flame.
Explosive in the form of vapor when
exposed to heat or flame. Can react
vigorously with oxidizing materials. To fight
fire, use CO_2, dry chemical. When heated to
decomposition it emits highly toxic fumes of
SO_x and may explode. See also SULFIDES.

TFQ000   **CAS: 79-19-6**   **HR: 3**
THIOSEMICARBAZIDE

mf: CH_5N_3S mw: 91.15

PROP: Needles from water. Mp: 182–184°.
Sol in water, alc.

SYNS: N-AMINOTHIOUREA □ HYDRAZINECARBO-
THIOAMIDE □ RCRA WASTE NUMBER P116 □
THIOCARBAMYLHYDRAZINE □ 3-THIOSEMI-
CARBAZIDE □ TSC □ USAF EK-1275

CONSENSUS REPORTS: EPA Extremely
Hazardous Substances List. Reported in
EPA TSCA Inventory.

SAFETY PROFILE: Poison by ingestion,
intraperitoneal, and intravenous routes.
Questionable carcinogen with experimental
tumorigenic data. Human mutation data
reported. When heated to decomposition it
emits very toxic fumes of NO_x and SO_x.

TFQ750   **CAS: 52-24-4**   **HR: 3**
THIOTRIETHYLENEPHOSPHORAMIDE

mf: $C_6H_{12}N_3PS$ mw: 189.24

PROP: Crystals from pentane, Et_2O, or
C_6H_6/pet ether. Mp: 52°. Very sol in EtOH;
sol in C_6H_6, Et_2O, and $CHCl_3$.

SYNS: CBC 806495 □ GIROSTAN □ NCI-C01649 □ NSC-
6396 □ ONCOTEPA □ ONCOTIOTEPA □ 1,1′,1″-PHOS-
PHINOTHIOYLIDYNETRISAZIRIDINE □ PHOS-
PHOROTHIOIC ACID TRIETHYLENETRIAMIDE □ SK
6882 □ TESPAMINE □ THIOFOZIL □ THIOPHOSPH-
AMIDE □ THIO-TEP □ TIOFOSFAMID □ TIOFOZIL □
TRIAZIRIDINYLPHOSPHINE SULFIDE □ N,N′,N″-TRI-
1,2-ETHANEDIYLPHOSPHOROTHIOIC TRIAMIDE □
N,N′,N″-TRI-1,2-ETHANEDIYLTHIOPHOSPHORAMIDE
□ TRI(ETHYLENEIMINO)THIOPHOSPHORAMIDE □
N,N′,N″-TRIETHYLENEPHOSPHOROTHIOIC TRIAMIDE
□ N,N′,N″-TRIETHYLENETHIOPHOSPHAMIDE □
N,N′,N″-TRIETHYLENETHIOPHOSPHORAMIDE □
TRIETHYLENETHIOPHOSPHOROTRIAMIDE □ TRIS(1-
AZIRIDINYL)PHOSPHINE SULFIDE □ TRIS(ETHYL-
ENIMINO)THIOPHOSPHATE □ TSPA

CONSENSUS REPORTS: NTP 10th Report
on Carcinogens. IARC Cancer Review:
Group 1 IMEMDT 50,123,90; Animal
Sufficient Evidence IMEMDT 9,85,75;
Animal Sufficient Evidence IMEMDT
50,123,90; Human Sufficient Evidence
IMEMDT 50,123,90; Human Inadequate
Evidence IMEMDT 9,85,75. NCI
Carcinogenesis Bioassay (ipr); Clear
Evidence: mouse, rat NCITR* NCI-CG-

TR-58,78. EPA Genetic Toxicology Program.

SAFETY PROFILE: Confirmed human carcinogen producing leukemia. Poison by ingestion, intraperitoneal, intravenous, and subcutaneous routes. Experimental teratogenic data. Human systemic effects by parenteral route: paresthesia, bone marrow changes, and leukemia. Experimental reproductive effects. Human mutation data reported. When heated to decomposition it emits very toxic fumes of PO_x, SO_x, and NO_x.

TFR250   CAS: 141-90-2   HR: 2
2-THIOURACIL
mf: $C_4H_4N_2OS$ mw: 128.16
PROP: Small crystals or prisms from water or alc; bitter taste. Mp: 340° (decomp). Practically insol in water, alc, ether, and acids; sol in alkalies.
SYNS: ANTAGOTHYROID □ ANTAGOTHYROIL □ DERACIL □ 2,3-DIHYDRO-2-THIOXO-4(1H)-PYRIMID- INONE □ 6-HYDROXY-2-MERCAPTOPYRIMIDINE □ 4- HYDROXY-2(1H)-PYRIMIDINETHIONE □ 2-MER- CAPTO-4-HYDROXYPYRIMIDINE □ 2-MERCAPTO-4- PYRIMIDINOL □ 2-MERCAPTO-4-PYRIMIDONE □ 2- MERCAPTOPYRIMID-4-ONE □ NOBILEN □ 2-THIO-6- OXYPYRIMIDINE □ 2-THIO-1,3-PYRIMIDIN-4-ONE □ THIOURACIL □ 6-THIOURACIL □ TIOURACYL - (POLISH) □ TU □ 2-TU

CONSENSUS REPORTS: IARC Cancer Review: Group 3 IMEMDT 7,56,87; Animal Sufficient Evidence IMEMDT 7,85,74. Reported in EPA TSCA Inventory. EPA Genetic Toxicology Program.

SAFETY PROFILE: Moderately toxic by ingestion. Human teratogenic effects by unspecified routes: developmental abnormalities of the central nervous system, craniofacial area, and endocrine system. Human reproductive effects by unspecified route: effects on newborn, including viability index changes. Experimental teratogenic effects. Other experimental reproductive effects. Questionable carcinogen with experimental neoplastigenic and tumorigenic data. Mutation data reported. When heated to decomposition it emits very toxic fumes

of NO_x and SO_x. Used in the treatment of hyperthyroidism, angina pectoris, and congestive heart failure. See also MERCAPTANS and KETONES.

TFS350   CAS: 137-26-8   HR: 3
THIRAM
mf: $C_6H_{12}N_2S_4$ mw: 240.44
PROP: Crystals from $CHCl_3$/EtOH. Mp: 156°, d: 1.30, bp: 129° @ 20 mm. Insol in water; sol in alc, ether, acetone, and chloroform. IDLH 100 mg/m³.
SYNS: AATACK □ ACCELERATOR THIURAM □ ACETO TETD □ ARASAN □ AULES □ BIS((DIMETHYL- AMINO)CARBONOTHIOYL) DISULPHIDE □ BIS(DI- METHYL-THIOCARBAMOYL)-DISULFID (GERMAN) □ BIS(DIMETHYLTHIOCARBAMOYL) DISULFIDE □ CHIPCO THIRAM 75 □ CYURAM DS □ DISOLFURO DI TETRAMETILTIOURAME (ITALIAN) □ DISULFURE de TETRAMETHYLTHIOURAME (FRENCH) □ α,α'-DI- THIOBIS(DIMETHYLTHIO)FORMAMIDE □ 1,1'-DI- THIOBIS(N,N-DIMETHYLTHIO)FORMAMIDE □ N,N'-- (DITHIODICARBONOTHIOYL)BIS(N-METHYL- METHANAMINE) □ EKAGOM TB □ FALITIRAM □ FERMIDE □ FERNACOL □ FERNASAN □ FERNIDE □ FLO PRO T SEED PROTECTANT □ HERMAL □ HERMAT TMT □ HERYL □ HEXATHIR □ KREGASAN □ MERCURAM □ METHYL THIRAM □ METHYL THIURAMDISULFIDE □ METHYL TUADS □ NOBECUTAN □ NOMERSAN □ NORMERSAN □ PANORAM 75 □ POLYRAM ULTRA □ POMARSOL □ POMASOL □ PURALIN □ RCRA WASTE NUMBER U244 □ REZIFILM □ ROYAL TMTD □ SADOPLON □ SPOTRETE □ SQ 1489 □ TERSAN □ TETRAMETHYLDI- URANE SULPHITE □ TETRAMETHYLENETHIURAM DI- SULPHIDE □ TETRAMETHYLTHIOCARBAMOYLDI- SULPHIDE □ TETRAMETHYLTHIORAMDISULFIDE - (DUTCH) □ TETRAMETHYL-THIRAM DISULFID - (GERMAN) □ TETRAMETHYLTHIURAM BISULFIDE □ TETRAMETHYLTHIURAM DISULFIDE □ N,N,N',N'- TETRAMETHYLTHIURAM DISULFIDE □ N,N-TETRA- METHYLTHIURAM DISULPHIDE □ TETRAMETHYL THIURANE DISULFIDE □ TETRAPOM □ TETRASIPTON □ TETRATHIURAM DISULFIDE □ TETRATHIURAM DISULPHIDE □ THILLATE □ THIMER □ THIOSAN □ THIOTEX □ THIOTOX □ THIRAMAD □ THIRAME (FRENCH) □ THIRASAN □ THIULIX □ THIURAD □ THIURAM □ THIURAMIN □ THIURAMYL □ THYLATE □ TIRAMPA □ TIURAM - (POLISH) □ TIURAMYL □ TMTD □ TMTDS □ TRAMETAN □ TRIDIPAM □ TRIPOMOL □ TTD □ TUADS □ TUEX □ TULISAN □ USAF B-30 □ USAF EK- 2089 □ USAF P-5 □ VANCIDA TM-95 □ VANCIDE TM □

T

VUAGT-I-4 □ VULCAFOR TMTD □ VULKACIT MTIC □ VULKACIT THIURAM □ VULKACIT THIURAM/C
CONSENSUS REPORTS: IARC Cancer Review: Group 3 IMEMDT 7,56,87; Human Inadequate Evidence IMEMDT 12,225,76; Animal Inadequate Evidence IMEMDT 12,225,76. EPA Genetic Toxicology Program. Reported in EPA TSCA Inventory.
OSHA PEL: TWA 5 mg/m³
ACGIH TLV: TWA 1 mg/m³; Not Classifiable as a Human Carcinogen; - (Proposed: inhalable fraction and vapor 0.05 mg/m³; (sensitizer); Confirmed Animal Carcinogen with Unknown Revelance to Humans)
DFG MAK: 5 mg/m³
SAFETY PROFILE: Poison by ingestion and intraperitoneal routes. Questionable carcinogen with experimental tumorigenic and teratogenic data. Other experimental reproductive effects. Mutation data reported. Affects human pulmonary system. A mild allergen and irritant. Acute poisoning in experimental animals produced liver, kidney, and brain damage. Dangerous in a fire; see NITROGEN MONOXIDE and SULFUR DIOXIDE.

TFS750   CAS: 7440-29-1   HR: 3
THORIUM
DOT: UN 2975
af: Th   aw: 232.00
PROP: Silvery-white, air-stable, soft, ductile metal. Attacked vigorously by aq HCl, slowly by H_2O, rapidly by steam. Reacts with H_2 (3°), N_2 (6°), halogens, sulfur, carbon, phosphorus on heating. Alloys with many metals. Props of metal considerably affected by presence of small amount of ThO_2. D: 11.72, mp: 1750 ± 30°. A radioactive material.
SYNS: THORIUM-232 □ THORIUM METAL, pyrophoric - (DOT)
CONSENSUS REPORTS: Reported in EPA TSCA Inventory.
DOT CLASSIFICATION: 7; Label: RADI-OACTIVE, SPONT Combustible

SAFETY PROFILE: Suspected carcinogen. Taken internally as ThO_2, it has proven to be carcinogenic due to its radioactivity. On an acute basis it has caused dermatitis. Flammable in the form of dust when exposed to heat or flame, or by chemical reaction with oxidizers. The powder may ignite spontaneously in air. Potentially hazardous reactions with chlorine, fluorine, bromine, oxygen, phosphorus, silver, sulfur, air, nitryl fluoride, peroxyformic acid.

TFT000   CAS: 10026-08-1   HR: 3
THORIUM CHLORIDE
mf: Cl_4Th mw: 373.80
PROP: White, odorless crystals or deliquescent, white solid. D: 4.59, mp: 770°, bp: 928°. Sol in water (with hydrol) and alc.
SYNS: TETRACHLOROTHORIUM □ THORIUM TETRACHLORIDE
CONSENSUS REPORTS: Reported in EPA TSCA Inventory. EPA Genetic Toxicology Program.
SAFETY PROFILE: Poison by intravenous route. Moderately toxic by intraperitoneal and subcutaneous routes. When heated to decomposition it emits toxic fumes of Cl⁻. See also THORIUM.

TFT250   CAS: 15457-87-1   HR: 3
THORIUM HYDRIDE
mf: H_4Th mw: 236.07
PROP: Black, metallic crystals. D: 8.24, mp: decomp explosively @ red heat. Reacts with water.
SAFETY PROFILE: Suspected carcinogen. Explodes on heating in air. The powder ignites spontaneously in air. See also THORIUM and HYDRIDES.

TFT500   CAS: 13823-29-5   HR: 3
THORIUM(IV) NITRATE
DOT: UN 2976
mf: N_4O_{12}•Th   mw: 480.04
PROP: White, crystalline mass or deliquescent, white plates. Mp: 55° - (decomp). Very sol in water, alc.
SYNS: NITRIC ACID, THORIUM(4+) SALT □ THORIUM (4+) NITRATE □ THORIUM TETRANITRATE

CONSENSUS REPORTS: Reported in EPA TSCA Inventory.

DOT CLASSIFICATION: 7; Label: RADI-OACTIVE, Oxidizer

SAFETY PROFILE: Poison by intraperitoneal, intravenous, and intratracheal routes. Moderately toxic by ingestion. Experimental reproductive effects. Radioactive. An oxidizing material; when in contact with readily combustible substances will cause violent combustion or ignition. When heated to decomposition it emits toxic fumes of NO_x. See also THORIUM and NITRATES.

TFT750   CAS: 1314-20-1   HR: 3
THORIUM OXIDE

mf: O_2Th mw: 264.00

PROP: Heavy, white crystalline powder or solid. D: 9.7, mp: 3390°, bp: 4400°. Insol in H_2O, dil acid or alkali; sol in hot H_2SO_4.

SYNS: THORIA □ THORIUM DIOXIDE □ THOROTRAST □ THORTRAST □ UMBRATHOR

CONSENSUS REPORTS: NTP 10th Report on Carcinogens. Community Right-To-Know List. Reported in EPA TSCA Inventory.

SAFETY PROFILE: Confirmed human carcinogen producing angiosarcoma, liver and kidney tumors, lymphoma and other tumors of the blood system, and tumors at the application site. See also THORIUM.

TFU500   CAS: 299-75-2   HR: 3
l-THREITOL-1,4-BISMETHANE-SULFONATE

mf: $C_6H_{14}O_6S_2$ mw: 246.32

SYNS: CB 2562 □ 1,4-DIMETHANESULFONATE THREITOL □ (2s,3s)-1,4-DIMETHANESULFONATE TREITOL □ NSC-39069 □ TREOSULFAN □ TRESULFAN

CONSENSUS REPORTS: IARC Cancer Review: Group 1 IMEMDT 7,363,87; Human Sufficient Evidence IMEMDT 26,341,81. EPA Genetic Toxicology Program.

SAFETY PROFILE: Confirmed carcinogen. Poison by intravenous route. Human mutation data reported. When heated to decomposition it emits toxic fumes of SO_x.

TFU750   CAS: 72-19-5   HR: 2
l-THREONINE

mf: $C_4H_9NO_3$ mw: 119.14

PROP: An essential amino acid. Colorless crystals or white crystalline powder from alc (aq); slt sweet taste. Mp: 255–257° with decomp. Sol in water; very sol in hot water; insol in alc, chloroform, ether.

SYNS: l-2-AMINO-3-HYDROXYBUTYRIC ACID □ THREONINE

CONSENSUS REPORTS: Reported in EPA TSCA Inventory.

SAFETY PROFILE: Moderately toxic by intraperitoneal route. When heated to decomposition it emits toxic fumes of NO_x.

TFW000   CAS: 546-80-5   HR: 3
THUJONE

mf: $C_{10}H_{16}O$ mw: 152.26

PROP: A liquid. Bp: 78° @ 12 mm. A flavor constituent. A major component of wormwood oil, (*Artemisia absinthium L.*), which is the principal ingredient of absinthe, a liquor. Occurs as α, l, (−) or β, d, (+) called isothujone.

SYNS: (1S-1-α,4-α,5-α)-4-METHYL-1-(1-METHYLETH-YL)-BICYCLO(3.1.0)HEXAN-3-ONE □ (1S,4R,5R)-(−)-3-THUJANONE □ THUJON □ (−)-THUJONE □ α-THUJONE □ l-THUJONE

CONSENSUS REPORTS: Reported in EPA TSCA Inventory.

SAFETY PROFILE: Poison by intravenous, intraperitoneal, and subcutaneous routes. Moderately toxic by ingestion. Serious physiological consequences from abuse of absinthe (mainly in France) led to its abolition in 1915. Wormwood is still used in concentrations of less than 10 ppm in flavored wines. Thujone at 30 mg/kg causes convulsions associated with lesions of the cerebral cortex. Little is known of thujone metabolism. Both forms occur in wormwood oil, oak moss. The α form is major constituent of cedar leaf oil or oil of thuja, sage. The β form occurs in tansy, yarrow. When heated to decomposition it emits acrid smoke and irritating fumes.

T

TFW500 CAS: 13537-18-3 HR: 3
THULIUM CHLORIDE
mf: Cl_3Tm mw: 275.28
PROP: Hygroscopic pale-yellow solid. Mp: 824°.
CONSENSUS REPORTS: Reported in EPA TSCA Inventory.
SAFETY PROFILE: Poison by intraperitoneal route. Mildly toxic by ingestion. A skin and eye irritant. When heated to decomposition it emits toxic fumes of Cl^-. See also THULIUM and RARE EARTHS.

TFX500 CAS: 8007-46-3 HR: 2
THYME OIL
PROP: From distillation of flowering plant *Thymus vulgaris L.* (Fam. *Labiatae*). Colorless to reddish-brown liquid; pleasant odor, sharp taste. D: 0.930 @ 25°/25°, refr index: 1.495 @ 20°.
SYNS: OIL OF THYME □ THYMIAN OEL (GERMAN) □ THYM OIL
CONSENSUS REPORTS: Reported in EPA TSCA Inventory.
SAFETY PROFILE: Moderately toxic by ingestion. A severe skin irritant. Mutation data reported. An allergen. Combustible when exposed to heat or flame. When heated to decomposition it emits acrid smoke and irritating fumes.

TFX750 CAS: 8007-46-3 HR: 2
THYME OIL RED
PROP: Main constituents are thymol, carvacrol. Found in plants *Thymus vulgaris L.* and *Thymus zygis L.* (FCTXAV 12,807,74).
SYN: SPANISH THYME OIL
SAFETY PROFILE: Mildly toxic by ingestion. A severe skin irritant. When heated to decomposition it emits acrid smoke and irritating fumes.

TFX790 CAS: 50-89-5 HR: 2
THYMIDINE
mf: $C_{10}H_{14}N_2O_5$ mw: 242.26
PROP: Rosettes of needles. Mp: 186–187°. Sol in water, methanol, hot alc, hot acetone, hot ethyl acetate, pyridine, glacial acetic acid; sltly sol in hot chloroform.
SYNS: DEOXYTHYMIDINE □ 2'-DEOXYTHYMIDINE □ DT □ DTHYD □ 5-METHYLDEOXYURIDINE □ THYMIDIN □ THYMINEDEOXYRIBOSIDE □ THYMINE-2-DEOXYRIBOSIDE
CONSENSUS REPORTS: Reported in EPA TSCA Inventory. EPA Genetic Toxicology Program.
SAFETY PROFILE: Moderately toxic by intraperitoneal route. An experimental teratogen. Experimental reproductive effects. Human mutation data reported. When heated to decomposition it emits toxic fumes of NO_x.

TFX810 CAS: 89-83-8 HR: 3
THYMOL
mf: $C_{10}H_{14}O$ mw: 150.24
PROP: Colorless, translucent crystals or plates from EtOAc or AcOH or Me_2CO; pungent, caustic taste with odor of thyme. Mp: 51°, bp: 233°, d: 0.972, vap press: 1 mm @ 64°. Sol in water, alkali; very sol in alc, ether, chloroform.
SYNS: p-CYMEN-3-OL □ 3-p-CYMENOL □ 3-HYDROXY-p-CYMENE □ 3-HYDROXY-1-METHYL-4-ISOPROPYLBENZENE □ ISOPROPYL CRESOL □ 6-ISOPROPYL-m-CRESOL □ 2-ISOPROPYL-5-METHYLPHENOL □ 1-METHYL-3-HYDROXY-4-ISOPROPYLBENZENE □ 5-METHYL-2-ISOPROPYL-1-PHENOL □ 5-METHYL-2-(1-METHYLETHYL)PHENOL □ THYME CAMPHOR □ THYMIC ACID □ m-THYMOL
CONSENSUS REPORTS: Reported in EPA TSCA Inventory. EPA Genetic Toxicology Program.
SAFETY PROFILE: Poison by ingestion and intravenous routes. Moderately toxic by subcutaneous route. Experimental reproductive effects. Mutation data reported. An allergen. Incompatible with acetanilide. When heated to decomposition it emits acrid smoke and irritating fumes. An FDA over-the-counter drug used as an antibacterial and antifungal agent.

TGA700 CAS: 80-59-1 HR: 2
TIGLIC ACID
mf: $C_5H_8O_2$ mw: 100.13

PROP: Triclinic plates or rods or crystals from water; spicy odor. D: 0.972, mp: 63.5–64°, bp: 198.5°. Sltly sol in cold water; sol in hot water, alc, and eth.

SYNS: 2-BUTENOIC ACID, 2-METHYL-, (E)-(9CI) □ CEVADIC ACID □ CROTONIC ACID, 2-METHYL-, (E)- □ (E)-2,3-DIMETHYLACRYLIC ACID □ trans-α-β-DIMETHYLACRYLIC ACID □ trans-2,3-DIMETHYLACRYLIC ACID □ trans-2-METHYL-2-BUTENOIC ACID □ (E)-2-METHYLCROTONIC ACID □ trans-2-METHYLCROTONIC ACID □ TIGLINIC ACID

CONSENSUS REPORTS: Reported in EPA TSCA Inventory.

SAFETY PROFILE: Low toxicity by ingestion and skin contact. A severe skin irritant. When heated to decomposition it emits acrid smoke and irritating fumes.

TGB250　CAS: 7440-31-5　HR: 2
TIN
af: Sn　　aw: 118.71

PROP: Cubic, gray, crystalline metallic element. Mp: 231.9°, stabilizes @ <18°, d: 7.31, vap press: 1 mm @ 1492°, bp: 2625°. IDLH 100 mg/m³ (as Sn).

SYNS: SILVER MATT POWDER □ TIN (α) □ TIN FLAKE □ TIN POWDER □ ZINN (GERMAN)

CONSENSUS REPORTS: Reported in EPA TSCA Inventory.

OSHA PEL: Organic Compounds: TWA 0.1 mg(Sn)/m³ (skin); Inorganic Compounds - (except oxides): TWA 2 mg/m³

ACGIH TLV: TWA metal, oxide, and inorganic compounds (except SnH₄) as Sn 2 mg/m³; organic compounds TWA 0.1 mg-(Sn)/m³; STEL 0.2 mg(Sn)/m³ (skin)

DFG MAK: Inorganic 2 mg/m³, organic 0.1 mg/m³

NIOSH REL: (Organotin Compounds) TWA 0.1 mg(Sn)/m³

SAFETY PROFILE: An inhalation hazard. Questionable carcinogen with experimental tumorigenic data by implant route. Combustible in the form of dust when exposed to heat or by spontaneous chemical reaction with Br_2, BrF_3, Cl_2, ClF_3, $Cu(NO_3)$, K_2O_2, S. See also POWDERED METALS and TIN COMPOUNDS.

TGB475　CAS: 2398-96-1　HR: 1
TINACTIN
mf: $C_{19}H_{17}NOS$　　mw: 307.43

PROP: Crystals from alc. Mp: 110.5–111.5°. Insol in water; sparingly sol in methanol, ethanol; sol in chloroform (1:1.5), acetone - (1:7), CCl_4 (1:9).

SYNS: AFTATE □ CHINOFUNGIN □ DERMOXIN □ m,N-DIMETHYLTHIOCARBANILIC ACID-o-2 NAPHTHYL ESTER □ FOCUSAN □ FUNGISTOP □ HI-ALAZIN □ METHYL (3-METHYLPHENYL)CARBAMOTHIOIC ACID, o-2-NAPHTHALENYL ESTER □ NAPHTHIOMATE T □ o-2-NAPHTHYL m,N-DIMETHYLTHIOCARBANILATE □ 2-NAPHTHYL N-METHYL-N-(3-TOLYL)THIONO-CARBAMATE □ PITREX □ SORGOA □ SPORILINE □ TIMOPED □ TINADERM □ TOLNAFTATE □ TOLNAPHTHATE □ TOLSANIL □ TONOFTAL

CONSENSUS REPORTS: EPA Genetic Toxicology Program. Reported in EPA TSCA Inventory.

SAFETY PROFILE: Mildly toxic by ingestion. An experimental teratogen. When heated to decomposition it emits toxic fumes of SO_x and NO_x. A fungicide used to control athlete's foot. See also ESTERS and CARBAMATES.

TGB750　CAS: 7789-67-5　HR: 3
TIN(IV) BROMIDE (1:4)
mf: Br_4Sn　　mw: 438.33

PROP: White crystalline mass or hygroscopic crystals. Fumes in air; decomp in water. Mp: 31°, bp: 202°, d (liquid): 3.340 @ 35°, vap press: 10 mm @ 72.7°.

SYNS: STANNIC BROMIDE □ TIN PERBROMIDE □ TIN TETRABROMIDE

CONSENSUS REPORTS: Reported in EPA TSCA Inventory.

OSHA PEL: TWA 2 mg(Sn)/m³

ACGIH TLV: TWA 2 mg(Sn)/m³

SAFETY PROFILE: Poison by intravenous route. Violent reaction with NO_2Cl. When heated to decomposition it emits toxic fumes of Br^-. See also BROMIDES and TIN COMPOUNDS.

TGC000　CAS: 7772-99-8　HR: 3
TIN(II) CHLORIDE (1:2)
mf: Cl_2Sn　　mw: 189.59

T

PROP: Colorless or white crystals. D: 2.71, mp: 246°, bp: 623°. Sol in less than its own weight of water; very sol in hydrochloric acid (dilute or conc); sol in alc, ethyl acetate, glacial acetic acid, sodium hydroxide solution.
SYNS: C.I. 77864 □ NCI-C02722 □ STANNOUS CHLORIDE (FCC) □ TIN DICHLORIDE □ TIN PROTOCHLORIDE
CONSENSUS REPORTS: NTP Carcinogenesis Bioassay (feed); No Evidence: mouse, rat NTPTR* NTP-TR-231,82. Reported in EPA TSCA Inventory. EPA Genetic Toxicology Program.
OSHA PEL: TWA 2 mg(Sn)/m³
ACGIH TLV: TWA 2 mg(Sn)/m³
SAFETY PROFILE: Poison by ingestion, intraperitoneal, intravenous, and subcutaneous routes. Experimental reproductive effects. Human mutation data reported. Potentially explosive reaction with metal nitrates. Violent reactions with hydrogen peroxide, ethylene oxide, hydrazine hydrate, nitrates, K, Na. Ignition on contact with bromine trifluoride. A vigorous reaction with calcium acetylide is initiated by flame. When heated to decomposition it emits toxic fumes of Cl⁻. See also TIN COMPOUNDS.

TGC250   CAS: 7646-78-8   HR: 3
TIN(IV) CHLORIDE (1:4)
DOT: UN 1827
mf: Cl₄Sn   mw: 260.49
PROP: Colorless, fuming caustic liquid or crystals. Mp: −33°, bp: 114.1°, d: 2.232, vap press: 10 mm @ 10°. Sol in H₂O, CCl₄, and C₆H₆.
SYNS: ETAIN (TETRACHLORURE d') (FRENCH) □ LIBAVIUS FUMING SPIRIT □ STAGNO (TETRA-CLORURO di) (ITALIAN) □ STANNIC CHLORIDE, anhydrous (DOT) □ TIN CHLORIDE, fuming (DOT) □ TIN PERCHLORIDE (DOT) □ TIN TETRACHLORIDE, anhydrous (DOT) □ TINTETRACHLORIDE (DUTCH) □ Z-INNTETRACHLORID (GERMAN)
CONSENSUS REPORTS: Reported in EPA TSCA Inventory. EPA Genetic Toxicology Program.
OSHA PEL: TWA 2 mg(Sn)/m³
ACGIH TLV: TWA 2 mg(Sn)/m³

DOT CLASSIFICATION: 8; Label: Corrosive
SAFETY PROFILE: Poison by intraperitoneal route. Moderately toxic by inhalation. A corrosive irritant to skin, eyes, and mucous membranes. Combustible by chemical reaction. Upon contact with moisture, considerable heat is generated. Violent reaction with K, Na, turpentine, ethylene oxide, alkyl nitrates. Dangerous; hydrochloric acid is liberated on contact with moisture or heat. When heated to decomposition it emits toxic fumes of Cl⁻. See also HYDROCHLORIC ACID.

TGC280   CAS: 13940-16-4   HR: 3
TIN CHLORIDE IODIDE
mf: Cl₂I₂Sn   mw: 443.39
PROP: Red mobile liquid. D: 3.287 @ 15°, bp: 297°. Sol in C₆H₆, CS₂, CHCl₃, and cold H₂O; decomp in hot H₂O.
SYN: STANNIC DICHLORIDE DIIODIDE
OSHA PEL: TWA 2 mg(Sn)/m³
ACGIH TLV: TWA 2 mg(Sn)/m³; (Proposed: (inhalable fraction) 0.1 mg/m³; Not Classifiable as a Human Carcinogen)
SAFETY PROFILE: Poison by intravenous route. When heated to decomposition it emits toxic fumes of Sn, I⁻, and Cl⁻.

TGC282   CAS: 10026-06-9   HR: 3
TIN(IV) CHLORIDE, PENTAHYDRATE - (1:4:5)
DOT: UN 2440
mf: Cl₄Sn•5H₂O   mw: 350.59
SYNS: STANNIC CHLORIDE PENTAHYDRATE □ STANNIC CHLORIDE, pentahydrate (DOT) □ TETRACHLOROSTANNANE PENTAHYDRATE
OSHA PEL: TWA 2 mg(Sn)/m³
ACGIH TLV: TWA 2 mg(Sn)/m³
DOT CLASSIFICATION: 8; Label: Corrosive
SAFETY PROFILE: A poison by intraperitoneal and intravenous routes. Mutation data reported. A corrosive liquid. When heated to decomposition it emits toxic vapors of tin and Cl⁻.

TGC500   **HR: D**
TIN COMPOUNDS
OSHA PEL: Organic Compounds: TWA 0.1 mg(Sn)/m³ (skin); Inorganic Compounds - (except oxides): TWA 2 mg/m³
ACGIH TLV: TWA metal, oxide, and inorganic compounds (except SnH₄) as Sn 2 mg/m³; organic compounds TWA 0.1 mg-(Sn)/m³; STEL 0.2 mg(Sn)/m³ (skin).
DFG MAK: Inorganic 2 mg/m³, organic 0.1 mg/m³
NIOSH REL: (Organotin Compounds) TWA 0.1 mg(Sn)/m³
SAFETY PROFILE: Variable toxicity. Elemental tin and inorganic tin compounds have low toxicity and are poorly absorbed when ingested. Some inorganic tin salts are irritating or can liberate toxic fumes on decomposition. The latter is particularly true of tin halogens. Tin hydride is highly toxic with effects similar to those of arsenic hydride. Inhalation of tin dusts over a period of years may cause pneumoconiosis. Some of the organic tin compounds are strong poisons. Short-chain alkyl tin compounds - (e.g., ethyl and methyl compounds) are particularly toxic. Generally alkyl tin compounds are more toxic than aryl compounds and short chain compounds are more toxic than long-chain compounds. The toxicity increases with the number of alkyl groups. Tetramethyl tin chloride and triethyl tin chloride are very toxic to the nervous system. They are lipid soluble and can be absorbed through the skin. Symptoms recede slowly. The concentration of tin in condensed milk from cans may reach 160 ppm and could become hazardous for babies. Some alkyl tin compounds have high ecotoxicity. They have been used in marine paints to prevent growth on boat hulls, but may have too much environmental effect for this purpose. See also TIN and specific compounds.

TGD000   **CAS: 58-14-0**   **HR: 3**
TINDURIN
mf: $C_{12}H_{13}ClN_4$ mw: 248.74

PROP: A solid. Mp: 233–234° (capillary), mp: 240–242° (Cu block).
SYNS: CD □ CHLORIDIN □ CHLORIDINE □ 5-(4'-CHLOROPHENYL)-2,4-DIAMINO-6-ETHYLPYRIMIDINE □ 5-(4-CHLOROPHENYL)-6-ETHYL-2,4-PYRIMIDINEDI-AMINE □ DARACLOR □ DARAPRAM □ DARAPRIM □ DARAPRIME □ 2,4-DIAMINO-5-p-CHLOROPHENYL-6-ETHYLPYRIMIDINE □ 2,4-DIAMINO-5-(4-CHLORO-PHENYL)-6-ETHYLPYRIMIDINE □ DIAMINOPYRITAM-IN □ ERBAPRELINA □ KHLORIDIN □ MALACID □ MALOCID □ MALOCIDE □ MALOPRIM □ NCI-C01683 □ NSC-3061 □ PIRIMECIDAN □ PIRIMETAMINA -(SPANISH) □ 4753 R.P. □ WR 2978
CONSENSUS REPORTS: IARC Cancer Review: Group 3 IMEMDT 7,56,87; Animal Limited Evidence IMEMDT 13,233,77. NCI Carcinogenesis Bioassay (feed); Inadequate Studies: mouse NCITR* NCI-CG-TR-77,78; (feed); No Evidence: rat NCITR* NCI-CG-TR-77,78. EPA Genetic Toxicology Program.
SAFETY PROFILE: Poison by ingestion, subcutaneous, and intraperitoneal routes. Experimental teratogenic and reproductive effects. Questionable carcinogen. Human mutation data reported. When heated to decomposition it emits very toxic fumes of Cl⁻ and NO_x. Used as an antimalarial drug for humans and to treat toxoplasmosis in hogs.

TGD100   **CAS: 7783-47-3**   **HR: 3**
TIN FLUORIDE
mf: F_2Sn mw: 156.69
PROP: Monoclinic, lamellar plates or colorless hygroscopic prisms. Mp: 213°, d: - (25°) 4.57, sublimes @ 7°. Sol in water - (about 30%). Forms an oxyfluoride, $SnOF_2$, on exposure to air. Prac insol in MeOH, Et_2O, and $CHCl_3$.
SYNS: FLUORISTAN □ STANNOUS FLUORIDE □ TIN BIFLUORIDE □ TIN DIFLUORIDE
CONSENSUS REPORTS: IARC Cancer Review: Group 3 IMEMDT 7,208,87. EPA Genetic Toxicology Program. Reported in EPA TSCA Inventory.
OSHA PEL: TWA 2 mg(Sn)/m³; 2.5 mg(F)-/m³
ACGIH TLV: TWA 2 mg(Sn)/m³; TWA 2.5 mg(F)/m³; BEI: 3 mg/g creatinine of

fluorides in urine prior to shift; 10 mg/g creatinine of fluorides in urine at end of shift.

NIOSH REL: TWA 2.5 mg(F)/m^3

SAFETY PROFILE: Poison by ingestion and intraperitoneal routes. Questionable carcinogen. Mutation data reported. When heated to decomposition it emits toxic fumes of F$^-$. See also TIN COMPOUNDS.

TGD500   CAS: 10294-70-9   HR: 3
TIN(II) IODIDE

mf: I$_2$Sn   mw: 372.49

PROP: Powder or red needles. Mp: 320°, bp: 720°, d: 5.21. Sol in water.

SYN: STANNOUS IODIDE

CONSENSUS REPORTS: Reported in EPA TSCA Inventory.

OSHA PEL: TWA 2 mg(Sn)/m^3

ACGIH TLV: TWA 2 mg(Sn)/m^3; (Proposed: (inhalable fraction) 0.1 mg/m^3; Not Classifiable as a Human Carcinogen)

SAFETY PROFILE: Poison by intravenous route. When heated to decomposition it emits toxic fumes of I$^-$. See also IODIDES and TIN COMPOUNDS.

TGD750   CAS: 7790-47-8   HR: 3
TIN(IV) IODIDE (1:4)

mf: I$_4$Sn   mw: 626.29

PROP: Red cubic crystals. Hydrolyzes completely in H$_2$O. Mp: 144.5°, bp: 364°, d: 4.473 @ 0°. Sol in CCl$_4$, CHCl$_3$, C$_6$H$_6$, CS$_2$, EtOH, and Et$_2$O.

SYNS: STANNIC IODIDE □ TIN TETRAIODIDE

CONSENSUS REPORTS: Reported in EPA TSCA Inventory.

OSHA PEL: TWA 2 mg(Sn)/m^3

ACGIH TLV: TWA 2 mg(Sn)/m^3; (Proposed: (inhalable fraction) 0.1 mg/m^3; Not Classifiable as a Human Carcinogen)

SAFETY PROFILE: Poison by intravenous route. Strong reaction with NO$_2$Cl, (K + S), (Na + S). When heated to decomposition it emits toxic fumes of I$^-$. See also TIN COMPOUNDS and IODIDES.

TGE100   CAS: 13863-31-5   HR: 1
TINOPAL 5BM

mf: C$_{38}$H$_{36}$N$_{12}$O$_8$S$_2$•2Na   mw: 898.96

SYN: 2,2'-STILBENEDISULFONIC ACID, 4,4'-BIS((4-ANILINO-6-((2-HYDROXYETHYL)METHYLAMINO)-s-TRIAZIN- 2-YL)AMINO)-, DISODIUM SALT

CONSENSUS REPORTS: Reported in EPA TSCA Inventory.

SAFETY PROFILE: Experimental reproductive effects. An eye irritant. When heated to decomposition it emits toxic fumes of NO$_x$ and SO$_x$.

TGE150   CAS: 27344-41-8   HR: 2
TINOPAL CBS

mf: C$_{28}$H$_{20}$O$_6$S$_2$•2Na   mw: 562.58

SYNS: 2,2'-((1,1'-BIPHENYL)-4,4'-DIYLDI-2,1-ETHENEDIYL)BIS-BENZENESULFONIC ACID DISODIUM SALT □ DISODIUM-4,4'-BIS(2-SULFOSTYRYL)-BIPHENYL □ STILBENE 3 □ TINOPAL CBS-X

CONSENSUS REPORTS: EPA Genetic Toxicology Program. Reported in EPA TSCA Inventory.

SAFETY PROFILE: Mildly toxic by ingestion. A skin and severe eye irritant. Experimental reproductive effects. When heated to decomposition it emits toxic fumes of SO$_x$ and Na$_2$O.

TGE250   CAS: 814-94-8   HR: 2
TIN(II) OXALATE

mf: C$_2$H$_2$O$_4$•Sn   mw: 208.73

PROP: Powder. Mp: 280° (decomp), d: 3.56. Insol in H$_2$O.

SYNS: ETHANEDIOIC ACID, TIN(2+) SALT (1:1) (9CI) □ OXALIC ACID, TIN(2+) SALT (1:1) (8CI) □ STANNOUS OXALATE □ STAVELAN CINATY (CZECH) □ TIN OXALATE □ TIN(2+) OXALATE

CONSENSUS REPORTS: Reported in EPA TSCA Inventory.

OSHA PEL: TWA 2 mg(Sn)/m^3

ACGIH TLV: TWA 2 mg(Sn)/m^3

DOT CLASSIFICATION: 6.1; Label: KEEP AWAY FROM FOOD

SAFETY PROFILE: Moderately toxic by ingestion. When heated to decomposition it emits acrid smoke and irritating fumes. See also TIN COMPOUNDS and OXALATES.

TGE300   CAS: 1332-29-2   HR: 2
TIN OXIDE
mf: O$_2$Sn   mw: 150.69
SYNS: MESA □ STANNOXYL
CONSENSUS REPORTS: Reported in EPA
TSCA Inventory.
OSHA PEL: TWA 2 mg/m^3
ACGIH TLV: TWA 2 mg(Sn)/m^3
SAFETY PROFILE: When heated to
decomposition it emits acrid smoke and
irritating fumes. See also TIN
COMPOUNDS.

TGE500   CAS: 25324-56-5   HR: 3
TIN(IV) PHOSPHIDE
DOT: UN 1433
mf: PSn   mw: 149.66
PROP: Silver-white crystals or solid with a
dull metallic luster. D: 6.56.
SYN: STANNIC PHOSPHIDE (DOT)
CONSENSUS REPORTS: Reported in EPA
TSCA Inventory.
OSHA PEL: TWA 2 mg(Sn)/m^3
ACGIH TLV: TWA 2 mg(Sn)/m^3
DOT CLASSIFICATION: 4.3; Label:
Dangerous When Wet, Poison
SAFETY PROFILE: A flammable solid.
Reacts with moisture or acid fumes to
liberate highly toxic phosphine gas. When
heated to decomposition it emits toxic
fumes of PO$_x$. See also PHOSPHINE,
PHOSPHIDES, and TIN COMPOUNDS.

TGF250   CAS: 7440-32-6   HR: 3
TITANIUM
af: Ti   aw: 47.90
PROP: Dark-gray amorphous powder or
hard, lustrous white metal. Not attacked by
alkalies or cold mineral acids (except HF).
Resists corrosion (oxide layer). D: 4.5 @
20°, autoign temp: 1200° for solid metal in
air, 250° for powder, mp: 1667°, bp: 3285°.
SYNS: CONTIMET 30 □ C.P. TITANIUM □ IMI 115 □
NCI-C04251 □ OREMET □ T 40 □ TITANATE □
TITANIUM 50A □ TITANIUM ALLOY □ VT 1
CONSENSUS REPORTS: Reported in EPA
TSCA Inventory.
SAFETY PROFILE: Questionable
carcinogen with experimental tumorigenic

data. Experimental reproductive effects.
The dust may ignite spontaneously in air.
Flammable when exposed to heat or flame
or by chemical reaction. Titanium can burn
in an atmosphere of carbon dioxide,
nitrogen, or air. Also reacts violently with
BrF$_3$, CuO, PbO, (Ni + KClO$_3$), metaloxy
salts, halocarbons, halogens, CO$_2$, metal
carbonates, Al, water, AgF, O$_2$, nitryl
fluoride, HNO$_3$, O$_2$, KClO$_3$, KNO$_3$,
KMnO$_4$, steam @ 704°, trichloroethylene,
trichlorotrifluoroethane. Ordinary
extinguishers are often ineffective against
titanium fires. Such fires require special
extinguishers designed for metal fires. In
airtight enclosures, titanium fires can be
controlled by the use of argon or helium.
Titanium, in the absence of moisture, burns
slowly, but evolves much heat. The
application of water to burning titanium can
cause an explosion. Finely divided titanium
dust and powders, like most metal powders,
are potential explosion hazards when
exposed to sparks, open flame, or high-heat
sources. See also TITANIUM
COMPOUNDS, POWDERED METALS,
and MAGNESIUM.

TGF500   CAS: 7440-32-6   HR: D
TITANIUM (wet powder)
SYN: TITANIUM METAL POWDER, WET (DOT)
CONSENSUS REPORTS: Reported in EPA
TSCA Inventory.
SAFETY PROFILE: An experimental
teratogen. See also TITANIUM
COMPOUNDS and TITANIUM.

TGG250   CAS: 7705-07-9   HR: 3
TITANIUM CHLORIDE
DOT: UN 2441/UN 2869
mf: Cl$_3$Ti   mw: 154.25
PROP: Colorless to light-yellow liquid;
fumes in moist air. þ&□ -form is a red-violet
solid, β-form is a brown solid. The β or þ&□
transformation occurs at 250–300° or at 40°
in an inert solv containing TiCl$_4$.
Disproportionates to TiCl$_2$ and TiCl$_4$ above
4°. Readily oxidized in air. Mp: −30°, bp:

T

136.4°, d: 1.772 @ 25°/25°, vap press: 10 mm @ 21.3°.

SYNS: TAC 121 □ TAC 131 □ TITANIUM(III) CHLORIDE □ TITANIUM TRICHLORIDE MIXTURES - (UN 2869) (DOT) □ TITANIUM TRICHLORIDE MIXTURES, pyrophoric (UN 2441) (DOT) □ TITANIUM TRICHLORIDE, pyrophoric (UN 2441) (DOT) □ TITANOUS CHLORIDE □ TRICHLOROTITANIUM

CONSENSUS REPORTS: Reported in EPA TSCA Inventory.

DOT CLASSIFICATION: 8; Label: Corrosive (UN 2869); DOT Class: 4.2; Label: Spontaneously Combustible, Corrosive (UN 2441)

SAFETY PROFILE: A corrosive irritant to skin, eyes, and mucous membranes. A severe corrosive because it liberates heat and hydrochloric acid upon contact with moisture. If spilled on skin, wipe off with dry cloth before applying water. May ignite spontaneously in air. Flammable when exposed to heat or flame. Reacts violently with K, HF. Experimental reproductive effects. When heated to decomposition it emits toxic fumes of Cl⁻. See also TITANIUM COMPOUNDS.

TGG500 HR: D
TITANIUM COMPOUNDS

SAFETY PROFILE: These compounds are generally considered to be physiologically inert. There are no reported cases in the literature where titanium as such has caused human intoxication. The dusts of titanium or most titanium compounds such as titanium oxide may be placed in the nuisance category. Titanium tetrachloride, however, is an irritant and corrosive material, because, when exposed to moisture, it hydrolyzes to hydrogen chloride. See also TITANIUM and specific compounds.

TGG760 CAS: 13463-67-7 HR: 1
TITANIUM DIOXIDE

mf: O_2Ti mw: 79.90

PROP: White amorphous powder or white solid with very high refractive index. Loses O_2 in air to form $TiO_{1.985}$ which melts at *ca.*

18°. Mp: 1860° (decomp), d: 4.26. Insol in water, hydrochloric acid, dil sulfuric acid, and alc; sol in hot concentrated H_2SO_4 and HF. IDLH 5000 mg/m³.

SYNS: A-FIL CREAM □ ATLAS WHITE TITANIUM DI-OXIDE □ AUSTIOX □ BAYERITIAN □ BAYERTITAN □ BAYTITAN □ CALCOTONE WHITE T □ C.I. 77891 □ C.I. PIGMENT WHITE 6 □ COSMETIC WHITE C47-5175 □ C-WEISS 7 (GERMAN) □ FLAMENCO □ HOMBITAN □ HORSE HEAD A-410 □ KH 360 □ KRONOS TITANIUM DIOXIDE □ LEVANOX WHITE RKB □ NCI-C04240 □ RAYOX □ RUNA RH20 □ RUTILE □ TIOFINE □ TIOXIDE □ TITANDIOXID (SWEDEN) □ TITANIUM OXIDE □ TRIOXIDE(S) □ TRONOX □ UNITANE O-110 □ 1700 WHITE □ ZOPAQUE

CONSENSUS REPORTS: NCI Carcinogenesis Bioassay (feed); No Evidence: mouse, rat NCITR* NCI-CG-TR-97,79. Reported in EPA TSCA Inventory. EPA Genetic Toxicology Program.

OSHA PEL: TWA Total Dust: 10 mg/m³; Respirable Fraction: 5 mg/m³

ACGIH TLV: TWA (nuisance particulate) 10 mg/m³ of total dust (when toxic impurities are not present, e.g., quartz <1%); Not Classifiable as a Human Carcinogen

DFG MAK: 1.5 mg/m³

SAFETY PROFILE: A nuisance dust. A human skin irritant. Questionable carcinogen with experimental carcinogenic, neoplastigenic, and tumorigenic data. Violent or incandescent reaction with metals at high temperatures (e.g., aluminum, calcium, magnesium, potassium, sodium, zinc, lithium). See also TITANIUM COMPOUNDS.

TGH250 CAS: 13825-74-6 HR: 2
TITANIUM SULFATE

DOT: NA 1760

mf: $H_2O_4S•1/2Ti$

PROP: Colorless solid, decomp to TiO_2 + SO_3. Colorless crystals.

SYNS: SULFURIC ACID, TITANIUM(4+) SALT (2:1) □ TITANIUM DISULFATE □ TITANIUM SULFATE SOLUTION (DOT)

CONSENSUS REPORTS: Reported in EPA TSCA Inventory.

DOT CLASSIFICATION: 8; Label: Corrosive

SAFETY PROFILE: A corrosive irritant to skin, eyes, and mucous membranes. When heated to decomposition it emits toxic fumes of SO_x. See also TITANIUM COMPOUNDS and SULFATES.

TGH252   CAS: 13693-11-3   HR: 2
TITANIUM SULFATE SOLUTION
DOT: NA 1760
mf: $O_8S_2 \cdot Ti$ mw: 240.02
SYNS: SULFURIC ACID, TITANIUM(4+) SALT (2:1) □ TITANIUM DISULFATE
CONSENSUS REPORTS: Reported in EPA TSCA Inventory.
DOT CLASSIFICATION: 8; Label: Corrosive
SAFETY PROFILE: A corrosive to skin and eyes. When heated to decomposition it emits corrosive and toxic vapors of SO_x and Ti.

TGH350   CAS: 7550-45-0   HR: 3
TITANIUM TETRACHLORIDE
DOT: UN 1838
mf: Cl_4Ti mw: 189.70
PROP: Colorless, readily hydrolyzed liquid; penetrating acid odor. D: 1.726, mp: −24.1°, bp: 136.4°. Sol in cold water and alc; decomp in hot water.
SYNS: TETRACHLORURE de TITANE (FRENCH) □ TITAANTETRACHLORID (DUTCH) □ TITANE - (TETRACHLORURE de) (FRENCH) □ TITANIO TETRACHLORURO di (ITALIAN) □ TITANIUM CHLORIDE □ TITANTETRACHLORID (GERMAN)
CONSENSUS REPORTS: EPA Extremely Hazardous Substances List. Community Right-To-Know List. Reported in EPA TSCA Inventory.
DOT CLASSIFICATION: 8; Label: Corrosive, Poison
SAFETY PROFILE: Poison by inhalation. A corrosive irritant to skin, eyes, and mucous membranes. When heated to decomposition it emits toxic fumes of Cl^-. See also TITANIUM COMPOUNDS.

TGI100   HR: 3
TOBACCO PLANT
PROP: Large annual or perennial shrubs with leaves that are often broad, hairy, and sticky. The trumpet-shaped flowers are white, yellow, green-yellow, or red. The seed capsule holds many small seeds. *N. tabacum* is the principal commercial tobacco in the western countries. *N. rustica*, native to South America, is found sporadically across the United States and is the most widely cultivated tobacco in the Orient. *N. longiflora* is commonly cultivated as a garden ornamental. *N. attenuata* grows in the region bounded by Idaho, Baja California, and Texas. *N. glauca* is native to South America and now grows in the southwestern United States, Hawaii, Mexico, and the West Indies.
SYNS: NICOTIANA ATTENUATA □ NICOTIANA GLAUCA □ NICOTIANA LONGIFLORA □ NICOTIANA RUSTICA □ NICOTIANA TABACUM □ PAKA (HAWAII) □ TABAC (FRENCH) □ TABACO (SPANISH)
SAFETY PROFILE: Confirmed human carcinogen by several routes. The whole plant contains poisonous nicotine and other chemically related alkaloids. The primary alkaloid in *N. tabacum* is nicotine. The primary alkaloid in *N. glauca* is anabasine. Ingestion of any part of the plant can cause salivation, nausea, vomiting, distorted perceptions, convulsions, vasomotor collapse, and respiratory failure. Most serious poisonings result from ingestion of the leaves in salad, use of infusions as enemas, or skin absorption of alkaloids during commercial harvesting. See also SMOKELESS TOBACCO, NICOTINE, and other tobacco entries.

TGJ050   CAS: 58-95-7   HR: D
d-α-TOCOPHERYL ACETATE
mf: $C_{31}H_{52}O_3$ mw: 472.83
PROP: From vacuum steam distillation and acetylation of edible vegetable oil products. - (FCC III) Colorless to yellow oil; odorless. Mp: 26.5–27.5°, bp: 205° @ 0.02 mm - (bath). Insol in water, sol in alc; misc with acetone, chloroform, ether, vegetable oil.
SYNS: ALFACOL □ COMBINAL E □ CONTOPHERON □ ECOFROL □ ECON □ E-FEROL □ ENDO E DOMPE □ EPHYNAL ACETATE □ EPSILAN-M □ EREVIT □ E-TOPLEX □ EVIPHEROL □ FERTILVIT □ GEVEX □ JUVELA □ TOCOPHEREX □ (+)-α-TOCOPHEROL

T

ACETATE □ α-TOCOPHEROL ACETATE □ d-α-TOCO-
PHEROL ACETATE □ (+)-α-TOCOPHERYL ACETATE □
α-TOCOPHERYL ACETATE □ d-α-TOCOPHERYL
ACETATE □ (R,R,R)-α-TOCOPHERYL ACETATE □ -
(2R,4'R,8'R)-α-TOCOPHERYL ACETATE □ TOCOPHRIN
□ TOFAXIN □ TOKOFEROL ACETATE □ VITAMIN E
ACETATE

SAFETY PROFILE: When heated to
decomposition it emits acrid smoke and
irritating fumes.

TGJ060   CAS: 4345-03-3   HR: D
dl-α-TOCOPHERYL ACID SUCCINATE

mf: $C_{33}H_{54}O_5$ mw: 530.79
PROP: From vacuum steam distillation and
succinylation of edible vegetable oil
products. (FCC III) Colorless to white
crystalline powder or needles from pet ether;
odorless and tasteless. Mp: 76–77°. Insol in
water; very sol in chloroform, sol in acetone,
alc, ether, vegetable oil.
SAFETY PROFILE: When heated to
decomposition it emits acrid smoke and
irritating fumes.

TGJ750   CAS: 119-93-7   HR: 3
o-TOLIDINE

mf: $C_{14}H_{16}N_2$ mw: 212.32
PROP: White to reddish crystals or leaflets.
Mp: 129–131°. Very sltly sol in water; sol in
alc, ether, acetic acid.
SYNS: BIANISIDINE □ 4,4'-BI-o-TOLUIDINE □ C.I.
37230 □ C.I. AZOIC DIAZO COMPONENT 113 □ (4,4'-DI-
AMINE)-3,3'-DIMETHYL(1,1'-BIPHENYL) □ 4,4'-DIAM-
INO-3,3'-DIMETHYLBIPHENYL □ 4,4'-DIAMINO-3,3'-DI-
METHYLDIPHENYL □ DIAMINODITOLYL □ 3,3'-DI-
METHYLBENZIDIN □ 3,3'-DIMETHYLBENZIDINE □
3,3'-DIMETHYL-4,4'-BIPHENYLDIAMINE □ 3,3'-DI-
METHYLBIPHENYL-4,4'-DIAMINE □ 3,3'-DIMETHYL--
(1,1'-BIPHENYL)-4,4'-DIAMINE □ 3,3'-DIMETHYL-4,4'-DI-
PHENYLDIAMINE □ 3,3'-DIMETHYLDIPHENYL-4,4'-DI-
AMINE □ 4,4'-DI-o-TOLUIDINE □ FAST DARK BLUE
BASE R □ RCRA WASTE NUMBER U095 □ o-TOLIDIN □
2-TOLIDIN (GERMAN) □ 2-TOLIDINA (ITALIAN) □
TOLIDINE □ o,o'-TOLIDINE □ 2-TOLIDINE □ 3,3'-
TOLIDINE
CONSENSUS REPORTS: NTP 10th Report
on Carcinogens. IARC Cancer Review:
Group 2B IMEMDT 7,56,87; Animal
Limited Evidence IMEMDT 1,87,72. EPA
Genetic Toxicology Program. Community

Right-To-Know List. Reported in EPA
TSCA Inventory.
ACGIH TLV: Animal Carcinogen
DFG MAK: Animal Carcinogen, Suspected
Human Carcinogen
NIOSH REL: (o-Toluidine) CL 0.02
mg/m³/60M; avoid skin contact
SAFETY PROFILE: Confirmed carcinogen
with experimental carcinogenic and
tumorigenic data. Poison by intraperitoneal
route. Moderately toxic by ingestion.
Human mutation data reported. When
heated to decomposition it emits toxic
fumes of NO_x.

TGK750   CAS: 108-88-3   HR: 3
TOLUENE

DOT: UN 1294
mf: C_7H_8 mw: 92.15
PROP: Colorless liquid; benzol-like odor.
Mp: −95 to −94.5°, fp: −95°, bp: 110.4°,
flash p: 40°F (CC), ULC: 75–80, lel: 1.27%,
uel: 7%, d: 0.866 @ 20°/4°, autoign temp:
996°F, vap press: 36.7 mm @ 30°, vap d:
3.14. Insol in water; sol in acetone; misc in
abs alc, ether, chloroform. IDLH 500 ppm.
SYNS: ANTISAL 1a □ BENZENE, METHYL- □
METHACIDE □ METHANE, PHENYL- □ METHYL-
BENZENE □ METHYLBENZOL □ NCI-C07272 □ PHEN-
YLMETHANE □ RCRA WASTE NUMBER U220 □
TOLUEEN (DUTCH) □ TOLUEN (CZECH) □ TOLUOL -
(DOT) □ TOLUOLO (ITALIAN) □ TOLU-SOL
CONSENSUS REPORTS: Community
Right-To-Know List. Reported in EPA
TSCA Inventory. EPA Genetic Toxicology
Program.
OSHA PEL: TWA 100 ppm; STEL 150 ppm
ACGIH TLV: TWA 20 ppm; Not Classifiable
as a Human Carcinogen; BEI: 1.6 g/g
creatinine of hippuric acid in urine at end of
shift; 0.05 mg/L toluene in venous blood
prior to last shift of workweek; BEI: BEI:
0.05 mg/L toluene in venous blood prior to
last shift of work ek; 0.5 mg/L o-cresol in
urine at end of shift
DFG MAK: 50 ppm (190 mg/m³); BAT: 340
µg/dL in blood at end of shift
NIOSH REL: (Toluene) TWA 100 ppm; CL
200 ppm/10M

DOT CLASSIFICATION: 3; Label: Flammable Liquid

SAFETY PROFILE: Poison by intraperitoneal route. Moderately toxic by intravenous and subcutaneous routes. Mildly toxic by inhalation. An experimental teratogen. Human systemic effects by inhalation: CNS recording changes, hallucinations or distorted perceptions, motor activity changes, antipsychotic, psychophysiological test changes, and bone marrow changes. Experimental reproductive effects. Mutation data reported. A human eye irritant. An experimental skin and severe eye irritant.

Toluene is derived from coal tar, and commercial grades usually contain small amounts of benzene as an impurity. Inhalation of 200 ppm of toluene for 8 hours may cause impairment of coordination and reaction time; with higher concentrations (up to 800 ppm) these effects are increased and are observed in a shorter time. In the few cases of acute toluene poisoning reported, the effect has been that of a narcotic, the workman passing through a stage of intoxication into one of coma. Recovery following removal from exposure has been the rule. An occasional report of chronic poisoning describes an anemia and leukopenia, with biopsy showing a bone marrow hypoplasia. These effects, however, are less common in people working with toluene, and they are not as severe. At 200–500 ppm, headache, nausea, eye irritation, loss of appetite, a bad taste, lassitude, impairment of coordination and reaction time are reported, but are not usually accompanied by any laboratory or physical findings of significance. With higher concentrations, the above complaints are increased and in addition, anemia, leukopenia, and enlarged liver may be found in rare cases. A common air contaminant, emitted from modern building materials - (CENEAR 69,22,91). Used in production of drugs of abuse.

Flammable liquid. A very dangerous fire hazard when exposed to heat, flame, or oxidizers. Explosive in the form of vapor when exposed to heat or flame. Explosive reaction with 1,3-dichloro-5,5-dimethyl-2,4-imidazolididione, dinitrogen tetraoxide, concentrated nitric acid, H_2SO_4 + HNO_3, N_2O_4, $AgClO_4$, BrF_3, UF_6, sulfur dichloride. Forms an explosive mixture with tetranitromethane. Can react vigorously with oxidizing materials. To fight fire, use foam, CO_2, dry chemical. When heated to decomposition it emits acrid smoke and irritating fumes.

TGL500   CAS: 25376-45-8   HR: 3
TOLUENEDIAMINE

mf: $C_7H_{10}N_2$ mw: 122.19

SYNS: BENZENEDIAMINE, ar-METHYL- □ DIAMINOTOLUENE □ METHYLPHENYLENEDIAMINE □ RCRA WASTE NUMBER U221

CONSENSUS REPORTS: Community Right-To-Know List. Reported in EPA TSCA Inventory.

SAFETY PROFILE: A poison by ingestion, intraperitoneal, and intravenous routes. When heated to decomposition it emits toxic fumes of NO_x. See also other toluene diamine entries and AROMATIC AMINES.

TGL750   CAS: 95-80-7   HR: 3
TOLUENE-2,4-DIAMINE

DOT: UN 1709

mf: $C_7H_{10}N_2$ mw: 122.19

PROP: Needles from water or prisms from alc. Mp: 99°, bp: 292°, bp: 148–150° @ 8 mm, vap press: 1 mm @ 106.5°.

SYNS: 3-AMINO-p-TOLUIDINE □ 5-AMINO-o-TOLUIDINE □ AZOGEN DEVELOPER H □ BENZOFUR MT □ C.I. 76035 □ C.I. OXIDATION BASE □ C.I. OXIDATION BASE 20 □ C.I. OXIDATION BASE 35 □ C.I. OXIDATION BASE 200 □ DEVELOPER 14 □ DEVELOPER B □ DEVELOPER DB □ DEVELOPER DBJ □ DEVELOPER H □ DEVELOPER MC □ DEVELOPER MT □ DEVELOPER MT-CF □ DEVELOPER MTD □ DEVELOPER T □ 1,3-DIAMINO-4-METHYLBENZENE □ 2,4-DIAMINO-1-METHYL-BENZENE □ 2,4-DIAMINOTOLUEN (CZECH) □ DIAMINOTOLUENE □ 2,4-DIAMINOTOLUENE □ 2,4-DIAMINO-1-TOLUENE □ 2,4-DIAMINOTOLUOL □ EUCAN-

INE GB □ FOURAMINE □ FOURAMINE J □ FOURRINE 94 □ FOURRINE M □ META TOLUYLENE DIAMINE □ 4-METHYL-1,3-BENZENEDIAMINE □ 4-METHYL-m-PHENYLENEDIAMINE □ MTD □ NAKO TMT □ NCI-C02302 □ PELAGOL GREY J □ PELAGOL J □ PONTAMINE DEVELOPER TN □ RCRA WASTE NUMBER U221 □ RENAL MD □ TDA □ TERTRAL G □ 2,4-TOLAMINE □ m-TOLUENEDIAMINE □ 2,4-TOLUENEDIAMINE □ 2,4-TOLUENEDIAMINE (DOT) □ m-TOLUYLENDIAMIN -(CZECH) □ m-TOLUYLENEDIAMINE □ 2,4-TOLUYL-ENEDIAMINE □ 2,4-TOLUYLENEDIAMINE (DOT) □ m-TOLYENEDIAMINE □ m-TOLYLENEDIAMINE □ 2,4-TOLYLENEDIAMINE □ 4-m-TOLYLENEDIAMINE □ ZOBA GKE □ ZOGEN DEVELOPER H

CONSENSUS REPORTS: NTP 10th Report on Carcinogens. IARC Cancer Review: Group 2B IMEMDT 7,56,87; Animal Sufficient Evidence IMEMDT 16,83,78. NCI Carcinogenesis Bioassay (feed); Clear Evidence: mouse, rat NCITR* NCI-CG-TR-162,79. Community Right-To-Know List. Reported in EPA TSCA Inventory. EPA Genetic Toxicology Program.
DFG MAK: Animal Carcinogen, Suspected Human Carcinogen
DOT CLASSIFICATION: 6.1; Label: KEEP AWAY FROM FOOD
SAFETY PROFILE: Confirmed carcinogen with experimental carcinogenic data. Poison by intraperitoneal and subcutaneous routes. Experimental reproductive effects. Human mutation data reported. A skin and eye irritant. This material has a marked toxic action upon the liver and can cause fatty degeneration of that organ. When heated to decomposition it emits toxic fumes of NO_x. See also other toluenediamine entries and AROMATIC AMINES.

**TGM000   CAS: 95-70-5   HR: 3
TOLUENE-2,5-DIAMINE**
mf: $C_7H_{10}N_2$ mw: 122.19
PROP: Colorless, crystalline tablets or plates from C_6H_6. Mp: 64°, bp: 274°.
SYNS: 4-AMINO-2-METHYLANILINE □ C.I. 76042 □ 2,5-DIAMINOTOLUENE □ 2-METHYL-1,4-BENZENEDI-AMINE □ 2-METHYL-p-PHENYLENEDIAMINE □ p-TOLUENEDIAMINE □ p-TOLUYLENDIAMINE □ TOLUYLENE-2,5-DIAMINE □ p,m-TOLYLENEDIAMINE

CONSENSUS REPORTS: IARC Cancer Review: Group 3 IMEMDT 7,56,87; Animal Inadequate Evidence IMEMDT 16,97,78. Reported in EPA TSCA Inventory. EPA Genetic Toxicology Program.
SAFETY PROFILE: Poison by ingestion and subcutaneous routes. A skin irritant. Mutation data reported. Questionable carcinogen. Has a toxic action upon the liver and can cause fatty degeneration of that organ. Its total effect upon the body seems to take place three different ways. It is toxic to the central nervous system, producing jaundice by action on the liver and spleen, and anemia by destruction of the red blood cells. This action is quite similar to that of aniline, although by no means identical with it. Its high boiling point and the fact that the material is solid at room temperature make it somewhat less hazardous than aniline, particularly at ordinary working temperatures. The literature contains a reference to a permanent injury to an eye due to the use of this material as an eyelash dye. It is considered to be an irritating dye material. When heated to decomposition it emits toxic fumes of NO_x. See also other toluene diamine entries and AROMATIC AMINES.

**TGM400   CAS: 6369-59-1   HR: 2
p-TOLUENEDIAMINE SULFATE**
mf: $C_7H_{10}N_2 \cdot 7H_2O_4S$ mw: 808.75
SYNS: C.I. 76043 □ C.I. OXIDATION BASE 4 □ FOURAMINE STD □ 2-METHYL-1,4-BENZENEDIAMINE □ NCI-C01832 □ 2,5-TDS □ 2,5-TOULENEDIAMINE SULFATE
CONSENSUS REPORTS: IARC Cancer Review: Animal Inadequate Evidence IMEMDT 16,97,78. EPA Genetic Toxicology Program. Reported in EPA TSCA Inventory.
SAFETY PROFILE: An experimental teratogen. Experimental reproductive effects. Questionable carcinogen with experimental tumorigenic data. Mutation data reported. When heated to decomposition it emits toxic fumes of SO_x

and NO$_x$. See also AROMATIC AMINES and SULFATES.

TGM740   CAS: 26471-62-5   HR: 3
TOLUENE-1,3-DIISOCYANATE
DOT: UN 2078
mf: C$_9$H$_6$N$_2$O$_2$ mw: 174.17
SYNS: BENZENE-, 1,3-DIISOCYANATOMETHYL- □ DESMODUR T100 □ DIISOCYANATOMETHYL-BENZENE □ DIISOCYANATOTOLUENE □ HYLENE-T □ ISOCYANIC ACID, METHYLPHENYLENE ESTER □ -METHYL-m-PHENYLENE DIISOCYANATE □ METHYL-PHENYLENE ISOCYANATE □ MONDUR-TD □ MON-DUR-TD-80 □ NACCONATE-100 □ NIAX ISOCYANATE TDI □ RCRA WASTE NUMBER U223 □ RUBINATE TDI □ RUBINATE TDI 80/20 □ T 100 □ TDI □ TDI-80 □ TDI 80-20 □ TOLUENE DIISOCYANATE □ TOLYLENE DI-ISOCYANATE □ TOLYLENE ISOCYANATE
CONSENSUS REPORTS: NTP 10th Report on Carcinogens. IARC Cancer Review: Group 2B IMEMDT 7,56,87, Animal Sufficient Evidence IMEMDT 39,287,86; Human Inadequate Evidence IMEMDT 39,287,86. NTP Carcinogenesis Studies - (gavage); Clear Evidence: mouse, rat NTPTR* NTP-TR-251,86. Reported in EPA TSCA Inventory.
NIOSH REL: (TDI) 10H TWA 0.005 ppm; CL 0.02 ppm/10M
DOT CLASSIFICATION: 6.1; Label: Poison; DOT Class: 6.1; Label: Poison, Flammable Liquid; DOT Class: 3; Label: Flammable Liquid, Poison
SAFETY PROFILE: Confirmed carcinogen with experimental carcinogenic and neoplastigenic data. Poison by inhalation. Moderately toxic by ingestion. Severe skin irritant. Human mutation data reported. Capable of producing severe dermatitis and bronchial spasm. A common air contaminant. A flammable liquid when exposed to heat or flame. Explosive in the form of vapor when exposed to heat or flame. To fight fire, use dry chemical, CO$_2$. Potentially violent polymerization reaction with bases or acyl chlorides. Reaction with water releases carbon dioxide. Storage in polyethylene containers is hazardous due to absorption of water through the plastic.

When heated to decomposition it emits highly toxic fumes of NO$_x$. See also ISOCYANATES.

TGM750   CAS: 584-84-9   HR: 3
TOLUENE-2,4-DIISOCYANATE
DOT: UN 2206/UN 2207/UN 2478/UN 3080
mf: C$_9$H$_6$N$_2$O$_2$ mw: 174.17
PROP: Clear, faintly yellow liquid; sharp, pungent odor. Mp: 19.5-21.5°, d: (liquid) 1.2244 @ 20°/4°, bp: 124-126° @ 18 mm, flash p: 270°F (OC), vap d: 6.0, lel: 0.9%, uel: 9.5%. Misc with alc (decomp), ether, acetone, carbon tetrachloride, benzene, chlorobenzene, kerosene, olive oil. IDLH 2.5 ppm.
SYNS: CRESORCINOL DIISOCYANATE □ DESMODUR T80 □ DI-ISOCYANATE de TOLUYLENE □ DI-ISO-CYANATOLUENE □ 2,4-DIISOCYANATO-1--METHYLBENZENE (9CI) □ 2,4-DIISOCYANATO-TOLUENE □ DIISOCYANAT-TOLUOL □ HYLENE T □ HYLENE TCPA □ HYLENE TLC □ HYLENE TM □ HYL-ENE TM-65 □ HYLENE TRF □ ISOCYANIC ACID, -METHYLPHENYLENE ESTER □ ISOCYANIC ACID, 4--METHYL-m-PHENYLENE ESTER □ 4-METHYL-PHEN-YLENE DIISOCYANATE □ 4-METHYL-PHENYLENE ISOCYANATE □ MONDUR TD □ MONDUR TD-80 □ MONDUR TDS □ NACCONATE 100 □ NCI-C50533 □ NIAX TDI □ NIAX TDI-P □ RCRA WASTE NUMBER U223 □ RUBINATE TDI 80/20 □ TDI (OSHA) □ 2,4-TDI □ TDI-80 □ TOLUEEN-DIISOCYANAAT □ TOLUEN-DI-SOCIANATO □ TOLUENE DIISOCYANATE □ 2,4-TOLUENEDIISOCYANATE □ TOLUILENODWUIZO-CYJANIAN □ TULUYLENDIISOCYANAT □ TOLUYL-ENE-2,4-DIISOCYANATE □ m-TOLYLENE DI-ISOCYANATE □ TOLYLENE-2,4-DIISOCYANATE □ 2,4-TOLYLENEDIISOCYANATE
CONSENSUS REPORTS: IARC Cancer Review: Group 2B IMEMDT 7,56,87; Human Inadequate Evidence IMEMDT 39,287,86; Animal Sufficient Evidence IMEMDT 39,287,86. Community Right-To-Know List. EPA Extremely Hazardous Substances List. Reported in EPA TSCA Inventory.
OSHA PEL: TWA 0.005 ppm; STEL 0.02 ppm
ACGIH TLV: TWA 0.005 ppm; STEL 0.02 ppm; Not Classifiable as a Human Carcinogen; (Proposed: TWA 0.005 ppm;

T

STEL 0.02 ppm (sensitizer); Not Classifiable as a Human Carcinogen); (Proposed: TWA inhalable fraction and vapor 0.001 ppm; STEL 0.003; skin; sensitizer; Not Classifiable as a Human Carcinogen)
DFG MAK: 0.01 ppm (0.072 mg/m^3)
NIOSH REL: (Diisocyanates) TWA 0.005 ppm; CL 0.02 ppm/10M
DOT CLASSIFICATION: 6.1; Label: KEEP AWAY FROM FOOD (UN 2207)
SAFETY PROFILE: Confirmed carcinogen. Poison by ingestion, inhalation, and intravenous routes. Human systemic effects by inhalation: unspecified changes to the eyes and sense of smell, respiratory obstruction, cough, sputum, and other pulmonary and gastrointestinal changes. Mutation data reported. A severe skin and eye irritant. Capable of producing severe dermatitis and bronchial spasm. A common air contaminant. Combustible when exposed to heat or flame. Explosive in the form of vapor when exposed to heat or flame. To fight fire, use dry chemical, CO$_2$. Potentially violent polymerization reaction with bases or acyl chlorides. Reaction with water releases carbon dioxide. Storage in polyethylene containers is hazardous due to absorption of water through the plastic. When heated to decomposition it emits highly toxic fumes of NO$_x$. See also ISOCYANATES.

TGM800   CAS: 91-08-7   HR: 3
TOLUENE-2,6-DIISOCYANATE
DOT: UN 2207
mf: C$_9$H$_6$N$_2$O$_2$ mw: 174.17
PROP: A liquid. Bp: 129–133° @ 18 mm.
SYNS: 2,6-DIISOCYANATO-1-METHYLBENZENE □ 2,6-DIISOCYANATOTOLUENE □ HYLENE TM □ 2--METHYL-m-PHENYLENE ESTER, ISOCYANIC ACID □ 2-METHYL-m-PHENYLENE ISOCYANATE □ NIAX TDI □ 2,6-TDI □ 2,6-TOLUENE DIISOCYANATE □ m-TOLYLENE DIISOCYANATE □ TOLYLENE-2,6-DI-ISOCYANATE
CONSENSUS REPORTS: IARC Cancer Review: Group 2B IMEMDT 7,56,87; Human Inadequate Evidence IMEMDT 39,287,86; Animal Sufficient Evidence

IMEMDT 39,287,86. Reported in EPA TSCA Inventory. Community Right-To-Know List. EPA Hazardous Substances List. EPA Extremely Hazardous Substances List.
ACGIH TLV: TWA 0.005 ppm; STEL 0.02 ppm; Not Classifiable as a Human Carcinogen; (Proposed: TWA 0.005 ppm; STEL 0.02 ppm (sensitizer); Not Classifiable as a Human Carcinogen); (Proposed: TWA inhalable fraction and vapor 0.001 ppm; STEL 0.003; skin; sensitizer; Not Classifiable as a Human Carcinogen)
DFG MAK: 0.01 ppm (0.072 mg/m^3)
NIOSH REL: (Diisocyanates) TWA 0.005 ppm; CL 0.02 ppm/10M
DOT CLASSIFICATION: 6.1; Label: KEEP AWAY FROM FOOD (UN 2207); DOT Class: 6.1; Label: Poison (UN 2206); DOT Class: 6.1; Label: Poison, Flammable Liquid (UN 3080); DOT Class: 3; Label: Flammable Liquid, Poison (UN 2478)
SAFETY PROFILE: Confirmed carcinogen. Poison by ingestion and inhalation. Human systemic effects by inhalation: olfactory, eye, and pulmonary changes. Flammable liquid. When heated to decomposition it emits toxic fumes of NO$_x$. See also CYANATES.

TGN250   CAS: 88-19-7   HR: 3
o-TOLUENESULFONAMIDE
mf: C$_7$H$_9$NO$_2$S   mw: 171.23
PROP: Tetragonal prisms. Mp: 156°. Sol in water, alc.
SYNS: o-METHYLBENZENESULFONAMIDE □ 2--METHYLBENZENESULFONAMIDE □ ONCO-CARBIDE □ ORTHO-TOLUOL-SULFONAMID (GERMAN) □ OTS □ OXYUREA □ TOLUENE-2-SULFONAMIDE
CONSENSUS REPORTS: IARC Cancer Review: Group 2B IMEMDT 7,334,87; Animal Limited Evidence IMEMDT 22,111,80. Reported in EPA TSCA Inventory. EPA Genetic Toxicology Program.
SAFETY PROFILE: Suspected carcinogen with experimental tumorigenic data. Mildly toxic by ingestion. Experimental reproductive effects. Mutation data

reported. An eye irritant. When heated to decomposition it emits very toxic fumes of NO_x and SO_x. Used as a chemical intermediate in the production of saccharin.

TGO750   **CAS: 100-53-8**   **HR: 3**
α-TOLUENETHIOL
DOT: UN 1228/UN 3071
mf: C_7H_8S mw: 124.21
PROP: A water-white, mobile liquid; strong odor. Bp: 194.8°, d: 1.058 @ 20°, vap d: 4.28.
SYNS: BENZYLHYDROSULFIDE □ BENZYL MERCAPTAN □ BENZYLTHIOL □ (MERCAPTOMETH-YL)BENZENE □ α-MERCAPTOTOLUENE □ METHANETHIOL, PHENYL- □ PHENYLMETHAN-ETHIOL □ PHENYLMETHYL MERCAPTAN □ THIOBENZYL ALCOHOL □ α-TOLYL MERCAPTAN □ USAF EK-1509
CONSENSUS REPORTS: Reported in EPA TSCA Inventory.
DOT CLASSIFICATION: 3; Label: Flammable Liquid, Poison (UN 1228); DOT Class: 6.1; Label: Poison, Flammable Liquid (UN 3071)
SAFETY PROFILE: Poison by inhalation and intraperitoneal routes. Moderately toxic by ingestion. An eye irritant. Questionable carcinogen with experimental tumorigenic data. Flammable when exposed to heat or flame. Can react vigorously with oxidizing materials. To fight fire, use foam, CO_2, dry chemical, water spray, mist, fog. When heated to decomposition and on contact with acid or acid fumes it emits highly toxic fumes of SO_x. See also SULFIDES and MERCAPTANS.

TGP000   **CAS: 137-06-4**   **HR: 3**
o-TOLUENETHIOL
DOT: UN 1228/UN 3071
mf: C_7H_8S mw: 124.21
PROP: Leaves or liquid. Mp: 15°, bp: 106° @ 50 mm. Insol in H_2O; sol in common org solvs.
SYNS: o-MERCAPTOTOLUENE □ o-METHYL-BENZENETHIOL □ 2-METHYLBENZENETHIOL □ o--METHYLTHIOPHENOL □ 2-METHYLTHIOPHENOL □ o-THIOCRESOL □ 2-TOLUENETHIOL □ o-TOLYL MERCAPTAN □ USAF EK-2676

CONSENSUS REPORTS: Reported in EPA TSCA Inventory.
DOT CLASSIFICATION: 3; Label: Flammable Liquid, Poison (UN 1228); DOT Class: 6.1; Label: Poison, Flammable Liquid (UN 3071)
SAFETY PROFILE: Poison by intraperitoneal route. A flammable liquid. When heated to decomposition it emits toxic fumes of SO_x. See also α-TOLUENETHIOL and MERCAPTANS.

TGP250   **CAS: 106-45-6**   **HR: 3**
p-TOLUENETHIOL
DOT: UN 1228/UN 3071
mf: C_7H_8S mw: 124.21
PROP: Leaflets from EtOH (aq). Mp: 43–44°, bp: 195°. Sol in AcOH.
SYNS: p-MERCAPTOTOLUENE □ p-METHYLBEN-ZENETHIOL □ 4-METHYLBENZENETHIOL □ p-METH-YLPHENYLMERCAPTAN □ 4-METHYLPHENYLMER-CAPTAN □ p-METHYLTHIOPHENOL □ 4-METHYL-THIOPHENOL □ p-THIOCRESOL □ 4-THIOCRESOL □ 4-TOLUENETHIOL □ p-TOLYL MERCAPTAN □ p-TOL-YLTHIOL □ USAF EK-510
CONSENSUS REPORTS: Reported in EPA TSCA Inventory.
DOT CLASSIFICATION: 3; Label: Flammable Liquid, Poison (UN 1228); DOT Class: 6.1; Label: Poison, Flammable Liquid (UN 3071)
SAFETY PROFILE: Poison by intraperitoneal route. A flammable liquid. When heated to decomposition it emits toxic fumes of SO_x. See also α-TOLUENETHIOL and MERCAPTANS.

TGQ500   **CAS: 108-44-1**   **HR: 3**
m-TOLUIDINE
DOT: UN 1708
mf: C_7H_9N mw: 107.17
PROP: Colorless liquid. Mp: −43.6°, bp: 203.3°, d: 0.989 @ 20°/4°, vap press: 1 mm @ 41°, vap d: 3.90. Sltly sol in water; sol in alc, ether.
SYNS: 3-AMINO-1-METHYLBENZENE □ 3-AM-INOPHENYLMETHANE □ 3-AMINOTOLUEN (CZECH) □ m-AMINOTOLUENE □ 3-AMINOTOLUENE □ m--METHYLANILINE □ 3-METHYLANILINE □ m-METH-

T

YLBENZENAMINE □ 3-METHYLBENZENAMINE □ m-TOLUIDIN (CZECH) □ 3-TOLUIDINE □ m-TOLYLAMINE

CONSENSUS REPORTS: Reported in EPA TSCA Inventory.
OSHA PEL: TWA 2 ppm (skin)
ACGIH TLV: TWA 2 ppm (skin); Not Classifiable as a Human Carcinogen
DOT CLASSIFICATION: 6.1; Label: Poison
SAFETY PROFILE: Poison by ingestion and intraperitoneal routes. A skin and eye irritant. Flammable when exposed to heat or flame. Can react vigorously on contact with oxidizing materials. To fight fire, use foam, CO_2, dry chemical. When heated to decomposition it emits highly toxic fumes of NO_x. See also ANILINE and o-TOLUIDINE.

TGQ750   CAS: 95-53-4   HR: 3
o-TOLUIDINE
DOT: UN 1708
mf: C_7H_9N mw: 107.17
PROP: Colorless liquid. Mp: −16.3° - (dimorph), bp: 200–202°, ULC: 20–25, flash p: 185° (CC), d: 1.004 @ 20°/4°, autoign temp: 900°F, vap press: 1 mm @ 44°, vap d: 3.69. Sltly sol in water, dil acid; sol in alc and ether.
SYNS: 1-AMINO-2-METHYLBENZENE □ 2-AMINO-1--METHYLBENZENE □ o-AMINOTOLUENE □ 2-AMINOTOLUENE □ C.I. 37077 □ 1-METHYL-2-AMINOBENZENE □ 2-METHYL-1-AMINOBENZENE □ o--METHYLANILINE □ 2-METHYLANILINE □ o-METHYL-BENZENAMINE □ 2-METHYLBENZENAMINE □ o-TOLUIDIN (CZECH) □ 2-TOLUIDINE □ o-TOLUIDYNA (POLISH) □ o-TOLYLAMINE
CONSENSUS REPORTS: NTP 10th Report on Carcinogens. IARC Cancer Review: Group 2B IMEMDT 7,362,87; Human Inadequate Evidence IMEMDT 16,349,78; Human Limited Evidence IMEMDT 27,155,82; Animal Inadequate Evidence IMEMDT 16,349,78. EPA Genetic Toxicology Program. Community Right-To-Know List. Reported in EPA TSCA Inventory.
OSHA PEL: TWA 5 ppm (skin)
ACGIH TLV: TWA 2 ppm (skin); Animal Carcinogen

DFG MAK: Animal Carcinogen, Suspected Human Carcinogen
DOT CLASSIFICATION: 6.1; Label: Poison
SAFETY PROFILE: Confirmed carcinogen with experimental neoplastigenic and tumorigenic data. Poison by ingestion and intraperitoneal routes. Moderately toxic by skin contact. Human systemic effects by inhalation: urine volume increase, hematuria, and blood methemoglobinemia-carboxyhemoglobinemia. An experimental teratogen. Human mutation data reported. A skin and severe eye irritant. Human mucous membrane effects. Can produce severe systemic disturbances. The main portal of entry into the body is the respiratory tract, particularly in cases of industrial exposure. The symptoms produced are headache, weakness, difficulty in breathing, air hunger, psychic disturbances, and marked irritation of the kidneys and bladder. The literature does not yield any good data for comparing the toxicity of the o-, m-, and p-isomers. Their behavior is generally comparable to that of aniline. It has been determined experimentally that a concentration of about 100 ppm is the maximum endurable for 1 hour without serious consequences and that 6–23 ppm is endurable for several hours without serious disturbances.

Flammable when exposed to heat or flame. Hypergolic reaction with red fuming nitric acid. Can react with oxidizing materials. To fight fire, use foam, CO_2, dry chemical. When heated to decomposition it emits highly toxic fumes of NO_x. See also ANILINE.

TGR000   CAS: 106-49-0   HR: 3
p-TOLUIDINE
DOT: UN 1708
mf: C_7H_9N mw: 107.17
PROP: Colorless leaflets. Mp: 44.5° (anhyd), mp: 42°, bp: 82.2° @ 10 mm, flash p: 188°F (CC), d: 1.046 @ 20°/4°, autoign temp: 900°F, vap press: 1 mm @ 42°, vap d: 3.90. Sol in water, dil acid, CS_2; very sol in alc, ether.

SYNS: 4-AMINO-1-METHYLBENZENE □ 4-AM-INOTOLUEN (CZECH) □ p-AMINOTOLUENE □ 4-AM-INOTOLUENE □ C.I. 37107 □ C.I. AZOIC COUPLING COMPONENT 107 □ p-METHYLANILINE □ 4-METHYL-ANILINE □ p-METHYLBENZENAMINE □ 4-METHYL-BENZENAMINE □ NAPHTOL AS-KG □ NAPHTOL AS-KGLL □ p-TOLUIDIN (CZECH) □ 4-TOLUIDINE □ TOL-YLAMINE □ p-TOLYLAMINE

CONSENSUS REPORTS: Reported in EPA TSCA Inventory. EPA Genetic Toxicology Program.

OSHA PEL: TWA 2 ppm (skin)

ACGIH TLV: TWA 2 ppm (skin); Animal Carcinogen

DFG MAK: Confirmed Animal Carcinogen with Unknown Relevance to Humans

DOT CLASSIFICATION: 6.1; Label: Poison

SAFETY PROFILE: Confirmed carcinogen. Poison by ingestion and intraperitoneal routes. Mutation data reported. A severe skin and eye irritant. Flammable when exposed to heat, flame, or oxidizers. Can react vigorously on contact with oxidizing materials. To fight fire, use foam, CO_2, dry chemical. When heated to decomposition it emits highly toxic fumes of NO_x. See also o-TOLUIDINE and ANILINE.

TGS500   CAS: 636-21-5   HR: 3
o-TOLUIDINE HYDROCHLORIDE
mf: $C_7H_9N \cdot ClH$ mw: 143.63
PROP: Monoclinic prisms. Mp: 218–220°, bp: 242°. Sol in water; sltly sol in alc.
SYNS: 1-AMINO-2-METHYLBENZENE HYDRO-CHLORIDE □ 2-AMINO-1-METHYLBENZENE HYDRO-CHLORIDE □ o-AMINOTOLUENE HYDROCHLORIDE □ 2-AMINOTOLUENE HYDROCHLORIDE □ 1-METH-YL-2-AMINOBENZENE HYDROCHLORIDE □ 2-METH-YL-1-AMINOBENZENE HYDROCHLORIDE □ o-METH-YLANILINE HYDROCHLORIDE □ 2-METHYLANILINE HYDROCHLORIDE □ o-METHYLBENZENAMINE HYDROCHLORIDE □ 2-METHYLBENZENAMINE HYDROCHLORIDE □ NCI-C02335 □ RCRA WASTE NUMBER U222 □ 2-TOLUIDINE HYDROCHLORIDE □ o-TOLYLAMINE HYDROCHLORIDE

CONSENSUS REPORTS: NTP 10th Report on Carcinogens. IARC Cancer Review: Animal Sufficient Evidence IMEMDT 27,155,82. NCI Carcinogenesis Bioassay - (feed); Clear Evidence: mouse, rat NCITR* NCI-CG-TR-153,79. EPA Genetic

Toxicology Program. Community Right-To-Know List. Reported in EPA TSCA Inventory.

SAFETY PROFILE: Confirmed carcinogen with experimental carcinogenic data. Poison by intraperitoneal route. Moderately toxic by ingestion. Mutation data reported. When heated to decomposition it emits very toxic fumes of HCl and NO_x. See also o-TOLUIDINE.

TGS750   CAS: 540-23-8   HR: 3
p-TOLUIDINE HYDROCHLORIDE
mf: $C_7H_9N \cdot ClH$ mw: 143.63
PROP: Needles or crystals from acetic ether. Mp: 243°, bp: 257.5°. Sol in water, alc; insol in ether, benzene.
SYNS: 4-AMINOTOLUENE HYDROCHLORIDE □ 4--METHYLANILINE HYDROCHLORIDE □ 4-METHYL-BENZENAMINE HYDROCHLORIDE □ p-TOLUID-INIUM CHLORIDE

CONSENSUS REPORTS: Reported in EPA TSCA Inventory. EPA Genetic Toxicology Program.

SAFETY PROFILE: Poison by intraperitoneal route. Moderately toxic by ingestion. Questionable carcinogen with experimental carcinogenic and tumorigenic data. When heated to decomposition it emits very toxic fumes of NO_x and HCl. See also p-TOLUIDINE.

TGV100   CAS: 102395-95-9 HR: 3
4-TOLYLARSENOUS ACID
mf: C_7H_7AsO mw: 182.06
SYNS: p-ARSENOSOTOLUENE □ TOLUENE, p-ARSENOSO-

OSHA PEL: TWA 0.5 mg(As)/m³

SAFETY PROFILE: Poison by intravenous route. When heated to decomposition it emits toxic fumes of As.

TGW000   CAS: 2646-17-5   HR: 3
1-(o-TOLYLAZO)-2-NAPHTHOL
mf: $C_{17}H_{14}N_2O$ mw: 262.33
PROP: Red needles from AcOH. Mp: 132–133°.
SYNS: A.F.ORANGE No. 2 □ AIZEN FOOD ORANGE No. 2 □ ATUL OIL ORANGE T □ C.I. 12100 □ C.I.

SOLVENT ORANGE 2 □ D&C ORANGE No. 2 □ DOLKWAL ORANGE SS □ EXTRACT D&C ORANGE No. 4 □ FAT ORANGE II □ HEXACOL OIL ORANGE SS □ LACQUER ORANGE V □ 1-((2-METHYLPHENYL)AZO)-2-NAPHTHALENOL □ OIL ORANGE O'PEL □ OIL ORANGE SS □ OLEAL ORANGE SS □ ORANGE 3R SOLUBLE IN GREASE □ ORGANOL ORANGE 2R □ TOLUENE-2-AZONAPHTHOL-2 □ o-TOLUENO-AZO-β-NAPHTHOL □ 1-(o-TOLYLAZO)-β-NAPHTHOL

CONSENSUS REPORTS: IARC Cancer Review: Group 2B IMEMDT 7,56,87; Animal Sufficient Evidence IMEMDT 8,165,75. Reported in EPA TSCA Inventory. EPA Genetic Toxicology Program.

SAFETY PROFILE: Confirmed carcinogen with experimental carcinogenic and neoplastigenic data. Poison by intravenous route. Mildly toxic by ingestion. Mutation data reported. When heated to decomposition it emits toxic fumes of NO_x. Used to color cosmetics, varnishes, oils, fats and waxes, petroleum products.

TGX100   CAS: 614-34-6   HR: 2
p-TOLYL BENZOATE
mf: $C_{14}H_{12}O_2$ mw: 212.26
PROP: Solid. Mp: 70–71°.
SYNS: BENZOIC ACID, 4-METHYLPHENYL ESTER □ BENZOIC ACID, p-TOLYL ESTER □ p-CRESYL BENZOATE □ 4-METHYLPHENYL BENZOATE

CONSENSUS REPORTS: Reported in EPA TSCA Inventory.

SAFETY PROFILE: Moderately toxic by ingestion. A skin irritant. When heated to decomposition it emits acrid smoke and irritating fumes.

TGY075   CAS: 106-43-4   HR: 3
p-TOLYL CHLORIDE
DOT: UN 2238
mf: C_7H_7Cl mw: 126.59
PROP: Liquid. Bp: 162.4°, d: (20°/4°) 1.0697, mp: 7.5°. Sltly sol in water; sol in alc, benzene, chloroform, ether.
SYNS: BENZENE, 1-CHLORO-4-METHYL- □ 4-CHLORO-1-METHYLBENZENE □ p-CHLOROTOLUENE □ 4-CHLOROTOLUENE □ p-CHLOROTOLUENE (DOT)

CONSENSUS REPORTS: EPA Genetic Toxicology Program. Reported in EPA TSCA Inventory.

DOT CLASSIFICATION: 3; Label: Flammable Liquid

SAFETY PROFILE: Moderately toxic by ingestion. Mildly toxic by inhalation. Flammable when exposed to heat or flame. When heated to decomposition it emits toxic fumes of Cl^-. See also TOLYL CHLORIDE and CHLORINATED HYDROCARBONS, AROMATIC.

THA250   CAS: 103-93-5   HR: 2
p-TOLYL ISOBUTYRATE
mf: $C_{11}H_{14}O_2$ mw: 178.25
PROP: Colorless liquid; characteristic odor. D: 0.990–0.996, refr index: 1.485, flash p: 212°F. Sol in alc; insol in water.
SYNS: p-CRESYL ISOBUTYRATE □ FEMA No. 3075 □ ISOBUTYRIC ACID, p-TOLYL ESTER □ PARACRESYL ISOBUTYRATE

CONSENSUS REPORTS: Reported in EPA TSCA Inventory.

SAFETY PROFILE: Moderately toxic by ingestion and skin contact. Combustible liquid. When heated to decomposition it emits acrid smoke and irritating fumes.

THD850   CAS: 607-88-5   HR: 2
p-TOLYL SALICYLATE
mf: $C_{14}H_{12}O_3$ mw: 228.26
SYNS: BENZOIC ACID, 2-HYDROXY-, 4-METHYL-PHENYL ESTER (9CI) □ p-CRESYL SALICYLATE □ 4--METHYLPHENYL 2-HYDROXYBENZOATE □ 4-METH-YLPHENYL SALICYLATE □ SALICYLIC ACID, p-TOLYL ESTER

CONSENSUS REPORTS: Reported in EPA TSCA Inventory.

SAFETY PROFILE: Moderately toxic by ingestion. A skin irritant. When heated to decomposition it emits acrid smoke and irritating fumes.

THF310   CAS: 17766-75-5   HR: 3
1-(p-TOLYL)-4-(3,4,5-TRIMETHOXY-
**   BENZOYL)PIPERAZINE**
mf: $C_{21}H_{26}N_2O_4$ mw: 370.49
SYNS: KETONE, 4-(p-TOLYL)PIPERAZINYL 3,4,5-TRIMETHOXYPHENYL □ PIPERAZINE, 1-(p-TOLYL)-4--

(3,4,5-TRIMETHOXYBENZOYL)- □ 4-(p-TOLYL)-PIPERAZINYL 3,4,5-TRIMETHOXYPHENYL KETONE

DOT CLASSIFICATION: 3; Label: Flammable Liquid

SAFETY PROFILE: A poison by intraperitoneal route. A flammable liquid. When heated to decomposition it emits toxic vapors of NO_x.

THG300   CAS: 17605-83-3   HR: 3
α-TOMATINE HYDROCHLORIDE

mf: $C_{50}H_{83}NO_{21} \cdot ClH$ mw: 1070.80

SYNS: TOMATINE HYDROCHLORIDE □ α-TOMATINE, HYDROCHLORIDE

SAFETY PROFILE: A poison by intravenous route. An eye irritant. When heated to decomposition it emits toxic vapors of NO_x, HCl, and Cl^-.

THH355   CAS: 31984-14-2   HR: 3
N-TOSYL-β-ALANINE CHLOROMETHYL KETONE

mf: $C_{11}H_{14}ClNO_3$ mw: 243.71

SYNS: N-(4-CHLORO-3-OXOBUTYL)-p-TOLUENESULFONAMIDE □ p-TOLUENESULFONAMIDE, N-(4-CHLORO-3-OXOBUTYL)-

DOT CLASSIFICATION: 3; Label: Flammable Liquid

SAFETY PROFILE: A poison by an unspecified route. A flammable liquid. When heated to decomposition it emits toxic vapors of NO_x and Cl^-.

THH360   CAS: 31982-00-0   HR: 3
N-TOSYL-I-ALANINE CHLOROMETHYL KETONE

mf: $C_{11}H_{14}ClNO_3$ mw: 275.77

SYNS: l-N-(3-CHLORO-1-METHYLACETONYL)-p-TOLUENESULFONAMIDE □ p-TOLUENESULFONAMIDE, N-(3-CHLORO-1-METHYL-ACETONYL)-, l-

DOT CLASSIFICATION: 3; Label: Flammable Liquid

SAFETY PROFILE: A poison by an unspecified route. A flammable liquid. When heated to decomposition it emits toxic vapors of NO_x, SO_x, and Cl^-.

THH375   CAS: 32065-38-6   HR: 3
N-TOSYL-β-ALANINE DIAZOMETHYL KETONE

mf: $C_{11}H_{13}N_3O_3S$ mw: 267.33

SYNS: N-(4-DIAZO-3-OXOBUTYL)-p-TOLUENESULFONAMIDE □ p-TOLUENESULFONAMIDE, N-(4-DIAZO-3-OXOBUTYL)-

DOT CLASSIFICATION: 3; Label: Flammable Liquid

SAFETY PROFILE: A poison by an unspecified route. A flammable liquid. When heated to decomposition it emits toxic vapors of NO_x and SO_x.

THH380   CAS: 31981-99-4   HR: 3
N-TOSYL-I-ALANINE DIAZOMETHYL KETONE

mf: $C_{11}H_{13}N_3O_3S$ mw: 267.33

SYNS: N-(3-DIAZO-1-METHYLACETONYL)-p-TOLUENESULFONAMIDE □ p-TOLUENESULFONAMIDE, N-(3-DIAZO-1-METHYLACETONYL)-

DOT CLASSIFICATION: 3; Label: Flammable Liquid

SAFETY PROFILE: A poison by an unspecified route. A flammable liquid. When heated to decomposition it emits toxic vapors of NO_x and SO_x.

THH450   CAS: 402-71-1   HR: D
I-1-TOSYLAMIDO-2-PHENYLETHYL CHLOROMETHYL KETONE

mf: $C_{17}H_{18}ClNO_3S$ mw: 351.87

SYNS: BENZENESULFONAMIDE, N-(3-CHLORO-2-OXO-1-(PHENYLMETHYL)PROPYL)-4-METHYL-,(S)- □ l-N-(α-(CHLOROACETYL)PHENETHYL)-p-TOLUENESULFONAMIDE □ p-TOLUENESULFONAMIDE, N-(α--(CHLOROACETYL)PHENETHYL)-, (–)-

CONSENSUS REPORTS: Reported in EPA TSCA Inventory.

DOT CLASSIFICATION: 3; Label: Flammable Liquid

SAFETY PROFILE: Experimental reproductive effects. A flammable liquid. When heated to decomposition it emits toxic fumes of NO_x, SO_x, and Cl^-.

THH460   CAS: 72676-77-8   HR: 3
6-(N-TOSYL)AMINOCAPROIC ACID DI-AZOMETHYL KETONE

mf: $C_{14}H_{19}N_3O_3S$ mw: 309.42

SYNS: N-(7-DIAZO-6-OXOHEPTYL)-p-TOLUENESULONAMIDE □ p-TOLUENESULFONAMIDE, N-(7-DIAZO-6-OXOHEPTYL)-

T

DOT CLASSIFICATION: 3; Label: Flammable Liquid

SAFETY PROFILE: A poison by an unspecified route. A flammable liquid. When heated to decomposition it emits toxic vapors of NO_x and Cl^-.

THH470   CAS: 72676-78-9   HR: 3
N-TOSYL-d,l-ISOLEUCINE CHLORO-
METHYL KETONE

mf: $C_{14}H_{20}ClNO_3$ mw: 285.80

SYNS: N-(3-CHLORO-1-sec-BUTYLACETONYL)-p-TOLUENESULFONAMIDE □ p-TOLUENESULFONAMIDE, N-(3-CHLORO-1-sec-BUTYL-ACETONYL)-

DOT CLASSIFICATION: 3; Label: Flammable Liquid

SAFETY PROFILE: A poison by an unspecified route. A flammable liquid. When heated to decomposition it emits toxic vapors of NO_x and Cl^-.

THH480   CAS: 72676-74-5   HR: 3
N-TOSYL-d,l-ISOLEUCINE DIAZOMETH-
YL KETONE

mf: $C_{14}H_{18}N_3O_3S$ mw: 308.41

SYNS: N-(1-(DIAZOACETYL)-2-METHYLBUTYL)-p-TOLUENESULFONAMIDE □ p-TOLUENESULFON-AMIDE, N-(1-(DIAZOACETYL)-2-METHYLBUTYL)-

DOT CLASSIFICATION: 3; Label: Flammable Liquid

SAFETY PROFILE: A poison by an unspecified route. A flammable liquid. When heated to decomposition it emits toxic vapors of NO_x and SO_x.

THH490   CAS: 72676-73-4   HR: 3
N-TOSYL-l-LEUCINE DIAZOMETHYL
KETONE

mf: $C_{14}H_{19}N_3O_3S$ mw: 309.42

SYNS: N-(3-DIAZO-1-ISOBUTYLACETONYL)-p-TOLUENESULFONAMIDE □ p-TOLUENESULFONAMIDE, N-(3-DIAZO-1-ISOBUTYL-ACETONYL)-

DOT CLASSIFICATION: 3; Label: Flammable Liquid

SAFETY PROFILE: A poison by an unspecified route. A flammable liquid. When heated to decomposition it emits toxic vapors of NO_x and SO_x.

THH550   CAS: 102516-65-4   HR: 3
TOSYL-l-PHENYLALANYLCHLORO-
METHYL KETONE

mf: $C_{18}H_{18}ClNO_5S$ mw: 395.88

SYNS: N-(CHLOROACETYL)-3-PHENYL-N-(p-TOLYL-SULFONYL)ALANINE □ N-TOSYL-l-PHENYLALANINE CHLOROMETHYL KETONE

DOT CLASSIFICATION: 3; Label: Flammable Liquid

SAFETY PROFILE: Poison by intravaginal route. Experimental reproductive effects. When heated to decomposition it emits toxic fumes of Cl^-, SO_x, and NO_x.

THH555   CAS: 26020-35-9   HR: 3
N-TOSYL-l-VALINE CHLOROMETHYL
KETONE

mf: $C_{13}H_{18}ClNO_3S$ mw: 303.83

SYNS: N-(3-CHLORO-1-ISOPROPYLACETONYL)-p-TOLUENESULFONAMIDE □ p-TOLUENESULFON-AMIDE, N-(3-CHLORO-1-ISOPROPYLACETONYL)-

DOT CLASSIFICATION: 3; Label: Flammable Liquid

SAFETY PROFILE: A poison by an unspecified route. A flammable liquid. When heated to decomposition it emits toxic vapors of NO_x, SO_x, and Cl^-.

THI255   CAS: 85079-48-7   HR: 3
TOXIN T-17

SYN: T-17 TOXIN

SAFETY PROFILE: A poison by ingestion, intraperitoneal, and intravenous routes. When heated to decomposition it emits acrid smoke and irritating vapors.

THI500   CAS: 23031-36-9   HR: 3
d,d-T80-PRALLETHRIN

mf: $C_{19}H_{24}O_3$ mw: 300.43

SYNS: AI3-29750 □ CYCLOPROPANECARBOXYLIC ACID, 2,2-DIMETHYL-3-(2-METHYLPROPENYL)-, ESTER WITH 4-HYDROXY-3-METHYL-2-(2-PROPYNYL)-2-CYCLOPENTEN-1-ONE, trans-(+−)- □ ETOC □ PRALLETHRIN □ S 4068 □ S-4068 SF

SAFETY PROFILE: A poison by ingestion and inhalation. Low toxicity by skin contact. When heated to decomposition it emits acrid smoke and irritating vapors.

THJ250　　CAS: 9000-65-1　　HR: 2
TRAGACANTH GUM
PROP: From the shrub *Astragalus gummifier* Labillardiere. Powder is white, pieces are white to pale yellow, translucent, and horny; odorless with mucilaginous taste.
SYNS: GUM TRAGACANTH □ TRAGACANTH
CONSENSUS REPORTS: Reported in EPA TSCA Inventory.
SAFETY PROFILE: Mildly toxic by ingestion. A mild allergen. A skin and eye irritant. Combustible when exposed to heat or flame. When heated to decomposition it emits acrid smoke and irritating fumes.

THL600　　　　　　　　　　HR: D
TRENBOLONE
mf: $C_{18}H_{22}O_2$　　mw: 270.38
PROP: Crystals. Mp: 186°.
SYNS: 4,9,11,-ESTRATRIEN-17β-OL-3-ONE □ 17β-HYDROXYESTRA-4,9,11-TRIEN-3-ONE □ TRIENBOLONE □ TRIENOLONE
SAFETY PROFILE: When heated to decomposition it emits acrid smoke and irritating fumes.

THM500　　CAS: 102-76-1　　HR: 3
TRIACETYL GLYCERIN
mf: $C_9H_{14}O_6$　　mw: 218.23
PROP: Colorless oily liquid; slt fatty odor and taste. Mp: −78°, bp: 258°, flash p: 280°F (COC), d: 1.161, autoign temp: 812°F, vap d: 7.52. Sol in water; misc with alc, ether, chloroform.
SYNS: ENZACTIN □ FEMA No. 2007 □ FUNGACETIN □ GLYCERINE TRIACETATE □ GLYCEROL TRIACETATE □ GLYCERYL TRIACETATE □ GLYPED □ KESSCOFLEX TRA □ KODAFLEX TRIACETIN □ 1,2,3-PROPANETRIOL TRIACETATE □ TRIACETIN - (FCC) □ VANAY
CONSENSUS REPORTS: Reported in EPA TSCA Inventory.
SAFETY PROFILE: Poison by ingestion. Moderately toxic by intraperitoneal, subcutaneous, and intravenous routes. An eye irritant. Combustible when exposed to heat, flame, or powerful oxidizers. To fight fire, use alcohol foam, water, CO_2, dry chemical. When heated to decomposition it emits acrid smoke and irritating fumes.

THN000　　CAS: 102-70-5　　HR: 3
TRIALLYLAMINE
DOT: UN 2610
mf: $C_9H_{15}N$　　mw: 137.25
PROP: Oily liquid with unpleasant odor. D: 0.800 @ 20°/4°, mp: <−70°, bp: 150–151°, flash p: 103°F (TOC).
SYN: N-N-DI-2-PROPENYL-2-PROPEN-1-AMINE
CONSENSUS REPORTS: Reported in EPA TSCA Inventory.
DOT CLASSIFICATION: 3; Label: Flammable Liquid
SAFETY PROFILE: Poison by skin contact and intraperitoneal routes. Moderately toxic by ingestion and inhalation. An eye and severe skin irritant. Human systemic effects by inhalation: structural or functional changes in trachea or bronchi. Flammable liquid when exposed to heat, flame or oxidizers. To fight fire, use foam, alcohol foam, fog. When heated to decomposition it emits toxic fumes of NO_x. See also AMINES and ALLYL COMPOUNDS.

THN500　　CAS: 101-37-1　　HR: 3
TRIALLYL CYANURATE
mf: $C_{12}H_{15}N_3O_3$　　mw: 243.24
PROP: Bp: 120° @ 5 mm, fp: 27.3°, flash p: >176°F (TOC), d: 1.1133 @ 30°, vap press: 1 mm @ 100°.
SYNS: TRIPROPARGYL CYANURATE □ 2,4,6-TRIPROP-2-YNYLOXY-s-TRIAZINE □ 2,4,6-TRIS(ALLYL-OXY)TRIAZINE
CONSENSUS REPORTS: Reported in EPA TSCA Inventory.
SAFETY PROFILE: Poison by intravenous route. Flammable when exposed to heat, flame, or oxidizers. To fight fire, use spray, foam, dry chemical. When heated to decomposition or on contact with acid or acid fumes it emits highly toxic fumes of CN^- and NO_x. See also ESTERS and ALLYL COMPOUNDS.

T

THN800   CAS: 4000-16-2   HR: 2
TRIAMINOGUANIDINE NITRATE
mf: $CH_8N_6 \cdot NO_3$ mw: 166.16
SYNS: CARBONOHYDRAZONIC DIHYDRAZIDE, MONONITRATE (9CI) □ TAGN
CONSENSUS REPORTS: Reported in EPA TSCA Inventory.
SAFETY PROFILE: Moderately toxic by intravenous route. An experimental teratogen. Experimental reproductive effects. Human mutation data reported. Decomposes violently at 230°C. When heated to decomposition it emits toxic fumes of NO_x. See also NITRATES and AMINES.

THP250   CAS: 17168-85-3   HR: 3
TRIAMMINEDIPEROXOCHROMIUM(IV)
mf: $CrH_9N_3O_4$ mw: 167.09
SAFETY PROFILE: Suspected carcinogen. Chromium compounds are generally poisons. May explode with heat or shock. May explode at 120°C. An oxidizer. When heated to decomposition it emits toxic fumes of NO_x. See also CHROMIUM COMPOUNDS, PEROXIDES, and AMINES.

THQ500   CAS: 7784-19-2   HR: 3
TRIAMMONIUM HEXAFLUOROALUMINATE
mf: $AlF_6 \cdot 3H_4N$ mw: 195.13
PROP: White hygroscopic powder. Insol in EtOH.
SYNS: AMMONIUM ALUMINUM FLUORIDE □ AMMONIUM CRYOLITE □ AMMONIUM FLUOALUMINATE □ AMMONIUM HEXAFLUOROALUMINATE □ TRIAMMONIUM ALUMINUM HEXAFLUORIDE
CONSENSUS REPORTS: Reported in EPA TSCA Inventory.
OSHA PEL: TWA 2.5 mg(F)/m³
ACGIH TLV: TWA 2 mg(Al)/m³; TWA 2.5 mg(F)/m³; BEI: 3 mg/g creatinine of fluorides in urine prior to shift; 10 mg/g creatinine of fluorides in urine at end of shift.
NIOSH REL: TWA 2.5 mg(F)/m³
SAFETY PROFILE: Poison by intravenous route. When heated to decomposition it emits very toxic fumes of F⁻, NO_x, and NH_3. See also ALUMINUM COMPOUNDS and FLUORIDES.

THR750   CAS: 461-89-2   HR: 3
s-TRIAZINE-3,5(2H,4H)-DIONE
mf: $C_3H_3N_3O_2$ mw: 113.09
PROP: Crystals from H_2O. Mp: 278–280° - (sinters).
SYNS: 4(6)-AZAURACIL □ 6-AZAURACIL □ NSC-3425 □ USAF CB-30
CONSENSUS REPORTS: Reported in EPA TSCA Inventory.
SAFETY PROFILE: Poison by intraperitoneal route. Experimental reproductive effects. Questionable carcinogen with experimental tumorigenic data. Mutation data reported. When heated to decomposition it emits toxic fumes of NO_x.

THT280   CAS: 274-98-6   HR: 3
(1,2,4)TRIAZOLO(4,3-a)PYRIMIDINE
mf: $C_5H_4N_4$ mw: 120.11
SAFETY PROFILE: A poison by ingestion. When heated to decomposition it emits toxic vapors of NO_x.

THV000   CAS: 1329-86-8   HR: 3
TRIBROMOETHANOL
mf: $C_2H_3Br_4O$ mw: 282.78
PROP: Crystals; ethereal odor, aromatic taste. Bp: 92° @ 10 mm, mp: 79–82°, decomp @ 70°. Sltly water sol; sol in alc, org solvs.
SYNS: AVERTIN □ BROMETHOL □ ETHOBROM □ NARCOLAN □ NARKOLAN □ TRIBROMETHANOL □ TRIBROMOETHYL ALCOHOL
SAFETY PROFILE: Poison by ingestion, intravenous, intraperitoneal, and rectal routes. Experimental reproductive effects. When heated to decomposition it emits toxic fumes of Br⁻. An anesthetic drug.

THV500   CAS: 73941-35-2   HR: 3
2,4,5-TRIBROMOIMIDAZOLE CADMIUM SALT (2:1)
mf: $C_6Br_6N_4 \cdot Cd$ mw: 719.96
SYN: CADMIUM salt of 2,4-5-TRIBROMOIMIDAZOLE

CONSENSUS REPORTS: Cadmium and its compounds are on the Community Right-To-Know List.

OSHA PEL: TWA 5 µg(Cd)/m³

ACGIH TLV: TWA 0.002 mg(Cd)/m³ - (respirable dust), Suspected Human Carcinogen); BEI: 5 µg/g creatinine in urine; 5 µg/L in blood

NIOSH REL: (Cadmium) Reduce to lowest feasible level

SAFETY PROFILE: Confirmed human carcinogen. Poison by intravenous route. When heated to decomposition it emits very toxic fumes of Br⁻, Cd, and NOₓ. See also BROMIDES and CADMIUM COMPOUNDS.

THX250   CAS: 102-82-9   HR: 3
TRIBUTYLAMINE
DOT: UN 2542

mf: $C_{12}H_{27}N$ mw: 185.40

PROP: A colorless, hygroscopic liquid with characteristic odor. Mp: −70°, bp: 216–217°, flash p: 187°F (OC), d: 0.78–0.79, vap d: 6.38. Sparingly sol in water; sol in alc, ether.

SYNS: TRI-n-BUTYLAMINE ☐ TRIS-N-BUTYLAMINE

CONSENSUS REPORTS: Reported in EPA TSCA Inventory.

DOT CLASSIFICATION: 8; Label: Corrosive

SAFETY PROFILE: Poison by ingestion, inhalation, skin contact, and subcutaneous routes. A central nervous system stimulant, irritant, and sensitizer. A corrosive irritant to skin, eyes, and mucous membranes. Flammable when exposed to heat, flame, or oxidizers. Can react with oxidizing materials. To fight fire, use foam, CO₂, dry chemical. When heated to decomposition it emits toxic fumes of NOₓ. See also AMINES.

THX500   CAS: 122-56-5   HR: 3
TRI-n-BUTYL BORANE
mf: $C_{12}H_{27}B$ mw: 182.20

PROP: Colorless pyrophoric liquid. Mp: 34°, bp: 170° @ 222 mm, d: 0.747 @ 25°, vap press: 1 mm @ 20°, flash p: −32°F. Insol in water; sol in most org solvs.

SYNS: BORIC ACID, TRIBUTYL ESTER ☐ TBB ☐ TRIBUTYLBORINE

CONSENSUS REPORTS: Reported in EPA TSCA Inventory.

SAFETY PROFILE: Poison by intravenous route. Moderately toxic by ingestion. A very dangerous fire hazard when exposed to heat or flame; can ignite spontaneously. When heated to decomposition it emits acrid smoke and irritating fumes. See also BORANES.

THX750   CAS: 688-74-4   HR: 3
TRI-n-BUTYL BORATE
mf: $C_{12}H_{27}BO_3$ mw: 230.20

PROP: Colorless, mobile, moisture-sensitive liquid; odor like n-butanol. Bp: 230°, fp: <−70°, flash p: 200°F (COC), d: 0.847 @ 28°, vap d: 7.95.

SYNS: BORESTER 2 ☐ BORIC ACID, TRI-sec-BUTYL ESTER ☐ BUTYL BORATE ☐ n-BUTYL BORATE ☐ TRIBUTOXYBORANE ☐ TRI-n-BUTOXYBORANE ☐ TRIBUTYL BORATE

CONSENSUS REPORTS: Reported in EPA TSCA Inventory.

SAFETY PROFILE: Moderately toxic by ingestion and intraperitoneal routes. An eye irritant. Flammable when exposed to heat, flame, or oxidizers. To fight fire, use foam, CO₂, dry chemical. When heated to decomposition or on contact with acid or acid fumes it can emit toxic fumes; on contact with oxidizing materials it can react vigorously. See also BORANES and BORON COMPOUNDS.

THY750   CAS: 681-99-2   HR: 3
TRIBUTYLISOCYANATOSTANNANE
DOT: UN 2207/UN 2206/UN 3080/UN 2478

mf: $C_{13}H_{27}NOSn$ mw: 332.10

SYNS: STANNANE, (ISOCYANATO)TRIBUTYL- ☐ STANNANE, TRIBUTYLISOCYANATO- ☐ TIN, TRIBUTYL-, ISOTHIOCYANATE ☐ TRIBUTYLSTANN-YL ISOCYANATE ☐ TRIBUTYLTIN ISOCYANATE ☐ TRI-n-BUTYLTIN ISOCYANATE ☐ TRIBUTYLTIN ISOTHIOCYANATE

OSHA PEL: TWA 0.1 mg(Sn)/m³ (skin)

ACGIH TLV: TWA 0.1 mg(Sn)/m³; STEL 0.2 mg(Sn)/m³ (skin).

NIOSH REL: (Organotin Compounds) TWA 0.1 mg(Sn)/m³
DOT CLASSIFICATION: 6.1; Label: KEEP AWAY FROM FOOD (UN 2207); Class: 6.1; Label: Poison (UN 2206); Class: 6.1; Label: Poison, Flammable Liquid (UN 3080); Class: 3; Label: Flammable Liquid, Poison (UN 2478).
SAFETY PROFILE: A poison. Flammable liquid. When heated to decomposition it emits toxic fumes of NO$_x$. See also TIN COMPOUNDS and ISOCYANATES.

**THZ000   CAS: 2155-70-6   HR: 3
TRIBUTYL(METHACRYLOXY)-
       STANNANE**
mf: C$_{16}$H$_{32}$O$_2$Sn   mw: 375.17
SYNS: TRIBUTYL(METHACRYLOYLOXY)STANNANE □ TRIBUTYL((2-METHYL-1-OXO-2-PROPENYL)OXY)-STANNANE □ TRIBUTYLSTANNYL METHACRYLATE □ TRIBUTYLTIN METHACRYLATE
CONSENSUS REPORTS: Reported in EPA TSCA Inventory.
OSHA PEL: TWA 0.1 mg(Sn)/m³ (skin)
ACGIH TLV: TWA 0.1 mg(Sn)/m³; STEL 0.2 mg(Sn)/m³ (skin).
DFG MAK: 0.0021 ppm (0.05 mg/m³)
NIOSH REL: (Organotin Compounds) TWA 0.1 mg(Sn)/m³
SAFETY PROFILE: Poison by ingestion and intravenous routes. When heated to decomposition it emits acrid smoke and irritating fumes. See also TIN COMPOUNDS.

**TIA250   CAS: 126-73-8   HR: 3
TRIBUTYL PHOSPHATE**
mf: C$_{12}$H$_{27}$O$_4$P   mw: 266.36
PROP: Colorless odorless liquid. Bp: 289° - (decomp), mp: <−80°, flash p: 295°F - (COC), d: 0.982 @ 20°, vap d: 9.20. Sol in water; misc in alc and ether. IDLH 30 ppm.
SYNS: CELLUPHOS 4 □ TBP □ TRIBUTILFOSFATO - (ITALIAN) □ TRIBUTYLE (PHOSPHATE de) (FRENCH) □ TRIBUTYLFOSFAAT (DUTCH) □ TRIBUTYLPHOSPHAT - (GERMAN) □ TRI-n-BUTYL PHOSPHATE
CONSENSUS REPORTS: Reported in EPA TSCA Inventory.
OSHA PEL: TWA 0.2 ppm

ACGIH TLV: TWA 0.2 ppm
SAFETY PROFILE: Poison by intraperitoneal and intravenous routes. Moderately toxic by ingestion, inhalation, and subcutaneous routes. Experimental reproductive effects. A skin, eye, and mucous membrane irritant. Combustible when exposed to heat or flame. To fight fire, use CO$_2$, dry chemical, fog, mist. When heated to decomposition it emits toxic fumes of PO$_x$.

**TIB000   CAS: 5488-45-9   HR: 3
TRIBUTYL(8-QUINOLINOLATO)TIN**
mf: C$_{21}$H$_{33}$NOSn   mw: 434.24
SYN: (8-QUINOLINOLATO)TRIBUTYLSTANNANE
OSHA PEL: TWA 0.1 mg(Sn)/m³ (skin)
ACGIH TLV: TWA 0.1 mg(Sn)/m³; STEL 0.2 mg(Sn)/m³ (skin).
NIOSH REL: (Organotin Compounds) TWA 0.1 mg(Sn)/m³
SAFETY PROFILE: Poison by intravenous route. When heated to decomposition it emits toxic fumes of NO$_x$. See also TIN COMPOUNDS.

**TIB500   CAS: 688-73-3   HR: 2
TRI-n-BUTYLSTANNANE HYDRIDE**
mf: C$_{12}$H$_{28}$Sn   mw: 291.09
PROP: A liquid. D: 1.103 @ 20°, bp: 112.5–113.5° @ 8 mm.
SYNS: TRIBUTYLSTANNIC HYDRIDE □ TRIBUTYLTIN HYDRIDE □ TRI-n-BUTYLTIN HYDRIDE
CONSENSUS REPORTS: Reported in EPA TSCA Inventory.
OSHA PEL: TWA 0.1 mg(Sn)/m³ (skin)
ACGIH TLV: TWA 0.1 mg(Sn)/m³; STEL 0.2 mg(Sn)/m³ (skin).
NIOSH REL: (Organotin Compounds) TWA 0.1 mg(Sn)/m³
SAFETY PROFILE: Moderately toxic by inhalation. When heated to decomposition it emits acrid smoke and irritating fumes. See also TIN COMPOUNDS.

TID100   CAS: 2669-35-4   HR: 3
**TRIBUTYLTIN CYCLOHEXANECARBO-
    XYLATE**
mf: $C_{19}H_{38}O_2Sn$ mw: 417.26
SYNS: CYCLOHEXANECARBOXYLIC ACID, TRIBUT-
YLSTANNYL ESTER □ ((CYCLOHEXYLCARBONYL)O-
XY)TRIBUTYLSTANNANE
OSHA PEL: TWA 0.1 mg(Sn)/m³ (skin)
ACGIH TLV: TWA 0.1 mg(Sn)/m³; STEL
0.2 mg/m³ (skin)
NIOSH REL: (Organotin Compound) TWA
0.1 mg(Sn)/m³
SAFETY PROFILE: Poison by intravenous
route. When heated to decomposition it
emits toxic fumes of Sn.

TID150   CAS: 20369-63-5   HR: 3
**TRIBUTYLTIN DIMETHYLDITHIO-
    CARBAMATE**
mf: $C_{15}H_{33}NS_2Sn$ mw: 410.30
SYNS: N,N-DIMETHYLDITHIOCARBAMIC ACID S-
TRIBUTYLSTANNYL ESTER □ STANNANE, ((DIMETH-
YLTHIOCARBAMOYL)THIO)TRIBUTYL-
OSHA PEL: TWA 0.1 mg(Sn)/m³ (skin)
ACGIH TLV: TWA 0.1 mg(Sn)/m³; STEL
0.2 mg/m³ (skin)
NIOSH REL: (Organotin Compound) TWA
0.1 mg(Sn)/m³
SAFETY PROFILE: Poison by intravenous
route. When heated to decomposition it
emits toxic fumes of NO_x, SO_x, and Sn.

TIE600   CAS: 53404-82-3   HR: 3
TRIBUTYLTIN ISOPROPYLSUCCINATE
mf: $C_{19}H_{38}O_4Sn$ mw: 449.26
SYNS: STANNANE, (ISOPROPYLSUCCINYLOXY)
TRIBUTYL- □ SUCCINIC ACID, O-ISOPROPYL-O'-
TRIBUTYLSTANNYL ESTER
OSHA PEL: TWA 0.1 mg(Sn)/m³ (skin)
ACGIH TLV: TWA 0.1 mg(Sn)/m³; STEL
0.2 mg/m³ (skin)
NIOSH REL: (Organotin Compound) TWA
0.1 mg(Sn)/m³
SAFETY PROFILE: Poison by intravenous
route. When heated to decomposition it
emits toxic fumes of Sn.

TIF250   CAS: 28801-69-6   HR: 3
TRIBUTYLTIN NEODECANOATE
mf: $C_{22}H_{46}O_2Sn$ mw: 461.37

SYNS: 4,4-DIMETHYLOCTANOIC ACID, TRIBUTYL-
STANNYL ESTER □ (4,4-DIMETHYLOCTANOYLOXY)-
TRIBUTYLSTANNANE □ HYDROXYTRIBUTYL-
STANNANE-4,4-DIMETHYLOCTANOATE □ TRIBUTYL-
(NEODECANOYLOXY)STANNANE
OSHA PEL: TWA 0.1 mg(Sn)/m³
ACGIH TLV: TWA 0.1 mg(Sn)/m³; STEL
0.2 mg(Sn)/m³ (skin).
NIOSH REL: (Organotin Compounds) TWA
0.1 mg(Sn)/m³
SAFETY PROFILE: Poison by intravenous
route. Moderately toxic by ingestion. When
heated to decomposition it emits acrid
smoke and irritating fumes. See also TIN
COMPOUNDS.

TIG250   CAS: 150-50-5   HR: 3
S,S,S-TRIBUTYL TRITHIOPHOSPHITE
mf: $C_{12}H_{27}PS_3$ mw: 298.54
PROP: Colorless to pale-yellow liquid with a
mild characteristic odor. Bp: 142–145° @
4.5 mm, flash p: 295°F (COC), d: 0.987 @
20°/4°.
SYNS: CHEMAGRO B-1776 □ DELEAF DEFOLIANT □
EASY OFF-D □ FOLEX □ MERPHOS □
PHOSPHOROTRITHIOUS ACID, S,S,S-TRIBUTYL ESTER
□ TRIBUTYL PHOSPHOROTRITHIOITE □ S,S,S-TRIBUT-
YL PHOSPHOROTRITHIOITE
SAFETY PROFILE: Poison by
intraperitoneal route. Moderately toxic by
ingestion and skin contact. A cholinesterase
inhibitor. Combustible when exposed to
heat or flame. Can react vigorously with
oxidizing materials. When heated to
decomposition it emits highly toxic fumes of
PO_x and SO_x. Used as a defoliant. See also
PARATHION.

TIG750   CAS: 60-01-5   HR: 3
TRIBUTYRIN
mf: $C_{15}H_{26}O_6$ mw: 302.41
PROP: Colorless, oily liquid; bitter taste.
Mp: −75°, d: 1.0356 @ 20°/20°, bp:
305–310°, flash p: 212°F. Insol in water;
very sol in alc, ether, chloroform.
SYNS: BUTANOIC ACID, 1,2,3-PROPANETRIYL ESTER
□ BUTYRIC ACID TRIESTER with GLYCERIN □ BUTYR-
YL TRIGLYCERIDE □ FEMA No. 2223 □ GLYCEROL
TRIBUTYRATE □ KODAFLEX □ TRIBUTYROIN

T

CONSENSUS REPORTS: Reported in EPA TSCA Inventory.

SAFETY PROFILE: Poison by intravenous route. Low toxicity by ingestion. Questionable carcinogen with experimental tumorigenic data. Combustible liquid. When heated to decomposition it emits acrid smoke and irritating fumes. See also ESTERS.

TIH000   CAS: 12380-95-9   HR: 3
TRICADMIUM DINITRIDE

mf: Cd_3N_2 mw: 365.21

PROP: Moisture- and air-sensitive black powder.

SYN: CADMIUM NITRIDE

CONSENSUS REPORTS: Cadmium compounds are on the Community Right-To-Know List.

OSHA PEL: TWA 5 μg(Cd)/m³

ACGIH TLV: TWA 0.002 mg(Cd)/m³ - (respirable dust), Suspected Human Carcinogen); BEI: 5 μg/g creatinine in urine; 5 μg/L in blood

NIOSH REL: (Cadium) Reduce to lowest feasible level

SAFETY PROFILE: Confirmed human carcinogen. Many cadmium compounds are poisons. Explodes violently on shock or heating. Explodes on contact with water, acids, or bases. When heated to decomposition it emits very toxic fumes of NO_x and Cd. See also NITRIDES and CADMIUM COMPOUNDS.

TII250   CAS: 76-03-9   HR: 3
TRICHLOROACETIC ACID

DOT: UN 1839/UN 2564

mf: $C_2HCl_3O_2$ mw: 163.38

PROP: Colorless, rhombic, deliq crystals. Bp: 197.5°, fp: 57.7°, flash p: none, mp: 57–58°, d: 1.6298 @ 61°/4°, vap press: 1 mm @ 51.0°. Sol in water and alc.

SYNS: ACETO-CAUSTIN □ ACIDE TRICHLO-RACETIQUE (FRENCH) □ ACIDO TRICLOROACETICO - (ITALIAN) □ AMCHEM GRASS KILLER □ DOW SODI-UM TCA SOLUTION □ KONESTA □ KYSELINA TRICHLOROCTOVA □ NA TA □ SODIUM TCA SOLUTION □ TCA □ TRICHLOORAZIJNZUUR (DUTCH)

□ TRICHLORESSIGSAEURE (GERMAN) □ TRICHLOROACETIC ACID (UN 1839) (DOT) □ TRICHLOROACETIC ACID, solution (UN 2564) (DOT) □ TRICHLOROETHANOIC ACID □ VARITOX

CONSENSUS REPORTS: Reported in EPA TSCA Inventory. EPA Genetic Toxicology Program.

OSHA PEL: TWA 1 ppm

ACGIH TLV: TWA 1 ppm; Confirmed Animal Carcinogen with Unknown Relevance to Humans

DOT CLASSIFICATION: 8; Label: Corrosive

SAFETY PROFILE: Poison by ingestion and subcutaneous routes. Moderately toxic by intraperitoneal route. Questionable carcinogen with experimental carcinogenic data. Experimental reproductive effects. Mutation data reported. A corrosive irritant to skin, eyes, and mucous membranes. When heated to decomposition it emits toxic fumes of Cl^- and Na_2O. Used as an herbicide.

TII750   CAS: 545-06-2   HR: 3
TRICHLOROACETONITRILE

mf: C_2Cl_3N mw: 144.38

PROP: Crystals or liquid; odor of chloral and hydrogen cyanide. Mp: 44°, bp: 85.7°, d: 1.44° @ 25°/4°.

SYNS: CYANOTRICHLOROMETHANE □ NITRILE TRICHLORACETIQUE (FRENCH) □ TRICHLOR-ACETONITRIL (GERMAN) □ TRICHLOROMETHYL CYANIDE □ TRICHLOROMETHYLNITRILE □ TRICHLOURACETONITRIL (DUTCH) □ TRITOX

CONSENSUS REPORTS: Cyanide and its compounds are on the Community Right-To-Know List. Reported in EPA TSCA Inventory.

SAFETY PROFILE: Poison by ingestion and intravenous routes. Moderately toxic by inhalation and skin contact. Human mutation data reported. A skin and severe eye irritant. An experimental teratogen. Other experimental reproductive effects. When heated to decomposition or in reaction with water, steam, acid, or acid fumes it produces toxic fumes of CN^-, Cl^-, and NO_x. Used as an insecticide. See also NITRILES and CYANIDE.

TIJ150   **CAS: 76-02-8**   **HR: 2**
TRICHLOROACETYL CHLORIDE
DOT: UN 2442
mf: C_2Cl_4O mw: 181.82
PROP: Mp: $-146°$, bp: 114–116°, refr index: 1.4700, d: 1.629, flash p: none.
SYNS: TRICHLOROACETIC ACID CHLORIDE □ TRICHLOROACETOCHLORIDE
CONSENSUS REPORTS: EPA Extremely Hazardous Substances List. Reported in EPA TSCA Inventory.
DOT CLASSIFICATION: 8; Label: Corrosive, Poison
SAFETY PROFILE: Moderately toxic by inhalation and ingestion. A corrosive irritant to skin, eyes, and mucous membranes. When heated to decomposition it emits toxic fumes of Cl^-.

TIK250   **CAS: 120-82-1**   **HR: 3**
1,2,4-TRICHLOROBENZENE
mf: $C_6H_3Cl_3$ mw: 181.44
PROP: Colorless liquid. Mp: 17°, bp: 213°, flash p: 230°F (CC), d: 1.454 @ 25°/25°, vap press: 1 mm @ 38.4°, vap d: 6.26. Sol in water.
SYNS: HOSTETEX L-PEC □ unsym-TRICHLORO-BENZENE □ 1,2,5-TRICHLOROBENZENE □ 1,3,4-TRICHLOROBENZENE □ 1,2,4-TRICHLOROBENZENE - (ACGIH,OSHA) □ 1,2,4-TRICHLOROBENZOL □ TROJCHLOROBENZEN □ TROJCHLOROBENZEN - (POLISH)
CONSENSUS REPORTS: Community Right-To-Know List. Reported in EPA TSCA Inventory.
OSHA PEL: CL 5 ppm
ACGIH TLV: CL 5 ppm
DFG MAK: Confirmed Animal Carcinogen with Unknown Relevance to Humans
SAFETY PROFILE: Poison by ingestion. Moderately toxic by intraperitoneal route. An experimental teratogen. Experimental reproductive effects. Mutation data reported. A skin irritant. Combustible when exposed to heat or flame. Can react vigorously with oxidizing materials. To fight fire, use water, foam, CO_2, dry chemical. When heated to decomposition it emits toxic fumes of Cl^-. See also CHLOR-

INATED HYDROCARBONS, AROMATIC.

TIK300   **CAS: 108-70-3**   **HR: 2**
1,3,5-TRICHLOROBENZENE
mf: $C_6H_3Cl_3$ mw: 181.44
SYNS: BENZENE, 1,3,5-TRICHLORO- □ s-TRI-CHLOROBENZENE □ sym-TRICHLOROBENZENE
CONSENSUS REPORTS: Reported in EPA TSCA Inventory.
DFG MAK: 5 ppm
SAFETY PROFILE: Moderately toxic by ingestion and intraperitoneal routes. Mutation data reported. When heated to decomposition it emits toxic vapors of Cl^-.

TIL360   **CAS: 2431-50-7**   **HR: 3**
2,3,4-TRICHLOROBUTENE-1
mf: $C_4H_5Cl_3$ mw: 159.44
SYN: 1-BUTENE, 2,3,4-TRICHLORO-
CONSENSUS REPORTS: Reported in EPA TSCA Inventory.
DFG MAK: Animal Carcinogen; Suspected Human Carcinogen
SAFETY PROFILE: Confirmed carcinogen. Poison by ingestion. When heated to decomposition it emits toxic fumes of Cl^-.

TIN000   **CAS: 79-00-5**   **HR: 3**
1,1,2-TRICHLOROETHANE
mf: $C_2H_3Cl_3$ mw: 133.40
PROP: Nonflammable, mobile liquid with a pleasant odor. Bp: 114°, fp: $-35°$, d: 1.4416 @ 20°/4°, vap press: 40 mm @ 35.2°. Insol in H_2O; misc in most org solvs. IDLH 100 ppm.
SYNS: ETHANE TRICHLORIDE □ NCI-C04579 □ RCRA WASTE NUMBER U227 □ β-T □ 1,1,2-TRICHLO-RETHANE □ β-TRICHLOROETHANE □ 1,2,2-TRI-CHLOROETHANE □ TROJCHLOROETAN(1,1,2) - (POLISH) □ VINYL TRICHLORIDE
CONSENSUS REPORTS: IARC Cancer Review: Group 3 IMEMDT 7,56,87; Animal Limited Evidence IMEMDT 20,533,79. NCI Carcinogenesis Bioassay - (gavage); No Evidence: rat NCITR* NCI-CG-TR-74,78; (gavage); Clear Evidence: mouse NCITR* NCI-CG-TR-74,78.

T

Community Right-To-Know List. Reported in EPA TSCA Inventory.

OSHA PEL: TWA 10 ppm (skin)
ACGIH TLV: TWA 10 ppm (skin); Not Classifiable as a Human Carcinogen
DFG MAK: 10 ppm (55 mg/m³); Confirmed Animal Carcinogen with Unknown Relevance to Humans
SAFETY PROFILE: Suspected carcinogen with experimental carcinogenic data. Poison by ingestion, intravenous, and subcutaneous routes. Moderately toxic by inhalation, skin contact, and intraperitoneal routes. Experimental reproductive effects. Mutation data reported. An eye and severe skin irritant. Has narcotic properties and acts as a local irritant to the eyes, nose, and lungs. It may also be injurious to the liver and kidneys. Incompatible with potassium. When heated to decomposition it emits toxic fumes of Cl⁻. See also CHLOR-INATED HYDROCARBONS, ALIPHATIC; and other trichloroethane entries.

TIN750   CAS: 75-94-5   HR: 3
TRICHLOROETHENYLSILANE
DOT: UN 1305
mf: C₂H₃Cl₃Si   mw: 161.49
PROP: Fuming liquid. Bp: 90.6°, d: 1.265 @ 25/25°, flash p: 16°F.
SYNS: SILANE, VINYL TRICHLORO 1-150 □ TRICHLORO(VINYL)SILANE □ TRICHLOROVINYL SILICANE □ UNION CARBIDE A-150 □ VINYLSILICON TRICHLORIDE □ VINYL TRICHLOROSILANE (DOT) □ VINYL TRICHLOROSILANE, INHIBITED (DOT)
CONSENSUS REPORTS: Reported in EPA TSCA Inventory.
DOT CLASSIFICATION: 3; Label: Flammable Liquid, Corrosive
SAFETY PROFILE: Moderately toxic by ingestion, inhalation, and skin contact. A corrosive irritant to skin, eyes, and mucous membranes. A very dangerous fire hazard when exposed to heat or flame. Reacts violently with water, moist air, or steam to produce toxic and corrosive fumes. When heated to decomposition it emits toxic fumes of Cl⁻. See also CHLOROSILANES.

TIO750   CAS: 79-01-6   HR: 3
TRICHLOROETHYLENE
DOT: UN 1710
mf: C₂HCl₃   mw: 131.38
PROP: Clear, colorless, nonflammable, mobile liquid; characteristic sweet odor of chloroform. D: 1.4649 @ 20°/4°, bp: 86.7°, mp: −84°, fp: −86.8°, autoign temp: 788°F, vap press: 100 mm @ 32°, vap d: 4.53, refr index: 1.477 @ 20°. Immisc with water; misc with alc, ether, acetone, carbon tetrachloride. Insol in H₂O; sol in most org solvs. IDLH 1000 ppm.
SYNS: ACETYLENE TRICHLORIDE □ ALGYLEN □ ANAMENTH □ BENZINOL □ BLACOSOLV □ CECOLENE □ 1-CHLORO-2,2-DICHLOROETHYLENE □ CHLORYLEA □ CHORYLEN □ CIRCOSOLV □ CRAWHASPOL □ DENSINFLUAT □ 1,1-DICHLORO-2-CHLOROETHYLENE □ DOW-TRI □ DUKERON □ ETH-INYL TRICHLORIDE □ ETHYLENE TRICHLORIDE □ FLECK-FLIP □ FLUATE □ GERMALGENE □ LANADIN □ LETHURIN □ NARCOGEN □ NARKOSOID □ NCI-C04546 □ NIALK □ PERM-A-CHLOR □ PETZINOL □ RCRA WASTE NUMBER U228 □ THRETHYLENE □ TRIAD □ TRIASOL □ TRICHLOORETHEEN (DUTCH) □ TRICHLOORETHYLEEN (DUTCH) □ TRICHLORAE-THEN (GERMAN) □ TRICHLORAETHYLEN (GERMAN) □ TRICHLORAN □ TRICHLORETHENE (FRENCH) □ TRICHLORETHYLENE (FRENCH) □ TRICHLOROETH-ENE □ 1,2,2-TRICHLOROETHYLENE □ TRI-CLENE □ TRICLORETENE (ITALIAN) □ TRICLOROETILENE - (ITALIAN) □ TRIELINA (ITALIAN) □ TRILENE □ TRIMAR □ TRI-PLUS □ VESTROL □ VITRAN □ WESTROSOL
CONSENSUS REPORTS: NTP 10th Report on Carcinogens. IARC Cancer Review: Group 3 IMEMDT 7,364,87; Animal Limited Evidence IMEMDT 20,545,79; Human Inadequate Evidence IMEMDT 20,545,79; Animal Sufficient Evidence IMEMDT 11,263,76. NCI Carcinogenesis Bioassay (gavage); No Evidence: rat NCITR* NCI-CG-TR-2,76; (gavage); Clear Evidence: mouse NCITR* NCI-CG-TR-2,76. Community Right-To-Know List. Reported in EPA TSCA Inventory. EPA Genetic Toxicology Program.
OSHA PEL: TWA 50 ppm; STEL 200 ppm
ACGIH TLV: TWA 10 ppm; 25 STEL; Suspected Human Carcinogen; BEI: 100

mg(trichloroacetic acid)/g creatinine in urine at end of workweek

DFG MAK: Confirmed Human Carcinogen; BAT: 500 µg/dL in blood at end of shift or workweek

NIOSH REL: (Trichloroethylene) TWA 250 ppm; (Waste Anesthetic Gases) CL 2 ppm/1H

DOT CLASSIFICATION: 6.1; Label: KEEP AWAY FROM FOOD

SAFETY PROFILE: Confirmed carcinogen with experimental carcinogenic, tumorigenic, and teratogenic data. Experimental poison by intravenous and subcutaneous routes. Moderately toxic experimentally by ingestion and intraperitoneal routes. Mildly toxic to humans by ingestion and inhalation. Mildly toxic experimentally by inhalation. Human systemic effects by ingestion and inhalation: eye effects, somnolence, hallucinations or distorted perceptions, gastrointestinal changes, and jaundice. Experimental reproductive effects. Human mutation data reported. An eye and severe skin irritant. Inhalation of high concentrations causes narcosis and anesthesia. A form of addiction has been observed in exposed workers. Prolonged inhalation of moderate concentrations causes headache and drowsiness. Fatalities following severe, acute exposure have been attributed to ventricular fibrillation resulting in cardiac failure. There is damage to liver and other organs from chronic exposure. A common air contaminant.

Nonflammable, but high concentrations of trichloroethylene vapor in high-temperature air can be made to burn mildly if plied with a strong flame. Though such a condition is difficult to produce, flames or arcs should not be used in closed equipment that contains any solvent residue or vapor. Reacts with alkali, epoxides, e.g., 1-chloro-2,3-epoxypropane, 1,4-butanediol mono-2,3-epoxypropylether, 1,4-butanediol di-2,3-epoxypropylether, 2,2-bis[(4(2',3'-epoxypropoxy)phenyl)propane] to form the spontaneously flammable gas

dichloroacetylene. Can react violently with Al, Ba, N_2O_4, Li, Mg, liquid O_2, O_3, KOH, KNO_3, Na, NaOH, Ti. Reacts with water under heat and pressure to form HCl gas. When heated to decomposition it emits toxic fumes of Cl⁻. See also CHLORINATED HYDROCARBONS, ALIPHATIC.

TIP500   CAS: 75-69-4   HR: 2
TRICHLOROFLUOROMETHANE
mf: CCl_3F mw: 137.36

PROP: Colorless liquid. Mp: −111°, bp: 24.1°, d: 1.484 @ 17.2°.

SYNS: ALGOFRENE TYPE 1 □ ARCTON 9 □ ELECTRO-CF 11 □ ESKIMON 11 □ FLUOROCARBON No. 11 □ FLUOROTRICHLOROMETHANE (OSHA) □ FLUOROTROJCHLOROMETAN (POLISH) □ FREON 11 □ FREON MF □ FRIGEN 11 □ GENETRON 11 □ HALOCARBON 11 □ ISCEON 131 □ ISOTRON 11 □ LEDON 11 □ MONOFLUOROTRICHLOROMETHANE □ NCI-C04637 □ RCRA WASTE NUMBER U121 □ TRICHLOROMONOFLUOROMETHANE □ UCON REFRIGERANT 11

CONSENSUS REPORTS: NCI Carcinogenesis Bioassay (gavage); No Evidence: mouse NCITR* NCI-CG-TR-106,78; (gavage); Inadequate Studies: rat NCITR* NCI-CG-TR-106,78. Reported in EPA TSCA Inventory.

OSHA PEL: CL 1000 ppm

ACGIH TLV: CL 1000 ppm; Not Classifiable as a Human Carcinogen

DFG MAK: 1000 ppm (5700 mg/m³)

SAFETY PROFILE: High concentrations cause narcosis and anesthesia in humans. Human systemic effects by inhalation: conjunctiva irritation, fibrosing alveolitis, and liver changes. Experimental poison by inhalation. Moderately toxic by intraperitoneal route. Reacts violently with aluminum, barium, or lithium. When heated to decomposition it emits highly toxic fumes of F⁻ and Cl⁻. Used as an aerosol propellant, refrigerant, and blowing agent for polymeric foams. See also CHLORINATED HYDROCARBONS, ALIPHATIC; and FLUORIDES.

T

TIQ250   CAS: 52-68-6   HR: 3
(2,2,2-TRICHLORO-1-HYDROXYETHYL)
DIMETHYLPHOSPHONATE
DOT: NA 2783
mf: $C_4H_8Cl_3O_4P$ mw: 257.44
PROP: Crystals. Sol in C_6H_6, EtOH; mod
sol in H_2O; spar sol in Et_2O and pet ether.
SYNS: AEROL 1 (pesticide) □ AGROFOROTOX □
ANTHON □ BAY 15922 □ BAYER 15922 □ BAYER L
13/59 □ BILARCIL □ BOVINOX □ BRITON □ BRITTEN
□ CEKUFON □ CHLORAK □ CHLORFOS □
CHLOROFOS □ CHLOROFTALM □ CHLOROPHOS □
CHLOROPHTHALM □ CHLOROXYPHOS □ CICLOSOM
□ CLOROFOS (RUSSIAN) □ COMBOT EQUINE □
DANEX □ DEP (pesticide) □ DEPTHON □ DETF □ DI-
METHOXY-2,2,2-TRICHLORO-1-HYDROXY-ETHYL-
PHOSPHINE OXIDE □ O,O-DIMETHYL-(1-HYDROXY-
2,2,2-TRICHLORAETHYL)-PHOSPHAT (GERMAN) □
O,O-DIMETHYL-(1-HYDROXY-2,2,2-TRICHLORAETHYL-
)PHOSPHONSAEURE ESTER (GERMAN) □ O,O-DI-
METHYL-(1-HYDROXY-2,2,2-TRICHLORO)ETHYL
PHOSPHATE □ DIMETHYL-1-HYDROXY-2,2,2-TRI-
CHLOROETHYL PHOSPHONATE □ O,O-DIMETHYL-(1-
HYDROXY-2,2,2-TRICHLOROETHYL)PHOSPHONATE □
O,O-DIMETHYL-1-OXY-2,2,2-TRICHLOROETHYL
PHOSPHONATE □ O,O-DIMETHYL-(2,2,2-TRICHLOOR-
1-HYDROXY-ETHYL)-FOSFONAAT (DUTCH) □ O,O-DI-
METHYL-(2,2,2-TRICHLOR-1-HYDROXY-AETHYL)-
PHOSPHONAT (GERMAN) □ DIMETHYLTRICHLORO-
HYDROXYETHYL PHOSPHONATE □ DIMETHYL-2,2,2-
TRICHLORO-1-HYDROXYETHYLPHOSPHONATE □
O,O-DIMETHYL-2,2,2-TRICHLORO-1-HYDROXYETHYL
PHOSPHONATE □ O,O-DIMETIL-(2,2,2-TRICLORO-1-
IDROSSI-ETIL)-FOSFONATO (ITALIAN) □ DIMETOX □
DIPTERAX □ DIPTEREX □ DIPTEREX 50 □ DIPTEVUR
□ DITRIFON □ DYLOX □ DYLOX-METASYSTOX-R □
DYREX □ DYVON □ ENT 19,763 □ EQUINO-ACID □
EQUINO-AID □ FLIBOL E □ FLIEGENTELLER □
FOROTOX □ FOSCHLOR □ FOSCHLOREM (POLISH) □
FOSCHLOR R-50 □ 1-HYDROXY-2,2,2-TRICHLOROETH-
YLPHOSPHONIC ACID DIMETHYL ESTER □ HYPO-
DERMACID □ LEIVASOM □ LOISOL □ MASOTEN □
MAZOTEN □ METHYL CHLOROPHOS □ METIFONATE
□ METRIFONATE □ METRIPHONATE □ NCI-C54831 □
NEGUVON □ NEGUVON A □ PHOSCHLOR R50 □
POLFOSCHLOR □ PROXOL □ RICIFON □ RITSIFON □
SATOX 20WSC □ SOLDEP □ SOTIPOX □ TRICHLO-
ORFON (DUTCH) □ TRICHLORFON (USDA, ACGIH) □
2,2,2-TRICHLORO-1-HYDROXYETHYL-PHOSPHONATE,
DIMETHYL ESTER □ (2,2,2-TRICHLORO-1-HYDRO-
XYETHYL)PHOSPHONIC ACID DIMETHYL ESTER □
TRICHLOROPHON □ TRICHLORPHENE □ TRI-
CHLORPHON □ TRICHLORPHON FN □ TRINEX □
TUGON □ TUGON FLY BAIT □ TUGON STABLE SPRAY
□ VERMICIDE BAYER 2349 □ VOLFARTOL □ VOTEXIT
□ WEC 50 □ WOTEXIT

CONSENSUS REPORTS: IARC Cancer
Review: Group 3 IMEMDT 7,56,87; Animal
Inadequate Evidence IMEMDT 30,207,83.
Community Right-To-Know List. EPA
Genetic Toxicology Program.
ACGIH TLV: TWA 1 mg/m³; Not
Classifiable as a Human Carcinogen
SAFETY PROFILE: Poison by ingestion,
inhalation, intraperitoneal, subcutaneous,
intravenous, and intramuscular routes.
Moderately toxic by skin contact. Human
systemic effects: true cholinesterase.
Experimental teratogenic and reproductive
effects. Questionable carcinogen with
experimental carcinogenic and tumorigenic
data. Human mutation data reported. An eye
irritant. When heated to decomposition it
emits very toxic fumes of Cl^- and PO_x.

TIQ750   CAS: 87-90-1   HR: 2
N,N',N''-TRICHLOROISOCYANURIC
ACID
DOT: UN 2468
mf: $C_3Cl_3N_3O_3$ mw: 232.41
PROP: White crystals; chlorine odor. Mp:
248°. Moderately sol in water; sol in
chlorinated solvs, very polar solvs.
SYNS: ACL 85 □ CBD 90 □ FICHLOR 91 □ FI CLOR 91
□ ISOCYANURIC CHLORIDE □ KYSELINA
TRICHLOISOKYANUROVA (CZECH) □ NSC-405124 □
SYMCLOSEN □ SYMCLOSENE □ TRICHLORINATED
ISOCYANURIC ACID □ TRICHLOROCYANURIC ACID □
TRICHLOROISOCYANIC ACID □ TRICHLOROISOCY-
ANURIC ACID □ 1,3,5-TRICHLOROISOCYANURIC ACID
□ TRICHLORO-s-TRIAZINETRIONE □ 1,3,5-TRI-
CHLORO-1,3,5-TRIAZINETRIONE □ TRICHLORO-s-
TRIAZINE-2,4,6(1H,3H,5H)-TRIONE □ 1,3,5-TRICHLORO-
2,4,6-TRIOXOHEXAHYDRO-s-TRIAZINE
CONSENSUS REPORTS: Reported in EPA
TSCA Inventory.
DOT CLASSIFICATION: 5.1; Label: Oxidizer
SAFETY PROFILE: Moderately toxic to
humans and experimentally by ingestion.
Mildly toxic experimentally by skin contact.
Human systemic effects by ingestion:
ulceration or bleeding from stomach. A
severe skin and eye irritant. Toxicity
symptoms include emaciation, lethargy,
weakness, and delayed death. Autopsy

shows inflammation of gastrointestinal tract, liver discoloration, and kidney hyperemia.

A powerful oxidizer. Forms an explosive product with cyanuric acid + sodium hydroxide. Potentially violent reaction with combustible materials. When heated to decomposition it emits very toxic fumes of Cl⁻ and NO$_x$. Used to chlorinate swimming pools.

TIR900   CAS: 5216-25-1   HR: 3
p-TRICHLOROMETHYLCHLORO- BENZENE

mf: $C_7H_4Cl_4$ mw: 229.91

SYNS: BENZENE, 1-CHLORO-4-(TRICHLOROMETH-YL)-(9CI) □ p-CHLOROBENZOTRICHLORIDE □ 4-CHLOROBENZOTRICHLORIDE □ p-CHLOROPHENYL-TRICHLOROMETHANE □ 1-CHLORO-4-(TRICHLORO-METHYL)BENZENE □ p,α-α-α-TETRACHLOROTOLU-ENE □ α-α-α-4-TETRACHLOROTOLUENE □ TOLU-ENE, α-α-α-p-TETRACHLORO-

CONSENSUS REPORTS: Reported in EPA TSCA Inventory.

DFG MAK: Animal Carcinogen, Suspected Human Carcinogen

SAFETY PROFILE: Suspected carcinogen. A poison by inhalation. Moderately toxic by ingestion and skin contact. When heated to decomposition it emits toxic vapors of Cl⁻.

TIR920   CAS: 503-38-8   HR: 2
TRICHLOROMETHYL CHLOROFORMATE

mf: $C_2Cl_4O_2$ mw: 197.82

SYNS: CARBONOCHLORIDIC ACID TRICHLORO-METHYL ESTER □ DIFOSGEN □ DIPHOSGEN □ DI-PHOSGENE □ FORMIC ACID, CHLORO-, TRICHLORO-METHYL ESTER □ METHANOL, TRICHLORO-, CHLOROFORMATE □ TRICHLORMETHYLESTER KYSELINY CHLORMRAVENCI

CONSENSUS REPORTS: Reported in EPA TSCA Inventory.

DOT CLASSIFICATION: 6.1; Label: Poison, Corrosive

SAFETY PROFILE: Low toxicity by inhalation. A corrosive liquid. When heated to decomposition it emits toxic vapors of Cl⁻.

TIT250   CAS: 133-07-3   HR: 3
N-(TRICHLOROMETHYLTHIO)PHTHAL-IMIDE

mf: $C_9H_4Cl_3NO_2S$ mw: 296.55

PROP: Crystals. Mp: 177°.

SYNS: FOLPAN □ FOLPET □ FTALAN □ ORTHO-PHALTAN □ PHALTAN □ PHTHALTAN □ THIOPHAL □ N-(TRICHLOR-METHYLTHIO)-PHTHALAMID -(GERMAN) □ N-(TRICHLOROMETHYLMERCAPTO)-PHTHALIMIDE □ 2-((TRICHLOROMETHYL)THIO)-1H-ISOINDOLE-1,3(2H)-DIONE □ TROYSAN ANTI-MILDEW O

CONSENSUS REPORTS: Reported in EPA TSCA Inventory. EPA Genetic Toxicology Program.

SAFETY PROFILE: Poison by intraperitoneal route. Moderately toxic by ingestion. Questionable carcinogen with experimental tumorigenic and teratogenic data. Experimental reproductive effects. Human mutation data reported. When heated to decomposition it emits very toxic fumes of Cl⁻, NO$_x$, and SO$_x$. Used as a fungicide.

TIT500   CAS: 1321-65-9   HR: 3
TRICHLORONAPHTHALENE

mf: $C_{10}H_5Cl_3$ mw: 231.50

PROP: A white solid.

SYNS: HALOWAX □ NIBREN WAX □ SEEKAY WAX

CONSENSUS REPORTS: Reported in EPA TSCA Inventory.

OSHA PEL: TWA 5 mg/m³ (skin)

ACGIH TLV: TWA 5 mg/m³ (skin)

DFG MAK: 5 mg/m³

SAFETY PROFILE: A poison. The chlorinated naphthalenes have toxic effects on the skin and liver. See also CHLOR-INATED HYDROCARBONS, ALIPHATIC; and POLYCHLORINATED BIPHENYLS.

TIV750   CAS: 95-95-4   HR: 3
2,4,5-TRICHLOROPHENOL

mf: $C_6H_3Cl_3O$ mw: 197.44

PROP: Colorless needles or gray flakes from pet ether; strong phenolic odor. Mp: 68°, bp: 252°, d: 1.678 @ 25°/4°, vap press: 1

mm @ 72.0°. Insol in water; sol in CCl₄, alc, benzene, and ether.

SYNS: COLLUNOSOL □ DOWICIDE 2 □ DOWICIDE B □ NCI-C61187 □ NURELLE □ PREVENTOL I □ RCRA WASTE NUMBER U230

CONSENSUS REPORTS: IARC Cancer Review: Human Limited Evidence IMEMDT 41,319,86; Animal Inadequate Evidence IMEMDT 20,349,79. Chlorophenol compounds are on the Community Right-To-Know List. Reported in EPA TSCA Inventory.

SAFETY PROFILE: Suspected carcinogen with experimental neoplastigenic data. Poison by intraperitoneal and intravenous routes. Moderately toxic by ingestion and subcutaneous routes. Experimental reproductive effects. Mutation data reported. When heated to decomposition it emits toxic fumes of Cl⁻ and explodes. See also CHLOROPHENOLS.

TIW000   CAS: 88-06-2   HR: 3
2,4,6-TRICHLOROPHENOL

mf: C₆H₃Cl₃O   mw: 197.44

PROP: Colorless needles from AcOH or yellow solid; strong phenolic odor. Mp: 68°, bp: 244.5°, fp: 62°, d: 1.490 @ 75°/4°, vap press: 1 mm @ 76.5°. Sol in water; very sol in alc and ether.

SYNS: DOWICIDE 2S □ NCI-C02904 □ OMAL □ PHENACHLOR □ RCRA WASTE NUMBER U231 □ 2,4,6-TRICHLORFENOL (CZECH)

CONSENSUS REPORTS: NTP 10th Report on Carcinogens. IARC Cancer Review: Animal Inadequate Evidence IMEMDT 20,349,79; Human Limited Evidence IMEMDT 41,319,86. NCI Carcinogenesis Bioassay (feed); Clear Evidence: mouse, rat NCITR* NCI-CG-TR-155,79. Chlorophenol compounds are on the Community Right-To-Know List. Reported in EPA TSCA Inventory. EPA Genetic Toxicology Program.

SAFETY PROFILE: Confirmed carcinogen with experimental carcinogenic data. Poison by intraperitoneal route. Moderately toxic by ingestion and skin contact. A skin and severe eye irritant. Experimental

reproductive effects. Mutation data reported. When heated to decomposition it emits toxic fumes of Cl⁻. Used as a germicide and preservative. See also CHLOROPHENOLS.

TIX500   CAS: 93-72-1   HR: 3
α-(2,4,5-TRICHLOROPHENOXY)-
PROPIONIC ACID

mf: C₉H₇Cl₃O₃   mw: 269.51

PROP: Crystals. Mp: 182°. Sltly water-sol.

SYNS: ACIDE 2-(2,4,5-TRICHLORO-PHENOXY) PROPIONIQUE (FRENCH) □ ACIDO 2-(2,4,5-TRICLORO-FENOSSI)-PROPIONICO (ITALIAN) □ AMCHEM 2,4,5-TP □ AQUA-VEX □ COLOR-SET □ DED-WEED □ DOUBLE STRENGTH □ FENOPROP □ FENORMONE □ FRUITONE T □ HERBICIDES, SILVEX □ KURAN □ KURON □ KUROSAL □ MILLER NU SET □ PROPON □ RCRA WASTE NUMBER U233 □ SILVEX (USDA) □ SILVI-RHAP □ STA-FAST □ 2,4,5-TC □ 2,4,5-TCPPA □ 2,4,5-TP □ 2-(2,4,5-TRICHLOOR-FENOXY)-PROPIONZUUR -(DUTCH) □ 2-(2,4,5-TRICHLOROPHENOXY)PROPIONIC ACID □ 2,4,5-TRICHLOROPHENOXY-α-PROPIONIC ACID □ 2-(2,4,5-TRICHLOR-PHENOXY)-PROPION-SAEURE (GERMAN) □ WEED-B-GON

CONSENSUS REPORTS: IARC Cancer Review: Human Limited Evidence IMEMDT 41,357,86.

SAFETY PROFILE: A suspected carcinogen. Poison by ingestion. An experimental teratogen. When heated to decomposition it emits toxic fumes of Cl⁻.

TIY800   CAS: 4511-19-7   HR: 2
2,4,6-TRICHLOROPHENYL CHLORO-
FORMATE

mf: C₇H₂Cl₄O₂   mw: 259.89

SYNS: CARBONOCHLORIDIC ACID, 2,4,6-TRICHLO-ROPHENYL ESTER (9CI) □ FORMIC ACID, CHLORO-, 2,4,6-TRICHLOROPHENYL ESTER □ PHENOL, 2,4,6-TRICHLORO-, CHLOROFORMATE □ TL 399

DOT CLASSIFICATION: 6.1; Label: Poison, Corrosive

SAFETY PROFILE: Low toxicity by inhalation. A corrosive liquid. When heated to decomposition it emits toxic vapors of Cl⁻.

TJA750   CAS: 98-13-5   HR: 3
TRICHLOROPHENYLSILANE

DOT: UN 1804

mf: $C_6H_5Cl_3Si$ mw: 211.55
PROP: Liquid. Bp: 201°, d: 1.321, refr index: 1.5247.
SYNS: PHENYLSILICON TRICHLORIDE □ PHENYL TRICHLOROSILANE (DOT) □ SILICON PHENYL TRICHLORIDE
CONSENSUS REPORTS: EPA Extremely Hazardous Substances List. Reported in EPA TSCA Inventory.
DOT CLASSIFICATION: 8; Label: Corrosive
SAFETY PROFILE: Poison by inhalation and intravenous routes. Moderately toxic by ingestion and skin contact. A corrosive irritant to skin, eyes, and mucous membranes. When heated to decomposition it emits toxic fumes of Cl⁻. See also CHLOROSILANES.

TJB600   CAS: 96-18-4   HR: 3
1,2,3-TRICHLOROPROPANE
mf: $C_3H_5Cl_3$ mw: 147.43
PROP: Bp: 158°, d: 1.414 @ 20°/20°, flash p: 180°F (OC).
SYNS: ALLYL TRICHLORIDE □ GLYCEROL TRICHLOROHYDRIN □ GLYCERYL TRICHLOROHYDR-IN □ NCI-C60220 □ TRICHLOROHYDRIN
CONSENSUS REPORTS: NTP 10th Report on Carcinogens. Reported in EPA TSCA Inventory.
OSHA PEL: TWA 10 ppm
ACGIH TLV: TWA 10 ppm (skin); Animal Carcinogen
DFG MAK: Confirmed Animal Carcinogen, Suspected Human Carcinogen
SAFETY PROFILE: Confirmed carcinogen. Poison by ingestion. Moderately toxic by inhalation and skin contact. Experimental reproductive effects. A skin and severe eye irritant. Mutation data reported. Moderately flammable by heat, flames (sparks), or powerful oxidizers. See also ALLYL COMPOUNDS and CHLORINATED HYDROCARBONS, ALIPHATIC. When heated to decomposition it yields highly toxic Cl⁻. To fight fire, use water (as a blanket), spray, mist, dry chemical.

TJD500   CAS: 10025-78-2   HR: 3
TRICHLOROSILANE

DOT: UN 1295
mf: Cl_3HSi mw: 135.45
PROP: Colorless, very volatile fuming liquid. Mp: −126.5°, bp: 31.8°, flash p: −18.4°F -(OC), d: 1.35 @ 0°, vap press: 400 mm @ 14.5°, vap d: 4.7, autoign temp: 219°F. Sol in benzene, carbon disulfide, chloroform, carbon tetrachloride. Fumes in air. Decomp in water.
SYNS: SILICI-CHLOROFORME (FRENCH) □ SILI-CIUMCHLOROFORM (GERMAN) □ SILICO-CHLOROFORM □ TRICHLOORSILAAN (DUTCH) □ TRICHLOROMONOSILANE □ TRICHLORSILAN - (GERMAN) □ TRICLOROSILANO (ITALIAN)
CONSENSUS REPORTS: Reported in EPA TSCA Inventory.
DOT CLASSIFICATION: 4.3; Label: Dangerous When Wet, Flammable Liquid, Corrosive
SAFETY PROFILE: Moderately toxic by ingestion and inhalation. A corrosive irritant to skin, eyes, and mucous membranes. A very dangerous fire hazard when exposed to heat, flame, or by chemical reaction. May be ignited by spark or impact. Spontaneously flammable in air. Explosive reaction with acetonitrile + diphenyl sulfoxide. Will react with water or steam to produce heat and toxic and corrosive fumes. Can react vigorously with oxidizing materials. To fight fire, use CO₂, dry chemical. When heated to decomposition it emits toxic fumes of Cl⁻. See also CHLOROSILANES.

TJD750   CAS: 108-77-0   HR: 3
2,4,6-TRICHLOROTRIAZINE
DOT: UN 2670
mf: $C_3Cl_3N_3$ mw: 184.41
PROP: Monoclinic, colorless crystals from C_6H_6; pungent odor. Mp: 154°, bp: 190°, d: 1.32 @ 20°/4°, vap press: 2 mm @ 70°, vap d: 6.36. Contains 96.9% cyanuric chloride; the remainder is cyanuric acid -(VOONAW 12(4),78,66).
SYNS: CHLOROTRIAZINE □ CYANURCHLORIDE □ CYANURIC ACID CHLORIDE □ CYANURIC CHLORIDE (DOT) □ CYANURIC TRICHLORIDE (DOT) □ CYANUR-YL CHLORIDE □ KYANURCHLORID (CZECH) □ s-TRIAZINE TRICHLORIDE □ TRICHLOROCYANIDINE

□ TRICHLORO-s-TRIAZINE □ sym-TRICHLOROTRIAZ-INE □ 1,3,5-TRICHLOROTRIAZINE □ 2,4,6-TRICHLORO-s-TRIAZINE □ 2,4,6-TRICHLORO-1,3,5-TRIAZINE □ sym-TRICHLOTRIAZIN (CZECH) □ TRICYANOGEN CHLORIDE
CONSENSUS REPORTS: Reported in EPA TSCA Inventory.
DOT CLASSIFICATION: 8; Label: Corrosive
SAFETY PROFILE: Poison by ingestion, inhalation, and intravenous routes. Questionable carcinogen with experimental tumorigenic data. Experimental reproductive effects. A corrosive. A skin and severe eye irritant. An allergen. Has been reported as causing irritation of mucous membranes and heart rhythm disturbances in humans. Violent reaction with water - (above 30°C), acetone + water, methanol, methanol + sodium hydrogen carbonate, 2-ethoxyethanol, dimethyl formamide, 3-butanone + sodium hydroxide + water, allyl alcohol + sodium hydroxide + water (at 28°C). When heated to decomposition it emits toxic fumes of Cl⁻ and NOₓ. See also CHLORIDES.

TJF250   CAS: 540-09-0   HR: 3
12-TRICOSANONE
mf: $C_{23}H_{46}O$ mw: 338.69
PROP: Scales or plates. D: 0.809, mp: 69°. Insol in water; sol in alc.
SYN: DI-n-UNDECYL KETONE
CONSENSUS REPORTS: Reported in EPA TSCA Inventory.
DOT CLASSIFICATION: 3; Label: Flammable Liquid
SAFETY PROFILE: Poison by intravenous route. A flammable liquid. When heated to decomposition it emits acrid smoke and irritating fumes. See also KETONES.

TJJ400   CAS: 7774-82-5   HR: 1
2-TRIDECENAL
mf: $C_{13}H_{24}O$ mw: 196.37
PROP: White to yellow liquid; oily, citrus odor. D: 0.842–0.862, refr index: 1.457. Sol in alc, fixed oils; insol in water.
SYN: FEMA No. 3082

CONSENSUS REPORTS: Reported in EPA TSCA Inventory.
SAFETY PROFILE: Low toxicity by ingestion and skin contact. When heated to decomposition it emits acrid smoke and irritating fumes.

TJL700   CAS: 5870-82-6   HR: 3
1,1,3-TRIETHOXYBUTANE
mf: $C_{10}H_{22}O_3$ mw: 190.32
SYNS: BUTANE, 1,1,3-TRIETHOXY- □ BUTYRALDEHYDE, 3-ETHOXY-, DIETHYL ACETAL
SAFETY PROFILE: A poison by ingestion and skin contact. Low toxicity by inhalation. A mild skin irritant. When heated to decomposition it emits acrid smoke and irritating vapors.

TJN750   CAS: 97-93-8   HR: 3
TRIETHYLALUMINUM
mf: $C_6H_{15}Al$ mw: 114.19
PROP: Fp: −52.5°, d: 0.837 @ 20°, vap press: 4 mm @ 83°, flash p: <−63°F, bp: 194°.
SYNS: ALUMINUM, TRIETHYL- □ TEA □ TRIETHYL-ALUMINUM
CONSENSUS REPORTS: Reported in EPA TSCA Inventory.
ACGIH TLV: TWA 2 mg(Al)/m³
SAFETY PROFILE: Extremely destructive to living tissue. A very dangerous fire hazard when exposed to heat or flame. Ignites spontaneously in air. Explodes violently in water. To fight fire, use CO_2, dry sand, dry chemical. Do not use water, foam, or halogenated fire-fighting agents. Explosive reaction with alcohols (e.g., methanol, ethanol, propanol), carbon tetrachloride, N,N-dimethylformamide + heat. Incompatible with halogenated hydrocarbons; triethyl borane. When heated to decomposition it emits acrid smoke and irritating fumes. See also ALUMINUM COMPOUNDS and ORGANOMETALS.

TJO000   CAS: 121-44-8   HR: 3
TRIETHYLAMINE
DOT: UN 1296
mf: $C_6H_{15}N$ mw: 101.22

PROP: Colorless liquid with fishy or ammonia odor. Mp: −114.8°, bp: 89.5°, flash p: 20°F (OC), d: 0.7255 @ 25°/4°, vap d: 3.48, lel: 1.2%, uel: 8.0%. Misc in water, alc, ether. IDLH 200 ppm.
SYNS: (DIETHYLAMINO)ETHANE □ N,N-DIETHYL-ETHANAMINE □ TEN □ TRIAETHYLAMIN (GERMAN) □ TRIETILAMINA (ITALIAN)
CONSENSUS REPORTS: Reported in EPA TSCA Inventory.
OSHA PEL: TWA 10 ppm; STEL 15 ppm
ACGIH TLV: TWA 1 ppm; STEL 3 ppm - (skin); Not Classifiable as a Human Carcinogen
DFG MAK: 1 ppm (4.2 mg/m^3)
DOT CLASSIFICATION: 3; Label: Flammable Liquid
SAFETY PROFILE: Moderately toxic by ingestion and skin contact. Mildly toxic by inhalation. Human systemic effects: visual field changes. Experimental reproductive effects. Mutation data reported. A skin and severe eye irritant. Can cause kidney and liver damage. A very dangerous fire hazard when exposed to heat, flame, or oxidizers. Explosive in the form of vapor when exposed to heat or flame. Complex with dinitrogen tetraoxide explodes below 0°C when undiluted with solvent. Exothermic reaction with maleic anhydride above 150°C. Can react with oxidizing materials. Incompatible with N_2O_4. To fight fire, use CO_2, dry chemical, alcohol foam. When heated to decomposition it emits toxic fumes of NO_x.

TJP250   CAS: 97-94-9   HR: 3
TRIETHYLBORANE
mf: $C_6H_{15}B$ mw: 98.02
PROP: Colorless, fuming liquid. Mp: −93°, d: 0.6961 @ 23°, bp: 95°.
SYN: TRIETHYLBORINE
CONSENSUS REPORTS: Reported in EPA TSCA Inventory.
SAFETY PROFILE: Poison by ingestion and intraperitoneal routes. Mildly toxic by inhalation. Animal experiments show that the vapor is a poison which causes

pulmonary irritation and convulsions. A very dangerous fire hazard by spontaneous chemical reaction with oxidizers. Spontaneously flammable in air. Explodes in oxygen atmospheres. Hypergolic reaction with triethylaluminum. Ignites on contact with chlorine, bromine, or other halogens. Will react with water or steam to produce toxic and flammable vapors. To fight fire, do NOT use halogenated extinguishing agents. When heated to decomposition or upon contact with air it emits toxic acrid smoke and irritating fumes. See also BORANES and BORON COMPOUNDS.

TJP750   CAS: 77-93-0   HR: 2
TRIETHYL CITRATE
mf: $C_{12}H_{20}O_7$ mw: 276.32
PROP: Colorless oily liquid; odorless. Bp: 294°, flash p: 303°F (COC), d: 1.136 @ 25°, vap press: 1 mm @ 107.0°. Sltly sol in water; misc in alc, ether.
SYNS: CITROFLEX 2 □ ETHYL CITRATE □ 2-HYDRO-XY,1,2,3-PROPANETRICARBOXYLIC ACID, TRIETHYL ESTER □ TEC
CONSENSUS REPORTS: Reported in EPA TSCA Inventory.
SAFETY PROFILE: Moderately toxic by intraperitoneal route. Mildly toxic by ingestion and inhalation. Combustible liquid when exposed to heat or flame. To fight fire, use dry chemical, CO_2. When heated to decomposition it emits acrid smoke and irritating fumes. See also ESTERS and CITRIC ACID.

TJP775   CAS: 12075-68-2   HR: 3
TRIETHYLDIALUMINUM TRICHLORIDE
mf: $C_6H_{15}Al_2Cl_3$ mw: 247.51
PROP: Yellow liquid. Bp: 114.5–116.5° @ 50 mm.
SYNS: ETHYLALUMINUM SESQUICHLORIDE □ SESQUIETHYLALUMINUM CHLORIDE □ TRICHLOROTRIETHYLDIALUMINIUM □ TRICHLOROTRIETHYLDIALUMINUM □ TRIETHYL-ALUMINUM SESQUICHLORIDE □ TRIETHYL-TRICHLORODIALUMINUM
CONSENSUS REPORTS: Reported in EPA TSCA Inventory.

ACGIH TLV: TWA 2 mg(Al)/m^3
SAFETY PROFILE: Mixtures with carbon tetrachloride explode at room temperature. When heated to decomposition it emits toxic fumes of Cl$^-$. See also ALUMINUM COMPOUNDS.

TJQ000   CAS: 112-27-6   HR: 3
TRIETHYLENE GLYCOL
mf: C$_6$H$_{14}$O$_4$ mw: 150.20
PROP: Odorless, colorless liquid; hygroscopic. Fp: −7.3°, flash p: 350°F, mp: −4.3°, d: 1.122 @ 25°/25°, lel: 0.9%, uel: 9.2%, autoign temp: 700°F, vap press: 1 mm @ 114°, vap d: 5.17, bp: 285°. Misc in water, alc, benzene; insol in pet ether; very sltly sol in ether.
SYNS: DI-β-HYDROXYETHOXYETHANE □ 3,6-DI-OXAOCTANE-1,8-DIOL □ 2,2'-(1,2-ETHANEDIYLBIS(O-XY))BISETHANOL □ 2,2'-ETHYLENEDIOXYDI-ETHANOL □ 2,2'-ETHYLENEDIOXYETHANOL □ ETH-YLENE GLYCOL-BIS-(2-HYDROXYETHYL ETHER) □ ETHYLENE GLYCOL DIHYDROXYDIETHYL ETHER □ GLYCOL BIS(HYDROXYETHYL) ETHER □ TEG □ TRIGEN □ TRIGLYCOL
CONSENSUS REPORTS: Glycol ether compounds are on the Community Right-To-Know List. Reported in EPA TSCA Inventory.
SAFETY PROFILE: Poison by intravenous route. Mildly toxic to humans by ingestion. Experimental reproductive effects. An eye and skin irritant. Many glycol ether compounds have dangerous human reproductive effects. Combustible when exposed to heat or flame. Can react with oxidizing materials. Explosive in the form of vapor when exposed to heat, flame, or spark. To fight fire, use alcohol foam, dry chemical. When heated to decomposition it emits acrid smoke and irritating fumes. See also ESTERS and GLYCOL ETHERS.

TJQ100   CAS: 1680-21-3   HR: 2
TRIETHYLENE GLYCOL DIACRYLATE
mf: C$_{12}$H$_{18}$O$_6$ mw: 258.30
SYNS: ACRYLIC ACID, DIESTER with TRIETHYLENE GLYCOL □ 2-PROPENOIC ACID, 1,2-ETHANEDIYLBIS-(OXY-2,1-ETHANEDIYL) ESTER (9CI)

CONSENSUS REPORTS: Reported in EPA TSCA Inventory.
SAFETY PROFILE: Moderately toxic by ingestion and skin contact. Questionable carcinogen with experimental tumorigenic data. Severe skin and eye irritant. When heated to decomposition it emits acrid smoke and irritating fumes.

TJR000   CAS: 112-24-3   HR: 3
TRIETHYLENETETRAMINE
DOT: UN 2259
mf: C$_6$H$_{18}$N$_4$ mw: 146.28
PROP: Moderately viscous, yellowish liquid or oil. Bp: 272°, mp: 12°, flash p: 275°F, d: 0.982, vap press: <0.01 mm @ 20°, autoign temp: 640°F. Very sol in water and ether.
SYNS: ARALDITE HARDENER HY 951 □ ARALDITE HY 951 □ N,N'-BIS(2-AMINOETHYL)-1,2-DIAM-INOETHANE □ N,N'-BIS(2-AMINOETHYL)ETHYL-ENEDIAMINE □ N,N'-BIS(2-AMINOETHYL)-1,2-ETHYL-ENEDIAMINE □ DEH 24 □ 3,6-DIAZAOCTANE-1,8-DI-AMINE □ HY 951 □ TECZA □ TETA □ 1,4,7,10-TETRAAZADECANE □ TRIEN □ TRIENTINE
CONSENSUS REPORTS: Reported in EPA TSCA Inventory.
DOT CLASSIFICATION: 8; Label: Corrosive
SAFETY PROFILE: Poison by intravenous route. Moderately toxic by ingestion and skin contact. An experimental teratogen. Experimental reproductive effects. Mutation data reported. A corrosive irritant to skin, eyes, and mucous membranes. Causes skin sensitization. Combustible when exposed to heat or flame. Ignites on contact with cellulose nitrate of high surface area. Can react with oxidizing materials. To fight fire, use CO$_2$, dry chemical, alcohol foam. When heated to decomposition it emits toxic fumes of NO$_x$.

TJS250   CAS: 1067-14-7   HR: 3
TRIETHYL LEAD CHLORIDE
mf: C$_6$H$_{15}$ClPb   mw: 329.85
PROP: White crystals. Mp: 172° (decomp).
SYN: TRIETHYLCHLOROPLUMBANE
CONSENSUS REPORTS: Lead and its compounds are on the Community Right-To-Know List. Reported in EPA TSCA

Inventory. EPA Genetic Toxicology Program.

SAFETY PROFILE: Poison by intraperitoneal, subcutaneous, parenteral, and intravenous routes. An experimental teratogen. Experimental reproductive effects. Mutation data reported. When heated to decomposition it emits very toxic fumes of Cl⁻ and Pb. See also LEAD COMPOUNDS and CHLORIDES.

TJT750   CAS: 78-40-0   HR: 2
TRIETHYL PHOSPHATE
mf: $C_6H_{15}O_4P$ mw: 182.18
PROP: Pleasant smelling liquid. Mp: $-56.5°$, flash p: 240°F (OC), d: 1.067-1.072 @ 20°/20°, vap press: 1 mm @ 39.6°, vap d: 6.28, bp: 215-216°. Sol in most org solvs, water, alc, ether.
SYNS: ETHYL PHOSPHATE □ TEP
CONSENSUS REPORTS: Reported in EPA TSCA Inventory.
SAFETY PROFILE: Moderately toxic by ingestion, intraperitoneal, and intravenous routes. Experimental reproductive effects. Mutation data reported. Causes cholinesterase inhibition, but to a lesser extent than parathion. May be expected to cause nerve injury similar to that of other phosphate esters. Combustible when exposed to heat or flame. Can react vigorously with oxidizing materials. To fight fire, use CO_2, dry chemical, alcohol foam. When heated to decomposition it emits toxic fumes of PO_x. See also PARATHION.

TJT800   CAS: 122-52-1   HR: 3
TRIETHYL PHOSPHITE
DOT: UN 2323
mf: $C_6H_{15}O_3P$ mw: 166.18
PROP: A liquid with characteristic, obnoxious, phosphite odor. Bp: 156°, refr index: 1.4130, d: 0.969, flash p: 130° F.
SYN: FOSFORYN TROJETYLOWY (CZECH)
CONSENSUS REPORTS: Reported in EPA TSCA Inventory.
DOT CLASSIFICATION: 3; Label: Flammable Liquid

SAFETY PROFILE: Moderately toxic by ingestion. A skin and eye irritant. Flammable liquid when exposed to heat, sparks, or flame. When heated to decomposition it emits toxic fumes of PO_x.

TJU850   CAS: 73926-90-6   HR: 3
TRIETHYLTIN BROMIDE-2-PIPECOLINE
mf: $C_6H_{15}BrSn•C_6H_{13}N$ mw: 385.01
SYNS: BROMOTRIETHYLSTANNANE compounded with 2-PIPECOLINE (1:1) □ STANNANE, BROMOTRIETHYL-, compounded with 2-PIPECOLINE (1:1)
OSHA PEL: TWA 0.1 mg(Sn)/m³ (skin)
ACGIH TLV: TWA 0.1 mg(Sn)/m³; STEL 0.2 mg/m³ (skin)
NIOSH REL: (Organotin Compound) 10H TWA 0.1 mg(Sn)/m³
SAFETY PROFILE: Poison by intravenous route. When heated to decomposition it emits toxic fumes of NO_x, Sn, and Br⁻.

TJX500   CAS: 354-32-5   HR: 2
TRIFLUOROACETYL CHLORIDE
DOT: UN 3057
SYN: PERFLUOROACETYL CHLORIDE
CONSENSUS REPORTS: Reported in EPA TSCA Inventory.
DOT CLASSIFICATION: 2.2; Label: Nonflammable Gas, Corrosive
SAFETY PROFILE: Corrosive to skin, eyes, and materials. When heated to decomposition it emits very toxic fumes of F⁻ and Cl⁻. See also FLUORIDES and CHLORIDES.

TJY100   CAS: 75-63-8   HR: 1
TRIFLUOROBROMOMETHANE
DOT: UN 1009
mf: $CBrF_3$ mw: 148.92
PROP: A gas. D: 1.58, fp: $-168°$, bp: $-57.8°$. IDLH 40,000 ppm.
SYNS: BROMOFLUOROFORM □ BROMOTRIFLUOROMETHANE □ F-13B1 □ FREON 13B1 □ HALON 1301 □ R13B1 (DOT) □ TRIFLUOROMONOBROMOMETHANE
CONSENSUS REPORTS: Reported in EPA TSCA Inventory.
OSHA PEL: TWA 1000 ppm
ACGIH TLV: TWA 1000 ppm

DFG MAK: 1000 ppm (6200 mg/m³)
DOT CLASSIFICATION: 2.2; Label:
Nonflammable Gas
SAFETY PROFILE: Mildly toxic by
inhalation. Incompatible with aluminum.
When heated to decomposition it emits
toxic fumes of F⁻ and Br⁻. See also
BROMIDES and FLUORIDES.

TJY175   CAS: 75-88-7   HR: 3
2,2,2-TRIFLUOROCHLOROETHANE
mf: $C_2H_2ClF_3$ mw: 118.49
PROP: A liquid. D: 1.389 @ 0°/4°, fp:
−105.5°, bp: 6.1°.
SYNS: CFC 133a □ 1-CHLORO-2,2,2-TRIFLUORO-
ETHANE □ 2-CHLORO-1,1,1-TRIFLUOROETHANE □ FC
133a □ FREON 133a □ GENETRON 133a □ R 133a □ 1,1,1-
TRIFLUORO-2-CHLOROETHANE □ 1,1,1-TRIFLUORO-
ETHYL CHLORIDE
CONSENSUS REPORTS: IARC Cancer
Review: Group 3 IMEMDT 7,56,87; Animal
Limited Evidence IMEMDT 41,253,86.
Reported in EPA TSCA Inventory.
SAFETY PROFILE: A poison by inhalation.
Experimental reproductive effects.
Questionable carcinogen with experimental
carcinogenic data. When heated to
decomposition it emits toxic fumes of F⁻.

TJY500   CAS: 306-83-2   HR: 3
1,1,1-TRIFLUORO-2,2-DICHLORO-
   ETHANE
mf: $C_2HCl_2F_3$ mw: 152.93
PROP: A liquid. D: 1.475 @ 15°/4°, mp:
−107°, bp: 28.7°.
SYNS: 2,2-DICHLORO-1,1,1-TRIFLUOROETHANE □
FC 123 □ FREON 123 □ R 123
CONSENSUS REPORTS: Reported in EPA
TSCA Inventory.
DFG MAK: Confirmed Animal Carcinogen
with Unknown Relevance to Humans
SAFETY PROFILE: Suspected carcinogen.
Moderately toxic by inhalation. When
heated to decomposition it emits very toxic
fumes of F⁻ and Cl⁻. See also CHLOR-
INATED HYDROCARBONS,
ALIPHATIC; and FLUORIDES.

TKA250   CAS: 76-05-1   HR: 3
TRIFLUOROETHANOIC ACID
DOT: UN 2699
mf: $C_2HF_3O_2$ mw: 114.03
PROP: Colorless liquid; strong pungent
odor. Mp: −15.25°, bp: 71.1° @ 734 mm, d:
1.535 @ 0°. Misc in water.
SYNS: KYSELINA TRIFLUOROCTOVA □ PERFLU-
OROACETIC ACID □ TRIFLUORACETIC ACID □
TRIFLUOROACETIC ACID (DOT)
CONSENSUS REPORTS: Reported in EPA
TSCA Inventory. EPA Genetic Toxicology
Program.
DOT CLASSIFICATION: 8; Label: Corrosive
SAFETY PROFILE: Poison by ingestion and
intraperitoneal routes. Moderately toxic by
intravenous route. Mildly toxic by
inhalation. A corrosive irritant to skin, eyes,
and mucous membranes. When heated to
decomposition it emits toxic fumes of F⁻.
Used as a strong organic acid catalyst.

TKA350   CAS: 75-89-8   HR: 3
2,2,2-TRIFLUOROETHANOL
mf: $C_2H_3F_3O$ mw: 100.05
PROP: Liquid. Bp: 103–105° @ 742 mm, d:
1.288, refr index: <1.3000.
SYN: TFE
CONSENSUS REPORTS: Reported in EPA
TSCA Inventory. EPA Genetic Toxicology
Program.
SAFETY PROFILE: Poison by ingestion,
intravenous, and intraperitoneal routes.
Moderately toxic by inhalation and skin
contact. Experimental reproductive effects.
A severe skin and eye irritant. When heated
to decomposition it emits toxic fumes of F⁻.

TKB250   CAS: 406-90-6   HR: 3
2,2,2-TRIFLUOROETHYL VINYL ETHER
mf: $C_4H_5F_3O$ mw: 126.09
PROP: A liquid. D: 1.14 @ 20°/4°, bp:
42.5° @ 751 mm.
SYNS: FLOROXENE □ FLUOOXENE □ FLUOROMAR
□ FLUOROXENE □ FLUORXENE □ FLURXENE □ -
(2,2,2-TRIFLUOROETHOXY)ETHENE
CONSENSUS REPORTS: Reported in EPA
TSCA Inventory. EPA Genetic Toxicology
Program.

NIOSH REL: (Waste Anesthetic Gases) CL 2 ppm/1H
SAFETY PROFILE: Poison by inhalation. Moderately toxic by intraperitoneal route. An experimental teratogen. Human systemic effects by inhalation: jaundice and liver function tests impaired. Experimental reproductive effects. Mutation data reported. When heated to decomposition it emits toxic fumes of F⁻. Used as an anesthetic. See also FLUORIDES and ETHERS.

TKB285   CAS: 101931-68-4 HR: 3
(3,3,3-TRIFLUORO-2-HYDROXY-2---(TRIFLUOROMETHYL))PROPYL BENZYL KETONE
mf: $C_{12}H_{10}F_6O_2$ mw: 300.22
SYN: 2-PENTANONE, 4-HYDROXY-1-PHENYL-5,5,5-TRIFLUORO-4-(TRIFLUOROMETHYL)-
DOT CLASSIFICATION: 3; Label: Flammable Liquid
SAFETY PROFILE: A poison by intravenous route. A flammable liquid. When heated to decomposition it emits toxic vapors of F⁻.

TKB310   CAS: 1493-13-6   HR: 3
TRIFLUOROMETHANE SULFONIC ACID
mf: CHF_3O_3S mw: 150.08
PROP: A liquid. Hygroscopic. Bp: 162°, refr index: 1.3270, d: 1.696, flash p: none.
SYN: TRIFLIC ACID
SAFETY PROFILE: A corrosive irritant to the skin, eyes, and mucous membranes. A strong acid. Violent reaction with acyl chlorides or aromatic hydrocarbons evolves toxic hydrogen chloride gas. When heated to decomposition it emits toxic fumes of F⁻ and SO$_x$. See also FLUORIDES.

TKB800   CAS: 312-73-2   HR: 3
2-TRIFLUOROMETHYL BENZIMID-AZOLE
mf: $C_8H_5F_3N_2$ mw: 186.15
SYNS: 1H-BENZIMIDAZOLE, 2-(TRIFLUOROMETHYL-)- □ BENZIMIDAZOLE, 2-TRIFLUOROMETHYL-
SAFETY PROFILE: A poison by ingestion and intravenous routes. When heated to decomposition it emits toxic vapors of NO$_x$ and F⁻.

TKH250   CAS: 59544-89-7   HR: 3
TRIFLUORO SELENIUM HEXAFLUORO ARSENATE
mf: AsF_9Se mw: 324.9
CONSENSUS REPORTS: Selenium and its compounds, as well as arsenic and its compounds, are on the Community Right-To-Know List.
OSHA PEL: TWA 0.01 mg(As)/m³; Cancer Hazard; TWA 0.2 mg(Se)/m³
ACGIH TLV: BEI: 35 μ (As)/L inorganic arsenic and methylated metabolites in urine; TWA 0.2 mg(Se)/m³
DFG MAK: DFG TRK: 0.2 mg/m³ calculated as arsenic in that portion of dust that can possibly be inhaled
NIOSH REL: CL 2 μg(As)/m³
SAFETY PROFILE: Arsenic compounds are poisons. Violent reaction with water. When heated to decomposition it emits very toxic fumes of As, F⁻, and Se. See also FLUORIDES, ARSENIC COMPOUNDS, and SELENIUM COMPOUNDS.

TKH300   CAS: 101913-67-1 HR: 3
TRIFLUOROSTANNITE HEXADECYL-AMINE
mf: $C_{16}H_{35}N•F_3HSn$ mw: 418.22
SYN: TRIFLUOROSTANNITE OF HEXADECYLAMINE
OSHA PEL: TWA 2 mg(Sn)/m³
ACGIH TLV: TWA 2 mg(Sn)/m³
SAFETY PROFILE: Poison by intravenous route. When heated to decomposition it emits toxic fumes of NO$_x$, Sn, and F⁻.

TKJ250   CAS: 329-01-1   HR: 3
(α,α,α-TRIFLUORO-m-TOLYL) ISOCYANATE
DOT: UN 2206/UN 2207/UN 2478/UN 3080
mf: $C_8H_4F_3NO$ mw: 187.13
PROP: Bp: 54° @ 11 mm, refr index: 1.4700, d: 1.359, flash p: 138° F.
SYNS: ISOCYANIC ACID, (m-TRIFLUOROMETHYL-PHENYL) ESTER □ TIC

CONSENSUS REPORTS: Reported in EPA TSCA Inventory.

DOT CLASSIFICATION: 6.1; Label: KEEP AWAY FROM FOOD (UN 2207); DOT Class: 6.1; Label: Poison (UN 2206); DOT Class: 6.1; Label: Poison, Flammable Liquid (UN 3080); DOT Class: 3; Label: Flammable Liquid, Poison (UN 2478)

SAFETY PROFILE: Moderately toxic by ingestion, inhalation, and intraperitoneal routes. A lachrymator. Flammable liquid when exposed to heat, sparks, or flame. When heated to decomposition it emits very toxic fumes of NO_x and F^-. See also ISOCYANATES and FLUORIDES.

TKK250   CAS: 440-17-5   HR: 3
TRIFLUPERAZINE DIHYDROCHLORIDE
mf: $C_{21}H_{24}F_3N_3S \bullet 2ClH$ mw: 480.46

PROP: Crystals from EtOH. Mp: 242–243°. Sol in H_2O.

SYNS: ESKAZINE □ ESKAZINE DIHYDROCHLORIDE □ FLUOPERAZINE □ JATRONEURAL □ 10-(3-(4-METH-YL-1-PIPERAZINYL)PROPYL)-2-TRIFLUOROMETHYL-PHENOTHIAZINE DIHYDROCHLORIDE □ SKF 5019 □ STELAZINE □ STELAZINE DIHYDROCHLORIDE □ TERFLUZINE □ TERFLUZINE DIHYDROCHLORIDE □ TRIFLORPERAZINE DIHYDROCHLORIDE □ TRIFLUO-PERAZINE HYDROCHLORIDE □ TRIFLUOROPERAZ-INE DIHYDROCHLORIDE □ TRIFLUOROPYRAZIN DI-HYDROCHLORIDE □ TRIFTAZIN □ TRIPHTHAZINE □ TRIPHTHAZINE DIHYDROCHLORIDE □ TRYPTAZINE DIHYDROCHLORIDE

CONSENSUS REPORTS: Reported in EPA TSCA Inventory. EPA Genetic Toxicology Program.

SAFETY PROFILE: Poison by intravenous and intraperitoneal routes. Moderately toxic by ingestion. Experimental reproductive effects. Human mutation data reported. When heated to decomposition it emits very toxic fumes of F^-, NO_x, SO_x, and HCl. See also FLUORIDES.

TKK500   CAS: 749-13-3   HR: 3
TRIFLUPERIDOL
mf: $C_{22}H_{23}F_4NO_2$ mw: 409.46

SYNS: 4'-FLUORO-4-(4-HYDROXY-4-(α,α,α-TRIFLU-ORO-m-TOLYL)PIPERIDINO)BUTYROPHENONE □ 4-FLUORO-4,4-IDROSSI-4-(m-TRIFLUOROMETIL-FENIL)-PIPERIDINO-BUTIRROFENONE (ITALIAN) □ 1-(4-

FLUOROPHENYL)-4-(4-HYDROXY-4-(3-(TRIFLUORO-METHYL)PHENYL)-1-PIPERIDINYL)-1-BUTANONE □ MCN-JR-2498 □ PSICOPERIDOL-R □ PSYCHOPERIDOL □ R-2498 □ TRIFLUPERIDOLO (ITALIAN) □ TRIPERIDOL

SAFETY PROFILE: Poison by ingestion, subcutaneous, and intraperitoneal routes. An experimental teratogen. Experimental reproductive effects. When heated to decomposition it emits very toxic fumes of F^- and NO_x. See also FLUORIDES and TRIFLUPERIDOL HYDROCHLORIDE.

TKL000   CAS: 146-54-3   HR: 3
TRIFLUPROMAZINE
mf: $C_{18}H_{19}F_3N_2S$ mw: 352.45

PROP: A viscous oil. Bp: 176°, refr index: 1.5780.

SYNS: 10-(3-(DIMETHYLAMINO)PROPYL)-2-(TRI-FLUOROMETHYL) PHENOTHIAZINE □ N,N-DIMETH-YL-2-(TRIFLUOROMETHYL)-10H-PHENOTHIAZINE-10-PROPANAMINE □ VESPRIN

CONSENSUS REPORTS: EPA Genetic Toxicology Program.

SAFETY PROFILE: Poison by ingestion, intravenous, and intraperitoneal routes. Experimental reproductive effects. Mutation data reported. When heated to decomposition it emits very toxic fumes of F^-, NO_x, and SO_x. See also FLUORIDES.

TKL100   CAS: 26644-46-2   HR: 1
TRIFORINE
mf: $C_{10}H_{14}Cl_6N_4O_2$ mw: 434.98

PROP: White crystals. Mp: 155°. Sol in water, CMF, DMSO.

SYNS: BIFORMYCHLORAZIN □ BIFORMYLCHLO-RAZIN □ N,N'-BIS(1-FORMAMIDO-2,2,2-TRICHLORO-ETHYL)PIPERAZINE □ 1,4-BIS(1-FORMAMIDO-2,2,2-TRICHLOROETHYL)PIPERAZINE □ CA 70203 □ CELA 50 □ CELA W 524 □ CME 74770 □ COMPOUND W □ CW 524 □ FORMAMIDE, N,N'-(1,4-PIPERAZINEDIYLBIS-(2,2,2-TRICHLOROETHYLIDENE))BIS-(8CI,9CI) □ FUNG-INEX □ N,N'-(PIPERAZINEDIYLBIS(2,2,2-TRICHLORO-ETHYLIDENE)) BIS(FORMAMIDE) □ SAPROL □ W 524

SAFETY PROFILE: Low toxicity by ingestion, inhalation, and skin contact. Human systemic effects: change in taste function. When heated to decomposition emits toxic fumes of NO_x and Cl^-.

TKO250   CAS: 1421-63-2   HR: 3
2′,4′,5′-TRIHYDROXYBUTYROPHENONE
mf: $C_{10}H_{12}O_4$ mw: 196.22
PROP: Yellow-tan crystals. Mp: 149–153°,
d: 6.0 lb/gal @ 20°. Very sltly sol in water;
sol in alc, propylene glycol.
SYNS: THBP □ 2,4,5-TRIHYDROXYBUTYROPHEN-
ONE □ USAF EK
CONSENSUS REPORTS: Reported in EPA
TSCA Inventory.
SAFETY PROFILE: Poison by
intraperitoneal route. Mutation data
reported. When heated to decomposition it
emits acrid smoke and irritating fumes. See
also KETONES.

TKP500   CAS: 102-71-6   HR: 2
TRIHYDROXYTRIETHYLAMINE
mf: $C_6H_{15}NO_3$ mw: 149.22
PROP: Hygroscopic, pale-yellow viscous
liquid. Mp: 21.6°, bp: 360°, flash p: 355°F -
(CC), d: 1.1258 @ 20°/20°, vap press: 10
mm @ 205°, vap d: 5.14.
SYNS: DALTOGEN □ NITRILO-2,2′,2″-TRIETHANOL □
2,2′,2″-NITRILOTRIETHANOL □ STEROLAMIDE □
THIOFACO T-35 □ TRIAETHANOLAMIN-NG □
TRIETHANOLAMIN □ TRIETHANOLAMINE (ACGIH) □
TRIETHYLOLAMINE □ TRI(HYDROXYETHYL)AMINE
□ 2,2′,2″-TRIHYDROXYTRIETHYLAMINE □ TRIS(2-
HYDROXYETHYL)AMINE □ TROLAMINE
CONSENSUS REPORTS: Reported in EPA
TSCA Inventory. EPA Genetic Toxicology
Program.
ACGIH TLV: TWA 0.5 mg/m³
SAFETY PROFILE: Moderately toxic by
intraperitoneal route. Mildly toxic by
ingestion. Liver and kidney damage have
been demonstrated in animals from chronic
exposure. A human and experimental skin
irritant. An eye irritant. Questionable
carcinogen with experimental carcinogenic
data. Combustible liquid when exposed to
heat or flame; can react vigorously with
oxidizing materials. To fight fire, use alcohol
foam, CO_2, dry chemical. When heated to
decomposition it emits toxic fumes of NO_x
and CN^-.

TKR050   CAS: 13904-39-7   HR: 3
TRIIODOPROPYLGERMANE
mf: $C_3H_7GeI_3$ mw: 496.39
SYNS: GERMANE, TRIIODOPROPYL- □ GERMANE,
PROPYLTRIIODO- □ PROPYLTRIIODOGERMANE □
TRIJOD-PROPYLGERMAN
SAFETY PROFILE: A poison by ingestion.
When heated to decomposition it emits
toxic vapors of Ge and I^-.

TKR500   CAS: 100-99-2   HR: 3
TRIISOBUTYLALUMINUM
mf: $(C_4H_9)_3Al$ mw: 198.3
PROP: Clear, colorless liquid. D: 0.7859 @
20°, mp: 6°, vap press: 1 mm @ 47°, flash p:
<4°, fp: 4.3°, bp: 86° @ 10 mm.
SYNS: ALUMINUM, TRIS(2-METHYLPROPYL)-(9CI) □
TRIISOBUTYLALANE □ TRIISOBUTYLALUMINIUM □
TRIS(2-METHYLPROPYL)ALUMINUM
CONSENSUS REPORTS: Reported in EPA
TSCA Inventory.
ACGIH TLV: TWA 2 mg(Al)/m³
DOT CLASSIFICATION: Flammable Solid;
Label: Spontaneously Combustible
SAFETY PROFILE: A poison. Extremely
destructive to living tissue. A very dangerous
fire hazard; ignites on exposure to air.
Incompatible with moisture, acids, air,
alcohols, amines, halogens. To fight fire, use
CO_2, dry sand, dry chemical. Do not use
water, foam, or halogenated extinguishing
agents. When heated to decomposition it
emits acrid smoke and irritating fumes.

TKT750   CAS: 19464-55-2   HR: 3
TRIISOPROPYLTIN ACETATE
mf: $C_{11}H_{24}O_2Sn$ mw: 307.04
SYN: ACETOXYTRIISOPROPYLSTANNANE
OSHA PEL: TWA 0.1 mg(Sn)/m³ (skin)
ACGIH TLV: TWA 0.1 mg(Sn)/m³; STEL
0.2 mg(Sn)/m³ (skin).
NIOSH REL: (Organotin Compounds)
TWA 0.1 mg(Sn)/m³
SAFETY PROFILE: Poison by ingestion
and intravenous routes. When heated to
decomposition it emits acrid smoke and
irritating fumes. See also TIN
COMPOUNDS and ESTERS.

T

TKT850　CAS: 73928-00-4　HR: 3
TRIISOPROPYLTIN UNDECYLENATE
mf: $C_{20}H_{42}O_2Sn$　　mw: 433.31
SYNS: STANNANE, TRIISOPROPYL(UNDECANOYLO-
XY)- □ UNDECANOIC ACID, TRIISOPROPYLSTANNYL
ESTER
OSHA PEL: TWA 0.1 mg(Sn)/m³ (skin)
ACGIH TLV: TWA 0.1 mg(Sn)/m³; STEL
0.2 mg/m³ (skin)
NIOSH REL: (Organotin Compound) 10H
TWA 0.1 mg(Sn)/m³
SAFETY PROFILE: Poison by intravenous
route. When heated to decomposition it
emits toxic fumes of Sn.

TKV000　CAS: 552-30-7　HR: 2
TRIMELLITIC ANHYDRIDE
mf: $C_9H_4O_5$　　mw: 192.13
PROP: Crystals or needles. Mp: 162°, bp:
240–245° @ 14 mm. Sol in acetone, ethyl
acetate, dimethylformamide.
SYNS: ANHYDROTRIMELLIC ACID □ 1,2,4-BENZ-
ENETRICARBOXYLIC ACID ANHYDRIDE □ 1,2,4-
BENZENETRICARBOXYLIC ACID, CYCLIC 1,2-
ANHYDRIDE □ 1,2,4-BENZENETRICARBOXYLIC
ANHYDRIDE □ 4-CARBOXYPHTHALIC ANHYDRIDE □
1,3-DIHYDRO-1,3-DIOXO-5-ISOBENZOFURANCARBOX-
YLIC ACID □ 1,3-DIOXO-5-PHTHALANCARBOXYLIC
ACID □ DIPHENYLMETHANE-4,4'-DIISOCYANATE-
TRIMELLIC ANHYDRIDE-ETHOMID HT POLYMER □
NCI-C56633 □ TMA □ TMAN □ TRIMELLIC ACID
ANHYDRIDE □ TRIMELLIC ACID-1,2-ANHYDRIDE □
TRIMELLITIC ACID CYCLIC-1,2-ANHYDRIDE
CONSENSUS REPORTS: Reported in EPA
TSCA Inventory.
OSHA PEL: TWA 0.005 ppm
ACGIH TLV: TWA CL 0.04 mg/m³; -
(Proposed: inhalable fraction and vapor;
0.0005 mg/m³; STEL 0.002 mg/m³; skin;
sensitizer.
DFG MAK: 0.04 mg/m³
NIOSH REL: (Trimellitic Anhydride): handle
as extremely toxic
SAFETY PROFILE: Moderately toxic by
ingestion. Has caused pulmonary edema
from inhalation. Irritant to lungs and air
passages. May be a powerful allergen.
Typical attack consists of breathlessness,
wheezing, cough, running nose,
immunological sensitization, and asthma
symptoms. When heated to decomposition

it emits acrid smoke and irritating fumes.
See also ANHYDRIDES.

TKW000　CAS: 12136-15-1　HR: 3
TRIMERCURY DINITRIDE
mf: Hg_3N_2　　mw: 629.78
PROP: A solid. IDLH 10 mg/m³ (as Hg).
SYN: MERCURY NITRIDE
CONSENSUS REPORTS: Mercury and its
compounds are on the Community Right-
To-Know List.
ACGIH TLV: TWA 0.1 mg(Hg)/m³ (skin);
BEI: 35 µg/g creatinine total inorganic
mercury in urine preshift; 15 µg/g creatinine
total inorganic mercury in blood at end of
shift at end of workweek.
DOT CLASSIFICATION: Forbidden
SAFETY PROFILE: Mercury compounds
are poisons. An explosive sensitive to
friction, impact, heating, or contact with
sulfuric acid. Incompatible with sulfuric
acid. When heated to decomposition it emits
very toxic fumes of Hg and NO_x. See also
MERCURY COMPOUNDS and
NITRIDES.

TLA600　CAS: 6163-73-1　HR: 1
TRI-(2-METHOXYETHANOL)PHOSPH-
　　ATE
mf: $C_9H_{21}O_7P$　　mw: 272.27
SYNS: ETHANOL, 2-METHOXY-, PHOSPHATE (3:1) □
2-METHOXYETHANOL PHOSPHATE (3:1) □ TRIS-(2-
METHOXYETHYL)FOSFAT
CONSENSUS REPORTS: Reported in EPA
TSCA Inventory.
SAFETY PROFILE: A skin and eye irritant.
When heated to decomposition it emits
toxic fumes of PO_x.

TLA650　CAS: 315706-65-1　HR: 3
3,4,5-TRIMETHOXY-N-(4-PROPYL-
　　CYCLOHEXYL)BENZAMIDE
mf: $C_{19}H_{29}NO_4$　　mw: 335.44
SAFETY PROFILE: A poison by ingestion.
When heated to decomposition it emits
toxic vapors of NO_x.

TLB750 CAS: 2487-90-3 HR: 2
TRIMETHOXY SILANE
DOT: NA 9269
mf: $C_3H_{10}O_3Si$ mw: 122.22
PROP: A liquid. Mp: −115°, bp: 84°, refr
index: 1.3580, d: 0.960.
SYNS: TRIMETHOXYSILANE (DOT) □ SILANE,
TRIMETHOXY-
CONSENSUS REPORTS: Reported in EPA
TSCA Inventory.
DOT CLASSIFICATION: 6.1; Label: Poison,
Flammable Liquid
SAFETY PROFILE: Moderately toxic by
inhalation. Mildly toxic by ingestion and skin
contact. When heated to decomposition it
emits acrid smoke and irritating fumes.

TLC600 CAS: 25147-91-5 HR: 3
N-(3-(TRIMETHOXYSILYL)PROPYL)-1,3-
PROPANEDIAMINE
mf: $C_9H_{24}N_2O_3Si$ mw: 236.44
SYNS: (N-(3-AMINOPROPYL)-3-AMINOPROPYL)-
TRIMETHOXYSILANE □ 1,3-PROPANEDIAMINE, N-(3--
(TRIMETHOXYSILYL)PROPYL)- □ SILANE 40-47
SAFETY PROFILE: A poison by ingestion
and skin contact. A mild skin and severe eye
irritant. When heated to decomposition it
emits toxic vapors of NO_x.

TLD500 CAS: 75-50-3 HR: 3
TRIMETHYLAMINE
DOT: UN 1083/UN 1297
mf: C_3H_9N mw: 59.13
PROP: Volatile liquid with fishy odor, or
colorless gas with pungent, ammonia-like
odor; saline taste. Bp: 2.87°, lel: 2%, uel:
11.6%, fp: −117.1°, mp: −117.2°, d: 0.662
@ −5°, autoign temp: 374°F, vap d: 2.0,
flash p: 20°F (CC). Misc with alc; sol in
ether, benzene, toluene, xylene, chloroform.
SYNS: TMA □ TRIMETHYLAMINE, anhydrous (UN 1083)
(DOT) □ TRIMETHYLAMINE, aqueous solutions not >50%
trimethylamine, by weight (UN 1297) (DOT)
CONSENSUS REPORTS: Reported in EPA
TSCA Inventory.
OSHA PEL: TWA 10 ppm; STEL 15 ppm
ACGIH TLV: TWA 5 ppm; STEL 15 ppm
DOT CLASSIFICATION: 2.1; Label:
Flammable Gas (UN 1083); DOT Class: 3;
Label: Flammable Liquid (UN 1297)

SAFETY PROFILE: Poison by intravenous
route. Moderately toxic by subcutaneous
and rectal routes. Mildly toxic by inhalation.
A very dangerous fire hazard when exposed
to heat or flame. Self-reactive. Moderately
explosive in the form of vapor when
exposed to heat or flame. Can react with
oxidizing materials. To fight fire, stop flow
of gas. Potentially explosive reaction with
bromine + heat, ethylene oxide,
triethynylaluminum. When heated to
decomposition it emits toxic fumes of NO_x.
See also AMINES.

TLG250 CAS: 137-17-7 HR: 3
2,4,5-TRIMETHYLANILINE
mf: $C_9H_{13}N$ mw: 135.23
PROP: Needles in H_2O. Mp: 68°, bp:
234–235°.
SYNS: 1-AMINO-2,4,5-TRIMETHYLBENZENE □ psi-
CUMIDINE □ NCI-C02299 □ PSEUDOCUMIDINE □
1,2,4-TRIMETHYL-5-AMINOBENZENE □ 2,4,5-TRI-
METHYLANILIN (CZECH) □ 2,4,5-TRIMETHYL-
BENZENAMINE
CONSENSUS REPORTS: IARC Cancer
Review: Group 3 IMEMDT 7,56,87; Animal
Limited Evidence IMEMDT 27,177,82.
NCI Carcinogenesis Bioassay (feed); Clear
Evidence: mouse, rat NCITR* NCI-CG-
TR-160,79.
DFG MAK: Animal Carcinogen, Suspected
Human Carcinogen
SAFETY PROFILE: Confirmed carcinogen
with experimental carcinogenic and
tumorigenic data. Moderately toxic by
ingestion. Mutation data reported. When
heated to decomposition it emits toxic
fumes of NO_x. Used as a dye, pigment, and
printing ink. See also ANILINE DYES.

TLG500 CAS: 88-05-1 HR: 3
2,4,6-TRIMETHYLANILINE
mf: $C_9H_{13}N$ mw: 135.23
PROP: A liquid. Mp: 233°, d: 0.96, refr
index: 1.5510, bp: 232–233°.
SYNS: AMINOMESITYLENE □ 2-AMINOMESITYL-
ENE □ 1-AMINO-2,4,6-TRIMETHYLBENZEN (CZECH) □
2-AMINO-1,3,5-TRIMETHYLBENZENE □ MESIDIN -

T

(CZECH) □ MESIDINE □ MESITYLAMINE □ MEZID-
INE □ 2,4,6-TRIMETHYLBENZENAMINE
CONSENSUS REPORTS: IARC Cancer
Review: Group 3 IMEMDT 7,56,87; Animal
Inadequate Evidence IMEMDT 27,177,82.
EPA Extremely Hazardous Substances List.
Reported in EPA TSCA Inventory.
SAFETY PROFILE: Poison by inhalation.
Moderately toxic by ingestion. A skin and
severe eye irritant. Questionable carcinogen
with experimental carcinogenic data.
Mutation data reported. When heated to
decomposition it emits toxic fumes of NO_x.

TLG750   CAS: 21436-97-5   HR: 3
2,4,5-TRIMETHYLANILINE HYDROCHLORIDE
mf: $C_9H_{13}N \cdot ClH$ mw: 171.69
SYNS: 1-AMINO-2,4,5-TRIMETHYLBENZENE HYDRO-
CHLORIDE □ psi-CUMIDINE HYDROCHLORIDE □
PSEUDOCUMIDINE HYDROCHLORIDE □ 1,2,4-TRI-
METHYL-5-AMINOBENZENE HYDROCHLORIDE □
2,4,5-TRIMETHYLBENZENAMINE HYDROCHLORIDE
CONSENSUS REPORTS: IARC Cancer
Review: Animal Inadequate Evidence
IMEMDT 27,177,82.
SAFETY PROFILE: Poison by
intraperitoneal route. Moderately toxic by
ingestion. Questionable carcinogen with
experimental carcinogenic, neoplastigenic,
and tumorigenic data. When heated to
decomposition it emits very toxic fumes of
NO_x and HCl.

TLH000   CAS: 6334-11-8   HR: 3
2,4,6-TRIMETHYLANILINE HYDRO-CHLORIDE
mf: $C_9H_{13}N \cdot ClH$ mw: 171.69
SYNS: AMINOMESITYLENE HYDROCHLORIDE □ 2-
AMINOMESITYLENE HYDROCHLORIDE □ 2-AMINO-
1,3-5-TRIMETHYLBENZENE HYDROCHLORIDE □
MESIDINE HYDROCHLORIDE □ MESITYLAMINE
HYDROCHLORIDE □ 2,4,6-TRIMETHYLBENZENAMINE
HYDROCHLORIDE
CONSENSUS REPORTS: IARC Cancer
Review: Animal Inadequate Evidence
IMEMDT 27,177,82.
SAFETY PROFILE: Suspected carcinogen
with experimental carcinogenic and
neoplastigenic data. Poison by

intraperitoneal route. Moderately toxic by
ingestion. When heated to decomposition it
emits very toxic fumes of NO_x and HCl.

TLN125   CAS: 63884-77-5   HR: 3
β,γ,γ-TRIMETHYLCAPROALDEHYDE THIOSEMICARBAZONE
SYN: CAPROALDEHYDE, β,γ,γ-TRIMETHYL-,
THIOSEMICARBAZONE
SAFETY PROFILE: A poison by ingestion.
When heated to decomposition it emits
toxic vapors of NO_x and SO_x.

TLH250   CAS: 41262-21-9   HR: 3
TRIMETHYLARSINE SELENIDE
mf: C_3H_9AsSe mw: 199.00
PROP: Thin needles from EtOH. Mp: *ca* 1°
decomp.
SYN: ARSINE SELENIDE, TRIMETHYL-
OSHA PEL: TWA 0.5 mg(As)/m^3; 0.2 mg-
(Se)/m^3
ACGIH TLV: BEI: 35 μ (As)/L inorganic
arsenic and methylated metabolites in urine;
TWA 0.2 mg(Se)/m^3
SAFETY PROFILE: Poison by intravenous
route. When heated to decomposition it
emits toxic fumes of As and Se.

TLK750   CAS: 13345-64-7   HR: 3
7,8,12-TRIMETHYLBENZ(a)ANTHRACENE
mf: $C_{21}H_{18}$ mw: 270.39
SYNS: 7,8,12-TMBA □ 5:9:10-TRIMETHYL-1:2-
BENZANTHRACENE
SAFETY PROFILE: Poison by intravenous
route. Experimental reproductive effects.
Questionable carcinogen with experimental
carcinogenic and tumorigenic data. Mutation
data reported. When heated to
decomposition it emits acrid smoke and
irritating fumes.

TLL250   CAS: 25551-13-7   HR: 3
TRIMETHYL BENZENE
mf: C_9H_{12} mw: 120.21
SYN: TRIMETHYL BENZENE (mixed isomers)
CONSENSUS REPORTS: Reported in EPA
TSCA Inventory.
OSHA PEL: TWA 25 ppm

ACGIH TLV: TWA 25 ppm
SAFETY PROFILE: Mildly toxic by ingestion. A skin and eye irritant. Flammable when exposed to heat, flame, and oxidizers. When heated to decomposition it emits acrid smoke and irritating fumes. See also individual trimethyl benzene isomers.

TLL500   CAS: 526-73-8   HR: 3
1,2,3-TRIMETHYL BENZENE
mf: C_9H_{12} mw: 120.21
PROP: Mp: −25.4°, bp: 176.1°, d: 0.894, refr index: 1.5139, vap d: 4.15, flash p: 119° F, autoign temp: 878° F.
SYNS: HEMIMELLITENE □ TRIMETHYL BENZENE
CONSENSUS REPORTS: Reported in EPA TSCA Inventory.
OSHA PEL: TWA 25 ppm
ACGIH TLV: TWA 25 ppm
DFG MAK: 20 ppm
SAFETY PROFILE: Mildly toxic by ingestion. Flammable liquid when exposed to heat, sparks, or flame. To fight fire, use water spray, mist, dry chemical, CO_2, foam. When heated to decomposition it emits acrid smoke and irritating fumes.

TLL750   CAS: 95-63-6   HR: 3
1,2,4-TRIMETHYL BENZENE
mf: C_9H_{12} mw: 120.21
PROP: A liquid. Mp: −44°, d: 0.888 @ 4°, fp: −61°, bp: 168.89°, flash p: 130°F, autoign temp: 959°F. Insol in water; sol in alc, benzene, and ether.
SYNS: ASYMMETRICAL TRIMETHYL BENZENE □ psi-CUMENE □ PSEUDOCUMENE □ PSEUDOCUMOL □ as-TRIMETHYL BENZENE □ 1,2,5-TRIMETHYL BENZENE
CONSENSUS REPORTS: Reported in EPA TSCA Inventory. Community Right-To-Know List.
OSHA PEL: TWA 25 ppm
ACGIH TLV: TWA 25 ppm
DFG MAK: 20 ppm
SAFETY PROFILE: Moderately toxic by intraperitoneal route. Mildly toxic by inhalation. Can cause central nervous system depression, anemia, bronchitis. Flammable liquid when exposed to heat, sparks, or

flame. To fight fire, use foam, alcohol foam, mist. Emitted from modern building materials (CENEAR 69,22,91). When heated to decomposition it emits acrid smoke and irritating fumes.

TLM050   CAS: 108-67-8   HR: 3
1,3,5-TRIMETHYL BENZENE
DOT: UN 2325
mf: C_9H_{12} mw: 120.21
PROP: A liquid; peculiar odor. Mp: −44.8°, d: 0.8637 @ 20°/4°, bp: 164.7°, autoign temp: 1022°F. Insol in water; misc in alc, benzene, and ether.
SYNS: BENZENE, 1,3,5-TRIMETHYL- □ FLEET-X □ MESITYLENE □ TMB □ TRIMETHYL BENZENE -(ACGIH) □ sym-TRIMETHYLBENZENE □ TRIMETHYL BENZOL
CONSENSUS REPORTS: Reported in EPA TSCA Inventory.
OSHA PEL: TWA 25 ppm
ACGIH TLV: TWA 25 ppm
DFG MAK: 20 ppm
DOT CLASSIFICATION: 3; Label: Flammable Liquid
SAFETY PROFILE: Poison by inhalation. Moderately toxic by intraperitoneal route. Human systemic effects by inhalation: sensory changes involving peripheral nerves, somnolence (general depressed activity), and structural or functional change in trachea or bronchi. Reports of leukopenia and thrombocytopenia in experimental animals. A mild skin and eye irritant. A flammable liquid when exposed to heat or flame; can react vigorously with oxidizing materials. Violent reaction with HNO_3. To fight fire, use water spray, fog, foam, CO_2. Emitted from modern building materials (CENEAR 69,22,91). When heated to decomposition it emits acrid smoke and irritating fumes.

TLN000   CAS: 121-43-7   HR: 3
TRIMETHYL BORATE
DOT: UN 2416
mf: $C_3H_9BO_3$ mw: 103.93
PROP: Colorless, moisture-sensitive liquid; fumes in air. Decomp in water; misc in alc, ether. Mp: −29°, bp: 68°, flash p: <73°F, d:

T

0.92 @ 20°, vap d: 3.59. Sol in polar non-hydroxylic solvs.

SYNS: BORESTER O □ METHYL BORATE □ TRIMETHOXYBORINE □ TRIMETHYLESTER KYSEL-INY BORITE

CONSENSUS REPORTS: Reported in EPA TSCA Inventory.

DOT CLASSIFICATION: 3; Label: Flammable Liquid

SAFETY PROFILE: Moderately toxic by ingestion, skin contact, and intraperitoneal routes. An eye irritant. A very dangerous fire hazard when exposed to heat, flame, or oxidizers. Moderately explosive when exposed to flame. Will react with water or steam to produce toxic and flammable vapors. To fight fire, use dry chemical, CO_2, spray, foam. When heated to decomposition it emits acrid smoke and irritating fumes. See also ESTERS and BORON COMPOUNDS.

TLN250   CAS: 75-77-4   HR: 3
TRIMETHYL CHLOROSILANE
DOT: UN 1298

mf: C_3H_9ClSi mw: 108.66

PROP: Colorless liquid. Bp: 57°, mp: −40°, d: 0.854 @ 25°/25°, flash p: −18°F. Sol in benzene, ether, perchloroethylene.

SYNS: CHLOROTRIMETHYLSILANE □ SILANE, TRI-METHYLCHLORO- □ SILICANE, CHLOROTRIMETHYL- □ TL 1163 □ TRIMETHYLCHLOROSILANE (DOT)

CONSENSUS REPORTS: EPA Extremely Hazardous Substances List. Reported in EPA TSCA Inventory.

DOT CLASSIFICATION: 3; Label: Flammable Liquid, Corrosive; DOT Class: 4.3; Label: Dangerous When Wet, Flammable Liquid

SAFETY PROFILE: Poison by ingestion and skin contact. Moderately toxic by inhalation and intraperitoneal routes. A corrosive irritant to skin, eyes, and mucous membranes. Questionable carcinogen with experimental neoplastigenic data. Mutation data reported. A flammable liquid and very dangerous fire hazard when exposed to heat or flame. Violent reaction with water or hexafluoroisopropylideneamino lithium. A

preparative hazard. To fight fire, use foam, alcohol foam, fog. When heated to decomposition it emits toxic fumes of Cl^-. An intermediate in the production of silicones. See also CHLOROSILANES.

TLP500   CAS: 147-47-7   HR: 2
2,2,4-TRIMETHYL-1,2-DIHYDROQU-INOLINE
mf: $C_{12}H_{15}N$ mw: 173.28

PROP: A solid or liquid. Mp: 26–27°, bp: 255–260° @ 743 mm.

SYNS: ACETONANIL □ ACETONANYL □ ACETONE ANIL □ AGERITE RESIN D □ 1,2-DIHYDRO-2,2,4-TRI-METHYLQUINOLINE □ FLECTOL A □ FLECTOL H □ FLECTOL PASTILLES □ NCI-C60902 □ 2,2,4-TRIMETH-YL-1,2-DIHYDROCHINOLIN □ TRIMETHYL-1,2-DI-HYDROQUINOLINE □ VULKANOX HS/LG □ VULKANOX HS/POWDER

CONSENSUS REPORTS: Reported in EPA TSCA Inventory.

SAFETY PROFILE: Moderately toxic by ingestion. When heated to decomposition it emits toxic fumes of NO_x.

TLP800   CAS: 316172-59-5   HR: 3
1,7,7-TRIMETHYL-o-(3-(2,6-DIMETHYL-4-MORPHOLINYL)-2-HYDRO-XYPROPYL)OXIME (1R,4R)-BICYCLO(2.2.1)HEPTAN-2-ONE
mf: $C_{19}H_{34}N_2O_3$ mw: 338.49

SAFETY PROFILE: A poison by ingestion. When heated to decomposition it emits toxic vapors of NO_x.

TLR500   CAS: 544-13-8   HR: 3
1,3-TRIMETHYLENEDINITRILE
mf: $C_5H_6N_2$ mw: 94.13

PROP: Colorless liquid. D: 0.989 @ 15°/4°, mp: −29°, bp: 144–147° @ 13 mm. Sol in water; insol in ether.

SYNS: 1,3-DICYANOPROPANE □ GLUTARIC ACID D-INITRILE □ GLUTARODINITRILE □ GLUTARO-NITRILE □ PENTANEDINITRILE □ PYROTARTARIC ACID NITRILE

CONSENSUS REPORTS: Cyanide and its compounds are on the Community Right-To-Know List. Reported in EPA TSCA Inventory.

SAFETY PROFILE: Poison by subcutaneous route. Moderately toxic by ingestion. When heated to decomposition it emits very toxic fumes of NO_x and CN^-. See also NITRILES.

TLR675 CAS: 2825-82-3 HR: 3
exo-TRIMETHYLENENORBORNANE
mf: $C_{10}H_{16}$ mw: 136.26
SYNS: exo-HEXAHYDRO-4,7-METHANOINDAN □ JP-10 □ exo-TETRAHYDROBICYCLOPENTADIENE □ exo-TETRAHYDRODI(CYCLOPENTADIENE) □ exo-TRICYCLO(5.2.1.0^{2,6})DECANE □ exo-5,6-TRIMETHYL-ENENORBORNANE
CONSENSUS REPORTS: Reported in EPA TSCA Inventory.
SAFETY PROFILE: Moderately toxic by ingestion. Mildly toxic by inhalation. An experimental teratogen. Experimental reproductive effects. Questionable carcinogen with experimental carcinogenic and tumorigenic data. Mutation data reported. Used as a major component of cruise missile fuel. When heated to decomposition it emits acrid smoke and irritating fumes.

TLT150 CAS: 314238-35-2 HR: 3
1,7,7-TRIMETHYL-o-(3-(HEXAHYDRO-1H-AZEPIN-1-YL)-2-HYDROXYPROP-YL)OXIME (1R,4R)-BICYCLO(2.2.1)-HEPTAN-2-ONE
mf: $C_{19}H_{34}N_2O_2$ mw: 322.49
SAFETY PROFILE: A poison by ingestion. When heated to decomposition it emits toxic vapors of NO_x.

TLT750 CAS: 60597-20-8 HR: 3
TRIMETHYLHYDRAZINE HYDRO-CHLORIDE
mf: $C_3H_{10}N_2 \cdot ClH$ mw: 110.61
SAFETY PROFILE: Suspected carcinogen with experimental carcinogenic data. When heated to decomposition it emits very toxic fumes of HCl and NO_x.

TLT757 CAS: 314238-34-1 HR: 3
1,7,7-TRIMETHYL-o-(2-HYDROXY-3-(4-MORPHOLINYL)PROPYL)OXIME -

(1R,4R)-BICYCLO(2.2.1)HEPTAN-2-ONE
mf: $C_{17}H_{30}N_2O_3$ mw: 310.44
SAFETY PROFILE: A poison by ingestion. When heated to decomposition it emits toxic vapors of NO_x.

TLT763 CAS: 314238-32-9 HR: 3
1,7,7-TRIMETHYL-o-(2-HYDROXY-3-(1-PYRROLIDINYL)PROPYL)OXIME - (1R,4R)-BICYCLO(2.2.1)HEPTAN-2-ONE
mf: $C_{17}H_{30}N_2O_2$ mw: 294.44
SAFETY PROFILE: A poison by ingestion. When heated to decomposition it emits toxic vapors of NO_x.

TLU750 CAS: 3475-63-6 HR: 3
1,1,3-TRIMETHYL-3-NITROSOUREA
mf: $C_4H_9N_3O_2$ mw: 131.16
SYNS: N-NITROSO-TRIMETHYLHARNSTOFF -(GERMAN) □ NITROSOTRIMETHYLUREA □ N-NITROSOTRIMETHYLUREA □ TRIMETHYL-NITROSOHARNSTOFF (GERMAN) □ N-TRIMETHYL-N-NITROSOUREA
CONSENSUS REPORTS: EPA Genetic Toxicology Program.
SAFETY PROFILE: Poison by ingestion and intravenous routes. Experimental teratogenic effects. Questionable carcinogen with experimental tumorigenic data. Mutation data reported. Many N-nitroso compounds are carcinogens. When heated to decomposition it emits toxic fumes of NO_x. See also N-NITROSO COMPOUNDS.

TLX600 CAS: 149-73-5 HR: 3
TRIMETHYL ORTHOFORMATE
mf: $C_4H_{10}O_3$ mw: 106.14
PROP: Colorless liquid; pungent odor. Vap d: 3.67, fp: 15°, bp: 103–105°, flash p: 59°F.
SYNS: METHYLESTER KYSELINY ORTHOMRAVENCI (CZECH) □ METHYL ORTHOFORMATE □ ORTHO-FORMIC ACID, TRIMETHYL ESTER □ ORTHOMRAVEN-CAN METHYLNATY (CZECH) □ TRIMETHO-XYMETHANE
CONSENSUS REPORTS: Reported in EPA TSCA Inventory.

T

SAFETY PROFILE: A skin and eye irritant. A very dangerous fire hazard when exposed to heat or flame; can react with oxidizing materials. Hazardous to prepare. To fight fire, use CO_2, fog, haze. When heated to decomposition it emits acrid smoke and irritating fumes. See also ESTERS.

TLX800   HR: D
2,4,5-TRIMETHYL Δ-3-OXAZOLINE
mf: $C_6H_{11}NO$ mw: 113.16
PROP: Yellow-orange liquid; powerful, musty, nut-like odor. D: 0.911–0.932, refr index: 1.414–1.435. Sol in alc, propylene glycol, water; insol in fixed oils.
SYN: FEMA No. 3525
SAFETY PROFILE: When heated to decomposition emits toxic fumes of NO_x.

TLY000   CAS: 64047-30-9   HR: 3
TRIMETHYL-2-OXEPANONE (mixed isomers)
mf: $C_9H_{16}O_2$ mw: 156.25
SYN: TRIMETHYL-ε-LACTONE (mixed isomers)
CONSENSUS REPORTS: IARC Cancer Review: Animal Sufficient Evidence IMEMDT 19,303,79.
SAFETY PROFILE: Confirmed carcinogen. Mildly toxic by ingestion and skin contact. When heated to decomposition it emits acrid smoke and irritating fumes. See also KETONES.

TLY500   CAS: 540-84-1   HR: 3
2,2,4-TRIMETHYLPENTANE
mf: C_8H_{18} mw: 114.26
PROP: Clear liquid; odor of gasoline. Bp: 99.2°, mp: −107.5°, fp: −116°, flash p: 10°F, d: 0.692 @ 20°/4°, autoign temp: 779°F, vap press: 40.6 mm @ 21°, vap d: 3.93, lel: 1.1%, uel: 6.0%.
SYNS: ISOBUTYLTRIMETHYLETHANE □ ISOOCTANE (DOT)
CONSENSUS REPORTS: Reported in EPA TSCA Inventory.
NIOSH REL: TWA (Alkanes) 350 mg/m^3
SAFETY PROFILE: Mutation data reported. High concentrations can cause narcosis. A very dangerous fire hazard when exposed to heat, flame, oxidizers. Can react vigorously with reducing materials. Explosive in the form of vapor when exposed to heat or flame. To fight fire, use CO_2, dry chemical. When heated to decomposition it emits acrid smoke and irritating fumes. See also ALKANES.

TMD250   CAS: 512-56-1   HR: 3
TRIMETHYL PHOSPHATE
mf: $C_3H_9O_4P$ mw: 140.09
PROP: Pleasant smelling liquid. D: 1.97 @ 19.5°/0°, bp: 197.2°. Sol in alc, water, ether, and org solvs.
SYNS: METHYL PHOSPHATE □ NCI-C03781 □ PHOSPHORIC ACID, TRIMETHYL ESTER □ TMP □ O,O,O-TRIMETHYL PHOSPHATE
CONSENSUS REPORTS: NCI Carcinogenesis Bioassay (gavage); Clear Evidence: mouse, rat NCITR* NCI-CG-TR-81,78. Reported in EPA TSCA Inventory. EPA Genetic Toxicology Program.
DFG MAK: Confirmed Animal Carcinogen with Unknown Relevance to Humans
SAFETY PROFILE: Suspected carcinogen with experimental carcinogenic, neoplastigenic, tumorigenic, and teratogenic data. Moderately toxic by ingestion, skin contact, intraperitoneal, and intravenous routes. Experimental reproductive effects. Human mutation data reported. Explodes when heat distilled. When heated to decomposition it emits toxic fumes of PO_x. See also ESTERS.

TMD400   CAS: 20819-54-9   HR: 3
TRIMETHYLPHOSPHINE SELENIDE
mf: C_3H_9PSe mw: 155.05
PROP: Needles from EtOH. Mp: 140.5–141°.
SYN: PHOSPHINE SELENIDE, TRIMETHYL-
OSHA PEL: TWA 0.2 mg(Se)/m^3
ACGIH TLV: TWA 0.2 mg(Se)/m^3
SAFETY PROFILE: Poison by intravenous route. When heated to decomposition it emits toxic fumes of PO_x and Se.

TMD500   CAS: 121-45-9   HR: 3

TRIMETHYL PHOSPHITE
DOT: UN 2329
mf: $C_3H_9O_3P$ mw: 124.09
PROP: Air-sensitive, colorless liquid with powerful sickly odor. D: 1.046 @ 20°/4°, mp: −78°, bp: 111–112°, vap d: 4.3, bp: 232–234°F, flash p: 130°F (OC). Insol in water; sol in hexane, benzene, acetone, alc, ether, carbon tetrachloride, kerosene.
SYNS: FOSFORYN TROJMETYLOWY (CZECH) □ -METHYL PHOSPHITE □ PHOSPHORUS ACID, TRI-METHYL ESTER □ TRIMETHOXYFOSFIN □ TRI-METHOXYPHOSPHINE □ TRIMETHYLFOSFIT □ TRI-METHYL PHOSPHITE
CONSENSUS REPORTS: Reported in EPA TSCA Inventory.
OSHA PEL: TWA 2 ppm
ACGIH TLV: TWA 2 ppm
DOT CLASSIFICATION: 3; Label: Flammable Liquid
SAFETY PROFILE: Moderately toxic by ingestion and skin contact. An experimental teratogen. A severe skin and eye irritant. Flammable liquid when exposed to heat, flame, or oxidizers. To fight fire, use water, foam, fog, CO_2. Violent explosive reaction on contact with magnesium perchlorate or trimethyl platinum(IV) azide tetramer. When heated to decomposition it emits toxic fumes of PO_x. An intermediate in the production of pesticides, fire retardants, and organic phosphorus additives. See also ESTERS.

TMD650   CAS: 816-80-8   HR: 3
TRIMETHYL PHOSPHOROTRITHIOATE
mf: $C_3H_9PS_3$ mw: 172.27
SYN: PHOSPHOROTRITHIOUS ACID, TRIMETHYL ESTER
SAFETY PROFILE: A poison by ingestion, intraperitoneal, and intravenous routes. Moderately toxic by skin contact. A severe eye irritant. When heated to decomposition it emits toxic vapors of SO_x and PO_x.

TME270   CAS: 14667-55-1   HR: 2
2,3,5-TRIMETHYLPYRAZINE
mf: $C_7H_{10}N_2$ mw: 122.19

PROP: Colorless to sltly yellow liquid; baked potato, peanut odor. D: 0.960–0.990 @ 20°, bp: 171–172° @ 735 mm, refr index: 1.503, flash p: 153°F. Sol in water and org solvs.
SYNS: FEMA No. 3244 □ TRIMETHYLPYRAZINE
CONSENSUS REPORTS: Reported in EPA TSCA Inventory.
SAFETY PROFILE: Moderately toxic by ingestion. Combustible liquid. When heated to decomposition emits toxic fumes of NO_x.

TME500   CAS: 25930-79-4   HR: 3
TRIMETHYLSELENONIUM
mf: C_3H_9Se mw: 124.08
SYNS: SELENONIUM, TRIMETHYL- □ TRIMETHYL-SELENONIUM ION
OSHA PEL: TWA 0.2 mg(Se)/m³
ACGIH TLV: TWA 0.2 mg(Se)/m³
SAFETY PROFILE: Poison by subcutaneous route. When heated to decomposition it emits toxic fumes of Se.

TME600   CAS: 18987-38-7   HR: 3
TRIMETHYLSELENONIUM CHLORIDE
mf: $C_3H_9Se•Cl$ mw: 159.53
SYN: SELENONIUM, TRIMETHYL-, CHLORIDE
OSHA PEL: TWA 0.2 mg(Se)/m³
ACGIH TLV: TWA 0.2 mg(Se)/m³
SAFETY PROFILE: Poison by intraperitoneal route. When heated to decomposition it emits toxic fumes of Se and Cl⁻.

TMH750   CAS: 2489-77-2   HR: 3
1,1,3-TRIMETHYL-2-THIOUREA
mf: $C_4H_{10}N_2S$ mw: 118.22
PROP: Prisms from C_6H_6/ligroin. Mp: 87–88°. Trimethylthiourea tested in NCITR* NCI-CG-TR-129 contained 15% 1,3-dimethyl-2-thiourea and 5% Zeolex 80.
SYNS: NCI-C02186 □ TRIMETHYLTHIOUREA □ N,N,N'-TRIMETHYLTHIOUREA
CONSENSUS REPORTS: NCI Carcinogenesis Bioassay (feed); No Evidence: mouse NCITR* NCI-CG-TR-129,79. Reported in EPA TSCA Inventory.
SAFETY PROFILE: Poison by ingestion. Questionable carcinogen with experimental carcinogenic data. Mutation data reported.

T

When heated to decomposition it emits very toxic fumes of NO_x and SO_x. See also ISOTHIOUREA.

TMI000   CAS: 1118-14-5   HR: 3
TRIMETHYLTIN ACETATE
mf: $C_5H_{12}O_2Sn$ mw: 222.86
PROP: White crystals. Mp: 196–197°. Spar sol in $CHCl_3$ and CCl_4.
SYN: ACETOXYTRIMETHYLSTANNANE
OSHA PEL: TWA 0.1 mg(Sn)/m³ (skin)
ACGIH TLV: TWA 0.1 mg(Sn)/m³; STEL 0.2 mg(Sn)/m³ (skin).
NIOSH REL: (Organotin Compounds) TWA 0.1 mg(Sn)/m³
SAFETY PROFILE: Poison by ingestion. When heated to decomposition it emits acrid smoke and irritating fumes. See also T-IN COMPOUNDS.

TMI100   CAS: 73940-86-0   HR: 3
TRIMETHYLTIN CYANATE
mf: C_4H_9NOSn mw: 205.83
SYNS: CYANIC ACID, TRIMETHYLSTANNYL ESTER □ STANNANE, CYANATOTRIMETHYL-
OSHA PEL: TWA 0.1 mg(Sn)/m³ (skin)
ACGIH TLV: TWA 0.1 mg(Sn)/m³; STEL 0.2 mg/m³ (skin)
NIOSH REL: (Organotin Compound) 10H TWA 0.1 mg(Sn)/m³
SAFETY PROFILE: Poison by intravenous route. When heated to decomposition it emits toxic fumes of NO_x and Sn.

TMI500   CAS: 63869-87-4   HR: 3
TRIMETHYLTIN SULPHATE
mf: $C_3H_{10}O_4SSn$ mw: 260.88
SYN: TRIMETHYLSTANNANE SULPHATE
OSHA PEL: TWA 0.1 mg(Sn)/m³ (skin)
ACGIH TLV: TWA 0.1 mg(Sn)/m³; STEL 0.2 mg(Sn)/m³ (skin).
NIOSH REL: (Organotin Compounds) TWA 0.1 mg(Sn)/m³
SAFETY PROFILE: Poison by ingestion and intraperitoneal routes. When heated to decomposition it emits toxic fumes of SO_x. See also SULFATES and TIN COMPOUNDS.

TMJ000   CAS: 1709-70-2   HR: 2
1,3,5-TRIMETHYL-2,4,6-TRIS(3,5-DI-tert-BUTYL-4-HYDROXYBENZYL) BENZENE
mf: $C_{54}H_{78}O_3$ mw: 775.32
PROP: Crystals from $CH_2Cl_2/2,3,3$-trimethylpentane. Mp: 244°.
SYNS: AHYDOL (RUSSIAN) □ ANTIOXIDANT 330 □ AO-40 □ ETHANOX 330 □ SANTOQUIN EMULSION □ SANTOQUIN MIXTURE 6
CONSENSUS REPORTS: Reported in EPA TSCA Inventory.
SAFETY PROFILE: Moderately toxic by ingestion. An experimental teratogen. When heated to decomposition it emits acrid smoke and irritating fumes.

TMJ750   CAS: 86-21-5   HR: 3
TRIMETON
mf: $C_{16}H_{20}N_2$ mw: 240.38
SYNS: p-AMINOSALICYLSAEURES SALZ (GERMAN) □ 2-(α-(2-DIMETHYLAMINOETHYL)BENZYL)PYRIDINE □ 2-(3-DIMETHYLAMINO-1-PHENYLPROPYL)PYRIDINE □ N,N-DIMETHYL-3-PHENYL-3-(2-PYRIDYL)PROPYL-AMINE □ NCI-C60695 □ 1-PHENYL-1-(2-PYRIDYL)-3-DI-METHYLAMINOPROPANE □ 3-PHENYL-3-(2-PYRIDYL)-N,N-DIMETHYLPROPYLAMINE
SAFETY PROFILE: Poison by intravenous route. Human systemic effects by ingestion: central nervous system effects. When heated to decomposition it emits toxic fumes of NO_x.

TMK100   CAS: 3564-66-7   HR: 3
(+)-TRIMIPRAMINE
mf: $C_{20}H_{26}N_2$ mw: 294.48
SYNS: 5H-DIBENZ(B,F)AZEPINE, 10,11-DIHYDRO-5-(3-(DIMETHYLAMINO)-2-METHYLPROPYL)-, (+)- □ 5H-DI-BENZ(B,F)AZEPINE-5-PROPANAMINE, 10,11-DIHYDRO-N,N,β-TRIMETHYL-, (+)- □ 5H-DIBENZ(B,F)AZEPINE, 5-(3-(DIMETHYLAMINO)-2-METHYLPROPYL)-10,11-DI-HYDRO-, (+)- □ 10633 RP
SAFETY PROFILE: A poison by ingestion, intraperitoneal, and intravenous route. When heated to decomposition it emits toxic vapors of NO_x.

TMK250   CAS: 630-72-8   HR: 3
TRINITROACETONITRILE
mf: $C_2N_4O_6$ mw: 176.05

PROP: Crystal mass with camphoraceous odor. Mp: 41.5°.
CONSENSUS REPORTS: Cyanide and its compounds are on the Community Right-To-Know List.
DOT CLASSIFICATION: Forbidden
SAFETY PROFILE: An explosive sensitive to friction, impact, or rapid heating to 220°C. When heated to decomposition it emits toxic fumes of CN^- and NO_x. See also NITRILES and NITRO COMPOUNDS.

TMK500   CAS: 99-35-4   HR: 3
1,3,5-TRINITROBENZENE
DOT: UN 0214/UN 1354
mf: $C_6H_3N_3O_6$ mw: 213.12
PROP: Yellow crystals or dimorphic crystals from EtOH or HNO_3. Mp: 122°, bp: decomp, d: 1.760 @ 20°/4°.
SYNS: RCRA WASTE NUMBER U234 □ TNB □ TRINITROBENZEEN □ TRINITROBENZENE □ TRINITROBENZENE, dry or wetted with <30% water, by weight (UN 0214) (DOT) □ TRINITROBENZENE, wetted with not <30% water, by weight (UN 1354) (DOT) □ TRINITROBENZOL (GERMAN)
CONSENSUS REPORTS: Reported in EPA TSCA Inventory.
DOT CLASSIFICATION: EXPLOSIVE 1.1D; Label: EXPLOSIVE 1.1D (UN 0214); DOT Class: 4.1; Label: Flammable Solid - (UN 1354)
SAFETY PROFILE: Poison by ingestion and intravenous routes. Mutation data reported. A severe explosion hazard when shocked or exposed to heat. Trinitrobenzene is considered a powerful high explosive and has more shattering power than TNT. Although it is less sensitive to impact than TNT, it is not used much because it is difficult to produce. The complex with potassium trimethyl stannate explodes at room temperature. Forms heat-sensitive explosive complexes with alkyl or aryl metallates (e.g., lithium or potassium salts of trimethyl-, triethyl-, or triphenyl-germanate, -silanate, or -stannate). Can react vigorously with reducing materials. When heated to decomposition it emits highly toxic fumes of NO_x and explodes. See also NITRO

COMPOUNDS of AROMATIC HYDROCARBONS.

TML000   CAS: 129-66-8   HR: 3
TRINITROBENZOIC ACID (dry)
DOT: UN 0215/UN 1355
mf: $C_7H_3N_3O_8$ mw: 257.13
PROP: Orthorhombic or rhombohedral crystals from H_2O. Mp: 228.7°. Sol @ 25° - (2.05% in water, 26.6% in alc, 14.7% in ether), sol in methanol; sltly sol in benzene.
SYNS: TRINITROBENZOIC ACID, dry or wetted with <30% water, by weight (UN 0215) (DOT) □ TRINITROBENZOIC ACID, wetted with not <30% water, by weight (UN 1355) (DOT)
DOT CLASSIFICATION: EXPLOSIVE 1.1D; Label: EXPLOSIVE 1.1D (UN 0215); DOT Class: 4.1; Label: Flammable Solid - (UN 1355)
SAFETY PROFILE: An explosive. A hazard in preparation. Reacts with heavy metals to form heat- or impact-sensitive explosive salts. When heated to decomposition it emits toxic fumes of NO_x. See also NITRO COMPOUNDS of AROMATIC HYDROCARBONS and EXPLOSIVES, HIGH.

TML100   CAS: 5029-46-9   HR: 3
4,4,4-TRINITROBUTYRIC ACID
mf: $C_4H_5N_3O_8$ mw: 223.12
SYNS: BA 2759 □ BUTANOIC ACID, 4,4,4-TRINITRO- □ BUTYRIC ACID, 4,4,4-TRINITRO- □ γ,γ,γ-TRINITROBUTYRIC ACID □ USAF SE-3
SAFETY PROFILE: A poison by inhalation and intraperitoneal routes. Moderately toxic by ingestion. When heated to decomposition it emits toxic vapors of NO_x.

TML325   CAS: 28260-61-9   HR: 3
TRINITROCHLOROBENZENE
DOT: UN 0155
mf: $C_6H_2ClN_3O_6$ mw: 247.56
SYN: PICRYL CHLORIDE (DOT)
DOT CLASSIFICATION: EXPLOSIVE 1.1D; Label: EXPLOSIVE 1.1D
SAFETY PROFILE: Mutation data reported. An explosive. When heated to decomposition it emits toxic fumes of Cl^-

T

and NO$_x$. See also NITRO COMPOUNDS of AROMATIC HYDROCARBONS and EXPLOSIVES, HIGH.

TML750   CAS: 81-15-2   HR: 3
2,4,6-TRINITRO-1,3-DIMETHYL-5-tert-BUTYLBENZENE
DOT: UN 2956
mf: $C_{12}H_{15}N_3O_6$ mw: 297.30
PROP: Plates or needles from EtOH with strong musk odor. Mp: 112–113°.
SYNS: BENZENE, 1-tert-BUTYL-3,5-DIMETHYL-2,4,6-TRINITRO- □ 5-tert-BUTYL-2,4,6-TRINITROXYLENE □ 5-tert-BUTYL-2,4,6-TRINITRO-m-XYLENE (DOT) □ MUSK XYLENE □ MUSK XYLENE (DOT) □ MUSK XYLOL □ 2,4,6-TRINITRO-3,5-DIMETHYL-tert-BUTYLBENZENE □ m-XYLENE, 5-tert-BUTYL-2,4,6-TRINITRO- □ XYLENE MUSK
CONSENSUS REPORTS: Reported in EPA TSCA Inventory.
DOT CLASSIFICATION: 4.1; Label: Flammable Solid
SAFETY PROFILE: Low oral toxicity. A human skin irritant. A flammable solid. When heated to decomposition it emits toxic fumes of NO$_x$. See also NITRO COMPOUNDS of AROMATIC HYDROCARBONS.

TMM000   CAS: 918-54-7   HR: 3
2,2,2-TRINITROETHANOL
mf: $C_2H_3N_3O_7$ mw: 181.08
PROP: Long needles. Mp: 72°, bp: 60–62° @ 2 mm.
SYN: TRINITROETHANOL (DOT)
DOT CLASSIFICATION: Forbidden
SAFETY PROFILE: Poison by intraperitoneal route. A shock-sensitive explosive. When heated to decomposition it emits toxic fumes of NO$_x$. See also NITRO COMPOUNDS.

TMM250   CAS: 129-79-3   HR: 3
2,4,7-TRINITROFLUOREN-9-ONE
mf: $C_{13}H_5N_3O_7$ mw: 315.21
PROP: Pale-yellow needles from AcOH or C_6H_6. Mp: 176°.
SYNS: 2,4,7-TRINITRO-9-FLUORENONE □ 2,4,7-TR-INITROFLUORENONE (MAK)

CONSENSUS REPORTS: Reported in EPA TSCA Inventory.
DFG MAK: Confirmed Animal Carcinogen with Unknown Relevance to Humans
SAFETY PROFILE: Suspected carcinogen with experimental tumorigenic data. Mildly toxic by ingestion. Human mutation data reported. A skin and eye irritant. When heated to decomposition it emits highly toxic fumes of NO$_x$. See also NITRO COMPOUNDS of AROMATIC HYDROCARBONS and KETONES.

TMM500   CAS: 517-25-9   HR: 3
TRINITROMETHANE
mf: CHN_3O_6 mw: 151.05
PROP: Crystals. Mp: 15°, d: 1.469, bp: decomp >25°. Sol in water.
SYN: NITROFORM
CONSENSUS REPORTS: Reported in EPA TSCA Inventory.
DOT CLASSIFICATION: Forbidden
SAFETY PROFILE: Poison by ingestion and intraperitoneal routes. Moderately toxic by inhalation. Irritating to skin, eyes, and mucous membranes. Inhalation can cause headache and nausea. Causes mild narcosis. A very dangerous explosion hazard; explodes when heated rapidly. Dissolution is exothermic and solutions of more than 50% can explode. Mixtures of 90% trinitromethane + 10% isopropyl alcohol in polyethylene bottles have exploded. Frozen mixtures with 2-propanol (10%) explode when thawed. Can explode during distillation. Mixtures with divinyl ketone can explode at 4°C. When heated to decomposition it emits toxic fumes of NO$_x$. See also NITRO COMPOUNDS.

TMN000   CAS: 75321-19-6   HR: 3
1,3,6-TRINITROPYRENE
mf: $C_{16}H_7N_3O_6$ mw: 337.26
SYN: TRINITROPYRENE
DFG MAK: Confirmed Animal Carcinogen with Unknown Relevance to Humans
SAFETY PROFILE: Suspected carcinogen. Mutation data reported. When heated to

decomposition it emits toxic fumes of NO$_x$. See also NITRO COMPOUNDS of AROMATIC HYDROCARBONS.

TMN490   CAS: 118-96-7   HR: 3
2,4,6-TRINITROTOLUENE
DOT: UN 0209/UN 1356
mf: $C_7H_5N_3O_6$ mw: 227.15
PROP: Colorless, monoclinic, rhombohedral crystals from EtOH. Mp: 82°, bp: 240° - (explodes), flash p: explodes, d: 1.654. Sol in hot water, alc, ether. IDLH 500 mg/m^3
SYNS: BENZENE, 2-METHYL-1,3,5-TRINITRO- □ ENTSUFON □ NCI-C56155 □ TNT (OSHA) □ α-TNT □ TNT, dry or wetted with <30% water, by weight (UN 0209) - (DOT) □ TNT-TOLITE (FRENCH) □ TOLIT □ TOLITE □ 2,4,6-TRINITROTOLUEEN (DUTCH) □ TRINITRO-TOLUENE □ TRINITROTOLUENE (UN 0209) (DOT) □ TRINITROTOLUENE, wetted with not <30% water, by weight (UN 1356) (DOT) □ s-TRINITROTOLUENE □ sym-TR-INITROTOLUENE □ 2,4,6-TRINITROTOLUENE □ 2,4,6-TRINITROTOLUENE (ACGIH,OSHA) □ s-TRINITRO-TOLUOL □ sym-TRINITROTOLUOL □ 2,4,6-TRINITRO-TOLUOL (GERMAN) □ TRITOL □ TROJNITROTOLUEN (POLISH) □ TROTYL □ TROTYL OIL
CONSENSUS REPORTS: Reported in EPA TSCA Inventory. EPA Genetic Toxicology Program.
OSHA PEL: TWA 0.5 mg/m^3 (skin)
ACGIH TLV: TWA 0.1 ppm
DFG MAK: 0.011 ppm (0.1 mg/m^3); Confirmed Animal Carcinogen with Unknown Relevance to Humans
DOT CLASSIFICATION: EXPLOSIVE 1.1D; Label: EXPLOSIVE 1.1D (UN 0209); DOT Class: 4.1; Label: Flammable Solid - (UN 1356)
SAFETY PROFILE: Suspected carcinogen. Poison by subcutaneous route. Moderately toxic by ingestion. Human systemic effects by ingestion: hallucinations or distorted perceptions, cyanosis, and gastrointestinal changes. Experimental reproductive effects. Mutation data reported. A skin irritant. Has been implicated in aplastic anemia. Can cause headache, weakness, anemia, liver injury. May be absorbed through skin.

Flammable or explosive when exposed to heat or flame. Moderate explosion hazard; will detonate under strong shock. It detonates at around 240°C but can be distilled safely under reduced pressure. It is a comparatively insensitive explosive. In small quantities it will burn quietly if not confined. However, sudden heating of any quantity will cause it to detonate; the accumulation of heat when large quantities are burning will cause detonation. In other respects it is one of the most stable of all high explosives, and there are but a few restrictions for its handling. It is for this reason, from the military standpoint, that TNT is quantitatively the most used. It requires a fall of 130 cm for a 2 kg weight to detonate it. It is one of the most powerful high explosives. It can be detonated by the usual detonators and blasting caps (at least a No. 6). For full efficiency, the use of a high-velocity initiator, such as tetryl, is required. TNT is one of those explosives containing an oxygen deficiency. In other words, the addition of products that are oxygen rich can enhance its explosive power. Also mono- and dinitrotoluene may be added for reduction of the temperature of the explosion and to make the explosion flashless. Various materials are added to TNT to make what are known as permissible explosives. TNT may be regarded as the equivalent of 40% dynamite and can be used underwater. It is also used in the manufacture of a detonator fuse known as cordeau detonant. For the military, TNT finds use in all types of bursting charges, including armor-piercing types, although it is somewhat too sensitive to be ideal for this purpose and has since been replaced to a great extent by ammonium picrate. It is a relatively expensive explosive and does not compete seriously with dynamite for general commercial use.

Highly dangerous; explodes with shock or heating to 297°C. Various materials can reduce the explosive temperature: red lead - (to 192°C), sodium carbonate (to 218°C), potassium hydroxide (to 192°C). Mixtures with sodium dichromate + sulfuric acid may

T

ignite spontaneously. Reacts with nitric acid + metals (e.g., lead or iron) to form explosive products more sensitive to shock, friction, or contact with nitric or sulfuric acids. Reacts with potassium hydroxide dissolved in methanol to form explosive aci-nitro salts. Bases (e.g., sodium hydroxide, potassium iodide, tetramethyl ammonium octahydrotriborate) induce deflagration in molten TNT. Can react vigorously with reducing materials. When heated to decomposition it emits highly toxic fumes of NO_x. See also NITRO COMPOUNDS of AROMATIC HYDROCARBONS and EXPLOSIVES, HIGH.

TMO000   CAS: 538-23-8   HR: 3
TRIOCTANOIN

mf: $C_{27}H_{50}O_6$ mw: 470.77

PROP: Crystals from $Me_2CO/EtOH$. Mp: 10°.

SYNS: CAPRYLIC ACID TRIGLYCERIDE □ GLYCEROL TRICAPRYLATE □ GLYCEROL TRIOCTANOATE □ GLYCERYL TRIOCTANOATE □ MCT □ OCTANOIC ACID, 1,2,3-PROPANETRIYL ESTER □ OCTANOIC ACID TRIGLYCERIDE □ RATO □ TRICAPRYLIC GLYCERIDE □ TRICAPRYLIN □ TRIOCTANOYLGLYCEROL

CONSENSUS REPORTS: Reported in EPA TSCA Inventory. EPA Genetic Toxicology Program.

SAFETY PROFILE: Poison by intraperitoneal route. Moderately toxic by intravenous route. Mildly toxic by ingestion. Experimental reproductive effects. When heated to decomposition it emits acrid smoke and irritating fumes. See also ESTERS.

TMO550   CAS: 2467-12-1   HR: 2
TRI-n-OCTYL BORATE

mf: $C_{24}H_{51}BO_3$ mw: 398.56

PROP: Colorless moisture-sensitive liquid; odor of octyl alc. Bp: 182–184° @ 0.5 mm, flash p: 370°F (COC), d: 0.846 @ 23°, vap d: 13.7. Sol in non-hydroxylic solvs.

SYN: BORIC ACID, TRI-n-OCTYL ESTER

CONSENSUS REPORTS: Reported in EPA TSCA Inventory.

SAFETY PROFILE: Moderately toxic by ingestion. An eye irritant. Combustible when exposed to heat or flame; can react with oxidizing materials. To fight fire, use foam, CO_2, dry chemical. When heated to decomposition it emits acrid smoke and irritating fumes. See also ESTERS and BORON COMPOUNDS.

TMO600   CAS: 78-30-8   HR: 3
TRIORTHOCRESYL PHOSPHATE

mf: $C_{21}H_{21}O_4P$ mw: 368.39

PROP: Colorless liquid. Mp: −25 to −30°, bp: 410° (slt decomp), flash p: 437°F, d: 1.17, autoign temp: 725°F, vap d: 12.7. Insol in water; sol in alc and ether. IDLH 40 mg/m^3.

SYNS: o-CRESYL PHOSPHATE □ PHOSFLEX 179-C □ PHOSPHORIC ACID, TRI-o-CRESYL ESTER □ PHOSPHORIC ACID, TRIS(2-METHYLPHENYL) ESTER □ TOCP □ TOFK □ o-TOLYL PHOSPHATE □ TOTP □ TRICRESYL PHOSPHATE □ TRI-o-CRESYL PHOSPHATE □ o-TRIKRESYLPHOSPHATE (GERMAN) □ TRI-2--METHYLPHENYL PHOSPHATE □ TRIS(o-CRESYL)-PHOSPHATE □ TRIS(o-METHYLPHENYL)PHOSPHATE □ TRIS(o-TOLYL)-PHOSPHATE □ TRI-o-TOLYL PHOSPHATE □ TRI-2-TOLYL PHOSPHATE □ TROJKREZYLU FOSFORAN (POLISH)

CONSENSUS REPORTS: Reported in EPA TSCA Inventory.

OSHA PEL: TWA 0.1 mg/m^3 (skin)

ACGIH TLV: TWA 0.1 mg/m^3 (skin); Not Classifiable as a Human Carcinogen

SAFETY PROFILE: Poison by subcutaneous, intramuscular, intravenous, and intraperitoneal routes. Moderately toxic by ingestion. Most of the cases of tri-o-cresyl phosphate poisoning have followed its ingestion. In 1930, some 15,000 persons were affected in the United States, and of these, 10 died. The responsible material was found to be an alcoholic drink known as Jamaica ginger, or "jake." This beverage had been adulterated with about 2% of tri-o-cresyl phosphate. The affected persons developed a polyneuritis, which progressed, in many cases, with degeneration of the peripheral motor nerves, the anterior horn cells, and the pyramidal tracts. Sensory changes were absent. Since 1930 there have been several other outbreaks of poisoning following ingestion of the material. Tri-o-cresyl phosphate is more toxic than the m-form, and much more so than tri-p-cresyl phosphate or triphenyl phosphate. Experimental reproductive effects.

Combustible when exposed to heat or flame. Can react with oxidizing materials. To fight fire, use CO_2, dry chemical. When heated to decomposition it emits highly toxic fumes of PO_x. See also PHOSPHATES.

TMP000   CAS: 110-88-3   HR: 3
s-TRIOXANE
mf: $C_3H_6O_3$ mw: 90.09

PROP: Crystals or solid from Et_2O. Stable, cyclic trimer of formaldehyde, having characteristic ethanol- and chloroform-like odors. Mp: 64°, bp: 114.5°, subl readily, lel: 3.6%, uel: 28.7%, flash p: 113°F (OC), d: 1.17 @ 65°, autoign temp: 777°F, vap press: 13 mm @ 25°, vap d: 3.1. Very sol in water, alc, ketones, ether, acetone, chlorinated and aromatic hydrocarbons, org solvs; sltly sol in pentane, pet ether.

SYNS: POLYOXYMETHYLENE □ TRIOSSIMETELENE (ITALIAN) □ TRIOXANE □ sym-TRIOXANE □ 1,3,5-TRIOXANE □ TRIOXYMETHYLEEN (DUTCH) □ TRIO-XYMETHYLEN (GERMAN) □ TRIOXYMETHYLENE

CONSENSUS REPORTS: Reported in EPA TSCA Inventory.

SAFETY PROFILE: Mutation data reported. Can evolve toxic formaldehyde fumes when heated strongly or in contact with strong acids or acid fumes. Flammable liquid when exposed to heat, flame, or oxidizers. May explode when heated. Explosive in the form of vapor when exposed to heat or flame. Explodes on impact, possibly due to peroxide contamination. Mixtures with hydrogen peroxide are explosives sensitive to heat, shock, or contact with lead. Mixtures with liquid oxygen are highly explosive. Incompatible with oxidizing materials. To fight fire, use foam, CO_2, or dry chemical. When heated to decomposition it emits acrid smoke and

irritating fumes. See also FORMALDEHYDE.

TMP750   CAS: 91-81-6   HR: 3
TRIPELENNAMINE
mf: $C_{16}H_{21}N_3$ mw: 255.40

PROP: Oily liquid; amine odor. Bp: 185–190° @ 0.1 mm. Freely sol in water and alc; sltly sol in ether; practically insol in benzene and chloroform.

SYNS: BENZOXALE □ 2-(BENZYL(2-DIMETHYL AM-INOETHYL)AMINO)PYRIDINE □ N-BENZYL-N',N'-DI-METHYL-N-2-PYRIDYLETHYLENE DIAMINE □ BENZ-YL-(α-PYRIDYL)-DIMETHYLAETHYLENDIAMIN - (GERMAN) □ CIZARON □ DEHISTIN □ β-DIMETHYL-AMINO ETHYL-2-PYRIDYLAMINOTOLUENE □ β-DI-METHYLAMINOETHYL-2-PYRIDYLBENZYLAMINE □ N,N-DIMETHYL-N'-BENZYL-N'-(α-PYRIDYL)ETHYL-ENEDIAMINE □ NCI-C60662 □ PBZ □ PIRIBENZIL □ PYRIBENZAMINE □ PYRINAMINE BASE □ RESISTAM-INE □ TONARIL □ TRIPELENAMINE □ TRIPELENNAMINA (ITALIAN)

SAFETY PROFILE: Poison by ingestion and intraperitoneal routes. Human mutation data reported. Has been implicated in aplastic anemia. Used as an antihistamine. Addicts have added it to paregoric to make "blue velvet," which can cause euphoria by injection. When heated to decomposition it emits toxic fumes of NO_x.

TMQ250   CAS: 6304-33-2   HR: 3
2,3,3-TRIPHENYLACRYLONITRILE
mf: $C_{21}H_{15}N$ mw: 281.37

SYNS: α,β-DIPHENYLCINNAMONITRILE □ α-(DI-PHENYLMETHYLENE)BENZENEACETIC ACID □ TRIPHENYLACRYLONITRILE □ α,β,β-TRIPHENYLACR-YLONITRILE □ TRIPHENYLCYANOETHYLENE

CONSENSUS REPORTS: Cyanide and its compounds are on the Community Right-To-Know List.

SAFETY PROFILE: Poison by ingestion and intravenous routes. Questionable carcinogen with experimental carcinogenic data. When heated to decomposition it emits toxic fumes of NO_x and CN^-. See also NITRILES.

T

TMQ500　　**CAS: 603-34-9**　　**HR: 2**
TRIPHENYLAMINE
mf: $C_{18}H_{15}N$　　mw: 245.34
PROP: Monoclinic crystals from EtOAc. D: 0.774 @ 0°/0°, mp: 127°, bp: 195–205° @ 10–22 mm.
SYN: N,N-DIPHENYLANILINE
CONSENSUS REPORTS: Reported in EPA TSCA Inventory.
OSHA PEL: TWA 5 mg/m³
SAFETY PROFILE: Moderately toxic by ingestion. When heated to decomposition it emits toxic fumes of NO_x. See also AROMATIC AMINES.

TMQ550　　**CAS: 4756-75-6**　　**HR: 3**
TRIPHENYLANTIMONY OXIDE
mf: $C_{18}H_{15}OSb$　　mw: 369.08
PROP: Crystals.
SYN: STIBINE OXIDE, TRIPHENYL-
OSHA PEL: TWA 0.5 mg(Sb)/m³
ACGIH TLV: TWA 0.5 mg(Sb)/m³
NIOSH REL: (Antimony) 10H TWA 0.5 mg-(Sb)/m³
SAFETY PROFILE: Poison by intravenous route. When heated to decomposition it emits toxic fumes of Sb.

TMQ600　　**CAS: 3958-19-8**　　**HR: 3**
TRIPHENYL ANTIMONY SULFIDE
mf: $C_{18}H_{15}SSb$　　mw: 385.14
PROP: Crystals or monoclinic needles from EtOH. Mp: 120°. Very sol in C_6H_6, $CHCl_3$, HOAc; sol in EtOH; sltly sol in Et_2O and pet ether.
SYN: STIBINE SULFIDE, TRIPHENYL-
OSHA PEL: TWA 0.5 mg(Sb)/m³
ACGIH TLV: TWA 0.5 mg(Sb)/m³
NIOSH REL: (Antimony) 10H TWA 0.5 mg-(Sb)/m³
SAFETY PROFILE: Poison by intravenous route. When heated to decomposition it emits toxic fumes of SO_x and Sb.

TMS250　　**CAS: 58-72-0**　　**HR: 2**
TRIPHENYLETHYLENE
mf: $C_{20}H_{16}$　　mw: 256.36

PROP: Crystals from EtOH or AcOH. Mp: 72–73°, bp: 220–221° @ 14 mm.
SYN: 1,1,2-TRIPHENYLETHYLENE
CONSENSUS REPORTS: Reported in EPA TSCA Inventory.
SAFETY PROFILE: Experimental reproductive effects. Questionable carcinogen with experimental tumorigenic data. Human mutation data reported. When heated to decomposition it emits acrid smoke and irritating fumes.

TMT750　　**CAS: 115-86-6**　　**HR: 3**
TRIPHENYL PHOSPHATE
mf: $C_{18}H_{15}O_4P$　　mw: 326.30
PROP: Colorless, odorless, crystalline solid or prisms from EtOH or EtOH/pet ether. Mp: 49–50°, bp: 245° @ 11 mm, flash p: 428°F (CC), d: 1.268 @ 60°, vap press: 1 mm @ 193.5°. Insol in water; sol in alc, benzene, ether, chloroform, and acetone. IDLH 1000 mg/m³.
SYNS: CELLUFLEX TPP □ PHOSPHORIC ACID, TRIPHENYL ESTER □ TPP
CONSENSUS REPORTS: Reported in EPA TSCA Inventory.
OSHA PEL: TWA 3 mg/m³
ACGIH TLV: TWA 3 mg/m³; Not Classifiable as a Human Carcinogen
SAFETY PROFILE: Poison by subcutaneous route. Moderately toxic by ingestion. Absorbed slowly, particularly by skin contact. Not a potent cholinesterase inhibitor. Combustible when exposed to heat or flame. To fight fire, use CO_2, dry chemical. When heated to decomposition it emits toxic fumes of PO_x. See also TRITOLYL PHOSPHATE.

TMU000　　**CAS: 603-35-0**　　**HR: 2**
TRIPHENYLPHOSPHINE
mf: $C_{18}H_{15}P$　　mw: 262.30
PROP: Odorless crystals, plates, or prisms from Et_2O. Mp: 79°, bp: >360°, d: 1.194, flash p: 356°F (OC), vap d: 9.0. Insol in water; sol in HCl, benzene; sltly sol in alc; very sol in ether.

CONSENSUS REPORTS: Reported in EPA TSCA Inventory.

SAFETY PROFILE: Moderately toxic by ingestion. Mildly toxic by inhalation. A skin and eye irritant. Combustible when exposed to heat or flame. Slight explosion hazard in the form of vapor when exposed to flame. Can react vigorously with oxidizing materials. To fight fire, use dry chemical, fog, CO_2. When heated to decomposition it emits highly toxic fumes of phosphine and PO_x. See also PHOSPHINE and PHENOL.

TMU250   CAS: 101-02-0   HR: 3
TRIPHENYL PHOSPHITE
mf: $C_{18}H_{15}O_3P$ mw: 310.30

PROP: Water-white to pale-yellow solid or oily liquid; clean and pleasant odor. D: 1.184 @ 25°/25°, mp: 21–23°, bp: 183–184° @ 0.1 mm, flash p: 425°F (OC). Insol in water.

SYNS: EFED □ PHOSPHOROUS ACID, TRIPHENYL ESTER □ TRIFENOXYFOSFIN (CZECH) □ TRIFENYL-FOSFIT (CZECH)

CONSENSUS REPORTS: Reported in EPA TSCA Inventory.

SAFETY PROFILE: Poison by intraperitoneal and subcutaneous routes. Moderately toxic by ingestion. An experimental eye and severe human skin irritant. Combustible when exposed to heat or flame. To fight fire, use CO_2, mist, dry chemical. When heated to decomposition it emits toxic fumes of PO_x. See also PHENOL.

TMV250   CAS: 603-36-1   HR: 3
TRIPHENYL STIBINE
mf: $C_{18}H_{15}Sb$ mw: 353.08

PROP: Crystals or prisms from EtOH, pet ether, or Me_2CO (aq). Mp: 53–57°, d: 1.4343 @ 25°, bp: >360°. Water-insol; sol in org solvs.

SYN: TRIPHENYLANTIMONY

CONSENSUS REPORTS: Antimony and its compounds are on the Community Right-To-Know List. Reported in EPA TSCA Inventory.

OSHA PEL: TWA 0.5 mg(Sb)/m^3

ACGIH TLV: TWA 0.5 mg(Sb)/m^3

NIOSH REL: (Antimony) TWA 0.5 mg(Sb)-/m^3

SAFETY PROFILE: Poison by ingestion and intraperitoneal routes. Flammable when exposed to heat or flame. Can react vigorously with oxidizing materials. To fight fire, use water, foam, mist. When heated to decomposition it emits toxic fumes of Sb. See also ANTIMONY COMPOUNDS.

TMV800   CAS: 2847-65-6   HR: 3
TRIPHENYLTIN p-ACETAMIDO-
BENZOATE
mf: $C_{27}H_{23}NO_3Sn$ mw: 528.20

SYN: STANNANE, ((p-ACETAMIDOBENZOYL)OXY)-TRIPHENYL-

OSHA PEL: TWA 0.1 mg(Sn)/m^3 (skin)

ACGIH TLV: TWA 0.1 mg(Sn)/m^3; STEL 0.2 mg/m^3 (skin)

NIOSH REL: (Organotin Compound) 10H TWA 0.1 mg(Sn)/m^3

SAFETY PROFILE: Poison by intravenous route. When heated to decomposition it emits toxic fumes of NO_x and Sn.

TMV825   CAS: 73927-89-6   HR: 3
TRIPHENYLTIN CYANOACETATE
mf: $C_{21}H_{17}NO_2Sn$ mw: 434.08

SYNS: ACETIC ACID, CYANO-, TRIPHENYLSTANNYL ESTER □ STANNANE, (CYANOACETOXY)TRIPHENYL-

OSHA PEL: TWA 0.1 mg(Sn)/m^3 (skin)

ACGIH TLV: TWA 0.1 mg(Sn)/m^3; STEL 0.2 mg/m^3 (skin)

NIOSH REL: (Organotin Compound) 10H TWA 0.1 mg(Sn)/m^3

SAFETY PROFILE: Poison by intravenous route. When heated to decomposition it emits toxic fumes of NO_x and Sn.

TMW600   CAS: 67410-20-2   HR: 3
TRIPHENYLTIN PROPIOLATE
mf: $C_{21}H_{16}O_2Sn$ mw: 419.06

SYNS: (ACETYLENECARBONYLOXY)TRIPHENYLTIN □ PROPIOLIC ACID, TRIPHENYLSTANNYL ESTER □ STANNANE, (ACETYLENECARBONYLOXY)TRIPHEN-YL-

OSHA PEL: TWA 0.1 mg(Sn)/m^3 (skin)

ACGIH TLV: TWA 0.1 mg(Sn)/m^3; STEL 0.2 mg/m^3 (skin)

T

NIOSH REL: (Organotin Compound) 10H TWA 0.1 mg(Sn)/m³
SAFETY PROFILE: Poison by intravenous route. When heated to decomposition it emits toxic fumes of Sn.

TMX350   CAS: 68541-88-8   HR: 3
TRIPIPERIDINOPHOSPHINE SELENIDE
mf: $C_{15}H_{30}N_3PSe$ mw: 362.41
SYN: PHOSPHINE SELENIDE, TRIPIPERIDINO-
OSHA PEL: TWA 0.2 mg(Se)/m³
ACGIH TLV: TWA 0.2 mg(Se)/m³
SAFETY PROFILE: Poison by intravenous route. When heated to decomposition it emits toxic fumes of NO_x, PO_x, and Se.

TMX600   CAS: 14023-90-6   HR: 3
TRIPOTASSIUM HEXACYANO-
MANGANATE(3-)
mf: C_6MnN_6•3K   mw: 328.36
PROP: Air-stable dark red-brown needles from 10% KCN/EtOH.
SYNS: MANGANATE(3-), HEXACYANO-, TRIPOTASSIUM □ MANGANATE(3-), HEXAKIS(CYANO-C)-, TRIPOTASSIUM, (OC-6-11)- □ POTASSIUM MANGANOCYANIDE
OSHA PEL: CL 5 mg(Mn)/m³
ACGIH TLV: TWA 5 mg(Mn)/m³
SAFETY PROFILE: Poison by ingestion. When heated to decomposition it emits toxic fumes of NO_x and Mn.

TMY100   CAS: 102-67-0   HR: 3
TRIPROPYLALUMINUM
mf: $C_9H_{21}Al$ mw: 156.28
PROP: Air- and moisture-sensitive liquid. Mp: −107°, bp: 82–84°, d: 0.823.
SYNS: ALUMINUM, TRIPROPYL- □ TRIPROPYLALUM-INUM (DOT)
CONSENSUS REPORTS: Reported in EPA TSCA Inventory.
ACGIH TLV: TWA 2 mg(Al)/m³
SAFETY PROFILE: Pyrophoric, moisture-sensitive, flammable solid. Danger from spontaneous combustion. When heated to decomposition it emits toxic fumes of Al.

TMY250   CAS: 102-69-2   HR: 3
TRI-N-PROPYLAMINE

DOT: UN 2260
mf: $C_9H_{21}N$ mw: 143.31
PROP: Liquid. Mp: −93°, bp: 156°, flash p: 105°F (OC), d: 0.75, vap d: 4.9. Very sltly sol in water.
SYNS: N,N-DIPROPYL-1-PROPANAMINE □ TRIPROP-YLAMINE (DOT)
CONSENSUS REPORTS: Reported in EPA TSCA Inventory.
DOT CLASSIFICATION: 3; Label: Flammable Liquid, Corrosive
SAFETY PROFILE: Poison by ingestion. Moderately toxic by skin contact and inhalation. A corrosive irritant to skin, eyes, and mucous membranes. Flammable when exposed to heat, flame, or oxidizers. Can react with oxidizing materials. To fight fire, use foam, CO_2, dry chemical. When heated to decomposition it emits toxic fumes of NO_x. See also AMINES.

TMY850   CAS: 67445-50-5   HR: 3
TRIPROPYL(BUTYLTHIO)STANNANE
mf: $C_{13}H_{30}SSn$ mw: 337.18
SYNS: (BUTYLTHIO)TRIPROPYLSTANNANE □ STANNANE, (BUTYLTHIO)TRIPROPYL-
OSHA PEL: TWA 0.1 mg(Sn)/m³ (skin)
ACGIH TLV: TWA 0.1 mg(Sn)/m³; STEL 0.2 mg/m³ (skin)
NIOSH REL: (Organotin Compound) 10H TWA 0.1 mg(Sn)/m³
SAFETY PROFILE: Poison by intraperitoneal route. When heated to decomposition it emits toxic fumes of SO_x and Sn.

TNC500   CAS: 126-72-7   HR: 3
TRIS
mf: $C_9H_{15}Br_6O_4P$ mw: 697.67
PROP: Crystals. D: 2.24, flash p: >112°.
SYNS: ANFRAM 3PB □ APEX 462-5 □ BROMKAL P 67-6HP □ 2,3-DIBROMO-1-PROPANOL PHOSPHATE □ 2,3-DIBROMO-1-PROPANOL, PHOSPHATE (3:1) □ (2,3-DI-BROMOPROPYL) PHOSPHATE □ FIREMASTER T23P-LV □ FLACAVON R □ FLAMMEX AP □ FYROL HB32 □ NCI-C03270 □ PHOSPHORIC ACID, TRIS(2,3-DI-BROMOPROPYL) ESTER □ RCRA WASTE NUMBER U235 □ TDBP (CZECH) □ TRIS (flame retardant) □ TRIS(DI-BROMOPROPYL)PHOSPHATE □ TRIS(2,3-DI-BROMOPROPYL) PHOSPHATE □ TRIS(2,3-DI-

BROMOPROPYL) PHOSPHORIC ACID ESTER □ TRIS-2,3-DIBROMPROPYL ESTER KYSELINY FOSFORECNE - (CZECH) □ USAF DO-41 □ ZETIFEX ZN

CONSENSUS REPORTS: NTP 10th Report on Carcinogens. IARC Cancer Review: Group 2A IMEMDT 7,341,87; Animal Sufficient Evidence IMEMDT 20,575,79; Human Limited Evidence IMEMDT 20,575,79. NCI Carcinogenesis Bioassay - (feed); Clear Evidence: mouse, rat NCITR* NCI-CG-TR-76,78. Community Right-To-Know List. Reported in EPA TSCA Inventory. EPA Genetic Toxicology Program.

SAFETY PROFILE: Confirmed carcinogen with experimental carcinogenic, neoplastigenic, tumorigenic, and teratogenic data. Poison by intraperitoneal route. Moderately toxic by ingestion. Experimental reproductive effects. Human mutation data reported. An eye and severe skin irritant. Can cause testicular atrophy and sterility. Once used as a flame retardant additive to synthetic textiles and plastics, particularly in children's sleepwear. Use discontinued because it can be absorbed by human skin, or chewed or sucked off sleepwear by infants. May be flammable when exposed to heat or flame. When heated to decomposition it emits very toxic fumes of Br^- and PO_x.

TND000   CAS: 68-76-8   HR: 3
TRIS(1-AZIRIDINYL)-p-BENZOQUINONE
mf: $C_{12}H_{13}N_3O_2$ mw: 231.28
PROP: Purple crystals from EtOAc. Mp: 162.5–163°.
SYNS: BAYER 3231 □ 1,1',1"-(3,6-DIOXO-1,4-CYCLOHEXADIENE-1,2,4-TRIYL)TRISAZIRIDINE □ NSC-29215 □ ONCOVEDEX □ PRENIMON □ RIKER 601 □ 10257 R.P. □ TEIB □ TRENIMON □ TRIAZICHON - (GERMAN) □ TRIAZIQUINONE □ TRIAZIQUONE □ 2,3,5-TRI-(1-AZIRIDINYL)-p-BENZOQUINONE □ 2,3,5-TRIETHYLENEIMINO-1,4-BENZOQUINONE □ TRIETHYLENIMINOBENZOQUINONE □ TRISAETHYL-ENIMINOBENZOCHINON (GERMAN) □ 2,3,5-TRIS-(AZIRIDINO)-1,4-BENZOQUINONE □ 2,3,5-TRIS(1-AZIRIDINO)-p-BENZOQUINONE □ TRIS(AZIRIDINYL)-p-BENZOQUINONE □ 2,3,5-TRIS(1-AZIRIDINYL)-p-BENZOQUINONE □ 2,3,5-TRIS(AZIRIDINYL)-1,4-

BENZOQUINONE □ 2,3,5-TRIS(1-AZIRIDINYL)-2,5-CYL-OHEXADIENE-1,4-DIONE □ 2,3,5-TRISETHYLENEIM-INOBENZOQUINONE □ TRISETHYLENEIMINOQU-INONE □ 2,3,5-TRIS(ETHYLENIMINO)BENZOQU-INONE □ 2,3,5-TRIS(ETHYLENIMINO)-p-BENZOQU-INONE □ 2,3,5-TRIS(ETHYLENIMINO)-1,4-BENZOQU-INONE

CONSENSUS REPORTS: IARC Cancer Review: Group 3 IMEMDT 7,367,87; Animal Sufficient Evidence IMEMDT 9,67,75; Human Inadequate Evidence IMEMDT 9,67,75. Community Right-To-Know List. EPA Genetic Toxicology Program.

SAFETY PROFILE: Poison by intraperitoneal, intravenous, and parenteral routes. Experimental teratogenic and reproductive effects. Questionable carcinogen with experimental carcinogenic data. Human mutation data reported. When heated to decomposition it emits toxic fumes of NO_x. Used as a drug for the treatment of neoplastic diseases.

TND250   CAS: 545-55-1   HR: 3
TRIS-(1-AZIRIDINYL)PHOSPHINE OXIDE
DOT: UN 2501
mf: $C_6H_{12}N_3OP$ mw: 173.18
PROP: Colorless crystals or solid. Mp: 41°, bp: 90° @ 23 mm. Sol in water, alc, ether.
SYNS: APHOXIDE □ APO □ 1-AZIRIDINYL PHOSPH-INE OXIDE (TRIS) (DOT) □ CBC 906288 □ ENT 24,915 □ IMPERON FIXER T □ NSC 9717 □ 1,1',1"-PHOSPHINYL-IDYNETRISAZIRIDINE □ PHOSPHORIC ACID TRIETH-YLENE IMIDE □ PHOSPHORIC ACID TRIETHYL-ENEIMINE (DOT) □ SK-3818 □ TEF □ TEPA □ TRIA-ETHYLENPHOSPHORSAEUREAMID (GERMAN) □ TRIAZIRIDINOPHOSPHINE OXIDE □ TRI(AZIRIDIN-YL)PHOSPHINE OXIDE □ TRI(-1-AZIRIDINYL)PHOSPH-INE OXIDE □ N,N',N"-TRI-1,2-ETHANEDIYL PHOSPHORIC TRIMIDE □ TRIETHYLENEPHOS-PHOROTRIAMIDE □ TRIS(1-AZIRIDINE)PHOSPHINE OXIDE □ TRIS(N-ETHYLENE)PHOSPHOROTRIAMID-ATE
CONSENSUS REPORTS: IARC Cancer Review: Group 3 IMEMDT 7,56,87; Animal Inadequate Evidence IMEMDT 9,75,75. EPA Genetic Toxicology Program.
DOT CLASSIFICATION: 6.1; Label: Poison
SAFETY PROFILE: Poison by ingestion, skin contact, intravenous, and

T

intraperitoneal routes. Experimental teratogenic and reproductive effects. Questionable carcinogen with experimental carcinogenic and neoplastigenic data. Human mutation data reported. A corrosive irritant to the skin, eyes, and mucous membranes. When heated to decomposition it emits very toxic fumes of PO_x and NO_x. Used as an acaricide and in the permanent-press treatment of cotton.

TND500   CAS: 51-18-3   HR: 3
TRISAZIRIDINYLTRIAZINE
mf: $C_9H_{12}N_6$ mw: 204.27
PROP: Small crystals, or powder from $CHCl_3$. Mp: 139° (decomp). Water-sol.
SYNS: DRP 859025 □ ENT 25,296 □ M-9500 □ NSC 9706 □ PERSISTOL □ R-246 □ SEM (cytostatic) □ SK1133 □ TRETAMINE □ TRIAETHYLENMELAMIN (GERMAN) □ TRIAMELIN □ 1,1',1''-s-TRIAZINE-2,4,6-TRIYLTRISAZI-RIDINE □ TRIAZIRIDINYL TRIAZINE □ TRIETHANO-MELAMINE □ 2,4,6-TRI(ETHYLENEIMINO)-1,3,5-TRIAZ-INE □ 2,4,6-TRIETHYLENEIMINO-s-TRIAZINE □ TRIETHYLENEMELAMINE □ 2,4,6-TRIETHYLENIM-INO-s-TRIAZINE □ 2,4,6-TRIETHYLENIMINO-1,3,5-TRIAZINE □ 2,4,6-TRIS(1-AZIRIDINYL)-s-TRIAZINE □ 2,4,6-TRIS(1'-AZIRIDINYL)-1,3,5-TRIAZINE □ TRIS(ETH-YLENEIMINO)TRIAZINE □ TRISETHYLENEIMINO-1,3,5-TRIAZINE □ 2,4,6-TRIS(ETHYLENEIMINO)-s-TRIAZINE □ 2,4,6-TRIS(ETHYLENEIMINO)-s-TRIAZINE
CONSENSUS REPORTS: IARC Cancer Review: Group 3 IMEMDT 7,56,87; Animal Sufficient Evidence IMEMDT 9,95,75. EPA Genetic Toxicology Program.
SAFETY PROFILE: Poison by ingestion, intraperitoneal, intramuscular, intravenous, and subcutaneous routes. Experimental teratogenic and reproductive effects. Questionable carcinogen with experimental neoplastigenic and tumorigenic data. Human mutation data reported. Can cause gastrointestinal tract disturbances and bone marrow depression. When heated to decomposition it emits highly toxic fumes of NO_x. Used as an antineoplastic agent and as an insect sterilant.

TNF500   CAS: 817-09-4   HR: 3
TRIS(2-CHLOROETHYL)AMMONIUM CHLORIDE

mf: $C_6H_{12}Cl_3N \cdot ClH$ mw: 241.00
PROP: Crystals from $EtOH/Me_2CO$. Mp: 131–132.2°.
SYNS: LEKAMIN □ NSC-30211 □ R-47 □ SINALOST □ SK-100 □ TRICHLORMETHINE □ TRICHLORMETH-INIUM CHLORIDE □ TRI(β-CHLOROETHYL)AMINE HYDROCHLORIDE □ TRI-(2-CHLOROETHYL)AMINE HYDROCHLORIDE □ 2,2',2''-TRICHLOROTRIETHYLAM-INE HYDROCHLORIDE □ TRICHLOR-TRIAETHYLAM-IN-HYDROCHLORID (GERMAN) □ TRILLEKAMIN □ TRIMITAN □ TRIMUSTINE □ TRIMUSTINE HYDRO-CHLORIDE □ TRIS(β-CHLOROETHYL)AMINE HYDROCHLORIDE □ TRIS(2-CHLOROETHYL)AMINE HYDROCHLORIDE □ TRIS(2-CHLOROETHYL)AMINE MONOHYDROCHLORIDE □ TRIS-N-LOST □ TS-160
CONSENSUS REPORTS: IARC Cancer Review: Group 3 IMEMDT 7,56,87; Animal Inadequate Evidence IMEMDT 9,229,75. EPA Genetic Toxicology Program.
SAFETY PROFILE: Poison by ingestion, subcutaneous, intravenous, and intraperitoneal routes. Human systemic effects by ingestion and intravenous routes: somnolence, anorexia, headache, thrombosis distant from injection site, nausea or vomiting, and leukopenia. Experimental reproductive effects. Mutation data reported. Questionable carcinogen with experimental carcinogenic data. When heated to decomposition it emits very toxic fumes of Cl^-, NH_3, and NO_x. Used as an antineoplastic agent.

TNG050   CAS: 427-45-2   HR: 3
TRIS(p-CHLOROPHENYL)TIN FLUORIDE
mf: $C_{18}H_{12}Cl_3FSn$ mw: 472.34
SYN: STANNANE, FLUOROTRIS(p-CHLOROPHENYL)-
OSHA PEL: TWA 0.1 mg(Sn)/m³ (skin)
ACGIH TLV: TWA 0.1 mg(Sn)/m³; STEL 0.2 mg/m³ (skin)
NIOSH REL: (Organotin Compound) 10H TWA 0.1 mg(Sn)/m³
SAFETY PROFILE: Poison by intravenous route. When heated to decomposition it emits toxic fumes of Sn, Cl^-, and F^-.

TNH850   CAS: 6939-83-9   HR: 3
TRIS(DODECYLTHIO)ANTIMONY
mf: $C_{36}H_{75}S_3Sb$ mw: 726.04

SYN: STIBINE, TRIS(DODECYLTHIO)-
ACGIH TLV: TWA 0.5 mg(Sb)/m³
OSHA PEL: TWA 0.5 mg(Sb)/m³
NIOSH REL: (Antimony) 10H TWA 0.5 mg-(Sb)/m³
SAFETY PROFILE: Poison by intravenous route. When heated to decomposition it emits toxic fumes of SO_x and Sb.

TNK250   CAS: 57-39-6   HR: 3
TRIS(1-METHYLETHYLENE)PHOS-PHORIC TRIAMIDE
mf: $C_9H_{18}N_3OP$ mw: 215.27
PROP: Amber-colored liquid; amine odor. Bp: 118–125° @ 1 mm, d: 1.079 @ 25°/25°. Misc with water and all org solvs.
SYNS: C 3172 □ ENT 50,003 □ MAPO □ METEPA □ METHAPHOXIDE □ METHYL APHOXIDE □ 1,1',1"-PHOSPHINYLIDYNETRIS(2-METHYL)AZRIDINE □ TRIS(2-METHYL-1-AZIRIDINYL)PHOSPHINE OXIDE □ TRIS(2-METHYLAZIRIDIN-1-YL)PHOSPHINE OXIDE □ N,N',N"-TRIS(1-METHYLETHYLENE)PHOSPHORAMIDE
CONSENSUS REPORTS: IARC Cancer Review: Group 3 IMEMDT 7,56,87; Animal Inadequate Evidence IMEMDT 9,107,75. Reported in EPA TSCA Inventory. EPA Genetic Toxicology Program.
SAFETY PROFILE: Poison by ingestion, skin contact, intraperitoneal, and subcutaneous routes. Experimental teratogenic and reproductive effects. Questionable carcinogen with experimental carcinogenic data. Animal experiments suggest cholinesterase inhibition, possibly due to metabolic products of this material in the body. When heated to decomposition it emits very toxic fumes of NO_x and PO_x.

TNK400   CAS: 38668-83-6   HR: 3
TRIS(OCTAMETHYLPYROPHOSPHOR-AMIDE)MANGANESE(2+), DI-PERCHLORATE
mf: $C_{24}H_{72}MnN_{12}O_9P_6 \cdot 2ClO_4$ mw: 1112.74
SYNS: MANGANESE(2+), TRIS(OCTAMETHYLDI-PHOSPHORAMIDE-Op,Op')-, (OC-6-11)-, DIPERCHL-ORATE □ PERCHLORIC ACID, MANGANESE(2+) SALT, compounded with 3 mols. of OCTAMETHYL-PYROPHOSPHORAMIDE
OSHA PEL: CL 5 mg(Mn)/m³

ACGIH TLV: TWA 5 mg(Mn)/m³
SAFETY PROFILE: Poison by intraperitoneal route. When heated to decomposition it emits toxic fumes of NO_x, Mn, PO_x, and ClO^-.

TNL250   CAS: 150-38-9   HR: 3
TRISODIUM EDETATE
mf: $C_{10}H_{13}N_2O_8 \cdot 3Na$ mw: 358.22
PROP: Crystals from H_2O. Mp: >300°.
SYNS: EDETATE TRISODIUM □ EDTA TRISODIUM SALT □ N,N'-1,2-ETHANEDIYLBIS(N-CARBOXYMETH-YL)GLYCINE, TRISODIUM SALT □ ETHYLENEDIAM-INEACETIC ACID TRISODIUM SALT □ ETHYLENEDI-AMINETETRAACETIC ACID, TRISODIUM SALT □ NCI-C03974 □ NEVANAID-B POWDER □ PERMA KLEER 50, TRISODIUM SALT □ SEQUESTRENE Na3 □ SEQUES-TRENE TRISODIUM □ SEQUESTRENE TRISODIUM SALT □ TRILON AO □ TRISODIUM EDTA □ TRISODI-UM ETHYLENEDIAMINETETRAACETATE □ TRISODI-UM HYDROGEN ETHYLENEDIAMINETETRAACETATE □ TRISODIUM HYDROGEN (ETHYLENEDINITRILO)-TETRAACETATE □ TRISODIUM VERSENATE □ VERSENE 9
CONSENSUS REPORTS: Reported in EPA TSCA Inventory.
SAFETY PROFILE: Poison by intraperitoneal route. Moderately toxic by ingestion. When heated to decomposition it emits toxic fumes of NO_x and Na_2O.

TNN250   CAS: 21679-31-2   HR: 2
TRIS(2,4-PENTANEDIONATO)-CHROMIUM
mf: $C_{15}H_{21}CrO_6$ mw: 349.36
PROP: IDLH 25 mg/m³ [as Cr(III)].
SYNS: CHROMIC ACETYLACETONATE □ CHRO-MIUM ACETYLACETONATE □ CHROMIUM(3+) ACET-YLACETONATE □ CHROMIUM(III) ACETYLACETON-ATE □ CHROMIUM TRIACETYLACETONATE □ CHROMIUM TRIS(ACETYLACETONATE) □ CHROMIUM TRIS(2,4-PENTANEDIONATE) □ TRIS(ACETYL-ACETONATO)CHROMIUM □ TRIS(ACETYLACETON-ATO)CHROMIUM(III) □ TRIS(2,4-PENTANEDI-ONATO)CHROMIUM(3+)
CONSENSUS REPORTS: Chromium and its compounds are on the Community Right-To-Know List. Reported in EPA TSCA Inventory.
OSHA PEL: TWA 0.5 mg(Cr)/m³
ACGIH TLV: TWA 0.5 mg(Cr)/m³; Not Classifiable as a Carcinogen

SAFETY PROFILE: Moderately toxic by ingestion. Mildly toxic by skin contact. When heated to decomposition it emits acrid smoke and irritating fumes. See also CHROMIUM COMPOUNDS.

TNN500   CAS: 16432-36-3   HR: 3
TRIS(1-PHENYL-1,3-BUTANEDIONO)-CHROMIUM(III)

mf: $C_{30}H_{27}CrO_6$ mw: 535.57

PROP: Olive-drab crystals. Mp: 237–238°. Sol in C_6H_6 and $CHCl_3$. IDLH 25 mg/m³ [as Cr(III)].

SYNS: CHROMIUM TRIS(BENZOYLACETONATE) □ TRIS(BENZOYLACETONATO)CHROMIUM □ TRIS(1-PHENYL-1,3-BUTANEDIONATO)CHROMIUM □ TRIS(1-PHENYL-1,3-BUTANEDIONATO)CHROMIUM(3+) □ TRIS(1-PHENYL-1,3-BUTANEDIONATO-O,O')-CHROMIUM

CONSENSUS REPORTS: Chromium and its compounds are on the Community Right-To-Know List. Reported in EPA TSCA Inventory.

OSHA PEL: TWA 0.5 mg(Cr)/m³

ACGIH TLV: TWA 0.5 mg(Cr)/m³; Not Classifiable as a Carcinogen

SAFETY PROFILE: Poison by intravenous route. When heated to decomposition it emits acrid smoke and irritating fumes. See also CHROMIUM COMPOUNDS.

TNN760   CAS: 5910-77-0   HR: 3
TRIS(TRIDECYL)AMINE

mf: $C_{39}H_{81}N$ mw: 564.21

SYNS: 1-TRIDECANAMINE, N,N-DITRIDECYL- □ TRI-N-TRIDECYLAMINE □ TRITRIDECYL AMINE

SAFETY PROFILE: A poison by ingestion and skin contact. When heated to decomposition it emits toxic vapors of NO_x.

TNP250   CAS: 786-19-6   HR: 3
TRITHION

mf: $C_{11}H_{16}ClO_2PS_3$ mw: 342.87

PROP: Amber liquid. Bp: 82° @ 0.1 mm, d: 1.29 @ 20°. Essentially insol in water; misc in common solvents.

SYNS: ACARITHION □ AKARITHION □ CARBOFENOTHION (DUTCH) □ S-((p-CHLOROPHENYLTHIO)METHYL)-O,O-DIETHYL PHOSPHORODITHIOATE □ S-(4-CHLOROPHENYLTHIOMETHYL)DIETHYL PHOSPHOROTHIOLOTHIONATE □ DAGADIP □ O,O-DIAETHYL-S-((4-CHLOR-PHENYL-THIO)-METHYL)DITHIOPHOSPHAT (GERMAN) □ O,O-DIETHYL-S-((4-CHLOOR-FENYL-THIO)-METHYL)-DITHIOFOSFAAT - (DUTCH) □ O,O-DIETHYL-S-p-CHLORFENYLTHIO-METHYLESTER KYSELINY DITHIOFOSFORECNE - (CZECH) □ O,O-DIETHYL-S-p-CHLOROLPHENYLTHIO-METHYL DITHIOPHOSPHATE □ O,O-DIETHYL-P-CHLOROPHENYLMERCAPTOMETHYL DITHIOPHOSPHATE □ O,O-DIETHYL-S-(4-CHLOROPHENYLTHIOMETHYL) DITHIOPHOSPHATE □ O,O-DIETHYL-S-(p-CHLOROPHENYLTHIOMETHYL) PHOSPHORODITHIOATE □ O,O-DIETHYL-DITHIOPHOSPHORIC ACID, p-CHLOROPHENYLTHIOMETHYL ESTER □ O,O-DIETIL-S-((4-CLORO-FENIL-TIO)-METILE)-DITIOFOSFATO (ITALIAN) □ DITHIOPHOSPHATE de O,O-DIETHYLE et de (4-CHLORO-PHENYL) THIOMETHYLE (FRENCH) □ ENDYL □ ENT 23,708 □ GARRATHION □ LETHOX □ NEPHOCARP □ OLEOAKARITHION □ R-1303 □ STAUFFER R-1,303 □ TRITHION MITICIDE

CONSENSUS REPORTS: EPA Farm Worker Field Reentry FEREAC 39,16888,74. EPA Extremely Hazardous Substances List.

SAFETY PROFILE: Poison by ingestion, skin contact, and intraperitoneal routes. Moderately toxic by subcutaneous route. Mutation data reported. A cholinesterase inhibitor. When heated to decomposition it emits very toxic fumes of SO_x, PO_x, and Cl⁻. See also PARATHION, ESTERS, and MERCAPTANS.

TNP500   CAS: 1330-78-5   HR: 2
TRITOLYL PHOSPHATE

DOT: UN 2574

mf: $C_{21}H_{21}O_4P$ mw: 368.39

PROP: Oily, flame-resistant liquid. D: 1.16, bp: 265°, pour point: 28°, flash p: 410°F. Insol in water; misc with all common org solvs and thinners, linseed oil, chinawood oil, and castor oil.

SYNS: CELLUFLEX 179C □ CRESYL PHOSPHATE □ DISFLAMOLL TKP □ DURAD □ FLEXOL PLASTICIZER TCP □ FYRQUEL 150 □ IMOL S 140 □ KRONITEX □ L-INDOL □ NCI-C61041 □ PHOSPHATE de TRICRESYLE - (FRENCH) □ PHOSPHORIC ACID, TRITOLYL ESTER □ TRICRESILFOSFATI (ITALIAN) □ TRICRESYL-FOSFATEN (DUTCH) □ TRICRESYL PHOSPHATE □ TRICRESYLPHOSPHATE, with more than 3% ortho isomer - (DOT) □ TRIKRESYLPHOSPHATE (GERMAN) □ TRIS-(TOLYLOXY)PHOSPHINE OXIDE

CONSENSUS REPORTS: Reported in EPA TSCA Inventory.

DOT CLASSIFICATION: 6.1; Label: Poison
SAFETY PROFILE: Moderately toxic by ingestion and skin contact. Human systemic effects by ingestion: flaccid paralysis without anesthesia, motor activity changes, and muscle weakness. An experimental teratogen. Experimental reproductive effects. An eye and skin irritant. Combustible. When heated to decomposition it emits toxic fumes of PO_x.

TNT500   CAS: 22089-22-1   HR: 3
TROPHOSPHAMIDE
mf: $C_9H_{18}Cl_3N_2O_2P$ mw: 323.61
SYNS: A-4828 □ ASTA Z 4828 □ 2-(BIS(2-CHLOROETH-YL)AMINO)-3-(2-CHLOROETHYL)TETRAHYDRO-2H-1,3,2-OXAPHOSPHORINE-2-OXIDE □ 3-(2-CHLORO-ETHYL)-2-(BIS(2-CHLOROETHYL)AMINO)PERHYDRO-2H-1,3,2-OXAZAPHOSPHORINE-2-OXIDE □ CYCLO-PHOSPHAMIDE-N-MONOCHLOROETHYL derivative □ IXOTEN □ NSC 109723 □ TFF □ TRIFOSFAMIDE □ TRILOFOSFAMIDA □ TRILOPHOSPHAMIDE □ N,N,N'-TRIS(2-CHLORAETHYL)-N',O-PROPYLEN-PHOSPHORSAEUREESTER-DIAMID (GERMAN) □ N,N,N'-TRIS(2-CHLOROETHYL)-N',O-PROPYLENE PHOSPHORIC ACID ESTER DIAMIDE □ N,N,3-TRIS(2-CHLOROETHYL)TETRAHYDRO-2H-1,3,2-OXAPHOS-PHORIN-2-AMINE-2-OXIDE □ TRISFOSFAMIDE □ TRISPHOSPHAMIDE □ TROFOSFAMID □ TROPHOSPHAMID □ Z 4828
CONSENSUS REPORTS: EPA Genetic Toxicology Program.
SAFETY PROFILE: Poison by intraperitoneal, subcutaneous, and intravenous routes. Moderately toxic by ingestion. Human mutation data reported. Human systemic effects by unspecified routes: hematuria, leukopenia, and thrombocytopenia. When heated to decomposition it emits very toxic fumes of Cl^-, NO_x, and PO_x.

TNU000   CAS: 132-17-2   HR: 3
TROPINE BENZOHYDRYL ETHER
METHANESULFONATE
mf: $C_{21}H_{25}NO \cdot CH_4O_3S$ mw: 403.58
PROP: Crystals from Me_2CO/Et_2O. Mp: 143°.
SYNS: BENZATROPINE METHANESULFONATE □ BENZOTROPINE MESYLATE □ BENZOTROPINE METHANESULFONATE □ BENZTROPINE MESYLATE □ BENZTROPINE METHANESULFONATE □ 3-DI-PHENYLMETHOXYTROPANE MESYLATE □ 3-DIPHEN-YLMETHOXYTROPANE METHANESULFONATE
SAFETY PROFILE: Poison by ingestion, intravenous, subcutaneous, and intraperitoneal routes. Human systemic effects by ingestion: psychotropic effects. Mutation data reported. When heated to decomposition it emits very toxic fumes of NO_x and SO_x. See also ETHERS.

TNW500   CAS: 54-12-6   HR: 2
dl-TRYPTOPHAN
mf: $C_{11}H_{12}N_2O_2$ mw: 204.25
PROP: Plates from EtOH (aq). Mp: 275–282°. White crystals or crystalline powder; odorless. Sol in water, dil acids, alkalies; sltly sol in alc. Optically inactive.
CONSENSUS REPORTS: Reported in EPA TSCA Inventory. EPA Genetic Toxicology Program.
SAFETY PROFILE: Experimental reproductive effects. Questionable carcinogen with experimental carcinogenic data. When heated to decomposition it emits toxic fumes of NO_x.

TNX000   CAS: 73-22-3   HR: 2
l-TRYPTOPHAN
mf: $C_{11}H_{12}N_2O_2$ mw: 204.25
PROP: Leaflets or plates from dil alc. An essential amino acid occurring in isomeric forms. Mp: decomp @ 289°. The l and dl forms: white crystals or crystalline powder; slt bitter taste. dl Form: sltly sol in water. l Form: sol in water, hot alc, and alkali hydroxides; insol in chloroform.
SYNS: α'-AMINO-3-INDOLEPROPRIONIC ACID □ α-AMINO-INDOLE-3-PROPRIONIC ACID □ l-α-AMINO-3--INDOLEPROPRIONIC ACID □ 2-AMINO-3-INDOL-3-YL-PROPRIONIC ACID □ EH 121 □ INDOLE-3-ALANINE □ 1-β-3-INDOLYLALANINE □ NCI-C01729 □ (−)-TRYPTOPHAN □ l-TRYPTOPHAN (FCC) □ TRYPTOPHANE
CONSENSUS REPORTS: NCI Carcinogenesis Bioassay (feed); No Evidence: mouse, rat NCITR* NCI-CG-TR-71,78. Reported in EPA TSCA Inventory.

T

SAFETY PROFILE: Moderately toxic by intraperitoneal route. Experimental teratogenic and reproductive effects. Human mutation data reported. Questionable carcinogen with experimental tumorigenic data. When heated to decomposition it emits toxic fumes of NO_x.

TNX275   CAS: 62450-06-0   HR: 3
TRYPTOPHAN P1

mf: $C_{13}H_{13}N_3$ mw: 211.29

SYNS: 3-AMINO-1,4-DIMETHYL-γ-CARBOLINE □ 3-AMINO-1,4-DIMETHYL-5H-PYRIDO(4,3-b)INDOLE □ 1,4-DIMETHYL-5H-PYRIDO(4,3-b)INDOL-3-AMINE □ TRP-P-1 □ dl-TRYPTOPHAN, pyrolyzate 1

CONSENSUS REPORTS: IARC Cancer Review: Group 2B IMEMDT 7,56,87; Animal Sufficient Evidence IMEMDT 31,247,83. EPA Genetic Toxicology Program.

SAFETY PROFILE: Confirmed carcinogen with experimental carcinogenic and neoplastigenic data. Poison by ingestion. Human mutation data reported. When heated to decomposition it emits toxic fumes of NO_x.

TOA000   CAS: 57-94-3   HR: 3
TUBOCURARINE HYDROCHLORIDE

mf: $C_{38}H_{44}N_2O_6 \cdot 2Cl$ mw: 694.74

SYNS: AMERIZOL □ CURARIN-HAF □ DELACURARINE □ DEXTROTUBOCURARINE CHLORIDE □ d-7',12'-DIHYDROXY-6,6'-DIMETHOXY-2,2',2'-TRIMETHYL-TUBOCURARANIUM CHLORIDE □ INTOCOSTRIN □ d-PARACURARINE CHLORIDE □ TUBADIL □ TUBARINE □ TUBOCURARINE CHLORIDE □ (+)-TUBOCURARINE CHLORIDE □ d-TUBOCURARINE CHLORIDE □ TUBOCURARINE, CHLORIDE, HYDROCHLORIDE, (+)- - (8CI) □ d-TUBOCURARINE DICHLORIDE □ d-TUBOCURARINE HYDROCHLORIDE □ (+)-TUBOCURARINE HYDROCHLORIDE

CONSENSUS REPORTS: EPA Genetic Toxicology Program.

SAFETY PROFILE: Poison by ingestion, intravenous, intraperitoneal, and subcutaneous routes. Human toxicity: Large doses and overdoses may cause respiratory paralysis and hypotension. When heated to decomposition it emits very toxic fumes of NO_x and Cl^-. Used as a muscle relaxant.

TOA500   HR: 2
TUNG NUT MEALS

SAFETY PROFILE: Toxic by ingestion. Contact causes dermatitis. Ingestion causes nausea, vomiting, cramps, diarrhea and tenesmus, thirst, dizziness, lethargy, and disorientation. Large doses can cause fever, tachycardia, and respiratory effects. Combustible in the form of dust when exposed to heat or flame. Processed material must be cooled thoroughly before storage so as not to overdry; can react with oxidizing materials. See also SAPONIN and TUNG NUT.

TOA510   HR: 2
TUNG NUT OIL

PROP: Pale-yellow liquid; characteristic disagreeable odor. Sol in chloroform, ether, carbon disulfide, and oils. Polymerized product is practically insol in org solvs.

SYN: CHINAWOOD OIL

SAFETY PROFILE: Toxic by ingestion. Contact causes dermatitis. Ingestion causes nausea, vomiting, cramps, diarrhea and tenesmus, thirst, dizziness, lethargy, and disorientation. Large doses can cause fever, tachycardia, and respiratory effects. Combustible when exposed to heat or flame. Can react with oxidizing materials.

TOA750   CAS: 7440-33-7   HR: 3
TUNGSTEN

af: W   aw: 183.85

PROP: A steely-gray to white, cuttable, forgeable, and spinnable metallic element. Fairly soft when pure. Mp: 3410°, d: 19.3 @ 20°, bp: 5900°.

SYN: WOLFRAM

CONSENSUS REPORTS: Reported in EPA TSCA Inventory.

OSHA PEL: TWA (Insoluble compounds) 5 mg(W)/m³; STEL 10 mg(W)/m³; (soluble compounds) 1 mg(W)/m³; STEL 3 mg(W)-/m³

ACGIH TLV: TWA (Insoluble compounds) 5 mg(W)/m³; STEL 10 mg(W)/m³; (soluble compounds) 1 mg(W)/m³; STEL 3 mg(W)-/m³

NIOSH REL: (Tungsten, Insoluble) TWA 5 mg(W)/m³

SAFETY PROFILE: An inhalation hazard. Mildly toxic by an unspecified route. An experimental teratogen. Experimental reproductive effects. A skin and eye irritant. Flammable in the form of dust when exposed to flame. The powdered metal may ignite on contact with air or oxidants (e.g., bromine pentafluoride, bromine, chlorine trifluoride, potassium perchlorate, potassium dichromate, nitryl fluoride, fluorine, oxygen difluoride, iodine pentafluoride, hydrogen sulfide, sodium peroxide, lead(IV) oxide). See also TUNGSTEN COMPOUNDS and POWDERED METALS.

TOB500   CAS: 12070-12-1   HR: 3
TUNGSTEN CARBIDE
mf: WC   mw: 195.9.

PROP: Solid with metallic luster. Electrical conductor. Mp: >2755° (decomp). Insol.

CONSENSUS REPORTS: Reported in EPA TSCA Inventory.

OSHA PEL: TWA 5 mg(W)/m³; STEL 10 mg(W)/m³

ACGIH TLV: TWA 5 mg(W)/m³; STEL 10 mg(W)/m³

NIOSH REL: (Tungsten, insoluble) TWA 5 mg(W)/m³

SAFETY PROFILE: Chronic inhalation causes lung damage in humans. Ignites at 600°C in nitrogen oxide atmospheres. Violent reaction with F_2, ClF_3, NO_x, IF_5, PbO_2, NO_2, N_2O. See also TUNGSTEN COMPOUNDS.

TOC500   HR: 2
TUNGSTEN COMPOUNDS
OSHA PEL: TWA (insoluble compounds) 5 mg(W)/m³; STEL 10 mg(W)/m³; (soluble compounds) 1 mg(W)/m³; STEL 3 mg(W)-/m³

ACGIH TLV: TWA (insoluble compounds) 5 mg(W)/m³; STEL 10 mg(W)/m³; (soluble compounds) 1 mg(W)/m³; STEL 3 mg(W)-/m³

SAFETY PROFILE: Tungsten compounds are considered somewhat more toxic than those of molybdenum. However, industrially, this element does not constitute an important health hazard. Exposure is related chiefly to the dust arising from the crushing and milling of the two chief ores of tungsten, namely, scheelite and wolframite. The feeding of 2, 5, and 10% of diet as tungsten metal over a period of 70 days has shown no marked effect upon the growth of rats, as measured in terms of gain in weight. Sodium tungstate (Na_2WO_4), the most soluble salt, is moderately toxic by ingestion. Large overdoses cause central nervous system disturbances, diarrhea, respiratory failure, and death in experimental animals. Ammonium-p-tungstate has been found to be much less toxic to rats upon ingestion than either tungstic oxide or sodium tungstate. Tungsten carbide (WC) is chronically toxic to humans by inhalation although the effect may be due to contamination by cobalt. Heavy exposure to the dust or the ingestion of large amounts of the soluble compounds produces changes in body weight, behavior, blood cells, cholinesterase activity, and sperm in experimental animals. See also specific compounds.

TOC550   CAS: 7783-82-6   HR: 3
TUNGSTEN HEXAFLUORIDE
DOT: UN 2196

mf: F_6W mw: 297.85

PROP: Colorless gas or pale-yelow liquid - (orthorhombic crystals when solid). Hexane solutions often strongly colored. Mp: 2.3°, bp: 17.5°. Misc in many solvs (e.g., cyclohexane, CCl_4).

SYN: TUNGSTEN FLUORIDE

CONSENSUS REPORTS: Reported in EPA TSCA Inventory.

OSHA PEL: TWA 2.5 mg(F)/m³

ACGIH TLV: TWA 2.5 mg(F)/m³; BEI: 3 mg/g creatinine of fluorides in urine prior to shift; 10 mg/g creatinine of fluorides in urine at end of shift; 5 mg(W)/m³; STEL 10 mg(W)/m³

DOT CLASSIFICATION: 2.3; Label: Poison Gas

SAFETY PROFILE: A poison and corrosive liquid or gas.

TOC750   CAS: 1314-35-8   HR: 2
TUNGSTEN OXIDE

mf: O₃W   mw: 231.85

PROP: Heavy, yellow powder. Insol in water; sol in caustic alkalies; very sltly sol in acids.

SYNS: C.I. 77901 □ TUNGSTEN BLUE □ TUNGSTEN TRIOXIDE □ TUNGSTIC ANHYDRIDE □ TUNGSTIC OXIDE □ WOLFRAMITE

CONSENSUS REPORTS: Reported in EPA TSCA Inventory.

OSHA PEL: TWA 5 mg(W)/m³

ACGIH TLV: TWA 5 mg(W)/m³

NIOSH REL: TWA 5 mg(W)/m³

SAFETY PROFILE: Moderately toxic by ingestion. Can react violently with ClF₃, Li, Cl₂. See also TUNGSTEN COMPOUNDS.

TOD625   HR: D
TURMERIC

PROP: From solvent extraction of dried ground rhizome of *Curcuma ionga L.* Bright-yellow powder or yellow-orange to brown liquid; mustard taste. Misc in water.

SYN: OLEORESIN TUMERIC

SAFETY PROFILE: Human mutation data reported. When heated to decomposition it emits acrid smoke and irritating fumes.

TOD750   CAS: 8006-64-2   HR: 3
TURPENTINE

DOT: UN 1299/UN 1300

PROP: Colorless liquid; characteristic odor. Bp: 154–170°, lel: 0.8%, flash p: 95°F (CC), d: 0.854–0.868 @ 25°/25°, autoign temp: 488°F, vap d: 4.84, ULC: 40–50. IDLH 800 ppm.

SYNS: OIL of TURPENTINE □ OIL of TURPENTINE, RECTIFIED □ SPIRIT of TURPENTINE □ SPIRITS of TURPENTINE □ TEREBENTHINE □ TERPENTIN OEL - (GERMAN) □ TURPENTINE (UN 1299) (DOT) □

TURPENTINE OIL □ TURPENTINE OIL, RECTIFIER □ TURPENTINE STEAM DISTILLED □ TURPENTINE SUBSTITUTE (UN 1300) (DOT)

CONSENSUS REPORTS: Reported in EPA TSCA Inventory.

OSHA PEL: TWA 100 ppm

ACGIH TLV: TWA 20 ppm (sensitizer); Not Classifiable as a Human Carcinogen

DFG MAK: 100 ppm (560 mg/m³)

DOT CLASSIFICATION: 3; Label: Flammable Liquid

SAFETY PROFILE: An experimental poison by intravenous route. Moderately toxic to humans by ingestion. Mildly toxic experimentally by ingestion and inhalation. Human systemic effects by ingestion and inhalation: conjunctiva irritation, other olfactory and eye effects, hallucinations or distorted perceptions, antipsychotic, headache, pulmonary, and kidney changes. A human eye irritant. Irritating to skin and mucous membranes. Can cause serious irritation of kidneys. Questionable carcinogen with experimental tumorigenic data. A common air contaminant. A very dangerous fire hazard when exposed to heat or flame; can react vigorously with oxidizing materials. Avoid impregnation of combustibles with turpentine. Keep cool and ventilated. Spontaneous heating is possible. Moderate explosion hazard in the form of vapor when exposed to flame; can react violently with Ca(OCl)₂, Cl₂, CrO₃, Cr-(OCl)₂, SnCl₄, hexachloromelamine, trichloromelamine. To fight fire, use foam, CO₂, dry chemical. When heated to decomposition it emits acrid smoke and irritating fumes.

TOE600   CAS: 1401-69-0   HR: 3
TYLOSIN

mf: C₄₅H₇₇NO₁₇   mw: 904.23

PROP: Amorphous solid or crystals from water. Mp: 128–132°. Sol in water at 25°: 5 mg/mL. Sol in lower alc, esters, and ketones, in chlorinated hydrocarbons, benzene, ether.

SYNS: TYLAN □ TYLON

SAFETY PROFILE: Poison by intravenous route. Moderately toxic by ingestion and intraperitoneal routes. When heated to decomposition it emits toxic fumes of NO_x. See also TYLOSIN HYDROCHLORIDE.

TOE750   CAS: 11032-12-5   HR: 3
TYLOSIN HYDROCHLORIDE
mf: $C_{45}H_{77}NO_{17} \cdot ClH$ mw: 940.69
PROP: Crystals. Mp: 141–145°.
SAFETY PROFILE: Poison by intravenous route. When heated to decomposition it emits very toxic fumes of NO_x and HCl.

TOE810   HR: D
TYLOSIN and SULFAMETHAZINE
SAFETY PROFILE: When heated to decomposition it emits acrid smoke and irritating fumes.

TOG300   CAS: 60-18-4   HR: D
l-TYROSINE
mf: $C_9H_{11}NO_3$ mw: 181.21
PROP: Colorless, silky needles or white crystalline powder from water. Mp: 290–295° (decomp) (slow heat). Almost insol in water, dil mineral acids, alkaline solutions; sltly sol in alc.
SYNS: l-β-(p-HYDROXYPHENYL)ALANINE □ TYROS-INE □ p-TYROSINE □ l-p-TYROSINE
CONSENSUS REPORTS: Reported in EPA TSCA Inventory.
SAFETY PROFILE: An experimental teratogen. Experimental reproductive effects. When heated to decomposition it emits acrid smoke and irritating fumes.

T

U

UJJ000   CAS: 112-44-7   HR: 1
1-UNDECANAL
mf: $C_{11}H_{22}O$ mw: 170.33
PROP: Colorless to sltly yellow liquid; sweet, fatty, floral odor. Mp: −4°, bp: 117° @ 18 mm, flash p: 235°F (COC), d: 0.830 @ 20°/4°, refr index: 1.430, vap press: 0.04 mm @ 20°, vap d: 5.94. Sol in fixed oils, propylene glycol; insol in glycerin, water @ 223°. Reported in lemon and mandarin oils (FCTXAV 11,477,73).
SYNS: ALDEHYDE-14 □ 1-DECYL ALDEHYDE □ FEMA No. 3092 □ HENDECANAL □ HENDECAN-ALDEHYDE □ UNDECANAL □ n-UNDECANAL □ UNDECANALDEHYDE □ UNDECYL ALDEHYDE □ n-UNDECYL ALDEHYDE □ UNDECYLIC ALDEHYDE
CONSENSUS REPORTS: Reported in EPA TSCA Inventory.
SAFETY PROFILE: Low toxicity by ingestion and skin contact. A skin irritant. Combustible liquid when exposed to heat or flame. To fight fire, use CO_2, dry chemical. When heated to decomposition it emits acrid smoke and irritating fumes. See also ALDEHYDES.

UJS000   CAS: 1120-21-4   HR: 3
UNDECANE
DOT: UN 2330
mf: $C_{11}H_{24}$ mw: 156.35
PROP: Colorless liquid. D: 0.7402 @ 20°/4°, fp: −25.75°, bp: 195.6°, flash p: 149°F (OC), vap d: 5.4. Insol in water.
SYNS: HENDECANE □ n-UNDECANE
CONSENSUS REPORTS: Reported in EPA TSCA Inventory.
DOT CLASSIFICATION: 3; Label: Flammable Liquid
SAFETY PROFILE: Moderately toxic by intravenous route. Flammable liquid when exposed to heat, sparks, flame, or oxidizers.

To fight fire, use foam, mist, dry chemical. Emitted from modern building materials (CENEAR 69,22,91). When heated to decomposition it emits acrid smoke and irritating fumes. See also ALKANES.

UKS000   CAS: 112-12-9   HR: 2
2-UNDECANONE
mf: $C_{11}H_{22}O$ mw: 170.33
PROP: Colorless liquid. Mp: 12°, fp: 15°, bp: 223°, flash p: 192°F (CC), d: 0.829 @ 30°, vap d: 5.9. Insol in water.
SYNS: 2-HENDECANONE □ METHYL NONYL KETONE □ METHYL-n-NONYL KETONE □ MGK DOG AND CAT REPELLENT □ NONYL METHYL KETONE
CONSENSUS REPORTS: Reported in EPA TSCA Inventory.
DOT CLASSIFICATION: 3; Label: Flammable Liquid
SAFETY PROFILE: Moderately toxic by ingestion. Combustible when exposed to heat or flame; can react with oxidizing materials. To fight fire, use CO_2, dry chemical. When heated to decomposition it emits acrid smoke and irritating fumes. See also KETONES.

ULA000   CAS: 927-49-1   HR: 3
6-UNDECANONE
mf: $C_{11}H_{22}O$ mw: 170.33
PROP: A liquid. Mp: 14–15°, bp: 226°.
SYNS: AMYL KETONE □ DIAMYL KETONE □ DIPENTYL KETONE □ 6-OXOUNDECANE □ PENTYL KETONE □ UNDECAN-6-ONE
CONSENSUS REPORTS: Reported in EPA TSCA Inventory.
DOT CLASSIFICATION: 3; Label: Flammable Liquid
SAFETY PROFILE: Poison by intravenous route. Moderately toxic by ingestion. When heated to decomposition it emits acrid

U

smoke and irritating fumes. See also KETONES.

ULJ000   CAS: 112-45-8   HR: 2
10-UNDECENAL

mf: $C_{11}H_{20}O$ mw: 168.31

PROP: Colorless to light yellow liquid; rose odor. D: 0.840–0.850, bp: 101–103° @ 10 mm, refr index: 1.441–1.447, flash p: 212°F. Sol in fixed oils, propylene glycol; insol in water @ 235°, glycerin.

SYNS: ALDEHYDE C-11, UNDECYLENIC □ FEMA No. 3095 □ HENDECENAL □ 1-UNDECEN-10-AL □ UNDEC-YLENALDEHYDE □ 10-UNDECYLENEALDEHYDE □ UNDECYLENIC ALDEHYDE

CONSENSUS REPORTS: Reported in EPA TSCA Inventory.

SAFETY PROFILE: Low toxicity by ingestion and skin contact. A skin irritant. Combustible liquid. When heated to decomposition it emits acrid smoke and irritating fumes. See also ALDEHYDES.

ULS400   CAS: 5760-50-9   HR: 1
9-UNDECENOIC ACID, METHYL ESTER

mf: $C_{12}H_{22}O_2$ mw: 198.34

PROP: Fragrance.

SYNS: METHYL 9-UNDECENOATE □ METHYL 10-UNDECENOATE □ METHYL UNDECYLENATE

CONSENSUS REPORTS: Reported in EPA TSCA Inventory.

SAFETY PROFILE: A skin irritant. When heated to decomposition it emits acrid smoke and irritating fumes.

ULS875   HR: D
2-UNDECENOL

mf: $C_{11}H_{22}O$ mw: 170.30

PROP: White to sltly yellow liquid; oily, sweet, floral odor. D: 0.847, refr index: 1.450 @ 22°. Insol in water.

SAFETY PROFILE: When heated to decomposition it emits acrid smoke and irritating fumes.

UNA000   CAS: 112-42-5   HR: 2
UNDECYL ALCOHOL

mf: $C_{11}H_{24}O$ mw: 172.35

PROP: Colorless liquid; fatty-floral odor. D: 0.820–0.840, refr index: 1.437–1.443, mp: 19°, bp: 131° @ 15 mm, flash p: 234°F. Sol in fixed oils; insol in water.

SYNS: ALCOHOL C-11 □ FEMA No. 3097 □ HENDEC-ANOIC ALCOHOL □ 1-HENDECANOL □ HENDECYL ALCOHOL □ n-HENDECYLENIC ALCOHOL □ n-UNDECANOL

CONSENSUS REPORTS: Reported in EPA TSCA Inventory.

SAFETY PROFILE: Moderately toxic by ingestion. A skin irritant. Combustible liquid. Mutation data reported. When heated to decomposition it emits acrid smoke and irritating fumes. See also ALCOHOLS.

UNA100   CAS: 67785-74-4   HR: 1
UNDECYLENIC ALDEHYDE DIGERAN-YL ACETAL

mf: $C_{31}H_{54}O_2$ mw: 458.85

PROP: Fragrance and flavor.

SYNS: 11,11-BIS-((3,7-DIMETHYL-2,6-OCTADIEN-YL)OXY)-1-UNDECENE □ 11,11-DIGERANYLOXY-1-UNDECENE □ 10-UNDECENAL DIGERANYL ACETAL □ 1-UNDECENE, 11,11-BIS((3,7-DIMETHYL-2,6-OCTA-DIENYL)OXY)-

CONSENSUS REPORTS: Reported in EPA TSCA Inventory.

SAFETY PROFILE: Low toxicity by ingestion and skin contact. A skin irritant. When heated to decomposition it emits acrid smoke and irritating fumes.

UNJ800   CAS: 66-22-8   HR: 2
URACIL

mf: $C_4H_4N_2O_2$ mw: 112.10

PROP: Needles from water. Mp: 335° with effervescence. Freely sol in hot water; sparingly sol in cold water (100 parts of water at 25° dissolves 0.358 part of uracil); almost insol in alc, ether; sol in ammonia water and in other alkalies.

SYNS: 2,4-DIHYDROXYPYRIMIDINE □ 2,4-DIOXO-PYRIMIDINE □ HYBAR X □ PIROD □ 2,4-PYRIMI-DINEDIOL □ 2,4-PYRIMIDINEDIONE □ 2,4(1H,3H)-PYRIMIDINEDIONE (9CI) □ PYROD

CONSENSUS REPORTS: EPA Genetic Toxicology Program. Reported in EPA TSCA Inventory.

SAFETY PROFILE: Moderately toxic by intraperitoneal route. An experimental teratogen. Experimental reproductive effects. Questionable carcinogen with experimental tumorigenic data. Mutation data reported. When heated to decomposition it emits toxic fumes of NO_x.

UNS000   **CAS: 7440-61-1**   **HR: 3**
URANIUM
DOT: UN 2979
af: U   aw: 238.00
PROP: A heavy, silvery-white, malleable, ductile, softer-than-steel, metallic element. Tarnishes in air. α- and β-forms are brittle, the γ-form softer and more malleable. Mp: 1132°, bp: 3818°, d: 18.95. Radioactive material. IDLH 10 mg/m³ (as U).
SYN: URANIUM METAL, pyrophoric (DOT)
CONSENSUS REPORTS: Reported in EPA TSCA Inventory.
OSHA PEL: TWA Soluble Compounds: 0.05 mg(U)/m³; Insoluble Compounds 0.2 mg(U)/m³; STEL 0.6 mg(U)/m³
ACGIH TLV: TWA 0.2 mg(U)/m³; STEL 0.6 mg(U)/m³
DFG MAK: 0.25 mg/m³
DOT CLASSIFICATION: 7; Label: RADIOACTIVE, SPONT Combustible
SAFETY PROFILE: A highly toxic element on an acute basis. The permissible levels for soluble compounds are based on chemical toxicity, whereas the permissible body level for insoluble compounds is based on radiotoxicity. The high chemical toxicity of uranium and its salts is largely shown in kidney damage, which may not be reversible. Acute arterial lesions may occur after acute exposures. The most soluble uranium compounds are UF_6, $UO_2(NO_3)_2$, UO_2Cl_2, UO_2F_2, and uranyl acetates, sulfates, and carbonates. Some moderately soluble compounds are UF_4, UO_2, UO_4, $(NH_4)_2 U_2O_7$, UO_3, and uranyl nitrates. The rapid passage of soluble uranium compounds through the body tends to allow relatively large amounts to be absorbed. Soluble uranium compounds may be absorbed through the skin. The least soluble compounds are high-fired UO_2, U_3O_8, and uranium hydrides and carbides. The high toxicity effect of insoluble compounds is largely due to lung irradiation by inhaled particles. This material is transferred from the lungs of animals quite slowly.

A very dangerous fire hazard in the form of a solid or dust when exposed to heat or flame. It can react violently with air, Cl_2, F_2, HNO_3, NO, Se, S, water, NH_3, BrF_3, trichloroethylene, nitryl fluoride. During storage it may form a pyrophoric surface due to effects of air and moisture. Depleted uranium (the [238]U by-product of the uranium enrichment process, with relatively low radioactivity) is used in armor-piercing shells, ship or aircraft ballast, and counterbalances. Uranium is also used in making colored ceramic glazes.

UOJ000   **CAS: 7783-81-5**   **HR: 3**
URANIUM FLUORIDE (fissile)
DOT: UN 2977/UN 2978
mf: F_6U mw: 352.00
PROP: Colorless crystals, rapidly hydrolyzed by H_2O. Containing more than 1% U-235 (DOT). Mp: 64°, sublimes @ 56°. Decomp by H_2O, EtOH, and Et_2O. Sol in CCl_4 and $CHCl_3$. IDLH 10 mg/m³ (as U).
SYNS: URANIUM HEXAFLUORIDE, fissile (containing >1% U-235) (UN 2977) (DOT) □ URANIUM HEXAFLUOR-IDE, fissile excepted or non-fissile (UN 2978) (DOT)
OSHA PEL: TWA Soluble Compounds: 0.05 mg(U)/m³
ACGIH TLV: TWA 0.2 mg(U)/m³; STEL 0.6 mg(U)/m³; 2.5 mg(F)/m³
DOT CLASSIFICATION: 7; Label: RADIOACTIVE, Corrosive
SAFETY PROFILE: Radioactive poison. A corrosive irritant to skin, eyes, and mucous membranes. Violent reaction with hydroxy compounds (e.g., ethanol, water). Vigorous reaction with aromatic hydrocarbons (e.g., benzene, toluene, xylene). When heated to decomposition it emits toxic fumes of F^-. See also FLUORIDES and URANIUM.

U

UPA000   CAS: 13598-56-6   HR: 3
URANIUM(III) HYDRIDE
mf: H$_3$U   mw: 241.06
PROP: Fine black powder. Insol in EtOH,
Me$_2$O, and H$_2$O. IDLH 10 mg/m^3 (as U).
SAFETY PROFILE: A radioactive material.
The powder ignites spontaneously in air or
on contact with water. Potentially explosive
reaction with halocarbons. See also
HYDRIDES and URANIUM.

UPS000   CAS: 541-09-3   HR: 3
URANIUM OXYACETATE
mf: C$_4$H$_6$O$_6$U•2H$_2$O   mw: 424.19
PROP: Mp: loses 2H$_2$O @ 110°, bp: 275°
(decomp), d: 2.893 @ 15°. IDLH 10 mg/m^3
(as U).
SYNS: URANIUM ACETATE □ URANYL ACETATE
CONSENSUS REPORTS: Reported in EPA
TSCA Inventory.
OSHA PEL: TWA 0.05 mg(U)/m^3
ACGIH TLV: TWA 0.2 mg(U)/m^3
SAFETY PROFILE: Poison by
intraperitoneal route. A radioactive material.
See also URANIUM.

UQA000   CAS: 13536-84-0   HR: 3
URANIUM OXYFLUORIDE
mf: F$_2$O$_2$U   mw: 308.00
PROP: Hygroscopic light yellow solid, air-
stable to 4°. Very sol in H$_2$O; insol in org
solvs. IDLH 10 mg/m^3 (as U).
SYNS: URANIUM FLUORIDE OXIDE □ URANYL
FLUORIDE
CONSENSUS REPORTS: Reported in EPA
TSCA Inventory.
OSHA PEL: TWA 0.05 mg(U)/m^3; 2.5
mg(F)/m^3
ACGIH TLV: TWA 0.2 mg(U); TWA 2.5
mg(F)/m^3; BEI: 3 mg/g creatinine of
fluorides in urine prior to shift; 10 mg/g
creatinine of fluorides in urine at end of
shift.
NIOSH REL: TWA 2.5 mg(F)/m^3
SAFETY PROFILE: Poison by intravenous
route. When heated to decomposition it
emits toxic fumes of F$^-$. See also
FLUORIDES and URANIUM.

UQJ000   CAS: 10026-10-5   HR: 3
URANIUM TETRACHLORIDE
mf: Cl$_4$U   mw: 379.80
PROP: Cubic, dark green-gray deliquescent
crystals or solid. Mp: 590°, bp: 791°, d:
4.725 @ 25°/4°. Freely sol in water
(decomp); insol in hydrocarbons, ethyl
ether. Should be stored in sealed ampules.
IDLH 10 mg/m^3 (as U).
SYN: URANIUM(IV) CHLORIDE
CONSENSUS REPORTS: Reported in EPA
TSCA Inventory.
OSHA PEL: TWA 0.05 mg(U)/m^3
ACGIH TLV: TWA 0.2 mg(U)/m^3; STEL 0.6
mg(U)/m^3
SAFETY PROFILE: Probably a poison.
When heated to decomposition it emits
toxic fumes of Cl$^-$. See also URANIUM.

URA000   CAS: 7791-26-6   HR: 3
URANYL CHLORIDE
mf: Cl$_2$O$_2$U   mw: 340.90
PROP: Bright yellow, deliq crystals. Mp:
578°. Very hygroscopic. Volatile above 775°.
Very sol in water; sol in alc, acetone; insol in
benzene. Unstable in aq solutions. IDLH 10
mg/m^3 (as U).
CONSENSUS REPORTS: Reported in EPA
TSCA Inventory.
OSHA PEL: TWA 0.05 mg(U)/m^3
ACGIH TLV: TWA 0.2 mg(U)/m^3; STEL 0.6
mg(U)/m^3
SAFETY PROFILE: Poison by
intraperitoneal route. A radioactive material.
When heated to decomposition it emits
toxic fumes of Cl$^-$. See also URANIUM.

URA200   CAS: 10102-06-4   HR: 3
URANYL NITRATE (solid)
DOT: UN 2981
mf: N$_2$O$_8$U   mw: 394.02
PROP: Sol in H$_2$O and Et$_2$O, many org
solvs. IDLH 10 mg/m^3 (as U).
SYN: BIS(NITRATO-O,O')DIOXO URANIUM (solid)
CONSENSUS REPORTS: Reported in EPA
TSCA Inventory.
OSHA PEL: TWA 0.2 mg(U)/m^3; STEL 0.6
mg(U)/m^3

ACGIH TLV: TWA 0.2 mg(U)/m³; STEL 0.6 mg(U)/m³
DOT CLASSIFICATION: 7; Label: Radioactive, Oxidizer
SAFETY PROFILE: Poison by inhalation. Moderately toxic by ingestion. Human mutation data reported. A corrosive irritant to skin, eyes, and mucous membranes. A radioactive material. A powerful explosive and oxidizer. Incompatible with cellulose. Ether solutions in sunlight may explode. When heated to decomposition it emits toxic fumes of NOₓ. See also URANYL NITRATE HEXAHYDRATE and URANIUM.

URS000   CAS: 13520-83-7   HR: 3
URANYL NITRATE HEXAHYDRATE
DOT: UN 2980
mf: N₂O₈U•6H₂O   mw: 502.14
PROP: Rhombic, deliquescent, yellow crystals. Mp: 60.2°, bp: 118°, decomp @ 100°, d: 2.807 @ 13°. Sol in H₂O and common org solvs forming range of 1:1 and 1:2 complexes. IDLH 10 mg/m³ (as U).
SYNS: BIS(NITRATO)DIOXOURANIUM HEXA-HYDRATE □ DINITRATODIOXOURANIUM, HEXA-HYDRATE □ URANYL NITRATE HEXAHYDRATE, solution (DOT)
OSHA PEL: TWA 0.05 mg(U)/m³
ACGIH TLV: TWA 0.2 mg(U)/m³; STEL 0.6 mg(U)/m³
DOT CLASSIFICATION: 7; Label: RADIOACTIVE, Corrosive
SAFETY PROFILE: Poison by ingestion, subcutaneous, intravenous, and intraperitoneal routes. Mutation data reported. A corrosive irritant to skin, eyes, and mucous membranes. A radioactive material. When heated to decomposition it emits toxic fumes of NOₓ. See also URANIUM.

USS000   CAS: 57-13-6   HR: 2
UREA
mf: CH₄N₂O   mw: 60.07
PROP: White crystals or tetragonal needles or prisms with faint salty taste from H₂O or EtOH. Mp: 132.7°, bp: decomp, d: (solid)

1.335. Very sol in H₂O; sol in MeOH and EtOH; insol in CHCl₃ and C₆H₆.
SYNS: CARBAMIDE □ CARBAMIDE RESIN □ CARBAMIMIDIC ACID □ CARBONYL DIAMIDE □ CARBONYLDIAMINE □ ISOUREA □ NCI-C02119 □ PRESPERSION, 75 UREA □ PSEUDOUREA □ SUPERCEL 3000 □ UREAPHIL □ UREOPHIL □ UREVERT □ VARIOFORM II
CONSENSUS REPORTS: Reported in EPA TSCA Inventory. EPA Genetic Toxicology Program.
SAFETY PROFILE: Moderately toxic by intravenous and subcutaneous routes. Human reproductive effects by intraplacental route: fertility effects. Experimental reproductive effects. Human mutation data reported. A human skin irritant. Questionable carcinogen with experimental carcinogenic and neoplastigenic data. Reacts with sodium hypochlorite or calcium hypochlorite to form the explosive nitrogen trichloride. Incompatible with NaNO₂, P₂Cl₅, nitrosyl perchlorate. Preparation of the ¹⁵N-labeled urea is hazardous. When heated to decomposition it emits toxic fumes of NOₓ.

UTJ000   CAS: 124-47-0   HR: 3
UREA NITRATE (wet)
DOT: UN 0220/UN 1357
mf: CH₅N₃O₄   mw: 123.09
PROP: Colorless minerals or prisms. Mp: 152° decomp. Very sltly sol in hot water; sol in alc; insol in HNO₃.
SYNS: ACIDOGEN NITRATE □ UREA, MONO-NITRATE (8CI,9CI) □ UREA NITRATE □ UREA NITRATE, dry or wetted with <20% water, by weight (UN 0220) (DOT) □ UREA NITRATE, wetted with not <20% water, by weight (UN 1357) (DOT)
DOT CLASSIFICATION: EXPLOSIVE 1.1D; Label: EXPLOSIVE 1.1D (UN 0220); DOT Class: 4.1; Label: Flammable Solid (UN 1357)
SAFETY PROFILE: A mild irritant. Flammable when exposed to heat or flame. The dry nitrate may explode when heated. The presence of heavy metals (e.g., lead, iron) catalyzes the thermal decomposition of urea nitrate. When heated to decomposition it emits toxic fumes of NOₓ.

UVA000   CAS: 51-79-6   HR: 3
URETHANE
mf: $C_3H_7NO_2$ mw: 89.11
PROP: Colorless, odorless crystals, prisms from C_6H_6 or toluene with cooling, saline taste. Mp: 49°, bp: 103° @ 54 mm, d: 1.107, vap press: 10 mm @ 77.8°, vap d: 3.07. Very sol in H_2O, EtOH, Et_2O, $CHCl_3$, and C_6H_6; spar sol in ligroin.
SYNS: A 11032 □ AETHYLCARBAMAT (GERMAN) □ AETHYLURETHAN (GERMAN) □ CARBAMIC ACID, ETHYL ESTER □ CARBAMIDSAEURE-AETHYLESTER (GERMAN) □ ESTANE 5703 □ ETHYL CARBAMATE □ ETHYLURETHAN □ ETHYL URETHANE □ o-ETHYL-URETHANE □ LEUCETHANE □ LEUCOTHANE □ NSC 746 □ PRACARBAMIN □ PRACARBAMINE □ RCRA WASTE NUMBER U238 □ U-COMPOUND □ URETAN ETYLOWY (POLISH) □ URETHAN
CONSENSUS REPORTS: NTP 10th Report on Carcinogens. IARC Cancer Review: Group 2B IMEMDT 7,56,87; Animal Sufficient Evidence IMEMDT 7,111,74. Community Right-To-Know List. Reported in EPA TSCA Inventory. EPA Genetic Toxicology Program.
DFG MAK: Animal Carcinogen, Suspected Human Carcinogen
SAFETY PROFILE: Confirmed carcinogen with experimental carcinogenic, neoplastigenic, and tumorigenic data. A transplacental carcinogen. Moderately toxic by ingestion, intraperitoneal, subcutaneous, intramuscular, parenteral, and intravenous routes. An experimental teratogen. Experimental reproductive effects. Human mutation data reported. Causes depression of bone marrow and occasionally focal degeneration in the brain. Can also produce central nervous system depression, nausea and vomiting. Has been found in over 1000 beverages sold in the United States. The most heavily contaminated liquors are bourbons, sherries, and fruit brandies (some had 1000 to 12,000 ppb urethane). Many whiskeys, table and dessert wines, brandies, and liqueurs contain potentially hazardous amounts of urethane. The allowable limit for urethane in alcoholic beverages is 125 ppb. It is formed as a side product during processing.

Hot aqueous acids or alkalies decompose urethane to ethanol, carbon dioxide, and ammonia. Reacts with phosphorus pentachloride to form an explosive product. When heated it emits toxic fumes of NO_x. Used as an intermediate in the manufacture of pharmaceuticals, pesticides, and fungicides. See also CARBAMATES.

UVA400   CAS: 69-93-2   HR: D
URIC ACID
mf: $C_5H_4N_4O_3$ mw: 168.13
PROP: Odorless, tasteless, rhombic prisms or plates. Sol in alkalies and glycerol; spar sol in mineral acids; very spar sol in H_2O; insol in EtOH and Et_2O.
SYNS: LITHIC ACID □ 1H-PURINE-2,6,8(3H)-TRIONE, 7,9-DIHYDRO- (9CI) □ 2,6,8-TRIHYDROXYPURINE □ 2,6,8-TRIOXOPURINE □ 2,6,8-TRIOXYPURINE
CONSENSUS REPORTS: Reported in EPA TSCA Inventory.
SAFETY PROFILE: Experimental reproductive effects. Mutation data reported. When heated to decomposition it emits toxic fumes of NO_x.

UVS500   CAS: 9039-53-6   HR: D
UROKINASE
SYNS: UROKINASE (ENZYME-ACTIVATING) □ WIN 22005 □ WIN-KINASE
CONSENSUS REPORTS: Reported in EPA TSCA Inventory.
SAFETY PROFILE: An experimental teratogen. Experimental reproductive effects. Used in the treatment of diseases caused by blood clots.

V

VAG000 **CAS: 110-62-3** **HR: 3**
n-VALERALDEHYDE
DOT: UN 2058
mf: $C_5H_{10}O$ mw: 86.15
PROP: Liquid. Flash p: 53.6°F, fp: −92°, bp:
102–103°, d: 0.8095 @ 20°/4°. Very sltly sol
in water; misc with org solvs.
SYNS: AMYL ALDEHYDE □ BUTYL FORMAL □
PENTANAL □ n-PENTANAL □ VALERAL □
VALERIANIC ALDEHYDE □ VALERIC ACID
ALDEHYDE □ VALERIC ALDEHYDE □
VALERYLALDEHYDE
CONSENSUS REPORTS: Reported in EPA
TSCA Inventory.
OSHA PEL: TWA 50 ppm
ACGIH TLV: TWA 50 ppm
DOT CLASSIFICATION: 3; Label:
Flammable Liquid
SAFETY PROFILE: Moderately toxic by
ingestion. Mildly toxic by inhalation and skin
contact. A severe eye and skin irritant. A
very dangerous fire hazard when exposed to
heat or flame. When heated to
decomposition it emits acrid smoke and
irritating fumes. See also ALDEHYDES.

VAQ000 **CAS: 109-52-4** **HR: 2**
VALERIC ACID
mf: $C_5H_{10}O_2$ mw: 102.15
PROP: Colorless, mobile liquid with
unpleasant, penetrating, rancid odor. D:
0.940 @ 20°/4°, refr index: 1.405–1.14 @
25°, mp: −34.5°, fp: −34.5°, bp: 186.4°,
flash p: 203°F. Sol in EtOH and Et2O; spar
sol in H2O.
SYNS: BUTANECARBOXYLIC ACID □ 1-BUTANE-
CARBOXYLIC ACID □ FEMA No. 3101 □ PENTANOIC
ACID □ n-PENTANOIC ACID □ PROPYLACETIC ACID
□ VALERIANIC ACID □ n-VALERIC ACID
CONSENSUS REPORTS: Reported in EPA
TSCA Inventory.
SAFETY PROFILE: Moderately toxic by
ingestion, intravenous, and subcutaneous

routes. Mildly toxic by inhalation. A
corrosive irritant to skin, eyes, and mucous
membranes. Combustible liquid. When
heated to decomposition it emits acrid
smoke and irritating fumes. Used in
perfumes.

VAV000 **CAS: 108-29-2** **HR: 2**
4-VALEROLACTONE
mf: $C_5H_8O_2$ mw: 100.13
PROP: Colorless, mobile liquid; sweet,
herbaceous odor. Mp: −31°, bp: 205–206.5°,
flash p: 205°F (COC), d: 1.047–1.054, refr
index: 1.43, vap d: 3.45. Misc in alc, fixed
oils, water.
SYNS: FEMA No. 3103 □ 4-HYDROXYPENTANOIC
ACID LACTONE □ 4-HYDROXYVALERIC ACID
LACTONE □ γ-METHYL-γ-BUTYROLACTONE □ 4-
METHYL-γ-BUTYROLACTONE □ γ-PENTALACTONE □
4-PENTANOLIDE □ γ-VALEROLACTONE (FCC)
CONSENSUS REPORTS: Reported in EPA
TSCA Inventory.
SAFETY PROFILE: Moderately toxic by
ingestion. A skin irritant. Mutation data
reported. Combustible liquid when exposed
to heat or flame; can react with oxidizing
materials. To fight fire, use water, foam,
CO2, dry chemical. When heated to
decomposition it emits acrid smoke and
irritating fumes.

VBA000 **CAS: 638-29-9** **HR: 2**
VALERYL CHLORIDE
DOT: UN 2502
mf: C_5H_9ClO mw: 120.59
PROP: Fp: 32°, bp: 125–127°, d: 1.016.
CONSENSUS REPORTS: Reported in EPA
TSCA Inventory.
DOT CLASSIFICATION: 8; Label:
Corrosive
SAFETY PROFILE: A corrosive irritant to
skin, eyes, and mucous membranes. When

V

heated to decomposition it emits toxic fumes of Cl⁻.

VBA100   CAS: 1701-73-1   HR: 3
4-VALERYLPYRIDINE
mf: $C_{10}H_{13}NO$ mw: 163.24

SYNS: BUTYL 4-PYRIDYL KETONE □ KETONE, BUTYL 4-PYRIDYL □ 1-PENTANONE, 1-(4-PYRIDYL)- □ PYRIDINE, 4-VALERYL- □ 1-(4-PYRIDYL)-1-PENTANONE

DOT CLASSIFICATION: 3; Label: Flammable Liquid

SAFETY PROFILE: A poison by intraperitoneal route. A flammable liquid. When heated to decomposition it emits toxic vapors of NO_x.

VBK000   CAS: 90-22-2   HR: 3
VALETHAMATE BROMIDE
mf: $C_{19}H_{32}NO_2 \bullet Br$ mw: 386.43

PROP: Crystals from ethanol and ether or acetone. Mp: 118–120°. Freely sol in water and alc; practically insol in ether.

SYNS: 2-DIETHYLAMINOETHYL-3-METHYL-2-PHENYLVALERATE METHYLBROMIDE □ 2-DIETHYL-AMINOETHYL-2-PHENYL-3-METHYLVALERATE METHYL BROMIDE □ DIETHYL(2-HYDROXYETH-YL)METHYLAMMONIUM-3-METHYL-2-PHENYL-VALERATE BROMIDE □ EDIPOSIN □ EPIDOSIN □ EPIDOZIN □ 3-METHYL-2-PHENYLVALERIC ACID-2-DIETHYLAMINOETHYL ESTER METHYL BROMIDE □ 3-METHYL-2-PHENYLVALERIC ACID DIETHYL(2-HYDROXYETHYL)METHYLAMMONIUM BROMIDE ESTER □ 2-((3-METHYL-2-PHENYLVALERYL)OXY)-N,N-DIETHYL-N-METHYLETHANAMINIUM BROMIDE □ MUREL □ PHENYLMETHYLVALERIANSAEURE-β-DIAETHYLAMINOAETHYLESTER-BROMMETHYLAT (GERMAN) □ RESITAN □ VALETHAMATE

SAFETY PROFILE: Poison by ingestion, subcutaneous, and intravenous routes. See also ESTERS and BROMIDES. When heated to decomposition it emits very toxic fumes of NO_x, NH_3, and Br⁻.

VBP000   CAS: 72-18-4   HR: 1
VALINE
mf: $C_5H_{11}NO_2$ mw: 117.17

PROP: White, crystalline solid; characteristic taste. Mp (dl): 298° (decomp), mp (l): 315°, d (l): 1.230. Sol in water; very sltly sol in alc; insol in ether. An essential amino acid.

SYNS: l-(+)-α-AMINOISOVALERIC ACID □ l-VALINE (FCC)

CONSENSUS REPORTS: Reported in EPA TSCA Inventory.

SAFETY PROFILE: Mutation data reported. When heated to decomposition it emits toxic fumes of NO_x.

VCA100   CAS: 4420-67-1   HR: 3
VALLAROSOLANOSIDE
mf: $C_{32}H_{48}O_{10}$ mw: 592.80

SYN: CARD-20(22)-ENOLIDE, 16-(ACETYLOXY)-3-((6-DEOXY-3-o-METHYL-α-l-ALTROPYRANOSYL)OXY)-14-HYDROXY-, (3-β,5-β,16-β)-

SAFETY PROFILE: A poison by intravenous route. When heated to decomposition it emits acrid smoke and irritating vapors.

VCP000   CAS: 7440-62-2   HR: 3
VANADIUM
af: V   aw: 50.94

PROP: A bright, white, soft, ductile metal; sltly radioactive. Corrosion resistant (oxide film). Resistant to fused alkalies, attacked by hot concentrated mineral acids. Bp: 3380°, d: 6.11 @ 18.7°, mp: 1917°. Insol in water.

CONSENSUS REPORTS: Reported in EPA TSCA Inventory.

OSHA PEL: Respirable Dust and Fume: TWA 0.05 mg(V_2O_5)/m³

NIOSH REL: TWA 1.0 mg(V)/m³

SAFETY PROFILE: An inhalation hazard. Poison by subcutaneous route. Questionable carcinogen with experimental tumorigenic data. Flammable in dust form from heat, flame, or sparks. Violent reaction with BrF_3, Cl_2, lithium, nitryl fluoride, oxidants. When heated to decomposition it emits toxic fumes of VO_x. See also VANADIUM COMPOUNDS.

VCZ000   HR: D
VANADIUM COMPOUNDS
NIOSH REL: (Vanadium Compounds) CL 0.05 mg(V)/m³/15M

SAFETY PROFILE: Variable toxicity. Vanadium compounds act chiefly as an irritant to the conjunctiva and respiratory tract. Acute and chronic exposure can give rise to conjunctivitis, rhinitis, reversible irritation of the respiratory tract, and to bronchitis, bronchospasms, and asthma-like diseases in more severe cases. There is still some controversy as to the effects of industrial exposure on other systems of the body. Responses are mostly acute, seldom chronic. The first report of human vanadium poisoning described rather widespread systemic effects, consisting of polycythemia, followed by red blood cell destruction and anemia, loss of appetite, pallor and emaciation, albuminuria and hematuria, gastrointestinal disorders, nervous complaints, and cough, sometimes severe enough to cause hemoptysis. More recent reports describe symptoms that, for the most part, are restricted to the conjunctiva and respiratory system, no evidence being found of disturbances of the gastrointestinal tract, kidneys, blood, or central nervous system. Vanadate (VO_3^-) is a potent inhibitor of the sodium pump, an enzyme universally present in eukaryotic organisms. The absorption of V_2O_5 by inhalation is nearly 100%. Though certain workers believe that it is only the pentoxide that is harmful, other investigators have found that patronite dust (chiefly vanadium sulfide) is quite toxic to animals, causing acute pulmonary edema. Acute poisoning in animals by ingestion of vanadium compounds causes nervous disturbances, paralysis of legs, respiratory failure, convulsions, bloody diarrhea, and death. Poisoning by inhalation causes bleeding of the nose and acute bronchitis. Some compounds have reported mutation effects. VF_5 and the oxyhalogenides of pentavalent vanadium (VOF_3, $VOCl_3$, $VOBr_3$) are volatile. Vanadium compounds are common air contaminants. The fumes are highly toxic. The major use of vanadium and its alloys is in the steel industry. When heated to decomposition they emit toxic fumes of VO_x. See also specific compounds.

VDP000   CAS: 7727-18-6   HR: 3
VANADIUM OXYTRICHLORIDE
DOT: UN 2443
mf: Cl_3OV mw: 173.29

PROP: Lemon yellow liquid, freezing to deep orange solid. Mp: −77°, bp: 126.7°, d: 1.811 @ 32°.
SYNS: TRICHLOROOXOVANADIUM □ VANADIUM TRICHLORIDE OXIDE □ VANADYL TRICHLORIDE
CONSENSUS REPORTS: Reported in EPA TSCA Inventory.
ACGIH TLV: TWA 0.05 mg(V_2O_5)/m^3
NIOSH REL: (Vanadium Compounds) CL 0.05 mg(V)/m^3/15M
DOT CLASSIFICATION: 8; Label: Corrosive
SAFETY PROFILE: Poison by ingestion. A corrosive irritant to skin, eyes, and mucous membranes. Explosive reaction with sodium. Violently hygroscopic. Violent reaction with rubidium (at 60°C), potassium. When heated to decomposition it emits toxic fumes of VO_x and Cl^-. See also VANADIUM COMPOUNDS and HYDROCHLORIC ACID.

VDU000   CAS: 1314-62-1   HR: 3
VANADIUM PENTOXIDE (dust)
DOT: UN 2862
mf: O_5V_2 mw: 181.88
PROP: Yellow to red crystalline powder or orange solid. Loses oxygen reversibly on heating. Amphoteric. Mild oxidizing agent. Mp: 677°, bp: decomp @ 1750°, d: 3.357 @ 18°. Insol. IDLH 35 mg/m^3 (as V).
SYNS: ANHYDRIDE VANADIQUE (FRENCH) □ C.I. 77938 □ RCRA WASTE NUMBER P120 □ VANADIC ANHYDRIDE □ VANADIO, PENTOSSIDO di (ITALIAN) □ VANADIUM DUST and FUME (ACGIH) □ VANADIUM(V) OXIDE □ VANADIUM PENTAOXIDE □ VANADIUMPENTOXID (GERMAN) □ VANADIUM PENTOXIDE, non-fused form (DOT) □ VANADIUM-PENTOXYDE (DUTCH) □ VANADIUM, PENTOXYDE de (FRENCH) □ WANADU PIECIOTLENEK (POLISH)
CONSENSUS REPORTS: Reported in EPA TSCA Inventory. EPA Genetic Toxicology Program.
OSHA PEL: Respirable Dust and Fume:
ACGIH TLV: TWA 0.05 mg(V_2O_5)/m^3; Not Classifiable as a Human Carcinogen; BEI: 50 µg/g creatinine of vanadium in urine at end of shift at end of workweek.; (Proposed: inhalable fraction TWA 0.02

mg(V_2O_5)/m³; Confirmed Animal Carcinogen with Unknown Revelance to Humans)

DFG MAK: (fine dust) 0.05 mg/m³

NIOSH REL: (Vanadium Compounds) CL 0.05 mg(V)/m³/15M

DOT CLASSIFICATION: 6.1; Label: Poison

SAFETY PROFILE: Poison by ingestion, inhalation, intraperitoneal, subcutaneous, intratracheal, and intravenous routes. An experimental teratogen. Human systemic effects by inhalation: bronchiolar constriction, including asthma, cough, dyspnea, sputum, and conjunctiva irritation. Experimental reproductive effects. Mutation data reported. A respiratory irritant; causes skin pallor, greenish-black tongue, chest pain, cough, dyspnea, palpitation, lung changes. When ingested it causes gastrointestinal tract disturbances. May also cause a papular skin rash. Mixtures with calcium + sulfur + water may ignite spontaneously. The absorption of V_2O_5 by inhalation is nearly 100%. Incompatible with ClF_3, Li, peroxyformic acid. When heated to decomposition it emits acrid smoke and irritating fumes of VO_x. See also VANADIUM COMPOUNDS.

VDZ000   CAS: 1314-62-1   HR: 3
VANADIUM PENTOXIDE (fume)
mf: O_5V_2 mw: 181.88

PROP: IDLH 35 mg/m³ (as V).

SYNS: RCRA WASTE NUMBER P120 □ VANADIUM DUST and FUME (ACGIH) □ VANADIUM (OSHA) □ VANADIUM PENTOXIDE, nonfused form (DOT)

CONSENSUS REPORTS: EPA Extremely Hazardous Substances List. Reported in EPA TSCA Inventory. EPA Genetic Toxicology Program.

OSHA PEL: Respirable Dust and Fume: TWA 0.05 mg(V_2O_5)/m³

ACGIH TLV: TWA 0.05 mg(V_2O_5)/m³; Not Classifiable as a Human Carcinogen; BEI: 50 µg/g creatinine of vanadium in urine at end of shift at end of workweek.; (Proposed: inhalable fraction TWA 0.02 mg(V_2O_5)/m³; Confirmed Animal Carcinogen with Unknown Revelance to Humans)

NIOSH REL: (Vanadium Compound) CL 0.05 mg(V)/m³/15M

DOT CLASSIFICATION: 6.1; Label: Poison

SAFETY PROFILE: A poison by several routes. Can react violently with (Ca + S + H_2O), ClF_3, Li. When heated to decomposition it emits toxic fumes of VO_x. See also VANADIUM PENTOXIDE (dust).

VEA000   CAS: 1314-34-7   HR: 3
VANADIUM SESQUIOXIDE
mf: O_3V_2 mw: 149.88

PROP: Black crystals or solid. Not amphoteric. Shows unusual electrical properties with a tenfold change in resistance between 225 and 4°. There is a metal-insulator transition at 155K. Mp: 1970°, d: 4.87 @ 18°.

SYNS: VANADIC OXIDE □ VANADIUM OXIDE □ VANADIUM TRIOXIDE

CONSENSUS REPORTS: Reported in EPA TSCA Inventory.

ACGIH TLV: TWA 0.05 mg(V_2O_5)/m³

NIOSH REL: (Vanadium Compound) CL 0.05 mg(V)/m³/15M

SAFETY PROFILE: Poison by ingestion, subcutaneous, and intratracheal routes. Ignites when heated in air. When heated to decomposition it emits toxic fumes of VO_x. See also VANADIUM COMPOUNDS.

VEF000   CAS: 7632-51-1   HR: 3
VANADIUM TETRACHLORIDE
DOT: UN 2444
mf: Cl_4V mw: 192.74

PROP: Reddish-brown liquid. Readily hydrolyzes; decomp slowly to VCl_3 + Cl_2 at room temp. Mp: −28°, bp: 148.5°, d: 1.816 @ 30°. Sol in CCl_4 and donor solvs.

SYN: VANADIUM CHLORIDE

CONSENSUS REPORTS: Reported in EPA TSCA Inventory.

ACGIH TLV: TWA 0.05 mg(V_2O_5)/m³

NIOSH REL: (Vanadium Compounds) CL 0.05 mg(V)/m³/15M

DOT CLASSIFICATION: 6.1; Label: Poison

SAFETY PROFILE: Poison by ingestion. A corrosive irritant to skin, eyes, and mucous membranes. When heated to decomposition it emits toxic fumes of VO$_x$ and Cl$^-$. See also VANADIUM COMPOUNDS and HYDROCHLORIC ACID.

VEK000   CAS: 13470-26-3   HR: 3
VANADIUM TRIBROMIDE
mf: Br$_3$V   mw: 290.67
PROP: Green-black deliq crystals or solid, violet vapor. Disproportionates to VBr$_4$ + VBr$_2$; thermal decomp to VBr$_2$ +Br$_2$. Mp: decomp. Sol in donor solvs.
SYN: VANADIUM BROMIDE
CONSENSUS REPORTS: Reported in EPA TSCA Inventory.
ACGIH TLV: TWA 0.05 mg(V$_2$O$_5$)/m^3
NIOSH REL: CL 0.05 mg(V)/m^3/15M
SAFETY PROFILE: Poison by subcutaneous route. When heated to decomposition it emits toxic fumes of VO$_x$ and Br$^-$. See also VANADIUM COMPOUNDS and BROMIDES.

VEP000   CAS: 7718-98-1   HR: 3
VANADIUM TRICHLORIDE
DOT: UN 2475
mf: Cl$_3$V   mw: 157.29
PROP: Pink crystals or violet very hygroscopic solid; disproportionates @ >4° to VCl$_4$ + VCl$_2$. Mp: decomp, d: 3.00 @ 18°. Sol in aq HCl and most donor solvs.
SYN: VANADIUM(III) CHLORIDE
CONSENSUS REPORTS: Reported in EPA TSCA Inventory.
ACGIH TLV: TWA 0.05 mg(V$_2$O$_5$)/m^3
NIOSH REL: (Vanadium Compounds) CL 0.05 mg(V)/m^3/15M
DOT CLASSIFICATION: 8; Label: Corrosive
SAFETY PROFILE: Poison by ingestion and subcutaneous routes. A corrosive irritant to skin, eyes, and mucous membranes. Extremely violent reaction with methyl magnesium iodide and other Grignard reagents. When heated to decomposition it emits toxic fumes of VO$_x$ and Cl$^-$. See also VANADIUM COMPOUNDS and HYDROCHLORIC ACID.

VEZ000   CAS: 27774-13-6   HR: 3
VANADYL SULFATE

DOT: UN 2931
mf: O$_5$SV   mw: 163.00
PROP: Blue crystals or solid.
SYNS: C.I. 77940 □ OXYSULFATOVANADIUM
CONSENSUS REPORTS: Reported in EPA TSCA Inventory.
ACGIH TLV: TWA 0.05 mg(V$_2$O$_5$)/m^3
NIOSH REL: (Vanadium Compounds) CL 0.05 mg(V)/m^3/15M
DOT CLASSIFICATION: 6.1; Label: Poison
SAFETY PROFILE: A poison and an inhalation hazard. Poison by intravenous, intraperitoneal, and subcutaneous routes. Mutation data reported. When heated to decomposition it emits toxic fumes of VO$_x$ and SO$_x$. See also SULFATES and VANADIUM COMPOUNDS.

VFK000   CAS: 121-33-5   HR: 2
VANILLIN
mf: C$_8$H$_8$O$_3$ mw: 152.16
PROP: White, crystalline needles from water; vanilla odor. D: 1.056, bp: 285°, mp: 80–81°. Sol in 125 parts water, 20 parts glycerin, 2 parts 95% alc, chloroform, ether.
SYNS: FEMA No. 3107 □ 4-HYDROXY-m-ANISALDE-HYDE □ 4-HYDROXY-3-METHOXYBENZALDEHYDE □ LIOXIN □ 3-METHOXY-4-HYDROXYBENZALDEHYDE □ METHYLPROTOCATECHUALDEHYDE □ VANILLA □ VANILLALDEHYDE □ VANILLIC ALDEHYDE □ p-VANILLIN □ ZIMCO
CONSENSUS REPORTS: Reported in EPA TSCA Inventory.
SAFETY PROFILE: Moderately toxic by ingestion, intraperitoneal, subcutaneous, and intravenous routes. Experimental reproductive effects. Human mutation data reported. Can react violently with Br$_2$, HClO$_4$, potassium-tert-butoxide, tert-chlorobenzene + NaOH, formic acid + thallium nitrate. When heated to decomposition it emits acrid smoke and irritating fumes. See also ALDEHYDES.

VFP100   CAS: 122-48-5   HR: 2
VANILLYL ACETONE
mf: C$_{11}$H$_{14}$O$_3$ mw: 194.25
PROP: Crystals from pet ether. Mp: 40–41°, bp: 187–188°. Sltly sol in water, pet ether; sol in ether.

V

SYNS: 2-BUTANONE, 4-(4-HYDROXY-3-METHOXY-PHENYL)- □ GINGERONE □ 4-(4-HYDROXY-3-METHOXYPHENYL)-2-BUTANONE □ (4-HYDROXY-3-METHOXYPHENYL)ETHYL METHYL KETONE □ 3-METHOXY-4-HYDROXY-BENZYLACETONE □ (0)-PARADOL □ ZINGERONE □ ZINGIBERONE

CONSENSUS REPORTS: Reported in EPA TSCA Inventory.

DOT CLASSIFICATION: 3; Label: Flammable Liquid

SAFETY PROFILE: Moderately toxic by ingestion. A skin irritant. A flammable liquid. When heated to decomposition it emits acrid smoke and irritating fumes.

VGP000   CAS: 51-43-4   HR: 3
VASOTONIN

mf: $C_9H_{13}NO_3$ mw: 183.23

PROP: White powder. Mp: 216° (decomp).

SYNS: ADNEPHRINE □ ADRENAL □ 1-ADRENALIN □ ADRENALIN-MEDIHALER □ ADRENAMINE □ ADRENAN □ ADRENAPAX □ ADRENASOL □ ADREN-ATRATE □ ADRENODIS □ ADRENOHORMA □ ADRENUTOL □ ADRINE □ ASMATANE MIST □ ASTHMA METER MIST □ ASTMAHALIN □ BALMA-DREN □ BERNARENIN □ BIORENINE □ BOSMIN □ BREVIRENIN □ BRONKAID MIST □ CHELAFRIN □ CORISOL □ 3,4-DIHYDROXY-α-((METHYLAMINO)-METHYL)BENZYL ALCOHOL □ 1-1-(3,4-DIHYDROXY-PHENYL)-2-METHYLAMINOETHANOL □ DRENAMIST □ DYLEPHRIN □ DYSPNE-INHAL □ EPIFRIN □ EPINEPHRAN □ EPINEPHRINE □ (–)-EPINEPHRINE □ (R)-EPINEPHRINE □ 1-EPINEPHRINE □ 1-EPINEPHR-INE (synthetic) □ EPIRENAMINE □ EPIRENAN □ EPITRATE □ ESPHYGMOGENINA □ EXADRIN □ GLYCIRENAN □ HAEMOSTASIN □ HEKTALIN □ HEMISINE □ HEMOSTASIN □ (R)-4-(1-HYDROXY-2-(METHYLAMINO)ETHYL)-1,2-BENZENEDIOL (9CI) □ HYPERNEPHRIN □ HYPORENIN □ INTRANEFRIN □ KIDOLINE □ LEVORENIN □ LYOPHRIN □ MEDIHAL-ER-EPI □ METANEPHRIN □ METHYLARTERENOL □ MUCIDRINA □ MYOSTHENINE □ MYTRATE □ NEPHRIDINE □ NIERALINE □ PARANEPHRIN □ PRIMATENE MIST □ RCRA WASTE NUMBER P042 □ RENAGLADIN □ RENALEPTINE □ RENALINA □ RENOFORM □ RENOSTYPRICIN □ RENOSTYPTIN □ SCURENALINE □ SINDRENINA □ SOLADREN □ SPHYGMOGENIN □ STRYPTIRENAL □ SUPRACAPSUL-IN □ SUPRADIN □ SUPRANEPHRANE □ SUPRAN-EPHRINE □ SUPRANOL □ SUPRARENIN □ SUPREL □ SURENINE □ SUSPHRINE □ SYMPATHIN I □ TAKAM-INA □ TOKAMINA □ TONOGEN □ VAPONEFRIN □ VASOCONSTRICTINE □ VASOCONSTRICTOR □ VASODRINE □ VASOTON

CONSENSUS REPORTS: Reported in EPA TSCA Inventory. EPA Genetic Toxicology Program.

SAFETY PROFILE: Human poison by subcutaneous route. Experimental poison by ingestion, skin contact, subcutaneous, intraperitoneal, intravenous, and intramuscular routes. Human systemic effects: cardiomyopathy including infarction, arrhythmias. An experimental teratogen. Experimental reproductive effects. Mutation data reported. When heated to decomposition it emits toxic fumes of NO_x. Used as an adrenergic, sympathomimetic, vasoconstrictor, bronchodilator, and cardiac stimulant.

VGU200   CAS: 68956-68-3   HR: 1
VEGETABLE OIL

SYNS: VEGETABLE OIL MIST (OSHA) □ VISCOLEO OIL

CONSENSUS REPORTS: Reported in EPA TSCA Inventory.

OSHA PEL: TWA 15 mg/m³, total dust; TWA 5 mg/m³, respirable fraction

SAFETY PROFILE: A nuisance mist. When heated to decomposition it emits acrid smoke and irritating fumes.

VLA000   CAS: 143-67-9   HR: 3
VINCALEUKOBLASTINE SULFATE (1:1) (SALT)

mf: $C_{46}H_{58}N_4O_9 \cdot H_2O_4S$ mw: 909.16

PROP: A solid. Mp: 284–285°.

SYNS: EXAL □ 29060 LE □ NSC 49842 □ VELBAN □ VELBE □ VINBLASTINE SULFATE □ VINCALEUKO-BLASTINE SULFATE □ VLB MONOSULFATE

CONSENSUS REPORTS: IARC Cancer Review: Group 3 IMEMDT 7,371,87; Animal Inadequate Evidence IMEMDT 26,349,81; Human Inadequate Evidence IMEMDT 26,349,81. EPA Genetic Toxicology Program.

SAFETY PROFILE: Poison by ingestion, intraperitoneal, and intravenous routes. An experimental teratogen. Human systemic effects by intravenous route: blood leukopenia and hair changes. Experimental reproductive effects. Questionable carcinogen. Human mutation data reported.

When heated to decomposition it emits very toxic fumes of NO_x and SO_x. See also VINCALEUKOBLASTINE and SULFATES.

VLU200   CAS: 83768-87-0   HR: 3
VINTHIONINE
mf: $C_6H_{11}NO_2S$ mw: 161.24
SYNS: S-ETHENYL-dl-HOMOCYSTEINE □ S-VINYL-dl-HOMOCYSTEINE
SAFETY PROFILE: Experimental reproductive effects. Suspected carcinogen with experimental carcinogenic data. Mutation data reported. When heated to decomposition it emits toxic fumes of SO_x and NO_x.

VLU250   CAS: 108-05-4   HR: 3
VINYL ACETATE
DOT: UN 1301
mf: $C_4H_6O_2$ mw: 86.10
PROP: Colorless, mobile liquid; polymerizes to solid on exposure to light. Mp: $-92.8°$, fp: $-100°$, bp: $73°$, flash p: $18°F$, d: 0.9335 @ $20°$, autoign temp: $800°F$, vap press: 100 mm @ $21.5°$, lel: 2.6%, uel: 13.4%, vap d: 3.0. Misc in alc, ether. Somewhat sol in water.
SYNS: ACETATE de VINYLE □ ACETIC ACID, ETHENYL ESTER □ ACETIC ACID, ETHYLENE ETHER □ ACETIC ACID VINYL ESTER □ 1-ACETOXYETHYLENE □ ETHANOIC ACID, ETHENYL ESTER □ ETHENYL ACETATE □ ETHENYL ETHANOATE □ OCTAN WINYLU (POLISH) □ VAC □ VINILE (ACETATO di) (ITALIAN) □ VINYLACETAAT (DUTCH) □ VINYLACETAT (GERMAN) □ VINYL ACETATE, inhibited (DOT) □ VINYL ACETATE H.Q. □ VINYL A MONOMER □ VINYLE (ACETATE de) (FRENCH) □ VINYLESTER KYSELINY OCTOVE □ VINYL ETHANOATE □ VYAC □ ZESET T
CONSENSUS REPORTS: IARC Cancer Review: Group 3 IMEMDT 7,56,87; Animal Inadequate Evidence IMEMDT 19,341,79; IMEMDT 39,113,86; Human Inadequate Evidence IMEMDT 39,113,86. Reported in EPA TSCA Inventory. Community Right-To-Know List. EPA Extremely Hazardous Substances List.
OSHA PEL: TWA 10 ppm; STEL 20 ppm
ACGIH TLV: 10 ppm, STEL: 15 ppm; Animal Carcinogen

DFG MAK: 10 ppm (35 mg/m³); Confirmed Animal Carcinogen with Unknown Relevance to Humans
NIOSH REL: (Vinyl Acetate) CL 15 mg/m³/15M
DOT CLASSIFICATION: 3; Label: Flammable Liquid
SAFETY PROFILE: Confirmed carcinogen with experimental carcinogenic and tumorigenic data. Moderately toxic by ingestion, inhalation, and intraperitoneal routes. A skin and eye irritant. Experimental reproductive effects. Human mutation data reported. Highly dangerous fire hazard when exposed to heat, flame, or oxidizers. A storage hazard, it may undergo spontaneous exothermic polymerization. Reaction with air or water to form peroxides that catalyze an exothermic polymerization reaction has caused several large industrial explosions. Reaction with hydrogen peroxide forms the explosive peracetic acid. Reacts with oxygen above $50°C$ to form an unstable explosive peroxide. Reacts with ozone to form the explosive vinyl acetate ozonide. Solution polymerization of the acetate dissolved in toluene has resulted in large industrial explosions. Polymerization reaction with dibenzoyl peroxide + ethyl acetate may release ignitable and explosive vapors. The vapor may react vigorously with desiccants (e.g., silica gel or alumina). Incompatible (explosive) with 2-amino ethanol, chlorosulfonic acid, ethylenediamine, ethyleneimine, HCl, HF, HNO_3, oleum, peroxides, H_2SO_4. See also ESTERS.

VMP000   CAS: 593-60-2   HR: 3
VINYL BROMIDE
DOT: UN 1085
mf: C_2H_3Br mw: 106.96
PROP: A gas or liquid. Mp: $-138°$, bp: $15.6°$, d: 1.51. Insol in water; misc in alc, ether.
SYNS: BROMOETHENE □ BROMOETHYLENE □ BROMURE de VINYLE (FRENCH) □ VINILE (BROMURO di) (ITALIAN) □ VINYLBROMID (GERMAN) □ VINYL BROMIDE, inhibited (DOT) □ VINYLE (BROMURE de) (FRENCH)

V

CONSENSUS REPORTS: NTP 10th Report on Carcinogens. IARC Cancer Review: Group 2A IMEMDT 7,56,87; Animal Sufficient Evidence IMEMDT 39,133,86; Animal Inadequate Evidence IMEMDT 19,367,79. Community Right-To-Know List. Reported in EPA TSCA Inventory. EPA Genetic Toxicology Program.
OSHA PEL: TWA 5 ppm
ACGIH TLV: TWA 0.5 ppm; Suspected Human Carcinogen
DFG MAK: Human Carcinogen
NIOSH REL: (Vinyl Bromide) Lowest Detectable Level
DOT CLASSIFICATION: 2.1; Label: Flammable Gas
SAFETY PROFILE: Confirmed carcinogen with experimental carcinogenic, neoplastigenic, and tumorigenic data. Moderately toxic by ingestion. Mutation data reported. A very dangerous fire hazard when exposed to heat or flame. Can react violently with oxidizing materials. May polymerize in sunlight. To fight fire, use CO_2, dry chemical, or water spray. When heated to decomposition it emits toxic fumes of Br^-. See also BROMIDES and VINYL CHLORIDE.

VMZ000   CAS: 111-34-2   HR: 3
VINYL BUTYL ETHER
DOT: UN 2352
mf: $C_6H_{12}O$ mw: 100.18
PROP: Liquid. Mp: $-112.7°$, bp: $94.2°$, flash p: $-9°$, d: 0.7803 @ $20°/20°$, vap d: 3.45.
SYNS: BUTOXYETHENE □ BUTYL VINYL ETHER □ BUTYL VINYL ETHER (inhibited) □ 1-(ETHENYLOXY) BUTANE □ VINYL-n-BUTYL ETHER
CONSENSUS REPORTS: Reported in EPA TSCA Inventory.
DOT CLASSIFICATION: 3; Label: Flammable Liquid
SAFETY PROFILE: Mildly toxic by ingestion, skin contact, and inhalation. A skin and eye irritant. A very dangerous fire hazard when exposed to heat or flame. To fight fire, use foam, CO_2, dry chemical, alcohol foam. Moderately explosive by spontaneous chemical reaction. Can react

with oxidizing materials. When heated to decomposition it emits acrid smoke and irritating fumes. See also ETHERS.

VNF000   CAS: 123-20-6   HR: 3
VINYL BUTYRATE
DOT: UN 2838
mf: $C_6H_{10}O_2$ mw: 114.16
PROP: D: 0.9, vap d: 4.0, bp: $116°$, flash p: $68°F$ (OC), lel: 1.4%, uel: 8.8%.
SYNS: BUTYRIC ACID, VINYL ESTER □ VINYL BUTYRATE, INHIBITED (DOT) □ VINYLESTER KYSELINY MASELNE
DOT CLASSIFICATION: 3; Label: Flammable Liquid
SAFETY PROFILE: Mildly toxic by inhalation and ingestion. A skin and eye irritant. A very dangerous fire hazard when exposed to heat, flame, or oxidizers. Explosive in the form of vapor when exposed to heat or flame. To fight fire, use alcohol foam, fog, mist, CO_2. When heated to decomposition it emits acrid smoke and irritating fumes. See also ESTERS.

VNK000   CAS: 15805-73-9   HR: 3
VINYL CARBAMATE
mf: $C_3H_5NO_2$ mw: 87.09
PROP: White crystalline powder. Mp: $54°-55°$. Mod sol in water.
SYN: CARBAMIC ACID, VINYL ESTER
SAFETY PROFILE: Poison by intraperitoneal route. Questionable carcinogen with experimental neoplastigenic data. Human mutation data reported. When heated to decomposition it emits toxic fumes of NO_x. See also ESTERS and CARBAMATES.

VNP000   CAS: 75-01-4   HR: 3
VINYL CHLORIDE
DOT: UN 1086
mf: C_2H_3Cl mw: 62.50
PROP: Colorless liquid or gas (when inhibited); faintly sweet odor. Mp: $-160°$, bp: $-13.9°$, lel: 4%, uel: 22%, flash p: $17.6°F$ (COC), fp: $-159.7°$, d (liquid): 0.9195 @ $15°/4°$, vap press: 2600 mm @ $25°$, vap d: 2.15, autoign temp: $882°F$. Sltly sol in water; sol in alc; very sol in ether.

SYNS: CHLORETHENE □ CHLORETHYLENE □ CHLOROETHENE □ CHLOROETHYLENE □ CHLORURE de VINYLE (FRENCH) □ CLORURO di VINILE (ITALIAN) □ ETHYLENE MONOCHLORIDE □ MONOCHLOROETHENE □ MONOCHLOROETHYL-ENE (DOT) □ RCRA WASTE NUMBER U043 □ TROVIDUR □ VC □ VCM □ VINILE (CLORURO di) (ITALIAN) □ VINYLCHLORID (GERMAN) □ VINYL CHLORIDE MONOMER □ VINYL C MONOMER □ VINYLE (CHLORURE de) (FRENCH) □ WINYLU CHLOREK (POLISH)

CONSENSUS REPORTS: NTP 10th Report on Carcinogens. IARC Cancer Review: Group 1 IMEMDT 7,373,87; Animal Sufficient Evidence IMEMDT 19,377,79; IMEMDT 7,291,74; Human Limited Evidence IMEMDT 7,291,74; Human Sufficient Evidence IMEMDT 19,377,79. Community Right-To-Know List. Reported in EPA TSCA Inventory. EPA Genetic Toxicology Program.

OSHA PEL: Cancer Suspect Agent

ACGIH TLV: TWA 1 ppm; Confirmed Human Carcinogen

DFG MAK: DFG TRK: Confirmed Human Carcinogen

NIOSH REL: (Vinyl Chloride) Lowest Detectable Level

DOT CLASSIFICATION: 2.1; Label: Flammable Gas

SAFETY PROFILE: Confirmed human carcinogen producing liver and blood tumors. Moderately toxic by ingestion. Experimental teratogenic data. Experimental reproductive effects. Human reproductive effects by inhalation: changes in spermatogenesis. Human mutation data reported. A severe irritant to skin, eyes, and mucous membranes. Causes skin burns by rapid evaporation and consequent freezing. In high concentration it acts as an anesthetic. Chronic exposure has produced liver injury. Circulatory and bone changes in the fingertips have been reported in workers handling unpolymerized materials.

A very dangerous fire hazard when exposed to heat, flame, or oxidizers. Large fires of this material are practically inextinguishable. A severe explosion hazard in the form of vapor when exposed to heat or flame. Long-term exposure to air may result in formation of peroxides that can initiate explosive polymerization of the chloride. Can react vigorously with oxidizing materials. Can explode on contact with oxides of nitrogen. Obtain instructions for its use from the supplier before storing or handling this material. To fight fire, stop flow of gas. When heated to decomposition it emits highly toxic fumes of Cl^-. See also CHLORINATED HYDROCARBONS, ALIPHATIC.

VOA000   CAS: 106-87-6   HR: 3
VINYL CYCLOHEXENE DIOXIDE
mf: $C_8H_{12}O_2$ mw: 140.20
PROP: Colorless liquid. D: 1.098 @ 20°/20°, bp: 227°, flash p: 230°F, mp: -55°. Very sol in water.

SYNS: CHISSONOX 206 □ EP-206 □ 1,2-EPOXY-4-(EPOXYETHYL)CYCLOHEXANE □ 1-EPOXYETHYL-3,4-EPOXYCYCLOHEXANE □ 3-(EPOXYETHYL)-7-OXABI-CYCLO(4.1.0)HEPTANE □ 3-(1,2-EPOXYETHYL)-7-OXABICYCLO(4.1.0)HEPTANE □ 4-(EPOXYETHYL)-7-OXABICYCLO(4.1.0)HEPTANE □ 4-(1,2-EPOXYETHYL)-7-OXABICYCLO(4.1.0)HEPTANE □ ERLA-2270 □ ERLA-2271 □ 1-ETHYLENEOXY-3,4-EPOXYCYCLOHEXANE □ NCI-C60139 □ 3-OXIRANYL-7-OXABICYCLO(4.1.0)HEPT-ENE □ UCET TEXTILE FINISH 11-74 (OBS.) □ UNOX EPOXIDE 206 □ VINYL CYCLOHEXENE DIEPOXIDE □ 4-VINYLCYCLOHEXENE DIEPOXIDE □ 4-VINYL-1-CYCLOHEXENE DIEPOXIDE □ 4-VINYL-1,2-CYCLO-HEXENE DIEPOXIDE □ 1-VINYL-3-CYCLOHEXENE DIOXIDE □ 4-VINYLCYCLOHEXENE DIOXIDE □ 4-VINYL-1-CYCLOHEXENE DIOXIDE (MAK)

CONSENSUS REPORTS: NTP 10th Report on Carcinogens. IARC Cancer Review: Group 3 IMEMDT 7,56,87; Animal Sufficient Evidence IMEMDT 11,141,76. Reported in EPA TSCA Inventory.

OSHA PEL: TWA 10 ppm (skin)

ACGIH TLV: TWA 0.1 ppm; Animal Carcinogen

DFG MAK: Animal Carcinogen, Suspected Human Carcinogen

SAFETY PROFILE: Confirmed carcinogen with experimental carcinogenic and tumorigenic data. Poison by unspecified route. Moderately toxic by ingestion and skin contact. Mildly toxic by inhalation. Experimental reproductive effects. Mutation data reported. A severe skin irritant.

V

Combustible when exposed to heat or flame. To fight fire, use water, foam, dry chemical. When heated to decomposition it emits acrid smoke and irritating fumes.

VOP000   CAS: 109-93-3   HR: 3
VINYL ETHER
DOT: UN 1167
mf: C_4H_6O mw: 70.10
PROP: Colorless liquid; very volatile with characteristic odor. Bp: 39°, ULC: 100, lel: 1.7%, uel: 27%, flash p: <−22°F (CC), d: 0.774 @ 20°/20°, autoign temp: 680°F, vap d: 2.41. Very sltly sol in water; misc in alc, ether.
SYNS: DIVINYL ETHER (DOT) □ DIVINYL ETHER, inhibited (DOT) □ DIVYNYL OXIDE □ ETHENYL-OXYETHENE □ 1,1'-OXYBISETHENE □ VINESTHENE □ VINESTHESIN □ VINETHEN □ VINETHENE □ VINETHER □ VINIDYL □ VINYDAN
DOT CLASSIFICATION: 3; Label: Flammable Liquid
SAFETY PROFILE: Mildly toxic by inhalation. Mutation data reported. Prolonged exposure causes liver injury. A very dangerous fire hazard when exposed to heat or flame; can react vigorously with oxidizing materials. A severe explosion hazard in the form of vapor when exposed to heat or flame. Forms peroxides when exposed to air or oxygen. Hypergolic reaction with concentrated nitric acid. To fight fire, use CO_2, dry chemical. When heated to decomposition it emits acrid smoke and irritating fumes. Used as an inhalation anesthetic. See also ETHERS.

VPA000   CAS: 75-02-5   HR: 3
VINYL FLUORIDE
DOT: UN 1860
mf: CH_2:CHF   mw: 46
PROP: Colorless gas. Mp: −160.5°, bp: −51°, fp: −160.5°, lel: 2.6%, uel: 21.7%. Insol in water; sol in alc, ether.
SYNS: ETHENE, FLUORO- □ ETHYLENE, FLUORO- (8CI) □ FLUOROETHENE □ FLUOROETHYLENE □ MONOFLUOROETHYLENE □ VINYL FLUORIDE, inhibited (DOT)

CONSENSUS REPORTS: NTP 10th Report on Carcinogens. Reported in EPA TSCA Inventory.
ACGIH TLV: TWA 1 ppm; Suspected Human Carcinogen
NIOSH REL: (Vinyl Chloride) TWA 1 ppm; CL 5 ppm/15M
DOT CLASSIFICATION: 2.1; Label: Flammable Gas
SAFETY PROFILE: Confirmed carcinogen. A poison. Mutation data reported. A very dangerous fire hazard. To fight fire, stop flow of gas. When heated to decomposition it emits toxic fumes of F^-. See also FLUORIDES.

VPK000   CAS: 75-35-4   HR: 3
VINYLIDENE CHLORIDE
DOT: UN 1303
mf: $C_2H_2Cl_2$ mw: 96.94
PROP: Colorless, volatile liquid. Bp: 31.6°, lel: 7.3%, uel: 16.0%, fp: −122°, flash p: 0°F (OC), d: 1.213 @ 20°/4°, autoign temp: 1058°F.
SYNS: CHLORURE de VINYLIDENE (FRENCH) □ 1-1-DCE □ 1,1-DICHLOROETHENE □ 1,1-DICHLORO-ETHYLENE □ NCI-C54262 □ RCRA WASTE NUMBER U078 □ SCONATEX □ VDC □ VINYLIDENE CHLORIDE (II) □ VINYLIDENE DICHLORIDE □ VINYLIDINE CHLORIDE
CONSENSUS REPORTS: IARC Cancer Review: Group 3 IMEMDT 7,376,87; Human Inadequate Evidence IMEMDT 39,195,86, IMEMDT 19,439,79; Animal Limited Evidence IMEMDT 39,195,86; Animal Sufficient Evidence IMEMDT 19,439,79. EPA Genetic Toxicology Program. Reported in EPA TSCA Inventory. Community Right-To-Know List.
OSHA PEL: TWA 1 ppm
ACGIH TLV: TWA 5 ppm; Not Classifiable as a Human Carcinogen
DFG MAK: 2 ppm (8 mg/m³); Confirmed Animal Carcinogen with Unknown Relevance to Humans
NIOSH REL: (Vinyl Halides) TWA reduce to lowest detectable level
DOT CLASSIFICATION: 3; Label: Flammable Liquid

SAFETY PROFILE: Suspected carcinogen with experimental carcinogenic, neoplastigenic, tumorigenic, and teratogenic data. Poison by inhalation, ingestion, and intravenous routes. Moderately toxic by subcutaneous route. Human systemic effects by inhalation: general anesthesia, liver and kidney changes. Experimental reproductive effects. Mutation data reported. See also VINYL CHLORIDE. A very dangerous fire hazard when exposed to heat or flame. Moderately explosive in the form of gas when exposed to heat or flame. It forms explosive peroxides upon exposure to air. Potentially explosive reaction with chlorotrifluoroethylene at 180°C. Reaction with ozone forms dangerous products. Explosive reaction with perchloryl fluoride when heated above 100°C. Also can explode spontaneously. Reacts violently with chlorosulfonic acid, HNO_3, oleum. Can react vigorously with oxidizing materials. To fight fire, use alcohol foam, CO_2, dry chemical. When heated to decomposition it emits toxic fumes of Cl^-. See also CHLORINATED HYDROCARBONS, ALIPHATIC.

VPP000   CAS: 75-38-7   HR: 3
VINYLIDENE FLUORIDE
DOT: UN 1959
mf: $C_2H_2F_2$ mw: 64.04
PROP: Odorless, colorless gas. Bp: <−70°, fp: −144°, lel: 5.5%, uel: 21.3%.
SYNS: 1,1-DIFLUOROETHENE □ 1,1-DIFLUORO-ETHYLENE (DOT, MAK) □ ETHENE, 1,1-DIFLUORO- □ HALOCARBON 1132A □ NCI-C60208 □ R1132a (DOT) □ VDF □ VINYLIDENE DIFLUORIDE
CONSENSUS REPORTS: IARC Cancer Review: Group 3 IMEMDT 7,56,87; Animal Inadequate Evidence IMEMDT 39,227,86. Reported in EPA TSCA Inventory.
ACGIH TLV: TWA 500 ppm; Not Classifiable as a Human Carcinogen
DFG MAK: Confirmed Animal Carcinogen with Unknown Relevance to Humans
DOT CLASSIFICATION: 2.1; Label: Flammable Gas

NIOSH REL: (Vinyl Halides) TWA reduce to lowest detectable level
SAFETY PROFILE: Suspected carcinogen with experimental neoplastigenic data. Mildly toxic by inhalation. Mutation data reported. A very dangerous fire hazard when exposed to heat, flame, or oxidizers. Explosive in the form of vapor when exposed to heat or flame. Violent reaction with hydrogen chloride when heated under pressure. To fight fire, stop flow of gas. When heated to decomposition it emits toxic fumes of F^-. See also FLUORIDES.

VPP100   CAS: 25232-42-2   HR: 3
1-VINYLIMIDAZOLE HOMOPOLYMER
mf: $(C_5H_6N_2)_x$
SYNS: 1H-IMIDAZOLE, 1-ETHENYL-, HOMO-POLYMER □ IMIDAZOLE, 1-VINYL-, POLYMERS □ LUFIXAN □ POLY(VINYLIMIDAZOLE) □ POLY(N-VINYLIMIDAZOLE) □ POLY(1-VINYLIMIDAZOLE) □ N-VINYLIMIDAZOLE HOMOPOLYMER □ N-VINYL-IMIDAZOLE POLYMER
SAFETY PROFILE: A poison by intravenous route. When heated to decomposition it emits toxic vapors of NO_x.

VQK000   CAS: 105-38-4   HR: 3
VINYL PROPIONATE
mf: $C_5H_8O_2$ mw: 100.13
PROP: Liquid. D: 0.9173 @ 20°/20°, bp: 95°, fp: −81.1°, flash p: 34°F (OC), vap d: 3.3. Almost insol in water.
SYN: PROPANOIC ACID, ETHENYL ESTER
CONSENSUS REPORTS: Reported in EPA TSCA Inventory.
SAFETY PROFILE: Mildly toxic by ingestion and inhalation. A skin and eye irritant. Mutation data reported. A very dangerous fire hazard when exposed to heat or flame. To fight fire, use alcohol foam, mist, fog. When heated to decomposition it emits acrid smoke and irritating fumes. See also ESTERS.

VQK650   CAS: 25013-15-4   HR: 3
VINYL TOLUENE
DOT: UN 2618
mf: C_9H_{10} mw: 118.19

PROP: Clear colorless liquid. Mp: 75 C, bp: 170°. Sol in water: <1 mg/mL @ 24.5°. IDLH 400 ppm.

SYNS: METHYLSTYRENE □ NCI-C56406 □ TOLUENE, VINYL (mixed isomers) □ VINYL TOLUENE, inhibited mixed isomers (DOT) □ 3- and 4-VINYL TOLUENE (mixed isomers)

CONSENSUS REPORTS: Reported in EPA TSCA Inventory.

OSHA PEL: TWA 100 ppm

ACGIH TLV: TWA 50 ppm; STEL 100 ppm; Not Classifiable as a Human Carcinogen

DFG MAK: 100 ppm (490 mg/m³)

DOT CLASSIFICATION: 3; Label: Flammable Liquid

SAFETY PROFILE: Moderately toxic by ingestion and inhalation. An experimental teratogen. Human systemic effects by inhalation: eye and olfactory effects. Experimental reproductive effects. Mutation data reported. A skin and eye irritant. Flammable when exposed to heat or flame; can react vigorously with oxidizing materials. When heated to decomposition it emits acrid smoke and irritating fumes.

VRF000   CAS: 11006-76-1   HR: 2
VIRGINIAMYCIN

PROP: White powder. Decomp @ 138–140°. Sltly sol in water and dil acid; sol in methanol, ethanol, acetone, benzene; almost insol in ligroin.

SYNS: ANTIBIOTIC No. 899 □ ESKALIN V □ MIKAMYCIN □ OSTREOGRYCIN □ PATRICIN □ PRISTINAMYCIN □ PYOSTACINE □ RP7293 □ SKF 7988 □ STAFAC □ STAPHYLOMYCIN □ STAPYOCINE □ STREPTOGRAMIN □ VERNAMYCIN □ VIRGIMYCIN

SAFETY PROFILE: Moderately toxic by ingestion, intraperitoneal, and subcutaneous routes. Used as an antibiotic.

VSK600   CAS: 68-26-8   HR: 3
VITAMIN A

mf: $C_{20}H_{30}O$ mw: 286.50

PROP: Yellow crystals or light-yellow to red oil; mild fishy odor. Mp: 63–64°, bp: 137–138° @ 0.000001 mm. Very sol in chloroform, ether; sol in abs alc, vegetable oil; insol in glycerin, water.

SYNS: ACON □ AFAXIN □ AGIOLAN □ ALPHALIN □ ALPHASTEROL □ ANATOLA □ ANTI-INFECTIVE VITAMIN □ ANTIXEROPHTHALMIC VITAMIN □ AORAL □ APEXOL □ AQUASYNTH □ AVIBON □ AVITA □ AVITOL □ BIOSTEROL □ CHOCOLA A □ 3,7-DIMETHYL-9-(2,6,6-TRIMETHYL-1-CYCLOHEXEN-1-YL)-2,4,6,8-NONATETRAEN-1-OL □ DISATABS TABS □ DOFSOL □ EPITELIOL □ HI-A-VITA □ LARD FACTOR □ MYVPACK □ OLEOVITAMIN A □ OPHTHALAMIN □ PREPALIN □ RETINOL □ all-trans RETINOL □ RETRO-VITAMIN A □ TESTAVOL □ VAFLOL □ VI-ALPHA □ VITAMIN A1 □ VITAMIN A1 ALCOHOL □ all-trans-VITAMIN A ALCOHOL □ VITAVEL-A □ VITPEX □ VOGAN □ VOGAN-NEU

CONSENSUS REPORTS: Reported in EPA TSCA Inventory. EPA Genetic Toxicology Program.

SAFETY PROFILE: Moderately toxic by ingestion. Human teratogenic effects by ingestion: developmental abnormalities of the craniofacial area and urogenital system. An experimental teratogen. Experimental reproductive effects. Human mutation data reported. When heated to decomposition it emits acrid smoke and irritating fumes.

VSK900   CAS: 127-47-9   HR: 2
VITAMIN A ACETATE

mf: $C_{22}H_{32}O_2$ mw: 328.54

PROP: Crystals from MeOH. Mp: 57–58°.

SYNS: CRYSTALETS □ MYVAK □ MYVAX □ RETINOL ACETATE □ RETINYL ACETATE □ all-trans-RETINYL ACETATE □ trans-VITAMIN A ACETATE □ VITAMIN A ALCOHOL ACETATE

CONSENSUS REPORTS: Reported in EPA TSCA Inventory.

SAFETY PROFILE: Moderately toxic by ingestion. Experimental teratogenic and reproductive effects. Questionable carcinogen with experimental neoplastigenic data. Mutation data reported. When heated to decomposition it emits acrid smoke and irritating fumes. See also VITAMIN A.

VSK950   CAS: 302-79-4   HR: 3
VITAMIN A ACID

mf: $C_{20}H_{28}O_2$ mw: 300.48

PROP: Crystals from MeOH. Mp: 180–182°.

SYNS: ABEREL □ 3,7-DIMETHYL-9-(2,6,6-TRIMETHYL-1-CYCLOHEXEN)-1-YL-2,4,6,8-NONATETRAENOIC ACID □ NSC-122758 □ β-RA □ RETIN-A □ RETINOIC ACID □ β-RETINOIC ACID □ all-trans-RETINOIC ACID □ TRETINOIN

CONSENSUS REPORTS: Reported in EPA TSCA Inventory. EPA Genetic Toxicology Program.

SAFETY PROFILE: Poison by ingestion, intraperitoneal, subcutaneous, and intravenous routes. Experimental reproductive effects. Questionable carcinogen with experimental neoplastigenic and teratogenic data. Human mutation data reported. A human skin irritant. When heated to decomposition it emits acrid smoke and irritating fumes. Used to treat acne and other skin problems.

VSK955   CAS: 4759-48-2   HR: 3
13-cis-VITAMIN A ACID

mf: $C_{20}H_{28}O_2$ mw: 300.48

PROP: Crystals from EtOH. Mp: 189–190°.

SYNS: ISOTRETINOIN □ NEOVITAMIN A ACID □ 13-RA □ 13-cis-RETINOIC ACID □ RO-4-3780

CONSENSUS REPORTS: Reported in EPA TSCA Inventory.

SAFETY PROFILE: Poison by intraperitoneal route. Moderately toxic by ingestion. A human teratogen by ingestion with fetal developmental abnormalities of the skin and appendages and other postnatal effects. Human reproductive effects. Human systemic effects: decreased immune response, diarrhea, hypermotility, irritative dermatitis, sweating. Human mutation data reported. An experimental teratogen. Other experimental reproductive effects. When heated to decomposition it emits acrid smoke and irritating fumes.

VSK975   CAS: 514-85-2   HR: D
9-cis-VITAMIN A ALDEHYDE

mf: $C_{20}H_{28}O$ mw: 284.48

SYNS: 9-cis-3,7-DIMETHYL-9-(2,6,6-TRIMETHYL-1-CYCLOHEXEN-1-YL)-2,4,6,8-NONATETRAENAL □ ISORETINENE a □ 9-cis-RETINAL □ 9-cis-RETINALDEHYDE

CONSENSUS REPORTS: Reported in EPA TSCA Inventory.

SAFETY PROFILE: Experimental reproductive effects. When heated to decomposition it emits acrid smoke and irritating fumes.

VSP000   CAS: 79-81-2   HR: 1
VITAMIN A PALMITATE

mf: $C_{36}H_{60}O_2$ mw: 524.96

PROP: A solid or liquid. Mp: 28–29°.

SYNS: AQUASOL □ AROVIT □ RETINOL PALMITATE □ RETINYL PALMITATE

CONSENSUS REPORTS: Reported in EPA TSCA Inventory. EPA Genetic Toxicology Program.

SAFETY PROFILE: Mildly toxic by ingestion. An experimental teratogen. Experimental reproductive effects. Human mutation data reported. When heated to decomposition it emits acrid smoke and irritating fumes.

VSU000   CAS: 65-22-5   HR: 3
VITAMIN B₆ HYDROCHLORIDE

mf: $C_8H_9NO_3 \cdot ClH$ mw: 203.64

SYNS: 3-HYDROXY-5-(HYDROXYMETHYL)-2-METHYLISONICOTINALDEHYDE, HYDROCHLORIDE □ 2-METHYL-3-HYDROXY-4-FORMYL-5-HYDROXYMETHYLPYRIDINE HYDROCHLORIDE □ PYRIDOXAL HYDROCHLORIDE

CONSENSUS REPORTS: Reported in EPA TSCA Inventory.

SAFETY PROFILE: Poison by intramuscular, intravenous, and intraperitoneal routes. Moderately toxic by ingestion and subcutaneous routes. When heated to decomposition it emits very toxic fumes of NO_x and HCl. See also ALDEHYDES.

VSU100   CAS: 58-85-5   HR: D
VITAMIN B₇

mf: $C_{10}H_{16}N_2O_3S$ mw: 244.34

SYNS: BIOEPIDERM □ BIOS II □ BIOTIN □ (+)-BIOTIN □ d-BIOTIN □ d-(+)-BIOTIN □ COENZYME R □ FACTOR S □ FACTOR S (vitamin) □ 1H-THIENO(3,4-d)IMIDAZOLE-4-PENTANOIC ACID, HEXAHYDRO-2-OXO-, (3aS-(3a-α-4-β, 6a-α))- □ VITAMIN H

CONSENSUS REPORTS: Reported in EPA TSCA Inventory.

SAFETY PROFILE: An experimental teratogen. Experimental reproductive effects. When heated to decomposition it emits toxic fumes of NO_x and SO_x.

V

VSZ000   **CAS: 68-19-9**   **HR: 3**
VITAMIN B₁₂ COMPLEX

mf: $C_{63}H_{88}CoN_{14}O_{14}P$ mw: 1355.55

PROP: Red needles. Mp: @ >300°. The anti-pernicious-anemia vitamin. All vitamin B₁₂ compounds contain the cobalt atom in its trivalent state. There are at least three active forms: cyanocobalamin, hydroxycobalamin, and nitrocobalamin. Dark-red crystals or crystalline powder. Very hygroscopic; sltly sol in water; sol in alc; insol in acetone, chloroform, ether.

SYNS: ANACOBIN □ B-12 □ BERUBIGEN □ BETALIN 12 CRYSTALLINE □ BEVATINE-12 □ BEVIDOX □ BYLADOCE □ CABADON M □ COBADOCE FORTE □ COBALIN □ COBAMIN □ COBIONE □ COTEL □ COVIT □ CRYSTAMIN □ CRYSTWEL □ CYANO-B12 □ CYANOCOBALAMIN □ CYCOLAMIN □ CYKOBEMIN-ET □ CYREDIN □ CYTACON □ CYTAMEN □ CYTOBION □ DEPINAR □ DIMETHYLBENZIMID-AZOLYCOBAMIDE □ 5,6-DIMETHYLBENZIMIDAZOLY-COBAMIDE CYANIDE □ DISTIVIT (B12 PEPTIDE) □ DOBETIN □ DOCEMINE □ DOCIBIN □ DOCIGRAM □ DODECABEE □ DODECAVITE □ DODEX □ DUCO-BEE □ DUODECIBIN □ EMBIOL □ EMOCICLINA □ ERITRONE □ ERYCYTOL □ ERYTHROTIN □ EUHAEMON □ EXTRINSIC FACTOR □ FACTOR II (VITAMIN) □ FRESMIN □ HEMO-B-DOZE □ HEMOMIN □ HEPAGON □ HEPAVIS □ HEPCOVITE □ LACTOBACILLUS LACTIS DORNER FACTOR □ LLD FACTOR □ MACRABIN □ MEGABION □ MEGALOVEL □ MILBEDOCE □ NAGRAVON □ NORMOCYTIN □ PERNAEMON □ PERNAEVIT □ PERNIPURON □ PLECYAMIN □ POYAMIN □ REBRAMIN □ REDAMINA □ REDISOL □ RHODACRYST □ RUBESOL □ RUBRAMIN □ RUBRIPCA □ RUBROCITOL □ SYTOBEX □ VIBALT □ VIBISONE □ VIRUBRA □ VITAMIN B12 (FCC) □ VITARUBIN □ VITA-RUBRA □ VITRAL □ VI-TWEL

CONSENSUS REPORTS: Cobalt and its compounds are on the Community Right-To-Know List. Reported in EPA TSCA Inventory. EPA Genetic Toxicology Program.

NIOSH REL: (Cobalt) Insufficient evidence for recommending limit

SAFETY PROFILE: Poison by subcutaneous route. Moderately toxic by intraperitoneal route. An experimental teratogen. Experimental reproductive effects. When heated to decomposition it emits very toxic fumes of PO$_x$ and NO$_x$. See also COBALT COMPOUNDS.

VSZ100   **CAS: 50-14-6**   **HR: 3**
VITAMIN D2

mf: $C_{28}H_{44}O$ mw: 396.72

PROP: White crystals or prisms from Me₂CO; odorless. Mp: 115–118°. Insol in water; sol in alc, chloroform, ether, and fatty oils.

SYNS: d-ARTHIN □ CALCIFEROL □ CALCIFERON 2 □ CONDACAPS □ CONDOCAPS □ CONDOL □ CRTRON □ CRYSTALLINA □ DARAL □ DAVITAMON D □ DAVITIN □ DECAPS □ DEE-OSTEROL □ DEE-RON □ DEE-RONAL □ DEE-ROUAL □ DELTALIN □ DERATOL □ DETALUP □ DIACTOL □ DIVIT URTO □ DORAL □ DRISDOL □ ERGOCALCIFEROL □ ERGORONE □ ERGOSTEROL, activated □ ERGOSTER-OL, irradiated □ ERTRON □ FORTODYL □ GELTABS □ HI-DERATOL □ INFRON □ IRRADIATED ERGOSTA-5,7,22-TRIEN-3-β-OL □ METADEE □ MULSIFEROL □ MYKOSTIN □ OLEOVITAMIN D □ OSTELIN □ RADIOSTOL □ RADSTERIN □ 9,10,SECOERGOSTA-5,7,10(19),22-TETRAEN-3-β-OL □ SHOCK-FEROL □ STEROGYL □ VIGANTOL □ VIOSTEROL □ VITAVEL-D

CONSENSUS REPORTS: EPA Extremely Hazardous Substances List.

SAFETY PROFILE: Poison by ingestion, intraperitoneal, intravenous, and intramuscular routes. An experimental teratogen. Human systemic effects by ingestion: anorexia, nausea or vomiting, and weight loss. Experimental reproductive effects. When heated to decomposition it emits acrid smoke and irritating fumes.

VSZ450   **CAS: 59-02-9**   **HR: D**
VITAMIN E

mf: $C_{29}H_{50}O_2$ mw: 430.79

PROP: dl-Form: Sltly viscous, pale-yellow oil; d-form: red liquid; odorless. Natural α-tocopherol has been crystallized. Mp: 2.5–3.5°, d: (25°/4°) 0.950, bp: (0.1 mm Hg) 200–220°. Practically insol in water; freely sol in oils, fats, acetone, alc, chloroform, ether, other fat solvents. Gradually darkens on exposure to light.

SYNS: ALMEFROL □ ANTISTERILITY VITAMIN □ COVI-OX □ DENAMONE □ EMIPHEROL □ ENDO E □ EPHYNAL □ EPROLIN □ EPSILAN □ ESORB □ ETAMICAN □ ETAVIT □ EVION □ EVITAMINUM □ ILITIA □ PHYTOGERMINE □ PROFECUNDIN □ SPAVIT □ SYNTOPHEROL □ d-α-TOCOPHEROL (FCC) □ dl-α-TOCOPHEROL (FCC) □ α-TOCOPHEROL □ (R,R,R)-α-TOCOPHEROL □ (2R,4'R,8'R)-α-TOCOPHEROL

□ TOKOPHARM □ 5,7,8-TRIMETHYLTOCOL □ VASCUALS □ VERROL □ VITAPLEX E □ VITAYONON □ VITEOLIN

CONSENSUS REPORTS: Reported in EPA TSCA Inventory.

SAFETY PROFILE: Experimental reproductive effects. Mutation data reported. When heated to decomposition it emits acrid smoke and irritating fumes.

VSZ500   CAS: 12001-79-5   HR: 2
VITAMIN K
CONSENSUS REPORTS: Reported in EPA TSCA Inventory.
SAFETY PROFILE: Moderately toxic by subcutaneous route. An experimental teratogen. When heated to decomposition it emits acrid smoke and irritating fumes.

VTF000   CAS: 595-33-5   HR: 3
VOLIDAN
mf: $C_{24}H_{32}O_4$ mw: 384.56
PROP: Crystals from MeOH (aq). Mp: 214–216°.
SYNS: 17-α-ACETOXY-6-DEHYDRO-6-METHYL-PROGESTERONE □ 17-ACETOXY-6-METHYLPREGNA-4,6-DIENE-3,20-DIONE □ 17-α-ACETOXY-6-METHYL-PREGNA-4,6-DIENE-3,20-DIONE □ 17-α-ACETOXY-6-METHYL-4,6-PREGNADIENE-3,20-DIONE □ BDH 1298 □ 6-DEHYDRO-6-METHYL-17-α-ACETOXYPROGESTER-ONE □ DMAP □ 17-HYDROXY-6-METHYLPREGNA-4,6-DIENE-3,20-DIONE ACETATE □ MEGACE □ MEG-ESTROL ACETATE □ MEGESTRYL ACETATE □ 6-METHYL-17-α-ACETOXYPREGNA-4,6-DIENE-3,20-DIONE □ 6-METHYL-6-DEHYDRO-17-α-ACETOXY-PROGESTERONE □ 6-METHYL-6-DEHYDRO-17-α-ACETYLPROGESTERONE □ 6-METHYL-17-α-HYDROXY-Δ^6-PROGESTERONE ACETATE □ 6-METHYL-$\Delta^{4,6}$-PREGNADIEN-17-α-OL-3,20-DIONE ACETATE □ NSC-71423 □ OVABAN □ SC10363

CONSENSUS REPORTS: IARC Cancer Review: Animal Limited Evidence IMEMDT 21,431,79.

SAFETY PROFILE: Suspected carcinogen with experimental carcinogenic and teratogenic data. Poison by intravenous route. Human reproductive effects by ingestion and implant routes: effects on ovaries and fallopian tubes, menstrual cycle changes, and female fertility index changes. Mutation data reported. Experimental reproductive effects. When heated to decomposition it emits acrid smoke and irritating fumes. An FDA proprietary drug used to treat endometriosis and breast cancer. A steroid.

V

W

WAT200　CAS: 81-81-2　HR: 3
WARFARIN
mf: $C_{19}H_{16}O_4$　mw: 308.35

PROP: Colorless, odorless, tasteless crystals. Mp: 161°. Sol in acetone, dioxane; sltly sol in methanol, ethanol; very sol in alkaline aqueous sol; insol in water and benzene. IDLH 100 mg/m³.

SYNS: 3-(α-ACETONYLBENZYL)-4-HYDROXY-COUMARIN □ ARAB RAT DETH □ ATHROMBINE-K □ BRUMIN □ COMPOUND 42 □ d-CON □ CO-RAX □ COUMADIN □ COUMAFENE □ DETHMORE □ EASTERN STATES DUOCIDE □ 4-HYDROXY-3-(3-OXO-1-FENYL-BUTYL) CUMARINE (DUTCH) □ 4-HYDROXY-3-(3-OXO-1-PHENYL-BUTYL)-CUMARIN (GERMAN) □ 4-IDROSSI-3-(3-OXO-)-(FENIL-BUTIL)-CUMARINE (ITALIAN) □ KUMADER □ LIQUA-TOX □ MOUSE PAK □ 3-(α-PHENYL-β-ACETYLETHYL)-4-HYDRO-XYCOUMARIN □ 3-(1'-PHENYL-2'-ACETYLETHYL)-4-HYDROXYCOUMARIN □ (PHENYL-1 ACETYL-2 ETHYL)-3-HYDROXY-4 COUMARINE (FRENCH) □ PROTHROMADIN □ RAT-A-WAY □ RAT-B-GON □ RAT-GARD □ RAT & MICE BAIT □ RATS-NO-MORE □ RCRA WASTE NUMBER P001 □ RO-DETH □ ROUGH & READY MOUSE MIX □ SOLFARIN □ SPRAY-TROL BRANCH RODEN-TROL □ TWIN LIGHT RAT AWAY □ WARFARINE (FRENCH) □ ZOOCOUMARIN (RUSSIAN)

CONSENSUS REPORTS: Reported in EPA TSCA Inventory. EPA Extremely Hazardous Substances List.

OSHA PEL: TWA 0.1 mg/m³
ACGIH TLV: TWA 0.1 mg/m³
DFG MAK: 0.5 mg/m³

SAFETY PROFILE: A human poison by ingestion. Poison by inhalation and intravenous routes. Moderately toxic by skin contact, subcutaneous, and intraperitoneal routes. Human systemic effects by ingestion: hemorrhage, ulceration or bleeding from small intestine, blood clotting factor change. Human reproductive effects by ingestion and intramuscular routes: fetal death and physical abnormalities at birth. Human teratogenic effects include developmental abnormalities of the craniofacial area, musculoskeletal system, and respiratory system. An experimental teratogen. Other experimental reproductive effects. Used as an oral anticoagulant and as a rodenticide. When heated to decomposition it emits acrid smoke and fumes.

WAT220　CAS: 129-06-6　HR: 3
WARFARIN SODIUM
mf: $C_{19}H_{15}O_4 \cdot Na$　mw: 330.33

PROP: Crystalline solid.

SYNS: 3-(α-ACETONYLBENZYL)-4-HYDROXY-COUMARIN SODIUM SALT □ ATHROMBIN □ COUMADIN SODIUM □ 4-HYDROXY-3-(3-OXO-1-PHENYLBUTYL)-2H-1-BENZOPYRAN-2-ONE SODIUM SALT (9CI) □ MAREVAN (SODIUM SALT) □ PANWARFIN □ PROTHROMBIN □ RATSUL SOLUBLE □ SODIUM COUMADIN □ SODIUM WARFARIN □ TINTORANE □ VARFINE □ WARAN □ WARCOUMIN □ WARFILONE

CONSENSUS REPORTS: Reported in EPA TSCA Inventory. EPA Extremely Hazardous Substances List.

SAFETY PROFILE: Poison to humans by ingestion. Experimental poison by ingestion and intravenous routes. Human systemic effects by ingestion: dermatitis. Human reproductive effects by ingestion: fetotoxicity, abnormal condition of newborn at birth, other newborn physical effects, and teratogenic effects including developmental abnormalities of the eye and ear, craniofacial area, skin and appendages, musculoskeletal system, cardiovascular system, and gastrointestinal system of the fetus. An experimental teratogen. Other experimental reproductive effects. Mutation data reported. An anticoagulant drug. When heated to decomposition it emits toxic fumes of Na_2O.

WAT230   CAS: 133743-71-2   HR: 3
WATANIDIPINE HYDROCHLORIDE
SYNS: AE0047 □ PYRIDINE-3,5-CARBOXYLIC ACID, 1,4-DIHYDRO-2,6-DIMETHYL-4-(3-NITROPHENYL)-, 2-(4-(4-BENZHYDRYLPIPERAZIN-1-YL)PHENYL)ETHYL METHYL ESTER, (+−)-

SAFETY PROFILE: A poison by ingestion, intraperitoneal, and subcutaneous routes. When heated to decomposition it emits toxic vapors of NO_x.

WBJ000   HR: 3
WELDING FUMES
OSHA PEL: TWA 5 mg(Cd)/m³
ACGIH TLV: TWA 5 mg(Cd)/m³

SAFETY PROFILE: When welding is done on a surface coated with cadmium, toxic and carcinogenic fumes of cadmium are evolved. When zinc-coated surfaces are welded, toxic quantities of zinc oxide may be liberated. When painted surfaces are welded, lead or other pigment fumes may be liberated. And when fluoride fluxes are used in welding, very toxic fluoride fumes are evolved. When oily surfaces are welded, offensive and toxic fumes can be liberated, and, when the welding torch is improperly ignited, carbon monoxide, which is very toxic, may be evolved. Also, NO_x is formed. It is therefore considered hazardous to inhale excessive amounts of welding fumes. It is also possible to inhale sufficient quantities of iron oxide from welding to cause siderosis. Metal fume fever is a common reaction. It is characterized by chills, fever, sweating, and leukocytosis coming on several hours after exposure. Recovery is usually complete in 24–48 hours and there are no significant after effects. Safety goggles are required to protect against spatter. Light-filtering goggles are required to shield the eyes against the intense UV light from the arc. See also specific metals and their compounds (e.g., CADMIUM and CADMIUM COMPOUNDS).

WBJ700   CAS: 68917-73-7   HR: 1
WHEAT GERM OIL
PROP: Bland yellow oil. Misc with chloroform, ether, pet ether, and benzene; sltly sol in alc.
SYNS: BRAN ABSOLUTE □ CAV-ECOL □ MERIT □ MYOPONE □ OILS, WHEAT GERM □ UNIDERM WGO □ WHEAT HUSK OIL

CONSENSUS REPORTS: Reported in EPA TSCA Inventory.

SAFETY PROFILE: A skin and eye irritant. When heated to decomposition it emits acrid smoke and irritating fumes.

WBL100   CAS: 8002-80-0   HR: D
WHEAT GLUTEN
PROP: Natural protein portion of grain. White powder.

SAFETY PROFILE: When heated to decomposition it emits acrid smoke and irritating fumes.

WBL155   HR: D
WHEY, PROTEIN CONCENTRATE
PROP: White powder.

SAFETY PROFILE: When heated to decomposition it emits acrid smoke and irritating fumes.

WBL165   HR: D
WHEY, REDUCED MINERALS
SYN: REDUCED MINERALS WHEY

SAFETY PROFILE: When heated to decomposition it emits acrid smoke and irritating fumes.

WBS000   HR: 3
WHISKEY
PROP: Light yellow-amber liquid. Pleasant to fruity odor. D: 0.923–0.935 @ 15.56°; 47–53% of ethanol, by volume, flash p: 80.0°F (CC). Made by distillation of fermented malted grains, e.g., corn, rye, or barley. After distillation, whiskey is aged in wooden containers for up to several years. The aging extracts such components as acids and esters from the wood and promotes oxidation of components of raw whiskey and some reactions between organic components to form new flavors.

SAFETY PROFILE: The carcinogen urethane is sometimes found in whiskey.

The whiskey equivalent of 1 ounce of pure ethanol per capita per day has been cited as healthful to adults to relieve stress and promote relaxation. However, it is often abused, which can lead to habituation with consequent liver damage, malnutrition, and a wide variety of other physical and mental problems, including the development of cancer. A fire hazard when exposed to heat or flame. To fight fire, use water, water spray, alcohol foam, CO_2, dry chemical. See also ETHANOL and URETHANE.

WCA000   HR: 2
WINE
PROP: An alcoholic beverage made from the fermented juice of grapes, other fruits, or plants. Contains 7–20% ethanol by volume. Concentrations of alcohol higher than those produced naturally are obtained by fortifying with pure ethanol. The distinctive colors, tastes, bouquets of wines are sometimes produced by adding coloring matter, sugar, acetic acid, salts, and higher fatty acids.
SAFETY PROFILE: Some wines contain the carcinogen urethane. The wine equivalent of 1 ounce of pure ethanol per capita per day has been cited as healthful to adults to relieve stress and promote relaxation. However, it is often abused, which can lead to habituation with consequent liver damage, malnutrition, and a wide variety of

other physical and mental problems, including the development of cancer. Some of the additives to wines have been known to cause allergic reactions in humans. See also ETHANOL and URETHANE.

WCJ100   HR: 3
WOOD DUST
CONSENSUS REPORTS: NTP 10th Report on Carcinogens.
ACGIH TLV: Hardwoods & softwoods (nonallergenic) TWA 5 mg/m³ Not Classifiable as a Human Carcinogen; western red cedar TWA 0.5 mg/m³ (skin, sensitizer) Not Classifiable as a Human Carcinogen; beech and oak (skin, sensitizer) TWA 5 mg/m³ Confirmed Human Carcinogen; birch, mahogany, teak, walnut TWA 5 mg/m³ (skin, sensitizer) Suspected Human Carcinogen; (Proposed: (nonallergenic and noncarcinogenic) TWA 1 mg/m³; respiratory allergenic 0.5 mg/m³ (sensitizer); (Birch, mahogany, teak, walnut) Suspected Human Carcinogen; (oak and beech) Confirmed Human Carcinogen; all other wood dusts ; Not Classifiable as a Human Carcinogen)
SAFETY PROFILE: Vary from confirmed human carcinogens to noncarcinogenic depending on wood species. Some are sensitizers. When heated to decomposition it emits acrid smoke and irritating fumes.

W

X

XAA500   **CAS: 130209-82-4**   **HR: 3**
XALATAN
mf: C$_{26}$H$_{40}$O$_5$ mw: 432.66
SYNS: 5-HEPTENOIC ACID, 7-(3,5-DIHYDROXY-2-(3-HYDROXY-5-PHENYLPENTYL)CYCLOPENTYL)-,1-METHYLETHYL ESTER, (1R-(1-α(Z),2-β(R*),3-α,5-α))- □ LATANOPROST □ PHXA 41 □ XA 41
SAFETY PROFILE: A poison by ingestion and intravenous route. Human systemic effects. When heated to decomposition it emits acrid smoke and irritating vapors.

XAK800   **CAS: 11138-66-2**   **HR: D**
XANTHAN GUM
PROP: Produced by fermentation of a carbohydrate with *Xanthomonas campestris*. Cream-colored powder. Sol in hot or cold water.
SAFETY PROFILE: When heated to decomposition it emits acrid smoke and irritating fumes.

XDJ000   **CAS: 298-81-7**   **HR: 3**
XANTHOTOXIN
mf: C$_{12}$H$_8$O$_4$ mw: 216.20
PROP: Crystals from EtOH (aq). Mp: 148°.
SYNS: AMMOIDIN □ 6-HYDROXY-7-METHOXY-5-BENZOFURANACRYLIC ACID Δ-LACTONE □ MELADININ □ MELADININE □ MELOXINE □ METHOXA-DOME □ METHOXSALEN □ 8-METHOXY-(FURANO-3'.2':6.7-COUMARIN) □ 9-METHOXY-7H-FURO(3,2-g)BENZOPYRAN-7-ONE □ 8-METHOXY-2',3',6,7-FURO-COUMARIN □ 8-METHOXY-4',5',6,7-FUROCOUMARIN □ 8-METHOXYPSORALEN □ 9-METHOXYPSORALEN □ 8-MOP □ 8-MP □ NCI-C55903 □ OXSORALEN □ OXY-PSORALEN □ PRORALONE-MOP
CONSENSUS REPORTS: NTP 10th Report on Carcinogens. IARC Cancer Review: Group 1 IMEMDT 7,243,87; Human Inadequate Evidence IMEMDT 24,101,80; Animal Inadequate Evidence IMEMDT 24,101,80. Reported in EPA TSCA Inventory. EPA Genetic Toxicology Program.
SAFETY PROFILE: Confirmed carcinogen. Poison by intraperitoneal route. Moderately toxic by ingestion and subcutaneous routes. Human mutation data reported. When heated to decomposition it emits acrid smoke and irritating fumes. A drug used to treat skin diseases.

XDS000   **CAS: 7440-63-3**   **HR: 1**
XENON
DOT: UN 2036/UN 2591
af: Xe   aw: 131.29
PROP: Colorless, odorless, tasteless, monatomic gaseous element. Reacts with fluorine and very powerful fluorinating agents. D (gas): 5.8878 g/L, d (liq): 3.52 @ −109°, mp: −112°, bp: −107°. Sol in H$_2$O.
SYNS: XENON (UN 0236) (DOT) □ XENON, refrigerated liquid (cryogenic liquids) (UN 2591) (DOT)
CONSENSUS REPORTS: Reported in EPA TSCA Inventory.
DOT CLASSIFICATION: 2.2; Label: Nonflammable Gas
SAFETY PROFILE: An inert gas that acts as a simple asphyxiant. For a discussion of toxicity effects, see ARGON. A common air contaminant.

XGS000   **CAS: 1330-20-7**   **HR: 3**
XYLENE
DOT: UN 1307
mf: C$_8$H$_{10}$ mw: 106.18
PROP: A clear liquid. Bp: 138.5°, flash p: 100°F (TOC), d: 0.864 @ 20°/4°, vap press: 6.72 mm @ 21°. Composition: as nonaromatics 0.07%, toluene 14%, ethyl benzene 19.27%, p-xylene 7.84%, m-xylene 65.01%, o-xylene 7.63%, C9 and aromatics 0.04% (TXAPA9 33,543,75).

SYNS: DIMETHYLBENZENE □ KSYLEN (POLISH) □ METHYL TOLUENE □ NCI-C55232 □ RCRA WASTE NUMBER U239 □ VIOLET 3 □ XILOLI (ITALIAN) □ XYLENEN (DUTCH) □ XYLOL (DOT) □ XYLOLE (GERMAN)

CONSENSUS REPORTS: Reported in EPA TSCA Inventory. EPA Genetic Toxicology Program. Community Right-To-Know List.

OSHA PEL: TWA 100 ppm; STEL 150 ppm

ACGIH TLV: TWA 100 ppm; STEL 150 ppm; BEI: methyl hippuric acids in urine at end of shift 1.5 g/g creatinine; Not Classifiable as a Human Carcinogen

DFG MAK: (all isomers) 100 ppm (440 mg/m^3); BAT: 150 μg/dL in blood at end of shift

NIOSH REL: (Xylene) TWA 100 ppm; CL 200 ppm/10M

DOT CLASSIFICATION: 3; Label: Flammable Liquid

SAFETY PROFILE: Moderately toxic by intraperitoneal and subcutaneous routes. Mildly toxic by ingestion and inhalation. An experimental teratogen. Human systemic effects by inhalation: olfactory changes, conjunctiva irritation, and pulmonary changes. Experimental reproductive effects. Mutation data reported. A human eye irritant. An experimental skin and severe eye irritant. Some temporary corneal effects are noted, as well as some conjunctival irritation by instillation (adding drops to the eyes one drop at a time). Irritation can start @ 200 ppm. A very dangerous fire hazard when exposed to heat or flame; can react with oxidizing materials. To fight fire, use foam, CO_2, dry chemical. When heated to decomposition it emits acrid smoke and irritating fumes. See also other xylene entries.

XHA000   CAS: 108-38-3   HR: 3
m-XYLENE
mf: C_8H_{10} mw: 106.18
PROP: Colorless, mobile liquid. Mp: −47.9°, bp: 139°, lel: 1.1%, uel: 7.0%, flash p: 77°F, d: 0.864 @ 20°/4°, vap press: 10 mm @ 28.3°, vap d: 3.66, autoign temp: 986°F.

Insol in water; misc with alc, ether, and some org solvs. IDLH 900 ppm.

SYNS: m-DIMETHYLBENZENE □ 1,3-DIMETHYLBENZENE □ 1,3-XYLENE □ m-XYLOL (DOT)

CONSENSUS REPORTS: Community Right-To-Know List. Reported in EPA TSCA Inventory.

OSHA PEL: TWA 100 ppm; STEL 150 ppm

ACGIH TLV: TWA 100 ppm; STEL 150 ppm; BEI: methyl hippuric acids in urine at end of shift 1.5 g/g creatinine; Not Classifiable as a Human Carcinogen

NIOSH REL: (Xylene) TWA 100 ppm; CL 200 ppm/10M

DOT CLASSIFICATION: 3; Label: Flammable Liquid

SAFETY PROFILE: Moderately toxic by intraperitoneal route. Mildly toxic by ingestion, skin contact, and inhalation. An experimental teratogen. Human systemic effects by inhalation: motor activity changes, ataxia, and irritability. Experimental reproductive effects. A severe skin irritant. A common air contaminant. A very dangerous fire hazard when exposed to heat or flame; can react with oxidizing materials. Explosive in the form of vapor when exposed to heat or flame. To fight fire, use foam, CO_2, dry chemical. Emitted from modern building materials (CENEAR 69,22,91). When heated to decomposition it emits acrid smoke and irritating fumes. See also other xylene entries.

XHJ000   CAS: 95-47-6   HR: 3
o-XYLENE
mf: C_8H_{10} mw: 106.18
PROP: Colorless, mobile liquid. D: 0.880 @ 20°/4°, mp: −25.2°, bp: 144.4°, flash p: 62.6°F, lel: 1.0%, uel: 6.0%. Insol in water; misc in abs alc, ether. IDLH 900 ppm.

SYNS: o-DIMETHYLBENZENE □ 1,2-DIMETHYLBENZENE □ o-METHYLTOLUENE □ 1,2-XYLENE □ o-XYLOL

CONSENSUS REPORTS: Community Right-To-Know List. Reported in EPA TSCA Inventory.

OSHA PEL: TWA 100 ppm; STEL 150 ppm

ACGIH TLV: TWA 100 ppm; STEL 150 ppm; BEI: methyl hippuric acids in urine at end of shift 1.5 g/g creatinine; Not Classifiable as a Human Carcinogen
NIOSH REL: (Xylene) TWA 100 ppm; CL 200 ppm/10M
DOT CLASSIFICATION: 3; Label: Flammable Liquid
SAFETY PROFILE: Moderately toxic by intraperitoneal route. Mildly toxic by ingestion and inhalation. An experimental teratogen. A common air contaminant. A very dangerous fire hazard when exposed to heat or flame. Explosive in the form of vapor when exposed to heat or flame. To fight fire, use foam, CO_2, dry chemical. Incompatible with oxidizing materials. When heated to decomposition it emits acrid smoke and irritating fumes. Emitted from modern building materials (CENEAR 69,22,91). See also other xylene entries.

XHS000   CAS: 106-42-3   HR: 3
p-XYLENE
mf: C_8H_{10} mw: 106.18
PROP: Clear plates, prisms, or liquid. Bp: 138.3°, lel: 1.1%, uel: 7.0%, flash p: 77°F (CC), mp: 13–14°, d: 0.8611 @ 20°/4°, vap press: 10 mm @ 27.3°, vap d: 3.66, autoign temp: 986°F. Insol in water; sol in alc, ether, org solvs. IDLH 900 ppm.
SYNS: CHROMAR □ p-DIMETHYLBENZENE □ 1,4-DIMETHYLBENZENE □ p-METHYLTOLUENE □ SCINTILLAR □ 1,4-XYLENE □ p-XYLOL (DOT)
CONSENSUS REPORTS: Community Right-To-Know List. Reported in EPA TSCA Inventory.
OSHA PEL: TWA 100 ppm; STEL 150 ppm
ACGIH TLV: TWA 100 ppm; STEL 150 ppm; BEI: methyl hippuric acids in urine at end of shift 1.5 g/g creatinine; Not Classifiable as a Human Carcinogen
NIOSH REL: (Xylene) TWA 100 ppm; CL 200 ppm/10M
SAFETY PROFILE: Moderately toxic by intraperitoneal route. Mildly toxic by ingestion and inhalation. An experimental teratogen. Experimental reproductive effects. May be narcotic in high concentrations. Chronic toxicity not established, but is less toxic than benzene. A very dangerous fire hazard when exposed to heat or flame; can react with oxidizing materials. Explosive in the form of vapor when exposed to heat or flame. To fight fire, use foam, CO_2, dry chemical. Potentially explosive reaction with acetic acid + air, 1,3-dichloro-5,5-dimethyl-2,4-imidazolidindione, nitric acid + pressure. When heated to decomposition it emits acrid smoke and irritating fumes. See also other xylene entries.

XHS800   CAS: 1477-55-0   HR: 2
m-XYLENE-α,α'-DIAMINE
mf: $C_8H_{12}N_2$ mw: 136.22
PROP: A liquid. Bp: 245–248°.
SYNS: 1,3-BIS-AMINOMETHYLBENZEN (CZECH) □ MXDA □ m-PHENYLENEBIS(METHYLAMINE) □ m-XYLYLENDIAMIN (CZECH)
CONSENSUS REPORTS: Reported in EPA TSCA Inventory.
OSHA PEL: TWA CL 0.1 mg/m³ (skin)
ACGIH TLV: TWA CL 0.1 mg/m³ (skin)
SAFETY PROFILE: Moderately toxic by skin contact and ingestion. Mildly toxic by inhalation. A severe skin and eye irritant. When heated to decomposition it emits toxic fumes of NO_x. Used to make polyamide fibers and resins and as a curing agent.

XIJ000   CAS: 3634-83-1   HR: 3
m-XYLENE DIISOCYANATE
DOT: UN 2207/UN 3080
mf: $C_{10}H_8N_2O_2$ mw: 188.20
SYNS: BENZENE, 1,3-BIS(ISOCYANATOMETHYL)-(9CI) □ 1,3-BIS(ISOCYANATOMETHYL)BENZENE □ 1,3-BIS-(ISOKYANATOMETHYL)BENZEN □ m-PHENYLENEDIMETHYLENE ISOCYANATE □ TAKENATE □ TAKENATE 500 □ m-XDI □ m-XYLIDENE DIISOCYANATE □ m-XYLYLENDIISOKYANAT □ m-XYLYLENE DIISOCYANATE □ XYLYLENDIISOKYANAT (CZECH)
CONSENSUS REPORTS: Reported in EPA TSCA Inventory.
NIOSH REL: (Diisocyanates) TWA 0.005 ppm; CL 0.02 ppm/10M

X

DOT CLASSIFICATION: 6.1; Label: KEEP AWAY FROM FOOD (UN 2207); DOT Class: 6.1; Label: Poison (UN 2206); DOT Class: 6.1; Label: Poison, Flammable Liquid (UN 3080); DOT Class: 3; Label: Flammable Liquid, Poison (UN 2478)
SAFETY PROFILE: Moderately toxic by ingestion. A severe skin and eye irritant. A sensitizer. A flammable liquid. When heated to decomposition it emits very toxic fumes of NO_x. See also ISOCYANATES.

XJJ000   CAS: 88-61-9   HR: 2
2,4-XYLENESULFONIC ACID
mf: $C_8H_{10}O_3S$ mw: 186.24
PROP: Plates or prisms from H_2O. Mp: 61–62°.
SYNS: 2,4-DIMETHYLBENZENESULFONIC ACID □ m-XYLENESULFONIC ACID □ m-XYLENE-4-SULFONIC ACID
CONSENSUS REPORTS: Reported in EPA TSCA Inventory.
DOT CLASSIFICATION: 8; Label: Corrosive
SAFETY PROFILE: Moderately toxic by intraperitoneal route. A corrosive. When heated to decomposition it emits toxic fumes of SO_x.

XKA000   CAS: 1300-71-6   HR: 3
XYLENOL
DOT: UN 2261
mf: $C_8H_{10}O$ mw: 122.18
PROP: The six isomers of xylenol are sltly sol in water; very sol in alc, chloroform, ether, benzene; sol in NaOH soln.
SYNS: DIMETHYLPHENOL □ PHENOL, DIMETHYL- □ STERICOL □ XILENOLI (ITALIAN) □ XYLENOLEN (DUTCH) □ XYLENOLS (DOT)
CONSENSUS REPORTS: Reported in EPA TSCA Inventory.
DOT CLASSIFICATION: 6.1; Label: Poison
SAFETY PROFILE: A poison. When heated to decomposition it emits acrid smoke and irritating fumes. See also other xylenol entries.

XKJ000   CAS: 526-75-0   HR: 3
2,3-XYLENOL
mf: $C_8H_{10}O$ mw: 122.18

PROP: Needles or crystals from EtOH (aq). Mp: 75°, bp: 218°. Sol in water, alc.
SYNS: 2,3-DIMETHYLPHENOL □ PHENOL, 2,3-DIMETHYL- □ o-XYLENOL
CONSENSUS REPORTS: Reported in EPA TSCA Inventory.
SAFETY PROFILE: Poison by intravenous route. When heated to decomposition it emits acrid smoke and irritating fumes. See also other xylenol entries.

XKJ500   CAS: 105-67-9   HR: 3
2,4-XYLENOL
mf: $C_8H_{10}O$ mw: 122.18
PROP: Needles from EtOH (aq) or C_6H_6. Mp: 26°, bp: 97–98° @ 14 mm. Sol in water and alc.
SYNS: 2,4-DIMETHYLPHENOL □ 4,6-DIMETHYL-PHENOL □ 1-HYDROXY-2,4-DIMETHYLBENZENE □ RCRA WASTE NUMBER U101 □ m-XYLENOL
CONSENSUS REPORTS: Reported in EPA TSCA Inventory.
SAFETY PROFILE: Poison by intravenous and intraperitoneal routes. Moderately toxic by ingestion and skin contact. Questionable carcinogen with experimental carcinogenic data. When heated to decomposition it emits acrid smoke and irritating fumes. See also other xylenol entries.

XKS000   CAS: 95-87-4   HR: 3
2,5-XYLENOL
mf: $C_8H_{10}O$ mw: 122.18
PROP: Crystals from EtOH. Mp: 74.5°, bp: 211.5–213.5°.
SYNS: 2,5-DIMETHYLPHENOL □ 3,6-DIMETHYL-PHENOL □ 2,5-DMP □ 6-METHYL-m-CRESOL □ p-XYLENOL □ 1,2,5-XYLENOL
CONSENSUS REPORTS: Reported in EPA TSCA Inventory.
SAFETY PROFILE: Poison by ingestion. Moderately toxic by an unspecified route. When heated to decomposition it emits acrid smoke and irritating fumes. Questionable carcinogen with experimental tumorigenic data. Used in disinfectants, solvents, pharmaceuticals, plasticizers, and wetting agents. See also other xylenol entries.

XLS000   CAS: 108-68-9   HR: 3
3,5-XYLENOL
mf: $C_8H_{10}O$ mw: 122.18
PROP: White crystals from water. Mp: 64°,
bp: 219.5°, d: 1.0362, vap press: 1 mm @
62°. Sltly sol in water; sol in alc.
SYNS: 3,5-DIMETHYLPHENOL □ 3,5-DMP □ 1,3,5-
XYLENOL
CONSENSUS REPORTS: Reported in EPA
TSCA Inventory. EPA Genetic Toxicology
Program.
SAFETY PROFILE: Poison by
intraperitoneal route. Moderately toxic by
ingestion. A severe eye irritant. Questionable
carcinogen with experimental tumorigenic
data. When heated to decomposition it
emits acrid smoke and irritating fumes. See
also other xylenol entries.

XMA000   CAS: 1300-73-8   HR: 3
XYLIDINE
mf: $C_8H_{11}N$ mw: 121.20
PROP: Usually liquid (except for o-4-
xylidine). Bp: 213–226°, flash p: 206° (CC),
d: 0.97–0.99, vap d: 4.17. Sltly sol in water;
sol in alc. IDLH 50 ppm.
SYNS: ACID LEATHER BROWN 2G □ ACID ORANGE
24 □ AMINODIMETHYLBENZENE □ 11460 BROWN □
DIMETHYLANILINE □ DIMETHYLPHENYLAMINE □
RESORCINE BROWN J □ RESORCINE BROWN R □
XILIDINE (ITALIAN) □ XYLIDINEN (DUTCH)
CONSENSUS REPORTS: Reported in EPA
TSCA Inventory.
OSHA PEL: TWA 0.2 ppm (skin)
ACGIH TLV: TWA 0.5 ppm (skin); Animal
Carcinogen
DFG MAK: (all isomers except 2,4-xylidene)
5 ppm (25 mg/m³)
SAFETY PROFILE: Confirmed carcinogen.
Poison by intravenous route. Moderately
toxic by ingestion. This material, which so
closely resembles aniline in the character of
its toxic effects, is actually twice as toxic as
aniline. It can cause injury to the blood and
the liver. It does not necessarily give any
alarm or warning, such as cyanosis,
headache, and dizziness, which characterize
aniline poisoning. Thus, it may be
considered a more insidious poison than

aniline, and severe and possibly fatal
intoxication may come about through skin
absorption. Combustible when exposed to
heat or flame. Can react vigorously with
oxidizing materials. To fight fire, use foam,
CO_2, dry chemical. When heated to
decomposition it emits toxic fumes of NO_x.
See also ANILINE and other xylidine
entries.

XMJ000   CAS: 87-59-2   HR: 3
2,3-XYLIDINE
mf: $C_8H_{11}N$ mw: 121.20
PROP: Liquid. D: 0.991 @ 15°, mp: <−15°,
bp: 220°. Very sltly sol in water; sol in alc,
ether.
SYNS: 2,3-DIMETHYLANILINE □ 2,3-DIMETHYL-
BENZENAMINE □ 2,3-DIMETHYLPHENYLAMINE □ o-
XYLIDINE □ 2,3-XYLYLAMINE
CONSENSUS REPORTS: Reported in EPA
TSCA Inventory.
DFG MAK: (all isomers except 2,4-xylidene)
5 ppm (25 mg/m³)
SAFETY PROFILE: A poison. Moderately
toxic by ingestion. Mutation data reported.
When heated to decomposition it emits
toxic fumes of NO_x. See also other xylidine
entries.

XMS000   CAS: 95-68-1   HR: 3
2,4-XYLIDINE
mf: $C_8H_{11}N$ mw: 121.20
PROP: Liquid. Bp: 214°, mp: 16°, d: 0.978
@ 19.6°/4°. Very sltly sol in water.
SYNS: 1-AMINO-2,4-DIMETHYLBENZENE □ 4-
AMINO-1,3-DIMETHYLBENZENE □ 4-AMINO-3-
METHYLTOLUENE □ 4-AMINO-1,3-XYLENE □ 2,4-
DIMETHYLANILINE □ 2,4-DIMETHYLBENZENAMINE
□ 2,4-DIMETHYLPHENYLAMINE □ 2-METHYL-p-
TOLUIDINE □ 4-METHYL-o-TOLUIDINE □ 2,4-
XYLIDENE (MAK) □ m-XYLIDINE □ m-4-XYLIDINE
CONSENSUS REPORTS: IARC Cancer
Review: Group 3 IMEMDT 7,56,87; Animal
Inadequate Evidence IMEMDT 16,367,78.
Reported in EPA TSCA Inventory.
DFG MAK: Animal Carcinogen, Suspected
Human Carcinogen
SAFETY PROFILE: Suspected carcinogen.
Poison by ingestion. Mutation data reported.

X

When heated to decomposition it emits toxic fumes of NO_x. See also other xylidine entries.

XNA000   CAS: 95-78-3   HR: 3
2,5-XYLIDINE
mf: $C_8H_{11}N$ mw: 121.20
PROP: Colorless oil or pale-yellow leaflets. Bp: 214°, d: 0.979 @ 21°/4°, mp: 155°. Very sltly sol in water.
SYNS: 1-AMINO-2,5-DIMETHYLBENZENE □ 3-AMINO-1,4-DIMETHYLBENZENE □ 2-AMINO-1,4-XYLENE □ 2,5-DIMETHYLANILINE □ 2,5-DIMETHYL-BENZENAMINE □ 2,5-DIMETHYLPHENYLAMINE □ 5-METHYL-o-TOLUIDINE □ 6-METHYL-m-TOLUIDINE □ p-XYLIDINE (DOT)
CONSENSUS REPORTS: IARC Cancer Review: Group 3 IMEMDT 7,56,87; Animal Inadequate Evidence IMEMDT 16,377,78. Reported in EPA TSCA Inventory.
DFG MAK: (all isomers except 2,4-xylidene) Confirmed Animal Carcinogen with Unknown Relevance to Humans
SAFETY PROFILE: Suspected carcinogen. A poison. Moderately toxic by ingestion. Questionable carcinogen. Mutation data reported. When heated to decomposition it emits toxic fumes of NO_x. See also other xylidine entries.

XNJ000   CAS: 87-62-7   HR: 3
2,6-XYLIDINE
mf: $C_8H_{11}N$ mw: 121.20
PROP: Liquid. D: 0.980 @ 15°, mp: 10–12°, bp: 216–217°.
SYNS: 2,6-DIMETHYLANILINE □ 2,6-DIMETHYL-BENZENAMINE □ NCI-C56188 □ o-XYLIDINE □ 2,6-XYLYLAMINE
CONSENSUS REPORTS: IARC Cancer Review: Group 2B IMEMDT 57,323,93; Animal Sufficient Evidence IMEMDT 57,323,93; Human Inadequate Evidence IMEMDT 57,323,93. Community Right-To-Know List. Reported in EPA TSCA Inventory.
DFG MAK: Animal Carcinogen, Suspected Human Carcinogen
SAFETY PROFILE: Suspected carcinogen. Moderately toxic by ingestion. Mutation data

reported. Questionable carcinogen with experimental carcinogenic data. When heated to decomposition it emits toxic fumes of NO_x. See also other xylidine entries.

XNS000   CAS: 95-64-7   HR: 3
3,4-XYLIDINE
mf: $C_8H_{11}N$ mw: 121.20
PROP: Crystals from pet ether. Mp: 51°, bp: 226°. Insol in water; sol in petroleum ether.
SYNS: 3,4-DIMETHYLAMINOBENZENE □ 3,4-DIMETHYLANILINE □ 3,4-DIMETHYLPHENYLAMINE □ 3,4-XYLYLAMINE
CONSENSUS REPORTS: Reported in EPA TSCA Inventory.
DFG MAK: (all isomers except 2,4-xylidene) Confirmed Animal Carcinogen with Unknown Relevance to Humans
SAFETY PROFILE: Suspected carcinogen. Poison by ingestion. Mutation data reported. When heated to decomposition it emits toxic fumes of NO_x. See also other xylidine entries.

XOA000   CAS: 108-69-0   HR: 3
3,5-XYLIDINE
mf: $C_8H_{11}N$ mw: 121.20
PROP: An oily liquid. D: 0.972 @ 20°/4°, bp: 221–222°.
SYNS: 3,5-DIMETHYLANILINE □ 3,5-DIMETHYL-BENZENAMINE □ 3,5-DIMETHYLPHENYLAMINE □ 3,5-XYLYLAMINE
CONSENSUS REPORTS: Reported in EPA TSCA Inventory.
DFG MAK: (all isomers except 2,4-xylidene) Confirmed Animal Carcinogen with Unknown Relevance to Humans
SAFETY PROFILE: Suspected carcinogen. Moderately toxic by ingestion. Mutation data reported. When heated to decomposition it emits toxic fumes of NO_x. See also other xylidine entries.

XOS500   CAS: 23420-61-3   HR: 3
3,4-XYLIDINO-2-OXAZOLINE
mf: $C_{11}H_{14}N_2O$ mw: 190.27
SYNS: 2-(3,4-DIMETHYLANILINO)-2-OXAZOLINE □ 2-OXAZOLAMINE, N-(3,4-DIMETHYLPHENYL)-4,5-

DIHYDRO- □ 2-OXAZOLINE, 3,4-XYLIDINO- □ 2-OXAZOLINE, 2-(3,4-XYLIDINO)- □ A 5160 □ 3,4-XYLIDINE, N-(2-OXAZOLIN-2-YL)-

SAFETY PROFILE: A poison by intravenous route. When heated to decomposition it emits toxic vapors of NO_x.

XPJ000 **CAS: 87-99-0** **HR: 1**
XYLITOL
mf: $C_5H_{12}O_5$ mw: 152.17
PROP: White crystals or crystalline powder; sweet taste with cooling sensation. Two forms: metastable, rhombic crystals and stable, monoclinic crystals. Mp: 61–61.5° (metastable), mp: 93–94.5° (stable). Sol in water; sltly sol in alc.
SYNS: EUTRIT □ KANNIT □ KLINIT □ NEWTOL □ TORCH □ XYLITE □ XYLITE (SUGAR) □ XYLITON
CONSENSUS REPORTS: Reported in EPA TSCA Inventory.
SAFETY PROFILE: Very low toxicity by ingestion. When heated to decomposition it emits acrid smoke and irritating fumes. A sugar.

XRA000 **CAS: 3118-97-6** **HR: 2**
1-(2,4-XYLYLAZO)-2-NAPHTHOL
mf: $C_{18}H_{16}N_2O$ mw: 276.36
SYNS: A.F. RED No. 5 □ AIZEN FOOD RED No. 5 □ BRASILAZINA OIL SCARLET 6G □ BRILLIANT OIL SCARLET B □ CALCO OIL SCARLET BL □ CERES ORANGE RR □ CERISOL SCARLET G □ CEROTIN-SCHARLACH G □ C.I. 12140 □ C.I. SOLVENT ORANGE 7 □ 1-((2,4-DIMETHYLPHENYL)AZO)-2-NAPHTHALENOL □ EXTRACT D&C RED No. 14 □ FAST OIL ORANGE II □ FAT RED (YELLOWISH) □ FAT SCARLET 2G □ FETTORANGE B □ GRASAN ORANGE 3R □ LACQUER ORANGE VR □ MOTIROT G □ OIL ORANGE KB □ OIL ORANGE N EXTRA □ OIL ORANGE R □ OIL ORANGE 2R □ OIL ORANGE X □ OIL ORANGE XO □ OIL RED GRO □ OIL RED O □ OIL RED RO □ OIL RED XO □ OIL SCARLET □ OIL SCARLET 371 □ OIL SCARLET APYO □ OIL SCARLET BL □ OIL SCARLET 6G □ OIL SCARLET L □ OIL SCARLET YS □ ORANGE INSOLUBLE OLG □ ORANGE INSOLUBLE RR □ ORANGE OIL KB □ PONCEAU INSOLUBLE OLG □ PYRONALROT R □ RED B □ RED No. 5 □ RESIN SCARLET 2R □ RESOFORM ORANGE R □ ROT B □ ROT GG FETTLOESLICH □ SOMALIA ORANGE A2R □ SOMALIA ORANGE 2R □ SOUDAN II □ SUDAN AX □ SUDAN ORANGE □ SUDAN ORANGE RPA □ SUDAN ORANGE RRA □ SUDAN RED □ SUDAN SCARLET 6G

□ SUDAN X □ WAXAKOL VERMILION L □ 1-XYLYLAZO-2-NAPHTHOL □ 1-(o-XYLYLAZO)-2-NAPHTHOL
CONSENSUS REPORTS: IARC Cancer Review: Group 3 IMEMDT 7,56,87; Animal Sufficient Evidence IMEMDT 8,233,75. Reported in EPA TSCA Inventory. Community Right-To-Know List. EPA Genetic Toxicology Program.
SAFETY PROFILE: Questionable carcinogen with experimental carcinogenic data. Mutation data reported. When heated to decomposition it emits toxic fumes of NO_x.

XRS000 **CAS: 35884-77-6** **HR: 3**
XYLYL BROMIDE
DOT: UN 1701
mf: C_8H_9Br mw: 185.08
PROP: Colorless liquid. Bp: 212–215° (slt decomp), d: 1.371 @ 23°. Almost insol in water; sol in alc, ether.
SYN: BROMURE de XYLYLE (FRENCH)
DOT CLASSIFICATION: 6.1; Label: Poison
SAFETY PROFILE: A poison. A powerful irritant. When heated to decomposition it emits toxic fumes of Br^-.

XSS260 **CAS: 1014-98-8** **HR: 3**
p-XYLYLENE ISOCYANATE
DOT: UN 2206/UN 2207/UN 2478/UN 3080
mf: $C_{10}H_6N_2O_2$ mw: 186.18
SYN: ISOCYANIC ACID, p-XYLYLENE ESTER
DOT CLASSIFICATION: 6.1; Label: KEEP AWAY FROM FOOD (UN 2207); Class: 6.1; Label: Poison (UN 2206); Class: 6.1; Label: Poison, Flammable Liquid (UN 3080); Class: 3; Label: Flammable Liquid, Poison (UN 2478)
SAFETY PROFILE: A poison. A flammable liquid. When heated to decomposition it emits toxic vapors of CN^-.

X

XWS100   CAS: 77248-44-3   HR: 3
((2,6-XYLYL)SULFINYL)METHYLCARB-
AMIC ACID-2,3-DIHYDRO-2,2-
DIMETHYL-7-BENZOFURANYL
ESTER

mf: $C_{20}H_{23}NO_5S$ mw: 389.50

SYNS: CARBAMIC ACID, ((2,6-DIMETHYLPHENOXY)-SULFINYL)METHYL-, 2,3-DIHYDRO-2,2-DIMETHYL-7-BENZOFURANYL ESTER □ 2,3-DIHYDRO-2,2-DIMETHYLBENZOFURANYL-7-(METHYL)(2,6-DIMETHYLPHENOXYSULFINYL)CARBAMATE

SAFETY PROFILE: A poison by ingestion. When heated to decomposition it emits toxic vapors of NO_x and SO_x.

Y

YBS500   CAS: 75444-63-2   HR: 3
β-YOHIMBINE HYDROCHLORIDE
mf: $C_{21}H_{26}N_2O_3 \cdot ClH$ mw: 390.95
SYNS: YOHIMBAN-16-CARBOXYLIC ACID, 17-
HYDROXY-, METHYL ESTER, MONOHYDRO-
CHLORIDE, (16α,17β)- □ β-YOHIMBIN HYDRO-
CHLORIDE
SAFETY PROFILE: A poison by ingestion
and intravenous routes. When heated to
decomposition it emits toxic vapors of NO_x,
HCl, and Cl^-.

YDA000   CAS: 7440-64-4   HR: 2
YTTERBIUM
af: Yb   aw: 173.04
PROP: A bright, silvery, lustrous, soft,
malleable, ductile, and fairly stable element;
somewhat air- and moisture-sensitive. It is
attacked by dil and conc mineral acids. Mp:
824°, bp: 1193°, d: 6.977. A rare earth.
CONSENSUS REPORTS: Reported in EPA
TSCA Inventory.
SAFETY PROFILE: As a lanthanon it may
have an anticoagulant action on blood.
Questionable carcinogen with experimental
tumorigenic data. Flammable in the form of
dust when reacted with air, halogens. See
also LANTHANUM and RARE EARTHS.

YDJ000   CAS: 10361-91-8   HR: 3
YTTERBIUM CHLORIDE
mf: Cl_3Yb mw: 279.39
PROP: Hexahydrate, deliq needles, crystals,
or hygroscopic white solid. D: 2.575, mp:
150–155°. Sol in water.
SYN: YTTERBIUM TRICHLORIDE
CONSENSUS REPORTS: Reported in EPA
TSCA Inventory.
SAFETY PROFILE: Poison by
intraperitoneal route. Mildly toxic by
ingestion. An experimental teratogen. A skin
and eye irritant. When heated to

decomposition it emits toxic fumes of Cl^-.
See also YTTERBIUM, RARE EARTHS,
and CHLORIDES.

YDS800   CAS: 13768-67-7   HR: 3
YTTERBIUM NITRATE
mf: $N_3O_9 \cdot Yb$ mw: 235.05
PROP: Hygroscopic colorless solid or
crystals. Sol in water and alc.
SYN: NITRIC ACID, YTTERBIUM(3+) SALT
CONSENSUS REPORTS: Reported in EPA
TSCA Inventory.
SAFETY PROFILE: Poison by ingestion and
intraperitoneal routes. Experimental
reproductive effects. When heated to
decomposition it emits toxic fumes of NO_x.
See also YTTERBIUM, RARE EARTHS,
and NITRATES.

YEJ000   CAS: 7440-65-5   HR: 3
YTTRIUM
af: Y   aw: 88.9059
PROP: Hexagonal, silvery-metallic colored
element. Reasonably air-stable; gray-black
when finely divided. Brittle, rather harder
than zinc. Reacts slowly with cold H_2O,
rapidly with dil acid. Burns easily. Reacts
with Cl_2 at 2° and O_2 at 4°. Mp: 1522°, bp:
3338°, d: 4.469. IDLH 500 mg/m³ (as Y).
SYN: YTTRIUM-89
CONSENSUS REPORTS: Reported in EPA
TSCA Inventory.
OSHA PEL: TWA 1 mg(Y)/m³
ACGIH TLV: TWA 1 mg(Y)/m³
DFG MAK: 5 mg(Y)/m³
SAFETY PROFILE: It may have an
anticoagulant effect on the blood.
Flammable in the form of dust when reacted
with air, halogens.

YES000 **CAS: 10361-92-9** **HR: 3**
YTTRIUM CHLORIDE
mf: Cl_3Y mw: 195.26
PROP: Hexahydrate, colorless, deliq crystals.
Mp: 721°, bp: 1507°. Sol in water, alc, Py,
DMSO; sltly sol in THF and DMF.
SYN: YTTRIUM TRICHLORIDE
CONSENSUS REPORTS: Reported in EPA
TSCA Inventory.
ACGIH TLV: TWA 1 mg(Y)/m³
SAFETY PROFILE: Poison by
intraperitoneal route. When heated to
decomposition it emits toxic fumes of Cl⁻.
See also YTTRIUM and RARE EARTHS.

YFJ000 **CAS: 10361-93-0** **HR: 3**
YTTRIUM(III) NITRATE (1:3)
mf: $N_3O_9 \cdot Y$ mw: 274.94
PROP: Hexahydrate, deliq crystals. Sol in
water.
SYN: NITRIC ACID, YTTRIUM(3+) SALT
CONSENSUS REPORTS: Reported in EPA
TSCA Inventory.
ACGIH TLV: TWA 1 mg(Y)/m³

SAFETY PROFILE: Poison by
intraperitoneal route. Moderately toxic by
intravenous route. Experimental
reproductive effects. Questionable
carcinogen with experimental tumorigenic
data. A skin and eye irritant. Mutation data
reported. When heated to decomposition it
emits toxic fumes of NO_x. See also
NITRATES, YTTRIUM, and RARE
EARTHS.

YGA000 **CAS: 1314-36-9** **HR: 3**
YTTRIUM OXIDE
mf: O_3Y_2 mw: 225.82
PROP: White powder or solid. D: 4.84, mp:
2410°. Insol in water.
SYN: YTTRIA
CONSENSUS REPORTS: Reported in EPA
TSCA Inventory.
ACGIH TLV: TWA 1 mg(Y)/m³
SAFETY PROFILE:
A poison by intraperitoneal route. See also
YTTRIUM and RARE EARTHS.

Z

ZAT100   CAS: 9010-66-6   HR: D
ZEIN
PROP: Powder. Insol in water, alc; sol in glycols, glycol ethers.
SAFETY PROFILE: When heated to decomposition it emits acrid smoke and irritating fumes.

ZBJ000   CAS: 7440-66-6   HR: 3
ZINC
DOT: UN 1435/UN 1436
af: Zn   aw: 65.37
PROP: Bluish-white, lustrous, metallic element. Not perceptibly attacked by pure H_2O. Mp: 419.8°, bp: 908°, d: 7.14 @ 25°, vap press: 1 mm @ 487°. Stable in dry air.
SYNS: BLUE POWDER □ C.I. 77945 □ C.I. PIGMENT BLACK 16 □ C.I. PIGMENT METAL 6 □ EMANAY ZINC DUST □ GRANULAR ZINC □ JASAD □ MERRILLITE □ PASCO □ ZINC ASHES (UN 1435) (DOT) □ ZINC DUST □ ZINC DUST (DOT) □ ZINC POWDER □ ZINC POWDER (DOT)
CONSENSUS REPORTS: Zinc and its compounds are on the Community Right-To-Know List. Reported in EPA TSCA Inventory. EPA Genetic Toxicology Program.
DOT CLASSIFICATION: 4.3; Label: Dangerous When Wet, Spontaneously Combustible; DOT Class: 4.3; Label: Dangerous When Wet (UN 1435)
SAFETY PROFILE: Human systemic effects by ingestion: cough, dyspnea, and sweating. A human skin irritant. Pure zinc powder, dust, and fume are relatively nontoxic to humans by inhalation. The difficulty arises from oxidation of zinc fumes immediately prior to inhalation or presence of impurities such as Cd, Sb, As, Pb. Inhalation may cause sweet taste, throat dryness, cough, weakness, generalized aches, chills, fever, nausea, vomiting.

Flammable in the form of dust when exposed to heat or flame. May ignite spontaneously in air when dry. Explosive in the form of dust when reacted with acids. Incompatible with NH_4NO_3, BaO_2, $Ba(NO_3)_2$, Cd, CS_2, chlorates, Cl_2, ClF_3, CrO_3, (ethyl acetoacetate + tribromoneopentyl alcohol), F_2, hydrazine mononitrate, hydroxylamine, $Pb(N_3)_2$, $(Mg + Ba(NO_3)_2 + BaO_2)$, $MnCl_2$, HNO_3, performic acid, $KClO_3$, KNO_3, K_2O_2, Se, $NaClO_3$, Na_2O_2, S, Te, H_2O, $(NH_4)_2S$, As_2O_3, CS_2, $CaCl_2$, NaOH, chlorinated rubber, catalytic metals, halocarbons, o-nitroanisole, nitrobenzene, nonmetals, oxidants, paint primer base, pentacarbonyliron, transition metal halides, seleninyl bromide. To fight fire, use special mixtures of dry chemical. When heated to decomposition it emits toxic fumes of ZnO. See also ZINC COMPOUNDS.

ZDJ000   CAS: 1303-39-5   HR: 3
ZINC ARSENATE
mf: $As_4O_{15} \cdot 5Zn$ mw: 866.53
PROP: White, odorless powder.
SYNS: ARSENIC ACID, ZINC SALT □ ZINC ARSENATE, BASIC
CONSENSUS REPORTS: Arsenic and its compounds, as well as zinc and its compounds, are on the Community Right-To-Know List.
OSHA PEL: TWA 0.01 mg(As)/m³; Cancer Hazard
ACGIH TLV: BEI: 35 μ (As)/L inorganic arsenic and methylated metabolites in urine
NIOSH REL: CL 0.002 mg(As)/m³/15M
DOT CLASSIFICATION: 6.1; Label: Poison
SAFETY PROFILE: Confirmed human carcinogen. A poison. When heated to decomposition it emits toxic fumes of As and ZnO. See also ARSENIC COMPOUNDS and ZINC

COMPOUNDS.

ZDS000   CAS: 10326-24-6   HR: 3
ZINC-m-ARSENITE
DOT: UN 1712
mf: $AsHO_2 \cdot 1/2Zn$ mw: 140.61
PROP: A white powder.
SYNS: ARSENIOUS ACID, ZINC SALT (9CI) □ ZINC
ARSENITE, solid (DOT) □ ZINC METAARSENITE □
ZINC METHARSENITE □ ZMA
CONSENSUS REPORTS: Arsenic and its
compounds, as well as zinc and its
compounds, are on the Community Right-
To-Know List.
OSHA PEL: TWA 0.01 mg(As)/m³; Cancer
Hazard
ACGIH TLV: TWA 0.01 mg/m³; Confirmed
Human Carcinogen; BEI: 35 µ (As)/L
inorganic arsenic and methylated
metabolites in urine
NIOSH REL: (Inorganic Arsenic) CL 0.002
mg(As)/m³/15M
DOT CLASSIFICATION: 6.1; Label: Poison
SAFETY PROFILE: Confirmed human
carcinogen. Poison by ingestion and
intraperitoneal routes. Moderately toxic by
skin contact. When heated to decomposition
it emits toxic fumes of As and ZnO. See
also ARSENIC COMPOUNDS and ZINC
COMPOUNDS.

ZEJ050   CAS: 3486-35-9   HR: D
ZINC CARBONATE (1:1)
mf: $CO_3 \cdot Zn$ mw: 125.38
PROP: Rhombohedral colorless crystals.
Practically insol in H_2O.
SYN: CARBONIC ACID, ZINC SALT (1:1)
CONSENSUS REPORTS: Reported in EPA
TSCA Inventory.
SAFETY PROFILE: An experimental
teratogen. When heated to decomposition it
emits toxic fumes of CO and Zn.

ZES000   CAS: 10361-95-2   HR: 3
ZINC CHLORATE
DOT: UN 1513
mf: $Cl_2O_6 \cdot Zn$ mw: 232.27
PROP: Colorless, very deliq crystals.
CONSENSUS REPORTS: Zinc and its
compounds are on the Community Right-

To-Know List. Reported in EPA TSCA
Inventory.
DOT CLASSIFICATION: 5.1; Label: Oxidizer
SAFETY PROFILE: A powerful oxidizer.
Probably a skin, eye, and mucous membrane
irritant. The tetrahydrated salt explodes at
60°C. Explosive reaction with copper(II)
sulfide. Can react violently with Al, Sb_2S_3,
As, C, charcoal, Cu, MnO_2, metal sulfides,
dibasic organic acids, organic matter, P, S,
H_2SO_4. Incandescent reaction with
antimony(III) sulfide, arsenic(III) sulfide,
tin(II) sulfide, tin(IV) sulfide. When heated
to decomposition it emits toxic fumes of Cl⁻
and ZnO. See also CHLORATES and
ZINC COMPOUNDS.

ZFA000   CAS: 7646-85-7   HR: 3
ZINC CHLORIDE
DOT: UN 1840/UN 2331
mf: Cl_2Zn mw: 136.27
PROP: Odorless, colorless, cubic, white,
highly deliq crystals. Mp: 290°, bp: 732°, d:
2.91 @ 25°, vap press: 1 mm @ 428°. Sol in
MeOH, EtOH, Et_2O, and Me_2O; very sol in
H_2O. IDLH 50 mg/m³.
SYNS: BUTTER of ZINC □ CHLORURE de ZINC
(FRENCH) □ ZINC BUTTER □ ZINC CHLORIDE
(ACGIH, OSHA) □ ZINC CHLORIDE, anhydrous (UN 2331)
(DOT) □ ZINC CHLORIDE, solution (UN 1840) (DOT) □
ZINC (CHLORURE de) (FRENCH) □ ZINC DICHLORIDE
□ ZINC MURIATE, solution (DOT) □ ZINCO (CLORURO
di) (ITALIAN) □ ZINKCHLORID (GERMAN) □
ZINKCHLORIDE (DUTCH)
CONSENSUS REPORTS: Zinc and its
compounds are on the Community Right-
To-Know List. Reported in EPA TSCA
Inventory. EPA Genetic Toxicology
Program.
OSHA PEL: Fume: TWA 1 mg/m³; STEL 2
mg/m³
ACGIH TLV: TWA 1 mg/m³; STEL 2
mg/m³ (fume)
DOT CLASSIFICATION: 8; Label: Corrosive
SAFETY PROFILE: Poison by ingestion,
intravenous, and intraperitoneal routes.
Human systemic effects by inhalation: pulm-
onary changes. An experimental teratogen.
Experimental reproductive effects.
Questionable carcinogen with experimental

tumorigenic data. Human mutation data reported. A corrosive irritant to skin, eyes, and mucous membranes. Exposure to $ZnCl_2$ fumes or dusts can cause dermatitis, boils, conjunctivitis, gastrointestinal tract upsets. The fumes are highly toxic. Incompatible with potassium. Mixtures of the powdered chloride and powdered zinc are flammable. When heated to decomposition it emits toxic fumes of Cl^- and ZnO. See also ZINC COMPOUNDS and CHLORIDES.

ZFJ100   CAS: 13530-65-9   HR: 3
ZINC CHROMATE
mf: $CrH_2O_4 \cdot Zn$ mw: 183.39
PROP: Lemon-yellow prisms. Mp: 316°. Sol in H_2O. IDLH Ca [15 mg/m³ {as Cr(VI)}].
SYNS: BASIC ZINC CHROMATE □ BUTTERCUP YELLOW □ CHROMIC ACID, ZINC SALT □ CHROMIUM ZINC OXIDE □ C.I. 77955 □ C.I. PIGMENT YELLOW 36 □ CITRON YELLOW □ C.P. ZINC YELLOW X-883 □ PRIMROSE YELLOW □ PURE ZINC CHROME □ ZINC CHROMATE(VI) HYDROXIDE □ ZINC CHROME YELLOW □ ZINC CHROMIUM OXIDE □ ZINC HYDROXYCHROMATE □ ZINC TETRAOXY-CHROMATE 76A □ ZINC YELLOW
CONSENSUS REPORTS: NTP 10th Report on Carcinogens. IARC Cancer Review: Group 1 IMEMDT 7,165,87; Human Sufficient Evidence IMEMDT 23,205,80; Animal Sufficient Evidence IMEMDT 23,205,80. EPA Genetic Toxicology Program. Reported in EPA TSCA Inventory. Zinc and chromium and their compounds are on the Community Right-To-Know List.
OSHA PEL: CL 0.1 mg(CrO₃)/m³
ACGIH TLV: TWA 0.01 mg(Cr)/M³; Confirmed Human Carcinogen
DFG MAK: DFG TRK: 0.1 mg/m³; Human Carcinogen
NIOSH REL: (Chromium(VI)) TWA 0.001 mg(Cr(VI))/m³
SAFETY PROFILE: Confirmed human carcinogen producing lung tumors. A poison via intravenous route. Human mutation data reported. See also CHROMIUM COMPOUNDS and ZINC COMPOUNDS.

ZFJ122   HR: 3
ZINC CHROMATE, POTASSIUM DICHROMATE, and ZINC HYDROXIDE (3:1:1)
mf: $CrK_2O_4 \cdot 3CrO_4Zn \cdot H_2O_2Zn$ mw: 837.70
SYN: POTASSIUM DICHROMATE, ZINC CHROMATE, and ZINC HYDROXIDE (1:3:1)
OSHA PEL: CL 0.1 mg(CrO₃)/m³
ACGIH TLV: TWA 0.01 mg(Cr)/M³; Confirmed Human Carcinogen
DFG MAK: DFG TRK: 0.1 mg/m³; Human Carcinogen
NIOSH REL: (Chromium (VI)) TWA 0.001 mg(Cr(VI))/m³
SAFETY PROFILE: Confirmed human carcinogen with experimental carcinogenic data.

ZFJ125   CAS: 37300-23-5   HR: 3
ZINC CHROMATE with ZINC HYDROX-IDE and CHROMIUM OXIDE (9:1)
mf: $CrO_4 \cdot Zn \cdot H_4O_2Zn \cdot CrO_3$ mw: 183.39
PROP: Yellow chromate pigment.
SYN: ZINC YELLOW
CONSENSUS REPORTS: Reported in EPA TSCA Inventory.
OSHA PEL: CL 0.1 mg(CrO₃)/m³
ACGIH TLV: TWA 0.01 mg(Cr)/M³; Confirmed Human Carcinogen
DFG MAK: DFG TRK: 0.1 mg/m³; Human Carcinogen
NIOSH REL: (Chromium (VI)) TWA 0.001 mg(Cr(VI))/m³
SAFETY PROFILE: Confirmed human carcinogen producing lung tumors. Mutation data reported. See also CHROMIUM COMPOUNDS and ZINC COMPOUNDS.

ZFJ250   CAS: 546-46-3   HR: D
ZINC CITRATE
mf: $C_6H_5O_7 \cdot 3/2Zn$ mw: 290.18
PROP: Crystals from H_2O. Sltly sol in H_2O; insol in $CHCl_3$, EtOH, Et_2O, and Me_2CO.
SYNS: CITRIC ACID, ZINC SALT (2:3) □ 1,2,3-PROPANETRICARBOXYLIC ACID, 2-HYDROXY-, ZINC SALT (2:3) (9CI)
CONSENSUS REPORTS: Reported in EPA TSCA Inventory.

SAFETY PROFILE: Experimental reproductive effects. When heated to decomposition it emits acrid smoke and irritating fumes.

ZFS000   HR: D
ZINC COMPOUNDS

CONSENSUS REPORTS: Zinc and its compounds are on the Community Right-To-Know List.

SAFETY PROFILE: Variable toxicity, but generally of low toxicity. However, zinc salts, such as chromates and arsenates, are experimental carcinogens. Zinc is not inherently a toxic element. However, when heated, it evolves a fume of zinc oxide, which, when inhaled fresh, can cause a disease known as "brass founders," "ague," or "brass chills," resulting in a sweet taste, throat dryness, cough, weakness, generalized aching, fever, nausea, and vomiting. It is possible for people to become immune to it, but this immunity can be broken by cessation of exposure of only a few days. Zinc oxide dust that is not freshly formed is virtually innocuous. There is no cumulative effect from the inhalation of zinc fumes. Exposure to zinc chloride fumes can cause damage to the mucous membranes of the nasopharnyx and respiratory tract, and give rise to a pale gray cyanosis; fatalities have resulted. Soluble salts of zinc have a harsh metallic taste; small doses can cause nausea and vomiting, whereas larger doses cause violent vomiting and purging. Some cases of intoxication have been reported due to drinking liquids stored in galvanized containers, and in dialysis patients, using a dialyzate prepared with water that had been stored in a galvanized tank. In general, the continued administration of zinc salts in small doses has no effect in humans except those of disordered digestion and constipation. Workers in zinc refining have been reported to suffer from a variety of non-specific intestinal, respiratory, and nervous symptoms. Ulceration of the nasal septum and eczematous dermatosis are also reported. It has been stated that zinc oxide or zinc stearate dust can block the ducts of the sebaceous glands and give rise to a papular, pustular eczema in workers engaged in packing these compounds into barrels. Sensitivity to zinc oxide in humans is extremely rare. Zinc chloride and zinc sulfate, because of caustic action, can cause ulceration of the fingers, hands, and forearms of those who use these compounds as a flux in soldering or other industrial use. This condition has even been observed in men who handle railway ties that have been impregnated with this material. Common air contaminants. When heated to decomposition it emits toxic fumes of ZnO.

ZGA000   CAS: 557-21-1   HR: 3
ZINC CYANIDE

DOT: UN 1713

mf: C_2N_2Zn mw: 117.41

PROP: Rhombic, colorless crystals. Mp: decomp @ 800°. Insol in water; sol in solns of alkali cyanides; decomp by dil mineral acid.

SYNS: CYANURE de ZINC (FRENCH) □ RCRA WASTE NUMBER P121 □ ZINC DICYANIDE

CONSENSUS REPORTS: Zinc and its compounds, as well as cyanide and its compounds, are on the Community Right-To-Know List. Reported in EPA TSCA Inventory.

DOT CLASSIFICATION: 6.1; Label: Poison

OSHA PEL: TWA 5 mg(CN)/m³

ACGIH TLV: CL 5 mg(CN)/m³ (skin)

DFG MAK: 5 mg/m³

NIOSH REL: (Cyanide) CL 5 mg(CN)/m³/10M

SAFETY PROFILE: Poison by intraperitoneal route. Can react violently with Mg. When heated to decomposition it emits toxic fumes of CN⁻, ZnO, and NOₓ. Used in electroplating operations. See also CYANIDE and ZINC COMPOUNDS.

ZGJ100   CAS: 7779-86-4   HR: 3
ZINC DITHIONITE

DOT: UN 1931

mf: $O_4S_2•Zn$ mw: 193.49

SYNS: DITHIONOUS ACID, ZINC SALT (1:1) □ ZINC HYDROSULFITE □ ZINC HYDROSULFITE (DOT)
CONSENSUS REPORTS: Reported in EPA TSCA Inventory.
DOT CLASSIFICATION: 9; Label: None
SAFETY PROFILE: Possibly a poison. When heated to decomposition it emits toxic vapors of zinc and SO_x.

ZHS000 CAS: 7783-49-5 HR: 3
ZINC FLUORIDE
mf: F_2Zn mw: 103.37
PROP: Tetragonal needles or white crystalline mass. D: 5.00 @ 25°, mp: 872°, bp: 1500°, vap press: 1 mm @ 970°. Sltly sol in aq HF; sol in HCl, HNO_3, and NH_4OH.
SYN: ZINC FLUORURE (FRENCH)
CONSENSUS REPORTS: Zinc and its compounds are on the Community Right-To-Know List. Reported in EPA TSCA Inventory.
OSHA PEL: TWA 2.5 mg(F)/m³
ACGIH TLV: TWA 2.5 mg(F)/m³; BEI: 3 mg/g creatinine of fluorides in urine prior to shift; 10 mg/g creatinine of fluorides in urine at end of shift.
NIOSH REL: (Fluorides, Inorganic) TWA 2.5 mg(F)/m³
SAFETY PROFILE: Poison by subcutaneous route. Can react violently with potassium. A fluorination agent. When heated to decomposition it emits toxic fumes of F^- and ZnO. See also FLUORIDES and ZINC COMPOUNDS.

ZIA000 CAS: 16871-71-9 HR: 3
ZINC FLUOSILICATE
DOT: UN 2855
mf: $F_6Si•Zn$ mw: 207.46
PROP: Hexagonal white crystals. Sol in water and inorganic acid.
SYNS: FUNGOL □ FUNGONIT GF 2 □ SILICON ZINC FLUORIDE □ ZINC FLUOROSILICATE □ ZINC FLUOROSILICATE (DOT) □ ZINC HEXAFLUOROSILICATE
CONSENSUS REPORTS: Zinc and its compounds are on the Community Right-To-Know List. Reported in EPA TSCA Inventory.
OSHA PEL: TWA 2.5 mg(F)/m³

ACGIH TLV: TWA 2.5 mg(F)/m³; BEI: 3 mg/g creatinine of fluorides in urine prior to shift; 10 mg/g creatinine of fluorides in urine at end of shift.
NIOSH REL: TWA 2.5 mg(F)/m³
DOT CLASSIFICATION: 6.1; Label: KEEP AWAY FROM FOOD
SAFETY PROFILE: Poison by ingestion and subcutaneous routes. When heated to decomposition it emits toxic fumes of F^- and ZnO. See also ZINC COMPOUNDS.

ZIA750 CAS: 4468-02-4 HR: 1
ZINC GLUCONATE
mf: $C_{12}H_{22}O_4Zn$ mw: 295.71
PROP: White granular or crystalline powder. Sol in water; very sltly sol in alc.
SYN: ZINC,((BIS(d-GLUCONATO-O¹),O²))- (9CI)
CONSENSUS REPORTS: Reported in EPA TSCA Inventory.
SAFETY PROFILE: Experimental reproductive effects. When heated to decomposition it emits toxic fumes of ZnO.

ZIJ100 CAS: 7779-86-4 HR: 3
ZINC HYDROSULFITE
DOT: UN 1931
mf: $O_4S_2•Zn$ mw: 193.49
PROP: White amorphous. Sol in water. Wood pulp bleach.
SYNS: DITHIONOUS ACID, ZINC SALT (1:1) □ ZINC DITHIONITE □ ZINC DITHIONITE (DOT) □ ZINC HYDROSULFITE
CONSENSUS REPORTS: Reported in EPA TSCA Inventory.
DOT CLASSIFICATION: 9; Label: None
SAFETY PROFILE: Probably a poison. When heated to decomposition it emits toxic vapors of zinc and SO_x.

ZJA000 CAS: 22323-45-1 HR: 3
ZINC MERCURY CHROMATE COMPLEX
mf: $7ZnO•2HgO•2CrO_3•7H_2O$ mw: 1328.91
SYNS: CHROMIC ACID, MERCURY ZINC COMPLEX □ EXPERIMENTAL FUNGICIDE 224 (UNION CARBIDE) □ MERCURY ZINC CHROMATE COMPLEX
CONSENSUS REPORTS: Zinc, mercury, chromium, and their compounds are on the Community Right-To-Know List.

Z

OSHA PEL: CL 0.1 mg(CrO$_3$)/m^3
ACGIH TLV: TWA 0.05 mg(Cr)/m^3;
Confirmed Human Carcinogen
DFG MAK: Animal Carcinogen, Suspected
Human Carcinogen
NIOSH REL: (Chromium): TWA 0.001
mg(Cr)/m^3; (Mercury, Aryl and Inorganic)
CL 0.1 mg/m^3 (skin)
SAFETY PROFILE: Confirmed carcinogen.
Moderately toxic by ingestion. When heated
to decomposition it emits very toxic fumes
of Hg and ZnO. See also MERCURY
COMPOUNDS, ZINC COMPOUNDS,
and CHROMIUM COMPOUNDS.

ZJJ000   CAS: 7779-88-6   HR: 3
ZINC NITRATE
DOT: UN 1514
mf: N$_2$O$_6$•Zn   mw: 189.39
PROP: A: needles; B: tetragonal, colorless
crystals; A: trihydrate; B: hexahydrate; d: (B)
2.065 @ 14°; mp: (A) 42.5°; mp: (B) 36.4°;
bp: (B) loses 6H$_2$O @ 105–131°. Very sol in
alc; sol in water.
SYNS: NITRATE de ZINC (FRENCH) □ NITRIC ACID,
ZINC SALT
CONSENSUS REPORTS: Zinc and its
compounds are on the Community Right-
To-Know List. Reported in EPA TSCA
Inventory.
DOT CLASSIFICATION: 5.1; Label: Oxidizer
SAFETY PROFILE: A powerful oxidizer.
Can react violently with C, Cu, metal
sulfides, organic matter, P, S. When heated
to decomposition it emits toxic fumes of
NO$_x$ and ZnO. See also NITRATES and
ZINC COMPOUNDS.

ZJS400   CAS: 7779-90-0   HR: 2
ZINC ORTHOPHOSPHATE
mf: H$_3$O$_4$P•3/$_2$Zn   mw: 386.05
SYNS: BONDERITE 40 □ BONDERITE 880 □ C.I.
77964 □ C.I. PIGMENT WHITE 32 □ DELAPHOS □
DELAPHOS 2M □ FLECK'S EXTRAORDINARY CEMENT
□ GRANODINE 16NC □ GRANODINE 80 □
HEUCOPHOS ZP 10 □ J 0852 □ LF BOWSEI PW 2 □ LF-
PW 2 □ MICROPHOS 90 □ NEUTRAL ZINC PHOSPHATE
□ PHOSPHINOX PZ 06 □ PIGMENT WHITE 32 □ SICOR
ZNP/M □ SICOR ZNP/S □ TRIBASIC ZINC PHOSPHATE
□ TRIZINC DIPHOSPHATE □ VIRCHEM 931 □

WEATHER COAT 1000 □ ZINC ACID PHOSPHATE □
ZINC PHOSPHATE (3:2) □ ZP-SB □ ZPF
CONSENSUS REPORTS: Reported in EPA
TSCA Inventory.
SAFETY PROFILE: Moderately toxic by
intraperitoneal route. When heated to
decomposition it emits acrid smoke and
irritating fumes.

ZKA000   CAS: 1314-13-2   HR: 3
ZINC OXIDE
mf: OZn   mw: 81.37
PROP: Odorless, white or yellowish powder.
Hexagonal white crystals. Mp: >1800°, d:
5.47. Insol in water and alc; sol in dil acetic
or mineral acids, ammonia. IDLH 500
mg/m^3.
SYNS: AKRO-ZINC BAR 85 □ AMALOX □ AZO-33 □
AZODOX-55 □ CALAMINE (spray) □ CHINESE WHITE □
C.I. 77947 □ C.I. PIGMENT WHITE 4 □ CYNKU TLENEK
(POLISH) □ EMANAY ZINC OXIDE □ EMAR □ FEL-
LING ZINC OXIDE □ FLOWERS of ZINC □ GREEN
SEAL-8 □ HUBBUCK'S WHITE □ KADOX-25 □ K-ZINC
□ OZIDE □ OZLO □ PASCO □ PERMANENT WHITE □
PHILOSOPHER'S WOOL □ PROTOX TYPE 166 □ RED-
SEAL-9 □ SNOW WHITE □ WHITE SEAL-7 □ ZINCITE
□ ZINCOID □ ZINC OXIDE FUME (MAK) □ ZINC
WHITE
CONSENSUS REPORTS: Zinc and its
compounds are on the Community Right-
To-Know List. Reported in EPA TSCA
Inventory.
OSHA PEL: Fume: TWA 5 mg/m^3; STEL
10 mg/m^3; Dust: TWA Total Dust: 10
mg/m^3; Respirable Fraction: 5 mg/m^3
ACGIH TLV: TWA 2 mg/m^3; STEL 10
mg/m^3; respirable fraction
DFG MAK: 1.5 mg/m^3
NIOSH REL: TWA (Zinc Oxide) 5 mg/m^3;
CL 15 mg/m^3/15M
SAFETY PROFILE: Moderately toxic to
humans by ingestion. Poison experimentally
by intraperitoneal route. An experimental
teratogen. Other experimental reproductive
effects. Human systemic effects by
inhalation of freshly formed fumes: metal
fume fever with chills, fever, tightness of
chest, cough, dyspnea, and other pulmonary
changes. Mutation data reported. A skin and
eye irritant. Has exploded when mixed with
chlorinated rubber. Violent reaction with

Mg, linseed oil. When heated to decomposition it emits toxic fumes of ZnO. See also ZINC COMPOUNDS.

ZKS100   CAS: 10025-64-6   HR: 3
ZINC PERCHLORATE HEXAHYDRATE
mf: $Cl_2O_8 \cdot Zn \cdot 6H_2O$ mw: 372.39
SYN: PERCHLORIC ACID, ZINC SALT, HEXAHYDRATE
DOT CLASSIFICATION: 5.1; Label: Oxidizer
SAFETY PROFILE: A poison by intraperitoneal route. An oxidizer. When heated to decomposition it emits toxic vapors of zinc and Cl^-.

ZLA000   CAS: 23414-72-4   HR: 3
ZINC PERMANGANATE
DOT: UN 1515
mf: $Mn_2O_8 \cdot Zn$ mw: 303.25
PROP: Violet-brown or black, hygroscopic crystals.
CONSENSUS REPORTS: Zinc, manganese, and their compounds are on the Community Right-To-Know List.
ACGIH TLV: TWA 0.03 mg(Mn)/m^3
DOT CLASSIFICATION: 5.1; Label: Oxidizer
SAFETY PROFILE: Probably a skin, eye, and mucous membrane irritant. Flammable by chemical reaction with reducing agents. A powerful oxidizing agent. When heated to decomposition it emits toxic fumes of ZnO. Used as an antiseptic. See also MANGANESE COMPOUNDS and ZINC COMPOUNDS.

ZLJ000   CAS: 1314-22-3   HR: 3
ZINC PEROXIDE
DOT: UN 1516
mf: O_2Zn mw: 97.37
PROP: Odorless, fairly unstable, yellow-white powder. D: 1.571 (theoretical). Decomp @ >150°. Sol in dil acids.
SYN: ZINC SUPEROXIDE
CONSENSUS REPORTS: Zinc and its compounds are on the Community Right-To-Know List. Reported in EPA TSCA Inventory.
DOT CLASSIFICATION: 5.1; Label: Oxidizer
SAFETY PROFILE: Systemic toxicity is similar to zinc oxide. Flammable when

exposed to heat or by chemical reaction with reducing materials. Finely divided powder is slightly soluble in water, decomposes rapidly at 150°. A powerful oxidizer and dangerous when mixed with highly combustible materials. A very dangerous explosion hazard when exposed to heat. Explodes at 212°. Can react violently with Al and Zn. Very dangerous, will react with water or steam to produce heat. Vigorous reaction with reducing materials. When heated to decomposition it emits toxic fumes of ZnO. See also PEROXIDES and ZINC COMPOUNDS.

ZLS000   CAS: 1314-84-7   HR: 3
ZINC PHOSPHIDE
DOT: UN 1714
mf: P_2Zn_3 mw: 258.05
PROP: Cubic, dark-gray, tetragonal crystals or powder with faint phosphorus odor. Stable when dry. Mp: 420°, bp: 1100°, d: 4.55 @ 13°. Insol in water, alc; sol in benzene, carbon disulfide.
SYNS: BLUE-OX □ KILRAT □ MOUS-CON □ PHOSPHURE de ZINC (FRENCH) □ PHOSVIN □ RCRA WASTE NUMBER P122 □ RUMETAN □ ZINCO (FOSFURO di) (ITALIAN) □ ZINC (PHOSPHURE de) (FRENCH) □ ZINC-TOX □ ZINKFOSIDE (DUTCH) □ ZINKPHOSPHID (GERMAN) □ ZP
CONSENSUS REPORTS: EPA Extremely Hazardous Substances List. Zinc and its compounds are on the Community Right-To-Know List. Reported in EPA TSCA Inventory.
DOT CLASSIFICATION: 4.3; Label: Dangerous When Wet, Poison
SAFETY PROFILE: Human poison by ingestion causing nausea, vomiting, death. Flammable when exposed to heat or flame. This material is stable while kept dry. In moist air, it decomposes slowly. Reacts violently with acids or acid fumes to emit the highly toxic and flammable phosphine. Violent reaction with concentrated sulfuric acid, nitric acid, and oxidizing materials. Incompatible with HCl, H_2SO_4. When heated to decomposition it emits toxic fumes of PO$_x$ and ZnO. Used as an acute rodenticide. See also PHOSPHIDES and

Z

ZINC COMPOUNDS.

ZMJ000   CAS: 13463-41-7   HR: 3
ZINC PYRIDINE-2-THIOL-1-OXIDE
mf: $C_{10}H_8N_2O_2S_2 \cdot Zn$ mw: 317.69

PROP: White solid. Mp: 262°. Sol in DMSO, DMF, and $CHCl_3$.

SYNS: BIS(1-HYDROXY-2(1H)-PYRIDINETHIONATO)-ZINC □ BIS(2-PYRIDYLTHIO)ZINC, 1,1'-DIOXIDE □ OM-1563 □ OMADINE ZINC □ 2-PYRIDINETHIOL-1-OXIDE, ZINC SALT □ PYRITHIONE ZINC □ VANCIDE P □ ZINC OMADINE □ ZINCPOLYANEMINE □ ZINC PT □ ZINC PYRIDINETHIONE □ ZINC PYRION □ ZINC PYRITHIONE

CONSENSUS REPORTS: Zinc and its compounds are on the Community Right-To-Know List. Reported in EPA TSCA Inventory.

SAFETY PROFILE: Poison by ingestion, skin contact, intraperitoneal, and intravenous routes. Moderately toxic by subcutaneous route. An experimental teratogen. Experimental reproductive effects. An eye irritant. When heated to decomposition it emits very toxic fumes of NO_x, SO_x, and ZnO. Used as an anti-dandruff agent in shampoos. See also ZINC COMPOUNDS and SULFIDES.

ZMS000   CAS: 557-05-1   HR: 3
ZINC STEARATE
mf: $Zn(C_{18}H_{35}O_2)_2$ mw: 632.30

PROP: White powder. Mp: 130°, flash p: 530°F (OC), autoign temp: 790°F. Insol in water, alc, ether; sol in benzene. Decomp in dil acids.

SYNS: DIBASIC ZINC STEARATE □ OCTADECANOIC ACID, ZINC SALT □ STEARIC ACID, ZINC SALT □ ZINC DISTEARATE □ ZINC OCTADECANOATE

CONSENSUS REPORTS: Zinc and its compounds are on the Community Right-To-Know List. Reported in EPA TSCA Inventory.

OSHA PEL: TWA Total Dust: 10 mg/m³; Respirable Fraction: 5 mg/m³

ACGIH TLV: TWA 10 mg/m³ of total dust when toxic impurities are not present, e.g., quartz <1%

SAFETY PROFILE: Poison by intratracheal route. Inhalation of zinc stearate has been reported as causing pulmonary fibrosis. A nuisance dust. Combustible when exposed to heat or flame. To fight fire, use water, foam, CO_2, dry chemical. When heated to decomposition it emits toxic fumes of ZnO. See also ZINC COMPOUNDS.

ZNA000   CAS: 7733-02-0   HR: 3
ZINC SULFATE
mf: $O_4S \cdot Zn$ mw: 161.43

PROP: Rhombic, colorless crystals or crystalline powder. Mp: decomp @ 740°, d: 3.74 @ 15°. Sol in water; almost insol in alc.

SYNS: BONAZEN □ BUFOPTO ZINC SULFATE □ OP-THAL-ZIN □ SULFATE de ZINC (FRENCH) □ SULFURIC ACID, ZINC SALT (1:1) □ VERAZINC □ WHITE COPPERAS □ WHITE VITRIOL □ ZINC SULPHATE □ ZINC VITRIOL □ ZINKOSITE

CONSENSUS REPORTS: Zinc and its compounds are on the Community Right-To-Know List. Reported in EPA TSCA Inventory. EPA Genetic Toxicology Program.

SAFETY PROFILE: Poison by ingestion, intraperitoneal, subcutaneous, and intravenous routes. Human systemic effects by ingestion: acute pulmonary edema, agranulocytosis, blood pressure decrease, diarrhea and other gastrointestinal changes, hypermotility, increased pulse rate without blood pressure decrease, level changes for metals other than Na/K/Fe/Ca/P/Cl, microcytosis with or without anemia, normocytic anemia. Experimental teratogenic and reproductive effects. Questionable carcinogen with experimental tumorigenic data. Human mutation data reported. An eye irritant. When heated to decomposition it emits toxic fumes of SO_x and ZnO. See also SULFATES and ZINC COMPOUNDS.

ZNJ000   CAS: 7446-20-0   HR: 3
ZINC SULFATE HEPTAHYDRATE (1:1:7)
mf: $O_4SZn \cdot 7H_2O$ mw: 287.57

PROP: Colorless, odorless, orthorhombic crystals or crystalline powder from water. D: 1.97, mp: 100°. Decomp @ >500°. Insol in alc, glycerin.

SYNS: SULFURIC ACID, ZINC SALT (1:1), HEPTAHYDRATE □ WHITE VITRIOL □ ZINC SULFATE

☐ ZINC SULFATE (1:1) HEPTAHYDRATE ☐ ZINC VITRIOL

CONSENSUS REPORTS: Zinc and its compounds are on the Community Right-To-Know List.

SAFETY PROFILE: Human poison by an unspecified route. Poison experimentally by subcutaneous, intravenous, and intraperitoneal routes. Moderately toxic by ingestion. Experimental reproductive effects. When heated to decomposition it emits toxic fumes of SO_x and ZnO. See also ZINC SULFATE.

ZOA000   CAS: 7440-67-7   HR: 3
ZIRCONIUM
DOT: UN 1358/UN 1932/UN 2008/UN 2009/UN 2858
af: Zr   aw: 91.224
PROP: A grayish-white, lustrous, metallic element; very sltly radioactive. Very resistant to corrosion but embrittled by N, O, and C. Oxidizes rapidly at 6°. Nitrided slowly at 700°. Mp: 1852°, bp: 4200°, d: 6.506 @ 20°. IDLH 50 mg/m³ (as Zr).
SYNS: ZIRCAT ☐ ZIRCONIUM (ACGIH,OSHA) ☐ ZIRCONIUM, dry, coiled wire, finished metal sheets, strip (UN 2858) (DOT) ☐ ZIRCONIUM, dry, finished sheets, strip or coiled wire (UN 2009) (DOT) ☐ ZIRCONIUM METAL, dry, chemically produced, finer than 20 mesh particle size (UN 2008) ☐ ZIRCONIUM POWDER, dry (UN 2008) (DOT) ☐ ZIRCONIUM POWDER, wetted with not <25% water (UN 1358) (DOT) ☐ ZIRCONIUM SCRAP (UN 1932) (DOT)
CONSENSUS REPORTS: Reported in EPA TSCA Inventory.
OSHA PEL: TWA 5 mg(Zr)/m³; STEL 10 mg(Zr)/m³; Not Classifiable as a Human Carcinogen
ACGIH TLV: TWA 5 mg(Zr)/m³; STEL 10 mg(Zr)/m³; Not Classifiable as a Human Carcinogen
DFG MAK: 1 mg(Zr)/m³
DOT CLASSIFICATION: 4.1; Label: Flammable Solid (UN 2858, UN 1358); DOT Class: 4.2; Label: Spontaneously Combustible (UN 2008, UN 2009, UN 1932)
SAFETY PROFILE: A very dangerous fire hazard in the form of dust when exposed to heat or flame or by chemical reaction with oxidizers. May ignite spontaneously. A dangerous explosion hazard in the form of dust by chemical reaction with air, alkali hydroxides, alkali metal chromates, dichromates, molybdates, sulfates, tungstates, borax, CCl_4, CuO, Pb, PbO, P, $KClO_3$, KNO_3, nitrylfluoride. Explosive range: 0.16 g/L in air. To fight fire, use special mixtures, dry chemical, salt, or dry sand. See also ZIRCONIUM COMPOUNDS.

ZPA000   CAS: 10026-11-6   HR: 3
ZIRCONIUM CHLORIDE
DOT: UN 2503
mf: Cl_4Zr mw: 233.02
PROP: White solid; easily hydrolyzed; lustrous crystals. Mp: subl @ 437°, bp: 331°, d: 2.80, vap press: 1 mm @ 190°. IDLH 50 mg/m³ (as Zr).
SYNS: ZIRCONIUM(IV) CHLORIDE (1:4) ☐ ZIRCONIUM TETRACHLORIDE (DOT) ☐ ZIRCONIUM TETRACHLORIDE, solid (DOT)
CONSENSUS REPORTS: Reported in EPA TSCA Inventory.
OSHA PEL: TWA 5 mg(Zr)/m³; STEL 10 mg(Zr)/m³; Not Classifiable as a Human Carcinogen
ACGIH TLV: TWA 5 mg(Zr)/m³; STEL 10 mg(Zr)/m³
DFG MAK: 1 mg(Zr)/m³
DOT CLASSIFICATION: 8; Label: Corrosive
SAFETY PROFILE: Moderately toxic by ingestion. A corrosive irritant to skin, eyes, and mucous membranes. Ignites spontaneously in air. When heated to decomposition it emits toxic fumes of Cl⁻. See also ZIRCONIUM COMPOUNDS and HYDROCHLORIC ACID.

ZPJ000   CAS: 10119-31-0   HR: 1
ZIRCONIUM CHLORIDE HYDROXIDE
mf: ClHOZr   mw: 143.68
PROP: IDLH 50 mg/m³ (as Zr).
SYNS: ZIRCONIUM CHLOROHYDRATE ☐ ZIRCONIUM HYDROXYCHLORIDE ☐ ZIRCONYL HYDROXYCHLORIDE
CONSENSUS REPORTS: Reported in EPA TSCA Inventory.
OSHA PEL: TWA 5 mg(Zr)/m³; STEL 10 mg(Zr)/m³

Z

ACGIH TLV: TWA 5 mg(Zr)/m³; STEL 10 mg(Zr)/m³; Not Classifiable as a Human Carcinogen
DFG MAK: 1 mg(Zr)/m³
SAFETY PROFILE: A human skin irritant. When heated to decomposition it emits toxic fumes of Cl⁻. Used as an antiperspirant. See also ZIRCONIUM COMPOUNDS and CHLORIDES.

ZQA000   HR: 2
ZIRCONIUM COMPOUNDS
PROP: IDLH 50 mg/m³ (as Zr).
OSHA PEL: TWA 5 mg(Zr)/m³; STEL 10 mg(Zr)/m³
ACGIH TLV: TWA 5 mg(Zr)/m³; STEL 10 mg(Zr)/m³; Not Classifiable as a Human Carcinogen
DFG MAK: (insoluble compounds) 1 mg(Zr)/m³
SAFETY PROFILE: Zirconium is not an important industrial poison; however, poisoning may occur due to excessive exposure to zirconium salts. Deaths in rabbits have been caused by intravenous injection of 150 mg/kg of body weight. Inhalation of ZrCl₄ (6 mg Zr/m³) for 60 days produces slight decreases in hemo-globin and red blood cell count in dogs and increases mortality in rats and guinea pigs. Most zirconium compounds in common use are insoluble and considered inert. Pulmonary granuloma in zirconium workers has been reported and sodium zirconium lactate has been held responsible for skin granulomas. Avoid inhalation of Zr-containing aerosols, which can cause lung granulomas. Zirconium-containing drugs or cosmetic products are being controlled by the FDA.

ZQS000   CAS: 7783-64-4   HR: 3
ZIRCONIUM FLUORIDE
mf: F₄Zr   mw: 167.22
PROP: Refractive crystals or white solid which readily sublimes; water-sol. D: 4.6 @ 16°, subl @ 600°. Very sol in HF. IDLH 50 mg/m³ (as Zr).
SYN: ZIRCONIUM TETRAFLUORIDE

CONSENSUS REPORTS: Reported in EPA TSCA Inventory.
OSHA PEL: TWA 5 mg(Zr)/m³; STEL 10 mg(Zr)/m³
ACGIH TLV: TWA 5 mg(Zr)/m³; STEL 10 mg(Zr)/m³; Not Classifiable as a Human Carcinogen; TWA 2.5 mg(F)/m³; BEI: 3 mg/g creatinine of fluorides in urine prior to shift; 10 mg/g creatinine of fluorides in urine at end of shift.
DFG MAK: 1 mg(Zr)/m³
NIOSH REL: (Fluorides, Inorganic) 10H TWA 2.5 mg(F)/m³
SAFETY PROFILE: Poison by intravenous route. When heated to decomposition it emits toxic fumes of F⁻. See also ZIRCONIUM COMPOUNDS and FLUORIDES.

ZQS100   CAS: 70983-41-4   HR: 3
ZIRCONIUM GLUCONATE
PROP: IDLH 50 mg/m³ (as Zr).
CONSENSUS REPORTS: Reported in EPA TSCA Inventory.
OSHA PEL: TWA 5 mg(Zr)/m³; STEL 10 mg(Zr)/m³
ACGIH TLV: TWA 5 mg(Zr)/m³; STEL 10 mg(Zr)/m³; Not Classifiable as a Human Carcinogen
SAFETY PROFILE: Poison by intraperitoneal route. When heated to decomposition it emits toxic fumes of Zr.

ZRA000   CAS: 7704-99-6   HR: 3
ZIRCONIUM HYDRIDE
DOT: UN 1437
mf: H₂Zr   mw: 93.24
PROP: Metallic dark-gray to black powder. D: 5.6, autoign temp: 270° (in air). IDLH 50 mg/m³ (as Zr).
CONSENSUS REPORTS: Reported in EPA TSCA Inventory.
OSHA PEL: TWA 5 mg(Zr)/m³; STEL 10 mg(Zr)/m³
ACGIH TLV: TWA 5 mg(Zr)/m³; STEL 10 mg(Zr)/m³; Not Classifiable as a Human Carcinogen
DFG MAK: 1 mg(Zr)/m³

DOT CLASSIFICATION: 4.1; Label: Flammable Solid
SAFETY PROFILE: A powerful reducing agent. Flammable when dry or wet. Very dangerous to handle; can explode. Incandesces when heated in air. See also HYDRIDES and ZIRCONIUM COMPOUNDS.

ZSA000   CAS: 13746-89-9   HR: 2
ZIRCONIUM NITRATE
DOT: UN 2728
mf: $N_4O_{12} \cdot Zr$ mw: 339.26
PROP: White hygroscopic crystals. Very sol in water; sol in alc. IDLH 50 mg/m^3 (as Zr).
SYN: DUSICNAN ZIRKONICITY (CZECH)
CONSENSUS REPORTS: Reported in EPA TSCA Inventory.
OSHA PEL: TWA 5 mg(Zr)/m^3; STEL 10 mg(Zr)/m^3
ACGIH TLV: TWA 5 mg(Zr)/m^3; STEL 10 mg(Zr)/m^3; Not Classifiable as a Human Carcinogen
DFG MAK: 1 mg(Zr)/m^3
DOT CLASSIFICATION: 5.1; Label: Oxidizer
SAFETY PROFILE: Moderately toxic by inhalation and ingestion. A powerful oxidizer. When heated to decomposition it emits toxic fumes of NO$_x$. See also NITRATES and ZIRCONIUM COMPOUNDS.

ZSJ000   CAS: 7699-43-6   HR: 3
ZIRCONIUM OXYCHLORIDE
mf: Cl_2OZr mw: 178.12
PROP: Crystals or white solid. Forms adducts by direct reaction in EtOH. D: 1.91. Very sol in water, alc. IDLH 50 mg/m^3 (as Zr).
SYNS: BASIC ZIRCONIUM CHLORIDE □ CHLOROZIRCONYL □ DICHLOROOXOZIRCONIUM □ NCI-C60811 □ ZIRCONYL CHLORIDE
CONSENSUS REPORTS: Reported in EPA TSCA Inventory.
OSHA PEL: TWA 5 mg(Zr)/m^3; STEL 10 mg(Zr)/m^3
ACGIH TLV: TWA 5 mg(Zr)/m^3; STEL 10 mg(Zr)/m^3; Not Classifiable as a Human Carcinogen
DFG MAK: 1 mg(Zr)/m^3

SAFETY PROFILE: Poison by intraperitoneal route. Moderately toxic by ingestion and subcutaneous routes. Questionable carcinogen with experimental neoplastigenic data. When heated to decomposition it emits toxic fumes of Cl$^-$. Used as an antiperspirant. See also ZIRCONIUM COMPOUNDS and CHLORIDES.

ZSS000   CAS: 14940-68-2   HR: 2
ZIRCONIUM(IV) SILICATE (1:1)
mf: O_4SiZr mw: 183.31
PROP: Usually shades of brown, green, gray, yellow, and red, tetragonal, bipyramidal crystals. D: 4.56, mp: 2550°. IDLH 50 mg/m^3 (as Zr).
SYNS: HYACINTH □ SILICIC ACID, ZIRCONIUM(4+) SALT (1:1) □ ZIRCON
CONSENSUS REPORTS: Reported in EPA TSCA Inventory.
OSHA PEL: TWA 5 mg(Zr)/m^3; STEL 10 mg(Zr)/m^3
ACGIH TLV: TWA 5 mg(Zr)/m^3; STEL 10 mg(Zr)/m^3; Not Classifiable as a Human Carcinogen
DFG MAK: 1 mg(Zr)/m^3
SAFETY PROFILE: See SILICATES and ZIRCONIUM COMPOUNDS.

ZTJ000   CAS: 14644-61-2   HR: 3
ZIRCONIUM(IV) SULFATE (1:2)
DOT: NA 9163
mf: $O_8S_2 \cdot Zr$ mw: 283.34
PROP: Tetrahydrate, crystalline solid. IDLH 50 mg/m^3 (as Zr).
SYNS: DISULFATOZIRCONIC ACID □ SULFURIC ACID, ZIRCONIUM(4+) SALT (2:1) □ ZIRCONYL SULFATE
CONSENSUS REPORTS: Reported in EPA TSCA Inventory.
OSHA PEL: TWA 5 mg(Zr)/m^3; STEL 10 mg(Zr)/m^3
ACGIH TLV: TWA 5 mg(Zr)/m^3; STEL 10 mg(Zr)/m^3; Not Classifiable as a Human Carcinogen
DFG MAK: 1 mg(Zr)/m^3
DOT CLASSIFICATION: 8; Label: Corrosive
SAFETY PROFILE: Poison by

Z

intraperitoneal route. Moderately toxic by ingestion and subcutaneous routes. A corrosive. Experimental reproductive effects. Mutation data reported. When heated to decomposition it emits toxic fumes of SO_x. See also SULFATES and ZIRCONIUM COMPOUNDS.

ZTK400   CAS: 1291-32-3   HR: 3
ZIRCONOCENE, DICHLORIDE
mf: $C_{10}H_{10}Cl_2Zr$ mw: 292.32
PROP: Colorless crystals. Mp: 248°. IDLH 50 mg/m³ (as Zr).
SYN: ZIRCONIUM, DICHLORO-DI-pi-CYCLOPENTADIENYL-
CONSENSUS REPORTS: Reported in EPA TSCA Inventory.
OSHA PEL: TWA 5 mg(Zr)/m³; STEL 10 mg(Zr)/m³
ACGIH TLV: TWA 5 mg(Zr)/m³; STEL 10 mg(Zr)/m³; Not Classifiable as a Human Carcinogen
SAFETY PROFILE: Poison by intraperitoneal route. Mutation data reported. When heated to decomposition it emits toxic fumes of Zr and Cl⁻.

ZUS000   CAS: 22144-77-0   HR: 3
ZYGOSPORIN A
mf: $C_{30}H_{37}NO_6$ mw: 507.68
PROP: Needles from Me_2CO/pet ether. Mp: 268–271°.
SYNS: 3-BENZYL-3,3-α,4,5,6,6-α,9,10,12,15-DECAHYDRO-6,12,15-TRIHYDROXY-4,10,12-TRIMETHYL-5-METHYLENE-1H-CYCLOUNDEC(d)ISOINDOLE-1,11(2H)-DIONE, 15-ACETATE □ CYTOCHALASIN D
SAFETY PROFILE: Poison by ingestion, subcutaneous, and intraperitoneal routes. An

experimental teratogen. Experimental reproductive effects. Human mutation data reported. When heated to decomposition it emits toxic fumes of NO_x.

ZVJ000   CAS: 315-30-0   HR: 3
ZYLOPRIM
mf: $C_5H_4N_4O$ mw: 136.13
PROP: Crystals from H_2O. Mp: 383–384°.
SYNS: ADENOCK □ AL-100 □ ALLOPURINOL □ ALLOZYM □ ALLURAL □ ALOSITOL □ ALULINE □ ANOPROLIN □ ANZIEF □ APURIN □ APUROL □ BLEMINOL □ BLOXANTH □ BW 56-158 □ CAPLENAL □ CELLIDRIN □ DABROSIN □ 1,5-DIHYDRO-4H-PYRAZOLO(3,4-d)PYRIMIDIN-4-ONE □ EMBARIN □ EPIDROPAL □ FOLIGAN □ GICHTEX □ HPP □ 4'-HYDROXYPYRAZOLOL(3,4-d)PYRIMIDINE □ 4-HYDRO-XY-1H-PYRAZOLO(3,4-d)PYRIMIDINE □ 4-HYDROXY-3,4-PYRAZOLOPYRIMIDINE □ 4-HYDROXYPYRAZO-LO(3,4-d)PYRIMIDINE □ 4-HYDROXYPYRAZOLYL(3,4-d)PYRIMIDINE □ KETANRIFT □ KETOBUN-A □ LOPURIN □ LYSURON □ MINIPLANOR □ MONARCH □ NEKTROHAN □ NSC-1390 □ 4H-PYRAZOLO(3,4-d)PYRIMIDIN-4-ONE □ REMID □ RIBALL □ SUSPEND-OL □ TAKANARUMIN □ URBOL □ URICEMIL □ URITAS □ UROBENYL □ UROSIN □ XANTURAT □ ZYLORIC
CONSENSUS REPORTS: Reported in EPA TSCA Inventory.
SAFETY PROFILE: Human poison by ingestion. Poison experimentally by intraperitoneal and subcutaneous routes. An experimental teratogen. Human systemic effects by ingestion: blood leukopenia, dermatitis, jaundice, muscle weakness, thrombocytopenia. When heated to decomposition it emits toxic fumes of NO_x. An FDA proprietary drug used as a xanthine oxidase inhibitor.

CAS Number Cross-Index

75-00-3	see EHH000	75-87-6	see CDN550	78-81-9	see IIM000
75-01-4	see VNP000	75-88-7	see TJY175	78-82-0	see IJX000
75-02-5	see VPA000	75-89-8	see TKA350	78-83-1	see IIL000
75-04-7	see EFU400	75-91-2	see BRM250	78-84-2	see IJS000
75-05-8	see ABE500	75-94-5	see TIN750	78-85-3	see MGA250
75-07-0	see AAG250	75-98-9	see PJA500	78-86-4	see CEU250
75-08-1	see EMB100	75-99-0	see DGI400	78-87-5	see PNJ400
75-09-2	see MJP450	76-01-7	see PAW500	78-88-6	see DGH400
75-12-7	see FMY000	76-02-8	see TIJ150	78-89-7	see CKR500
75-15-0	see CBV500	76-03-9	see TII250	78-90-0	see PMK250
75-16-1	see MLE000	76-05-1	see TKA250	78-92-2	see BPW750
75-18-3	see TFP000	76-06-2	see CKN500	78-93-3	see MKA400
75-19-4	see CQD750	76-11-9	see TBP000	78-94-4	see BOY500
75-20-7	see CAN750	76-12-0	see TBP050	78-95-5	see CDN200
75-21-8	see EJN500	76-13-1	see FOO000	78-97-7	see LAQ000
75-25-2	see BNL000	76-14-2	see FOO509	79-00-5	see TIN000
75-26-3	see BNY000	76-15-3	see CJI500	79-01-6	see TIO750
75-27-4	see BND500	76-22-2	see CBA750	79-03-8	see PMW500
75-28-5	see MOR750	76-25-5	see AQX500	79-04-9	see CEC250
75-29-6	see CKQ000	76-42-6	see PCG500	79-06-1	see ADS250
75-31-0	see INK000	76-44-8	see HAR000	79-07-2	see CDY850
75-33-2	see IMU000	76-49-3	see BMD100	79-08-3	see BMR750
75-34-3	see DFF809	76-87-9	see HON000	79-09-4	see PMU750
75-35-4	see VPK000	76-99-3	see MDO750	79-10-7	see ADS750
75-36-5	see ACF750	77-06-5	see GEM000	79-11-8	see CEA000
75-37-6	see ELN500	77-07-6	see LFG000	79-16-3	see MFT750
75-38-7	see VPP000	77-09-8	see PDO750	79-19-6	see TFQ000
75-39-8	see AAG500	77-21-4	see DYC800	79-20-9	see MFW100
75-43-4	see DFL000	77-47-4	see HCE500	79-21-0	see PCL500
75-44-5	see PGX000	77-58-7	see DDV600	79-22-1	see MIG000
75-45-6	see CFX500	77-65-6	see BNK000	79-24-3	see NFY500
75-46-7	see CBY750	77-73-6	see DGW000	79-27-6	see ACK250
75-47-8	see IEP000	77-78-1	see DUD100	79-29-8	see DQT400
75-50-3	see TLD500	77-81-6	see EIF000	79-31-2	see IJU000
75-52-5	see NHM500	77-83-8	see ENC000	79-34-5	see TBQ100
75-54-7	see DFS000	77-89-4	see ADD750	79-36-7	see DEN400
75-55-8	see PNL400	77-90-7	see ADD400	79-38-9	see CLQ750
75-56-9	see PNL600	77-92-9	see CMS750	79-39-0	see MDN500
75-57-0	see TDK000	77-93-0	see TJP750	79-40-3	see DXO200
75-59-2	see TDK500	78-00-2	see TCF000	79-41-4	see MDN250
75-60-5	see HKC000	78-06-8	see DEJ200	79-42-5	see TFK250
75-61-6	see DKG850	78-10-4	see EPF550	79-43-6	see DEL000
75-62-7	see BOH750	78-11-5	see PBC250	79-44-7	see DQY950
75-63-8	see TJY100	78-18-2	see CPC300	79-46-9	see NIY000
75-64-9	see BPY250	78-20-6	see DEF200	79-57-2	see HOH500
75-65-0	see BPX000	78-30-8	see TMO600	79-69-6	see IGW500
75-66-1	see MOS000	78-34-2	see DVQ709	79-78-7	see AGI500
75-68-3	see CFX250	78-35-3	see LGB000	79-92-5	see CBA500
75-69-4	see TIP500	78-40-0	see TJT750	80-05-7	see BLD500
75-71-8	see DFA600	78-46-6	see DDV800	80-08-0	see SOA500
75-72-9	see CLR250	78-48-8	see BSH250	80-10-4	see DFF000
75-73-0	see CBY250	78-51-3	see BPK250	80-15-9	see IOB000
75-74-1	see TDR500	78-59-1	see IMF400	80-17-1	see BBS300
75-75-2	see MDR250	78-62-6	see DHG000	80-26-2	see TBE250
75-77-4	see TLN250	78-63-7	see DRJ800	80-40-0	see EPW500
75-78-5	see DFE259	78-67-1	see ASL750	80-43-3	see DGR600
75-79-6	see MQC500	78-70-6	see LFX000	80-48-8	see MLL250
75-80-9	see ARW250	78-76-2	see BMX750	80-51-3	see OPE000
75-83-2	see DQT200	78-77-3	see BNR750	80-56-8	see PIH250
75-85-4	see PBV000	78-78-4	see EIK000	80-59-1	see TGA700
75-86-5	see MLC750	78-79-5	see IMS000	80-62-6	see MLH750

80-71-7 see HMB500	87-19-4 see IJN000	90-80-2 see GFA200
81-07-2 see BCE500	87-20-7 see IME000	90-87-9 see HII600
81-13-0 see PAG200	87-25-2 see EGM000	90-89-1 see DIW000
81-15-2 see TML750	87-29-6 see API750	90-94-8 see MQS500
81-64-1 see DMH000	87-31-0 see DUR800	91-08-7 see TGM800
81-81-2 see WAT200	87-33-2 see CCK125	91-10-1 see DOJ200
82-21-3 see DVW100	87-44-5 see CCN000	91-15-6 see PHY000
82-28-0 see AKP750	87-51-4 see ICN000	91-17-8 see DAE800
82-33-7 see DBY700	87-56-9 see MRU900	91-20-3 see NAJ500
82-43-9 see DEO750	87-59-2 see XMJ000	91-22-5 see QMJ000
82-44-0 see CEI000	87-60-5 see CLK200	91-23-6 see NER000
82-45-1 see AIA750	87-62-7 see XNJ000	91-40-7 see PEG500
82-46-2 see DEO700	87-63-8 see CLK227	91-51-0 see LFT100
82-66-6 see DVV600	87-65-0 see DFY000	91-53-2 see SAV000
82-68-8 see PAX000	87-66-1 see PPQ500	91-56-5 see ICR000
82-71-3 see SMP500	87-68-3 see HCD250	91-59-8 see NBE500
83-26-1 see PIH175	87-69-4 see TAF750	91-60-1 see NAP500
83-34-1 see MKV750	87-78-5 see MAW100	91-64-5 see CNV000
83-43-2 see MOR500	87-85-4 see HEC000	91-66-7 see DIS700
83-44-3 see DAQ400	87-86-5 see PAX250	91-71-4 see TFD750
83-66-9 see BRU500	87-89-8 see IDE300	91-81-6 see TMP750
83-67-0 see TEO500	87-90-1 see TIQ750	91-93-0 see DCJ400
83-73-8 see DNF600	87-99-0 see XPJ000	91-94-1 see DEQ600
83-86-3 see PIB250	88-04-0 see CLW000	91-95-2 see BGK500
83-88-5 see RIK000	88-05-1 see TLG500	91-97-4 see DQS000
83-89-6 see ARQ250	88-06-2 see TIW000	92-06-8 see TBC620
84-15-1 see TBC640	88-09-5 see DHI400	92-13-7 see PIF000
84-17-3 see DAL600	88-10-8 see DIW400	92-15-9 see ABA500
84-65-1 see APK250	88-12-0 see EEG000	92-26-2 see DBT200
84-66-2 see DJX000	88-15-3 see ABI500	92-48-8 see MIP750
84-69-5 see DNJ400	88-19-7 see TGN250	92-52-4 see BGE000
84-74-2 see DEH200	88-24-4 see MJN250	92-53-5 see PFS750
84-75-3 see DKP600	88-29-9 see ACL750	92-54-6 see PFX000
85-00-7 see DWX800	88-61-9 see XJJ000	92-62-6 see DBN600
85-01-8 see PCW250	88-69-7 see IQX100	92-67-1 see AJS100
85-41-6 see PHX000	88-72-2 see NMO525	92-69-3 see BGJ500
85-43-8 see TDB000	88-73-3 see CJB750	92-82-0 see PDB500
85-44-9 see PHW750	88-74-4 see NEO000	92-84-2 see PDP250
85-68-7 see BEC500	88-75-5 see NIF010	92-87-5 see BBX000
85-70-1 see BQP750	88-89-1 see PID000	92-93-3 see NFQ000
85-83-6 see SBC500	88-99-3 see PHW250	92-94-4 see TBC750
85-84-7 see FAG130	89-00-9 see QOJ200	93-05-0 see DJV200
85-86-9 see OHI200	89-72-5 see BSE000	93-08-3 see ABC500
85-91-6 see MGQ250	89-78-1 see MCF750	93-15-2 see AGE250
85-98-3 see DJC400	89-80-5 see MCG275	93-16-3 see IKR000
86-21-5 see TMJ750	89-83-8 see TFX810	93-23-2 see LBW000
86-29-3 see DVX200	89-84-9 see DMG400	93-28-7 see EQS000
86-30-6 see DWI000	89-86-1 see HOE600	93-29-8 see AAX750
86-50-0 see ASH500	89-98-5 see CEI500	93-46-9 see NBL000
86-51-1 see DNZ200	90-00-6 see PGR250	93-53-8 see COF000
86-54-4 see HGP495	90-01-7 see HMK100	93-55-0 see EOL500
86-56-6 see DSU400	90-03-9 see CHW675	93-58-3 see MHA750
86-57-7 see NHQ000	90-04-0 see AOV900	93-65-2 see CIR500
86-65-7 see NBE850	90-05-1 see GKI000	93-68-5 see ABA000
86-72-6 see CBN100	90-15-3 see NAW500	93-71-0 see CFK000
86-74-8 see CBN000	90-16-4 see BDH000	93-72-1 see TIX500
86-85-1 see MLH000	90-22-2 see VBK000	93-89-0 see EGR000
86-87-3 see NAK500	90-33-5 see MKP500	94-09-7 see EFX000
86-88-4 see AQN635	90-43-7 see BGJ250	94-11-1 see IOY000
87-13-8 see EEV200	90-64-2 see MAP000	94-13-3 see HNU500
87-18-3 see BSH100	90-65-3 see PAP750	94-17-7 see BHM750

94-30-4 see AOV000	96-88-8 see SBB000	99-35-4 see TMK500
94-36-0 see BDS000	97-00-7 see CGM000	99-55-8 see NMP500
94-41-7 see CDH000	97-02-9 see DUP600	99-56-9 see ALL500
94-46-2 see IHP100	97-11-0 see POO000	99-59-2 see NEQ500
94-58-6 see DMD600	97-16-5 see DFY400	99-65-0 see DUQ200
94-59-7 see SAD000	97-17-6 see DFK600	99-71-8 see BSE250
94-62-2 see PIV600	97-23-4 see MJM500	99-76-3 see HJL500
94-63-3 see POS750	97-39-2 see DXP200	99-83-2 see MCC000
94-74-6 see CIR250	97-44-9 see ABX500	99-85-4 see MCB750
94-86-0 see IRY000	97-53-0 see EQR500	99-86-5 see MLA250
94-96-2 see EKV000	97-54-1 see IKQ000	99-87-6 see CQI000
95-06-7 see CDO250	97-56-3 see AIC250	99-89-8 see IQZ000
95-13-6 see IBX000	97-62-1 see ELS000	99-91-2 see CEB250
95-14-7 see BDH250	97-63-2 see EMF000	99-92-3 see AHR240
95-33-0 see CPI250	97-64-3 see LAJ000	99-99-0 see NMO550
95-41-0 see HFO700	97-72-3 see IJW000	100-00-5 see NFS525
95-45-4 see DBH000	97-74-5 see BJL600	100-01-6 see NEO500
95-47-6 see XHJ000	97-77-8 see DXH250	100-02-7 see NIF000
95-48-7 see CNX000	97-85-8 see IIW000	100-06-1 see MDW750
95-49-8 see CLK100	97-86-9 see IIY000	100-07-2 see AOY250
95-50-1 see DEP600	97-88-1 see MHU750	100-17-4 see NER500
95-51-2 see CEH670	97-93-8 see TJN750	100-20-9 see TAV250
95-53-4 see TGQ750	97-94-9 see TJP250	100-21-0 see TAN750
95-54-5 see PEY250	97-95-0 see EGW000	100-25-4 see DUQ600
95-55-6 see ALT000	97-96-1 see DHI000	100-36-7 see DJI400
95-57-8 see CJK250	98-00-0 see FPU000	100-37-8 see DHO500
95-63-6 see TLL750	98-01-1 see FPQ875	100-39-0 see BEC000
95-64-7 see XNS000	98-02-2 see FPM000	100-40-3 see CPD750
95-68-1 see XMS000	98-05-5 see BBL750	100-41-4 see EGP500
95-69-2 see CLK220	98-06-6 see BQJ250	100-42-5 see SMQ000
95-70-5 see TGM000	98-07-7 see BFL250	100-44-7 see BEE375
95-74-9 see CLK215	98-08-8 see BDH500	100-47-0 see BCQ250
95-78-3 see XNA000	98-09-9 see BBS750	100-50-5 see FNK025
95-79-4 see CLK225	98-11-3 see BBS250	100-51-6 see BDX500
95-80-7 see TGL750	98-12-4 see CPR250	100-52-7 see BAY500
95-83-0 see CFK125	98-13-5 see TJA750	100-53-8 see TGO750
95-85-2 see CEH250	98-16-8 see AID500	100-56-1 see PFM500
95-87-4 see XKS000	98-29-3 see BSK000	100-57-2 see PFN100
95-88-5 see CLD750	98-46-4 see NFJ500	100-61-8 see MGN750
95-92-1 see DJT200	98-47-5 see NFB500	100-63-0 see PFI000
95-95-4 see TIV750	98-50-0 see ARA250	100-65-2 see PFJ250
96-08-2 see LFV000	98-51-1 see BSP500	100-66-3 see AOX750
96-09-3 see EBR000	98-54-4 see BSE500	100-73-2 see ADR500
96-11-7 see GGG000	98-56-6 see CEM825	100-74-3 see ENL000
96-12-8 see DDL800	98-72-6 see NIJ500	100-75-4 see NLJ500
96-14-0 see MNI500	98-73-7 see BQK500	100-79-8 see DVR600
96-18-4 see TJB600	98-77-1 see PIY500	100-86-7 see DQQ200
96-22-0 see DJN750	98-82-8 see COE750	100-88-9 see CPQ625
96-23-1 see DGG400	98-83-9 see MPK250	100-97-0 see HEI500
96-24-2 see CDT750	98-85-1 see PDE000	100-99-2 see TKR500
96-27-5 see MRM750	98-86-2 see ABH000	101-02-0 see TMU250
96-29-7 see EMU500	98-87-3 see BAY300	101-14-4 see MJM200
96-31-1 see DUM200	98-88-4 see BDM500	101-25-7 see DVF400
96-32-2 see MHR250	98-92-0 see NCR000	101-31-5 see HOU000
96-33-3 see MGA500	98-94-2 see DRF709	101-37-1 see THN500
96-34-4 see MIF775	98-95-3 see NEX000	101-39-3 see MIO000
96-37-7 see MIU500	98-96-4 see POL500	101-41-7 see MHA500
96-45-7 see IAQ000	99-03-6 see AHR500	101-48-4 see PDX000
96-48-0 see BOV000	99-06-9 see HJI100	101-61-1 see MJN000
96-69-5 see TFC600	99-08-1 see NMO500	101-68-8 see MJP400
96-80-0 see DNP000	99-09-2 see NEN500	101-77-9 see MJQ000

101-80-4 see OPM000	104-45-0 see PNE250	106-35-4 see EHA600
101-83-7 see DGT600	104-46-1 see PMQ750	106-38-7 see BOG255
101-84-8 see PFA850	104-50-7 see OCE000	106-42-3 see XHS000
101-86-0 see HFO500	104-51-8 see BQI750	106-43-4 see TGY075
101-90-6 see REF000	104-53-0 see HHP000	106-44-5 see CNX250
101-96-2 see DEG200	104-54-1 see CMQ740	106-45-6 see TGP250
101-97-3 see EOH000	104-55-2 see CMP969	106-46-7 see DEP800
102-01-2 see AAY000	104-57-4 see BEP250	106-47-8 see CEH680
102-06-7 see DWC600	104-61-0 see CNF250	106-48-9 see CJK750
102-08-9 see DWN800	104-65-4 see CMR500	106-49-0 see TGR000
102-09-0 see DVZ000	104-68-7 see PEQ750	106-50-3 see PEY500
102-20-5 see PDI000	104-75-6 see EKS500	106-51-4 see QQS200
102-22-7 see GDM400	104-76-7 see EKQ000	106-63-8 see IIK000
102-29-4 see RDZ900	104-78-9 see DIY800	106-68-3 see ODI000
102-36-3 see IKH099	104-90-5 see EOS000	106-72-9 see DSD775
102-50-1 see MGO500	104-91-6 see NLF200	106-75-2 see OPO000
102-54-5 see FBC000	104-93-8 see MGP000	106-87-6 see VOA000
102-62-5 see DBF600	104-94-9 see AOW000	106-88-7 see BOX750
102-67-0 see TMY100	105-11-3 see DVR200	106-89-8 see EAZ500
102-69-2 see TMY250	105-13-5 see MED500	106-90-1 see ECH500
102-70-5 see THN000	105-14-6 see. NOB800	106-92-3 see AGH150
102-71-6 see TKP500	105-21-5 see HBA550	106-93-4 see EIY500
102-76-1 see THM500	105-30-6 see AOK750	106-94-5 see BNX750
102-77-2 see BDG000	105-36-2 see EGV000	106-95-6 see AFY000
102-79-4 see BQM000	105-37-3 see EPB500	106-96-7 see PMN500
102-81-8 see DDU600	105-38-4 see VQK000	106-97-8 see BOR500
102-82-9 see THX250	105-39-5 see EHG500	106-99-0 see BOP500
103-23-1 see AEO000	105-40-8 see EMQ500	107-00-6 see EFS500
103-26-4 see MIO500	105-45-3 see MFX250	107-02-8 see ADR000
103-27-5 see PFO000	105-46-4 see BPV000	107-03-9 see PML500
103-28-6 see IJV000	105-53-3 see EMA500	107-05-1 see AGB250
103-29-7 see BFX500	105-54-4 see EHE000	107-06-2 see EIY600
103-33-3 see ASL250	105-55-5 see DKC400	107-07-3 see EIU800
103-34-4 see BKU500	105-56-6 see EHP500	107-10-8 see PND250
103-36-6 see EHN000	105-57-7 see AAG000	107-11-9 see AFW000
103-37-7 see BED000	105-58-8 see DIX200	107-12-0 see PMV750
103-38-8 see ISW000	105-60-2 see CBF700	107-13-1 see ADX500
103-41-3 see BEG750	105-64-6 see DNR400	107-14-2 see CDN500
103-45-7 see PFB250	105-67-9 see XKJ500	107-15-3 see EEA500
103-48-0 see PDF750	105-74-8 see LBR000	107-16-4 see HIM500
103-50-4 see BEO250	105-75-9 see DEC600	107-18-6 see AFV500
103-54-8 see CMQ730	105-76-0 see DED600	107-19-7 see PMN450
103-56-0 see CMR850	105-82-8 see AAG850	107-20-0 see CDY500
103-60-6 see PDS900	105-83-9 see BGU750	107-21-1 see EJC500
103-61-7 see CMQ800	105-85-1 see CMT750	107-22-2 see GIK000
103-65-1 see IKG000	105-86-2 see GCY000	107-25-5 see MQL750
103-69-5 see EGK000	105-87-3 see DTD800	107-27-7 see CHC500
103-71-9 see PFK250	105-90-8 see GDM450	107-29-9 see AAH250
103-72-0 see ISQ000	105-99-7 see AEO750	107-30-2 see CIO250
103-75-3 see EER500	106-11-6 see HKJ000	107-31-3 see MKG750
103-82-2 see PDY850	106-19-4 see DWQ875	107-35-7 see TAG750
103-83-3 see DQP800	106-21-8 see DTE600	107-37-9 see AGU250
103-84-4 see AAQ500	106-22-9 see CMT250	107-41-5 see HFP875
103-85-5 see PGN250	106-23-0 see CMS845	107-44-8 see IPX000
103-89-9 see ABJ250	106-24-1 see DTD000	107-49-3 see TCF250
103-90-2 see HIM000	106-25-2 see DTD200	107-59-5 see BQR100
103-93-5 see THA250	106-27-4 see IHP400	107-66-4 see DEG700
103-95-7 see COU500	106-30-9 see EKN050	107-70-0 see MEX250
104-01-8 see MFE250	106-31-0 see BSW550	107-71-1 see BSC250
104-12-1 see CKB000	106-32-1 see ENY000	107-72-2 see PBY750
104-20-1 see MFF580	106-33-2 see ELY700	107-75-5 see CMS850

107-81-3 see BNU500	109-06-8 see MOY000	110-58-7 see PBV505
107-82-4 see BNP250	109-08-0 see MOW750	110-60-1 see BOS000
107-83-5 see IKS600	109-09-1 see CKW000	110-61-2 see SNE000
107-87-9 see PBN250	109-19-3 see ISX000	110-62-3 see VAG000
107-88-0 see BOS500	109-20-6 see GDK000	110-63-4 see BOS750
107-89-1 see AAH750	109-21-7 see BQM500	110-65-6 see BST500
107-92-6 see BSW000	109-27-3 see TEF500	110-66-7 see PBM000
107-98-2 see PNL250	109-42-2 see BSS100	110-68-9 see MHV000
107-99-3 see CGW000	109-43-3 see DEH600	110-69-0 see BSU500
108-01-0 see DOY800	109-44-4 see BJO225	110-71-4 see DOE600
108-03-2 see NIX500	109-52-4 see VAQ000	110-73-6 see EGA500
108-05-4 see VLU250	109-53-5 see IJQ000	110-74-7 see PNM500
108-09-8 see DQU600	109-55-7 see AJQ100	110-75-8 see CHI250
108-10-1 see HFG500	109-56-8 see INN400	110-78-1 see PNP000
108-11-2 see MKW600	109-59-1 see INA500	110-80-5 see EES350
108-18-9 see DNM200	109-60-4 see PNC250	110-82-7 see CPB000
108-20-3 see IOZ750	109-61-5 see PNH000	110-83-8 see CPC579
108-21-4 see INE100	109-65-9 see BMX500	110-85-0 see PIJ000
108-22-5 see MQK750	109-66-0 see PBK250	110-86-1 see POP250
108-23-6 see IOL000	109-69-3 see BQQ750	110-87-2 see DMC200
108-24-7 see AAX500	109-70-6 see BNA825	110-88-3 see TMP000
108-29-2 see VAV000	109-73-9 see BPX750	110-89-4 see PIL500
108-30-5 see SNC000	109-74-0 see BSX250	110-91-8 see MRP750
108-31-6 see MAM000	109-76-2 see PMK500	110-93-0 see MKK000
108-38-3 see XHA000	109-77-3 see MAO250	110-96-3 see DNH400
108-39-4 see CNW750	109-78-4 see HGP000	111-12-6 see MND275
108-42-9 see CEH675	109-79-5 see BRR900	111-13-7 see ODG000
108-43-0 see CJK500	109-83-1 see MGG000	111-15-9 see EES400
108-44-1 see TGQ500	109-84-2 see HHC000	111-17-1 see BHM000
108-45-2 see PEY000	109-86-4 see EJH500	111-26-2 see HFK000
108-46-3 see REA000	109-87-5 see MGA850	111-27-3 see HFJ500
108-50-9 see DTU800	109-89-7 see DHJ200	111-30-8 see GFQ000
108-57-6 see DXQ745	109-90-0 see ELS500	111-31-9 see HES000
108-58-7 see REA100	109-92-2 see EQF500	111-34-2 see VMZ000
108-60-1 see BII250	109-93-3 see VOP000	111-36-4 see BRQ500
108-62-3 see TDW500	109-94-4 see EKL000	111-40-0 see DJG600
108-64-5 see ISY000	109-95-5 see ENN000	111-42-2 see DHF000
108-65-6 see PNL265	109-97-7 see PPS250	111-43-3 see PNM000
108-67-8 see TLM050	109-99-9 see TCR750	111-44-4 see DFJ050
108-68-9 see XLS000	110-00-9 see FPK000	111-46-6 see DJD600
108-69-0 see XOA000	110-01-0 see TDC730	111-49-9 see HDG000
108-70-3 see TIK300	110-02-1 see TFM250	111-55-7 see EJD759
108-73-6 see PGR000	110-12-3 see MKW450	111-60-4 see EJM500
108-77-0 see TJD750	110-13-4 see HEQ500	111-65-9 see OCU000
108-78-1 see MCB000	110-15-6 see SMY000	111-68-2 see HBL600
108-82-7 see DNH800	110-16-7 see MAK900	111-69-3 see AER250
108-83-8 see DNI800	110-17-8 see FOU000	111-70-6 see HBL500
108-84-9 see HFJ000	110-18-9 see TDQ750	111-71-7 see HBB500
108-86-1 see PEO500	110-19-0 see IIJ000	111-75-1 see BQC000
108-87-2 see MIQ740	110-22-5 see ACV500	111-76-2 see BPJ850
108-88-3 see TGK750	110-26-9 see MJL500	111-77-3 see DJG000
108-89-4 see MOY250	110-27-0 see IQN000	111-84-2 see NMX000
108-90-7 see CEJ125	110-38-3 see EHE500	111-87-5 see OEI000
108-91-8 see CPF500	110-40-7 see DJY600	111-90-0 see CBR000
108-93-0 see CPB750	110-41-8 see MQI550	111-91-1 see BID750
108-94-1 see CPC000	110-43-0 see MGN500	111-92-2 see DDT800
108-95-2 see PDN750	110-44-1 see SKU000	111-94-4 see BIQ500
108-98-5 see PFL850	110-45-2 see IHS000	112-04-9 see OBI000
108-99-6 see PIB920	110-46-3 see IMB000	112-05-0 see NMY000
109-01-3 see MOD250	110-49-6 see EJJ500	112-07-2 see BPM000
109-02-4 see MMA250	110-54-3 see HEN000	112-12-9 see UKS000

112-14-1	see OEG000	116-15-4	see HDF000	121-25-5	see AOD175
112-15-2	see CBQ750	116-16-5	see HCL500	121-29-9	see POO100
112-23-2	see HBO500	116-29-0	see CKM000	121-32-4	see EQF000
112-24-3	see TJR000	116-45-0	see SNH875	121-33-5	see VFK000
112-27-6	see TJQ000	116-54-1	see DEM800	121-39-1	see EOK600
112-30-1	see DAI600	116-85-8	see AKE250	121-43-7	see TLN000
112-31-2	see DAG000	117-10-2	see DMH400	121-44-8	see TJO000
112-31-2	see DAG200	117-39-5	see QCA000	121-45-9	see TMD500
112-32-3	see OEY100	117-79-3	see AIB000	121-46-0	see NNG000
112-34-5	see DJF200	117-80-6	see DFT000	121-54-0	see BEN000
112-36-7	see DIW800	117-81-7	see DVL700	121-59-5	see CBJ000
112-42-5	see UNA000	117-82-8	see DOF400	121-66-4	see ALQ000
112-44-7	see UJJ000	117-84-0	see DVL600	121-69-7	see DQF800
112-45-8	see ULJ000	118-00-3	see GLS000	121-73-3	see CJB250
112-50-5	see EFL000	118-02-5	see DVF300	121-75-5	see MAK700
112-53-8	see DXV600	118-29-6	see HMP100	121-79-9	see PNM750
112-54-9	see DXT000	118-48-9	see IHN200	121-82-4	see CPR800
112-55-0	see LBX000	118-52-5	see DFE200	121-86-8	see CJG800
112-57-2	see TCE500	118-55-8	see PGG750	122-00-9	see MFW250
112-60-7	see TCE250	118-58-1	see BFJ750	122-03-2	see COE500
112-61-8	see MJW000	118-71-8	see MAO350	122-04-3	see NFK100
112-62-9	see OHW000	118-74-1	see HCC500	122-09-8	see DTJ400
112-73-2	see DDW200	118-92-3	see API500	122-10-1	see SOY000
112-80-1	see OHU000	118-93-4	see HIN500	122-11-2	see SNN300
112-90-3	see OHM700	118-96-7	see TMN490	122-14-5	see DSQ000
112-92-5	see OAX000	119-12-0	see POP000	122-19-0	see DTC600
112-96-9	see OBG000	119-27-7	see DUP800	122-34-9	see BJP000
113-18-8	see CHG000	119-34-6	see NEM480	122-39-4	see DVX800
113-38-2	see EDR000	119-36-8	see MPI000	122-40-7	see AOG500
113-45-1	see MNQ000	119-38-0	see DSK200	122-42-9	see CBM000
113-48-4	see OES000	119-53-9	see BCP250	122-43-0	see BQJ350
113-52-0	see DLH630	119-60-8	see DGV600	122-48-5	see VFP100
113-92-8	see TAI500	119-61-9	see BCS250	122-51-0	see ENY500
114-03-4	see HOA575	119-65-3	see IRX000	122-52-1	see TJT800
114-03-4	see HON800	119-84-6	see HHR500	122-56-5	see THX500
114-07-8	see EDH500	119-90-4	see DCJ200	122-57-6	see SMS500
114-49-8	see HOT500	119-93-7	see TGJ750	122-60-1	see PFF360
114-80-7	see POD000	120-02-5	see CBI250	122-62-3	see BJS250
115-02-6	see ASA500	120-08-1	see DRS800	122-66-7	see HHG000
115-07-1	see PMO500	120-12-7	see APG500	122-67-8	see IIQ000
115-09-3	see MDD750	120-20-7	see DOE200	122-72-5	see HHP500
115-10-6	see MJW500	120-36-5	see DGB000	122-74-7	see HHQ550
115-11-7	see IIC000	120-40-1	see BKE500	122-78-1	see BBL500
115-21-9	see EPY500	120-46-7	see PFU300	122-88-3	see CJN000
115-25-3	see CPS000	120-47-8	see HJL000	122-97-4	see HHP050
115-26-4	see BJE750	120-51-4	see BCM000	122-98-5	see AOR750
115-28-6	see CDS000	120-57-0	see PIW250	122-99-6	see PER000
115-29-7	see EAQ750	120-58-1	see IRZ000	123-00-2	see AMF250
115-31-1	see IHZ000	120-61-6	see DUE000	123-03-5	see CCX000
115-32-2	see BIO750	120-71-8	see MGO750	123-05-7	see BRI000
115-38-8	see ENB500	120-72-9	see ICM000	123-07-9	see EOE100
115-58-2	see PBS250	120-78-5	see BDE750	123-11-5	see AOT530
115-77-5	see PBB750	120-80-9	see CCP850	123-19-3	see DWT600
115-86-6	see TMT750	120-82-1	see TIK250	123-20-6	see VNF000
115-90-2	see FAQ800	120-83-2	see DFX800	123-23-9	see SNC500
115-93-5	see CQL250	120-92-3	see CPW500	123-25-1	see SNB000
115-95-7	see LFY600	120-94-5	see MPB250	123-28-4	see TFD500
115-96-8	see CGO500	120-95-6	see DCI000	123-29-5	see ENW000
116-06-3	see CBM500	121-14-2	see DVH000	123-30-8	see ALT250
116-09-6	see ABC000	121-17-5	see NFS700	123-31-9	see HIH000
116-14-3	see TCH500	121-19-7	see HMY000	123-32-0	see DTU600

123-33-1 see DMC600	127-20-8 see DGI600	134-62-3 see DKC800
123-35-3 see MRZ150	127-25-3 see MFT500	134-71-4 see EAX500
123-38-6 see PMT750	127-33-3 see MIJ500	134-72-5 see EAY500
123-39-7 see MKG500	127-47-9 see VSK900	134-84-9 see MHF750
123-42-2 see DBF750	127-58-2 see SJW475	135-02-4 see AOT525
123-51-3 see IHP000	127-65-1 see CDP000	135-12-6 see CJD600
123-54-6 see ABX750	127-69-5 see SNN500	135-19-3 see NAX000
123-61-5 see BBP000	127-79-7 see ALF250	135-20-6 see ANO500
123-62-6 see PMV500	127-85-5 see ARA500	135-51-3 see ROF300
123-63-7 see PAI250	127-91-3 see POH750	135-87-5 see BCI500
123-66-0 see EHF000	128-04-1 see SGM500	135-88-6 see PFT500
123-68-2 see AGA500	128-08-5 see BOF500	135-98-8 see BQJ000
123-72-8 see BSU250	128-37-0 see BFW750	136-23-2 see BIX000
123-73-9 see COB260	128-44-9 see SJN700	136-30-1 see SGF500
123-75-1 see PPS500	128-46-1 see DME000	136-35-6 see DWO800
123-77-3 see ASM300	128-53-0 see MAL250	136-40-3 see PDC250
123-86-4 see BPU750	128-58-5 see JAT000	136-60-7 see BQK250
123-88-6 see MEP250	128-80-3 see BLK000	136-77-6 see HFV500
123-91-1 see DVQ000	128-93-8 see BNN550	136-92-5 see DJD400
123-92-2 see IHO850	129-00-0 see PON250	137-05-3 see MIQ075
123-93-3 see MCM750	129-06-6 see WAT220	137-06-4 see TGP000
124-02-7 see DBI600	129-15-7 see MMG000	137-07-5 see AIF500
124-04-9 see AEN250	129-16-8 see MCV000	137-08-6 see CAU750
124-05-0 see EIQ000	129-49-7 see MQP500	137-17-7 see TLG250
124-06-1 see ENL850	129-66-8 see TML000	137-26-8 see TFS350
124-07-2 see OCY000	129-74-8 see BOM250	137-30-4 see BJK500
124-09-4 see HEO000	129-79-3 see TMM250	137-32-6 see MHS750
124-13-0 see OCO000	130-15-4 see NBA500	137-40-6 see SJL500
124-16-3 see BPL500	130-26-7 see CHR500	137-58-6 see DHK400
124-17-4 see BQP500	130-80-3 see DKB000	137-97-3 see DXP600
124-18-5 see DAG400	130-89-2 see QJS000	138-22-7 see BRR600
124-19-6 see NMW500	130-95-0 see QHJ000	138-85-2 see SHU000
124-22-1 see DXW000	131-11-3 see DTR200	138-86-3 see MCC250
124-30-1 see OBC000	131-14-6 see APK850	138-89-6 see DSY600
124-38-9 see CBU250	131-16-8 see DWV500	139-02-6 see SJF000
124-40-3 see DOQ800	131-17-9 see DBL200	139-05-9 see SGC000
124-41-4 see SIK450	131-18-0 see AON300	139-06-0 see CAR000
124-43-6 see HIB500	131-52-2 see SJA000	139-07-1 see BEM000
124-47-0 see UTJ000	131-56-6 see DMI600	139-08-2 see TCA500
124-48-1 see CFK500	131-57-7 see MES000	139-13-9 see AMT500
124-65-2 see HKC500	131-70-4 see MRF525	139-25-3 see MJN750
124-87-8 see PIE500	131-73-7 see HET500	139-33-3 see EIX500
125-04-2 see HHR000	131-74-8 see ANS500	139-40-2 see PMN850
125-12-2 see IHX600	131-79-3 see FAG135	139-65-1 see TFI000
125-28-0 see DKW800	132-17-2 see TNU000	139-70-8 see CMU050
125-33-7 see DBB200	132-27-4 see BGJ750	139-94-6 see ENV500
126-07-8 see GKE000	132-32-1 see AJV000	140-11-4 see BDX000
126-17-0 see POJ000	132-53-6 see NLB500	140-26-1 see PDF775
126-52-3 see EEH000	132-69-4 see BBW500	140-29-4 see PEA750
126-64-7 see LFZ000	133-06-2 see CBG000	140-31-8 see AKB000
126-72-7 see TNC500	133-07-3 see TIT250	140-39-6 see MNR250
126-73-8 see TIA250	133-10-8 see SEP000	140-40-9 see ABY900
126-75-0 see DAP200	133-14-2 see BIX750	140-49-8 see CEC000
126-85-2 see CFA500	133-18-6 see APJ500	140-56-7 see DOU600
126-92-1 see TAV750	133-32-4 see ICP000	140-57-8 see SOP500
126-96-5 see SGE400	133-55-1 see DRO400	140-67-0 see AFW750
126-98-7 see MGA750	133-58-4 see NHK900	140-79-4 see DVF200
126-99-8 see NCI500	134-03-2 see ARN125	140-82-9 see DHQ100
127-09-3 see SEG500	134-20-3 see APJ250	140-88-5 see EFT000
127-18-4 see PCF275	134-29-2 see AOX250	141-00-4 see CAI750
127-19-5 see DOO800	134-32-7 see NBE700	141-01-5 see FBJ100

327-98-0 see EPY000	431-03-8 see BOT500	499-75-2 see CCM000
328-38-1 see LER000	434-07-1 see PAN100	500-38-9 see NBR000
329-01-1 see TKJ250	438-41-5 see MDQ250	500-92-5 see CKB250
330-54-1 see DXQ500	438-67-5 see EDV600	501-53-1 see BEF500
330-55-2 see DGD600	439-14-5 see DCK759	502-39-6 see MLF250
333-18-6 see EIW000	440-17-5 see TKK250	502-55-6 see BJU000
333-20-0 see PLV750	443-48-1 see MMN250	502-85-2 see HJS500
333-25-5 see DEW000	443-79-8 see IKX010	503-01-5 see ILK000
333-29-9 see DXN600	446-72-0 see GCM350	503-17-3 see COC500
333-41-5 see DCM750	446-86-6 see ASB250	503-30-0 see OMW000
334-22-5 see BHN750	451-40-1 see PEB000	503-38-8 see TIR920
334-48-5 see DAH400	457-60-3 see NCJ500	503-74-2 see ISU000
334-88-3 see DCP800	460-07-1 see ACB250	504-15-4 see MPH500
335-76-2 see PCG725	460-19-5 see COO000	504-20-1 see PGW250
337-47-3 see SOX500	461-72-3 see HGO600	504-24-5 see AMI500
339-43-5 see BSM000	461-89-2 see THR750	504-29-0 see AMI000
340-56-7 see MDT250	462-06-6 see FGA000	505-57-7 see HFA500
341-69-5 see OJW000	462-08-8 see AMI250	505-60-2 see BIH250
342-69-8 see MPU000	462-27-1 see FIH100	505-66-8 see HGI900
350-03-8 see ABI000	462-95-3 see EFT500	506-61-6 see PLS250
352-32-9 see FMC000	463-04-7 see AOL500	506-63-8 see DQR200
352-93-2 see EPH000	463-51-4 see KEU000	506-64-9 see SDP000
353-16-2 see FHC200	463-58-1 see CCC000	506-68-3 see COO500
353-36-6 see FIB000	463-71-8 see TFN500	506-77-4 see COO750
353-42-4 see BMH000	463-82-1 see NCH000	506-78-5 see COP000
353-50-4 see CCA500	464-10-8 see NMQ000	506-82-1 see DQW800
353-59-3 see BNA250	464-41-5 see BMD300	506-85-4 see FOS050
354-32-5 see TJX500	464-45-9 see NCQ820	506-87-6 see ANE000
357-57-3 see BOL750	465-42-9 see CBF760	506-93-4 see GLA000
359-06-8 see FFR000	465-73-6 see IKO000	506-96-7 see ACD750
359-83-1 see DOQ400	466-24-0 see PIC250	507-02-8 see ACO500
360-89-4 see OBO000	469-59-0 see JCS000	507-09-5 see TFA500
361-09-1 see SFW000	469-62-5 see DAB879	507-19-7 see BQM250
363-03-1 see PEL750	469-79-4 see KFK000	507-20-0 see BQR000
363-24-6 see DVJ200	470-90-6 see CDS750	507-28-8 see TEA300
366-18-7 see BGO500	471-46-5 see OLO000	507-70-0 see BMD000
366-70-1 see PME500	477-30-5 see MIW500	509-09-1 see PBF250
366-93-8 see BHN000	478-43-3 see RHZ700	509-14-8 see TDY250
367-51-1 see SKH500	478-84-2 see BNM250	510-15-6 see DER000
368-43-4 see BBT250	479-13-0 see COF350	511-55-7 see PEM750
368-68-3 see DKG100	479-20-9 see ARQ600	512-24-3 see CCZ000
368-97-8 see DKI400	479-45-8 see TEG250	512-48-1 see DJU200
370-81-0 see COF675	479-50-5 see DHU000	512-56-1 see TMD250
371-40-4 see FFY000	479-92-5 see INY000	512-85-6 see ARM500
371-62-0 see FIE000	480-16-0 see MRN500	513-37-1 see IKE000
371-86-8 see PHF750	480-54-6 see RFP000	513-42-8 see IMW000
373-02-4 see NCX000	485-50-7 see CBF550	513-48-4 see IEH000
375-22-4 see HAX500	488-17-5 see DNE000	513-77-9 see BAJ250
379-79-3 see EDC500	489-98-5 see PIC800	513-78-0 see CAD800
389-08-2 see EID000	490-79-9 see GCU000	513-85-9 see BOT000
399-24-6 see FKQ100	492-18-2 see SIH500	513-86-0 see ABB500
402-71-1 see THH450	492-41-1 see NNM000	514-73-8 see DJT800
404-82-0 see PDM250	492-80-8 see IBB000	514-78-3 see CBE800
406-90-6 see TKB250	493-52-7 see CCE500	514-85-2 see VSK975
407-99-8 see FLR100	494-03-1 see BIF250	516-95-0 see EBA100
408-35-5 see SIZ025	494-38-2 see BJF000	517-16-8 see EME500
409-21-2 see SCQ000	495-48-7 see ASO750	517-25-9 see TMM500
420-04-2 see COH500	496-67-3 see BNP750	518-47-8 see FEW000
420-12-2 see EJP500	497-18-7 see CBS500	520-45-6 see MFW500
427-45-2 see TNG050	497-19-8 see SFO000	520-52-5 see PHU500
428-59-1 see HDF050	497-56-3 see DUT000	522-00-9 see DIR000

522-23-6 see MDU750	541-59-3 see MAM750	557-48-2 see NMV760
523-87-5 see DYE600	541-69-5 see PEY750	557-98-2 see CKS000
524-40-3 see RJK100	541-73-1 see DEP599	557-99-3 see ACM000
525-02-0 see BEM750	541-85-5 see EGI750	558-13-4 see CBX750
526-73-8 see TLL500	541-91-3 see MIT625	563-12-2 see EEH600
526-75-0 see XKJ000	542-46-1 see CMU850	563-25-7 see DDY800
527-07-1 see SHK800	542-55-2 see IIR000	563-41-7 see SBW500
527-09-3 see CNM100	542-56-3 see IJD000	563-45-1 see MHT250
528-29-0 see DUQ400	542-62-1 see BAK750	563-46-2 see MHT000
529-65-7 see EFQ500	542-63-2 see DIV000	563-47-3 see CIU750
530-43-8 see CDP700	542-75-6 see DGG950	563-68-8 see TEI250
531-18-0 see HDY000	542-88-1 see BIK000	563-80-4 see MLA750
531-76-0 see BHT750	542-90-5 see EPP000	564-00-1 see DHB800
531-82-8 see AAL750	542-92-7 see CPU500	568-75-2 see HMF000
531-85-1 see BBX750	543-49-7 see HBE500	569-57-3 see CLO750
531-86-2 see BBY000	543-59-9 see PBW500	569-58-4 see AGW750
532-27-4 see CEA750	543-80-6 see BAH500	569-61-9 see RMK020
532-28-5 see MAP250	543-81-7 see BFP000	569-65-3 see HGC500
532-32-1 see SFB000	543-82-8 see ILM000	572-48-5 see DXO000
532-34-3 see BRT000	543-90-8 see CAD250	573-58-0 see SGQ500
532-43-4 see TET500	544-13-8 see TLR500	577-11-7 see DJL000
532-82-1 see PEK000	544-16-1 see BRV500	578-54-1 see EGK500
533-73-3 see BBU250	544-17-2 see CAS250	578-94-9 see PDB000
533-74-4 see DSB200	544-25-2 see COY000	581-89-5 see NHQ500
534-07-6 see BIK250	544-63-8 see MSA250	582-25-2 see PKW760
534-13-4 see DSK900	544-92-3 see CNL000	582-61-6 see BDL750
534-15-6 see DOO600	545-06-2 see TII750	583-03-9 see BQJ500
534-17-8 see CDC750	545-55-1 see TND250	583-15-3 see MCX500
534-22-5 see MKH000	546-46-3 see ZFJ250	583-39-1 see BCC500
534-52-1 see DUS700	546-80-5 see TFW000	583-58-4 see LJB000
536-17-4 see DOT800	546-89-4 see LGO100	583-60-8 see MIR500
536-29-8 see DFX400	548-00-5 see BKA000	583-63-1 see BDC250
536-59-4 see PCI550	548-62-9 see AOR500	584-02-1 see IHP010
536-69-6 see BSI000	549-18-8 see EAI000	584-03-2 see BOS250
537-00-8 see CCY500	551-08-6 see BRQ100	584-08-7 see PLA000
538-07-8 see BID250	551-11-1 see POC500	584-79-2 see AFR250
538-23-8 see TMO000	551-74-6 see MAW750	584-84-9 see TGM750
538-28-3 see BEU500	552-30-7 see TKV000	585-54-6 see AQZ900
538-71-6 see DXX000	553-84-4 see PCI750	586-62-9 see TBE000
538-93-2 see IIN000	553-97-9 see MHI250	587-85-9 see DWD800
539-17-3 see DPO200	554-12-1 see MOT000	587-98-4 see MDM775
539-88-8 see EFS600	554-13-2 see LGZ000	588-59-0 see SLR000
539-90-2 see BSW500	554-84-7 see NIE600	589-16-2 see EGL000
540-09-0 see TJF250	555-30-6 see DNA800	589-98-0 see OCY100
540-18-1 see AOG000	555-43-1 see GGU400	590-00-1 see PLS750
540-23-8 see TGS750	555-84-0 see NDY000	590-01-2 see BSJ500
540-42-1 see PMV250	556-24-1 see ITC000	590-21-6 see PMR750
540-51-2 see BNI500	556-52-5 see GGW500	590-28-3 see PLC250
540-54-5 see CKP750	556-56-9 see AGI250	590-92-1 see BOB250
540-59-0 see DFI210	556-61-6 see ISE000	591-27-5 see ALS990
540-63-6 see EEB000	556-64-9 see MPT000	591-60-6 see BPV250
540-67-0 see EMT000	556-88-7 see NHA500	591-78-6 see HEV000
540-72-7 see SIA500	556-89-8 see NMQ500	591-87-7 see AFU750
540-73-8 see DSF600	557-04-0 see MAJ030	591-89-9 see PLU500
540-84-1 see TLY500	557-05-1 see ZMS000	592-01-8 see CAQ500
540-88-5 see BPV100	557-17-5 see MOU830	592-04-1 see MDA250
541-09-3 see UPS000	557-18-6 see DJO100	592-05-2 see LCU000
541-25-3 see CLV000	557-19-7 see NDB500	592-31-4 see BSS250
541-41-3 see EHK500	557-20-0 see DKE600	592-35-8 see BQP250
541-42-4 see IQQ000	557-21-1 see ZGA000	592-41-6 see HFB000
541-53-7 see DXL800	557-40-4 see DBK000	592-42-7 see HCR500

592-62-1 see MGS750	613-37-6 see PEG250	627-93-0 see DOQ300
592-76-7 see HBJ000	613-94-5 see BBV250	628-02-4 see HEM500
592-84-7 see BRK000	614-00-6 see MMU250	628-28-4 see BRU780
592-85-8 see MCU250	614-34-6 see TGX100	628-32-0 see EPC125
593-29-3 see PLS775	614-45-9 see BSC500	628-52-4 see CMJ000
593-53-3 see FJK000	614-68-6 see IKG725	628-63-7 see AOD725
593-60-2 see VMP000	614-95-9 see NKE500	628-73-9 see HER500
593-70-4 see CHI900	615-05-4 see DBO000	628-81-9 see EHA500
593-74-8 see DSM450	615-15-6 see MHC250	628-83-1 see BSN500
593-82-8 see DSG000	615-45-2 see DCE200	628-85-3 see DWU000
593-89-5 see DFP200	615-50-9 see DCE600	628-86-4 see MDC000
594-27-4 see TDV750	615-53-2 see MMX250	628-94-4 see AEN000
594-31-0 see DGO800	615-65-6 see CLK210	628-96-6 see EJG000
594-42-3 see PCF300	615-66-7 see CEG600	629-13-0 see DCL600
594-72-9 see DFU000	615-67-8 see CHM000	629-14-1 see EJE500
595-33-5 see VTF000	616-23-9 see DGG450	629-15-2 see EJF000
596-03-2 see DDO200	616-38-6 see MIF000	629-17-4 see EJC035
596-51-0 see GIC000	616-91-1 see ACH000	629-35-6 see DEE000
597-64-8 see TCF750	617-79-8 see EHA000	630-08-0 see CBW750
598-14-1 see DFH200	617-89-0 see FPW000	630-10-4 see SBV000
598-31-2 see BNZ000	618-25-7 see CBJ750	630-20-6 see TBQ000
598-55-0 see MHZ000	619-01-2 see DKV150	630-60-4 see OKS000
598-57-2 see NHN500	619-15-8 see DVH200	630-72-8 see TMK250
598-58-3 see MMF500	621-64-7 see NKB700	630-93-3 see DNU000
598-63-0 see LCP000	621-82-9 see CMP975	631-60-7 see MDE250
598-73-2 see BOJ000	622-44-6 see PFJ400	631-61-8 see ANA000
598-74-3 see AOE200	622-45-7 see CPF000	632-22-4 see TDX250
598-78-7 see CKS750	622-62-8 see EFA100	632-69-9 see RMP175
600-14-6 see PBL350	622-78-6 see BEU250	633-03-4 see BAY750
600-25-9 see CJE000	623-07-4 see CHW750	634-66-2 see TBN740
600-40-8 see DUV710	623-26-7 see BBP250	635-22-3 see CJA185
601-77-4 see NKA000	623-42-7 see MHY000	635-65-4 see HAO900
602-01-7 see DVG800	623-68-7 see COB900	636-21-5 see TGS500
602-38-0 see DUX710	623-70-1 see COB750	636-23-7 see DCE000
602-87-9 see NEJ500	623-73-4 see DCN800	637-03-6 see PEG750
603-34-9 see TMQ500	624-46-4 see MKA750	638-21-1 see PFV250
603-35-0 see TMU000	624-54-4 see AON350	638-23-3 see CBR675
603-36-1 see TMV250	624-61-3 see DDJ800	638-29-9 see VBA000
605-65-2 see DPN200	624-74-8 see DNE500	638-38-0 see MAQ000
605-69-6 see DUX800	624-83-9 see MKX250	638-49-3 see AOJ500
605-71-0 see DUX700	624-91-9 see MMF750	639-58-7 see CLU000
606-20-2 see DVH400	624-92-0 see DRQ400	640-15-3 see PHI500
606-23-5 see IBS000	625-22-9 see DEC000	640-19-7 see FFF000
606-37-1 see DUX650	625-45-6 see MDW275	641-16-7 see TDY600
607-35-2 see NJD500	625-52-5 see EQD875	642-65-9 see DBF200
607-57-8 see NGB000	625-53-6 see EPR600	644-06-4 see AEX850
607-88-5 see THD850	625-55-8 see IPC000	644-31-5 see ACC250
608-71-9 see PAU250	625-58-1 see ENM500	644-35-9 see PNS250
608-73-1 see BBP750	625-84-3 see DTV300	644-97-3 see DGE400
608-93-5 see PAV500	626-17-5 see PHX550	645-48-7 see PGM750
609-93-8 see DUT600	626-38-0 see AOD735	646-06-0 see DVR800
609-99-4 see HKE600	626-48-2 see MQI750	657-24-9 see DQR600
610-39-9 see DVH600	626-67-5 see MOG500	657-27-2 see LJO000
611-13-2 see MKH600	626-82-4 see BRK900	659-70-1 see ITB000
611-95-0 see BDL860	627-03-2 see EEK500	665-66-7 see AED250
612-12-4 see DGP400	627-11-2 see CGU199	673-06-3 see PEC500
612-64-6 see NKD000	627-12-3 see PNG250	673-31-4 see PGA750
612-82-8 see DQM000	627-13-4 see PNQ500	674-81-7 see NKH000
612-83-9 see DEQ800	627-30-5 see CKP600	674-82-8 see KFA000
613-13-8 see APG100	627-44-1 see DJO400	676-83-5 see MOC250
613-35-4 see BFX000	627-63-4 see FOY000	676-97-1 see MOB399

1125-27-5 see DFJ800	1304-82-1 see BKY000	1317-65-3 see CAO000
1125-88-8 see DOG700	1305-62-0 see CAT225	1317-70-0 see OBU100
1126-78-9 see BQH850	1305-78-8 see CAU500	1319-77-3 see CNW500
1126-79-0 see BSF750	1305-79-9 see CAV500	1320-37-2 see DGL600
1128-05-8 see MPQ900	1305-99-3 see CAW250	1321-31-9 see EEL100
1129-41-5 see MIB750	1306-19-0 see CAH500	1321-64-8 see PAW750
1132-20-3 see CEC100	1306-23-6 see CAJ750	1321-65-9 see TIT500
1132-39-4 see PGH250	1307-86-4 see CNC233	1321-74-0 see DXQ740
1137-41-3 see AIR250	1307-96-6 see CND125	1322-98-1 see DAJ000
1138-80-3 see CBR125	1308-04-9 see CND825	1327-53-3 see ARI750
1139-30-6 see CCN100	1308-14-1 see CMH260	1329-86-8 see THV000
1141-88-4 see DXJ800	1308-31-2 see CMI500	1330-20-7 see XGS000
1145-73-9 see DUB800	1308-38-9 see CMJ900	1330-43-4 see DXG035
1155-38-0 see MGZ000	1309-32-6 see COE000	1330-78-5 see TNP500
1162-65-8 see AEU250	1309-37-1 see IHC450	1331-22-2 see MIR250
1163-19-5 see PAU500	1309-42-8 see MAG750	1332-21-4 see ARM250
1165-39-5 see AEV000	1309-48-4 see MAH500	1332-29-2 see TGE300
1166-52-5 see DXX200	1309-60-0 see LCX000	1332-37-2 see IHG100
1177-87-3 see DBC400	1309-64-4 see AQF000	1332-58-7 see KBB600
1178-29-6 see MQR200	1310-53-8 see GDS000	1333-39-7 see HJH500
1184-57-2 see MLG000	1310-53-8 see GEC000	1333-74-0 see HHW500
1187-00-4 see BKM500	1310-58-3 see PLJ500	1333-82-0 see CMK000
1187-59-3 see MGA300	1310-65-2 see LHI100	1333-83-1 see SHQ500
1189-85-1 see BQV000	1310-73-2 see SHS000	1333-86-4 see CBT750
1191-15-7 see DNI600	1310-73-2 see SHS500	1335-31-5 see MDA500
1191-16-8 see DOQ350	1310-82-3 see RPZ000	1335-32-6 see LCH000
1191-50-0 see SIO000	1312-73-8 see PLT250	1335-87-1 see HCK500
1191-79-3 see BAI800	1313-13-9 see MAS000	1335-88-2 see TBR000
1191-80-6 see MDF250	1313-27-5 see MRE000	1336-21-6 see ANK250
1192-75-2 see BJP450	1313-60-6 see SJC500	1336-36-3 see PJL750
1193-54-0 see DFN800	1313-82-2 see SJY500	1336-80-7 see FBC100
1197-16-6 see CCI550	1313-85-5 see SJT000	1338-02-9 see NAS000
1197-40-6 see FPX028	1313-97-9 see NCC000	1338-23-4 see MKA500
1198-77-2 see SKO575	1313-99-1 see NDF500	1338-24-5 see NAR000
1208-52-2 see MJP750	1314-06-3 see NDH500	1338-39-2 see SKV000
1215-16-3 see BHO500	1314-13-2 see ZKA000	1338-41-6 see SKV150
1218-34-4 see ADE075	1314-18-7 see SMK500	1338-43-8 see SKV100
1229-55-6 see CMS238	1314-20-1 see TFT750	1341-24-8 see CDN505
1239-31-2 see HJB225	1314-22-3 see ZLJ000	1341-49-7 see ANJ000
1239-45-8 see DBV400	1314-32-5 see TEL050	1343-90-4 see MAJ000
1241-94-7 see DWB800	1314-34-7 see VEA000	1344-00-9 see SEM000
1264-72-8 see PKD300	1314-35-8 see TOC750	1344-28-1 see AHE250
1271-19-8 see DGW200	1314-36-9 see YGA000	1344-40-7 see LCV100
1271-28-9 see NDA500	1314-41-6 see LDS000	1344-43-0 see MAT250
1271-54-1 see BGY700	1314-56-3 see PHS250	1344-67-8 see CNJ950
1271-55-2 see ABA750	1314-60-9 see AQF750	1344-95-2 see CAW850
1291-32-3 see ZTK400	1314-61-0 see TAF500	1345-04-6 see AQL500
1300-71-6 see XKA000	1314-62-1 see VDU000	1390-65-4 see CCK590
1300-73-8 see XMA000	1314-62-1 see VDZ000	1397-89-3 see AOC500
1302-52-9 see BFO500	1314-80-3 see PHS000	1400-61-9 see NOH500
1302-78-9 see BAV750	1314-84-7 see ZLS000	1401-55-4 see CDM250
1303-00-0 see GBK000	1314-85-8 see PHS500	1401-55-4 see MQV250
1303-18-0 see ARJ750	1314-87-0 see LDZ000	1401-55-4 see TAD750
1303-28-2 see ARH500	1314-96-1 see SMM000	1401-69-0 see TOE600
1303-33-9 see ARI000	1315-04-4 see AQF500	1402-68-2 see AET750
1303-39-5 see ZDJ000	1317-34-6 see MAT500	1403-17-4 see LFF000
1303-86-2 see BMG000	1317-35-7 see MAV550	1403-66-3 see GCO000
1303-96-4 see SFF000	1317-36-8 see LDN000	1404-04-2 see NCE000
1304-28-5 see BAO000	1317-39-1 see CNO000	1405-10-3 see NCG000
1304-29-6 see BAO250	1317-42-6 see CNE200	1405-41-0 see GCS000
1304-56-9 see BFT250	1317-60-8 see HAO875	1405-86-3 see GIG000

1405-87-4 see BAC250	1707-95-5 see ONY000	2094-99-7 see IKG800
1406-05-9 see PAQ000	1708-39-0 see BBA000	2104-64-5 see EBD700
1406-11-7 see PKC000	1709-70-2 see TMJ000	2114-33-2 see ABU800
1406-65-1 see CKN000	1712-64-7 see IQP000	2130-56-5 see BFX250
1407-03-0 see AMY700	1719-53-5 see DEY800	2155-70-6 see THZ000
1421-63-2 see TKO250	1722-62-9 see CBR250	2162-74-5 see BJE550
1455-77-2 see DCF200	1733-25-1 see IRN100	2163-80-6 see MRL750
1456-28-6 see DTA000	1738-25-6 see DPU000	2164-17-2 see DUK800
1461-22-9 see CLP500	1746-01-6 see TAI000	2169-75-7 see EAG100
1461-25-2 see TBM250	1746-77-6 see IOJ000	2179-59-1 see AGR500
1464-43-3 see DWY800	1746-81-2 see CKD500	2191-10-8 see CAD750
1464-53-5 see BGA750	1754-58-1 see PEV500	2198-61-0 see IHU100
1467-79-4 see DRF600	1762-95-4 see ANW750	2210-25-5 see INH000
1476-23-9 see AGJ000	1763-23-1 see HAS075	2216-51-5 see MCG250
1476-53-5 see NOB000	1777-84-0 see NEL000	2218-96-4 see DUR425
1477-19-6 see BBJ500	1779-25-5 see CGB500	2223-93-0 see OAT000
1477-55-0 see XHS800	1789-58-8 see DFK000	2234-13-1 see OAP000
1492-93-9 see BHP750	1797-74-6 see PMS500	2238-07-5 see DKM200
1493-13-6 see TKB310	1836-75-5 see DFT800	2243-62-1 see NAM000
1498-40-4 see EOQ000	1837-57-6 see EDW500	2244-11-3 see MDL500
1498-51-7 see EOR000	1838-59-1 see AGH000	2244-16-8 see CCM100
1500-94-3 see ACI550	1866-31-5 see AGC000	2244-21-5 see PLD000
1516-32-1 see BSO500	1867-66-9 see CKD750	2273-43-0 see BSL500
1528-74-1 see DUS000	1885-14-9 see CBX109	2273-45-2 see DTH400
1538-09-6 see BFC750	1888-71-7 see HCM000	2274-11-5 see EIP000
1563-66-2 see CBS275	1892-29-1 see DXM600	2275-14-1 see PDC750
1569-69-3 see CPB625	1893-33-0 see FHG000	2277-19-2 see NNA325
1570-45-2 see ELU000	1897-45-6 see TBQ750	2278-50-4 see CMU475
1582-09-8 see DUV600	1907-13-7 see ABW750	2279-64-3 see PFP500
1592-23-0 see CAX350	1912-24-9 see ARQ725	2279-76-7 see CLU250
1596-84-5 see DQD400	1912-28-3 see MJW250	2294-47-5 see DCL125
1600-27-7 see MCS750	1918-00-9 see MEL500	2303-16-4 see DBI200
1606-83-3 see BST900	1918-02-1 see PIB900	2310-17-0 see BDJ250
1609-47-8 see DIZ100	1918-13-4 see DGM600	2312-35-8 see SOP000
1615-80-1 see DJL400	1929-82-4 see CLP750	2312-76-7 see DUU600
1622-79-3 see FKI000	1934-21-0 see FAG140	2318-18-5 see DMX200
1623-24-1 see PHE500	1936-15-8 see HGC000	2321-07-5 see FEV000
1624-01-7 see OCM100	1937-37-7 see AQP000	2338-05-8 see FAW100
1624-02-8 see BLS750	1942-78-5 see EMR100	2338-12-7 see NFJ000
1628-58-6 see DQC400	1943-83-5 see IKH000	2349-07-7 see HFQ550
1629-58-9 see PBR250	1948-33-0 see BRM500	2373-98-0 see DMI400
1632-16-2 see EKR500	1949-20-8 see OOE000	2385-85-5 see MQW500
1633-83-6 see BOU250	1951-25-3 see AJK750	2386-25-6 see ACI500
1634-04-4 see MHV859	1955-45-9 see DTH000	2386-52-9 see SDR500
1638-22-8 see BSE450	1972-08-3 see TCM250	2386-90-5 see BJN250
1639-09-4 see HBD500	1976-28-9 see AHD650	2392-39-4 see DAE525
1639-60-7 see PNA500	1977-10-2 see DCS200	2393-53-5 see EDV500
1642-54-2 see DIW200	1983-10-4 see FME000	2398-96-1 see TGB475
1649-08-7 see DFA000	1984-23-2 see IKH780	2401-85-6 see DUS600
1649-18-9 see FLU000	2016-57-1 see DAG600	2409-55-4 see BQV750
1675-54-3 see BLD750	2032-65-7 see DST000	2425-06-1 see CBF800
1679-07-8 see CPW300	2035-99-6 see IHP500	2425-74-3 see BRJ750
1679-09-0 see MHS550	2039-87-4 see CLE750	2426-08-6 see BRK750
1680-21-3 see TJQ100	2050-46-6 see CCP900	2429-74-5 see CMO500
1689-82-3 see HJF000	2050-92-2 see DCH200	2431-50-7 see TIL360
1689-84-5 see DDP000	2058-46-0 see HOI000	2432-99-7 see AMW000
1694-09-3 see FAG120	2058-52-8 see CIS750	2435-76-9 see DCQ600
1698-60-8 see PEE750	2068-78-2 see LEZ000	2436-90-0 see CMT050
1701-71-9 see PNV755	2078-54-8 see DNR800	2439-10-3 see DXX400
1701-73-1 see VBA100	2079-89-2 see AMB750	2440-45-1 see BJT250
1707-14-8 see MNV750	2092-16-2 see CAY250	2463-45-8 see CBW400

3772-26-7 see IRG100	4484-72-4 see DYA800	5329-14-6 see SNK500
3775-90-4 see BQD250	4485-12-5 see LHQ100	5332-73-0 see MFM000
3778-73-2 see IMH000	4511-19-7 see TIY800	5340-36-3 see MND050
3794-64-7 see HAW000	4525-46-6 see BFM750	5344-27-4 see POR500
3810-74-0 see SLY500	4549-40-0 see NKY000	5392-40-5 see DTC800
3810-81-9 see BKS810	4549-43-3 see MMT500	5410-29-7 see NEX500
3811-04-9 see PLA250	4549-44-4 see EHC000	5410-78-6 see AOR640
3811-73-2 see MCQ750	4564-87-8 see CBT250	5412-01-1 see IOJ500
3817-11-6 see HJQ350	4568-81-4 see DSC100	5415-07-6 see OOK200
3819-00-9 see PII500	4568-82-5 see DNF500	5419-55-6 see IOI000
3825-26-1 see ANP625	4568-83-6 see EID250	5421-46-5 see ANM500
3844-63-1 see NKL000	4584-46-7 see DRC000	5421-48-7 see AAS250
3848-24-6 see HEQ200	4587-15-9 see BHX300	5430-13-7 see FGA100
3913-02-8 see BSA500	4591-46-2 see TDY075	5432-28-0 see NKT500
3913-71-1 see DAI350	4593-81-1 see DDY000	5459-93-8 see EHT000
3922-90-5 see OHM900	4621-04-9 see IOO300	5471-51-2 see RBU000
3926-62-3 see SFU500	4638-44-2 see ASI300	5486-03-3 see BOO632
3942-54-9 see CKF000	4657-20-9 see DWC650	5488-45-9 see TIB000
3953-10-4 see EGZ000	4662-17-3 see DNF450	5522-43-0 see NJA000
3958-19-8 see TMQ600	4682-94-4 see FQL200	5536-17-4 see AQQ900
3982-91-0 see TFO000	4685-14-7 see PAI990	5538-94-3 see DTF820
3999-01-7 see LGF900	4691-65-0 see DXE500	5550-12-9 see GLS800
4000-16-2 see THN800	4756-45-0 see HLS500	5566-34-7 see CDR575
4016-11-9 see EBQ700	4756-75-6 see TMQ550	5593-70-4 see BSP250
4016-11-9 see EKM200	4759-48-2 see VSK955	5598-13-0 see CMA250
4016-14-2 see IPD000	4795-29-3 see TCS500	5606-24-6 see BRR800
4028-32-4 see BHB950	4808-30-4 see HCA700	5634-39-9 see IEL800
4044-65-9 see PFA500	4812-22-0 see NHE000	5683-33-0 see DQB400
4062-60-6 see DEC100	4822-44-0 see MCK000	5687-22-9 see ARA100
4075-79-0 see PDY500	4824-78-6 see EGV500	5714-22-7 see SOQ450
4075-81-4 see CAW400	4826-62-4 see DXU280	5716-15-4 see DHF200
4080-31-3 see CEG550	4940-11-8 see EMA600	5743-04-4 see CAD275
4097-89-6 see NEI800	4964-27-6 see DIS850	5743-27-1 see CAM600
4098-71-9 see IMG000	4985-15-3 see DPH600	5760-50-9 see ULS400
4109-96-0 see DGK300	4988-64-1 see MCQ500	5798-79-8 see BMW250
4164-06-1 see ARJ800	4998-76-9 see CPA775	5800-19-1 see MQQ000
4164-07-2 see ARJ760	5001-51-4 see CAT650	5870-82-6 see TJL700
4164-28-7 see DSV200	5029-46-9 see TML100	5894-60-0 see HCQ000
4170-30-3 see COB250	5034-77-5 see IAN000	5902-79-4 see MLF500
4213-45-0 see QDS000	5064-31-3 see SIP500	5902-95-4 see CAM000
4224-87-7 see MIF762	5117-17-9 see DJY800	5903-13-9 see MME809
4230-97-1 see AGM500	5124-30-1 see MJM600	5905-52-2 see LAL000
4253-22-9 see DEI200	5131-60-2 see CJY120	5910-77-0 see TNN760
4261-68-1 see CGV600	5137-55-3 see MQH000	5910-85-0 see HAV450
4288-84-0 see CHS250	5160-02-1 see CHP500	5910-87-2 see NMV775
4342-03-4 see DAB600	5185-71-7 see BIA300	5910-89-4 see DTU400
4342-36-3 see BDR750	5185-76-2 see DIS775	5967-09-9 see BGQ000
4345-03-3 see TGJ060	5185-77-3 see BHW300	5967-73-7 see MDP770
4350-09-8 see HOA600	5185-78-4 see DIP100	5970-32-1 see MCU000
4350-09-8 see HOO000	5185-80-8 see ARJ770	5989-27-5 see LFU000
4356-33-6 see SKS150	5188-42-1 see BJW825	5989-54-8 see MCC500
4362-40-7 see CIL800	5205-11-8 see MHU150	6011-14-9 see GHI100
4368-28-9 see FOQ000	5208-87-7 see BCJ000	6012-97-1 see TBV750
4387-13-7 see BJW800	5216-25-1 see TIR900	6018-89-9 see NCX500
4418-26-2 see SGD000	5234-68-4 see CCC500	6029-87-4 see FOT000
4420-67-1 see VCA100	5275-02-5 see ILG200	6032-29-7 see PBM750
4431-24-7 see EIQ200	5275-69-4 see ACT250	6055-19-2 see CQC500
4435-53-4 see MHV750	5283-66-9 see OGE000	6080-56-4 see LCJ000
4439-24-1 see IIP000	5283-67-0 see NNE000	6104-30-9 see IIV000
4465-94-5 see CQN000	5285-87-0 see TFF600	6109-97-3 see AJV250
4468-02-4 see ZIA750	5307-14-2 see ALL750	6117-91-5 see BOY000

6147-53-1 see CNA500	7128-68-9 see DDV225	7440-37-1 see AQW250
6163-73-1 see TLA600	7149-24-8 see COU510	7440-38-2 see ARA750
6164-98-3 see CJJ250	7157-29-1 see MRR760	7440-39-3 see BAH250
6217-24-9 see DVY100	7177-48-2 see AOD125	7440-41-7 see BFO750
6283-24-5 see ABQ000	7203-92-1 see DSN600	7440-42-8 see BMD500
6285-05-8 see CKT500	7207-97-8 see MGQ775	7440-43-9 see CAD000
6285-34-3 see DAJ450	7209-38-3 see BGV000	7440-44-0 see CBT500
6296-45-3 see CHF500	7220-81-7 see AEU750	7440-45-1 see CCY250
6304-33-2 see TMQ250	7227-91-0 see DTP000	7440-46-2 see CDC000
6305-43-7 see DDK600	7227-92-1 see DSX400	7440-47-3 see CMI750
6317-18-6 see MJT500	7235-40-7 see CCK685	7440-48-4 see CNA250
6318-57-6 see HJE400	7280-37-7 see PIK450	7440-50-8 see CNI000
6325-54-8 see CIG250	7287-19-6 see BKL250	7440-57-5 see GIS000
6332-68-9 see BAL275	7289-52-3 see MIW075	7440-59-7 see HAM500
6334-11-8 see TLH000	7300-34-7 see BGU600	7440-61-1 see UNS000
6358-53-8 see DOK200	7310-87-4 see DJQ300	7440-62-2 see VCP000
6368-72-5 see EOJ500	7320-37-8 see EBX500	7440-63-3 see XDS000
6369-59-1 see TGM400	7328-05-4 see HEW200	7440-64-4 see YDA000
6376-26-7 see DHP200	7339-53-9 see MKN250	7440-65-5 see YEJ000
6423-43-4 see PNL000	7361-61-7 see DMW000	7440-66-6 see ZBJ000
6453-98-1 see FQO050	7361-89-9 see PNI850	7440-67-7 see ZOA000
6459-94-5 see CMM330	7392-96-3 see DDZ000	7440-69-9 see BKU750
6484-52-2 see ANN000	7399-02-2 see DEH650	7440-74-6 see ICF000
6485-34-3 see CAM750	7400-08-0 see CNU825	7446-07-3 see TAJ750
6485-40-1 see CCM120	7411-49-6 see BGK750	7446-08-4 see SBQ500
6512-83-0 see SBU150	7414-83-7 see DXD400	7446-09-5 see SOH500
6533-73-9 see TEJ000	7417-67-6 see MMT000	7446-11-9 see SOR500
6542-37-6 see OMM300	7424-00-2 see FAM100	7446-14-2 see LDY000
6649-23-6 see LFA000	7428-48-0 see LDX000	7446-18-6 see TEM000
6696-47-5 see OHO000	7429-90-5 see AGX000	7446-20-0 see ZNJ000
6707-60-4 see OKW110	7439-89-6 see IGK800	7446-27-7 see LDU000
6746-59-4 see ENK500	7439-91-0 see LAV000	7446-34-6 see SBT000
6795-23-9 see AEW000	7439-92-1 see LCF000	7446-70-0 see AGY750
6804-07-5 see FOI000	7439-93-2 see LGO000	7447-39-4 see CNK500
6834-92-0 see SJU000	7439-95-4 see MAC750	7447-40-7 see PLA500
6837-24-7 see CPQ275	7439-96-5 see MAP750	7447-41-8 see LHB000
6842-15-5 see PMP750	7439-97-6 see MCW250	7452-79-1 see EMP600
6843-30-7 see DOU700	7439-98-7 see MRC250	7487-88-9 see MAJ250
6863-58-7 see BRH760	7440-00-8 see NBX000	7487-94-7 see MCY475
6884-59-9 see SAG100	7440-01-9 see NCG500	7488-56-4 see SBR000
6901-97-9 see IFW000	7440-02-0 see NCW500	7492-70-8 see BQP000
6902-91-6 see GDO200	7440-03-1 see NDZ000	7493-74-5 see AGQ750
6915-15-7 see MAN000	7440-04-2 see OKE000	7495-93-4 see CAD500
6923-22-4 see MRH209	7440-06-4 see PJD500	7496-02-8 see NFT400
6923-52-0 see AQJ750	7440-09-7 see PKT250	7521-80-4 see BSR000
6939-83-9 see TNH850	7440-09-7 see PKT500	7526-26-3 see MDQ825
6959-48-4 see CIV000	7440-16-6 see RHF000	7546-30-7 see MCW000
7011-83-8 see MIW050	7440-17-7 see RPA000	7550-45-0 see TGH350
7027-11-4 see IMG500	7440-18-8 see RRU000	7553-56-2 see IDM000
7046-61-9 see NHI500	7440-21-3 see SCP000	7558-63-6 see MRF000
7047-84-9 see AHA250	7440-22-4 see SDI500	7558-79-4 see SJH090
7060-74-4 see OHO200	7440-23-5 see see500	7558-80-7 see SJH100
7068-83-9 see MHW500	7440-23-5 see SEF500	7568-37-8 see MHY550
7081-44-9 see SLJ000	7440-25-7 see TAE750	7572-29-4 see DEN600
7083-24-1 see BEL525	7440-28-0 see TEI000	7578-36-1 see BMG750
7085-85-0 see EHP700	7440-29-1 see TFS750	7580-67-8 see LHH000
7090-25-7 see NBJ500	7440-31-5 see TGB250	7601-54-9 see SJH200
7101-57-7 see ADJ550	7440-32-6 see TGF250	7601-89-0 see PCE750
7101-58-8 see PIU100	7440-32-6 see TGF500	7601-90-3 see PCD250
7101-64-6 see TCO100	7440-33-7 see TOA750	7616-94-6 see PCF750
7101-65-7 see MRU077	7440-36-0 see AQB750	7631-86-9 see SCI000

7631-89-2 see ARD750	7722-84-1 see HIB050	7782-49-2 see SBP000
7631-90-5 see SFE000	7722-88-5 see TEE500	7782-50-5 see CDV750
7631-95-0 see DXE800	7723-14-0 see PHO500	7782-63-0 see FBO000
7631-98-3 see DXZ000	7723-14-0 see PHP010	7782-64-1 see MAS750
7631-99-4 see SIO900	7726-95-6 see BMP000	7782-65-2 see GEI100
7632-00-0 see SIQ500	7727-15-3 see AGX750	7782-77-6 see NMR000
7632-50-0 see ANF800	7727-18-6 see VDP000	7782-78-7 see NMJ000
7632-51-1 see VEF000	7727-21-1 see DWQ000	7782-79-8 see HHG500
7637-07-2 see BMG700	7727-37-9 see NGP500	7782-89-0 see LGT000
7645-25-2 see ARC750	7727-43-7 see BAP000	7782-92-5 see SEN000
7646-69-7 see SHO500	7727-54-0 see ANR000	7782-99-2 see SOO500
7646-78-8 see TGC250	7733-02-0 see ZNA000	7783-00-8 see SBO000
7646-79-9 see CNB599	7738-94-5 see CMH250	7783-06-4 see HIC500
7646-85-7 see ZFA000	7757-79-1 see PLL500	7783-07-5 see HIC000
7646-93-7 see PKX750	7757-81-5 see SJV000	7783-08-6 see SBN500
7647-01-0 see HHL000	7757-82-6 see SJY000	7783-20-2 see ANU750
7647-01-0 see HHX000	7757-83-7 see SJZ000	7783-28-0 see ANR500
7647-10-1 see PAD500	7757-93-9 see CAW100	7783-30-4 see MDC750
7647-14-5 see SFT000	7758-01-2 see PKY300	7783-33-7 see NCP500
7647-15-6 see SFG500	7758-02-3 see PKY500	7783-35-9 see MDG500
7647-17-8 see CDD000	7758-05-6 see PLK250	7783-40-6 see MAF500
7647-18-9 see AQD000	7758-09-0 see PLM500	7783-41-7 see ORA000
7647-19-0 see PHR750	7758-16-9 see DXF800	7783-42-8 see TFL250
7650-84-2 see PNI600	7758-19-2 see SFT500	7783-46-2 see LDF000
7660-25-5 see LFI000	7758-88-5 see CDA750	7783-47-3 see TGD100
7664-38-2 see PHB250	7758-94-3 see FBI000	7783-48-4 see SMI500
7664-39-3 see HHU500	7758-95-4 see LCQ000	7783-49-5 see ZHS000
7664-41-7 see AMY500	7758-97-6 see LCR000	7783-50-8 see FAX000
7664-93-9 see SOI500	7758-98-7 see CNP250	7783-54-2 see NGW000
7664-93-9 see SOI530	7761-88-8 see SDS000	7783-55-3 see PHQ500
7664-98-4 see DWV000	7764-50-3 see DKV175	7783-56-4 see AQE000
7665-72-7 see BRK800	7772-76-1 see ANR750	7783-60-0 see SOR000
7672-94-8 see CLH800	7772-98-7 see SKI000	7783-61-1 see SDF650
7681-11-0 see PLK500	7772-99-8 see TGC000	7783-64-4 see ZQS000
7681-38-1 see SEG800	7773-01-5 see MAR000	7783-66-6 see IDT000
7681-49-4 see SHF500	7773-06-0 see ANU650	7783-70-2 see AQF250
7681-52-9 see SHU500	7774-29-0 see MDD000	7783-71-3 see TAF250
7681-53-0 see SHV000	7774-29-0 see MDD250	7783-79-1 see SBS000
7681-57-4 see SII000	7774-41-6 see ARC500	7783-80-4 see TAK250
7681-82-5 see SHW000	7774-82-5 see TJJ400	7783-81-5 see UOJ000
7681-93-8 see PIF750	7775-09-9 see SFS000	7783-82-6 see TOC550
7683-59-2 see DMV600	7775-11-3 see DXC200	7783-95-1 see SDQ500
7693-26-7 see PLJ250	7775-14-6 see SHR500	7784-08-9 see SDM100
7697-37-2 see NED500	7775-27-1 see SJE000	7784-18-1 see AHB000
7697-37-2 see NEE500	7778-18-9 see CAX500	7784-19-2 see THQ500
7698-91-1 see MGL600	7778-39-4 see ARB250	7784-21-6 see AHB500
7699-41-4 see SCL000	7778-43-0 see ARC000	7784-30-7 see PHB500
7699-43-6 see ZSJ000	7778-44-1 see ARB750	7784-33-0 see ARF250
7704-34-9 see SOD500	7778-50-9 see PKX250	7784-34-1 see ARF500
7704-99-6 see ZRA000	7778-54-3 see HOV500	7784-35-2 see ARI250
7705-07-9 see TGG250	7778-66-7 see PLK000	7784-37-4 see MDF350
7705-08-0 see FAU000	7778-74-7 see PLO500	7784-40-9 see LCK000
7718-54-9 see NDH000	7778-80-5 see PLT000	7784-41-0 see ARD250
7718-98-1 see VEP000	7779-41-1 see AFJ700	7784-42-1 see ARK250
7719-09-7 see TFL000	7779-86-4 see ZGJ100	7784-44-3 see DCG800
7719-12-2 see PHT275	7779-86-4 see ZIJ100	7784-45-4 see ARG750
7720-78-7 see FBN100	7779-88-6 see ZJJ000	7784-46-5 see SEY500
7721-01-9 see TAF000	7779-90-0 see ZJS400	7785-84-4 see SKM500
7722-06-7 see AMN300	7782-41-4 see FEZ000	7785-87-7 see MAU250
7722-64-7 see PLP000	7782-44-7 see OQW000	7786-29-0 see MNC175
7722-84-1 see HIB010	7782-49-2 see SBO500	7786-34-7 see MQR750

7786-67-6 see MCE750	7803-51-2 see PGY000	8007-56-5 see HHM000
7786-81-4 see NDK500	7803-52-3 see SLQ000	8007-70-3 see AOU250
7787-32-8 see BAM000	7803-55-6 see ANY250	8007-75-8 see BFO000
7787-36-2 see PCK000	7803-62-5 see SDH575	8007-80-5 see CCO750
7787-47-5 see BFQ000	7803-63-6 see ANJ500	8007-87-2 see COE175
7787-49-7 see BFR500	7803-68-1 see TAI750	8008-20-6 see KEK000
7787-52-2 see BFR750	8000-25-7 see RMU000	8008-26-2 see OGM850
7787-56-6 see BFU500	8000-26-8 see PIH400	8008-45-5 see NOG500
7787-69-1 see CDC500	8000-26-8 see PIH500	8008-51-3 see CBB500
7787-71-5 see BMQ325	8000-28-0 see LCD000	8008-52-4 see CNR735
7788-97-8 see CMJ560	8000-42-8 see CBG500	8008-56-8 see LEI000
7788-98-9 see ANF500	8000-46-2 see GDA000	8008-57-9 see OGY000
7789-00-6 see PLB250	8000-48-4 see EQQ000	8008-79-5 see SKY000
7789-04-0 see CMK300	8000-66-6 see CCJ625	8012-74-6 see LIC000
7789-06-2 see SMH000	8000-68-8 see PAL750	8012-89-3 see BAU000
7789-09-5 see ANB500	8001-25-0 see OIQ000	8012-95-1 see MQV750
7789-17-5 see CDE000	8001-26-1 see LGK000	8013-75-0 see FQT000
7789-18-6 see CDE250	8001-29-4 see CNU000	8013-76-1 see BLV500
7789-20-0 see HAK000	8001-30-7 see CNS000	8014-13-9 see COF325
7789-21-1 see FLZ000	8001-31-8 see CNR000	8014-19-5 see PAE000
7789-23-3 see PLF500	8001-35-2 see CDV100	8014-95-7 see SOI520
7789-24-4 see LHF000	8001-50-1 see TBC500	8015-01-8 see MBU500
7789-27-7 see TEK000	8001-58-9 see CMY825	8015-12-1 see EEH520
7789-29-9 see PKU250	8001-61-4 see CNH792	8015-14-3 see LJE000
7789-29-9 see PKU500	8001-79-4 see CCP250	8015-19-8 see DNX500
7789-30-2 see BMQ000	8001-85-2 see BMA750	8015-30-3 see EAP000
7789-38-0 see SFG000	8001-88-5 see BGO750	8015-64-3 see AOO760
7789-40-4 see TEI750	8002-03-7 see PAO000	8015-73-4 see BAR250
7789-42-6 see CAD600	8002-05-9 see PCR250	8015-79-0 see OGK000
7789-47-1 see MCY000	8002-05-9 see PCS250	8015-88-1 see CCL750
7789-59-5 see PHU000	8002-09-3 see PIH750	8015-91-6 see CMQ510
7789-60-8 see PHT250	8002-26-4 see TAC000	8015-92-7 see CDH750
7789-61-9 see AQK000	8002-66-2 see CDH500	8015-97-2 see CMY500
7789-67-5 see TGB750	8002-68-4 see JEA000	8016-14-6 see HHW560
7789-69-7 see PHR250	8002-74-2 see PAH750	8016-20-4 see GJU000
7789-75-5 see CAS000	8002-75-3 see PAE500	8016-31-7 see LII000
7789-80-2 see CAT500	8002-80-0 see WBL100	8016-45-3 see PIG730
7789-82-4 see CAT750	8003-19-8 see DGG000	8016-68-0 see SBA000
7790-30-9 see TEK500	8004-13-5 see PFA860	8016-85-1 see TAD500
7790-47-8 see TGD750	8006-28-8 see see000	8016-88-4 see TAF700
7790-58-1 see PLU000	8006-39-1 see TBD500	8021-27-0 see AAC250
7790-59-2 see PLR750	8006-61-9 see GBY000	8021-29-2 see FBV000
7790-78-5 see CAE425	8006-64-2 see TOD750	8021-39-4 see BAT850
7790-79-6 see CAG250	8006-75-5 see DNU400	8022-15-9 see LCA000
7790-84-3 see CAJ250	8006-77-7 see PIG740	8022-37-5 see ARL250
7790-86-5 see CCY750	8006-78-8 see LBK000	8024-37-1 see COG000
7790-91-2 see CDX750	8006-80-2 see OHI000	8028-73-7 see ARE500
7790-92-3 see HOV000	8006-82-4 see BLW250	8028-73-7 see ARE750
7790-93-4 see CDU000	8006-84-6 see FAP000	8028-89-5 see CBG125
7790-94-5 see CLG500	8006-87-9 see OGY220	8030-30-6 see NAH600
7790-98-9 see PCD500	8006-90-4 see PCB250	8032-32-4 see PCT250
7790-99-0 see IDS000	8007-01-0 see RNA000	8042-47-5 see MQV875
7791-10-8 see SMF500	8007-02-1 see LEH000	8047-67-4 see IHG000
7791-11-9 see RPF000	8007-08-7 see GEQ000	8048-52-0 see DBX400
7791-12-0 see TEJ250	8007-11-2 see OJO000	8049-17-0 see FBG000
7791-20-0 see NDA000	8007-12-3 see OGQ100	8050-07-5 see OIM000
7791-23-3 see SBT500	8007-20-3 see CCQ500	8050-09-7 see RNU100
7791-25-5 see SOT000	8007-45-2 see CMY800	8050-89-3 see OJK340
7791-26-6 see URA000	8007-45-2 see CMY805	8052-41-3 see SLU500
7791-27-7 see PPR500	8007-46-3 see TFX500	8052-42-4 see ARO500
7803-49-8 see HLM500	8007-46-3 see TFX750	8052-42-4 see PCR500

10124-56-8	see	SHM500	10595-95-6	see	MKB000	12125-02-9	see ANE500

10124-56-8 see SHM500
10137-69-6 see CPE500
10137-74-3 see CAO500
10138-41-7 see ECX500
10138-62-2 see HGG000
10141-00-1 see PLB500
10141-05-6 see CNC500
10161-85-0 see DJA325
10168-80-6 see ECY500
10190-99-5 see SEY050
10192-29-7 see ANE250
10192-30-0 see ANB600
10210-68-1 see CNB500
10217-52-4 see HGU500
10222-01-2 see DDM000
10241-05-1 see MRD500
10265-92-6 see DTQ400
10290-12-7 see CNN500
10294-33-4 see BMG400
10294-34-5 see BMG500
10294-40-3 see BAK250
10294-70-9 see TGD500
10309-79-2 see MHN750
10311-84-9 see DBI099
10318-26-0 see DDJ000
10325-94-7 see CAH000
10326-24-6 see ZDS000
10361-03-2 see SII500
10361-37-2 see BAK000
10361-44-1 see BKW250
10361-79-2 see PLX750
10361-80-5 see PLY250
10361-82-7 see SAR500
10361-84-9 see SBC000
10361-91-8 see YDJ000
10361-92-9 see YES000
10361-93-0 see YFJ000
10361-95-2 see ZES000
10377-60-3 see MAH000
10377-66-9 see MAS900
10380-28-6 see BLC250
10415-75-5 see MDE750
10415-87-9 see PFR200
10421-48-4 see FAY200
10421-48-4 see IHB900
10428-19-0 see TBL500
10431-47-7 see SBO100
10453-86-8 see BEP500
10465-27-7 see SAR000
10519-11-6 see DAF100
10519-12-7 see DAF150
10540-29-1 see NOA600
10543-95-0 see HDA000
10544-63-5 see EHO200
10544-72-6 see NGU500
10545-99-0 see SOG500
10546-24-4 see MME500
10563-26-5 see TBI700
10584-98-2 see DDY600
10588-01-9 see SGI000
10589-74-9 see PBX500

10595-95-6 see MKB000
10605-21-7 see MHC750
11006-76-1 see VRF000
11015-37-5 see MRA250
11024-24-1 see DKL400
11032-12-5 see TOE750
11041-12-6 see CME400
11054-70-9 see LBF500
11056-06-7 see BLY000
11069-19-5 see DEV200
11071-15-1 see AQH000
11096-82-5 see PJN250
11097-69-1 see PJN000
11100-14-4 see PJN750
11103-86-9 see PLW500
11104-28-2 see PJM000
11114-46-8 see FBD000
11120-29-9 see PJO250
11133-98-5 see CNI600
11135-81-2 see PLS500
11138-49-1 see AHG000
11138-66-2 see XAK800
11141-16-5 see PJM250
12001-26-2 see MQS250
12001-28-4 see ARM275
12001-29-5 see ARM268
12001-79-5 see VSZ500
12001-89-7 see DGR200
12002-03-8 see COF500
12002-19-6 see MCV250
12005-86-6 see SHM000
12007-97-5 see MRC650
12009-21-1 see BAP750
12010-12-7 see BFQ750
12011-76-6 see DAC450
12012-50-9 see PLW200
12014-28-7 see CAI125
12018-18-7 see NDA100
12030-88-5 see PLE260
12031-80-0 see LHO000
12034-12-7 see SJZ100
12035-39-1 see NDL500
12035-52-8 see NCY100
12035-72-2 see NDJ500
12037-82-0 see PHQ750
12039-52-0 see TEL500
12042-91-0 see AHA000
12054-48-7 see NDE000
12057-74-8 see MAI000
12058-85-4 see SJI500
12060-00-3 see LED000
12067-99-1 see PHU750
12068-85-8 see IGV000
12070-12-1 see TOB500
12075-68-2 see TJP775
12079-65-1 see CPV000
12089-29-1 see BGY720
12108-13-3 see MAV750
12111-24-9 see CAY500
12124-99-1 see ANJ750
12125-01-8 see ANH250

12125-02-9 see ANE500
12125-03-0 see PLS760
12125-56-3 see NDE010
12125-77-8 see COY100
12126-59-9 see ECU750
12126-59-9 see PMB000
12136-15-1 see TKW000
12137-13-2 see NDJ475
12142-88-0 see NDL425
12161-82-9 see BFO250
12164-94-2 see ANA750
12165-69-4 see PHT750
12172-73-5 see ARM262
12174-11-7 see PAE750
12192-57-3 see ART250
12206-14-3 see SKF000
12232-67-6 see BFR000
12244-57-4 see GJC000
12245-39-5 see CPR840
12255-10-6 see NCY125
12255-80-0 see NDJ399
12256-33-6 see NDJ400
12263-85-3 see MGC225
12298-43-0 see HAF375
12298-68-9 see PLW285
12380-95-9 see TIH000
12400-16-7 see BFR250
12401-86-4 see SIN500
12427-38-2 see MAS500
12510-42-8 see EDC650
12540-13-5 see CNI500
12542-85-7 see MGC230
12604-53-4 see FBE000
12604-58-9 see FBP000
12607-70-4 see NCY600
12623-78-8 see AEC000
12640-89-0 see SBT200
12645-50-0 see IHB800
12656-85-8 see MRC000
12672-29-6 see PJM750
12709-98-7 see LDM000
12737-87-0 see PJO750
12758-40-6 see CCF125
12770-50-2 see BFP250
12788-93-1 see ADF250
12789-03-6 see CDR760
12789-46-7 see PBW750
13007-92-6 see HCB000
13010-47-4 see CGV250
13055-82-8 see DNA600
13065-64-0 see DQA710
13092-75-6 see SDJ000
13106-47-3 see BFP750
13106-76-8 see ANM750
13121-70-5 see CQH650
13121-71-6 see ABW600
13138-45-9 see NDG000
13147-25-6 see ELG500
13183-79-4 see MPQ250
13194-48-4 see EIN000
13256-07-0 see AOL000

15181-46-1	see	HIC600	16672-87-0	see	CDS125	17766-75-5	see	THF310
15191-85-2	see	SCN500	16680-47-0	see	ECW520	17773-41-0	see	HMR550
15194-98-6	see	CAM300	16721-80-5	see	SHR000	17804-35-2	see	BAV575
15284-15-8	see	MDP240	16731-55-8	see	PLR250	17822-74-1	see	DHQ800
15356-70-4	see	MCG000	16752-77-5	see	MDU600	17831-71-9	see	ADT050
15421-84-8	see	DIO200	16812-54-7	see	NDL100	17861-62-0	see	NDG550
15442-77-0	see	BIX500	16842-03-8	see	CNC230	17902-23-7	see	FLZ050
15451-93-1	see	DIE350	16853-85-3	see	LHS000	18048-06-1	see	DJA330
15457-87-1	see	TFT250	16870-90-9	see	HOP259	18252-65-8	see	DEU115
15467-20-6	see	DXF000	16871-71-9	see	ZIA000	18312-12-4	see	IDZ100
15468-32-3	see	SCK000	16871-90-2	see	PLH750	18431-36-2	see	HFR100
15503-86-3	see	RFU000	16872-11-0	see	FDD125	18454-12-1	see	LCS000
15529-90-5	see	CLQ500	16872-11-0	see	HHS600	18461-55-7	see	AAU250
15535-79-2	see	DVL200	16881-77-9	see	MJE900	18472-87-2	see	ADG250
15546-11-9	see	BKO250	16893-85-9	see	DXE000	18497-13-7	see	DLO400
15546-16-4	see	BHK250	16919-27-0	see	PLI000	18507-89-6	see	DAI495
15571-58-1	see	DVM800	16919-58-7	see	ANF250	18662-53-8	see	NEI000
15598-34-2	see	PPC100	16921-30-5	see	PLR000	18771-38-5	see	BJT800
15611-84-4	see	PLN050	16923-95-8	see	PLG500	18810-58-7	see	BAI000
15630-89-4	see	SJB400	16924-00-8	see	PLH000	18868-43-4	see	MRD250
15652-38-7	see	DAE600	16925-39-6	see	CAX250	18883-66-4	see	SMD000
15663-27-1	see	PJD000	16940-66-2	see	SFF500	18897-36-4	see	CAI350
15686-63-2	see	EDV700	16940-81-1	see	HDE000	18917-91-4	see	AHC750
15687-27-1	see	IIU000	16941-12-1	see	CKO750	18917-93-6	see	LAL100
15699-18-0	see	NCY050	16949-15-8	see	LHT000	18972-56-0	see	MAG250
15702-65-5	see	BKJ275	16949-65-8	see	MAF600	18987-38-7	see	TME600
15707-23-0	see	ENF200	16961-83-4	see	SCO500	19005-95-9	see	ADE050
15721-02-5	see	TBO000	16962-07-5	see	AHG875	19010-66-3	see	LCW000
15721-33-2	see	DNO200	16962-40-6	see	ANI250	19010-79-8	see	CAI400
15770-21-5	see	PPY300	17010-21-8	see	CAG500	19049-40-2	see	BFT500
15805-73-9	see	VNK000	17014-71-0	see	PLP250	19089-92-0	see	HFM600
15825-70-4	see	MAW250	17029-22-0	see	PLH500	19287-45-7	see	DDI450
15829-53-5	see	MDF750	17031-32-2	see	CNO500	19441-09-9	see	ANT300
15879-93-3	see	GFA000	17040-19-6	see	DAP600	19464-55-2	see	TKT750
15922-78-8	see	HOC000	17068-78-9	see	ARM266	19473-49-5	see	MRK500
15930-94-6	see	CMK500	17090-79-8	see	MRE225	19525-20-3	see	TER500
15954-91-3	see	CAF750	17125-80-3	see	BAO750	19526-81-9	see	EAJ500
15972-60-8	see	CFX000	17168-85-3	see	THP250	19624-22-7	see	PAT750
16018-21-6	see	MNP450	17185-68-1	see	AQQ125	19910-65-7	see	BSD000
16033-21-9	see	PFS500	17256-39-2	see	CGJ280	19992-69-9	see	PPL500
16037-91-5	see	AQH800	17297-82-4	see	ORI400	20198-77-0	see	DFJ400
16039-55-7	see	CAG750	17372-87-1	see	BNH500	20236-55-9	see	BAO900
16066-38-9	see	DWV400	17380-19-7	see	COP550	20240-98-6	see	MNT500
16069-36-6	see	DGV100	17380-21-1	see	COP525	20241-03-6	see	DSR200
16071-86-6	see	CMO750	17381-88-3	see	DBJ400	20246-69-9	see	CAD550
16078-34-5	see	ACO320	17476-04-9	see	LGS000	20265-96-7	see	CJR200
16091-18-2	see	DVK200	17501-44-9	see	PBL750	20265-97-8	see	AOX500
16111-62-9	see	DJK800	17563-48-3	see	BRA550	20325-40-0	see	DOA800
16219-75-3	see	ELO500	17597-95-4	see	HFG700	20333-40-8	see	BRF550
16219-99-1	see	NKQ100	17605-83-3	see	THG300	20369-63-5	see	TID150
16291-96-6	see	CDI250	17608-59-2	see	NKC000	20398-06-5	see	EEE000
16301-26-1	see	ASP000	17617-23-1	see	FMQ000	20667-12-3	see	SDU500
16339-07-4	see	NKW500	17639-93-9	see	CKT000	20738-78-7	see	DWX200
16339-16-5	see	CIQ500	17650-98-5	see	CAK285	20740-05-0	see	HFN500
16409-45-3	see	MCG500	17661-50-6	see	TCB100	20777-39-3	see	LCA100
16432-36-3	see	TNN500	17692-34-1	see	HHK050	20777-49-5	see	DKV160
16478-59-4	see	HAP100	17702-41-9	see	DAE400	20816-12-0	see	OKK000
16532-79-9	see	BNV750	17766-63-1	see	PFX600	20819-54-9	see	TMD400
16543-55-8	see	NLD500	17766-66-4	see	CKJ100	20830-75-5	see	DKN400
16568-02-8	see	AAH000	17766-68-6	see	MFH760	20830-81-3	see	DAC000
16595-80-5	see	LFA020	17766-70-0	see	MFH770	20859-73-8	see	AHE750

64741-58-8 see GBW025	68071-23-8 see EBU100	73747-53-2 see ICW100
64741-59-9 see DXG840	68085-85-8 see GJU600	73747-54-3 see MQE100
64741-61-3 see DXG810	68133-73-3 see IHS100	73791-39-6 see ALV100
64741-80-6 see RDK100	68308-34-9 see COD750	73791-40-9 see BRQ800
64741-88-4 see MQV850	68334-30-5 see DHE900	73791-41-0 see DGA425
64741-89-5 see MQV855	68411-30-3 see LGF825	73791-42-1 see CKD800
64741-96-4 see MQV845	68476-30-2 see DHE800	73791-43-2 see IPS100
64741-97-5 see MQV852	68476-33-5 see FOP200	73791-44-3 see MNU050
64742-03-6 see MQV860	68476-34-6 see DHE850	73791-45-4 see MOT800
64742-04-7 see MQV859	68476-85-7 see LGM000	73816-43-0 see BKJ500
64742-05-8 see MQV862	68527-78-6 see AOH100	73825-85-1 see SNS100
64742-10-5 see MQV863	68527-79-7 see ICS100	73926-81-5 see DDG600
64742-11-6 see MQV857	68541-88-8 see TMX350	73926-85-9 see BIV900
64742-17-2 see MQV872	68555-58-8 see PMB600	73926-87-1 see CET000
64742-18-3 see MQV760	68603-42-9 see CNF330	73926-90-6 see TJU850
64742-19-4 see MQV770	68808-54-8 see AJR500	73927-89-6 see TMV825
64742-20-7 see MQV765	68833-55-6 see MCW349	73927-90-9 see QSJ800
64742-21-8 see MQV775	68848-64-6 see LHP000	73928-00-4 see TKT850
64742-44-5 see PCS260	68917-43-1 see HGA100	73940-86-0 see TMI100
64742-45-6 see PCS270	68917-50-0 see SOY100	73940-87-1 see BLS900
64742-47-8 see KEK100	68917-73-7 see WBJ700	73941-35-2 see THV500
64742-52-5 see MQV790	68952-98-7 see BMO825	73953-53-4 see HHA100
64742-53-6 see MQV800	68955-53-3 see CAY710	73973-02-1 see PDB300
64742-54-7 see MQV795	68955-54-4 see PMC400	74037-18-6 see DEU125
64742-55-8 see MQV805	68956-68-3 see VGU200	74038-78-1 see DSK300
64742-56-9 see MQV840	68956-82-1 see CNE000	74252-25-8 see IDA100
64742-63-8 see MQV820	69011-63-8 see MAF000	74278-22-1 see KHU000
64742-64-9 see MQV835	69012-64-2 see SCH001	75198-31-1 see NGI800
64742-65-0 see MQV825	69029-52-3 see LDC000	75321-19-6 see TMN000
64742-68-3 see MQV776	69226-45-5 see DVL800	75321-20-9 see DVD400
64742-69-4 see MQV777	69382-20-3 see BGC250	75444-63-2 see YBS500
64742-70-7 see MQV778	69521-64-8 see HKY650	75464-11-8 see BSY400
64742-71-8 see MQV779	69853-15-2 see DUO500	75625-24-0 see HCA650
65089-17-0 see CLW500	69929-16-4 see EES100	75965-74-1 see MFB400
65405-73-4 see GDM100	70084-70-7 see TAC200	76180-96-6 see AKT600
65996-92-1 see CMY900	70134-26-8 see DEU100	76706-99-5 see CEV840
65996-93-2 see CMZ100	70145-55-0 see DWZ100	77094-11-2 see AJQ600
65997-15-1 see PKS750	70288-86-7 see ITD875	77248-44-3 see XWS100
66104-24-3 see BFP500	70303-47-8 see BSO200	77267-50-6 see MKM800
66408-78-4 see CMS324	70536-17-3 see AFI850	77276-08-5 see INE062
66499-61-4 see MCS600	70983-41-4 see ZQS100	77405-29-9 see EIM100
66637-25-0 see MID860	71016-15-4 see MHW350	77500-04-0 see AJQ675
66637-32-9 see MPU600	71108-04-8 see DRV300	77536-66-4 see ARM260
66637-35-2 see DUG550	71108-06-0 see DSP710	77536-67-5 see ARM264
66733-21-9 see EDC700	71751-41-2 see ARW200	77536-68-6 see ARM280
67239-28-5 see PMM300	71752-69-7 see NKO400	77824-43-2 see AQE320
67410-20-2 see TMW600	72117-72-7 see COW780	77824-44-3 see AQE305
67445-50-5 see TMY850	72432-14-5 see AOY270	78246-54-5 see HLX925
67465-26-3 see PNV760	72589-96-9 see CAE375	79622-59-6 see CEX800
67465-28-5 see EPI400	72676-73-4 see THH490	80387-97-9 see BJK600
67465-66-1 see DIR100	72676-74-5 see THH480	81861-89-4 see EEA700
67590-56-1 see EQD100	72676-77-8 see THH460	81861-90-7 see PML270
67590-57-2 see IRQ100	72676-78-9 see THH470	81861-94-1 see EEB050
67730-10-3 see DWW700	72869-73-9 see TAD250	81862-00-2 see DPL300
67730-11-4 see AKS250	73128-65-1 see MDF050	81877-66-9 see BKS640
67774-32-7 see FBU509	73160-32-4 see MQM150	83768-87-0 see VLU200
67785-74-4 see UNA100	73245-91-7 see DAB630	84837-04-7 see SLB500
67874-81-1 see CCR525	73263-81-7 see EIB600	85079-48-7 see THI255
68000-78-2 see IHN300	73419-42-8 see CAK250	86073-23-6 see ENF050
68006-83-7 see ALD750	73688-85-4 see DPO275	88208-15-5 see NLY750
68037-57-0 see EHC900	73747-22-5 see BEL550	88671-89-0 see MRW775

89213-87-6 see HGL680	107097-80-3 see LII100	193551-21-2 see HKA123
90034-98-3 see TCU110	107133-36-8 see PCJ230	200398-40-9 see FGA200
90035-12-4 see CKA030	107359-69-3 see BNV800	205943-18-6 see BGD088
90035-14-6 see HOI245	107359-74-0 see CFA800	214899-21-5 see ASA600
90293-48-4 see DRL425	107359-76-2 see BNX035	219959-86-1 see ALX120
90293-50-8 see MES550	107746-52-1 see MEL100	296269-48-2 see DQA720
90293-54-2 see DOV870	108171-26-2 see PAH800	296269-49-3 see DPY700
90466-79-8 see BJA200	108171-27-3 see PAH810	296269-51-7 see DPW630
91297-11-9 see DBL300	108944-67-8 see NBW100	296269-52-8 see PGE550
91465-08-6 see LAS200	109581-93-3 see TAA900	296269-53-9 see MNX420
91724-16-2 see SME500	110147-48-3 see MMA600	296269-54-0 see EOM650
91845-41-9 see PCR000	110690-43-2 see BDN600	301644-18-8 see DLU650
92065-91-3 see ALQ640	112885-41-3 see MRU253	301644-21-3 see DMA500
93334-51-1 see SLO100	112945-52-5 see SCH002	301644-24-6 see CJR220
93763-70-3 see PCJ400	116425-35-5 see AET600	301644-25-7 see CJR210
93780-95-1 see DMS410	117568-24-8 see ACI640	301644-27-9 see BNV752
94948-59-1 see HGL920	117929-15-4 see DVD900	302542-42-3 see IEI700
96811-96-0 see DPI750	120373-24-2 see IRR050	302542-44-5 see DSQ810
98271-51-3 see MOL300	122322-19-4 see CFC600	302542-49-0 see MNX310
99071-30-4 see HNK575	122322-20-7 see BOC600	302542-50-3 see DSQ830
99591-73-8 see DVO920	122322-22-9 see BNA350	302542-60-5 see DSQ840
100482-34-6 see PFI600	122322-24-1 see CKW330	302542-63-8 see IEI740
101043-37-2 see AQV990	125276-72-4 see DCS821	302959-32-6 see DSU300
101652-13-5 see BMT150	126268-14-2 see MPF300	314238-32-9 see TLT763
101670-78-4 see BFC200	128758-36-1 see CJT800	314238-34-1 see TLT757
101831-65-6 see MOH290	130209-82-4 see XAA500	314238-35-2 see TLT150
101913-67-1 see TKH300	133743-71-2 see WAT230	315706-65-1 see TLA650
101931-68-4 see TKB285	136572-09-3 see IGJ550	315706-69-5 see DGI630
102107-61-9 see NHE600	143390-89-0 see MLI900	315706-70-8 see PNH522
102280-93-3 see MLH100	143563-20-6 see DOS400	315706-71-9 see MOU820
102395-95-9 see TGV100	143621-35-6 see AMI600	315706-72-0 see NIY600
102488-99-3 see AEB750	149950-60-7 see EAN525	315706-73-1 see PNH533
102489-70-3 see TCI100	153049-45-7 see DXI480	315706-75-3 see FLL100
102492-24-0 see SDL000	153436-22-7 see DFW730	315706-76-4 see BNX330
102504-71-2 see DPS700	153857-27-3 see DDC810	315706-78-6 see PGB850
102516-61-0 see AKI900	159081-23-9 see IAT275	315706-79-7 see MNW790
102516-65-4 see THH550	171599-83-0 see SCF600	316172-59-5 see TLP800
103426-96-6 see HCK550	174175-11-2 see MKQ600	326800-76-4 see FKK035
103426-97-7 see HCK600	189624-85-9 see NDY550	326800-80-0 see DLI630
105650-23-5 see AKZ200	190133-94-9 see BCP690	

Synonym Cross-Index

1080 see SHG500
γ-6480 see BOV000
3A see SMQ500
A 00 see AGX000
A 21 see DMV600
A-36 see DAE600
A 71 see CHG000
A 95 see AGX000
A 99 see AGX000
A-20D see GLU000
A-310 see ALF250
A 361 see ARQ725
A 3-80 see SMQ500
A-502 see SNJ000
A 688 see PDT250
688A see DDG800
A 884 see DBA800
A 995 see AGX000
A 999 see AGX000
1212A see GLU000
A 1530 see AQF000
A 1582 see AQF000
1A-4OA see AKE250
A-4760 see EDM000
A-4828 see TNT500
A 4942 see IMH000
A 5160 see XOS500
A 10846 see DUD800
A 11032 see UVA000
A 3823A see MRE225
A 60-20R see PJS750
A 60-70R see PJS750
A-91033 see DQA400
A 1 (sorbent) see AHE250
A 15 (polymer) see AAX175
A 100 (pharmaceutical) see IGS000
AA-9 see DXW200
AA 1099 see AGX000
AA1199 see AGX000
AAB see PEI000
AACAPTAN see CBG000
AACIFEMINE see EDU500
AAF see FDR000
2-AAF see FDR000
AAFERTIS see FAS000
AALINDAN see BBQ500
AAMANGAN see MAS500
AAN see AAY000
AAPROTECT see BJK500
AAT see PAK000
AAT see AIC250
o-AAT see AIC250
AATACK see TFS350
AATP see PAK000
AATREX see ARQ725
AATREX 4L see ARQ725
AATREX 80W see ARQ725
AATREX NINE-O see ARQ725
AAVOLEX see BJK500
AAZIRA see BJK500
2-AB see BPY000

ABAMECTIN see ARW200
ABAR see LEN000
ABATE see TAL250
ABATHION see TAL250
ABAVIT see PFP500
ABBOCILLIN see BDY669
ABBOTT 40566 see TCM250
ABCID see SNN300
ABENSANIL see HIM000
ABEREL see VSK950
ABESON NAM see DXW200
ABESTA see RDK000
ABICEL see CCU150
ABICOL see RDK000
ABIES ALBA OIL see AAC250
ABIETIC ACID, METHYL ESTER see MFT500
ABIGUANIL see AHO250
ABIOL see HJL500
"A" BLASTING POWDER see ERF500
ABMINTHIC see DJT800
ABRACOL S.L.G see OAV000
ABRAREX see AHE250
ABRIAL LAVANDIN OIL see LCA000
ABROVAL see BNP750
ABSINTHIUM see ARL250
ABSOLUTE ETHANOL see EFU000
ABSORBABLE GELATIN SPONGE see PCU360
ABSTENSIL see DXH250
ABSTINYL see DXH250
5-AC see ARY000
AC 8 see PJS750
AC 394 see PJS750
AC 680 see PJS750
AC 1198 see PMO250
AC 1220 see PJS750
AC 3422 see EEH600
AC 5223 see DXX400
AC 5230 see ADA725
AC-12682 see DSP400
AC 18133 see EPC500
AC 26,691 see CQL250
AC-43064 see DXN600
AC 47031 see PGW750
AC 47470 see DHH400
AC 52160 see TAL250
ACACIA see AQQ500
ACACIA DEALBATA GUM see AQQ500
ACACIA GUM see AQQ500
ACACIA MOLLISSIMA TANNIN see MQV250
ACACIA SENEGAL see AQQ500
ACACIA SYRUP see AQQ500
ACADYL see BJZ000
ACAMOL see HIM000
ACAR see DER000
ACARABEN 4E see DER000
ACARACIDE see SOP500
ACARIN see BIO750
ACARITHION see TNP250
ACARON see CJJ250
ACAVYL see BJZ000
ACCELERATOR CZ see CPI250

ACCELERATOR L see BJK500
ACCELERATOR THIURAM see TFS350
ACCELERINE see DSY600
ACCEL R see BKU500
ACCENT see MRL500
ACCICURE HBS see CPI250
ACCO FAST RED KB BASE see CLK225
ACCOTHION see DSQ000
ACCUSAND see SCK600
ACCUZOLE see SNN500
AC-DI-SOL NF see SFO500
ACECOLINE see ABO000
ACEDOXIN see DKL800
ACEDRON see BBK500
ACE-E 50 see EHP700
ACE-EE see EHP700
ACENTERINE see ADA725
ACEOTHION see DSQ000
ACEPHAT (GERMAN) see DOP600
ACEPHATE see DOP600
ACEPHATE-MET see DTQ400
ACEPRAMINE see AJD000
ACEPROMAZINA see ABH500
ACEPROMAZINE see ABH500
ACEPROMIZINA see ABH500
ACERDOL see CAV250
ACESAL see ADA725
ACETAAL (DUTCH) see AAG000
ACETACID RED B see HJF500
ACETACID RED J see FMU070
ACETAGESIC see HIM000
ACETAL see AAG000
ACETAL see ADA725
ACETALDEHYD (GERMAN) see AAG250
ACETALDEHYDE see AAG250
ACETALDEHYDE, 4-(9-ACRIDINYL)-2-METHYL-3-
THIOSEMICARBAZONE see MKA300
ACETALDEHYDE, AMINE SALT see AAG500
ACETALDEHYDE AMMONIA see AAG500
ACETALDEHYDE DIMETHYL ACETAL see DOO600
ACETALDEHYDE, ((3,7-DIMETHYL-2,6-
OCTADIENYL)OXY)-, (E)- see GDM100
ACETALDEHYDE, DIPROPYL ACETAL see AAG850
ACETALDEHYDE-DI-n-PROPYL ACETAL see AAG850
ACETALDEHYDE-N-FORMYL-N-
METHYLHYDRAZONE see AAH000
ACETALDEHYDE, (HEXYLOXY)-, DIMETHYL
ACETAL see HFG700
ACETALDEHYDE-N-METHYL-N-
FORMYLHYDRAZONE see AAH000
ACETALDEHYDE OXIME see AAH250
ACETALDEHYDE, TETRAMER see TDW500
ACETALDEHYDE, TRICHLORO-(9CI) see CDN550
ACETALDEHYDE, TRIMER see PAI250
ACETAL DIETHYLIQUE (FRENCH) see AAG000
ACETALDOL see AAH750
ACETALDOXIME see AAH250
ACETALE (ITALIAN) see AAG000
ACETALGIN see HIM000
ACETAMIDE see AAI000
ACETAMIDE, N-(4-((2-HYDROXY-5-
METHYLPHENYL)AZO)PHENYL)- see AAQ250
ACETAMIDE, N-(5-NITRO-2-THIAZOLYL)- see
ABY900
ACETAMIDE, N-PHENYL- see AAQ500
5-ACETAMIDE-1,3,4-THIADIAZOLE-2-
SULFONAMIDE see AAI250
ACETAMIDOBENZENE see AAQ500

4-ACETAMIDOBIPHENYL see PDY500
7-ACETAMIDO-6,7-DIHYDRO-1,2,3,10-
TETRAMETHOXY-BENZO(a)HEPTALEN-9(5H)-ONE
see CNG938
4-ACETAMIDO-2-ETHOLXBENZOIC ACID METHYL
ESTER see EEK100
1-ACETAMIDO-4-ETHOXYBENZENE see ABG750
2-ACETAMIDOFLUORENE see FDR000
1-(N-ACETAMIDOFLUOROMETHYL)-
NAPHTHALENE see MME809
4-ACETAMIDO-2'-HYDROXY-5'-
METHYLAZOBENZENE see AAQ250
3-ACETAMIDO-4-HYDROXY-PHENYLARSONIC
ACID see ABX500
l-α-ACETAMIDO-β-MERCAPTOPROPIONIC ACID see
ACH000
2-ACETAMIDO-4-(5-NITRO-2-FURYL)THIAZOLE see
AAL750
p-ACETAMIDOPHENACYL CHLORIDE see CEC000
4-ACETAMIDOPHENOL see HIM000
p-ACETAMIDOPHENOL see HIM000
2-ACETAMIDO-5-SULFONAMIDO-1,3,4-
THIADIAZOLE see AAI250
ACETAMIDOTHIADIAZOLESULFONAMIDE see
AAI250
p-ACETAMIDOTOLUENE see ABJ250
ACETAMINE DIAZO BLACK RD see DCJ200
ACETAMINE YELLOW CG see AAQ250
p-ACETAMINOFENYL-2-HYDROXYETHYLSULFON
see HLB500
p-ACETAMINOFENYL-β-HYDROXYETHYLSULFON
see HLB500
2-ACETAMINOFLUORENE see FDR000
2-ACETAMINO-4-(5-NITRO-2-FURYL)THIAZOLE see
AAL750
4-ACETAMINO-2-NITROPHENETOLE see NEL000
ACETAMINOPHEN see HIM000
p-ACETAMINOPHENOL see HIM000
ACETAMOX see AAI250
ACETANHYDRIDE see AAX500
ACETANIL see AAQ500
ACETANILID see AAQ500
ACETANILIDE see AAQ500
ACETANILIDE, 2-ACETYL- see AAY000
ACETANILIDE, 4'-(2-HYDROXYETHYLSULFONYL)-
see HLB500
ACETANISOLE (FCC) see MDW750
ACETARSOL see ABX500
ACETARSONE see ABX500
ACETATE d'AMYLE (FRENCH) see AOD725
ACETATE BLUE G see TBG700
ACETATE de BUTYLE (FRENCH) see BPU750
ACETATE de BUTYLE SECONDAIRE (FRENCH) see
BPV000
ACETATE C-8 see OEG000
ACETATE de CELLOSOLVE (FRENCH) see EES400
ACETATE de CUIVRE (FRENCH) see CNI250
ACETATE de l'ETHER MONOETHYLIQUE de
l'ETHYLENE-GLYCOL (FRENCH) see EES400
ACETATE de l'ETHER MONOMETHYLIQUE de
l'ETHYLENE-GLYCOL (FRENCH) see EJJ500
ACETATE d'ETHYLGLYCOL (FRENCH) see EES400
ACETATE FAST ORANGE R see AKP750
ACETATE FAST RED 2B see AKE250
ACETATE FAST YELLOW G see AAQ250
ACETATE d'ISOBUTYLE (FRENCH) see IIJ000
ACETATE d'ISOPROPYLE (FRENCH) see INE100

ACETATE of LIME see CAL750
ACETATE de METHYLE (FRENCH) see MFW100
ACETATE de METHYLE GLYCOL (FRENCH) see EJJ500
ACETATE P.A. see AGQ750
ACETATE PHENYLMERCURIQUE (FRENCH) see ABU500
ACETATE de PLOMB (FRENCH) see LCV000
ACETATE de PROPYLE NORMAL (FRENCH) see PNC250
ACETATE de TRIPHENYL-ETAIN (FRENCH) see ABX250
ACETATE de VINYLE see VLU250
ACETATO(2-AMINO-5-NITROPHENYL)MERCURY see ABQ250
(ACETATO)(p-AMINOPHENYL)MERCURY see ABQ000
ACETATO di CELLOSOLVE (ITALIAN) see EES400
(ACETATO)(DIETHOXYPHOSPHINYL)MERCURY see AAS250
ACETATO(2-METHOXYETHYL)MERCURY see MEO750
ACETATO di METIL CELLOSOLVE (ITALIAN) see EJJ500
(ACETATO)PHENYLMERCURY see ABU500
(ACETATO)PHENYL-MERCURY mixed with CHLOROETHYL-MERCURY (19:1) see PFO500
ACETATO di STAGNO TRIFENILE (ITALIAN) see ABX250
(ACETATO)(2,3,5,6-TETRAMETHYLPHENYL)MERCURY see AAS500
(ACETATO)(TRIMETAARSENITO)DICOPPER see COF500
ACETATOTRIPHENYLSTANNANE see ABX250
ACETAZINE see ABH500
ACETAZOLAMID see AAI250
ACETAZOLAMIDE see AAI250
ACETAZOLEAMIDE see AAI250
ACETDIMETHYLAMIDE see DOO800
ACETEIN see ACH000
ACETENE see EIO000
ACETETHYLANILIDE see EFQ500
ACETEUGENOL see EQS000
ACETHYDRAZIDE see ACM750
ACETHYLPROMAZIN see ABH500
ACETIC ACID see AAT250
ACETIC ACID (aqueous solution) (DOT) see AAT250
ACETIC ACID ALLYL ESTER see AFU750
ACETIC ACID AMIDE see AAI000
ACETIC ACID, AMINO-sec-BUTYL- see IKX000
ACETIC ACID, AMMONIUM SALT see ANA000
ACETIC ACID, AMYL ESTER see AOD725
ACETIC ACID, AMYL ESTER see AOD750
ACETIC ACID, ANHYDRIDE (9CI) see AAX500
ACETIC ACID ANILIDE see AAQ500
ACETIC ACID, BARIUM SALT see BAH500
ACETIC ACID BENZYL ESTER see BDX000
ACETIC ACID, (((3,5-BIS(1,1-DIMETHYLETHYL)-4-HYDROXYPHENYL)METHYL)THIO)-, 2-ETHYLHEXYL ESTER see BJK600
ACETIC ACID, compd. with BORON FLUORIDE (BF3) (8CI) see BMG750
ACETIC ACID-2-BUTOXY ESTER see BPV000
ACETIC ACID n-BUTYL ESTER see BPU750
ACETIC ACID-tert-BUTYL ESTER see BPV100
ACETIC ACID, CADMIUM SALT see CAD250
ACETIC ACID, CADMIUM SALT, DIHYDRATE see CAD275

ACETIC ACID CHLORIDE see ACF750
ACETIC ACID, CHLORO-, tert-BUTYL ESTER see BQR100
ACETIC ACID, CHLORO-, 1,1-DIMETHYLETHYL ESTER see BQR100
ACETIC ACID, CHROMIUM (2+) SALT (8CI, 9CI) see CMJ000
ACETIC ACID, CINNAMYL ESTER see CMQ730
ACETIC ACID, CITRONELLYL ESTER see AAU000
ACETIC ACID, COBALT(2+) SALT see CNC000
ACETIC ACID, COBALT(2+) SALT, TETRAHYDRATE see CNA500
ACETIC ACID, CUPRIC SALT see CNI250
ACETIC ACID, CYANO-, TRIPHENYLSTANNYL ESTER see TMV825
ACETIC ACID, (1,2-CYCLOHEXYLIDENEDINITRILO)TETRA-, CADMIUM COMPLEX, TRANS- see CAD900
ACETIC ACID, 9-DECENYL ESTER see DAI450
ACETIC ACID, 2,2'-((DIBUTYLSTANNYLENE)BIS(THIO))BIS-, DINONYL ESTER see DEH650
ACETIC ACID, ((DIBUTYLSTANNYLENE)DITHIO)DI-, DINONYL ESTER (8CI) see DEH650
ACETIC ACID, DIMETHYL- see IJU000
ACETIC ACID DIMETHYLAMIDE see DOO800
ACETIC ACID-1,3-DIMETHYLBUTYL ESTER see HFJ000
ACETIC ACID-1,1-DIMETHYLETHYL ESTER see BPV100
ACETIC ACID-3,7-DIMETHYL-6-OCTEN-1-YL ESTER see AAU000
ACETIC ACID-4,6-DINITRO-o-CRESYL ESTER see AAU250
ACETIC ACID, ETHENYL ESTER see VLU250
ACETIC ACID ETHENYL ESTER HOMOPOLYMER see AAX250
ACETIC ACID ETHENYL ESTER POLYMER with CHLORETHENE (9CI) see AAX175
ACETIC ACID, ETHOXY- see EEK500
ACETIC ACID-2-ETHOXYETHYL ESTER see EES400
ACETIC ACID, ETHYLENE ETHER see VLU250
ACETIC ACID GERANIOL ESTER see DTD800
ACETIC ACID, GLACIAL see AAT250
ACETIC ACID HEXYL ESTER see HFI500
ACETIC ACID, (o-((2-HYDROXY-3-HYDROXYMERCURI)PROPYL)CARBAMOYL)PHENOXY- see NCM800
ACETIC ACID, IRON(2+) SALT see FBH000
ACETIC ACID, ISOBUTYL ESTER see IIJ000
ACETIC ACID, ISOPENTYL ESTER see IHO850
ACETIC ACID, ISOPROPENYL ESTER see MQK750
ACETIC ACID ISOPROPYL ESTER see INE100
ACETIC ACID LEAD(2+) SALT see LCV000
ACETIC ACID, LEAD(2+) SALT TRIHYDRATE see LCJ000
ACETIC ACID LINALOOL ESTER see LFY600
ACETIC ACID, LITHIUM SALT see LGO100
ACETIC ACID MANGANESE(II) SALT (2:1) see MAQ000
ACETIC ACID, MERCURY(2+) SALT see MCS750
ACETIC ACID-3-METHOXYBUTYL ESTER see MHV750
ACETIC ACID, 2-METHOXY-1-METHYLETHYL ESTER see PNL265
ACETIC ACID METHYL ESTER see MFW100

ACETIC ACID-1-METHYLETHYL ESTER (9CI) see INE100

ACETIC ACID METHYLNITROSAMINOMETHYL ESTER see AAW000

ACETIC ACID, (METHYL-ONN-AZOXY)METHYL ESTER see MGS750

ACETIC ACID, α-METHYL-PHENETHYL ESTER see ABU800

ACETIC ACID-4-METHYLPHENYL ESTER see MNR250

ACETIC ACID-2-METHYLPROPYL ESTER see IIJ000

ACETIC ACID-1-METHYLPROPYL ESTER (9CI) see BPV000

ACETIC ACID, NICKEL(2+) SALT see NCX000

ACETIC ACID, NICKEL(+2) SALT, TETRAHYDRATE see NCX500

ACETIC ACID, NITRILOTRI-, MERCURY(II) COMPLEX see MDF050

ACETIC ACID, OCTYL ESTER see OEG000

ACETIC ACID, ((OCTYLSTANNYLIDYNE)TRITHIO)TRI-, TRIS(2-ETHYLHEXYL) ESTER see OGI000

ACETIC ACID, PHENYL-, BUTYL ESTER see BQJ350

ACETIC ACID, PHENYL-, 3,7-DIMETHYL-2,6-OCTADIENYL ESTER, (E)-(8CI) see GDM400

ACETIC ACID, PHENYL-, 3,7-DIMETHYL-6-OCTENYL ESTER see CMU050

ACETIC ACID-2-PHENYLETHYL ESTER see PFB250

ACETIC ACID, PHENYLMERCURY DERIV. see ABU500

ACETIC ACID PHENYLMETHYL ESTER see BDX000

ACETIC ACID-2-PROPENYL ESTER see AFU750

ACETIC ACID, n-PROPYL ESTER see PNC250

ACETIC ACID, SAMARIUM SALT see SAR000

ACETIC ACID, SODIUM SALT see SEG500

ACETIC ACID, TRIANHYDRIDE with ANTIMONIC ACID see AQJ750

ACETIC ACID, glacial or acetic acid solution, >80% acid, by weight (UN 2790) (DOT) see AAT250

ACETIC ACID solution, >10% but not >80% acid, by weight (UN 2790) (DOT) see AAT250

ACETIC ACID VINYL ESTER see VLU250

ACETIC ACID, VINYL ESTER, POLYMER with CHLOROETHYLENE see AAX175

ACETIC ACID VINYL ESTER POLYMERS see AAX250

ACETIC ALDEHYDE see AAG250

ACETIC ANHYDRIDE see AAX500

ACETIC CHLORIDE see ACF750

ACETIC ETHER see EFR000

ACETIC OXIDE see AAX500

ACETIC PEROXIDE see PCL500

ACETICYL see ADA725

ACETIDIN see EFR000

ACETILE DIAZO BLACK N see DPO200

ACETILSALICILICO see ADA725

ACETILUM ACIDULATUM see ADA725

ACETIMIDIC ACID see AAI000

ACETISAL see ADA725

ACETISOEUGENOL see AAX750

ACETOACETAMIDOBENZENE see AAY000

ACETOACETANILID see AAY000

ACETOACETANILIDE see AAY000

o-ACETOACETANISIDE see ABA500

ACETOACET-o-ANISIDIN (CZECH) see ABA500

ACETOACETIC ACID ANILIDE see AAY000

ACETOACETIC ACID-o-ANISIDIDE see ABA500

ACETOACETIC ACID BUTYL ESTER see BPV250

ACETOACETIC ACID, ETHYL ESTER see EFS000

ACETOACETIC ACID, TRIESTER WITH 2-ETHYL-2-(HYDROXYMETHYL)-1,3-PROPANEDIOL see ELJ600

ACETOACETIC ACID, 1,1,1-TRIHYDROXY-METHYLPROPANE TRIESTER see ELJ600

ACETOACETIC ANILIDE see AAY000

ACETOACETIC ESTER see EFS000

ACETOACETIC METHYL ESTER see MFX250

ACETOACETONE see ABX750

ACETOACET-o-TOLUIDIDE see ABA000

2-ACETOACETYLAMINOANISOLE see ABA500

((ACETOACETYL)AMINO)BENZENE see AAY000

2-ACETOACETYLAMINOTOLUENE see ABA000

ACETOACETYLANILINE see AAY000

ACETOACETYL-o-ANISIDE see ABA500

ACETOACETYL-o-ANISIDINE see ABA500

ACETOACETYL-o-ANISINE see ABA500

ACETOACETYL-2-METHYLANILIDE see ABA000

ACETOAMINOFLUORENE see FDR000

ACETOANILIDE see AAQ500

ACETOARSENITE de CUIVRE (FRENCH) see COF500

ACETO AZIB see ASL750

ACETO-CAUSTIN see TII250

ACETOCID see SNQ710

ACETO DIPP see NBL000

ACETO DNPT 40 see DVF400

ACETO DNPT 80 see DVF400

ACETO DNPT 100 see DVF400

ACETOFERROCENE see ABA750

ACETOHEXAMIDE see ABB000

ACETO HMT see HEI500

ACETOHYDRAZIDE see ACM750

ACETOIN see ABB500

ACETOL see ADA725

ACETOL (1) see ABC000

ACETONANIL see TLP500

ACETONANYL see TLP500

1-ACETONAPHTHALENE see ABC475

β-ACETONAPHTHALENE see ABC500

ACETONAPHTHONE see ABC500

1-ACETONAPHTHONE see ABC475

2-ACETONAPHTHONE see ABC500

1'-ACETONAPHTHONE see ABC475

2'-ACETONAPHTHONE see ABC500

α-ACETONAPHTHONE see ABC475

β-ACETONAPHTHONE see ABC500

ACETONCIANHIDRINEI (ROUMANIAN) see MLC750

ACETONCIANIDRINA (ITALIAN) see MLC750

ACETONCYAANHYDRINE (DUTCH) see MLC750

ACETONCYANHYDRIN (GERMAN) see MLC750

ACETON (GERMAN, DUTCH, POLISH) see ABC750

ACETONE see ABC750

ACETONE ANIL see TLP500

ACETONE CHLOROFORM see ABD000

ACETONECYANHYDRINE (FRENCH) see MLC750

ACETONE CYANOHYDRIN (ACGIH,DOT) see MLC750

ACETONE, HEXACHLORO- see HCL500

ACETONE OILS (DOT) see ABC750

ACETONE PEROXIDE see ABE000

ACETONIC ACID see LAG000

ACETONITRIL (GERMAN, DUTCH) see ABE500

ACETONITRILE see ABE500

ACETONITRILE, AMINO-, MONOHYDROCHLORIDE (9CI) see GHI100

ACETONITRILE, BENZOYLPHENYL- see OOK200

ACETONKYANHYDRIN (CZECH) see MLC750

ACETONYL see ADA725
ACETONYL ACETONE see HEQ500
3-(α-ACETONYLBENZYL)-4-HYDROXYCOUMARIN
see WAT200
3-(α-ACETONYLBENZYL)-4-HYDROXY-COUMARIN
SODIUM SALT see WAT220
ACETONYL BROMIDE see BNZ000
ACETONYL CHLORIDE see CDN200
ACETO PBN see PFT500
ACETOPHEN see ADA725
ACETO-p-PHENALIDE see ABG750
p-ACETOPHENETIDE see ABG750
p-ACETOPHENETIDIDE see ABG750
ACETO-p-PHENETIDIDE see ABG750
ACETOPHENETIDIN see ABG750
ACETOPHENETIDINE see ABG750
ACETO-4-PHENETIDINE see ABG750
ACETOPHENETIN see ABG750
ACETOPHENONE see ABH000
ACETOPHENONE, 2'-HYDROXY-(8CI) see HIN500
ACETOPHENONE, 4'-HYDROXY-, o-
(ISOPROPYLCARBAMOYL)OXIME, o-ESTER WITH
o,o-DIETHYL PHOSPHOROTHIO see DJP520
ACETOPHENONE, 2-HYDROXY-2-PHENYL- see
BCP250
ACETOPROMAZINE see ABH500
3-ACETOPYRIDINE see ABI000
ACETOQUAT CPC see CCX000
ACETOQUAT CTAB see HCQ500
ACETOQUINONE BLUE L see TBG700
ACETOQUINONE BLUE R see TBG700
ACETOQUINONE LIGHT GOOSEBERRY RL see
AKE250
ACETOQUINONE LIGHT ORANGE JL see AKP750
ACETOQUINONE LIGHT YELLOW see AAQ250
ACETOQUINONE LIGHT YELLOW 4JLZ see AAQ250
ACETOSAL see ADA725
ACETOSALIC ACID see ADA725
ACETOSALIN see ADA725
ACETO SDD 40 see SGM500
ACETOSPAN see AQX500
ACETOSULFAMIN see SNQ710
ACETO TETD see TFS350
2-ACETOTHIENONE see ABI500
ACETOTHIOAMIDE see TFA000
2-ACETOTHIOPHENE see ABI500
ACETO TMTM see BJL600
4-ACETOTOLUIDE see ABJ250
p-ACETOTOLUIDE see ABJ250
p-ACETOTOLUIDIDE see ABJ250
2-ACETOXYBENZOIC ACID see ADA725
o-ACETOXYBENZOIC ACID see ADA725
17-ACETOXY-6-CHLORO-6-
DEHYDROPROGESTERONE see CBF250
17-α-ACETOXY-6-CHLORO-6-
DEHYDROPROGESTERONE see CBF250
17-α-ACETOXY-6-CHLORO-6,7-
DEHYDROPROGESTERONE see CBF250
17-α-ACETOXY-6-CHLOROPREGNA-4,6-DIENE-3,20-
DIONE see CBF250
17-α-ACETOXY-6-CHLORO-4,6-PREGNADIENE-3,20-
DIONE see CBF250
17-α-ACETOXY-6-DEHYDRO-6-
METHYLPROGESTERONE see VTF000
ACETOXYDIETHYLPHENYLSTANNANE see DJV800
α-ACETOXY DIMETHYLNITROSAMINE see AAW000

ACETOXYDIPHENYLETHYLSTANNANE see EIM100
ACETOXYETHANE see EFR000
1-ACETOXYETHYLENE see VLU250
2-ACETOXYETHYLTRIMETHYLAMMONIUM
CHLORIDE see ABO000
ACETOXYL see BDS000
p-(ACETOXYMERCURI)ANILINE see ABQ000
(ACETOXYMERCURI)BENZENE see ABU500
2-(ACETOXYMERCURI)-4-NITROANILINE see
ABQ250
1-ACETOXY-2-METHOXY-4-ALLYLBENZENE see
EQS000
4-ACETOXY-3-METHOXY-1-PROPENYLBENZENE
see AAX750
ACETOXYMETHYL-METHYL-NITROSAMIN
(GERMAN) see AAW000
ACETOXYMETHYL METHYLNITROSAMINE see
AAW000
N-α-ACETOXYMETHYL-N-METHYLNITROSAMINE
see AAW000
17-ACETOXY-6-METHYLPREGNA-4,6-DIENE-3,20-
DIONE see VTF000
17-α-ACETOXY-6-METHYLPREGNA-4,6-DIENE-3,20-
DIONE see VTF000
17-α-ACETOXY-6-METHYL-4,6-PREGNADIENE-3,20-
DIONE see VTF000
17-α-ACETOXY-6-α-METHYLPREGN-4-ENE-3,20-
DIONE see MCA000
17-ACETOXY-6-α-METHYLPROGESTERONE see
MCA000
1-ACETOXY-N-NITROSODIMETHYLAMINE see
AAW000
17-β-ACETOXY-19-NOR-17-α-PREGN-4-EN-20-YN-3-
ONE see ABU000
17-ACETOXY-19-NOR-17-α-PREGN-4-EN-20-YN-3-
ONE see ABU000
2-ACETOXYPENTANE see AOD735
3-ACETOXYPHENOL see RDZ900
ACETOXYPHENYLMERCURY see ABU500
2-ACETOXY-1-PHENYLPROPANE see ABU800
1-ACETOXYPROPANE see PNC250
2-ACETOXYPROPANE see INE100
3-ACETOXYPROPENE see AFU750
4-ACETOXYTOLUENE see MNR250
p-ACETOXYTOLUENE see MNR250
α-ACETOXYTOLUENE see BDX000
ACETOXYTRICYCLOHEXYLSTANNANE see ABW600
ACETOXYTRIETHYLSTANNANE see ABW750
ACETOXYTRIETHYLTIN see ABW750
ACETOXYTRIHEXYLSTANNANE see ABX000
ACETOXYTRIHEXYLTIN see ABX000
ACETOXYTRIISOPROPYLSTANNANE see TKT750
ACETOXYTRIMETHYLSTANNANE see TMI000
ACETOXY-TRIPHENYL-STANNAN (GERMAN) see
ABX250
ACETOXYTRIPHENYLSTANNANE see ABX250
ACETOXY-TRIPHENYL-STANNANE see ABX250
ACETOXYTRIPHENYLTIN see ABX250
ACETOZALAMIDE see AAI250
ACETO ZDBD see BIX000
ACETO ZDED see BJK500
ACETO ZDMD see BJK500
ACET-p-PHENALIDE see ABG750
ACETPHENARSINE see ABX500
ACETPHENETIDIN see ABG750
ACET-p-PHENETIDIN see ABG750

p-ACETPHENETIDIN see ABG750
ACET-THEOCIN see TEP000
ACETYLACETANILIDE see AAY000
α-ACETYLACETANILIDE see AAY000
ACETYLACETONATE-1,5-CYCLOOCTADIENE
RHODIUM see CPR840
ACETYL ACETONE see ABX750
N-(ACETYLACETYL)ANILINE see AAY000
ACETYLADRIAMYCIN see DAC000
ACETYLAMINOBENZENE see AAQ500
N-ACETYL-4-AMINOBENZENESULFONAMIDE see
SNQ710
4-ACETYLAMINOBIPHENYL see PDY500
S-(2-(ACETYLAMINO)ETHYL)-O,O-DIMETHYL
PHOSPHORODITHIOATE see DOP200
2-ACETYLAMINO-FLUOREN (GERMAN) see FDR000
4-ACETYLAMINOFLUOREN (GERMAN) see ABY000
4-ACETYLAMINOFLUORENE see ABY000
N-ACETYL-2-AMINOFLUORENE see FDR000
2-ACETYLAMINOFLUORENE (OSHA) see FDR000
3-ACETYLAMINO-4-HYDROXYPHENYLARSONIC
ACID see ABX500
(3-(ACETYLAMINO)-4-
HYDROXYPHENYL)ARSONINE (9CI) see ABX500
2-ACETYLAMINO-4-(5-NITRO-2-FURYL)THIAZOLE
see AAL750
2-ACETYLAMINO-5-NITROTHIAZOLE see ABY900
p-(ACETYLAMINO)PHENACYL CHLORIDE see
CEC000
p-ACETYLAMINOPHENOL see HIM000
N-ACETYL-p-AMINOPHENOL see HIM000
p-ACETYLAMINOPHENYL DERIVATIVE of
NITROGEN MUSTARD see BHO500
N¹-ACETYL-4-AMINOPHENYLSULFONAMIDE see
SNQ710
2-ACETYLAMINO-1,3,4-THIADIAZOLE-5-
SULFONAMIDE see AAI250
4-(ACETYLAMINO)TOLUENE see ABJ250
ACETYL ANHYDRIDE see AAX500
ACETYLANILINE see AAQ500
3-ACETYLANILINE see AHR500
4-ACETYLANILINE see AHR240
m-ACETYLANILINE see AHR500
N-ACETYLANILINE see AAQ500
4-ACETYLANISOLE see MDW750
p-ACETYLANISOLE see MDW750
ACETYLATED MONO- and DIGLYCERIDES see
ACA900
ACETYLATED MONOGLYCERIDES see ACA900
1-ACETYLAZIRIDINE see ACB250
ACETYLBENZENE see ABH000
1-(p-ACETYLBENZENESULFONYL)-3-
CYCLOHEXYLUREA see ABB000
ACETYL BENZOYL PEROXIDE (solid) see ACC250
ACETYL BROMIDE see ACD750
ACETYLBUTYRYL see HEQ200
ACETYL CHLORIDE see ACF750
ACETYLCHOLINE CHLORIDE see ABO000
ACETYLCHOLINE HYDROCHLORIDE see ABO000
ACETYLCHOLINIUM CHLORIDE see ABO000
4-ACETYL-N-((CYCLOHEXYLAMINO)CARBONYL)-
BENZENESULFONAMIDE see ABB000
ACETYLCYSTEINE see ACH000
N-ACETYLCYSTEINE see ACH000
N-ACETYL-l-CYSTEINE see ACH000
N-ACETYL-N-CYSTEINE see ACH000
N-ACETYL-l-CYSTEINE (9CI) see ACH000

2-ACETYLDIBENZOTHIOPHENE see ACH090
3-ACETYL-10-(3-
DIMETHYLAMINOPROPYL)PHENOTHIAZINE see
ABH500
3-ACETYL-2,4-DIMETHYL-PYRROLE see ACI500
3-ACETYL-2,5-DIMETHYL-PYRROLE see ACI550
10-ACETYLDITHRANOL see ACI640
ACETYLEN see ACI750
ACETYLENE see ACI750
ACETYLENE, dissolved (DOT) see ACI750
ACETYLENE BLACK see CBT750
(ACETYLENECARBONYLOXY)TRIPHENYLTIN see
TMW600
ACETYLENE COMPOUNDS and ALKYNES see ACJ125
ACETYLENE DICHLORIDE see DFI210
trans-ACETYLENE DICHLORIDE see ACK000
ACETYLENE, METHYL- see MFX590
ACETYLENE TETRABROMIDE see ACK250
ACETYLENE TETRACHLORIDE see TBQ100
ACETYLENE TRICHLORIDE see TIO750
ACETYL ENHEPTIN see ABY900
ACETYLENOGEN see CAN750
ACETYL ETHER see AAX500
ACETYL ETHYLENE see BOY500
ACETYLETHYLENEIMINE see ACB250
ACETYL ETHYL TETRAMETHYL TETRALIN see
ACL750
ACETYLETHYL TETRAMETHYLTETRALIN see
ACL750
ACETYLEUGENOL see EQS000
ACETYLFERROCENE see ABA750
1-ACETYLFERROCENE see ABA750
N-ACETYL-N-9H-FLUOREN-2-YL-ACETAMIDE see
DBF200
ACETYL FLUORIDE see ACM000
ACETYL HYDRAZIDE see ACM750
N-ACETYLHYDRAZINE see ACM750
ACETYL HYDROPEROXIDE see PCL500
N-ACETYL-4-HYDROXY-m-ARSANILIC ACID see
ABX500
2-ACETYL-7-(2-HYDROXY-3-sec-
BUTYLAMINOPROPOXY)BENZOFURAN see ACN310
2-ACETYL-4-(2-HYDROXY-3-tert-
BUTYLAMINOPROPOXY)BENZOFURAN see ACN300
2-ACETYL-7-(2-HYDROXY-3-tert-
BUTYLAMINOPROPOXY)BENZOFURAN see ACN320
2-ACETYL-10-(3-(4-(β-
HYDROXYETHYL)PIPERIDINO)PROPYL)PHENOTH
IAZINE see PII500
2-ACETYL-5-HYDROXY-3-OXO-4-HEXENOIC ACID
Δ-LACTONE see MFW500
2-ACETYL-7-(2-HYROXY-3-
ISOPROPYLAMINOPROPOXY)BENZOFURAN see
HLK600
ACETYLIDES see ACO000
ACETYLIN see ADA725
ACETYL-3-INDOLE see ICY100
3-ACETYLINDOLE see ICY100
3-ACETYLINDOLE-5-CARBONITRILE see COP550
5-ACETYLINDOLINE see ACO320
ACETYL IODIDE see ACO500
ACETYLISOEUGENOL see AAX750
ACETYLISOPENTANOYL see MKL300
ACETYL ISOVALERYL see MKL300
ACETYL MERCAPTAN see TFA500
N-ACETYL-3-MERCAPTOALANINE see ACH000
ACETYLMETHIONINE see ACQ275

N-ACETYLMETHIONINE see ACQ275
N-ACETYL-l-METHIONINE see ACQ275
4-(N-ACETYL-N-METHYL)AMINO-4'-(N',N'-
DIMETHYLAMINO)AZOBENZENE see DPQ200
N'-ACETYL-N'-METHYL-4'-AMINO-N,N-DIMETHYL-
4-AMINOAZOBENZENE see DPQ200
ACETYL METHYL BROMIDE see BNZ000
ACETYLMETHYLCARBAMIC ACID 2,2-DIMETHYL-
1,3-BENZODIOXOL-4-YL ESTER see BCJ005
ACETYL METHYL CARBINOL see ABB500
ACETYL-METHYL-NITROSO-HARNSTOFF
(GERMAN) see ACR400
ACETYLMETHYLNITROSOUREA see ACR400
N'-ACETYL-METHYLNITROSOUREA see ACR400
3-ACETYL-10-(3'-N-METHYL-PIPERAZINO-N'-
PROPYL)PHENOTHIAZIN see ACR500
3-ACETYL-6-METHYL-2,4-PYRANDIONE see MFW500
3-ACETYL-6-METHYLPYRANDIONE-2,4 see MFW500
3-ACETYL-6-METHYL-2H-PYRAN-2,4(3H)-DIONE see
MFW500
3-ACETYL-10-(3'-
MORPHOLINOPROPYL)PHENOTHIAZIN see
MMA600
1-ACETYLNAPHTHALENE see ABC475
2-ACETYLNAPHTHALENE see ABC500
β-ACETYLNAPHTHALENE see ABC500
2-ACETYL-5-NITROFURAN see ACT250
ACETYLON FAST BLUE G see TBG700
ACETYLON FAST PINK B see AKE250
ACETYL OXIDE see AAX500
(3-β,5-α)-3-(ACETYLOXY)ANDROSTAN-17-ONE see
HJB225
2-(ACETYLOXY)BENZOIC ACID see ADA725
17-(ACETYLOXY)-6-CHLOROPREGNA-4,6-DIENE-
3,20-DIONE see CBF250
10-(ACETYLOXY)-1,8-DIHYDROXY-9(10H)-
ANTHRACENONE see ACI640
4-ACETYLOXY-12,13-EPOXY-3,7,15-TRIHYDROXY-(3-
α,4-β,7-β)-TRICHOTHEC-9-EN-8-ONE see FQR000
17-(ACETYLOXY)-6-METHYL-16-
METHYLENEPREGNA-4,6-DIENE-3,20-DIONE (9CI)
see MCB380
(6-α)-17-(ACETYLOXY)-6-METHYLPREG-4-ENE-3,20-
DIONE see MCA000
(17-α)-17-(ACETYLOXY)-19-NORPREGN-4-EN-20-YN-
3-ONE see ABU000
17-ACETYLOXY(17-α)-19-NORPREGN-4-ESTREN-17-
β-OL-ACETATE-3-ONE see ABU000
2-(ACETYLOXY)-N,N,N-
TRIMETHYLETHANAMINIUM CHLORIDE see
ABO000
(ACETYLOXY)TRIPHENYL-STANNANE (9CI) see
ABX250
ACETYL PEROXIDE see ACV500
ACETYL PEROXIDE, solid, or >25% in solution (DOT)
see ACV500
ACETYL PEROXIDE, not >25% in solution (UN 2084)
(DOT) see ACV500
ACETYLPHENETIDIN see ABG750
N-ACETYL-p-PHENETIDINE see ABG750
2-ACETYLPHENOL see HIN500
o-ACETYLPHENOL see HIN500
ACETYLPHOSPHORAMIDOTHIOIC ACID-O,S-
DIMETHYL ESTER see DOP600
ACETYL PHTHALYL CELLULOSE see CCU050
ACETYLPROMAZINE see ABH500
ACETYLPROPIONYL see PBL350

3-ACETYLPYRIDINE see ABI000
β-ACETYLPYRIDINE see ABI000
ACETYLRESORCINOL see RDZ900
4-ACETYLRESORCINOL see DMG400
ACETYLSAL see ADA725
ACETYLSALICYLIC ACID see ADA725
ACETYLSALICYLSAEURE (GERMAN) see ADA725
N-ACETYL-SARCOLYSIL VALINE ETHYL ETHER see
ARM000
N'-ACETYLSULFANILAMIDE see SNQ710
N¹-ACETYLSULFANILAMIDE see SNQ710
N-ACETYLSULFANILAMINE see SNQ710
6-ACETYL-1,1,4,4-TETRAMETHYL-7-ETHYL-1,2,3,4,-
TETRALIN see ACL750
7-ACETYL-1,1,4,4-TETRAMETHYL-1,2,3,4-
TETRAHYDRONAPHTHALENE see ACL750
7-α-ACETYLTHIO-3-OXO-17-α-PREGN-4-ENE-21,17-
β-CARBOLACTONE see AFJ500
7-α-ACETYLTHIO-3-OXO-17-β-PREGN-4-ENE-21,17-
β-CARBOLACTONE see AFJ500
2-ACETYLTHIOPHENE see ABI500
p-ACETYLTOLUENE see MFW250
N-ACETYL-p-TOLUIDIDE see ABJ250
ACETYL-p-TOLUIDINE see ABJ250
ACETYL TRIBUTYL CITRATE see ADD400
ACETYL TRIETHYL CITRATE see ADD750
N-ACETYL TRIMETHYLCOLCHICINIC ACID
METHYL ETHER see CNG938
3-ACETYL-2,4,5-TRIMETHYL-PYRROLE see ADE050
ACETYL-l-TRP see ADE075
ACETYLTRYPTOPHAN see ADE075
ACETYL-l-TRYPTOPHAN see ADE075
N-ACETYLTRYPTOPHAN see ADE075
N-ACETYL-l-TRYPTOPHAN see ADE075
(S)-N-ACETYLTRYPTOPHAN see ADE075
ACETYLZIRCONIUM, ACETONATE see PBL750
AC GA see PJS750
ACH CHLORIDE see ABO000
ACHIOTE see APE100
ACHROCIDIN see ABG750
ACHROMYCIN see TBX250
ACHROMYCIN HYDROCHLORIDE see TBX250
ACID see DJO000
ACIDAL BRIGHT PONCEAU 3R see FMU080
ACIDAL FAST ORANGE see HGC000
ACIDAL PONCEAU G see FMU070
ACID AMIDE see NCR000
ACID AMMONIUM CARBONATE see ANB250
ACID AMMONIUM FLUORIDE see ANJ000
ACID AMMONIUM SULFATE see ANJ500
ACID BLUE 9 see FMU059
ACID BRILLIANT RUBINE 2G see HJF500
ACID BRILLIANT SCARLET 3R see FMU080
ACID BUTYL PHOSPHATE see ADF250
ACID CHROME BLUE BA see HJF500
ACID COPPER ARSENITE see CNN500
ACIDE ACETIQUE (FRENCH) see AAT250
ACIDE ACETYLSALICYLIQUE (FRENCH) see ADA725
ACIDE ANISIQUE (FRENCH) see MPI000
ACIDE ARSENIEUX see ARI750
ACIDE ARSENIQUE LIQUIDE (FRENCH) see ARB250
ACIDE BENZOIQUE (FRENCH) see BCL750
ACIDE BROMACETIQUE (FRENCH) see BMR750
ACIDE BROMHYDRIQUE (FRENCH) see HHJ000
ACIDE CACODYLIQUE (FRENCH) see HKC000
ACIDE CARBOLIQUE (FRENCH) see PDN750
ACIDE CHLORACETIQUE (FRENCH) see CEA000

ACIDE CHLORHYDRIQUE (FRENCH) see HHL000
ACIDE 2-(4-CHLORO-2-METHYL-PHENOXY)PROPIONIQUE (FRENCH) see CIR500
ACIDE CHROMIQUE (FRENCH) see CMH250
ACIDE CRESYLIQUE (FRENCH) see CNW500
ACIDE CYANHYDRIQUE (FRENCH) see HHS000
ACIDE-2-(2,4-DICHLORO-PHENOXY)PROPIONIQUE (FRENCH) see DGB000
ACIDE DIMETHYLARSINIQUE (FRENCH) see HKC000
ACIDE DISELINO SALICYLIQUE see SBU150
ACIDE ETHYLENEDIAMINETETRACETIQUE (FRENCH) see EIX000
ACIDE 1-ETIL-7-METIL-1,8-NAFTIRIDIN-4-ONE-3-CARBOSSILICO (ITALIAN) see EID000
ACIDE FLUORHYDRIQUE (FRENCH) see HHU500
ACIDE FLUOROSILICIQUE (FRENCH) see SCO500
ACIDE FLUOSILICIQUE (FRENCH) see SCO500
ACIDE FORMIQUE (FRENCH) see FNA000
ACIDE (ISOBUTYL-4 PHENYL)-2 PROPIONIQUE (FRENCH) see IIU000
ACIDE METHYL-o-BENZOIQUE (FRENCH) see MPI000
ACIDE MONOCHLORACETIQUE (FRENCH) see CEA000
ACIDE-MONOFLUORACETIQUE (FRENCH) see FIC000
ACIDE NALIDIXICO (ITALIAN) see EID000
ACIDE NALIDIXIQUE (FRENCH) see EID000
ACIDE NICOTINIQUE (FRENCH) see NCQ900
ACIDE NITRIQUE (FRENCH) see NED500
l'ACIDE OLEIQUE (FRENCH) see OHU000
ACIDE OXALIQUE (FRENCH) see OLA000
ACIDE PERACETIQUE (FRENCH) see PCL500
ACIDE PHOSPHORIQUE (FRENCH) see PHB250
ACIDE PHTHALIQUE (FRENCH) see PHW250
ACIDE PICRIQUE (FRENCH) see PID000
ACIDE PROPIONIQUE (FRENCH) see PMU750
ACIDE SULFHYDRIQUE (FRENCH) see HIC500
ACIDE SULFURIQUE (FRENCH) see SOI500
d'ACIDE TANNIQUE (FRENCH) see TAD750
ACIDE TEREPHTHALIQUE (FRENCH) see TAN750
ACIDE THIOGLYCOLIQUE (FRENCH) see TFJ100
ACIDE TRICHLORACETIQUE (FRENCH) see TII250
ACIDE 2-(2,4,5-TRICHLORO-PHENOXY)PROPIONIQUE (FRENCH) see TIX500
ACID FAST ORANGE EGG see HGC000
ACID FAST RED FB see HJF500
ACIDIC METANIL YELLOW see MDM775
ACID LEAD ARSENATE see LCK000
ACID LEAD ORTHOARSENATE see LCK000
ACID LEATHER BROWN 2G see XMA000
ACID LEATHER ORANGE PGW see HGC000
ACID LEATHER RED BG see CMM330
ACID LEATHER RED KPR see FMU070
ACID LEATHER YELLOW PRW see MDM775
ACID LEATHER YELLOW R see MDM775
ACID LEATHER YELLOW T see FAG140
ACID LIGHT ORANGE G see HGC000
ACID METANIL YELLOW see MDM775
ACIDO ACETICO (ITALIAN) see AAT250
ACIDO o-ACETIL-BENZOICO (ITALIAN) see ADA725
ACIDO ACETILSALICILICO (ITALIAN) see ADA725
ACIDO p-AMINOBENZOICO see AIH600
ACIDO BROMIDRICO (ITALIAN) see HHJ000
ACIDO CIANIDRICO (ITALIAN) see HHS000
ACIDO CLORIDRICO (ITALIAN) see HHL000

ACIDO 2-(4-CLORO-2-METIL-FENOSSI)-PROPIONICO (ITALIAN) see CIR500
ACIDO-2-(2,4-DICLORO-FENOSSI)-PROPIONICO (ITALIAN) see DGB000
ACIDO (3,6-DICLORO-2-METOSSI)-BENZOICO (ITALIAN) see MEL500
ACIDO-5-FENIL-5-ETILBARBITURICO (ITALIAN) see EOK000
ACIDO FLUORIDRICO (ITALIAN) see HHU500
ACIDO FLUOSILICICO (ITALIAN) see SCO500
ACIDO FORMICO (ITALIAN) see FNA000
ACIDO FOSFORICO (ITALIAN) see PHB250
ACIDOGEN NITRATE see UTJ000
ACIDO-m-IDROSSIBENZOICO (ITALIAN) see HJI100
ACIDO MANDELICO see MAP000
ACIDOMONOCLOROACETICO (ITALIAN) see CEA000
ACIDO MONOFLUOROACETIO (ITALIAN) see FIC000
ACIDO NITRICO (ITALIAN) see NED500
ACIDO OSSALICO (ITALIAN) see OLA000
ACIDO PICRICO (ITALIAN) see PID000
ACID ORANGE 10 see HGC000
ACID ORANGE 11 see DDO200
ACID ORANGE 24 see XMA000
ACIDO SALICILICO (ITALIAN) see SAI000
ACIDO SOLFORICO (ITALIAN) see SOI500
ACIDO TRICLOROACETICO (ITALIAN) see TII250
ACIDO 2-(2,4,5-TRICLORO-FENOSSI)-PROPIONICO (ITALIAN) see TIX500
ACID PONCEAU 4R see FMU080
ACID PONCEAU R see FMU070
ACID POTASSIUM SULFATE see PKX750
ACID QUININE HYDROCHLORIDE see QIJ000
ACID RED 18 see FMU080
ACID RED 26 see FMU070
ACID RED 92 see ADG250
ACID RED 114 see CMM330
ACID RUBINE see HJF500
ACID SCARLET see FMU070
ACID SCARLET 3R see FMU080
ACID-SPAR see CAS000
ACID-TREATED HEAVY NAPHTHENIC DISTILLATE see MQV760
ACID-TREATED LIGHT NAPHTHENIC DISTILLATE see MQV770
ACID-TREATED LIGHT PARAFFINIC DISTILLATE see MQV775
ACID-TREATED RESIDUAL OIL see MQV872
ACIDUM ACETYLSALICYLICUM see ADA725
ACIDUM NICOTINICUM see NCQ900
ACID VIOLET see FAG120
ACID YELLOW 23 see FAG140
ACID YELLOW 36 see MDM775
ACID YELLOW T see FAG140
ACID YELLOW TRA see FAG150
ACIFLOCTIN see AEN250
ACIGENA see HCL000
ACILAN ORANGE GX see HGC000
ACILAN PONCEAU RRL see FMU070
ACILAN SCARLET V3R see FMU080
ACILAN TURQUOISE BLUE AE see FMU059
ACILAN YELLOW GG see FAG140
ACILETTEN see CMS750
ACILLIN see AIV500
ACIMETION see MDT740
ACIMETTEN see ADA725
ACINETTEN see AEN250

ACINITRAZOL see ABY900
ACINITRAZOLE see ABY900
ACINTENE A see PIH250
ACINTENE DP see MCC250
ACINTENE DP DIPENTENE see MCC250
ACINTENE O see PMQ750
ACISAL see ADA725
ACL-59 see PLD000
ACL 60 see SGG500
ACL 70 see DGN200
ACL 85 see TIQ750
ACNEGEL see BDS000
ACNESTROL see DKA600
ACOCANTHERIN see OKS000
ACON see VSK600
ACP 6 see PJS750
AC 8 (POLYMER) see PJS750
ACQUINITE see ADR000
ACQUINITE see CKN500
ACRALDEHYDE see ADR000
ACRICHINE see ARQ250
ACRIDINE see ADJ500
ACRIDINE, 9-(2,2-BIS(2-
CHLOROETHYL)HYDRAZINO)-,
MONOHYDROCHLORIDE see BID800
ACRIDINE-9-CARBOXAMIDE, N,N-DIETHYL-1,2,3,4-
TETRAHYDRO- see ADJ550
ACRIDINE-9-CARBOXAMIDE, 1,2,3,4-TETRAHYDRO-
N,N-DIETHYL- see ADJ550
ACRIDINE, 2-CHLORO-9-(2,2-
DIMETHYLHYDRAZINO)- see CGH800
3,6-ACRIDINEDIAMINE see DBN600
3,6-ACRIDINEDIAMINE, MONOHYDROCHLORIDE
(9CI) see PMH250
ACRIDINE, 9-(2,2-DIMETHYLHYDRAZINO)-,
MONOHYDROCHLORIDE see DSG330
ACRIDINE, 9-(MORPHOLINOAMINO)-,
MONO(METHYL SULFATE) see MRR115
ACRIDINE, 9-(MORPHOLINOCARBONYL)-1,2,3,4-
TETRAHYDRO- see MRR760
ACRIDINE ORANGE see BJF000
ACRIDINE ORANGE FREE BASE see BJF000
ACRIDINE, 9-(PIPERIDINOAMINO)- see PIN200
ACRIDINE, 1,2,3,4-TETRAHYDRO-9-
(MORPHOLINOCARBONYL)- see MRR760
ACRIDINE, 1,2,3,4-TETRAHYDRO-9-
(PIPERIDINOCARBONYL)- see PIU100
ACRIDINE YELLOW BASE see DBT200
4-(9-ACRIDINYL)-2-METHYL-3-
THIOSEMICARBAZONE ACETONE see ADQ600
ACRIFLAVIN see DBX400
ACRIFLAVINE mixture with PROFLAVINE see DBX400
ACRIFLAVINIUM CHLORIDE see DBX400
ACRIFLAVINIUM CHLORIDUM see DBX400
ACRIFLAVON see DBX400
ACRILAFIL see ADY500
ACRINAMINE see ARQ250
ACRINOL see EDW500
ACRIQUINE see ARQ250
ACRITET see ADX500
ACROART see PJS750
ACROLACTINE see EDW500
ACROLEIC ACID see ADS750
ACROLEIN see ADR000
ACROLEINA (ITALIAN) see ADR000
ACROLEIN ACETAL see DHH800
ACROLEIN DIMER see ADR500
ACROLEIN DIMER, stabilized (DOT) see ADR500

ACROLEINE (DUTCH, FRENCH) see ADR000
ACROLEIN, 2-METHYL- see MGA250
ACROMONA see MMN250
ACRONIZE see CMA750
ACRYLALDEHYD (GERMAN) see ADR000
ACRYLALDEHYDE see ADR000
ACRYLALDEHYDE DIETHYL ACETAL see DHH800
ACRYLAMIDE see ADS250
ACRYLAMIDE, N-METHYL- see MGA300
ACRYLATE-ACRYLAMIDE RESINS see ADS400
ACRYLATE d'ETHYLE (FRENCH) see EFT000
ACRYLATE de METHYLE (FRENCH) see MGA500
ACRYLIC ACID see ADS750
ACRYLIC ACID, inhibited (DOT) see ADS750
ACRYLIC ACID BUTYL ESTER see BPW100
ACRYLIC ACID n-BUTYL ESTER (MAK) see BPW100
ACRYLIC ACID, 2-CYANO-, ETHYL ESTER
(6CI,7CI,8CI) see EHP700
ACRYLIC ACID, DIESTER with TETRAETHYLENE
GLYCOL see ADT050
ACRYLIC ACID, DIESTER with TRIETHYLENE
GLYCOL see TJQ100
ACRYLIC ACID, ETHYLENE ESTER see EIP000
ACRYLIC ACID, ETHYLENE GLYCOL DIESTER see
EIP000
ACRYLIC ACID ETHYL ESTER see EFT000
ACRYLIC ACID, GLACIAL see ADS750
ACRYLIC ACID-2-HYDROXYPROPYL ESTER see
HNT600
ACRYLIC ACID ISOBUTYL ESTER see IIK000
ACRYLIC ACID-2-METHOXYETHYL ESTER see
MIF750
ACRYLIC ACID, 2-METHYL- see MDN250
ACRYLIC ACID METHYL ESTER (MAK) see MGA500
ACRYLIC ACID, 2-METHYL-, METHYL ESTER see
MLH750
ACRYLIC ACID,
OXYBIS(ETHYLENEOXYETHYLENE) ESTER see
ADT050
ACRYLIC ALDEHYDE see ADR000
ACRYLIC AMIDE see ADS250
ACRYLITE see PKB500
ACRYLNITRIL (GERMAN, DUTCH) see ADX500
ACRYLON see ADX500
ACRYLONITRILE see ADX500
ACRYLONITRILE, inhibited (DOT) see ADX500
2-ACRYLONITRILE, 1-HYDROXYETHYL- see HKY650
ACRYLONITRILE MONOMER see ADX500
ACRYLONITRILE POLYMER with STYRENE see
ADY500
ACRYLONITRILE-STYRENE COPOLYMER see
ADY500
ACRYLONITRILE-STYRENE POLYMER see ADY500
ACRYLONITRILE-STYRENE RESIN see ADY500
ACRYLOPHENONE see PMQ250
ACRYLSAEUREAETHYLESTER (GERMAN) see
EFT000
ACRYLSAEUREMETHYLESTER (GERMAN) see
MGA500
ACRYPET see PKB500
ACS see ADY500
ACS see AJD000
ACTEDRON see BBK000
ACTELIC see DIN800
ACTELLIC see DIN800
ACTELLIFOG see DIN800
ACTICARBONE see CBT500
ACTICEL see SCH002

ACTI-CHLORE see CDP000
ACTINOLITE ASBESTOS see ARM260
ACTINOMYCIN 1048A see AEC000
ACTINOMYCIN 2104L see AEB750
ACTINOMYCIN L see AEB750
ACTINOMYCIN S see AEC000
ACTIVATED ALUMINUM OXIDE see AHE250
ACTIVATED ATTAPULGITE see PAE750
ACTIVATED CARBON see CBT500
ACTIVATED CARBON see CDI000
ACTIVE ACETYL ACETATE see EFS000
ACTIVE DICUMYL PEROXIDE see DGR600
ACTIVOL see ALT250
ACTOR Q see DVR200
ACTOZINE see PGA750
ACTOZINE see BCA000
AC-TRY see ADE075
ACTYBARYTE see BAP000
ACTYLOL see LAJ000
ACYLPYRIN see ADA725
ACYTOL see LAJ000
5-ACZ see ARY000
AD 1 see AGX000
AD1M see AGX000
ADAB see DPO200
ADALIN see BNK000
ADAMANTANAMINE HYDROCHLORIDE see
AED250
1-ADAMANTANAMINE HYDROCHLORIDE see
AED250
ADAMANTINE HYDROCHLORIDE see AED250
ADAMANTYLAMINE HYDROCHLORIDE see AED250
1-ADAMANTYLAMINE HYDROCHLORIDE see
AED250
ADAMSITE see PDB000
ADAMYCIN see HOH500
ADANON see MDO750
ADC AURAMINE O see IBA000
ADC BRILLIANT GREEN CRYSTALS see BAY750
ADCHEM GMO see GGR200
ADDISOMNOL see BNK000
ADDUKT HEXACHLORCYKLOPENTADIENU S
CYKLOPENTADIENEM (CZECH) see HCN000
ADELFAN see RDK000
ADELPHANE see RDK000
ADELPHIN see RDK000
ADELPHIN-ESIDREX-K see RDK000
ADENINE see AEH000
ADENINE ARABINOSIDE see AQQ900
ADENINIMINE see AEH000
ADENOCK see ZVJ000
ADENOSINE-5'-MONOPHOSPHATE see AOA125
ADENOSINE-5-MONOPHOSPHORIC ACID see
AOA125
ADENOSINE-5'-MONOPHOSPHORIC ACID see
AOA125
ADENOSINE PHOSPHATE see AOA125
ADENOSINE-5'-PHOSPHATE see AOA125
ADENOSINE-5'-PHOSPHORIC ACID see AOA125
ADENOVITE see AOA125
ADENYL see AOA125
ADENYLIC ACID see AOA125
tert-ADENYLIC ACID see AOA125
ADEPSINE OIL see MQV750
ADERGON see AOR500
ADERMINE see PPK250
ADERMINE HYDROCHLORIDE see PPK500
ADHERE see MIQ075

ADHESIVE 502 see EHP700
ADIAZINE see PPP500
ADILACTETTEN see AEN250
ADINE 0102 see PCC480
ADIPAMIDE see AEN000
ADIPAN see BBK000
ADIPEX see MDQ500
ADIPEX see MDT600
ADIPIC ACID see AEN250
ADIPIC ACID, BIS(2-ETHOXYETHYL) ESTER see
BJO225
ADIPIC ACID BIS(2-ETHYLHEXYL) ESTER see
AEO000
ADIPIC ACID DIAMIDE see AEN000
ADIPIC ACID DIBUTYL ESTER see AEO750
ADIPIC ACID DINITRILE see AER250
ADIPIC ACID NITRILE see AER250
ADIPIC ACID, POLYMER with 1,4-BUTANEDIOL and
METHYLENEDI-p-PHENYLENE ISOCYANATE see
PKM250
ADIPIC ACID, POLYMER with 1,4-BUTANEDIOL,
METHYLENEDI-p-PHENYLENE ISOCYANATE and
2,2'-(p-PHENYLENEDIOXY)DIETHANOL see PKM500
ADIPIC ACID, POLYMER with ETHYLENE GLYCOL
and METHYLENEDI-p-PHENYLENE ISOCYANATE
see PKL750
ADIPIC DIAMIDE see AEN000
ADIPIC KETONE see CPW500
ADIPINIC ACID see AEN250
ADIPLON see PPO000
ADIPODINITRILE see AER250
ADIPOL 2EH see AEO000
ADIPONITRILE see AER250
ADJUDETS see BBK500
(−)-ADLUMIDINE see CBF550
l-ADLUMIDINE see CBF550
l-ADM see AES750
ADMER PB 02 see PMP500
ADM HYDROCHLORIDE see HKA300
ADMUL see OAV000
ADNEPHRINE see VGP000
ADO see AGX000
ADOBACILLIN see AIV500
ADOGENEN 142 see OBC000
ADOL see OAX000
ADOL see HCP000
ADOL 68 see OAX000
ADONAL see EOK000
ADORM see TDA500
ADR see HKA300
ADRAN see IIU000
ADRENAL see VGP000
1-ADRENALIN see VGP000
ADRENALIN-MEDIHALER see VGP000
ADRENAMINE see VGP000
ADRENAN see VGP000
ADRENAPAX see VGP000
ADRENASOL see VGP000
ADRENATRATE see VGP000
ADRENODIS see VGP000
ADRENOHORMA see VGP000
ADRENOR see NNO500
ADRENUTOL see VGP000
ADRIACIN see HKA300
ADRIAMYCIN see AES750
ADRIAMYCIN see HKA300
ADRIAMYCIN-HCl see AES750

ADRIAMYCIN, HYDROCHLORIDE see HKA300
ADRIAMYCIN SEMIQUINONE see AES750
ADRIANOL see SPC500
ADRIBLASTIN see HKA300
ADRIBLASTINA see AES750
ADRIBLASTINE see HKA300
ADRINE see VGP000
ADRIXINE see BBK500
ADROIDIN see PAN100
ADRONAL see CPB750
ADROYD see PAN100
ADRUCIL see FMM000
ADULSIN see MIF760
ADVASTAB 401 see BFW750
ADVASTAB 800 see TFD500
ADVASTAB 17 MO see BKK750
ADVAWAX 140 see OAV000
AE see AGX000
AE0047 see WAT230
(S)-AE0047 see NDY550
(S)-(+)-AE 0047 see NDY550
AED see CQJ750
AENH (GERMAN) see ENV000
AERO see MCB000
AERO-CYANAMID see CAQ250
AERO CYANAMID GRANULAR see CAQ250
AERO CYANAMID SPECIAL GRADE see CAQ250
AERO CYANATE see PLC250
AERO liquid HCN see HHS000
AEROL 1 (pesticide) see TIQ250
AEROSEB-DEX see SOW000
AEROSEB-HC see CNS750
AEROSIL see SCH002
AEROSOL GPG see DJL000
AEROSOL of THERMOVACUUM CADMIUM see
CAK000
AEROTEX GLYOXAL 40 see GIK000
AEROTHENE MM see MJP450
AEROTHENE TT see MIH275
AERUGIDIOL see AET600
AESCIN SODIUM SALT see EDM000
AESCULETIN DIMETHYL ETHER see DRS800
AESCUSAN SODIUM SALT see EDM000
AETHALDIAMIN (GERMAN) see EEA500
AETHANETHIOL (GERMAN) see EMB100
AETHANOL (GERMAN) see EFU000
AETHANOLAMIN (GERMAN) see EEC600
AETHER see EJU000
AETHIONIN see EEI000
AETHON see ENY500
AETHOPROPAZIN see DIR000
2-AETHOXY-AETHYLACETAT (GERMAN) see EES400
2-AETHOXY-6,9-DIAMINOACRIDINLACTAT
(GERMAN) see EDW500
p-AETHOXYPHYLHARNSTOFF (GERMAN) see
EFE000
AETHYLACETAT (GERMAN) see EFR000
AETHYLACRYLAT (GERMAN) see EFT000
AETHYL-AETHANOL-NITROSOAMIN (GERMAN) see
ELG500
AETHYLALKOHOL (GERMAN) see EFU000
AETHYLAMINE (GERMAN) see EFU400
2-AETHYLAMINO-4-CHLOR-6-ISOPROPYLAMINO-
1,3,5-TRIAZIN (GERMAN) see ARQ725
2-AETHYLAMINO-4-ISOPROPYLAMINO-6-CHLOR-
1,3,5-TRIAZIN (GERMAN) see ARQ725
AETHYLANILIN (GERMAN) see EGK000
AETHYLBENZOL (GERMAN) see EGP500

AETHYLBUTYLKETON (GERMAN) see EHA600
AETHYL-N-BUTYL-NITROSOAMIN (GERMAN) see
EHC000
AETHYLCARBAMAT (GERMAN) see UVA000
AETHYLCHLORID (GERMAN) see EHH000
AETHYL-CHLORVYNOL see CHG000
AETHYL-2-(3',5'-DIJOD-4'-OXYBENZOYL)-3
CUMARON see EID200
S-AETHYL-N,N-DIPROPYLTHIOLCARBAMAT
(GERMAN) see EIN500
AETHYLENBROMID (GERMAN) see EIY500
AETHYLENCHLORID (GERMAN) see EIY600
AETHYLENECHLORHYDRIN (GERMAN) see EIU800
AETHYLENEDIAMIN (GERMAN) see EEA500
AETHYLENGLYKOLAETHERACETAT (GERMAN)
see EES400
AETHYLENGLYKOLMETHYLAETHERACETAT
(GERMAN) see EJJ500
AETHYLENGLYKOL-MONOMETHYLAETHER
(GERMAN) see EJH500
AETHYLENIMIN (GERMAN) see EJM900
AETHYLENOXID (GERMAN) see EJN500
AETHYLENSULFID (GERMAN) see EJP500
AETHYLFORMIAT (GERMAN) see EKL000
AETHYLHARNSTOFF und NATRIUMNITRIT
(GERMAN) see EQE000
AETHYLHARNSTOFF und NITRIT (GERMAN) see
EQE000
1-AETHYLHEXANOL (GERMAN) see EKQ000
AETHYLIDENCHLORID (GERMAN) see DFF809
AETHYLIS see EHH000
AETHYLIS CHLORIDUM see EHH000
AETHYLMERCAPTAN (GERMAN) see EMB100
AETHYLMETHYLKETON (GERMAN) see MKA400
O-AETHYL-O-(3-METHYL-4-
METHYLTHIOPHENYL)-ISOPROPYLAMIDO-
PHOSPHORSAEURE ESTER (GERMAN) see FAK000
O-AETHYL-O-n(4-NITROPHENYL)-PHENYL-
MONOTHIOPHOSPHONAT (GERMAN) see EBD700
AETHYLNITROSO-HARNSTOFF (GERMAN) see
ENV000
AETHYLNITROSOURETHAN (GERMAN) see NKE500
5-AETHYL-5-PENTYL-(2')-BARBITURSAEURE
(GERMAN) see PBS250
O-AETHYL-S-PHENYL-AETHYL-
DITHIOPHOSPHONAT (GERMAN) see FMU045
5-AETHYL-5-PHENYL-HEXAHYDROPYRIMIDIN-4,6-
DION (GERMAN) see DBB200
AETHYL-4-PICOLYLNITROSAMIN (GERMAN) see
NLH000
N-AETHYLPIPERIDIN (GERMAN) see EOS500
AETHYLRHODANID (GERMAN) see EPP000
O-AETHYL-O-(2,4,5-TRICHLORPHENYL)-
AETHYLTHIONOPHOSPHONAT (GERMAN) see
EPY000
AETHYLURETHAN (GERMAN) see UVA000
AETHYL-VINYL-NITROSOAMIN (GERMAN) see
NKF000
AETT see ACL750
AF 101 see DXQ500
AF 260 see AHC000
AF 864 see BBW500
AF-2 (preservative) see FQN000
AFASTOGEN BLUE 5040 see DNE400
AFATIN see BBK500
AFAXIN see VSK600
AFBI see AEU250
A.F. BLUE No. 1 see FMU059

AFCOLAC B 101 see SMR000
AFCOLENE see SMQ500
AFCOLENE 666 see SMQ500
AFCOLENE S 100 see SMQ500
AFESIN see CKD500
AFFIRM see ARW200
AFIBRIN see AJD000
AFICIDE see BBQ500
A-FIL CREAM see TGG760
AFI-TIAZIN see PDP250
AFL see AEW500
AFL 1081 see FFF000
AFLATOXICOL see AEW500
AFLATOXICOL NATURAL EPIMER see AEW500
AFLATOXIN see AET750
AFLATOXIN B see AEU250
AFLATOXIN B1 see AEU250
AFLATOXIN B2 see AEU750
AFLATOXIN G1 see AEV000
AFLATOXIN G1 mixed with AFLATOXIN B1 see
AEV250
AFLATOXIN M1 see AEW000
AFLATOXIN Ro see AEW500
AFLIX see DRR200
AFLON see TAI250
AFLUON see MAF500
AFNOR see CJJ000
A.F.ORANGE No. 2 see TGW000
A.F. RED No. 1 see FAG018
A.F. RED No. 5 see XRA000
AFRICAN COFFEE TREE see CCP000
AFTATE see TGB475
A.F. VIOLET No 1 see FAG120
A.F YELLOW No. 2 see FAG130
A.F. YELLOW No. 3 see FAG135
A.F. YELLOW NO. 4 see FAG140
A.F. YELLOW NO. 5 see FAG150
AG 3 see CBT500
AG 5 see CBT500
AG 3 (ADSORBENT) see CBT500
AG 5 (ADSORBENT) see CBT500
AGALITE see TAB750
AGALLO FORTE see MEP250
AGALLOL see MEP250
AGALLOLAT see MEP250
AGALOL see MEP250
AGAR see AEX250
AGAR-AGAR see AEX250
AGAR AGAR FLAKE see AEX250
AGAR-AGAR GUM see AEX250
AGATE see SCI500
AGATE see SCJ500
AGC see GFA000
AGE see AGH150
AGEDAL see DPH600
AGEFLEX BGE see BRK750
AGEFLEX FM-1 see DPG600
AGEFLEX FM-4 see BQD250
AGELFLEX FM-10 see IKM000
AGENAP see NAR000
AGENE see NGQ500
AGENT 504 see DAI600
AGENT AT 717 see PKQ250
AGENT BLUE see HKC000
AGERATOCHROMENE see AEX850
AGERITE see BLE500
AGERITEDPPD see BLE500
AGERITE POWDER see PFT500

AGERITE RESIN D see TLP500
AGERITE WHITE see NBL000
AGIDOL see BFW750
AGIDOL 7 see MJN250
AGILENE see PJS750
AGIOLAN see VSK600
AGI TALC, BC 1615 see TAB750
AGLICID see BSQ000
AGOFOLLIN see EDR000
AGORAL see PDO750
AGOSTILBEN see DKA600
AGOVIRIN see TBG000
AGRAZINE see PDP250
AGREFLAN see DUV600
AGRIA 1050 see DSQ000
AGRIBON see SNN300
AGRICIDE MAGGOT KILLER (F) see CDV100
AGRICULTURAL LIMESTONE see CAO000
AGRIDIP see CNU750
AGRIFLAN 24 see DUV600
AGRIMEK see ARW200
AGRI-MYCIN see SLY500
AGRIMYCIN 17 see SLW500
AGRION see SGH500
AGRISIL see EPY000
AGRISOL G-20 see BBQ500
AGRISTREP see SLY500
AGRITAN see DAD200
AGRITOX see CIR250
AGRITOX see EPY000
AGRIYA 1050 see DSQ000
A-GRO see MNH000
AGROCERES see HAR000
AGROCIDE see BBQ500
AGROFOROTOX see TIQ250
AGRONAA see NAK500
AGRONEXIT see BBQ500
AGROSAN see ABU500
AGROSAND see ABU500
AGROSAN GN 5 see ABU500
AGROSOL see MLF250
AGROSOL S see CBG000
AGROTHION see DSQ000
AGROXONE see CIR250
AGROXONE 3 see SIL500
AGROX 2-WAY and 3-WAY see CBG000
AGRYPNAL see EOK000
AGSTONE see CAO000
AH see DBM800
AHCOCID FAST SCARLET R see FMU070
AHCO DIRECT BLACK GX see AQP000
AH-289 HYDROCHLORIDE see CDR250
AHYDOL (RUSSIAN) see TMJ000
A 66 HYDROCHLORIDE see MNV750
A-HYDROCORT see HHR000
AHYPNON see MKA250
AI 3-22542 see DKC800
AI3-25722 see BHL800
AI3-29158 see AHJ750
AI3-29750 see THI500
AI3-35966 see ISZ000
AI 3-51408 see DCD050
AIBN see ASL750
TRANS-AID see ANW750
AIMAX see MLJ500
D-P-A INJECTION see PAG200
AIP see AHE750
AIR, refrigerated liquid see AFG250

AIRBRON see ACH000
AIREDALE BLACK ED see AQP000
AIREDALE BLUE D see CMO500
AIREDALE BLUE 2BD see CMO000
AIREDALE BLUE FFD see CMN750
AIREDALE CARMOISINE see HJF500
AIREDALE YELLOW T see FAG140
AIR-FLO GREEN see CNN500
AIRLOCK see CAU500
AIR, compressed (UN 1002) (DOT) see AFG250
AIR, refrigerated liquid (cryogenic liquid) (UN 1003) (DOT)
see AFG250
AIR, refrigerated liquid (cryogenic liquid) non-pressurized
(UN 1003) (DOT) see AFG250
AISELAZINE see HGP500
AISEMIDE see CHJ750
AITC see AGJ250
AIZEN ACID PHLOXINE PB see ADG250
AIZEN AURAMINE see IBA000
AIZEN BRILLIANT BLUE FCF see FMU059
AIZEN BRILLIANT SCARLET 3RH see FMU080
AIZEN CRYSTAL VIOLET see AOR500
AIZEN CRYSTAL VIOLET EXTRA PURE see AOR500
AIZEN DIAMOND GREEN GH see BAY750
AIZEN DIRECT BLUE 2BH see CMO000
AIZEN DIRECT DEEP BLACK EH see AQP000
AIZEN DIRECT DEEP BLACK GH see AQP000
AIZEN DIRECT DEEP BLACK RH see AQP000
AIZEN DIRECT SKY BLUE 5BH see CMO500
AIZEN EOSINE GH see BNH500
AIZEN FOOD ORANGE No. 2 see TGW000
AIZEN FOOD RED No. 5 see XRA000
AIZEN FOOD VIOLET No 1 see FAG120
AIZEN FOOD YELLOW NO. 5 see FAG150
AIZEN METANIL YELLOW see MDM775
AIZEN METHYLENE BLUE BH see BJI250
AIZEN PONCEAU RH see FMU070
AIZEN PRIMULA BROWN BRLH see CMO750
AIZEN RHODAMINE 6GCP see RGW000
AIZEN TARTRAZINE see FAG140
AIZEN URANINE see FEW000
AJAX GMO see GGR200
AJINOMOTO see MRL500
AK-33X see MAV750
AK (ADSORBENT) see CBT500
AKAR see DER000
AKARITHION see TNP250
AKARITOX see CKM000
AKERSTOX see AQV990
AKF-94 see DNU330
AKLOMIDE see AFH400
AKLOMIX see AFH400
AKLOMIX-3 see HMY000
AKNADI EXTRACT see SLO100
AKOTIN see NCQ900
AKRICHIN see ARQ250
AKROCHEM ETU-22 see IAQ000
AKROLEIN (CZECH) see ADR000
AKROLEINA (POLISH) see ADR000
AKRO-MAG see MAH500
AKRO-ZINC BAR 85 see ZKA000
AKRYLAMID (CZECH) see ADS250
AKRYLONITRYL (POLISH) see ADX500
AKTAMIN see NNO500
AKTIKON see ARQ725
AKTIKON PK see ARQ725
AKTINIT A see ARQ725
AKTINIT PK see ARQ725

AKTINIT S see BJP000
AKTIVEX see CCX000
AKTIVIN see CDP000
AKULON see PJY500
AKZO CHEMIE MANEB see MAS500
AL-100 see ZVJ000
ALABASTER see CAX750
ALABASTER NO. 3 see FAG150
ALACHLOR (USDA) see CFX000
ALACINE see PFC750
ALAIXOL II see OAT000
AL-ALCHILI (ITALIAN) see DNI600
ALAMINE 4 see DXW000
ALAMINE 6 see HCO500
ALAMINE 7 see OBC000
ALAMINE 11 see OHM700
ALAMINE 7D see OBC000
ALANE see AHB500
ALANEX see CFX000
ALANINE see AFH625
l-ALANINE see AFH625
(S)-ALANINE see AFH625
l-(+)-ALANINE see AFH625
α-ALANINE see AFH625
l-α-ALANINE see AFH625
ALANINE, 3-(((3-AMINO-4-
HYDROXYPHENYL)PHENYLARSINO)THIO)- see
AKI900
ALANINE, 3-(p-CHLOROPHENYL)-, dl- see FAM100
DL-ALANINE, N-(2,6-DIMETHYLPHENYL)-N-
(METHOXYACETYL)-, METHYL ESTER (9CI) see
MDM100
ALANINE NITROGEN MUSTARD see PED750
ALANINE, PHENYL- see PEC750
ALANINE, 3-PHENYL- see PEC750
l-ALANINE, PHENYL- see PEC750
β-ALANINOL see PMM250
β-ALANYL-l-HISTIDINE see CCK665
ALAR see DQD400
ALAR-85 see DQD400
ALATHON see PJS750
ALATHON 14 see PJS750
ALATHON 15 see PJS750
ALATHON 5B see PJS750
ALATHON 1560 see PJS750
ALATHON 6600 see PJS750
ALATHON 7026 see PJS750
ALATHON 7040 see PJS750
ALATHON 7050 see PJS750
ALATHON 7140 see PJS750
ALATHON 7511 see PJS750
ALATHON 71XHN see PJS750
ALAUN (GERMAN) see AGX000
ALAZIN see PFC750
ALAZINE see PFC750
ALBAGEL PREMIUM USP 4444 see BAV750
ALBAMINE see SNQ710
ALBAMYCIN see NOB000
ALBAMYCIN SODIUM see NOB000
ALBEMAP see BBK500
ALBEXAN see SNM500
ALBIGEN A see PKQ250
ALBIOTIC see LGD000
ALBOLINE see MQV750
ALBON see SNN300
ALBONE see HIB050
ALBONE 35 see HIB010

ALBONE 50 see HIB010
ALBONE 70 see HIB010
ALBONE 35CG see HIB010
ALBONE 50CG see HIB010
ALBONE 70CG see HIB010
ALBORAL see DCK759
ALBOSAL see SNM500
ALBUCID see SNQ710
ALBUMIN see AFI850
ALBUMIN MACRO AGGREGATES see AFI850
ALCALASE see BAC000
ALCHLOQUIN see CHR500
ALCIDE see CDW450
ALCIDE LD see SFT500
ALCOA 331 see AHC000
ALCOA F 1 see AHE250
ALCOA SODIUM FLUORIDE see SHF500
ALCOBAM NM see SGM500
ALCOBAM ZM see BJK500
ALCOHOL see EFU000
ALCOHOL, anhydrous see EFU000
ALCOHOL, dehydrated see EFU000
ALCOHOL C-8 see OEI000
ALCOHOL C-9 see NNB500
ALCOHOL C-10 see DAI600
ALCOHOL C-11 see UNA000
ALCOHOL C-12 see DXV600
ALCOHOL C-16 see HCP000
ALCOHOL, DENATURED see AFJ000
ALCOHOLS, n.o.s. (UN 1987) (DOT) see EFU000
ALCOHOLS, toxic, n.o.s. (UN 1986) (DOT) see EFU000
ALCOOL ALLILCO (ITALIAN) see AFV500
ALCOOL ALLYLIQUE (FRENCH) see AFV500
ALCOOL AMILICO (ITALIAN) see IHP000
ALCOOL AMYLIQUE (FRENCH) see AOE000
ALCOOL BUTYLIQUE (FRENCH) see BPW500
ALCOOL BUTYLIQUE SECONDAIRE (FRENCH) see BPW750
ALCOOL BUTYLIQUE TERTIAIRE (FRENCH) see BPX000
ALCOOL ETHYLIQUE (FRENCH) see EFU000
ALCOOL ETILICO (ITALIAN) see EFU000
l'ALCOOL n-HEPTYLIQUE PRIMAIRE (FRENCH) see HBL500
ALCOOL ISOAMYLIQUE (FRENCH) see IHP000
ALCOOL ISOBUTYLIQUE (FRENCH) see IIL000
ALCOOL ISOPROPILICO (ITALIAN) see INJ000
ALCOOL ISOPROPYLIQUE (FRENCH) see INJ000
ALCOOL METHYL AMYLIQUE (FRENCH) see MKW600
ALCOOL METHYLIQUE (FRENCH) see MGB150
ALCOOL METILICO (ITALIAN) see MGB150
ALCOOL PROPILICO (ITALIAN) see PND000
ALCOOL PROPYLIQUE (FRENCH) see PND000
ALCOPAN-250 see PAG200
ALCOPHOBIN see DXH250
ALCOPOL O see DJL000
ALCOTEX 88/05 see PKP750
ALCOTEX 88/10 see PKP750
ALCOWAX 6 see PJS750
ALDACOL Q see PKQ250
ALDACTAZIDE see AFJ500
ALDACTIDE see AFJ500
ALDACTONE see AFJ500
ALDACTONE A see AFJ500
ALDECARB see CBM500
ALDEHYDE-14 see UJJ000
ALDEHYDE ACETIQUE (FRENCH) see AAG250

ALDEHYDE ACRYLIQUE (FRENCH) see ADR000
ALDEHYDE AMMONIA see AAG500
ALDEHYDE B see COU500
ALDEHYDE BUTYRIQUE (FRENCH) see BSU250
ALDEHYDE C-6 see HEM000
ALDEHYDE C-8 see OCO000
ALDEHYDE C-9 see NMW500
ALDEHYDE C-10 see DAG000
ALDEHYDE C-18 see CNF250
ALDEHYDE C-10 DIMETHYLACETAL see AFJ700
ALDEHYDE C-12, MNA see MQI550
ALDEHYDECOLLIDINE see EOS000
ALDEHYDE CROTONIQUE (FRENCH) see COB260
ALDEHYDE C-11, UNDECYLENIC see ULJ000
ALDEHYDE-2-ETHYLBUTYRIQUE (FRENCH) see DHI000
ALDEHYDE FORMIQUE (FRENCH) see FMV000
ALDEHYDE M.N.A. see MQI550
ALDEHYDE PROPIONIQUE (FRENCH) see PMT750
ALDEHYDES see AFJ800
ALDEHYDINE see EOS000
ALDEHYDODICHLOROMALEIC ACID see MRU900
ALDEIDE ACETICA (ITALIAN) see AAG250
ALDEIDE ACRILICA (ITALIAN) see ADR000
ALDEIDE BUTIRRICA (ITALIAN) see BSU250
ALDEIDE FORMICA (ITALIAN) see FMV000
ALDICARBE (FRENCH) see CBM500
ALDICARB (USDA) see CBM500
ALDIFEN see DUZ000
AL-DIISOBUTYL see DNI600
ALDINAMID see POL500
ALDO-28 see OAV000
ALDO 40 see GGR200
ALDO-72 see OAV000
ALDO HMS see OAV000
ALDOL see AAH750
ALDOMET see DNA800
ALDOMETIL see DNA800
ALDOMIN see DNA800
ALDO MO-FG see GGR200
ALDO MS see OAV000
ALDO MSA see OAV000
ALDO MSLG see OAV000
ALDOMYCIN see NGE500
ALDOXIME see AAH250
ALDREX see AFK250
ALDREX 30 see AFK250
ALDRICH see DOJ200
ALDRIN see AFK250
ALDRIN, cast solid (DOT) see AFK250
ALDRINE (FRENCH) see AFK250
ALDRITE see AFK250
ALDROSOL see AFK250
ALDYL A see PJS750
ALENTIN see BSM000
ALEPSIN see DNU000
ALERYL see BBV500
ALESTEN see SNQ710
ALEUDRIN see DMV600
ALEVIATIN see DKQ000
ALFACOL see TGJ050
ALFAMAT see GFA000
ALFANAFTILAMINA (ITALIAN) see NBE700
ALFA-NAFTYLOAMINA (POLISH) see NBE700
ALFA-TOX see DCM750
ALFENOL 3 see PKF500
ALFENOL 9 see PKF500
ALFICETYN see CDP250

ALFIMID see DYC800
ALFOL 8 see OEI000
ALFOL 12 see DXV600
ALFUCIN see NGE500
ALGAE MEAL, DRIED see AFK925
ALGAFAN see PNA500
ALGAROBA see LIA000
ALGEON 22 see CFX500
ALGIL see DAM700
K'-ALGILINE see SEH000
ALGIMYCIN see ABU500
ALGIN see SEH000
ALGIN see CAM200
ALGIN (polysaccharide) see SEH000
ALGINATE KMF see SEH000
ALGIN GUM see CAO250
ALGINIC ACID see AFL000
ALGIPON L-1168 see SEH000
ALGISTAT see DFT000
ALGLOFLON see TAI250
ALGOCOR see EID200
ALGOFLON SV see TAI250
ALGOFRENE 22 see CFX500
ALGOFRENE TYPE 1 see TIP500
ALGOFRENE TYPE 2 see DFA600
ALGOFRENE TYPE 5 see DFL000
ALGOFRENE TYPE 6 see CFX500
ALGOFRENE TYPE 67 see ELN500
ALGOSEDIV see TEH500
ALGOTROPYL see HIM000
ALGRAIN see EFU000
ALGYLEN see TIO750
ALICYANATE see PLC250
ALIDOCHLOR see CFK000
ALINDOR see BRF500
ALIPHATIC and AROMATIC EPOXIDES see AFM250
ALIPHATIC CHLORINATED HYDROCARBONS see
CDV250
ALIQUAT 336 see MQH000
ALIQUAT 336N see MQH000
ALIQUAT 336-PTC see MQH000
ALITHON 7050 see PJS750
ALIZARINE CYANINE GREEN BASE see BLK000
ALKALIES see AFM500
ALKALOID H 3, from COLCHICUM ANTUMNALE see
MIW500
ALKAMID see PJY500
ALKANES see AFN250
ALKAPOL PEG-200 see PJT000
ALKAPOL PEG-300 see PJT000
ALKAPOL PEG-600 see PJT000
ALKAPOL PEG-6000 see PJT000
ALKAPOL PEG-8000 see PJT000
ALKAPOL PPG-1200 see PKI500
ALKARAU see RDK000
ALKARSODYL see HKC500
ALKASERP see RDK000
ALKATHENE see PJS750
ALKATHENE 200 see PJS750
ALKATHENE 22 300 see PJS750
ALKATHENE 17/04/00 see PJS750
ALKATHENE ARN 60 see PJS750
ALKATHENE RXDG33 see TAI250
ALKATHENE WJG 11 see PJS750
ALKATHENE WNG 14 see PJS750
ALKATHENE XDG 33 see PJS750
ALKATHENE XJK 25 see PJS750
ALK-AUBS see DXH250

ALK-ENZYME see BAC000
ALKERAN see PED750
ALKIRON see MPW500
ALKOHOL (GERMAN) see EFU000
ALKOHOLU ETYLOWEGO (POLISH) see EFU000
ALKOTEX see PKP750
ALKOVERT see PIB250
ALLBRI ALUMINUM PASTE and POWDER see AGX000
ALLBRI NATURAL COPPER see CNI000
ALLEDRYL see BBV500
(+)-ALLELRETHONYL (+)-cis,trans-
CHRYSANTHEMATE see AFR250
ALLERCLOR see TAI500
ALLERGAN 211 see DAS000
ALLERGAN B see BBV500
ALLERGEVAL see BBV500
ALLERGICAL see BBV500
ALLERGIN see TAI500
ALLERGIN see BBV500
ALLERGINA see BBV500
ALLERGISAN see TAI500
ALLERGIVAL see BBV500
ALLERON see PAK000
ALLETHRIN see AFR250
d-ALLETHRIN see AFR250
d-trans ALLETHRIN see AFR250
ALLETHRIN I see AFR250
ALLEVIATE see AFR250
ALLIDOCHLOR see CFK000
ALLIED PE 617 see PJS750
ALLILE (CLORURO di) (ITALIAN) see AGB250
ALLIL-GLICIDIL-ETERE (ITALIAN) see AGH150
1-ALLILOSSI-2,3 EPOSSIPROPANO (ITALIAN) see
AGH150
ALLILOWY ALKOHOL (POLISH) see AFV500
ALLOCAINE see AIT250
ALLODENE see BBK000
ALLO-GLAUCOTOXIGENIN see AFS800
ALLOGLAUCOTOXIGENIN see AFS800
ALLOGLUCOTOXIGENIN see AFS800
ALLOMALEIC ACID see FOU000
ALLO-OCIMENOL see LFX000
ALLOPURINOL see ZVJ000
ALLOXAN see AFT750
ALLOXAN MONOHYDRATE see MDL500
ALLOZYM see ZVJ000
ALLTEX see CDV100
ALLTOX see CDV100
ALLUMINIO(CLORURO DI) (ITALIAN) see AGY750
ALLUMINIO DIISOBUTIL-MONOCLORURO
(ITALIAN) see CGB500
ALLURAL see ZVJ000
ALLUVAL see BNP750
ALLYL ACETATE see AFU750
ALLYL AL see AFV500
ALLYL ALCOHOL see AFV500
ALLYL ALDEHYDE see ADR000
ALLYLALKOHOL (GERMAN) see AFV500
ALLYLAMINE see AFW000
p-ALLYLANISOLE see AFW750
5-ALLYL-1,3-BENZODIOXOLE see SAD000
ALLYL BROMIDE see AFY000
ALLYL CAPROATE see AGA500
ALLYL CAPRYLATE see AGM500
ALLYLCATECHOL METHYLENE ETHER see SAD000
ALLYLCHLORID (GERMAN) see AGB250
ALLYL CHLORIDE see AGB250
ALLYL CHLOROCARBONATE see AGB500

ALLYL CHLOROFORMATE see AGB500
ALLYL CHLOROFORMATE (DOT) see AGB500
ALLYL CINERIN see AFR250
ALLYL CINNAMATE see AGC000
ALLYL CYCLOHEXANEPROPIONATE see AGC500
3-ALLYLCYCLOHEXYL PROPIONATE see AGC500
1-ALLYL-3,4-DIMETHOXYBENZENE see AGE250
4-ALLYL-1,2-DIMETHOXYBENZENE see AGE250
ALLYLDIOXYBENZENE METHYLENE ETHER see SAD000
ALLYLE (CHLORURE d') (FRENCH) see AGB250
ALLYL ENANTHATE see AGH250
ALLYLENE see MFX590
ALLYL-2,3-EPOXYPROPYL ETHER see AGH150
ALLYLESTER KYSELINY CHLORMRAVENCI see AGB500
ALLYLESTER KYSELINY MRAVENCI see AGH000
ALLYLETHER see DBK000
ALLYL FORMATE see AGH000
ALLYLGLYCIDAETHER (GERMAN) see AGH150
ALLYL GLYCIDYL ETHER see AGH150
4-ALLYLGUAIACOL see EQR500
ALLYL HEPTANOATE see AGH250
ALLYL HEPTOATE see AGH250
ALLYL HEPTYLATE see AGH250
ALLYL HEXAHYDROPHENYLPROPIONATE see AGC500
ALLYL HEXANOATE (FCC) see AGA500
ALLYL HOMOLOG of CINERIN I see AFR250
4-ALLYL-1-HYDROXY-2-METHOXYBENZENE see EQR500
d,l-2-ALLYL-4-HYDROXY-3-METHYL-2-CYCLOPENTEN-1-ONE-d,l-CHRYSANTHEMUM MONOCARBOXYLATE see AFR250
ALLYLHYDROXYPHENYLARSINE OXIDE see AGQ775
ALLYLIC ALCOHOL see AFV500
ALLYL IODIDE see AGI250
ALLYL α-IONONE see AGI500
ALLYL ISOCYANATE see AGJ000
ALLYL ISORHODANIDE see AGJ250
ALLYL ISOSULFOCYANATE see AGJ250
ALLYL ISOTHIOCYANATE see AGJ250
ALLYL ISOTHIOCYANATE, stabilized (DOT) see AGJ250
ALLYL ISOVALERATE see ISV000
ALLYL ISOVALERIANATE see ISV000
3-ALLYL-4-KETO-2-METHYLCYCLOPENTENYL CHRYSANTHEMUMMONOCARBOXYLATE see AFR250
ALLYL MERCAPTAN see AGJ500
4-ALLYL-1-METHOXYBENZENE see AFW750
4-ALLYL-2-METHOXYPHENOL see EQR500
4-ALLYL-2-METHOXYPHENOL ACETATE see EQS000
5-ALLYL-5-(1-METHYLBUTYL)-2-THIOBARBITURATE SODIUM see SOX500
5-ALLYL-5-(1-METHYLBUTYL)-2-THIO-BARBITURIC ACID SODIUM SALT see SOX500
ALLYL 3-METHYLBUTYRATE see ISV000
1-ALLYL-3,4-METHYLENEDIOXYBENZENE see SAD000
4-ALLYL-1,2-METHYLENEDIOXYBENZENE see SAD000
5-ALLYL-1-METHYL-5-(1-METHYL-2-PENTYNYL)BARBITURIC ACID SODIUM SALT see MDU500

3-ALLYL-2-METHYL-4-OXO-2-CYCLOPENTEN-1-YL CHRYSANTHEMATE see AFR250
dl-3-ALLYL-2-METHYL-4-OXOCYCLOPENT-2-ENYL dl-cis trans CHRYSANTHEMATE see AFR250
ALLYL MUSTARD OIL see AGJ250
ALLYL OCTANOATE see AGM500
(±)-1-(β-(ALLYLOXY)-2,4-DICHLOROPHENETHYL)IMIDAZOLE see FPB875
1-ALLYLOXY-2,3-EPOXY-PROPAAN (DUTCH) see AGH150
1-ALLYLOXY-2,3-EPOXYPROPAN (GERMAN) see AGH150
1-(ALLYLOXY)-2,3-EPOXYPROPANE see AGH150
ALLYL PHENOXYACETATE see AGQ750
ALLYL PHENYLACETATE see PMS500
ALLYL-3-PHENYLACRYLATE see AGC000
ALLYL PHENYL ARSINIC ACID see AGQ775
ALLYL PROPYL DISULFIDE see AGR500
m-ALLYLPYROCATECHIN METHYLENE ETHER see SAD000
4-ALLYLPYROCATECHOL FORMALDEHYDE ACETAL see SAD000
ALLYLPYROCATECHOL METHYLENE ETHER see SAD000
ALLYLRETHRONYL dl-cis-trans-CHRYSANTHEMATE see AFR250
ALLYLSENFOEL (GERMAN) see AGJ250
ALLYL SEVENOLUM see AGJ250
ALLYL THIOCARBONIMIDE see AGJ250
ALLYL TRICHLORIDE see TJB600
ALLYL TRICHLOROSILANE see AGU250
ALLYLTRICHLOROSILANE, stabilized (DOT) see AGU250
4-ALLYLVERATROLE see AGE250
ALMEDERM see HCL000
ALMEFRIN see SPC500
ALMEFROL see VSZ450
ALMITE see AHE250
ALMOCARPINE see PIF000
ALMOCARPINE see PIF250
ALMOND ARTIFICIAL ESSENTIAL OIL see BAY500
ALMOND OIL BITTER, FFPA (FCC) see BLV500
ALOCHLOR see CFX000
ALODAN (GEROT) see DAM700
ALON see AHE250
ALOSITOL see ZVJ000
ALPEN see AIV500
ALPEN-N see SEQ000
ALPEROX C see LBR000
ALPHALIN see VSK600
ALPHA MEDOPA see DNA800
ALPHANAPHTHYL THIOUREA see AQN635
ALPHANAPHTYL THIOUREE (FRENCH) see AQN635
ALPHASOL OT see DJL000
ALPHASPRA see NAK500
ALPHASTEROL see VSK600
ALPHAZURINE see FMU059
ALPHEX FIT 221 see PJS750
AL-PHOS see AHE750
ALPINE TALC USP, BC 127 see TAB750
ALPINE TALC USP, BC 141 see TAB750
ALPINE TALC USP, BC 662 see TAB750
ALPINYL see HIM000
ALQOVERIN see BRF500
ALRATO see AQN635
ALRHEUMAT see BDU500
ALRHEUMUM see BDU500

ALSERIN see RDK000
ALTADIOL see EDS100
ALTAN see DMH400
ALTAX see BDE750
ALTCO SPERSE FAST YELLOW GFN NEW see AAQ250
ALTOWHITES see KBB600
ALTOX see AFK250
ALTRAD see EDO000
ALUDRINE see DMV600
ALULINE see ZVJ000
ALUM see AHG750
ALUMIGEL see AHC000
ALUMINA see AHE250
β-ALUMINA see AHE250
β-ALUMINA see AHG000
γ-ALUMINA see AHE250
β''-ALUMINA see AHG000
ALUMINA FIBRE see AGX000
ALUMINA HYDRATE see AHC000
ALUMINA HYDRATED see AHC000
α-ALUMINA (OSHA) see AHE250
ALUMINATE(1−), TETRAHYDRO-, SODIUM, (T-4)-(9CI) see SEM500
ALUMINA TRIHYDRATE see AHC000
α-ALUMINA TRIHYDRATE see AHC000
ALUMINIC ACID see AHC000
ALUMINIUM BRONZE see AGX000
ALUMINON see AGW750
ALUMINOPHOSPHORIC ACID see PHB500
ALUMINUM see AGX000
ALUMINUM 27 see AGX000
ALUMINUM A00 see AGX000
ALUMINUM ACID PHOSPHATE see PHB500
ALUMINUM ALLOY, Al,Be see BFP250
ALUMINUM ALUM see AHG750
ALUMINUM AMMONIUM SULFATE see AGX250
ALUMINUM BERYLLIUM ALLOY see BFP250
ALUMINUM BOROHYDRIDE (DOT) see AHG875
ALUMINUM BOROHYDRIDE in devices (DOT) see AHG875
ALUMINUM BROMIDE see AGX750
ALUMINUM BROMIDE, anhydrous (UN 1725) (DOT) see AGX750
ALUMINUM BROMIDE, solution (UN 2580) (DOT) see AGX750
ALUMINUM CHLORHYDRATE see AHA000
ALUMINUM CHLORHYDROL see AHA000
ALUMINUM CHLORHYDROXIDE see AHA000
ALUMINUMCHLORID (GERMAN) see AGY750
ALUMINUM CHLORIDE see AGY750
ALUMINUM CHLORIDE (1:3) see AGY750
ALUMINUM CHLORIDE, solution (DOT) see AGY750
ALUMINUM CHLORIDE, anhydrous (DOT) see AGY750
ALUMINUM CHLORIDE HYDROXIDE see AHA000
ALUMINUM CHLOROHYDROXIDE see AHA000
ALUMINUM DEHYDRATED see AGX000
ALUMINUM DEXTRAN see AHA250
ALUMINUM FLAKE see AGX000
ALUMINUM FLUORIDE see AHB000
ALUMINUM FLUORURE (FRENCH) see AHB000
ALUMINUM FOSFIDE (DUTCH) see AHE750
ALUMINUM HYDRATE see AHC000
ALUMINUM HYDRIDE see AHB500
ALUMINUM HYDROBORATE see AHG875
ALUMINUM HYDROXIDE see AHC000
ALUMINUM(III) HYDROXIDE see AHC000

ALUMINUM HYDROXIDE CHLORIDE see AHA000
ALUMINUM HYDROXIDE GEL see AHC000
ALUMINUM HYDROXIDE OXIDE see AHC250
ALUMINUM HYDROXYCHLORIDE see AHA000
ALUMINUM LACTATE see AHC750
ALUMINUM LITHIUM HYDRIDE see LHS000
ALUMINUM MAGNESIUM PHOSPHIDE see AHD250
ALUMINUM METAHYDROXIDE see AHC250
ALUMINUM METAL (OSHA) see AGX000
ALUMINUM MONOPHOSPHATE see PHB500
ALUMINUM MONOPHOSPHIDE see AHE750
ALUMINUM MONOSTEARATE see AHA250
ALUMINUM, molten (NA 9260) (DOT) see AGX000
ALUMINUM NICOTINATE see AHD650
ALUMINUM NITRATE (DOT) see AHD750
ALUMINUM(III) NITRATE (1:3) see AHD750
ALUMINUM OXIDE see AHE250
ALUMINUM OXIDE (2:3) see AHE250
α-ALUMINUM OXIDE see AHE250
β-ALUMINUM OXIDE see AHE250
γ-ALUMINUM OXIDE see AHE250
ALUMINUM OXIDE HYDRATE see AHC000
ALUMINUM OXIDE TRIHYDRATE see AHC000
ALUMINUM PHOSPHATE see PHB500
ALUMINUM PHOSPHATE (1:1) see PHB500
ALUMINUM PHOSPHATE, solution (DOT) see PHB500
ALUMINUM PHOSPHIDE see AHE750
ALUMINUM POWDER see AGX000
ALUMINUM POWDER, coated (UN 1309) (DOT) see AGX000
ALUMINUM POWDER, uncoated (UN 1396) (DOT) see AGX000
ALUMINUM PYRO POWDERS (OSHA) see AGX000
ALUMINUM SESQUIOXIDE see AHE250
ALUMINUM SODIUM FLUORIDE see SHF000
ALUMINUM SODIUM HYDRIDE see SEM500
ALUMINUM SODIUM OXIDE see AHG000
ALUMINUM SODIUM SULFATE see AHG500
ALUMINUM STEARATE (ACGIH) see AHA250
ALUMINUM SULFATE see AHG750
ALUMINUM SULFATE (2:3) see AHG750
ALUMINUM SULPHATE see AHG750
ALUMINUM TETRAHYDROBORATE see AHG875
ALUMINUM TRIBROMIDE see AGX750
ALUMINUM, TRIBROMOTRIMETHYLDI- see MGC225
ALUMINUM TRICHLORIDE see AGY750
ALUMINUM, TRICHLOROTRIMETHYLDI- see MGC230
ALUMINUM, TRIETHYL- see TJN750
ALUMINUM TRIFLUORIDE see AHB000
ALUMINUM TRIHYDRAT see AHC000
ALUMINUM TRIHYDRIDE see AHB500
α-ALUMINUM TRIHYDRIDE see AHB500
ALUMINUM TRIHYDROXIDE see AHC000
ALUMINUM TRINITRATE see AHD750
ALUMINUM, TRIPROPYL- see TMY100
ALUMINUM, TRIS(2-HYDROXYPROPANOATO-O¹),O²- (9CI) see AHC750
ALUMINUM, TRIS(LACTATO)- see AHC750
ALUMINUM, TRIS(2-METHYLPROPYL)-(9CI) see TKR500
ALUMINUM TRISULFATE see AHG750
ALUMINUM WELDING FUMES (OSHA) see AGX000
ALUMITE see AHE250
A 21 LUNDBECK see KFK000
ALUNDUM see AHE250
ALUNEX see TAI500

ALUNITINE see AHD650
ALUPHOS see PHB500
ALURAL see BNP750
ALURENE see CLH750
ALUSAL see AHC000
ALUTOR M 70 see PKB500
ALUZINE see CHJ750
ALVEDON see HIM000
ALVINOL see CHG000
ALVIT see DHB400
ALVYL see PKP750
ALYSINE see SJO000
ALZODEF see CAQ250
AMABEVAN see CBJ000
AMACEL BLUE GG see TBG700
AMACEL DEVELOPED NAVY SD see DCJ200
AMACEL PINK B see AKE250
AMACEL PURE BLUE B see TBG700
AMACEL YELLOW G see AAQ250
AMACID BLUE FG CONC see FMU059
AMACID CHROME BLUE R see HJF500
AMACID LAKE SCARLET 2R see FMU070
AMACID MILLING RED PRS see CMM330
AMACID YELLOW M see MDM775
AMACID YELLOW T see FAG140
AMADIL see HIM000
AMAIZO W 13 see SLJ500
AMALOX see ZKA000
AMANIL BLACK GL see AQP000
AMANIL BLACK WD see AQP000
AMANIL BLUE 2BX see CMO000
AMANIL SKY BLUE see CMO250
AMANIL SKY BLUE see CMO500
AMANIL SKY BLUE 6B see CMN750
AMANIL SUPRA BROWN LBL see CMO750
AMANTADINE HYDROCHLORIDE see AED250
AMAPLAST GREEN OZ see BLK000
AMARSAN see ABX500
AMARTHOL FAST RED TR BASE see CLK220
AMARTHOL FAST RED TR BASE see CLK235
AMARTHOL FAST RED TR SALT see CLK235
AMARTHOL FAST SCARLET G BASE see NMP500
AMARTHOL FAST SCARLET G SALT see NMP500
AMATIN see HCC500
AMAX see BDG000
AMAZOLON see AED250
AMB see AOC500
AMBEN see AIH600
AMBENYL see BAU750
AMBER ACID see SMY000
AMBEROL ST 140F see AHC000
AMBESIDE see SNM500
AMBLOSIN see AIV500
AMBOCHLORIN see CDO500
AMBOCLORIN see CDO500
AMBOFEN see CDP250
AMBRACYN see TBX250
AMBUSH see AHJ750
AMBUSH see CBM500
AMBYTHENE see PJS750
AMCAP see AOD125
AMCHEM 68-250 see CDS125
AMCHEM GRASS KILLER see TII250
AMCHEM R 14 see PKL750
AMCHEM 2,4,5-TP see TIX500
AMCHLOR see ANE500
AMCIDE see ANU650
AMCILL see AIV500

AMCILL see AOD125
AMCILL-S see SEQ000
AMCO see PMP500
AMD see DNA800
AMDEX see BBK500
AMDON GRAZON see PIB900
AMDRAM see DBA800
AMEBAN see CBJ000
AMEBARSONE see CBJ000
AMEBICIDE see EAN000
AMEBIL see CHR500
AMEDRINE see DBA800
AMEISENATOD see BBQ500
AMEISENMITTEL MERCK see BBQ500
AMEISENSAEURE (GERMAN) see FNA000
AMEPROMAT see MQU750
AMERCIAN CYANAMID 18133 see EPC500
AMERCIDE see CBG000
AMERFIL see PMP500
AMERICAINE see EFX000
AMERICAN CL-26691 see CQL250
AMERICAN CYANAMID 4,049 see MAK700
AMERICAN CYANAMID 5223 see DXX400
AMERICAN CYANAMID 12,503 see POP000
AMERICAN CYANAMID 12880 see DSP400
AMERICAN CYANAMID 47031 see PGW750
AMERICAN CYANAMID AC 43,064 see DXN600
AMERICAN CYANAMID AC 52,160 see TAL250
AMERICAN CYANAMID CL-26,691 see CQL250
AMERICAN CYANAMID CL-47,300 see DSQ000
AMERICAN CYANAMID CL-47470 see DHH400
AMERICAN PENICILLIN see BFD250
AMERICAN PENNYROYAL OIL see PAR500
AMERIZOL see TOA000
AMEROL see AMY050
AMETHOPTERIN see MDV500
AMETHYST see SCI500
AMETHYST see SCJ500
AMETOTERINA see ABY900
AMETYCIN see AHK500
d-AMFETASUL see BBK500
AMFIPEN see AIV500
AM-FOL see AMY500
AMIANTHUS see ARM250
AMIBIARSON see CBJ000
AMICAR see AJD000
AMICIDE see ANU650
AMIDAZOPHEN see DOT000
AMIDE PP see NCR000
AMIDES see AHL750
AMIDES, COCO, N,N-BIS(HYDROXYETHYL) see
CNF330
AMIDINE BLUE 4B see CMO250
4-AMIDINO-1-(NITROSAMINOAMIDINO)-1-
TETRAZENE see TEF500
4-AMIDINO-1-(NITROSOAMINOAMIDINO)-1-
TETRAZENE see TEF500
N^1-AMIDINOSULFANILAMIDE see AHO250
AMID KYSELINY AKRYLOVE see ADS250
AMID KYSELINY OCTOVE (POLISH) see AAI000
AMID KYSELINY STAVELOVE (CZECH) see OLO000
o-AMIDOAZOTOLUOL (GERMAN) see AIC250
o-AMIDOBENZOIC ACID see API500
AMIDOCYANOGEN see COH500
AMIDOFEBRIN see DOT000
AMIDO-G-ACID see NBE850
AMIDONE see MDO750
AMIDOPHEN see DOT000

AMIDOPHENAZONE see DOT000
AMIDOPYRAZOLINE see DOT000
AMIDOPYRIN see DOT000
AMIDOSULFONIC ACID see SNK500
AMIDOSULFURIC ACID see SNK500
AMIDOUREA HYDROCHLORIDE see SBW500
AMIDOXAL see SNN500
AMIDRINE see ILM000
AMIDRYL see BBV500
AMIFUR see NGE500
AMILAN CM 1001 see PJY500
AMIMYCIN see OHM900
AMINARSON see CBJ000
AMINARSONE see CBJ000
AMINAZIN MONOHYDROCHLORIDE see CKP500
AMINE AB see OBC000
AMINE BB see DXW000
AMINES see AHP750
AMINIC ACID see FNA000
AMINICOTIN see NCR000
AMINITROZOL see ABY900
AMINITROZOLE see ABY900
AMINOACETIC ACID see GHA000
AMINOACETONITRILE HYDROCHLORIDE see
GHI100
m-AMINOACETOPHENONE see AHR500
p-AMINOACETOPHENONE see AHR240
3'-AMINOACETOPHENONE see AHR500
4'-AMINOACETOPHENONE see AHR240
β-AMINOACETOPHENONE see AHR500
m-AMINOACETYLBENZENE see AHR500
p-AMINOACETYLBENZENE see AHR240
AMINOADAMANTANE HYDROCHLORIDE see
AED250
1-AMINOADAMANTENE HYDROCHLORIDE see
AED250
2-AMINOAETHANOL (GERMAN) see EEC600
4-AMINO-N-
(AMINOIMINOMETHYL)BENZENESULFONAMIDE
see AHO250
2-AMINOANILINE see PEY250
3-AMINOANILINE see PEY000
4-AMINOANILINE see PEY500
m-AMINOANILINE see PEY000
p-AMINOANILINE see PEY500
3-AMINOANILINE DIHYDROCHLORIDE see PEY750
m-AMINOANILINE DIHYDROCHLORIDE see PEY750
2-AMINOANISOLE see AOV900
4-AMINOANISOLE see AOW000
o-AMINOANISOLE see AOV900
p-AMINOANISOLE see AOW000
2-AMINOANISOLE HYDROCHLORIDE see AOX250
o-AMINOANISOLE HYDROCHLORIDE see AOX250
2-AMINOANTHRACENE see APG100
β-AMINOANTHRACENE see APG100
1-AMINO-9,10-ANTHRACENEDIONE see AIA750
2-AMINO-9,10-ANTHRACENEDIONE see AIB000
1-AMINOANTHRACHINON (CZECH) see AIA750
1-AMINOANTHRAQUINONE see AIA750
2-AMINOANTHRAQUINONE see AIB000
1-AMINO-9,10-ANTHRAQUINONE see AIA750
2-AMINO-9,10-ANTHRAQUINONE see AIB000
α-AMINOANTHRAQUINONE see AIA750
β-AMINOANTHRAQUINONE see AIB000
2-AMINO-4-ARSENOSOPHENOL see OOK100
AMINOARSON see CBJ000
AMINOAZOBENZENE see PEI000

4-AMINOAZOBENZENE see PEI000
p-AMINOAZOBENZENE see PEI000
4-AMINO-1,1'-AZOBENZENE see PEI000
4-AMINOAZOBENZOL see PEI000
p-AMINOAZOBENZOL see PEI000
2-AMINO-5-AZOTOLUENE see AIC250
4'-AMINO-2,3'-AZOTOLUENE see AIC250
4'-AMINO-2:3'-AZOTOLUENE see AIC250
AMINOAZOTOLUENE (indicator) see AIC250
o-AMINOAZOTOLUENE (MAK) see AIC250
o-AMINOAZOTOLUENO (SPANISH) see AIC250
o-AMINOAZOTOLUOL see AIC250
m-AMINOBENZAL FLUORIDE see AID500
AMINOBENZENE see AOQ000
4-AMINOBENZENEARSONIC ACID see ARA250
p-AMINOBENZENEARSONIC ACID see ARA250
p-AMINOBENZENEAZODIMETHYLANILINE see
DPO200
(S)-α-AMINOBENZENEPROPANOIC ACID see
PEC750
p-AMINOBENZENESULFAMIDE see SNM500
p-AMINOBENZENESULFONACETAMIDE see
SNQ710
4-AMINOBENZENESULFONAMIDE see SNM500
p-AMINOBENZENESULFONAMIDE see SNM500
5-(p-AMINOBENZENESULFONAMIDO)-3,4-
DIMETHYLISOOXALE see SNN500
5-(p-AMINOBENZENESULFONAMIDO)-3,4-
DIMETHYLISOXAZOLE see SNN500
2-(p-AMINOBENZENESULFONAMIDO)-4,6-
DIMETHYLPYRIMIDINE see SNJ000
2-p-
AMINOBENZENESULFONAMIDOQUINOXALINE
see QTS000
2-(p-AMINOBENZENESULFONAMIDO)THIAZOLE
see TEX250
N-(4-AMINOBENZENESULFONYL)-N'-BUTYLUREA
see BSM000
p-AMINOBENZENESULFONYLGUANIDINE see
AHO250
5-(p-AMINOBENZENESULPHONAMIDE)-3,4-
DIMETHYLISOXAZOLE see SNN500
5-(p-AMINOBENZENESULPHONAMIDO)-3,4-
DIMETHYLISOXAZOLE see SNN500
2-p-
AMINOBENZENESULPHONAMIDOQUINOXALINE
see QTS000
2-(p-AMINOBENZENESULPHONAMIDO)THIAZOLE
see TEX250
N-p-AMINOBENZENESULPHONYLGUANIDINE
MONOHYDRATE see AHO250
2-AMINOBENZENETHIOL see AIF500
2-AMINOBENZIMIDAZOLE see AIG000
2-AMINO-6-BENZIMIDAZOLYL PHENYLKETONE
see AIH000
AMINOBENZOIC ACID see AIH600
2-AMINOBENZOIC ACID see API500
4-AMINOBENZOIC ACID see AIH600
o-AMINOBENZOIC ACID see API500
p-AMINOBENZOIC ACID see AIH600
γ-AMINOBENZOIC ACID see AIH600
4-AMINOBENZOIC ACID 2-
(DIETHYLAMINO)ETHYL ESTER,
HYDROCHLORIDE see AIT250
p-AMINOBENZOIC ACID-2-DIETHYLAMINOETHYL
ESTER, HYDROCHLORIDE see AIT250
4-AMINOBENZOIC ACID ETHYL ESTER see EFX000

o-AMINOBENZOIC ACID, ETHYL ESTER see EGM000
p-AMINOBENZOIC ACID ETHYL ESTER see EFX000
2-AMINOBENZOIC ACID METHYL ESTER see APJ250
o-AMINOBENZOIC ACID METHYL ESTER see APJ250
2-AMINOBENZOIC ACID-3-PHENYL-2-PROPENYL
ESTER see API750
6-(4'-AMINOBENZOL-SULFONAMIDO)-2,4-
DIMETHYLPYRIMIDIN (GERMAN) see SNJ000
(p-AMINOBENZOLSULFONYL)-2-AMINO-4,6-
DIMETHYLPYRIMIDIN (GERMAN) see SNJ000
(p-AMINOBENZOLSULFONYL)-2-AMINO-4-
METHYLPYRIMIDIN (GERMAN) see ALF250
p-AMINOBENZOPHENONE see AIR250
3-AMINOBENZOTRIFLUORIDE see AID500
m-AMINOBENZOTRIFLUORIDE see AID500
2-AMINO-5-BENZOYLBENZIMIDAZOLE see AIH000
p-AMINOBENZOYLDIETHYLAMINOETHANOL
HYDROCHLORIDE see AIT250
o-AMINOBENZOYLFORMIC ANHYDRIDE see ICR000
4-(4-AMINOBENZYL)ANILINE see MJQ000
AMINOBENZYLPENICILLIN see AIV500
d-(−)-α-AMINOBENZYLPENICILLIN see AIV500
d-(−)-α-AMINOBENZYLPENICILLIN SODIUM SALT
see SEQ000
AMINOBENZYLPENICILLIN TRIHYDRATE see
AOD125
α-AMINOBENZYLPENICILLIN TRIHYDRATE see
AOD125
4-AMINOBIPHENYL see AJS100
p-AMINOBIPHENYL see AJS100
5-AMINO-1-BIS(DIMETHYLAMIDE)PHOSPHORYL-3-
PHENYL-1,2,4-TRIAZOLE see AIX000
5-AMINO-1-BIS(DIMETHYLAMIDO)PHOSPHORYL-3-
PHENYL-1,2,4-TRIAZOLE see AIX000
5-AMINO-1-(BIS(DIMETHYLAMINO)PHOSPHINYL)-
3-PHENYL-1,2,4-TRIAZOLE see AIX000
AMINOBIS(PROPYLAMINE) see AIX250
AMINOBIS(PROPYLAMINE) see AIX250
4-AMINO-N-(5-BROMO-4,6-DIMETHYL-2-
PYRIMIDINYL)BENZENESILFANILAMIDE see
SNH875
1-AMINO-BUTAAN (DUTCH) see BPX750
1-AMINOBUTAN (GERMAN) see BPX750
1-AMINOBUTANE see BPX750
2-AMINOBUTANE see BPY000
(S)-AMINOBUTANEDIOIC ACID see ARN850
4-AMINO-N-
((BUTYLAMINO)CARBONYL)BENZENESULFONAMI
DE see BSM000
4-AMINO-6-tert-BUTYL-3-METHYLTHIO-as-TRIAZIN-
5-ONE see MQR275
4-AMINO-6-tert-BUTYL-3-(METHYLTHIO)-1,2,4-
TRIAZIN-5-ONE see MQR275
AMINOCAINE see AIT250
AMINOCAPROIC ACID see AJD000
6-AMINOCAPROIC ACID see AJD000
ε-AMINOCAPROIC ACID see AJD000
ω-AMINOCAPROIC ACID see AJD000
AMINOCAPROIC LACTAM see CBF700
AMINO-α-CARBOLINE see AJD750
2-AMINO-α-CARBOLINE see AJD750
2,2'-((4-
((AMINOCARBONYL)AMINO)PHENYL)ARSINIDENE
)BIS(THIO)BISACETIC ACID see CBI250
2,2'-(((4-
((AMINOCARBONYL)AMINO)PHENYL)ARSINIDENE
BIS(THIO))BIS) BENZOIC ACID see TFD750

(4-((AMINOCARBONYL)AMINO)PHENYL)ARSONIC
ACID see CBJ000
N-(AMINOCARBONYL)-2-BROMO-2-
ETHYLBUTANAMIDE see BNK000
N-(AMINOCARBONYL)-2-BROMO-3-
METHYLBUTANAMIDE see BNP750
1-AMINO-2-CARBOXYBENZENE see API500
1-AMINO-4-CARBOXYBENZENE see AIH600
3-AMINO-N-(α-
CARBOXYPHENETHYL)SUCCINAMIC ACID N-
METHYL ESTER, stereoisomer see ARN825
m-AMINOCHLOROBENZENE see CEH675
1-AMINO-2-CHLOROBENZENE see CEH670
1-AMINO-3-CHLOROBENZENE see CEH675
1-AMINO-4-CHLOROBENZENE see CEH680
1-AMINO-4-CHLOROBENZENE HYDROCHLORIDE
see CJR200
5-AMINO-4-CHLORO-2,3-DIHYDRO-3-OXO-2-
PHENYLPYRIDAZINE see PEE750
1-AMINO-2-CHLORO-6-METHYLBENZENE see
CLK200
1-AMINO-3-CHLORO-2-METHYLBENZENE see
CLK200
1-AMINO-3-CHLORO-4-METHYLBENZENE see
CLK215
1-AMINO-3-CHLORO-6-METHYLBENZENE see
CLK225
2-AMINO-4-CHLOROPHENOL (DOT) see CEH250
5-AMINO-4-CHLORO-2-PHENYL-3(2H)-
PYRIDAZINONE see PEE750
2-AMINO-3-CHLOROTOLUENE see CLK227
2-AMINO-4-CHLOROTOLUENE see CLK225
2-AMINO-5-CHLOROTOLUENE see CLK220
2-AMINO-6-CHLOROTOLUENE see CLK200
4-AMINO-2-CHLOROTOLUENE see CLK215
2-AMINO-5-CHLOROTOLUENE HYDROCHLORIDE
see CLK235
3-AMINO-p-CRESOL METHYL ESTER see MGO750
m-AMINO-p-CRESOL, METHYL ESTER see MGO750
AMINOCYCLOHEXANE see CPF500
AMINOCYCLOHEXANE HYDROCHLORIDE see
CPA775
4-AMINO-DAB see DPO200
AMINODARONE see AJK750
1-AMINODECANE see DAG600
4-AMINO-4-DEOXY-N¹⁰-
METHYLPTEROYLGLUTAMATE see MDV500
4-AMINO-4-DEOXY-N¹⁰-
METHYLPTEROYLGLUTAMIC ACID see MDV500
4-AMINO-4-DEOXYPTEROYLGLUTAMATE see
AMG750
2-AMINO-9-(2-DEOXY-β-d-RIBOFURANOSYL)-9H-
PURINE-6-THIOL HYDRATE see TFJ250
4-AMINO-N-
(DIAMINOMETHYLENE)BENZENESULFONAMIDE
see AHO250
2-AMINO-4-DICHLOROARSINOPHENOL
HYDROCHLORIDE see DFX400
1-AMINO-3-(DIETHYLAMINO)PROPANE see DIY800
p-AMINODIETHYLANILINE see DJV200
4-AMINODIFENIL (SPANISH) see AJS100
l-2-AMINO-1-(3,4-DIHYDROXYPHENYL)ETHANOL
see NNO500
2-AMINO-3-(3,4-DIHYDROXYPHENYL)PROPANOIC
ACID see DNA200
2-AMINO-1-(2,5-DIMETHOXYPHENYL)-1-
PROPANOL HYDROCHLORIDE see MDW000

4-AMINO-N-(2,6-DIMETHOXY-4-PYRIMIDINYL)BENZENESULFONAMIDE see SNN300

4-AMINO-4'-DIMETHYLAMINOAZOBENZENE see DPO200

4'-AMINO-N,N-DIMETHYL-4-AMINOAZOBENZENE see DPO200

3-AMINO-7-(DIMETHYLAMINO)-2-METHYL-PHENOSELENAZIN-5-IUM CHLORIDE see SBU950

1-AMINO-3-DIMETHYLAMINOPROPANE see AJQ100

4-AMINO-2',3-DIMETHYLAZOBENZENE see AIC250

4'-AMINO-2,3'-DIMETHYLAZOBENZENE see AIC250

AMINODIMETHYLBENZENE see XMA000

1-AMINO-2,4-DIMETHYLBENZENE see XMS000

1-AMINO-2,5-DIMETHYLBENZENE see XNA000

3-AMINO-1,4-DIMETHYLBENZENE see XNA000

4-AMINO-1,3-DIMETHYLBENZENE see XMS000

3-AMINO-1,4-DIMETHYL-γ-CARBOLINE see TNX275

4-AMINO-6-(1,1-DIMETHYLETHYL)-3-(METHYLTHIO)-1,2,4-TRIAZIN-5(4H)-ONE see MQR275

2-AMINO-3,4-DIMETHYLIMIDAZO(4,5-f)QUINOLINE see AJQ600

2-AMINO-3,8-DIMETHYLIMIDAZO(4,5-f)QUINOXALINE see AJQ675

2-AMINO-3,8-DIMETHYL-3H-IMIDAZO(4,5-f)QUINOXALINE see AJQ675

4-AMINO-N-(3,4-DIMETHYL-5-ISOXAZOLYL)BENZENESULPHONAMIDE see SNN500

3-AMINO-1,4-DIMETHYL-5H-PYRIDO(4,3-b)INDOLE see TNX275

3-AMINO-1,4-DIMETHYL-5H-PYRIDO(4,3-b)INDOLE ACETATE see AJR500

4-AMINODIPHENYL see AJS100

p-AMINODIPHENYL see AJS100

p-AMINODIPHENYLIMIDE see PEI000

2-AMINODIPYRIDO(1,2-a:3',2'-d)-IMIDAZOLE see DWW700

1-AMINODODECANE see DXW000

2-AMINOETANOLO (ITALIAN) see EEC600

AMINOETHANE see EFU400

1-AMINOETHANE see EFU400

2-AMINO-ETHANESELENOL HYDROCHLORIDE see AJS900

2-AMINOETHANESELENOSULFURIC ACID see AJS950

2-AMINOETHANESULFONIC ACID see TAG750

2-AMINOETHANETHIOL see AJT250

1-AMINOETHANOL see AAG500

2-AMINOETHANOL (MAK) see EEC600

4-AMINOETHOXYBENZENE see PDK790

2-AMINOETHOXYETHANOL see AJU250

2-(2-AMINOETHOXY)ETHANOL see AJU250

4-AMINO-N-(6-ETHOXY-3-PYRIDAZINYL)BENZENESULFONAMIDE see SNJ100

α-AMINOETHYL ALCOHOL see AAG500

β-AMINOETHYL ALCOHOL see EEC600

o-AMINOETHYLBENZENE see EGK500

1-AMINO-4-ETHYLBENZENE see EGL000

β-AMINOETHYLBENZENE see PDE250

α-(1-AMINOETHYL)BENZENEMETHANOL HYDROCHLORIDE see PMJ500

α-(1-AMINOETHYL)-BENZYL ALCOHOL see NNM000

α-(1-AMINOETHYL)BENZYL ALCOHOL HYDROCHLORIDE see PMJ500

3-(2-AMINOETHYL)-1-BENZYL-5-METHOXY-2-METHYLINDOLE HYDROCHLORIDE see BEM750

3-AMINO-9-ETHYLCARBAZOLE see AJV000

3-AMINO-N-ETHYLCARBAZOLE see AJV000

3-AMINO-9-ETHYLCARBAZOLEHYDROCHLORIDE see AJV250

4-(2-AMINOETHYL)DIETHYLENETRIAMINE see NEI800

α-(1-AMINOETHYL)-2,5-DIMETHOXYBENZYL ALCOHOL HYDROCHLORIDE see MDW000

2-AMINOETHYL DISULFIDE DIHYDROCHLORIDE see CQJ750

AMINOETHYLENE see EJM900

AMINOETHYLETHANEDIAMINE see DJG600

N-(2-AMINOETHYL)ETHYLENEDIAMINE see DJG600

1-AMINO-2-ETHYLHEXAN (CZECH) see EKS500

4-(2-AMINOETHYL)IMIDAZOLE BIS(DIHYDROGEN PHOSPHATE) see HGE000

4-(2-AMINOETHYL)IMIDAZOLE DI-ACID PHOSPHATE see HGE000

2-AMINOETHYL MERCAPTAN see AJT250

AMINOETHYLPIPERAZINE see AKB000

N-AMINOETHYLPIPERAZINE see AKB000

1-(2-AMINOETHYL)PIPERAZINE see AKB000

N-(2-AMINOETHYL)PIPERAZINE see AKB000

N-(β-AMINOETHYL)PIPERAZINE see AKB000

4-(2-AMINOETHYL)PYROCATECHOL see DYC400

2-AMINOETHYLSULFONIC ACID see TAG750

2-AMINO-4-(ETHYLTHIO)BUTYRIC ACID see EEI000

dl-2-AMINO-4-(ETHYLTHIO)BUTYRIC ACID see EEI000

AMINOFENAZONE (ITALIAN) see DOT000

p-AMINOFENETOL see PDK790

5-AMINO-3-FENIL-1-BIS(-DIMETILAMINO)-FOSFORIL-1,2,4-TRIAZOLO (ITALIAN) see AIX000

m-AMINOFENOL (CZECH) see ALS990

p-AMINOFENOL (CZECH) see ALT250

5-AMINO-3-FENYL-1-BIS(DIMETHYL-AMINO)-FOSFORYL-1,2,4-TRIAZOOL (DUTCH) see AIX000

AMINOFLUOREN (GERMAN) see FDI000

2-AMINOFLUORENE see FDI000

AMINOFORM see HEI500

2-AMINOGLUTARAMIC ACID see GFO050

l-2-AMINOGLUTARAMIDIC ACID see GFO050

l-2-AMINOGLUTARIC ACID see GFO000

α-AMINOGLUTARIC ACID see GFO000

1-AMINOHEPTANE see HBL600

AMINOHEXAHYDROBENZENE see CPF500

AMINOHEXAHYDROBENZENE HYDROCHLORIDE see CPA775

1-AMINOHEXANE see HFK000

ω-AMINOHEXANOIC ACID see AJD000

6-AMINOHEXANOIC ACID CYCLIC LACTAM see CBF700

6-AMINOHEXANOIC ACID HOMOPOLYMER see PJY500

α-AMINOHYDROCINNAMIC ACID see PEC750

1-AMINO-4-HYDROXY-9,10-ANTHRACENEDIONE see AKE250

1-AMINO-4-HYDROXYANTHRAQUINONE see AKE250

2-AMINO-1-HYDROXYBENZENE see ALT000

3-AMINO-1-HYDROXYBENZENE see ALS990

4-AMINO-1-HYDROXYBENZENE see ALT250

4-AMINO-2-HYDROXYBENZENEARSONIC ACID see HJE400

α-AMINO-p-HYDROXYBENZYLPENICILLIN
TRIHYDRATE see AOA100
l-2-AMINO-3-HYDROXYBUTYRIC ACID see TFU750
(R)-4-(2-AMINO-1-HYDROXYETHYL)-1,2-
BENZENEDIOL see NNO500
(2S-(2-α,5-α,6-β(S*)))-6-((AMINO(4-
HYDROXYPHENYL)ACETYL)AMINO)-3,3-
DIMETHYL-7-OXO-4-THIA-1-
AZABICYCLO(3.2.0)HEPTANE-2-CARBOXYLIC ACID
TRIHYDRATE see AOA100
((5-(3-AMINO-4-HYDROXYPHENYL)ARSENO)-2-
HYDROXYANILINO)METHANOL SULFOXYLATE
SODIUM see NCJ500
(3-AMINO-4-HYDROXYPHENYL)ARSENOUS ACID
see OOK100
(3-AMINO-4-HYDROXYPHENYL)ARSONOUS
DICHLORIDE MONOHYDROCHLORIDE see DFX400
3-AMINO-4-HYDROXYPHENYL DICHLORARSINE
HYDROCHLORIDE see DFX400
(3-AMINO-4-HYDROXYPHENYL)DICHLOROARSINE
HYDROCHLORIDE see DFX400
3-(((3-AMINO-4-
HYDROXYPHENYL)PHENYLARSINO)THIO)ALANIN
E see AKI900
l-N-(p-(((-2-AMINO-4-HYDROXY-6-
PTERIDINYL)METHYL)AMINO)BENZOYL)GLUTAM
IC ACID see FMT000
2-AMINOHYPOXANTHINE see GLI000
α-AMINO-INDOLE-3-PROPRIONIC ACID see TNX000
α'-AMINO-3-INDOLEPROPRIONIC ACID see TNX000
l-α-AMINO-3-INDOLEPROPRIONIC ACID see TNX000
2-AMINO-3-INDOL-3-YL-PROPRIONIC ACID see
TNX000
2-AMINOISOBUTANE see BPY250
α-AMINOISOCAPROIC ACID see LES000
l-(+)-α-AMINOISOVALERIC ACID see VBP000
AMINOKAPRON see AJD000
AMINOMERCURIC CHLORIDE see MCW500
AMINOMESITYLENE see TLG500
2-AMINOMESITYLENE see TLG500
AMINOMESITYLENE HYDROCHLORIDE see TLH000
2-AMINOMESITYLENE HYDROCHLORIDE see
TLH000
AMINOMETHANE see MGC250
1-AMINO-2-METHOXYBENZENE see AOV900
1-AMINO-4-METHOXYBENZENE see AOW000
1-AMINO-2-METHOXY-5-METHYLBENZENE see
MGO750
7-AMINO-9-α-METHOXYMITOSANE see AHK500
3-AMINO-4-METHOXYNITROBENZENE see NEQ500
2-AMINO-1-METHOXY-4-NITROBENZENE see
NEQ500
3-AMINO-4-METHOXYTOLUENE see MGO750
4-AMINO-2-METHYLANILINE see TGM000
2-AMINO-4-METHYLANISOLE see MGO750
1-AMINO-2-METHYL-9,10-ANTHRACENEDIONE see
AKP750
1-AMINO-2-METHYLANTHRAQUINONE see AKP750
1-AMINO-2-METHYLBENZENE see TGQ750
2-AMINO-1-METHYLBENZENE see TGQ750
3-AMINO-1-METHYLBENZENE see TGQ500
4-AMINO-1-METHYLBENZENE see TGR000
1-AMINO-2-METHYLBENZENE HYDROCHLORIDE
see TGS500
2-AMINO-1-METHYLBENZENE HYDROCHLORIDE
see TGS500

2-AMINO-3-METHYL-α-CARBOLINE see ALD750
3-AMINO-1-METHYL-γ-CARBOLINE see ALD500
l-α-(AMINOMETHYL)-3,4-DIHYDROXYBENZYL
ALCOHOL see NNO500
2-AMINO-6-METHYLDIPYRIDO(1,2-a:3',2'-
d)IMIDAZOLE see AKS250
4-AMINO-10-METHYLFOLIC ACID see MDV500
2-AMINO-6-METHYLHEPTANE see ILM000
6-AMINO-2-METHYLHEPTANE see ILM000
2-AMINO-3-METHYLIMIDAZO(4,5-f)QUINOLINE see
AKT600
4-AMINO-N-(5-METHYL-3-
ISOXAZOLYL)BENZENESULFONAMIDE see SNK000
l-α-AMINO-γ-METHYLMERCAPTOBUTYRIC ACID see
MDT750
1-AMINO-2-METHYL-5-NITROBENZENE see NMP500
2-AMINO-3-METHYLPENTANOIC ACID see IKX000
2-AMINO-4-METHYLPENTANOIC ACID see LES000
2-AMINO-1-METHYL-6-PHENYLIMIDAZO(4,5-
B)PYRIDINE see AKZ200
1-AMINO-2-METHYLPROPANE see IIM000
2-AMINO-2-METHYLPROPANE see BPY250
(−)-α-(AMINOMETHYL)PROTOCATECHUYL
ALCOHOL see NNO500
2-AMINO-3-METHYL-9H-PYRIDO(2,3-b)INDOLE see
ALD750
3-AMINO-1-METHYL-5H-PYRIDO(4,3-b)INDOLE see
ALD500
4-AMINO-N-(4-METHYL-2-PYRIMIDINYL)-
BENZENESULFONAMIDE see ALF250
4-AMINO-N-(4-METHYL-2-PYRIMIDINYL)-
BENZENESULFONAMIDE MONOSODIUM SALT see
SJW475
3-(((4-AMINO-2-METHYL-5-PYRIMIDINYL)METHYL)-
5-(2-HYDROXYETHYL)-4-METHYLTHIAZOLIUM
CHLORIDE see TES750
3-(4-AMINO-2-METHYLPYRIMIDYL-5-METHYL)-4-
METHYL-5,β-HYDROXYETHYLTHIAZOLIUM
NITRATE see TET500
4-AMINO-N-METHYL-1,2,5-SELENADIAZOLE-3-
CARBOXAMIDE see MGL600
2-AMINO-4-(METHYLTHIO)BUTYRIC ACID see
MDT750
l(−)-AMINO-γ-METHYLTHIOBUTYRIC ACID see
MDT750
4-AMINO-3-METHYLTOLUENE see XMS000
2-AMINO-4-METHYLVALERIC ACID see LES000
l,2-AMINO-4-METHYLVALERIC ACID see LES000
dl-2-AMINO-3-METHYLVALERIC ACID see IKX010
dl-2-AMINO-4-METHYLVALERIC ACID see LER000
α-AMINO-β-METHYLVALERIC ACID see IKX000
α-AMINO-γ-METHYLVALERIC ACID see LES000
1-AMINONAFTALEN (CZECH) see NBE700
2-AMINONAFTALEN (CZECH) see NBE500
1-AMINONAPHTHALENE see NBE700
2-AMINONAPHTHALENE see NBE500
7-AMINO-1,3-NAPHTHALENEDISULFONIC ACID see
NBE850
2-AMINO-4-NITROANILINE see ALL500
4-AMINO-2-NITROANILINE see ALL750
2-AMINO-4-NITROANISOLE see NEQ500
m-AMINONITROBENZENE see NEN500
p-AMINONITROBENZENE see NEO500
1-AMINO-2-NITROBENZENE see NEO000
1-AMINO-3-NITROBENZENE see NEN500
1-AMINO-4-NITROBENZENE see NEO500

2-AMINO-5-(5-NITRO-2-FURYL)-1,3,4-THIADIAZOLE see NGI500

5-AMINO-2-(5-NITRO-2-FURYL)-1,3,4-THIADIAZOLE see NGI500

4-AMINO-2-NITROPHENOL see NEM480

2-((4-AMINO-2-NITROPHENYL)AMINO)ETHANOL see ALO750

AMINONITROTHIAZOLE see ALQ000

2-AMINO-5-NITROTHIAZOLE see ALQ000

AMINONITROTHIAZOLUM see ALQ000

2-AMINO-4-NITROTOLUENE see NMP500

AMINONUCLEOSIDE see ALQ625

AMINONUCLEOSIDE PUROMYCIN see ALQ625

4-AMINO-N[10]-METHYLPTEROYLGLUTAMIC ACID see MDV500

1-AMINOOCTADECANE see OBC000

2-AMINOOCTANE see OEK010

l-α-AMINO-4(OR 5)-IMIDAZOLEPROPIONIC ACID see HGE700

l-α-AMINO-4(OR 5)-IMIDAZOLEPROPIONIC ACID MONOHYDROCHLORIDE see HGE800

(4-(2-AMINO-2-OXOETHYL)AMINO)PHENYL)ARSONIC ACID see CBJ750

2-AMINO-2-OXOETHYL-2,2-DIMETHYL-N-(((METHYLAMINO)CARBONYL)OXY)PROPANIMIDO THIOATE see ALQ640

1-AMINO-4-OXYANTHRAQUINONE see AKE250

d-(−)-α-AMINOPENICILLIN see AIV500

1-AMINOPENTANE see PBV505

2-AMINOPENTANE see DHJ200

2-AMINOPENTANEDIOIC ACID see GFO000

4-AMINO-PGA see AMG750

AMINOPHEN see AOQ000

AMINOPHENAZONE see DOT000

4-AMINOPHENETOLE see PDK790

p-AMINOPHENETOLE see PDK790

2-AMINOPHENOL see ALT000

3-AMINOPHENOL see ALS990

4-AMINOPHENOL see ALT250

m-AMINOPHENOL see ALS990

o-AMINOPHENOL see ALT000

m-AMINOPHENOL (DOT) see ALS990

p-AMINOPHENOL (DOT) see ALT250

AMINOPHENUROBUTANE see BSM000

6-(d(−)-α-AMINOPHENYLACETAMIDO)PENICILLANIC ACID see AIV500

AMINOPHENYLARSINE ACID see ARA250

p-AMINOPHENYLARSINE ACID see ARA250

p-AMINOPHENYLARSINE OXIDE DIHYDRATE see ALV100

p-AMINOPHENYLARSINIC ACID see ARA250

4-AMINOPHENYLARSONIC ACID see ARA250

(4-AMINOPHENYL)ARSONIC ACID SODIUM SALT see ARA500

4-((4-AMINOPHENYL)AZO)-N,N-DIMETHYLBENZENAMINE see DPO200

5-AMINO-3-PHENYL-1-BIS (DIMETHYL-AMINO)-PHOSPHORYLE-1,2,4-TRIAZOLE (FRENCH) see AIX000

5-AMINO-3-PHENYL-1-BIS(DIMETHYLAMINO)-PHOSPHORYL-1H-1,2,4-TRIAZOL (GERMAN) see AIX000

1-AMINO-2-PHENYLETHANE see PDE250

4-AMINOPHENYL ETHER see OPM000

p-AMINOPHENYL ETHER see OPM000

4-((4-AMINOPHENYL)(4-IMINO-2,5-CYCLOHEXADIEN-1-YLIDENE)METHYL), MONOCHLORIDE see RMK020

p-AMINOPHENYLMERCURIC ACETATE see ABQ000

3-AMINOPHENYLMETHANE see TGQ500

(AMINOPHENYLMETHYL)-PENICILLIN see AIV500

AMINOPHENYLNORHARMAN see ALX120

d-2-AMINO-1-PHENYLPROPANE see AOA500

2-AMINO-1-PHENYL-1-PROPANOL see NNM000

(±)-2-AMINO-1-PHENYL-1-PROPANOL HYDROCHLORIDE see PMJ500

1-(4-AMINOPHENYL)-1-PROPANONE see AMC000

α-AMINO-β-PHENYLPROPIONIC ACID see PEC750

dl-α-AMINO-β-PHENYLPROPIONIC ACID see PEC500

9-(4'-AMINOPHENYL)-9H-PYRIDO(3,4-B)INDOLE see ALX120

4-AMINOPHENYLSULFONAMIDE see SNM500

p-AMINOPHENYLSULFONAMIDE see SNM500

5-(4-AMINOPHENYLSULFONAMIDO)-3,4-DIMETHYLISOXAZOLE see SNN500

N-((4-AMINOPHENYL)SULFONYL)ACETAMIDE see SNQ710

5-(4-AMINOPHENYLSULPHONAMIDO)-3,4-DIMETHYLISOXAZOLE see SNN500

3-(p-AMINOPHENYLSULPHONAMIDO)-5-METHYLISOXAZOLE see SNK000

5-AMINO-3-PHENYL-1,2,4-TRIAZOLE-1-YL-N,N,N',N'-TETRAMETHYLPHOSPHODIAMIDE see AIX000

5-AMINO-3-PHENYL-1,2,4-TRIAZOLYL-1-BIS(DIMETHYLAMIDO)PHOSPHATE see AIX000

5-AMINO-3-PHENYL-1,2,4-TRIAZOLYL-N,N,N'N'-TETRAMETHYL-PHOSPHONAMIDE see AIX000

p-(5-AMINO-3-PHENYL-1H-1,2,4-TRIAZOL-1-YL)-N,N,N'-TETRAMETHYL PHOSPHONIC DIAMIDE see AIX000

AMINOPHYLLINE see TEP000

2-AMINO-PROPAAN (DUTCH) see INK000

2-AMINOPROPAN (GERMAN) see INK000

1-AMINOPROPANE see PND250

2-AMINOPROPANE see INK000

1-AMINOPROPANE-1,3-DICARBOXYLIC ACID see GFO000

2-AMINO-PROPANO (ITALIAN) see INK000

l-2-AMINOPROPANOIC ACID see AFH625

(S)-2-AMINOPROPANOIC ACID see AFH625

3-AMINOPROPANOL see PMM250

3-AMINO-1-PROPANOL see PMM250

γ-AMINOPROPANOL see PMM250

3-AMINOPROPENE see AFW000

α-AMINOPROPIONIC ACID see AFH625

3-AMINOPROPIONITRILE see AMB500

β-AMINOPROPIONITRILE see AMB500

β-AMINOPROPIONITRILE FUMARATE see AMB750

p-AMINOPROPIOPHENONE see AMC000

3-AMINOPROPYL ALCOHOL see PMM250

N-(2-(3-AMINOPROPYLAMINO)ETHYL)-1,3-PROPANEDIAMINE see TBI700

N-(3-AMINOPROPYL)-N'-(3-((3-AMINOPROPYL)AMINO)PROPYL)-1,3-PROPANEDIAMINE see TED600

(N-(3-AMINOPROPYL)-3-AMINOPROPYL)TRIMETHOXYSILANE see TLC600

3-AMINOPROPYLENE see AFW000

3-(2-AMINOPROPYL)INDOLE see AME500

N-(3-AMINOPROPYL)MORFOLIN see AMF250

4-AMINOPROPYLMORPHOLINE see AMF250

N-(3-AMINOPROPYL)MORPHOLINE see AMF250

1-[(4-AMINO-2-PROPYL-5-PYRIMIDINYL)METHYL]-2-METHYLPYRIDINIUM CHLORIDE see AOD175
1-(4-AMINO-2-n-PROPYL-5-PYRIMIDINYLMETHYL)-2-PICOLINIUM CHLORIDE see AOD175
AMINOPTERIDINE see AMG750
AMINOPTERIN see AMG750
4-AMINOPTEROYLGLUTAMIC ACID see AMG750
6-AMINOPURINE see AEH000
6-AMINO-1H-PURINE see AEH000
6-AMINO-3H-PURINE see AEH000
6-AMINO-9H-PURINE see AEH000
2-AMINOPYRIDINE see AMI000
AMINO-2-PYRIDINE see AMI000
3-AMINOPYRIDINE see AMI250
AMINO-3-PYRIDINE see AMI250
4-AMINOPYRIDINE see AMI500
AMINO-4-PYRIDINE see AMI500
o-AMINOPYRIDINE see AMI000
p-AMINOPYRIDINE see AMI500
m-AMINOPYRIDINE (DOT) see AMI250
α-AMINOPYRIDINE see AMI000
γ-AMINOPYRIDINE see AMI500
3-AMINO-PYRIDINE-2-CARBOXALDEHYDE see AMI600
3-AMINOPYRIDINE-2-CARBOXALDEHYDE THIOSEMICARBAZONE see AMI600
2-((3-AMINO-2-PYRIDINYL)METHYLENE)HYDRAZINECARBOTHIOAMIDE see AMI600
2-AMINO-9H-PYRIDO(2,3-B)INDOLE see AJD750
4-AMINO-N-2-PYRIMIDINYLBENZENESULFONAMIDE see PPP500
AMINOPYRINE see DOT000
AMINOPYRINE SODIUM SULFONATE see AMK500
4-AMINO-1-β-d-RIBOFURANOSYL-d-TRIAZIN-2(1H)-ONE see ARY000
4-AMINO-1-β-d-RIBOFURANOSYL-1,3,5-TRIAZIN-2(1H)-ONE see ARY000
p-AMINOSALICYLATE SODIUM see SEP000
p-AMINOSALICYLIC ACID SODIUM SALT see SEP000
p-AMINOSALICYLSAEURES SALZ (GERMAN) see TMJ750
4-AMINO-1,2,5-SELENADIAZOLE-3-CARBOXAMIDE see AMN300
4-AMINOSEMICARBAZIDE see CBS500
(AMINOSUBERIC ACID 1,7)-EEL CALCITONIN see CBR300
l-α-AMINOSUCCINAMIC ACID see ARN810
l-AMINOSUCCINIC ACID see ARN850
AMINOSULFONIC ACID see SNK500
O-(4-(AMINOSULFONYL)PHENYL) O,O-DIMETHYL PHOSPHOROTHIOATE see CQL250
N-(5-(AMINOSULFONYL)-1,3,4-THIADIAZOL-2-YL)ACETAMIDE see AAI250
5-(AMINOSULFURANYL)-4-CHLORO-2-((2-FURNAYLMETHYL)AMINO)BENZOIC ACID see CHJ750
5-AMINO-2,2,4,4-TETRAKIS(TRIFLUOROMETHYL)IMIDAZOLIDINE see AMQ500
4-AMINO-2,2,5,5-TETRAKIS(TRIFLUOROMETHYL)-3-IMIDAZOLINE see AMQ500
2-AMINO-1,3,4-THIADIAZOLEHYDROCHLORIDE see AMR500
2-AMINO-1,3,4-THIADIAZOLE, MONOHYDROCHLORIDE see AMR500

4-AMINO-N-2-THIAZOLYLBENZENESULFONAMIDE see TEX250
2-AMINOTHIOPHENOL see AIF500
o-AMINOTHIOPHENOL see AIF500
N-AMINOTHIOUREA see TFQ000
3-AMINOTOLUEN (CZECH) see TGQ500
4-AMINOTOLUEN (CZECH) see TGR000
2-AMINOTOLUENE see TGQ750
3-AMINOTOLUENE see TGQ500
4-AMINOTOLUENE see TGR000
m-AMINOTOLUENE see TGQ500
o-AMINOTOLUENE see TGQ750
p-AMINOTOLUENE see TGR000
2-AMINOTOLUENE HYDROCHLORIDE see TGS500
4-AMINOTOLUENE HYDROCHLORIDE see TGS750
o-AMINOTOLUENE HYDROCHLORIDE see TGS500
3-AMINO-p-TOLUIDINE see TGL750
5-AMINO-o-TOLUIDINE see TGL750
AMINOTRIACETIC ACID see AMT500
AMINOTRIAZOLE see AMY050
2-AMINOTRIAZOLE see AMY050
3-AMINOTRIAZOLE see AMY050
3-AMINO-s-TRIAZOLE see AMY050
2-AMINO-1,3,4-TRIAZOLE see AMY050
3-AMINO-1,2,4-TRIAZOLE see AMY050
3-AMINO-1H-1,2,4-TRIAZOLE see AMY050
3-AMINO-1,2,4-TRIAZOLE (ACGIH) see AMY050
AMINOTRIAZOLE (PLANT REGULATOR) see AMY050
AMINO TRIAZOLE WEEDKILLER 90 see AMY050
AMINOTRIAZOL-SPRITZPULVER see AMY050
4-AMINO-3,5,6-TRICHLOROPICOLINIC ACID see PIB900
4-AMINO-3,5,6-TRICHLORO-2-PICOLINIC ACID see PIB900
4-AMINO-3,5,6-TRICHLORPICOLINSAEURE (GERMAN) see PIB900
1-AMINO-2,4,6-TRIMETHYLBENZEN (CZECH) see TLG500
1-AMINO-2,4,5-TRIMETHYLBENZENE see TLG250
2-AMINO-1,3,5-TRIMETHYLBENZENE see TLG500
1-AMINO-2,4,5-TRIMETHYLBENZENE HYDROCHLORIDE see TLG750
2-AMINO-1,3,5-TRIMETHYLBENZENE HYDROCHLORIDE see TLH000
AMINOUNDECANOIC ACID see AMW000
11-AMINOUNDECANOIC ACID see AMW000
11-AMINOUNDECYLIC ACID see AMW000
AMINOURACIL MUSTARD see BIA250
AMINOUREA see HGU000
AMINOUREA HYDROCHLORIDE see SBW500
2-AMINO-1,4-XYLENE see XNA000
4-AMINO-1,3-XYLENE see XMS000
AMINOZIDE see DQD400
1,2-AMINOZOPHENYLENE see BDH250
AMINUTRIN see LJM700
AMINZOL SOLUBLE see ALQ000
AMIODARONE see AJK750
AMIOYL see BCA000
AMIP 15m see TAI250
AMIPENIX S see AIV500
AMIPHOS see DOP200
AMI-PILO see PIF250
AMIPROL see DCK759
AMIRAL see CJO250
AMISTURA P see PIF250
AMISYL see BCA000
AMITAKON see BCA000

AMITID see EAI000
AMITOL see AMY050
AMITRENE see BBK500
AMITRIL see AMY050
AMITRIL see EAI000
AMITRIL T.L. see AMY050
AMITRIPTYLINE CHLORIDE see EAI000
AMITROL see AMY050
AMITROL 90 see AMY050
AMITROLE see AMY050
AMITROL-T see AMY050
AMITRYPTYLINE HYDROCHLORIDE see EAI000
AMIXICOTYN see NCR000
AMIZIL HYDROCHLORIDE see BCA000
AMIZOL see AMY050
AMIZOL D see AMY050
AMIZOL DP NAU see AMY050
AMIZOL F see AMY050
AMMAT see ANU650
AMMATE see ANU650
AMMELIDE see MCB000
AMMN see AAW000
AMMOFORM see HEI500
AMMOIDIN see XDJ000
AMMONERIC see ANE500
AMMONIA see AMY500
AMMONIA, anhydrous, liquefied (DOT) see AMY500
AMMONIA ANHYDROUS see AMY500
AMMONIA AQUEOUS see ANK250
AMMONIAC (FRENCH) see AMY500
AMMONIACA (ITALIAN) see AMY500
AMMONIA GAS see AMY500
AMMONIAK (GERMAN) see AMY500
AMMONIA SOLUTIONS, relative density <0.880 at 15
degrees C in water, with >50% ammonia (DOT) see
AMY500
AMMONIA SOLUTIONS, with >10% but not >35%
ammonia (UN 2672) (DOT) see ANK250
AMMONIA SOLUTIONS, with >35% but not >50%
ammonia (UN 2073) (DOT) see ANK250
AMMONIATED GLYCYRRHIZIN see AMY700
AMMONIATED GLYCYRRHIZIN see GIE100
AMMONIATED MERCURY see MCW500
AMMONIA WATER 29% see ANK250
AMMONIO (DICROMATO DI) (ITALIAN) see ANB500
AMMONIOFORMALDEHYDE see HEI500
AMMONIUM ACETATE see ANA000
AMMONIUM ACID ARSENATE see DCG800
AMMONIUM ACID SULFATE see ANJ500
AMMONIUM ALUMINUM FLUORIDE see THQ500
AMMONIUM AMIDOSULFONATE see ANU650
AMMONIUM AMIDOSULPHATE see ANU650
AMMONIUM AMINOFORMATE see AND750
AMMONIUM ARSENATE, solid (DOT) see DCG800
AMMONIUM AURINTRICARBOXYLATE see AGW750
AMMONIUM AZIDE see ANA750
AMMONIUM BICARBONATE (1:1) see ANB250
AMMONIUMBICHROMAAT (DUTCH) see ANB500
AMMONIUM BICHROMATE see ANB500
AMMONIUM BIFLUORIDE see ANJ000
AMMONIUM BISULFATE see ANJ500
AMMONIUM BISULFIDE see ANJ750
AMMONIUM BISULFITE see ANB600
AMMONIUM BOROFLUORIDE see ANH000
AMMONIUM CADMIUM CHLORIDE see AND250
AMMONIUM CARBAMATE see AND750
AMMONIUM CARBAZOATE see ANS500
AMMONIUMCARBONAT (GERMAN) see ANE000

AMMONIUM CARBONATE see ANB250
AMMONIUM CARBONATE see ANE000
AMMONIUM CHLORATE see ANE250
AMMONIUMCHLORID (GERMAN) see ANE500
AMMONIUM CHLORIDE see ANE500
AMMONIUM CHLOROPLATINATE see ANF250
AMMONIUM CHROMATE see ANF500
AMMONIUM CHROMATE(VI) see ANF500
AMMONIUM CHROMIC SULFATE see ANF750
AMMONIUM CITRATE see ANF800
AMMONIUM CITRATE, DIBASIC (DOT) see ANF800
AMMONIUM CRYOLITE see THQ500
AMMONIUMDICHROMAAT (DUTCH) see ANB500
AMMONIUMDICHROMAT (GERMAN) see ANB500
AMMONIUM DICHROMATE see ANB500
AMMONIUM DICHROMATE(VI) see ANB500
AMMONIUM DIFLUORIDE see ANJ000
AMMONIUM, DIMETHYLDIOCTYL-, CHLORIDE see
DTF820
AMMONIUM DISULFATONICKELATE(II) see NCY050
AMMONIUM FLUOALUMINATE see THQ500
AMMONIUM FLUOBORATE see ANH000
AMMONIUM FLUORIDE see ANH250
AMMONIUM FLUORIDE comp. with HYDROGEN
FLUORIDE (1:1) see ANJ000
AMMONIUM FLUOROBERYLLATE see ANH300
AMMONIUM FLUOROBORATE see ANH000
AMMONIUM FLUOROSILICATE (DOT) see COE000
AMMONIUM FLUORURE (FRENCH) see ANH250
AMMONIUM FLUOSILICATE see COE000
AMMONIUMGLUTAMINAT (GERMAN) see MRF000
AMMONIUM GLYCYRRHIZINATE see GIE100
AMMONIUM HEXACHLOROPLATINATE(IV) see
ANF250
AMMONIUM HEXAFLUOROALUMINATE see
THQ500
AMMONIUM HEXAFLUOROSILICATE see COE000
AMMONIUM HEXAFLUOROTITANATE see ANI250
AMMONIUM HEXAFLUOROVANADATE see ANI500
AMMONIUM HYDROFLUORIDE see ANJ000
AMMONIUM HYDROGEN BIFLUORIDE see ANJ000
AMMONIUM HYDROGEN CARBONATE see ANB250
AMMONIUM HYDROGEN DIFLUORIDE see ANJ000
AMMONIUM HYDROGEN FLUORIDE see ANJ000
AMMONIUM HYDROGEN FLUORIDE, solid (UN
1727) (DOT) see ANJ000
AMMONIUM HYDROGEN FLUORIDE, solution (UN
2817) (DOT) see ANJ000
AMMONIUM HYDROGEN SULFATE see ANJ500
AMMONIUM HYDROGEN SULFIDE see ANJ750
AMMONIUM HYDROGEN SULFITE see ANB600
AMMONIUM HYDROSULFIDE see ANJ750
AMMONIUM HYDROSULFIDE, solution (DOT) see
ANJ750
AMMONIUM HYDROXIDE see ANK250
AMMONIUM ISETHIONATE see ANL100
AMMONIUM MAGNESIUM ARSENATE see MAD025
AMMONIUM MAGNESIUM ARSENATE DIHYDRATE
see MAD025
AMMONIUM MAGNESIUM CHROMATE see ANM000
AMMONIUM MERCAPTAN see ANJ750
AMMONIUM MERCAPTOACETATE see ANM500
AMMONIUM METAVANADATE (DOT) see ANY250
AMMONIUM MOLYBDATE see ANM750
AMMONIUM MONOHYDROGEN SULFATE see
ANJ500
AMMONIUM MONOSULFITE see ANB600
AMMONIUM MURIATE see ANE500

AMMONIUM NICKEL SULFATE see NCY050
AMMONIUM NITRATE see ANN000
AMMONIUM(I) NITRATE(1:1) see ANN000
AMMONIUM NITRATE, liquid (hot concentrated solution) (UN 2426) (DOT) see ANN000
AMMONIUM NITRATE, with >0.2% combustible substances (UN 0222) (DOT) see ANN000
AMMONIUM NITRATE, with not >0.2% of combustible substances (UN 1942) (DOT) see ANN000
AMMONIUM-N-NITROSOPHENYLHYDROXYLAMINE see ANO500
AMMONIUM OXALATE see ANO750
AMMONIUM PARAMOLYBDATE see ANM750
AMMONIUM PENTADECAFLUOROOCTANATE see ANP625
AMMONIUM PENTA PEROXODICHROMATE see ANP000
AMMONIUM PERCHLORATE (DOT) see PCD500
AMMONIUM PERFLUOROCAPRILATE see ANP625
AMMONIUM PERFLUOROCAPRYLATE see ANP625
AMMONIUM PERFLUOROOCTANOATE see ANP625
AMMONIUM PEROXYCHROMATE see ANQ750
AMMONIUM PEROXYDISULFATE see ANR000
AMMONIUM PERSULFATE see ANR000
AMMONIUM PHOSPHATE see ANR500
AMMONIUM PHOSPHATE, DIBASIC see ANR500
AMMONIUM PHOSPHATE, MONOBASIC see ANR750
AMMONIUM PICRATE see ANS500
AMMONIUM PICRATE, dry or wetted with <10% water, by weight (UN 0004) (DOT) see ANS500
AMMONIUM PICRATE, wetted with not <10% water, by weight (UN 1310) (DOT) see ANS500
AMMONIUM PICRONITRATE see ANS500
AMMONIUM PLATINIC CHLORIDE see ANF250
AMMONIUM POLYSULFIDE (solution) see ANT000
AMMONIUM POLYSULFIDE, solution (DOT) see ANT000
AMMONIUM POTASSIUM HYDROGEN PHOSPHATE see ANT100
AMMONIUM REINECKATE HYDRATE see ANT300
AMMONIUM RHODANATE see ANW750
AMMONIUM RHODANIDE see ANW750
AMMONIUM SALTPETER see ANN000
AMMONIUMSALZ der AMIDOSULFONSAEURE (GERMAN) see ANU650
AMMONIUM SILICOFLUORIDE see COE000
AMMONIUM SILICON FLUORIDE see COE000
AMMONIUM SULFAMATE see ANU650
AMMONIUM SULFATE (2:1) see ANU750
AMMONIUM SULFHYDRATE see ANJ750
AMMONIUM SULFIDE, solution, red see ANT000
AMMONIUM SULFIDE (POLY-) see ANT000
AMMONIUM SULFOCYANATE see ANW750
AMMONIUM SULFOCYANIDE see ANW750
AMMONIUM SULPHAMATE see ANU650
AMMONIUM SULPHATE see ANU750
AMMONIUM-d-TARTRATE see DCH000
AMMONIUM TARTRATE (DOT) see DCH000
AMMONIUM TELLURATE see ANV750
AMMONIUM TETRACHLOROPLATINATE see ANV800
AMMONIUM TETRAFLUOROBERYLLATE see ANH300
AMMONIUM TETRAFLUOROBORATE see ANH000
AMMONIUM TETRAFLUOROBORATE(1-) see ANH000
AMMONIUM, TETRAMETHYL-, HYDROXIDE see TDK500

AMMONIUM TETRAPEROXO CHROMATE see ANW500
AMMONIUM THIOCYANATE see ANW750
AMMONIUM THIOGLYCOLATE see ANM500
AMMONIUM THIOGLYCOLLATE see ANM500
AMMONIUM TRISULFIDE see ANT000
AMMONIUM VANADATE see ANY250
AMMONIUM VANADO-ARSENATE see ANY750
AMMONIUMYL, DIBUTYL-, HEXACHLOROSTANNATE(2-) (2:1) see BIV900
AMMONYX 4 see DTC600
AMMONYX CA SPECIAL see DTC600
AMMONYX CPC see CCX000
AMN see AOL000
AMNESTROGEN see ECU750
AMNICOTIN see NCR000
AMNOSED see TDA500
AMNUCOL see SEH000
AMOCO 1010 see PMP500
AMOCO 610A4 see PJS750
AMOEBAL see ABX500
AMOENOL see CHR500
AMOGLANDIN see POC500
AMOIL see AON300
AMONIAK (POLISH) see AMY500
A1-MORIN see MRN500
AMORPHOUS CROCIDOLITE ASBESTOS see ARM275
AMORPHOUS QUARTZ see SCK600
AMORPHOUS SILICA see SCK600
AMORPHOUS SILICA see DCJ800
AMORPHOUS SILICA DUST see SCH002
AMORPHOUS SILICA FUME see SCH001
AMOSENE see MQU750
AMOSITE ASBESTOS see ARM262
AMOSITE (OBS.) see ARM250
AMOSYT see DYE600
AMOX see NGS500
AMOXICILLIN TRIHYDRATE see AOA100
AMP see AOA125
5-AMP see AOA125
5'-AMP see AOA125
AMP (nucleotide) see AOA125
AMPACET E/C see EHG100
AMPERIL see AIV750
AMPERIL see AOD125
AMPHAETEX see BBK500
AMPHEDRINE see BBK500
AMPHEDROXY see DBA800
AMPHEDROXYN see DBA800
AMPHENICOL see CDP250
AMPHEREX see BBK500
(+)-AMPHETAMINE see AOA500
d-AMPHETAMINE see AOA500
dl-AMPHETAMINE see BBK000
(+)-AMPHETAMINE SULFATE see BBK500
d-AMPHETAMINE SULFATE see BBK500
AMPHIBOLE see ARM250
AMPHICOL see CDP250
AMPHOJEL see AHC000
AMPHOMORONAL see AOC500
AMPHOTERICIN beta see AOC500
AMPHOTERICIN B see AOC500
AMPHOTERICIN B DEOXYCHOLATE see FPC200
AMPHOTERICIN B, MIXT. WITH (3-α,5-β,12-α)-3,12-DIHYDROXYCHOLAN-24-OIC ACIDMONOSODIUM SALT see FPC200

AMPHOTERICIN B SODIUM DESOXYCHOLATE see FPC200
AMPHOTERICINE B see AOC500
AMPHOZONE see AOC500
AMPI-BOL see AIV500
AMPICHEL see AOD125
d-AMPICILLIN see AIV500
d-(−)-AMPICILLIN see AIV500
AMPICILLIN A see AIV500
AMPICILLIN ACID see AIV500
AMPICILLIN ANHYDRATE see AIV500
AMPICILLIN SODIUM see SEQ000
AMPICILLIN SODIUM SALT see SEQ000
AMPICILLIN TRIHYDRATE see AOD125
AMPICILLIN (USDA) see AIV500
AMPICIN see AIV500
AMPIKEL see AIV500
AMPIKEL see AOD125
AMPIMED see AIV500
AMPINOVA see AOD125
AMPIPENIN see AIV500
AMPLIACTIL MONOHYDROCHLORIDE see CKP500
AMPLIN see AOD125
AMPLISOM see AIV500
AMPLITAL see AIV500
AMPLIVIX see EID200
AMPROLENE see EJN500
AMPROLIUM see AOD175
AMPY-PENYL see AIV500
AMS see ANU650
AMSCO H-J see NAH600
AMSCO H-SB see NAH600
AMSECLOR see CDP250
AMSPEC-KR see AQF000
AMSUSTAIN see AOA500
AMSUSTAIN see BBK500
AMTHIO see ANW750
AMUDANE see GKE000
AMUNO see IDA000
AMYGDALIC ACID see MAP000
AMYGDALINIC ACID see MAP000
AMYGDALONITRILE see MAP250
n-AMYL ACETATE see AOD725
AMYL ACETATE (DOT) see AOD725
sec-AMYL ACETATE see AOD735
AMYL ACETATE (mixed isomers) see AOD750
AMYL ACETIC ESTER see AOD725
AMYL ACID PHOSPHATE (DOT) see PBW750
AMYL ALCOHOL see AOE000
N-AMYL ALCOHOL see AOE000
sec-AMYL ALCOHOL see PBM750
tert-AMYL ALCOHOL (DOT) see PBV000
AMYL ALCOHOL, NORMAL see AOE000
AMYL ALDEHYDE see VAG000
N-AMYLALKOHOL (CZECH) see AOE000
AMYLAMINE see PBV505
n-AMYLAMINE see PBV505
AMYLAMINE (DOT) see PBV505
iso-AMYLAMINE see AOE200
AMYLAMINE (mixed isomers) see PBV500
AMYLAMINES (DOT) see PBV500
AMYLAZETAT (GERMAN) see AOD725
AMYL BENZOATE see IHP100
AMYL BUTYRATE see AOG000
n-AMYL BUTYRATE see AOG000
γ-N-AMYLBUTYROLACTONE see CNF250
AMYLCARBINOL see HFJ500

n-AMYL CHLORIDE see PBW500
AMYL CHLORIDE (DOT) see PBW500
α-AMYL CINNAMALDEHYDE see AOG500
AMYL CINNAMATE see AOG600
α-AMYL CINNAMIC ALDEHYDE see AOG500
AMYL CINNAMYLIDENE METHYL ANTHRANILATE see AOH100
AMYLDICHLORARSINE see AOI200
N-AMYLDICHLORARSINE see AOI200
AMYLENE see AOI800
tert-AMYLENE see IHR220
AMYLENE HYDRATE see PBV000
n-AMYLENE PENTENE see AOI800
AMYLENES, MIXED see AOJ000
AMYLESTER KYSELINY DUSICNE see AOL250
AMYLESTER KYSELINY OCTOVE see AOD725
2-AMYLESTER KYSELINY OCTOVE see AOD735
sek.AMYLESTER KYSELINY OCTOVE see AOD735
AMYL ETHER see PBX000
n-AMYL ETHER see PBX000
AMYLETHYLCARBINOL see OCY100
AMYL ETHYL KETONE see ODI000
AMYLETHYLMETHYLCARBINOL see MND050
AMYL FORMATE see AOJ500
n-AMYL FORMATE see AOJ500
AMYL HEXANOATE see IHU100
AMYL HYDRIDE (DOT) see PBK250
AMYL HYDROSULFIDE see PBM000
AMYL KETONE see ULA000
n-AMYL MERCAPTAN see PBM000
AMYL MERCAPTAN (DOT) see PBM000
tert-AMYLMERCAPTAN see MHS550
AMYL METHYL ALCOHOL see AOK750
AMYL METHYL CARBINOL see HBE500
AMYL-METHYL-CETONE (FRENCH) see MGN500
n-AMYL METHYL KETONE see MGN500
AMYL METHYL KETONE (DOT) see MGN500
n-AMYL-N-METHYLNITROSAMINE see AOL000
AMYL NITRATE see AOL250
n-AMYL NITRITE see AOL500
AMYL NITRITE (DOT) see AOL500
n-AMYLNITROSOUREA see PBX500
1-AMYL-1-NITROSOUREA see PBX500
AMYLOFENE see EOK000
AMYLOGLUCOSIDASE see AOM125
AMYLOMAIZE VII see SLJ500
β-AMYLOSE see CCU150
AMYLOWY ALKOHOL (POLISH) see IHP000
4-n-AMYLPHENOL see AOM250
2-sec-AMYLPHENOL see AOM500
α-AMYL-β-PHENYLACROLEIN see AOG500
AMYL PHTHALATE see AON300
AMYL PROPANOATE see AON350
AMYL PROPIONATE see AON350
n-AMYL PROPIONATE see AON350
AMYL SULFHYDRATE see PBM000
AMYL THIOALCOHOL see PBM000
tert-AMYLTHIOL see MHS550
AMYL TRICHLOROSILANE see PBY750
AMYLUM see SLJ500
AMYL-Δ-VALEROLACTONE see DAF200
AMYLVINYLCARBINOL see ODW000
AMYL ZIMATE see BJK500
AMYTAL SODIUM see AON750
AN see AAQ500
ANA see NAK500
ANAC 110 see CNI000

ANACARDONE see DJS200
ANACETIN see CDP250
ANACOBIN see VSZ000
ANACORDONE see DJS200
ANADOLOR see AIT250
ANADOMIS GREEN see CMJ900
ANADROL see PAN100
ANADROYD see PAN100
ANAESTHETIC ETHER see EJU000
ANAFEBRINA see DOT000
ANAFLON see HIM000
ANAGIARDIL see MMN250
ANAMENTH see TIO750
ANANASE see BMO000
ANAPAC see ABG750
ANAPOLON see PAN100
ANASTERON see PAN100
ANASTERONAL see PAN100
ANASTERONE see PAN100
ANASTRESS see MQU750
ANATASE see OBU100
ANATHYLMON see MQU750
ANATOLA see VSK600
ANATRAN see ABH500
ANAUTINE see DYE600
ANAYODIN see SHW000
ANCAMINE TL see MJQ000
ANCHRED STANDARD see IHC450
ANCILLIN see AOD125
ANCOLAN see HGC500
ANCOLAN DIHYDROCHLORIDE see MBX250
ANCOR EN 80/150 see IGK800
ANCORTONE see PLZ000
ANCYLOL see DNG000
ANDAKSIN see MQU750
ANDAXIN see MQU750
ANDRAMINE see DYE600
ANDRANE see CCR510
ANDREZ see SMR000
ANDROGEN see TBG000
ANDROLIN see TBF500
ANDROMETH see MPN500
ANDRONAQ see TBF500
ANDROSAN see MPN500
ANDROSAN see TBG000
ANDROSAN (tablets) see MPN500
ANDROSTAN-17-ONE, 3-(ACETYLOXY)-, (3-β,5-α)- see HJB225
ANDROSTEN see MPN500
4-ANDROSTENE-17-α-METHYL-17-β-OL-3-ONE see MPN500
Δ⁴-ANDROSTENE-17-β-PROPIONATE-3-ONE see TBG000
ANDROST-4-EN-17β-OL-3-ONE see TBF500
Δ⁴-ANDROSTEN-17(β)-OL-3-ONE see TBF500
ANDROSTEN-17-ONE, 3-β-HYDROXY-, ACETATE see HJB225
ANDROTESTON see TBG000
ANDROTEST P see TBG000
ANDRUSOL see TBF500
ANDRUSOL-P see TBG000
ANDUR see PKL500
ANELIX see HIM000
ANELMID see DJT800
ANERGAN see ABH500
ANERTAN see MPN500
ANERTAN see TBG000

ANERTAN (tablets) see MPN500
ANERVAL see BRF500
ANESTACON see DHK400
ANESTACON HYDROCHLORIDE see DHK600
ANESTHENYL see MGA850
ANESTHESIA ETHER see EJU000
ANESTHESIN see EFX000
ANESTHESOL see AIT250
ANESTHETIC COMPOUND No. 347 see EAT900
ANESTHETIC ETHER see EJU000
ANESTHONE see EFX000
ANESTIL see AIT250
ANETHOLE (FCC) see PMQ750
ANEURAL see MQU750
ANEURINE see TES750
ANEUXRAL see MQU750
ANEXOL see CDP000
ANFLAGEN see IIU000
ANFRAM 3PB see TNC500
ANGELICA OIL, root see AOO760
ANGELICA ROOT OIL see AOO760
ANGELICA SEED OIL see AOO790
ANGELIKA OEL see AOO760
ANGIBID see NGY000
ANGICAP see PBC250
ANGIFLAN see DBX400
ANGININE see NGY000
ANGIOLINGUAL see NGY000
ANGITET see PBC250
ANGLISLITE see LDY000
ANGORIN see NGY000
ANGUIFUGAN see DJT800
ANHIBA see HIM000
ANHYDRIDE ACETIQUE (FRENCH) see AAX500
ANHYDRIDE ARSENIEUX see ARI750
ANHYDRIDE ARSENIQUE (FRENCH) see ARH500
ANHYDRIDE CARBONIQUE (FRENCH) see CBU250
ANHYDRIDE CARBONIQUE et OXYDE d'ETHYLENE MELANGES (FRENCH) see EJO000
ANHYDRIDE CHROMIQUE (FRENCH) see CMK000
ANHYDRIDE PHTHALIQUE (FRENCH) see PHW750
ANHYDRIDES see AOP500
ANHYDRIDE VANADIQUE (FRENCH) see VDU000
ANHYDRID KYSELINY KROTONOVE see COB900
ANHYDRID KYSELINY MASELNE see BSW550
ANHYDRID KYSELINY OCTOVE see AAX500
ANHYDRID KYSELINY TETRAHYDROFTALOVE (CZECH) see TDB000
3,6-ANHYDRO-d-GALACTAN see CCL250
ANHYDRO-d-GLUCITOL MONOOCTADECANOATE see SKV150
ANHYDROGLUCOCHLORAL see GFA000
ANHYDROL see EFU000
ANHYDRONE see PCE000
ANHYDROSORBITOL STEARATE see SKV150
ANHYDRO-o-SULFAMINEBENZOIC ACID see BCE500
ANHYDROTRIMELLIC ACID see TKV000
ANHYDROUS AMMONIA see AMY500
ANHYDROUS BORAX see DXG035
ANHYDROUS CALCIUM SULFATE see CAX500
ANHYDROUS CHLORAL see CDN550
ANHYDROUS CHLOROBUTANOL see ABD000
ANHYDROUS HYDRIODIC ACID see HHI500
ANHYDROUS HYDROBROMIC ACID see HHJ000
ANHYDROUS HYDROCHLORIC ACID see HHL000
ANHYDROUS IRON OXIDE see IHC450
ANHYDROUS OXIDE of IRON see IHC450

ANHYDROUS SODIUM ARSANILATE see ARA500
ANICON KOMBI see CIR250
ANICON M see CIR250
ANIDRIDE ACETICA (ITALIAN) see AAX500
ANIDRIDE CROMICA (ITALIAN) see CMK000
ANIDRIDE CROMIQUE (FRENCH) see CMJ900
ANIDRIDE FTALICA (ITALIAN) see PHW750
ANILID KYSELINY ACETOCTOVE see AAY000
ANILIN (CZECH) see AOQ000
ANILINA (ITALIAN, POLISH) see AOQ000
ANILINE see AOQ000
ANILINE, N-ACETYL- see AAQ500
ANILINE, p-ARSENOSO- see ARJ755
ANILINE, p-ARSENOSO-N,N-BIS(2-CHLOROETHYL)-
see ARJ760
ANILINE, p-ARSENOSO-N,N-BIS(2-
HYDROXYETHYL)- see ARJ770
ANILINE, p-ARSENOSO-N,N-DIETHYL- see ARJ800
ANILINE, p-ARSENOSO-, DIHYDRATE see ALV100
p-ANILINEARSONIC ACID see ARA250
ANILINE CHLORIDE see BBL000
ANILINE, p-CHLORO-, HYDROCHLORIDE see CJR200
ANILINE, 4-CHLORO-3-NITRO- see CJA185
ANILINE, p-DICHLOROARSINO-,
HYDROCHLORIDE see AOR640
ANILINE DYES see AOQ500
ANILINE, p-ETHOXY- see PDK790
ANILINE, o-ETHYL-(8CI) see EGK500
ANILINE GREEN see BAY750
ANILINE HYDROCHLORIDE (DOT) see BBL000
ANILINE, 4-NITRO- see NEO500
ANILINE OIL see AOQ000
"ANILINE SALT" see BBL000
p-ANILINESULFONAMIDE see SNM500
ANILINE-p-SULFONIC AMIDE see SNM500
ANILINE, 2,4,6-TRINITRO- see PIC800
ANILINE VIOLET see AOR500
ANILINE VIOLET PYOKTANINE see AOR500
ANILINE YELLOW see PEI000
ANILINIUM CHLORIDE see BBL000
ANILINOBENZENE see DVX800
2-ANILINOBENZOIC ACID see PEG500
o-ANILINOBENZOIC ACID see PEG500
4-ANILINODICHLOROARSINE, HYDROCHLORIDE
see AOR640
ANILINOETHANE see EGK000
2-ANILINOETHANOL see AOR750
ANILINOMETHANE see MGN750
ANILINONAPHTHALENE see PFT500
2-ANILINONAPHTHALENE see PFT500
ANIMAG see MAH500
ANIMAL OIL see BMA750
2-ANISALDEHYDE see AOT525
o-ANISALDEHYDE see AOT525
p-ANISALDEHYDE see AOT530
ANISE ALCOHOL see MED500
ANISE CAMPHOR see PMQ750
ANISEED OIL see AOU250
ANISENE see CLO750
ANISE OIL see AOU250
o-ANISIC ACID see MPI000
p-ANISIC ACID, ETHYL ESTER see AOV000
ANISIC ACID HYDRAZIDE see AOV500
p-ANISIC ACID, HYDRAZIDE see AOV500
ANISIC ALCOHOL see MED500
ANISIC ALDEHYDE see AOT530
ANISIC HYDRAZIDE see AOV500
2-ANISIDINE see AOV900

4-ANISIDINE see AOW000
o-ANISIDINE see AOV900
p-ANISIDINE see AOW000
o-ANISIDINE HYDROCHLORIDE see AOX250
p-ANISIDINE HYDROCHLORIDE see AOX500
o-ANISIDINE NITRATE see NEQ500
ANIS OEL (GERMAN) see AOU250
p-ANISOL ALCOHOL see MED500
ANISOLE see AOX750
ANISOLE, 2,4-DIAMINO-, HYDROGEN SULFATE see
DBO400
ANISOLE, 2,4-DIAMINO-, SULFATE see DBO400
ANISOPYRADAMINE see DBM800
ANISOYL CHLORIDE see AOY250
N-ANISOYL-GABA see AOY270
ANISOYLHYDRAZINE see AOV500
p-ANISOYLHYDRAZINE see AOV500
ANISYL ACETATE see AOY400
2-(p-ANISYL)ACETIC ACID see MFE250
ANISYLACETONE see MFF580
ANISYL ALCOHOL (FCC) see MED500
o-ANISYLAMINE see AOV900
p-ANISYLAMINE see AOW000
o-ANISYLAMINE HYDROCHLORIDE see AOX250
ANISYL FORMATE see MFE250
p-ANISYL 4-PYRIDYL KETONE see MFH930
p-ANISYOL CHLORIDE see AOY250
ANKILOSTIN see PCF275
ANN (GERMAN) see AAW000
ANNALINE see CAX750
ANNANOX CK see SCQ000
ANNATTO EXTRACT see APE100
(6)ANNULENE see BBL250
ANODYNON see EHH000
ANOFEX see DAD200
ANOL see CPB750
ANOPROLIN see ZVJ000
ANOREXIDE see BBK000
ANOVLAR 21 see EEH520
ANOXOMER see APE300
ANOZOL see DJX000
ANPROLENE see EJN500
ANPROLINE see EJN500
ANQUIL see RDK000
ANSAR see HKC000
ANSAR 160 see HKC500
ANSAR 170 see MRL750
ANSAR 560 see HKC500
ANSEPRON see PGA750
ANSIACAL see MDQ250
ANSIATAN see MQU750
ANSIBASE RED KB see CLK225
ANSIL see MQU750
ANSIOLISINA see DCK759
ANSIOWAS see MQU750
ANSUL ETHER 181AT see PBO500
ANTABUS see DXH250
ANTABUSE see DXH250
ANTADIX see DXH250
ANTADOL see BRF500
ANTAENYL see DXH250
ANTAETHAN see DXH250
ANTAETHYL see DXH250
ANTAETIL see DXH250
ANTAGE W 500 see MJN250
ANTAGONATE see TAI500
ANTAGOTHYROID see TFR250
ANTAGOTHYROIL see TFR250

ANTAK see DAI600
ANTALCOL see DXH250
ANTALVIC see PNA500
ANTAROX A-200 see PKF500
ANTENE see BJK500
ANTETAN see DXH250
ANTETHYL see DXH250
ANTETIL see DXH250
ANTEYL see DXH250
ANTHER see IHV050
ANTHIO see DRR200
ANTHIOLIMINE see LGU000
ANTHIOMALINE see LGU000
ANTHION see DWQ000
ANTHIPHEN see MJM500
ANTHISAN MALEATE see DBM800
ANTHIUM DIOXIDE see CDW450
ANTHON see TIQ250
ANTHOPHYLITE see ARM264
ANTHRACEN (GERMAN) see APG500
2-ANTHRACENAMINE see APG100
ANTHRACENE see APG500
2-ANTHRACENECARBOXYLIC ACID, 9,10-
DIHYDRO-4,5-DIHYDROXY-9,10-DIOXO- see RHZ700
9,10-ANTHRACENEDIONE see APK250
9,10-ANTHRACENEDIONE, 1-AMINO-4-HYDROXY-
(9CI) see AKE250
9,10-ANTHRACENEDIONE, 1-BROMO-4-
(METHYLAMINO)- see BNN550
9,10-ANTHRACENEDIONE, 2,6-DIAMINO- see
APK850
9,10-ANTHRACENEDIONE, 1,4-DIAMINO-5-NITRO-
(9CI) see DBY700
9,10-ANTHRACENEDIONE, 1,5-DICHLORO- see
DEO700
9,10-ANTHRACENEDIONE, 1,8-DICHLORO- see
DEO750
9,10-ANTHRACENEDIONE, 1,5-DIPHENOXY- see
DVW100
9(10H)-ANTHRACENONE, 10-(ACETYLOXY)-1,8-
DIHYDROXY- see ACI640
9(10H)-ANTHRACENONE, 1,8-DIHYDROXY-10-(1-
OXOBUTYL)- see BSY400
9(10H)-ANTHRACENONE, 1,8-DIHYDROXY-10-(1-
OXOETHYL)- see ACI640
ANTHRACIN see APG500
ANTHRACITE PARTICLES see CMY760
2-ANTHRACYLAMINE see APG100
ANTHRADIONE see APK250
2-ANTHRAMINE see APG100
ANTHRANILIC ACID see API500
ANTHRANILIC ACID, N-(2-
BENZYLIDENEHEPTYLIDENE)-, METHYL ESTER
see AOH100
ANTHRANILIC ACID, N-(3-(p-tert-BUTYLPHENYL)-2-
METHYLPROPYLIDENE)-, METHYL ESTER see
LFT100
ANTHRANILIC ACID, CINNAMYL ESTER see API750
ANTHRANILIC ACID, METHYL ESTER see APJ250
ANTHRANILIC ACID, PHENETHYL ESTER see
APJ500
ANTHRAPOLE AZ see BQK250
ANTHRAQUINONE see APK250
9,10-ANTHRAQUINONE see APK250
ANTHRAQUINONE, 1-AMINO-4-HYDROXY- see
AKE250
ANTHRAQUINONE, 2-BROMO-1,5-DIAMINO-4,8-
DIHYDROXY- see BNC800

ANTHRAQUINONE, 1-BROMO-4-(METHYLAMINO)-
see BNN550
ANTHRAQUINONE, 2,6-DIAMINO- see APK850
α-ANTHRAQUINONYLAMINE see AIA750
β-ANTHRAQUINONYLAMINE see AIB000
2,6-ANTHRAQUINONYLDIAMINE see APK850
ANTHRASORB see CBT500
2-ANTHROIC ACID, 9,10-DIHYDRO-4,5-
DIHYDROXY-9,10-DIOXO- see RHZ700
2-ANTHRYLAMINE see APG100
ANTIAETHAN see DXH250
ANTIBASON see MPW500
ANTIBIOTIC A-5283 see PIF750
ANTIBIOTIC FN 1636 see PEC750
ANTIBIOTIC No. 899 see VRF000
ANTIBIOTIC PA-105 see OHM900
ANTIBIOTIC S 7481F1 see CQH100
ANTIBIOTIC SF 2052 SULFATE see DAB630
ANTIBIOTIC TM 25 see HOH500
ANTIBIOTIC U 18496 see ARY000
ANTIBIOTIC X 537 see LBF500
ANTIBULIT see SHF500
ANTICANITIC VITAMIN see AIH600
ANTICARIE see HCC500
ANTIDEPRIN HYDROCHLORIDE see DLH630
ANTIDUROL see DAM700
ANTIEGENE MB see BCC500
ANTIETANOL see DXH250
ANTIETIL see DXH250
ANTIFEBRIN see AAQ500
ANTIFOAM FD 62 see SCR400
ANTIFOLAN see MDV500
ANTIFORMIN see SHU500
ANTIGESTIL see DKA600
ANTIHELMYCIN see AQB000
ANTIHIST see DBM800
ANTIKNOCK-33 see MAV750
ANTIKOL see DXH250
ANTILEPSIN see DNU000
ANTIMALARINA see ARQ250
ANTIMIGRANT C 45 see SEH000
ANTIMILACE see TDW500
ANTIMIT see BIE500
ANTIMOINE FLUORURE (FRENCH) see AQE000
ANTIMOINE (TRICHLORURE d') see AQC500
ANTIMOL see SFB000
ANTIMONIAL SAFFRON see AQF500
ANTIMONIC "ACID" see AQF750
ANTIMONIC CHLORIDE see AQD000
ANTIMONIC OXIDE see AQF750
ANTIMONIC SULFIDE see AQF500
ANTIMONIO (PENTACLORURO DI) (ITALIAN) see
AQD000
ANTIMONIO (TRICLORURO di) see AQC500
ANTIMONIOUS OXIDE see AQF000
ANTIMONOUS CHLORIDE see AQC500
ANTIMONOUS CHLORIDE (DOT) see AQC500
ANTIMONOUS FLUORIDE see AQE000
ANTIMONOUS SULFIDE see AQL500
ANTIMONPENTACHLORID (GERMAN) see AQD000
ANTIMONTRICHLORID see AQC500
ANTIMONWASSERSTOFFES (GERMAN) see SLQ000
ANTIMONY see AQB750
ANTIMONY(III) ACETATE see AQJ750
ANTIMONY BLACK see AQB750
ANTIMONY BUTTER see AQC500
ANTIMONY CHLORIDE see AQC500

ANTIMONY(V) CHLORIDE see AQD000
ANTIMONY CHLORIDE (DOT) see AQC500
ANTIMONY(III) CHLORIDE see AQC500
ANTIMONY COMPOUNDS see AQD500
ANTIMONY EMETINE IODIDE see EAM000
ANTIMONY FLUORIDE see AQF250
ANTIMONY(V) FLUORIDE see AQF250
ANTIMONY(III) FLUORIDE (1:3) see AQE000
ANTIMONY GLANCE see AQL500
ANTIMONY HYDRIDE see SLQ000
ANTIMONY LACTATE see AQE250
ANTIMONY LACTATE, solid (DOT) see AQE250
ANTIMONYL-2,4-DIHYDROXY PYRIMIDINE see
AQE305
ANTIMONYL-7-FORMYL-8-HYDROXYQUINOLINE-
5-SULPHONATE see AQE320
ANTIMONYL POTASSIUM TARTRATE see AQG250
ANTIMONY, compounded with NICKEL (1:1) see
NCY100
ANTIMONY ORANGE see AQL500
ANTIMONY OXIDE see AQF000
ANTIMONY(3+) OXIDE see AQF000
ANTIMONY PENTACHLORIDE see AQD000
ANTIMONY PENTACHLORIDE (DOT) see AQD000
ANTIMONY(V) PENTAFLUORIDE see AQF250
ANTIMONY PENTAFLUORIDE (DOT) see AQF250
ANTIMONY PENTAOXIDE see AQF750
ANTIMONY PENTASULFIDE see AQF500
ANTIMONY PENTOXIDE see AQF750
ANTIMONY PERCHLORIDE see AQD000
ANTIMONY PEROXIDE see AQF000
ANTIMONY POTASSIUM TARTRATE see AQG250
l-ANTIMONY POTASSIUM TARTRATE see AQH000
ANTIMONY POWDER (DOT) see AQB750
ANTIMONY RED see AQF500
ANTIMONY REGULUS see AQB750
ANTIMONY SESQUIOXIDE see AQF000
ANTIMONY SESQUISULFIDE see AQL500
ANTIMONY SODIUM GLUCONATE see AQH800
ANTIMONY SODIUM OXIDE-l-(+)-TARTRATE see
AQI750
ANTIMONY SODIUM TARTRATE see AQI750
ANTIMONY SULFIDE see AQF500
ANTIMONY SULFIDE see AQL500
ANTIMONY TRIACETATE see AQJ750
ANTIMONY TRIBROMIDE see AQK000
ANTIMONY TRIBROMIDE see AQK000
ANTIMONY TRIBROMIDE, solid or solution (DOT) see
AQK000
ANTIMONY TRICHLORIDE see AQC500
ANTIMONY TRICHLORIDE, solid (DOT) see AQC500
ANTIMONY TRICHLORIDE, liquid (DOT) see AQC500
ANTIMONY TRICHLORIDE, solution (DOT) see
AQC500
ANTIMONY TRIFLUORIDE see AQE000
ANTIMONY TRIFLUORIDE, solid or solution (DOT) see
AQE000
ANTIMONY TRIHYDRIDE see SLQ000
ANTIMONY TRIOXIDE see AQF000
ANTIMONY TRIPHENYLDICHLORIDE see DGO800
ANTIMONY TRISULFIDE see AQL500
ANTIMONY TRISULFIDE COLLOID see AQL500
ANTIMONY VERMILION see AQL500
ANTIMONY WHITE see AQF000
ANTIMOONPENTACHLORIDE (DUTCH) see AQD000
ANTIMOONTRICHLORIDE see AQC500
ANTIMUCIN WDR see ABU500
ANTINONIN see DUS700

ANTIO see DRR200
ANTIOXIDANT 29 see BFW750
ANTIOXIDANT 116 see PFT500
ANTIOXIDANT 330 see TMJ000
ANTIOXIDANT 425 see MJN250
ANTIOXIDANT AS see TFD500
ANTIOXIDANT DBPC see BFW750
ANTIOXIDANT LTDP see TFD500
ANTIOXIDANT MB (CZECH) see BCC500
ANTIOXIDANT PBN see PFT500
ANTI-PELLAGRA VITAMIN see NCQ900
ANTIPHEN see MJM500
ANTIPYONIN see SFF000
ANTIPYRINE see AQN000
(ANTIPYRINYLMETHYLAMINO)METHANESULFON
IC ACID SODIUM SALT see AMK500
ANTIREN see PIJ000
ANTI-RUST see SIQ500
ANTISACER see DKQ000
ANTISACER see DNU000
ANTISAL 1a see TGK750
ANTISEPTOL see BEN000
ANTISOL 1 see PCF275
ANTISTERILITY VITAMIN see VSZ450
ANTISTOMINUM see BBV500
ANTISTREPT see SNM500
ANTITROMBOSIN see BJZ000
ANTIVERM see PDP250
ANTIVITIUM see DXH250
ANTIXEROPHTHALMIC VITAMIN see VSK600
ANTOL see EGV000
ANTOMIN see BBV500
ANTORA see PBC250
ANTOX see AQF000
ANTOXYLIC ACID see ARA250
ANTRANCINE 12 see BQI000
ANTRAPUROL see DMH400
ANTRIOPEPTIN (HUMAN α-COMPONENT) see
HGL680
ANTU see AQN635
ANTURAT see AQN635
ANTYMON (POLISH) see AQB750
ANTYMONOWODOR (POLISH) see SLQ000
ANTYWYLEGACZ see CMF400
ANU see PBX500
ANURAL see MQU750
ANUSPIRAMIN see BRF500
ANXIETIL see MQU750
ANZIEF see ZVJ000
ANZON-TMS see AQF000
AO 29 see BFW750
AO-40 see TMJ000
AO 4K see BFW750
AO 425 see MJN250
AO A1 see AGX000
AOM see ASP250
AOMB see BCC500
AORAL see VSK600
4-AP see AMI500
AP 50 see AQF000
A5MP see AOA125
A1-0109 P see AHE250
A 1588LP see AQF000
APACHLOR see CDS750
APADODINE see DXX400
APADON see HIM000
APADRIN see MRH209

APAMIDE see HIM000
APAMINE see DBA800
APAP see HIM000
APARSIN see BBQ500
APASCIL see MQU750
APATATE DRAPE see TES750
APAURIN see DCK759
APAVAP see DGP900
APAVINPHOS see MQR750
APC see ABG750
APCO 2330 see PEY000
APELAGRIN see NCQ900
APETAIN see BBK500
APEX 462-5 see TNC500
APEXOL see VSK600
APFO see ANP625
APGA see AMG750
APHAMITE see PAK000
APHENYLBARBIT see EOK000
APHOSAL see GFA000
APHOXIDE see TND250
APHTIRIA see BBQ500
API No. 2 FUEL OIL see DHE800
APLIDAL see BBQ500
APL-LUSTER see TEX000
β-APN see AMB750
APO see TND250
APO see AQO300
APOCAROTENAL see AQO300
β-APO-8'-CAROTENAL see AQO300
APOCID ORANGE 2G see HGC000
APOMINE BLACK GX see AQP000
APOPLON see RDK000
A-POXIDE see MDQ250
APOZEPAM see DCK759
APPA see PHX250
APPL-SET see NAK500
APPRESINUM see HGP500
APPRESSIN see HGP495
APRELAZINE see HGP500
APRESAZIDE see HGP500
APRESINE see HGP500
APRESOLIN see HGP495
APRESOLIN see HGP500
APRESOLINE-ESIDRIX see HGP500
APRESOLINE HYDROCHLORIDE see HGP500
APREZOLIN see HGP495
APREZOLIN see HGP500
APROBIT see DQA400
APROL 160 see MND100
APROL 161 see MND050
APRON see MDM100
APRON 2E see MDM100
APRON FL see MDM100
AP-S see ANT000
APSICAL see RDK000
APTAL see CFD990
APURIN see ZVJ000
APUROL see ZVJ000
APV see CBR000
APYONINE AURAMINE BASE see IBB000
AQUA AMMONIA see ANK250
AQUACAL see CAX350
AQUACAT see CNA250
AQUA CERA see HKJ000
AQUACHLORAL see CDO000
AQUACIDE see DWX800

AQUACRINE see EDV000
AQUA-1,2-
DIAMINOPROPANEDIPEROXOCHROMIUM(IV)
DIHYDRATE see AQQ125
AQUAFIL see SCH002
AQUA FORTIS see NED500
AQUAKAY see MMD500
AQUALINE see ADR000
AQUAMOLLIN see EIV000
AQUAMYCETIN see CDP250
AQUAPEL (POLYSACCHARIDE) see SLJ500
AQUAPLAST see SFO500
AQUA REGIA see HHM000
AQUAREX METHYL see SIB600
AQUARILLS see CFY000
AQUARIUS see CFY000
AQUASOL see VSP000
AQUASYNTH see VSK600
AQUATHOL see EAR000
AQUATIN see CLU000
AQUA-VEX see TIX500
AQUAVIRON see TBG000
AQUAZINE see BJP000
AQUINONE see MMD500
AR2 see AGX000
AR 3 see CBT500
AR 12008 see DIO200
AR 81242 see DEC100
ARABIC GUM see AQQ500
9-β-d-ARABINO FURANOSYL ADENINE see AQQ900
ARABINOGALACTAN see AQR800
(+)-ARABINOGALACTAN see AQR800
ARABINOSYLADENINE see AQQ900
9-ARABINOSYLADENINE see AQQ900
β-d-ARABINOSYLADENINE see AQQ900
d-ARABOASCORBIC ACID see EDE600
ARAB RAT DETH see WAT200
ARACET APV see PKP750
ARACHIS OIL see PAO000
ARACIDE see SOP500
ARAGONITE see CAO000
ARALDITE ACCELERATOR 062 see DQP800
ARALDITE ERE 1359 see REF000
ARALDITE HARDENER 972 see MJQ000
ARALDITE HARDENER HY 951 see TJR000
ARALDITE HY 951 see TJR000
ARALO see PAK000
ARAMITE see SOP500
ARAMITEARARAMITE-15W see SOP500
ARANCIO CROMO (ITALIAN) see LCS000
ARASAN see TFS350
ARATAN see MEP250
ARATHANE see AQT500
ARATRON see SOP500
ARBITEX see BBQ500
ARBOCEL see CCU150
ARBOCEL BC 200 see CCU150
ARBOCELL B 600/30 see CCU150
ARBOGAL see DSQ000
ARBOROL see DUS700
ARBOTECT see TEX000
ARBUZ see PAG500
ARCADINE see BCA000
ARCHIDYN see RKP000
ARCOBAN see MQU750
ARCOSOLV see DWT200
ARCOTRATE see PBC250

ARCTON see CBY750
ARCTON 0 see CBY250
ARCTON 3 see CLR250
ARCTON 4 see CFX500
ARCTON 6 see DFA600
ARCTON 7 see DFL000
ARCTON 9 see TIP500
ARCTON 22 see CFX500
ARCTON 33 see FOO509
ARCTON 63 see FOO000
ARCTON 114 see FOO509
ARCTUVIN see HIH000
ARCUM R-S see RDK000
ARDALL see SJO000
ARDEX see BBK500
ARECA CATECHU see BFW000
ARECA CATECHU Linn., nut extract see BFW000
ARECA CATECHU Linn., fruit extract see BFW000
ARECAIDINE METHYL ESTER see AQT750
ARECOLINE see AQT750
ARECOLINE BASE see AQT750
AREDION see CKM000
AREGINAL see EKL000
ARESIN see CKD500
ARETAN see MEP250
ARETAN 6 see MEP250
ARETAN-NIEUW see BKS810
AREZIN see CKD500
AREZINE see CKD500
ARFICIN see RKP000
ARGAMINE see AQW000
ARGENT FLUORURE (FRENCH) see SDQ500
ARGENTIC FLUORIDE see SDQ500
ARGENTIUM CREDE see SDI750
ARGENTOUS OXIDE see SDU500
ARGENTUM see SDI500
ARGEZIN see ARQ725
5-l-ARGININECYANOGINOSIN LA see AQV990
ARGININE HYDROCHLORIDE see AQW000
l-ARGININE HYDROCHLORIDE see AQW000
ARGININE MONOHYDROCHLORIDE see AQW000
l-ARGININE MONOHYDROCHLORIDE see AQW000
ARGIVENE see AQW000
ARGO BRAND CORN STARCH see SLJ500
ARGON see AQW250
ARHEOL see OGY220
ARILAT see CBM750
ARILATE see BAV575
ARILATE see CBM750
ARIOTOX see TDW500
ARISTOCORT ACETONIDE see AQX500
ARISTODERM see AQX500
ARISTOGEL see AQX500
ARISTOLOCHIC ACID see AQY250
ARISTOLOCHIC ACID I SODIUM SALT see SEY050
ARISTOLOCHIC ACID, SODIUM SALT see SEY050
ARISTOLOCHINE see AQY250
ARIZOLE see PIH750
ARIZOLE see PMQ750
ARKLONE P see FOO000
ARKOPAL N-090 see PKF000
ARKOTINE see DAD200
ARKOZAL see BSQ000
ARLACEL 60 see SKV150
ARLACEL 80 see SKV100
ARLACEL 161 see OAV000
ARLACEL 169 see OAV000
ARLOSOL GREEN B see BLK000

ARMCO IRON see IGK800
ARMEEN 18 see OBC000
ARMEEN 12D see DXW000
ARMEEN 16D see HCO500
ARMEEN 18D see OBC000
ARMEEN O see OHM700
ARMENIAN BOLE see IHC450
ARMODOUR see PKQ059
ARMOFILM see OBC000
ARMOFOS see SKN000
ARMOSTAT 801 see OAV000
ARMOTAN MO see SKV100
ARMOTAN MS see SKV150
ARMOTAN PML-20 see PKG000
ARMOTAN PMO-20 see PKL100
ARNAUDON'S GREEN see CMK300
ARNAUDON'S GREEN (HEMIHEPTAHYDRATE) see
CMK300
ARNOSULFAN see SNN300
ARO see CBT750
AROALL see SJO000
AROCHLOR 1221 see PJM000
AROCHLOR 1242 see PJM500
AROCHLOR 1254 see PJN000
AROCHLOR 1260 see PJN250
AROCLOR see PJL750
AROCLOR 1016 see PJL750
AROCLOR 1221 see PJL750
AROCLOR 1232 see PJL750
AROCLOR 1232 see PJM250
AROCLOR 1242 see PJL750
AROCLOR 1242 see PJM500
AROCLOR 1248 see PJL750
AROCLOR 1248 see PJM750
AROCLOR 1254 see PJL750
AROCLOR 1254 see PJN000
AROCLOR 1260 see PJL750
AROCLOR 1260 see PJN250
AROCLOR 1262 see PJL750
AROCLOR 1262 see PJN500
AROCLOR 1268 see PJL750
AROCLOR 1268 see PJN750
AROCLOR 2565 see PJL750
AROCLOR 2565 see PJO000
AROCLOR 4465 see PJL750
AROCLOR 4465 see PJO250
AROCLOR 5442 see PJL750
AROFLOW see CBT750
AROGEN see CBT750
AROMATIC AMINES see AQY750
AROMATIC CASTOR OIL see CCP250
AROMEX see CBT750
ARON ALPHA 402X see EHP700
ARON ALPHA D see EHP700
ARON COMPOUND HW see PKQ059
AROSOL see PER000
AROTONE see CBT750
AROVEL see CBT750
AROVIT see VSP000
ARQUAD DM14B-90 see TCA500
ARQUAD DM18B-90 see DTC600
ARRESIN see CKD500
ARROW see CBT750
ARROWROOT STARCH see SLJ500
ARSACETIN see AQZ900
ARSACETIN SODIUM SALT see AQZ900
9-ARSAFLUORENINIC ACID see ARA100
ARSAFLUORINIC ACID see ARA100

ARSAMBIDE see CBJ000
ARSAMIN see ARA500
ARSAN see HKC000
ARSANILIC ACID see ARA250
4-ARSANILIC ACID see ARA250
p-ARSANILIC ACID see ARA250
ARSANILIC ACID, N-ACETYL-, SODIUM SALT see
AQZ900
p-ARSANILIC ACID, N,N-BIS(2-CHLOROETHYL)- see
BIA300
p-ARSANILIC ACID, N,N-DIETHYL- see DIS775
ARSANILIC ACID, MONOSODIUM SALT see ARA500
ARSANILIC ACID SODIUM SALT see ARA500
ARSECLOR see DFX400
ARSECODILE see HKC500
ARSEN (GERMAN, POLISH) see ARA750
ARSENATE see ARB250
ARSENATE of IRON, FERRIC see IGN000
ARSENATE of IRON, FERROUS see IGM000
ARSENATE of LEAD see LCK000
ARSENENOUS ACID, CALCIUM SALT (2:1) see
CAM300
ARSENENOUS ACID, POTASSIUM SALT see PKV500
ARSENIATE de CALCIUM see ARB750
ARSENIATE de MAGNESIUM (FRENCH) see ARD000
ARSENIATE de PLOMB (FRENCH) see ARC750
ARSENIC see ARA750
ARSENIC-75 see ARA750
ARSENIC, metallic (DOT) see ARA750
ARSENIC ACID see ARH500
m-ARSENIC ACID see ARB000
o-ARSENIC ACID see ARB250
ARSENIC ACID, solid (DOT) see ARB250
ARSENIC ACID, solid (DOT) see ARC500
ARSENIC ACID, liquid (DOT) see ARB250
ARSENIC ACID, AMMONIUM MAGNESIUM SALT,
HYDRATE (1:1:1:2) see MAD025
ARSENIC ACID ANHYDRIDE see ARH500
ARSENIC ACID, CALCIUM SALT see CAM222
ARSENIC ACID, CALCIUM SALT (2:3) see ARB750
ARSENIC ACID, DISODIUM SALT see ARC000
ARSENIC ACID, DISODIUM SALT, HEPTAHYDRATE
see ARC250
ARSENIC ACID (H3-AS-O4), AMMONIUM
MAGNESIUM SALT, (1:1:1) see MAD025
ARSENIC ACID, HEMIHYDRATE see ARC500
ARSENIC ACID, LEAD SALT see ARC750
ARSENIC ACID, LEAD(2+) SALT (2:3) see LCK100
ARSENIC ACID, MAGNESIUM SALT see ARD000
ARSENIC ACID, METHYLPHENYL-(9CI) see HMK200
ARSENIC ACID, MONOPOTASSIUM SALT see ARD250
ARSENIC ACID, MONOSODIUM SALT see ARD500
ARSENIC ACID, MONOSODIUM SALT see ARD600
ARSENIC ACID, SODIUM SALT see ARD750
ARSENIC ACID, SODIUM SALT (9CI) see ARD500
ARSENIC ACID, TRICESIUM SALT see CDC375
ARSENIC(V) ACID, TRISODIUM SALT,
HEPTAHYDRATE (1:3:7) see ARE000
ARSENIC ACID, ZINC SALT see ZDJ000
ARSENICAL DUST see ARE500
ARSENICAL FLUE DUST see ARE500
ARSENICAL FLUE DUST see ARE750
ARSENICALS see ARA750
ARSENICALS see ARF750
ARSENIC ANHYDRIDE see ARH500
ARSENIC BLACK see ARA750
ARSENIC BLANC see ARI750
ARSENIC(III) BROMIDE see ARF250

ARSENIC BUTTER see ARF500
ARSENIC CHLORIDE see ARF500
ARSENIC(III) CHLORIDE see ARF500
ARSENIC COMPOUNDS see ARF750
ARSENIC DICHLOROETHANE see DFH200
ARSENIC DISULFIDE see ARJ100
ARSENIC FLUORIDE see ARI250
ARSENIC HEMISELENIDE see ARG500
ARSENIC HYDRID see ARK250
ARSENIC HYDRIDE see ARK250
ARSENIC IODIDE see ARG750
ARSENIC OXIDE see ARH500
ARSENIC OXIDE see ARI750
ARSENIC(V) OXIDE see ARH500
ARSENIC(III) OXIDE see ARI750
ARSENIC PENTASULFIDE see ARH250
ARSENIC PENTOXIDE see ARH500
ARSENIC SESQUIOXIDE see ARI750
ARSENIC SESQUISULFIDE see ARI000
ARSENIC SULFIDE see ARI000
ARSENIC SULFIDE (DOT) see ARJ100
ARSENIC SULFIDE YELLOW see ARI000
ARSENIC SULPHIDE see ARI000
ARSENIC TERSULPHIDE see ARI000
ARSENIC TRIBROMIDE see ARF250
ARSENIC TRIFLUORIDE see ARI250
ARSENIC TRIHYDRIDE see ARK250
ARSENIC TRIIODIDE see ARG750
ARSENIC TRIOXIDE see ARI750
ARSENIC TRIOXIDE mixed with SELENIUM DIOXIDE
(1:1) see ARJ000
ARSENIC TRISULFIDE see ARI000
ARSENIC TRISULFIDE see ARJ100
ARSENIC TRISULFIDE (DOT) see ARI000
ARSENICUM ALBUM see ARI750
ARSENIC YELLOW see ARI000
ARSENIDES see ARJ250
ARSENIGEN SAURE see ARI750
ARSENIOUS ACID see ARI750
ARSENIOUS ACID, CALCIUM SALT see CAM500
ARSENIOUS ACID (H3AsO3), TRISODIUM SALT (8CI)
see SEY200
ARSENIOUS ACID, SODIUM SALT see SEY500
ARSENIOUS ACID, SODIUM SALT see ARJ500
ARSENIOUS ACID, SODIUM SALT POLYMERS see
ARJ500
ARSENIOUS ACID, STRONTIUM SALT see SME500
ARSENIOUS ACID, TRISILVER(1+) SALT see SDM100
ARSENIOUS ACID, ZINC SALT (9CI) see ZDS000
ARSENIOUS CHLORIDE see ARF500
ARSENIOUS OXIDE see ARI750
ARSENIOUS SULPHIDE see ARI000
ARSENIOUS TRIOXIDE see ARI750
ARSENITE see ARI750
ARSENITE de POTASSIUM (FRENCH) see PKV500
ARSENITE de SODIUM (FRENCH) see SEY500
ARSENIURETTED HYDROGEN see ARK250
ARSENOLITE see ARI750
ARSENOMARCASITE see ARJ750
ARSENOPYRITE see ARJ750
p-ARSENOSOANILINE see ARJ755
4-ARSENOSOANILINE, DIHYDRATE see ALV100
ARSENOSOBENZENE see PEG750
p-ARSENOSO-N,N-BIS(2-CHLOROETHYL)ANILINE
see ARJ760
p-ARSENOSO-N,N-BIS(2-HYDROXYETHYL)ANILINE
see ARJ770
p-ARSENOSO-N,N-DIETHYLANILINE see ARJ800

4-ARSENOSO-2-NITROPHENOL see NHE600
p-ARSENOSOTOLUENE see TGV100
ARSENOUS ACID see ARI750
ARSENOUS ACID ANHYDRIDE see ARI750
ARSENOUS ACID, SODIUM SALT (9CI) see SEY500
ARSENOUS ACID, TRISILVER(1+) SALT (9CI) see
SDM100
ARSENOUS ACID, TRISODIUM SALT see SEY200
ARSENOUS ANHYDRIDE see ARI750
ARSENOUS BROMIDE see ARF250
ARSENOUS CHLORIDE see ARF500
ARSENOUS FLUORIDE see ARI250
ARSENOUS HYDRIDE see ARK250
ARSENOUS IODIDE see ARG750
ARSENOUS OXIDE see ARI750
ARSENOUS OXIDE ANHYDRIDE see ARI750
ARSENOUS SULFIDE see ARI000
ARSENOUS TRIBROMIDE see ARF250
ARSENOUS TRICHLORIDE (9CI) see ARF500
ARSENOUS TRIIODIDE (9CI) see ARG750
ARSENOWODOR (POLISH) see ARK250
ARSENOXIDE SODIUM see ARJ900
ARSENTRIOXIDE see ARI750
ARSENWASSERSTOFF (GERMAN) see ARK250
ARSEVAN see NCJ500
ARSINE see ARK250
ARSINE, (p-AMINOPHENYL)DICHLORO-,
HYDROCHLORIDE see AOR640
ARSINE, (p-AMINOPHENYL)OXO-, DIHYDRATE see
ALV100
ARSINE, AMYLDICHLORO- see AOI200
ARSINE, sec-BUTYLDICHLORO- see BQY300
ARSINE, CHLORO(2-CHLOROVINYL)PHENYL- see
PER600
ARSINE, DICHLOROHEPTYL- see HBN600
ARSINE, DICHLOROHEXYL- see HFP600
ARSINE, DICHLOROPENTYL- see AOI200
ARSINE, DIIODOMETHYL- see MGQ775
ARSINE, DIPHENYLHYDROXY- see DVY100
ARSINE, ETHYLENEBIS(DIPHENYL)- see EIQ200
ARSINE, HYDROXYDIPHENYL- see DVY100
ARSINE OXIDE, ALLYLHYDROXYPHENYL- see
AGQ775
ARSINE OXIDE, BUTYLHYDROXYISOPROPYL- see
BRQ800
ARSINE OXIDE, (o-CHLOROPHENYL)(3-(2,4-
DICHLOROPHENOXY)-2-
HYDROXYPROPYL)HYDROXY- see DGA425
ARSINE OXIDE, (m-CHLOROPHENYL)HYDROXY(β-
HYDROXYPHENETHYL)- see CKA575
ARSINE OXIDE, (p-
CHLOROPHENYL)HYDROXYMETHYL- see CKD800
ARSINE OXIDE, DIBUTYLHYDROXY- see DDV250
ARSINE OXIDE, DIETHYLHYDROXY- see DIS850
ARSINE OXIDE, HYDROXY(2-
HYDROXYPROPYL)PHENYL- see HNX600
ARSINE OXIDE, HYDROXYISOBUTYLISOPROPYL-
see IPS100
ARSINE OXIDE, HYDROXYMETHYLPHENETHYL-
see MNU050
ARSINE OXIDE, HYDROXYMETHYLPHENYL- see
HMK200
ARSINE OXIDE, HYDROXYMETHYLPROPYL- see
MOT800
ARSINE, OXO(4-CARBOXY)PHENYL- see CCI550
ARSINE SELENIDE, TRIMETHYL- see TLH250
ARSINE SULFIDE, DIMETHYLDI- see DQG700

ARSINETTE see LCK000
ARSINETTE see LCK100
ARSINIC ACID, DIBUTYL-(9CI) see DDV250
ARSINIC ACID, DIMETHYL-(9CI) see HKC000
ARSINIC ACID, DIMETHYL-, SODIUM SALT (9CI) see
HKC500
ARSINOSOLVIN see ARA500
ARSODENT see ARI750
ARSONATE liquid see MRL750
ARSONIC ACID see ABX500
ARSONIC ACID, (4-(ACETYLAMINO)PHENYL)-,
MONOSODIUM SALT (9CI) see AQZ900
ARSONIC ACID, (4-AMINOPHENYL)-,
MONOSODIUM SALT (9CI) see ARA500
ARSONIC ACID, COPPER(2+) SALT (1:1) (9CI) see
CNN500
ARSONIC ACID, (4-HYDROXYPHENYL)-, polymer with
FORMALDEHYDE see BCJ150
ARSONIC ACID, POTASSIUM SALT see PKV500
ARSONIC ACID, SODIUM SALT (9CI) see ARJ500
ARSONIUM, (3-
HYDROXYPHENYL)DIETHYLMETHYL-, IODIDE,
METHYLCARBAMATE see MID900
ARSONIUM, TETRAPHENYL-, CHLORIDE see
TEA300
4-ARSONOPHENYLGLYCINAMIDE see CBJ750
p-ARSONOPHENYLUREA see CBJ000
ARSONOUS DICHLORIDE, ETHYL-(9CI) see DFH200
ARSONOUS DICHLORIDE, METHYL-(9CI) see DFP200
ARSONOUS DICHLORIDE, (1-METHYLPROPYL)-
(9CI) see BQY300
ARSONOUS DICHLORIDE, PHENYL-(9CI) see
DGB600
ARSONOUS DIIODIDE, METHYL-(9CI) see MGQ775
ARSPHEN see ABX500
ARSPHENAMINE METHYLENESULFOXYLIC ACID
SODIUM SALT see NCJ500
ARSYCODILE see HKC500
ART 2 see CBT500
ARTEGODAN see PAH250
ARTEMISIA OIL see ARL250
ARTEMISIA OIL (WORMWOOD) see ARL250
ARTERENOL see NNO500
l-ARTERENOL see NNO500
ARTEROCOLINE see ABO000
ARTEROCYN see PLV750
ARTERODY see BBJ750
d-ARTHIN see VSZ100
ARTHO LM see MLH000
ARTIC see MIF765
ARTIFICIAL ALMOND OIL see BAY500
ARTIFICIAL ANT OIL see FPQ875
ARTIFICIAL BARITE see BAP000
ARTIFICIAL CINNAMON OIL see CCO750
ARTIFICIAL GUM see DBD800
ARTIFICIAL HEAVY SPAR see BAP000
ARTIFICIAL MUSTARD OIL see AGJ250
ARTIFICIAL SWEETENING SUBSTANZ GENDORF
450 see SJN700
ARTISIL BLUE SAP see TBG700
ARTISIL DIRECT RED 3BP see AKE250
ARTISIL DIRECT YELLOW G see AAQ250
ARTISIL ORANGE 3RP see AKP750
ARTISIL RED 3BP see AKE250
ARTISIL YELLOW G see AAQ250
ARTISIL YELLOW 2GN see AAQ250
ARTIZIN see BRF500
ARTOLON see MQU750

ARTOMYCIN see TBX250
ARTOSIN see BSQ000
ARTOZIN see BSQ000
ARTRACIN see IDA000
ARTRIL 300 see IIU000
ARTRINOVO see IDA000
ARTRIVIA see IDA000
ARTRIZONE see BRF500
ARTROPAN see BRF500
ARUMEL see FMM000
ARVYNOL see CHG000
ARWOOD COPPER see CNI000
ARYLAM see CBM750
ARZENE see PEG750
AS see CCU250
AS-15 see SLY500
AS-17665 see NDY500
ASA see ADA725
A.S.A. see ADA725
ASA COMPOUND see ABG750
A.S.A. EMPIRIN see ADA725
ASAGRAN see ADA725
ASAHISOL 1527 see AAX250
ASALIN see ARM000
ASALINE see ARM000
ASATARD see ADA725
ASAZOL see MRL750
ASB 516 see AAX250
ASBEST (GERMAN) see ARM250
ASBESTINE see TAB750
ASBESTOS see ARM250
7-45 ASBESTOS see ARM268
ASBESTOS (ACGIH) see ARM260
ASBESTOS (ACGIH) see ARM262
ASBESTOS (ACGIH) see ARM264
ASBESTOS (ACGIH) see ARM268
ASBESTOS (ACGIH) see ARM275
ASBESTOS (ACGIH) see ARM280
ASBESTOS, ACTINOLITE see ARM260
ASBESTOS, AMOSITE see ARM262
ASBESTOS, ANTHOPHYLITE see ARM264
ASBESTOS, ANTHOPHYLLITE see ARM266
ASBESTOS, CHRYSOTILE see ARM268
ASBESTOS, CROCIDOLITE see ARM275
ASBESTOS FIBER see ARM250
ASBESTOS, TREMOLITE see ARM280
ASC 66825 see CEX800
ASC 67178 see CEX800
ASCABIN see BCM000
ASCABIOL see BCM000
ASCARIDOL see ARM500
ASCARIDOLE see ARM500
ASCARIDOLE (organic peroxide) (DOT) see ARM500
ASCARISIN see ARM500
ASCARYL see HFV500
ASCEPTICHROME see MCV000
ASCORBIC ACID see ARN000
l-ASCORBIC ACID see ARN000
l(+)-ASCORBIC ACID see ARN000
ASCORBIC ACID SODIUM SALT see ARN125
l-ASCORBIC ACID SODIUM SALT see ARN125
ASCORBICIN see ARN125
ASCORBIN see ARN125
ASCORBUTINA see ARN000
ASCORBYL PALMITATE see ARN150
ASCOSERP see RDK000
ASCOSERPINA see RDK000
ASECRYL see GIC000

ASEPTICHROME see MCV000
ASEPTOFORM see HJL500
ASEPTOFORM E see HJL000
ASEPTOFORM P see HNU500
ASEX see SFS000
ASIDON 3 see TEH500
ASIPRENOL see DMV600
AS 61CL see ADY500
ASMADION see TEH500
ASMALAR see DMV600
ASMATANE MIST see VGP000
ASMAVAL see TEH500
ASM MB see BCC500
ASP 47 see SOD100
ASPALON see ADA725
ASPARAGIC ACID see ARN850
l-ASPARAGIC ACID see ARN850
l-ASPARAGINE see ARN810
ASPARAGINIC ACID see ARN850
l-ASPARAGINIC ACID see ARN850
ASPARTAME see ARN825
ASPARTIC ACID see ARN850
l-ASPARTIC ACID see ARN850
(l)-ASPARTIC ACID see ARN850
(S)-ASPARTIC ACID see ARN850
l-(+)-ASPARTIC ACID see ARN850
l-ASPARTIC ACID, N-ACETYL-, DILITHIUM SALT see DNU330
l-ASPARTIC ACID, COMPD. WITH 18-DECARBOXY-40-DEMETHYL-3,7-DIDEOXO-N^3)-((DIMETHYLAMINO) ACETYL)-18-(((2-(DIMETHYLAMINO)ETHYL)AMINO)CARBONYL)-3,7-DIHYDROXY-N^{47})-METHYL-5-OXOCANDICIDIN D, CYCLIC 15,19-HEMIACETAL (2:1) see DOS400
ASPARTYLPHENYLALANINE METHYL ESTER see ARN825
N-l-α-ASPARTYL-l-PHENYLALANINE 1-METHYL ESTER (9CI) see ARN825
ASPERGUM see ADA725
ASPHALT see ARO500
ASPHALT, at or above its Fp (DOT) see ARO500
ASPHALT FUMES (ACGIH) see ARO500
ASPHALT, PETROLEUM see PCR500
ASPHALT, PETROLEUM see ARO500
ASPHALTUM see ARO500
ASPIRDROPS see ADA725
ASPIRIN see ADA725
ASPIRINE see ADA725
ASPON-CHLORDANE see CDR750
ASPRO see ADA725
ASSIFLAVINE see DBX400
ASSIPRENOL see DMV600
ASSUGRIN see SGC000
ASSUGRIN FEINUSS see SGC000
ASSUGRIN VOLLSUSS see SGC000
ASTA see CQC650
ASTA B518 see CQC650
ASTA Z 4828 see TNT500
ASTA Z 4942 see IMH000
ASTERIC see ADA725
ASTHENTHILO see DKL800
ASTHMA METER MIST see VGP000
ASTMAHALIN see VGP000
A-STOFF see CDN200
ASTOMIN see MLP250
ASTRA CHRYSOIDINE R see PEK000
ASTRA DIAMOND GREEN GX see BAY750

ASTRALON see PKQ059
ASTRAZOLO see SNN500
ASTRIDINE see CCK125
ASTRINGEN see AHA000
ASTROBAIN see OKS000
ASTROBOT see DGP900
ASTROCAR see DJS200
ASTRUMAL see PLO500
ASUCCIN see SMY000
ASUGRYN see SGC000
ASUNTHOL see CNU750
ASYMMETRICAL TRIMETHYL BENZENE see TLL750
AT see AMY050
3,A-T see AMY050
AT 7 see HCL000
o-AT see AIC250
AT-90 see AMY050
AT-290 see PED750
AT 717 see PKQ250
ATA see AMY050
ATABRINE see ARQ250
ATACTIC POLYPROPYLENE see PMP500
ATACTIC POLYSTYRENE see SMQ500
ATACTIC POLY(VINYL CHLORIDE) see PKQ059
ATALCO C see HCP000
ATALCO S see OAX000
ATARA see CJR909
ATARAX see CJR909
ATARAXOID see CJR909
ATARAZOID see CJR909
ATAZINA see CJR909
ATAZINAX see ARQ725
ATCC No. 20034 see GAV050
ATCP see PIB900
ATDA HYDROCHLORIDE see AMR500
ATEC see ADD750
ATENSINE see DCK759
ATERAX see CJR909
ATERIAN see AHO250
ATEROCYN see PLV750
ATGARD see DGP900
ATHAPROPAZINE see DIR000
ATHOPROPAZIN see DIR000
ATHROMBIN see WAT220
ATHROMBINE-K see WAT200
ATHYLEN (GERMAN) see EIO000
ATHYLENGLYKOL (GERMAN) see EJC500
ATHYLENGLYKOL-MONOATHYLATHER
(GERMAN) see EES350
ATHYL-GUSATHION see EKN000
ATILEN see DCK759
ATIRAN see MEP250
ATLACIDE see SFS000
ATLANTIC see CBT750
ATLANTIC BLACK BD see AQP000
ATLANTIC BLACK C see AQP000
ATLANTIC BLACK E see AQP000
ATLANTIC BLACK EA see AQP000
ATLANTIC BLACK GAC see AQP000
ATLANTIC BLACK GG see AQP000
ATLANTIC BLACK GXCW see AQP000
ATLANTIC BLACK GXOO see AQP000
ATLANTIC BLACK SD see AQP000
ATLANTIC BLUE 2B see CMO000
ATLANTIC CONGO RED see SGQ500
ATLANTIC RESIN FAST BLUE see CMN750
ATLANTIC RESIN FAST BROWN BRL see CMO750
ATLANTIC SKY BLUE A see CMO500

ATLAS "A" see SEY500
ATLAS G 2146 see HKJ000
ATLAS WHITE TITANIUM DIOXIDE see TGG760
AT LIQUID see AMY050
ATLOX 1087 see PKL100
AuTM see GJC000
ATMONIL see AOR500
ATMOS 150 see OAV000
ATMUL 67 see OAV000
ATMUL 84 see OAV000
ATMUL 124 see OAV000
ATOMIT see CAO000
ATOREL see IDE000
ATOSIL see DQA400
ATOXAN see CBM750
ATOXICOCAINE see AIT250
ATOXYL see ARA500
ATOXYLIC ACID see ARA250
AT-17 PHOSPHATE see MLP250
ATRANEX see ARQ725
ATRANORIN see ARQ600
ATRASINE see ARQ725
ATRATOL see SFS000
ATRATOL A see ARQ725
ATRAVET see ABH500
ATRAXINE see MQU750
ATRAZIN see ARQ725
ATRAZINE see ARQ725
ATRED see ARQ725
ATREX see ARQ725
α-hmn ATRIAL NATRIURETIC HORMONE see
HGL680
ATRIOPEPTIN-28 (HUMAN) see HGL680
α-ATRIOPEPTIN (HUMAN) see HGL680
ATRIOPEPTIN-33(RAT), 1-DE-l-LEUCINE-2-DE-l-
ALANINE-3-DEGLYCINE-4-DE-l-PROLINE-5-DE-l-
ARGININE-1 7-l-METHIONINE- see HGL680
ATRIVYL see MMN250
ATROCHIN see SBG000
ATROPIN (GERMAN) see ARR000
ATROPINE see ARR000
(−)-ATROPINE see HOU000
ATROPINE METHONITRATE see MGR500
ATROPINE METHYL NITRATE see MGR500
ATROPINE SULFATE (2:1) see ARR500
ATROPIN SIRAN (CZECH) see ARR500
ATROPINSULFAT (GERMAN) see ARR500
ATROQUIN see SBG000
ATSETOZIN see ABH500
ATTAC 6 see CDV100
ATTAC 6-3 see CDV100
ATTACLAY see PAE750
ATTACLAY X 250 see PAE750
ATTACOTE see PAE750
ATTAGEL see PAE750
ATTAGEL 40 see PAE750
ATTAGEL 50 see PAE750
ATTAGEL 150 see PAE750
ATTAPULGITE see PAE750
ATTAR ROSE see RNA000
ATTAR of ROSE see RNA000
ATTASORB see PAE750
ATUL ACID CRYSTAL ORANGE G see HGC000
ATUL ACID SCARLET 3R see FMU080
ATUL CONGO RED see SGQ500
ATUL CRYSTAL RED F see HJF500
ATUL DIRECT BLACK E see AQP000

ATUL DIRECT BLUE 2B see CMO000
ATUL DIRECT SKY BLUE see CMO500
ATUL FAST YELLOW R see DOT300
ATUL OIL ORANGE T see TGW000
ATUL OIL RED G see OHI200
ATUL ORANGE R see PEJ500
ATUL SUNSET YELLOW FCF see FAG150
ATUL TARTRAZINE see FAG140
AU 3 see CBT500
AUBYGEL GS see CCL250
AUBYGUM DM see CCL250
AULES see TFS350
AULIGEN see BJU000
AURAMINE BASE see IBB000
AURAMINE HYDROCHLORIDE see IBA000
AURAMINE (MAK) see IBA000
AURAMINE (MAK) see IBB000
AURAMINE O (BIOLOGICAL STAIN) see IBA000
AURAMINE YELLOW see IBA000
AURANILE see DKQ000
AURANILE see DNU000
AURANTIA see HET500
AURANTICA see MRN500
AUREOCINA see CMA750
AUREOMYCIN see CMA750
AUREOMYCIN A-377 see CMA750
AUREOMYKOIN see CMA750
AUREOTAN see ART250
AURIC CHLORIDE see GIW176
AURINE-TRICARBOXYLATE d'AMMONIUM
(FRENCH) see AGW750
AURINTRICARBOXYLIC ACID AMMONIUM SALT see
AGW750
AURIPIGMENT see ARI000
AUROMYOSE see ART250
AUROPAN see NBU000
AURORA YELLOW see CAJ750
AUROTAN see ART250
1-AUROTHIO-d-GLUCOPYRANOSE see ART250
AUROTHIOGLUCOSE see ART250
AURUMINE see ART250
AUSTIOX see TGG760
AUSTRACIL see CDP250
AUSTRACOL see CDP250
AUSTRALIAN GUM see AQQ500
AUSTRALOL see IQZ000
AUSTRAPEN see AIV500
AUSTRAPINE see RDK000
AUSTRIAN CINNABAR see LCS000
AUSTROMINAL see EOK000
AUSTROVIT PP see NCR000
AUTAN see DKC800
AUTHRON see ART250
AUTOLYZED YEAST EXTRACT see BAD400
AUTOMIN see BBV500
AUTOMOTIVE DIESEL OIL see DHE900
AUTOMOTIVE DIESEL OIL see FOP000
AUXINUTRIL see PBB750
AV00 see AGX000
AV000 see AGX000
AVADEX see DBI200
AVERMECTIN B(SUB 1) see ARW200
AVERMECTIN B(SUB 1) TECHNICAL GRADE see
ARW200
AVERMIN see AOR500
AVERSAN see DXH250
AVERTIN see THV000
AVERTIN see ARW250

AVERZAN see DXH250
AVIBEST C see ARM268
AVIBON see VSK600
AVICEL see CCU150
AVICEL 101 see CCU150
AVICEL 102 see CCU150
AVICEL PH 101 see CCU150
AVICEL PH 105 see CCU150
AVICOL see PAX000
AVID EC see ARW200
AVIOMARIN see DYE600
AVIROL 118 CONC see SIB600
AVISUN see PMP500
AVITA see VSK600
AVITOL see VSK600
AVITROL see AMI500
AVLON see DBX400
AVLOSULPHONE see SOA500
AVLOTANE see HCI000
AVOGODRITE see PKY000
AVOLIN see DTR200
AVOMEC see ARW200
AVOMINE see DQA400
AVON GREEN A-4379 see BAY750
AW-15'1129 see CKK050
AWPA #1 see CMY825
AXOLE see FPK000
AXURIS see AOR500
AY-5406 see BCA000
AY-6108 see AIV500
AY 9944 see BHN000
AY 24034 see LIU370
AY-57,062 see ABH500
AY-61122 see MLJ500
AYAA see AAX250
AYAF see AAX250
AYERMATE see MQU750
AYFIVIN see BAC250
9-AZAANTHRACENE see ADJ500
10-AZAANTHRACENE see ADJ500
5-AZA-10-ARSENAANTHRACENE CHLORIDE see
PDB000
AZABENZENE see POP250
3-AZABICYCLO(3.2.2)NONANE, 3-
(CHLOROACETYL)- see CEC100
3-AZABICYCLO(3.2.2)NONANE, 3-(IODOACETYL)-
see IDZ100
AZACITIDINE see ARY000
AZACYCLOHEPTANE see HDG000
1-AZACYCLOHEPTANE see HDG000
2-AZACYCLOHEPTANONE see CBF700
AZACYCLOHEXANE see PIL500
1-AZA-2,4-CYCLOPENTADIENE see PPS250
AZACYCLOPENTANE see PPS500
AZACYCLOPROPANE see EJM900
AZACYTIDINE see ARY000
5-AZACYTIDINE see ARY000
5'-AZACYTIDINE see ARY000
7-AZADIBENZ(a,h)ANTHRACENE see DCS400
7-AZADIBENZ(a,j)ANTHRACENE see DCS600
7-AZA-7H-DIBENZO(c,g)FLUORENE see DCY000
9-AZAFLUORENE see CBN000
1-AZAINDENE see ICM000
3-AZAINDOLE see BCB750
AZALINE see ARM000
1-AZANAPHTHALENE see QMJ000
2-AZANAPHTHALENE see IRX000
AZANIL RED SALT TRD see CLK235

AZANIN see ASB250
3-AZAPENTANE-1,5-DIAMINE see DJG600
AZAPERONE (USDA) see FLU000
AZAPLANT see AMY050
AZAPLANT KOMBI see AMY050
AZASERIN see ASA500
AZASERINE see ASA500
l-AZASERINE see ASA500
AZASPIRACID see ASA600
AZATHIOPRINE see ASB250
AZATIOPRIN see ASB250
6-AZAURACIL see THR750
4(6)-AZAURACIL see THR750
AZBOLEN ASBESTOS see ARM264
AZBOLEN ASBESTOS see ARM266
AZDEL see PMP500
AZDID see BRF500
AZEPERONE see FLU000
AZEPROMAZINE see ABH500
AZETYLAMINOFLUOREN (GERMAN) see FDR000
AZIDE see SFA000
AZIDINE BLUE 3B see CMO250
AZIJNZUUR (DUTCH) see AAT250
AZIJNZUURANHYDRIDE (DUTCH) see AAX500
AZIMETHYLENE see DCP800
AZIMIDOBENZENE see BDH250
AZIMINOBENZENE see BDH250
AZINDOLE see BCB750
AZINE see POP250
AZINE DEEP BLACK EW see AQP000
AZINE SKY BLUE 5B see CMO500
AZINFOS-ETHYL (DUTCH) see EKN000
AZINFOS-METHYL (DUTCH) see ASH500
AZINOS see EKN000
AZINPHOS-AETHYL (GERMAN) see EKN000
AZINPHOS ETHYL see EKN000
AZINPHOS-ETILE (ITALIAN) see EKN000
AZINPHOS METHYL see ASH500
AZINPHOS METHYL, liquid (DOT) see ASH500
AZINPHOS-METILE (ITALIAN) see ASH500
AZIRANE see EJM900
AZIRIDIN (GERMAN) see EJM900
AZIRIDINE see EJM900
AZIRIDINE, 1-(3-(BIS(2-CHLOROETHYL)AMINO-p-
TOLUOYL))- see BHQ760
AZIRIDINE, 1,1'-CARBONYLBIS- see BJP450
1-AZIRIDINE ETHANOL see ASI000
1-AZIRIDINYL m-(BIS(2-
CHLOROETHYL)AMINO)PHENYL KETONE see
ASI300
2-(1-AZIRIDINYL)ETHANOL see ASI000
1-(1-AZIRIDINYL)-N-(p-
METHOXYPHENYL)FORMAMIDE see MFF250
1-AZIRIDINYL PHOSPHINE OXIDE (TRIS) (DOT) see
TND250
AZIUM see SFA000
AZIUM see SOW000
AZO-33 see ZKA000
AZOAMINE RED ZH see NEO500
AZOAMINE SCARLET see NEQ500
AZOBASE MNA see NEN500
AZOBENZEEN (DUTCH) see ASL250
AZOBENZENE see ASL250
AZOBENZENE OXIDE see ASO750
AZOBENZIDE see ASL250
AZOBENZOL see ASL250
AZOBISBENZENE see ASL250
AZOBISISOBUTYLONITRILE see ASL750

α,α'-AZOBISISOBUTYLONITRILE see ASL750
AZOBISISOBUTYRONITRILE see ASL750
2,2'-AZOBIS(ISOBUTYRONITRILE) see ASL750
2,2'-AZOBIS(2-METHYLPROPIONITRILE) see ASL750
AZOCARD BLACK EW see AQP000
AZOCARD BLUE 2B see CMO000
AZOCARD RED CONGO see SGQ500
AZODIBENZENE see ASL250
AZODIBENZENEAZOFUME see ASL250
AZODICARBONAMIDE see ASM300
AZODIISOBUTYRONITRILE see ASL750
2,2'-AZODIISOBUTYRONITRILE see ASL750
AZODIISOBUTYRONITRILE (DOT) see ASL750
α,α'-AZODIISOBUTYRONITRILE see ASL750
AZODINE see PDC250
AZODIUM see PDC250
AZODOX-55 see ZKA000
AZODRIN-71 see MRH209
AZODRIN (OSHA) see MRH209
AZODRIN PESTICIDE see MRH209
AZODYNE see PDC250
AZOENE FAST BLUE BASE see DCJ200
AZOENE FAST ORANGE GR BASE see NEO000
AZOENE FAST RED KB BASE see CLK225
AZOENE FAST RED TR BASE see CLK220
AZOENE FAST RED TR SALT see CLK235
AZOENE FAST SCARLET GC BASE see NMP500
AZOENE FAST SCARLET GC SALT see NMP500
AZOFENE see BDJ250
AZOFIX BLUE B SALT see DCJ200
AZOFIX SCARLET G SALT see NMP500
AZOFOS see MNH000
AZO-GANTANOL see SNK000
AZO GANTRISIN see PDC250
AZO GANTRISIN see SNN500
AZO GASTANOL see PDC250
AZOGEN DEVELOPER A see NAX000
AZOGEN DEVELOPER H see TGL750
AZOGENE ECARLATE R see NEQ500
AZOGENE FAST ORANGE GR see NEO000
AZOGENE FAST RED TR see CLK220
AZOGENE FAST RED TR see CLK235
AZOGENE FAST SCARLET G see NMP500
AZOGNE FAST BLUE B see DCJ200
AZOIC DIAZO COMPONENT 6 see NEO000
AZOIC DIAZO COMPONENT 12 see NMP500
AZOIC DIAZO COMPONENT 32 see CLK225
AZOIC DIAZO COMPONENT 37 see NEO500
AZOIC DIAZO COMPONENT 46 see CLK200
AZOIC DIAZO COMPONENT 11 BASE see CLK220
AZOIC DIAZO COMPONENT 11 BASE see CLK235
AZOIC DIAZO COMPONENT 13 BASE see NEQ500
AZOIC RED 36 see MGO750
AZOIMIDE see HHG500
AZOLAN see AMY050
AZOLE see PPS250
AZOLE see AMY050
AZOLID see BRF500
AZOLMETAZIN see SNJ000
AZO-MANDELAMINE see PDC250
AZOMINE see PDC250
AZOMINE BLACK EWO see AQP000
AZOMINE BLUE 2B see CMO000
AZOPHENYLENE see PDB500
AZOPHOS see MNH000
AZORUBIN see HJF500
AZOSEPTALE see TEX250

AZOSSIBENZENE (ITALIAN) see ASO750
AZO-STANDARD see PDC250
AZO-STAT see PDC250
AZOSULFIZIN see SNN500
AZOTE (FRENCH) see NGR500
AZOTHIOPRINE see ASB250
AZOTIC ACID see NED500
AZOTO (ITALIAN) see NGR500
AZOTOWY KWAS (POLISH) see NED500
AZOTOX see DAD200
AZOTOYPERITE see BIE500
AZOTREX see PDC250
AZOTURE de SODIUM (FRENCH) see SFA000
AZOTU TLENKI (POLISH) see NGT500
AZOXYAETHAN (GERMAN) see ASP000
AZOXYBENZEEN (DUTCH) see ASO750
AZOXYBENZENE see ASO750
AZOXYBENZIDE see ASO750
AZOXYBENZOL (GERMAN) see ASO750
AZOXYDIBENZENE see ASO750
AZOXYETHANE see ASP000
AZOXYMETHANE see ASP250
AZS see ASA500
AZTEC BPO see BDS000
AZULENE, 1,2,3,4,5,6,7,8-OCTAHYDRO-1,4-
DIMETHYL-7-(1-METHYLETHYLIDENE)-,
MONOEPOXIDE see EBU100
6(1H)-AZULENONE, 2,3,3A,7,8,8A-HEXAHYDRO-1,3A-
DIHYDROXY-1,4-DIMETHYL-7-(1-
METHYLETHYLIDENE)-, (1S,3AR,8AR)- see AET600
AZZURRO DIRETTO 3B see CMO250
B10 see SFO500
B-12 see VSZ000
B32 see HCL000
B-71 see PKC000
B 75 see IGS000
B-500 see QMJ000
B 518 see CQC650
B 995 see DQD400
B-1,776 see BSH250
B 4992 see BKS810
B 77488 see BAT750
B 859-35 see NDY550
Ba 2797 see CAY500
Bi 3411 see CDO000
B 8509-035 see NDY550
B 10 (polysaccharide) see SFO500
BA see BBC250
BA 2726 see DBF800
BA 2759 see TML100
BA5968 see HGP495
BA 5968 see HGP500
BA 30,803 see BCH750
BA 51-090462 see DCY400
BABROCID see NGE500
BACIGUENT see BAC250
BACI-JEL see BAC250
BACILIQUIN see BAC250
BACILLOL see CNW500
BACILLOPEPTIDASE A see BAC000
BACILLOPEPTIDASE B see BAC000
BACILLUS SUBTILIS CARLSBERG see BAC000
BACITEK OINTMENT see BAC250
BACITRACIN see BAC250
BACITRACIN METHYLENE DISALICYLATE see
BAC260
BACL VITAMIN H1 see AIH600
BACO AF 260 see AHC000

BACTERAMID see SNM500
BACTESULF see SNN500
BACTOL see CHR500
BACTOLATEX see SMQ500
BACTOPEN see SLJ000
BACTRIM see SNK000
BACTROL see BGJ750
BACTROVET see SNN300
BACTYLAN see SEP000
BADIL see AOR500
BAGALOL see MEP250
BAGAODRYL see BBV500
BAGOLAX see MIF760
BAKELITE AYAA see AAX250
BAKELITE DFD 330 see PJS750
BAKELITE DHDA 4080 see PJS750
BAKELITE DYNH see PJS750
BAKELITE LP 70 see AAX175
BAKELITE LP 90 see AAX250
BAKELITE RMD 4511 see ADY500
BAKELITE SMD 3500 see SMQ500
BAKELITE VLFV see AAX175
BAKELITE VMCC see AAX175
BAKELITE VYNS see AAX175
BAKER'S ANTIFOL see DKC800
BAKER'S P AND S LIQUID and OINTMENT see
PDN750
BAKERS YEAST EXTRACT see BAD400
BAKERS YEAST GLYCAN see BAD400
BAKING SODA see SFC500
BAKONTAL see BAP000
BAKTOL see CFD990
BAKTOLAN see CFD990
BAL see BAD750
BALFON 7000 see TAI250
BALMADREN see VGP000
BALS 3A see RNU100
BALSAM CAPTIVI see CNH792
BALSAM FIR OIL see FBU850
BALSAM FIR, OREGON see OJK340
BALSAM, OREGON see OJK340
BALSAM of PERU see BAE750
BALSAM PERU OIL (FCC) see BAE750
BALSAMS, COPAIBA see CNH792
BAMBERMYCIN see MRA250
BAMD 400 see MQU750
B-AMIN see TES750
BANANA OIL see IHO850
BANANOTE see MDW750
BANASIL see RDK000
BANDIS G100 see RNU100
BANEX see MEL500
BANGTON see CBG000
BAN-HOE see CBM000
BANISIL see RDK000
BANLEN see MEL500
BANMINTH see TCW750
BANOCIDE see DIW200
BANTENOL see MHL000
BANTHIONINE see MDT740
BANVEL see MEL500
BANVEL HERBICIDE see MEL500
BAPN see AMB500
BAPN FUMARATE see AMB750
BARACOUMIN see BJZ000
BARAMINE see BBV500
BARAZAE see SNN500
BARBAPIL see EOK000

BARBENYL see EOK000
BARBILEHAE (BARBILETTAE) see EOK000
BARBIPHENYL see EOK000
BARBITA see EOK000
BARBITAL Na see BAG250
BARBITAL SODIUM see BAG250
BARBITAL SOLUBLE see BAG250
BARBITONE SODIUM see BAG250
BARBONAL see EOK000
BARBOPHEN see EOK000
BARDIOL see EDO000
BARECO POLYWAX 2000 see PJS750
BARECO WAX C 7500 see PJS750
BARIDIUM see PDC250
BARIDOL see BAP000
BARIO (PEROSSIDO di) (ITALIAN) see BAO250
BARITE see BAP000
BARITOP see BAP000
BARITRATE see PBC250
BARIUM see BAH250
BARIUM ACETATE see BAH500
BARIUM AZIDE see BAI000
BARIUM AZIDE, dry or wetted with <50% water, by
weight (UN 0224) (DOT) see BAI000
BARIUM AZIDE, wetted with not <50% water, by weight
(UN 1571) (DOT) see BAI000
BARIUM BINOXIDE see BAO250
BARIUM BROMATE see BAI750
BARIUM CADMIUM STEARATE see BAI800
BARIUM CARBONATE see BAJ250
BARIUM CARBONATE (1:1) see BAJ250
BARIUM CHLORATE see BAJ500
BARIUM CHLORIDE see BAK000
BARIUM CHROMATE (1:1) see BAK250
BARIUM CHROMATE(VI) see BAK250
BARIUM CHROMATE OXIDE see BAK250
BARIUM COMPOUNDS (soluble) see BAK500
BARIUM CYANIDE see BAK750
BARIUM CYANIDE, solid (DOT) see BAK750
BARIUM DIACETATE see BAH500
BARIUM DIBENZYLPHOSPHATE see BAL275
BARIUM DICHLORIDE see BAK000
BARIUM DICYANIDE see BAK750
BARIUM DINITRATE see BAN250
BARIUM DIOXIDE see BAO250
BARIUM FLUORIDE see BAM000
BARIUM FLUOROSILICATE see BAO750
BARIUM FLUOSILICATE see BAO750
BARIUM HEXAFLUOROSILICATE see BAO750
BARIUM HEXAFLUOROSILICATE(2-) see BAO750
BARIUM MONOXIDE see BAO000
BARIUM NITRATE (DOT) see BAN250
BARIUM(II) NITRATE (1:2) see BAN250
BARIUM OXIDE see BAO000
BARIUM PERCHLORATE (DOT) see PCD750
BARIUM PERMANGANATE see PCK000
BARIUMPEROXID (GERMAN) see BAO250
BARIUM PEROXIDE see BAO250
BARIUMPEROXYDE (DUTCH) see BAO250
BARIUMPOLYSULFID see BAO300
BARIUM POLYSULFIDE see BAO300
BARIUM PROTOXIDE see BAO000
BARIUMSILICOFLUORID see BAO750
BARIUM SILICOFLUORIDE see BAO750
BARIUM SILICON FLUORIDE see BAO750
BARIUM STYPHNATE see BAO900
BARIUM SULFATE see BAP000
BARIUM SULFIDE see BAO300

BARIUM SULFIDE see BAP250
BARIUM SUPEROXIDE see BAO250
BARIUM ZIRCONATE see BAP750
BARIUM ZIRCONIUM OXIDE see BAP750
BARIUM ZIRCONIUM(IV) OXIDE see BAP750
BARIUM ZIRCONIUM TRIOXIDE see BAP750
BARIZON see MOV000
BAROS CAMPHOR see BMD000
BAROSPERSE see BAP000
BAROTRAST see BAP000
BARPENTAL see NBU000
BARQUAT SB-25 see DTC600
BARQUINOL see CHR500
BARSEB HC see CNS750
BARTOL see EOK000
BARYTA see BAO000
BARYTA WHITE see BAP000
BARYTA YELLOW see BAK250
BARYTES see BAP000
BARYUM FLUORURE (FRENCH) see BAM000
BAS see BEM750
BAS-3460 see MHC750
BAS 67054 see MHC750
BASAGRAN see MJY500
BASAMID see DSB200
BASAMID G see DSB200
BASAMID-GRANULAR see DSB200
BASAMID P see DSB200
BASAMID-PUDER see DSB200
BASCHEM 12 see MAG750
BASCOREZ see AAX250
BASE 661 see SMR000
BASECIL see MPW500
BASE OIL see PCR250
BASETHYRIN see MPW500
BAS 352 F see RMA000
BASFAPON see DGI400
BASFAPON B see DGI400
BASFAPON B see DGI600
BASFAPON/BASFAPON N see DGI400
BASF III see SMQ500
BASF-MANEB SPRITZPULVER see MAS500
BASF URSOL 3GA see ALT000
BASF URSOL D see PEY500
BASF URSOL EG see ALS990
BASF URSOL ERN see NAW500
BASF URSOL P BASE see ALT250
BAS 351-H see MJY500
BASIC ALUMINUM CHLORATE see AHA000
BASIC BLUE 9 see BJI250
BASIC BRIGHT GREEN see BAY750
BASIC CHROMIUM CARBONATE see CMJ100
BASIC LEAD ACETATE see LCH000
BASIC LEAD CHROMATE see LCS000
BASIC NICKEL CARBONATE see NCY500
BASIC NICKEL(II) CARBONATE see NCY600
BASIC ORANGE 3RN see BJF000
BASIC PARAFUCHSINE see RMK020
BASIC RED 1 see RGW000
BASIC RHODAMINE YELLOW see RGW000
BASIC RHODAMINIC YELLOW see RGW000
BASIC VIOLET 3 see AOR500
BASIC VIOLET BN see AOR500
BASIC ZINC CHROMATE see ZFJ100
BASIC ZIRCONIUM CHLORIDE see ZSJ000
BASIL OIL see BAR250
BASIL OIL, EUROPEAN TYPE (FCC) see BAR250
BASIL OIL, SWEET see BAR250

BASINEX see DGI400
BASLE GREEN see COF500
BASOLAN see MCO500
BASSA see MOV000
BASUDIN see DCM750
BASUDIN 10 G see DCM750
BATASAN see ABX250
BATAZINA see BJP000
BATRILEX see PAX000
BAU see CBT500
BAUXITE RESIDUE see IHC450
BAVISTIN see MHC750
BAX see BAU750
BAX 1400Z see PMO250
BAY 1470 see DMW000
BAY 1518 see DJB460
BAY 1521 see DPH600
BAY 2353 see DFV400
BAY 5621 see BAT750
BAY 5821 see DJY200
BAY 9026 see DST000
BAY 9027 see ASH500
BAY 11405 see MNH000
BAY 15203 see DAO800
BAY 15922 see TIQ250
BAY 16225 see EKN000
BAY 18436 see DAP400
BAY 19149 see DGP900
BAY 19639 see DXH325
BAY 21097 see DAP000
BAY 23129 see PHI500
BAY 25141 see FAQ800
BAY 30130 see DGI000
BAY 39731 see MIA250
BAY 41637 see MOV000
BAY 41831 see DSQ000
BAY 61597 see MQR275
BAY 68138 see FAK000
BAY 70143 see CBS275
BAY 71628 see DTQ400
BAY 77049 see DJY200
BAY 77488 see BAT750
BAY 105807 see MIA250
BAYCARB see MOV000
BAYCOVIN see DIZ100
BAY DIC 1468 see MQR275
BAY-E-393 see SOD100
BAY E-601 see MNH000
BAY E-605 see PAK000
BAYER 73 see DFV400
BAYER 2353 see DFV400
BAYER 3231 see TND000
BAYER 5072 see DOU600
BAYER 5081 see EPY000
BAYER 5360 see MMN250
BAYER 15922 see TIQ250
BAYER 16259 see EKN000
BAYER 17147 see ASH500
BAYER 19564 see DKB170
BAYER 19639 see DXH325
BAYER 20315 see DAP600
BAYER 37289 see EPY000
BAYER 37344 see DST000
BAYER 39731 see MIA250
BAYER 41637 see MOV000
BAYER 41831 see DSQ000
BAYER 6159H see MQR275
BAYER 6443H see MQR275

BAYER 71628 see DTQ400
BAYER 94337 see MQR275
BAYER 21/199 see CNU750
BAYER 25/154 see DAP400
BAYER 41367C see MOV000
BAYER-E 393 see SOD100
BAYER E-605 see PAK000
BAYERITIAN see TGG760
BAYER 1440 L see DNA800
BAYER L 13/59 see TIQ250
BAYER S767 see FAQ800
BAYER S 4400 see EPY000
BAYER S 5660 see DSQ000
BAYERTITAN see TGG760
BAY 6681 F see CJO250
BAY LEAF OIL see BAT500
BAY LEAF OIL see LBK000
BAYLETON see CJO250
BAYLUSCID see DFV400
BAY-MEB-6447 see CJO250
BAYMIX 50 see CNU750
BAY-NTN-9306 see SOU625
BAY OIL see BAT500
BAY OIL see LBK000
BAYOL F see MQV750
BAYPRESOL see DNA800
BAYRE 77488 see BAT750
BAYRITES see BAP000
BAYRUSIL see DJY200
BAYTAN see MEP250
BAYTHION see BAT750
BAYTITAN see TGG760
BAY VA 1470 see DMW000
BAZUDEN see DCM750
BBC see BAV575
BBC see BMW250
BBC 12 see DDL800
BBCE see BIQ500
BBH see BBQ500
"B" BLASTING POWDER see ERF500
BBN see BMW250
BBN see HJQ350
BBNOH see HJQ350
BBP see BEC500
BCM see MHC750
BCME see BIK000
BCNU see BIF750
BiCNU see BIF750
BCPN see BQQ250
BCS COPPER FUNGICIDE see CNP250
BDCM see BND500
BDH 1298 see VTF000
BDH 29-790 see SMQ500
BDH 29-801 see TAI250
BDMA see DQP800
BD(a,h)P see DCY200
BDU see BNC750
5-BDU see BNC750
BE see BNI500
BEACILLIN see BFC750
BEAMETTE see PMP500
BEAN SEED PROTECTANT see CBG000
BECAPTAN see AJT250
BECILAN see PPK500
BECOREL see PAN100
BEECHWOOD CRESOATE see BAT850
BEESIX see PPK250
BEESWAX see BAU000

BEESWAX, WHITE see BAU000
BEESWAX, YELLOW see BAU000
BEET-KLEEN see CBM000
BEET SUGAR see SNH000
BEFLAVINE see RIK000
BEFUNOLOL see HLK600
BEHA see AEO000
BEHP see DVL700
BEIVON see TES750
BEK see BJU000
BELAMINE BLACK GX see AQP000
BELAMINE BLUE 2B see CMO000
BELAMINE SKY BLUE A see CMO500
BELAMINE SKY BLUE FF see CMN750
BELCOMYCIN see PKD300
BELDAVRIN see HOT500
BELLASTHMAN see DMV600
BELL CML(E) see CAU500
BELL MINE see CAT225
BELL MINE PULVERIZED LIMESTONE see CAO000
BELMARK see FAR100
BELT see CDR750
BELUSTINE see CGV250
BEMEGRIDE see MKA250
BENA see BAU750
BENA see BBV500
BENACHLOR see BBV500
BENACTIZINE HYDROCHLORIDE see BCA000
BENACTYZIN (CZECH) see BCA000
BENACTYZINE CHLORIDE see BCA000
BENACTYZINE HYDROCHLORIDE see BCA000
BENADON see PPK500
BENADON see BBV500
BENADRIN see BBV500
BENADRYL see BAU750
BENADRYL see BBV500
BENADRYL HYDROCHLORIDE see BAU750
BENAKTIN see BCA000
BENALGIN see BBW500
BEN-ALLERGIN see BBV500
BENANSERIN HYDROCHLORIDE see BEM750
BENAPON see BBV500
BENASPIR see ADA725
BENAZYL see RDK000
BENCARBATE see DQM600
BENCIDAL BLACK E see AQP000
BENCIDAL BLUE 2B see CMO000
BENCIDAL BLUE 3B see CMO250
BENDEX see BLU000
BENDIGON see RDK000
BENDIOCARB see MHZ000
BENDIOCARB see DQM600
BENDIOXIDE see MJY500
BENDOPA see DNA200
BENDYLATE see BAU750
BENFOS see DGP900
BENGAL GELATIN see AEX250
BENGAL ISINGLASS see AEX250
BEN-HEX see BBQ500
BENICOT see NCR000
BENKFURAN see NGE000
BENLATE 50 see BAV575
BENOCTEN see BAU750
BENODAINE HYDROCHLORIDE see BCI500
BENODIN see BBV500
BENODINE see BBV500
BENOMYL see BAV575
BENOMYL 50W see BAV575

BENOVOCYLIN see EDP000
BENOXYL see BDS000
BEN-P see BFC750
BENSULFOID see SOD500
BENSYLYT see DDG800
BENSYLYTE see PDT250
BENT see MDQ250
BENTAZON see MJY500
BENTONE see KBB600
BENTONITE see BAV750
BENTONITE 2073 see BAV750
BENTONITE MAGMA see BAV750
BENTOX 10 see BBQ500
BEN-U-RON see HIM000
(5R,6R)-BENXYLPENICILLIN see BDY669
BENYLAN see BBV500
BENYLATE see BCM000
BENZAC see BDS000
1,2-BENZACENAPHTHENE see FDF000
BENZ(e)ACEPHENANTHRYLENE see BAW250
3,4-BENZ(e)ACEPHENANTHRYLENE see BAW250
BENZACILLIN see BFC750
BENZACONINE see PIC250
BENZADONE GREY M see CMU475
BENZAHEX see BBQ750
BENZAKNEW see BDS000
BENZALACETON (GERMAN) see SMS500
BENZALACETONE see SMS500
2-BENZALACETOPHENONE see CDH000
BENZAL ALCOHOL see BDX500
BENZAL-(BENZYL-CYANID) (GERMAN) see DVX600
BENZAL CHLORIDE see BAY300
BENZALDEHYDE see BAY500
BENZALDEHYDE CYANOHYDRIN see MAP250
BENZALDEHYDE, DIMETHYL ACETAL see DOG700
BENZALDEHYDE GLYCERYL ACETAL (FCC) see BBA000
BENZALDEHYDE GREEN see BAY750
BENZALDEHYDE, 2-METHOXY-(9CI) see AOT525
BENZALDEHYDKYANHYDRIN (CZECH) see MAP250
BENZAL GLYCERYL ACETAL see BBA000
BENZALIN see DLY000
BENZAMIDE, 4-AMINO-5-CHLORO-2-ETHOXY-N-((4-((4-FLUOROPHENYL)METHYL)-2-MORPHOLINYL)MET HYL)- see MRU253
BENZAMIL BLACK E see AQP000
BENZAMIL SUPRA BROWN BRLL see CMO750
BENZAMINE BLUE see CMO250
BENZANIL BLUE 2B see CMO000
BENZANIL SKY BLUE see CMO500
BENZANTHRACENE see BBC250
BENZ(a)ANTHRACENE see BBC250
1,2-BENZANTHRACENE see BBC250
1,2-BENZ(a)ANTHRACENE see BBC250
1,2:5,6-BENZANTHRACENE see DCT400
1,2-BENZANTHRAZEN (GERMAN) see BBC250
BENZANTHRENE see BBC250
1,2-BENZANTHRENE see BBC250
BENZANTINE see BBV500
BENZARONE see BBJ500
5H-BENZ(B)ARSINDOLE, 5-HYDROXY-, 5-OXIDE see ARA100
BENZATHINE BENZYLPENICILLIN see BFC750
BENZATHINE PENICILLIN see BFC750
BENZATHINE PENICILLIN G see BFC750
BENZATROPINE METHANESULFONATE see TNU000
BENZAZIDE see BDL750

BENZAZIMIDE see BDH000
BENZAZIMIDONE see BDH000
1-BENZAZINE see QMJ000
2-BENZAZINE see IRX000
1-BENZAZOLE see ICM000
BENZAZOLINE HYDROCHLORIDE see BBJ750
BENZBROMARON see DDP200
BENZBROMARONE see DDP200
BENZ-o-CHLOR see DER000
15,16-BENZDEHYDROCHOLANTHRENE see DCR400
BENZEDRINE see BBK000
(±)-BENZEDRINE see BBK000
dl-BENZEDRINE see BBK000
d-BENZEDRINE SULFATE see BBK500
BENZEEN (DUTCH) see BBL250
BENZEHIST see BAU750
BENZEN (POLISH) see BBL250
BENZENACETIC ACID see PDY850
BENZENAMINE see AOQ000
BENZENAMINE, N-BUTYL-(9CI) see BQH850
BENZENAMINE, 4-CHLORO-, HYDROCHLORIDE see CJR200
BENZENAMINE, 2-CHLORO-4-METHYL- see CLK210
BENZENAMINE, N,N-DIETHYL-(9CI) see DIS700
BENZENAMINE, ar-ETHOXY- see EEL100
BENZENAMINE, 4-ETHOXY-(9CI) see PDK790
BENZENAMINE, 2-ETHYL-(9CI) see EGK500
BENZENAMINE, 4-FLUORO-(9CI) see FFY000
BENZENAMINE HYDROCHLORIDE see BBL000
BENZENAMINE, N,N'-METHANETETRAYLBIS(2,6-BIS(1-METHYLETHYL)- see BJE550
BENZENAMINE, 2-METHOXY-(9CI) see AOV900
BENZENAMINE, 2-METHOXY-, HYDROCHLORIDE (9CI) see AOX250
BENZENAMINE, N-METHYL-(9CI) see MGN750
BENZENAMINE, 4,4'-METHYLENEBIS- see MJQ000
BENZENAMINE, 4,4'-METHYLENEBIS-, DIHYDROCHLORIDE see MJQ100
BENZENAMINE, 2-METHYL-5-NITRO- see NMP500
BENZENAMINE, N-(2-METHYL-2-NITROPROPYL)-p-NITROSO-(9CI) see NHK800
BENZENAMINE, N-METHYL-N,2,4,6-TETRANITRO-(9CI) see TEG250
BENZENAMINE, N,N,-DIMETHYL-(9CI) see DQF800
BENZENAMINE, 4-NITRO-(9CI) see NEO500
BENZENAMINE, 4-(9H-PYRIDO(3,4-B)INDOL-9-YL)- see ALX120
BENZENAMINE, N,2,4,6-TETRANITRO- see TDY075
BENZENAMINIUM, 3-(((DIMETHYLAMINO)CARBONYL)OXY)-N,N,N-TRIMETHYL-, BROMIDE (9CI) see POD000
BENZENE see BBL250
BENZENEACETALDEHYDE see BBL500
BENZENEACETIC ACID see PDY850
BENZENEACETIC ACID, BUTYL ESTER (9CI) see BQJ350
BENZENEACETIC ACID, 3,7-DIMETHYL-2,6-OCTADIENYL ESTER, (E)- see GDM400
BENZENEACETIC ACID, 3,7-DIMETHYL-6-OCTENYL ESTER (9CI) see CMU050
BENZENEACETIC ACID, ETHYL ESTER (9CI) see EOH000
BENZENEACETIC ACID, 3-HEXENYL ESTER, (Z)- see HFE625
BENZENEACETIC ACID, α-(METHOXYIMINO)-2-((2-METHYLPHENOXY)METHYL)-, METHYL ESTER, (α-E)- see MLI900

BENZENEACETIC ACID, METHYL ESTER see MHA500
BENZENEACETIC ACID, 2-PHENYLETHYL ESTER see PDI000
BENZENEACETIC ACID, 2-PROPENYL ESTER see PMS500
BENZENEACETONITRILE see PEA750
BENZENE, (ACETOXYMERCURI)- see ABU500
BENZENE, (ACETOXYMERCURIO)- see ABU500
BENZENEARSONIC ACID see BBL750
BENZENEARSONIC ACID, 4-AMINO-2-HYDROXY- see HJE400
BENZENEARSONIC ACID, 3,4-DIFLUORO- see DKG100
BENZENEARSONIC ACID, 4-(p-DIMETHYLAMINOPHENYLAZO)-, HYDROCHLORIDE see DPO275
BENZENEARSONIC ACID, p-FLUORO- see FGA100
BENZENE AZIMIDE see BDH250
4-BENZENEAZOANILINE see PEI000
BENZENEAZOBENZENE see ASL250
BENZENEAZOBENZENEAZO-β-NAPHTHOL see OHI200
BENZENEAZODIMETHYLANILINE see DOT300
BENZENE-1-AZO-2-NAPHTHOL see PEJ500
BENZENEAZO-β-NAPHTHOL see PEJ500
1-BENZENEAZO-2-NAPHTHYLAMINE see FAG130
1-BENZENE-AZO-β-NAPHTHYLAMINE see FAG130
p-BENZENEAZOPHENOL see HJF000
BENZENE, 1,3-BIS(ISOCYANATOMETHYL)-(9CI) see XIJ000
BENZENE, 1-tert-BUTYL-3,5-DIMETHYL-2,4,6-TRINITRO- see TML750
BENZENECARBALDEHYDE see BAY500
BENZENECARBINOL see BDX500
BENZENECARBONAL see BAY500
BENZENECARBONYL CHLORIDE see BDM500
BENZENECARBOXYLIC ACID see BCL750
BENZENE CHLORIDE see CEJ125
BENZENE, 4-CHLORO-1-(4-CHLOROPHENOXY)-2-NITRO- see CJD600
BENZENE, 1-CHLORO-4-METHYL- see TGY075
BENZENE, CHLOROMETHYL-(9CI) see CLK130
BENZENE, 1-CHLORO-4-(TRICHLOROMETHYL)-(9CI) see TIR900
m-BENZENEDIAMINE see PEY000
o-BENZENEDIAMINE see PEY250
p-BENZENEDIAMINE see PEY500
1,2-BENZENEDIAMINE see PEY250
1,3-BENZENEDIAMINE see PEY000
1,4-BENZENEDIAMINE see PEY500
m-BENZENEDIAMINE DIHYDROCHLORIDE see PEY750
1,3-BENZENEDIAMINE HYDROCHLORIDE see PEY750
1,3-BENZENEDIAMINE, 4-METHOXY, SULFATE (1:1) (9CI) see DBO400
BENZENEDIAMINE, ar-METHYL- see TGL500
BENZENE, 1,4-DIAZIDO- see DCL125
BENZENEDIAZONIUM, 4-(HYDROXYMETHYL)-, TETRAFLUOROBORATE(1-) see HLX925
BENZENE, DIBROMO- see DDJ900
1,3-BENZENEDICARBONITRILE see PHX550
1,4-BENZENEDICARBONYL CHLORIDE see TAV250
1,4-BENZENEDICARBONYL DICHLORIDE see TAV250

1,4-BENZENEDICARBOXAMIDE, N,N'-DIMETHYL-
N,N'-DINITROSO-(9CI) see DRO400
o-BENZENEDICARBOXYLIC ACID see PHW250
p-BENZENEDICARBOXYLIC ACID see TAN750
BENZENE-1,2-DICARBOXYLIC ACID see PHW250
1,2-BENZENEDICARBOXYLIC ACID see PHW250
1,4-BENZENEDICARBOXYLIC ACID see TAN750
1,2-BENZENEDICARBOXYLIC ACID ANHYDRIDE
see PHW750
1,2-BENZENEDICARBOXYLIC ACID BI(2-
METHOXYETHYL)ESTER (9CI) see DOF400
1,2-BENZENEDICARBOXYLIC ACID, BUTYL
PHENYLMETHYL ESTER see BEC500
o-BENZENEDICARBOXYLIC ACID, DIBUTYL ESTER
see DEH200
BENZENE-o-DICARBOXYLIC ACID DI-n-BUTYL
ESTER see DEH200
1,2-BENZENEDICARBOXYLIC ACID, DIETHYL
ESTER see DJX000
1,2-BENZENEDICARBOXYLIC ACID, DIHEPTYL
ESTER (9CI) see HBP400
1,2-BENZENEDICARBOXYLIC ACID DIHEXYL
ESTER see DKP600
1,2-BENZENEDICARBOXYLIC ACID, DIISOOCTYL
ESTER see ILR100
1,2-BENZENEDICARBOXYLIC ACID DIMETHYL
ESTER see DTR200
1,4-BENZENE DICARBOXYLIC ACID DIMETHYL
ESTER (9CI) see DUE000
o-BENZENEDICARBOXYLIC ACID DIOCTYL ESTER
see DVL600
1,2-BENZENEDICARBOXYLIC ACID DIOCTYL
ESTER see DVL600
1,2-BENZENEDICARBOXYLIC ACID, DIPENTYL
ESTER see AON300
1,2-BENZENEDICARBOXYLIC ACID, DIPROPYL
ESTER see DWV500
BENZENE, 1,2-DICHLORO- see DEP600
BENZENE, o-DIETHOXY- see CCP900
BENZENE, 1,2-DIETHOXY-(9CI) see CCP900
BENZENE, p-DIHYDROXY- see HIH000
BENZENE-1,3-DIISOCYANATE see BBP000
BENZENE-1,3-DIISOCYANATE see BBP000
BENZENE, 1,3-DIISOCYANATO- see BBP000
BENZENE-, 1,3-DIISOCYANATOMETHYL- see
TGM740
p-BENZENEDINITRILE see BBP250
m-BENZENEDIOL see REA000
o-BENZENEDIOL see CCP850
p-BENZENEDIOL see HIH000
1,2-BENZENEDIOL see CCP850
1,3-BENZENEDIOL see REA000
1,4-BENZENEDIOL see HIH000
1,3-BENZENEDIOL, DIACETATE see REA100
BENZENE-1,3-DIOL, 2,4-DINITROSO- see DVF300
1,3-BENZENEDIOL, MONOACETATE see RDZ900
1,4-BENZENEDIOL, 2,3,5-TRIMETHYL- (9CI) see
POG300
1,3-BENZENEDIOL, 2,4,6-TRINITRO-, BARIUM SALT,
HYDRATE (2:1:1) see BAO900
BENZENE, DIVINYL- see DXQ740
BENZENE, 1,1'-(1,2-ETHENEDIYL)BIS-(9CI) see
SLR000
BENZENE, ETHENYL-, HOMOPOLYMER (9CI) see
SMQ500
BENZENEFORMIC ACID see BCL750
BENZENE HEXACHLORIDE see BBP750

α-BENZENEHEXACHLORIDE see BBQ000
β-BENZENEHEXACHLORIDE see BBR000
Δ-BENZENEHEXACHLORIDE see BFW500
γ-BENZENE HEXACHLORIDE see BBQ500
BENZENEHEXACHLORIDE (mixed isomers) see
BBQ750
trans-α-BENZENEHEXACHLORIDE see BBR000
BENZENE HEXACHLORIDE-α-isomer see BBQ000
BENZENE HEXACHLORIDE-γ-isomer see BBQ500
BENZENE, 1-ISOCYANATO-2-METHYL- see IKG725
BENZENE, 1-(1-ISOCYANATO-1-METHYLETHYL)-3-
(1-METHYLETHENYL)- see IKG800
BENZENE, 1-ISOCYANO-4-NITRO- see IKH780
BENZENE ISOPROPYL see COE750
BENZENE-1-ISOTHIOCYANATE see ISQ000
BENZENEMETHANOIC ACID see BCL750
BENZENEMETHANOL see BDX500
BENZENEMETHANOL, α-ETHYNYL-, CARBAMATE
(9CI) see PGE000
BENZENEMETHANOL, 2-HYDROXY- (9CI) see
HMK100
BENZENEMETHANOL, α-METHYL- see PDE000
BENZENEMETHANOL, α-(1-
(METHYLAMINO)ETHYL)-, HYDROCHLORIDE, (R-
(R*,S*))- see EAW500
BENZENE, METHOXY see AOX750
BENZENE, METHYL- see TGK750
BENZENE, (2-(3-METHYLBUTOXY)ETHYL)- see
IHV050
BENZENE, METHYLDINITRO- see DVG600
BENZENE, 2-METHYL-1,3,5-TRINITRO- see TMN490
BENZENENITRILE see BCQ250
BENZENE, 1,1'-OXYBIS-, HEXACHLORO derivatives
(9CI) see CDV175
BENZENEPROPANAL see HHP000
BENZENEPROPANENITRILE, β-OXO-α-PHENYL-
(9CI) see OOK200
3-BENZENEPROPANOL see HHP050
BENZENEPROPANOL CARBAMATE see PGA750
BENZENEPROPANOL, PROPANOATE (9CI) see
HHQ550
BENZENESULFOHYDRAZIDE see BBS300
BENZENESULFONAMIDE, 4-AMINO-N-
(DIAMINOMETHYLENE)- see AHO250
BENZENESULFONAMIDE, N-CHLORO-4-METHYL-,
SODIUM SALT (9CI) see CDP000
BENZENESULFONAMIDE, N-(3-CHLORO-2-OXO-1-
(PHENYLMETHYL)PROPYL)-4-METHYL-,(S)- see
THH450
BENZENESULFONAMIDE, p-HYDROXY-, O-ESTER
with O,O-DIMETHYL PHOSPHOROTHIOATE see
CQL250
BENZENE SULFONCHLORIDE see BBS750
BENZENESULFONIC ACID see BBS250
BENZENESULFONIC (ACID) CHLORIDE see BBS750
BENZENESULFONIC ACID, DODECYL- see LBU100
BENZENESULFONIC ACID, HYDRAZIDE see BBS300
BENZENESULFONIC ACID, OXYBIS-,
DIHYDRAZIDE (9CI) see OPE000
BENZENESULFONIC HYDRAZIDE see BBS300
BENZENESULFONOHYDRAZIDE see BBS300
BENZENESULFONYL CHLORIDE see BBS750
BENZENESULFONYL HYDRAZIDE see BBS300
BENZENESULFONYL HYDRAZINE see BBS300
BENZENE SULPHONOHYDRAZIDE see BBS300

BENZENE SULPHONYL CHLORIDE (DOT) see
BBS750
BENZENESULPHONYL FLUORIDE see BBT250
BENZENETETRAHYDRIDE see CPC579
BENZENETHIOL, 2,4-DINITRO- see DUR425
1,2,4-BENZENETRICARBOXYLIC ACID ANHYDRIDE
see TKV000
1,2,4-BENZENETRICARBOXYLIC ACID, CYCLIC 1,2-
ANHYDRIDE see TKV000
1,2,4-BENZENETRICARBOXYLIC ANHYDRIDE see
TKV000
BENZENE, 1,3,5-TRICHLORO- see TIK300
BENZENE, 1,3,5-TRIMETHYL- see TLM050
BENZENE-s-TRIOL see PGR000
1,2,3-BENZENETRIOL see PPQ500
1,2,4-BENZENETRIOL see BBU250
BENZENE-1,3,5-TRIOL see PGR000
1,3,5-BENZENETRIOL see PGR000
BENZENOL see PDN750
BENZENOSULFOCHLOREK (POLISH) see BBS750
BENZENOSULPHOCHLORIDE see BBS750
BENZENYL CHLORIDE see BFL250
BENZENYL FLUORIDE see BDH500
BENZENYL TRICHLORIDE see BFL250
BENZETHACIL see BFC750
BENZETHONIUM CHLORIDE see BEN000
BENZETONIUM CHLORIDE see BEN000
BENZEX see BBQ750
2,3-BENZFLUORANTHENE see BAW250
3,4-BENZFLUORANTHENE see BAW250
10,11-BENZFLUORANTHENE see BCJ250
BENZ(j)FLUOROANTHRENE see BCJ250
BENZHORMOVARINE see EDP000
BENZHYDRAMINE see BBV500
BENZHYDRAMINE HYDROCHLORIDE see BAU750
BENZHYDRAMINUM see BBV500
BENZHYDRAZIDE see BBV250
BENZHYDRIL see BBV500
BENZHYDRYL see BBV500
o-BENZHYDRYLDIMETHYLAMINOETHANOL see
BBV500
o-BENZHYDRYLDIMETHYLAMINOETHANOL-8-
CHLOROTHEOPHYLLINATE see DYE600
2-(BENZHYDRYLOXY)-N,N-
DIMETHYLETHYLAMINE see BBV500
2-(BENZHYDRYLOXY)-N,N-
DIMETHYLETHYLAMINE with 8-
CHLOROTHEOPHYLLINE see DYE600
2-(BENZHYDRYLOXY)-N,N-
DIMETHYLETHYLAMINEHYDROCHLORIDE see
BAU750
BENZIDAMINE HYDROCHLORIDE see BBW500
BENZIDIN (CZECH) see BBX000
BENZIDINA (ITALIAN) see BBX000
BENZIDINE see BBX000
3,3'-BENZIDINEDICARBOXYLIC ACID see BFX250
BENZIDINE HYDROCHLORIDE see BBX750
BENZIDINE SULFATE see BBY000
BENZIDINE SULPHATE and HYDRAZINE-BENZENE
see BBY300
BENZILAN see DER000
BENZILATE DU DIETHYLAMINO-ETHANOL
CHLORHYDRATE (FRENCH) see BCA000
BENZILE (CLORURO di) (ITALIAN) see BEE375
BENZILIC ACID-β-DIETHYLAMINOETHYL ESTER
HYDROCHLORIDE see BCA000
BENZIL, MONOOXIME see BCA300

α-BENZIL MONOOXIME see BCA300
BENZIL, MONOXIME see BCA300
α-BENZIL MONOXIME see BCA300
BENZIL, β-MONOXIME see BCA300
BENZIL, OXIME see BCA300
BENZIMIDAZOLE see BCB750
o-BENZIMIDAZOLE see BCB750
1H-BENZIMIDAZOLE (9CI) see BCB750
BENZIMIDAZOLE, 2-AMINO-5-BENZOYL- see
AIH000
2-BENZIMIDAZOLECARBAMIC ACID, 5-BENZOYL-,
METHYL ESTER see MHL000
BENZIMIDAZOLE-2-CARBAMIC ACID, METHYL
ESTER see MHC750
1H-BENZIMIDAZOLE, 2-(2-CHLOROPHENYL)- see
CJR550
2-BENZIMIDAZOLETHIOL see BCC500
BENZIMIDAZOLE, 2-TRIFLUOROMETHYL- see
TKB800
1H-BENZIMIDAZOLE, 2-(TRIFLUOROMETHYL)- see
TKB800
N-2-(BENZIMIDAZOLYL) CARBAMATE see MHC750
1H-BENZIMIDAZOL-2-YLCARBAMIC ACID METHYL
ESTER see MHC750
4-(2-BENZIMIDAZOLYL)THIAZOLE see TEX000
BENZIMINAZOLE see BCB750
BENZIN B70 see NAH600
BENZINDAMINE HYDROCHLORIDE see BBW500
1H-BENZ(6,7)INDAZOLO(2,3,4-
fgh)NAPHTH(2",3":6',7')INDOLO(3',2':5,6)ANTHR
A(2,1,9-mna) ACRIDINE-5,8,13,25-TETRAONE see
CMU475
1-BENZINE see QMJ000
BENZINE (LIGHT PETROLEUM DISTILLATE) see
PCT250
BENZINE (OBS.) see BBL250
BENZIN (OBS.) see BBL250
BENZINOFORM see CBY000
BENZINOL see TIO750
BENZIODARON see EID200
BENZIODARONE see EID200
3-BENZISOTHIAZOLINONE-1,1-DIOXIDE see
BCE500
1,2-BENZISOTHIAZOL-3(2H)-ONE-1,1-DIOXIDE see
BCE500
1,2-BENZISOTHIAZOL-3(2H)-ONE-1,1-DIOXIDE,
CALCIUM SALT see CAM750
BENZISOTRIAZOLE see BDH250
BENZOANTHRACENE see BBC250
BENZO(a)ANTHRACENE see BBC250
1,2-BENZOANTHRACENE see BBC250
BENZOATE see BCL750
BENZOATE d'OESTRADIOL (FRENCH) see EDP000
BENZOATE d'OESTRONE (FRENCH) see EDV500
BENZOATE of SODA see SFB000
BENZOATE SODIUM see SFB000
1-BENZOAZO-2-NAPHTHOL see PEJ500
BENZO BLUE see CMO250
BENZO BLUE GS see CMO000
BENZOCAINE see EFX000
BENZO-CHINON (GERMAN) see QQS200
BENZO(d,e,f)CHRYSENE see BCS750
BENZO CONGO RED see SGQ500
BENZOCTAMINE HYDROCHLORIDE see BCH750
BENZO DEEP BLACK E see AQP000
BENZODIAPIN see MDQ250
1,3-BENZODIAZOLE see BCB750

1,2-BENZODIHYDROPYRONE (FCC) see HHR500
BENZODIOXANE HYDROCHLORIDE see BCI500
1-(1,4-BENZODIOXAN-2-
YLMETHYL)PIPERIDINEHYDROCHLORIDE see
BCI500
3,4-BENZODIOXOLE-5-CARBOXALDEHYDE see
PIW250
1,3-BENZODIOXOLE-5-(2-PROPEN-1-OL) see BCJ000
1,3-BENZODIOXOL-4-OL, 2,2-DIMETHYL-,
ACETYLMETHYLCARBAMATE see BCJ005
1-(5-(1,3-BENZODIOXOL-5-YL)-1-OXO-2,4-
PENTADIENYL)PIPERIDINE (E,E)- (9CI) see PIV600
1,3-BENZODIOXOL-5-YL-OXO-2,4-PENTADIENYL-
PIPERINE see PIV600
BENZODOL see BCJ150
BENZOEPIN see EAQ750
BENZOESAEURE (GERMAN) see BCL750
BENZOESAEURE (NA-SALZ) (GERMAN) see SFB000
BENZOESTROFOL see EDP000
BENZO(1)FLUORANTHENE see BCJ250
BENZO(b)FLUORANTHENE see BAW250
BENZO(e)FLUORANTHENE see BAW250
BENZO(j)FLUORANTHENE see BCJ250
BENZO(k)FLUORANTHENE see BCJ280
2,3-BENZOFLUORANTHENE see BAW250
3,4-BENZOFLUORANTHENE see BAW250
7,8-BENZOFLUORANTHENE see BCJ250
8,9-BENZOFLUORANTHENE see BCJ280
11,12-BENZOFLUORANTHENE see BCJ280
11,12-BENZO(k)FLUORANTHENE see BCJ280
2,3-BENZOFLUORANTHRENE see BAW250
BENZO(jk)FLUORENE see FDF000
BENZOFOLINE see EDP000
BENZOFORM BLACK BCN-CF see AQP000
BENZOFURAN see BCK250
BENZO(b)FURAN see BCK250
2,3-BENZOFURAN see BCK250
BENZOFURAN, 3-(p-(2-
(DIETHYLAMINO)ETHOXY)BENZOYL)-2-ETHYL-
see EDV700
BENZOFURAN, 3-(3,5-DIIODO-4-
HYDROXYBENZOYL)-2-ETHYL- see EID200
BENZOFURAN, (2-ETHYL-3-(4'-
HYDROXYBENZOYL)) see BBJ500
BENZOFUR D see PEY500
BENZOFURFURAN see BCK250
BENZOFUR GG see ALT000
BENZOFUR MT see TGL750
6H-BENZOFURO(3,2-c)(1)BENZOPYRAN-6-ONE, 3,9-
DIHDYROXY- see COF350
BENZOFUROLINE see BEP500
BENZOFUR P see ALT250
BENZO-GYNOESTRYL see EDP000
BENZO(a)HEPTALEN-9(5H)-ONE, 6,7-DIHYDRO-
1,2,3,10-TETRAMETHOXY-7-(METHYLAMINO)-, (S)-
see MIW500
BENZOHYDRAZIDE see BBV250
BENZOHYDRAZINE see BBV250
BENZOHYDROQUINONE see HIH000
2-(BENZOHYDRYLOXY)-N,N-
DIMETHYLETHYLAMINE see BBV500
BENZOIC ACID see BCL750
BENZOIC ACID (DOT) see BCL750
BENZOIC ACID, 4-AMINO- see AIH600
BENZOIC ACID, 2-AMINO-, 2-PHENYLETHYL
ESTER see APJ500
BENZOIC ACID, 4-ARSENOSO- see CCI550
BENZOIC ACID AZIDE see BDL750

BENZOIC ACID, 4-BENZOYL- see BDL860
BENZOIC ACID, BENZYL ESTER see BCM000
BENZOIC ACID-n-BUTYL ESTER see BQK250
BENZOIC ACID, CHLORIDE see BDM500
BENZOIC ACID, 2,4-DIHYDROXY- (9CI) see HOE600
BENZOIC ACID,2-((3-(4-(1,1-
DIMETHYLETHYL)PHENYL)-2-
METHYLPROPYLIDENE)AMINO)-, METHYLESTER
see LFT100
BENZOIC ACID, 2,2'-DISELENOBIS- see SBU150
BENZOIC ACID ESTRADIOL see EDP000
BENZOIC ACID, 3-((3-(ETHOXYMETHYL)-5-
FLUORO-3,6-DIHYDRO-2,6-DIOXO-1(2H)-
PYRIMIDINYL)CARB ONYL)-, see BDN600
BENZOIC ACID, 3-FORMYL-2,4-DIHYDROXY-6-
METHYL-, 3-HYDROXY-4-(METHOXYCARBONYL)-
2,5-DIMETHYLPHENYL ESTER see ARQ600
BENZOIC ACID, 2-HYDROXY-, 4-(1,1-
DIMETHYLETHYL)PHENYL ESTER see BSH100
BENZOIC ACID, 2-HYDROXY-3,5-DINITRO-(9CI) see
HKE600
BENZOIC ACID, 2-HYDROXY-, 3-METHYL-2-
BUTENYL ESTER see PMB600
BENZOIC ACID, 2-HYDROXY-, 4-METHYLPHENYL
ESTER (9CI) see THD850
BENZOIC ACID, 3-METHYL-2-BUTENYL ESTER see
MHU150
BENZOIC ACID, 1-(3-METHYL)BUTYL ESTER see
IHP100
BENZOIC ACID, 4-METHYLPHENYL ESTER see
TGX100
BENZOIC ACID, o-(METHYLTELLURO)-, SODIUM
SALT see MPN275
BENZOIC ACID NITRILE see BCQ250
BENZOIC ACID, PEROXIDE see BDS000
BENZOIC ACID, o-(PHENYLHYDROXYARSINO)- see
PFI600
BENZOIC ACID, PHENYLMETHYL ESTER see
BCM000
BENZOIC ACID, SODIUM SALT see SFB000
BENZOIC ACID, p-TOLYL ESTER see TGX100
BENZOIC ALDEHYDE see BAY500
BENZOIC ETHER see EGR000
BENZOIC HYDRAZIDE see BBV250
o-BENZOIC SULPHIMIDE see BCE500
BENZOIC TRICHLORIDE see BFL250
BENZOIMIDAZOLE see BCB750
BENZOIN see BCP250
BENZOKETCTRIAZINE see BDH000
BENZOL (DOT) see BBL250
BENZOLE see BBL250
BENZO LEATHER BLACK E see AQP000
BENZOLENE see BBL250
BENZOLINE see PCT250
BENZOLO (ITALIAN) see BBL250
5H-BENZO(d)NAPHTH(2,1-B)AZEPIN-12-OL, 11-
CHLORO-6,6A,7,8,9,13b-HEXAHYDRO-7-METHYL-,
HYDROCHLORIDE, (6as,13br)- see BCP690
BENZONE see BRF500
BENZONITRILE see BCQ250
BENZONITRILE (DOT) see BCQ250
BENZONITRILE, p-ISOPROPYL- see IOD050
BENZONITRILE, 4-(1-METHYLETHYL)- see IOD050
BENZOPENICILLIN see BDY669
BENZO(rst)PENTAPHENE see BCQ500
BENZOPEROXIDE see BDS000
BENZO(a)PHENANTHRENE see BBC250
BENZO(a)PHENANTHRENE see CML810

BENZO(b)PHENANTHRENE see BBC250
1,2-BENZOPHENANTHRENE see CML810
2,3-BENZOPHENANTHRENE see BBC250
BENZO(def)PHENANTHRENE see PON250
BENZOPHENONE see BCS250
BENZOPHENONE-3 see MES000
p-BENZOPHENONE, METHYL- see MHF750
BENZOPHOSPHATE see BDJ250
3,4-BENZOPIRENE (ITALIAN) see BCS750
2H-1-BENZOPYRAN, 6,7-DIMETHOXY-2,2-
DIMETHYL- see AEX850
2H-1-BENZOPYRAN-2-ONE see CNV000
BENZO(a)PYRENE see BCS750
3,4-BENZOPYRENE see BCS750
6,7-BENZOPYRENE see BCS750
BENZO(a)PYRENE-6-METHANOL see BCV250
BENZO(b)PYRIDINE see QMJ000
BENZO(c)PYRIDINE see IRX000
1,2-BENZOPYRONE see CNV000
BENZOPYRROLE see ICM000
2,3-BENZOPYRROLE see ICM000
1,4-BENZOQUINE see QQS200
BENZOQUINOL see HIH000
BENZO(b)QUINOLINE see ADJ500
2,3-BENZOQUINOLINE see ADJ500
o-BENZOQUINONE see BDC250
p-BENZOQUINONE see QQS200
1,2-BENZOQUINONE see BDC250
1,4-BENZOQUINONE see QQS200
BENZOQUINONE (DOT) see QQS200
BENZOQUINONE (DOT) see BDC250
1,4-BENZOQUINONE DIOXINE see DVR200
p-BENZOQUINONE, 2-PHENYL- see PEL750
2,1,3-BENZOSELENADIAZOLE, 5,6-DIMETHYL- see
DQO650
2,1,3-BENZOSELENADIAZOLE, 5-METHYL- see
MHI300
BENZOSELENAZOLIUM, 3-ETHYL-2-(3-(3-ETHYL-2-
BENZOSELENAZOLINYLIDENE)-2-
METHYLPROPENYL)-, IODIDE see DJQ300
BENZO SKY BLUE A-CF see CMO500
BENZO SKY BLUE S see CMO500
o-BENZOSULFIMIDE see BCE500
BENZOSULPHIMIDE see BCE500
BENZO-2-SULPHIMIDE see BCE500
BENZOTHIAZOLE DISULFIDE see BDE750
2-BENZOTHIAZOLETHIOL see BDF000
2-BENZOTHIAZOLETHIOL, ZINC SALT (2:1) see
BHA750
BENZOTHIAZOLE-2-THIONE see BDF000
2(3H)-BENZOTHIAZOLETHIONE see BDF000
BENZOTHIAZOLYL DISULFIDE see BDE750
2-BENZOTHIAZOLYL DISULFIDE see BDE750
2-BENZOTHIAZOLYL MERCAPTAN see BDF000
2-BENZOTHIAZOLYL-N-MORPHOLINOSULFIDE see
BDG000
2-BENZOTHIAZOLYLSULFENYL MORPHOLINE see
BDG000
4-(2-BENZOTHIAZOLYLTHIO)MORPHOLINE see
BDG000
BENZOTHIAZYL-2-CYCLOHEXYLSULFENAMIDE
see CPI250
BENZOTRIAZINEDITHIOPHOSPHORIC ACID
DIMETHOXY ESTER see ASH500
BENZOTRIAZINE derivative of an ETHYL
DITHIOPHOSPHATE see EKN000
BENZOTRIAZINE derivative of a METHYL
DITHIOPHOSPHATE see ASH500

1,2,3-BENZOTRIAZIN-4(1H)-ONE see BDH000
3H-1,2,3-BENZOTRIAZIN-4-ONE see BDH000
1H-BENZOTRIAZOLE see BDH250
1,2,3-BENZOTRIAZOLE see BDH250
1H-BENZOTRIAZOLE, 6-NITRO- see NFJ000
BENZOTRICHLORIDE (DOT, MAK) see BFL250
BENZOTRIFLUORIDE see BDH500
2,5,8-BENZOTRIOXACYCLOUNDECIN-1,9-DIONE,
3,4,6,7-TETRAHYDRO-(9CI) see DJD700
BENZOTROPINE MESYLATE see TNU000
BENZOTROPINE METHANESULFONATE see
TNU000
BENZOXALE see TMP750
2H-3,1-BENZOXAZINE-2,4(1H)-DIONE see IHN200
2H-1,3-BENZOXAZINE-2,4(3H)-DIONE, 6-CHLORO-2-
THIO- see CLH800
2H-1,3-BENZOXAZINE-2,4(3H)-DIONE, 6,8-
DIBROMO-2-THIO- see DDM820
2H-1,3-BENZOXAZINE-2,4(3H)-DIONE, 6,8-
DICHLORO-2-THIO- see DFC300
2H-1,3-BENZOXAZINE-2,4(3H)-DIONE, 6-METHYL-2-
THIO- see MPS600
2H-1,3-BENZOXAZINE-2,4(3H)-DIONE, 8-NITRO-2-
THIO- see NFW210
2H-1,3-BENZOXAZINE-2,4(3H)-DIONE, 2-THIO- see
TFC570
S-((3-BENZOXAZOLINYL-6-CHLORO-2-
OXO)METHYL) O,O-
DIETHYLPHOSPHORODITHIOATE see BDJ250
BENZ(h)OXIRENO(5,6)BENZ(1,2-A)ACRIDINE-2,3-
DIOL, 1A,2,3,13C-TETRAHYDRO-, (1A-α,2-β,3-α,13C-
α)-(+−)- see DCS821
BENZ(c)OXIRENO(5,6)BENZ(1,2-H)ACRIDINE-2,3-
DIOL, 1A,2,3,13C-TETRAHYDRO-, (1AS-(1A-α,2-β,3-
α,13C-α))- see DMS410
3-BENZOXY-1-(2-METHYLPIPERIDINO)PROPANE
see PIV750
BENZOYL see BDS000
2-BENZOYLACETOPHENONE see PFU300
BENZOYLACONINE see PIC250
BENZOYL ALCOHOL see BDX500
BENZOYL AZIDE see BDL750
BENZOYLBENZENE see BCS250
N-2 (5-BENZOYL-BENZIMIDAZOLE) CARBAMATE
de METHYLE see MHL000
5-BENZOYL-2-BENZIMIDAZOLECARBAMIC ACID
METHYL ESTER see MHL000
N-(BENZOYL-5, BENZIMIDAZOLYL)-2, CARBAMATE
de METHYLE see MHL000
(5-BENZOYL-1H-BENZIMIDAZOL-2-YL)-CARBAMIC
ACID METHYL ESTER see MHL000
p-BENZOYLBENZOIC ACID see BDL860
α-BENZOYLBENZYL CYANIDE see OOK200
BENZOYL CHLORIDE see BDM500
BENZOYL CHLORIDE (DOT) see BDM500
BENZOYL CHLORIDE, METHOXY-(9CI) see AOY250
BENZOYL CYANIDE-o-
(DIETHOXYPHOSPHINOTHIOYL)OXIME see BAT750
3-BENZOYLFURAN see FQO050
3-BENZOYLHYDRATROPIC ACID see BDU500
m-BENZOYLHYDRATROPIC ACID see BDU500
BENZOYL HYDRAZIDE see BBV250
BENZOYL METHIDE see ABH000
BENZOYLMETHYLECGONINE see CNE750
BENZOYL-γ-(2-METHYLPIPERIDINO)PROPANOL see
PIV750

BENZYLIDENEACETALDEHYDE see CMP969
BENZYLIDENE ACETONE see SMS500
2-BENZYLIDENEACETOPHENONE see CDH000
BENZYLIDENE CHLORIDE see BAY300
BENZYLIDENE CHLORIDE (DOT) see BAY300
BENZYLIDENE GLYCEROL see BBA000
BENZYLIDENEPHENYLACETONITRILE see DVX600
BENZYLIDYNE CHLORIDE see BFL250
BENZYLIDYNE FLUORIDE see BDH500
BENZYLIMIDAZOLINE HYDROCHLORIDE see
BBJ750
2-BENZYL-2-IMIDAZOLINE
MONOHYDROCHLORIDE see BBJ750
1-BENZYL-2-INDOLYL HYDROXYMETHYL
KETONE see BES300
BENZYL ISOBUTYRATE (FCC) see IJV000
BENZYL-ISOTHIOCYANATE see BEU250
BENZYLISOTHIOUREA HYDROCHLORIDE see
BEU500
BENZYLISOTHIOURONIUM CHLORIDE see BEU500
2-BENZYLISOTHIOURONIUM CHLORIDE see
BEU500
BENZYL ISOVALERATE (FCC) see ISW000
BENZYLKYANID see PEA750
BENZYL MERCAPTAN see TGO750
BENZYL METHANOATE see BEP250
1-BENZYL-2-METHYL-3-(2-AMINOETHYL)-5-
METHOXYINDOLE HYDROCHLORIDE see BEM750
BENZYL-3-METHYLBUTANOATE see ISW000
BENZYL-3-METHYL BUTYRATE see ISW000
BENZYLMETHYLCARBINYL ACETATE see ABU800
1-BENZYL-2-METHYLHYDRAZINE see MHN750
1-BENZYL-2-METHYL-5-METHOXYTRYPTAMINE
HYDROCHLORIDE see BEM750
BENZYL-2-METHYL PROPIONATE see IJV000
BENZYL MUSTARD OIL see BEU250
BENZYL MUSTARD OIL see BFL000
BENZYL NITRILE see PEA750
BENZYL OXIDE (CZECH) see BEO250
7-(BENZYLOXY)-6-N-BUTYL-1,4-DIHYDRO-4-OXO-3-
QUINOLINECARBOXYLIC ACID METHYL ESTER see
NCN600
7-(BENZYLOXY)-6-N-BUTYL-4-HYDROXY-3-
QUINOLINECARBOXYLIC ACID METHYL see
NCN600
BENZYLOXYCARBONYL CHLORIDE see BEF500
BENZYLOXYCARBONYLGLYCINE see CBR125
N-BENZYLOXYCARBONYLGLYCINE see CBR125
5-BENZYLOXY-3-ISONIPECOTOYLINDOLE see
BFC200
BENZYLPENCILLINDIBENZYLETHYLENEDIAMIN
E SALT see BFC750
BENZYLPENICILLIN see BDY669
BENZYLPENICILLIN BENZATHINE see BFC750
BENZYLPENICILLIN G see BDY669
BENZYLPENICILLINIC ACID see BDY669
BENZYL PENICILLINIC ACID SODIUM SALT see
BFD250
BENZYLPENICILLIN SODIUM see BFD250
N-BENZYL-N-PHENOXYISOPROPYL-β-
CHLORETHYLAMINE HYDROCHLORIDE see
DDG800
BENZYL PHENYLACETATE see BFD400
BENZYL γ-PHENYLACRYLATE see BEG750
BENZYL PHENYLFORMATE see BCM000
BENZYL PHENYL KETONE see PEB000

BENZYLPHOSPHONIC ACID DIBUTYL ESTER see
BFD760
BENZYL PROPIONATE see BFD800
BENZYL-(α-PYRIDYL)-
DIMETHYLAETHYLENDIAMIN (GERMAN) see
TMP750
BENZYL RED BR see CMM330
BENZYL SALICYLATE see BFJ750
BENZYLSENFOEL (GERMAN) see BEU250
BENZYLSTEARYLDIMETHYLAMMONIUM
CHLORIDE see DTC600
BENZYLT see PDT250
BENZYL THIOCYANATE see BFL000
BENZYLTHIOL see TGO750
BENZYL THIOPSEUDOUREA HYDROCHLORIDE see
BEU500
2-BENZYL-2-THIO-PSEUDOUREA
HYDROCHLORIDE see BEU500
BENZYLTHIURONIUM CHLORIDE see BEU500
S-BENZYLTHIURONIUM CHLORIDE see BEU500
BENZYL TRICHLORIDE see BFL250
BENZYL TRIMETHYL AMMONIUM IODIDE see
BFM750
BENZYL VIOLET see FAG120
BENZYL VIOLET 3B see FAG120
BENZYLYT see DDG800
3,4-BENZYPYRENE see BCS750
BENZYRIN see BBW500
BENZYTOL see CLW000
BEOSIT see EAQ750
BEPANTHEN see PAG200
BEPANTHENE see PAG200
BEPANTOL see PAG200
BERCEMA FERTAM 50 see FAS000
BERCEMA NMC50 see CBM750
BERELEX see GEM000
BERGAMIOL see LFY600
BERGAMOT OIL rectified see BFO000
BERGAMOTTE OEL (GERMAN) see BFO000
BERKENDYL see CDP000
BERKFLAM B 10 see PCC480
BERKFLAM B 10E see PAU500
BERKFURIN see NGE000
BERKMYCEN see HOH500
BERKOMINE see DLH630
BERMAT see CJJ250
BERNARENIN see VGP000
BERNICE see CNE750
BERNIES see CNE750
BERNOCAINE see AIT250
BERNSTEINSAEURE (GERMAN) see SMY000
BERNSTEINSAEURE-ANHYDRID (GERMAN) see
SNC000
BERNSTEINSAEURE-2,2-DIMETHYLHYDRAZID
(GERMAN) see DQD400
BEROL 478 see DJL000
BERONALD see CHJ750
BERTHOLITE see CDV750
BERTHOLLET SALT see PLA250
BERTRANDITE see BFO250
BERUBIGEN see VSZ000
BERYL see BFO500
BERYLLATE(2-), TETRAFLUORO-, DIAMMONIUM see
ANH300
BERYLLATE(2-), TETRAFLUORO-, DIAMMONIUM,
(T-4)- see ANH300
BERYLLIA see BFT250

BERYLLIUM see BFO750
BERYLLIUM-9 see BFO750
BERYLLIUM ACETATE see BFP000
BERYLLIUM ACETATE, BASIC see BFT500
BERYLLIUM ACETATE, NORMAL see BFP000
BERYLLIUM ALUMINOSILICATE see BFO500
BERYLLIUM ALUMINUM ALLOY see BFP250
BERYLLIUM ALUMINUM SILICATE see BFO500
BERYLLIUM CARBONATE see BFP500
BERYLLIUM CARBONATE (1:1) see BFP750
BERYLLIUM CARBONATE, BASIC see BFP500
BERYLLIUM CHLORIDE see BFQ000
BERYLLIUM COMPOUND with NIOBIUM (12:1) see BFQ750
BERYLLIUM COMPOUNDS see BFQ500
BERYLLIUM COMPOUNDS, n.o.s. (UN 1566) (DOT) see BFO750
BERYLLIUM COMPOUND with TITANIUM (12:1) see BFR000
BERYLLIUM COMPOUND with VANADIUM (12:1) see BFR250
BERYLLIUM-COPPER ALLOY see CNI600
BERYLLIUM-COPPER-COBALT ALLOY see CNK700
BERYLLIUM DICHLORIDE see BFQ000
BERYLLIUM DIFLUORIDE see BFR500
BERYLLIUM DIHYDROXIDE see BFS250
BERYLLIUM DINITRATE see BFT000
BERYLLIUM FLUORIDE see BFR500
BERYLLIUM HYDRATE see BFS250
BERYLLIUM HYDRIDE see BFR750
BERYLLIUM HYDROGEN PHOSPHATE (1:1) see BFS000
BERYLLIUM HYDROXIDE see BFS250
BERYLLIUM LACTATE see LAH000
BERYLLIUM MANGANESE ZINC SILICATE see BFS750
BERYLLIUM MONOXIDE see BFT250
BERYLLIUM NITRATE see BFT000
BERYLLIUM ORTHOSILICATE see SCN500
BERYLLIUM OXIDE see BFT250
BERYLLIUM OXIDE ACETATE see BFT500
BERYLLIUMOXIDE CARBONATE see BFP500
BERYLLIUM OXYACETATE see BFT500
BERYLLIUM OXYFLUORIDE see BFT750
BERYLLIUM PERCHLORATE see BFU000
BERYLLIUM PHOSPHATE see BFS000
BERYLLIUM SILICATE see SCN500
BERYLLIUM SILICATE HYDRATE see BFO250
BERYLLIUM SILICIC ACID see SCN500
BERYLLIUM SULFATE (1:1) see BFU250
BERYLLIUM SULFATE TETRAHYDRATE (1:1:4) see BFU500
BERYLLIUM SULPHATE TETRAHYDRATE see BFU500
BERYLLIUM TETRAHYDROBORATE see BFU750
BERYLLIUM TETRAHYDROBORATETRIMETHYLAMINE see BFV000
BERYLLIUM, powder (UN 1567) (DOT) see BFO750
BERYLLIUM ZINC SILICATE see BFV250
BERYL ORE see BFO500
BESANTIN see MHL000
BETABION see TES750
BETACIDE P see HNU500
BETADID see HJS850
BETAFEDRINA see BBK500
BETAFEDRINE see BBK500
BETALIN 12 CRYSTALLINE see VSZ000

BETALIN S see TES750
BETA-NAFTYLOAMINA (POLISH) see NBE500
d-BETAPHEDRINE see BBK500
BETAPRONE see PMT100
BETAPYRIMIDUM see DJS200
BETARUNDUM see SCQ000
BETARUNDUM ST-S see SCQ000
BETARUNDUM UF see SCQ000
BETARUNDUM ULTRAFINE see SCQ000
BETAXIN see TES750
BETAXINA see EID000
BETAZED see BRF500
BETEL NUT see BFW000
BETEL QUID EXTRACT see BFW125
BETEL TOBACCO EXTRACT see BFW135
BETHIAMIN see TES750
BETRAMIN see BBV500
BETULA OIL see MPI000
BEVATINE-12 see VSZ000
BEVIDOX see VSZ000
BEWON see TES750
BEXIDE see BJU000
BEXOL see BBQ500
BEXON see MMN250
BEXT see BJU000
BEXTRENE XL 750 see SMQ500
B(b)F see BAW250
B(j)F see BCJ250
BF 5930 see OJW000
BFP see BJE750
BFPO see BJE750
BFV see FMV000
BG 5930 see OJW000
BG 6080 see CBT500
BGE see BRK750
t-BGE see BRK800
BGE (OSHA) see BRK750
BHA (FCC) see BQI000
BHBN see HJQ350
BHC see BBQ500
BHC see BJZ000
α-BHC see BBQ000
β-BHC see BBR000
Δ-BHC see BFW500
γ-BHC see BBQ500
BHC (USDA) see BBP750
BH DALAPON see DGI400
B-HERBATOX see SFS000
BHIMSAIM CAMPHOR see BMD000
BH MCPA see CIR250
BH MECOPROP see CIR500
BHP see DNB200
BHT (food grade) see BFW750
BI-58 see DSP400
4',4'''-BIACETANILIDE see BFX000
BIACETYL see BOT500
BIALFLAVINA see DBX400
BIALLYL see HCR500
BIALMINAL see EOK000
BIALPIRINIA see ADA725
BIALZEPAM see DCK759
p,p-BIANILINE see BBX000
4,4'-BIANILINE see BBX000
N,N'-BIANILINE see HHG000
BIANISIDINE see TGJ750
5,5'-BIANTHRANILIC ACID see BFX250
BIBENZAL see SLR000

BIBENZENE see BGE000
BIBENZYL see BFX500
BIBENZYLIDENE see SLR000
BIBENZYLIDINE see SLR000
BIBESOL see DGP900
BIC see BRQ500
BIC see IAN000
BICAM ULV see DQM600
BICA-PENICILLIN see BFC750
BICARBONATE of SODA see SFC500
BICARBURET of HYDROGEN see BBL250
BICARBURETTED HYDROGEN see EIO000
BICHLORACETIC ACID see DEL000
BICHLORENDO see MQW500
BICHLORIDE of MERCURY see MCY475
BICHLORURE d'ETHYLENE (FRENCH) see EIY600
BICHLORURE de MERCURE (FRENCH) see MCY475
BICHLORURE de PROPYLENE (FRENCH) see PNJ400
BICHROMATE d'AMMONIUM (FRENCH) see ANB500
BICHROMATE OF POTASH see PKX250
BICHROMATE of SODA see SGI000
BICHROMATE de SODIUM (FRENCH) see SGI000
BICILLIN see BFC750
BICKIE-MOL see HIM000
BICOLASTIC A 75 see SMQ500
BICOLENE C see PJS750
BICOLENE H see SMQ500
BICOLENE P see PMP500
BICORTONE see PLZ000
BICYCLO(4.4.0)DECANE see DAE800
BICYCLO(2.2.1)HEPTADIENE see NNG000
BICYCLO(2.2.1)HEPTANE, 2-CHLORO-1,7,7-TRIMETHYL-, endo- see BMD300
BICYCLO(2.2.1)HEPTANE, 2,2-DIMETHYL-3-METHYLENE-(9CI) see CBA500
BICYCLO(2.2.1)HEPTAN-2-OL, 1,7,7-TRIMETHYL-, endo-(9CI) see BMD000
BICYCLO(2.2.1)HEPTAN-2-OL, 1,7,7-TRIMETHYL-, ACETATE, EXO- see IHX600
BICYCLO(2.2.1)HEPTENE-2-DICARBOXYLIC ACID, 2-ETHYLHEXYLIMIDE see OES000
BICYCLOPENTADIENE see DGW000
BIDIRL see DGQ875
BIDRIN see DGQ875
BIEBRICH SCARLET BPC see SBC500
BIEBRICH SCARLET RED see SBC500
BIEBRICH SCARLET R MEDICINAL see SBC500
BIETHYLENE see BOP500
1,1'-BI(ETHYLENE OXIDE) see BGA750
BIETHYLXANTHOGENTRISULFIDE see BJU000
BIFLUORIDEN (DUTCH) see FEZ000
BIFLUORURE de POTASSIUM (FRENCH) see PKU250
BIFORMYCHLORAZIN see TKL100
BIFORMYLCHLORAZIN see TKL100
BIFURON see NGG500
BIG DIPPER see DVX800
BIGUANIDINE, CHROMATE see BJW825
BIGUMAL see CKB250
BILARCIL see TIQ250
BILCOLIC see MKP500
BILEVON see HCL000
BILICANTE see MKP500
BILINE-8,12-DIPROPIONIC ACID, 1,10,19,22,23,24-HEXAHYDRO-2,7,13,17-TETRAMETHYL-1,19-DIOXO-3,18-DIVINYL- see HAO900
BILIRUBIN see HAO900
BILIRUBIN IX-α see HAO900

BILOBORN see MRH209
BILOBRAN see MRH209
BIMETHYL see EDZ000
2,3,1',8'-BINAPHTHYLENE see BCJ280
BINDON see ONY000
BINDON ATHYLATHER see BGC250
BINDON ETHYL ETHER see BGC250
BINITROBENZENE see DUQ200
BINOTAL see AIV500
BINOTAL SODIUM see SEQ000
BIO 5,462 see EAQ750
BIOACRIDIN see DBX400
BIOALLETHRIN see AFR250
BIOALTRINA see AFR250
BIOBAMAT see MQU750
BIOBAN-C see CAW400
BIO-BEADS S-S 2 see SMQ500
BIOCALC see CAT225
BIOCETIN see CDP250
BIOCIDE see ADR000
BIOCOLINA see CMF750
BIO-DES see DKA600
BIODOPA see DNA200
BIOEPIDERM see VSU100
BIOEPIDERM see BGD100
BIOFANAL see NOH500
BIOFUREA see NGE500
BIOGRISIN-FP see GKE000
BIOMET TBTO see BLL750
BIOMITSIN see CMA750
BIOMYCIN see CMA750
BIOMYDRIN see SPC500
BIONIC see NCQ900
BIO-PERGE see ISD066
BIOPHEDRIN see EAW000
BIOPHENICOL see CDP250
BIOPRASE see BAC000
BIOQUIN see QPA000
BIOQUIN see BLC250
BIOQUIN 1 see BLC250
BIORENINE see VGP000
BIOREX see BGD088
BIOSEDAN see NBU000
BIOSEPT see CCX000
BIOSERPINE see RDK000
BIOS II see VSU100
BIOS II see BGD100
BIO-SOFT D-40 see DXW200
BIO-SOFT S 100 see LBU100
BIOSOL VETERINARY see NCG000
BIOSTAT see HOH500
BIOSTAT PA see HOH500
BIOSTEROL see VSK600
SYMBIO see SNN300
BIO-TESTICULINA see TBG000
BIOTHION see TAL250
BIOTIN see VSU100
BIOTIN see BGD100
(+)-BIOTIN see VSU100
(+)-BIOTIN see BGD100
d-BIOTIN see VSU100
d-BIOTIN see BGD100
D-BIOTIN see BGD100
d-(+)-BIOTIN see VSU100
D-(+)-BIOTIN see BGD100
BIOXIRANE see BGA750
2,2'-BIOXIRANE see BGA750
(S-(R*,R*))-2,2'-BIOXIRANE see BOP750

(R*,S*)-2,2'-BIOXIRANE see DHB800
BIOXYDE d'AZOTE (FRENCH) see NEG100
BIOXYDE de PLOMB (FRENCH) see LCX000
BIPHENYL see BGE000
1,1'-BIPHENYL see BGE000
4-BIPHENYLACETAMIDE see PDY500
N-4-BIPHENYLACETAMIDE see PDY500
BIPHENYLAMINE see AJS100
4-BIPHENYLAMINE see AJS100
p-BIPHENYLAMINE see AJS100
(1,1'-BIPHENYL)-4-AMINE see AJS100
BIPHENYL, mixed with BIPHENYL OXIDE (3:7) see
PFA860
BIPHENYL, DECABROMO- see PCC480
1,1'-BIPHENYL, 2,2',3,3',4,4',5,5',6,6'-DECABROMO- see
PCC480
4,4'-BIPHENYLDIAMINE see BBX000
(1,1'-BIPHENYL)-4,4'-DIAMINE (9CI) see BBX000
(1,1'-BIPHENYL)-4,4'-DIAMINE, DIHYDROCHLORIDE
see BBX750
(1,1'-BIPHENYL)-4,4'-DIAMINE SULFATE (1:1) see
BBY000
BIPHENYL-DIPHENYL ETHER mixture see PFA860
N,N'-(1,1'-BIPHENYL)-4,4'-DIYLBIS-ACETAMIDE 4',4'''-
BIACETANILIDE see BFX000
2,2'-(BIPHENYL)-4,4'-DIYLBIS(2-HYDROXY-4,4-
DIMETHYLMORPHOLINIUM) see HAP100
2,2'-(1,1'-BIPHENYL)-4,4'-DIYLBIS(2-HYDROXY-4,4-
DIMETHYL)-MORPHOLINIUM DIBROMIDE see
HAQ000
2,2'-((1,1'-BIPHENYL)-4,4'-DIYLDI-2,1-
ETHENEDIYL)BIS-BENZENESULFONIC ACID
DISODIUM SALT see TGE150
2,2'-(4,4'-BIPHENYLENE)BIS(2-HYDROXY-4,4-
DIMETHYLMORPHOLINIUM) see HAP100
4,4'-BIPHENYLENEDIAMINE see BBX000
N-(p-BIPHENYLMETHYL)-ATROPINIUM BROMIDE
see PEM750
N,4-BIPHENYL-METHYL-dl-TROPEYL-α-
TROPINIUMBROMID (GERMAN) see PEM750
p-BIPHENYLMETHYL-(dl-TROPYL-α-
TROPINIUM)BROMIDE (GERMAN) see PEM750
1,1-BIPHENYL, NONABROMO- see NMV735
2-BIPHENYLOL see BGJ250
4-BIPHENYLOL see BGJ500
o-BIPHENYLOL see BGJ250
(1,1'-BIPHENYL)-2-OL see BGJ250
2-BIPHENYLOL, SODIUM SALT see BGJ750
(1,1'-BIPHENYL)-2-OL, SODIUM SALT see BGJ750
BIPHENYL OXIDE see PFA850
1,1'-BIPHENYL, mixed with 1,1'-OXYBIS(BENZENE) see
PFA860
BIPHENYL, POLYCHLORO- see PJL750
BIPHENYL SELENIUM see PGH250
3,3',4,4'-BIPHENYLTETRAMINE see BGK500
3,3',4,4'-BIPHENYLTETRAMINE
TETRAHYDROCHLORIDE see BGK750
N-(4-BIPHENYLYL)ACETAMIDE see PDY500
N,N'-4,4'-BIPHENYLYLENEBISACETAMIDE see
BFX000
4-BIPHENYLYL ETHYLKETONE see BGM100
(1,4'-BIPIPERIDINE)-1'-CARBOXYLIC ACID, 4,11-
DIETHYL-3,4,12,14-TETRAHYDRO-4-HYDROXY-3,14-
DIOXO-1H-PYRANO(3',4':6,7)INDOLIZINO(1,2-
B)QUINOLIN-9-YL ESTER,
MONOHYDROCHLORIDE,TRIHYDRATE, (S)- see
IGJ550

BIPOTASSIUM CHROMATE see PLB250
BIPYRIDINE see BGO500
2,2'-BIPYRIDINE see BGO500
α,α'-BIPYRIDINE see BGO500
(2,4'-BIPYRIDINE)-3',5'-DICARBOXYLIC ACID, 1',4'-
DIHYDRO-2',6'-DIMETHYL-,-DIETHYL ESTER see
DJB460
2,2'-BIPYRIDYL see BGO500
α,α'-BIPYRIDYL see BGO500
BIRCH TAR OIL see BGO750
BIRCH TAR OIL, RECTIFIED (FCC) see BGO750
BIRLANE see CDS750
BIRNENOEL see AOD725
BIRTHWORT see AQY250
2,7-BIS(ACETAMIDO)FLUORENE see BGP250
BIS(ACETATO)TETRAHYDROXYTRILEAD see
LCH000
BIS(ACETATO)TRIHYDROXYTRILEAD see LCJ000
BIS(ACETO)DIHYDROXYTRILEAD see LCH000
BIS(ACETOXY)CADMIUM see CAD250
BIS(ACETOXYDIBUTYLSTANNANE) OXIDE see
BGQ000
BIS(ACETYLOXY)DIBUTYLSTANNANE see DBF800
BIS(ACETYLOXY)MERCURY see MCS750
S-(1,2-BIS(AETHOXY-CARBONYL)-AETHYL)-O,O-
DIMETHYL-DITHIOPHOSPHAT (GERMAN) see
MAK700
2,4-BIS(AETHYLAMINO)-6-CHLOR-1,3,5-TRIAZIN
(GERMAN) see BJP000
BIS AMINE see MJM200
BIS(4-AMINO-3-CHLOROPHENYL) ETHER see
BGT000
BIS(2-AMINOETHYL)AMINE see DJG600
BIS(β-AMINOETHYL)AMINE see DJG600
N,N'-BIS(2-AMINOETHYL)-1,2-DIAMINOETHANE see
TJR000
N,N-BIS(2-AMINOETHYL)-1,2-ETHANEDIAMINE see
NEI800
N,N'-BIS(2-AMINOETHYL)ETHYLENEDIAMINE see
TJR000
N,N'-BIS(2-AMINOETHYL)-1,2-ETHYLENEDIAMINE
see TJR000
BIS-p-AMINOFENYLMETHAN see MJQ000
4,4'-BIS(1-AMINO-8-HYDROXY-2,4-DISULFO-7-
NAPHTHYLAZO)-3,3'-BITOLYL, TETRASODIUM
SALT see BGT250
4,4'-BIS(7-(1-AMINO-8-HYDROXY-2,4-
DISULFO)NAPHTHYLAZO)-3,3'-BITOLYL,
TETRASODIUM SALT see BGT250
4,4'-BIS(1-AMINO-8-HYDROXY-2,4-DISULPHO-7-
NAPHTHYLAZO)-3,3'-BITOLYL, TETRASODIUM
SALT see BGT250
1,3-BIS-AMINOMETHYLBENZEN (CZECH) see
XHS800
BIS-4-AMINO-3-METHYLFENYLMETHAN (CZECH)
see MJO250
BIS(2-AMINOPHENYL)DISULFIDE see DXJ800
BIS(o-AMINOPHENYL)DISULFIDE see DXJ800
1,1'-BIS(2-AMINOPHENYL)DISULFIDE see DXJ800
BIS(4-AMINOPHENYL)ETHER see OPM000
BIS(p-AMINOPHENYL)ETHER see OPM000
BIS(4-AMINOPHENYL)METHANE see MJQ000
BIS(p-AMINOPHENYL)METHANE see MJQ000
2',4-BIS(AMINOPHENYL)METHANE see MJP750
BIS(4-AMINOPHENYL) SULFIDE see TFI000
BIS(p-AMINOPHENYL)SULFIDE see TFI000
BIS(4-AMINOPHENYL) SULFONE see SOA500

BIS(p-AMINOPHENYL) SULFONE see SOA500
BIS(4-AMINOPHENYL) SULPHIDE see TFI000
BIS(p-AMINOPHENYL)SULPHIDE see TFI000
BIS(4-AMINOPHENYL)SULPHONE see SOA500
BIS(p-AMINOPHENYL)SULPHONE see SOA500
1,4-BIS(3-AMINOPROPOXY)BUTANE see BGU600
1,4-BIS(γ-AMINOPROPOXY)BUTANE see BGU600
BIS-(3-AMINOPROPYL)AMINE see AIX250
N,N'-BIS(3-AMINOPROPYL)DIAMINOETHANE see
TBI700
N,N'-BIS(3-AMINOPROPYL)ETHYLENEDIAMINE see
TBI700
BIS(3-AMINOPROPYL)METHYLAMINE see BGU750
N,N-BIS(3-AMINOPROPYL)METHYLAMINE see
BGU750
BIS(γ-AMINOPROPYL)METHYLAMINE see BGU750
BIS(ω-AMINOPROPYL)METHYLAMINE see BGU750
N,N-BIS(γ-AMINOPROPYL)METHYLAMINE see
BGU750
1,4-BIS(AMINOPROPYL)PIPERAZINE see BGV000
BIS(AMINOPROPYL)PIPERAZINE (DOT) see BGV000
4,4'-BIS(((4-ANILINO-6-METHOXY-s-TRIAZIN-2-
YL)AMINO)-2,2'-STILBENEDISULFONIC ACID)
DISODIUM SALT see BGW100
2,2-BIS(p-ANISYL)-1,1,1-TRICHLOROETHANE see
MEI450
BIS(1-AZIRIDINYL)KETONE see BJP450
BIS-BENZENE CHROMIUM see BGY700
BIS(BENZENE)CHROMIUM IODIDE see BGY720
BIS(BENZENE)CHROMIUM(1+)IODIDE see BGY720
BIS(1,3)BENZODIOXOLO(5,6-A:4',5'-
G)QUINOLIZINIUM, 6,7-DIHYDRO- see CNR100
BIS(BENZOTHIAZOLYL)DISULFIDE see BDE750
BIS(2-BENZOTHIAZOLYLTHIO)ZINC see BHA750
BIS(2-BENZOTHIAZYL) DISULFIDE see BDE750
BIS(2-BENZOYLBENZOATO)BIS(3-(1-METHYL-2-
PYRROLIDINYL)PYRIDINE) NICKEL TRIHYDRATE
see BHB000
BIS(BISDIMETHYLAMINOPHOSPHONOUS)ANHYDR
IDE see OCM000
4,4'-BIS((4-BIS((2-HYDROXYETHYL)AMINO)-6-
CHLORO-s-TRIAZIN-2-YL)AMINO)-2,2'-
STILBENEDISULFONIC ACID, DISODIUM SALT see
BHB950
BIS(BUTOXYETHYL) ETHER see DDW200
BIS(2-BUTOXYETHYL) ETHER see DDW200
BIS(BUTOXYMALEOYLOXY)DIBUTYLSTANNANE
see BHK250
BIS(BUTOXYMALEOYLOXY)DIOCTYLSTANNANE
see BHK500
BIS(2-BUTYL)ETHER see BRH760
BIS(3-tert-BUTYL-4-HYDROXY-6-METHYLPHENYL)
SULFIDE see TFC600
α-α'-BIS(tert-
BUTYLPEROXY)DIISOPROPYLBENZENE see BHL100
BIS(n-BUTYL)SEBACATE see DEH600
S-(1,2-BIS(CARBETHOXY)ETHYL)-O,O-DIMETHYL
DITHIOPHOSPHATE see MAK700
2,3-
BIS(CARBOMETHOXYMERCAPTO)QUINOXALINE
see BHL800
BIS(CARBONATO(2-))DIHYDROXYTRIBERYLLIUM
see BFP500
(Z,Z)-BIS((3-CARBOXYACRYLOYL)OXY)DIOCTYL-
STANNANE DIISOOCTYL ESTER (8CI) see BKL000
BIS-β-CARBOXYETHYLGERMANIUM SESQUIOXIDE
see CCF125

BIS(2-CARBOXYETHYL) SULFIDE see BHM000
3,6-BIS(CARBOXYMETHYL)-3,5-
DIAZOOCTANEDIOIC ACID see EIX000
N,N-BIS(CARBOXYMETHYL)GLYCINE see AMT500
N,N-BIS(CARBOXYMETHYL)GLYCINE DISODIUM
SALT see DXF000
N,N-BIS(CARBOXYMETHYL)GLYCINE SODIUM
SALT see SEP500
N,N-BIS(CARBOXYMETHYL)GLYCINE TRISODIUM
SALT MONOHYDRATE see NEI000
BIS(CARBOXYMETHYLMERCAPTO)(p-
UREIDOPHENYL)ARSINE see CBI250
BIS(CARBOXYMETHYLTHIO)(p-
UREIDOPHENYL)ARSINE see CBI250
p-(BIS(o-CARBOXYPHENYLMERCAPTO)-ARSINO)-
PHENYLUREA see TFD750
BIS(3-CARBOXYPROPIONYL) PEROXIDE see SNC500
N,N-BIS-(β-CHLORAETHYL)-AMIN (GERMAN) see
BHN750
N,N-BIS-(β-CHLORAETHYL)-N',O-PROPYLEN-
PHOSPHORSAEURE-ESTER-DIAMID (GERMAN) see
CQC650
BIS(5-CHLOR-2-HYDROXYPHENYL)-METHAN see
MJM500
BIS(p-CHLOROBENZOYL) PEROXIDE see BHM750
trans-N,N'-BIS(2-CHLOROBENZYL)-1,4-
CYCLOHEXANEBIS(METHYLAMINE)
DIHYDROCHLORIDE see BHN000
1,3-BIS((p-
CHLOROBENZYLIDENE)AMINO)GUANIDINE see
RLK890
1,2-BIS((CHLOROCARBONYL)OXY)ETHANE see
EIQ000
BIS(2-CHLOROETHOXY)METHANE see BID750
BIS-β-CHLOROETHYLAMINE see BHN750
BIS(2-CHLOROETHYL)AMINE HYDROCHLORIDE
see BHO250
N,N-BIS(2-CHLOROETHYL)AMINE
HYDROCHLORIDE see BHO250
BIS(β-CHLOROETHYL)AMINE HYDROCHLORIDE
see BHO250
4'-(BIS(2-CHLOROETHYL)AMINO)ACETANILIDE see
BHO500
4-(BIS(2-
CHLOROETHYL)AMINO)BENZENEBUTANOIC
ACID see CDO500
1-((BIS(2-
CHLOROETHYL)AMINO)BENZOYL)PIPERIDINE see
BHP150
2-(BIS(2-CHLOROETHYL)AMINO)-3-(2-
CHLOROETHYL)TETRAHYDRO-2H-1,3,2-
OXAPHOSPHORINE-2-OXIDE see TNT500
1,6-BIS-(CHLOROETHYLAMINO)-1,6-DESOXY-d-
MANNITOLDIHYDROCHLORIDE see MAW750
1,6-BIS-(CHLOROETHYLAMINO)-1,6-DIDEOXY-d-
MANNITOLDIHYDROCHLORIDE see MAW750
4'-(BIS(2-CHLOROETHYL)AMINO)-2-FLUORO
ACETANILIDE see BHP750
1-(3-(BIS(2-CHLOROETHYL)AMINO-4-
METHYLBENZOYL)AZIRIDINE) see BHQ760
1-(3-(BIS(2-CHLOROETHYL)AMINO)-4-
METHYLBENZOYL)MORPHOLINE see BHR400
9-(4-BIS(2-CHLOROETHYL)AMINO-1-
METHYLBUTYLAMINO)-6-CHLORO-2-
METHOXYACRIDINE DIHYDROCHLORIDE see
QDS000

2-BIS(2-CHLOROETHYL)AMINONAPHTHALENE see BIF250

1-BIS(2-CHLOROETHYL)AMINO-1-OXA-2-AZA-5-OXAPHOSPHORIDINE MONOHYDRATE see CQC500

2-(BIS(2-CHLOROETHYL)AMINO)-1-OXA-3-AZA-2-PHOSPHOCYCLOHEXANE 2-OXIDE MONOHYDRATE see CQC500

2-(BIS(2-CHLOROETHYL)AMINO)-2H-1,3,2-OXAAZAPHOSPHORINE 2-OXIDE see CQC650

(p-(BIS(2-CHLOROETHYL)AMINO)PHENYL)ACETATE CHOLESTEROL see CME250

(p-(BIS(2-CHLOROETHYL)AMINO)PHENYL)ACETIC ACID CHOLESTEROL ESTER see CME250

(4-(BIS(2-CHLOROETHYL)AMINO)PHENYL)ACETIC ACID CHOLESTERYL ESTER see CME250

3-(p-(BIS(2-CHLOROETHYL)AMINO)PHENYL)ALANINE see BHT750

4-(BIS(2-CHLOROETHYL)AMINO)-l-PHENYLALANINE see PED750

4-(BIS(2-CHLOROETHYL)AMINO)-dl-PHENYLALANINE see BHT750

l-3-(p-(BIS(2-CHLOROETHYL)AMINO)PHENYL)ALANINE see PED750

p-N-BIS(2-CHLOROETHYL)AMINO-l-PHENYLALANINE see PED750

3-(p-(p-(BIS(2-CHLOROETHYL)AMINO)PHENYL))-l-ALANINE see PED750

dl-3-(p-(BIS(2-CHLOROETHYL)AMINO)PHENYL)ALANINE see BHT750

4-(p-(BIS(2-CHLOROETHYL)AMINO)PHENYL)BUTYRIC ACID see CDO500

γ-(p-BIS(2-CHLOROETHYL)AMINOPHENYL)BUTYRIC ACID see CDO500

4-(p-BIS(β-CHLOROETHYL)AMINOPHENYL)BUTYRIC ACID see CDO500

2-(p-BIS(2-CHLOROETHYL)AMINOPHENYL)-1,3,2-DITHIARSENOLANE see BHW300

m-(BIS(2-CHLOROETHYL)AMINO)PHENYL MORPHOLINO KETONE see BHX300

N,N-BIS(β-CHLOROETHYL)-AMINO-N'-O-PROPYLENE-PHOSPHORIC ACID ESTER DIAMIDE see IMH000

5-(BIS(2-CHLOROETHYL)AMINO)-2,4(1H,3H)PYRIMIDINEDIONE see BIA250

2-(BIS(2-CHLOROETHYL)AMINO)TETRAHYDROOXAZAPHOSPHORINE CYCLOHEXYLAMINE SALT see CQN000

(BIS(CHLORO-2-ETHYL)AMINO)-2-TETRAHYDRO-3,4,5,6-OXAZAPHOSPHORINE-1,3,2-OXIDE-2-MONOHYDRATE see CQC500

1-(3-(BIS(2-CHLOROETHYL)AMINO)-p-TOLUOYL)PIPERIDINE see BIA100

3-(BIS(2-CHLOROETHYL)AMINO)-p-TOLYL PIPERIDYL KETONE see BIA100

5-(BIS(2-CHLOROETHYL)AMINO)URACIL see BIA250

5-N,N-BIS(2-CHLOROETHYL)AMINOURACIL see BIA250

BIS(2-CHLOROETHYL)AMMONIUM CHLORIDE see BHO250

N,N-BIS(2-CHLOROETHYL)-p-ARSANILIC ACID see BIA300

O,O-BIS(2-CHLOROETHYL)-O-(3-CHLORO-4-METHYL-7-COUMARINYL) PHOSPHATE see DFH600

2,3-(N,N(1)-BIS(2-CHLOROETHYL)DIAMIDO)-1,3,2-OXAZAPHOSPHORIDINOXY see IMH000

BIS(2-CHLOROETHYL) ETHER see DFJ050

BIS(β-CHLOROETHYL) ETHER see DFJ050

BIS(2-CHLOROETHYL)ETHYLAMINE see BID250

BIS(2-CHLOROETHYL)FORMAL see BID750

BIS(β-CHLOROETHYL)FORMAL see BID750

9-(2',2'-BIS-β-CHLORO-ETHYL-HYDRAZINO)ACRIDINE HYDROCHLORIDE see BID800

9-(2,2-BIS(2-CHLOROETHYL)HYDRAZINO)ACRIDINE MONOHYDROCHLORIDE see BID800

N,N-BIS(2-CHLOROETHYL)-N'-(3-HYDROXYPROPYL)PHOSPHORODIAMIDATE, CYCLOHEXYLAMMONIUM SALT see CQN000

N,N-BIS(2-CHLOROETHYL)-N'-(3-HYDROXYPROPYL)PHOSPHORODIAMIDIC ACID intramol. ESTER see CQC650

BIS(2-CHLOROETHYL)METHYLAMINE see BIE250

N,N-BIS(2-CHLOROETHYL)METHYLAMINE see BIE250

BIS(β-CHLOROETHYL)METHYLAMINE see BIE250

BIS(2-CHLOROETHYL)METHYLAMINE HYDROCHLORIDE see BIE500

N,N-BIS(2-CHLOROETHYL)-2-NAPHTHYLAMINE see BIF250

BIS(2-CHLOROETHYL)-β-NAPHTHYLAMINE see BIF250

BISCHLOROETHYLNITROSOUREA see BIF750

BIS(2-CHLOROETHYL)NITROSOUREA see BIF750

1,3-BIS-(2-CHLOROETHYL)-1-NITROSOUREA see BIF750

N,N'-BIS(2-CHLOROETHYL)-N-NITROSOUREA see BIF750

1,3-BIS(β-CHLOROETHYL)-1-NITROSOUREA see BIF750

N,N-BIS(β-CHLOROETHYL)-N',O-PROPYLENEPHOSPHORIC ACID ESTER AMINE MONOHYDRATE see CQC500

N,N-BIS(2-CHLOROETHYL)-N',O-PROPYLENEPHOSPHORIC ACID ESTER DIAMIDE see CQC650

N,N-BIS(β-CHLOROETHYL)-N',O-TRIMETHYLENEPHOSPHORIC ACID ESTER DIAMIDE MONOHYDRATE see CQC500

N,N-BIS(β-CHLOROETHYL)-N',O-TRIMETHYLENEPHOSPHORIC ACID ESTER DIAMIDE see CQC650

BIS(2-CHLOROETHYL)PHOSPHORAMIDE CYCLIC PROPANOLAMIDE ESTER MONOHYDRATE see CQC500

BIS(2-CHLOROETHYL)PHOSPHORAMIDE-CYCLIC PROPANOLAMIDE ESTER see CQC650

N,N-BIS(2-CHLOROETHYL)-N'-3-PHOSPHORODIAMIDIC ACID HYDROXYLPROPYLCYCLOHEXYLAMINE SALT see CQN000

BIS(2-CHLOROETHYL)SULFIDE see BIH250

BIS(β-CHLOROETHYL)SULFIDE see BIH250

BIS(2-CHLOROETHYLSULFONYLMETHYL)ETHER see OPG100

BIS(DIMETHYLAMINO)FLUOROPHOSPHATE see BJE750

BISDIMETHYLAMINOFLUOROPHOSPHINE OXIDE see BJE750

3,7-BIS(DIMETHYL AMINO)PHENAZA THIONIUM CHLORIDE see BJI250

3,7-BIS(DIMETHYLAMINO)PHENOTHIAZIN-5-IUM CHLORIDE see BJI250

BIS(p-(N,N-DIMETHYLAMINO)PHENYL)KETONE see MQS500

BIS(p-DIMETHYLAMINOPHENYL)METHANE see MJN000

BIS(p-(N,N-DIMETHYLAMINO)PHENYL)METHANE see MJN000

p,p'-BIS(N,N-DIMETHYLAMINOPHENYL)METHANE see MJN000

BIS(4-(DIMETHYLAMINO)PHENYL)METHANONE see MQS500

BIS(p-DIMETHYLAMINOPHENYL)METHYLENEIMINE see IBB000

1,1-BIS(p-DIMETHYLAMINOPHENYL)METHYLENIMINEHYDROCHLORIDE see IBA000

BIS(DIMETHYLAMINO)PHOSPHONOUS ANYHYDRIDE see OCM000

BIS(DIMETHYLAMINO)PHOSPHORIC ANHYDRIDE see OCM000

BIS(α,α-DIMETHYLBENZYL)PEROXIDE see DGR600

BIS(DIMETHYLCARBAMODITHIOATO-S,S')LEAD see LCW000

BIS(DIMETHYLCARBAMODITHIOATO-S,S')ZINC see BJK500

BIS(DIMETHYLDITHIOCARBAMATE de ZINC) (FRENCH) see BJK500

BIS(DIMETHYLDITHIOCARBAMATO)ZINC see BJK500

BIS(DIMETHYLDITHIOCARBAMIATO)LEAD see LCW000

N,N'-BIS(1,1-DIMETHYLETHYL)-1,2-ETHANEDIAMINE see DEC100

3-((3,5-BIS(1,1-DIMETHYLETHYL)-4-HYDROXYPHENYL)METHYLENE)-1-METHOXY-2-PYRROLIDINONE see MEL100

((3,5-BIS(1,1-DIMETHYLETHYL)-4-HYDROXYPHENYL)METHYL)PHOSPHONIC ACID, MONOETHYL ESTER, NICKEL(2+) SALT (2:1) see BJK560

(((3,5-BIS(1,1-DIMETHYLETHYL)-4-HYDROXYPHENYL)METHYL)THIO)ACETIC ACID 2-ETHYLHEXYL ESTER see BJK600

2,6-BIS(1,1-DIMETHYLETHYL)-4-METHYLPHENOL see BFW750

11,11-BIS-((3,7-DIMETHYL-2,6-OCTADIENYL)OXY)-1-UNDECENE see UNA100

1,1'-BIS(DIMETHYLOCTOXYSILYL)FERROCENE see BJK780

1,1'-BIS(DIMETHYL(OCTYLOXY)SILYL)FERROCENE see BJK780

BIS(DIMETHYL-THIOCARBAMOYL)-DISULFID (GERMAN) see TFS350

BIS(DIMETHYLTHIOCARBAMOYL) DISULFIDE see TFS350

BIS(DIMETHYLTHIOCARBAMOYL)SULFIDE see BJL600

BIS(DIMETHYLTHIOCARBAMYL) MONOSULFIDE see BJL600

BIS(N,N-DIMETIL-DITIOCARBAMMATO) DI ZINCO (ITALIAN) see BJK500

(±)-1,2-BIS(3,5-DIOXOPIPERAZINE-1-YL)PROPANE see PIK250

(±)-1,2-BIS(3,5-DIOXOPIPERAZINYL)PROPANE see PIK250

BIS(1,3-DITHIOCYANATO-1,1,3,3-TETRABUTYLDISTANNOXANE) see BJM700

BIS(DITHIOPHOSPHATE de O,O-DIETHYLE) de S,S'-(1,4-DIOXANNE-2,3-DIYLE) (FRENCH) see DVQ709

BIS(DODECANOLOXY)DIOCTYLSTANNANE see DVJ800

BIS(DODECANOYLOXY)DI-n-BUTYLSTANNANE see DDV600

BIS(DODECYLOXYCARBONYLETHYL) SULFIDE see TFD500

BIS(2,3-EPOXYCYCLOPENTYL) ETHER see BJN250

m-BIS(2,3-EPOXYPROPOXY)BENZENE see REF000

1,3-BIS(2,3-EPOXYPROPOXY)BENZENE see REF000

BIS(2,3-EPOXYPROPYL)ETHER see DKM200

2,2-BIS(4-(2,3-EPOXYPROPYLOXY)PHENYL)PROPANE see BLD750

S-(1,2-BIS(ETHOXY-CARBONYL)-ETHYL)-O,O-DIMETHYL-DITHIOFOSFAAT (DUTCH) see MAK700

S-(1,2-BIS(ETHOXYCARBONYL)ETHYL)-O,O-DIMETHYL PHOSPHORODITHIOATE see MAK700

S-1,2-BIS(ETHOXYCARBONYL)ETHYL-O,O-DIMETHYL THIOPHOSPHATE see MAK700

BIS(2-ETHOXYETHYL) ADIPATE see BJO225

BIS(2-ETHOXYETHYL)ETHER see DIW800

2,4-BIS(ETHYLAMINO)-6-CHLORO-s-TRIAZINE see BJP000

BIS(ETHYLENEDIAMINE)(MERCURICTETRATHIOCYANATO)COPPER see BJP425

BISETHYLENEUREA see BJP450

BIS(2-ETHYLHEXANOYLOXY)DIBUTYL STANNANE see BJQ250

BIS(2-ETHYLHEXYL) ADIPATE see AEO000

BIS(2-ETHYLHEXYL)-1,2-BENZENEDICARBOXYLATE see DVL700

BIS(2-ETHYLHEXYL) ESTER PHOSPHOROUS ACID CADMIUM SALT see CAD500

BIS(ETHYLHEXYL) ESTER of SODIUM SULFOSUCCINIC ACID see DJL000

BIS(2-ETHYLHEXYL) FUMARATE see DVK600

BIS(2-ETHYLHEXYL)HYDROGEN PHOSPHATE see BJR750

BIS(2-ETHYLHEXYL)ORTHOPHOSPHORIC ACID see BJR750

BIS(2-ETHYLHEXYLOXYCARBONYLMETHYLTHIO)DIBUTYLSTANNANE see BKK250

BIS(2-ETHYLHEXYLOXYCARBONYLMETHYLTHIO)DIBUTYLSTANNANE see DDY600

BIS(2-ETHYLHEXYLOXYCARBONYLMETHYLTHIO)DIMETHYLSTANNANE see BKK500

BIS(2-ETHYLHEXYL) PHOSPHATE see BJR750

BIS(2-ETHYLHEXYL)PHOSPHORIC ACID see BJR750

BIS(2-ETHYLHEXYL)PHTHALATE see DVL700

BIS(2-ETHYLHEXYL) SEBACATE see BJS250

BIS(2-ETHYLHEXYL)SODIUM SULFOSUCCINATE see DJL000

BIS(2-ETHYLHEXYL)-S-SODIUM SULFOSUCCINATE see DJL000

1,4-BIS(2-ETHYLHEXYL) SODIUM SULFOSUCCINATE see DJL000

1,4-BIS(2-ETHYLHEXYL)SULFOBUTANEDIOIC ACID ESTER, SODIUM SALT see DJL000

BIS(2-ETHYLHEXYLTHIOGLYCOLATE)DIBUTYLTIN see DDY600

BIS(2-ETHYLHEXYLTHIOGLYCOLATE)DIOCTYLTIN see DVM800

BIS(ETHYLMERCURI) PHOSPHATE see BJT250

1,1-BIS(p-ETHYLPHENYL)-2,2-DICHLOROETHANE see DJC000

2,2-BIS(p-ETHYLPHENYL)-1,1-DICHLOROETHANE see DJC000

BIS(N-ETHYL-N-PHENYL)UREA see DJC400

BIS(ETHYLTHIO)METHYLENE MALONONITRILE see BJT800

(BIS(ETHYLTHIO)METHYLENE)PROPANEDINITRILE see BJT800

BIS(ETHYLXANTHIC)DISULFIDE see BJU000

BIS(ETHYLXANTHOGEN) DISULFIDE see BJU000

S-(1,2-BIS(ETOSSI-CARBONIL)-ETIL)-O,O-DIMETIL-DITIOFOSFATO (ITALIAN) see MAK700

BISFEROL A (GERMAN) see BLD500

BIS(3-FLUOROSALICYLALDEHYDE)-ETHYLENEDIIMINE-COBALT see EIS000

1,4-BIS(1-FORMAMIDO-2,2,2-TRICHLOROETHYL)PIPERAZINE see TKL100

N,N'-BIS(1-FORMAMIDO-2,2,2-TRICHLOROETHYL)PIPERAZINE see TKL100

BIS(FORMYLMETHYL) MERCURY see BJW800

m-BIS(GLYCIDYLOXY)BENZENE see REF000

BIS(4-GLYCIDYLOXYPHENYL)DIMETHYLAMETHANE see BLD750

2,2-BIS(p-GLYCIDYLOXYPHENYL)PROPANE see BLD750

BIS(GUANIDINIUM) CHROMATE see BJW825

BIS(l-HISTIDINATO)MANGANESE TETRAHYDRATE see BJX800

BIS(HYDROGEN MALEATO)DIOCTYLTIN BIS(2-ETHYLHEXYL) ESTER see DVM600

BIS(HYDROXYAETHYL)-AETHER-DINITRAT (GERMAN) see DJE400

BIS(β-HYDROXYAETHYL)NITROSAMIN (GERMAN) see NKM000

BIS(2-HYDROXYBENZOATO-O^{1}O^2-), (T-4)-CADMIUM (9CI) see CAI400

BIS(2-HYDROXY-3-tert-BUTYL-5-ETHYLPHENYL)METHANE see MJN250

BIS(4-HYDROXY-5-tert-BUTYL-2-METHYLPHENYL) SULFIDE see TFC600

BIS-2-HYDROXY-5-CHLORFENYLMETHAN see MJM500

BIS(2-HYDROXY-5-CHLOROPHENYL)METHANE see MJM500

BISHYDROXYCOUMARIN see BJZ000

BIS(4-HYDROXY-3-COUMARIN) ACETIC ACID ETHYL ESTER see BKA000

BIS-3,3'-(4-HYDROXYCOUMARINYL)ACETIC ACID ETHYL ESTER see BKA000

BIS-(4-HYDROXY-3-COUMARINYL)ETHYL ACETATE see BKA000

BIS(4-HYDROXYCOUMARIN-3-YL)METHANE see BJZ000

BIS(2-HYDROXYETHYL)AMINE see DHF000

BIS(2-HYDROXYETHYL)CARBAMODITHIOIC ACID, MONOPOTASSIUM SALT see PKX500

N,N-BIS(2-HYDROXYETHYL)COCOAMIDE see CNF330

N,N-BIS(2-HYDROXYETHYL)COCONUT FATTY ACID AMIDE see CNF330

N,N-BIS(2-HYDROXYETHYL)COCONUT OIL AMIDE see CNF330

BIS(2-HYDROXYETHYL)DITHIOCARBAMIC ACID, MONOPOTASSIUM SALT see PKX500

BIS(2-HYDROXYETHYL)DITHIOCARBAMIC ACID, POTASSIUM SALT see PKX500

N,N-BIS(2-HYDROXYETHYL)DODECAN AMIDE see BKE500

BIS(2-HYDROXYETHYL) ETHER see DJD600

BIS(2-HYDROXYETHYL)LAURAMIDE see BKE500

N,N-BIS(HYDROXYETHYL)LAURAMIDE see BKE500

N,N-BIS(2-HYDROXYETHYL)LAURAMIDE see BKE500

N,N-BIS(β-HYDROXYETHYL)LAURAMIDE see BKE500

N',N'-BIS(2-HYDROXYETHYL)-N-METHYL-2-NITRO-p-PHENYLENEDIAMINE see BKF250

BIS(β-HYDROXYETHYL)NITROSAMINE see NKM000

BIS[HYDROXYETHYLPOLY(ETHYLENEOXY)ETHYL PROPYLENEGLYCOL see PJK150

2,2-BIS-4'-HYDROXYFENYLPROPAN (CZECH) see BLD500

BIS(HYDROXYLAMINE) SULFATE see OLS000

3-BIS(HYDROXYMETHYL)AMINO-6-(5-NITRO-2-FURYLETHENYL)-1,2,4-TRIAZINE see BKH500

BIS(HYDROXYMETHYL)FURATRIZINE see BKH500

BIS(HYDROXYMETHYL)PHOSPHINE OXIDE see DSG600

2,2-BIS(HYDROXYMETHYL)-1,3-PROPANEDIOL see PBB750

2,2-BIS(HYDROXYMETHYL)-1,3-PROPANEDIOL TETRANITRATE see PBC250

BIS(4-HYDROXY-3-NITROPHENYL)MERCURY see MCS600

BIS(4-HYDROXY-2-OXO-2H-1-BENZOPYRAN-3-YL)ACETIC ACID ETHYL ESTER see BKA000

BIS(4-HYDROXYPHENYL) DIMETHYLMETHANE see BLD500

BIS(4-HYDROXYPHENYL)DIMETHYLMETHANE DIGLYCIDYL ETHER see BLD750

3,4-BIS(4-HYDROXYPHENYL)-2,4-HEXADIENE see DAL600

3,4-BIS(p-HYDROXYPHENYL)-2,4-HEXADIENE see DAL600

3,4-BIS(p-HYDROXYPHENYL)-3-HEXENE see DKA600

3,3-BIS(p-HYDROXYPHENYL)PHTHALIDE see PDO750

BIS(4-HYDROXYPHENYL)PROPANE see BLD500

2,2-BIS(4-HYDROXYPHENYL)PROPANE see BLD500

2,2-BIS(p-HYDROXYPHENYL)PROPANE see BLD500

2,2-BIS(4-HYDROXYPHENYL)PROPANE, DIGLYCIDYL ETHER see BLD750

2,2-BIS(p-HYDROXYPHENYL)PROPANE, DIGLYCIDYL ETHER see BLD750

N-BIS(2-HYDROXYPROPYL)NITROSAMINE see DNB200

2,2'-BISHYDROXYPROPYLNITROSAMINE see DNB200

BIS(1-HYDROXY-2(1H)-PYRIDINETHIONATO)ZINC see ZMJ000

BIS(8-HYDROXYQUINOLINE-5-SULFONIC ACID) MANGANESE(II) see BKJ275

BIS(2-HYDROXY-3,5,6-TRICHLOROPHENYL)METHANE see HCL000
BIS(3-INDOLEMETHYLENEMORPHOLINIUM)HEXACHLOROSTANNATE see BKJ500
BIS(ISOBUTYL)ALUMINUM CHLORIDE see CGB500
BIS(ISOBUTYL)HYDROALUMINUM see DNI600
BIS(4-ISOCYANATOCYCLOHEXYL)METHANE see MJM600
1,3-BIS(ISOCYANATOMETHYL)BENZENE see XIJ000
BIS(4-ISOCYANATOPHENYL)METHANE see MJP400
BIS(p-ISOCYANATOPHENYL)METHANE see MJP400
BIS(1,4-ISOCYANATOPHENYL)METHANE see MJP400
1,3-BIS-(ISOKYANATOMETHYL)BENZEN see XIJ000
BIS(ISOOCTYLOXYCARBONYLMETHYLTHIO)DIBUTYL STANNANE see BKK250
BIS(ISOOCTYLOXYCARBONYLMETHYLTHIO)DIMETHYLSTANNANE see BKK500
BIS(ISOOCTYLOXYCARBONYLMETHYLTHIO)DIOCTYL STANNANE see BKK750
BIS(ISOOCTYLOXYMALEOYLOXY)DIOCTYLSTANNANE see BKL000
BIS(ISOPROPYLAMIDO) FLUOROPHOSPHATE see PHF750
2,4-BIS(ISOPROPYLAMINO)-6-CHLORO-s-TRIAZINE see PMN850
2,4-BIS(ISOPROPYLAMINO)-6-METHYLMERCAPTO-s-TRIAZINE see BKL250
4,6-BIS(ISOPROPYLAMINO)-2-METHYLMERCAPTO-s-TRIAZINE see BKL250
2,4-BIS(ISOPROPYLAMINO)-6-METHYLTHIO-s-TRIAZINE see BKL250
2,4-BIS(ISOPROPYLAMINO)-6-METHYLTHIO-1,3,5-TRIAZINE see BKL250
BIS(ISOPROPYLBENZENE)CHROMIUM see DGR200
BIS(LACTATO)MAGNESIUM see LAL100
BIS(LAUROYLOXY)DIBUTYLSTANNANE see DDV600
BIS(LAUROYLOXY)DI(n-BUTYL)STANNANE see DDV600
BIS(LAUROYLOXY)DIOCTYLSTANNANE see DVJ800
1,3-BISMALEIMIDO BENZENE see BKL750
BISMATE see BKW000
BIS(MERCAPTOACETATE)DIOCTYLTIN BIS(2-ETHYLHEXYL) ESTER see DVM800
BIS(MERCAPTOACETATE)DIOCTYL-TIN BIS(ISOOCTYL) ESTER see BKK750
BIS(MERCAPTOBENZOTHIAZOLATO)ZINC see BHA750
BIS(MERCAPTO)DIOCTYLTIN BIS(DODECYL) ESTER see DVM400
1,4-BIS(METHANESULFONOXY)BUTANE see BOT250
BIS(METHANE SULFONYL)-d-MANNITOL see BKM500
(1,4-BIS(METHANESULFONYLOXY)BUTANE) see BOT250
BIS(2-(2-METHOXYETHOXY)ETHYL) ETHER see PBO500
BIS(2-METHOXYETHYL)ETHER see PBO500
BIS(METHOXYETHYL) PHTHALATE see DOF400
BIS(2-METHOXYETHYL) PHTHALATE see DOF400
BIS(METHOXYMALEOYLOXY)DIBUTYLSTANNANE see BKO250
1,1-BIS(p-METHOXYPHENYL)-2,2,2-TRICHLOROETHANE see MEI450
2,2-BIS(p-METHOXYPHENYL)-1,1,1-TRICHLOROETHANE see MEI450

1,2:5,6-BIS-o-(1-METHYLETHYLIDENE)-α-d-GLUCOFURANOSE, ((((2-(DIMETHYLAMINO)-2-OXO-1-(METHYLTHIO)ETHYLIDENE)AMINO)OXY)CARBONYL)METHYLAMIDOSULFITE see BKS640
N,N'-BIS(1-METHYLETHYL)-6-METHYL-THIO-1,3,5-TRIAZINE-2,4-DIAMINE see BKL250
2,6-BIS(1-METHYLETHYL)PHENOL see DNR800
BIS(6-METHYLHEPTYL)ESTER of PHTHALIC ACID see ILR100
BIS(METHYLMERCURIC)SULFATE see BKS810
BIS-(METHYLMERCURY)-SULFATE see BKS810
BIS-(METHYLMERKURI)SULFAT see BKS810
N,N'-BIS(2-METHYLPHENYL)THIOUREA see DXP600
N-BISMETHYLPTEROYLGLUTAMIC ACID see MDV500
N,N-BIS(METHYLQUECKSILBER)-p-TOLUOL-SULFAMID see MLH100
1,6-BIS-o-METHYLSULFONYL-d-MANNITOL see BKM500
N,N'-BIS(2-METHYLSULFONYL-2-METHYLPROPIONALDEHYDE-o-(N-METHYLCARBAMOYL)OXIME)SULFIDE see BKT300
N,N'-BIS(1-METHYLTHIO-1-(N,N-DIMETHYLCARBONYL)FORMALDEHYDE-o-(N-METHYLCARBAMOYL)OXIME)SULFIDE see BKU120
BIS(MONOISOPROPYLAMINO)FLUOROPHOSPHATE see PHF750
BIS(MONOISOPROPYLAMINO)FLUOROPHOSPHINE OXIDE see PHF750
N,N'-BISMORPHOLINE DISULFIDE see BKU500
BISMORPHOLINO DISULFIDE see BKU500
BISMUTH see BKU750
BISMUTH-209 see BKU750
BISMUTH COMPOUNDS see BKV750
BISMUTH DIMETHYL DITHIOCARBAMATE see BKW000
BISMUTH NITRATE see BKW250
BISMUTH SESQUITELLURIDE see BKY000
BISMUTH TELLURIDE see BKY000
BISMUTH TELLURIDE, UNDOPED see BKY000
BISMUTH VIOLET see AOR500
BIS(NITRATO)DIOXOURANIUM HEXAHYDRATE see URS000
BIS(NITRATO-O,O')DIOXO URANIUM (solid) see URA200
BIS(NITRATO-O)OXOZIRCONIUM see BLA000
BISODIUM TARTRATE see BLC000
BISOFLEX 81 see DVL700
BISOFLEX DOA see AEO000
BISOFLEX DOP see DVL700
BISOFLEX DOS see BJS250
BIS(OLEOYLOXY)DIBUTYLSTANNANE see DEJ000
BISOLVOMYCIN see HOI000
BISOMEL see IQN000
BIS(1-OXODODECYL)PEROXIDE see LBR000
BIS-(2-OXOPROPYL)-N-NITROSAMINE see NJN000
BIS(1-OXOPROPYL)PEROXIDE see DWQ800
BIS(8-OXYQUINOLINE)COPPER see BLC250
BISPENTAFLUOROSULFUR OXIDE see BLD000
BISPHENOL A see BLD500
BISPHENOL A DIGLYCIDYL ETHER see BLD750
1,4-BIS(PHENYL AMINO)BENZENE see BLE500
BIS(PHENYLMERCURI)METHYLENEDINAPHTHALENESULFONATE see PFN000
2,6-BIS(PICRYLAMINO)-3,5-DINITROPYRIDINE see PQC525

2,4-BIS(PROPYLAMINO)-6-CHLOR-1,3,5-TRIAZIN (GERMAN) see PMN850
BIS(2-PYRIDYLTHIO)ZINC, 1,1'-DIOXIDE see ZMJ000
BIS(8-QUINOLINATO)COPPER see BLC250
BIS(8-QUINOLINOLATO)COPPER see BLC250
BIS(8-QUINOLINOLATO-N¹,O⁸)-COPPER see BLC250
BIS(5-SULFO-8-QUINOLINOLATO-N¹,O⁸) MANGANESE(II) see BKJ275
BISTERIL see PDC250
BIS-N,N,N',N'-TETRAMETHYLPHOSPHORODIAMIDIC ANHYDRIDE see OCM000
1,2-BIS(THIOCYANATO)ETHANE see EJC035
BIS(THIOCYANATO)-MERCURY see MCU250
1,4-BIS(p-TOLYLAMINO)ANTHRAQUINONE see BLK000
BIS-1,4-p-TOLYLAMINOANTHRCHINON (CZECH) see BLK000
1,3-BIS(o-TOLYL)-2-THIOUREA see DXP600
BIS-(TRI-N-BUTYLCIN)OXID (CZECH) see BLL750
BIS(TRIBUTYLOXIDE) of TIN see BLL750
BIS(TRI-N-BUTYLPHOSPHINE)DICHLORONICKEL see BLS250
BIS(TRIBUTYLSTANNYL)OXIDE see BLL750
BIS(TRIBUTYL TIN)OXIDE see BLL750
BIS(TRIBUTYLTIN)SULFIDE see HCA700
BIS(TRI-N-BUTYLZINN)-OXYD (GERMAN) see BLL750
BIS-2,3,5-TRICHLOR-6-HYDROXYFENYLMETHAN (CZECH) see HCL000
BIS(3,5,6-TRICHLORO-2-HYDROXYPHENYL)METHANE see HCL000
1:4-BIS-TRICHLOROMETHYL BENZENE see HCM500
BIS(TRIETHYL TIN) SULFATE see BLN500
BIS(TRIFLUOROMETHYLTHIO)MERCURY see BLQ525
BIS(TRIISOBUTYLSTANNANE) see HDY100
BIS(2,4,6-TRINITRO-PHENYL)-AMIN (GERMAN) see HET500
BIS(TRIPHENYLPHOSPHINE)DICHLORONICKEL see BLS250
BIS(TRIPHENYL SILYL)CHROMATE see BLS750
BIS(TRIPHENYLTIN)ACETYLENEDICARBOXYLATE see BLS900
BIS(TRIPROPYLTIN)OXIDE see BLT300
BIS(TRIS(p-DIMETHYLAMINOPHENYL)PHOSPHINE OXIDE)STANNIC CHLORIDE COMPLEX see BLT775
BIS(TRIS(β,β-DIMETHYLPHENETHYL)TIN)OXIDE see BLU000
BIS(TRIS(2-METHYL-2-PHENYLPROPYL)TIN)OXIDE see BLU000
BISULFAN see BOT250
BISULFITE see SOH500
BISULFITE see HIC600
BISULFITE de SODIUM (FRENCH) see SFE000
BISULPHANE see BOT250
BISULPHITE see HIC600
BITEMOL see BJP000
BITEMOL S 50 see BJP000
BITHION see TAL250
BITIRAZINE see DIW000
4,4'-BI-o-TOLUIDINE see TGJ750
(m,o'-BITOLYL)-4-AMINE see BLV250
BITOSCANATE see PFA500
BITTER ALMOND OIL see BLV500
BITTER ALMOND OIL CAMPHOR see BCP250
BITTER FENNEL OIL see FAP000

BITUMEN (MAK) see ARO500
BIVERM see PDP250
BIVINYL see BOP500
BIXA ORELLANA see APE100
B-K LIQUID see SHU500
B-K POWDER see HOV500
γ-BL see BOV000
BLA see LCH000
BLACAR 1716 see PKQ059
BLACK AND WHITE BLEACHING CREAM see HIH000
BLACK ANTIMONY see AQL500
BLACK BIRCH OIL see SOY100
BLACK BLASTING POWDER see ERF500
BLACK LEAD see CBT500
BLACK LEAF see NDN000
BLACK MANGANESE OXIDE see MAS000
BLACK MAX see EHP700
BLACK 2EMBL see AQP000
BLACK 4EMBL see AQP000
BLACK OXIDE of IRON see IHC450
BLACK PEARLS see CBT750
BLACK PEPPER OIL see BLW250
BLACK POWDER, compressed (DOT) see PLL750
BLACK POWDER, compressed (DOT) see ERF500
BLACK POWDER, granular or as a meal (UN 0027) (DOT) see PLL750
BLACK POWDER, granular or as a meal (UN 0027) (DOT) see ERF500
BLACK POWDER, in pellets (UN 0028) (DOT) see PLL750
BLACK POWDER, in pellets (UN 0028) (DOT) see ERF500
BLACOSOLV see TIO750
BLADAFUM see SOD100
BLADAFUME see SOD100
BLADAFUN see SOD100
BLADAN see PAK000
BLADAN see TCF250
BLADAN see EEH600
BLADAN see HCY000
BLADAN BASE see HCY000
BLADAN-M see MNH000
BLADEX see BLW750
BLADEX 80WP see BLW750
BLAETTERALKOHOL see HFE000
BLANC FIXE see BAP000
BLANDLUBE see MQV750
BLANOSE BWM see SFO500
BLASTING GELATIN (DOT) see NGY000
BLASTING OIL see NGY000
BLASTING POWDER see PLL750
BLASTING POWDER see ERF500
BLAUES PYOKTANIN see AOR500
BLAUSAEURE (GERMAN) see HHS000
BLAUWZUUR (DUTCH) see HHS000
BLEACHING POWDER see HOV500
BLEACHING POWDER, containing 39% or less chlorine (DOT) see HOV500
BLEIACETAT (GERMAN) see LCV000
BLEIAZETAT (GERMAN) see LCJ000
BLEIPHOSPHAT (GERMAN) see LDU000
BLEISTEARAT (GERMAN) see LDX000
BLEISULFAT (GERMAN) see LDY000
BLEKIT EVANSA (POLISH) see BGT250
BLEMINOL see ZVJ000
BLENDED RED OXIDES of IRON see IHC450

BLENOXANE see BLY000
BLEO see BLY000
BLEOCIN see BLY000
BLEOMYCIN see BLY000
BLEPH-10 see SNQ710
BLEU BRILLIANT FCF see FMU059
BLEU DIAMINE see CMO250
BLEX see DIN800
BLM see BLY000
BLO see BOV000
BLOAT GUARD see PJK150
BLOC see FAK100
BLOCADREN see DDG800
BLON see BOV000
BLOODSTONE see HAO875
BLOTIC see MKA000
BLOXANTH see ZVJ000
BLUE 2B see CMO000
11388 BLUE see FMU059
BLUE ASBESTOS (DOT) see ARM275
BLUE BN BALSE see DCJ200
BLUE CHAMOMILE OIL see CDH500
BLUE COPPER see CNP250
BLUE CROSS see CGN000
BLUE EMB see CMO250
BLUE OIL see AOQ000
BLUE OIL see COD750
BLUE-OX see ZLS000
BLUE POWDER see ZBJ000
BLUE STAR see AQF000
BLUE STONE see CNP250
BLUE VITRIOL see CNP250
BLUTON see IIU000
BM 06011 see SFX730
BMC see MHC750
BMD see BAC260
BMOO see BRT000
BN see BFW000
B-NINE see DQD400
BNM see BAV575
BNU see BSA250
BOEA see BKA000
B.O.E.A. see BKA000
BOF-A2 see BDN600
BOG MANGANESE see MAS000
BOH see HHC000
BOIS D'ARC (FRENCH) see MRN500
BOIS D'INDE see LBK000
BOIS d'INDE see BAT500
BOIS de ROSE OIL see BMA550
BOL see BNM250
BOL-148 see BNM250
BOLATRON see PKQ059
BOLETIC ACID see FOU000
BOLINAN see PKQ250
BOLLS-EYE see HKC000
BOLLS-EYE see HKC500
BOLSTAR see SOU625
BOMBITA see DBA800
BOMYL see SOY000
BONADETTES see HGC500
BONADOXIN see HGC500
BONAID see BOO632
BONAMID see PJY500
BONAMINE see HGC500
BONAPICILLIN see AIV500
BONAZEN see ZNA000
BONBRAIN see TEH500

BOND CH 18 see AAX250
BONDERITE 40 see ZJS400
BONDERITE 880 see ZJS400
BONDING AGENT M 3 see OMM300
BONE OIL see BMA750
BONIBAL see DXH250
BONIDE BLUE DEATH RAT KILLER see PHP010
BONIDE KRAB CRABGRASS KILLER see PLC250
BONIDE TOPZOL RAT BAITS and KILLING SYRUP see RCF000
BONLOID see PKQ059
BONOFORM see TBQ100
BONOMOLD OE see HJL000
BONOMOLD OP see HNU500
BOOKSAVER see AAX250
BOOMER-RID see SMN500
BOP see NJN000
BORACIC ACID see BMC000
BORACSU see SFF000
BORANE, COMPD. WITH N2H4 see HGU025
BORANE with DIMETHYLAMINE (1:1) see DOR200
BORANE, TRIFLUORO-, DIHYDRATE see BMG800
BORATES, TETRA, SODIUM SALT, anhydrous (OSHA) see DXG035
BORATES, TETRA, SODIUM SALT, anhydrous (OSHA, ACGIH) see SFF000
BORATE(1-), TETRAFLUORO-, CADMIUM (2:1) (9CI) see CAG000
BORATE(1-), TETRAFLUORO-, HYDROGEN see FDD125
BORATE(1-), TETRAFLUORO-, HYDROGEN see HHS600
BORATE(1-), TETRAHYDRO-, ALUMINUM (3:1) (9CI) see AHG875
BORAX (8CI) see SFF000
BORAX DECAHYDRATE see SFF000
BORAX GLASS see DXG035
BORDEN 2123 see AAX250
BORDERMASTER see CIR250
BOREA see BMM650
BORER SOL see EIY600
BORESTER 2 see THX750
BORESTER O see TLN000
BORIC ACID see BMC000
BORIC ACID, DISODIUM SALT see DXG035
BORIC ACID, ETHYL ESTER see BMC250
BORIC ACID, TRIBUTYL ESTER see THX500
BORIC ACID, TRI-sec-BUTYL ESTER see THX750
BORIC ACID, TRIISOPROPYL ESTER see IOI000
BORIC ACID, TRI-n-OCTYL ESTER see TMO550
BORIC ANHYDRIDE see BMG000
BORICIN see SFF000
BORNANE, 2-CHLORO-, endo- see BMD300
1-2-BORNANOL see NCQ820
2-BORNANOL, endo- see BMD000
2-BORNANONE see CBA750
BORNATE see IHZ000
BORNEO CAMPHOR see BMD000
(−)-BORNEOL see NCQ820
BORNEOL see BMD000
BORNEOL (DOT) see BMD000
trans-BORNEOL see BMD000
(1S,2R,4S)-(−)-1-BORNEOL see NCQ820
BORNYL ACETATE see BMD100
l-BORNYL ACETATE see BMD100
BORNYL ALCOHOL see BMD000
1-BORNYL ALCOHOL see NCQ820
BORNYL CHLORIDE see BMD300

2-BORNYL CHLORIDE see BMD300
BOROETHANE see DDI450
BOROFAX see BMC000
BOROFLUORIC ACID see FDD125
BOROFLUORIC ACID see HHS600
BOROHYDRURE de POTASSIUM (FRENCH) see PKY250
BOROHYDRURE de SODIUM (FRENCH) see SFF500
BOROLIN see PIB900
BORON see BMD500
BORON BROMIDE see BMG400
BORON CHLORIDE see BMG500
BORON COMPOUNDS see BME500
BORON FLUORIDE see BMG700
BORON FLUORIDE, compd. with ACETIC ACID see BMG750
BORON FLUORIDE DIHYDRATE see BMG800
BORON, (HYDRAZINE-KAPPAN)TRIHYDRO-, (T-4)-see HGU025
BORON, (HYDRAZINE-N)TRIHYDRO-, (T-4)- see HGU025
BORON HYDRIDE see DDI450
BORON OXIDE see BMG000
BORON SESQUIOXIDE see BMG000
BORON TRIBROMIDE see BMG400
BORON TRICHLORIDE see BMG500
BORON TRIFLUORIDE see BMG700
BORON TRIFLUORIDE–ACETIC ACID COMPLEX see BMG750
BORON TRIFLUORIDE DIHYDRATE see BMG800
BORON TRIFLUORIDE DIHYDRATE (DOT) see BMG800
BORON TRIFLUORIDE-DIMETHYL ETHER see BMH000
BORON TRIFLUORIDE DIMETHYL ETHERATE (DOT) see BMH000
BORON TRIOXIDE see BMG000
BORSAEURE (GERMAN) see BMC000
BORSIL P see SCK600
BOSAN SUPRA see DAD200
BOSMIN see VGP000
BOTRYODIPLODIN see BMK290
(−)-BOTRYODIPLODIN see BMK290
BOURBONAL see EQF000
BOV see SOI500
BOVIDERMOL see DAD200
BOVINOX see TIQ250
BOVIZOLE see TEX000
BOVOFLAVIN see DBX400
B(a)P see BCS750
BPE-I see PJS750
BP-KLP see SMQ500
BPL see PMT100
BPMC see MOV000
BPPS see SOP000
BR-931 see CLW500
BRACKEN FERN, DRIED see BML000
BRALEN KB 2-11 see PJS750
BRALEN RB 03-23 see PJS750
BRAN ABSOLUTE see WBJ700
BRASILAMINA BLACK GN see AQP000
BRASILAMINA BLUE 2B see CMO000
BRASILAMINA BLUE 3B see CMO250
BRASILAMINA CONGO 4B see SGQ500
BRASILAN AZO RUBINE 2NS see HJF500
BRASILAN METANIL YELLOW see MDM775
BRASILAN ORANGE 2G see HGC000

BRASILAZET BLUE GR see TBG700
BRASILAZINA OIL RED B see SBC500
BRASILAZINA OIL SCARLET see OHI200
BRASILAZINA OIL SCARLET 6G see XRA000
BRASILAZINA OIL YELLOW G see PEI000
BRASILAZINA OIL YELLOW R see AIC250
BRASILAZINA ORANGE Y see PEK000
BRASSICOL see PAX000
BRAUNSTEIN (GERMAN) see MAS000
BRAVO see TBQ750
BRAVO 6F see TBQ750
BRAVO-W-75 see TBQ750
BRECOLANE NDG see DJD600
BRELLIN see GEM000
BREMIL see CFY000
BRENOL see ART250
BRENTAMINE FAST BLUE B BASE see DCJ200
BRENTAMINE FAST ORANGE GR BASE see NEO000
BRENTAMINE FAST RED TR BASE see CLK220
BRENTAMINE FAST RED TR SALT see CLK235
BREON see PKQ059
BREON 351 see AAX175
BRESTAN see ABX250
BRESTANOL see CLU000
BREVIMYTAL see MDU500
BREVINYL see DGP900
BREVIRENIN see VGP000
BREVITAL SODIUM see MDU500
BRICK OIL see CMY825
BRIETAL SODIUM see MDU500
BRIGHT RED see CHP500
BRILLIANT ACRIDINE ORANGE E see BJF000
BRILLIANT BLUE see FMU059
BRILLIANT BLUE R see BMM500
BRILLIANT CHROME LEATHER BLACK H see AQP000
BRILLIANT CRIMSON RED see HJF500
BRILLIANT FAST YELLOW see DOT300
BRILLIANT FAT SCARLET R see CMS238
BRILLIANT GREEN SULFATE see BAY750
BRILLIANT OIL ORANGE R see PEJ500
BRILLIANT OIL ORANGE Y BASE see PEK000
BRILLIANT OIL SCARLET B see XRA000
BRILLIANT OIL YELLOW see IBB000
BRILLIANT PONCEAU 3R see FMU080
BRILLIANT PONCEAU G see FMU070
BRILLIANT RED see CHP500
BRILLIANT SCARLET see CHP500
BRILLIANT SCARLET see FMU080
BRILLIANT TONER Z see CHP500
BRILLIANT VIOLET 5B see AOR500
BRIMONIDINE see BMM575
BRIMSTONE see SOD500
BRINDERDIN see RDK000
BRISERINE see RDK000
BRISTACYCLINE see TBX250
BRITACIL see AIV500
BRITISH ALUMINUM AF 260 see AHC000
BRITISH ANTILEWISITE see BAD750
BRITON see TIQ250
BRITTEN see TIQ250
BRITTOX see DDP000
BRL see AIV500
BRL 1341 see AIV500
BRL-1621 see SLJ000
BRL 2333 TRIHYDRATE see AOA100
BR 55N see PAU500
BRN 4729620 see SKS150

BROBAMATE see MQU750
BROCADISIPAL see OJW000
BROCADOPA see DNA200
BROCASIPAL see OJW000
BROCIDE see EIY600
BROCKMANN, ALUMINUM OXIDE see AHE250
BRODAN see CMA100
BROGDEX 555 see SGM500
BROM (GERMAN) see BMP000
BROMACETOCARBAMIDE see BNK000
BROMACIL see BMM650
BROMADAL see BNK000
BROMADEL see BNK000
BROMALLYLENE see AFY000
BROMARAL see BNP750
BROMAT see HCQ500
BROMATES see BMN500
BROMATE de SODIUM (FRENCH) see SFG000
BROMAZIL see BMM650
BROMBENZYL CYANIDE see BMW250
BROMCARBAMIDE see BNP750
d-2-BROM-DIETHYLAMIDE of LYSERGIC ACID see
BNM250
BROME (FRENCH) see BMP000
BROMELAIN see BMO000
BROMELAINS see BMO000
BROMELIN see BMO000
BROMETHOL see THV000
BROMETHOL see ARW250
BROMEX see DFK600
BROMIC ACID, POTASSIUM SALT see PKY300
BROMIC ACID, SODIUM SALT see SFG000
BROMIC ETHER see EGV400
BROMIDES see BMO750
BROMIDE SALT OF POTASSIUM see PKY500
BROMIDE SALT of SODIUM see SFG500
BROMID UHLICITY see CBX750
BROMINAL see DDP000
BROMINAL M & PLUS see CIR250
BROMINATED VEGETABLE (SOYBEAN) OIL see
BMO825
BROMINE see BMP000
BROMINE, solution (DOT) see BMP000
BROMINE AZIDE see BMP250
BROMINE CYANIDE see COO500
BROMINE NITRIDE see BMP250
BROMINE PENTAFLUORIDE see BMQ000
BROMINE TRIFLUORIDE see BMQ325
BROMINEX see DDP000
BROMINIL see DDP000
BROMISOVAL see BNP750
BROMISOVALERYLUREA see BNP750
α-BROMISOVALERYLUREA see BNP750
BROMISOVALUM see BNP750
BROMIZOVAL see BNP750
BROMKAL 80 see BMQ800
BROMKAL 80-9D see NMV735
BROMKAL 83-10DE see PAU500
BROMKAL 82-ODE see PAU500
BROMKAL P 67-6HP see TNC500
BROM LSD see BNM250
BROMLYSERGAMIDE see BNM250
2-BROM-d-LYSERGIC ACID DIETHYLAMINE see
BNM250
BROM-METHAN (GERMAN) see MHR200
BROMNATRIUM (GERMAN) see SFG500
BROMO (ITALIAN) see BMP000

BROMOACETIC ACID see BMR750
α-BROMOACETIC ACID see BMR750
BROMOACETIC ACID, solid or solution (DOT) see
BMR750
BROMOACETIC ACID, ETHYL ESTER see EGV000
BROMOACETIC ACID METHYL ESTER see MHR250
BROMOACETONE see BNZ000
BROMOACETONE (DOT) see BNZ000
BROMOACETONE, liquid (DOT) see BNZ000
1-BROMOACETYL-α-α-DIPHENYL-4-
PIPERIDINEMETHANOL see BMS300
BROMO ACID see BNH500
γ-BROMOALLYLENE see PMN500
3-BROMOALLYL ISOCYANATE see BMT150
BROMOBENZENE (DOT) see PEO500
4-BROMOBENZENEACETONITRILE see BNV750
4-BROMOBENZYLCYANIDE see BNV750
p-BROMOBENZYL CYANIDE see BNV750
α-BROMOBENZYL CYANIDE see BMW250
BROMOBENZYLNITRILE see BMW250
α-BROMOBENZYLNITRILE see BMW250
2-BROMO-2-(BROMOMETHYL)GLUTARONITRILE
see DDM500
1-BROMOBUTANE see BMX500
2-BROMOBUTANE see BMX750
5-BROMO-3-sec-BUTYL-6-METHYLURACIL see
BMM650
BROMOCARBAMIDE see BNP750
BROMOCHLORODIFLUOROMETHANE see BNA250
7-BROMO-6-CHLOROFEBRIFUGINE
HYDROBROMIDE see HAF600
7-BROMO-6-CHLORO-3-[3-(3-HYDROXY-2-
PIPERDINYL)-2-OXOPROPYL]-4(3H)-
QUINAZOLINONE HYDROBROMIDE see HAF600
3-BROMO-N-(2-
CHLOROMERCURICYCLOHEXYL)PROPIONAMIDE
see CET000
BROMOCHLOROMETHANE see CES650
2-(3-BROMO-4-CHLOROPHENYL)-4-CHLORO-5-((6-
CHLORO-3-PYRIDINYL)METHOXY)-3(2H)-
PYRIDAZINONE see BNA350
O-(4-BROMO-2-CHLOROPHENYL)-O-ETHYL-S-
PROPYL PHOSPHOROTHIOATE see BNA750
1-BROMO-3-CHLOROPROPANE see BNA825
3-BROMO-1-CHLOROPROPENE see BNA880
BROMOCYAN see COO500
BROMOCYANOGEN see COO500
BROMODEOXYURIDINE see BNC750
5-BROMODEOXYURIDINE see BNC750
5-BROMO-2-DEOXYURIDINE see BNC750
5-BROMO-2'-DEOXYURIDINE see BNC750
5-BROMODESOXYURIDINE see BNC750
2-BROMO-1,5-DIAMINO-4,8-
DIHYDROXYANTHRAQUINONE see BNC800
BROMODICHLOROMETHANE see BND500
4-BROMO-2,5-DICHLOROPHENOL-o-ESTER with
O,O-DIETHYL PHOSPHOROTHIOATE see EGV500
O-(4-BROMO-2,5-DICHLOROPHENYL)-O,O-
DIETHYL PHOSPHOROTHIOATE see EGV500
O-(4-BROMO-2,5-DICHLOROPHENYL)-O,O-
DIETHYLPHOSPHOROTHIONATE see EGV500
O-(4-BROMO-2,5-DICHLOROPHENYL)-O-METHYL
PHENYLPHOSPHONOTHIOATE see LEN000
2-BROMO-9,10-DIDEHYDRO-N,N-DIETHYL-6-
METHYLERGOLINE-8-β-CARBOXAMIDE see
BNM250
BROMODIETHYLACETYLCARBAMIDE see BNK000

BROMODIETHYLACETYLUREA see BNK000
BROMODIETHYLGOLD see DJJ850
α-BROMO-β-DIMETHYLPROPANOYLUREA see BNP750
N¹-(5-BROMO-4,6-DIMETHYL-2-PYRIMIDINYL)BENZENESULFONAMIDE see SNH875
BROMODIPHENYLMETHANE see BNG750
BROMODIPHENYLMETHANE (solution) see BNH000
BROMOEOSINE see BNH500
3-BROMO-1,2-EPOXYPROPANE see BNI000
BROMOETHANE see EGV400
BROMOETHANOIC ACID see BMR750
α-BROMOETHANOIC ACID see BMR750
BROMOETHANOL see BNI500
2-BROMO ETHANOL see BNI500
BROMOETHENE see VMP000
2-BROMO-2-ETHYLBUTYRLUREA see BNK000
(α-BROMO-α-ETHYLBUTYRYL)CARBAMIDE see BNK000
1-BROMO-ETHYL-BUTYRYL-UREA see BNK000
2-BROMO-2-ETHYLBUTYRYLUREA see BNK000
(α-BROMO-α-ETHYLBUTYRYL)UREA see BNK000
BROMOETHYLENE see VMP000
BROMOFLOR see CDS125
BROMOFLUORESCEIC ACID see BNH500
BROMO FLUORESCEIN see BNH500
BROMOFLUOROFORM see TJY100
BROMOFORM see BNL000
BROMOFORME (FRENCH) see BNL000
BROMOFORMIO (ITALIAN) see BNL000
BROMOFOS-ETHYL see EGV500
BROMOFUME see EIY500
BROMO-O-GAS see MHR200
BROMOHEXYLMERCURY see HFR100
BROMO(2-HYDROXYETHYL)MERCURY AMMONIA SALT see BNL275
5-BROMO-6-(2-IMIDAZOLIN-2-YLAMINO)QUINOXALINE see BMM575
2-BROMOISOBUTANE see BQM250
α-BROMOISOVALERIC ACID UREIDE see BNP750
α-BROMOISOVALEROYLUREA see BNP750
(α-BROMOISOVALERYL)UREA see BNP750
2-BROMO-d-LYSERGIC ACID DIETHYLAMIDE see BNM250
BROMOLYSERGIDE see BNM250
2-(BROMOMERCURI) ETHANOL-AMMONIA (1:0.8 moles) compound see BNL275
BROMOMETANO (ITALIAN) see MHR200
BROMOMETHANE see MHR200
BROMOMETHANE mixed with DIBROMOETHANE see BNM750
1-BROMO-4-(METHYLAMINO)ANTHRAQUINONE see BNN550
(BROMOMETHYL)BENZENE see BEC000
1-BROMO-3-METHYL BUTANE see BNP250
2-BROMO-3-METHYLBUTYRYLUREA see BNP750
BROMOMETHYL METHYL KETONE see BNZ000
5-BROMO-6-METHYL-3-(1-METHYLPROPYL)-2,4(1H,3H)-PYRIMIDINEDIONE see BMM650
5-BROMO-6-METHYL-3-(1-METHYLPROPYL)URACIL see BMM650
p-(BROMOMETHYL)NITROBENZENE see BEC000
1-BROMO-2-METHYLPROPANE see BNR750
2-BROMO-2-METHYLPROPANE (DOT) see BQM250

8-β-((5-BROMONICOTINOYLOXY)METHYL)-1,6-DIMETHYL-10-α-METHOXYERGOLINE see NDM000
2-BROMO-2-NITROPANE-1,3-DIOL see BNT250
2-BROMO-2-NITROPROPAN-1,3-DIOL see BNT250
2-BROMO-2-NITRO-1,3-PROPANEDIOL see BNT250
β-BROMO-β-NITROTRIMETHYLENEGLYCOL see BNT250
2-BROMOPENTANE see BNU500
4-BROMOPHENYLACETONITRILE see BNV750
p-BROMOPHENYLACETONITRILE see BNV750
2-(4-BROMOPHENYL)ACETONITRILE see BNV750
α-BROMOPHENYLACETONITRILE see BMW250
4-((3-(((4-BROMOPHENYL)AMINO)-4,5-DIHYDRO-2H-BENZ(gNDAZOL-2-YL)ACETYL)MORPHOLINE see BNV752
2-(4-BROMOPHENYL)-4-CHLORO-5-((4-CHLOROPHENYL)METHOXY)-3(2H)-PYRIDAZINONE see BNV800
BROMOPHENYLMETHANE see BEC000
5-((4-BROMOPHENYL)METHOXY)-4-CHLORO-2-(4-CHLORO-2-FLUOROPHENYL)-3(2H)-PYRIDAZINONE see BNX035
3-(4-BROMOPHENYL)-N-(4-PROPYLCYCLOHEXYL)-2-PROPENAMIDE see BNX330
BROMOPHOSETHYL see EGV500
1-BROMOPROPANE see BNX750
2-BROMOPROPANE see BNY000
1-BROMOPROPANE (DOT) see BNX750
BROMO-2-PROPANONE see BNZ000
1-BROMO-2-PROPANONE see BNZ000
3-BROMOPROPENE see AFY000
1-BROMO-2-PROPENE see AFY000
3-BROMOPROPIONIC ACID see BOB250
β-BROMOPROPIONIC ACID see BOB250
3-BROMOPROPYL CHLORIDE see BNA825
3-BROMOPROPYLENE see AFY000
3-BROMO-1-PROPYNE see PMN500
3-BROMOPROPYNE (DOT) see PMN500
5-((6-BROMO-3-PYRIDINYL)METHOXY)-4-CHLORO-2-(4-CHLOROPHENYL)-3(2H)-PYRIDAZINONE see BOC600
1-BROMO-2,5-PYRROLIDINEDIONE see BOF500
BROMO SELTZER see ABG750
BROMOSILANE see BOE750
N-BROMOSUCCIMIDE see BOF500
N-BROMOSUCCINIMIDE see BOF500
5-BROMOSULFAMETHAZINE see SNH875
p-BROMOTOLUENE see BOG255
ω-BROMOTOLUENE see BEC000
α-BROMOTOLUENE (DOT) see BEC000
α-BROMO-α-TOLUNITRILE see BMW250
BROMOTRICHLOROMETHANE see BOH750
BROMOTRIETHYLSTANNANE see BOI750
BROMOTRIETHYLSTANNANE compounded with 2-PIPECOLINE (1:1) see TJU850
BROMOTRIFLUOROETHENE see BOJ000
BROMO TRIFLUOROETHYLENE see BOJ000
BROMOTRIFLUOROMETHANE see TJY100
5-BROMOURACIL see BOL000
BROMOURACIL DEOXYRIBOSIDE see BNC750
5-BROMOURACIL DEOXYRIBOSIDE see BNC750
5-BROMOURACIL-2-DEOXYRIBOSIDE see BNC750
BROMOVAL see BNP750
BROMOVALEROCARBAMIDE see BNP750
BROMOVALERYLUREA see BNP750
BROMOWODOR (POLISH) see HHJ000

BROMOXIL see BNP750
BROMOXYNIL see DDP000
BROMURAL see BNP750
BROMURE de CYANOGEN (FRENCH) see COO500
BROMURE d'ETHYLE see EGV400
BROMURE de METHYLE (FRENCH) see MHR200
BROMURE de VINYLE (FRENCH) see VMP000
BROMURE de XYLYLE (FRENCH) see XRS000
BROMURO di ETILE (ITALIAN) see EIY500
BROMURO di METILE (ITALIAN) see MHR200
BROMUVAN see BNP750
BROMVALERYLUREA see BNP750
BROMVALETONE see BNP750
BROMVALETONUM see BNP750
BROMVALUREA see BNP750
BROMWASSERSTOFF (GERMAN) see HHJ000
BROMYL see BNP750
BRONCHODIL see DNA600
BRONCHOLYSIN see ACH000
BRONCHOSPASMIN see DNA600
BRONKAID MIST see VGP000
BRONKEPHRINE see DMV600
BRONOCOT see BNT250
BRONOPOL see BNT250
BRONOSOL see BNT250
BRONTIN see EAG100
BRONTINA see EAG100
BRONTINE see EAG100
BRONTISOL see EAG100
BRONZE BROMO see BNH500
BRONZE POWDER see CNI000
BRONZE RED RO see CHP500
BRONZE SCARLET see CHP500
BROOM (DUTCH) see BMP000
BROOMMETHAAN (DUTCH) see MHR200
BROOMWATERSTOF (DUTCH) see HHJ000
B ROSE LIQUID see CCK590
BROSERPINE see RDK000
BROVALIN see BNP750
BROVALUREA see BNP750
BROVARIN see BNP750
11460 BROWN see XMA000
BROWN ACETATE see CAL750
BROWN ASBESTOS (DOT) see ARM275
BROWN COPPER OXIDE see CNO000
BROXURIDINE see BNC750
BROXYNIL see DDP000
BRUCINA (ITALIAN) see BOL750
BRUCINE see BOL750
(−)-BRUCINE see BOL750
BRUCINE (DOT) see BOL750
BRUDR see BNC750
BRUFANEUXOL see DOT000
BRUFANIC see IIU000
BRUFEN see IIU000
BRUINSTEEN (DUTCH) see MAS000
BRUMIN see WAT200
BRUSH BUSTER see MEL500
BS 5930 see OJW000
BS 6987 see EAG100
BSB-S 40 see SMQ500
BSB-S-E see SMQ500
BSC-REFINE D see BBS750
B-SELEKTONON M see CIR250
BTKH see BEU500
BTO see BLL750
BUBURONE see IIU000
BUCACID AZURE BLUE see FMU059

BUCACID BRILLIANT SCARLET 3R see FMU080
BUCACID FAST ORANGE G see HGC000
BUCACID METANIL YELLOW see MDM775
BUCACID TARTRAZINE see FAG140
BUCARBAN see BSM000
BUCB see DJF200
BUCCALSONE see HHR000
BUCETIN see HJS850
BUCLIZINE DIHYDROCHLORIDE see BOM250
BUCLODIN see BOM250
BUCROL see BSM000
BUCS see BPJ850
BUCTRIL see DDP000
BUCTRIL INDUSTRIAL see DDP000
BUDOFORM see CHR500
BUDR see BNC750
5-BUDR see BNC750
BU2AE see DDU600
BUENO see MRL750
BUFAPTO METHALOSE see MIF760
BUFEN see ABU500
BUFF-A-COMP see ABG750
BUFON see DKA600
BUFOPTO ZINC SULFATE see ZNA000
BUG MASTER see CBM750
BUKARBAN see BSM000
BUKS see BFW750
BULEN A see PJS750
BULEN A 30 see PJS750
BULKALOID see MIF760
BULPUR see PLC250
BUNSENITE see NDF500
BUNT-CURE see HCC500
BUNT-NO-MORE see HCC500
BUQUINOLATE see BOO632
BURCOL see BSM000
BURESE see CNE750
BUREX (CZECH) see PEE750
BURITAL SODIUM see SOX500
BURNISH GOLD see GIS000
BURNOL see DBX400
BURNTISLAND RED see IHC450
BURNT LIME see CAU500
BURNT SIENNA see IHC450
BURNT UMBER see IHC450
BUROFLAVIN see DBX400
BURONIL see FKI000
BURTOLIN see DMC600
BURTONITE 44 see FPQ000
BURTONITE V-7-E see GLU000
BURTONITE-V-40-E see CCL250
BUSONE see BRF500
BUSTREN see SMQ500
BUSTREN K 500 see SMQ500
BUSTREN K 525-19 see SMQ500
BUSTREN U 825 see SMQ500
BUSTREN U 825E11 see SMQ500
BUSTREN Y 825 see SMQ500
BUSTREN Y 3532 see SMQ500
BUTACIDE see PIX250
BUTACOMPREN see BRF500
BUTACOTE see BRF500
BUTADIEEN (DUTCH) see BOP500
BUTA-1,3-DIEEN (DUTCH) see BOP500
BUTADIEN (POLISH) see BOP500
BUTA-1,3-DIEN (GERMAN) see BOP500
BUTADIENDIOXYD (GERMAN) see BGA750
BUTADIENE see BOP100

1,3-BUTADIENE see BOP500
BUTA-1,3-DIENE see BOP500
α-γ-BUTADIENE see BOP500
BUTADIENE DIEPOXIDE see BGA750
l-BUTADIENE DIEPOXIDE see BOP750
1,3-BUTADIENE DIEPOXIDE see BGA750
BUTADIENE DIMER see CPD750
BUTADIENE DIOXIDE see BGA750
1,3-BUTADIENE, 2-ETHYL- see EGV600
BUTADIENE MONOEPOXIDE see EBJ500
BUTADIENE MONOXIDE see EBJ500
BUTADIENES, inhibited (DOT) see BOP100
1,3-BUTADIENE-STYRENE COPOLYMER see SMR000
BUTADIENE-STYRENE POLYMER see SMR000
1,3-BUTADIENE-STYRENE POLYMER see SMR000
BUTADIENE-STYRENE RESIN see SMR000
BUTADIENE-STYRENE RUBBER (FCC) see SMR000
BUTADIONE see BOT500
BUTAFUME see BPY000
BUTAKON 85-71 see SMR000
BUTAL see BSU250
BUTALAN see BRF500
BUTALDEHYDE see BSU250
BUTALGINA see BRF500
BUTALIDON see BRF500
BUTALYDE see BSU250
BUTAMID see BSQ000
BUTANAL see BSU250
n-BUTANAL (CZECH) see BSU250
BUTANAL OXIME see BSU500
BUTANAMIDE, N-(4-ETHOXYPHENYL)-3-
HYDROXY- see HJS850
BUTANAMIDE, 3-OXO-N-PHENYL-(9CI) see AAY000
1-BUTANAMINE see BPX750
2-BUTANAMINE see BPY000
2-BUTANAMINE, 3-METHYL-(9CI) see AOE200
1,3-BUTANDIOL (GERMAN) see BOS500
BUTANE see BOR500
n-BUTANE (DOT) see BOR500
BUTANECARBOXYLIC ACID see VAQ000
1-BUTANECARBOXYLIC ACID see VAQ000
1,4-BUTANEDIAMINE see BOS000
1,4-BUTANEDIAMINE, 2-METHYL-, POLYMER with
α-HYDRO-ω-HYDROXYPOLY(OXY-1,4-
BUTANEDIYL) and 1,1'-METHYLENEBIS(4-
ISOCYANATOCYCLOHEXANE) see PKN500
1,4-BUTANEDICARBOXAMIDE see AEN000
1,4-BUTANEDICARBOXYLIC ACID see AEN250
BUTANE DIEPOXIDE see BGA750
1,4-BUTANEDINITRILE see SNE000
BUTANEDIOIC ACID see SMY000
BUTANEDIOIC ACID, 2,3-BIS(BENZOYLOXY)-, (R-
(R*,R*))- see DDE300
BUTANEDIOIC ACID, DIETHYL ESTER see SNB000
BUTANEDIOIC ACID, 2,3-DIHYDROXY- see TAF750
BUTANEDIOIC ACID, HYDROXY-(9CI) see MAN000
BUTANEDIOIC ACID MONO(2,2-
DIMETHYLHYDRAZIDE) see DQD400
BUTANEDIOIC ANHYDRIDE see SNC000
1,2-BUTANEDIOL see BOS250
1,3-BUTANEDIOL see BOS500
BUTANE-1,3-DIOL see BOS500
1,4-BUTANEDIOL see BOS750
BUTANE-1,4-DIOL see BOS750
2,3-BUTANEDIOL see BOT000
1,4-BUTANEDIOL BIS(3-AMINOPROPYL) ETHER see
BGU600

1,4-BUTANEDIOL DIMETHANESULPHONATE see
BOT250
1,4-BUTANEDIOL DIMETHYL SULFONATE see
BOT250
1,4-BUTANEDIOL POLYMER with 1,1'-
METHYLENEBIS(4-ISOCYANATOBENZENE) see
PKP000
2,3-BUTANEDIONE see BOT500
BUTANEDIONE (DOT) see BOT500
BUTANE, 1-METHOXY-(9CI) see BRU780
BUTANE MIXTURES (DOT) see BOR500
BUTANEN (DUTCH) see BOR500
BUTANENITRILE see BSX250
n-BUTANENITRILE see BSX250
BUTANENITRILE, 2-HYDROXY-4-(METHYLTHIO)-
see HMR550
BUTANE, 2,2'-OXYBIS-(9CI) see BRH760
BUTANESULFONE see BOU250
BUTANE SULTONE see BOU250
Δ-BUTANE SULTONE see BOU250
1,4-BUTANESULTONE (MAK) see BOU250
5H,6H-6,5A,13A,14-
(1,2,3,4)BUTANETETRACYCLOOCTA(1,2-B:5,6-
B')DINAPHTHALENE see LIV000
tert-BUTANETHIOL see MOS000
2-BUTANETHIOL, 2-METHYL- see MHS550
BUTANETHIOL (OSHA) see BRR900
BUTANE, 1,1,3-TRIETHOXY- see TJL700
BUTANI (ITALIAN) see BOR500
BUTANOIC ACID see BSW000
BUTANOIC ACID, ANHYDRIDE (9CI) see BSW550
BUTANOIC ACID-2-BUTOXY-1-METHYL-2-
OXOETHYL ESTER (9CI) see BQP000
BUTANOIC ACID ETHYL ESTER see EHE000
BUTANOIC ACID, 4-HYDROXY-, MONOLITHIUM
SALT see LHM800
BUTANOIC ACID, 4-((4-
METHOXYBENZOYL)AMINO)- see AOY270
BUTANOIC ACID, 3-METHYLBUTYL ESTER (9CI) see
IHP400
BUTANOIC ACID, 3-OXO-, 2-((1,3-
DIOXOBUTOXY)METHYL)-2-ETHYL-1,3-
PROPANEDIYL ESTER see ELJ600
BUTANOIC ACID PENTYL ESTER see AOG000
BUTANOIC ACID, 1,2,3-PROPANETRIYL ESTER see
TIG750
BUTANOIC ACID, 4,4,4-TRINITRO- see TML100
BUTANOIC ANHYDRIDE see BSW550
BUTAN-1-OL see BPW500
1-BUTANOL see BPW500
BUTAN-2-OL see BPW750
2-BUTANOL see BPW750
n-BUTANOL see BPW500
BUTANOL (DOT) see BPW500
tert-BUTANOL see BPX000
BUTANOL (FRENCH) see BPW500
sec-BUTANOL (DOT) see BPW750
2-BUTANOL ACETATE see BPV000
3-BUTANOLAL see AAH750
BUTANOL (4)-BUTYL-NITROSAMINE see HJQ350
BUTANOLEN (DUTCH) see BPW500
4-BUTANOLIDE see BOV000
BUTANOLO (ITALIAN) see BPW500
2-BUTANOL-3-ONE see ABB500
2,3-BUTANOLONE see ABB500
BUTANOL SECONDAIRE (FRENCH) see BPW750
BUTANOL TERTIAIRE (FRENCH) see BPX000

BUTANONE 2 (FRENCH) see MKA400
2-BUTANONE, O,O'-
(ETHENYLMETHYLSILYLENE)DIOXIME, (E,Z)- see
MQM150
2-BUTANONE, O,O'-
(ETHENYLMETHYLSILYLENE)DIOXIME, (2E,2'Z)-
see MQM150
2-BUTANONE, 4-(4-HYDROXY-3-
METHOXYPHENYL)- see VFP100
2-BUTANONE, 4-(4-HYDROXYPHENYL)- see RBU000
2-BUTANONE, 4-(p-METHOXYPHENYL)-(6CI,7CI,8CI)
see MFF580
2-BUTANONE (OSHA) see MKA400
2-BUTANONE, OXIME see EMU500
2-BUTANONE, 4-PHENYL-, 3-
THIOSEMICARBAZONE see PEO600
1-BUTANONE, 1-(4-PYRIDYL)- see PNV755
2-BUTANONE, SEMICARBAZONE see MKA750
BUTANOX LPT see MKA500
BUTANOX M 50 see MKA500
BUTANOX M 105 see MKA500
BUTANTRONE see BSY400
BUTAPIRAZOL see BRF500
BUTAPYRAZOLE see BRF500
BUTARECBON see BRF500
BUTARTRINA see BRF500
BUTAZATE see BIX000
BUTAZATE 50-D see BIX000
BUTAZINA see BRF500
BUTAZONA see BRF500
BUTAZONE see BRF500
2-BUTENAL see COB250
(E)-2-BUTENAL see COB260
trans-2-BUTENAL see COB260
1-BUTENE see BOW250
BUTENE, DICHLORO- see DEV200
1-BUTENE, 3,4-DICHLORO- see DEV100
(E)-BUTENEDIOIC ACID see FOU000
(Z)-BUTENEDIOIC ACID see MAK900
cis-BUTENEDIOIC ACID see MAK900
trans-BUTENEDIOIC ACID see FOU000
2-BUTENEDIOIC ACID BIS(2-ETHYLHEXYL) ESTER
see DVK600
2-BUTENEDIOIC ACID, DIBUTYL ESTER see DED600
cis-BUTENEDIOIC ANHYDRIDE see MAM000
3-BUTENE-2-ONE see BOY500
1-BUTENE OXIDE see BOX750
1-BUTENE, 2,3,4-TRICHLORO- see TIL360
2-BUTENOIC ACID see COB500
α-BUTENOIC ACID see COB500
2-BUTENOIC ACID, ANHYDRIDE (9CI) see COB900
2-BUTENOIC ACID, 2,3-DICHLOR-4-OXO-, (Z)-(9CI)
see MRU900
2-BUTENOIC ACID-3-(DIETHOXY
PHOSPHINOTHIOYL)ETHYL ESTER see EIB600
2-BUTENOIC ACID, 3-
((DIETHYLPHOSPHINOTHIOYL)OXY)-, ETHYL
ESTER see EIB600
2-BUTENOIC ACID, ETHYL ESTER see EHO200
2-BUTENOIC ACID, ETHYL ESTER, (E)-(9CI) see
COB750
trans-2-BUTENOIC ACID ETHYL ESTER see COB750
2-BUTENOIC ACID, 3-
((ETHYL(PROPYLAMINO)PHOSPHINOTHIOYL)OXY
)-, METHYL ESTER see MKB320
2-BUTENOIC ACID, HEXYL ESTER see HFM600
2-BUTENOIC ACID, 2-METHYL-, (E)-(9CI) see TGA700

2-BUTENOIC ACID, 2-METHYL, 1-ISOPROPYL
ESTER (E)- see IRN100
2-BUTENOL see BOY000
2-BUTEN-1-OL see BOY000
3-BUTENO-β-LACTONE see KFA000
2-BUTEN-1-OL, 3-METHYL-, ACETATE see DOQ350
2-BUTEN-1-OL, 3-METHYL-, BENZOATE see MHU150
2-BUTEN-1-OL, 3-METHYL-, SALICYLATE see PMB600
3-BUTEN-2-ONE see BOY500
3-BUTEN-2-ONE, 4-(2,5,6,6-TETRAMETHYL-2-
CYCLOHEXEN-1-YL)-(9CI) see IGW500
2-BUTENYL ALCOHOL see BOY000
(2-BUTENYLIDENE)ACETIC ACID see SKU000
BUTIDIONA see BRF500
BUTIFOS see BSH250
n-BUTILAMINA (ITALIAN) see BPX750
BUTILCHLOROFOS see DDP000
BUTILE (ACETATI di) (ITALIAN) see BPU750
BUTIL METACRILATO (ITALIAN) see MHU750
BUTINOX see BLL750
BUTIPHOS see BSH250
BUTISERPAZIDE-25 see RDK000
BUTISERPAZIDE-50 see RDK000
BUTISERPINE see RDK000
BUTISULFINA see BSM000
BUTOBEN see DTC800
BUTOCIDE see PIX250
BUTOKSYETYLOWY ALKOHOL (POLISH) see BPJ850
BUTONE see BRF500
BUTOPYRONOXYL see BRT000
2-BUTOSSI-ETANOLO (ITALIAN) see BPJ850
BUTOXIDE see PIX250
2-BUTOXY-AETHANOL (GERMAN) see BPJ850
1-BUTOXYBUTANE see BRH750
t-BUTOXYCARBONYL AZIDE see BQI250
tert-BUTOXYCARBONYL AZIDE (DOT) see BQI250
BUTOXYDIETHYLENE GLYCOL see DJF200
BUTOXYDIGLYCOL see DJF200
BUTOXYETHANOL see BPJ850
2-BUTOXYETHANOL see BPJ850
n-BUTOXYETHANOL see BPJ850
2-BUTOXY-1-ETHANOL see BPJ850
2-BUTOXYETHANOL ACETATE see BPM000
2-BUTOXYETHANOL PHOSPHATE see BPK250
BUTOXYETHENE see VMZ000
2-(2-BUTOXYETHOXY)ETHANOL see DJF200
2-(2-BUTOXYETHOXY)ETHANOL ACETATE see
BQP500
α-(2-(2-BUTOXYETHOXY)ETHOXY)-4,5-
METHYLENEDIOXY-2-PROPYLTOLUENE see PIX250
α-(2-(2-n-BUTOXYETHOXY)-ETHOXY)-4,5-
METHYLENEDIOXY-2-PROPYLTOLUENE see PIX250
5-((2-(2-BUTOXYETHOXY)ETHOXY)METHYL)-6-
PROPYL-1,3-BENZODIOXOLE see PIX250
2-(2-BUTOXYETHOXY)ETHYL ACETATE see BQP500
1-BUTOXY ETHOXY-2-PROPANOL see BPL500
1-(2-BUTOXYETHOXY)-2-PROPANOL see BPL500
2-BUTOXYETHYL ACETATE see BPM000
2-BUTOXYETHYL ESTER ACETIC ACID see BPM000
BUTOXYL see MHV750
BUTOXYPHENYL see BSF750
BUTOZ see BRF500
BUTTER of ANTIMONY see AQC500
BUTTER of ANTIMONY see AQD000
BUTTERCUP YELLOW see PLW500
BUTTERCUP YELLOW see ZFJ100
BUTTERCUP YELLOW see CMK500

BUTTERSAEURE (GERMAN) see BSW000
BUTTER YELLOW see AIC250
BUTTER YELLOW see DOT300
BUTTER of ZINC see ZFA000
BUTYLACETAT (GERMAN) see BPU750
BUTYL ACETATE see BPU750
1-BUTYL ACETATE see BPU750
2-BUTYL ACETATE see BPV000
n-BUTYL ACETATE see BPU750
sec-BUTYL ACETATE see BPV000
sec-BUTYL ACETATE see BPV000
tert-BUTYL ACETATE see BPV100
BUTYLACETATEN (DUTCH) see BPU750
BUTYLACETIC ACID see HEU000
BUTYL ACETOACETATE see BPV250
n-BUTYL ACID PHOSPHATE see ADF250
BUTYL ACRYLATE see BPW100
n-BUTYL ACRYLATE see BPW100
BUTYLACRYLATE, INHIBITED (DOT) see BPW100
BUTYL ADIPATE see AEO750
2-BUTYL ALCOHOL see BPW750
n-BUTYL ALCOHOL see BPW500
BUTYL ALCOHOL (DOT) see BPW500
sec-BUTYL ALCOHOL see BPW750
tert-BUTYL ALCOHOL see BPX000
sec-BUTYL ALCOHOL ACETATE see BPV000
n-BUTYL ALDEHYDE see BSU250
n-BUTYLAMIN (GERMAN) see BPX750
n-BUTYLAMINE see BPX750
sec-BUTYLAMINE see BPY000
tert-BUTYLAMINE see BPY250
BUTYLAMINE, tertiary see BPY250
BUTYLAMINE (OSHA) see BPX750
N-((BUTYLAMINO)CARBONYL)-4-
METHYLBENZENESULFONAMIDE see BSQ000
2-BUTYLAMINOETHANOL see BQC000
tert-BUTYL AMINO ETHYL METHACRYLATE see
BQD250
2-(tert-BUTYLAMINO)ETHYL METHACRYLATE see
BQD250
N-BUTYLANILINE see BQH850
N-(n-BUTYL)ANILINE see BQH850
N-n-BUTYLANILINE (DOT) see BQH850
BUTYLATED HYDROXYANISOLE see BQI000
BUTYLATED HYDROXYMETHYLPHENOL see
BQI050
BUTYLATED HYDROXYTOLUENE see BFW750
tert-BUTYL AZIDOFORMATE see BQI250
N-BUTYLBENZENAMINE (9CI) see BQH850
n-BUTYLBENZENE see BQI750
sec-BUTYLBENZENE see BQJ000
tert-BUTYLBENZENE see BQJ250
BUTYLBENZENEACETATE see BQJ350
α-BUTYLBENZENEMETHANOL see BQJ500
BUTYL BENZOATE see BQK250
n-BUTYL BENZOATE see BQK250
2-BUTYL-3-BENZOFURANYL p-((2-
DIETHYLAMINO)ETHOXY)-m,m-DIIODOPHENYL
KETONE see AJK750
p-tert-BUTYL BENZOIC ACID see BQK500
α-BUTYLBENZYL ALCOHOL see BQJ500
1-(p-tert-BUTYLBENZYL)-4-(p-
CHLORODIPHENYLMETHYL)PIPERAZINE
DIHYDROCHLORIDE see BOM250
1-(p-tert-BUTYLBENZYL-4-p-CHLORO-α-
PHENYLBENZYL)PIPERAZINE
DIHYDROCHLORIDE see BOM250

BUTYL BENZYL PHTHALATE see BEC500
n-BUTYL BENZYL PHTHALATE see BEC500
N-BUTYL-N,N-BIS(HYDROXY ETHYL)AMINE see
BQM000
BUTYL BORATE see THX750
n-BUTYL BORATE see THX750
1-BUTYL BROMIDE see BNR750
BUTYL BROMIDE (DOT) see BMX500
iso-BUTYL BROMIDE see BNR750
sec-BUTYL BROMIDE see BMX750
n-BUTYL BROMIDE (DOT) see BMX500
tert-BUTYL BROMIDE see BQM250
3-sek.BUTYL-5-BROM-6-METHYLURACIL (GERMAN)
see BMM650
n-BUTYL-1-BUTANAMINE see DDT800
n-BUTYL n-BUTANOATE see BQM500
BUTYL-BUTANOL(4)-NITROSAMIN see HJQ350
BUTYL-BUTANOL-NITROSAMINE see HJQ350
S-sec-BUTYL S-tert-BUTYL o-ETHYL
PHOSPHORODITHIOATE see ENF050
n-BUTYL BUTYRATE see BQM500
n-BUTYL n-BUTYRATE see BQM500
BUTYL BUTYRATE (FCC) see BQM500
BUTYL BUTYROLACTATE see BQP000
γ-n-BUTYL-γ-BUTYROLACTONE see OCE000
BUTYL BUTYRYL LACTATE see BQP000
BUTYL CAPROATE see BRK900
BUTYL CARBAMATE see BQP250
1-(BUTYLCARBAMOYL)-2-
BENZIMIDAZOLECARBAMIC ACID, METHYL
ESTER see BAV575
1-(BUTYLCARBAMOYL)-2-BENZIMIDAZOL-
METHYLCARBAMAT (GERMAN) see BAV575
1-(N-BUTYLCARBAMOYL)-2-(METHOXY-
CARBOXAMIDO)-BENZIMIDAZOL (GERMAN) see
BAV575
N'-(BUTYLCARBAMOYL)SULFANILAMIDE see
BSM000
N¹-(BUTYLCARBAMOYL)SULFANILAMIDE see
BSM000
N-BUTYLCARBINOL see AOE000
dl-sec-BUTYLCARBINOL see MHS750
BUTYL CARBITOL see DJF200
BUTYL CARBITOL ACETATE see BQP500
BUTYL CARBITOL 6-PROPYLPIPERONYL ETHER see
PIX250
BUTYL-CARBITYL (6-PROPYLPIPERONYL) ETHER
see PIX250
BUTYL CARBOBUTOXYMETHYL PHTHALATE see
BQP750
1-((tert-BUTYLCARBONYL-4-
CHLOROPHENOXY)METHYL)-1H-1,2,4-TRIAZOLE
see CJO250
N-BUTYL-(3-CARBOXY PROPYL)NITROSAMINE see
BQQ250
4-tert-BUTYLCATECHOL see BSK000
BUTYL CELLOSOLVE see BPJ850
BUTYL CELLOSOLVE ACETATE see BPM000
n-BUTYL CHLORIDE see BQQ750
BUTYL CHLORIDE (DOT) see BQQ750
sec-BUTYL CHLORIDE see CEU250
tert-BUTYL CHLORIDE see BQR000
tert-BUTYL CHLOROACETATE see BQR100
α-BUTYL-α(4-CHLOROPHENYL)-1H-1,2,4-
THIAZOLE-1-PROPANENITRILE see MRW775
tert-BUTYL CHROMATE see BQV000
2-tert-BUTYL-p-CRESOL see BQV750

sec-BUTYLDICHLORARSINE see BQY300
sec-BUTYLDICHLOROARSINE see BQY300
N-BUTYLDIETHANOLAMINE see BQM000
o-BUTYL DIETHYLENE GLYCOL see DJF200
n-BUTYLDIETHYLTIN IODIDE see BRA550
BUTYL DIGLYME see DDW200
BUTYL-3,4-DIHYDRO-2,2-DIMETHYL-4-OXO-2H-PYRAN-6-CARBOXYLATE see BRT000
2-BUTYL-3-(3,5-DIIODO-4-(2-DIETHYLAMINOETHOXY)BENZOYL)BENZOFURAN see AJK750
2-N-BUTYL-3',5'-DIIODO-4'-N-DIETHYLAMINOETHOXY-3-BENZOYLBENZOFURAN see AJK750
BUTYL DIOXITOL see DJF200
4-BUTYL-1,2-DIPHENYL-3,5-DIOXO PYRAZOLIDINE see BRF500
4-BUTYL-1,2-DIPHENYLPYRAZOLIDINE-3,5-DIONE see BRF500
BUTYL DISELENIDE see BRF550
BUTYLE (ACETATE de) (FRENCH) see BPU750
BUTYLENE see BOW250
α-BUTYLENE see BOW250
γ-BUTYLENE see IIC000
BUTYLENEDIAMINE see BOS000
1,4-BUTYLENEDIAMINE see BOS000
1,2-BUTYLENE GLYCOL see BOS250
1,4-BUTYLENE GLYCOL see BOS750
2,3-BUTYLENE GLYCOL see BOT000
β-BUTYLENE GLYCOL see BOS500
1,3-BUTYLENE GLYCOL (FCC) see BOS500
BUTYLENE HYDRATE see BPW750
BUTYLENE OXIDE see TCR750
BUTYLENE OXIDE see BOX750
1,2-BUTYLENE OXIDE see BOX750
1,2-BUTYLENE OXIDE, stabilized (DOT) see BOX750
1,4-BUTYLENE SULFONE see BOU250
BUTYLENIN see IIU000
n-BUTYL ESTER of 3,4-DIHYDRO-2,2-DIMETHYL-4-OXO-2H-PYRAN-6-CARBOXYLIC ACID see BRT000
BUTYLESTER KYSELINY MRAVENCI see BRK000
terc.BUTYLESTER KYSELINY PEROXYBENZOOVE (CZECH) see BSC500
BUTYL ETHANOATE see BPU750
n-BUTYL ETHER see BRH750
BUTYL ETHER (DOT) see BRH750
sec-BUTYL ETHER see BRH760
BUTYL ETHYL ACETALDEHYDE see BRI000
BUTYL ETHYL ACETIC ACID see BRI250
BUTYL ETHYLENE see HFB000
o-BUTYL ETHYLENE GLYCOL see BPJ850
n-BUTYL ETHYL KETONE see EHA600
2-sec.-BUTYLFENOL (CZECH) see BSE000
p-tert-BUTYLFENOL (CZECH) see BSE500
2-sek.BUTYLFENYLESTER KYSELINY METHYLKARBAMINOVE (CZECH) see MOV000
p-terc.BUTYLFENYLESTER KYSELINY SALICYLOVE see BSH100
BUTYL FORMAL see VAG000
tert-BUTYL FORMAMIDE see BRJ750
n-BUTYL FORMATE see BRK000
BUTYL FORMATE (DOT) see BRK000
BUTYL GLYCIDYL ETHER see BRK750
n-BUTYL GLYCIDYL ETHER see BRK750
t-BUTYL GLYCIDYL ETHER see BRK800
BUTYL GLYCOL see BPJ850
BUTYLGLYCOL (FRENCH, GERMAN) see BPJ850

BUTYL HEXANOATE see BRK900
n-BUTYL HEXANOATE see BRK900
BUTYL HYDROGEN PHTHALATE see MRF525
terc.BUTYLHYDROPEROXID (CZECH) see BRM250
tert-BUTYLHYDROPEROXIDE see BRM250
tert-BUTYLHYDROQUINONE see BRM500
BUTYL HYDROXIDE see BPW500
tert-BUTYL HYDROXIDE see BPX000
BUTYLHYDROXYANISOLE see BQI000
tert-BUTYLHYDROXYANISOLE see BQI000
tert-BUTYL-4-HYDROXYANISOLE see BQI000
2(3)-tert-BUTYL-4-HYDROXYANISOLE see BQI000
BUTYL p-HYDROXYBENZOATE see DTC800
n-BUTYL-(4-HYDROXYBUTYL)NITROSAMINE see HJQ350
N-BUTYL-N-(4-HYDROXYBUTYL)NITROSAMINE see HJQ350
BUTYLHYDROXYISOPROPYLARSINE OXIDE see BRQ800
BUTYLHYDROXYOXOSTANNANE see BSL500
BUTYL α-HYDROXYPROPIONATE see BRR600
BUTYLHYDROXYTOLUENE see BFW750
BUTYLIDENE PHTHALIDE see BRQ100
3-BUTYLIDENE PHTHALIDE see BRQ100
n-BUTYLIDENE PHTHALIDE see BRQ100
N-BUTYL-2,2'-IMINODIETHANOL see BQM000
sec-BUTYL IODIDE see IEH000
BUTYL ISOBUTYRATE see BRQ350
n-BUTYL ISOCYANATE see BRQ500
n-BUTYL ISOPENTANOATE see ISX000
BUTYL(ISOPROPYL)ARSINIC ACID see BRQ800
1-BUTYL ISOVALERATE see ISX000
n-BUTYL ISOVALERATE see ISX000
BUTYL ISOVALERIANATE see ISX000
2-tert-BUTYL-p-KRESOL (CZECH) see BQV750
BUTYL LACTATE see BRR600
n-BUTYL LACTATE see BRR600
N-BUTYLMELAMINE see BRR800
BUTYL MERCAPTAN see BRR900
n-BUTYL MERCAPTAN see BRR900
tert-BUTYL MERCAPTAN see MOS000
n-BUTYL MERCAPTAN (ACGIH,DOT) see BRR900
n-BUTYL MESITYL OXIDE OXALATE see BRT000
n-BUTYLMESITYLOXID OXALATE see BRT000
BUTYLMETHACRYLAAT (DUTCH) see MHU750
BUTYL-2-METHACRYLATE see MHU750
N-BUTYL METHACRYLATE see MHU750
BUTYL 3-METHYLBUTYRATE see ISX000
6-tert-BUTYL-3-METHYL-2,4-DINITRO ANISOLE see BRU500
BUTYL METHYL ETHER (DOT) see BRU780
BUTYL METHYL KETONE see HEV000
n-BUTYL METHYL KETONE see HEV000
2-tert-BUTYL-4-METHYLPHENOL see BQV750
1-BUTYL-3-(p-METHYLPHENYLSULFONYL)UREA see BSQ000
BUTYL-2-METHYL-2-PROPENOATE see MHU750
BUTYL NAMATE see SGF500
n-BUTYL NITRITE see BRV500
BUTYL NITRITE (DOT) see BRV500
sec-BUTYL NITRITE see BRV750
4-(BUTYLNITROSAMINO)-1-BUTANOL see HJQ350
4-(n-BUTYLNITROSAMINO)-1-BUTANOL see HJQ350
4-(BUTYLNITROSOAMINO)BUTANOIC ACID see BQQ250
n-BUTYL-N-NITROSO-1-BUTAMINE see BRY500
BUTYLNITROSOHARNSTOFF (GERMAN) see BSA250

n-BUTYLNITROSOUREA see BSA250
1-BUTYL-1-NITROSOUREA see BSA250
N-n-BUTYL-N-NITROSOUREA see BSA250
2-BUTYL-1-OCTANOL see BSA500
2-BUTYLOCTYL ALCOHOL see BSA500
BUTYLOHYDROKSYANIZOL (POLISH) see BQI000
BUTYLONE see NBU000
BUTYLOWY ALKOHOL (POLISH) see BPW500
BUTYL OXITOL see BPJ850
tert-BUTYLOXYCARBONYL AZIDE see BQI250
t-BUTYL PERACETATE see BSC250
tert-BUTYL PERACETATE see BSC250
terc.BUTYLPERBENZOAN (CZECH) see BSC500
t-BUTYL PERBENZOATE see BSC500
tert-BUTYL PERBENZOATE see BSC500
t-BUTYL PEROXYACETATE see BSC250
tert-BUTYL PEROXYACETATE, >76% in solution (DOT) see BSC250
t-BUTYL PEROXY BENZOATE see BSC500
sec-BUTYL PEROXYDICARBONATE see BSD000
t-BUTYL PEROXYPIVALATE see BSD250
tert-BUTYL PEROXYPIVALATE see BSD250
tert-BUTYL PERPIVALATE see BSD250
BUTYLPHEN see BSE500
4-n-BUTYLPHENOL see BSE450
4-t-BUTYLPHENOL see BSE500
4-sec BUTYL PHENOL see BSE250
o-sec-BUTYLPHENOL see BSE000
p-sec-BUTYLPHENOL see BSE250
p-tert-BUTYLPHENOL (MAK) see BSE500
2-(p-tert-BUTYLPHENOXY)CYCLOHEXYL PROPARGYL SULFITE see SOP000
2-(p-tert-BUTYLPHENOXY)CYCLOHEXYL 2-PROPYNYL SULFITE see SOP000
BUTYLPHENOXYISOPROPYL CHLOROETHYL SULFITE see SOP500
2-(p-BUTYLPHENOXY)ISOPROPYL 2-CHLOROETHYL SULFITE see SOP500
2-(4-tert-BUTYLPHENOXY)ISOPROPYL-2-CHLOROETHYL SULFITE see SOP500
2-(p-tert-BUTYLPHENOXY)ISOPROPYL 2'-CHLOROETHYL SULPHITE see SOP500
2-(p-tert-BUTYLPHENOXY)-1-METHYLETHYL 2-CHLOROETHYL ESTER of SULPHUROUS ACID see SOP500
2-(p-BUTYLPHENOXY)-1-METHYLETHYL 2-CHLOROETHYL SULFITE see SOP500
2-(p-tert-BUTYLPHENOXY)-1-METHYLETHYL-2-CHLOROETHYL SULFITE ESTER see SOP500
2-(p-tert-BUTYLPHENOXY)-1-METHYLETHYL 2'-CHLOROETHYL SULPHITE see SOP500
2-(p-tert-BUTYLPHENOXY)-1-METHYLETHYL SULPHITE of 2-CHLOROETHANOL see SOP500
1-(p-tert-BUTYLPHENOXY)-2-PROPANOL-2-CHLOROETHYL SULFITE see SOP500
BUTYL PHENYL ACETATE see BBA000
BUTYL PHENYLACETATE see BQJ350
n-BUTYL PHENYLACETATE see BQJ350
BUTYL PHENYL ETHER see BSF750
o-sec-BUTYLPHENYL METHYLCARBAMATE see MOV000
2-sec-BUTYLPHENYL N-METHYLCARBAMATE see MOV000
p-tert-BUTYLPHENYL SALICYLATE see BSH100
BUTYL PHOSPHORIC ACID see ADF250
BUTYL PHOSPHOROTRITHIOATE see BSH250
n-BUTYL PHTHALATE (DOT) see DEH200

BUTYL PHTHALATE BUTYL GLYCOLATE see BQP750
BUTYL PHTHALYL BUTYL GLYCOLATE see BQP750
5-BUTYL PICOLINIC ACID see BSI000
BUTYL PROPANOATE see BSJ500
BUTYL-2-PROPENOATE see BPW100
BUTYL PROPIONATE see BSJ500
n-BUTYL PROPIONATE see BSJ500
5-BUTYL-2-PYRIDINECARBOXYLIC ACID see BSI000
BUTYL 4-PYRIDYL KETONE see VBA100
BUTYLPYRIN see BRF500
4-tert-BUTYLPYROCATECHOL see BSK000
p-tert-BUTYLPYROCATECHOL see BSK000
4-tert-BUTYLPYROKATECHIN (CZECH) see BSK000
n-BUTYL RHODANATE see BSN500
BUTYL STANNOIC ACID see BSL500
N-BUTYLSULFANILYLUREA see BSM000
1-BUTYL-3-SULFANILYL UREA see BSM000
n-BUTYL THIOCYANATE see BSN500
(BUTYLTHIO)TRIOCTYLSTANNANE see BSO200
(BUTYLTHIO)TRIPROPYLSTANNANE see TMY850
n-BUTYL THIOUREA see BSO500
BUTYL TITANATE see BSP250
p-tert-BUTYLTOLUENE see BSP500
n-BUTYL-N'-p-TOLUENESULFONYLUREA see BSQ000
1-BUTYL-3-(p-TOLYL SULFONYL)UREA see BSQ000
1-BUTYL-3-TOSYLUREA see BSQ000
N-n-BUTYL-N'-TOSYLUREA see BSQ000
BUTYLTRICHLOROSILANE see BSR000
BUTYL TRICHLORO STANNANE see BSR250
tert-BUTYL TRIMETHYLPEROXYACETATE see BSD250
5-tert-BUTYL-2,4,6-TRINITROXYLENE see TML750
5-tert-BUTYL-2,4,6-TRINITRO-m-XYLENE (DOT) see TML750
BUTYLTRIS(2-ETHYLHEXYLOXYCARBONYLMETHYLTHIO)STANNANE see BSS000
BUTYLTRIS(ISOOCTYLOXYCARBONYLMETHYLTHIO)STANNANE see BSS000
BUTYL 10-UNDECENOATE see BSS100
BUTYL UNDECYLENATE see BSS100
N-BUTYLUREA see BSS250
1-BUTYLUREA and SODIUM NITRITE (2:1) see BSS500
BUTYL VINYL ETHER see VMZ000
BUTYL VINYL ETHER (inhibited) see VMZ000
BUTYL ZIMATE see BIX000
BUTYL ZIRAM see BIX000
1-BUTYNE see EFS500
2-BUTYNE see COC500
2-BUTYNE, 1-CHLORO-4-MERCAPTO-, S-ESTER WITH DIPHENYLPHOSPHINOTHIOATE see CEV840
2-BUTYNE-1,4-DIOL see BST500
1,4-BUTYNEDIOL (DOT) see BST500
BUTYNOIC ACID, 3-PHENYL-2-PROPENYL ESTER see CMQ800
BUTYNORATE see DDV600
1,1'-(2-BUTYNYLENEDIOXY)BIS(3-CHLORO)-2-PROPANOL) see BST900
BUTYRAL see BSU250
BUTYRALDEHYD (GERMAN) see BSU250
n-BUTYRALDEHYDE see BSU250
BUTYRALDEHYDE (CZECH) see BSU250
BUTYRALDEHYDE, 3-ETHOXY-, DIETHYL ACETAL see TJL700
n-BUTYRALDEHYDE OXIME see BSU500
N-BUTYRALDOXIME see BSU500

BUTYRALDOXIME (DOT) see BSU500
BUTYRANHYDRID see BSW550
BUTYRANILIDE, 4'-ETHOXY-3-HYDROXY- see HJS850
BUTYRHODANID (GERMAN) see BSN500
n-BUTYRIC ACID see BSW000
BUTYRIC ACID ANHYDRIDE see BSW550
n-BUTYRIC ACID ANHYDRIDE see BSW550
BUTYRIC ACID, CINNAMYL ESTER see CMQ800
BUTYRIC ACID, α-α-DIMETHYLPHENETHYL ESTER see BEL850
BUTYRIC ACID ESTER with BUTYL LACTATE see BQP000
BUTYRIC ACID, HEXYL ESTER see HFM700
BUTYRIC ACID ISOBUTYL ESTER see BSW500
BUTYRIC ACID LACTONE see BOV000
BUTYRIC ACID NITRILE see BSX250
BUTYRIC ACID TRIESTER with GLYCERIN see TIG750
BUTYRIC ACID, 4,4,4-TRINITRO- see TML100
BUTYRIC ACID, VINYL ESTER see VNF000
BUTYRIC ALDEHYDE see BSU250
BUTYRIC ANHYDRIDE see BSW550
n-BUTYRIC ANHYDRIDE see BSW550
BUTYRIC ETHER see EHE000
BUTYRIC or NORMAL PRIMARY BUTYL ALCOHOL see BPW500
α-BUTYROLACTONE see BOV000
β-BUTYROLACTONE see BSX000
γ-BUTYROLACTONE (FCC) see BOV000
BUTYRONE (DOT) see DWT600
BUTYRONITRILE see BSX250
BUTYRONITRILE (DOT) see BSX250
BUTYRONITRILE, 4-(DIETHOXYMETHYLSILYL)- see COR500
BUTYRONITRILE, 4-(TRIETHOXYSILYL)- see COR800
BUTYROPHENONE, 4-(4-(p-CHLOROPHENYL)-4-HYDROXYPIPERIDINO)-4'-(DIMETHYLAMINO)- see CKA580
BUTYRYL DITHRANOL see BSY400
10-BUTYRYLDITHRANOL see BSY400
10-BUTYRYL DITHRANOL see BSY400
BUTYRYL LACTONE see BOV000
BUTYRYL OXIDE see BSW550
4-BUTYRYLPYRIDINE see PNV755
BUTYRYL TRIGLYCERIDE see TIG750
BUVETZONE see BRF500
2-n-BUYTLAMINOETHANOL see BQC000
BUZON see BRF500
BUZULFAN see BOT250
BVU see BNP750
B-W see SJU000
BW 56-158 see ZVJ000
BW 57-322 see ASB250
BW-21-Z see AHJ750
BY 935 see NDY550
BYLADOCE see VSZ000
2,2'-BYPYRIDIN see BGO500
B-3-Zh see CMS212
BZ 55 see BSQ000
BZCF see BEF500
BZF-60 see BDS000
BZI see BCB750
BZT see BEN000
C-56 see HCE500
C6E3 see HFT550
C 709 see DGQ875

C 1414 see MRH209
C 2018 see CCU250
C 2059 see DUK800
C 3172 see TNK250
C-5068 see HGP495
C 5968 see HGP495
C 6379 see DBA800
C 6866 see BIE500
C 8514 see CJJ250
C-10015 see CDS750
C-12669 see MIW500
C 13963 see DXN830
C-13963 see DXN830
8057HC see DSQ000
CA see BMW250
CA 3 see EHP700
CA 33 see CAM200
CA 8-3A see EHP700
CA 80-15 see CCU250
CA 70203 see TKL100
CA 3 (ADHESIVE) see EHP700
CABADON M see VSZ000
CAB-O-GRIP see AHE250
CAB-O-GRIP II see SCH002
CABRONAL see EOK000
CAB-O-SIL see SCH002
CAB-O-SPERSE see SCH002
CACHALOT L-50 see DXV600
CACHALOT C-50 see HCP000
C-8 ACID see OCY000
CACODYLATE de SODIUM (FRENCH) see HKC500
CACODYLIC ACID (DOT) see HKC000
CACODYLIC ACID SODIUM SALT see HKC500
CACP see PJD000
CADCO 0115 see SMQ500
CADDY see CAE250
CADET see BDS000
CADMINATE see CAI750
CADMIUM see CAD000
CADMIUM(II) ACETATE see CAD250
CADMIUM ACETATE (DOT) see CAD250
CADMIUM ACETATE DIHYDRATE see CAD275
CADMIUM(II) ACETATE, MONOHYDRATE see CAE800
CADMIUM AMIDE see CAD325
CADMIUM AZIDE see CAD350
CADMIUM BARIUM STEARATE see BAI800
CADMIUM BIS(N-AMYLDITHIOCARBAMATE) see CAD550
CADMIUM BIS(2-ETHYLHEXYL) PHOSPHITE see CAD500
CADMIUM, BIS(1-HYDROXY-2(1H)-PYRIDINETHIONATO)- see CAI350
CADMIUM BIS(PENTYLDITHIOCARBAMATE) see CAD550
CADMIUM, BIS(PENTYLDITHIOCARBAMATO)- see CAD550
CADMIUM, BIS(SALICYLATO)- see CAI400
CADMIUM BROMIDE see CAD600
CADMIUM CAPRYLATE see CAD750
CADMIUM CARBONATE see CAD800
CADMIUM CATION see CAG600
CADMIUM CDTA see CAD900
CADMIUM CHLORATE see CAE000
CADMIUM CHLORIDE see CAE250
CADMIUM CHLORIDE, DIHYDRATE see CAE375
CADMIUM CHLORIDE, HYDRATE (2:5) see CAE425
CADMIUM CHLORIDE, MONOHYDRATE see CAE500

CADMIUM COMPOUNDS see CAE750
CADMIUM DIACETATE see CAD250
CADMIUM DIACETATE DIHYDRATE see CAD275
CADMIUM DIACETATE MONOHYDRATE see CAE800
CADMIUM DIAMIDE see CAD325
CADMIUM DIAMYL DITHIOCARBAMATE see CAD550
CADMIUM DIAZIDE see CAD350
CADMIUM DIBROMIDE see CAD600
CADMIUM DICHLORIDE see CAE250
CADMIUM DICYANIDE see CAF500
CADMIUM DIETHYL DITHIOCARBAMATE see BJB500
CADMIUM DILAURATE see CAG775
CADMIUM DINITRATE see CAH000
CADMIUM DODECANOATE see CAG775
CADMIUM(II) EDTA COMPLEX see CAF750
CADMIUM FLUOBORATE see CAG000
CADMIUM FLUORIDE see CAG250
CADMIUM FLUOROBORATE see CAG000
CADMIUM FLUOROSILICATE see CAG500
CADMIUM FLUORURE (FRENCH) see CAG250
CADMIUM FLUOSILICATE see CAG500
CADMIUM FUME see CAH750
CADMIUM GOLDEN see CMS212
CADMIUM GOLDEN 366 see CAJ750
CADMIUM HEXAFLUOROSILICATE (7CI) see CAG500
CADMIUM ION see CAG600
CADMIUM, ION (Cd^{2+}) see CAG600
CADMIUM LACTATE see CAG750
CADMIUM LAURATE see CAG775
CADMIUM LEMON see CMS212
CADMIUM LEMON YELLOW 527 see CAJ750
CADMIUM MONOCARBONATE see CAD800
CADMIUM MONOSULFIDE see CAJ750
CADMIUM MONOXIDE see CAH500
CADMIUM NITRATE see CAH000
CADMIUM(II) NITRATE see CAH000
CADMIUM(II) NITRATE TETRAHYDRATE (1:2:4) see CAH250
CADMIUM NITRIDE see TIH000
CADMIUM OCTADECANOATE see OAT000
CADMIUM ORANGE see CAJ750
CADMIUM OXIDE see CAH500
CADMIUM OXIDE FUME see CAH750
CADMIUM PHOSPHATE see CAI000
CADMIUM PHOSPHIDE see CAI125
CADMIUM PRIMROSE see CMS212
CADMIUM PRIMROSE 819 see CAJ750
CADMIUM PROPIONATE see CAI250
CADMIUM PT see CAI350
CADMIUM 2-PYRIDINETHIONE see CAI350
CADMIUM SALICYLATE see CAI400
CADMIUM SELENIDE see CAI500
CADMIUM SILICON FLUORIDE see CAG500
CADMIUM STEARATE see OAT000
CADMIUM(II) STEARATE see OAT000
CADMIUM SUCCINATE see CAI750
CADMIUM SULFATE see CAJ000
CADMIUM SULFATE (1:1) see CAJ000
CADMIUM SULFATE (1:1) HYDRATE (3:8) see CAJ250
CADMIUM SULFATE OCTAHYDRATE see CAJ250
CADMIUM SULFATE TETRAHYDRATE see CAJ500
CADMIUM SULFIDE see CAJ750
CADMIUM SULFIDE (AMORPHOUS) see CAJ760
CADMIUM SULFIDE mixed with ZINC SULFIDE (1:1) see CMS212

CADMIUM SULFIDE mixed with ZINC SULFIDE (5:95) see CAJ770
CADMIUM SULFIDE mixed with ZINC SULFIDE (8:92) see CAJ772
CADMIUM SULPHATE see CAJ000
CADMIUM SULPHIDE see CAJ750
CADMIUM TETRAFLUOROBORATE (7CI) see CAG000
CADMIUM THERMOVACUUM AEROSOL see CAK000
CADMIUM-THIONEINE see CAK250
CADMIUM salt of 2,4-5-TRIBROMOIMIDAZOLE see THV500
CADMIUM YELLOW see CAJ750
CADMIUM YELLOW 000 see CAJ750
CADMIUM YELLOW 892 see CAJ750
CADMIUM YELLOW 10G CONC. see CAJ750
CADMIUM YELLOW CONC. DEEP see CAJ750
CADMIUM YELLOW CONC. GOLDEN see CAJ750
CADMIUM YELLOW CONC. LEMON see CAJ750
CADMIUM YELLOW CONC. PRIMROSE see CAJ750
CADMIUM YELLOW OZ DARK see CAJ750
CADMIUM YELLOW PRIMROSE 47-4100 see CAJ750
CADMIUM(2Cd^{2+}) see CAG600
CADMOPUR GOLDEN YELLOW N see CAJ750
CADMOPUR YELLOW see CAJ750
CADOX see MKA500
CADOX see BDS000
CADOX TBH see BRM250
CADOX TS see BIX750
CADOX TS 40,50 see BIX750
CADPX PS see BHM750
CAERULEIN see CAK285
CAESIUM HYDROXIDE, solid (UN 2682) (DOT) see CDD750
CAESIUM HYDROXIDE, solution (UN 2681) (DOT) see CDD750
CAF see CDP250
CAF see CEA750
CAFFEIN see CAK500
CAFFEINE see CAK500
CAFRON see BCA000
CAID see CJJ000
CAIROX see PLP000
CAJEPUTENE see MCC250
CAKE ALUM see AHG750
CALAMINE (spray) see ZKA000
CALAMUS OIL see OGK000
CALAR see CAM000
CALCIA see CAU500
CALCICOL see CAS750
CALCID see CAQ500
CALCIFEROL see VSZ100
CALCIFERON 2 see VSZ100
CALCINED BARYTA see BAO000
CALCINED BRUCITE see MAH500
CALCINED DIATOMITE see SCJ000
CALCINED MAGNESIA see MAH500
CALCINED MAGNESITE see MAH500
CALCINED SODA see SIN500
CALCIOFON see CAS750
CALCIPUR see CAS750
CALCITE see CAO000
CALCIUM ACETATE see CAL750
CALCIUM ACETYLIDE see CAN750
CALCIUM ACID METHANEARSONATE see CAM000
CALCIUM ACID METHYL ARSONATE see CAM000
CALCIUM ALGINATE see CAM200
CALCIUMARSENAT see ARB750
CALCIUM ARSENATE see CAM222

CALCIUM TRISODIUM DIETHYLENE TRIAMINE PENTAACETATE see CAY500
CALCIUM TRISODIUM DTPA see CAY500
CALCIUM TRISODIUM PENTETATE see CAY500
CALCIUM TRISODIUM SALT of DIETHYLENETRIAMINEPENTAACETIC ACID see CAY500
CALCOCID BLUE EG see FMU059
CALCOCID BRILLIANT SCARLET 3RN see FMU080
CALCOCID FAST LIGHT ORANGE 2G see HGC000
CALCOCID 2RIL see FMU070
CALCOCID URANINE B4315 see FEW000
CALCOCID VIOLET 4BNS see FAG120
CALCOCID YELLOW MCG see FAG140
CALCOCID YELLOW MXXX see MDM775
CALCOCID YELLOW XX see FAG140
CALCODUR BROWN BRL see CMO750
CALCODUR RESIN FAST BLUE see CMN750
CALCOGAS ORANGE NC see PEJ500
C 10 ALCOHOL see DAI600
CALCOLAKE SCARLET 2R see FMU070
CALCOMINE BLACK see AQP000
CALCOMINE BLACK EXL see AQP000
CALCOMINE BLUE 2B see CMO000
CALCO OIL ORANGE 7078 see PEJ500
CALCO OIL RED D see SBC500
CALCO OIL SCARLET BL see XRA000
CALCOSYN PINK B see AKE250
CALCOSYN YELLOW GC see AAQ250
CALCOSYN YELLOW GCN see AAQ250
CALCOTONE RED see IHC450
CALCOTONE WHITE T see TGG760
CALCOZINE BLUE ZF see BJI250
CALCOZINE BRILLIANT GREEN G see BAY750
CALCOZINE CHRYSOIDINE Y see PEK000
CALCOZINE MAGENTA N see RMK020
CALCOZINE ORANGE YS see PEK000
CALCOZINE RED 6G see RGW000
CALCOZINE RHODAMINE 6GX see RGW000
CALCOZINE VIOLET C see AOR500
CALCOZINE VIOLET 6BN see AOR500
CALCOZINE YELLOW OX see IBA000
CALCYAN see CAQ500
CALCYANIDE see CAQ500
C-8 ALDEHYDE see OCO000
C-9 ALDEHYDE see NMW500
C-10 ALDEHYDE see DAG000
C-16 ALDEHYDE see ENC000
C-12 ALDEHYDE, LAURIC see DXT000
CALDOPENTAMINE see TED600
CALEDON GREY M see CMU475
CALFLO E see CAW850
CALGINATE see CAM200
CALGLUCOL see CAS750
CALGLUCON see CAS750
CALGON see SHM500
CALICO YELLOW see MRN500
CALIDRIA RG 100 see ARM268
CALIDRIA RG 144 see ARM268
CALIDRIA RG 600 see ARM268
C12-14-tert-ALKYL AMINES see CAY710
CALMADIN see MQU750
CALMATHION see MAK700
CALMAX see MQU750
CALMINAL see EOK000
CALMIREN see MQU750
CALMOCITENE see DCK759
CALMODEN see MDQ250

CALMONAL see HGC500
CALMORE see TEH500
CALMOREX see TEH500
CALMOTIN see BNP750
CALOCAIN see MNQ000
CALOCHLOR see MCY475
CALO-CLOR see CAY950
CALOGREEN see MCW000
CALOMEL see MCW000
CALOMELANO (ITALIAN) see MCW000
CALOSAN see MCW000
CALOXOL CP2 see CAU500
CALOXOL W3 see CAU500
CALPANATE see CAU750
CALPLUS see CAO750
CALPOL see HIM000
CALSIL see CAW850
CALSMIN see DLY000
CALSOFT F-90 see DXW200
CALSOFT LAS 99 see LBU100
CALSOL see EIV000
CALSTAR see CAX350
CALTAC see CAO750
CALVIT see CAT225
CALX see CAU500
CALXYL see CAU500
CAM see CDP250
CAMA see CAM000
CAMCOLIT see LGZ000
CAMITE see BMW250
CAMOMILE OIL, ENGLISH TYPE (FCC) see CDH750
CAMOMILE OIL GERMAN see CDH500
CAMPAPRIM A 1544 see AMY050
CAMPBELLINE OIL ORANGE see PEJ500
CAMPHANE, 2-HYDROXY- see BMD000
2-CAMPHANOL see BMD000
1-2-CAMPHANOL see NCQ820
2-CAMPHANONE see CBA750
CAMPHECHLOR see CDV100
CAMPHENE see CBA500
CAMPHOCHLOR see CDV100
CAMPHOCLOR see CDV100
CAMPHOFENE HUILEUX see CDV100
CAMPHOGEN see CQI000
CAMPHOL see BMD000
CAMPHOR see CBA750
CAMPHOR-natural see CBA750
CAMPHOR, synthetic (ACGIH, DOT) see CBA750
CAMPHOR OIL see CBB500
CAMPHOR OIL, RECTIFIED see CBB500
CAMPHOR OIL WHITE see CBB500
CAMPHOR OIL YELLOW see CBB500
CAMPHOR TAR see NAJ500
CAMPHOZONE see DJS200
CAMPILIT see COO500
CAMPOSAN see CDS125
CAMPOVITON 6 see PPK500
CANACERT SUNSET YELLOW FCF see FAG150
CANACERT TARTRAZINE see FAG140
CANADOL see PCT250
CANANGA OIL see CBC100
CANARY CHROME YELLOW 40-2250 see LCR000
CANCARB see CBT750
CANDAMIDE see LGZ000
CANDASEPTIC see CFD990
CANDEPTIN see LFF000
CANDEREL see ARN825
CANDEX see NOH500

CANDEX see ARQ725
CANDIDA LIPOLYTICA see CBC400
CANDIMON see LFF000
CANDIO-HERMAL see NOH500
CANDLE SCARLET 2B see SBC500
CANDLE SCARLET B see SBC500
CANDLE SCARLET G see SBC500
CANE SUGAR see SNH000
CANOGARD see DGP900
CANQUIL-400 see MQU750
CANTABILINE see MKP500
CANTHA see CBE800
CANTHAXANTHIN see CBE800
CANTREX see KAL000
CAO 1 see BFW750
CAO 3 see BFW750
CAP see CBF250
CAP see CDP250
CAP see CEA750
CAPAROL see BKL250
CAPISTEN see BDU500
CAPLENAL see ZVJ000
CAPMUL see PKG000
CAPMUL see PKL030
CAPMUL POE-O see PKL100
CAPNOIDINE see CBF550
(−)-CAPNOIDINE see CBF550
l-CAPNOIDINE see CBF550
CAPORIT see HOV500
CAP-P see CDP700
CAP-PALMITATE see CDP700
CAPRALDEHYDE see DAG000
CAPRALENSE see AJD000
CAPRAMOL see AJD000
CAPRAN 80 see PJY500
CAPRIC ACID see DAH400
n-CAPRIC ACID see DAH400
CAPRIC ACID ETHYL ESTER see EHE500
CAPRIC ALCOHOL see DAI600
CAPRIC ALDEHYDE see DAG000
CAPRIN see ADA725
CAPRINALDEHYDE see DAG000
CAPRINIC ACID see DAH400
CAPRINIC ALCOHOL see DAI600
CAPRINIC ALDEHYDE see DAG000
CAPROALDEHYDE see HEM000
CAPROALDEHYDE, β,γ,γ-TRIMETHYL-,
THIOSEMICARBAZONE see TLN125
CAPROAMIDE see HEM500
CAPROAMIDE POLYMER see PJY500
CAPROCID see AJD000
CAPROIC ACID see HEU000
n-CAPROIC ACID see HEU000
CAPROIC ALDEHYDE see HEM000
CAPROKOL see HFV500
CAPROLACTAM see CBF700
6-CAPROLACTAM see CBF700
ω-CAPROLACTAM (MAK) see CBF700
CAPROLACTAM OLIGOMER see PJY500
ε-CAPROLACTAM POLYMERE (GERMAN) see PJY500
CAPROLATTAME (FRENCH) see CBF700
CAPROLIN see CBM750
CAPROLISIN see AJD000
CAPRON see PJY500
CAPRONALDEHYDE see HEM000
CAPRONAMIDE see HEM500
CAPRONIC ACID see HEU000

CAPRONITRILE see HER500
CAPROYL ALCOHOL see HFJ500
n-CAPROYLALDEHYDE see HEM000
CAPRYL ALCOHOL see OEI000
CAPRYLAMINE see OEK010
CAPRYLDINITROPHENYL CROTONATE see AQT500
2-CAPRYL-4,6-DINITROPHENYL CROTONATE see
AQT500
CAPRYLIC ACID see OCY000
n-CAPRYLIC ACID see OCY000
CAPRYLIC ACID TRIGLYCERIDE see TMO000
CAPRYLIC ALCOHOL see OEI000
CAPRYLYL ACETATE see OEG000
CAPRYNIC ACID see DAH400
CAPSANTHIN see CBF760
CAPSEBON see CAJ750
CAPSINE see DUS700
CAPTAF see CBG000
CAPTAFOL see CBF800
CAPTAN see CBG000
CAPTANCAPTENEET 26,538 see CBG000
CAPTANE see CBG000
CAPTAN-STREPTOMYCIN 7.5-0.1 POTATO SEED
PIECE PROTECTANT see CBG000
CAPTAX see BDF000
CAPTEX see CBG000
CAPTOFOL see CBF800
CAP-O-TRAN see MQU750
CAPUT MORTUUM see IHC450
CAP-WAKO see CCU050
CARADATE 30 see MJP400
CARAMEL see CBG125
CARAMEL COLOR see CBG125
CARASTAY see CCL250
CARASTAY G see CCL250
CARAWAY OIL see CBG500
CARBACRYL see ADX500
CARBADOX (USDA) see FOI000
CARBAMALDEHYDE see FMY000
CARBAMATE see FAS000
CARBAMATE de l'ETHINYLCYCLOHEXANOL
(FRENCH) see EEH000
CARBAMATES see CBH750
CARBAMAZINE see DIW000
CARBAMIC ACID, ACETYLMETHYL-, 2,2-
DIMETHYL-1,3-BENZODIOXOL-4-YL ESTER see
BCJ005
CARBAMIC ACID, ACETYLMETHYL-, 2,3-
(ISOPROPYLIDENEDIOXY)PHENYL ESTER see
BCJ005
CARBAMIC ACID, ((p-
BROMOPHENYL)THIO)METHYL-, 2,3-DIHYDRO-2,2-
DIMETHYL-7-BENZOFURANYL ESTER see DRL100
CARBAMIC ACID, BUTYL ESTER see BQP250
CARBAMIC ACID, (BUTYLTHIO)METHYL-, 2,3-
DIHYDRO-2,2-DIMETHYL-7-BENZOFURANYL
ESTER see DLH830
CARBAMIC ACID, DIMETHYL-(9CI) see DQY950
CARBAMIC ACID, DIMETHYLDITHIO-,
ANHYDROSULFIDE see BJL600
CARBAMIC ACID, DIMETHYL-, 2-(1,3-DITHIOLAN-2-
YL)PHENYL ESTER see DXN830
CARBAMIC ACID, DIMETHYL-, o-(1,3-DITHIOLAN-2-
YL)PHENYL ESTER see DXN830
CARBAMIC ACID, DIMETHYLDITHIO-, ZINC SALT
(2:1) see BJK500

CARBAMIC ACID, ((((1,1-DIMETHYLETHOXY)SULFINYL)METHYL)-, 2-(1-METHYLETHOXY)PHENYL ESTER see INE062

CARBAMIC ACID, DIMETHYL-, ester with (m-HYDROXYPHENYL)TRIMETHYLAMMONIUM BROMIDE see POD000

CARBAMIC ACID, ((2,6-DIMETHYLPHENOXY)SULFINYL)METHYL-, 2,3-DIHYDRO-2,2-DIMETHYL-7-BENZOFURANYL ESTER see XWS100

CARBAMIC ACID, (((((1,4-DITHIAN-2-YLIDENEAMINO)OXY)CARBONYL)METHYLAMINO)THIO)METHYL-,ETHYL ESTER see MLL100

CARBAMIC ACID, ESTER WITH SALICYLALDEHYDE DIMETHYL ACETAL see SAG100

CARBAMIC ACID, ETHYLENEBIS(DITHIO)-, MANGANESE SALT see MAS500

CARBAMIC ACID, ETHYL ESTER see UVA000

CARBAMIC ACID, ((4-FLUOROPHENYL)THIO)METHYL-, 2,3-DIHYDRO-2,2-DIMETHYL-7-BENZOFURANYL ESTER see DLH850

CARBAMIC ACID HYDRAZIDE see HGU000

CARBAMIC ACID, ISOPROPYL ESTER see IOJ000

CARBAMIC ACID, N-METHYL-, 3-DIETHYLARSINOPHENYL ESTER, METHIODIDE see MID900

CARBAMIC ACID-1-METHYLETHYL ESTER see IOJ000

CARBAMIC ACID, METHYL-, o-ISOPROPYLPHENYL ESTER see MIA250

CARBAMIC ACID, METHYL-, 2-(1-METHYLETHYL)PHENYL ESTER see MIA250

CARBAMIC ACID, METHYL-, 3-METHYLPHENYL ESTER (9CI) see MIB750

CARBAMIC ACID, METHYL((4-NITROPHENYL)THIO)-, 2,3-DIHYDRO-2,2-DIMETHYL-7-BENZOFURANYL ESTER see DLH870

CARBAMIC ACID, METHYL-, SPIRO(1,3-BENZODIOXOLE-2,1'-CYCLOHEXAN)-4-YL ESTER see SLD550

CARBAMIC ACID, METHYL((1,1,2,2-TETRACHLOROETHYL)THIO)-, 2,3-DIHYDRO-2,2-DIMETHYL-7-BENZOFURANYL ESTER see DLH890

CARBAMIC ACID, METHYL-, 3-TOLYL ESTER see MIB750

CARBAMIC ACID-3-PHENYLPROPYL ESTER see PGA750

CARBAMIC ACID, 1-PHENYL-2-PROPYNYL ESTER see PGE000

CARBAMIC ACID, PROPYL ESTER see PNG250

CARBAMIC ACID, VINYL ESTER see VNK000

CARBAMIDAL see DJS200

CARBAMIDE see USS000

CARBAMIDE PEROXIDE see HIB500

CARBAMIDE RESIN see USS000

p-CARBAMIDOBENZENEARSONIC ACID see CBJ000

4-CARBAMIDOPHENYL BIS(CARBOXYMETHYLTHIO)ARSENITE see CBI250

p-CARBAMIDOPHENYL-BIS(2-CARBOXYPHENYLMERCAPTO)ARSINE see TFD750

4-CARBAMIDOPHENYL BIS(o-CARBOXYPHENYLTHIO)ARSENITE see TFD750

p-CARBAMIDOPHENYL-DI(1'-CARBOXYPHENYL-2')THIOARSENITE see TFD750

CARBAMIDSAEURE-AETHYLESTER (GERMAN) see UVA000

CARBAMIMIDIC ACID see USS000

CARBAMINE see CBM750

CARBAMINOPHENYL-p-ARSONIC ACID see CBJ000

p-CARBAMINO PHENYL ARSONIC ACID see CBJ000

CARBAMMATO di FENILETINILCARBINOLO see PGE000

CARBAMONITRILE see COH500

(p-CARBAMOYLAMINO)PHENYLARSINOBIS(2-THIO-ACETIC ACID) see CBI250

N-CARBAMOYLARSANILIC ACID see CBJ000

1-CARBAMOYLFORMIMIDIC ACID see OLO000

CARBAMOYLHYDRAZINE see HGU000

p-(((CARBAMOYLMETHYL)AMINO)-BENZENEARSONIC ACID see CBJ750

N-(CARBAMOYLMETHYL)ARSANILIC ACID see CBJ750

1-CARBAMOYLOXY-3-PHENYLPROPANE see PGA750

4-CARBAMYLAMINOPHENYLARSONIC ACID see CBJ000

N-CARBAMYL ARSANILIC ACID see CBJ000

CARBAMYL CHLORIDE, N,N-DIMETHYL- see DQY950

CARBAMYLHYDRAZINE see HGU000

CARBAMYLHYDRAZINE HYDROCHLORIDE see SBW500

2-CARBAMYL PYRAZINE see POL500

CARBANIL see PFK250

CARBANILIC ACID ISOPROPYL ESTER see CBM000

CARBANOCHLORIDIC ACID, NAPHTHYL ESTER see NBH200

CARBANOLATE see CBM500

CARBARSONE (USDA) see CBJ000

CARBARYL see CBM750

CARBARYL (ACGIH,DOT,OSHA) see CBM750

CARBASONE see CBJ000

CARBATOX see CBM750

CARBATOX-60 see CBM750

CARBATOX-75 see CBM750

CARBAVUR see CBM750

CARBAX see BIO750

CARBAZAMIDE see HGU000

CARBAZIC ACID HYDRAZIDE see CBS500

CARBAZIDE see CBS500

CARBAZINC see BJK500

CARBAZOLE see CBN000

9H-CARBAZOLE see CBN000

CARBAZOLE, 3-(p-HYDROXYANILINO)- see CBN100

4-(3-CARBAZOLYLAMINO)PHENOL see CBN100

CARBAZONE, DIPHENYLTHIO- see DWN200

CARBAZOTIC ACID see PID000

CARBENDAZIM see MHC750

CARBENDAZIME see MHC750

CARBENDAZIM and SODIUM NITRITE (5:1) see CBN375

CARBENDAZOL see MHC750

CARBENDAZOLE see MHC750

CARBENDAZYM see MHC750

CARBETHOXYACETIC ESTER see EMA500

CARBETHOXY MALATHION see MAK700

p-CARBETHOXYPHENOL see HJL000

CARBETOVUR see MAK700

CARBETOX see MAK700

CARBICRON see DGQ875

CARBIDE 6-12 see EKV000

CARBIDE BLACK E see AQP000

CARBILAZINE see DIW000

CARBIMIDE see COH500

CARBINAMINE see MGC250

CARBINOL see MGB150

CARBINOXAMIDE MALEATE see TAI500
CARBITOL see CBR000
CARBITOL see DJD600
CARBITOL ACETATE see CBQ750
CARBITOL CELLOSOLVE see CBR000
CARBITOL SOLVENT see CBR000
CARBOBENZOXY CHLORIDE see BEF500
CARBOBENZOXYLGLYCINE see CBR125
CARBOBENZOYL GLYCINE see CBR125
N-CARBOBENZOYLGLYCINE see CBR125
CARBOBENZYLOXY CHLORIDE see BEF500
CARBOBENZYLOXYGLYCINE see CBR125
N-CARBOBENZYLOXYGLYCINE see CBR125
2-CARBO-n-BUTOXY-6,6-DIMETHYL-5,6-DIHYDRO-1,4-PYRONE see BRT000
CARBOCAINE see SBB000
CARBOCAINE HYDROCHLORIDE see CBR250
CARBOCALCITONIN see CBR300
CARBOCISTEINE see CBR675
CARBOCIT see CBR675
CARBO-CORT see CMY800
CARBOCYSTEINE see CBR675
CARBO D see BJE550
CARBODIHYDRAZIDE see CBS500
CARBODIIMIDE, BIS(2,6-DIISOPROPYLPHENYL)- see BJE550
CARBODIS see CBT750
CARBOFENOTHION (DUTCH) see TNP250
CARBOFOS see MAK700
CARBOFRAX M see SCQ000
CARBOFURAN see CBS275
CARBOGEN (8CI) see CBV250
CARBOHYDRASE see AOM125
CARBOHYDRASE, ASPERGILLUS see CBS400
CARBOHYDRASE and PROTEASE, mixed see CBS410
CARBOHYDRAZIDE see CBS500
CARBOLAC see CBT750
CARBOLAC 1 see CBT750
CARBOLIC ACID see PDN750
CARBOLITH see LGZ000
CARBOLON see SCQ000
CARBOLSAEURE (GERMAN) see PDN750
CARBOMAL see BNK000
CARBOMATE see CBM750
CARBOMET see CBT750
CARBOMETHENE see KEU000
2-CARBOMETHOXYANILINE see APJ250
o-CARBOMETHOXYANILINE see APJ250
2-β-CARBOMETHOXY-3-β-BENZOXYTROPANE see CNE750
4'-CARBOMETHOXY-2,3'-DIMETHYLAZOBENZENE see CBS750
4'-CARBOMETHOXY-2,3'-DIMETHYLAZOBENZOL see CBS750
α-2-CARBOMETHOXY-1-METHYLVINYL DIMETHYL PHOSPHATE see MQR750
2-CARBOMETHOXY-1-METHYLVINYL-N-PROPYL ETHYLPHOSPHONOAMIDOTHIOATE see MKB320
2-CARBOMETHOXY-1-PROPEN-2-YL DIMETHYL PHOSPHATE see MQR750
CARBOMYCIN see CBT250
CARBOMYCIN A see CBT250
CARBON see CBT500
CARBON-12 see CBT500
CARBON, activated (DOT) see CBT500
CARBON, animal or vegetable origin (DOT) see CBT500
CARBONA see CBY000

CARBON, ACTIVATED see CDI000
CARBONAZIDIC ACID, 1,1-DIMETHYLETHYL ESTER see BQI250
CARBON BICHLORIDE see PCF275
CARBON BISULFIDE (DOT) see CBV500
CARBON BISULPHIDE see CBV500
CARBON BLACK see CBT750
CARBON BLACK, ACETYLENE see CBT750
CARBON BLACK BV and V see CBT750
CARBON BLACK, CHANNEL see CBT750
CARBON BLACK, FURNACE see CBT750
CARBON BLACK, LAMP see CBT750
CARBON BLACK, THERMAL see CBT750
CARBON BROMIDE see CBX750
CARBON CHLORIDE see CBY000
CARBON CHLOROSULFIDE see TFN500
CARBON D see DXD200
CARBON DICHLORIDE see PCF275
CARBON DIFLUORIDE OXIDE see CCA500
CARBON DIOXIDE see CBU250
CARBON DIOXIDE, mixture with NITROGEN OXIDE (N$_2$O) see CBV000
CARBON DIOXIDE mixed with NITROUS OXIDE see CBV000
CARBON DIOXIDE–NITROUS OXIDE mixture (DOT) see CBV000
CARBON DIOXIDE mixed with OXYGEN see CBV250
CARBON DIOXIDE-OXYGEN mixture (DOT) see CBV250
CARBON DIOXIDE, solid (UN 1845) (DOT) see CBU250
CARBON DIOXIDE, refrigerated liquid (UN 2187) (DOT) see CBU250
CARBON DISULFIDE see CBV500
CARBON DISULPHIDE see CBV500
CARBONE (OXYCHLORURE de) (FRENCH) see PGX000
CARBONE (OXYDE de) (FRENCH) see CBW750
CARBONE (SUFURE de) (FRENCH) see CBV500
CARBON FERROCHROMIUM see FBD000
CARBON FLUORIDE see CBY250
CARBON FLUORIDE OXIDE see CCA500
CARBON HEXACHLORIDE see HCI000
CARBON HYDRIDE NITRIDE (CHN) see HHS000
CARBONIC ACID, AMMONIUM SALT see ANE000
CARBONIC ACID ANHYDRIDE see CBU250
CARBONIC ACID, BARIUM SALT (1:1) see BAJ250
CARBONIC ACID BERYLLIUM SALT (1:1) see BFP750
CARBONIC ACID, CADMIUM SALT see CAD800
CARBONIC ACID, CALCIUM SALT (1:1) see CAO000
CARBONIC ACID, CHROMIUM SALT see CMJ100
CARBONIC ACID, CYCLIC 3-CHLOROPROPYLENE ESTER see CBW400
CARBONIC ACID, DIAMMONIUM SALT see ANE000
CARBONIC ACID, DICESIUM SALT see CDC750
CARBONIC ACID DIHYDRAZIDE see CBS500
CARBONIC ACID, DILITHIUM SALT see LGZ000
CARBONIC ACID, DIPHENYL ESTER see DVZ000
CARBONIC ACID, DIPOTASSIUM SALT see PLA000
CARBONIC ACID, DISODIUM SALT see SFO000
CARBONIC ACID, DITHALLIUM(1+) SALT see TEJ000
CARBONIC ACID, DITHIODI-, o,o'-DIMETHYL S,S'-(2,3-QUINOXALINEDIYL) ESTER see BHL800
CARBONIC ACID GAS see CBU250
CARBONIC ACID, LEAD(2+) SALT (1:1) see LCP000
CARBONIC ACID LITHIUM SALT see LGZ000
CARBONIC ACID METHYL-4-(o-TOLYLAZO)-o-TOLYL ESTER see CBS750

CARBONIC ACID, MONOAMMONIUM SALT see ANB250

CARBONIC ACID MONOSODIUM SALT see SFC500

CARBONIC ACID, NICKEL SALT (1:1) see NCY500

CARBONIC ACID, NICKEL SALT, BASIC see NCY600

CARBONIC ACID, THIO-, o,o'-DIMETHYL S,S'-2,3-QUINOXALINEDIYL ESTER see BHL800

CARBONIC ACID, THIO-, o-METHYL ESTER, S,S-DIESTER WITH 2,3-QUINOXALINEDITHIOL see BHL800

CARBONIC ACID, ZINC SALT (1:1) see ZEJ050

CARBONIC ANHYDRASE INHIBITOR NO. 6063 see AAI250

CARBONIC ANHYDRIDE see CBU250

CARBONIC DIFLUORIDE see CCA500

CARBONIC DIHYDRAZIDE see CBS500

CARBONIC OXIDE see CBW750

CARBONIMIDIC DICHLORIDE, PHENYL-(9CI) see PFJ400

CARBONIMIDIC DIHYDRAZIDE, BIS((4-CHLOROPHENYL)METHYLENE)- see RLK890

4,4'-CARBONIMIDOYLBIS(N,N-DIMETHYLBENZENAMINE) see IBB000

4,4'-CARBONIMIDOYLBIS(N,N-DIMETHYLBENZENAMINE)MONOHYDROCHLORIDE see IBA000

CARBONIO (OSSICLORURO di) (ITALIAN) see PGX000

CARBONIO (OSSIDO di) (ITALIAN) see CBW750

CARBONIO (SOLFURO di) (ITALIAN) see CBV500

CARBON MONOXIDE see CBW750

CARBON MONOXIDE (ACGIH,OSHA) see CBW750

CARBON MONOXIDE, refrigerated liquid (cryogenic liquid) (NA 9202) (DOT) see CBW750

CARBON MONOXIDE (UN 1016) (DOT) see CBW750

CARBON NITRIDE see COO000

CARBON NITRIDE ION (CN^{1-}) see COI500

CARBONOCHLORIDE ACID, 1,2-ETHANEDIYL ESTER see EIQ000

CARBONOCHLORIDE ACID-1-METHYL ESTER see IOL000

CARBONOCHLORIDIC ACID, 9H-FLUOREN-9-YLMETHYL ESTER see FEI100

CARBONOCHLORIDIC ACID, 1-NAPHTHALENYL ESTER see NBH200

CARBONOCHLORIDIC ACID, OXYDI-2,1-ETHANEDIYL ESTER see OPO000

CARBONOCHLORIDIC ACID PHENYL ESTER see CBX109

CARBONOCHLORIDIC ACID, PROPYL ESTER see PNH000

CARBONOCHLORIDIC ACID TRICHLOROMETHYL ESTER see TIR920

CARBONOCHLORIDIC ACID, 2,4,6-TRICHLOROPHENYL ESTER (9CI) see TIY800

CARBONOHYDRAZIDE see CBS500

CARBONOHYDRAZONIC DIHYDRAZIDE, MONONITRATE (9CI) see THN800

CARBON OIL see BBL250

CARBONOTHIOIC ACID, S,S'-2,3-QUINOXALINEDIYL o,o'-DIMETHYL ESTER see BHL800

CARBONOTHIOIC DICHLORIDE (9CI) see TFN500

CARBON OXIDE see CBU250

CARBON OXIDE (CO) see CBW750

CARBON OXIDE SULFIDE see CCC000

CARBON OXYCHLORIDE see PGX000

CARBON OXYFLUORIDE see CCA500

CARBON OXYSULFIDE see CCC000

CARBON S see SGM500

CARBON SILICIDE see SCQ000

CARBON SULFIDE see CBV500

CARBON SULPHIDE (DOT) see CBV500

CARBON TET see CBY000

CARBON TETRABROMIDE see CBX750

CARBON TETRACHLORIDE see CBY000

CARBON TETRAFLUORIDE see CBY250

CARBON TRIFLUORIDE see CBY750

CARBON TRIFLUORIDE see CBY750

CARBONYLBIS(AZIRIDINE) see BJP450

CARBONYLBIS(1-AZIRIDINE) see BJP450

CARBONYLCHLORID (GERMAN) see PGX000

CARBONYL CHLORIDE see PGX000

CARBONYL CHLORIDE, THIO- see TFN500

CARBONYL DIAMIDE see USS000

CARBONYLDIAMINE see USS000

CARBONYL DIFLUORIDE see CCA500

CARBONYLDIHYDRAZINE see CBS500

CARBONYL FLUORIDE see CCA500

CARBONYL IRON see IGK800

CARBONYLS see CCB609

CARBONYL SULFIDE see CCC000

CARBONYL SULFIDE-^{32}S see CCC000

CARBOPHOS see MAK700

CARBOPOL EXTRA see CBT500

CARBOPOL M see CBT500

CARBOPOL Z 4 see CBT500

CARBOPOL Z EXTRA see CBT500

CARBORAFFIN see CDI000

CARBORAFINE see CDI000

CARBORUNDEUM see SCQ000

CARBORUNDUM see SCQ000

CARBOSIEVE see CBT500

CARBOSORBIT R see CBT500

CARBOSPOL see AGJ250

CARBOTHIALDIN see DSB200

CARBOTHIALDINE see DSB200

CARBOWAX see PJT000

CARBOWAX see PJT200

CARBOWAX 1000 see PJT250

CARBOWAX 1500 see PJT500

CARBOWAX 4000 see PJT750

CARBOWAX 6000 see PJU000

5-CARBOXANILIDO-2,3-DIHYDRO-6-METHYL-1,4-OXATHIIN see CCC500

CARBOXIDE see CAT225

CARBOXINE see CCC500

CARBOXIN (USDA) see CCC500

CARBOXYANILINE see API500

2-CARBOXYANILINE see API500

4-CARBOXYANILINE see AIH600

o-CARBOXYANILINE see API500

p-CARBOXYANILINE see AIH600

3-(α-CARBOXY-o-ANISAMIDO)-2-METHOXYPROPYL HYDROXYMERCURY, MONOSODIUM SALT see SIH500

CARBOXYBENZENE see BCL750

2-CARBOXY-4'-(DIMETHYLAMINO)AZOBENZENE see CCE500

2-CARBOXYDIPHENYLAMINE see PEG500

2-CARBOXYDIPHENYLARSINOUS ACID see PFI600

CARBOXYETHANE see PMU750

CARBOXYETHYLGERMANIUM SESQUIOXIDE see CCF125

2-CARBOXYETHYLGERMASESQUIOXANE see CCF125

3-CARBOXY-1-ETHYL-7-METHYL-1,8-NAPHTHIDIN-4-ONE see EID000

2'-CARBOXY-2-HYDROXY-4-METHOXYBENZOPHENONE(o-(2-HYDROXY-p-ANISOYL)BENZOIC ACID) see HLS500

3-CARBOXY-5-HYDROXY-1-p-SULFOPHENYL-4-p-SULFOPHENYLAZOPYRAZOLE TRISODIUM SALT see FAG140

(4-(CARBOXY METHOXY)-3-CHLOROPHENYL)(5,5-DIETHYL-2,4,6(1H,3H,5H)-PYRIMIDINETRIONATO)-O²-MERCURY, MONOSODIUM SALT see CCG500

CARBOXYMETHYL CELLULOSE see SFO500

CARBOXYMETHYL CELLULOSE, SODIUM see SFO500

CARBOXYMETHYL CELLULOSE, SODIUM SALT see SFO500

l-CARBOXYMETHYLCYSTEINE see CBR675

S-(CARBOXYMETHYL)CYSTEINE see CBR675

3,3'-(CARBOXYMETHYLENE)BIS(4-HYDROXYCOUMARIN) ETHYL ESTER see BKA000

((CARBOXYMETHYLIMINO)BIS(ETHYLENENITRILO))TETRAACETIC ACID see DJG800

N-(γ-CARBOXYMETHYLMERCAPTOMERCURI-β-METHOXY)PROPYLCAMPHORAMIC ACID DISODIUM SALT see TFK270

1-CARBOXYMETHYL-1-METHYLPYRROLIDINIUM IODIDE METHYL ESTER see CCH300

(CARBOXYMETHYLTHIO)ACETIC ACID see MCM750

3-(CARBOXYMETHYLTHIO)ALANINE see CBR675

l-3-((CARBOXYMETHYL)THIO)ALANINE see CBR675

3-CARBOXYPHENOL see HJI100

o-CARBOXYPHENYL ACETATE see ADA725

p-CARBOXYPHENYLAMINE see AIH600

p-CARBOXY PHENYLARSENOXIDE see CCI550

(p-CARBOXYPHENYL)CHLOROMERCURY see CHU500

9-(o-CARBOXYPHENYL)-6-HYDROXY-3-ISOXANTHENONE see FEV000

9-o-CARBOXYPHENYL-6-HYDROXY-3-ISOXANTHONE, DISODIUM SALT see FEW000

9-(o-CARBOXYPHENYL)-6-HYDROXY-3H-XANTHEN-3-ONE see FEV000

((o-CARBOXYPHENYL)THIO)ETHYLMERCURY SODIUM SALT see MDI000

4-CARBOXYPHTHALIC ANHYDRIDE see TKV000

3-CARBOXYPYRIDINE see NCQ900

4-CARBOXYRESORCINOL see HOE600

CARBRITAL see NBU000

CARBUTAMID see BSM000

CARBUTAMIDE see BSM000

CARDAMINE see DJS200

CARDAMIST see NGY000

CARDAMON see CCJ625

CARDAMON OIL see CCJ625

CARD-20(22)-ENOLIDE, 16-(ACETYLOXY)-3-((6-DEOXY-3-o-METHYL-α-l-ALTROPYRANOSYL)OXY)-14-HYDROXY-, (3-β,5-β,16-β)- see VCA100

CARD-20(22)-ENOLIDE, 3-((6-DEOXY-3-o-METHYL-α-l-MANNOPYRANOSYL)OXY)-14-HYDROXY-, (3-β, 5-β)- see SKS150

CARD-20(22)-ENOLIDE, 19-OXO-3,14,15-TRIHYDROXY-, (3-β,5-α,15-β)- see AFS800

5-α-CARD-20(22)-ENOLIDE, 3-β,14,15-β-TRIHYDROXY-19-OXO- see AFS800

CARDIAGEN see DJS200

CARDIAMID see DJS200

CARDIAMINA see DJS200

CARDIAMINE see DJS200

CARDIDIGIN see DKL800

CARDIGIN see DKL800

CARDIMON see DJS200

CARDIO see CCK125

CARDIO-GREEN see CCK000

CARDIOMONE see AOA125

CARDIOSERPIN see RDK000

CARDIS see CCK125

CARDITIVO see RDK000

CARDITOXIN see DKL800

CARDIVIX see EID200

CARDOVERINA see PAH250

CARFENE see ASH500

CARFIMAT see PGE000

CARFIMATE see PGE000

CARICIDE see DIW000

CARICIDE see DIW200

CARINA see PKQ059

CARINEX GP see SMQ500

CARINEX HR see SMQ500

CARINEX HRM see SMQ500

CARINEX SB 59 see SMQ500

CARINEX SB 61 see SMQ500

CARINEX SL 273 see SMQ500

CARINEX TGX/MF see SMQ500

CARITROL see DIW200

CARLONA 900 see PJS750

CARLONA 58-030 see PJS750

CARLONA 18020 FA see PJS750

CARLONA P see PMP500

CARLONA PXB see PJS750

CARMETHOSE see SFO500

CARMINAPH see PEJ500

CARMINE see CCK590

CARMINIC ACID see CCK590

CARMINOMYCIN see KBU000

CARMOISIN (GERMAN) see HJF500

CARMOISINE ALUMINUM LAKE see HJF500

CARMOISINE SUPRA see HJF500

CARMUBRIS see BIF750

CARMUSTIN see BIF750

CARMUSTINE see BIF750

CARNOSINE see CCK665

l-CARNOSINE see CCK665

CAROB BEAN GUM see LIA000

CAROB FLOUR see LIA000

CAROFAM see EID200

CAROID see PAG500

CAROLYSINE see BIE500

CAROTENE see CCK685

β-CAROTENE see CCK685

β-CAROTENE-4,4'-DIONE see CBE800

CARPENE see DXX400

CARPERITIDE see HGL680

CARPOLIN see CBM750

CARRAGEEN see CCL250

CARRAGEENAN, CALCIUM(II) SALT see CAO250

CARRAGEENAN, DEGRADED see CCL500

CARRAGEENAN (FCC) see CCL250

CARRAGEENAN GUM see CCL250

CARRAGHEANIN see CCL250

CARRAGHEEN see CCL250

CARRAGHEENAN see CCL250

CARREL-DAKIN SOLUTION see SHU500
CARROT SEED OIL see CCL750
CARRSERP see RDK000
CARRTIME see BBK500
CARSONOL SLS see SIB600
CARSONON N-9 see PKF000
CARSONON PEG-4000 see PJT750
CARSOQUAT SDQ-25 see DTC600
CARSTAB DLTDP see TFD500
CARTOSE see GFG000
CARVACROL see CCM000
CARVANIL see CCK125
CARVASIN see CCK125
CARVIL see MOV000
(−)-CARVONE see CCM120
(+)-CARVONE see CCM100
1-CARVONE see CCM120
d-CARVONE see CCM100
l(−)-CARVONE see CCM120
(R)-CARVONE see CCM120
(S)-CARVONE see CCM100
d(+)-CARVONE see CCM100
(S)-(+)-CARVONE see CCM100
CARYLDERM see CBM750
CARYOLYSIN see BIE250
CARYOLYSINE see BIE500
CARYOLYSINE HYDROCHLORIDE see BIE500
CARYOPHYLLENE see CCN000
CARYOPHYLLENE EPOXIDE see CCN100
β-CARYOPHYLLENE EPOXIDE see CCN100
β-CARYOPHYLLENE (FCC) see CCN000
CARYOPHYLLENE OXIDE see CCN100
(−)-CARYOPHYLLENE OXIDE see CCN100
β-CARYOPHYLLENE OXIDE see CCN100
CARYOPHYLLIC ACID see EQR500
CARZOL see CJJ250
CARZOL SP see DSO200
CARZONAL see FLZ050
CARZONAL see FMM000
CASALIS GREEN see CMJ900
CASCARILLA OIL see CCO500
CASEIN and CASEINATE SALTS (FCC) see SFQ000
CASEIN-SODIUM see SFQ000
CASEIN, SODIUM COMPLEX see SFQ000
CASEINS, SODIUM COMPLEXES see SFQ000
CASPAN see MDD750
CASSEL BROWN see MAT500
CASSEL GREEN see MAT250
CASSIA ALDEHYDE see CMP969
CASSIA OIL see CCO750
CASSIAR AK see ARM268
CASSIC ACID see RHZ700
CASTANEA SATIVA MILL TANNIN see CDM250
CASTOR BEAN see CCP000
CASTOR BEANS (DOT) see CCP000
CASTOR FLAKE (DOT) see CCP000
CASTOR MEAL (DOT) see CCP000
CASTOR OIL see CCP250
CASTOR OIL AROMATIC see CCP250
CASTOR OIL PLANT see CCP000
CASTOR POMACE (DOT) see CCP000
CAT (herbicide) see BJP000
CATACIDE see DIW000
CATALIN CAO-3 see BFW750
CATALOID see SCH002
CATALYTIC-DEWAXED HEAVY NAPHTHENIC
DISTILLATE see MQV776

CATALYTIC-DEWAXED HEAVY PARAFFINIC
DISTILLATE see MQV778
CATALYTIC-DEWAXED LIGHT NAPHTHENIC
DISTILLATE see MQV777
CATALYTIC-DEWAXED LIGHT PARAFFINIC
DISTILLATE see MQV779
CATECHIN see CCP850
CATECHOL see CCP850
CATECHOL DIETHYL ETHER see CCP900
CATHOMYCIN SODIUM see NOB000
CATHOMYCIN SODIUM LYOVAC see NOB000
CATILAN see CDP250
CAUSOIN see DKQ000
CAUSTIC POTASH see PLJ500
CAUSTIC POTASH, dry, solid, flake, bead, or granular
(DOT) see PLJ500
CAUSTIC POTASH, liquid or solution (DOT) see PLJ500
CAUSTIC SODA see SHS000
CAUSTIC SODA, dry (DOT) see SHS000
CAUSTIC SODA, bead (DOT) see SHS000
CAUSTIC SODA, flake (DOT) see SHS000
CAUSTIC SODA, solid (DOT) see SHS000
CAUSTIC SODA, solution see SHS500
CAUSTIC SODA, liquid (DOT) see SHS000
CAUSTIC SODA, granular (DOT) see SHS000
CAUSTIC SODA, solution (DOT) see SHS000
CAVALITE BRILLIANT BLUE R see BMM500
CAV-ECOL see WBJ700
CAVI-TROL see SHF500
CAVONLY see TDA500
4-CB see TBO700
CB 1348 see CDO500
CB 1506 see CHC750
1522 CB see ABH500
C.B. 2041 see BOT250
CB 2511 see BKM500
CB 2562 see TFU500
CB 3025 see PED750
CB-3307 see BHT750
CB-4564 see CQC500
CB 4564 see CQC650
CB-4835 see BIA250
CBC 806495 see TFQ750
CBC 906288 see TND250
CBD 90 see TIQ750
CBS see CPI250
(CBZ)GLY see CBR125
CC 914 see CBI250
CC 11511 see DYC800
"C" CARRIE see CNE750
CCC see CAQ250
CCC PLANT GROWTH REGULANT see CMF400
CCH see HOV500
CCHO see CPD000
C.C. No. 914 see CBI250
CCNU see CGV250
CCS 203 see BPW500
CCS 301 see BPW750
CCUCOL see ASB250
CD see TGD000
CD 2 see LFK000
CD 68 see CDR750
CDA 101 see CNI000
CDA 102 see CNI000
CDA 110 see CNI000
CDA 122 see CNI000
CDAA see CFK000
CDAAT see CFK000

CDB 60 see DGN200
CDB 63 see SGG500
CDBM see CFK500
CDDP see PJD000
CDEC see CDO250
CDM see CJJ250
CDP see LFK000
CDT see BJP000
CE see TEG250
CEBETOX see MIW250
CEBITATE see ARN125
CEBROGEN see GFO050
CEBRUM see MDQ250
CECALGINE TBV see SEH000
CECARBON see CBT500
CE CE CE CE see CMF400
CECENU see CGV250
CECIL see CNE750
CECOLENE see TIO750
CEDAD see BCA000
CEDAR LEAF OIL see CCQ500
CEDOCARD see CCK125
CEDRAMBER see CCR525
CEDRANE, 8,9-EPOXIDE see CCR510
CEDR-8-ENE EPOXIDE see CCR510
CEDROL FORMATE see CCR524
CEDROL METHYL ETHER see CCR525
CEDRO OIL see LEI000
CEDRYL FORMATE see CCR524
CEE see PMB000
CEE DEE see HCQ500
CEENU see CGV250
CEEPRYN see CCX000
CEEPRYN CHLORIDE see CCX000
CEFAPIRIN (GERMAN) see CCX500
CEFATIN see OAV000
CEFRACYCLINE TABLETS see TBX250
CEFTIOFUR see CCS575
CEGLUTION see LGZ000
CEKIURON see DXQ500
CEKUBARYL see CBM750
CEKU C.B. see HCC500
CEKUDAZIM see MHC750
CEKUDIFOL see BIO750
CEKUFON see TIQ250
CEKUGIB see GEM000
CEKUMETA see TDW500
CEKUMETHION see MNH000
CEKUSAN see BJP000
CEKUSAN see DGP900
CEKUSIL see ABU500
CEKUSIL UNIVERSAL A see MEO750
CEKUSIL UNIVERSAL C see MEP250
CEKUTHOATE see DSP400
CEKUTROTHION see DSQ000
CEKUZINA-S see BJP000
CEKUZINA-T see ARQ725
CELA 50 see TKL100
CELA S-2225 see EGV500
CELA A-36 see DAE600
CELACOL M see MIF760
CELACOL M20 see MIF760
CELACOL M 20P see MIF760
CELACOL M450 see MIF760
CELACOL MM see MIF760
CELACOL MM 10P see MIF760
CELANEX see BBQ500
CELANOL DOS 75 see DJL000

CELANTHRENE PURE BLUE BRS see TBG700
CELANTHRENE RED 3BN see AKE250
CELA W 524 see TKL100
CELEX see CCU250
CELGARD 2500 see PMP500
CELINHOL -A see OAV000
CELITE see DCJ800
CELLACETATE see CCU050
CELLAPRET see MIF760
CELLEX MX see CCU150
CELLIDRIN see ZVJ000
CELLITAZOL B see DCJ200
CELLITON BLUE G see TBG700
CELLITON DISCHARGE YELLOW GL see AAQ250
CELLITON FAST PINK BA-CF see AKE250
CELLITON FAST PINK BN see AKE250
CELLITON FAST VIOLET B see DBY700
CELLITON FAST VIOLET BA-CF see DBY700
CELLITON FAST YELLOW G see AAQ250
CELLITON FAST YELLOW GA see AAQ250
CELLITON FAST YELLOW GA-CF see AAQ250
CELLITON ORANGE R see AKP750
CELLITON VIOLET B see DBY700
CELLITON YELLOW G see AAQ250
CELLMIC S see OPE000
CELLOFAS see SFO500
CELLOFOR (CZECH) see DWO800
CELLOGEL C see SFO500
CELLOGRAN see MIF760
CELLOIDIN see CCU250
CELLON see TBQ100
CELLOPHANE see CCT250
CELLOSOLVE (DOT) see EES350
CELLOSOLVE ACETATE (DOT) see EES400
CELLOSOLVE SOLVENT see EES350
CELLOTHYL see MIF760
CELLPRO see SFO500
CELLUFIX FF 100 see SFO500
CELLUFLEX see CGO500
CELLUFLEX 179C see TNP500
CELLUFLEX DOP see DVL600
CELLUFLEX DPB see DEH200
CELLUFLEX TPP see TMT750
CELLUGEL see SFO500
CELLULOSE 248 see CCU150
α-CELLULOSE see CCU150
CELLULOSE, ACETATE HYDROGEN 1,2-
BENZENEDICARBOXYLATE (9CI) see CCU050
CELLULOSE ACETATE MONOPHTHALATE see
CCU050
CELLULOSE, ACETATE PHTHALATE see CCU050
CELLULOSE ACETOPHTHALATE see CCU050
CELLULOSE ACETYLPHTHALATE see CCU050
CELLULOSE (ACGIH,OSHA) see CCU150
CELLULOSE CRYSTALLINE see CCU150
CELLULOSE ETHYL see EHG100
CELLULOSE ETHYLATE see EHG100
CELLULOSE GEL see CCU100
CELLULOSE GLYCOLIC ACID, SODIUM SALT see
SFO500
CELLULOSE GUM see SFO500
CELLULOSE METHYL see MIF760
CELLULOSE METHYLATE see MIF760
CELLULOSE, MICROCRYSTALLINE see CCU100
CELLULOSE NITRATE see CCU250
CELLULOSE, NITRATE (9CI) see CCU250
CELLULOSE, POWDERED see CCU150

CELLULOSE SODIUM GLYCOLATE see SFO500
CELLULOSE TETRANITRATE see CCU250
CELLUMETH see MIF760
CELLUPHOS 4 see TIA250
CELLU-QUIN see BLC250
CELMER see MEP250
CELMER see ABU500
CELMIDE see EIY500
CELMONE see NAK500
CELOGEN BSH see BBS300
CELOGEN OT see OPE000
CELON A see EIX000
CELON ATH see EIX000
CELON E see EIV000
CELON H see EIV000
CELON IS see EIV000
CELPHIDE see AHE750
CELPHOS see PGY000
CELPHOS see AHE750
CELTHIGN see MAK700
CELUFI see CCU150
CELUTATE PINK B see AKE250
CELUTATE PINK BN see AKE250
CELUTATE PINK BY see AKE250
CELUTATE YELLOW GH see AAQ250
CEMEDINE 3000RP see EHP700
CEMEDINE 3000RP TYPE-II see EHP700
CEMEDINE 3000RS see EHP700
CEMEDINE 3000RS TYPE-II see EHP700
CEMENT (rubber) see CCW250
CEMENT BLACK see MAS000
CEMENT, PORTLAND see PKS750
CEMENT, RUBBER see CCW250
CENITRON OB see OPE000
CENOLATE see ARN125
CENSTIM see DLH630
CENSTIN see DLH630
CENTEDEIN see MNQ000
CENTIMIDE see HCQ500
CENTRALGIN see DAM700
CENTRALINE BLUE 3B see CMO250
CENTREDIN see MNQ000
CENTURY 1240 see SLK000
CENTURY CD FATTY ACID see OHU000
CEP see CDS125
2-CEPA see CDS125
CEPACILINA see BFC750
CEPACILLINA see BFC750
CEPACOL CHLORIDE see CCX000
CEPAVERIN see PAH250
CEPHA see CDS125
CEPHAPIRIN see CCX500
CEPHA 10LS see CDS125
CEPHROL see CMT250
CEPO see CCU150
CEPO CFM see CCU150
CEPO S 20 see CCU150
CEPO S 40 see CCU150
CEPRIM see CCX000
CERAPHYL 368 see OFG100
CERAPHYL 375 see ISC550
CERASINE YELLOW GG see DOT300
CERASIN RED see OHI200
CERASINROT see OHI200
CERASYNT see HKJ000
CERASYNT 1000-D see OAV000
CERASYNT S see OAV000
CERASYNT SD see OAV000

CERASYNT SE see OAV000
CERASYNT WM see OAV000
CERAZOL (suspension) see TEX250
CERCINE see DCK759
CERECHLOR 54 see PAH780
CERECLOR see PAH780
CERECLOR 30 see PAH780
CERECLOR 42 see PAH780
CERECLOR 48 see PAH780
CERECLOR 52 see PAH780
CERECLOR 54 see PAH780
CERECLOR 70 see PAH780
CERECLOR 51L see PAH780
CERECLOR 56L see PAH780
CERECLOR 63L see PAH780
CERECLOR 65L see PAH780
CERECLOR 70L see PAH780
CERECLOR S 42 see PAH780
CERECLOR S52 see PAH780
CERECLOR S70 see PAH780
CERECLOR 50LV see PAH780
CERELOSE see GFG000
CEREPAP see DNA200
CERESAN see ABU500
CERESAN see CHC500
CERESAN M see EME500
CERESAN UNIVERSAL see ABU500
CERESAN UNIVERSAL-FEUCHTBEIZE see BKS810
CERESAN-UNIVERSAL NASSBEIZE see MEP250
CERESAN UNIVERSAL NAZBEIZE see MEP250
CERESOL see ABU500
CERES ORANGE R see PEJ500
CERES ORANGE RR see XRA000
CERESPAN see PAH250
CERES RED 7B see EOJ500
CERES RED BB see SBC500
CERES RED G see CMS238
CERES RED G 102 see CMS238
CERES YELLOW R see PEI000
CEREWET see BKS810
CERIC DISULFATE see CDB400
CERIC SULFATE see CDB400
CERIC SULPHATE see CDB400
CERISOL SCARLET G see XRA000
CERISOL YELLOW AB see FAG130
CERISOL YELLOW TB see FAG135
CERIUM see CCY250
CERIUM ACETATE see CCY500
CERIUM CHLORIDE see CCY750
CERIUM(III) CHLORIDE see CCY750
CERIUM CITRATE see CCZ000
CERIUM(III) CITRATE see CCZ000
CERIUM COMPOUNDS see CDA250
CERIUM DISULFATE see CDB400
CERIUM FLUORIDE see CDA750
CERIUM FLUORURE (FRENCH) see CDA750
CERIUM NITRATE see CDB000
CERIUM(3+) NITRATE see CDB000
CERIUM(III) NITRATE see CDB000
CERIUM SULFATE see CDB400
CERIUM(4+) SULFATE see CDB400
CERIUM(IV) SULFATE see CDB400
CERIUM TRIACETATE see CCY500
CERIUM TRICHLORIDE see CCY750
CERIUM TRIFLUORIDE see CDA750
CERIUM TRINITRATE see CDB000
CERN KYPOVA 8 see CMU475
CERN PRIMA 38 see AQP000

CEROTINE PONCEAU 3B see SBC500
CEROTINORANGE G see PEJ500
CEROTINSCHARLACH G see XRA000
CEROTINSCHARLACH R see OHI200
CEROUS ACETATE see CCY500
CEROUS CHLORIDE see CCY750
CEROUS CITRATE see CCZ000
CEROUS FLUORIDE see CDA750
CEROUS NITRATE see CDB000
CERTICOL CARMOISINE S see HJF500
CERTICOL ORANGE GS see HGC000
CERTICOL PONCEAU MXS see FMU070
CERTICOL PONCEAU 4RS see FMU080
CERTICOL SUNSET YELLOW CFS see FAG150
CERTICOL TARTRAZOL YELLOW S see FAG140
CERTINAL see ALT250
CERTIQUAL EOSINE see BNH500
CERTIQUAL FLUORESCEINE see FEW000
CERTIQUAL OIL RED see OHI200
CERTOLAKE SUNSET YELLOW see FAG150
CERTOX see SMN500
CERUBIDIN see DAC000
CERULEIN see CAK285
CERUSSETE see LCP000
CERVEN DISPERZNI 15 see AKE250
CERVEN KOSENILOVA A see FMU080
CERVEN KUMIDINOVA see FAG018
CERVEN KYSELA 26 see FMU070
CERVEN KYSELA 114 see CMM330
CERVEN ROZPOUSTEDLOVA 23 see OHI200
CERVEN ROZPOUSTEDLOVA 24 see SBC500
CERVEN ZASADITA 1 see RGW000
CERVICUNDIN see EQJ500
CERVOLIDE see OKW110
CES see SOP500
CES see ECU750
CESIUM see CDC000
CESIUM-133 see CDC000
CESIUM ARSENATE see CDC375
CESIUM BROMIDE see CDC500
CESIUM CARBONATE see CDC750
CESIUM CHLORIDE see CDD000
CESIUM FLUORIDE see CDD500
CESIUM HYDRATE see CDD750
CESIUM HYDROXIDE see CDD750
CESIUM HYDROXIDE (ACGIH, OSHA) see CDD750
CESIUM HYDROXIDE DIMER see CDD750
CESIUM IODIDE see CDE000
CESIUM MONOCHLORIDE see CDD000
CESIUM MONOFLUORIDE see CDD500
CESIUM MONOIODIDE see CDE000
CESIUM NITRATE (DOT) see CDE250
CESIUM(I) NITRATE (1:1) see CDE250
CET see BJP000
CETAB see HCQ500
CETACORT see CNS750
CETADOL see HIM000
CETAFFINE see HCP000
CETAIN see AIT250
CETAL see HCP000
CETALOL CA see HCP000
CETAMIUM see CCX000
CETAROL see HCQ500
CETAVLON see HCQ500
CETHYLOSE see MIF760
CETHYTIN see MIF760
CETIL LIGHT ORANGE GG see HGC000
CETOBEMIDON see KFK000

CETOBEMIDONE see KFK000
CETRIMIDE see HCQ500
CETRIMONIUM BROMIDE see HCQ500
CETYL ALCOHOL see HCP000
CETYLAMIN (GERMAN) see HCO500
CETYLAMINE see HCO500
CETYLAMINE see HCQ500
CETYL 2-ETHYLHEXANOATE see HCP550
CETYLIC ACID see PAE250
CETYLIC ALCOHOL see HCP000
CETYLOL see HCP000
CETYLPYRIDINIUM CHLORIDE see CCX000
1-CETYLPYRIDINIUM CHLORIDE see CCX000
N-CETYLPYRIDINIUM CHLORIDE see CCX000
CETYLTRIMETHYLAMMONIUM BROMIDE see
HCQ500
N-CETYLTRIMETHYLAMMONIUM BROMIDE see
HCQ500
CEVADIC ACID see TGA700
CEVANOL see BCA000
CEVIAN A 678 see AAX250
CEVIAN HL see ADY500
CEVITAMIC ACID see ARN000
CEVITAMIN see ARN000
CEYLON CINNAMON BARK OIL see CMQ510
CEYLON ISINGLASS see AEX250
CEYLON-ZIMT OEL see CMQ510
CF 8 see CBT500
CFC see PGE000
CFC 22 see CFX500
CFC 31 see CHI900
CFC-112 see TBP050
CFC-112a see TBP000
CFC 133a see TJY175
CFC 142b see CFX250
CF 8 (CARBON) see CBT500
C.F.S. see TEM000
CFV see CDS750
CG 117 see MDM100
CG-1283 see MQW500
CGA 15324 see BNA750
CGA 26351 see CDS750
CGA 48988 see MDM100
C-GREEN 10 see BLK000
CH see SBW500
CHA see CPF500
CHALCEDONY see SCI500
CHALCEDONY see SCJ500
CHALCONE see CDH000
CHALCONE, 4-METHYL-(6CI,7CI,8CI) see MIF762
CHALCONE, 2′,3,4′-TRIHYDROXY-4′,6′-DIMETHOXY-,
4′-(6-O-(6-DEOXY-α-l-MANNOPYRANOSYL)-β-d-
GLUCOPYRANOSIDE) see HBU400
CHALK see CAO000
CHALOXYD MEKP-HA 1 see MKA500
CHALOXYD MEKP-LA 1 see MKA500
CHAMBER CRYSTALS see NMJ000
CHAMELEON MINERAL see PLP000
CHAMOMILE-GERMAN OIL see CDH500
CHAMOMILE OIL see CDH500
CHAMOMILE OIL (ROMAN) see CDH750
CHANNEL BLACK see CBT750
CHANNING'S SOLUTION see NCP500
CHAPCO Cu-NAP see NAS000
CHARCOAL see CDI250
CHARCOAL, ACTIVATED (DOT) see CDI000
CHARCOAL (BRIQUETTES) see CDI250

CHARCOAL SCREENINGS (DOT) see CDI250
CHARCOAL (SHELL) see CDJ000
CHARCOAL, SHELL (DOT) see CDJ000
CHARCOAL WOOD (DOT) see CDI250
CHAVICOL METHYL ETHER see AFW750
1,3-CHBP see BNA825
CHEELOX BF see EIV000
CHEELOX BF ACID see EIX000
CHEELOX BR-33 see EIV000
CHEL 300 see AMT500
CHEL 330 see DJG800
CHEL 330 ACID see DJG800
CHELADRATE see EIX500
CHELAFER see FBC100
CHELAFRIN see VGP000
CHELAPLEX III see EIX500
CHELATON III see EIX500
CHEL DTPA see DJG800
CHELEN see EHH000
CHEL-IRON see FBC100
CHELON 100 see EIV000
CHEMAGRO 1,776 see BSH250
CHEMAGRO 2353 see DFV400
CHEMAGRO 4965 see SNT100
CHEMAGRO 25141 see FAQ800
CHEMAGRO 37289 see EPY000
CHEMAGRO B-1776 see TIG250
CHEMAGRO B-1776 see BSH250
CHEMAID see HKC500
CHEMANOX 11 see BFW750
CHEMANOX 22 see MJN250
CHEMATHION see MAK700
CHEM BAM see DXD200
CHEMCARB see CAD800
CHEMCOCCIDE see RLK890
CHEMCOLOX 200 see EIV000
CHEMCOLOX 340 see EIX000
CHEMCOR see PJS750
CHEMETRON FIRE SHIELD see AQF000
CHEMFORM see MEI450
CHEMFORM see SLW500
CHEMFORM see DKC800
CHEMFORM see DMC600
CHEM-HOE see CBM000
CHEMICAL 109 see AQN635
CHEMICAL MACE see CEA750
CHEMICETIN see CDP250
CHEMICETINA see CDP250
CHEMI-CHARL see SHM500
CHEMIFLUOR see SHF500
CHEMIOFURAN see NGE000
CHEMLINK 160 see PER250
CHEMLON see PJY500
CHEM NEB see MAS500
CHEMOCCIDE see RLK890
CHEMOFURAN see NGE500
CHEMOSEPT see TEX250
CHEMOUAG see SNN500
CHEMOX PE see DUZ000
CHEM PELS C see SEY500
CHEM-PHENE see CDV100
CHEMPLEX 3006 see PJS750
CHEMRAT see PIH175
CHEM RICE see DGI000
CHEMSECT DNOC see DUS700
CHEM-SEN 56 see SEY500
CHEM-TOL see PAX250
CHEQUE see MQS225

CHERTS see SCI500
CHERTS see SCJ500
CHESTNUT TANNIN see CDM250
CHEVRON 9006 see DTQ400
CHEVRON ACETONE see ABC750
CHEVRON ORTHO 9006 see DTQ400
CHEVRON RE 12,420 see DOP600
CHEWING TOBACCO see SED400
CHEXMATE see HKC000
CHICAGO BLUE 6B see CMN750
CHICLIDA see HGC500
CHILE SALTPETER see SIO900
CHIMCOCCIDE see RLK890
CHIMOREPTIN see DLH630
CHINALPHOS see DJY200
CHINAWOOD OIL see TOA510
CHINESE ISINGLASS see AEX250
CHINESE RED see LCS000
CHINESE SEASONING see MRL500
CHINESE WHITE see ZKA000
CHINIDIN (GERMAN) see QFS000
CHININ (GERMAN) see QHJ000
CHININDIHYDROCHLORID (GERMAN) see QIJ000
CHINOFER see IGS000
CHINOFORM see CHR500
CHINOFUNGIN see TGB475
CHINOIN see EID000
CHINOLEINE see QMJ000
CHINOLIN see QMJ000
CHINOLINE see QMJ000
p-CHINON (GERMAN) see QQS200
CHINON (DUTCH, GERMAN) see QQS200
CHINONE see QQS200
CHIP-CAL see ARB750
CHIP-CAL GRANULAR see ARB750
CHIPCO 26019 see GIA000
CHIPCO BUCTRIL see DDP000
CHIPCO CRAB-KLEEN see DDP000
CHIPCO THIRAM 75 see TFS350
CHIPCO TURF HERBICIDE MCPP see CIR500
CHIPMAN 11974 see BDJ250
CHIPTOX see CIR250
l-CHIRO-INOSITOL, 4-AMINO-1,4-DIDEOXY-3-o-(2,6-
DIAMINO-2,3,4,6,7-PENTADEOXY-β-l-LYXO-
HEPTOPYRANOSYL)-6-o-METHYL-1-(2-
(FORMIMIDOYLAMINO)-N-METHYLACETAMIDO)-,
SULFATE (1:2),HYDRATE see DAB630
CHISSO 507B see PMP500
CHISSONOX 206 see VOA000
CHKHZ 18 see DVF400
CHLODRONATE SODIUM see SFX730
CHLOFENVINPHOS see CDS750
CHLOMIN see CDP250
CHLOMYCOL see CDP250
CHLOOR (DUTCH) see CDV750
3-CHLOORANILINEN (DUTCH) see CEH675
2-CHLOORBENZALDEHYDE (DUTCH) see CEI500
o-CHLOORBENZALDEHYDE (DUTCH) see CEI500
CHLOORBENZEEN (DUTCH) see CEJ125
2-CHLOOR-1,3-BUTADIEEN (DUTCH) see NCI500
CHLOORDAAN (DUTCH) see CDR750
O-2-CHLOOR-1-(2,4-DICHLOOR-FENYL)-VINYL-O,O-
DIETHYLFOSFAAT (DUTCH) see CDS750
1-CHLOOR-2,4-DINITROBENZEEN (DUTCH) see
CGM000
1-CHLOOR-2,3-EPOXY-PROPAAN (DUTCH) see
EAZ500

CHLOORETHAAN (DUTCH) see EHH000
2-CHLOORETHANOL (DUTCH) see EIU800
CHLOORFACINON (DUTCH) see CJJ000
3-(4-CHLOOR-FENYL)-1,1-DIMETHYLUREUM (DUTCH) see CJX750
2(2-(4-CHLOOR-FENYL-2-FENYL)-ACETYL)-INDAAN-1,3-DION (DUTCH) see CJJ000
CHLOOR-METHAAN (DUTCH) see MIF765
2-(4-CHLOOR-2-METHYL-FENOXY)-PROPIONZUUR (DUTCH) see CIR500
1-CHLOOR-4-NITROBENZEEN (DUTCH) see NFS525
CHLOORPIKRINE (DUTCH) see CKN500
CHLOORWATERSTOF (DUTCH) see HHL000
CHLOPHEN see PJL750
CHLOR (GERMAN) see CDV750
CHLORACETAMID (GERMAN) see CDY850
CHLORACETIC ACID see CEA000
CHLORACETONE see CDN200
CHLORACETONE see CDN200
CHLORACETONITRILE see CDN500
CHLORACETOPHENONE see CDN505
CHLORACETOPHENONE see CEA750
CHLORACETYL CHLORIDE see CEC250
CHLORACTIL see CKP500
2-CHLOROETHANOL (GERMAN) see EIU800
N-(2-CHLORAETHYL)-N'-(2 CHLOROETHYL)-N'-o-PROPYLEN-PHOSPHORSAEUREESTER-DIAMID (GERMAN) see IMH000
2-CHLORAETHYL-PHOSPHONSAEURE (GERMAN) see CDS125
2-CHLORAETHYL-TRIMETHYLAMMONIUMCHLORID see CMF400
CHLORAK see TIQ250
CHLORAL see CDN550
CHLORAL, anhydrous, inhibited (DOT) see CDN550
CHLORALDURAT see CDO000
CHLORAL HYDRATE see CDO000
CHLORALLYL DIETHYLDITHIOCARBAMATE see CDO250
2-CHLORALLYL DIETHYLDITHIOCARBAMATE see CDO250
CHLORALLYLENE see AGB250
CHLORALONE see CDP000
CHLORALOSANE see GFA000
α-CHLORALOSE see GFA000
CHLORAMBUCIL see CDO500
CHLORAMEISENSAEUREAETHYLESTER (GERMAN) see EHK500
CHLORAMEISENSAEURE METHYLESTER (GERMAN) see MIG000
CHLORAMEX see CDP250
CHLORAMFICIN see CDP250
CHLORAMFILIN see CDP250
CHLORAMIN see BIE500
CHLORAMINE see BIE500
CHLORAMINE BLACK C see AQP000
CHLORAMINE BLACK EC see AQP000
CHLORAMINE BLACK ERT see AQP000
CHLORAMINE BLACK EX see AQP000
CHLORAMINE BLACK EXR see AQP000
CHLORAMINE BLACK XO see AQP000
CHLORAMINE BLUE see CMO250
CHLORAMINE BLUE 2B see CMO000
CHLORAMINE CARBON BLACK S see AQP000
CHLORAMINE CARBON BLACK SJ see AQP000
CHLORAMINE CARBON BLACK SN see AQP000
CHLORAMINE FAST BROWN BRL see CMO750

CHLORAMINE SKY BLUE 4B see CMO500
CHLORAMINE SKY BLUE A see CMO500
CHLORAMINE T see CDP000
CHLORAMIN HYDROCHLORIDE see BIE500
CHLORAMINOPHEN see CDO500
CHLORAMINOPHENE see CDO500
CHLORAMP (RUSSIAN) see PIB900
CHLORAMPHENICOL see CDP250
d-CHLORAMPHENICOL see CDP250
d-threo-CHLORAMPHENICOL see CDP250
CHLORAMPHENICOL MONOPALMITATE see CDP700
CHLORAMPHENICOL PALMITATE see CDP700
CHLORAMSAAR see CDP250
CHLORANAUTINE see DYE600
4-CHLORANILIN (CZECH) see CEH680
m-CHLORANILINE see CEH675
o-CHLORANILINE see CEH670
p-CHLORANILINE see CEH680
1-CHLORANTHRACHINON (CZECH) see CEI000
CHLORARSOL see DFX400
CHLORASAN see CDP000
CHLORASEN see DFX400
CHLORASEPTINE see CDP000
CHLORASOL see CDP250
CHLORA-TABS see CDP250
CHLORATE de CALCIUM (FRENCH) see CAO500
CHLORATE of POTASH (DOT) see PLA250
CHLORATE de POTASSIUM (FRENCH) see PLA250
CHLORATES see CDQ000
CHLORATE SALT of SODIUM see SFS000
CHLORATE of SODA (DOT) see SFS000
CHLORAX see SFS000
CHLORAZAN see CDP000
CHLORAZENE see CDP000
CHLORAZIN see CKP500
CHLORAZOL BLACK E see AQP000
CHLORAZOL BLACK EA see AQP000
CHLORAZOL BLACK E (BIOLOGICAL STAIN) see AQP000
CHLORAZOL BLACK EN see AQP000
CHLORAZOL BLUE 3B see CMO250
CHLORAZOL BLUE B see CMO000
CHLORAZOL BURL BLACK E see AQP000
CHLORAZOL LEATHER BLACK ENP see AQP000
CHLORAZOL SILK BLACK G see AQP000
CHLORAZOL SKY BLUE FF see BGT250
CHLORAZOL SKY BLUE FF see CMN750
CHLORAZONE see CDP000
2-CHLORBENZALDEHYD (GERMAN) see CEI500
CHLORBENZENE see CEJ125
1-p-CHLORBENZHYDRYL-m-METHYLBENZYLPIPERAZINE DIHYDROCHLORIDE see MBX250
CHLORBENZILATE see DER000
CHLORBENZOL see CEJ125
o-CHLORBENZONITRIL (CZECH) see CEM000
N-p-CHLORBENZOYL-5-METHOXY-2-METHYLINDOLE-3-ACETIC ACID see IDA000
CHLORBICYCLENE (FRENCH) see DAM700
2-CHLOR-1,3-BUTADIEN (GERMAN) see NCI500
CHLORBUTANOL see ABD000
4-CHLORBUTAN-1-OL (GERMAN) see CEU500
CHLORBUTOL see ABD000
CHLORCHOLINCHLORID see CMF400
CHLORCHOLINE CHLORIDE see CMF400
CHLORCOSANE see PAH780
p-CHLOR-m-CRESOL see CFD990

CHLORCYAN see COO750
CHLORCYCLIZINE HYDROCHLORIDE see CDR250
CHLORCYCLIZINIUM CHLORIDE see CDR250
CHLORDAN see CDR750
trans-CHLORDAN see CDR575
γ-CHLORDAN see CDR575
γ-CHLORDAN see CDR750
CHLORDANE see CDR750
CHLORDANE see CDR760
CHLORDANE, liquid (DOT) see CDR750
γ(trans)-CHLORDANE see CDR575
CHLORDANE, TECHNICAL see CDR760
CHLORDECONE see KEA000
CHLORDENE see HCN000
CHLORDIAZACHEL see MDQ250
CHLORDIAZEPOXIDE see LFK000
CHLORDIAZEPOXIDE HYDROCHLORIDE see
MDQ250
CHLORDIAZEPOXIDE MONOHYDROCHLORIDE
see MDQ250
O-2-CHLOR-1-(2,4-DICHLOR-PHENYL)-VINYL-O,O-
DIAETHYLPHOSPHAT (GERMAN) see CDS750
CHLORDIMEFORM see CJJ250
CHLORDIMETHYLETHER (CZECH) see CIO250
1-CHLOR-2,4-DINITROBENZENE see CGM000
CHLORE (FRENCH) see CDV750
CHLORENDIC ACID see CDS000
CHLOREPIN see CIR750
1-CHLOR-2,3-EPOXY-PROPAN (GERMAN) see EAZ500
CHLORESENE see BBQ500
CHLORESSIGSAEURE-N-(METHOXYMETHYL)-2,6-
DIAETHYLANILID (GERMAN) see CFX000
CHLORESTROLO see CLO750
CHLORETHAMINACIL see BIA250
CHLORETHAMINE see BIE500
CHLOR-ETHAMINE see EIW000
2-CHLORETHANOL (GERMAN) see EIU800
CHLORETHAZINE see BIE500
CHLORETHENE see VNP000
CHLORETHEPHON see CDS125
CHLORETHYL see EHH000
CHLORETHYLENE see VNP000
2-CHLORETHYLESTER KYSELINY
CHLORMRAVENCI see CGU199
2-CHLORETHYLISOKYANAT see IKH000
2-CHLORETHYLPHOSPHONIC ACID see CDS125
2-CHLORETHYL VINYL ETHER see CHI250
CHLORETONE see ABD000
CHLOREX see DFJ050
CHLOREXTOL see PJL750
CHLOREZ 700 see PAH780
CHLOREZ 700HMP see PAH780
CHLORFACINON (GERMAN) see CJJ000
CHLORFENAMIDINE see CJJ250
CHLORFENIDIM see CJX750
p-CHLORFENOL (CZECH) see CJK750
CHLORFENVINFOS see CDS750
CHLORFENVINFOS see CDS750
CHLORFENVINPHOS see CDS750
3-CHLOR-p-FENYLENDIAMIN (CZECH) see CEG600
m-CHLORFENYLISOKYANAT see CKA750
p-CHLORFENYLISOKYANAT (CZECH) see CKB000
CHLORFOS see TIQ250
CHLOR-N-(2-FURYLMETHYL)-5-
SULFAMYLANTHRANILSAEURE (GERMAN) see
CHJ750
CHLORGUANIDE see CKB250

CHLORHEXIDIN (CZECH) see BIM250
CHLORHEXIDINE see BIM250
6-CHLORHEXYLISOKYANAT see CHL250
CHLORHYDRATE d'ANILINE (FRENCH) see BBL000
CHLORHYDRATE de 4-CHLOROORTHOTOLUIDINE
(FRENCH) see CLK235
CHLORHYDRATE de NICOTINE (FRENCH) see
NDP400
CHLORHYDRATE de PAPAVERINE (FRENCH) see
PAH250
CHLORHYDRATE de TETRACYCLINE (FRENCH) see
TBX250
CHLORHYDRIN see CDT750
α-CHLORHYDRIN see CDT750
CHLORHYDROL see AHA000
CHLORHYDROL, GRANULAR see AHA000
CHLORHYDROL, IMPALPABLE see AHA000
CHLORIAZID see CLH750
CHLORIC ACID see CDU000
CHLORIC ACID, solution, containing not more than 10%
acid (DOT) see CDU000
CHLORIC ACID, AMMONIUM SALT see ANE250
CHLORIC ACID, BARIUM SALT see BAJ500
CHLORIC ACID, COPPER SALT see CNJ900
CHLORIC ACID, STRONTIUM SALT see SMF500
CHLORIC ACID, THALLIUM(1+) SALT see TEJ100
CHLORICOL see CDP250
CHLORID AMONNY (CZECH) see ANE500
CHLORID ANILINU (CZECH) see BBL000
CHLORID ANTIMONITY see AQC500
CHLORIDAZON see PEE750
CHLORID-N-BUTYLCINICITY (CZECH) see BSR250
CHLORID DI-n-BUTYLCINICITY (CZECH) see
DDY200
CHLORID DRASELNY (CZECH) see PLA500
CHLORIDEAZEPOXIDE HYDROCHLORIDE see
MDQ250
CHLORIDE de CHOLINE (FRENCH) see CMF750
CHLORIDE of LIME (DOT) see HOV500
CHLORIDE of PHOSPHORUS see PHT275
CHLORIDES see CDU250
CHLORIDE of SULFUR see SOG500
CHLORIDE of SULFUR (DOT) see SON510
CHLORID FENYLRTUTNATY (CZECH) see PFM500
CHLORIDIAZEPIDE see LFK000
CHLORIDIAZEPOXIDE see LFK000
CHLORIDIN see TGD000
CHLORIDINE see TGD000
CHLORID KREMICITY (CZECH) see SCQ500
CHLORID KYSELINY CHLOROCTOVE see CEC250
CHLORID KYSELINY DICHLOROCTOVE see DEN400
CHLORID KYSELINY DIMETHYLKARBAMINOVE
see DQY950
CHLORID RTUTNATY (CZECH) see MCY475
CHLORID TRI-n-BUTYLCINICITY (CZECH) see
CLP500
CHLORIDUM see EHH000
CHLORIERTE BIPHENYLE, CHLORGEHALT 42%
(GERMAN) see PJM250
CHLORIERTE BIPHENYLE, CHLORGEHALT 54%
(GERMAN) see PJN000
CHLORINATED BIPHENYL see PJL750
CHLORINATED CAMPHENE see CDV100
CHLORINATED DIPHENYL see PJL750
CHLORINATED DIPHENYLENE see PJL750
CHLORINATED DIPHENYL OXIDE see CDV175
CHLORINATED HC, ALIPHATIC see CDV250

CHLORINATED HC AROMATIC see CDV500
CHLORINATED HYDROCARBONS, ALIPHATIC see CDV250
CHLORINATED HYDROCARBONS, AROMATIC see CDV500
CHLORINATED HYDROCHLORIC ETHER see DFF809
CHLORINATED LIME (DOT) see HOV500
CHLORINATED NAPHTHALENES see CDV575
CHLORINATED PARAFFINS see PAH780
CHLORINATED PARAFFINS (C$_{12}$, 60% CHLORINE) see PAH800
CHLORINATED PARAFFINS (C$_{23}$, 43% CHLORINE) see PAH810
CHLORINATED POLYETHER POLYURETHAN see CDV625
CHLORINDAN see CDR750
CHLORINE see CDV750
CHLORINE AZIDE see CDW000
CHLORINE CYANIDE see COO750
CHLORINE DIOXIDE see CDW450
CHLORINE DIOXIDE, not hydrated (DOT) see CDW450
CHLORINE FLUORIDE see CDX750
CHLORINE FLUORIDE (ClF$_5$) see CDX250
CHLORINE FLUORIDE OXIDE see PCF750
CHLORINE MOL. see CDV750
CHLORINE NITRIDE see NGQ500
CHLORINE OXIDE see CDW450
CHLORINE(IV) OXIDE see CDW450
CHLORINE OXYFLUORIDE see PCF750
CHLORINE PENTAFLUORIDE see CDX250
CHLORINE PENTAFLUORIDE (DOT) see CDX250
CHLORINE PEROXIDE see CDW450
CHLORINE SULFIDE see SOG500
CHLORINE TRIFLUORIDE see CDX750
CHLORITES see CDY250
5-CHLOR-7-JOD-8-HYDROXY-CHINOLIN (GERMAN) see CHR500
CHLOR KIL see CDR750
CHLORKU LITU (POLISH) see LHB000
CHLORMADINON ACETATE see CBF250
CHLORMADINONE ACETATE see CBF250
CHLORMADINONU (POLISH) see CBF250
CHLORMENE see TAI500
CHLORMEQUAT see CMF400
CHLORMEQUAT CHLORIDE see CMF400
CHLOR-METHAN (GERMAN) see MIF765
CHLORMETHINE see BIE250
CHLORMETHINE HYDROCHLORIDE see BIE500
CHLORMETHINE-N-OXIDE HYDROCHLORIDE see CFA750
CHLORMETHINUM see BIE500
2-(4-CHLOR-2-METHYL-PHENOXY)-PROPIONSAEURE (GERMAN) see CIR500
3-CHLOR-2-METHYL-PROP-1-EN (GERMAN) see CIU750
CHLORNAFTINA see BIF250
CHLORNAPHAZIN see BIF250
CHLORNAPHTHIN see BIF250
1-CHLOR-4-NITROBENZOL (GERMAN) see NFS525
CHLORNITROMYCIN see CDP250
CHLOROACETALDEHYDE see CDY500
2-CHLOROACETALDEHYDE see CDY500
CHLOROACETALDEHYDE MONOMER see CDY500
CHLOROACETAMIDE see CDY850
2-CHLOROACETAMIDE see CDY850
α-CHLOROACETAMIDE see CDY850

CHLOROACETIC ACID see CEA000
α-CHLOROACETIC ACID see CEA000
CHLOROACETIC ACID tert-BUTYL ESTER see BQR100
CHLOROACETIC ACID CHLORIDE see CEC250
CHLOROACETIC ACID, ETHYL ESTER see EHG500
CHLOROACETIC ACID METHYL ESTER see MIF775
CHLOROACETIC ACID SODIUM SALT see SFU500
CHLOROACETIC ACID, solid (UN 1751) (DOT) see CEA000
CHLOROACETIC ACID, liquid (UN 1750) (DOT) see CEA000
CHLOROACETIC CHLORIDE see CEC250
CHLOROACETONE see CDN200
CHLOROACETONE, stabilized (DOT) see CDN200
2-CHLOROACETONITRILE see CDN500
CHLOROACETONITRILE (DOT) see CDN500
α-CHLOROACETONITRILE see CDN500
1-CHLOROACETOPHENONE see CEA750
4-CHLOROACETOPHENONE see CEB250
p-CHLOROACETOPHENONE see CEB250
4'-CHLOROACETOPHENONE see CEB250
α-CHLOROACETOPHENONE see CEA750
ω-CHLOROACETOPHENONE see CEA750
CHLOROACETOPHENONE, liquid or solid (DOT) see CEA750
6-CHLORO-17-α-ACETOXY-4,6-PREGNADIENE-3,20-DIONE see CBF250
Δ^6-6-CHLORO-17-α-ACETOXYPROGESTERONE see CBF250
4'-CHLOROACETYL ACETANILIDE see CEC000
4'-(CHLOROACETYL)ACETANILIDE see CEC000
N-(CHLOROACETYL)-3-AZABICYCLO(3.2.1)NONANE see CEC100
CHLOROACETYL CHLORIDE see CEC250
1-CHLOROACETYL-α-α-DIPHENYL-4-PIPERIDINEMETHANOL see CEC300
2-CHLOROACETYLFLUORENE see CEC700
l-N-(α-(CHLOROACETYL)PHENETHYL)-p-TOLUENESULFONAMIDE see THH450
N-(CHLOROACETYL)-3-PHENYL-N-(p-TOLYLSULFONYL)ALANINE see THH550
CHLOROACRYLONITRILE see CEE750
2-CHLOROACRYLONITRILE see CEE750
α-CHLOROACRYLONITRILE see CEE750
CHLOROAETHAN (GERMAN) see EHH000
α-CHLOROALLYL CHLORIDE see DGG950
γ-CHLOROALLYL CHLORIDE see DGG950
2-CHLOROALLYL DIETHYLDITHIOCARBAMATE see CDO250
2-CHLOROALLYL-N,N-DIETHYLDITHIOCARBAMATE see CDO250
CHLOROALLYLENE see AGB250
1-(3-CHLOROALLYL)-3,5,7-TRIAZA-1-AZONIAADAMANTANE CHLORIDE see CEG550
CHLOROALONIL see TBQ750
CHLOROALOSANE see GFA000
CHLOROAMBUCIL see CDO500
3-CHLORO-4-AMINOANILINE see CEG600
p-CHLORO-o-AMINOPHENOL see CEH250
2-CHLORO-4-AMINOTOLUENE see CLK215
3-CHLORO-2-AMINOTOLUENE see CLK227
4-CHLORO-2-AMINOTOLUENE see CLK225
5-CHLORO-2-AMINOTOLUENE see CLK220
5-CHLORO-2-AMINOTOLUENE HYDROCHLORIDE see CLK235

2-CHLOROANILINE see CEH670
3-CHLOROANILINE see CEH675
4-CHLOROANILINE see CEH680
m-CHLOROANILINE see CEH675
o-CHLOROANILINE see CEH670
p-CHLOROANILINE see CEH680
m-CHLOROANILINE, solid see CEH675
o-CHLOROANILINE, solid see CEH670
p-CHLOROANILINE, solid see CEH680
m-CHLOROANILINE, liquid see CEH675
o-CHLOROANILINE, liquid see CEH670
p-CHLOROANILINE, liquid see CEH680
3-CHLOROANILINE (ITALIAN) see CEH675
4-CHLOROANILINE HYDROCHLORIDE see CJR200
p-CHLOROANILINE HYDROCHLORIDE see CJR200
p-CHLOROANILINIUM CHLORIDE see CJR200
1-CHLORO-9,10-ANTHRACENEDIONE see CEI000
1-CHLOROANTHRAQUINONE see CEI000
1-CHLORO-9,10-ANTHRAQUINONE see CEI000
α-CHLOROANTHRAQUINONE see CEI000
CHLOROBEN see DEP600
CHLOROBENZAL see BAY300
2-CHLOROBENZALDEHYDE see CEI500
o-CHLOROBENZALDEHYDE see CEI500
α-CHLOROBENZALDEHYDE see BDM500
2-CHLOROBENZAL MALONONITRILE see CEQ600
o-CHLOROBENZAL MALONONITRILE see CEQ600
CHLOROBENZEN (POLISH) see CEJ125
3-CHLOROBENZENAMINE see CEH675
4-CHLOROBENZENAMINE see CEH680
2-CHLORO-BENZENAMINE (9CI) see CEH670
4-CHLOROBENZENAMINE HYDROCHLORIDE see CJR200
CHLOROBENZENE see CEJ125
4-CHLOROBENZENEAMINE see CEH680
o-CHLOROBENZENECARBOXALDEHYDE see CEI500
2-CHLORO-1,4-BENZENEDIAMINE see CEG600
4-CHLORO-1,3-BENZENEDIAMINE see CJY120
1-(p-CHLOROBENZHYDRYL)-4-(p-tert-BUTYLBENZYL)DIETHYLENEDIAMINE DIHYDROCHLORIDE see BOM250
1-p-CHLOROBENZHYDRYL-4-p-(tert)-BUTYLBENZYLPIPERAZINE DIHYDROCHLORIDE see BOM250
1-(p-CHLOROBENZHYDRYL)-4-(2-(2-HYDROXYETHOXY)ETHYL)DIETHYLENEDIAMINE see CJR909
N-(4-CHLOROBENZHYDRYL)-N'-(HYDROXYETHOXYETHYL)PIPERAZINE see CJR909
1-(p-CHLOROBENZHYDRYL)-4-(2-(2-HYDROXYETHOXY)ETHYL)PIPERAZINE see CJR909
1-(p-CHLOROBENZHYDRYL)-4-(m-METHYLBENZYL)DIETHYLENEDIAMINE see HGC500
1-p-CHLOROBENZHYDRYL-4-m-METHYLBENZYLPIPERAZINE see HGC500
1-(p-CHLOROBENZHYDRYL)-4-METHYLPIPERAZINE HYDROCHLORIDE see CDR250
CHLOROBENZOL (DOT) see CEJ125
o-CHLOROBENZONITRILE see CEM000
6-CHLORO-2H-1,2,4-BENZOTHIADIAZINE-7-SULFONAMIDE-1,1-DIOXIDE see CLH750
4-CHLOROBENZOTRICHLORIDE see TIR900
p-CHLOROBENZOTRICHLORIDE see TIR900
p-CHLOROBENZOTRIFLUORIDE see CEM825

1-(p-CHLOROBENZOYL)-5-METHOXY-2-METHYLINDOLE-3-ACETIC ACID see IDA000
1-(p-CHLOROBENZOYL)-2-METHYL-5-METHOXYINDOLE-3-ACETIC ACID see IDA000
1-(p-CHLOROBENZOYL)-2-METHYL-5-METHOXY-3-INDOLE-ACETIC ACID see IDA000
α-(1-(p-CHLOROBENZOYL)-2-METHYL-5-METHOXY-3-INDOLYL)ACETIC ACID see IDA000
2-(p-CHLOROBENZOYL)-1-(2-MORPHOLINOETHYL)PYRROLE MONOHYDROCHLORIDE see CEO100
p-CHLOROBENZOYL PEROXIDE (DOT) see BHM750
CHLOROBENZYLATE see DER000
o-CHLOROBENZYLIDENE MALONITRILE see CEQ600
2-CHLOROBENZYLIDENE MALONONITRILE see CEQ600
o-CHLOROBENZYLIDENE MALONONITRILE see CEQ600
CHLORO BIPHENYL see PJL750
CHLORO 1,1-BIPHENYL see PJL750
2-CHLORO-4,6-BIS(ETHYLAMINO)-s-TRIAZINE see BJP000
1-CHLORO-3,5-BISETHYLAMINO-2,4,6-TRIAZINE see BJP000
2-CHLORO-4,6-BIS(ETHYLAMINO)-1,3,5-TRIAZINE see BJP000
CHLOROBIS(2-METHYLPROPYL)ALUMINUM see CGB500
CHLOROBLE M see MAS500
2-CHLOROBMN see CEQ600
CHLOROBROMOMETHANE see CES650
1-CHLORO-3-BROMOPROPANE (DOT) see BNA825
ω-CHLOROBROMOPROPANE see BNA825
trans-CHLORO(2-(3-BROMOPROPIONAMIDO)CYCLOHEXYL)MERCURY see CET000
CHLOROBUTADIENE see NCI500
2-CHLOROBUTA-1,3-DIENE see NCI500
2-CHLORO-1,3-BUTADIENE see NCI500
2-CHLOROBUTANE see CEU250
1-CHLOROBUTANE (DOT) see BQQ750
4-CHLORO-1-BUTANE-OL see CEU500
CHLOROBUTANOL see ABD000
4-CHLOROBUTANOL see CEU500
4-CHLORO-1-BUTANOL see CEU500
CHLOROBUTIN see CDO500
CHLOROBUTINE see CDO500
N-(3-CHLORO-1-sec-BUTYLACETONYL)-p-TOLUENESULFONAMIDE see THH470
S-(4-CHLORO-2-BUTYNYL) DIPHENYLPHOSPHINOTHIOATE see CEV840
CHLOROCAIN see CBR250
CHLOROCAINE see AIT250
2-CHLOROCAMPHANE see BMD300
CHLOROCAMPHENE see CDV100
CHLOROCAPS see CDP250
CHLOROCARBONATE D'ETHYLE (FRENCH) see EHK500
CHLOROCARBONATE de METHYLE (FRENCH) see MIG000
CHLOROCARBONIC ACID METHYL ESTER see MIG000
3-CHLOROCHLORDENE see HAR000
3-CHLORO-N-(5-CHLORO-2,6-DINITRO-4-TRIFLUOROMETHYLPHENYL)-5-TRIFLUOROMETHYL-2-PYRIDINAMINE see CEX800

2-CHLORO-9-(2,2-DIMETHYLHYDRAZINO)ACRIDINE see CGH800
4-CHLORO-3,5-DIMETHYLPHENOL see CLW000
((4-CHLORO-6-((2,3-DIMETHYLPHENYL)AMINO)-2-PYRIMIDINYL)THIO)ACETIC ACID see CLW250
1-CHLORO-N,N-DIMETHYL-2-PROPANAMINE HYDROCHLORIDE see CGJ280
CHLORODINITROBENZENE see CGL750
1-CHLORO-2,4-DINITROBENZENE see CGM000
4-CHLORO-1,3-DINITROBENZENE see CGM000
6-CHLORO-1,3-DINITROBENZENE see CGM000
CHLORODINITROBENZENE (mixed isomers) (DOT) see CGL750
1-CHLORO-2,4-DINITROBENZOL (GERMAN) see CGM000
1-CHLORO-2,4-DINITRONAPHTHALENE see DUS600
CHLORODIPHENYLARSINE see CGN000
CHLORODIPHENYL (21% Cl) see PJM000
CHLORODIPHENYL (32% Cl) see PJM250
CHLORODIPHENYL (48% Cl) see PJM750
CHLORODIPHENYL (60% Cl) see PJN250
CHLORODIPHENYL (62% Cl) see PJN500
CHLORODIPHENYL (68% Cl) see PJN750
CHLORODIPHENYL (42% Cl) (OSHA) see PJM500
CHLORODIPHENYL (54% Cl) (OSHA) see PJN000
1-(p-CHLORODIPHENYLMETHYL)-4-(2-(2-HYDROXYETHOXY)ETHYL)PIPERAZINE see CJR909
2-CHLORO-N,N-DI-2-PROPENYLACETAMIDE see CFK000
1-CHLORO-2,3-EPOXYPROPANE see EAZ500
3-CHLORO-1,2-EPOXYPROPANE see EAZ500
CHLOROETENE see MIH275
2-CHLOROETHANAL see CDY500
2-CHLORO-1-ETHANAL see CDY500
2-CHLOROETHANAMIDE see CDY850
CHLOROETHANE see EHH000
2-CHLOROETHANEPHOSPHONIC ACID see CDS125
CHLOROETHANOIC ACID see CEA000
Δ-CHLOROETHANOL see EIU800
2-CHLOROETHANOL-2-(p-tert-BUTYLPHENOXY)-1-METHYLETHYL SULFITE see SOP500
2-CHLOROETHANOL ESTER with 2-(p-tert-BUTYLPHENOXY)-1-METHYLETHYL SULFITE see SOP500
2-CHLOROETHANOL HYDROGEN PHOSPHATE ESTER with 3-CHLORO-7-HYDROXY-4-METHYLCOUMARIN see DFH600
2-CHLOROETHANOL (MAK) see EIU800
2-CHLOROETHANOL PHOSPHATE see CGO500
2-CHLOROETHANOL PHOSPHATE DIESTER ESTER with 3-CHLORO-7-HYDROXY-4-METHYLCOUMARIN see DFH600
CHLOROETHENE see MIH275
CHLOROETHENE see VNP000
CHLOROETHENE HOMOPOLYMER see PKQ059
(2-CHLOROETHENYL) ARSONOUS DICHLORIDE see CLV000
1,1',1''-(1-CHLORO-1-ETHENYL-2-YLIDENE)-TRIS(4-METHOXYBENZENE) see CLO750
(2-CHLOROETHOXY)ETHENE see CHI250
2-CHLORO-1-(3-ETHOXY-4-NITROPHENOXY)-4-(TRIFLUOROMETHYL)BENZENE see OQU100
2-CHLOROETHYL ALCOHOL see EIU800
β-CHLOROETHYL ALCOHOL see EIU800
2-CHLORO-4-ETHYLAMINEISOPROPYLAMINE-s-TRIAZINE see ARQ725

2-CHLORO-4-ETHYLAMINO-6-(1-CYANO-1-METHYL)ETHYLAMINO-s-TRIAZINE see BLW750
1-CHLORO-3-ETHYLAMINO-5-ISOPROPYLAMINO-s-TRIAZINE see ARQ725
2-CHLORO-4-ETHYLAMINO-6-ISOPROPYLAMINO-s-TRIAZINE see ARQ725
1-CHLORO-3-ETHYLAMINO-5-ISOPROPYLAMINO-2,4,6-TRIAZINE see ARQ725
2-CHLORO-4-ETHYLAMINO-6-ISOPROPYLAMINO-1,3,5-TRIAZINE see ARQ725
2-(4-CHLORO-6-ETHYLAMINO-s-TRIAZINE-2-YLAMINO)-2-METHYL-PROPIONITRILE see BLW750
2-(4-CHLORO-6-ETHYLAMINO-1,3,5-TRIAZINE-2-YLAMINO)-2-METHYLPROPIONITRILE see BLW750
2-((4-CHLORO-6-(ETHYLAMINO)-1,3,5-TRIAZIN-2-YL)AMINO)-2-METHYL-PROPANENITRILE see BLW750
2-((4-CHLORO-6-(ETHYLAMINO)-s-TRIAZIN-2-YL)AMINO)-2-METHYLPROPIONITRILE see BLW750
3-(2-CHLOROETHYL)-2-(BIS(2-CHLOROETHYL)AMINO)PERHYDRO-2H-1,3,2-OXAZAPHOSPHORINE-2-OXIDE see TNT500
β-CHLOROETHYL-β'-(p-tert-BUTYLPHENOXY)-α'-METHYLETHYL SULFITE see SOP500
β-CHLOROETHYL-β-(p-tert-BUTYLPHENOXY)-α-METHYLETHYL SULPHITE see SOP500
3-(2-CHLOROETHYL)-2-((2-CHLOROETHYL)AMINO)PERHYDRO-2H-1,3,2-OXAZAPHOSPHORINE OXIDE see IMH000
3-(2-CHLOROETHYL)-2-((2-CHLOROETHYL)AMINO)TETRAHYDRO-2H-1,3,2-OXAZAPHOSPHORINE-2-OXIDE see IMH000
N-(2-CHLOROETHYL)-N'-(2-CHLOROETHYL)-N',O-PROPYLENEPHOSPHORIC ACID DIAMIDE see IMH000
N-(2-CHLOROETHYL)-N'-(2-CHLOROETHYL)-N',O-PROPYLENEPHOSPHORIC ACID ESTER DIAMIDE see IMH000
2-CHLOROETHYL CHLOROFORMATE see CGU199
CHLOROETHYLCYCLOHEXYLNITROSOUREA see CGV250
(CHLORO-2-ETHYL)-1-CYCLOHEXYL-3-NITROSOUREA see CGV250
1-(2-CHLOROETHYL)-3-CYCLOHEXYL-1-NITROSOUREA see CGV250
N-(2-CHLOROETHYL)-N'-CYCLOHEXYL-N-NITROSOUREA see CGV250
N-(2-CHLOROETHYL)DIBENZYLAMINE see DCT050
2-CHLOROETHYLDIISOPROPYLAMINE HYDROCHLORIDE see CGV600
N-(CHLOROETHYL)DIISOPROPYLAMINE HYDROCHLORIDE see CGV600
(β-CHLOROETHYL)DIISOPROPYLAMINE HYDROCHLORIDE see CGV600
(2-CHLOROETHYL)DIMETHYLAMINE see CGW000
N-(2-CHLOROETHYL)DIMETHYLAMINE see CGW000
β-CHLOROETHYLDIMETHYLAMINE see CGW000
CHLOROETHYLENE see VNP000
CHLOROETHYLENE POLYMER see PKQ059
CHLOROETHYLENEVINYL ACETATE POLYMER see AAX175
CHLOROETHYL ETHER see DFJ050
CHLOROETHYL ETHYL SULFIDE see CGY750
2-CHLOROETHYL ETHYL SULFIDE see CGY750
2-CHLOROETHYL ETHYL THIOETHER see CGY750
1-(2-CHLOROETHYL)-3-(d-GLUCOPYRANOS-2-YL)-1-NITROSOUREA see CLX000

CHLOROETHYLIDENE FLUORIDE see CFX250

α-CHLOROETHYLIDENE FLUORIDE see CFX250

2-CHLOROETHYL ISOCYANATE see IKH000

CHLOROETHYL MERCURY see CHC500

2-CHLOROETHYL METHANESULFONATE see CHC750

β-CHLOROETHYLMETHANESULFONATE see CHC750

CHLOROETHYL METHANESULPHONATE see CHC750

2-CHLOROETHYL 1-METHYL-2-(p-tert-BUTYLPHENOXY)ETHYL SULPHITE see SOP500

1-(2-CHLOROETHYL)-3-(4-METHYLCYCLOHEXYL)-1-NITROSOUREA see CHD250

1-(2-CHLOROETHYL)-3-(trans-4-METHYLCYCLOHEXYL)-1-NITROSOUREA see CHD250

N-(2-CHLOROETHYL)-N'-(trans-4-METHYLCYCLOHEXYL)-N-NITROSOUREA see CHD250

N-(2-CHLOROETHYL)-N-(1-METHYLETHYL)-2-PROPANAMINE HYDROCHLORIDE see CGV600

6-CHLORO-N-ETHYL-N'-(1-METHYLETHYL)-1,3,5-TRIAZINE-2,4-DIAMINE (9CI) see ARQ725

N-(2-CHLOROETHYL)-N-(1-METHYL-2-PHENOXYETHYL)BENZENEMETHANAMINE see PDT250

N-(2-CHLOROETHYL)-N-(1-METHYL-2-PHENOXYETHYL)BENZENEMETHANAMINE HYDROCHLORIDE see DDG800

N-(2-CHLOROETHYL)-N-(1-METHYL-2-PHENOXYETHYL)BENZYLAMINE see PDT250

N-(2-CHLOROETHYL)-N-(1-METHYL-2-PHENOXYETHYL)BENZYLAMINE HYDROCHLORIDE see DDG800

4-(2-CHLOROETHYL)MORPHOLINE HYDROCHLORIDE see CHE000

2-((((2-CHLOROETHYL)NITROSOAMINO)CARBONYL)AMINO)-2-DEOXY-d-GLUCOPYRANOSE see CLX000

2-((((2-CHLOROETHYL)NITROSOAMINO)CARBONYL)AMINO)-2-DEOXY-d-GLUCOSE see CLX000

N-(2-CHLOROETHYL)-N-NITROSOETHYLCARBAMATE see CHF500

2-(3-(2-CHLOROETHYL)-3-NITROSOUREIDO)-2-DEOXY-d-GLUCOSOPYRANOSE see CLX000

2-(3-(2-CHLOROETHYL)-3-NITROSOUREIDO)-d-GLUCO-PYRANOSE see CLX000

N-(β-CHLOROETHYL)-N-NITROSOURETHAN see CHF500

2-CHLOROETHYL-N-NITROSOURETHANE see CHF500

CHLOROETHYLOWY ALKOHOL (POLISH) see EIU800

1-CHLORO-3-ETHYL-1-PENTEN-4-YN-3-OL see CHG000

2-CHLOROETHYL SULFUROUS ACID-2-(4-(1,1-DIMETHYLETHYL)PHENOXY)-1-METHYLETHYL ESTER see SOP500

2-CHLOROETHYL SULPHITE of 1-(p-tert-BUTYLPHENOXY)-2-PROPANOL see SOP500

1-CHLORO-2-(ETHYLTHIO)ETHANE see CGY750

(2-CHLOROETHYL)TRIMETHYLAMMONIUM CHLORIDE see CMF400

(β-CHLOROETHYL)TRIMETHYLAMMONIUM CHLORIDE see CMF400

2-CHLOROETHYL VINYL ETHER see CHI250

CHLOROFENVINPHOS see CDS750

CHLOROFLUOROMETHANE see CHI900

CHLORO FLUORO SULFONE see SOT500

CHLOROFOLIN see CKN000

CHLOROFORM see CHJ500

CHLOROFORME (FRENCH) see CHJ500

CHLOROFORMIC ACID ALLYL ESTER see AGB500

CHLOROFORMIC ACID BENZYL ESTER see BEF500

CHLOROFORMIC ACID 2-CHLOROETHYL ESTER see CGU199

CHLOROFORMIC ACID DIMETHYLAMIDE see DQY950

CHLOROFORMIC ACID, ESTER with 1-NAPHTHOL see NBH200

CHLOROFORMIC ACID ETHYL ESTER see EHK500

CHLOROFORMIC ACID 2-FLUOROETHYL ESTER see FIH100

CHLOROFORMIC ACID ISOPROPYL ESTER see IOL000

CHLOROFORMIC ACID METHYL ESTER see MIG000

CHLOROFORMIC ACID 1-NAPHTHYL ESTER see NBH200

CHLOROFORMIC ACID PHENYL ESTER see CBX109

CHLOROFORMIC ACID PROPYL ESTER see PNH000

CHLOROFORMIC DIGITALIN see DKN400

CHLOROFORMYL CHLORIDE see PGX000

CHLOROFOS see TIQ250

CHLOROFTALM see TIQ250

4-CHLORO-N-FURFURYL-5-SULFAMOYLANTHRANILIC ACID see CHJ750

4-CHLORO-N-(2-FURYLMETHYL)-5-SULFAMOYLANTHRANILIC ACID see CHJ750

CHLOROFYL see CKN000

CHLOROGUANIDE see CKB250

CHLOROHEXYL ISOCYANATE see CHL250

CHLOROHYDRIC ACID see HHL000

epi-CHLOROHYDRIN see EAZ500

α-CHLOROHYDRIN see CDT750

CHLOROHYDROL see AHA000

CHLOROHYDROQUINONE see CHM000

2-CHLORO-10-3-(1-(2-HYDROXYETHYL)-4-PIPERAZINYL)PROPYL PHENOTHIAZINE see CJM250

5-CHLORO-8-HYDROXY-7-IODOQUINOLINE see CHR500

3-CHLORO-7-HYDROXY-4-METHYLCOUMARIN BIS(2-CHLOROETHYL)PHOSPHATE see DFH600

3-CHLORO-7-HYDROXY-4-METHYL-COUMARIN-O,O-DIETHYL PHOSPHOROTHIOATE see CNU750

3-CHLORO-7-HYDROXY-4-METHYL-COUMARIN-O-ESTER with O,O-DIETHYL PHOSPHOROTHIOATE see CNU750

(−)-N-((5-CHLORO-8-HYDROXY-3-METHYL-1-OXO-7-ISOCHROMANYL)CARBONYL)-3-PHENYLALANINE see CHP250

5-CHLORO-2-((2-HYDROXY-1-NAPHTHALENYL)AZO)-4-METHYLBENZENE SULFONIC ACID, BARIUM SALT (2:1) see CHP500

5-CHLORO-2-((2-HYDROXY-1-NAPHTHALENYL)AZO)-4-METHYLBENZENE SULPHONIC ACID, BARIUM SALT see CHP500

5-CHLORO-2-((2-HYDROXY-1-NAPHTHYL)AZO)-p-TOLUENE SULFONIC ACID, BARIUM SALT see CHP500

CHLORO(o-HYDROXYPHENYL)MERCURY see CHW675

CHLORO(p-HYDROXYPHENYL)MERCURY see CHW750

6-CHLORO-17-α-HYDROXYPREGNA-4,6-DIENE-3,20-DIONE ACETATE see CBF250

2-CHLORO-HYDROXYTOLUENE see CFD990

6-CHLORO-3-HYDROXYTOLUENE see CFD990

6-CHLORO-17-α-HYDROXY-Δ⁶-PROGESTERONE ACETATE see CBF250

5-CHLORO-7-IODO-8-HYDROXYQUINOLINE see CHR500

CHLOROIODOQUINE see CHR500

5-CHLORO-7-IODO-8-QUINOLINOL see CHR500

2-CHLOROISOBUTANE see BQR000

α-CHLOROISOBUTYLENE see IKE000

γ-CHLOROISOBUTYLENE see CIU750

1-CHLORO-2-ISOPROPOXY-2-PROPANOL see CHS250

N-(3-CHLORO-1-ISOPROPYLACETONYL)-p-TOLUENESULFONAMIDE see THH555

(β-CHLOROISOPROPYL)DIMETHYLAMINE HYDROCHLORIDE see CGJ280

CHLOROJECT L see CDP250

CHLOROMADINONE ACETATE see CBF250

CHLOROMAX see CDP250

S-(6-CHLORO-3-(MERCAPTOMETHYL)-2-BENZOXAZOLINONE)-O,O-DIETHYL PHOSPHORODITHIOATE see BDJ250

(CHLOROMERCURI)BENZENE see PFM500

p-(CHLOROMERCURI)BENZOIC ACID see CHU500

p-CHLOROMERCURIC BENZOIC ACID see CHU500

o-CHLOROMERCURIPHENOL see CHW675

p-CHLOROMERCURIPHENOL see CHW750

p-(CHLOROMERCURI)PHENOL see CHW750

CHLOROMETHANE see MIF765

CHLOROMETHANE SULFONATE d'ETHYLE (FRENCH) see CHC750

CHLOROMETHOXY ETHANE see CIM000

CHLORO(2-METHOXYETHYL)MERCURY see MEP250

3-CHLORO-7-METHOXY-9-(1-METHYL-4-DIETHYLAMINOBUTYLAMINO)ACRIDINE see ARQ250

l-N-(3-CHLORO-1-METHYLACETONYL)-p-TOLUENESULFONAMIDE see THH360

7-CHLORO-2-METHYLAMINO-5-PHENYL-3H-1,4-BENZODIAZEPINE 4-OXIDE see LFK000

7-CHLORO-2-METHYLAMINO-5-PHENYL-3H-1,4-BENZODIAZEPIN 4-OXIDE see LFK000

7-CHLORO-2-METHYLAMINO-5-PHENYL-3H-1,4-BENZODIAZEPIN, 4-OXIDE, HYDROCHLORIDE see MDQ250

4-CHLORO-5-(METHYLAMINO)-2-(α,α,α-TRIFLUORO-m-TOLYL)-3(2H)-PYRIDAZINONE see NNQ100

2-CHLORO-4-METHYLANILINE see CLK210

3-CHLORO-2-METHYLANILINE see CLK200

3-CHLORO-4-METHYLANILINE see CLK215

3-CHLORO-6-METHYLANILINE see CLK225

4-CHLORO-2-METHYLANILINE see CLK220

4-CHLORO-6-METHYLANILINE see CLK220

5-CHLORO-2-METHYLANILINE see CLK225

6-CHLORO-2-METHYLANILINE see CLK227

4-CHLORO-2-METHYLANILINE HYDROCHLORIDE see CLK235

4-CHLORO-6-METHYLANILINE HYDROCHLORIDE see CLK235

CHLOROMETHYLATED AMINATED STRYENE-DIVINYLBENZENE RESIN see CIF775

7-CHLOROMETHYL BENZ(a)ANTHRACENE see CIG250

CHLOROMETHYLBENZENE see BEE375

CHLOROMETHYLBENZENE see CLK130

4-CHLORO-1-METHYLBENZENE see TGY075

2-CHLORO-1-METHYLBENZENE (9CI) see CLK100

4-CHLORO-2-METHYLBENZENEAMINE see CLK220

4-CHLORO-2-METHYLBENZENEAMINE HYDROCHLORIDE see CLK235

7-CHLORO-1-METHYL-5-3H-1,4-BENZODIAZEPIN-2(1H)-ONE see DCK759

3-CHLORO-4-METHYL-7-COUMARINYL DIETHYLPHOSPHATE see CIK750

3-CHLORO-4-METHYL-7-COUMARINYL DIETHYL PHOSPHOROTHIOATE see CNU750

O-3-CHLORO-4-METHYL-7-COUMARINYL-O,O-DIETHYL PHOSPHOROTHIOATE see CNU750

CHLOROMETHYL CYANIDE see CDN500

4-(CHLOROMETHYL)-2,2-DIMETHYL-1,3-DIOXOLANE see CIL800

1-CHLOROMETHYLETHYLENE GLYCOL CYCLIC SULFITE see CKQ750

(CHLOROMETHYL)ETHYLENE OXIDE see EAZ500

CHLOROMETHYL ETHYL ETHER see CIM000

(2-CHLORO-1-METHYLETHYL) ETHER see BII250

3-CHLORO-4-METHYL-7-HYDROXYCOUMARIN DIETHYL THIOPHOSPHORIC ACID ESTER see CNU750

CHLOROMETHYLMERCURY see MDD750

7-CHLOROMETHYL-12-METHYL BENZ(a)ANTHRACENE see CIN750

CHLOROMETHYL METHYL ETHER see CIO250

2-CHLORO-2-METHYL-N-NITROSOETHANAMINE see CIQ500

2-CHLORO-N-METHYL-N-NITROSOETHYLAMINE see CIQ500

CHLOROMETHYLOXIRANE see EAZ500

2-(CHLOROMETHYL)OXIRANE see EAZ500

7-CHLORO-1-METHYL-2-OXO-5-PHENYL-3H-1,4-BENZODIAZEPINE see DCK759

4-CHLORO-3-METHYLPHENOL see CFD990

(4-CHLORO-2-METHYLPHENOXY)ACETIC ACID see CIR250

4-CHLORO-2-METHYLPHENOXYACETIC ACID SODIUM SALT see SIL500

2-(4-CHLORO-2-METHYLPHENOXY)PROPIONIC ACID see CIR500

4-CHLORO-2-METHYLPHENOXY-α-PROPIONIC ACID see CIR500

(+)-α-(4-CHLORO-2-METHYLPHENOXY) PROPIONIC ACID see CIR500

7-CHLORO-N-METHYL-5-PHENYL-3H-1,4-BENZODIAZEPIN-2-AMINE-4-OXIDE see LFK000

7-CHLORO-1-METHYL-5-PHENYL-1H-1,5-BENZODIAZEPINE-2,4(3H,5H)-DIONE see CIR750

7-CHLORO-1-METHYL-5-PHENYL-2H-1,4-BENZODIAZEPIN-2-ONE see DCK759

7-CHLORO-1-METHYL-5-PHENYL-1,3-DIHYDRO-2H-1,4-BENZODIAZEPIN-2-ONE see DCK759

N'-(4-CHLORO-2-METHYLPHENYL)-N,N-DIMETHYLMETHANIMIDAMIDE see CJJ250

7-CHLORO-N-METHYL-5-PHENYL-EH-1,4-BENZODIAZEPIN-2-AMINE-4-OXIDE, MONOHYDROCHLORIDE see MDQ250

S-(4-CHLORO-3-METHYLPHENYL) o-ETHYL ETHYLPHOSPHONODITHIOATE see EMR100

CHLOROMETHYL PHENYL KETONE see CEA750

2-CHLORO-11-(4-
METHYLPIPERAZINO)DIBENZO(b,f)(1,4)THIAZEPIN
E see CIS750
2-CHLORO-11-(4-METHYL-1-PIPERAZINYL)-
DIBENZO(b,f)(1,4)OXAZEPINE see DCS200
2-CHLORO-11-(4-METHYL-1-PIPERAZINYL)-
DIBENZO(b,f)(1,4)OXOAZEPINE see DCS200
2-CHLORO-11-(4-METHYL-1-
PIPERAZINYL)DIBENZO(b,f)(1,4)THIAZEPINE see
CIS750
2-CHLORO-10-(3-(1-METHYL-4-PIPERAZINYL)-
PROPYL)-PHENOTHIAZINE see PMF500
2-CHLORO-10-(3-(4-METHYL-1-
PIPERAZINYL)PROPYL)PHENOTHIAZINE see
PMF500
3-CHLORO-10-(3-(1-METHYL-4-
PIPERAZINYL)PROPYL)PHENOTHIAZINE see
PMF500
CHLORO-3 (N-METHYLPIPERAZINYL-3 PROPYL)-10
PHENOTHIAZINE (FRENCH) see PMF500
2-CHLORO-2-METHYLPROPANE see BQR000
1-CHLORO-2-METHYLPROPENE see IKE000
3-CHLORO-2-METHYLPROPENE see CIU750
1-CHLORO-2-METHYL-1-PROPENE see IKE000
3-CHLORO-2-METHYL-1-PROPENE see CIU750
3-(CHLOROMETHYL) PYRIDINE HYDROCHLORIDE
see CIV000
3-CHLORO-4-METHYL-UMBELLIFERONE BIS(2-
CHLOROETHYL)PHOSPHATE see DFH600
3-CHLORO-4-METHYLUMBELLIFERONE-O-ESTER
with O,O-DIETHYL PHOSPHOROTHIOATE see
CNU750
CHLOROMROWCZAN 1-NAFTYLU (CZECH) see
NBH200
CHLOROMYCETIN see CDP250
CHLORONAFTINA see BIF250
CHLORONAPHTHINE see BIF250
CHLORONITRIN see CDP250
4-CHLORO-3-NITROANILINE see CJA185
2-CHLORO-4-NITROBENZAMIDE see AFH400
CHLORONITROBENZENE see CJA950
2-CHLORONITROBENZENE see CJB750
4-CHLORONITROBENZENE see NFS525
CHLORO-m-NITROBENZENE see CJB250
m-CHLORONITROBENZENE see CJB250
CHLORO-o-NITROBENZENE see CJB750
o-CHLORONITROBENZENE see CJB750
p-CHLORONITROBENZENE see NFS525
1-CHLORO-2-NITROBENZENE see CJB750
1-CHLORO-3-NITROBENZENE see CJB250
1-CHLORO-4-NITROBENZENE see NFS525
2-CHLORO-1-NITROBENZENE see CJB750
4-CHLORO-1-NITROBENZENE see NFS525
m-CHLORONITROBENZENE (DOT) see CJB250
o-CHLORONITROBENZENE, liquid (DOT) see CJB750
CHLORONITROBENZENE, ortho, liquid (DOT) see
CJA950
2-CHLORO-4-NITROPHENYLAMIDE-6-
CHLOROSALICYLIC ACID see DFV400
4-CHLORO-2-NITROPHENYL p-CHLOROPHENYL
ETHER see CJD600
N-(2-CHLORO-4-NITROPHENYL)-5-
CHLOROSALICYLAMIDE see DFV400
CHLORONITROPROPANE see CJE000
1-CHLORO-1-NITROPROPANE see CJE000
2-CHLORO-4-NITROTOLUENE see CJG800

4-CHLORO-3-NITRO-α,α,α-TRIFLUOROTOLUENE
see NFS700
3-(6-CHLORO-2-OXOBENZOXAZOLIN-3-
YL)METHYL-O,O-DIETHYL
PHOSPHOROTHIOLOTHIONATE see BDJ250
N-(4-CHLORO-3-OXOBUTYL)-p-
TOLUENESULFONAMIDE see THH355
CHLOROPARAFFINE 40G see PAH780
CHLOROPENTAFLUOROETHANE see CJI500
CHLOROPENTAHYDROXYDIALUMINUM see
AHA000
1-CHLOROPENTANE see PBW500
1-CHLORO-3-(PENTYLOXY)-2-PROPANOL see
CHS250
CHLOROPEROXYL see CDW450
CHLOROPHACINONE see CJJ000
CHLOROPHEN see PAX250
CHLOROPHENAMADIN see CJJ250
CHLOROPHENAMIDINE see CJJ250
4-CHLOROPHENE-1,3-DIAMINE see CJY120
2-CHLOROPHENOL see CJK250
3-CHLOROPHENOL see CJK500
4-CHLOROPHENOL see CJK750
m-CHLOROPHENOL see CJK500
o-CHLOROPHENOL see CJK250
p-CHLOROPHENOL see CJK750
o-CHLOROPHENOL, solid see CJK250
o-CHLOROPHENOL, liquid see CJK250
CHLOROPHENOLS see CJL000
CHLOROPHENOTHAN see DAD200
CHLOROPHENOTHANE see DAD200
4-(3-(2-CHLOROPHENOTHIAZIN-10-YL)PROPYL)-1-
PIPERAZINEETHANOL see CJM250
CHLOROPHENOTOXUM see DAD200
10-CHLOROPHENOXARSINE see CJM750
10-CHLORO-10H-PHENOXARSINE see CJM750
(4-CHLOROPHENOXY)ACETIC ACID see CJN000
p-CHLOROPHENOXYACETIC ACID see CJN000
10-CHLOROPHENOXYARSINE see CJM750
1-(4-CHLOROPHENOXY)-3,3-DIMETHYL-1-(1,2,4-
TRIAZOL-1-YL)-2-BUTAN-2-ONE see CJO250
1-(4-CHLOROPHENOXY)-3,3-DIMETHYL-1-(1H-1,2,4-
TRIAZOL-1-YL)-2-BUTANONE see CJO250
2-(4-CHLOROPHENOXY-2-METHYL)PROPIONIC
ACID see CIR500
2-(α-p-CHLOROPHENYLACETYL)INDANE-1,3-
DIONE see CJJ000
dl-4-CHLOROPHENYLALANINE see FAM100
dl-p-CHLOROPHENYLALANINE see FAM100
3-CHLOROPHENYLAMINE see CEH675
4-CHLOROPHENYLAMINE see CEH680
m-CHLOROPHENYLAMINE see CEH675
4-CHLOROPHENYLAMINE HYDROCHLORIDE see
CJR200
p-CHLOROPHENYLAMINE HYDROCHLORIDE see
CJR200
N-(((4-CHLOROPHENYL)AMINO)CARBONYL)-2,6-
DIFLUOROBENZAMIDE see CJV250
4-((3-((4-CHLOROPHENYL)AMINO)-4,5-DIHYDRO-
2H-BENZ(G)INDAZOL-2-YL)ACETYL)MORPHOLINE
see CJR210
3-((4-CHLOROPHENYL)AMINO)-4,5-DIHYDRO-N-
(PHENYLMETHYL)-2H-BENZ(g)INDAZOLE-2-
ACETAMIDE see CJR220
2-(o-CHLOROPHENYL)BENZIMIDAZOLE see CJR550
2-(2-CHLOROPHENYL)-1H-BENZIMIDAZOLE see
CJR550

1-(p-CHLORO-α-PHENYLBENZYL)-4-(2-((2-HYDROXYETHOXY)ETHYL)PIPERAZINE) see CJR909

1-(p-CHLORO-α-PHENYLBENZYL)-4-(m-METHYLBENZYL)PIPERAZINE see HGC500

2-(2-(4-(p-CHLORO-α-PHENYLBENZYL)-1-PIPERAZINYL)ETHOXY)ETHANOL see CJR909

p-CHLOROPHENYL CHLORIDE see DEP800

2-(4-CHLOROPHENYL)-5-((4-CHLOROPHENYL)METHOXY)-4-IODO-3(2H)-PYRIDAZINONE see CJT800

(2-CHLOROPHENYL)-α-(4-CHLOROPHENYL)-5-PYRIMIDINEMETHANOL see FAK100

α-(2-CHLOROPHENYL)-α-(4-CHLOROPHENYL)-5-PYRIMIDINEMETHANOL see FAK100

5-(4'-CHLOROPHENYL)-2,4-DIAMINO-6-ETHYLPYRIMIDINE see TGD000

(o-CHLOROPHENYL)(3-(2,4-DICHLOROPHENOXY)-2-HYDROXYPROPYL)HYDROXYARSINEOXIDE see DGA425

1-(4-CHLOROPHENYL)-3-(2,6-DIFLUOROBENZOYL)UREA see CJV250

1-(p-CHLOROPHENYL)-3,3-DIMETHYLUREA see CJX750

3-(4-CHLOROPHENYL)-1,1-DIMETHYLUREA see CJX750

3-(p-CHLOROPHENYL)-1,1-DIMETHYLUREA see CJX750

N'-(4-CHLOROPHENYL)-N,N-DIMETHYLUREA see CJX750

N-(p-CHLOROPHENYL)-N',N'-DIMETHYLUREA see CJX750

1-(4-CHLORO PHENYL)-3,3-DIMETHYLUREE (FRENCH) see CJX750

2-CHLORO-p-PHENYLENEDIAMINE see CEG600

4-CHLORO-m-PHENYLENEDIAMINE see CJY120

4-CHLORO-o-PHENYLENEDIAMINE see CFK125

o-CHLORO-p-PHENYLENEDIAMINE see CEG600

p-CHLORO-o-PHENYLENEDIAMINE see CFK125

4-CHLORO-1,2-PHENYLENEDIAMINE see CFK125

4-CHLOROPHENYLENE-1,3-DIAMINE see CJY120

4-CHLORO-1,3-PHENYLENEDIAMINE see CJY120

1-(4-CHLOROPHENYL)ETHANONE see CEB250

2-CHLORO-1-PHENYLETHANONE see CEA750

3-(3-(4-(2-(4-CHLOROPHENYL)ETHYL)PHENYL)-1,2,3,4-TETRAHYDRO-1-NAPHTHALENYL)-4-HYDROXY2H-1-BENZOPYRAN-2-ONE see CKA030

5-(4-CHLOROPHENYL)-6-ETHYL-2,4-PYRIMIDINEDIAMINE see TGD000

(m-CHLOROPHENYL)HYDROXY(β-HYDROXYPHENETHYL)ARSINE OXIDE see CKA575

(p-CHLOROPHENYL)HYDROXYMETHYLARSINE OXIDE see CKD800

4-(4-(p-CHLOROPHENYL)-4-HYDROXYPIPERIDINO)-4'-(DIMETHYLAMINO)BUTYROPHENO NE see CKA580

m-CHLOROPHENYL ISOCYANATE see CKA750

p-CHLOROPHENYL ISOCYANATE see CKB000

1-(p-CHLOROPHENYL)-5-ISOPROPYLBIGUANIDE see CKB250

CHLOROPHENYLMETHANE see BEE375

3-(4-CHLOROPHENYL)-1-METHOXY-1-METHYLUREA see CKD500

N'-(4-CHLOROPHENYL)-N-METHOXY-N-METHYLUREA see CKD500

N-(4-CHLOROPHENYL)-N'-METHOXY-N-METHYLUREA see CKD500

2-(o-CHLOROPHENYL)-2-(METHYLAMINO)CYCLOHEXANONE HYDROCHLORIDE see CKD750

(4-CHLOROPHENYL)METHYLARSINIC ACID see CKD800

o-CHLOROPHENYL METHYLCARBAMATE see CKF000

2-CHLOROPHENYL-N-METHYLCARBAMATE see CKF000

6-(3-(o-CHLOROPHENYL)-5-METHYL-4-ISOXAZOLECARBOXAMIDEO)-3,3-DIMETHYL-7-OXO-4-THIA-1-AZABICYCLO(3.2.0)HEPTANE-2-CARBOXYLIC ACID, SODIUM SALT, MONOHYDRATE see SLJ000

1-(p-CHLOROPHENYL)-3-METHYL-3-NITROSOUREA see MMW775

3-(p-CHLOROPHENYL)-1-METHYL-1-NITROSOUREA see MMW775

4-CHLORO-2-(PHENYLMETHYL)PHENOL SODIUM SALT see SFB200

2-((p-CHLOROPHENYL)PHENYLACETYL)-1,3-INDANDIONE see CJJ000

2(2-(4-CHLOROPHENYL)-2-PHENYLACETYL)INDAN-1,3-DIONE see CJJ000

2-((4-CHLOROPHENYL)PHENYLACETYL)-1H-INDENE-1,3(2H)-DIONE see CJJ000

8-(4-(4-CHLOROPHENYLPHENYLMETHYL)PIPERAZINYL)-3,6-DIOXAOCTANOL see HHK050

4-(p-CHLOROPHENYL)PIPERAZINYL 3,4,5-TRIMETHYOXYPHENYL KETONE see CKJ100

1-p-CHLOROPHENYL-1-(2-PYRIDYL)-3-DIMETHYLAMINOPROPANE MALEATE see TAI500

5-(p-CHLOROPHENYL)-2,3,5,6-TETRAHYDROIMIDAZO(1,2-c)QUINAZOLIN see CKK050

((p-CHLOROPHENYL)THIO)METHANETHIOL-S-ESTER with O,O-DIMETHYL PHOSPHORODITHIOATE see MQH750

S-((p-CHLOROPHENYLTHIO)METHYL)-O,O-DIETHYL PHOSPHORODITHIOATE see TNP250

S-(4-CHLOROPHENYLTHIOMETHYL)DIETHYL PHOSPHOROTHIOLOTHIONATE see TNP250

S-(((4-CHLOROPHENYL)THIO)METHYL) O,O-DIMETHYLPHOSPHORODITHIOATE see MQH750

S-(((p-CHLOROPHENYL)THIO)METHYL) O,O-DIMETHYL PHOSPHORODITHIOATE see MQH750

2-P-CHLOROPHENYL-2-(1H-1,2,4-TRIAZOL-1-YLMETHYL)HEXANENITRILE see MRW775

p-CHLOROPHENYLTRICHLOROMETHANE see TIR900

4-CHLOROPHENYL-2,4,5-TRICHLOROPHENYL SULFONE see CKM000

p-CHLOROPHENYL-2,4,5-TRICHLOROPHENYL SULFONE see CKM000

p-CHLOROPHENYL-2,4,5-TRICHLOROPHENYL SULPHONE see CKM000

CHLOROPHENYLTRICHLOROSILANE see CKM250

(p-CHLOROPHENYL)TRIFLUOROMETHANE see CEM825

1-(p-CHLOROPHENYL)-4-(3,4,5-TRIMETHOXYBENZOYL)PIPERAZINE see CKJ100

CHLOROPHOS see TIQ250

CHLORO-PHOSPHONOTHIOIC ACID-O,O-DIETHYL ESTER see DJW600

CHLOROPHOSPHORIC ACID DIETHYL ESTER see DIY000

4-CHLORO-o-TOLUIDINE HYDROCHLORIDE see CLK235

4-CHLORO-o-TOLUIDINE HYDROCHLORIDE (DOT) see CLK235

N'-(4-CHLORO-o-TOLYL)-N,N-DIMETHYLFORMAMIDINE see CJJ250

((4-CHLORO-o-TOLYL)OXY)ACETIC ACID see CIR250

(p-CHLORO-o-TOLYLOXY)ACETIC ACID SODIUM SALT see SIL500

2-(p-CHLORO-o-TOLYLOXY)PROPIONIC ACID see CIR500

CHLOROTRIANISENE see CLO750

CHLOROTRIANIZEN see CLO750

CHLOROTRIAZINE see TJD750

CHLOROTRIBUTYLSTANNANE see CLP500

1-CHLORO-4-(TRICHLOROMETHYL)BENZENE see TIR900

2-CHLORO-6-(TRICHLOROMETHYL)PYRIDINE see CLP750

2-CHLORO-1-(2,4,5-TRICHLOROPHENYL)VINYL DIMETHYL PHOSPHATE see TBW100

2-CHLORO-1-(2,4,5-TRICHLOROPHENYL)VINYL PHOSPHORIC ACID DIMETHYL ESTER see TBW100

CHLORO(TRIETHYLPHOSPHINE)GOLD see CLQ500

CHLOROTRIFLUORIDE see CDX750

1-CHLORO-2,2,2-TRIFLUOROETHANE see TJY175

2-CHLORO-1,1,1-TRIFLUOROETHANE see TJY175

2-CHLORO-1,1,2-TRIFLUOROETHYL DIFLUOROMETHYL ETHER see EAT900

CHLOROTRIFLUOROETHYLENE see CLQ750

1-CHLORO-1,2,2-TRIFLUOROETHYLENE see CLQ750

2-CHLORO-1,1,2-TRIFLUOROETHYLENE see CLQ750

CHLOROTRIFLUOROMETHANE see CLR250

CHLOROTRIFLUOROMETHANE mixed with TRIFLUOROMETHANE see FOO562

CHLOROTRIFLUOROMETHANE and TRIFLUOROMETHANE AZEOTROPIC MIXTURE (DOT) see FOO562

4-CHLOROTRIFLUOROMETHYLBENZENE see CEM825

p-CHLOROTRIFLUOROMETHYLBENZENE see CEM825

4-(4-CHLORO-α,α,α-TRIFLUORO-m-TOLYL)-1-(4,4-BIS(p-FLUOROPHENYL)BUTYL)-4-PIPERIDINOL see PAP250

2-CHLORO-α-α-α-TRIFLUORO-p-TOLYL-3-ETHOXY-4-NITROPHENYL ETHER see OQU100

7-CHLORO-4,6,2'-TRIMETHOXY-6'-METHYLGRIS-2'-EN-3,4'-DIONE see GKE000

1-CHLORO-4-(TRIMETHYL)-BENZENE (9CI) see CEM825

endo-2-CHLORO-1,7,7-TRIMETHYLBICYCLO(2.2.1)HEPTANE see BMD300

2-CHLORO-N,N,N-TRIMETHYLETHANAMINIUM CHLORIDE see CMF400

CHLOROTRIMETHYLSILANE see TLN250

CHLOROTRIMETHYLSTANNANE see CLT000

CHLOROTRIMETHYLTIN see CLT000

CHLOROTRIPHENYLSTANNANE see CLU000

CHLOROTRIPHENYLTIN see CLU000

CHLOROTRIPROPYLSTANNANE see CLU250

CHLOROTRISIN see CLO750

CHLOROTRIS(p-METHOXYPHENYL)ETHYLENE see CLO750

CHLORO(TRIVINYL)STANNANE see CLU500

CHLOROUS ACID, SODIUM SALT (8CI,9CI) see SFT500

CHLOROVINYLARSINE DICHLORIDE see CLV000

β-CHLOROVINYLBICHLOROARSINE see CLV000

2-CHLOROVINYLDICHLOROARSINE see CLV000

(2-CHLOROVINYL)DICHLOROARSINE see CLV000

β-CHLOROVINYL ETHYLETHYNYL CARBINOL see CHG000

3-(β-CHLOROVINYL)-1-PENTYN-3-OL see CHG000

CHLOROVULES see CDP250

CHLOROWAX see PAH780

CHLOROWAX 40 see PAH780

CHLOROWAX 50 see PAH780

CHLOROWAX 70 see PAH780

CHLOROWAX 70S see PAH780

CHLOROWAX 500C see PAH780

CHLOROWAX S 70 see PAH780

CHLOROWODOR (POLISH) see HHL000

CHLOROX see SHU500

CHLORO-XYLENOL see CLW000

p-CHLORO-m-XYLENOL see CLW000

4-CHLORO-3,5-XYLENOL see CLW000

(4-CHLORO-6-(2,3-XYLIDINO)-2-PYRIMIDINYLTHIO)ACETIC ACID see CLW250

2-((4-CHLORO-6-(2,3-XYLIDINO)-2-PYRIMIDINYL)THIO)-N-(2-HYDROXYETHYL)ACETAMIDE see CLW500

CHLOROXYPHOS see TIQ250

CHLOROZIRCONYL see ZSJ000

CHLOROZONE see CDP000

CHLOROZOTOCIN see CLX000

6-CHLORO-Δ⁶-17-ACETOXYPROGESTERONE see CBF250

6-CHLORO-Δ⁶-(17-α)ACETOXYPROGESTERONE see CBF250

6-CHLORO-Δ⁶-DEHYDRO-17-ACETOXYPROGESTERONE see CBF250

6-CHLORO-Δ⁴,⁶-PREGNADIENE-17-α-OL-3,20-DIONE-17-ACETATE see CBF250

CHLORPERAZINE see PMF500

CHLORPHACINON (ITALIAN) see CJJ000

CHLORPHENAMIDINE see CJJ250

CHLORPHENIRAMINE MALEATE see TAI500

o-CHLORPHENOL (GERMAN) see CJK250

CHLORPHENVINFOS see CDS750

CHLORPHENVINPHOS see CDS750

(±)-p-CHLORPHENYLALANINE see FAM100

3-(4-CHLOR-PHENYL)-1,1-DIMETHYL-HARNSTOFF (GERMAN) see CJX750

3-(4-CHLORPHENYL)-1-METHOXY-1-METHYLHARNSTOFF (GERMAN) see CKD500

((4-CHLORPHENYL)-1-PHENYL)-ACETYL-1,3-INDANDION (GERMAN) see CJJ000

1-(4-CHLORPHENYL)-1-PHENYL-ACETYL-INDAN-1,3-DION (GERMAN) see CJJ000

2(2-(4-CHLOR-PHENYL-2-PHENYL)ACETYL)INDAN-1,3-DION (GERMAN) see CJJ000

CHLOR-O-PIC see CKN500

CHLORPIKRIN (GERMAN) see CKN500

3-CHLORPROPAN-1-OL see CKP600

3-CHLORPROPEN (GERMAN) see AGB250

CHLORPYRIFOS see CMA100

CHLORPYRIFOS-ETHYL see CMA100

CHLORPYRIFOS-METHYL see CMA250

CHLORPYRIPHOS see CMA100

CHLORPYRIPHOS-ETHYL see CMA100

CHLORSAEURE (GERMAN) see SFS000

CHLORSAL see CLH750

CHLORSEPTOL see CDP000

CHLORSULFONAMIDO
DIHYDROBENZOTHIADIAZINE DIOXIDE see
CFY000
CHLORTEN see MIH275
CHLORTETRACYCLINE see CMA750
CHLORTETRACYCLINE, 6-DEMETHYL- see MIJ500
CHLORTHALONIL (GERMAN) see TBQ750
CHLORTHIAZIDE see CLH750
CHLORTHIEPIN see EAQ750
2-CHLOR-4-TOLUIDIN (CZECH) see CLK210
3-CHLOR-2-TOLUIDIN (CZECH) see CLK200
α-CHLORTOLUOL (GERMAN) see BEE375
N'-(4-CHLOR-o-TOLYL)-N,N-
DIMETHYLFORMAMIDIN (GERMAN) see CJJ250
CHLORTOX see CDR750
CHLORTRIANISEN see CLO750
CHLORTRIFLUORAETHYLEN (GERMAN) see CLQ750
CHLOR-TRIMETON see TAI500
CHLOR-TRIMETON MALEATE see TAI500
CHLOR-TRIPOLON see TAI500
CHLORURE d'ALUMINUM (FRENCH) see AGY750
CHLORURE ANTIMONIEUX see AQC500
CHLORURE d'ARSENIC (FRENCH) see ARF500
CHLORURE ARSENIEUX (FRENCH) see ARF500
CHLORURE de BENZENYLE (FRENCH) see BFL250
CHLORURE de BENZYLE (FRENCH) see BEE375
CHLORURE de BENZYLIDENE see BAY300
CHLORURE de BORE (FRENCH) see BMG500
CHLORURE de BUTYLE (FRENCH) see BQQ750
CHLORURE de CHLORACETYLE (FRENCH) see
CEC250
CHLORURE de CHROMYLE (FRENCH) see CML125
CHLORURE de CYANOGENE see COO750
CHLORURE de DICHLORACETYLE (FRENCH) see
DEN400
CHLORURE d'ETHYLE (FRENCH) see EHH000
CHLORURE d'ETHYLENE (FRENCH) see EIY600
CHLORURE d'ETHYLIDENE (FRENCH) see DFF809
CHLORURE de FUMARYLE (FRENCH) see FOY000
CHLORURE de LITHIUM (FRENCH) see LHB000
CHLORURE MERCUREUX (FRENCH) see MCW000
CHLORURE MERCURIQUE (FRENCH) see MCY475
CHLORURE de METHALLYLE (FRENCH) see CIU750
CHLORURE de METHYLE (FRENCH) see MIF765
CHLORURE de METHYLENE (FRENCH) see MJP450
CHLORURE PERRIQUE see FAU000
CHLORURE de VINYLE (FRENCH) see VNP000
CHLORURE de VINYLIDENE (FRENCH) see VPK000
CHLORURE de ZINC (FRENCH) see ZFA000
CHLORURIT see CLH750
CHLORVINPHOS see DGP900
CHLORWASSERSTOFF (GERMAN) see HHL000
CHLORYL see EHH000
CHLORYL ANESTHETIC see EHH000
CHLORYLEA see TIO750
CHLORYL RADICAL see CDW450
CHLORZIDE see CFY000
CHLOTRIDE see CLH750
CHLZ see CLX000
CHOCOLA A see VSK600
CHOLAXINE see SKV200
CHOLECALCIFEROL see CMC750
CHOLEIC ACID see DAQ400
CHOLEREBIC see DAQ400
epi-CHOLESTANOL see EBA100
CHOLESTAN-3-OL, (3-α-5α-)-(9CI) see EBA100
5-α-CHOLESTAN-3-α-OL (8CI) see EBA100

α-CHOLESTANOL (7CI) see EBA100
CHOLEST-5-EN-3-β-OL see CMD750
5-CHOLESTEN-3-β-OL see CMD750
5:6-CHOLESTEN-3-β-OL see CMD750
Δ⁵-CHOLESTEN-3-β-OL see CMD750
5-CHOLESTEN-3-β-OL 3-(p-(BIS(2-
CHLOROETHYL)AMINO)PHENYL)ACETATE see
CME250
CHOLESTERIN see CMD750
CHOLESTEROL see CMD750
CHOLESTEROL BASE H see CMD750
CHOLESTERYL ALCOHOL see CMD750
CHOLESTERYL-p-BIS(2-CHLOROETHYL)AMINO
PHENYLACETATE see CME250
CHOLESTRIN see CMD750
CHOLESTROL see CMD750
CHOLESTYRAMINE see CME400
CHOLESTYRAMINE CHLORIDE see CME400
CHOLESTYRAMINE RESIN see CME400
CHOLIC ACID, MONOSODIUM SALT see SFW000
CHOLIFLAVIN see DBX400
CHOLINE BITARTRATE see CMF300
CHOLINE CHLORHYDRATE see CMF750
CHOLINE CHLORIDE ACETATE see ABO000
CHOLINE CHLORIDE (FCC) see CMF750
CHOLINE DICHLORIDE see CMF400
CHOLINE HYDROCHLORIDE see CMF750
CHOLINIUM CHLORIDE see CMF750
CHOLLY see CNE750
CHOLOREBIC see DAQ400
CHONDRUS see CCL250
CHONDRUS EXTRACT see CCL250
CHORYLEN see TIO750
CHOT see PBC250
CHR 9 see HFT550
CHRISTENSENITE see SCK000
CHROMALUM HEXAHYDRATE see CMG800
2-CHROMANONE see HHR500
CHROMAR see XHS000
CHROMARGYRE see MCV000
CHROMATE(1-),
DIAMMINETETRAKIS(ISOTHIOCYANATO)-,
AMMONIUM, HYDRATE see ANT300
CHROMATE OF POTASSIUM see PLB250
CHROMATE de PLOMB (FRENCH) see LCR000
CHROMATE of SODA see DXC200
CHROME see CMI750
CHROME ALUM see PLB500
CHROMEDIA CC 31 see CCU150
CHROMEDIA CF 11 see CCU150
CHROME FAST BLUE 2R see HJF500
CHROME FERROALLOY see FBD000
CHROME FLUORURE see CMJ560
CHROME GREEN see CMJ900
CHROME GREEN see LCR000
CHROME LEATHER BLACK E see AQP000
CHROME LEATHER BLACK EC see AQP000
CHROME LEATHER BLACK EM see AQP000
CHROME LEATHER BLACK G see AQP000
CHROME LEATHER BLUE 2B see CMO000
CHROME LEATHER BLUE 3B see CMO250
CHROME LEATHER BRILLIANT BLACK ER see
AQP000
CHROME LEATHER BROWN BRLL see CMO750
CHROME LEATHER PURE BLUE see CMO500
CHROME LEATHER SKY BLUE see CMN750
CHROME LEMON see LCR000

CHROME OCHER see CMJ900
CHROME ORANGE see LCS000
CHROME ORE see CMI500
CHROME OXIDE see CMJ900
CHROME OXIDE GREEN see CMJ900
CHROME POTASH ALUM see PLB500
CHROME (TRIOXYDE de) (FRENCH) see CMK000
CHROME VERMILION see MRC000
CHROME YELLOW see LCR000
CHROMIA see CMJ900
CHROMIC ACETATE see CMH000
CHROMIC ACETATE(III) see CMH000
CHROMIC ACETYLACETONATE see TNN250
CHROMIC ACID see CMH250
CHROMIC ACID see CMJ900
CHROMIC ACID see CMK000
CHROMIC(VI) ACID see CMH250
CHROMIC(VI) ACID see CMK000
CHROMIC(III) ACID see CMH260
CHROMIC ACID, BARIUM SALT (1:1) see BAK250
CHROMIC ACID, BIS(TRIPHENYLSILYL) ESTER see BLS750
CHROMIC ACID, CALCIUM SALT (1:1) see CAP500
CHROMIC ACID, CALCIUM SALT (1:1), DIHYDRATE see CAP750
CHROMIC ACID, CHROMIUM(3+) SALT (3:2) see CMI250
CHROMIC ACID, DIAMMONIUM SALT see ANF500
CHROMIC ACID, DI-tert-BUTYL ESTER see BQV000
CHROMIC ACID, DIPOTASSIUM SALT see PKX250
CHROMIC ACID, DISODIUM SALT see SGI000
CHROMIC ACID, DISODIUM SALT, DECAHYDRATE see SFW500
CHROMIC ACID GREEN see CMJ900
CHROMIC ACID, LEAD and MOLYBDENUM SALT see LDM000
CHROMIC ACID, LEAD(2+) SALT (1:1) see LCR000
CHROMIC ACID LEAD SALT with LEAD MOLYBDATE see LDM000
CHROMIC ACID, MERCURY ZINC COMPLEX see ZJA000
CHROMIC ACID, solid (NA 1463) (DOT) see CMK000
CHROMIC ACID, POTASSIUM ZINC SALT (2:2:1) see PLW500
CHROMIC ACID, STRONTIUM SALT (1:1) see SMH000
CHROMIC ACID, solution (UN 1755) (DOT) see CMK000
CHROMIC ACID, ZINC SALT see ZFJ100
CHROMIC ACID, ZINC SALT (1:2) see CMK500
CHROMIC ANHYDRIDE see CMK000
CHROMIC CHLORIDE see CMJ250
CHROMIC CHROMATE see CMI250
CHROMIC FLUORIDE see CMJ560
CHROMIC FLUORIDE, solid (UN1756) (DOT) see CMJ560
CHROMIC FLUORIDE, solution (UN1757) (DOT) see CMJ560
CHROMIC (III) HYDROXIDE see CMH260
CHROMIC NITRATE see CMJ600
CHROMIC OXIDE see CMJ900
CHROMIC OXYCHLORIDE see CML125
CHROMIC PHOSPHATE see CMK300
CHROMIC POTASSIUM SULFATE see PLB500
CHROMIC POTASSIUM SULPHATE see PLB500
CHROMIC TRIFLUORIDE see CMJ560
CHROMIC TRIOXIDE see CMK000
CHROMITE see CMI500
CHROMITE (mineral) see CMI500
CHROMITE ORE see CMI500

CHROMIUM see CMI750
CHROMIUM ACETATE see CMH000
CHROMIUM(2+) ACETATE see CMJ000
CHROMIUM(II) ACETATE see CMJ000
CHROMIUM(III) ACETATE see CMH000
CHROMIUM ACETATE HYDRATE see CMJ000
CHROMIUM ACETYLACETONATE see TNN250
CHROMIUM(3+) ACETYLACETONATE see TNN250
CHROMIUM(III) ACETYLACETONATE see TNN250
CHROMIUM ALLOY, BASE, Cr,C,Fe,N,Si (FERROCHROMIUM) see FBD000
CHROMIUM ALLOY, Cr,C,Fe,N,Si see FBD000
CHROMIUM, BIS(BENZENE)-(8CI) see BGY700
CHROMIUM(1+), BIS(BENZENE)-, IODIDE (8CI) see BGY720
CHROMIUM, BIS(BENZENE)IODO- see BGY720
CHROMIUM, BIS(eta⁶)-BENZENE)-(9CI) see BGY700
(CHROMIUM(1+), BIS(eta⁶)-BENZENE)-, IODIDE (9CI) see BGY720
CHROMIUM CARBONATE see CMJ100
CHROMIUM CARBONYL (MAK) see HCB000
CHROMIUM CARBONYL (OC-6-11) (9CI) see HCB000
CHROMIUM CHLORIDE see CMJ250
CHROMIUM(III) CHLORIDE (1:3) see CMJ250
CHROMIUM CHLORIDE, anhydrous see CMJ250
CHROMIUM CHLORIDE, HEXAUREA see HEZ800
CHROMIUM CHLORIDE OXIDE see CML125
CHROMIUM CHROMATE (MAK) see CMI250
CHROMIUM COMPOUNDS see CMJ500
CHROMIUM DIACETATE see CMJ000
CHROMIUM DICHLORIDE DIOXIDE see CML125
CHROMIUM DIOXIDE DICHLORIDE see CML125
CHROMIUM(VI) DIOXYCHLORIDE see CML125
CHROMIUM(II), DIPHENYL- see BGY700
CHROMIUM(III), DIPHENYL-, IODIDE see BGY720
CHROMIUM DISODIUM OXIDE see DXC200
CHROMIUM(III) FLUORIDE see CMJ560
CHROMIUM HEXACARBONYL see HCB000
CHROMIUM(3+), HEXAKIS(UREA-O)-, TRICHLORIDE, (OC-6-11)-(9CI) see HEZ800
CHROMIUM(3+), HEXAKIS(UREA)-, TRICHLORIDE (8CI) see HEZ800
CHROMIUM(III) HEXA-UREA CHLORIDE see HEZ800
CHROMIUM(III) HYDROXIDE see CMH260
CHROMIUM LEAD OXIDE see LCS000
CHROMIUM METAL (OSHA) see CMI750
CHROMIUM MONOPHOSPHATE see CMK300
CHROMIUM NICKEL OXIDE see NDA100
CHROMIUM NITRATE see CMJ600
CHROMIUM (3+) NITRATE see CMJ600
CHROMIUM NITRATE (DOT) see CMJ600
CHROMIUM(III) NITRATE see CMJ600
CHROMIUM ORTHOPHOSPHATE see CMK300
CHROMIUM OXIDE see CMJ900
CHROMIUM OXIDE see CMK000
CHROMIUM(3+) OXIDE see CMJ900
CHROMIUM(VI) OXIDE see CMK000
CHROMIUM(III) OXIDE see CMJ900
CHROMIUM(VI) OXIDE (1:3) see CMK000
CHROMIUM(III) OXIDE (2:3) see CMJ900
CHROMIUM OXIDE, NICKEL OXIDE, and IRON OXIDE FUME see IHE000
CHROMIUM OXYCHLORIDE see CML125
CHROMIUM PHOSPHATE see CMK300
CHROMIUM POTASSIUM SULFATE (1:1:2) see PLB500
CHROMIUM POTASSIUM SULPHATE see PLB500
CHROMIUM POTASSIUM ZINC OXIDE see CMK400
CHROMIUM SESQUIOXIDE see CMJ900

CHROMIUM SODIUM OXIDE see SGI000
CHROMIUM SODIUM OXIDE see DXC200
CHROMIUM(III) SULFATE, HEXAHYDRATE (2:3:6) see CMG800
CHROMIUM SULFATE, PENTADECAHYDRATE see CMK425
CHROMIUM TRIACETATE see CMH000
CHROMIUM TRIACETYLACETONATE see TNN250
CHROMIUM TRICHLORIDE see CMJ250
CHROMIUM TRIFLUORIDE see CMJ560
CHROMIUM TRIHYDROXIDE see CMH260
CHROMIUM TRINITRATE see CMJ600
CHROMIUM TRIOXIDE see CMK000
CHROMIUM(3+) TRIOXIDE see CMJ900
CHROMIUM(6+) TRIOXIDE see CMK000
CHROMIUM TRIOXIDE, anhydrous (DOT) see CMK000
CHROMIUM TRIOXIDE, anhydrous (UN 1463) (DOT) see CMK000
CHROMIUM TRIS(ACETYLACETONATE) see TNN250
CHROMIUM TRIS(BENZOYLACETONATE) see TNN500
CHROMIUM TRIS(2,4-PENTANEDIONATE) see TNN250
CHROMIUM YELLOW see LCR000
CHROMIUM ZINC OXIDE see ZFJ100
CHROMIUM(6+)ZINC OXIDE HYDRATE (1:2:6:1) see CMK500
CHROMOFLAVINE see DBX400
CHROMOSMON see BJI250
CHROMOSORB T see TAI250
CHROMOSULFURIC ACID (UN 2240) (DOT) see CLG500
CHROMOTRICHIA FACTOR see AIH600
ANTI-CHROMOTRICHIA FACTOR see AIH600
anti-CHROMOTRICHIA FACTOR see AIH600
CHROMO (TRIOSSIDO di) (ITALIAN) see CMK000
CHROMOUS ACETATE see CMJ000
CHROMOUS ACETATE MONOHYDRATE see CMJ000
CHROMOXYCHLORID (GERMAN) see CML125
CHROMSAEUREANHYDRID (GERMAN) see CMK000
CHROMTRIOXID (GERMAN) see CMK000
CHROMYLCHLORID (GERMAN) see CML125
CHROMYL CHLORIDE see CML125
CHROOMOXYLCHLORIDE (DUTCH) see CML125
CHROOMTRIOXYDE (DUTCH) see CMK000
CHROOMZUURANHYDRIDE (DUTCH) see CMK000
CHRYSANTHEMUMDICARBOXYLIC ACID MONOMETHYL ESTER PYRETHROLONE ESTER see POO100
CHRYSAZIN see DMH400
CHRYSAZIN-3-CARBOXYLIC ACID see RHZ700
CHRYSENE see CML810
CHRYSOIDIN see PEK000
CHRYSOIDINE see PEK000
CHRYSOIDINE(II) see PEK000
CHRYSOIDINE A see PEK000
CHRYSOIDINE B see PEK000
CHRYSOIDINE C CRYSTALS see PEK000
CHRYSOIDINE G see PEK000
CHRYSOIDINE GN see PEK000
CHRYSOIDINE HR see PEK000
CHRYSOIDINE J see PEK000
CHRYSOIDINE M see PEK000
CHRYSOIDINE ORANGE see PEK000
CHRYSOIDINE PRL see PEK000
CHRYSOIDINE PRR see PEK000
CHRYSOIDINE SL see PEK000

CHRYSOIDINE SPECIAL (biological stain and indicator) see PEK000
CHRYSOIDINE SS see PEK000
CHRYSOIDINE Y see PEK000
CHRYSOIDINE Y BASE NEW see PEK000
CHRYSOIDINE Y CRYSTALS see PEK000
CHRYSOIDINE Y EX see PEK000
CHRYSOIDINE YGH see PEK000
CHRYSOIDINE YL see PEK000
CHRYSOIDINE YN see PEK000
CHRYSOIDINE Y SPECIAL see PEK000
CHRYSOIDIN FB see PEK000
CHRYSOIDIN Y see PEK000
CHRYSOIDIN YN see PEK000
CHRYSOMYKINE see CMA750
CHRYSON see BEP500
CHRYSOTILE ASBESTOS see ARM268
CHRYSRON see BEP500
CHRYTEMIN see DLH630
CHRYZOIDYNA F.B. (POLISH) see PEK000
CHWASTOKS see SIL500
CHWASTOX see SIL500
CHWASTOX see CIR250
CI-2 see MAV750
C.I. 23 see OHI200
C.I. 27 see HGC000
C.I. 79 see FMU070
C.I. 185 see FMU080
C.I. 258 see SBC500
CI-337 see ASA500
CI-406 see PAN100
CI 581 see CKD750
CI-588 see BGO500
C.I. 640 see FAG140
C.I. 671 see FMU059
C.I. 766 see FEW000
C.I. 1956 see CKN000
C.I. 10305 see PID000
C.I. 10315 see DUX800
C.I. 10355 see DVX800
C.I. 10360 see HET500
C.I. 11000 see PEI000
C.I. 11020 see DOT300
C.I. 11025 see DPO200
C.I. 11050 see DHM500
C.I. 11160 see AIC250
C.I. 11270 see PEK000
C.I. 11380 see FAG130
C.I. 11390 see FAG135
C.I. 11855 see AAQ250
C.I. 12055 see PEJ500
C.I. 12100 see TGW000
C.I. 12140 see XRA000
C.I. 12150 see CMS238
C.I. 12156 see DOK200
C.I. 13020 see CCE500
C.I. 13065 see MDM775
C.I. 14720 see HJF500
C.I. 15985 see FAG150
C.I. 16150 see FMU070
C.I. 16155 see FAG018
C.I. 16255 see FMU080
C.I. 19140 see FAG140
C.I. 22120 see SGQ500
C.I. 22610 see CMO000
C.I. 23060 see DEQ600
C.I. 23635 see CMM330
C.I. 23850 see CMO250

C.I. 23860 see BGT250
C.I. 24110 see DCJ200
C.I. 24400 see CMO500
C.I. 24410 see CMN750
C.I. 26050 see EOJ500
C.I. 26100 see OHI200
C.I. 26105 see SBC500
C.I. 30145 see CMO750
C.I. 30235 see AQP000
C.I. 37025 see NEO000
C.I. 37030 see NEN500
C.I. 37035 see NEO500
C.I. 37077 see TGQ750
C.I. 37085 see CLK235
C.I. 37105 see NMP500
C.I. 37107 see TGR000
C.I. 37115 see AOX250
C.I. 37130 see NEQ500
C.I. 37225 see BBX000
C.I. 37230 see TGJ750
C.I. 37270 see NBE500
C.I. 37275 see AIA750
C.I. 37500 see NAX000
C.I. 41000 see IBA000
C.I. 42040 see BAY750
C.I. 42090 see FMU059
C.I. 42500 see RMK020
C.I. 42555 see AOR500
C.I. 42640 see FAG120
C.I. 45160 see RGW000
C.I. 45330 see FEV000
C.I. 45380 see BNH500
C.I. 45410 see ADG250
C.I. 46005 see BJF000
C.I. 47031 see PGW750
C.I. 59825 see JAT000
C.I. 60700 see AKP750
C.I. 60710 see AKE250
C.I. 61200 see BMM500
C.I. 61565 see BLK000
C.I. 62030 see DBY700
C.I. 64500 see TBG700
C.I. 71000 see CMU475
C.I. 75300 see COG000
C.I. 75660 see MRN500
C.I. 75670 see QCA000
C.I. 76000 see AOQ000
C.I. 76010 see PEY250
C.I. 76020 see ALL500
C.I. 76025 see PEY000
C.I. 76027 see CJY120
C.I. 76035 see TGL750
C.I. 76042 see TGM000
C.I. 76043 see TGM400
C.I. 76043 see DCE600
C.I. 76050 see DBO000
C.I. 76051 see DBO400
C.I. 76060 see PEY500
C.I. 76065 see CEG600
C.I. 76070 see ALL750
C.I. 76500 see CCP850
C.I. 76505 see REA000
C.I. 76515 see PPQ500
C.I. 76520 see ALT000
C.I. 76545 see ALS990
C.I. 76555 see NEM480
C.I. 76605 see NAW500
C.I. 77000 see AGX000

C.I. 77002 see AHC000
C.I. 77050 see AQB750
C.I. 77052 see AQF000
C.I. 77056 see AQC500
C.I. 77060 see AQL500
C.I. 77061 see AQF500
C.I. 77086 see ARI000
C.I. 77099 see BAJ250
C.I. 77103 see BAK250
C.I. 77120 see BAP000
C.I. 77180 see CAD000
C.I. 77185 see CAD250
C.I. 77199 see CAJ750
C.I. 77205 see CMS212
C.I. 77223 see CAP500
C.I. 77223 see CAP750
C.I. 77231 see CAX750
C.I. 77265 see CBT500
C.I. 77266 see CBT750
C.I. 77288 see CMJ900
C.I. 77295 see CMJ250
C.I. 77320 see CNA250
C.I. 77322 see CND125
C.I. 77323 see CND825
C.I. 77400 see CNI000
C.I. 77402 see CNO000
C.I. 77410 see COF500
C.I. 77480 see GIS000
C.I. 77491 see IHC450
C.I. 77575 see LCF000
C.I. 77577 see LDN000
C.I. 77578 see LDS000
C.I. 77580 see LCX000
C.I. 77600 see LCR000
C.I. 77601 see LCS000
C.I. 77605 see MRC000
C.I. 77610 see LCU000
C.I. 77620 see LCV100
C.I. 77622 see LDU000
C.I. 77630 see LDY000
C.I. 77640 see LDZ000
C.I. 77718 see TAB750
C.I. 77726 see MAT250
C.I. 77727 see MAT500
C.I. 77728 see MAS000
C.I. 77755 see PLP000
C.I. 77760 see MCT500
C.I. 77764 see MCW000
C.I. 77775 see NCW500
C.I. 77777 see NDF500
C.I. 77779 see NCY500
C.I. 77795 see PJD500
C.I. 77805 see SBO500
C.I. 77820 see SDI500
C.I. 77847 see SMM000
C.I. 77864 see TGC000
C.I. 77891 see TGG760
C.I. 77901 see TOC750
C.I. 77938 see VDU000
C.I. 77940 see VEZ000
C.I. 77945 see ZBJ000
C.I. 77947 see ZKA000
C.I. 77955 see ZFJ100
C.I. 77964 see ZJS400
C.I. 11160B see AIC250
C.I. 45370:1 see DDO200
C.I. 3/11855 see AAQ250
C.I. 52015 (CZECH) see BJI250

C.I. ACID BLUE 9, DIAMMONIUM SALT see FMU059
C.I. ACID ORANGE 10 see HGC000
C.I. ACID RED 2 see CCE500
C.I. ACID RED 18 see FMU080
C.I. ACID RED 26 see FMU070
C.I. ACID RED 92 see ADG250
C.I. ACID RED 114 see CMM330
C.I. ACID RED 14, DISODIUM SALT see HJF500
C.I. ACID RED 26, DISODIUM SALT see FMU070
C.I. ACID RED 114, DISODIUM SALT see CMM330
C.I. ACID YELLOW 23 see FAG140
C.I. ACID YELLOW 36 see MDM775
C.I. ACID YELLOW 73 see FEW000
C.I. ACID YELLOW 36 MONOSODIUM SALT see MDM775
C.I. ACID YELLOW 23, TRISODIUM SALT see FAG140
CIANURO di SODIO (ITALIAN) see SGA500
CIANURO di VINILE (ITALIAN) see ADX500
C.I. AZOIC COUPLING COMPONENT 1 see NAX000
C.I. AZOIC COUPLING COMPONENT 107 see TGR000
C.I. AZOIC DIAZO COMPONENT 6 see NEO000
C.I. AZOIC DIAZO COMPONENT 7 see NEN500
C.I. AZOIC DIAZO COMPONENT 11 see CLK235
C.I. AZOIC DIAZO COMPONENT 12 see NMP500
C.I. AZOIC DIAZO COMPONENT 13 see NEQ500
C.I. AZOIC DIAZO COMPONENT 37 see NEO500
C.I. AZOIC DIAZO COMPONENT 48 see DCJ200
C.I. AZOIC DIAZO COMPONENT 112 see BBX000
C.I. AZOIC DIAZO COMPONENT 113 see TGJ750
C.I. AZOIC DIAZO COMPONENT 114 see NBE700
C.I. AZOIC RED 83 see MGO750
C.I. 41000B see IBB000
CIBA 709 see DGQ875
CIBA 1414 see MRH209
CIBA 2059 see DUK800
CIBA 5968 see HGP495
CIBA 5968 see HGP500
CIBA 7115 see KFK000
CIBA 8514 see CJJ250
CIBA 8514 see KEA000
CIBA 12669A see MIW500
CIBACET BRILLIANT VIOLET 3B see DBY700
CIBACETE DIAZO NAVY BLUE 2B see DCJ200
CIBACETE RED 3B see AKE250
CIBACETE YELLOW GBA see AAQ250
CIBACET RED 3B see AKE250
CIBACET RED E3B see AKE250
CIBACET SAPPHIRE BLUE G see TBG700
CIBACET YELLOW 2GC see AAQ250
CIBACET YELLOW GBA see AAQ250
CIBA-GEIGY GS 13005 see DSO000
C.I. BASIC BLUE 9 see BJI250
C.I. BASIC GREEN 1, SULFATE (1:1) see BAY750
C.I. BASIC ORANGE 2 see PEK000
C.I. BASIC ORANGE 3 see PEK000
C.I. BASIC ORANGE 14 see BJF000
C.I. BASIC ORANGE 2, MONOHYDROCHLORIDE see PEK000
C.I. BASIC RED 1 see RGW000
C.I. BASIC RED 1, MONOHYDROCHLORIDE see RGW000
C.I. BASIC RED 9, MONOHYDROCHLORIDE see RMK020
C.I. BASIC VIOLET 3 see AOR500
C.I. BASIC YELLOW 2 see IBA000
C.I. BASIC YELLOW 2, FREE BASE see IBB000
C.I. BASIC YELLOW 2, MONOHYDROCHLORIDE see IBA000

CIC see IKH000
CICLOESANO (ITALIAN) see CPB000
CICLOESANOLO (ITALIAN) see CPB750
CICLOESANONE (ITALIAN) see CPC000
CICLORAL see BSM000
CICLOSOM see TIQ250
CICLOSPORIN see CQH100
CIDALON see IHZ000
CIDAMEX see AAI250
CIDANDOPA see DNA200
C.I. DEVELOPER 4 see REA000
C.I. DEVELOPER 5 see NAX000
C.I. DEVELOPER 13 see PEY500
C.I. DEVELOPER 17 see NEO500
CIDEX see GFQ000
C.I. DIRECT BLACK 38 see AQP000
C.I. DIRECT BLACK 38, DISODIUM SALT see AQP000
C.I. DIRECT BLUE 1 see CMN750
C.I. DIRECT BLUE 14 see CMO250
C.I. DIRECT BLUE 15 see CMO500
C.I. DIRECT BLUE 53 see BGT250
C.I. DIRECT BLUE 1, TETRASODIUM SALT see CMN750
C.I. DIRECT BLUE 6, TETRASODIUM SALT see CMO000
C.I. DIRECT BLUE 14, TETRASODIUM SALT see CMO250
C.I. DIRECT BLUE 15, TETRASODIUM SALT see CMO500
C.I. DIRECT BROWN see CMO750
C.I. DIRECT BROWN 78, DIAMMONIUM SALT see FMU059
C.I. DIRECT RED 28 see SGQ500
C.I. DIRECT RED 28, DISODIUM SALT see SGQ500
C.I. 45350 DISODIUM SALT see FEW000
C.I. DISPERSE BLACK 3 see DPO200
C.I. DISPERSE BLACK 6 see DCJ200
C.I. DISPERSE BLACK 6 DIHYDROCHLORIDE see DOA800
C.I. DISPERSE BLUE 1 see TBG700
C.I. DISPERSE ORANGE 11 see AKP750
C.I. DISPERSE RED 15 see AKE250
C.I. DISPERSE VIOLET 8 see DBY700
C.I. DISPERSE YELLOW 3 see AAQ250
CIDOCETINE see CDP250
CIDREX see CFY000
C.I. FOOD BLUE 2 see FMU059
C.I. FOOD ORANGE 4 see HGC000
C.I. FOOD RED 3 see HJF500
C.I. FOOD RED 5 see FMU070
C.I. FOOD RED 6 see FAG018
C.I. FOOD RED 7 see FMU080
C.I. FOOD RED 16 see CMS238
C.I. FOOD RED 6, DISODIUM SALT see FAG018
C.I. FOOD VIOLET 2 see FAG120
C.I. FOOD YELLOW 3 see FAG150
C.I. FOOD YELLOW 4 see FAG140
C.I. FOOD YELLOW 10 see FAG130
C.I. FOOD YELLOW 11 see FAG135
C.I. FOOD YELLOW 3, DISODIUM SALT see FAG150
C.I. 45350 (FREE ACID) see FEV000
CIGARETTE REFINED TAR see CMP800
CIGARETTE SMOKE CONDENSATE see SEC000
CIGARETTE TAR see CMP800
CILAG 61 see HDY000
CILEFA ORANGE S see FAG150
CILEFA PONCEAU 4R see FMU080
CILEFA YELLOW T see FAG140

CILLA BLUE EXTRA see TBG700
CILLA FAST PINK BN see AKE250
CILLA FAST VIOLET B see DBY700
CILLA FAST YELLOW G see AAQ250
CILLA ORANGE R see AKP750
CILLENTA see BFC750
CILLORAL see BDY669
CILOPEN see BDY669
CIMEXAN see MAK700
C.I. MORDANT VIOLET 39, TRIAMMONIUM SALT (8CI) see AGW750
C.I. No. 77278 see CMJ900
C.I. No. 46005:1 see BJF000
C.I. NATURAL BROWN 8 see MAT500
C.I. NATURAL RED 1 see QCA000
C.I. NATURAL YELLOW 8 see MRN500
C.I. NATURAL YELLOW 10 see QCA000
C.I. NATURAL YELLOW 11 see MRN500
CINENE see MCC250
CINERIN I ALLYL HOMOLOG see AFR250
CINNAMAL see CMP969
CINNAMALDEHYDE see CMP969
CINNAMEIN see BEG750
CINNAMENE see SMQ000
CINNAMENOL see SMQ000
CINNAMIC ACID see CMP975
trans-CINNAMIC ACID BENZYL ESTER see BEG750
CINNAMIC ACID, ISOBUTYL ESTER see IIQ000
CINNAMIC ALCOHOL see CMQ740
CINNAMON BARK OIL see CCO750
CINNAMON BARK OIL, CEYLON TYPE (FCC) see CCO750
CINNAMON LEAF OIL see CMQ510
CINNAMON LEAF OIL, CEYLON see CMQ510
CINNAMON OIL see CCO750
CINNAMON OIL, CEYLON see CMQ510
CINNAMOPHENONE see CDH000
CINNAMYL ACETATE see CMQ730
CINNAMYL ALCOHOL see CMQ740
CINNAMYL ALCOHOL ANTHRANILATE see API750
CINNAMYL ALCOHOL, FORMATE see CMR500
CINNAMYL ALCOHOL, SYNTHETIC see CMQ740
CINNAMYL ALDEHYDE see CMP969
CINNAMYL-2-AMINOBENZOATE see API750
CINNAMYL-o-AMINOBENZOATE see API750
CINNAMYL ANTHRANILATE (FCC) see API750
CINNAMYL BUTYRATE see CMQ800
CINNAMYL FORMATE see CMR500
CINNAMYL ISOVALERATE see CMR800
CINNAMYL METHANOATE see CMR500
CINNAMYL PROPIONATE see CMR850
CINNIMIC ALDEHYDE see CMP969
CIN-QUIN see QFS000
CINU see CGV250
C.I. OXIDATION BASE see TGL750
C.I. OXIDATION BASE 4 see TGM400
C.I. OXIDATION BASE 7 see ALS990
C.I. OXIDATION BASE 10 see PEY500
C.I. OXIDATION BASE 12 see DBO000
C.I. OXIDATION BASE 16 see PEY250
C.I. OXIDATION BASE 17 see ALT000
C.I. OXIDATION BASE 20 see TGL750
C.I. OXIDATION BASE 22 see ALL750
C.I. OXIDATION BASE 26 see CCP850
C.I. OXIDATION BASE 31 see REA000
C.I. OXIDATION BASE 32 see PPQ500
C.I. OXIDATION BASE 33 see NAW500
C.I. OXIDATION BASE 35 see TGL750

C.I. OXIDATION BASE 6A see ALT250
C.I. OXIDATION BASE 12A see DBO400
C.I. OXIDATION BASE 200 see TGL750
CIPE see PJS750
C.I. PIGMENT BLACK 6 see CBT750
C.I. PIGMENT BLACK 7 see CBT750
C.I. PIGMENT BLACK 10 see CBT500
C.I. PIGMENT BLACK 13 see CND125
C.I. PIGMENT BLACK 14 see MAS000
C.I. PIGMENT BLACK 16 see ZBJ000
C.I. PIGMENT BROWN 8 see MAS000
C.I. PIGMENT GREEN 17 see CMJ900
C.I. PIGMENT GREEN 21 (9CI) see COF500
C.I. PIGMENT METAL 2 see CNI000
C.I. PIGMENT METAL 3 see GIS000
C.I. PIGMENT METAL 4 see LCF000
C.I. PIGMENT METAL 6 see ZBJ000
C.I. PIGMENT ORANGE 20 see CAJ750
C.I. PIGMENT ORANGE 21 see LCS000
C.I. PIGMENT RED see CHP500
C.I. PIGMENT RED see LCS000
C.I. PIGMENT RED 101 see IHC450
C.I. PIGMENT RED 104 see MRC000
C.I. PIGMENT RED 104 see LDM000
C.I. PIGMENT RED 105 see LDS000
C.I. PIGMENT RED 107 see AQL500
C.I. PIGMENT WHITE 3 see LDY000
C.I. PIGMENT WHITE 4 see ZKA000
C.I. PIGMENT WHITE 6 see TGG760
C.I. PIGMENT WHITE 10 see BAJ250
C.I. PIGMENT WHITE 11 see AQF000
C.I. PIGMENT WHITE 21 see BAP000
C.I. PIGMENT WHITE 25 see CAX750
C.I. PIGMENT WHITE 32 see ZJS400
C.I. PIGMENT YELLOW see ARI000
C.I. PIGMENT YELLOW 31 see BAK250
C.I. PIGMENT YELLOW 32 see SMH000
C.I. PIGMENT YELLOW 33 see CAP500
C.I. PIGMENT YELLOW 33 see CAP750
C.I. PIGMENT YELLOW 34 see LCR000
C.I. PIGMENT YELLOW 35 see CMS212
C.I. PIGMENT YELLOW 36 see ZFJ100
C.I. PIGMENT YELLOW 37 see CAJ750
C.I. PIGMENT YELLOW 46 see LDN000
C.I. PIGMENT YELLOW 48 see LCU000
CIPLAMYCETIN see CDP250
CIPOVIOL W 72 see PKP750
CIRAM see BJK500
CIRCOSOLV see TIO750
C.I. REACTIVE BLUE 19 see BMM500
C.I. REACTIVE BLUE 19, DISODIUM SALT see BMM500
CIRPONYL see MQU750
CIRRASOL 185A see NMY000
CIRRASOL-OD see HCQ500
C.I. SOLVENT BLUE 7 see PEI000
C.I. SOLVENT BLUE 18 see TBG700
C.I. SOLVENT GREEN 3 see BLK000
C.I. SOLVENT ORANGE 2 see TGW000
C.I. SOLVENT ORANGE 3 see PEK000
C.I. SOLVENT ORANGE 7 see XRA000
C.I. SOLVENT ORANGE 15 see BJF000
C.I. SOLVENT RED see CMS238
C.I. SOLVENT RED 19 see EOJ500
C.I. SOLVENT RED 23 see OHI200
C.I. SOLVENT RED 24 see SBC500
C.I. SOLVENT RED 53 see AKE250
C.I. SOLVENT RED 72 see DDO200

C.I. SOLVENT RED 80 see DOK200
C.I. SOLVENT YELLOW 1 see PEI000
C.I. SOLVENT YELLOW 2 see DOT300
C.I. SOLVENT YELLOW 3 see AIC250
C.I. SOLVENT YELLOW 5 see FAG130
C.I. SOLVENT YELLOW 7 see HJF000
C.I. SOLVENT YELLOW 14 see PEJ500
C.I. SOLVENT YELLOW 34 see IBB000
C.I. SOLVENT YELLOW 92 see AAQ250
C.I. SOLVENT YELLOW 94 see FEV000
C.I. SOLVENT YELLOW 99 see AAQ250
CISPLATINO (SPANISH) see PJD000
CISPLATYL see PJD000
CISTEAMINA (ITALIAN) see AJT250
CITARIN L see LFA020
CITGRENILE see MNB500
CITOBARYUM see BAP000
CITOCOR see DJS200
CITOFUR see FLZ050
CITOL see ALT250
CITOMULGAN M see OAV000
CITOSULFAN see BOT250
CITOX see DAD200
CITRACETAL see CMS324
CITRA-FORT see ABG750
CITRAL ETHYLENE GLYCOL ACETAL see CMS324
CITRAL (FCC) see DTC800
CITRETTEN see CMS750
CITRIC ACID see CMS750
CITRIC ACID, anhydrous see CMS750
CITRIC ACID, ACETYL TRIETHYL ESTER see ADD750
CITRIC ACID, AMMONIUM SALT see ANF800
CITRIC ACID, COPPER(2+) SALT (8CI) see CNK625
CITRIC ACID, TRIPOTASSIUM SALT see PLB750
CITRIC ACID, ZINC SALT (2:3) see ZFJ250
CITRO see CMS750
CITROFLEX 2 see TJP750
CITROFLEX A 2 see ADD750
CITRONELLAL see CMS845
CITRONELLAL HYDRATE see CMS850
CITRONELLENE see CMT050
CITRONELLOL see CMT250
α-CITRONELLOL see DTF400
α-CITRONELLYL ACETATE see RHA000
CITRONELLYL ACETATE (FCC) see AAU000
CITRONELLYL BUTYRATE see CMT600
CITRONELLYL ETHYL ETHER see EES100
CITRONELLYL FORMATE see CMT750
CITRONELLYL ISOBUTYRATE see CMT900
CITRONELLYL PHENYLACETATE see CMU050
CITRONELLYL PROPIONATE see CMU100
CITRON YELLOW see PLW500
CITRON YELLOW see ZFJ100
CITRULLAMON see DKQ000
CITRULLAMON see DNU000
CITRUS RED No. 2 see DOK200
CITTERAL see SEQ000
C.I. VAT BLACK 8 see CMU475
CIVETONE see CMU850
cis-CIVETONE see CMU850
CIZARON see TMP750
CK3 see CBT750
ChKhZ 9 see BBS300
CL 337 see ASA500
CL 369 see CKD750
CL 9140 see QOJ200

CL 10304 see AJD000
CL 12503 see POP000
CL 12,625 see PIF750
CL 12880 see DSP400
CL-14377 see MDV500
CL 18133 see EPC500
CL 26691 see CQL250
CL-43,064 see DXN600
CL 47300 see DSQ000
CL-47,470 see DHH400
CL 52160 see TAL250
CL-62362 see DCS200
CL-71563 see DCS200
CL 19217 4090L 7-5525 see EGI000
CLAFEN see CQC500
CLAFEN see CQC650
CLAIRSIT see PCF300
CLAODICAL see CCK125
CLAPHENE see CQC650
CLARK I see CGN000
CLARO 5591 see SLJ500
CLARY OIL see CMU900
CLARY SAGE OIL see CMU900
CLAUDELITE see ARI750
CLAUDETITE see ARI750
CLEARASIL BENZOYL PEROXIDE LOTION see BDS000
CLEARASIL BP ACNE TREATMENT see BDS000
CLEAREL see SLJ500
CLEARJEL see SLJ500
CLESTOL see DJL000
CLF II see CBT500
CLIFT see MCI750
CLIMATERINE see DKA600
CLIMESTRONE see ECU750
CLIN see SJO000
CLINDROL 101CG see BKE500
CLINDROL 200CGN see CNF330
CLINDROL 202CGN see CNF330
CLINDROL SDG see HKJ000
CLINDROL SEG see EJM500
CLINDROL SUPERAMIDE 100L see BKE500
CLINDROL SUPERAMIDE 100CG see CNF330
CLINESTROL see DKB000
CLIOQUINOL see CHR500
CLIQUINOL see CHR500
CLIRADON see KFK000
CLIRADONE see KFK000
CLIXODYNE see HIM000
CLM see SFY500
CLOBAZAM see CIR750
CLODRONATE SODIUM see SFX730
CLOFENOTANE see DAD200
CLONITRALID see DFV400
CLONT see MMN250
CLOPHEN see PJL750
CLOPHEN A60 see PJN250
CLOPIDOL see CMX850
CLOPOXIDE see LFK000
CLOPROSTENOL see CMX880
CLORAFIN see PAH780
CLORALIO see CDN550
CLORAMIDINA see CDP250
CLORAMIN see BIE250
CLORARSEN see DFX400
CLOR CHEM T-590 see CDV100
CLORDAN (ITALIAN) see CDR750
CLORDIAZEPOSSIDO (ITALIAN) see LFK000

CLORDION see CBF250
CLOREPIN see CIR750
CLORESTROLO see CLO750
CLOREX see DFJ050
CLORINA see CDP000
CLORNAPHAZINE see BIF250
CLORO (ITALIAN) see CDV750
CLOROAMFENICOLO (ITALIAN) see CDP250
CLOROBEN see DEP600
2-CLOROBENZALDEIDE (ITALIAN) see CEI500
CLOROBENZENE (ITALIAN) see CEJ125
1-p-CLORO-BENZOIL-5-METOXI-2-METILINDOL-3-
ACIDO ACETICO (SPANISH) see IDA000
2-CLORO-1,3-BUTADIENE (ITALIAN) see NCI500
CLORODANE see CDR750
O-2-CLORO-1-(2,4-DICLORO-FENIL)-VINYL-O,O-
DIETILFOSFATO (ITALIAN) see CDS750
CLORODIFENILI, CLORO 42% (ITALIAN) see PJM500
CLORODIFENILI, CLORO 54% (ITALIAN) see PJN000
1-CLORO-2,4-DINITROBENZENE (ITALIAN) see
CGM000
1-CLORO-2,3-EPOSSIPROPANO (ITALIAN) see
EAZ500
CLOROETANO (ITALIAN) see EHH000
2-CLOROETANOLO (ITALIAN) see EIU800
(CLORO-2-ETIL)-1-CICLOESIL-3-NITROSOUREA
(ITALIAN) see CGV250
3-(4-CLORO-FENIL)-1,1-DIMETIL-UREA (ITALIAN)
see CJX750
2(2-(4-CLORO-FENIL-2-FENIL)-ACETIL)INDAN-1,3-
DIONE (ITALIAN) see CJJ000
CLOROFORMIO (ITALIAN) see CHJ500
CLOROFOS (RUSSIAN) see TIQ250
CLOROMETANO (ITALIAN) see MIF765
7-CLORO-2-METILAMINO-5-FENIL-3H-1,4-
BENZOIDIAZEPINA 4-OSSIDO (ITALIAN) see LFK000
3-CLORO-2-METIL-PROP-1-ENE (ITALIAN) see
CIU750
CLOROMISAN see CDP250
1-CLORO-4-NITROBENZENE (ITALIAN) see NFS525
CLOROPICRINA (ITALIAN) see CKN500
CLOROPIRIL see TAI500
CLOROPRENE (ITALIAN) see NCI500
CLOROSAN see CDP000
CLOROSINTEX see CDP250
CLOROTRISIN see CLO750
CLOROX see SHU500
CLORSULON see CMX010
CLORTRAN see ABD000
CLORURO DI ETILE (ITALIAN) see EHH000
CLORURO di ETHENE (ITALIAN) see EIY600
CLORURO di ETILIDENE (ITALIAN) see DFF809
CLORURO di MERCURIO (ITALIAN) see MCY475
CLORURO MERCUROSO (ITALIAN) see MCW000
CLORURO di METALLILE (ITALIAN) see CIU750
CLORURO di METILE (ITALIAN) see MIF765
CLORURO di VINILE (ITALIAN) see VNP000
CLOTRIDE see CLH750
CLOVE LEAF OIL see CMY500
CLOVE LEAF OIL MADAGASCAR see CMY500
CLOXACILLIN SODIUM MONOHYDRATE see SLJ000
CLOXAPEN see SLJ000
CLOXAZEPINE see DCS200
CLOXYPEN see SLJ000
CLYSAR see PMP500
CMA see CBF250
CMB 50 see CBT500
CMB 200 see CBT500

CMC see SFO500
CMC 7H see SFO500
CM-CELLULOSE Na SALT see SFO500
CMC SODIUM SALT see SFO500
CMDP see MQR750
CME 74770 see TKL100
C-METON see TAI500
CML 21 see CAU500
CML 31 see CAU500
CMME see CIO250
CMPP see CIR500
CMU see CJX750
CMW BONE CEMENT see PKB500
CN see CEA750
CN 2 see EHP700
CN 4 see EHP700
CN 447 see DEJ000
CN 8676 see EEI000
CN-15,757 see ASA500
CN-52,372-2 see CKD750
CNC see NAS000
CO 12 see DXV600
CO-1214 see DXV600
CO-1670 see HCP000
CO-1895 see OAX000
CO-1897 see OAX000
COAL CONVERSION MATERIALS, SRC-II HEAVY
DISTILLATE see CMY750
COAL DUST see CMY760
COAL FACINGS see CMY760
COAL GAS (UN 1023) (DOT) see HHJ500
COAL, GROUND BITUMINOUS (DOT) see CMY760
COAL LIQUID see PCR250
COAL-MILLED see CMY760
COAL NAPHTHA see BBL250
COAL OIL see PCR250
COAL OIL see KEK000
COAL SLAG-MILLED see CMY760
COAL TAR see CMY800
COAL TAR, AEROSOL see CMY800
COAL TAR, AEROSOL see CMY805
COAL TAR CREOSOTE see CMY825
COAL TAR DISTILLATE see CMY900
COAL TAR DISTILLATES see CMY900
COAL TAR DISTILLATES, flammable (DOT) see
CMY900
COAL TAR DYE see CMY920
COAL TAR OIL see CMY825
COAL TAR OIL (DOT) see CMY825
COAL TAR PITCH VOLATILES see CMZ100
COAL TAR PITCH VOLATILES: PHENANTHRENE see
PCW250
COAL TAR SOLUTION USP see CMY800
COAPT see MIQ075
COATHYLENE HA 1671 see PJS750
COATHYLENE PF 0548 see PMP500
COBADEX see CNS750
COBADOCE FORTE see VSZ000
COBALIN see VSZ000
COBALT see CNA250
COBALT-59 see CNA250
COBALT ACETATE see CNC000
COBALT(2+) ACETATE see CNC000
COBALT(II) ACETATE see CNC000
COBALT ACETATE TETRAHYDRATE see CNA500
COBALT BIS(NITRATE) see CNC500
COBALT BLACK see CND125
COBALT CARBONYL see CNB500

COBALT(II) CHLORIDE see CNB599
COBALT COMPOUNDS see CNB850
COBALT DIACETATE see CNC000
COBALT DIACETATE TETRAHYDRATE see CNA500
COBALT DICHLORIDE see CNB599
COBALT DIFLUORIDE see CNC100
COBALT, DI-MU-CARBONYLHEXACARBONYLDI-, (CO-CO) see CNB500
COBALT DINITRATE see CNC500
COBALT DIPERCHLORATE HEXAHYDRATE see CND900
COBALT(II) FLUORIDE see CNC100
COBALT HYDROCARBONYL see CNC230
COBALT HYDROXIDE see CNC233
COBALT(III) HYDROXIDE see CNC233
COBALTIC HYDROXIDE see CNC233
COBALTIC OXIDE see CND825
COBALT LINOLEATE see CNC245
COBALT MOLYBDATE see CNC250
COBALT(2+) MOLYBDATE see CNC250
COBALT MOLYBDENUM OXIDE see CNC250
COBALT MONOOXIDE see CND125
COBALT MONOSULFIDE see CNE200
COBALT MONOXIDE see CND125
COBALT MURIATE see CNB599
COBALT NAPHTHENATE, POWDER (DOT) see NAR500
COBALT NITRATE see CNC500
COBALT(2+) NITRATE see CNC500
COBALT(II) NITRATE see CNC500
COBALT OCTACARBONYL see CNB500
COBALTOUS ACETATE TETRAHYDRATE see CNA500
COBALTOUS CHLORIDE see CNB599
COBALTOUS DIACETATE see CNC000
COBALTOUS DICHLORIDE see CNB599
COBALTOUS FLUORIDE see CNC100
COBALTOUS MOLYBDATE see CNC250
COBALTOUS NITRATE see CNC500
COBALTOUS OXIDE see CND125
COBALTOUS PERCHLORATE, HEXAHYDRATE see CND900
COBALTOUS SULFATE see CNE125
COBALTOUS SULFIDE see CNE200
COBALT OXIDE see CND125
COBALT(2+) OXIDE see CND125
COBALT(3+) OXIDE see CND825
COBALT(II) OXIDE see CND125
COBALT(III) OXIDE see CND825
COBALT OXIDE (8CI,9CI) see CND825
COBALT PERCHLORATE HEXAHYDRATE see CND900
COBALT(II) PERCHLORATE, HEXAHYDRATE see CND900
COBALT PEROXIDE see CND825
COBALT RESINATE, precipitated see CNE000
COBALT SESQIOXIDE see CND825
COBALT SESQUIOXIDE see CND825
COBALT SULFATE see CNE125
COBALT SULFATE (1:1) see CNE125
COBALT (2+) SULFATE see CNE125
COBALT(II) SULFATE (1:1) see CNE125
COBALT SULFIDE see CNE200
COBALT(II) SULFIDE see CNE200
COBALT SULFIDE (amorphous) see CNE200
COBALT(II) SULPHATE see CNE125
COBALT TALLATE see CNE240
COBALT TETRACARBONYL see CNB500

COBALT TETRACARBONYL DIMER see CNB500
COBALT TRIHYDROXIDE see CNC233
COBALT TRIOXIDE see CND825
COBAMIN see VSZ000
COBEX (polymer) see PKQ059
COBIONE see VSZ000
COBRATEC #99 see BDH250
COCAFURIN see NGE500
COCAINE see CNE750
(−)-COCAINE see CNE750
l-COCAINE see CNE750
β-COCAINE see CNE750
COCAMIDE DEA see CNF330
CO CAP IMIPRAMINE 25 see DLH630
COCCIDINE A see DUP300
COCCIDIOSTAT C see CMX850
COCCIDOT see DUP300
COCCINE see FMU080
COCCOCLASE see PPO000
COCCULIN see PIE500
COCCULUS see PIE500
COCCULUS solid (DOT) see PIE500
COCHENILLEROT A see FMU080
COCHINEAL RED A see FMU080
COCOA FATTY ACIDS, POTASSIUM SALTS see CNF175
COCO-DIAZINE see PPP500
COCO DIETHANOLAMIDE see BKE500
COCONUT ALDEHYDE see CNF250
COCONUT BUTTER see CNR000
COCONUT DIETHANOLAMIDE see CNF330
COCONUT DIETHANOLAMINE see CNF330
COCONUT MEAL PELLETS, containing 6–13% moisture and no more than 10% residual fat (DOT) see CNR000
COCONUT OIL ACID DIETHANOLAMINE see CNF330
COCONUT OIL ACID DIETHANOLAMINE see CNF330
COCONUT OIL ACID DIETHANOLAMINE CONDENSATE see CNF330
COCONUT OIL AMIDE of DIETHANOLAMINE see BKE500
COCONUT OIL (FCC) see CNR000
COCONUT PALM OIL see CNR000
CODECHINE see BBQ500
CODELCORTONE see PMA000
CODEMPIRAL see ABG750
CODHYDRINE see DKW800
CODIAZINE see PPP500
CODIBARBITA see EOK000
COENZYME R see VSU100
COENZYME R see BGD100
CO-ESTRO see ECU750
COFFEIN (GERMAN) see CAK500
COFFEINE see CAK500
COGNAC OIL see EKN050
COHASAL-1H see SEH000
CO-HYDELTRA see PMA000
COHYDRIN see DKW800
COIR DEEP BLACK C see AQP000
COKE see CNE750
COKE OVEN EMISSIONS see CNG929
COKE OVEN EMISSIONS (OSHA) see CNG929
COKE POWDER see CBT500
COLACE see DJL000
COLACID PONCEAU 4R see FMU080
COLACID PONCEAU SPECIAL see FMU070

COLAMINE see EEC600
COLCEMID see MIW500
COLCEMIDE see MIW500
COLCHAMIN see MIW500
COLCHAMINE see MIW500
COLCHICIN (GERMAN) see CNG938
COLCHICINA (ITALIAN) see CNG938
COLCHICINE see CNG938
7-α-H-COLCHICINE see CNG938
COLCHICINE, 7-DEACETAMIDO-7-
(METHYLAMINO)- see MIW500
COLCHICINE, DEACETYL-N-METHYL- see MIW500
COLCHINEOS see CNG938
COLCHISOL see CNG938
COLCIN see CNG938
COLCOTHAR see IHC450
COLEBENZ see BCM000
COLECALCIFEROL see CMC750
COLEMID see MIW500
COLESTYRAMIN see CME400
COL-EVAC see SFC500
COLFARIT see ADA725
COLIMYCIN M see SFY500
COLIMYCIN SULFATE see PKD300
COLISONE see PLZ000
COLISTIMETHATE SODIUM see SFY500
COLISTINASE see BAC000
COLISTIN SODIUM METHANESULFONATE see
SFY500
COLISTIN SULFAT see PKD300
COLISTIN SULFATE see PKD300
COLISTIN, SULFATE (SALT) see PKD300
COLISTIN SULFOMETHAT see SFY500
COLISTIN SULFOMETHATE SODIUM see SFY500
COLISTRIMETHATE SODIUM see SFY500
COLLARGOL see SDI750
COLLIDINE, ALDEHYDECOLLIDINE see EOS000
COLLIRON I.V. see IHG000
COLLOCARB see CBT750
COLLODION see CCU250
COLLODION COTTON see CCU250
COLLODION WOOL see CCU250
COLLOID 775 see CCL250
COLLOIDAL ARSENIC see ARA750
COLLOIDAL CADMIUM see CAD000
COLLOIDAL FERRIC OXIDE see IHC450
COLLOIDAL GOLD see GIS000
COLLOIDAL MANGANESE see MAP750
COLLOIDAL MERCURY see MCW250
COLLOIDAL SELENIUM see SBO500
COLLOIDAL SILICA see SCH002
COLLOIDAL SILICON DIOXIDE see SCH002
COLLOIDAL SULFUR see SOD500
COLLOKIT see SOD500
COLLOMIDE see SNM500
COLLOWELL see SFO500
COLLOXYLIN see CCU250
COLLUNOSOL see TIV750
COLLUNOVAR see NCJ500
COLLUNOVER see NCJ500
COLOGNE EARTH see MAT500
COLOGNE SPIRIT see EFU000
COLOGNE UMBER see MAT500
COLOGNE YELLOW see LCR000
COLOMBIAN BLACK TOBACCO CIGARETTE
REFINED TAR see CMP800
COLOMYCIN SYRUP see PKD300

COLONATRAST see BAP000
COLONIAL SPIRIT see MGB150
COLOR-SET see TIX500
COLPOVISTER see EDU500
COLSALOID see CNG938
COLSUL see SOD500
COLSULANYDE see SNM500
COLTS FOOT see PCR000
COLUMBIA BLACK EP see AQP000
COLUMBIA CARBON see CBT750
COLUMBIA LCK see CBT500
COLUMBIAN SPIRITS (DOT) see MGB150
COLUMBIUM see NDZ000
COLUMBIUM PENTACHLORIDE see NEA000
COLY-MYCIN INJECTABLE see SFY500
COLYSTINMETHANSULFONAT (GERMAN) see
SFY500
COLZA OIL see RBK200
COMBINACE see CAM200
COMBINAL E see TGJ050
COMBOT EQUINE see TIQ250
COMBUSTION IMPROVER-2 see MAV750
COMESTROL see DKA600
COMESTROL ESTROBENE see DKA600
COMFREY, RUSSIAN see RRP000
COMITAL see DKQ000
COMITE see SOP000
COMMERCIAL DIESEL FUEL NO. 2 see DHE850
COMMON SALT see SFT000
COMMON SENSE COCKROACH and RAT
PREPARATIONS see PHP010
COMMOTIONAL see ABG750
COMPALOX see AHE250
COMPAZINE see PMF500
COMPERLAN LD see BKE500
COMPITOX see CIR500
COMPLEMIX see DJL000
COMPLEXONE see EIV000
COMPLEXON I see AMT500
COMPLEXON II see EIX000
COMPLEXON III see EIX500
COMPOCILLIN G see BDY669
COMPOUND 42 see WAT200
COMPOUND 118 see AFK250
COMPOUND 269 see EAT500
COMPOUND 338 see DER000
COMPOUND 347 see EAT900
COMPOUND 497 see DHB400
COMPOUND 604 see DFT000
COMPOUND-666 see BBP750
COMPOUND 711 see IKO000
COMPOUND 889 see DVL700
COMPOUND 88R see SOP500
COMPOUND 923 see DFY400
COMPOUND M-81 see PHI500
COMPOUND 1081 see FFF000
COMPOUND 1189 see KEA000
COMPOUND 2046 see MQR750
COMPOUND 3422 see PAK000
COMPOUND 3956 see CDV100
COMPOUND 4049 see MAK700
COMPOUND 4072 see CDS750
COMPOUND-4992 see BKS810
COMPOUND 7744 see CBM750
COMPOUND 01748 see DJT800
COMPOUND 33355 see MKB750
COMPOUND 33,828 see MLJ500
COMPOUND B DICAMBA see MEL500

COMPOUND F see CNS750
COMPOUND-1452-F see EME500
COMPOUND G-11 see HCL000
COMPOUND 6-12 INSECT REPELLENT see EKV000
COMPOUND No. 1080 see SHG500
COMPOUND W see TKL100
COMYCETIN see CDP250
d-CON see WAT200
CONAC A see CPI250
CONAC S see CPI250
CONCHININ see QFS000
CONCO AAS-35 see DXW200
CONCO NI-90 see PKF000
CONCO NIX-100 see PKF500
CONCO SULFATE WA see SIB600
CONDACAPS see VSZ100
CONDENSATE PL see BKE500
CONDENSATES (PETROLEUM), VACUUM TOWER
(9CI) see MQV755
CONDITION see DCK759
CONDOCAPS see VSZ100
CONDOL see VSZ100
CONDUCTEX see CBT500
CONDUCTEX see CBT750
CONDYLON see CNG938
CONDY'S CRYSTALS see PLP000
CONEST see ECU750
CONESTORAL see EDV600
CONESTRON see ECU750
CONFECTIONER'S SUGAR see SNH000
CONFORTID see IDA000
CONGOBLAU 3B see CMO250
CONGO BLUE see CMO250
CONGO RED see SGQ500
CONIGON BC see EIV000
CONJES see ECU750
CONJUGATED EQUINE ESTROGEN see PMB000
CONJUGATED ESTROGENS see ECU750
CONJUTABS see ECU750
CONOCO C-50 see DXW200
CONOTRANE see PFN000
CONOVID see EAP000
CONOVID E see EAP000
CONQUININE see QFS000
CONSDRIN see SPC500
CONSDRIN HYDROCHLORIDE see SPC500
CONSTONATE see DJL000
CONT see MMN250
CONTAVERM see PDP250
CONTERGAN see TEH500
CONTIMET 30 see TGF250
CONTINAL see NBU000
CONTINENTAL see CBT750
CONTINENTAL see KBB600
CONTINEX see CBT750
CONTIZELL see PKQ059
CONTOPHERON see TGJ050
CONTRA CREME see ABU500
CONTRADOL see ABG750
CONTRALIN see DXH250
CONTRAPOT see DXH250
CONTRHEUMA RETARD see ADA725
CONTROVLAR see EEH520
CONVUL see DKQ000
COOMASSIE VIOLET see FAG120
CO-OP HEXA see HCC500
COPAGEL PB 25 see SFO500
COPAIBA BALSAM see CNH792

COPAIBA OIL see CNH792
COPAIBA OLEORESIN see CNH792
COPAL Z see SMQ500
COPAROGIN see FLZ050
COPHARCILIN see AIV500
COPOX see CNO000
COPPER see CNI000
COPPER-8 see BLC250
COPPER ACETATE see CNI250
COPPER(2+) ACETATE see CNI250
COPPER(II) ACETATE see CNI250
COPPER ACETOARSENITE (DOT) see COF500
COPPER ACETOARSENITE, solid (DOT) see COF500
COPPER ACETYLIDE see CNI500
COPPER-AIRBORNE see CNI000
COPPER ALLOY, Cu, Be see CNI600
COPPER ALLOY, Cu, Be, Co see CNK700
COPPER ARSENITE, solid (DOT) see CNN500
COPPERAS see FBN100
COPPERAS see FBO000
COPPER-BERYLLIUM ALLOY see CNI600
COPPER BICHLORIDE see CNK500
COPPER,
BIS(ETHYLENEDIAMINE)(MERCURICTETRATHIOC
YANATO)- see BJP425
COPPER(2+), BIS(ETHYLENEDIAMINE)-,
TETRAKIS(THIOCYANATO)MERCURATE(2-),
POLYMERS see BJP425
COPPER BRONZE see CNI000
COPPER CARBIDE see CNI500
COPPER CHLORATE see CNJ900
COPPER CHLORATE (DOT) see CNJ900
COPPER CHLORIDE see CNJ950
COPPER(2+) CHLORIDE see CNK500
COPPER(II) CHLORIDE see CNK500
COPPER CHLORIDE (DOT) see CNJ950
COPPER(II) CHLORIDE (1:2) see CNK500
COPPER CITRATE see CNK625
COPPER(I) CITRATE see CNK625
COPPER-COBALT-BERYLLIUM see CNK700
COPPER COMPOUNDS see CNK750
COPPER CYANAMIDE see CNL250
COPPER CYANIDE see CNL000
COPPER(I) CYANIDE see CNL000
COPPER(II) CYANIDE see CNL250
COPPER CYANIDE (DOT) see CNL250
COPPER DIACETATE see CNI250
COPPER(2+) DIACETATE see CNI250
COPPER DINITRATE see CNM750
COPPER-ETHYLENEDIAMINE COMPLEX see
DBU800
COPPER GLUCONATE see CNM100
COPPER HYDROXYQUINOLATE see BLC250
COPPER-8-HYDROXYQUINOLATE see BLC250
COPPER-8-HYDROXYQUINOLINATE see BLC250
COPPER-8-HYDROXYQUINOLINE see BLC250
COPPER-MILLED see CNI000
COPPER MONOSULFATE see CNP250
COPPER NAPHTHENATE see NAS000
COPPER(2+) NITRATE see CNM750
COPPER(II) NITRATE see CNM750
COPPER NORDOX see CNO000
COPPER ORTHOARSENITE see CNN500
COPPER(I) OXIDE see CNO000
COPPER OXINATE see BLC250
COPPER (2+) OXINATE see BLC250
COPPER OXINE see BLC250
COPPER OXYQUINOLATE see BLC250

COPPER OXYQUINOLINE see BLC250
COPPER(II) PERCHLORATE, DIHYDRATE see CNO500
COPPER(I) POTASSIUM CYANIDE see PLC175
COPPER QUINOLATE see BLC250
COPPER-8-QUINOLATE see BLC250
COPPER-8-QUINOLINOL see BLC250
COPPER QUINOLINOLATE see BLC250
COPPER-8-QUINOLINOLATE see BLC250
COPPER SALT 2-HYDROXY-1,2,3-PROPANETRICARBOXYLIC ACID (1:2) see CNK625
COPPER SARDEX see CNO000
COPPER SLAG-AIRBORNE see CNI000
COPPER SLAG-MILLED see CNI000
COPPER SODIUM CYANIDE see SFZ100
COPPER SULFATE see CNP250
COPPER(II) SULFATE (1:1) see CNP250
COPPER UVERSOL see NAS000
COPRA (DOT) see CNR000
COPRA (OIL) see CNR000
COPRA PELLETS (DOT) see CNR000
COPREN see GFW000
COPROL see DJL000
COPTICIDE see SNM500
COPTISINE see CNR100
COQUES DU LEVANT (FRENCH) see PIE500
CORACON see DJS200
CORAETHAMIDE see DJS200
CORAETHAMIDUM see DJS200
CORALEPT see DJS200
CORAMINE see DJS200
CORAVITA see DJS200
CORAX see MDQ250
CO-RAX see WAT200
CORAX see CBT750
CORAX P see CBT750
CORAZONE see DJS200
CORDIAMID see DJS200
CORDIAMIN see DJS200
CORDIAMINE see DJS200
CORDITON see DJS200
CORDULAN see CMD750
CORDYNIL see DJS200
COREDIOL see DJS200
COREINE see CCL250
CORESPIN see DJS200
CORETHAMIDE see DJS200
CORETONE see DJS200
CORFLEX 880 see ILR100
CORIAL EM FINISH F see CCU250
CORIANDER OIL see CNR735
CORICIDIN see ABG750
CORID see AOD175
CORIFORTE see ABG750
CORINE see CNE750
CORISOL see VGP000
CORIZIUM see NGG500
CORLAN see HHR000
CORLUTIN see PMH500
CORLUVITE see PMH500
CORMED see DJS200
CORMID see DJS200
CORMOTYL see DJS200
CORN ENDOSPERM OIL see CNR850
CORNMINT OIL, PARTIALLY DEMENTHOLIZED see MCB625
CORN OIL see CNS000
CORNOTONE see DJS200

CORNOX-M see CIR250
CORNOX RD see DGB000
CORNOX RK see DGB000
CORN PRODUCTS see SLJ500
CORN SILK and CORN SILK EXTRACT see CNS100
CORN SUGAR see GFG000
CORODINOC see DUU600
CORONA COROZATE see BJK500
CORONAL-CRINOS see EID200
CORONATE MR 200 see PKB100
COROPHOS see MRH209
COROSORBIDE see CCK125
COROSUL D AND S see SOD500
COROTHION see PAK000
COROTONIN see DJS200
COROVIT see DJS200
COROVLISS see CCK125
COROXON see CIK750
COROZATE see BJK500
CORPORIN see PMH500
CORPS PRALINE see MAO350
CORPUS LUTEUM HORMONE see PMH500
CORRONAROBETIN see TEH500
CORROSIVE MERCURY CHLORIDE see MCY475
CORROSIVE SUBLIMATE see MCY475
CORRY'S SLUG DEATH see TDW500
CORSONE see SOW000
CORTAN see PLZ000
CORTANCYL see PLZ000
CORT-DOME see CNS750
Δ-CORTELAN see PLZ000
CORTHION see PAK000
CORTHIONE see PAK000
CORTIDELT see PLZ000
CORTILAN-NEU see CDR750
CORTISOL see CNS750
Δ¹-CORTISOL see PMA000
CORTISOL ALCOHOL see CNS750
CORTISOL HEMISUCCINATE SODIUM SALT see HHR000
CORTISOL SODIUM HEMISUCCINATE see HHR000
CORTISOL SODIUM SUCCINATE see HHR000
CORTISOL-21-SODIUM SUCCINATE see HHR000
CORTISOL SUCCINATE, SODIUM SALT see HHR000
Δ-CORTISONE see PLZ000
Δ¹-CORTISONE see PLZ000
CORTISPRAY see CNS750
Δ-CORTONE see PLZ000
CORUNDUM FUME see CNT250
CORVIC 55/9 see PKQ059
CORVIC 236581 see AAX175
CORVITAN see DJS200
CORVITIN see DBA800
CORVITOL see DJS200
CORVITONE see DJS200
CORYBAN-D see ABG750
CORYLON see HMB500
CORYLONE see HMB500
CORYWAS see DJS200
COSAN see SOD500
COSDEN 550 see SMQ500
COSDEN 945E see SMQ500
COSMETIC CORAL RED KO BLUISH see CHP500
COSMETIC WHITE C47-5175 see TGG760
COSMETOL see CCP250
COSMOPEN see BDY669
COTEL see VSZ000

COTNION-ETHYL see EKN000
COTNION METHYL see ASH500
COTOFILM see HCL000
COTONE see PLZ000
COTORAN see DUK800
COTORAN MULTI 50WP see DUK800
CO-TRIMOXAZOLE see SNK000
COTTON AIDE HC see HKC000
COTTON DUST see CNT750
COTTONEX see DUK800
COTTON RED L see SGQ500
COTTONSEED, MODIFIED PRODUCTS see CNT950
COTTONSEED OIL (unhydrogenated) see CNU000
COUMADIN see WAT200
COUMADIN SODIUM see WAT220
COUMAFENE see WAT200
COUMAPHOS see CNU750
COUMAPHOS-O-ANALOG see CIK750
COUMAPHOS OXYGEN ANALOG (USDA) see CIK750
4-COUMARIC ACID see CNU825
p-COUMARIC ACID see CNU825
COUMARIN see CNV000
COUMARIN 4 see MKP500
cis-o-COUMARINIC ACID LACTONE see CNV000
COUMARINIC ANHYDRIDE see CNV000
COUMARONE see BCK250
COUMESTROL see COF350
COURLENE-X3 see PJS750
COURLOSE A 590 see SFO500
COVI-OX see VSZ450
COVIT see VSZ000
COVOL see PKP750
COVOL 971 see PKP750
COXISTAT see NGE500
COYDEN see CMX850
COZYME see PAG200
CP see CQC650
4-CP see CJN000
CP 34 see TFM250
CP 4517 see HMR550
CP 4572 see CDO250
CP 6,343 see CFK000
CP 10,188 see FAM100
CP 15,336 see DBI200
CP 25017 see NHK800
CP 26890 see BJT800
CP 47114 see DSQ000
CP 49674 see DOP200
CP 50144 see CFX000
CP 53926 see DRR200
CP 78601 see TBV300
CP 10423-18 see TCW750
CP-15467-61 see LGZ000
CPA see CJN000
CPA see CQC650
C-PAL see FAM100
CP BASIC SULFATE see CNP250
CPC 3005 see SLJ500
CPC 6448 see SLJ500
CPCA see BIO750
C.P. CHROME LIGHT 2010 see LCS000
C.P. CHROME ORANGE DARK 2030 see LCS000
C.P. CHROME ORANGE MEDIUM 2020 see LCS000
C.P. CHROME YELLOW LIGHT see LCR000
4-Cl-m-PD see CJY120
4-Cl-o-PD see CFK125
CPDC see PJD000
CPDD see PJD000

CPE see PJS750
CPE 16 see PJS750
CPE 25 see PJS750
CPH see CDP250
CPIRON see FBJ100
CPMC see CKF000
CPT see CLK215
CdPT see CAI350
C.P. TITANIUM see TGF250
CPZ see CKP500
C.P. ZINC YELLOW X-883 see ZFJ100
C-QUENS see CBF250
CR see DDE200
CR 200 see PKB100
CR 409 see BJE750
CR 1505 see LII100
CR 3029 see MAS500
CRADEX see BBK500
CRAG 85W see DSB200
CRAG 974 see DSB200
CRAG FUNGICIDE 974 see DSB200
CRAG NEMACIDE see DSB200
CRAG SEVIN see CBM750
CRATECIL see EOK000
CRAWHASPOL see TIO750
CRECHLOR S 45 see PAH780
C RED 2 see CMS238
CREDO see SHF500
CREMODIAZINE see PPP500
CREMOMERAZINE see ALF250
CREMOMETHAZINE see SNJ000
CREOSOTE see CMY825
CREOSOTE, from COAL TAR see CMY825
CREOSOTE OIL see CMY825
CREOSOTE P1 see CMY825
CREOSOTUM see CMY825
CRESIDINE see MGO750
m-CRESIDINE see MGO500
p-CRESIDINE see MGO750
CRESOATE, WOOD see BAT850
CRESOL see CNW500
2-CRESOL see CNX000
3-CRESOL see CNW750
4-CRESOL see CNX250
m-CRESOL see CNW750
o-CRESOL see CNX000
p-CRESOL see CNX250
p-CRESOL ACETATE see MNR250
o-CRESOL, DINITRO-, SODIUM SALT see SGP550
CRESOLI (ITALIAN) see CNW500
p-CRESOL METHYL ETHER see MGP000
CRESORCINOL DIISOCYANATE see TGM750
CRESOTINE BLUE 2B see CMO000
CRESOTINE BLUE 3B see CMO250
CRESOTINE PURE BLUE see CMO500
CRESOTOL see DUU600
CRESTANIL see MQU750
CRESTOXO see CDV100
p-CRESYL ACETATE (FCC) see MNR250
p-CRESYL BENZOATE see TGX100
CRESYLIC ACID see CNW500
m-CRESYLIC ACID see CNW750
o-CRESYLIC ACID see CNX000
p-CRESYLIC ACID see CNX250
CRESYLIC CREOSOTE see CMY825
p-CRESYL ISOBUTYRATE see THA250
m-CRESYL METHYLCARBAMATE see MIB750
p-CRESYL METHYL ETHER see MGP000

CRESYL PHOSPHATE see TNP500
o-CRESYL PHOSPHATE see TMO600
p-CRESYL SALICYLATE see THD850
CRILL 3 see SKV150
CRILL 10 see PKL100
CRILL K 3 see SKV150
CRILLON L.D.E. see BKE500
CRIMSON ANTIMONY see AQL500
CRIMSON EMBL see HJF500
CRIMSON SX see FMU080
CRINOTHENE see PKB500
CRISALIN see DUV600
CRISAPON see DGI400
CRISATRINA see ARQ725
CRISAZINE see ARQ725
CRISODIN see MRH209
CRISODRIN see MRH209
CRISPATINE see FOT000
CRISTALLOSE see SJN700
CRISTALLOVAR see EDV000
CRISTAPURAT see DKL800
CRISTERONE T see TBF500
CRISTOBALITE see SCI500
CRISTOBALITE see SCJ000
CRISTOXO 90 see CDV100
CRISULFAN see EAQ750
CRISURON see DXQ500
CROCIDOLITE (DOT) see ARM275
CROCIDOLITE ASBESTOS see ARM275
CROCOITE see LCR000
CROCUS MARTIS ADSTRINGENS see IHC450
CRODACID see MSA250
CRODACOL-CAS see HCP000
CRODACOL-S see OAX000
CRODAMINE 1.18D see OBC000
CRODAMOL IPM see IQN000
CROFLEX see CBT750
CROLAC see CBT750
CROLEAN see ADR000
CROMILE, CLORURO di (ITALIAN) see CML125
CROMOCI see CQM325
CROMO, OSSICLORURO di (ITALIAN) see CML125
CRONETAL see DXH250
CROSPOVIDONE see PKQ150
CROTALINE see MRH000
CROTONAL see COB260
CROTONALDEHYDE see COB250
CROTONALDEHYDE see COB260
(E)-CROTONALDEHYDE see COB260
CROTONALDEHYDE, stabilized (DOT) see COB250
CROTONATE de 2,4-DINITRO 6-(1-METHYL-HEPTYL)-PHENYLE (FRENCH) see AQT500
CROTONATE d'ETHYLE (FRENCH) see COB750
CROTONIC ACID see COB500
CROTONIC ACID, solid see COB500
α-CROTONIC ACID see COB500
CROTONIC ACID ANHYDRIDE see COB900
CROTONIC ACID, ETHYL ESTER see EHO200
α-CROTONIC ACID ETHYL ESTER see COB750
CROTONIC ACID, 2-METHYL-, (E)- see TGA700
CROTONIC ALDEHYDE see COB250
CROTONIC ALDEHYDE see COB260
CROTONIC ANHYDRIDE see COB900
CROTONYL ALCOHOL see BOY000
CROTONYLENE see COC500
CROTYL ALCOHOL see BOY000
CROTYLIDENE ACETIC ACID see SKU000

CROYSULFONE see SOA500
CRTRON see VSZ100
CRUDE ARSENIC see ARI750
CRUDE COAL TAR see CMY800
CRUDE OIL see PCR250
CRUDE PETROLEUM see PCR250
CRUDE SHALE OILS see COD750
CRUNCH see CBM750
CRYOFLUORAN see FOO509
CRYOFLUORANE see FOO509
CRYOLITE see SHF000
CRYOPOLYTHENE see PJS750
CRYPTOCRYSTALLINE QUARTZ see SCK600
CRYPTOGIL OL see PAX250
CRYPTOHALITE see COE000
CRYSALBA see CAX500
CRYSTAL CHROME ALUM see PLB500
CRYSTALETS see VSK900
CRYSTALLINA see VSZ100
CRYSTALLINE DEHYDROXY SODIUM ALUMINUM, CARBONATE see DAC450
CRYSTALLINE DIGITALIN see DKL800
CRYSTALLIZED VERDIGRIS see CNI250
CRYSTALLOSE see SJN700
CRYSTAL O see CCP250
CRYSTAL ORANGE 2G see HGC000
CRYSTAL PROPANIL-4 see DGI000
CRYSTALS of VENUS see CNI250
CRYSTAL VIOLET see AOR500
CRYSTAL VIOLET 6B see AOR500
CRYSTAL VIOLET 10B see AOR500
CRYSTAL VIOLET AO see AOR500
CRYSTAL VIOLET AON see AOR500
CRYSTAL VIOLET BASE see AOR500
CRYSTAL VIOLET BP see AOR500
CRYSTAL VIOLET BPC see AOR500
CRYSTAL VIOLET CHLORIDE see AOR500
CRYSTAL VIOLET EXTRA PURE see AOR500
CRYSTAL VIOLET EXTRA PURE APN see AOR500
CRYSTAL VIOLET EXTRA PURE APNX see AOR500
CRYSTAL VIOLET FN see AOR500
CRYSTAL VIOLET HL2 see AOR500
CRYSTAL VIOLET O see AOR500
CRYSTAL VIOLET 5BO see AOR500
CRYSTAL VIOLET 6BO see AOR500
CRYSTAL VIOLET PURE DSC see AOR500
CRYSTAL VIOLET PURE DSC BRILLIANT see AOR500
CRYSTAL VIOLET SS see AOR500
CRYSTAL VIOLET TECHNICAL see AOR500
CRYSTAL VIOLET USP see AOR500
CRYSTAMET see SJU000
CRYSTAMIN see VSZ000
CRYSTAPEN see BFD250
CRYSTAR see SCQ000
CRYSTAR see ADA725
CRYSTEX see SOD500
CRYSTHION 2L see ASH500
CRYSTHYON see ASH500
CRYSTODIGIN see DKL800
CRYSTOGEN see EDV000
CRYSTOIDS see HFV500
CRYSTOL CARBONATE see SFO000
CRYSTOLON 37 see SCQ000
CRYSTOLON 39 see SCQ000
CRYSTOSERPINE see RDK000
CRYSTOSOL see MQV750
CRYSTWEL see VSZ000
CRY-O-VAC L see PJS750

CS see CEQ600
C-Sn-9 see BLL750
50-CS-46 see EME050
60-CS-16 see CMF400
CSAC see EES400
CSC see SEC000
CSF-GIFTWEIZEN see TEM000
CS LAFARGE see CAW850
CT see CLH750
CTA see CLO750
CTAB see HCQ500
CTC see CMA750
CTFE see CLQ750
CTR 6669 see MHC750
CTX see CQC650
CUBEB OIL see COE175
CUBES see DJO000
CUBIC NITER see SIO900
CUCUMBER ALCOHOL see NMV780
CUCUMBER ALDEHYDE see NMV760
CUCUMBER DUST see ARB750
CUEMID see CME400
CULLEN EARTH see MAT500
CULMINAL K 42 see MIF760
CUM see COE750
CUMA see BJZ000
CUMAFOS (DUTCH) see CNU750
CUMALDEHYDE see COE500
CUMAN see BJK500
CUMAN L see BJK500
p-CUMARIC ACID see CNU825
CUMEEN (DUTCH) see COE750
CUMEENHYDROPEROXYDE (DUTCH) see IOB000
CUMENE see COE750
psi-CUMENE see TLL750
CUMENE ALDEHYDE see COF000
CUMENE HYDROPEROXIDE (DOT) see IOB000
CUMENE HYDROPEROXIDE, TECHNICALLY PURE
(DOT) see IOB000
CUMENE PEROXIDE see DGR600
p-CUMENOL see IQZ000
CUMENT HYDROPEROXIDE see IOB000
CUMENYL HYDROPEROXIDE see IOB000
o-CUMENYL METHYLCARBAMATE see MIA250
p-CUMIC ALDEHYDE see COE500
CUMID see BJZ000
psi-CUMIDINE see TLG250
psi-CUMIDINE HYDROCHLORIDE see TLG750
CUMINALDEHYDE see COE500
CUMINIC ALDEHYDE (FCC) see COE500
CUMIN OIL see COF325
CUMINYL ALDEHYDE see COE500
CUMINYL NITRILE see IOD050
CUMMIN see COF325
CUMOESTEROL see COF350
ψ-CUMOHYDROQUINONE see POG300
CUMOLHYDROPEROXID (GERMAN) see IOB000
CUMOSTROL see COF350
CUMYL HYDROPEROXIDE see IOB000
α-CUMYL HYDROPEROXIDE see IOB000
CUMYL HYDROPEROXIDE, TECHNICAL PURE
(DOT) see IOB000
CUMYL PEROXIDE see DGR600
CUNAPSOL see NAS000
CUNILATE see BLC250
CUNILATE 2472 see BLC250
CUPFERRON see ANO500

CUPPER OXIDE (RUSSIAN) see CNO000
CUPRAL see SGJ000
CUPRATE(1-), DICYANO-, POTASSIUM see PLC175
CUPRATE(2-), TRIS(CYANO-C)-, DISODIUM see
SFZ100
CUPRENIL see MCR750
CUPRIC ACETATE see CNI250
CUPRIC ACETOARSENITE see COF500
CUPRIC ARSENITE see CNN500
CUPRIC CHLORIDE see CNK500
CUPRIC CITRATE see CNK625
CUPRIC CYANIDE (DOT) see CNL250
CUPRIC DIACETATE see CNI250
CUPRIC DICHLORIDE see CNK500
CUPRIC DINITRATE see CNM750
CUPRIC DIPERCHLORATE TETRAHYDRATE see
CNO500
CUPRICELLULOSE see CCU150
CUPRIC GREEN see CNN500
CUPRIC-8-HYDROXYQUINOLATE see BLC250
CUPRICIN see CNL000
CUPRIC NITRATE (DOT) see CNM750
CUPRIC-8-QUINOLINOLATE see BLC250
CUPRIC SULFATE see CNP250
CUPRIETHYLENE DIAMINE see DBU800
CUPRIETHYLENEDIAMINE, solution (DOT) see
DBU800
CUPRIMINE see MCR750
CUPRINOL see NAS000
CUPRIZANE see COF675
CUPRIZONE see COF675
CUPROCITROL see CNK625
CUPROUS CYANIDE see CNL000
CUPROUS OXIDE see CNO000
CUPROUS POTASSIUM CYANIDE see PLC175
CURACRON see BNA750
CURALIN M see MJM200
CURARIN-HAF see TOA000
CURATERR see CBS275
CURAX see CPI250
CURCUMA OIL see COG000
CURCUMIN see COG000
CURCUMINE see COG000
CURENE see PKL500
CURENE 442 see MJM200
CURESAN see MEP250
CURETAN see MEP250
CURETARD A see DWI000
CURITAN see DXX400
CURITHANE see MJQ000
CURITHANE 103 see MGA500
CURITHANE C126 see DEQ600
CUROL BRIGHT RED 4R see FMU080
CURON FAST YELLOW 5G see FAG140
CURZATE M see MAS500
CUSTOS see MHC750
CUTICURA ACNE CREAM see BDS000
CUTTING OILS see COH000
CUZ 3 see CBT500
CVP see CDS750
CW 524 see TKL100
C-WEISS 7 (GERMAN) see TGG760
CWN 2 see CBT500
CY see CQC650
CY-39 see PHU500
CY 116 see AJD000
CYAANWATERSTOF (DUTCH) see HHS000
CYALANE see DXN600

CYAMOPSIS GUM see GLU000
CYANACETATE ETHYLE (GERMAN) see EHP500
CYANAMIDE see CAQ250
CYANAMIDE see COH500
CYANAMIDE CALCIQUE (FRENCH) see CAQ250
CYANAMIDE, CALCIUM SALT (1:1) see CAQ250
CYANAMID GRANULAR see CAQ250
CYANAMID SPECIAL GRADE see CAQ250
CYANASET see MJM200
CYANAZINE see BLW750
CYANHYDRINE d'ACETONE (FRENCH) see MLC750
CYANIC ACID, POTASSIUM SALT see PLC250
CYANIC ACID, SODIUM SALT see COI250
CYANIC ACID, TRIMETHYLSTANNYL ESTER see TMI100
CYANIDE see COI500
CYANIDE ANION see COI500
CYANIDE(CN^{1-}) see COI500
CYANIDE(CN^{1-}) ION see COI500
CYANIDE ION see COI500
CYANIDELONON 1522 see QCA000
CYANIDE of POTASSIUM see PLC500
CYANIDE of SODIUM see SGA500
CYANIDE SOLUTIONS (DOT) see COI500
CYANIDES (OSHA) see PLC500
CYANIDE, dry (UN 1588) see COI500
CYANINE GREEN G BASE see BLK000
CYANOACETIC ACID ETHYL ESTER see EHP500
CYANOACETIC ESTER see EHP500
CYANOACETONITRILE see MAO250
2-CYANOACRYLIC ACID, METHYL ESTER see MIQ075
α-CYANOACRYLIC ACID METHYL ESTER see MIQ075
CYANOAMINE see COH500
CYANO-B12 see VSZ000
CYANOBENZENE see BCQ250
4-CYANOBENZONITRILE see BBP250
α-CYANOBENZYL PHENYL KETONE see OOK200
CYANOBOND W100 see EHP700
CYANOBOND W300 see EHP700
CYANOBRIK see SGA500
CYANOBROMIDE see COO500
CYANOCOBALAMIN see VSZ000
p-CYANOCUMENE see IOD050
2-CYANO-N-(2-CYANOETHYL)ETHANAMINE see BIQ500
α-CYANODEOXYBENZOIN see OOK200
α-CYANODIPHENYLMETHANE see DVX200
CYANOETHANE see PMV750
2-CYANOETHANOL see HGP000
2-CYANOETHYL ALCOHOL see HGP000
β-CYANOETHYLAMINE see AMB500
CYANOETHYLENE see ADX500
CYANO(4-FLUORO-3-PHENOXYPHENYL)METHYL-3-(2,2-DICHLOROETHENYL)-2,2-DIMETHYLCYCLOPROPANECARBOXYLATE see CON750
CYANOGAS see CAQ500
CYANOGEN see COO000
CYANOGENAMIDE see COH500
CYANOGEN BROMIDE see COO500
CYANOGEN CHLORIDE see COO750
CYANOGEN CHLORIDE, inhibited (DOT) see COO750
CYANOGEN CHLORIDE (ACGIH,OSHA) see COO750
CYANOGENE (FRENCH) see COO000
CYANOGEN GAS (DOT) see COO000

CYANOGEN IODIDE see COP000
CYANOGEN MONOBROMIDE see COO500
CYANOGEN NITRIDE see COH500
CYANOGINOSIN LA, 5-l-ARGININE- see AQV990
CYANOGINOSIN LR see AQV990
CYANOGRAN see SGA500
2-CYANO-10-(3-(4-HYDROXYPIPERIDINO)PROPYL)PHENOTHIAZINE see PIW000
2-CYANO-10-(3-(4-HYDROXY-1-PIPERIDYL)PROPYL)PHENOTHIAZINE see PIW000
CYANO-3-((HYDROXY-4-PIPERIDYL-1)-3 PROPYL)-10-PHENOTHIAZINE (FRENCH) see PIW000
5-CYANO-3-INDOLYL ISOPROPYL KETONE see COP525
5-CYANO-3-INDOLYLMETHYL KETONE see COP550
α-CYANOISOPROPYLAMIDE OF THE O,O-DIETHYLTHIOPHOSPHORYL ACETIC ACID see PHK250
CYANOLIT see MIQ075
CYANOMETHANE see ABE500
CYANOMETHANOL see HIM500
(CYANOMETHYL)BENZENE see PEA750
N-(CYANOMETHYL)DIMETHYLAMINE see DOS200
S-N-(1-CYANO-1-METHYLETHYL)CARBAMOYLMETHYL DIETHYL PHOSPHOROTHIOLATE see PHK250
S-(((1-CYANO-1-METHYL-ETHYL)CARBAMOYL)METHYL) O,O-DIETHYL PHOSPHOROTHIOATE see PHK250
3-(CYANOMETHYL)INDOLE see ICW000
CYANO(METHYLMERCURI)GUANIDINE see MLF250
CYANON S see EHP700
CYANON 5MSP see EHP700
α-CYANO-3-PHENOXYBENZYL-2-(4-CHLOROPHENYL)ISOVALERATE PYDRIN see FAR100
α-CYANO-3-PHENOXYBENZYL-2-(4-CHLOROPHENYL)-3-METHYLBUTYRATE see FAR100
CYANO(3-PHENOXYPHENYL)METHYL 4-CHLORO-α-(1-METHYLETHYL)BENZENEACETATE see FAR100
(+)CYANO(3-PHENOXYPHENYL)METHYL(±)-1-(DIFLUOROMETHOXY)-α-(1-METHYLETHYL)BENZENEACETATE see COQ390
CYANOPHOS see DGP900
1-CYANOPROPANE see BSX250
2-CYANOPROPANE see IJX000
2-CYANOPROPENE-1 see MGA750
(3-CYANOPROPYL)DIETHOXY(METHYL) SILANE see COR500
(3-CYANOPROPYL) TRIETHOXYSILANE see COR800
CYANOSIN see ADG250
CYANOSIN (ACID DYE) see ADG250
CYANOSINE see ADG250
α-CYANOSTILBENE see DVX600
α-CYANOTOLUENE see PEA750
ω-CYANOTOLUENE see PEA750
CYANOTRICHLOROMETHANE see TII750
CYANOX 425 see MJN250
CYANOX LTDP see TFD500
CYANSAN see COI250
CYANTIN see NGE000
CYANURAMIDE see MCB000
CYANURCHLORIDE see TJD750
CYANURE see COI500

CYANURE d'ARGENT (FRENCH) see SDP000
CYANURE de CALCIUM (FRENCH) see CAQ500
CYANURE de CUIVRE (FRENCH) see CNL250
CYANURE de MERCURE (FRENCH) see MDA250
CYANURE de METHYL (FRENCH) see ABE500
CYANURE de PLOMB (FRENCH) see LCU000
CYANURE de POTASSIUM (FRENCH) see PLC500
CYANURE de SODIUM (FRENCH) see SGA500
CYANURE de VINYLE (FRENCH) see ADX500
CYANURE de ZINC (FRENCH) see ZGA000
CYANURIC ACID CHLORIDE see TJD750
CYANURIC CHLORIDE (DOT) see TJD750
CYANURIC TRIAMIDE see MCB000
CYANURIC TRICHLORIDE (DOT) see TJD750
CYANUROTRIAMIDE see MCB000
CYANUROTRIAMINE see MCB000
CYANURYL CHLORIDE see TJD750
CYANWASSERSTOFF (GERMAN) see HHS000
CYASORB UV 9 see MES000
CYAZIN see ARQ725
CYBIS see EID000
CYCASIN see COU000
CYCASIN ACETATE see MGS750
CYCAS REVOLUTA GLUCOSIDE see COU000
CYCLADIENE see DAL600
CYCLAL CETYL ALCOHOL see HCP000
CYCLALIA see CMS850
CYCLAMAL see COU500
CYCLAMATE see SGC000
CYCLAMATE see CPQ625
CYCLAMATE CALCIUM see CAR000
CYCLAMATE, CALCIUM SALT see CAR000
CYCLAMATE SODIUM see SGC000
CYCLAMEN ALDEHYDE see COU500
CYCLAMEN ALDEHYDE DIETHYL ACETAL see COU510
CYCLAMIC ACID see CPQ625
CYCLAMIC ACID SODIUM SALT see SGC000
CYCLAMIDE see ABB000
CYCLAN see CAR000
CYCLIC ETHYLENE(DIETHOXYPHOSPHINOTHIOYL)DITHIOIMIDOCARBONATE see PGW750
CYCLIC ETHYLENE (DIETHOXYPHOSPHINOTHIOYL)DITHIOIMIDOCARBONATE see DXN600
CYCLIC ETHYLENE P,P-DIETHYL PHOSPHONODITHIOIMIDOCARBONATE see PGW750
CYCLIC ETHYLENE ESTER of (DIETHOXYPHOSPHINOTHIOYL)DITHIOIMIDOCARBONIC ACID see DXN600
CYCLIC (HYDROXYMETHYL)ETHYLENE ACETAL ACETONE see DVR600
CYCLIC N',O-PROPYLENE ESTER of N,N-BIS(2-CHLOROETHYL)PHOSPHORODIAMIDIC ACID MONOHYDRATE see CQC500
CYCLIC PROPYLENE (DIETHOXYPHOSPHINYL)DITHIOIMIDOCARBONATE see DHH400
CYCLIC SOSO see DVO920
CYCLIC-SOSO see DVO920
CYCLOBARBITAL see TDA500
CYCLOBARBITOL see TDA500
CYCLOBARBITONE see TDA500
CYCLOCEL see CMF400
CYCLOCHEM GMS see OAV000
CYCLOCHEM INEO see ISC550

α-CYCLOCITRYLIDENEACETONE see IFW000
β-CYCLOCITRYLIDENEACETONE see IFX000
α-CYCLOCITRYLIDENE-4-METHYLBUTAN-3-ONE see COW780
CYCLODAN see EAQ750
3,7-CYCLODECADIEN-1-ONE, 3,7-DIMETHYL-10-(1-METHYLETHYLIDENE)-, (3E,7E)- see GDO200
6-CYCLODECENE-1,4-DIONE, 6,10-DIMETHYL-3-(1-METHYLETHYL)-, (3R,6E,10S)- see NBW100
CYCLODISONE see DVO920
CYCLODORM see TDA500
CYCLOGEST see PMH500
9-CYCLOHEPTADECEN-1-ONE see CMU850
9-CYCLOHEPTADECEN-1-ONE, (Z)-(8CI,9CI) see CMU850
6H-CYCLOHEPTA(b)QUINOLINE-11-CARBOXAMIDE, N,N-DIETHYL-7,8,9,10-TETRAHYDRO- see TCO100
6H-CYCLOHEPTA(b)QUINOLINE, 7,8,9,10-TETRAHYDRO-11-(MORPHOLINOCARBONYL)- see MRU077
1,3,5-CYCLOHEPTATRIENE see COY000
CYCLOHEPTATRIENE (DOT) see COY000
CYCLOHEPTATRIENE MOLYBDENUM TRICARBONYL see COY100
CYCLOHEXAAN (DUTCH) see CPB000
CYCLOHEXADIENEDIONE see QQS200
1,4-CYCLOHEXADIENEDIONE see QQS200
2,5-CYCLOHEXADIENE-1,4-DIONE see QQS200
3,5-CYCLOHEXADIENE-1,2-DIONE see BDC250
2,5-CYCLOHEXADIENE-1,4-DIONE DIOXIME see DVR200
2,5-CYCLOHEXADIENE-1,4-DIONE, 2-PHENYL-(9CI) see PEL750
1,4-CYCLOHEXADIENE DIOXIDE see QQS200
CYCLOHEXAMETHYLENE CARBAMIDE see CPB050
CYCLOHEXAMETHYLENIMINE see HDG000
CYCLOHEXAN (GERMAN) see CPB000
CYCLOHEXANAMIDE see CPB050
CYCLOHEXANAMINE see CPF500
CYCLOHEXANAMINE HYDROCHLORIDE see CPA775
CYCLOHEXANE see CPB000
CYCLOHEXANECARBONITRILE, 5-OXO-1,3,3-TRIMETHYL- see IMG500
CYCLOHEXANECARBOXAMIDE see CPB050
CYCLOHEXANECARBOXYLIC ACID, LEAD SALT see NAS500
CYCLOHEXANECARBOXYLIC ACID, TRIBUTYLSTANNYL ESTER see TID100
CYCLOHEXANEFORMAMIDE see CPB050
cis-1,2,3,5-trans-4,6-CYCLOHEXANEHEXOL see IDE300
CYCLOHEXANE, ISOCYANATO-(9CI) see CPN500
CYCLOHEXANE, 5-ISOCYANATO-1-(ISOCYANATOMETHYL)-1,3,3-TRIMETHYL-(9CI) see IMG000
CYCLOHEXANE OXIDE see CPD000
CYCLOHEXANESULFAMIC ACID, CALCIUM SALT see CAR000
CYCLOHEXANESULFAMIC ACID, MONOSODIUM SALT see SGC000
CYCLOHEXANESULPHAMIC ACID see CPQ625
CYCLOHEXANESULPHAMIC ACID, MONOSODIUM SALT see SGC000
CYCLOHEXANETHIOL see CPB625
CYCLOHEXANE, 1-(TRICHLOROSILYL)- see CPR250
CYCLOHEXANOL see CPB750

CYCLOPHOSPHAMIDE see CQC650
CYCLOPHOSPHAMIDE HYDRATE see CQC500
CYCLOPHOSPHAMIDE-N-MONOCHLOROETHYL
derivative see TNT500
CYCLOPHOSPHAMIDE MONOHYDRATE see CQC500
CYCLOPHOSPHAMIDUM see CQC500
CYCLOPHOSPHAMIDUM see CQC650
CYCLOPHOSPHAN see CQC500
CYCLOPHOSPHAN see CQC650
CYCLOPHOSPHANE see CQC500
CYCLOPHOSPHANUM see CQC500
CYCLOPHOSPHORAMIDE see CQC650
CYCLOPROPANE see CQD750
CYCLOPROPANE, liquefied (DOT) see CQD750
CYCLOPROPANE, 1-(N-AMINO)CARBAMOYL-2-
METHYL- see MIV300
CYCLOPROPANECARBOXYLIC ACID, 3-(2-CHLORO-
3,3,3-TRIFLUORO-1-PROPENYL)-2,2-DIMETHYL-,
CYANO(3-PHENOXYPHENYL)METHYL ESTER see
GJU600
CYCLOPROPANECARBOXYLIC ACID, 3-(2-CHLORO-
3,3,3-TRIFLUORO-1-PROPENYL)-2,2-DIMETHYL-
,CYANO(3-PHENOXYPHENYL)METHYL ESTER, (1-
α(S*),3-α(Z))-(+−)- see LAS200
CYCLOPROPANECARBOXYLIC ACID, 3-(2,2-
DICHLOROETHENYL)-2,2-DIMETHYL-, CYANO(3-
PHENOXYPHENYL) METHYL ESTER, MIXED WITH
ACETIC ACID ANHYDRIDE, 5-((2-(2-
BUTOXYETHOXY)ETHOXY)METHYL)-6-PROPYL-
1,3-BENZODIOXOLE, DIMETHYLBENZENE AND 1-
METHYL 2-PYRROLIDINONE see BGD088
CYCLOPROPANECARBOXYLIC ACID, 2,2-
DIMETHYL-3-(2-METHYLPROPENYL)-, ESTER WITH
4-HYDROXY-3-METHYL-2-(2-PROPYNYL)-2-
CYCLOPENTEN-1-ONE, trans-(+−)- see THI500
CYCLORYL 21 see SIB600
CYCLOSAN see MCW000
CYCLOSIA see CMS850
CYCLOSPORIN see CQH100
CYCLOSPORIN A see CQH100
CYCLOSPORINE see CQH100
CYCLOSPORINE A see CQH100
CYCLOSTIN see CQC650
CYCLOTEN see HMB500
CYCLOTETRAMETHYLENE OXIDE see TCR750
CYCLOTETRAMETHYLENE TETRANITRAMINE see
CQH250
CYCLOTETRAMETHYLENETETRANITRAMINE (dry
or unphlegmatized) (DOT) see CQH250
CYCLOTETRAMETHYLENETETRANITRAMINE,
wetted (UN 0226) (DOT) see CQH250
CYCLOTETRAMETHYLENETETRANITRAMINE,
desensitized (UN 0483) (DOT) see CQH250
CYCLOTETRASILOXANE, 2,6-DIPHENYL-2,4,4,6,8,8-
HEXAMETHYL- see DWC650
CYCLOTON V see HCQ500
CYCLOTRIMETHYLENENITRAMINE see CPR800
CYCLOTRIMETHYLENETRINITRAMINE see CPR800
CYCLOTRIMETHYLENETRINITRAMINE, wetted (UN
0072) (DOT) see CPR800
CYCLOTRIMETHYLENETRINITRAMINE, desensitized
(UN 0483) (DOT) see CPR800
CYCOCEL see CMF400
CYCOCEL-EXTRA see CMF400
CYCOGAN see CMF400
CYCOGAN EXTRA see CMF400
CYCOLAMIN see VSZ000

CYCTEINAMINE see AJT250
CYFEN see DSQ000
CYFLEE see CQL250
CYFOS see IMH000
CYGON see DSP400
CYGON INSECTICIDE see DSP400
CYHALOTHRIN see GJU600
CYHALOTHRINE see GJU600
CYHALOTHRIN K see LAS200
CYHEXATIN see CQH650
CYJANOWODOR (POLISH) see HHS000
CYKAZINE see COU000
CYKLOHEKSAN (POLISH) see CPB000
CYKLOHEKSANOL (POLISH) see CPB750
CYKLOHEKSANON (POLISH) see CPC000
CYKLOHEKSEN (POLISH) see CPC579
CYKLOHEXANTHIOL see CPB625
CYKLOHEXYLMERKAPTAN (CZECH) see CPB625
CYKLONIT see CPR800
CYKOBEMINET see VSZ000
CY-L 500 see CAQ250
CYLAN see PGW750
CYLAN see CAR000
CYLAN see DXN600
CYLPHENICOL see CDP250
CYMAG see SGA500
CYMATE see BJK500
CYMBI see AIV500
CYMBI see AOD125
CYMEL see MCB000
CYMENE see CQI000
p-CYMENE see CQI000
2-p-CYMENOL see CCM000
3-p-CYMENOL see TFX810
p-CYMEN-3-OL see TFX810
CYMETHION see MDT750
CYMETOX see MIW250
CYMIDON see KFK000
CYMOL see CQI000
CYMONIC ACID see FIC000
CYNARON see MDT740
CYNEM see EPC500
CYNKU TLENEK (POLISH) see ZKA000
CYNOGAN see BMM650
CYOCEL see CMF400
CYOLAN see DXN600
CYOLANE see PGW750
CYOLANE INSECTICIDE see PGW750
CYOLANE INSECTICIDE see DXN600
CYPENTIL see PIL500
CYPIP see DIW000
CYPONA see DGP900
CYPREX see DXX400
CYPREX 65W see DXX400
CYPRON see MQU750
CYRAL see DBB200
CYREDIN see VSZ000
CYREN see DKA600
CYREN B see DKB000
CYRSTHION see EKN000
CYSTAMIN see HEI500
CYSTAMINE DIHYDROCHLORIDE see CQJ750
CYSTAMINE "MCCLUNG" see PDC250
CYSTEAMIDE see AJT250
CYSTEAMINE see AJT250
CYSTEIN see CQK000
CYSTEINE see CQK000
l-CYSTEINE see CQK000

l-(+)-CYSTEINE see CQK000
CYSTEINE CHLORHYDRATE see CQK250
CYSTEINE DISULFIDE see CQK325
CYSTEINE HYDROCHLORIDE see CQK250
l-CYSTEINE HYDROCHLORIDE see CQK250
l-CYSTEINE HYDROCHLORIDE see CQK250
l-CYSTEINE MONOHYDROCHLORIDE (FCC) see CQK250
CYSTIN see CQK325
(−)-CYSTINE see CQK325
l-CYSTINE see CQK325
CYSTINE ACID see CQK325
CYSTOGEN see HEI500
CYSTOIDS ANTHELMINTIC see HFV500
CYSTOPYRIN see PDC250
CYSTORELIN see LIU305
CYSTORELIN see LIU370
CYSTURAL see PDC250
CYTACON see VSZ000
CYTAMEN see VSZ000
CYTEL see DSQ000
CYTEN see DSQ000
CYTHIOATE see CQL250
CYTHION see MAK700
CYTOBION see VSZ000
CYTOCHALASIN D see ZUS000
CYTOCHROME C see CQM325
CYTOPHOSPHAN see CQC500
CYTOPHOSPHAN see CQC650
CYTOREST see CQM325
CYTOSINE DEOXYRIBOSIDE see DAQ850
CYTOXAL ALCOHOL see CQN000
CYTOXAN see CQC500
CYTOXAN see CQC650
CYTOXYL ALCOHOL CYCLOHEXYLAMMONIUM SALT see CQN000
CYTROL see AMY050
CYTROL AMITROLE-T see AMY050
CYTROLANE see DHH400
CYTROLE see AMY050
CYURAM DS see TFS350
CYZINE PREMIX see ABY900
CZT see CLX000
CZTEROCHLOREK WEGLA (POLISH) see CBY000
2,3,7,8-CZTEROCHLORODWUBENZO-p-DWUOKSYNY (POLISH) see TAI000
1,1,2,2-CZTEROCHLOROETAN (POLISH) see TBQ100
CZTEROCHLOROETYLEN (POLISH) see PCF275
CZTEROETHLEK OLOWIU (POLISH) see TCF000
D-D see DGG000
D-50 see POL500
D 50 see AAX250
D 212 see DIR100
D 704 see ORI400
D 735 see CCC500
D 860 see BSQ000
D 1221 see CBS275
D-1410 see DSP600
DA see CGN000
DA see DNA200
DA 737S see EHP700
2,4-DAA see DBO000
DAAB see DWO800
DAAE see DCN800
2,4-DAA SULFATE see DBO400
DAB see DOT300
DABICYCLINE see HOH500
DAB-N-OXIDE see DTK600

DABROSIN see ZVJ000
DABYLEN see BAU750
DABYLEN see BBV500
DAC 649 see TBV300
DAC 2797 see TBQ750
DACARBAZINE see DAB600
DACONATE 6 see MRL750
DACONIL see TBQ750
DACONIL 2787 FLOWABLE FUNGICIDE see TBQ750
DACORTIN see PLZ000
DACOSOIL see TBQ750
DACOVIN see PKQ059
DACTIMICIN SULFATE see DAB630
DACTIN see DFE200
DADEX see BBK500
DADOX d-CITRAMINE see BBK500
DADPE see OPM000
DADPM see MJQ000
DADPS see SOA500
DAEP see DOP200
DAF 68 see DVL700
DAGADIP see TNP250
DAGENAN see PPO000
DAGUTAN see SJN700
DAI CARI XBN see BQK250
DAICEL 1150 see SFO500
DAIFLON see CLQ750
DAIFLON 22 see CFX500
DAIFLON S 3 see FOO000
DAILON see DXQ500
DAINICHI CHROME ORANGE R see LCS000
DAINICHI CHROME YELLOW G see LCR000
DAINICHI FAST SCARLET G BASE see NMP500
DAINICHI LAKE RED C see CHP500
DAISHIKI BRILLIANT SCARLET 3R see FMU080
DAISOLAC see PJS750
DAITO ORANGE BASE R see NEN500
DAITO RED BASE TR see CLK220
DAITO RED SALT TR see CLK235
DAITO SCARLET BASE G see NMP500
DAKINS SOLUTION see SHU500
DAKTIN see DFE200
DALAPON see DGI600
DALAPON 85 see DGI400
DALAPON SODIUM see DGI600
DALAPON SODIUM SALT see DGI600
DALAPON (USDA) see DGI400
DAL-E-RAD see MRL750
DALF see MNH000
DALTOGEN see TKP500
DAMILEN HYDROCHLORIDE see EAI000
DAMINOZIDE (USDA) see DQD400
DANA see NJW500
DANAMID see PJY500
DANAMINE see DJS200
DANANTIZOL see MCO500
DANEX see TIQ250
DANFIRM see AAX250
DANIZOL see MMN250
DANSYL see DPN200
DANSYL CHLORIDE see DPN200
DANTAFUR see NGE000
DANTEN see DKQ000
DANTEN see DNU000
DANTHION see PAK000
DANTHRON see DMH400
DANTINAL see DKQ000
DANTOIN see DFE200

DANTOIN see DNU000
DANTOINAL KLINOS see DKQ000
DANTOINE see DKQ000
DANTRON see DMH400
DANUVIL 70 see PKQ059
DAP see DOT000
DAPA see DOU600
DAPAZ see MQU750
DAPHENE see DSP400
DAPLEN see PJS750
DAPLEN AD see PMP500
DAPLEN 1810 H see PJS750
DAPM see MJQ000
DAPON 35 see DBL200
DAPON R see DBL200
DAPRISAL see ABG750
DAPSONE see SOA500
DARACLOR see TGD000
DARAL see VSZ100
DARAMIN see CAM750
DARAMMON see ANE500
DARAPRAM see TGD000
DARAPRIM see TGD000
DARAPRIME see TGD000
DARATAK see AAX250
DAR-CHEM 14 see SLK000
DARCO see CBT500
DAREBON see RDK000
DARID QH see SEH000
DARILOID QH see SEH000
DAROPERVAMIN see DBA800
DARVIC 110 see PKQ059
DARVIS CLEAR 025 see PKQ059
DARVON see DAB879
DARVON COMPOUND see ABG750
DARVON HYDROCHLORIDE see PNA500
DAS see DOU600
DASANIT see FAQ800
DASIKON see ABG750
DASKIL see NCQ900
DATC see DBI200
DATHROID see LFG050
DATRIL see HIM000
DATURINE see HOU000
DAUCUS OIL see CCL750
DAUNAMYCIN see DAC000
DAUNOMYCIN see DAC000
DAUNOMYCIN, 4-DEMETHOXY-,
   HYDROCHLORIDE see DAN100
DAUNORUBICIN see DAC000
DAUNORUBICINE see DAC000
DAVISON SG-67 see SCH002
DAVITAMON D see VSZ100
DAVITAMON PP see NCQ900
DAVITIN see VSZ100
DAWE'S DESTROL see DKA600
DAWSON 100 see MHR200
DAWSONITE see DAC450
DAZOMET see DSB200
DAZZEL see DCM750
2NDB see ALL750
4NDB see ALL500
DB 133 see FQL200
DB 134 see DNF500
DB 135 see DSC100
DB 136 see DNF450
DB 138 see EID250
DBA see DCT400

DBA see DQJ200
DB(a,h)A see DCT400
1,2,5,6-DBA see DCT400
DB(a,h)AC see DCS400
DB(a,j)AC see DCS600
7H-DB(c,g)C see DCY000
DBCP see DDL800
DBD see ASH500
DBD see DDJ000
DBDPO see PAU500
DBE see EIY500
DBED DIPENCILLIN G see BFC750
DBED PENICILLIN see BFC750
DBF see DEC400
DBH see BBP750
DBH see BBQ500
DBM see DED600
DBMP see BFW750
DBN see BRY500
DBNA see BRY500
DBNPA see DDM000
DBOT see DEF400
DBP see DEH200
DB(a,e)P see NAT500
DB(a,i)P see BCQ500
DB(a,l)P see DCY400
DBPC (technical grade) see BFW750
D.B.T.C. see DDY200
DBTL see DDV600
DC 360 see SCR400
DCA see DEL000
DCA see DFE200
DCA 70 see AAX250
DCB see DEP600
DCB see DEQ600
DCB see DEV000
1,4-DCB see DEV000
D&C BLUE No. 4 see FMU059
D&C BLUE NUMBER 1 see BJI250
DCBN see DGM600
2,3-DCDT see DBI200
1-1-DCE see VPK000
1,2-DCE see EIY600
DCEE see DFJ050
D&C GREEN No. 6 see BLK000
DCHA see DGT600
DCIP see BII250
DCIP (nematocide) see BII250
DCM see MJP450
DCMO see CCC500
DCMU see DXQ500
DCNU see CLX000
D&C ORANGE No. 3 see HGC000
D&C ORANGE No. 2 see TGW000
D&C ORANGE NO. 5 see DDO200
DCP see DFX800
2,4-DCP see DFX800
DCPA see DGI000
D&C RED No. 5 see FMU070
D&C RED No. 9 see CHP500
D&C RED No. 22 see BNH500
D & C RED NO. 17 see OHI200
D and C RED NO. 28 see ADG250
D.C.S. see BGJ750
dCYD see DAQ850
D&C YELLOW No. 7 see FEV000
D&C YELLOW No. 8 see FEW000
D and C YELLOW NO. 5 see FAG140

DECASPRAY see SOW000
n-DECATYL ALCOHOL see DAI600
DECCOTANE see BPY000
DECCOX see DAI495
DECELITH H see PKQ059
DECEMTHION P-6 see PHX250
2-DECENAL see DAI350
cis-4-DECENAL see DAI360
trans-2-DECEN-1-AL see DAI350
DECENALDEHYDE see DAI350
cis-4-DECEN-1-AL (FCC) see DAI360
9-DECEN-1-OL, ACETATE see DAI450
DECENTAN see CJM250
DECENYL ACETATE see DAI450
9-DECENYL ACETATE see DAI450
DECHAN see DGU200
DECHLORANE 4070 see MQW500
DECHLORANE A-O see AQF000
DECLOMYCIN see MIJ500
DECOFOL see BIO750
n-DECOIC ACID see DAH400
DECOQUINATE see DAI495
DECORPA see GLU000
DECORTANCYL see PLZ000
DECORTIN see PLZ000
DECORTIN H see PMA000
DECORTISYL see PLZ000
DECTAN see DBC400
DECTANCYL see SOW000
DE-CUT see DMC600
DECYL ALCOHOL see DAI600
n-DECYL ALCOHOL see DAI600
DECYL ALDEHYDE see DAG000
1-DECYL ALDEHYDE see UJJ000
1-DECYL ALDEHYDE see DAG000
n-DECYL ALDEHYDE see DAG000
DECYLALDEHYDE DMA see AFJ700
DECYLAMINE see DAG600
DECYL BENZENE SODIUM SULFONATE see DAJ000
DECYLIC ACID see DAH400
n-DECYLIC ACID see DAH400
DECYLIC ALCOHOL see DAI600
DECYLIC ALDEHYDE see DAG000
DECYL METHYL ETHER see MIW075
DECYL OCTYL ALCOHOL see OAX000
6-DECYLOXY-7-ETHOXY-4-HYDROXY-3-
QUINOLINECARBOXYLID ACID ETHYL ESTER see
DAI495
4-DECYLOXY-2-HYDROXYPHENYL 4-
DECYLOXYPHENYL KETONE see DAJ450
DECYLTRIPHENYLPHOSPHONIUM
BROMOCHLOROTRIPHENYLSTANNATE see DAE600
(DECYL-TRIPHENYL-PHOSPHONIUM)-TRIPHENYL-
BROM-CHLOR-STANNAT (GERMAN) see DAE600
DEDC see SGJ000
DEDELO see DAD200
DEDEVAP see DGP900
DEDK see SGJ000
DED-WEED see TIX500
DED-WEED see CIR250
DED-WEED see DGI400
DED-WEED CRABGRASS KILLER see PLC250
DEE-OSTEROL see VSZ100
DEEP LEMON YELLOW see SMH000
DEE-RON see VSZ100
DEE-RONAL see VSZ100
DEE-ROUAL see VSZ100
DEET see DKC800

DEF see BSH250
DEF DEFOLIANT see BSH250
DE-FEND see DSP400
DEFILIN see DJL000
DEFILTRAN see AAI250
DEFLAMON-WIRKSTOFF see MMN250
DEFOAMER S-10 see SCP000
DE-FOL-ATE see SFS000
DEFOLIT see TEX600
DEG see DJD600
DEGALAN S 85 see PKB500
DEGALOL see DAQ400
DEGRASSAN see DUS700
DE-GREEN see BSH250
DEGUSSA see CBT750
D.E.H. 20 see DJG600
DEH 24 see TJR000
D.E.H. 26 see TCE500
DEHA see AEO000
DEHA see DJN000
DEHACODIN see DKW800
DEHISTIN see TMP750
DEHP see DVL700
DEHPA EXTRACTANT see BJR750
DEHYDRACETIC ACID see MFW500
DEHYDRATIN see AAI250
DEHYDRITE see PCE000
DEHYDROACETIC ACID (FCC) see MFW500
DEHYDROACETIC ACID, SODIUM SALT see SGD000
6-DEHYDRO-6-CHLORO-17-α-
ACETOXYPROGESTERONE see CBF250
7-DEHYDROCHOLESTROL, ACTIVATED see CMC750
Δ^1-DEHYDROCORTISOL see PMA000
1-DEHYDROCORTISONE see PLZ000
Δ-1-DEHYDROCORTISONE see PLZ000
DEHYDROEPIANDROSTERONE ACETATE see
HJB225
1-DEHYDROHYDROCORTISONE see PMA000
Δ^1-DEHYDROHYDROCORTISONE see PMA000
DEHYDROLINALOOL see LFY333
DEHYDRO-β-LINALOOL see LFY333
6-DEHYDRO-6-METHYL-17-α-
ACETOXYPROGESTERONE see VTF000
6-DEHYDRO-16-METHYLENE-6-METHYL-17-
ACETOXYPROGESTERONE see MCB380
1-DEHYDRO-16-α-METHYL-9-α-
FLUOROHYDROCORTISONE see SOW000
DEHYDRORETRONECINE see DAL400
DEHYDROSTILBESTROL see DAL600
DEHYQUART STC-25 see DTC600
DEINAIT see CJR909
DEIQUAT see DWX800
DEJO see DJT800
DEK see DJN750
DE-KALIN see DAE800
DEKALINA (POLISH) see DAE800
DEKORTIN see PLZ000
DEKRYSIL see DUS700
DEKSONAL see DOU600
DELAC J see DWI000
DELAC S see CPI250
DELACURARINE see TOA000
DELADIOL see EDS100
DELAHORMONE UNIMATIC see EDS100
DELAPHOS see ZJS400
DELAPHOS 2M see ZJS400
DELATESTRYL see MPN500

DELCORTOL see PMA000
DELEAF DEFOLIANT see TIG250
DELESTROGEN see EDS100
DELESTROGEN 4X see EDS100
DELGESIC see ADA725
DELICIA see PGY000
DELICIA see AHE750
DELLIPSOIDS see BBK500
DELMOFULVINA see GKE000
DELNAV see DVQ709
DELONIN AMIDE see NCR000
DELOWAS S see TBD000
DELOWAX OM see TBD000
DELPET 50M see PKB500
DELPHENE see DKC800
m-DELPHENE see DKC800
DELPHINIC ACID see ISU000
DELSENE see MHC750
DELSENE M see MAS500
DELSTEROL see CMC750
DELTA see CJJ000
DELTA-CORTEF see PMA000
DELTACORTELAN see PLZ000
DELTACORTENOL see PMA000
DELTACORTISONE see PLZ000
DELTACORTONE see PLZ000
DELTACORTRIL see PMA000
DELTA-DOME see PLZ000
DELTA F see PMA000
DELTAFLUORENE see SOW000
DELTAGLUCONOLACTONE see GFA200
DELTALIN see VSZ100
DELTA-MVE see MKB750
DELTAMYCIN A see CBT250
DELTAN see DUD800
DELTA-STAB see PMA000
DELTATHIONE see GFW000
DELTAZINA see PPP500
DELTISONE see PLZ000
DELTRATE-20 see PBC250
DELTYLEXTRA see IQN000
DELUSSA BLACK FW see CBT750
DELVEX see DJT800
DELYSID see DJO000
DEMA see BIE500
DEMASORB see DUD800
DEMAVET see DUD800
DEMECLOCYCLINE see MIJ500
DEMECOLCIN see MIW500
DEMECOLCINE see MIW500
DEMEPHION see MIW250
DEMEROL see DAM700
DEMEROL HYDROCHLORIDE see DAM700
DEMESO see DUD800
4-DEMETHOXYDAUNORUBICIN
HYDROCHLORIDE see DAN100
DEMETHYLCHLOROTETRACYCLINE see MIJ500
6-DEMETHYLCHLOROTETRACYCLINE see MIJ500
6-DEMETHYL-7-CHLOROTETRACYCLINE see MIJ500
DEMETHYLCHLORTETRACYCLIN see MIJ500
DEMETHYLCHLORTETRACYCLINE see MIJ500
6-DEMETHYLCHLORTETRACYCLINE see MIJ500
6-DEMETHYL-7-CHLORTETRACYCLINE see MIJ500
DEMETHYLCHLORTETRACYCLINE BASE see MIJ500
o-DEMETHYLDAUNOMYCIN see KBU000
DEMETHYLDOPAN see BIA250
DEMETHYL-EPIODOPHYLLOTOXIN ETHYLIDENE
GLUCOSIDE see EAV500

4-DEMETHYLEPIODODPHYLLOTOXIN-β,d-
ETHYLIDENEGLUCOSIDE see EAV500
4'-DEMETHYLEPIPODOPHYLLOTOXIN-9-(4,6-O-
ETHYLIDENE-β-d-GLUCOPYRANOSIDE see EAV500
4'-DEMETHYLEPIPODOPHYLLOTOXIN
ETHYLIDENE-β,d-GLUCOSIDE see EAV500
4-DEMETHYL-EPIPODOPHYLLOTOXIN-β,d-
ETHYLIDEN-GLUCOSIDE see EAV500
4'-DEMETHYLEPIPODOPHYLLOTOXIN-9-(4,6-O-2-
THENYLIDENE-β-d-GLUCOPYRANOSIDE see
EQP000
4'-DEMETHYL-EPIPODOPHYLLOTOXIN-β-d-
THENYLIDENE-GLUCOSIDE see EQP000
4'-O-DEMETHYL-1-O-(4,6-O-ETHYLIDENE-β,d-
GLUCOPYRANOSYL)EPIPODOPHYLLOTOXIN see
EAV500
4'-DEMETHYL 1-O-(4,6-O,O-(2-THENYLIDENE)-β-d-
GLUCOPYRANOSYL)EPIPODOPHYLLOTOXIN see
EQP000
6-DEMETIL-7-CLOROTETRACICLINA see MIJ500
DEMETON-O-METHYL see DAO800
DEMETON-S-METHYL see DAP400
DEMETON-S-METHYLSULFON (GERMAN) see
DAP600
DEMETON-S-METHYLSULFONE see DAP600
DEMETON-S-METHYL-SULFOXID (GERMAN) see
DAP000
DEMETON-O-METHYL SULFOXIDE see DAP000
DEMETON-S-METHYL SULFOXIDE see DAP000
DEMETON-S-METHYL-SULPHONE see DAP600
DEMETON-METHYL SULPHOXIDE see DAP000
DEMETON-O-METILE (ITALIAN) see DAO800
DEMETON-S-METILE (ITALIAN) see DAP400
DEMETON-S see DAP200
DEMOS-L40 see DSP400
DEMSODROX see DUD800
DEN see NJW500
DENA see NJW500
DENAMONE see VSZ450
DENAPON see CBM750
DENAPON, NITROSATED (JAPANESE) see NBJ500
DENATURED ALCOHOL (DOT) see AFJ000
DENATURED SPIRITS see AFJ000
DENDREPAR see PEM750
DENDRID see DAS000
DENDRITIS see SFT000
DENKA F 90 see SCK600
DENKA FB 44 see SCK600
DENKALAC 61 see AAX175
DENKA QP3 see SMQ500
DENKA VINYL SS 80 see PKQ059
DENSIC C 500 see SCQ000
DENSINFLUAT see TIO750
DENYL see DKQ000
DENYL see DNU000
DENYLSODIUM see DNU000
DEOBASE see KEK000
DEODOPHYLL see CKN000
DEODORIZED WINTERIZED COTTONSEED OIL see
CNU000
DEOFED see MDT600
DEOFED see DBA800
DEORLENE GREEN JJO see BAY750
DEOSAN see SHU500
DEOVAL see DAD200
2-DEOXY-3-ARABINO-HEXOSE see DAR600

2-DEOXY-d-ARABINO-HEXOSE see DAR600
DEOXYBENZOIN see PEB000
DEOXYCHOLATE SODIUM see SGE000
DEOXYCHOLATIC ACID see DAQ400
7-α-DEOXYCHOLIC ACID see DAQ400
DEOXYCHOLIC ACID (FCC) see DAQ400
DEOXYCHOLIC ACID SODIUM SALT see SGE000
DEOXYCYTIDINE see DAQ850
2'-DEOXYCYTIDINE see DAQ850
14-DEOXY-14-((2-DIETHYLAMINOETHYL-THIO)-
ACETOXY)MUTILINE see TET800
DEOXYEPHEDRINE see DBA800
d-DEOXYEPHEDRINE HYDROCHLORIDE see
MDT600
9-DEOXY-12,13-EPOXY-9-OXOLEUCOMYCIN V 3-
ACETATE 4ᴮ-(3-METHYLBUTANOATE) see CBT250
DEOXYFLUOROURIDINE see DAR400
2'-DEOXY-5-FLUOROURIDINE see DAR400
2-DEOXYGLUCOSE see DAR600
2-DEOXY-d-GLUCOSE see DAR600
d-2-DEOXYGLUCOSE see DAR600
2-DEOXY-d-GLUCOSE (FRENCH) see DAR600
2'-DEOXY-5-IODOURIDINE see DAS000
2-DEOXY-2-
(((METHYLNITROSOAMINO)CARBONYL)AMINO)-d-
GLUCOPYRANOSE see SMD000
2-DEOXY-2-(3-METHYL-3-NITROSOUREIDO)-d-
GLUCOPYRANOSE see SMD000
2-DEOXY-2-(3-METHYL-3-NITROSOUREIDO)-α(and
β)-d-GLUCOPYRANOSE see SMD000
DEOXYNOREPHEDRINE see BBK000
2-DEOXYPHENOBARBITAL see DBB200
1-β-d-2'-DEOXYRIBOFURANOSYL-5-
FLUOROURACIL see DAR400
1-(2-DEOXY-β-d-RIBOFURANOSYL)-5-IODOURACIL
see DAS000
1-β-d-2'-DEOXYRIBOFURANOSYL-5-IODOURACIL
see DAS000
DEOXYRIBONUCLEOSIDE CYTOSINE see DAQ850
DEOXYRIBOSE CYTIDINE see DAQ850
DEOXYTETRARIC ACID see MAN000
4-DEOXYTETRONIC ACID see BOV000
β-DEOXYTHIOGUANOSINE see TFJ250
β-2'-DEOXY-6-THIOGUANOSINE MONOHYDRATE
see TFJ250
DEOXYTHYMIDINE see TFX790
2'-DEOXYTHYMIDINE see TFX790
DEP (pesticide) see TIQ250
DEPALLETHRIN see AFR250
DEPARAL see CMC750
DEPC see DIZ100
DEPEN see MCR750
DEPHADREN see AOA500
DEPHADREN see BBK500
DEPINAR see VSZ000
DEPOESTRADIOL see DAZ115
DEPOESTRADIOL CYPIONATE see DAZ115
DEPOFEMIN see DAZ115
DEPO-PROVERA see MCA000
DEPOSUL see SNN300
DEPOXIN see DBA800
DEPRANCOL see PNA500
DEPRELIN see SNE000
DEPREX see EAI000
DEPRINOL see DLH630
DEPROMIC see PNA500

DEPTHON see TIQ250
DEPTRIN see EAG100
DEPTROPINE CITRATE see EAG100
DEQUEST 2010 see HKS780
DEQUEST 2015 see HKS780
DEQUEST Z 010 see HKS780
DE 83R see PAU500
D.E.R. 332 see BLD750
DERACIL see TFR250
DERATOL see VSZ100
DEREUMA see DOT000
DERGRAMIN see SOW000
DERIBAN see DGP900
DERIZENE see SPC500
DERIZENE see DNU000
DERMACORT see CNS750
DERMADEX see HCL000
DERMA FAST BROWN W-GL see CMO750
DERMAFOSU (POLISH) see RMA500
DERMAGINE see OAV000
DERMAPHOS see RMA500
DERMASORB see DUD800
DERMATON see CDS750
DERMISTINE see BBV500
DERMODRIN see BBV500
DERMOFURAL see NGE500
DERMOXIN see TGB475
DERONIL see SOW000
DEROSAL see MHC750
DERRIBANTE see DGP900
DES (synthetic estrogen) see DKA600
DESACETYLMETHYLCOLCHICINE see MIW500
N-DESACETYLMETHYLCOLCHICINE see MIW500
N-DESACETYL-N-METHYLCOLCHICINE see MIW500
DESADRENE see SOW000
DESAMETASONE see SOW000
DESAMINE see DBA800
DESD see DKB000
DESDEMIN see CHJ750
DESENTOL see BBV500
DESERPINE see RDK000
DESERTALC 57 see TAB750
DESERT RED see CHP500
DESFEDRIN see DBA800
DESICAL P see CAU500
DESICCANT L-10 see ARB250
DESINFECT see CDP000
DESIPRAMINE HYDROCHLORIDE see DLS600
DESMA see DKA600
DESMECOLCHINE see MIW500
DESMECOLCINE see MIW500
DESMETHYLDOPAN see BIA250
DESMETHYLIMIPRAMINE HYDROCHLORIDE see
DLS600
DESMODUR 44 see MJP400
DESMODUR 44V20 see PKB100
DESMODUR H see DNJ800
DESMODUR N see DNJ800
DESMODUR PU 1520A20 see PKB100
DESMODUR T80 see TGM750
DESMODUR T100 see TGM740
DESOLET see SFS000
DESORMONE see DGB000
14-DESOSSI-14-((2-DIETILAMINOETIL)MERCAPTO-
ACETOSSI)MUTILIN IDROGENO FUMARATO
(ITALIAN) see TET800
DESOSSIEFEDRINA see DBA800
DES-OXA-D see DBA800

DESOXEDRINE see DBA800
DESOXIN see DBA800
DESOXO-5 see MDT600
DESOXO-5 see DBA800
DESOXON 1 see PCL500
DESOXYBENZOIN see PEB000
DESOXYCHOLATE AMPHOTERICIN B see FPC200
DESOXYCHOLIC ACID see DAQ400
DESOXYCHOLSAEURE (GERMAN) see DAQ400
DESOXYCYTIDIN (GERMAN) see DAQ850
14-DESOXY-14-((DIETHYLAMINOETHYL)-MERCAPTO ACETOXYL)-MUTILIN HYDROGEN FUMARATE see TET800
DESOXYEPHEDRINE HYDROCHLORIDE see DBA800
d-DESOXYEPHEDRINE HYDROCHLORIDE see MDT600
l-DESOXYEPHEDRINE HYDROCHLORIDE see MDQ500
DESOXYFED see MDT600
DESOXYFED see DBA800
DESOXYN see MDT600
DESOXYN see BBK500
DESOXYN see DBA800
DESOXYNE see MDT600
(±)-DESOXYNOREPHEDRINE see BBK000
racemic-DESOXYNOREPHEDRINE see BBK000
DESOXYPHED see DBA800
2-DESOXYPHENOBARBITAL see DBB200
DESOXYPHENOBARBITONE see DBB200
DESPHEN see CDP250
DE-SPROUT see DMC600
DESSON see CLW000
DESTENDO see BCA000
DESTIM see MDT600
DESTIM see DBA800
DESTRIOL see EDU500
DESTROL see DKA600
DESTRONE see EDV000
DESTRUXOL APPLEX see DXE000
DESTRUXOL BORER-SOL see EIY600
DESTRUXOL ORCHID SPRAY see NDN000
DESURIC see DDP200
DESYPHED see MDT600
DET see DKC800
m-DET see DKC800
DETA see DJG600
m-DETA see DKC800
DETAL see DUS700
DETALUP see VSZ100
DETAMIDE see DKC800
DETERGENT 66 see SIB600
DETERGENT HD-90 see DXW200
DETF see TIQ250
DETHMORE see WAT200
DETIA GAS EX-B see PGY000
DETIA GAS EX-B see AHE750
DETICENE see DAB600
DETMOL-EXTRAKT see BBQ500
DETMOL MA see MAK700
DETMOL MA 96% see MAK700
DETMOL U.A. see CMA100
DETOX see DAD200
DETOX 25 see BBQ500
DETOXAN see DAD200
DETOXARGIN see AQW000
DETREOMYCINE see CDP250
DETREOPAL see CDP700

DETREX see DBA800
DETTOL see CLW000
DEUSLON-A see EDU500
DEUTERIUM OXIDE see HAK000
DEVAL RED K see CLK220
DEVAL RED TR see CLK220
DEVEGAN see ABX500
DEVELIN see PNA500
DEVELOPER 11 see PEY000
DEVELOPER 13 see PEY500
DEVELOPER 14 see TGL750
DEVELOPER A see NAX000
DEVELOPER AMS see NAX000
DEVELOPER B see TGL750
DEVELOPER BN see NAX000
DEVELOPER DB see TGL750
DEVELOPER DBJ see TGL750
DEVELOPER H see TGL750
DEVELOPER MC see TGL750
DEVELOPER MT see TGL750
DEVELOPER MT-CF see TGL750
DEVELOPER MTD see TGL750
DEVELOPER P see NEO500
DEVELOPER PF see PEY500
DEVELOPER R see REA000
DEVELOPER SODIUM see NAX000
DEVELOPER T see TGL750
DEVICARB see CBM750
DEVIGON see DSP400
DEVIKOL see DGP900
DEVIPON see DGI400
DEVISULPHAN see EAQ750
DEVITHION see MNH000
DEVOL ORANGE B see NEO000
DEVOL ORANGE R see NEN500
DEVOL RED GG see NEO500
DEVOL RED K see CLK235
DEVOL RED TA SALT see CLK235
DEVOL RED TR see CLK235
DEVOL SCARLET B see NMP500
DEVOL SCARLET G SALT see NMP500
DEVORAN see BBQ500
DEVOTON see MFW100
DEX see BJU000
DEXA see SOW000
DEXACORT see SOW000
DEXACORT see DAE525
DEXA-CORTIDELT see SOW000
DEXA-CORTIDELT HOSTACORTIN H see PMA000
DEXADELTONE see SOW000
DEXADRESON see DAE525
DEXAGRO see DAE525
DEXAIME see BBK500
DEXALINE see BBK500
DEXALME see BBK500
DEXAMED see BBK500
DEXAMETH see SOW000
DEXAMETHASONE ACETATE see DBC400
DEXAMETHASONE ALCOHOL see SOW000
DEXAMETHASONE DISODIUM PHOSPHATE see DAE525
DEXAMETHASONE SODIUM PHOSPHATE see DAE525
DEXAMETHAZONE SODIUM PHOSPHATE see DAE525
DEXAMINE see BBK500
DEXAMPHAMINE see BBK500
DEXAMPHETAMINE see AOA500

DEXAMPHETAMINE see BBK500
DEXAMPHETAMINE SULFATE see BBK500
DEXAMYL see BBK500
DEXEDRINA see BBK500
DEXEDRINE see AOA500
DEXEDRINE SULFATE see BBK500
DEXIES see BBK500
DEXNIGULDIPINE HYDROCHLORIDE see NDY550
DEXON see DOU600
DEXONE see SOW000
DEXON E 117 see PMP500
DEXOPHRINE see DBA800
DEXOVAL see MDT600
DEXOVAL see DBA800
DEXPANTHENOL (FCC) see PAG200
DEXTELAN see SOW000
DEXTIM see MDT600
DEXTRAN see DBD700
DEXTRAN 1 see DBC800
DEXTRAN 2 see DBD000
DEXTRAN 5 see DBD200
DEXTRAN 10 see DBD400
DEXTRAN 11 see DBD600
DEXTRAN 70 see DBD700
DEXTRAN ION COMPLEX see IGS000
DEXTRANS see DBD800
DEXTRAVEN see DBD700
DEXTRINS see DBD800
DEXTROAMPHETAMINE SULFATE see BBK500
DEXTRO CALCIUM PANTOTHENATE see CAU750
DEXTROFER 75 see IGS000
DEXTRO-α-METHYLPHENETHYLAMINE SULFATE
see BBK500
DEXTROMYCETIN see CDP250
DEXTRONE see DWX800
DEXTRO-1-PHENYL-2-AMINOPROPANE SULFATE
see BBK500
DEXTRO-β-PHENYLISOPROPYLAMINE SULFATE
see BBK500
DEXTROPROPOXYPHENE see DAB879
DEXTROPROPOXYPHENE HYDROCHLORIDE see
PNA500
DEXTROPROXYPHEN HYDROCHLORIDE see
PNA500
DEXTROPUR see GFG000
DEXTROSE, anhydrous see GFG000
DEXTROSE (FCC) see GFG000
DEXTROSOL see GFG000
DEXTROTUBOCURARINE CHLORIDE see TOA000
DEZIBARBITUR see EOK000
DEZONE see SOW000
DF 118 see DKW800
DF 493 see NOB800
DFA see DVX800
DFD 0173 see PJS750
DFD 0188 see PJS750
DFD 2005 see PJS750
DFD 6005 see PJS750
DFD 6032 see PJS750
DFD 6040 see PJS750
DFDJ 5505 see PJS750
DFM see DHE750
DFP see IRF000
DFT see DWN800
2-DG see DAR600
DGE see DKM200
DGNB 3825 see PJS750

DHA see MFW500
DHA-SODIUM see SGD000
2,4-DHBA see HOE600
2,5-DHBA see GCU000
DHMS see DME000
DHNT see BKH500
DHPN see DNB200
DHS see MFW500
DI-μ-(THIOCYANATODI-n-
BUTYLSTANNYLOXO)BIS(THIOCYANATODI-n-
BUTYLTIN) see BJM700
DIABASE SCARLET G see NMP500
DIABEN see BSQ000
DIABENYL see BBV500
DIABETAMID see BSQ000
DIABETOL see BSQ000
DIABORAL see BSM000
DIABUTAL see NBU000
DIABUTON see BSQ000
DIABYLEN see BBV500
DIACARB see AAI250
DIACELLITON FAST BLUE R see TBG700
DIACELLITON FAST GREY G see DCJ200
DIACELLITON FAST PINK B see AKE250
DIACELLITON FAST VIOLET B see DBY700
DIACELLITON FAST YELLOW G see AAQ250
DIACEL NAVY DC see DCJ200
DIACEPAN see DCK759
2-DIACETAMIDOFLUORENE see DBF200
2,7-DIACETAMIDOFLUORENE see BGP250
1,3-DIACETATE GLYCEROL see DBF600
1,2-DIACETATE 1,2,3-PROPANETRIOL see DBF600
DIACETIC ETHER see EFS000
DIACETIN see DBF600
1,2-DI-ACETIN see DBF600
1,3-DIACETIN see DBF600
2,3-DIACETIN see DBF600
DIACETONALCOHOL (DUTCH) see DBF750
DIACETONALCOOL (ITALIAN) see DBF750
DIACETONALKOHOL (GERMAN) see DBF750
DIACETONE see DBF750
DIACETONE ALCOHOL see DBF750
DIACETONE-ALCOOL (FRENCH) see DBF750
DIACETONE PEROXIDES, solid, or >25% in solution
(DOT) see ACV500
1,3-DIACETOXYBENZENE see REA100
DIACETOXYBUTYLTIN see DBF800
DIACETOXYDIBUTYL STANNANE see DBF800
DIACETOXYDIBUTYLTIN see DBF800
3-β,17-β-DIACETOXY-17-α-ETHYNYL-4-OESTRENE
see EQJ500
DIACETOXYMERCURIPHENOL see HNK575
DIACETOXYMERCURY see MCS750
4,15-DIACETOXY-8-(3-METHYLBUTYRYLOXY)-12,13-
EPOXY-Δ-9-TRICHOTHECEN-3-OL see FQS000
4-β,15-DIACETOXY-8-α-(3-METHYLBUTYRYLOXY)-3-
α-HYDROXY-12,13-EPOXYTRICHOTHEC-9-ENE see
FQS000
3-β,17-β-DIACETOXY-19-NOR-17-α-PREGN-4-EN-20-
YNE see EQJ500
DIACETOXYTETRABUTYLDISTANNOXANE see
BGQ000
4,4'-DIACETYLAMINOBIPHENYL see BFX000
2-DIACETYLAMINOFLUORENE see DBF200
2,7-DIACETYLAMINOFLUORENE see BGP250
N-DIACETYL-2-AMINOFLUORENE see DBF200
N,N-DIACETYL-2-AMINOFLUORENE see DBF200

4,4'-DIACETYLBENZIDINE see BFX000
N,N'-DIACETYL BENZIDINE see BFX000
DIACETYL DIOXIME see DBH000
1,2-DIACETYLETHANE see HEQ500
α,β-DIACETYLETHANE see HEQ500
DIACETYL (FCC) see BOT500
N,N-DIACETYL-2-FLUORENAMINE see DBF200
DIACETYL GLYCERINE see DBF600
DIACETYLMANGANESE see MAQ000
DIACETYLMETHANE see ABX750
DIACETYL PEROXIDE (MAK) see ACV500
DIACETYL TARTARIC ACID ESTERS of MONO- and
DIGLYCERIDES see DBH700
DIACID see BNK000
DIACID METANIL YELLOW see MDM775
DIACOTTON BLUE BB see CMO000
DIACOTTON CONGO RED see SGQ500
DIACOTTON DEEP BLACK see AQP000
DIACOTTON DEEP BLACK RX see AQP000
DIACOTTON SKY BLUE 5B see CMO500
DIACOTTON SKY BLUE 6B see CMN750
DIACRID see DBX400
* DIACTOL see VSZ100
DIACYCINE see TBX250
DIADEM CHROME BLUE R see HJF500
DI-ADRESON F see PMA000
DIAETHANOLAMIN (GERMAN) see DHF000
DIAETHANOLNITROSAMIN (GERMAN) see NKM000
1,1-DIAETHOXY-AETHAN (GERMAN) see AAG000
DIAETHYLACETAL (GERMAN) see AAG000
DIAETHYLAETHER (GERMAN) see EJU000
O,O-DIAETHYL-S-(2-AETHYLTHIO-AETHYL)-
DITHIOPHOSPHAT (GERMAN) see DXH325
O,O-DIAETHYL-S-(2-AETHYLTHIO-AETHYL)-
MONOTHIOPHOSPHAT (GERMAN) see DAP200
O,O-DIAETHYL-S-(AETHYLTHIO-METHYL)-
DITHIOPHOSPHAT (GERMAN) see PGS000
DIAETHYLALLYLACETAMIDE (GERMAN) see
DJU200
DIAETHYLAMIN (GERMAN) see DHJ200
DIAETHYLAMINOAETHANOL (GERMAN) see
DHO500
o-DIAETHYLAMINOAETHOXY-BENZANILID
(GERMAN) see DHP200
DIAETHYLANILIN (GERMAN) see DIS700
O,O-DIAETHYL-O-(4-BROM-2,5-DICHLOR)-PHENYL-
MONOTHIOPHOSPHAT (GERMAN) see EGV500
DIAETHYLCARBONAT (GERMAN) see DIX200
O,O-DIAETHYL-O-(CHINOXALYL-(2))-
MONOTHIOPHOSPHAT (GERMAN) see DJY200
O,O-DIAETHYL-O-(3-CHLOR-4-METHYL-CUMARIN-
7-YL)-MONOTHIOPHOSPHAT (GERMAN) see
CNU750
O,O-DIAETHYL-S-(6-CHLOR-2-OXO-BEN(b)-1,3-
OXALIN-3-YL)-METHYL-DIT HIOPHOSPHAT
(GERMAN) see BDJ250
O,O-DIAETHYL-S-((4-CHLOR-PHENYL-THIO)-
METHYL)DITHIOPHOSPHAT (GERMAN) see TNP250
O,O-DIAETHYL-o-(α-CYANBENZYLIDEN-AMINO)-
THIONPHOSPHAT (GERMAN) see BAT750
O,O-DIAETHYL-o-(α-CYANO-
BENZYLIDENAMINO)-MONOTHIOPHOSPHAT
(GERMAN) see BAT750
O,O-DIAETHYL-O-(2,5-DICHLOR-4-BROMPHENYL)-
THIONOPHOSPHAT (GERMAN) see EGV500
O,O-DIAETHYL-O-1-(4,5-DICHLORPHENYL)-2-
CHLOR-VINYL-PHOSPHAT (GERMAN) see CDS750

O,O-DIAETHYL-O-2,4-DICHLOR-PHENYL-
MONOTHIOPHOSPHAT (GERMAN) see DFK600
O,O-DIAETHYL-S((2,5-DICHLOR-PHENYL-THIO)-
METHYL)-DITHIOPHOSPHAT (GERMAN) see PDC750
O,O-DIAETHYL-O-2,4-DICHLORPHENYL-
THIONOPHOSPHAT (GERMAN) see DFK600
1,2-DIAETHYLHYDRAZINE (GERMAN) see DJL400
O,O-DIAETHYL-O-(2-ISOPROPYL-4-METHYL-
PYRIMIDIN-6-YL)-MONOTHIOPHOSPHAT
(GERMAN) see DCM750
O,O-DIAETHYL-O-(2-ISOPROPYL-4-METHYL)-6-
PYRIMIDYL-THIONOPHOSPHAT (GERMAN) see
DCM750
O,O-DIAETHYL-S-(1-METHYLAETHYL)-
CARBAMOYL-METHYL-MONOTHIOPHOSPHAT
(GERMAN) see PHK250
O,O-DIAETHYL-O-4-METHYLSULFINYL-PHENYL-
MONOTHIOPHOSPHAT (GERMAN) see FAQ800
DIAETHYL-NICOTINAMID (GERMAN) see DJS200
O,O-DIAETHYL-O-(4-NITROPHENYL)-
MONOTHIOPHOSPHAT (GERMAN) see PAK000
DIAETHYLNITROSAMIN (GERMAN) see NJW500
O,O-DIAETHYL-S-(4-OXOBENZOTRIAZIN-3-
METHYL)-DITHIOPHOSPHAT (GERMAN) see
EKN000
O,O-DIAETHYL-S-((4-OXO-3H-1,2,3-BENZOTRIAZIN-
3-YL)-METHYL)-DITHIOPHOSPHAT (GERMAN) see
EKN000
O,O-DIAETHYL-O-(PYRAZIN-2YL)-
MONOTHIOPHOSPHAT (GERMAN) see EPC500
O,O-DIAETHYL-O-(2-PYRAZINYL)-
THIONOPHOSPHAT (GERMAN) see EPC500
DIAETHYLSULFAT (GERMAN) see DKB110
O,O-DIAETHYL-S-(3-THIA-PENTYL)-
DITHIOPHOSPHAT (GERMAN) see DXH325
DIAETHYLTHIOPHOSPHORSAEUREESTER des
AETHYLTHIOGLYKOL (GERMAN) see DAP200
O,O-DIAETHYL-O-3,5,6-TRICHLOR-2-
PYRIDYLMONOTHIOPHOSPHAT see CMA100
DIAETHYLZINNDICHLORID (GERMAN) see DEZ000
DIAFURON see NGG500
DIAGRABROMYL see BNP750
DIAKARB see AAI250
DIAKARMON see SKV200
DIAKON see MLH750
DIAKON see PKB500
DIAL-A-GESIC see HIM000
DIALIFOR see DBI099
DIALLAAT (DUTCH) see DBI200
DIALLAT (GERMAN) see DBI200
DIALLATE see DBI200
DIALLYL see HCR500
DIALLYLAMINE see DBI600
DIALLYLCHLOROACETAMIDE see CFK000
N,N-DIALLYLCHLOROACETAMIDE see CFK000
N,N-DIALLYL-2-CHLOROACETAMIDE see CFK000
N,N-DIALLYL-α-CHLOROACETAMIDE see CFK000
DIALLYLDIBROMO STANNANE see DBJ400
DIALLYL ETHER see DBK000
DIALLYL MALEATE see DBK200
DIALLYL PHTHALATE see DBL200
DIALLYL SELENIDE see DBL300
DIALLYLTIN DIBROMIDE see DBJ400
DIALUMINUM SULPHATE see AHG750
DIALUMINUM TRIOXIDE see AHE250
DIALUMINUM TRISULFATE see AHG750
DIALUX see ADY500

DIAMARIN see DYE600
DIAMET see SIL500
DIAMIDAFOS see PEV500
DIAMIDE see HGS000
DIAMIDFOS see PEV500
DIAMINE see HGS000
2,4-DIAMINEANISOLE see DBO000
DIAMINE BLUE 2B see CMO000
DIAMINE BLUE 3B see CMO250
DIAMINE DEEP BLACK EC see AQP000
(4,4'-DIAMINE)-3,3'-DIMETHYL(1,1'-BIPHENYL) see
TGJ750
DIAMINE DIPENICILLIN G see BFC750
DIAMINE DIRECT BLACK E see AQP000
DIAMINE SKY BLUE CI see CMO500
DIAMINE SKY BLUE FF see BGT250
DIAMINIDE MALEATE see DBM800
2,8-DIAMINOACRIDINE see DBN600
3,6-DIAMINOACRIDINE see DBN600
3,6-DIAMINOACRIDINE mixture with 3,6-DIAMINO-
10-METHYLACRIDINIUM CHLORIDE see DBX400
3,6-DIAMINOACRIDINE MONOHYDROCHLORIDE
see PMH250
2,8-DIAMINOACRIDINIUM see DBN600
3,6-DIAMINOACRIDINIUM see DBN600
3,6-DIAMINOACRIDINIUM CHLORIDE see PMH250
3,6-DIAMINOACRIDINIUM CHLORIDE
HYDROCHLORIDE see PMH250
2,8-DIAMINOACRIDINIUM CHLORIDE
MONOHYDROCHLORIDE see PMH250
1,2-DIAMINOAETHAN (GERMAN) see EEA500
2,4-DIAMINOANISOL see DBO000
2,4-DIAMINOANISOLE see DBO000
2,4-DIAMINOANISOLE BASE see DBO000
m-DIAMINOANISOLE 1,3-DIAMINO-4-
METHOXYBENZENE see DBO000
2,4-DIAMINOANISOLE SULPHATE see DBO400
2,4-DIAMINOANISOLE SULPHATE see DBO400
2,4-DIAMINO-ANISOL SULPHATE see DBO400
2,6-DIAMINOANTHRACHINON see APK850
2,6-DIAMINOANTHRAQUINONE see APK850
2,6-DIAMINO-9,10-ANTHRAQUINONE see APK850
1,5-DIAMINOANTHRARUFIN see DBP909
4,8-DIAMINOANTHRARUFIN see DBP909
3,7-DIAMINO-5-AZAANTHRACENE see DBN600
2,4-DIAMINOAZOBENZENE HYDROCHLORIDE see
PEK000
m-DIAMINOBENZENE see PEY000
o-DIAMINOBENZENE see PEY250
p-DIAMINOBENZENE see PEY500
1,2-DIAMINOBENZENE see PEY250
1,3-DIAMINOBENZENE see PEY000
1,4-DIAMINOBENZENE see PEY500
m-DIAMINOBENZENE DIHYDROCHLORIDE see
PEY750
1,3-DIAMINOBENZENE DIHYDROCHLORIDE see
PEY750
3,3'-DIAMINOBENZIDENE see BGK500
3,3'-DIAMINOBENZIDINE
TETRAHYDROCHLORIDE see BGK750
6,6'-DIAMINO-m,m'-BIPHENOL see DMI400
4,4'-DIAMINOBIPHENYL see BBX000
p,p'-DIAMINOBIPHENYL see BBX000
4,4'-DIAMINO-1,1'-BIPHENYL see BBX000
4,4'-DIAMINOBIPHENYL-3,3'-DICARBOXYLIC ACID
see BFX250
4,4'-DIAMINO-3,3'-BIPHENYLDICARBOXYLIC ACID
see BFX250

4,4'-DIAMINO-3,3'-BIPHENYLDIOL see DMI400
4,4'-DIAMINOBIPHENYLOXIDE see OPM000
1,4-DIAMINOBUTANE see BOS000
2,6-DIAMINO-4-BUTYLAMINO-s-TRIAZINE see
BRR800
α,ε-DIAMINOCAPROIC ACID see LJM700
3,4-DIAMINOCHLOROBENZENE see CFK125
1,2-DIAMINO-4-CHLOROBENZENE see CFK125
3,4-DIAMINO-1-CHLOROBENZENE see CFK125
2,4-DIAMINO-5-(4-CHLOROPHENYL)-6-
ETHYLPYRIMIDINE see TGD000
2,4-DIAMINO-5-p-CHLOROPHENYL-6-
ETHYLPYRIMIDINE see TGD000
DI(-4-AMINO-3-CHLOROPHENYL)METHANE see
MJM200
DIAMINOCILLIAN see BFC750
DI-(4-AMINO-3-CLOROFENIL)METANO (ITALIAN)
see MJM200
4,4'-DIAMINO-3,3'-DICHLOROBIPHENYL see DEQ600
4,4'-DIAMINO-3,3'-DICHLORODIPHENYL see DEQ600
4,4'-DIAMINO-3,3'-DICHLORODIPHENYLMETHANE
see MJM200
2,2'-DIAMINODIETHYLAMINE see DJG600
DIAMINODIFENILSULFONA (SPANISH) see SOA500
p,p'-DIAMINODIFENYLMETHAN see MJQ000
1,5-DIAMINO-4,8-DIHYDROXY-9,10-
ANTHRACENEDIONE see DBP909
1,5-DIAMINO-4,8-DIHYDROXYANTHRAQUINONE
see DBP909
4,8-DIAMINO-1,5-DIHYDROXYANTHRAQUINONE
see DBP909
leuco-1,5-DIAMINO-4,8-
DIHYDROXYANTHRAQUINONE see DBP909
3,3'-DIAMINO-4,4'-DIHYDROXY ARSENOBENZENE
METHYLENESULFOXYLATE SODIUM see NCJ500
2,8-DIAMINO-3,7-DIMETHYLACRIDINE see DBT200
3,6-DIAMINO-2,7-DIMETHYLACRIDINE see DBT200
4,4'-DIAMINO-3,3'-DIMETHYLBIPHENYL see TGJ750
4,4'-DIAMINO-3,3'-DIMETHYLBIPHENYL
DIHYDROCHLORIDE see DQM000
4,4'-DIAMINO-3,3'-DIMETHYLDIPHENYL see TGJ750
1,12-DIAMINO-4,9-DIOXADODECANE see BGU600
p-DIAMINODIPHENYL see BBX000
4,4'-DIAMINODIPHENYL see BBX000
O,O'-DIAMINO DIPHENYL DISULFIDE see DXJ800
DIAMINODIPHENYL ETHER see OPM000
4,4-DIAMINODIPHENYL ETHER see OPM000
p,p'-DIAMINODIPHENYL ETHER see OPM000
4,4'-DIAMINODIPHENYLMETHAN see MJQ000
2,4'-DIAMINODIPHENYLMETHAN (GERMAN) see
MJP750
DIAMINODIPHENYLMETHANE see MJQ000
2,4'-DIAMINODIPHENYLMETHANE see MJP750
4,4'-DIAMINODIPHENYLMETHANE see MJQ000
o,p'-DIAMINODIPHENYLMETHANE see MJP750
p,p'-DIAMINODIPHENYLMETHANE see MJQ000
4,4'-DIAMINODIPHENYLMETHANE (DOT) see
MJQ000
4,4'-DIAMINODIPHENYL OXIDE see OPM000
4,4'-DIAMINODIPHENYL SULFIDE see TFI000
p,p'-DIAMINODIPHENYL SULFIDE see TFI000
DIAMINO-4,4'-DIPHENYL SULFONE see SOA500
4,4'-DIAMINODIPHENYL SULFONE see SOA500
p,p'-DIAMINODIPHENYL SULFONE see SOA500
p,p'-DIAMINODIPHENYL SULPHIDE see TFI000
p,p-DIAMINODIPHENYL SULPHONE see SOA500
DIAMINO-4,4'-DIPHENYL SULPHONE see SOA500

DI-n-AMYL ETHER see PBX000
DIAMYL KETONE see ULA000
DI-tert-AMYLPHENOL see DCI000
2,4-DI-tert-AMYLPHENOL see DCI000
DIAMYL PHTHALATE see AON300
DIAN see BLD500
DIANABOL see MPN500
DIANAT (RUSSIAN) see MEL500
DIANATE see MEL500
1,2:3,4-DIANHYDROERYTHRITOL see DHB800
1,4:3,6-DIANHYDROSORBITOL-2,5-DINITRATE see CCK125
DIANILBLAU see CMO250
DIANIL BLUE see CMO250
p,p'-DIANILINE see BBX000
DIANILINOMETHANE see MJQ000
o-DIANISIDIN (CZECH, GERMAN) see DCJ200
o-DIANISIDINA (ITALIAN) see DCJ200
o-DIANISIDINE see DCJ200
3,3'-DIANISIDINE see DCJ200
o-DIANISIDINE DIHYDROCHLORIDE see DOA800
DIANISIDINE DIISOCYANATE see DCJ400
DIANISYLTRICHLORETHANE see MEI450
2,2-DI-p-ANISYL-1,1,1-TRICHLOROETHANE see MEI450
DIANIX FAST VIOLET B see DBY700
DIANON see DCM750
DIANTIMONY PENTOXIDE see AQF750
DIANTIMONY TRIOXIDE see AQF000
DIANTIMONY TRISULFIDE see AQL500
DIAPADRIN see DGQ875
DIAPAM see DCK759
DIAPHENYLSULFONE see SOA500
DIAPHENYLSULPHON see SOA500
DIAPHENYLSULPHONE see SOA500
DIAPHTAMINE BLACK V see AQP000
DIAPHTAMINE BLUE BB see CMO000
DIAPHTAMINE PURE BLUE see CMO500
DIAPP see BEN000
DIAQUONE see DMH400
DIAREX 43G see SMQ500
DIAREX 600 see SMR000
DIAREX HF 55 see SMQ500
DIAREX HF 77 see SMQ000
DIAREX HF 77 see SMQ500
DIAREX HF 55-247 see SMQ500
DIAREX HS 77 see SMQ500
DIAREX HT 88 see SMQ500
DIAREX HT 90 see SMQ500
DIAREX HT 190 see SMQ500
DIAREX HT 500 see SMQ500
DIAREX HT 88A see SMQ500
DIAREX YH 476 see SMQ500
DIARSENIC ACID, TETRASODIUM SALT see SEY100
DIARSENIC PENTOXIDE see ARH500
DIARSENIC TRIOXIDE see ARI750
DIARSENIC TRISULFIDE see ARI000
DIARSENIC TRISULPHIDE see ARI000
DIASTATIN see NOH500
DIASTYL see DKA600
DIASULFA see SNN300
DIASULFYL see SNN300
DIATER see DXQ500
DIATERR-FOS see DCM750
DIATHESIN see HMK100
DIATO BLUE BASE B see DCJ200
DIATOMACEOUS EARTH see DCJ800
DIATOMACEOUS EARTH, NATURAL see DCJ800

DIATOMACEOUS SILICA see DCJ800
DIATOMITE see DCJ800
DIAZACHEL (OBS.) see MDQ250
1,4-DIAZACYCLOHEPTANE see HGI900
1,6-DIAZA-3,4,8,9,12,13-HEXAOXABICYCLO(4.4.4)TETRADECANE see DCK700
1,3-DIAZAINDENE see BCB750
2,3-DIAZAINDOLE see BDH250
DIAZAJET see DCM750
3,6-DIAZAOCTANE-1,8-DIAMINE see TJR000
4,5-DIAZAPHENANTHRENE see PCY250
DIAZATOL see DCM750
DIAZEPAM see DCK759
1H-1,4-DIAZEPINE, HEXAHYDRO- see HGI900
DIAZETARD see DCK759
DIAZIDE see DCM750
1,4-DIAZIDOBENZENE see DCL125
1,4-DIAZIDOBENZENE see DCL125
p-DIAZIDOBENZENE (DOT) see DCL125
1,2-DIAZIDOETHANE see DCL600
DIAZINE BLACK E see AQP000
DIAZINE BLUE 2B see CMO000
DIAZINE BLUE 3B see CMO250
DIAZINE DIRECT BLACK E see AQP000
DIAZINE DIRECT BLACK G see AQP000
DIAZINON see DCM750
DIAZINONE see DCM750
DIAZIRINE see DCP800
DIAZITOL see DCM750
DI-AZO see PDC250
DIAZO see DUR800
DIAZOACETATE (ESTER)-l-SERINE see ASA500
l-DIAZOACETATE (ESTER) SERINE see ASA500
DIAZO-ACETIC ACID ESTER with SERINE see ASA500
DIAZOACETIC ACID, ETHYL ESTER see DCN800
DIAZOACETIC ESTER see DCN800
N-DIAZOACETILGLICINA-IDRAZIDE (ITALIAN) see DCO800
DIAZOACETYLGLYCINE HYDRAZIDE see DCO800
N-(DIAZOACETYL)GLYCINE HYDRAZINE see DCO800
N-DIAZOACETYL GLYCYLHYDRAZIDE see DCO800
N-(1-(DIAZOACETYL)-2-METHYLBUTYL)-p-TOLUENESULFONAMIDE see THH480
o-DIAZOACETYL-l-SERINE see ASA500
DIAZOAMINOBENZEN (CZECH) see DWO800
DIAZOAMINOBENZENE see DWO800
p-DIAZOAMINOBENZENE see DWO800
DIAZOAMINOBENZOL (GERMAN) see DWO800
DIAZOBENZENE see ASL250
DIAZOBLEU see BGT250
DIAZOCARD CHRYSOIDINE G see PEK000
2-DIAZO-4,6-DINITROBENZENE-1-OXIDE see DUR800
DIAZODINITROPHENOL (dry) (DOT) see DUR800
DIAZODINITROPHENOL, wetted with not <40% H_2O or mixture of alcohol & H_2O (UN 0074) (DOT) see DUR800
DIAZOESSIGSAEURE-AETHYLESTER (GERMAN) see DCN800
DIAZO FAST ORANGE GR see NEO000
DIAZO FAST ORANGE R see NEN500
DIAZO FAST RED AL see AIA750
DIAZO FAST RED GG see NEO500
DIAZO FAST RED TR see CLK235
DIAZO FAST RED TRA see CLK220
DIAZO FAST RED TRA see CLK235

DIAZO FAST SCARLET G see NMP500
DIAZOIMIDE see HHG500
N-(3-DIAZO-1-ISOBUTYLACETONYL)-p-
TOLUENESULFONAMIDE see THH490
DIAZOL see DCM750
DIAZOL BLACK 2V see AQP000
DIAZOL BLUE 2B see CMO000
1,2-DIAZOLE see POM500
DIAZOLONE see PPP500
DIAZOL PURE BLUE 4B see CMO500
DIAZOL PURE BLUE FF see BGT250
DIAZOMETHANE see DCP800
N-(3-DIAZO-1-METHYLACETONYL)-p-
TOLUENESULFONAMIDE see THH380
DIAZO NERO MICROSETILE G see DPO200
N-(4-DIAZO-3-OXOBUTYL)-p-
TOLUENESULFONAMIDE see THH375
N-(7-DIAZO-6-OXOHEPTYL)-p-
TOLUENESULONAMIDE see THH460
5-DIAZOPYRIMIDINE-2,4(3H)-DIONE see DCQ600
5-DIAZO-2,4(1H,3H)-PYRIMIDINEDIONE see DCQ600
DIAZOTIZING SALTS see SIQ500
DIAZOURACIL see DCQ600
5-DIAZOURACIL see DCQ600
DIAZYL see PPP500
DIAZYL see SNJ000
DIBA see DNH125
DIBAM see SGM500
DI-BAPN FUMARATE see AMB750
DIBASIC AMMONIUM ARSENATE see DCG800
DIBASIC AMMONIUM PHOSPHATE see ANR500
DIBASIC LEAD ACETATE see LCV000
DIBASIC LEAD ARSENATE see LCK000
DIBASIC LEAD CARBONATE see LCP000
DIBASIC LEAD METAPHOSPHATE see LCV100
DIBASIC LEAD PHOSPHITE see LCV100
DIBASIC SODIUM PHOSPHATE see SJH090
DIBASIC ZINC STEARATE see ZMS000
DIBEKACIN see DCQ800
DIBENCIL see BFC750
DIBENCILLIN see BFC750
DIBENYLIN see PDT250
DIBENYLINE see PDT250
DIBENZ(a,j)ACEANTHRYLENE see DCR400
DIBENZACEPIN see DCS200
DIBENZ(a,d)ACRIDINE see DCS400
DIBENZ(a,f)ACRIDINE see DCS600
DIBENZ(a,h)ACRIDINE see DCS400
DIBENZ(a,j)ACRIDINE see DCS600
1,2,5,6-DIBENZACRIDINE see DCS400
1,2,7,8-DIBENZACRIDINE see DCS600
3,4,5,6-DIBENZACRIDINE see DCS600
DIBENZ(a,h)ACRIDINE 3,4-DIOL-1,2-EPOXIDE see
DCS821
DIBENZAMINE see DCT050
1,2,5,6-DIBENZANTHRACEEN (DUTCH) see DCT400
DIBENZ(a,h)ANTHRACENE see DCT400
1,2:5,6-DIBENZANTHRACENE see DCT400
1,2:5,6-DIBENZ(a)ANTHRACENE see DCT400
DIBENZARSENOLE, 5-HYDROXY-, 5-OXIDE see
ARA100
DIBENZARSENOLIC ACID see ARA100
5H-DIBENZARSOLE, 5-HYDROXY-, 5-OXIDE see
ARA100
5H-DIBENZ(b,f)AZEPINE, 3,7-DICHLORO-10,11-
DIHYDRO-5-(3-(DIMETHYLAMINO)PROPYL)- see
DEL400

5H-DIBENZ(b,f)AZEPINE,-3,7-DICHLORO-5-(3-
(DIMETHYLAMINO)PROPYL)-10,11-DIHYDRO- see
DEL400
5H-DIBENZ(b,f)AZEPINE,-10,11-DIHYDRO-3,7-
DICHLORO-5-(3-(DIMETHYLAMINO)PROPYL)- see
DEL400
5H-DIBENZ(B,F)AZEPINE, 10,11-DIHYDRO-5-(3-
(DIMETHYLAMINO)-2-METHYLPROPYL)-, (+)- see
TMK100
5H-DIBENZ(B,F)AZEPINE, 5-(3-(DIMETHYLAMINO)-
2-METHYLPROPYL)-10,11-DIHYDRO-, (+)- see
TMK100
5H-DIBENZ(B,F)AZEPINE-5-PROPANAMINE, 10,11-
DIHYDRO-N,N,β-TRIMETHYL-, (+)- see TMK100
3,4,5,6-DIBENZCARBAZOL see DCY000
3,4,5,6-DIBENZCARBAZOLE see DCY000
DIBENZENECHROMIUM see BGY700
DIBENZENECHROMIUM IODIDE see BGY720
DIBENZHEPTROPINE see EAG100
DIBENZHEPTROPINE CITRATE see EAG100
DIBENZO(a,j)ACRIDINE see DCS600
1,2,5,6-DIBENZOACRIDINE see DCS400
DIBENZO(a,h)ANTHRACENE see DCT400
1,2:5,6-DIBENZOANTHRACENE see DCT400
DIBENZOAZEPINE see DCS200
3,4,5,6-DIBENZOCARBAZOLE see DCY000
7H-DIBENZO(c,g)CARBAZOLE see DCY000
DIBENZO(b,def)CHRYSENE see DCY200
DIBENZO(def,p)CHRYSENE see DCY400
DIBENZO(def,p)CHRYSENE-11,12-DIOL, 11,12-
DIHYDRO-, (11S,12S)-REL- see DDC810
DIBENZO-p-DIOXIN, 1,2,3,4,6,7,8-HEPTACHLORO-
see HAR100
DIBENZO(B,E)(1,4)DIOXIN, 1,2,3,4,6,7,8-
HEPTACHLORO- see HAR100
DIBENZO(a,jk)FLUORENE see BCJ250
DIBENZO(b,jk)FLUORENE see BCJ280
DIBENZOFURAN, DICHLORO- see DEX200
1,2,5,6-DIBENZONAPHTHALENE see CML810
DIBENZOPARADIAZINE see PDB500
DIBENZOPARATHIAZINE see PDP250
DIBENZO PQD see DVR200
DIBENZOPYRAZINE see PDB500
DIBENZO(a,d)PYRENE see DCY400
DIBENZO(a,e)PYRENE see NAT500
DIBENZO(a,h)PYRENE see DCY200
DIBENZO(a,i)PYRENE see BCQ500
DIBENZO(a,l)PYRENE see DCY400
DIBENZO(b,h)PYRENE see BCQ500
1,2:3,4-DIBENZOPYRENE see DCY400
1,2:4,5-DIBENZOPYRENE see NAT500
1,2:6,7-DIBENZOPYRENE see DCY200
1,2:7,8-DIBENZOPYRENE see BCQ500
2,3:4,5-DIBENZOPYRENE see DCY400
3,4:8,9-DIBENZOPYRENE see DCY200
1,2:9,10-DIBENZOPYRENE see DCY400
3,4:9,10-DIBENZOPYRENE see BCQ500
(+−)-DIBENZO(a,l)PYRENE-11,12-DIHYDRODIOL see
DDC810
DIBENZO(b,e)PYRIDINE see ADJ500
DIBENZOPYRROLE see CBN000
DIBENZO(b,d)PYRROLE see CBN000
DIBENZOTHIAZEPINE see CIS750
DIBENZOTHIAZINE see PDP250
DIBENZO-1,4-THIAZINE see PDP250
DI-2-BENZOTHIAZOLYLDISULFIDE see BDE750
DIBENZOTHIAZYL DISULFIDE see BDE750

2,2'-DIBENZOTHIAZYLDISULFIDE see BDE750
DIBENZOTHIEN-2-YL METHYL KETONE see ACH090
DIBENZ(b,f)(1,4)OXAZEPINE see DDE200
DIBENZOYLMETHANE see PFU300
DIBENZOYLPEROXID (GERMAN) see BDS000
DIBENZOYL PEROXIDE (MAK) see BDS000
DIBENZOYLPEROXYDE (DUTCH) see BDS000
DIBENZOYLTARTARIC ACID see DDE300
DIBENZOYLTHIAZYL DISULFIDE see BDE750
DIBENZ(a,i)PYRENE see BCQ500
1,2,3,4-DIBENZPYRENE see DCY400
1,2:7,8-DIBENZPYRENE see BCQ500
3,4,8,9-DIBENZPYRENE see DCY200
4,5,6,7-DIBENZPYRENE see DCY400
3,4:9,10-DIBENZPYRENE see BCQ500
DIBENZTHIAZYL DISULFIDE see BDE750
DIBENZYL see BFX500
DIBENZYL CHLORETHYLAMINE see DCT050
N,N-DIBENZYL-β-CHLOROETHYLAMINE see DCT050
DIBENZYLENE see DDG800
DIBENZYLETHER (CZECH) see BEO250
N,N'-DIBENZYLETHYLENEDIAMINE BIS(BENZYL PENICILLIN) see BFC750
DIBENZYLETHYLENEDIAMINE-DI-PENICILLIN G see BFC750
N,N'-DIBENZYLETHYLENEDIAMINE, compounded with PENICILLIN G (1:2) see BFC750
DIBENZYLETHYLSULFONIUM IODIDE MERCURIC IODIDE see DDG600
DIBENZYLETHYLSULFONIUM IODIDE with MERCURY IODIDE (1:1) see DDG600
DIBENZYLIN see DDG800
DIBENZYLINE see PDT250
DIBENZYLINE HYDROCHLORIDE see DDG800
DIBENZYLMERCURY see DDH000
DIBENZYRAN see DDG800
DIBESTIL see DKB000
DIBESTROL see DKA600
DIBONDRIN see BBV500
DIBORANE see DDI450
DIBORANE MIXTURES (NA 1911) see DDI450
DIBORON HEXAHYDRIDE see DDI450
DIBOVAN see DAD200
DIBP see DNJ400
DIBROLUUR see BNP750
1,2-DIBROMAETHAN (GERMAN) see EIY500
DIBROMCHLORPROPAN (GERMAN) see DDL800
1,2-DIBROM-3-CHLOR-PROPAN (GERMAN) see DDL800
DIBROMOACETONITRILE see DDJ400
DIBROMOACETYLENE see DDJ800
DIBROMOBENZENE see DDJ900
2,2'-DIBROMOBIACETYL see DDK600
α,α'-DIBROMOBIACETYL see DDK600
DIBROMOCHLOROMETHANE see CFK500
DIBROMOCHLOROPROPANE see DDL800
1,2-DIBROMO-3-CHLOROPROPANE see DDL800
1,2-DIBROMO-3-CLORO-PROPANO (ITALIAN) see DDL800
DIBROMOCYANOACETAMIDE see DDM000
α,α-DIBROMO-α-CYANOACETAMIDE see DDM000
2,6-DIBROMO-4-CYANOPHENOL see DDP000
DIBROMODIBUTYLSTANNANE see DDM400
DIBROMODIBUTYLTIN see DDM400
1,2-DIBROMO-2,4-DICYANOBUTANE see DDM500

1,6-DIBROMODIDEOXYDULCITOL see DDJ000
1,6-DIBROMO-1,6-DIDEOXYDULCITOL see DDJ000
1,6-DIBROMO-1,6-DIDEOXYGALACTITOL see DDJ000
1,6-DIBROMO-1,6-DIDEOXY-d-GALACTITOL see DDJ000
DIBROMODIFLUOROMETHANE see DKG850
6,8-DIBROMO-DIHYDRO-1,3-BENZOXAZINE-2-THIONE-4-ONE see DDM820
DIBROMODIMETHYL STANNANE see DUG800
DIBROMODULCITOL see DDJ000
DIBROMODULCITOL see DDJ000
1,6-DIBROMODULCITOL see DDJ000
1,2-DIBROMOETANO (ITALIAN) see EIY500
sym-DIBROMOETHANE see EIY500
α,β-DIBROMOETHANE see EIY500
1,2-DIBROMOETHANE (MAK) see EIY500
DIBROMOFLUORESCEIN see DDO200
4',5'-DIBROMOFLUORORESCEIN see DDO200
3,5-DIBROMO-4-HYDROXYBENZONITRILE see DDP000
3-(3,5-DIBROMO-4-HYDROXYBENZOYL)-2-ETHYLBENZOFURAN see DDP200
2,7-DIBROMO-4-HYDROXYMERCURIFLUORESCEINE DISODIUM SALT see MCV000
3,5-DIBROMO-4-HYDROXYPHENYLCYANIDE see DDP000
3,5-DIBROMO-4-HYDROXYPHENYL-2-ETHYL-3-BENZOFURANYL KETONE see DDP200
(3,5-DIBROMO-4-HYDROXYPHENYL)(2-ETHYL-3-BENZOFURANYL)METHANONE see DDP200
DIBROMOMETHANE see DDP800
2,2-DIBROMO-3-NITRILOPROPIONAMIDE see DDM000
DIBROMOPHENYLARSINE see DDR200
2,3-DIBROMO-1-PROPANOL PHOSPHATE see TNC500
2,3-DIBROMO-1-PROPANOL, PHOSPHATE (3:1) see TNC500
(2,3-DIBROMOPROPYL) PHOSPHATE see TNC500
6,8-DIBROMO-2-THIO-2H-1,3-BENZOXAZINE-2,4(3H)-DIONE see DDM820
DIBROMURE d'ETHYLENE (FRENCH) see EIY500
1,2-DIBROOM-3-CHLOORPROPAAN (DUTCH) see DDL800
1,2-DIBROOMETHAAN (DUTCH) see EIY500
DIBUTIL see DIR000
2,2'-DIBUTOXYETHYL ETHER see DDW200
DIBUTYL ACID PHOSPHATE see DEG700
DIBUTYL ADIPATE see AEO750
DI-N-BUTYL ADIPATE see AEO750
DIBUTYL ADIPINATE see AEO750
DIBUTYLAMID KYSELINY MRAVENCI see DEC400
n-DIBUTYLAMINE see DDT800
DI-n-BUTYLAMINE see DDT800
DI(n-BUTYL)AMINE (DOT) see DDT800
DIBUTYLAMINE, HEXACHLOROSTANNANE (2:1) see BIV900
DIBUTYLAMINE, HEXAFLUOROARSENATE(1-) see DDV225
DIBUTYLAMINE, 4-HYDROXY-N-NITROSO- see HJQ350
DIBUTYLAMINOETHANOL see DDU600
2-DIBUTYLAMINOETHANOL see DDU600
2-N-DIBUTYLAMINOETHANOL see DDU600
2-DI-n-BUTYLAMINOETHANOL see DDU600

N,N-DI-n-BUTYLAMINOETHANOL (DOT) see DDU600

β-N-DIBUTYLAMINOETHYL ALCOHOL see DDU600

DI-n-BUTYLAMMONIUM HEXAFLUOROARSENATE see DDV225

DIBUTYLARSINIC ACID see DDV250

DIBUTYLATED HYDROXYTOLUENE see BFW750

DIBUTYL-1,2-BENZENEDICARBOXYLATE see DEH200

DI-N-BUTYL BENZYLPHOSPHONATE see BFD760

DIBUTYLBIS((3-CARBOXYACRYLOYL)OXY)-STANNANE DIMETHYL ESTER (Z,Z) (8CI) see BKO250

DIBUTYLBIS((2-ETHYLHEXANOYL)OXY)-STANNANE see BJQ250

DIBUTYLBIS((2-ETHYL-1-OXOHEXYL)OXY)-STANNANE (9CI) see BJQ250

DIBUTYLBIS(LAUROYLOXY)STANNANE see DDV600

DIBUTYLBIS(LAUROYLOXY)TIN see DDV600

DIBUTYLBIS(OLEOYLOXY)STANNANE see DEJ000

DIBUTYLBIS((1-OXO-9-OCTADECENYL)OXY)STANNANE (Z,Z) see DEJ000

DIBUTYL BUTANEPHOSPHONATE see DDV800

DIBUTYL BUTYLPHOSPHONATE see DDV800

DIBUTYL CARBITOL see DDW200

DIBUTYL-o-(o-CARBOXYBENZOYL) GLYCOLATE see BQP750

DIBUTYL-o-CARBOXYBENZOYLOXYACETATE see BQP750

2,6-DI-tert-BUTYL-p-CRESOL (OSHA, ACGIH) see BFW750

DIBUTYLDICHLOROGERMANE see DDY000

DIBUTYLDICHLOROSTANNANE see DDY200

DIBUTYLDICHLOROTIN see DDY200

DIBUTYLDI(2-ETHYLHEXYLOXYCARBONYLMETHYLTHIO)STANNANE see DDY600

DIBUTYLDIFLUOROSTANNANE see DDY800

DIBUTYL(DIFORMYLOXY)STANNANE see DDZ000

2,2-DIBUTYLDIHYDRO-6H-1,3,2-OXATHIASTANNIN-6-ONE see DEJ200

DIBUTYLDIPENTANOYLOXYSTANNANE see DEA600

DIBUTYL DISELENIDE see BRF550

DI-n-BUTYL-DISELENIDE see BRF550

DIBUTYLDISELENIUM see BRF550

DIBUTYLDITHIOCARBAMIC ACID, NICKEL SALT see BIW750

DIBUTYLDITHIOCARBAMIC ACID SODIUM SALT see SGF500

DIBUTYLDITHIO-CARBAMIC ACID ZINC COMPLEX see BIX000

DIBUTYLDITHIOCARBAMIC ACID ZINC SALT see BIX000

DIBUTYLESTER KYSELINY FUMAROVE see DEC600

DIBUTYL ESTER SULFURIC ACID see DEC000

N,N-DIBUTYLETHANOLAMINE see DDU600

DI-sec-BUTYL ETHER see BRH760

DI-n-BUTYL ETHER (DOT) see BRH750

N,N-DI-tert-BUTYLETHYLENEDIAMINE see DEC100

N,N-DI-n-BUTYLFORMAMIDE see DEC400

N,N-DI-n-BUTYLFORMAMIDE see DEC400

DIBUTYL FUMARATE see DEC600

DI-n-BUTYLGERMANEDICHLORIDE see DDY000

DIBUTYL HEXANEDIOATE see AEO750

DIBUTYL HYDROGEN PHOSPHATE see DEG700

N,N-DIBUTYL-N-(2-HYDROXYETHYL)AMINE see DDU600

2,6-DI-tert-BUTYL-1-HYDROXY-4-METHYLBENZENE see BFW750

3,5-DI-tert-BUTYL-4-HYDROXYTOLUENE see BFW750

2,6-DI-terc.BUTYL-p-KRESOL (CZECH) see BFW750

DIBUTYL MALEATE see DED600

DIBUTYL(3-MERCAPTOPROPIONATO(2-))TIN see DEJ200

DIBUTYLMERCURY see DEE000

DI-sec-BUTYLMERCURY see DEE200

2,6-DI-tert-BUTYL-4-METHYLPHENOL see BFW750

2,6-DI-tert-BUTYL-p-METHYLPHENOL see BFW750

DI-n-BUTYLNITROSAMIN (GERMAN) see BRY500

DI-n-BUTYLNITROSAMINE see BRY500

N,N-DI-n-BUTYLNITROSAMINE see BRY500

DIBUTYLNITROSOAMINE see BRY500

N,N-DIBUTYLNITROSOAMINE see BRY500

2,2-DIBUTYL-1-OXA-2-STANNA-3-THIACYCLOHEXAN-6-ONE see DEJ200

2,2-DIBUTYL-1,3,2-OXATHIASTANNOLANE see DEF150

2,2-DIBUTYL-1,3,2-OXATHIASTANNOLANE-5-OXIDE see DEF200

DIBUTYL OXIDE see BRH750

DIBUTYLOXIDE of TIN see DEF400

DIBUTYLOXOSTANNANE see DEF400

DIBUTYLOXOTIN see DEF400

DI-sec-BUTYL PEROXYDICARBONATE see BSD000

DI-sec-BUTYL PEROXYDICARBONATE, not more than 52% in solution (DOT) see BSD000

DI-sec-BUTYL PEROXYDICARBONATE, technically pure (DOT) see BSD000

N,N'-DI-sec-BUTYL-p-PHENYLENEDIAMINE see DEG200

DIBUTYL PHENYL PHOSPHATE see DEG600

DIBUTYL PHOSPHATE see DEG700

DIBUTYL PHOSPHATE see DEG700

DI-n-BUTYL PHOSPHATE see DEG700

DIBUTYL PHTHALATE see DEH200

DI-n-BUTYL PHTHALATE see DEH200

DIBUTYLRTUT see DEE000

DIBUTYL SEBACATE see DEH600

DI-n-BUTYL SEBACATE see DEH600

DIBUTYLSTANNANE OXIDE see DEF400

2,2'-((DIBUTYLSTANNYLENE)BIS(THIO))BISACETIC ACID DINONYL ESTER see DEH650

DI-n-BUTYLSULFAT (GERMAN) see DEC000

DIBUTYL SULFATE see DEC000

DIBUTYL(THIOACETOXY)STANNANE see DEF200

DIBUTYLTHIOXOSTANNANE see DEI200

DIBUTYLTIN BIS(2-ETHYLHEXANOATE) see BJQ250

DIBUTYLTIN BIS(α-ETHYLHEXANOATE) see BJQ250

DIBUTYL-TIN BIS(ISOOCTYLTHIOGLYCOLLATE) see BKK250

DIBUTYLTIN BIS(METHYL MALEATE) see BKO250

DIBUTYLTIN BIS(MONOMETHYL MALEATE) see BKO250

DIBUTYLTIN CHLORIDE see DDY200

DIBUTYL TIN DIACETATE see DBF800

DIBUTYL TIN DIBROMIDE see DDM400

DIBUTYLTIN DICHLORIDE see DDY200

DI-n-BUTYLTIN DICHLORIDE see DDY200

DI-n-BUTYLTIN DI(DODECANOATE) see DDV600

DIBUTYLTIN DI(2-ETHYLHEXANOATE) see BJQ250

DI-n-BUTYLTIN DI-2-ETHYLHEXANOATE see BJQ250

DIBUTYLTIN DI(2-ETHYLHEXOATE) see BJQ250
DI-n-BUTYLTIN DI-2-ETHYLHEXYLTHIOGLYCOLATE see DDY600
DIBUTYLTIN DIFLUORIDE see DDY800
DI-n-BUTYLTIN DIFORMATE see DDZ000
DIBUTYLTIN DILAURATE (USDA) see DDV600
DI-N-BUTYLTIN DI(MONOBUTYL)MALEATE see BHK250
DIBUTYLTIN DIOLEATE see DEJ000
DI-n-BUTYLTIN DIPENTANOATE see DEA600
DIBUTYLTIN LAURATE see DDV600
DIBUTYLTIN MERCAPTOPROPIONATE see DEJ200
DIBUTYLTIN-O,S-MERCAPTOPROPIONATE see DEJ200
DIBUTYLTIN-S,O-3-MERCAPTOPROPIONATE see DEJ200
DIBUTYLTIN-S,O-β-MERCAPTOPROPIONATE see DEJ200
DIBUTYLTIN METHYL MALEATE see BKO250
DIBUTYLTIN OXIDE see DEF400
DI-n-BUTYLTIN OXIDE see DEF400
DIBUTYLTIN SULFIDE see DEI200
6,6-DIBUTYL-4,8,11-TRIOXO-5,7,12-TRIOXA-6-STANNATRIDECA-2,9-DIENOIC ACID METHYL ESTER see BKO250
DIBUTYLZINN-S,S'-BIS(ISOOCTYLTHIOGLYCOLAT) (GERMAN) see BKK250
DI-n-BUTYL-ZINN DI-2-AETHYLHEXYL THIOGLYKOLAT (GERMAN) see DDY600
DI-n-BUTYL-ZINN-DICHLORID (GERMAN) see DDY200
DIBUTYL-ZINN-DILAURAT (GERMAN) see DDV600
DI-N-BUTYL-ZINN-DI(MONOBUTYL)MALEINAT (GERMAN) see BHK250
DI-n-BUTYLZINN-DIMONOMETHYLMALEINAT (GERMAN) see BKO250
DI-n-BUTYL-ZINN-OXYD (GERMAN) see DEF400
DI-n-BUTYLZINN THIOGLYKOLAT (GERMAN) see DEF200
DIC see DAB600
DIC 1468 see MQR275
DICALCIUM PHOSPHATE see CAW100
DICALITE see SCH002
DICAMBA (DOT) see MEL500
DICANDIOL see MQU750
DICAPTOL see BAD750
1,7-DICARBACALCITONIN (sal), 1-BUTANOIC ACID-26-l-ASPARTIC ACID-27-l-VALINE-29-l-ALANINE- see CBR300
1,7-DICARBACALCITONIN (EEL), 1-BUTANOIC ACID- see CBR300
DICARBAM see CBM750
2,2-DI(CARBAMOYLOXYMETHYL)PENTANE see MQU750
S-(1,2-DICARBETHOXYETHYL)-O,O-DIMETHYLDITHIOPHOSPHATE see MAK700
DICARBETHOXYMETHANE see EMA500
DICARBOETHOXYETHYL-O,O-DIMETHYL PHOSPHORODITHIOATE see MAK700
DICARBONIC ACID DIETHYL ESTER see DIZ100
o-DICARBOXYBENZENE see PHW250
3,3'-DICARBOXYBENZIDINE see BFX250
((1,2-DICARBOXYETHYL)THIO)GOLD DISODIUM SALT see GJC000
DICAROCIDE see DIW200
DICARZOL see DSO200
DICESIUM CARBONATE see CDC750

DICESIUM DICHLORIDE see CDD000
DICESIUM DIFLUORIDE see CDD500
DICESIUM DIIODIDE see CDE000
DICESTAL see MJM500
DICHA see DGT600
DICHAN (CZECH) see DGU200
DICHAPETULUM CYMOSUM (HOOK) ENGL see PLG000
DICHLOFENTHION see DFK600
DICHLOFENTION see DFK600
DICHLONE (DOT) see DFT000
p-DICHLOORBENZEEN (DUTCH) see DEP800
1,4-DICHLOORBENZEEN (DUTCH) see DEP800
1,1-DICHLOOR-2,2-BIS(4-CHLOOR FENYL)-ETHAAN (DUTCH) see BIM500
1,1-DICHLOORETHAAN (DUTCH) see DFF809
1,2-DICHLOORETHAAN (DUTCH) see EIY600
2,2'-DICHLOORETHYLETHER (DUTCH) see DFJ050
DICHLOORFEEN see MJM500
2-(2,4-DICHLOOR-FENOXY)-PROPIONZUUR (DUTCH) see DGB000
3-(3,4-DICHLOOR-FENYL)-1,1-DIMETHYLUREUM (DUTCH) see DXQ500
3-(3,4-DICHLOOR-FENYL)-1-METHOXY-1-METHYLUREUM (DUTCH) see DGD600
3,6-DICHLOOR-2-METHOXY-BENZOEIZUUR (DUTCH) see MEL500
1,1-DICHLOOR-1-NITROETHAAN (DUTCH) see DFU000
(2,2-DICHLOOR-VINYL)-DIMETHYL-FOSFAAT (DUTCH) see DGP900
DICHLOORVO (DUTCH) see DGP900
DICHLORACETIC ACID see DEL000
DICHLORACETYL CHLORIDE see DEN400
1,1-DICHLORAETHAN (GERMAN) see DFF809
1,2-DICHLOR-AETHAN (GERMAN) see EIY600
1,2-DICHLOR-AETHEN (GERMAN) see DFI210
p-DI-(2-CHLORAETHYL)-AMINO-dl-PHENYL-ALANIN (GERMAN) see BHT750
S-(2,3-DICHLOR-ALLYL)-N,N-DIISOPROPYL-MONOTHIOCARBAMAAT (DUTCH) see DBI200
2,3-DICHLORALLYL-N,N-(DIISOPROPYL)-THIOCARBAMAT (GERMAN) see DBI200
DICHLORAMINE see BIE250
1,5-DICHLORANTHRACHINON see DEO700
1,8-DICHLORANTHRACHINON see DEO750
DICHLORANTIN see DFE200
3,3'-DICHLORBENZIDIN (CZECH) see DEQ600
4,4'-DICHLORBENZILSAEUREAETHYLESTER (GERMAN) see DER000
o-DICHLOR BENZOL see DEP600
p-DICHLORBENZOL (GERMAN) see DEP800
1,4-DICHLOR-BENZOL (GERMAN) see DEP800
1,1-DICHLOR-2,2-BIS(4-CHLOR-PHENYL)-AETHAN (GERMAN) see BIM500
2,2'-DICHLOR-DIAETHYLAETHER (GERMAN) see DFJ050
3,3'-DICHLOR-4,4'-DIAMINO-DIPHENYLAETHER (GERMAN) see BGT000
3,3'-DICHLOR-4,4'-DIAMINODIPHENYLMETHAN (GERMAN) see MJM200
DICHLOR-DIFENYLSILAN see DFF000
DICHLORDIMETHYLAETHER (GERMAN) see BIK000
DICHLOREMULSION see EIY600
DICHLOREN see BIE500
DICHLOREN (GERMAN) see BIE250
DICHLOREN HYDROCHLORIDE see BIE500
DICHLORETHANOIC ACID see DEL000

2,4-DICHLORO-4'-NITRODIPHENYL ETHER see DFT800

DICHLORONITROETHANE see DFU000

1,1-DICHLORO-1-NITROETHANE see DFU000

2,4-DICHLORO-1-(4-NITROPHENOXY)BENZENE see DFT800

2',5-DICHLORO-4'-NITROSALICYLANILIDE see DFV400

2,3-DICHLORO-4-OXO-2-BUTENOIC ACID see MRU900

4,6-DICHLORO-3-((1E)-3-OXO-3-(PHENYLAMINO)-1-PROPENYL)-1H-INDOLE-2-CARBOXYLIC ACID, see DFW730

DICHLOROOXOZIRCONIUM see ZSJ000

DICHLOROPENTANE see DFX000

DICHLOROPENTANES (DOT) see DFX000

DICHLOROPENTYLARSINE see AOI200

DICHLOROPHEN see MJM500

DICHLOROPHENARSINE HYDROCHLORIDE see DFX400

DICHLOROPHEN B see MJM500

DICHLOROPHENE see MJM500

2,4-DICHLOROPHENOL see DFX800

2,6-DICHLOROPHENOL see DFY000

2,4-DICHLOROPHENOL BENZENESULFONATE see DFY400

3-(3,4-DICHLOROPHENOL)-1,1-DIMETHYLUREA see DXQ500

2,4-DICHLORO-PHENOL-O-ESTER with O,O-DIETHYL PHOSPHOROTHIOATE see DFK600

(2,4-DICHLOROPHENOXY)ACETIC ACID, ISOPROPYL ESTER see IOY000

(2-4-DICHLOROPHENOXY)ACETIC ACID-1-METHYLETHYL ESTER (9CI) see IOY000

2,4-DICHLOROPHENOXYACETIC ACID, SODIUM SALT see SGH500

3-(2,4-DICHLOROPHENOXY)-2-HYDROXYPROPYL-o-CHLOROPHENYL ARSINIC ACID see DGA425

4-(2,4-DICHLOROPHENOXY)NITROBENZENE see DFT800

2-(2,4-DICHLOROPHENOXY) PROPIONIC ACID see DGB000

α-(2,4-DICHLOROPHENOXY) PROPIONIC ACID see DGB000

DICHLOROPHENYLARSINE see DGB600

2,4-DICHLOROPHENYL BENZENESULFONATE see DFY400

2,4-DICHLOROPHENYL BENZENESULPHONATE see DFY400

O-2,4-DICHLOROPHENYL-O,O-DIETHYL PHOSPHOROTHIOATE see DFK600

2,4-DICHLORO-PHENYL DIETHYL PHOSPHOROTHIONATE see DFK600

1,6-DI(4'-CHLOROPHENYLDIGUANIDO)HEXANE see BIM250

N'-(3,4-DICHLOROPHENYL)-N,N-DIMETHYLUREA see DXQ500

1-(3,4-DICHLOROPHENYL)-3,3-DIMETHYLUREE (FRENCH) see DXQ500

2,4-DICHLOROPHENYL ESTER of BENZENESULFONIC ACID see DFY400

2,4-DICHLOROPHENYL ESTER BENZENESULPHONIC ACID see DFY400

3-(3,5-DICHLOROPHENYL)-5-ETHENYL-5-METHYL-2,4-OXAZOLIDINEDIONE see RMA000

3,4-DICHLOROPHENYL ISOCYANATE see IKH099

3-(3,4-DICHLOROPHENYL)-1-METHOXYMETHYLUREA see DGD600

3-(3,4-DICHLOROPHENYL)-1-METHOXY-1-METHYLUREA see DGD600

N'-(3,4-DICHLOROPHENYL)-N-METHOXY-N-METHYLUREA see DGD600

1-(3,4-DICHLOROPHENYL)3-METHOXY-3-METHYLUREE (FRENCH) see DGD600

3-(3,5-DICHLOROPHENYL)-N-(1-METHYLETHYL)-2,4-DIOXO-1-IMIDAZOLIDINECARBOXAMIDE see GIA000

N-(3,4-DICHLOROPHENYL)-N'-METHYL-N'-METHOXYUREA see DGD600

3-(3,5-DICHLOROPHENYL)-5-METHYL-5-VINYL-2,4-OXAZOLIDINEDIONE see RMA000

2,4-DICHLOROPHENYL-4-NITROPHENYL ETHER see DFT800

2,4-DICHLOROPHENYL-p-NITROPHENYL ETHER see DFT800

DICHLOROPHENYLPHOSPHINE see DGE400

DICHLOROPHENYLPHOSPHINE SULFIDE see PFW200

DICHLOROPHENYLPHOSPHINE SULFIDE see PFW210

N-(3,4-DICHLOROPHENYL)PROPANAMIDE see DGI000

1-(2-(2,4-DICHLOROPHENYL)-2-(2-PROPENYLOXY)ETHYL)-1H-IMIDAZOLE see FPB875

N-(3,4-DICHLOROPHENYL)PROPIONAMIDE see DGI000

((2,5-DICHLOROPHENYL)THIO)METHYLCARBAMIC ACID, 2,3-DIHYDRO-2,2-DIMETHYL-7-BENZOFURANYL ESTER see DGF100

2,5-DICHLOROPHENYLTHIOMETHYL O,O-DIETHYL PHOSPHORODITHIOATE see PDC750

S-(2,5-DICHLOROPHENYLTHIOMETHYL) O,O-DIETHYL PHOSPHORODITHIOATE see PDC750

S-(2,5-DICHLOROPHENYLTHIOMETHYL) DIETHYL PHOSPHOROTHIOLOTHIONATE see PDC750

DI-(p-CHLOROPHENYL)TRICHLOROMETHYLCARBINOL see BIO750

(DICHLOROPHENYL)TRICHLOROSILANE see DGF200

DICHLOROPHENYLTRICHLOROSILANE (DOT) see DGF200

DICHLOROPHOS see DGP900

DICHLOROPHOSPHORIC ACID, ETHYL ESTER see EOR000

DICHLOROPROP see DGB000

1,2-DICHLOROPROPANE see PNJ400

1,3-DICHLOROPROPANE see DGF800

α,β-DICHLOROPROPANE see PNJ400

DICHLOROPROPANE-DICHLOROPROPENE MIXTURE see DGG000

2,3-DICHLOROPROPANOL see DGG450

1,2-DICHLOROPROPANOL-3 see DGG450

1,2-DICHLORO-3-PROPANOL see DGG450

1,3-DICHLORO-2-PROPANOL see DGG400

2,3-DICHLORO-1-PROPANOL see DGG450

1,3-DICHLOROPROPANOL-2 (DOT) see DGG400

1,3-DICHLORO-2-PROPANOL PHOSPHATE (3:1) see FQU875

1,3-DICHLORO-2-PROPANONE see BIK250

DICHLOROPROPENE see DGG700

1,3-DICHLOROPROPENE see DGG950

2,3-DICHLOROPROPENE see DGH400

2,3-DICHLORO-1-PROPENE see DGH400
DICHLOROPROPENE (DOT) see DGG950
(Z)-1,3-DICHLOROPROPENE see DGH200
cis-1,3-DICHLOROPROPENE see DGH200
1,3-DICHLOROPROPENE and 1,2-
DICHLOROPROPANE MIXTURE see DGG000
2,3-DICHLORO-2-PROPENE-1-THIOL
DIISOPROPYLCARBAMATE see DBI200
S-(2,3-DICHLORO-2-PROPENYL)ESTER, BIS(1-
METHYLETHYL) CARBAMOTHIOIC ACID see DBI200
DICHLOROPROPIONANILIDE see DGI000
3,4-DICHLOROPROPIONANILIDE see DGI000
3',4'-DICHLOROPROPIONANILIDE see DGI000
2,2-DICHLOROPROPIONIC ACID see DGI400
α-DICHLOROPROPIONIC ACID see DGI400
α,α-DICHLOROPROPIONIC ACID see DGI400
2,2-DICHLOROPROPIONIC ACID, SODIUM SALT see
DGI600
α,α-DICHLOROPROPIONIC ACID SODIUM SALT see
DGI600
1,3-DICHLOROPROPYENE-1 see DGG950
2,4-DICHLORO-N-(4-
PROPYLCYCLOHEXYL)BENZAMIDE see DGI630
DICHLOROPROPYLENE see DGG700
DICHLOROPROPYLENE see DGG950
DICHLOROPROPYLENE see DGI700
1,3-DICHLOROPROPYLENE see DGG950
2,3-DICHLOROPROPYLENE see DGH400
cis-1,3-DICHLOROPROPYLENE see DGH200
α,γ-DICHLOROPROPYLENE see DGG950
3,4-DICHLORO-2,5-PYRROLIDINEDIONE see DFN800
DICHLOROSAL see CFY000
3,5-DICHLOROSALICYLIC ACID see DGK200
DICHLOROSILANE see DGK300
DICHLOROSILANE see DGK300
DICHLOROSULFANE see SOG500
DICHLOROTETRAFLUOROETHANE see DGL600
sym-DICHLOROTETRAFLUOROETHANE see FOO509
1,2-DICHLORO-1,1,2,2-TETRAFLUOROETHANE
(MAK) see FOO509
DICHLOROTETRAFLUOROETHANE (OSHA, ACGIH)
see FOO509
2,6-DICHLOROTHIOBENZAMIDE see DGM600
6,8-DICHLORO-2-THIO-2H-1,3-BENZOXAZINE-
2,4(3H)-DIONE see DFC300
DICHLOROTHIOCARBONYL see TFN500
DICHLOROTITANOCENE see DGW200
α-α-DICHLOROTOLUENE see BAY300
1,3-DICHLORO-s-TRIAZINE-2,4,6(1H,3H,5H)-TRIONE
see DGN200
DICHLORO-S-TRIAZINE-2,4,6(1H,3H,5H)-TRIONE
POTASSIUM DERIV see PLD000
1,3-DICHLORO-s-TRIAZINE-2,4,6(1H,3H,5H)TRIONE
POTASSIUM SALT see PLD000
4,4'-DICHLORO-α-
(TRICHLOROMETHYL)BENZHYDROL see BIO750
2,2'-DICHLOROTRIETHYLAMINE see BID250
2,2-DICHLORO-1,1,1-TRIFLUOROETHANE see TJY500
DICHLOROTRIPHENYLANTIMONY see DGO800
DICHLOROTRIPHENYLSTIBINE see DGO800
DICHLOROVAS see DGP900
2,2-DICHLOROVINYL ALCOHOL, DIMETHYL
PHOSPHATE see DGP900
DICHLOROVINYLARSINE CHLORIDE see BIQ250
DICHLOROVINYLCHLOROARSINE (DOT) see
BIQ250

2,2-DICHLOROVINYL DIMETHYL PHOSPHATE see
DGP900
2,2-DICHLOROVINYL DIMETHYL PHOSPHORIC
ACID ESTER see DGP900
DICHLOROVOS see DGP900
α,α'-DICHLORO-o-XYLENE see DGP400
DICHLORPHEN see MJM500
2-(2,4-DICHLOR-PHENOXY)-PROPIONSAEURE
(GERMAN) see DGB000
3-(3,4-DICHLOR-PHENYL)-1,1-DIMETHYL-
HARNSTOFF (GERMAN) see DXQ500
3-(3,4-DICHLOR-PHENYL)-1-METHOXY-1-METHYL-
HARNSTOFF (GERMAN) see DGD600
3-(4,5-DICHLORPHENYL)-1-METHOXY-1-
METHYLHARNSTOFF (GERMAN) see DGD600
2,4,-DICHLORPHENYL-4-NITROPHENYLAETHER
(GERMAN) see DFT800
1-(2-(2,4-DICHLORPHENYL)-2-
(PROPENYLOXY)AETHYL)-1H-IMIDAZOLE see
FPB875
DICHLORPHOS see DGP900
DICHLORPROP see DGB000
DICHLORPROPAN-DICHLORPROPENGEMISCH
(GERMAN) see DGG000
DICHLOR-s-TRIAZIN-2,4,6(1H,3H,5H)TRIONE
POTASSIUM see PLD000
(2,2-DICHLOR-VINYL)-DIMETHYL-PHOSPHAT
(GERMAN) see DGP900
O-(2,2-DICHLORVINYL)-O,O-DIMETHYLPHOSPHAT
(GERMAN) see DGP900
DICHLORVOS see DGP900
DICHLOSALE see DFV400
DICHLOTIAZID see CFY000
DICHLOTRIDE see CFY000
DICHROMIUM TRIOXIDE see CMJ900
DI(2-CIANOETIL)AMMINA (ITALIAN) see BIQ500
DICK (GERMAN) see DFH200
S-(2,3-DICLORO-ALLIL)-N,N-DIISOPROPIL-
MONOTIOCARBAMMATO (ITALIAN) see DBI200
p-DICLOROBENZENE (ITALIAN) see DEP800
1,4-DICLOROBENZENE (ITALIAN) see DEP800
1,1-DICLORO-2,2-BIS(4-CLORO-FENIL)-ETANO
(ITALIAN) see BIM500
3,3'-DICLORO-4,4'-DIAMINODIFENILMETANO
(ITALIAN) see MJM200
1,1-DICLOROETANO (ITALIAN) see DFF809
1,2-DICLOROETANO (ITALIAN) see EIY600
2,2'-DICLOROETILETERE (ITALIAN) see DFJ050
3-(3,4-DICLORO-FENYL)-1,1-DIMETIL-UREA
(ITALIAN) see DXQ500
1,1-DICLORO-1-NITROETANO (ITALIAN) see DFU000
(2,2-DICLORO-VINIL)DIMETILFOSFATO (ITALIAN)
see DGP900
DICLOTRIDE see CFY000
DICOBALT CARBONYL see CNB500
DICOBALT OCTACARBONYL see CNB500
DICOBALT OXIDE see CND825
DICOBALT TRIOXIDE see CND825
DICOFOL see BIO750
DICOL see DJD600
DICONIRT D see SGH500
DICOPHANE see DAD200
DICOPPER MONOXIDE see CNO000
DICOPUR-M see CIR250
DICORTOL see PMA000
DICORVIN see DKA600
DICOTEX see CIR250

DICOTEX 80 see SIL500
DICOUMARIN see BJZ000
DICOUMAROL see BJZ000
DICRESYL see MIB750
DICRESYL N-METHYLCARBAMATE see MIB750
DICROTOFOS (DUTCH) see DGQ875
DICROTOPHOS see DGQ875
DICUMACYL see BKA000
DICUMAN see BJZ000
DICUMARINE see BJZ000
DICUMENE CHROMIUM see DGR200
DICUMENYLCHROMIUM see DGR200
DICUMYL PEROXIDE (DOT) see DGR600
DI-α-CUMYL PEROXIDE see DGR600
DI-CUP see DGR600
DI-CUP 40 KF see DGR600
DI-CUPR see DGR600
DICUPRAL see DXH250
2,2'-DICYANO-2,2'-AZOPROPANE see ASL750
m-DICYANOBENZENE see PHX550
o-DICYANOBENZENE see PHY000
p-DICYANOBENZENE see BBP250
1,2-DICYANOBENZENE see PHY000
1,3-DICYANOBENZENE see PHX550
1,4-DICYANOBENZENE see BBP250
1,4-DICYANOBUTANE see AER250
β,β-DICYANO-o-CHLOROSTYRENE see CEQ600
2,2'-DICYANODIETHYLAMINE see BIQ500
s-DICYANOETHANE see SNE000
1,2-DICYANOETHANE see SNE000
DI-(2-CYANOETHYL)AMINE see BIQ500
DICYANOGEN see COO000
DICYANOMETHANE see MAO250
1,3-DICYANOPROPANE see TLR500
1,3-DICYANOTETRACHLOROBENZENE see TBQ750
DICYCLOHEXANO-18-CROWN-6 see DGV100
DICYCLOHEXYL ADIPATE see DGT500
N,N-DICYCLOHEXYLAMINE see DGT600
DICYCLOHEXYLAMINE (DOT) see DGT600
DICYCLOHEXYLAMINE NITRITE see DGU200
DICYCLOHEXYLAMINONITRITE see DGU200
DICYCLOHEXYLAMMONIUM NITRITE see DGU200
DICYCLOHEXYL-18-CROWN-6 see DGV100
DICYCLOHEXYL KETONE see DGV600
DICYCLOHEXYLMETHANE-4,4'-DIISOCYANATE see MJM600
DICYCLOHEXYLTIN OXIDE see DGV900
DICYCLOPENTADIENE see DGW000
DICYCLOPENTADIENYLDICHLOROTITANIUM see DGW200
DI-2,4-CYCLOPENTADIEN-1-YL IRON see FBC000
DICYCLOPENTADIENYL IRON (OSHA, ACGIH) see FBC000
DI-pi-CYCLOPENTADIENYLNICKEL see NDA500
DICYCLOPENTADIENYLTITANIUMDICHLORIDE see DGW200
DICYKLOHEXYLAMIN (CZECH) see DGT600
DICYKLOHEXYLAMIN NITRIT (CZECH) see DGU200
DICYKLOPENTADIEN (CZECH) see DGW000
DICYNIT (CZECH) see DGU200
DICYSTEINE see CQK325
DID 95 see CJM750
DIDAKENE see PCF275
DIDANDIN see DVV600
DIDAN-TDC-250 see DKQ000
DIDECANOYLTRIETHYLENE GLYCOL ESTER (mixed isomers) see DAH450

9,10-DIDEHYDRO-N,N-DIETHYL-2-BROMO-6-METHYLERGOLINE-8-β-CARBOXAMIDE see BNM250
9,10-DIDEHYDRO-N,N-DIETHYL-6-METHYL-ERGOLINE-8-β-CARBOXAMIDE see DJO000
9,10-DIDEHYDRO-N,N-DIETHYL-6-METHYL-ERGOLINE-8-β-CARBOXAMIDE-d-TARTRATE with METHANOL (1:2) see LJG000
7,8-DIDEHYDRO-4,5-α-EPOXY-3-ETHOXY-17-METHYLMORPHINAN-6-α-OL HYDROCHLORIDE DIHYDRATE see ENK500
7,8-DIDEHYDRO-4,5-α-EPOXY-17-METHYLMORPHINAN-3,6-α-DIOL HYDROCHLORIDE see MRO750
7,8-DIDEHYDRO-4,5-α-EPOXY-17-METHYLMORPHINE HYDROCHLORIDE see MRO750
3,8-DIDEHYDRORETRONECINE see DAL400
1,6-DIDEOXY-1,6-DI(2-CHLOROETHYLAMINO)-d-MANNITOLDIHYDROCHLORIDE see MAW750
DIDEOXYKANAMYCIN B see DCQ800
3',4'-DIDEOXYKANAMYCIN B see DCQ800
DIDEUTERIUM OXIDE see HAK000
DI-2,4-DICHLOROBENZOYL PEROXIDE, >75% with water (DOT) see BIX750
DIDIGAM see DAD200
DIDIMAC see DAD200
3,6-DI(DIMETHYLAMINO)ACRIDINE see BJF000
1,2-DI-(DIMETHYLAMINO)ETHANE (DOT) see TDQ750
DIDOC see AAI250
DIDODECANOYLOXYDIOCTYLSTANNANE see DVJ800
DIDODECYL 3,3'-THIODIPROPIONATE see TFD500
DIDRATE see DKW800
DIDRONEL R see DXD400
DIDROXAN see MJM500
DIDROXANE see MJM500
DIELDREX see DHB400
DIELDRIN see DHB400
DIELDRINE (FRENCH) see DHB400
DIELDRITE see DHB400
DIELTAMID see DKC800
DIENESTROL see DAL600
DIENOESTROL see DAL600
β-DIENOESTROL see DAL600
DIENOL see DAL600
DIENOL S see SMR000
DIENPAX see DCK759
DIEPOXYBUTANE see BGA750
l-DIEPOXYBUTANE see BOP750
2,4-DIEPOXYBUTANE see BGA750
1,2:3,4-DIEPOXYBUTANE see BGA750
(2S,3S)-DIEPOXYBUTANE see BOP750
meso-DIEPOXYBUTANE see DHB800
(R*,S*)-DIEPOXYBUTANE see DHB800
l-1,2:3,4-DIEPOXYBUTANE see BOP750
(2S,3S)-1,2:3,4-DIEPOXYBUTANE see BOP750
meso-1,2,3,4-DIEPOXYBUTANE see DHB800
1,2,8,9-DIEPOXYLIMONENE see LFV000
1,2:8,9-DIEPOXYMENTHANE see LFV000
1,2:8,9-DIEPOXY-p-MENTHANE see LFV000
DI(2,3-EPOXYPROPYL) ETHER see DKM200
DIESEL EXHAUST see DHE485
DIESEL FUEL (DOT) see DHE900
DIESEL FUEL (DOT) see FOP000
DIESEL FUEL MARINE see DHE750

DIESEL FUEL MARINE see DHE800
DIESEL FUEL NO. 2 see DHE850
DIESEL FUEL NO. 4 see DHE750
DIESEL FUELS see DHE900
DIESEL OIL (PETROLEUM) see DHE900
DIESEL OIL (PETROLEUM) see FOP000
DIESEL OILS see DHE900
DIESEL OILS see FOP000
DIESEL TEST FUEL see DHE900
DIESEL TEST FUEL see FOP000
S,S-DIESTER with DITHIO-p-
UREIDOBENZENEARSONOUS ACID o-
MERCAPTOBENZOIC ACID see TFD750
DIESTER with o-MERCAPTOBENZOIC ACID
DITHIO-p-UREIDOBENZENEARSONOUS ACID see
TFD750
DI-ESTRYL see DKA600
DIETADIONE (ITALIAN) see DJT400
DIETHADION see DJT400
DIETHADIONE see DJT400
DIETHANOLAMIN (CZECH) see DHF000
DIETHANOLAMINE see DHF000
DIETHANOLAMMONIUM MALEIC HYDRAZIDE see
DHF200
DIETHANOLLAURAMIDE see BKE500
N,N-DIETHANOLLAURAMIDE see BKE500
N,N-DIETHANOLLAURIC ACID AMIDE see BKE500
DIETHANOLNITROSOAMINE see NKM000
DIETHION see EEH600
o-DIETHOXYBENZENE see CCP900
1,2-DIETHOXYBENZENE see CCP900
1,2-DI(ETHOXYCARBONYL)ETHYL-O,O-DIMETHYL
PHOSPHORODITHIOATE see MAK700
S-(1,2-DI(ETHOXYCARBONYL)ETHYL) DIMETHYL
PHOSPHOROTHIOLOTHIONATE see MAK700
DIETHOXYDIMETHYLSILANE see DHG000
1,1-DIETHOXY-ETHAAN (DUTCH) see AAG000
1,1-DIETHOXYETHANE see AAG000
1,2-DIETHOXYETHANE see EJE500
DIETHOXY ETHYL ADIPATE see BJO225
DIETHOXY-3-KYANPROPYL-METHYLSILAN see
COR500
DIETHOXYMETHANE (DOT) see EFT500
α-
(((DIETHOXYPHOSPHINOTHIOYL)OXY)IMINO)BE
NZENEACETONITRILE see BAT750
(DIETHOXYPHOSPHINYL)DITHIOIMIDOCARBONI
C ACID CYCLIC ETHYLENE ESTER see PGW750
2-(DIETHOXYPHOSPHINYLIMINO)-1,3-
DITHIOLANE see PGW750
2-(DIETHOXYPHOSPHINYLIMINO)-1,3-
DITHIOLANE see DXN600
2-(DIETHOXYPHOSPHINYLIMINO)-4-METHYL-1,3-
DITHIOLANE see DHH400
(DIETHOXY-PHOSPHINYL)MERCURY ACETATE see
AAS250
DIETHOXYPHOSPHORUS OXYCHLORIDE see
DIY000
3,3-DIETHOXYPROPENE see DHH800
3,3-DIETHOXY-1-PROPENE see DHH800
(DIETHOXY-THIOPHOSPHORYLOXYIMINO)-
PHENYL ACETONITRILE see BAT750
DIETHQUINALPHION see DJY200
DIETHQUINALPHIONE see DJY200
DIETHYL see BOR500
DIETHYL ACETAL see AAG000
DIETHYL ACETALDEHYDE see DHI000

N,N-DIETHYLACETAMIDE see DHI200
DIETHYLACETIC ACID see DHI400
O,O-DIETHYL-S-N-(A-
CYANOISOPROPYL)CARBOMOYLMETHYL
PHOSPHOROTHIOATE see PHK250
DIETHYLAMINE see DHJ200
N,N-DIETHYLAMINE see DHJ200
DIETHYLAMINE, 2,2'-DICHLORO-N-METHYL-,
OXIDE see CFA500
(DIETHYLAMINO)ACETONITRILE see DHJ600
N,N-DIETHYLAMINOACETONITRILE see DHJ600
DIETHYLAMINOACETO-2,6-XYLIDIDE see DHK400
2-(DIETHYLAMINO)-2',6'-ACETOXYLIDIDE see
DHK400
α-DIETHYLAMINOACETO-2,6-XYLIDIDE see
DHK400
α-DIETHYLAMINO-2,6-ACETOXYLIDIDE see
DHK400
2-(DIETHYLAMINO)-2',6'-ACETOXYLIDIDE
HYDROCHLORIDE see DHK600
2-(DIETHYLAMINO)-2',6'-ACETOXYLIDIDE
MONOHYDROCHLORIDE see DHK600
α-DIETHYLAMINO-2,5-ACETOXYLIDINE
HYDROCHLORIDE see DHK600
DIETHYLAMINOACET-2,6-XYLIDIDE see DHK400
4-(DIETHYLAMINO)ANILINE see DJV200
p-(DIETHYLAMINO)ANILINE see DJV200
N,N-DIETHYLAMINOBENZENE see DIS700
α-DIETHYLAMINO-2,6-DIMETHYLACETANILIDE
see DHK400
ω-DIETHYLAMINO-2,6-DIMETHYLACETANILIDE
see DHK400
ω-DIETHYLAMINO-2,6-DIMETHYLACETANILIDE
HYDROCHLORIDE see DHK600
3-(DIETHYLAMINO)-7-((p-
(DIMETHYLAMINO)PHENYL)AZO)-5-
PHENYLPHENAZINIUM CHLORIDE see DHM500
2-(DIETHYLAMINO)-N-(2,6-
DIMETHYLPHENYL)ACETAMIDE
MONOHYDROCHLORIDE see DHK600
(DIETHYLAMINO)ETHANE see TJO000
DIETHYLAMINOETHANOL see DHO500
2-DIETHYLAMINOETHANOL see DHO500
2-(DIETHYLAMINO)ETHANOL see DHO500
N-DIETHYLAMINOETHANOL see DHO500
2-N-DIETHYLAMINOETHANOL see DHO500
DIETHYLAMINOETHANOL (DOT) see DHO500
β-DIETHYLAMINOETHANOL see DHO500
DIETHYLAMINOETHANOL-4-AMINOBENZOATE
HYDROCHLORIDE see AIT250
o-(DIETHYLAMINOETHOXY)BENZANILIDE see
DHP200
2-(2-(DIETHYLAMINO)ETHOXY)BENZANILIDE see
DHP200
DIETHYLAMINOETHOXYETHANOL see DHQ100
2-(2-(DIETHYLAMINO)ETHOXY)ETHANOL see
DHQ100
2-(β-(DIETHYLAMINO)ETHOXY)ETHANOL see
DHQ100
2-(2-(DIETHYLAMINO)ETHOXY)-3-
METHYLBENZANILIDE see DHQ800
p-(2-(DIETHYLAMINO)ETHOXY)PHENYL 2-ETHYL-
3-BENZOFURANYL KETONE see EDV700
β-DIETHYLAMINOETHYL ALCOHOL see DHO500
2-DIETHYLAMINOETHYL-p-AMINOBENZOATE
HYDROCHLORIDE see AIT250

1-((2-(DIETHYLAMINO)ETHYL)AMINO)-4-
(HYDROXYMETHYL)THIOXANTHEN-9-ONE see
LIM000
1-((2-(DIETHYLAMINO)ETHYL)AMINO)-4-
(HYDROXYMETHYL)9H-THIOXANTHEN-9-ONE see
LIM000
1-(2'-DIETHYLAMINO)ETHYLAMINO-4-
METHYLTHIOXANTHENONE see DHU000
1-((2-(DIETHYLAMINO)ETHYL)AMINO)-4-METHYL-
9H-THIOXANTHEN-9-ONE see DHU000
2-DIETHYLAMINOETHYL BENZILATE
HYDROCHLORIDE see BCA000
β-DIETHYLAMINOETHYL BENZILATE
HYDROCHLORIDE see BCA000
2-DIETHYLAMINOETHYL-2-CYCLOPENTYL-2-(2-
THIENYL)HYDROXYACETATE METHOBROMIDE
see PBS000
2-DIETHYLAMINOETHYL-α-CYCLOPENTYL-2-
THIOPHENEGLYCOLATE METHOBROMIDE see
PBS000
2-DIETHYLAMINOETHYL DIPHENYLGLYCOLATE
HYDROCHLORIDE see BCA000
1-(2-(DIETHYLAMINO)ETHYL)-2-(p-
ETHOXYBENZYL)-5-BENZIMIDAZOLYL METHYL
KETONE see DHZ050
2-DIETHYLAMINOETHYL-3-METHYL-2-
PHENYLVALERATE METHYLBROMIDE see VBK000
1-(2-(DIETHYLAMINO)ETHYL)-2-p-PHENETIDINO-
5-BENZIMIDAZOLYL METHYL KETONE see DIE350
2-DIETHYLAMINOETHYL-2-PHENYL-3-
METHYLVALERATE METHYL BROMIDE see VBK000
5-β-DIETHYLAMINOETHYL-3-PHENYL-1,2,4-
OXADIAZOLE CITRATE see OOE000
10-(2-DIETHYLAMINO-2-
METHYLETHYL)PHENOTHIAZINE see DIR000
2-DIETHYLAMINO-6-METHYLPYRIMIDIN-4-YL
DIMETHYL PHOSPHOROTHIONATE see DIN800
O-(2-(DIETHYLAMINO-6-METHYLPYRIMIDIN-4-YL)-
O,O-DIMETHYL PHOSPHOROTHIOATE see DIN800
O-(2-(DIETHYLAMINO)-6-METHYL-4-
PYRIMIDINYL)-O,O-DIMETHYL
PHOSPHOROTHIOATE see DIN800
7-DIETHYLAMINO-5-METHYL-s-TRIAZOLO(1,5-
a)PYRIMIDINE see DIO200
1-(DIETHYLAMINO)-4-PENTANONE see NOB800
5-DIETHYLAMINO-2-PENTANONE see NOB800
2-(p-(DIETHYLAMINOPHENYL))-1,3,2-
DITHIARSENOLANE see DIP100
10-DIETHYLAMINOPROPIONYL-3-
TRIFLUOROMETHYL PHENOTHIAZINE
HYDROCHLORIDE see FDE000
N-(3-DIETHYLAMINOPROPYL)AMINE see DIY800
N,N-DIETHYLAMINOPROPYLAMINE see DIY800
3-(DIETHYLAMINO)PROPYLAMINE (DOT) see
DIY800
2-DIETHYLAMINO-1-PROPYL-N-
DIBENZOPARATHIAZINE see DIR000
10-(2-DIETHYLAMINOPROPYL)PHENOTHIAZINE
see DIR000
10-(2-(DIETHYLAMINO)PROPYL)-10H-PYRIDO(3,2-
B)(1,4)BENZOTHIAZINE see DIR100
DIETHYLAMINOTRIMETHYLENAMINE see DIY800
N,N-DIETHYLANILIN (CZECH) see DIS700
DIETHYLANILINE see DIS700
N,N-DIETHYLANILINE see DIS700
N,N-DIETHYL-p-ARSANILIC ACID see DIS775
DIETHYL ARSINIC ACID see DIS850

DIETHYLBARBITURATE MONOSODIUM see BAG250
5,5-DIETHYLBARBITURIC ACID SODIUM deriv. see
BAG250
N,N-DIETHYLBENZENAMINE see DIS700
DIETHYL BENZENE see DIU000
m-DIETHYLBENZENE see DIU200
DIETHYLBENZENE (DOT) see DIU000
DIETHYLBENZOL see DIU000
DIETHYLBERYLLIUM see DIV000
DIETHYLBIS(OCTANOYLOXY)STANNANE see
DIV600
DIETHYLBIS(1-OXOOCTYL)OXY)STANNANE see
DIV600
O,O-DIETHYL-O-(4-BROOM-2,5-DICHLOOR-FENYL)-
MONOTHIOFOSFAAT (DUTCH) see EGV500
DIETHYLCADMIUM see DIV800
DIETHYLCARBAMAZANE CITRATE see DIW200
DIETHYLCARBAMAZINE see DIW000
DIETHYLCARBAMAZINE ACID CITRATE see DIW200
DIETHYLCARBAMAZINE CITRATE see DIW200
DIETHYLCARBAMAZINE HYDROGEN CITRATE see
DIW200
DIETHYLCARBAMIC CHLORIDE see DIW400
DIETHYLCARBAMIDOYL CHLORIDE see DIW400
DIETHYLCARBAMODITHIOIC ACID 2-CHLORO-2-
PROPENYL ESTER see CDO250
DIETHYLCARBAMODITHIOIC ACID, SODIUM SALT
see SGJ000
DIETHYLCARBAMOYL CHLORIDE see DIW400
N,N-DIETHYLCARBAMOYL CHLORIDE see DIW400
1-DIETHYLCARBAMOYL-4-METHYLPIPERAZINE see
DIW000
1-DIETHYLCARBAMOYL-4-METHYLPIPERAZINE
DIHYDROGEN CITRATE see DIW200
DIETHYLCARBAMYL CHLORIDE see DIW400
1-DIETHYLCARBAMYL-4-METHYLPIPERZINE see
DIW000
N,N-DIETHYLCARBANILIDE see DJC400
DIETHYL CARBINOL see IHP010
DIETHYLCARBINOL see IHP010
DIETHYL CARBITOL see DIW800
DIETHYL CARBONATE see DIX200
DIETHYL CARBONATE (DOT) see DIX200
DIETHYL CELLOSOLVE (DOT) see EJE500
DIETHYLCETONE (FRENCH) see DJN750
O,O-DIETHYL-O-(2-
CHINOXALYL)PHOSPHOROTHIOATE see DJY200
O,O-DIETHYL-S-((4-CHLOOR-FENYL-THIO)-
METHYL)-DITHIOFOSFAAT (DUTCH) see TNP250
O,O-DIETHYL-O-(3-CHLOOR-4-METHYL-CUMARIN-
7-YL)MONOTHIOFOSFAAT (DUTCH) see CNU750
O,O-DIETHYL-S-((6-CHLOOR-2-OXO-
BENZOXAZOLIN-3-YL)-METHYL)-DITHIO
FOSFAAT (DUTCH) see BDJ250
O,O-DIETHYL-S-p-
CHLORFENYLTHIOMETHYLESTER KYSELINY
DITHIOFOSFORECNE (CZECH) see TNP250
O,O-DIETHYL-S-(6-CHLOROBENZOXAZOLINYL-3-
METHYL)DITHIOPHOSPHATE see BDJ250
O,O-DIETHYL-O-(2-CHLORO-1-(2',4'-
DICHLOROPHENYL)VINYL) PHOSPHATE see
CDS750
O,O-DIETHYL-S-p-CHLOROLPHENYLTHIOMETHYL
DITHIOPHOSPHATE see TNP250
DIETHYL-3-CHLORO-4-METHYL-7-COUMARINYL
PHOSPHATE see CIK750
O,O-DIETHYL-O-(3-CHLORO-4-
METHYLCOUMARIN-7-YL)PHOSPHATE see CIK750

O,O-DIETHYL-O-(3-CHLORO-4-METHYL-7-COUMARINYL)PHOSPHOROTHIOATE see CNU750

O,O-DIETHYL-O-(3-CHLORO-4-METHYLCOUMARINYL-7) THIOPHOSPHATE see CNU750

O,O-DIETHYL-O-(3-CHLORO-4-METHYL-2-OXO-2H-BENZOPYRAN-7-YL)PHOSPHOROTHIOATE see CNU750

O,O-DIETHYL-3-CHLORO-4-METHYL-7-UMBELLIFERONE THIOPHOSPHATE see CNU750

O,O-DIETHYL-O-(3-CHLORO-4-METHYLUMBELLIFERYL)PHOSPHOROTHIOATE see CNU750

DIETHYL-3-CHLORO-4-METHYLUMBELLIFERYL THIONOPHOSPHATE see CNU750

O,O-DIETHYL-S-((6-CHLORO-2-OXOBENZOXAZOLIN-3-YL)METHYL)PHOSPHORODITHIOATE see BDJ250

O,O-DIETHYL-S-(6-CHLORO-2-OXO-BENZOXAZOLIN-3-YL)METHYL-PHOSPHORO THIOLOTHIONATE see BDJ250

O,O-DIETHYL-P-CHLOROPHENYLMERCAPTOMETHYL DITHIOPHOSPHATE see TNP250

O,O-DIETHYL-S-(4-CHLOROPHENYLTHIOMETHYL) DITHIOPHOSPHATE see TNP250

O,O-DIETHYL-S-(p-CHLOROPHENYLTHIOMETHYL) PHOSPHORODITHIOATE see TNP250

DIETHYL CHLOROPHOSPHATE see DIY000

O,O-DIETHYL-S-(2-CHLORO-1-PHTHALIMIDOETHYL)PHOSPHORODITHIOATE see DBI099

DIETHYLCHLOROTHIOPHOSPHATE see DJW600

DIETHYLCHLORTHIOFOSFAT (CZECH) see DJW600

O,O-DIETHYL-S-((2-CYAAN-2-METHYL-ETHYL)-CARBAMOYL)-METHYL-MONOTHIOFOSFAAT (DUTCH) see PHK250

P,P-DIETHYL CYCLIC ETHYLENE ESTER OF PHOSPHONODITHIOIMIDOCARBONIC ACID see PGW750

p,p-DIETHYL CYCLIC PROPYLENE ESTER of PHOSPHONODITHIOIMIDOCARBONIC ACID see DHH400

DIETHYL DECANEDIOATE see DJY600

DIETHYL-1,10-DECANEDIOATE see DJY600

N,N-DIETHYL-1,3-DIAMINOPROPANE see DIY800

DIETHYLDIAZENE-1-OXIDE see ASP000

DIETHYL DICARBONATE see DIZ100

O,O-DIETHYL-O-(2,4-DICHLOOR-FENYL)-MONOTHIOFOSFAAT (DUTCH) see DFK600

O,O-DIETHYL O-2,5-DICHLORO-4-BROMOPHENYL-PHOSPHOROTHIOATE see EGV500

O,O-DIETHYL O-(2,5-DICHLORO-4-BROMOPHENYL) THIOPHOSPHATE see EGV500

O,O-DIETHYL-O-(2,4-DICHLOROPHENYL)PHOSPHOROTHIOATE see DFK600

DIETHYL 2,4-DICHLOROPHENYL PHOSPHOROTHIONATE see DFK600

O,O-DIETHYL-S-(2,5-DICHLOROPHENYLTHIOMETHYL)DITHIOPHOSPHATE see PDC750

O,O-DIETHYL-S-(2,5-DICHLOROPHENYLTHIOMETHYL)DITHIOPHOSPHORAN see PDC750

O,O-DIETHYL-S-(2,5-DICHLOROPHENYLTHIOMETHYL)PHOSPHORODITHIOATE see PDC750

O,O-DIETHYL-S-(2,5-DICHLOROPHENYLTHIOMETHYL)PHOSPHOROTHIOLOTHIONATE see PDC750

O,O-DIETHYL-O-2,4-DICHLOROPHENYL THIOPHOSPHATE see DFK600

DIETHYLDICHLOROSILANE (DOT) see DEY800

DIETHYLDICHLOROSTANNANE see DEZ000

DIETHYL (2-(DIETHOXYMETHYLSILYL)ETHYL)PHOSPHONATE see DJA330

5,5-DIETHYLDIHYDRO-2H-1,3-OXAZINE-2,4(3H)-DIONE see DJT400

O,O-DIETHYL O-(2,3-DIHYDRO-3-OXO-2-PHENYL-6-PYRIDAZINYL)PHOSPHOROTHIOATE see POP000

DIETHYLDIIODOSTANNANE see DJB000

DIETHYL (DIMETHOXYPHOSPHINOTHIOYLTHIO) BUTANEDIOATE see MAK700

DIETHYL (DIMETHOXYPHOSPHINOTHIOYLTHIO)SUCCINATE see MAK700

DIETHYL 2,6-DIMETHYL-4(2-PYRIDYL)-1,4-DIHYDRO-3,5-PYRIDINE-DICARBOXYLATE see DJB460

DIETHYLDIPHENYL DICHLOROETHANE see DJC000

sym-DIETHYLDIPHENYLUREA see DJC400

1,3-DIETHYL-1,3-DIPHENYLUREA see DJC400

N,N-DIETHYL-N,N-DIPHENYLUREA see DJC400

DIETHYLDITHIO BIS(THIONOFORMATE) see BJU000

DIETHYLDITHIOCARBAMATE SODIUM see SGJ000

DIETHYLDITHIOCARBAMIC ACID-2-CHLOROALLYL ESTER see CDO250

DIETHYLDITHIOCARBAMIC ACID SELENIUM(II) SALT see DJD400

DIETHYLDITHIOCARBAMIC ACID SODIUM see SGJ000

DIETHYLDITHIOCARBAMIC ACID, SODIUM SALT see SGJ000

DIETHYLDITHIOCARBAMIC ACID TELLURIUM SALT see EPJ000

DIETHYLDITHIOCARBAMIC ACID ZINC SALT see BJC000

O,O-DIETHYL 1,3-DITHIOLAN-2-YLIDENEPHOSPHORAMIDOTHIOATE see DXN600

DIETHYL-N-1,3-DITHIOLANYL-2-IMINO PHOSPHATE see DXN600

O,O-DIETHYL-DITHIOPHOSPHORIC ACID, p-CHLOROPHENYLTHIOMETHYL ESTER see TNP250

3-DIETHYLDITHIOPHOSPHORYLMETHYL-6-CHLOROBENZOXAZOLONE-2 see BDJ250

DIETHYL DIXANTHOGEN see BJU000

DIETHYL EMME see EEV200

1,4-DIETHYLENEDIAMINE see PIJ000

N,N-DIETHYLENE DIAMINE (DOT) see PIJ000

DIETHYLENE DIOXIDE see DVQ000

1,4-DIETHYLENE DIOXIDE see DVQ000

DIETHYLENE ETHER see DVQ000

DIETHYLENE GLYCOL see DJD600

DIETHYLENE GLYCOL, BISCHLOROFORMATE see OPO000

DIETHYLENE GLYCOL BISPHTHALATE see DJD700

DIETHYLENE GLYCOL-n-BUTYL ETHER see DJF200

DIETHYLENE GLYCOL BUTYL ETHER ACETATE see BQP500

DIETHYLENEGLYCOL DIBUTYL ETHER see DDW200

O,O-DIETHYLPHOSPHOROCHLORIDOTHIOATE see DJW600

O,O-DIETHYL PHOSPHORODITHIOATE S-ester with 3-(MERCAPTOMETHYL)-1,2,3-BENZOTRIAZIN-4(3H)-ONE see EKN000

O,O-DIETHYLPHOSPHOROTHIOATE, O-ESTER with 6-HYDROXY-2-PHENYL-3(2H)-PYRIDAZINONE see POP000

O,O-DIETHYL PHOSPHOROTHIOATE, o-ESTER with PHENYLGLYOXYLONITRILE OXIME see BAT750 4-(o-(o,o-DIETHYLPHOSPHOROTHIOYL))BENZALDOXIMIN

O-N-METHYLCARBAMATE see DJW890

DIETHYL PHTHALATE see DJX000

DIETHYL-o-PHTHALATE see DJX000

DIETHYL PROPANEDIOATE see EMA500

DIETHYL-O-2-PYRAZINYL PHOSPHOROTHIONATE see EPC500

O,O-DIETHYL-O-2-PYRAZINYL PHOSPHOTHIONATE see EPC500

O,O-DIETHYL-O-PYRAZINYL THIOPHOSPHATE see EPC500

N,N-DIETHYL-3-PYRIDINECARBOXAMIDE see DJS200

3,3-DIETHYL-1-(m-PYRIDYL)TRIAZENE see DJY000

DIETHYL PYROCARBONATE see DIZ100

DIETHYL PYROCARBONIC ACID see DIZ100

O,O-DIETHYL-O-QUINOXALIN-2-YL PHOSPHOROTHIOATE see DJY200

O,O-DIETHYL-O-(2-QUINOXALINYL) PHOSPHOROTHIOATE see DJY200

O,O-DIETHYL-O-(2-QUINOXALYL) PHOSPHOROTHIOATE see DJY200

O,O-DIETHYL-O-2-QUINOXALYLTHIOPHOSPHATE see DJY200

O,O-DIETHYL Se-(2-DIETHYLAMINOETHYL)PHOSPHOROSELENOATE see DJA325

DIETHYL SEBACATE see DJY600

N,N-DIETHYLSELENOUREA see DJY800

1,1-DIETHYL-2-SELENOUREA see DJY800

DIETHYL SODIUM DITHIOCARBAMATE see SGJ000

DIETHYLSTANNIUM DIIODIDE see DJB000

DIETHYLSTANNIUMDIJODID (GERMAN) see DJB000

DIETHYLSTANNYL DICHLORIDE see DEZ000

2,2'-DIETHYL-4,4'-STILBENEDIOL see DKA600

α,α'-DIETHYLSTILBENEDIOL see DKA600

α,α'-DIETHYL-(E)-4,4'-STILBENEDIOL see DKA600

α,α'-DIETHYL-4,4'-STILBENEDIOL see DKA600

trans-α,α'-DIETHYL-4,4'-STILBENEDIOL see DKA600

α,α'-DIETHYL-4,4'-STILBENEDIOL, DIPROPIONATE see DKB000

α,α'-DIETHYL-4,4'-STILBENEDIOL trans-DIPROPIONATE see DKB000

trans-α,α'-DIETHYL-4,4'-STILBENEDIOL DIPROPIONATE see DKB000

α,α'-DIETHYL-4,4'-STILBENEDIOL DIPROPIONYL ESTER see DKB000

DIETHYLSTILBENE DIPROPIONATE see DKB000

DIETHYLSTILBESTEROL see DKA600

trans-DIETHYLSTILBESTEROL see DKA600

DIETHYLSTILBESTEROL DIPROPIONATE see DKB000

DIETHYLSTILBESTROL see DKA600

trans-DIETHYLSTILBESTROL see DKA600

DIETHYLSTILBESTROL DIPROPIONATE see DKB000

DIETHYLSTILBESTROL PROPIONATE see DKB000

DIETHYLSTILBOESTEROL see DKA600

trans-DIETHYLSTILBOESTEROL see DKA600

DIETHYL SUCCINATE (FCC) see SNB000

DIETHYL SULFATE see DKB110

DIETHYLSULFID (CZECH) see EPH000

DIETHYL SULFIDE (DOT) see EPH000

DIETHYL SULFIDE-2,2'-DICARBOXYLIC ACID see BHM000

o,o-DIETHYL o-TETRAHYDROFURFURYL ESTER PHOSPHOROTHIOIC ACID see DKB170

5,5-DIETHYLTETRAHYDRO-2H-1,3-OXAZINE-2,4(3H)-DIONE see DJT400

O,O-DIETHYL-O-(7,8,9,10-TETRAHYDRO-6-OXOBENZO(C)CHROMAN-3-YL)PHOSPHOROTHIOATE see DXO000

O,O-DIETHYL-O-(7,8,9,10-TETRAHYDRO-6-OXO-6H-DIBENZO(b,d)PYRAN-3-YL)PHOSPHOROTHIOATE see DXO000

O,O-DIETHYL-O-(3,4-TETRAMETHYLENECOUMARINYL-7) THIOPHOSPHATE see DXO000

DIETHYL TETRAOXOSULFATE see DKB110

DIETHYLTHIADICARBOCYANINE IODIDE see DJT800

3,3'-DIETHYLTHIADICARBOCYANINE IODIDE see DJT800

N,N-DIETHYLTHIOCARBAMIDE see DKC400

DIETHYLTHIOETHER see EPH000

DIETHYL THIOPHOSPHORIC ACIDESTER of 3-CHLORO-4-METHYL-7-HYDROXYCOUMARIN see CNU750

DIETHYLTHIOPHOSPHORYL CHLORIDE (DOT) see DJW600

1,3-DIETHYLTHIOUREA see DKC400

1,3-DIETHYL-2-THIOUREA see DKC400

N,N-DIETHYLTHIOUREA see DKC400

DIETHYLTIN CHLORIDE see DEZ000

DIETHYLTIN DICAPRYLATE see DIV600

DIETHYLTIN DICHLORIDE see DEZ000

DIETHYLTIN DIIODIDE see DJB000

DIETHYLTIN DIOCTANOATE see DIV600

DIETHYLTOLUAMIDE see DKC800

DIETHYL-m-TOLUAMIDE see DKC800

N,N-DIETHYL-m-TOLUAMIDE see DKC800

DIETHYL TRIAZENE see DKD200

DIETHYL-TRIAZENE see DKD200

1,3-DIETHYLTRIAZENE see DKD200

1,3-DIETHYL-1-TRIAZENE see DKD200

N,N-DIETHYLTRIAZENE see DKD200

1,3-DIETHYLTRIAZINE see DKD200

O,O-DIETHYL O-3,5,6-TRICHLORO-2-PYRIDYL PHOSPHOROTHIOATE see CMA100

DIETHYL XANTHOGENATE see BJU000

DIETHYLXANTHOGEN DISULFIDE see BJU000

DIETHYLZINC see DKE600

DIETIL see CAR000

DIETILAMIDE-CARBOPIRIDINA see DJS200

DIETILAMINA (ITALIAN) see DHJ200

α-DIETILAMINO-2,6-DIMETILACETANILIDE (ITALIAN) see DHK400

O,O-DIETIL-O-(4-BROMO-2,5-DICLORO-FENIL)-MONOTIOFOSFATO (ITALIAN) see EGV500

O,O-DIETIL-S-((2-CIAN-2-METIL-ETIL)-CARBAMOIL)-METIL-MONOTIOFOSFATO (ITALIAN) see PHK250

O,O-DIETIL-S-((4-CLORO-FENIL-TIO)-METILE)-
DITIOFOSFATO (ITALIAN) see TNP250
O,O-DIETIL-O-(3-CLORO-4-METIL-CUMARIN-7-IL-
MONOTIOFOSFATO) (ITALIAN) see CNU750
O,O-DIETIL-S-((6-CLORO-2-OXO-BENZOSSAZOLIN-
3-IL)-METIL)-DITIOFOSFATO (ITALIAN) see BDJ250
O,O-DIETIL-O-(2,4-DICLORO-FENIL)-
MONOTIOFOSFATO (ITALIAN) see DFK600
5,5-DIETILDIIDRO-1,3-OSSAZIN-2,4-DIONE
(ITALIAN) see DJT400
DIETILESTILBESTROL (SPANISH) see DKA600
O,O-DIETIL-S-(2-ETILTIO-ETIL)-DITIOFOSFATO
(ITALIAN) see DXH325
O,O-DIETIL-S-(2-ETILTIO-ETIL)-
MONOTIOFOSFATO (ITALIAN) see DAP200
O,O-DIETIL-S-(ETILTIO-METIL)-DITIOFOSFATO
(ITALIAN) see PGS000
O,O-DIETIL-O-(2-ISOPROPIL-4-METIL-PIRIMIDIN-6-
IL)-MONOTIOFOSFATO (ITALIAN) see DCM750
O,O-DIETIL-O-(4-NITRO-FENIL)-
MONOTIOFOSFATO (ITALIAN) see PAK000
O,O-DIETIL-S-((4-OXO-3H-1,2,3-BENZOTRIAZIN-3-
IL)-METIL)-DITIOFOSFATO (ITALIAN) see EKN000
1,1-DIETOSSIETANO (ITALIAN) see AAG000
DIETROXINE see DJT400
O,O-DIETYL-S-2-ETYLMERKAPTOETYLTIOFOSFAT
(CZECH) see DAP200
DIFEDRYL see BBV500
DIFENHYDRAMIN see BBV500
DIFENHYDRAMINE HYDROCHLORIDE see BAU750
DIFENIDRAMINA (ITALIAN) see BBV500
DIFENILHIDANTOINA (SPANISH) see DKQ000
DIFENIL-METAN-DIISOCIANATO (ITALIAN) see
MJP400
DIFENIN see DKQ000
DIFENIN see DNU000
1,5-DIFENOXYANTHRACHINON see DVW100
DIFENTHOS see TAL250
N,N'-DIFENYL-p-FENYLENDIAMIN (CZECH) see
BLE500
2-(DIFENYL-HYDROXYACETOXY)ETHYL-
DIETHYLAMMONIUMCHLORID (CZECH) see
BCA000
DIFENYLMETHAAN-DIISSOCYANAAT (DUTCH) see
MJP400
DIFETOIN see DNU000
DIFFLAM see BBW500
DIFFOLLISTEROL see EDP000
DIFHYDAN see DKQ000
DIFHYDAN see DNU000
DIFLUBENZURON see CJV250
DIFLUNISAL see DKI600
3,4-DIFLUOROBENZENEARSONIC ACID see DKG100
1,1-DIFLUORO-1-CHLOROETHANE see CFX250
DIFLUOROCHLOROETHANES (DOT) see CFX250
DIFLUOROCHLOROMETHANE see CFX500
DIFLUORODIBROMOMETHANE see DKG850
DIFLUORODICHLOROMETHANE see DFA600
DIFLUORODIMETHYLSTANNANE see DKH200
DIFLUOROETHANE see ELN500
1,1-DIFLUOROETHANE see ELN500
1,1-DIFLUOROETHENE see VPP000
1,1-DIFLUOROETHYLENE (DOT, MAK) see VPP000
DIFLUOROFORMALDEHYDE see CCA500
2',4'-DIFLUORO-4-HYDROXY-3-
BIPHENYLCARBOXYLIC ACID see DKI600
2',4'-DIFLUORO-4-HYDROXY-(1,1'-BIPHENYL)-3-
CARBOXYLIC ACID see DKI600

2',4'-DIFLUORO-4-HYDROXY-(1',1-DIPHENYL)-3-
CARBOXYLIC ACID see DKI600
DIFLUOROMONOCHLOROMETHANE see CFX500
DIFLUOROPHENYLARSINE see DKI400
5-(2,4-DIFLUOROPHENYL)SALICYLIC ACID see
DKI600
1,1-DIFLUORO-1,2,2,2-TETRACHLOROETHANE see
TBP000
1,2-DIFLUORO-1,1,2,2-TETRACHLOROETHANE see
TBP050
DIFLUPYL see IRF000
DIFLURON see CJV250
DIFLUROPHATE see IRF000
DIFO see BJE750
DIFOLATAN see CBF800
DIFOLLICULINE see EDP000
DIFONATE see FMU045
DIFOSAN see CBF800
DIFOSGEN see TIR920
DIFURYLMETHANE see FPX028
DI-2-FURYLMETHANE see FPX028
2,2'-DIFURYLMETHANE see FPX028
DI-α-FURYLMETHANE see FPX028
DIGACIN see DKN400
DIGENEA SIMPLEX MUCILAGE see AEX250
11,11-DIGERANYLOXY-1-UNDECENE see UNA100
DIGERMIN see DUV600
DIGIBUTINA see BRF500
DIGILONG see DKL800
DIGIMED see DKL800
DIGIMERCK see DKL800
DIGISIDIN see DKL800
DIGITALIN see DKL800
DIGITALINE (FRENCH) see DKL800
DIGITALINE CRISTALLISEE see DKL800
DIGITALINE NATIVELLE see DKL800
DIGITALINUM VERUM see DKL800
DIGITALIS GLYCOSIDE see DKN400
DIGITIN see DKL400
DIGITONIN see DKL400
DIGITOPHYLLIN see DKL800
DIGITOXIGENIN-TRIDIGITOXOSID (GERMAN) see
DKL800
DIGITOXIGENIN TRIDIGITOXOSIDE see DKL800
DIGITOXIN see DKL800
DIGLYCERIDE ACETIC ACID see DBF600
DIGLYCIDYL BISPHENOL A ETHER see BLD750
DIGLYCIDYL ETHER see DKM200
DIGLYCIDYL ETHER of 2,2-BIS(4-
HYDROXYPHENYL)PROPANE see BLD750
DIGLYCIDYL ETHER of 2,2-BIS(p-
HYDROXYPHENYL)PROPANE see BLD750
DIGLYCIDYL ETHER of BISPHENOL A see BLD750
DIGLYCIDYL ETHER of 4,4'-
ISOPROPYLIDENEDIPHENOL see BLD750
1,3-DIGLYCIDYLOXYBENZENE see REF000
DIGLYCIDYL RESORCINOL ETHER see REF000
DIGLYCOL see DJD600
DIGLYCOLAMINE see AJU250
DIGLYCOLDINITRAAT (DUTCH) see DJE400
DIGLYCOL (DINITRATE de) (FRENCH) see DJE400
DIGLYCOL MONOBUTYL ETHER see DJF200
DIGLYCOL MONOBUTYL ETHER ACETATE see
BQP500
DIGLYCOL MONOETHYL ETHER see CBR000
DIGLYCOL MONOETHYL ETHER ACETATE see
CBQ750

DIGLYCOL MONOMETHYL ETHER see DJG000
DIGLYCOL MONOSTEARATE see HKJ000
DIGLYCOL STEARATE see HKJ000
DIGLYKOLDINITRAT (GERMAN) see DJE400
DIGOXIGENIN-TRIDIGITOXOSID (GERMAN) see DKN400
DIGOXIN see DKN400
DIGOXINE see DKN400
DIHDYROPYRONE see BRT000
DIHEPTYLMERCURY see DKO000
DIHEPTYL PHTHALATE see HBP400
DI-n-HEPTYL PHTHALATE see HBP400
DIHEXYLAMINE see DKO600
DI-N-HEXYLAMINE see DKO600
DIHEXYL PHTHALATE see DKP600
DI-n-HEXYL PHTHALATE see DKP600
DIHEXYLTIN DICHLORIDE see DFC200
DIHIDRAL see BBV500
DIHIDROCLORURO de BENZIDINA (SPANISH) see BBX750
DIHYCON see DKQ000
DI-HYDAN see DKQ000
DI-HYDAN see DNU000
DIHYDANTOIN see DKQ000
DIHYDANTOIN see DNU000
DIHYDRIN see DKW800
DIHYDROAFLATOXIN B1 see AEU750
DIHYDROANETHOLE see PNE250
DIHYDROAZIRENE see EJM900
DIHYDRO-1H-AZIRINE see EJM900
2,3-DIHYDRO-1,3-BENZOXAZINE-4H-2-THIONE-4-ONE see TFC570
2,3-DIHYDRO-5-CARBOXANILIDO-6-METHYL-1,4-OXATHIIN see CCC500
DIHYDROCARVEOL see DKV150
1,6-DIHYDROCARVEOL see DKV150
DIHYDROCARVEOL ACETATE see DKV160
DIHYDROCARVEYL ACETATE see DKV160
d-DIHYDROCARVONE see DKV175
DIHYDROCARVYL ACETATE see DKV160
DIHYDROCHLORIDE SALT OF DIETHYLENEDIAMINE see PIK000
3,4-DIHYDRO-6-CHLORO-7-SULFAMYL-1,2,4-BENZOTHIADIAZINE-1,1-DIOXIDE see CFY000
DIHYDROCHLOROTHIAZID see CFY000
DIHYDROCHLOROTHIAZIDE see CFY000
3,4-DIHYDROCHLOROTHIAZIDE see CFY000
DIHYDROCINNAMALDEHYDE see HHP000
DIHYDROCITRONELLOL see DTE600
DIHYDROCODEINE see DKW800
7,8-DIHYDROCODEINE see DKW800
DIHYDROCOUMARIN see HHR500
3,4-DIHYDROCOUMARIN see HHR500
DIHYDROCUMINYL ALCOHOL see PCI550
9,10-DIHYDRO-8a,10,-DIAZONIAPHENANTHRENE DIBROMIDE see DWX800
9,10-DIHYDRO-8a,10a-DIAZONIAPHENANTHRENE(1,1'-ETHYLENE-2,2'-BIPYRIDYLIUM)DIBROMIDE see DWX800
3-α-((10,11-DIHYDRO-5H-DIBENZO(A,D)CYCLOHEPTEN-5-YL)OXY)TROPAN DIHYDROGEN CITRATE see EAG100
3-α-(10,11-DIHYDRO-5H-DIBENZO(A,D)CYCLOHEPTEN-5-YLOXY)TROPANE CITRATE see EAG100
DIHYDRO-5,5-DIETHYL-2H-1,3-OXAZINE-2,4(3H)-DIONE see DJT400

9,10-DIHYDRO-4,5-DIHYDROXY-9,10-DIOXO-2-ANTHRACENECARBOXYLIC ACID see RHZ700
14,19-DIHYDRO-12,13-DIHYDROXY(13-α,14-α)-20-NORCROTALANAN-11,15-DIONE see MRH000
1,4-DIHYDRO-1,4-DIKETONAPHTHALENE see NBA500
10,11-DIHYDRO-5-(3-(DIMETHYLAMINO)PROPYL)-5H-DIBENZ(b,f)AZEPINE HYDROCHLORIDE see DLH630
2,3-DIHYDRO-2,2-DIMETHYL-7-BENZOFURANYL (BUTYLTHIO)METHYLCARBAMATE see DLH830
2,3-DIHYDRO-2,2-DIMETHYL-7-BENZOFURANYL((4-FLUOROPHENYL)THIO)METHYLCARBAMATE see DLH850
2,3-DIHYDRO-2,2-DIMETHYL-7-BENZOFURANYL METHYLCARBAMATE see CBS275
2,3-DIHYDRO-2,2-DIMETHYLBENZOFURANYL-7-N-METHYLCARBAMATE see CBS275
2,3-DIHYDRO-2,2-DIMETHYLBENZOFURANYL-7-(METHYL)(2,6-DIMETHYLPHENOXYSULFINYL)CARBAMATE see XWS100
2,3-DIHYDRO-2,2-DIMETHYL-7-BENZOFURANYLMETHYL((4-NITROPHENYL)THIO)CARBAMATE see DLH870
2,3-DIHYDRO-2,2-DIMETHYL-7-BENZOFURANYLMETHYL(1,1,2,2-TETRACHLOROETHYL)THIOCARBAMATE see DLH890
10,11-DIHYDRO-N,N-DIMETHYL-5H-DIBENZ(b,f)AZEPINE-5-PROPANAMINE MONOHYDROCHLORIDE see DLH630
10,11-DIHYDRO-N,N-DIMETHYL-5H-DIBENZO(a,d)-CYCLOHEPTENE-Δ^{5,γ}-PROPYLAMINE HCL see EAI000
1,2-DIHYDRO-1,5-DIMETHYL-4-((1-METHYLETHYL)AMINO)-2-PHENYL-3H-PYRAZOL-3-ONE see INY000
2,5-DIHYDRO-1,2-DIMETHYL-3-(2-NAPHTHALENYL)-1H-PYRROLE, REL-(2R,3R)-2,3-DIHYDROXYBUTANEDIOATE (1:1) see DLI630
3,4-DIHYDRO-2,2-DIMETHYL-4-OXO-2H-PYRAN-6-CARBOXYLIC ACID-n-BUTYL ESTER see BRT000
3,7-DIHYDRO-1,3-DIMETHYL-1H-PURINE-2,6-DIONE see TEP000
3,7-DIHYDRO-3,7-DIMETHYL-1H-PURINE-2,6-DIONE see TEO500
DIHYDRO-2,2-DIOCTYL-6H-1,3,2-OXATHIASTANNIN-6-ONE see DVN800
1,3-DIHYDRO-1,3-DIOXO-5-ISOBENZOFURANCARBOXYLIC ACID see TKV000
5,6-DIHYDRO-DIPYRIDO(1,2a;2,1c)PYRAZINIUM DIBROMIDE see DWX800
DIHYDROESTRIN BENZOATE see EDP000
1,2-DIHYDRO-6-ETHOXY-2,2,4-TRIMETHYLQUINOLINE see SAV000
1,4-DIHYDRO-1-ETHYL-7-METHYL-4-OXO-1,8-NAPHTHYRIDINE-3-CARBOXYLIC ACID see EID000
DIHYDROFOLLICULAR HORMONE see EDO000
DIHYDROFOLLICULIN see EDO000
DIHYDROFOLLICULIN BENZOATE see EDP000
DIHYDRO-2,5-FURANDIONE see SNC000
2,5-DIHYDROFURAN-2,5-DIONE see MAM000
DIHYDRO-2(3H)-FURANONE see BOV000
DIHYDROGEN DIOXIDE see HIB050
DIHYDROGEN HEXACHLOROPLATINATE see CKO750

DIHYDROGEN HEXACHLOROPLATINATE(2-) see CKO750

DIHYDROGEN HEXACHLOROPLATINATE HEXAHYDRATE see DLO400

(DIHYDROGEN MERCAPTOSUCCINATO)GOLD DISODIUM SALT see GJC000

DIHYDROHYDROXYCODEINONE see PCG500

DIHYDRO-14-HYDROXYCODEINONE see PCG500

3a,12c-DIHYDRO-8-HYDROXY-6-METHOXY-7H-FURO(3',2' :4,5)FURO(2,3-C)XANTHEN-7-ONE see SLP000

4,5-DIHYDRO-2-HYDROXYMETHYLENE-17-α-METHYLTESTOSTERONE see PAN100

(R)-2,3-DIHYDRO-1-HYDROXY-1H-PYRROLIZINE-7-METHANOL see DAL400

4,5-DIHYDROIMIDAZOLE-2(3H)-THIONE see IAQ000

1,6-DIHYDRO-6-IMINOPURINE see AEH000

3,6-DIHYDRO-6-IMINOPURINE see AEH000

DIHYDRO-ISOJASMONE see HFO700

12,β,13,α-DIHYDROJERVINE see DLQ800

1,2-DIHYDRO-2-KETOBENZISOSULFONAZOLE see BCE500

1,2-DIHYDRO-2-KETOBENZISOSULPHONAZOLE see BCE500

13,14-DIHYDRO-15-KETO-20-ETHYL-PGF2 see IRR050

DIHYDROMENFORMON see EDO000

1,2-DIHYDRO-4-METHOXY-1-METHYL-2-OXONICOTINONITRILE see RJK100

S-(2,3-DIHYDRO-5-METHOXY-2-OXO-1,3,4-THIADIAZOL-3-METHYL) see DSO000

10,11-DIHYDRO-5-(3-(METHYLAMINO)PROPYL)-5H-DIBENZ(b,f)AZEPINE HYDROCHLORIDE see DLS600

1,2-DIHYDRO-3-METHYL-BENZ(j)ACEANTHRYLENE see MIJ750

5,6-DIHYDRO-2-METHYL-3-CARBOXANILIDO-1,4-OXATHIIN (GERMAN) see CCC500

15,16-DIHYDRO-11-METHYLCYCLOPENTA(a)PHENANTHREN-17-ONE see MJE500

15,16-DIHYDRO-11-METHYL-17H-CYCLOPENTA(a)PHENANTHREN-17-ONE see MJE500

4,5-DIHYDRO-N-(1-METHYLETHYL)-3-(PHENYLAMINO)-2H-BENZ(g)INDAZOLE-2-ACETAMIDE see DLU650

2,3-DIHYDRO-6-METHYL-1,4-OXATHIIN-5-CARBOXANILIDE see CCC500

5,6-DIHYDRO-2-METHYL-1,4-OXATHIIN-3-CARBOXANILIDE see CCC500

5,6-DIHYDRO-2-METHYL-N-PHENYL-1,4-OXATHIIN-3-CARBOXAMIDE see CCC500

(S)-(+)-5,6-DIHYDRO-6-METHYL-2H-PYRAN-2-ONE see PAJ500

2,3-DIHYDRO-6-METHYL-2-THIOXO-4(1H)-PYRIMIDINONE see MPW500

DIHYDROMORPHINONE HYDROCHLORIDE see DNU300

DIHYDROMYRCENE see CMT050

DIHYDRONEOPINE see DKW800

1,2-DIHYDRO-5-NITRO-ACENAPHTHYLENE see NEJ500

1,3-DIHYDRO-7-NITRO-5-PHENYL-2H-1,4-BENZODIAZEPIN-2-ONE see DLY000

DIHYDRONORGUAIARETIC ACID see NBR000

DIHYDROOXIRENE see EJN500

2,3-DIHYDRO-3-OXOBENZISOSULFONAZOLE see BCE500

2,3-DIHYDRO-3-OXOBENZISOSULPHONAZOLE see BCE500

3,4-DIHYDRO-4-OXO-3-BENZOTRIAZINYLMETHYL O,O-DIETHYL PHOSPHORODITHIOATE see EKN000

S-(3,4-DIHYDRO-4-OXO-1,2,3-BENZOTRIAZIN-3-YLMETHYL) O,O-DIETHYL PHOSPHORODITHIOATE see EKN000

S-(3,4-DIHYDRO-4-OXO-1,2,3-BENZOTRIAZIN-3-YLMETHYL)-O,O-DIMETHYL PHOSPHORODITHIOATE see ASH500

S-(3,4-DIHYDRO-4-OXO-BENZO(α)(1,2,3)TRIAZIN-3-YLMETHYL)-O,O-DIMETHYL PHOSPHORODITHIOATE see ASH500

O-(1,6)-(DIHYDRO-6-OXO-1-PHENYLPYRIDAZIN-3-YL), O,O-DIETHYL PHOSPHOROTHIOATE see POP000

4-((4,5-DIHYDRO-3-(PHENYLAMINO)-2H-BENZ(G)INDAZOL-2-YL)ACETYL)MORPHOLINE see DMA500

2,3-DIHYDRO-6-PROPYL-2-THIOXO-4(1H)-PYRIMIDINONE see PNX000

DIHYDROPSEUDOIONONE see GDE400

α-β-DIHYDROPSEUDOIONONE see GDE400

1,7-DIHYDRO-6H-PURINE-6-THIONE see POK000

1,7-DIHYDRO-6H-PURIN-6-ONE see DMC000

1,2-DIHYDRO-3,6-PYRADIZINEDIONE see DMC600

DIHYDROPYRAN see DMC200

3,4-DIHYDROPYRAN see DMC200

2H-3,4-DIHYDROPYRAN see DMC200

Δ²-DIHYDROPYRAN see DMC200

3,4-DIHYDRO-2H-PYRAN-2-CARBOXALDEHYDE see ADR500

2,3-DIHYDRO-1,4-PYRAN-2-KARBOXALDEHYD see ADR500

1,5-DIHYDRO-4H-PYRAZOLO(3,4-d)PYRIMIDIN-4-ONE see ZVJ000

1,2-DIHYDROPYRIDAZINE-3,6-DIONE see DMC600

1,2-DIHYDRO-3,6-PYRIDAZINEDIONE see DMC600

6,7-DIHYDROPYRIDO(1,2a;2',1'-c)PYRAZINEDIUM DIBROMIDE see DWX800

DIHYDROSAFROLE see DMD600

DIHYDROSTREPTOMYCIN see DME000

2,3-DIHYDROSUCCINIC ACID see TAF750

6,7-DIHYDRO-1,2,3,10-TETRAMETHOXY-7-(METHYLAMINO)-BENZO(α)HEPTALEN-9 (5H)-ONE see MIW500

DIHYDROTHEELIN see EDO000

N-(5,6-DIHYDRO-4H-1,3-THIAZINYL)-2,6-XYLIDINE see DMW000

2,3-DIHYDROTHIIRENE see EJP500

2,3-DIHYDRO-2-THIOXO-4(1H)-PYRIMIDINONE see TFR250

1,2-DIHYDRO-2,2,4-TRIMETHYL-6-ETHOXYQUINOLINE see SAV000

3,7-DIHYDRO-1,3,7-TRIMETHYL-1H-PURINE-2,6-DIONE see CAK500

1,2-DIHYDRO-2,2,4-TRIMETHYLQUINOLINE see TLP500

2,4-DIHYDROXYACETOPHENONE see DMG400

2',4'-DIHYDROXYACETOPHENONE see DMG400

1,8-DIHYDROXY-9,10-ANTHRACENEDIONE see DMH400

1,4-DIHYDROXYANTHRACHINON (CZECH) see DMH000

1,8-DIHYDROXYANTHRACHINON (CZECH) see DMH400

1,4-DIHYDROXYANTHRAQUINONE see DMH000

1,8-DIHYDROXYANTHRAQUINONE see DMH400
1,4-DIHYDROXY-9,10-ANTHRAQUINONE see
DMH000
3,4-DIHYDROXYBENZALDEHYDE METHYLENE
KETAL see PIW250
1,4-DIHYDROXY-BENZEEN (DUTCH) see HIH000
1,4-DIHYDROXYBENZEN (CZECH) see HIH000
DIHYDROXYBENZENE see HIH000
m-DIHYDROXYBENZENE see REA000
o-DIHYDROXYBENZENE see CCP850
p-DIHYDROXYBENZENE see HIH000
1,2-DIHYDROXYBENZENE see CCP850
1,3-DIHYDROXYBENZENE see REA000
1,4-DIHYDROXYBENZENE see HIH000
1,3-DIHYDROXYBENZENE DIACETATE see REA100
DIHYDROXYBENZENE (OSHA) see HIH000
3,3'-DIHYDROXYBENZIDINE see DMI400
2,4-DIHYDROXYBENZOFENON (CZECH) see DMI600
2,4-DIHYDROXYBENZOIC ACID see HOE600
2,5-DIHYDROXYBENZOIC ACID see GCU400
1,4-DIHYDROXY-BENZOL (GERMAN) see HIH000
2,4-DIHYDROXYBENZOPHENONE see DMI600
2,6-DIHYDROXY-5-BIS(2-
CHLOROETHYL)AMINOPYRAMIDINE see BIA250
1,3-DIHYDROXYBUTANE see BOS500
1,4-DIHYDROXYBUTANE see BOS750
2,3-DIHYDROXYBUTANE see BOT000
2,3-DIHYDROXYBUTANEDIOIC ACID see TAF750
2,3-DIHYDROXYBUTANEDIOIC ACID,
DIAMMONIUM SALT see DCH000
DIHYDROXYCHLOROTHIAZIDUM see CFY000
DIHYDROXY 3-12 CHOLANATE de Na (FRENCH) see
SGE000
3,12-DIHYDROXYCHOLANIC ACID see DAQ400
3-α,12-α-DIHYDROXYCHOLANIC ACID see DAQ400
3-α,12-α-DIHYDROXY-5-β-CHOLANOIC ACID see
DAQ400
3-α,12-α-DIHYDROXY-5-β-CHOLAN-24-OIC ACID see
DAQ400
(3-α,5-β,12-α)-3,12-DIHYDROXY-CHOLAN-24-OIC
ACID MONOSODIUM SALT see SGE000
3-α,12-α-DIHYDROXY-5-β-CHOLAN-24-OIC ACID
SODIUM SALT see SGE000
3-α,12-α-DIHYDROXYCHOLANSAEURE (GERMAN)
see DAQ400
DI-(4-HYDROXY-3-COUMARINYL)METHANE see
BJZ000
1,5-DIHYDROXY-4,8-DIAMINOANTHRACHINON
(CZECH) see DBP909
1,5-DIHYDROXY-4,8-DIAMINOANTHRAQUINONE
see DBP909
2,2'-DIHYDROXY-5,5'-
DICHLORODIPHENYLMETHANE see MJM500
2,2'-DIHYDROXYDIETHYLAMINE see DHF000
DIHYDROXYDIETHYL ETHER see DJD600
β,β'-DIHYDROXYDIETHYL ETHER see DJD600
4,4'-DIHYDROXYDIETHYLSTILBENE see DKA600
4,4'-DIHYDROXY-α,β-DIETHYLSTILBENE see
DKA600
DIHYDROXYDIETHYLSTILBENE DIPROPIONATE
see DKB000
4,4'-DIHYDROXY-α,β-DIETHYLSTILBENE
DIPROPIONATE see DKB000
d-7',12'-DIHYDROXY-6,6'-DIMETHOXY-2,2',2'-
TRIMETHYLTUBOCURARANIUM CHLORIDE see
TOA000

N-(2,4-DIHYDROXY-3,3-DIMETHYLBUTYRYL)-β-
ALANINE CALCIUM see CAU750
4,4'-DIHYDROXYDIPHENYLDIMETHYLMETHANE
see BLD500
p,p'-DIHYDROXYDIPHENYLDIMETHYLMETHANE
see BLD500
4,4'-DIHYDROXYDIPHENYLDIMETHYLMETHANE
DIGLYCIDYL ETHER see BLD750
p,p'-DIHYDROXYDIPHENYLDIMETHYLMETHANE
DIGLYCIDYL ETHER see BLD750
4,4'-DIHYDROXYDIPHENYLPROPANE see BLD500
p,p'-DIHYDROXYDIPHENYLPROPANE see BLD500
2,2-(4,4'-DIHYDROXYDIPHENYL)PROPANE see
BLD500
4,4'-DIHYDROXYDIPHENYL-2,2-PROPANE see
BLD500
4,4'-DIHYDROXY-2,2-DIPHENYLPROPANE see
BLD500
2,2'-DIHYDROXY-DI-n-PROPYLNITROSOAMINE see
DNB200
(+/-)-3α,4β-DIHYDROXY-1α,2α-EPOXY-1,2,3,4-
TETRAHYDRODIBEN-Z(a,h)ACRIDINE see DCS821
(+)-(1R,2S,3S,4R)-3,4-DIHYDROXY-1,2-EPOXY-1,2,3,4-
TETRAHYDRODIBENZ(c,h)ACRIDINE see DMS410
(1R,2S,3S,4R)-3,4-DIHYDROXY-1,2-EPOXY-1,2,3,4-
TETRAHYDRODIBENZ(c,h)ACRIDINE see DMS410
3,17-β-DIHYDROXYESTRA-1,3,5(10)-TRIENE see
EDO000
3,17-β-DIHYDROXY-1,3,5(10)-ESTRATRIENE see
EDO000
DIHYDROXYESTRIN see EDO000
1,2-DIHYDROXYETHANE see EJC500
DI-β-HYDROXYETHOXYETHANE see TJQ000
DI(2-HYDROXYETHYL)AMINE see DHF000
2,2'-DIHYDROXYETHYL ETHER see DJD600
DI(HYDROXYETHYL) ETHER DINITRATE see
DJE400
3,17-β-DIHYDROXY-17-α-ETHYNYL-1,3,5(10)-
ESTRATRIENE see EEH500
3,17-β-DIHYDROXY-17-α-ETHYNYL-1,3,5(10)-
OESTRATRIENE see EEH500
3',6'-DIHYDROXYFLUORAN see FEV000
DIHYDROXYFLUORANE see FEV000
2,2'-DIHYDROXY-3,3',5,5',6,6'-
HEXACHLORODIPHENYLMETHANE see HCL000
2,2'-DIHYDROXY-3,5,6,3',5',6'-
HEXACHLORODIPHENYLMETHANE see HCL000
7-(3,5-DIHYDROXY-2-(3-HYDROXY-1-
OCTENYL)CYCLOPENTYL)-5-HEPTENOIC ACID see
POC500
7-(3,5-DIHYDROXY-2-(3-HYDROXY-1-
OCTENYL)CYCLOPENTYL)-5-HEPTENOIC ACID,
THAM see POC750
7-(3,5-DIHYDROXY-2-(3-HYDROXY-1-
OCTENYL)CYCLOPENTYL)-5-HEPTENOIC ACID,
TRIMETHAMINE SALT see POC750
d-(+)-2,4-DIHYDROXY-N-(3-HYDROXYPROPYL)-3,3-
DIMETHYLBUTYRAMIDE see PAG200
3,4-DIHYDROXY-α-
((ISOPROPYLAMINO)METHYL)BENZYL ALCOHOL
see DMV600
β,β'-DIHYDROXYISOPROPYL CHLORIDE see CDT750
5,6-DIHYDRO-2-(2,6-XYLIDINO)-4H-1,3-THIAZINE
see DMW000

3,4-DIHYDROXY-α-
((METHYLAMINO)METHYL)BENZYL ALCOHOL see
VGP000
3-DI(HYDROXYMETHYL)AMINO-6-(5-NITRO-2-
FURYLETHENYL)-1,2,4-TRIAZINE see BKH500
3-DI(HYDROXYMETHYL)AMINO-6-(2-(5-NITRO-2-
FURYL)VINYL)-1,2,4-TRIAZINE see BKH500
1,3-DIHYDROXY-5-METHYLBENZENE see MPH500
2,12-DIHYDROXY-4-METHYL-11,16-
DIOXOSENECIONANIUM see DMX200
DI-4-HYDROXY-3,3'-METHYLENEDICOUMARIN see
BJZ000
DIHYDROXYMETHYL FURATRIZINE see BKH500
2,4-DIHYDROXY-2-METHYLPENTANE see HFP875
d-threo-N-(1,1'-DIHYDROXY-1-p-
NITROPHENYLISOPROPYL)DICHLOROACETAMID
E see CDP250
2,2'-DIHYDROXY-N-NITROSODIETHYLAMINE see
NKM000
3,17-β-DIHYDROXYOESTRA-1,3,5-TRIENE see
EDO000
3,17-β-DIHYDROXY-1,3,5(10)-OESTRATRIENE see
EDO000
DIHYDROXYOESTRIN see EDO000
(5Z,11-α,13E,15S)-11,15-DIHYDROXY-9-OXOPROSTA-
5,13-DIEN-1-OIC ACID see DVJ200
3,5-DIHYDROXYPHENOL see PGR000
DIHYDROXY-l-PHENYLALANINE see DNA200
3,4-DIHYDROXYPHENYLALANINE see DNA200
(−)-3,4-DIHYDROXYPHENYLALANINE see DNA200
l-DIHYDROXYPHENYL-l-ALANINE see DNA200
3,4-DIHYDROXYPHENYL-l-ALANINE see DNA200
3,4-DIHYDROXY-l-PHENYLALANINE see DNA200
l-3,4-DIHYDROXYPHENYLALANINE see DNA200
(−)-3-(3,4-DIHYDROXYPHENYL)-l-ALANINE see
DNA200
3-(3,4-DIHYDROXYPHENYL)-l-ALANINE see DNA200
l-α-DIHYDROXYPHENYLALANINE see DNA200
β-(3,4-DIHYDROXYPHENYL)-l-ALANINE see DNA200
l-β-(3,4-DIHYDROXYPHENYL)ALANINE see DNA200
l-3,4-DIHYDROXYPHENYL-α-ALANINE see DNA200
β-(3,4-DIHYDROXYPHENYL)-α-ALANINE see
DNA200
l-1-(3,4-DIHYDROXYPHENYL)-2-AMINOETHANOL
see NNO500
l-3,4-DIHYDROXYPHENYLETHANOLAMINE see
NNO500
DIHYDROXYPHENYLETHANOLISOPROPYLAMINE
see DMV600
1-(2,4-DIHYDROXYPHENYL)ETHANONE see
DMG400
3,4'(4,4'-DIHYDROXYPHENYL)HEX-3-ENE see
DKA600
7-(3-(2-(3,5-DIHYDROXYPHENYL-2-HYDROXY-
ETHYLAMINO)PROPYL))THEOPHYLLINE
HYDROCHLORIDE see DNA600
1-(3,4-DIHYDROXYPHENYL)-2-
ISOPROPYLAMINOETHANOL see DMV600
l-(−)-3-(3,4-DIHYDROXYPHENYL)-2-
METHYLALANINE see DNA800
l(−)-β-(3,4-DIHYDROXYPHENYL)-α-
METHYLALANINE see DNA800
1-1-(3,4-DIHYDROXYPHENYL)-2-
METHYLAMINOETHANOL see VGP000
α-DI(p-HYDROXYPHENYL)PHTHALIDE see PDO750
2,2-DI(4-HYDROXYPHENYL)PROPANE see BLD500

β-DI-p-HYDROXYPHENYLPROPANE see BLD500
2-(3,4-DIHYDROXYPHENYL)-3,5,7-TRIHYDROXY-4H-
1-BENZOPYRAN-4-ONE see QCA000
DIHYDROXYPHTHALOPHENONE see PDO750
17,21-DIHYDROXYPREGNA-1,4-DIENE-3,11,20-
TRIONE see PLZ000
17-α,21-DIHYDROXY-14-α-PREGN-4-ENE-3,20-
DIONE 21-IODOACETATE see DNA850
1,2-DIHYDROXYPROPANE see PML000
17R,21-α-DIHYDROXY-4-PROPYLAJMALANIUM
HYDROGEN TARTRATE see DNB000
DI(2-HYDROXY-n-PROPYL)AMINE see DNB200
2,3-DIHYDROXYPROPYL CHLORIDE see CDT750
N,N-DI-(2-HYDROXYPROPYL)NITROSAMINE see
DNB200
2,4-DIHYDROXYPYRIMIDINE see UNJ800
2,3-DIHYDROXY-(R-(R*,R*))-BUTANEDIOIC ACID
DISODIUM SALT (9CI) see BLC000
8,8'-DIHYDROXY-RUGULOSIN see LIV000
12,18-DIHYDROXY-SENECIONAN-11,16-DIONE see
RFP000
3',6'-DIHYDROXYSPIRO(ISOBENZOFURAN-
1(3H),9'(9H)-XANTHEN)-3-ONE see FEV000
2,3-DIHYDROXYTOLUENE see DNE000
3,5-DIHYDROXYTOLUENE see MPH500
α-2-DIHYDROXYTOLUENE see HMK100
1,3-DIHYDROXY-2,4,6-TRINITROBENZENE see
SMP500
2,4-DIHYDROXY-1,3,5-TRINITROBENZENE see
SMP500
6-(6,10-DIHYDROXYUNDECYL)-β-RESORCYLIC
ACID-μ-LACTONE see RBF100
1-α,24(R)-DIHYDROXYVITAMIN D3 see SBL600
1,4-DIIDROBENZENE (ITALIAN) see HIH000
DIIDRO-5,5-DIETIL-2H-1,3-OSSAZIN-2,4(3H)-DIONE
(ITALIAN) see DJT400
1,3-DIIMINOISOINDOLIN (CZECH) see DNE400
1,3-DIIMINOISOINDOLINE see DNE400
DIIODOACETYLENE see DNE500
DIIODOETHYNE see DNE500
DIIODO-3,3 HYDROXY-4 BENZOYL 2 FURANNE see
DNF500
3,5-DIIODO-4-HYDROXYPHENYL 2,5-DIMETHYL-3-
FURYL KETONE see DNF450
3,5-DIIODO-4-HYDROXYPHENYL 2-ETHYL-3-
BENZOFURANYL KETONE see EID200
3,5-DIIODO-4-HYDROXYPHENYL 5-ETHYL-2-FURYL
KETONE see EID250
3,5-DIIODO-4-HYDROXYPHENYL 2-FURYL
KETONE see DNF500
DIIODOHYDROXYQUIN see DNF600
DIIODOHYDROXYQUIN see DNF600
DIIODOHYDROXYQUINOLINE see DNF600
5,7-DIIODO-8-HYDROXYQUINOLINE see DNF600
DIIODOMETHYLARSINE see MGQ775
2,6-DIIODO-4-NITROPHENOL see DNG000
5,7-DIIODO-OXINE see DNF600
DIIODOQUIN see DNF600
5,7-DIIODO-8-QUINOLINOL see DNF600
DIIRON TRISULFATE see FBA000
DIISOBUTILCHETONE (ITALIAN) see DNI800
DIISOBUTYL ADIPATE see DNH125
DIISOBUTYLALUMINIUM HYDRIDE see DNI600
DIISOBUTYLALUMINUM CHLORIDE see CGB500
DIISOBUTYLALUMINUM HYDRIDE see DNI600
DIISOBUTYLALUMINUM MONOCHLORIDE see
CGB500

DIISOBUTYLAMINE see DNH400
DIISOBUTYL CARBINOL see DNH800
DI-ISOBUTYLCETONE (FRENCH) see DNI800
DIISOBUTYLCHLOROALUMINUM see CGB500
DIISOBUTYLESTER KYSELINY FTALOVE see DNJ400
DIISOBUTYLHYDROALUMINUM see DNI600
DIISOBUTYLKETON (DUTCH, GERMAN) see DNI800
DIISOBUTYL KETONE see DNI800
DIISOBUTYLOXOSTANNANE see DNJ000
DIISOBUTYLPHENOXYETHOXYETHYLDIMETHYL
BENZYL AMMONIUM CHLORIDE see BEN000
DIISOBUTYL PHTHALATE see DNJ400
DIISOBUTYLTIN OXIDE see DNJ000
DIISOBUTYRYL PEROXIDE see DNJ600
4-4'-DIISOCYANATE de DIPHENYLMETHANE
(FRENCH) see MJP400
DI-ISOCYANATE de TOLUYLENE see TGM750
1,3-DIISOCYANATOBENZENE see BBP000
4,4'-DIISOCYANATO-3,3'-DIMETHOXY-1,1'-
BIPHENYL see DCJ400
4,4'-DIISOCYANATO-3,3'-DIMETHYL-1,1'-BIPHENYL
see DQS000
4,4'-DIISOCYANATODIPHENYLMETHANE see
MJP400
1,6-DIISOCYANATOHEXANE see DNJ800
DI-ISO-CYANATOLUENE see TGM750
DIISOCYANATOMETHYLBENZENE see TGM740
2,6-DIISOCYANATO-1-METHYLBENZENE see
TGM800
2,4-DIISOCYANATO-1-METHYLBENZENE (9CI) see
TGM750
1,5-DIISOCYANATONAPHTHALENE see NAM500
DIISOCYANATOTOLUENE see TGM740
2,4-DIISOCYANATOTOLUENE see TGM750
2,6-DIISOCYANATOTOLUENE see TGM800
DIISOCYANAT-TOLUOL see TGM750
2,3-DIISONITROSOBUTANE see DBH000
DIISOOCTYL ACID PHOSPHATE see DNK800
DIISOOCTYL
((DIOCTYLSTANNYLENE)DITHIO)DIACETATE see
BKK750
DIISOOCTYL PHOSPHATE (DOT) see DNK800
DIISOOCTYL PHTHALATE see ILR100
DIISOPHENOL see DNG000
DIISOPROPANOLNITROSAMINE see DNB200
N,N'-DIISOPROPIL-FOSFORODIAMMIDO-
FLUORURO (ITALIAN) see PHF750
DIISOPROPOXYPHOSPHORYL FLUORIDE see
IRF000
s-DIISOPROPYLACETONE see DNI800
DI(ISOPROPYLAMIDO)PHOSPHORYLFLUORIDE see
PHF750
DIISOPROPYLAMINE see DNM200
2-DIISOPROPYLAMINOETHANOL see DNP000
2-(DIISOPROPYLAMINO)ETHYL CHLORIDE
HYDROCHLORIDE see CGV600
S-(2-DIISOPROPYLAMINOETHYL)-O-ETHYL
METHYL PHOSPHONOTHIOLATE see EIG000
DIISOPROPYLBENZENE HYDROPEROXIDE, not
more than 72% in solution (DOT) see DNS000
DIISOPROPYLBENZENE PEROXIDE see DGR600
DIISOPROPYLBERYLLIUM see DNO200
N,N'-DIISOPROPYL-DIAMIDO-FOSFORZUUR-
FLUORIDE (DUTCH) see PHF750
N,N'-DIISOPROPYL-DIAMIDO-PHOSPHORSAEURE-
FLUORID (GERMAN) see PHF750
N,N'-DIISOPROPYLDIAMIDOPHOSPHORYL
FLUORIDE see PHF750

DIISOPROPYL ETHANOLAMINE see DNP000
N,N-DIISOPROPYL ETHANOLAMINE see DNP000
DIISOPROPYL ETHER see IOZ750
DIISOPROPYL FLUOROPHOSPHATE see IRF000
O,O-DIISOPROPYL FLUOROPHOSPHATE see IRF000
DIISOPROPYL FLUOROPHOSPHONATE see IRF000
DIISOPROPYLFLUOROPHOSPHORIC ACID ESTER
see IRF000
DIISOPROPYLFLUORPHOSPHORSAEUREESTER
(GERMAN) see IRF000
DIISOPROPYLIDENE ACETONE see PGW250
sym-DIISOPROPYLIDENE ACETONE see PGW250
DIISOPROPYLMERCURY see DNQ800
DIISOPROPYLNITROSAMIN (GERMAN) see NKA000
DIISOPROPYL OXIDE see IOZ750
DIISOPROPYLOXOSTANNANE see DNR200
DIISOPROPYL PERDICARBONATE see DNR400
DIISOPROPYL PEROXYDICARBONATE see DNR400
2,6-DIISOPROPYLPHENOL see DNR800
DIISOPROPYLPHENYLHYDROPEROXIDE (solution)
see DNS000
DIISOPROPYL PHOSPHOFLUORIDATE see IRF000
N,N'-DIISOPROPYLPHOSPHORODIAMIDIC
FLUORIDE see PHF750
DIISOPROPYL PHOSPHOROFLUORIDATE see IRF000
O,O'-DIISOPROPYL PHOSPHORYL FLUORIDE see
IRF000
DI-ISOPROPYLTHIOLOCARBAMATE de S-(2,3-
DICHLOROALLYLE) (FRENCH) see DBI200
DIISOPROPYLTIN DICHLORIDE see DNT000
DIISOPROPYLTIN OXIDE see DNR200
1,4-DIISOTHIOCYANATOBENZENE see PFA500
DIKETENE see KFA000
DIKETENE, inhibited (DOT) see KFA000
2,3-DIKETOBUTANE see BOT500
4,4'-DIKETO-β-CAROTENE see CBE800
2,5-DIKETOHEXANE see HEQ500
1,3-DIKETOHYDRINDENE see IBS000
2,3-DIKETOINDOLINE see ICR000
DIKETONE ALCOHOL see DBF750
2,5-DIKETOTETRAHYDROFURAN see SNC000
DIKONIT see SGG500
DIKOTEKS see SIL500
DIKOTEX 30 see SIL500
DILAFURANE see EID200
DILANGIL see MAW250
DILANTIN see DKQ000
DILANTIN see DNU000
DILANTIN DB see DEP600
DILANTINE see DKQ000
DILANTIN SODIUM see DNU000
DILATIN DB see DEP600
DILAUDID see DNU300
DILAUDID HYDROCHLORIDE see DNU300
DILAUROYL PEROXIDE see LBR000
DILAUROYL PEROXIDE, TECHNICAL PURE (DOT)
see LBR000
DILAURYLESTER KYSELINY β',β'-
THIODIPROPIONOVE see TFD500
DILAURYL THIODIPROPIONATE see TFD500
DILAURYL 3,3'-THIODIPROPIONATE see TFD500
DILAURYL β-THIODIPROPIONATE see TFD500
DILAURYL β',β'-THIODIPROPIONATE see TFD500
DILA-VASAL see EID200
DILEAD(II) LEAD(IV) OXIDE see LDS000
DI-LEN see DNU000
DILENE see BIM500

DILIC see HKC000
DILITHIUM N-ACETYL-l-ASPARTATE see DNU330
DILITHIUM CARBONATE see LGZ000
DILL FRUIT OIL see DNU400
DILL HERB OIL see DNU400
DILL HERB OIL, AMERICAN TYPE see DNU390
DILL OIL see DNU390
DILL OIL see DNU400
DILL SEED OIL see DNU400
DILL SEED OIL, AMERICAN TYPE see DNU390
DILL SEED OIL, EUROPEAN TYPE see DNU400
DILL WEED OIL see DNU400
DILOMBRIN see DJT800
DILOSPAN S see PGR000
DILUEX see PAE750
DILURAN see AAI250
1,3-DIMALEIMIDOBENZENE see BKL750
DIMANGANESE TRIOXIDE see MAT500
DIMANIN C see SGG500
DIMAPP see DQA400
DIMAPYRIN see DOT000
DIMAS see DQD400
DIMATE 267 see DSP400
DIMAZINE see DSF400
DIMEDROL see BBV500
DIMEDRYL see BBV500
DIMEFOX see BJE750
DIMEHYPO see DXC900
DIMEHYPO JUMBO see DXC900
DIMELIN see ABB000
DIMELOR see ABB000
DIMEMORFAN PHOSPHATE see MLP250
DIMENFORMON see EDO000
DIMENFORMON BENZOATE see EDP000
DIMENFORMON DIPROPIONATE see EDR000
DIMENFORMONE see EDP000
DIMENFORMON PROLONGATUM see EDO000
DIMENHYDRINATE see DYE600
DIMERCAPROL PROPANOL see BAD750
1,2-DIMERCAPTOETHANE see EEB000
DIMERCAPTOL see BAD750
2,3-DIMERCAPTOL-1-PROPANOL see BAD750
DIMERCAPTOPROPANOL see BAD750
2,3-DIMERCAPTOPROPANOL see BAD750
2,3-DIMERCAPTOPROPAN-1-OL see BAD750
DIMERCUROUS METHANE ARSONATE see DNW000
DIMER CYKLOPENTADIENU (CZECH) see DGW000
1,6-DIMESYL-d-MANNITOL see BKM500
1,4-DIMESYLOXYBUTANE see BOT250
DIMETACRINE BITARTRATE see DRM000
DIMETACRIN HYDROGENTARTRATE see DRM000
DIMETATE see DSP400
DIMETAZINA see SNN300
DIMETHACRINE TARTRATE see DRM000
DIMETHADIONE see PMO250
1,6-DIMETHANESULFONATE-d-MANNITOL see BKM500
1,4-DIMETHANESULFONATE THREITOL see TFU500
(2s,3s)-1,4-DIMETHANESULFONATE TREITOL see TFU500
1,4-DIMETHANESULFONOXYBUTANE see BOT250
1,6-DIMETHANE-SULFONOXY-d-MANNITOL see BKM500
1,4-DI(METHANESULFONYLOXY)BUTANE see BOT250
1,6-DIMETHANESULPHONOXY-1,6-DIDEOXY-d-MANNITOL see BKM500

1,4-DIMETHANESULPHONYLOXYBUTANE see BOT250
DIMETHISTERONE and ETHINYL ESTRADIOL see DNX500
DIMETHOAAT (DUTCH) see DSP400
DIMETHOAT (GERMAN) see DSP400
DIMETHOATE (USDA) see DSP400
DIMETHOAT TECHNISCH 95% see DSP400
DIMETHOGEN see DSP400
1,2-DIMETHOXY-4-ALLYLBENZENE see AGE250
2,6-DIMETHOXY-4-(p-AMINOBENZENESULFONAMIDO)PYRIMIDINE see SNN300
(trans)-2,5-DIMETHOXY-4'-AMINOSTILBENE see DON400
2,3-DIMETHOXYBENZALDEHYDE see DNZ200
2,5-DIMETHOXYBENZENEAZO-β-NAPHTHOL see DOK200
3,3'-DIMETHOXYBENZIDIN (CZECH) see DCJ200
3,3'-DIMETHOXYBENZIDINE see DCJ200
3,3'-DIMETHOXYBENZIDINE DIHYDROCHLORIDE see DOA800
3,3'-DIMETHOXYBENZIDINE-4,4'-DIISOCYANATE see DCJ400
6,7-DIMETHOXYBENZOPYRAN-2-ONE see DRS800
3,3-DIMETHOXY-(1,1'-BIPHENYL)-4,4'-DIAMINE DIHYDROCHLORIDE see DOA800
3,3'-DIMETHOXY-4,4'-BIPHENYLENE DIISOCYANATE see DCJ400
6,7-DIMETHOXYCOUMARIN see DRS800
DIMETHOXY-DDT see MEI450
1,1-DIMETHOXYDECANE see AFJ700
10,10-DIMETHOXYDECANE see AFJ700
6,7-DIMETHOXY-2,2-DIMETHYL-2H-BENZO(b)PYRAN see AEX850
p,p'-DIMETHOXYDIPHENYLTRICHLOROETHANE see MEI450
3,4-DIMETHOXYDOPAMINE see DOE200
DIMETHOXY-DT see MEI450
DIMETHOXYETHANE see DOE600
1,2-DIMETHOXYETHANE see DOE600
1,1-DIMETHOXYETHANE (DOT) see DOO600
1,2-DIMETHOXYETHANE (DOT) see DOE600
α,β-DIMETHOXYETHANE see DOE600
(2,2-DIMETHOXYETHYL)-BENZENE (9CI) see PDX000
DIMETHOXY ETHYL PHTHALATE see DOF400
DI(2-METHOXYETHYL)PHTHALATE see DOF400
DIMETHOXYMETHANE see MGA850
DIMETHOXYMETHYLBENZENE see DOG700
DIMETHOXYMETHYLSILANE see MJE900
3,4-DIMETHOXYPHENETHYLAMINE see DOE200
3,4-DIMETHOXY-β-PHENETHYLAMINE see DOE200
4-(2,5-DIMETHOXYPHENETHYL)ANILINE see DON400
2,6-DIMETHOXYPHENOL see DOJ200
1-((2,5-DIMETHOXYPHENYL)AZO)-2-NAPHTHALENOL see DOK200
1-((2,5-DIMETHOXYPHENYL)AZO)-2-NAPHTHOL see DOK200
2,5-DIMETHOXY-1-(PHENYLAZO)-2-NAPHTHOL see DOK200
1-(1-(2,5-DIMETHOXYPHENYL)AZO)-2-NAPHTHOL see DOK200
1,1-DIMETHOXY-2-PHENYLETHANE see PDX000
DIMETHOXYPHENYLETHYLAMINE see DOE200
3,4-DIMETHOXYPHENYLETHYLAMINE see DOE200

2-(3,4-DIMETHOXYPHENYL)ETHYLAMINE see DOE200

3,4-DIMETHOXYPHENYLETHYLAMINE (base) see DOE200

β-(3,4-DIMETHOXYPHENYL)ETHYLAMINE see DOE200

3,4-DIMETHOXY-β-PHENYLETHYLAMINE see DOE200

4-(2-(2,5-DIMETHOXYPHENYL)ETHYL)BENZENAMINE see DON400

β-(2,5-DIMETHOXYPHENYL)-β-HYDROXYISOPROPYLAMINE HYDROCHLORIDE see MDW000

DIMETHOXYPHENYLMETHANE see DOG700

1-((3,4-DIMETHOXYPHENYL)METHYL)-6,7-DIMETHOXYISOQUINOLINE see PAH000

1,1-DIMETHOXY-2-PHENYLPROPANE see HII600

1-(3,4-DIMETHOXYPHENYL)-2-PROPENE see AGE250

2,2-DI-(p-METHOXYPHENYL)-1,1,1-TRICHLOROETHANE see MEI450

DI(p-METHOXYPHENYL)-TRICHLOROMETHYL METHANE see MEI450

DIMETHOXYPHOSPHINE OXIDE see DSG600

((DIMETHOXYPHOSPHINOTHIOYL)THIO)BUTANE DIOIC ACID DIETHYL ESTER see MAK700

3-((DIMETHOXYPHOSPHINYL)OXY)-2-BUTENOIC ACID METHYL ESTER see MQR750

3-(DIMETHOXYPHOSPHINYLOXY)-N,N-DIMETHYL-cis-CROTONAMIDE see DGQ875

3-(DIMETHOXYPHOSPHINYLOXY)-N,N-DIMETHYLISOCROTONAMIDE see DGQ875

3-(DIMETHOXYPHOSPHINYLOXY)N-METHYL-cis-CROTONAMIDE see MRH209

DIMETHOXY POLYETHYLENE GLYCOL see DOM100

1,2-DIMETHOXY-4-PROPENYLBENZENE see IKR000

N¹-(2,6-DIMETHOXY-4-PYRIMIDINYL)SULFANILAMIDE see SNN300

4-(2,5-DIMETHOXY)STILBENAMINE see DON400

2',5'-DIMETHOXYSTILBENAMINE see DON400

2,5-DIMETHOXY-4'-STILBENAMINE see DON400

2,3-DIMETHOXYSTRYCHNIDIN-10-ONE see BOL750

2,3-DIMETHOXYSTRYCHNINE see BOL750

DIMETHOXY STRYCHNINE (DOT) see BOL750

DIMETHOXYSULFADIAZINE see SNN300

2,4-DIMETHOXY-6-SULFANILAMIDO-1,3-DIAZINE see SNN300

2,6-DIMETHOXY-4-SULFANILAMIDOPYRIMIDINE see SNN300

DIMETHOXYTETRAETHYLENE GLYCOL see PBO500

DIMETHOXYTETRAGLYCOL see PBO500

DIMETHOXY-2,2,2-TRICHLORO-1-HYDROXY-ETHYL-PHOSPHINE OXIDE see TIQ250

6,7-DIMETHOXY-1-VERATRYLISOQUINOLINE see PAH000

6,7-DIMETHOXY-1-VERATRYLISOQUINOLINE HYDROCHLORIDE see PAH250

DIMETHOXYVIOLANTHRONE see JAT000

16,17-DIMETHOXYVIOLANTHRONE see JAT000

DIMETHYL see EDZ000

DIMETHYLACETAL see DOO600

DIMETHYLACETAMIDE see DOO800

N,N-DIMETHYLACETAMIDE see DOO800

O,O-DIMETHYL-S-(2-ACETAMIDOETHYL) ESTER PHOSPHORODITHIOIC ACID see DOP200

DIMETHYLACETIC ACID see IJU000

DIMETHYLACETONE see DJN750

DIMETHYLACETONE AMIDE see DOO800

DIMETHYLACETONITRILE see IJX000

O,O-DIMETHYL-S-(2-(ACETYLAMINO)ETHYL) DITHIOPHOSPHATE see DOP200

O,O-DIMETHYL-S-(2-ACETYLAMINOETHYL) PHOSPHORODITHIOATE see DOP200

DIMETHYLACETYLENE see COC500

O,S-DIMETHYLACETYLPHOSPHOROAMIDOTHIOATE see DOP600

DIMETHYL ACID PHOSPHITE see DSG600

(E)-2,3-DIMETHYLACRYLIC ACID see TGA700

trans-2,3-DIMETHYLACRYLIC ACID see TGA700

trans-α-β-DIMETHYLACRYLIC ACID see TGA700

DIMETHYL ADIPATE see DOQ300

DIMETHYLAETHANOLAMIN (GERMAN) see DOY800

O,O-DIMETHYL-S-(2-AETHYLSULFINYL-AETHYL)-THIOLPHOSPHAT (GERMAN) see DAP000

O,O-DIMETHYL-S-(2-AETHYLSULFONYL-AETHYL)-THIOLPHOSPHAT (GERMAN) see DAP600

O,O-DIMETHYL-S-(2-AETHYLTHIO-AETHYL)-DITHIO PHOSPHAT (GERMAN) see PHI500

O,O-DIMETHYL-O-(2-AETHYLTHIO-AETHYL) MONOTHIOPHOSPHAT (GERMAN) see DAO800

O,O-DIMETHYL-S-(2-AETHYLTHIO-AETHYL)-MONOTHIOPHOSPHAT (GERMAN) see DAP400

DIMETHYL ALDEHYDE see DOO600

DIMETHYLALLYL ACETATE see DOQ350

3,3-DIMETHYLALLYL ACETATE see DOQ350

γ,γ-DIMETHYLALLYL ACETATE see DOQ350

2-(3,3-DIMETHYLALLYL)CYCLAZOCINE see DOQ400

2-DIMETHYLALLYL-5,9-DIMETHYL-2'-HYDROXYBENZOMORPHAN see DOQ400

2-(3,3-DIMETHYLALLYL)-2',2'-HYDROXY-5,9-DIMETHYL-6,7-BENZOMORPHAN see DOQ400

DIMETHYLAMIDE ACETATE see DOO800

DIMETHYLAMID KYSELINY CHLORMRAVENCI see DQY950

DIMETHYLAMIDOETHOXYPHOSPHORYL CYANIDE see EIF000

DIMETHYLAMINE see DOQ800

DIMETHYLAMINE, solution (DOT) see DOQ800

DIMETHYLAMINE, anhydrous (DOT) see DOQ800

DIMETHYLAMINE, aqueous solution (DOT) see DOQ800

DIMETHYLAMINE BENZHYDRYL ESTER HYDROCHLORIDE see BAU750

DIMETHYLAMINE BORANE see DOR200

DIMETHYLAMINE-EPICHLOROHYDRIN COPOLYMER see DOR500

4-(DIMETHYLAMINE)-3,5-XYLYL-N-METHYLCARBAMATE see DOS000

DIMETHYLAMINOACETONITRILE see DOS200

2-DIMETHYLAMINOACETONITRILE (DOT) see DOS200

N',N'-DIMETHYL-4'-AMINO-N-ACETYL-N-MONOMETHYL-4-AMINOAZOBENZENE see DPQ200

N'-DIMETHYLAMINOACETYLPARTRICIN A DIMETHYLAMINOETHYLAMIDE DIASPARTAT see DOS400

DIMETHYLAMINOAETHANOL (GERMAN) see DOY800

β-DIMETHYLAMINO-AETHYL-BENZHYDRYL-AETHER (GERMAN) see BBV500

5-(DIMETHYLAMINOAETHYL-OXYIMINO)-5H-DIBENZO(a,d)CYCLOHEPTA-1,4-DIENHYDROCHLORID (GERMAN) see DPH600

DIMETHYLAMINO-ANALGESINE see DOT000

DIMETHYLAMINOANTIPYRINE see DOT000

4-(DIMETHYLAMINO)ANTIPYRINE see DOT000

p-DIMETHYLAMINOAZOBENZEN (CZECH) see DOT300

DIMETHYLAMINOAZOBENZENE see DOT300

4-DIMETHYLAMINOAZOBENZENE see DOT300

p-DIMETHYLAMINOAZOBENZENE see DOT300

4-(N,N-DIMETHYLAMINO)AZOBENZENE see DOT300

N,N-DIMETHYL-4-AMINOAZOBENZENE see DOT300

N,N-DIMETHYL-p-AMINOAZOBENZENE see DOT300

2',3-DIMETHYL-4-AMINOAZOBENZENE see AIC250

4-DIMETHYLAMINOAZOBENZENE AMINE-N-OXIDE see DTK600

p-(DIMETHYLAMINO)AZOBENZENE-o-CARBOXYLIC ACID see CCE500

4'-DIMETHYLAMINOAZOBENZENE-2-CARBOXYLIC ACID see CCE500

N,N-DIMETHYLAMINOAZOBENZENE-N-OXIDE see DTK600

DIMETHYLAMINOAZOBENZOL see DOT300

4-DIMETHYLAMINOAZOBENZOL see DOT300

p-DIMETHYLAMINO-AZOBENZOL (GERMAN) see DOT300

DIMETHYLAMINOAZOPHENE see DOT000

p-DIMETHYLAMINOBENZALRHODANINE see DOT800

5-(p-DIMETHYLAMINOBENZAL)RHODANINE see DOT800

p-(DIMETHYLAMINO)BENZAL-5-RHODANINE see DOT800

(DIMETHYLAMINO)BENZENE see DQF800

3,4-DIMETHYLAMINOBENZENE see XNS000

p-DIMETHYLAMINOBENZENE DIAZO SODIUM SULFONATE see DOU600

p-DIMETHYLAMINOBENZENEDIAZOSODIUM SULPHONATE see DOU600

p-(DIMETHYLAMINO)BENZENEDIAZOSULFONATE see DOU600

4-DIMETHYLAMINOBENZENEDIAZOSULFONIC ACID, SODIUM SALT see DOU600

p-DIMETHYLAMINOBENZENEDIAZOSULFONIC ACID, SODIUM SALT see DOU600

p-(DIMETHYLAMINO)BENZENEDIAZOSULPHONATE see DOU600

4-DIMETHYLAMINOBENZENEDIAZOSULPHONIC ACID, SODIUM SALT see DOU600

p-(DIMETHYLAMINO)BENZENEDIAZOSULPHONIC ACID, SODIUM SALT see DOU600

p-DIMETHYLAMINOBENZOLDIAZOSULFONAT (NATRIUMSALZ) (GERMAN) see DOU600

4,4'-DIMETHYLAMINOBENZOPHENONIMIDE see IBB000

5-(p-DIMETHYLAMINOBENZOYLIDENE)RHODANINE see DOT800

5-DIMETHYLAMINO-3-BENZOYLINDOLE see DOU700

p-DIMETHYLAMINOBENZYLIDENE RHODAMINE see DOT800

2',3-DIMETHYL-4-AMINOBIPHENYL see BLV250

3,2'-DIMETHYL-4-AMINOBIPHENYL see BLV250

N-DIMETHYL AMINO-β-CARBAMYL PROPIONIC ACID see DQD400

(DIMETHYLAMINO)CARBONYL CHLORIDE see DQY950

N,N-DIMETHYLAMINOCARBONYL CHLORIDE see DQY950

8-((DIMETHYLAMINO)CARBONYL)-5-OXO-2,4,9-TRIMETHYL-6,11-DIOXA-3-THIA-2,4,7,10-TETRAAZADODECA-7,9-DIENOIC ACID, 2,3-DIHYDRO-2,2-DIMETHYL-7-BENZOFURANYL ESTER see DOV870

1-(N,N-DIMETHYLAMINO)-3-(p-CHLOROPHENYL-3-α-PYRIDYL)PROPANE MALEATE see TAI500

DIMETHYLAMINOCYANPHOSPHORSAEUREAETHYLESTER (GERMAN) see EIF000

(DIMETHYLAMINO)CYCLOHEXANE see DRF709

N,N-DIMETHYLAMINOCYCLOHEXANE see DRF709

4-(DIMETHYLAMINO)-1,2-DIHYDRO-1,5-DIMETHYL-2-PHENYL-3H-PYRAZOL-3-ONE see DOT000

4-(DIMETHYLAMINO)-3,5-DIMETHYLPHENOL METHYLCARBAMATE (ESTER) see DOS000

4-(DIMETHYLAMINO)-3,5-DIMETHYLPHENYL ESTER, METHYLCARBAMIC ACID see DOS000

4-(DIMETHYLAMINO)-3,5-DIMETHYLPHENYL-N-METHYLCARBAMATE see DOS000

4-DIMETHYLAMINO-2,3-DIMETHYL-1-PHENYL-3-PYRAZOLIN-5-ONE see DOT000

4-DIMETHYLAMINO-2,3-DIMETHYL-1-PHENYL-5-PYRAZOLONE see DOT000

3,2'-DIMETHYL-4-AMINODIPHENYL see BLV250

1-6-DIMETHYLAMINO-4,4-DIPHENYL-3-HEPTANONE HYDROCHLORIDE see MDP770

p,p-DIMETHYLAMINODIPHENYLMETHANE see MJN000

α-(+)-4-DIMETHYLAMINO-1,2-DIPHENYL-3-METHYL-2-BUTANOL PROPIONATE ESTER see DAB879

DIMETHYLAMINOETHANOL see DOY800

2-(DIMETHYLAMINO)ETHANOL see DOY800

N-DIMETHYLAMINOETHANOL see DOY800

N,N-DIMETHYLAMINOETHANOL see DOY800

β-DIMETHYLAMINOETHANOL see DOY800

β-DIMETHYLAMINOETHANOL DIPHENYLMETHYL ETHER see BBV500

2-(DIMETHYLAMINO)ETHANOL METHACRYLATE see DPG600

α-(2-DIMETHYLAMINOETHOXY)DIPHENYLMETHANE see BBV500

cis-1-(p-(2-(N,N-DIMETHYLAMINO)ETHOXY)PHENYL)-1,2-DIPHENYLBUT-1-ENE see NOA600

trans-1-(p-β-DIMETHYLAMINOETHOXYPHENYL)-1,2-DIPHENYLBUT-1-ENE CITRATE see TAD175

β-DIMETHYLAMINOETHYL ALCOHOL see DOY800

β-DIMETHYLAMINOETHYL BENZHYDRYL ESTER HYDROCHLORIDE see BAU750

β-DIMETHYLAMINOETHYLBENZHYDRYLETHER see BBV500

2-(α-(2-DIMETHYLAMINOETHYL)BENZYL)PYRIDINE see TMJ750

DIMETHYLAMINOETHYL CHLORIDE see CGW000

2-DIMETHYLAMINOETHYLCHLORIDE see CGW000

β-(DIMETHYLAMINO)ETHYL CHLORIDE see CGW000

2-(DIMETHYLAMINO)ETHYL ESTER METHACRYLIC ACID see DPG600

3-(2-(DIMETHYLAMINO)ETHYL)-1H-INDOL-4-OL DIHYDROGEN PHOSPHATE ESTER see PHU500

3-2'-DIMETHYLAMINOETHYLINDOL-4-PHOSPHATE see PHU500

3-(2-DIMETHYLAMINOETHYL)INDOL-4-YL DIHYDROGEN PHOSPHATE see PHU500

DIMETHYLAMINOETHYL METHACRYLATE see DPG600

2-(DIMETHYLAMINO)ETHYL METHACRYLATE see DPG600

N,N-DIMETHYLAMINOETHYL METHACRYLATE see DPG600

β-DIMETHYLAMINOETHYL METHACRYLATE see DPG600

N-DIMETHYLAMINOETHYL-N-p-METHOXY-α-AMINOPYRIDINE MALEATE see DBM800

2-((2-(DIMETHYLAMINO)ETHYL)(p-METHOXYBENZYL)AMINO)PYRIDINE BIMALEATE see DBM800

2-((2-(DIMETHYLAMINO)ETHYL)(p-METHOXYBENZYL)AMINO)PYRIDINE MALEATE see DBM800

2-DIMETHYLAMINOETHYL-2-METHYLBENZHYDRYL ETHERHYDROCHLORIDE see OJW000

5-DIMETHYLAMINOETHYLOXYIMINO-5H-DIBENZO(a,d)CYCLOHEPTA-1,4-DIENE HYDROCHLORIDE see DPH600

β-DIMETHYLAMINO ETHYL-2-PYRIDYLAMINOTOLUENE see TMP750

β-DIMETHYLAMINOETHYL-2-PYRIDYLBENZYLAMINE see TMP750

2-((2-(DIMETHYLAMINO)ETHYL)(SELENOPHENE-2-YLMETHYL)AMINO)PYRIDINE see DPI750

10-(3-DIMETHYLAMINOISOPROPYL)PHENOTHIAZINE HYDROCHLORIDE see PMI750

DIMETHYLAMINO-ISOPROPYL-PHENTHIAZIN (GERMAN) see DQA400

2-(DIMETHYLAMINO)-N-(((METHYLAMINO)CARBONYL)OXY)-2-OXOETHANIMIDOTHIOIC ACID METHYL ESTER see DSP600

4-(N,N-DIMETHYLAMINO)-3'-METHYLAZOBENZENE see DUH600

d-4-DIMETHYLAMINO-3-METHYL-1,2-DIPHENYL-2-BUTANOL PROPIONATE HYDROCHLORIDE see PNA500

m-(((DIMETHYLAMINO)METHYLENE)AMINO)PHENYLMETHYL CARBAMATE,HYDROCHLORIDE see DSO200

3-DIMETHYLAMINOMETHYLENEIMINOPHENYL-N-METHYLCARBAMATE, HYDROCHLORIDE see DSO200

(2-DIMETHYLAMINO-2-METHYL)ETHYL-N-DIBENZOPARATHIAZINE see DQA400

10-(2-(DIMETHYLAMINO)-2-METHYLETHYL)PHENOTHIAZINE see DQA400

N-(2'-DIMETHYLAMINO-2'-METHYL)ETHYLPHENOTHIAZINE see DQA400

N-(2'-DIMETHYLAMINO-2'-METHYL)ETHYLPHENOTHIAZINE HYDROCHLORIDE see PMI750

s-α-(2-(DIMETHYLAMINO)-1-METHYLETHYL)-α-PHENYLBENZENEETHANOL PROPIOATE HYDROCHLORIDE see PNA500

N-DIMETHYLAMINO-2-METHYLETHYL THIODIPHENYLAMINE see DQA400

trans-2-((DIMETHYLAMINO)METHYLIMINO)-5-(2-(5-NITRO-2-FURYL)VINYL)-1,3,4-OXADIAZOLE see DPL000

3-(DIMETHYLAMINO)-1-METHYL-3-OXO-1-PROPENYL DIMETHYL PHOSPHATE see DGQ875

2-(DIMETHYLAMINO)-N-(((METHYL(((2-PHENYL-1,3-DIOXAN-5-YL)METHOXY)SULFINYL)AMINO)CARBONYL)OXY)-2-OXO-ETHANIMIDOTHIOIC ACID, METHYL ESTER see DPL300

2-DIMETHYLAMINO-1-(METHYLTHIO)GLYOXAL-o-METHYLCARBAMOYLMONOXIME see DSP600

1-DIMETHYLAMINONAPHTHALENE see DSU400

DIMETHYLAMINONAPHTHALENESULFONYL CHLORIDE see DPN200

1-DIMETHYLAMINONAPHTHALENE-5-SULFONYL CHLORIDE see DPN200

1-(DIMETHYLAMINO)-5-NAPHTHALENESULFONYLCHLORIDE see DPN200

5-(DIMETHYLAMINO)-1-NAPHTHALENESULFONYL CHLORIDE see DPN200

5-DIMETHYLAMINONAPHTHYL-5-SULFONYL CHLORIDE see DPN200

4-(DIMETHYLAMINO)NITROSOBENZENE see DSY600

p-(DIMETHYLAMINO)NITROSOBENZENE see DSY600

5-(DIMETHYLAMINOOXYIMINO)-5H-DIBENZO(a,b)CYCLOHEPTA-1,4-DIENE HYDROCHLORIDE see DPH600

DIMETHYLAMINOPHENAZON (GERMAN) see DOT000

DIMETHYLAMINOPHENAZONE see DOT000

4-DIMETHYLAMINOPHENAZONE see DOT000

4-(p-DIMETHYLAMINOPHENYLAZO)ANILINE see DPO200

4-DIMETHYLAMINOPHENYLAZOBENZENE see DOT300

4-(p-DIMETHYLAMINOPHENYLAZO)-BENZENEARSONIC ACID HYDROCHLORIDE see DPO275

2-((4-DIMETHYLAMINO)PHENYLAZO)BENZOIC ACID see CCE500

o-((p-(DIMETHYLAMINO)PHENYL)AZO)BENZOIC ACID see CCE500

4-((p-(DIMETHYLAMINO)PHENYL)AZO)-N-METHYLACETANILIDE see DPQ200

N-(4-((4-(DIMETHYLAMINO)PHENYL)AZO)PHENYL)-N-METHYLACETAMIDE see DPQ200

4-(DIMETHYLAMINO)PHENYL-3-(4-(4-CHLOROPHENYL)-4-HYDROXYPIPERIDINO)-PROPYL KETONE see CKA580

DIMETHYL ARSINIC SULFIDE see DQG700
((DIMETHYLARSINO)OXY)SODIUM-As-OXIDE see
HKC500
N,N-DIMETHYL-p-AZOANILINE see DOT300
2,3'-DIMETHYLAZOBENZENE-4'-
METHYLCARBONATE see CBS750
DIMETHYLBENZANTHRACENE see DQJ200
DIMETHYLBENZ(a)ANTHRACENE see DQJ200
7,12-DIMETHYLBENZANTHRACENE see DQJ200
9,10-DIMETHYL-BENZANTHRACENE see DQJ200
7,12-DIMETHYLBENZ(a)ANTHRACENE see DQJ200
9,10-DIMETHYLBENZ(a)ANTHRACENE see DQJ200
9,10-DIMETHYL-1,2-BENZANTHRACENE see DQJ200
9,10-DIMETHYL-1,2-BENZANTHRAZEN (GERMAN)
see DQJ200
DIMETHYLBENZANTHRENE see DQJ200
O,O-DIMETHYL-S-(BENZAZIMINOMETHYL)
DITHIOPHOSPHATE see ASH500
2,2-DIMETHYL-1,3-BENZDIOXOL-4-YL-N-
METHYLCARBAMATE see DQM600
α,α-DIMETHYLBENZEETHANAMINE see DTJ400
2,3-DIMETHYLBENZENAMINE see XMJ000
2,4-DIMETHYLBENZENAMINE see XMS000
2,5-DIMETHYLBENZENAMINE see XNA000
2,6-DIMETHYLBENZENAMINE see XNJ000
3,5-DIMETHYLBENZENAMINE see XOA000
DIMETHYLBENZENE see XGS000
m-DIMETHYLBENZENE see XHA000
o-DIMETHYLBENZENE see XHJ000
p-DIMETHYLBENZENE see XHS000
1,2-DIMETHYLBENZENE see XHJ000
1,3-DIMETHYLBENZENE see XHA000
1,4-DIMETHYLBENZENE see XHS000
N,N-DIMETHYLBENZENEAMINE see DQF800
DIMETHYL-1,2-BENZENEDICARBOXYLATE see
DTR200
DIMETHYL-1,4-BENZENE DICARBOXYLATE see
DUE000
N,N-DIMETHYLBENZENEMETHANAMINE see
DQP800
DIMETHYL BENZENEORTHODICARBOXYLATE see
DTR200
2,4-DIMETHYLBENZENESULFONIC ACID see XJJ000
3,3'-DIMETHYLBENZIDIN see TGJ750
3,3'-DIMETHYLBENZIDINE see TGJ750
3,3'-DIMETHYLBENZIDINE DIHYDROCHLORIDE
see DQM000
DIMETHYLBENZIMIDAZOLYCOBAMIDE see VSZ000
5,6-DIMETHYLBENZIMIDAZOLYCOBAMIDE
CYANIDE see VSZ000
7,12-DIMETHYLBENZO(a)ANTHRACENE see DQJ200
2,2-DIMETHYL-1,3-BENZODIOX-4-OL
METHYLCARBAMATE see DQM600
2,2-DIMETHYLBENZO-1,3-DIOXOL-4-YL
METHYLCARBAMATE see DQM600
5,6-DIMETHYL-2,1,3-BENZOSELENODIAZOLE see
DQO650
O,O-DIMETHYL-S-(1,2,3-BENZOTRIAZINYL-4-
KETO)METHYL PHOSPHORODITHIOATE see
ASH500
1,4-DIMETHYL-2,3-BENZPHENANTHRENE see
DQJ200
N,N-DIMETHYLBENZYLAMINE see DQP800
N,N-DIMETHYLBENZYLAMINE
HEXAFLUOROARSENATE see BEL550
DIMETHYL BENZYL CARBINOL see DQQ200

DIMETHYL BENZYL CARBINYL ACETATE see
DQQ375
DIMETHYLBENZYLCARBINYL BUTYRATE see
BEL850
DIMETHYL BENZYL CARBINYL BUTYRATE see
DQQ380
α,α-DIMETHYLBENZYL HYDROPEROXIDE (MAK)
see IOB000
DIMETHYLBENZYLOCTADECYLAMMONIUM
CHLORIDE see DTC600
N,N-DIMETHYL-N'-BENZYL-N'-(α-
PYRIDYL)ETHYLENEDIAMINE see TMP750
DIMETHYL BERYLLIUM see DQR200
DIMETHYLBERYLLIUM-1,2-DIMETHOXYETHANE
see DQR289
1,1-DIMETHYLBIGUANIDE see DQR600
N,N-DIMETHYLBIGUANIDE see DQR600
3,2'-DIMETHYL-4-BIPHENYLAMINE see BLV250
3,3'-DIMETHYLBIPHENYL-4,4'-BIPHENYLDIAMINE
DIHYDROCHLORIDE see DQM000
3,3'-DIMETHYL-4,4'-BIPHENYLDIAMINE see TGJ750
3,3'-DIMETHYLBIPHENYL-4,4'-DIAMINE see TGJ750
3,3'-DIMETHYL-(1,1'-BIPHENYL)-4,4'-DIAMINE see
TGJ750
2,3'-DIMETHYLBIPHENYL-4,4'-DIAMINE
DIHYDROCHLORIDE see DQM000
3,3'-DIMETHYL-4,4'-BIPHENYLENE DIISOCYANATE
see DQS000
1,1'-DIMETHYL-4,4'-BIPYRIDINIUM see PAI990
DIMETHYL 1,3-BIS(CARBOMETHOXY)-1-PROPEN-2-
YL PHOSPHATE see SOY000
β,γ-DIMETHYL-α,Δ-BIS(3,4-
DIHYDROXYPHENYL)BUTANE see NBR000
O,O-DIMETHYL-S-(1,2-
BIS(ETHOXYCARBONYL)ETHYL)DITHIOPHOSPHA
TE see MAK700
DIMETHYL BIS(p-HYDROXYPHENYL)METHANE see
BLD500
1,2-DIMETHYLBUTANE see IKS600
2,2-DIMETHYLBUTANE see DQT200
2,3-DIMETHYLBUTANE see DQT400
1,3-DIMETHYL BUTANOL see AOK750
1,3-DIMETHYLBUTYL ACETATE see HFJ000
1,3-DIMETHYL BUTYLAMINE see DQU600
1,3-DIMETHYLBUTYLAMINE (DOT) see DQU600
DIMETHYLCADMIUM see DQW800
DIMETHYLCARBAMATE-d'l-ISOPROPYL-3-METHYL-
5-PYRAZOLYLE (FRENCH) see DSK200
DIMETHYLCARBAMIC ACID CHLORIDE see DQY950
N,N-DIMETHYLCARBAMIC ACID CHLORIDE see
DQY950
DIMETHYLCARBAMIC ACID 3-METHYL-1-(1-
METHYLETHYL)-1H-PYRAZOL-5-YL ESTER see
DSK200
DIMETHYLCARBAMIC CHLORIDE see DQY950
DIMETHYLCARBAMIDOYL CHLORIDE see DQY950
N,N-DIMETHYLCARBAMIDOYL CHLORIDE see
DQY950
DIMETHYLCARBAMODITHIOIC ACID, IRON
COMPLEX see FAS000
DIMETHYLCARBAMODITHIOIC ACID, IRON(3+)
SALT see FAS000
DIMETHYLCARBAMODITHIOIC ACID, ZINC
COMPLEX see BJK500
DIMETHYLCARBAMODITHIOIC ACID, ZINC SALT
see BJK500
DIMETHYLCARBAMOYL CHLORIDE see DQY950

N,N-DIMETHYLCARBAMOYL CHLORIDE see
DQY950
DIMETHYL CARBAMOYL CHLORIDE (ACGIH,DOT)
see DQY950
cis-2-DIMETHYLCARBAMOYL-1-METHYLVINYL
DIMETHYLPHOSPHATE see DGQ875
DIMETHYLCARBAMYL CHLORIDE see DQY950
N,N-DIMETHYLCARBAMYL CHLORIDE see DQY950
DIMETHYLCARBINOL see INJ000
2,2-DIMETHYL-6-CARBOBUTOXY-2,3-DIHYDRO-4-
PYRONE see BRT000
α,α-DIMETHYL-α'-CARBOBUTOXY-DIHYDRO-γ-
PYRONE see BRT000
O,O-DIMETHYL-O-(2-CARBOMETHOXY-1-
METHYLVINYL) PHOSPHATE see MQR750
DIMETHYL-1-CARBOMETHOXY-1-PROPEN-2-YL
PHOSPHATE see MQR750
DIMETHYL CARBONATE see MIF000
O,O-DIMETHYL-S-
(CARBONYLMETHYLMORPHOLINO)
PHOSPHORODITHIOATE see PHI500
DIMETHYLCELLOSOLVE see DOE600
DIMETHYLCHLOROETHER see CIO250
DIMETHYL(2-CHLOROETHYL)AMINE see CGW000
DIMETHYL(2-CHLOROETHYL)AMINE
HYDROCHLORIDE see DRC000
DIMETHYL-β-CHLOROETHYLAMINE
HYDROCHLORIDE see DRC000
DIMETHYLCHLOROFORMAMIDE see DQY950
DIMETHYL-p-CHLOROPHENYLTHIOMETHYL
DITHIOPHOSPHATE see MQH750
O,O-DIMETHYL-S-(p-
CHLOROPHENYLTHIOMETHYL)PHOSPHORODITH
IOATE see MQH750
1,1-DIMETHYL-3-(p-CHLOROPHENYL)UREA see
CJX750
N,N-DIMETHYL-N'-(4-CHLOROPHENYL)UREA see
CJX750
DIMETHYL CHLOROTHIOPHOSPHATE (DOT) see
DTQ600
DIMETHYLCHLORTHIOFOSAT (CZECH) see DTQ600
O,O-DIMETHYL-O-2-CHLOR-1-(2,4,5-
TRICHLORPHENYL)-VINYL-PHOSPHAT (GERMAN)
see TBW100
2,2-DIMETHYL-7-COUMARANYL-N-
METHYLCARBAMATE see CBS275
DIMETHYLCYANAMIDE see DRF600
N,N-DIMETHYLCYCLOHEXANAMINE see DRF709
DIMETHYLCYCLOHEXYLAMINE see DRF709
N,N-DIMETHYLCYCLOHEXYLAMINE (DOT) see
DRF709
DIMETHYLCYSTEINE see MCR750
β,β-DIMETHYLCYSTEINE see MCR750
DIMETHYL DIALKYL AMMONIUM CHLORIDE see
DRI500
3,3'-DIMETHYL-4,4'-DIAMINODIPHENYLMETHANE
see MJO250
N,N-DIMETHYL-1,3-DIAMINOPROPANE see AJQ100
DIMETHYLDIARSINE SULFIDE see DQG700
2,3-DIMETHYL-1,4-DIAZINE see DTU400
2,5-DIMETHYL-1,4-DIAZINE see DTU600
2,5-DIMETHYL-2,5-DI(t-BUTYLPEROXY)HEXANE see
DRJ800
2,5-DIMETHYL-2,5-DI(tert-BUTYLPEROXY)HEXANE
see DRJ800
O,O-DIMETHYL-S-1,2-(DICARBAETHOXYAETHYL)-
DITHIOPHOSPHAT (GERMAN) see MAK700

O,O-DIMETHYL-S-(1,2-DICARBETHOXYETHYL)
DITHIOPHOSPHATE see MAK700
O,O-DIMETHYL-S-(1,2-
DICARBETHOXYETHYL)PHOSPHORODITHIOATE
see MAK700
O,O-DIMETHYL-S-(1,2-DICARBETHOXYETHYL)
THIOTHIONOPHOSPHATE see MAK700
DIMETHYL-1,3-DI(CARBOMETHOXY)-1-PROPEN-2-
YL PHOSPHATE see SOY000
DIMETHYL-2,2-DICHLOROETHENYL PHOSPHATE
see DGP900
DIMETHYL-1,1'-DICHLOROETHER see BIK000
1,1-DIMETHYL-3-(3,4-DICHLOROPHENYL)UREA see
DXQ500
DIMETHYLDICHLOROSILANE (DOT) see DFE259
DIMETHYL DICHLOROVINYL PHOSPHATE see
DGP900
DIMETHYL-2,2-DICHLOROVINYL PHOSPHATE see
DGP900
O,O-DIMETHYL DICHLOROVINYL PHOSPHATE see
DGP900
O,O-DIMETHYL-O-2,2-DICHLOROVINYL
PHOSPHATE see DGP900
DIMETHYL-DICHLORSILAN see DFE259
O,O-DIMETHYL-O-(2,2-DICHLOR-VINYL)-
PHOSPHAT (GERMAN) see DGP900
O,O-DIMETHYL-S-1,2-
DI(ETHOXYCARBAMYL)ETHYL
PHOSPHORODITHIOATE see MAK700
DIMETHYL-DIETHOXYSILAN (CZECH) see DHG000
DIMETHYLDIETHOXYSILANE (DOT) see DHG000
N,N-DIMETHYLDIGUANIDE see DQR600
2,2-DIMETHYL-2,3-DIHYDROBENZOFURAN-7-YL-N-
(4-BROMOPHENYLTHIO)-N-METHYLCARBAMATE
see DRL100
2,2-DIMETHYL-2,3-DIHYDROBENZOFURAN-7-YL
ESTER, METHYLCARBAMIC ACID see CBS275
2,2-DIMETHYL-2,3-DIHYDRO-7-BENZOFURANYL-N-
METHYLCARBAMATE see CBS275
O,O-DIMETHYL-S-(3,4-DIHYDRO-4-KETO-1,2,3-
BENZOTRIAZINYL-3-METHYL)
DITHIOPHOSPHATE see ASH500
2,5-DIMETHYL-2,5-DIHYDROPEROXYHEXANE,
>82% with water (DOT) see DSE800
O,O-DIMETHYL-S-1,2-
DIKARBETOXYLETHYLDITIOFOSFAT (CZECH) see
MAK700
DIMETHYL DIKETONE see BOT500
DIMETHYL 3-
(DIMETHOXYPHOSPHINYLOXY)GLUTACONATE
see SOY000
N,5-DIMETHYL-4-((DIMETHYLAMINO)CARBONYL)-
N-((4-(1,1-DIMETHYLETHYL)PHENYL)THIO)-2,7-
DIOXA-3,6-DIAZAOCTA-3,5-DIENAMIDE see DRL425
3-keto-1,5-DIMETHYL-4-DIMETHYLAMINO-2-
PHENYL-2,3-DIHYDROPYRAZOLE see DOT000
1,5-DIMETHYL-4-DIMETHYLAMINO-2-PHENYL-3-
PYRAZOLONE see DOT000
2,3-DIMETHYL-4-DIMETHYLAMINO-1-PHENYL-5-
PYRAZOLONE see DOT000
9,9-DIMETHYL-10-
DIMETHYLAMINOPROPYLACRIDAN HYDROGEN
TARTRATE see DRM000
9,9-DIMETHYL-10-(3-
DIMETHYLAMINO)PROPYLACRIDINE TARTRATE
see DRM000
3,3-DIMETHYL-4-(DIMETHYLAMINO)-4-(o-
TOLYL)BUTYL o-TOLYL KETONE see DRM110

O,O-DIMETHYL-O-(2-DIMETHYL-CARBAMOYL-1-METHYL-VINYL)PHOSPHAT (GERMAN) see DGQ875

O,O-DIMETHYL-O-(N,N-DIMETHYLCARBAMOYL-1-METHYLVINYL) PHOSPHATE see DGQ875

O,O-DIMETHYL-O-(1,4-DIMETHYL-3-OXO-4-AZA-PENT-1-ENYL)FOSFAAT (DUTCH) see DGQ875

O,O-DIMETHYL-O-(1,4-DIMETHYL-3-OXO-4-AZA-PENT-1-ENYL)PHOSPHATE see DGQ875

N,N'-DIMETHYL-N,N'-DINITROSO-1,4-BENZENEDICARBOXAMIDE see DRO400

N,N'-DIMETHYL-N,N'-DINITROSOTEREPHTHALAMIDE see DRO400

DIMETHYLDIOCTYLAMMONIUM CHLORIDE see DTF820

2,2-DIMETHYL-1,3-DIOXOLANE-4-METHANOL see DVR600

cis-N,N-DIMETHYL-2-(p-(1,2-DIPHENYL-1-BUTENYL)PHENOXY)ETHYLAMINE see NOA600

3,3'-DIMETHYL-4,4'-DIPHENYLDIAMINE see TGJ750

3,3'-DIMETHYLDIPHENYL-4,4'-DIAMINE see TGJ750

3,3'-DIMETHYLDIPHENYLMETHANE-4,4'-DIISOCYANATE see MJN750

DIMETHYL DISULFIDE see DRQ400

DIMETHYLDITHIOCARBAMATE ZINC SALT see BJK500

DIMETHYLDITHIOCARBAMIC ACID, IRON SALT see FAS000

DIMETHYLDITHIOCARBAMIC ACID, IRON(3+) SALT see FAS000

DIMETHYLDITHIOCARBAMIC ACID, LEAD SALT see LCW000

DIMETHYLDITHIOCARBAMIC ACID, SODIUM SALT see SGM500

N,N-DIMETHYLDITHIOCARBAMIC ACID S-TRIBUTYLSTANNYL ESTER see TID150

DIMETHYLDITHIOCARBAMIC ACID, ZINC SALT see BJK500

O,O-DIMETHYLDITHIOPHOSPHATE DIETHYLMERCAPTOSUCCINATE see MAK700

DIMETHYLDITHIOPHOSPHORIC ACID N-METHYLBENZAZIMIDE ESTER see ASH500

O,O-DIMETHYL DITHIOPHOSPHORYLACETIC ACID-N-METHYL-N-FORMYLAMIDE see DRR200

O,O-DIMETHYLDITHIOPHOSPHORYLACETIC ACID-N-MONOMETHYLAMIDE SALT see DSP400

O,O-DIMETHYL-DITHIOPHOSPHORYLESSIGSAEURE MONOMETHYLAMID (GERMAN) see DSP400

DIMETHYLENEDIAMINE see EEA500

DIMETHYLENE GLYCOL see BOT000

DIMETHYLENEIMINE see EJM900

DIMETHYLENE OXIDE see EJN500

DIMETHYLENIMINE see EJM900

6,7-DIMETHYLESCULETIN see DRS800

O,S-DIMETHYL ESTER AMIDE of AMIDOTHIOATE see DTQ400

O,O-DIMETHYLESTER KYSELINY CHLORTHIOFOSFORECNE (CZECH) see DTQ600

DIMETHYLESTER KYSELINY FOSFORITE (CZECH) see DSG600

DIMETHYLESTER KYSELINY SIROVE (CZECH) see DUD100

DIMETHYL ESTER PHOSPHORIC ACID ESTER with METHYL 3-HYDROXYCROTONATE see MQR750

1,1-DIMETHYLETHANOL see BPX000

DIMETHYLETHANOLAMINE see DOY800

N,N-DIMETHYLETHANOLAMINE see DOY800

DIMETHYLETHANOLAMINE (DOT) see DOY800

DIMETHYL ETHER (DOT) see MJW500

DIMETHYLETHOXYSILANE (ACGIH) see EES200

O,O-DIMETHYL-S-(2-ETHSULFONYLETHYL)PHOSPHOROTHIOATE see DAP600

DIMETHYL-S-(2-ETHSULFONYLETHYL)THIOPHOSPHATE see DAP600

O,O-DIMETHYL-S-(2-ETHTHIOETHYL)PHOSPHOROTHIOATE see DAP400

DIMETHYL-S-(2-ETHTHIOETHYL)THIOPHOSPHATE see DAP400

O,O-DIMETHYL-S-(2-ETHTHIONYLETHYL)PHOSPHOROTHIOATE see DAP000

DIMETHYL-S-(2-ETHTHIONYLETHYL)THIOPHOSPHATE see DAP000

1,1-DIMETHYLETHYLAMINE see BPY250

1-(4-(3-((1,1-DIMETHYLETHYL)AMINO)-2-HYDROXYPROPOXY)-2-BENZOFURANYL)ETHANONE see ACN300

1-(7-(3-((1,1-DIMETHYLETHYL)AMINO)-2-HYDROXYPROPOXY)-2-BENZOFURANYL)ETHANONE see ACN320

4-(1,1-DIMETHYLETHYL)-1,2-BENZENEDIOL see BSK000

DIMETHYLETHYLCARBINOL see PBV000

1,1-DIMETHYLETHYL CHLOROACETATE see BQR100

1,1-DIMETHYLETHYL GLYCIDYL ETHER see BRK800

1,1-DIMETHYLETHYL HYDROPEROXIDE see BRM250

5,5-DIMETHYL-2-(ETHYLIMINO)-1,3-DITHIOLAN-4-ONE-o-((METHYLAMINO)CARBONYL)OXIME see DRV300

O,O-DIMETHYL-S-(2-ETHYLMERCAPTOETHYL) DITHIOPHOSPHATE see PHI500

O,O-DIMETHYL-O-ETHYLMERCAPTOETHYL THIOPHOSPHATE see DAO800

O,O-DIMETHYL-S-ETHYLMERCAPTOETHYL THIOPHOSPHATE see DAP400

O,O-DIMETHYL-S-ETHYLMERCAPTOETHYL THIOPHOSPHATE, THIOLO ISOMER see DAP400

O,O-DIMETHYL 2-ETHYLMERCAPTOETHYL THIOPHOSPHATE, THIONO ISOMER see DAO800

O,O-DIMETHYL-S-2-ETHYLMERKAPTOETHYLESTER KYSELINY DITHIOFOSFORECNE (CZECH) see PHI500

4-(1,1-DIMETHYLETHYL)PHENOL see BSE500

2-(4-(1,1-DIMETHYLETHYL)PHENOXY)CYCLOHEXYL 2-PROPYNYL ESTER, SULFUROUS ACID see SOP000

2-(4-(1,1-DIMETHYLETHYL)PHENOXY)CYCLOHEXYL 2-PROPYNYL SULFITE see SOP000

O,O-DIMETHYL-S-(2-ETHYLSULFINYL-ETHYL)-MONOTHIOFOSFAAT (DUTCH) see DAP000

O,O-DIMETHYL-S-(2-(ETHYLSULFINYL)ETHYL) PHOSPHOROTHIOATE see DAP000

O,O-DIMETHYL-S-(2-ETHYLSULFINYL)ETHYL THIOPHOSPHATE see DAP000

O,O-DIMETHYL-S-ETHYL-2-SULFONYLETHYL PHOSPHOROTHIOLATE see DAP600

O,O-DIMETHYL-S-ETHYLSULPHINYLETHYL PHOSPHOROTHIOLATE see DAP000

O,O-DIMETHYL-S-ETHYLSULPHONYLETHYL PHOSPHOROTHIOLATE see DAP600

O,O-DIMETHYL-S-(2-ETHYLTHIO-ETHYL)-DITHIOFOSFAAT (DUTCH) see PHI500

O,O-DIMETHYL-O-(2-ETHYL-THIO-ETHYL)-
MONOTHIOFOSFAAT (DUTCH) see DAO800
O,O-DIMETHYL-S-(2-ETHYLTHIO-ETHYL)-
MONOTHIOFOSFAAT (DUTCH) see DAP400
O,O-DIMETHYL S-(2-(ETHYLTHIO)ETHYL)
PHOSPHORODITHIOATE see PHI500
O,O-DIMETHYL-O-2-(ETHYLTHIO)ETHYL
PHOSPHOROTHIOATE see DAO800
O,O-DIMETHYL-S-(2-
(ETHYLTHIO)ETHYL)PHOSPHOROTHIOATE see
DAP400
DIMETHYL FORMAL see MGA850
DIMETHYLFORMALDEHYDE see ABC750
DIMETHYLFORMAMID (GERMAN) see DSB000
DIMETHYLFORMAMIDE see DSB000
N,N-DIMETHYL FORMAMIDE see DSB000
N,N-DIMETHYLFORMAMIDE (DOT) see DSB000
DIMETHYLFORMOCARBOTHIALDINE see DSB200
O,O-DIMETHYL-S-(N-FORMYL-N-
METHYLCARBAMOYLMETHYL)
PHOSPHORODITHIOATE see DRR200
DIMETHYLFOSFIT see DSG600
DIMETHYLFOSFONAT see DSG600
2,5-DIMETHYL-3-FURYL p-HYDROXYPHENYL
KETONE see DSC100
N,N-DIMETHYLGLYCINONITRILE see DOS200
DIMETHYLGLYOXAL see BOT500
DIMETHYLGLYOXIME see DBH000
N,N'-DIMETHYLHARNSTOFF (GERMAN) see
DUM200
2,6-DIMETHYL-2,5-HEPTADIEN-4-ONE see PGW250
2,6-DIMETHYL HEPTANOL-4 see DNH800
2,6-DIMETHYL-4-HEPTANOL see DNH800
2,6-DIMETHYL-HEPTAN-4-ON (DUTCH, GERMAN)
see DNI800
2,6-DIMETHYLHEPTAN-4-ONE see DNI800
2,6-DIMETHYL-4-HEPTANONE see DNI800
2,6-DIMETHYL-5-HEPTENAL see DSD775
DIMETHYLHEXANE DIHYDROPEROXIDE (dry) see
DSE800
DIMETHYL HEXANEDIOATE see DOQ300
1,5-DIMETHYLHEXYLAMINE see ILM000
α,ε-DIMETHYLHEXYLAMINE see ILM000
1,2-DIMETHYLHYDRAZIN (GERMAN) see DSF600
DIMETHYLHYDRAZINE see DSF400
1,1-DIMETHYLHYDRAZINE see DSF400
1,2-DIMETHYLHYDRAZINE see DSF600
1,2-DIMETHYL-HYDRAZINE see DSF600
N,N-DIMETHYLHYDRAZINE see DSF400
N,N'-DIMETHYLHYDRAZINE see DSF600
sym-DIMETHYLHYDRAZINE see DSF600
uns-DIMETHYLHYDRAZINE see DSF400
asym-DIMETHYLHYDRAZINE see DSF400
unsym-DIMETHYLHYDRAZINE see DSF400
1,1-DIMETHYLHYDRAZINE (GERMAN) see DSF400
DIMETHYLHYDRAZINE, symmetrical (DOT) see
DSF600
DIMETHYLHYDRAZINE, unsymmetrical (DOT) see
DSF400
1,2-DIMETHYLHYDRAZINE DIHYDROCHLORIDE
see DSF800
N,N'-DIMETHYLHYDRAZINE DIHYDROCHLORIDE
see DSF800
sym-DIMETHYLHYDRAZINE DIHYDROCHLORIDE
see DSF800
1,1-DIMETHYLHYDRAZINE HYDROCHLORIDE see
DSG000

1,2-DIMETHYLHYDRAZINE HYDROCHLORIDE see
DSG200
sym-DIMETHYLHYDRAZINE HYDROCHLORIDE see
DSG200
9-(2,2-DIMETHYLHYDRAZINO)ACRIDINE
MONOHYDROCHLORIDE see DSG330
2-(2,2-DIMETHYLHYDRAZINO)-4-(5-NITRO-2-
FURYL)THIAZOLE see DSG400
DIMETHYL HYDROGEN PHOSPHITE see DSG600
DIMETHYLHYDROGENPHOSPHITE see DSG600
DIMETHYL-2,5 (HYDROXY 4 BENZOYL) 3
FURANNE see DSC100
N,N-DIMETHYL-2-HYDROXYETHYLAMINE see
DOY800
N,N-DIMETHYL-N-(2-HYDROXYETHYL)AMINE see
DOY800
DIMETHYL 3-HYDROXYGLUTACONATE
DIMETHYL PHOSPHATE see SOY000
2,2-DIMETHYL-5-HYDROXYMETHYL-1,3-
DIOXOLANE see DVR600
2',3-DIMETHYL-4-(2-
HYDROXYNAPHTHYLAZO)AZOBENZENE see
SBC500
3,7-DIMETHYL-7-HYDROXYOCTANAL see CMS850
1,1-DIMETHYL-3-HYDROXYPYRROLIDINIUM
BROMIDE-α-CYCLOPENTYLMANDELATE see
GIC000
O,O-DIMETHYL-(1-HYDROXY-2,2,2-
TRICHLORAETHYL)-PHOSPHAT (GERMAN) see
TIQ250
O,O-DIMETHYL-(1-HYDROXY-2,2,2-
TRICHLORAETHYL)PHOSPHONSAEURE ESTER
(GERMAN) see TIQ250
O,O-DIMETHYL-(1-HYDROXY-2,2,2-
TRICHLORO)ETHYL PHOSPHATE see TIQ250
DIMETHYL-1-HYDROXY-2,2,2-TRICHLOROETHYL
PHOSPHONATE see TIQ250
O,O-DIMETHYL-(1-HYDROXY-2,2,2-
TRICHLOROETHYL)PHOSPHONATE see TIQ250
3,4-DIMETHYL-3H-IMIDAZO(4,5-f)QUINOLIN-2-
AMINE see AJQ600
3,8-DIMETHYL-3H-IMIDAZO(4,5-f)QUINOXALIN-2-
AMINE see AJQ675
DIMETHYLIMIPRAMINE HYDROCHLORIDE see
DLS600
DIMETHYLIONONE see COW780
1,3-DIMETHYL-α-IONONE see COW780
α-α-DIMETHYL-m-ISOPROPENYL BENZYL
ISOCYANATE see IKG800
DIMETHYL-5-(1-ISOPROPYL-3-
METHYLPYRAZOLYL)CARBAMATE see DSK200
4,4-DIMETHYL-1-ISOPROPYL-2-NONYL-2-
IMIDAZOLINE see DSK300
1,3-DIMETHYLISOTHIOUREA see DSK900
3,4-DIMETHYLISOXALE-5-SULFANILAMIDE see
SNN500
3,4-DIMETHYLISOXALE-5-SULPHANILAMIDE see
SNN500
N[1]-(3,4-DIMETHYL-5-
ISOXAZOLYL)SULFANILAMIDE see SNN500
N'-(3,4)DIMETHYLISOXAZOL-5-YL-
SULPHANILAMIDE see SNN500
N[1]-(3,4-DIMETHYL-5-
ISOXAZOLYL)SULPHANILAMIDE see SNN500
DIMETHYLKARBAMOYLCHLORID see DQY950
DIMETHYLKETAL see ABC750
DIMETHYLKETOL see ABB500

DIMETHYL KETONE see ABC750
DIMETHYLMAGNESIUM see DSL600
DIMETHYLMALEIC ANHYDRIDE see DSM000
α,β-DIMETHYLMALEIC ANHYDRIDE see DSM000
DIMETHYL MERCURY see DSM450
DIMETHYLMESCALINE see DOE200
DIMETHYLMETHANE see PMJ750
N,N-DIMETHYL-N'-(4-METHOXYBENZYL)-N'-(2-
PYRIDYL)ETHYLENEDIAMINE MALEATE see
DBM800
O,O-DIMETHYL-O-2-METHOXYCARBONYL-1-
METHYL-VINYL-PHOSPHAT (GERMAN) see MQR750
DIMETHYL 2-METHOXYCARBONYL-1-
METHYLVINYL PHOSPHATE see MQR750
DIMETHYL METHOXYCARBONYLPROPENYL
PHOSPHATE see MQR750
DIMETHYL (1-METHOXYCARBOXYPROPEN-2-
YL)PHOSPHATE see MQR750
3,3-DIMETHYL-1-p-METHOXYPHENYLTRIAZENE
see DSN600
O,O-DIMETHYL-S-(5-METHOXY-1,3,4-
THIADIAZOLINYL-3-METHYL) DITHIOPHOSPHATE
see DSO000
O,O-DIMETHYL-S-(2-METHOXY-1,3,4-THIADIAZOL-
5-(4H)-ONYL-(4)-METHYL)-DITHIOPHOSPHAT
(GERMAN) see DSO000
O,O-DIMETHYL-S-(2-METHOXY-1,3,4-THIADIAZOL-
5(4H)-ONYL-(4)-METHYL) PHOSPHORODITHIOATE
see DSO000
O,O-DIMETHYL-S-((2-METHOXY-1,3,4 (4H)-
THIODIAZOL-5-ON-4-YL)-
METHYL)DITHIOFOSFAAT (DUTCH) see DSO000
(O,O-DIMETHYL)-S-(-2-METHOXY-Δ²-1,3,4-
THIADIAZOLIN-5-ON-4-
YLMETHYL)DITHIOPHOSPHATE DIMETHYL
PHOSPHOROTHIOLOTHIONATE see DSO000
N,N-DIMETHYL-N'-
(((METHYLAMINO)CARBONYL)OXY)PHENYLMETH
ANIMIDAMIDE MONOHYDROCHLORIDE see
DSO200
2,2-DIMETHYL-4-(N-
METHYLAMINOCARBOXYLATO)-1,3-
BENZODIOXOLE see DQM600
O,O-DIMETHYL-S-(2-(METHYLAMINO)-2-
OXOETHYL) PHOSPHORODITHIOATE see DSP400
2,2-DIMETHYL-4-(N-METHYLCARBAMATO)-1,3-
BENZODIOXOLE see DQM600
O,O-DIMETHYL-S-(N-METHYL-CARBAMOYL)-
METHYL-DITHIOFOSFAAT (DUTCH) see DSP400
(O,O-DIMETHYL-S-(N-METHYL-CARBAMOYL-
METHYL)-DITHIOPHOSPHAT) (GERMAN) see
DSP400
O,O-DIMETHYL-S-(N-
METHYLCARBAMOYLMETHYL)
DITHIOPHOSPHATE see DSP400
O,O-DIMETHYL METHYLCARBAMOYLMETHYL
PHOSPHORODITHIOATE see DSP400
O,O-DIMETHYL-S-(N-
METHYLCARBAMOYLMETHYL)
PHOSPHORODITHIOATE see DSP400
O,O-DIMETHYL-O-(2-N-METHYLCARBAMOYL-1-
METHYL-VINYL)-FOSFAAT (DUTCH) see MRH209
O,O-DIMETHYL-O-(2-N-METHYLCARBAMOYL-1-
METHYL)-VINYL-PHOSPHAT (GERMAN) see MRH209
O,O-DIMETHYL-O-(2-N-METHYLCARBAMOYL-1-
METHYL-VINYL) PHOSPHATE see MRH209

N,N-DIMETHYL-α-
METHYLCARBAMOYLOXYIMINO-α-
(METHYLTHIO)ACETAMIDE see DSP600
N',N'-DIMETHYL-N-((METHYLCARBAMOYL)OXY)-1-
METHYLTHIOOXAMIMIDIC ACID see DSP600
N',N'-DIMETHYL-N-((METHYLCARBAMOYL)OXY)-1-
THIOOXAMIMIDIC ACID METHYL ESTER see
DSP600
O,O-DIMETHYL-S-(N-METHYLCARBAMYLMETHYL)
THIOTHIONOPHOSPHATE see DSP400
O,O-DIMETHYL O-(1-METHYL-2-CARBOXYVINYL)
PHOSPHATE see MQR750
N,N-DIMETHYL-N'-(2-METHYL-4-CHLOROPHENYL)-
FORMAMIDINE see CJJ250
N,N-DIMETHYL-N'-(2-METHYL-4-CHLORPHENYL)-
FORMADIN (GERMAN) see CJJ250
O,O-DIMETHYL-S-(3-METHYL-2,4-DIOXO-3-AZA-
BUTYL)-DITHIOFOSFAAT (DUTCH) see DRR200
O,O-DIMETHYL-S-(3-METHYL-2,4-DIOXO-3-AZA-
BUTYL)-DITHIOPHOSPHAT (GERMAN) see DRR200
6,6-DIMETHYL-2-
METHYLENEBICYCLO(3.1.1)HEPTANE see POH750
DIMETHYLMETHYLENE-p,p'-DIPHENOL see BLD500
5,5-DIMETHYL-2-((1-METHYLETHYL)IMINO)1,3-
DITHIOLAN-4-ONE, o-
((METHYL((TRICHLOROMETHYL)THIO)AMINO)CAR
BONYL)OXIME see DSP710
O,O-DIMETHYL-S-(N-METHYL-N-FORMYL-
CARBAMOYLMETHYL)-DITHIOPHOSPHAT see
DRR200
O,O-DIMETHYL-S-(N-METHYL-N-
FORMYLCARBAMOYLMETHYL)PHOSPHORODITHI
OATE see DRR200
O,O-DIMETHYL O-(4-
METHYLMERCAPTOPHENYL)PHOSPHATE see
PHD250
(E)-DIMETHYL 1-METHYL-3-(METHYLAMINO)-3-
OXO-1-PROPENYL see MRH209
DIMETHYL-1-METHYL-2-
(METHYLCARBAMOYL)VINYLPHOSPHATE, cis
PHOSPHATE see MRH209
O,O-DIMETHYL-O-(3-METHYL-4-NITROFENYL)-
MONOTHIOFOSFAAT (DUTCH) see DSQ000
O,O-DIMETHYL-O-(3-METHYL-4-NITRO-PHENYL)-
MONOTHIOPHOSPHAT (GERMAN) see DSQ000
O,O-DIMETHYL-O-(3-METHYL-4-NITROPHENYL)
PHOSPHOROTHIOATE see DSQ000
DIMETHYL-3-METHYL-4-
NITROPHENYLPHOSPHOROTHIONATE see DSQ000
O,O-DIMETHYL-O-(3-METHYL-4-NITROPHENYL)
THIOPHOSPHATE see DSQ000
N,N-DIMETHYL-p-(3'-
METHYLPHENYLAZO)ANILINE see DUH600
N,N-DIMETHYL-4-((3-
METHYLPHENYL)AZO)BENZENAMINE see DUH600
N,N-DIMETHYL-2-(o-METHYL-α-
PHENYLBENZYLOXY)ETHYLAMINE
HYDROCHLORIDE see OJW000
3,5-DIMETHYL-N-(2-METHYLPHENYL)-4-NITRO-1H-
PYRAZOLE-1-ACETAMIDE see DSQ810
3,5-DIMETHYL-N-(3-METHYLPHENYL)-1H-
PYRAZOLE-1-ACETAMIDE see DSQ830
3,5-DIMETHYL-N-(4-METHYLPHENYL)-1H-
PYRAZOLE-1-ACETAMIDE see DSQ840
3,3-DIMETHYL-1-(m-METHYLPHENYL)TRIAZENE
see DSR200

3,3-DIMETHYL-1-(o-METHYLPHENYL)TRIAZENE see MNT500

DIMETHYL METHYLPHOSPHONATE see DSR400

O,O-DIMETHYL-O-(3-METHYL) PHOSPHOROTHIOATE see DSQ000

DIMETHYL-3-(2-METHYL-1-PROPENYL)CYCLOPROPANECARBOXYLATE see BEP500

3,5-DIMETHYL-4-(METHYLTHIO)PHENOL METHYLCARBAMATE see DST000

3,5-DIMETHYL-4-METHYL-THIOPHENYL-N-CARBAMAT (GERMAN) see DST000

3,5-DIMETHYL-4-METHYLTHIOPHENYL-N-METHYLCARBAMATE see DST000

DIMETHYL-p-(METHYLTHIO)PHENYL PHOSPHATE see PHD250

O,O-DIMETHYL-S-(N-MONOMETHYL)-CARBAMYL METHYLDITHIOPHOSPHATE see DSP400

DIMETHYL MONOSULFATE see DUD100

3,17-DIMETHYL-9-α,13-α,14-α-MORPHINAN PHOSPHATE see MLP250

(9-α,13-α,14-α)-3,17-DIMETHYLMORPHINAN PHOSPHATE see MLP250

1,2-DIMETHYL-3-(2-NAPHTHALENYL)(2R,3S)-REL-3-PYRROLIDINOL DROCHLORIDE see DSU300

N,N-DIMETHYL-1-NAPHTHYLAMINE see DSU400

DIMETHYL-α-NAPHTHYLAMINE see DSU400

α-DIMETHYLNAPHTHYLAMINE see DSU400

N,N-DIMETHYL-α-NAPHTHYLAMINE see DSU400

DIMETHYLNITRAMIN (GERMAN) see DSV200

DIMETHYLNITRAMINE see DSV200

DIMETHYLNITROAMINE see DSV200

O,O-DIMETHYL-O-p-NITROFENYLESTER KYSELINY THIOFOSFORECNE (CZECH) see MNH000

O,O-DIMETHYL-O-(4-NITROFENYL)-MONOTHIOFOSFAAT (DUTCH) see MNH000

DIMETHYLNITROMETHANE see NIY000

O,O-DIMETHYL-O-(4-NITRO-3-METHYLPHENYL)THIOPHOSPHATE see DSQ000

O,O-DIMETHYL-O-(4-NITRO-PHENYL)-MONOTHIOPHOSPHAT (GERMAN) see MNH000

DIMETHYL p-NITROPHENYL MONOTHIOPHOSPHATE see MNH000

O,O-DIMETHYL-O-(4-NITROPHENYL) PHOSPHOROTHIOATE see MNH000

O,O-DIMETHYL-O-(p-NITROPHENYL) PHOSPHOROTHIOATE see MNH000

DIMETHYL 4-NITROPHENYL PHOSPHOROTHIONATE see MNH000

O,O-DIMETHYL-O-(4-NITROPHENYL)-THIONOPHOSPHAT (GERMAN) see MNH000

O,O-DIMETHYL-O-(p-NITROPHENYL)-THIONOPHOSPHAT (GERMAN) see MNH000

DIMETHYL-p-NITROPHENYL THIONPHOSPHATE see MNH000

DIMETHYL p-NITROPHENYL THIOPHOSPHATE see MNH000

O,O-DIMETHYL-O-p-NITROPHENYL THIOPHOSPHATE see MNH000

3,3-DIMETHYL-1-(p-NITROPHENYL)TRIAZENE see DSX400

DIMETHYLNITROSAMIN (GERMAN) see NKA600

DIMETHYLNITROSAMINE see NKA600

N,N-DIMETHYLNITROSAMINE see NKA600

DIMETHYLNITROSOAMINE see NKA600

N,N-DIMETHYL-p-NITROSOANILINE see DSY600

DIMETHYL-p-NITROSOANILINE (DOT) see DSY600

N,N-DIMETHYL-4-NITROSOBENZENAMINE see DSY600

DIMETHYLNITROSOHARNSTOFF (GERMAN) see DTB200

DIMETHYLNITROSOMORPHOLINE see DTA000

2,6-DIMETHYLNITROSOMORPHOLINE see DTA000

2,6-DIMETHYL-N-NITROSOMORPHOLINE see DTA000

DIMETHYL(p-NITROSOPHENYL)AMINE see DSY600

1,3-DIMETHYLNITROSOUREA see DTB200

1,3-DIMETHYL-N-NITROSOUREA see DTB200

N,N'-DIMETHYLNITROSOUREA see DTB200

O,O-DIMETHYL-O-4-NITRO-m-TOLYL PHOSPHOROTHIOATE see DSQ000

3,7-DIMETHYL-2,6-OCTADADIEN-1-YL PROPIONATE see GDM450

DIMETHYLOCTADECYLBENZYLAMMONIUM CHLORIDE see DTC600

3,7-DIMETHYL-2,6-OCTADIENAL see DTC800

3,7-DIMETHYL-1,6-OCTADIENE see CMT050

2,6-DIMETHYL-2,7-OCTADIENE-6-OL see LFX000

2,6-DIMETHYLOCTA-2,7-DIEN-6-OL see LFX000

3,7-DIMETHYLOCTA-1,6-DIEN-3-OL see LFX000

3,7-DIMETHYL-1,6-OCTADIEN-3-OL see LFX000

3,7-DIMETHYL-(E)-2,6-OCTADIEN-1-OL see DTD000

3,7-DIMETHYL-(Z)-2,6-OCTADIEN-1-OL see DTD200

2-cis-3,7-DIMETHYL-2,6-OCTADIEN-1-OL see DTD200

2,6-DIMETHYL-trans-2,6-OCTADIEN-8-OL see DTD000

3,7-DIMETHYL-trans-2,6-OCTADIEN-1-OL see DTD000

3,7-DIMETHYL-1,6-OCTADIEN-3-OL ACETATE see LFY600

trans-3,7-DIMETHYL-2,6-OCTADIEN-1-OL ACETATE see DTD800

3,7-DIMETHYL-1,6-OCTADIEN-3-OL BENZOATE see LFZ000

trans-3,7-DIMETHYL-2,6-OCTADIEN-1-OL FORMATE see GCY000

3,7-DIMETHYL-1,6-OCTADIEN-3-OL ISOBUTYRATE see LGB000

(E)-3,7-DIMETHYL-2,6-OCTADIEN-1-OL PROPIONATE see GDM450

3,7-DIMETHYL-1,6-OCTADIEN-3-YL ACETATE see LFY600

3,7-DIMETHYL-2-trans-6-OCTADIENYL ACETATE see DTD800

trans-3,7-DIMETHYL-2,6-OCTADIEN-1-YL ACETATE see DTD800

3,7-DIMETHYL-1,6-OCTADIEN-3-YL BENZOATE see LFZ000

3,7-DIMETHYL-2,6-OCTADIEN-1-YL BENZOATE see GDE800

3,7-DIMETHYL-2,6-OCTADIENYL ESTER FORMIC ACID (E) see GCY000

trans-2,6-DIMETHYL-2,6-OCTADIEN-8-YL ETHANOATE see DTD800

3,7-DIMETHYL-1,6-OCTADIEN-3-YL FORMATE see LGA050

trans-3,7-DIMETHYL-2,6-OCTADIEN-1-YL FORMATE see GCY000

3,7-DIMETHYL-1,6-OCTADIEN-3-YL ISOBUTYRATE see LGB000

trans-3,7-DIMETHYL-2,6-OCTADIENYL ISOPENTANOATE see GDK000

3,7-DIMETHYL-2,6-OCTADIEN-1-YL PHENYLACETATE see GDM400

trans-3,7-DIMETHYL-2,6-OCTADIEN-1-YL PHENYLACETATE see GDM400

N-(2,6-DIMETHYLPHENYL)-N-(METHOXYACETYL)-DL-ALANINE METHYL ESTER see MDM100
N,N-DIMETHYL-3((1-PHENYLMETHYL)-1H-INDAZOL-3-YL)OXY-1-PROPANAMINE HYDROCHLORIDE see BBW500
N-(2,6-DIMETHYLPHENYL)-1-METHYL-2-PIPERIDINECARBOXAMIDE-MONOHYDROCHLORIDE see CBR250
2,3-DIMETHYL-1-PHENYL-3-PYRAZOLIN-5-ONE see AQN000
2,3-DIMETHYL-1-PHENYL-5-PYRAZOLONE see AQN000
N,N-DIMETHYL-3-PHENYL-3-(2-PYRIDYL)PROPYLAMINE see TMJ750
3,3-DIMETHYL-1-PHENYLTRIAZENE see DTP000
3,3-DIMETHYL-1-PHENYL-1-TRIAZENE see DTP000
DIMETHYLPHOSPHATE ESTER with 3-HYDROXY-N,N-DIMETHYL-cis-CROTONAMIDE see DGQ875
DIMETHYL PHOSPHATE ESTER of 3-HYDROXY-N-METHYL-cis-CROTONAMIDE see MRH209
DIMETHYL PHOSPHATE of 3-HYDROXY-N,N-DIMETHYL-cis-CROTONAMIDE see DGQ875
DIMETHYL PHOSPHATE of 3-HYDROXY-N-METHYL-cis-CROTONAMINE see MRH209
DIMETHYL PHOSPHITE see DSG600
DIMETHYL PHOSPHONATE see DSG600
DIMETHYLPHOSPHORAMIDOCYANIDIC ACID, ETHYL ESTER see EIF000
O,S-DIMETHYL PHOSPHORAMIDOTHIOATE see DTQ400
O,O-DIMETHYLPHOSPHOROCHLORIDOTHIOATE see DTQ600
DIMETHYL PHOSPHOROCHLORIDOTHIOATE (DOT) see DTQ600
O,O-DIMETHYL PHOSPHORODITHIOATE N-FORMYL-2-MERCAPTO-N-METHYLACETAMIDE-S-ESTER see DRR200
N-((O,O-DIMETHYLPHOSPHORODITHIOYL)ETHYL)ACETAMIDE see DOP200
O,O-DIMETHYL PHOSPHOROTHIOATE-O,O-DIESTER with 4,4'-THIODIPHENOL see TAL250
DIMETHYL PHOSPHOROUS ACID see DSG600
DIMETHYL PHTHALATE see DTR200
(O,O-DIMETHYL-PHTHALIMIDOMETHYL-DITHIOPHOSPHATE) see PHX250
O,O-DIMETHYL S-(N-PHTHALIMIDOMETHYL) DITHIOPHOSPHATE
/SSS± □ O,O-DIMETHYL S-PHTHALIMIDOMETHYL PHOSPHORODITHIOATE see PHX250
DIMETHYLPOLYSILOXANE see DTR850
1,2-DIMETHYLPROPANAMINE see AOE200
2,2-DIMETHYLPROPANE see NCH000
2,2-DIMETHYLPROPANE, other than pentane and isopentane (DOT) see NCH000
N,N-DIMETHYL-1,3-PROPANEDIAMINE see AJQ100
2,2-DIMETHYLPROPANOIC ACID see PJA500
2,2-DIMETHYLPROPANOIC ACID ISOOCTADECYL ESTER see ISC550
2,2-DIMETHYLPROPANOYL CHLORIDE see DTS400
DIMETHYL PROPIOLACTONE see DTH000
3,3-DIMETHYL-β-PROPIOLACTONE see DTH000
2,2-DIMETHYLPROPIONIC ACID see PJA500
α,α-DIMETHYLPROPIONIC ACID see PJA500
2,2-DIMETHYLPROPIONYL CHLORIDE see DTS400
1,2-DIMETHYLPROPYLAMINE see AOE200

N,N-DIMETHYL-1,3-PROPYLENEDIAMINE see AJQ100
2,3-DIMETHYLPYRAZINE see DTU400
2,5-DIMETHYLPYRAZINE see DTU600
2,6-DIMETHYLPYRAZINE see DTU800
3,4-DIMETHYLPYRIDINE see LJB000
1,4-DIMETHYL-5H-PYRIDO(4,3-b)INDOL-3-AMINE see TNX275
1,4-DIMETHYL-5H-PYRIDO(4,3-b)INDOL-3-AMINE ACETATE see AJR500
1,4-DIMETHYL-5H-PYRIDO(4,3-b)INDOL-3-AMINE MONOACETATE see AJR500
(3,3-DIMETHYL-1-(m-PYRIDYL-N-OXIDE))TRIAZENE see DTV200
N¹-(4,6-DIMETHYL-2-PYRIMIDINYL)SULFANILAMIDE see SNJ000
N-(4,6-DIMETHYL-2-PYRIMIDYL)SULFANILAMIDE see SNJ000
1,3-DIMETHYL PYROGALLATE see DOJ200
2,5-DIMETHYLPYRROLE see DTV300
N,N-DIMETHYL-3-(PYRROLIDIN-1-YL)PROPIONAMIDE see DTV330
2,4-DIMETHYLPYRROL-3-YL METHYL KETONE see ACI500
2,5-DIMETHYLPYRROL-3-YL METHYL KETONE see ACI550
o,o'-DIMETHYL S,S'-2,3-QUINOXALINEDIYL THIOCARBONATE see BHL800
6,7-DIMETHYL-9-d-RIBITYLISOALLOXAZINE see RIK000
7,8-DIMETHYL-10-d-RIBITYLISOALLOXAZINE see RIK000
7,8-DIMETHYL-10-(d-RIBO-2,3,4,5-TETRAHYDROXYPENTYL)ISOALLOXAZINE see RIK000
DIMETHYL SILICONE see DTR850
N,N-DIMETHYL-4-STILBENAMINE see DUB800
(E)-N,N-DIMETHYL-4-STILBENAMINE see DUC000
trans-N,N-DIMETHYL-4-STILBENAMINE see DUC000
N,N-DIMETHYL-p-STYRYLANILINE see DUB800
DIMETHYLSULFAAT (DUTCH) see DUD100
O,O-DIMETHYL O-p-SULFAMOYLPHENYL PHOSPHOROTHIOATE see CQL250
3,4-DIMETHYL-5-SULFANILAMIDOISOXAZOLE see SNN500
4,6-DIMETHYL-2-SULFANILAMIDOPYRIMIDINE see SNJ000
DIMETHYLSULFAT (CZECH) see DUD100
DIMETHYL SULFATE see DUD100
DIMETHYLSULFID (CZECH) see TFP000
DIMETHYL SULFIDE (DOT) see TFP000
DIMETHYLSULFIDE-α,α'-DICARBOXYLIC ACID see MCM750
1,4-DIMETHYLSULFONOXYBUTANE see BOT250
DIMETHYL SULFOXIDE see DUD800
3,4-DIMETHYL-5-SULPHANILAMIDOISOXAZOLE see SNN500
as-DIMETHYL SULPHATE see MLH500
DIMETHYL SULPHIDE (ACGIH) see TFP000
3,4-DIMETHYL-5-SULPHONAMIDOISOXAZOLE see SNN500
DIMETHYL SULPHOXIDE see DUD800
DIMETHYL TEREPHTHALATE see DUE000
N,N-DIMETHYL-N-TETRADECYLBENZENEMETHANAMINIUM, CHLORIDE (9CI) see TCA500

3,5-DIMETHYLTETRAHYDRO-1,3,5-THIADIAZINE-2-THIONE see DSB200
3,5-DIMETHYLTETRAHYDRO-1,3,5-2H-THIADIAZINE-2-THIONE see DSB200
3,5-DIMETHYL-1,3,5-2H-TETRAHYDROTHIADIAZINE-2-THIONE see DSB200
3,5-DIMETHYLTETRAHYDRO-2H-1,3,5-THIADIAZINE-2-THIONE see DSB200
3,5-DIMETHYL-1,2,3,5-TETRAHYDRO-1,3,5-THIADIAZINETHIONE-2 see DSB200
4,4'-(2,3-DIMETHYLTETRAMETHYLENE)DIPYROCATECHOL see NBR000
O,O-DIMETHYL-S-(3-THIA-PENTYL)-MONOTHIOPHOSPHAT (GERMAN) see DAP400
DIMETHYLTHIENYLCETONE see DUG425
DIMETHYLTHIOCARBAMIDE see DSK900
N,N'-DIMETHYLTHIOCARBAMIDE see DSK900
2,2'-DIMETHYLTHIOCARBANILIDE see DXP600
3,3-DIMETHYL-2,5-THIOMORPHOLINEDIONE 2-(o-((METHYLAMINO)CARBONYL)OXIME) see MID860
3,4-DIMETHYL-2,5-THIOMORPHOLINEDIONE, 2-(o-((METHYL((TRICHLOROMETHYL)THIO)AMINO)CARBONYL)OXIME) see DUG550
3,5-DIMETHYL-2-THIONOTETRAHYDRO-1,3,5-THIADIAZINE see DSB200
O,O-DIMETHYLTHIOPHOSPHORIC ACID, p-CHLOROPHENYL ESTER see MQH750
2-(O,S-DIMETHYLTHIOPHOSPHORYLIMINO)-3-METHYLTHIAZOLIDINE see MEY200
1,3-DIMETHYLTHIOUREA see DSK900
sym-DIMETHYLTHIOUREA see DSK900
DIMETHYL-TIN BIS(ISOOCTYLTHIOGLYCOLLATE) see BKK500
DIMETHYLTIN DIBROMIDE see DUG800
DIMETHYLTIN DIFLUORIDE see DKH200
DIMETHYLTIN FLUORIDE see DKH200
DIMETHYLTIN OXIDE see DTH400
N,N-DIMETHYL-p-((m-TOLYL)AZO)ANILINE see DUH600
3,3-DIMETHYL-1-(m-TOLYL)TRIAZENE see DSR200
3,3-DIMETHYL-1-(o-TOLYL)TRIAZENE see MNT500
(DIMETHYLTRIAZENO)IMIDAZOLECARBOXAMIDE see DAB600
4-(DIMETHYLTRIAZENO)IMIDAZOLE-5-CARBOXAMIDE see DAB600
5-(DIMETHYLTRIAZENO)IMIDAZOLE-4-CARBOXAMIDE see DAB600
5-(3,3-DIMETHYLTRIAZENO)IMIDAZOLE-4-CARBOXAMIDE see DAB600
4-(3,3-DIMETHYL-1-TRIAZENO)IMIDAZOLE-5-CARBOXAMIDE see DAB600
5-(3,3-DIMETHYL-1-TRIAZENO)IMIDAZOLE-4-CARBOXAMIDE see DAB600
4-(5)-(3,3-DIMETHYL-1-TRIAZENO)IMIDAZOLE-5(4)-CARBOXAMIDE see DAB600
3-(3',3'-DIMETHYLTRIAZENO)PYRIDINE-N-OXIDE see DTV200
3-(3',3'-DIMETHYLTRIAZENO)-PYRIDIN-N-OXID (GERMAN) see DTV200
5-(3,3-DIMETHYL-1-TRIAZENYL)-1H-IMIDAZOLE-4-CARBOXAMIDE see DAB600
O,O-DIMETHYL-(2,2,2-TRICHLOOR-1-HYDROXY-ETHYL)-FOSFONAAT (DUTCH) see TIQ250
O,O-DIMETHYL-(2,2,2-TRICHLOR-1-HYDROXY-AETHYL)PHOSPHONAT (GERMAN) see TIQ250
DIMETHYLTRICHLOROHYDROXYETHYL PHOSPHONATE see TIQ250

DIMETHYL-2,2,2-TRICHLORO-1-HYDROXYETHYLPHOSPHONATE see TIQ250
O,O-DIMETHYL-2,2,2-TRICHLORO-1-HYDROXYETHYL PHOSPHONATE see TIQ250
O,O-DIMETHYL-O-2,4,5-TRICHLOROPHENYL PHOSPHOROTHIOATE see RMA500
DIMETHYL TRICHLOROPHENYL THIOPHOSPHATE see RMA500
O,O-DIMETHYL-O-(2,4,5-TRICHLOROPHENYL)THIOPHOSPHATE see RMA500
O,O-DIMETHYL-O-(3,5,6-TRICHLORO-2-PYRIDYL)PHOSPHOROTHIOATE see CMA250
O,O-DIMETHYL-O-(2,4,5-TRICHLORPHENYL)-THIONOPHOSPHAT (GERMAN) see RMA500
N,N-DIMETHYL-2-(TRIFLUOROMETHYL)-10H-PHENOTHIAZINE-10-PROPANAMINE see TKL000
1,1-DIMETHYL-3-(3-TRIFLUOROMETHYLPHENYL)UREA see DUK800
N,N-DIMETHYL-N'-(3-TRIFLUOROMETHYLPHENYL)UREA see DUK800
1,1-DIMETHYL-3-(α,α,α-TRIFLUORO-m-TOLYL)UREA see DUK800
9-cis-3,7-DIMETHYL-9-(2,6,6-TRIMETHYL-1-CYCLOHEXEN-1-YL)-2,4,6,8-NONATETRAENAL see VSK975
3,7-DIMETHYL-9-(2,6,6-TRIMETHYL-1-CYCLOHEXEN)-1-YL-2,4,6,8-NONATETRAENOIC ACID see VSK950
3,7-DIMETHYL-9-(2,6,6-TRIMETHYL-1-CYCLOHEXEN-1-YL)-2,4,6,8-NONATETRAEN-1-OL see VSK600
6,10-DIMETHYL-UNDECA-5,9-DIEN-2-ONE see GDE400
1,3-DIMETHYLUREA see DUM200
N,N'-DIMETHYLUREA see DUM200
sym-DIMETHYLUREA see DUM200
β,β-DIMETHYLVINYL CHLORIDE see IKE000
1,5-DIMETHYL-1-VINYL-4-HEXEN-1-OL BENZOATE see LFZ000
1,5-DIMETHYL-1-VINYL-4-HEXEN-1-YL BENZOATE see LFZ000
1,5-DIMETHYL-1-VINYL-4-HEXENYL ESTER, ISOBUTYRIC ACID see LGB000
DIMETHYL VIOLOGEN see PAI990
1,3-DIMETHYLXANTHINE see TEP000
3,7-DIMETHYLXANTHINE see TEO500
DIMETHYL YELLOW see DOT300
DIMETHYL YELLOW-N,N-DIMETHYLANILINE see DOT300
DIMETHYLZINN-S,S'-BIS(ISOOCTYLTHIOGLYCOLAT) (GERMAN) see BKK500
DIMETHYOXYDOPAMINE see DOE200
10,11-DIMETHYSTRYCHNINE see BOL750
5-(DIMETILAMINOETILOSIMINO-5H-DIBENZO(a,d)CICLOEPTA-1,4-DIENE) CLORIDRATO (ITALIAN) see DPH600
N-(γ-DIMETILAMINOPROPIL)-IMINODIBENZILE CLORIDRATO (ITALIAN) see DLH630
O,O-DIMETIL-O-(1,4-DIMETIL-3-OXO-4-AZA-PENT-1-ENIL)-FOSFATO (ITALIAN) see DGQ875
2,6-DIMETIL-EPTAN-4-ONE (ITALIAN) see DNI800
O,O-DIMETIL-S-(2-ETILITIO-ETIL)-MONOTIOFOSFATO (ITALIAN) see DAP400
O,O-DIMETIL-S-(2-ETIL-SOLFINIL-ETIL)-MONOTIOFOSFATO (ITALIAN) see DAP000

O,O-DIMETIL-S-(ETILTIO-ETIL)-DITIOFOSFATO (ITALIAN) see PHI500
O,O-DIMETIL-O-(2-ETILTIO-ETIL)-MONOTIOFOSFATO (ITALIAN) see DAO800
DIMETILFORMAMIDE (ITALIAN) see DSB000
O,O-DIMETIL-S-(N-FORMIL-N-METIL-CARBAMOIL-METIL)-DITIOFOSFATO (ITALIAN) see DRR200
O,O-DIMETIL-S-(N-METIL-CARBAMOIL-METIL)-DITIOFOSFATO (ITALIAN) see DSP400
O,O-DIMETIL-O-(2-N-METILCARBAMOIL-1-METIL-VINIL)-FOSFATO (ITALIAN) see MRH209
O,O-DIMETIL-O-(3-METIL-4-NITRO-FENIL)-MONOTIOFOSFATO (ITALIAN) see DSQ000
O,O-DIMETIL-S-((2-METOSSI-1,3,4-(4H)-TIADIZAOL-5-ON-4-IL)-METIL)-DITIFOSFATO (ITALIAN) see DSO000
O,O-DIMETIL-O-(4-NITRO-FENIL)-MONOTIOFOSFATO (ITALIAN) see MNH000
O,O-DIMETIL-S-((4-OXO-3H-1,2,3-BENZOTRIAZIN-3-IL)-METIL)-DITIOFOSFATO (ITALIAN) see ASH500
3,5-DIMETIL-PERIDRO-1,3,5-THIADIAZIN-2-TIONE (ITALIAN) see DSB200
DIMETILSOLFATO (ITALIAN) see DUD100
O,O-DIMETIL-(2,2,2-TRICLORO-1-IDROSSI-ETIL)-FOSFONATO (ITALIAN) see TIQ250
DIMETON see DSP400
3,3'-DIMETOSSIBENZODINA (ITALIAN) see DCJ200
DIMETOX see TIQ250
DIMETYLFORMAMIDU (CZECH) see DSB000
DIMEVUR see DSP400
DIMEXIDE see DUD800
DIMEZATHINE see SNJ000
DIMILIN see CJV250
DIMIPRESSIN see DLH630
DIMITAN see BIE500
DIMORPHOLINE DISULFIDE see BKU500
DIMORPHOLINIUM HEXACHLOROSTANNATE see DUO500
DIMORPHOLINO DISULFIDE see BKU500
DIMPEA see DOE200
DIMPYLATE see DCM750
DI-MU-CARBONYLHEXACARBONYLDICOBALT see CNB500
DINACORYL see DJS200
3,4,5,6-DINAPHTHACARBAZOLE see DCY000
1,2,5,6-DINAPHTHACRIDINE see DCS400
3,4,6,7-DINAPHTHACRIDINE see DCS600
DI-β-NAPHTHYL-p-PHENYLDIAMINE see NBL000
DI-β-NAPHTHYL-p-PHENYLENEDIAMINE see NBL000
N,N'-DI-β-NAPHTHYL-p-PHENYLENEDIAMINE see NBL000
sym-DI-β-NAPHTHYL-p-PHENYLENEDIAMINE see NBL000
DINATRIUM-AETHYLENBISDITHIOCARBAMAT (GERMAN) see DXD200
DINATRIUM-(N,N'-AETHYLEN-BIS(DITHIOCARBAMAT)) (GERMAN) see DXD200
DINATRIUM-(N,N'-ETHYLEEN-BIS(DITHIOCARBAMAAT)) (DUTCH) see DXD200
DINATRIUMPYROPHOSPHAT (GERMAN) see DXF800
DINEODYMIUM TRIOXIDE see NCC000
DINICKEL TRIOXIDE see NDH500
DINIL see PFA860
DINILE see SNE000
DINITOLMID see DUP300
DINITOLMIDE see DUP300

2,4-DINITRANILINE see DUP600
DINITRATE de DIETHYLENE-GLYCOL (FRENCH) see DJE400
1,3-DINITRATO-2,2-BIS(NITRATOMETHYL)PROPANE see PBC250
DINITRATODIOXOURANIUM, HEXAHYDRATE see URS000
DINITRILE of ISOPHTHALIC ACID see PHX550
2,4-DINITROANILIN (GERMAN) see DUP600
2,4-DINITROANILINA (ITALIAN) see DUP600
2,4-DINITROANILINE see DUP600
2,4-DINITROANISOL see DUP800
2,4-DINITROANISOLE see DUP800
α-DINITROANISOLE see DUP800
3,5-DINITROBENZAMIDE see DUQ150
2,4-DINITROBENZENAMIME see DUP600
DINITROBENZENE see DUQ180
m-DINITROBENZENE see DUQ200
o-DINITROBENZENE see DUQ400
p-DINITROBENZENE see DUQ600
1,2-DINITROBENZENE see DUQ400
1,3-DINITROBENZENE see DUQ200
2,4-DINITROBENZENE see DUQ200
DINITROBENZENE, solution (DOT) see DUQ180
2,4-DINITROBENZENETHIOL see DUR425
1,3-DINITROBENZOL see DUQ200
DINITROBENZOL, solid (DOT) see DUQ180
5,7-DINITRO-1,2,3-BENZOXADIAZOLE see DUR800
4,4'-DINITROBIFENYL (CZECH) see DUS000
4,4'-DINITROBIPHENYL see DUS000
3,5-DINITRO-N,N'-BIS(2,4,6-TRINITROPHENYL)-2,6-PYRIDINEDIAMINE see PQC525
4,6-DINITRO-2-CAPRYLPHENYL CROTONATE see AQT500
4,6-DINITRO-2-(2-CAPRYL)PHENYL CROTONATE see AQT500
DINITROCHLOROBENZENE see CGL750
2,4-DINITROCHLOROBENZENE see CGM000
1,3-DINITRO-4-CHLOROBENZENE see CGM000
2,4-DINITRO-1-CHLOROBENZENE see CGM000
DINITROCHLOROBENZENE (DOT) see CGL750
DINITROCHLOROBENZOL see CGM000
2,4-DINITRO-1-CHLORO-NAPHTHALENE see DUS600
DINITROCRESOL see DUS700
DINITRO-o-CRESOL see DUS700
DINITRO-p-CRESOL see DUT600
2,4-DINITRO-o-CRESOL see DUS700
2,6-DINITRO-p-CRESOL see DUT600
3,5-DINITRO-o-CRESOL see DUT000
4,6-DINITRO-o-CRESOL see DUS700
4,6-DINITRO-o-CRESOLO (ITALIAN) see DUS700
DINITRO-o-CRESOL SODIUM SALT see SGP550
DINITRO-o-CRESOL SODIUM SALT see DUU600
3,5-DINITRO-o-CRESOL SODIUM SALT see DUU600
4,6-DINITRO-o-CRESOL SODIUM SALT see DUU600
DINITRODENDTROXAL see DUS700
DINITRODIGLICOL (ITALIAN) see DJE400
DINITRODIGLYKOL (CZECH) see DJE400
2,6-DINITRO-N,N-DIPROPYL-4-(TRIFLUOROMETHYL)BENZENAMINE see DUV600
2,6-DINITRO-N,N-DI-N-PROPYL-α,α,α-TRIFLURO-p-TOLUIDINE see DUV600
1,1-DINITROETHANE see DUV710
1,1-DINITROETHANE (dry) (DOT) see DUV710
2,4-DINITROFENOL (DUTCH) see DUZ000
DINITROFENOLO (ITALIAN) see DUZ000
2,4-DINITROFLUOROBENZENE see DUW400

2,4-DINITRO-1-FLUOROBENZENE see DUW400
DINITROGEN MONOXIDE see NGU000
DINITROGEN TETRAFLUORIDE see TCI000
DINITROGEN TETROXIDE see NGU500
DINITROGEN TETROXIDE, liquefied (DOT) see
NGU500
DINITROGLICOL (ITALIAN) see EJG000
DINITROGLYCOL see EJG000
3,5-DINITRO-2-HYDROXYTOLUENE see DUS700
4,6-DINITROKRESOL (DUTCH) see DUS700
4,6-DINITRO-o-KRESOL (CZECH) see DUS700
4,6-DINITRO-o-KRESYLESTER KYSELINY OCTOVE
(CZECH) see AAU250
DINITROL see DUS700
2,6-DINITRO-3-METHOXY-4-tert-BUTYLTOLUENE
see BRU500
DINITROMETHYL CYCLOHEXYLTRIENOL see
DUS700
DINITRO(1-METHYLHEPTYL)PHENYL
CROTONATE see AQT500
2,4-DINITRO-6-(1-METHYLHEPTYL)PHENYL
CROTONATE see AQT500
2,4-DINITRO-6-METHYLPHENOL see DUS700
2,4-DINITRO-6-METHYLPHENOL SODIUM SALT see
DUU600
2,4-DINITRO-1-NAFTOL see DUX800
1,3-DINITRONAPHTHALENE see DUX650
1,5-DINITRONAPHTHALENE see DUX700
1,5-DINITRONAPHTHALENE see DUX700
1,8-DINITRONAPHTHALENE see DUX710
2,4-DINITRO-1-NAPHTHOL see DUX800
2-4 DINITRO-α-NAPHTOL see DUX800
2-4 DINITRO-α-NAPHTOL (FRENCH) see DUX800
2,4-DINITRO-6-(2-OCTYL)PHENYL CROTONATE see
AQT500
DINITROPHENOL see DUY600
DINITROPHENOL see DUY600
2,4-DINITROPHENOL see DUZ000
α-DINITROPHENOL see DUZ000
DINITROPHENOL SOLUTIONS (UN 1599) (DOT) see
DUY600
DINITROPHENOL, dry or wetted with <15% water, by
weight (UN 0076) (DOT) see DUY600
DINITROPHENOL, wetted with not <15% water, by
weight (UN 1320) (DOT) see DUY600
DINITROPHENYLMETHANE see DVG600
2,4-DINITROPHENYLMETHYL ETHER see DUP800
DINITROPYRENE see DVD400
DINITROPYRENE see DVD600
DINITROPYRENE see DVD800
1,3-DINITROPYRENE see DVD400
1,6-DINITROPYRENE see DVD600
1,8-DINITROPYRENE see DVD800
2,7-DINITROPYRENE see DVD900
2,4-DINITRORESORCINOL (heavy metal salts of) (dry)
(DOT) see DVF300
3,5-DINITROSALICYLIC ACID see HKE600
N,N'-DINITROSO-N,N'-
DIMETHYLTEREPHTALSAUREAMID see DRO400
N,N'-DINITROSO-N,N'-
DIMETHYLTEREPHTHALAMIDE, not >72% as a paste
(DOT) see DRO400
DINITROSOPENTAMETHYLENETETRAMINE see
DVF400
N,N-DINITROSOPENTAMETHYLENETETRAMINE
see DVF400

3,4-DI-N-NITROSOPENTAMETHYLENETETRAMINE
see DVF400
3,7-DI-N-NITROSOPENTAMETHYLENETETRAMINE
see DVF400
N¹,N³-DINITROSOPENTAMETHYLENETETRAMINE
see DVF400
DINITROSOPIPERAZIN (GERMAN) see DVF200
DINITROSOPIPERAZINE see DVF200
1,4-DINITROSOPIPERAZINE see DVF200
N,N'-DINITROSOPIPERAZINE see DVF200
DINITROSORBIDE see CCK125
2,4-DINITROSO-m-RESORCINOL see DVF300
3,7-DINITROSO-1,3,5,7-
TETRAAZABICYCLO[3.3.1]NONANE see DVF400
3,5-DINITRO-o-TOLUAMIDE see DUP300
DINITROTOLUENE see DVG600
2,3-DINITROTOLUENE see DVG800
2,4-DINITROTOLUENE see DVH000
2,5-DINITROTOLUENE see DVH200
2,6-DINITROTOLUENE see DVH400
3,4-DINITROTOLUENE see DVH600
DINITROTOLUENES, molten (DOT) see PGN100
DINITROTOLUENES, liquid or solid (DOT) see DVG600
2,4-DINITROTOLUOL see DVH000
2,6-DINITRO-4-TRIFLUORMETHYL-N,N-
DIPROPYLANILIN (GERMAN) see DUV600
DINKUM OIL see EQQ000
DINOC see DUS700
DINOC see DUU600
DINOLEINE see DNF600
DINOPOL NOP see DVL600
DINOPROST see POC500
DINOPROSTONE see DVJ200
DINOPROST TROMETHAMINE (USDA) see POC750
DINOSOL see SNN300
DINOVEX see DAL600
DINTOIN see DKQ000
DINTOINA see DNU000
DINURANIA see DUS700
DINYL see PFA860
DIOCTLYN see DJL000
DIOCTYL ADIPATE see AEO000
DIOCTYLAL see DJL000
DIOCTYL-o-BENZENEDICARBOXYLATE see DVL600
DIOCTYLBIS(LAUROYLOXY)STANNANE see DVJ800
DIOCTYLDIDODECANOYLOXYSTANNANE see
DVJ800
DIOCTYLDI(LAUROYLOXY)STANNANE see DVJ800
DIOCTYLDIMETHYLAMMONIUM CHLORIDE see
DTF820
2,2-DIOCTYL-1,3-DIOXA-2-STANNA-7-THIADECAN-
4,10-DIONE see DVN909
2,2-DIOCTYL-1,3,2-DIOXASTANNEPIN-4,7-DIONE see
DVK200
DIOCTYL ESTER of SODIUM SULFOSUCCINATE see
DJL000
DIOCTYL ESTER of SODIUM SULFOSUCCINIC ACID
see DJL000
DIOCTYL FUMARATE see DVK600
DIOCTYL-MEDO FORTE see DJL000
2,2-DIOCTYL-1,3,2-OXATHIASTANNOLANE-5-
OXIDE see DVL200
DIOCTYLOXOSTANNANE see DVL400
DIOCTYL PHTHALATE see DVL600
DIOCTYL PHTHALATE see DVL700
n-DIOCTYL PHTHALATE see DVL600
DI-sec-OCTYL PHTHALATE see DVL700

DIOCTYL(1,2-PROPYLENEDIOXYBIS(MALEOYLDIOXY))STANNANE see DVL800
DIOCTYL SEBACATE see BJS250
DIOCTYL SODIUM SULFOSUCCINATE (FCC) see DJL000
(Z,Z)-4,4'-((DIOCTYLSTANNYLENE)BIS(OXY))BIS(4-OXO-2-BUTANOIC ACID) DIISOOCTYL ESTER see BKL000
DIOCTYLSTANNYLENE MALEATE see DVK200
DIOCTYL SULFOSUCCINATE SODIUM SALT see DJL000
DIOCTYLTHIOACETOXYSTANNANE see DVL200
DIOCTYLTHIOXOSTANNANE see DVM000
DI-N-OCTYLTIN BIS(BUTYL MALEATE) see BHK500
DI-n-OCTYLTIN BIS(DODECYL MERCAPTIDE) see DVM400
DI-n-OCTYLTIN BIS(2-ETHYLHEXYL MALEATE) see DVM600
DI-n-OCTYLTIN BIS(2-ETHYLHEXYL) MERCAPTOACETATE see DVM800
DIOCTYLTINBIS(ISOOCTYL MALEATE) see BKL000
DIOCTYLTIN BIS(ISOOCTYL MERCAPTOACETATE) see BKK750
DIOCTYLTIN-S,S'-BIS(ISOOCTYL MERCAPTOACETATE) see BKK750
DIOCTYLTIN BIS(ISOOCTYL THIOGLYCOLATE) see BKK750
DIOCTYL-TIN BIS(ISOOCTYLTHIOGLYCOLLATE) see BKK750
DI-n-OCTYLTIN DIISOOCTYL THIOGLYCOLATE see BKK750
DIOCTYLTIN DILAURATE see DVJ800
DI-n-OCTYLTIN DILAURATE see DVJ800
DI-N-OCTYLTIN DIMONOBUTYLMALEATE see BHK500
DI-n-OCTYLTIN DI(1,2-PROPYLENEGLYCOLMALEATE) see DVL800
DI-N-OCTYLTIN-2-ETHYLHEXYLDIMERCAPTOETHANOATE see DVM800
DIOCTYLTIN MALEATE see DVK200
DI-n-OCTYLTIN MALEATE see DVK200
DIOCTYLTIN-β-MERCAPTOPROPIONATE see DVN800
DI-n-OCTYLTIN β-MERCAPTOPROPIONATE see DVN800
DIOCTYLTIN OXIDE see DVL400
DI-n-OCTYLTIN OXIDE see DVL400
DI-n-OCTYLTIN SULFIDE see DVM000
DIOCTYLTIN-3,3'-THIODIPROPIONATE see DVN909
DIOCTYLTIN THIOGLYCOLATE see DVL200
DI-n-OCTYLTIN THIOGLYCOLATE see DVL200
DI-N-OCTYLTIN-THIOGLYCOLIC ACID 2-ETHYLHEXYL ESTER see DVM800
DI-n-OCTYL-ZINN-BIS(2-AETHYLHEXYLMALEINAT) (GERMAN) see DVM600
DI-n-OCTYL-ZINN-DI-ISOOCTYLTHIOGLYKOLAT (GERMAN) see BKK750
DI-n-OCTYL-ZINN DILAURAT (GERMAN) see DVJ800
DI-N-OCTYLZINN-DIMONOBUTYLMALEINAT (GERMAN) see BHK500
DI-n-OCTYL-ZINN-DI-(1,2-PROPYLENGLYKOLMALEINAT)(GERMAN) see DVL800
DI-n-OCTYLZINN MALEINAT see DVK200

DI-n-OCTYL-ZINN β-MERCAPTOPROPIONAT (GERMAN) see DVN800
DI-n-OCTYL-ZINN OXYD (GERMAN) see DVL400
DI-n-OCTYL-ZINN THIOGLYKOLAT (GERMAN) see DVL200
DIODOQUIN see DNF600
DIODOXYLIN see DNF600
DIOFORM see DFI210
DIOGYN see EDO000
DIOGYN B see EDP000
DIOGYNETS see EDO000
DIOKAN see DVQ000
DIOKSAN (POLISH) see DVQ000
DIOKSYNY (POLISH) see TAI000
DIOLAMINE see DHF000
DIOLANE see HFP875
DIOLICE see CNU750
DIOMEDICONE see DJL000
DI-ON see DXQ500
DIONE 21-ACETATE see RKP000
DIONIN see ENK500
DIONONE see DMH400
DIORTHOTOLYLGUANIDINE see DXP200
1,4-DIOSSAN-2,3-DIYL-BIS(O,O-DIETIL-DITIOFOSFATO) (ITALIAN) see DVQ709
DIOSSANO-1,4 (ITALIAN) see DVQ000
1,4-DIOSSIBENZENE (ITALIAN) see QQS200
2,4-DIOSSI-5-DIAZOPIRIMIDINA (ITALIAN) see DCQ600
DIOSSIDONE see BRF500
DIOSUCCIN see DJL000
DIOTHENE see PJS750
DIOTILAN see DJL000
DIOVAC see DJL000
DIOVOCYCLIN see EDR000
DIOVOCYLIN see EDR000
DIOXAAN-1,4 (DUTCH) see DVQ000
1,4-DIOXAAN-2,3-DIYL-BIS(O,O-DIETHYL-DITHIOFOSFAAT) (DUTCH) see DVQ709
2,3-DIOXABICYCLO(2.2.2)OCT-5-ENE, 1-ISOPROPYL-4-METHYL- see ARM500
1,6-DIOXACYCLOHEPTADECAN-17-ONE see OKW110
1,7-DIOXACYCLOHEPTADECAN-17-ONE see OKW100
1,3-DIOXACYCLOPENTANE see DVR800
4,9-DIOXA-1,12-DIAMINODODECANE see BGU600
6,9-DIOXA-3,12-DIAZATETRADECANEDIOIC ACID, 3,12-BIS(CARBOXYMETHYL)-(9CI) see EIT000
4,8-DIOXA-3,9-DITHIA-2,10-DIAZAUNDECANEDIOIC ACID, 2,10-DIMETHYL-, BIS(2,3-DIHYDRO-2,2-DIMETHYL-7-BENZOFURANYL) ESTER, 3,9-DIOXIDE see PML270
1,5,2,4-DIOXADITHIEPANE-2,2,4,4-TETRAOXIDE see DVO920
1,5,2,4-DIOXADITHIEPANE, 2,2,4,4-TETRAOXIDE see DVO920
4,9-DIOXADODECANE-1,12-DIAMINE see BGU600
2,5-DIOXAHEXANE see DOE600
p-DIOXAN (CZECH) see DVQ000
DIOXAN-1,4 (GERMAN) see DVQ000
2,3-p-DIOXANDITHIOL S,S-BIS(O,O-DIETHYL PHOSPHORODITHIOATE) see DVQ709
1,4-DIOXAN-2,3-DIYL-BIS(O,O-DIAETHYL-DITHIOPHOSPHAT) (GERMAN) see DVQ709

1,4-DIOXAN-2,3-DIYL-BIS(O,O-DIETHYLPHOSPHOROTHIOLOTHIONATE) see DVQ709
1,4-DIOXAN-2,3-DIYL-O,O,O',O'-TETRAETHYL DI(PHOSPHORODITHIOATE) see DVQ709
DIOXANE see DVQ000
p-DIOXANE see DVQ000
2,3-p-DIOXANE-S,S-BIS(O,O-DIETHYLPHOSPHORODITHIOATE) see DVQ709
p-DIOXANE-2,3-DITHIOL-S,S-DIESTER with O,O-DIETHYL PHOSPHORODITHIOATE see DVQ709
p-DIOXANE-2,3-DIYL ETHYL PHOSPHORODITHIOATE see DVQ709
1,4-DIOXANE (MAK) see DVQ000
DIOXANNE (FRENCH) see DVQ000
3,6-DIOXAOCTANE-1,8-DIOL see TJQ000
3,6-DIOXAOCTANOL, 8-(4-(4-CHLOROPHENYLPHENYLMETHYL)PIPERAZINYL)- see HHK050
5,11-DIOXA-9-THIA-4,7,9,12-TETRAAZAPENTADECA-3,12-DIENEDINITRILE, 6,10-DIOXO-2,2,7,9,14,14-HEXAMETHYL- see BIQ660
DIOXATHION see DVQ709
DIOXIME-p-BENZOQUINONE see DVR200
DIOXIME-1,4-CYCLOHEXADIENEDIONE see DVR200
DIOXIME-2,5-CYCLOHEXADIENE-1,4-DIONE see DVR200
DIOXIN (herbicide contaminant) see TAI000
DIOXINE see TAI000
DIOXITOL see CBR000
9,10-DIOXOANTHRACENE see APK250
p-DIOXOBENZENE see HIH000
1,1',1''-(3,6-DIOXO-1,4-CYCLOHEXADIENE-1,2,4-TRIYL)TRISAZIRIDINE see TND000
2,6-DIOXO-5-DIAZOPYRIMIDINE see DCQ600
DIOXODICHLOROCHROMIUM see CML125
3,3'-(DIOXODIGERMOXANYLENE) DIPROPANOIC ACID see CCF125
3,5-DIOXO-1,2-DIPHENYL-4-N-BUTYLPYRAZOLIDENE see BRF500
DI-OXO-DI-N-PROPYLNITROSAMINE see NJN000
2,2'-DIOXO-DI-N-PROPYLNITROSAMINE see NJN000
2,3-DIOXOINDOLINE see ICR000
DIOXOLAN see DVR600
1,3-DIOXOLAN see DVR800
1,3-DIOXOLANE see DVR800
DIOXOLANE (DOT) see DVR600
1,3-DIOXOLANE, 2-(2,6-DIMETHYL-1,5-HEPTADIENYL)- see CMS324
1,3-DIOXOLANE-4-METHANOL, 2-(1-IODOETHYL)- see IEL800
1,3-DIOXOLAN-2-ONE, 4-(CHLOROMETHYL)- see CBW400
1,3-DIOXOLO(4,5-H)ISOQUINOLINE see MJR775
2,6-DIOXO-4-METHYL-4-ETHYLPIPERIDINE see MKA250
DIOXONE see DJT400
2,2'-DIOXO-N-NITROSODIPROPYLAMINE see NJN000
1,3-DIOXOPHTHALAN see PHW750
1,3-DIOXO-5-PHTHALANCARBOXYLIC ACID see TKV000
2,6-DIOXO-3-PHTHALIMIDOPIPERIDINE see TEH500
2-(2-6-DIOXO-3-PIPERIDINYL)1H-ISOINDOLE-1,3(2H)-DIONE see TEH500
N-(2,6-DIOXO-3-PIPERIDYL)PHTHALIMIDE see TEH500

N,N-DI(2-OXOPROPYL)NITROSAMINE see NJN000
2,2'-DIOXOPROPYL-N-PROPYLNITROSAMINE see NJN000
2,4-DIOXOPYRIMIDINE see UNJ800
1,4-DIOXYANTHRAQUINONE (RUSSIAN) see DMH000
m-DIOXYBENZENE see REA000
o-DIOXYBENZENE see CCP850
p-DIOXYBENZENE see HIH000
1,4-DIOXYBENZENE see QQS200
3,3'-DIOXYBENZIDINE see DMI400
1,4-DIOXY-BENZOL (GERMAN) see QQS200
DIOXYBUTADIENE see BGA750
DIOXYDE de BARYUM (FRENCH) see BAO250
DIOXYDEMETON-S-METHYL see DAP600
3,3'-(DIOXYDICARBONYL)DIPROPIONIC ACID see SNC500
DIOXYETHYLENE ETHER see DVQ000
DIOXYMETHYLENE-PROTOCATECHUIC ALDEHYDE see PIW250
DI(p-OXYPHENYL)-2,4-HEXADIENE see DAL600
DIOZOL see BRF500
DIPA see DNM200
DIPAM see DCK759
DIPAN see DVX200
DIPANOL see MCC250
DIPARALENE HYDROCHLORIDE see CDR250
DIPAV see PAH250
DIPAXIN see DVV600
DIPEGYL see NCR000
DIPELARGONYL PEROXIDE see NNA100
DI(PENTANOYLOXY)DIBUTYLSTANNANE see DEA600
DIPENTENE see MCC250
DIPENTENE DIOXIDE see LFV000
DIPENTYLAMINE see DCH200
DIPENTYL ETHER see PBX000
DIPENTYL KETONE see ULA000
2,4-DI-tert-PENTYLPHENOL see DCI000
DIPENTYL PHTHALATE see AON300
DI-n-PENTYLPHTHALATE see AON300
DIPENTYLTIN DICHLORIDE see DVV200
DIPEPTIDE SWEETENER see ARN825
DIPHACIN see DVV600
DIPHACINONE see DVV600
DIPHANTINE see BBV500
DIPHANTOIN see DKQ000
DIPHANTOINE SODIUM see DNU000
M-DIPHAR see MAS500
DIPHEBUZOL see BRF500
DIPHEDAL see DKQ000
DIPHEDAN see DNU000
DIPHENACIN see DVV600
DIPHENADIONE see DVV600
DIPHENATE see DNU000
DIPHENATRILE see DVX200
DIPHENHYDRINATE see DYE600
DIPHENIN see DNU000
DIPHENINE see DKQ000
DIPHENINE SODIUM see DNU000
o-DIPHENOL see CCP850
1,5-DIPHENOXYANTHRAQUINONE see DVW100
DIPHENTHANE 70 see MJM500
DIPHENTOIN see DKQ000
DIPHENTOIN see DNU000
DIPHENYLACETONITRILE see DVX200
2-DIPHENYLACETYL-1,3-DIKETOHYDRINDENE see DVV600

2-(DIPHENYLACETYL)INDAN-1,3-DIONE see DVV600
2-DIPHENYLACETYL-1,3-INDANDIONE see DVV600
2-(DIPHENYLACETYL)-1H-INDENE-1,3(2H)-DIONE see DVV600
2,3-DIPHENYLACRYLONITRILE see DVX600
α,β-DIPHENYLACRYLONITRILE see DVX600
DIPHENYLAMINE see DVX800
N,N-DIPHENYLAMINE see DVX800
DIPHENYLAMINE-2-CARBOXYLIC ACID see PEG500
DIPHENYLAMINECHLORARSINE see PDB000
DIPHENYLAMINECHLOROARSINE (DOT) see PDB000
DIPHENYLAMINE, HEXANITRO- see HET500
DIPHENYLAN see DKQ000
N,N-DIPHENYLANILINE see TMQ500
DIPHENYLAN SODIUM see DNU000
DIPHENYLARSINOUS ACID see DVY100
DIPHENYLARSINOUS CHLORIDE see CGN000
DIPHENYLBENZENE see TBD000
m-DIPHENYLBENZENE see TBC620
p-DIPHENYLBENZENE see TBC750
1,2-DIPHENYLBENZENE see TBC640
1,4-DIPHENYLBENZENE see TBC750
DIPHENYL BLUE 2B see CMO000
DIPHENYL BLUE 3B see CMO250
DIPHENYL BRILLIANT BLUE see CMO500
DIPHENYL BRILLIANT BLUE FF see CMN750
DIPHENYLBUTAZONE see BRF500
(Z)-2-(p-(1,2-DIPHENYL-1-BUTENYL)-PHENOXY)-N,N-DIMETHYLETHYLAMINE see NOA600
1,2-DIPHENYL-4-BUTYL-3,5-DIOXOPYRAZOLIDINE see BRF500
DIPHENYL CARBONATE see DVZ000
DIPHENYLCHLOORARSINE (DUTCH) see CGN000
DIPHENYLCHLOROARSINE (DOT) see CGN000
DIPHENYLCHROMIUM see BGY700
DIPHENYLCHROMIUM(III) IODIDE see BGY720
α,β-DIPHENYLCINNAMONITRILE see TMQ250
DIPHENYL-α-CYANOMETHANE see DVX200
DIPHENYL DEEP BLACK G see AQP000
DIPHENYLDIAZENE see ASL250
1,2-DIPHENYLDIAZENE see ASL250
DIPHENYL DICHLOROSILANE (DOT) see DFF000
DIPHENYLDIIMIDE see ASL250
DIPHENYL mixed with DIPHENYL OXIDE see PFA860
DIPHENYLE CHLORE, 42% de CHLORE (FRENCH) see PJM500
DIPHENYLE CHLORE, 54% de CHLORE (FRENCH) see PJN000
4,4'-DIPHENYLENEDIAMINE see BBX000
DIPHENYLENEIMINE see CBN000
DIPHENYLENIMIDE see CBN000
DIPHENYLENIMINE see CBN000
1,2-DIPHENYLETHANE see BFX500
1,2-DIPHENYLETHANEDIONE MONOXIME see BCA300
1,2-DIPHENYLETHANONE see PEB000
DIPHENYL ETHER see PFA850
1,2-DIPHENYLETHYLENE see SLR000
α-β-DIPHENYLETHYLENE see SLR000
DIPHENYL-2-ETHYLHEXYL PHOSPHATE see DWB800
DIPHENYL FAST BROWN BRL see CMO750
DIPHENYLGLYCOLLIC ACID-2-(DIETHYLAMINO)ETHYL ESTER HYDROCHLORIDE see BCA000

DIPHENYLGLYOXAL PEROXIDE see BDS000
DIPHENYLGUANIDINE see DWC600
1,3-DIPHENYLGUANIDINE see DWC600
N,N'-DIPHENYLGUANIDINE see DWC600
2,6-DIPHENYL-2,4,6,6,8,8-HEXAMETHYLCYCLOTETRASILOXANE see DWC650
DIPHENYLHYDANTOIN see DKQ000
5,5-DIPHENYLHYDANTOIN see DKQ000
DIPHENYLHYDANTOINE (FRENCH) see DKQ000
DIPHENYLHYDANTOIN SODIUM see DNU000
5,5-DIPHENYLHYDANTOIN SODIUM see DNU000
DIPHENYLHYDRAMINE see BBV500
DIPHENYLHYDRAMINE HYDROCHLORIDE see BAU750
1,2-DIPHENYLHYDRAZINE see HHG000
sym-DIPHENYLHYDRAZINE see HHG000
DIPHENYLHYDROXYARSINE see DVY100
5,5-DIPHENYLIMIDAZOLIDIN-2,4-DIONE see DKQ000
5,5-DIPHENYL-2,4-IMIDAZOLIDINEDIONE see DKQ000
5,5-DIPHENYL-2,4-IMIDAZOLIDINE-DIONE, MONOSODIUM SALT see DNU000
DIPHENYL KETONE see BCS250
DIPHENYLMERCURY see DWD800
DIPHENYLMETHAN-4,4'-DIISOCYANAT (GERMAN) see MJP400
2,4'-DIPHENYLMETHANEDIAMINE see MJP750
4,4'-DIPHENYLMETHANEDIAMINE see MJQ000
DIPHENYL METHANE DIISOCYANATE see MJP400
4,4'-DIPHENYLMETHANE DIISOCYANATE see MJP400
p,p'-DIPHENYLMETHANE DIISOCYANATE see MJP400
DIPHENYLMETHANE 4,4'-DIISOCYANATE (DOT) see MJP400
DIPHENYLMETHANE-4,4'-DIISOCYANATE-TRIMELLIC ANHYDRIDE-ETHOMID HT POLYMER see TKV000
DIPHENYL METHANEPHOSPHONATE see MDQ825
DIPHENYLMETHANONE see BCS250
2-(DIPHENYLMETHOXY)-N,N-DIMETHYL-ETHANAMINE HYDROCHLORIDE see BAU750
2-(DIPHENYLMETHOXY)-N,N-DIMETHYLETHYLAMINE see BBV500
2-DIPHENYLMETHOXY-N,N-DIMETHYLETHYLAMINE HYDROCHLORIDE see BAU750
3-DIPHENYLMETHOXYTROPANE MESYLATE see TNU000
3-DIPHENYLMETHOXYTROPANE METHANESULFONATE see TNU000
DIPHENYLMETHYL BROMIDE (DOT) see BNG750
DIPHENYL METHYL BROMIDE, solution (DOT) see BNH000
DIPHENYLMETHYLCYANIDE see DVX200
α-(DIPHENYLMETHYLENE)BENZENEACETIC ACID see TMQ250
DIPHENYL METHYLPHOSPHONATE see MDQ825
1-(3-(4-(DIPHENYLMETHYL)-1-PIPERAZINYL)PROPYL)-2-BENZIMIDAZOLINONE see OMG000
1-(3-(4-(DIPHENYLMETHYL)-1-PIPERAZINYL)PROPYL)-1,3-DIHYDRO-2H-BENZIMIDAZOL-2-ONE see OMG000
DIPHENYLMETHYLSILANOL see DWH550

4-DIPHENYLMETHYLTROPYLTROPINIUM BROMIDE see PEM750
4-DIPHENYLMETHYL-dl-TROPYLTROPINIUM BROMIDE see PEM750
DIPHENYLNITROSAMIN (GERMAN) see DWI000
DIPHENYLNITROSAMINE see DWI000
N,N-DIPHENYLNITROSAMINE see DWI000
DIPHENYL N-NITROSOAMINE see DWI000
o-DIPHENYLOL see BGJ250
2,2-DI(4-PHENYLOL)PROPANE see BLD500
DIPHENYL (OSHA) see BGE000
DIPHENYL OXIDE see PFA850
DIPHENYLOXIDE-4,4'-DISULFOHYDRAZIDE (DOT) see OPE000
DIPHENYL-p-PHENYLENEDIAMINE see BLE500
N,N'-DIPHENYL-p-PHENYLENEDIAMINE see BLE500
α,α-DIPHENYL-2-PIPERIDINEMETHANOL HYDROCHLORIDE see PII750
1,3-DIPHENYL-1,3-PROPANEDIONE see PFU300
1,3-DIPHENYL-1-PROPEN-3-ONE see CDH000
(+)-1,2-DIPHENYL-2-PROPIONOXY-3-METHYL-4-DIMETHYLAMINOBUTANE HYDROCHLORIDE see PNA500
DIPHENYLPROPYLPHOSPHINE see PNI600
1,1-DIPHENYL-2-PROPYN-1-OL CYCLOHEXANECARBAMATE see DWL400
1,1-DIPHENYL-2-PROPYNYL-N-CYCLOHEXYLCARBAMATE see DWL400
1,1-DIPHENYL-2-PROPYNYL ESTER CYCLOHEXANECARBAMIC ACID see DWL400
DIPHENYL SELENIDE see PGH250
DIPHENYL SKY BLUE 6B see CMO500
1,3-DIPHENYL-1,1,3,3-TETRAMETHYLDISILOXANE see DWN150
3,3',4,4'-DIPHENYLTETRAMINE see BGK500
N,N'-DIPHENYLTHIOCARBAMIDE see DWN800
sym-DIPHENYLTHIOCARBAMIDE see DWN800
DIPHENYLTHIOCARBAZONE see DWN200
DIPHENYLTHIOUREA see DWN800
1,3-DIPHENYLTHIOUREA see DWN800
1,3-DIPHENYL-2-THIOUREA see DWN800
N,N'-DIPHENYLTHIOUREA see DWN800
sym-DIPHENYLTHIOUREA see DWN800
1,3-DIPHENYLTRIAZENE see DWO800
DIPHENYLTRICHLOROETHANE see DAD200
DIPHERGAN see PMI750
DI-PHETINE see DKQ000
DI-PHETINE see DNU000
DIPHONE see SOA500
DIPHOSGEN see TIR920
DIPHOSGENE see PGX000
DIPHOSGENE see TIR920
DIPHOSPHORIC ACID, DISODIUM SALT see DXF800
DIPHOSPHORIC ACID TETRAETHYL ESTER see TCF250
DIPHOSPHORUS PENTOXIDE see PHS250
DIPHYL see PFA860
DIPICRYLAMINE see HET500
DIPICRYLAMINE (DOT) see HET500
DIPIGYL see NCR000
DIPIKRYLAMIN see HET500
DIPIPERAL see FHG000
DIPIPERON see FHG000
DIPIPERONE see FHG000
DIPIRARTRIL-TROPICO see DUD800
DIPIRIN see DOT000

DIPN see DNB200
DIPOFENE see DCM750
DIPOLYOXYETHYLATEDPOLYPROPYLENEGLYCOL ETHER see PJK150
DIPO-SAFT see BFC750
DIPOTASSIUM CHROMATE see PLB250
DIPOTASSIUM DICHLORIDE see PLA500
DIPOTASSIUM DICHROMATE see PKX250
DIPOTASSIUM MONOCHROMATE see PLB250
DIPOTASSIUM NICKEL TETRACYANIDE see NDI000
DIPOTASSIUM PERSULFATE see DWQ000
DIPOTASSIUM SELENITE see SBO100
DIPOTASSIUM TETRACYANONICKELATE see NDI000
DIPOTASSIUM TRICHLORONITROPLATINATE see PLN050
DIPPEL'S OIL see BMA750
DIPPING ACID see SOI500
DIPRAM see DGI000
DIPRAZINE see DQA400
DIPRIVAN see DNR800
DIPRON see SNM500
DIPROPANOIC ACID GERMANIUM SESQUIOXIDE see CCF125
DI-2-PROPENYLAMINE see DBI600
DI-2-PROPENYL ESTER, 1,2-BENZENEDICARBOXYLIC ACID see DBL200
N-N-DI-2-PROPENYL-2-PROPEN-1-AMINE see THN000
DIPROPIONATE d'OESTRADIOL (FRENCH) see EDR000
DIPROPIONATO de ESTILBENE (SPANISH) see DKB000
p,p'-DIPROPIONOXY-trans-α,β-DIETHYLSTILBENE see DKB000
DIPROPIONYL PEROXIDE see DWQ800
DIPROPIONYL PEROXIDE, >28% in solution (DOT) see DWQ800
1,1-DIPROPOXYETHANE see AAG850
DIPROPYL ACETAL see AAG850
DIPROPYL ADIPATE see DWQ875
DI-n-PROPYL ADIPATE see DWQ875
DIPROPYLAMINE see DWR000
DI-n-PROPYLAMINE see DWR000
n-DIPROPYLAMINE see DWR000
DIPROPYLAMINE, 3,3'-BIS((3-(AMINOPROPYL))AMINO)- see TED600
4-(DI-N-PROPYLAMINO)-3,5-DINITRO-1-TRIFLUOROMETHYLBENZENE see DUV600
DIPROPYLCARBAMOTHIOIC ACID-S-ETHYL ESTER see EIN500
N,N-DI-N-PROPYL-2,6-DINITRO-4-TRIFLUOROMETHYLANILINE see DUV600
DI-n-PROPYL DISELENIDE see PNI850
DIPROPYLENE GLYCOL METHYL ETHER see DWT200
DIPROPYLENE GLYCOL MONOMETHYL ETHER see DWT200
DIPROPYLENETRIAMINE see AIX250
DIPROPYL ETHER see PNM000
DIPROPYL KETONE see DWT600
DIPROPYL MERCURY see DWU000
DIPROPYL METHANE see HBC500
DI-n-PROPYLNITROSAMINE see NKB700
DIPROPYLNITROSOAMINE see NKB700
DIPROPYL OXIDE see PNM000
DIPROPYLOXOSTANNANE see DWV000

DI-n-PROPYL PEROXYDICARBONATE see DWV400
DIPROPYL PHTHALATE see DWV500
DI-n-PROPYL PHTHALATE see DWV500
N,N-DIPROPYL-1-PROPANAMINE see TMY250
N,N-DIPROPYLTHIOCARBAMIC ACID-S-ETHYL
ESTER see EIN500
DIPROPYLTIN CHLORIDE see DFF400
DIPROPYLTIN DICHLORIDE see DFF400
DI-n-PROPYLTIN DICHLORIDE see DFF400
DIPROPYLTIN OXIDE see DWV000
N,N-DIPROPYL-4-TRIFLUOROMETHYL-2,6-
DINITROANILINE see DUV600
DIPROSTRON see EDR000
DIPROZIN see DQA400
DIPTERAX see TIQ250
DIPTEREX see TIQ250
DIPTEREX 50 see TIQ250
DIPTEVUR see TIQ250
DIPYRIDO(1,2-a:3',2'-d)IMIDAZOL-2-AMINE see
DWW700
2,2'-DIPYRIDYL see BGO500
α,α'-DIPYRIDYL see BGO500
DI-3-PYRIDYLMERCURY see DWX200
DIPYRIN see DOT000
cis-DIPYRROLIDINEDICHLOROPLATINUM(II) see
DEU200
DI-1H-PYRROL-2-YL KETONE see PPY300
DIQUAT see DWX800
DIQUAT DIBROMIDE see DWX800
DI-QUINOL see DNF600
DIRAX see AQN635
DIRECT BLACK 3 see AQP000
DIRECT BLACK 38 see AQP000
DIRECT BLACK A see AQP000
DIRECT BLACK BRN see AQP000
DIRECT BLACK CX see AQP000
DIRECT BLACK CXR see AQP000
DIRECT BLACK E see AQP000
DIRECT BLACK EW see AQP000
DIRECT BLACK EX see AQP000
DIRECT BLACK FR see AQP000
DIRECT BLACK GAC see AQP000
DIRECT BLACK GW see AQP000
DIRECT BLACK GX see AQP000
DIRECT BLACK GXR see AQP000
DIRECT BLACK JET see AQP000
DIRECT BLACK META see AQP000
DIRECT BLACK METHYL see AQP000
DIRECT BLACK N see AQP000
DIRECT BLACK RX see AQP000
DIRECT BLACK SD see AQP000
DIRECT BLACK WS see AQP000
DIRECT BLACK Z see AQP000
DIRECT BLUE 6 see CMO000
DIRECT BLUE 14 see CMO250
DIRECT BLUE 15 see CMO500
DIRECT BLUE 10G see CMO500
DIRECT BLUE HH see CMO500
DIRECT BRILLIANT BLUE FF see CMN750
DIRECT BROWN 95 see CMO750
DIRECT BROWN BR see PEY000
DIRECT DEEP BLACK E see AQP000
DIRECT DEEP BLACK EAC see AQP000
DIRECT DEEP BLACK EA-CF see AQP000
DIRECT DEEP BLACK E EXTRA see AQP000
DIRECT DEEP BLACK EW see AQP000
DIRECT DEEP BLACK EX see AQP000

DIRECT PURE BLUE see CMO500
DIRECT PURE BLUE M see CMO500
DIRECT RED 28 see SGQ500
DIRECT SKY BLUE A see CMO500
DIREKTAN see NCQ900
DIREMA see CFY000
DIREX 4L see DXQ500
DIREXIODE see DNF600
DIRIDONE see PDC250
DIROX see HIM000
DISALUNIL see CFY000
DISATABS TABS see VSK600
DISELENIDE, BIS(2,2-DIETHOXYETHYL)- see BJA200
DISELENIDE, DIBUTYL-(9CI) see BRF550
DISELENIDE, DIPROPYL-(9CI) see PNI850
3,3'-DISELENODIALANINE see DWY800
β,β'-DISELENODIPROPIONIC ACID, SODIUM SALT
see DWZ100
DISELENO SALICYLIC ACID see SBU150
DISETIL see DXH250
DISFLAMOLL TKP see TNP500
DISILANE see DXA000
DISILOXANE, 1,3-DIPHENYL-1,1,3,3-
TETRAMETHYL- see DWN150
DISILVER OXIDE see SDU500
DISILYN see BEN000
DISIPAL HYDROCHLORIDE see OJW000
DISODIUM ARSENATE see ARC000
DISODIUM ARSENATE, HEPTAHYDRATE see
ARC250
DISODIUM ARSENIC ACID see ARC000
DISODIUM AUROTHIOMALATE see GJC000
DISODIUM-4,4'-BIS((4-ANILINO-6-METHOXY-s-
TRIAZIN-2-YL)AMINO)STILBENE-2,2'-
DISULFONATE see BGW100
DISODIUM-4,4'-BIS(2-SULFOSTYRYL)BIPHENYL see
TGE150
DISODIUM CARBONATE see SFO000
DISODIUM-N-(3-
(CARBOXYMETHYLTHIOMERCURI)-2-
METHOXYPROPYL)-α-CAMPHORAMATE see TFK270
DISODIUM CHROMATE see DXC200
DISODIUM CITRATE see DXC400
DISODIUM CLODRONATE see SFX730
DISODIUM DEXAMETHASONE PHOSPHATE see
DAE525
DISODIUM DIACID
ETHYLENEDIAMINETETRAACETATE see EIX500
DISODIUM-2,7-DIBROM-4-HYDROXY-MERCURI-
FLUORESCEIN see MCV000
DISODIUM-2',7'-DIBROMO-4'-
(HYDROXYMERCURY)FLUORESCEIN see MCV000
DISODIUM
(DICHLOROMETHYLENE)BISPHOSPHONATE see
SFX730
DISODIUM DICHROMATE see SGI000
DISODIUM DIFLUORIDE see SHF500
DISODIUM DIHYDROGEN
ETHYLENEDIAMINETETRAACETATE see EIX500
DISODIUM
DIHYDROGEN(ETHYLENEDINITRILO)TETRAACET
ATE see EIX500
DISODIUM DIHYDROGEN-(1-
HYDROXYETHYLIDENE)DIPHOSPHONATE see
DXD400
DISODIUM DIHYDROGEN PYROPHOSPHATE see
DXF800

DISODIUM S,S'-(2-DIMETHYLAMINO-1,3-PROPANEDIYL)BIS(THIOSULFATE) see DXC900

DISODIUM (2,4-DIMETHYLPHENYLAZO)-2-HYDROXYNAPHTHALENE-3,6-DISULFONATE see FMU070

DISODIUM (2,4-DIMETHYLPHENYLAZO)-2-HYDROXYNAPHTHALENE-3,6-DISULPHONATE see FMU070

DISODIUM DIOXIDE see SJC500

DISODIUM DIPHOSPHATE see DXF800

DISODIUM DISULFITE see SII000

DISODIUM EDATHAMIL see EIX500

DISODIUM EDETATE see EIX500

DISODIUM EDTA (FCC) see EIX500

DISODIUM EOSIN see BNH500

DISODIUM ETHANOL-1,1-DIPHOSPHONATE see DXD400

DISODIUM ETHYDRONATE see DXD400

DISODIUM ETHYLENEBIS(DITHIOCARBAMATE) see DXD200

DISODIUM ETHYLENE-1,2-BISDITHIOCARBAMATE see DXD200

DISODIUM ETHYLENEDIAMINETETRAACETATE see EIX500

DISODIUM ETHYLENEDIAMINETETRAACETIC ACID see EIX500

DISODIUM (ETHYLENEDINITRILO)TETRAACETATE see EIX500

DISODIUM (ETHYLENEDINITRILO)TETRAACETIC ACID see EIX500

DISODIUM ETIDRONATE see DXD400

DISODIUM GMP see GLS800

DISODIUM-5'-GMP see GLS800

DISODIUM-5'-GUANYLATE see GLS800

DISODIUM GUANYLATE (FCC) see GLS800

DISODIUM HEXAFLUOROSILICATE see DXE000

(2-)-DISODIUM HEXAFLUOROSILICATE see DXE000

DISODIUM HYDROGEN ARSENATE see ARC000

DISODIUM HYDROGEN CITRATE see DXC400

DISODIUM HYDROGEN NITRILOTRIACETATE see DXF000

DISODIUM HYDROGEN ORTHOARSENATE see ARC000

DISODIUM HYDROGEN PHOSPHATE see SJH090

DISODIUM-6-HYDROXY-3-OXO-9-XANTHENE-o-BENZOATE see FEW000

DISODIUM-3-HYDROXY-4-((2,4,5-TRIMETHYLPHENYL)AZO)-2,7-NAPHTHALENEDISULFONATE see FAG018

DISODIUM-3-HYDROXY-4-((2,4,5-TRIMETHYLPHENYL)AZO)-2,7-NAPHTHALENEDISULFONIC ACID see FAG018

DISODIUM-3-HYDROXY-4-((2,4,5-TRIMETHYLPHENYL)AZO)-2,7-NAPHTHALENEDISULPHONATE see FAG018

DISODIUM-3-HYDROXY-4-((2,4,5-TRIMETHYLPHENYL)AZO)-2,7-NAPHTHALENEDISULPHONIC ACID see FAG018

DISODIUM IMP see DXE500

DISODIUM INOSINATE see DXE500

DISODIUM-5'-INOSINATE see DXE500

DISODIUM INOSINE-5'-MONOPHOSPHATE see DXE500

DISODIUM INOSINE-5'-PHOSPHATE see DXE500

DISODIUM METASILICATE see SJU000

DISODIUM MOLYBDATE see DXE800

DISODIUM MONOHYDROGEN ARSENATE see ARC000

DISODIUM MONOHYDROGEN PHOSPHATE see SJH090

DISODIUM MONOSELENIDE see SJT000

DISODIUM MONOSILICATE see SJU000

DISODIUM MONOXIDE see SIN500

DISODIUM NITRILOTRIACETATE see DXF000

DISODIUM NITROSYLPENTACYANOFERRATE see SIU500

DISODIUM ORTHOPHOSPHATE see SJH090

DISODIUM OXIDE see SIN500

DISODIUM PEROXIDE see SJC500

DISODIUM PHOSPHATE see SJH090

DISODIUM PHOSPHORIC ACID see SJH090

DISODIUM PYROPHOSPHATE see DXF800

DISODIUM PYROSULFITE see SII000

DISODIUM SALT of EDTA see EIX500

DISODIUM SALT of 2-(4-SULPHO-1-NAPHTHYLAZO)-1-NAPHTHOL-4-SULPHONIC ACID see HJF500

DISODIUM SALT of 1-p-SULPHOPHENYLAZO-2-NAPHTHOL-6-SULPHONIC ACID see FAG150

DISODIUM SALT of 1-(2,4-XYLYLAZO)-2-NAPHTHOL-3,6-DISULFONIC ACID see FMU070

DISODIUM SALT of 1-(2,4-XYLYLAZO)-2-NAPHTHOL-3,6-DISULPHONIC ACID see FMU070

DISODIUM SELENATE see DXG000

DISODIUM SELENITE see SJT500

DISODIUM SEQUESTRENE see EIX500

DISODIUM SILICOFLUORIDE see DXE000

DISODIUM 2-(4-STYRYL-3-SULFOPHENYL)-7-SULFO-2H-NAPHTHO(1,2-d)TRIAZOLE see DXG025

DISODIUM SULFATE see SJY000

DISODIUM SULFITE see SJZ000

DISODIUM-2-(4-SULFO-1-NAPHTHYLAZO)-1-NAPHTHOL-4-SULFONATE see HJF500

DISODIUM-2-(4-SULPHO-1-NAPHTHYLAZO)-1-NAPHTHOL-4-SULPHONATE see HJF500

DISODIUM TARTRATE see BLC000

DISODIUM l-(+)-TARTRATE see BLC000

DISODIUM TETRABORATE see DXG035

DISODIUM TETRACEMATE see EIX500

DISODIUM TETRAOXATUNGSTATE (2-) see SKN500

DISODIUM TETRAOXOTUNGSTATE (2-) see SKN500

DISODIUM TUNGSTATE see SKN500

DISODIUM VERSENATE see EIX500

DISODIUM VERSENE see EIX500

DISOFEN see DNG000

DISOLFURO DI TETRAMETILTIOURAME (ITALIAN) see TFS350

DISOPHENOL see DNG000

2,4-D ISOPROPYL ESTER see IOY000

DISOQUIN see DNF600

DISORLON see CCK125

DISPADOL see DAM700

DISPAL see AHE250

DISPAMIL see PAH250

DISPARICIDA see ABX500

DISPASOL M see PKB500

DISPERMINE see PIJ000

DISPERSE BLUE NO 1 see TBG700

DISPERSED ORANGE 11348 see FAG150

DISPERSED VIOLET 12197 see FAG120

DISPERSED YELLOW 12116 see FAG150

DISPERSE FAST PINK B see AKE250

DISPERSE FAST YELLOW G see AAQ250

DISPERSE MB-61 see TFC600

DISPERSE ORANGE see AKP750

DISPERSE RED 15 see AKE250

DISPERSE RED 25 see AKE250
DISPERSE VIOLET 2S see DBY700
DISPERSE YELLOW 3 see AAQ250
DISPERSE YELLOW G see AAQ250
DISPERSE YELLOW Z see AAQ250
DISPERSIVE YELLOW 3T see AAQ250
DISPERSOL FAST YELLOW G see AAQ250
DISPERSOL ORANGE D-G see AKE250
DISPERSOL PRINTING YELLOW G see AAQ250
DISPERSOL RED PP see SBC500
DISPERSOL YELLOW A-G see AAQ250
DISPERSOL YELLOW PP see PEJ500
DISSOLVANT APV see DJD600
DISTANNANE, HEXAISOPROPYL- see HDY100
DISTANNATHIANE, HEXABUTYL-(9CI) see HCA700
DISTANNOXANE, BIS(1,3-DITHIOCYANATO-1,1,3,3-TETRABUTYL)- see BJM700
DISTANNOXANE, HEXAETHYL- see HCX050
DISTANNOXANE, 1,1,1,3,3,3-HEXAPROPYL- see BLT300
DISTANNTHIANE, HEXABUTYL- see HCA700
DISTANNTHIANE, HEXAETHYL- see HCX100
DISTANNTHIANE, HEXAPROPYL- see HEW200
DISTAVAL see TEH500
DISTAXAL see TEH500
DISTEARIN see OAV000
DISTEARYL THIODIPROPIONATE see DXG650
DISTILBENE see DKA600
DISTILBENE see DKB000
DISTILLATE FUEL, MARINE, PETROLEUM DERIV. see DHE750
DISTILLATES (COAL TAR) see CMY900
DISTILLATES (PETROLEUM), ACID-TREATED HEAVY NAPHTHENIC (9CI) see MQV760
DISTILLATES (PETROLEUM), ACID-TREATED LIGHT NAPHTHENIC (9CI) see MQV770
DISTILLATES (PETROLEUM), ACID-TREATED LIGHT PARAFFINIC (9CI) see MQV775
DISTILLATES (PETROLEUM), HEAVY CATALYTIC CRACKED see DXG810
DISTILLATES (PETROLEUM), HEAVY NAPHTHENIC (9CI) see MQV780
DISTILLATES (PETROLEUM), HEAVY PARAFFINIC (9CI) see MQV785
DISTILLATES (PETROLEUM), HYDROTREATED (mild) HEAVY NAPHTHENIC (9CI) see MQV790
DISTILLATES (PETROLEUM), HYDROTREATED (mild) HEAVY PARAFFINIC (9CI) see MQV795
DISTILLATES (PETROLEUM), HYDROTREATED (mild) LIGHT NAPHTHENIC (9CI) see MQV800
DISTILLATES (PETROLEUM), HYDROTREATED (mild) LIGHT PARAFFINIC (9CI) see MQV805
DISTILLATES (PETROLEUM), LIGHT CATALYTIC CRACKED see DXG840
DISTILLATES (PETROLEUM), LIGHT NAPHTHENIC (9CI) see MQV810
DISTILLATES (PETROLEUM), LIGHT PARAFFINIC (9CI) see MQV815
DISTILLATES (PETROLEUM), SOLVENT-DEWAXED HEAVY NAPHTHENIC (9CI) see MQV820
DISTILLATES (PETROLEUM), SOLVENT-DEWAXED HEAVY PARAFFINIC (9CI) see MQV825
DISTILLATES (PETROLEUM), SOLVENT-DEWAXED LIGHT NAPHTHENIC (9CI) see MQV835
DISTILLATES (PETROLEUM), SOLVENT-DEWAXED LIGHT PARAFFINIC (9CI) see MQV840
DISTILLATES (PETROLEUM), SOLVENT-REFINED (mild) HEAVY NAPHTHENIC (9CI) see MQV845

DISTILLATES (PETROLEUM), SOLVENT-REFINED (mild) HEAVY PARAFFINIC (9CI) see MQV850
DISTILLATES (PETROLEUM), SOLVENT-REFINED (mild) LIGHT NAPHTHENIC (9CI) see MQV852
DISTILLATES (PETROLEUM), SOLVENT-REFINED (mild) LIGHT PARAFFINIC (9CI) see MQV855
DISTILLED LIME OIL see OGM850
DISTILLED MUSTARD see BIH250
DISTIVIT (B12 PEPTIDE) see VSZ000
DISTOKAL see HCI000
DISTOL 8 see EIV000
DISTOPAN see HCI000
DISTOPIN see HCI000
DISTOVAL see TEH500
DISULFAN see DXH250
DISULFATOZIRCONIC ACID see ZTJ000
DISULFIRAM see DXH250
DISULFOTON see DXH325
DISULFURAM see DXH250
DISULFUR DICHLORIDE see SON510
DISULFURE de TETRAMETHYLTHIOURAME (FRENCH) see TFS350
DISULFUROUS ACID, DISODIUM SALT see SII000
DISULFUR PENTOXYDICHLORIDE see PPR500
DISULFURYL CHLORIDE see PPR500
DISULONE see SOA500
DISULPHINE LAKE BLUE EG see FMU059
DISULPHURAM see DXH250
DISULPHURIC ACID see SOI520
DISYNFORMON see EDV000
DI-SYSTON see DXH325
DITAVEN see DKL800
DITHALLIUM CARBONATE see TEJ000
DITHALLIUM SULFATE see TEM000
DITHALLIUM(1+) SULFATE see TEM000
DITHALLIUM TRIOXIDE see TEL050
DITHANE D-14 see DXD200
DITHANE A-4 see DUQ600
DITHANE A-40 see DXD200
DITHANE M 22 see MAS500
DITHANE M 22 SPECIAL see MAS500
DITHIADENOXIDE see DXI480
DITHIADENOXID HYDROGEN MALEATE see DXI480
1,3,2-DITHIARSENOLANE, 2-(p-BIS(2-CHLOROETHYL)AMINOPHENYL)- see BHW300
1,3,2-DITHIARSENOLANE, 2-(p-(DIETHYLAMINO)PHENYL)- see DIP100
1,3,2-DITHIARSENOLE, 2-CHLORODIHYDRO- see EIU900
1,3,2-DITHIARSOLANE, 2-CHLORO- see EIU900
DITHIAZANINE IODIDE see DJT800
DITHIAZANIN IODIDE see DJT800
DITHIAZININE see DJT800
DITHIO see SOD100
O,O-DITHIO-BIS-ANILINE see DXJ800
2,2'-DITHIOBISANILINE see DXJ800
2,2'-DITHIOBIS(BENZOTHIAZOLE) see BDE750
1,1'-DITHIOBIS(N,N-DIETHYLTHIOFORMAMIDE) see DXH250
1,1'-DITHIOBIS(N,N-DIMETHYLTHIO)FORMAMIDE see TFS350
α,α'-DITHIOBIS(DIMETHYLTHIO)FORMAMIDE see TFS350
2,2'-DITHIO-BIS-(ETHYLAMINE) DIHYDROCHLORIDE see CQJ750
DITHIOBISMORPHOLINE see BKU500

DIVINYLENE OXIDE see FPK000
DIVINYLENE SULFIDE see TFM250
DIVINYLENIMINE see PPS250
DIVINYL ETHER (DOT) see VOP000
DIVINYL ETHER, inhibited (DOT) see VOP000
DIVIPAN see DGP900
DIVIT URTO see VSZ100
DIVULSAN see DNU000
DIVYNYL OXIDE see VOP000
DIXANTHOGEN see BJU000
DIXIBEN see EID000
DIXIE see CBT750
DIXIE see KBB600
DIXIECELL see CBT750
DIXIEDENSED see CBT750
DIXITHERM see CBT750
DIXON 164 see TAI250
DIXOPAK see PJS750
DIZENE see DEP600
DIZINON see DCM750
DKB see DCQ800
DKC 1347 see CLK215
DKD see DIZ100
DLP-87 see PPP750
DLP 787 see PPP750
DLT see TFD500
DLTDP see TFD500
DLTP see TFD500
DLX-6000 see TAI250
DM see PDB000
DM see DAC000
DMA see DOO800
DMA see DOQ800
DMAA see HKC000
DMAB see DOR200
DMAB see DOT300
3,2'-DMAB see BLV250
DMAC see DOO800
DMAE see DOY800
DMAEE see BJH750
DMAP see VTF000
DMASA see DQD400
DMBA see DQJ200
7,12-DMBA see DQJ200
DMBC see DQQ200
DMCC see DQY950
DMCT see MIJ500
DMDJ 4309 see PJS750
DMDJ 5140 see PJS750
DMDJ 7008 see PJS750
DMDK see SGM500
DMDT see MEI450
p,p'-DMDT see MEI450
DMEP see DOF400
DMF see BJE750
DMF see DSB000
DMFA see DSB000
DMH see DSF400
DMH see DSF600
DMH see DSF800
DMH see DSG200
DMI HYDROCHLORIDE see DLS600
DMM see BKM500
DMMP see DSR400
DMN see NKA600
DMNA see NKA600
DMNM see DSV200
DMNM see DTA000

DMNO see DSV200
DMN-OAC see AAW000
DMNT see DSG400
DMO see PMO250
DMP see DTR200
2,5-DMP see XKS000
3,5-DMP see XLS000
DMPE see DOE200
DMPEA see DOE200
DMPT see DTP000
DMPTP see TFD500
DMS see TFP000
DMS see DUD100
DMS-70 see DUD800
DMS-90 see DUD800
DMSA see DQD400
DMS(METHYL SULFATE) see DUD100
DMSO see DUD800
DMSP see FAQ800
DMTP (JAPAN) see DSO000
DMTT see DSB200
DMU see DXQ500
DNA see DUP600
DNCB see CGM000
DN-DRY MIX No. 2 see DUS700
2,4-DNFB see DUW400
DNOC-SODIUM see SGP550
DNOC SODIUM SALT see DUU600
DNOK (CZECH) see DUS700
DNOK-ACETAT (CZECH) see AAU250
DNOP see DVL600
DNP see DNG000
2,4-DNP see DUZ000
DNPC see DUT600
DNPD see NBL000
DNPMT see DVF400
DNPT see DVF400
DNPZ see DVF200
DNT see DVH000
2,3-DNT see DVG800
2,4-DNT see DVH000
2,5-DNT see DVH200
2,6-DNT see DVH400
3,4-DNT see DVH600
DNTP see PAK000
DO 14 see SOP000
DOA see AEO000
DOBANIC ACID 83 see LBU100
DOBANIC ACID JN see LBU100
DOBENDAN see CCX000
DOBETIN see VSZ000
DOCEMINE see VSZ000
DOCIBIN see VSZ000
DOCIGRAM see VSZ000
DOCTAMICINA see CDP250
DOCUSATE SODIUM see DJL000
DODAT see DAD200
DODECABEE see VSZ000
DODECACHLOROOCTAHYDRO-1,3,4-METHENO-
2H-CYCLOBUTA(c,d)PENTALENE see MQW500
1,1a,2,2,3,3a,4,5,5,5a,5b,6-
DODECACHLOROOCTAHYDRO-1,3,4-METHENO-
1H-CYCLOBUTA(c,d)PENTALENE see MQW500
DODECACHLOROPENTACYCLODECANE see
MQW500
DODECACHLOROPENTACYCLO(3,2,2,0^{2,6},0^{3,9},0^{5,10})DE
CANE see MQW500
DODECAHYDRODIPHENYLAMINE see DGT600

DODECAHYDROPHENYLAMINE NITRITE see DGU200

Δ-DODECALACTONE see DXS700

Δ-DODECALACTONE see HBP450

1-DODECANAL see DXT000

1-DODECANAMINE (9CI) see DXW000

1-DODECANETHIOL see LBX000

tert-DODECANETHIOL see DXT800

DODECANOIC ACID see LBL000

DODECANOIC ACID, CADMIUM SALT (9CI) see CAG775

1-DODECANOL see DXV600

n-DODECANOL see DXV600

DODECANOYL PEROXIDE see LBR000

1,6,10-DODECATRIEN-3-OL, 3,7,11-TRIMETHYL-, ACETATE, (S-(Z))- see NCN800

DODECAVITE see VSZ000

2-DODECENAL see DXU280

DODECENE EPOXIDE see DXU400

DODECENYLSUCCINIC ANHYDRIDE see DXV000

DODECOIC ACID see LBL000

DODECYL ALCOHOL see DXV600

n-DODECYL ALCOHOL see DXV600

DODECYL ALCOHOL, HYDROGEN SULFATE, SODIUM SALT see SIB600

1-DODECYL ALDEHYDE see DXT000

DODECYLAMINE see DXW000

1-DODECYLAMINE see DXW000

n-DODECYLAMINE see DXW000

DODECYL BENZENE SODIUM SULFONATE see DXW200

n-DODECYLBENZENESULFONIC ACID see LBU100

DODECYLBENZENESULFONIC ACID (DOT) see LBU100

DODECYLBENZENESULFONIC ACID SODIUM SALT see DXW200

DODECYLBENZENESULPHONATE, SODIUM SALT see DXW200

DODECYLBENZENESULPHONIC ACID see LBU100

DODECYLBENZENSULFONAN SODNY (CZECH) see DXW200

DODECYL DIMETHYL BENZYLAMMONIUM CHLORIDE see BEM000

DODECYLDIMETHYL(2-PHENOXYETHYL)AMMONIUM BROMIDE see DXX000

DODECYLESTER KYSELINY GALLOVE see DXX200

DODECYL GALLATE see DXX200

N-DODECYLGUANIDINACETAT (GERMAN) see DXX400

DODECYLGUANIDINE ACETATE see DXX400

N-DODECYLGUANIDINE ACETATE see DXX400

2-DODECYLISOQUINOLINIUM BROMIDE see LBW000

1-DODECYL MERCAPTAN see LBX000

m-DODECYL MERCAPTAN see LBX000

tert-DODECYLMERCAPTAN see DXT800

DODECYL MERCAPTAN (ACGIH) see LBX000

tert. DODECYLMERKAPTAN (CZECH) see DXT800

DODECYLPHENOL (mixed isomers) see DXY600

N-DODECYLSARCOSINE SODIUM SALT see DXZ000

DODECYL SODIUM SULFATE see SIB600

DODECYL SULFATE, SODIUM SALT see SIB600

tert-DODECYLTHIOL see DXT800

DODECYLTRICHLOROSILANE see DYA800

DODEX see VSZ000

DODGUADINE see DXX400

DODIGEN 2617 see DTF820

DODINE see DXX400

DODINE ACETATE see DXX400

DODINE, mixture with GLYODIN see DXX400

DOF see DVK600

DOFSOL see VSK600

DOGQUADINE see DXX400

DOJYOPICRIN see CKN500

DOKIRIN see BLC250

DOKTACILLIN see AIV500

DOL see BBQ750

DOLANTAL see DAM700

DOLANTIN see DAM700

DOLANTIN HYDROCHLORIDE see DAM700

DOLANTOL see DAM700

DOLAREN see DAM700

DOLARGAN see DAM700

DOLCO MOUSE CEREAL see SMN500

DOLCONTRAL see DAM700

DOLCYMENE see CQI000

DOLEAN pH 8 see ADA725

DOLENAL see DAM700

DOLENE see PNA500

DOLENE see DAB879

DOLENOL see DAM700

DOLEN-PUR see HCD250

DOLESTAN see BAU750

DOLESTINE see DAM700

DOLGIN see IIU000

DOL GRANULE see BBQ500

DOLICUR see DUD800

DOLIGUR see DUD800

DOLIN see DAM700

DOLIPOL see BSQ000

DOLIPRANE see HIM000

DOLKWAL ORANGE SS see TGW000

DOLKWAL PONCEAU 3R see FAG018

DOLKWAL SUNSET YELLOW see FAG150

DOLKWAL TARTRAZINE see FAG140

DOLKWAL YELLOW AB see FAG130

DOLKWAL YELLOW OB see FAG135

DOLMIX see BBQ750

DOLOBID see DKI600

DOLOBIL see DKI600

DOLOBIS see DKI600

DOLOCAP see PNA500

DOLOCHLOR see CKN500

DOLOGAL see DAM700

DOLOMITE see CAO000

DOLONEURINE see DAM700

DOLONIL see PDC250

DOLOPETHIN see DAM700

DOLOPHINE see MDO750

d-DOLOPHINE HYDROCHLORIDE see MDP240

DOLOSAL see DAM700

DOLOVIN see IDA000

DOLOXENE see PNA500

DOLOXENE see DAB879

DOLVANOL see DAM700

DOMALIUM see DCK759

DOMATOL see AMY050

DOMATOL 88 see AMY050

DOMESTROL see DKA600

DOMF see MCV000

DOMICAL see EAI000

DOMICILLIN see SEQ000

DOMOSO see DUD800

DONMOX see AAI250

DOP see DVL700
(−)-DOPA see DNA200
l-DOPA see DNA200
DOPAFLEX see DNA200
DOPAL see DNA200
DOPAMET see DNA800
DOPAMINE see DYC400
DOPARKINE see DNA200
DOPASOL see DNA200
DOPEGYT see DNA800
DOPIDRIN see DBA800
DOPN see NJN000
DOPRIN see DNA200
DOPTAEC see DNA800
DORAL see VSZ100
DORAPHEN see PNA500
DORBANE see DMH400
DORBANEX see DMH400
DORIDEN see DYC800
DORIDEN-SED see DYC800
DORINAMIN see BBW500
DORISUL see SNN300
DORLYL see PKQ059
DORMABROL see MQU750
DORMAL see CDO000
DORME see PMI750
DORMIGENE see BNP750
DORMIRAL see EOK000
DORMITURIN see BNK000
DORSULFAN see SNN500
DORVICIDE A see BGJ750
DORVON see SMQ500
DORVON FR 100 see SMQ500
DOS see BJS250
D.O.T. see DUP300
DOTG see BKK750
DOTG ACCELERATOR see DXP200
DOTMENT 324 see AHE250
DOTYCIN see EDH500
DOUBLE STRENGTH see TIX500
DOUGLAS FIR OIL see OJK340
DOW 209 see SMR000
DOW 360 see SMQ500
DOW 456 see SMQ500
DOW 665 see SMQ500
DOW 860 see SMQ500
DOW 1683 see SMQ500
DOWANOL see CBR000
DOWANOL 33B see PNL250
DOWANOL-50B see DWT200
DOWANOL DB see DJF200
DOWANOL DE see CBR000
DOWANOL DM see DJG000
DOWANOL DPM see DWT200
DOWANOL EB see BPJ850
DOWANOL EE see EES350
DOWANOL EIPAT see INA500
DOWANOL EM see EJH500
DOWANOL EP see PER000
DOWANOL EPH see PER000
DOWANOL PM see PNL250
DOWANOL (R) PMA GLYCOL ETHER ACETATE see PNL265
DOWANOL PM GLYCOL ETHER see PNL250
DOWANOL TE see EFL000
DOWCHLOR see CDR750
DOWCIDE 1 see BGJ250
DOWCIDE 7 see PAX250

DOWCIDE 1 ANTIMICROBIAL see BGJ250
DOWCO 139 see DOS000
DOWCO-163 see CLP750
DOWCO 169 see PEV500
DOWCO 179 see CMA100
DOWCO 184 see CEG550
DOWCO 186 see HON000
DOWCO-213 see CQH650
DOWCO 217 see CMA250
DOW DEFOLIANT see SFU500
DOW DORMANT FUNGICIDE see SJA000
DOW ET 14 see RMA500
DOW ET 57 see RMA500
DOWFLAKE see CAO750
DOWFROST see PML000
DOWFUME see MHR200
DOWFUME 40 see EIY500
DOWFUME EDB see EIY500
DOWFUME MC-2 SOIL FUMIGANT see MHR200
DOWFUME N see DGG000
DOWFUME W-8 see EIY500
DOWICIDE see BGJ750
DOWICIDE 2 see TIV750
DOWICIDE 6 see TBT000
DOWICIDE 7 see PAX250
DOWICIDE 2S see TIW000
DOWICIDE A see BGJ750
DOWICIDE A & A FLAKES see BGJ750
DOWICIDE B see TIV750
DOWICIDE EC-7 see PAX250
DOWICIDE G see PAX250
DOWICIDE G-ST see SJA000
DOWICIDE Q see CEG550
DOWICIL 75 see CEG550
DOWICIL 100 see CEG550
DOWIZID A see BGJ750
DOW LATEX 612 see SMR000
DOWLEX FILM see PJS750
DOW MCP AMINE WEED KILLER see CIR250
DOW MX 5514 see SMQ500
DOW MX 5516 see SMQ500
DOW PENTACHLOROPHENOL DP-2
ANTIMICROBIAL see PAX250
DOW-PER see PCF275
DOWPON see DGI400
DOWPON see DGI600
DOWPON M see DGI400
DOW SODIUM TCA SOLUTION see TII250
DOWTHERM see PFA860
DOWTHERM 209 see PNL250
DOWTHERM A see PFA860
DOWTHERM E see DEP600
DOWTHERM SR 1 see EJC500
DOW-TRI see TIO750
DOWZENE DHC see PIK000
D-OX see SHR500
DOXCIDE 50 see CDW450
DOXEPHRIN see DBA800
DOX HYDROCHLORIDE see HKA300
DOXINATE see DJL000
DOXOL see DJL000
DOXORUBICIN see AES750
DOXORUBICIN see HKA300
DOXORUBICIN HYDROCHLORIDE see HKA300
DOXYFED see MDT600
DOXYFED see DBA800
2,4-DP see DGB000
2-(2,4-DP) see DGB000

DPA see DGI000
DPA see DVX800
DPA see HET500
2,2-DPA see DGI600
DPBS see DFY400
DPD see DJV200
D & P DOUBLE O CRABGRASS KILLER see PLC250
DPG see DWC600
DPG ACCELERATOR see DWC600
DPH see DKQ000
DPH see DNU000
DPN see NKB700
DPNA see NKB700
DPP see PAK000
DPP see AON300
DPPD see BLE500
DP X 1410 see MME809
DPX 1410 see DSP600
DPX 3674 see HFA300
DQDA 1868 see PJS750
DQUIGARD see DGP900
DQWA 0355 see PJS750
DRABET see BSQ000
DRACYLIC ACID see BCL750
DRAGOCAL see CAS750
DRAKEOL see MQV750
DRAKEOL see MQV875
DRALZINE see HGP500
DRAMAMIN see DYE600
DRAMAMINE see DYE600
DRAMARIN see DYE600
DRAMYL see DYE600
DRAT see CJJ000
DRAZA see DST000
DRC 1339 see CLK215
DRC 3341 see MIB750
DREFT see SIB600
DRENAMIST see VGP000
DRENOL see CFY000
DRENUSIL-R see RDK000
DREWMULSE POE-SMO see PKL100
DREWMULSE TP see OAV000
DREWMULSE V see OAV000
DREWSORB 60 see SKV150
DREXEL see DXQ500
DREXEL DEFOL see SFS000
DREXEL DIURON 4L see DXQ500
DREXEL METHYL PARATHION 4E see MNH000
DREXEL PARATHION 8E see PAK000
DREXEL-SUPER P see DMC600
DRI-DIE PESTICIDE 67 see SCH002
DRIERITE see CAX500
DRILL TOX-SPEZIAL AGLUKON see BBQ500
DRINALFA see MDT600
DRINALFA see DBA800
DRINOX see AFK250
DRINOX see HAR000
DRISDOL see VSZ100
DRI-TRI see SJH200
DROCODE see DKW800
DROMILAC see DBV400
DROMISOL see DUD800
levo-DROMORAN see LFG000
DROMYL see DYE600
DROPCILLIN see BDY669
DROP LEAF see SFS000
DROPP see TEX600
DROXOL see MDQ250

DROXOLAN see DAQ400
DRP 859025 see TND500
DRUMULSE AA see OAV000
DRUPINA 90 see BJK500
DRY AND CLEAR see BDS000
DRY ICE see CBU250
DRY ICE (UN 1845) (DOT) see CBU250
DRYISTAN see BBV500
DRYLISTAN see BBV500
DRYOBALANOPS CAMPHOR see BMD000
DRYPTAL see CHJ750
DS see DKB110
DSE see DXD200
DS-Na see SFW000
2,4-D SODIUM SALT see SGH500
DSP see SJH090
DSS see SOA500
DSS see DJL000
DST see DME000
DST 50 see SMR000
DT see TFX790
DTB see DXL800
DTHYD see TFX790
DTIC see DAB600
DTIC-DOME see DAB600
DTMC see BIO750
DTPA see DJG800
DTPA CALCIUM TRISODIUM SALT see CAY500
DU see DCQ600
DU 112307 see CJV250
DU-A 1 see SCQ000
DU-A 2 see SCQ000
DU-A 3 see SCQ000
DU-A 4 see SCQ000
DU-A 3C see SCQ000
DUATOK see TEX250
DUBRONAX see SOA500
DUCKALGIN see SEH000
DUCOBEE see VSZ000
DUFALONE see BJZ000
DUGERASE see AIT250
DUKERON see TIO750
DUKSEN see DCK759
DULCIDOR see GFA000
DULCINE see EFE000
DULL 704 see PJY500
DULSIVAC see DJL000
DULZOR-ETAS see SGC000
DUMASIN see CPW500
DUMITONE see SOA500
DUMOCYCIN see TBX250
DUMOGRAN see MPN500
DUNCAINE see DHK400
DUNCAINE HYDROCHLORIDE see DHK600
DUNERYL see EOK000
DUNKELGELB see PEJ500
DUODECIBIN see VSZ000
DUODECYL ALCOHOL see DXV600
DUODECYLIC ACID see LBL000
DUODECYLIC ALDEHYDE see DXT000
DUO-KILL see DGP900
DUOLAX see DMH400
DUOMYCIN see CMA750
DUOSOL see DJL000
DUOTRATE see PBC250
DUPHAR see CKM000
DUPONOL see SIB600
DU PONT 326 see DGD600

DU PONT 1991 see BAV575
DU PONT HERBICIDE 326 see DGD600
DU PONT HERBICIDE 976 see BMM650
DU PONT INSECTICIDE 1179 see MDU600
DU PONT PC CRABGRASS KILLER see PLC250
DURABIOTIC see BFC750
DURAD see TNP500
DURA-ESTRADIOL see EDS100
DURAFUR BLACK R see PEY500
DURAFUR BROWN see ALL750
DURAFUR BROWN 2R see ALL750
DURAFUR BROWN MN see DBO400
DURAFUR BROWN RB see ALT250
DURAFUR DEVELOPER C see CCP850
DURAFUR DEVELOPER D see NAW500
DURAFUR DEVELOPER G see REA000
DURAMAX see ADA725
DURAN see DXQ500
DURANIT see SMR000
DURANITRAT see CCK125
DURANOL BRILLIANT BLUE CB see TBG700
DURANOL BRILLIANT BLUE VIOLET BR see DBY700
DURANOL BRILLIANT VIOLET BR see DBY700
DURANOL ORANGE G see AKP750
DURANOL RED 2B see AKE250
DURA-PENITA see BFC750
DURAPHOS see MQR750
DURASORB see DUD800
DURATOX see DAP400
DURAVOS see DGP900
DURAX see CPI250
DUR-EM 204 see GGR200
DURETHAN BK see PJY500
DURETTER see FBN100
DUREX see CBT750
DURFAX 80 see PKL100
DURGACET YELLOW G see AAQ250
DUROFERON see FBN100
DUROFOL P see PKQ059
DUROID 5870 see TAI250
DUROMINE see DTJ400
DUROPENIN see BFC750
DUROSPERSE YELLOW G see AAQ250
DUROTOX see PAX250
DURSBAN see CMA100
DURSBAN F see CMA100
DURSBAN METHYL see CMA250
DURTAN 60 see SKV150
DUSICNAN BARNATY (CZECH) see BAN250
DUSICNAN CERITY (CZECH) see CDB000
DUSICNAN KADEMNATY (CZECH) see CAH250
DUSICNAN ZIRKONICITY (CZECH) see ZSA000
DUSITAN DICYKLOHEXYLAMINU (CZECH) see DGU200
DUSITAN SODNY (CZECH) see SIQ500
DUSOLINE see CMD750
DUSORAN see CMD750
DUTCH LIQUID see EIY600
DUTCH OIL see EIY600
DUTCH-TREAT see HKC500
DU-TER see HON000
DUTION see DXH325
DUVILAX BD 20 see AAX250
DV see DAL600
DV 400 see PKB500
D3-VIGANTOL see CMC750
DW3418 see BLW750
DWARF PINE NEEDLE OIL see PIH400

DWUBROMOETAN (POLISH) see EIY500
DWUCHLOROCZTEROFLUOROETAN (POLISH) see DGL600
DWUCHLORODWUETYLOWY ETER (POLISH) see DFJ050
DWUCHLORODWUFLUOROMETAN (POLISH) see DFA600
DWUCHLOROFLUOROMETAN (POLISH) see DFL000
DWUCHLOROPROPAN (POLISH) see PNJ400
DWUETYLOAMINA (POLISH) see DHJ200
DWUETYLOWY ETER (POLISH) see EJU000
DWUFENYLOGUANIDYNA (POLISH) see DWC600
DWUMETYLOFORMAMID (POLISH) see DSB000
DWUMETYLOANILINA (POLISH) see DQF800
DWUMETYLOSULFOTLENKU (POLISH) see MAO250
DWUMETYLOWY SIARCZAN (POLISH) see DUD100
DWU-β-NAFTYLO-p-FENYLODWUAMINA (POLISH) see NBL000
DWUNITROBENZEN (POLISH) see DUQ200
DWUNITRO-o-KREZOL (POLISH) see DUS700
3,3'-DWUOKSYBENZYDYNA (POLISH) see DMI400
DWUSIARCZEK DWUBENZOTIAZYLU (POLISH) see BDE750
DX see AES750
DXEWMULSE POE-SML see PKG000
DXM 100 see PJS750
DXMS see SOW000
DYALL see PJS750
DYANACIDE see ABU500
DYAZIDE see CFY000
DYCARB see DQM600
DYDELTRONE see PMA000
DYE EVANS BLUE see BGT250
DYE FDC YELLOW LAKE 6 see FAG150
DYE FD & C YELLOW LAKE 6 see FAG150
DYE FDC YELLOW NO. 6 see FAG150
DYE FD & C YELLOW NO. 6 see FAG150
DYE FD and C YELLOW NO. 5 see FAG140
DYE GS see ALL750
DYESTROL see DKA600
DYE SUNSET YELLOW see FAG150
DYETONE see SFG000
DYFLOS see IRF000
DYFONATE see FMU045
DYKANOL see PJL750
DYKOL see DAD200
DYLAMON see BBV500
DYLAN see PJS750
DYLAN SUPER see PJS750
DYLAN WPD 205 see PJS750
DYLARK 250 see SMQ500
DYLENE see SMQ500
DYLENE 8 see SMQ500
DYLENE 9 see SMQ500
DYLENE 8G see SMQ500
DYLEPHRIN see VGP000
DYLITE F 40 see SMQ500
DYLITE F 40L see SMQ500
DYLOX see TIQ250
DYLOX-METASYSTOX-R see TIQ250
DYMADON see HIM000
DYMEL 22 see CFX500
DYMELOR see ABB000
DYMEX see ABH000
DYNA-CARBYL see CBM750
DYNACORYL see DJS200
DYNADUR see PKQ059

DYNALIN INJECTABLE see TET800
DYNAMICARDE see DJS200
DYNAMITE see DYG000
DYNAMUTILIN see TET800
DYNARSAN see ABX500
DYNASTEN see PAN100
DYNA-ZINA see DLH630
DYNAZONE see NGE500
DYNEX see DXQ500
DYNH see PJS750
DYNK 2 see PJS750
DYNOSOL see DUU600
DYNOVAS see PAH250
DYODIN see DNF600
DYPERTANE COMPOUND see RDK000
DYP-97 F see LBR000
DYPHONATE see FMU045
DYPRIN see MDT740
DYREX see TIQ250
DYSPNE-INHAL see VGP000
DYSPROSIUM CHLORIDE see DYG600
DYTHOL see CMD750
DYTOL M-83 see OEI000
DYTOL S-91 see DAI600
DYTOL E-46 see OAX000
DYTOL F-11 see HCP000
DYTOL J-68 see DXV600
DYVON see TIQ250
DYZOL see DCM750
E6 see PKO500
E 62 see PKQ059
E 102 see FAG140
E 110 see FAG150
E 140 see CKN000
E 158 see DAP600
E393 see SOD100
E 534 see PKB100
583E see POC750
E 66P see PKQ059
686E see SMQ500
E 1440 see CCU250
E 3314 see HAR000
E 5110 see MEL100
E 7256 see LBU100
EA 3547 see DDE200
EAA see EFS000
EACA see AJD000
EACA KABI see AJD000
EACS see AJD000
EAGLE GERMANTOWN see CBT750
EAK see ODI000
EARTHCIDE see PAX000
EARTHNUT OIL see PAO000
EASEPTOL see HJL000
EASTBOND M 5 see PMP500
EASTERN STATES DUOCIDE see WAT200
EAST INDIAN SANDALWOOD OIL see OGY220
EASTMAN 910 see MIQ075
EASTMAN 7663 see DJT800
EASTMAN 910 ADHESIVE see MIQ075
EASTMAN INHIBITOR DHPB see DMI600
EASTMAN INHIBITOR HPT see HEK000
EASTMAN 910 MONOMER see MIQ075
EASTONE YELLOW GN see AAQ250
EASY OFF-D see TIG250
EATAN see DLY000
EB see BGT250
EB see EGP500

EBECRYL 110 see PER250
E-D-BEE see EIY500
EBERPINE see RDK000
EBERSPINE see RDK000
EBONTA see EIT000
EBSERPINE see RDK000
EBUCIN see CAS750
EBZ see EDP000
E.C. 3.4.4.16 see BAC000
E.C. 3.4.4.24 see BMO000
E.C. 3.4.21.14 see BAC000
ECATOX see PAK000
ECCOTHAL see TEM000
ECF see EHK500
ECGONINE, METHYL ESTER, BENZOATE (ESTER)
see CNE750
ECH see EAZ500
ECIPHIN see EAW000
ECLORIL see CDO500
ECM see ADA725
ECOFROL see TGJ050
ECON see TGJ050
ECONOCHLOR see CDP250
ECOPRO see TAL250
ECOTRIN see ADA725
ECP see DAZ115
ECP see DFK600
ECTIBAN see AHJ750
ECTILURAN see TEH500
ECTORAL see RMA500
ECTRIN see FAR100
ECUANIL see MQU750
ECZECIDIN see CHR500
ED see DFH200
EDA see DCN800
EDATHAMIL see EIX000
EDATHAMIL DISODIUM see EIX500
EDATHANIL TETRASODIUM see EIV000
EDB see EIY500
EDB-85 see EIY500
EDC see EIY600
EDCO see MHR200
EDEMOX see AAI250
EDEN see LFK000
EDENAL see MQU750
EDETATE DISODIUM see EIX500
EDETATE SODIUM see EIV000
EDETATE TRISODIUM see TNL250
EDETIC ACID see EIX000
EDETIC ACID TETRASODIUM SALT see EIV000
EDICOL PONCEAU RS see FMU070
EDICOL SUPRA BLUE E6 see FMU059
EDICOL SUPRA CARMOISINE WS see HJF500
EDICOL SUPRA PONCEAU 4R see FMU080
EDICOL SUPRA TARTRAZINE N see FAG140
EDICOL SUPRA YELLOW FC see FAG150
EDIPOSIN see VBK000
EDISOL M see MIF760
EDISTIR RB see SMQ500
EDISTIR RB 268 see SMR000
EDTA (chelating agent) see EIX000
EDTA ACID see EIX000
d'E.D.T.A. DISODIQUE (FRENCH) see EIX500
EDTA, DISODIUM SALT see EIX500
EDTA, SODIUM SALT see EIV000
EDTA TETRASODIUM SALT see EIV000
EDTA TRISODIUM SALT see TNL250
EEC No. E924 see PKY300

(E,E)-2,4-DECADIENAL see DAE450
EENA see ELG500
EENKAPTON (DUTCH) see PDC750
EEREX GRANULAR WEED KILLER see BMM650
EEREX WATER SOLUBLE CONCENTRATE WEED
KILLER see BMM650
EF 10 see SCK600
EFACIN see NCQ900
EF CORLIN see CNS750
EFED see TMU250
EFEDRIN see EAW000
E-FEROL see TGJ050
EFEROX see LFG050
EFFEMOLL DOA see AEO000
EFFLUDERM (free base) see FMM000
EFFROXINE see DBA800
EFFUSAN see DUS700
EFLORAN see MMN250
EFRICEL see SPC500
EFROXINE see MDT600
EFUDEX see FMM000
EFUDIX see FMM000
EFURANOL see DLH630
EFV 250/400 see IGK800
EGBE see BPJ850
EGDME see DOE600
EGDN see EJG000
EGG YELLOW A see FAG140
EGITOL see HCI000
EGM see EJH500
EGME see EJH500
EGTA see EIT000
EH 121 see TNX000
EHDP see HKS780
EHEN see ELG500
EHRLICH 594 see ABX500
EI see EJM900
EI-12880 see DSP400
EI 38,555 see CMF400
EI 47031 see PGW750
EI 47300 see DSQ000
EI-47470 see DHH400
EI 52160 see TAL250
EICOSAHYDRO
DIBENZO(b,k)(1,4,7,10,13,16)HEXAOXACYCLOOCTA
DECIN see DGV100
EIRENAL see PGA750
EISENDEXTRAN (GERMAN) see IGS000
EISENDIMETHYLDITHIOCARBAMAT (GERMAN) see
FAS000
EISENOXYD see IHC450
EISEN(III)-TRIS(N,N-DIMETHYLDITHIOCARBAMAT)
(GERMAN) see FAS000
EITDRONATE DISODIUM see DXD400
EK 54 see DUU600
EK 1700 see PEY250
EKAGOM CBS see CPI250
EKAGOM TB see TFS350
EKAGOM TEDS see DXH250
EKALUX see DJY200
EKATIN see PHI500
EKATIN AEROSOL see PHI500
EKATINE-25 see PHI500
EKATIN TD see DXH325
EKATIN ULV see PHI500
EKATOX see PAK000
EKAVYL SD 2 see PKQ059
EKKO CAPSULES see DKQ000

EKOMINE see MGR500
EKTAFOS see DGQ875
EKTASOLVE de ACETATE see CBQ750
EKTASOLVE DB see DJF200
EKTASOLVE DB ACETATE see BQP500
EKTASOLVE EB see BPJ850
EKTASOLVE EB ACETATE see BPM000
EKTASOLVE EE see EES350
EKTASOLVE EE ACETATE SOLVENT see EES400
EKTASOLVE EIB see IIP000
EKTASOLVE EP see PNG750
EKVACILLIN see SLJ000
EK 1108GY-A see TAI250
EL 222 see FAK100
EL 4049 see MAK700
ELAIOMYCIN see EAG000
ELALDEHYDE see PAI250
ELAMOL see EAG100
ELANCOBAN see MRE225
ELANCOLAN see DUV600
ELAOL see DEH200
ELARGIN see EAG100
ELARGYL see EAG100
ELASTONON see BBK000
ELAVIL see EAI000
ELAVIL HYDROCHLORIDE see EAI000
ELAYL see EIO000
ELCACID MILLING FAST RED RS see CMM330
ELCATONIN see CBR300
ELCEMA F 150 see CCU150
ELCEMA G 250 see CCU150
ELCEMA P 050 see CCU150
ELCEMA P 100 see CCU150
ELCIDE 75 see MDI000
ELCORIL see CDO500
EL-CORTELAN SOLUBLE see HHR000
ELCOZINE CHRYSOIDINE Y see PEK000
ELCOZINE RHODAMINE 6GDN see RGW000
ELDADRYL see BAU750
ELDEZOL see NGE500
ELDIATRIC C see CAK500
ELDODRAM see DYE600
ELDOPAL see DNA200
ELDOPAQUE see HIH000
ELDOQUIN see HIH000
ELECTRO-CF 11 see TIP500
ELECTRO-CF 12 see DFA600
ELECTRO-CF 22 see CFX500
ELECTRONIC E-2 see HIC000
ELEMENTAL SELENIUM see SBO500
ELENIUM see MDQ250
ELENIUM see LFK000
ELEPSINDON see DKQ000
ELEUDRON see TEX250
ELF see CBT750
ELFANEX see RDK000
ELFAN WA SULPHONIC ACID see LBU100
ELFTEX see CBT750
ELGETOL see DUS700
ELGETOL see DUU600
ELICIDE see MDI000
ELIPOL see DUS700
ELITONE see DJS200
ELIXICON see TEP000
ELIXOPHYLLIN see TEP000
ELIXOPHYLLINE see TEP000
ELJON LAKE RED C see CHP500
ELJON PINK TONER see RGW000

ELMASIL see AMY050
ELMEDAL see BRF500
ELMER'S GLUE ALL see AAX250
ELOBROMOL see DDJ000
EL P.E.T.N. see PBC250
ELPON see PMP500
ELRODORM see DYC800
ELSERPINE see RDK000
ELTEX see PJS750
ELTEX 6037 see PJS750
ELTEX A 1050 see PJS750
ELTROXIN see LFG050
ELVACITE see PKB500
ELVANOL see PKP750
ELVANOL 50-42 see PKP750
ELVANOL 52-22 see PKP750
ELVANOL 70-05 see PKP750
ELVANOL 71-30 see PKP750
ELVANOL 90-50 see PKP750
ELVANOL 522-22 see PKP750
ELVANOL 73125G see PKP750
ELYZOL see MMN250
EM see EDH500
EM 3 see RNU100
910EM see EHP700
EM 923 see DFY400
EMAFORM see CHR500
EMAGRIN see SPC500
EMANAY ATOMIZED ALUMINUM POWDER see
AGX000
EMANAY ZINC DUST see ZBJ000
EMANAY ZINC OXIDE see ZKA000
EMANIL see DAS000
EMAR see ZKA000
EMATHLITE see KBB600
EMAZOL RED B see EAJ500
EMBACETIN see CDP250
EMBAFUME see MHR200
EMBANOX see BQI000
EMBARIN see ZVJ000
EMBATHION see EEH600
EMBECHINE see BIE500
EMBEQUIN see DNF600
EMBICHIN see BIE250
EMBICHIN see BIE500
EMBICHIN HYDROCHLORIDE see BIE500
EMBIKHINE see BIE500
EMBINAL see BAG250
EMBIOL see VSZ000
EMBUTAL see NBU000
EMC see CHC500
EMCEPAN see CIR250
EMCOL CA see OAV000
EMCOL DS-50 CAD see HKJ000
EMCOL-IM see IQN000
EMCOL MSK see OAV000
EMCOL O see GGR200
EMEDAN see BSM000
EMERALD GREEN see BAY750
EMERALD GREEN see COF500
EMERESSENCE 1160 see PER000
EMEREST 2301 see OHW000
EMEREST 2314 see IQN000
EMEREST 2350 see EJM500
EMEREST 2400 see OAV000
EMEREST 2401 see OAV000
EMEREST 2801 see OHW000
EMERSAL 6400 see SIB600

EMERSAL 6465 see TAV750
EMERSOL 120 see SLK000
EMERSOL 140 see PAE250
EMERSOL 143 see PAE250
EMERSOL 210 see OHU000
EMERSOL 213 see OHU000
EMERSOL 6321 see OHU000
EMERSOL 233LL see OHU000
EMERSOL 221 LOW TITER WHITE OLEIC ACID see
OHU000
EMERSOL 220 WHITE OLEIC ACID see OHU000
EMERY 655 see MSA250
EMERY 2218 see MJW000
EMERY 2219 see OHW000
EMERY 2310 see OHW000
EMERY 5791 see MCN250
EMERY 6705 see PER000
EMERY OLEIC ACID ESTER 2221 see GGR200
EMERY OLEIC ACID ESTER 2301 see OHW000
EMETINE ANTIMONY IODIDE see EAM000
EMETINE, DIHYDROCHLORIDE see EAN000
(−)-EMETINE DIHYDROCHLORIDE see EAN000
1-EMETINE DIHYDROCHLORIDE see EAN000
EMETINE HYDROCHLORIDE see EAN000
EMETIQUE (FRENCH) see AQG250
EMETREN see CDP250
EMFAC 1202 see NMY000
EMI-CORLIN see HHR000
EMID 6511 see BKE500
EMID 6541 see BKE500
EMIPHEROL see VSZ450
EMISAN 6 see MEP250
EMISOL see AMY050
EMISOL 50 see AMY050
EMISOL F see AMY050
EMITEFUR see BDN600
EMIVIRINE see EAN525
EMMATOS see MAK700
EMMATOS EXTRA see MAK700
EMMI see EME050
EMOCICLINA see VSZ000
EMODIN see IIU000
EMO-NIK see NDN000
EMPAL see CIR250
EMPG see ENC000
EMPILAN 2848 see EJM500
EMPIRIN see ADA725
EMPIRIN COMPOUND see ABG750
EMPLETS POTASSIUM CHLORIDE see PLA500
EMQ see SAV000
EMRITE 6009 see GGR200
EMS see EMF500
EMSORB 2500 see SKV100
EMSORB 2505 see SKV150
EMSORB 2515 see SKV000
EMSORB 6900 see PKL100
EMSORB 6907 see SKV195
EMSORB 6915 see PKG000
EMT 25,299 see MDV500
EMTAL 596 see TAB750
EMTEXATE see MDV500
EMTS see EME500
EMUL P.7 see OAV000
EMULSIFIABLE OIL see MQV855
EMULSIFIER No. 104 see SIB600
EMULSION 212 see FQU875
EMULSIPHOS 440/660 see SJH200
E-MYCIN see EDH500

EN 18133 see EPC500
ENALLYNYMAL SODIUM see MDU500
ENAMEL WHITE see BAP000
ENANTHAL see HBB500
ENANTHALDEHYDE see HBB500
ENANTHIC ALCOHOL see HBL500
ENANTHOLE see HBB500
ENANTHYLIC ETHER see EKN050
ENARMON see TBG000
ENAVID see EAP000
ENCORTON see PLZ000
ENDEP see EAI000
ENDOBION see NCR000
ENDOCEL see EAQ750
ENDO E see VSZ450
ENDO E DOMPE see TGJ050
ENDOFOLLICOLINA D.P. see EDR000
ENDOFOLLICULINA see EDV000
ENDOLAT see DAM700
ENDOMETHYLENETETRAHYDROPHTHALIC ACID,
N-2-ETHYLHEXYL IMIDE see OES000
3,6-ENDOOXOHEXAHYDROPHTHALIC ACID see
EAR000
ENDOSOL see EAQ750
ENDOSULFAN see EAQ750
ENDOSULPHAN see EAQ750
ENDOTHAL see EAR000
ENDOTHALL see EAR000
ENDOTHAL TECHNICAL see EAR000
ENDOXAN see CQC650
ENDOXANA see CQC500
ENDOXANAL see CQC650
ENDOXAN-ASTA see CQC500
ENDOXAN MONOHYDRATE see CQC500
ENDOXAN R see CQC500
3,6-ENDOXOHEXAHYDROPHTHALIC ACID see
EAR000
ENDRATE see EIX000
ENDRATE DISODIUM see EIX500
ENDRATE TETRASODIUM see EIV000
ENDREX see EAT500
ENDRIN see EAT500
ENDRINE (FRENCH) see EAT500
ENDRIN KETONE see KFK200
ENDUXAN see CQC500
ENDYDOL see ADA725
ENDYL see TNP250
ENELFA see HIM000
ENERIL see HIM000
ENFENEMAL see ENB500
ENFLURANE see EAT900
ENGLISH RED see IHC450
ENHEPTIN see ALQ000
ENHEPTIN A see ABY900
ENIACID BRILLIANT RUBINE 3B see HJF500
ENIACID LIGHT ORANGE G see HGC000
ENIACID METANIL YELLOW GN see MDM775
ENIACID SUNSET YELLOW see FAG150
ENIAL ORANGE I see PEJ500
ENIAL RED IV see SBC500
ENIAL YELLOW 2G see DOT300
ENIANIL BLACK CN see AQP000
ENIANIL BLUE 2BN see CMO000
ENIANIL BRILLIANT BLUE FF see CMN750
ENIANIL PURE BLUE AN see CMO500
ENICOL see CDP250
ENIDREL see EAP000
ENILOCONAZOL (SP) see FPB875

ENIPRESSER see RDK000
ENJAY CD 460 see PMP500
ENKEFAL see DNU000
ENKELFEL see DKQ000
ENORDEN see MQU750
ENOVID see EAP000
ENOVID-E see EAP000
ENPHENEMAL see ENB500
ENPROMATE see DWL400
ENSEAL see PLA500
ENSODORM see EOK000
ENSURE see EAQ750
ENS-ZEM WEEVIL BAIT see DXE000
ENT 9 see BRT000
17-ENT see ABU000
ENT 38 see PDP250
ENT 54 see ADX500
ENT 92 see IHZ000
ENT 154 see DUS700
ENT 262 see DTR200
ENT 375 see EKV000
ENT 884 see COF500
ENT 988 see BJK500
ENT 1,501 see DXE000
ENT 1,506 see DAD200
ENT 1,656 see EIY600
ENT 1,716 see MEI450
ENT 1,860 see PCF275
ENT 2,435 see NDR500
ENT 3,424 see NDN000
ENT 3,776 see DFT000
ENT 4,225 see BIM500
ENT 4,504 see DFJ050
ENT 4,705 see CBY000
ENT 7,543 see POO100
ENT 7,796 see BBQ500
ENT 8,184 see OES000
ENT 8,420 see DGG000
ENT 8,601 see BBP750
ENT 9,232 see BBQ000
ENT 9,233 see BBR000
ENT 9,234 see BFW500
ENT 9,735 see CDV100
ENT 9,932 see CDR750
ENT 14,250 see PIX250
ENT 14,611 see ASL250
ENT 14,689 see FAS000
ENT 14,875 see MAS500
ENT 15,108 see PAK000
ENT 15,152 see HAR000
ENT 15,349 see EIY500
ENT 15,406 see PNJ400
ENT 15,949 see AFK250
ENT 16,225 see DHB400
ENT 16,273 see SOD100
ENT 16,391 see KEA000
ENT 16,436 see DXX400
ENT 16,519 see SOP500
ENT 17,034 see MAK700
ENT 17,251 see EAT500
ENT 17,291 see OCM000
ENT 17,292 see MNH000
ENT 17,470 see DFK600
ENT 17,510 see AFR250
ENT 17,798 see EBD700
ENT 17,956 see CNU750
ENT 18,596 see DER000
ENT 18,771 see TCF250

ENT 18,862 see DAO800
ENT 18,870 see DMC600
ENT 19,060 see DSK200
ENT 19,109 see BJE750
ENT 19,244 see IKO000
ENT 19,442 see TBC500
ENT 19,507 see DCM750
ENT 19,763 see TIQ250
ENT 20,218 see DKC800
ENT 20,279 see MFF580
ENT 20,738 see DGP900
ENT 20,852 see DDP000
ENT 21,040 see AGE250
ENT 22,014 see EKN000
ENT 22,374 see MQR750
ENT 22,542 see DKC800
ENT 22,897 see DVQ709
ENT 22,952 see POO000
ENT 23,233 see ASH500
ENT 23,284 see RMA500
ENT 23,437 see DXH325
ENT 23,444 see DKB170
ENT 23,648 see BIO750
ENT 23,708 see TNP250
ENT 23,737 see CKM000
ENT 23,968 see POP000
ENT 23,969 see CBM750
ENT 23,979 see EAQ750
ENT 24,042 see PGS000
ENT 24,105 see EEH600
ENT 24,482 see DGQ875
ENT 24,650 see DSP400
ENT 24,727 see AQT500
ENT 24,833 see SOY000
ENT 24,915 see TND250
ENT 24,945 see FAQ800
ENT 24,964 see DAP000
ENT 24,969 see CDS750
ENT 24,979 see BLL750
ENT 24,984 see SHF000
ENT 24,986 see DXO000
ENT 25,208 see ABX250
ENT 25,294 see BIE250
ENT 25,296 see TND500
ENT 25,445 see AMY050
ENT 25,550 see SCH002
ENT 25,580 see EPC500
ENT 25,584 see EBW500
ENT 25,585 see PDC750
ENT 25,599 see MQH750
ENT 25,640 see CQL250
ENT 25,670 see MIA250
ENT 25,705 see PHX250
ENT 25,712 see EPY000
ENT 25,715 see DSQ000
ENT 25,719 see MQW500
ENT 25,726 see DST000
ENT 25,734 see PHD250
ENT 25,764 see TBV750
ENT 25,766 see DOS000
ENT 25,796 see FMU045
ENT 25,809 see DXN600
ENT 25,823 see DFV400
ENT 25,830 see PGW750
ENT 25,991 see DHH400
ENT 26,079 see AMG750
ENT 26,263 see EJN500
ENT 26,396 see EMF500

ENT 26,538 see CBG000
ENT 26,592 see BGA750
ENT 27,045 see EMR100
ENT 27,093 see CBM500
ENT 27,129 see MRH209
ENT 27,163 see BDJ250
ENT 27,164 see CBS275
ENT 27,165 see TAL250
ENT 27,193 see DSO000
ENT 27,221 see SNT100
ENT 27,223 see AIX000
ENT 27,226 see SOP000
ENT 27,257 see DRR200
ENT 27,258 see EGV500
ENT 27,311 see CMA100
ENT 27,318 see EIN000
ENT 27,320 see DBI099
ENT 27,335 see CJJ250
ENT 27,341 see MDU600
ENT 27,346 see DOP200
ENT 27,394 see DJY200
ENT 27,396 see DTQ400
ENT 27,474 see BEP500
ENT 27,488 see BAT750
ENT 27,520 see CMA250
ENT 27,566 see DSO200
ENT 27,567 see CJJ250
ENT 27,572 see FAK000
ENT 27,738 see BLU000
ENT 27,822 see DOP600
ENT 27,989 see MKA000
ENT 28,009 see HON000
ENT 29,054 see CJV250
ENT 33,348 see HFE520
ENT 50,003 see TNK250
ENT 50,146 see RDK000
ENT 50,324 see EJM900
ENT 50,434 see AQG250
ENT 50,439 see BIA250
ENT 50,882 see HEK000
ENT 51,799 see MJQ500
ENT 61,241 see ACM750
ENT 27,699GC see DIN800
ENTERICIN see ADA725
ENTERO-BIO FORM see CHR500
ENTEROMYCETIN see CDP250
ENTEROPHEN see ADA725
ENTEROQUINOL see CHR500
ENTEROSALICYL see SJO000
ENTEROSALIL see SJO000
ENTEROSARINE see ADA725
ENTEROSEDIV see TEH500
ENTEROSEPT see DNF600
ENTEROSEPTOL see CHR500
ENTEROTOXON see NGG500
ENTERO-VIOFORM see CHR500
ENTEROZOL see CHR500
ENTERUM LOCORTEN see CHR500
ENTIZOL see MMN250
ENTOMOXAN see BBQ500
ENTRAMIN see ALQ000
ENTROKIN see CHR500
ENTROPHEN see ADA725
ENTSUFON see TMN490
ENTUSIL see SNN500
ENT 25,552-X see CDR750
ENT 27,395-X see CQH650
ENU see NKE500

ENU see ENV000
ENZACTIN see THM500
ENZAMIN see BBW500
ENZAPROST see POC500
ENZAPROST F see POC500
E.O. see EJN500
EO 5A see IGK800
EOSIN BLUE see ADG250
EOSINE see BNH500
EOSINE BLUE see ADG250
EOSINE BLUISH see ADG250
EOSINE SODIUM SALT see BNH500
EOSINE YELLOWISH see BNH500
EOSIN GELBLICH (GERMAN) see BNH500
EP 30 see PAX250
EP 160 see PKP750
EP-205 see BJN250
EP-206 see VOA000
EP-332 see DSO200
EP-333 see CJJ250
EP-411 see PFC750
EP 1463 see AAX250
EP-161E see ISE000
EPAL 6 see HFJ500
EPAL 8 see OEI000
EPAL 10 see DAI600
EPAL 12 see DXV600
EPAL 16NF see HCP000
E-PAM see DCK759
EPAMIN see DKQ000
EPAMIN see DNU000
EPANUTIN see DKQ000
EPANUTIN see DNU000
EPASMIR "5" see DKQ000
EPDANTOINE SIMPLE see DKQ000
EPE see EAV500
EPELIN see DKQ000
EPELIN see DNU000
EPHEDRAL see EAW000
EPHEDRATE see EAW000
EPHEDREMAL see EAW000
EPHEDRIN see EAW000
EPHEDRINE see EAW000
l-EPHEDRINE see EAW000
l(−)-EPHEDRINE see EAW000
EPHEDRINE HYDROCHLORIDE see EAW500
EPHEDRINE HYDROCHLORIDE see EAW500
EPHEDRINE HYDROCHLORIDE see EAX000
(−)-EPHEDRINE HYDROCHLORIDE see EAX000
d-EPHEDRINE HYDROCHLORIDE see EAW995
l-EPHEDRINE, HYDROCHLORIDE see EAW500
l-EPHEDRINE HYDROCHLORIDE see EAX000
dl-EPHEDRINE HYDROCHLORIDE see EAX500
1-EPHEDRINE SULFATE see EAY500
EPHEDRITAL see EAW000
EPHEDROL see EAW000
EPHEDROSAN see EAW000
EPHEDROTAL see EAW000
EPHEDSOL see EAW000
EPHENDRONAL see EAW000
EPHETONIN see EAX500
EPHETONINE see EAX500
EPHORRAN see DXH250
EPHOXAMIN see EAW000
EPHYNAL see VSZ450
EPHYNAL ACETATE see TGJ050
EPIANDROSTERONE, DEHYDRO-, ACETATE see HJB225

EPIBENZALIN see DLY000
EPIBLOC see CDT750
EPIBROMHYDRIN see BNI000
EPIBROMOHYDRIN (DOT) see BNI000
EPIBROMOHYDRINE see BNI000
EPICHLOORHYDRINE (DUTCH) see EAZ500
EPICHLORHYDRIN (GERMAN) see EAZ500
EPICHLORHYDRINE (FRENCH) see EAZ500
EPICHLOROHYDRIN see EAZ500
α-EPICHLOROHYDRIN see EAZ500
(dl)-α-EPICHLOROHYDRIN see EAZ500
EPICHLOROHYDRIN-DIMETHYLAMINE COPOLYMER see DOR500
EPICHLOROHYDRYNA (POLISH) see EAZ500
EPICHLOROPHYDRIN see EAZ500
EPICHOLESTANOL see EBA100
EPI-CLEAR see BDS000
EPICLORIDRINA (ITALIAN) see EAZ500
EPICUR see MQU750
EPICURE DDM see MJQ000
EPIDEHYDROCHOLESTERIN see EBA100
EPIDIAN 5 see IPO000
3,17-EPIDIHYDROXYESTRATRIENE see EDO000
3,17-EPIDIHYDROXYOESTRATRIENE see EDO000
EPIDORM see EOK000
EPIDOSIN see VBK000
EPIDOZIN see VBK000
EPIDROPAL see ZVJ000
EPIFENYL see DKQ000
EPIFENYL see DNU000
EPIFRIN see VGP000
EPIHYDAN see DKQ000
EPIHYDAN see DNU000
EPIHYDRIN ALCOHOL see GGW500
EPIHYDRINALDEHYDE see GGW000
EPIHYDRINE ALDEHYDE see GGW000
EPIKURE DDM see MJQ000
EPILAN see MKB250
EPILAN see DKQ000
EPILAN-D see DNU000
EPILANTIN see DKQ000
EPILANTIN see DNU000
EPINAT see DKQ000
EPINAT see DNU000
EPINELBON see DLY000
EPINEPHRAN see VGP000
EPINEPHRINE see VGP000
(−)-EPINEPHRINE see VGP000
1-EPINEPHRINE see VGP000
(R)-EPINEPHRINE see VGP000
1-EPINEPHRINE (synthetic) see VGP000
EPINEPHRINE ISOPROPYL HOMOLOG see DMV600
EPINOVAL see DJU200
EPIRENAMINE see VGP000
EPIRENAN see VGP000
EPI-REZ 508 see BLD750
EPI-REZ 510 see BLD750
EPIROTIN see BBW500
EPISED see DKQ000
EPISEDAL see EOK000
EPITELIOL see VSK600
EPITRATE see VGP000
EPN see EBD700
EPOBRON see IIU000
EPOLENE C see PJS750
EPOLENE C 10 see PJS750
EPOLENE C 11 see PJS750

EPOLENE E see PJS750
EPOLENE E 10 see PJS750
EPOLENE E 12 see PJS750
EPOLENE M 5K see PMP500
EPOLENE N see PJS750
EPON 828 see BLD750
EPON 828 see IPO000
EPORAL see SOA500
EPOXIDE 269 see LFV000
EPOXIDE A see BLD750
EPOXIDIZED SOYBEAN OIL see EBH525
1,2-EPOXYAETHAN (GERMAN) see EJN500
1,4-EPOXY-1,3-BUTADIENE see FPK000
EPOXYBUTANE see BOX750
1,2-EPOXYBUTANE see BOX750
1,4-EPOXYBUTANE see TCR750
1,2-EPOXYBUTENE-3 see EBJ500
3,4-EPOXY-1-BUTENE see EBJ500
EPOXYCARYOPHYLLENE see CCN100
1,2-EPOXY-3-CHLOROPROPANE see EAZ500
1,2-EPOXYCYCLOHEXANE see CPD000
3,6-endo-EPOXY-1,2-CYCLOHEXANEDICARBOXYLIC
ACID see EAR000
(−)-EPOXYDIHYDROCARYOPHYLLENE see CCN100
5,6-EPOXY-5,6-DIHYDRO-7-METHYLBENZ(A)
ANTHRACENE see MGZ000
1,2-EPOXYDODECANE see DXU400
1,2-EPOXY-4-(EPOXYETHYL)CYCLOHEXANE see
VOA000
EPOXYETHANE see EJN500
1,2-EPOXYETHANE see EJN500
1,2-EPOXY-3-ETHOXYPROPANE see EBQ700
1,2-EPOXY-3-ETHOXYPROPANE see EKM200
1,2-EPOXY-3-ETHOXYPROPANE (DOT) see EKM200
1,2-EPOXYETHYLBENZENE see EBR000
EPOXYETHYLBENZENE (8CI) see EBR000
1-EPOXYETHYL-3,4-EPOXYCYCLOHEXANE see
VOA000
3-(EPOXYETHYL)-7-OXABICYCLO(4.1.0)HEPTANE
see VOA000
4-(EPOXYETHYL)-7-OXABICYCLO(4.1.0)HEPTANE
see VOA000
3-(1,2-EPOXYETHYL)-7-
OXABICYCLO(4.1.0)HEPTANE see VOA000
4-(1,2-EPOXYETHYL)-7-
OXABICYCLO(4.1.0)HEPTANE see VOA000
EPOXYGUAIENE see EBU100
EPOXYHEPTACHLOR see EBW500
1,2-EPOXYHEXADECANE see EBX500
4,5-α-EPOXY-3-HYDROXY-17-
METHYLMORPHINAN-6-ONE HYDROCHLORIDE
see DNU300
1,2-EPOXY-3-ISOPROPOXYPROPANE see IPD000
4-(1,2-EPOXY-1-METHYLETHYL)-1-METHYL-7-
OXABICYCLO(4.1.0)HEPTANE see LFV000
α-β-EPOXY-β-METHYLHYDROCINNAMIC ACID,
ETHYL ESTER see ENC000
1,2-EPOXY-3-PHENOXYPROPANE see PFF360
2,3-EPOXYPROPANAL see GGW000
2,3-EPOXY-1-PROPANAL see GGW000
EPOXYPROPANE see PNL600
1,2-EPOXYPROPANE see PNL600
1,3-EPOXYPROPANE see OMW000
2,3-EPOXYPROPANE see PNL600
2,3-EPOXYPROPANOL see GGW500
2,3-EPOXY-1-PROPANOL see GGW500
2,3-EPOXY-1-PROPANOL ACRYLATE see ECH500

2,3-EPOXY-1-PROPANOL (OSHA) see GGW500
2,3-EPOXYPROPIONALDEHYDE see GGW000
2,3-EPOXYPROPYL ACRYLATE see ECH500
2,3-EPOXYPROPYL BUTYL ETHER see BRK750
2,3-EPOXYPROPYL CHLORIDE see EAZ500
2,3-EPOXYPROPYL ESTER ACRYLIC ACID see
ECH500
2,3-EPOXYPROPYL ISOPROPYL ETHER see IPD000
2,3-EPOXYPROPYLPHENYL ETHER see PFF360
(2,3-EPOXYPROPYL)TRIMETHYLAMMONIUM
CHLORIDE see GGY200
EPOXY RESINS, CURED see ECL500
EPOXY RESINS, UNCURED see ECM500
EPOXYSTYRENE see EBR000
α,β-EPOXYSTYRENE see EBR000
6-β,7-β-EPOXY-3-α-TROPANYL S-(−)-TROPATE see
SBG000
EPOXYTROPINE TROPATE see SBG000
EPRAZIN see POL500
EPROFIL see TEX000
EPROLIN see VSZ450
EPSAMON see AJD000
EPSICAPRON see AJD000
EPSILAN see VSZ450
EPSILAN-M see TGJ050
EPSOM SALTS see MAJ250
EPSYLON KAPROLAKTAM (POLISH) see CBF700
EPT see EQP000
EPTAC 1 see BJK500
EPTACLORO (ITALIAN) see HAR000
1,4,5,6,7,8,8-EPTACLORO-3a,4,7,7a-TETRAIDRO-4,7-
endo-METANO-INDENE (ITALIAN) see HAR000
EPTAL see DKQ000
EPTAM see EIN500
EPTANI (ITALIAN) see HBC500
EPTAN-3-ONE (ITALIAN) see EHA600
EPTC see EIN500
EPTOIN see DKQ000
EPTOIN see DNU000
E-PVC see PKQ059
EQ see SAV000
EQUAL see ARN825
EQUANIL SUSPENSION see MQU750
EQUIBRAL see MDQ250
EQUI BUTE see BRF500
EQUIGEL see DGP900
EQUIGYNE see ECU750
EQUIGYNE see ECU750
EQUILIN SODIUM SULFATE see ECW520
EQUILIN, SULFATE, SODIUM SALT (6CI) see ECW520
EQUILIUM see MQU750
EQUILIUM see PGE000
EQUINIL see MQU750
EQUINO-ACID see TIQ250
EQUINO-AID see TIQ250
EQUIPOISE see CJR909
EQUITDAZIN see MHC750
EQUIZOLE see TEX000
ERADEX see CMA100
ERAMIDE see CDR250
ERANTIN see PNA500
ERASE see HKC000
ERASOL see BIE500
ERASOL HYDROCHLORIDE see BIE500
ERASOL-IDO see BIE500
ERBAPLAST see CDP250
ERBAPRELINA see TGD000

ERBIUM CHLORIDE see ECX500
ERBIUM(III) NITRATE (1:3) see ECY500
ERBIUM TRICHLORIDE see ECX500
ERCO-FER see FBJ100
ERCOFERRO see FBJ100
ERE 1359 see REF000
EREVIT see TGJ050
ERGADENYLIC ACID see AOA125
ERGAM see EDC500
ERGATE see EDC500
ERGOCALCIFEROL see VSZ100
ERGOMAR see EDC500
ERGOPLAST ADC see DGT500
ERGOPLAST AdDO see AEO000
ERGOPLAST FDO see DVL700
ERGORONE see VSZ100
ERGOSTAT see EDC500
ERGOSTEROL, activated see VSZ100
ERGOSTEROL, irradiated see VSZ100
ERGOTAMINE BITARTRATE see EDC500
ERGOTAMINE TARTRATE see EDC500
ERGOTARTRATE see EDC500
ERIBUTAZONE see BRF500
ERIDAN see DCK759
ERIE BLACK B see AQP000
ERIE BLACK BF see AQP000
ERIE BLACK GAC see AQP000
ERIE BLACK GXOO see AQP000
ERIE BLACK JET see AQP000
ERIE BLACK NUG see AQP000
ERIE BLACK RXOO see AQP000
ERIE BRILLIANT BLACK S see AQP000
ERIE CONGO 4B see SGQ500
ERIE FIBRE BLACK VP see AQP000
ERINA see MQU750
ERINIT see PBC250
ERINITRIT see SIQ500
ERIO FAST ORANGE AS see HGC000
ERIOGLAUCINE see FMU059
ERION see ARQ250
ERIONITE see EDC650
ERIONITE (CAKNA (AL2SI7O18)2.14H2O) see EDC700
ERIONYL RED RS see CMM330
ERIOSKY BLUE see FMU059
ERIO TARTRAZINE see FAG140
ERITRONE see VSZ000
ERITROXILINA see CNE750
ERL-2774 see BLD750
ERLA-2270 see VOA000
ERLA-2271 see VOA000
ERR 4205 see BJN250
ERROLON see CHJ750
ERSERINE see PIA500
ERTALON 6SA see PJY500
ERTILEN see CDP250
ERTRON see VSZ100
ERYCIN see EDH500
ERYCYTOL see VSZ000
ERYSAN see BIF250
ERYTHORBIC ACID see EDE600
ERYTHRENE see BOP500
ERYTHRITOL ANHYDRIDE see BGA750
ERYTHRITOL ANHYDRIDE see DHB800
ERYTHROCIN see EDH500
ERYTHROGRAN see EDH500
ERYTHROGUENT see EDH500
ERYTHROMYCIN see EDH500
ERYTHROMYCIN A see EDH500

ERYTHROTIN see VSZ000
ERYTROXYLIN see CNE750
ESACHLOROBENZENE (ITALIAN) see HCC500
ESAIDRO-1,3,5-TRINITRO-1,3,5-TRIAZINA (ITALIAN)
see CPR800
ESAMETILENTETRAMINA (ITALIAN) see HEI500
ESANI (ITALIAN) see HEN000
ESANITRODIFENILAMINA (ITALIAN) see HET500
ESBRITE see SMQ500
ESBRITE 2 see SMQ500
ESBRITE 4 see SMQ500
ESBRITE 8 see SMQ500
ESBRITE 4-62 see SMQ500
ESBRITE G 10 see SMQ500
ESBRITE G-P 2 see SMQ500
ESBRITE LBL see SMQ500
ESBRITE 500HM see SMQ500
ESCAMBIA 2160 see PKQ059
ESCASPERE see RDK000
ESCIN, SODIUM SALT see EDM000
ESCOPARONE see DRS800
ESCOREZ 7404 see SMQ500
ESCULETIN DIMETHYL ETHER see DRS800
ESDRAGOL see AFW750
ESEN see PHW750
ESERINE see PIA500
ESERINE SALICYLATE see PIA750
ESEROLEIN, METHYLCARBAMATE (ESTER) see
PIA500
ESERPINE see RDK000
ESIDREX see CFY000
ESIDRIX see CFY000
ESKABARB see EOK000
ESKADIAZINE see PPP500
ESKALIN V see VRF000
ESKALITH see LGZ000
ESKASERP see RDK000
ESKAZINE see TKK250
ESKAZINE DIHYDROCHLORIDE see TKK250
ESKIMON 11 see TIP500
ESKIMON 12 see DFA600
ESKIMON 22 see CFX500
ESOPHOTRAST see BAP000
ESORB see VSZ450
ESPADOL see CLW000
ESPENAL see DXH250
ESPERAL see DXH250
ESPERFOAM FR see MKA500
ESPEROX 10 see BSC500
ESPEROX 31M see BSD250
ESPHYGMOGENINA see VGP000
ESSENCE of MIRBANE see NEX000
ESSENCE of MYRBANE see NEX000
ESSENCE of NIOBE see EGR000
ESSENCE OF NIOBE see MHA750
ESSENCE of ROSE see RNA000
ESSEX see CBT750
ESSEX 1360 see HLB400
ESSEX GUM 1360 see HLB400
ESSIGESTER (GERMAN) see EFR000
ESSIGSAEURE (GERMAN) see AAT250
ESSIGSAEUREANHYDRID (GERMAN) see AAX500
ESSO FUNGICIDE 406 see CBG000
ESTABEX U 18 see DVK200
ESTANE 5703 see UVA000
ESTAR see CMY800
ESTASIL see MQU750
ESTERASE-LIPASE see EDN100

ESTERE CIANOACETICO see EHP500
O-ESTER-p-NITROPHENOL with O-ETHYL PHENYL
PHOSPHONOTHIOATE see EBD700
ESTERON 44 see IOY000
ESTERONE see EDV000
ESTEROQUINONE LIGHT YELLOW 4JL see AAQ250
α-ESTER PALMITIC ACID with D-threo-(−)-2,2-
DICHLORO-N-(β-HYDROXY-α-
(HYDROXYMETHYL)-p-
NITROPHENETHYL)ACETAMIDE see CDP700
ESTERS see EDN500
ESTEVE see BRF500
ESTIBOGLUCONATO SODICO see AQH800
ESTILBEN see DKA600
ESTILBEN see DKB000
ESTILBIN see DKB000
ESTIMULEX see DBA800
ESTINERVAL see PFC750
ESTOL 603 see OAV000
ESTOL 1550 see DJX000
ESTONATE see DAD200
ESTON-B see EDP000
ESTONE YELLOW GN see AAQ250
ESTONOX see CDV100
ESTOSTERIL see PCL500
ESTRADEP see DAZ115
ESTRADIOL see EDO000
d-ESTRADIOL see EDO000
cis-ESTRADIOL see EDO000
α-ESTRADIOL see EDO000
β-ESTRADIOL see EDO000
ESTRADIOL-17-β see EDO000
17-β-ESTRADIOL see EDO000
3,17-β-ESTRADIOL see EDO000
d-3,17-β-ESTRADIOL see EDO000
ESTRADIOL BENZOATE see EDP000
ESTRADIOL-3-BENZOATE see EDP000
β-ESTRADIOL BENZOATE see EDP000
β-ESTRADIOL-3-BENZOATE see EDP000
ESTRADIOL-17-β-BENZOATE see EDP000
17-β-ESTRADIOL BENZOATE see EDP000
ESTRADIOL-17-β-3-BENZOATE see EDP000
17-β-ESTRADIOL-3-BENZOATE see EDP000
ESTRADIOL CYCLOPENTYLPROPIONATE see
DAZ115
ESTRADIOL-17-CYCLOPENTYLPROPIONATE see
DAZ115
ESTRADIOL-17-β-CYCLOPENTYLPROPIONATE see
DAZ115
ESTRADIOL-CYPIONATE see DAZ115
ESTRADIOL-17-CYPIONATE see DAZ115
ESTRADIOL-17-β-CYPIONATE see DAZ115
ESTRADIOL DIPROPIONATE see EDR000
ESTRADIOL-3,17-DIPROPIONATE see EDR000
β-ESTRADIOL DIPROPIONATE see EDR000
17-β-ESTRADIOL DIPROPIONATE see EDR000
β-ESTRADIOL-3,17-DIPROPIONATE see EDR000
3,17-β-ESTRADIOL DIPROPIONATE see EDR000
ESTRADIOL MONOBENZOATE see EDP000
17-β-ESTRADIOL MONOBENZOATE see EDP000
ESTRADIOL PHOSPHATE POLYMER see EDS000
ESTRADIOL POLYESTER with PHOSPHORIC ACID
see EDS000
ESTRADIOL VALERATE see EDS100
ESTRADIOL-17-VALERATE see EDS100

ESTRADIOL 17-β-VALERATE see EDS100
ESTRADIOL VALERIANATE see EDS100
ESTRADURIN see EDS000
ESTRAGARD see DAL600
ESTRAGON OIL see TAF700
ESTRALDINE see EDO000
ESTRATAB see ECU750
ESTRA-1,3,5(10),7-TETRAEN-17-ONE, 3-HYDROXY-,
HYDROGEN SULFATE SODIUM SALT (8CI) see
ECW520
ESTRA-1,3,5(10),7-TETRAEN-17-ONE, 3-(SULFOOXY)-,
SODIUM SALT see ECW520
1,3,5-ESTRATRIENE-3,17-β-DIOL see EDO000
ESTRA-1,3,5(10)-TRIENE-3,17-β-DIOL see EDO000
17-β-ESTRA-1,3,5(10)-TRIENE-3,17-DIOL see EDO000
ESTRA-1,3,5(10)-TRIENE-3,17-β-DIOL, 3-BENZOATE
see EDP000
1,3,5(10)-ESTRATRIENE-3,17-β-DIOL 3-BENZOATE
see EDP000
ESTRA-1,3,5(10)-TRIENE-3,17-DIOL (17-β)-3-
BENZOATE see EDP000
(17-β)-ESTRA-1,3,5(10)-TRIENE-3,17-DIOL 17-
CYCLOPENTANEPROPANOATE (9CI) see DAZ115
1,3,5(10)-ESTRATRIENE-3,17-β-DIOL DIPROPIONATE
see EDR000
ESTRA-1,3,5(10)-TRIENE-3,17-DIOL (17-β)-
DIPROPIONATE see EDR000
(17-β)-ESTRA-1,3,5(10)-TRIENE-3,17-DIOL-17-
PENTANOATE (9CI) see EDS100
(17-β)-ESTRA-1,3,5(10)-TRIENE-3,17-DIOL POLYMER
with PHOSPHORIC ACID see EDS000
ESTRA-1,3,5(10)-TRIENE-3,16-α,17-β-TRIOL see
EDU500
1,3,5-ESTRATRIENE-3-β,16-α,17-β-TRIOL see EDU500
(16-α,17-β)-ESTRA-1,3,5(10)-TRIENE-3,16,17-TRIOL see
EDU500
1,3,5-ESTRATRIEN-3-OL-17-ONE see EDV000
1,3,5(10)-ESTRATRIEN-3-OL-17-ONE see EDV000
4,9,11,-ESTRATRIEN-17β-OL-3-ONE see THL600
Δ-1,3,5-ESTRATRIEN-3-β-OL-17-ONE see EDV000
ESTRA-1,3,5(10)-TRIEN-17-ONE, 3-(SULFOOXY)-,
compounded with PIPERAZINE (1:1) see PIK450
ESTRA-1,3,5(10)-TRIEN-17-ONE, 3-(SULFOXY)-,
SODIUM SALT (9CI) see EDV600
ESTRATRIOL see EDU500
ESTRAVEL see EDS100
ESTREPTOCIDA see SNM500
ESTRIFOL see ECU750
ESTRIL see DKA600
ESTRIN see EDV000
ESTRIOL see EDU500
16-α,17-β-ESTRIOL see EDU500
3,16-α,17-β-ESTRIOL see EDU500
ESTRIOLO (ITALIAN) see EDU500
ESTROATE see ECU750
ESTROBEN see DKB000
ESTROBENE see DKA600
ESTROBENE see DKB000
ESTROCON see ECU750
ESTRODIENOL see DAL600
ESTROGEN see DKA600
ESTROGEN see EEH500
ESTROGENIN see DKB000
ESTROGENS, CONJUGATES see PMB000
ESTROICI see EDR000

ESTROL see EDV000
ESTROMED see ECU750
ESTROMENIN see DKA600
ESTRON see EDV000
ESTRONA (SPANISH) see EDV000
ESTRONE see EDV000
ESTRONE-A see EDV000
ESTRONE BENZOATE see EDV500
ESTRONE, HYDROGEN SULFATE, compounded with
PIPERAZINE (1:1) (8CI) see PIK450
ESTRONE, HYDROGEN SULFATE, SODIUM SALT see
EDV600
ESTRONE SODIUM SULFATE see EDV600
ESTRONE SULFATE SODIUM see EDV600
ESTRONE SULFATE SODIUM SALT see EDV600
ESTRONE-3-SULFATE SODIUM SALT see EDV600
ESTRONEX see EDR000
ESTROPAN see ECU750
ESTROPIPATE see PIK450
ESTRORAL see DAL600
ESTROSEL see DGP900
ESTROSOL see DGP900
ESTROSTILBEN see DKB000
ESTROSYN see DKA600
ESTROVITE see EDO000
ESTRUGENONE see EDV000
ESTRUMATE see CMX880
ESTRUSOL see EDV000
ESTYRENE 4-62 see SMQ500
ESTYRENE AS see ADY500
ESTYRENE G 15 see SMQ500
ESTYRENE G 20 see SMQ500
ESTYRENE G-P 4 see SMQ500
ESTYRENE H 61 see SMQ500
ESTYRENE 500SH see SMQ500
ETs see EHG100
ET 14 see RMA500
ET 57 see RMA500
ETABENZARONE see EDV700
ETABUS see DXH250
ETAFOS see EEE200
ETAIN (TETRACHLORURE d') (FRENCH) see TGC250
ETAMICAN see VSZ450
ETAMINAL SODIUM see NBU000
ETANAUTINE see BBV500
ETANOLAMINA (ITALIAN) see EEC600
ETANOLO (ITALIAN) see EFU000
ETANTIOLO (ITALIAN) see EMB100
ETAPERAZIN see CJM250
ETAPERAZINE see CJM250
ETAPHOS see EEE200
ETAVIT see VSZ450
ETCHLORVINOLO see CHG000
ETERE ETILICO (ITALIAN) see EJU000
ETH see EEI000
ETHAANTHIOL (DUTCH) see EMB100
ETHACRIDINE LACTATE see EDW500
ETHAL see HCP000
ETHAMINAL SODIUM see NBU000
ETHAN see PII500
ETHANAL see AAG250
ETHANAL OXIME see AAH250
ETHANAMIDE see AAI000
ETHANAMINE see EFU400
ETHANAMINIUM, 2-CHLORO-N,N,N-TRIMETHYL-,
CHLORIDE (9CI) see CMF400
ETHANE see EDZ000
ETHANECARBOXYLIC ACID see PMU750

ETHANE, CHLOROPENTAFLUORO-, mixt. with
CHLORODIFLUOROMETHANE see FOO560
ETHANEDIAMIDE see OLO000
1,2-ETHANEDIAMINE see EEA500
1,2-ETHANEDIAMINE, N,N-BIS(2-AMINOETHYL)- see
NEI800
1,2-ETHANEDIAMINE, N,N'-BIS(1,1-
DIMETHYLETHYL)- see DEC100
1,2-ETHANEDIAMINE, DIHYDROCHLORIDE see
EIW000
ETHANE, 1,2-DIAMINO-, COPPER COMPLEX see
DBU800
ETHANE, 1,2-DIAZIDO- see DCL600
1,2-ETHANEDICARBOXYLIC ACID see SMY000
ETHANE DICHLORIDE see EIY600
ETHANEDINITRILE see COO000
ETHANE, 1,1-DINITRO- see DUV710
ETHANEDIOIC ACID see OLA000
ETHANEDIOIC ACID BIS(CYCLOHEXYLIDENE
HYDRAZIDE) see COF675
ETHANEDIOIC ACID DIAMMONIUM SALT see
ANO750
ETHANEDIOIC ACID, TIN(2+) SALT (1:1) (9CI) see
TGE250
1,2-ETHANEDIOL see EJC500
1,2-ETHANEDIOL DIACETATE see EJD759
ETHANEDIOL DINITRATE see EJG000
1,2-ETHANEDIOL DIPROPANOATE (9CI) see COB260
ETHANEDIONE, DIPHENYL-, MONOOXIME see
BCA300
ETHANEDIONIC ACID see OLA000
N,N'-(1,2-ETHANEDIOXYSULFINYL)BIS(S-METHYL-
N-METHYLCARBAMOYLOXYTHIOACETIMIDATE)
see EEA700
ETHANEDITHIOAMIDE see DXO200
1,2-ETHANEDITHIOL see EEB000
1,2-ETHANEDITHIOL, CYCLIC ESTER with P,P-
DIETHYL PHOSPHONODITHIOIMIDOCARBONATE
see PGW750
1,2-ETHANEDITHIOL, CYCLIC S,S-ESTER with
PHOSPHONODITHIOIMIDOCARBONIC ACID P,P-
DIETHYL ESTER see PGW750
N,N'-(1,2-ETHANEDITHIOSULFINYL)BIS(S-METHYL-
N-METHYLCARBAMOYLOXYTHIOACETIMIDATE)
see EEB050
1,2-ETHANEDIYLBIS(CARBAMODITHIOATO)(2−)-
MANGANESE see MAS500
1,2-ETHANEDIYLBISCARBAMODITHIOIC ACID
DISODIUM SALT see DXD200
1,2-ETHANEDIYLBISCARBAMODITHIOIC ACID
MANGANESE COMPLEX see MAS500
1,2-ETHANEDIYLBISCARBAMODITHIOIC ACID,
MANGANESE(2+) SALT (1:1) see MAS500
N,N'-1,2-ETHANEDIYLBIS(N-
(CARBOXYMETHYL))GLYCINE see EIX000
N,N'-1,2-ETHANEDIYLBIS(N-
(CARBOXYMETHYL)GLYCINE) DISODIUM SALT see
EIX500
N,N'-1,2-ETHANEDIYLBIS(N-
(CARBOXYMETHYL))GLYCINE TETRASODIUM
SALT see EIV000
N,N'-1,2-ETHANEDIYLBIS(N-
CARBOXYMETHYL)GLYCINE, TRISODIUM SALT see
TNL250
1,2-ETHANEDIYLBISMANEB, MANGANESE(2+)
SALT (1:1) see MAS500
2,2'-(1,2-ETHANEDIYLBIS(OXY))BISETHANOL see
TJQ000

1,2-ETHANEDIYL THIOCYANATE see EJC035
ETHANE HEXACHLORIDE see HCI000
ETHANE, HEXANITRO- see HET675
ETHANEHYDRAZONIC ACID see ACM750
ETHANE-1-HYDROXY-1,1-DIPHOSPHONATE see HKS780
ETHANE-1-HYDROXY-1,1-DIPHOSPHONIC ACID DISODIUM SALT see DXD400
ETHANENITRILE see ABE500
ETHANE, 1,1'-(OXYBIS(METHYLENESULFONYL))BIS(2-CHLORO- see OPG100
ETHANE PENTACHLORIDE see PAW500
ETHANEPEROXOIC ACID see PCL500
ETHANEPEROXOIC ACID, 1,1-DIMETHYLETHYL ESTER see BSC250
ETHANESELENOL, 2-AMINO-, HYDROCHLORIDE see AJS900
ETHANE, 1,1'-SULFINYLBIS(1,2-DICHLORO- see SNT100
ETHANESULFONIC ACID, 2-AMINO- see TAG750
ETHANESULFONIC ACID, 2-HYDROXY-, AMMONIUM SALT see ANL100
ETHANE, 1,1,2,2-TETRACHLORO-1,2-DIFLUORO- see TBP050
ETHANETHIOAMIDE see TFA000
ETHANETHIOIC ACID see TFA500
ETHANETHIOL see EMB100
ETHANETHIOLIC ACID see TFA500
ETHANE TRICHLORIDE see TIN000
1,1,1-ETHANETRIOL DIPHOSPHONATE see HKS780
ETHANE, compressed (UN 1035) (DOT) see EDZ000
ETHANE, refrigerated liquid (UN 1961) (DOT) see EDZ000
ETHANIMIDOTHIOIC ACID, N,N'-(1,2-ETHANEDIYLBIS(OXYSULFINYL(METHYLIMINO)CARBONYLOXY)BIS-, DIMETHYL ESTER see EEA700
ETHANIMIDOTHIOIC ACID, N,N'-(1,2-ETHANEDIYLBIS(THIOSULFINYL(METHYLIMINO)CARBONYLOXY))BIS-, DIMETHYL ESTER see EEB050
ETHANIMIDOTHIOIC ACID, N-((((((HEXYLOXY)SULFINYL)METHYLAMINO)CARBONYL)OXY)-, METHYL ESTER see MKM800
ETHANIMIDOTHIOIC ACID, N,N'-(THIOBIS((METHYLAMINO)CARBONYLOXY))BIS(2-(DIMETHYLAMINO)-2-OXO-, DIMETHYL ESTER see BKU120
ETHANOIC ACID see AAT250
ETHANOIC ACID, ETHENYL ESTER see VLU250
ETHANOIC ANHYDRATE see AAX500
ETHANOLAMINE see EEC600
β-ETHANOLAMINE see EEC600
ETHANOLAMINE, solution (DOT) see EEC600
ETHANOL, 1-AMINO-(8CI,9CI) see AAG500
ETHANOL, 2-((4-AMINO-2-NITROPHENYL)AMINO)- see ALO750
ETHANOL, 2-(2-(2-(4-(p-CHLORO-α-PHENYLBENZYL)-1-PIPERAZINYL)ETHOXY)ETHOXY)- see HHK050
ETHANOL,- 2-(2-(2-(4-((4-CHLOROPHENYL)PHENYLMETHYL)-1-PIPERAZINYL)ETHOXY)ETHOXY)- see HHK050
ETHANOL, 2-(2-(DIETHYLAMINO)ETHOXY)- see DHQ100
ETHANOL, 2-(ETHYL(3-METHYL-4-NITROSOPHENYL)AMINO)- see ELG100

ETHANOL, 2-(2-(2-(HEXYLOXY)ETHOXY)ETHOXY)- see HFT550
ETHANOL, 2-(2-ISOPROPOXYETHOXY)- see IOJ500
ETHANOLISOPROPYLAMINE see INN400
ETHANOL, 2-(ISOPROPYLAMINO)- see INN400
ETHANOL (MAK) see EFU000
ETHANOL, 2-METHOXY-, PHOSPHATE (3:1) see TLA600
ETHANOL, 2-(2-(1-METHYLETHOXY)ETHOXYL)- see IOJ500
ETHANOL, 2-((1-METHYLETHYL)AMINO)- (9CI) see INN400
ETHANOL, 1-PHENYL- see PDE000
ETHANOL, 2-PHENYL- see PDD750
ETHANOL 200 PROOF see EFU000
4-ETHANOLPYRIDINE see POR500
ETHANOL SOLUTIONS (UN 1170) (DOT) see EFU000
ETHANOL THALLIUM(1+) SALT see EEE000
1-ETHANOL-2-THIOL see MCN250
ETHANONE, 2-CHLORO-1-(9H-FLUOREN-2-YL)- see CEC700
ETHANONE, 2-CHLORO-1-PHENYL- see CEA750
ETHANONE, 2-((4-(DICHLOROACETYL)PHENYL)AMINO)-2-HYDROXY-1-(4-PHENOXYPHENYL)- see DEN880
ETHANONE, 1-(4-(3-((1,1-DIMETHYLETHYL)AMINO)-2-HYDROXYPROPOXY)-2-BENZOFURANYL)- see ACN300
ETHANONE, 1-(7-(3-((1,1-DIMETHYLETHYL)AMINO)-2-HYDROXYPROPOXY)-2-BENZOFURANYL)- see ACN320
ETHANONE, 1-(2,5-DIMETHYL-3-THIENYL)- see DUG425
ETHANONE, 1-(2-ETHYL-7-(2-HYDROXY-3-((1-METHYLETHYL)AMINO)PROPOXY)-4-BENZOFURANYL)- see ELI600
ETHANONE-1-(3-ETHYL-5,6,7,8-TETRAHYDRO-5,5,8,8-TETRAMETHYL-2-NAPHTHALENYL)(9CI) see ACL750
ETHANONE, 2-HYDROXY-1,2-DIPHENYL- see BCP250
ETHANONE, 1-(7-(2-HYDROXY-3-((1-METHYLETHYL)AMINO)PROPOXY)-2-BENZOFURANYL)-(9CI) see HLK600
ETHANONE, 1-(7-(2-HYDROXY-3-((1-METHYLPROPYL)AMINO)PROPOXY)-2-BENZOFURANYL)- see ACN310
ETHANONE, 1-(2-HYDROXYPHENYL)-(9CI) see HIN500
ETHANONE, 1-(5-METHOXY-1H-INDOL-3-YL)-2-(4-PIPERIDINYL)-, MONOHYDROCHLORIDE (9CI) see MES900
ETHANONE, 1-(4-METHOXYPHENYL)-(9CI) see MDW750
ETHANONE, 1-(4-METHYLPHENYL)-(9CI) see MFW250
ETHANONE, 1-(5-METHYL-2-THIENYL)- see MPR300
ETHANONE, 1-(1-NAPHTHALENYL)-(9CI) see ABC475
ETHANONE, 1-PHENYL-, MONOCHLORO DERIV. see CDN505
ETHANOX see EEH600
ETHANOX 330 see TMJ000
ETHANOYL CHLORIDE see ACF750
ETHAPERAZINE see CJM250
ETHAPHOS see EEE200
ETHAVAN see EQF000
ETHAZATE see BJC000
ETHCHLOROVYNOL see CHG000

ETHCHLORVINYL see CHG000
ETHCLORVYNOL see CHG000
ETHEFON see CDS125
ETHEL see CDS125
ETHENE see EIO000
ETHENE, 1,1-DIFLUORO- see VPP000
1,1'-(1,2-ETHENEDIYL)DIBENZENE see SLR000
ETHENE, FLUORO- see VPA000
ETHENE OXIDE see EJN500
ETHENE POLYMER see PJS750
ETHENE, 1,1'-SULFINYLBIS(2,2-DICHLORO- see SNT100
ETHENOL HOMOPOLYMER (9CI) see PKP750
ETHENONE see KEU000
ETHENYL ACETATE see VLU250
ETHENYLBENZENE see SMQ000
ETHENYLBENZENE HOMOPOLYMER see SMQ500
ETHENYLBENZENE POLYMER with 1,3-BUTADIENE see SMR000
4-ETHENYL-BENZENESULFONIC ACID SODIUM SALT, HOMOPOLYMER (9CI) see SJK375
4-ETHENYL-1-CYCLOHEXENE see CPD750
ETHENYL ETHANOATE see VLU250
S-ETHENYL-dl-HOMOCYSTEINE see VLU200
1-(ETHENYLOXY) BUTANE see VMZ000
ETHENYLOXYETHENE see VOP000
1-ETHENYL-2-PYRROLIDINONE see EEG000
1-ETHENYL-2-PYRROLIDINONE HOMOPOLYMER see PKQ250
1-ETHENYL-2-PYRROLIDINONE POLYMERS see PKQ250
ETHEPHON see CDS125
ETHER see EJU000
ETHER, BIS(2-(2-METHOXYETHOXY)ETHYL) see PBO500
ETHER, BUTYL 2,3-EPOXYPROPYL see BRK750
ETHER, BUTYL GLYCIDYL see BRK750
ETHER BUTYLIQUE (FRENCH) see BRH750
ETHER, BUTYL METHYL see BRU780
ETHER CHLORATUS see EHH000
ETHER, 4-CHLOROPHENYL (4'-CHLORO-2'-NITRO)PHENYL see CJD600
ETHER CYANATUS see PMV750
ETHER, DECYL METHYL see MIW075
ETHER DICHLORE (FRENCH) see DFJ050
ETHER, DI-n-PENTYL- see PBX000
ETHER, DI-n-PROPYL- see PNM000
ETHER ETHYLBUTYLIQUE (FRENCH) see EHA500
ETHER ETHYLIQUE (FRENCH) see EJU000
ETHER, HEXACHLOROPHENYL see CDV175
ETHER HYDROCHLORIC see EHH000
ETHERIN see PJS750
ETHER ISOPROPYLIQUE (FRENCH) see IOZ750
ETHER METHYLIQUE MONOCHLORE (FRENCH) see CIO250
ETHER, METHYL PHENYL see AOX750
ETHER, METHYL PROPYL see MOU830
ETHER MONOETHYLIQUE de l'ETHYLENE-GLYCOL (FRENCH) see EES350
ETHER MONOETHYLIQUE de l'HYDROQUINONE see EFA100
ETHER MONOMETHYLIQUE de l'ETHYLENE-GLYCOL (FRENCH) see EJH500
ETHER MURIATIC see EHH000
ETHEROL E see PJS750
ETHERON see PKL500
ETHERON SPONGE see PKL500
ETHER, PENTACHLOROPHENYL see PAW250

ETHERS see EEG500
ETHEVERSE see CDS125
ETHICON PTFE see TAI250
ETHIDE see DFU000
ETHIDIUM BROMIDE see DBV400
ETHINAMATE see EEH000
ETHINE see ACI750
ETHINODIOL DIACETATE see EQJ500
1-ETHINYLCYCLOHEXYL CARBAMATE see EEH000
1-ETHINYLCYCLOHEXYL CARBONATE see EEH000
17-α-ETHINYL-3,17-DIHYDROXY-$\Delta^{1,3,5}$-ESTRATRIENE see EEH500
17-α-ETHINYL-3,17-DIHYDROXY-$\Delta^{1,3,5}$-OESTRATRIENE see EEH500
ETHINYL ESTRADIOL see EEH500
17-ETHINYLESTRADIOL see EEH500
17-ETHINYL-3,17-ESTRADIOL see EEH500
17-α-ETHINYLESTRADIOL see EEH500
17-α-ETHINYL-17-β-ESTRADIOL see EEH500
ETHINYL ESTRADIOL and DIMETHISTERONE see DNX500
ETHINYLESTRADIOL-3-METHYL ETHER see MKB750
17-α-ETHINYL ESTRADIOL 3-METHYL ETHER see MKB750
ETHINYLESTRADIOL-3-METHYL ETHER and NORETHYNODRED (1:50) see EAP000
ETHINYL ESTRADIOL and NORETHINDRONE ACETATE see EEH520
17-ETHINYL-5(10)-ESTRAENEOLONE see EEH550
17-α-ETHINYL-ESTRA(5,10)ENEOLONE see EEH550
17-α-ETHINYLESTRA-4-EN-17-β-OL-3-ONE see NNP500
17-α-ETHINYLESTRA-1,3,5(10)-TRIENE-3,17-β-DIOL see EEH500
17-α-ETHINYL-5,10-ESTRENOLONE see EEH550
ETHINYLESTRIOL see EEH500
17-α-ETHINYL-17-β-HYDROXY-Δ:4-ESTREN-3-ONE see NNP500
17-α-ETHINYL-17-β-HYDROXY-$\Delta^{5(10)}$-ESTREN-3-ONE see EEH550
17-α-ETHINYL-19-NORTESTOSTERONE see NNP500
17-α-ETHINYL-19-NORTESTOSTERONE ACETATE see ABU000
17-α-ETHINYL-19-NORTESTOSTERONE-17-β-ACETATE see ABU000
ETHINYLOESTRADIOL see EEH500
17-ETHINYL-3,17-OESTRADIOL see EEH500
ETHINYLOESTRADIOL-3-METHYL ETHER see MKB750
17-α-ETHINYL OESTRADIOL-3-METHYL ETHER see MKB750
ETHINYL OESTRADIOL mixed with NORETHISTERONE ACETATE see EEH520
ETHINYL-OESTRANOL see EEH500
17-α-ETHINYLOESTRA-1,3,5(10)-TRIENE-3,17-β-DIOL see EEH500
ETHINYLOESTRIOL see EEH500
ETHINYL TRICHLORIDE see TIO750
17-α-ETHINYL-$\Delta^{5,10-19}$-NORTESTOSTERONE see EEH550
17-α-ETHINYL-$\Delta^{1,3,5(10)}$OESTRATRIENE-3,17-β-DIOL see EEH500
ETHIOL see EEH600
ETHIOLACAR see MAK700

ETHION see EEH600
ETHION, dry see CMA100
ETHIONIN see EEI000
ETHIONINE see EEI000
(±)-ETHIONINE see EEI000
dl-ETHIONINE see EEI000
ETHLON see PAK000
ETHOBROM see THV000
ETHOBROM see ARW250
ETHOCAINE see AIT250
ETHOCEL see EHG100
ETHOCEL 150 see EHG100
ETHOCEL 890 see EHG100
ETHOCEL E7 see EHG100
ETHOCEL E50 see EHG100
ETHOCEL MED see EHG100
ETHOCEL N7 see EHG100
ETHOCEL N10 see EHG100
ETHOCEL N200 see EHG100
ETHOCEL STD see EHG100
ETHOCHLORVYNOL see CHG000
ETHODAN see EEH600
ETHODIN see EDW500
ETHODRYL see DIW000
ETHODRYL CITRATE see DIW200
ETHOHEXADIOL see EKV000
ETHOL see HCP000
ETHOMEEN C/15 see EEJ000
ETHONE see ENY500
ETHOPABATE see EEK100
ETHOPROP see EIN000
ETHOPROPAZINE see DIR000
ETHOPROPHOS see EIN000
ETHOVAN see EQF000
2-ETHOXY-4-ACETAMIDOBENZOID ACID METHYL ESTER see EEK100
4-ETHOXYACETANILIDE see ABG750
p-ETHOXYACETANILIDE see ABG750
ETHOXY ACETATE see EES400
ETHOXYACETIC ACID see EEK500
2-ETHOXYACETIC ACID see EEK500
ETHOXYANILINE see EEL100
4-ETHOXYANILINE see PDK790
p-ETHOXYANILINE see PDK790
6-ETHOXY-m-ANOL see IRY000
4-ETHOXYBENZENAMINE see PDK790
ETHOXYCARBONYLDIAZOMETHANE see DCN800
ETHOXYCARBONYLETHYLENE see EFT000
ETHOXYCARBONYLMETHYL BROMIDE see EGV000
ETHOXY CHLOROMETHANE see CIM000
2-ETHOXY-6,9-DIAMINOACRIDINE LACTATE see EDW500
2-ETHOXY-6,9-DIAMINOACRIDINE LACTATE HYDRATE see EDW500
2-ETHOXY-6,9-DIAMINOACRIDINIUM LACTATE see EDW500
ETHOXY DIGLYCOL see CBR000
2-ETHOXY DIHYDROPYRAN see EER500
2-ETHOXY-3,4-DIHYDRO-1,2-PYRAN see EER500
2-ETHOXY-3,4-DIHYDRO-2H-PYRAN see EER500
2-ETHOXY-2,3-DIHYDRO-γ-PYRAN see EER500
6-ETHOXY-1,2-DIHYDRO-2,2,4-TRIMETHYLQUINOLINE see SAV000
1-ETHOXY-3,7-DIMETHYL-2,6-OCTADIENE see GDG100
8-ETHOXY-2,6-DIMETHYLOCTENE-2 see EES100
ETHOXYDIMETHYLSILANE see EES200

ETHOXYETHANE see EJU000
2-ETHOXYETHANOL see EES350
2-ETHOXYETHANOL ACETATE see EES400
2-ETHOXYETHANOL, ESTER with ACETIC ACID see EES400
ETHOXY ETHENE see EQF500
2-(2-ETHOXYETHOXY)ETHANOL see CBR000
2-(2-ETHOXYETHOXY)ETHANOL ACETATE see CBQ750
1-ETHOXY-2-(β-ETHOXYETHOXY)ETHANE see DIW800
2-(2-(2-ETHOXYETHOXY)ETHOXY)ETHANOL see EFL000
2-ETHOXY-ETHYLACETAAT (DUTCH) see EES400
ETHOXYETHYL ACETATE see EES400
2-ETHOXYETHYL ACETATE see EES400
β-ETHOXYETHYL ACETATE see EES400
2-ETHOXYETHYLE, ACETATE de (FRENCH) see EES400
ETHOXYFORMIC ANHYDRIDE see DIX200
3-ETHOXY-4-HYDROXYBENZALDEHYDE see EQF000
4'-ETHOXY-3-HYDROXYBUTYRANILIDE see HJS850
1-ETHOXY-2-HYDROXY-4-PROPENYLBENZENE see IRY000
2-(3-ETHOXY-1-INDANYLIDENE)-1,3-DINDANDIONE see BGC250
ETHOXYLATED MONO- and DIGLYCERIDES see EEU100
ETHOXYLATED SORBITAN MONOOLEATE see PKL100
ETHOXYMETHANE see EMT000
ETHOXY METHYL CHLORIDE see CIM000
ETHOXYMETHYLENEMALONIC ACID, ETHYL ESTER see EEV200
1-(ETHOXYMETHYL)-5-(1-METHYLETHYL)-6-(PHENYLMETHYL)-2,4(1H,3H)-PYRIMIDINEDIONE see EAN525
(ETHOXYMETHYL)OXIRANE see EBQ700
(ETHOXYMETHYL)OXIRANE see EKM200
ETHOXY-4-NITROPHENOXYPHENYLPHOSPHINE SULFIDE see EBD700
N-(4-ETHOXY-3-NITRO)PHENYLACETAMIDE see NEL000
4-ETHOXYPHENOL see EFA100
p-ETHOXYPHENOL see EFA100
N-(4-ETHOXYPHENYL)ACETAMIDE see ABG750
N-p-ETHOXYPHENYLACETAMIDE see ABG750
4-ETHOXY-7-PHENYL-3,5-DIOXA-6-AZA-4-PHOSPHAOCT-6-ENE-8-NITRILE-4-SULFIDE see BAT750
N-(4-ETHOXYPHENYL)-3'-NITROACETAMIDE see NEL000
4-ETHOXYPHENYLUREA see EFE000
p-ETHOXYPHENYLUREA see EFE000
N-(4-ETHOXYPHENYL)UREA see EFE000
1-ETHOXYPROPANE see EPC125
N¹-(6-ETHOXY-3-PYRIDAZINYL)SULFANILAMIDE see SNJ100
ETHOXYQUINE see SAV000
ETHOXYQUIN (FCC) see SAV000
6-ETHOXY-3-SULFANILAMIDOPYRIDAZINE see SNJ100
ETHOXYTRIETHYLENE GLYCOL see EFL000
ETHOXYTRIGLYCOL see EFL000
6-ETHOXY-2,2,4-TRIMETHYL-1,2-DIHYDROQUINOLINE see SAV000

ETHRANE see EAT900
ETHREL see CDS125
ANTI-ETHYL see DXH250
ETHYLACETAAT (DUTCH) see EFR000
ETHYLACETANILIDE see EFQ500
N-ETHYLACETANILIDE see EFQ500
ETHYL ACETATE see EFR000
ETHYLACETIC ACID see BSW000
ETHYL ACETIC ESTER see EFR000
ETHYL ACETOACETATE (FCC) see EFS000
ETHYL ACETONE see PBN250
ETHYL ACETYL ACETATE see EFS000
ETHYL ACETYLACETONATE see EFS000
ETHYL ACETYLENE see EFS500
ETHYL ACETYLENE, INHIBITED see EFS500
2-ETHYL-4-ACETYL-7-(2-HYDROXY-3-
ISOPROPYLAMINOPROPOXY)BENZOFURAN see
ELI600
ETHYL 3-ACETYLPROPIONATE see EFS600
ETHYLACRYLAAT (DUTCH) see EFT000
ETHYL ACRYLATE see EFT000
ETHYLAENE GLYCOL FORMAL see DVR800
ETHYLAKRYLAT (CZECH) see EFT000
ETHYLAL see EFT500
ETHYL ALCOHOL see EFU000
ETHYLALCOHOL (DUTCH) see EFU000
ETHYL ALCOHOL, anhydrous see EFU000
ETHYL ALCOHOL SOLUTIONS (UN 1170) (DOT) see
EFU000
ETHYL ALCOHOL THALLIUM (I) see EEE000
ETHYL ALDEHYDE see AAG250
ETHYLALUMINUM SESQUICHLORIDE see TJP775
ETHYLAMINE see EFU400
ETHYLAMINE, 2-CHLORO-N,N-TRIMETHYL-,
HYDROCHLORIDE see CGJ280
ETHYLAMINE-2-(DIPHENYLMETHOXY)-N,N-
DIMETHYL, compound with 8-
CHLOROTHEOPHYLLINE (1:1) see DYE600
ETHYLAMINE (UN 1036) (DOT) see EFU400
ETHYLAMINE, aqueous solution with not <50% but not
>70% ethylamine (UN 2270) (DOT) see EFU400
N-ETHYLAMINOBENZENE see EGK000
ETHYL AMINOBENZOATE see EFX000
ETHYL-4-AMINOBENZOATE see EFX000
ETHYL-o-AMINOBENZOATE see EGM000
ETHYL-p-AMINOBENZOATE see EFX000
2-ETHYLAMINOETHANOL see EGA500
2-(ETHYLAMINO)ETHANOL see EGA500
ETHYL-p-AMINOPHENYL KETONE see AMC000
2-ETHYLAMINOTHIADIAZOLE see EGI000
2-ETHYLAMINO-1,3,4-THIADIAZOLE see EGI000
ETHYLAMYLCARBINOL see OCY100
ETHYL-n-AMYLCARBINOL see OCY100
ETHYL AMYL KETONE see ODI000
ETHYL AMYL KETONE see EGI750
ETHYL sec-AMYL KETONE see EGI750
ETHYLAN see DJC000
ETHYLANILINE see EGK000
2-ETHYLANILINE see EGK500
2-ETHYL ANILINE see EGK500
4-ETHYLANILINE see EGL000
N-ETHYLANILINE see EGK000
p-ETHYLANILINE see EGL000
2-ETHYLANILINE (DOT) see EGK500
ETHYL ANISATE see AOV000
ETHYL-p-ANISATE (FCC) see AOV000
ETHYLAN LD see CNF330
ETHYLAN MLD see BKE500

ETHYL ANTHRANILATE see EGM000
ETHYLARSONOUS DICHLORIDE see DFH200
ETHYLBENZEEN (DUTCH) see EGP500
2-ETHYLBENZENAMINE see EGK500
N-ETHYLBENZENAMINE see EGK000
N-ETHYLBENZENAMINO see EGK000
ETHYL BENZENE see EGP500
ETHYL BENZENEACETATE see EOH000
ETHYL BENZOATE see EGR000
2-ETHYL-3-BENZOFURANYL p-HYDROXYPHENYL
KETONE see BBJ500
ETHYLBENZOL see EGP500
ETHYL BENZYL ACETOACETATE see EFS000
ETHYLBIS(2-CHLOROETHYL)AMINE see BID250
ETHYLBIS(β-CHLOROETHYL)AMINE see BID250
ETHYL BISCOUMACETATE see BKA000
ETHYL BIS(4-HYDROXYCOUMARINYL)ACETATE
see BKA000
ETHYL BIS(4-HYDROXY-3-COUMARINYL)ACETATE
see BKA000
ETHYL BORATE (DOT) see BMC250
ETHYL BROMACETATE see EGV000
ETHYL BROMIDE see EGV400
ETHYL BROMOACETATE see EGV000
ETHYL-α-BROMOACETATE see EGV000
ETHYL BROMOPHOS see EGV500
2-ETHYLBUTADIENE see EGV600
2-ETHYL-1,3-BUTADIENE see EGV600
2-ETHYLBUTANAL see DHI000
2-ETHYL-1-BUTANAMINE see EHA000
ETHYL BUTANOATE see EHE000
2-ETHYL BUTANOIC ACID see DHI400
2-ETHYLBUTANOL see EGW000
2-ETHYLBUTANOL-1 see EGW000
2-ETHYL-1-BUTANOL see EGW000
2-ETHYL-1-BUTENE see EGW500
ETHYL (E)-2-BUTENOATE see COB750
ETHYLBUTYLACETALDEHYDE see BRI000
ETHYL BUTYLACETATE (DOT) see EHF000
2-ETHYLBUTYLACRYLATE see EGZ000
2-ETHYLBUTYL ALCOHOL see EGW000
2-ETHYLBUTYLAMINE see EHA000
ETHYLBUTYLCETONE (FRENCH) see EHA600
2-ETHYLBUTYL ESTER, ACRYLIC ACID see EGZ000
2-ETHYLBUTYLESTER KYSELINY AKRYLOVE see
EGZ000
ETHYL BUTYL ETHER see EHA500
ETHYLBUTYLKETON (DUTCH) see EHA600
ETHYL BUTYL KETONE see EHA600
ETHYL-N-BUTYLNITROSAMINE see EHC000
2-ETHYLBUTYL SILICATE see EHC900
ETHYL BUTYRALDEHYDE see DHI000
ETHYL BUTYRALDEHYDE (DOT) see DHI000
α-ETHYLBUTYRALDEHYDE see DHI000
2-ETHYLBUTYRALDEHYDE (DOT,FCC) see DHI000
ETHYL n-BUTYRATE see EHE000
ETHYL BUTYRATE (DOT,FCC) see EHE000
α-ETHYLBUTYRIC ACID see DHI400
2-ETHYLBUTYRIC ACID (FCC) see DHI400
2-ETHYLBUTYRIC ALDEHYDE see DHI000
ETHYL CADMATE see BJB500
ETHYL CAPRATE see EHE500
ETHYL CAPRINATE see EHE500
α-ETHYLCAPROALDEHYDE see BRI000
ETHYL CAPROATE see EHF000
α-ETHYLCAPROIC ACID see BRI250
ETHYL CAPRYLATE see ENY000

ETHYL CARBAMATE see UVA000
ETHYL CARBINOL see PND000
ETHYL CARBITOL see CBR000
ETHYL CARBONATE see DIX200
ETHYL CELLOSOLVE see EES350
ETHYL CELLOSOLVE ACETAAT (DUTCH) see
EES400
ETHYLCELLULOSE see EHG100
ETHYLCHLOORFORMIAAT (DUTCH) see EHK500
ETHYL CHLORACETATE see EHG500
ETHYL CHLORIDE see EHH000
ETHYL CHLOROACETATE see EHG500
ETHYL-α-CHLOROACETATE see EHG500
ETHYL CHLOROCARBONATE (DOT) see EHK500
ETHYL 4-CHLORO-α-(4-CHLOROPHENYL)-α-
HYDROXYBENZENEACETATE see DER000
ETHYL CHLOROETHANOATE see EHG500
9-((3-ETHYL-2-
CHLOROETHYL)AMINOPROPYLAMINO)-4-
METHOXYACRIDINE DIHYDROCHLORIDE see
EHJ500
ETHYL-N-(β-CHLOROETHYL)-N-
NITROSOCARBAMATE see CHF500
ETHYL-2-CHLOROETHYL SULFIDE see CGY750
ETHYL-β-CHLOROETHYL SULFIDE see CGY750
ETHYL CHLOROFORMATE see EHK500
ETHYL CHLOROTHIOFORMATE (DOT) see CLJ750
ETHYL-β-CHLOROVINYLETHYNYL CARBINOL see
CHG000
ETHYLCHLORVYNOL see CHG000
ETHYL-trans-CINNAMATE see EHN000
ETHYL CINNAMATE (FCC) see EHN000
ETHYL CITRATE see TJP750
ETHYLCROTONATE see COB750
ETHYL CROTONATE see EHO200
ETHYL (E)-CROTONATE see COB750
ETHYL CROTONATE (DOT) see COB750
ETHYL trans-CROTONATE see COB750
ETHYL CYANIDE see PMV750
ETHYL CYANOACETATE see EHP500
ETHYL CYANOACETATE see EHP500
ETHYL CYANOACRYLATE see EHP700
ETHYL 2-CYANOACRYLATE see EHP700
ETHYL α-CYANOACRYLATE see EHP700
ETHYL CYANOETHANOATE see EHP500
ETHYL 2-CYANO-2-PROPENOATE see EHP700
5-ETHYL-5-CYCLOHEXENYLBARBITURIC ACID see
TDA500
N-ETHYL(CYCLOHEXYL)AMINE see EHT000
N-ETHYL-CYCLOHEXYLAMINE see EHT000
ETHYL CYMATE see BJC000
p,p-ETHYL DDD see DJC000
p,p'-ETHYL-DDD see DJC000
ETHYL DECANOATE (FCC) see EHE500
ETHYL DECYLATE see EHE500
ETHYL 6-(N-DECYLOXY)-7-ETHOXY-4-
HYDROXYQUINOLINE-3-CARBOXYLATE see
DAI495
ETHYL DIAZOACETATE see DCN800
ETHYL 2,3-DIBROMOPROPANOATE see EHY050
ETHYL 2,3-DIBROMOPROPIONATE see EHY050
ETHYL α,β-DIBROMOPROPIONATE see EHY050
ETHYL-4,4'-DICHLOROBENZILATE see DER000
ETHYL-p,p'-DICHLOROBENZILATE see DER000
ETHYL-4,4'-DICHLORODIPHENYL GLYCOLLATE see
DER000
N-ETHYL-DICHLOROMALEINIMIDE see DFJ400

ETHYL-4,4'-DICHLOROPHENYL GLYCOLLATE see
DER000
ETHYL DICHLOROSILANE (DOT) see DFK000
ETHYLDICOUMAROL see BKA000
ETHYLDICOUMAROL ACETATE see BKA000
ETHYL DIETHYLENE GLYCOL see CBR000
ETHYL 3-((DIETHYLPHOSPHINOTHIOYL)OXY)-2-
BUTENOATE see EIB600
ETHYL DIGLYME see DIW800
5-ETHYLDIHYDRO-5-(1-METHYLBUTYL)-2-
THIOXO-4,6(1H,5H)-PYRIMIDINEDIONE,
MONOSODIUM SALT see PBT500
(3S-cis)-3-ETHYLDIHYDRO-4-((1-METHYL-1H-
IMIDAZOL-5-YL)METHYL)-2(3H)-FURANONE see
PIF000
1-ETHYL-1,4-DIHYDRO-7-METHYL-4-OXO-1,8-
NAPHTHYRIDINE-3-CARBOXYLIC ACID see EID000
5-ETHYLDIHYDRO-5-PHENYL-4,6(1H,5H)-
PYRIMIDINEDIONE see DBB200
ETHYL-4,4'-DIHYDROXYDICOUMARINYL-3,3'-
ACETATE see BKA000
2-ETHYL-3-(3',5'-DIIODO-4'-HYDROXYBENZOYL)-
CUMARONE see EID200
ETHYL-2-(DIIODO-3,5 HYDROXY-4 BENZOYL)5-
FURANNE see EID250
ETHYL-6,7-DIISOBUTOXY-4-
HYDROXYQUINOLINE-3-CARBOXYLATE see
BOO632
O-ETHYL-S-2-DIISOPROPYLAMINOETHYL
METHYLPHOSPHONOTHIOTE see EIG000
ETHYL-S-DIISOPROPYLAMINOETHYL
METHYLTHIOPHOSPHONATE see EIG000
ETHYL DIMETHYLAMIDOCYANOPHOSPHATE see
EIF000
ETHYL N,N-DIMETHYLAMINO CYANOPHOSPHATE
see EIF000
ETHYL-S-DIMETHYLAMINOETHYL
METHYLPHOSPHONOTHIOLATE see EIG000
ETHYLDIMETHYLMETHANE see EIK000
ETHYL DIMETHYLPHOSPHORAMIDOCYANIDATE
see EIF000
ETHYL-N,N-
DIMETHYLPHOSPHORAMIDOCYANIDATE see
EIF000
2-ETHYL-3,5(6)-DIMETHYLPYRAZINE see EIL100
10-ETHYL-4,4-DIOCTYL-7-OXO-8-OXA-3,5-DITHIA-4-
STANNATETRADECANOIC ACID-2-ETHYLHEXYL
ESTER see DVM800
ETHYLDIOL ACRILATE (RUSSIAN) see EIP000
ETHYLDIPHENYLTIN ACETATE see EIM100
O-ETHYL-S,S-DIPROPYL ESTER,
PHOSPHORODITHIOIC ACID see EIN000
O-ETHYL-S,S-DIPROPYLPHOSPHORODITHIOATE
see EIN000
S-ETHYL-N,N-DIPROPYLTHIOCARBAMATE see
EIN500
S-ETHYL-N,N-DI-N-PROPYLTHIOCARBAMATE see
EIN500
ETHYL DI-N-PROPYLTHIOLCARBAMATE see EIN500
ETHYL-N,N-DIPROPYLTHIOLCARBAMATE see
EIN500
ETHYL-N,N-DI-N-PROPYLTHIOLCARBAMATE see
EIN500
ETHYLDITHIOURAME see DXH250
ETHYLDITHIURAME see DXH250
ETHYL DODECANOATE see ELY700
ETHYLE (ACETATE d') (FRENCH) see EFR000

ETHYLE, CHLOROFORMIAT d' (FRENCH) see EHK500
ETHYLEEN-CHLOORHYDRINE (DUTCH) see EIU800
ETHYLEENDIAMINE (DUTCH) see EEA500
ETHYLEENDICHLORIDE (DUTCH) see EIY600
ETHYLEENIMINE (DUTCH) see EJM900
ETHYLEENOXIDE (DUTCH) see EJN500
ETHYLE (FORMIATE d') (FRENCH) see EKL000
ETHYL ENANTHATE see EKN050
ETHYLENE see EIO000
ETHYLENE, compressed (DOT) see EIO000
ETHYLENE, refrigerated liquid (DOT) see EIO000
ETHYLENE ACETATE see EJD759
ETHYLENE ACRYLATE see EIP000
ETHYLENE ALCOHOL see EJC500
ETHYLENE ALDEHYDE see ADR000
1,1'-ETHYLENE-2,2'-BIPYRIDYLIUM DIBROMIDE see DWX800
ETHYLENE BIS(CHLOROFORMATE) see EIQ000
ETHYLENEBIS-(DIPHENYLARSINE) see EIQ200
ETHYLENEBIS(DITHIOCARBAMATE) DISODIUM SALT see DXD200
ETHYLENEBISDITHIOCARBAMATE MANGANESE see MAS500
N,N'-ETHYLENE BIS(DITHIOCARBAMATE MANGANEUX) (FRENCH) see MAS500
N,N'-ETHYLENE BIS(DITHIOCARBAMATE de SODIUM) (FRENCH) see DXD200
ETHYLENEBIS(DITHIOCARBAMATO) MANGANESE see MAS500
ETHYLENEBIS(DITHIOCARBAMATO)MANGANESE and ZINC ACETATE (50:1) see EIQ500
ETHYLENEBIS(DITHIOCARBAMIC ACID) DISODIUM SALT see DXD200
ETHYLENEBIS(DITHIOCARBAMIC ACID) MANGANESE SALT see MAS500
ETHYLENEBIS(DITHIOCARBAMIC ACID) MANGANOUS SALT see MAS500
N,N'-ETHYLENE BIS(3-FLUOROSALICYLIDENEIMINATO)COBALT(II) see EIS000
ETHYLENEBIS(IMINODIACETIC ACID) DISODIUM SALT see EIX500
ETHYLENEBIS(IMINODIACETIC ACID) TETRASODIUM SALT see EIV000
(ETHYLENEBIS(OXYETHYLENENITRILO))TETRAACETIC ACID see EIT000
ETHYLENE BROMIDE see EIY500
ETHYLENEBROMOHYDRIN see BNI500
ETHYLENECARBOXAMIDE see ADS250
ETHYLENECARBOXYLIC ACID see ADS750
ETHYLENE CHLORIDE see EIY600
ETHYLENE CHLOROFORMATE see EIQ000
ETHYLENE CHLOROHYDRIN see EIU800
ETHYLENE CHLOROTHIOARSENATE(III) see EIU900
ETHYLENE CYANIDE see SNE000
ETHYLENE CYANOHYDRIN see HGP000
ETHYLENE DIACRYLATE see EIP000
1,2-ETHYLENEDIAMINE see EEA500
ETHYLENE-DIAMINE (FRENCH) see EEA500
ETHYLENEDIAMINEACETIC ACID TRISODIUM SALT see TNL250
N,N'-ETHYLENEDIAMINEDIACETIC ACID TETRASODIUM SALT see EIV000
ETHYLENEDIAMINE HYDROCHLORIDE see EIW000
ETHYLENEDIAMINE (OSHA) see EEA500
ETHYLENEDIAMINETETRAACETATE see EIX000

ETHYLENEDIAMINETETRAACETATE DISODIUM SALT see EIX500
ETHYLENEDIAMINETETRAACETIC ACID see EIX000
ETHYLENEDIAMINE-N,N,N',N'-TETRAACETIC ACID see EIX000
ETHYLENEDIAMINETETRAACETIC ACID, DISODIUM SALT see EIX500
ETHYLENEDIAMINETETRAACETIC ACID, TETRASODIUM SALT see EIV000
ETHYLENEDIAMINETETRAACETIC ACID, TRISODIUM SALT see TNL250
ETHYLENEDIAMMONIUM CHLORIDE see EIW000
1,2-ETHYLENE DIBROMIDE see EIY500
(E)1,2-ETHYLENEDICARBOXYLIC ACID see FOU000
cis-1,2-ETHYLENEDICARBOXYLIC ACID see MAK900
trans-1,2-ETHYLENEDICARBOXYLIC ACID see FOU000
ETHYLENE DICHLORIDE see EIY600
1,2-ETHYLENE DICHLORIDE see EIY600
ETHYLENE DICYANIDE see SNE000
ETHYLENE DIGLYCOL see DJD600
ETHYLENE DIGLYCOL MONOETHYL ETHER see CBR000
ETHYLENE DIGLYCOL MONOMETHYL ETHER see DJG000
ETHYLENE DIHYDRATE see EJC500
ETHYLENE DIMERCAPTAN see EEB000
α-ETHYLENE DIMERCAPTAN see EEB000
ETHYLENE DIMETHYL ETHER see DOE600
ETHYLENE DINITRATE see EJG000
ETHYLENEDINITRILOTETRAACETIC ACID see EIX000
(ETHYLENEDINITRILO)TETRAACETIC ACID CADMIUM(II) COMPLEX see CAF750
(ETHYLENEDINITRILO)-TETRAACETIC ACID DISODIUM SALT see EIX500
ETHYLENEDIOXYBIS(ETHYLENEAMINO)TETRAACETIC ACID see EIT000
2,2'-ETHYLENEDIOXYDIETHANOL see TJQ000
2,2'-ETHYLENEDIOXYETHANOL see TJQ000
ETHYLENE DIPYRIDYLIUM DIBROMIDE see DWX800
1,1-ETHYLENE 2,2-DIPYRIDYLIUM DIBROMIDE see DWX800
1,1'-ETHYLENE-2,2'-DIPYRIDYLIUM DIBROMIDE see DWX800
ETHYLENEDITHIOCYANATE see EJC035
ETHYLENE DITHIOGLYCOL see EEB000
ETHYLENEDITHIOL see EEB000
1,2-ETHYLENEDIYLBIS(CARBAMODITHIOATO)MANGANESE see MAS500
ETHYLENE EPISULFIDE see EJP500
ETHYLENE EPISULPHIDE see EJP500
ETHYLENE FLUORIDE see ELN500
ETHYLENE, FLUORO-(8CI) see VPA000
ETHYLENE FORMATE see EJF000
ETHYLENE GLYCOL see EJC500
ETHYLENE GLYCOL ACETATE see EJD759
ETHYLENE GLYCOL BIS(AMINOETHYL ETHER)TETRAACETATE see EIT000
ETHYLENE GLYCOL BIS(β-AMINOETHYL ETHER)TETRAACETATE see EIT000
ETHYLENE GLYCOL BIS(2-AMINOETHYL ETHER)TETRAACETIC ACID see EIT000

ETHYLENE GLYCOL BIS(2-AMINOETHYL ETHER)-
N,N,N',N'-TETRAACETIC ACID see EIT000
ETHYLENE GLYCOL BIS(β-AMINOETHYL ETHER)-
N,N'-TETRAACETIC ACID see EIT000
ETHYLENE GLYCOL, BISCHLOROFORMATE see
EIQ000
ETHYLENE GLYCOL BIS(CHLOROMETHYL)ETHER
see BIJ250
ETHYLENE GLYCOL-BIS-(2-HYDROXYETHYL
ETHER) see TJQ000
ETHYLENE GLYCOL-n-BUTYL ETHER see BPJ850
ETHYLENE GLYCOL, CHLOROHYDRIN see EIU800
ETHYLENE GLYCOL DIACETATE see EJD759
ETHYLENE GLYCOL DIACRYLATE see EIP000
ETHYLENE GLYCOL DI(CHLOROFORMATE) see
EIQ000
ETHYLENE GLYCOL DIETHYL ETHER see EJE500
ETHYLENE GLYCOL DIFORMATE see EJF000
ETHYLENE GLYCOL DIHYDROXYDIETHYL
ETHER see TJQ000
ETHYLENE GLYCOL DIMETHYL ETHER see
DOE600
ETHYLENE GLYCOL DINITRATE see EJG000
ETHYLENE GLYCOL DIPROPIONATE (8CI) see
COB260
ETHYLENE GLYCOL ETHYL ETHER see EES350
ETHYLENE GLYCOL ETHYL ETHER ACETATE see
EES400
ETHYLENE GLYCOL ISOPROPYL ETHER see INA500
ETHYLENE GLYCOL METHYL ETHER see EJH500
ETHYLENE GLYCOL METHYL ETHER ACETATE see
EJJ500
ETHYLENE GLYCOL MONOBUTYL ETHER
ACETATE (MAK) see BPM000
ETHYLENE GLYCOL MONOBUTYL ETHER (MAK,
DOT) see BPJ850
ETHYLENE GLYCOL MONOETHYL ETHER see
EES350
ETHYLENE GLYCOL MONOETHYL ETHER (DOT)
see EES350
ETHYLENE GLYCOL MONOETHYL ETHER
ACETATE (MAK, DOT) see EES400
ETHYLENE GLYCOL MONOISOBUTYL ETHER see
IIP000
ETHYLENE GLYCOL, MONOISOPROPYL ETHER see
INA500
ETHYLENE GLYCOL MONOMETHYL ETHER
ACETATE see EJJ500
ETHYLENE GLYCOL MONOMETHYL ETHER
ACRYLATE see MIF750
ETHYLENE GLYCOL MONOMETHYL ETHER (MAK,
DOT) see EJH500
ETHYLENE GLYCOL MONOPHENYL ETHER see
PER000
ETHYLENE GLYCOL-MONO-PROPYL ETHER see
PNG750
ETHYLENE GLYCOL-MONO-n-PROPYL ETHER see
PNG750
ETHYLENE GLYCOL, MONOSTEARATE see EJM500
ETHYLENE GLYCOL PHENYL ETHER see PER000
ETHYLENE GLYCOL STEARATE see EJM500
ETHYLENE HEXACHLORIDE see HCI000
ETHYLENE HOMOPOLYMER see PJS750
ETHYLENEIMINE see EJM900
ETHYLENE IMINE, INHIBITED (DOT) see EJM900
ETHYLENE MONOCHLORIDE see VNP000
ETHYLENE NITRATE see EJG000

ETHYLENENITROSOUREA see NKL000
ETHYLENE OXIDE see EJN500
ETHYLENE OXIDE, mixed with CARBON DIOXIDE
see EJO000
ETHYLENE OXIDE and CARBON DIOXIDE
MIXTURES (DOT) see EJO000
ETHYLENE OXIDE, ETHYL- see BOX750
ETHYLENE OXIDE POLYMER see EJO025
ETHYLENE OXIDE and PROPYLENE OXIDE BLOCK
POLYMER see PJK200
ETHYLENE (OXYDE d') (FRENCH) see EJN500
1-ETHYLENEOXY-3,4-EPOXYCYCLOHEXANE see
VOA000
ETHYLENE POLYMER see PJS750
ETHYLENE POLYMERS (8CI) see PJS750
ETHYLENE PROPIONATE see COB260
ETHYLENESUCCINIC ACID see SMY000
ETHYLENE SULFIDE see EJP500
ETHYLENE SULPHIDE see EJP500
ETHYLENE TETRACHLORIDE see PCF275
ETHYLENE THIOCYANATE see EJC035
ETHYLENE THIOUREA see IAQ000
1,3-ETHYLENE-2-THIOUREA see IAQ000
N,N'-ETHYLENETHIOUREA see IAQ000
ETHYLENETHIOUREA mixed with SODIUM NITRITE
see IAR000
l'ETHYLENE THIOUREE (FRENCH) see IAQ000
ETHYLENE TRICHLORIDE see TIO750
ETHYLENGLYKOLDINITRAT (CZECH) see EJG000
ETHYLENIMINE see EJM900
ETHYL-α,β-EPOXYHYDROCINNAMATE see EOK600
ETHYL α,β-EPOXY-β-METHYLHYDROCINNAMATE
see ENC000
ETHYL 2,3-EPOXY-3-METHYL-3-
PHENYLPROPIONATE see ENC000
ETHYL-α,β-EPOXY-α-PHENYLPROPIONATE see
EOK600
ETHYL ESTER of N-ACETYL-dl-SARCOSYLYL-dl-
VALINE see ARM000
ETHYL ESTER of 4,4'-DICHLOROBENZILIC ACID see
DER000
ETHYL ESTER of 2,3-EPOXY-3-PHENYLBUTANOIC
ACID see ENC000
ETHYLESTER KYSELINY DUSITE see ENN000
ETHYLESTER KYSELINY N-ETHYL-N-
NITROSOKARBAMINOVE see NKE500
ETHYLESTER KYSELINY KROTONOVE see EHO200
ETHYLESTER KYSELINY KYANOCTOVE see EHP500
ETHYLESTER KYSELINY METHYLKARBAMINOVE
see EMQ500
ETHYLESTER KYSELINY MLECNE see LAJ000
ETHYLESTER KYSELINY ORTHOMRAVENCI
(CZECH) see ENY500
ETHYL ESTER of METHANESULFONIC ACID see
EMF500
ETHYL ESTER of METHYLNITROSO-CARBAMIC
ACID see MMX250
ETHYL ESTER of METHYLSULFONIC ACID see
EMF500
ETHYL ESTER of METHYLSULPHONIC ACID see
EMF500
ETHYL ESTER of MONOACETIC ACID see MFW100
N-ETHYL-ETHANAMINE see DHJ200
ETHYL ETHANOATE see EFR000
ETHYL ETHER see EJU000
3-ETHYL-2-(5-(3-ETHYL-2-
BENZOTHIAZOLINYLIDENE)-1,3-

PENTADIENYL)BENZOTHIAZOLIUM IODIDE see DJT800
ETHYL ETHYLENE OXIDE see BOX750
ETHYL N-ETHYLNITROSOCARBAMATE see NKE500
O,O-ETHYL S-2(ETHYLTHIO)ETHYL PHOSPHORODITHIOATE see DXH325
ETHYLETHYNE see EFS500
ETHYLEX GUM 2020 see HLB400
ETHYL FLUORIDE (DOT) see FIB000
ETHYL FORMATE see EKL000
ETHYLFORMIAAT (DUTCH) see EKL000
ETHYLFORMIC ACID see PMU750
ETHYL FORMIC ESTER see EKL000
ETHYL GERANYL ETHER see GDG100
ETHYL GLYCIDYL ETHER see EBQ700
ETHYL GLYCIDYL ETHER see EKM200
ETHYLGLYKOLACETAT (GERMAN) see EES400
ETHYL GLYME see EJE500
ETHYL GREEN see BAY750
ETHYL GUSATHION see EKN000
ETHYL GUTHION see EKN000
ETHYL HEPTANOATE see EKN050
ETHYL HEPTANOATE see EKN050
ETHYL n-HEPTANOATE see EKN050
ETHYL HEPTOATE see EKN050
ETHYL HEPTYLATE see EKN050
ETHYLHEXABITAL see TDA500
5-ETHYLHEXAHYDRO-4,6-DIOXO-5-PHENYLPHYIMIDINE see DBB200
5-ETHYLHEXAHYDRO-5-PHENYLPYRIMIDINE-4,6-DIONE see DBB200
2-ETHYLHEXALDEHYDE see BRI000
ETHYLHEXALDEHYDE (DOT) see BRI000
2-ETHYLHEXANAL see BRI000
ETHYL HEXANEDIOL see EKV000
2-ETHYLHEXANEDIOL-1,3 see EKV000
2-ETHYLHEXANE-1,3-DIOL see EKV000
2-ETHYL-1,3-HEXANEDIOL see EKV000
ETHYL HEXANOATE (FCC) see EHF000
2-ETHYLHEXANOIC ACID see BRI250
2-ETHYLHEXANOL see EKQ000
2-ETHYL-1-HEXANOL see EKQ000
2-ETHYL-1-HEXANOL ESTER with DIPHENYL PHOSPHATE see DWB800
2-ETHYL-1-HEXANOL HYDROGEN PHOSPHATE see BJR750
2-ETHYL-1-HEXANOL HYDROGEN SULFATE, SODIUM SALT see TAV750
2-ETHYL-1-HEXANOL SULFATE SODIUM SALT see TAV750
2-ETHYL-1-HEXENE see EKR500
2-ETHYL HEXENE-1 see EKR500
2-ETHYLHEXOIC ACID see BRI250
2-ETHYLHEXYL ALCOHOL see EKQ000
2-ETHYL HEXYLAMINE see EKS500
N-(2-ETHYLHEXYL)BICYCLO-(2,2,1)-HEPT-5-ENE-2,3-DICARBOXIMIDE see OES000
2-ETHYLHEXYL DIPHENYL ESTER PHOSPHORIC ACID see DWB800
2-ETHYLHEXYL DIPHENYLPHOSPHATE see DWB800
ETHYL HEXYLENE GLYCOL see EKV000
2-ETHYLHEXYL FUMARATE see DVK600
N-2-ETHYLHEXYLIMIDEENDOMETHYLENETETRAHYDROPHTHALIC ACID see OES000
N-(2-ETHYLHEXYL)-5-NORBORNENE-2,3-DICARBOXIMIDE see OES000

2-ETHYLHEXYL PALMITATE see OFG100
ETHYLHEXYL PHTHALATE see DVL700
2-ETHYLHEXYL PHTHALATE see DVL700
2-ETHYLHEXYL SEBACATE see BJS250
2-ETHYLHEXYL SODIUM SULFATE see TAV750
2-ETHYLHEXYL SULFOSUCCINATE SODIUM see DJL000
2-(2-ETHYLHEXYL)-3a,4,7,7a-TETRAHYDRO-4,7-METHANO-1H-ISOINDOLE-1,3(2H)-DIONE see OES000
S-ETHYL-HOMOCYSTEINE see EEI000
S-ETHYL-dl-HOMOCYSTEINE see EEI000
ETHYL HYDRATE see EFU000
ETHYL HYDRIDE see EDZ000
ETHYL HYDROGEN PEROXIDE see ELD000
ETHYL HYDROPEROXIDE see ELD000
ETHYL HYDROPERSULFIDE see EEB000
ETHYL HYDROSULFIDE see EMB100
ETHYL HYDROXIDE see EFU000
ETHYL-p-HYDROXYBENZOATE see HJL000
2-ETHYL-3-(p-HYDROXYBENZOYL)BENZOFURAN see BBJ500
2-ETHYL-4'-HYDROXY-3-BENZOYLBENZOFURAN see BBJ500
ETHYL-2 (HYDROXY-4 BENZOYL)-3 BENZOFURANNE see BBJ500
ETHYL-2-HYDROXY-2,2-BIS(4-CHLOROPHENYL)ACETATE see DER000
N-ETHYL-N-(2-HYDROXYETHYL)-3-METHYL-4-NITROSOANILINE see ELG100
ETHYL-2-HYDROXYETHYLNITROSAMINE see ELG500
N-ETHYL-N-HYDROXYETHYLNITROSAMINE see ELG500
1-(2-ETHYL-7-(2-HYDROXY-3-((1-METHYLETHYL)AMINO)PROPOXY)-4-BENZOFURANYL) ETHANONE see ELI600
α-ETHYL-β-(HYDROXYMETHYL)-1-METHYL-IMIDAZOLE-5-BUTYRIC ACID, γ-LACTONE see PIF000
2-ETHYL-2-(HYDROXYMETHYL)-1,3-PROPANEDIOL TRIACETOACETATE see ELJ600
ETHYL-p-HYDROXYPHENYL KETONE see ELL500
ETHYL (4-(m-HYDROXYPHENYL)-1-METHYL)-4-PIPERIDYL KETONE see KFK000
ETHYL 2-HYDROXYPROPIONATE see LAJ000
ETHYL α-HYDROXYPROPIONATE see LAJ000
2-ETHYL-3-HYDROXY-4H-PYRAN-4-ONE see EMA600
ETHYLIC ACID see AAT250
5-ETHYLIDENEBICYCLO(2.2.1)HEPT-2-ENE see ELO500
ETHYLIDENE CHLORIDE see DFF809
ETHYLIDENE DICHLORIDE see DFF809
ETHYLIDENE DIETHYL ETHER see AAG000
ETHYLIDENE DIFLUORIDE see ELN500
ETHYLIDENE DIMETHYL ETHER see DOO600
ETHYLIDENE FLUORIDE see ELN500
ETHYLIDENE GYROMITRIN see AAH000
ETHYLIDENEHYDROXYLAMINE see AAH250
trans-15-ETHYLIDENE-12-β-HYDROXY-4,12-α,13-β-TRIMETHYL 8-OXO-4,8 SECOSENEC-1-ENINE see DMX200
ETHYLIDENELACTIC ACID see LAG000
ETHYLIDENE NORBORNENE see ELO500
5-ETHYLIDENE-2-NORBORNENE see ELO500
ETHYLIDICHLORARSINE see DFH200
ETHYLIDICHLOROARSINE (DOT) see DFH200

ETHYLIMINE see EJM900
ETHYL ISOBUTANOATE see ELS000
ETHYL ISOBUTYRATE see ELS000
ETHYLISOBUTYRATE (DOT) see ELS000
ETHYL ISOCYANATE see ELS500
ETHYL ISOCYANATE (DOT) see ELS500
ETHYL ISONICOTINATE see ELU000
5-ETHYL-5-ISOPENTYLBARBITURIC ACID SODIUM
SALT see AON750
ETHYL ISOVALERATE (FCC) see ISY000
ETHYL KETOVALERATE see EFS600
ETHYL 4-KETOVALERATE see EFS600
ETHYL LACTATE (DOT,FCC) see LAJ000
ETHYL LAEVULINATE see EFS600
ETHYL LAURATE see ELY700
ETHYL LEVULATE see EFS600
N-ETHYLMALEIMIDE see MAL250
ETHYL MALONATE see EMA500
ETHYL MALTOL see EMA600
ETHYLMERCAPTAAN (DUTCH) see EMB100
ETHYL MERCAPTAN see EMB100
β-ETHYLMERCAPTOETHYL DIMETHYL
THIONOPHOSPHATE see DAO800
ETHYLMERCURIC CHLORIDE see CHC500
ETHYLMERCURICHLORENDIMIDE see EME050
ETHYLMERCURIC PHOSPHATE see BJT250
N-(ETHYLMERCURI)-1,4,5,6,7,7-
HEXACHLOROBICYCLO(2.2.1)HEPT-5-ENE-2,3-
DICARBOXIMIDE see EME050
N-ETHYLMERCURI-3,4,5,6,7,7-HEXACHLORO-3,6-
ENDOMETHYLENE-1,2,3,6-
TETRAHYDROPHTHALIMIDE see EME050
N-ETHYLMERCURI-N-PHENYL-p-
TOLUENESULFONAMIDE see EME500
N-ETHYLMERCURI-1,2,3,6-TETRAHYDRO-3,6-
ENDOMETHANO-3,4,5,6,7,7-
HEXACHLOROPHTHALIMIDE see EME050
o-(ETHYLMERCURITHIO)BENZOIC ACID SODIUM
SALT see MDI000
ETHYLMERCURITHIOSALICYLIC ACID SODIUM
SALT see MDI000
N-(ETHYLMERCURI)-p-TOLUENESULFONANILIDE
see EME500
N-(ETHYLMERCURI)-p-
TOLUENESULPHONANILIDE see EME500
ETHYLMERCURY CHLORIDE see CHC500
ETHYLMERCURY PHOSPHATE see BJT250
ETHYLMERCURY p-TOLUENESULFANILIDE see
EME500
ETHYLMERCURY-p-TOLUENE SULFONAMIDE see
EME500
ETHYLMERCURY-p-TOLUENESULFONANILIDE see
EME500
ETHYLMERKAPTAN (CZECH) see EMB100
β-ETHYLMERKAPTOETHYLCHLORID (CZECH) see
CGY750
ETHYL METHACRYLATE see EMF000
ETHYL METHACRYLATE, INHIBITED (DOT) see
EMF000
ETHYL METHANESULFONATE see EMF500
ETHYL METHANESULPHONATE see EMF500
ETHYL METHANOATE see EKL000
ETHYL METHANSULFONATE see EMF500
ETHYL METHANSULPHONATE see EMF500
ETHYL-4-METHOXYBENZOATE see AOV000
ETHYL-p-METHOXYBENZOATE see AOV000
ETHYL-2-METHYLACRYLATE see EMF000

ETHYL-α-METHYL ACRYLATE see EMF000
ETHYL-p-METHYL BENZENESULFONATE see
EPW500
5-ETHYL-5-(3-METHYLBUTYL)BARBITURIC ACID,
SODIUM DERIVATIVE see AON750
5-ETHYL-5-(1-METHYLBUTYL)BARBITURIC ACID
SODIUM SALT see NBU000
5-ETHYL-5-(1-METHYLBUTYL)-2,4,6(1H,3H,5H)-
PYRIMIDINETRIONE MONOSODIUM SALT (9CI) see
NBU000
5-ETHYL-5-(1-METHYLBUTYL)-2-THIOBARBITURIC
ACID MONOSODIUM see PBT500
ETHYL 2-METHYLBUTYRATE see EMP600
ETHYL METHYLCARBAMATE see EMQ500
ETHYL-N-METHYLCARBAMATE see EMQ500
ETHYLMETHYL CARBINOL see BPW750
ETHYL METHYL CETONE (FRENCH) see MKA400
o-ETHYL-S-(3-METHYL-4-CHLOROPHENYL)ETHYL
PHOSPHONODITHIOATE see EMR100
1-ETHYL-7-METHYL-1,4-DIHYDRO-1,8-
NAPHTHYRIDINE-4-ONE-3-CARBOXYLIC ACID see
EID000
1-ETHYL-7-METHYL-1,4-DIHYDRO-1,8-
NAPHTHYRIDIN-4-ONE-3-CARBOXYLIC ACID see
EID000
4-ETHYL-4-METHYL-2,6-DIOXOPIPERIDINE see
MKA250
ETHYL METHYLENE PHOSPHORODITHIOATE see
EEH600
ETHYL METHYL ETHER see EMT000
ETHYL METHYL ETHER (DOT) see EMT000
3-ETHYL-3-METHYLGLUTARIMIDE see MKA250
β-ETHYL-β-METHYLGLUTARIMIDE see MKA250
7-ETHYL-2-METHYL-4-HEXADECANOL SULFATE
SODIUM SALT see SIO000
ETHYLMETHYLKETON (DUTCH) see MKA400
ETHYL METHYL KETONE (DOT) see MKA400
ETHYL METHYL KETONE OXIME see EMU500
ETHYL METHYL KETONE PEROXIDE see MKA500
ETHYL-METHYLKETONOXIM see EMU500
ETHYL METHYL KETOXIME see EMU500
O-ETHYL-O-(4-(METHYLMERCAPTO)PHENYL)-S-N-
PROPYLPHOSPHOROTHIONOTHIOLATE see
SOU625
ETHYL-3-METHYL-4-(METHYLTHIO)PHENYL(1-
METHYLETHYL)PHOSPHORAMIDATE see FAK000
ETHYLMETHYLNITROSAMINE see MKB000
2-(ETHYL(3-METHYL-4-
NITROSOPHENYL)AMINO)ETHANOL see ELG100
1-ETHYL-7-METHYL-4-OXO-1,4-DIHYDRO-1,8-
NAPHTHYRIDINE-3-CARBOXYLIC ACID see EID000
N-ETHYLMETHYLPHENYLBARBITURIC ACID see
ENB500
5-ETHYL-1-METHYL-5-PHENYLBARBITURIC ACID
see ENB500
5-ETHYL-N-METHYL-5-PHENYLBARBITURIC ACID
see ENB500
ETHYL METHYLPHENYLGLYCIDATE see ENC000
5-ETHYL-3-METHYL-5-PHENYLHYDANTOIN see
MKB250
5-ETHYL-3-METHYL-5-PHENYL-2,4(3H,5H)-
IMIDAZOLEDIONE see MKB250
5-ETHYL-3-METHYL-5-PHENYLIMIDAZOLIDIN-2,4-
DIONE see MKB250
ETHYL-1-METHYL-4-PHENYLISONIPECOTATE
HYDROCHLORIDE see DAM700

ETHYL-1-METHYL-4-PHENYLPIPERIDINE-4-
CARBOXYLATE HYDROCHLORIDE see DAM700
ETHYL-1-METHYL-4-PHENYLPIPERIDYL-4-
CARBOXYLATE HYDROCHLORIDE see DAM700
5-ETHYL-1-METHYL-5-PHENYL-2,4,6(1H,3H,5H)-
PYRIMIDINETRIONE see ENB500
4-ETHYL-4-METHYL-2,6-PIPERIDINEDIONE see
MKA250
ETHYL-2-METHYLPROPANOATE see ELS000
ETHYL-2-METHYL-2-PROPENOATE see EMF000
ETHYL-2-METHYLPROPIONATE see ELS000
o-ETHYL S-1-METHYLPROPYL S-1,1-
DIMETHYLETHYL PHOSPHORODITHIOATE see
ENF050
2-ETHYL-3-METHYLPYRAZINE see ENF200
3-ETHYL-6-METHYLPYRIDINE see EOS000
5-ETHYL-2-METHYLPYRIDINE see EOS000
O-ETHYL-O-(4-
(METHYLTHIO)PHENYL)PHOSPHORODITHIOIC
ACID-S-PROPYL ESTER see SOU625
O-ETHYL-O-(4-(METHYLTHIO)PHENYL) S-PROPYL
PHOSPHORODITHIOATE see SOU625
ETHYL-4-(METHYLTHIO)-m-TOLYL ISOPROPYL
PHOSPHORAMIDATE see FAK000
N-ETHYL-α-METHYL-m-
TRIFLUOROMETHYLPHENETHYLAMINE see
PDM250
N-ETHYL-α-METHYL-m-
(TRIFLUOROMETHYL)PHENETHYLAMINE
HYDROCHLORIDE see PDM250
ETHYL MONOBROMOACETATE see EGV000
ETHYL MONOCHLORACETATE see EHG500
ETHYL MONOCHLOROACETATE see EHG500
ETHYL MONOSULFIDE see EPH000
ETHYLMORPHINE HYDROCHLORIDE see ENK500
ETHYL MORPHINE HYDROCHLORIDE
DIHYDRATE see ENK500
4-ETHYLMORPHOLINE see ENL000
N-ETHYLMORPHOLINE see ENL000
ETHYL MYRISTATE see ENL850
3-ETHYLNIRVANOL see MKB250
ETHYL NITRATE see ENM500
ETHYL NITRILE see ABE500
ETHYL NITRITE see ENN000
ETHYL NITRITE see ENN000
ETHYL NITRITE SOLUTIONS (DOT) see ENN000
O-ETHYL-O-((4-NITROFENYL)-FENYL)-
MONOTHIOFOSFONAAT (DUTCH) see EBD700
ETHYL-p-NITROPHENYL
BENZENETHIONOPHOSPHONATE see EBD700
O-ETHYL O-(4-
NITROPHENYL)BENZENETHIONOPHOSPHONATE
see EBD700
ETHYL-p-NITROPHENYL
BENZENETHIOPHOSPHATE see EBD700
ETHYL-p-NITROPHENYL
BENZENETHIOPHOSPHONATE see EBD700
ETHYL-p-NITROPHENYL
PHENYLPHOSPHONOTHIOATE see EBD700
O-ETHYL-O-(4-NITROPHENYL)
PHENYLPHOSPHONOTHIOATE see EBD700
O-ETHYL-O-p-NITROPHENYL
PHENYLPHOSPHONOTHIOLATE see EBD700
O-ETHYL-O-p-NITROPHENYL
PHENYLPHOSPHOROTHIOATE see EBD700
ETHYL-p-NITROPHENYL
THIONOBENZENEPHOSPHATE see EBD700

ETHYL-p-NITROPHENYL
THIONOBENZENEPHOSPHONATE see EBD700
2-(ETHYLNITROSAMINO)ETHANOL see ELG500
4-((ETHYLNITROSAMINO)METHYL)PYRIDINE see
NLH000
ETHYLNITROSOANILINE see NKD000
N-ETHYL-N-NITROSOBENZENAMINE see NKD000
N-ETHYL-N-NITROSOBUTYLAMINE see EHC000
ETHYLNITROSOCARBAMIC ACID, ETHYL ESTER
see NKE500
N-ETHYL-N-NITROSOCARBAMIC ACID ETHYL
ESTER see NKE500
N-ETHYL-N-NITROSOCARBAMIDE see ENV000
N-ETHYL-N-NITROSO-ETHANAMINE see NJW500
N-ETHYL-N-NITROSOETHENAMINE see NKF000
N-ETHYL-N-NITROSOETHENYLAMINE see NKF000
ETHYLNITROSOUREA see ENV000
1-ETHYL-1-NITROSOUREA see ENV000
N-ETHYL-N-NITROSO-UREA see ENV000
ETHYL NITROSOURETHAN see NKE500
N-ETHYL-N-NITROSOURETHAN see NKE500
ETHYL NITROSOURETHANE see NKE500
N-ETHYL-N-NITROSOURETHANE see NKE500
N-ETHYL-N-NITROSOVINYLAMINE see NKF000
1-ETHYL-3-(5-NITRO-2-THIAZOLYL) UREA see
ENV500
N-ETHYL-N'-(5-NITRO-2-THIAZOLYL)UREA see
ENV500
ETHYL NONANOATE see ENW000
ETHYL NONYLATE see ENW000
ETHYL OCTANOATE see ENY000
ETHYL OCTYLATE see ENY000
ETHYL OENANTHATE see EKN050
ETHYL OENANTHYLATE see EKN050
ETHYLOLAMINE see EEC600
1-(β-ETHYLOL)-2-METHYL-5-NITRO-3-
AZAPYRROLE see MMN250
ETHYL ORTHOFORMATE see ENY500
ETHYL ORTHOSILICATE see EPF550
ETHYL OXALATE see DJT200
ETHYL OXALATE (DOT) see DJT200
ETHYLOXIRANE see BOX750
ETHYL-3-OXOBUTANOATE see EFS000
ETHYL-3-OXOBUTYRATE see EFS000
ETHYL 4-OXOPENTANOATE see EFS600
ETHYL 4-OXOVALERATE see EFS600
4-ETHYLOXYPHENOL see EFA100
ETHYL PABATE see EEK100
ETHYL PARABEN see HJL000
ETHYL PARASEPT see HJL000
ETHYL PARATHION see PAK000
ETHYL PELARGONATE see ENW000
5-ETHYL-5-PENTYLBARBITURIC ACID see PBS250
ETHYL PERCHLORATE see EOD000
ETHYL PEROXYCARBONATE see DJU600
ETHYL PHENACETATE see EOH000
2-ETHYLPHENOL see PGR250
4-ETHYLPHENOL see EOE100
o-ETHYLPHENOL see PGR250
ETHYL PHENYLACETATE see EOH000
ETHYL-β-PHENYLACRYLATE see EHN000
ETHYLPHENYLAMINE see EGK000
N-ETHYL-1-((4-(PHENYLAZO)PHENYL)AZO)-2-
NAPHTHALENAMINE see EOJ500
N-ETHYL-1-((p-(PHENYLAZO)PHENYL)AZO)-2-
NAPHTHALENAMINE see EOJ500

N-ETHYL-1-((4-(PHENYLAZO)PHENYL)AZO)-2-NAPHTHYLAMINE see EOJ500
N-ETHYL-1-((p-(PHENYLAZO)PHENYL)AZO)-2-NAPHTHYLAMINE see EOJ500
5-ETHYL-5-PHENYLBARBITURIC ACID see EOK000
5-ETHYL-5-PHENYLBARBITURIC ACID SODIUM see SID000
5-ETHYL-5-PHENYLBARBITURIC ACID SODIUM SALT see SID000
ETHYL PHENYL DICHLOROSILANE (DOT) see DFJ800
3-ETHYL-3-PHENYL-2,6-DIKETOPIPERIDINE see DYC800
3-ETHYL-3-PHENYL-2,6-DIOXOPIPERIDINE see DYC800
ETHYL-2-PHENYLETHANOATE see EOH000
O-ETHYL-S-PHENYL ETHYLDITHIOPHOSPHONATE see FMU045
O-ETHYL-S-PHENYL ETHYLPHOSPHONODITHIOATE see FMU045
2-ETHYL-2-PHENYLGLUTARIMIDE see DYC800
α-ETHYL-α-PHENYLGLUTARIMIDE see DYC800
ETHYL PHENYLGLYCIDATE see EOK600
ETHYL-3-PHENYLGLYCIDATE see EOK600
5-ETHYL-5-PHENYLHEXAHYDROPYRIMIDINE-4,6-DIONE see DBB200
ETHYL PHENYL KETONE see EOL500
5-ETHYL-5-PHENYL-N-METHYLBARBITURIC ACID see ENB500
O-ETHYL-PHENYL-p-NITROPHENYL THIOPHOSPHONATE see EBD700
3-ETHYL-3-PHENYL-2,6-PIPERIDINEDIONE see DYC800
ETHYL-3-PHENYLPROPENOATE see EHN000
α-(5-ETHYL-1-PHENYL-1H-PYRAZOL-4-YL)-1-PIPERIDINEBUTANOL see EOM650
5-ETHYL-5-PHENYL-2,4,6-(1H,3H,5H)PYRIMIDINETRIONE see EOK000
5-ETHYL-5-PHENYL-2,4,6-(1H,3H,5H)PYRIMIDINETRIONE MONOSODIUM SALT see SID000
ETHYL(N-PHENYL-p-TOLUENESULFONAMIDATO)MERCURY see EME500
ETHYL(N-PHENYL-p-TOLUENESULFONAMIDO)MERCURY see EME500
ETHYL PHOSPHATE see TJT750
ETHYL PHOSPHONOTHIOIC DICHLORIDE see EOP600
ETHYL PHOSPHONOTHIOIC DICHLORIDE, anhydrous (DOT) see EOP600
ETHYLPHOSPHONOTHIONIC DICHLORIDE see EOP600
ETHYL PHOSPHONOTHIOYL DICHLORIDE see EOP600
ETHYL PHOSPHONOUS DICHLORIDE see EOQ000
ETHYL PHOSPHONOUS DICHLORIDE, anhydrous (DOT) see EOQ000
ETHYL PHOSPHORODICHLORIDATE see EOR000
ETHYL PHTHALATE see DJX000
ETHYLPHTHALYL ETHYL GLYCOLATE see EOR525
5-ETHYL-2-PICOLINE see EOS000
5-ETHYL-α-PICOLINE see EOS000
1-ETHYLPIPERIDINE see EOS500
ETHYL PROPENOATE see EFT000
ETHYL-2-PROPENOATE see EFT000
ETHYL PROPIONATE see EPB500
ETHYL PROPYL ETHER see EPC125

ETHYL n-PROPYL ETHER see EPC125
2-ETHYL-3-PROPYL-1,3-PROPANEDIOL see EKV000
ETHYLPROTAL see EQF000
ETHYL PTS see EPW500
ETHYL PYRAZINYL PHOSPHOROTHIOATE see EPC500
ETHYL PYROCARBONATE see DIZ100
2-ETHYL PYROMECONIC ACID see EMA600
N-ETHYLPYRROLIDINONE see EPC700
1-ETHYL-2-PYRROLIDINONE see EPC700
N-ETHYLPYRROLIDONE see EPC700
ETHYL RHODANATE see EPP000
ETHYL-S see BID250
ETHYL SEBACATE see DJY600
ETHYL SELENAC see DJD400
ETHYL SILICATE see EPF550
ETHYL SILICON TRICHLORIDE see EPY500
ETHYL SUCCINATE see SNB000
ETHYL SULFATE see DKB110
ETHYL SULFHYDRATE see EMB100
ETHYL SULFIDE see EPH000
S-(2-(ETHYLSULFINYL)ETHYL)-O,O-DIMETHYL PHOSPHOROTHIOATE see DAP000
ETHYL SULFOCYANATE see EPP000
3-(3-ETHYLSULFONYL)PENTYL PIPERIDINO KETONE see EPI400
ETHYL TELLURAC see EPJ000
3'-ETHYL-5',6',7',8'-TETRAHYDRO-5',5',8'-TETRAMETHYL-2'-ACETONAPHTHONE see ACL750
1-(3-ETHYL-5,6,7,8-TETRAHYDRO-5,5,8,8-TETRAMETHYL-2-NAPHTHALENYL)-ETHANONE see ACL750
ETHYL TETRAPHOSPHATE see HCY000
ETHYL TETRAPHOSPHATE, HEXA- see HCY000
ETHYL THIOALCOHOL see EMB100
2-(ETHYLTHIO)CHLOROETHANE see CGY750
ETHYL THIOCYANATE see EPP000
ETHYLTHIOETHANE see EPH000
2-(ETHYLTHIO)-ETHANETHIOL S-ESTER with O,O-DIETHYL PHOSPHOROTHIOATE see DAP200
ETHYL THIOETHER see EPH000
2-ETHYLTHIOETHYL CHLORIDE see CGY750
S-2-(ETHYLTHIO)ETHYL O,O-DIETHYL ESTER OF PHOSPHORODITHIOIC ACID see DXH325
2-ETHYLTHIOETHYL O,O-DIMETHYL PHOSPHORODITHIOATE see PHI500
S-(2-(ETHYLTHIO)ETHYL) O,O-DIMETHYLPHOSPHORODITHIONATE see PHI500
O-(2-(ETHYLTHIO)ETHYL)-O,O-DIMETHYL PHOSPHOROTHIOATE see DAO800
S-(2-(ETHYLTHIO)ETHYL)-O,O-DIMETHYL PHOSPHOROTHIOATE see DAP400
S-(2-(ETHYLTHIO)ETHYL)DIMETHYL PHOSPHOROTHIOLATE see DAP400
S-(2-(ETHYLTHIO)ETHYL)DIMETHYL PHOSPHOROTHIOLOTHIONATE see PHI500
2-(ETHYLTHIO)ETHYL DIMETHYL PHOSPHOROTHIONATE see DAO800
S-(2-(ETHYLTHIO)ETHYL)-O,O-DIMETHYL THIOPHOSPHATE see DAP400
ETHYL THIOMETON see DXH325
ETHYLTHIOMETON B see DXH325
ETHYLTHIONOPHOSPHONYL DICHLORIDE see EOP600
ETHYLTHIOPHOSPHONIC DICHLORIDE see EOP600
ETHYL THIOPYROPHOSPHATE see SOD100
ETHYL THIOUREA see EPR600

1-ETHYLTHIOUREA see EPR600
ETHYL THIRAM see DXH250
ETHYL THIUDAD see DXH250
ETHYL THIURAD see DXH250
ETHYL-α-TOLUATE see EOH000
ETHYL(p-TOLUENESULFONANILIDATO)MERCURY see EME500
ETHYL-p-TOLUENESULFONATE see EPW500
ETHYL TOSYLATE see EPW500
ETHYL-p-TOSYLATE see EPW500
ETHYL TRICHLOROPHENYLETHYLPHOSPHONOTHIOATE see EPY000
O-ETHYL-O-2,4,5-TRICHLOROPHENYL ETHYLPHOSPHONOTHIOATE see EPY000
ETHYL TRICHLOROSILANE see EPY500
ETHYLTRICHLOROSILANE (DOT) see EPY500
4-ETHYL-2,6,7-TRIOXA-1-ARSABICYCLO(2.2.2)OCTANE see EQD100
ETHYL TUADS see BJB500
ETHYL TUADS see DXH250
ETHYL TUEX see DXH250
ETHYLUREA see EQD875
1-ETHYLUREA see EQD875
N-ETHYLUREA see EQD875
ETHYLUREA and SODIUM NITRITE (2:1) see EQE000
ETHYLURETHAN see UVA000
ETHYL URETHANE see UVA000
o-ETHYLURETHANE see UVA000
ETHYL VANILLIN see EQF000
ETHYL VINYL ETHER see EQF500
ETHYL VINYL KETONE see PBR250
ETHYLVINYLNITROSAMINE see NKF000
ETHYL XANTHOGEN DISULFIDE see BJU000
ETHYL ZIMATE see BJC000
ETHYL ZIRUM see BJC000
ETHYNE see ACI750
ETHYNE, DICHLORO-(9CI) see DEN600
ETHYNE, DIIODO- see DNE500
ETHYNODIOL ACETATE see EQJ500
ETHYNODIOL DIACETATE see EQJ500
β-ETHYNODIOL DIACETATE see EQJ500
17-α-ETHYNYL-17-β-ACETOXY-19-NORANDROST-4-EN-3-ONE see ABU000
α-ETHYNYLBENZENEMETHANOL CARBAMATE see PGE000
α-ETHYNYLBENZYL CARBAMATE see PGE000
ETHYNYLCARBINOL see PMN450
1-ETHYNYLCYCLOHEXANOL CARBAMATE see EEH000
1-ETHYNYLCYCLOHEXYL CARBAMATE see EEH000
1-ETHYNYLCYCLOHEXYL ESTER CARBAMIC ACID see EEH000
17-α-ETHYNYL-3,17-DIHYDROXY-4-ESTRENE DIACETATE see EQJ500
17-ETHYNYL-3,17-DIHYDROXY-1,3,5-OESTRATRIENE see EEH500
ETHYNYLENEBIS(CARBONYLOXY)BIS(TRIPHENYL STANNANE) see BLS900
ETHYNYLESTRADIOL see EEH500
17-α-ETHYNYLESTRADIOL see EEH500
17-α-ETHYNYLESTRADIOL-17-β see EEH500
ETHYNYLESTRADIOL-3-METHYL ETHER see MKB750
17-ETHYNYLESTRADIOL-3-METHYL ETHER see MKB750

17-α-ETHYNYLESTRADIOL-3-METHYL ETHER see MKB750
17-α-ETHYNYL-1,3,5(10)-ESTRATRIENE-3,17-β-DIOL see EEH500
17-α-ETHYNYLESTRA-1,3,5(10)-TRIENE-3,17-β-DIOL see EEH500
17-α-ETHYNYLESTR-4-ENE-3-β,17-β-DIOL ACETATE see EQJ500
17-α-ETHYNYL-4-ESTRENE-3-β,17-β-DIOL DIACETATE see EQJ500
17-α-ETHYNYL-4-ESTRENE-3-β,17-β-DIOL DIACETATE see EQJ500
17-α-ETHYNYL-4-ESTREN-17-OL-3-ONE see NNP500
17-α-ETHYNYLESTR-5(10)-EN-17-β-OL-3-ONE see EEH550
17-α-ETHYNYL-5(10)-ESTREN-17-OL-3-ONE see EEH550
17-α-ETHYNYL-ESTR-5(10)-EN-3-ON-17-β-OL see EEH550
17-α-ETHYNYL-17-HYDROXY-4-ESTREN-3-ONE see NNP500
17-α-ETHYNYL-17-β-HYDROXY-5(10)-ESTREN-3-ONE see EEH550
17-α-ETHYNYL-17-β-HYDROXYESTR-5(10)-EN-3-ONE see EEH550
17-α-ETHYNYL-17-HYDROXYESTR-5(10)-EN-3-ONE see EEH550
17-α-ETHYNYL-17-HYDROXY-5(10)-ESTREN-3-ONE see EEH550
17-α-ETHYNYL-17-HYDROXYESTR-4-EN-3-ONE ACETATE see ABU000
(+)-17-α-ETHYNYL-17-β-HYDROXY-3-METHOXY-1,3,5(10)-ESTRATRIENE see MKB750
(+)-17-α-ETHYNYL-17-β-HYDROXY-3-METHOXY-1,3,5(10)-OESTRATRIENE see MKB750
17-α-ETHYNYL-17-β-HYDROXY-19-NORANDROST-4-EN-3-ONE see NNP500
17-α-ETHYNYL-17-β-HYDROXY-3-OXO-Δ5(10)-ESTRENE see EEH550
17-α-ETHYNYL-17-β-HYDROXY-Δ5(10)-ESTREN-3-ONE see EEH550
ETHYNYLMETHANOL see PMN450
17-α-ETHYNYL-3-METHOXY-1,3,5(10)-ESTRATRIEN-17-β-OL see MKB750
17-ETHYNYL-3-METHOXY-1,3,5(10)-ESTRATRIEN-17-β-OL see MKB750
17-α-ETHYNYL-3-METHOXY-17-β-HYDROXY-Δ-1,3,5(10)-ESTRATRIENE see MKB750
17-α-ETHYNYL-3-METHOXY-17-β-HYDROXY-Δ-1,3,5(10)-OESTRATRIENE see MKB750
17-ETHYNYL-3-METHOXY-1,3,5(10)-OESTRATIEN-17-β-OL see MKB750
17-α-ETHYNYL-19-NORANDROST-4-ENE-3-β,17-β-DIOL DIACETATE see EQJ500
17-α-ETHYNYL-19-NORANDROST-4-EN-17-β-OL-3-ONE see NNP500
17-α-ETHYNYL-19-NOR-4-ANDROSTEN-17-β-OL-3-ONE see NNP500
17-α-ETHYNYL-19-NOR-5(10)-ANDROSTEN-17-β-OL-3-ONE see EEH550
17-α-ETHYNYL-19-NORTESTOSTERONE see NNP500
17-α-ETHYNYL-19-NORTESTOSTERONE ACETATE see ABU000

ETHYNYLOESTRADIOL see EEH500
17-ETHYNYLOESTRADIOL see EEH500
17-α-ETHYNYLOESTRADIOL see EEH500
17-α-ETHYNYL-17-β-OESTRADIOL see EEH500
17-α-ETHYNYLOESTRADIOL-17-β see EEH500
ETHYNYLOESTRADIOL METHYL ETHER see MKB750
17-ETHYNYLOESTRADIOL-3-METHYL ETHER see MKB750
17-α-ETHYNYLOESTRADIOL METHYL ETHER see MKB750
17-α-ETHYNYLOESTRADIOL-3-METHYL ETHER see MKB750
17-α-ETHYNYL-1,3,5(10)-OESTRATRIENE-3,17-β-DIOL see EEH500
17-α-ETHYNYLOESTRA-1,3,5(10)-TRIENE-3,17-β-DIOL see EEH500
17-α-ETHYNYL-1,3,5-OESTRATRIENE-3,17-β-DIOL see EEH500
17-ETHYNYLOESTRA-1,3,5(10)-TRIENE-3,17-β-DIOL see EEH500
ETIDRONIC ACID see HKS780
ETIL ACRILATO (ITALIAN) see EFT000
ETILACRILATULUI (ROMANIAN) see EFT000
ETILAMINA (ITALIAN) see EFU400
ETILBENZENE (ITALIAN) see EGP500
ETILBUTILCHETONE (ITALIAN) see EHA600
ETIL CLOROCARBONATO (ITALIAN) see EHK500
ETIL CLOROFORMIATO (ITALIAN) see EHK500
ETILE (ACETATO di) (ITALIAN) see EFR000
ETILE (FORMIATO di) (ITALIAN) see EKL000
N,N'-ETILEN-BIS(DITIOCARBAMMATO) di MANGANESE (ITALIAN) see MAS500
N,N'-ETILEN-BIS(DITIOCARBAMMATO) di SODIO (ITALIAN) see DXD200
ETILENE (OSSIDO di) (ITALIAN) see EJN500
ETILENIMINA (ITALIAN) see EJM900
ETILFEN see EOK000
ETILMERCAPTANO (ITALIAN) see EMB100
O-ETIL-O-((4-NITRO-FENIL)-FENIL)-MONOTIOFOSFONATO (ITALIAN) see EBD700
ETIN see EDC500
ETINAMATE see EEH000
ETIOL see MAK700
ETO see EJN500
ETOC see THI500
ETODROXINE see HHK050
ETODROXYZINE see HHK050
ETOKSYETYLOWY ALKOHOL (POLISH) see EES350
E-TOPLEX see TGJ050
ETOPOSIDE see EAV500
ETOPROPEZINA see DIR000
2-ETOSSIETIL-ACETATO (ITALIAN) see EES400
ETP see EQP000
ETs (POLYSACCHARIDE) see EHG100
ETROFOL see CKF000
ETROFOLAN see MIA250
ETROLENE see RMA500
ETU see IAQ000
ETYLENU TLENEK (POLISH) see EJN500
ETYLOAMINA (POLISH) see EFU400
ETYLOBENZEN (POLISH) see EGP500
ETYLOWY ALKOHOL (POLISH) see EFU000
ETYLU BROMEK (POLISH) see EGV400
ETYLU CHLOREK (POLISH) see EHH000
ETYLU KRZEMIAN (POLISH) see EPF550

EUBASIN see PPO000
EUBASINUM see PPO000
EUCALMYL see FLU000
EUCALYPTUS OIL see EQQ000
EUCANINE GB see TGL750
EUCHESSINA see PDO750
EUCHEUMA SPINOSUM GUM see CCL250
EUCHRYSINE see BJF000
EUCISTEN see EID000
EUCISTIN see PDC250
EUCLORINA see CDP000
EUCORAN see DJS200
EUFIN see DIX200
EUFLAVINE see DBX400
EUFODRIANL see MDT600
EUGENIC ACID see EQR500
EUGENOL see EQR500
EUGENOL ACETATE see EQS000
1,3,4-EUGENOL ACETATE see EQS000
1,3,4-EUGENOL METHYL ETHER see AGE250
EUGENYL ACETATE see EQS000
EUGENYL METHYL ETHER see AGE250
EUHAEMON see VSZ000
EUKALYPTUS OEL (GERMAN) see EQQ000
EUKRATON see MKA250
EULAVA SM see MAG250
EUMICTON see AAI250
EUMIDRINA see MGR500
EUMIN see MMN250
EUMYDRIN see MGR500
EUNATROL see OIA000
EUNERPAN see FKI000
EUNOCTIN see DLY000
EUPHODRIN see DBA800
EUPRACTONE see PMO250
EUPRAMIN see DLH630
EURECOR see CCK125
EURESOL see RDZ900
EUROCERT AZORUBINE see HJF500
EUROCERT COCHINEAL RED A see FMU080
EUROCERT ORANGE FCF see FAG150
EUROCERT TARTRAZINE see FAG140
EURODOPA see DNA200
EUROGALE see MKP500
EUROPEN see MGR500
EUROPHAN see PKQ059
EUROPIC CHLORIDE see ERA500
EUROPIUM CHLORIDE see ERA500
EUSAPRIM see SNK000
EUSCOPOL see HOT500
EUSTIDIL see DFH600
EUSTIGMIN BROMIDE see POD000
EUTENSIN see CHJ750
EUTHATAL see NBU000
EUTHYROX see LFG050
EUTRIT see XPJ000
EUVESTIN see DKB000
EVABLIN see BGT250
EVAC-Q-KIT see PDO750
EVAC-Q-KWIK see PDO750
EVAC-U-GEN see PDO750
EVANS BLUE DYE see BGT250
EVAU-SUPER see SFS000
EVE see EQF500
EVERCYN see HHS000
EVEX see ECU750
EVEX see EDV600
EVION see VSZ450

EVIPHEROL see TGJ050
EVIPLAST 80 see DVL700
EVIPLAST 81 see DVL700
EVITAMINUM see VSZ450
EVOLA see DEP800
EWEISS see BAP000
EX 4355 see DLS600
EXACT-S see TFP000
EXADRIN see VGP000
EXAGAMA see BBQ500
EXAL see VLA000
EXCELSIOR see CBT750
EXD see BJU000
EXDOL see HIM000
EXHAUST GAS see CBW750
EXHORAN see DXH250
EXHORRAN see DXH250
EXITELITE see AQF000
EXMIGRA see EDC500
EXMIN see AHJ750
EXOFENE see HCL000
EXON 450 see AAX175
EXON 454 see AAX175
EXON 605 see PKQ059
EXONAL see FLZ050
EXOTHERM see TBQ750
EXOTHERM TERMIL see TBQ750
EXP 338 see AMQ500
EXP 105-1 see AED250
EXPANDEX see DBD700
EXPERIMENTAL FUNGICIDE 5223 see DXX400
EXPERIMENTAL FUNGICIDE 224 (UNION
CARBIDE) see ZJA000
EXPERIMENTAL INSECTICIDE 711 see IKO000
EXPERIMENTAL INSECTICIDE 4049 see MAK700
EXPERIMENTAL INSECTICIDE 7744 see CBM750
EXPERIMENTAL INSECTICIDE 12,880 see DSP400
EXPERIMENTAL INSECTICIDE 52160 see TAL250
EXPERIMENTAL NEMATOCIDE 18,133 see EPC500
EXPERIMENTAL TICK REPELLENT 3 see AEO750
EXPLOSION ACETYLENE BLACK see CBT750
EXPLOSION BLACK see CBT750
EXPLOSIVE D see ANS500
EXPLOSIVES, HIGH see ERF000
EXPLOSIVES, LOW see ERF500
EXPLOSIVES, PERMITTED see ERG000
EXSEL see SBR000
EXSICATED SODIUM SULFITE see SJZ000
EXSICCATED FERROUS SULFATE see FBN100
EXSICCATED FERROUS SULPHATE see FBN100
EXSICCATED SODIUM PHOSPHATE see SJH090
EXTACOL see PGA750
EXT. D&C RED No. 15 see FAG018
EXT D&C YELLOW No. 1 see MDM775
EXT. D&C YELLOW No. 9 see FAG130
EXT. D&C YELLOW No. 10 see FAG135
EXTENCILLINE see BFC750
EXTENICILLINE see BFC750
EXTERMATHION see MAK700
EXTERNAL BLUE 1 see BJI250
EXTHRIN see AFR250
EXTRACT D&C ORANGE No. 4 see TGW000
EXTRACT D&C RED No. 10 see HJF500
EXTRACT D&C RED No. 14 see XRA000
EXTRACTS (PETROLEUM), HEAVY NAPHTHENIC
DISTILLATE SOLVENT (9CI) see MQV857
EXTRACTS (PETROLEUM), HEAVY PARAFFINIC
DISTILLATE SOLVENT (9CI) see MQV859

EXTRACTS (PETROLEUM), LIGHT NAPHTHENIC
DISTILLATE SOLVENT (9CI) see MQV860
EXTRACTS (PETROLEUM), LIGHT PARAFFINIC
DISTILLATE SOLVENT (9CI) see MQV862
EXTRACTS (PETROLEUM), RESIDUAL OIL
SOLVENT (9CI) see MQV863
EXTRA FINE 200 SALT see SFT000
EXTRA FINE 325 SALT see SFT000
EXTRANASE see BMO000
EXTRAR see DUS700
EXTREMA see SCQ500
EXTREMA see AQF000
EXTREMA see EPF550
EXTREMA see GDY000
EXTREN see ADA725
EXTRINSIC FACTOR see VSZ000
EXURATE see DDP200
EYEULES see ARR000
E-Z-OFF D see BSH250
E-Z-PAQUE see BAP000
E² see EDO000
F 12 see DFA600
F 13 see CLR250
F 14 see CBY250
16 F see ARM266
F 22 see CFX500
F 44 see SCK600
F-112 see TBP050
F 114 see FOO509
F-115 see CJI500
F 125 see SCK600
F 190 see ABX500
190 F see ABX500
F 735 see CCC500
F 933 see BCI500
F 1162 see SNM500
1358F see SOA500
F 1358 see SOA500
F-13B1 see TJY100
F 2387 see DVX600
Fa 100 see EQR500
23F203 see PJS750
F 1 (complexon) see EIX500
F 10 (pesticide) see MAS500
FA see FMV000
FAA see FDR000
FAA see FFF000
FAA see FIC000
2-FAA see FDR000
2,7-FAA see BGP250
F-diAA see DBF200
FABRITONE PE see PJS750
FAC 5273 see PIX250
FACTITIOUS AIR see NGU000
FACTOR II (VITAMIN) see VSZ000
FACTOR PP see NCR000
FACTOR S see VSU100
FACTOR S see BGD100
FACTOR S (vitamin) see VSU100
FACTOR S (VITAMIN) see BGD100
FAIR 30 see DMC600
FAIR PS see DMC600
FALISAN see MEP250
FALITHION see DSQ000
FALITIRAM see TFS350
FALKITOL see HCI000
FALL see SFS000
FAM see MME809

FANAL PINK B see RGW000
FANAL PINK GFK see RGW000
FANAL RED 25532 see RGW000
FANFT see NGM500
FANNOFORM see FMV000
FANODORMO see TDA500
FANTERRIN see HOH500
FARBRUSS see CBT750
FARGAN see PMI750
FARGAN see DQA400
FARINEX 100 see SLJ500
FARLUTIN see MCA000
FARMCO ATRAZINE see ARQ725
FARMCO DIURON see DXQ500
FARMCO PROPANIL see DGI000
FARMICETINA see CDP250
FARMIN 80 see OBC000
FARMOTAL see PBT500
FARTOX see PAX000
FAS-CILE see MQU750
FASCIOLIN see CBY000
FASCIOLIN see HCI000
FASCO-TERPENE see CDV100
FASERTON see AHE250
FASTBALLS see BBK500
FAST BLUE B BASE see DCJ200
FAST CORINTH BASE B see BBX000
FAST DARK BLUE BASE R see TGJ750
FAST GARNET GBC BASE see AIC250
FAST GREEN JJO see BAY750
FAST LIGHT ORANGE GA see HGC000
FASTOGEN BLUE FP-3100 see DNE400
FASTOGEN BLUE SH-100 see DNE400
FAST OIL ORANGE see PEJ500
FAST OIL ORANGE II see XRA000
FAST OIL RED B see SBC500
FAST OIL SCARLET III see OHI200
FAST OIL YELLOW see AIC250
FAST OIL YELLOW B see DOT300
FAST ORANGE see PEJ500
FAST ORANGE BASE GR see NEO000
FAST ORANGE BASE JR see NEO000
FAST ORANGE GC BASE see CEH675
FAST ORANGE GR BASE see NEO000
FAST ORANGE O BASE see NEO000
FAST ORANGE R SALT see NEN500
FAST RED 2G BASE see NEO500
FAST RED BASE 2J see NEO500
FAST RED BASE GG see NEO500
FAST RED BASE TR see CLK220
FAST RED BB see SBC500
FAST RED BB BASE see AOX250
FAST RED GG BASE see NEO500
FAST RED KB AMINE see CLK225
FAST RED KB BASE see CLK225
FAST RED KB SALT see CLK225
FAST RED KB SALT SUPRA see CLK225
FAST RED KBS SALT see CLK225
FAST RED MP BASE see NEO500
FAST RED P BASE see NEO500
FAST RED R see OHI200
FAST RED SALT TR see CLK235
FAST RED SALT TRA see CLK235
FAST RED SALT TRN see CLK235
FAST RED SG BASE see NMP500
FAST RED 5CT BASE see CLK220
FAST RED TR see CLK220
FAST RED TR11 see CLK220

FAST RED TR BASE see CLK220
FAST RED TRO BASE see CLK220
FAST RED TR SALT see CLK235
FAST RED 5CT SALT see CLK235
FAST SCARLET BASE B see NBE500
FAST SCARLET BASE G see NMP500
FAST SCARLET BASE J see NMP500
FAST SCARLET G see NMP500
FAST SCARLET G BASE see NMP500
FAST SCARLET GC BASE see NMP500
FAST SCARLET G SALT see NMP500
FAST SCARLET J SALT see NMP500
FAST SCARLET M4NT BASE see NMP500
FAST SCARLET R see NEQ500
FAST SCARLET T BASE see NMP500
FAST SCARLET TR BASE see CLK200
FAST SPIRIT YELLOW AAB see PEI000
FASTUM see BDU500
FAST WHITE see LDY000
FAST YELLOW AT see AIC250
FAST YELLOW B see AIC250
FAST YELLOW GC BASE see CEH670
FAT ORANGE II see TGW000
FAT PONCEAU R see SBC500
FAT RED 2B see SBC500
FAT RED 7B see EOJ500
FAT RED B see SBC500
FAT RED BB see SBC500
FAT RED BG see CMS238
FAT RED (BLUISH) see OHI200
FAT RED BS see SBC500
FAT RED G see OHI200
FAT RED G see CMS238
FAT RED HRR see OHI200
FAT RED R see OHI200
FAT RED RS see OHI200
FAT RED RS see CMS238
FAT RED TS see SBC500
FAT RED (YELLOWISH) see XRA000
FAT SCARLET 2G see XRA000
FAT SCARLET LB see OHI200
FATSCO ANT POISON see ARD750
FAT SOLUBLE GREEN ANTHRAQUINONE see BLK000
FAT SOLUBLE RED S see CMS238
FAT SOLUBLE RED ZH see OHI200
FATTY ACIDS see FAB850
FAT YELLOW see DOT300
FAUSTAN see DCK759
FAVISTAN see MCO500
FB/2 see DWX800
FB 217 see PJS750
FBC CMPP see CIR500
FBHC see BBQ750
FB SODIUM PERCARBONATE see SJB400
FC 12 see DFA600
FC 14 see CBY250
FC-21 see DFL000
FC 22 see CFX500
FC 31 see CHI900
FC 112 see TBP050
FC 114 see FOO509
FC 123 see TJY500
FC-143 see ANP625
FC-1318 see OBO000
FC 133a see TJY175
FC142b see CFX250
FC 152a see ELN500

FC 4648 see PKQ059
FC-C 318 see CPS000
FC-MY 5450 see SMQ500
FDA see DGB600
FDA 0101 see SHF500
FDA 0345 see BIF750
FDA 1446 see AFR250
FDA 1541 see EIN500
FD & C NO. 6 see FAG150
FD and C NO. 6 see FAG150
FD&C RED No. 1 see FAG018
FD&C VIOLET No. 1 see FAG120
FD and C YELLOW 6 see FAG150
FD and C YELLOW LAKE NO. 6 see FAG150
FD&C YELLOW No. 3 see FAG130
FD&C YELLOW No. 4 see FAG135
FD&C YELLOW No. 5 see FAG140
FD&C YELLOW No. 6 see FAG150
FDC YELLOW NO. 6 see FAG150
FD and C YELLOW NO. 6 see FAG150
FD & C YELLOW NO. 6 ALUMINIUM LAKE see
FAG150
FD & C YELLOW NO. 5 TARTRAZINE see FAG140
FDUR see DAR400
FEBRILIX see HIM000
FEBRININA see DOT000
FEBRO-GESIC see HIM000
FEBROLIN see HIM000
FEBRON see DOT000
FECAMA see DGP900
FECTO see CBT750
FECTRIM see SNK000
FEDACIN see NGE500
FEDAL-UN see PCF275
Fe-DEXTRAN see IGS000
FEDRIN see EAW000
FEEN-A-MINT GUM see PDO750
FEENO see PDP250
FEGLOX see DWX800
FEINALMIN see DLH630
FEKABIT see PLA250
FELAZINE see PFC750
FELBEN see BAU750
α-FELLANDRENE see MCC000
FELLING ZINC OXIDE see ZKA000
FELLOZINE see PMI750
FELMANE see FMQ000
FELSULES see CDO000
FEMACOID see ECU750
FEMADOL see PNA500
FEMA No. 2003 see AAG250
FEMA No. 2005 see MDW750
FEMA No. 2006 see AAT250
FEMA No. 2007 see THM500
FEMA No. 2008 see ABB500
FEMA No. 2009 see ABH000
FEMA No. 2011 see AEN250
FEMA No. 2026 see AGC500
FEMA No. 2031 see AGH250
FEMA No. 2032 see AGA500
FEMA No. 2033 see AGI500
FEMA No. 2034 see AGJ250
FEMA No. 2037 see AGM500
FEMA No. 2045 see ISV000
FEMA No. 2055 see IHO850
FEMA No. 2058 see IHP100
FEMA No. 2060 see IHP400

FEMA No. 2061 see AOG500
FEMA No. 2063 see AOG600
FEMA No. 2069 see IHS000
FEMA No. 2075 see IHU100
FEMA No. 2082 see AON350
FEMA No. 2084 see IME000
FEMA No. 2085 see ITB000
FEMA No. 2086 see PMQ750
FEMA No. 2097 see AOX750
FEMA No. 2098 see AOY400
FEMA No. 2099 see MED500
FEMA No. 2109 see ARN000
FEMA No. 2127 see BAY500
FEMA No. 2134 see BCS250
FEMA No. 2135 see BDX000
FEMA No. 2137 see BDX500
FEMA No. 2138 see BCM000
FEMA No. 2140 see BED000
FEMA No. 2141 see IJV000
FEMA No. 2142 see BEG750
FEMA No. 2149 see BFD400
FEMA No. 2150 see BFD800
FEMA No. 2151 see BFJ750
FEMA No. 2152 see ISW000
FEMA No. 2159 see BMD100
FEMA No. 2160 see IHX600
FEMA No. 2170 see MKA400
FEMA No. 2174 see BPU750
FEMA No. 2175 see IIJ000
FEMA No. 2178 see BPW500
FEMA No. 2179 see IIL000
FEMA No. 2183 see BQI000
FEMA No. 2184 see BFW750
FEMA No. 2186 see BQM500
FEMA No. 2187 see BSW500
FEMA No. 2188 see BRQ350
FEMA No. 2190 see BQP000
FEMA No. 2193 see IIQ000
FEMA No. 2203 see DTC800
FEMA No. 2209 see BBA000
FEMA No. 2210 see IJF400
FEMA No. 2213 see IJN000
FEMA No. 2218 see ISX000
FEMA No. 2219 see BSU250
FEMA No. 2220 see IJS000
FEMA No. 2221 see BSW000
FEMA No. 2222 see IJU000
FEMA No. 2223 see TIG750
FEMA No. 2224 see CAK500
FEMA No. 2229 see CBA500
FEMA No. 2245 see CCM000
FEMA No. 2249 see CCM100
FEMA No. 2249 see CCM120
FEMA No. 2252 see CCN000
FEMA No. 2286 see CMP969
FEMA No. 2288 see CMP975
FEMA No. 2293 see CMQ730
FEMA No. 2294 see CMQ740
FEMA No. 2295 see API750
FEMA No. 2299 see CMR500
FEMA No. 2301 see CMR850
FEMA No. 2302 see CMR800
FEMA No. 2306 see CMS750
FEMA No. 2307 see CMS845
FEMA No. 2309 see CMT250
FEMA No. 2311 see AAU000
FEMA No. 2312 see CMT600
FEMA No. 2313 see CMT900

FEMA No. 2857 see PFB250
FEMA No. 2858 see PDD750
FEMA No. 2862 see PDF750
FEMA No. 2866 see PDI000
FEMA No. 2868 see PDK200
FEMA No. 2871 see PDF775
FEMA No. 2873 see PDS900
FEMA No. 2874 see BBL500
FEMA No. 2876 see PDX000
FEMA No. 2878 see PDY850
FEMA No. 2885 see HHP050
FEMA No. 2886 see COF000
FEMA No. 2887 see HHP000
FEMA No. 2888 see PGA800
FEMA No. 2890 see HHP500
FEMA No. 2902 see PIH250
FEMA No. 2903 see POH750
FEMA No. 2911 see PIW250
FEMA No. 2922 see IRY000
FEMA No. 2923 see PMT750
FEMA No. 2926 see INE100
FEMA No. 2930 see PNE250
FEMA No. 2962 see MCE750
FEMA No. 2980 see CMT250
FEMA No. 2981 see RHA000
FEMA No. 2981 see DTF400
FEMA No. 2984 see RHA100
FEMA No. 3006 see OGY220
FEMA No. 3007 see SAU400
FEMA No. 3045 see TBD500
FEMA No. 3047 see TBE250
FEMA No. 3053 see TBE600
FEMA No. 3060 see TCU600
FEMA No. 3073 see MNR250
FEMA No. 3075 see THA250
FEMA No. 3082 see TJJ400
FEMA No. 3092 see UJJ000
FEMA No. 3095 see ULJ000
FEMA No. 3097 see UNA000
FEMA No. 3101 see VAQ000
FEMA No. 3102 see ISU000
FEMA No. 3103 see VAV000
FEMA No. 3107 see VFK000
FEMA No. 3135 see DAE450
FEMA No. 3149 see EIL100
FEMA No. 3155 see ENF200
FEMA No. 3164 see HAV450
FEMA No. 3183 see MEX350
FEMA No. 3212 see NMV775
FEMA No. 3213 see NNA300
FEMA No. 3215 see ODQ800
FEMA No. 3237 see TDU800
FEMA No. 3244 see TME270
FEMA No. 3264 see DAI360
FEMA No. 3271 see DTU400
FEMA No. 3272 see DTU600
FEMA No. 3273 see DTU800
FEMA No. 3289 see HBI800
FEMA No. 3291 see BOV000
FEMA No. 3302 see MFN285
FEMA No. 3309 see MOW750
FEMA No. 3317 see NMV760
FEMA No. 3326 see ABC750
FEMA No. 3354 see HFM600
FEMA No. 3355 see HKC600
FEMA No. 3386 see PPS250
FEMA No. 3406 see MLA300
FEMA No. 3432 see IIN300

FEMA No. 3465 see NNA530
FEMA No. 3497 see HFE550
FEMA No. 3498 see ISZ000
FEMA No. 3499 see HFR200
FEMA No. 3500 see HFQ600
FEMA No. 3525 see TLX800
FEMA No. 3558 see MLA250
FEMA No. 3559 see MCB750
FEMA No. 3565 see DKV175
FEMA No. 3581 see OCY100
FEMA No. 3583 see OEG100
FEMA No. 3587 see ODW030
FEMA No. 3632 see PDF790
FEMA No. 7071 see DTV300
FEMERGIN see EDC500
FEMEST see ECU750
FEMESTRAL see EDO000
FEMESTRONE see EDP000
FEMESTRONE INJECTION see EDV000
FEM H see ECU750
FEMIDYN see EDV000
FEMMA see ABU500
FEMOGEN see ECU750
FEMOGEN see EDO000
FEMOGEX see EDS100
FEMPROPAZINE see DIR000
FEMULEN see EQJ500
FENACET BLUE G see TBG700
FENACET FAST PINK B see AKE250
FENACET FAST VIOLET B see DBY700
FENACET FAST YELLOW G see AAQ250
FENACETINA see ABG750
FENACET YELLOW G see AAQ250
FENAFOR RED PB see CMM330
FENAMIC ACID see PEG500
FENAMIN see ARQ725
FENAMIN BLACK E see AQP000
FENAMIN BLUE 2B see CMO000
FENAMINE see AMY050
FENAMINE see ARQ725
FENAMINOSULF see DOU600
FENAMIN SKY BLUE see CMO500
FENAMIN SKY BLUE 3F see CMN750
FENAMIPHOS see FAK000
FENANTOIN see DKQ000
FENANTOIN see DNU000
FENARIMOL see FAK100
FENAROL see CJR909
FENARSONE see CBJ000
FENARTIL see BRF500
FENASAL see DFV400
FENATE see IGS000
FENATROL see ARQ725
FENAVAR see AMY050
FENAZIL see PMI750
FENAZIL see DQA400
FENAZO BLUE XR see FMU059
FENAZO EOSINE XG see BNH500
FENAZO RED C see HJF500
FENAZO SCARLET 2R see FMU070
FENAZO SCARLET 3R see FMU080
FENAZO YELLOW M see MDM775
FENAZO YELLOW T see FAG140
FENBITAL see EOK000
FENBUTATIN OXIDE see BLU000
FENCAL see ARB750
FENCHEL OEL (GERMAN) see FAP000
FENCHLONINE see FAM100

FENCHLOORFOS (DUTCH) see RMA500
FENCHLORFOS see RMA500
FENCHLORFOSU (POLISH) see RMA500
FENCHLOROPHOS see RMA500
FENCHLORPHOS see RMA500
FENCLONIN see FAM100
FENCLONINE see FAM100
FENCLOR see PJL750
FENCLOR 42 see PJL750
FENDON see HIM000
FENELZIN see PFC750
FENERGAN see PMI750
FENERGAN see DQA400
FENESTERIN see CME250
FENESTRIN see CME250
FENETAZINA see DQA400
β-FENETHYLALKOHOL see PDD750
p-FENETIDIN see PDK790
FENFLUORAMINE HYDROCHLORIDE see PDM250
FENIBUTAZONA see BRF500
FENIBUTOL see BRF500
FENICOL see CDP250
FENIDANTOIN "S" see DKQ000
FENILBUTINE see BRF500
FENILDICLOROARSINA (ITALIAN) see DGB600
FENILFAR see SPC500
FENILIDINA see BRF500
FENILIDRAZINA (ITALIAN) see PFI000
2-FENILPROPANO (ITALIAN) see COE750
FENILPROPANOLAMINA (ITALIAN) see NNM000
FENIPENTOL see BQJ500
FENITOIN see DNU000
FENITOX see DSQ000
FENITROTHION see DSQ000
FENITROTION (HUNGARIAN) see DSQ000
FENNEL OIL see FAP000
FENNOSAN see QPA000
FENNOSAN B 100 see DSB200
FENOBARBITAL see EOK000
FENOBCARB see MOV000
FENOBUCARB see MOV000
FENOL (DUTCH, POLISH) see PDN750
FENOLFTALEIN see PDO750
FENOLO (ITALIAN) see PDN750
FENOLOVO see HON000
FENOLOVO ACETATE see ABX250
FENOPHOSPHON see EPY000
FENOPROP see TIX500
FENORMONE see TIX500
FENOTHIAZINE (DUTCH) see PDP250
FENOTIAZINA (ITALIAN) see PDP250
FENOTONE see BRF500
FENOVERM see PDP250
FENOX see SPC500
FENOXYBENZAMIN see DDG800
2-FENOXYETHANOL (CZECH) see PER000
FENOXYL CARBON N see DUZ000
FENPROBAMATO see PGA750
FENPROPAZINA see DIR000
FENSULFOTHION see FAQ800
FENTAL see FLZ050
FENTANEST see PDW750
FENTANYL CITRATE see PDW750
FENTAZIN see CJM250
FENTIAZIN see PDP250
FENTIN ACETAAT (DUTCH) see ABX250
FENTIN ACETAT (GERMAN) see ABX250

FENTIN ACETATE see ABX250
FENTIN CHLORIDE see CLU000
FENTINE ACETATE (FRENCH) see ABX250
FENTIN HYDROXIDE see HON000
FENVALERATE see FAR100
1-FENYL-4-AMINO-5-CHLOR-6-PYRIDAZINON (CZECH) see PEE750
FENYLBUTAZON see BRF500
FENYL-CELLOSOLVE (CZECH) see PER000
FENYLDICHLORARSIN see DGB600
1-FENYL-3,3-DIMETHYLTRIAZIN see DTP000
m-FENYLENDIAMIN (CZECH) see PEY000
FENYLENODWUAMINA (POLISH) see PEY500
FENYLEPSIN see DKQ000
FENYLESTER KYSELINY CHLORMRAVENCI (CZECH) see CBX109
FENYLESTER KYSELINY SALICYLOVE see PGG750
1-FENYLETHANOL see PDE000
β-FENYLETHANOL see PDD750
FENYLETTAE see EOK000
FENYLFOSFIN see PFV250
FENYL-GLYCIDYLETHER (CZECH) see PFF360
FENYLHIST see BAU750
FENYLHYDRAZINE (DUTCH) see PFI000
FENYL-α-HYDROXYBENZYLKETON see BCP250
FENYLISOKYANAT see PFK250
FENYLKYANID see BCQ250
FENYLMERCURIACETAT (CZECH) see ABU500
FENYLMERCURICHLORID (CZECH) see PFM500
FENYLMERKURINITRAT see MCU750
FENYL-METHYLKARBINOL see PDE000
2-FENYLOTIOMOCZNIK (POLISH) see DWN800
1-FENYLPIPERAZIN see PFX000
2-FENYL-PROPAAN (DUTCH) see COE750
FENYPRIN see DBA800
FENYTOINE see DKQ000
FENYTOINE see DNU000
FENZEN (CZECH) see BBL250
FEOJECTIN see IHG000
FEOSOL see FBN100
FEOSOL see FBO000
FEOSPAN see FBN100
FEOSTAT see FBJ100
FERBAM see FAS000
FERBAM 50 see FAS000
FERBAM, IRON SALT see FAS000
FERBECK see FAS000
FERDEX 100 see IGS000
FERGON see FBK000
FERGON PREPARATIONS see FBK000
FER-IN-SOL see FBN100
FER-IN-SOL see FBO000
FERKETHION see DSP400
FERLUCON see FBK000
FERMATE FERBAM FUNGICIDE see FAS000
FERMENICIDE LIQUID see SOH500
FERMENICIDE POWDER see SOH500
FERMENTATION ALCOHOL see EFU000
FERMENTATION AMYL ALCOHOL see IHP000
FERMENTATION BUTYL ALCOHOL see IIL000
FERMIDE see TFS350
FERMINE see DTR200
FERMOCIDE see FAS000
FERNACOL see TFS350
FERNASAN see TFS350
FERNIDE see TFS350
FERNISOLONE see PMA000

FERNOXENE see SGH500
FERO-GRADUMET see FBN100
FERO-GRADUMET see FBO000
FEROTON see FBJ100
FER PENTACARBONYLE (FRENCH) see IHG500
FERRADOW see FAS000
FERRALYN see FBN100
FERRATE(4-), HEXACYANO-, TETRAPOTASSIUM see TEC500
FERRATE(4-), HEXAKIS(CYANO-C)-, TETRAPOTASSIUM, (OC-6-11)- see TEC500
FERRIAMICIDE see MQW500
FERRIC ARSENATE, solid (DOT) see IGN000
FERRIC ARSENITE, solid (DOT) see IGO000
FERRIC ARSENITE, BASIC see IGO000
FERRIC CHLORIDE see FAU000
FERRIC CHLORIDE (UN 1733) (DOT) see FAU000
FERRIC CHLORIDE, solution (UN 2582) (DOT) see FAU000
FERRIC CHOLINE CITRATE see FBC100
FERRIC CITRATE see FAW100
FERRIC DEXTRAN see IGS000
FERRIC DIMETHYLDITHIOCARBAMATE see FAS000
FERRIC FLUORIDE see FAX000
FERRIC NITRATE see FAY200
FERRIC NITRATE see IHB900
FERRIC NITRATE (DOT) see FAY200
FERRIC NITRATE (DOT) see IHB900
FERRICON see ROF300
FERRIC OXIDE see IHC450
FERRIC OXIDE, SACCHARATED see IHG000
FERRIC PYROPHOSPHATE see FAZ525
FERRIC SACCHARATE IRON OXIDE (MIX.) see IHG000
FERRIC SULFATE see FBA000
FERRICYTOCHROME C see CQM325
FERRIDEXTRAN see IGS000
FERRIVENIN see IHG000
FERROANTHOPHYLLITE see ARM264
FERROCENE see FBC000
FERROCENE, ACETYL- see ABA750
FERROCENE, 1,1'-BIS(DIMETHYL(OCTYLOXY)SILYL)- see BJK780
FERROCHOLINATE see FBC100
FERROCHROME see FBD000
FERROCHROME (exothermic) see FBD000
exothermic FERROCHROME see FBD000
FERROCHROMIUM see FBD000
FERROCYTOCHROME C see CQM325
FERRODEXTRAN see IGS000
FERROFLUKIN 75 see IGS000
FERROFOS 510 see HKS780
FERROFUME see FBJ100
FERROGLUCIN see IGS000
FERROGLUKIN 75 see IGS000
FERRO-GRADUMET see FBN100
FERRO LEMON YELLOW see CAJ750
FERROLIP see FBC100
FERROMANGANESE (exothermic) see FBE000
exothermic FERROMANGANESE (DOT) see FBE000
FERRONAT see FBJ100
FERRONE see FBJ100
FERRONICUM see FBK000
FERRO ORANGE YELLOW see CAJ750
FERROSILICON see FBG000
FERROSILICON, containing more than 30% but less than 90% SILICON (DOT) see FBG000
FERROSULFAT (GERMAN) see FBN100

FERROSULFATE see FBN100
FERROTEMP see FBJ100
FERRO-THERON see FBN100
FERROUS ACETATE see FBH000
FERROUS ARSENATE (DOT) see IGM000
FERROUS ARSENATE, solid (DOT) see IGM000
FERROUS CHLORIDE see FBI000
FERROUS CHLORIDE, solid (NA 1759) (DOT) see FBI000
FERROUS CHLORIDE, solution (NA 1760) (DOT) see FBI000
FERROUS FERRITE see IHG100
FERROUS FUMARATE see FBJ100
FERROUS GLUCONATE see FBK000
FERROUS LACTATE see LAL000
FERROUS SULFATE see FBN100
FERROUS SULFATE (FCC) see FBO000
FERROUS SULFATE HEPTAHYDRATE see FBO000
FERROVAC E see IGK800
FERROVANADIUM DUST see FBP000
FERRO YELLOW see CAJ750
FERRUGO see IHC450
FERRUM see FBJ100
FERSAMAL see FBJ100
FERSOLATE see FBN100
FERTENE see PJS750
FERTILVIT see TGJ050
FERTIRAL see LIU370
FESOFOR see FBO000
FESOTYME see FBO000
FETTORANGE B see XRA000
FETTORANGE R see PEJ500
FETTPONCEAU G see OHI200
FETTROT see OHI200
FETTSCHARLACH see OHI200
FETTSCHARLACH LB see OHI200
FF see FQN000
FFB 32 see PHB500
F-5-FU see FLZ050
FG 834 see SMQ500
FG 5111 see FKI000
FHCH see BBQ750
F.I 106 see AES750
FI 106 see HKA300
F.I. 6145 see PIW000
FI6339 see DAC000
FI 6714 see NDM000
FI 6804 see HKA300
FIBERGLASS see FBQ000
FIBERS, REFRACTORY CERAMIC see RCK725
FIBRE BLACK VF see AQP000
FIBRENE C 400 see TAB750
FIBROTAN see PFN000
FIBROUS CROCIDOLITE ASBESTOS see ARM275
FIBROUS GLASS see FBQ000
FIBROUS GLASS DUST (ACGIH) see FBQ000
FIBROUS GRUNERITE see ARM250
FIBROUS TREMOLITE see ARM280
FICAM see DQM600
FICHLOR 91 see TIQ750
FICIN see FBS000
FI CLOR 71 see DGN200
FI CLOR 91 see TIQ750
FI CLOR 60S see SGG500
FICUS PROTEASE see FBS000
FICUS PROTEINASE see FBS000
FILARIOL see EGV500
FILARSEN see DFX400

FILMERINE see SIQ500
FILTRASORB see CBT500
FILTRASORB 200 see CBT500
FILTRASORB 400 see CBT500
FINE GUM HES see SFO500
FINEMEAL see BAP000
FINIMAL see HIM000
FINTIN ACETATO (ITALIAN) see ABX250
FINTINE HYDROXYDE (FRENCH) see HON000
FINTIN HYDROXID (GERMAN) see HON000
FINTIN HYDROXYDE (DUTCH) see HON000
FINTIN IDROSSIDO (ITALIAN) see HON000
FIORINAL see ABG750
FIR BALSAM OREGON see OJK340
FIRE DAMP see MDQ750
FIREMASTER BP-6 see FBU000
FIREMASTER FF-1 see FBU509
FIREMASTER T23P-LV see TNC500
FIR NEEDLE OIL, CANADIAN TYPE see FBU850
FIR NEEDLE OIL, SIBERIAN see FBV000
FIRON see FBJ100
FISH BERRY see PIE500
FISONS NC 2964 see DSO000
FITROL see KBB600
FITROL DESICCATE 25 see KBB600
FIXANOL BLACK E see AQP000
FIXANOL BLUE 2B see CMO000
FIXOL see CMS850
6FK see HCZ000
FK 506 see TAA900
FLAC see ARB750
FLACAVON R see TNC500
FLAGEMONA see MMN250
FLAGESOL see MMN250
FLAGIL see MMN250
FLAGYL see MMN250
FLAMENCO see TGG760
FLAMMEX AP see TNC500
FLAMMEX B 10 see PCC480
FLAMOLIN MF 15711 see PJS750
FLAMRUSS see CBT750
FLAMYCIN see CMA750
FLANOGEN ELA see CCL250
FLAVACRIDINUM HYDROCHLORICUM see DBX400
FLAVAXIN see RIK000
FLAVAZONE see NGE500
FLAVINE see DBX400
FLAVIOFORM see DBX400
FLAVIPIN see DBX400
FLAVISEPT see DBX400
FLAVITROL see EDW500
FLAVOMYCELIN see LIV000
FLAVOMYCIN see MRA250
FLAVOPHOSPHOLIPOL see MRA250
FLAVUROL see MCV000
FLEBOCORTID see HHR000
FLECK-FLIP see TIO750
FLECK'S EXTRAORDINARY CEMENT see ZJS400
FLECTOL A see TLP500
FLECTOL H see TLP500
FLECTOL PASTILLES see TLP500
FLEET-X see TLM050
FLEXAMINE G see BLE500
FLEXAZONE see BRF500
FLEXCHLOR see PAH780
FLEXIBLE COLLODION see CCU250
FLEXICHEM see CAX350
FLEXICHEM CS see CAX350

FLEXIMEL see DVL700
FLEXOL A 26 see AEO000
FLEXOL DOP see DVL700
FLEXOL PLASTICIZER DIP see ILR100
FLEXOL PLASTICIZER DOP see DVL700
FLEXOL PLASTICIZER TCP see TNP500
FLEXO RED 482 see RGW000
FLIBOL E see TIQ250
FLIEGENTELLER see TIQ250
FLINDIX see CCK125
FLINT see SCI500
FLINT see SCJ500
FLIT 406 see CBG000
FLOCOOL 180 see SJC500
FLOCOR see PKQ059
FLO-MOR see PAI000
FLO PRO T SEED PROTECTANT see TFS350
FLO PRO V SEED PROTECTANT see CCC500
FLORALTONE see GEM000
FLORAQUIN see DNF600
FLORDIMEX see CDS125
FLOREL see CDS125
FLORES MARTIS see FAU000
FLORIDINE see SHF500
FLORITE R see CAW850
FLOROCID see SHF500
FLOROL see PGR250
FLOROPIPAMIDE see FHG000
FLOROPRYL see IRF000
FLOROXENE see TKB250
FLOTHENE see PJS750
FLOVACIL see DKI600
FLOWERS of ANTIMONY see AQF000
FLOWERS of SULPHUR (DOT) see SOD500
FLOWERS of ZINC see ZKA000
FLOXURIDIN see DAR400
FLOXURIDINE see DAR400
FLOZENGES see SHF500
FLUATE see TIO750
FLUAZINAM see CEX800
FLUCYTHRINATE see COQ390
FLUE DUST, ARSENIC CONTAINING see ARE500
FLUE GAS see CBW750
FLUGENE 22 see CFX500
FLUGEX 12B1 see BNA250
FLUIFORT see CBR675
FLUIMUCETIN see ACH000
FLUIMUCIL see ACH000
FLUKOIDS see CBY000
FLUMAMINE see DQR600
FLUMEN see CLH750
FLUMICIL see ACH000
FLUNIGET see DKI600
FLUOBORIC ACID see FDD125
FLUOBORIC ACID see HHS600
FLUOBORIC ACID (DOT) see FDD125
FLUOBORIC ACID (DOT) see HHS600
FLUO-KEM see TAI250
FLUOMETURON see DUK800
FLUOMINE see EIS000
FLUOMINE DUST see EIS000
FLUON see TAI250
FLUOOXENE see TKB250
FLUOPERAZINE see TKK250
FLUOPERIDOL see FLU000
FLUOPHOSGENE see CCA500
FLUOPHOSPHORIC ACID DI(DIMETHYLAMIDE) see BJE750

FLUOPHOSPHORIC ACID, DIISOPROPYL ESTER see IRF000
FLUOR (DUTCH, FRENCH, GERMAN, POLISH) see FEZ000
FLUORACETATO di SODIO (ITALIAN) see SHG500
5-FLUORACIL (GERMAN) see FMM000
FLUORACIZINE see FDE000
FLUORAKIL 100 see FFF000
FLUORAL see SHF500
3,6-FLUORANDIOL see FEV000
3',6'-FLUORANDIOL see FEV000
FLUORANE 114 see FOO509
4-FLUORANILIN see FFY000
FLUORANTHENE see FDF000
FLUORAQUIN see DNF600
FLUOREN-2-AMINE see FDI000
2-FLUORENAMINE see FDI000
2-FLUORENEAMINE see FDI000
FLUORENE, 2-(CHLOROACETYL)- see CEC700
2-FLUORENYLACETAMIDE see FDR000
N-FLUOREN-2-YL ACETAMIDE see FDR000
N-2-FLUORENYLACETAMIDE see FDR000
N-FLUOREN-4-YLACETAMIDE see ABY000
N-4-FLUORENYLACETAMIDE see ABY000
FLUORENYL-2-ACETHYDROXAMIC ACID see HIP000
N-FLUOREN-2-YL ACETOHYDROXAMIC ACID see HIP000
N-2-FLUORENYL ACETOHYDROXAMIC ACID see HIP000
2,7-FLUORENYLBISACETAMIDE see BGP250
N,N'-FLUOREN-2,7-YLBISACETAMIDE see BGP250
2-FLUORENYLDIACETAMIDE see DBF200
N-FLUOREN-2-YLDIACETAMIDE see DBF200
N-2-FLUORENYLDIACETAMIDE see DBF200
N,N'-FLUOREN-2,7-YLENEBISACETAMIDE see BGP250
N,N'-2,7-FLUORENYLENEBISACETAMIDE see BGP250
N,N'-(FLUOREN-2,7-YLENE)BIS(ACETYLAMINE) see BGP250
N,N'-2,7-FLUORENYLENEDIACETAMIDE see BGP250
9-FLUORENYLMETHYL CHLOROFORMATE see FEI100
FLUORESCEIN see FEV000
FLUORESCEIN, soluble see FEW000
FLUORESCEIN, 4',5'-DIBROMO- see DDO200
FLUORESCEINE see FEV000
FLUORESCEIN MERCURIACETATE see FEV100
FLUORESCEIN MERCURIC ACETATE see FEV100
FLUORESCEIN MERCURY ACETATE see FEV100
FLUORESCEIN SODIUM see FEW000
FLUORESCEIN SODIUM B.P see FEW000
FLUORESSIGSAEURE (GERMAN) see SHG500
2-FLUORETHYLESTER KYSELINY CHLORMRAVENCI see FIH100
FLUORID BORITY-DIMETHYLETHER (1:1) see BMH000
FLUORIDENT see SHF500
FLUORIDES see FEY000
FLUORID HLINITY (CZECH) see AHB000
FLUORID KYSELINY OCTOVE see ACM000
FLUORID SODNY (CZECH) see SHF500
FLUORIGARD see SHF500
FLUORINE see FEZ000
FLUORINE, compressed (DOT) see FEZ000
FLUORINEED see SHF500
FLUORINE MONOXIDE see ORA000
FLUORINE OXIDE see ORA000

FLUORINSE see SHF500
FLUORISTAN see TGD100
FLUOR-I-STRIP A.T. see FEW000
FLUORITAB see SHF500
FLUORITE see CAS000
FLUOR-O-KOTE see SHF500
FLUORO (ITALIAN) see FEZ000
FLUOROACETAMIDE see FFF000
2-FLUOROACETAMIDE see FFF000
FLUOROACETATE see FIC000
FLUOROACETIC ACID see FIC000
2-FLUOROACETIC ACID see FIC000
FLUOROACETIC ACID (DOT) see FIC000
FLUOROACETIC ACID AMIDE see FFF000
FLUOROACETIC ACID, SODIUM SALT see SHG500
p-FLUOROACETYLAMINOPHENYL DERIVATIVE of NITROGEN MUSTARD see BHP750
FLUOROACETYL CHLORIDE see FFR000
4-FLUOROANILINE see FFY000
p-FLUOROANILINE see FFY000
4-FLUOROBENZENAMINE see FFY000
FLUOROBENZENE see FGA000
4-FLUOROBENZENEARSONIC ACID see FGA100
1-(2-(4-(6-FLUORO-1,2-BENZISOXAZOL-3-YL)-1-PIPERIDINYL)ETHYL)-3-PHENYL-2-IMIDAZOLIDINONE see FGA200
6-FLUOROBENZO(a)PYRENE see FGI100
1'-(3-(p-FLUOROBENZOYL)PROPYL)(1,4'-BIPIPERIDINE)-4'-CARBOXAMIDE see FHG000
1-(3-(p-FLUOROBENZOYL)PROPYL)-4-PIPERIDINOISONIPACOTAMIDE see FHG000
1-(3-(4-FLUOROBENZOYL)PROPYL)-4-(2-PYRIDYL)PIPERAZINE see FLU000
FLUOROBISISOPROPYLAMINO-PHOSPHINE OXIDE see PHF750
FLUOROBLASTIN see FMM000
3-FLUOROBUTYL ISOCYANATE see FHC200
FLUOROBUTYROPHENONE see FHG000
FLUOROCARBON-12 see DFA600
FLUOROCARBON-22 see CFX500
FLUOROCARBON 113 see FOO000
FLUOROCARBON 114 see FOO509
FLUOROCARBON-115 see CJI500
FLUOROCARBON 1211 see BNA250
FLUOROCARBON FC142b see CFX250
FLUOROCARBON No. 11 see TIP500
FLUOROCHROME see MCV000
FLUORODEOXYURIDINE see DAR400
5-FLUORODEOXYURIDINE see DAR400
5-FLUORO-2-DEOXYURIDINE see DAR400
5-FLUORO-2'-DEOXYURIDINE see DAR400
β-5-FLUORO-2'-DEOXYURIDINE see DAR400
FLUORODICHLOROMETHANE see DFL000
9-α-FLUORO-11-β,21-DIHYDROXY-16-α-ISOPROYLIDENEDIOXY-1,4-PREGNADIENE, 3,20-DIONE see AQX500
FLUORODIISOPROPYL PHOSPHATE see IRF000
1-FLUORO-2,4-DINITROBENZENE see DUW400
1,2,4-FLUORODINITROBENZENE see DUW400
FLUOROETHANE see FIB000
FLUOROETHANOIC ACID see FIC000
2-FLUOROETHANOL see FIE000
β-FLUOROETHANOL see FIE000
FLUOROETHENE see VPA000
2-FLUOROETHYL CHLOROFORMATE see FIH100
FLUOROETHYLENE see VPA000
FLUOROFLEX see TAI250

FLUOROFORM see CBY750
FLUOROFORMYL FLUORIDE see CCA500
FLUOROFUR see FLZ050
9-α-FLUORO-16-HYDROXYPREDNISOLONE
ACETONIDE see AQX500
4'-FLUORO-4-(4-HYDROXY-4-(α,α,α-TRIFLUORO-m-TOLYL)PIPERIDINO)BUTYROPHENONE see TKK500
4-FLUORO-4,4-IDROSSI-4-(m-TRIFLUOROMETIL-FENIL)-PIPERIDINO-BUTIRROFENONE (ITALIAN)
see TKK500
FLUOROISOPROPOXYMETHYLPHOSPHINE OXIDE
see IPX000
9-α-FLUORO-16-α-17-α-ISOPROPYLEDENE DIOXY
PREDNISOLONE see AQX500
9-α-FLUORO-16-α-17-α-ISOPROPYLIDENEDIOXY-Δ-1-HYDROCORTISONE see AQX500
FLUOROLON 4 see TAI250
FLUOROMAR see TKB250
FLUOROMETHANE see FJK000
Δ¹-9-α-FLUORO-16-α-METHYLCORTISOL see SOW000
2-FLUORO-N-METHYL-N-1-NAPHTHALENYLACETAMIDE see MME809
2-FLUORO-N-METHYL-N-1-NAPHTHYLACETAMIDE
see MME809
4'-FLUORO-4-(4-METHYLPIPERIDINO)BUTYROPHENONE
HYDROCHLORIDE see FKI000
9-α-FLUORO-16-α-METHYLPREDNISOLONE see
SOW000
9-α-FLUORO-16-α-METHYL-1,4-PREGNADIENE-11-β,17-α,21-TRIOL-3,20-DIONE see SOW000
4-α-FLUORO-16-α-METHYL-11-β,17,21-TRIHYDROXYPREGNA-1,4-DIENE-3,20-DIONE see
SOW000
3-(6-FLUORO-2-NAPHTHALENYL)-2,5-DIHYDRO-1,2-DIMETHYL-1H-PYRROLE, EL-(2R,3R)-2,3-DIHYDROXYBUTANEDIOATE (1:1) see FKK035
9-FLUORONONYL PHENYL KETONE see FKQ100
8-FLUOROOCTYL PHENYL KETONE see FKT050
FLUOROPAK 80 see TAI250
p-FLUOROPHENYLAMINE see FFY000
1-(4-FLUOROPHENYL)-4-(4-HYDROXY-4-(3-(TRIFLUOROMETHYL)PHENYL)-1-PIPERIDINYL)-1-BUTANONE see TKK500
3-(4-FLUOROPHENYL)-N-(4-PROPYLCYCLOHEXYL)-2-PROPENAMIDE see FLL100
1-(4-FLUOROPHENYL)-4-(4-(2-PYRIDINYL)-1-PIPERAZINYL)-1-BUTANONE see FLU000
FLUOROPHOSGENE see CCA500
FLUOROPHOSPHORIC ACID, anhydrous see PHJ250
4'-FLUORO-4-(4-N-PIPERIDINO-4-CARBAMIDOPIPERIDINO)BUTYROPHENONE see
FHG000
p-FLUORO-γ-(4-PIPERIDINO-4-CARBAMOYLPIPERIDINO)BUTYROPHENONE see
FHG000
FLUOROPLAST 3 see CLQ750
FLUOROPLAST 4 see TCH500
FLUOROPLEX see FMM000
3-FLUOROPROPYL ISOCYANATE see FLR100
FLUOROPRYL see IRF000
4'-FLUORO-4-(4-(2-PYRIDYL)-1-PIPERAZINYL)BUTYROPHENONE see FLU000
5-FLUORO-2,4-PYRIMIDINEDIONE see FMM000
5-FLUORO-2,4(1H,3H)-PYRIMIDINEDIONE see
FMM000

FLUOROSILICIC ACID see SCO500
FLUOROSULFONIC ACID (DOT) see FLZ000
FLUOROSULFONYL CHLORIDE see SOT500
FLUOROSULFURIC ACID see FLZ000
5-FLUORO-1-(TETRAHYDRO-2-FURANYL)-2,4-PYRIMIDINEDIONE see FLZ050
5-FLUORO-1-(TETRAHYDRO-2-FURANYL)-2,4(1H,3H)-PYRIMIDINEDIONE see FLZ050
5-FLUORO-1-(TETRAHYDROFURAN-2-YL)URACIL
see FLZ050
5-FLUORO-1-(TETRAHYDRO-3-FURYL)URACIL see
FLZ050
p-FLUOROTOLUENE see FMC000
FLUOROTRIBUTYLSTANNANE see FME000
FLUOROTRICHLOROMETHANE (OSHA) see TIP500
9-α-FLUORO-11-β,17-α,21-TRIHYDROXY-16-α-METHYLPREGNA-1,4-DIENE-3,20-DIONE see
SOW000
9-FLUORO-11-β,17,21-TRIHYDROXY-16-α-METHYLPREGNA-1,4-DIENE-3,20-DIONE see
SOW000
9-FLUORO-11-β,17,21-TRIHYDROXY-16-α-METHYLPREGNA-1,4-DIENE-3,20-DIONE ACETATE
see DBC400
9-FLUORO-11-β,17,21-TRIHYDROXY-16-α-METHYLPREGNA-1,4-DIENE-3,20-DIONE-21-(DIHYDROGEN PHOSPHATE) DISODIUM SALT see
DAE525
FLUOROTROJCHLOROMETAN (POLISH) see TIP500
FLUOROURACIL see FMM000
5-FLUOROURACIL see FMM000
5-FLUOROURACIL DEOXYRIBOSIDE see DAR400
5-FLUOROURACIL-2'-DEOXYRIBOSIDE see DAR400
FLUOROWODOR (POLISH) see HHU500
FLUOROXENE see TKB250
FLUORPLAST 4 see TAI250
5-FLUORPROPYPYRIMIDINE-2,4-DIONE see FMM000
FLUORSPAR see CAS000
5-FLUORURACIL (GERMAN) see FMM000
FLUORURE de BORE (FRENCH) see BMG700
FLUORURE de N,N'-DIISOPROPYLE
PHOSPHORODIAMIDE (FRENCH) see PHF750
FLUORURE de POTASSIUM (FRENCH) see PLF500
FLUORURES ACIDE (FRENCH) see FEZ000
FLUORURE de SODIUM (FRENCH) see SHF500
FLUORURE de SULFURYLE (FRENCH) see SOU500
FLUORURE de N,N,N',N'-TETRAMETHYLE
PHOSPHORO-DIAMIDE (FRENCH) see BJE750
FLUORURE de THIONYLE (FRENCH) see TFL250
FLUORURI ACIDI (ITALIAN) see FEZ000
FLUORURIDINE DEOXYRIBOSE see DAR400
FLUORWASSERSTOFF (GERMAN) see HHU500
FLUORWATERSTOF (DUTCH) see HHU500
FLUORXENE see TKB250
FLUORYL see CBY750
FLUOSILICATE de AMMONIUM (FRENCH) see
COE000
FLUOSILICATE de MAGNESIUM (FRENCH) see
MAG250
FLUOSILICATE de SODIUM see DXE000
FLUOSILICIC ACID see SCO500
FLUOSTIGMINE see IRF000
FLUOSULFONIC ACID (DOT) see FLZ000
FLUOTITANATE de POTASSIUM (FRENCH) see
PLI000
FLURACIL see FMM000
FLURA-GEL see SHF500

FLURAZEPAM see FMQ000
FLURCARE see SHF500
FLURI see FMM000
FLURIL see FMM000
FLUROXENE see TKB250
FLUTONE see AQX500
FLUVIN see CFY000
FLUX MAAG see NDN000
FLY BAIT GRITS see SOY000
FLY-DIE see DGP900
FLY FIGHTER see DGP900
FLYPEL see DKC800
FM 510 see PJS750
FMA see ABU500
FMA see FEV100
FMA (analytical reagent) see FEV100
FMC 249 see AFR250
FMC-1240 see EEH600
FMC 5273 see PIX250
FMC 5462 see EAQ750
FMC 5488 see CKM000
FMC 10242 see CBS275
FMC 17370 see BEP500
FMC 33297 see AHJ750
FMC 41655 see AHJ750
FM-NTS see CCU250
FNT see NDY500
FOBEX see BCA000
FOCUSAN see TGB475
FOLACIN see FMT000
FOLAN RED B see CMM330
FOLATE see FMT000
FOLBEX see DER000
FOLBEX SMOKE-STRIPS see DER000
FOLCID see CBF800
FOLCYSTEINE see FMT000
FOLETHION see DSQ000
FOLEX see TIG250
FOLIC ACID see FMT000
FOLIC ACID, 4-AMINO- see AMG750
FOLIDOL see PAK000
FOLIDOL M see MNH000
FOLIGAN see ZVJ000
FOLIKRIN see EDV000
FOLIONE see MND275
FOLIPEX see EDV000
FOLISAN see EDV000
FOLLESTRINE see EDV000
FOLLICORMON see EDP000
FOLLICULAR HORMONE see EDV000
FOLLICULAR HORMONE HYDRATE see EDU500
FOLLICULIN see EDV000
FOLLICULINE BENZOATE see EDV000
FOLLICUNODIS see EDV000
FOLLICYCLIN P see EDR000
FOLLIDIENE see DAL600
FOLLIDIENE see DKA600
FOLLIDRIN see EDP000
FOLLIDRIN see EDV000
FOLLORMON see DAL600
FOLOSAN see PAX000
FOLPAN see TIT250
FOLPET see TIT250
FOMAC see HCL000
FOMAC 2 see PAX000
FOMREZ SUL-3 see DBF800
FOMREZ SUL-4 see DDV600
FONATOL see DKA600

FONOFOS see FMU045
FONOLINE see MQV750
FONTARSOL see DFX400
FONURIT see AAI250
FOOD BLUE 1 see FMU059
FOODCOL SUNSET YELLOW FCF see FAG150
FOOD DYE RED No. 104 see ADG250
FOOD RED 5 see HJF500
FOOD RED 6 see FMU080
FOOD RED 7 see FMU080
FOOD RED 16 see CMS238
FOOD RED COLOR No. 105, SODIUM SALT see RMP175
FOOD RED No. 101 see FMU070
FOOD RED No. 102 see FMU080
FOOD RED No. 104 see ADG250
FOOD RED No. 105, SODIUM SALT see RMP175
FOOD STARCH, MODIFIED see FMU100
FOOD YELLOW 3 see FAG150
FOOD YELLOW 4 see FAG140
FOOD YELLOW 5 see FAG140
FOOD YELLOW 6 see FAG150
FOOD YELLOW NO. 4 see FAG140
FORAAT (DUTCH) see PGS000
FORANE 22 see CFX500
FORLIN see BBQ500
FORMAGENE see PAI000
FORMAL see MAK700
FORMAL see MGA850
FORMALDEHYD (CZECH, POLISH) see FMV000
FORMALDEHYDE see FMV000
FORMALDEHYDE, solution (DOT) see FMV000
FORMALDEHYDE BIS(β-CHLOROETHYL) ACETAL see BID750
FORMALDEHYDE CYANOHYDRIN see HIM500
FORMALDEHYDE DIMETHYLACETAL see MGA850
FORMAL GLYCOL see DVR800
FORMALIN see FMV000
FORMALIN 40 see FMV000
FORMALIN (DOT) see FMV000
FORMALINA (ITALIAN) see FMV000
FORMALINE (GERMAN) see FMV000
FORMALINE BLACK C see AQP000
FORMALIN-LOESUNGEN (GERMAN) see FMV000
FORMALITH see FMV000
FORMAMIDE see FMY000
FORMAMIDE, N,N'-(1,4-PIPERAZINEDIYLBIS(2,2,2-TRICHLOROETHYLIDENE))BIS-(8CI,9CI) see TKL100
FORMAMINE see HEI500
FORMATRIX see ECU750
3-FORMAZANTHIOL, 1,5-DIPHENYL- see DWN200
FORMETANATE HYDROCHLORIDE see DSO200
FORMIATE de METHYLE (FRENCH) see MKG750
FORMIATE de PROPYLE (FRENCH) see PNM500
FORMIC ACID see FNA000
FORMIC ACID, ALLYL ESTER see AGH000
FORMIC ACID, AZIDO-, tert-BUTYL ESTER see BQI250
FORMIC ACID, CALCIUM SALT see CAS250
FORMIC ACID, CHLORO-, 2-CHLOROETHYL ESTER see CGU199
FORMIC ACID, CHLORO-, FLUOREN-9-YLMETHYL ESTER see FEI100
FORMIC ACID, CHLORO-, 2-FLUOROETHYL ESTER see FIH100
FORMIC ACID, CHLORO-, OXYDIETHYLENE ESTER see OPO000

FORMIC ACID, CHLORO-, TRICHLOROMETHYL ESTER see TIR920
FORMIC ACID, CHLORO-, 2,4,6-TRICHLOROPHENYL ESTER see TIY800
FORMIC ACID, CINNAMYL ESTER see CMR500
FORMIC ACID, CITRONELLYL ESTER see CMT750
FORMIC ACID-3,7-DIMETHYL-6-OCTEN-1-YL ESTER see CMT750
FORMIC ACID, ETHYL ESTER see EKL000
FORMIC ACID, GERANIOL ESTER see GCY000
FORMIC ACID, HEPTYL ESTER see HBO500
FORMIC ACID, ISOBUTYL ESTER see IIR000
FORMIC ACID, ISOPENTYL ESTER see IHS000
FORMIC ACID, ISOPROPYL ESTER see IPC000
FORMIC ACID, METHYLHYDRAZIDE see FNW000
FORMIC ACID, OCTYL ESTER see OEY100
FORMIC ALDEHYDE see FMV000
FORMIC ANAMMONIDE see HHS000
FORMIC BLACK BA see AQP000
FORMIC BLACK C see AQP000
FORMIC BLACK CW see AQP000
FORMIC BLACK MTG see AQP000
FORMIC BLACK TG see AQP000
FORMIC ETHER see EKL000
FORMIC 2-(4-(5-NITROFURYL)-2-THIAZOLYL)HYDRAZIDE see NDY500
FORMIN see HEI500
FORMOL see FMV000
FORMONITRILE see HHS000
FORMOSA CAMPHOR see CBA750
FORMOSA CAMPHOR OIL see CBB500
FORMOSE OIL OF CAMPHOR see CBB500
FORMOSULFACETAMIDE see SNQ710
FORMOSULFATHIAZOLE see TEX250
FORMOTHION see DRR200
FORMVAR 1285 see AAX250
2-FORMYLAMINO-4-(5-NITRO-2-FURYL)THIAZOLE see NGM500
4-FORMYLCYCLOHEXENE see FNK025
2-FORMYL-3,4-DIHYDRO-2H-PYRAN see ADR500
N-FORMYLDIMETHYLAMINE see DSB000
α-FORMYLETHYLBENZENE see COF000
2-(2-FORMYLHYDRAZINO)-4-(5-NITRO-2-FURYL)THIAZOLE see NDY500
FORMYLIC ACID see FNA000
S-(2-(FORMYLMETHYLAMINO)-2-OXOETHYL)-O,O-DIMETHYLPHOSPHORODITHIOATE see DRR200
N-FORMYL-N-METHYLCARBAMOYLMETHYL-O,O-DIMETHYL PHOSPHORODITHIOATE see DRR200
S-(N-FORMYL-N-METHYLCARBAMOYLMETHYL)-O,O-DIMETHYL PHOSPHORODITHIOATE see DRR200
S-(N-FORMYL-N-METHYLCARBAMOYLMETHYL) DIMETHYL PHOSPHOROTHIOLOTHIONATE see DRR200
1-FORMYL-1-METHYLHYDRAZINE see FNW000
N-FORMYL-N-METHYLHYDRAZINE see FNW000
2-FORMYL-1-METHYLPYRIDINIUM IODIDE OXIME see POS750
2-FORMYL-N-METHYLPYRIDINIUM OXIME IODIDE see POS750
2-FORMYLQUINOXALINE-1,4-DIOXIDE CARBOMETHOXYHYDRAZONE see FOI000
FORMYL TRICHLORIDE see CHJ500
FORMYL VIOLET S4BN see FAG120
FOROTOX see TIQ250
FOR-SYN see BEP500

FORTALGESIC see DOQ400
FORTALIN see DOQ400
FORTECORTIN see SOW000
FORTHION see MAK700
FORTIFLEX 6015 see PJS750
FORTIFLEX A 60/50 see PJS750
FORTIGRO see FOI000
FORTION NM see DSP400
FORTODYL see VSZ100
FORTRACIN see BAC250
FORTRACIN (BACITRACIN-MD) see BAC260
FORTRAL see DOQ400
FORTROL see BLW750
FORTURF see TBQ750
FOSCHLOR see TIQ250
FOSCHLOREM (POLISH) see TIQ250
FOSCHLOR R-50 see TIQ250
FOSDRIN see MQR750
FOS-FALL "A" see BSH250
FOSFAMID see DSP400
FOSFERMO see PAK000
FOSFEX see PAK000
FOSFIVE see PAK000
FOSFONO 50 see EEH600
FOSFORAN TROJ-(1,3-DWUCHLOROIZOPROPYLOWY) (POLISH) see FQU875
FOSFORO BIANCO (ITALIAN) see PHP010
FOSFORO(PENTACHLORURO di) (ITALIAN) see PHR500
FOSFORO(TRICLORURO di) (ITALIAN) see PHT275
FOSFOROWODOR (POLISH) see PGY000
FOSFOROXYCHLORID see PHQ800
FOSFORPENTACHLORIDE (DUTCH) see PHR500
FOSFORTHIOCHLORID see TFO000
FOSFORTRICHLORIDE (DUTCH) see PHT275
FOSFORYN TROJETYLOWY (CZECH) see TJT800
FOSFORYN TROJMETYLOWY (CZECH) see TMD500
FOSFORZUUROPLOSSINGEN (DUTCH) see PHB250
FOSFOTHION see MAK700
FOSFOTION see MAK700
FOSFOTOX see DSP400
FOSFURI di ALLUMINIO (ITALIAN) see AHE750
FOSFURI di MAGNESIO (ITALIAN) see MAI000
FOSGEEN (DUTCH) see PGX000
FOSGEN (POLISH) see PGX000
FOSGENE (ITALIAN) see PGX000
FOSOVA see PAK000
FOSSIL FLOUR see SCH002
FOSTER GRANT 834 see SMQ500
FOSTERN see PAK000
FOSTEX see BDS000
FOSTION MM see DSP400
FOSTOX see PAK000
FOSTRIL see HCL000
FOSVEL see LEN000
FOSVEX see TCF250
FOURAMIEN 2R see ALL750
FOURAMINE see TGL750
FOURAMINE BA see DBO400
FOURAMINE BROWN AP see PPQ500
FOURAMINE D see PEY500
FOURAMINE EG see ALS990
FOURAMINE ERN see NAW500
FOURAMINE J see TGL750
FOURAMINE OP see ALT000
FOURAMINE P see ALT250
FOURAMINE PCH see CCP850

FOURAMINE RS see REA000
FOURAMINE STD see TGM400
FOURNEAU 190 see ABX500
FOURNEAU 933 see BCI500
FOURNEAU 1162 see SNM500
FOURRINE 1 see PEY500
FOURRINE 36 see ALL750
FOURRINE 57 see NEM480
FOURRINE 65 see ALS990
FOURRINE 68 see CCP850
FOURRINE 76 see DBO400
FOURRINE 79 see REA000
FOURRINE 84 see ALT250
FOURRINE 94 see TGL750
FOURRINE 99 see NAW500
FOURRINE BROWN 2R see ALL750
FOURRINE BROWN PR see NEM480
FOURRINE BROWN PROPYL see NEM480
FOURRINE D see PEY500
FOURRINE EG see ALS990
FOURRINE ERN see NAW500
FOURRINE M see TGL750
FOURRINE P BASE see ALT250
FOURRINE PG see PPQ500
FOURRINE SLA see DBO400
FOUR THOUSAND FORTY-NINE see MAK700
FOZALON see BDJ250
FP 4 see PJS750
FPA see FHG000
FR 28 see DXG035
FR 222 see MKA500
FR 300 see PAU500
FRACINE see NGE500
FRADIOMYCIN SULFATE see NCG000
FRAESEOL see ENC000
FRAGIVIX see BBJ500
FRAMBINONE see RBU000
FRAMED see BJP000
FRANKINCENSE GUM see OIM000
FRANKLIN see CAO000
FRANOCIDE see DIW200
FRANOZAN see DIW200
FRANROZE see FLZ050
FRATOL see SHG500
FREE BENZYLPENICILLIN see BDY669
FREE COCONUT OIL see CNR000
FREEMANS WHITE LEAD see LDY000
FRENANTOL see ELL500
FRENCH GREEN see COF500
FRENOHYPON see ELL500
FRENOLON DIFUMARATE see MDU750
FRENTIROX see MCO500
FREON see CFX500
FREON 11 see TIP500
FREON 12 see DFA600
FREON 13 see CLR250
FREON 14 see CBY250
FREON 21 see DFL000
FREON 22 see CFX500
FREON 23 see CBY750
FREON 30 see MJP450
FREON 31 see CHI900
FREON 41 see FJK000
FREON 112 see TBP050
FREON 113 see FOO000
FREON 114 see FOO509
FREON 115 see CJI500
FREON 123 see TJY500

FREON 142 see CFX250
FREON 152 see ELN500
FREON 500 see DFB400
FREON 502 see FOO560
FREON 503 see FOO562
FREON 12B1 see BNA250
FREON 133a see TJY175
FREON 13B1 see TJY100
FREON 142b see CFX250
FREON 12-B2 see DKG850
FREON C-318 see CPS000
FREON F-12 see DFA600
FREON F-23 see CBY750
FREON MF see TIP500
FREON R 112 see TBP050
FREON 113TR-T see FOO000
FRESENIUS D 6 see CCU150
FRESMIN see VSZ000
FRIDERON see RBF100
FRIGEN see CFX500
FRIGEN 11 see TIP500
FRIGEN 12 see DFA600
FRIGEN 22 see CFX500
FRIGEN 114 see FOO509
FRIGEN 113a see FOO000
FRIGIDERM see FOO509
FRISIUM see CIR750
FROWNCIDE see CEX800
FRP 53 see PAU500
FRUCOTE see BPY000
FRUCTOSE (FCC) see LFI000
FRUITDO see BLC250
FRUITONE see NAK500
FRUITONE T see TIX500
FRUIT RED A EXTRA YELLOWISH GEIGY see HJF500
FRUIT SUGAR see LFI000
FRUMIN AL see DXH325
FRUMIN G see DXH325
FRUSEMIDE see CHJ750
FRUSEMIN see CHJ750
FRUSID see CHJ750
FRUSTAN see DCK759
FRUTABS see LFI000
FS 74 see SCK600
FTAALZUURANHYDRIDE (DUTCH) see PHW750
F1-TABS see SHF500
FTAFLEX DIBA see DNH125
FTALAN see TIT250
FTALOPHOS see PHX250
FTALOWY BEZWODNIK (POLISH) see PHW750
FTORAFUR see FLZ050
FTORLON 4 see TAI250
FTOROPLAST 4 see TAI250
5-FU see FMM000
p-FUCHSIN see RMK020
FUCHSINE DR-001 see RMK020
FUCHSINE SPC see RMK020
FUCLASIN see BJK500
FUCLASIN ULTRA see BJK500
FUDR see DAR400
5-FUDR see DAR400
FUEL OIL see FOP000
FUEL OIL NO. 2 (DOT) see DHE800
FUEL OIL, RESIDUAL see FOP200
FUELS, DIESEL see DHE900
FUELS, DIESEL see FOP000
FUELS, DIESEL, NO. 2 see DHE850
FUGU POISON see FOQ000

FUKI-NO-TOH (JAPANESE) see PCR000
FUKLASIN see BJK500
FUKLASIN ULTRA see FAS000
FULAID see FLZ050
FULCIN see GKE000
FULCINE see GKE000
FULFEEL see FLZ050
FUL-GLO see FEW000
FULLY HYDROGENATED RAPESEED OIL see
RBK200
FULMINATE of MERCURY see MDC000
FULMINATE of MERCURY (dry) (DOT) see MDC000
FULMINATES see FOS000
FULMINATING MERCURY (DOT) see MDC000
FULMINIC ACID see FOS050
FULSIX see CHJ750
FULUVAMIDE see CHJ750
FULVICAN GRISACTIN see GKE000
FULVICIN see GKE000
FULVINA see GKE000
FULVINE see FOT000
FULVISTATIN see GKE000
FUMAFER see FBJ100
FUMAGON see DDL800
FUMAR-F see FBJ100
FUMARIC ACID see FOU000
FUMARIC ACID, DIBUTYL ESTER see DEC600
FUMAROYL CHLORIDE see FOY000
FUMARYLCHLORID (CZECH) see FOY000
FUMARYL CHLORIDE see FOY000
FUMAZONE see DDL800
FUMED SILICA see SCH002
FUMED SILICON DIOXIDE see SCH002
FUMETOBAC see NDN000
FUMIGANT-1 (OBS.) see MHR200
FUMIGRAIN see ADX500
FUMING LIQUID ARSENIC see ARF500
FUMING SULFURIC ACID see SOI520
FUMIRON see FBJ100
FUMITOXIN see AHE750
FUMO-GAS see EIY500
FUNDAL see CJJ250
FUNDAL 500 see CJJ250
FUNDASOL see BAV575
FUNDEX see CJJ250
FUNDUSCEIN see FEW000
FUNGACETIN see THM500
FUNGAFLOR see FPB875
FUNGICIDE 1991 see BAV575
FUNGICIDE FX see MJM500
FUNGICLOR see PAX000
FUNGIFEN see PAX250
FUNGILIN see AOC500
FUNGIMAR see CNO000
FUNGINEX see TKL100
FUNGISONE see AOC500
FUNGISTOP see TGB475
FUNGITOX OR see ABU500
FUNGIVIN see GKE000
FUNGIZONE see AOC500
FUNGIZONE INTRAVENOUS see FPC200
FUNGOL see ZIA000
FUNGOL B see SHF500
FUNGONIT GF 2 see ZIA000
FUNGOSTOP see BJK500
FUNGUS BAN TYPE II see CBG000
FURACILLIN see NGE500
FURACIN see SPC500

FURACINETTEN see NGE500
FURACOCCID see NGE500
FURACORT see NGE500
FURACYCLINE see NGE500
FURADAN see CBS275
FURADANTIN see NGE000
FURADONIN see NGE000
FURAFLUOR see FLZ050
FURAL see FPQ875
2-FURALDEHYDE see FPQ875
FURALDON see NGE500
FURALE see FPQ875
l-FURALTADONE HYDROCHLORIDE see FPI150
FURAN see FPK000
FURAN (DOT) see FPK000
2-FURANALDEHYDE see FPQ875
2-FURANCARBINOL see FPU000
2-FURANCARBONAL see FPQ875
2-FURANCARBOXALDEHYDE see FPQ875
FURAN-α-CARBOXYLIC ACID METHYL ESTER see
MKH600
2,5-FURANDIONE see MAM000
2,5-FURANDIONE, 3-(DODECENYL)DIHYDRO- see
DXV000
FURANIDINE see TCR750
FURANIUM see FEW000
2-FURANMETHANETHIOL see FPM000
2-FURANMETHANOL see FPU000
2-FURANMETHYLAMINE see FPW000
FURAN, 2,2′-METHYLENEBIS- see FPX028
FURAN, 2,2′-METHYLENEDI- see FPX028
FURAN, 2-(2-NITROVINYL)- see FQM100
FURAN-OFTENO see NGE500
FURANTHRIL see CHJ750
FURANTHRYL see CHJ750
FURANTOIN see NGE000
FURANTRIL see CHJ750
1-(3-FURANYL)-4-METHYL-1-PENTANONE see
PCI750
3-FURANYLPHENYLMETHANONE see FQO050
FURAPLAST see NGE500
FURASEPTYL see NGE500
FURATOL see SHG500
FURATONE see BKH500
FURATONE-S see BKH500
FURAXONE see NGG500
FURAZOL see NGG500
FURAZOLIDON see NGG500
FURAZOLIDONE (USDA) see NGG500
FURAZON see NGG500
FURAZONE see NGE500
FUR BLACK 41867 see PEY500
FUR BROWN 41866 see PEY500
FURCELLERAN GUM see FPQ000
FURESIS see CHJ750
FURESOL see NGE500
FURFURAL see FPQ875
2-FURFURAL see FPQ875
FURFURAL ALCOHOL see FPU000
FURFURALDEHYDE see FPQ875
FURFURALE (ITALIAN) see FPQ875
FURFURAN see FPK000
FURFURIN see NGE500
FURFUROL see FPQ875
FURFUROLE see FPQ875
FURFURYL ALCOHOL see FPU000
2-FURFURYLALKOHOL (CZECH) see FPU000

FURFURYLAMINE see FPW000
2-(2-FURFURYL)FURAN see FPX028
FURFURYL MERCAPTAN see FPM000
FURIDIAZINE see NGI500
FURIDON see NGG500
2-FURIL-METANALE (ITALIAN) see FPQ875
FURMETHONOL see FPI150
FURNAL see CBT750
FURNEX see CBT750
FURNEX N 765 see CBT750
FUROBACTINA see NGE000
FURODAN see CBS275
FURO(3,4-E)-1,3-BENZODIOXOL-8(6H)-ONE, 6-
(5,6,7,8-TETRAHYDRO-6-METHYL-1,3-DIOXOLO(4,5-
G) ISOQUINOLIN-5-YL)-, (R-(R*,R*))- see CBF550
FUROFUTRAN see FLZ050
2-FUROIC ACID, METHYL ESTER see MKH600
FUROLE see FPQ875
α-FUROLE see FPQ875
FUROSEDON see CHJ750
FUROSEMID see CHJ750
FUROSEMIDE see CHJ750
FUROSEMIDE "MITA" see CHJ750
FUROVAG see NGG500
FUROX see NGG500
FUROXAL see NGG500
FUROXANE see NGG500
FUROXONE SWINE MIX see NGG500
FUROZOLIDINE see NGG500
FURRO D see PEY500
FURRO EG see ALS990
FURRO ER see NAW500
FURRO L see DBO000
FURRO P BASE see ALT250
FURRO SLA see DBO400
FURSEMID see CHJ750
FURSEMIDE see CHJ750
FUR YELLOW see PEY500
FURYL ALCOHOL see FPU000
FURYLAMIDE see FQN000
2-FURYLCARBINOL see FPU000
α-FURYLCARBINOL see FPU000
FURYLFURAMIDE see FQN000
2-FURYL p-HYDROXYPHENYL KETONE see FQL200
β-FURYL ISOAMYL KETONE see PCI750
2-FURYL-METHANAL see FPQ875
(2-FURYL)METHANOL see FPU000
1-(2-FURYL)METHYLAMINE see FPW000
1-(3-FURYL)-4-METHYL-1-PENTANONE see PCI750
2-FURYL-1-NITROETHENE see FQM100
α-2-FURYL-5-NITRO-2-FURANACYRLAMIDE see
FQN000
2-(2-FURYL)-3-(5-NITRO-2-FURYL)ACRYLAMIDE see
FQN000
2-(2-FURYL)-3-(5-NITRO-2-FURYL)ACRYLIC ACID
AMIDE see FQN000
α-(FURYL)-β-(5-NITRO-2-FURYL)ACRYLIC AMIDE see
FQN000
3-FURYL PHENYL KETONE see FQO050
FUSARENONE X see FQR000
FUSARIC ACID see BSI000
FUSARINIC ACID see BSI000
FUSARIOTOXIN T 2 see FQS000
FUSED BORAX see DXG035
FUSED BORIC ACID see BMG000
FUSED QUARTZ see SCK600
FUSED SILICA see SCK600

FUSELEX see SCK600
FUSELEX RD 120 see SCK600
FUSELEX RD 40-60 see SCK600
FUSELEX ZA 30 see SCK600
FUSELOEL (GERMAN) see FQT000
FUSEL OIL see FQT000
FUSEL OIL, REFINED (FCC) see FQT000
FUSID see CHJ750
FUSSOL see FFF000
FUTRAFUL see FLZ050
FUTRAMINE D see PEY500
FUTRAMINE EG see ALS990
FUVACILLIN see NGE500
FUXAL see SNN300
FW 293 see BIO750
FW 734 see DGI000
FW 925 see DFT800
FYDALIN see BNK000
FYDE see FMV000
FYFANON see MAK700
FYROL CEF see CGO500
FYROL FR 2 see FQU875
FYROL HB32 see TNC500
FYRQUEL 150 see TNP500
FYTIC ACID see PIB250
G-0 see FQM100
G 0 see HDG000
G 1 see HIM000
G 4 see MJM500
G-11 see HCL000
G 50 see EHG100
G 200 see EHG100
G 301 see DCM750
G 338 see DER000
G 572 see CJR550
G 996 see CDS125
G-1029 see AIH000
G 22150 see DLH630
G 22355 see DLH630
G 23992 see DER000
G-24480 see DCM750
G 28364 see DEL400
G-29288 see MQH750
G 30027 see ARQ725
G 34161 see BKL250
G 35020 see DLS600
GA see EIF000
GA see GEM000
GADEXYL see MQU750
GAFCOL EB see BPJ850
GAFCOTE see PKQ059
GALACTASOL see GLU000
GALACTASOL A see SLJ500
GALACTICOL see DDJ000
α-GALACTOSIDASE see GAV050
4-(β-d-GALACTOSIDO)-d-GLUCOSE see LAR000
GALECRON see CJJ250
GALENA see LDZ000
GALFER see FBJ100
GALLIC ACID, DODECYL ESTER see DXX200
GALLIC ACID, LAURYL ESTER see DXX200
GALLIC ACID, PROPYL ESTER see PNM750
GALLIUM ARSENIDE see GBK000
GALLIUM MONOARSENIDE see GBK000
GALLIUM-NICKEL ALLOY see NDD500
GALLOCHROME see MCV000
GALLOGAMA see BBQ500

GALLOTANNIC ACID see TAD750
GALLOTANNIN see TAD750
GALLOTOX see ABU500
GALLOXON see DFH600
GALOFAK see BDY669
GALOXANE see DFH600
GALOZONE see CCL250
GALVATOL 1-60 see PKP750
GAMACID see BBQ500
GAMAPHEX see BBQ500
GAMAQUIL see PGA750
GAMASERPIN see RDK000
GAMASOL 90 see DUD800
GAMENE see BBQ500
GAMISO see BBQ500
GAMMA-COL see BBQ500
GAMMACORTEN see SOW000
GAMMAHEXA see BBQ500
GAMMAHEXANE see BBQ500
GAMMALIN see BBQ500
GAMMA OH see HJS500
GAMMASERPINE see RDK000
GAMMEXANE see BBP750
GAMMOPAZ see BBQ500
GAMONIL see CBM750
GAMOPHENE see HCL000
GANEAKE see ECU750
GANEX P 804 see PKQ250
GANIDAN see AHO250
GANOZAN see CHC500
GANPHEN see PMI750
GANSIL see CDP000
GANTANOL see SNK000
GANTRISINE see SNN500
GARAMYCIN see GCO000
GARAMYCIN see GCS000
GARANTOSE see BCE500
GARDENAL SODIUM see SID000
GARDENTOX see DCM750
GARDEPANYL see EOK000
GARNITAN see DGD600
GAROX see BDS000
GARRATHION see TNP250
GARVOX see DQM600
GAS-FURNACE BLACK see CBT750
GAS OIL see DHE800
GAS OIL (DOT) see DHE800
GAS OILS (petroleum), light vacuum see GBW025
GASOLINE see GBY000
GASOLINE, UNLEADED see GCE100
GASTEX see CBT750
GASTRINIDE see TEH500
GASTRIPON see PEM750
GASTRODYN see GIC000
GAULTHERIA OIL, ARTIFICIAL see MPI000
GB see IPX000
GBL see DWT600
GBS see SEG800
GC-1106 see HCL500
GC 3707 see SOY000
GC 4072 see CDS750
GC 6936 see ABX250
GC 8993 see CLU000
GC 10000 see SCQ000
GC 3944-3-4 see PAX000
GDUE see OMM300
Ge 132 see CCF125
GEARPHOS see MNH000

GEARPHOS see PAK000
GECHLOREERDEDIFENYL (DUTCH) see PJM500
GEDEX see SMQ500
GEIGY 338 see DER000
GEIGY 13005 see DSO000
GEIGY 24480 see DCM750
GEIGY 27,692 see BJP000
GEIGY 30,027 see ARQ725
GEIGY-BLAU 536 see BGT250
GEIGY G-23611 see DSK200
GEIGY G-28029 see PDC750
GEIGY G-29288 see MQH750
GELACILLIN see BDY669
GELAN I see EIF000
GELATIN see PCU360
GELATINE see PCU360
GELATINE DYNAMITE see DYG000
GELATIN FOAM see PCU360
GELATINS see PCU360
L-GELB 2 see FAG140
GELBER PHOSPHOR (GERMAN) see PHP010
GELBIN see CAP500
GELBIN YELLOW ULTRAMARINE see CAP750
GELBORANGE-S see FAG150
GELCARIN see CCL250
GELCARIN HMR see CCL250
G-ELEVEN see HCL000
GELFOAM see PCU360
GEL II see SHF500
GELOCATIL see HIM000
GELOSE see AEX250
GELOZONE see CCL250
GELSTAPH see SLJ000
GELTABS see VSZ100
GELUCYSTINE see CQK325
GELUTION see SHF500
GELVA CSV 16 see AAX250
GELVATOL see PKP750
GELVATOL 1-30 see PKP750
GELVATOL 1-90 see PKP750
GELVATOL 3-91 see PKP750
GELVATOL 20-30 see PKP750
GELVATOL 2090 see PKP750
GENACORT see CNS750
GENACRON YELLOW G see AAQ250
GENAZO RED KB SOLN see CLK225
GENDRIV 162 see GLU000
GENEP EPTC see EIN500
GENERAL CHEMICALS 1189 see KEA000
GENERAL CHEMICALS 3707 see SOY000
GENERAL CHEMICALS 8993 see CLU000
GENETRON 11 see TIP500
GENETRON 12 see DFA600
GENETRON 13 see CLR250
GENETRON 21 see DFL000
GENETRON 22 see CFX500
GENETRON 23 see CBY750
GENETRON 100 see ELN500
GENETRON 101 see CFX250
GENETRON 112 see TBP050
GENETRON 113 see FOO000
GENETRON 114 see FOO509
GENETRON 115 see CJI500
GENETRON 316 see FOO509
GENETRON 1113 see CLQ750
GENETRON 133a see TJY175
GENETRON 142b see CFX250
GENIPHENE see CDV100

GENISIS see ECU750
GENISTEIN see GCM350
GENISTEOL see GCM350
GENISTERIN see GCM350
GENITE see DFY400
GENITHION see PAK000
GENITOL see DFY400
GENITOX see DAD200
GENITRON BSH see BBS300
GENO-CRISTAUZ GREMY see TBF500
GENOPTIC see GCS000
GENOPTIC S.O.P. see GCS000
GENOTHERM see PKQ059
GENOXAL see CQC500
GENOXAL see CQC650
GENTAMYCIN see GCO000
GENTAMYCIN see GCO000
GENTAMYCIN-CREME (GERMAN) see GCO000
GENTAMYCIN SULFATE see GCS000
GENTERSAL see AOR500
GENTIANAVIOLETT see AOR500
GENTIAN VIOLET see AOR500
GENTIAVERM see AOR500
GENTICID see AOR500
GENTIOLETTEN see AOR500
GENTISATE see GCU000
GENTISIC ACID see GCU000
GENTRAN see DBD700
GENTRON 142B see CFX250
GENU see CCL250
GENUGEL see CCL250
GENUGEL CJ see CCL250
GENUGOL RLV see CCL250
GENUINE ACETATE CHROME ORANGE see LCS000
GENUINE ORANGE CHROME see LCS000
GENUINE PARIS GREEN see COF500
GENUVISCO J see CCL250
GENVIS see SLJ500
GEOCARB 50EC see MOV000
GEOMYCIN see HOH500
GEON see PKQ059
GEON 135 see AAX175
GEON LATEX 151 see PKQ059
GERANIOL ACETATE see DTD800
GERANIOL ALCOHOL see DTD000
GERANIOL EXTRA see DTD000
GERANIOL (FCC) see DTD000
GERANIOL FORMATE see GCY000
GERANIOL TETRAHYDRIDE see DTE600
GERANIUM CRYSTALS see PFA850
GERANIUM OIL see GDA000
GERANIUM OIL ALGERIAN TYPE see GDA000
GERANIUM OIL, EAST INDIAN TYPE see PAE000
GERANIUM OIL, TURKISH TYPE see PAE000
GERANOXY ACETALDEHYDE see GDM100
GERANYL ACETATE (FCC) see DTD800
GERANYL ACETONE see GDE400
GERANYL ALCOHOL see DTD000
GERANYL BENZOATE see GDE800
GERANYL ETHYL ETHER see GDG100
GERANYL FORMATE (FCC) see GCY000
GERANYL ISOVALERATE see GDK000
GERANYL OXYACETALDEHYDE see GDM100
GERANYL PHENYLACETATE see GDM400
GERANYL PROPIONATE see GDM450
GERANYL α-TOLUATE see GDM400
GERFIL see PMP500

GERISON see SNM500
ANTI-GERM 77 see BEN000
GERMACRONE see GDO200
GERMAIN'S see CBM750
GERMALGENE see TIO750
GERMA-MEDICA see HCL000
GERMANATE(2-), BIS(2-CARBOXYLATOETHYL)TRIOXODI-, DIHYDROGEN (9CI) see CCF125
GERMAN CHAMOMILE OIL see CDH500
GERMANE (DOT) see GEI100
GERMANE, PROPYLTRIIODO- see TKR050
GERMANE, TRIIODOPROPYL- see TKR050
GERMANIA see GEC000
GERMANIC ACID see GEC000
GERMANIC OXIDE (crystalline) see GDS000
GERMANIUM BROMIDE see GDW000
GERMANIUM CHLORIDE see GDY000
GERMANIUM COMPOUNDS see GEA000
GERMANIUM DIOXIDE see GEC000
GERMANIUM HYDRIDE see GEI100
GERMANIUM OXIDE see GEC000
GERMANIUM OXIDE (GeO_2) see GEC000
GERMANIUM TETRABROMIDE see GDW000
GERMANIUM TETRACHLORIDE see GDY000
GERMANIUM TETRAHYDRIDE see GEI100
GEROBIT see DBA800
GEROT-EPILAN see MKB250
GEROT-EPILAN-D see DKQ000
GEROVIT see DBA800
GEROX see SLW500
GERTLEY BORATE see SFF000
GERVOT see MDT600
GESAFID see DAD200
GESAGARD see BKL250
GESAMIL see PMN850
GESAPON see DAD200
GESAPRIM see ARQ725
GESARAN see BJP000
GESAREX see DAD200
GESAROL see DAD200
GESATOP see BJP000
GESFID see MQR750
GESOPRIM see ARQ725
GESTID see MQR750
GETTYSOLVE-B see HEN000
GETTYSOLVE-C see HBC500
GEVEX see TGJ050
GH see MJM500
GH 20 see PKP750
GHA 331 see AHC000
GIALLO CROMO (ITALIAN) see LCR000
GIARDIL see NGG500
GIARLAM see NGG500
GIATRICOL see MMN250
GIBBERELLIC ACID see GEM000
GIBBERELLIN see GEM000
GIBBREL see GEM000
GIBS see CAX500
GIB-SOL see GEM000
GIB-TABS see GEM000
GICHTEX see ZVJ000
GIE see DVR600
GIFBLAAR POISON see FIC000
GIHITAN see DCK759
GILOTHERM OM 2 see TBD000
GILUCARD see RDK000
GILUCOR NITRO see NGY000

GIMID see DYC800
GINARSOL see ABX500
GINEFLAVIR see MMN250
GINGER OIL see GEQ000
GINGERONE see VFP100
GIRACID see PDC250
GIRL see CNE750
GIROSTAN see TFQ750
GL 02 see PKP750
GL 03 see PKP750
GLACIAL ACETIC ACID see AAT250
GLACIAL ACRYLIC ACID see ADS750
GLANDUBOLIN see EDV000
GLANDUCORPIN see PMH500
GLASS see FBQ000
GLASS FIBERS see FBQ000
GLAUPAX see AAI250
GLAURAMINE see IBB000
GLAZD PENTA see PAX250
GLEBOFOS see DXH325
GLEEM see SHF500
GLENTONIN-RETARD see CCK125
GLICOL MONOCLORIDRINA (ITALIAN) see EIU800
GLIKOCEL TA see SFO500
GLIMID see DYC800
GLO 5 see PKP750
GLOBENICOL see CDP250
GLOBOID see ADA725
GLOMAX see KBB600
GLONOIN see NGY000
GLONSEN see SHK800
GLOROUS see CDP250
GLOSSO STERANDRYL see MPN500
GLOVER see LCF000
GLUCAL see CAS750
GLUCID see BCE500
GLUCIDORAL see BSM000
GLUCINIUM see BFO750
GLUCINUM see BFO750
GLUCITOL see SKV200
d-GLUCITOL see SKV200
GLUCOBIOGEN see CAS750
GLUCOCHLORAL see GFA000
GLUCOCHLORALOSE see GFA000
α-d-GLUCOCHLORALOSE see GFA000
GLUCODIGIN see DKL800
GLUCO-FERRUM see FBK000
GLUCOFREN see BSM000
GLUCOLIN see GFG000
GLUCOMYLASE see AOM125
GLUCONATE de CALCIUM (FRENCH) see CAS750
GLUCONATO di CALCIO see CAS750
GLUCONATO di SODIO (ITALIAN) see SHK800
d-GLUCONIC ACID, CALCIUM SALT (2:1) (9CI) see CAS750
d-GLUCONIC ACID, CYCLIC ESTER with ANTIMONIC ACID (H₈Sb₂O₉) (2:1),TRISODIUM SALT, NONAHYDRATE see AQH800
d-GLUCONIC ACID, MONOPOTASSIUM SALT (9CI) see PLG800
d-GLUCONIC ACID, 2,4:2',4'-O-(OXYDISTIBYLIDYNE)BIS-, Sb,Sb'-DIOXIDE, TRISODIUM SALT, NONAHYDRATE see AQH800
GLUCONIC ACID POTASSIUM SALT see PLG800
GLUCONIC ACID SODIUM SALT see SHK800
d-GLUCONIC Δ-LACTONE see GFA200
GLUCONOLACTONE see GFA200

GLUCONO-Δ-LACTONE see GFA200
GLUCONSAN K see PLG800
GLUCOPHAGE see DQR600
GLUCOPHAGE LA 6023 see DQR600
4-(α-d-GLUCOPYRANOSIDO)-α-GLUCOPYRANOSE see MAO500
α-D-GLUCOPYRANOSIDURONIC ACID, (3-β,20-β)-20-CARBOXY-11-OXO-30-NOROLEAN-12-EN-3-YL 2-O-β-D-GLUCOPYRANURONOSYL-, AMMONIATE see GIE100
α-d-GLUCOPYRANOSYL β-d-FRUCTOFURANOSIDE see SNH000
(d-GLUCOPYRANOSYLTHIO)GOLD see ART250
GLUCOSE see GFG000
d-GLUCOSE see GFG000
d-GLUCOSE, anhydrous see GFG000
GLUCOSE LIQUID see GFG000
(α-d-GLUCOSIDO)-β-d-FRUCTOFURANOSIDE see SNH000
4-(α-d-GLUCOSIDO)-d-GLUCOSE see MAO500
N-d-GLUCOSYL(2)-N'-NITROSOMETHYLHARNSTOFF (GERMAN) see SMD000
N-d-GLUCOSYL-(2)-N'-NITROSOMETHYLUREA see SMD000
β-d-GLUCOSYLOXYAZOXYMETHANE see COU000
(1-d-GLUCOSYLTHIO)GOLD see ART250
GLUEOPHOGE see DQR600
GLU-P-I see AKS250
GLUMIN see GFO050
GLU-P-2 see DWW700
GLUPAN see TEH500
GLUPAX see AAI250
GLUSATE see GFO000
GLUSIDE see BCE500
GLUTACID see GFO000
GLUTACYL see MRL500
GLUTAMIC ACID see GFO000
l-GLUTAMIC ACID see GFO000
α-GLUTAMIC ACID see GFO000
GLUTAMIC ACID AMIDE see GFO050
GLUTAMIC ACID-5-AMIDE see GFO050
l-GLUTAMIC ACID, MONOPOTASSIUM SALT see MRK500
GLUTAMIC ACID, SODIUM SALT see MRL500
d-GLUTAMIENSUUR see GFO000
GLUTAMINE see GFO050
γ-GLUTAMINE see GFO050
l-GLUTAMINE (9CI, FCC) see GFO050
GLUTAMINIC ACID see GFO000
l-GLUTAMINIC ACID see GFO000
GLUTAMINOL see GFO000
GLUTAMMATO MONOSODICO (ITALIAN) see MRL500
GLUTANON see TEH500
GLUTARAL see GFQ000
GLUTARALDEHYD (CZECH) see GFQ000
GLUTARALDEHYDE see GFQ000
GLUTARDIALDEHYDE see GFQ000
GLUTARIC ACID DINITRILE see TLR500
GLUTARIC DIALDEHYDE see GFQ000
GLUTARODINITRILE see TLR500
GLUTARONITRILE see TLR500
GLUTARONITRILE, 2-BROMO-2-(BROMOMETHYL)- see DDM500
GLUTATHIMID see DYC800

GLUTATHIONE see GFW000
GLUTATHIONE (reduced) see GFW000
GLUTATIOL see GFW000
GLUTATIONE see GFW000
GLUTATON see GFO000
GLUTAVENE see MRL500
GLUTETHIMID see DYC800
GLUTETHIMIDE see DYC800
GLUTETIMIDE see DYC800
GLUTIDE see GFW000
GLUTINAL see GFW000
GLYBUTAMIDE see BSM000
GLYCERIN see GGA000
GLYCERIN, anhydrous see GGA000
GLYCERIN, synthetic see GGA000
GLYCERINE see GGA000
GLYCERINE MONOOLEATE see GGR200
GLYCERINE TRIACETATE see THM500
GLYCERIN-α-MONOCHLORHYDRIN see CDT750
GLYCERIN MONOOLEATE see GGR200
GLYCERIN MONOSTEARATE see OAV000
GLYCERINTRINITRATE (CZECH) see NGY000
GLYCERITE see TAD750
GLYCERITOL see GGA000
GLYCEROL see GGA000
GLYCEROLACETONE see DVR600
GLYCEROL CHLOROHYDRIN see CDT750
GLYCEROL-α-CHLOROHYDRIN see CDT750
GLYCEROL DIACETATE see DBF600
sym-GLYCEROL DICHLOROHYDRIN see DGG400
GLYCEROL-α,β-DICHLOROHYDRIN see DGG450
GLYCEROL α,γ-DICHLOROHYDRIN see DGG400
GLYCEROL DIMETHYLKETAL see DVR600
GLYCEROL EPICHLORHYDRIN see EAZ500
GLYCEROL ESTER of PARTIALLY DIMERIZED
ROSIN see GGA850
GLYCEROL ESTER of POLYMERIZED ROSIN see
GGA865
GLYCEROL ESTER of WOOD ROSIN see GGA875
GLYCEROL-LACTO PALMITATE see GGA900
GLYCEROL-α-MONOCHLOROHYDRIN (DOT) see
CDT750
GLYCEROL MONOOLEATE see GGA925
GLYCEROL MONOOLEATE see GGR200
GLYCEROL MONOSTEARATE see OAV000
GLYCEROL, NITRIC ACID TRIESTER see NGY000
GLYCEROL OLEATE see GGR200
GLYCEROL TRIACETATE see THM500
GLYCEROL TRIBROMOHYDRIN see GGG000
GLYCEROL TRIBUTYRATE see TIG750
GLYCEROL TRICAPRYLATE see TMO000
GLYCEROL TRICHLOROHYDRIN see TJB600
GLYCEROL (TRI(CHLOROMETHYL))ETHER see
GGI000
GLYCEROLTRINITRAAT (DUTCH) see NGY000
GLYCEROL TRINITRATE see NGY000
GLYCEROL (TRINITRATE de) (FRENCH) see NGY000
GLYCEROL TRIOCTANOATE see TMO000
GLYCERYL-α-CHLOROHYDRIN see CDT750
GLYCERYL-1,3-DIACETATE see DBF600
GLYCERYL MONOOLEATE see GGR200
GLYCERYL MONOSTEARATE see OAV000
GLYCERYL NITRATE see NGY000
GLYCERYL OLEATE see GGR200
GLYCERYL TRIACETATE see THM500
GLYCERYL TRIBROMOHYDRIN see GGG000
GLYCERYL TRICHLOROHYDRIN see TJB600

GLYCERYL TRINITRATE see NGY000
GLYCERYL TRIOCTANOATE see TMO000
GLYCERYL TRISTEARATE see GGU400
GLYCIDAL see GGW000
GLYCIDALDEHYDE see GGW000
GLYCIDE see GGW500
GLYCIDOL see GGW500
GLYCIDYL ACRYLATE see ECH500
GLYCIDYL ALCOHOL see GGW500
GLYCIDYLALDEHDYE see GGW000
GLYCIDYL BUTYL ETHER see BRK750
GLYCIDYL ISOPROPYL ETHER see IPD000
GLYCIDYL PHENYL ETHER see PFF360
GLYCIDYL PROPENATE see ECH500
GLYCIDYL-TRIMETHYL-AMMONIUM CHLORIDE
see GGY200
GLYCINE see GHA000
GLYCINE, N,N-BIS(CARBOXYMETHYL)-(9CI) see
AMT500
GLYCINE, N,N-BIS(CARBOXYMETHYL)-, DISODIUM
SALT (9CI) see DXF000
GLYCINE, N,N-BIS(CARBOXYMETHYL)-, SODIUM
SALT (9CI) see SEP500
GLYCINE, N-METHYL-N-NITROSO- see NLR500
GLYCINOL see EEC600
GLYCINONITRILE HYDROCHLORIDE see GHI100
GLYCINONITRILE, MONOHYDROCHLORIDE see
GHI100
GLYCIRENAN see VGP000
GLYCO-FLAVINE see DBX400
GLYCOL see EJC500
GLYCOL ALCOHOL see EJC500
GLYCOL BIS(HYDROXYETHYL) ETHER see TJQ000
GLYCOL BROMIDE see EIY500
GLYCOL BROMOHYDRIN see BNI500
GLYCOL BUTYL ETHER see BPJ850
GLYCOL CHLOROHYDRIN see EIU800
GLYCOL CYANOHYDRIN see HGP000
GLYCOL DIACETATE see EJD759
GLYCOL DIBROMIDE see EIY500
GLYCOL DICHLORIDE see EIY600
GLYCOL DIFORMATE see EJF000
GLYCOL DIMETHYL ETHER see DOE600
GLYCOLDINITRAAT (DUTCH) see EJG000
GLYCOL DINITRATE see EJG000
GLYCOL (DINITRATE DE) (FRENCH) see EJG000
GLYCOL ETHER see DJD600
GLYCOL ETHER de ACETATE see CBQ750
GLYCOL ETHER DB see DJF200
GLYCOL ETHER DB ACETATE see BQP500
GLYCOL-ETHERDIAMINETETRAACETIC ACID see
EIT000
GLYCOL ETHER EB see BPJ850
GLYCOL ETHER EB ACETATE see BPJ850
GLYCOL ETHER EE see EES350
GLYCOL ETHER EE ACETATE see EES400
GLYCOL ETHER EM see EJH500
GLYCOL ETHER EM ACETATE see EJJ500
GLYCOL ETHER PM see PNL250
GLYCOL ETHYLENE ETHER see DVQ000
GLYCOL ETHYL ETHER see DJD600
GLYCOL ETHYL ETHER see EES350
GLYCOL FORMAL see DVR800
GLYCOLIC ACID, PHENYL- see MAP000
GLYCOLIC ACID, THIO- see TFJ100
GLYCOLIC ACID, 2-THIO- see TFJ100
GLYCOLIC NITRILE see HIM500
GLYCOLIXIR see GHA000

GLYCOL METHYL ETHER see EJH500
GLYCOL MONOBUTYL ETHER see BPJ850
GLYCOL MONOBUTYL ETHER ACETATE see
BPM000
GLYCOLMONOCHLOORHYDRINE (DUTCH) see
EIU800
GLYCOL MONOCHLOROHYDRIN see EIU800
GLYCOL MONOETHYL ETHER see EES350
GLYCOL MONOETHYL ETHER ACETATE see
EES400
GLYCOL MONOMETHYL ETHER see EJH500
GLYCOL MONOMETHYL ETHER ACETATE see
EJJ500
GLYCOL MONOMETHYL ETHER ACRYLATE see
MIF750
GLYCOL MONOPHENYL ETHER see PER000
GLYCOL MONOSTEARATE see EJM500
GLYCOLONITRILE see HIM500
GLYCOLS, POLYETHYLENE, DIMETHYL ETHER see
DOM100
GLYCOL STEARATE see EJM500
GLYCOLYLUREA see HGO600
GLYCOMONOCHLORHYDRIN see EIU800
GLYCOMUL O see SKV100
GLYCOMUL S see SKV150
GLYCON S-70 see SLK000
GLYCON DP see SLK000
GLYCONITRILE see HIM500
GLYCON RO see OHU000
GLYCON TP see SLK000
GLYCON WO see OHU000
GLYCOPHEN see GIA000
GLYCOPHENE see GIA000
GLYCOPYRROLATE see GIC000
GLYCOPYRROLATE BROMIDE see GIC000
GLYCOPYRRONIUM BROMIDE see GIC000
GLYCOSOLVE DIP see IOJ500
GLYCOSPERSE L-20 see PKG000
GLYCOSPERSE O-20 see PKL100
GLYCOSPERSE L-20X see PKG000
GLYCOSPERSE L-20X see PKL000
GLYCOSPERSE TS 20 see SKV195
GLYCO STEARIN see HKJ000
GLYCYL ALCOHOL see GGA000
GLYCYRON see GIG000
GLYCYRRHETINIC ACID GLYCOSIDE see GIG000
GLYCYRRHIZIC ACID see GIG000
GLYCYRRHIZIC ACID (8CI) see GIG000
18-β-GLYCYRRHIZIC ACID see GIG000
GLYCYRRHIZIC ACID, AMMONIUM SALT see GIE100
GLYCYRRHIZIN see GIG000
β-GLYCYRRHIZIN see GIG000
GLYCYRRHIZINIC ACID see GIG000
GLYESTRIN see ECU750
GLYKOLDINITRAT (GERMAN) see EJG000
GLYME see DOE600
GLYME-23 see DOM100
GLYMOL see MQV750
GLYODEX 3722 see CBG000
GLYOXAL see GIK000
GLYOXAL, DIMETHYL- see BOT500
GLYOXALINE-5-ALANINE see HGE700
GLYOXALINE-5-ALANINE
MONOHYDROCHLORIDE see HGE800
GLY-OXIDE see HIB500
GLYPED see THM500
GLYPHOSATE see PHA500

GLYSANOL B see ART250
GLYSOLETTEN see EOK000
GLYTAC see GGY200
GLYTAC A 100 see GGY200
GM 14 see PKP750
G-MAC see GGY200
G-M-F see DVR200
GMI see DLS600
GMO 8903 see GGR200
GMP DISODIUM SALT see GLS800
5'-GMP DISODIUM SALT see GLS800
GMP SODIUM SALT see GLS800
GM SULFATE see GCS000
Gn-RH see LIU370
GOAL see OQU100
GOHSENOL see PKP750
GOHSENOL AH 22 see PKP750
GOHSENOL GH see PKP750
GOHSENOL GH 17 see PKP750
GOHSENOL GH 20 see PKP750
GOHSENOL GH 23 see PKP750
GOHSENOL GL 03 see PKP750
GOHSENOL GL 05 see PKP750
GOHSENOL GL 08 see PKP750
GOHSENOL GM 14 see PKP750
GOHSENOL GM 94 see PKP750
GOHSENOL GM 14L see PKP750
GOHSENOL KH 17 see PKP750
GOHSENOL NH 05 see PKP750
GOHSENOL NH 17 see PKP750
GOHSENOL NH 18 see PKP750
GOHSENOL NH 20 see PKP750
GOHSENOL NH 26 see PKP750
GOHSENOL NK 114 see PKP750
GOHSENOL NL 05 see PKP750
GOHSENOL NM 14 see PKP750
GOHSENYL E 50 Y see AAX250
GOLD see GIS000
1721 GOLD see CNI000
GOLD BOND see CCP250
GOLD BRONZE see CNI000
GOLD CHLORIDE see GIW176
GOLD(III) CHLORIDE see GIW176
GOLD DUST see CNE750
GOLDEN ANTIMONY SULFIDE see AQF500
GOLDEN YELLOW see DUX800
GOLD FLAKE see GIS000
GOLD LEAF see GIS000
GOLD ORANGE MP see DOU600
GOLD POWDER see GIS000
GOLD SATINOBRE see LDS000
GOLD SODIUM THIOMALATE see GJC000
GOLD THIOGLUCOSE see ART250
GOLD TRICHLORIDE see GIW176
GOMBARDOL see SNM500
GONACRINE see DBX400
GONADORELIN see LIU370
GONADORELIN ACETATE see LIU305
GONADOTROPIN-RELEASING FACTOR see LIU370
GONADOTROPIN RELEASING HORMONE see
LIU370
GOODRITE 1800X73 see SMR000
GOOD-RITE GP 264 see DVL700
GOPHER BAIT see SMN500
GOPHER-GITTER see SMN500
GORE-TEX see TAI250
GOTAMINE TARTRATE see EDC500
GOTHNION see ASH500

G 2 (OXIDE) see AHE250
GOYL see ABX500
GP 7I see SCK600
GP 11I see SCK600
2100 GP see PJS750
GP-40-66:120 see HCD250
GPKh see HAR000
G 50 (POLYSACCHARIDE) see EHG100
GR see GLS000
GRAAFINA see EDP000
de GRAAFINA see EDP000
GRAFESTROL see DKA600
GRAHAM'S SALT see SII500
GRAIN ALCOHOL see EFU000
GRAIN SORGHUM HARVEST-AID see SFS000
GRAMEVIN see DGI400
GRAMEVIN see DGI600
GRAMISAN see MEP250
GRAMOXONE S see PAI990
GRAMPENIL see AIV500
GRANEX O see SFS000
GRANMAG see MAH500
GRANODINE 80 see ZJS400
GRANODINE 16NC see ZJS400
GRANOSAN see CHC500
GRANOSAN M see EME500
GRANOX NM see HCC500
GRANOX PPM see CBG000
GRANULAR ZINC see ZBJ000
GRANULATED SUGAR see SNH000
GRANUTOX see PGS000
GRAPE COLOR EXTRACT see GJS300
GRAPEFRUIT OIL see GJU000
GRAPEFRUIT OIL, expressed see GJU000
GRAPEFRUIT OIL, coldpressed see GJU000
GRAPE SUGAR see GFG000
GRAPHITE see CBT500
GRAPHITE SYNTHETIC (ACGIH,OSHA) see CBT500
GRAPHLOX see HCE500
GRASAL BRILLIANT RED B see SBC500
GRASAL BRILLIANT RED G see OHI200
GRASAL BRILLIANT YELLOW see DOT300
GRASAL YELLOW see FAG130
GRASAN BRILLIANT RED B see SBC500
GRASAN ORANGE 3R see XRA000
GRASAN ORANGE R see PEJ500
GRASCIDE see DGI000
GRASEX see CDN550
GRASOL BLUE 2GS see TBG700
GRATIBAIN see OKS000
GRATUS STROPHANTHIN see OKS000
GRAVIDOX see PPK250
GRAVINOL see DYE600
GRAVOCAIN see DHK400
GRAVOCAIN HYDROCHLORIDE see DHK600
GRAVOL see DYE600
GRAY ACETATE see CAL750
11091 GREEN see BLK000
11661 GREEN see CMJ900
GREEN CHROME OXIDE see CMJ900
GREEN CHROMIC OXIDE see CMJ900
GREEN CINNABAR see CMJ900
GREEN CROSS CRABGRASS KILLER see PLC250
GREEN DENSIC see SCQ000
GREEN DENSIC GC 800 see SCQ000
GREEN No. 2 see BLK000
GREEN NICKEL OXIDE see NDF500
GREENOCKITE see CAJ750

GREEN OIL see APG500
GREEN OIL see COD750
GREEN ROUGE see CMJ900
GREEN SEAL-8 see ZKA000
GREEN VITRIOL see FBN100
GREEN VITROL see FBO000
GRENADE see GJU600
GRENADES, empty primed (NA0349) (DOT) see GJU600
GRENADES, hand or rifle, with bursting charge (UN0284, UN0285, UN0292, UN0293) (DOT) see GJU600
GRENADES, practice, hand or rifle (UN0452, UN0110, UN0318, UN0372) (DOT) see GJU600
GREOSIN see GKE000
GRESFEED see GKE000
G-RESINS see PJS750
GREX see PJS750
GREX PP 60-002 see PJS750
GREY ARSENIC see ARA750
GRICIN see GKE000
GRIFFEX see ARQ725
GRIFFIN MANEX see MAS500
GRIFULVIN see GKE000
GRILON see PJY500
GRIPPEX see TEH500
GRISACTIN see GKE000
GRISCOFULVIN see GKE000
GRISEFULINE see GKE000
GRISEO see GKE000
(+)-GRISEOFULVIN see GKE000
GRISEOFULVIN-FORTE see GKE000
GRISEOFULVINUM see GKE000
GRISETIN see GKE000
GRISOFULVIN see GKE000
GRISOL see TCF250
GRISOLEN see PJS750
GRISOVIN see GKE000
GRIS-PEG see GKE000
GROCEL see GEM000
GROCO see LGK000
GROCO 2 see OHU000
GROCO 4 see OHU000
GROCO 54 see SLK000
GROCO 5L see OHU000
GROCOLENE see GGA000
GROCOR 5500 see OAV000
GROCOR 6000 see OAV000
GROSAFE see CBT500
GROUNDNUT OIL see PAO000
GROUND VOCLE SULPHUR see SOD500
GRUNDIER ARBEZOL see PAX250
GRYSIO see GKE000
GS 6 see IGK800
G 339 S see CAX350
GS 6244 see FOI000
GSH see GFW000
G-STROPHANTHIN see OKS000
GT see PCU360
GT41 see BOT250
GT-1012 see DNB000
GT 2041 see BOT250
GTG see ART250
GTN see NGY000
GUAIACOL see GKI000
m-GUAIACOL see REF050
GUAICOL see GKI000
GUAMIDE see AHO250
GUANAZOLE see DCF200
GUANICIL see AHO250

GUANIDAN see AHO250
GUANIDINE, CYANO-, METHYLMERCURY deriv. see MLF250
GUANIDINE, N-METHYL-N'-NITRO-N-NITROSO-(9CI) see MMP000
GUANIDINE MONONITRATE see GLA000
GUANIDINE NITRATE (DOT) see GLA000
GUANIDINE, NITRO- see NHA500
GUANIDINE, SULFANILYL- see AHO250
GUANINE see GLI000
GUANINE, 9-β-D-RIBOFURANOSYL- see GLS000
GUANINE RIBOSIDE see GLS000
GUANIOL see DTD000
GUANOSINE see GLS000
GUANYLIC ACID SODIUM SALT see GLS800
GUANYL NITROSAMINO GUANYL TETRAZENE see TEF500
1-GUANYL-4-NITROSAMINOGUANYLTETRAZENE see TEF500
GUANYL NITROSAMINOGUANYLTETRAZENE (dry) (DOT) see TEF500
GUANYL NITROSAMINOGUANYLTETRAZENE, wetted or tetrazene, wetted with not <30% water or (DOT) see TEF500
N₁-GUANYLSULFANILAMIDE see AHO250
GUAR see GLU000
GUARAN see GLU000
GUARANINE see CAK500
GUAR FLOUR see GLU000
GUAR GUM see GLU000
GUATEMALA LEMONGRASS OIL see LEH000
GUDAKHU (INDIA) see SED400
GUESAPON see DAD200
GUESAROL see DAD200
GUICITRINA see AIV500
GUICITRINE see AIV500
GUIGNER'S GREEN see CMJ900
GULITOL see SKV200
l-GULITOL see SKV200
GUM ARABIC see AQQ500
GUM CAMPHOR see CBA750
GUM CARRAGEENAN see CCL250
GUM CHON 2 see CCL250
GUM CHROND see CCL250
GUM CYAMOPSIS see GLU000
GUM GHATTI see GLY000
GUM GUAIAC see GLY100
GUM GUAR see GLU000
GUM OPIUM see OJG000
GUM OVALINE see AQQ500
GUM SENEGAL see AQQ500
GUM STERCULIA see KBK000
GUM TRAGACANTH see THJ250
GUNCOTTON see CCU250
GUNPOWDER see PLL750
GUNPOWDER see ERF500
GUNPOWDER, compressed (UN 0028) (DOT) see PLL750
GUNPOWDER, compressed (UN 0028) (DOT) see ERF500
GUNPOWDER, granular or as a meal (UN 0027) (DOT) see PLL750
GUNPOWDER, granular or as a meal (UN 0027) (DOT) see ERF500
GUNPOWDER, in pellets (UN 0028) (DOT) see PLL750
GUNPOWDER, in pellets (UN 0028) (DOT) see ERF500
GUSATHION see ASH500

GUSATHION A see EKN000
GUSERVIN see GKE000
GUSTAFSON CAPTAN 30-DD see CBG000
GUTHION (DOT) see ASH500
GUTHION (ETHYL) see EKN000
GUTTAGENA see PKQ059
GYNAESAN see EDU500
GYN-ANOVLAR see EEH520
GYNECLORINA see CDP000
GYNECORMONE see EDP000
GYNEFOLLIN see DAL600
GYNERGEN see EDC500
GYNERGON see EDO000
GYNESTREL see EDO000
GYNFORMONE see EDP000
GYNOCHROME see MCV000
GYNOESTRYL see EDO000
GYNOFON see ABY900
GYNOLETT see DKB000
GYNONLAR 21 see EEH520
GYNOPHARM see DKA600
GYNOPLIX see ABX500
GY-PHENE see CDV100
GYPSINE see LCK000
GYPSINE see LCK100
GYPSUM see CAX750
GYPSUM STONE see CAX750
GYROMITRIN see AAH000
GYRON see DAD200
H-34 see HAR000
H 46 see AHC000
H 321 see DST000
H-365 see ELL500
454H see SMQ500
H 520 see RDK000
H-4723 see CIR750
96H60 see DFH600
HACHI-SUGAR see SGC000
HAEMATITE see HAO875
HAEMOFORT see FBO000
HAEMOSTASIN see VGP000
HAFFKININE see ARQ250
HAIMASED see SIA500
HAITIN see HON000
HALAMID see CDP000
HALANE see DFE200
HALARSOL see DFX400
HALBMOND see BAU750
HALF-CYSTEINE see CQK000
HALF-CYSTINE see CQK000
HALF-MUSTARD GAS see CGY750
HALF-MYLERAN see EMF500
HALITE see SFT000
HALIZAN see TDW500
HALLOYSITE see HAF375
HALLTEX see SEH000
HALOCARBON 11 see TIP500
HALOCARBON 14 see CBY250
HALOCARBON 23 see CBY750
HALOCARBON 112 see TBP050
HALOCARBON 113 see FOO000
HALOCARBON 114 see FOO509
HALOCARBON 115 see CJI500
HALOCARBON 112a see TBP000
HALOCARBON 152A see ELN500
HALOCARBON 1132A see VPP000
HALOCARBON C-138 see CPS000
HALOCARBON 13/UCON 13 see CLR250

HALOFUGINONE HYDROBROMIDE see HAF600
HALOMYCETIN see CDP250
HALON see DFA600
HALON 14 see CBY250
HALON 1001 see MHR200
HALON 1011 see CES650
HALON 1202 see DKG850
HALON 1211 see BNA250
HALON 1301 see TJY100
HALON 2001 see EGV400
HALON 10001 see MKW200
HALON TFEG 180 see TAI250
HALOWAX see TBR000
HALOWAX see TIT500
HALOWAX 1014 see HCK500
HALOXON see DFH600
HALSO 99 see CLK100
HALTRON 22 see CFX500
HALVIC 223 see PKQ059
HAMIDOP see DTQ400
HAMILTON RED see CHP500
HAMP-ENE 100 see EIV000
HAMP-ENE 215 see EIV000
HAMP-ENE 220 see EIV000
HAMP-ENE ACID see EIX000
HAMP-ENE Na4 see EIV000
HAMP-EX ACID see DJG800
HAMPSHIRE GLYCINE see GHA000
HAMPSHIRE NTA see SIP500
HAMPSHIRE NTA ACID see AMT500
HANANE see BJE750
HANSACOR see DJS200
HANSAMID see NCR000
HAPLOS see EOK000
HAPPY DUST see CNE750
HAPTOCIL see PPO000
HARE-RID see SMN500
HARMAR see PNA500
HARMOGEN see PIK450
HARMONIN see MQU750
HAROWAX L 9 see GGR200
HARRICAL see CCK125
HARTOL see MQU750
HARVATRATE see MGR500
HARVEN see SGD000
HARVEST-AID see SFS000
HAS (GERMAN) see HLB400
HASETHROL see PBC250
HATCOL DIBP see DNJ400
HATCOL DOP see DVL700
HAVERO-EXTRA see DAD200
HAVIDOTE see EIX000
HAZODRIN see MRH209
m-HBA see HJI100
HBB see HCA500
HBBN see HJQ350
HC see CNS750
HC-3 see HAQ000
HC-58 see CBR300
HCA see HCL500
HCA WEEDKILLER see HCL500
HCB see HCC500
HCBD see HCD250
HC BLUE 1 see BKF250
HCCH see BBP750
HCCH see BBQ500
HCCPD see HCE500
HCE see EBW500

HCH see BBQ500
α-HCH see BBQ000
β-HCH see BBR000
γ-HCH see BBQ500
HCP see ABD000
HCP see HCL000
HC RED NO. 3 see ALO750
HCS 3260 see CDR750
HCTZ see CFY000
HCZ see CFY000
HDEHP see BJR750
HDMTX see MDV500
HD PONCEAU 4R see FMU080
HD SUNSET YELLOW FCF see FAG150
HD SUNSET YELLOW FCF SUPRA see FAG150
HD TARTRAZINE see FAG140
HD TARTRAZINE SUPRA see FAG140
HEARTS see BBK500
HEAVENLY BLUE see DJO000
HEAVY CATALYTICALLY CRACKED DISTILLATE
see DXG810
HEAVY NAPHTHENIC DISTILLATE see MQV780
HEAVY NAPHTHENIC DISTILLATE SOLVENT
EXTRACT see MQV857
HEAVY NAPHTHENIC DISTILLATES (PETROLEUM)
see MQV780
HEAVY OIL see CMY825
HEAVY PARAFFINIC DISTILLATE see MQV785
HEAVY PARAFFINIC DISTILLATE, SOLVENT
EXTRACT see MQV859
HEAVY WATER see HAK000
HEAVY WATER-d2 see HAK000
HEAZLEWOODITE see NDJ500
HEBABIONE HYDROCHLORIDE see PPK500
HEBANIL see CKP500
HEB-CORT see CNS750
HECLOTOX see BBQ500
HECTOGRAPH VIOLET SR see AOR500
HECTO VIOLET R see AOR500
HEDAPUR M 52 see CIR250
HEDEX see HIM000
HEDOLIT see DUS700
HEDONAL see DGB000
HEDONAL DP see DGB000
HEDONAL MCPP see CIR500
HEDP see HKS780
HEKSAN (POLISH) see HEN000
HEKSOGEN (POLISH) see CPR800
HEKTALIN see VGP000
HELFO DOPA see DNA200
HELFOSERPIN see RDK000
HELICON see ADA725
HELIOGEN see CDP000
HELIO RED TONER LCLL see CHP500
HELIOSTABLE BRILLIANT PINK B EXTRA see
RGW000
HELIOTRIDINE ESTER with LASIOCARPUM and
ANGELIC ACID see LBG000
HELIOTROPIN see PIW250
HELIOTROPYL ACETATE see PIX000
HELIUM see HAM500
HELIUM, compressed (UN 1046) (DOT) see HAM500
HELIUM, refrigerated liquid (cryogenic liquid) (UN 1963)
(DOT) see HAM500
HELMETINA see PDP250
HELMIRANE see DFH600
HELMIRON see DFH600

HELMIRONE see DFH600
HELOTHION see SOU625
HEMATIN-PROTEIN see CQM325
HEMATITE see HAO875
HEMATOIDIN see HAO900
HEMETOIDIN see HAO900
HEMICHOLINE see HAQ000
HEMICHOLINIUM-3 see HAQ000
HEMICHOLINIUM see HAP100
HEMICHOLINIUM BROMIDE see HAQ000
HEMICHOLINIUM-3-BROMIDE see HAQ000
HEMICHOLINIUM DIBROMIDE see HAQ000
HEMICHOLINIUM-3-DIBROMIDE see HAQ000
HEMIMELLITENE see TLL500
HEMISINE see VGP000
HEMO-B-DOZE see VSZ000
HEMOCAPROL see AJD000
HEMODAL see MMD500
HEMODESIS see PKQ250
HEMODEX see DBD700
HEMODEZ see PKQ250
HEMOFURAN see NGE500
HEMOMIN see VSZ000
HEMOPAR see AJD000
HEMOSTASIN see VGP000
HEMOSTYPTANON see EDU500
HEMOTON see FBJ100
HEMPA see HEK000
HENDECANAL see UJJ000
HENDECANALDEHYDE see UJJ000
HENDECANE see UJS000
HENDECANOIC ALCOHOL see UNA000
1-HENDECANOL see UNA000
2-HENDECANONE see UKS000
HENDECENAL see ULJ000
HENDECYL ALCOHOL see UNA000
n-HENDECYLENIC ALCOHOL see UNA000
HENNOLETTEN see EOK000
HENU see HKW500
HEOD see DHB400
HEPACHOLINE see CMF750
HEPAGON see VSZ000
HEPAVIS see VSZ000
HEPCOVITE see VSZ000
HEPIN see AJD000
HEPT see TCF250
HEPTACHLOOR (DUTCH) see HAR000
1,4,5,6,7,8,8-HEPTACHLOOR-3a,4,7,7a-TETRAHYDRO-4,7-endo-METHANO-INDEEN (DUTCH) see HAR000
HEPTACHLOR see HAR000
HEPTACHLORE (FRENCH) see HAR000
HEPTACHLOR EPOXIDE (USDA) see EBW500
HEPTACHLORODIBENZO-p-DIOXIN see HAR100
1,2,3,4,6,7,8-HEPTACHLORODIBENZODIOXIN see HAR100
1,2,3,4,6,7,8-HEPTACHLORODIBENZO-p-DIOXIN see HAR100
3,4,5,6,7,8,8-HEPTACHLORODICYCLOPENTADIENE see HAR000
3,4,5,6,7,8,8a-HEPTACHLORODICYCLOPENTADIENE see HAR000
1,4,5,6,7,8,8-HEPTACHLORO-2,3-EPOXY-2,3,3a,4,7,7a-HEXAHYDRO-4,7-METHANOINDENE see EBW500
1,4,5,6,7,8,8-HEPTACHLORO-2,3-EPOXY-3a,4,7,7a-TETRAHYDRO-4,7-METHANOINDAN see EBW500
2,3,4,5,6,7,7-HEPTACHLORO-1a,1b,5,5a,6,6a-HEXAHYDRO-2,5-METHANO-2H-INDENO(1,2-b)OXIRENE see EBW500

1,4,5,6,7,8,8-HEPTACHLORO-3a,4,7,7a-TETRAHYDRO-4,7-ENDOMETHANOINDENE see HAR000
1,4,5,6,7,10,10-HEPTACHLORO-4,7,8,9,-TETRAHYDRO-4,7-ENDOMETHYLENEINDENE see HAR000
1,4,5,6,7,8,8a-HEPTACHLORO-3a,4,7,7a-TETRAHYDRO-4,7-METHANOINDANE see HAR000
1(3a),4,5,6,7,8,8-HEPTACHLORO-3a(1),4,7,7a-TETRAHYDRO-4,7-METHANOINDENE see HAR000
1,4,5,6,7,8,8-HEPTACHLORO-3a,4,7,7a-TETRAHYDRO-4,7-METHANOINDENE see HAR000
1,4,5,6,7,8,8-HEPTACHLORO-3a,4,7,7a-TETRAHYDRO-4,7-METHANOL-1H-INDENE see HAR000
1,4,5,6,7,8,8-HEPTACHLORO-3a,4,7,7,7a-TETRAHYDRO-4,7-METHYLENE INDENE see HAR000
1,4,5,6,7,8,8-HEPTACHLOR-3a,4,7,7,7a-TETRAHYDRO-4,7-endo-METHANO-INDEN (GERMAN) see HAR000
1,1,2,2,3,3,4,4,5,5,6,6,7,7,8,8,8-HEPTADECAFLUORO-1-OCTANESULFONIC ACID see HAS075
1-HEPTADECANECARBOXYLIC ACID see SLK000
2,4-HEPTADIENAL see HAV450
HEPTADIENAL-2,4 see HAV450
trans,trans-2,4-HEPTADIENAL see HAV450
HEPTADONE see MDO750
HEPTAFLUORJODPROPAN see HAY300
HEPTAFLUORMASELNAN STRIBRNY (CZECH) see HAW000
HEPTAFLUOROBUTANOIC ACID, SILVER SALT see HAW000
HEPTAFLUOROBUTYRIC ACID see HAX500
HEPTAFLUOROIODOPROPANE see HAY300
HEPTAGRAN see HAR000
γ-HEPTALACTONE see HBA550
HEPTALDEHYDE see HBB500
n-HEPTALDEHYDE see HBB500
HEPTAMUL see HAR000
HEPTAN (POLISH) see HBC500
HEPTANAL see HBB500
1-HEPTANAMINE see HBL600
HEPTANE see HBC500
n-HEPTANE see HBC500
1-HEPTANECARBOXYLIC ACID see OCY000
HEPTANEN (DUTCH) see HBC500
1-HEPTANETHIOL see HBD500
1-HEPTANOL see HBL500
2-HEPTANOL see HBE500
HEPTANOL-2 see HBE500
n-HEPTANOL see HBL500
n-HEPTANOL-1 (FRENCH) see HBL500
HEPTANOL, FORMATE see HBO500
HEPTANOLIDE-1,4 see HBA550
HEPTANOLIDE-4,1 see HBA550
HEPTANON see MDO750
HEPTAN-3-ON (DUTCH, GERMAN) see EHA600
2-HEPTANONE see MGN500
HEPTAN-3-ONE see EHA600
3-HEPTANONE see EHA600
HEPTAN-4-ONE see DWT600
4-HEPTANONE see DWT600
3-HEPTANONE, 5-METHYL- see EGI750
1-HEPTANONE, 1-(4-PYRIDYL)- see HBF600
4-HEPTANOYLPYRIDINE see HBF600
2,4-HEPTDAIENAL see HAV450
4-HEPTENAL see HBI800
cis-4-HEPTEN-1-AL see HBI800
n-HEPTENE see HBJ000
1-n-HEPTENE see HBJ000

5-HEPTENOIC ACID, 7-(3,5-DIHYDROXY-2-(3-HYDROXY-5-PHENYLPENTYL)CYCLOPENTYL)-,1-METHYLETHYL ESTER, (1R-(1-α(Z),2-β(R*),3-α,5-α))- see XAA500
5-HEPTENOIC ACID, 7-(3,5-DIHYDROXY-2-(3-OXODECYL)CYCLOPENTYL)-, 1-METHYLETHYL ESTER,(1R-(1α(Z),2β,3α,5α))- see IRR050
5-HEPTEN-2-ONE, 6-METHYL- see MKK000
HEPTENYL ACROLEIN see DAE450
HEPTYL ALCOHOL see HBL500
HEPTYLAMINE see HBL600
1-HEPTYLAMINE see HBL600
n-HEPTYLAMINE see HBL600
HEPTYLAMINE, 1-METHYL- see OEK010
HEPTYL CARBINOL see OEI000
HEPTYLDICHLORARSINE see HBN600
1-HEPTYLENE see HBJ000
HEPTYL FORMATE see HBO500
HEPTYL HYDRIDE see HBC500
n-HEPTYL p-HYDROXYBENZOATE see HBP300
HEPTYL MERCAPTAN see HBD500
n-HEPTYLMERCAPTAN see HBD500
n-HEPTYL METHANOATE see HBO500
HEPTYLPARABEN see HBP300
HEPTYL PHTHALATE see HBP400
HEPTYL 4-PYRIDYL KETONE see HBF600
n-HEPTYL-Δ-VALEROLACTONE see HBP450
HEPZIDE see ENV500
HERB-ALL see MRL750
HERBAN M see MRL750
HERBATOX see DXQ500
HERBAX TECHNICAL see DGI000
HERBAZIN see BJP000
HERBEX see BJP000
HERBICIDE 326 see DGD600
HERBICIDE 976 see BMM650
HERBICIDE C-2059 see DUK800
HERBICIDE M see CIR250
HERBICIDES, MONURON see CJX750
HERBICIDES, SILVEX see TIX500
HERBIDAL TOTAL see AMY050
HERBIZOLE see AMY050
HERBOXY see BJP000
HERCOFLAT 135 see PMP500
HERCOFLEX 260 see DVL700
HERCO PRILLS see ANN000
HERCULES 3956 see CDV100
HERCULES 6937 see DJJ393
HERCULES 14503 see DBI099
HERCULES P6 see PBB750
HERCULES TOXAPHENE see CDV100
HERCULON see PMP500
HERKAL see DGP900
HERMAL see TFS350
HERMAT TMT see TFS350
HERMAT ZDM see BJK500
HERMAT Zn-MBT see BHA750
HERMESETAS see BCE500
HEROPON see DBA800
HERPESIL see DAS000
HERPIDU see DAS000
HERPLEX see DAS000
HERPLEX LIQUIFILM see DAS000
HERYL see TFS350
HES see HLB400
HESPANDER see HLB400
HESPANDER INJECTION see HLB400

HESPERIDIN METHYLCHALCONE see HBU400
HET see HCY000
HET ACID see CDS000
HETAMIDE ML see BKE500
HETEROAUXIN see ICN000
HETEROFOS see HBU415
HETEROPHOS see HBU415
HETP see HCY000
HETRAZAN see DIW200
HEUCOPHOS ZP 10 see ZJS400
HEWETEN 10 see CCU150
HEXA see BBP750
HEXABALM see HCL000
HEXABETALIN see PPK500
HEXABROMOBIPHENYL see HCA500
2,4,5,2',4',5'-HEXABROMOBIPHENYL see FBU509
HEXABROMOBIPHENYL (technical grade) see FBU000
1,2,3,4,6,7-HEXABROMONAPHTHALENE see HCA650
HEXABUTYLDISTANNOXANE see BLL750
HEXABUTYLDISTANNTHIANE see HCA700
1,1,1,3,3,3-HEXABUTYLDISTANNTHIANE see HCA700
HEXABUTYLDITIN see BLL750
HEXACAP see CBG000
HEXACARBONYLCHROMIUM see HCB000
HEXACARBONYL CHROMIUM see HCB000
HEXA C.B. see HCC500
HEXACERT YELLOW NO. 5 see FAG140
HEXACHLOR see BBP750
HEXACHLOR-AETHAN (GERMAN) see HCI000
HEXACHLORAN see BBP750
HEXACHLORAN see BBQ500
γ-HEXACHLORAN see BBQ500
α-HEXACHLORANE see BBQ000
γ-HEXACHLORANE see BBQ500
HEXACHLORBENZOL (GERMAN) see HCC500
HEXACHLOR-1,3-BUTADIEN (CZECH) see HCD250
HEXACHLORCYCLOHEXAN (GERMAN) see BBQ000
HEXACHLORCYKLOPENTADIEN (CZECH) see HCE500
1,2,3,4,6,7-HEXACHLORINATED NAPHTHALANE see HCK550
HEXACHLORNAFTALEN see HCK500
HEXACHLOROACETONE (DOT) see HCL500
HEXACHLOROBENZENE see HCC500
β-HEXACHLOROBENZENE see BBR000
γ-HEXACHLOROBENZENE see BBQ500
1,2,3,4,7,7-HEXACHLOROBICYCLO(2.2.1)HEPTEN-5,6-BIOXYMETHYLENESULFITE see EAQ750
α,β-1,2,3,4,7,7-HEXACHLOROBICYCLO(2.2.1)-2-HEPTENE-5,6-BISOXYMETHYLENE SULFITE see EAQ750
HEXACHLOROBUTADIENE see HCD250
1,1,2,3,4,4-HEXACHLORO-1,3-BUTADIENE see HCD250
HEXACHLORO-1,3-BUTADIENE (MAK) see HCD250
HEXACHLOROCYCLOHEXANE see BBP750
1,2,3,4,5,6-HEXACHLOROCYCLOHEXANE see BBP750
α-HEXACHLOROCYCLOHEXANE see BBQ000
β-HEXACHLOROCYCLOHEXANE see BBR000
Δ-HEXACHLOROCYCLOHEXANE see BFW500
Δ-1,2,3,4,5,6-HEXACHLOROCYCLOHEXANE see BFW500
1-α,2-α,3-β,4-α,5-β,6-β-HEXACHLOROCYCLOHEXANE see BBQ000
1-α,2-α,3-β,4-α,5-α,6-β-HEXACHLOROCYCLOHEXANE see BBQ500

1-α,2-α,3-α,4-β,5-α,6-β-
HEXACHLOROCYCLOHEXANE see BFW500
1-α,2-β,3-α,4-β,5-α,6-β-
HEXACHLOROCYCLOHEXANE see BBR000
1,2,3,4,5,6-HEXACHLOROCYCLOHEXANE (mixture of
isomers) see BBQ750
1,2,3,4,5,6-HEXACHLOROCYCLOHEXANE, γ-ISOMER
see BBQ500
γ-HEXACHLOROCYCLOHEXANE (MAK) see BBQ500
α-1,2,3,4,5,6-HEXACHLOROCYCLOHEXANE (MAK)
see BBQ000
β-1,2,3,4,5,6-HEXACHLOROCYCLOHEXANE (MAK)
see BBR000
HEXACHLOROCYCLOPENTADIENE see HCE500
HEXACHLORO-1,3-CYCLOPENTADIENE see HCE500
1,2,3,4,5,5-HEXACHLORO-1,3-CYCLOPENTADIENE
see HCE500
HEXACHLOROCYCLOPENTADIENE
(ACGIH,DOT,OSHA) see HCE500
HEXACHLOROCYCLOPENTADIENEDIMER see
MQW500
1,2,3,4,5,5-HEXACHLORO-1,3-CYCLOPENTADIENE
DIMER see MQW500
1,2,3,6,7,8-HEXACHLORODIBENZO-p-DIOXIN mixed
with 1,2,3,7,8,9-HEXACHLORODIBENZO-p-DIOXIN
see HCF500
1,2,3,7,8,9-HEXACHLORODIBENZO-p-DIOXIN mixed
with 1,2,3,6,7,8-HEXACHLORODIBENZO-p-DIOXIN
see HCF500
2,2',3,3',5,5'-HEXACHLORO-6,6'-
DIHYDROXYDIPHENYLMETHANE see HCL000
HEXACHLORODIPHENYL ETHER see CDV175
HEXACHLORODIPHENYL OXIDE see CDV175
HEXACHLOROEPOXYOCTAHYDRO-endo,exo-
DIMETHANONAPHTHALENE see DHB400
HEXACHLOROEPOXYOCTAHYDRO-endo,endo-
DIMETHANONAPHTHALENE see EAT500
HEXACHLOROETHANE see HCI000
1,1,1,2,2,2-HEXACHLOROETHANE see HCI000
HEXACHLOROETHYLENE see HCI000
1,4,5,6,77-HEXACHLORO-N-(ETHYLMERCURI)-5-
NORBORNENE-2,3-DICARBOXIMIDE see EME050
HEXACHLOROFEN (CZECH) see HCL000
HEXACHLOROHEXAHYDRO-endo-exo-
DIMETHANONAPHTHALENE see AFK250
1,2,3,4,10,10-HEXACHLORO-1,4,4a,5,8,8a-
HEXAHYDRO-1,4,5,8-DIMETHANONAPHTHALENE
see AFK250
1,2,3,4,10,10-HEXACHLORO-1,4,4a,5,8,8a-
HEXAHYDRO-1,4-endo-exo-5,8-
DIMETHANONAPHTHALENE see AFK250
1,2,3,4,10,10-HEXACHLORO-1,4,4a,5,8,8a-
HEXAHYDRO-exo-1,4,-endo-5,8-
DIMETHANONAPHTHALENE see AFK250
1,2,3,4,10,10-HEXACHLORO-1,4,4a,5,8,8a-
HEXAHYDRO-1,4-endo,endo-5,8-
DIMETHANONAPHTHALENE see IKO000
1,2,3,4,10,10-HEXACHLORO-1,4,4a,5,8,8a-
HEXAHYDRO-1,4,5,8-endo,endo-
DIMETHANONAPHTHALENE see IKO000
HEXACHLOROHEXAHYDROMETHANO-2,4,3-
BENZODIOXATHIEPIN-3-OXIDE see EAQ750
6,7,8,9,10,10-HEXACHLORO-1,5,5a,6,9,9a-
HEXAHYDRO-6,9-METHANO-2,4,3-
BENZODIOXATHIEPIN-3-OXIDE see EAQ750
HEXACHLORONAPHTHALENE see HCK500

1,2,3,4,6,7-HEXACHLORONAPHTHALENE see
HCK550
1,2,3,5,6,7-HEXACHLORONAPHTHALENE see
HCK600
1,4,5,6,7,7-HEXACHLORO-5-NORBORNENE-2,3-
DICARBOXYLIC ACID see CDS000
1,4,5,6,7,7-HEXACHLORO-5-NORBORNENE-2,3-
DIMETHANOL CYCLIC SULFITE see EAQ750
3,4,5,6,9,9-HEXACHLORO-1a,2,2a,3,6,6a,7,7a-
OCTAHYDRO-2,7:3,6-DIMETHANONAPHTH(2,3-
b)OXIRENE see DHB400
3,4,5,6,9,9-HEXACHLORO-1a,2,2a,3,6,6a,7,7a-
OCTAHYDRO-2,7:3,6-DIMETHANONAPHTH(2,3-
b)OXIRENE see EAT500
HEXACHLOROPHANE see HCL000
HEXACHLOROPHEN see HCL000
HEXACHLOROPHENE see HCL000
HEXACHLOROPHENE (DOT) see HCL000
HEXACHLOROPLATINATE(2-) DIPOTASSIUM see
PLR000
HEXACHLOROPLATINIC ACID see CKO750
HEXACHLOROPLATINIC(IV) ACID see CKO750
HEXACHLOROPLATINIC(4+) ACID, HYDROGEN-
see CKO750
HEXACHLORO-2-PROPANONE see HCL500
1,1,1,3,3,3-HEXACHLORO-2-PROPANONE see HCL500
HEXACHLOROPROPENE see HCM000
HEXACHLOROPROPYLENE see HCM000
4,5,6,7,8,8-HEXACHLORO-3a,4,7,7a-TETRAHYDRO-4,7-
METHANOINDENE see HCN000
α,α,α,α',α',α'-HEXACHLORO-p-XYLENE see HCM500
α,α'-HEXACHLOROXYLENE see HCM500
4,5,6,7,8,8-HEXACHLOR-Δ^{1,5}-TETRAHYDRO-4,7-
METHANOINDEN see HCN000
HEXACID 698 see HEU000
HEXACID 898 see OCY000
HEXACID 1095 see DAH400
HEXACID C-9 see NMY000
HEXACOL CARMOISINE see HJF500
HEXACOL OIL ORANGE SS see TGW000
HEXACOL ORANGE GG CRYSTALS see HGC000
HEXACOL PONCEAU 4R see FMU080
HEXACOL PONCEAU MX see FMU070
HEXACOL SUNSET YELLOW FCF see FAG150
HEXACOL SUNSET YELLOW FCF SUPRA see FAG150
HEXACOL SUNSET YELLOW FCP see FAG150
HEXACOL SUNSET YELLOW F & F SUPRA see
FAG150
HEXACOL TARTRAZINE see FAG140
HEXACYANOFERRATE(3-) TRIPOTASSIUM see
PLF250
HEXADECADROL see SOW000
1-HEXADECANAMINE see HCO500
HEXADECANOIC ACID see PAE250
HEXADECANOIC ACID, 2-ETHYLHEXYL ESTER
(9CI) see OFG100
HEXADECANOL see HCP000
1-HEXADECANOL see HCP000
HEXADECAN-1-OL see HCP000
n-HEXADECANOL see HCP000
HEXADECENE EPOXIDE see EBX500
2-HEXADECEN-1-OL, 3,7,11,15-TETRAMETHYL-, (R-
(R*,R*-(E)))-(9CI) see PIB600
n-HEXADECOIC ACID see PAE250
HEXADECYL ALCOHOL see HCP000
n-HEXADECYL ALCOHOL see HCP000
N-HEXADECYLAMINE see HCO500

HEXADECYL 2-ETHYLHEXANOATE see HCP550
HEXADECYLIC ACID see PAE250
HEXADECYLPYRIDINIUM CHLORIDE see CCX000
1-HEXADECYLPYRIDINIUM CHLORIDE see CCX000
n-HEXADECYLPYRIDINIUM CHLORIDE see CCX000
HEXADECYLTRICHLOROSILANE see HCQ000
HEXADECYLTRIMETHYLAMMONIUM BROMIDE
see HCQ500
(1-HEXADECYL)TRIMETHYLAMMONIUM BROMIDE
see HCQ500
N-HEXADECYLTRIMETHYLAMMONIUM BROMIDE
see HCQ500
N-HEXADECYL-N,N,N-TRIMETHYLAMMONIUM
BROMIDE see HCQ500
HEXADIENE see HCQ600
1,5-HEXADIENE see HCR500
HEXA-1,5-DIENE see HCR500
HEXADIENIC ACID see SKU000
HEXADIENOIC ACID see SKU000
2,4-HEXADIENOIC ACID see SKU000
trans-trans-2,4-HEXADIENOIC ACID see SKU000
2,4-HEXADIENOIC ACID POTASSIUM SALT see
PLS750
HEXADIONA see DBB200
HEXADRIN see CQC650
HEXADRIN see EAT500
HEXADROL see SOW000
HEXAETHYLDISTANNOXANE see HCX050
1,1,1,3,3,3-HEXAETHYLDISTANNOXANE see HCX050
HEXAETHYLDISTANNTHIANE see HCX100
1,1,1,3,3,3-HEXAETHYLDISTANNTHIANE see HCX100
HEXAETHYLTETRAFOSFAT see HCY000
HEXAETHYL TETRAPHOSPHATE see HCY000
HEXAETHYL TETRAPHOSPHATE, liquid or solid
(DOT) see HCY000
HEXAFEN see HCL000
HEXAFERB see FAS000
HEXAFLUOROACETONE see HCZ000
HEXAFLUOROACETONE HYDRATE see HDA000
HEXAFLUOROEPOXYPROPANE see HDF050
HEXAFLUORO-1,2-EPOXYPROPANE see HDF050
HEXAFLUOROISOPROPANOL see HDC500
HEXAFLUOROKIESELSAEURE (GERMAN) see
SCO500
HEXAFLUOROKIEZELZUUR (DUTCH) see SCO500
HEXAFLUOROPHOSPHORIC ACID see HDE000
1,1,1,3,3,3-HEXAFLUORO-2-PROPANOL see HDC500
HEXAFLUORO-2-PROPANONE HYDRATE see
HDA000
HEXAFLUOROPROPENE see HDF000
HEXAFLUOROPROPENE EPOXIDE see HDF050
HEXAFLUOROPROPENE OXIDE see HDF050
HEXAFLUOROPROPYLENE (DOT) see HDF000
HEXAFLUOROPROPYLENE OXIDE (DOT) see
HDF050
HEXAFLUOROSILICATE(2-) DIHYDROGEN see
SCO500
HEXAFLUOROSILICATE (2-1) LEAD(II) SALT
DIHYDRATE see LDG000
HEXAFLUOROSILICATE(2-) MAGNESIUM (1:1) see
MAF600
HEXAFLUOROSILICATE (2−), NICKEL see NDD000
HEXAFLUORO VANADATE (3-) TRIAMMONIUM
SALT see ANI500
HEXAFLUORURE de SOUFRE (FRENCH) see SOI000
HEXAFLUOSILICIC ACID see SCO500
HEXAFLURATE see PLH500
HEXAFORM see HEI500

HEXAHYDROANILINE see CPF500
HEXAHYDROANILINE HYDROCHLORIDE see
CPA775
HEXAHYDROAZEPINE see HDG000
HEXAHYDRO-1H-AZEPINE see HDG000
HEXAHYDRO-2-AZEPINONE see CBF700
HEXAHYDRO-2H-AZEPIN-2-ONE see CBF700
HEXAHYDRO-2H-AZEPIN-2-ONE HOMOPOLYMER
see PJY500
HEXAHYDROBENZENAMINE see CPF500
HEXAHYDROBENZENE see CPB000
HEXAHYDROBENZOIC ACID AMIDE see CPB050
HEXAHYDROCRESOL see MIQ745
HEXAHYDRO-1,4-DIAZEPINE see HGI900
HEXAHYDRO-1,4-DIAZINE see PIJ000
1,2,3,4,5,6-HEXAHYDRO-6,11-DIMETHYL-3-(3-
METHYL-2-BUTENYL)-2,6-METHANO-3-
BENZAZOCINE see DOQ400
exo-HEXAHYDRO-4,7-METHANOINDAN see TLR675
HEXAHYDROMETHYLPHENOL see MIQ745
HEXAHYDRO-1-NITROSO-1H-AZEPINE see NKI000
HEXAHYDRO-3,6-endo-OXYPHTHALIC ACID see
EAR000
HEXAHYDROPHENOL see CPB750
HEXAHYDROPYRAZINE see PIJ000
HEXAHYDROPYRIDINE see PIL500
HEXAHYDROTHYMOL see MCF750
HEXAHYDROTOLUENE see MIQ740
HEXAHYDRO-1,3,5-TRINITRO-1,3,5-TRIAZIN
(GERMAN) see CPR800
HEXAHYDRO-1,3,5-TRINITRO-1,3,5-TRIAZINE see
CPR800
HEXA(HYDROXYMETHYL)MELAMINE see HDY000
HEXAISOBUTYLDITIN see HDY100
HEXAKIS(μ-ACETATO)-μ⁴-OXOTETRABERYLLIUM
see BFT500
HEXAKIS(μ-ACETATO-O:O')-μ⁴-
OXOTETRABERYLLIUM see BFT500
HEXAKIS(DIHYDROGEN PHOSPHATE) MYO-
INOSITOL see PIB250
HEXAKIS(β,β-
DIMETHYLPHENETHYL)DISTANNOXANE see
BLU000
HEXAKIS(HYDROXYMETHYL)MELAMINE see
HDY000
HEXAKIS(HYDROXYMETHYL)-1,3,5-TRIAZINE-2,4,6-
TRIAMINE see HDY000
HEXAKIS(2-METHYL-2-
PHENYLPROPYL)DISTANNOXANE see BLU000
HEXALDEHYDE (DOT) see HEM000
HEXALIN see CPB750
HEXAMETAPHOSPHATE, SODIUM SALT see SHM500
HEXAMETAPOL see HEK000
HEXAMETHYLBENZENE see HEC000
HEXAMETHYLENAMINE see HEI500
HEXAMETHYLENDIISOKYANAT see DNJ800
HEXAMETHYLENE see CPB000
HEXAMETHYLENEAMINE see HEI500
1,1'-HEXAMETHYLENEBIS(5-(p-
CHLOROPHENYL)BIGUANIDE) see BIM250
HEXAMETHYLENEDIAMINE see HEO000
1,6-HEXAMETHYLENEDIAMINE see HEO000
HEXAMETHYLENEDIAMINE, solid (UN 2280) (DOT)
see HEO000
HEXAMETHYLENEDIAMINE, solution (UN 1783)
(DOT) see HEO000
HEXAMETHYLENE DIISOCYANATE see DNJ800

HEXAMETHYLENE-1,6-DIISOCYANATE see DNJ800
1,6-HEXAMETHYLENE DIISOCYANATE see DNJ800
HEXAMETHYLENE DIISOCYANATE (DOT) see
DNJ800
HEXAMETHYLENE IMINE (DOT) see HDG000
HEXAMETHYLENETETRAAMINE see HEI500
HEXAMETHYLENETETRAMINE see HEI500
HEXAMETHYLENETRIPEROXYDIAMINE see
DCK700
HEXAMETHYLENIMINE see HDG000
HEXAMETHYLENTETRAMIN (GERMAN) see HEI500
2,3,3,4,4,5-HEXAMETHYL-2-HEXANETHIOL see
DXT800
HEXAMETHYLOLMELAMIN (CZECH) see HDY000
HEXAMETHYLOLMELAMINE see HDY000
HEXAMETHYLPARAOSANILINE CHLORIDE see
AOR500
HEXAMETHYLPHOSPHORAMIDE see HEK000
HEXAMETHYLPHOSPHORIC ACID TRIAMIDE
(MAK) see HEK000
HEXAMETHYLPHOSPHORIC TRIAMIDE see HEK000
N,N,N,N,N,N-HEXAMETHYLPHOSPHORIC
TRIAMIDE see HEK000
HEXAMETHYLPHOSPHOROTRIAMIDE see HEK000
HEXAMETHYLPHOSPHOTRIAMIDE see HEK000
HEXAMETHYL-p-ROSANILINE CHLORIDE see
AOR500
HEXAMETHYL-p-ROSANILINE HYDROCHLORIDE
see AOR500
HEXAMETHYL VIOLET see AOR500
HEXAMIC ACID see CPQ625
HEXAMIDINE see DBB200
HEXAMIDINE (the antispasmodic) see DBB200
HEXAMINE (DOT) see HEI500
HEXAMITE see TCF250
HEXAMOL SLS see SIB600
HEXANAL see HEM000
1-HEXANAL see HEM000
HEXANAMIDE see HEM500
1-HEXANAMINE see HFK000
HEXANAPHTHENE see CPB000
n-HEXANE see HEN000
HEXANE (DOT) see HEN000
HEXANEDIAMIDE (9CI) see AEN000
1,6-HEXANEDIAMINE see HEO000
HEXANE, 2,5-DIMETHYL-, 2,5-DIHYDROPEROXIDE
see DSE800
HEXANEDINITRILE see AER250
1,6-HEXANEDIOIC ACID see AEN250
HEXANEDIOIC ACID, BIS(2-ETHYLHEXYL) ESTER
see AEO000
HEXANEDIOIC ACID—DIBUTYL ESTER see AEO750
HEXANEDIOIC ACID, DICYCLOHEXYL ESTER (9CI)
see DGT500
HEXANEDIOIC ACID DINITRILE see AER250
HEXANEDIOIC ACID, DIOCTYL ESTER see AEO000
HEXANEDIOIC ACID, POLYMER with 1,4-
BUTANEDIOL and 1,1'-METHYLENEBIS(4-
ISOCYANATOBENZENE) see PKM250
HEXANEDIOIC ACID, POLYMER with 1,3-
ETHANEDIOL and 1,1'-METHYLENEBIS(4-
ISOCYANATOBENZENE) see PKL750
1,2-HEXANEDIOL see HFP875
1,6-HEXANEDIOL DIISOCYANATE see DNJ800
2,3-HEXANEDIONE see HEQ200
2,5-HEXANEDIONE see HEQ500
2,3-HEXANEDIONE, 5-METHYL- see MKL300

6-HEXANELACTAM see CBF700
HEXA-NEMA see DFK600
HEXANEN (DUTCH) see HEN000
HEXANENITRILE see HER500
HEXANES (FCC) see HEN000
1-HEXANETHIOL see HES000
2,2',4,4',6,6'-HEXANITRODIFENYLAMIN see HET500
HEXANITRODIFENYLAMINE (DUTCH) see HET500
HEXANITRODIPHENYLAMINE see HET500
HEXANITRODIPHENYLAMINE (FRENCH) see
HET500
2,2',4,4',6,6'-HEXANITRODIPHENYLAMINE see
HET500
2,4,6,2',4',6'-HEXANITRODIPHENYLAMINE see
HET500
2,4,6,2',4',6'-HEXANITRODIPHENYLAMINE see
HET500
HEXANITROETHANE see HET675
HEXANITROL see MAW250
HEXANOIC ACID see HEU000
n-HEXANOIC ACID see HEU000
HEXANOIC ACID, BIS(2-ETHOXYETHYL) ESTER see
BJO225
HEXANOIC ACID, BUTYL ESTER see BRK900
HEXANOIC ACID, 2-ETHYL-, HEXADECYL ESTER
see HCP550
HEXANOL see HFJ500
1-HEXANOL see HFJ500
n-HEXANOL see HFJ500
sec-HEXANOL (DOT) see EGW000
tert-HEXANOL (9CI, DOT) see HFJ600
HEXANON see CPC000
2-HEXANONE see HEV000
HEXANONE-2 see HEV000
HEXANONE ISOXIME see CBF700
1-HEXANONE, 1-(4-PYRIDYL)- see PBX800
1-HEXANONE, 1-(4-PYRIDYL)- see HEW050
HEXANONISOXIM (GERMAN) see CBF700
4-HEXANOYLPYRIDINE see PBX800
4-HEXANOYLPYRIDINE see HEW050
HEXAPLAS M/1B see DNJ400
HEXAPLAS M/B see DEH200
HEXAPLAS M/O see ILR100
HEXAPLIN see RDK000
1,1,1,3,3,3-HEXAPROPYLDISTANNOXANE see BLT300
HEXAPROPYLDISTANNTHIANE see HEW200
1,1,1,3,3,3-HEXAPROPYLDISTANNTHIANE see
HEW200
HEXASUL see SOD500
HEXATHIR see TFS350
HEXATOX see BBQ500
HEXATYPE CARMINE B see EOJ500
HEXAUREA CHROMIC CHLORIDE see HEZ800
HEXAVIBEX see PPK500
HEXAVIN see CBM750
HEXAZANE see PIL500
HEXAZINONE see HFA300
HEXAZIR see BJK500
HEXEMAL see TDA500
2-HEXENAL see HFA500
HEX-2-ENAL see HFA500
HEX-2-EN-1-AL see HFA500
trans-2-HEXEN-1-AL see HFA525
5-HEXENAL, 2,6-EPOXY- see ADR500
2-HEXEN-1-AL, 2-ISOPROPYL-5-METHYL- see IKM100
2-HEXEN-1-AL, 5-METHYL-2-(1-METHYLETHYL)- see
IKM100

3-HEXENAMIDE, 2-(HYDROXYIMINO)-6-METHOXY-4-METHYL-5-NITRO-, (2E,3E)- see MKQ600
HEXENE see HFB000
1-HEXENE see HFB000
4-HEXENOIC ACID, 2-ACETYL-5-HYDROXY-3-OXO, Δ-LACTONE, SODIUM derivative see SGD000
2-HEXENOL see HFD500
2-HEXEN-1-OL, (E)- see HFD500
cis-3-HEXENOL see HFE000
trans-2-HEXENOL see HFD500
β-γ-HEXENOL see HFE000
2-HEXEN-1-OL ACETATE see HFE100
γ-HEXENOLACTONE see PAJ500
cis-3-HEXEN-1-OL (FCC) see HFE000
trans-2-HEXEN-1-OL (FCC) see HFD500
2-HEXEN-5,1-OLIDE see PAJ500
D"-HEXENOLLACTONE see PAJ500
4-HEXEN-1-OL, 5-METHYL-2-(1-METHYLETHENYL)-, ACETATE see LCA100
3-HEXEN-1-OL, PROPANOATE (Z)- see HFE650
HEX-2-ENYL ACETATE see HFE100
2-HEXENYL ACETATE see HFE100
2-HEXEN-1-YL-ACETATE see HFE100
(E)-2-HEXENYL ACETATE see HFE100
trans-2-HEXENYL ACETATE see HFE100
β,γ-HEXENYL ISOBUTANOATE see HFE520
cis-3-HEXENYL ISOBUTYRATE see HFE520
cis-3-HEXENYL ISOVALERATE (FCC) see ISZ000
cis-3-HEXENYL 2-METHYLBUTYRATE see HFE550
cis-3-HEXENYL PHENYLACETATE see HFE625
β,γ-HEXENYL PROPANOATE see HFE650
cis-3-HEXENYL PROPIONATE see HFE650
β,γ-HEXENYL α-TOLUATE see HFE625
HEXERMIN see PPK500
HEXICIDE see BBQ500
HEXIDE see HCL000
HEXILMETHYLENAMINE see HEI500
HEXMETHYLPHOSPHORAMIDE see HEK000
HEXOBION see PPK500
HEXOGEEN (DUTCH) see CPR800
HEXOGEN see CPR800
HEXOGEN 5W see CPR800
HEXOGEN (Explosive) see CPR800
HEXOGEN, wetted (UN 0072) (DOT) see CPR800
HEXOGEN, desensitized (UN 0483) (DOT) see CPR800
n-HEXOIC ACID see HEU000
1,6-HEXOLACTAM see CBF700
HEXOLITE see CPR800
HEXOLITE, dry or wetted with <15% water, by weight (UN 0118) (DOT) see CPR800
HEXON (CZECH) see HFG500
HEXONE see HFG500
HEXOPHENE see HCL000
HEXOSAN see HCL000
HEXOXYACETALDEHYDE DIMETHYLACETAL see HFG700
2-HEXOXYACETALDEHYDE DIMETHYLACETAL see HFG700
β-HEXOXYACETALDEHYDE DIMETHYLACETAL see HFG700
HEXYCLAN see BBQ750
HEXYL (GERMAN, DUTCH) see HET500
HEXYL ACETATE see HFI500
1-HEXYL ACETATE see HFI500
sec-HEXYL ACETATE see HFJ000
n-HEXYL ACETATE (FCC) see HFI500

β-HEXYLACROLEIN see NNA300
HEXYL ALCOHOL see HFJ500
n-HEXYL ALCOHOL see HFJ500
sec-HEXYL ALCOHOL see EGW000
tert-HEXYL ALCOHOL see HFJ600
HEXYL ALCOHOL, ACETATE see HFI500
HEXYLAMINE see HFK000
N-HEXYLAMINE see HFK000
HEXYLAN see BBP750
4-HEXYL-1,3-BENZENEDIOL see HFV500
HEXYL BUTANOATE see HFM700
n-HEXYL BUTANOATE see HFM700
n-HEXYL n-BUTANOATE see HFM700
HEXYL-2-BUTENOATE see HFM600
n-HEXYL 2-BUTENOATE see HFM600
HEXYL BUTYRATE see HFM700
1-HEXYL BUTYRATE see HFM700
n-HEXYL BUTYRATE see HFM700
n-HEXYL CARBORANE see HFN500
HEXYL CINNAMALDEHYDE see HFO500
α-HEXYLCINNAMALDEHYDE (FCC) see HFO500
HEXYL CINNAMIC ALDEHYDE see HFO500
α-HEXYLCINNAMIC ALDEHYDE see HFO500
HEXYL CROTONATE see HFM600
2-n-HEXYL-2-CYCLOPENTEN-1-ONE see HFO700
HEXYLDICARBADODECABORANE(12) see HFN500
HEXYLDICHLORARSINE see HFP600
4-HEXYL-1,3-DIHYDROXYBENZENE see HFV500
HEXYLENE see HFB000
HEXYLENE GLYCOL see HFP875
HEXYLENIC ALDEHYDE see HFA500
HEXYL ETHANOATE see HFI500
HEXYL ISOBUTANOATE see HFQ550
n-HEXYL ISOBUTANOATE see HFQ550
HEXYL ISOBUTYRATE see HFQ550
1-HEXYL ISOBUTYRATE see HFQ550
n-HEXYL ISOBUTYRATE see HFQ550
HEXYL ISOVALERATE see HFQ600
HEXYL MERCAPTAN see HES000
HEXYLMERCURIC BROMIDE see HFR100
n-HEXYLMERCURIC BROMIDE see HFR100
HEXYL MERCURY BROMIDE see HFR100
HEXYL 2-METHYLBUTYRATE see HFR200
2-(2-(2-(HEXYLOXY)ETHOXY)ETHOXY)ETHANOL see HFT550
α-n-HEXYL-β-PHENYLACROLEIN see HFO500
HEXYLRESORCIN (GERMAN) see HFV500
4-HEXYLRESORCINE see HFV500
HEXYLRESORCINOL see HFV500
4-HEXYLRESORCINOL see HFV500
p-HEXYLRESORCINOL see HFV500
4-n-HEXYLRESORCINOL see HFV500
HEXYLTRICHLOROSILANE see HFX500
HEYDEFLON see TAI250
HF 10 see SMQ500
HF 55 see SMQ500
HF 77 see SMQ500
HF3170 see DCS200
HFA see FIC000
H-35-F 87 (BVM) see DSQ000
HFDB 4201 see PJS750
HFIP see HDC500
81723 HFU see TET800
Hg 532 see PGA750
H.G. BLENDING see SFT000
HGI see BBQ500
HH 102 see SMQ500

HHDN see AFK250
HHI 11 see SMQ500
HI-ALAZIN see TGB475
HI-A-VITA see VSK600
HIBANIL see CKP500
HIBAWOOD OIL see HGA100
HIBERNA see DQA400
HIBERNAL see CKP500
HIBESTROL see DKA600
HIBISCOLIDE see OKW110
HIBITANE see BIM250
HICO CCC see CMF400
HICOPHOR PR see MCB000
HIDACID AZO RUBINE see HJF500
HIDACID AZURE BLUE see FMU059
HIDACID DIBROMO FLUORESCEIN see BNH500
HIDACID FAST ORANGE G see HGC000
HIDACID FAST SCARLET 3R see FMU080
HIDACID FLUORESCEIN see FEV000
HIDACID METANIL YELLOW see MDM775
HIDACID SCARLET 2R see FMU070
HIDACID URANINE see FEW000
HIDACO BRILLIANT CRYSTAL VIOLET see AOR500
HIDACO BRILLIANT GREEN see BAY750
HIDACO CRYSTAL VIOLET see AOR500
HIDACO METHYLENE BLUE SALT FREE see BJI250
HIDACO OIL ORANGE see PEJ500
HIDACO OIL RED see SBC500
HIDACO OIL YELLOW see AIC250
HIDAN see DKQ000
HIDANTILO see DKQ000
HIDANTINA SENOSIAN see DKQ000
HIDANTINA VITORIA see DKQ000
HIDANTOMIN see DKQ000
HIDAZID TARTRAZINE see FAG140
HI-DERATOL see VSZ100
HIDRALAZIN see HGP495
HIDRALAZIN see HGP500
HIDRIL see CFY000
HIDROCHLORTIAZID see CFY000
HIDRO-COLISONA see CNS750
HIDROESTRON see EDP000
HIDRORONOL see CFY000
HIDROTIAZIDA see CFY000
HI-DRY see TCE250
HI-ENTEROL see CHR500
HIESTRONE see EDV000
HI-FAX see PJS750
HI-FAX 1900 see PJS750
HI-FAX 4401 see PJS750
HI-FAX 4601 see PJS750
HI-FLASH NAPHTHA see NAH600
HIFOL see BIO750
HIGHROSIN see RNU100
HIGILITE see AHC000
HIGOSAN see MEP250
HIGUERETA (CUBA, PUERTO RICO) see CCP000
HIGUERILLA (MEXICO) see CCP000
HIGUEROXYL DELABARRE see FBS000
HI-JEL see BAV750
HILDAN see EAQ750
HILDIT see DAD200
HILITE 60 see DGN200
HILTHION see MAK700
HILTHION 25WDP see MAK700
HILTONIL FAST BLUE B BASE see DCJ200
HILTONIL FAST ORANGE GR BASE see NEO000
HILTONIL FAST ORANGE R BASE see NEN500

HILTONIL FAST RED KB BASE see CLK225
HILTONIL FAST SCARLET G BASE see NMP500
HILTONIL FAST SCARLET GC BASE see NMP500
HILTONIL FAST SCARLET G SALT see NMP500
HILTOSAL FAST BLUE B SALT see DCJ200
HINDASOL BLUE B SALT see DCJ200
HINDASOL RED TR SALT see CLK235
HIOXYL see HIB050
HIPNAX see DLY000
HIPOFTALIN see HGP495
HIPOFTALIN see HGP500
HI-POINT 90 see MKA500
HI-POINT 180 see MKA500
HI-POINT PD-1 see MKA500
HIPOSERPIL see RDK000
HIPPUZON see TEH500
HIPSAL see DLY000
HISERPIA see RDK000
HISHIREX 502 see PKQ059
HISPACET FAST YELLOW G see AAQ250
HISPACID BRILLIANT SCARLET 3RF see FMU080
HISPACID FAST ORANGE 2G see HGC000
HISPACID FAST YELLOW T see FAG140
HISPACID YELLOW MG see MDM775
HISPAMIN BLACK EF see AQP000
HISPAMIN BLUE 2B see CMO000
HISPAMIN BLUE 3BX see CMO250
HISPAMIN CONGO 4B see SGQ500
HISPAMIN SKY BLUE 3B see CMO500
HISPAMIN SKY BLUE 6B see CMN750
HISPAVIC 229 see PKQ059
HISPERSE YELLOW G see AAQ250
HISTABID see SPC500
HISTABUTYZINE DIHYDROCHLORIDE see BOM250
HISTADUR see TAI500
HISTADUR DURA-TABS see TAI500
HISTALEN see TAI500
HISTAMETHINE see HGC500
HISTAMETHIZINE see HGC500
HISTAMETIZINE see HGC500
HISTAMETIZYNE see HGC500
HISTAMINE ACID PHOSPHATE see HGE000
HISTAMINE DICHLORIDE see HGD500
HISTAMINE DIHYDROCHLORIDE see HGD500
HISTAMINE DIPHOSPHATE see HGE000
HISTAMINE PHOSPHATE (1:2) see HGE000
HISTAPAN see TAI500
HISTARGAN see DQA400
HISTATEX see DBM800
HISTAXIN see BBV500
HISTIDINE see HGE700
l-HISTIDINE, N-β-ALANYL- see CCK665
l-HISTIDINE (FCC) see HGE700
HISTIDINE MONOHYDROCHLORIDE see HGE800
HISTOCARB see CBJ000
HISTYRENE S 6F see SMR000
HI-STYROL see SMQ500
HITACERAM SC 101 see SCQ000
HI-YIELD DESICCANT H-10 see ARB250
HIZEX see PJS750
HIZEX 5000 see PJS750
HIZEX 5100 see PJS750
HIZEX 1091J see PJS750
HIZEX 1291J see PJS750
HIZEX 1300J see PJS750
HIZEX 2100J see PJS750
HIZEX 2200J see PJS750

HIZEX 3300F see PJS750
HIZEX 3300S see PJS750
HIZEX 6100P see PJS750
HIZEX 7300F see PJS750
HIZEX 3000B see PJS750
HIZEX 7000F see PJS750
HIZEX 2100LP see PJS750
HIZEX 5100LP see PJS750
HIZEX 3000S see PJS750
HIZEX 5000S see PJS750
H.K. FORMULA No. K. 7117 see FMU059
HL-331 see ABU500
HL 2447 see DFV400
HL 8700 see PMI750
HLS 831 see BSQ000
HMB see HFR100
HMBD see HLX925
HMD see PAN100
HMDA see HEO000
HMDI see DNJ800
h-MG see CGY750
7-HM-12-MBA see HMF000
HMP see SHM500
HMPA see HEK000
HMPT see HEK000
HMT see HEI500
HMX see CQH250
HMX (dry or unphlegmatized) (DOT) see CQH250
HMX, wetted (UN 0226) (DOT) see CQH250
beta HMY see CQH250
HN1 see BID250
HN 1 see CGW000
HN2 see BIE250
HN₂ AMINE OXIDE see CFA500
HN2.HCl see BIE500
HN2 HYDROCHLORIDE see BIE500
HN₂ OXIDE HYDROCHLORIDE see CFA750
HN₂ OXIDE MUSTARD see CFA500
HNU see HKW500
HOCA see DXH250
HOCH see FMV000
HODAG GMS see OAV000
HODAG PSML-20 see PKG000
HODAG SMS see SKV150
HODAG SVO 9 see PKL100
HOE 2,671 see EAQ750
HOE 2747 see CKD500
HOE 2810 see DGD600
HOE-2824 see ABX250
HOE 2872 see CLU000
HOE 17411 see MHC750
HOECHST 10720 see KFK000
HOECHST PA 190 see PJS750
HOECHST WAX PA 520 see PJS750
HOE-S 2617 see DTF820
HOGGAR see BNK000
HOKKO-MYCIN see SLW500
HOKMATE see FAS000
HOLBAMATE see MQU750
HOLIN see EDU500
HOLMIUM CHLORIDE see HGG000
HOLOXAN see IMH000
HOMANDREN see MPN500
HOMANDREN (amps) see TBG000
HOMBITAN see TGG760
HOME HEATING OIL No. 2 see DHE800
#2 HOME HEATING OILS see DHE800
HOMIDIUM BROMIDE see DBV400

HOMOANISIC ACID see MFE250
HOMOLLE'S DIGITALIN see DKN400
HOMOOLAN see HIM000
HOMOPIPERAZINE see HGI900
HOMOPIPERIDINE see HDG000
HOMOSTERONE see TBF500
HOMOVERATRYLAMINE see DOE200
HONEY DIAZINE see PPP500
HONG KIEN see ABU500
HONGKONG ROSIN WW see RNU100
HOOKER NO. 1 CHRYSOTILE ASBESTOS see ARM268
HOPCIDE see CKF000
HOPCIN see MOV000
HOPS OIL see HGK800
HORFEMINE see DKB000
HORMALE see MPN500
HORMATOX see DGB000
HORMEX ROOTING POWDER see ICP000
HORMIT see SGH500
HORMOCEL-2CCC see CMF400
HORMODIN see ICP000
HORMOFEMIN see DAL600
HORMOFLAVEINE see PMH500
HORMOFOLLIN see EDV000
HORMOGYNON see EDP000
HORMOLUTON see PMH500
HORMOMED see EDU500
HORMONIN see EDU500
HORMONISENE see CLO750
HORMOTESTON see TBG000
HORMOTUHO see CIR250
HORMOVARINE see EDV000
HORSE ATRIAL NATRIURETIC PEPTIDE-28 see HGL680
HORSE-CYTOCHROME C see CQM325
HORSE HEAD A-410 see TGG760
HORSE HEART CYTOCHROME C see CQM325
HORTFENICOL see CDP250
HORTICULTURAL SPRAY OIL see MQV855
HOSTACORTIN see PLZ000
HOSTACORTIN see PMA000
HOSTAFLEX VP 150 see AAX175
HOSTAFLON see TAI250
HOSTALEN see PJS750
HOSTALEN GD 620 see PJS750
HOSTALEN GD 6250 see PJS750
HOSTALEN GF 4760 see PJS750
HOSTALEN GF 5750 see PJS750
HOSTALEN GM 5010 see PJS750
HOSTALEN GUR see PJS750
HOSTALEN HDPE see PJS750
HOSTALEN PP see PMP500
HOSTALIT see PKQ059
HOSTAQUICK see ABU500
HOSTETEX L-PEC see TIK250
HOSTYREN N see SMQ500
HOSTYREN N 4000 see SMQ500
HOSTYREN N 7001 see SMQ500
HOSTYREN N 4000V see SMQ500
HOSTYREN S see SMQ500
HOWFLEX GBP see DJD700
β-HPN see HGP000
HPOP see HNX500
HPP see ZVJ000
2-HPPN see NLM500
β-HPPN see NLM500
HPT see HEK000

HR 376 see CIR750
HRS 1276 see MQW500
HRS 1655 see HCE500
HRW 13 see SLJ500
HS see HGW500
HS-119-1 see PEE750
HT 88 see SMQ500
HT 88A see SMQ500
HT 91-1 see SMQ500
HT 972 see MJQ000
HT-400 E 1/8' see CNC250
HT-F 76 see SMQ500
HTH see HOV500
HTP see HCY000
5-HTP see HOO100
l-5-HTP see HOA600
l-5-HTP see HOO000
1-α-H,5-α-H-TROPANE, 3-α-((10,11-DIHYDRO-5H-
DIBENZO(AD)CYCLOHEPTEN-5-YL)OXY)-,
CITRATE (1:1) see EAG100
1-α-H,5-α-H-TROPAN-3-α-OL (±)-TROPATE (ESTER)
see ARR000
1-α-H,5-α-H-TROPAN-3-α-OL (±)-TROPATE (ESTER),
SULFATE (2:1) SALT see ARR500
HUBBUCK'S WHITE see ZKA000
HUBER see CBT750
HUILE d'ANILINE (FRENCH) see AOQ000
HUILE de CAMPHRE (FRENCH) see CBA750
HUILE de FUSEL (FRENCH) see FQT000
HUILE H50 see TFM250
HULS P 6500 see PMP500
HUMAN ATRIAL NATRIURETIC FACTOR (99-126) see
HGL680
HUMAN ATRIAL NATRIURETIC PEPTIDE (99-126) see
HGL680
HUMAN ATRIAL NATRIURETIC PEPTIDE (1-28) (99-
126) see HGL680
HUMAN ATRIOPEPTIN(1-28) see HGL680
HUMAN ATRIOPEPTIN(99-126) see HGL680
HUMAN RECOMBINANT TUMOR NECROSIS
FACTOR-α see HGL920
HUMENEGRO see CBT750
HUMIFEN WT 27G see DJL000
HUNGARIAN CHAMOMILE OIL see CDH500
HUNGAZIN see ARQ725
HUNGAZIN DT see BJP000
HUNGAZIN PK see ARQ725
HUSEPT EXTRA see CLW000
HVA 2 see BKL750
HVA-2 CURING AGENT see BKL750
HW 4 see CQH250
HW 920 see DXQ500
HX 3/5 see CCU250
HXR see IDE000
HY 951 see TJR000
HYACINTH see ZSS000
HYACINTHAL see COF000
HYACINTHIN see BBL500
HYADRINE see BBV500
HYADUR see DUD800
HYAMINE see BEN000
HYAMINE 1622 see BEN000
HYBAR X see UNJ800
HYBERNAL see CKP500
HYCANTHON see LIM000
HYCANTHONE see LIM000
HYCAR LX 407 see SMR000

HY-CHLOR see HOV500
HYCHOTINE see CJR909
HYCLORITE see SHU500
HYCORACE see HHR000
HYDAN see DFE200
HYDAN (antiseptic) see DFE200
HYDANTAL see DKQ000
HYDANTIN SODIUM see DNU000
HYDANTOIN see DKQ000
HYDANTOIN see HGO600
HYDANTOIN SODIUM see DNU000
HYDELTRA see PMA000
HYDELTRONE see PMA000
HYDOUT see EAR000
HYDOXIN see PPK250
HYDRACRYLIC ACID β-LACTONE see PMT100
HYDRACRYLONITRILE see HGP000
HYDRAL see CDO000
HYDRAL 705 see AHC000
HYDRALAZINE see HGP495
HYDRALAZINE CHLORIDE see HGP500
HYDRALAZINE HYDROCHLORIDE see HGP500
HYDRALAZINE MONOHYDROCHLORIDE see
HGP500
HYDRAL de CHLORAL see CDO000
HYDRALIN see CPB750
HYDRALLAZINE see HGP495
HYDRALLAZINE HYDROCHLORIDE see HGP500
HYDRAPHEN see PFN000
HYDRAPRESS see HGP500
HYDRARGAPHEN see PFN000
HYDRARGYRUM BIJODATUM (GERMAN) see
MDD000
HYDRATED ALUMINA see AHC250
HYDRATED LIME see CAT225
HYDRATROP ALDEHYDE see COF000
HYDRATROPIC ALDEHYDE see COF000
HYDRATROPIC ALDEHYDE DIMETHYL ACETAL see
PGA800
HYDRAZIDE BSG see BBS300
HYDRAZID KYSELINY MALEINOVE see DMC600
HYDRAZINE see HGS000
HYDRAZINE, anhydrous (DOT) see HGS000
HYDRAZINE AQUEOUS SOLUTIONS with >64%
hydrazine, by weight (DOT) see HGS000
HYDRAZINE AQUEOUS SOLUTIONS, with not >64%
hydrazine, by weight (DOT) see HGU500
HYDRAZINE-BENZENE see PFI000
HYDRAZINE-BENZENE and BENZIDINE SULFATE
see BBY300
HYDRAZINE BORANE (1:1) see HGU025
HYDRAZINECARBOTHIOAMIDE see TFQ000
HYDRAZINECARBOTHIOAMIDE, 2-((3-AMINO-2-
PYRIDINYL)METHYLENE)- see AMI600
HYDRAZINECARBOTHIOAMIDE, 2-PHENYL-(9CI)
see PGM750
HYDRAZINE CARBOXAMIDE see HGU000
HYDRAZINECARBOXAMIDE
MONOHYDROCHLORIDE see SBW500
HYDRAZINE, COMPD. WITH BORANE (1:1) see
HGU025
HYDRAZINE DIFLUORIDE see HGU100
HYDRAZINE, DIHYDROFLUORIDE see HGU100
HYDRAZINE HYDRATE see HGU500
HYDRAZINE HYDRATE, with not >64% hydrazine, by
weight (DOT) see HGU500
HYDRAZINE, HYDROCHLORIDE see HGV000

HYDRAZINE HYDROGEN SULFATE see HGW500
(T-4)-(HYDRAZINE-KAPPAN)TRIHYDROBORON see
HGU025
HYDRAZINE, N-((2-
METHYLCYCLOPROPYL)CARBONYL)- see MIV300
HYDRAZINE MONOCHLORIDE see HGV000
HYDRAZINE MONOSULFATE see HGW500
HYDRAZINE SULFATE (1:1) see HGW500
HYDRAZINE SULPHATE see HGW500
HYDRAZINE, TRIFLUOROSTANNITE see HHA100
HYDRAZINE YELLOW see FAG140
HYDRAZINIUM CHLORIDE see HGV000
HYDRAZINIUM MONOCHLORIDE see HGV000
HYDRAZINIUM SULFATE see HGW500
HYDRAZINIUM TRIFLUOROSTANNITE see HHA100
HYDRAZINOBENEZENE see PFI000
2-HYDRAZINOETHANOL see HHC000
1-HYDRAZINO-2-PHENYLETHANE see PFC500
1-HYDRAZINO-2-PHENYLETHANE HYDROGEN
SULPHATE see PFC750
HYDRAZINOPHTHALAZINE see HGP495
1-HYDRAZINOPHTHALAZINE see HGP495
1-HYDRAZINOPHTHALAZINE HYDROCHLORIDE
see HGP500
1-HYDRAZINOPHTHALAZINE
MONOHYDROCHLORIDE see HGP500
HYDRAZOBENZEN (CZECH) see HHG000
HYDRAZOBENZENE see HHG000
HYDRAZODIBENZENE see HHG000
HYDRAZOETHANE see DJL400
HYDRAZOIC ACID see HHG500
HYDRAZOMETHANE see MKN000
HYDRAZOMETHANE see DSF600
HYDRAZONIUM SULFATE see HGW500
HYDRAZYNA (POLISH) see HGS000
HYDRIODIC ACID see HHI500
HYDRIODIC ACID, solution (UN 1787) (DOT) see
HHI500
HYDRIODIDE-ENTROL see CHR500
HYDRITE see KBB600
HYDRO-AQUIL see CFY000
HYDROAZOETHANE see DJL400
HYDROBIS(2-METHYLPROPYL)ALUMINUM see
DNI600
HYDROBROMIC ACID see HHJ000
HYDROBROMIC ACID SOLUTION, >49% hydrobromic
acid (UN 1788) (DOT) see HHJ000
HYDROBROMIC ACID SOLUTION, not >49%
hydrobromic acid (UN 1788) (DOT) see HHJ000
HYDROBROMIC ETHER see EGV400
HYDROCARBON GAS see HHJ500
HYDROCARBON GASES, COMPRESSED, N.O.S. (UN
1964) (DOT) see HHJ500
HYDROCARBON GASES, LIQUEFIED, N.O.S. (UN
1965) (DOT) see HHJ500
HYDROCARBON GASES MIXTURES, COMPRESSED,
N.O.S. (UN 1964) (DOT) see HHJ500
HYDROCARBON GASES MIXTURES, LIQUEFIED,
N.O.S. (UN 1965) (DOT) see HHJ500
HYDROCERIN see CMD750
HYDROCHINON (CZECH, POLISH) see HIH000
HYDROCHLORBENZETHYLAMINE see HHK050
HYDROCHLORIC ACID see HHL000
HYDROCHLORIC ACID, mixed with NITRIC ACID (3:1)
see HHM000
HYDROCHLORIC ACID, solution (UN 1789) (DOT) see
HHL000
HYDROCHLORIC ETHER see EHH000

HYDROCHLORIDE see HHL000
l-HYDROCHLORIDE ARGININE see AQW000
HYDROCHLOROFLUOROCARBON 142b see CFX250
HYDROCHLORTHIAZID see CFY000
HYDROCINNAMALDEHYDE see HHP000
HYDROCINNAMALDEHYDE, p-ISOPROPYL-α-
METHYL-, DIETHYL ACETAL see COU510
HYDROCINNAMIC ACID, α-AMINO- see PEC750
HYDROCINNAMIC ALCOHOL see HHP050
HYDROCINNAMIC ALDEHYDE see HHP000
HYDROCINNAMYL ACETATE see HHP500
HYDROCINNAMYL ALCOHOL see HHP050
HYDROCINNAMYL PROPIONATE see HHQ550
HYDROCODIN see DKW800
Δ¹-HYDROCORTISONE see PMA000
11-β-HYDROCORTISONE see CNS750
HYDROCORTISONE FREE ALCOHOL see CNS750
HYDROCORTISONE SODIUM SUCCINATE see
HHR000
HYDROCORTISONE-21-SODIUM SUCCINATE see
HHR000
HYDROCORTISYL see CNS750
HYDROCORTONE see CNS750
HYDROCOUMARIN see HHR500
HYDROCYANIC ACID see HHS000
HYDROCYANIC ACID, aqueous solutions <5% HCN
(NA 1613) (DOT) see HHS000
HYDROCYANIC ACID, ION(CN¹⁻) see COI500
HYDROCYANIC ACID, POTASSIUM SALT see PLC500
HYDROCYANIC ACID (PRUSSIC), unstabilized (DOT)
see HHS000
HYDROCYANIC ACID, SODIUM SALT see SGA500
HYDROCYANIC ACID, aqueous solutions not >20%
hydrocyanic acid (UN 1613) (DOT) see HHS000
HYDROCYANIC ETHER see PMV750
HYDROCYCLIN see HOI000
HYDRODARCO see CBT500
HYDRODELTALONE see PMA000
HYDRODELTISONE see PMA000
HYDRODIISOBUTYLALUMINUM see DNI600
HYDRODIURETIC see CFY000
HYDRO-DIURIL see CFY000
HYDROFLUOBORIC ACID see FDD125
HYDROFLUOBORIC ACID see HHS600
HYDROFLUORIC ACID see HHU500
HYDROFLUORIC ACID mixed with SULFURIC ACID
see HHV000
HYDROFLUORIC ACID, solution, >60% strength (UN
1790) (DOT) see HHU500
HYDROFLUORIC ACID, solution, not >60% strength
(UN 1790) (DOT) see HHU500
HYDROFLUORIC and SULFURIC ACIDS, MIXTURE
(DOT) see HHV000
HYDROFLUORIDE see HHU500
HYDROFLUOSILICIC ACID see SCO500
HYDROFOL see PAE250
HYDROFOL ACID 1255 see LBL000
HYDROFOL ACID 1495 see MSA250
HYDROFOL ACID 1655 see SLK000
HYDROFURAN see TCR750
HYDROGEN see HHW500
HYDROGEN (DOT) see HHW500
HYDROGEN, compressed (DOT) see HHW500
HYDROGEN, refrigerated liquid (DOT) see HHW500
HYDROGEN ANTIMONIDE see SLQ000
HYDROGEN ARSENIDE see ARK250
HYDROGENATED FISH OIL see HHW560

HYDROGENATED MDI see MJM600
HYDROGENATED TERPHENYLS see HHW800
HYDROGEN AZIDE see HHG500
HYDROGEN BROMIDE (ACGIH,OSHA,MAK) see HHJ000
HYDROGEN BROMIDE, anhydrous (UN 1048) (DOT) see HHJ000
HYDROGEN CARBOXYLIC ACID see FNA000
HYDROGEN CHLORIDE see HHX000
HYDROGEN CHLORIDE, anhydrous (UN 1050) (DOT) see HHL000
HYDROGEN CHLORIDE, refrigerated liquid (UN 2186) (DOT) see HHL000
HYDROGEN CYANAMIDE see COH500
HYDROGEN CYANIDE see HHS000
HYDROGEN CYANIDE (ACGIH,OSHA) see HHS000
HYDROGEN CYANIDE, anhydrous, stabilized (UN 1051) (DOT) see HHS000
HYDROGEN CYANIDE, anhydrous, stabilized, absorbed in a porous inert material (UN 1614) (DOT) see HHS000
HYDROGEN DIMETHYL PHOSPHITE see DSG600
HYDROGEN DIOXIDE see HIB050
HYDROGENE SULFURE (FRENCH) see HIC500
HYDROGEN FLUORIDE, anhydrous (UN 1052) (DOT) see HHU500
HYDROGEN HEXACHLOROPLATINATE(4+) see CKO750
HYDROGEN HEXAFLUOROPHOSPHATE see HDE000
HYDROGEN HEXAFLUOROSILICATE see SCO500
HYDROGEN IODIDE see HHI500
HYDROGEN IODIDE, anhydrous (UN 2197) (DOT) see HHI500
HYDROGEN NITRATE see NED500
HYDROGEN PEROXIDE, 30% see HIB010
HYDROGEN PEROXIDE, 90% see HIB050
HYDROGEN PEROXIDE, 8% to 20% see HIB005
HYDROGEN PEROXIDE, solution, 30% see HIB010
HYDROGEN PEROXIDE, solution, 8% to 20% (DOT) see HIB005
HYDROGEN PEROXIDE, stabilized with >60% hydrogen peroxide (DOT) see HIB050
HYDROGEN PEROXIDE CARBAMIDE see HIB500
HYDROGEN PEROXIDE & PEROXYACETIC ACID MIXT., with acids, H$_2$O not >5% peroxyacetic acid (DOT) see PCL500
HYDROGEN PEROXIDE, aqueous solutions with not <8% but <20% hydrogen peroxide (UN 2984) (DOT) see HIB010
HYDROGEN PEROXIDE, aqueous solutions with >40%, not >60% hydrogen peroxide (UN 2014) see HIB010
HYDROGEN PEROXIDE with UREA (1:1) see HIB500
HYDROGEN PHOSPHIDE see PGY000
HYDROGEN POTASSIUM FLUORIDE see PKU250
HYDROGEN SELENIDE see HIC000
HYDROGEN SELENIDE, anhydrous (DOT) see HIC000
21-(HYDROGEN SUCCINATE)CORTISOL, MONOSODIUM SALT see HHR000
HYDROGEN SULFIDE see HIC500
HYDROGEN SULFITE see HIC600
HYDROGEN SULFITE SODIUM see SFE000
HYDROGEN SULFURIC ACID see HIC500
HYDROGEN TETRAFLUOROBORATE see FDD125
HYDROGEN TETRAFLUOROBORATE see HHS600
α-HYDRO-omega-HYDROXYPOLY(OXY-1,2-ETHANEDIYL) see PJT000

α-HYDRO-ω-HYDROXY-POLY(OXY-1,2-ETHANEDIYL) see PJT500
α-HYDRO-ω-HYDROXY-POLY(OXYRTHYLENE)-POLY(OXYPROPYLENE)(51-57 MOLES)POLY(OXYETHYLENE) BLOCK POLYMER see PJK200
HYDROLIN see SHR500
HYDROLOSE see MIF760
HYDROL SW see PKF500
HYDRO-MAG MA see MAG750
HYDROMIREX see MRI750
HYDROMORPHONE HYDROCHLORIDE see DNU300
HYDROMOX R see RDK000
HYDRONITRIC ACID see HHG500
HYDROOT see SOI500
HYDROPERIT see HIB500
HYDROPEROXIDE see HIB050
HYDROPEROXIDE, ACETYL see PCL500
HYDROPEROXIDE, ETHYL see ELD000
1-HYDROPEROXYCYCLOHEXYL-1-HYDROXYCYCLOHEXYL PEROXIDE see CPC300
HYDROPEROXYDE de BUTYLE TERTIAIRE (FRENCH) see BRM250
HYDROPEROXYDE de CUMENE (FRENCH) see IOB000
HYDROPEROXYDE de CUMYLE (FRENCH) see IOB000
2-HYDROPEROXY-2-METHYLPROPANE see BRM250
HYDROPHENOL see CPB750
HYDROPRES see RDK000
HYDROPRES KA see RDK000
HYDROQUINOL see HIH000
HYDROQUINOLE see HIH000
HYDROQUINONE see HIH000
m-HYDROQUINONE see REA000
o-HYDROQUINONE see CCP850
p-HYDROQUINONE see HIH000
α-HYDROQUINONE see HIH000
HYDROQUINONE, liquid or solid (DOT) see HIH000
HYDROQUINONECARBOXYLIC ACID see GCU000
HYDROQUINONE MONOETHYL ETHER see EFA100
HYDROQUINONE MONOMETHYL ETHER see MFC700
HYDROQUINONE, TRIMETHYL- see POG300
HYDRO-RAPID see CHJ750
HYDRORETROCORTIN see PMA000
HYDRORUBEANIC ACID see DXO200
HYDROSALURIC see CFY000
HYDROSCINE HYDROBROMIDE see HOT500
HYDROSILICOFLUORIC ACID see SCO500
HYDROSULFITE ANION see HIC600
HYDROTHAL-47 see EAR000
HYDROTHIDE see CFY000
HYDROTREATED (mild) HEAVY NAPHTHENIC DISTILLATE see MQV790
HYDROTREATED (mild) HEAVY NAPHTHENIC DISTILLATES (PETROLEUM) see MQV790
HYDROTREATED (mild) HEAVY PARAFFINIC DISTILLATE see MQV795
HYDROTREATED KEROSENE see KEK100
HYDROTREATED (mild) LIGHT NAPHTHENIC DISTILLATE see MQV800
HYDROTREATED (mild) LIGHT NAPHTHENIC DISTILLATES (PETROLEUM) see MQV800
HYDROTREATED (mild) LIGHT PARAFFINIC DISTILLATE see MQV805
HYDROTREATED NAPHTHA see NAH600

HYDROTROPALDEHYDE DIMETHYL ACETAL see HII600
HYDROTROPIC ALDEHYDE DIMETHYL ACETAL see HII600
HYDROXINE see CJR909
HYDROXINE YELLOW L see FAG140
N-HYDROXY-AAF see HIP000
2-(N-HYDROXYACETAMIDO)FLUORENE see HIP000
N-HYDROXY-2-ACETAMIDOFLUORENE see HIP000
4-HYDROXYACETANILIDE see HIM000
p-HYDROXYACETANILIDE see HIM000
4'-HYDROXYACETANILIDE see HIM000
HYDROXYACETONE see ABC000
HYDROXYACETONITRILE see HIM500
2-HYDROXYACETONITRILE see HIM500
o-HYDROXYACETOPHENONE see HIN500
2'-HYDROXYACETOPHENONE see HIN500
N-HYDROXY-2-ACETYLAMINOFLUORENE see HIP000
N-HYDROXY-N-ACETYL-2-AMINOFLUORENE see HIP000
2-(HYDROXYACETYL)INDOLE see HIS100
2-(HYDROXYACETYL)-1-METHYLINDOLE see HIS120
2-(HYDROXYACETYL)-3-METHYLINDOLE see HIS130
3-(HYDROXYACETYL)-1-METHYLINDOLE see HIS140
2-(2-HYDROXYACETYL)-1-METHYLINDOLE see HIS120
3-(2-HYDROXYACETYL)-2-METHYLINDOLE see HIS150
4-HYDROXYAFLATOXIN B1 see AEW000
1-HYDROXY-4-AMINOANTHRAQUINONE see AKE250
α-HYDROXY-β-AMINOPROPYLBENZENE HYDROCHLORIDE see PMJ500
7-β-HYDROXYANDROST-4-EN-3-ONE see TBF500
17-β-HYDROXYANDROST-4-EN-3-ONE see TBF500
17-β-HYDROXY-4-ANDROSTEN-3-ONE see TBF500
17-HYDROXY-(17-β)-ANDROST-4-EN-3-ONE see TBF500
3-β-HYDROXYANDROSTEN-17-ONE ACETATE see HJB225
2-HYDROXYANILINE see ALT000
3-HYDROXYANILINE see ALS990
4-HYDROXYANILINE see ALT250
o-HYDROXYANILINE see ALT000
p-HYDROXYANILINE see ALT250
4-HYDROXY-m-ANISALDEHYDE see VFK000
2-HYDROXYANISOLE see GKI000
3-HYDROXYANISOLE see REF050
m-HYDROXYANISOLE see REF050
o-HYDROXYANISOLE see GKI000
4-HYDROXY-1-ANTHRAQUINONYLAMINE see AKE250
2-HYDROXY-p-ARSANILIC ACID see HJE400
HYDROXYATHYLSTARKE (GERMAN) see HLB400
4-HYDROXYAZOBENZENE see HJF000
p-HYDROXYAZOBENZENE see HJF000
4-HYDROXY-3,4'-AZODI-1-NAPHTHALENESULFONIC ACID, DISODIUM SALT see HJF500
4-HYDROXY-3,4'-AZODI-1-NAPHTHALENESULPHONIC ACID, DISODIUM SALT see HJF500
HYDROXYBENZENE see PDN750

2-HYDROXYBENZENEMETHANOL see HMK100
HYDROXYBENZENESULFONIC ACID see HJH500
3-HYDROXYBENZISOTHIAZOL-S,S-DIOXIDE see BCE500
2-HYDROXYBENZOIC ACID see SAI000
3-HYDROXYBENZOIC ACID see HJI100
m-HYDROXYBENZOIC ACID see HJI100
o-HYDROXYBENZOIC ACID see SAI000
p-HYDROXYBENZOIC ACID ETHYL ESTER see HJL000
2-HYDROXYBENZOIC ACID METHYL ESTER see MPI000
o-HYDROXYBENZOIC ACID, METHYL ESTER see MPI000
p-HYDROXYBENZOIC ACID METHYL ESTER see HJL500
2-HYDROXYBENZOIC ACID MONOSODIUM SALT see SJO000
4-HYDROXYBENZOIC ACID PROPYL ESTER see HNU500
p-HYDROXYBENZOIC ACID PROPYL ESTER see HNU500
p-HYDROXYBENZOIC ETHYL ESTER see HJL000
o-HYDROXYBENZOIC SODIUM SALT see SJO000
HYDROXYBENZOPYRIDINE see QPA000
HYDROXY-4 BENZOYL-2-FURANNE see FQL200
p-HYDROXYBENZYL ACETONE see RBU000
2-HYDROXYBENZYL ALCOHOL see HMK100
o-HYDROXYBENZYL ALCOHOL see HMK100
α-HYDROXYBENZYL PHENYL KETONE see BCP250
2-HYDROXYBIFENYL (CZECH) see BGJ250
2-HYDROXYBIPHENYL see BGJ250
4-HYDROXYBIPHENYL see BGJ500
o-HYDROXYBIPHENYL see BGJ250
p-HYDROXYBIPHENYL see BGJ500
2-HYDROXYBIPHENYL SODIUM SALT see BGJ750
4-HYDROXY-6,7-BIS(2-METHYLPROPOXY)-3-QUINOLINECARBOXYLIC ACID ETHER ESTER see BOO632
3-HYDROXYBUTANAL see AAH750
1-HYDROXYBUTANE see BPW500
2-HYDROXYBUTANE see BPW750
HYDROXYBUTANEDIOIC ACID see MAN000
4-HYDROXYBUTANOIC ACID LACTONE see BOV000
3-HYDROXYBUTANOIC ACID-β-LACTONE see BSX000
4-HYDROXYBUTANOIC ACID MONOLITHIUM SALT see LHM800
3-HYDROXY-2-BUTANONE see ABB500
1-HYDROXY-4-tert-BUTYLBENZENE see BSE500
4-HYDROXYBUTYLBUTYLNITROSAMINE see HJQ350
3-HYDROXYBUTYRALDEHYDE see AAH750
β-HYDROXYBUTYRALDEHYDE see AAH750
γ-HYDROXYBUTYRATE SODIUM SALT see HJS500
γ-HYDROXYBUTYRIC ACID CYCLIC ESTER see BOV000
HYDROXYBUTYRIC ACID LACTONE see BSX000
3-HYDROXYBUTYRIC ACID LACTONE see BSX000
4-HYDROXYBUTYRIC ACID γ-LACTONE see BOV000
β-HYDROXYBUTYRIC ACID-p-PHENETIDIDE see HJS850
4-HYDROXYBUTYRIC ACID SODIUM SALT see HJS500
γ-HYDROXYBUTYROLACTONE see BOV000
3-HYDROXY-p-BUTYROPHENETIDIDE see HJS850

2-HYDROXYCAMPHANE see BMD000
HYDROXYCELLULOSE see CCU150
8-HYDROXY-CHINOLIN (GERMAN) see QPA000
2-HYDROXY-5-CHLORO-N-(2-CHLORO-4-NITROPHENYL)BENZAMIDE see DFV400
3-β-HYDROXYCHOLEST-5-ENE see CMD750
HYDROXYCINE see CJR909
4-HYDROXYCINNAMIC ACID see CNU825
p-HYDROXYCINNAMIC ACID see CNU825
4'-HYDROXYCINNAMIC ACID see CNU825
o-HYDROXYCINNAMIC ACID LACTONE see CNV000
7-HYDROXYCITRONELLAL see CMS850
HYDROXYCITRONELLAL DIMETHYL ACETAL see HJV700
HYDROXYCITRONELLAL (FCC) see CMS850
HYDROXYCITRONELLAL-INDOLE (SCHIFF BASE) see ICS100
HYDROXYCITRONELLYLIDENE-INDOLE see ICS100
17-HYDROXYCORTICOSTERONE see CNS750
11-β-HYDROXYCORTISONE see CNS750
3-HYDROXYCROTONIC ACID METHYL ESTER DIMETHYL PHOSPHATE see MQR750
3-HYDROXYCYCLOHEXADIEN-1-ONE see REA000
HYDROXYCYCLOHEXANE see CPB750
trans-1-(4-HYDROXYCYCLOHEXYL)-4-(4-FLUOROPHENYL)-5-(2-METHOXYPYRIMIDIN-4-YL)IMIDAZOLE see HKA123
2-HYDROXY-p-CYMENE see CCM000
3-HYDROXY-p-CYMENE see TFX810
14-HYDROXYDAUNOMYCIN see AES750
14'-HYDROXYDAUNOMYCIN see AES750
14-HYDROXYDAUNORUBICINE see AES750
HYDROXYDAUNORUBICIN HYDROCHLORIDE see HKA300
HYDROXYDE de POTASSIUM (FRENCH) see PLJ500
HYDROXYDE de SODIUM (FRENCH) see SHS000
HYDROXYDE de TETRAMETHYLAMMONIUM (FRENCH) see TDK500
HYDROXYDE de TRIPHENYL-ETAIN (FRENCH) see HON000
3-HYDROXY-4,15-DIACETOXY-8-(3-METHYLBUTYRYLOXY)-12,13-EPOXY-Δ⁹-TRICHOTHECENE see FQS000
4-HYDROXY-3,5-DIBROMOBENZONITRILE see DDP000
4-HYDROXY-3,5-DI-tert-BUTYLTOLUENE see BFW750
2-(HYDROXY)-5-(2,4-DIFLUOROPHENYL)BENZOIC ACID see DKI600
14-HYDROXYDIHYDROCODEINONE see PCG500
8-HYDROXY-5,7-DIIODOQUINOLINE see DNF600
4-HYDROXY-6,7-DIISOBUTOXY-3-QUINOLINECARBOXYLIC ACID ETHYL ESTER see BOO632
β-HYDROXY-β-(2,5-DIMETHOXYPHENYL)-ISOPROPYLAMINE HYDROCHLORIDE see MDW000
HYDROXYDIMETHYLARSINE OXIDE see HKC000
HYDROXYDIMETHYLARSINE OXIDE, SODIUM SALT see HKC500
HYDROXYDIMETHYLARSINE OXIDE, SODIUM SALT see HKC500
1-HYDROXY-2,4-DIMETHYLBENZENE see XKJ500
3-HYDROXYDIMETHYL CROTONAMIDE DIMETHYL PHOSPHATE see DGQ875
3-HYDROXY-N,N-DIMETHYL-cis-CROTONAMIDE DIMETHYL PHOSPHATE see DGQ875

2'-HYDROXY-5,9-DIMETHYL-2-(3,3-DIMETHYLALLYL)-6,7-BENZOMORPHAN see DOQ400
dl-2'-HYDROXY-5,9-DIMETHYL-2-(3,3-DIMETHYLALLYL)-6,7-BENZOMORPHAN see DOQ400
17-β-HYDROXY-7-α,17-DIMETHYLESTR-4-EN-3-ONE see MQS225
(7-α,17-β)-17-HYDROXY-7,17-DIMETHYL-ESTR-4-EN-3-ONE (9CI) see MQS225
7-HYDROXY-3,7-DIMETHYL OCTANAL see CMS850
7-HYDROXY-3,7-DIMETHYLOCTAN-1-AL see CMS850
7-HYDROXY-3,7-DIMETHYL OCTANAL:ACETAL see HJV700
6-HYDROXY-3,7-DIMETHYLOCTANOIC ACID LACTONE see HKC600
3-HYDROXY-4,5-DIMETHYLOL-α-PICOLINE see PPK250
3-HYDROXY-4,5-DIMETHYLOL-α-PICOLINE HYDROCHLORIDE see PPK500
1-HYDROXY-2,4-DINITROBENZENE see DUZ000
2-HYDROXY-3,5-DINITROBENZOIC ACID see HKE600
2-HYDROXYDIPHENYL see BGJ250
4-HYDROXYDIPHENYL see BGJ500
o-HYDROXYDIPHENYL see BGJ250
p-HYDROXYDIPHENYL see BGJ500
4,4'-HYDROXY-γ,Δ-DIPHENYL-β,Δ-HEXADIENE see DAL600
α-HYDROXYDIPHENYLMETHANE-β-DIMETHYLAMINOETHYL ETHER HYDROCHLORIDE see BAU750
2-HYDROXYDIPHENYL SODIUM see BGJ750
2-HYDROXYDIPHENYL, SODIUM SALT see BGJ750
1-HYDROXY-1,1-DIPHOSPHONOETHANE see HKS780
5-HYDROXYDODECANOIC ACID LACTONE see HBP450
5-HYDROXYDODECANOIC ACID Δ-LACTONE see HBP450
3-HYDROXY-1,2-EPOXYPROPANE see GGW500
16-α-HYDROXYESTRADIOL see EDU500
3-HYDROXYESTRA-1,3,5(10)-TRIEN-17-ONE see EDV000
17β-HYDROXYESTRA-4,9,11-TRIEN-3-ONE see THL600
3-HYDROXYESTRA-1,3,5(10)-TRIEN-17-ONE BENZOATE see EDV500
HYDROXYESTRIN BENZOATE see EDP000
1-HYDROXYETHANECARBOXYLIC ACID see LAG000
HYDROXYETHANEDIPHOSPHONIC ACID see HKS780
1-HYDROXYETHANEDIPHOSPHONIC ACID see HKS780
2-HYDROXYETHANESULFONIC ACID AMMONIUM SALT see ANL100
2-HYDROXY-1-ETHANETHIOL see MCN250
HYDROXY ETHER see EES350
17-β-HYDROXY-17-α-ETHINYL-5(10)-ESTREN-3-ONE see EEH550
4-HYDROXY-3-ETHOXYBENZALDEHYDE see EQF000
2-(2-HYDROXYETHOXY)ETHYL ESTER STEARIC ACID see HKJ000
2-(1-HYDROXYETHYL)ACRYLONITRILE see HKY650

2-HYDROXYETHYLAMINE see EEC600
β-HYDROXYETHYLAMINE see EEC600
4-(2-HYDROXYETHYL)AMINO-3-NITROANILINE see ALO750
2-HYDROXY-1-ETHYLAZIRIDINE see ASI000
N-(2-HYDROXYETHYL)AZIRIDINE see ASI000
β-HYDROXY-1-ETHYLAZIRIDINE see ASI000
N-(β-HYDROXYETHYL)AZIRIDINE see ASI000
β-HYDROXYETHYLBENZENE see PDD750
1-(2-HYDROXYETHYL)-4-(3-(2-CHLORO-10-PHENOTHIAZINYL)PROPYL)PIPERAZINE see CJM250
β-HYDROXYETHYLDIMETHYLAMINE see DOY800
2-HYDROXYETHYL ESTER STEARIC ACID see EJM500
N-HYDROXYETHYL ETHYLENE IMINE see ASI000
1-(2-HYDROXYETHYL)ETHYLENIMINE see ASI000
N-(2-HYDROXYETHYL)ETHYLENIMINE see ASI000
HYDROXYETHYL HYDRAZINE see HHC000
N-(2-HYDROXYETHYL)HYDRAZINE see HHC000
β-HYDROXYETHYLHYDRAZINE see HHC000
1-HYDROXYETHYLIDENE-1,1-DIPHOSPHONIC ACID see HKS780
(1-HYDROXYETHYLIDENE)DIPHOSPHONIC ACID DISODIUM SALT see DXD400
3-(1-HYDROXYETHYLIDENE)-6-METHYL-2H-PYRAN-2,4(3H)-DIONE, SODIUM SALT see SGD000
(N-HYDROXYETHYL)ISOPROPYLAMINE see INN400
β-HYDROXYETHYL ISOPROPYL ETHER see INA500
2-HYDROXYETHYL MERCAPTAN see MCN250
1-HYDROXYETHYL METHYL KETONE see ABB500
1-HYDROXYETHYL-2-METHYL-5-NITROIMIDAZOLE see MMN250
1-(2-HYDROXYETHYL)-2-METHYL-5-NITROIMIDAZOLE see MMN250
1-(2-HYDROXY-1-ETHYL)-2-METHYL-5-NITROIMIDAZOLE see MMN250
1-(β-HYDROXYETHYL)-2-METHYL-5-NITROIMIDAZOLE see MMN250
N¹-(2-HYDROXYETHYL)-2-NITRO-p-PHENYLENEDIAMINE see ALO750
1-(2-HYDROXYETHYL)-1-NITROSOUREA see HKW500
N-(2-HYDROXYETHYL)PHENYLAMINE see AOR750
β-HYDROXYETHYL PHENYL ETHER see PER000
γ-(4-(β-HYDROXYETHYL)PIPERAZIN-1-YL)PROPYL-2-CHLOROPHENOTHIAZINE see CJM250
10-(3'-(1''-β-HYDROXYETHYL-4''-PIPERAZINYL)-PROPYL)-THIOPHENYLPYRIDYAMINE HYDROCHLORIDE see ORI400
10-(3-(4-(2-HYDROXYETHYL)PIPERIDINO)PROPYL)PHENOTHIAZIN-2-YL METHYL KETONE see PII500
(2-HYDROXYETHYL)-2-PROPENENITRILE see HKY650
4-(2-HYDROXYETHYL)PYRIDINE see POR500
4-(β-HYDROXYETHYL)PYRIDINE see POR500
3-HYDROXY-2-ETHYL-4-PYRONE see EMA600
HYDROXYETHYL STARCH see HLB400
2-HYDROXYETHYL STARCH see HLB400
o-(HYDROXYETHYL)STARCH see HLB400
o-(2-HYDROXYETHYL)STARCH see HLB400
2-HYDROXYETHYL STARCH ETHER see HLB400
4'-(2-HYDROXYETHYLSULFONYL)ACETANILIDE see HLB500

(2-HYDROXYETHYL)TRIMETHYLAMMONIUM BITARTRATE see CMF300
(2-HYDROXYETHYL)TRIMETHYLAMMONIUM CHLORIDE see CMF750
(2-HYDROXYETHYL)TRIMETHYLAMMONIUM CHLORIDE ACETATE see ABO000
N-HYDROXY-2-FAA see HIP000
3-(4'-HYDROXYFENYL)AMINOKARBAZOL see CBN100
o-HYDROXYFENYLMERKURICHLORID see CHW675
N-HYDROXY-N-(2-FLUORENYL)ACETAMIDE see HIP000
3-HYDROXYGLUTACONIC ACID, DIMETHYL ESTER, DIMETHYL PHOSPHATE see SOY000
1-HYDROXYHEPTANE see HBL500
2-HYDROXYHEPTANE see HBE500
4-HYDROXYHEPTANOIC ACID LACTONE see HBA550
4-HYDROXYHEPTANOIC ACID, γ-LACTONE see HBA550
1-HYDROXYHEXANE see HFJ500
5-HYDROXY-2-HEXENOIC ACID LACTONE see PAJ500
3-β-HYDROXY-1-α-H,5-α-H-TROPANE-2-β-CARBOXYLIC ACID METHYL ESTER, BENZOATE see CNE750
o-HYDROXY-HYDROCINNAMIC ACID-Δ-LACTONE see HHR500
1-HYDROXY-1-HYDROPEROXYDICYCLOHEXYL PEROXIDE see CPC300
1-HYDROXY-1'-HYDROPEROXYDICYCLOHEXYL PEROXIDE see CPC300
HYDROXYHYDROQUINONE see BBU250
(o-((2-HYDROXY-3-(HYDROXYMERCURY)PROPYL)CARBAMOYL)PHENOXY)ACETIC ACID see NCM800
17-β-HYDROXY-2-HYDROXYMETHYLENE-17-α-METHYL-3-ANDROSTANONE see PAN100
17-β-HYDROXY-2-(HYDROXYMETHYLENE)-17-α-METHYL-5-α-ANDROSTAN-3-ONE see PAN100
17-β-HYDROXY-2-(HYDROXYMETHYLENE)-17-METHYL-5-α-ANDROSTAN-3-ONE see PAN100
17-HYDROXY-2-(HYDROXYMETHYLENE)-17-METHYL-5-α-17-β-ANDROSTAN-3-ONE see PAN100
3-HYDROXY-5-(HYDROXYMETHYL)-2-METHYLISONICOTINALDEHYDE, HYDROCHLORIDE see VSU000
HYDROXY(4-HYDROXY-3-NITROPHENYL)MERCURY see HLO400
7-(3-HYDROXY-2-(3-HYDROXY-1-OCTENYL)-5-OXOCYCLOPENTYL)-5-HEPTENOIC ACID see DVJ200
HYDROXY(2-HYDROXYPROPYL)PHENYLARSINE OXIDE see HNX600
2-HYDROXYIMINOMETHYL-1-METHYLPYRIDINIUM IODIDE see POS750
HYDROXYISOBUTYLISOPROPYLARSINE OXIDE see IPS100
α-HYDROXYISOBUTYRONITRILE see MLC750
7-(2-HYDROXY-3-(ISOPROPYLAMINO)PROPOXY)-2-BENZOFURANYL METHYL KETONE see HLK600
3-HYDROXY-17-KETO-ESTRA-1,3,5-TRIENE see EDV000
4-HYDROXY-2-KETO-4-METHYLPENTANE see DBF750

3-HYDROXY-17-KETO-OESTRA-1,3,5-TRIENE see EDV000
HYDROXYLAMINE see HLM500
HYDROXYLAMINE NEUTRAL SULFATE see OLS000
HYDROXYLAMINE SULFATE see OLS000
HYDROXYLAMINE SULFATE (2:1) see OLS000
HYDROXYLAMMONIUM SULFATE see OLS000
HYDROXYLATED LECITHIN see HLN700
6-HYDROXY-2-MERCAPTOPYRIMIDINE see TFR250
o-((3-HYDROXYMERCURI-2-METHOXYPROPYL)CARBAMOYL)PHENOXYACETIC ACID MONOSODIUM SALT see SIH500
N-(γ-HYDROXYMERCURI-β-METHOXYPROPYL)SALICYLAMIDE-o-ACETIC ACID SODIUM SALT see SIH500
HYDROXYMERCURI-o-NITROPHENOL see HLO400
HYDROXYMERCURIPROPANOLAMIDE of o-CARBOXYPHENOXYACETIC ACID see NCM800
4-HYDROXY-3-METHOXYALLYLBENZENE see EQR500
1-HYDROXY-2-METHOXY-4-ALLYLBENZENE see EQR500
4-HYDROXY-3-METHOXYBENZALDEHYDE see VFK000
1-HYDROXY-2-METHOXYBENZENE see GKI000
6-HYDROXY-7-METHOXY-5-BENZOFURANACRYLIC ACID Δ-LACTONE see XDJ000
2-HYDROXY-4-METHOXYBENZOPHENONE see MES000
o-(2-HYDROXY-4-METHOXYBENZOYL)BENZOIC ACID see HLS500
6-HYDROXY-3-METHOXY-N-METHYL-4,5-EPOXYMORPHINAN see DKW800
4-(4-HYDROXY-3-METHOXYPHENYL)-2-BUTANONE see VFP100
(4-HYDROXY-3-METHOXYPHENYL)ETHYL METHYL KETONE see VFP100
(2-HYDROXY-4-METHOXYPHENYL)PHENYLMETHANONE see MES000
1-HYDROXY-2-METHOXY-4-PROPENYLBENZENE see IKQ000
4-HYDROXY-3-METHOXY-1-PROPENYLBENZENE see IKQ000
1-HYDROXY-2-METHOXY-4-PROP-2-ENYLBENZENE see EQR500
2-HYDROXY-3-METHYL-4-ACETYLTETRAHYDROFURANE see BMK290
N-(HYDROXYMETHYL)ACRYLAMIDE see HLU500
α-HYDROXY-β-METHYL AMINE PROPYLBENZENE see EAW000
(R)-4-(1-HYDROXY-2-(METHYLAMINO)ETHYL)-1,2-BENZENEDIOL (9CI) see VGP000
(–)-α-HYDROXY-β-(METHYLAMINO)ETHYL-α-(3-HYDROXYBENZENE) HYDROCHLORIDE see SPC500
1-α-HYDROXY-β-METHYLAMINO-3-HYDROXY-1-ETHYLBENZENE see NCL500
(R)-3-HYDROXY-α-((METHYLAMINO)METHYL)BENZENEMETHANOL see NCL500
(R)-3-HYDROXY-α-((METHYLAMINO)METHYL)BENZENEMETHANOL HYDROCHLORIDE see SPC500

(–)-m-HYDROXY-α-(METHYLAMINOMETHYL)BENZYL ALCOHOL see NCL500
1-m-HYDROXY-α-((METHYLAMINO)METHYL)-BENZYL ALCOHOL see NCL500
1-m-HYDROXY-α-(METHYLAMINOMETHYL)BENZYL ALCOHOL HYDROCHLORIDE see SPC500
1-HYDROXY-2-METHYLAMINO-1-PHENYLPROPANE see EAW000
17-β-HYDROXY-17-METHYLANDROST-4-EN-3-ONE see MPN500
HYDROXY METHYL ANETHOL see IRY000
5-(HYDROXYMETHYL)-1-AZA-3,7-DIOXABICYCLO(3.3.0)OCTANE see OMM300
1-HYDROXY-2-METHYLBENZENE see CNX000
1-HYDROXY-3-METHYLBENZENE see CNW750
1-HYDROXY-4-METHYLBENZENE see CNX250
4-(HYDROXYMETHYL)BENZENEDIAZONIUM TETRAFLUOROBORATE see HLX925
6-HYDROXYMETHYLBENZO(a)PYRENE see BCV250
7-HYDROXY-4-METHYLCOUMARIN see MKP500
3-HYDROXY-N-METHYL-cis-CROTONAMIDE DIMETHYL PHOSPHATE see MRH209
2-HYDROXY-3-METHYL-2-CYCLOPENTEN-1-ONE see HMB500
4-HYDROXYMETHYL-2,6-DI-tert-BUTYLPHENOL see BQI050
4-HYDROXYMETHYL-2,2-DIMETHYL-1,3-DIOXOLANE see DVR600
2-HYDROXYMETHYLENE-17-α-METHYL-5-α-ANDROSTAN-17-β-OL-3-ONE see PAN100
2-HYDROXYMETHYLENE-17-α-METHYL-DIHYDROTESTOSTERONE see PAN100
2-(HYDROXYMETHYLENE)-17-α-METHYLDIHYDROTESTOSTERONE see PAN100
2-HYDROXYMETHYLENE-17-α-METHYL-17-β-HYDROXY-3-ANDROSTANONE see PAN100
4-(1-HYDROXY-2-((1-METHYLETHYL)AMINO)ETHYL)-1,2-BENZENEDIOL see DMV600
1-(7-(2-HYDROXY-3-((1-METHYLETHYL)AMINO)PROPOXY)-2-BENZOFURANYL)ETHANONE see HLK600
(HYDROXYMETHYL)ETHYLENE ACETATE see DBF600
2-HYDROXYMETHYLFURAN see FPU000
3-HYDROXYMETHYL-n-HEPTAN-4-OL see EKV000
3-α-HYDROXY-8-METHYL-1-α-H,5-α-H-TROPANIUM NITRATE (±)-TROPATE (ESTER) see MGR500
HYDROXYMETHYL 2-INDOLYL KETONE see HIS100
HYDROXYMETHYLINITRILE see HIM500
3-HYDROXY-1-METHYL-4-ISOPROPYLBENZENE see TFX810
HYDROXYMETHYLMERCURY see MLG000
7-HYDROXYMETHYL-12-METHYLBENZ(a)ANTHRACENE see HMF000
17-HYDROXY-6-METHYL-16-METHYLENEPREGNA-4,6-DIENE-3,20-DIONE, ACETATE see MCB380
HYDROXYMETHYL 1-METHYL-2-INDOLYL KETONE see HIS120
HYDROXYMETHYL 3-METHYL-2-INDOLYL KETONE see HIS130
(–)-3-HYDROXY-N-METHYLMORPHINAN see LFG000

7-HYDROXY-4-METHYL-2-OXO-2H-1-BENZOPYRAN see MKP500

3-(HYDROXYMETHYL)-8-OXO-7-(2-(4-PYRIDYLTHIO)ACETAMIDO)-5-THIA-1-AZABICYCLO(4.2.0)OCT-2-ENE-2-CARBOXYLIC ACID, ACETATE (ESTER) see CCX500

4-HYDROXY-4-METHYL-PENTAN-2-ON (GERMAN, DUTCH) see DBF750

4-HYDROXY-4-METHYLPENTANONE-2 see DBF750

4-HYDROXY-4-METHYL PENTAN-2-ONE see DBF750

4-HYDROXY-4-METHYL-2-PENTANONE see DBF750

2-HYDROXYMETHYLPHENOL see HMK100

o-(HYDROXYMETHYL)PHENOL see HMK100

HYDROXYMETHYLPHENYLARSINE OXIDE see HMK200

N-(4-((2-HYDROXY-5-METHYLPHENYL)AZO)PHENYL)ACETAMIDE see AAQ250

HYDROXYMETHYLPHTHALIMIDE see HMP100

N-(HYDROXYMETHYL)PHTHALIMIDE see HMP100

17-HYDROXY-6-METHYLPREGNA-4,6-DIENE-3,20-DIONE ACETATE see VTF000

17-HYDROXY-6-α-METHYLPREGN-4-ENE-3,20-DIONE ACETATE see MCA000

17-α-HYDROXY-6-α-METHYLPREGN-4-ENE-3,20-DIONE ACETATE see MCA000

17-α-HYDROXY-6-α-METHYLPROGESTERONE ACETATE see MCA000

1-HYDROXYMETHYLPROPANE see IIL000

N-(HYDROXYMETHYL)-2-PROPENAMIDE see HLU500

2-HYDROXY-2-METHYLPROPIONITRILE see MLC750

1-(HYDROXYMETHYL)PROPYLAMIDE of 1-METHYL-(+)-LYSERGIC ACID HYDROGEN MALEATE see MQP500

3-HYDROXY-2-METHYL-4H-PYRAN-4-ONE see MAO350

5-HYDROXY-6-METHYL-3,4-PYRIDINEDICARBINOL HYDROCHLORIDE see PPK500

5-HYDROXY-6-METHYL-3,4-PYRIDINEDIMETHANOL see PPK250

5-HYDROXY-6-METHYL-3,4-PYRIDINEDIMETHANOL HYDROCHLORIDE see PPK500

3-HYDROXY-2-METHYL-4-PYRONE see MAO350

3-HYDROXY-2-METHYL-γ-PYRONE see MAO350

12-HYDROXY-4-METHYL-4,8-SECOSENECIONAN-8,11,16-TRIONE see DMX200

2-HYDROXY-4-(METHYLTHIO)BUTANENITRILE see HMR550

3-HYDROXY-α-METHYL-l-TYROSINE see DNA800

1-HYDROXYNAPHTHALENE see NAW500

2-HYDROXYNAPHTHALENE see NAX000

α-HYDROXYNAPHTHALENE see NAW500

β-HYDROXYNAPHTHALENE see NAX000

4-HYDROXY-3-NITROANILINE see NEM480

2-HYDROXYNITROBENZENE see NIF010

3-HYDROXYNITROBENZENE see NIE600

4-HYDROXYNITROBENZENE see NIF000

m-HYDROXYNITROBENZENE see NIE600

4-HYDROXY-3-NITROBENZENEARSONIC ACID see HMY000

4-HYDROXY-3-NITROPHENYLARSONIC ACID see HMY000

N-HYDROXY-N-NITROSO-BENZENAMINE, AMMONIUM SALT see ANO500

4-HYDROXYNONANOIC ACID, γ-LACTONE see CNF250

(17-α)-17-HYDROXY-19-NORPREGN-4-EN-20-YN-3-ONE see NNP500

(17-α)-17-HYDROXY-19-NORPREGN-5(10)-EN-20-YN-3-ONE see EEH550

17-β-HYDROXY-19-NORPREGN-4-EN-20-YN-3-ONE see NNP500

17-HYDROXY(17-α)-19-NORPREGN-5(10)-EN-20-YN-3-ONE see EEH550

17-HYDROXY-19-NOR-17-α-PREGN-4-EN-20-YN-3-ONE see NNP500

17-HYDROXY-19-NOR-17-α-PREGN-5(10)-EN-20-YN-3-ONE see EEH550

17-β-HYDROXY-19-NOR-17-α-PREGN-4-EN-20-YN-3-ONE ACETATE see ABU000

17-HYDROXY-19-NOR-17-α-PREGN-4-EN-20-YN-3-ONE ACETATE see ABU000

1-HYDROXYOCTANE see OEI000

5-HYDROXYOCTANOIC ACID LACTONE see OCE000

16-α-HYDROXYOESTRADIOL see EDU500

3-HYDROXY-OESTRA-1,3,5(10)-TRIEN-17-ONE see EDV000

3-HYDROXY-1,3,5(10)-OESTRATRIEN-17-ONE see EDV000

16-HYDROXY-11-OXAHEXADECANOIC ACID, ω-LACTONE see OKW100

16-HYDROXY-12-OXAHEXADECANOIC ACID, ω-LACTONE see OKW110

γ-HYDROXY-β-OXOBUTANE see ABB500

4-HYDROXY-3-(3-OXO-1-FENYL-BUTYL) CUMARINE (DUTCH) see WAT200

4-HYDROXY-3-(3-OXO-1-PHENYLBUTYL)-2H-1-BENZOPYRAN-2-ONE SODIUM SALT (9CI) see WAT220

4-HYDROXY-3-(3-OXO-1-PHENYL-BUTYL)-CUMARIN (GERMAN) see WAT200

2'-HYDROXYPELARGIDENOLON 1522 see MRN500

3-HYDROXY-2-PENTANEDIOIC ACID, DIMETHYL ESTER, DIMETHYL PHOSPHATE see SOY000

4-HYDROXYPENTANOIC ACID LACTONE see VAV000

p-HYDROXYPHENETOLE see EFA100

2-HYDROXYPHENOL see CCP850

3-HYDROXYPHENOL see REA000

m-HYDROXYPHENOL see REA000

o-HYDROXYPHENOL see CCP850

p-HYDROXYPHENOL see HIH000

1-HYDROXY-2-PHENOXYETHANE see PER000

N-(4-HYDROXYPHENYL)ACETAMIDE see HIM000

m-HYDROXYPHENYL ACETATE see RDZ900

α-HYDROXYPHENYLACETIC ACID see MAP000

HYDROXYPHENYLACETONITRILE see MAP250

2-HYDROXY-2-PHENYLACETOPHENONE see BCP250

α-HYDROXY-α-PHENYLACETOPHENONE see BCP250

p-HYDROXYPHENYLACRYLIC ACID see CNU825

β-(4-HYDROXYPHENYL)ACRYLIC ACID see CNU825

l-β-(p-HYDROXYPHENYL)ALANINE see TOG300

(4-HYDROXYPHENYL)ARSONIC ACID polymer with FORMALDEHYDE see BCJ150

7-HYDROXY-8-(PHENYLAZO)-1,3-NAPHTHALENEDISULFONIC ACID, DISODIUM SALT see HGC000

6-HYDROXY-5-((p-SULPHOPHENYL)AZO)-2-
NAPHTHALENESULPHONIC ACID, DISODIUM
SALT see FAG150
5-HYDROXYTETRACYCLINE see HOH500
5-HYDROXYTETRACYCLINE HYDROCHLORIDE see
HOI000
4-HYDROXY-3-(1,2,3,4-TETRAHYDRO-3-(4-(4-
(TRIFLUOROMETHYL)PHENOXY) PHENYL)-1-
NAPHTHALENYL)2H-1-BENZOPYRAN-2-ONE see
HOI245
7-HYDROXY-3,4-TETRAMETHYLENECOUMARIN-
O,O-DIETHYL THIOPHOSPHATE see DXO000
HYDROXYTHIOSPASMIN see HOJ150
HYDROXYTOLUENE see BDX500
4-HYDROXYTOLUENE see CNX250
m-HYDROXYTOLUENE see CNW750
o-HYDROXYTOLUENE see CNX000
p-HYDROXYTOLUENE see CNX250
α-HYDROXYTOLUENE see BDX500
α-HYDROXY-α-TOLUIC ACID see MAP000
HYDROXYTOLUOLE (GERMAN) see CNW500
4'-((6-HYDROXY-m-TOLYL)AZO)ACETANILIDE see
AAQ250
HYDROXYTRIBUTYLSTANNANE-4,4-
DIMETHYLOCTANOATE see TIF250
β-HYDROXYTRICARBALLYLIC ACID see CMS750
1-HYDROXY-2,2,2-TRICHLOROETHYLPHOSPHONIC
ACID DIMETHYL ESTER see TIQ250
2-HYDROXYTRIETHYLAMINE see DHO500
3-HYDROXY-4-((2,4,5-TRIMETHYLPHENYL)AZO)-2,7-
NAPHTHALENEDISULFONIC ACID, DISODIUM
SALT see FAG018
3-HYDROXY-4-((2,4,5-TRIMETHYLPHENYL)AZO)-2,7-
NAPHTHALENEDISULPHONIC ACID, DISODIUM
SALT see FAG018
2-HYDROXY-1,3,5-TRINITROBENZENE see PID000
3-HYDROXY-2,4,6-TRINITROPHENOL see SMP500
HYDROXYTRIPHENYLSTANNANE see HON000
HYDROXYTRIPHENYLTIN see HON000
HYDROXYTRYPTOPHAN see HOO100
5-HYDROXYTRYPTOPHAN see HOA575
5-HYDROXYTRYPTOPHAN see HON800
5-HYDROXYTRYPTOPHAN see HOO100
5-HYDROXY-l-TRYPTOPHAN see HOA600
5-HYDROXY-l-TRYPTOPHAN see HOO000
dl-HYDROXYTRYPTOPHAN see HOA575
dl-HYDROXYTRYPTOPHAN see HON800
l-5-HYDROXYTRYPTOPHAN see HOA600
l-5-HYDROXYTRYPTOPHAN see HOO000
(±)-5-HYDROXYTRYPTOPHAN see HOA575
(±)-5-HYDROXYTRYPTOPHAN see HON800
dl-5-HYDROXYTRYPTOPHAN see HOA575
5-HYDROXYTRYPTOPHANE see HOO100
3-HYDROXYTYRAMINE see DYC400
3-HYDROXY-l-TYROSINE see DNA200
l-o-HYDROXYTYROSINE see DNA200
4-HYDROXYVALERIC ACID LACTONE see VAV000
7-HYDROXYXANTHINE see HOP259
3-HYDROXY-4-(2,4-XYLYLAZO)-3,7-
NAPHTHALENEDISULFONIC ACID, DISODIUM
SALT see FMU070
3-HYDROXY-4-(2,4-XYLYLAZO)-3,7-
NAPHTHALENEDISULFONIC ACID, DISODIUM
SALT see FMU070
o-HYDROXYZIMTSAEURE-LACTON (GERMAN) see
CNV000
HYDROXYZINE see CJR909

17-β-HYDROXY-Δ⁴-ANDROSTEN-3-ONE see TBF500
HYDRURE de LITHIUM (FRENCH) see LHH000
HYFLAVIN see RIK000
HYGROMIX-8 see AQB000
HYGROMYCIN B (USDA) see AQB000
HYGROTON-RESERPINE see RDK000
HYLEMOX see EEH600
HYLENE M50 see MJP400
HYLENE-T see TGM740
HYLENE T see TGM750
HYLENE TCPA see TGM750
HYLENE TLC see TGM750
HYLENE TM see TGM750
HYLENE TM see TGM800
HYLENE TM-65 see TGM750
HYLENE TRF see TGM750
HYMECROMONE see MKP500
HYMORPHAN see DNU300
HYONIC PE-250 see PKF500
HYOSAN see MJM500
HYOSCINE see SBG000
(−)-HYOSCINE see SBG000
HYOSCINE BROMIDE see HOT500
HYOSCINE F HYDROBROMIDE see HOT500
HYOSCINE HYDROBROMIDE see HOT500
(−)-HYOSCINE HYDROBROMIDE see HOT500
1-HYOSCINE HYDROBROMIDE see HOT500
(−)-HYOSCYAMINE see HOU000
HYOSCYAMINE see HOU000
1-HYOSCYAMINE see HOU000
dl-HYOSCYAMINE see ARR000
dl-HYOSCYAMINE METHYLNITRATE see MGR500
HYOSCYINE HYDROBROMIDE see HOT500
HYOSOL see SBG000
dl-HYOSYAMINE METHYLNITRATE see MGR500
HYPERAZIN see HGP500
HYPERCAL B see RDK000
HYPERNEPHRIN see VGP000
HYPEROL see HIB500
HYPERPAX see DNA800
HYPERTANE FORTE see RDK000
HYPERTENAIN see MAW250
HYPERTENSAN see RDK000
HY-PHI 1055 see OHU000
HY-PHI 1088 see OHU000
HY-PHI 1199 see SLK000
HY-PHI 2066 see OHU000
HY-PHI 2088 see OHU000
HY-PHI 2102 see OHU000
HYPNOGEN see EOK000
HYPNONE see ABH000
HYPNOREX see LGZ000
HYPNOSTAN see PBT500
HYPNO-TABLINETTEN see EOK000
HYPO see SKI000
HYPOCHLORITES see HOU500
HYPOCHLORITE SOLUTIONS containing >7% available
chlorine by wt. see SHU500
HYPOCHLORITE SOLUTIONS with >5% but <16%
available chlorine by wt. (DOT) see SHU500
HYPOCHLORITE SOLUTIONS with 16% or more
available chlorine by wt. (UN 1791) see SHU500
HYPOCHLOROUS ACID see HOV000
HYPOCHLOROUS ACID, CALCIUM SALT see HOV500
HYPOCHLOROUS ACID, POTASSIUM SALT see
PLK000
HYPODERMACID see TIQ250

HYPONITROUS ACID see HOW500
HYPONITROUS ACID ANHYDRIDE see NGU000
HYPOPHENON see ELL500
HYPOPHTHALIN see HGP495
HYPOPHTHALIN see HGP500
HYPORENIN see VGP000
HYPOS see HGP500
HYPOTHIAZIDE see CFY000
HYPOXANTHINE see DMC000
HYPOXANTHINE NUCLEOSIDE see IDE000
HYPOXANTHINE RIBONUCLEOSIDE see IDE000
HYPOXANTHINE RIBOSIDE see IDE000
HYPOXANTHINE-d-RIBOSIDE see IDE000
HYPOXANTHOSINE see IDE000
HYRE see RIK000
HYSCO see HOT500
HYSCYLENE P see PDX000
HYSTRENE 80 see SLK000
HYSTRENE 8016 see PAE250
HYSTRENE 9014 see MSA250
HYSTRENE 9512 see LBL000
HYTONE LOTION see CNS750
HYTOX see MIA250
HYVAR see BMM650
HYVAREX see BMM650
HYVAR X see BMM650
HYVAR X BROMACIL see BMM650
HYVAR X WEED KILLER see BMM650
I 337A see CDP250
I 7840 see IKG900
IA see IDZ000
IA see IHN200
IAA see ICN000
IAB see AOC500
IA-PRAM see DLH630
IBA see ICP000
IBD see CCK125
IBDU see IIV000
IBENZMETHYZINE HYDROCHLORIDE see PME500
IBENZMETHYZIN HYDROCHLORIDE see PME500
IBIODRAL see BBV500
IBIOFURAL see NGE500
IBIOSUC see SGC000
IBIOTON see TAI500
IBIOTYZIL see BCA000
IBN see IJD000
IBUFEN see IIU000
IBUPROCIN see IIU000
IBUPROFEN see IIU000
IBZ see PME500
IC 6002 see PIW000
ICG see CCK000
ICI 543 see PMP500
I.C.I. 33,828 see MLJ500
ICI 35868 see DNR800
ICI 46,474 see NOA600
ICI 46,474 see TAD175
ICI 59118 see RCA375
ICI 80996 see CMX880
ICI 146814 see GJU600
racemic-ICI 80,996 see CMX880
ICIG 1109 see CGV250
ICIG 1110 see CHD250
I.C.I. LTD. COMPOUND NUMBER 80996 see CMX880
ICI-PP 557 see AHJ750
ICI-PP 563 see GJU600
ICON see LAS200
ICR 1θ see QDS000

ICR 377 see EHJ500
ICR 451 see CIG250
ICRF-159 see PIK250
ICRF 159 see RCA375
ICTALIS SIMPLE see DKQ000
IDANTOIL see DNU000
IDANTOIN see DKQ000
IDANTOINAL see DNU000
IDEXUR see DAS000
IDOCYL NOVUM see SJO000
IDOMETHINE see IDA000
IDOSERP see RDK000
IDOXENE see DAS000
IDOXURIDIN see DAS000
IDOXURIDINE see DAS000
IDPN see BIQ500
IDRAGIN see ADA725
IDRALAZINA (ITALIAN) see HGP495
IDRAZINA SOLFATO (ITALIAN) see HGW500
IDRIANOL see SPC500
IDROCHINONE (ITALIAN) see HIH000
IDROESTRIL see DKA600
IDROGENO SOLFORATO (ITALIAN) see HIC500
IDROPEROSSIDO di CUMENE (ITALIAN) see IOB000
IDROPEROSSIDO di CUMOLO (ITALIAN) see IOB000
IDROSSIDO DI STAGNO TRIFENILE (ITALIAN) see HON000
1',1-(2-IDROSSIETIL)4-(3-(2-CLORO-10-FENOTIAZIL)PROPILPIPERAZINA (ITALIAN) see CJM250
4-IDROSSI-4-METIL-PENTAN-2-ONE (ITALIAN) see DBF750
4-IDROSSI-3-(3-OXO-)-(FENIL-BUTIL)-CUMARINE (ITALIAN) see WAT200
IDROSSIZINA see CJR909
IDROTIAZIDE see CFY000
IDRYL see FDF000
IDSOSERP see RDK000
IDU see DAS000
IDUCHER see DAS000
IDULEA see DAS000
IDUOCULOS see DAS000
IDUR see DAS000
IDURIDIN see DAS000
I-EBU see EAN525
IERGIGAN see DQA400
IF (fumigant) see TBV750
IFC see CBM000
IFIBRIUM see LFK000
IFOSFAMID see IMH000
IFOSFAMIDE see IMH000
IGE see IPD000
IGELITE F see PKQ059
IGE (OSHA) see IPD000
IGEPAL CA-63 see PKF500
IGEPAL CO-630 see PKF000
IGNOTINE see CCK665
IGROSIN see SIH500
II-C-2 see DOQ400
IKF 1216 see CEX800
IKURIN see ANU650
ILITIA see VSZ450
ILLOXOL see DHB400
ILOPAN see PAG200
ILOTYCIN see EDH500
IMAVATE see DLH630
IMAVEROL see FPB875
IMAZALIL see FPB875

IMBRILON see IDA000
IMFERON see IGS000
IMI 115 see TGF250
IMIDA-LAB see TEH500
IMIDALINE HYDROCHLORIDE see BBJ750
IMIDAN see PHX250
IMIDAN (PEYTA) see TEH500
1H-IMIDAZOLE-4-ETHANAMINE PHOSPHATE (1:2) see HGE000
1H-IMIDAZOLE, 1-ETHENYL-, HOMOPOLYMER see VPP100
IMIDAZOLE MUSTARD see IAN000
IMIDAZOLE, 1-VINYL-, POLYMERS see VPP100
2,4-IMIDAZOLIDINEDIONE (9CI) see HGO600
2-IMIDAZOLIDINETHIONE see IAQ000
2-IMIDAZOLIDINETHIONE mixed with SODIUM NITRITE see IAR000
2-IMIDAZOLINE, 4,4-DIMETHYL-1-ISOPROPYL-2-NONYL- see DSK300
2-(1H-IMIDAZOL-4-YLMETHYL)-8H-INDENO(1,2-D)THIAZOLE MONOFUMARATE see IAT275
1H-IMIDAZO(4,5-B)PYRIDIN-2-AMINE, 1-METHYL-6-PHENYL-(9CI) see AKZ200
IMIDAZO(4,5-B)PYRIDINE, 2-AMINO-1-METHYL-6-PHENYL- see AKZ200
1H-IMIDAZO(4,5-B)PYRIDINE, 2-AMINO-1-METHYL-6-PHENYL- see AKZ200
IMIDAZO(1,2-c)QUINAZOLINE, 5-(4-CHLOROPHENYL)-2,3,5,6-TETRAHYDRO- see CKK050
IMIDAZO(1,2-c)QUINAZOLINE, 2,3,5,6-TETRAHYDRO-5-(p-CHLOROPHENYL)- see CKK050
IMIDAZO(2,1-β)THIAZOLE MONOHYDROCHLORIDE see LFA020
IMIDENE see TEH500
IMIDOBENZYLE see DLH630
4,4'-(IMIDOCARBONYL)BIS(N,N-DIMETHYLAMINE) MONOHYDROCHLORIDE see IBA000
4,4'-(IMIDOCARBONYL)BIS(N,N-DIMETHYLANILINE) see IBB000
IMIDOL see DLH630
IMIDOLE see PPS250
IMILANYLE see DLH630
2,2'-IMINOBISETHANOL see DHF000
2,2'-IMINOBISETHYLAMINE see DJG600
3,3'-IMINOBISPROPANENITRILE see BIQ500
IMINOBIS(PROPYLAMINE) see AIX250
3,3'-IMINOBIS(PROPYLAMINE) see AIX250
4,4'-((4-IMINO-2,5-CYCLOHEXADIEN-1-YLIDENE)METHYLENE)DIANILINE MONOHYDROCHLORIDE-o-TOLUIDINE see RMK020
2,2'-IMINODIETHANOL see DHF000
2,2'-IMINODI-ETHANOL with 1,2-DIHYDRO-3,6-PYRIDAZINEDIONE (1:1) see DHF200
2,2'-IMINODI-N-NITROSOETHANOL see NKM000
3,3'-IMINODIPROPIONITRILE see BIQ500
β,β-IMINODIPROPIONITRILE see BIQ500
IMINO-β,β'-DIPROPIONITRILE see BIQ500
β,β'-IMINODIPROPIONITRILE see BIQ500
IMINOPHOSPHATE see DXN600
2(3H)-IMINO-9-β-D-RIBOFURANOSYL-9H-PURIN-6(1H)-ONE see GLS000
IMIPRAMINA (ITALIAN) see DLH630
IMIPRAMINE see DLH630
IMIPRAMINEDEMETHYL HYDROCHLORIDE see DLS600

IMIPRAMINE HYDROCHLORIDE see DLH630
IMIPRAMINE MONOHYDROCHLORIDE see DLH630
IMIPRIN see DLH630
IMOL S 140 see TNP500
IMOTRYL see BBW500
IMP DISODIUM SALT see DXE500
5'-IMP DISODIUM SALT see DXE500
IMPEDEX see SJL500
IMPERIAL GREEN see COF500
IMPERON FIXER T see TND250
IMPERVOTAR see CMY800
IMPF see IPX000
IMP HYDROCHLORIDE see DLH630
IMPINGEMENT BLACK see CBT750
IMPOSIL see IGS000
IMPROVED WILT PRUF see PKQ059
IMPRUVOL see BFW750
IMP SODIUM SALT see DXE500
IMS see IPY000
IMURAN see ASB250
IMUREK see ASB250
IMUREL see ASB250
IMVITE I.G.B.A. see BAV750
IMWITOR 191 see OAV000
IMWITOR 900K see OAV000
INACID see IDA000
INACTIVE LIMONENE see MCC250
INAKOR see ARQ725
INAMYCIN see NOB000
INBUTON see BSM000
INCIDOL see BDS000
INCRECEL see CMF400
INDALCA AG see GLU000
INDALCA AG-BV see GLU000
INDALCA AG-HV see GLU000
INDALONE see BRT000
1,3-INDANDIONE see IBS000
INDANTHREN GREY M see CMU475
INDANTHREN GREY MG see CMU475
INDENE see IBX000
1H-INDENE-1,3(2H)-DIONE see IBS000
INDENO(1,2,3-cd)PYRENE see IBZ000
8H-INDENO(1,2-D)THIAZOLE, 2-(1H-IMIDAZOL-4-YLMETHYL)-, (E)-2-BUTENEDIOATE (1:1) see IAT275
INDIAN BERRY see PIE500
INDIAN GUM see AQQ500
INDIAN GUM see GLY000
INDIAN RED see IHC450
INDIAN RED see LCS000
INDIGENOUS PEANUT OIL see PAO000
INDIGO BLUE 2B see CMO000
INDISULFAT (GERMAN) see ICJ000
INDIUM see ICF000
INDIUM CHLORIDE see ICK000
INDIUM MONOPHOSPHIDE see ICI300
INDIUM NITRATE see ICI000
INDIUM PHOSPHIDE see ICI300
INDIUM SULFATE see ICJ000
INDIUM TRICHLORIDE see ICK000
INDOCID see IDA000
INDOCIN I.V. see IDA100
INDOCYANINE GREEN see CCK000
INDOCYBIN see PHU500
INDOL (GERMAN) see ICM000
3-INDOLACETONITRILE see ICW000
INDOLE see ICM000
3-INDOLEACETIC ACID see ICN000
1H-INDOLE-3-ACETIC ACID see ICN000

β-INDOLEACETIC ACID see ICN000
β-INDOLE-3-ACETIC ACID see ICN000
1H-INDOLE-3-ACETIC ACID, 1-(4-CHLOROBENZOYL)-5-METHOXY-2-METHYL-, SODIUM SALT, TRIHYDRATE see IDA100
INDOLEACETONITRILE see ICW000
INDOLE-3-ACETONITRILE see ICW000
1H-INDOLE-3-ACETONITRILE see ICW000
INDOLE-3-ALANINE see TNX000
INDOLE, 5-BENZYLOXY-3-ISONIPECOTOYL- see BFC200
1H-INDOLE-3-BUTANOIC ACID see ICP000
INDOLE BUTYRIC see ICP000
INDOLE BUTYRIC ACID see ICP000
3-INDOLEBUTYRIC ACID see ICP000
β-INDOLEBUTYRIC ACID see ICP000
γ-(INDOLE-3)-BUTYRIC ACID see ICP000
INDOLE-5-CARBONITRILE, 3-ACETYL- see COP550
INDOLE-5-CARBONITRILE, 3-(2-METHYLPROPIONYL)- see COP525
1H-INDOLE-2-CARBOXYLIC ACID, 1-(2-((1-(ETHOXYCARBONYL)BUTYL)AMINO)-1-OXOPROPYL)OCTAHYDRO-,(2S-(1(R*(R*)),2α,3Aβ,7Aβ))-, COMPD. WITH 2-METHYL-2-PROPANAMINE (1:1) see PCJ230
INDOLE, 5-(DIMETHYLAMINO)-3-(PIPERIDINOACETYL)- see DPT300
INDOLE-2,3-DIONE see ICR000
INDOLE, 3-(HYDROXYACETYL)-1-METHYL see HIS140
INDOLE, 3-(HYDROXYACETYL)-2-METHYL see HIS150
INDOLENE see ICS100
INDOLE, 1-PROPIONYL- see ICW100
INDOLIN see BBW500
INDOLINE, 5-ACETYL- see ACO320
2,3-INDOLINEDIONE see ICR000
INDOLYLACETIC ACID see ICN000
INDOLYL-3-ACETIC ACID see ICN000
3-INDOLYLACETIC ACID see ICN000
β-INDOLYLACETIC ACID see ICN000
α-INDOL-3-YL-ACETIC ACID see ICN000
INDOLYLACETONITRILE see ICW000
3-INDOLYLACETONITRILE see ICW000
1-β-3-INDOLYLALANINE see TNX000
INDOLYL-3-BUTYRIC ACID see ICP000
4-(INDOLYL)BUTYRIC ACID see ICP000
4-(INDOL-3-YL)BUTYRIC ACID see ICP000
4-(3-INDOLYL)BUTYRIC ACID see ICP000
γ-(INDOL-3-YL)BUTYRIC ACID see ICP000
γ-(3-INDOLYL)BUTYRIC ACID see ICP000
3-INDOLYL-γ-BUTYRIC ACID see ICP000
INDOL-1-YL ETHYL KETONE see ICW100
2-INDOLYL METHOXYMETHYL KETONE see ICW200
INDOL-3-YL METHYL KETONE see ICY100
INDOLYL-3-MORPHOLINOMETHYL KETONE see ICZ150
INDOLYL-3-PIPERIDINOMETHYL KETONE see ICZ200
INDOMECOL see IDA000
INDOMED see IDA000
INDOMETHACIN see IDA000
INDOMETHACIN SODIUM TRIHYDRATE see IDA100
INDOMETHAZINE see IDA000
INDOMETICINA (SPANISH) see IDA000

INDONAPHTHENE see IBX000
INDOPAN see AME500
INDOPTIC see IDA000
INDO-RECTOLMIN see IDA000
INDO-TABLINEN see IDA000
INDUSTRENE 105 see OHU000
INDUSTRENE 205 see OHU000
INDUSTRENE 206 see OHU000
INDUSTRENE 4516 see PAE250
INDUSTRENE 5016 see SLK000
INERTEEN see PJL750
INEXIT see BBQ500
ANTI-INFECTIVE VITAMIN see VSK600
INFILTRINA see DUD800
INFLAMEN see BMO000
ANTI-INFLAMMATORY HORMONE see CNS750
INFLAZON see IDA000
INFRON see VSZ100
INFUSORIAL EARTH see DCJ800
INHIBINE see HIB050
INHIBISOL see MIH275
INICARDIO see DJS200
INITIATING EXPLOSIVE DIAZODINITROPHENOL (DOT) see DUR800
INITIATING EXPLOSIVE LEAD MONONITRORESORCINATE (DOT) see LDP000
INITIATING EXPLOSIVE NITROSOGUANIDINE see NKH000
INITIATING EXPLOSIVE PENTAERYTHRITE TETRANITRATE (DOT) see PBC250
INK ORANGE JSN see HGC000
INNOXALON see EID000
INO see IDE000
INOSIE see IDE000
INOSINE see IDE000
β-INOSINE see IDE000
INOSINE, 2-AMINO- see GLS000
INOSINE-5'-MONOPHOSPHATE DISODIUM see DXE500
INOSIN-5'-MONOPHOSPHATE DISODIUM see DXE500
INOSITHEXAPHOSPHORSAEURE (GERMAN) see PIB250
INOSITOL see IDE300
i-INOSITOL see IDE300
meso-INOSITOL see IDE300
INOSITOL HEXAPHOSPHATE see PIB250
INOVITAN PP see NCR000
INSARIOTOXIN see FQS000
INSECTICIDE 1,179 see MDU600
INSECTICIDE No. 497 see DHB400
INSECTICIDE No. 4049 see MAK700
INSECTICIDE-NEMATICIDE 1410 see DSP600
INSECTOPHENE see EAQ750
6-12-INSECT REPELLENT see EKV000
INSOLUBLE SACCHARINE see BCE500
IN-SONE see PLZ000
INSPIR see ACH000
INSULAMINA see DNA200
INSULTON see MKB250
INTALBUT see BRF500
INTALPRAM see DLH630
INTEBAN SP see IDA000
INTERCHEM ACETATE DEVELOPED BLACK see DPO200
INTERCHEM ACETATE PINK BLF see AKE250
INTERCHEM ACETATE YELLOW G see AAQ250

INTERCHEM DIRECT BLACK Z see AQP000
INTERCHEM DISPERSE YELLOW GH see AAQ250
INTERCHEM HISPERSE PINK BH see AKE250
INTERFLO see PJS750
INTERNATIONAL ORANGE 2221 see LCS000
INTEROX see HIB010
INTERPINA see RDK000
INTEXAN SB-85 see DTC600
INTEXSAN CPC see CCX000
INTEXSAN LQ75 see LBW000
INTOCOSTRIN see TOA000
INTRACID FAST ORANGE G see HGC000
INTRACORT see HHR000
INTRADEX see DBD700
INTRAMYCETIN see CDP250
INTRANEFRIN see VGP000
INTRASPERSE YELLOW GBA see AAQ250
INTRASPERSE YELLOW GBA EXTRA see AAQ250
INTRATHION see PHI500
INTRATION see PHI500
INTRAVAL SODIUM see PBT500
INTRAZONE RED BR see CMM330
INVENOL see BSM000
IODATES see IDJ700
IODE (FRENCH) see IDM000
IODENTEROL see CHR500
IODIC ACIODIC ACID, POTASSIUM SALT see PLK250
IODIDES see IDL000
IODINATED CASEIN see IDL100
IODINATED GLYCEROL see IEL800
IODINE see IDM000
IODINE AZIDE see IDN000
IODINE(I) AZIDE see IDN000
IODINE AZIDE (dry) (DOT) see IDN000
IODINE CHLORIDE see IDS000
IODINE CRYSTALS see IDM000
IODINE CYANIDE see COP000
IODINE MONOCHLORIDE see IDS000
IODINE PENTAFLUORIDE see IDT000
IODINE-POTASSIUM IODIDE see PLW285
IODINE SUBLIMED see IDM000
IODIO (ITALIAN) see IDM000
IODOACETAMIDE see AOC500
IODOACETAMIDE see IDW000
2-IODOACETAMIDE see IDW000
α-IODOACETAMIDE see IDW000
IODOACETATE see IDZ000
IODOACETATE SODIUM SALT see SIN000
IODOACETIC ACID see IDZ000
IODOACETIC ACID SODIUM SALT see SIN000
N-(IODOACETYL)-3-AZABICYCLO(3.2.2)NONANE see
IDZ100
1-IODOACETYL-α-α-DIPHENYL-4-
PIPERIDINEMETHANOL see IDZ200
IODOAZIDE see IDN000
2-IODOBUTANE see IEH000
IODOCHLORHYDROXYQUINOL see CHR500
IODOCHLORHYDROXYQUINOLINE see CHR500
7-IODO-5-CHLORO-8-HYDROXYQUINOLINE see
CHR500
7-IODO-5-CHLOROXINE see CHR500
5-IODODEOXYURIDINE see DAS000
5-IODO-2'-DEOXYURIDINE see DAS000
4-IODO-3,5-DIMETHYL-N-(2-METHYLPHENYL)-1H-
PYRAZOLE-1-ACETAMIDE see IEI700
4-IODO-3,5-DIMETHYL-N-(4-METHYLPHENYL)-1H-
PYRAZOLE-1-ACETAMIDE see IEI740

IODOENTEROL see CHR500
2-(1-IODOETHYL)-1,3-DIOXOLANE-4-METHANOL
see IEL800
IODOFORM see IEP000
IODOHEPTAFLUOROPROPANE see HAY300
IODOMETANO (ITALIAN) see MKW200
IODOMETHANE see MKW200
3-IODOPROPENE see AGI250
3-IODO-1-PROPENE see AGI250
3-IODOPROPIONIC ACID see IEY000
3-IODOPROPYLENE see AGI250
IODOPROPYLIDENE GLYCEROL see IEL800
IODOQUINOL see DNF600
5-IODOURACIL DEOXYRIBOSIDE see DAS000
IODURE de MERCURE (FRENCH) see MDC750
IODURE de METHYLE (FRENCH) see MKW200
IODURIL see SHW000
IOMESAN see DFV400
IOMEZAN see DFV400
IONET S-80 see SKV100
IONET S 60 see SKV150
IONOL see BFW750
IONOL (antioxidant) see BFW750
α-IONONE see IFW000
β-IONONE see IFX000
IOPEZITE see PKX250
IOQUIN SUSPENSION see DNF600
IP-82 see IIU000
IPA see DMV600
IPDI see IMG000
IPHOSPHAMIDE see IMH000
IPN see PHX550
IPO 8 see TBW100
IPOGLICONE see BSQ000
IPOGNOX 89 see TFD500
IPOLINA see HGP500
IPPC see CBM000
IPRAL SODIUM see NBU000
IPRODIONE see GIA000
IPROGEN see DLH630
IPRONIDAZOLE (USDA) see IGH000
IPROPRAN see IGH000
IPSILON see AJD000
IPSOFLAME see BRF500
IPSOTIAN see MQU750
IR 125 see CCK000
IRADICAV see SHF500
IRAMIL see DLH630
IRAX see PJS750
IRC 453 see CIN750
IRCON see FBJ100
IRENAL see PLO500
IRENAT see PLO500
IRENE see DLS600
IRGACHROME ORANGE OS see LCS000
IRGALITE 1104 see CBT500
IRGALITE BRONZE RED CL see BNH500
IRGALITE RED CBN see CHP500
IRGALON see EIV000
IRGANOX PS 800 see TFD500
IRGASTAB 2002 see BJK560
IRGASTAB 2002 HT see BJK560
IRIFAN see TDA500
IRINOTECTAN HYDROCHLORIDE HYDRATE see
IGJ550
IRISH GUM see CCL250
IRISH MOSS EXTRACT see CCL250

IRISH MOSS GELOSE see CCL250
IRIUM see SIB600
IROCAINE see AIT250
IRO-JEX see IGS000
IROMIN see FBK000
IRON see IGK800
IRON(2+) ACETATE see FBH000
IRON(II) ACETATE see FBH000
IRON ARSENATE (DOT) see IGM000
IRON(II) ARSENATE (3:2) see IGM000
IRON(III) ARSENATE (1:1) see IGN000
IRON(III)-o-ARSENITE PENTAHYDRATE see IGO000
IRONATE see FBO000
IRON BIS(CYCLOPENTADIENE) see FBC000
IRON CAPRYLATE see IGQ050
IRON CARBONYL see IHG500
IRON, CARBONYL (FCC) see IGK800
IRON CHLORIDE see FAU000
IRON(III) CHLORIDE see FAU000
IRON(II) CHLORIDE (1:2) see FBI000
IRON CHOLINE CITRATE COMPLEX see FBC100
IRON CHROMITE see CMI500
IRON(III) CITRATE see FAW100
IRON DEXTRAN see IGS000
IRON-DEXTRAN COMPLEX see IGS000
IRON DEXTRAN INJECTION see IGS000
IRON DIACETATE see FBH000
IRON DICHLORIDE see FBI000
IRON DICYCLOPENTADIENYL see FBC000
IRON DIMETHYLDITHIOCARBAMATE see FAS000
IRON DISULFIDE see IGV000
IRON DUST see IGW000
α-IRONE see IGW500
IRON, ELECTROLYTIC see IGK800
IRON, ELEMENTAL see IGK800
IRON FLUORIDE see FAX000
IRON FUMARATE see FBJ100
IRON GLUCONATE see FBK000
IRON HYDROGENATED DEXTRAN see IGS000
IRON(2+) LACTATE see LAL000
IRON MONOSULFATE see FBN100
IRON NAPHTHENATE see IHB700
IRON NICKEL SULFIDE see NDE500
IRON NICKEL ZINC OXIDE see IHB800
IRON NITRATE see FAY200
IRON NITRATE see IHB900
IRON (III) NITRATE, ANHYDROUS see FAY200
IRON (III) NITRATE, ANHYDROUS see IHB900
IRON ORE see HAO875
IRONORM INJECTION see IGS000
IRON OXIDE see IHC450
IRON OXIDE see IHG100
IRON(III) OXIDE see IHC450
IRON OXIDE, spent see IHG100
IRON OXIDE, CHROMIUM OXIDE, and NICKEL OXIDE FUME see IHE000
IRON OXIDE RED see IHC450
IRON OXIDE RED 130B see IHG100
IRON OXIDE, SACCHARATED see IHG000
IRON PENTACARBONYL see IHG500
IRON PERSULFATE see FBA000
IRON PROTOCHLORIDE see FBI000
IRON PROTOSULFATE see FBN100
IRON PYRITES see IGV000
IRON PYROPHOSPHATE see FAZ525
IRON, REDUCED (FCC) see IGK800
IRON SACCHARATE see IHG000

IRON SESQUIOXIDE see IHC450
IRON SESQUISULFATE see FBA000
IRON SPONGE, spent obtained from coal gas purification (DOT) see IHG100
IRON SUGAR see IHG000
IRON SULFATE (2:3) see FBA000
IRON(III) SULFATE see FBA000
IRON(II) SULFATE (1:1) see FBN100
IRON(II) SULFATE (1:1), HEPTAHYDRATE see FBO000
IRON SULFIDE see IGV000
IRON TERSULFATE see FBA000
IRON TRICHLORIDE see FAU000
IRON TRIFLUORIDE see FAX000
IRON TRINITRATE see FAY200
IRON TRINITRATE see IHB900
IRON VITRIOL see FBN100
IRON VITROL see FBO000
IROSPAN see FBN100
IROSUL see FBN100
IROSUL see FBO000
IROX (GADOR) see FBK000
IRRADIATED ERGOSTA-5,7,22-TRIEN-3-β-OL see VSZ100
IRRATHENE R see PJS750
IRTRAN 1 see MAF500
IRTRAN 3 see CAS000
2341 I.S. see BIQ500
ISACONITINE see PIC250
ISATIC ACID LACTAM see ICR000
ISATIDINE see RFU000
ISATIN see ICR000
ISATINIC ACID ANHYDRIDE see ICR000
ISATOIC ACID ANHYDRIDE see IHN200
ISATOIC ANHYDRIDE see IHN200
ISCEON 22 see CFX500
ISCEON 113 see FOO000
ISCEON 122 see DFA600
ISCEON 131 see TIP500
ISCOBROME see MHR200
ISCOBROME D see EIY500
ISCOVESCO see DKA600
ISDIN see CCK125
I-SEDRIN see EAW000
ISEPAMICIN DISULFATE see IHN300
ISETHION see GFW000
ISICAINA see DHK400
ISICAINE HYDROCHLORIDE see DHK600
ISKIA-C see ARN125
ISMICETINA see CDP250
ISMIPUR see POK000
ISOACETOPHORONE see IMF400
ISOAMYCIN see BBK000
ISOAMYL ACETATE see IHO850
ISOAMYL ALCOHOL see IHP000
ISOAMYL ALCOHOL see IHP010
ISOAMYL ALKOHOL (CZECH) see IHP000
ISO-AMYLALKOHOL (GERMAN) see IHP000
ISOAMYL BENZOATE see IHP100
ISOAMYL BROMIDE see BNP250
ISOAMYL BUTANOATE see IHP400
ISOAMYL BUTYLATE see IHP400
ISOAMYL BUTYRATE see IHP400
ISOAMYL-n-BUTYRATE see IHP400
ISOAMYL CAPROATE see IHU100
ISOAMYL CAPRYLATE see IHP500
ISOAMYL CINNAMATE see AOG600

ISOAMYLENE see IHR220
ISOAMYL ETHANOATE see IHO850
5-ISOAMYL-5-ETHYLBARBITURIC ACID, SODIUM
DERIVATIVE see AON750
ISOAMYL FORMATE see IHS000
ISOAMYL GERANATE see IHS100
ISOAMYL HEXANOATE see IHU100
ISOAMYL HEXANOATE see IHU100
ISOAMYLHYDRIDE see EIK000
ISOAMYL o-HYDROXYBENZOATE see IME000
ISOAMYL ISOVALERATE (FCC) see ITB000
ISOAMYL METHANOATE see IHS000
ISOAMYL METHYL KETONE see MKW450
ISOAMYL NITRITE see IMB000
ISOAMYL OCTANOATE see IHP500
ISOAMYLOL see IHP000
ISOAMYL 3-PENTYL PROPENATE see AOG600
ISOAMYL PHENYLETHYL ETHER see IHV050
ISOAMYL PROPIONATE see AON350
ISOAMYL SALICYLATE (FCC) see IME000
ISOANETHOLE see AFW750
ISOBAC 20 see HCL000
ISOBARB see NBU000
1,3-ISOBENZOFURANDIONE see PHW750
1(3H)-ISOBENZOFURANONE, 3,3-BIS(4-
HYDROXYPHENYL)- see PDO750
1(3H)-ISOBENZOFURANONE, 3-BUTYLIDENE-(9CI)
see BRQ100
ISO-BID see CCK125
ISOBIND 100 see PKB100
ISOBORNEOL THIOCYANATOACETATE see IHZ000
ISOBORNYL ACETATE see IHX600
ISOBORNYL ACETATE see IHX600
ISOBORNYL THIOCYANATOACETATE see IHZ000
ISOBORNYL THIOCYANOACETATE see IHZ000
ISOBROMYL see BNP750
ISOBUTANAL see IJS000
ISOBUTANE see MOR750
ISOBUTANE (DOT) see MOR750
ISOBUTANE MIXTURES (DOT) see MOR750
ISOBUTANOL (DOT) see IIL000
ISOBUTENAL see MGA250
ISOBUTENE see IIC000
ISOBUTENYL CHLORIDE see CIU750
ISOBUTENYL METHYL KETONE see MDJ750
2-ISOBUTOXYETHANOL see IIP000
ISOBUTYL ACETATE see IIJ000
ISOBUTYL ACRYLATE see IIK000
ISOBUTYL ACRYLATE, inhibited (DOT) see IIK000
ISOBUTYL ADIPATE see DNH125
ISOBUTYL ALCOHOL see IIL000
ISOBUTYLALDEHYDE see IJS000
ISOBUTYL ALDEHYDE (DOT) see IJS000
ISOBUTYLALKOHOL (CZECH) see IIL000
ISOBUTYLAMINE see IIM000
ISOBUTYLBENZENE see IIN000
ISOBUTYL BROMIDE see BNR750
ISOBUTYL BUTANOATE see BSW500
ISOBUTYL-2-BUTENOATE see IIN300
ISOBUTYL BUTYRATE (FCC) see BSW500
ISOBUTYLCARBINOL see IHP000
ISOBUTYL CELLOSOLVE see IIP000
ISOBUTYL CINNAMATE see IIQ000
ISOBUTYLDIUREA see IIV000
ISOBUTYLENE (DOT) see IIC000
ISOBUTYLENEDIUREA see IIV000
ISOBUTYLESTER KYSELINY METHAKRYLOVE see
IIY000

ISOBUTYLESTER KYSELINY MRAVENCI see IIR000
ISOBUTYLESTER KYSELINY OCTOVE see IIJ000
ISOBUTYL FORMATE see IIR000
ISO-BUTYL FORMATE see IIR000
4-ISOBUTYLHYDRATROPIC ACID see IIU000
p-ISOBUTYLHYDRATROPIC ACID see IIU000
ISOBUTYL-o-HYDROXYBENZOATE see IJN000
1,1'-ISOBUTYLIDENEBISUREA see IIV000
ISOBUTYLIDENEDIUREA see IIV000
ISOBUTYL ISOBUTYRATE see IIW000
ISOBUTYLISOBUTYRATE (DOT) see IIW000
ISOBUTYL KETONE see DNI800
ISOBUTYL METHACRYLATE see IIY000
ISOBUTYL-α-METHACRYLATE see IIY000
ISOBUTYL METHYL CARBINOL see MKW600
ISOBUTYL-METHYLKETON (CZECH) see HFG500
ISOBUTYL METHYL KETONE see HFG500
ISOBUTYLMETHYLMETHANOL see MKW600
ISOBUTYL NITRITE see IJD000
ISOBUTYL PHENYLACETATE see IJF400
2-(4-ISOBUTYLPHENYL)PROPANOIC ACID see
IIU000
2-(p-ISOBUTYLPHENYL)PROPIONIC ACID see IIU000
α-(4-ISOBUTYLPHENYL)PROPIONIC ACID see IIU000
α-p-ISOBUTYLPHENYLPROPIONIC ACID see IIU000
ISOBUTYL PROPENOATE see IIK000
ISOBUTYL-2-PROPENOATE see IIK000
ISOBUTYL PROPIONATE (DOT) see PMV250
ISOBUTYL SALICYLATE see IJN000
ISOBUTYLTRIMETHYLETHANE see TLY500
ISOBUTYL VINYL ETHER see IJQ000
ISOBUTYRALDEHYD (CZECH) see IJS000
ISOBUTYRALDEHYDE see IJS000
ISOBUTYRIC ACID see IJU000
ISOBUTYRIC ACID (DOT) see IJU000
ISOBUTYRIC ACID, BENZYL ESTER see IJV000
ISOBUTYRIC ACID, ETHYL ESTER see ELS000
ISOBUTYRIC ACID, HEXYL ESTER see HFQ550
ISOBUTYRIC ACID, ISOBUTYL ESTER see IIW000
ISOBUTYRIC ACID, 2-PHENOXYETHYL ESTER
(6CI,8CI) see PDS900
ISOBUTYRIC ACID, p-TOLYL ESTER see THA250
ISOBUTYRIC ALDEHYDE see IJS000
ISOBUTYRIC ANHYDRIDE see IJW000
ISOBUTYRONITRILE see IJX000
ISOCAINE-ASID see AIT250
ISOCAINE BASE see PIV750
ISOCAINE-HEISLER see AIT250
ISOCINCHOMERONIC ACID, 1,4-DIHYDRO-2,6-
DIMETHYL-4-(2-PYRIDYL)-, DIETHYL ESTER see
DJB460
ISO-CORNOX see CIR500
ISOCROTYL CHLORIDE see IKE000
ISOCUMENE see IKG000
ISOCYANATE 580 see PKB100
ISOCYANATE de METHYLE (FRENCH) see MKX250
ISOCYANATES see IKG349
ISOCYANATOCYCLOHEXANE see CPN500
ISOCYANATOETHANE see ELS500
2-ISOCYANATOETHYL METHACRYLATE see IKG700
β-ISOCYANATOETHYL METHACRYLATE see IKG700
ISO-CYANATOMETHANE see MKX250
1-ISOCYANATO-2-METHYLBENZENE see IKG725
1-(1-ISOCYANATO-1-METHYLETHYL)-3-(1-
METHYLETHENYL)BENZENE see IKG800
3-ISOCYANATOMETHYL-3,5,5-
TRIMETHYLCYCLOHEXYLISOCYANATE see IMG000

1-ISOCYANATOPROPANE see PNP000
3-ISOCYANATOPROPYLTRIETHOXYSILANE see IKG900
γ-ISOCYANATOPROPYLTRIETHOXYSILANE see IKG900
ISOCYANIC ACID, ALLYL ESTER see AGJ000
ISOCYANIC ACID, 3-BROMOALLYL ESTER see BMT150
ISOCYANIC ACID, BUTYL ESTER see BRQ500
ISOCYANIC ACID-2-CHLOROETHYL ESTER see IKH000
ISOCYANIC ACID, 6-CHLOROHEXYL ESTER see CHL250
ISOCYANIC ACID-m-CHLOROPHENYL ESTER see CKA750
ISOCYANIC ACID-p-CHLOROPHENYL ESTER see CKB000
ISOCYANIC ACID, CYCLOHEXYL ESTER see CPN500
ISOCYANIC ACID-3,4-DICHLOROPHENYL ESTER see IKH099
ISOCYANIC ACID, DIESTER with 1,6-HEXANEDIOL see DNJ800
ISOCYANIC ACID, 3,3'-DIMETHYL-4,4'-BIPHENYLENE ESTER see DQS000
ISOCYANIC ACID, ESTER with DI-o-TOLUENEMETHANE see MJN750
ISOCYANIC ACID, ETHYL ESTER see ELS500
ISOCYANIC ACID, 4-FLUOROBUTYL ESTER see FHC200
ISOCYANIC ACID, 3-FLUOROPROPYL ESTER see FLR100
ISOCYANIC ACID, HEXAMETHYLENE ESTER see DNJ800
ISOCYANIC ACID, m-ISOPROPENYL-α-α-DIMETHYL BENZYL ESTER see IKG800
ISOCYANIC ACID, METHYLENEDI-p-PHENYLENE ESTER, POLYMER with 1,4-BUTANEDIOL see PKP000
ISOCYANIC ACID, METHYL ESTER see MKX250
ISOCYANIC ACID, METHYLPHENYLENE ESTER see TGM740
ISOCYANIC ACID, METHYLPHENYLENE ESTER see TGM750
ISOCYANIC ACID, 4-METHYL-m-PHENYLENE ESTER see TGM750
ISOCYANIC ACID-1,5-NAPHTHYLENE ESTER see NAM500
ISOCYANIC ACID, OCTADECYL ESTER see OBG000
ISOCYANIC ACID, PHENYL ESTER see PFK250
ISOCYANIC ACID, PROPYL ESTER see PNP000
ISOCYANIC ACID, o-TOLYL ESTER see IKG725
ISOCYANIC ACID, 3-(TRIETHOXYSILYL)PROPYL ESTER see IKG900
ISOCYANIC ACID, (m-TRIFLUOROMETHYLPHENYL) ESTER see TKJ250
ISOCYANIC ACID, p-XYLYLENE ESTER see XSS260
ISOCYANIDE see COI500
1-ISOCYANO-4-NITROBENZENE see IKH780
ISOCYANURIC ACID, DICHLORO- see DGN200
ISOCYANURIC ACID, DICHLORO-, POTASSIUM SALT see PLD000
ISOCYANURIC CHLORIDE see TIQ750
ISOCYANURIC DICHLORIDE see DGN200
ISODECYL METHACRYLATE see IKM000
ISODEMETON see DAP200
ISODIENESTROL see DAL600
ISODIHYDROLAVANDULYL ALDEHYDE see IKM100
ISODIPHENYLBENZENE see TBC620

ISODRIN see IKO000
ISODUR see IIV000
ISOENDOXAN see IMH000
ISOESTRAGOLE see PMQ750
ISOEUGENOL see IKQ000
ISOEUGENOL ACETATE see AAX750
1,3,4-ISOEUGENOL METHYL ETHER see IKR000
ISOEUGENYL ACETATE (FCC) see AAX750
ISOEUGENYL METHYL ETHER see IKR000
ISOFEDROL see EAW000
ISOFEDROL see EAY500
ISOFLAV BASE see DBN600
ISOFLUOROPHATE see IRF000
ISOFLUROPHATE see IRF000
ISOFORON see IMF400
ISOFORONE (ITALIAN) see IMF400
ISOFOSFAMIDE see IMH000
ISOFTALODINITRIL (CZECH) see PHX550
ISOHEXANE see IKS600
ISOHEXYL ALCOHOL see AOK750
ISOHOL see INJ000
ISOHOMOGENOL see IKR000
ISOINDOLE-1,3-DIONE see PHX000
1,3-ISOINDOLEDIONE see PHX000
1H-ISOINDOLE-1,3(2H)-DIONE, 2-(HYDROXYMETHYL)- see HMP100
1,3-ISOINDOLINEDIONE see PHX000
ISO-K see BDU500
ISOKET see CCK125
ISOL see HFP875
ISOLAN see DSK200
ISOLANE (FRENCH) see DSK200
ISOLEUCINE see IKX000
ISOLEUCINE see IKX000
dl-ISOLEUCINE see IKX010
l-ISOLEUCINE (FCC) see IKX000
ISOL LAKE RED LCS 12527 see CHP500
ISOMACK see CCK125
ISOMELAMINE see MCB000
β-ISOMER see BBR000
ISOMETASYSTOX see DAP400
ISOMETASYSTOX SULFONE see DAP600
ISOMETHEPTENE see ILK000
ISOMETHYLSYSTOX see DAP400
ISOMETHYLSYSTOX SULFONE see DAP600
ISOMETHYLSYSTOX SULFOXIDE see DAP000
ISOMIN see TEH500
ISOMYN see BBK000
ISOMYST see IQN000
ISONAL see ENB500
ISONAL (ROUSSEL) see ENB500
ISONAPHTHOL see NAX000
ISONATE see MJP400
ISONATE 390P see PKB100
ISONICOTINIC ACID, ETHYL ESTER see ELU000
ISONIPECAINE HYDROCHLORIDE see DAM700
3-ISONIPECOTYLINDOLE see ILG200
ISONITOX see MIW250
ISONITROPROPANE see NIY000
ISONORENE see DMV600
ISONYL see ILK000
ISOOCTANE (DOT) see TLY500
ISOOCTANOL see ILL000
ISOOCTYL ALCOHOL see ILL000
2-ISOOCTYL AMINE see ILM000
ISOOCTYL PHTHALATE see ILR100
ISOPENTANE (DOT) see EIK000

ISOPENTANOIC ACID (DOT) see ISU000
ISOPENTANOIC ACID, PHENYLMETHYL ESTER see ISW000
ISOPENTANOL see IHP000
ISOPENTENE see IHR220
ISOPENTENES (DOT) see IHR220
ISOPENT-2-ENYL ACETATE see DOQ350
ISOPENTYL ACETATE see IHO850
ISOPENTYL ALCOHOL see IHP000
ISOPENTYL ALCOHOL ACETATE see IHO850
ISOPENTYL ALCOHOL, FORMATE see IHS000
ISOPENTYL ALCOHOL NITRITE see IMB000
ISOPENTYL BENZOATE see IHP100
ISOPENTYL BROMIDE see BNP250
ISOPENTYL BUTANOATE see IHP400
ISOPENTYL BUTYRATE see IHP400
ISOPENTYL FORMATE see IHS000
ISOPENTYL HEXANOATE see IHU100
ISOPENTYL-n-HEXANOATE see IHU100
ISOPENTYL-2-HYDROXYPHENYL METHANOATE see IME000
ISOPENTYL ISOVALERATE see ITB000
ISOPENTYL METHYL KETONE see MKW450
ISOPENTYL NITRITE see IMB000
3-((ISOPENTYL)NITROSOAMINO)-2-BUTANONE see MHW350
ISOPENTYL OCTANOATE see IHP500
ISOPENTYL SALICYLATE see IME000
ISOPESTOX see PHF750
ISOPHEN see MDT600
ISOPHEN see DBA800
ISOPHENERGAN see DQA400
ISOPHENICOL see CDP250
ISOPHORONE see IMF400
ISOPHORONE DIAMINE DIISOCYANATE see IMG000
ISOPHORONE DIISOCYANATE see IMG000
ISOPHORONEDIISOCYANATE, solution, 70%, by weight (DOT) see IMG000
ISOPHORONENITRILE see IMG500
ISOPHOSPHAMIDE see IMH000
ISOPHRIN see NCL500
ISOPHRINE see SPC500
ISOPHRIN HYDROCHLORIDE see SPC500
ISOPHTHALODINITRILE see PHX550
ISOPHTHALONITRILE see PHX550
ISOPRENALINE see DMV600
ISOPRENE see IMS000
ISOPRENE, INHIBITED (DOT) see IMS000
ISOPROCARB see MIA250
ISOPROCARBE see MIA250
ISOPROMETHAZINE see DQA400
ISOPROPANETHIOL see IMU000
ISOPROPANOL (DOT) see INJ000
ISOPROPENE CYANIDE see MGA750
ISOPROPENIL-BENZOLO (ITALIAN) see MPK250
ISOPROPENYL ACETATE see MQK750
ISOPROPENYL ACETATE (DOT) see MQK750
ISOPROPENYL-BENZEEN (DUTCH) see MPK250
ISOPROPENYLBENZENE see MPK250
ISOPROPENYL-BENZOL (GERMAN) see MPK250
ISOPROPENYL CARBINOL see IMW000
4-ISOPROPENYL-CYCLOHEX-1-ENE-1-METHANOL see PCI550
ISOPROPENYLESTER KYSELINY OCTOVE see MQK750
(+)-4-ISOPROPENYL-1-METHYLCYCLOHEXENE see LFU000

ISOPROPENYLNITRILE see MGA750
ISOPROPHYL METHYLPHOSPHONOFLUORIDATE see IPX000
ISOPROPILAMINA (ITALIAN) see INK000
ISOPROPILBENZENE (ITALIAN) see COE750
ISOPROPILE (ACETATO di) (ITALIAN) see INE100
ISOPROPIL-N-FENIL-CARBAMMATO (ITALIAN) see CBM000
(1-ISOPROPIL-3-METIL-1H-PIRAZOL-5-IL)-N,N-DIMETIL-CARBAMMATO (ITALIAN) see DSK200
(3)-O-2-ISOPROPOXY-CARBONYL-1-METHYLVINYL-O-METHYL ETHYLPHOSPHORAMIDOTHIOATE see MKA000
2-ISOPROPOXYETHANOL see INA500
(ISOPROPOXYMETHYL)OXIRANE see IPD000
ISOPROPOXYMETHYLPHORYL, FLUORIDE see IPX000
2-ISOPROPOXYPHENYL (METHYL)(T-BUTOXYSULFINYL)CARBAMATE see INE062
2-ISOPROPOXYPROPANE see IOZ750
ISOPROPYDRIN see DMV600
ISOPROPYLACETAAT (DUTCH) see INE100
ISOPROPYLACETAT (GERMAN) see INE100
ISOPROPYL ACETATE see INE100
ISOPROPYL (ACETATE d') (FRENCH) see INE100
ISOPROPYLACETIC ACID see ISU000
ISOPROPYL ACETIC ACID, BENZYL ESTER see ISW000
ISOPROPYLACETONE see HFG500
ISOPROPYL ACID PHOSPHATE solid see PHE500
ISOPROPYL ACRYLAMIDE see INH000
N-ISOPROPYLACRYLAMIDE see INH000
ISOPROPYLADRENALINE see DMV600
ISOPROPYL ALCOHOL see INJ000
ISO-PROPYLALKOHOL (GERMAN) see INJ000
ISOPROPYLAMID KYSELINY AKRYLOVE see INH000
ISOPROPYLAMINE see INK000
ISOPROPYLAMINOETHANOL see INN400
2-ISOPROPYLAMINOETHANOL see INN400
N-ISOPROPYLAMINOETHANOL see INN400
ISOPROPYLAMINO-O-ETHYL-(4-METHYLMERCAPTO-3-METHYLPHENYL)PHOSPHATE see FAK000
ISOPROPYLAMINOMETHYL-3,4-DIHYDROXYPHENYL CARBINOL see DMV600
α-(ISOPROPYLAMINOMETHYL)PROTOCATECHUYL ALCOHOL see DMV600
N-ISOPROPYLANILINE see INX000
ISOPROPYLANTIPYRIN see INY000
ISOPROPYLANTIPYRINE see INY000
4-ISOPROPYLANTIPYRINE see INY000
ISOPROPYLARTERENOL see DMV600
4-ISOPROPYLBENZALDEHYDE see COE500
p-ISOPROPYLBENZALDEHYDE see COE500
ISOPROPYLBENZEEN (DUTCH) see COE750
ISOPROPYL BENZENE see COE750
p-ISOPROPYLBENZENECARBOXALDEHYDE see COE500
ISOPROPYLBENZENE HYDROPEROXIDE see IOB000
ISOPROPYLBENZENE PEROXIDE see DGR600
ISOPROPYLBENZOL see COE750
ISOPROPYL-BENZOL (GERMAN) see COE750
p-ISOPROPYLBENZONITRILE see IOD050
3-ISOPROPYL-2,1,3-BENZOTHIADIAZINON-(4)-2,2-DIOXID (GERMAN) see MJY500

3-ISOPROPYL-1H-2,1,3-BENZOTHIADIAZIN-4(3H)-ONE-2,2-DIOXIDE see MJY500
ISOPROPYL BORATE see IOI000
ISOPROPYL BROMIDE see BNY000
ISOPROPYL CARBAMATE see IOJ000
1-(p-ISOPROPYLCARBAMOYLBENZYL)-2-METHYLHYDRAZINE HYDROCHLORIDE see PME500
2-(p-(ISOPROPYLCARBAMOYL)BENZYL)-1-METHYLHYDRAZINE HYDROCHLORIDE see PME500
1-ISOPROPYL CARBAMOYL-3-(3,5-DICHLOROPHENYL)-HYDANTOIN see GIA000
ISOPROPYL CARBANILATE see CBM000
ISOPROPYL CARBANILIC ACID ESTER see CBM000
ISOPROPYLCARBINOL see IIL000
ISOPROPYL CARBITOL see IOJ500
ISOPROPYL CELLOSOLVE see INA500
ISOPROPYL CHLORIDE see CKQ000
ISOPROPYL CHLOROCARBONATE see IOL000
ISOPROPYL CHLOROFORMATE see IOL000
ISOPROPYL CHLOROMETHANOATE see IOL000
ISOPROPYL CITRATE see IOO222
ISOPROPYL CRESOL see TFX810
ISOPROPYL-o-CRESOL see CCM000
6-ISOPROPYL-m-CRESOL see TFX810
ISOPROPYL CYANIDE see IJX000
4-ISOPROPYLCYCLOHEXANOL see IOO300
p-ISOPROPYLCYCLOHEXANOL see IOO300
ISOPROPYL-2,4-D ESTER see IOY000
N-ISOPROPYL-β-DIHYDROXYPHENYL-β-HYDROXYETHYLAMINE see DMV600
ISOPROPYL DIMETHYL CARBINOL see AOK750
4-ISOPROPYL-2,3-DIMETHYL-1-PHENYL-3-PYRAZOLIN-5-ONE see INY000
ISOPROPYL EPOXYPROPYL ETHER see IPD000
ISOPROPYLESTER KYSELINY DUSITE see IQQ000
ISOPROPYLESTER KYSELINY KARBAMINOVE see IOJ000
ISOPROPYLESTER KYSELINY OCTOVE see INE100
N-ISOPROPYLETHANOLAMINE see INN400
ISOPROPYL ETHER see IOZ750
ISOPROPYL-N-FENYL-CARBAMAAT (DUTCH) see CBM000
ISOPROPYL FLUOPHOSPHATE see IRF000
ISOPROPYL FORMATE see IPC000
ISOPROPYLFORMIC ACID see IJU000
ISOPROPYL GLYCIDYL ETHER see IPD000
ISOPROPYL GLYCOL see INA500
ISOPROPYLIDENEACETONE see MDJ750
4,4'-ISOPROPYLIDENEBISPHENOL see BLD500
p,p'-ISOPROPYLIDENEBISPHENOL see BLD500
2,3-ISOPROPYLIDENEDIOXYPHENYL METHYLCARBAMATE see DQM600
p,p'-ISOPROPYLIDENEDIPHENOL see BLD500
4,4'-ISOPROPYLIDENEDIPHENOL DIGLYCIDYL ETHER see BLD750
4,4'-ISOPROPYLIDENEDIPHENOL, POLYMER with 1-CHLORO-2,3-EPOXYPROPANE see IPO000
ISOPROPYLIDENE GLYCEROL see DVR600
1,2-o-ISOPROPYLIDENE GLYCEROL see DVR600
ISOPROPYL ISOBUTYL ARSINIC ACID see IPS100
ISOPROPYLKYANID see IJX000
ISOPROPYL MERCAPTAN (DOT) see IMU000
ISOPROPYL MESYLATE see IPY000
ISOPROPYL METHANEFLUOROPHOSPHONATE see IPX000

ISOPROPYLMETHANESULFONATE see IPY000
ISOPROPYL METHANE SULPHONATE see IPY000
4-ISOPROPYL-1-METHYLBENZENE see CQI000
ISOPROPYL 2-METHYL-2-BUTENOATE see IRN100
ISOPROPYL α-METHYL CROTONATE see IRN100
4-ISOPROPYL-1-METHYL-1,5-CYCLOHEXADIENE see MCC000
5-ISOPROPYL-2-METHYL-1,3-CYCLOHEXADIENE see MCC000
2-ISOPROPYL-5-METHYLCYCLOHEXANOL see MCF750
4-ISOPROPYL-1-METHYLCYCLOHEXAN-3-OL see MCG000
l-2-ISOPROPYL-5-METHYL-CYCLOHEXAN-1-OL ACETATE see MCG750
2-ISOPROPYL-5-METHYL-CYCLOHEXAN-1-ONE, racemic see MCE250
ISOPROPYL METHYLFLUOROPHOSPHATE see IPX000
2-ISOPROPYL-5-METHYL-2-HEXEN-1-AL see IKM100
N-ISOPROPYL-p-(2-METHYLHYDRAZINOMETHYL)BENZAMIDEHYDROCHLORIDE see PME500
N-ISOPROPYL-α-(2-METHYLHYDRAZINO)-p-TOLUAMIDE HYDROCHLORIDE see PME500
p-ISOPROPYL-α-METHYLHYDROCINNAMIC ALDEHYDE see COU500
ISOPROPYL METHYL KETONE see MLA750
2-ISOPROPYL-1-METHYL-5-NITROIMIDAZOLE see IGH000
2-ISOPROPYL-5-METHYLPHENOL see TFX810
5-ISOPROPYL-2-METHYLPHENOL see CCM000
p-ISOPROPYL-α-METHYLPHENYLPROPYL ALDEHYDE see COU500
ISOPROPYL METHYLPHOSPHONOFLUORIDATE see IPX000
O-ISOPROPYL METHYLPHOSPHONOFLUORIDATE see IPX000
ISOPROPYL-METHYL-PHOSPHORYL FLUORIDE see IPX000
(1-ISOPROPYL-3-METHYL-1H-PYRAZOL-5-YL)-N,N-DIMETHYLCARBAMAAT (DUTCH) see DSK200
(1-ISOPROPYL-3-METHYL-1H-PYRAZOL-5-YL)-N,N-DIMETHYL-CARBAMAT (GERMAN) see DSK200
ISOPROPYLMETHYLPYRAZOLYL DIMETHYLCARBAMATE see DSK200
1-ISOPROPYL-3-METHYL-5-PYRAZOLYL DIMETHYLCARBAMATE see DSK200
1-ISOPROPYL-3-METHYLPYRAZOLYL-(5)-DIMETHYLCARBAMATE see DSK200
O-2-ISOPROPYL-4-METHYLPYRIMIDYL-O,O-DIETHYL PHOSPHOROTHIOATE see DCM750
ISOPROPYLMETHYLPYRIMIDYL DIETHYL THIOPHOSPHATE see DCM750
ISOPROPYL MYRISTATE see IQN000
ISOPROPYL NITRATE see IQP000
ISOPROPYL NITRILE see IJX000
ISOPROPYL NITRITE see IQQ000
ISOPROPYL NORADRENALINE see DMV600
1-ISOPROPYLNORADRENALINE see DMV600
N-ISOPROPYLNORADRENALINE see DMV600
ISOPROPYL OILS see IQU000
3-ISOPROPYLOXYPROPYLENE OXIDE see IPD000
ISOPROPYL PERCARBONATE see DNR400
ISOPROPYL PEROXYDICARBONATE see DNR400
ISOPROPYLPHENAZONE see INY000
2-ISOPROPYLPHENOL see IQX100

4-ISOPROPYLPHENOL see IQZ000
o-ISOPROPYLPHENOL see IQX100
p-ISOPROPYLPHENOL see IQZ000
o-ISOPROPYLPHENOL METHYLCARBAMATE see MIA250
ISOPROPYL-N-PHENYL-CARBAMAT (GERMAN) see CBM000
ISOPROPYL PHENYLCARBAMATE see CBM000
ISOPROPYL-N-PHENYLCARBAMATE see CBM000
o-ISOPROPYL-N-PHENYL CARBAMATE see CBM000
2-ISOPROPYLPHENYL N-METHYLCARBAMATE see MIA250
o-ISOPROPYLPHENYL N-METHYLCARBAMATE see MIA250
ISOPROPYL-N-PHENYLURETHAN (GERMAN) see CBM000
ISOPROPYL PHOSPHORIC ACID see PHE500
ISOPROPYL PHOSPHOROFLUORIDATE see IRF000
N-ISOPROPYLPYRROLIDINONE see IRG100
1-ISOPROPYL-2-PYRROLIDINONE see IRG100
ISOPROPYL TETRADECANOATE see IQN000
ISOPROPYLTHIOL see IMU000
ISOPROPYL TIGLATE see IRN100
p-ISOPROPYLTOLUENE see CQI000
4-ISOPROPYL-2,6,7-TRIOXA-1-ARSABICYCLO(2.2.2)OCTANE see IRQ100
ISOPROPYL UNOPROSTONE see IRR050
(+)-ISOPROPYL, Z-7-((1R,2R,3R,5S)-3,5-DIHYDROXY-2-(3-OXODECYL)CYCLOPENTYL)HEPT-5-ENOATE see IRR050
ISOPROTERENOL see DMV600
l-ISOPROTERENOL see DMV600
ISOPTO-CARPINE see PIF250
ISOPTO CETAMIDE see SNQ710
ISOPTO FENICOL see CDP250
ISOPTO HYOSCINE see SBG000
ISOPULEGOL (FCC) see MCE750
ISO-PUREN see CCK125
ISOPYRINE see INY000
ISOQUINOLINE see IRX000
ISOQUINOLINE, 6,7-DIMETHOXY-1-VERATRYL-, HYDROCHLORIDE see PAH250
ISORBID see CCK125
ISORDIL see CCK125
ISORDIL TEMBIDS see CCK125
ISORENIN see DMV600
ISORETINENE a see VSK975
ISOSAFROEUGENOL see IRY000
ISOSAFROLE see IRZ000
ISOSCOPIL see HOT500
ISOSET CX 11 see PKB100
ISOSORBIDE DINITRATE see CCK125
ISOSTEARYL NEOPENTANOATE see ISC550
ISOSTENASE see CCK125
ISOTACTIC POLYPROPYLENE see PMP500
ISOTAZIN see DIR000
ISOTHAN see LBW000
ISOTHAZINE see DIR000
ISOTHIAZINE see DIR000
ISOTHIAZOLINONE CHLORIDE see ISD066
3(2H)-ISOTHIAZOLONE, 5-CHLORO-2-METHYL-, MIXT. WITH 2-METHYL-3(2H)-ISOTHIAZOLONE see ISD066
ISOTHIOCYANATE d'ALLYLE (FRENCH) see AGJ250
ISOTHIOCYANATE de METHYLE (FRENCH) see ISE000
ISOTHIOCYANATOBENZENE see ISQ000
ISOTHIOCYANATOMETHANE see ISE000

3-ISOTHIOCYANATO-1-PROPENE see AGJ250
ISOTHIOCYANIC ACID BENZYL ESTER see BEU250
ISOTHIOCYANIC ACID, METHYL ESTER see ISE000
ISOTHIOCYANIC ACID-p-PHENYLENE ESTER see PFA500
ISOTHIOCYANIC ACID, PHENYL ESTER see ISQ000
ISOTHIOUREA see ISR000
ISOTHIOURONIUM CHLORIDE, BENZYL see BEU500
ISOTHYMOL see CCM000
ISOTIOCIANATO di METILE (ITALIAN) see ISE000
ISOTONIL see DRM000
ISOTOX see BBQ500
ISOTRATE see CCK125
ISOTRETINOIN see VSK955
ISOTRON 11 see TIP500
ISOTRON 12 see DFA600
ISOTRON 22 see CFX500
ISOUREA see USS000
ISOURETHANE see PKL500
ISOVAL see BNP750
8-ISOVALERATE see FQS000
ISOVALERIANIC AICD see ISU000
ISOVALERIC ACID see ISU000
ISOVALERIC ACID, ALLYL ESTER see ISV000
ISOVALERIC ACID, BENZYL ESTER see ISW000
ISOVALERIC ACID, BUTYL ESTER see ISX000
(E)-ISOVALERIC ACID-3,7-DIMETHYL-2,6-OCTADIENYL ESTER see GDK000
ISOVALERIC ACID, ETHYL ESTER see ISY000
(Z)-ISOVALERIC ACID-3-HEXENYL see ISZ000
ISOVALERIC ACID, ISOPENTYL ESTER see ITB000
ISOVALERIC ACID, METHYL ESTER see ITC000
ISOVALERONE see DNI800
ISOXAMIN see SNN500
ISRAVIN see DBX400
ISTIN see DMH400
ISTONYL see DRM000
ISUPREL see DMV600
ISUPREN see DMV600
IT 40 see SMQ500
ITAMID see PJY500
ITAMIDONE see DOT000
ITINEROL see HGC500
ITOPAZ see EEH600
ITURAN see NGE000
IUDR see DAS000
5-IUDR see DAS000
IVALON see PKP750
IVALON see FMV000
IVAUGAN see CFY000
IVE see IJQ000
IVERMECTIN see ITD875
IVIRON see IHG000
IVORAN see DAD200
IVORIT see EJM500
IXODEX see DAD200
IXOTEN see TNT500
IXPER 25M see MAH750
IZOFORON (POLISH) see IMF400
IZOPROPYLOWY ETER (POLISH) see IOZ750
IZOSYSTOX (CZECH) see DAP200
J 242 see AHG000
J 2Fp see GLU000
J 400 see PMP500
J 0852 see ZJS400
JACUTIN see BBP750
JACUTIN see BBQ500
JADE GREEN BASE see JAT000

JADO see CBT500
JA-FA IPM see IQN000
JAFFNA TOBACCO see BFW135
JAGUAR see GLU000
JAGUAR 6000 see GLU000
JAGUAR A 20D see GLU000
JAGUAR A 40F see GLU000
JAGUAR A 20 B see GLU000
JAGUAR GUM A-20-D see GLU000
JAGUAR No. 124 see GLU000
JAGUAR PLUS see GLU000
JAIKIN see SFF000
JANIMINE see DLH630
JANUPAP see HIM000
JANUS GREEN B see DHM500
JANUS GREEN V see DHM500
JAPAN AGAR see AEX250
JAPAN CAMPHOR see CBA750
JAPANESE CAMPHOR OIL see CBB500
JAPANESE OIL OF CAMPHOR see CBB500
JAPAN ISINGLASS see AEX250
JAPAN RED 104 see ADG250
JASAD see ZBJ000
JASMINALDEHYDE see AOG500
JATRONEURAL see TKK250
JAUNE AB see FAG130
JAUNE de BEURRE (FRENCH) see DOT300
JAUNE OB see FAG135
JAUNE ORANGE S see FAG150
JAUNE SOLEIL see FAG150
JAVA METANIL YELLOW G see MDM775
JAVA ORANGE 2G see HGC000
JAVA PONCEAU 2R see FMU070
JAVA RUBINE N see HJF500
JAVA SCARLET 3R see FMU080
JAVEX see SHU500
JAYSOL see EFU000
JAYSOL S see EFU000
JB 8181 see DLS600
177 J.D. see AJD000
JEFFAMINE AP-20 see MJQ000
JEFFERSOL DB see DJF200
JEFFERSOL EB see BPJ850
JEFFERSOL EE see EES350
JEFFERSOL EM see EJH500
JEFFOX see PJT000
JEFFOX see PJT200
JEFFOX see PKI500
JEN-DIRIL see CFY000
JERVINE see JCS000
JESUIT'S BALSAM see CNH792
JEWELER'S ROUGE see IHC450
JIFFY GROW see ICP000
JISC 3108 see AGX000
JISC 3110 see AGX000
J-LIBERTY see MDQ250
JOD (GERMAN, POLISH) see IDM000
2-JODBUTAN see IEH000
JODCYAN see COP000
JODDEOXIURIDIN see DAS000
JODID SODNY see SHW000
JOD-METHAN (GERMAN) see MKW200
JOLT see EIN000
JONIT see PFA500
JOOD (DUTCH) see IDM000
JOODMETHAAN (DUTCH) see MKW200
JORCHEM 400 ML see PJT000
JP-10 see TLR675

J SOFT C 4 see DTC600
J-SUL see SNN500
JUDEAN PITCH see ARO500
JULIN'S CARBON CHLORIDE see HCC500
JUNIPER BERRY OIL see JEA000
JUNIPER OIL see JEA000
JUSONIN see SFC500
JUVASON see PLZ000
JUVELA see TGJ050
JUVOCAINE see AIT250
JZF see BLE500
K 17 see TEH500
K 52 see OHU000
K-82 see CAJ770
K-83 see CAJ772
K 257 see CBT500
K 525 see SMQ500
K6-30 see ARM268
K 4710 see KFK000
K62-105 see LEN000
K25 (polymer) see PKQ250
KADMIUM (GERMAN) see CAD000
KADMIUMCHLORID (GERMAN) see CAE250
KADMIUMSTEARAT (GERMAN) see OAT000
KADMU TLENEK (POLISH) see CAH500
KADOX-25 see ZKA000
KAERGONA see MMD500
KAFAR COPPER see CNI000
KAISER CHEMICALS 11 see FOO000
KAISER CHEMICALS 12 see DFA600
KAISER NCO 20 see PKB100
KAKO BLUE B SALT see DCJ200
KAKODYLAN DODNY see HKC500
KAKO RED TR BASE see CLK220
KAKO TARTRAZINE see FAG140
KALCIT see CAD800
KALEX see EIV000
KALGAN see PFC750
KALITABS see PLA500
KALIUMARSENIT (GERMAN) see PKV500
KALIUM-BETA see PLG800
KALIUMCARBONAT (GERMAN) see PLA000
KALIUMCHLORAAT (DUTCH) see PLA250
KALIUMCHLORAT (GERMAN) see PLA250
KALIUMCYANAT (GERMAN) see PLC250
KALIUM-CYANID (GERMAN) see PLC500
KALIUMDICHROMAT (GERMAN) see PKX250
KALIUMHYDROXID (GERMAN) see PLJ500
KALIUMHYDROXYDE (DUTCH) see PLJ500
KALIUMNITRAT (GERMAN) see PLL500
KALIUMPERMANGANAAT (DUTCH) see PLP000
KALIUMPERMANGANAT (GERMAN) see PLP000
KALIUMPERMANGANAT (GERMAN) see CAV250
KALKHYDRATE see CAT225
KALLOCRYL K see PKB500
KALLODENT CLEAR see PKB500
KALMETTUMSOMNIFERUM see GFA000
KALMOCAPS see MDQ250
KALMOCAPS see LFK000
KALMUS OEL (GERMAN) see OGK000
KALO see ARB750
KALOMEL (GERMAN) see MCW000
KALPREN see CAS750
KALTOSTAT see CAM200
KALZIUMARSENIAT (GERMAN) see ARB750
KALZIUMZYKLAMATE (GERMAN) see CAR000
KAM 1000 see SLK000
KAM 2000 see SLK000

KAM 3000 see SLK000
KAMAVER see CDP250
KAMBAMINE RED TR see CLK220
KAMFOCHLOR see CDV100
KAMILLENOEL see CDH500
KAMPFER (GERMAN) see CBA750
KAMPOSAN see CDS125
KAMPSTOFF "LOST" see BIH250
KANAMICINA (ITALIAN) see KAL000
KANAMYCIN see KAL000
KANAMYCIN A see KAL000
KANAMYTREX see KAL000
KANDISET see BCE500
KANECHLOR see PJL750
KANECHLOR 300 see PJL750
KANECHLOR 300 see PJO500
KANECHLOR 400 see PJL750
KANECHLOR 400 see PJO750
KANECHLOR 500 see PJP000
KANNIT see XPJ000
KANONE see MMD500
KANTREX see KAL000
KAOCHLOR see PLA500
KAOLIN see KBB600
KAON see PLG800
KAON-Cl see PLA500
KAON ELIXIR see PLG800
KAOPAOUS see KBB600
KAOPHILLS-2 see KBB600
β,KAPPA-CAROTEN-6'-ONE, 3,3'-DIHYDROXY-,
(3R,3'S,5'R)- see CBF760
KAPPAXAN see MMD500
e-KAPROLAKTAM (CZECH) see CBF700
KAPROLIT see PJY500
KAPROLON see PJY500
KAPROMIN see PJY500
KAPRON see PJY500
KAPTAN see CBG000
KAPTAX see BDF000
KARATE see LAS200
KARAYA GUM see KBK000
KARBAM BLACK see FAS000
KARBAM WHITE see BJK500
KARBANIL see PFK250
KARBARYL (POLISH) see CBM750
KARBASPRAY see CBM750
KARBATOX see CBM750
KARBATOX 75 see CBM750
KARBATOX ZAWIESINOWY see CBM750
KARBOFOS see MAK700
KARBORAFIN see CDI000
KARBOSEP see CBM750
KARBROMAL see BNK000
KARCON see MMD500
KARDIAMID see DJS200
KARDONYL see DJS200
KARENZU DK2 see PFU300
KAREON see MMD500
KARIDIUM see SHF500
KARIGEL see SHF500
KARION see SKV200
KARI-RINSE see SHF500
KARLAN see RMA500
KARMESIN see HJF500
KARMEX see DXQ500
KARMEX DIURON HERBICIDE see DXQ500
KARMEX DW see DXQ500

KARMEX MONURON HERBICIDE see CJX750
KARMEX W. MONURON HERBICIDE see CJX750
KARMINOMYCIN see KBU000
KARNOZZN see CCK665
KARSAN see FMV000
KARTRYL see BNK000
KASSIA OEL (GERMAN) see CCO750
KASTONE see HIB010
KATAMINE AB see DTC600
KATCHUNG OIL see PAO000
KATHON 886 see ISD066
KATHON BIOCIDE see ISD066
KATHON CG see ISD066
KATHON LP PRESERVATIVE see OFE000
KATHON LX see ISD066
KATHON RH 886 see ISD066
KATHON SP 70 see OFE000
KATHON 886 W see ISD066
KATHON 886MW see ISD066
KATHON WT see ISD066
KATHRO see CMD750
KATIV-G see MMD500
KATLEX see CHJ750
KATORIN see PLG800
KATRON see BCA000
KAUTSCHIN see MCC250
KAYAFUME see MHR200
KAYAKU ACID BRILLIANT SCARLET 3R see FMU080
KAYAKU BLUE B BASE see DCJ200
KAYAKU CONGO RED see SGQ500
KAYAKU DIRECT see CMO000
KAYAKU DIRECT DEEP BLACK EX see AQP000
KAYAKU DIRECT DEEP BLACK GX see AQP000
KAYAKU DIRECT DEEP BLACK S see AQP000
KAYAKU DIRECT LEATHER BLACK EX see AQP000
KAYAKU DIRECT SKY BLUE 5B see CMO500
KAYAKU DIRECT SKY BLUE 6B see CMN750
KAYAKU DIRECT SPECIAL BLACK AAX see AQP000
KAYAKU FOOD COLOUR YELLOW NO. 4 see
FAG140
KAYAKU SCARLET G BASE see NMP500
KAYAKU TARTRAZINE see FAG140
KAYALON FAST BLUE BR see TBG700
KAYALON FAST VIOLET BR see DBY700
KAYALON FAST YELLOW G see AAQ250
KAYANOL MILLING RED RS see CMM330
KAYASET YELLOW G see AAQ250
KAYAZINON see DCM750
KAYAZOL see DCM750
KAY CIEL see PLA500
KAYDOL see MQV750
KAYDOL see MQV875
KAYEXALATE see SJK375
KAYKLOT see MMD500
KAYLITE see PKQ059
KAYQUINONE see MMD500
KAYTRATE see PBC250
KAZOE see SFA000
KB see ISD066
KB (POLYMER) see SMQ500
K-BRITE see SHR500
KC-400 see PJO750
KC-500 see PJP000
KCA ACETATE FAST YELLOW G see AAQ250
KCA FOODCOL SUNSET YELLOW FCF see FAG150
KCA FOODCOL TARTRAZINE PF see FAG140
KCA TARTRAZINE PF see FAG140
K 55E see SMR000

KE 709 see RNU100
KEDAVON see TEH500
KEESTAR see SLJ500
KEFENID see BDU500
KELACID see AFL000
KELAMERAZINE see ALF250
KELCO GEL LV see SEH000
KELCOLOID see PNJ750
KELCOSOL see SEH000
KELENE see EHH000
KELGIN see SEH000
KELGUM see SEH000
KELOFORM see EFX000
KELP see KDK700
KELSET see SEH000
KELSIZE see SEH000
KELTANE see BIO750
KELTEX see SEH000
KELTHANE (DOT) see BIO750
p,p'-KELTHANE see BIO750
KELTHANE DUST BASE see BIO750
KELTHANETHANOL see BIO750
KELTONE see SEH000
KEMAMINE P690 see DXW000
KEMAMINE P 989 see OHM700
KEMAMINE P990 see OBC000
KEMDAZIN see MHC750
KEMESTER 105 see OHW000
KEMESTER 115 see OHW000
KEMESTER 205 see OHW000
KEMESTER 213 see OHW000
KEMICETINE see CDP250
KEMIKAL see CAT225
KEMITRACIN 10 see BAC260
KEMODRIN see DBA800
KEMOLATE see PHX250
KENACHROME BLUE 2R see HJF500
KENACORT-A see AQX500
KENALOG see AQX500
KENAPON see DGI400
KENDALL'S COMPOUND F see CNS750
KEPMPLEX 100 see EIV000
KEPONE see KEA000
KERALYT see SAI000
KERASALICYL see SJO000
KERECID see DAS000
KEROCAINE see AIT250
KEROSAL see SJO000
KEROSENE see KEK000
KEROSENE (PETROLEUM), hydrotreated see KEK100
KEROSINE see KEK000
KEROSINE (petroleum) see KEK000
KESSAR see TAD175
KESSCO 40 see OAV000
KESSCOFLEX MCP see DOF400
KESSCOFLEX TRA see THM500
KESSCO ISOPROPYL MYRISTATE see IQN000
KESSCOMIR see IQN000
KESSOBAMATE see MQU750
KESSODANTEN see DKQ000
KESSODRATE see CDO000
KESTREL (Pesticide) see AHJ750
KESTRIN see ECU750
KESTRONE see EDV000
KETAJECT see CKD750
KETALAR see CKD750
KETALGIN see MDO750
KETAMINE see CKD750

KETAMINE HYDROCHLORIDE see CKD750
KETANEST see CKD750
KETANRIFT see ZVJ000
KETASET see CKD750
KETAVET see CKD750
KETENE see KEU000
KETENE DIMER see KFA000
KETJENBLACK EC see CBT750
Δ-KETO 153 see KFK200
3-(3-KETO-7-α-ACETYLTHIO-17-β-HYDROXY-4-
ANDROSTEN-17-α-YL)PROPIONIC ACID LACTONE
see AFJ500
KETOBEMIDONE see KFK000
4-KETOBENZOTRIAZINE see BDH000
KETOBUN-A see ZVJ000
β-KETOBUTYRANILIDE see AAY000
KETOCYCLOPENTANE see CPW500
KETODESTRIN see EDV000
Δ-KETOENDRIN see KFK200
KETO-ETHYLENE see KEU000
3-KETO-l-GULOFURANOLACTONE see ARN000
KETOHEXAMETHYLENE see CPC000
2-KETOHEXAMETHYLENIMINE see CBF700
KETOHYDROXY-ESTRATRIENE see EDV000
KETOHYDROXYESTRIN see EDV000
KETOHYDROXYESTRIN BENZOATE see EDV500
KETOHYDROXYOESTRIN see EDV000
2,3-KETOINDOLINE see ICR000
KETOLAR see CKD750
KETOLE see ICM000
γ-KETO-β-METHOXY-Δ-METHYLENE-Δ^α-
HvEXENOIC ACID see PAP750
KETONE, 2-AMINO-5-BENZIMIDAZOLYL PHENYL
see AIH000
KETONE, p-ANISYL 4-PYRIDYL see MFH930
KETONE, 3-AZABICYCLO(3.2.2)NONYL
CHLOROMETHYL see CEC100
KETONE, 3-AZABICYCLO(3.2.2)NONYL
IODOMETHYL see IDZ100
KETONE, 1-AZIRIDINYL 3-(BIS(2-
CHLOROETHYL)AMINO)-p-TOLYL see BHQ760
KETONE, 1-BENZYL-2-INDOLYL
HYDROXYMETHYL- see BES300
KETONE, 5-BENZYLOXY-3-INDOLYL 4-PIPERIDYL
see BFC200
KETONE, 4-BIPHENYL ETHYL see BGM100
KETONE, m-(BIS(2-
CHLOROETHYL)AMINO)PHENYL PIPERIDINO see
BHP150
KETONE, 3-(BIS(2-CHLOROETHYL)AMINO)-p-
TOLYL MORPHOLINO- see BHR400
KETONE, 3-(BIS(2-CHLOROETHYL)AMINO)-p-
TOLYL PIPERIDYL- see BIA100
KETONE, BROMOMETHYL 4-
(DIPHENYLHYDROXYMETHYL)PIPERIDINO see
BMS300
KETONE, 7-(3-(sec-BUTYLAMINO)-2-
HYDROXYPROPOXY)-2-BENZOFURANYL METHYL
see ACN310
KETONE, 4-(3-(tert-BUTYLAMINO)-2-
HYDROXYPROPOXY)-2-BENZOFURANYL METHYL
see ACN300
KETONE, 7-(3-(tert-BUTYLAMINO)-2-
HYDROXYPROPOXY)-2-BENZOFURANYL METHYL
see ACN320
KETONE, BUTYL 4-PYRIDYL see VBA100

KETONE, CHLOROMETHYL 4-(DIPHENYLHYDROXYMETHYL)PIPERIDINO see CEC300

KETONE, CHLOROMETHYL 2-FLUORENYL see CEC700

KETONE, p-CHLOROPHENYL 1-(2-MORPHOLINOETHYL)PYRROL-2-YL, HYDROCHLORIDE see CEO100

KETONE, 4-(o-CHLOROPHENYL)-1-PIPERAZINYLMETHYL 3-PYRIDYL- see KGK120

KETONE, 4-(p-CHLOROPHENYL)PIPERAZINYL 3,4,5-TRIMETHOXYPHENYL see CKJ100

KETONE, CYCLOHEXYL 4-PYRIDYL see CPI375

KETONE, CYCLOPROPYL 4-((2-HYDROXY-3-(1,2-DIHYDRO-2-IMINO-4-METHYLPYRIDINO)PROPOXY)PHENYL) see PMM300

KETONE, 2-DIBENZOTHIENYL METHYL see ACH090

KETONE, 2-((4-(DICHLOROACETYL)PHENYL)AMINO)-2-HYDROXY-1-(4-PHENOXYPHENYL)- see DEN880

KETONE, 4,5-DICHLOROPYRROL-2-YL 2,6-DIHYDROXYPHENYL see POK400

KETONE, p-(2-(DIETHYLAMINO)ETHOXY)PHENYL 2-ETHYL-3-BENZOFURANYL see EDV700

KETONE, 1-(2-(DIETHYLAMINO)ETHYL)-2-(p-ETHOXYBENZYL)-5-BENZIMIDAZOLYL METHYL see DHZ050

KETONE, 1-(2-(DIETHYLAMINO)ETHYL)-2-p-PHENETIDINO-5-BENZIMIDAZOLYL METHYL see DIE350

KETONE, DIETHYLAMINO(1,2,3,4-TETRAHYDRO-9-ACRIDINYL) see ADJ550

KETONE, DIETHYLAMINO(7,8,9,10-TETRAHYDRO-11-6H-CYCLOHEPTA(b)QUINOLINYL) see TCO100

KETONE, 3,5-DIIODO-4-HYDROXYPHENYL 2,5-DIMETHYL-3-FURYL see DNF450

KETONE, 3,5-DIIODO-4-HYDROXYPHENYL 2-ETHYL-3-BENZOFURANYL see EID200

KETONE, 3,5-DIIODO-4-HYDROXYPHENYL 5-ETHYL-2-FURYL see EID250

KETONE, 3,5-DIIODO-4-HYDROXYPHENYL 2-FURYL see DNF500

KETONE, DIMETHYL see ABC750

KETONE, 5-DIMETHYLAMINO-3-INDOLYL PHENYL see DOU700

KETONE, 3,3-DIMETHYL-4-(DIMETHYLAMINO)-4-(o-TOLYL)BUTYL o-TOLYL see DRM110

KETONE, 2,5-DIMETHYL-3-FURYL p-HYDROXYPHENYL see DSC100

KETONE, 2,5-DIMETHYLPYRROL-3-YL METHYL see ACI550

KETONE, 2,5-DIMETHYL-3-THIENYL METHYL see DUG425

KETONE, 4-(DIPHENYLHYDROXYMETHYL)PIPERIDINO IODOMETHYL see IDZ200

KETONE, DI-1H-2-PYRROLYL see PPY300

KETONE, 2-ETHYL-7-(2-HYDROXY-3-(ISOPROPYLAMINO)PROPOXY)-4-BENZOFURANYL METHYL see ELI600

KETONE, ETHYL 4-(m-HYDROXYPHENYL)-1-METHYLPIPERIDYL see KFK000

KETONE, ETHYL (3-(3-ETHYLSULFONYL)PENTYL PIPERIDINO see EPI400

KETONE, ETHYL VINYL see PBR250

KETONE, 2-FURYL p-HYDROXYPHENYL see FQL200

KETONE, 3-FURYL PHENYL see FQO050

KETONE, HEPTYL 4-PYRIDYL see HBF600

KETONE, α-HYDROXYBENZYL PHENYL see BCP250

KETONE, 7-(2-HYDROXY-3-(ISOPROPYLAMINO)PROPOXY)-2-BENZOFURANYL METHYL see HLK600

KETONE, HYDROXYMETHYL 2-INDOLYL- see HIS100

KETONE, HYDROXYMETHYL 1-METHYL-2-INDOLYL- see HIS120

KETONE, HYDROXYMETHYL 1-METHYL-3-INDOLYL see HIS140

KETONE, HYDROXYMETHYL 2-METHYL-3-INDOLYL see HIS150

KETONE, HYDROXYMETHYL 3-METHYL-2-INDOLYL- see HIS130

KETONE, 5-INDOLINYL METHYL see ACO320

KETONE, 2-INDOLYL METHOXYMETHYL- see ICW200

KETONE, 3-INDOLYL MORPHOLINOMETHYL see ICZ150

KETONE, 3-INDOLYL PIPERIDINOMETHYL see ICZ200

KETONE, 3-INDOLYL 4-PIPERIDYL see ILG200

KETONE, METHOXYMETHYL 1-METHYL-2-INDOLYL see MDX300

KETONE, METHOXYMETHYL 3-METHYL-2-INDOLYL- see MDX310

KETONE, 4-(o-METHOXYPHENYL)-1-PIPERAZINYLMETHYL 3-PYRIDYL- see KGK130

KETONE, 4-(o-METHOXYPHENYL)PIPERAZINYL 3,4,5-TRIMETHOXYPHENYL see MFH760

KETONE, 4-(p-METHOXYPHENYL)PIPERAZINYL 3,4,5-TRIMETHOXYPHENYL see MFH770

KETONE, (p-METHOXYPHENYL) 4-PYRIDYL see MFH930

KETONE, 2-METHYLCYCLOPROPYL HYDRAZINO see MIV300

KETONE, METHYL ISOAMYL see MKW450

KETONE, METHYL 10-(3-(4-METHYL-1-PIPERAZINYL)PROPYL)PHENOTHIAZIN-2-YL see ACR500

KETONE, METHYL 5-METHYL-1-(2-QUINOLYL)-4-PYRAZOLYL see MLW600

KETONE, METHYL 5-METHYL-1-(2-QUINOXALINYL)-4-PYRAZOLYL see MLW630

KETONE, METHYL 5-METHYL-2-THIENYL see MPR300

KETONE, METHYL 10-(3-MORPHOLINOPROPYL)PHENOTHIAZIN-2-YL see MMA600

KETONE METHYL PHENYL see ABH000

KETONE, 2-METHYLPIPERIDINO 2-NAPHTHYL see MOH290

KETONE, METHYL 10-(3-PIPERIDINOPROPYL)PHENOTHIAZIN-2-YL see MOL300

KETONE, METHYL 1,4,5,6-TETRAHYDRO-2-METHYLCYCLOPENTA(b)PYRROL-3-YL see MPO800

KETONE, (α-METHYL-m-TRIFLUOROMETHYLPHENETHYLAMINOMETHYL) PIPERIDINO see MQE100

KETONE, METHYL 2,4,5-TRIMETHYLPYRROL-3-YL see ADE050

KETONE, MORPHOLINO(1,2,3,4-TETRAHYDRO-9-ACRIDINYL) see MRR760

KETONE, PENTYL 4-PYRIDYL see PBX800

KETONE, PENTYL 4-PYRIDYL see HEW050

KETONE, 1-PHENETHYL-3-PIPERIDYL PIPERIDINO see PDI550
KETONE, 4-PHENYL-1-PIPERAZINYLMETHYL-3-PYRIDYL- see NDW520
KETONE, 4-PHENYLPIPERAZINYL 3,4,5-TRIMETHOXYPHENYL see PFX600
KETONE, PHENYL (1-PIPERIDINOCYCLOHEXYL) see PFY200
KETONE, PIPERIDINO(1,2,3,4-TETRAHYDRO-9-ACRIDINYL) see PIU100
KETONE PROPANE see ABC750
KETONE, PROPYL 4-PYRIDYL see PNV755
KETONES see KGA000
KETONE, 4-(p-TOLYL)PIPERAZINYL 3,4,5-TRIMETHOXYPHENYL see THF310
KETONOX see MKA500
KETOPENTAMETHYLENE see CPW500
KETOPROFEN see BDU500
KETOPRON see BDU500
β-KETOPROPANE see ABC750
1-(2-KETO-2-(3'-PYRIDYL)ETHYL)-4-(2'-CHLOROPHENYL)PIPERAZINE see KGK120
1-(2-KETO-2-(3'-PYRIDYL)ETHYL)-4-(2'-METHOXYPHENYL)PIPERAZINE see KGK130
1-(2-KETO-2-(3'-PYRIDYL)ETHYL)-4-(PHENYL)PIPERAZINE see NDW520
l-3-KETOTHREOHEXURONIC ACID LACTONE see ARN000
2-KETO-1,7,7-TRIMETHYLNORCAMPHANE see CBA750
KEVADON see TEH500
KEY-SERPINE see RDK000
KF-1820 see DOQ400
K-GRAN see PLA000
KH 360 see TGG760
KHAINI (INDIA) see SED400
KHALADON 22 see CFX500
KHAROPHEN see ABX500
KHE 0145 see MIA250
KHIMCOCCID see RLK890
KHIMCOECID see RLK890
KHIMKOKTSID see RLK890
KHIMKOKTSIDE see RLK890
KHLADON 113 see FOO000
KHLADON 744 see CBU250
KHLORIDIN see TGD000
KHLORTRIANIZEN see CLO750
KHP 2 see AHE250
K-IAO see PLG800
KIATRIUM see DCK759
KIDOLINE see VGP000
KIEFERNADEL OEL (GERMAN) see PIH500
KIESELGUHR see DCJ800
KIESELSAEURE (GERMAN) see SCL000
KIEZELFLUORWATERSTOFZUUR (DUTCH) see SCO500
K III see DUS700
KILDIP see DGB000
KILL-ALL see SEY500
KILLAX see TCF250
KILLEEN see CCL250
KILL KANTZ see AQN635
KILMAG see ARB750
KILMITE 40 see TCF250
KILPROP see CIR500
KILRAT see ZLS000
KILSEM see CIR250

KINGCOT see CCU150
KING'S GOLD see ARI000
KING'S GREEN see COF500
KING'S YELLOW see ARI000
KING'S YELLOW see LCR000
KIPCA see MMD500
KIRESUTO B see EIX500
KIRESUTO NTB see DXF000
KIRKSTIGMINE BROMIDE see POD000
KIRTICOPPER see CNJ950
KITINE see RDK000
KITON CRIMSON 2R see HJF500
KITON FAST ORANGE G see HGC000
KITON ORANGE MNO see MDM775
KITON PONCEAU R see FMU070
KITON PURE BLUE L see FMU059
KITON SCARLET 4R see FMU080
KITON YELLOW MS see MDM775
KITON YELLOW T see FAG140
KIWAM (INDIA) see SED400
KIWI LUSTR 277 see BGJ250
KKM 43 see ISD066
KLAVI KORDAL see NGY000
KLEER-LOT see AMY050
KLEGECELL see PKQ059
KLIMANOSID see RDK000
KLIMORAL see EDU500
KLINGTITE see NAK500
KLINIT see XPJ000
KLION see MMN250
K-LOR see PLA500
KLORAMIN see BIE500
KLORAMIN see CDP000
KLORAMINE-T see CDP000
KLOREX see SFS000
KLOROCIN see SHU500
KLORPROMAN see CKP500
KLORPROMEX see CKP500
KLORT see MQU750
KLOTRIX see PLA500
KLOTTONE see MMD500
KLUCEL see HNV000
KM see SMQ500
KM see KAL000
4K-2M see CIR250
KM (the antibiotic) see KAL000
KMH see DMC600
KM (POLYMER) see SMQ500
KMTS 212 see SFO500
K1-N see PLK500
KNEE PINE OIL see PIH400
KNOCKMATE see FAS000
KNOLLIDE see PLK500
KO 08 see SCR400
KOAXIN see MMD500
KOBALT (GERMAN, POLISH) see CNA250
KOBALT CHLORID (GERMAN) see CNB599
KOBU see PAX000
KOBUTOL see PAX000
KOCHINEAL RED A FOR FOOD see FMU080
KOCIDE see SOD500
KODAFLEX see TIG750
KODAFLEX DBS see DEH600
KODAFLEX DIBP see DNJ400
KODAFLEX DOA see AEO000
KODAFLEX DOP see DVL700
KODAFLEX TRIACETIN see THM500
KODAK LR 115 see CCU250

KOFFEIN (GERMAN) see CAK500
KOHLENDIOXYD (GERMAN) see CBU250
KOHLENDISULFID (SCHWEFELKOHLENSTOFF)
(GERMAN) see CBV500
KOHLENMONOXID (GERMAN) see CBW750
KOHLENOXYD (GERMAN) see CBW750
KOHLENSAEURE (GERMAN) see CBU250
KOKAIN see CNE750
KOKAN see CNE750
KOKAYEEN see CNE750
KOKOTINE see BBQ500
KOLCHAMIN see MIW500
KOLCHICIN see MIW500
KOLI (HAWAII) see CCP000
KOLKAMIN see MIW500
KOLKLOT see MMD500
KOLLIDON see PKQ250
KOLOFOG see SOD500
KOLOSPRAY see SOD500
KOLPHOS see PAK000
KOLPON see EDV000
KOLTAR see OQU100
KOMEEN see DBU800
KOMPLEXON I see AMT500
KOMPLXON see EIV000
KONDREMUL see MQV750
KONESTA see TII250
KONLAX see DJL000
KOOLMONOXYDE (DUTCH) see CBW750
KOOLSTOFDISULFIDE (ZWAVELKOOLSTOF)
(DUTCH) see CBV500
KOOLSTOFOXYCHLORIDE (DUTCH) see PGX000
KOPFUME see EIY500
KOPLEN 2 see SMQ500
KOPLEX AQUATIC HERBICIDE see DBU800
KOP MITE see DER000
KOPOLYMER BUTADIEN STYRENOVY (CZECH) see
SMR000
KOPROL see PDO750
KOPSOL see DAD200
KOP-THIODAN see EAQ750
KOP-THION see MAK700
KORAD see PKB500
KORDIAMIN see DJS200
KORIUM see MJM500
KORLAN see RMA500
KORLANE see RMA500
KORODIL see CCK125
KOROSEAL see PKQ059
KORO-SULF see SNN500
KORUM see HIM000
KOSATE see DJL000
KOSMINK see CBT750
KOSMOBIL see CBT750
KOSMOLAK see CBT750
KOSMOS see CBT750
KOSMOTHERM see CBT750
KOSMOVAR see CBT750
KOSTIL see ADY500
KOTION see DSQ000
KOTOL see BBQ750
KP 2 see PAX000
KP 140 see BPK250
K-PIN see PIB900
K-PRENDE-DOME see PLA500
K PREPARATION see BJU000
KR 2537 see SMQ500
KRASTEN 1.4 see SMQ500

KRASTEN 052 see SMQ500
KRASTEN SB see SMQ500
KRATEDYN see EAW000
KRECALVIN see DGP900
KREGASAN see TFS350
KRENITE (OBS.) see DUS700
KRENITE (OBS.) see DUU600
KRESAMONE see DUS700
KRESIDIN see MGO750
m-KRESOL see CNW750
p-KRESOL see CNX250
o-KRESOL (GERMAN) see CNX000
KRESOLE (GERMAN) see CNW500
KRESOLEN (DUTCH) see CNW500
KRESOXIM-METHYL see MLI900
KRESOXIM-METHYL TECHNICAL see MLI900
KREZIDINE see MGO750
KREZOL (POLISH) see CNW500
KREZONE see CIR250
KREZONITE see DUU600
KREZOTOL 50 see DUS700
KRISTALLOSE see SJN700
KRISTALL-VIOLETT see AOR500
KRO 1 see SMR000
KROKYDOLITH (GERMAN) see ARM275
KROLOR ORANGE RKO 786D see MRC000
KROMAD see KHU000
KROMON GREEN B see CLK235
KRONITEX see TNP500
KRONITEX KP-140 see BPK250
KRONOS TITANIUM DIOXIDE see TGG760
KROTENAL see DXH250
KROTONALDEHYD (CZECH) see COB250
KROVAR II see BMM650
KRYOLITH (GERMAN) see SHF000
KRYPTOCUR see LIU370
KRYSID see AQN635
KSYLEN (POLISH) see XGS000
K-THROMBYL see MMD500
KUBARSOL see ABX500
KUEMMEL OIL (GERMAN) see CBG500
KUMADER see WAT200
KUMIAI see MIB750
KUMORAN see BJZ000
KUMULUS see SOD500
KUPFEROXYDUL (GERMAN) see CNO000
KUPFERRON (CZECH) see ANO500
KUPFERSULFAT (GERMAN) see CNP250
KURALON VP see PKP750
KURAN see TIX500
KURARE OM 100 see AAX250
KURARE POVAL 1700 see PKP750
KURARE PVA 205 see PKP750
KURATE POVAL 120 see PKP750
KURON see TIX500
KUROSAL see TIX500
KUSA-TOHRU see SFS000
KUSATOL see SFS000
KUSNARIN see EID000
K-VITAN see MMD500
KW-125 see AES750
KW-4354 see OMG000
KWAS BENZYDYNODWUKAROKSYLOWY (POLISH)
see BFX250
KWAS METANIOWY (POLISH) see FNA000
KWELL see BBQ500
KWELLS see HOT500
KWIK (DUTCH) see MCW250

KWIK-KIL see SMN500
KWIKSAN see ABU500
KWIT see EEH600
KW-2-LE-T see LFA020
KYANID SODNY (CZECH) see SGA500
KYANID STRIBRNY (CZECH) see SDP000
KYANOSTRIBRNAN DRASELNY (CZECH) see PLS250
KYLAR see DQD400
KYOCRISTINE see LEZ000
KYONATE see PLV750
KYPCHLOR see CDR750
KYPFOS see MAK700
KYPMAN 80 see MAS500
KYPTHION see PAK000
KYSELINA ADIPOVA (CZECH) see AEN250
KYSELINA AKRYLOVA see ADS750
KYSELINA AMIDOSULFONOVA (CZECH) see SNK500
KYSELINA p-AMINOBENZOOVA see AIH600
KYSELINA BENZOOVA (CZECH) see BCL750
KYSELINA BROMOCTOVA see BMR750
KYSELINA CHLOROCTOVA see CEA000
KYSELINA CITRONOVA (CZECH) see CMS750
KYSELINA DICHLORISOKYANUROVA (CZECH) see DGN200
KYSELINA DICHLOROCTOVA see DEL000
KYSELINA DI-(2-ETHYLHEXYL)FOSFORECNA see BJR750
KYSELINA 2,3-DIHYDROXYBUTANDIOVA see TAF750
KYSELINA DUSICNE see NED500
KYSELINA DUSITE see NMR000
KYSELINA 3,6-ENDOMETHYLEN-3,4,5,6,7,7-HEXACHLOR-Δ^4-TETRAHYDROFTALOVA (CZECH) see CDS000
KYSELINA 2-FENYL-2-HYDROXYETHANOVA see MAP000
KYSELINA FUMAROVA (CZECH) see FOU000
KYSELINA HET (CZECH) see CDS000
KYSELINA HYDROXYBUTANDIOVA (CZECH) see MAN000
KYSELINA 2-HYDROXYPROPANOVA see LAG000
KYSELINA ISOMASELNA see IJU000
KYSELINA JABLECNA (CZECH) see MAN000
KYSELINA JANTAROVA see SMY000
KYSELINA KAKODYLOVA see HKC000
KYSELINA MANDLOVA see MAP000
KYSELINA MERKAPTOOCTOVA see TFJ100
KYSELINA METHAKRYLOVA see MDN250
KYSELINA METHANSULFONOVA (CZECH) see MDR250
KYSELINA MLECNA (CZECH) see LAG000
KYSELINA MUKOCHLOROVA see MRU900
KYSELINA NITRILOTRIOCTOVA see AMT500
KYSELINA NITROBENZEN-m-SULFONOVA (CZECH) see NFB500
KYSELINA PEROXYOCTOVA see PCL500
KYSELINA PIKROVA see PID000
KYSELINA PROPIONOVA see PMU750
KYSELINA STAVELOVA (CZECH) see OLA000
KYSELINA SULFAMINOVA (CZECH) see SNK500
KYSELINA TERFTALOVA (CZECH) see TAN750
KYSELINA-β,β'-THIODIPROPIONOVA (CZECH) see BHM000
KYSELINA THIOGLYKOLOVA see TFJ100
KYSELINA THIOOCTOVA see TFA500

KYSELINA TRICHLOISOKYANUROVA (CZECH) see TIQ750
KYSELINA TRICHLOROCTOVA see TII250
KYSELINA TRIFLUOROCTOVA see TKA250
KYSELINA VINNA see TAF750
KYSLICNIK DI-n-BUTYLCINICITY (CZECH) see DEF400
KYSLICNIK DIISOBUTYLCINICITY (CZECH) see DNJ000
KYSLICNIK DIISOPROPYLCINICITY (CZECH) see DNR200
KYSLICNIK DI-N-PROPYLCINICITY (CZECH) see DWV000
KYSLICNIK TRI-N-BUTYLCINICITY (CZECH) see BLL750
KZ 3M see SCQ000
KZ 5M see SCQ000
KZ 7M see SCQ000
K-ZINC see ZKA000
L-2 see BIQ250
L16 see AGX000
84L see DIW000
L-310 see LGK000
L-395 see DSP400
L 1811 see DJT400
L2214 see DDP200
L 2329 see EID200
L. 3428 see AJK750
L-5103 see RKP000
Le-100 see EIF000
L-01748 see DJT800
L-36352 see DUV600
LA 1 see DLY000
LA 01 see CCU150
LA 6023 see DQR600
LA'AU-'AILA (HAWAII) see CCP000
LABAZ see AJK750
2329 LABAZ see EID200
LABDANOL see IIQ000
LABICAN see MDQ250
LABOPAL see DKQ000
LACOLIN see LAM000
LACQREN 506 see SMQ500
LACQREN 550 see SMQ500
LACQTEN 1020 see PJS750
LACQUER ORANGE V see TGW000
LACQUER ORANGE VG see PEJ500
LACQUER ORANGE VR see XRA000
LACQUER RED V see SBC500
LACQUER RED V 2G see CMS238
LACQUER RED V3B see EOJ500
LACQUER RED VS see SBC500
LACTASE ENZYME PREPARATIONS from KLUYVEROMYCES LACTIS see LAE350
LACTATE d'ETHYLE (FRENCH) see LAJ000
LACTIC ACID see LAG000
dl-LACTIC ACID see LAG000
LACTIC ACID, ANTIMONY SALT see AQE250
LACTIC ACID, BERYLLIUM SALT see LAH000
LACTIC ACID, BUTYL ESTER see BRR600
LACTIC ACID, BUTYL ESTER, BUTYRATE see BQP000
LACTIC ACID, CADMIUM SALT see CAG750
LACTIC ACID, ETHYL ESTER see LAJ000
LACTIC ACID, IRON(2+) SALT (2:1) see LAL000
LACTIC ACID LITHIUM SALT see LHL000
LACTIC ACID, MAGNESIUM SALT see LAL100
LACTIC ACID, MONOSODIUM SALT see LAM000
LACTIC ACID SODIUM SALT see LAM000

LACTIN see LAR000
LACTOBACILLUS LACTIS DORNER FACTOR see VSZ000
LACTOBARYT see BAP000
LACTOBIOSE see LAR000
LACTOCAINE see AIT250
LACTOFLAVIN see RIK000
LACTOFLAVINE see RIK000
LACTONITRILE see LAQ000
LACTOSE see LAR000
d-LACTOSE see LAR000
LACTYLATED FATTY ACID ESTERS of GLYCEROL and PROPYLENE GLYCOL see LAR400
LADAKAMYCIN see ARY000
LAEVORAL see LFI000
LAEVOSAN see LFI000
LAEVOXIN see LFG050
LAKE BLUE B BASE see DCJ200
LAKE PONCEAU see FMU070
LAKE RED C see CHP500
LAKE RED KB BASE see CLK225
LAKE SCARLET G BASE see NMP500
LAKE YELLOW see FAG140
LAMAR see FLZ050
LAMBDA-CYHALOTHRIN see LAS200
LAMBDA-CYHALOTHRIN see LAS200
LAMBDA-CYHALOTHRIN TECHNICAL see LAS200
LAMBETH see PMP500
LAMBRATEN see AJT250
LAMDIOL see EDO000
LAMIDON see IIU000
LAMITEX see SEH000
LAMORYL see GKE000
LANADIN see TIO750
LANATOXIN see DKL800
LANAZINE see DBA800
LANDALGINE see AFL000
LANDISAN see MEO750
LAND PLASTER see CAX750
LANDRAX see CQM325
LANETTE WAX-S see SIB600
LANEX see DUK800
LANGFORD see KBB600
LANGORAN see CCK125
LANICOR see DKN400
LANNATE see MDU600
LANODOXIN see DNF600
LANOL see CMD750
LANOLIN, anhydrous see LAU550
LANOPHYLLIN see TEP000
LANOXIN see DKN400
LANTHANACETAT (GERMAN) see LAW000
LANTHANUM see LAV000
LANTHANUM ACETATE see LAW000
LANTHANUM CHLORIDE see LAX000
LANTHANUM NITRATE see LBA000
LANTHANUM TRIACETATE see LAW000
LAPAV see PAH250
LARCH GUM see AQR800
LARD FACTOR see VSK600
LARGACTIL MONOHYDROCHLORIDE see CKP500
LARGAKTYL see CKP500
LARIXIC ACID see MAO350
LARIXINIC ACID see MAO350
LARODON see INY000
LARODOPA see DNA200
LARTEN see MQU750
LARVACIDE see CKN500

LASALOCID see LBF500
LASERDIL see CCK125
LASEX see CHJ750
LASIOCARPINE see LBG000
LASIX see CHJ750
LAS-Na see LGF825
LASSO see CFX000
LAS, SODIUM SALT see LGF825
LATANOPROST see XAA500
LATEXOL SCARLET R see CHP500
LATKA 666 see BBP750
LATKA 7744 see CBM750
LATSCHENKIEFEROEL see PIH400
LAUDRAN DI-n-BUTYLCINICITY (CZECH) see DDV600
LAUGHING GAS see NGU000
LAURAMIDE DEA see BKE500
LAUREL CAMPHOR see CBA750
LAUREL LEAF OIL see BAT500
LAUREL LEAF OIL see LBK000
LAURIC ACID see LBL000
LAURIC ACID, CADMIUM SALT (2:1) see CAG775
LAURIC ACID, DIBUTYLSTANNYLENE derivative see DDV600
LAURIC ACID, DIBUTYLSTANNYLENE SALT see DDV600
LAURIC ACID DIETHANOLAMIDE see BKE500
LAURIC ALCOHOL see DXV600
LAURIC DIETHANOLAMIDE see BKE500
LAURINAMINE see DXW000
LAURINE see CMS850
LAURINIC ALCOHOL see DXV600
LAUROSTEARIC ACID see LBL000
LAUROX see LBR000
LAUROYL DIETHANOLAMIDE see BKE500
LAUROYL PEROXIDE see LBR000
LAUROYL PEROXIDE, TECHNICALLY PURE (DOT) see LBR000
LAURYDOL see LBR000
LAURYL 24 see DXV600
LAURYL ALCOHOL (FCC) see DXV600
n-LAURYL ALCOHOL, PRIMARY see DXV600
LAURYL ALDEHYDE (FCC) see DXT000
LAURYLAMINE see DXW000
n-LAURYLAMINE see DXW000
LAURYLBENZENESULFONIC ACID see LBU100
LAURYL DIETHANOLAMIDE see BKE500
LAURYL GALLATE see DXX200
LAURYLGUANIDINE ACETATE see DXX400
LAURYLISOQUINOLINIUM BROMIDE see LBW000
LAURYL MERCAPTAN see LBX000
m-LAURYL MERCAPTAN see LBX000
LAURYL SODIUM SULFATE see SIB600
LAURYL SULFATE, SODIUM SALT see SIB600
LAURYL 3,3'-THIODIPROPIONATE see TFD500
LAUSIT see IDA000
LAUXTOL see PAX250
LAUXTOL A see PAX250
LAV see CMY800
LAVANDIN ABSOLUTE see LCA000
LAVANDIN BENZOL ABSOLUTE see LCA000
LAVANDIN OIL see LCA000
LAVANDULYL ACETATE see LCA100
LAVATAR see CMY800
LAVENDEL OEL (GERMAN) see LCD000
LAVENDER OIL see LCD000
LAVENDER OIL, SPIKE see SLB500
LAVOFLAGIN see ABY900

LAXANORM see DMH400
LAXANTHREEN see DMH400
LAXINATE see DJL000
LAXIPUR see DMH400
LAXIPURIN see DMH400
LAXOGEN see PDO750
LAYOR CARANG see AEX250
LAZO see CFX000
LB 502 see CHJ750
"L," CARPSERP see RDK000
LCR see LEY000
LD 400 see PJS750
LD 600 see PJS750
LDA see BKE500
LDE see BKE500
LDPE 4 see PJS750
LD RUBBER RED 16913 see CHP500
29060 LE see VLA000
LEA-COV see SHF500
LEAD see LCF000
LEAD ACETATE see LCV000
LEAD(2+) ACETATE see LCV000
LEAD(II) ACETATE see LCV000
LEAD ACETATE, BASIC see LCH000
LEAD ACETATE TRIHYDRATE see LCJ000
LEAD ACETATE(II), TRIHYDRATE see LCJ000
LEAD ACID ARSENATE see LCK000
LEAD ARSENATE see ARC750
LEAD ARSENATE see LCK000
LEAD ARSENATE see LCK100
LEAD ARSENATE, solid (DOT) see LCK000
LEAD ARSENATE (standard) see LCK000
LEAD(II) ARSENITE see LCL000
LEAD ARSENITES (DOT) see LCL000
LEAD(II) AZIDE see LCM000
LEAD AZIDE (dry) (DOT) see LCM000
LEAD AZIDE, wetted with not <20% water or mixture of
alcohol and water, by weight (DOT) see LCM000
LEAD BOTTOMS see LDY000
LEAD BROWN see LCX000
LEAD CARBONATE see LCP000
LEAD(2+) CARBONATE see LCP000
LEAD CHLORIDE see LCQ000
LEAD(2+) CHLORIDE see LCQ000
LEAD(II) CHLORIDE see LCQ000
LEAD CHROMATE see LCR000
LEAD CHROMATE(VI) see LCR000
LEAD CHROMATE, BASIC see LCS000
LEAD CHROMATE MOLYBDATE SULFATE RED see
MRC000
LEAD CHROMATE OXIDE (MAK) see LCS000
LEAD CHROMATE, RED see LCS000
LEAD CHROMATE, SULPHATE and MOLYBDATE see
LDM000
LEAD COMPOUNDS see LCT000
LEAD(II) CYANIDE see LCU000
LEAD CYANIDE (DOT) see LCU000
LEAD DIACETATE see LCV000
LEAD DIACETATE TRIHYDRATE see LCJ000
LEAD DIBASIC ACETATE see LCV000
LEAD DIBASIC PHOSPHITE see LCV100
LEAD DICHLORIDE see LCQ000
LEAD DIFLUORIDE see LDF000
LEAD DIMETHYLDITHIOCARBAMATE see LCW000
LEAD DINITRATE see LDO000
LEAD DIOXIDE see LCX000
LEAD DIPERCHLORATE see LDS499
LEAD DISODIUM EDTA see LDB000

LEAD DISODIUM
ETHYLENEDINITRILOTETRACETATE see LDB000
LEAD DROSS see LDC000
LEAD DROSS (DOT) see LDY000
LEAD(2-),
((ETHYLENEDINITRILO)TETRAACETATO)-,
DISODIUM see LDB000
LEAD FLAKE see LCF000
LEAD FLUOBORATE see LDE000
LEAD(II) FLUORIDE see LDF000
LEAD FLUORIDE (DOT) see LDF000
LEAD(II) FLUOROSILICATE see LDG000
LEAD-MOLYBDENUM CHROMATE see LDM000
LEAD MONONITRORESORCINATE (DRY) (DOT) see
LDP000
LEAD MONOSUBACETATE see LCH000
LEAD MONOXIDE see LDN000
LEAD NAPHTHENATE see NAS500
LEAD NITRATE see LDO000
LEAD(2+) NITRATE see LDO000
LEAD(II) NITRATE see LDO000
LEAD(II) NITRATE (1:2) see LDO000
LEAD NITRORESORCINATE see LDP000
LEAD(II) OLEATE (1:2) see LDQ000
LEAD ORTHOPHOSPHATE see LDU000
LEAD ORTHOPLUMBATE see LDS000
LEAD OXIDE see LDN000
LEAD(II) OXIDE see LDN000
LEAD(IV) OXIDE see LCX000
LEAD OXIDE BROWN see LCX000
LEAD OXIDE PHOSPHONATE, HEMIHYDRATE see
LCV100
LEAD OXIDE RED see LDS000
LEAD OXIDE YELLOW see LDN000
LEAD PERCHLORATE see LDS499
LEAD(2+) PERCHLORATE see LDS499
LEAD(II) PERCHLORATE see LDS499
LEAD PERCHLORATE, solid or solution (DOT) see
LDS499
LEAD(II) PERCHLORATE, HEXAHYDRATE (1:2:6) see
LDT000
LEAD PEROXIDE (DOT) see LCX000
LEAD PHOSPHATE see LDU000
LEAD(2+) PHOSPHATE see LDU000
LEAD PHOSPHATE (3:2) see LDU000
LEAD(II) PHOSPHATE (3:2) see LDU000
LEAD PHOSPHITE, dibasic (DOT) see LCV100
LEAD PROTOXIDE see LDN000
LEAD S2 see LCF000
LEAD SCRAP see LDC000
LEAD SILICATE see LDW000
LEAD STEARATE see LDX000
LEAD STYPHNATE see LEE000
LEAD STYPHNATE (dry) (DOT) see LEE000
LEAD STYPHNATE, wetted or lead trinitroresorcinate,
wetted with not <20% water or mixt. (DOT) see LEE000
LEAD SUBACETATE see LCH000
LEAD(II) SULFATE (1:1) see LDY000
LEAD SULFATE, solid, containing more than 3% free acid
(DOT) see LDY000
LEAD SULFIDE see LDZ000
LEAD SUPEROXIDE see LCX000
LEAD(II) TARTRATE (1:1) see LEA000
LEAD, TETRAETHYL- see TCF000
LEAD TETRAOXIDE see LDS000
LEAD TITANATE see LED000
LEAD TRINITRORESORCINATE see LEE000
LEAF ALCOHOL see HFE000

LEAF ALDEHYDE see HFA500
LEAF GREEN see CMJ900
LEATHER FAST RED B see CMM330
LEATHER ORANGE HR see PEK000
LEATHER PURE BLUE HB see BJI250
LE CAPTANE (FRENCH) see CBG000
LECITHIN see LEF180
LEDERMYCIN see MIJ500
LE DINITROCRESOL-4,6 (FRENCH) see DUS700
LEDON 11 see TIP500
LEDON 12 see DFA600
LEDON 114 see FOO509
LEDOSTEN see DJT400
LEFEBAR see EOK000
LEGUMEX DB see CIR250
LEHYDAN see DKQ000
LEINOLEIC ACID see LGG000
LEIPZIG YELLOW see LCR000
LEIVASOM see TIQ250
LEKAMIN see TNF500
LEMAC 1000 see AAX250
LEMBROL see DCK759
LEMISERP see RDK000
LEMOFLUR see SHF500
LEMOL see PKP750
LEMOL 5-88 see PKP750
LEMOL 5-98 see PKP750
LEMOL 12-88 see PKP750
LEMOL 16-98 see PKP750
LEMOL 24-98 see PKP750
LEMOL 30-98 see PKP750
LEMOL 51-98 see PKP750
LEMOL 60-98 see PKP750
LEMOL 75-98 see PKP750
LEMOL GF-60 see PKP750
LEMON CHROME see BAK250
LEMONENE see BGE000
LEMONGRASS OIL WEST INDIAN see LEH000
LEMON OIL see LEI000
LEMON OIL, desert type, coldpressed see LEI025
LEMON OIL, COLDPRESSED (FCC) see LEI000
LEMON OIL, EXPRESSED see LEI000
LEMONOL see DTD000
LEMON YELLOW see BAK250
LEMON YELLOW see LCR000
LEMON YELLOW A see FAG140
LEMON YELLOW A GEIGY see FAG140
LENDINE see BBQ500
LENITRAL see NGY000
LENOCYCLINE see HOH500
LENTIZOL see EAI000
LENTOCILLIN see BFC750
LENTOPENIL see BFC750
LENTOTRAN see MDQ250
LENTOX see BBQ500
LEO 72a see BHO250
LEOMYPEN see BFC750
LEOPENTAL see PBT500
LEOSTESIN see DHK400
LEOSTESIN HYDROCHLORIDE see DHK600
LEOSTIGMINE BROMIDE see POD000
LEPASEN see SEP000
LEPENIL see MQU750
LEPETOWN see MQU750
LEPHEBAR see EOK000
LEPIMIDIN see DBB200
LEPINAL see EOK000
LEPITOIN see DKQ000

LEPITOIN see DNU000
LEPITOIN SODIUM see DNU000
LEPSIN see DKQ000
LEPSIRAL see DBB200
LEPTAMIN see DJS200
LEPTANAL see PDW750
LEPTON see DJT400
LEPTOPHOS see LEN000
LERBEK see CMX850
LERCIGAN see DQA400
LERGIGAN see PMI750
LERGIGAN see DQA400
LERTUS see BDU500
LESAN see DOU600
LESTEMP see HIM000
LETHALAIRE G-52 see TCF250
LETHALAIRE G-54 see PAK000
LETHALAIRE G-57 see SOD100
LETHALAIRE G-59 see OCM000
LETHELMIN see PDP250
LETHOX see TNP250
LETHURIN see TIO750
LETTER see LFG050
LETYL see MQU750
LEUCARSONE see CBJ000
LEUCETHANE see UVA000
LEUCIDIL see BCA000
LEUCIN (GERMAN) see LES000
LEUCINE see LES000
l-LEUCINE see LES000
dl-LEUCINE see LER000
ε-LEUCINE see AJD000
LEUCO-4 see AEH000
LEUCOL see QMJ000
LEUCOLINE see QMJ000
LEUCOLINE see IRX000
LEUCOSULFAN see BOT250
LEUCOTHANE see UVA000
LEUCOVYL PA 1302 see AAX175
LEUKAEMOMYCIN C see DAC000
LEUKERAN see POK000
LEUKERAN see CDO500
LEUKERSAN see CDO500
LEUKOL see QMJ000
LEUKOMYAN see CDP250
LEUKORAN see CDO500
LEUNA M see CIR250
LEUPURIN see POK000
LEUROCRISTINE see LEY000
LEUROCRISTINE SULFATE (1:1) see LEZ000
LEVADONE see MDP770
LEVAMISOLE see LFA000
LEVAMISOLE see LFA020
LEVAMISOLE HYDROCHLORIDE see LFA020
LEVANOL RED GG see CMM330
LEVANOX GREEN GA see CMJ900
LEVANOX RED 130A see IHC450
LEVANOX WHITE RKB see TGG760
LEVARGIN see AQW000
LEVARTERENOL see NNO500
LEVAXIN see LFG050
LEV HYDROCHLORIDE see LFA020
LEVIUM see DCK759
LEVOARTERENOL see NNO500
LEVOGLUTAMID see GFO050
LEVOGLUTAMIDE see GFO050
LEVOMYCETIN see CDP250

LEVOMYSOL HYDROCHLORIDE see LFA020
LEVONORADRENALINE see NNO500
LEVONOREPINEPHRINE see NNO500
LEVOPHED see NNO500
LEVORENIN see VGP000
LEVORIN see LFF000
LEVOROXINE see LFG050
LEVORPHAN see LFG000
LEVORPHANOL see LFG000
LEVOTHROID see LFG050
LEVOTHYL see MDP770
LEVOTHYROX see LFG050
LEVOTHYROXINE SODIUM see LFG050
LEVOTHYROXINE SODIUM see LFG050
LEVUGEN see LFI000
LEVULINIC ACID, ETHYL ESTER see EFS600
LEVULOSE see LFI000
LEWISITE see CLV000
LEWISITE (ARSENIC COMPOUND) see CLV000
LEWISITE II see BIQ250
LEWISITE I OXIDE see DEW000
LEWIS-RED DEVIL LYE see SHS000
LEXATOL see SPC500
LEXONE see MQR275
LEYSPRAY see CIR250
LEYTOSAN see PFP500
LEYTOSAN see ABU500
LFA 2043 see GIA000
LF BOWSEI PW 2 see ZJS400
LF-PW 2 see ZJS400
L-GRUEN No. 1 see CKN000
L-GRUEN No. 1 (GERMAN) see CKN000
LGYCOSPERSE S-20 see PKL030
LH RELEASING FACTOR see LIU370
LH-RELEASING HORMONE see LIU370
LH-RF see LIU370
LHRH see LIU370
LH-RH see LIU370
LH-RH/FSH-RH see LIU370
LIBAVIUS FUMING SPIRIT see TGC250
LIBERETAS see DCK759
LIBIOLAN see MQU750
LIBRAX see LFK000
LIBRININ see LFK000
LIBRITABS see LFK000
LIBRIUM see MDQ250
LIBRIUM see LFK000
LIBRIUM HYDROCHLORIDE see MDQ250
LICAREOL ACETATE see LFY600
LICHENIC ACID see FOU000
LIDA-MANTLE see DHK400
LIDAMYCIN CREME see NCG000
LIDENAL see BBQ500
LIDOCAINE see DHK400
LIDOCAINE HYDROCHLORIDE see DHK600
LIDOL see DAM700
LIDOTHESIN HYDROCHLORIDE see DHK600
LIFEAMPIL see AIV500
LIFEAMPIL see AOD125
LIFRIL see FLZ050
LIGHT CAMPHOR OIL see CBB500
LIGHT CATALYTICALLY CRACKED DISTILLATE see DXG840
LIGHT CATALYTICALLY CRACKED NAPHTHA see NAQ540
LIGHT ESTER PO-A see PER250
LIGHT GAS OIL see GBW025
LIGHTHOUSE CHROME BLUE 2R see HJF500

LIGHT NAPHTHENIC DISTILLATE see MQV810
LIGHT NAPHTHENIC DISTILLATE, SOLVENT EXTRACT see MQV860
LIGHT NAPHTHENIC DISTILLATES (PETROLEUM) see MQV810
LIGHT OIL OF CAMPHOR see CBB500
LIGHT ORANGE CHROME see LCS000
LIGHT PARAFFINIC DISTILLATE see MQV815
LIGHT PARAFFINIC DISTILLATE, SOLVENT EXTRACT see MQV862
LIGHT RED see IHC450
LIGHT SPAR see CAX750
LIGHT STRAIGHT-RUN NAPHTHA see NAQ560
LIGNALOE OIL see BMA550
LIGNASAN FUNGICIDE see BJT250
LIGNASAN-X see BJT250
LIGNOCAINE see DHK400
LIGNOCAINE HYDROCHLORIDE see DHK600
LIGROIN see PCT250
LIHOCIN see CMF400
LIKUDEN see GKE000
LILIAL-METHYLANTHRANILATE, Schiff's base see LFT100
LILLY 1516 see DQA400
LILLY 01516 see DQA400
LILLY 22451 see MDU500
LILLY 36,352 see DUV600
LILLY 37231 see LEZ000
LILO see PDO000
LILYL ALDEHYDE see CMS850
LIMARSOL MALAGRIDE see ABX500
LIMAS see LGZ000
LIMBUX see CAT225
LIME see CAU500
LIME ACETATE see CAL750
LIME, BURNED see CAU500
LIME CHLORIDE see HOV500
LIMED ROSIN see CAW500
LIME MILK see CAT225
LIME-NITROGEN (DOT) see CAQ250
LIME OIL see OGM850
LIME OIL, distilled (FCC) see OGM850
LIME PYROLIGNITE see CAL750
LIMESTONE (FCC) see CAO000
LIME, UNSLAKED (DOT) see CAU500
LIME WATER see CAT225
LIMONENE see MCC250
1-LIMONENE see MCC500
d-LIMONENE see LFU000
(+)-R-LIMONENE see LFU000
d-(+)-LIMONENE see LFU000
dl-LIMONENE see MCC250
LIMONENE DIOXIDE see LFV000
(−)-LIMONENE (FCC) see MCC500
LINALOL see LFX000
LINALOL ACETATE see LFY600
LINALOOL ACETATE see LFY600
LINALOOL, DEHYDRO- see LFY333
LINALOOL ISOBUTYRATE see LGB000
LINALYL ACETATE see LFY600
LINALYL ALCOHOL see LFX000
LINALYL BENZOATE see LFZ000
LINALYL FORMATE see LGA050
LINALYL ISOBUTYRATE see LGB000
LINARODIN see MDW750
LINASEN see EOK000
LINCOCIN see LGD000

LINCOLCINA see LGD000
LINCOLNENSIN see LGD000
LINCOMYCIN see LGD000
LINCOMYCINE (FRENCH) see LGD000
LINDAGRAIN see BBQ500
LINDAN see DGP900
α-LINDANE see BBQ000
β-LINDANE see BBR000
Δ-LINDANE see BFW500
LINDANE (ACGIH, DOT, USDA) see BBQ500
LINDEROL see NCQ820
LINDOL see TNP500
LINEAR ALKYLBENZENE SULFONATE, SODIUM SALT see LGF825
LINEX 4L see DGD600
LINFOLIZIN see CDO500
LINFOLYSIN see CDO500
LINGEL see BRF500
LINGRAINE see EDC500
LINGRAN see EDC500
LINGUSORBS see PMH500
LINOLEAMIDE see LGF900
LINOLEIC ACID see LGG000
9,12-LINOLEIC ACID see LGG000
LINOLEIC ACID AMIDE see LGF900
LINORMONE see CIR250
LINOROX see DGD600
LINSEED OIL see LGK000
LINTOX see BBQ500
LINUREX see DGD600
LINURON see DGD600
LINURON (herbicide) see DGD600
LIOXIN see VFK000
LIPAN see DUS700
LIPARITE see CAS000
LIPHADIONE see CJJ000
LIPID CRIMSON see SBC500
LIPO-DIAZINE see PPP500
LIPO EGMS see EJM500
LIPO GMS 410 see OAV000
LIPO GMS 450 see OAV000
LIPO GMS 600 see OAV000
LIPO-LEVAZINE see PPP500
LIPO-LUTIN see PMH500
LIPOPILL see DTJ400
LIPOSORB L-20 see PKG000
LIPOSORB O-20 see PKL100
LIPOSORB O-20 see SKV100
LIPOSORB S-20 see PKL030
LIPOSORB S-20 see SKV150
LIPOSORB O see SKV100
LIPOSORB S see SKV150
LIPOTRIL see CMF750
LIPTAN see IIU000
LIQUACILLIN see BDY669
LIQUADIAZINE see PPP500
LIQUAGESIC see HIM000
LIQUAMYCIN INJECTABLE see HOI000
LIQUAMYCIN LA 200 see HOH500
LIQUA-TOX see WAT200
LIQUEFIED CARBON DIOXIDE see LGL000
LIQUEFIED PETROLEUM GAS see LGM000
LIQUEFIED PETROLEUM GAS (DOT) see IIC000
LIQUIBARINE see BAP000
LIQUID BRIGHT PLATINUM see PJD500
LIQUID CAMPHOR see CBB500
LIQUID CARBONIC GAS see LGL000

LIQUID ETHYLENE see EIO000
LIQUIDOW see CAO750
LIQUID PITCH OIL see CMY825
LIQUID ROSIN see TAC000
LIQUIGEL see AHC000
LIQUIMETH see MDT750
LIQUIPHENE see ABU500
LIQUI-SAN see MLH000
LIQUI-STIK see NAK500
LIQUITAL see EOK000
LIQUOPHYLLINE see TEP000
LIQUORICE see GIG000
LIRANOX see CIR500
LIROBETAREX see CJX750
LIROHEX see TCF250
LIROMATIN see ABX250
LIROPON see DGI400
LIROPREM see PAX250
LIROSTANOL see ABX250
LIROTHION see PAK000
LISACORT see PLZ000
LISERGAN see ABH500
LISKONUM see LGZ000
LISSOLAMINE see HCQ500
LITAC see ADY500
LITEX CA see SMR000
LITHALURE see LHQ100
LITHAMIDE see LGT000
LITHANE see LGZ000
LITHARGE see LDN000
LITHARGE YELLOW L-28 see LDN000
LITHIC ACID see UVA400
LITHICARB see LGZ000
LITHINATE see LGZ000
LITHIUM see LGO000
LITHIUM ACETATE see LGO100
LITHIUM ALANATE see LHS000
LITHIUM ALUMINOHYDRIDE see LHS000
LITHIUM ALUMINUM HYDRIDE (DOT) see LHS000
LITHIUM ALUMINUM HYDRIDE, ETHEREAL (DOT) see LHS000
LITHIUM ALUMINUM TETRAHYDRIDE see LHS000
LITHIUM ALUMINUM TRI-tert-BUTOXYHYDRIDE see LGS000
LITHIUM AMIDE see LGT000
LITHIUM AMIDE, POWDERED see LGT000
LITHIUM ANTIMONIOTHIOMALATE see LGU000
LITHIUM ANTIMONY THIOMALATE see LGU000
LITHIUM BOROHYDRIDE (DOT) see LHT000
LITHIUM CARBONATE see LGZ000
LITHIUM CARBONATE (2:1) see LGZ000
LITHIUM CHLORIDE see LHB000
LITHIUM FERRO SILICON see LHK000
LITHIUM FLUORIDE see LHF000
LITHIUM FLUORURE (FRENCH) see LHF000
LITHIUM HYDRIDE see LHH000
LITHIUM HYDRIDE (UN 1414) (DOT) see LHH000
LITHIUM HYDRIDE, fused solid (UN 2805) (DOT) see LHH000
LITHIUM HYDROXIDE see LHI100
LITHIUM HYDROXIDE, solution (DOT) see LHI100
LITHIUM HYDROXIDE, monohydrate or lithium hydroxide, solid (DOT) see LHI100
LITHIUM HYDROXIDE (Li(OH)) (9CI) see LHI100
LITHIUM HYDROXYBUTYRATE see LHM800
LITHIUM γ-HYDROXYBUTYRATE see LHM800
LITHIUM HYPOCHLORITE see LHJ000

LITHIUM HYPOCHLORITE COMPOUND, dry, containing more than 39% available chlorine (DOT) see LHJ000
LITHIUM IRON SILICON see LHK000
LITHIUM LACTATE see LHL000
LITHIUM METAL (DOT) see LGO000
LITHIUM METAL, IN CARTRIDGES (DOT) see LGO000
LITHIUM NITRIDE see LHM000
LITHIUM OCTADECANOATE see LHQ100
LITHIUM OXYBUTYRATE see LHM800
LITHIUM PEROXIDE see LHO000
LITHIUM SILICON see LHP000
LITHIUM STEARATE see LHQ100
LITHIUM TETRAHYDROALUMINATE see LHS000
LITHIUM TETRAHYDROBORATE see LHT000
LITHOBID see LGZ000
LITHOGRAPHIC STONE see CAO000
LITHOLITE see LHQ100
LITHONATE see LGZ000
LITHOSOL ORANGE R BASE see NMP500
LITHOTABS see LGZ000
LITICON see DOQ400
LIV 1176 see DVK200
L.J. 206 see CBR675
LK 14304-18 see BMM575
L 2642-LABAZ see EDV700
LLD FACTOR see VSZ000
LM 91 see CJJ000
LM-2717 see CIR750
LM SEED PROTECTANT see MLH000
LOBAMINE see MDT740
LO-BAX see HOV500
LOCRON EXTRA see AHA000
LOCRON FLAKES see AHA000
LOCRON POWDER see AHA000
LOCRON SOLUTION see AHA000
LOCUST BEAN GUM see LIA000
LODRONATE see SFX730
LOFEPRAMINE see DLH630
LOHA see IGK800
LOISOL see TIQ250
LOMBRISTOP see TEX000
LO MICRON TALC 1 see TAB750
LO MICRON TALC, BC 1621 see TAB750
LO MICRON TALC USP, BC 2755 see TAB750
LOMPER see MHL000
LOMUPREN see DMV600
LOMUSTINE see CGV250
S-LON see PKQ059
LON 41 see DTM600
LONAMIN see DTJ400
LONARID see HIM000
LONDON PURPLE see LIC000
LONGACILIAN see BFC750
LONGICIL see BFC750
LONGIFENE see BOM250
LONGIFENE see HGC500
β-LONGILOBINE see RFP000
LONIN 3 see BHO500
LONOCOL M see MAS500
LONZA G see PKQ059
LOPRESS see HGP500
LOPURIN see ZVJ000
L-ORANGE 2 see FAG150
L. ORANGE Z2010 see FAG150
LOREX see DGD600

LORINAL see CDO000
LORMIN see CBF250
LOROL see DXV600
LOROL 20 see OEI000
LOROL 22 see DAI600
LOROL 24 see HCP000
LOROL 28 see OAX000
LOROMISIN see CDP250
LOROX see DGD600
LOROXIDE see BDS000
LOROX LINURON WEED KILLER see DGD600
LORPHEN see TAI500
LORSBAN see CMA100
LOSANTIN see HOV500
N-LOST see BIE500
S-LOST see BIH250
N-LOST (GERMAN) see BIE250
LOSUNGSMITTEL APV see CBR000
LOVAGE OIL see LII000
LOVISCOL see CBR675
LOVOSA see SFO500
LOW ERUCIC ACID RAPESEED OIL see RBK200
LOWESERP see RDK000
LOWETRATE see PBC250
LOWPSTRON see CHJ750
LOXANOL K see HCP000
LOXAPINE see DCS200
LOXIGLUMIDE see LII100
LOXIOL G 10 see GGR200
LOXON see DFH600
LOXURAN see DIW200
LPG see BFC750
LPG see LGM000
LPG ETHYL MERCAPTAN 1010 see EMB100
L.P.G. (OSHA, ACGIH) see LGM000
LPT see PKB500
LR 115 see CCU250
LRF see LIU370
LRH see LIU370
LS 061A see SMQ500
L.S. 3394 see BLL750
LS 4442 see CLU000
LS 1028E see SMQ500
LSD see DJO000
d-LSD see DJO000
LSD-25 see DJO000
LSD TARTRATE see LJG000
LTAN see DMH400
LUBERGAL see EOK000
LUCALOX see AHE250
LUCANTHON see DHU000
LUCANTHONE see DHU000
LUCANTHONE METABOLITE see LIM000
LUCEL (polysaccharide) see SFO500
LUCIDOL see BDS000
LUCIDOL DELTAX see MKA500
LUCITE see PKB500
LUCOFLEX see PKQ059
LUCORTEUM SOL see PMH500
LUCOVYL PE see PKQ059
LUDOX see SCH002
LUFIXAN see VPP100
LUGOL'S IODINE see PLW285
LUGOL'S SOLUTION see PLW285
LULAMIN see TEH500
LULIBERIN see LIU370
LUMBRICAL see PIJ000
LUMEN see EOK000

LUMESETTES see EOK000
LUMICREASE BLUE 4GL see CMN750
LUMINAL see EOK000
LUMINAL SODIUM see SID000
LUMOFRIDETTEN see EOK000
LUNAR CAUSTIC see SDS000
LUPAREEN see PMP500
LUPERCO see BDS000
LUPERCO see DGR600
LUPERCO CST see BIX750
LUPEROX see DGR600
LUPEROX 500R see DGR600
LUPEROX 500T see DGR600
LUPEROX FL see BDS000
LUPERSOL see MKA500
LUPERSOL 11 see BSD250
LUPERSOL 70 see BSC250
LUPERSOL DDA 30 see MKA500
LUPERSOL DDM see MKA500
LUPERSOL DNF see MKA500
LUPERSOL DSW see MKA500
LUPERSOL Δ-X see MKA500
LUPOLEN 1010H see PJS750
LUPOLEN 1800H see PJS750
LUPOLEN 1800S see PJS750
LUPOLEN 1810H see PJS750
LUPOLEN 4261A see PJS750
LUPOLEN 6011H see PJS750
LUPOLEN 6011L see PJS750
LUPOLEN 6042D see PJS750
LUPOLEN KR 1032 see PJS750
LUPOLEN KR 1051 see PJS750
LUPOLEN KR 1257 see PJS750
LUPOLEN L 6041D see PJS750
LUPOLEN N see PJS750
LUPRANATE M 10 see PKB100
LUPRANATE M 70 see PKB100
LUPRANATE M 20S see PKB100
LUPRINATE M 20 see PKB100
LUPROLITE ACETATE see LIU305
LUPROSIL see PMU750
LURAMIN see EOK000
LURAN see ADY500
LURAZOL BLACK BA see AQP000
LURGO see DSP400
LURIDE see SHF500
LUSIL see SNM500
LUSMIT see TFD500
LUSTRAN see ADY500
LUSTREX see SMQ500
LUSTREX H 77 see SMQ500
LUSTREX HH 101 see SMQ500
LUSTREX HH 101 see SMQ500
LUSTREX HP 77 see SMQ500
LUSTREX HT 88 see SMQ500
LUTALYSE see POC750
LUTEAL HORMONE see PMH500
LUTEINIZING HORMONE-RELEASING FACTOR see LIU370
LUTEINIZING HORMONE-RELEASING FACTOR (PIG), ACETATE (SALT) see LIU305
LUTEINIZING HORMONE-RELEASING FACTOR (SWINE), ACETATE (SALT) see LIU305
LUTEINIZING HORMONE-RELEASING HORMONE see LIU370
LUTEOHORMONE see PMH500
LUTEOSAN see PMH500

LUTEOSKYRIN see LIV000
(−)-LUTEOSKYRIN see LIV000
LUTEOSTIMULIN see LIU370
LUTETIA RED CLN see CHP500
LUTETIUM CHLORIDE see LIW000
LUTETIUM(III) NITRATE (1:3) see LIY000
LUTEX see PMH500
3,4-LUTIDINE see LJB000
LUTINYL see CBF250
LUTOCYCLIN see PMH500
LUTOFAN see PKQ059
LUTO-METRODIOL see EQJ500
LUTOSOL see INJ000
LUTRELEF see LIU370
LUTREPULSE see LIU305
LUTROL see PJT000
LUTROL-9 see EJC500
LUTROMONE see PMH500
LUVISKOL see PKQ250
LUXISTELM see PHI500
LUXON see DFH600
LW 3170 see DCS200
LX 14-0 see CQH250
LXON see DFH600
LYCEDAN see AOA125
LYCOID DR see GLU000
LYDOL see DAM700
LYE see PLJ500
LYE (DOT) see SHS000
LYE, solution see SHS500
LYGOMME CDS see CCL250
LYMPHOSCAN see AQL500
LYNDIOL see LJE000
LYNESTRENOL mixed with MESTRANOL see LJE000
LYNESTROL see EEH550
LYNESTROL mixed with MESTRANOL see LJE000
LYNOESTRENOL mixed with MESTRANOL see LJE000
LYOPHRIN see VGP000
LYP 97 see LBR000
LYSERGAMID see DJO000
LYSERGAURE DIETHYLAMID see DJO000
d-LYSERGIC ACID DIETHYLAMIDE see DJO000
LYSERGIC ACID DIETHYLAMIDE-25 see DJO000
LYSERGIC ACID DIETHYLAMIDE TARTRATE see LJG000
d-LYSERGIC ACID DIETHYLAMIDE TARTRATE see LJG000
LYSERGIDE see DJO000
LYSERGSAEUREDIAETHYLAMID see DJO000
LYSINE see LJM700
l-(+)-LYSINE see LJM700
l-LYSINE (9CI) see LJM700
LYSINE ACID see LJM700
l-LYSINE HYDROCHLORIDE see LJO000
LYSINE MONOHYDROCHLORIDE see LJO000
l-LYSINE MONOHYDROCHLORIDE see LJO000
LYSIVANE see DIR000
LYSOCOCCINE see SNM500
LYSOFON see AGW750
LYSOFORM see FMV000
LYSURON see ZVJ000
LYTECA SYRUP see HIM000
LYTRON 5202 see SMR000
M 40 see CIR250
M-74 see DXH325
M 81 see PHI500
N-2-M see CCK665
M 140 see CDR750

M 176 see DUD800
M 410 see CDR750
M 3180 see CQH650
M-4209 see CBT250
M 5055 see CDV100
M-9500 see TND500
M 13/20 see PKP750
M 3/158 see DAP600
MA-1214 see DXV600
MAA see AFI850
MAAC see HFJ000
MABLIN see BOT250
MA 100 (CARBON) see CBT500
MACASIROOL see CHJ750
MACE (lachrymator) see CEA750
MACE OIL see OGQ100
MACH-NIC see NDN000
MACOCYN see HOH500
MACQUER'S SALT see ARD250
MACRABIN see VSZ000
MACRODANTIN see NGE000
MACRODIOL see EDO000
MACROGOL 1000 see PJT250
MACROGOL 4000 see PJT750
MACROGOL 400 BPC see EJC500
MACROL see EDO000
MACROPAQUE see BAP000
MACROSE see DBD700
MACROSPHERICAL 95 see AHA000
MACULOTOXIN see FOQ000
MADAGASCAR LEMONGRASS OIL see LEH000
MADHURIN see SJN700
MADIOL see MQU750
MADRIBON see SNN300
MADRIGID see SNN300
MADRINE see DBA800
MADRIQID see SNN300
MADROXIN see SNN300
MADROXINE see SNN300
MAFU see DGP900
MAG see MRF000
MAGBOND see BAV750
MAGCAL see MAH500
MAGCHEM 100 see MAH500
MAGECOL see CBT750
MAGLITE see MAH500
MAGNACIDE H see ADR000
MAGNAMYCIN see CBT250
MAGNAMYCIN A see CBT250
MAGNESIA see MAH500
MAGNESIA MAGMA see MAG750
MAGNESIA USTA see MAH500
MAGNESIA WHITE see CAX750
MAGNESIO (ITALIAN) see MAC750
MAGNESIUM see MAC750
MAGNESIUM ALLOYS, powder (UN 1418) (DOT) see
MAC750
MAGNESIUM ALLOYS with >50% magnesium in pellets,
turnings or ribbons (UN 1869) (DOT) see MAC750
MAGNESIUM ALUMINUM PHOSPHIDE (DOT) see
AHD250
MAGNESIUM AMMONIUM ARSENATE see MAD025
MAGNESIUM AMMONIUM ARSENATE DIHYDRATE
see MAD025
MAGNESIUM ARSENATE see ARD000
MAGNESIUM ARSENATE PHOSPHOR see ARD000
MAGNESIUM, BIS(2-HYDROXYPROPANOATO-
O(1),O(2))-, (T-4)-(9CI) see LAL100

MAGNESIUM, BIS(LACTATO)-(8CI) see LAL100
MAGNESIUM CLIPPINGS see MAC750
MAGNESIUM COMPOUNDS see MAE750
MAGNESIUM DIPROPYLACETATE see MAK275
MAGNESIUM DROSS, wet or hot (DOT) see MAF000
MAGNESIUM DROSS (HOT) see MAF000
MAGNESIUM FLUORIDE see MAF500
MAGNESIUM FLUOROSILICATE (DOT) see MAF600
MAGNESIUM FLUORURE (FRENCH) see MAF500
MAGNESIUM FLUOSILICATE see MAF600
MAGNESIUM FLUOSILICATE see MAG250
MAGNESIUMFOSFIDE (DUTCH) see MAI000
MAGNESIUM GLYCEROPHOSPHATE see MAG100
MAGNESIUM GOLD PURPLE see GIS000
MAGNESIUM GRANULES, coated particle size not <149
microns (UN 2950) (DOT) see MAC750
MAGNESIUM HEXAFLUOROSILICATE see MAG250
MAGNESIUM HYDRATE see MAG750
MAGNESIUM HYDROXIDE see MAG750
MAGNESIUM LACTATE see LAL100
MAGNESIUM NITRATE (DOT) see MAH000
MAGNESIUM(II) NITRATE (1:2) see MAH000
MAGNESIUM OXIDE see MAH500
MAGNESIUM OXIDE FUME (ACGIH) see MAH500
MAGNESIUM PELLETS see MAC750
MAGNESIUM PERCHLORATE see PCE000
MAGNESIUM PEROXIDE see MAH750
MAGNESIUM PEROXIDE, solid (DOT) see MAH750
MAGNESIUM PHOSPHIDE see MAI000
MAGNESIUM POWDERED see MAC750
MAGNESIUM RIBBONS see MAC750
MAGNESIUM SILICATE HYDRATE see MAJ000
MAGNESIUM SILICOFLUORIDE see MAG250
MAGNESIUM STEARATE see MAJ030
MAGNESIUM STEARATE see MAJ030
MAGNESIUM STEARATE (ACGIH) see MAJ030
MAGNESIUM SULFATE (1:1) see MAJ250
MAGNESIUM SULFATE adduct of 2,2-DITHIO-BIS-
PYRIDINE 1-OXIDE see OIU850
MAGNESIUM SULPHATE see MAJ250
MAGNESIUM TURNINGS (DOT) see MAC750
MAGNESIUM (UN 1869) (DOT) see MAC750
MAGNESIUM, powder (UN 1418) (DOT) see MAC750
MAGNESIUM VALPROATE see MAK275
MAGNETIC 70, 90, and 95 see SOD500
MAGNEZU TLENEK (POLISH) see MAH500
MAGNOLIONE see OOO100
MAGOX see MAG750
MAGOX see MAH500
MAGOX 85 see MAH500
MAGOX 90 see MAH500
MAGOX 95 see MAH500
MAGOX 98 see MAH500
MAGOX OP see MAH500
MAGSALYL see SJO000
MAH see DMC600
MAINTAIN 3 see DMC600
MAIPEDOPA see DNA200
MAIZENA see SLJ500
MAJOL PLX see SFO500
MAJSOLIN see DBB200
MAKAROL see DKA600
MALACHITE GREEN G see BAY750
MALACID see TGD000
MALACIDE see MAK700
MALAFOR see MAK700
MALAGRAN see MAK700
MALAKILL see MAK700

MALAMAR see MAK700
MALAMAR 50 see MAK700
MALAPHELE see MAK700
MALAPHOS see MAK700
MALASOL see MAK700
MALASPRAY see MAK700
MALATHION see MAK700
MALATHION ULV CONCENTRATE see MAK700
MALATHIOZOO see MAK700
MALATHON see MAK700
MALATHYL LV CONCENTRATE & ULV
CONCENTRATE see MAK700
MALATION (POLISH) see MAK700
MALATOL see MAK700
MALATOX see MAK700
MALAYAN CAMPHOR see BMD000
MALAZIDE see DMC600
MALDISON see MAK700
MALEALDEHYDIC ACID, DICHLORO-4-OXO-, (Z)-
(9CI) see MRU900
MALEIC ACID see MAK900
MALEIC ACID ANHYDRIDE (MAK) see MAM000
MALEIC ACID, DIALLYL ESTER see DBK200
MALEIC ACID, DIBUTYL ESTER see DED600
MALEIC ACID-N-ETHYLIMIDE see MAL250
MALEIC ACID N-ETHYLIMIDE see MAL250
MALEIC ACID HYDRAZIDE see DMC600
MALEIC ANHYDRIDE see MAM000
MALEIC ANHYDRIDE adduct of BUTADIENE see
TDB000
MALEIC HYDRAZIDE see DMC600
MALEIC HYDRAZIDE 30% see DMC600
MALEIC HYDRAZIDE DIETHANOLAMINE SALT see
DHF200
MALEIC HYDRAZINE see DMC600
MALEIMIDE see MAM750
MALEIMIDE, N-ETHYL- see MAL250
MALEIN 30 see DMC600
MALEINIC ACID see MAK900
MALEINIMIDE see MAM750
MALEINSAEUREHYDRAZID see DMC600
MALENIC ACID see MAK900
N,N-MALEOYLHYDRAZINE see DMC600
MALESTRONE see MPN500
MALESTRONE (AMPS) see TBF500
MALGESIC see BRF500
MALIC ACID see MAN000
MALIC ACID, 3-HYDROXY- see TAF750
MALIPUR see CBG000
MALIX see EAQ750
MALLOFEEN see PDC250
MALLOPHENE see PDC250
MALMED see MAK700
MALOCID see TGD000
MALOCIDE see TGD000
MALOGEN see MPN500
MALONIC ACID, DIETHYL ESTER see EMA500
MALONIC ACID, (ETHOXYMETHYLENE)-,
DIETHYL ESTER see EEV200
MALONIC ACID ETHYL ESTER NITRILE see EHP500
MALONIC DINITRILE see MAO250
MALONIC ESTER see EMA500
MALONONITRILE see MAO250
MALONONITRILE, CARBONYL-, DIETHYL
MERCAPTOLE see BJT800
MALOPRIM see SOA500
MALOPRIM see TGD000
MALPHOS see MAK700

MALTOBIOSE see MAO500
MALTODEXTRIN see MAO300
MALTOL see MAO350
MALTOSE see MAO500
d-MALTOSE see MAO500
MALTOX see MAK700
MALTOX MLT see MAK700
MALT SUGAR see MAO500
α-MALT SUGAR see MAO500
MALYSOL see MKA250
MALZID see DMC600
MAM AC see MGS750
MAM ACETATE see MGS750
MAMALLET-A see DOT000
MAMBNA see MHW350
MAMMEX see NGE500
MAMN see AAW000
MANADRIN see EAW000
MANAM see MAS500
MANCHESTER YELLOW see DUX800
MANDARIN OIL, COLDPRESSED see MAO900
MANDELIC ACID see MAP000
MANDELIC ACID NITRILE see MAP250
MANDRIN see EAW000
MANEB see MAS500
MANEB 80 see MAS500
MANEBA see MAS500
MANEBE 80 see MAS500
MANEBE (FRENCH) see MAS500
MANEBGAN see MAS500
MANEB PREPARATIONS, stabilized against self-heating
(UN 2968) (DOT) see MAS500
MANEB PREPARATIONS with not <60% maneb (UN
2210) (DOT) see MAS500
MANEB (UN 2210) (DOT) see MAS500
MANEB, stabilized (UN 2968) (DOT) see MAS500
MANEB plus ZINC ACETATE (50:1) see EIQ500
MANEB ZL4 see MAS500
MANESAN see MAS500
MANEX see MAS500
MANEXIN see MAW250
MANGAANBIOXYDE (DUTCH) see MAS000
MANGAANDIOXYDE (DUTCH) see MAS000
MANGAAN(II)-(N,N'-ETHYLEEN-
BIS(DITHIOCARBAMAAT)) (DUTCH) see MAS500
MANGACAT see MAP750
MANGAN (POLISH) see MAP750
MANGAN(II)-(N,N'-AETHYLEN-
BIS(DITHIOCARBAMAT)) (GERMAN) see MAS500
MANGANATE(3-), HEXACYANO-, TRIPOTASSIUM
see TMX600
MANGANATE(3-), HEXAKIS(CYANO-C)-,
TRIPOTASSIUM, (OC-6-11)- see TMX600
MANGANDIOXID (GERMAN) see MAS000
MANGANESE see MAP750
MANGANESE ACETATE see MAQ000
MANGANESE(2+) ACETATE see MAQ000
MANGANESE(II) ACETATE see MAQ000
MANGANESE ACETYLACETONATE see MAQ500
MANGANESE BINOXIDE see MAS000
MANGANESE (BIOSSIDO di) (ITALIAN) see MAS000
MANGANESE (BIOXYDE de) (FRENCH) see MAS000
MANGANESE, BIS(l-HISTIDINATO-N,O)-,
TETRAHYDRATE see BJX800
MANGANESE, BIS(l-HISTIDINATO)-,
TETRAHYDRATE see BJX800

MANGANESE, BIS(5-SULFO-8-QUINOLINOLATO)-
see BKJ275
MANGANESE BLACK see MAS000
MANGANESE(II) CHLORIDE (1:2) see MAR000
MANGANESE CITRATE see MAR260
MANGANESE COMPOUNDS see MAR500
MANGANESE CYCLOPENTADIENYL
TRICARBONYL see CPV000
MANGANESE DIACETATE see MAQ000
MANGANESE DICHLORIDE see MAR000
MANGANESE DINITRATE see MAS900
MANGANESE (DIOSSIDO di) (ITALIAN) see MAS000
MANGANESE DIOXIDE see MAS000
MANGANESE (DIOXYDE de) (FRENCH) see MAS000
MANGANESE ETHYLENE-1,2-
BISDITHIOCARBAMATE see MAS500
MANGANESE(II)
ETHYLENEBIS(DITHIOCARBAMATE) see MAS500
MANGANESE(II) ETHYLENE
DI(DITHIOCARBAMATE) see MAS500
MANGANESE FLUORIDE see MAS750
MANGANESE(II) FLUORIDE see MAS750
MANGANESE FLUORURE (FRENCH) see MAS750
MANGANESE GLYCEROPHOSPHATE see MAS810
MANGANESE GREEN see MAT250
MANGANESE LINOLEATE see MAS818
MANGANESE MANGANATE see MAT500
MANGANESE,
(METHYLCYCLOPENTADIENYL)TRICARBONYL- see
MAV750
MANGANESE MONOXIDE see MAT250
MANGANESE NITRATE see MAS900
MANGANESE(2+) NITRATE see MAS900
MANGANESE(II) NITRATE see MAS900
MANGANESE NITRATE (DOT) see MAS900
MANGANESE (II) NITRATE, ANHYDROUS see
MAS900
MANGANESE OXIDE see MAS000
MANGANESE OXIDE see MAV550
MANGANESE(II) OXIDE see MAT250
MANGANESE(IV) OXIDE see MAS000
MANGANESE(III) OXIDE see MAT500
MANGANESE PEROXIDE see MAS000
MANGANESE SESQUIOXIDE see MAT500
MANGANESE(II) SULFATE (1:1) see MAU250
MANGANESE SUPEROXIDE see MAS000
MANGANESE TALLATE see MAV100
MANGANESE,
TETRACARBONYL(TRIFLUOROMETHYLTHIO)-,
dimer see TBN200
MANGANESE TETROXIDE see MAV550
MANGANESE TRICARBONYL
METHYLCYCLOPENTADIENYL see MAV750
MANGANESE TRIOXIDE see MAT500
MANGANESE(2+),
TRIS(OCTAMETHYLDIPHOSPHORAMIDE-Op,Op')-,
(OC-6-11)-, DIPERCHLORATE see TNK400
MANGANESE ZINC BERYLLIUM SILICATE see
BFS750
MANGANIC OXIDE see MAT500
MANGAN NITRIDOVANY (CZECH) see MAP750
MANGANOMANGANIC OXIDE see MAV550
MANGANOUS ACETATE see MAQ000
MANGANOUS ACETYLACETONATE see MAQ500
MANGANOUS CHLORIDE see MAR000
MANGANOUS DINITRATE see MAS900
MANGANOUS NITRATE see MAS900
MANGANOUS OXIDE see MAT250

MANGANOUS SULFATE see MAU250
MAN-GRO see MAU250
MANHEXIN see MAW250
MANICOLE see MAW250
MANITE see MAW250
MANNEX see MAW250
MANNITOL see MAW100
d-MANNITOL BUSULFAN see BKM500
MANNITOL HEXANITRATE see MAW250
d-MANNITOL HEXANITRATE see MAW250
MANNITOL HEXANITRATE (dry) (DOT) see MAW250
MANNITOL HEXANITRATE, wetted with not <40%
water, by weight or mixture (NA 0133) (DOT) see MAW250
MANNITOL MUSTARD DIHYDROCHLORIDE see
MAW750
MANNITOL MYLERAN see BKM500
MANNITRIN see MAW250
MANNOGRANOL see BKM500
MANNOMUSTINE DIHYDROCHLORIDE see
MAW750
MANOLENE 6050 see PJS750
MANOXAL OT see DJL000
MAN'S MOTHERWORT see CCP000
MANTADAN see AED250
MANUCOL see SEH000
MANUCOL DM see SEH000
MANUFACTURED IRON OXIDES see IHC450
MANUTEX see SEH000
MANZATE see MAS500
MANZATE 200 see MAS500
MANZATE D see MAS500
MANZATE MANEB FUNGICIDE see MAS500
MANZEB see MAS500
MANZIN see MAS500
MAOH see MKW600
MAO-REM see PFC750
MAPLE BRILLIANT BLUE FCF see FMU059
MAPLE LACTONE see HMB500
MAPLE PONCEAU 3R see FAG018
MAPLE SUNSET YELLOW FCF see FAG150
MAPLE TARTRAZOL YELLOW see FAG140
MAPO see TNK250
MAPOLOSE 60SH50 see MIF760
MAPOLOSE M25 see MIF760
MAPP (OSHA) see MFX600
MAPROFIX 563 see SIB600
MAPROFIX WAC-LA see SIB600
MARALATE see MEI450
MARANTA see SLJ500
MARANYL F 114 see PJY500
MAR BATE see MQU750
MARBLE see CAO000
MARBON 9200 see SMR000
MAREVAN (SODIUM SALT) see WAT220
MAREX see HGC500
MARGONIL see MQU750
MARICAINE see DHK400
MARIMET 45 see CAW850
MARINCO H see MAG750
MARINE DIESEL FUEL see DHE750
MARISILAN see AIV500
MARITUS YELLOW see DUX800
MARJORAM OIL, SPANISH see MBU500
MARKS 4-CPA see CJN000
MARLATE see MEI450
MARLEX 9 see PJS750
MARLEX 50 see PJS750
MARLEX 60 see PJS750

MARLEX 960 see PJS750
MARLEX 6003 see PJS750
MARLEX 6009 see PJS750
MARLEX 6015 see PJS750
MARLEX 6050 see PJS750
MARLEX 6060 see PJS750
MARLEX 9400 see PMP500
MARLEX EHM 6001 see PJS750
MARLEX M 309 see PJS750
MARLEX TR 704 see PJS750
MARLEX TR 880 see PJS750
MARLEX TR 885 see PJS750
MARLEX TR 906 see PJS750
MARLON AS 3 see LBU100
MARLOPHEN 820 see PKF500
MARMAG see MAH500
MARMER see DXQ500
MARNITENSION SIMPLE see RDK000
MAROXOL-50 see DUZ000
MARS BROWN see IHC450
MARSH GAS see MDQ750
MARSIN see MNV750
MARS RED see IHC450
MARTRATE-45 see PBC250
MARVEX see DGP900
MARVINAL see PKQ059
MASCHITT see CFY000
MASENATE see TBG000
MASENONE see MPN500
MASEPTOL see HJL500
MASHERI (INDIA) see SED400
MASOTEN see TIQ250
MASSICOT see LDN000
MASSICOTTE see LDN000
MASTESTONA see MPN500
MASTIPHEN see CDP250
MATCH see MLJ500
MATENON see MQS225
MATRICARIA CAMPHOR see CBA750
MATROMYCIN see OHO200
MATSUTAKE ALCOHOL (JAPANESE) see ODW000
MATTING ACID (DOT) see SOI500
MATULANE see PME500
MAURYLENE see PMP500
MAVISERPIN see RDK000
MAXATASE see BAC000
MAXIDEX see SOW000
MAXITATE see MAW250
MAXULVET see SNN300
MAYCOR see CCK125
MAYSERPINE see RDK000
MAZIDE 30 see DHF200
MAZOTEN see TIQ250
MB see MHR200
M+B 695 see PPO000
M+B 760 see TEX250
MB 10064 see DDP000
MBA see BIE250
MBA HYDROCHLORIDE see BIE500
MBAO see CFA500
MBAO HYDROCHLORIDE see CFA750
MBC see MHC750
MBC see BAV575
MBCP see LEN000
MBDZ see MHL000
MBH see PME500
MBK see HEV000
MBNA see MHW500

MBOCA see MJM200
MBOT see MJO250
MBP see MRF525
MBT see BDF000
MBTS see BDE750
MBTS RUBBER ACCELERATOR see BDE750
MBX see MHR200
6-MC see MIP750
2M-4C see CIR250
MC6897 see DQM600
MCA see CEA000
3-MCA see MIJ750
MeCsAc see EJJ500
MCB see CEJ125
MCE see EIF000
MCF see MIG000
MCN-A 2833-109 see PCJ230
MCNAMEE see KBB600
MCN-JR-2498 see TKK500
MCN-JR-3345 see FHG000
MCN-JR-4263-49 see PDW750
MCO 8000 see MIF760
M1 (COPPER) see CNI000
M2 (COPPER) see CNI000
MCP see CIR250
2M-4CP see CIR500
MC 4000 cP see MIF760
MCPA see CIR250
MCPA SODIUM SALT see SIL500
MCPP see CIR500
2-MCPP see CIR500
MCPP-D-4 see CIR500
MCPP 2,4-D see CIR500
MCPP-K-4 see CIR500
MC 20000S see MIF760
MCT see TMO000
MCT see CPV000
M 3 (CURING AGENT) see OMM300
l-(α-MD) see DNA800
MDA see MJQ000
MDAB see DUH600
3'-MDAB see DUH600
MDBA see MEL500
MD BACITRACIN see BAC260
MDI see MJP400
MDI-CR see PKB100
MDI-CR 100 see PKB100
MDI-CR 200 see PKB100
MDI-CR 300 see PKB100
2-ME see MCN250
ME-1700 see BIM500
MEA see AJT250
MEA see EEC600
MEADOW GREEN see COF500
MEARLMAID see GLI000
MEASURIN see ADA725
MEAVERIN see CBR250
MEB see MAS500
MEB 6447 see CJO250
MEBACID see ALF250
MEBARAL see ENB500
MEBENDAZOLE (USDA) see MHL000
MEBENVET see MHL000
MEBEREL see ENB500
MEBICHLORAMINE see BIE500
MEBR see MHR200
ME4 BROMINAL see DDP000

MEBUBARBITAL SODIUM see NBU000
MEBUMAL NATRIUM see NBU000
MEBUMAL SODIUM see NBU000
MECADOX see FOI000
MECB see DJG000
ME-CCNU see CHD250
MECHLORETHAMINE see BIE250
MECHLORETHAMINE HYDROCHLORIDE see
BIE500
MECHLORETHAMINE OXIDE see CFA500
MECHLORETHAMINE OXIDE HYDROCHLORIDE
see CFA750
MECLIZINE see HGC500
MECLIZINE DIHYDROCHLORIDE see MBX250
MECLOZINE see HGC500
MECODIN see MDO750
MECODRIN see BBK000
MECOMEC see CIR500
MECOPEOP see CIR500
MECOPER see CIR500
MECOPEX see CIR500
MECOPROP see CIR500
MECOTURF see CIR500
MECPROP see CIR500
MECRAMINE see AJT250
MECRILAT see MIQ075
MECRYLATE see MIQ075
MECS see EJH500
MEDAMYCIN see TBX250
MEDARON see NGG500
MEDEMANOL see MAW250
MEDIAMID see DJS200
MEDIAMYCETINE see CDP250
MEDIBEN see MEL500
MEDI-CALGON see SHM500
MEDIDRYL see BBV500
MEDIFLAVIN see DBX400
MEDIHALER-EPI see VGP000
MEDILLA see MKP500
MEDINAL see BAG250
MEDOMET see DNA800
MEDOPREN see DNA800
MEDROL see MOR500
MEDROL DOSEPAK see MOR500
MEDRONE see MOR500
MEDROXYPROGESTERONE ACETATE see MCA000
MEE see EDP000
MEFEDINA see DAM700
M.E.G. see EJC500
MEGABION see VSZ000
MEGACE see VTF000
MEGACILLIN SUSPENSION see BFC750
MEGACORT see DAE525
MEGADIURIL see CFY000
MEGALOVEL see VSZ000
MEGAPHEN see CKP500
MEGATOX see FFF000
MEGESTROL ACETATE see VTF000
MEGESTRYL ACETATE see VTF000
MEGIMIDE see MKA250
6-ME-GLU-P-2 see AKS250
MeIQ see AJQ600
MEISEI TERYL DIAZO BLACK CR see DPO200
MEISEI TERYL DIAZO BLUE HR see DCJ200
MEK see MKA400
MEK-OXIME see EMU500
MEK PEROXIDE see MKA500
MEKP (OSHA) see MKA500

MELABON see ABG750
MELADININ see XDJ000
MELADININE see XDJ000
MELAMINE see MCB000
MELAMINE, BUTYL- see BRR800
MELANILINE see DWC600
MELBIN see DQR600
MELDONE see CNU750
MELENGESTROL ACETATE see MCB380
MELETIN see QCA000
MELILOTAL see MFW250
MELILOTIN see HHR500
MELILOTOL see HHR500
MELINITE see PID000
MELIPAN see MCB500
MELIPAX see CDV100
MELIPRAMIN see DLH630
MELIPRAMINE see DLH630
MELIPRAMINE HYDROCHLORIDE see DLH630
MELIPRAMIN HYDROCHLORIDE see DLH630
MELITOXIN see BJZ000
MELLITE 131 see DEH650
MELLITE 825 see DVK200
MELLOSE see MIF760
MELOGEL see SLJ500
MELOXINE see XDJ000
MELPHALAN see PED750
MELPREX see DXX400
MELSEDIN see MDT250
MELUNA see SLJ500
MEMA see MEO750
MEMA see MLF250
MEMC see MEP250
MEMCOZINE see SNN300
ME-MDA see MJO250
MEMMI see MLF500
MEMPA see HEK000
MENADION see MMD500
MENADIONE see MMD500
MENAGEN see EDV000
MENAPHTAM see CBM750
MENAPHTHON see MMD500
MENAPHTONE see MMD500
MENDEL see MQU750
MENDIAXON see MKP500
MENDRIN see EAT500
MENEST see ECU750
MENFORMON see EDV000
MENHYDRINATE see DYE600
MENIPHOS see MQR750
MENITE see MQR750
Me$_2$NMOR see DTA000
MENOGEN see ECU750
MENOMYCIN see MRA250
MENOSTILBEEN see DKA600
MENOTAB see ECU750
MENOTROL see ECU750
MENTA-BAL see ENB500
MENTHA ARVENSIS, OIL see MCB625
MENTHA ARVENSIS OIL, PARTIALLY
DEMENTHOLIZED (FCC) see MCB625
p-MENTHA-1,3-DIENE see MLA250
p-MENTHA-1,4-DIENE see MCB750
p-MENTHA-1,5-DIENE see MCC000
p-MENTHA-1,8-DIENE see MCC250
p-MENTHA-1,8-DIENE see LFU000
1,8(9)-p-MENTHADIENE see MCC250
d-p-MENTHA-1,8-DIENE see LFU000

(S)-(−)-p-MENTHA-1,8-DIENE see MCC500
p-MENTHA-1,8-DIEN-7-OL see PCI550
(R)-(−)-p-MENTHA-6,8-DIEN-2-ONE see CCM120
1-6,8(9)-p-MENTHADIEN-2-ONE see CCM120
d-p-MENTHA-6,8,(9)-DIEN-2-ONE see CCM100
p-MENTHAN-3-OL see MCF750
dl-3-p-MENTHANOL see MCG000
l-p-MENTHAN-3-ONE see MCG275
p-MENTHAN-3-ONE racemic see MCE250
8-p-MENTHEN-2-OL see DKV150
p-MENTH-1-EN-8-OL see TBD500
p-MENTH-8-EN-3-OL see MCE750
8(9)-p-MENTHEN-3-OL see MCE750
p-MENTH-8-EN-2-OL, ACETATE see DKV160
8-p-MENTHEN-2-ONE see DKV175
p-MENTH-8-EN-2-ONE see DKV175
p-MENTH-8-EN-2-YL ACETATE see DKV160
MENTHENYL KETONE see MCF525
1-(p-MENTHEN-6-YL)-1-PROPANONE see MCF525
MENTHEN-1-YL-8 PROPIONATE see TBE600
MENTHOL see MCF750
l-MENTHOL see MCF750
l-MENTHOL see MCG250
3-p-MENTHOL see MCG000
dl-MENTHOL see MCG000
MENTHOL racemic see MCG000
MENTHOL racemique (FRENCH) see MCG000
MENTHOL, ACETATE (8CI) see MCG500
MENTHONE see MCG275
p-MENTHONE see MCG275
trans-MENTHONE see MCG275
MENTHONE, racemic see MCE250
l-MENTHONE (FCC) see MCG275
MENTHYL ACETATE see MCG500
(−)-MENTHYL ACETATE see MCG750
dl-MENTHYL ACETATE see MCG500
1-p-MENTH-3-YL ACETATE see MCG750
l-p-MENTH-3-YL ACETATE see MCG750
MENTHYL ACETATE racemic see MCG500
l-MENTHYL ACETATE (FCC) see MCG750
(−)-MENTHYL ALCOHOL see MCG250
p-MENTH-3-YL ESTER-dl-ACETIC ACID see MCG500
MEONINE see MDT740
MEP see EOS000
MEPACRINE see ARQ250
MEPADIN see DAM700
MEPAMTIN see MQU750
ME-PARATHION see MNH000
MEPATON see MNH000
MEPAVLON see MQU750
MEPERIDINE HYDROCHLORIDE see DAM700
MEPHABUTAZONE see BRF500
MEPHACYCLIN see TBX250
MEPHADRYL see BBV500
MEPHANAC see CIR250
MEPHASERPIN see RDK000
MEPHEDINE see DAM700
MEPHENAMINE HYDROCHLORIDE see OJW000
MEPHENAMIN HYDROCHLORIDE see OJW000
MEPHENYTOIN see MKB250
MEPHOBARBITAL see ENB500
MEPHOBARBITONE see ENB500
MEPHOSFOLAN see DHH400
MEPHYTAL see ENB500
MEPIOSINE see MQU750
MEPIVACAINE see SBB000
dl-MEPIVACAINE see SBB000
MEPIVACAINE HYDROCHLORIDE see CBR250

dl-MEPIVACAINE HYDROCHLORIDE see CBR250
MEPIVASTESIN see CBR250
MEPOSED see MQU750
MEP (Pesticide) see DSQ000
MEPRANIL see MQU750
MEPRO see CIR500
MEPROBAM see MQU750
MEPROBAMAT (GERMAN) see MQU750
MEPROBAMATE see MQU750
MEPROBAMATO (ITALIAN) see MQU750
MEPROCOMPREN see MQU750
MEPROCON CMC see MQU750
MEPRODIL see MQU750
MEPROFEN see BDU500
MEPROLEAF see MQU750
MEPROSAN see MQU750
MEPROSCILLARIN see MCI750
MEPROTABS see MQU750
MEPROZINE see MQU750
MEPTOX see MNH000
MEPTRAN see MQU750
MEPYRAMINE MALEATE see DBM800
MEQUINOL see MFC700
MERAKLON see PMP500
MERBAPHEN see CCG500
MERBENTUL see CLO750
MERBROMIN see MCV000
MERCALEUKIN see POK000
MERCAMINE see AJT250
MERCAPTAMINE see AJT250
MERCAPTAN AMYLIQUE (FRENCH) see PBM000
MERCAPTAN METHYLIQUE (FRENCH) see MLE650
MERCAPTAN METHYLIQUE PERCHLORE
(FRENCH) see PCF300
MERCAPTANS see MCJ500
MERCAPTAZOLE see MCO500
2-MERCAPTOACETANILIDE see MCK000
α-MERCAPTOACETANILIDE see MCK000
MERCAPTOACETATE see TFJ100
MERCAPTOACETIC ACID see TFJ100
2-MERCAPTOACETIC ACID see TFJ100
α-MERCAPTOACETIC ACID see TFJ100
MERCAPTOACETIC ACID, DIESTER with DITHIO-p-
UREIDOBENZENEARSONOUS ACID see CBI250
MERCAPTOACETIC ACID SODIUM SALT see SKH500
β-MERCAPTOALANINE see CQK000
o-MERCAPTOANILINE see AIF500
2-MERCAPTOBENZIMIDAZOLE see BCC500
o-MERCAPTOBENZOIC ACID, DIESTER with
DITHIO-p-UREIDOBENZENEARSONOUS ACID see
TFD750
MERCAPTOBENZOIMIDAZOLE see BCC500
2-MERCAPTOBENZOIMIDAZOLE see BCC500
MERCAPTOBENZOTHIAZOLE see BDF000
2-MERCAPTOBENZOTHIAZOLE see BDF000
2-MERCAPTOBENZOTHIAZOLEDISULFIDE see
BDE750
2-MERCAPTOBENZOTHIAZOLE SODIUM
DERIVATIVE see SIG500
2-MERCAPTOBENZOTHIAZOLE SODIUM SALT see
SIG500
2-MERCAPTOBENZOTHIAZOLE ZINC SALT see
BHA750
2-MERCAPTOBENZOTHIAZYLDISULFIDE see
BDE750
(MERCAPTOBUTANEDIOATO(1-))GOLD DISODIUM
SALT see GJC000

MERCAPTOCYCLOPENTANE see CPW300
MERCAPTODIACETIC ACID see MCM750
MERCAPTODIMETHUR (DOT) see DST000
1-MERCAPTODODECANE see LBX000
MERCAPTOETHANOL see MCN250
2-MERCAPTOETHANOL see MCN250
β-MERCAPTOETHANOL see MCN250
(2-MERCAPTOETHYL)AMINE see AJT250
β-MERCAPTOETHYLAMINE see AJT250
1-MERCAPTOGLYCEROL see MRM750
2-MERCAPTO-4-HYDROXY-6-METHYLPYRIMIDINE see MPW500
2-MERCAPTO-4-HYDROXY-6-N-PROPYLPYRIMIDINE see PNX000
2-MERCAPTO-4-HYDROXYPYRIMIDINE see TFR250
2-MERCAPTOIMIDAZOLINE see IAQ000
MERCAPTOMERIN SODIUM see TFK270
(MERCAPTOMETHYL)BENZENE see TGO750
3-(MERCAPTOMETHYL)-1,2,3-BENZOTRIAZIN-4(3H)-ONE-O,O-DIMETHYL PHOSPHORODITHIOATE see ASH500
3-(MERCAPTOMETHYL)-1,2,3-BENZOTRIAZIN-4(3H)-ONE-O,O-DIMETHYL PHOSPHORODITHIOATE-S-ESTER see ASH500
2-MERCAPTO-1-METHYLIMIDAZOLE see MCO500
N-(MERCAPTOMETHYL)PHTHALIMIDE S-(O,O-DIMETHYL PHOSPHORODITHIOATE) see PHX250
2-MERCAPTO-6-METHYLPYRIMID-4-ONE see MPW500
2-MERCAPTO-6-METHYL-4-PYRIMIDONE see MPW500
2-MERCAPTONAPHTHALENE see NAP500
β-MERCAPTONAPHTHALENE see NAP500
1-MERCAPTOPROPANE see PML500
2-MERCAPTOPROPANE see IMU000
1-MERCAPTO-2,3-PROPANEDIOL see MRM750
3-MERCAPTO-1,2-PROPANEDIOL see MRM750
α-MERCAPTOPROPANOIC ACID see TFK250
2-MERCAPTOPROPIONIC ACID see TFK250
α-MERCAPTOPROPIONIC ACID see TFK250
MERCAPTOPROPIONIC ACID, DIBUTYLTIN SALT see DEJ200
2-MERCAPTO-6-PROPYL-4-PYRIMIDONE see PNX000
2-MERCAPTO-6-PROPYLPYRIMID-4-ONE see PNX000
6-MERCAPTOPURIN see POK000
MERCAPTOPURIN (GERMAN) see POK000
6-MERCAPTOPURINE see POK000
MERCAPTOPURINE RIBONUCLEOSIDE see MCQ500
6-MERCAPTOPURINE RIBOSIDE see MCQ500
2-MERCAPTOPYRIDINE-N-OXIDE SODIUM SALT see MCQ750
2-MERCAPTO-4-PYRIMIDINOL see TFR250
2-MERCAPTO-4-PYRIMIDONE see TFR250
2-MERCAPTOPYRIMID-4-ONE see TFR250
MERCAPTOSUCCINIC ACID ANTIMONATE(III) HEXALITHIUM SALT see LGU000
MERCAPTOSUCCINIC ACID-S-ANTIMONY DERIVATIVE LITHIUM SALT see LGU000
MERCAPTOSUCCINIC ACID DIETHYL ESTER see MAK700
MERCAPTOSUCCINIC ACID, GOLD SODIUM SALT see GJC000
MERCAPTOSUCCINIC ACID, THIOANTIMONATE(III), DILITHIUM SALT see LGU000
7-MERCAPTO-1,3,4,6-TETRAZAINDENE see POK000
MERCAPTOTHION see MAK700

MERCAPTOTION (SPANISH) see MAK700
o-MERCAPTOTOLUENE see TGP000
p-MERCAPTOTOLUENE see TGP250
α-MERCAPTOTOLUENE see TGO750
d-MERCAPTOVALINE see MCR750
d,3-MERCAPTOVALINE see MCR750
MERCAPTURIC ACID see ACH000
(R)-MERCAPTURIC ACID see ACH000
MERCAPURIN see POK000
MERCAZOLYL see MCO500
MERCHLORATE see MEP250
MERCHLORETHANAMINE see BIE500
MERCKOGEN 6000 see AAX250
MERCOL 25 see DXW200
MERCUFENOL CHLORIDE see CHW675
MERCURAM see TFS350
MERCURAMIDE see SIH500
MERCURAN see MEO750
MERCURANINE see MCV000
MERCURATE(4-), BIS(N,N-BIS(CARBOXYMETHYL)GLYCINATO(3-)-N,O,O',O")-, TETRAHYDROGEN see MDF050
MERCURE (FRENCH) see MCW250
MERCURIACETATE see MCS750
MERCURIALIN see MGC250
MERCURIBIS-o-NITROPHENOL see MCS600
MERCURIC ACETATE see MCS750
MERCURIC AMMONIUM CHLORIDE, solid see MCW500
MERCURIC ARSENATE see MDF350
MERCURIC BENZOATE see MCX500
MERCURIC BENZOATE, solid (DOT) see MCX500
MERCURIC BROMIDE see MCY000
MERCURIC BROMIDE, solid see MCY000
MERCURIC CHLORIDE (DOT) see MCY475
MERCURIC CHLORIDE, AMMONIATED see MCW500
MERCURIC CYANIDE, solid (DOT) see MDA250
MERCURIC DIACETATE see MCS750
MERCURIC IODIDE see MDD000
MERCURIC IODIDE, solid see MDD000
MERCURIC IODIDE, solution see MDD000
MERCURIC IODIDE, RED see MDD000
MERCURIC NITRATE see MDF050
MERCURIC OLEATE, solid (DOT) see MDF250
MERCURIC OXIDE see MCT500
MERCURIC OXIDE, solid (DOT) see MCT500
MERCURIC OXIDE, RED see MCT500
MERCURIC OXIDE, YELLOW see MCT500
MERCURIC OXYCYANIDE see MDA500
MERCURIC OXYCYANIDE, solid (desensitized) (DOT) see MDA500
MERCURIC POTASSIUM CYANIDE (DOT) see PLU500
MERCURIC POTASSIUM CYANIDE, solid (DOT) see PLU500
MERCURIC POTASSIUM IODIDE see NCP500
MERCURIC POTASSIUM IODIDE, solid (DOT) see NCP500
MERCURIC SALICYLATE see MCU000
MERCURIC SALICYLATE, solid (DOT) see MCU000
MERCURIC SULFATE, solid see MDG500
MERCURIC SULFOCYANATE see MCU250
MERCURIC SULFOCYANATE, solid (DOT) see MCU250
MERCURIC SULFOCYANIDE see MCU250
MERCURIC THIOCYANATE see MCU250
MERCURIC THIOCYANATE, solid (DOT) see MCU250
MERCURIDIACETALDEHYDE see BJW800
MERCURIO (ITALIAN) see MCW250

MERCURIPHENYL ACETATE see ABU500
MERCURIPHENYL CHLORIDE see PFM500
MERCURIPHENYL NITRATE see MCU750
MERCURISALICYLIC ACID see MCU000
MERCURITAL see SIH500
MERCUROCHLORID (DUTCH) see MCW000
MERCUROCHROME see MCV000
MERCUROCHROME-220 SOLUBLE see MCV000
MERCUROCOL see MCV000
MERCUROL see MCV250
MERCUROME see MCV000
MERCUROPHAGE see MCV000
MERCUROTHIOLATE see MDI000
MERCUROUS ACETATE see MDE250
MERCUROUS ACETATE, solid (DOT) see MDE250
MERCUROUS AZIDE (DOT) see MCX000
MERCUROUS BROMIDE, solid (DOT) see MCX750
MERCUROUS CHLORIDE see MCW000
MERCUROUS GLUCONATE see MDC500
MERCUROUS GLUCONATE, solid (DOT) see MDC500
MERCUROUS IODIDE see MDC750
MERCUROUS NITRATE, solid (DOT) see MDE750
MERCUROUS OXIDE, BLACK, solid (DOT) see MDF750
MERCURY see MCW250
MERCURY ACETATE see MCS750
MERCURY ACETATE see MDE250
MERCURY(2+) ACETATE see MCS750
MERCURY(II) ACETATE see MCS750
MERCURY(II) ACETATE, PHENYL- see ABU500
MERCURY, ACETOXY(2-METHOXYETHYL)- see MEO750
MERCURY, ACETOXYPHENYL- see ABU500
MERCURY ACETYLIDE see MCW349
MERCURY ACETYLIDE (DOT) see MCW349
MERCURY AMIDE CHLORIDE see MCW500
MERCURY AMINE CHLORIDE see MCW500
MERCURY AMMONIATED see MCW500
MERCURY AZIDE see MCX000
MERCURY(I) AZIDE see MCX000
MERCURY(II) BENZOATE see MCX500
MERCURY BICHLORIDE see MCY475
MERCURY BINIODIDE see MDD000
MERCURY, BIS(ACETATO)(mu-(3',6'-DIHYDROXY-2',7'-FLUORANDIYL))DI- see FEV100
MERCURY, BIS(4-HYDROXY-3-NITROPHENYL)- see MCS600
MERCURY, BIS(TRIFLUOROMETHYLTHIO)- see BLQ525
MERCURY BISULFATE see MDG500
MERCURY(I) BROMIDE (1:1) see MCX750
MERCURY(II) BROMIDE (1:2) see MCY000
MERCURY, BROMOHEXYL see HFR100
MERCURY, BROMO(2-HYDROXYETHYL)-, compound with AMMONIA (1:0.8 moles) see BNL275
MERCURY, (3-(α-CARBOXY-o-ANISAMIDO)-2-HYDROXYPROPYL)HYDROXY- see NCM800
MERCURY(I) CHLORIDE see MCW000
MERCURY(II) CHLORIDE see MCY475
MERCURY (E)-CHLORO(2-(3-BROMOPROPIONAMIDO)CYCLOHEXYL) see CET000
MERCURY, CHLORO(2-HYDROXYPHENYL)- see CHW675
MERCURY COMPOUNDS, INORGANIC see MCZ000
MERCURY COMPOUNDS, ORGANIC see MDA000
MERCURY(II) CYANIDE see MDA250
MERCURY CYANIDE OXIDE see MDA500
MERCURY DIACETATE see MCS750

MERCURY, DIMETHYL see DSM450
MERCURY DITHIOCYANATE see MCU250
MERCURY(II) FULMINATE see MDC000
MERCURY FULMINATE, wetted with not <20% water, or mixture (UN 0135) (DOT) see MDC000
MERCURY(I) GLUCONATE see MDC500
MERCURY, HYDROXY(4-HYDROXY-3-NITROPHENYL)- see HLO400
MERCURY, HYDROXYPHENYL- see PFN100
MERCURY, (4-HYDROXY-m-PHENYLENE)BIS(ACETATO)- see HNK575
MERCURY(I) IODIDE see MDC750
MERCURY(II) IODIDE see MDD000
MERCURY IODIDE (DOT) see MDC750
MERCURY(II) IODIDE (solution) see MDD250
MERCURY IODIDE, solution (DOT) see MDC750
MERCURYL ACETATE see MCS750
MERCURY, METALLIC (DOT) see MCW250
MERCURY METHYLCHLORIDE see MDD750
MERCURY, (2-METHYL-5-NITROPHENOLATO(2-)-C^6,O^1)-(9CI) see NHK900
MERCURY MONOACETATE see MDE250
MERCURY MONOCHLORIDE see MCW000
MERCURY NITRATE see MDF000
MERCURY(I) NITRATE (1:1) see MDE750
MERCURY(II) NITRATE (1:2) see MDF000
MERCURY NITRIDE see TKW000
MERCURY, NITRILOTRIACETATE see MDF050
MERCURY NUCLEATE, solid (DOT) see MCV250
MERCURY OLEATE see MDF250
MERCURY(II) ORTHOARSENATE see MDF350
MERCURY(I) OXIDE see MDF750
MERCURY(II) OXIDE see MCT500
MERCURY OXYCYANIDE see MDA500
MERCURY PERCHLORIDE see MCY475
MERCURY PERNITRATE see MDF000
MERCURY PERSULFATE see MDG500
MERCURY(II) POTASSIUM IODIDE see NCP500
MERCURY PROTOCHLORIDE see MCW000
MERCURY PROTOIODIDE see MDC750
MERCURY SALICYLATE see MCU000
MERCURY SUBSALICYLATE see MCU000
MERCURY(II) SULFATE (1:1) see MDG500
MERCURY, SULFATOBIS(METHYL- see BKS810
MERCURY(II) THIOCYANATE see MCU250
MERCURY THIOCYANATE (DOT) see MCU250
MERCURY, (((p-TOLYL)SULFAMOYL)IMINO)BIS(METHYL)- see MLH100
MERCURY ZINC CHROMATE COMPLEX see ZJA000
MERCUSAL see SIH500
MEREX see KEA000
MERFALAN see BHT750
MERFAMIN see MDI000
MERFAZIN see PFM500
MERGAMMA see ABU500
MERGE see MRL750
MERIDIL see MNQ000
MERIT see WBJ700
MERIZONE see BRF500
MERKAPTOBENZIMIDAZOL (CZECH) see BCC500
2-MERKAPTOBENZOTIAZOL see BDF000
2-MERKAPTOBENZTHIAZOL see BDF000
2-MERKAPTOIMIDAZOLIN (CZECH) see IAQ000
MERKAZIN see BKL250
MERMETH see SNJ000
MERN see POK000
MERONIDAL see MMN250

MEROXYL see AOR500
MEROXYLAN see AOR500
MEROXYLAN-WANDER see AOR500
MEROXYL-WANDER see AOR500
MERPAN see CBG000
MERPHALAN see BHT750
o-MERPHALAN see BHT750
MERPHENYL NITRATE see MCU750
MERPHOS see TIG250
MERPOL see EJN500
MERRILLITE see ZBJ000
MERSALIN see SIH500
MERSALYL see SIH500
MERSOLITE see ABU500
MERSOLITE 1 see PFN100
MERSOLITE 2 see PFM500
MERSOLITE 7 see MCU750
MERSOLITE 8 see ABU500
MERTEC see TEX000
MERTESTATE see TBF500
MERTHIOLATE see MDI000
MERTHIOLATE SALT see MDI000
MERTHIOLATE SODIUM see MDI000
MERTIONIN see MDT740
MERTORGAN see MDI000
MERVAMINE see DJL000
MERZONIN SODIUM see MDI000
MES see MJW250
MESA see TGE300
MESAMATE see MRL750
MESAMATE CONCENTRATE see MRL750
MESANTOIN see MKB250
MESATON see NCL500
MESCALINE see MDI500
MESIDIN (CZECH) see TLG500
MESIDINE see TLG500
MESIDINE HYDROCHLORIDE see TLH000
MESITYLAMINE see TLG500
MESITYLAMINE HYDROCHLORIDE see TLH000
MESITYLENE see TLM050
MESITYLOXID (GERMAN) see MDJ750
MESITYL OXIDE see MDJ750
MESITYLOXYDE (DUTCH) see MDJ750
MESOMILE see MDU600
MESOXALYLCARBAMIDE see AFT750
MESOXALYLCARBAMIDE MONOHYDRATE see
MDL500
MESOXALYLUREA see AFT750
MESOXALYLUREA MONOHYDRATE see MDL500
MESTERONE see MPN500
MESTRANOL see MKB750
MESTRANOL mixed with LYNESTRENOL see LJE000
MESTRANOL mixed with LYNESTROL see LJE000
MESTRANOL mixed with NORETHYNODREL see
EAP000
MESTRENOL see MKB750
MESULFA see ALF250
MESURAL see LFK000
MESUROL see DST000
META see TDW500
METAARSENIC ACID see ARB000
META BLACK see AQP000
METABOLITE C see SOA500
METACE see CLO750
METACEN see IDA000
METACETALDEHYDE see TDW500
METACETONE see DJN750
METACETONIC ACID see PMU750

METACHLOR see CFX000
METACIDE see MNH000
METACIL see MPW500
METACORTANDRACIN see PLZ000
METACORTANDRALONE see PMA000
METACRATE see MIB750
METADEE see VSZ100
METADELPHENE see DKC800
METAFOS see MNH000
METAFOS see SII500
METAFUME see MHR200
METAISOSEPTOX see DAP400
METAISOSYSTOX see DAP400
METAISOSYSTOX-SOLFON 20 315 see DAP600
METAISOSYSTOXSULFOXIDE see DAP000
METAKRYLAN METYLU (POLISH) see MLH750
METALAXIL see MDM100
METALAXYL see MDM100
METALCAPTASE see MCR750
METALDEHYD (GERMAN) see TDW500
METALDEHYDE (DOT) see TDW500
METALDEIDE (ITALIAN) see TDW500
METALLIBURE see MLJ500
METALLIC ARSENIC see ARA750
METALLIC OSMIUM see OKE000
METAMFETAMINA see DBA800
METAMID see PJY500
METAMIDOFOS ESTRELLA see DTQ400
METAMPHETAMIN see DBA800
METAMPHETAMINE HYDROCHLORIDE see MDT600
METANA ALUMINUM PASTE see AGX000
METANDREN see MPN500
METANEPHRIN see VGP000
METANEX see CBT750
METANFETAMINA see DBA800
METANILE YELLOW O see MDM775
METANIL YELLOW see MDM775
METANIL YELLOW 1955 see MDM775
METANIL YELLOW C see MDM775
METANIL YELLOW E see MDM775
METANIL YELLOW EXTRA see MDM775
METANIL YELLOW F see MDM775
METANIL YELLOW G see MDM775
METANIL YELLOW GRIESBACH see MDM775
METANIL YELLOW K see MDM775
METANIL YELLOW KRSU see MDM775
METANIL YELLOW M3X see MDM775
METANIL YELLOW O see MDM775
METANIL YELLOW PL see MDM775
METANIL YELLOW S see MDM775
METANIL YELLOW SUPRA P see MDM775
METANIL YELLOW VS see MDM775
METANIL YELLOW WS see MDM775
METANIL YELLOW Y see MDM775
METANIL YELLOW YK see MDM775
METANITE see MGR500
METANOLO (ITALIAN) see MGB150
METANTIOLO (ITALIAN) see MLE650
METAOKSEDRIN see SPC500
METAOXEDRIN see NCL500
METAOXEDRIN see SPC500
METAOXEDRINUM see SPC500
METAPHEN see NHK900
METAPHENYLENEDIAMINE see PEY000
METAPHOR see MNH000
METAPHOS see MNH000
METAPHOSPHORIC ACID, CALCIUM SODIUM SALT
see CAX260

METAPLEX NO see PKB500
METAQUEST A see EIX000
METAQUEST B see EIX500
METAQUEST C see EIV000
METARTRIL see IDA000
METASILICIC ACID see SCL000
METASOL see MLH000
METASOL 30 see ABU500
METASOL P-6 see PFO000
METASOL TK-100 see TEX000
METASON see TDW500
METASYMPATOL see NCL500
METASYNEPHRINE see NCL500
METASYSTEMOX see DAP000
METASYSTOX FORTE see DAP400
METASYSTOX-R see DAP000
METATENSIN see RDK000
METATHIONE see DSQ000
METATION see DSQ000
META TOLUYLENE DIAMINE see TGL750
METATOLYLENEDIAMINE DIHYDROCHLORIDE
see DCE000
METAUPON see OHU000
METAXANIN see MDM100
METAXITE see ARM268
METAXON see CIR250
METAXONE see SIL500
METAZIN see SNJ000
METAZOL see MLH000
METAZOLO see MCO500
METEPA see TNK250
METERFER see FBJ100
METERFOLIC see FBJ100
METFORMIN see DQR600
"METH" see MDQ500
METHAANTHIOL (DUTCH) see MLE650
METHABOL see PAN100
METHACETONE see DJN750
METHACHLOR see CFX000
METHACIDE see TGK750
METHACRALDEHYDE (DOT) see MGA250
METHACROLEIN see MGA250
METHACRYLATE de BUTYLE (FRENCH) see MHU750
METHACRYLATE de METHYLE (FRENCH) see
MLH750
METHACRYLIC ACID see MDN250
METHACRYLIC ACID, inhibited (DOT) see MDN250
METHACRYLIC ACID (ACGIH,OSHA) see MDN250
METHACRYLIC ACID AMIDE see MDN500
METHACRYLIC ACID-DIVINYLBENZENE
COPOLYMER see MDN525
METHACRYLIC ACID, ISOBUTYL ESTER see IIY000
METHACRYLIC ACID, ISODECYL ESTER see IKM000
METHACRYLIC ACID, METHYL ESTER (MAK) see
MLH750
METHACRYLIC ACID METHYL ESTER POLYMERS
see PKB500
METHACRYLIC ALDEHYDE see MGA250
METHACRYLIC AMIDE see MDN500
METHACRYLONITRILE, inhibited see MGA750
METHACRYLOYLOXYETHYL ISOCYANATE see
IKG700
METHACRYLSAEUREBUTYLESTER (GERMAN) see
MHU750
METHACRYLSAEUREMETHYL ESTER (GERMAN) see
MLH750
METHADONE see MDO750
d-METHADONE HYDROCHLORIDE see MDP240

l-METHADONE HYDROCHLORIDE see MDP770
METHAFORM see ABD000
METHAKRYLALDEHYD see MGA250
METHALLIBURE see MLJ500
METHALLYL ALCOHOL (DOT) see IMW000
METHALLYL CHLORIDE see CIU750
α-METHALLYL CHLORIDE see CIU750
METHAMIDOPHOS see DTQ400
METHAMIN see HEI500
METHAMINODIAZEPINE HYDROCHLORIDE see
MDQ250
METHAMINODIAZEPOXIDE see LFK000
METHAMINODIAZEPOXIDE HYDROCHLORIDE see
MDQ250
(+)-METHAMPHETAMINE CHLORIDE see MDT600
METHAMPHETAMINE HYDROCHLORIDE see
MDT600
METHAMPHETAMINE HYDROCHLORIDE see
DBA800
(+)-METHAMPHETAMINE HYDROCHLORIDE see
MDT600
d-METHAMPHETAMINE HYDROCHLORIDE see
MDT600
l-METHAMPHETAMINE HYDROCHLORIDE see
MDQ500
METHAMPHETAMINIUM CHLORIDE see MDT600
METHANAL see FMV000
METHANAMIDE see FMY000
METHANAMINE (9CI) see MGC250
METHANAMINE, N-NITRO-(9CI) see NHN500
METHANE see MDQ750
METHANEARSONIC ACID DIMERCURY SALT see
DNW000
METHANE BASE see MJN000
METHANECARBONITRILE see ABE500
METHANECARBOTHIOLIC ACID see TFA500
METHANECARBOXAMIDE see AAI000
METHANECARBOXYLIC ACID see AAT250
METHANE, CHLOROTRIFLUORO-, mixt. with
TRIFLUOROMETHANE (9CI) see FOO562
METHANE, CYANO- see ABE500
METHANEDICARBOXYLIC ACID, DIETHYL ESTER
see EMA500
METHANE DICHLORIDE see MJP450
METHANEDITHIOL-S,S-DIESTER with O,O-DIETHYL
ESTER PHOSPHORODITHIOIC ACID see EEH600
METHANE, PHENYL- see TGK750
METHANEPHOSPHONIC ACID, DIPHENYL ESTER
see MDQ825
METHANESULFONIC ACID see MDR250
METHANESULFONIC ACID CHLOROETHYL ESTER
see CHC750
METHANESULFONIC ACID, METHYLENE ESTER
see MJQ500
METHANESULFONIC ACID-1-METHYLETHYL
ESTER see IPY000
METHANESULFONIC ACID, SILVER SALT see
SDR500
METHANESULFONIC ACID, SILVER(1+) SALT see
SDR500
METHANESULFONIC ACID TETRAMETHYLENE
ESTER see BOT250
METHANESULPHONIC ACID ETHYL ESTER see
EMF500
METHANESULPHONIC ACID METHYL ESTER see
MLH500
METHANE, TETRABROMIDE see CBX750

METHANE, TETRABROMO- see CBX750
METHANE TETRACHLORIDE see CBY000
METHANE, TETRAFLUORO- see CBY250
METHANE TETRAMETHYLOL see PBB750
N,N'-METHANETETRAYLBIS(2,6-BIS(1-METHYLETHYL)BENZENAMINE) see BJE550
METHANETHIOL see MLE650
METHANETHIOL, PHENYL- see TGO750
METHANE TRICHLORIDE see CHJ500
METHANE, TRIFLUORO-, mixt. with CHLOROTRIFLUOROMETHANE see FOO562
METHANE, compressed (UN 1971) (DOT) see MDQ750
METHANE, refrigerated liquid (cryogenic liquid) (UN 1972) (DOT) see MDQ750
1H-3a,7-METHANOAZULENE, OCTAHYDRO-6-METHOXY-3,6,8,8-TETRAMETHYL-,(3R-(3-α-3a-β, 6-α-7-β,8aα-))- see CCR525
1H-3-α-7-METHANOAZULEN-6-OL, OCTAHYDRO-3,6,8,8-TETRAMETHYL-, FORMATE, (3R-(3-α-3a-β,6-α-7-β,8aα-))- see CCR524
METHANOIC ACID see FNA000
4,7-METHANOINDAN, 2,2,4,5,6,7,8,8-OCTACHLORO-3a,4,7,7a-TETRAHYDRO- see CDR575
4,7-METHANOINDENE, 4,5,6,7,8,8-HEXACHLORO-3a,4,7,7a-TETRAHYDRO- see HCN000
4,7-METHANO-1H-INDENE, 2,2,4,5,6,7,8,8-OCTACHLORO-2,3,3a,4,7,7a-HEXAHYDRO- (9CI) see CDR575
4,7-METHANOINDEN, 1,2,4,5,6,7,8,9-OCTACHLORO-3A,4,7,7A-TETRAHYDRO- see CDR760
METHANOL see MGB150
METHANOLACETONITRILE see HGP000
N-METHANOLACRYLAMIDE see HLU500
METHANOL, BENZYL- see PDD750
METHANOL, OXIRANYL- see GGW500
METHANOL, SODIUM SALT see SIK450
METHANOL, TRICHLORO-, CHLOROFORMATE see TIR920
METHANONE, (2-AMINO-1H-BENZIMIDAZOL-5-YL)PHENYL- see AIH000
METHANONE, (4,5-DICHLORO-1H-PYRROL-2-YL)(2,6-DIHYDROXYPHENYL)- see POK400
METHANONE, (4-(2-(DIETHYLAMINO)ETHOXY)PHENYL)(2-ETHYL-3-BENZOFURANYL)-(9CI) see EDV700
METHANONE, 3-FURFANYLPHENYL-(9CI) see FQO050
4,7-METHANOSELENOPHENE, OCTAHYDRO-2,2,3,3-TETRAFLUORO- see TCI100
METHANTHIOL (GERMAN) see MLE650
METHAPHOXIDE see TNK250
METHAQUALONE HYDROCHLORIDE see MDT250
METHASAN see BJK500
METHAZATE see BJK500
METHAZINE see IDA000
METHEDRINE see MDT600
METHEDRINE see DBA800
METHEDRINE HYDROCHLORIDE see MDT600
METHEDRINE HYDROCHLORIDE see DBA800
METHENAMINE see HEI500
2,5,7-METHENO-3H-CYCLOPENTA(A)PENTALEN-3-ONE, 3B,4,5,6,6,6A-HEXACHLORODECAHYDRO-, (2-α-3A-β,3B-β,4-β,5-β,6A-β,7-α-7A-β,8R*)- see KFK200
N,N'-METHENYL-o-PHENYLENEDIAMINE see BCB750
METHENYL TRIBROMIDE see BNL000
METHENYL TRICHLORIDE see CHJ500

METHIACIL see MPW500
METHIAMAZOLE see MCO500
METHIDATHION see DSO000
METHILANIN see MDT740
METHIOCARB see DST000
METHIOCIL see MPW500
METHIONAL see MPV400
METHIONAL see TET900
METHIONAMINE see ACQ275
METHIONINE see MDT750
l-METHIONINE see MDT750
l-(−)-METHIONINE see MDT750
(±)-METHIONINE see MDT740
dl-METHIONINE see MDT740
METHOCEL 10 see MIF760
METHOCEL 15 see MIF760
METHOCEL 181 see MIF760
METHOCEL 400 see MIF760
METHOCEL 4000 see MIF760
METHOCEL A see MIF760
METHOCEL CHG see MIF760
METHOCEL 4000CPS see MIF760
METHOCEL HG see HNX000
METHOCEL MC see MIF760
METHOCEL MC 25 see MIF760
METHOCEL MC4000 see MIF760
METHOCEL MC 8000 see MIF760
METHOCEL 400CPS see MIF760
METHOCEL SM 100 see MIF760
METHOCILLIN-S see SLJ000
METHOGAS see MHR200
METHOHEXITAL SODIUM see MDU500
METHOHEXITONE SODIUM see MDU500
METHOIN see MKB250
METHOLCARB see MIB750
METHOLENE 2218 see MJW000
METHOMYL see MDU600
METHOPHENAZATE ACID FUMARATE see MDU750
METHOPHENAZINE DIFUMARATE see MDU750
METHOPLAIN see DNA800
METHOPTERIN see MDV500
METHOTEXTRATE see MDV500
METHOTREXATE see MDV500
METHOXA-DOME see XDJ000
METHOXAMINE HYDROCHLORIDE see MDW000
METHOXCIDE see MEI450
METHOXO see MEI450
METHOXONE see SIL500
METHOXONE see CIR250
METHOXONE see CIR500
METHOXSALEN see XDJ000
METHOXYACETIC ACID see MDW275
2-METHOXYACETIC ACID see MDW275
2-METHOXYACETOACETANILIDE see ABA500
o-METHOXYACETOACETANILIDE see ABA500
2'-METHOXYACETOACETANILIDE see ABA500
4-METHOXYACETOFENON see MDW750
4-METHOXYACETOPHENONE see MDW750
p-METHOXYACETOPHENONE see MDW750
4'-METHOXYACETOPHENONE see MDW750
1-METHOXY-2-ACETOXYPROPANE see PNL265
2-(METHOXYACETYL)INDOLE see ICW200
2-(METHOXYACETYL)-1-METHYLINDOLE see MDX300
2-(METHOXYACETYL)-3-METHYLINDOLE see MDX310
2-METHOXY-AETHANOL (GERMAN) see EJH500
2-METHOXYAETHYLACETAT (GERMAN) see EJJ500

1-METHOXY-AETHYL-METHYLNITROSAMIN (GERMAN) see MEP500
METHOXYAETHYLQUECKSILBERCHLORID see MEP250
p-METHOXYALLYLBENZENE see AFW750
2-METHOXY-4-ALLYLPHENOL see EQR500
2-METHOXY-1-AMINOBENZENE see AOV900
2-METHOXY-1-AMINOBENZENE HYDROCHLORIDE see AOX250
2-METHOXYANILINE see AOV900
4-METHOXYANILINE see AOW000
o-METHOXYANILINE see AOV900
p-METHOXYANILINE see AOW000
2-METHOXYANILINE HYDROCHLORIDE see AOX250
o-METHOXYANILINE HYDROCHLORIDE see AOX250
2-METHOXYBENZALDEHYDE see AOT525
4-METHOXYBENZALDEHYDE see AOT530
6-METHOXYBENZALDEHYDE see AOT525
o-METHOXYBENZALDEHYDE see AOT525
p-METHOXYBENZALDEHYDE (FCC) see AOT530
2-METHOXYBENZENAMINE see AOV900
4-METHOXYBENZENAMINE see AOW000
2-METHOXYBENZENAMINE HYDROCHLORIDE see AOX250
METHOXYBENZENE see AOX750
4-METHOXYBENZENEACETIC ACID see MFE250
4-METHOXYBENZENEAMINE see AOW000
2-METHOXYBENZENEAMINE HYDROCHLORIDE see AOX250
2-METHOXYBENZENECARBOXALDEHYDE see AOT525
4-METHOXY-1,3-BENZENEDIAMINE see DBO000
4-METHOXY-1,3-BENZENEDIAMINE SULFATE see DBO400
4-METHOXY-1,3-BENZENEDIAMINE SULFATE (1:1) see DBO400
4-METHOXY-1,3-BENZENEDIAMINE SULPHATE see DBO400
4-METHOXYBENZENEMETHANOL see MED500
2-METHOXYBENZOIC ACID see MPI000
o-METHOXYBENZOIC ACID see MPI000
4-METHOXYBENZOIC ACID HYDRAZIDE see AOV500
p-METHOXYBENZOIC ACID HYDRAZIDE see AOV500
p-METHOXYBENZOIC HYDRAZIDE see AOV500
METHOXYBENZOYL CHLORIDE see AOY250
4-METHOXYBENZOYL HYDRAZIDE see AOV500
4-METHOXYBENZOYLHYDRAZINE see AOV500
(p-METHOXYBENZOYL)HYDRAZINE see AOV500
4-(4-METHOXYBENZOYL)PYRIDINE see MFH930
p-METHOXYBENZYL ACETATE see AOY400
4-METHOXYBENZYLACETONE see MFF580
4-METHOXYBENZYL ALCOHOL see MED500
p-METHOXYBENZYL ALCOHOL see MED500
N-p-METHOXYBENZYL-N'-N'-DIMETHYL-N-α-PYRIDYLETHYLENEDIAMINE MALEATE see DBM800
p-METHOXYBENZYL FORMATE see MFE250
4-METHOXYBIPHENYL see PEG250
p-METHOXYBIPHENYL see PEG250
1-METHOXYBUTANE see BRU780
α-METHOXYBUTANE see BRU780
3-METHOXYBUTYL ACETATE see MHV750

3-METHOXYBUTYLESTER KYSELINY OCTOVE see MHV750
2-(METHOXY-CARBONYLAMINO)-BENZIMIDAZOL see MHC750
2-(METHOXYCARBONYLAMINO)-BENZIMIDAZOLE see MHC750
2-(METHOXYCARBONYL)ANILINE see APJ250
3-METHOXYCARBONYL-6-N-BUTYL-7-BENZYLOXY-4-OXOQUINOLINE see NCN600
METHOXYCARBONYL CHLORIDE see MIG000
METHOXYCARBONYLETHYLENE see MGA500
(2-METHOXYCARBONYL-1-METHYL-VINYL)-DIMETHYL-FOSFAAT (DUTCH) see MQR750
(2-METHOXYCARBONYL-1-METHYL-VINYL)-DIMETHYL-PHOSPHAT (GERMAN) see MQR750
2-METHOXYCARBONYL-1-METHYLVINYL DIMETHYLPHOSPHATE see MQR750
1-METHOXYCARBONYL-1-PROPEN-2-YL DIMETHYL PHOSPHATE see MQR750
(1-METHOXYCARBOXYPROPEN-2-YL)PHOSPHORIC ACID, DIMETHYL ESTER see MQR750
METHOXYCHLOR see MEI450
p,p'-METHOXYCHLOR see MEI450
2-METHOXY-6-CHLORO-9-(4-BIS(2-CHLOROETHYL)AMINO-1-METHYLBUTYLAMINO)ACRIDINE DIHYDROCHLORIDE see QDS000
2-METHOXY-6-CHLORO-9-DIETHYLAMINOPENTYLAMINOACRIDINE see ARQ250
2-METHOXY-6-CHLORO-9-(3-(ETHYL-2-CHLOROETHYL)AMINOPROPYLAMINO)ACRIDINE DIHYDROCHLORIDE see QDS000
6'-METHOXYCINCHONAN-9-OL see QFS000
(8-α,9R)-6'-METHOXYCINCHONAN-9-OL see QHJ000
6'-METHOXYCINCHONAN-9-OL DIHYDROCHLORIDE see QIJ000
6-METHOXYCINCHONINE see QHJ000
METHOXY-DDT see MEI450
1-METHOXYDECANE see MIW075
N-METHOXY-3-(3,5-DI-tert-BUTYL-4-HYDROXYBENZYLIDENE)-2-PYRROLIDONE see MEL100
2-METHOXY-3,6-DICHLOROBENZOIC ACID see MEL500
METHOXYDIGLYCOL see DJG000
10-METHOXY-1,6-DIMETHYLERGOLINE-8-METHANOL-5-BROMO-3-PYRIDINECARBOXYLATE (ester) see NDM000
10-METHOXY-1,6-DIMETHYL-ERGOLIN-8-β-METHANOL-(5-BROMNICOTINAT) (GERMAN) see NDM000
1-METHOXY-2,4-DINITROBENZENE see DUP800
METHOXYDIURON see DGD600
METHOXYETHANE see EMT000
2-METHOXYETHANOL (ACGIH) see EJH500
2-METHOXYETHANOL, ACETATE see EJJ500
2-METHOXYETHANOL, ACRYLATE see MIF750
2-METHOXYETHANOL PHOSPHATE (3:1) see TLA600
METHOXYETHENE see MQL750
METHOXY ETHER of PROPYLENE GLYCOL see PNL250
3-METHOXY-17-α-ETHINYLESTRADIOL see MKB750
3-METHOXY-17-α-ETHINYLOESTRADIOL see MKB750
2-(2-METHOXYETHOXY)ETHANOL see DJG000
2-METHOXY-ETHYL ACETAAT (DUTCH) see EJJ500

2-METHOXYETHYL ACETATE (ACGIH) see EJJ500
4-METHOXY-9-(3-(ETHYL-2-CHLOROETHYL) see EHJ500
2-METHOXYETHYLE, ACETATE de (FRENCH) see EJJ500
METHOXYETHYL MERCURIC ACETATE see MEO750
METHOXYETHYL MERCURIC ACETATE see MEO750
METHOXYETHYL MERCURIC CHLORIDE see MEP250
2-METHOXYETHYLMERCURIC CHLORIDE see MEP250
(β-METHOXYETHYL)MERCURIC CHLORIDE see MEP250
METHOXYETHYLMERCURY ACETATE see MEO750
METHOXYETHYLMERCURY CHLORIDE see MEP250
2-METHOXYETHYLMERCURY CHLORIDE see MEP250
β-METHOXYETHYLMERCURY CHLORIDE see MEP250
2-METHOXYETHYLMERKURIACETAT see MEO750
2-METHOXYETHYLMERKURICHLORID see MEP250
1-METHOXY ETHYL METHYLNITROSAMINE see MEP500
2-METHOXYETHYL PHTHALATE see DOF400
3-METHOXY-17-α-ETHYNOESTRADIOL see MKB750
3-METHOXYETHYNYLESTRADIOL see MKB750
3-METHOXY-17-α-ETHYNYLESTRADIOL see MKB750
3-METHOXY-17-α-ETHYNYL-1,3,5(10)-ESTRATRIEN-17-β-OL see MKB750
3-METHOXYETHYNYLOESTRADIOL see MKB750
3-METHOXY-17-ETHYNYLOESTRADIOL-17-β see MKB750
3-METHOXY-17-α-ETHYNYL-1,3,5(10)-OESTRATRIEN-17-β-OL see MKB750
1-p-METHOXYFENYL-3,3-DIMETHYLTRIAZEN (CZECH) see DSN600
8-METHOXY-(FURANO-3'.2':6.7-COUMARIN) see XDJ000
9-METHOXY-7H-FURO(3,2-g)BENZOPYRAN-7-ONE see XDJ000
8-METHOXY-2',3',6,7-FUROCOUMARIN see XDJ000
8-METHOXY-4',5',6,7-FUROCOUMARIN see XDJ000
3-METHOXY-4-HYDROXYBENZALDEHYDE see VFK000
4-METHOXY-2-HYDROXYBENZOPHENONE see MES000
3-METHOXY-4-HYDROXY-BENZYLACETONE see VFP100
β-METHOXY-β'-HYDROXYDIETHYL ETHER see DJG000
METHOXYHYDROXYETHANE see EJH500
9-(1-(METHOXYIMINO)ETHYL)-6-OXO-N,N,2,2,5-PENTAMETHYL-7-OXA-3,4-DITHIA-5,8-DIAZADEC-8-EN-10-AMIDE see MES550
5-METHOXY-1H-INDOL-3-YL 4-PIPERIDYLMETHYL KETONE MONOHYDROCHLORIDE see MES900
(5-METHOXY-3-INDOLYL)-(4-PIPERIDYL-METHYL-)-KETON HYDROCHLORID see MES900
2-METHOXY-5-METHYLANILINE see MGO750
4-METHOXY-2-METHYLANILINE see MGO500
4-METHOXY-2-METHYLBENZENAMINE see MGO500
2-METHOXY-5-METHYL-BENZENAMINE (9CI) see MGO750
1-METHOXY-1-METHYL-3-(3,4-DICHLOROPHENYL)UREA see DGD600

METHOXYMETHYL 1-METHYL-2-INDOLYL KETONE see MDX300
METHOXYMETHYL 3-METHYL-2-INDOLYL KETONE see MDX310
3-METHOXY-5-METHYL-4-OXO-2,5-HEXADIENOIC ACID see PAP750
4-METHOXY-4-METHYL-2-PENTANONE see MEX250
4-METHOXY-4-METHYLPENTAN-2-ONE (DOT) see MEX250
2-METHOXY-2-METHYLPROPANE see MHV859
2-METHOXY-3(5)-METHYLPYRAZINE see MEX350
p-METHOXY-β-METHYLSTYRENE see PMQ750
2-(METHOXY(METHYLTHIO)PHOSPHINYLIMINO)-3-METHYL-1,3-THIAZOLINE see MEY200
METHOXYN see DBA800
2-METHOXY-5-NITROANILINE see NEQ500
2-METHOXY-5-NITROBENZENAMINE see NEQ500
2-METHOXYNITROBENZENE see NER000
4-METHOXYNITROBENZENE see NER500
p-METHOXYNITROBENZENE see NER500
1-METHOXY-2-NITROBENZENE see NER000
1-METHOXY-4-NITROBENZENE see NER500
7-METHOXY-2-NITRONAPHTHO(2,1-b)FURAN see MFB400
8-METHOXY-6-NITROPHENANTHOL-(3,4-d)-1,3-DIOXOLE-5-CARBOXYLIC ACID see AQY250
8-METHOXY-6-NITROPHENANTHRO(3,4-D)-1,3-DIOXOLE-5-CARBOXYLIC ACID SODIUM SALT see SEY050
3-METHOXY-17-α-19-NORPREGNA-K,3,5(10)-TRIEN-20-YN-17-OL see MKB750
(17-α)-3-METHOXY-19-NORPREGNA-1,3,5(10)-TRIEN-20-YN-17-OL see MKB750
3-METHOXY-19-NOR-17-α-PREGNA-1,3,5(10)-TRIEN-10-YN-17-OL see MKB750
d-threo-METHOXY-3-(1-OCTENYL-O,N,N-AZOXY)-2-BUTANOL see EAG000
S-((5-METHOXY-2-OXO-1,3,4-THIADIAZOL-3(2H)-YL)METHYL)-O,O-DIMETHYL PHOSPHORODITHIOATE see DSO000
2-METHOXYPHENOL see GKI000
3-METHOXYPHENOL see REF050
4-METHOXYPHENOL see MFC700
m-METHOXYPHENOL see REF050
o-METHOXYPHENOL see GKI000
p-METHOXYPHENOL see MFC700
4-METHOXYPHENYLACETIC ACID see MFE250
p-METHOXYPHENYLACETIC ACID see MFE250
o-METHOXYPHENYLAMINE see AOV900
p-METHOXYPHENYLAMINE see AOW000
o-METHOXYPHENYLAMINE HYDROCHLORIDE see AOX250
N-(p-METHOXYPHENYL)-1-AZIRIDINECARBOXAMIDE see MFF250
p-METHOXYPHENYLBUTANONE see MFF580
4-p-METHOXYPHENYL-2-BUTANONE see MFF580
4-(p-METHOXYPHENYL)-2-BUTANONE see MFF580
p-METHOXYPHENYL-N-CARBAMOYLAZIRIDINE see MFF250
1-(p-METHOXYPHENYL)-3,3-DIMETHYLTRIAZENE see DSN600
4-METHOXY-m-PHENYLENEDIAMINE see DBO000
p-METHOXY-m-PHENYLENEDIAMINE see DBO000
4-METHOXY-m-PHENYLENEDIAMINE SULFATE see DBO400
4-METHOXY-m-PHENYLENEDIAMINE SULPHATE see DBO400

p-METHOXY-m-PHENYLENEDIAMINE SULPHATE see DBO400
4-METHOXYPHENYL METHYL KETONE see MDW750
p-METHOXYPHENYL METHYL KETONE see MDW750
4-(o-METHOXYPHENYL)PIPERAZINYL 3,4,5-TRIMETHOXYPHENYL KETONE see MFH760
4-(p-METHOXYPHENYL)PIPERAZINYL 3,4,5-TRIMETHOXYPHENYL KETONE see MFH770
1-(p-METHOXYPHENYL)PROPENE see PMQ750
p-METHOXYPHENYL 4-PYRIDYL KETONE see MFH930
1-(o-METHOXYPHENYL)-4-(3,4,5-TRIMETHOXYBENZOYL)PIPERAZINE see MFH760
1-(p-METHOXYPHENYL)-4-(3,4,5-TRIMETHOXYBENZOYL)PIPERAZINE see MFH770
α-ω-METHOXYPOLY(ETHYLENE OXIDE) see DOM100
1-METHOXYPROPANE see MOU830
α-METHOXY PROPANE see MOU830
1-METHOXY-2-PROPANOL see PNL250
4-METHOXYPROPENYLBENZENE see PMQ750
1-METHOXY-4-PROPENYLBENZENE see PMQ750
1-METHOXY-4-(2-PROPENYL)BENZENE see AFW750
2-METHOXY-4-PROPENYLPHENOL see IKQ000
2-METHOXY-4-PROP-2-ENYLPHENOL see EQR500
2-METHOXY-4-(2-PROPENYL)PHENOL see EQR500
2-METHOXY-4-PROPENYLPHENYL ACETATE see AAX750
3-METHOXYPROPYLAMINE see MFM000
1-METHOXY-4-PROPYLBENZENE see PNE250
8-METHOXYPSORALEN see XDJ000
9-METHOXYPSORALEN see XDJ000
2-METHOXYPYRAZINE see MFN285
α-(6-METHOXY-4-QUINOLYL)-5-VINYL-2-QUINUCLIDINEMETHANOL see QFS000
α-(6-METHOXY-4-QUINOYL)-5-VINYL-2-QUINCLIDINEMETHANOL see QHJ000
4-METHOXYTOLUENE see MGP000
p-METHOXYTOLUENE see MGP000
4-METHOXY-m-TOLUIDINE see MGO750
6-METHOXY-α-(5-VINYL-2-QUINUCLIDINYL)-4-QUINOLINEMETHANOL see QFS000
METHULOSE see MIF760
12-METHYBENZ(a)ANTHRACENE-7-METHANOL see HMF000
METHYL ABIETATE see MFT500
METHYLACETAAT (DUTCH) see MFW100
METHYLACETALDEHYDE see PMT750
METHYLACETAMIDE see MFT750
N-METHYLACETAMIDE see MFT750
METHYL 4-ACETAMIDO-2-ETHOXYBENZOATE see EEK250
4-METHYLACETANILIDE see ABJ250
p-METHYLACETANILIDE see ABJ250
4'-METHYLACETANILIDE see ABJ250
METHYLACETAT (GERMAN) see MFW100
METHYL ACETATE see MFW100
METHYL ACETIC ACID see PMU750
METHYLACETIC ANHYDRIDE see PMV500
2'-METHYLACETOACETANILIDE see ABA000
METHYLACETOACETATE see MFX250
METHYL ACETONE (DOT) see MKA400
p-METHYL ACETOPHENONE see MFW250
4'-METHYL ACETOPHENONE see MFW250
METHYLACETOPYRONONE see MFW500

METHYL(ACETOXYMETHYL)NITROSAMINE see AAW000
6-METHYL-17-α-ACETOXYPREGNA-4,6-DIENE-3,20-DIONE see VTF000
6-α-METHYL-17-α-ACETOXYPREGN-4-ENE-3,20-DIONE see MCA000
6-α-METHYL-17-α-ACETOXYPROGESTERONE see MCA000
METHYL ACETYLACETATE see MFX250
METHYL ACETYLACETONATE see MFX250
1-METHYL-4-ACETYLBENZENE see MFW250
METHYL ACETYLENE see MFX590
METHYL ACETYLENE-PROPADIENE MIXTURE see MFX600
METHYL ACETYLENE and PROPADIENE MIXTURES, stabilized (DOT) see MFX600
2-METHYLACROLEIN see MGA250
α-METHYLACROLEIN see MGA250
β-METHYLACROLEIN see COB250
β-METHYL ACROLEIN see COB260
METHYLACRYLAAT (DUTCH) see MGA500
METHYLACRYLALDEHYDE see MGA250
METHYLACRYLALDEHYDE see MGA250
2-METHYLACRYLAMIDE see MDN500
N-METHYLACRYLAMIDE see MGA300
METHYL-ACRYLAT (GERMAN) see MGA500
METHYL ACRYLATE see MGA500
METHYL ACRYLATE, INHIBITED (DOT) see MGA500
3-METHYLACRYLIC ACID see COB500
α-METHYLACRYLIC ACID see MDN250
β-METHYLACRYLIC ACID see COB500
α-METHYL ACRYLIC AMIDE see MDN500
METHYLACRYLONITRILE see MGA750
α-METHYLACRYLONITRILE see MGA750
METHYL ADIPATE see DOQ300
METHYLAETHYLNITROSAMIN (GERMAN) see MKB000
METHYLAL see MGA850
METHYL ALCOHOL see MGB150
METHYL ALDEHYDE see FMV000
1-METHYL-2-ALDOXIMINOPYRIDINIUM IODIDE see POS750
METHYLALKOHOL (GERMAN) see MGB150
2-METHYL-ALLYLCHLORID (GERMAN) see CIU750
2-METHYLALLYL CHLORIDE see CIU750
METHYL ALLYL CHLORIDE (DOT) see CIU750
β-METHYLALLYL CHLORIDE see CIU750
1-METHYL-5-ALLYL-5-(1-METHYL-2-PENTYNYL)BARBITURIC ACID SODIUM SALT see MDU500
METHYLALLYLNITROSAMIN (GERMAN) see MMT500
METHYLALLYLNITROSAMINE see MMT500
N-((1-METHYLALLYL)THIOCARBAMOYL)-N'-(METHYLTHIOCARBAMOYL)HYDRAZINE see MLJ500
1-α-METHYLALLYLTHIOCARBAMOYL-2-METHYLTHIOCARBAMOYLHYDRAZINE see MLJ500
METHYL ALUMINIUM SESQUIBROMIDE see MGC225
METHYL ALUMINIUM SESQUICHLORIDE see MGC230
METHYL ALUMINUM SESQUIBROMIDE see MGC225
METHYL ALUMINUM SESQUIBROMIDE see MGC225
METHYL ALUMINUM SESQUICHLORIDE see MGC230
METHYLAMINE see MGC250

METHYLAMINE (ACGIH,OSHA) see MGC250
METHYLAMINEN (DUTCH) see MGC250
METHYLAMINE, anhydrous (UN 1061) (DOT) see
MGC250
METHYLAMINE, aqueous solution (UN 1235) (DOT) see
MGC250
4-METHYL-2-AMINOANISOLE see MGO750
METHYLAMINOANTIPYRINE SODIUM
METHANESULFONATE see AMK500
(METHYLAMINO)BENZENE see MGN750
N-METHYLAMINOBENZENE see MGN750
1-METHYL-2-AMINOBENZENE see TGQ750
2-METHYL-1-AMINOBENZENE see TGQ750
1-METHYL-2-AMINOBENZENE HYDROCHLORIDE
see TGS500
2-METHYL-1-AMINOBENZENE HYDROCHLORIDE
see TGS500
METHYL 2-AMINOBENZOATE see APJ250
METHYL o-AMINOBENZOATE see APJ250
1-METHYLAMINO-4-BROMANTHRACHINON see
BNN550
1-(METHYLAMINO)-4-BROMOANTHRAQUINONE
see BNN550
4-METHYLAMINO-1,5-DIMETHYL-2-PHENYL-3-
PYRAZOLONE SODIUM METHANESULFONATE see
AMK500
2-METHYLAMINOETHANOL see MGG000
N-METHYLAMINOETHANOL see MGG000
β-(METHYLAMINO)ETHANOL see MGG000
m-METHYLAMINOETHANOLPHENOL see NCL500
m-METHYLAMINOETHANOLPHENOL
HYDROCHLORIDE see SPC500
(R-(R*,S*))-α-(1-
(METHYLAMINO)ETHYL)BENZENEMETHANOL
HYDROCHLORIDE see EAX000
(−)-α-(1-METHYLAMINOETHYL)BENZYL ALCOHOL
see EAW000
1-α-(1-METHYLAMINOETHYL)BENZYL ALCOHOL
see EAW000
dl-α-(1-(METHYLAMINO)ETHYL) BENZYL
ALCOHOL HYDROCHLORIDE see EAX500
1-α-(1-(METHYLAMINO)ETHYL)BENZYL ALCOHOL
SULFATE see EAY500
2-METHYL-6-AMINOHEPTANE see ILM000
2-METHYLAMINO METHYL BENZOATE see MGQ250
1-
METHYLAMINOMETHYLDIBENZO(b,c)BICYCLO(2,2
,2)OCTADIENE HYDROCHLORIDE see BCH750
6-METHYLAMINO-2-METHYLHEPTENE see ILK000
METHYLAMINOPHENYLDIMETHYLPYRAZOLONE
METHANESULFONATE SODIUM see AMK500
1-2-METHYLAMINO-1-PHENYLPROPANOL see
EAW000
N-(γ-METHYLAMINOPROPYL)IMINODIBENZYL
HYDROCHLORIDE see DLS600
METHYLAMINOPTERIN see MDV500
1-METHYL-3-AMINO-5H-PYRIDO(4,3-b)INDOLE see
ALD500
N-METHYL-4-AMINO-1,2,5-SELENADIAZOLE-3-
CARBOXAMIDE see MGL600
METHYLAMPHETAMINE HYDROCHLORIDE see
MDT600
METHYLAMPHETAMINE HYDROCHLORIDE see
DBA800
d-METHYLAMPHETAMINE HYDROCHLORIDE see
MDT600

N-METHYLAMPHETAMINE HYDROCHLORIDE see
MDT600
METHYLAMYL ACETATE see HFJ000
METHYL AMYL ACETATE (DOT) see HFJ000
METHYL AMYL ALCOHOL see MKW600
METHYLAMYL ALCOHOL see AOK750
METHYL AMYL CARBINOL see HBE500
METHYL-AMYL-CETONE (FRENCH) see MGN500
METHYL n-AMYL KETONE see MGN500
METHYL AMYL KETONE (DOT) see MGN500
METHYLAMYLNITROSAMIN (GERMAN) see AOL000
METHYLAMYLNITROSAMINE see AOL000
METHYL-N-AMYLNITROSAMINE see AOL000
METHYLANILINE see MGN750
2-METHYLANILINE see TGQ750
3-METHYLANILINE see TGQ500
4-METHYLANILINE see TGR000
m-METHYLANILINE see TGQ500
N-METHYLANILINE see MGN750
o-METHYLANILINE see TGQ750
p-METHYLANILINE see TGR000
N-METHYLANILINE (ACGIH,DOT) see MGN750
2-METHYLANILINE HYDROCHLORIDE see TGS500
4-METHYLANILINE HYDROCHLORIDE see TGS750
o-METHYLANILINE HYDROCHLORIDE see TGS500
2-METHYL-p-ANISIDINE see MGO500
5-METHYL-o-ANISIDINE see MGO750
p-METHYL ANISOLE see MGP000
9-METHYLANTHRACENE see MGP750
METHYL ANTHRANILATE (FCC) see APJ250
N-METHYLANTHRANILIC ACID, METHYL ESTER
see MGQ250
2-METHYL-1-ANTHRAQUINONYLAMINE see
AKP750
METHYL APHOXIDE see TNK250
METHYLARSENIC ACID, SODIUM SALT see MRL750
METHYLARSINE DICHLORIDE see DFP200
METHYLARSINE DIIODIDE see MGQ775
METHYLARSONOUS DICHLORIDE see DFP200
METHYLARTERENOL see VGP000
METHYL ASPARTYLPHENYLALANATE see ARN825
1-METHYL N-l-α-ASPARTYL-l-PHENYLALANINE see
ARN825
METHYL ATROPINE NITRATE see MGR500
N-METHYLATROPINE NITRATE see MGR500
8-METHYLATROPINIUM NITRATE see MGR500
N-METHYLATROPINIUM NITRATE see MGR500
2-METHYLAZACYCLOPROPANE see PNL400
METHYLAZINPHOS see ASH500
2-METHYLAZIRIDINE see PNL400
METHYLAZOXYMETHANOL ACETATE see MGS750
METHYLAZOXYMETHANOL GLUCOSIDE see
COU000
METHYLAZOXYMETHANOL-β-d-GLUCOSIDE see
COU000
METHYLAZOXYMETHYL ACETATE see MGS750
METHYLAZOXYMETHYLESTER KYSELINY
OCTOVE (CZECH) see MGS750
METHYLBEN see HJL500
7-METHYLBENZ(a)ANTHRACENE-5,6-OXIDE see
MGZ000
N-METHYLBENZAZIMIDE,
DIMETHYLDITHIOPHOSPHORIC ACID ESTER see
ASH500
METHYLBENZEDRIN see DBA800
METHYL 1H-BENZEMEDAZOL-2-YLCARBAMATE
see MHC750

3-METHYL BUTANOL see IHP000
2-METHYL BUTANOL-1 see MHS750
2-METHYL BUTANOL-2 see PBV000
2-METHYL-2-BUTANOL see PBV000
2-METHYL-4-BUTANOL see IHP000
3-METHYLBUTAN-1-OL see IHP000
3-METHYLBUTAN-3-OL see PBV000
3-METHYL-1-BUTANOL (CZECH) see IHP000
3-METHYLBUTANOL NITRITE see IMB000
3-METHYL-2-BUTANONE see MLA750
3-METHYL BUTAN-2-ONE (DOT) see MLA750
METHYLBUTENE see IHR220
2-METHYLBUTENE see IHR220
2-METHYL-1-BUTENE see MHT000
3-METHYL-1-BUTENE see MHT250
trans-2-METHYL-2-BUTENOIC ACID see TGA700
3-METHYL-3-BUTEN-2-ON (GERMAN) see MKY500
2-METHYL-1-BUTEN-3-ONE see MKY500
3-METHYL-2-BUTENYL ACETATE see DOQ350
3-METHYL-2-BUTENYL BENZOATE see MHU150
3-(3-METHYL-2-BUTENYL)-1,2,3,4,5,6-HEXAHYDRO-6,11-DIMETHYL-2,6-METHANO-3-BENZAZOCIN-8-OL see DOQ400
3-METHYL-2-BUTENYL SALICYLATE see PMB600
(2-(3-METHYLBUTOXY)ETHYL)BENZENE see IHV050
3'-METHYLBUTTERGELB (GERMAN) see DUH600
1-METHYLBUTYL ACETATE see AOD735
3-METHYLBUTYL ACETATE see IHO850
3-METHYL-1-BUTYL ACETATE see IHO850
2-METHYL-BUTYLACRYLAAT (DUTCH) see MHU750
2-METHYL-BUTYLACRYLAT (GERMAN) see MHU750
2-METHYL BUTYLACRYLATE see MHU750
METHYLBUTYLAMINE see MHV000
N-(METHYL) BUTYL AMINE see MHV000
N-METHYL-n-BUTYLAMINE see MHV000
p-METHYL-tert-BUTYLBENZENE see BSP500
1-METHYL-4-tert-BUTYLBENZENE see BSP500
1-(3-METHYL)BUTYL BENZOATE see IHP100
3-METHYLBUTYL BROMIDE see BNP250
3-METHYLBUTYL BUTYRATE see IHP400
METHYL-1-(BUTYLCARBAMOYL)-2-BENZIMIDAZOLYLCARBAMATE see BAV575
METHYL-1,3-BUTYLENE GLYCOL ACETATE see MHV750
3-METHYLBUTYL ETHANOATE see IHO850
METHYL BUTYL ETHER see BRU780
METHYL n-BUTYL ETHER see BRU780
METHYL tert-BUTYL ETHER see MHV859
METHYL tert-BUTYL ETHER (DOT) see MHV859
3-METHYLBUTYL FORMATE see IHS000
3-METHYLBUTYL 2-HYDROXYBENZOATE see IME000
METHYL n-BUTYL KETONE (ACGIH) see HEV000
N-3-METHYLBUTYL-N-1-METHYL ACETONYLNITROSAMINE see MHW350
METHYL-N-(p-tert-BUTYL-α-METHYLHYDROCINNAMYLIDENE) ANTHRANILATE see LFT100
3-METHYLBUTYL NITRITE see IMB000
METHYL-BUTYL-NITROSAMIN (GERMAN) see MHW500
METHYLBUTYLNITROSAMINE see MHW500
METHYL-N-BUTYLNITROSAMINE see MHW500
METHYL BUTYRATE see MHY000
METHYL-n-BUTYRATE see MHY000
3-METHYLBUTYRIC ACID see ISU000
β-METHYLBUTYRIC ACID see ISU000

3-METHYLBUTYRIC ACID, ALLYL ESTER see ISV000
(E)-3-METHYLBUTYRIC ACID-3,7-DIMETHYL-2,6-OCTADIENYL ESTER see GDK000
3-METHYLBUTYRIC ACID, ETHYL ESTER see ISY000
4-METHYL-γ-BUTYROLACTONE see VAV000
γ-METHYL-γ-BUTYROLACTONE see VAV000
8-(3-METHYLBUTYRYLOXY)-DIACETOXYSCIRPENOL see FQS000
METHYL CADMIUM AZIDE see MHY550
METHYL-CALMINAL see ENB500
METHYL CARBAMATE see MHZ000
METHYLCARBAMATE-1-NAPHTHALENOL see CBM750
METHYLCARBAMATE-1-NAPHTHOL see CBM750
N-METHYLCARBAMATE de 1-NAPHTYLE see CBM750
N-METHYLCARBAMATE de 1-NAPHTYLE (FRENCH) see CBM750
METHYLCARBAMIC ACID o-sec-BUTYLPHENYL ESTER see MOV000
METHYLCARBAMIC ACID-o-CUMENYL ESTER see MIA250
METHYL CARBAMIC ACID 2,3-DIHYDRO-2,2-DIMETHYL-7-BENZOFURANYL ESTER see CBS275
METHYL-CARBAMIC ACID, ESTER with ESEROLINE see PIA500
METHYLCARBAMIC ACID, ETHYL ESTER see EMQ500
METHYLCARBAMIC ACID-2,3-(ISOPROPYLIDENEDIOXY)PHENYL ESTER see DQM600
METHYL CARBAMIC ACID-4-(METHYLTHIO)-3,5-XYLYL ESTER see DST000
METHYLCARBAMIC ACID-1-NAPHTHYL ESTER see CBM750
METHYLCARBAMIC ACID m-TOLYL ESTER see MIB750
S-METHYLCARBAMOYLMETHYL-O,O-DIMETHYL PHOSPHORODITHIOATE see DSP400
2-(o-(METHYLCARBAMOYL)OXIMINO)-3,3-DIMETHYLTETRAHYDRO-1,4-THIAZIN-5-ONE see MID860
2-(o-(METHYLCARBAMOYL)OXIMINO)-3-METHYLTETRAHYDRO-1,4-THIAZIN-5-ONE see MPU600
(3-(N-METHYLCARBAMOYLOXY)PHENYL)TRIMETHYL-ARSONIUM IODIDE see MID900
1-METHYL-4-CARBETHOXY-4-PHENYLPIPERIDINE HYDROCHLORIDE see DAM700
METHYLCARBINOL see EFU000
METHYL CARBITOL see DJG000
METHYL-4-CARBOMETHOXY BENZOATE see DUE000
METHYL CARBONATE see MIF000
METHYLCARBONYL FLUORIDE see ACM000
METHYLCARBOPHENOTHION see MQH750
17-β-(1-METHYL-3-CARBOXYPROPYL)-ETIOCHOLANE-3-α,12-α-DIOL see DAQ400
METHYLCATECHOL see GKI000
3-METHYLCATECHOL see DNE000
METHYL-CCNU see CHD250
trans-METHYL-CCNU see CHD250
METHYL CEDRYL ETHER see CCR525
METHYL CELLOSOLVE ACETATE (OSHA, DOT) see EJJ500
METHYL CELLOSOLVE ACRYLATE see MIF750
METHYL CELLOSOLVE (OSHA, DOT) see EJH500

METHYL CELLOSOLYE ACETAAT (DUTCH) see
EJJ500
METHYL CELLULOSE see MIF760
METHYL CELLULOSE-A see MIF760
METHYL CELLULOSE ETHER see MIF760
4-METHYLCHALCONE see MIF762
p-METHYLCHALCONE see MIF762
METHYL CHAVICOL see AFW750
METHYL CHEMOSEPT see HJL500
METHYLCHLOORFORMIAT (DUTCH) see MIG000
METHYL-2-CHLORAETHYLNITROSAMIN
(GERMAN) see CIQ500
METHYLCHLORID (GERMAN) see MIF765
METHYL CHLORIDE see MIF765
METHYL CHLOROACETATE see MIF775
METHYL CHLOROACETATE (DOT) see MIF775
2-METHYL-4-CHLOROANILINE see CLK220
4-METHYL-2-CHLOROANILINE see CLK210
2-METHYL-4-CHLOROANILINE HYDROCHLORIDE
see CLK235
2-METHYLCHLOROBENZENE see CLK100
1-METHYL-2-CHLOROBENZENE see CLK100
METHYL CHLOROCARBONATE see MIG000
METHYL(2-CHLOROETHYL)NITROSAMINE see
CIQ500
METHYL CHLOROFORM see MIH275
METHYL CHLOROFORMATE (DOT) see MIG000
METHYLCHLOROMETHYL ETHER (DOT) see CIO250
METHYL CHLOROMETHYL ETHER, anhydrous (DOT)
see CIO250
3-METHYL-4-CHLOROPHENOL see CFD990
2-METHYL-4-CHLOROPHENOXYACETIC ACID see
CIR250
(2-METHYL-4-CHLOROPHENOXY)ACETIC ACID,
SODIUM SALT see SIL500
2-(2-METHYL-4-CHLOROPHENOXY)PROPIONIC
ACID see CIR500
α-(2-METHYL-4-CHLOROPHENOXY)PROPIONIC
ACID see CIR500
2-METHYL-4-CHLOROPHENOXY-α-PROPIONIC
ACID see CIR500
N'-(2-METHYL-4-CHLOROPHENYL)-N,N-
DIMETHYLFORMAMIDINE see CJJ250
N-METHYL-N'-(p-CHLOROPHENYL)-N-
NITROSOUREA see MMW775
METHYL CHLOROPHOS see TIQ250
METHYLCHLOROPINDOL see CMX850
METHYL-2-CHLOROPROPIONATE (DOT) see CKT000
2-METHYL-4-CHLORPHENOXYESSIGSAEURE
(GERMAN) see CIR250
2-(2-METHYL-4-CHLORPHENOXY)-
PROPIONSAEURE (GERMAN) see CIR500
N'-(2-METHYL-4-CHLORPHENYL)-FORMAMIDIN-
HYDROCHLORID (GERMAN) see CJJ250
METHYLCHLORPINDOL see CMX850
METHYL CHLORPYRIFOS see CMA250
METHYLCHLORTETRACYCLINE see MIJ500
METHYLCHOLANTHRENE see MIJ750
3-METHYLCHOLANTHRENE see MIJ750
20-METHYLCHOLANTHRENE see MIJ750
5-METHYLCHRYSENE see MIN500
α-METHYLCINNAMALDEHYDE see MIO000
METHYL CINNAMATE see MIO500
METHYL CINNAMIC ALDEHYDE see MIO000
α-METHYLCINNAMIC ALDEHYDE see MIO000
METHYL CINNAMYLATE see MIO500
α-METHYLCINNIMAL see MIO000

METHYLCISTOX see DAO800
METHYLCOLCHICINE see MIW500
6-METHYLCOUMARIN see MIP750
6-METHYLCOUMARINIC ANHYDRIDE see MIP750
6-METHYL-m-CRESOL see XKS000
(E)-2-METHYLCROTONIC ACID see TGA700
trans-2-METHYLCROTONIC ACID see TGA700
p-METHYL-CUMENE see CQI000
METHYL CYANIDE see ABE500
METHYL CYANOACRYLATE see MIQ075
METHYL 2-CYANOACRYLATE see MIQ075
METHYL α-CYANOACRYLATE see MIQ075
METHYLCYCLOHEXANE see MIQ740
METHYLCYCLOHEXANOL see MIQ745
METHYLCYCLOHEXANOL (ACGIH,DOT,OSHA) see
MIQ745
METHYL CYCLOHEXANOLS, Fp not >60.5 degrees C
(DOT) see MIQ745
2-METHYL-CYCLOHEXANON (GERMAN, DUTCH)
see MIR500
METHYLCYCLOHEXANONE see MIR250
2-METHYLCYCLOHEXANONE see MIR500
o-METHYLCYCLOHEXANONE see MIR500
1-METHYLCYCLOHEXAN-2-ONE see MIR500
METHYLCYCLOHEXYLNITROSAMIN (GERMAN) see
NKT500
METHYLCYCLOHEXYLNITROSAMINE see NKT500
3-METHYLCYCLOPENTADECANONE see MIT625
3-METHYL-1-CYCLOPENTADECANONE see MIT625
METHYLCYCLOPENTADIENYL MANGANESE
TRICARBONYL see MAV750
2-METHYLCYCLOPENTADIENYL
MANGANESETRICARBONYL see MAV750
2-METHYLCYCLOPENTADIENYL MANGANESE
TRICARBONYL (ACGIH) see MAV750
METHYLCYCLOPENTADIENYL MANGANESE
TRICARBONYL (OSHA) see MAV750
METHYLCYCLOPENTANE see MIU500
METHYL CYCLOPENTANE (DOT) see MIU500
3-METHYLCYCLOPENTANE-1,2-DIONE see HMB500
METHYL CYCLOPENTENOLONE (FCC) see HMB500
METHYLCYCLOPROPANECARBONYLHYDRAZINE
see MIV300
METHYLCYKLOPENTADIENTRIKARBONYLMANG
ANIUM see MAV750
3'-METHYL-DAB see DUH600
N-METHYL-N-DEACETYLCOLCHICINE see MIW500
α-METHYL DECALACTONE see MIW050
4-METHYLDECANOLIDE see MIW050
METHYL DECYL ETHER see MIW075
METHYL n-DECYL ETHER see MIW075
6-METHYL-6-DEHYDRO-17-α-
ACETOXYPROGESTERONE see VTF000
6-METHYL-6-DEHYDRO-17-α-
ACETYLPROGESTERONE see VTF000
N-METHYLDEMECOLCINE see MIW500
O-METHYLDEMETON see DAO800
METHYL DEMETON METHYL see MIW250
METHYL-DEMETON-O see DAO800
METHYL DEMETON-O-SULFOXIDE see DAP000
METHYL DEMETON THIOESTER see DAP400
5-METHYLDEOXYURIDINE see TFX790
N-METHYL-N-DESACETYLCOLCHICINE see MIW500
N-METHYL-N-DESACETYLCOLCHICINE see MIW500
METHYL DIAZEPINONE see DCK759
METHYLDIBROMOGLUTARONITRILE see DDM500

4-METHYL-2,6-DI-terc.BUTYLFENOL (CZECH) see BFW750

METHYL-DI-tert-BUTYLPHENOL see BFW750

4-METHYL-2,6-DI-tert-BUTYLPHENOL see BFW750

METHYLDICHLORARSINE see DFP200

METHYL DICHLOROACETATE (DOT) see DEM800

METHYLDICHLOROARSINE (DOT) see DFP200

O-METHYL-O-2,5-DICHLORO-4-BROMOPHENYL PHENYLTHIOPHOSPHONATE see LEN000

N-METHYL-2,2'-DICHLORODIETHYLAMINE see BIE250

N-METHYL-2,2'-DICHLORODIETHYLAMINE HYDROCHLORIDE see BIE500

N-METHYL-2,2'-DICHLORODIETHYLAMINE-N-OXIDE HYDROCHLORIDE see CFA750

METHYL DICHLOROETHANOATE see DEM800

METHYLDI(2-CHLOROETHYL)AMINE see BIE250

METHYLDI(2-CHLOROETHYL)AMINE HYDROCHLORIDE see BIE500

N-METHYL-DI-2-CHLOROETHYLAMINE HYDROCHLORIDE see BIE500

METHYLDI(β-CHLOROETHYL)AMINE HYDROCHLORIDE see BIE500

N-METHYL-DI-2-CHLOROETHYLAMINE-N-OXIDE see CFA500

METHYLDI(2-CHLOROETHYL)AMINE-N-OXIDE HYDROCHLORIDE see CFA750

N-METHYLDICHLOROMALEINIMIDE see DFP800

METHYL DICHLOROSILANE (DOT) see DFS000

METHYL-DICHLORSILAN (CZECH) see DFS000

5-METHYL-7-DIETHYLAMINO-s-TRIAZOLO-(1,5-a)PYRIMIDINE see DIO200

3-METHYL-N,N-DIETHYLBENZAMIDE see DKC800

1-METHYL-4-DIETHYLCARBAMOYLPIPERAZINE CITRATE see DIW200

1-METHYL-4-DIETHYLCARBAMYLPIPERAZINE see DIW000

11-METHYL-15,16-DIHYDRO-17H-CYCLOPENTA(a)PHENANTHREN-17-ONE see MJE500

11-METHYL-15,16-DIHYDRO-17-OXOCYCLOPENTA(a)PHENANTHRENE see MJE500

α-METHYL-l-3,4-DIHYDROXYPHENYLALANINE see DNA800

l-α-METHYL-3,4-DIHYDROXYPHENYLALANINE see DNA800

α-METHYL-β-(3,4-DIHYDROXYPHENYL)-l-ALANINE see DNA800

l-(−)-α-METHYL-β-(3,4-DIHYDROXYPHENYL)ALANINE see DNA800

METHYLDIIODOARSINE see MGQ775

METHYL-3-(DIMETHOXYPHOSPHINYLOXY)CROTONATE see MQR750

METHYLDIMETHOXYSILANE see MJE900

3'-METHYL-4-DIMETHYLAMINOAZOBENZEN (CZECH) see DUH600

3'-METHYL-4-DIMETHYLAMINOAZOBENZENE see DUH600

M'-METHYL-p-DIMETHYLAMINOAZOBENZENE see DUH600

3'-METHYL-N,N-DIMETHYL-4-AMINOAZOBENZENE see DUH600

3'-METHYLDIMETHYLAMINOAZOBENZOL (GERMAN) see DUH600

METHYL-2-(DIMETHYLAMINO)-N-(((METHYLAMINO)CARBONYL)OXY)-2-OXOETHANIMIDOTHIOATE see DSP600

METHYL-4-DIMETHYLAMINO-3,5-XYLYL CARBAMATE see DOS000

METHYL-4-DIMETHYLAMINO-3,5-XYLYL ESTER of CARBAMIC ACID see DOS000

METHYL-1-(DIMETHYLCARBAMOYL)-N-(METHYLCARBAMOYLOXY)THIOFORMIMIDATE see DSP600

S-METHYL-1-(DIMETHYLCARBAMOYL)-N-((METHYLCARBAMOYL)OXY)THIOFORMIMIDATE see DSP600

METHYL 1,1-DIMETHYLETHYL ETHER see MHV859

METHYL-N',N-DIMETHYL-N-((METHYLCARBAMOYL)OXY)-1-THIOOXAMIMIDATE see DSP600

2-METHYL-3,5-DINITROBENZAMIDE see DUP300

METHYLDINITROBENZENE see DVG600

1-METHYL-2,4-DINITROBENZENE see DVH000

2-METHYL-1,3-DINITROBENZENE see DVH400

2-METHYL-1,4-DINITROBENZENE see DVH200

4-METHYL-1,2-DINITROBENZENE see DVH600

1-METHYL-2,3-DINITRO-BENZENE (9CI) see DVG800

2-METHYL-4,6-DINITROPHENOL see DUS700

2-METHYLDINITROPHENOL SODIUM SALT see SGP550

2-METHYL-4,6-DINITROPHENOL SODIUM SALT see DUU600

N-METHYL-N,N-DIOCTYL-1-OCTANAMINIUM CHLORIDE see MQH000

6-METHYL DIPYRIDO(1,2-a:3',2'-d)IMIDAZOL-2-AMINE see AKS250

(4-METHYL-1,3-DITHIOLAN-2-YLIDENE)PHOSPHORAMIDIC ACID, DIETHYL ESTER see DHH400

METHYLDOPA see DNA800

α-METHYL-l-DOPA see DNA800

l-α-METHYLDOPA see DNA800

METHYL DURSBAN see CMA250

METHYL-E 605 see MNH000

METHYLE (ACETATE de) (FRENCH) see MFW100

O,O-METHYLEEN-BIS(4-CHLOORFENOL) see MJM500

METHYLEEN-S,S'-BIS(O,O-DIETHYL-DITHIOFOSFAAT) (DUTCH) see EEH600

3,3'-METHYLEEN-BIS(4-HYDROXY-CUMARINE) (DUTCH) see BJZ000

METHYLE (FORMIATE de) (FRENCH) see MKG750

S,S'-METHYLEN-BIS(O,O-DIAETHYL-DITHIOPHOSPHAT) (GERMAN) see EEH600

3,3'-METHYLEN-BIS(4-HYDROXY-CUMARIN) (GERMAN) see BJZ000

3,4-METHYLENDIOXY-6-PROPYLBENZYL-n-BUTYL-DIAETHYLENGLYKOLAETHER (GERMAN) see PIX250

METHYLENDIRHODANID (CZECH, GERMAN) see MJT500

METHYLENE ACETONE see BOY500

METHYLENE BICHLORIDE see MJP450

METHYLENEBISACRYLAMIDE see MJL500

N,N'-METHYLENEBIS(ACRYLAMIDE) see MJL500

METHYLENEBIS(ANILINE) see MJQ000

2,4'-METHYLENEBIS(ANILINE) see MJP750

4,4'-METHYLENEBISANILINE see MJQ000

4,4'-METHYLENEBIS(BENZENEAMINE) see MJQ000

2,2'-METHYLENEBIS(6-tert-BUTYL-4-ETHYLPHENOL) see MJN250
4,4'-METHYLENE(BIS)-CHLOROANILINE see MJM200
4,4'-METHYLENE BIS(2-CHLOROANILINE) see MJM200
METHYLENE-4,4'-BIS(o-CHLOROANILINE) see MJM200
4,4'-METHYLENEBIS(o-CHLOROANILINE) see MJM200
p,p'-METHYLENEBIS(o-CHLOROANILINE) see MJM200
p,p'-METHYLENEBIS(α-CHLOROANILINE) see MJM200
4,4'-METHYLENEBIS-2-CHLOROBENZENAMINE see MJM200
2,2'-METHYLENEBIS(4-CHLOROPHENOL) see MJM500
2,2'-METHYLENEBIS(4-CHLOROPHENOL) see MJM500
METHYLENE BIS(4-CYCLOHEXYLISOCYANATE) see MJM600
METHYLENE BIS(4-CYCLOHEXYLISOCYANATE) (ACGIH,OSHA) see MJM600
METHYLENE-S,S'-BIS(O,O-DIAETHYL-DITHIOPHOSPHAT) (GERMAN) see EEH600
4,4'-METHYLENE BIS(N,N'-DIMETHYLANILINE) see MJN000
4,4'-METHYLENEBIS(N,N-DIMETHYL)BENZENAMINE see MJN000
2,2'-METHYLENEBIS(4-ETHYL-6-tert-BUTYLPHENOL) see MJN250
2,2'-METHYLENEBISFURAN see FPX028
3,3'-METHYLENEBIS(4-HYDROXY-1,2-BENZOPYRONE) see BJZ000
3,3'-METHYLENEBIS(4-HYDROXYCOUMARIN) see BJZ000
3,3'-METHYLENE-BIS(4-HYDROXYCOUMARINE) (FRENCH) see BJZ000
METHYLENEBIS(4-ISOCYANATOBENZENE) see MJP400
1,1-METHYLENEBIS(4-ISOCYANATOBENZENE) see MJP400
5,5'-METHYLENEBIS(2-ISOCYANATO)TOLUENE see MJN750
METHYLENE BIS(METHANESULFONATE) see MJQ500
4,4'-METHYLENEBIS(2-METHYLANILINE) see MJO250
4,4'-METHYLENEBIS(2-METHYLBENZENAMINE) see MJO250
METHYLENE-BIS-ORTHOCHLOROANILINE see MJM200
1,1'-(METHYLENEBIS(OXY)BIS(2-CHLOROETHANE)) see BID750
METHYLENEBIS(4-PHENYLENE ISOCYANATE) see MJP400
METHYLENEBIS(p-PHENYLENE ISOCYANATE) see MJP400
METHYLENE BISPHENYL ISOCYANATE see MJP400
METHYLENEBIS(4-PHENYL ISOCYANATE) see MJP400
METHYLENEBIS(p-PHENYL ISOCYANATE) see MJP400
4,4'-METHYLENEBIS(PHENYL ISOCYANATE) see MJP400
p,p'-METHYLENEBIS(PHENYL ISOCYANATE) see MJP400

2,2'-METHYLENEBIS(3,4,6-TRICHLOROPHENOL) see HCL000
METHYLENE BLUE see BJI250
METHYLENE BLUE (medicinal) see BJI250
METHYLENE BLUE A see BJI250
METHYLENE BLUE BB see BJI250
METHYLENE BLUE BB ZINC FREE see BJI250
METHYLENE BLUE CHLORIDE see BJI250
METHYLENE BLUE CHLORIDE (biological stain) see BJI250
METHYLENE BLUE D see BJI250
METHYLENE BLUE I (medicinal) see BJI250
METHYLENE BLUE NF (medicinal) see BJI250
METHYLENE BLUE POLYCHROME see BJI250
METHYLENE BLUE USP (medicinal) see BJI250
METHYLENE BLUE USP XII (medicinal) see BJI250
METHYLENE BROMIDE see DDP800
METHYLENE CHLORIDE see MJP450
METHYLENE CHLOROBROMIDE see CES650
METHYLENE CYANIDE see MAO250
N,N'-METHYLENEDIACRYLAMIDE see MJL500
METHYLENEDIANILINE see MJQ000
2,4'-METHYLENEDIANILINE see MJP750
4,4'-METHYLENEDIANILINE see MJQ000
p,p'-METHYLENEDIANILINE see MJQ000
4,4-METHYLENEDIANILINE (ACGIH) see MJQ000
4,4'-METHYLENEDIANILINE DIHYDROCHLORIDE see MJQ100
p,p'-METHYLENEDIANILINE DIHYDROCHLORIDE see MJQ100
METHYLENE DIBROMIDE see DDP800
METHYLENE DICHLORIDE see MJP450
3,4-METHYLENE-DIHYDROXYBENZALDEHYDE see PIW250
METHYLENE DIMETHANESULFONATE see MJQ500
METHYLENE DIMETHYL ETHER see MGA850
METHYLENEDINAPHTHALENESULFONIC ACID BISPHENYLMERCURI SALT see PFN000
1,5-METHYLENE-3,7-DINITROSO-1,3,5,7-TETRAAZACYCLOOCTAINE see DVF400
1,5-METHYLENE-3,7-DINITROSO-1,3,5,7-TETRAAZACYCLOOCTANE see DVF400
3,4-METHYLENEDIOXY-ALLYLBENZENE see SAD000
1,2-METHYLENEDIOXY-4-ALLYLBENZENE see SAD000
3,4-METHYLENEDIOXYBENZALDEHYDE see PIW250
3,4-METHYLENEDIOXYBENZYL ACETATE see PIX000
1,2-METHYLENEDIOXY-4-(1-HYDROXYALLYL)BENZENE see BCJ000
7,8-METHYLENEDIOXYISOQUINOLINE see MJR775
1,2-METHYLENEDIOXY-4-PROPENYLBENZENE see IRZ000
3,4-METHYLENEDIOXY-1-PROPENYL BENZENE see IRZ000
1,2-(METHYLENEDIOXY)-4-PROPYLBENZENE see DMD600
(3,4-METHYLENEDIOXY-6-PROPYLBENZYL)(BUTYL)DIETHYLENE GLYCOL ETHER see PIX250
3,4-METHYLENEDIOXY-6-PROPYLBENZYL-n-BUTYL DIETHYLENEGLYCOL ETHER see PIX250
4,4'-METHYLENEDIPHENYL DIISOCYANATE see MJP400
METHYLENEDI-p-PHENYLENE DIISOCYANATE see MJP400

METHYLENEDI-p-PHENYLENE ISOCYANATE see MJP400

4,4'-METHYLENEDIPHENYLENE ISOCYANATE see MJP400

METHYLENE DI(PHENYLENE ISOCYANATE) (DOT) see MJP400

4,4'-METHYLENEDIPHENYL ISOCYANATE see MJP400

METHYLENE DITHIOCYANATE see MJT500

4,4'-METHYLENE DI-o-TOLUIDINE see MJO250

METHYLENE GLYCOL see FMV000

3-METHYLENE-7-METHYL-1,6-OCTADIENE see MRZ150

4-METHYLENE-2-OXETANONE see KFA000

METHYLENE OXIDE see FMV000

S,S'-METHYLENE O,O,O',O'-TETRAETHYL PHOSPHORODITHIOATE see EEH600

8-METHYLENE-4,11,11-(TRIMETHYL)BICYCLO(7.2.0)UNDEC-4-ENE see CCN000

METHYLENIUM CERULEUM see BJI250

METHYL-18-EPIRESERPATE METHYL ETHER HYDROCHLORIDE see MQR200

METHYL ESTER of p-HYDROXYBENZOIC ACID see HJL500

METHYLESTER KISELINY OCTOVE (CZECH) see MFW100

METHYLESTER KYSELINY ANTHRANILOVE see APJ250

METHYLESTER KYSELINY BENZOOVE see MHA750

METHYLESTER KYSELINY BROMOCTOVE see MHR250

METHYLESTER KYSELINY CHLOROCTOVE see MIF775

METHYLESTER KYSELINY METHAKRYLOVE see MLH750

METHYLESTER KYSELINY ORTHOMRAVENCI (CZECH) see TLX600

METHYLESTER KYSELINY p-TOLUENSULFONOVE (CZECH) see MLL250

METHYL ESTER of METHANESULFONIC ACID see MLH500

METHYL ESTER of METHANESULPHONIC ACID see MLH500

METHYL ESTER STEARIC ACID see MJW000

METHYL ESTER of WOOD ROSIN see MFT500

METHYL ESTER of WOOD ROSIN, partially hydrogenated (FCC) see MFT500

METHYLE (SULFATE de) (FRENCH) see DUD100

4,4'-(1-METHYL-1,2-ETHANEDIYL)BIS-2,6-PIPERAZINEDIONE see PIK250

METHYL ETHANE SULFONATE see MJW250

METHYL ETHANE SULPHONATE see MJW250

1-METHYLETHANETHIOL see IMU000

N-METHYLETHANOANTHRACENE-9-(10H)-METHYLAMINE HYDROCHLORIDE see BCH750

METHYL ETHANOATE see MFW100

N-METHYLETHANOLAMINE see MGG000

METHYLETHENE see PMO500

METHYL ETHER see MJW500

METHYL ETHOXOL see EJH500

2-(2-(1-METHYLETHOXY)ETHOXYL)ETHANOL see IOJ500

((1-METHYLETHOXY)METHYL)OXIRANE see IPD000

2-(1-METHYLETHOXY)PHENYL (((1,1-DIMETHYLETHOXY)SULFINYL)METHYL)CARBAMATE see INE062

1-METHYLETHYLAMINE see INK000

4-(1-METHYLETHYL)-BENZALDEHYDE (9CI) see COE500

4-(1-METHYLETHYL)BENZONITRILE see IOD050

3-(1-METHYLETHYL)-1H-2,1,3-BENZOTHIAZAIN-4(3H)-ONE-2,2-DIOXIDE see MJY500

METHYLETHYLBROMOMETHANE see BMX750

METHYLETHYLCARBINOL see BPW750

METHYL ETHYL CELLULOSE see MJY550

4-METHYL-4-ETHYL-2,6-DIOXOPIPERIDINE see MKA250

METHYLETHYLENE see PMO500

METHYLETHYLENE GLYCOL see PML000

METHYL ETHYLENE OXIDE see PNL600

METHYLETHYLENIMINE see PNL400

2-METHYLETHYLENIMINE see PNL400

METHYL ETHYL ETHER (DOT) see EMT000

(E)-1-METHYLETHYL-3-((ETHYLAMINO)METHOXYPHOSPHINOTHIOYL)OXY-2-BUTENOATE see MKA000

1-(METHYLETHYL)-ETHYL 3-METHYL-4-(METHYLTHIO)PHENYL PHOSPHORAMIDATE see FAK000

3-METHYL-3-ETHYLGLUTARIMIDE see MKA250

β-METHYL-β-ETHYLGLUTARIMIDE see MKA250

2,2'-((1-METHYLETHYLIDENE)BIS(4,1-PHENYLENEOXYMETHYLENE))BISOXIRANE see BLD750

9-(3'-METHYL-4'-ETHYLIDENE-THIOSEMICARBAZIDO)ACRIDINE see MKA300

METHYL ETHYL KETONE see MKA400

METHYL ETHYL KETONE HYDROPEROXIDE see MKA500

METHYL ETHYL KETONE PEROXIDE see MKA500

METHYL ETHYL KETONE PEROXIDE, in solution with >9% by weight active oxygen (DOT) see MKA500

METHYL ETHYL KETONE SEMICARBAZONE see MKA750

METHYLETHYLKETONHYDROPEROXIDE see MKA500

METHYL ETHYL KETOXIME see EMU500

METHYLETHYLMETHANE see BOR500

N-(1-METHYLETHYL)-4-((2-METHYLHYDRAZINO)METHYL)BENZAMIDE MONOHYDROCHLORIDE see PME500

METHYLETHYLNITROSAMINE see MKB000

N,N-METHYLETHYLNITROSAMINE see MKB000

METHYLETHYLOLAMINE see MGG000

2-(1-METHYLETHYL)PHENOL see IQX100

4-(1-METHYLETHYL)PHENOL see IQZ000

1-METHYL-5-ETHYL-5-PHENYLBARBITURIC ACID see ENB500

3-METHYL-5-ETHYL-5-PHENYLHYDANTOIN see MKB250

2-(1-METHYLETHYL)PHENYL METHYLCARBAMATE see MIA250

N-(1-METHYLETHYL)-2-PROPANAMINE see DNM200

N-(1-METHYLETHYL)-2-PROPENAMIDE see INH000

METHYL 3-((ETHYL(PROPYLAMINO)PHOSPHINOTHIOYL)OXY)-2-BUTENOATE see MKB320

2-METHYL-5-ETHYLPYRIDINE see EOS000

6-METHYL-3-ETHYLPYRIDINE see EOS000

METHYL ETHYL PYRIDINE (DOT) see EOS000

2-METHYL-5-ETHYLPYRIDINE (DOT) see EOS000

3-METHYLETHYNYLESTRADIOL see MKB750

3-METHYLETHYNYLOESTRADIOL see MKB750

METHYL EUGENOL (FCC) see AGE250

METHYL FLUORIDE see FJK000
16-α-METHYL-9-α-FLUORO-1-DEHYDROCORTISOL see SOW000
METHYLFLUOROPHOSPHORIC ACID, ISOPROPYL ESTER see IPX000
16-α-METHYL-9-α-FLUOROPREDNISOLONE see SOW000
16-α-METHYL-9-α-FLUORO-1,4-PREGNADIENE-11-β,17-α,21-TRIOL-3,20-DIONE see SOW000
16-α-METHYL-9-α-FLUORO-11-β,17-α,21-TRIHYDROXYPREGNA-1,4-DIENE-3,20-DIONE see SOW000
16-α-METHYL-9-α-FLUORO-Δ¹-HYDROCORTISONE see SOW000
METHYLFLUORPHOSPHORSAEUREISOPROPYLESTER (GERMAN) see IPX000
METHYLFLURETHER see EAT900
METHYLFORMAMIDE see MKG500
N-METHYLFORMAMIDE see MKG500
METHYL FORMATE see MKG750
METHYLFORMIAAT (DUTCH) see MKG750
METHYLFORMIAT (GERMAN) see MKG750
METHYL FORMYL see DOO600
N-METHYL-N-FORMYLHYDRAZINE see FNW000
N-METHYL-N-FORMYL HYDRAZONE of ACETALDEHYDE see AAH000
METHYL FOSFERNO see MNH000
METHYLFURAN see MKH000
2-METHYLFURAN see MKH000
METHYL 2-FURANCARBOXYLATE see MKH600
METHYL FUROATE see MKH600
METHYL 2-FUROATE see MKH600
METHYL GLYCOL see PML000
METHYL GLYCOL see EJH500
METHYL GLYCOL ACETATE see EJJ500
METHYL GLYCOL MONOACETATE see EJJ500
METHYL GLYKOL (GERMAN) see EJH500
METHYLGLYKOLACETAT (GERMAN) see EJJ500
METHYL GUTHION see ASH500
METHYL HEPTANETHIOL see MKJ250
3-METHYL-5-HEPTANONE see EGI750
5-METHYL-3-HEPTANONE see EGI750
5-METHYL-3-HEPTANONE (OSHA) see ODI000
METHYL HEPTENONE see MKK000
6-METHYL-5-HEPTEN-2-ONE see MKK000
METHYL HEPTINE CARBONATE see MND275
1-METHYLHEPTYLAMINE see OEK010
2-METHYL-2-HEPTYLAMINE see ILM000
6-METHYL-2-HEPTYLAMINE see ILM000
(6-(1-METHYL-HEPTYL)-2,4-DINITRO-FENYL)-CROTONAAT (DUTCH) see AQT500
(6-(1-METHYL-HEPTYL)-2,3-DINITRO-PHENYL)-CROTONAT (GERMAN) see AQT500
2-(1-METHYLHEPTYL)-4,6-DINITROPHENYL CROTONATE see AQT500
N-METHYLHEXAHYDRO-2-PICOLINIC ACID, 2,6-DIMETHYLANILIDE see SBB000
5-METHYL-2,3-HEXANEDIONE see MKL300
2-METHYL-5-HEXANONE see MKW450
5-METHYL-2-HEXANONE see MKW450
5-METHYLHEXAN-2-ONE (DOT) see MKW450
METHYLHEXYLACETALDEHYDE see MNC175
METHYL HEXYL KETONE (FCC) see ODG000
METHYL-N-(((((HEXYLOXY)SULFINYL)METHYLAMINO)CARBONYL)OXY)ETHANIMIDOTHIOATE see MKM800
METHYL HYDANTOIN see MKB250

METHYL HYDRAZINE see MKN000
1-METHYL HYDRAZINE see MKN000
METHYLHYDRAZINE (DOT) see MKN000
METHYLHYDRAZINE HYDROCHLORIDE see MKN250
METHYL HYDRIDE see MDQ750
Δ¹-6-α-METHYLHYDROCORTISONE see MOR500
METHYL HYDROXIDE see MGB150
1-METHYL-4-HYDROXYBENZENE see CNX250
METHYL-o-HYDROXYBENZOATE see MPI000
METHYL p-HYDROXYBENZOATE see HJL500
2-METHYL-3-HYDROXY-4,5-BIS(HYDROXYMETHYL)PYRIDINE see PPK250
2-METHYL-3-HYDROXY-4,5-BIS(HYDROXYMETHYL)PYRIDINE HYDROCHLORIDE see PPK500
4-METHYL-7-HYDROXYCOUMARIN see MKP500
2-METHYL-3-HYDROXY-4,5-DIHYDROXYMETHYL-PYRIDIN (GERMAN) see PPK250
2-METHYL-3-HYDROXY-4,5-DI(HYDROXYMETHYL)PYRIDINE see PPK250
METHYL(β-HYDROXYETHYL)AMINE see MGG000
2-METHYL-1-(2-HYDROXYETHYL)-5-NITROIMIDAZOLE see MMN250
2-METHYL-3-(2-HYDROXYETHYL)-4-NITROIMIDAZOLE see MMN250
2-METHYL-3-HYDROXY-4-FORMYL-5-HYDROXYMETHYLPYRIDINE HYDROCHLORIDE see VSU000
METHYL-3-β-HYDROXY-1-α-H,5-α-H-TROPANE-2-β-CARBOXYLATE BENZOATE (ESTER) see CNE750
1-METHYL-2-HYDROXYIMINOMETHYLPYRIDINIUM IODIDE see POS750
(+−)-(E)-METHYL-2-((E)-HYDROXYIMINO)-5-NITRO-6-METHOXY-3-HEXENEAMIDE see MKQ600
1-METHYL-3-HYDROXY-4-ISOPROPYLBENZENE see TFX810
17-α-METHYL-2-HYDROXYMETHYLENE-17-HYDROXY-5-α-ANDROSTAN-3-ONE see PAN100
METHYL (4-(m-HYDROXYPHENYL)-1-METHYL)-4-PIPERIDYL KETONE see HNK950
6-α-METHYL-17-α-HYDROXYPROGESTERONE ACETATE see MCA000
2-METHYL-3-HYDROXY-4-PYRONE see MAO350
6-METHYL-17-α-HYDROXY-Δ⁶-PROGESTERONE ACETATE see VTF000
N,N'-METHYLIDENEBISACRYLAMIDE see MJL500
1,1',1'-(METHYLIDYNETRIS(OXY))TRIS(ETHANE) see ENY500
1-METHYLIMIDAZOLE-2-THIOL see MCO500
2-METHYL-1,3-INDANDIONE see MKV500
α-METHYL-β-INDOLAETHYLAMINE (GERMAN) see AME500
3-METHYLINDOLE see MKV750
3-METHYL-1H-INDOLE see MKV750
β-METHYLINDOLE see MKV750
α-METHYL-β-INDOLEETHYLAMINE see AME500
METHYL IODIDE see MKW200
METHYLISOAMYL ACETATE see HFJ000
METHYL ISOAMYL KETONE see MKW450
METHYL ISOBUTENYL KETONE see MDJ750
METHYL ISOBUTYL CARBINOL see MKW600
METHYLISOBUTYL CARBINOL see MKW600
METHYL ISOBUTYL CARBINOL see AOK750
METHYLISOBUTYLCARBINOL ACETATE see HFJ000

METHYLISOBUTYLCARBINYL ACETATE see HFJ000
METHYL-ISOBUTYL-CETONE (FRENCH) see HFG500
METHYLISOBUTYLKETON (DUTCH, GERMAN) see HFG500
METHYL ISOBUTYL KETONE (ACGIH, DOT) see HFG500
METHYLISOCYANAAT (DUTCH) see MKX250
METHYL ISOCYANAT (GERMAN) see MKX250
METHYL ISOCYANATE see MKX250
METHYL ISOCYANATE, solutions (DOT) see MKX250
METHYL ISOEUGENOL (FCC) see IKR000
METHYLISOMIN see DBA800
METHYLISOMYN see MDT600
METHYLISOOCTENYLAMINE see ILK000
METHYL ISOPENTANOATE see ITC000
1-METHYL-4-ISOPROPENYLCYCLOHEXAN-3-OL see MCE750
1-METHYL-4-ISOPROPENYL-1-CYCLOHEXENE see MCC250
1-1-METHYL-4-ISOPROPENYL-6-CYCLOHEXEN-2-ONE see CCM120
d-1-METHYL-4-ISOPROPENYL-6-CYCLOHEXEN-2-ONE see CCM100
METHYL ISOPROPENYL KETONE see MKY500
METHYL ISOPROPENYL KETONE INHIBITED (DOT) see MKY500
p-METHYLISOPROPYL BENZENE see CQI000
1-METHYL-4-ISOPROPYLBENZENE see CQI000
1-METHYL-2-p-(ISOPROPYLCARBAMOYL)BENZOHYDRAZINE HYDROCHLORIDE see PME500
1-METHYL-2-(p-ISOPROPYLCARBAMOYLBENZYL)HYDRAZINE HYDROCHLORIDE see PME500
1-METHYL-4-ISOPROPYLCYCLOHEXADIENE-1,3 see MLA250
1-METHYL-4-ISOPROPYL-1,3-CYCLOHEXADIENE see MLA250
1-METHYL-4-ISOPROPYLCYCLOHEXADIENE-1,4 see MCB750
2-METHYL-5-ISOPROPYL-1,3-CYCLOHEXADIENE see MCC000
6-METHYL-3-ISOPROPYLCYCLOHEXANOL see DKV150
5-METHYL-2-ISOPROPYL-2-HEXENAL see MLA300
α-METHYL-p-ISOPROPYLHYDROCINNAMALDEHYDE see COU500
α-METHYL-p-ISOPROPYL HYDROCINNAMIC ALDEHYDE DIETHYL ACETAL see COU510
METHYL ISOPROPYL KETONE see MLA750
2-METHYL-5-ISOPROPYLPHENOL see CCM000
5-METHYL-2-ISOPROPYL-1-PHENOL see TFX810
2-METHYL-3-(p-ISOPROPYLPHENYL)PROPIONALDEHYDE see COU500
5-METHYL-2-ISOPROPYL-3-PYRAZOLYL DIMETHYLCARBAMATE see DSK200
METHYL ISOSYSTOX see DAP400
METHYLISOTHIOCYANAAT (DUTCH) see ISE000
METHYL-ISOTHIOCYANAT (GERMAN) see ISE000
METHYL ISOTHIOCYANATE (DOT) see ISE000
METHYL ISOVALERATE see ITC000
METHYLISOVALERATE (DOT) see ITC000
N'-(5-METHYL-3-ISOXAZOLE)SULFANILAMIDE see SNK000
N'-(5-METHYL-3-ISOXAZOLYL)SULFANILAMIDE see SNK000

N'-(5-METHYLISOXAZOL-3-YL)SULPHANILAMIDE see SNK000
N^1-(5-METHYL-3-ISOXAZOLYL)SULPHANILAMIDE see SNK000
METHYLJODID (GERMAN) see MKW200
METHYLJODIDE (DUTCH) see MKW200
METHYL KETONE see ABC750
METHYLKYANID see ABE500
2-METHYLLACTONITRILE see MLC750
METHYL LEADATE see LCW000
METHYL-LOMUSTINE see CHD250
N-METHYL-LOST see BIE250
1-METHYL-LUMILYSERGOL-8-(5-BROMONICOTINATE)-10-METHYL ETHER see NDM000
METHYLMAGNESIUM BROMIDE (ethyl ether solution) see MLE000
METHYL MAGNESIUM BROMIDE in ETHYL ETHER (DOT) see MLE000
METHYLMERCAPTAAN (DUTCH) see MLE650
METHYL MERCAPTAN see MLE650
2-METHYLMERCAPTO-4,6-BIS(ISOPROPYLAMINO)-s-TRIAZINE see BKL250
4-METHYLMERCAPTO-3,5-DIMETHYLPHENYL N-METHYLCARBAMATE see DST000
METHYL-MERCAPTOFOS TEOLOVY see DAP400
1-METHYL-2-MERCAPTOIMIDAZOLE see MCO500
METHYLMERCAPTOPHOS see DAO800
3-(METHYLMERCAPTO)PROPIONALDEHYDE see MPV400
3-(METHYLMERCAPTO)PROPIONALDEHYDE see TET900
β-(METHYLMERCAPTO)PROPIONALDEHYDE see MPV400
β-(METHYLMERCAPTO)PROPIONALDEHYDE see TET900
METHYLMERCAPTOPROPIONIC ALDEHYDE see MPV400
METHYLMERCAPTOPROPIONIC ALDEHYDE see TET900
6-METHYLMERCAPTOPURINE RIBONUCLEOSIDE see MPU000
6-METHYLMERCAPTOPURINE RIBOSIDE see MPU000
1-METHYL-5-MERCAPTO-1,2,3,4-TETRAZOLE see MPQ250
4-METHYLMERCAPTO-3,5-XYLYL METHYLCARBAMATE see DST000
METHYLMERCURIC CHLORIDE see MDD750
METHYLMERCURIC CYANOGUANIDINE see MLF250
METHYLMERCURIC DICYANDIAMIDE see MLF250
METHYLMERCURIC DICYANDIAMIDE see MLF250
METHYLMERCURICHLORENDIMIDE see MLF500
METHYLMERCURIC HYDROXIDE see MLG000
METHYLMERCURIC SULFATE see BKS810
N-(METHYLMERCURI)-1,4,5,6,7,7-HEXACHLOROBICYCLO(2.2.1)HEPT-5-ENE-2,3-DICARBOXIMIDE see MLF500
8-(METHYLMERCURIOXY)QUINOLINE see MLH000
N-METHYLMERCURI-1,2,3,6-TETRAHYDRO-3,6-ENDOMETHANO-3,4,5,6,7,7-HEXACHLOROPHTHALIMIDE see MLF500
N-METHYLMERCURI-1,2,3,6-TETRAHYDRO-3,6-METHANO-3,4,5,6,7,7-HEXACHLOROPHTHALIMIDE see MLF500
METHYLMERCURY see MLF550
METHYL-MERCURY(1+) (9CI) see MLF550

METHYLMERCURY(I) CATION see MLF550
METHYLMERCURY CHLORIDE see MDD750
METHYLMERCURY DICYANDIAMIDE see MLF250
METHYLMERCURY HYDROXIDE see MLG000
METHYLMERCURY β-HYDROXYQUINOLATE see MLH000
METHYLMERCURY 8-HYDROXYQUINOLINATE see MLH000
METHYLMERCURY ION see MLF550
METHYLMERCURY ION(1+) see MLF550
METHYLMERCURY OXINATE see MLH000
METHYLMERCURY OXYQUINOLINATE see MLH000
METHYLMERCURY QUINOLINOLATE see MLH000
METHYL-MERCURY TOLUENESULPHAMIDE see MLH100
METHYLMERKURIDIKYANDIAMID see MLF250
METHYL MESYLATE see MLH500
METHYLMETHACRYLAAT (DUTCH) see MLH750
METHYL-METHACRYLAT (GERMAN) see MLH750
METHYL METHACRYLATE see MLH750
METHYL METHACRYLATE HOMOPOLYMER see PKB500
METHYL METHACRYLATE MONOMER, INHIBITED (DOT) see MLH750
METHYL METHACRYLATE POLYMER see PKB500
METHYL METHACRYLATE RESIN see PKB500
N-METHYLMETHANAMINE see DOQ800
N-METHYLMETHANAMINE with BORANE (1:1) see DOR200
METHYLMETHANE see EDZ000
METHYL METHANESULFONATE see MLH500
METHYL METHANESULPHONATE see MLH500
METHYL METHANOATE see MKG750
METHYL METHANSULFONAT (GERMAN) see MLH500
METHYL METHANSULFONATE see MLH500
METHYL METHANSULPHONATE see MLH500
2-METHYL-4-METHOXYANILINE see MGO500
4-METHYL-1-METHOXYBENZENE see MGP000
METHYL (E)-2-METHOXYIMINO-(2-(o-TOLYLOXYMETHYL)PHENYL)ACETATE see MLI900
METHYL-α-METHYLACRYLATE see MLH750
1-METHYL-6-(1-METHYLALLYL)DITHIOBIUREA see MLJ500
1-METHYL-6-(1-METHYLALLYL)-2,5-DITHIOBIUREA see MLJ500
METHYL METHYLAMINOBENZOATE see MGQ250
((METHYL N-((METHYLAMINO)CARBONYL)OXY)ETHANIMIDO)THIOATE see MDU600
2-METHYL-6-METHYLAMINO-2-HEPTENE see ILK000
METHYL-N-METHYL ANTHRANILATE see MGQ250
METHYL-4-METHYLBENZENESULFONATE see MLL250
METHYL-p-METHYLBENZENESULFONATE see MLL250
METHYL 2-METHYLBUTANOATE see MLL600
METHYL-3-METHYLBUTANOATE see ITC000
METHYL 2-METHYLBUTYRATE see MLL600
METHYL-3-METHYLBUTYRATE see ITC000
METHYL-N-((METHYLCARBAMOYL)OXY)THIOACETIMIDATE see MDU600
S-METHYL N-[METHYLCARBAMOYLOXY]THIOACETIMIDATE see MDU600

cis-1-METHYL-2-METHYL CARBAMOYL VINYL PHOSPHATE see MRH209
d-2-METHYL-5-(1-METHYLENENYL)-CYCLOHEXANONE see DKV175
7-METHYL-3-METHYLENE-1,6-OCTADIENE see MRZ150
1-METHYL-4-(1-METHYLETHENYL)-(S)-CYCLOHEXENE see MCC500
(R)-1-METHYL-4-(1-METHYLETHENYL)-CYCLOHEXENE see LFU000
(S)-2-METHYL-5-(1-METHYLETHENYL)-2-CYCLOHEXEN-1-ONE see CCM100
(R)-2-METHYL-5-(1-METHYLETHENYL)-2-CYCLOHEXEN-1-ONE (9CI) see CCM120
2-METHYL-5-(1-METHYLETHENYL)CYCLOHEXYL ACETATE see DKV160
5-METHYL-2-(1-METHYLETHENYL)-4-HEXEN-1-OL ACETATE see LCA100
(1S-1-α,4-α,5-α)-4-METHYL-1-(1-METHYLETHYL)-BICYCLO(3.1.0)HEXAN-3-ONE see TFW000
5-METHYL-2-(1-METHYLETHYL)CYCLOHEXANOL see MCF750
(1R-(1-α,2-β,5-α))-5-METHYL-2-(1-METHYLETHYL)CYCLOHEXANOL see MCG250
5-METHYL-2-(1-METHYLETHYL)-CYCLOHEXANOL (1-α,2-β,5-α) see MCG000
trans-5-METHYL-2-(1-METHYLETHYL)-CYCLOHEXANONE see MCG275
1-METHYL-4-(1-METHYLETHYL)-2,3-DIOXABICYCLO(2.2.2)OCT-5-ENE see ARM500
1-METHYL-4-(1-METHYLETHYLIDENE)CYCLOHEXENE see TBE000
1-METHYL-2-(1-METHYLETHYL)-5-NITRO-1H-IMIDAZOLE see IGH000
5-METHYL-2-(1-METHYLETHYL)PHENOL see TFX810
METHYL((METHYL(((5-METHYL-1,3-OXATHIOLAN-4-YLIDENE)AMINO)OXY)CARBONYL)AMINO)THIO)CARBAMIC ACID, ETHYL ESTER see MLO900
d-3-METHYL-N-METHYLMORPHINAN PHOSPHATE see MLP250
2-METHYL-4-((2-METHYLPHENYL)AZO)BENZENAMINE see AIC250
1-((2-METHYL-4-((2-METHYLPHENYL)AZO)PHENYL)AZO)-2-NAPHTHALENOL see SBC500
METHYL (3-METHYLPHENYL)CARBAMOTHIOIC ACID, o-2-NAPHTHALENYL ESTER see TGB475
2-METHYL-11-(4-METHYL-1-PIPERAZINYL)-DIBENZO(b,f)(1,4)THIAZEPINE see MQQ000
METHYL 10-(3-(4-METHYL-1-PIPERAZINYL)PROPYL)PHENOTHIAZIN-2-YL KETONE see ACR500
METHYL-2-METHYL-2-PROPENOATE see MLH750
α-METHYL-4-(2-METHYLPROPYL)BENZENEACETIC ACID see IIU000
2-METHYL-N-(2-METHYLPROPYL)-1-PROPANAMINE see DNH400
METHYL 5-METHYL-1-(2-QUINOLYL)-4-PYRAZOLYL KETONE see MLW600
METHYL 5-METHYL-1-(2-QUINOXALINYL)-4-PYRAZOLYL KETONE see MLW630
2-METHYL-2-(METHYLTHIO)PROPANAL-O-((METHYLAMINO)CARBONYL)OXIME see CBM500
2-METHYL-2-(METHYLTHIO)PROPANOL-o-((N-METHYL-N-MORPHOLINOSULFENYL)CARBAMOYL)OXIME see MLX820

2-METHYL-2-(METHYLTHIO)PROPANOL-o-((METHYL(4-MORPHOLINYLTHIO)AMINO)CARBONYL)OXIME see MLX820

2-METHYL-2-(METHYLTHIO)PROPIONALDEHYDE-O-(METHYLCARBAMOYL)OXIME see CBM500

2-METHYL-2-(METHYLTHIO)PROPIONALDEHYDE-o-(N-METHYL-N-(4-MORPHOLINOSULFENYL)CARBAMOYL)OXIME see MLX820

2-METHYL-2-(METHYLTHIO)PROPIONALDEHYDE OXIME see CBM500

2-METHYL-2-METHYLTHIO-PROPIONALDEHYD-O-(N-METHYL-CARBAMOYL)-OXIM (GERMAN) see CBM500

METHYL MONOBROMOACETATE see MHR250

METHYL MONOCHLORACETATE see MIF775

METHYL MONOCHLOROACETATE see MIF775

4-METHYLMORPHOLINE see MMA250

N-METHYL MORPHOLINE see MMA250

METHYLMORPHOLINE (DOT) see MMA250

METHYL 10-(3-MORPHOLINOPROPYL)PHENOTHIAZIN-2-YL KETONE see MMA600

6-METHYL-MP-RIBOSIDE see MPU000

METHYL MUSTARD OIL see ISE000

N-METHYL-1-NAFTYL-CARBAMAAT (DUTCH) see CBM750

METHYL NAMATE see SGM500

2-METHYL-1,4-NAPHTHALENDIONE see MMD500

2-METHYL-1,4-NAPHTHALENEDIONE see MMD500

2-METHYL-1,4-NAPHTHOCHINON (GERMAN) see MMD500

2-METHYL-1,4-NAPHTHOQUINONE see MMD500

3-METHYL-1,4-NAPHTHOQUINONE see MMD500

3-METHYL-2-NAPHTHYLAMINE see MME500

N-METHYL-1-NAPHTHYL-CARBAMAT (GERMAN) see CBM750

N-METHYL-1-NAPHTHYL CARBAMATE see CBM750

N-METHYL-α-NAPHTHYLCARBAMATE see CBM750

N-METHYL-N-(1-NAPHTHYL)FLUOROACETAMIDE see MME809

METHYL 1-NAPHTHYL KETONE see ABC475

METHYL-2-NAPHTHYL KETONE see ABC500

METHYL α-NAPHTHYL KETONE see ABC475

α-METHYL NAPHTHYL KETONE see ABC475

β-METHYL NAPHTHYL KETONE see ABC500

METHYL-β-NAPHTHYL KETONE (FCC) see ABC500

N-METHYL-N-(1-NAPHTHYL)MONOFLUOROACETAMIDE see MME809

N-METHYL-α-NAPHTHYLURETHAN see CBM750

METHYL NIRAN see MNH000

METHYLNITRAMINE see NHN500

METHYL NITRAMINE (dry) (DOT) see NHN500

METHYL NITRATE see MMF500

METHYL NITRITE see MMF750

2-METHYL-5-NITROANILINE see NMP500

6-METHYL-3-NITROANILINE see NMP500

2-METHYL-1-NITRO-9,10-ANTHRACENEDIONE see MMG000

2-METHYL-1-NITROANTHRAQUINONE see MMG000

2-METHYLNITROBENZENE see NMO525

3-METHYLNITROBENZENE see NMO500

4-METHYLNITROBENZENE see NMO550

m-METHYLNITROBENZENE see NMO500

o-METHYLNITROBENZENE see NMO525

p-METHYL NITROBENZENE see NMO550

2-METHYL-5-NITRO-BENZENEAMINE see NMP500

2-METHYL-5-NITROIMIDAZOLE-1-ETHANOL see MMN250

METHYLNITROIMIDAZOLYLMERCAPTOPURINE see ASB250

6-(1'-METHYL-4'-NITRO-5'-IMIDAZOLYL)-MERCAPTOPURINE see ASB250

6-(METHYL-p-NITRO-5-IMIDAZOLYL)-THIOPURINE see ASB250

6-((1-METHYL-4-NITROIMIDAZOL-5-YL)THIO)PURINE see ASB250

6-(1-METHYL-4-NITROIMIDAZOL-5-YLTHIO)PURINE see ASB250

6-(1-METHYL-p-NITRO-5-IMIDAZOLYL)-THIOPURINE see ASB250

6-((1-METHYL-4-NITRO-1H-IMIDAZOL-5-YL)THIO)-1H-PURINE see ASB250

METHYLNITRONITROSOGUANIDINE see MMP000

1-METHYL-3-NITRO-1-NITROSOGUANIDINE see MMP000

N-METHYL-N'-NITRO-N-NITROSOGUANIDINE see MMP000

N-METHYL-N'-NITRO-N-NITROSOGUANIDINE see MMP000

5-METHYL-2-NITRO-7-OXA-8-MERCURABICYCLO(4.2.0)OCTA-1,3,5-TRIENE see NHK900

METHYLNITROPHOS see DSQ000

N-(2-METHYL-2-NITROPROPYL)-p-NITROSOANILINE see NHK800

N-(2-METHYL-2-NITROPROPYL)-4-NITROSOBENZAMINE see NHK800

3-METHYLNITROSAMINOPROPIONITRILE see MMS200

4-(N-METHYL-N-NITROSAMINO)-1-(3-PYRIDYL)-1-BUTANONE see MMS500

METHYLNITROSOACETAMID (GERMAN) see MMT000

METHYLNITROSOACETAMIDE see MMT000

N-METHYL-N-NITROSOACETAMIDE see MMT000

1-METHYL-1-NITROSOACETYLUREA see ACR400

N-METHYL-N-NITROSO-N'-ACETYLUREA see ACR400

N-METHYL-N-NITROSOALLYLAMINE see MMT500

2-(N-METHYL-N-NITROSO)AMINOACETONITRILE see MMT750

α-(1-(N-METHYL-N-NITROSOAMINO)ETHYL)BENZYL ALCOHOL see NKC000

2-(N-METHYL-N-NITROSOAMINO)-1-PHENYL-1-PROPANOL see NKC000

1-(METHYLNITROSOAMINO)2-PROPANONE see NKV000

N-METHYL-N-NITROSO-4-AMINOPYRIDINE see NKQ100

4-(N-METHYL-N-NITROSOAMINO)-4-(3-PYRIDYL.)-1-BUTANONE see MMS500

N-METHYL-N-NITROSOANILINE see MMU250

N-METHYL-N-NITROSOBENZENAMINE see MMU250

N-METHYL-N-NITROSOBENZYLAMINE see MHP250

N-METHYL-N-NITROSOBUTYLAMINE see MHW500

N-METHYL-N-NITROSOCARBAMIC ACID, ETHYL ESTER see MMX250

METHYL-NITROSOCARBAMIC ACID-1-NAPHTHYL ESTER see NBJ500

4-METHYL-3-PENTEN-2-ON (DUTCH, GERMAN) see MDJ750

2-METHYL-2-PENTEN-4-ONE see MDJ750

4-METHYL-3-PENTEN-2-ONE see MDJ750

4-METHYL-2-PENTYL ACETATE see HFJ000

METHYL N-(β-PENTYLCINNAMYLIDENE)ANTHRANILATE see AOH100

METHYL PENTYL KETONE see MGN500

METHYL-N-PENTYLNITROSAMINE see AOL000

METHYLPERONE HYDROCHLORIDE see FKI000

(±)-α-METHYLPHENETHYLAMINE see BBK000

dl-α-METHYLPHENETHYLAMINE see BBK000

α-METHYLPHENETHYLAMINE, d-FORM see AOA500

d-α-METHYLPHENETHYLAMINE SULFATE see BBK500

1-(α-METHYLPHENETHYL)-4-PHENYLPIPERAZINE DIHYDROCHLORIDE see MNP450

3-(α-METHYLPHENETHYL)SYDONE IMINE MONOHYDROCHLORIDE see SPA000

METHYLPHENIDAN see MNQ000

METHYL PHENIDATE see MNQ000

METHYL PHENIDYL ACETATE see MNQ000

METHYLPHENOBARBITAL see ENB500

1-METHYLPHENOBARBITAL see ENB500

N-METHYLPHENOBARBITAL see ENB500

N-METHYLPHENOBARBITOL see ENB500

METHYLPHENOBARBITONE see ENB500

2-METHYLPHENOL see CNX000

3-METHYLPHENOL see CNW750

4-METHYLPHENOL see CNX250

m-METHYLPHENOL see CNW750

o-METHYLPHENOL see CNX000

p-METHYLPHENOL see CNX250

4-METHYLPHENOL METHYL ETHER see MGP000

α-METHYL PHENYLACETALDEHYDE see COF000

4-METHYLPHENYL ACETATE see MNR250

p-METHYLPHENYL ACETATE see MNR250

METHYL PHENYLACETATE (FCC) see MHA500

METHYL(2-PHENYLAETHYL)NITROSAMIN (GERMAN) see MNU250

METHYLPHENYLAMINE see MGN750

N-METHYLPHENYLAMINE see MGN750

METHYLPHENYLARSINIC ACID see HMK200

METHYLPHENYLARSONIC ACID see HMK200

1-(2-METHYLPHENYL)AZO-2-NAPHTHALENAMINE see FAG135

1-((2-METHYLPHENYL)AZO)-2-NAPHTHALENAMINE see FAG135

1-((2-METHYLPHENYL)AZO)-2-NAPHTHALENOL see TGW000

1-(2-METHYLPHENYL)AZO-2-NAPHTHYLAMINE see FAG135

METHYLPHENYLBARBITURIC ACID see ENB500

4-METHYLPHENYL BENZOATE see TGX100

N-METHYL-4-PHENYL-4-CARBETHOXYPIPERIDINE HYDROCHLORIDE see DAM700

METHYLPHENYLCARBINOL see PDE000

METHYL PHENYLCARBINYL ACETATE see MNT075

1-METHYL-4-PHENYL-4-CARBOETHOXYPIPERIDINE HYDROCHLORIDE see DAM700

1-METHYL-5-PHENYL-7-CHLORO-1,3-DIHYDRO-2H-1,4-BENZODIAZEPIN-2-ONE see DCK759

METHYLPHENYLDICHLOROSILANE (DOT) see DFQ800

1-(o-METHYLPHENYL)-3,3-DIMETHYL-TRIAZEN (GERMAN) see MNT500

1-(2-METHYLPHENYL)-3,3-DIMETHYLTRIAZENE see MNT500

1-(3-METHYLPHENYL)-3,3-DIMETHYLTRIAZENE see DSR200

1-(m-METHYLPHENYL)-3,3-DIMETHYLTRIAZENE see DSR200

1-(o-METHYLPHENYL)-3,3-DIMETHYL-TRIAZENE see MNT500

METHYLPHENYLENEDIAMINE see TGL500

2-METHYL-p-PHENYLENEDIAMINE see TGM000

4-METHYL-m-PHENYLENEDIAMINE see TGL750

2-METHYL-p-PHENYLENEDIAMINE SULPHATE see DCE600

4-METHYL-PHENYLENE DIISOCYANATE see TGM750

METHYL-m-PHENYLENE DIISOCYANATE see TGM740

2-METHYL-m-PHENYLENE ESTER, ISOCYANIC ACID see TGM800

METHYLPHENYLENE ISOCYANATE see TGM740

4-METHYL-PHENYLENE ISOCYANATE see TGM750

2-METHYL-m-PHENYLENE ISOCYANATE see TGM800

METHYL PHENYL ETHER see AOX750

α-METHYL-β-PHENYLETHYL ACETATE see ABU800

METHYL(2-PHENYLETHYL)ARSINIC ACID see MNU050

N-METHYL-5-PHENYL-5-ETHYLBARBITAL see ENB500

1-METHYL-5-PHENYL-5-ETHYLBARBITURIC ACID see ENB500

as-METHYLPHENYLETHYLENE see MPK250

3-METHYL-5,5-PHENYLETHYLHYDANTOIN see MKB250

METHYL-PHENYLETHYL-NITROSAMINE see MNU250

3-METHYL-3-PHENYLGLYCIDIC ACID ETHYL ESTER see ENC000

4-METHYLPHENYL 2-HYDROXYBENZOATE see THD850

2-METHYLPHENYL ISOCYANATE see IKG725

o-METHYLPHENYL ISOCYANATE see IKG725

1-METHYL-4-PHENYLISONIPECOTIC ACID ETHYL ESTER HYDROCHLORIDE see DAM700

l-N-METHYL-β-PHENYLISOPROPYLAMINE HYDROCHLORIDE see MDQ500

N-METHYL-β-PHENYLISOPROPYLAMINHYDROCHLORID (GERMAN) see DBA800

METHYL PHENYL KETONE see ABH000

4-METHYLPHENYLMERCAPTAN see TGP250

p-METHYLPHENYLMERCAPTAN see TGP250

m-METHYLPHENYL METHYLCARBAMATE see MIB750

3-METHYLPHENYL N-METHYLCARBAMATE see MIB750

3-METHYL-2-PHENYLMORPHOLINE HYDROCHLORIDE see MNV750

METHYLPHENYLNITROSAMINE see MMU250

METHYLPHENYLNITROSOUREA see MMY500

N-METHYL-N'-PHENYL-N-NITROSOUREA see MMY500

3-(4-METHYLPHENYL)-1-PHENYL-2-PROPEN-1-ONE see MIF762

METHYL α-PHENYL-α-(2-PIPERIDYL)ACETATE see
MNQ000
2-METHYL-1-PHENYLPROPANE see IIN000
2-METHYL-2-PHENYLPROPANE see BQJ250
2-METHYL-3-PHENYL-2-PROPENAL see MIO000
METHYL-3-PHENYLPROPENOATE see MIO500
3-(4-METHYLPHENYL)-N-(4-PROPYLCYCLOHEXYL)-
2-PROPENAMIDE see MNW790
N-(3-METHYLPHENYL)-1H-PYRAZOLE-1-
ACETAMIDE see MNX310
α-(5-METHYL-1-PHENYL-1H-PYRAZOL-4-YL)-1-
PIPERIDINEBUTANOL see MNX420
4-METHYLPHENYL SALICYLATE see THD850
3-METHYL-1-PHENYL-3-(2-SULFOETHYL)TRIAZENE
SODIUM SALT see PEJ250
3-METHYL-2-PHENYLTETRAHYDRO-2H-1,4-
OXAZINE HYDROCHLORIDE see MNV750
3-METHYL-2-PHENYLVALERIC ACID-2-
DIETHYLAMINOETHYL ESTER METHYL BROMIDE
see VBK000
3-METHYL-2-PHENYLVALERIC ACID DIETHYL(2-
HYDROXYETHYL)METHYLAMMONIUM BROMIDE
ESTER see VBK000
2-((3-METHYL-2-PHENYLVALERYL)OXY)-N,N-
DIETHYL-N-METHYLETHANAMINIUM BROMIDE
see VBK000
METHYL PHOSPHATE see TMD250
METHYL PHOSPHITE see TMD500
METHYL PHOSPHONATE see DSG600
METHYLPHOSPHONIC ACID DIMETHYL ESTER see
DSR400
METHYL PHOSPHONIC DICHLORIDE see MOB399
METHYLPHOSPHONOFLUORIDIC ACID
ISOPROPYL ESTER see IPX000
METHYLPHOSPHONOFLUORIDIC ACID-1-
METHYLETHYL ESTER see IPX000
METHYLPHOSPHONOTHIOIC ACID-S-(2-
(BIS(METHYLETHYL)AMINO)ETHYL)o-ETHYL
ESTER see EIG000
METHYL PHOSPHONOTHIOIC DICHLORIDE,
pyrophoric liquid (DOT) see MOC000
METHYLPHOSPHONOUS DICHLORIDE see MOC250
METHYL PHTHALATE see DTR200
N-METHYL-2-PIPECOLIC ACID, 2,6-
DIMETHYLANILIDE see SBB000
N-METHYL-2-PIPECOLIC ACID, 2,6-XYLIDIDE see
SBB000
(±)-1-METHYL-2',6'-PIPECOLOXYLIDIDE see SBB000
1-METHYL-2',6'-PIPECOLOXYLIDIDE
HYDROCHLORIDE see CBR250
dl-1-METHYL-2',6'-PIPECOLOXYLIDIDE
HYDROCHLORIDE see CBR250
1-METHYLPIPERAZINE see MOD250
N-METHYLPIPERAZINE see MOD250
3-(4-METHYLPIPERAZINYLIMINOMETHYL)-
RIFAMYCIN SV see RKP000
8-(4-METHYLPIPERAZINYLIMINOMETHYL)
RIFAMYCIN SV see RKP000
8-(((4-METHYL-1-
PIPERAZINYL)IMINO)METHYL)RIFAMYCIN SV see
RKP000
N-(γ-(4'-METHYLPIPERAZINYL-1')PROPYL)-3-
CHLOROPHENOTHIAZINE see PMF500
10-(3-(4-METHYL-1-PIPERAZINYL)PROPYL)-2-
TRIFLUOROMETHYLPHENOTHIAZINE
DIHYDROCHLORIDE see TKK250
N-METHYLPIPERIDINE see MOG500

1-METHYLPIPERIDINE (DOT) see MOG500
(1-METHYL-dl-PIPERIDINE-2-CARBOXYLIC ACID)-
2,6-DIMETHYLANILIDE HYDROCHLORIDE see
CBR250
γ-(4-METHYLPIPERIDINE)-p-
FLUOROBUTYROPHENONE HYDROCHLORIDE see
FKI000
2-METHYLPIPERIDINE β-NAPHTHOAMIDE see
MOH290
2-METHYL-1-PIPERIDINOPROPANOL, BENZOATE
see PIV750
(2-METHYLPIPERIDINO)PROPYL BENZOATE see
PIV750
METHYL 10-(3-
PIPERIDINOPROPYL)PHENOTHIAZIN-2-YL
KETONE see MOL300
γ-(2-METHYLPIPERIDYL)PROPYL BENZOATE see
PIV750
METHYL PIRIMIPHOS see DIN800
METHYLPREDNISOLONE see MOR500
6-α-METHYLPREDNISOLONE see MOR500
6-α-METHYL-4-PREGNENE-3,20-DION-17-α-OL
ACETATE see MCA000
METHYLPROPAMINE see DBA800
2-METHYLPROPANAL see IJS000
2-METHYL-1-PROPANAL see IJS000
2-METHYLPROPANE see MOR750
2-METHYLPROPANENITRILE see IJX000
α-METHYLPROPANENITRILE see IJX000
2-METHYL-2-PROPANETHIOL see MOS000
METHYL PROPANOATE see MOT000
2-METHYL PROPANOL see IIL000
2-METHYLPROPAN-1-OL see IIL000
2-METHYL-1-PROPANOL see IIL000
2-METHYL-2-PROPANOL see BPX000
2-METHYLPROPENAL (CZECH) see MGA250
2-METHYLPROPENAMIDE see MDN500
METHYL PROPENATE see MGA500
2-METHYLPROPENE see IIC000
2-METHYLPROPENENITRILE see MGA750
METHYL PROPENOATE see MGA500
METHYL-2-PROPENOATE see MGA500
2-METHYLPROPENOIC ACID see MDN250
2-METHYL-2-PROPENOIC ACID, ETHYL ESTER see
EMF000
2-METHYL-2-PROPENOIC ACID METHYL ESTER see
MLH750
2-METHYL-2-PROPENOIC ACID METHYL ESTER
HOMOPOLYMER see PKB500
2-METHYL-2-PROPENOIC ACID METHYL ESTER,
POLYMER with TRIBUTYL((2-METHYL-1-OXO-2-
PROPENYL)OXY)STANNANE see OIY000
2-METHYL-2-PROPENOIC ACID-2-METHYLPROPYL
ESTER see IIY000
2-METHYL-2-PROPEN-1-OL see IMW000
2-METHYLPROPIONALDEHYDE see IJS000
METHYL PROPIONATE see MOT000
2-METHYLPROPIONIC ACID see IJU000
α-METHYLPROPIONIC ACID see IJU000
2-METHYLPROPIONIC ACID, ETHYL ESTER see
ELS000
2-METHYLPROPIONITRILE see IJX000
3-(2-METHYLPROPIONYL)-5-
INDOLECARBONITRILE see COP525
2-METHYLPROPYL ACETATE see IIJ000
2-METHYL-1-PROPYL ACETATE see IIJ000
2-METHYLPROPYL ALCOHOL see IIL000

1-METHYLPROPYLAMINE see BPY000
METHYLPROPYLARSINIC ACID see MOT800
METHYL PROPYLATE see MOT000
2-METHYLPROPYL BUTYRATE see BSW500
METHYL PROPYL CARBINOL see PBM750
METHYL-PROPYL-CETONE (FRENCH) see PBN250
4-METHYL-N-(4-
PROPYLCYCLOHEXYL)BENZAMIDE see MOU820
METHYL PROPYL DIKETONE see HEQ200
β-METHYLPROPYL ETHANOATE see IIJ000
2-METHYL-2-PROPYLETHANOL see AOK750
METHYL PROPYL ETHER see MOU830
METHYL n-PROPYL ETHER see MOU830
N,N''-(2-METHYLPROPYLIDENE)BISUREA (9CI) see
IIV000
2-METHYLPROPYL ISOBUTYRATE see IIW000
METHYL-n-PROPYL KETONE see PBN250
METHYL PROPYL KETONE (ACGIH, DOT) see
PBN250
2-METHYLPROPYL METHACRYLATE see IIY000
METHYL-N-PROPYLNITROSAMINE see MNA000
METHYLPROPYLNITROSOAMINE see MNA000
2-(1-METHYLPROPYL)PHENYL
METHYLCARBAMATE see MOV000
2-METHYL-2-N-PROPYL-1,3-PROPANEDIOL
DICARBAMATE see MQU750
2-METHYLPROPYLPROPANOIC ACID-2-
METHYLPROPYL ESTER (9CI) see IIW000
2-METHYLPROPYL PROPIONATE see PMV250
2-METHYL-2-PROPYLTRIMETHYLENE CARBAMATE
see MQU750
METHYLPROTOCATECHUALDEHYDE see VFK000
2-METHYLPYRAZINE see MOW750
2-METHYLPYRIDINE see MOY000
3-METHYLPYRIDINE see PIB920
4-METHYLPYRIDINE see MOY250
α-METHYLPYRIDINE see MOY000
N-METHYLPYRIDINE-2-ALDOXIME IODIDE see
POS750
N-METHYLPYRIDINIUM-2-ALDOXIME IODIDE see
POS750
METHYL PYRIDYL KETONE see ABI000
METHYL-3-PYRIDYL KETONE see ABI000
METHYL-β-PYRIDYL KETONE see ABI000
1-METHYL-2-(3-PYRIDYL)PYRROLIDINE see NDN000
l-1-METHYL-2-(3-PYRIDYL)-PYRROLIDINE SULFATE
see NDR500
METHYLPYRIMAL see ALF250
6-METHYL-2,4(1H,3H)-PYRIMIDINEDIONE see
MQI750
N¹-(4-METHYL-2-PYRIMIDINYL)SULFANILAMIDE
see ALF250
N¹-(4-METHYL-2-PYRIMIDINYL)SULFANILAMIDE
SODIUM SALT see SJW475
3-METHYLPYROCATECHOL see DNE000
2-METHYL PYROMECONIC ACID see MAO350
METHYL PYROMUCATE see MKH600
1-METHYLPYRROLIDINE see MPB250
N-METHYLPYRROLIDINONE see MPF200
1-METHYL-2-PYRROLIDINONE see MPF200
1-METHYL-5-PYRROLIDINONE see MPF200
N-METHYL-2-PYRROLIDINONE see MPF200
3-(N-METHYLPYRROLIDINO)PYRIDINE see NDN000
3-(1-METHYL-2-PYRROLIDINYL)PYRIDINE see
NDN000
(S)-3-(1-METHYL-2-PYRROLIDINYL)PYRIDINE (9CI)
see NDN000

(S)-3-(1-METHYL-2-PYRROLIDINYL-PYRIDINE (R)-
(R,R))-2,3-DIHYDROXYBUTANEDIOATE (1:2) see
NDS500
(S)-3-(1-METHYL-2-PYRROLIDINYL)PYRIDINE
SULFATE (2:1) see NDR500
METHYLPYRROLIDONE see MPF200
N-METHYLPYRROLIDONE see MPF200
1-METHYL-2-PYRROLIDONE see MPF200
N-METHYL-2-PYRROLIDONE see MPF200
(−)-3-(1-METHYL-2-PYRROLIDYL)PYRIDINE see
NDN000
l-3-(1-METHYL-2-PYRROLIDYL)PYRIDINE see
NDN000
l-3-(1-METHYL-2-PYRROLIDYL)PYRIDINE SULFATE
see NDR500
3-((1-METHYLPYRROL-2-YL)METHYLENEAMINO)-4-
(PIPERIDINOMETHYL)-2-OXAZOLIDONE see
MPF300
METHYL-QUECKSILBER-TOLUOLSULFAMID see
MLH100
METHYLQUINAZOLONE HYDROCHLORIDE see
MDT250
2-METHYL-1,4-QUINONE see MHI250
METHYL RED see CCE500
METHYL RESERPATE 3,4,5-TRIMETHOXYBENZOIC
ACID see RDK000
METHYL RESERPATE 3,4,5-TRIMETHOXYBENZOIC
ACID ESTER see RDK000
5-METHYLRESORCINOL see MPH500
5-METHYLRESORCINOL ORCINOL see MPH500
METHYLRHODANID (GERMAN) see MPT000
6-METHYL-9-RIBOFURANOSYLPURINE-6-THIOL see
MPU000
METHYLROSANILINCHLORID see AOR500
METHYLROSANILINE CHLORIDE see AOR500
METHYLROSANILINUM CHLORATUM see AOR500
METHYL SALICYLATE see MPI000
METHYL SELENAC see SBQ000
METHYLSENFOEL (GERMAN) see ISE000
METHYLSERGIDE BIMALEATE see MQP500
METHYL SILICATE see MPI750
METHYL STEARATE see MJW000
α-METHYLSTYREEN (DUTCH) see MPK250
METHYLSTYRENE see VQK650
α-METHYL STYRENE see MPK250
α-METHYL-STYROL (GERMAN) see MPK250
5-METHYL-3-SULFANILAMIDOISOXAZOLE see
SNK000
METHYL SULFATE (DOT) see DUD100
METHYLSULFAZIN see ALF250
METHYL SULFIDE (DOT) see TFP000
METHYLSULFINYLMETHANE see DUD800
METHYL SULFOCYANATE see MPT000
METHYLSULFONIC ACID, ETHYL ESTER see EMF500
METHYL SULFOXIDE see DUD800
5-METHYL-3-SULPHANIL-AMIDOISOXAZOLE see
SNK000
METHYL SULPHIDE see TFP000
METHYLSYSTOX see DAO800
o-(METHYLTELLURO)BENZOIC ACID see MPN275
17-METHYLTESTOSTERON see MPN500
METHYLTESTOSTERONE see MPN500
17-METHYLTESTOSTERONE see MPN500
17-α-METHYLTESTOSTERONE see MPN500
METHYL TETRAHYDRO-5-HYDROXY-4-METHYL-3-
FURYL KETONE see BMK290

METHYL 1,4,5,6-TETRAHYDRO-2-
METHYLCYCLOPENTA(b)PYRROL-3-YL KETONE see
MPO800
METHYL-1,2,5,6-TETRAHYDRO-1-
METHYLNICOTINATE see AQT750
2-METHYL-2-(4-(1,2,3,4-TETRAHYDRO-1-
NAPHTHALENYL)PHENOXY)PROPANOIC ACID see
MCB500
2-METHYL-2-(4-(1,2,3,4-TETRAHYDRO-1-
NAPHTHYL)PHENOXY)PROPANOIC ACID see
MCB500
2-METHYL-2-(p-(1,2,3,4-TETRAHYDRO-1-
NAPHTHYL)PHENOXY)PROPIONIC ACID see
MCB500
α-METHYL-α-(p-1,2,3,4-TETRAHYDRONAPHTH-1-
YLPHENOXY)PROPIONIC ACID see MCB500
N-METHYL-Δ-TETRAHYDRONICOTINIC ACID
METHYL ESTER see AQT750
N-METHYLTETRAHYDROPYRIDINE-β-
CARBOXYLIC ACID METHYL ESTER see AQT750
N-METHYLTETRAHYDROPYRROLE see MPB250
N-METHYL-N,2,4,6-TETRANITROANILINE see
TEG250
N-METHYL-N,2,4,6-TETRANITROBENZENAMINE see
TEG250
1-METHYL-1H-TETRAZOLE-5-THIOL see MPQ250
METHYLTHEOBROMIDE see CAK500
1-METHYLTHEOBROMINE see CAK500
7-METHYLTHEOPHYLLINE see CAK500
METHYL 3-THIANAPHTHENYL KETONE see
MPQ900
METHYLTHIENYLCETONE see MPR300
2-METHYLTHIO-ACETALDEHYD-O-
(METHYLCARBAMOYL)-OXIM (GERMAN) see
MDU600
l-γ-METHYLTHIO-α-AMINOBUTYRIC ACID see
MDT750
6-METHYL-2-THIO-2H-1,3-BENZOXAZINE-2,4(3H)-
DIONE see MPS600
2-METHYLTHIO-4,6-BIS(ISOPROPYLAMINO)-s-
TRIAZINE see BKL250
METHYL THIOCYANATE see MPT000
4-METHYLTHIO-3,5-DIMETHYLPHENYL
METHYLCARBAMATE see DST000
2-(METHYLTHIO)-ETHANETHIOL-O,O-DIMETHYL
PHOSPHOROTHIOATE see MIW250
2-(METHYLTHIO)-ETHANETHIOL-S-ESTER with
O,O-DIMETHYL PHOSPHOROTHIOATE see MIW250
METHYLTHIOINOSINE see MPU000
6-METHYLTHIOINOSINE see MPU000
METHYLTHIOKYANAT see MPT000
METHYLTHIOMETHANE see TFP000
3-METHYL-2,5-THIOMORPHOLINEDIONE 2-(o-
((METHYLAMINO)CARBONYL)OXIME) see MPU600
METHYLTHIONINE CHLORIDE see BJI250
METHYLTHIONIUM CHLORIDE see BJI250
2-METHYLTHIOPHENOL see TGP000
4-METHYLTHIOPHENOL see TGP250
o-METHYLTHIOPHENOL see TGP000
p-METHYLTHIOPHENOL see TGP250
4-METHYLTHIOPHENYLDIMETHYL PHOSPHATE
see PHD250
METHYLTHIOPHOS see MNH000
3-(METHYLTHIO)PROPANAL see MPV400
3-(METHYLTHIO)PROPANAL see TET900
3-(METHYLTHIO)PROPIONALDEHYDE see MPV400
3-(METHYLTHIO)PROPIONALDEHYDE see TET900

β-(METHYLTHIO)PROPIONALDEHYDE see MPV400
β-(METHYLTHIO)PROPIONALDEHYDE see TET900
2-METHYLTHIO-PROPIONALDEHYD-O-
(METHYLCARBAMOYL)-OXIM (GERMAN) see
MDU600
METHYLTHIOPROPIONIC CYANOHYDRIN see
HMR550
6-(METHYLTHIO)PURINE RIBONUCLEOSIDE see
MPU000
6-METHYLTHIOPURINE RIBOSIDE see MPU000
6-METHYL-2-THIO-2,4-(1H3H)PYRIMIDINEDIONE
see MPW500
METHYLTHIOURACIL see MPW500
6-METHYLTHIOURACIL see MPW500
4-METHYL-2-THIOURACIL see MPW500
6-METHYL-2-THIOURACIL see MPW500
4-(METHYLTHIO)-3,5-XYLENOL
METHYLCARBAMATE see DST000
4-(METHYLTHIO)-3,5-XYLYL METHYLCARBAMATE
see DST000
METHYL THIRAM see TFS350
METHYL THIURAMDISULFIDE see TFS350
METHYLTIN TRICHLORIDE see MQC750
METHYL-α-TOLUATE see MHA500
METHYL TOLUENE see XGS000
o-METHYLTOLUENE see XHJ000
p-METHYLTOLUENE see XHS000
METHYL TOLUENE-4-SULFONATE see MLL250
METHYL-p-TOLUENESULFONATE see MLL250
α-METHYL-α-TOLUIC ALDEHYDE see COF000
2-METHYL-p-TOLUIDINE see XMS000
4-METHYL-o-TOLUIDINE see XMS000
5-METHYL-o-TOLUIDINE see XNA000
6-METHYL-m-TOLUIDINE see XNA000
2-METHYL-3-TOLYLCHINAZOLON-4
HYDROCHLORIDE (GERMAN) see MDT250
METHYL-p-TOLYL ETHER see MGP000
METHYL-p-TOLYL KETONE see MFW250
2-METHYL-3-o-TOLYL-4(3H)-QUINAZOLINONE
HYDROCHLORIDE see MDT250
2-METHYL-3-(o-TOLYL)-4-QUINAZOLONE
HYDROCHLORIDE see MDT250
METHYL TOSYLATE see MLL250
METHYL-p-TOSYLATE see MLL250
METHYLTRICAPRYLYLAMMONIUM CHLORIDE see
MQH000
METHYL TRICHLORIDE see CHJ500
METHYLTRICHLOROMETHANE see MIH275
4-((N-METHYL-N-
TRICHLOROMETHANESULFENYL)CARBAMOYLOX
IMINO)-1,3-DITHIOLANE see MQC320
METHYLTRICHLOROSILANE see MQC500
METHYLTRICHLOROSTANNANE see MQC750
METHYLTRICHLOROTIN see MQC750
METHYL-TRICHLORSILAN (CZECH) see MQC500
METHYL TRIFLUORIDE see CBY750
(α-METHYL-m-
TRIFLUOROMETHYLPHENETHYLAMINOMETHYL)
PIPERIDINO KETONE see MQE100
METHYLTRIMETHYLENE GLYCOL see BOS500
METHYL 2,4,5-TRIMETHYLPYRROL-3-YL KETONE
see ADE050
METHYLTRIOCTYLAMMONIUM CHLORIDE see
MQH000
4-METHYL-2,6,7-TRIOXA-1-
ARSABICYCLO(2.2.2)OCTANE see MQH100

METHYLTRIS(2-ETHYLHEXYLOXYCARBONYLMETHYLTHIO)STANNANE see MQH500
METHYL TRITHION see MQH750
α-METHYLTRYPTAMINE see AME500
METHYL TUADS see TFS350
4-METHYLUMBELLIFERON (CZECH) see MKP500
4-METHYLUMBELLIFERONE see MKP500
β-METHYLUMBELLIFERONE see MKP500
2-METHYLUNDECANAL see MQI550
METHYL 9-UNDECENOATE see ULS400
METHYL 10-UNDECENOATE see ULS400
METHYL UNDECYLENATE see ULS400
4-METHYLURACIL see MPW500
4-METHYLURACIL see MQI750
6-METHYLURACIL see MQI750
METHYLURETHAN see MHZ000
N-METHYL URETHAN see EMQ500
METHYLURETHANE see MHZ000
METHYLURETHANE see EMQ500
N-METHYLURETHANE see EMQ500
METHYLVINYL ACETATE see MQK750
METHYL-VINYL-CETONE (FRENCH) see BOY500
METHYL VINYL ETHER see MQL750
METHYLVINYLKETON (GERMAN) see BOY500
METHYL VINYL KETONE see BOY500
METHYLVINYLNITROSAMIN (GERMAN) see NKY000
METHYLVINYLNITROSAMINE see NKY000
METHYL VINYL OXIMINO SILANE see MQM150
METHYL VIOLET 10B see AOR500
METHYL VIOLET 10BD see AOR500
METHYL VIOLET 10BK see AOR500
METHYL VIOLET 10BN see AOR500
METHYL VIOLET 5BNO see AOR500
METHYL VIOLET 10BNS see AOR500
METHYL VIOLET 5BO see AOR500
METHYL VIOLET 10BO see AOR500
METHYLVIOLETT see AOR500
METHYL VIOLOGEN (2+) see PAI990
METHYL YELLOW see DOT300
METHYL ZIMATE see BJK500
METHYL ZINEB see BJK500
METHYL ZIRAM see BJK500
6-METHYL-Δ⁴,⁶-PREGNADIEN-17-α-OL-3,20-DIONE ACETATE see VTF000
METHYPHENYLMETHANOL see PDE000
METHYSERGIDE DIMALEATE see MQP500
METIAPINE see MQQ000
METICORTELONE see PMA000
METI-DERM see PMA000
METIFEX see EDW500
METIFONATE see TIQ250
METILACRILATO (ITALIAN) see MGA500
METILAMIL ALCOHOL (ITALIAN) see MKW600
METILAMINE (ITALIAN) see MGC250
2-METIL-6-AMINO-EPTANO (ITALIAN) see ILM000
3-METIL-BUTANOLO (ITALIAN) see IHP000
METIL CELLOSOLVE (ITALIAN) see EJH500
2-METILCICLOESANONE (ITALIAN) see MIR500
METILCLOROFORMIATO (ITALIAN) see MIG000
METILCLORPINDOL see CMX850
METILE (ACETATO di) (ITALIAN) see MFW100
O,O-METILEN-BIS(4-CLORO-FENOLO) see MJM500
3,3'-METILEN-BIS(4-IDROSSI-CUMARINA) (ITALIAN) see BJZ000
4,4-METILENE-BIS-o-CLOROANILINA (ITALIAN) see MJM200

(6-(1-METIL-EPITL)-2,4-DINITRO-FENIL)-CROTONATO (ITALIAN) see AQT500
METILETILCHETONE (ITALIAN) see MKA400
METIL (FORMIATO di) (ITALIAN) see MKG750
METILISOBUTILCHETONE (ITALIAN) see HFG500
METIL ISOCIANATO (ITALIAN) see MKX250
METILMERCAPTANO (ITALIAN) see MLE650
METILMERCAPTOFOSOKSID see DAP000
METIL METACRILATO (ITALIAN) see MLH750
N-METIL-1-NAFTIL-CARBAMMATO (ITALIAN) see CBM750
METILPARATION (HUNGARIAN) see MNH000
4-METILPENTAN-2-OLO (ITALIAN) see MKW600
4-METILPENTAN-2-ONE (ITALIAN) see HFG500
4-METIL-3-PENTEN-2-ONE (ITALIAN) see MDJ750
α-METIL-STIROLO (ITALIAN) see MPK250
2-METIL-2-TIOMETIL-PROPIONALDEID-O-(N-METIL-CARBAMOIL)-OSSIMA (ITALIAN) see CBM500
6-METIL-TIOURACILE (ITALIAN) see MPW500
METILTRIAZOTION see ASH500
METINDOL see IDA000
METIONE see MDT740
METIPREGNONE see MCA000
METIZOL see MCO500
METMERCAPTURON see DST000
2-METOKSY-4-ALLILOFENOL (POLISH) see EQR500
METOKSYCHLOR (POLISH) see MEI450
METOKSYETYLOWY ALKOHOL (POLISH) see EJH500
METOLCARB see MIB750
METOLOSE 60SH see MIF760
METOLOSE 60SH400 see MIF760
METOLOSE MC 8000 see MIF760
METOLOSE SM 15 see MIF760
METOLOSE SM 100 see MIF760
METOLOSE SM 4000 see MIF760
METOMIL (ITALIAN) see MDU600
METOPRYL see MOU830
METOSERPATE HYDROCHLORIDE see MQR200
(2-METOSSICARBONIL-1-METIL-VINIL)-DIMETIL-FOSFATO (ITALIAN) see MQR750
2-METOSSIETANOLO (ITALIAN) see EJH500
2-METOSSIETILACETATO (ITALIAN) see EJJ500
METOTHYRINE see MCO500
METOX see MEI450
METOXAL see SNK000
METOXIDON see SNN300
METOXYDE see HJL500
METRACTYL see MQU750
METRANIL see PBC250
METRIBUZIN see MQR275
METRIFONATE see TIQ250
METRIPHONATE see TIQ250
METRISONE see MOR500
METRODIOL see EQJ500
METRODIOL DIACETATE see EQJ500
METROGEN RED FORMER KB SOLN see CLK225
METRON see MNH000
METRONE see MPN500
METRONIDAZ see MMN250
METRONIDAZOL see MMN250
METRONIDAZOLO see MMN250
METROPINE see MGR500
METRO TALC 4604 see TAB750
METRO TALC 4608 see TAB750
METRO TALC 4609 see TAB750
METROXEDRINE see SPC500

METSO 20 see SJU000
METSO BEADS 2048 see SJU000
METSO BEADS, DRYMET see SJU000
METSO PENTABEAD 20 see SJU000
MET-SPAR see CAS000
METYCAINE see PIV750
METYLAL (POLISH) see MGA850
METYLENO-BIS-FENYLOIZOCYJANIAN see DNJ800
METYLENU CHLOREK (POLISH) see MJP450
METYLESTER KYSELINY SALICYLOVE (CZECH) see MPI000
METYLFENEMAL see ENB500
METYLOAMINA (POLISH) see MGC250
METYLOCYKLOHEKSAN (POLISH) see MIQ740
METYLOCYKLOHEKSANOL (POLISH) see MIQ745
METYLOCYKLOHEKSANON (POLISH) see MIR250
METYLOETYLOKETON (POLISH) see MKA400
METYLOHYDRAZYNA (POLISH) see MKN000
METYLOIZOBUTYLOKETON (POLISH) see HFG500
1-METYLO-2-MERKAPTOIMIDAZOLEM (POLISH) see MCO500
N-METYLO-N'-NITRO-N-NITROZOGUANIDYNY see MMP000
METYLOPARATION (POLISH) see MNH000
METYLOPROPYLOKETON (POLISH) see PBN250
METYLOWY ALKOHOL (POLISH) see MGB150
METYLPARATION (CZECH) see MNH000
METYLU BROMEK (POLISH) see MHR200
METYLU CHLOREK (POLISH) see MIF765
METYLU JODEK (POLISH) see MKW200
METYNA see ENB500
MEVINFOS (DUTCH) see MQR750
MEVINPHOS see MQR750
MEXACARBATE (DOT) see DOS000
MEXENE see BJK500
MEXICO WEED see CCP000
MEXIDEX see SOW000
MEXOCINE see MIJ500
MEXYL see ABX500
MEYPRALGIN R/LV see SEH000
MEZATON see NCL500
MEZATON see SPC500
MEZCALINE see MDI500
MEZCLINE see MDI500
MEZENE see BJK500
MEZIDINE see TLG500
MEZOLIN see IDA000
MEZOTOX see DFT800
MFA see FIC000
MFH see FNW000
MFI see IPX000
MFNA see MME809
MG 18370 see DTJ400
MG 18570 see DTJ400
MGA see MCB380
MGA 100 (STEROID) see MCB380
M7-GIFTKOERNER see TEM000
MGK-264 see OES000
MGK DIETHYLTOLUAMIDE see DKC800
MGK DOG AND CAT REPELLENT see UKS000
MH see DMC600
MH 30 see DMC600
MH-40 see DMC600
MH-532 see PGA750
MHA NITRILE see HMR550
MH 36 BAYER see DMC600
3-MI see MKV750
MIA see IDZ000

MIAK see MKW450
MIARSENOL see NCJ500
MIBC see MKW600
MIBK see HFG500
MIBOLERON see MQS225
MIBOLERONE see MQS225
MIC see MKW600
MIC see MKX250
MIC see ISE000
3-MIC see MKW600
MICA see MQS250
MICALEX see AHD650
MICA SILICATE see MQS250
MICHLER'S BASE see MJN000
MICHLER'S HYDRIDE see MJN000
MICHLER'S KETONE see MQS500
p,p'-MICHLER'S KETONE see MQS500
MICHLER'S METHANE see MJN000
MICOCHLORINE see CDP250
MICOFUME see DSB200
MICOL see HCQ500
MICREST see DKA600
MICROCAL 160 see CAW850
MICROCAL ET see CAW850
MICRO-CEL see CAW850
MICRO-CEL A see CAW850
MICRO-CEL B see CAW850
MICRO-CEL C see CAW850
MICRO-CEL E see CAW850
MICRO-CEL T see CAW850
MICRO-CEL T26 see CAW850
MICRO-CEL T38 see CAW850
MICRO-CEL T41 see CAW850
MICROCETINA see CDP250
MICRO-CHECK 12 see CBG000
MICRO-CHEK 11 see OFE000
MICRO-CHEK SKANE see OFE000
MICROCRYSTALLINE QUARTZ see SCK600
MICROCRYSTALLINE WAX see PCT600
MICROCYSTIN-A see AQV990
MICROCYSTIN-LR see AQV990
MICROCYSTIS AERUGINOSA TOXIN see AQV990
MICRO DDT 75 see DAD200
MICRODIOL see EDO000
MICRO DRY see AHA000
MICROEST see DKA600
MICROFLOTOX see SOD500
MICROGRIT WCA see AHE250
MICROIODIDE see PLW285
MICRO-LEX GREEN 5B see BLK000
MICROLYSIN see CKN500
MICRONEX see CBT750
MICROPHOS 90 see ZJS400
MICROSETILE BLUE EB see TBG700
MICROSETILE DIAZO BLACK G see DPO200
MICROSETILE ORANGE RA see AKP750
MICROSETILE PINK BN see AKE250
MICROSETILE YELLOW GR see AAQ250
MICROTEX LAKE RED CR see CHP500
MICROTHENE see PJS750
MICROTHENE 510 see PJS750
MICROTHENE 704 see PJS750
MICROTHENE 710 see PJS750
MICROTHENE F see PJS750
MICROTHENE FN 500 see PJS750
MICROTHENE FN 510 see PJS750
MICROTHENE MN 754-18 see PJS750
MICROZUL see CJJ000

MIDLON RED PRS see CMM330
MIDONE see DBB200
MIELUCIN see BOT250
MIERENZUUR (DUTCH) see FNA000
MIGHTY 150 see NAJ500
MIH HYDROCHLORIDE see PME500
MIIKE 20 see CBT750
MIK see HFG500
MIKAMETAN see IDA000
MIKAMYCIN see VRF000
MIKEDIMIDE see MKA250
MIKETHRENE GREY M see CMU475
MIKETHRENE GREY MG see CMU475
MIKETON FAST BLUE see TBG700
MIKETON FAST VIOLET B see DBY700
MIKETON FAST YELLOW G see AAQ250
MIKETORIN see EAI000
MIKROKALCIT see CAD800
MIKROLOUR see PJS750
MILBAM see BJK500
MILBAN see BJK500
MILBAN F see TFD500
MIL-B-4394-B see CES650
MILBEDOCE see VSZ000
MILBOL see BIO750
MILBOL 49 see BBQ500
MILCHSAEURE (GERMAN) see LAG000
MILDMEN see MDQ250
MILDMEN see LFK000
MILD MERCURY CHLORIDE see MCW000
MIL-DU-RID see BGJ750
MILEPSIN see DBB200
MILESTROL see DKA600
MILK ACID see LAG000
MILK-CLOTTING ENZYME from BACILLUS CEREUS see MQU075
MILK-CLOTTING ENZYME from MUCOR MIEHEI see MQU120
MILK OF LIME see CAT225
MILK OF MAGNESIA see MAG750
MILK SUGAR see LAR000
MILK WHITE see LDY000
MILLER NU SET see TIX500
MILLER P.C. WEEDKILLER see PLC250
MILLER'S FUMIGRAIN see ADX500
MILLICORTEN see SOW000
MILLING FAST RED B see CMM330
MILLING RED B see CMM330
MILLING RED BB see CMM330
MILLING RED SWB see CMM330
MILLIONATE 300 see PKB100
MILLIONATE MR see PKB100
MILLIONATE MR 100 see PKB100
MILLIONATE MR 200 see PKB100
MILLIONATE MR 300 see PKB100
MILLIONATE MR 340 see PKB100
MILLIONATE MR 400 see PKB100
MILLIONATE MR 500 see PKB100
MILMER see BLC250
MILOGARD see PMN850
MILPREM see MQU750
MILPREM see ECU750
MILPREX see DXX400
MILTANN see MQU750
MILTON see SHU500
MILTOWN see MQU750
MIMOSA TANNIN see MQV250
MINERAL DUSTS see MQV500

MINERAL FIRE RED 5DDS see MRC000
MINERAL FIRE RED 5GS see MRC000
MINERAL GREEN see COF500
MINERAL NAPHTHA see BBL250
MINERAL OIL see MQV750
MINERAL OIL, PETROLEUM CONDENSATES, VACUUM TOWER see MQV755
MINERAL OIL, PETROLEUM DISTILLATES, ACID-TREATED HEAVY NAPHTHENIC (mild or no solvent-refining or hydrotreatment) see MQV760
MINERAL OIL, PETROLEUM DISTILLATES, ACID-TREATED HEAVY PARAFFINIC (severe solvent-refining and/or hydrotreatment) see MQV765
MINERAL OIL, PETROLEUM DISTILLATES, ACID-TREATED LIGHT NAPHTHENIC (mild or no solvent-refining or hydrotreatment) see MQV770
MINERAL OIL, PETROLEUM DISTILLATES, ACID-TREATED LIGHT PARAFFINIC (mild or no solvent-refining or hydrotreatment) see MQV775
MINERAL OIL, PETROLEUM DISTILLATES CATALYTIC DEWAXED HEAVY NAPHTHENIC (mild or no solvent-refining or hydrotreatment) see MQV776
MINERAL OIL, PETROLEUM DISTILLATES CATALYTIC DEWAXED HEAVY PARAFFINIC (mild or no solvent-refining or hydrotreatment) see MQV778
MINERAL OIL, PETROLEUM DISTILLATES CATALYTIC DEWAXED LIGHT NAPHTHENIC (mild or no solvent-refining or hydrotreatment) see MQV777
MINERAL OIL, PETROLEUM DISTILLATES CATALYTIC DEWAXED LIGHT PARAFFINIC (mild or no solvent-refining or hydrotreatment) see MQV779
MINERAL OIL, PETROLEUM DISTILLATES, HEAVY NAPHTHENIC see MQV780
MINERAL OIL, PETROLEUM DISTILLATES, HEAVY PARAFFINIC see MQV785
MINERAL OIL, PETROLEUM DISTILLATES, HYDROTREATED (mild) HEAVY NAPHTHENIC see MQV790
MINERAL OIL, PETROLEUM DISTILLATES, HYDROTREATED (mild) HEAVY PARAFFINIC see MQV795
MINERAL OIL, PETROLEUM DISTILLATES, HYDROTREATED (mild) LIGHT NAPHTHENIC see MQV800
MINERAL OIL, PETROLEUM DISTILLATES, HYDROTREATED (mild) LIGHT PARAFFINIC see MQV805
MINERAL OIL, PETROLEUM DISTILLATES, LIGHT NAPHTHENIC see MQV810
MINERAL OIL, PETROLEUM DISTILLATES, LIGHT PARAFFINIC see MQV815
MINERAL OIL, PETROLEUM DISTILLATES, SOLVENT-DEWAXED HEAVY NAPHTHENIC (mild or no solvent-refining or hydrotreatment) see MQV820
MINERAL OIL, PETROLEUM DISTILLATES, SOLVENT-DEWAXED HEAVY PARAFFINIC (mild or no solvent-refining or hydrotreatment) see MQV825
MINERAL OIL, PETROLEUM DISTILLATES, SOLVENT-DEWAXED LIGHT NAPHTHENIC (mild or no solvent-refining or hydrotreatment) see MQV835
MINERAL OIL, PETROLEUM DISTILLATES, SOLVENT-DEWAXED LIGHT PARAFFINIC (mild or no solvent-refining or hydrotreatment) see MQV840
MINERAL OIL, PETROLEUM DISTILLATES, SOLVENT-REFINED (mild) HEAVY NAPHTHENIC see MQV845

MINERAL OIL, PETROLEUM DISTILLATES, SOLVENT-REFINED (mild) HEAVY PARAFFINIC see MQV850
MINERAL OIL, PETROLEUM DISTILLATES, SOLVENT-REFINED (mild) LIGHT NAPHTHENIC see MQV852
MINERAL OIL, PETROLEUM DISTILLATES, SOLVENT-REFINED (mild) LIGHT PARAFFINIC see MQV855
MINERAL OIL, PETROLEUM EXTRACTS, HEAVY NAPHTHENIC DISTILLATE SOLVENT see MQV857
MINERAL OIL, PETROLEUM EXTRACTS, HEAVY PARAFFINIC DISTILLATE SOLVENT see MQV859
MINERAL OIL, PETROLEUM EXTRACTS, LIGHT NAPHTHENIC DISTILLATE SOLVENT see MQV860
MINERAL OIL, PETROLEUM EXTRACTS, LIGHT PARAFFINIC DISTILLATE SOLVENT see MQV862
MINERAL OIL, PETROLEUM EXTRACTS, RESIDUAL OIL SOLVENT see MQV863
MINERAL OIL, PETROLEUM RESIDUAL OILS, ACID-TREATED see MQV872
MINERAL OIL, WHITE see MQV875
MINERAL OIL, WHITE (FCC) see MQV750
MINERAL ORANGE see LDS000
MINERAL PITCH see ARO500
MINERAL RED see LDS000
MINERAL WHITE see CAX750
MINETOIN see DKQ000
MINETOIN see DNU000
MINIHIST see DBM800
MINIPLANOR see ZVJ000
MINIUM see LDS000
MINIUM NON-SETTING RL-95 see LDS000
MINOPHAGEN A see AQW000
MINORLAR see EEH520
MINOVLAR see EEH520
MINTAL see NBU000
MINTEZOL see TEX000
MINT-O-MAG see MAG750
MIN-U-GEL 200 see PAE750
MIN-U-GEL 400 see PAE750
MIN-U-GEL FG see PAE750
MINURIC see DDP200
MINUS see SEH000
MINZIL see CLH750
MINZOLUM see TEX000
MIO 40GN see IHG100
MIO-PRESSIN see RDK000
MIPAFOX see PHF750
MIPAX see DTR200
MIPC see MIA250
MIPCIN see MIA250
MIPCINE see MIA250
MI-PILO OPHTH SOL see PIF250
MIPK see MLA750
MIPSIN see MIA250
MIRACIL D see DHU000
MIRAMID WM 55 see PJY500
MIRAPRONT see DTJ400
MIRASON 9 see PJS750
MIRASON 16 see PJS750
MIRASON M 15 see PJS750
MIRASON M 50 see PJS750
MIRASON M 68 see PJS750
MIRASON NEO 23H see PJS750
MIRATHEN see PJS750
MIRATHEN 1313 see PJS750
MIRATHEN 1350 see PJS750

MIRBANE OIL see NEX000
MIRCOSULFON see PPP500
MIREX see MQW500
MIROISTONIL see DRM000
MIRREX MCFD 1025 see PKQ059
MISHERI (INDIA) see SED400
MISHRI (INDIA) see SED400
MISODINE see DBB200
MISOLYNE see DBB200
MISPICKEL see ARJ750
MISTRON 2SC see TAB750
MISTRON FROST P see TAB750
MISTRON RCS see TAB750
MISTRON STAR see TAB750
MISTRON SUPER FROST see TAB750
MISTRON VAPOR see TAB750
MISULBAN see BOT250
MIT see ISE000
MIT-C see AHK500
MITC see ISE000
MITENON see MMD500
MITIGAN see BIO750
MITION see CKM000
MITIS GREEN see COF500
MITO-C see AHK500
MITOCIN-C see AHK500
MITOLAC see DDJ000
MITOLACTOL see DDJ000
MITOMEN see CFA500
MITOMEN see CFA750
MITOMIN see CFA500
MITOMYCIN see AHK500
MITOMYCIN-C see AHK500
MITOMYCINUM see AHK500
MITOSTAN see BOT250
MITOXAN see CQC500
MITOXAN see CQC650
MITOXANA see IMH000
MITOXINE see BIE500
MITSUI AURAMINE O see IBA000
MITSUI BLUE B BASE see DCJ200
MITSUI BRILLIANT GREEN G see BAY750
MITSUI CONGO RED see SGQ500
MITSUI CRYSTAL VIOLET see AOR500
MITSUI DIRECT BLACK EX see AQP000
MITSUI DIRECT BLACK GX see AQP000
MITSUI DIRECT BLUE 2BN see CMO000
MITSUI DIRECT SKY BLUE 5B see CMO500
MITSUI METANIL YELLOW see MDM775
MITSUI METHYLENE BLUE see BJI250
MITSUI RED TR BASE see CLK220
MITSUI RHODAMINE see RGW000
MITSUI RHODAMINE 6GCP see RGW000
MITSUI SCARLET G BASE see NMP500
MITSUI TARTRAZINE see FAG140
MIXTURE of p-METHENOLS see TBD500
MIZODIN see DBB200
MIZOLIN see DBB200
MJ 10061 see DDP200
MJF 9325 see IMH000
MJF-12264 see FLZ050
MK 56 see POL500
MK 125 see SOW000
MK-188 see RBF100
MK 351 see DNA800
MK 360 see TEX000
MK 647 see DKI600
MK 936 see ARW200

MK. B51 see DNA800
MKC-442 see EAN525
2M-4KH see CIR250
2M 4KHP see CIR500
2M-4KH SODIUM SALT see SIL500
ML 33F see SKV100
ML 55F see SKV100
MLT see MAK700
MM see BKM500
MMA see MGQ250
MMC see MDD750
MMC see AHK500
MMD see MLF250
MME see MFC700
MME see MLH750
MMH see MKN000
4-MMPD see DBO000
4-MMPD SULPHATE see DBO400
MMS see MLH500
MMT see MAV750
MMTs-BTR see MIF760
MNA see MMU250
MNA see NEN500
MNBK see HEV000
MNC see MMX000
MN-CELLULOSE see CCU150
MNCO see MQY325
MNE see NDM000
MnEBD see MAS500
MNFA see MME809
MNG see MMP000
McN-JR 4263 see PDW750
McN-JR-16,341 see PAP250
MNNG see MMP000
MNPN see MMS200
MNQ see MMD500
MNT see NMO500
MNU see MMX250
MNU see MNA750
MO 709 see DRM000
MOB see MES000
MOBAY MRS see PKB100
MOBENOL see BSQ000
MOBILAN see IDA000
MOBILAWN see DFK600
MOBIL V-C 9-104 see EIN000
MOCA see MJM200
MOCAP see EIN000
MODANE see DMH400
MODANE SOFT see DJL000
MODENOL see RDK000
MODIFIED POLYACRYLAMIDE RESINS see MRA075
MODOCOLL 1200 see SFO500
MODR FRALOSTANOVA 3G (CZECH) see DNE400
MODR METHYLENOVA (CZECH) see BJI250
MODR OSTACETOVA LR see BNC800
MODR PRIMA 15 see CMO500
MODULEX see CBT750
MOEBIQUIN see DNF600
MOENOMYCIN see MRA250
MOENOMYCIN A see MRA250
MOGADAN see DLY000
MOGUL see CBT750
MOGUL L see CBT750
MOHICAN RED A-8008 see CHP500
MOLACCO see CBT750
MOLASSES ALCOHOL see EFU000
MOLATOC see DJL000

MOLCER see DJL000
MOLDAMIN see BFC750
MOLDEX see HJL500
MOLECULAR CHLORINE see CDV750
MOLE DEATH see SMN500
MOL-IRON see FBO000
MOLLAN O see DVL700
MOLOFAC see DJL000
MOLOL see MQV750
MOLTEN ADIPIC ACID see AEN250
MOLURAME see BJK500
MOLYBDATE see MRC250
MOLYBDATE, CALCIUM see CAT750
MOLYBDATE ORANGE see MRC000
MOLYBDATE ORANGE Y 786D see MRC000
MOLYBDATE ORANGE YE 421D see MRC000
MOLYBDATE ORANGE YE 698D see MRC000
MOLYBDATE RED see MRC000
MOLYBDATE RED AA3 see MRC000
MOLYBDEN RED see MRC000
MOLYBDENUM see MRC250
MOLYBDENUM BORIDE see MRC650
MOLYBDENUM COMPOUNDS see MRC750
MOLYBDENUM DIOXIDE see MRD250
MOLYBDENUM-LEAD CHROMATE see LDM000
MOLYBDENUM ORANGE see LDM000
MOLYBDENUM OXIDE see MRD250
MOLYBDENUM(IV) OXIDE see MRD250
MOLYBDENUM(VI) OXIDE see MRE000
MOLYBDENUM PENTACHLORIDE see MRD500
MOLYBDENUM RED see MRC000
MOLYBDENUM, TRICARBONYL(1,3,5-
CYCLOHEPTATRIENE)- see COY100
MOLYBDENUM TRIOXIDE see MRE000
MOLYBDIC ACID DIAMMONIUM SALT see ANM750
MOLYBDIC ACID, DISODIUM SALT see DXE800
MOLYBDIC ACID (H_2MoO_4), CALCIUM SALT (1:1) see
CAT750
MOLYBDIC ANHYDRIDE see MRE000
MOLYBDIC TRIOXIDE see MRE000
MOLYKOTE 522 see TAI250
MOMENTUM see HIM000
MON 0573 see PHA500
MON 2139 see PHA500
MONACETYLFERROCENE see ABA750
MONAGYL see MMN250
MONAMID 150-LW see BKE500
MONAQUEST see DJG800
MONARCH see ZVJ000
MONARCH see CBT750
MONARGAN see ABX500
MONASIRUP see PMQ750
MONATE see MRL750
MONAWET MD 70E see DJL000
MONCIDE see HKC000
MONDUR E 429 see PKB100
MONDUR E 441 see PKB100
MONDUR E 541 see PKB100
MONDUR MR see PKB100
MONDUR MR 200 see PKB100
MONDUR MRS see PKB100
MONDUR MRS 10 see PKB100
MONDUR P see PFK250
MONDUR-TD see TGM740
MONDUR TD see TGM750
MONDUR-TD-80 see TGM740
MONDUR TD-80 see TGM750
MONDUR TDS see TGM750

MONELAN see MRE225
MONELGIN see OAV000
MONENSIC ACID see MRE225
MONENSIN A see MRE225
MONENSIN (USDA) see MRE225
MONEX see BJL600
MONHYDRIN see PMJ500
MONITAN see PKL100
MONITOR see DTQ400
MONOACETYLHYDRAZINE see ACM750
MONOAETHANOLAMIN (GERMAN) see EEC600
MONOALLYLAMINE see AFW000
MONOALUMINUM PHOSPHATE see PHB500
MONOAMMONIUM CARBONATE see ANB250
MONOAMMONIUM GLUTAMATE see MRF000
MONOAMMONIUM l-GLUTAMATE see MRF000
MONOAMMONIUM GLYCYRRHIZINATE see AMY700
MONOAMMONIUM GLYCYRRHIZINATE see GIE100
MONOAMMONIUM SULFAMATE see ANU650
MONOAMMONIUM SULFATE see ANJ500
MONOAMMONIUM SULFIDE see ANJ750
MONOAMMONIUM SULFITE see ANB600
MONOAMYLAMINE see PBV505
MONOAZO see MDM775
MONOBASIC LEAD ACETATE see LCH000
MONOBROMESSIGSAEURE (GERMAN) see BMR750
MONOBROMOACETIC ACID see BMR750
MONOBROMOACETONE see BNZ000
MONOBROMOBENZENE see PEO500
MONOBROMOETHANE see EGV400
MONOBROMOISOVALERYLUREA see BNP750
2-MONOBROMOISOVALERYLUREA see BNP750
MONOBROMOMETHANE see MHR200
MONOBUTILAMINA see BPX750
MONOBUTYLAMINE see BPX750
MONO-n-BUTYLAMINE see BPX750
MONOBUTYL GLYCOL ETHER see BPJ850
MONO-tert-BUTYLHYDROQUINONE see BRM500
MONOBUTYL PHTHALATE see MRF525
MONO-n-BUTYL PHTHALATE see MRF525
MONOCALCIUM ARSENITE see CAM500
MONOCHLOORAZIJNZUUR (DUTCH) see CEA000
MONOCHLOORBENZEEN (DUTCH) see CEJ125
MONOCHLORACETIC ACID see CEA000
MONOCHLORACETONE see CDN200
MONOCHLORBENZENE see CEJ125
MONOCHLORBENZOL (GERMAN) see CEJ125
MONOCHLORESSIGSAEURE (GERMAN) see CEA000
MONOCHLORETHANE see EHH000
MONOCHLORHYDRIN see CDT750
MONOCHLORHYDRINE du GLYCOL (FRENCH) see EIU800
MONOCHLOROACETALDEHYDE see CDY500
MONOCHLOROACETIC ACID see CEA000
MONOCHLOROACETIC ACID METHYL ESTER see MIF775
MONOCHLOROACETONE see CDN200
MONOCHLOROACETONE, inhibited (DOT) see CDN200
MONOCHLOROACETONE, stabilized (DOT) see CDN200
MONOCHLOROACETONE, unstabilized (DOT) see CDN200
MONOCHLOROACETONITRILE see CDN500
MONOCHLOROACETYL CHLORIDE see CEC250
α-MONOCHLOROANTHRAQUINONE see CEI000

MONOCHLOROBENZENE see CEJ125
MONOCHLORODIFLUOROMETHANE see CFX500
MONOCHLORODIMETHYL ETHER (MAK) see CIO250
MONOCHLOROETHANOIC ACID see CEA000
2-MONOCHLOROETHANOL see EIU800
MONOCHLOROETHENE see VNP000
MONOCHLOROETHYLENE (DOT) see VNP000
MONOCHLOROHYDRIN see CDT750
α-MONOCHLOROHYDRIN see CDT750
MONOCHLOROMETHANE see MIF765
MONOCHLOROMETHYL CYANIDE see CDN500
MONO-CHLORO-MONO-BROMO-METHANE see CES650
MONOCHLOROMONOFLUOROMETHANE see CHI900
MONOCHLOROPENTAFLUOROETHANE (DOT) see CJI500
MONOCHLOROSULFURIC ACID see CLG500
MONOCHLOROTRIFLUOROETHYLENE see CLQ750
MONOCHLOROTRIFLUOROMETHANE (DOT) see CLR250
MONOCHROMIUM OXIDE see CMK000
MONOCHROMIUM TRIOXIDE see CMK000
MONOCIL see MRH209
MONOCIL 40 see MRH209
"MONOCITE" METHACRYLATE MONOMER see MLH750
MONOCLAIR see CCK125
MONOCLOROBENZENE (ITALIAN) see CEJ125
MONOCOBALT OXIDE see CND125
MONOCRATILIN see MRH000
MONOCRON see MRH209
MONOCROTALINE see MRH000
MONOCROTOPHOS see MRH209
MONOCROTOPHOS see MRH209
MONOCROTOPHOS (ACGIH,OSHA) see MRH209
MONODEHYDROSORBITOL MONOOLEATE see SKV100
MONO- and DIGLYCERIDES see MRH215
MONO and DIGLYCERIDES, SODIUM SULFOACETATE DERIVATIVES see SJZ050
MONO-, DI-, and TRIPOTASSIUM CITRATE see MRH225
MONO-, DI-, and TRISTEARYL CITRATE see MRH235
MONODODECYLAMINE see DXW000
MONODRAL see PBS000
MONODRAL BROMIDE see PBS000
MONODRIN see MRH209
MONOETHANOLAMINE see EEC600
MONOETHYLAMINE (DOT) see EFU400
MONOETHYLAMINE, anhydrous (DOT) see EFU400
2-N-MONOETHYLAMINOETHANOL see EGA500
MONOETHYLENE GLYCOL see EJC500
MONOETHYLENE GLYCOL DIMETHYL ETHER see DOE600
MONOETHYL ETHER of DIETHYLENE GLYCOL see CBR000
MONO(2-ETHYLHEXYL)SULFATE SODIUM SALT see TAV750
MONOFLUORAZIJNZUUR (DUTCH) see FIC000
MONOFLUORESSIGSAEURE (GERMAN) see FIC000
MONOFLUORESSIGSAEURE, NATRIUM (GERMAN) see SHG500
MONOFLUOROACETAMIDE see FFF000
MONOFLUOROACETATE see FIC000
MONOFLUOROACETIC ACID see FIC000

MONOFLUOROETHANE see FIB000
MONOFLUOROETHYLENE see VPA000
MONOFLUOROPHOSPHORIC ACID, anhydrous see PHJ250
MONOFLUOROTRICHLOROMETHANE see TIP500
MONOFURACIN see NGE500
MONOGERMANE see GEI100
MONOGLYCERYL OLEATE see GGR200
MONO-GLYCOCOARD see DKL800
MONOGLYME see DOE600
MONO-N-HEXYLAMINE see HFK000
8-MONOHYDRO MIREX see MRI750
MONOHYDROXYBENZENE see PDN750
MONOHYDROXYMETHANE see MGB150
MONOIODOACETAMIDE see IDW000
MONOIODOACETATE see IDZ000
MONOIODOACETIC ACID see IDZ000
MONOIODURO di METILE (ITALIAN) see MKW200
MONOISOBUTYLAMINE see IIM000
MONOISOPROPYLAMINE see INK000
MONOISOPROPYLAMINOETHANOL see INN400
MONOISOPROPYL CITRATE see MRI785
MONOISOPROPYL ETHER of ETHYLENE GLYCOL see INA500
MONOLINURON see CKD500
MONOMETHYLACETAMIDE see MFT750
N-MONOMETHYLAMIDE of O,O-DIMETHYLDITHIOPHOSPHORYLACETIC ACID see DSP400
MONOMETHYLAMINE see MGC250
MONOMETHYL-AMINOAETHANOL (GERMAN) see MGG000
MONOMETHYLAMINOETHANOL see MGG000
N-MONOMETHYLAMINOETHANOL see MGG000
MONOMETHYLANILINE see MGN750
N-MONOMETHYLANILINE see MGN750
MONOMETHYL ANILINE (OSHA) see MGN750
MONOMETHYL ETHER of ETHYLENE GLYCOL see EJH500
MONOMETHYL ETHER HYDROQUINONE see MFC700
MONOMETHYLFORMAMIDE see MKG500
MONOMETHYL HYDRAZINE see MKN000
MONOMETHYL MERCURY CHLORIDE see MDD750
MONOMETHYLOLACRYLAMIDE see HLU500
MONOMETHYLTIN TRICHLORIDE see MQC750
MONONICKEL MONOSULFIDE see NDL100
MONONITROCHLOROBENZENE see CJA950
MONONITRONAPHTHALENE see NHP990
MONOOCTADECYLAMINE see OBC000
MONO-n-OCTYLTIN TRICHLORIDE see OGG000
MONO-n-OCTYL-TIN-TRIS-(2-ETHYLHEXYLMERCAPTOACETATE) see OGI000
MONO-N-OCTYL-ZINN-TRICHLORID (GERMAN) see OGG000
MONOOLEIN see GGR200
MONOOLEOYLGLYCEROL see GGR200
MONOPENTEK see PBB750
MONOPERACETIC ACID see PCL500
MONOPHEN see PFC750
MONOPHENOL see PDN750
MONOPLEX DBS see DEH600
MONOPLEX DOA see AEO000
MONOPLEX DOS see BJS250
MONOPOTASSIUM ARSENATE see ARD250
MONOPOTASSIUM DIHYDROGEN ARSENATE see ARD250
MONOPOTASSIUM GLUTAMATE see MRK500

MONOPOTASSIUM l-GLUTAMATE (FCC) see MRK500
MONOPOTASSIUM SULFATE see PKX750
MONO-N-PROPYLAMINE see PND250
MONOPROPYLENE GLYCOL see PML000
MONOPROPYL ETHER of ETHYLENE GLYCOL see PNG750
MONOPYRROLE see PPS250
MONORHEIN see RHZ700
MONOSILANE see SDH575
MONOSODIOGLUTAMMATO (ITALIAN) see MRL500
MONOSODIUM ACID METHANEARSONATE see MRL750
MONOSODIUM ACID METHARSONATE see MRL750
MONOSODIUM (4-AMINOPHENYL)ARSONATE see ARA500
MONOSODIUM ARSENATE see ARD600
MONOSODIUM ASCORBATE see ARN125
MONOSODIUM CARBONATE see SFC500
MONOSODIUM DIHYDROGEN PHOSPHATE see SJH100
MONOSODIUM-5-ETHYL-5-(1-METHYLBUTYL) THIOBARBITURATE see PBT500
MONOSODIUM GLUCONATE see SHK800
MONOSODIUM GLUTAMATE see MRL500
α-MONOSODIUM GLUTAMATE see MRL500
MONOSODIUM-l-GLUTAMATE (FCC) see MRL500
MONOSODIUM METHANEARSONATE see MRL750
MONOSODIUM METHANEARSONIC ACID see MRL750
MONOSODIUM METHYLARSONATE see MRL750
MONOSODIUM NOVOBIOCIN see NOB000
MONOSODIUM PHOSPHATE see SJH100
MONOSODIUM THYROXINE see LFG050
MONOSODIUM URATE see SKO575
MONOSORB XP-4 see SJH100
MONOSTEARIN see OAV000
MONOSULFUR DICHLORIDE see SOG500
MONOTEN see PFC750
MONOTHIOETHYLENEGLYCOL see MCN250
MONOTHIOGLYCEROL see MRM750
α-MONOTHIOGLYCEROL see MRM750
MONO-THIURAD see BJL600
MONOTHIURAM see BJL600
MONOTRICHLOR-AETHYLIDEN-α-GLUCOSE (GERMAN) see GFA000
MONOXONE see SFU500
β-MONOXYNAPHTHALENE see NAX000
MONSANTO CP 47114 see DSQ000
MONSANTO CP-49674 see DOP200
MONSUR see CBM750
MONTAN 80 see SKV100
MONTANE 60 see SKV150
MONTANOX 80 see PKL100
MONTAR see PJL750
MONTAR see HKC000
MONTHYBASE see EJM500
MONTHYLE see EJM500
MONTMORILLONITE see BAV750
MONTROSE PROPANIL see DGI000
MONUREX see CJX750
MONURON see CJX750
MONUROX see CJX750
MONURUON see CJX750
MONUURON see CJX750
MOON see GGA000
MOP see NKV000
8-MOP see XDJ000

MOPA see MFE250
MOPARI see DGP900
MOPLEN see PMP500
MOPLEN RO-QG 6015 see PJS750
MORBOCID see FMV000
MORBUSAN see ENB500
MOREPEN see AOD125
MORESTIN see EDV600
MORFINA (ITALIAN) see MRO500
MORIN see MRN500
MORIN see MRN500
MORONAL see NOH500
MOROSAN see DCK759
MORPHIA see MRO500
MORPHINA see MRO500
(−)-MORPHINE see MRO500
MORPHINE see MRO500
MORPHINE CHLORHYDRATE see MRO750
MORPHINE CHLORIDE see MRO750
MORPHINE HYDROCHLORIDE see MRO750
MORPHINE SULFATE see MRP250
MORPHINE SULPHATE see MRP250
MORPHINISM see MRO500
MORPHINUM see MRO500
MORPHIUM see MRO500
MORPHOLINE see MRP750
MORPHOLINE, N-AMINOPROPYL- see AMF250
MORPHOLINE, AQUEOUS MIXTURE (DOT) see
MRP750
MORPHOLINE, 4-(3-(BIS(2-CHLOROETHYL)AMINO)-
p-TOLUOYL)- see BHR400
MORPHOLINE DISULFIDE see BKU500
MORPHOLINE, N-METHYL- see MMA250
MORPHOLINIUM, 2,2'-(1,1'-BIPHENYL)-4,4'-
DIYLBIS(2-HYDROXY-4,4-DIMETHYL- see HAP100
MORPHOLINIUM, 2,2'-(4,4'-BIPHENYLENE)BIS(2-
HYDROXY-4,4-DIMETHYL- see HAP100
MORPHOLINIUM, HEXACHLOROSTANNATE(2-)
(2:1) see DUO500
MORPHOLINIUM, (3-INDOLYLMETHYLENE)-,
HEXACHLOROSTANNATE(2-) (2:1) see BKJ500
9-(MORPHOLINOAMINO)ACRIDINE
MONO(METHYL SULFATE) see MRR115
MORPHOLINOBENZENE see PFS750
4-MORPHOLINOCARBONYL-2,3-
TETRAMETHYLENEQUINOLINE see MRR760
MORPHOLINODISULFIDE see BKU500
l-5-(MORPHOLINOMETHYL)-3-((5-
NITROFURFURYLIDENE)AMINO)-2-
OXAZOLIDINONEHYDROCHLORIDE see FPI150
MORPHOLINO(7,8,9,10-TETRAHYDRO-11-(6H-
CYCLOHEPTA(b)QUINOLINYL)) KETONE see
MRU077
2-(MORPHOLINOTHIO)BENZOTHIAZOLE see
BDG000
MORPHOLINYLMERCAPTOBENZOTHIAZOLE see
BDG000
2-(4-MORPHOLINYLTHIO)BENZOTHIAZOLE see
BDG000
MORSODREN see MLF250
MORTON EP-227 see MLF250
MORTON EP332 see DSO200
MORTON SOIL DRENCH see MLF250
MORTON SOIL-DRENCH-C see MLF250
MORTON WP-161E see ISE000
MORTOPAL see TCF250
MOSANON see SEH000
MOSAPRIDE see MRU253

MOSCARDA see MAK700
MOSCHUS KETONE see MIT625
MOSS GREEN see COF500
MOSTEN see PMP500
MOTH BALLS (DOT) see NAJ500
MOTH FLAKES see NAJ500
MOTILYN see PAG200
MOTIORANGE R see PEJ500
MOTIROT 2R see OHI200
MOTIROT G see XRA000
MOTOR BENZOL see BBL250
MOTOR SPIRIT (DOT) see GBY000
MOTOX see CDV100
MOTRIN see IIU000
MOTTENHEXE see HCI000
MOUNTAIN GREEN see COF500
MOUS-CON see ZLS000
MOUSE-NOTS see SMN500
MOUSE PAK see WAT200
MOUSE-RID see SMN500
MOUSE-TOX see SMN500
MOVINYL 100 see PKQ059
MOVINYL 114 see AAX250
MOWIOL see PKP750
MOWIOL N 30-88 see PKP750
MOWIOL N 50-98 see PKP750
MOWIOL N 70-98 see PKP750
MOXIE see MEI450
MP see POK000
8-MP see XDJ000
MP 12-50 see TAB750
MP 25-38 see TAB750
MP 45-26 see TAB750
3-MPA see MFM000
M.P. CHLORCAPS T.D. see TAI500
MPG see MRK500
MPK see PBN250
MPK 90 see PKQ250
MPN see MNA000
MPNU see MMY500
MR 84 see SCK600
MR 200 see PKB100
MR 2000 see PKB100
MRAVENCAN DI-n-BUTYLCINICITY (CZECH) see
DDZ000
MRAVENCAN SODNY see SHJ000
MRAVENCAN VAPENATY (CZECH) see CAS250
MRC 910 see GIA000
MROWCZAN ETYLU (POLISH) see EKL000
MS 33 see SKV150
MS 53 see SNK000
MS 33F see SKV150
MSG see MRL500
MSMA see MRL750
MSMED see ECU750
MSZYCOL see BBQ500
MTs see MIF760
MTBE see MHV859
MTBHQ see BRM500
MTD see TGL750
MTD see DTQ400
M.T.F. see DUV600
MTMC see MIB750
M.T. MUCORETTES see MPN500
MTQ HYDROCHLORIDE see MDT250
MTU see MPW500
MTX see MDV500
MUCICLAR see CBR675

MUCIDRINA see VGP000
MUCOCHLORIC ACID see MRU900
MUCOCIS see CBR675
MUCODYNE see CBR675
MUCOLASE see CBR675
MUCOLEX see CBR675
MUCOLYTICUM see ACH000
MUCOLYTICUM LAPPE see ACH000
MUCOMYST see ACH000
MUCOPRONT see CBR675
MUCORAMA see PMJ500
MUCOSOLVIN see ACH000
MUGAN see CBM750
MUIRAMID see AAI250
MUL F 66 see PKL750
MULHOUSE WHITE see LDY000
MULSIFEROL see VSZ100
MULTAMAT see DQM600
MULTICHLOR see CDP000
MULTIN see HIM000
MURACIL see MPW500
MURCIL see MDQ250
MUREL see VBK000
MUREX see GFA000
MURFOS see PAK000
MURFULVIN see GKE000
MURIATE of PLATINUM see PJE000
MURIATIC ACID see HHL000
MURIATIC ETHER see EHH000
MURIOL see CJJ000
MURVIN see CBM750
MUSCLE ADENYLIC ACID see AOA125
MUSCONE see MIT625
MUSK 36A see ACL750
MUSK AMBRETTE see BRU500
MUSKONE see MIT625
MUSK R 1 see OKW100
MUSK XYLENE see TML750
MUSK XYLENE (DOT) see TML750
MUSK XYLOL see TML750
N-MUSTARD (GERMAN) see BIE500
MUSTARD GAS see BIH250
MUSTARD HD see BIH250
MUSTARD OIL see AGJ250
MUSTARD VAPOR see BIH250
MUSTARGEN see BIE250
MUSTARGEN see BIE500
MUSTARGEN HYDROCHLORIDE see BIE500
MUSTINE see BIE250
MUSTINE HYDROCHLOR see BIE500
MUSTINE HYDROCHLORIDE see BIE500
MUSTRON see CFA750
MUSUET SYNTHETIC see CMS850
MUSUETTINE PRINCIPLE see CMS850
MUTAGEN see BIE250
MUTAMYCIN see AHK500
MUTAMYCIN (MITOMYCIN for INJECTION) see AHK500
MUTHMANN'S LIQUID see ACK250
MUTOXIN see DAD200
MV see MAK275
MVNA see NKY000
2M-4X see SIL500
MX 4500 see SMQ500
MX 5514 see SMQ500
MX 5516 see SMQ500
MX 5517-02 see SMQ500
MXDA see XHS800

MYACYNE see NCE000
MY-B-DEN see AOA125
MYCAIFRADIN SULFATE see NCG000
MYCARDOL see PBC250
MYCHEL see CDP250
MYCIFRADIN see NCE000
MYCIFRADIN-N see NCG000
MYCIGIENT see NCG000
MYCINOL see CDP250
MYCLOBUTANIL see MRW775
MYCOBAN see SJL500
MYCOFARM see BFD250
MYCOPHYT see PIF750
MYCOSHIELD TMQTHC 20 see HOH500
MYCOSTATIN see NOH500
MYCOSTATIN 20 see NOH500
MYCOZOL see TEX000
MYCRONIL see BJK500
MYDFRIN see SPC500
MYDRIATIN see NNM000
MYDRIATINE see PMJ500
MYELOLEUKON see BOT250
MYKOSTIN see VSZ100
MYLEPSIN see DBB200
MYLEPSINUM see DBB200
MYLERAN see BOT250
MYLON (CZECH) see DSB200
MYLONE see DSB200
MYLONE 85 see DSB200
MYLOSAR see ARY000
MYNOSEDIN see IIU000
MYOCHRYSINE see GJC000
MYOCON see NGY000
MYOCRISIN see GJC000
MYODIGIN see DKL800
MYOFER 100 see IGS000
MYOGLYCERIN see NGY000
MYOHEMATIN see CQM325
MYO-INOSISTOL HEXAKISPHOSPHATE see PIB250
MYO-INOSITOL HEXAPHOSPHATE see PIB250
MYOPONE see WBJ700
MYOREXON see CCK125
MYOSTHENINE see VGP000
MYOSTIBIN see AQH800
MYOSTON see AOA125
MYOTRATE "10" see PBC250
MYPROZINE see PIF750
MYRAFORM see PKQ059
MYRCENE see MRZ150
MYRCIA OIL see BAT500
MYRCIA OIL see LBK000
MYRICIA OIL see BAT500
MYRINGACAINE DROPS see CHW675
MYRISTIC ACID see MSA250
MYRISTIC ACID, ISOPROPYL ESTER see IQN000
MYRISTICA OIL see NOG500
MYRISTYL ALCOHOL (mixed isomers) see TBY500
MYRISTYL-γ-PICOLINIUM CHLORIDE see MSB500
MYRISTYL STEARATE see TCB100
MYRISTYL SULFATE, SODIUM SALT see SIO000
MYSEDON see DBB200
MYSOLINE see DBB200
MYSORITE see ARM262
MYSTECLIN-F see AOC500
MYSTOX WFA see BGJ750
MYTOMYCIN see AHK500
MYTRATE see VGP000

MYVAK see VSK900
MYVAX see VSK900
MYVPACK see VSK600
N 5 see EHG100
N-9 see NNB300
N 135 see EHP700
N-399 see PEM750
N 521 see DSB200
N 2790 see FMU045
N 4446 see EMR100
Ni 270 see NCW500
168N15 see SMQ500
Ni 4303T see NCW500
NA see NBE500
NA see EID000
NA-22 see IAQ000
NA-101 see TDX000
N-1544A see PFC750
NAA 800 see NAK500
Na-AESCINAT see EDM000
NABAC see HCL000
NABAM see DXD200
NABAME (FRENCH) see DXD200
NABOLIN see MPN500
NAC see ACH000
NAC see CBM750
NACARAT A EXPORT see HJF500
NACCANOL NR see DXW200
NACCONATE-100 see TGM740
NACCONATE 100 see TGM750
NACCONATE 300 see MJP400
NACCONATE 400 see BBP000
NACCONATE H 12 see MJM600
NACCONOL 98SA see LBU100
NACELAN BLUE G see TBG700
NACELAN FAST YELLOW CG see AAQ250
NACELAN PINK B see AKE250
NACM-CELLULOSE SALT see SFO500
NAC-TB see ACH000
NACYCLYL see EDR000
NADEINE see DKW800
NA-DESOXYCHOLAT (GERMAN) see SGE000
NADISAL see SJO000
NADISAN see BSM000
NADIZAN see BSM000
NADONE see CPC000
NADOZONE see BRF500
NAFEEN see SHF500
NAFENOIC ACID see MCB500
NAFENOPIN see MCB500
NAFTALEN (POLISH) see NAJ500
1-NAFTILAMINA (SPANISH) see NBE700
β-NAFTILAMINA (ITALIAN) see NBE500
1-NAFTIL-TIOUREA (ITALIAN) see AQN635
2-NAFTOL (DUTCH) see NAX000
β-NAFTOL (DUTCH) see NAX000
2-NAFTOLO (ITALIAN) see NAX000
β-NAFTOLO (ITALIAN) see NAX000
α-NAFTYLAMIN (CZECH) see NBE700
β-NAFTYLAMIN (CZECH) see NBE500
1-NAFTYLAMINE (DUTCH) see NBE700
2-NAFTYLAMINE (DUTCH) see NBE500
1-NAFTYLESTER KYSELINY
METHYLKARBAMINOVE see CBM750
α-NAFTYL-N-METHYLKARBAMAT see CBM750
β-NAFTYLOAMINA (POLISH) see NBE500
1-NAFTYLTHIOUREUM (DUTCH) see AQN635

NAFUSAKU see NAK500
NAGARSE see BAC000
NAGRAVON see VSZ000
NAH see NCQ900
NAH 80 see SHO500
NAKO BROWN R see ALT250
NAKO H see PEY500
NAKO TEG see ALS990
NAKO TGG see REA000
NAKO TMT see TGL750
NAKO TRB see NAW500
NAKO TSA see DBO400
NAKO YELLOW EGA see ALT000
NALCAST see SCK600
NALCO 680 see AHG000
NALCOAG see SCH002
NALCON 243 see DSB200
NALIDIC ACID see EID000
NALIDICRON see EID000
NALIDIXIC ACID see EID000
NALIDIXIN see EID000
NALITUCSAN see EID000
NALKIL see BMM650
NALOX see MMN250
NALUTRON see PMH500
NAM see NCR000
Na MCPA see SIL500
NAMEKIL see TDW500
N4 AMINE see TBI700
NAMURON see TDA500
5-NAN see NEJ500
NANCHOR see RMA500
NANDERVIT-N see NCR000
NANKER see RMA500
NANKOR see RMA500
NANSA 1042P see LBU100
NANSA SSA see LBU100
NAOTIN see NCQ900
NAPA see HIM000
NAPACETIN see IIU000
NAPCLOR-G see SJA000
NAPENTAL see NBU000
NAPHID see NAR000
NAPHTAMINE BLUE 2B see CMO000
NAPHTAMINE BLUE 2B see CMO250
NAPHTAMINE BLUE 10G see CMO500
NAPHTAMINE SKY BLUE DD see CMN750
NAPHTHA see NAH600
NAPHTHA see NAH600
NAPHTHA, hydrotreated see NAH600
NAPHTHA COAL TAR (OSHA) see NAH600
2-NAPHTHALAMINE see NBE500
NAPHTHALANE see DAE800
2-NAPHTHALENAMINE see NBE500
NAPHTHALENE see NAJ500
NAPHTHALENE, molten (DOT) see NAJ500
NAPHTHALENE, crude or refined (DOT) see NAJ500
1-NAPHTHALENEACETIC ACID see NAK500
NAPHTHALENE-1-ACETIC ACID see NAK500
α-NAPHTHALENEACETIC ACID see NAK500
1,5-NAPHTHALENEDIAMINE see NAM000
1,5-NAPHTHALENE DIISOCYANATE see NAM500
NAPHTHALENE, 1,3-DINITRO- see DUX650
NAPHTHALENE, 1,8-DINITRO- see DUX710
1,4-NAPHTHALENEDIONE see NBA500
1,3-NAPHTHALENEDISULFONIC ACID, 8-((3,3'-
DIMETHYL-4'-((4-(((4-

METHYLPHENYL)SULFONYL)OXY)
PHENYL)AZO)(1,1'-BIPHENYL)-4-YL)AZO)-7-
HYDROXY-, DISODIUM SALT see CMM330
1,3-NAPHTHALENEDISULFONIC ACID, 7-
HYDROXY-8-((4-SULFO-1-NAPHTHYL)AZO)-,
TRISODIUM SALT see FMU080
1,2-(1,8-NAPHTHALENEDIYL)BENZENE see FDF000
NAPHTHALENE FAST ORANGE 2GS see HGC000
NAPHTHALENE, 1,2,3,4,6,7-HEXABROMO- see
HCA650
NAPHTHALENE, 1,2,3,4,6,7-HEXACHLORO- see
HCK550
NAPHTHALENE, 1,2,3,5,6,7-HEXACHLORO- see
HCK600
NAPHTHALENE INK SCARLET 4R see FMU080
NAPHTHALENE LAKE SCARLET R see FMU070
NAPHTHALENE, MONONITRO- see NHP990
NAPHTHALENE, NITRO- see NHP990
NAPHTHALENE OIL see CMY825
2-NAPHTHALENETHIOL see NAP500
NAPHTHALENE-2-THIOL see NAP500
β-NAPHTHALENETHIOL see NAP500
1-NAPHTHALENOL see NAW500
2-NAPHTHALENOL see NAX000
2-NAPHTHALENOL, DECAHYDRO-, FORMATE see
DAF150
2-NAPHTHALENOL, 1-((2-METHOXYPHENYL)AZO)-
(9CI) see CMS238
1-NAPHTHALENOL, METHYLCARBAMATE (9CI) see
CBM750
1-(1-NAPHTHALENYL)ETHANONE see ABC475
1-(2-NAPHTHALENYL)ETHANONE see ABC500
1-NAPHTHALENYL METHYLCARBAMATE see
CBM750
α-NAPHTHALENYL METHYLCARBAMATE see
CBM750
1-NAPHTHALENYLTHIOUREA see AQN635
NAPHTHALIDINE see NBE700
NAPHTHALIN (DOT) see NAJ500
NAPHTHALINE see NAJ500
α-NAPHTHALTHIOHARNSTOFF (GERMAN) see
AQN635
NAPHTHANE see DAE800
NAPHTHANIL BLUE B BASE see DCJ200
NAPHTHANIL SCARLET G BASE see NMP500
NAPHTHANTHRACENE see BBC250
NAPHTHA (PETROLEUM), LIGHT CATALYTIC
CRACKED see NAQ540
NAPHTHA (PETROLEUM), LIGHT STRAIGHT-RUN
see NAQ560
NAPHTHA SAFETY SOLVENT see SLU500
NAPHTHA (UN2553) (DOT) see NAH600
NAPHTHA, solvent (UN1256) (DOT) see NAH600
NAPHTHA, petroleum (UN1255) (DOT) see NAH600
NAPHTHENATE de COBALT (FRENCH) see NAR500
NAPHTHENE see NAJ500
NAPHTHENIC ACID see NAR000
NAPHTHENIC ACID, COBALT SALT see NAR500
NAPHTHENIC ACID, COPPER SALT see NAS000
NAPHTHENIC ACID, LEAD SALT see NAS500
NAPHTHENIC BASE LUBE STOCK see MQV845
NAPHTHENIC OILS (PETROLEUM), CATALYTIC
DEWAXED HEAVY (9CI) see MQV776
NAPHTHENIC OILS (PETROLEUM), CATALYTIC
DEWAXED LIGHT (9CI) see MQV777
NAPHTHIOMATE T see TGB475
NAPHTHO(1,2,3,4-def)CHRYSENE see NAT500

1-NAPHTHOL see NAW500
2-NAPHTHOL see NAX000
α-NAPHTHOL see NAW500
β-NAPHTHOL see NAX000
NAPHTHOL B see NAX000
2-NAPHTHOL, DECAHYDRO-, ACETATE see DAF100
2-NAPHTHOL, DECAHYDRO-, FORMATE see DAF150
2-NAPHTHOL-3,6-DISULFONIC ACID SODIUM SALT
see ROF300
1-NAPHTHOL N-METHYLCARBAMATE see CBM750
NAPHTHOL YELLOW see DUX800
1,4-NAPHTHOQUINONE see NBA500
α-NAPHTHOQUINONE see NBA500
NAPHTHOSOL FAST RED KB BASE see CLK225
α-NAPHTHOTHIOUREA see AQN635
2H-NAPHTHO(1,2-d)TRIAZOLE, 2-(4-STYRYL-3-
SULFOPHENYL)-7-SULFO-, DISODIUM SALT see
DXG025
NAPHTHO(1,2-d)TRIAZOLE-7-SULFONIC ACID,2-(4-
(2-PHENYLETHENYL)-3-SULFOPHENYL)-,
DISODIUM see DXG025
α-NAPHTHYLACETIC see NAK500
NAPHTHYLACETIC ACID see NAK500
1-NAPHTHYLACETIC ACID see NAK500
α-NAPHTHYLACETIC ACID see NAK500
β-NAPHTHYL ALCOHOL see NAX000
1-NAPHTHYLAMIN (GERMAN) see NBE700
2-NAPHTHYLAMIN (GERMAN) see NBE500
β-NAPHTHYLAMIN (GERMAN) see NBE500
1-NAPHTHYLAMINE see NBE700
2-NAPHTHYLAMINE see NBE500
6-NAPHTHYLAMINE see NBE500
α-NAPHTHYLAMINE see NBE700
β-NAPHTHYLAMINE see NBE500
NAPHTHYLAMINE BLUE see CMO250
2-NAPHTHYLAMINE-6,8-DISULFONIC ACID see
NBE850
NAPHTHYLAMINE MUSTARD see BIF250
2-NAPHTHYLAMINE MUSTARD see NBE500
N-(2-NAPHTHYL)ANILINE see PFT500
2-NAPHTHYLBIS(2-CHLOROETHYL)AMINE see
BIF250
β-NAPHTHYL-BIS-(β-CHLOROETHYL)AMINE see
BIF250
1-NAPHTHYL CHLOROCARBONATE see NBH200
1-NAPHTHYL CHLOROFORMATE see NBH200
α-NAPHTHYL CHLOROFORMATE see NBH200
β-NAPHTHYL-DI-(2-CHLOROETHYL)AMINE see
BIF250
α-NAPHTHYLENEACETIC ACID see NAK500
1,2-(1,8-NAPHTHYLENE)BENZENE see FDF000
1,5-NAPHTHYLENEDIAMINE see NAM000
NAPHTHYLENE YELLOW see DUX800
α-NAPHTHYLESSIGSAEURE (GERMAN) see NAK500
β-NAPHTHYL HYDROXIDE see NAX000
2-NAPHTHYL MERCAPTAN see NAP500
β-NAPHTHYL MERCAPTAN see NAP500
1-NAPHTHYL METHYLCARBAMATE see CBM750
1-NAPHTHYL N-METHYLCARBAMATE see CBM750
α-NAPHTHYL METHYLCARBAMATE see CBM750
α-NAPHTHYL N-METHYLCARBAMATE see CBM750
1-NAPHTHYL-N-METHYL-KARBAMAT see CBM750
1-NAPHTHYL METHYL KETONE see ABC475
2-NAPHTHYL METHYL KETONE see ABC500
α-NAPHTHYL METHYL KETONE see ABC475

β-NAPHTHYL METHYL KETONE see ABC500
1-NAPHTHYL METHYLNITROSOCARBAMATE see NBJ500
1-NAPHTHYL-N-METHYL-N-NITROSOCARBAMATE see NBJ500
2-NAPHTHYL N-METHYL-N-(3-TOLYL)THIONOCARBAMATE see TGB475
o-2-NAPHTHYL m,N-DIMETHYLTHIOCARBANILATE see TGB475
2-NAPHTHYLPHENYLAMINE see PFT500
β-NAPHTHYLPHENYLAMINE see PFT500
2-NAPHTHYL-p-PHENYLENEDIAMINE see NBL000
α-NAPHTHYLTHIOCARBAMIDE see AQN635
1-NAPHTHYL-THIOHARNSTOFF (GERMAN) see AQN635
2-NAPHTHYL THIOL see NAP500
1-(1-NAPHTHYL)-2-THIOUREA see AQN635
N-(1-NAPHTHYL)-2-THIOUREA see AQN635
α-NAPHTHYLTHIOUREA see AQN635
α-NAPHTHYLTHIOUREA (DOT) see AQN635
1-NAPHTHYL THIOUREA (MAK) see AQN635
1-NAPHTHYL-THIOUREE (FRENCH) see AQN635
NAPHTOCARD YELLOW O see FAG140
NAPHTOELAN FAST SCARLET G BASE see NMP500
NAPHTOELAN FAST SCARLET G SALT see NMP500
NAPHTOELAN ORANGE R BASE see NEN500
NAPHTOELAN RED GG BASE see NEO500
2-NAPHTOL (FRENCH) see NAX000
β-NAPHTOL (GERMAN) see NAX000
NAPHTOL AS-KG see TGR000
NAPHTOL AS-KGLL see TGR000
NAPHTOX see AQN635
NAPHYL-1-ESSIGSAEURE (GERMAN) see NAK500
NAPOLONE see MIF760
NAPOTON see MDQ250
NAPOTON see LFK000
NAPRINOL see HIM000
NAPROPION see SJL500
NAQUIVAL see RDK000
NARCEOL see MNR250
NARCOGEN see TIO750
NARCOLAN see THV000
NARCOLAN see ARW250
NARCOTILE see EHH000
NARCYLEN see ACI750
NARDELZINE see PFC750
NARDIL see PFC500
NARDIL see PFC750
NARGOLINE see NDM000
NARIGIX see EID000
NARKOLAN see THV000
NARKOLAN see ARW250
NARKOSOID see TIO750
NASDOL see TBG000
NASOL see EAW000
NASS (IRAN) see SED400
NASTENON see PAN100
NASWAR (PAKISTAN and AFGHANISTAN) see SED400
NA TA see TII250
NATACYN see PIF750
NATAMYCIN see PIF750
NATASOL FAST RED TR SALT see CLK235
NATHULANE see PME500
NATIONAL 120-1207 see AAX250
NATIVE CALCIUM SULFATE see CAX750
NATRASCORB see ARN125

NATRASCORB INJECTABLE see ARN000
NATREEN see SGC000
NATREEN see BCE500
NATRI-C see ARN125
NATRINAL see BAG250
NATRIONEX see AAI250
NATRI-PAS see SEP000
NATRIPHENE see BGJ750
NATRIUM see SEE500
NATRIUMACETAT (GERMAN) see SEG500
NATRIUMALUMINUMFLUORID (GERMAN) see SHF000
NATRIUMANTIMONYLTARTRAT (GERMAN) see AQI750
NATRIUMARSENIT (GERMAN) see ARJ500
NATRIUMAZID (GERMAN) see SFA000
NATRIUMAZIDE (DUTCH) see SFA000
NATRIUMBARBITALS (GERMAN) see BAG250
NATRIUMBICHROMAAT (DUTCH) see SGI000
NATRIUMCHLORAAT (DUTCH) see SFS000
NATRIUMCHLORAT (GERMAN) see SFS000
NATRIUMCHLORID (GERMAN) see SFT000
NATRIUM CITRICUM (GERMAN) see DXC400
NATRIUMDICHROMAAT (DUTCH) see SGI000
NATRIUMDICHROMAT (GERMAN) see SGI000
NATRIUM-3-α,12-α-DIHYDROXYCHOLANAT (GERMAN) see SGE000
NATRIUMFLUORACETAAT (DUTCH) see SHG500
NATRIUMFLUORACETAT (GERMAN) see SHG500
NATRIUM FLUORIDE see SHF500
NATRIUMGLUTAMINAT (GERMAN) see MRL500
NATRIUMHEXAFLUOROALUMINATE (GERMAN) see SHF000
NATRIUMHYDROXID (GERMAN) see SHS000
NATRIUMHYDROXYDE (DUTCH) see SHS000
NATRIUMHYPOPHOSPHIT (GERMAN) see SHV000
NATRIUMJODID (GERMAN) see SHW000
NATRIUMMOLYBDAT (GERMAN) see DXE800
NATRIUM NITRIT (GERMAN) see SIQ500
NATRIUMPERCHLORAAT (DUTCH) see PCE750
NATRIUMPERCHLORAT (GERMAN) see PCE750
NATRIUMPHOSPHAT (GERMAN) see SJH090
NATRIUMPROPIONAT see SJL500
NATRIUMPYROPHOSPHAT see TEE500
NATRIUMRHODANID (GERMAN) see SIA500
NATRIUMSALZ DER 2,2-DICHLORPROPIONSAEURE see DGI600
NATRIUMSELENIAT (GERMAN) see DXG000
NATRIUMSELENIT (GERMAN) see SJT500
NATRIUMSILICOFLUORID (GERMAN) see DXE000
NATRIUMSULFAT (GERMAN) see SJY000
NATRIUMSULFID (GERMAN) see SJZ000
NATRIUMTRIPOLYPHOSPHAT (GERMAN) see SKN000
NATRIUMZYKLAMATE (GERMAN) see SGC000
NATULAN see PME500
NATULANAR see PME500
NATULAN HYDROCHLORIDE see PME500
NATURAL CALCIUM CARBONATE see CAO000
NATURAL GASOLINE (DOT) see GBY000
NATURAL GAS, compressed (with high methane content) (UN 1971) (DOT) see MDQ750
NATURAL GAS, refrigerated liquid (cryogenic liquid) (with high methane content) (UN 1972) (DOT) see MDQ750
NATURAL IRON OXIDES see IHC450
NATURAL LEAD SULFIDE see LDZ000
NATURAL RED OXIDE see IHC450

NATURAL WINTERGREEN OIL see MPI000
NAUCAINE see AIT250
NAUGARD TJB see DWI000
NAUGARD TKB see NKB500
NAUGATUCK D-014 see SOP000
NAUGATUCK DET see DKC800
NAULI "GUM" see PMQ750
NAUSEN see BBV500
NAVICALM see HGC500
NAVRON see FFF000
NAXAMIDE see IMH000
NAXOL see CPB750
NAYPER B and BO see BDS000
NB2B see CMO000
NBHA see HJQ350
N.B. MECOPROP see CIR500
NBN see BRV500
NBS 706 see SMQ500
NC5 see HER500
NC 26 see BHO250
NC 100 see IGK800
NC 150 see PDC250
NC-262 see DSP400
NC-6897, ACETYL DERIVATIVE see BCJ005
NCI-C00044 see AFK250
NCI-C00066 see ASH500
NCI-C00077 see CBG000
NCI-C00099 see CDR750
NCI-C00102 see TBQ750
NCI-C00113 see DGP900
NCI-C00124 see DHB400
NCI-C00135 see DSP400
NCI-C00157 see EAT500
NCI-C00168 see TBW100
NCI-C00180 see HAR000
NCI-C00191 see KEA000
NCI-C00204 see BBQ500
NCI-C00215 see MAK700
NCI-C00226 see PAK000
NCI-C00237 see PIB900
NCI-C00259 see CDV100
NCI-C00260 see HON000
NCI-C00395 see DVQ709
NCI-C00408 see DER000
NCI-C00419 see PAX000
NCI-C00420 see DFT800
NCI-C00442 see DUV600
NCI-C00453 see CDO250
NCI-C00464 see DAD200
NCI-C00475 see BIM500
NCI-C00486 see BIO750
NCI-C00497 see MEI450
NCI-C00500 see DDL800
NCI-C00511 see EIY600
NCI-C00522 see EIY500
NCI-C00533 see CKN500
NCI-C00544 see DOS000
NCI-C00555 see BIM750
NCI-C00566 see EAQ750
NCI-C00920 see EJC500
NCI-C01445 see NEI000
NCI-C01478 see LBG000
NCI-C01514 see AES750
NCI-C01558 see CME250
NCI-C01569 see ARY000
NCI-C01592 see BOT250
NCI-C01616 see IAN000
NCI-C01627 see PIK250

NCI-C01638 see IMH000
NCI-C01649 see TFQ750
NCI-C01661 see DDG800
NCI-C01672 see PDC250
NCI-C01683 see TGD000
NCI-C01707 see TFI000
NCI-C01718 see SOA500
NCI-C01729 see TNX000
NCI-C01730 see API500
NCI-C01785 see POL500
NCI-C01810 see PME500
NCI-C01821 see DSY600
NCI-C01832 see TGM400
NCI-C01832 see DCE600
NCI-C01843 see NMP500
NCI-C01854 see HHG000
NCI-C01865 see DVH000
NCI-C01876 see AIB000
NCI-C01901 see AKP750
NCI-C01923 see MMG000
NCI-C01934 see NEQ500
NCI-C01956 see NHQ000
NCI-C01967 see NEJ500
NCI-C01978 see NEL000
NCI-C01989 see DBO400
NCI-C01990 see MJN000
NCI-C02006 see MQS500
NCI-C02017 see PGN250
NCI-C02028 see DBF800
NCI-C02039 see CEH680
NCI-C02040 see CLK215
NCI-C02051 see CLK225
NCI-C02073 see EFE000
NCI-C02084 see SIQ500
NCI-C02095 see AEN000
NCI-C02108 see AAI000
NCI-C02119 see USS000
NCI-C02142 see HEM500
NCI-C02175 see DCJ400
NCI-C02186 see TMH750
NCI-C02200 see SMQ000
NCI-C02222 see ALL750
NCI-C02244 see NKB500
NCI-C02299 see TLG250
NCI-C02302 see TGL750
NCI-C02335 see TGS500
NCI-C02368 see CLK235
NCI-C02551 see CAH500
NCI-C02653 see HCL000
NCI-C02664 see PJN000
NCI-C02686 see CHJ500
NCI-C02711 see CAJ750
NCI-C02722 see TGC000
NCI-C02733 see CAK500
NCI-C02766 see AMT500
NCI-C02799 see FMV000
NCI-C02813 see PIX250
NCI-C02835 see SGJ000
NCI-C02846 see CJX750
NCI-C02857 see EPJ000
NCI-C02868 see DJC000
NCI-C02880 see DWI000
NCI-C02891 see LCW000
NCI-C02904 see TIW000
NCI-C02915 see PFT500
NCI-C02926 see ASL250
NCI-C02937 see CAQ250
NCI-C02959 see DXH250

NCI-C02960 see CMF400
NCI-C02971 see MNH000
NCI-C02982 see MGO750
NCI-C02993 see MGO500
NCI-C03010 see DOU600
NCI-C03021 see NAM000
NCI-C03043 see AJV250
NCI-C03054 see DFE200
NCI-C03065 see ALQ000
NCI-C03134 see EFU000
NCI-C03167 see SMD000
NCI-C03258 see ANO500
NCI-C03270 see TNC500
NCI-C03292 see CFK125
NCI-C03305 see CJY120
NCI-C03361 see BBX000
NCI-C03372 see IAQ000
NCI-C03474 see ASB250
NCI-C03485 see CDO500
NCI-C03510 see API750
NCI-C03521 see BDH250
NCI-C03554 see TBQ100
NCI-C03598 see BFW750
NCI-C03601 see PHW750
NCI-C03678 see OAJ000
NCI-C03689 see DVQ000
NCI-C03703 see HCF500
NCI-C03714 see TAI000
NCI-C03736 see AOQ000
NCI-C03736 see BBL000
NCI-C03747 see AOX250
NCI-C03758 see AOX500
NCI-C03770 see CEC000
NCI-C03781 see TMD250
NCI-C03792 see ENV500
NCI-C03805 see BNK000
NCI-C03816 see DKC400
NCI-C03827 see DQD400
NCI-C03838 see CIV000
NCI-C03850 see DVR200
NCI-C03918 see DQJ200
NCI-C03941 see ALL500
NCI-C03963 see NEM480
NCI-C03974 see TNL250
NCI-C03985 see DGG950
NCI-C04126 see DTH000
NCI-C04159 see BKF250
NCI-C04240 see TGG760
NCI-C04251 see TGF250
NCI-C04502 see DGW200
NCI-C04535 see DFF809
NCI-C04546 see TIO750
NCI-C04568 see IEP000
NCI-C04579 see TIN000
NCI-C04580 see PCF275
NCI-C04591 see CBV500
NCI-C04604 see HCI000
NCI-C04615 see AGB250
NCI-C04626 see MIH275
NCI-C04637 see TIP500
NCI-C04671 see MDV500
NCI-C04693 see DAC000
NCI-C04706 see AHK500
NCI-C04717 see DAB600
NCI-C04740 see CGV250
NCI-C04773 see BIF750
NCI-C04784 see MPU000
NCI-C04795 see DDJ000

NCI-C04819 see DCF200
NCI-C04820 see BIA250
NCI-C04853 see PED750
NCI-C04864 see LEY000
NCI-C04886 see POK000
NCI-C04897 see PLZ000
NCI-C04900 see CQC650
NCI-C04922 see CQN000
NCI-C04944 see BHT750
NCI-C04955 see CHD250
NCI-C05970 see REA000
NCI-C06008 see TAB750
NCI-C06111 see BDX500
NCI-C06155 see BQQ750
NCI-C06224 see EHH000
NCI-C06360 see BEE375
NCI-C06428 see MQW500
NCI-C06462 see SFA000
NCI-C06508 see BDX000
NCI-C07272 see TGK750
NCI-C08640 see CBM500
NCI-C08662 see CNU750
NCI-C08673 see DCM750
NCI-C08695 see DUK800
NCI-C08991 see ARM250
NCI-C08991 see ARM280
NCI-C09007 see ARM275
NCI-C50011 see BCP250
NCI-C50033 see SBT000
NCI-C50044 see BII250
NCI-C50055 see DUE000
NCI-C50077 see PMO500
NCI-C50088 see EJN500
NCI-C50099 see PNL600
NCI-C50102 see MJP450
NCI-C50124 see PDN750
NCI-C50135 see EIU800
NCI-C50146 see OPM000
NCI-C50157 see RDK000
NCI-C50168 see CNU000
NCI-C50191 see SIB600
NCI-C50204 see TAV750
NCI-C50351 see BGJ250
NCI-C50384 see EFT000
NCI-C50395 see GLU000
NCI-C50419 see LIA000
NCI-C50442 see BJK500
NCI-C50453 see EQR500
NCI-C50464 see AGJ250
NCI-C50475 see AEX250
NCI-C50533 see TGM750
NCI-C50602 see BOP500
NCI-C50613 see AMW000
NCI-C50635 see BLD500
NCI-C50646 see CBF700
NCI-C50657 see DBL200
NCI-C50668 see MJP400
NCI-C50680 see MLH750
NCI-C50715 see MCB000
NCI-C50748 see AQQ500
NCI-C52459 see TBQ000
NCI-C52733 see DVL700
NCI-C52904 see NAJ500
NCI-C53587 see PAH780
NCI-C53634 see HCA500
NCI-C53781 see AAQ250
NCI-C53792 see CHP500
NCI-C53838 see HGC000

NCI-C53849 see HJF500
NCI-C53894 see PAW500
NCI-C53907 see FAG150
NCI-C53929 see PEJ500
NCI-C54262 see VPK000
NCI-C54375 see BEC500
NCI-C54386 see AEO000
NCI-C54546 see SBW000
NCI-C54557 see AQP000
NCI-C54568 see CMO750
NCI-C54579 see CMO000
NCI-C54604 see MJQ100
NCI-C54626 see MRC000
NCI-C54660 see DHF200
NCI-C54706 see FEW000
NCI-C54717 see ISV000
NCI-C54728 see DTD800
NCI-C54739 see RMK020
NCI-C54773 see DSG600
NCI-C54795 see DHE750
NCI-C54808 see ARN000
NCI-C54819 see IKE000
NCI-C54820 see CIU750
NCI-C54831 see TIQ250
NCI-C54853 see EES350
NCI-C54886 see CEJ125
NCI-C54900 see TBG700
NCI-C54922 see ALO750
NCI-C54933 see PAX250
NCI-C54944 see DEP600
NCI-C54955 see DEP800
NCI-C54966 see REF000
NCI-C54977 see EBR000
NCI-C54988 see TCF000
NCI-C54999 see CPD750
NCI-C55005 see CPC000
NCI-C55072 see CDS000
NCI-C55107 see CEA750
NCI-C55118 see CEQ600
NCI-C55130 see BNL000
NCI-C55141 see PNJ400
NCI-C55152 see AQF000
NCI-C55163 see CCP250
NCI-C55174 see DHF000
NCI-C55196 see NGE000
NCI-C55209 see OLA000
NCI-C55221 see SHF500
NCI-C55232 see XGS000
NCI-C55243 see BND500
NCI-C55254 see CFK500
NCI-C55265 see TAI500
NCI-C55276 see BBL250
NCI-C55287 see PAU500
NCI-C55298 see QPA000
NCI-C55301 see POP250
NCI-C55312 see CNF330
NCI-C55323 see BKE500
NCI-C55345 see DFX800
NCI-C55367 see BPX000
NCI-C55378 see PAX250
NCI-C55425 see GFQ000
NCI-C55447 see MKA500
NCI-C55469 see IEL800
NCI-C55481 see EGV400
NCI-C55492 see PEO500
NCI-C55505 see SEM000
NCI-C55527 see BOX750
NCI-C55538 see EBX500

NCI-C55549 see GGW500
NCI-C55561 see TBX250
NCI-C55572 see LFU000
NCI-C55583 see NKM000
NCI-C55594 see MHZ000
NCI-C55607 see HCE500
NCI-C55618 see IMF400
NCI-C55641 see SPC500
NCI-C55652 see EAY500
NCI-C55685 see PDE000
NCI-C55696 see SNC000
NCI-C55709 see CDP250
NCI-C55721 see DNA800
NCI-C55743 see PBC250
NCI-C55765 see DKQ000
NCI-C55776 see PJD000
NCI-C55787 see HFV500
NCI-C55798 see PDO750
NCI-C55801 see HIM000
NCI-C55812 see MIP750
NCI-C55823 see GEM000
NCI-C55834 see HIH000
NCI-C55845 see QQS200
NCI-C55856 see CCP850
NCI-C55878 see BOV000
NCI-C55890 see HHR500
NCI-C55903 see XDJ000
NCI-C55925 see CFY000
NCI-C55936 see CHJ750
NCI-C55947 see TDY250
NCI-C55969 see AOR500
NCI-C55992 see NIF000
NCI-C56031 see DFI210
NCI-C56064 see NGE500
NCI-C56075 see BAU750
NCI-C56086 see AOD125
NCI-C56100 see BFC750
NCI-C56111 see CMP969
NCI-C56122 see RGW000
NCI-C56133 see BAY500
NCI-C56144 see IDA000
NCI-C56155 see TMN490
NCI-C56166 see BCK250
NCI-C56177 see FPQ875
NCI-C56188 see XNJ000
NCI-C56199 see EID000
NCI-C56202 see FPK000
NCI-C56224 see FPU000
NCI-C56246 see QFS000
NCI-C56279 see COB260
NCI-C56280 see MNQ000
NCI-C56291 see BSU250
NCI-C56326 see AAG250
NCI-C56348 see DTC800
NCI-C56359 see PAH250
NCI-C56360 see DBB200
NCI-C56382 see BIE500
NCI-C56393 see EGP500
NCI-C56406 see VQK650
NCI-C56417 see BMC000
NCI-C56428 see DQF800
NCI-C56439 see IPD000
NCI-C56440 see HCZ000
NCI-C56451 see PKL500
NCI-C56462 see MRH000
NCI-C56473 see HOH500
NCI-C56484 see OGQ100
NCI-C56508 see HMY000

NCI-C56519 see BDF000
NCI-C56531 see BRF500
NCI-C56553 see BRV500
NCI-C56586 see CHP250
NCI-C56597 see SNH000
NCI-C56600 see SNJ000
NCI-C56633 see TKV000
NCI-C56644 see HOO100
NCI-C56655 see PAX250
NCI-C56666 see AGH150
NCI-C56762 see DSR400
NCI-C60015 see COG000
NCI-C60048 see DJX000
NCI-C60071 see MRL750
NCI-C60082 see NEX000
NCI-C60093 see PFJ250
NCI-C60106 see QCA000
NCI-C60117 see TAJ000
NCI-C60128 see CGO500
NCI-C60139 see VOA000
NCI-C60173 see MCY475
NCI-C60184 see PAD500
NCI-C60195 see PEC500
NCI-C60208 see VPP000
NCI-C60219 see PGX000
NCI-C60220 see TJB600
NCI-C60231 see CEA000
NCI-C60286 see PKL100
NCI-C60297 see CNV000
NCI-C60311 see CNA250
NCI-C60333 see HLU500
NCI-C60344 see NDK500
NCI-C60388 see NER000
NCI-C60399 see MCW250
NCI-C60402 see EEA500
NCI-C60413 see DER000
NCI-C60537 see NMO550
NCI-C60560 see TCR750
NCI-C60571 see HEN000
NCI-C60582 see PKQ250
NCI-C60606 see TAG750
NCI-C60639 see DYE600
NCI-C60662 see TMP750
NCI-C60673 see DQA400
NCI-C60695 see TMJ750
NCI-C60753 see DUP600
NCI-C60786 see NEO500
NCI-C60797 see PKQ059
NCI-C60811 see ZSJ000
NCI-C60822 see ABE500
NCI-C60866 see BRR900
NCI-C60899 see DIX200
NCI-C60902 see TLP500
NCI-C60913 see DSB000
NCI-C60924 see DWC600
NCI-C60935 see LBX000
NCI-C60946 see AFW750
NCI-C60957 see MES000
NCI-C60968 see IJS000
NCI-C60979 see IKQ000
NCI-C60980 see BCC500
NCI-C61018 see NMW500
NCI-C61029 see PMT750
NCI-C61041 see TNP500
NCI-C61052 see IJD000
NCI-C61074 see BAK000
NCI C61096 see CMM330
NCI-C61109 see CMN750

NCI-C61143 see MAU250
NCI-C61176 see ARA500
NCI-C61187 see TIV750
NCI-C61289 see CMO250
NCI C61290 see CMO500
NCI-C61405 see HEO000
NCI-C61494 see BEN000
NCI-C60253A see ARM262
NCI-C61223A see ARM268
NCI-CO1763 see BSQ000
NCI-CO2131 see BSS250
NCI-CO4035 see CFK000
NCI-CO5210 see CKP500
NCI-CO03247 see ABB000
NCO 20 see PKB100
NDBA see BRY500
NDEA see NJW500
NDELA see NKM000
NDFDA see PCG725
NDGA see NBR000
m,N-DIMETHYLTHIOCARBANILIC ACID-o-2
NAPHTHYL ESTER see TGB475
NDMA see NKA600
NDMA see DSY600
NDPA see NKB700
NDPA see DWI000
NDPhA see DWI000
NDRC-143 see AHJ750
NEA see NKD000
NEANTINE see DJX000
NEASINA see SNJ000
NEAT OIL OF SWEET ORANGE see OGY000
NEAUFATIN see TEH500
NEAZINE see PPP500
NEBERK see FLZ050
NECARBOXYLIC ACID see AFR250
NECATORINA see CBY000
NECATORINE see CBY000
NECCANOL SW see DXW200
NEDCIDOL see DCM750
NEEDLE ANTIMONY see AQL500
NEFCO see NGE500
NEFRECIL see PDC250
NEFRIX see CFY000
NEFTIN see NGG500
NEFURTHIAZOLE see NDY500
NEFUSAN see DSB200
NEGANOX DLTP see TFD500
NEGRAM see EID000
NEGUVON see TIQ250
NEGUVON A see TIQ250
NEKAL WT-27 see DJL000
NEKLACID FAST ORANGE 2G see HGC000
NEKLACID RED 3R see FMU080
NEKLACID RED RR see FMU070
NEKLACID RUBINE W see HJF750
NEKLACID YELLOW T see FAG140
NEKTROHAN see ZVJ000
NELBON see DLY000
NELLITE see PEV500
NEM see MAL250
NEM see ENL000
NEMA see MKB000
NEMA see PCF275
NEMABROM see DDL800
NEMACIDE see DFK600
NEMACTIL see PIW000
NEMACUR see FAK000

NEMAFENE see DGG000
NEMAFOS see EPC500
NEMAFUME see DDL800
NEMAGON see DDL800
NEMAGONE see DDL800
NEMAGON SOIL FUMIGANT see DDL800
NEMALITE see MAG750
NEMAMORT see BII250
NEMANAX see DDL800
NEMAPAN see TEX000
NEMAPAZ see DDL800
NEMAPHOS see EPC500
NEMASET see DDL800
NEMATOCIDE see DDL800
NEMATOCIDE see EPC500
NEMATOLYT see PAG500
NEMATOX see DDL800
NEMAZENE see PDP250
NEMAZINE see PDP250
NEMAZON see DDL800
NEMBU-SERPIN see RDK000
NEMBUTAL SODIUM see NBU000
NEMICIDE see LFA020
NENDRIN see EAT500
NENESIN see BNK000
NEO see TEH500
NEOANTERGAN MALEATE see DBM800
NEO-ANTITTENSOL see RDK000
NEO-ANTITTERSOL see RDK000
NEOARSOLUIN see NCJ500
NEOARSPHENAMINE see NCJ500
NEOBAR see BAP000
NEOBIOTIC see NCG000
NEOBOR see SFF000
NEOCAINE see AIT250
NEO-CALMA see CJR909
NEOCID see DAD200
NEOCIDOL see DCM750
NEO-CLEANER see SHU500
NEOCOCCYL see SNM500
NEO-CODEMA see CFY000
NEOCOMPENSAN see PKQ250
NEO-COROVAS see PBC250
NEO-CULTOL see MQV750
NEOCURDIONE see NBW100
NEO-DEMA see CLH750
NEODICOUMARIN see BKA000
NEODICOUMAROL see BKA000
NEODICUMARINUM see BKA000
NEODRENAL see DMV600
NEODRINE see DBA800
NEODYMIA see NCC000
NEODYMIUM see NBX000
NEODYMIUM CHLORIDE see NBY000
NEODYMIUM(III) NITRATE (1:3) see NCB000
NEODYMIUM OXIDE see NCC000
NEODYMIUM(3+) OXIDE see NCC000
NEODYMIUM(III) OXIDE see NCC000
NEODYMIUM SESQUIOXIDE see NCC000
NEODYMIUM TRIOXIDE see NCC000
NEO-EPININE see DMV600
NEO-ERGOTIN see EDC500
NEOESERINE BROMIDE see POD000
NEO-ESTRONE see ECU750
NEO-FAT 8 see OCY000
NEO-FAT 10 see DAH400
NEO-FAT 12 see LBL000
NEO-FAT 18-61 see SLK000

NEO-FAT 90-04 see OHU000
NEO-FAT 92-04 see OHU000
NEO-FAT 18-S see SLK000
NEO-FERRUM see IHG000
NEOFLUMEN see CFY000
NEOFOLLIN see EDS100
NEO-FULCIN see GKE000
NEO-GILURYTMAL see DNB000
NEOGLAUCIT see IRF000
NEOHEXANE (DOT) see DQT200
NEO-HOMBREOL see TBG000
NEO-HOMBREOL-M see MPN500
NEO-ISTAFENE see HGC500
NEOLIN see BFC750
NEOLOID see CCP250
NEO-MANTLE CREME see NCG000
NEOMCIN see NCE000
NEOMIX see NCG000
NEOMYCIN see NCE000
NEOMYCINE SULFATE see NCG000
NEOMYCIN SULFATE see NCG000
NEOMYCIN SULPHATE see NCG000
NEON see NCG500
NEO-NAVIGAN see DYE600
NEON, compressed (UN 1065) (DOT) see NCG500
NEON, refrigerated liquid (cryogenic liquid) (UN 1913)
   (DOT) see NCG500
NEO-OESTRANOL 1 see DKA600
NEO-OESTRANOL II see DKB000
NEOOXEDRINE see SPC500
NEOPANTANOYL CHLORIDE see DTS400
NEOPENTANE see NCH000
NEOPENTANETETRAYL NITRATE see PBC250
NEOPENTANOIC ACID see PJA500
NEOPHARMEDRINE see DBA800
NEOPHRYN see SPC500
NEOPLATIN see PJD000
NEOPOLEN see PJS750
NEOPOLEN 30N see PJS750
NEOPRENE see NCI500
NEORESTAMIN see TAI500
NEO-SALICYL see SJO000
NEOSALVARSAN see NCJ500
NEOSAR see CQC650
NEO-SCABICIDOL see BBQ500
NEOSEDYN see TEH500
NEOSEPT see HCL000
NEOSEPTAL CL see SHU500
NEOSERFIN see RDK000
NEOSERINE BROMIDE see POD000
NEO-SERP see RDK000
NEOSETILE BLUE EB see TBG700
NEOSETILE PINK BN see AKE250
NEOS-HIDANTOINA see DKQ000
NEO SILOX D see SFT500
NEO-SINEFRINA see SPC500
NEOSLOWTEN see RDK000
NEO-SPECTRA see CBT750
NEO-SPECTRA II see CBT750
NEOSTIGMINE BROMIDE see POD000
NEOSTIGMINE METHYL BROMIDE see POD000
NEOSTREPAL see SNN300
NEOSTREPSAN see TEX250
NEO-SUPRIMAL see HGC500
NEO-SUPRIMEL see HGC500
NEOSYDYN see TEH500
NEOSYMPATOL see SPC500
NEOSYNEPHRINE see NCL500

NEOSYNEPHRINE see SPC500
NEOSYNEPHRINE HYDROCHLORIDE see SPC500
NEOSYNESINE see SPC500
NEO-TESTIS see TBF500
NEOTEX see CBT750
NEOTHESIN see PIV750
NEOTHYL see MOU830
NEO-TRAN see MQU750
NEO-TRIC see MMN250
NEOVITAMIN A ACID see VSK955
NEOXAZOL see SNN500
NEOZEPAM see DLY000
NEOZEX 4010B see PJS750
NEOZEX 45150 see PJS750
NEO-ZINE see MNV750
NEOZONE D see PFT500
NEPHENTINE see MQU750
NEPHIS see EIY500
NEPHOCARP see TNP250
NEPHRAMIDE see AAI250
NEPHRIDINE see VGP000
NEPTAL see NCM800
NEPTUNE BLUE BRA CONCENTRATION see FMU059
NEQUINATE see NCN600
NERACID see CBG000
NERAL see DTC800
NEREB see MAS500
NERKOL see DGP900
NEROL (FCC) see DTD200
NEROLIDYL ACETATE see NCN800
NEROLI OIL, ARTIFICAL see APJ250
NERONE see MCF525
NERVACTON see BCA000
NERVANAID B ACID see EIX000
NERVANAID B LIQUID see EIV000
NERVANID B see EIV000
NERVATIL see BCA000
NERVOSETON see BAG250
NESDONAL SODIUM see PBT500
NESOL see MCC250
NESPOR see MAS500
NESSLER REAGENT see NCP500
NESTON see MDT740
NETAZOL see CIR250
NETOCYD see DJT800
NETSUSARIN see DOT000
NEU see NKE500
NEU see ENV000
NEUCHLONIC see DLY000
NEUCOCCIN see FMU080
NEULACTIL see PIW000
NEULEPTIL see PIW000
NEURAKTIL see BCA000
NEURAZINE see CKP500
NEUROBARB see EOK000
NEUROBENZIL see BCA000
NEUROCAINE see CNE750
NEURODYN see TEH500
NEUROLEPTONE see BCA000
NEURONIKA see ADA725
NEUROSEDIN see TEH500
NEUROZINA see CJR909
NEUT see SFC500
NEUTHION see GFW000
NEUTRAL ACRIFLAVINE see DBX400
NEUTRAL AMMONIUM CHROMATE see ANF500
NEUTRAL AMMONIUM FLUORIDE see ANH250

NEUTRAL POTASSIUM CHROMATE see PLB250
NEUTRAL SODIUM CHROMATE see DXC200
NEUTRAL VERDIGRIS see CNI250
NEUTRAL ZINC PHOSPHATE see ZJS400
NEUTRAZYME see SIB600
NEUTRONYX 600 see PKF000
NEUTRONYX 605 see PKF500
NEUTROSEL NAVY BN see DCJ200
NEUTROSEL RED TRVA see CLK235
NEUWIED GREEN see COF500
NEVANAID-B POWDER see TNL250
NEVAX see DJL000
NEVIGRAMON see EID000
NEVRODYN see TEH500
NEW COCCIN see FMU080
NEWCOL 60 see SKV150
NEW GREEN see COF500
NEW IMPROVED CERESAN see BJT250
NEW IMPROVED GRANOSAN see BJT250
NEWPHRINE see SPC500
NEW PONCEAU 4R see FMU070
NEWTOL see XPJ000
NEXAGAN see EGV500
NEXIT see BBQ500
NF see NGE500
NF 246 see NDY000
NF-260 see FPI150
NFIP see NGI800
NaFPAK see SHF500
Na FRINSE see SHF500
NG see NGY000
NG-180 see NGG500
NGAI CAMPHOR see NCQ820
NH 18 see PKP750
NHC see HFN500
NH-LOST see BHN750
NHMI see OBY000
3N4HPA see HMY000
NIA 249 see AFR250
NIA 5273 see PIX250
NIA 5462 see EAQ750
NIA 5488 see CKM000
NIA-9241 see BDJ250
NIA 10242 see CBS275
NIA 17170 see BEP500
NIA 33297 see AHJ750
NIACEVIT see NCR000
NIACIDE see FAS000
NIACIN see NCQ900
NIACINAMIDE see NCR000
NIACINAMIDE ASCORBATE see NCR025
NIACINAMIDE ASCORBIC ACID COMPLEX see NCR025
NIAGARA 1240 see EEH600
NIAGARA 5,462 see EAQ750
NIAGARA 5943 see AIX000
NIAGARA 9241 see BDJ250
NIAGARA BLUE see CMO250
NIAGARA BLUE 2B see CMO000
NIAGARA BLUE 4B see CMO500
NIAGARAMITE see SOP500
NIAGARA P.A. DUST see NDN000
NIAGARA SKY BLUE see CMO500
NIAGARA-STIK see NAK500
NIAGRA 10242 see CBS275
NIALATE see EEH600
NIALK see TIO750
NIAMIDE see NCR000

NIAMINE see DJS200
NIAPROOF 4 see SIO000
NIA PROOF 08 see TAV750
NIAX AFPI see PKB100
NIAX CATALYST AL see BJH750
NIAX FLAME RETARDANT 3 CF see CGO500
NIAX ISOCYANATE TDI see TGM740
NIAX TDI see TGM750
NIAX TDI see TGM800
NIAX TDI-P see TGM750
NIBREN WAX see TIT500
NIBROL see TEH500
NICACID see NCQ900
NICAMIDE see NCR000
NICAMIDE see DJS200
NICAMIN see NCQ900
NICAMINA see NCR000
NICAMINDON see NCR000
NICANGIN see NCQ900
NICASIR see NCR000
NICEL see MIF760
NICELATE see EID000
NICERGOLIN (GERMAN) see NDM000
NICERGOLINE see NDM000
NICETAMIDE see DJS200
NICETHAMIDE see DJS200
NICHEL (ITALIAN) see NCW500
NICHEL TETRACARBONILE (ITALIAN) see NCZ000
NICKEL see NCW500
NICKEL 270 see NCW500
NICKEL(II) ACETATE (1:2) see NCX000
NICKEL ACETATE TETRAHYDRATE see NCX500
NICKEL(II) ACETATE TETRAHYDRATE see NCX500
NICKEL AMMONIUM SULFATE see NCY050
NICKEL ANTIMONIDE see NCY100
NICKEL ARSENIDE see NCY110
NICKEL ARSENIDE (As$_2$-Ni$_5$) see NDJ399
NICKEL ARSENIDE (As$_8$-Ni$_{11}$) see NDJ400
NICKEL ARSENIDE SULFIDE see NCY125
NICKEL BISCYCLOPENTADIENE see NDA500
NICKEL BLACK see NDE010
NICKEL BOROFLUORIDE see NDC000
NICKEL(II) CARBONATE (1:1) see NCY500
NICKEL CARBONATE HYDROXIDE see NCY600
NICKEL, (CARBONATO(2-))TETRAHYDROXYTRI- see NCY600
NICKEL CARBONYL see NCZ000
NICKEL CARBONYLE (FRENCH) see NCZ000
NICKEL CHLORIDE see NDH000
NICKEL(II) CHLORIDE (1:2) see NDH000
NICKEL(II) CHLORIDE HEXAHYDRATE (1:2:6) see NDA000
NICKEL CHROMATE see NDA100
NICKEL CHROMITE see NDA100
NICKEL CHROMIUM OXIDE see NDA100
NICKEL, COMPOUND with pi-CYCLOPENTADIENYL (1:2) see NDA500
NICKEL COMPOUNDS see NDB000
NICKEL CYANIDE (DOT) see NDB500
NICKEL CYANIDE (solid) see NDB500
NICKEL DIACETATE TETRAHYDRATE see NCX500
NICKEL DIBUTYLDITHIOCARBAMATE see BIW750
NICKEL DIFLUORIDE see NDC500
NICKEL DIHYDROXIDE see NDE000
NICKEL DINITRITE see NDG550
NICKEL (DUST) see NCW500
NICKEL(II) FLUOBORATE see NDC000
NICKEL(II) FLUORIDE (1:2) see NDC500

NICKEL FLUOROBORATE see NDC000
NICKEL(II) FLUOSILICATE (1:1) see NDD000
NICKEL-GALLIUM ALLOY see NDD500
NICKEL(II) HYDROXIDE see NDE000
NICKEL(III) HYDROXIDE see NDE010
NICKELIC HYDROXIDE see NDE010
NICKELIC OXIDE see NDH500
NICKEL IRON SULFIDE see NDE500
NICKEL-IRON SULFIDE MATTE see NDE500
NICKEL MONOANTIMONIDE see NCY100
NICKEL MONOARSENIDE see NCY110
NICKEL MONOSULFATE HEXAHYDRATE see NDL000
NICKEL MONOSULFIDE see NDL100
NICKEL MONOXIDE see NDF500
NICKEL NITRATE see NDG000
NICKEL(II) NITRATE (1:2) see NDG000
NICKEL(2$^+$) NITRATE, HEXAHYDRATE see NDG500
NICKEL(II) NITRATE, HEXAHYDRATE (1:2:6) see NDG500
NICKEL NITRITE see NDG550
NICKEL NITRITE see NDG550
NICKELOCENE see NDA500
NICKELOUS ACETATE see NCX000
NICKELOUS ACETATE TETRAHYDRATE see NCX500
NICKELOUS CARBONATE see NCY500
NICKELOUS CHLORIDE see NDH000
NICKELOUS FLUORIDE see NDC500
NICKELOUS HYDROXIDE see NDE000
NICKELOUS OXIDE see NDF500
NICKELOUS SULFATE see NDK500
NICKELOUS SULFIDE see NDL100
NICKELOUS TETRAFLUOROBORATE see NDC000
NICKEL OXIDE see NDH500
NICKEL(II) OXIDE (1:1) see NDF500
NICKEL OXIDE, IRON OXIDE, and CHROMIUM OXIDE FUME see IHE000
NICKEL OXIDE (MAK) see NDF500
NICKEL OXIDE PEROXIDE see NDH500
NICKEL PARTICLES see NCW500
NICKEL(2+) PERCHLORATE, HEXAHYDRATE see NDJ000
NICKEL PEROXIDE see NDH500
NICKEL POTASSIUM CYANIDE see NDI000
NICKEL PROTOXIDE see NDF500
NICKEL REFINERY DUST see NDI500
NICKEL(2+) SALT PERCHLORIC ACID HEXAHYDRATE see NDJ000
NICKEL SELENIDE see NDJ475
NICKEL SELENIDE (3:2) CRYSTALLINE see NDJ475
NICKEL SESQUIOXIDE see NDH500
NICKEL SPONGE see NCW500
NICKEL SUBARSENIDE see NDJ399
NICKEL SUBARSENIDE see NDJ400
NICKEL SUBSELENIDE see NDJ475
NICKEL SUBSULFIDE see NDJ500
NICKEL SUBSULPHIDE see NDJ500
NICKEL (II) SULFAMATE see NDK000
NICKEL SULFARSENIDE see NCY125
NICKEL SULFATE see NDK500
NICKEL SULFATE (1:1) see NDK500
NICKEL(II) SULFATE see NDK500
NICKEL(2$^+$)SULFATE (1:1) see NDK500
NICKEL(II) SULFATE (1:1) see NDK500
NICKEL SULFATE HEXAHYDRATE see NDL000
NICKEL (II) SULFATE HEXAHYDRATE see NDL000

NICKEL(II) SULFATE HEXAHYDRATE (1:1:6) see NDL000
NICKEL SULFIDE see NDJ500
NICKEL SULFIDE see NDL100
NICKEL(2+) SULFIDE see NDL100
NICKEL(II) SULFIDE see NDL100
α-NICKEL SULFIDE (3:2) CRYSTALLINE see NDJ500
NICKEL SULPHATE HEXAHYDRATE see NDL000
NICKEL SULPHIDE see NDJ500
NICKEL TELLURIDE see NDL425
NICKEL TETRACARBONYL see NCZ000
NICKEL TETRACARBONYLE (FRENCH) see NCZ000
NICKEL(II) TETRAFLUOROBORATE see NDC000
NICKEL-TITANATE see NDL500
NICKEL TITANIUM OXIDE see NDL500
NICKEL TRIOXIDE see NDH500
NICKEL TRITADISULPHIDE see NDJ500
NICKEL ZINC FERRATE see IHB800
NICKEL ZINC FERRITE see IHB800
NICLOFEN see DFT800
NICLOSAMIDE see DFV400
NICO see NCQ900
NICO-400 see NCQ900
NICOBID see NCQ900
NICOBION see NCR000
NICOCAP see NCQ900
NICOCIDE see NDN000
NICOCIDIN see NCQ900
NICOCRISINA see NCQ900
NICODAN see NCQ900
NICODELMINE see NCQ900
NICO-DUST see NDN000
NICOFORT see NCR000
NICO-FUME see NDN000
NICOGEN see NCR000
NICOLAR see NCQ900
NICOLEN see NGG500
NICOMIDOL see NCR000
NICONACID see NCQ900
NICONAT see NCQ900
NICONAZID see NCQ900
NICOR see DJS200
NICORDAMIN see DJS200
NICORINE see DJS200
NICOROL see NCQ900
NICOROL see CHJ750
NICORYL see DJS200
NICOSAN 2 see NCR000
NICOSIDE see NCQ900
NICO-SPAN see NCQ900
NICOSYL see NCQ900
NICOTA see NCR000
NICOTAMIDE see NCR000
NICOTAMIN see NCQ900
NICOTENE see NCQ900
NICOTERGOLINE see NDM000
NICOTIANA ATTENUATA see TGI100
NICOTIANA GLAUCA see TGI100
NICOTIANA LONGIFLORA see TGI100
NICOTIANA RUSTICA see TGI100
NICOTIANA TABACUM see TGI100
NICOTIL see NCQ900
NICOTILAMIDE see NCR000
NICOTILILAMIDO see NCR000
NICOTINA (ITALIAN) see NDN000
NICOTINE see NDN000
(−)-NICOTINE see NDN000

l-NICOTINE see NDN000
NICOTINE, solid (DOT) see NDN000
NICOTINE, liquid (DOT) see NDN000
NICOTINE ACID see NCQ900
NICOTINE ACID AMIDE see NCR000
NICOTINE ACID TARTRATE see NDS500
NICOTINE ALKALOID see NDN000
NICOTINE BITARTRATE see NDS500
NICOTINE, COMPOUND, with NICKEL(II)-o-BENZOYL BENZOATE TRIHYDRATE (2:1) see BHB000
NICOTINE HYDROCHLORIDE see NDP400
NICOTINE HYDROCHLORIDE (d,l) see NDP400
NICOTINE HYDROCHLORIDE, solution (DOT) see NDP400
NICOTINE HYDROGEN TARTRATE see NDS500
−)-NICOTINE HYDROGEN TARTRATE see NDS500
NICOTINE MONOSALICYLATE see NDR000
NICOTINE, 1'-NITROSO-1'-DEMETHYL- see NLD500
NICOTINE SALICYLATE (DOT) see NDR000
NICOTINE SULFATE see NDR500
NICOTINE SULFATE see NDR500
NICOTINE SULFATE, solid or solution (DOT) see NDR500
NICOTINE TARTRATE see NDS500
NICOTINE TARTRATE (1:2) see NDS500
NICOTINE TARTRATE (DOT) see NDS500
NICOTINIC ACID see NCQ900
NICOTINIC ACID, ALUMINUM SALT see AHD650
NICOTINIC ACID AMIDE see NCR000
NICOTINIC ACID DIETHYLAMIDE see DJS200
NICOTINIC AMIDE see NCR000
NICOTINIPCA see NCQ900
NICOTINONITRILE, 1,2-DIHYDRO-4-METHOXY-1-METHYL-2-OXO- see RJK100
NICOTINOYL HYDRAZINE see NCQ900
1-NICOTINOYLMETHYL-4-PHENYL-PIPERAZINE see NDW520
NICOTINSAEURE (GERMAN) see NCQ900
NICOTINSAEUREAMID (GERMAN) see NCR000
NICOTOL see NCR000
NICOTYLAMIDE see NCR000
NICOVASAN see NCQ900
NICOVASEN see NCQ900
NICOVEL see NCQ900
NICOVEL see NCR000
NICOVIT see NCR000
NICOVITOL see NCR000
NICOZYMIN see NCR000
NICYL see NCQ900
NIDA see MMN250
NIERALINE see VGP000
NIESYMETRYCZNA DWU METYLOHYDRAZYNA (POLISH) see DSF400
NIFLEX see SAV000
NIFOS T see TCF250
NIFULIDONE see NGG500
NIFURADENE see NDY000
NIFURAN see NGG500
NIFURTHIAZOLE see NDY500
NIFUZON see NGE500
(−)-NIGALDIPINE HYDROCHLORIDE see NDY550
(R)-NIGALDIPINE HYDROCHLORIDE see NDY550
NIGLYCON see NGY000
NIH 7958 see DOQ400
NIHONTHRENE GREY M see CMU475
NIKARDIN see DJS200

NIKA-TEMP see PKQ059
NIKAVINYL SG 700 see PKQ059
NIKETAMID see DJS200
NIKETHAROL see DJS200
NIKETHYL see DJS200
NIKETILAMID see DJS200
NIKKELTETRACARBONYL (DUTCH) see NCZ000
NIKKOL OTP 70 see DJL000
NIKKOL SO 10 see SKV100
NIKKOL SO-15 see SKV100
NIKKOL SO-30 see SKV100
NIKKOL SS 30 see SKV150
NIKKOL TO see PKL100
NIKORIN see DJS200
NIKO-TAMIN see NCR000
NIKOTIN (GERMAN) see NDN000
NIKOTINSAEUREAMID (GERMAN) see NCR000
NIKOTINSULFAT see NDR500
NIKOTYNA (POLISH) see NDN000
NILACID see ABX500
NILODIN see DHU000
NILOX PBNA see PFT500
NILSTAT see NOH500
NIMCO CHOLESTEROL BASE H see CMD750
NIMERGOLINE see NDM000
NIMITEX see TAL250
NIMITOX see TAL250
NINOL 4821 see BKE500
NINOL 2012E see CNF330
NINOL AA62 see BKE500
NINOL AA-62 EXTRA see BKE500
NINOL AA-62 EXTRA see LBL000
NIOBE OIL see MHA750
NIOBIUM see NDZ000
NIOBIUM-93 see NDZ000
NIOBIUM CHLORIDE see NEA000
NIOBIUM ELEMENT see NDZ000
NIOBIUM PENTACHLORIDE see NEA000
NIOCINAMIDE see NCR000
NIOFORM see CHR500
NIOMIL see DQM600
NIONATE see FBK000
NIONG see NGY000
NIOZYMIN see NCR000
NIP see DFT800
NIPA 49 see PNM750
NIPAGALLIN LA see DXX200
NIPAGALLIN P see PNM750
NIPAGIN see HJL500
NIPAGIN A see HJL000
NIPAM see INH000
NIPANTIOX 1-F see BQI000
NIPAR S-20 see NIY000
NIPAR S-20 SOLVENT see NIY000
NIPAR S-30 SOLVENT see NIY000
NIPASOL see HNU500
NIPAZIN A see HJL000
NIPELLEN see NCQ900
NIPEON A 21 see PKQ059
NIPERYT see PBC250
NIPERYTH see PBC250
NIPODAL see PMF500
NIPOL 407 see SMR000
NIPOL 576 see PKQ059
NIPPAS see SEP000
NIPPON BLUE BB see CMO000
NIPPON DEEP BLACK see AQP000
NIPPON DEEP BLACK GX see AQP000

NIPPON DIRECT SKY BLUE see CMO500
NIPRIDE see SIU500
NIPSAN see DCM750
NIQUETAMIDA see DJS200
NIRAN see PAK000
NIRAN see CDR750
NIRATIC HYDROCHLORIDE see LFA020
NIRATIC-PURON HYDROCHLORIDE see LFA020
NIRVOTIN see PGE000
Ni 0901-S see NCW500
NISETAMIDE see DJS200
NISSAN AMINE AB see OBC000
NISSAN AMINE BB see DXW000
NISSAN CATION M2-100 see TCA500
NISSAN CATION S2-100 see DTC600
NISSAN NONION SP 60 see SKV150
NISSOL EC see MME809
NITARSONE see NIJ500
NITAZOL see ABY900
NITAZOLE see ABY900
NITER see PLL500
NITHIAMIDE see ABY900
NITHIAZID see ENV500
NITHIAZIDE see ENV500
NITOBANIL see FLZ050
NITOFEN see DFT800
NITOL see BIE500
NITOL "TAKEDA" see BIE500
NITORA see NGY000
NITRADOR see DUS700
NITRADOS see DLY000
NITRAFEN see DFT800
NITRAMIN see ALQ000
NITRAMINE see TEG250
NITRAMINE see ALQ000
NITRAMINE, METHYL- see NHN500
NITRAN see DUV600
4-NITRANILINE see NEO500
m-NITRANILINE see NEN500
p-NITRANILINE see NEO500
NITRANITOL see MAW250
NITRAPHEN see DFT800
NITRAPYRIN (ACGIH) see CLP750
NITRATE d'AMYLE (FRENCH) see AOL250
NITRATE d'ARGENT (FRENCH) see SDS000
NITRATE de BARYUM (FRENCH) see BAN250
NITRATE MERCUREUX (FRENCH) see MDE750
NITRATE MERCURIQUE (FRENCH) see MDF000
NITRATE de PLOMB (FRENCH) see LDO000
NITRATE de PROPYLE NORMAL (FRENCH) see PNQ500
NITRATES see NED000
NITRATE de SODIUM (FRENCH) see SIO900
NITRATE de STRONTIUM (FRENCH) see SMK000
NITRATE de ZINC (FRENCH) see ZJJ000
NITRATINE see SIO900
NITRATION BENZENE see BBL250
NITRAZEPAM see DLY000
NITRAZOL CF EXTRA see NEO500
NITRE see PLL500
NITRE CAKE see SEG800
NITRENPAX see DLY000
NITRIC ACID see NED500
NITRIC ACID, over 40% (DOT) see NED500
NITRIC ACID other than red fuming with >70% nitric acid (DOT) see NED500
NITRIC ACID other than red fuming with not >70% nitric acid (DOT) see NED500

NITRIC ACID, ALUMINUM SALT see AHD750
NITRIC ACID, ALUMINUM(3+) SALT see AHD750
NITRIC ACID, AMMONIUM SALT see ANN000
NITRIC ACID, BARIUM SALT see BAN250
NITRIC ACID, BERYLLIUM SALT see BFT000
NITRIC ACID, BISMUTH(3+) SALT see BKW250
NITRIC ACID, CADMIUM SALT see CAH000
NITRIC ACID, CADMIUM SALT, TETRAHYDRATE see CAH250
NITRIC ACID, CALCIUM SALT (8CI,9CI) see CAU000
NITRIC ACID, CERIUM(3+) SALT (8CI, 9CI) see CDB000
NITRIC ACID, CESIUM SALT see CDE250
NITRIC ACID, CHROMIUM (3+) SALT see CMJ600
NITRIC ACID, COBALT(2+) SALT see CNC500
NITRIC ACID, ERBIUM (3+) SALT see ECY500
NITRIC ACID, ETHYL ESTER see ENM500
NITRIC ACID, FUMING (DOT) see NEE500
NITRIC ACID, IRON(3+) SALT see FAY200
NITRIC ACID, IRON(3+) SALT see IHB900
NITRIC ACID, ISOPROPYL ESTER see IQP000
NITRIC ACID, LEAD(2+) SALT see LDO000
NITRIC ACID, LUTETIUM(3+) SALT see LIY000
NITRIC ACID, MAGNESIUM SALT (2:1) see MAH000
NITRIC ACID, MANGANESE(2+) SALT see MAS900
NITRIC ACID, MERCURY(I) SALT see MDE750
NITRIC ACID, MERCURY(II) SALT see MDF000
NITRIC ACID METHYL ESTER see MMF500
NITRIC ACID, NEODYMIUM SALT see NCB000
NITRIC ACID, NICKEL(II) SALT see NDG000
NITRIC ACID, NICKEL(2+) SALT, HEXAHYDRATE see NDG500
NITRIC ACID, PHENYLMERCURY SALT see MCU750
NITRIC ACID, POTASSIUM SALT see PLL500
NITRIC ACID, PRASEODYMIUM(3+) SALT see PLY250
NITRIC ACID, PROPYL ESTER see PNQ500
NITRIC ACID (RED FUMING) see NEE500
NITRIC ACID, RED FUMING (DOT) see NEE500
NITRIC ACID, SILVER(1+) SALT see SDS000
NITRIC ACID, SODIUM SALT see SIO900
NITRIC ACID, STRONTIUM SALT see SMK000
NITRIC ACID, THALLIUM(1+) SALT see TEK750
NITRIC ACID, THORIUM(4+) SALT see TFT500
NITRIC ACID TRIESTER OF GLYCEROL see NGY000
NITRIC ACID, YTTERBIUM(3+) SALT see YDS800
NITRIC ACID, YTTRIUM(3+) SALT see YFJ000
NITRIC ACID, ZINC SALT see ZJJ000
NITRIC ETHER see ENM500
NITRIC OXIDE see NEG100
NITRIC OXIDE and NITROGEN TETROXIDE MIXTURES see NGT500
NITRILE ACRILICO (ITALIAN) see ADX500
NITRILE ACRYLIQUE (FRENCH) see ADX500
NITRILE ADIPICO (ITALIAN) see AER250
NITRILES see NEH500
NITRILE TRICHLORACETIQUE (FRENCH) see TII750
NITRIL KISELINY DIETHYLAMINOOCTOVE (CZECH) see DHJ600
NITRIL KYSELINY-o-CHLORBENZOOVE (CZECH) see CEM000
NITRIL KYSELINY ISOFTALOVE (CZECH) see PHX550
NITRIL KYSELINY MALONOVE (CZECH) see MAO250
NITRIL KYSELINY MANDLOVE (CZECH) see MAP250
NITRIL KYSELINY TEREFTALOVE (CZECH) see BBP250

NITRILOACETIC ACID TRISODIUM SALT MONOHYDRATE see NEI000
NITRILOACETONITRILE see COO000
NITRILOTRIACETIC ACID see AMT500
NITRILOTRIACETIC ACID, DISODIUM SALT see DXF000
NITRILOTRIACETIC ACID SODIUM SALT see SEP500
NITRILOTRIACETIC ACID, TRISODIUM SALT see SIP500
NITRILOTRIACETIC ACID TRISODIUM SALT MONOHYDRATE see NEI000
NITRILO-2,2',2''-TRIETHANOL see TKP500
2,2',2''-NITRILOTRIETHANOL see TKP500
NITRILOTRIS(ETHYLAMINE) see NEI800
2,2',2''-NITRILOTRIS(ETHYLAMINE) see NEI800
NITRIN see NGY000
NITRINE see NGY000
NITRINE-TDC see NGY000
NITRITES see NEJ000
NITRITE de SODIUM (FRENCH) see SIQ500
NITRITO see NGR500
5-NITROACENAPHTHENE see NEJ500
5-NITROACENAPHTHYLENE see NEJ500
5-NITROACENAPHTHENE see NEJ500
2-NITRO-4-ACETAMINOFENETOL (CZECH) see NEL000
5-NITRO-2-ACETILAMINOTIAZOLO see ABY900
3-NITRO-p-ACETOPHENETIDE see NEL000
3-NITRO-p-ACETOPHENETIDIDE see NEL000
5-NITRO-p-ACETOPHENETIDIDE see NEL000
3'-NITRO-p-ACETOPHENETIDIN see NEL000
NITRO ACID 100 percent see HMY000
NITRO ACID SULFITE see NMJ000
m-NITROAMINOBENZENE see NEN500
2-NITRO-4-AMINOPHENOL see NEM480
o-NITRO-p-AMINOPHENOL see NEM480
5-NITRO-2-AMINOTHIAZOLE see ALQ000
4-NITRO-2-AMINOTOLUENE see NMP500
p-NITROANILINA see NEO500
2-NITROANILINE see NEO000
3-NITROANILINE see NEN500
m-NITROANILINE see NEN500
o-NITROANILINE see NEO000
o-NITROANILINE see NEO000
p-NITROANILINE see NEO500
p-NITROANILINE see NEO500
5-NITRO-o-ANISIDINE see NEQ500
p-NITROANISOL see NER500
2-NITROANISOLE see NER000
4-NITROANISOLE see NER500
o-NITROANISOLE see NER000
p-NITROANISOLE see NER500
3-NITRO-3-AZAPENTANE-1,5-DIISOCYANATE see NHI500
NITROBENZEEN (DUTCH) see NEX000
NITROBENZEN (POLISH) see NEX000
3-NITROBENZENAMINE see NEN500
4-NITROBENZENAMINE see NEO500
NITROBENZENE see NEX000
NITROBENZENE, liquid (DOT) see NEX000
4-NITROBENZENEARSONIC ACID see NIJ500
o-NITROBENZENEARSONIC ACID see NEX500
2-NITRO-1,4-BENZENEDIAMINE see ALL750
4-NITRO-1,2-BENZENEDIAMINE see ALL500
3-NITROBENZENEDIAZONIUM PERCHLORATE see NFA500
m-NITROBENZENE DIAZONIUM PERCHLORATE (DOT) see NFA500

NITROFUROXON see NGG500
2-(5-NITRO-2-FURYL)-5-AMINO-1,3,4-THIADIAZOLE see NGI500
5-(5-NITRO-2-FURYL)-2-AMINO-1,3,4-THIADIAZOLE see NGI500
3-((5-NITROFURYLIDENE)AMINO)-2-OXAZOLIDONE see NGG500
3-(5-NITRO-2-FURYL)-IMIDAZO(1,2-a)PYRIDINE see NGI800
(5-NITRO-2-FURYL) METHYL KETONE see ACT250
N-(4-(5-NITRO-2-FURYL)-2-THIAZOLYL)ACETAMIDE see AAL750
N-(4-(5-NITRO-2-FURYL)THIAZOL-2-YL)ACETAMIDE see AAL750
N-(4-(5-NITRO-2-FURYL)-2-THIAZOLYL)FORMAMID (GERMAN) see NGM500
N-(4-(5-NITRO-2-FURYL)-2-THIAZOLYL)FORMAMIDE see NGM500
N-(4-(5-NITRO-2-FURYL)-2-THIAZOLYL)-2,2,2-TRIFLUOROACETAMIDE see NGN500
6-(5-NITRO-2-FURYLVINYL)-3-(DIHYDROXYDIMETHYLAMINO)-1,2,4-TRIAZENE see BKH500
((6-(2-(5-NITRO-2-FURYL)VINYL)-as-TRIAZIN-3-YL)IMINO)DIMETHANOL see BKH500
N-(6-(2-(5-NITRO-2-FURYL)VINYL)-1,2,4-TRIAZIN-3-YL)IMINODIMETHANOL see BKH500
NITROGEN see NGP500
NITROGEN BROMIDE see BMP250
NITROGEN CHLORIDE see NGQ500
NITROGEN CHLORIDE see CDW000
NITROGEN DIOXIDE see NGR500
NITROGEN DIOXIDE, DI- see NGU500
NITROGEN FLUORIDE see NGW000
NITROGEN FLUORIDE OXIDE see NGS500
NITROGEN GAS see NGP500
NITROGEN HALF MUSTARD see CGW000
NITROGEN IODIDE see NGW500
NITROGEN IODIDE see IDN000
NITROGEN LIME see CAQ250
NITROGEN MONOXIDE see NEG100
NITROGEN MONOXIDE, mixed with NITROGEN TETROXIDE see NGT500
NITROGEN MUSTARD see BIE250
NITROGEN MUSTARD HYDROCHLORIDE see BIE500
NITROGEN MUSTARD OXIDE see CFA500
NITROGEN MUSTARD OXIDE see CFA750
NITROGEN MUSTARD-N-OXIDE see CFA500
NITROGEN MUSTARD-N-OXIDE see CFA750
NITROGEN MUSTARD-N-OXIDE HYDROCHLORIDE see CFA750
NITROGEN OXIDE see NGU000
NITROGEN OXYCHLORIDE see NMH000
NITROGEN PEROXIDE see NGR500
NITROGEN TETROXIDE see NGU500
NITROGEN TETROXIDE-NITRIC OXIDE MIXTURE see NGT500
NITROGEN TRICHLORIDE see NGQ500
NITROGEN TRICHLORIDE (DOT) see NGQ500
NITROGEN TRIFLUORIDE see NGW000
NITROGEN TRIIODIDE see NGW500
NITROGEN, compressed (UN 1066) (DOT) see NGP500
NITROGEN, refrigerated liquid (cryogenic liquid) (UN 1977) (DOT) see NGP500
NITROGLICERINA (ITALIAN) see NGY000
NITROGLICERYNA (POLISH) see NGY000
NITROGLYCERIN see NGY000

NITROGLYCERIN, liquid, not desensitized (DOT) see NGY000
NITROGLYCERINE see NGY000
NITROGLYCERIN, SPIRITS OF see NGY000
NITROGLYCERIN, desensitized, not <40% non-volatile water insoluble phlegmatizer (UN 0143) (DOT) see NGY000
NITROGLYCERIN, solution in alcohol, with >1% but not >5% nitroglycerin (UN 3064) (DOT) see NGY000
NITROGLYCERIN, solution in alcohol, with >1% but not >10% nitroglycerin (UN 0144) (DOT) see NGY000
NITROGLYCERIN, solution in alcohol, with not >1% nitroglycerin (UN 1204) (DOT) see NGY000
NITROGLYCEROL see NGY000
NITROGLYCOL see EJG000
NITROGLYKOL (CZECH) see EJG000
NITROGLYN see NGY000
NITROGRANULOGEN see BIE500
NITROGRANULOGEN HYDROCHLORIDE see BIE500
NITROGUANIDINE see NHA500
2-NITROGUANIDINE see NHA500
α-NITROGUANIDINE see NHA500
NITROGUANIDINE, wetted (UN 1336) (DOT) see NHA500
NITROGUANIDINE, dry or wetted with <20% water, by weight (UN 0282) (DOT) see NHA500
3-NITRO-3-HEXENE see NHE000
NITROHYDROCHLORIC ACID (DOT) see HHM000
NITROHYDROCHLORIC ACID, diluted (DOT) see HHM000
3-NITRO-4-HYDROXYBENZENEARSONIC ACID see HMY000
2-NITRO-1-HYDROXYBENZENE-4-ARSONIC ACID see HMY000
4-NITRO-5-HYDROXYMERCURIORTHOCRESOL see NHK900
3-NITRO-4-HYDROXYPHENYLARSENOUS ACID see NHE600
3-NITRO-4-HYDROXYPHENYLARSONIC ACID see HMY000
1-NITRO-2-IMIDAZOLIDONE see NKL000
N-NITRO-2-IMIDAZOLIDONE see NKL000
NITROIMINODIETHYLENEDIISOCYANIC ACID see NHI500
NITROISOPROPANE see NIY000
NITRO KLEENUP see DUZ000
NITROL see NGY000
NITROL see NHK800
NITROLAN see NGY000
NITRO-LENT see NGY000
NITROLETTEN see NGY000
NITROLIME see CAQ250
NITROLINGUAL see NGY000
NITROLOWE see NGY000
NITROL (PROMOTER) see NHK800
NITRO MANNITE see MAW250
NITROMANNITE (dry) (DOT) see MAW250
NITROMANNITE, wetted with not <40% water, by weight or mixture (NA 0133) (DOT) see MAW250
NITROMANNITOL see MAW250
NITROMEL see NGY000
NITROMERSOL see NHK900
NITROMERSOL SOLUTION see NHK900
NITROMETAN (POLISH) see NHM500
N-NITROMETHANAMINE see NHN500
NITROMETHANE see NHM500
3-NITRO-6-METHOXYANILINE see NEQ500

5-NITRO-2-METHOXYANILINE see NEQ500

2-NITRO-7-METHOXYNAPHTHO(2,1-b)FURAN see MFB400

N-NITROMETHYLAMINE see NHN500

3-NITRO-6-METHYLANILINE see NMP500

5-NITRO-2-METHYLANILINE see NMP500

1-NITRO-2-METHYLANTHRAQUINONE see MMG000

NITROMIM see CFA750

NITROMIN see CFA500

NITROMIN HYDROCHLORIDE see CFA750

NITROMIN IDO see ALQ000

NITROMURIATIC ACID (DOT) see HHM000

NITRON see CCU250

NITRONAPHTHALENE see NHP990

NITRONAPHTHALENE see NHP990

1-NITRONAPHTHALENE see NHQ000

2-NITRONAPHTHALENE see NHQ500

NITRONAPHTHALENE (DOT) see NHP990

α-NITRONAPHTHALENE see NHQ000

β-NITRONAPHTHALENE see NHQ500

5-NITRONAPHTHALENE ETHYLENE see NEJ500

NITRONET see NGY000

NITRONG see NGY000

N'-NITRO-N-NITROSO-N-METHYLGUANIDINE see MMP000

NITRONIUM TETRAFLUOROBORATE(1−) see NHS500

NITRON (NITROCELLULOSE) see CCU250

5-NITRO-N-(2-OXO-3-OXAZOLIDINYL)-2-FURANMETHANIMINE see NGG500

NITROPENTA see PBC250

NITROPENTAERYTHRITE see PBC250

NITROPENTAERYTHRITOL see PBC250

1-NITROPENTANE see AOL500

NITROPHEN see DFT800

NITROPHENE see DFT800

2-NITROPHENOL see NIF010

3-NITROPHENOL see NIE600

4-NITROPHENOL see NIF000

o-NITROPHENOL see NIF010

m-NITROPHENOL (DOT) see NIE600

p-NITROPHENOL (DOT) see NIF000

NITROPHENOLARSONIC ACID see HMY000

p-NITROPHENOL, O-ESTER with O,O-DIETHYLPHOSPHOROTHIOATE see PAK000

p-NITROPHENOXYTRIBUTYLTIN see NII200

m-NITROPHENYLAMINE see NEN500

p-NITROPHENYLAMINE see NEO500

4-NITROPHENYLARSONIC ACID see NIJ500

p-NITROPHENYLARSONIC ACID see NIJ500

d-(−)-threo-1-p-NITROPHENYL-2-DICHLORACETAMIDO-1,3-PROPANEDIOL see CDP250

d-threo-1-(p-NITROPHENYL)-2-(DICHLOROACETYLAMINO)-1,3-PROPANEDIOL see CDP250

7-NITRO-5-PHENYL-2,3-DIHYDRO-1H-1,4-BENZODIAZEPIN-2-ONE see DLY000

p-NITROPHENYLDIMETHYLTHIONOPHOSPHATE see MNH000

1-(p-NITROPHENYL-3,3-DIMETHYL-TRIAZEN (GERMAN) see DSX400

1-(4-NITROPHENYL)-3,3-DIMETHYLTRIAZENE see DSX400

1-(p-NITROPHENYL)-3,3-DIMETHYL-TRIAZENE see DSX400

NITRO-p-PHENYLENEDIAMINE see ALL750

2-NITRO-p-PHENYLENEDIAMINE see ALL750

4-NITRO-o-PHENYLENE-DIAMINE see ALL500

p-NITRO-o-PHENYLENEDIAMINE see ALL500

2-NITRO-1,4-PHENYLENEDIAMINE see ALL750

4-NITRO-1,2-PHENYLENEDIAMINE see ALL500

o-NITRO-p-PHENYLENEDIAMINE (MAK) see ALL750

4-NITROPHENYL ISOCYANIDE see IKH780

p-NITROPHENYL ISOCYANIDE see IKH780

p-NITROPHENYL ISONITRILE see IKH780

o-NITROPHENYL METHYL ETHER see NER000

N-(4-NITROPHENYL)-N'-(3-PYRIDINYLMETHYL)UREA see PPP750

NITROPHOS see DSQ000

NITROPORE OBSH see OPE000

NITROPORE OBSH see BBS300

NITROPROPANE see NIY000

1-NITROPROPANE see NIX500

2-NITROPROPANE see NIY000

NITROPROPANE (DOT) see NIY000

β-NITROPROPANE see NIY000

4-NITRO-N-(4-PROPYLCYCLOHEXYL)BENZAMIDE see NIY600

NITROPRUSSIDNATRIUM (GERMAN) see SIU500

1-NITROPYRENE see NJA000

3-NITROPYRENE see NJA000

4-NITROPYRENE see NJA100

4-NITROPYRIDINE-1-OXIDE see NJA500

4-NITROPYRIDINE-N-OXIDE see NJA500

8-NITROQUINOLINE see NJD500

4-NITROQUINOLINE-1-OXIDE see NJF000

4-NITROQUINOLINE-N-OXIDE see NJF000

NITRORECTAL see NGY000

NITROSAMINES see NJH000

NITRO-SIL see AMY500

N-NITROSO-N-(ACETOXY)METHYL-N-METHYLAMINE see AAW000

N-NITROSOAETHYLAETHANOLAMIN (GERMAN) see ELG500

N-NITROSOALLYLMETHYLAMINE see MMT500

N-NITROSOAMINODIETHANOL see NKM000

4-(NITROSOAMINO-N-METHYL)-1-(3-PYRIDYL)-1-BUTANONE see MMS500

N-(p-NITROSOANILINOMETHYL)-2-NITROPROPANE see NHK800

N-NITROSOAZACYCLOHEPTANE see NKI000

N-NITROSOAZACYCLOOCTANE see OBY000

N-NITROSOBENZYLMETHYLAMINE see MHP250

N-NITROSOBIS(2-HYDROXYETHYL)AMINE see NKM000

N-NITROSOBIS(2-HYDROXYPROPYL)AMINE see DNB200

N-NITROSOBIS(2-OXOPROPYL)AMINE see NJN000

N-NITROSO-N-BUTYL-N-(3-CARBOXYPROPYL)AMINE see BQQ250

N-NITROSO-N-BUTYLETHYLAMINE see EHC000

N-NITROSO-n-BUTYL-(4-HYDROXYBUTYL)AMINE see HJQ350

N-NITROSO-N-BUTYLMETHYLAMINE see MHW500

N-NITROSOBUTYLUREA see BSA250

N-NITROSOCARBARYL see NBJ500

NITROSO COMPOUNDS see NJT500

N-NITROSO COMPOUNDS see NJT550

1'-NITROSO-1'-DEMETHYLNICOTINE see NLD500

N-NITROSODIAETHANOLAMIN (GERMAN) see NKM000

N-NITROSODIAETHYLAMIN (GERMAN) see NJW500

N-NITROSODIBUTYLAMINE see BRY500

N-NITROSOMETHYLPHENYLAMINE (MAK) see MMU250

N-NITROSO-N-METHYL-2-PHENYLETHYLAMINE see MNU250

NITROSOMETHYLPHENYLUREA see MMY500

N-NITROSO-N'-METHYLPIPERAZIN (GERMAN) see NKW500

1-NITROSO-4-METHYLPIPERAZINE see NKW500

N-NITROSO-N'-METHYLPIPERAZINE see NKW500

NITROSOMETHYLPROPYLAMINE see MNA000

NITROSOMETHYL-N-PROPYLAMINE see MNA000

NITROSOMETHYLUREA see MNA750

1-NITROSO-1-METHYLUREA see MNA750

N-NITROSO-N-METHYLUREA see MNA750

NITROSOMETHYLURETHAN (GERMAN) see MMX250

NITROSOMETHYLURETHANE see MMX250

N-NITROSO-N-METHYLURETHANE see MMX250

N-NITROSOMETHYLVINYLAMINE see NKY000

N-NITROSOMORPHOLIN (GERMAN) see NKZ000

NITROSOMORPHOLINE see NKZ000

4-NITROSOMORPHOLINE see NKZ000

N-NITROSOMORPHOLINE (MAK) see NKZ000

NITROSO-NAC see NBJ500

2-NITROSO-1-NAPHTHOL see NLB500

NITROSONIUM BISULFITE see NMJ000

N'-NITROSONORNICOTINE see NLD500

N-NITROSOOOXAZOLIDIN (GERMAN) see NLE000

3-NITROSOOXAZOLIDINE see NLE000

N-NITROSOOXAZOLIDINE see NLE000

N-NITROSO-1,3-OXAZOLIDINE see NLE000

NITROSOOXAZOLIDONE see NLE000

N-NITROSO-2-OXO-N-PROPYL-N-PROPYLAMINE see ORS000

1-NITROSO-1-PENTYLUREA see PBX500

N-NITROSOPERHYDROAZEPINE see NKI000

NITROSOPHENOL see NLF200

4-NITROSOPHENOL see NLF200

p-NITROSOPHENOL see NLF200

4-NITROSO-N-PHENYLANILINE see NKB500

N-NITROSO-N-PHENYLANILINE see DWI000

p-NITROSO-N-PHENYLANILINE see NKB500

4-NITROSO-N-PHENYLBENZENAMINE see NKB500

N-NITROSOPHENYLHYDROXYLAMIN AMMONIUM SALZ (GERMAN) see ANO500

N-NITROSOPHENYLHYDROXYLAMINE AMMONIUM SALT see ANO500

N-NITROSO-4-PICOLYLETHYLAMINE see NLH000

NITROSOPIPERIDIN see NLJ500

N-NITROSO-PIPERIDIN see NLJ500

1-NITROSOPIPERIDINE see NLJ500

N-NITROSOPIPERIDINE see NLJ500

1-(NITROSOPROPYLAMINO)-2-PROPANOL see NLM500

1-(NITROSOPROPYLAMINO)-2-PROPANONE see ORS000

N-NITROSO-N-PROPYLPROPANAMINE see NKB700

N-NITROSO-N-PROPYL-1-PROPANAMINE see NKB700

NITROSOPROPYLUREA see NLO500

NITROSO-N-PROPYLUREA see NLO500

N-NITROSO-N-PROPYLUREA see NLO500

1-NITROSO-2-(3-PYRIDYL)PYRROLIDINE see NLD500

N-NITROSOPYRROLIDIN (GERMAN) see NLP500

1-NITROSOPYRROLIDINE see NLP500

N-NITROSOPYRROLIDINE see NLP500

3-(1-NITROSO-2-PYRROLIDINYL)PYRIDINE see NLD500

NITROSORBID see CCK125

NITROSORBIDE see CCK125

NITROSORBON see CCK125

N-NITROSOSARCOSINE see NLR500

NITROSO SARKOSIN see NLR500

N-NITROSO-TRIMETHYLHARNSTOFF (GERMAN) see TLU750

NITROSO-3,4,5-TRIMETHYLPIPERAZINE see NLY750

1-NITROSO-3,4,5-TRIMETHYLPIPERAZINE see NLY750

N-NITROSO-3,4,5-TRIMETHYLPIPERAZINE see NLY750

NITROSOTRIMETHYLUREA see TLU750

N-NITROSOTRIMETHYLUREA see TLU750

NITROSTABILIN see NGY000

NITROSTARCH see NMB000

NITROSTARCH, dry or wetted with <20% water, by weight (UN 0146) (DOT) see NMB000

NITROSTARCH, wetted with not <20% water, by weight (UN 1337) (DOT) see NMB000

NITROSTAT see NGY000

NITROSTIGMIN (GERMAN) see PAK000

NITROSTIGMINE see PAK000

NITROSYL CHLORIDE see NMH000

NITROSYL ETHOXIDE see ENN000

NITROSYLFERRICYTOCHROME C see CQM325

NITROSYL HYDROGEN SULFATE see NMJ000

NITROSYL HYDROXIDE see NMR000

NITROSYL SULFATE see NMJ000

NITROSYLSULFURIC ACID see NMJ000

N-(5-NITRO-2-THIAZOLYL)ACETAMIDE see ABY900

5-NITRO-2-THIAZOLYLAMINE see ALQ000

8-NITRO-2-THIO-2H-1,3-BENZOXAZINE-2,4(3H)-DIONE see NFW210

2-NITROTOLUENE see NMO525

3-NITROTOLUENE see NMO500

4-NITROTOLUENE see NMO550

m-NITROTOLUENE see NMO500

o-NITROTOLUENE see NMO525

p-NITROTOLUENE see NMO550

5-NITRO-2-TOLUIDINE see NMP500

5-NITRO-o-TOLUIDINE see NMP500

3-NITROTOLUOL see NMO500

4-NITROTOLUOL see NMO550

NITROTRIAZOLONE see NMP620

3-NITRO-1,2,4-TRIAZOL-5-ONE see NMP620

NITROTRIBROMOMETHANE see NMQ000

NITROTRICHLOROMETHANE see CKN500

m-NITROTRIFLUOROTOLUENE see NFJ500

m-NITROTRIFLUORTOLUOL (GERMAN) see NFJ500

NITROUREA see NMQ500

1-NITROUREA see NMQ500

N-NITROUREA see NMQ500

NITROUS ACID see NMR000

NITROUS ACID-n-BUTYL ESTER see BRV500

NITROUS ACID-sec-BUTYL ESTER see BRV750

NITROUS ACID, ISOBUTYL ESTER see IJD000

NITROUS ACID-3-METHYL BUTYL ESTER see IMB000

NITROUS ACID, METHYL ESTER see MMF750

NITROUS ACID, 1-METHYLETHYL ESTER (9CI) see IQQ000

NITROUS ACID-1-METHYL PROPYL ESTER see BRV750

NITROUS ACID, 2-METHYLPROPYL ESTER see IJD000

NITROUS ACID, NICKEL(2+) SALT see NDG550

NITROUS ACID, PENTYL ESTER see AOL500

NITROUS ACID, POTASSIUM SALT see PLM500

NITROUS ACID, SODIUM SALT see SIQ500
NITROUS DIPHENYLAMIDE see DWI000
NITROUS ETHER see ENN000
NITROUS ETHYL ETHER see ENN000
NITROUS FUMES see NEE500
NITROUS OXIDE (DOT) see NGU000
NITROUS OXIDE, compressed (UN 1070) (DOT) see NGU000
NITROUS OXIDE, refrigerated liquid (UN 2201) (DOT) see NGU000
β-NITROVINYLFURAN see FQM100
NITROX see MNH000
NITROXANTHIC ACID see PID000
NITROXYLENE see NMS000
NITROXYLENE see NMS000
NITROXYLENES (o-, m-, p-) (DOT) see NMS000
NITROZELL RETARD see NGY000
NITROZONE see NGE500
NITRUMON see BIF750
NITRYL FLUORIDE see NMT500
NITTO DIRECT SKY BLUE 5B see CMO500
NIVALENOL-4-O-ACETATE see FQR000
NIVEMYCIN see NCE000
NIVITIN see SKV200
NIXON E/C see EHG100
NIXON N/C see CCU250
NIX-SCALD see SAV000
NK 136 see DJT800
NK 171 see EAV500
NK 711 see LEN000
NK-843 see NGY000
NM 11 see PKP750
NM 14 see PKP750
NMA see MMU250
1-N-2-MA (RUSSIAN) see MMG000
NMBA see MHW500
NMC 50 see CBM750
NMDDA see NKU000
NMEA see MKB000
p-(N'-METHYLHYDRAZINOMETHYL)-N-ISOPROPYLBENZAMIDE HYDROCHLORIDE see PME500
No. 907 METRO TALC see TAB750
NMH see MNA750
N-6-MI see NKI000
NMO see CFA500
NMOP see NKV000
NMOR see NKZ000
NMP see MPF200
NMU see MNA750
NMUM see MMX250
NMUT see MMX250
NMVA see NKY000
NNDG see DQR600
1000 NN FERRITE see IHB800
NNK see MMS500
No. 3 CONC. SCARLET see CHP500
NOBECUTAN see TFS350
NOBEDON see HIM000
NOBFELON see IIU000
NOBFEN see IIU000
NOBGEN see IIU000
NOBILEN see TFR250
NOBLEN see PMP500
NO BUNT see HCC500
NO BUNT 40 see HCC500
NO BUNT 80 see HCC500

NO BUNT LIQUID see HCC500
NOCBIN see DXH250
NOCCELER CZ see CPI250
NOCRAC NS 5 see MJN250
NOCTEC see CDO000
NOCTOSEDIV see TEH500
NOCTOSOM see FMQ000
NODAPTON see GIC000
NO. 2 DIESEL FUEL see DHE850
NO-DOZ see CAK500
NO-ETU see NKK500
NOFLAMOL see PJL750
NOGEDAL see DPH600
NOGEST see MCA000
NOGOS see DGP900
NOGRAM see EID000
NOHFAA see HIP000
NOLTAM see TAD175
NOLTRAN see CMA250
NOLVADEX see TAD175
NOLVASAN see BIM250
NOMERSAN see TFS350
NONABROMOBIPHENYL see NMV735
NONADECAFLUORODECANOIC ACID see PCG725
NONADECAFLUORO-n-DECANOIC ACID see PCG725
2,6-NONADIENAL see NMV760
trans-2,cis-6-NONADIENAL see NMV760
trans,cis-2,6-NONADIENAL see NMV760
trans,cis-2,6-NONADIENAL see NMV760
trans,trans-2,4-NONADIENAL see NMV775
NONADIENOL see NMV780
trans,cis-2,6-NONADIENOL see NMV780
2-trans-6-cis-NONADIEN-1-OL see NMV780
γ-NONALACTONE (FCC) see CNF250
1-NONALDEHYDE see NMW500
NONALOL see NNB500
1,4-NONALOLIDE see CNF250
1-NONANAL see NMW500
NONANE see NMX000
1-NONANECARBOXYLIC ACID see DAH400
NONANOIC ACID see NMY000
NONANOIC ACID, ETHYL ESTER see ENW000
NONAN-1-OL see NNB500
1-NONANOL see NNB500
NONANOPHENONE, 9-FLUORO- see FKT050
NONANOYL PEROXIDE see NNA100
2-NONENAL see NNA300
NON-2-ENAL see NNA300
2-NONEN-1-AL see NNA300
6-NONENAL, (Z)- see NNA325
(Z)-6-NONENAL see NNA325
cis-6-NONENAL see NNA325
cis-6-NONEN-1-AL see NNA325
trans-2-NONENAL (FCC) see NNA300
2-NONENENITRILE, 3-METHYL- see MNB500
cis-6-NONEN-1-OL see NNA530
α-NONENYL ALDEHYDE see NNA300
NONEX 411 see HKJ000
NONION OP80R see SKV100
NONION SP 60 see SKV150
NONION SP 60R see SKV150
n-NONOIC ACID see NMY000
NONOX CL see NBL000
NONOX D see PFT500
NONOX DPPD see BLE500
NONOX TBC see BFW750

NONOXYNOL-9 see NNB300
NONYL ACETATE see NNB400
NONYL ALCOHOL see NNB500
n-NONYL ALCOHOL see NNB500
sec-NONYL ALCOHOL see DNH800
1-NONYL ALDEHYDE see NMW500
NONYLCARBINOL see DAI600
n-NONYLIC ACID see NMY000
NONYL METHYL KETONE see UKS000
NONYLPHENOXYPOLY(ETHYLENEOXY)ETHANO
L see NNB300
NONYLTRICHLOROSILANE see NNE000
2-NONYNAL DIMETHYLACETAL see NNE100
2-NONYNAL, DIMETHYL ACETAL see NNE100
2-NONYN-1-AL DIMETHYLACETAL see NNE100
NO-Pip see NLJ500
NOPALMATE see PLH500
NOPCOCIDE see TBQ750
NOPCOTE C 104 see CAX350
4-NOPD see ALL500
NO-PEST see DGP900
NO-PEST STRIP see DGP900
NOPINEN see POH750
NOPINENE see POH750
NOPOL (POLYMER) see PJS750
NOPTIL see EOK000
NO-PYR see NLP500
NOR-1 see MKQ600
NORACYCLINE see LJE000
(−)-NORADREC see NNO500
NORADRENALIN see NNO500
NORADRENALINA (ITALIAN) see NNO500
NORADRENALINE see NNO500
(−)-NORADRENALINE see NNO500
d-(−)-NORADRENALINE see NNO500
l-NORADRENALINE see NNO500
NORADRENLINE see NNO500
NORAL ALUMINUM see AGX000
NORAL EXTRA FINE LINING GRADE see AGX000
NORAL INK GRADE ALUMINUM see AGX000
NORAL NON-LEAFING GRADE see AGX000
NOR-AM EP 332 see DSO200
NORAM O see OHM700
No. 156 ORANGE CHROME see LCS000
NORARTRINAL see NNO500
NORBORAL see BSM000
NORBORNADIENE see NNG000
2,5-NORBORNADIENE see NNG000
NORCAIN see EFX000
NORCOZINE see CKP500
NORDHAUSEN ACID (DOT) see SOI500
NORDICOL see EDO000
NORDICORT see HHR000
NORDIHYDROGUAIARETIC ACID see NBR000
NORDIHYDROGUAIRARETIC ACID see NBR000
NORDOPAN see BIA250
NOREPHEDRANE see BBK000
(−)-NOREPHEDRINE see NNM000
dl-NOREPHEDRINE HYDROCHLORIDE see PMJ500
NOREPINEPHRINE see NNO500
(−)-NOREPINEPHRINE see NNO500
l-NOREPINEPHRINE see NNO500
NOREPIRENAMINE see NNO500
NORETHANDROL see EAP000
NORETHINDRONE-17-ACETATE see ABU000
NORETHINDRONE ACETATE and
ETHINYLESTRADIOL see EEH520
NORETHINODREL see EEH550

19-NOR-ETHINYL-4,5-TESTOSTERONE see NNP500
19-NOR-ETHINYL-5,10-TESTOSTERONE see EEH550
NORETHINYNODREL see EEH550
19-NORETHISTERONE see NNP500
19-NORETHISTERONE ACETATE see ABU000
NORETHISTERONE ACETATE mixed with ETHINYL
OESTRADIOL see EEH520
NORETHYNODRAL see EEH550
NORETHYNODREL see EEH550
19-NORETHYNODREL see EEH550
NORETHYNODREL and ETHINYLESTRADIOL-3-
METHYL ETHER (50:1) see EAP000
NORETHYNODREL mixed with MESTRANOL see
EAP000
19-NOR-17-α-ETHYNYLANDROSTEN-17-β-OL-3-ONE
see NNP500
19-NOR-17-α-ETHYNYL-17-β-HYDROXY-4-
ANDROSTEN-3-ONE see NNP500
19-NOR-17-α-ETHYNYLTESTOSTERONE see NNP500
19-NORETHYNYLTESTOSTERONE ACETATE see
ABU000
NORETHYSTERONE ACETATE see ABU000
NORFLURAZON see NNQ100
NORFORMS see ABU500
NORGE SALTPETER see CAU000
NORGINE see AFL000
NOR-HN2 see BHO250
NOR-HN2 HYDROCHLORIDE see BHO250
NORILGAN-S see SNN500
NORIMYCIN V see CDP250
NORISODRINE see DMV600
NORIT see CBT500
NORKOOL see EJC500
NORLESTRIN see EEH520
NORLEUCAMINE see PBV505
ε-NORLEUCINE see AJD000
NOR-LOST HYDROCHLORID (GERMAN) see BHO250
NORLUTATE see ABU000
NORLUTIN see NNP500
NORLUTINE ACETATE see ABU000
NORMAL LEAD ACETATE see LCV000
NORMAL LEAD ORTHOPHOSPHATE see LDU000
NORMERSAN see TFS350
NORMOCYTIN see VSZ000
NORMONSON see CHG000
NORMOSAN see CHG000
NORMOSON see CHG000
NOR-NITROGEN MUSTARD see BHN750
NORNITROGEN MUSTARD HYDROCHLORIDE see
BHO250
NORODIN see DBA800
NORODIN HYDROCHLORIDE see MDT600
NOROX BZP-250 see BDS000
NORPRAMIN see DLS600
(17-α)-19-NORPREGNA-1,3,5(10)-TRIEN-20-YNE-
3,17,DIOL see EEH500
19-NOR-17-α-PREGNA-1,3,5(10)-TRIEN-2-YNE-3,17-
DIOL see EEH500
(3-β,17-α)-19-NORPREGN-4-EN-20-YNE-3,17-DIOL
DIACETATE see EQJ500
NOR-PRESS 25 see HGP500
NORSULFASOL see TEX250
NORSULFAZOLE see TEX250
NORTEC see CDO000
NORTIMIL see DLS600
NORVAL see DJL000
NORVALAMINE see BPX750

NORVALINE, 3-METHYL- see IKX000
NORVALINE, 4-METHYL- see LES000
NORVINYL see PKQ059
NORVINYL P 6 see AAX175
NORWAY SALTPETER see CAU000
NORWEGIAN SALTPETER see CAU000
NO SCALD see DVX800
NOSIM see CCK125
NOSTEL see CHG000
NOTENQUIL see ABH500
NOTENSIL see ABH500
NOTESIL see ABH500
NOTEZINE see DIW000
NOURALGINE see SEH000
NOURITHION see PAK000
NOVA see MRW775
NOVADELOX see BDS000
NOVADEX see NOA600
NOVALON YELLOW 2GN see AAQ250
NOVAMIDON see DOT000
NOVAMIN see PMF500
NOVAMIN see DYE600
NOVAMINE see DYE600
NOVAMONT 2030 see PMP500
NOVANTOINA see DKQ000
NOVANTOINA see DNU000
NOVA-PHENO see EOK000
NOVARSAN see NCJ500
NOVARSENOBENZOL see NCJ500
NOVARSENOBILLON see NCJ500
NOVASUROL see CCG500
NOVATEC JUO 80 see PJS750
NOVATEC JVO 80 see PJS750
NOVATHION see DSQ000
NOVATONE see MDW750
NOVA W see MRW775
NOVERME see MHL000
NOVICODIN see DKW800
NOVID see ADA725
NOVIGAM see BBQ500
NOVOBIOCIN MONOSODIUM see NOB000
NOVOBIOCIN, MONOSODIUM SALT see NOB000
NOVOBIOCIN, SODIUM derivative see NOB000
NOVOCAIN-CHLORHYDRAT (GERMAN) see AIT250
NOVOCAINE HYDROCHLORIDE see AIT250
NOVOCAIN HYDROCHLORID (GERMAN) see AIT250
NOVOCAL see CAS750
NOVOCHLOROCAP see CDP250
NOVOCILLIN see BFD250
NOVOCONESTRON see ECU750
NOVODIPHENYL see DNU000
NOVODRIN see DMV600
NOVOLEN see PMP500
NOVOL KETONE see NOB800
NOVOMYCETIN see CDP250
NOVON 712 see PKQ059
NOVONAL see DJU200
NOVONIDAZOL see MMN250
NOVOPHENICOL see CDP250
NOVOPHENYL see BRF500
NOVOPHONE see SOA500
NOVOSAXAZOLE see SNN500
NOVOSCABIN see BCM000
NOVOSED see MDQ250
NOVOX see BSC500
NOVYDRINE see BBK000
NOXAL see DXH250
NOXIPTILINE HYDROCHLORIDE see DPH600

NOXIPTILIN HYDROCHLORID (GERMAN) see DPH600
NOXIPTYLINE HYDROCHLORIDE see DPH600
NOXODYN see TEH500
NOXYRON see DYC800
1-NP see NIX500
NP 2 see NCW500
2-NP see NIY000
2-NP see ALL750
NP-9 see NNB300
NP 212 see CJR909
NPA see DNB000
NPH-1091 see BDJ250
NPIP see NLJ500
2-NPPD see ALL750
NaPSt see SMQ500
NPU see NLO500
NPYR see NLP500
4-NQO see NJF000
NR.C 2294 see DNA800
NRDC 104 see BEP500
NSAR see NLR500
NSC-339 see DVF200
NSC-739 see AMG750
NSC-740 see MDV500
NSC-742 see ASA500
NSC 746 see UVA000
NSC-750 see BOT250
NSC-751 see EEI000
NSC-755 see POK000
NSC-757 see CNG938
NSC-759 see BCB750
NSC-762 see BIE250
NSC-762 see BIE500
NSC-763 see DUD800
NSC-1390 see ZVJ000
NSC-1532 see DUZ000
NSC-1895 see DCF200
NSC-2066 see TEP000
NSC-2100 see NGE500
NSC-2101 see HMY000
NSC-2107 see NGE000
NSC-2666 see DDV800
NSC-2752 see FOU000
NSC-3051 see MKG500
NSC-3058 see BDH250
NSC-3060 see PKV500
NSC-3061 see TGD000
NSC-3069 see CDP250
NSC-3070 see DKA600
NSC-3072 see SFA000
NSC-3073 see FMT000
NSC-3088 see CDO500
NSC-3094 see DTP000
NSC-3096 see MIW500
NSC-3138 see DOO800
NSC-3425 see THR750
NSC-4170 see MMD500
NSC-4730 see EGI000
NSC-4911 see MCQ500
NSC-5354 see PEY250
NSC-5356 see DSB000
NSC-6091 see SOA500
NSC-6396 see TFQ750
NSC-6470 see NDY000
NSC-6738 see DGP900
NSC-7760 see POS750
NSC-7764 see LAQ000

NSC-7778 see MES000
NSC-8028 see IIM000
NSC-8260 see MGA250
NSC-8806 see PED750
NSC-8819 see ADR000
NSC-9166 see TBG000
NSC-9169 see HOI000
NSC-9369 see MMP000
NSC-9698 see MAW750
NSC-9701 see MPN500
NSC-9704 see PMH500
NSC 9706 see TND500
NSC 9717 see TND250
NSC-9895 see EDO000
NSC-10023 see PLZ000
NSC-10107 see CFA500
NSC-10107 see CFA750
NSC-10108 see CLO750
NSC-10483 see CNS750
NSC-10873 see BHN750
NSC-10873 see BHO250
NSC-12169 see EDU500
NSC-14083 see SLW500
NSC-14210 see BHT750
NSC-15193 see DAR600
NSC-15432 see EEH550
NSC-16895 see LGZ000
NSC-18016 see CHC750
NSC-19893 see FMM000
NSC-19987 see MOR500
NSC-20264 see AOA125
NSC-22314 see TCF000
NSC-23519 see DCQ600
NSC-23909 see MNA750
NSC-24890 see NKE500
NSC-26154 see AJD000
NSC-26198 see PAN100
NSC-26271 see CQC500
NSC-26271 see CQC650
NSC-26805 see EMF500
NSC-26980 see AHK500
NSC-27640 see DAR400
NSC-28693 see MRH000
NSC-29215 see TND000
NSC-30152 see PMO250
NSC-30211 see TNF500
NSC-32606 see BOP750
NSC-33669 see EAN000
NSC-34462 see BIA250
NSC-34533 see GKE000
NSC-34652 see MKB250
NSC-37448 see PDT250
NSC-37538 see BKM500
NSC-39069 see TFU500
NSC-39084 see ASB250
NSC-39661 see DAS000
NSC-40774 see MPU000
NSC-45388 see DAB600
NSC-45403 see ENV000
NSC 49842 see VLA000
NSC-50256 see MLH500
NSC-50364 see MMN250
NSC-50413 see AQY250
NSC-52695 see CQN000
NSC-58404 see DCO800
NSC-58775 see DLY000
NSC-60380 see CMA250
NSC-60520 see DGH400

NSC-62209 see BIF250
NSC-66847 see TEH500
NSC-67574 see LEY000
NSC-67574 see LEZ000
NSC-69536 see MLJ500
NSC-70731 see LGD000
NSC-71261 see TFJ250
NSC-71423 see VTF000
NSC-73438 see NKL000
NSC-77213 see PME500
NSC-77518 see DCK759
NSC-79037 see CGV250
NSC-82151 see DAC000
NSC-82174 see EID000
NSC-82196 see IAN000
NSC-82261 see GCS000
NSC-83653 see AED250
NSC-84963 see AMN300
NSC-85598 see SMD000
NSC-85998 see SMD000
NSC-87418 see IKH000
NSC-87419 see CPN500
NSC-89945 see DMX200
NSC-92338 see CBF250
NSC-93169 see MGL600
NSC-95441 see CHD250
NSC-102816 see ARY000
NSC-104469 see CME250
NSC-104800 see DDJ000
NSC-107430 see DOQ400
NSC 109723 see TNT500
NSC-109724 see IMH000
NSC-111180 see ACH000
NSC-113926 see RKP000
NSC-114900 see DLH630
NSC-114901 see DLS600
NSC-115944 see EAT900
NSC-119875 see PJD000
NSC-122758 see VSK950
NSC-122819 see EQP000
NSC-123127 see AES750
NSC-129943 see PIK250
NSC-134434 see LIM000
NSC-138780 see FQS000
NSC-141540 see EAV500
NSC-148958 see FLZ050
NSC-150014 see HGW500
NSC-177023 see LFA020
NSC-178248 see CLX000
NSC-190935 see CJJ250
NSC-190945 see DOP200
NSC-190986 see DJY200
NSC-190987 see DTQ400
NSC-195022 see BEP500
NSC-195106 see FAK000
NSC-271675 see DEU115
NSC-281278 see SNS100
NSC-348948 see DVO920
NSC-405124 see TIQ750
NSC-409962 see BIF750
NSC-525334 see NDY500
NSC-527017 see AOC500
NSC-528986 see AIV500
NSC-D 254157 see CLX000
NTA see SIP500
NTA see AMT500
NTA SODIUM HYDRATE see NEI000
NTG see NGY000

NTM see DTR200
NTO see NMP620
NTO (DOT) see NMP620
N,N,α-TRIMETHYL-10H-PHENOTHIAZINE-10-
ETHANAMINE MONOHYDROCHLORIDE see PMI750
N,N,3-TRIS(2-CHLOROETHYL)TETRAHYDRO-2H-
1,3,2-OXAPHOSPHORIN-2-AMINE-2-OXIDE see
TNT500
NTs 62 see CCU250
NTs 218 see CCU250
NTs 222 see CCU250
NTs 539 see CCU250
NTs 542 see CCU250
NUARSOL see ARA500
NU-BAIT II see MDU600
NUCHAR see CBT500
NUCHAR 722 see CDI000
NUCIDOL see DCM750
NUDRIN see MDU600
NU-FLOW M see MRW775
NUFLUOR see SHF500
NUISANCE DUSTS and AEROSOLS see NOF000
NUJOL see MQV750
NU-LAWN WEEDER see DDP000
NULLAPON B see EIV000
NULLAPON BF-78 see EIV000
NULLAPON BF ACID see EIX000
NULLAPON BFC CONC see EIV000
NU MAN see MPN500
NU-MANESE see MAT250
NUMBER 2 BURNER FUEL see DHE800
NUMBER 2 FUEL OIL see DHE800
NUOPLAZ DOP see DVL700
NURELLE see TIV750
NU REXFORM see LCK100
NUSYN-NOXFISH see PIX250
NUTINAL see BCA000
NUTMEG OIL see NOG500
NUTMEG OIL, EAST INDIAN see NOG500
NU-TONE see NAK500
NUTRASWEET see ARN825
NUTRIFOS STP see SJH200
NUTROSE see SFQ000
NUVA see DGP900
NUVACRON see MRH209
NUVACRON 20 see MRH209
NUVANOL see DSQ000
NUVAPEN see AIV500
NUX VOMICA see SMN500
N 4000V see SMQ500
NVC 9025 see PJS750
NYACOL see SCH002
NYACOL 830 see SCH002
NYACOL 1430 see SCH002
NYACOL A 1530 see AQF000
NYCOLINE see PJT200
NYCO Liquid RED GF see RGW000
NYCOTON see CDO000
NYCTAL see BNK000
NYLMERATE see ABU500
NYLOMINE ACID RED P4B see HJF500
NYLON-6 see PJY500
NYLOQUINONE BLUE 2J see TBG700
NYLOQUINONE LIGHT YELLOW 4JL see AAQ250
NYLOQUINONE ORANGE JR see AKP750
NYLOQUINONE YELLOW 4J see AAQ250
NYMCEL S see SFO500

NYSCONITRINE see NGY000
NYSTAN see NOH500
NYSTATIN see NOH500
NYSTATINE see NOH500
NYSTAVESCENT see NOH500
NYTAL see TAB750
O 250 see SKV100
OAAT see AIC250
OBELINE PICRATE see ANS500
OBSH see OPE000
OBSTON see DJL000
OCDD see OAJ000
OCHRATOXIN A see CHP250
OCHRE see IHC450
OCI 56 see SGG500
OCIMUM BASILICUM OIL see BAR250
OCTACARBONYLDICOBALT see CNB500
1,2,4,5,6,7,8,8-OCTACHLOOR-3a,4,7,7a-TETRAHYDRO-
4,7-endo-METHANO-INDAAN (DUTCH) see CDR750
OCTACHLOR see CDR750
OCTACHLOROCAMPHENE see CDV100
OCTACHLORODIBENZODIOXIN see OAJ000
OCTACHLORODIBENZO-p-DIOXIN see OAJ000
OCTACHLORODIBENZO(b,e)(1,4)DIOXIN see OAJ000
1,2,3,4,6,7,8,9-OCTACHLORODIBENZODIOXIN see
OAJ000
OCTACHLORODIHYDRODICYCLOPENTADIENE
see CDR750
1,2,4,5,6,7,8,8-OCTACHLORO-2,3,3a,4,7,7a-
HEXAHYDRO-4,7-METHANOINDENE see CDR750
1,2,4,5,6,7,8,8-OCTACHLORO-2,3,3a,4,7,7a-
HEXAHYDRO-4,7-METHANO-1H-INDENE see
CDR750
1,2,4,5,6,7,8,8-OCTACHLORO-3a,4,7,7a-HEXAHYDRO-
4,7-METHYLENE INDANE see CDR750
OCTACHLORO-4,7-METHANOHYDROINDANE see
CDR750
OCTACHLORO-4,7-
METHANOTETRAHYDROINDANE see CDR750
1,2,4,5,6,7,8,8-OCTACHLORO-4,7-METHANO-3a,4,7,7a-
TETRAHYDROINDANE see CDR750
OCTACHLORONAPHTHALENE see OAP000
1,2,4,5,6,7,8,8-OCTACHLORO-3a,4,7,7a-TETRAHYDRO-
4,7-METHANOINDAN see CDR750
2,2,4,5,6,7,8,8-OCTACHLORO-3a,4,7,7a-TETRAHYDRO-
4,7-METHANOINDAN see CDR575
1,2,4,5,6,7,8,8-OCTACHLORO-3a,4,7,7a-TETRAHYDRO-
4,7-METHANOINDANE see CDR750
1,2,4,5,6,7,10,10-OCTACHLORO-4,7,8,9-TETRAHYDRO-
4,7-METHYLENEINDANE see CDR750
1,2,4,5,6,7,8,8-OCTACHLOR-3a,4,7,7a-TETRAHYDRO-
4,7-endo-METHANO-INDAN (GERMAN) see CDR750
OCTACIDE 264 see OES000
9,12-OCTADECADIENOIC ACID see LGG000
cis-9,cis-12-OCTADECADIENOIC ACID see LGG000
cis,cis-9,12-OCTADECADIENOIC ACID see LGG000
OCTADECANOIC ACID see SLK000
OCTADECANOIC ACID, BARIUM CADMIUM SALT
(4:1:1) (9CI) see BAI800
OCTADECANOIC ACID, CADMIUM SALT see OAT000
OCTADECANOIC ACID, CALCIUM SALT see CAX350
OCTADECANOIC ACID, LITHIUM SALT see LHQ100
OCTADECANOIC ACID, MAGNESIUM SALT see
MAJ030
OCTADECANOIC ACID, METHYL ESTER see MJW000
OCTADECANOIC ACID, MONOESTER with 1,2,3-
PROPANETRIOL see OAV000
OCTADECANOIC ACID, SODIUM SALT see SJV500

OCTADECANOIC ACID, TETRADECYL ESTER (9CI) see TCB100

OCTADECANOIC ACID, ZINC SALT see ZMS000

OCTADECANOL see OAX000

1-OCTADECANOL see OAX000

n-OCTADECANOL see OAX000

9,10-OCTADECENOIC ACID see OHU000

cis-OCTADEC-9-ENOIC ACID see OHU000

cis-9-OCTADECENOIC ACID see OHU000

cis-Δ^9-OCTADECENOIC ACID see OHU000

9-OCTADECENOIC ACID CALCIUM SALT see CAU300

(Z)-9-OCTADECENOIC ACID METHYL ESTER see OHW000

9-OCTADECENOIC ACID (Z)-, MONOESTER with 1,2,3-PROPANETRIOL (9CI) see GGR200

9-OCTADECENOIC ACID (Z)-, POTASSIUM SALT see OHY000

cis-9-OCTADECENYLAMINE see OHM700

OCTA DECYL ALCOHOL see OAX000

n-OCTADECYL ALCOHOL see OAX000

OCTADECYLAMINE see OBC000

1-OCTADECYLAMINE see OBC000

n-OCTADECYLAMINE see OBC000

N-OCTADECYL-N-BENZYL-N,N-DIMETHYLAMMONIUMCHLORIDE see DTC600

OCTADECYLDIMETHYLBENZYLAMMONIUM CHLORIDE see DTC600

OCTADECYL ISOCYANATE see OBG000

OCTADECYLTRICHLOROSILANE see OBI000

1,6-OCTADIENE, 3,7-DIMETHYL- see CMT050

2,6-OCTADIENE, 1-ETHOXY-3,7-DIMETHYL- see GDG100

2,6-OCTADIENOIC ACID, 3,7-DIMETHYL-, ISOPENTYL ESTER, (E)- see IHS100

2,6-OCTADIEN-1-OL, 3,7-DIMETHYL-, PROPIONATE, (E)-(8CI) see GDM450

OCTAFLUOROBUTENE-2 see OBO000

OCTAFLUOROBUT-2-ENE (DOT) see OBO000

1,1,1,2,3,4,4,4-OCTAFLUORO-2-BUTENE see OBO000

OCTAFLUOROCYCLOBUTANE (DOT) see CPS000

OCTAHEDRITE (MINERAL) see OBU100

1,2,3,4,5,6,7,8-OCTAHYDRO-1,4-DIMETHYL-7-(1-METHYLETHYLIDENE)AZULENEMONOEPOXIDE see EBU100

OCTAHYDRO-1-NITROSOAZOCINE see OBY000

OCTA-KLOR see CDR750

γ-OCTALACTONE see OCE000

OCTALENE see AFK250

OCTALOX see DHB400

OCTAMETHYL-DIFOSFORZUUR-TETRAMIDE (DUTCH) see OCM000

OCTAMETHYLDIPHOSPHORAMIDE see OCM000

OCTAMETHYL-DIPHOSPHORSAEURE-TETRAMID (GERMAN) see OCM000

OCTAMETHYLPYROPHOSPHORAMIDE see OCM000

OCTAMETHYL PYROPHOSPHORTETRAMIDE see OCM000

OCTAMETHYLSILANETETRAMINE see OCM100

OCTAMETHYL TETRAMIDO PYROPHOSPHATE see OCM000

1-OCTANAL see OCO000

OCTANALDEHYDE see OCO000

OCTANAL, 2-METHYL- see MNC175

2-OCTANAMINE see OEK010

1-OCTANAMINIUM, N,N-DIMETHYL-N-OCTYL-, CHLORIDE see DTF820

1-OCTANAMINIUM, N-METHYL-N,N-DIOCTYL-, CHLORIDE (9CI) see MQH000

OCTAN AMYLU (POLISH) see AOD725

OCTAN ANTIMONITY (CZECH) see AQJ750

OCTAN BARNATY (CZECH) see BAH500

OCTAN n-BUTYLU (POLISH) see BPU750

OCTANE see OCU000

n-OCTANE see OCU000

1-OCTANECARBOXYLIC ACID see NMY000

tert-OCTANETHIOL see MKJ250

tert-OCTANETHIOL see OFE030

OCTAN ETOKSYETYLU (POLISH) see EES400

OCTAN ETYLU (POLISH) see EFR000

OCTAN FENYLRTUTNATY (CZECH) see ABU500

OCTANIL see ILK000

OCTAN KOBALTNATY (CZECH) see CNA500

OCTAN MANGANATY (CZECH) see MAQ000

OCTAN MEDNATY (CZECH) see CNI250

OCTAN METYLU (POLISH) see MFW100

OCTANOIC ACID see OCY000

OCTANOIC ACID ALLYL ESTER see AGM500

OCTANOIC ACID, CADMIUM SALT (2:1) see CAD750

OCTANOIC ACID, ETHYL ESTER see ENY000

OCTANOIC ACID, ISOPENTYL ESTER see IHP500

OCTANOIC ACID, 1,2,3-PROPANETRIYL ESTER see TMO000

OCTANOIC ACID-2-PROPENYL ESTER see AGM500

OCTANOIC ACID TRIGLYCERIDE see TMO000

OCTANOL see OEI000

3-OCTANOL see OCY100

OCTANOL-3 see OCY100

n-OCTANOL see OEI000

D-n-OCTANOL see OCY100

1-OCTANOL ACETATE see OEG000

1-OCTANOL (FCC) see OEI000

OCTANOLIDE-1,4 see OCE000

3-OCTANOL, 3-METHYL- see MND050

2-OCTANONE see ODG000

3-OCTANONE see ODI000

OCTAN PROPYLU (POLISH) see PNC250

OCTAN WINYLU (POLISH) see VLU250

n-OCTANYL ACETATE see OEG000

trans-2-OCTEN-1-AL see ODQ800

2-OCTENE, 8-ETHOXY-2,6-DIMETHYL- see EES100

1-OCTEN-3-OL see ODW000

7-OCTEN-2-OL, 2,6-DIMETHYL-8-(1H-INDOL-1-YL)- see ICS100

1-OCTEN-3-OL, 3-METHYL- see MND100

1-OCTEN-3-YL ACETATE see ODW030

6-OCTEN-1-YN-3-OL, 3,7-DIMETHYL- see LFY333

OCTHILINONE see OFE000

OCTIC ACID see OCY000

OCTILIN see OEI000

OCTIN see ILK000

OCTINUM see ILK000

OCTODRINE see ILM000

OCTOGEN see CQH250

OCTOGEN, desensitized (UN 0483) (DOT) see CQH250

OCTOGEN, wetted with not <15% water, by weight (UN 0226) (DOT) see CQH250

n-OCTOIC ACID see OCY000

OCTOIL see DVL700

OCTOIL S see BJS250

OCTON see ILK000

OCTOWY ALDEHYD (POLISH) see AAG250

OCTOWY BEZWODNIK (POLISH) see AAX500

OCTOWY KWAS (POLISH) see AAT250

OCTOXINOL see PKF500

OCTOXYNOL see PKF500
OCTOXYNOL 3 see PKF500
OCTOXYNOL 9 see PKF500
OCTYL ACETATE see OEG000
1-OCTYL ACETATE see OEG000
3-OCTYL ACETATE see OEG100
n-OCTYL ACETATE see OEG000
β-OCTYL ACROLEIN see DXU280
OCTYL ADIPATE see AEO000
OCTYL ALCOHOL see OEI000
OCTYL ALCOHOL ACETATE see OEG000
OCTYL ALCOHOL, NORMAL-PRIMARY see OEI000
n-OCTYL ALDEHYDE see OCO000
2-OCTYLAMINE see OEK010
N-OCTYL BICYCLOHEPTENE DICARBOXIMIDE see OES000
N-OCTYLBICYCLO-(2.2.1)-5-HEPTENE-2,3-DICARBOXIMIDE see OES000
OCTYL CARBINOL see NNB500
OCTYLENE GLYCOL see EKV000
n-OCTYL ESTER of 3,4,5-TRIHYDROXYBENZOIC ACID see OFA000
OCTYL FORMATE see OEY100
n-OCTYL FORMATE see OEY100
OCTYL GALLATE see OFA000
n-OCTYLIC ACID see OCY000
2-OCTYL-4-ISOTHIAZOLIN-3-ONE see OFE000
2-OCTYL-3(2H)-ISOTHIAZOLONE see OFE000
tert-OCTYLMERCAPTAN see MKJ250
tert-OCTYLMERCAPTAN (DOT) see OFE030
OCTYL PALMITATE see OFG100
OCTYL PHENOL see OFK000
OCTYL PHENOL CONDENSED with 12–13 MOLES ETHYLENE OXIDE see PKF500
p-tert-OCTYLPHENOXYETHOXYETHYLDIMETHYLBENZYL AMMONIUM CHLORIDE see BEN000
p-tert-OCTYLPHENOXYPOLYETHOXYETHANOL see PKF500
OCTYL PHTHALATE see DVL600
OCTYL PHTHALATE see DVL700
n-OCTYL PHTHALATE see DVL600
OCTYL SEBACATE see BJS250
OCTYLTRICHLOROSILANE see OGE000
OCTYLTRICHLOROSTANNANE see OGG000
OCTYLTRIS(2-ETHYLHEXYLOXYCARBONYLMETHYLTHIO)STANNANE see OGI000
OCUSEPTINE see SJL500
OCUSOL see SPC500
ODA see SJN700
ODB see DEP600
ODB see EDP000
ODCB see DEP600
ODORLESS LIGHT PETROLEUM HYDROCARBONS see OGI200
O-DUE see TAG750
OE3 see EDU500
OEKOLP see DKA600
OENANTHAL see HBB500
OENANTHALDEHYDE see HBB500
OENANTHIC ALDEHYDE see HBB500
OENANTHIC ETHER see EKN050
OENANTHOL see HBB500
OESTERGON see EDO000
OESTRADIOL see EDO000
d-OESTRADIOL see EDO000

cis-OESTRADIOL see EDO000
α-OESTRADIOL see EDO000
β-OESTRADIOL see EDO000
OESTRADIOL-17-β see EDO000
3,17-β-OESTRADIOL see EDO000
d-3,17-β-OESTRADIOL see EDO000
OESTRADIOL BENZOATE see EDP000
OESTRADIOL-3-BENZOATE see EDP000
β-OESTRADIOL BENZOATE see EDP000
β-OESTRADIOL-3-BENZOATE see EDP000
17-β-OESTRADIOL-3-BENZOATE see EDP000
OESTRADIOL DIPROPIONATE see EDR000
OESTRADIOL-3,17-DIPROPIONATE see EDR000
β-OESTRADIOL DIPROPIONATE see EDR000
17-β-OESTRADIOL DIPROPIONATE see EDR000
3,17-β-OESTRADIOL DIPROPIONATE see EDR000
OESTRADIOL MONOBENZOATE see EDP000
OESTRADIOL PHOSPHATE POLYMER see EDS000
OESTRADIOL POLYESTER with PHOSPHORIC ACID see EDS000
OESTRADIOL R see EDO000
OESTRAFORM (BDH) see EDP000
OESTRASID see DAL600
OESTRA-1,3,5(10)-TRIENE-3,17-β-DIOL see EDO000
17-β-OESTRA-1,3,5(10)-TRIENE-3,17-DIOL see EDO000
1,3,5(10)-OESTRATRIENE-3,17-β-DIOL 3-BENZOATE see EDP000
OESTRA-1,3,5(10)-TRIENE-3,16-α,17-β-TRIOL see EDU500
1,3,5-OESTRATRIENE-3-β-3,16-α,17-β-TRIOL see EDU500
(16-α,17-β)-OESTRA-1,3,5(10)-TRIENE-3,16,17-TRIOL see EDU500
1,3,5-OESTRATRIEN-3-OL-17-ONE see EDV000
1,3,5(10)-OESTRATRIEN-3-OL-17-ONE see EDV000
Δ-1,3,5-OESTRATRIEN-3-β-OL-17-ONE see EDV000
OESTRATRIOL see EDU500
OESTRILIN see ECU750
OESTRIN see EDV000
OESTRIOL see EDU500
16-α,17-β-OESTRIOL see EDU500
3,16-α,17-β-OESTRIOL see EDU500
OESTRODIENE see DAL600
OESTRODIENOL see DAL600
OESTRO-FEMINAL see ECU750
OESTROFORM see EDV000
OESTROGENINE see DKA600
OESTROGLANDOL see EDO000
OESTROGYNAEDRON see DKB000
OESTROGYNAL see EDO000
OESTROL VETAG see DKA600
OESTROMENIN see DKA600
OESTROMENSIL see DKA600
OESTROMENSYL see DKA600
OESTROMIENIN see DKA600
OESTROMON see DKA600
OESTRONBENZOAT (GERMAN) see EDV500
OESTRONE see EDV000
OESTRONE-3-SULPHATE SODIUM SALT see EDV600
OESTROPAK MORNING see ECU750
OESTROPEROS see EDV000
OESTRORAL see DAL600
OFF see DKC800
OFHC Cu see CNI000
OFNACK see POP000

OFNA-PERL SALT RRA see CLK235
OFTALENT see CDP250
OFTALFRINE see SPC500
OFUNACK see POP000
OG 1 see SEH000
OGEEN 515 see OAV000
OGEEN GRB see OAV000
OGEEN M see OAV000
OGEN see PIK450
OH-BBN see HJQ350
17-β-OH-ESTRADIOL see EDO000
OHIO 347 see EAT900
7-OHM-MBA see HMF000
7-OHM-12-MBA see HMF000
17-β-OH-OESTRADIOL see EDO000
3,3'-(GERMANOIC ANHYDRIDE) DIPROPANOIC
ACID see CCF125
OIL of ANISEED see PMQ750
OIL, ARTEMISIA see ARL250
OIL, BITTER ALMOND see BLV500
OIL CAMPHOR SASSAFRASSY see CBB500
OIL-FURNACE BLACK see CBT750
OIL GREEN see CMJ900
OIL, HIBAWOOD see HGA100
OIL of MIRBANE (DOT) see NEX000
OIL MIST, MINERAL (OSHA, ACGIH) see MQV750
OIL of MOUNTAIN PINE see PIH400
OIL of MYRBANE see NEX000
OIL of MYRISTICA see NOG500
OIL of NUTMEG see NOG500
OIL OF ABIES ALBA see AAC250
OIL OF ALLSPICE see PIG740
OIL OF ANISE see AOU250
OIL OF ARBOR VITAE see CCQ500
OIL OF BASIL see BAR250
OIL OF BAY see BAT500
OIL OF BAY see LBK000
OIL OF BERGAMOT, rectified see BFO000
OIL OF BERGAMOT, coldpressed see BFO000
OIL OF CALAMUS, GERMAN see OGK000
OIL OF CAMPHOR RECTIFIED see CBB500
OIL OF CAMPHOR WHITE see CBB500
OIL OF CARAWAY see CBG500
OIL OF CARDAMON see CCJ625
OIL OF CASSIA see CCO750
OIL OF CEDAR LEAF see CCQ500
OIL OF CHINESE CINNAMON see CCO750
OIL OF CINNAMON see CCO750
OIL OF CINNAMON, CEYLON see CCO750
OIL OF CINNAMON, CEYLON see CMQ510
OIL OF CORIANDER see CNR735
OIL OF CUBEB see COE175
OIL OF EUCALYPTUS see EQQ000
OIL OF FENNEL see FAP000
OIL OF FUR see AAC250
OIL OF GERANIUM see GDA000
OIL OF GRAPEFRUIT see GJU000
OIL OF HARTSHORN see BMA750
OIL OF JUNIPER BERRY see JEA000
OIL OF LAVANDIN see LCA000
OIL OF LAVANDIN, ABRIAL TYPE see LCA000
OIL OF LAVENDER see LCD000
OIL OF LEMON see LEI000
OIL OF LEMON, desert type, coldpressed see LEI025
OIL OF LEMONGRASS, WEST INDIAN see LEH000
OIL OF LIME, distilled see OGM850
OIL OF LIME OIL, COLDPRESSED see OGM800

OIL OF MACE see OGQ100
OIL OF MARJORAM, SPANISH see MBU500
OIL OF MUSCATEL see CMU900
OIL OF MUSTARD, artificial see AGJ250
OIL OF MYRCIA see BAT500
OIL OF MYRCIA see LBK000
OIL OF NIOBE see MHA750
OIL OF NUTMEG, expressed see OGQ100
OIL OF ONION see OJD200
OIL OF ORANGE see OGY000
OIL OF ORIGANUM see OJO000
OIL OF PALMA CHRISTI see CCP250
OIL OF PELARGONIUM see GDA000
OIL OF PIMENTA see PIG740
OIL OF PIMENTO see PIG740
OIL OF ROSE see RNA000
OIL OF ROSE BLOSSOM see RNA000
OIL OF ROSE BULGARIAN see RNA000
OIL OF ROSE GERANIUM see GDA000
OIL OF SANDALWOOD, EAST INDIAN see OGY220
OIL OF SANTAL see OGY220
OIL OF SASSAFRAS see OHI000
OIL OF SHADDOCK see GJU000
OIL OF SILVER FIR see AAC250
OIL OF SILVER PINE see AAC250
OIL OF SPEARMINT see SKY000
OIL OF SPIKE LAVENDER see SLB500
OIL OF SWEET FLAG see OGK000
OIL OF SWEET ORANGE see OGY000
OIL OF THUJA see CCQ500
OIL OF THYME see TFX500
OIL OF WHITE CEDAR see CCQ500
OIL OF WINTERGREEN see MPI000
OIL ORANGE see PEJ500
OIL ORANGE 2R see XRA000
OIL ORANGE KB see XRA000
OIL ORANGE N EXTRA see XRA000
OIL ORANGE O'PEL see TGW000
OIL ORANGE R see XRA000
OIL ORANGE SS see TGW000
OIL ORANGE X see XRA000
OIL ORANGE XO see XRA000
OIL of PALMAROSA see PAE000
OIL of PARSLEY see PAL750
OIL PIMENTA BERRIES see PIG740
OIL of PIMENTA LEAF see PIG730
OIL of PINE see PIH750
OIL PINK see CMS238
OIL RED see OHI200
OIL RED see CMS238
OIL RED 3 see SBC500
OIL RED 7 see SBC500
OIL RED 2B see SBC500
OIL RED 3B see OHI200
OIL RED 3B see SBC500
OIL RED 3G see OHI200
OIL RED 47 see SBC500
OIL RED 4B see SBC500
OIL RED 113 see CMS238
OIL RED 282 see SBC500
OIL RED 6566 see OHI200
OIL RED A see SBC500
OIL RED APT see SBC500
OIL RED AS see OHI200
OIL RED B see OHI200
OIL RED BB see SBC500
OIL RED BS see SBC500
OIL RED D see SBC500

OIL RED ED see SBC500
OIL RED F see SBC500
OIL RED G see OHI200
OIL RED GO see SBC500
OIL RED GRO see XRA000
OIL RED IV see SBC500
OIL RED O see OHI200
OIL RED O see XRA000
OIL RED OG see CMS238
OIL RED PEL see SBC500
OIL RED RC see SBC500
OIL RED RO see XRA000
OIL RED RR see SBC500
OIL RED S see SBC500
OIL RED TAX see SBC500
OIL RED XO see XRA000
OIL RED ZD see SBC500
OIL ROSE GERANIUM ALGERIAN see GDA000
OILS, ALLSPICE see PIG740
OILS, ANGELICA ROOT see AOO760
OILS, BASIL see BAR250
OIL SCARLET see OHI200
OIL SCARLET see SBC500
OIL SCARLET see XRA000
OIL SCARLET 48 see SBC500
OIL SCARLET 6G see XRA000
OIL SCARLET 371 see XRA000
OIL SCARLET 389 see CMS238
OIL SCARLET APYO see XRA000
OIL SCARLET AS see OHI200
OIL SCARLET BL see XRA000
OIL SCARLET G see OHI200
OIL SCARLET L see XRA000
OIL SCARLET YS see XRA000
OILS, CARROT see CCL750
OILS, CEDAR LEAF see CCQ500
OILS, CHAMOMILE, GERMAN see CDH500
OILS, CINNAMON see CCO750
OILS, CLOVE LEAF see CMY500
OILS, CORIANDER see CNR735
OILS, CUBEB see COE175
OILS, CUMIN see COF325
OILS, DOUGLAS FIR see OJK340
OILS, JUNIPER see JEA000
OILS, LIME see OGM850
OIL SOLUBLE ANILINE YELLOW see PEI000
OIL SOLUBLE RED S see CMS238
OILS, PALM see PAE500
OILS, PIMENTA see PIG740
OILS, PINE see PIH750
OILS, SANDALWOOD see OGY220
OILS, SWEET BIRCH see SOY100
OILS, WHEAT GERM see WBJ700
OIL THUJA see CCQ500
OIL of TURPENTINE see TOD750
OIL of TURPENTINE, RECTIFIED see TOD750
OIL VERMILION see CMS238
OIL VERMILION LP see CMS238
OIL VIOLET see EOJ500
OIL of VITRIOL (DOT) see SOI500
OIL YELLOW see AIC250
OIL YELLOW see DOT300
OIL YELLOW 21 see AIC250
OIL YELLOW 2R see AIC250
OIL YELLOW 2681 see AIC250
OIL YELLOW A see AIC250
OIL YELLOW A see FAG130
OIL YELLOW AAB see PEI000

OIL YELLOW AT see AIC250
OIL YELLOW C see AIC250
OIL YELLOW I see AIC250
OIL YELLOW OB see FAG135
OIL YELLOW T see AIC250
OKASA-MASCUL see TBG000
OKITEN G 23 see PJS750
OKO see DGP900
OK PRE-GEL see SLJ500
OKSISYKLIN see HOH500
OKTADECYLAMIN (CZECH) see OBC000
OKTAN (POLISH) see OCU000
OKTANEN (DUTCH) see OCU000
terc.OKTANTHIOL see MKJ250
terc.OKTANTHIOL see OFE030
OKTATERR see CDR750
OKTOGEN see CQH250
OKTYLESTER KYSELINY GALLOVE see OFA000
OKULTIN M see CIR250
OL 27-400 see CQH100
OLAMINE see EEC600
OLATE FLAKES see OIA000
OLEAL ORANGE R see PEJ500
OLEAL ORANGE SS see TGW000
OLEAL RED BB see SBC500
OLEAL RED G see CMS238
OLEAL YELLOW 2G see DOT300
OLEAMINE see OHM700
OLEANDOMYCIN see OHM900
OLEANDOMYCINE see OHM900
OLEANDOMYCIN HYDROCHLORIDE see OHO000
OLEANDOMYCIN MONOHYDROCHLORIDE see
OHO000
OLEANDOMYCIN PHOSPHATE see OHO200
OLEATE of MERCURY see MDF250
OLEFIANT GAS see EIO000
OLEFINS see OHS000
OLEIC ACID see OHU000
OLEIC ACID CALCIUM SALT see CAU300
OLEIC ACID GLYCEROL MONOESTER see GGR200
OLEIC ACID LEAD SALT see LDQ000
OLEIC ACID, LEAD(2+) SALT (2:1) see LDQ000
cis-OLEIC ACID, METHYL ESTER see OHW000
OLEIC ACID MONOGLYCERIDE see GGR200
OLEIC ACID, POTASSIUM SALT see OHY000
OLEIC ACID, SODIUM SALT see OIA000
OLEINAMINE see OHM700
OLEJ NAPEDOWY III see DHE900
OLEJ NAPEDOWY III see FOP000
OLEOAKARITHION see TNP250
OLEOCUIVRE see CNO000
OLEOFOS 20 see PAK000
OLEOGESAPRIM see ARQ725
OLEOMYCETIN see CDP250
OLEO NORDOX see CNO000
OLEOPARAPHENE see PAK000
OLEOPARATHION see PAK000
OLEOPHOSPHOTHION see MAK700
OLEORESIN TUMERIC see TOD625
OLEOSUMIFENE see DSQ000
OLEOVITAMIN A see VSK600
OLEOVITAMIN D see VSZ100
OLEOVITAMIN D3 see CMC750
OLEOVOFOTOX see MNH000
OLEOYLGLYCEROL see GGR200
OLETAC 100 see PMP500
OLEUM see SOI520
OLEUM ABIETIS see PIH750

OLEUM SINAPIS VOLATILE see AGJ250
OLEYLAMIN (GERMAN) see OHM700
OLEYL AMINE see OHM700
OLEYLMONOGLYCERIDE see GGR200
OLIBANUM GUM see OIM000
OLICINE see GGR200
OLITREF see DUV600
OLIVE OIL see OIQ000
OLOTHORB see PKL100
OLOW (POLISH) see LCF000
OLPISAN see PAX000
OLTITOX see CBM750
OM-1563 see ZMJ000
OMACIDE 24 see HOC000
OMADINE MDS see OIU850
OMADINE ZINC see ZMJ000
OMAFLORA see HHC000
OMAHA see LCF000
OMAHA & GRANT see LCF000
OMAIN see MIW500
OMAINE see MIW500
OMAIT see SOP000
OMAL see TIW000
OMCHLOR see DFE200
OMEGA 127 see MKP500
OMEGA-BENZOYLACETOPHENONE see PFU300
OMEGA CHROME BLUE FB see HJF500
α,OMEGA-DIAMINO-4,9-DIOXADODECANE see BGU600
OM-HYDANTOINE see DKQ000
OM-HYDANTOINE SODIUM see DNU000
OMITE see SOP000
OMNIBON see SNN300
OMNI-PASSIN see DJT800
OMNIPEN see AIV500
OMNIPEN-N see SEQ000
OMNIZOLE see TEX000
OMP-2 see OIY000
OMPA see OCM000
OMPACIDE see OCM000
OMPATOX see OCM000
OMPAX see OCM000
OMS 14 see DGP900
OMS-29 see CBM750
OMS-32 see MIA250
OMS 43 see DSQ000
OMS-47 see DOS000
OMS-93 see DST000
OMS 468 see AFR250
OMS 570 see EAQ750
OMS-659 see EGV500
OMS-771 see CBM500
OMS 834 see MRH209
OMS-0971 see CMA100
OMS-1155 see CMA250
OMS-1206 see BEP500
OMS 1328 see CDS750
OMS 1804 see CJV250
ONCB see CJB750
ONCO-CARBIDE see TGN250
ONCOTEPA see TFQ750
ONCOTIOTEPA see TFQ750
ONCOVEDEX see TND000
ONCOVIN see LEY000
ONCOVIN see LEZ000
ONE-IRON see FBJ100
ONGROVIL S 165 see PKQ059

ONION OIL see OJD200
ONKOTIN see DBD700
ONOZUKA P 500 see CCU150
ONT see NMO525
ONYX see SCI500
ONYX see SCJ500
ONYXOL 345 see BKE500
O,O'DIANISIDINE see DCJ200
OOS see TAB750
OPALON see PKQ059
OPALON 400 see AAX175
OPE 30 see PKF500
OPERIDINE see DAM700
OPHTHALAMIN see VSK600
OPHTHALMADINE see DAS000
OPIUM see OJG000
OPLOSSINGEN (DUTCH) see FMV000
OPP see BGJ250
OPP-Na see BGJ750
OPP-SODIUM see BGJ750
OP-SULFA 30 see SNQ710
OPTAL see PND000
OPTEF see CNS750
OPTENYL see PAH250
OP-THAL-ZIN see ZNA000
OPTHEL-S see SNQ710
OPTHOCHLOR see CDP250
OPTIMIL see MDT250
OPTIPHYLLIN see TEP000
OPTOCIL see SCK600
OPTOCIL (QUARTZ) see SCK600
8-OQ see QPA000
ORABET see BSQ000
ORACET RED 3B see AKE250
ORACET SAPPHIRE BLUE G see TBG700
ORACON see DNX500
ORADEXON see SOW000
ORAFURAN see NGE000
ORAGEST see MCA000
ORALCID see ABX500
ORALIN see BSQ000
ORALSONE see HHR000
ORANGE #10 see HGC000
ORANGE A l'HUILE see PEJ500
ORANGE B see OJK325
ORANGE BASE CIBA II see NEO000
ORANGE BASE IRGA I see NEN500
ORANGE BASE IRGA II see NEO000
ORANGE CHROME see LCS000
ORANGE CRYSTALS see ABC500
ORANGE G (indicator) see HGC000
ORANGE G (biological stain) see HGC000
ORANGE GC BASE see CEH675
ORANGE G DYE see HGC000
ORANGE II R see FAG150
ORANGE INSOLUBLE OLG see PEJ500
ORANGE INSOLUBLE OLG see XRA000
ORANGE INSOLUBLE RR see XRA000
ORANGE LEAD see LDS000
ORANGE NITRATE CHROME see LCS000
ORANGE OIL see OGY000
ORANGE OIL see PDD750
ORANGE OIL, coldpressed (FCC) see OGY000
ORANGE OIL KB see XRA000
ORANGE PAL see FAG150
ORANGE PEL see PEJ500
ORANGE RESENOLE No. 3 see PEJ500
ORANGE RGL CONC. SPECIALLY PURE see FAG150

ORANGES see BBK500
ORANGE SALT CIBA II see NEO000
ORANGE SALT IRGA II see NEO000
ORANGE SOLUBLE A l'HUILE see PEJ500
ORANGE 3R SOLUBLE IN GREASE see TGW000
ORANGE YELLOW S see FAG150
ORANGE YELLOW S.AF see FAG150
ORANGE YELLOW S.FQ see FAG150
ORANIL see BSM000
ORANYL see BSM000
ORANZ G (POLISH) see HGC000
ORARSAN see ABX500
ORASONE see PLZ000
ORASULIN see BSM000
ORATRAST see BAP000
ORAVIRON see MPN500
ORBENIN SODIUM HYDRATE see SLJ000
ORBICIN see DCQ800
ORBON see OAV000
ORCANON see MPW500
ORCED see DGN200
ORCHARD BRAND ZIRAM see BJK500
ORCHIOL see TBG000
ORCHISTIN see TBG000
ORCIN see MPH500
ORCINOL see MPH500
ORDIMEL see ABB000
ORDINARY AZOXYBENZENE see ASO750
ORDINARY LACTIC ACID see LAG000
OREGON BALSAM see OJK340
OREMET see TGF250
ORESTOL see DKB000
ORETIC see CFY000
ORETON see TBG000
ORETON-F see TBF500
ORETON-M see MPN500
ORETON METHYL see MPN500
ORETON PROPIONATE see TBG000
OREZAN see BSQ000
ORGA-414 see AMY050
ORGAMIDE see PJY500
ORGANEX see CAK500
ORGANIC GLASS E 2 see PKB500
ORGANIDIN see IEL800
ORGANOL BORDEAUX B see EOJ500
ORGANOL FAST GREEN J see BLK000
ORGANOL ORANGE see PEJ500
ORGANOL ORANGE 2R see TGW000
ORGANOL RED B see SBC500
ORGANOL RED BS see OHI200
ORGANOL SCARLET see OHI200
ORGANOL VERMILION see CMS238
ORGANOL YELLOW see PEI000
ORGANOL YELLOW 25 see AIC250
ORGANOL YELLOW ADM see DOT300
ORGANOMETALS see OJM000
ORGASEPTINE see SNM500
ORGASTYPTIN see EDU500
ORIENTAL BERRY see PIE500
ORIENT OIL ORANGE PS see PEJ500
ORIENT OIL RED OG see CMS238
ORIENT OIL RED RR see SBC500
ORIENT OIL YELLOW GG see DOT300
ORIENT WATER PINK 2 see ADG250
ORIGANUM OIL see OJO000
ORIMON see PDP250
ORINASE see BSQ000
ORINAZ see BSQ000

ORION BLUE 3B see CMO250
ORIZON see PJS750
ORIZON 805 see PJS750
ORLUTATE see ABU000
OROLEVOL see MQU750
ORONOL see ART250
OROXINE see LFG050
ORPHENADRINE HYDROCHLORIDE see OJW000
ORPHENOL see BGJ750
ORPIMENT see ARI000
ORQUISTERONE see TBF500
ORRIS ROOT OIL see OJW100
ORSIN see PEY500
ORTEDRINE see BBK000
ORTHAMINE see PEY250
ORTHENE see DOP600
ORTHENE-755 see DOP600
ORTHESIN see EFX000
ORTHO 5865 see CBF800
ORTHO 9006 see DTQ400
ORTHO 12420 see DOP600
ORTHOARSENIC ACID see ARB250
ORTHOARSENIC ACID HEMIHYDRATE see ARC500
ORTHOBORIC ACID see BMC000
ORTHO C-1 DEFOLIANT & WEED KILLER see SFS000
ORTHOCIDE see CBG000
ORTHOCRESOL see CNX000
ORTHODICHLOROBENZENE see DEP600
ORTHODICHLOROBENZOL see DEP600
ORTHO N-4 DUST see NDN000
ORTHO N-5 DUST see NDN000
ORTHO EARWIG BAIT see DXE000
ORTHOFORMIC ACID, ETHYL ESTER see ENY500
ORTHOFORMIC ACID, TRIETHYL ESTER see ENY500
ORTHOFORMIC ACID, TRIMETHYL ESTER see TLX600
ORTHO GRASS KILLER see CBM000
ORTHOHYDROXYBENZOIC ACID see SAI000
ORTHOHYDROXYDIPHENYL see BGJ250
ORTHO-KLOR see CDR750
ORTHO L10 DUST see LCK000
ORTHO L10 DUST see LCK100
ORTHO L40 DUST see LCK000
ORTHO-LM APPLE SPRAY see MLH000
ORTHO LM CONCENTRATE see MLH000
ORTHO LM SEED PROTECTANT see MLH000
ORTHO MALATHION see MAK700
ORTHO-MITE see SOP500
ORTHOMRAVENCAN ETHYLNATY (CZECH) see ENY500
ORTHOMRAVENCAN METHYLNATY (CZECH) see TLX600
ORTHO P-G BAIT see COF500
ORTHOPHALTAN see TIT250
ORTHOPHENANTHROLINE see PCY250
ORTHOPHENYLPHENOL see BGJ250
ORTHOPHOS see PAK000
ORTHO PHOSPHATE DEFOLIANT see BSH250
ORTHOPHOSPHORIC ACID see PHB250
ORTHOPHOSPHORUS ACID see PGZ899
ORTHOSAN MB see DTC600
ORTHOSERPINA see RDK000
ORTHOSIL see SJU000
ORTHOTELLURIC ACID see TAI750
ORTHO-TOLUOL-SULFONAMID (GERMAN) see TGN250
ORTHO WEEVIL BAIT see DXE000
ORTHOXENOL see BGJ250

ORTIZON see HIB500
ORTRAN see DOP600
ORTRIL see DOP600
ORTUDUR see PKQ059
ORUDIS see BDU500
ORUVAIL see BDU500
ORVAGIL see MMN250
ORVINYLCARBINOL see AFV500
ORVUS WA PASTE see SIB600
ORYZANIN see TES750
ORYZANINE see TES750
OS 1897 see DDL800
OS 2046 see MQR750
OSAGE ORANGE see MRN500
OSAGE ORANGE CRYSTALS see MRN500
OSAGE ORANGE EXTRACT see MRN500
OSARSAL see ABX500
OSARSOLE see ABX500
OSBAC see MOV000
OSBON AC see PCL500
OSCINE see SBG000
OSIREN see AFJ500
OSMIC ACID see OKK000
OSMIUM see OKE000
OSMIUM(VIII) OXIDE see OKK000
OSMIUM TETROXIDE see OKK000
OSMOSOL EXTRA see PND000
OSOCIDE see CBG000
OSSALIN see SHF500
OSSIAMINA see CFA750
OSSICHLORIN see CFA750
OSSIDO di MESITILE (ITALIAN) see MDJ750
OSSIN see SHF500
OSTACET YELLOW P2G see AAQ250
OSTAMER see CDV625
OSTANTHREN GREY M see CMU475
OSTELIN see VSZ100
OSTEOBOND SURGICAL BONE CEMENT see PKB500
OSTREOGRYCIN see VRF000
OSVARSAN see ABX500
OSYROL see AFJ500
OTBE (FRENCH) see BLL750
OTC see HOH500
OTERBEN see BSQ000
OTETRYN see HOI000
OTOBIOTIC see NCG000
OTOFURAN see NGE500
OTOPHEN see CDP250
OTS see TGN250
OTS 11 see DVM800
OTTAFACT see CFD990
OTTANE (ITALIAN) see OCU000
OTTASEPT see CLW000
OTTASEPT EXTRA see CLW000
1,2,4,5,6,7,8,8-OTTOCHLORO-3A,4,7,7A-TETRAIDRO-4,7-endo-METANO-INDANO (ITALIAN) see CDR750
OTTOMETIL-PIROFOSFORAMMIDE (ITALIAN) see OCM000
OTTO ROSE see RNA000
OTTO of ROSE see RNA000
OUABAGENIN-l-RHAMNOSID (GERMAN) see OKS000
OUABAGENIN-l-RHAMNOSIDE see OKS000
OUABAIN see OKS000
OUABAINE see OKS000
OU-B see CBT500
OUBAIN see OKS000
OUTFLANK see AHJ750

OUTFLANK-STOCKADE see AHJ750
OVABAN see VTF000
OVADOFOS see DSQ000
OVADZIAK see BBQ500
OVAHORMON see EDO000
OVAHORMON BENZOATE see EDP000
OVANON see LJE000
OVARELIN see LIU370
OVARIOSTAT (FRENCH) see LJE000
OVASTEROL see EDO000
OVASTEROL-B see EDP000
OVASTEVOL see EDO000
OVEST see ECU750
OVESTERIN see EDU500
OVESTIN see EDU500
OVESTINON see EDU500
OVESTRION see EDU500
OVEX see EDP000
OVEX see EDV000
OVIFOLLIN see EDV000
OVIN see DNX500
OVISOT see ABO000
OVITELMIN see MHL000
OVOCICLINA see EDO000
OVOCYCLIN see EDO000
OVOCYCLIN BENZOATE see EDP000
OVOCYCLIN DIPROPIONATE see EDR000
OVOCYCLINE see EDO000
OVOCYCLIN M see EDP000
OVOCYCLIN-MB see EDP000
OVOCYCLIN-P see EDR000
OVOCYLIN see EDO000
O-V STATIN see NOH500
OVULEN 50 see EQJ500
OWISPOL GF see SMQ500
OXAALZUUR (DUTCH) see OLA000
1-OXA-4-AZACYCLOHEXANE see MRP750
7-OXABICYCLO(4.1.0)HEPTANE see CPD000
7-OXABICYCLO(2.2.1)HEPTANE-2,3-DICARBOXYLIC ACID see EAR000
OXACYCLOBUTANE see OMW000
OXACYCLOPENTADIENE see FPK000
OXACYCLOPENTANE see TCR750
OXACYCLOPROPANE see EJN500
4H-1,3,5-OXADIAZINE-4-THIONE, TETRAHYDRO-3,5-DIMETHYL- see TCQ275
(1,2,3)OXADIAZOLO(5,4-d)PYRIMIDIN-5(4H)-ONE see DCQ600
8-OXA-3,5-DITHIA-4-STANNAHEPTADECANOIC ACID, 4,4-DIBUTYL-7-OXO-, NONYLESTER (9CI) see DEH650
OXAF see BHA750
OXAFURADENE see NDY000
3-OXA-1-HEPTANOL see BPJ850
11-OXAHEXADECANOLIDE see OKW100
12-OXAHEXADECANOLIDE see OKW110
OXALAMIDE see OLO000
OXALATES see OKY000
OXALIC ACID see OLA000
OXALIC ACID DIAMIDE see OLO000
OXALIC ACID, DIAMMONIUM SALT see ANO750
OXALIC ACID, DIETHYL ESTER see DJT200
OXALIC ACID DINITRILE see COO000
OXALIC ACID, TIN(2+) SALT (1:1) (8CI) see TGE250
OXALONITRILE see COO000
OXALSAEURE (GERMAN) see OLA000
OXALYL CYANIDE see COO000

7-OXA-8-MERCURABICYCLO(4.2.0)OCTA-1,3,5-
TRIENE, 5-METHYL-2-NITRO- see NHK900
OXAMID (CZECH) see OLO000
OXAMIDE see OLO000
OXAMIMIDIC ACID see OLO000
OXAMMONIUM see HLM500
OXAMMONIUM SULFATE see OLS000
OXAMYL see DSP600
OXANAL YELLOW T see FAG140
OXANE see EJN500
3-OXAPENTANE-1,5-DIOL see DJD600
3-OXA-1,5-PENTANEDIOL see DJD600
1,3,2-OXATHIASTANNOLANE, 2,2-DIBUTYL- see
DEF150
1,2-OXATHIOLANE-2,2-DIOXIDE see PML400
OXATIMIDE see OMG000
OXATOMIDA see OMG000
OXATOMIDE see OMG000
5-OXATRICYCLO(8.2.0.0^{4,6})DODECANE, 4,12,12-
TRIMETHYL-9-METHYLENE-, (1R,4R,6R,10S)- see
CCN100
2-H-1,3,2-OXAZAPHOSPHORINANE see CQC650
2-OXAZOLAMINE, N-(3,4-DIMETHYLPHENYL)-4,5-
DIHYDRO- see XOS500
OXAZOLIDINE, 3-METHYL- see MND700
OXAZOLIDINE T see OMM300
2-OXAZOLIDONE, 3-((1-METHYLPYRROL-2-
YL)METHYLENEAMINO)-4-(PIPERIDINOMETHYL)-
see MPF300
2-OXAZOLINE, 3,4-XYLIDINO- see XOS500
2-OXAZOLINE, 2-(3,4-XYLIDINO)- see XOS500
1H,3H,5H-OXAZOLO(3,4-C)OXAZOLE-7A(7H)-
METHANOL see OMM300
m-OXEDRINE see NCL500
(−)-m-OXEDRINE see NCL500
m-OXEDRINE see SPC500
OXETAN see OMW000
OXETANE see OMW000
OXIAMIN see IDE000
OXIDATE LE see MHA750
OXIDATION BASE 22 see ALL750
OXIDATION BASE 25 see NEM480
OXIDATION BASE 12A see DBO400
OXIDE of CHROMIUM see CMJ900
OXIDIZED l-CYSTEINE see CQK325
OXIDOETHANE see EJN500
α,β-OXIDOETHANE see EJN500
OXILAPINE see DCS200
OXIME COPPER see BLC250
OXIMETHOLONUM see PAN100
OXIMETOLONA see PAN100
1-OXINDENE see BCK250
OXINE see QPA000
OXINE COPPER see BLC250
OXINE CUIVRE see BLC250
2-OXI-PROPYL-PROPYLNITROSAMIN (GERMAN) see
ORS000
OXIRAAN (DUTCH) see EJN500
OXIRANE see EJN500
OXIRANE-CARBOXALDEHYDE see GGW000
OXIRANE, ((1,1-DIMETHYLETHOXY)METHYL)- see
BRK800
OXIRANE, (ETHOXYMETHYL)-(9CI) see EBQ700
OXIRANE, (ETHOXYMETHYL)-(9CI) see EKM200
OXIRANEMETHANAMINIUM, N,N,N-TRIMETHYL-,
CHLORIDE (9CI) see GGY200

OXIRANE, ((1-METHYLETHOXY)METHYL)-(9CI) see
IPD000
OXIRANE, TRIFLUORO(TRIFLUOROMETHYL)- see
HDF050
3-OXIRANYL-7-OXABICYCLO(4.1.0)HEPTENE see
VOA000
OXITETRACYCLIN see HOH500
OXITOL see EES350
OXITOSONA-50 see PAN100
OXIURAN see AOR500
OXLOPAR see HOI000
OXO see TAB750
3'-(3-OXO-7-α-ACETYLTHIO-17-β-
HYDROXYANDROST-4-EN-17-β-YL)PROPIONIC
ACID LACTONE see AFJ500
2-OXO-1,2-BENZOPYRAN see CNV000
2-OXOBORNANE see CBA750
3-OXO-BUTANOIC ACID BUTYL ESTER see BPV250
3-OXOBUTANOIC ACID ETHYL ESTER see EFS000
3-OXOBUTANOIC ACID METHYL ESTER see MFX250
γ-OXO-α-BUTYLENE see BOY500
2-OXOCHROMAN see HHR500
OXODIOCTYLSTANNANE see DVL400
α-OXODIPHENYLMETHANE see BCS250
3-OXO-l-GULOFURANOLACTONE see ARN000
2-OXOHEXAMETHYLENIMINE see CBF700
2-(3-OXO-1-INDANYLIDENE)-1,3-INDANDIONE see
ONY000
OXOLAMINE CITRATE see OOE000
OXOLANE see TCR750
OXOLE see FPK000
OXOMETHANE see FMV000
OXOPHENARSINE see OOK100
2-OXO-2-(PHENYLAMINO)ETHYL
SELENOCYANATE see OOK175
β-OXO-α-PHENYLBENZENEPROPANENITRILE see
OOK200
3-OXO-N-PHENYLBUTANAMIDE see AAY000
17-(1-OXOPROPOXY)-(17-β)-ANDROST-4-EN-3-ONE
see TBG000
3-(2-OXOPROPYL)-2-PENTYLCYCLOPENTANONE
see OOO100
2-OXO-PROPYL-PROPYLNITROSAMINE see ORS000
(2-OXOPROPYL)PROPYLNITROSOAMINE see ORS000
(3-β,5-α,15-β)-19-OXO-3,14,15-TRIHYDROXYCARD-
20(22)-ENOLIDE see AFS800
5-OXO-1,3,3-
TRIMETHYLCYCLOHEXANECARBONITRILE see
IMG500
6-OXOUNDECANE see ULA000
22-OXOVINCALEUKOBLASTINE see LEY000
OXSORALEN see XDJ000
OXY-5 see BDS000
OXY-10 see BDS000
10-(3-(4-OXYAETHYL-PIPERAZINO)PROPYL-(1))-4-
AZAPHENTHIAZIN DIHYDROCHLORID see ORI400
OXYAMINE see CFA750
OXYBENZENE see PDN750
p-OXYBENZOESAEUREAETHYLESTER (GERMAN)
see HJL000
p-OXYBENZOESAEUREMETHYLESTER (GERMAN)
see HJL500
p-OXYBENZOESAEUREPROPYLESTER (GERMAN)
see HNU500
OXYBENZONE see MES000
OXYBENZOPYRIDINE see QPA000
OXYBIS(4-AMINOBENZENE) see OPM000

4,4'-OXYBISANILINE see OPM000
p,p'-OXYBIS(ANILINE) see OPM000
4,4'-OXYBISBENZENAMINE see OPM000
p,p'-OXYBISBENZENE DISULFONYLHYDRAZIDE see OPE000
OXYBISBENZENESULFONIC ACID DIHYDRAZIDE see OPE000
OXYBIS(BENZENESULFONYL HYDRAZIDE) see OPE000
p,p'-OXYBIS(BENZENESULFONYL) HYDRAZINE see OPE000
1,1'-OXYBIS(BUTANE) see BRH750
2,2'-OXYBISBUTANE see BRH760
4,4'-OXYBIS(2-CHLOROANILINE) see BGT000
4,4'-OXYBIS(2-CHLORO-BENZENAMINE) see BGT000
1,1'-OXYBIS(2-CHLORO)ETHANE see DFJ050
OXYBIS(CHLOROMETHANE) see BIK000
2,2'-OXYBIS(1-CHLOROPROPANE) see BII250
1,1'-OXYBISETHANE see EJU000
1,1'-(OXYBIS(2,1-ETHANEDIYLOXY))BISBUTANE see DDW200
2,2'-OXYBISETHANOL see DJD600
1,1'-OXYBISETHENE see VOP000
2,2'-(OXYBIS(ETHYLENEOXY))DIETHANOL see TCE250
OXYBISMETHANE see MJW500
1,1'-(OXYBIS(METHYLENESULFONYL))BIS(2-CHLOROETHANE) see OPG100
2,2'-OXYBIS-6-OXABICYCLO-(3.1.0)HEXANE see BJN250
1,1'-OXYBIS(2,3,4,5,6-PENTABROMOBENZENE) (9CI) see PAU500
1,1'-OXYBISPENTANE see PBX000
1,1'-OXYBISPROPANE see PNM000
3,3'-OXYBIS(1-PROPENE) see DBK000
OXYBIS(TRIBUTYLTIN) see BLL750
OXYBUTANAL see AAH750
β-OXYBUTTERSAEURE-p-PHENETIDID see HJS850
OXYBUTYRIC ALDEHYDE see AAH750
OXYCARBON SULFIDE see CCC000
OXYCHINOLIN see QPA000
o-OXYCHINOLIN (GERMAN) see QPA000
OXYCHLORID FOSFORECNY see PHQ800
OXYCHLORURE CHROMIQUE (FRENCH) see CML125
OXYCIL see SFS000
OXYCODEINONE see PCG500
OXYCOLOR see AOR500
OXY DBCP see DDL800
OXYDE d'ALLYLE et de GLYCIDYLE (FRENCH) see AGH150
OXYDE de BARYUM (FRENCH) see BAO000
OXYDE de CALCIUM (FRENCH) see CAU500
OXYDE de CARBONE (FRENCH) see CBW750
OXYDE de CHLORETHYLE (FRENCH) see DFJ050
OXYDE d'ETHYLE (FRENCH) see EJU000
OXYDE de MERCURE (FRENCH) see MCT500
OXYDE de MESITYLE (FRENCH) see MDJ750
OXYDEMETONMETHYL see DAP000
OXYDEMETON-METILE (ITALIAN) see DAP000
OXYDE NITRIQUE (FRENCH) see NEG100
OXYDE de PROPYLENE (FRENCH) see PNL600
OXYDE de TRIBUTYLETAIN see BLL750
OXYDIANILINE see OPM000
4,4'-OXYDIANILINE see OPM000
p,p'-OXYDIANILINE see OPM000
2,2'-OXYDIETHANOL see DJD600

N-(OXYDIETHYLENE)BENZOTHIAZOLE-2-SULFENAMIDE see BDG000
OXYDIETHYLENE BIS(CHLOROFORMATE) see OPO000
OXYDIETHYLENE CHLOROFORMATE see OPO000
OXYDIFORMIC ACID DIETHYL ESTER see DIZ100
OXYDIMETHYLQUINAZINE see AQN000
4,4'-OXYDIPHENYLAMINE see OPM000
OXYDI-p-PHENYLENEDIAMINE see OPM000
N-OXYD-LOST see CFA500
N-OXYD-LOST see CFA750
N-OXYD-MUSTARD see CFA500
OXYDOL see HIB050
OXYDRENE see DBA800
OXYETHYLIDENEDIPHOSPHONIC ACID see HKS780
1-(β-OXYETHYL)-2-METHYL-5-NITROIMIDAZOLE see MMN250
OXYFED see DBA800
OXYFLUORFEN see OQU100
OXYFLUORFENE see OQU100
OXYFUME see EJN500
OXYFUME 12 see EJN500
OXYFUME 20 see EJO000
OXYFUME 30 see EJO000
OXYFURADENE see NDY000
OXYGEN see OQW000
OXYGEN DIFLUORIDE see ORA000
OXYGEN FLUORIDE see ORA000
OXYGEN, compressed (UN 1072) (DOT) see OQW000
OXYGEN, refrigerated liquid (cryogenic liquid) (UN 1073) (DOT) see OQW000
OXYHYDROCHINON (GERMAN) see BBU250
OXYHYDROQUINONE see BBU250
OXYJECT 100 see HOI000
OXYLAN see DKQ000
OXYLITE see BDS000
OXYMAG see MAH500
OXYMETHALONE see PAN100
OXYMETHENOLONE see PAN100
OXYMETHOLONE see PAN100
OXYMETHYLENE see FMV000
OXYMETHYLPHTHALIMIDE see HMP100
OXYMURIATE OF POTASH see PLA250
OXYMYCIN see HOH500
OXYMYKOIN see HOH500
OXY-NH2 see CFA500
OXYOZYL see AOR500
OXYPENDYL HYDROCHLORIDE see ORI400
OXYPER see SJB400
OXYPHENALON see RBU000
OXYPHENIC ACID see CCP850
p-OXYPROPIOPHENONE see ELL500
β-OXYPROPYLPROPYLNITROSAMINE see ORS000
OXYPSORALEN see XDJ000
OXYQUINOLINE see QPA000
8-OXYQUINOLINE see QPA000
OXYQUINOLINOLEATE de CUIVRE (FRENCH) see BLC250
5-OXYRESORCINOL see PGR000
OXYSONIUM IODIDE see HOJ150
OXYSULFATOVANADIUM see VEZ000
OXYTERRACIN see HOH500
OXYTERRACINE see HOH500
OXYTERRACYNE see HOH500
OXYTETRACYCLINE see HOH500
OXYTETRACYCLINE AMPHOTERIC see HOH500

OXYTETRACYCLINE HYDROCHLORIDE see HOI000
OXYTOL ACETATE see EES400
m-OXYTOLUENE see CNW750
o-OXYTOLUENE see CNX000
p-OXYTOLUENE see CNX250
OXYTRIL M see DDP000
OXYUREA see TGN250
OXY WASH see BDS000
OZIDE see ZKA000
OZLO see ZKA000
OZON (POLISH) see ORW000
OZONE see ORW000
P07 see PKO500
P-25 see SLJ000
P-33 see CBT750
P-40 see DXG000
P-50 see AIV500
P68 see CBT750
P-165 see ASA500
P1250 see CBT750
P1496 see RBF100
P 1531 see PFC750
6020P see PJS750
P 2010B see PJS750
P 2020T see PJS750
P 2050T see PJS750
P 2070P see PJS750
P 4007T see PJS750
P 4070L see PJS750
PA see PAP750
PA see PEG500
PA 130 see PJS750
PA 190 see PJS750
PA 520 see PJS750
PA 560 see PJS750
PA 775 see OHM900
PA 6 (polymer) see PJY500
PAA see BBL500
PA'AILA (HAWAII) see CCP000
PABA see AIH600
PABANOL see AIH600
PABESTROL see DKA600
PABESTROL see DKB000
PABS see SNM500
PACEMO see HIM000
PACIFAN see NBU000
PACITRAN see MQR200
PACITRAN see DCK759
PAD 522 see PJS750
PADOPHENE see PDP250
PAINTER'S NAPHTHA see PCT250
PAISLEY POLYMER see PMP500
PAKA (HAWAII) see TGI100
PAKHTARAN see DUK800
PAL see PEC750
PALACET YELLOW GN see AAQ250
PALACOS see PKB500
PALAFER see FBJ100
PALANIL VIOLET 3B see DBY700
PALANIL YELLOW G see AAQ250
PALAPENT see NBU000
PALATINOL A see DJX000
PALATINOL AH see DVL700
PALATINOL BB see BEC500
PALATINOL C see DEH200
PALATINOL IC see DNJ400
PALATINOL M see DTR200
PALATONE see MAO350

PALE ORANGE CHROME see LCS000
PALESTROL see DKA600
PALINUM see TDA500
PALLADIUM CHLORIDE see PAD500
PALLADIUM(2+) CHLORIDE see PAD500
PALLADOUS CHLORIDE see PAD500
PALLETHRINE see AFR250
PALLICID see ABX500
PALMA CHRISTI (HAITI) see CCP000
PALMAROSA OIL see PAE000
PALM BUTTER see PAE500
PALMITA see PGA750
PALMITAMIDE see PAE240
PALMITIC ACID see PAE250
PALMITIC ACID AMIDE see PAE240
PALMITIC ACID, 2-ETHYLHEXYL ESTER see OFG100
PALMITOYL, l-ASCORBIC ACID see ARN150
PALMITYL ALCOHOL see HCP000
PALMITYLAMINE see HCO500
PALM OIL see PAE500
PALM OIL (UNHYDROGENATED) see PAE300
PALOPAUSE see ECU750
PALTET see TBX250
PALUDRINE see CKB250
PALYGORSCITE see PAE750
PALYGORSKIT (GERMAN) see PAE750
l-PAM see PED750
PAM (CZECH) see POS750
PAMACEL YELLOW G-3 see AAQ250
PAMAZONE see CJR909
PAMEION see PAH250
2-PAM IODIDE see POS750
PAMISAN see ABU500
PAMISYL SODIUM see SEP000
PAMOLYN see OHU000
PANACELAN see POC500
PANACIDE see MJM500
PANADOL see HIM000
PANADON see PAG200
PANAM see CBM750
PANCAL see CAU750
PANCALMA see MQU750
PANCID see SNN500
PANCIL see OFE000
PANCORAL see BQJ500
PANDEX see PKM250
PANDRINOX see MLF250
PANESTIN see TBG000
PANETS see HIM000
PANEX see HIM000
PANFLAVIN see DBX400
PANFURAN-S see BKH500
PANGUL see TEH500
PANMYCIN HYDROCHLORIDE see TBX250
PANO-DRENCH 4 see MLF250
PANODRIN A-13 see MLF250
PANOFEN see HIM000
PANOGEN see MEO750
PANOGEN see MLF250
PANOGEN 15 see MLF250
PANOGEN 43 see MLF250
PANOGEN M see MEO750
PANOGEN METOX see MEO750
PANOGEN PX see MLF250
PANOGEN TURF FUNGICIDE see MLF250
PANOGEN TURF SPRAY see MLF250
PANORAM 75 see TFS350
PANORAM D-31 see DHB400

PANOSINE see MMD500
PANOSPRAY 30 see MLF250
PANOXYL see BDS000
PANTALGINE see DAM700
PANTASOTE R 873 see PKQ059
PANTELMIN see MHL000
PANTHENOL see PAG200
d-PANTHENOL see PAG200
d(+)-PANTHENOL (FCC) see PAG200
PANTHER CREEK BENTONITE see BAV750
PANTHION see PAK000
PANTHODERM see PAG200
PANTHOJECT see CAU750
PANTHOLIC-L see IDE000
PANTHOLIN see CAU750
PANTOL see PAG200
PANTOMICINA see EDH500
PANTONSILETTEN see DBX400
PANTOSEDIV see TEH500
PANTOTHENATE CALCIUM see CAU750
PANTOTHENIC ACID, CALCIUM SALT see CAU750
(+)-PANTOTHENIC ACID, CALCIUM SALT see CAU750
PANTOTHENOL see PAG200
d-PANTOTHENOL see PAG200
PANTOTHENYL ALCOHOL see PAG200
d-PANTOTHENYL ALCOHOL see PAG200
d(+)-PANTOTHENYL ALCOHOL see PAG200
PANTOVERNIL see CDP250
PAN-TRANQUIL see MQU750
PANURIN see CFY000
PANWARFIN see WAT220
PAP see PDC250
PAP see ALT250
PAP-1 see AGX000
PAPAIN see PAG500
PAPANERINE see PAH000
PAPAVARINE CHLORHYDRATE see PAH250
PAPAVERINA (ITALIAN) see PAH000
PAPAVERINE see PAH000
PAPAVERINE CHLOROHYDRATE see PAH250
PAPAVERINE CHLOROHYDRATE see PAH250
PAPAVERINE HYDROCHLORIDE see PAH250
PAPAVERINE MONOHYDROCHLORIDE see PAH250
PAPAVERIN-HCL (GERMAN) see PAH250
PAPAYOTIN see PAG500
PAPER BLACK BA see AQP000
PAPER BLACK T see AQP000
PAPER BLUE R see AOR500
PAPER DEEP BLACK C see AQP000
PAPER RED HRR see FMU070
PAP H see PAH250
PAPI see PKB100
PAPI 20 see PKB100
PAPI 27 see PKB100
PAPI 135 see PKB100
PAPI 580 see PKB100
PAPI 901 see PKB100
PAPP see AMC000
PAPRIKA OLEORESIN see PAH280
PARA see PEY500
PARAAMINODIPHENYL see AJS100
PARABAR 441 see BFW750
PARABEN see HJL500
PARABEN see HNU500
PARABIS see MJM500
PARABROMOTOLUENE see BOG255
PARACAIN see AIT250

PARACETALDEHYDE see PAI250
PARACETAMOLE see HIM000
PARACETAMOLO (ITALIAN) see HIM000
PARACETANOL see HIM000
PARACETOPHENETIDIN see ABG750
PARACHLORAMINE see HGC500
PARACHLOROCIDUM see DAD200
PARACHLOROPHENOL see CJK750
PARACIDE see DEP800
PARACODIN see DKW800
PARACODINE see DKW800
PARACORT see PLZ000
PARACORTOL see PMA000
PARACOTOL see PMA000
PARACRESYL ACETATE see MNR250
PARACRESYL ISOBUTYRATE see THA250
PARA CRYSTALS see DEP800
d-PARACURARINE CHLORIDE see TOA000
PARACYMENE see CQI000
PARACYMOL see CQI000
PARADI see DEP800
PARADICHLORBENZOL (GERMAN) see DEP800
PARADICHLOROBENZENE see DEP800
PARADICHLOROBENZOL see DEP800
PARA-DIEN see DAL600
(0)-PARADOL see VFP100
PARADONE GREY M see CMU475
PARADONE GREY MG see CMU475
PARADONE OLIVE GREEN B see ALT000
PARADOW see DEP800
PARADUST see PAK000
PARAFFIN see PAH750
PARAFFIN HYDROCARBONS see PAH770
PARAFFIN OIL see MQV750
PARAFFIN OILS (PETROLEUM), CATALYTIC DEWAXED HEAVY (9CI) see MQV778
PARAFFIN OILS (PETROLEUM), CATALYTIC DEWAXED LIGHT (9CI) see MQV779
PARAFFIN WAX see PAH750
PARAFFIN WAXES and HYDROCARBON WAXES, CHLORINATED see PAH780
PARAFFIN WAXES and HYDROCARBON WAXES, CHLORINATED (C_{12}, 60% CHLORINE) see PAH800
PARAFFIN WAXES and HYDROCARBON WAXES, CHLORINATED (C_{23}, 43% CHLORINE) see PAH810
PARAFFIN WAX FUME (ACGIH) see PAH750
PARAFORM see FMV000
PARAFORMALDEHYDE see PAI000
PARAFORSN see PAI000
PARAFUCHSIN (GERMAN) see RMK020
PARAGLAS see PKB500
PARAL see PAI250
PARALDEHYD (GERMAN) see PAI250
PARALDEHYDE see PAI250
PARALDEIDE (ITALIAN) see PAI250
PARA M see AKE250
PARA-MAGENTA see RMK020
PARAMAL see DBM800
PARAMANDELIC ACID see MAP000
PARAMAR see PAK000
PARAMETHYL PHENOL see CNX250
PARAMINE BLACK B see AQP000
PARAMINE BLACK E see AQP000
PARAMINE BLUE 2B see CMO000
PARAMINE BLUE 3B see CMO250
PARAMINOL see AIH600
PARAMINOPROPIOPHENONE see AMC000
PARAMINYL MALEATE see DBM800

PARAMOTH see DEP800
PARANAPHTHALENE see APG500
PARANATE see AIH600
PARANEPHRIN see VGP000
PARANITROFENOL (DUTCH) see NIF000
PARANITROFENOLO (ITALIAN) see NIF000
PARANITROPHENOL (FRENCH, GERMAN) see NIF000
PARANITROSODIMETHYLANILIDE see DSY600
PARANOL see ALT250
PARANUGGETS see DEP800
PARA ORANGE see FAG150
PARAPAN see HIM000
PARAPEST M-50 see MNH000
PARAPHENOLAZO ANILINE see PEI000
PARAPHENYLEN-DIAMINE see PEY500
PARAPHOS see PAK000
PARAPLEX P 543 see PKB500
PARAQUAT see PAI990
PARAQUAT DICATION see PAI990
PARAQUAT ION see PAI990
PARAROSANILINE see RMK020
PARAROSANILINE CHLORIDE see RMK020
PARAROSANILINE HYDROCHLORIDE see RMK020
PARAROSANILINE, N,N,N',N',N'',N''-HEXAMETHYL-, CHLORIDE see AOR500
PARASAN see BCA000
PARASCORBIC ACID see PAJ500
PARASEPT see HJL500
PARASEPT see HNU500
PARASORBIC ACID see PAJ500
(+)-PARASORBINSAEURE (GERMAN) see PAJ500
PARASPEN see HIM000
PARASTARIN see EJM500
PARATAF see MNH000
PARATHENE see PAK000
PARATHESIN see EFX000
PARATHION see PAK000
M-PARATHION see MNH000
PARATHION, liquid (DOT) see PAK000
PARATHION and compressed gas mixture (DOT) see PAK230
PARATHION-ETHYL see PAK000
PARATHION METHYL see MNH000
PARATHION-METILE (ITALIAN) see MNH000
PARATOX see MNH000
PARAWET see PAK000
PARAXENOL see BGJ500
PARAXIN see CDP250
PARAZENE see DEP800
PARBOCYL-REV see SJO000
PARCIDOL see DIR000
PARCLAY see KBB600
PARCLOID see PKQ059
PARDA see DNA200
PARDIDOL see DIR000
PARDISOL see DIR000
PARDROYD see PAN100
PARENTERAL see CJR909
PARENTRACIN see BAC250
PAREST see MDT250
PAR ESTRO see ECU750
PARFEZINE see DIR000
PARFURAN see NGE000
PARIDINE RED LCL see CHP500
PARIDOL see HJL500
PARIS GREEN see COF500
PARIS RED see LDS000

PARIS YELLOW see LCR000
PARKIBLEU see CMO250
PARKIN see DIR000
PARKIPAN see CMO250
PARKISOL see DIR000
PARKOPHYLLIN see TEP000
PARKOSED see BNK000
PARKOTAL see EOK000
PARLODION see CCU250
PARMAVERT see NNE100
PARMETOL see CFD990
PARMOL see HIM000
PARMONE see NAK500
PARNATE see PET750
PAROIL CHLOREZ see PAH780
PAROL see MQV750
PAROL see MQV875
PAROL see CFD990
PAROLEINE see MQV750
PAROXON see ELL500
PAROXYL see ABX500
PAROXYPROPIONE see ELL500
PAROZONE see SHU500
PARPHEZEIN see DIR000
PARPON see BCA000
PARROT GREEN see COF500
PARSIDOL see DIR000
PARSITAN see DIR000
PARSLEY HERB OIL (FCC) see PAL750
PARSLEY OIL see PAL750
PARSLEY SEED OIL (FCC) see PAL750
PARTEL see DJT800
PARTREX see TBX250
PARVOLEX see ACH000
PARZATE see DXD200
PARZONE see DKW800
PASADE see SEP000
PASALON-RAKEET see SEP000
PASCO see ZBJ000
PASCO see ZKA000
PASEPTOL see HNU500
PASEXON 100T see SHK800
PASNAL see SEP000
PASSODICO see SEP000
PATENT BLUE AE see FMU059
PATENT GREEN see COF500
PATRICIN see VRF000
PATTINA V 82 see PKQ059
PAVABID see PAH250
PAVAGRANT see PAH250
PAVAKEY see PAH250
PAVASED see PAH250
PAVATEST see PAH250
PAVISOID see PAN100
PAXATE see DCK759
PAXISTIL see CJR909
PAXISYN see DLY000
PAYZE see BLW750
PB see PIX250
PBB see FBU000
PBB see FBU509
PBNA see PFT500
PBS see SID000
PBX(AF) 108 see CPR800
PBXW 108(E) see CPR800
PBZ see TMP750
PC 1 see BQJ500
PC-1421 see PII500

PCA see MCR750
PCA see PEE750
PCB see PJL750
PCB see PJM000
PCB see PJM250
PCB see PJM500
PCB see PJM750
PCB see PJN000
PCB see PJN250
PCB see PJN500
PCB see PJN750
PCB see PJO000
PCB see PJO250
PCB see PJO500
PCB see PJO750
PCB see PJP000
PCBs see PJL750
PCB HYDROCHLORIDE see PME500
PCC see CDV100
P.C. 80 CRABGRASS KILLER see PLC250
PCHO see PAI250
PCL see HCE500
PCM see PCF300
PCMC see CFD990
PCMX see CLW000
PCNB see PAX000
PCP see PAX250
PCPA see CJN000
PCPI see CKB000
2-N-p-PDA see ALL750
P.D.A.B. see DOT300
PDB see DEP800
PDCB see DEP800
PDD 6040I see CJV250
PDMT see DTP000
m-PDN see PHX550
p-PDN see BBP250
PDP see PDC250
PDT see DTP000
PE see PBB750
PE 512 see PJS750
PE 617 see PJS750
PEA see PDD750
β-PEA see PDD750
PEACOCK BLUE X-1756 see FMU059
PEANUT OIL see PAO000
PEARL ASH see PLA000
PEARL ASH see PLA250
PEARLPUSS see CCL250
PEARL STEARIC see SLK000
PEARLY GATES see DJO000
PEAR OIL see AOD725
PEAR OIL see IHO850
PEARSALL see AGY750
PEB1 see DAD200
PECAN SHELL POWDER see PAO000
PECTA-DIAZINE, suspension see PPP500
PECTALGINE see SEH000
PECTOX see CBR675
PEDIAFLOR see SHF500
PEDIDENT see SHF500
PEDRACZAK see BBQ500
PEDRIC see HIM000
PEERAMINE BLACK E see AQP000
PEERAMINE BLACK GXOO see AQP000
PEERAMINE CONGO RED see SGQ500
PEERLESS see CBT750

PEERLESS see KBB600
PEG 200 see PJT200
PEG 300 see PJT225
PEG 400 see PJT230
PEG 600 see PJT240
PEG 1000 see PJT250
PEG 1500 see PJT500
PEG 4000 see PJT750
PEG 6000 see PJU000
PEG-9 NONYL PHENYL ETHER see PKF000
PEG-9 OCTYL PHENYL ETHER see PKF500
PELADOW see CAO750
PELAGOL 3GA see ALT000
PELAGOL BA see DBO400
PELAGOL D see PEY500
PELAGOL DA see DBO000
PELAGOL DR see PEY500
PELAGOL EG see ALS990
PELAGOL GREY see DBO400
PELAGOL GREY C see CCP850
PELAGOL GREY D see PEY500
PELAGOL GREY GG see ALT000
PELAGOL GREY J see TGL750
PELAGOL GREY L see DBO000
PELAGOL GREY P BASE see ALT250
PELAGOL GREY RS see REA000
PELAGOL GREY SLA see DBO400
PELAGOL J see TGL750
PELAGOL L see DBO000
PELAGOL P BASE see ALT250
PELAGOL SLA see DBO400
PELARGIC ACID see NMY000
PELARGOL see DTE600
PELARGON (RUSSIAN) see NMY000
PELARGONIC ACID see NMY000
PELARGONIC ALCOHOL see NNB500
PELARGONIC ALDEHYDE see NMW500
PELARGONIUM OIL see GDA000
PELARGONOYL PEROXIDE see NNA100
PELARGONYL PEROXIDE see NNA100
PELARGONYL PEROXIDE, technically pure (DOT) see NNA100
PELASPAN 333 see SMQ500
PELASPAN ESP 109s see SMQ500
PELENTAN see BKA000
PELIDORM see BNK000
PELIKAN C 11/1431a see CBT500
PELLAGRAMIN see NCQ900
PELLAGRA PREVENTIVE FACTOR see NCQ900
PELLAGRIN see NCQ900
PELLCAFS see BBK500
PELLCAP see BBK500
PELLCAPS see BBK500
PELLETEX see CBT750
PELLON 2506 see PMP500
PELLUGEL see CCL250
PELMIN see NCR000
PELMINE see NCR000
PELONIN see NCQ900
PELONIN AMIDE see NCR000
PELSON see DLY000
PELTOL D see PEY500
PEN 100 see PJS750
PEN A see AOD125
PEN-A-BRASIVE see BFD250
PENADUR see BFC750
PENADUR L-A see BFC750
d-PENAMINE see MCR750

PEN A/N see SEQ000
PENBAR see NBU000
PENBRISTOL see AIV500
PENBRITIN see AIV500
PENBRITIN PAEDIATRIC see AIV500
PENBRITIN-S see SEQ000
PENBRITIN SYRUP see AIV500
PENBROCK see AIV500
PENCAL see ARB750
PENCARD see PBC250
PENCHLOROL see PAX250
PENCILLIC ACID see PAP750
PENCOGEL see CCL250
PENDEPON see BFC750
PEN-DI-BEN see BFC750
PENDITAN see BFC750
PENDURAN see BFC750
PENETECK see MQV750
PENETECK see MQV875
PENFLURIDOL see PAP250
PENFORD 260 see HLB400
PENFORD 280 see HLB400
PENFORD 290 see HLB400
PENFORD GUM 380 see SLJ500
PENFORD P 208 see HLB400
PENIALMEN see SEQ000
PENICILLAMIN see MCR750
(S)-PENICILLAMIN see MCR750
PENICILLAMINE see MCR750
d-PENICILLAMINE see MCR750
PENICILLIC ACID see PAP750
PENICILLIN see PAQ000
PENICILLIN G see BDY669
PENICILLIN G, compounded with N,N'-DIBENZYLETHYLENEDIAMINE (2:1) see BFC750
PENICILLIN-G, MONOSODIUM SALT see BFD250
PENICILLIN G SALT of N,N'-DIBENZYLETHYLENEDIAMINE see BFC750
PENICILLIN G, SODIUM see BFD250
PENICILLIN G, SODIUM SALT see BFD250
PENICLINE see AIV500
PENIDURAL see BFC750
PENIDURE see BFC750
PENILARYN see BFD250
PENILENTE see BFC750
PENITE see SEY500
PENITRACIN see BAC250
PENIZILLIN (GERMAN) see PAQ000
PENNAC see SGF500
PENNAC CBS see CPI250
PENNAC CRA see IAQ000
PENNAC MBT POWDER see BDF000
PENNAC MS see BJL600
PENNAC ZT see BHA750
PENNCAP-M see MNH000
PENNFLOAT M see LBX000
PENNFLOAT S see LBX000
PENN SALT TD-183 see TBV750
PENNWALT C-4852 see DSQ000
PENNWHITE see SHF500
PENNYROYAL OIL see PAR500
PENNZONE E see DKC400
PENOTRANE see PFN000
PENPHENE see TBV750
PENPHENE see CDV100
PENRECO see MQV750
PENSYN see AOD125
PENTA see PAX250

1,4,7,10,13-PENTAAZATRIDECANE see TCE500
PENTABARBITAL SODIUM see NBU000
PENTABORANE(9) see PAT750
PENTABORANE (ACGIH,DOT,OSHA) see PAT750
PENTABROMFENOL see PAU250
PENTABROMOPHENOL see PAU250
PENTABROMOPHENYL ETHER see PAU500
PENTABROMO PHOSPHORANE see PHR250
PENTABROMO PHOSPHORUS see PHR250
PENTACARBONYLIRON see IHG500
PENT-ACETATE see AOD725
PENTACHLOORETHAAN (DUTCH) see PAW500
PENTACHLOORFENOL (DUTCH) see PAX250
PENTACHLORAETHAN (GERMAN) see PAW500
PENTACHLORETHANE (FRENCH) see PAW500
PENTACHLORIN see DAD200
PENTACHLORNITROBENZOL (GERMAN) see PAX000
PENTACHLOROACETOPHENONE see PAV250
2',3',4',5',6'-PENTACHLOROACETOPHENONE see PAV250
PENTACHLOROANTIMONY see AQD000
PENTACHLOROBENZENE see PAV500
PENTACHLORO DIPHENYL OXIDE see PAW250
PENTACHLOROETHANE see PAW500
PENTACHLOROFENOL see PAX250
PENTACHLORONAPHTHALENE see PAW750
PENTACHLORONITROBENZENE see PAX000
PENTACHLOROPHENATE see PAX250
PENTACHLOROPHENATE SODIUM see SJA000
PENTACHLOROPHENOL see PAX250
2,3,4,5,6-PENTACHLOROPHENOL see PAX250
PENTACHLOROPHENOL (GERMAN) see PAX250
PENTACHLOROPHENOL, DOWICIDE EC-7 see PAX250
PENTACHLOROPHENOL, DP-2 see PAX250
PENTACHLOROPHENOL, SODIUM SALT see SJA000
PENTACHLOROPHENOL, TECHNICAL see PAX250
PENTACHLOROPHENOXY SODIUM see SJA000
PENTACHLOROPHENYL CHLORIDE see HCC500
PENTACHLORURE d'ANTIMOINE (FRENCH) see AQD000
PENTACIN see CAY500
PENTACINE see CAY500
PENTACLOROETANO (ITALIAN) see PAW500
PENTACLOROFENOLO (ITALIAN) see PAX250
PENTACON see PAX250
1-PENTADECANECARBOXYLIC ACID see PAE250
1,3-PENTADIENE-1-CARBOXYLIC ACID see SKU000
PENTAERYTHRITE see PBB750
PENTAERYTHRITE TETRANITRATE see PBC250
PENTAERYTHRITE TETRANITRATE (DOT) see PBC250
PENTAERYTHRITE TETRANITRATE, dry (DOT) see PBC250
PENTAERYTHRITE TETRANITRATE, desensitized, wet (DOT) see PBC250
PENTAERYTHRITE TETRANITRATE, with not less than 7% wax (DOT) see PBC250
PENTAERYTHRITOL see PBB750
PENTAERYTHRITOL ESTER of PARTIALLY HYDROGENATED WOOD ROSIN see PBB800
PENTAERYTHRITOL TETRANITRATE see PBC250
PENTAERYTHRITOL TETRANITRATE, diluted see PBC250
PENTAFIN see PBC250
PENTAFLUOROANTIMONY see AQF250
PENTAFLUOROIODINE see IDT000

PENTAFLUOROPROPIONIC ACID SILVER SALT see PBF250

PENTAFLUORPROPIONAN STRIBRNY (CZECH) see PBF250

PENTAGEN see PAX000

PENTAGIN see DOQ400

2′,3,4′,5,7-PENTAHYDROXYFLAVONE see MRN500

2′,4′,3,5,7-PENTAHYDROXYFLAVONE see MRN500

3,5,7,2′,4′-PENTAHYDROXYFLAVONE see MRN500

3,5,7,3′,4′-PENTAHYDROXYFLAVONE see QCA000

3,5,7,2′,4′-PENTAHYDROXYFLAVONOL see MRN500

PENTA-KIL see PAX250

PENTAL see NBU000

γ-PENTALACTONE see VAV000

PENTALIN see PAW500

PENTAMETHYLENE see CPV750

PENTAMETHYLENEDITHIOCARBAMATE see PIY500

PENTAMETHYLENEIMINE see PIL500

PENTAMETHYLENETETRAZOL see PBI500

1,5-PENTAMETHYLENETETRAZOLE see PBI500

PENTAMETHYLENE-1,5-TETRAZOLE see PBI500

PENTAN (POLISH) see PBK250

PENTANAL see VAG000

n-PENTANAL see VAG000

1-PENTANAMINE see PBV505

n-PENTANE see PBK250

tert-PENTANE see NCH000

PENTANE (ITALIAN) see PBK250

3-PENTANECARBOXYLIC ACID see DHI400

1,5-PENTANEDIAL see GFQ000

PENTANEDINITRILE see TLR500

PENTANEDINITRILE, 2-BROMO-2-(BROMOMETHYL)- see DDM500

PENTANEDIONE see ABX750

1,5-PENTANEDIONE see GFQ000

2,3-PENTANEDIONE see PBL350

2,4-PENTANEDIONE (FCC) see ABX750

2,4-PENTANEDIONE, NICKEL(II) DERIVATIVE see PBL500

2,4-PENTANEDIONE, ZIRCONIUM COMPLEX see PBL750

PENTANEN (DUTCH) see PBK250

PENTANE, 1,1′-OXYBIS-(9CI) see PBX000

1-PENTANETHIOL see PBM000

2-PENTANETHIOL, 2,4,4-TRIMETHYL- see MKJ250

2-PENTANETHIOL, 2,4,4-TRIMETHYL- see OFE030

PENTANOIC ACID see VAQ000

n-PENTANOIC ACID see VAQ000

tert-PENTANOIC ACID see PJA500

PENTANOIC ACID, 4-((3,4-DICHLOROBENZOYL)AMINO)-5-((3-METHOXYPROPYL)PENTYLAMINO)-5-OXO-, (+−)-see LII100

PENTANOIC ACID, 4-OXO-, ETHYL ESTER (9CI) see EFS600

PENTANOIC ACID, 2-PROPYL-, MAGNESIUM SALT see MAK275

PENTANOL-1 see AOE000

PENTAN-1-OL see AOE000

2-PENTANOL see PBM750

PENTANOL-2 see PBM750

3-PENTANOL see IHP010

PENTANOL-3 see IHP010

PENTAN-3-OL see IHP010

N-PENTANOL see AOE000

tert-PENTANOL see PBV000

1-PENTANOL ACETATE see AOD725

2-PENTANOL, ACETATE see AOD735

4-PENTANOLIDE see VAV000

3-PENTANOL, 3-METHYL-1-PHENYL- see PFR200

2-PENTANONE see PBN250

PENTANONE-3 see DJN750

3-PENTANONE see DJN750

2-PENTANONE, 5-(DIETHYLAMINO)- see NOB800

2-PENTANONE, 4-HYDROXY-1-PHENYL-5,5,5-TRIFLUORO-4-(TRIFLUOROMETHYL)- see TKB285

1-PENTANONE, 1-(4-PYRIDYL)- see VBA100

PENTANTIN see DAM700

PENTANTIN HYDROCHLORIDE see DAM700

PENTANYL see PDW750

2,5,8,11,14-PENTAOXAPENTADECANE see PBO500

PENTAPHENATE see SJA000

PENTASODIUM COLISTINMETHANESULFONATE see SFY500

PENTASODIUM TRIPHOSPHATE see SKN000

PENTASOL see PAX250

PENTASOL see AOE000

PENTASULFURE de PHOSPHORE (FRENCH) see PHS000

PENTAZOCINE see DOQ400

PENTECH see DAD200

PENTEK see PBB750

1-PENTENE, 3-METHYLENE- see EGV600

1-PENTEN-3-ONE see PBR250

1-PENTEN-3-ONE, 1-(2,6,6-TRIMETHYL-2-CYCLOHEXEN-1-YL)-2-METHYL- see COW780

PENTESTAN-80 see PBC250

PENTETATE TRISODIUM CALCIUM see CAY500

PENTETRATE UNICELLES see PBC250

PENTHAMIL see CAY500

PENTHAMIL see DJG800

PENTHAZINE see PDP250

PENTHIENATE BROMIDE see PBS000

PENTHIOBARBITAL SODIUM see PBT500

PENTIFORMIC ACID see HEU000

PENTOBARBITAL see PBS250

PENTOBARBITONE SODIUM see NBU000

PENTOLE see CPU500

PENTOLITE see PBT050

PENTOLITE, dry or wetted with <15% water, by weight (DOT) see PBT050

PENTONAL see NBU000

PENTOSTAM see AQH800

PENTOTHAL SODIUM see PBT500

PENTRATE see PBC250

PENTREX see AIV500

PENTREXL see AIV500

PENTRIOL see PBC250

PENTRYATE 80 see PBC250

PENTYL see NBU000

PENTYL ACETATE see AOD725

1-PENTYL ACETATE see AOD725

2-PENTYL ACETATE see AOD735

n-PENTYL ACETATE see AOD725

PENTYL ALCOHOL see AOE000

sec-PENTYL ALCOHOL see PBM750

tert-PENTYL ALCOHOL see PBV000

PENTYLAMINE see PBV505

1-PENTYLAMINE see PBV505

n-PENTYLAMINE see PBV505

PENTYLAMINE (mixed isomers) see PBV500

PENTYL BUTYRATE see AOG000

PENTYLCARBINOL see HFJ500

3-PENTYLCARBINOL see EGW000

sec-PENTYLCARBINOL see EGW000

PENTYL CHLORIDE see PBW500
α-PENTYLCINNAMALDEHYDE see AOG500
PENTYLCYCLOPENTANONEPROPANONE see OOO100
PENTYLDICHLOROARSINE see AOI200
PENTYLENE see AOI800
PENTYLENETETRAZOL see PBI500
PENTYL ESTER PHOSPHORIC ACID see PBW750
PENTYL ETHER see PBX000
PENTYL FORMATE see AOJ500
n-PENTYL FORMATE see AOJ500
PENTYLFORMIC ACID see HEU000
PENTYL KETONE see ULA000
PENTYL MERCAPTAN see PBM000
tert-PENTYL MERCAPTAN see MHS550
PENTYL NITRITE see AOL500
n-PENTYLNITROSOUREA see PBX500
PENTYL PENTYLAMINE see DCH200
p-PENTYLPHENOL see AOM250
o-(sec-PENTYL) PHENOL see AOM500
PENTYL PROPANOATE see AON350
n-PENTYL PROPANOATE see AON350
PENTYL PROPIONATE see AON350
PENTYL 4-PYRIDYL KETONE see PBX800
PENTYL 4-PYRIDYL KETONE see HEW050
PENTYLTRICHLOROSILANE see PBY750
3-PENTYL-6,6,9-TRIMETHYL-6a,7,8,10a-
TETRAHYDRO-6H-DIBENZO(b,d)PYRAN-1-OL see
TCM250
PENWAR see PAX250
PENZYLPENICILLIN SODIUM SALT see BFD250
PEP see EDS000
PEP 211 see PJS750
PEPPERMINT CAMPHOR see MCF750
PEPPERMINT OIL see PCB250
PERACETIC ACID see PCL500
PERAGAL ST see PKQ250
PERANDREN see TBF500
PERANDREN see TBG000
PERATOX see PAX250
PERAWIN see PCF275
PERBENZOATE de BUTYLE TERTIAIRE (FRENCH)
see BSC500
PERBROMOBIPHENYL see PCC480
PERBUTYL H see BRM250
PERCARBAMIDE see HIB500
PERCELINE OIL see HCP550
PERCHLOORETHYLEEN, PER (DUTCH) see PCF275
PERCHLOR see PCF275
PERCHLORAETHYLEN, PER (GERMAN) see PCF275
PERCHLORATE ACID, LEAD SALT, HEXAHYDRATE
see LDT000
PERCHLORATE de MAGNESIUM (FRENCH) see
PCE000
PERCHLORATES see PCD000
PERCHLORATE de SODIUM (FRENCH) see PCE750
PERCHLORETHYLENE see PCF275
PERCHLORETHYLENE, PER (FRENCH) see PCF275
PERCHLORIC ACID see PCD250
PERCHLORIC ACID, >72% acid by weight (DOT) see
PCD250
PERCHLORIC ACID, AMMONIUM SALT see PCD500
PERCHLORIC ACID, BARIUM SALT 3H₂O see PCD750
PERCHLORIC ACID, COBALT(II) SALT,
HEXAHYDRATE see CND900
PERCHLORIC ACID, COPPER(II) SALT, DIHYDRATE
(8CI, 9CI) see CNO500

PERCHLORIC ACID, ETHYL ESTER see EOD000
PERCHLORIC ACID, MAGNESIUM SALT see PCE000
PERCHLORIC ACID, MANGANESE(2+) SALT,
compounded with 3 mols. of
OCTAMETHYLPYROPHOSPHORAMIDE see TNK400
PERCHLORIC ACID, SODIUM SALT see PCE750
PERCHLORIC ACID, >50% but not >72% acid, by weight
(UN 1873) (DOT) see PCD250
PERCHLORIC ACID, not >50% acid, by weight (UN 1802)
(DOT) see PCD250
PERCHLORIC ACID, ZINC SALT, HEXAHYDRATE see
ZKS100
PERCHLORIDE of MERCURY see MCY475
PERCHLORMETHYLMERKAPTAN (CZECH) see
PCF300
PERCHLOROBENZENE see HCC500
PERCHLOROBUTADIENE see HCD250
PERCHLOROCYCLOPENTADIENE see HCE500
PERCHLORODIHOMOCUBANE see MQW500
PERCHLOROETHANE see HCI000
PERCHLOROETHYLENE see PCF275
PERCHLOROMETHANE see CBY000
PERCHLOROMETHYL MERCAPTAN see PCF300
PERCHLORON see HOV500
PERCHLOROPENTACYCLODECANE see MQW500
PERCHLOROPENTACYCLO(5.2.1.0²·⁶.0³·⁹.0⁵·⁸)DECANE
see MQW500
PERCHLOROTHIOPHENE see TBV750
PERCHLORURE d'ANTIMOINE (FRENCH) see
AQD000
PERCHLORURE de FER see FAU000
PERCHLORYL FLUORIDE see PCF750
PERCLENE see PCF275
PERCLOROETILENE (ITALIAN) see PCF275
PERCOBARB see PCG500
PERCOBARB see ABG750
PERCOCCIDE see ALF250
PERCODAN see PCG500
PERCODAN see ABG750
PERCOLATE see PHX250
PERCORAL see DJS200
PERCOSOLVE see PCF275
PERCUTACRINE see PMH500
PERCUTACRINE ANDROGENIQUE see TBF500
PERCUTATRINE OESTROGENIQUE ISCOVESCO see
DKA600
PERDOX see SJB400
PEREMESIN see HGC500
PEREQUIL see MQU750
PERFECTA see MQV750
PERFECTHION see DSP400
PERFENAZINA (ITALIAN) see CJM250
PERFLUOROACETIC ACID see TKA250
PERFLUOROACETYL CHLORIDE see TJX500
PERFLUOROAMMONIUM OCTANOATE see ANP625
PERFLUOROBUT-2-ENE see OBO000
PERFLUORO-2-BUTENE (DOT) see OBO000
PERFLUOROCYCLOBUTANE see CPS000
PERFLUORODECANOIC ACID see PCG725
PERFLUORO-N-DECANOIC ACID see PCG725
PERFLUOROETHENE see TCH500
PERFLUOROETHYLENE see TCH500
PERFLUORO HYDRAZINE see TCI000
PERFLUOROMETHANE see CBY250
PERFLUORO(METHYLOXIRANE) see HDF050
PERFLUOROOCTANESULFONIC ACID see HAS075
PERFLUOROPROPENE see HDF000
PERFLUOROPROPYLENE see HDF000

PERFLUOROPROPYLENE OXIDE see HDF050
PERGACID VIOLET 2B see FAG120
PERGANTENE see SHF500
PERGITRAL see PBC250
PERGLOTTAL see NGY000
PER-GLYCERIN see LAM000
PERHYDRIT see HIB500
PERHYDROAZEPINE see HDG000
2-PERHYDROAZEPINONE see CBF700
PERHYDROGERANIOL see DTE600
PERHYDROL see HIB050
PERHYDROL-UREA see HIB500
PERHYDRONAPHTHALENE see DAE800
PERICIAZINE see PIW000
PERICLASE see MAH500
PERICYAZINE see PIW000
PERIDEX-LA see PBC250
PERILLA ALCOHOL see PCI550
PERILLA KETONE see PCI750
PERILLOL see PCI550
PERILLYL ALCOHOL see PCI550
PERINDOPRIL tert-BUTYLAMINE see PCJ230
PERINDOPRIL ERBUMINE see PCJ230
PERIODIN see PLO500
PERISTON see PKQ250
PERITRATE see PBC250
PERITYL see PBC250
PERK see PCF275
PERKE see BBK500
PERKLONE see PCF275
PERLATAN see EDV000
PERLITE see PCJ400
PERLITON BLUE B see TBG700
PERLITON ORANGE 3R see AKP750
PERLITON PINK 3B see AKE250
PERLITON VIOLET B see DBY700
PERLITON YELLOW G see AAQ250
PERLUTEX see MCA000
PERM-A-CHLOR see TIO750
PERMACIDE see PAX250
PERMAGARD see PAX250
PERMAGEL see PAE750
PERMA KLEER see DJG800
PERMA KLEER 50 ACID see EIX000
PERMA KLEER 50 CRYSTALS see EIV000
PERMA KLEER 50 CRYSTALS DISODIUM SALT see EIX500
PERMA KLEER TETRA CP see EIV000
PERMA KLEER 50, TRISODIUM SALT see TNL250
PERMANENT WHITE see ZKA000
PERMANENT WHITE see BAP000
PERMANENT YELLOW see BAK250
PERMANGANATE of POTASH (DOT) see PLP000
PERMANGANATE de POTASSIUM (FRENCH) see PLP000
PERMANGANATES see PCJ500
PERMANGANATE de SODIUM (FRENCH) see SJC000
PERMANGANIC ACID, BARIUM SALT see PCK000
PERMANGANIC ACID(HMnO₄), CALCIUM SALT (8CI,9CI) see CAV250
PERMANGANIC ACID, SODIUM SALT see SJC000
PERMAPEN see BFC750
PERMASAN see PAX250
PERMATOX DP-2 see PAX250
PERMATOX PENTA see PAX250
PERMEK N see MKA500
PERMETHRIN (USDA) see AHJ750
PERMETRINA (PORTUGUESE) see AHJ750

PERMETRIN (HUNGARIAN) see AHJ750
PERMICORT see CNS750
PERMITAL see DYE600
PERMITE see PAX250
PERNAEMON see VSZ000
PERNAEVIT see VSZ000
PERNIPURON see VSZ000
PERONE see HIB050
PERONE 30 see HIB010
PERONE 35 see HIB010
PERONE 50 see HIB010
PEROSSIDO di BENZOILE (ITALIAN) see BDS000
PEROSSIDO di IDROGENO (ITALIAN) see HIB050
PEROXAN see HIB050
PEROXIDE see HIB050
PEROXIDE, BIS(1-OXONONYL) see NNA100
PEROXIDE, BIS(1-OXOPROPYL) see DWQ800
PEROXIDE, 1-HYDROPEROXYCYCLOHEXYL 1-HYDROXYCYCLOHEXYL see CPC300
PEROXIDE, (PHENYLENEBIS(1-METHYLETHYLIDENE))BIS(1,1-DIMETHYLETHYL)- see BHL100
PEROXIDE, (PHENYLENEDIISOPROPYLIDENE)BIS(tert-BUTYL)- see BHL100
PEROXIDES, INORGANIC see PCL000
PEROXIDES, ORGANIC see PCL250
PEROXIDE, (1,1,4,4-TETRAMETHYL-1,4-BUTANEDIYL)BIS(1,1-DIMETHYLETHYL) see DRJ800
PEROXIDE, (1,1,4,4-TETRAMETHYLTETRAMETHYLENE)BIS(tert-BUTYL) see DRJ800
1,4-PEROXIDO-p-MENTHENE-2 see ARM500
PEROXOACETIC ACID see PCL500
PEROXYACETIC ACID see PCL500
PEROXYACETIC ACID, >43% and with >6% hydrogen peroxide (DOT) see PCL500
PEROXYDE de BARYUM (FRENCH) see BAO250
PEROXYDE de BENZOYLE (FRENCH) see BDS000
PEROXYDE d'HYDROGENE (FRENCH) see HIB050
PEROXYDE de LAUROYLE (FRENCH) see LBR000
PEROXYDE de PLOMB (FRENCH) see LCX000
PEROXYDICARBONATE d'ISOPROPYLE (FRENCH) see DNR400
PEROXYDICARBONIC ACID, BIS(2-ETHYLHEXYL) ESTER see DJK800
PEROXYDICARBONIC ACID, BIS(1-METHYLETHYL) ESTER see DNR400
PEROXYDICARBONIC ACID, DIETHYL ESTER see DJU600
PEROXYDICARBONIC ACID, DI(2-ETHYLHEXYL) ESTER see DJK800
PEROXYDICARBONIC ACID DIPROPYL ESTER see DWV400
PEROXYDISULFURIC ACID DIPOTASSIUM SALT see DWQ000
PEROXY SODIUM CARBONATE see SJB400
PERPHENAZIN see CJM250
PERPHENAZINE see CJM250
PERSADOX see BDS000
PERSAMINE see DLH630
PERSEC see PCF275
PERSIAN RED see LCS000
PERSIA-PERAZOL see DEP800
PERSISTEN see DJT400
PERSISTOL see TND500
PERSKLERAN see RDK000
PERSPEX see PKB500
PERSULFATE d'AMMONIUM (FRENCH) see ANR000

PERSULFATE de SODIUM (FRENCH) see SJE000
PERSULFEN see SNN300
PERTHANE see DJC000
PERTOFRAM see DLH630
PERTOFRAN see DLS600
PERTOFRANE see DLS600
PERUSCABIN see BCM000
PERUVIAN BALSAM see BAE750
PERVETRAL see ORI400
PERVITIN see MDT600
PERVITIN see DBA800
PES 100 see PJS750
PES 200 see PJS750
M-74 (PESTICIDE) see DXH325
PESTMASTER see EIY500
PESTMASTER EDB-85 see EIY500
PESTMASTER (OBS.) see MHR200
PESTON XV see PHF750
PESTOX see OCM000
PESTOX 3 see OCM000
PESTOX 14 see BJE750
PESTOX 15 see PHF750
PESTOX III see OCM000
PESTOX IV see BJE750
PESTOX PLUS see PAK000
PESTOX XIV see BJE750
PESTOX XV see PHF750
PETASITES JAPONICUS MAXIM see PCR000
PETERSILIENSAMEN OEL (GERMAN) see PAL750
PETHIDINE CHLORIDE see DAM700
PETHION see PAK000
PETIDIN see DAM700
PETITGRAIN OIL, PARAGUAY TYPE see PCR100
PETNAMYCETIN see CDP250
PETROGALAR see MQV750
PETROHOL see INJ000
PETROL see PCR250
PETROL (DOT) see GBY000
PETROLATUM, liquid see MQV750
PETROLEUM see PCR250
PETROLEUM ASPHALT see PCR500
PETROLEUM ASPHALT see ARO500
PETROLEUM BENZIN see NAH600
PETROLEUM BITUMEN see ARO500
PETROLEUM CRUDE see PCR250
PETROLEUM CRUDE OIL (DOT) see PCR250
PETROLEUM DERIVED DISTILLATE FUEL, MARINE see DHE750
PETROLEUM-DERIVED NAPHTHA see NAH600
PETROLEUM DISTILLATE see PCS250
PETROLEUM DISTILLATES, CLAY-TREATED HEAVY NAPHTHENIC see PCS260
PETROLEUM DISTILLATES, CLAY-TREATED LIGHT NAPHTHENIC see PCS270
PETROLEUM DISTILLATES, HYDROTREATED (mild) HEAVY NAPHTHENIC see MQV790
PETROLEUM DISTILLATES (NAPHTHA) see NAH600
PETROLEUM DISTILLATES, SOLVENT-DEWAXED HEAVY PARAFFINIC see MQV825
PETROLEUM ETHER see PCT250
PETROLEUM GASES, liquefied or liquefied petroleum gas (DOT) see LGM000
PETROLEUM GAS, LIQUEFIED see LGM000
PETROLEUM OIL (UN1270) (DOT) see NAH600
PETROLEUM PITCH see ARO500
PETROLEUM ROOFING TAR see PCR500
PETROLEUM ROOFING TAR see ARO500
PETROLEUM SPIRIT (DOT) see PCT250

PETROLEUM SPIRITS see PCT250
PETROLEUM WAX see PCT600
PETROLEUM WAX, SYNTHETIC (FCC) see PCT600
PETROL ORANGE Y see PEJ500
PETROL YELLOW WT see DOT300
PETROTHENE see PJS750
PETROTHENE LB 861 see PJS750
PETROTHENE LC 731 see PJS750
PETROTHENE LC 941 see PJS750
PETROTHENE NA 219 see PJS750
PETROTHENE NA 227 see PJS750
PETROTHENE XL 6301 see PJS750
PETZINOL see TIO750
PEVIKON D 61 see PKQ059
PEVITON see NCQ900
PEYRONE'S CHLORIDE see PJD000
PF-3 see IRF000
PF 38 see FQU875
PFDA see PCG725
PFEFFERMINZ OEL (GERMAN) see PCB250
PFIKLOR see PLA500
PFIZERPEN A see AIV500
PFOS see HAS075
PG 12 see PML000
PGA see AHC000
PGDN see PNL000
PGE see PFF360
PGE2 see DVJ200
PGF2-α see POC500
PGF2-α THAM see POC750
PGF2-α TRIS SALT see POC750
PGF2-α TROMETHAMINE see POC750
PH 60-40 see CJV250
PHALDRONE see CDO000
PHALTAN see TIT250
PHANAMIPHOS see FAK000
PHANANTIN see DKQ000
PHANODORM see TDA500
PHANODORN see TDA500
PHARGAN see DQA400
PHARLON see EDS100
PHARMAGEL A see PCU360
PHARMAGEL AdB see PCU360
PHARMAGEL B see PCU360
PHARMASORB-COLLOIDAL see PAE750
PHARMAZOID RED KB see CLK225
PHAROS 100.1 see SMR000
PHASOLON see BDJ250
PH BC see BQJ500
M-PHDM see BKL750
PHEBUZIN see BRF500
α-PHELLANDRENE (FCC) see MCC000
PHEMERIDE see BEN000
PHE-MER-NITE see MCU750
PHEMEROL CHLORIDE see BEN000
PHEMETONE see ENB500
PHEMITHYN see BEN000
PHEMITON see ENB500
PHEMITONE see ENB500
PHENACETALDEHYDE DIMETHYL ACETAL see PDX000
p-PHENACETIN see ABG750
PHENACHLOR see TIW000
PHENACIDE see CDV100
PHENACYL CHLORIDE see CEA750
PHENADONE see MDO750
PHENADOR-X see BGE000

PHENAEMAL see EOK000
PHENALCO see MCU750
PHENALGENE see AAQ500
PHENALGIN see AAQ500
PHENALZINE see PFC750
PHENALZINE DIHYDROGEN SULFATE see PFC750
PHENALZINE HYDROGEN SULPHATE see PFC750
PHENAMINE BLACK BCN-CF see AQP000
PHENAMINE BLACK CL see AQP000
PHENAMINE BLACK E see AQP000
PHENAMINE BLACK E 200 see AQP000
PHENAMINE BLUE BB see CMO000
PHENAMINE SKY BLUE A see CMO500
PHENANTHREN (GERMAN) see PCW250
PHENANTHRENE see PCW250
PHENANTHRO(3,4-D)-1,3-DIOXOLE-5-CARBOXYLIC
ACID, 8-METHOXY-6-NITRO-, SODIUM SALT see
SEY050
o-PHENANTHROLINE see PCY250
1,10-PHENANTHROLINE see PCY250
1,10-o-PHENANTHROLINE see PCY250
β-PHENANTHROLINE see PCY250
PHENANTOIN see MKB250
PHENANTRIN see PCW250
PHENARSAZINE CHLORIDE see PDB000
10-PHENARSAZINETHIOL, S-ESTER with O,O-
DIISOOCTYLPHOSPHORODITHIOATE see PDB300
S-(10-PHENARSAZINYL)-O,O-
DIISOOCTYLPHOSPHORODITHIOATE see PDB300
PHENARSEN see OOK100
PHENASAL see DFV400
PHENATOINE see DKQ000
PHENATOX see CDV100
PHENAZINE see PDB500
PHENAZO see PDC250
PHENAZODINE see PDC250
PHENAZONE (pharmaceutical) see AQN000
PHENAZOPYRIDINE HYDROCHLORIDE see PDC250
PHENAZOPYRIDINIUM CHLORIDE see PDC250
PHENBUTAZOL see BRF500
PHENCAPTON see PDC750
PHENCEN see PMI750
PHENE see BBL250
PHENEDRINE see BBK000
PHENEGIC see PDP250
PHENELZIN see PFC750
PHENELZINE see PFC500
PHENELZINE ACID SULFATE see PFC750
PHENELZINE BISULPHATE see PFC750
PHENELZINE SULFATE see PFC750
PHENEMALUM see SID000
PHENERGAN see DQA400
PHENERGAN HYDROCHLORIDE see PMI750
PHENESTERINE see CME250
PHENESTRIN see CME250
PHENETHANOL see PDD750
2-PHENETHYL ACETATE see PFB250
β-PHENETHYL ACETATE see PFB250
PHENETHYL ALCOHOL see PDD750
2-PHENETHYL ALCOHOL see PDD750
α-PHENETHYL ALCOHOL see PDE000
β-PHENETHYL ALCOHOL see PDD750
β-PHENETHYLAMINE see PDE250
β-PHENETHYL-o-AMINOBENZOATE see APJ500
PHENETHYL ANTHRANILATE see APJ500
PHENETHYLCARBAMID (GERMAN) see EFE000
PHENETHYLENE see SMQ000

PHENETHYLENE OXIDE see EBR000
PHENETHYL ESTER ISOVALERIC ACID see PDF775
PHENETHYLHYDRAZINE see PFC500
PHENETHYLHYDRAZINE SULFATE (1:1) see PFC750
PHENETHYL ISOBUTYRATE see PDF750
PHENETHYL ISOVALERATE see PDF775
2-PHENETHYL 2-METHYLBUTYRATE see PDF790
PHENETHYLMETHYLETHYLCARBINOL see PFR200
PHENETHYL PHENYLACETATE see PDI000
1-PHENETHYL-3-
(PIPERIDINOCARBONYL)PIPERIDINE see PDI550
N-(1-PHENETHYL-4-
PIPERIDINYL)PROPIONANILIDE DIHYDROGEN
CITRATE see PDW750
1-PHENETHYL-3-PIPERIDYL PIPERIDINO KETONE
see PDI550
N-(1-PHENETHYL-4-PIPERIDYL)PROPIONANILIDE
CITRATE see PDW750
N-(1-PHENETHYL-4-PIPERIDYL)PROPIONANILIDE
DIHYDROGEN CITRATE see PDW750
PHENETHYL SALICYLATE see PDK200
p-PHENETIDIN see PDK790
PHENETIDINE see PDK790
PHENETIDINE see EEL100
PHENETIDINES (DOT) see EEL100
p-PHENETOLCARBAMID (GERMAN) see EFE000
p-PHENETOLCARBAMIDE see EFE000
p-PHENETOLECARBAMIDE see EFE000
p-PHENETYLUREA see EFE000
PHENFLUORAMINE HYDROCHLORIDE see PDM250
PHENIC ACID see PDN750
PHENIDYLATE see MNQ000
PHENISTAN see SPC500
PHENITOL see MCU750
PHENITROTHION see DSQ000
PHENLINE see PFC750
PHENMAD see ABU500
PHENMERZYL NITRATE see MCU750
PHENMETHYL TRIMETHYLAMMONIUM IODIDE
see BFM750
PHENMETRAZINE HYDROCHLORIDE see MNV750
PHENOBAL see EOK000
PHENOBAL SODIUM see SID000
PHENOBARBITAL see EOK000
PHENOBARBITAL ELIXIR see SID000
PHENOBARBITAL Na see SID000
PHENOBARBITAL SODIUM see SID000
PHENOBARBITAL SODIUM SALT see SID000
PHENOBARBITONE see EOK000
PHENOBARBITONE SODIUM see SID000
PHENOBARBITONE SODIUM SALT see SID000
PHENOBARBITURIC ACID see EOK000
PHENO BLACK EP see AQP000
PHENO BLACK SGN see AQP000
PHENO BLUE 2B see CMO000
PHENOCHLOR see PJL750
PHENOCLOR see PJL750
PHENOCLOR DP6 see PJN250
PHENODODECINIUM BROMIDE see DXX000
PHENODYNE see PFC750
PHENOHEP see HCI000
PHENOL see PDN750
PHENOL, molten (DOT) see PDN750
PHENOL ALCOHOL see PDN750
PHENOL, 2-AMINO-4-ARSENOSO- see OOK100
PHENOL, 2-AMINO-4-ARSENOSO-, SODIUM SALT
see ARJ900
PHENOL, 4-ARSENOSO-2-NITRO- see NHE600

(PHENYL-1 ACETYL-2 ETHYL)-3-HYDROXY-4 COUMARINE (FRENCH) see WAT200
PHENYL ACETYL NITRILE see PEA750
PHENYLACROLEIN see CMP969
3-PHENYLACROLEIN see CMP969
PHENYLACRYLIC ACID see CMP975
3-PHENYLACRYLIC ACID see CMP975
tert-β-PHENYLACRYLIC ACID see CMP975
3-PHENYLACRYLOPHENONE see CDH000
β-PHENYLACRYLOPHENONE see CDH000
β-PHENYLAETHYLAMIN (GERMAN) see PDE250
PHENYLAETHYL-HYDRAZIN see PFC750
PHENYLALANINE see PEC750
3-PHENYLALANINE see PEC750
d-PHENYLALANINE see PEC500
l-PHENYLALANINE see PEC750
(S)-PHENYLALANINE see PEC750
PHENYL-α-ALANINE see PEC750
β-PHENYLALANINE see PEC750
d-β-PHENYLALANINE see PEC500
l-β-PHENYLALANINE see PEC750
β-PHENYL-α-ALANINE, l- see PEC750
dl-PHENYLALANINE (FCC) see PEC500
l-PHENYLALANINE MUSTARD see PED750
dl-PHENYLALANINE MUSTARD see BHT750
PHENYLALANINE NITROGEN MUSTARD see PED750
PHENYLALANIN-LOST (GERMAN) see BHT750
γ-PHENYLALLYL ACETATE see CMQ730
3-PHENYLALLYL ALCOHOL see CMQ740
γ-PHENYLALLYL ALCOHOL see CMQ740
PHENYLAMINE see AOQ000
PHENYLAMINE HYDROCHLORIDE see BBL000
1-PHENYL-2-AMINO-AETHAN (GERMAN) see PDE250
2-(PHENYLAMINO)BENZOIC ACID see PEG500
4-(PHENYLAMINO)BUTANE see BQH850
1-PHENYL-4-AMINO-5-CHLOROPYRIDAZON-(6) (GERMAN) see PEE750
1-PHENYL-4-AMINO-5-CHLOROPYRIDAZONE-6 see PEE750
1-PHENYL-4-AMINO-5-CHLORO-6-PYRIDAZONE see PEE750
1-PHENYL-4-AMINO-5-CHLORPYRIDAZ-6-ONE see PEE750
trans-2-PHENYL-1-AMINOCYCLOPROPANE see PET750
4-PHENYLAMINODIPHENYLAMINE see BLE500
p-PHENYLAMINODIPHENYLAMINE see BLE500
1-PHENYL-2-AMINOETHANE see PDE250
2-(PHENYLAMINO)ETHANOL see AOR750
d-1-PHENYL-2-AMINOPROPAN (GERMAN) see AOA500
d-1-PHENYL-2-AMINOPROPANE see AOA500
dl-1-PHENYL-2-AMINOPROPANE see BBK000
d-1-PHENYL-2-AMINOPROPANE SULFATE see BBK500
dl-1-PHENYL-2-AMINO-1-PROPANOL MONOHYDROCHLORIDE see PMJ500
3-PHENYL-5-AMINO-1,2,4-TRIAZOLYL-(1)-(N,N'-TETRAMETHYL) DIAMIDOPHOSPHONATE see AIX000
N-PHENYLANILINE see DVX800
p-PHENYLANILINE see AJS100
p-PHENYLANISOLE see PEG250
PHENYLANTHRANILIC ACID see PEG500

N-PHENYLANTHRANILIC ACID see PEG500
PHENYL ARSENIC ACID see BBL750
PHENYLARSENOXIDE see PEG750
PHENYLARSINEDICHLORIDE see DGB600
PHENYL ARSINE OXIDE see PEG750
PHENYLARSONIC ACID see BBL750
PHENYLARSONOUS DIBROMIDE see DDR200
PHENYLARSONOUS DICHLORIDE see DGB600
4-(PHENYLAZO)ANILINE see PEI000
N-(PHENYLAZO)ANILINE see DWO800
p-(PHENYLAZO)ANILINE see PEI000
4-(PHENYLAZO)BENZENAMINE see PEI000
4-(PHENYLAZO)-1,3-BENZENEDIAMINE MONOHYDROCHLORIDE see PEK000
PHENYLAZODIAMINOPYRIDINE HYDROCHLORIDE see PDC250
3-PHENYLAZO-2,6-DIAMINOPYRIDINE HYDROCHLORIDE see PDC250
β-PHENYLAZO-α,α'-DIAMINOPYRIDINE HYDROCHLORIDE see PDC250
PHENYLAZO-α,α'-DIAMINOPYRIDINE MONOHYDROCHLORIDE see PDC250
N-PHENYLAZO-N-METHYLTAURINE SODIUM SALT see PEJ250
1-(PHENYLAZO)-2-NAPHTHALENAMINE see FAG130
1-(PHENYLAZO)-2-NAPHTHALENOL see PEJ500
1-(PHENYLAZO)-2-NAPHTHOL see PEJ500
1-PHENYLAZO-β-NAPHTHOL see PEJ500
1-PHENYLAZO-2-NAPHTHOL-6,8-DISULFONIC ACID, DISODIUM SALT see HGC000
1-PHENYLAZO-2-NAPHTHOL-6,8-DISULPHONIC ACID, DISODIUM SALT see HGC000
1-(PHENYLAZO)-2-NAPHTHYLAMINE see FAG130
4-PHENYLAZOPHENOL see HJF000
p-PHENYLAZOPHENOL see HJF000
p-PHENYLAZOPHENYLAMINE see PEI000
1-(4-PHENYLAZO-PHENYLAZO)-2-ETHYLAMINONAPHTHALENE see EOJ500
(PHENYLAZO-4-PHENYLAZO)-1-ETHYLAMINO-2-NAPHTHALENE see EOJ500
1-((4-(PHENYLAZO)PHENYL)AZO)-2-NAPHTHALENOL see OHI200
1-((p-PHENYLAZO)PHENYL)AZO-2-NAPHTHOL see OHI200
4-PHENYLAZO-m-PHENYLENEDIAMINE see PEK000
4-(PHENYLAZO)-m-PHENYLENEDIAMINE MONOHYDROCHLORIDE see PEK000
3-(PHENYLAZO)-2,6-PYRIDINEDIAMINE, HYDROCHLORIDE see PDC250
PHENYLAZOPYRIDINE HYDROCHLORIDE see PDC250
PHENYLAZO TABLETS see PDC250
(PHENYLAZO)THIOFORMIC ACID, 2-PHENYLHYDRAZIDE see DWN200
N-PHENYLBENEZENAMINE see DVX800
PHENYLBENZENE see BGE000
PHENYLBENZOQUINONE see PEL750
2-PHENYLBENZOQUINONE see PEL750
o-PHENYLBENZOQUINONE see PEL750
PHENYL-p-BENZOQUINONE see PEL750
PHENYL-1,4-BENZOQUINONE see PEL750
1-PHENYL-2-BENZOYLETHYLENE see CDH000
8-(p-PHENYLBENZYL)ATROPINIUM BROMIDE see PEM750
α-PHENYLBENZYLCYANIDE see DVX200
PHENYL BENZYL KETONE see PEB000
4-PHENYLBIPHENYL see TBC750

PHENYL BROMIDE see PEO500
1-PHENYLBUTANE see BQI750
2-PHENYLBUTANE see BQJ000
4-PHENYL-2-BUTANONE 3-THIOSEMICARBAZONE see PEO600
PHENYLBUTAZON (GERMAN) see BRF500
PHENYLBUTAZONE see BRF500
4-PHENYL-3-BUTEN-2-ONE see SMS500
2-PHENYL-tert-BUTYLAMINE see DTJ400
PHENYLBUTYLCARBINOL see BQJ500
PHENYLBUTYRIC ACID NITROGEN MUSTARD see CDO500
N-PHENYLCARBAMATE d'ISOPROPYLE (FRENCH) see CBM000
PHENYLCARBAMIC ACID-1-METHYLETHYL ESTER see CBM000
PHENYLCARBIMIDE see PFK250
PHENYLCARBINOL see BDX500
PHENYL CARBITOL see PEQ750
PHENYL CARBONATE see DVZ000
PHENYL CARBONIMIDE see PFK250
PHENYLCARBONIMIDIC DICHLORIDE see PFJ400
N-PHENYLCARBONIMIDIC DICHLORIDE see PFJ400
PHENYL CARBOXYLIC ACID see BCL750
PHENYL CARBYLAMINE CHLORIDE see PFJ400
PHENYLCARBYLAMINE CHLORIDE (DOT) see PFJ400
PHENYL CELLOSOLVE see PER000
PHENYL CELLOSOLVE ACRYLATE see PER250
PHENYL CHLORIDE see CEJ125
PHENYL CHLOROCARBONATE see CBX109
PHENYL CHLOROFORM see BFL250
PHENYL CHLOROFORMATE see CBX109
PHENYLCHLOROFORMATE (DOT) see CBX109
PHENYL CHLOROMERCURY see PFM500
PHENYLCHLOROMETHYLKETONE see CEA750
2-(2-PHENYL-2-(4-CHLOROPHENYL)ACETYL)-1,3-INDANDIONE see CJJ000
PHENYL(β-CHLOROVINYL)CHLORARSINE see PER600
PHENYL(β-CHLOROVINYL)CHLOROARSINE see PER600
α-PHENYLCINNAMONITRILE see DVX600
PHENYL CYANIDE see BCQ250
2-PHENYL-2,5-CYCLOHEXADIENE-1,4-DIONE see PEL750
PHENYL CYCLOHEXYL KETONE see PES750
trans-2-PHENYLCYCLOPROPYLAMINE see PET750
PHENYLDIBROMOARSINE see DDR200
PHENYL DICHLORARSINE see DGB600
PHENYLDICHLOROARSINE see DGB600
PHENYLDICHLOROPHOSPHINE see DGE400
3-PHENYL-5-(β-(DIETHYLAMINO)ETHYL)-1,2,4-OXADIAZOLE CITRATE see OOE000
PHENYLDIFLUOROARSINE see DKI400
1-PHENYL-2,3-DIMETHYL-4-DIMETHYLAMINOPYRAZOLONE-5 see DOT000
1-PHENYL-2,3-DIMETHYL-4-DIMETHYLAMINOPYRAZOL-5-ONE see DOT000
1-PHENYL-2,3-DIMETHYL-4-ISOPROPYL-3-PYRAZOLIN-5-ONE see INY000
1-PHENYL-2,3-DIMETHYL-4-ISOPROPYLPYRAZOL-5-ONE see INY000
O-PHENYL-N,N'-DIMETHYL PHOSPHORODIAMIDATE see PEV500
1-PHENYL-2,3-DIMETHYLPYRAZOLE-5-ONE see AQN000

1-PHENYL-2,3-DIMETHYL-5-PYRAZOLONE see AQN000
1-PHENYL-2,3-DIMETHYL-5-PYRAZOLONE-4-METHYLAMINOMETHANESULFONATESODIUM see AMK500
1-PHENYL-2,3-DIMETHYLPYRAZOLONE-(5)-4-METHYLAMINOMETHANESULFONICACID SODIUM see AMK500
PHENYL DIMETHYL PYRAZOLON METHYL AMINOMETHANE SODIUM SULFONATE see AMK500
1-PHENYL-3,3-DIMETHYLTRIAZENE see DTP000
PHENYLDIMETHYLTRIAZINE see DTP000
2-PHENYL-m-DIOXAN-5-OL see BBA000
4-PHENYLDIPHENYL see TBC750
PHENYL-DRANE see SPC500
N,N'-(m-PHENYLENE)BISMALEIMIDE see BKL750
m-PHENYLENEBIS(METHYLAMINE) see XHS800
2,2'-(1,3-PHENYLENEBIS(OXYMETHYLENE))BISOXIRANE see REF000
1,1'-(m-PHENYLENE)BIS-1H-PYROLE-2,5-DIONE (9CI) see BKL750
m-PHENYLENEDIACETATE see REA100
m-PHENYLENEDIAMINE see PEY000
o-PHENYLENEDIAMINE see PEY250
p-PHENYLENEDIAMINE see PEY500
1,3-PHENYLENEDIAMINE see PEY000
1,4-PHENYLENEDIAMINE see PEY500
m-PHENYLENEDIAMINE (DOT) see PEY000
1,2-PHENYLENEDIAMINE (DOT) see PEY250
1,3-PHENYLENEDIAMINE DIHYDROCHLORIDE see PEY750
m-PHENYLENEDIAMINE HYDROCHLORIDE see PEY750
PHENYLENEDIAMINE, META, solid (DOT) see PEY000
PHENYLENEDIAMINE, PARA, solid (DOT) see PEY500
p-PHENYLENE DIAZIDE see DCL125
p-PHENYLENEDICARBONYL DICHLORIDE see TAV250
m-PHENYLENE DIISOCYANATE see BBP000
(PHENYLENEDIISOPROPYLIDENE)BIS(tert-BUTYLPEROXIDE) see BHL100
PHENYLENE-1,4-DIISOTHIOCYANATE see PFA500
1,4-PHENYLENEDIISOTHIOCYANIC ACID see PFA500
N,N'-(m-PHENYLENEDIMALEIMIDE) see BKL750
m-PHENYLENEDIMETHYLENE ISOCYANATE see XIJ000
o-PHENYLENEDIOL see CCP850
m-PHENYLENE ISOCYANATE see BBP000
2,3-PHENYLENEPYRENE see IBZ000
2,3-o-PHENYLENEPYRENE see IBZ000
1,10-(o-PHENYLENE)PYRENE see IBZ000
1,10-(1,2-PHENYLENE)PYRENE see IBZ000
PHENYLENE THIOCYANATE see PFA500
o-PHENYLENETHIOUREA see BCC500
PHENYLEPHRINE see NCL500
(−)-PHENYLEPHRINE see NCL500
PHENYLEPHRINE HYDROCHLORIDE see SPC500
d-(−)-PHENYLEPHRINE HYDROCHLORIDE see SPC500
1-PHENYL-1,2-EPOXYETHANE see EBR000
PHENYL-2,3-EPOXYPROPYL ETHER see PFF360
PHENYLETHANAL see BBL500
PHENYLETHANE see EGP500
PHENYLETHANOIC ACID BUTYL ESTER see BQJ350

1-PHENYLETHANOL see PDE000
2-PHENYLETHANOL see PDD750
β-PHENYLETHANOL see PDD750
PHENYL ETHANOLAMINE see AOR750
N-PHENYLETHANOLAMINE see AOR750
1-PHENYLETHANONE see ABH000
PHENYLETHENE see SMQ000
PHENYL ETHER see PFA850
PHENYL ETHER-BIPHENYL MIXTURE see PFA860
PHENYL ETHER, HEXACHLORO derivative (8CI) see CDV175
PHENYL ETHER PENTACHLORO see PAW250
PHENYL ETHER TETRACHLORO see TBP250
2-PHENYLETHYL ACETATE see PFB250
α-PHENYL ETHYL ACETATE see MNT075
β-PHENYLETHYL ACETATE see PFB250
PHENYLETHYL ALCOHOL see PDD750
2-PHENYLETHYL ALCOHOL see PDD750
β-PHENYLETHYL ALCOHOL see PDD750
PHENYLETHYLAMINE see PDE250
2-PHENYLETHYLAMINE see PDE250
ω-PHENYLETHYLAMINE see PDE250
2-PHENYLETHYL-o-AMINOBENZOATE see APJ500
PHENYLETHYL ANTHRANILATE see APJ500
2-PHENYLETHYL ANTHRANILATE see APJ500
PHENYLETHYLBARBITURATE see EOK000
PHENYL-ETHYL-BARBITURIC ACID see EOK000
5-PHENYL-5-ETHYLBARBITURIC ACID see EOK000
PHENYLETHYLBARBITURIC ACID, SODIUM SALT see SID000
3-PHENYL-3-ETHYL-2,6-DIKETOPIPERIDINE see DYC800
3-PHENYL-3-ETHYL-2,6-DIOXOPIPERIDINE see DYC800
PHENYLETHYLENE see SMQ000
PHENYLETHYLENE OXIDE see EBR000
2-PHENYL-2-ETHYLGLUTARIC ACID IMIDE see DYC800
α-PHENYL-α-ETHYLGLUTARIC ACID IMIDE see DYC800
α-PHENYL-α-ETHYLGLUTARIMIDE see DYC800
5-PHENYL-5-ETHYL-HEXAHYDROPYRIMIDINE-4,6-DIONE see DBB200
2-PHENYLETHYLHYDRAZINE see PFC500
β-PHENYLETHYLHYDRAZINE see PFC500
β-PHENYLETHYLHYDRAZINE DIHYDROGEN SULFATE see PFC750
2-PHENYLETHYLHYDRAZINE DIHYDROGEN SULPHATE see PFC750
β-PHENYLETHYLHYDRAZINE HYDROGEN SULPHATE see PFC750
β-PHENYLETHYLHYDRAZINE SULFATE see PFC750
PHENYLETHYLHYDRAZINE SULPHATE see PFC750
PHENYLETHYL ISOAMYL ETHER see IHV050
PHENYLETHYL ISOBUTYRATE see PDF750
2-PHENYLETHYL ISOBUTYRATE see PDF750
β-PHENYLETHYL ISOBUTYRATE see PDF750
PHENYLETHYL ISOVALERATE see PDF775
β-PHENYLETHYL ISOVALERATE see PDF775
PHENYL ETHYL KETONE see EOL500
PHENYLETHYLMALONYLUREA see EOK000
5-PHENYL-5-ETHYL-3-METHYLBARBITURIC ACID see ENB500
2-PHENYLETHYL-3-METHYLBUTYRATE see PDF775
PHENYLETHYL METHYL ETHYL CARBINOL see PFR200

PHENYLETHYLMETHYLHYDANTOIN see MKB250
2-PHENYLETHYL-2-METHYLPROPIONATE see PDF750
2-PHENYLETHYL PHENYLACETATE see PDI000
β-PHENYLETHYL PHENYLACETATE see PDI000
2-PHENYLETHYL-α-TOLUATE see PDI000
PHENYLETHYNYLCARBINOL CARBAMATE see PGE000
PHENYLETTEN see EOK000
PHENYL FLUORIDE see FGA000
PHENYLFLUOROFORM see BDH500
PHENYLFORMIC ACID see BCL750
PHENYL GLYCIDYL ETHER see PFF360
PHENYLGLYCOLIC ACID see MAP000
PHENYLGLYCOLONITRILE see MAP250
PHENYLGLYOXYLONITRILE OXIME-O,O-DIETHYL PHOSPHOROTHIOATE see BAT750
2-PHENYLHYDRACRYLIC ACID-3-α-TROPANYL ESTER see ARR000
PHENYL HYDRATE see PDN750
PHENYLHYDRAZIN (GERMAN) see PFI000
PHENYLHYDRAZINE see PFI000
2-PHENYLHYDRAZINECARBOTHIOAMIDE see PGM750
PHENYLHYDRAZINE HYDROCHLORIDE see PFI250
PHENYLHYDRAZINE MONOHYDROCHLORIDE see PFI250
PHENYLHYDRAZIN HYDROCHLORID (GERMAN) see PFI250
PHENYLHYDRAZINIUM CHLORIDE see PFI250
PHENYL HYDRIDE see BBL250
PHENYL HYDROXIDE see PDN750
PHENYLHYDROXYACETIC ACID see MAP000
o-(PHENYLHYDROXYARSINO)BENZOIC ACID see PFI600
2-PHENYL-2-HYDROXYETHYL, m-CHLOROPHENYL ARSINIC ACID see CKA575
N-PHENYLHYDROXYLAMINE see PFJ250
β-PHENYLHYDROXYLAMINE see PFJ250
PHENYL HYDROXYMERCURY see PFN100
1-PHENYL-1-HYDROXYPENTANE see BQJ500
PHENYLIC ACID see PDN750
PHENYLIC ALCOHOL see PDN750
PHENYL-IDIUM see PDC250
PHENYL-IDIUM 200 see PDC250
PHENYLIMIDOCARBONYL CHLORIDE see PFJ400
PHENYLIMIDOCARBONYL CHLORIDE see PFJ400
N-PHENYLIMIDOPHOSGENE see PFJ400
PHENYLIMINOCARBONYL DICHLORIDE see PFJ400
N-PHENYLIMINOCARBONYL DICHLORIDE see PFJ400
PHENYL ISOCYANATE see PFK250
PHENYL ISOCYANIDE, p-NITRO- see IKH780
PHENYLISONITRILE DICHLORIDE see PFJ400
d-β-PHENYLISOPROPYLAMINE SULFATE see BBK500
N-PHENYL ISOPROPYL CARBAMATE see CBM000
3-(β-PHENYLISOPROPYL)-SIDNONIMINE HYDROCHLORIDE see SPA000
PHENYL ISOTHIOCYANATE see ISQ000
PHENYL KETONE see BCS250
PHENYL MERCAPTAN see PFL850
PHENYLMERCURIACETATE see ABU500
PHENYL MERCURIC ACETATE see ABU500
PHENYLMERCURIC CHLORIDE see PFM500
PHENYLMERCURIC DINAPHTHYLMETHANEDISULFONATE see PFN000

PHENYL MERCURIC FIXTAN see PFN000
PHENYLMERCURIC HYDROXIDE see PFN100
PHENYLMERCURIC HYDROXIDE (DOT) see PFN100
PHENYLMERCURIC 3,3'-METHYLENEBIS(2-NAPHTHALENESULFONATE) see PFN000
PHENYLMERCURIC NITRATE see MCU750
PHENYLMERCURIC UREA see PFP500
PHENYLMERCURI PROPIONATE see PFO000
PHENYLMERCURIUREA see PFP500
PHENYLMERCURY ACETATE see ABU500
PHENYLMERCURY ACETATE 95% plus ETHYLMERCURY CHLORIDE 5% see PFO500
PHENYLMERCURY CHLORIDE see PFM500
PHENYLMERCURY HYDROXIDE see PFN100
PHENYLMERCURY METHYLENEDINAPHTHALENESULFONATE see PFN000
PHENYLMERCURY NITRATE see MCU750
PHENYLMERCURY PROPIONATE see PFO000
PHENYLMERCURY UREA see PFP500
PHENYLMETHANE see TGK750
PHENYLMETHANETHIOL see TGO750
PHENYLMETHANOL see BDX500
PHENYLMETHYL ALCOHOL see BDX500
N-PHENYLMETHYLAMINE see MGN750
1-PHENYL-2-METHYLAMINE-PROPANOL-1-SULFATE see EAY500
1-PHENYL-2-METHYLAMINOPROPANOL see EAW000
1-PHENYL-2-METHYLAMINOPROPANOL-1 see EAX500
PHENYLMETHYLCARBINOL see PDE000
1-PHENYL-5-METHYL-8-CHLORO-1,2,4,5-TETRAHYDRO-2,4-DIOXO-3H-1,5-BENZODIAZEPINE see CIR750
PHENYLMETHYLDICHLOROSILANE see DFQ800
N-(PHENYLMETHYL)DIMETHYLAMINE see DQP800
α-(PHENYLMETHYLENE)BENZENEACETONITRILE see DVX600
2-(PHENYLMETHYLENE)OCTANOL see HFO500
PHENYLMETHYL ESTER THIOCYANIC ACID (9CI) see BFL000
PHENYL METHYL ETHER see AOX750
PHENYL METHYL KETONE see ABH000
PHENYLMETHYL MERCAPTAN see TGO750
PHENYLMETHYLNITROSAMINE see MMU250
3-PHENYL-1-METHYL-1-NITROSOHARNSTOFF (GERMAN) see MMY500
(PHENYLMETHYL) PENICILLINIC ACID see BDY669
1-PHENYL-3-METHYL-3-PENTANOL see PFR200
PHENYL p-METHYLSTYRYL KETONE see MIF762
1-PHENYL-3-METHYL-3-(2-SULFOAETHYL) NATRIUM SALZ (GERMAN) see PEJ250
1-PHENYL-3-METHYL-3-(2-SULFOETHYL)TRIAZENE, SODIUM SALT see PEJ250
2-PHENYL-3-METHYLTETRAHYDRO-1,4-OXAZINE HYDROCHLORIDE see MNV750
PHENYLMETHYLVALERIANSAEURE-β-DIAETHYLAMINOAETHYLESTER-BROMMETHYLAT (GERMAN) see VBK000
PHENYLMONOGLYCOL ETHER see PER000
1-PHENYL-3-MONOMETHYLTRIAZENE see PFS500
PHENYL MORPHOLINE see PFS750
N-PHENYLMORPHOLINE see PFS750
PHENYL MUSTARD OIL see ISQ000
PHENYL-2-NAPHTHYLAMINE see PFT500
N-PHENYL-2-NAPHTHYLAMINE see PFT500

PHENYL-β-NAPHTHYLAMINE see PFT500
N-PHENYL-β-NAPHTHYLAMINE see PFT500
4-PHENYL-NITROBENZENE see NFQ000
p-PHENYL-NITROBENZENE see NFQ000
N-PHENYL-p-NITROSOANILINE see NKB500
PHENYLOXIRANE see EBR000
1-PHENYLOXIRANE see EBR000
2-PHENYLOXIRANE see EBR000
β-PHENYL-γ-OXYPROPIONSAEURE-TROPYL-ESTER (GERMAN) see ARR000
PHENYLPENTANOL see BQJ500
1-PHENYLPENTANOL see BQJ500
PHENYL PERCHLORYL see HCC500
PHENYL PHENACYL KETONE see PFU300
2-PHENYLPHENOL see BGJ250
4-PHENYLPHENOL see BGJ500
o-PHENYLPHENOL see BGJ250
p-PHENYLPHENOL see BGJ500
2-PHENYLPHENOL SODIUM SALT see BGJ750
o-PHENYLPHENOL, SODIUM SALT see BGJ750
α-PHENYLPHENYLACETONITRILE see DVX200
PHENYLPHOSPHINE see PFV250
PHENYLPHOSPHINE DICHLORIDE see DGE400
PHENYLPHOSPHONOTHIOIC ACID O-(4-BROMO-2,5-BROMO-2,5-DICHLOROPHENYL) O-METHYL ESTER see LEN000
PHENYLPHOSPHONOUS ACID DICHLORIDE see DGE400
PHENYLPHOSPHONOUS DICHLORIDE see DGE400
PHENYL PHOSPHORUS DICHLORIDE see DGE400
PHENYL PHOSPHORUS DICHLORIDE (DOT) see DGE400
PHENYL PHOSPHORUS THIODICHLORIDE see PFW200
PHENYL PHOSPHORUS THIODICHLORIDE see PFW210
1-PHENYLPIPERAZINE see PFX000
N-PHENYLPIPERAZINE see PFX000
4-PHENYLPIPERAZINYL 3,4,5-TRIMETHOXYPHENYL KETONE see PFX600
α-PHENYL-2-PIPERIDINEACETIC ACID METHYL ESTER see MNQ000
PHENYL (1-PIPERIDINOCYCLOHEXYL) KETONE see PFY200
2-PHENYLPROPANAL see COF000
3-PHENYLPROPANAL see HHP000
3-PHENYL-1-PROPANAL see HHP000
1-PHENYLPROPANE see IKG000
2-PHENYLPROPANE see COE750
3-PHENYLPROPANOL see HHP050
γ-PHENYLPROPANOL see HHP050
1-PHENYL-2-PROPANOL ACETATE see ABU800
3-PHENYL-1-PROPANOL ACETATE see HHP500
PHENYLPROPANOLAMINE see NNM000
PHENYLPROPANOLAMINE HYDROCHLORIDE see PMJ500
3-PHENYL-1-PROPANOL CARBAMATE see PGA750
3-PHENYL-1-PROPANOL (FCC) see HHP050
1-PHENYL-1-PROPANONE see EOL500
3-PHENYLPROPENAL see CMP969
3-PHENYL-2-PROPENAL see CMP969
2-PHENYLPROPENE see MPK250
β-PHENYLPROPENE see MPK250
3-PHENYLPROPENOIC ACID see CMP975
3-PHENYL-2-PROPENOIC ACID see CMP975
3-PHENYL-2-PROPENOIC ACID METHYL ESTER (9CI) see MIO500

3-PHENYL-2-PROPENOIC ACID, 2-METHYLPROPYL ESTER see IIQ000
3-PHENYL-2-PROPENOIC ACID PHENYLMETHYL ESTER (9CI) see BEG750
3-PHENYL-2-PROPEN-1-OL see CMQ740
3-PHENYL-2-PROPEN-1-YL ACETATE see CMQ730
3-PHENYL-2-PROPENYLANTHRANILATE see API750
3-PHENYL-2-PROPEN-1-YL ANTHRANILATE see API750
PHENYLPROPENYL n-BUTYRATE see CMQ800
3-PHENYL-2-PROPEN-1-YL FORMATE see CMR500
3-PHENYL-2-PROPENYL PROPIONATE see CMR850
3-PHENYL-2-PROPEN-1-YL PROPIONATE see CMR850
α-PHENYLPROPIONALDEHYDE see COF000
β-PHENYLPROPIONALDEHYDE see HHP000
2-PHENYLPROPIONALDEHYDE DIMETHYL ACETAL see PGA800
2-PHENYLPROPIONALDEHYDE DIMETHYL ACETAL see HII600
2-PHENYLPROPIONALDEHYDE (FCC) see COF000
3-PHENYLPROPIONALDEHYDE (FCC) see HHP000
PHENYL(PROPIONYLOXY)MERCURY see PFO000
4-PHENYLPROPIOPHENONE see BGM100
PHENYLPROPYL ACETATE see HHP500
3-PHENYL-1-PROPYL ACETATE see HHP500
3-PHENYLPROPYL ACETATE (FCC) see HHP500
PHENYLPROPYL ALCOHOL see HHP050
3-PHENYLPROPYL ALCOHOL see HHP050
γ-PHENYLPROPYL ALCOHOL see HHP050
3-PHENYLPROPYL ALDEHYDE see HHP000
γ-PHENYLPROPYLCARBAMAT (GERMAN) see PGA750
γ-PHENYLPROPYL CARBAMATE see PGA750
3-PHENYL-N-(4-PROPYLCYCLOHEXYL)-2-PROPENAMIDE see PGB850
2-PHENYLPROPYLENE see MPK250
β-PHENYLPROPYLENE see MPK250
PHENYLPROPYL PROPIONATE see HHQ550
3-PHENYLPROPYL PROPIONATE see HHQ550
β-PHENYLPROPYL PROPIONATE see HHQ550
1-PHENYL-2-PROPYNYL CARBAMATE see PGE000
α-(1-PHENYL-1H-PYRAZOL-4-YL)-1-PIPERIDINEBUTANOL see PGE550
1-PHENYL-1-(2-PYRIDYL)-3-DIMETHYLAMINOPROPANE see TMJ750
1-PHENYL-1-(2-PYRIDYL)-3-DIMETHYLAMINOPROPANE HYDROCHLORIDE see PMS900
3-PHENYL-3-(2-PYRIDYL)-N,N-DIMETHYLPROPYLAMINE see TMJ750
PHENYLQUECKSILBERACETAT (GERMAN) see ABU500
PHENYLQUECKSILBERCHLORID (GERMAN) see PFM500
PHENYLQUINONE see PEL750
PHENYLRHODANID see TFF600
PHENYL SALICYLATE see PGG750
3-PHENYLSALICYLIC ACID see PGH000
PHENYL SELENIDE see PGH250
PHENYLSENFOEL (GERMAN) see ISQ000
PHENYLSILICON TRICHLORIDE see TJA750
PHENYL STYRYL KETONE see CDH000
PHENYLSULFOHYDRAZIDE see BBS300
PHENYLSULFONIC ACID see BBS250
PHENYLSULFONYL CHLORIDE see BBS750

PHENYLSULFONYL HYDRAZIDE see BBS300
PHENYLSULFONYLHYDRAZINE see BBS300
6-PHENYL-2,3,5,6-TETRAHYDROIMIDAZO(2,1-b)THIAZOLE see LFA000
N-PHENYL-N'-1,2,3-THIADIAZOL-5-YL-UREA see TEX600
PHENYLTHIOCARBAMIDE see PGN250
PHENYL THIOCYANATE see TFF600
PHENYLTHIONOPHOSPHONIC DICHLORIDE see PFW200
PHENYLTHIONOPHOSPHONIC DICHLORIDE see PFW210
PHENYLTHIOPHOSPHONATE de O-ETHYLE et O-4-NITROPHENYLE (FRENCH) see EBD700
1-PHENYLTHIOSEMICARBAZIDE see PGM750
α-(PHENYLTHIO)-p-TOLUIDINE see PGN100
1-PHENYLTHIOUREA see PGN250
N-PHENYLTHIOUREA see PGN250
1-PHENYL-2-THIOUREA see PGN250
α-PHENYLTHIOUREA see PGN250
(N-PHENYL-p-TOLUENESULFONAMIDO)ETHYLMERCURY see EME500
PHENYL p-TOLYL KETONE see MHF750
PHENYLTRICHLOROMETHANE see BFL250
PHENYL TRICHLOROSILANE (DOT) see TJA750
1-PHENYL-4-(3,4,5-TRIMETHOXYBENZOYL)PIPERAZINE see PFX600
PHENYL UREA-p-DI(CARBOXYMETHYL) THIOARSENITE see CBI250
PHENYLVINYL KETONE see PMQ250
PHENYRAL see EOK000
PHENYTOIN SODIUM see DNU000
PHERMERNITE see MCU750
PHETADEX see BBK500
PHILBLACK see CBT750
PHILBLACK N 550 see CBT750
PHILBLACK N 765 see CBT750
PHILBLACK O see CBT750
PHILIPS-DUPHAR PH 60-40 see CJV250
PHILODORM see TDA500
PHILOPON see MDT600
PHILOPON see DBA800
PHILOSOPHER'S WOOL see ZKA000
PHILOSTIGMIN BROMIDE see POD000
PHIP see AKZ200
PHISODANV see HCL000
PHISOHEX see HCL000
PHIX see ABU500
PHIXIA see CMS850
PHLOROGLUCIN see PGR000
PHLOROGLUCINOL see PGR000
PHLOROL see PGR250
PHLOXIN B see ADG250
PHLOXINE B see ADG250
PHLOXINE P see ADG250
PHLOXINE TONER B see BNH500
PHLOX RED TONER X-1354 see BNH500
PHOB see EOK000
PHOBEX see BCA000
PHONURIT see AAI250
PHORAT (GERMAN) see PGS000
PHORATE see PGS000
PHORATE-10G see PGS000
PHORBYOL see CCP250
PHORON (GERMAN) see PGW250
PHORONE see PGW250

PHOSADEN see AOA125
PHOSALON see BDJ250
PHOSALONE see BDJ250
PHOSCHLOR R50 see TIQ250
PHOSDRIN (OSHA) see MQR750
PHOSFENE see MQR750
PHOSFLEX 179-C see TMO600
PHOSFLEX T-BEP see BPK250
PHOS-FLUR see SHF500
PHOSFOLAN see PGW750
PHOSFOLAN see DXN600
PHOSGEN (GERMAN) see PGX000
PHOSGENE see PGX000
PHOSGENE, THIO- see TFN500
PHOSKIL see PAK000
PHOSMET see PHX250
PHOSPHADEN see AOA125
PHOSPHALUGEL see PHB500
PHOSPHAMID see DSP400
(Z)-PHOSPHAMIDON see PGX275
cis-PHOSPHAMIDON see PGX275
PHOSPHATE de O,O-DIETHYLE et de O-2-CHLORO-1-(2,4-DICHLOROPHENYL) VINYLE (FRENCH) see CDS750
PHOSPHATE de DIMETHYLE et de 2,2-DICHLOROVINYLE (FRENCH) see DGP900
PHOSPHATE de DIMETHYLE et de 2-DIMETHYLCARBAMOYL-1-METHYL VINYL (FRENCH) see DGQ875
PHOSPHATE de DIMETHYLE et de 2-METHOXYCARBONYL-1 METHYLVINYLE (FRENCH) see MQR750
PHOSPHATE de DIMETHYLE et de 2-METHYLCARBAMOYL 1-METHYL VINYLE see MRH209
PHOSPHATES see PGX500
PHOSPHATE, SODIUM HEXAMETA see SHM500
PHOSPHATE de TRICRESYLE (FRENCH) see TNP500
PHOSPHEMOL see PAK000
PHOSPHENE (FRENCH) see MQR750
PHOSPHENOL see PAK000
PHOSPHENTASIDE see AOA125
PHOSPHIDES see PGX750
PHOSPHINE see PGY000
PHOSPHINE, DIPHENYLPROPYL- see PNI600
PHOSPHINE OXIDE, TRIS(p-DIMETHYLAMINOPHENYL)-, compounded with STANNIC CHLORIDE (2:1) see BLT775
PHOSPHINE SELENIDE, TRIMETHYL- see TMD400
PHOSPHINE SELENIDE, TRIPIPERIDINO- see TMX350
PHOSPHINOTHIOIC ACID, DIPHENYL-, S-(4-CHLORO-2-BUTYNYL) ESTER see CEV840
1,1',1''-PHOSPHINOTHIOYLIDYNETRISAZIRIDINE see TFQ750
PHOSPHINOX PZ 06 see ZJS400
1,1',1''-PHOSPHINYLIDYNETRISAZIRIDINE see TND250
1,1',1''-PHOSPHINYLIDYNETRIS(2-METHYL)AZRIDINE see TNK250
PHOSPHONIC ACID see PGZ899
PHOSPHONIC ACID, BENZYL-, DIBUTYL ESTER see BFD760
PHOSPHONIC ACID, (DICHLOROMETHYLENE)BIS-, DISODIUM SALT see SFX730
PHOSPHONIC ACID, (DICHLOROMETHYLENE)DI-, DISODIUM SALT see SFX730

PHOSPHONIC ACID, (2-(DIETHOXYMETHYLSILYL)ETHYL)-, DIETHYL ESTER see DJA330
PHOSPHONIC ACID, 1-HYDROXY-1,1-ETHANEDIYL ESTER see HKS780
PHOSPHONIC ACID, (1-HYDROXYETHYLIDENE)BIS- see HKS780
PHOSPHONIC ACID, METHYL-, DIPHENYL ESTER see MDQ825
PHOSPHONIC ACID, (PHENYLMETHYL)-, DIBUTYL ESTER see BFD760
PHOSPHONODITHIOIC ACID, ETHYL-, S-(4-CHLORO-3-METHYLPHENYL)o-ETHYL ESTER see EMR100
PHOSPHONODITHIOIC ACID, ETHYL-, S-(4-CHLORO-m-TOLYL) o-ETHYL ESTER see EMR100
N-(PHOSPHONOMETHYL)GLYCINE see PHA500
PHOSPHONOTHIOIC DICHLORIDE, ETHYL- see EOP600
PHOSPHONOTHIOIC DICHLORIDE, PHENYL- see PFW200
PHOSPHONOTHIOIC DICHLORIDE, PHENYL- see PFW210
PHOSPHORAMIDOTHIOIC ACID, (3-METHYL-2-THIAZOLIDINYLIDENE)-, O,S-DIMETHYL ESTER see MEY200
PHOSPHORE BLANC (FRENCH) see PHP010
PHOSPHORE(PENTACHLORURE de) (FRENCH) see PHR500
PHOSPHORE(TRICHLORURE de) (FRENCH) see PHT275
PHOSPHORIC ACID see PHB250
PHOSPHORIC ACID, ALUMINUM SALT (1:1), (solution) see PHB500
PHOSPHORIC ACID, BERYLLIUM SALT (1:1) see BFS000
PHOSPHORIC ACID, 2-CHLORO-3-(DIETHYLAMINO)-1-METHYL-3-OXO-1-PROPENYL-,DIMETHYL ESTER, (Z)- see PGX275
PHOSPHORIC ACID, 2-CHLORO-1-(2,4,5-TRICHLOROPHENYL)ETHENYL DIMETHYL ESTER see TBW100
PHOSPHORIC ACID CHROMIUM (III) SALT see CMK300
PHOSPHORIC ACID, CHROMIUM(3+) SALT (1:1) see CMK300
PHOSPHORIC ACID, DIBUTYL PHENYL ESTER see DEG600
PHOSPHORIC ACID-2,2-DICHLOROETHENYL DIMETHYL ESTER see DGP900
PHOSPHORIC ACID, DIETHYL ESTER, with 3-CHLORO-7-HYDROXY-4-METHYLCOUMARIN see CIK750
PHOSPHORIC ACID, DIMETHYL ESTER, ESTER with DIMETHYL 3-HYDROXYGLUTACONATE see SOY000
PHOSPHORIC ACID, DIMETHYL ESTER, ESTER with cis-3-HYDROXY-N-METHYLCROTONAMIDE see MRH209
PHOSPHORIC ACID, DIMETHYL ESTER, ESTER WITH 2-CHLORO-N,N-DIETHYL-3-HYDROXYCROTANAMIDE, (Z)- see PGX275
PHOSPHORIC ACID, DIMETHYL 1-METHYL-3-(METHYLAMINO)-3-OXO-1-PROPENYL ESTER, (E)- see MRH209
PHOSPHORIC ACID DIMETHYL-p-(METHYLTHIO)PHENYL ESTER see PHD250
PHOSPHORIC ACID, DISODIUM SALT see SJH090
PHOSPHORIC ACID, ISOPROPYL ESTER see PHE500

PHOSPHORIC ACID, LEAD(2+) SALT (2:3) see LDU000
PHOSPHORIC ACID, TRI-o-CRESYL ESTER see TMO600
PHOSPHORIC ACID TRIETHYLENE IMIDE see TND250
PHOSPHORIC ACID TRIETHYLENEIMINE (DOT) see TND250
PHOSPHORIC ACID, TRIMETHYL ESTER see TMD250
PHOSPHORIC ACID, TRIPHENYL ESTER see TMT750
PHOSPHORIC ACID, TRIS(2,3-DIBROMOPROPYL) ESTER see TNC500
PHOSPHORIC ACID TRIS(1,3-DICHLORO-2-PROPYL)ESTER see FQU875
PHOSPHORIC ACID, TRIS(2-METHYLPHENYL) ESTER see TMO600
PHOSPHORIC ACID, TRISODIUM SALT see SJH200
PHOSPHORIC ACID, TRITOLYL ESTER see TNP500
PHOSPHORIC BROMIDE see PHR250
PHOSPHORIC CHLORIDE see PHR500
PHOSPHORIC SULFIDE see PHS000
PHOSPHORIC TRIS(DIMETHYLAMIDE) see HEK000
PHOSPHOROCHLORIDOTHIOIC ACID-O,O-DIMETHYL ESTER see DTQ600
PHOSPHORODICHLORIDIC ACID, ETHYL ESTER see EOR000
PHOSPHORODIFLUORIDIC ACID see PHF250
PHOSPHORODI(ISOPROPYLAMIDIC) FLUORIDE see PHF750
PHOSPHORODITHIOIC ACID, S-(6-CHLORO-3,4-DIHYDRO-2H-1-BENZOTHIOPYRAN-4-YL) o,o-DIMETHYL ESTER see CLH820
PHOSPHORODITHIOIC ACID-S-(2-CHLORO-1-(1,3-DIHYDRO-1,3-DIOXO-2H-ISOINDOL-2-YL))ETHYL-O,O-DIETHYL ESTER see DBI099
PHOSPHORODITHIOIC ACID-S-((2-CHLORO-1-PHTHALIMIDOETHYL)-O,)-DIETHYL ESTER see DBI099
PHOSPHORODITHIOIC ACID, o,o-DIETHYL S-(2-(ETHYLTHIO)-6-METHYL-4-PYRIMIDINYL) ESTER see DJJ393
PHOSPHORODITHIOIC ACID, O,O-DIISOOCTYL S-(10-PHENARSAZINYL) ESTER see PDB300
PHOSPHORODITHIOIC ACID-O,O-DIMETHYL ESTER-S-ESTER with DIETHYL MERCAPTOSUCCINATE see MAK700
PHOSPHORODITHIOIC ACID, O,O-DIMETHYL ESTER, S-ESTER with N-(2-MERCAPTOETHYL)ACETAMIDE see DOP200
PHOSPHORODITHIOIC ACID, S-(1,1-DIMETHYLETHYL) o-ETHYL S-(1-METHYLPROPYL) ESTER see ENF050
PHOSPHORODITHIOIC ACID, O,O-DIMETHYL-S-(2-ETHYLTHIO)ETHYL ESTER see PHI500
PHOSPHORODITHIOIC ACID-O,O-DIMETHYL-S-(2-(METHYLAMINO)-2-OXOETHYL) ESTER see DSP400
PHOSPHORODITHIOIC ACID-S,S'-1,4-DIOXANE-2,3-DIYL-O,O,O',O'-TETRAETHYL ESTER see DVQ709
PHOSPHORODITHIONIC ACID, S-2-(ETHYLTHIO)ETHYL-O,O-DIETHYL ESTER see DXH325
PHOSPHOROFLUORIDIC ACID see PHJ250
PHOSPHOROFLUORIDIC ACID, DIISOPROPYL ESTER see IRF000
PHOSPHOROSELENOIC ACID, Se-(2-(DIETHYLAMINO)ETHYL) O,O-DIETHYL ESTER see DJA325

PHOSPHOROTHIOIC ACID, O-(4-(AMINOSULFONYL)PHENYL) O,O-DIMETHYL ESTER (9CI) see CQL250
PHOSPHOROTHIOIC ACID-S-(((1-CYANO-1-METHYL-ETHYL)CARBAMOYL)METHYL)-O,O-DIETHYL ESTER see PHK250
PHOSPHOROTHIOIC ACID, o-(2,4-DICHLOROPHENYL) o-ETHYL S-PROPYL ESTER see EEE200
PHOSPHOROTHIOIC ACID, o,o-DIETHYL o-(4-(((((METHYLAMINO)CARBONYL)OXY)IMINO)METHYL)PHENYL)ESTER see DJW890
PHOSPHOROTHIOIC ACID, O,O-DIETHYL-O-(4-NITROPHENYL) ESTER see PAK000
PHOSPHOROTHIOIC ACID-O,O-DIETHYL, O-(p-NITROPHENYL)ESTER, mixed with compressed gas see PAK230
PHOSPHOROTHIOIC ACID-O,O-DIETHYL-O-2-PYRAZINYL ESTER see EPC500
PHOSPHOROTHIOIC ACID, o,o-DIETHYL o-((TETRAHYDRO-2-FURANYL)METHYL) ESTER see DKB170
PHOSPHOROTHIOIC ACID, o-ETHYL o-PHENYL S-PROPYL ESTER see HBU415
PHOSPHOROTHIOIC ACID TRIETHYLENETRIAMIDE see TFQ750
PHOSPHOROTHIOIC TRICHLORIDE see TFO000
PHOSPHOROTHIONIC TRICHLORIDE see TFO000
PHOSPHOROTRITHIOUS ACID, S,S,S-TRIBUTYL ESTER see TIG250
PHOSPHOROTRITHIOUS ACID, TRIMETHYL ESTER see TMD650
PHOSPHOROUS ACID, BERYLLIUM SALT see BFS000
PHOSPHOROUS ACID DIMETHYL ESTER see DSG600
PHOSPHOROUS ACID, TRIPHENYL ESTER see TMU250
PHOSPHOROUS OXYBROMIDE see PHU000
PHOSPHOROUS SULFOCHLORIDE see TFO000
PHOSPHOROUS THIOCHLORIDE see TFO000
PHOSPHOROUS TRICHLORIDE SULFIDE see TFO000
PHOSPHOROUS TRIFLUORIDE see PHQ500
PHOSPHOROUS (WHITE) see PHP010
PHOSPHOROUS YELLOW see PHP010
PHOSPHORPENTACHLORID (GERMAN) see PHR500
PHOSPHORSAEURELOESUNGEN (GERMAN) see PHB250
PHOSPHORTRICHLORID (GERMAN) see PHT275
PHOSPHORUS (red) see PHO500
PHOSPHORUS (yellow) see PHP010
PHOSPHORUS, amorphous (DOT) see PHO500
PHOSPHORUS ACID, TRIMETHYL ESTER see TMD500
PHOSPHORUS BROMIDE (DOT) see PHT250
PHOSPHORUS CHLORIDE see PHT275
PHOSPHORUS COMPOUNDS, INORGANIC see PHQ000
PHOSPHORUS FLUORIDE see PHQ500
PHOSPHORUS HEPTASULFIDE see PHQ750
PHOSPHORUS HEPTASULFIDE, free from yellow or white phosphorus (DOT) see PHQ750
PHOSPHORUS(V) OXIDE see PHS250
PHOSPHORUS OXYCHLORIDE see PHQ800
PHOSPHORUS OXYTRICHLORIDE see PHQ800
PHOSPHORUS PENTABROMIDE see PHR250
PHOSPHORUS PENTACHLORIDE see PHR500
PHOSPHORUS PENTAFLUORIDE see PHR750
PHOSPHORUS PENTAOXIDE see PHS250
PHOSPHORUS PENTASULFIDE see PHS000

PHOSPHORUS PENTASULFIDE, free from yellow or white phosphorus (DOT) see PHS000
PHOSPHORUS PENTOXIDE see PHS250
PHOSPHORUS PERCHLORIDE see PHR500
PHOSPHORUS PERSULFIDE see PHS000
PHOSPHORUS SESQUISULFIDE see PHS500
PHOSPHORUS SESQUISULFIDE, free from yellow or white phosphorus (DOT) see PHS500
PHOSPHORUS (III) SULFIDE (IV) see PHS500
PHOSPHORUS TRIBROMIDE see PHT250
PHOSPHORUS TRICHLORIDE see PHT275
PHOSPHORUS TRIFLUORIDE see PHQ500
PHOSPHORUS TRIHYDRIDE see PGY000
PHOSPHORUS TRIHYDROXIDE see PGZ899
PHOSPHORUS TRISULFIDE see PHT750
PHOSPHORUS TRISULFIDE, free from yellow or white phosphorus (DOT) see PHT750
PHOSPHORUS, white or yellow, dry or under water or in solution (UN 1381) (DOT) see PHP010
PHOSPHORUS WHITE, molten (UN 2447) (DOT) see PHP010
PHOSPHORUS, YELLOW (ACGIH,OSHA) see PHP010
PHOSPHORWASSERSTOFF (GERMAN) see PGY000
PHOSPHORYL BROMIDE see PHU000
PHOSPHORYL CHLORIDE see PHQ800
PHOSPHORYL HEXAMETHYLTRIAMIDE see HEK000
O-PHOSPHORYL-4-HYDROXY-N,N-DIMETHYLTRYPTAMINE see PHU500
PHOSPHORYL TRIBROMIDE see PHU000
PHOSPHOSTIGMINE see PAK000
PHOSPHOTEX see TEE500
PHOSPHOTHION see MAK700
PHOSPHOTOX E see EEH600
PHOSPHOTUNGSTIC ACID see PHU750
PHOSPHURE de MAGNESIUM (FRENCH) see MAI000
PHOSPHURES d'ALUMINUM (FRENCH) see AHE750
PHOSPHURE de SODIUM (FRENCH) see SJI500
PHOSPHURE de ZINC (FRENCH) see ZLS000
PHOSTIL see HBU415
PHOSTOXIN see AHE750
PHOSVEL see LEN000
PHOSVIN see ZLS000
PHOSVIT see DGP900
PHOTOMIREX see MRI750
PHOTOPHOR see CAW250
PHOXIME see BAT750
PHOXIN see BAT750
PHOZALON see BDJ250
PHP see ELL500
PHPH see BGE000
PHRENOLAN see MDU750
1,3-PHTHALANDIONE see PHW750
1(2H)-PHTHALAZINONE HYDRAZONE see HGP495
1(2H)-PHTHALAZINONE HYDRAZONE HYDROCHLORIDE see HGP500
1(2H)-PHTHALAZINONE, HYDRAZONE, MONOHYDROCHLORIDE see HGP500
PHTHALIC ACID see PHW250
PHTHALIC ACID ANHYDRIDE see PHW750
PHTHALIC ACID BIS(2-METHOXYETHYL) ESTER see DOF400
PHTHALIC ACID, DIALLYL ESTER see DBL200
o-PHTHALIC ACID, DIALLYL ESTER see DBL200
PHTHALIC ACID, DIETHYL ESTER see DJX000
PHTHALIC ACID, DIHEPTYL ESTER see HBP400
PHTHALIC ACID DIHEXYL ESTER see DKP600
PHTHALIC ACID DINITRILE see PHY000
PHTHALIC ACID DIOCTYL ESTER see DVL700

PHTHALIC ACID, DIPENTYL ESTER see AON300
PHTHALIC ACID, DIPROPYL ESTER see DWV500
PHTHALIC ACID METHYL ESTER see DTR200
PHTHALIC ANHYDRIDE see PHW750
o-PHTHALIC IMIDE see PHX000
PHTHALIDE 3,3,-BIS(p-HYDROXYPHENYL)- see PDO750
PHTHALIDE, 3-BUTYLIDENE- see BRQ100
PHTHALIMETTEN see PDO750
PHTHALIMIDE see PHX000
PHTHALIMIDE, N-(HYDROXYMETHYL)- see HMP100
PHTHALIMIDIMIDE see DNE400
PHTHALIMIDO-O,O-DIMETHYL PHOSPHORODITHIOATE see PHX250
2-PHTHALIMIDOGLUTARIC ACID ANHYDRIDE see PHX100
2-PHTHALIMIDOGLUTARIMIDE see TEH500
3-PHTHALIMIDOGLUTARIMIDE see TEH500
α-PHTHALIMIDOGLUTARIMIDE see TEH500
α-(N-PHTHALIMIDO)GLUTARIMIDE see TEH500
PHTHALIMIDOMETHYL ALCOHOL see HMP100
PHTHALIMIDOMETHYL-O,O-DIMETHYL PHOSPHORODITHIOATE see PHX250
PHTHALOCYANINE BLUE 01206 see DNE400
PHTHALODINITRILE see PHY000
m-PHTHALODINITRILE see PHX550
o-PHTHALODINITRILE see PHY000
p-PHTHALODINITRILE see BBP250
PHTHALOGEN see DNE400
PHTHALOL see DJX000
PHTHALONITRILE see PHY000
PHTHALOPHOS see PHX250
p-PHTHALOYL CHLORIDE see TAV250
p-PHTHALOYL DICHLORIDE see TAV250
N-PHTHALOYLGLUTAMIMIDE see TEH500
PHTHALSAEUREANHYDRID (GERMAN) see PHW750
PHTHALSAEUREDIAETHYLESTER (GERMAN) see DJX000
PHTHALSAEUREDIMETHYLESTER (GERMAN) see DTR200
PHTHALTAN see TIT250
N-PHTHALYLGLUTAMIC ACID IMIDE see TEH500
N-PHTHALYL-GLUTAMINSAEURE-IMID (GERMAN) see TEH500
α-N-PHTHALYLGLUTARAMIDE see TEH500
PHXA 41 see XAA500
PHYBAN see MRL750
PHYGON see DFT000
PHYGON PASTE see DFT000
PHYGON SEED PROTECTANT see DFT000
PHYGON XL see DFT000
PHYMONE see NAK500
PHYSEPTONE see MDO750
PHYSOSTIGMINE see PIA500
PHYSOSTIGMINE SALICYLATE (1:1) see PIA750
PHYSOSTOL see PIA500
PHYSOSTOL SALICYLATE see PIA750
PHYTAR see HKC000
PHYTAR 138 see HKC000
PHYTAR 560 see HKC000
PHYTAR 560 see HKC500
PHYTAR 600 see HKC000
PHYTIC ACID see PIB250
PHYTOGERMINE see VSZ450
PHYTOL see PIB600
trans-PHYTOL see PIB600
PHYTOMYCIN see SLY500

PHYTOSOL see EPY000
PIANADALIN see BNK000
PIAPONON see PMH500
PIC-CLOR see CKN500
PICCOLASTIC see SMQ500
PICCOLASTIC D-100 see SMQ500
PICCOLASTIC A see SMQ500
PICCOLASTIC A 5 see SMQ500
PICCOLASTIC A 25 see SMQ500
PICCOLASTIC A 50 see SMQ500
PICCOLASTIC A 75 see SMQ500
PICCOLASTIC C 125 see SMQ500
PICCOLASTIC D see SMQ500
PICCOLASTIC D 125 see SMQ500
PICCOLASTIC D 150 see SMQ500
PICCOLASTIC E 75 see SMQ500
PICCOLASTIC E 100 see SMQ500
PICCOLASTIC E 200 see SMQ500
PICFUME see CKN500
PICHTOSIN see IHX600
PICHTOSINE see IHX600
PICIS CARBONIS see CMY800
PICLORAM see PIB900
2-PICOLINE see MOY000
3-PICOLINE see PIB920
4-PICOLINE see MOY250
m-PICOLINE see PIB920
o-PICOLINE see MOY000
p-PICOLINE see MOY250
α-PICOLINE see MOY000
β-PICOLINE see PIB920
γ-PICOLINE see MOY250
PICRACONITINE see PIC250
PICRAGOL see PID200
PICRAMIC ACID, SODIUM SALT see PIC500
PICRAMIC ACID, ZIRCONIUM SALT (WET) see PIC750
PICRAMIDE see PIC800
PICRAMIDE (DOT) see PIC800
PICRATOL see ANS500
PICRIC ACID see PID000
PICRIC ACID (ACGIH,OSHA) see PID000
PICRIC ACID, AMMONIUM SALT see ANS500
PICRIC ACID, wet, with not <10% water (NA 1344)
(DOT) see PID000
PICRIC ACID, SILVER(1+) SALT see PID200
PICRIC ACID, dry or wetted with <30% water, by weight
(UN 0154) (DOT) see PID000
PICRIDE see CKN500
PICRITE (the explosive) see NHA500
PICRITE, dry or wetted with <20% water, by weight (UN
0282) (DOT) see NHA500
PICRITE, wetted with not <20% water, by weight (UN
1336) (DOT) see NHA500
PICRONITRIC ACID see PID000
PICROTIN, compounded with PICROTOXININ (1:1) see
PIE500
PICROTOL see PID200
PICROTOXIN see PIE500
PICROTOXINE see PIE500
PICRYL CHLORIDE (DOT) see TML325
PICRYLMETHYLNITRAMINE see TEG250
PICRYLNITROMETHYLAMINE see TEG250
PID see DVV600
PIECIOCHLOREK FOSFORU (POLISH) see PHR500
PIED PIPER MOUSE SEED see SMN500
PIELIK E see SGH500
PIGLET PRO-GEN V see ARA500

PIGMENT BLACK 7 see CBT750
PIGMENT GREEN 15 see LCR000
PIGMENT PONCEAU R see FMU070
PIGMENT RED CD see CHP500
PIGMENT TRANSPARENT YELLOW 2K see PIE800
PIGMENT TRANSPARENT YELLOW O see PIE830
PIGMENT WHITE 32 see ZJS400
PIGMENT YELLOW 33 see CAP750
PIGMENT YELLOW TRANSPARENT 2K see PIE800
PIGMENT YELLOW TRANSPARENT O see PIE830
PIG-WRACK see CCL250
PIKRINEZUUR (DUTCH) see PID000
PIKRINSAEURE (GERMAN) see PID000
PIKRYNOWY KWAS (POLISH) see PID000
PILLARDRIN see MRH209
PILLARON see DTQ400
PILLARSTIN see MHC750
PILLARZO see CFX000
PILLS (INDIA) see SED400
PILOCARPINE see PIF000
PILOCARPINE HYDROCHLORIDE see PIF250
PILOCARPINE MONOHYDROCHLORIDE see PIF250
PILOCARPINE MURIATE see PIF250
PILOCARPOL see PIF000
PILOCEL see PIF250
PILOMIOTIN see PIF250
PILOT 447 see MKP500
PILOT HD-90 see DXW200
PILOT SF-40 see DXW200
PILOVISC see PIF250
PILPOPHEN see DQA400
PIMACOL-SOL see NAK500
PIMAFUCIN see PIF750
PIMARICIN see PIF750
PIMELIC KETONE see CPC000
PIMENTA BERRY OIL see PIG740
PIMENTA LEAF OIL see PIG730
PIMENTA LEAF OIL see PIG740
PIMENTA OIL see PIG740
PIMENTA RACEMOSA OIL see LBK000
PIMENTO OIL see PIG740
PIN see EBD700
PINAKON see HFP875
PINANG see BFW000
PINDON (DUTCH) see PIH175
PINDONE see PIH175
2-PINENE see PIH250
2(10)-PINENE see POH750
α-PINENE (DOT) see PIH250
PINE NEEDLE OIL see FBV000
PINE NEEDLE OIL, DWARF see PIH400
PINE NEEDLE OIL, SCOTCH see PIH500
β-PINENE (FCC) see POH750
PINE OIL see PIH750
PINOCARVEOL see ODW030
PINUS MONTANA OIL see PIH400
PINUS PUMILIO OIL see PIH400
PIOMBO TETRA-ETILE see TCF000
N-N-PIP see NLJ500
PIPAMPERONE see FHG000
PIPANEPERONE see FHG000
PIPERACETAZINE see PII500
PIPERADROL HYDROCHLORIDE see PII750
PIPERAZIDINE see PIJ000
PIPERAZIN (GERMAN) see PIJ000
PIPERAZINE see PIJ000
PIPERAZINE, anhydrous see PIJ000

PIPERAZINE, 1-(o-CHLOROPHENYL)-4-
NICOTINOYLMETHYL- see KGK120
PIPERAZINE, 1-(p-CHLOROPHENYL)-4-(3,4,5-
TRIMETHOXYBENZOYL)- see CKJ100
PIPERAZINE DIHYDROCHLORIDE see PIK000
PIPERAZINE, 1-((3-(4,7-DIHYDRO-1-METHYL-7-OXO-
3-PROPYL-1H-PYRAZOLO(4,3-D)PYRIMIDIN-5-Y L)-4-
ETHOXYPHENYL)SULFONYL)-4-METHYL-, 2-
HYDROXY-1,2,3-PROPANETRICARBOXYLATE (1:1)
see SCF600
2,6-PIPERAZINEDIONE-4,4'-PROPYLENE
DIOXOPIPERAZINE see PIK250
N,N'-(PIPERAZINEDIYLBIS(2,2,2-
TRICHLOROETHYLIDENE)) BIS(FORMAMIDE) see
TKL100
PIPERAZINE, compounded with ESTRONE
HYDROGEN SULFATE (1:1) (8CI) see PIK450
PIPERAZINE ESTRONE SULFATE see PIK450
1-PIPERAZINEETHANOL, 4-(3-(10H-PYRIDO(3,2-
B)(1,4)BENZOTHIAZIN-10-YL)PROPYL)-
,DIHYDROCHLORIDE see ORI400
PIPERAZINE HYDROCHLORIDE see PIK000
PIPERAZINE, 1-(o-METHOXYPHENYL)-4-
NICOTINOYLMETHYL- see KGK130
PIPERAZINE, 1-(o-METHOXYPHENYL)-4-(3,4,5-
TRIMETHOXYBENZOYL)- see MFH760
PIPERAZINE, 1-(p-METHOXYPHENYL)-4-(3,4,5-
TRIMETHOXYBENZOYL)- see MFH770
PIPERAZINE, 1-(α-METHYLPHENETHYL)-4-
PHENYL-, DIHYDROCHLORIDE see MNP450
PIPERAZINE, 1-PHENYL-4-(3,4,5-
TRIMETHOXYBENZOYL)- see PFX600
PIPERAZINE, compounded with 3-(SULFOOXY)ESTRA-
1,3,5(10)-TRIEN-17-ONE (1:1) (9CI) see PIK450
PIPERAZINE, 1-(p-TOLYL)-4-(3,4,5-
TRIMETHOXYBENZOYL)- see THF310
PIPERIDIN (GERMAN) see PIL500
PIPERIDINE see PIL500
PIPERIDINE, 1-(m-(BIS(2-
CHLOROETHYL)AMINO)BENZOYL)- see BHP150
PIPERIDINE, 1-(3-(BIS(2-CHLOROETHYL)AMINO)-p-
TOLUOYL)- see BIA100
1-PIPERIDINECARBODITHIOIC ACID, compounded
with PIPERIDINE see PIY500
4-PIPERIDINEMETHANOL, 1-BROMOACETYL-α-α-
DIPHENYL- see BMS300
4-PIPERIDINEMETHANOL, 1-CHLOROACETYL-α-α-
DIPHENYL- see CEC300
4-PIPERIDINEMETHANOL, 1-(IODOACETYL)-α-α-
DIPHENYL- see IDZ200
PIPERIDINE, 1-((1-(2-PHENYLETHYL)-3-
PIPERIDINYL)CARBONYL)- see PDI550
PIPERIDINIUM see PIY500
9-(PIPERIDINOAMINO)ACRIDINE see PIN200
α-(1-PIPERIDINO)-CYCLOHEXYL PHENYL KETONE
see PFY200
2-PIPERIDINOMETHYL-1,4-BENZODIOXAN
HYDROCHLORIDE see BCI500
α-(2-PIPERIDYL)BENZHYDROL HYDROCHLORIDE
see PII750
4-(1-PIPERIDYL)CARBONYL-2,3-
TETRAMETHYLENEQUINOLINE see PIU100
2-(1-PIPERIDYLMETHYL)-1,4-BENZODIOXAN
HYDROCHLORIDE see BCI500
PIPERIN see PIV600
PIPERINE see PIV600
PIPEROCAINE see PIV750

PIPEROCYANOMAZINE see PIW000
PIPERONAL see PIW250
PIPERONALDEHYDE see PIW250
PIPERONYL see FHG000
PIPERONYL ACETATE see PIX000
PIPERONYL ALDEHYDE see PIW250
PIPERONYL BUTOXIDE see PIX250
PIPEROXANE HYDROCHLORIDE see BCI500
1-PIPEROYLPIPERIDINE see PIV600
PIPOLPHEN see DQA400
PIP-PIP see PIY500
PIPRADOL HYDROCHLORIDE see PII750
PIPRADROL HYDROCHLORIDE see PII750
PIRACAPS see TBX250
PIRAMIDON see DOT000
PIRARREUMOL "B" see BRF500
PIREF see DIZ100
PIRIBENZIL see TMP750
PIRID see PDC250
PIRIDACIL see PDC250
PIRIDANE see CMA100
PIRIDINA (ITALIAN) see POP250
PIRIDISIR see PPP500
PIRIDOL see DOT000
PIRIDOSAL see DAM700
PIRIDROL HYDROCHLORIDE see PII750
PIRIEX see TAI500
PIRIMAL-M see ALF250
PIRIMECIDAN see TGD000
PIRIMETAMINA (SPANISH) see TGD000
PIRIMIFOS-METHYL see DIN800
PIRINIXIL see CLW500
PIRITON see TAI500
PIRMAZIN see SNJ000
PIROD see UNJ800
PIROFOS see SOD100
PIROMIDINA see DOT000
PIRYDYNA (POLISH) see POP250
PITAYINE see QFS000
PITC see ISQ000
PITCH see CMZ100
PITCH, COAL TAR see CMZ100
PITREX see TGB475
PITTCHLOR see HOV500
PITTCIDE see HOV500
PITTCLOR see HOV500
PITTSBURGH PX-138 see DVL700
PIVACIN see PIH175
PIVADORM see BNP750
PIVADORN see BNP750
PIVAL see PIH175
PIVALDION (ITALIAN) see PIH175
PIVALDIONE (FRENCH) see PIH175
PIVALIC ACID see PJA500
PIVALIC ACID CHLORIDE see DTS400
PIVALIC ACID LACTONE see DTH000
PIVALOLACTONE see DTH000
PIVALOLYL CHLORIDE see DTS400
PIVALOYL CHLORIDE see DTS400
2-PIVALOYL-INDAAN-1,3-DION (DUTCH) see PIH175
2-PIVALOYL-INDAN-1,3-DION (GERMAN) see PIH175
2-PIVALOYL-1,3-INDANDIONE see PIH175
2-PIVALOYLINDANE-1,3-DIONE see PIH175
PIVALYL CHLORIDE see DTS400
2-PIVALYL-1,3-INDANDIONE see PIH175
PIVALYL VALONE see PIH175
PIVALYN see PIH175
PIXALBOL see CMY800

PIX CARBONIS see CMY800
PIX LITHANTHRACIS see CMY800
PKhNB see PAX000
PLACIDIL see CHG000
PLACIDOL see CJR909
PLACIDOL E see DJX000
PLACIDON see MQU750
PLACIDYL see CHG000
PLANADALIN see BNK000
PLANIUM see PJS750
PLANOCAINE see AIT250
PLANOCHROME see MCV000
PLANOFIX see NAK500
PLANOMIDE see TEX250
PLANT DITHIO AEROSOL see SOD100
PLANTDRIN see MRH209
PLANTFUME 103 SMOKE GENERATOR see SOD100
PLANTIFOG 160M see MAS500
PLANTOMYCIN see SLY500
PLANT PROTEASE CONCENTRATE see BMO000
PLANT PROTECTION PP511 see DIN800
PLANTULIN see PMN850
PLASDONE see PKQ250
PLASKON 201 see PJY500
PLASKON PP 60-002 see PJS750
PLASMASTERIL see HLB400
PLASTANOX 425 ANTIOXIDANT see MJN250
PLASTANOX LTDP see TFD500
PLASTANOX LTDP ANTIOXIDANT see TFD500
PLASTAZOTE X 1016 see PJS750
PLASTER of PARIS see CAX500
PLASTIBEST 20 see ARM268
PLASTOMOLL DOA see AEO000
PLASTORESIN ORANGE F4A see PEJ500
PLASTORESIN RED F see SBC500
PLASTORESIN RED FR see CMS238
PLASTORESIN VIOLET 5BO see AOR500
PLASTRONGA see PJS750
PLASTYLENE MA 2003 see PJS750
PLASTYLENE MA 7007 see PJS750
PLATH-LYSE see MJM500
PLATIBLASTIN see PJD000
cis-PLATIN see PJD000
PLATIN (GERMAN) see PJD500
PLATINATE(2-), HEXACHLORO-, DIHYDROGEN,
HEXAHYDRATE see DLO400
PLATINATE(2-), TETRACHLORO-, DIAMMONIUM see
ANV800
PLATINATE(2-), NITROTRICHLORO-, DIPOTASSIUM
see PLN050
PLATINATE(1-), TRICHLOROETHYLENE-,
DIPOTASSIUM see PLW200
PLATINATE(2-), TRICHLORO(NITRITO-N-),
DIPOTASSIUM (SP-4-2)- see PLN050
PLATINEX see PJD000
PLATINIC AMMONIUM CHLORIDE see ANF250
PLATINIC CHLORIDE see CKO750
PLATINIC POTASSIUM CHLORIDE see PLR000
PLATINOL see PJD000
PLATINOL AH see DVL700
PLATINOL DOP see DVL700
PLATINOUS CHLORIDE see PJE000
cis-PLATINOUS DIAMMINE DICHLORIDE see PJD000
PLATINOUS POTASSIUM CHLORIDE see PJD250
PLATINUM see PJD500
PLATINUM(II), BIS(METHYL SELENIDE)DICHLORO-
, cis- see DEU115

PLATINUM (II), BIS(METHYL SELENIDE)SULFATO-,
HYDRATE see SNS100
PLATINUM BLACK see PJD500
PLATINUM CHLORIDE see PJE000
PLATINUM(IV) CHLORIDE see PJE250
PLATINUM COMPOUNDS see PJE500
cis-PLATINUM(II) DIAMMINEDICHLORIDE see
PJD000
trans-PLATINUM(II)DIAMMINEDICHLORIDE see
DEX000
PLATINUM, DICHLOROBIS(METHYL SELENIDE)-,
cis- see DEU115
PLATINUM,
DICHLOROBIS(SELENOBIS(METHANE))-(SP-4-2) see
DEU115
PLATINUM SPONGE see PJD500
PLATINUM TETRACHLORIDE see PJE250
PLATINUM (2-), NITROTRICHLORO-, DIPOTASSIUM
see PLN050
PLAVOLEX see DBD700
PLAXIDOL see CJR909
PLECYAMIN see VSZ000
PLEGECYL see ABH500
PLEGICIN see ABH500
PLEGOMAZIN see CKP500
PLENASTRIL see PAN100
PLENUR see LGZ000
PLEOCIDE see ABY900
PLESSY'S GREEN (HEMIHEPTAHYDRATE) see
CMK300
PLEXIGLAS see PKB500
PLEXIGUM M 920 see PKB500
PLICTRAN see CQH650
PLIDAN see DCK759
PLIMASINE see MNQ000
PLIOFLEX see SMR000
PLIOGRIP see PKL500
PLIOLITE see BOP100
PLIOLITE S5 see SMR000
PLIOVAC AO see AAX175
PLIOVIC see PKQ059
PLIVA see BIE500
PLIVAPHEN see ABH500
PLLETIA see PMI750
PLOMB FLUORURE (FRENCH) see LDF000
PLOYMANNURONIC ACID see AFL000
PLUCKER see NAK500
PLUMBAGO see CBT500
PLUMBOPLUMBIC OXIDE see LDS000
PLUMBOUS ACETATE see LCJ000
PLUMBOUS ACETATE see LCV000
PLUMBOUS CHLORIDE see LCQ000
PLUMBOUS CHROMATE see LCR000
PLUMBOUS FLUORIDE see LDF000
PLUMBOUS OXIDE see LDN000
PLUMBOUS PHOSPHATE see LDU000
PLUMBOUS SULFIDE see LDZ000
PLURACOL P-410 see PJT000
PLURACOL P-710 see PJT000
PLURACOL P-1010 see PJT000
PLURACOL P-2010 see PJT000
PLURACOL P-3010 see PJT000
PLURACOL P-4010 see PJT000
PLURACOL E see PJT200
PLURACOL E-200 see PJT000
PLURACOL E-300 see PJT000
PLURACOL E-400 see PJT000
PLURACOL E-600 see PJT000

PLURACOL E-1500 see PJT000
PLURACOL E-4000 see PJT000
PLURACOL E-6000 see PJT000
PLURAGARD see MCB000
PLURAGARD C 133 see MCB000
PLUTONIUM COMPOUNDS see PJI000
PLYCTRAN see CQH650
PLYMOUTM IPM see IQN000
PMA see ABU500
PMAC see ABU500
PMACETATE see ABU500
PMAL see ABU500
PMAS see ABU500
PMB see ECU750
PMC see PFM500
P.M.F. see PFN000
PMMA see PKB500
PMP see PHX250
PMP SODIUM GLUCONATE see SHK800
PMS 1.5 see SCR400
PMS 300 see SCR400
PMS 154A see SCR400
PMS 200A see SCR400
PMT see PFS500
PNA see NEO500
PNB see NFQ000
PNCB see NFS525
PNOT see NMP500
PNS 25 see SCR400
PNT see NMO550
PNU see NLO500
POA see PER250
PODOPHYLLIN see PJJ000
PODOPHYLLUM see PJJ000
PODOPHYLLUM RESIN see PJJ000
POE 20 SORBITAN MONOLAURATE see PKG000
POINT TWO see SHF500
POLAAX see OAX000
POLACARITOX see CKM000
POLAMIDON see MDP770
POLAMIDONE see MDO750
POLARONIL (GERMAN) see TAI500
POLAR RED RS see CMM330
POLCOMINAL see EOK000
POLEON see EID000
POLFOSCHLOR see TIQ250
POLICAPRAN see PJY500
POLIFEN see TAI250
POLIGOSTYRENE see SMQ500
POLINALIN see DOT000
POLISEPTIL see TEX250
POLISIN see BKL250
POLITEF see TAI250
POLITEN see PJS750
POLITEN I 020 see PJS750
POLIVAL see TEX000
POLIVINIT see PKQ059
POLLACID see HCQ500
POLONIUM see PJJ750
POLONIUM CARBONYL see PJK000
POLOPIRYNA see ADA725
POLOXALENE see PJK150
POLOXAL RED 2B see HJF500
POLOXAMER 331 see PJK200
POLY see SKN000
POLYACRYLAMIDE RESINS, MODIFIED see MRA075
POLYAETHYLEN see PJS750

POLYAETHYLENGLYCOLE 200 (GERMAN) see
PJT200
POLYAETHYLENGLYKOLE 300 (GERMAN) see
PJT225
POLYAETHYLENGLYKOLE 400 (GERMAN) see
PJT230
POLYAETHYLENGLYKOLE 600 (GERMAN) see
PJT240
POLYAETHYLENGLYKOLE 1500 (GERMAN) see
PJT500
POLYAETHYLENGLYKOLE 4000 (GERMAN) see
PJT750
POLYAETHYLENGLYKOLE 6000 (GERMAN) see
PJU000
POLYAETHYLENGLYKOLE #1000 (GERMAN) see
PJT250
POLYAMIDE 6 see PJY500
POLY(ε-AMINOCAPROIC ACID) see PJY500
POLYARABINOGALACTAN see AQR800
POLYBENZARSOL see BCJ150
POLYBIS(2-ETHYLBUTYL)SILOXANE see EHC900
POLYBOR see SFF000
POLYBROMINATED BIPHENYL see FBU509
POLYBROMINATED BIPHENYL see HCA500
POLYBROMINATED BIPHENYL (FF-1) see FBU509
POLYBROMINATED BIPHENYLS see FBU000
POLYBUTADIENE-POLYSTYRENE COPOLYMER see
SMR000
POLYCAPROAMIDE see PJY500
POLY(ε-CAPROAMIDE) see PJY500
POLYCAPROLACTAM see PJY500
POLY(ε-CAPROLACTAM) see PJY500
POLYCAT 8 see DRF709
POLYCHLORCAMPHENE see CDV100
POLYCHLORINATED BIPHENYL (AROCLOR 1221)
see PJM000
POLYCHLORINATED BIPHENYL (AROCLOR 1232)
see PJM250
POLYCHLORINATED BIPHENYL (AROCLOR 1242)
see PJM500
POLYCHLORINATED BIPHENYL (AROCLOR 1248)
see PJM750
POLYCHLORINATED BIPHENYL (AROCLOR 1254)
see PJN000
POLYCHLORINATED BIPHENYL (AROCLOR 1260)
see PJN250
POLYCHLORINATED BIPHENYL (AROCLOR 1262)
see PJN500
POLYCHLORINATED BIPHENYL (AROCLOR 1268)
see PJN750
POLYCHLORINATED BIPHENYL (AROCLOR 2565)
see PJO000
POLYCHLORINATED BIPHENYL (AROCLOR 4465)
see PJO250
POLYCHLORINATED BIPHENYL (KANECHLOR 300)
see PJO500
POLYCHLORINATED BIPHENYL (KANECHLOR 400)
see PJO750
POLYCHLORINATED BIPHENYL (KANECHLOR 500)
see PJP000
POLYCHLORINATED BIPHENYLS see PJL750
POLYCHLORINATED CAMPHENES see CDV100
POLYCHLOROBIPHENYL see PJL750
POLYCHLOROCAMPHENE see CDV100
POLY(CHLOROETHYLENE) see PKQ059
POLYCILLIN see AIV500
POLYCILLIN see AOD125

POLYCILLIN-N see SEQ000
POLYCIZER DBP see DEH200
POLYCIZER DBS see DEH600
POLYCLAR L see PKQ250
POLYCLENE see DGB000
POLYCO see SJK000
POLYCO 2410 see SMR000
POLYCO 220NS see SMQ500
POLYCRON see BNA750
POLYCYCLIC MUSK see ACL750
POLYCYCLINE HYDROCHLORIDE see TBX250
POLYDESIS see PKP750
POLYDEXTROSE SOLUTION see PJQ430
POLYDIMETHYLSILOXANE see DTR850
POLY-EM 12 see PJS750
POLY-EM 40 see PJS750
POLY-EM 41 see PJS750
POLY(ESTRADIOL PHOSPHATE) see EDS000
POLYETHYLENE see PJS750
POLYETHYLENE AS see PJS750
POLYETHYLENE GLYCOL see PJT000
POLYETHYLENE GLYCOL 200 see PJT200
POLYETHYLENE GLYCOL 300 see PJT225
POLYETHYLENE GLYCOL 400 see PJT230
POLYETHYLENE GLYCOL 600 see PJT240
POLYETHYLENE GLYCOL 1000 see PJT250
POLYETHYLENE GLYCOL 1500 see PJT500
POLYETHYLENE GLYCOL 4000 see PJT750
POLYETHYLENE GLYCOL 6000 see PJU000
POLYETHYLENE GLYCOL DIMETHYL ETHER see
DOM100
POLYETHYLENE GLYCOL DISTEARATE see PJU500
POLYETHYLENE GLYCOL 300 DISTEARATE see
PJU500
POLYETHYLENE GLYCOL 400 (DI) STEARATE see
PJU500
POLYETHYLENE GLYCOL 600 (DI) STEARATE see
PJU500
POLYETHYLENE GLYCOL MONOETHER with p-tert-
OCTYLPHENYL see PKF500
POLYETHYLENE GLYCOL MONO(4-
OCTYLPHENYL) ETHER see PKF500
POLYETHYLENE GLYCOL MONO(4-tert-
OCTYLPHENYL) ETHER see PKF500
POLYETHYLENE GLYCOL MONO(p-tert-
OCTYLPHENYL) ETHER see PKF500
POLYETHYLENE GLYCOL MONOSTEARATE see
PJV250
POLYETHYLENE GLYCOL MONO(p-(1,1,3,3-
TETRAMETHYLBUTYL)PHENYL) ETHER see PKF500
POLYETHYLENE GLYCOL 450 NONYL PHENYL
ETHER see PKF000
POLYETHYLENE GLYCOL OCTYLPHENOL ETHER
see PKF500
POLYETHYLENE GLYCOL p-OCTYLPHENYL
ETHER see PKF500
POLYETHYLENE GLYCOL 450 OCTYL PHENYL
ETHER see PKF500
POLYETHYLENE GLYCOL p-tert-OCTYLPHENYL
ETHER see PKF500
POLYETHYLENE GLYCOL p-1,1,3,3,-
TETRAMETHYLBUTYLPHENYL ETHER see PKF500
POLY(ETHYLENE OXIDE) see PJT000
POLYETHYLENE RESINS see PJS750
POLY(ETHYLENE TETRAFLUORIDE) see TAI250
POLYFENE see TAI250
POLYFER see IGS000
POLYFIBRON 120 see SFO500

POLYFLEX see SMQ500
POLYFLON see TAI250
POLYFOAM PLASTIC SPONGE see PKL500
POLYFOAM SPONGE see PKL500
POLY-G see PJT200
POLY G 400 see PJT230
POLY-GIRON see TEH500
POLYGLUCIN see DBD700
POLYGLYCERATE (60) see EEU100
POLYGLYCEROL ESTERS of FATTY ACIDS see
PJX875
POLYGLYCOL 1000 see PJT250
POLYGLYCOL 4000 see PJT750
POLYGLYCOL DISTEARATE see PJU500
POLYGLYCOL E see PJT200
POLYGLYCOL E1000 see PJT250
POLYGLYCOL E-4000 see PJT750
POLYGLYCOL E-4000 USP see PJT750
POLYGON see SKN000
POLYGRIPAN see TEH500
POLY-G SERIES see PJT000
POLY(IMINOCARBONYLPENTAMETHYLENE) see
PJY500
POLY(IMINO(1-OXO-1,6-HEXANEDIYL)) see PJY500
POLYMALEIC ACID see PJY850
POLYMERS of EPICHLOROHYDRIN and 2,2-BIS(4-
HYDROXYPHENYL)PIPERAZINE see ECM500
POLYMERS, WATER-INSOLUBLE see PKA850
POLYMERS, WATER-SOLUBLE see PKA860
POLYMETHYLENEPOLYPHENYL ISOCYANATE see
PKB100
POLYMETHYLMETHACRYLATE see PKB500
POLYMINE D see BEN000
POLYMIST A12 see PJS750
POLYMONE see DGB000
POLYMUL CS 81 see PJS750
POLYMYXIN see PKC000
POLYMYXIN E SULFATE see PKD300
POLYMYXIN E SULFATE (SALT) see PKD300
POLYOESTRADIOL PHOSPHATE see EDS000
POLYOX see PJT000
POLY(1-(2-OXO-1-PYRROLIDINYL)ETHYLENE) see
PKQ250
POLY(OXY-1,2-ETHANEDIYL), α-HYDRO-omega-
HYDROXY- see PJT000
POLY(OXY-1,2-ETHANEDIYL), α-METHYL-ω-
METHOXY-(9CI) see DOM100
POLYOXYETHYLENE (75) see PJT750
POLYOXYETHYLENE 1500 see PJT500
POLYOXYETHYLENE DIMETHYL ETHER see
DOM100
POLYOXYETHYLENE (20) MONO- and
DIGLYCERIDES of FATTY ACIDS see EEU100
POLYOXYETHYLENE MONO(OCTYLPHENYL)
ETHER see PKF500
POLYOXYETHYLENE-8-MONOSTEARATE see
PJV250
POLYOXYETHYLENE (9) NONYL PHENYL ETHER
see PKF000
POLYOXYETHYLENE (9) OCTYLPHENYL ETHER
see PKF500
POLYOXYETHYLENE (13) OCTYLPHENYL ETHER
see PKF500
POLY(OXYETHYLENE)-p-tert-OCTYLPHENYL
ETHER see PKF500
POLY(OXYETHYLENE)-POLY(OXYPROPYLENE)-
POLY(OXYETHYLENE) POLYMER see PJK150

POLYOXYETHYLENE (20) SORBITAN
MONOLAURATE see PKG000
POLYOXYETHYLENE (20) SORBITAN
MONOLAURATE see PKG000
POLYOXYETHYLENE (20) SORBITAN
MONOLAURATE see PKL000
POLYOXYETHYLENE SORBITAN MONOOLEATE
see PKL100
POLYOXYETHYLENE SORBITAN MONOSTEARATE
see PKL030
POLYOXYETHYLENE 20 SORBITAN
MONOSTEARATE see PKL030
POLYOXYETHYLENE SORBITAN OLEATE see
PKL100
POLYOXYETHYLENE(8)STEARATE see PJV250
POLYOXYMETHYLENE see TMP000
POLYOXYMETHYLENE GLYCOLS see FMV000
POLYPHOS see SHM500
POLYPRO 1014 see PMP500
POLYPROPENE see PMP500
POLYPROPYLENE see PMP500
POLYPROPYLENE GLYCOL see PKI500
POLYPROPYLENE GLYCOL 750 see PKI750
POLYPROPYLENGLYKOL (CZECH) see PKI500
POLYRAM M see MAS500
POLYRAM ULTRA see TFS350
POLYSION N 22 see PJS750
POLYSIZER 173 see PKP750
POLY(SODIUM p-STYRENESULFONATE) see SJK375
POLY-SOLV see CBR000
POLY-SOLV DB see DJF200
POLY-SOLV DM see DJG000
POLY-SOLV EB see BPJ850
POLY-SOLV EE see EES350
POLY-SOLV EE ACETATE see EES400
POLY-SOLV EM see EJH500
POLY-SOLVE MPM see PNL250
POLY-SOLV TE see EFL000
POLYSORBAN 80 see PKL100
POLYSORBATE 20 see PKL000
POLYSORBATE 60 see PKL030
POLYSORBATE 65 see PKL050
POLYSORBATE 65 see SKV195
POLYSORBATE 80 see PKL100
POLYSORBATE 80, U.S.P. see PKL100
POLYSTROL D see SMQ500
POLYSTYRENE see SMQ500
POLYSTYRENE-ACRYLONITRILE see ADY500
POLYSTYRENE BW see SMQ500
POLYSTYRENE LATEX see SMQ500
POLYSTYROL see SMQ500
POLYTAC see PMP500
POLYTAR BATH see CMY800
POLYTEF see TAI250
POLYTETRAFLUOROETHENE see TAI250
POLYTETRAFLUOROETHYLENE see TAI250
POLYTHENE see PJS750
POLYTHERM see PKQ059
POLYTOX see DGB000
POLYURETHANE A see PKL500
POLYURETHANE ESTER FOAM see PKL500
POLYURETHANE ETHER FOAM see PKL500
POLYURETHANE FOAM see PKL500
POLYURETHANE SPONGE see PKL500
POLYURETHANE Y-195 see PKL750
POLYURETHANE Y-217 see PKM000
POLYURETHANE Y-218 see PKM250
POLYURETHANE Y-221 see PKM500

POLYURETHANE Y-222 see PKM750
POLYURETHANE Y-223 see PKN000
POLYURETHANE Y-224 see PKN250
POLYURETHANE Y-225 see PKN500
POLYURETHANE Y-226 see PKN750
POLYURETHANE Y-227 see PKO000
POLYURETHANE Y-238 see CDV625
POLYURETHANE Y-290 see PKO500
POLYURETHANE Y-302 see PKP000
POLYURETHANE Y-304 see PKP250
POLYVIDONE see PKQ250
POLYVINOL see PKP750
POLYVINYL ACETATE (FCC) see AAX250
POLYVINYL ALCOHOL see PKP750
POLY(n-VINYLBUTYROLACTAM) see PKQ250
POLYVINYLCHLORID (GERMAN) see PKQ059
POLYVINYL CHLORIDE see PKQ059
POLYVINYL CHLORIDE−POLYVINYL ACETATE see
AAX175
POLY(VINYLIMIDAZOLE) see VPP100
POLY(1-VINYLIMIDAZOLE) see VPP100
POLY(N-VINYLIMIDAZOLE) see VPP100
POLYVINYLPOLYPYRROLIDONE see PKQ150
POLY(1-VINYL-2-PYRROLIDINONE)
HOMOPOLYMER see PKQ250
POLYVINYLPYRROLIDONE see PKQ250
POLYVINYL SULFATE POTASSIUM SALT see PKS250
POLYVIOL see PKP750
POLYVIOL M 13/140 see PKP750
POLYVIOL MO 5/140 see PKP750
POLYVIOL W 25/140 see PKP750
POLYVIOL W 40/140 see PKP750
POLYWAX 1000 see PJS750
POLY-ZOLE AZDN see ASL750
POMALUS ACID see MAN000
POMARSOL see TFS350
POMARSOL Z FORTE see BJK500
POMASOL see TFS350
POMEX see CBM750
PONCEAU 3R see FAG018
PONCEAU 4R see FMU080
PONCEAU 4R ALUMINUM LAKE see FMU080
PONCEAU BNA see FMU070
PONCEAU INSOLUBLE OLG see OHI200
PONCEAU INSOLUBLE OLG see XRA000
PONCEAU R (BIOLOGICAL STAIN) see FMU070
PONCEAU XYLIDINE (BIOLOGICAL STAIN) see
FMU070
PONCYL see GKE000
PONDERAL see PDM250
PONDERAX see PDM250
PONDIMIN see PDM250
PONECIL see AIV500
PONTACYL RUBINE R see HJF500
PONTACYL SCARLET RR see FMU080
PONTACYL SKY BLUE 4BX see CMO500
PONTALITE see PKB500
PONTAMINE BLACK E see AQP000
PONTAMINE BLACK EBN see AQP000
PONTAMINE BLUE BB see CMO000
PONTAMINE BLUE 3BX see CMO250
PONTAMINE DEVELOPER TN see TGL750
PONTAMINE SKY BLUE see CMN750
PONTAMINE SKY BLUE 5BX see CMO500
POOGIPHALAM, nut extract see BFW000
POP see ELL500
POPROLIN see PMP500

POROFOR 57 see ASL750
POROFOR BSH see BBS300
POROFOR-BSH-PULVER see BBS300
POROFOR CHKHC-18 see DVF400
POROFOR ChKhZ 9 see BBS300
POROLEN see PJS750
POROPHOR B see DVF400
PORTLAND CEMENT see PKS750
PORTLAND CEMENT SILICATE see PKS750
PORTLAND STONE see CAO000
POSTAFEN see HGC500
PO-SYSTOX see DAP200
POTALIUM see PLG800
POTASH see PLA000
POTASH CHLORATE (DOT) see PLA250
POTASORAL see PLG800
POTASSA see PLJ500
POTASSE CAUSTIQUE (FRENCH) see PLJ500
POTASSIO (CHLORATO di) (ITALIAN) see PLA250
POTASSIO (IDROSSIDO di) (ITALIAN) see PLJ500
POTASSIO (PERMANGANATO di) (ITALIAN) see PLP000
POTASSIUM see PKT250
POTASSIUM (liquid alloy) see PKT500
POTASSIUM, metal liquid alloy (DOT) see PKT500
POTASSIUM ACID ARSENATE see ARD250
POTASSIUM ACID FLUORIDE see PKU250
POTASSIUM ACID FLUORIDE (solution) see PKU500
POTASSIUM ACID SULFATE see PKX750
POTASSIUM ANTIMONYL TARTRATE see AQG250
POTASSIUM ANTIMONYL-d-TARTRATE see AQG250
POTASSIUM ANTIMONYL-l-TARTRATE see AQH000
POTASSIUM ANTIMONY TARTRATE see AQG250
POTASSIUM ARSENATE see ARD250
POTASSIUM ARSENATE, MONOBASIC see ARD250
POTASSIUM ARSENITE see PKV500
POTASSIUM BENZOATE see PKW760
POTASSIUM BICHROMATE see PKX250
POTASSIUM BIFLUORIDE see PKU250
POTASSIUM BIFLUORIDE, solution (DOT) see PKU500
POTASSIUM BIFLUORIDE, solid or solution (DOT) see PKU250
POTASSIUM BIS(2-HYDROXYETHYL)DITHIOCARBAMATE see PKX500
POTASSIUM BISULFATE see PKX750
POTASSIUM BISULPHATE see PKX750
POTASSIUM BOROFLUORIDE see PKY000
POTASSIUM BOROHYDRATE see PKY250
POTASSIUM BOROHYDRIDE (DOT) see PKY250
POTASSIUM BROMATE see PKY300
POTASSIUM BROMIDE see PKY500
POTASSIUM CARBONATE (2:1) see PLA000
POTASSIUM CHLORATE see PLA250
POTASSIUM CHLORATE (DOT) see PLA250
POTASSIUM (CHLORATE de) (FRENCH) see PLA250
POTASSIUM CHLORATE (UN 1485) (DOT) see PLA250
POTASSIUM CHLORATE, solution (UN 2427) (DOT) see PLA250
POTASSIUM CHLORIDE see PLA500
POTASSIUM CHLORIDE OXIDE see PLK000
POTASSIUM CHLOROPLATINATE see PLR000
POTASSIUM CHLOROPLATINITE see PJD250
POTASSIUM CHROMATE(VI) see PLB250
POTASSIUM CHROMIC SULFATE see PLB500
POTASSIUM CHROMIC SULPHATE see PLB500
POTASSIUM CHROMIUM ALUM see PLB500
POTASSIUM CITRATE see PLB750
POTASSIUM COPPER(I) CYANIDE see PLC175

POTASSIUM CUPROCYANIDE see PLC175
POTASSIUM CYANATE see PLC250
POTASSIUM CYANIDE see PLC500
POTASSIUM CYANIDE, solution (DOT) see PLC500
POTASSIUM CYANONICKELATE HYDRATE see TBW250
POTASSIUM DICHLOROISOCYANURATE see PLD000
POTASSIUM DICHLORO-s-TRIAZINETRIONE see PLD000
POTASSIUM DICHROMATE(VI) see PKX250
POTASSIUM DICHROMATE, ZINC CHROMATE, and ZINC HYDROXIDE (1:3:1) see ZFJ122
POTASSIUM DICYANOCUPRATE(1-) see PLC175
POTASSIUM DIHYDROGEN ARSENATE see ARD250
POTASSIUM DIOXIDE see PLE260
POTASSIUM DISULPHATOCHROMATE(III) see PLB500
POTASSIUM FERRICYANATE see PLF250
POTASSIUM FERRICYANIDE see PLF250
POTASSIUM FERROCYANATE see TEC500
POTASSIUM FERROCYANIDE see TEC500
POTASSIUM FLUOBORATE see PKY000
POTASSIUM FLUORIDE see PKU250
POTASSIUM FLUORIDE see PLF500
POTASSIUM FLUORIDE, solution (DOT) see PLF500
POTASSIUM FLUOROACETATE see PLG000
POTASSIUM FLUOROBORATE see PKY000
POTASSIUM FLUORURE (FRENCH) see PLF500
POTASSIUM FLUOSILICATE see PLH750
POTASSIUM FLUOTANTALATE see PLH000
POTASSIUM FLUOZIRCONATE see PLG500
POTASSIUM GIBBERELLATE see PLG775
POTASSIUM GLUCONATE see PLG800
POTASSIUM d-GLUCONATE see PLG800
POTASSIUM GLUTAMATE see MRK500
POTASSIUM GLUTAMINATE see MRK500
POTASSIUM HEPTAFLUOROTANTALATE see PLH000
POTASSIUM HEXACHLOROPLATINATE(IV) see PLR000
POTASSIUM HEXACYANOFERRATE see TEC500
POTASSIUM HEXACYANOFERRATE(II) see TEC500
POTASSIUM HEXACYANOFERRATE(III) see PLF250
POTASSIUM HEXAFLUOROARSENATE see PLH500
POTASSIUM HEXAFLUOROSILICATE see PLH750
POTASSIUM HEXAFLUOROTITANATE see PLI000
POTASSIUM HYDRATE (DOT) see PLJ500
POTASSIUM HYDRIDE see PLJ250
POTASSIUM HYDROGEN ARSENATE see ARD250
POTASSIUM HYDROGEN DIFLUORIDE see PKU250
POTASSIUM HYDROGEN FLUORIDE see PKU250
POTASSIUM HYDROGEN FLUORIDE, solution (DOT) see PKU500
POTASSIUM HYDROGEN SULFATE, solid (DOT) see PKX750
POTASSIUM HYDROXIDE see PLJ500
POTASSIUM HYDROXIDE, dry, solid, flake, bead, or granular (DOT) see PLJ500
POTASSIUM HYDROXIDE, liquid or solution (DOT) see PLJ500
POTASSIUM (HYDROXYDE de) (FRENCH) see PLJ500
POTASSIUM HYPERCHLORIDE see PLO500
POTASSIUM HYPOCHLORITE see PLK000
POTASSIUM IODATE see PLK250
POTASSIUM IODIDE see PLK500
POTASSIUM IODOHYDRARGYRATE see NCP500
POTASSIUM ISOCYANATE see PLC250
POTASSIUM ISOTHIOCYANATE see PLV750

POTASSIUM MANGANOCYANIDE see TMX600
POTASSIUM MERCURIC IODIDE see NCP500
POTASSIUM METAARSENITE see PKV500
POTASSIUM METABISULFITE (DOT, FCC) see PLR250
POTASSIUM METAVANADATE see PLK810
POTASSIUM METAVANADATE see PLK810
POTASSIUM MONOCHLORIDE see PLA500
POTASSIUM MONOHYDROGEN DIFLUORIDE see PKU250
POTASSIUM MONOSULFIDE see PLT250
POTASSIUM NITRATE see PLL500
POTASSIUM NITRATE mixed with CHARCOAL and SULFUR (15:3:2) see PLL750
POTASSIUM NITRATE mixed with SODIUM NITRITE see PLM000
POTASSIUM NITRATE mixed (fused) with SODIUM NITRITE (DOT) see PLM000
POTASSIUM NITRITE (1:1) see PLM500
POTASSIUM NITRITE (DOT) see PLM500
POTASSIUM NITROTRICHLOROPLATINATE see PLN050
POTASSIUM cis-9-OCTADECENOIC ACID see OHY000
POTASSIUM OLEATE see OHY000
POTASSIUM OXYMURIATE see PLA250
POTASSIUM PERCHLORATE see PLO500
POTASSIUM PERCHLORATE see PLO500
POTASSIUM PERCHLORATE, solid or solution (DOT) see PLO500
POTASSIUM PERMANGANATE see PLP000
POTASSIUM (PERMANGANATE de) (FRENCH) see PLP000
POTASSIUM PEROXIDE see PLP250
POTASSIUM PEROXYDISULFATE see DWQ000
POTASSIUM PEROXYDISULPHATE see DWQ000
POTASSIUM PERSULFATE (DOT) see DWQ000
POTASSIUM PLATINIC CHLORIDE see PLR000
POTASSIUM PLATINOCHLORIDE see PJD250
POTASSIUM POLY(VINYL SULFATE) see PKS250
POTASSIUM PYROSULFITE see PLR250
POTASSIUM RHODANATE see PLV750
POTASSIUM RHODANIDE see PLV750
POTASSIUM SELENATE see PLR750
POTASSIUM SELENITE (7CI) see SBO100
POTASSIUM SELENOCYANATE see PLS000
POTASSIUM SILICOFLUORIDE (DOT) see PLH750
POTASSIUM SILVER CYANIDE see PLS250
POTASSIUM SODIUM ALLOY see PLS500
POTASSIUM SORBATE see PLS750
POTASSIUM STANNATE TRIHYDRATE see PLS760
POTASSIUM STEARATE see PLS775
POTASSIUM SULFATE (2:1) see PLT000
POTASSIUM SULFIDE (2:1) see PLT250
POTASSIUM SULFIDE, anhydrous (UN 1382) (DOT) see PLT250
POTASSIUM SULFIDE, hydrated with not <30% water of crystallization (UN 1847) (DOT) see PLT250
POTASSIUM SULFIDE with <30% water of crystallization (UN 1382) (DOT) see PLT250
POTASSIUM SULFITE see PLT500
POTASSIUM SULFOCYANATE see PLV750
POTASSIUM SUPEROXIDE (DOT) see PLE260
POTASSIUM TELLURITE see PLU000
POTASSIUM TETRABROMOPLATINATE see PLU250
POTASSIUM TETRACHLOROPLATINATE(II) see PJD250
POTASSIUM TETRACYANOMERCURATE(II) see PLU500
POTASSIUM TETRACYANONICKELATE see NDI000

POTASSIUM TETRACYANONICKELATE(II) see NDI000
POTASSIUM TETRAIODOMERCURATE(II) see NCP500
POTASSIUM THIOCYANATE see PLV750
POTASSIUM THIOCYANIDE see PLV750
POTASSIUM TRICHLOROETHYLENEPLATINATE see PLW200
POTASSIUM TRIIODIDE see PLW285
POTASSIUM TROCLOSENE see PLD000
POTASSIUM, metal alloys (UN 1420) (DOT) see PKT250
POTASSIUM VANADIUM TRIOXIDE see PLK810
POTASSIUM ZINC CHROMATE see PLW500
POTASSIUM ZINC CHROMATE see CMK400
POTASSIUM ZINC CHROMATE HYDROXIDE see PLW500
POTASSURIL see PLG800
POTATO ALCOHOL see EFU000
POTAVESCENT see PLA500
POTCRATE see PLA250
POTENTIATED ACID GLUTARALDEHYDE see GFQ000
POTIDE see PLK500
POTOMAC RED see CHP500
POUNCE see AHJ750
POVAL 117 see PKP750
POVAL 120 see PKP750
POVAL 203 see PKP750
POVAL 205 see PKP750
POVAL 217 see PKP750
POVAL 1700 see PKP750
POVAL C 17 see PKP750
POVIDONE (USP XIX) see PKQ250
POWDER GREEN see COF500
POYAMIN see VSZ000
PP 192 see CEX800
PP 321 see LAS200
PP511 see DIN800
PP 557 see AHJ750
PP 563 see GJU600
PPA see NNM000
PPC 3 see MIA250
PPD see PEY500
PPE 2 see PJS750
PPE201 see PKO500
PP FACTOR see NCQ900
PP-FACTOR see NCR000
P.P. FACTOR-PELLAGRA PREVENTIVE FACTOR see NCQ900
P.P.G. 750 see PKI750
PPZEIDAN see DAD200
PQD see DVR200
PRACARBAMIN see UVA000
PRACARBAMINE see UVA000
PRADUPEN see BDY669
PRAECIRHEUMIN see BRF500
PRAJMALINE BITARTRATE see DNB000
PRAJMALINE HYDROGEN TARTRATE see DNB000
PRALIDOXIME IODIDE see POS750
PRALIDOXIME METHIODIDE see POS750
PRALLETHRIN see THI500
PRALUMIN see TDA500
PRAPARAT 5968 see HGP500
PRASEODYMIUM CHLORIDE see PLX750
PRASEODYMIUM(III) NITRATE (1:3) see PLY250
PRECEPTIN see PKF500
PRECIPITATED BARIUM SULPHATE see BAP000
PRECIPITATED CALCIUM SULFATE see CAX750

PRECIPITATED SILICA see SCL000
PRECIPITATED SULFUR see SOD500
PRECIPITE BLANC see MCW000
PRECOCENE 2 see AEX850
PRECOCENE II see AEX850
PRECORT see PLZ000
PRECORTANCYL see PMA000
PRECORTISYL see PMA000
PREDENT see SHF500
PREDNE-DOME see PMA000
PREDNELAN see PMA000
PREDNICEN-M see PLZ000
PREDNILONGA see PLZ000
PREDNIS see PMA000
PREDNI-SEDIV see TEH500
PREDNISOLONE see PMA000
PREDNISON see PLZ000
PREDNISONE see PLZ000
PREDNIZON see PLZ000
PREDONIN see PMA000
PREDONINE see PMA000
PREEGLONE see DWX800
PREFEMIN see CHJ750
PREFRIN see SPC500
1,4-PREGNADIENE-17-α,21-DIOL-3,11,20-TRIONE see PLZ000
1,4-PREGNADIENE-3,20-DIONE-11-β,17-α,21-TRIOL see PMA000
1,4-PREGNADIENE-11-β,17-α,21-TRIOL-3,20-DIONE see PMA000
3,20-PREGNENE-4 see PMH500
17-α-PREGN-4-ENE-21-CARBOXYLIC ACID, 1-HYDROXY-7-α-MERCAPTO-3-OXO-α-LACTONE see AFJ500
PREGNENEDIONE see PMH500
PREGNENE-3,20-DIONE see PMH500
PREGN-4-ENE-3,20-DIONE see PMH500
4-PREGNENE-3,20-DIONE see PMH500
Δ^4-PREGNENE-3,20-DIONE see PMH500
14-α-PREGN-4-ENE-3,20-DIONE, 17-α,21-DIHYDROXY-, 21-IODOACETATE see DNA850
4-PREGNENE-11-β,17-α,21-TRIOL 3,20-DIONE see CNS750
PRELUDIN HYDROCHLORIDE see MNV750
PREMALIN see CKD500
PREMALIN see DGD600
PREMALOX see CBM000
PREMARIN see PMB000
PREMARIN see ECU750
PREMAZINE see BJP000
PREMERGE PLUS see BQI000
PREMGARD see BEP500
PREMODRIN see DBA800
PRENIMON see TND000
PRENTOX see PIX250
PRENYL ACETATE see DOQ350
PRENYL BENZOATE see MHU150
PRENYL SALICYLATE see PMB600
PREPALIN see VSK600
PREPARATION 125 see DFT800
PREPARATION AF see HEI500
PRESAMINE see DLH630
PRESERVAL M see HJL500
PRESERVAL P see HNU500
PRESERV-O-SOTE see CMY825
PRESFERSUL see FBO000

PRESINOL see DNA800
PRESOLISIN see DNA800
PRESOMEN see ECU750
PRESPERSION, 75 UREA see USS000
PRESSIMEDIN see RDK000
PRESSOMIN HYDROCHLORIDE see MDW000
PRETONINE see HOA575
PREVANGOR see PBC250
PREVENTAL see MJM500
PREVENTOL see MJM500
PREVENTOL CMK see CFD990
PREVENTOL GD see MJM500
PREVENTOL GDC see MJM500
PREVENTOL I see TIV750
PREVENTOL O EXTRA see BGJ250
PREVENTOL-ON see BGJ750
PREVENTOL ON & ON EXTRA see BGJ750
PREZERVIT see DSB200
PRIADEL see LGZ000
PRIGLONE see PAI990
PRILEPSIN see DBB200
PRILTOX see PAX250
PRIMACIONE see DBB200
PRIMACLONE see DBB200
PRIMACOL see NAK500
PRIMACONE see DBB200
PRIMAKTON see DBB200
PRIMARY AMYL ACETATE see AOD725
PRIMARY AMYL ALCOHOL see AOE000
PRIMARY DECYL ALCOHOL see DAI600
PRIMARY OCTYL ALCOHOL see OEI000
PRIMARY SODIUM PHOSPHATE see SJH100
PRIMATENE MIST see VGP000
PRIMATOL see ARQ725
PRIMATOL P see PMN850
PRIMATOL Q see BKL250
PRIMATOL S see BJP000
PRIMAZE see ARQ725
PRIMAZIN see SNJ000
PRIMENE JM-T see PMC400
PRIMENE 81-R see CAY710
PRIMIDON see DBB200
PRIMIDONE see DBB200
PRIMIER DIESEL FUEL see DHE850
PRIMIN see DSK200
PRIMINE see PMI750
PRIMODOS see EEH520
PRIMOFOL see EDO000
PRIMOGYN B see EDP000
PRIMOGYN BOLEOSUM see EDP000
PRIMOGYN I see EDP000
PRIMOL 335 see MQV750
PRIMOTEST see TBF500
PRIMROSE YELLOW see ZFJ100
PRINCILLIN see AOD125
PRINCIPAL BILE PIGMENT see HAO900
PRINCIPEN see AIV500
PRINCIPEN/N see SEQ000
PRINTEL'S see SMQ500
PRINTEX see CBT750
PRINTEX 60 see CBT750
PRIODERM see MAK700
PRISCOL see BBJ750
PRISCOLINE HYDROCHLORIDE see BBJ750
PRIST see EJH500
PRISTACIN see CCX000
PRISTINAMYCIN see VRF000
(99-126)-hmn PROATRIOPEPTIN see HGL680

PROAZAIMINE see DQA400
PROAZAMINE see DQA400
PROBAN see CQL250
PRO-BAN M see TEH500
PROBEDRYL see BBV500
PROBESE-P HYDROCHLORIDE see MNV750
PROCAINE HYDROCHLORIDE see AIT250
PROCALM see BCA000
PROCALMIDOL see MQU750
PROCARBAZIN (GERMAN) see PME500
PROCARBAZINE HYDROCHLORIDE see PME500
PROCARDINE see DJS200
PROCASIL see PNX000
PROCENE UF 1.5 see PJS750
PROCHLOROPERAZINE see PMF500
PROCHLORPEMAZINE see PMF500
PROCHLORPERAZINE see PMF500
PROCHLORPROMAZINE see PMF500
PROCIT see DQA400
PROCORMAN see DJS200
PROCTIN see SEH000
PROCYTOX see CQC500
PROCYTOX see CQC650
PRODALUMNOL see SEY500
PRODALUMNOL DOUBLE see SEY500
PRODAN see DXE000
PRODARAM see BJK500
PRODHYBASE ETHYL see EJM500
PRODICTAZIN see DIR000
PRODIERAZINE see DIR000
PRODOX 131 see IQX100
PRODOX 133 see IQZ000
PRODOX 156 see DCI000
PRODUCT 308 see HCP000
PRODUCT No. 161 see SIB600
PROFAM see CBM000
PROFAMINA see BBK000
PROFARMIL see TEH500
PROFAX see PMP500
PROFAX A 60-008 see PJS750
PROFECUNDIN see VSZ450
PROFEMIN see CHJ750
PROFENAMINA (ITALIAN) see DIR000
PROFENAMINUM see DIR000
PROFENID see BDU500
PROFENOFOS see BNA750
PROFENONE see ELL500
PROFERRIN see IHG000
PROFLAVIN see DBN600
PROFLAVINE see DBN600
PROFLAVINE HYDROCHLORIDE see PMH250
PROFLAVINE MONOHYDROCHLORIDE see PMH250
PROFOLIOL see DBN600
PROFOLIOL see EDO000
PROFORMIPHEN see DBN600
PROFUME A see CKN500
PROFUME (OBS.) see MHR200
PROFUNDOL see DBN600
PROFURA see DBN600
PROGALLIN LA see DXX200
PROGALLIN P see PNM750
PROGARMED see DBN600
PROGEKAN see PMH500
PRO-GEN see DBN600
PRO-GEN SODIUM see ARA500
PROGESIC see DBN600
PROGESTEROL see PMH500
PROGESTERONE see PMH500

β-PROGESTERONE see PMH500
PROGESTERONUM see PMH500
PROGESTIN see PMH500
PROGESTONE see PMH500
PRO-GIBB see GEM000
PROGUANIL see CKB250
PROGYNON see EDO000
PROGYNON see EDS100
PROGYNON B see EDP000
PROGYNON BENZOATE see EDP000
PROGYNON-DEPOT see EDS100
PROGYNON-DH see EDO000
PROGYNON-DP see EDR000
PROGYNOVA see EDS100
PROHEPTADIEN MONOHYDROCHLORIDE see EAI000
PROKARBOL see DUS700
PROKAYVIT see MMD500
PROLATE see PHX250
PROLIDON see PMH500
l-PROLINE see PMH900
PROLONGAL see IGS000
PROMACID see CKP500
PROMANTINE see PMI750
PROMAPAR see CKP500
PROMAR see DVV600
PROMARIT see ECU750
PROMAXON P60 see CAW850
PROMAZINAMIDE see DQA400
PROMAZINE HYDROCHLORIDE see PMI500
PROMETASIN see DQA400
PROMETAZIN see DQA400
PROMETHAZINE N-(2'-DIMETHYLAMINO-2'-METHYLETHYL)PHENOTHIAZINE HYDROCHLORIDE see PMI750
PROMETHAZINE HYDROCHLORIDE see PMI750
PROMETHIAZIN (GERMAN) see PMI750
PROMETHIAZINE see DQA400
PROMETREX see BKL250
PROMETRIN see BKL250
PROMETRYN see BKL250
PROMETRYNE (USDA) see BKL250
PROMEZATHINE see DQA400
PROMIBEN see DLH630
PROMIDIONE see GIA000
PROMINAL see ENB500
PROMIT see DBD700
PROMOTESTON see TBF500
PROMPTONAL see EOK000
PROMUL 5080 see HKJ000
PROMYR see IQN000
PRONTALBIN see SNM500
PRONTOSIL I see SNM500
PROPACIL see PNX000
PROPADRINE see NNM000
PROPADRINE HYDROCHLORIDE see PMJ500
PROPAL see CIR500
PROPALDEHYDE see PMT750
PROPAMINE D see TDQ750
PROPANAL see PMT750
PROPANAL, 2-METHYL-2-(METHYLSULFONYL)-o,o'-(THIOBIS((METHYLIMINO)CARBONYL))DIOXIME see BKT300
PROPANAL, 3-(METHYLTHIO)-(9CI) see MPV400
PROPANAL, 3-(METHYLTHIO)-(9CI) see TET900
PROPANAMINE see PND250
2-PROPANAMINE see INK000

DIHYDRO-2-IMINO-4-METHYLPYRIDINO)- see PMM300
PROPANOLE (GERMAN) see PND000
PROPANOLEN (DUTCH) see PND000
PROPANOLI (ITALIAN) see PND000
PROPANOLIDE see PMT100
PROPANOL, 2-METHYL-2-(METHYLTHIO)-, o-((METHYL(4-MORPHOLINYLTHIO)AMINO)CARBONYL)OXIME see MLX820
2-PROPANOL NITRITE see IQQ000
1-PROPANOL, 3-PHENYL-, PROPIONATE see HHQ550
PROPANONE see ABC750
2-PROPANONE see ABC750
1-PROPANONE, 3-(DIMETHYLAMINO)-1-INDOL-3-YL- see DPT300
1-PROPANONE, 3-(DIMETHYLAMINO)-1-(1H-INDOL-3-YL)- see DPT300
2-PROPANONE, HEXACHLORO- see HCL500
1-PROPANONE, 1-(4-(3-HYDROXYPHENYL)-1-METHYL-4-PIPERIDINYL)-(9CI) see KFK000
1-PROPANONE, 1-(4-(m-HYDROXYPHENYL)-1-METHYL-4-PIPERIDYL)- see KFK000
1-PROPANONE, 1-p-MENTH-6-EN-2-YL- see MCF525
PROPANOYL CHLORIDE see PMW500
PROPAPHEN see CKP500
PROPAPHENIN HYDROCHLORIDE see CKP500
PROPARGITE (DOT) see SOP000
PROPARGYL ALCOHOL see PMN450
PROPARGYL BROMIDE see PMN500
PROPASIN see PMN850
PROPASOL SOLVENT M see PNL250
PROPASTE 6708 see TAV750
PROPATHENE see PMP500
PROPAZINE see PMN850
PROPAZONE see PMO250
PROPEL see LAG000
PROPELLANT 12 see DFA600
PROPELLANT 22 see CFX500
PROPELLANT 114 see FOO509
PROPELLANT C318 see CPS000
2-PROPENAL see ADR000
PROP-2-EN-1-AL see ADR000
PROPENAL (CZECH) see ADR000
PROPENAL DIETHYL ACETAL see DHH800
2-PROPENAL, 2-METHYL- see MGA250
PROPENAMIDE see ADS250
2-PROPENAMIDE see ADS250
2-PROPENAMIDE, N-METHYL- (9CI) see MGA300
2-PROPENAMIDE, N-(1-METHYLETHYL)-(9CI) see INH000
2-PROPENAMINE see AFW000
2-PROPEN-1-AMINE see AFW000
PROPENE see PMO500
1-PROPENE see PMO500
PROPENE ACID see ADS750
1-PROPENE, 3-BROMO-1-CHLORO- see BNA880
1-PROPENE, DICHLORO- see DGG700
1-PROPENE, DICHLORO- see DGI700
1-PROPENE, 3,3-DIETHOXY-(9CI) see DHH800
1-PROPENE HOMOPOLYMER (9CI) see PMP500
1-PROPENE, 3-IODO-(9CI) see AGI250
PROPENENITRILE see ADX500
2-PROPENENITRILE see ADX500
2-PROPENENITRILE, (2-HYDROXYETHYL)- see HKY650

2-PROPENENITRILE POLYMER with ETHENYLBENZENE see ADY500
PROPENE OXIDE see PNL600
PROPENE POLYMERS see PMP500
PROPENE TETRAMER see PMP750
2-PROPENE-1-THIOL see AGJ500
PROPENOIC ACID see ADS750
2-PROPENOIC ACID (9CI) see ADS750
2-PROPENOIC ACID, 2-CYANO-, ETHYL ESTER see EHP700
2-PROPENOIC ACID, 1,2-ETHANEDIYLBIS(OXY-2,1-ETHANEDIYL) ESTER (9CI) see TJQ100
2-PROPENOIC ACID-1,2-ETHANEDIYL ESTER see EIP000
2-PROPENOIC ACID-2-ETHYLBUTYL ESTER see EGZ000
2-PROPENOIC ACID, ETHYL ESTER (MAK) see EFT000
2-PROPENOIC ACID-2-HYDROXYPROPYL ESTER see HNT600
2-PROPENOIC ACID, 2-METHYL-(9CI) see MDN250
PROPENOIC ACID METHYL ESTER see MGA500
2-PROPENOIC ACID METHYL ESTER see MGA500
2-PROPENOIC ACID, 2-METHYL-, METHYL ESTER see MLH750
2-PROPENOIC ACID-2-METHYLPROPYL ESTER see IIK000
2-PROPENOIC ACID OXIRANYLMETHYL ESTER see ECH500
2-PROPENOIC ACID, OXYBIS(2,1-ETHANEDIYLOXY-2,1-ETHANEDIYL)ESTER see ADT050
2-PROPENOIC ACID, 2-PHENOXYETHYL ESTER see PER250
PROPENOL see AFV500
PROPEN-1-OL-3 see AFV500
1-PROPEN-3-OL see AFV500
2-PROPEN-1-OL see AFV500
2-PROPEN-1-ONE see ADR000
2-PROPEN-1-ONE, 3-(4-METHYLPHENYL)-1-PHENYL- see MIF762
2-PROPENOPHENONE see PMQ250
1-PROPEN-2-YL ACETATE see MQK750
2-PROPENYLACRYLIC ACID see SKU000
PROPENYL ALCOHOL see AFV500
2-PROPENYL ALCOHOL see AFV500
4-PROPENYLANISOLE see PMQ750
p-PROPENYLANISOLE see PMQ750
p-1-PROPENYLANISOLE see PMQ750
5-(1-PROPENYL)-1,3-BENZODIOXOLE see IRZ000
5-(2-PROPENYL)-1,3-BENZODIOXOLE see SAD000
4-PROPENYLCATECHOL METHYLENE ETHER see IRZ000
PROPENYL CHLORIDE see PMR750
2-PROPENYL CHLORIDE see AGB250
PROPENYL CINNAMATE see AGC000
PROPENYL ETHER see DBK000
PROPENYLGUAETHOL (FCC) see IRY000
4-PROPENYLGUAIACOL see IKQ000
2-PROPENYL HEPTANOATE see AGH250
2-PROPENYL-N-HEXANOATE see AGA500
2-PROPENYL ISOTHIOCYANATE see AGJ250
2-PROPENYL ISOVALERATE see ISV000
2-PROPENYL 3-METHYLBUTANOATE see ISV000
4-PROPENYL-1,2-METHYLENEDIOXYBENZENE see IRZ000
((2-PROPENYLOXY)METHYL)OXIRANE see AGH150
2-PROPENYL PHENYLACETATE see PMS500

p-PROPENYLPHENYL METHYL ETHER see PMQ750
N-2-PROPENYL-2-PROPEN-1-AMINE see DBI600
4-PROPENYL VERATROLE see IKR000
PROPERICIAZINE see PIW000
PROPETAMPHOS see MKA000
PROPHAM see CBM000
PROPHENPYRIDAMINE HYDROCHLORIDE see PMS900
PROPHOS see EIN000
PROPILTHIOURACIL see PNX000
6-PROPIL-TIOURACILE (ITALIAN) see PNX000
PROPINE see MFX590
γ-PROPIOBUTYROLACTONE see HBA550
PROPIOCINE see EDH500
PROPIOKAN see TBG000
PROPIOLACTONE see PMT100
3-PROPIOLACTONE see PMT100
1,3-PROPIOLACTONE see PMT100
β-PROPIOLACTONE see PMT100
PROPIOLIC ACID, TRIPHENYLSTANNYL ESTER see TMW600
PROPIONALDEHYDE see PMT750
PROPIONALDEHYDE, 3-(METHYLTHIO)- see MPV400
PROPIONALDEHYDE, 3-(METHYLTHIO)- see TET900
PROPIONAMIDE, N,N-DIMETHYL-3-(PYRROLIDIN-1-YL)- see DTV330
PROPIONAN SODNY see SJL500
PROPIONATE d'ETHYLE (FRENCH) see EPB500
PROPIONATE de METHYLE (FRENCH) see MOT000
PROPIONE see DJN750
PROPIONIC ACID see PMU750
PROPIONIC ACID (ACGIH,DOT,OSHA) see PMU750
PROPIONIC ACID ANHYDRIDE see PMV500
PROPIONIC ACID CHLORIDE see PMW500
PROPIONIC ACID, CINNAMYL ESTER see CMR850
PROPIONIC ACID, 2,3-DIBROMO-, ETHYL ESTER see EHY050
PROPIONIC ACID-3,4-DICHLOROANILIDE see DGI000
PROPIONIC ACID, 3,3'-DISELENODI-, SODIUM SALT see DWZ100
PROPIONIC ACID, ETHYL ESTER see EPB500
PROPIONIC ACID GRAIN PRESERVER see PMU750
PROPIONIC ACID, 2-HYDROXY- see LAG000
PROPIONIC ACID, ISOBUTYL ESTER see PMV250
PROPIONIC ACID, 2-METHYL- see IJU000
PROPIONIC ACID, 2-METHYLENE- see MDN250
PROPIONIC ACID, PENTYL ESTER (6CI,7CI,8CI) see AON350
PROPIONIC ACID, PHENYLMERCURY SALT see PFO000
PROPIONIC ACID, 3-SELENINO- see SBN510
PROPIONIC ALDEHYDE see PMT750
PROPIONIC ANHYDRIDE see PMV500
PROPIONIC CHLORIDE see PMW500
PROPIONIC ETHER see EPB500
PROPIONIC NITRILE see PMV750
PROPIONITRILE, 2-HYDROXY- see LAQ000
β-PROPIONOLACTONE see PMT100
PROPIONONITRILE see PMV750
PROPIONYLBENZENE see EOL500
PROPIONYL CHLORIDE see PMW500
N-PROPIONYLINDOLE see ICW100
PROPIONYL OXIDE see PMV500
PROPIONYL PEROXIDE (DOT) see DWQ800
p-PROPIONYLPHENOL see ELL500

PROPIOPHENONE see EOL500
PROPIOPHENONE, 4'-(DIMETHYLAMINO)-3-(4-PHENYL-1,2,3,6-TETRAHYDRO-1-PYRIDYL)- see DPS700
PROPIOPHENONE, 4'-PHENYL- BGM100
PROPISAMINE see BBK000
PROPITAN see FHG000
PROP-JOB see DGI000
PROPOFOL see DNR800
PROPOLIN see PMP500
PROPON see TIX500
PROPONEX-PLUS see CIR500
PROPOPHANE see PMP500
PROPOX see PNA500
PROPOXYCHEL see PNA500
2-PROPOXYETHANOL see PNG750
(+)-PROPOXYPHENE see DAB879
d-PROPOXYPHENE see DAB879
PROPOXYPHENE HYDROCHLORIDE see PNA500
(+)-PROPOXYPHENE HYDROCHLORIDE see PNA500
d-PROPOXYPHENE HYDROCHLORIDE see PNA500
α-PROPOXYPHENE HYDROCHLORIDE see PNA500
α-d-PROPOXYPHENE HYDROCHLORIDE see PNA500
d-PROPOXYPHENE MONOHYDROCHLORIDE see PNA500
β-PROPRIOLACTONE (OSHA) see PMT100
β-PROPROLACTONE see PMT100
PROPROP see DGI400
PROPYCIL see PNX000
n-PROPYL ACETAL see AAG850
PROPYL ACETATE see PNC250
1-PROPYL ACETATE see PNC250
2-PROPYL ACETATE see INE100
n-PROPYL ACETATE see PNC250
PROPYLACETIC ACID see VAQ000
N-PROPYLAJMALINE BITARTRATE see DNB000
N-PROPYLAJMALINE HYDROGEN TARTRATE see DNB000
N-PROPYLAJMALINIUM BITARTRATE see DNB000
N-PROPYLAJMALINIUMHYDROGENTARTRAT (GERMAN) see DNB000
N⁴-PROPYLAJMALINIUM HYDROGEN TARTRATE see DNB000
PROPYL ALCOHOL see PND000
1-PROPYL ALCOHOL see PND000
n-PROPYL ALCOHOL see PND000
sec-PROPYL ALCOHOL (DOT) see INJ000
PROPYL ALDEHYDE see PMT750
i-PROPYLALKOHOL (GERMAN) see INJ000
n-PROPYL ALKOHOL (GERMAN) see PND000
PROPYLAMINE see PND250
2-PROPYLAMINE see INK000
N-PROPYLAMINE see PND250
sec-PROPYLAMINE see INK000
PROPYLAMINE, 1,2-DIMETHYL- see AOE200
PROPYLAMINE, 3,3'-IMINOBIS- see AIX250
PROPYLAMINE, 3,3'-(TETRAMETHYLENEDIOXY)BIS- see BGU600
4-PROPYLANISOLE see PNE250
4-n-PROPYLANISOLE see PNE250
p-n-PROPYL ANISOLE see PNE250
n-PROPYLBENZENE see IKG000
PROPYL BENZENE (DOT) see IKG000
5-PROPYL-1,3-BENZODIOXOLE see DMD600
PROPYL BROMIDE see BNX750
PROPYL CARBAMATE see PNG250

N-PROPYL CARBAMATE see PNG250
PROPYLCARBINOL see BPW500
N-PROPYLCARBINYL CHLORIDE see BQQ750
PROPYL CELLOSOLVE see PNG750
N-PROPYL CHLORIDE see CKP750
PROPYL CHLOROCARBONATE see PNH000
PROPYL CHLOROFORMATE see PNH000
n-PROPYL CHLOROFORMATE (DOT) see PNH000
PROPYL CYANIDE see BSX250
N-(4-PROPYLCYCLOHEXYL)BENZAMIDE see PNH522
N-(4-PROPYLCYCLOHEXYL)-3-(3,4,5-TRIMETHOXYPHENYL)-2-PROPENAMIDE see PNH533
PROPYLDIPHENYLPHOSPHINE see PNI600
PROPYL DISELENIDE see PNI850
PROPYLENE (DOT, ACGIH) see PMO500
PROPYLENE ALDEHYDE see ADR000
PROPYLENE ALDEHYDE see COB260
PROPYLENE CHLORIDE see PNJ400
PROPYLENECHLOROHYDRIN see CKR500
PROPYLENEDIAMINE see PMK250
1,3-PROPYLENEDIAMINE see PMK500
PROPYLENE DIAMINE (DOT) see PMK250
PROPYLENE DICHLORIDE see PNJ400
α,β-PROPYLENE DICHLORIDE see PNJ400
4,4'-PROPYLENEDI-2,6-PIPERAZINEDIONE see RCA375
PROPYLENE EPOXIDE see PNL600
1,2-PROPYLENE GLYCOL see PML000
α-PROPYLENEGLYCOL see PML000
PROPYLENE GLYCOL ALGINATE see PNJ750
PROPYLENE GLYCOL DINITRATE see PNL000
PROPYLENE GLYCOL-1,2-DINITRATE see PNL000
1,2-PROPYLENE GLYCOL DINITRATE see PNL000
PROPYLENE GLYCOL (FCC) see PML000
PROPYLENE GLYCOL LACTOSTEARATE see LAR400
PROPYLENE GLYCOL METHYL ETHER see PNL250
PROPYLENE GLYCOL MONOACRYLATE see HNT600
PROPYLENE GLYCOL MONO- and DIESTERS see PNL225
PROPYLENE GLYCOL MONO- and DIESTERS of FATTY ACIDS see PNL225
PROPYLENE GLYCOL MONOMETHYL ETHER see PNL250
PROPYLENE GLYCOL MONOMETHYL ETHER see PNL250
α-PROPYLENE GLYCOL MONOMETHYL ETHER see PNL250
PROPYLENE GLYCOL MONOMETHYL ETHER ACETATE see PNL265
PROPYLENE GLYCOL MONOMETHYL ETHER (ACGIH,OSHA) see PNL250
PROPYLENE GLYCOL MONOSTEARATE see PNL225
PROPYLENE GLYCOL USP see PML000
PROPYLENE IMINE see PNL400
1,2-PROPYLENEIMINE see PNL400
PROPYLENE IMINE, INHIBITED (DOT) see PNL400
PROPYLENE OXIDE see PNL600
1,2-PROPYLENE OXIDE see PNL600
1,3-PROPYLENE OXIDE see OMW000
PROPYLENE OXIDE and ETHYLENE OXIDE BLOCK POLYMER see PJK200
PROPYLENE OXIDE HEXAFLUORIDE see HDF050
PROPYLENE POLYMER see PMP500
PROPYLENE TETRAMER see PMP750

PROPYLENGLYKOL-MONOMETHYLAETHER see PNL250
PROPYLESTER KYSELINY DUSICNE see PNQ500
PROPYLESTER KYSELINY MRAVENCI see PNM500
PROPYLESTER KYSELINY OCTOVE see PNC250
n-PROPYL ESTER of 3,4,5-TRIHYDROXYBENZOIC ACID see PNM750
PROPYL ETHER see PNM000
β-PROPYL-α-ETHYLACROLEIN see BRI000
PROPYL ETHYL ETHER see EPC125
n-PROPYL FORMATE see PNM500
PROPYL FORMATE (DOT) see PNM500
PROPYLFORMIC ACID see BSW000
PROPYL GALLATE see PNM750
n-PROPYL GALLATE see PNM750
PROPYL HYDRIDE see PMJ750
PROPYL p-HYDROXYBENZOATE see HNU500
n-PROPYL p-HYDROXYBENZOATE see HNU500
PROPYLIC ALCOHOL see PND000
PROPYLIC ALDEHYDE see PMT750
n-PROPYLIDENE BUTYRALDEHYDE see HBI800
PROPYL ISOCYANATE see PNP000
1-PROPYL ISOCYANATE see PNP000
m-PROPYL ISOCYANATE see PNP000
PROPYL KETONE see DWT600
PROPYL MERCAPTAN see PML500
2-PROPYL MERCAPTAN see IMU000
N-PROPYL MERCAPTAN see PML500
PROPYL METHANOATE see PNM500
PROPYLMETHANOL see BPW500
PROPYLMETHYLCARBINYLETHYL BARBITURIC ACID SODIUM SALT see NBU000
4-PROPYL-1,2-METHYLENEDIOXYBENZENE see DMD600
PROPYL NITRATE see PNQ500
n-PROPYL NITRATE see PNQ500
N-PROPYLNITROSOHARNSTOFF (GERMAN) see NLO500
N-PROPYLNITROSOUREA see NLO500
1-PROPYL-1-NITROSOUREA see NLO500
PROPYLOWY ALKOHOL (POLISH) see PND000
PROPYLPARABEN (FCC) see HNU500
PROPYLPARASEPT see HNU500
n-PROPYL PERCARBONATE see DWV400
o-PROPYLPHENOL see PNS250
6-(PROPYLPIPERONYL)-BUTYL CARBITYL ETHER see PIX250
6-PROPYLPIPERONYL BUTYL DIETHYLENE GLYCOL ETHER see PIX250
N-PROPYL-1-PROPANAMINE see DWR000
PROPYL 4-PYRIDYL KETONE see PNV755
n-PROPYLSELENINIC ACID see PNV760
PROPYLTHIOL see PML500
N-PROPYLTHIOL see PML500
6-PROPYL-2-THIO-2,4(1H,3H)PYRIMIDINEDIONE see PNX000
PROPYL-THIORIST see PNX000
PROPYLTHIOURACIL see PNX000
4-PROPYL-2-THIOURACIL see PNX000
6-PROPYL-2-THIOURACIL see PNX000
6-N-PROPYLTHIOURACIL see PNX000
6-N-PROPYL-2-THIOURACIL see PNX000
PROPYL-THYRACIL see PNX000
n-PROPYLTRICHLOROSILANE see PNX250
PROPYLTRICHLOROSILANE (DOT) see PNX250
n-PROPYL-3,4,5-TRIHYDROXYBENZOATE see PNM750

PROPYLTRIIODOGERMANE see TKR050
5-PROPYL-4-(2,5,8-TRIOXA-DODECYL)-1,3-
BENZODIOXOL (GERMAN) see PIX250
4-PROPYL-2,6,7-TRIOXA-1-
STIBABICYCLO(2.2.2)OCTANE see PNX500
PROPYL URETHANE see PNG250
1-PROPYNE-3-OL see PMN450
PROPYNE (OSHA) see MFX590
3-PROPYNOL see PMN450
2-PROPYN-1-OL see PMN450
2-PROPYNYL ALCOHOL see PMN450
PROPYPHENAZONE see INY000
PROPYTHIOURACIL see PNX000
PROQUANIL see MQU750
PRORALONE-MOP see XDJ000
PROREX see PMI750
PROREX see DQA400
PROSEPTINE see SNM500
PROSEPTOL see SNM500
PROSERINE see POD000
PROSERINE BROMIDE see POD000
PROSEVOR 85 see CBM750
PRO-SONIL see TDA500
PROSTAGLANDIN E2 see DVJ200
(−)-PROSTAGLANDIN E2 see DVJ200
(15S)-PROSTAGLANDIN E2 see DVJ200
PROSTAGLANDIN F2-α see POC500
PROSTAGLANDIN F2-α-THAM see POC750
PROSTAGLANDIN F2-α THAM SALT see POC750
PROSTAGLANDIN F2a TROMETHAMINE see POC750
PROSTALMON F see POC500
PROSTAPHLIN-A see SLJ000
PROSTARMON F see POC500
PROSTIGMIN BROMIDE see POD000
PROSTIGMINE BROMIDE see POD000
PROSTIN E2 see DVJ200
PROSTIN F2-α see POC500
PROSTRUMYL see MPW500
PROTABEN P see HNU500
PROTACELL 8 see SEH000
PROTACHEM 630 see PKF000
PROTACHEM GMS see OAV000
PROTACTINIUM see POD500
PROTAGENT see PKQ250
PROTANABOL see PAN100
PROTANAL see SEH000
PROTARS see CAM300
PROTASORB L-20 see PKG000
PROTASORB O-20 see PKL100
PROTATEK see SEH000
PROTAZINE see PMI750
PROTAZINE see DQA400
PROTECTONA see DKA600
PROTERNOL see DMV600
PROTEX (POLYMER) see AAX250
PROTHAZIN see DQA400
PROTHIOFOS-OXON see EEE200
PROTHIUCIL see PNX000
PROTHIURONE see PNX000
PROTHROMADIN see WAT200
PROTHROMBIN see WAT220
PROTHYCIL see PNX000
PROTHYRAN see PNX000
PROTIURAL see PNX000
PROTOCATECHUIC ALDEHYDE ETHYL ETHER see
EQF000

PROTOCATECHUIC ALDEHYDE METHYLENE
ETHER see PIW250
PROTOCHLORURE d'IODE (FRENCH) see IDS000
PROTOPAM IODIDE see POS750
PROTOPET see MQV750
PROTOTYPE III SOFT see PKQ059
PROTOX TYPE 166 see ZKA000
PROTOXYL see ARA500
PROVIGAN see DQA400
PROVITAMIN D see CMD750
PROXAGESIC see PNA500
PROXAGESIC see DAB879
PROXITANE 4002 see PCL500
PROXOL see TIQ250
PROZINEX see PMN850
PROZOIN see PMU750
PRULET see PDO750
PRUNETOL see GCM350
PRUNOLIDE see CNF250
PRUSSIAN BROWN see IHC450
PRUSSIC ACID see HHS000
PRUSSIC ACID, UNSTABILIZED see HHS000
PRUSSITE see COO000
200U/P-RVM see PAE750
PRX 1195 see SMQ500
PRYSOLINE see DBB200
PRZEDZIORKOFOS (POLISH) see PDC750
PS see CKN500
PS 1 see SMQ500
PS 1 see AHE250
PS 2 see SMQ500
PS 200 see SMQ500
PS 209 see SMQ500
PS 454H see SMQ500
PS-B see SMQ500
PSB-C see SMQ500
PSB-S-E see SMQ500
PSB-S see SMQ500
PSB-S 40 see SMQ500
PSC CO-OP WEEVIL BAIT see DXE000
PSEUDOACETIC ACID see PMU750
PSEUDOBUTYLBENZENE see BQJ250
PSEUDOCUMENE see TLL750
PSEUDOCUMIDINE see TLG250
PSEUDOCUMIDINE HYDROCHLORIDE see TLG750
PSEUDOCUMOHYDROQUINONE see POG300
PSEUDOCUMOL see TLL750
PSEUDOHEXYL ALCOHOL see EGW000
PSEUDOPINEN see POH750
PSEUDOPINENE see POH750
PSEUDOTHEOPHYLLINE see TEP000
PSEUDOTHIOUREA see ISR000
PSEUDOTHYMINE see MQI750
PSEUDOUREA see USS000
PSICHIAL see MDQ250
PSICOPERIDOL-R see TKK500
PSICOSAN see MDQ250
PSICOSAN see LFK000
PSILOCIN PHOSPHATE ESTER see PHU500
PSILOCIPIN see PHU500
PSILOTSIBIN see PHU500
PSL see LEN000
PSML see PKG000
PS 2 (POLYMER) see SMQ500
PS 5 (POLYMER) see SMQ500
PSV-L see SMQ500
PSV-L 1 see SMQ500
PSV-L 2 see SMQ500

PSV-L 1S see SMQ500
PSYCHAMINE A 66 HYDROCHLORIDE see MNV750
PSYCHEDRINE see BBK000
PSYCHODRINE see BBK500
PSYCHOLIQUID see TEH500
PSYCHOPERIDOL see TKK500
PSYCHOTABLETS see TEH500
PSYCHOZINE see CKP500
PSYMOD see PII500
Pt-93 see PLW200
PTC see PGN250
PTEGLU see FMT000
PTERIDIUM AQUILINUM see BML000
PTERIS AQUALINA see BML000
PTEROYLGLUTAMIC ACID see FMT000
PTEROYL-l-GLUTAMIC ACID see FMT000
PTEROYLMONOGLUTAMIC ACID see FMT000
PTEROYL-l-MONOGLUTAMIC ACID see FMT000
PTFE see TAI250
PTG see EQP000
PTR-E 3 see EHP700
PTR-E 40 see EHP700
PTS 2 see PJS750
PTU see PGN250
PTU (thyreostatic) see PNX000
P 4007EU see PJS750
PULMOCLASE see CBR675
PURADIN see NGG500
PURAGEL see PCU360
PURALIN see TFS350
PURAPURIDINE see POJ000
PURASAN-SC-10 see ABU500
PURATRONIC CHROMIUM CHLORIDE see CMJ250
PURATRONIC CHROMIUM TRIOXIDE see CMK000
PURATURF 10 see ABU500
PURE CHRYSOIDINE YBH see PEK000
PURE CHRYSOIDINE YD see PEK000
PURE EOSINE YY see BNH500
PURE LEMON CHROME L3GS see LCR000
PURE ORANGE CHROME M see LCS000
PURE QUARTZ see SCI500
PURE QUARTZ see SCJ500
PUREX see SFT000
PURE ZINC CHROME see ZFJ100
PURGA see PDO750
PURGEN see PDO750
PURGOPHEN see PDO750
PURIMETHOL see POK000
1H-PURIN-6-AMINE see AEH000
PURIN B see SHU500
PURINE-6-THIOL see POK000
6-PURINETHIOL see POK000
3H-PURINE-6-THIOL see POK000
PURINETHOL see POK000
1H-PURINE-2,6,8(3H)-TRIONE, 7,9-DIHYDRO- (9CI) see UVA400
1H-PURINE-2,6,8(3H)-TRIONE, 7,9-DIHYDRO-, MONOSODIUM SALT (9CI) see SKO575
9H-PURIN-6-OL see DMC000
6(1H)-PURINONE see DMC000
PURIN-6(3H)-ONE see DMC000
PURODIGIN see DKL800
PUROSTROPHAN see OKS000
PURPLE MINT PLANT EXTRACT see PCI750
PURPLE RED see FMU080
PURPURID see DKL800
PURSERPINE see RDK000
PURTALC USP see TAB750

PUTRESCIN see BOS000
PUTRESCINE see BOS000
PVA see PKP750
PVA 008 see PKP750
PVC CORDO see AAX175
PVC (MAK) see PKQ059
PVP 8T see PJS750
PVP (FCC) see PKQ250
PVPP see PKQ150
PVS 4 see PKP750
PVSK see PKS250
PX 104 see DEH200
PX-138 see DVL600
PX-238 see AEO000
PX 404 see DEH600
PX 438 see BJS250
PY 100 see PJS750
PY2763 see SMQ500
PYBUTHRIN see PIX250
PYDRIN see FAR100
PYDT see DJY000
PYLOSTROPIN see MGR500
PYMAFED see DBM800
PYNAMIN see AFR250
PYNAMIN-FORTE see AFR250
PYNDT see DTV200
PYNOSECT see BEP500
PYOKTANIN see AOR500
PYOLUTEORIN see POK400
PYOSTACINE see VRF000
PYOVERM see AOR500
N-N-PYR see NLP500
PYRACRYL ORANGE Y see PEK000
PYRADEX see DBI200
PYRADONE see DOT000
PYRALCID see ALF250
PYRALENE see PJL750
PYRALIN see CCU250
PYRA MALEATE see DBM800
PYRAMIDON see DOT000
PYRAMIDONE see DOT000
PYRAMINE see PEE750
PYRAMIN RB see PEE750
PYRAN ALDEHYDE see ADR500
2H-PYRAN-2,6(3H)-DIONE, DIHYDRO-2-PHTHALIMIDO- see PHX100
PYRANILAMINE MALEATE see DBM800
PYRANINYL see DBM800
PYRANISAMINE MALEATE see DBM800
PYRANOL see PJL750
2H-PYRAN-2-ONE, 6-HEPTYLTETRAHYDRO- see HBP450
PYRANTEL TARTRATE see TCW750
PYRANTON see DBF750
PYRAZINAMIDE see POL500
PYRAZINEAMIDE see POL500
PYRAZINECARBOXAMIDE see POL500
PYRAZINE CARBOXYLAMIDE see POL500
PYRAZINE HEXAHYDRIDE see PIJ000
PYRAZINOIC ACID AMIDE see POL500
PYRAZINOL-O-ESTER with O,O-DIETHYL PHOSPHOROTHIOATE see EPC500
PYRAZODINE see PDC250
PYRAZOFEN see PDC250
PYRAZOL BLUE 3B see CMO250
PYRAZOLE see POM500
PYRAZOL FAST BRILLIANT BLUE VP see CMN750
PYRAZOLIDIN see BRF500

4H-PYRAZOLO(3,4-d)PYRIMIDIN-4-ONE see ZVJ000
PYRAZON see PEE750
PYRAZONE see PEE750
PYRAZONL see PEE750
PYREDAL see PDC250
PYREN (GERMAN) see PON250
PYRENE see PON250
PYRENE, 2,7-DINITRO- see DVD900
PYRENONE 606 see PIX250
PYREQUAN TARTRATE see TCW750
PYRESIN see AFR250
PYRESYN see AFR250
PYRETHERM see BEP500
PYRETHIA see DQA400
PYRETHIAZINE see DQA400
PYRETHRIN see POO000
PYRETHRIN see POO100
PYRETHRIN II see POO100
PYRETHROLONE CHRYSANTHEMUM
DICARBOXLIC ACID METHYL ESTER see POO100
PYRETHROLONE ESTER of
CHRYSANTHEMUMDICARBOXYLIC ACID
MONOMETHYL ESTER see POO100
(+)-PYRETHRONYL (+)-PYRETHRATE see POO100
PYRETRIN II see POO100
PYRIBENZAMINE see TMP750
PYRICAROYL see DJS200
PYRIDACIL see PDC250
PYRIDAFENTHION see POP000
PYRIDAMAL-100 see TAI500
PYRIDAPHENTHION see POP000
3(2H)-PYRIDAZINONE, 2-(3-BROMO-4-
CHLOROPHENYL)-4-CHLORO-5-((6-CHLORO-3-
PYRIDINYL)METHOXY)- see BNA350
3(2H)-PYRIDAZINONE, 2-(4-BROMOPHENYL)-4-
CHLORO-5-((4-CHLOROPHENYL)METHOXY)- see
BNV800
3(2H)-PYRIDAZINONE, 5-((4-
BROMOPHENYL)METHOXY)-4-CHLORO-2-(4-
CHLORO-2-FLUOROPHENYL)- see BNX035
3(2H)-PYRIDAZINONE, 5-((6-BROMO-3-
PYRIDINYL)METHOXY)-4-CHLORO-2-(4-
CHLOROPHENYL)- see BOC600
3(2H)-PYRIDAZINONE, 4-CHLORO-2-(4-CHLORO-2-
FLUOROPHENYL)-5-((4-
CHLOROPHENYL)METHOXY)- see CFA800
3(2H)-PYRIDAZINONE, 4-CHLORO-2-(4-
CHLOROPHENYL)-5-((6-IODO-3-
PYRIDINYL)METHOXY)- see CFC600
3(2H)-PYRIDAZINONE, 2-(4-CHLOROPHENYL)-5-((4-
CHLOROPHENYL)METHOXY)-4-IODO- see CJT800
3(2H)-PYRIDAZINONE, 5-((6-CHLORO-3-
PYRIDINYL)METHOXY)-2-(3,4-DICHLOROPHENYL)-
4-IODO- see CKW330
PYRIDENAL see PDC250
PYRIDENE see PDC250
PYRIDIATE see PDC250
PYRIDIN (GERMAN) see POP250
2-PYRIDINALDOXIM METHOJODID (GERMAN) see
POS750
PYRIDIN-2-ALDOXIN (CZECH) see POS750
2-PYRIDINAMIDE, N,N-DIMETHYL- see DQB400
3-PYRIDINAMINE see AMI250
4-PYRIDINAMINE see AMI500
α-PYRIDINAMINE see AMI000

2-PYRIDINAMINE, 3-CHLORO-N-(3-CHLORO-2,6-
DINITRO-4-(TRIFLUOROMETHYL)PHENYL)-5-
(TRIFLUOROMETHYL)- see CEX800
4-PYRIDINAMINE, N-METHYL-N-NITROSO- see
NKQ100
PYRIDINE see POP250
PYRIDINE, 3-ACETYL- see ABI000
2-PYRIDINE ALDOXIME IODOMETHYLATE see
POS750
PYRIDINE-2-ALDOXIME METHIODIDE see POS750
PYRIDINE-2-ALDOXIME METHYL IODIDE see
POS750
PYRIDINE, 4-BUTYRYL- see PNV755
PYRIDINE-3-CARBONIC ACID see NCQ900
3-PYRIDINECARBONITRILE, 1,2-DIHYDRO-4-
METHOXY-1-METHYL-2-OXO- see RJK100
PYRIDINE-3-CARBOXYDIETHYLAMIDE see DJS200
PYRIDINE-3-CARBOXYLIC ACID see NCQ900
3-PYRIDINECARBOXYLIC ACID see NCQ900
PYRIDINE-β-CARBOXYLIC ACID see NCQ900
3-PYRIDINECARBOXYLIC ACID, ALUMINUM SALT
see AHD650
PYRIDINE-3-CARBOXYLIC ACID AMIDE see NCR000
3-PYRIDINECARBOXYLIC ACID AMIDE see NCR000
PYRIDINE-3-CARBOXYLIC ACID DIETHYLAMIDE
see DJS200
PYRIDINE-3,5-CARBOXYLIC ACID, 1,4-DIHYDRO-
2,6-DIMETHYL-4-(3-NITROPHENYL)-, 2-(4-(4-
BENZHYDRYLPIPERAZIN-1-YL)PHENYL)ETHYL
METHYL ESTER, (+−)- see WAT230
4-PYRIDINECARBOXYLIC ACID, ETHYL ESTER see
ELU000
PYRIDINE-CARBOXYLIQUE-3 (FRENCH) see NCQ900
PYRIDINE, 4-(CYCLOHEXYLCARBONYL)- see CPI375
2,6-PYRIDINEDIAMINE, 3,5-DINITRO-N,N'-BIS(2,4,6-
TRINITROPHENYL)- see PQC525
2,3-PYRIDINEDICARBOXYLIC ACID see QOJ200
3,5-PYRIDINEDICARBOXYLIC ACID, 1,4-DIHYDRO-
2,6-DIMETHYL-4-(3-NITROPHENYL)-, 2-(4-(4-
(DIPHENYLMETHYL)-1-
PIPERAZINYL)PHENYL)ETHYL METHYL ESTER,
(S)-(+)- see NDY550
3,5-PYRIDINEDICARBOXYLIC ACID, 1,4-DIHYDRO-
2,6-DIMETHYL-4-(2-PYRIDYL)-, DIETHYL ESTER see
DJB460
PYRIDINE, 2-DIMETHYLAMINO- see DQB400
PYRIDINE, 2-((2-
(DIMETHYLAMINO)ETHYL)(SELENOPHENE-2-
YLMETHYL)AMINO)- see DPI750
4-PYRIDINEETHANOL see POR500
PYRIDINE, 4-HEPTANOYL- see HBF600
PYRIDINE, HEXAHYDRO-N-NITROSO- see NLJ500
PYRIDINE, 4-HEXANOYL- see PBX800
PYRIDINE, 4-HEXANOYL- see HEW050
PYRIDINE, 4-(4-METHOXYBENZOYL)- see MFH930
PYRIDINE, 4-(METHYLNITROSAMINO)- see NKQ100
PYRIDINE, 3-(1-METHYL-2-PYRROLIDINYL)-, (S)-,
SULFATE (2:1) see NDR500
PYRIDINE, 4-NITROSOMETHYLAMINO- see NKQ100
PYRIDINE, 3-(1-NITROSO-2-PYRROLIDINYL)-, (S)- see
NLD500
PYRIDINE, 3-(TETRAHYDRO-1-METHYLPYRROL-2-
YL) see NDN000
2-PYRIDINETHIOL-1-OXIDE SODIUM SALT see
MCQ750
2-PYRIDINETHIOL-1-OXIDE, ZINC SALT see ZMJ000
PYRIDINE, 4-VALERYL- see VBA100

PYRIDINIUM-2-ALDOXIME-N-METHYLIODIDE see POS750

PYRIDINIUM PERCHLORATE see PPC100

2-PYRIDINOL, 3,5,6-TRICHLORO-, O-ESTER with O,O-DIETHYL PHOSPHOROTHIOATE see CMA100

PYRIDIPCA see PPK500

PYRIDIUM see PDC250

PYRIDIVITE see PDC250

10H-PYRIDO(3,2-B)(1,4)BENZOTHIAZINE, 10-(2-(DIETHYLAMINO)PROPYL)- see DIR100

PYRIDOXAL HYDROCHLORIDE see VSU000

PYRIDOXINE see PPK250

PYRIDOXINE HYDROCHLORIDE (FCC) see PPK500

PYRIDOXINIUM CHLORIDE see PPK500

PYRIDOXINUM HYDROCHLORICUM (HUNGARIAN) see PPK500

PYRIDOXOL see PPK250

PYRIDOXOL HYDROCHLORIDE see PPK500

PYRIDROL see PII750

3-PYRIDYLAMINE see AMI250

4-PYRIDYLAMINE see AMI500

α-PYRIDYLAMINE see AMI000

1-(4-PYRIDYL)-1-BUTANONE see PNV755

1-(PYRIDYL-3-)-3,3-DIAETHYL-TRIAZEN (GERMAN) see DJY000

m-PYRIDYL-DIETHYL-TRIAZENE see DJY000

1-PYRIDYL-3,3-DIETHYLTRIAZENE see DJY000

1-(PYRIDYL-3)-3,3-DIETHYLTRIAZENE see DJY000

1-(3-PYRIDYL)-3,3-DIETHYLTRIAZENE see DJY000

1-(PYRIDYL-3)-3,3-DIMETHYL-TRIAZEN (GERMAN) see PPL500

1-(PYRIDYL-3)-3,3-DIMETHYL TRIAZENE see PPL500

1-(m-PYRIDYL)-3,3-DIMETHYL-TRIAZENE see PPL500

2-(4-PYRIDYL)ETHANOL see POR500

2-(γ-PYRIDYL)ETHANOL see POR500

1-(4-PYRIDYL)-1-HEPTANONE see HBF600

1-(4-PYRIDYL)-1-HEXANONE see PBX800

1-(4-PYRIDYL)-1-HEXANONE see HEW050

1-(3-PYRIDYLMETHYL)-3-(4-NITROPHENYL)UREA see PPP750

N-3-PYRIDYLMETHYL-N'-p-NITROPHENYLUREA see PPP750

β-PYRIDYL-α-N-METHYLPYRROLIDINE see NDN000

1-(PYRIDYL-3-N-OXID)-3,3-DIMETHYL-TRIAZEN (GERMAN) see DTV200

1-(PYRIDYL-3-N-OXIDE)-3,3-DIMETHYLTRIAZENE see DTV200

1-(4-PYRIDYL)-1-PENTANONE see VBA100

N-2-PYRIDYLSULFANILAMIDE see PPO000

N^1-2-PYRIDYLSULFANILAMIDE see PPO000

PYRILAMINE MALEATE see DBM800

PYRIMAL see PPP500

PYRIMAL M see ALF250

PYRIMIDINE-DEOXYRIBOSE N1-2'-FURANIDYL-5-FLUOROURACIL see FLZ050

2,4-PYRIMIDINEDIOL see UNJ800

2,4-PYRIMIDINEDIONE see UNJ800

2,4(1H,3H)-PYRIMIDINEDIONE (9CI) see UNJ800

2,4(1H,3H)-PYRIMIDINEDIONE, 1-(ETHOXYMETHYL)-5-(1-METHYLETHYL)-6-(PHENYLMETHYL)- see EAN525

PYRIMIDINE PHOSPHATE see DIN800

PYRIMIDINE, 3,4,5,6-TETRAHYDRO-2-(α-ETHOXYBENZYL)-5-ETHYL-5-METHYL- see TCQ700

2,4,5,6(1H,3H)-PYRIMIDINETETRONE see AFT750

2,4,5,6(1H,3H)-PYRIMIDINETETRONE HYDRATE see MDL500

2,4,6(1H,3H,5H)-PYRIMIDINETRIONE, 5,5-DIETHYL-, MONOSODIUM SALT (9CI) see BAG250

2,4,5,6-PYRIMIDINTETRON (CZECH) see AFT750

N^1-2-PYRIMIDINYL-SULFANILAMIDE see PPP500

PYRIMIDONE MEDI-PETS see DBB200

PYRIMINIL see PPP750

PYRIMINYL see PPP750

PYRIMIPHOS METHYL see DIN800

PYRINAMINE BASE see TMP750

PYRINAZINE see HIM000

β-PYRINE see PON250

PYRINEX see CMA100

PYRINURON see PPP750

PYRIPYRIDIUM see PDC250

PYRISEPT see CCX000

PYRISTAN see SPC500

PYRITHIONE ZINC see ZMJ000

PYRIZIN see PDC250

PYRLEUGAN see DLH630

PYROACETIC ACID see ABC750

PYROACETIC ETHER see ABC750

PYROBENZOL see BBL250

PYROBENZOLE see BBL250

PYROCARBONATE d'ETHYLE (FRENCH) see DIZ100

PYROCARBONIC ACID, DIETHYL ESTER see DIZ100

PYROCATECHIN see CCP850

PYROCATECHINIC ACID see CCP850

PYROCATECHOL see CCP850

PYROCATECHUIC ACID see CCP850

PYROCELLULOSE see CCU150

PYROCHOL see DAQ400

PYROD see UNJ800

PYRODONE see OES000

PYRODOXIN see PPK250

PYROGALLIC ACID see PPQ500

PYROGALLOL see PPQ500

PYROGALLOL DIMETHYLETHER see DOJ200

PYROGALLOL-1,3-DIMETHYL ETHER see DOJ200

PYROGUAIAC ACID see GKI000

PYROKOHLENSAEURE DIAETHYL ESTER (GERMAN) see DIZ100

M-PYROL see MPF200

PYROLUSITE BROWN see MAS000

PYROMUCIC ACID METHYL ESTER see MKH600

PYROMUCIC ALDEHYDE see FPQ875

PYRONALORANGE see PEJ500

PYRONALROT B see OHI200

PYRONALROT R see XRA000

PYROPENTYLENE see CPU500

PYROPHOSPHATE see TEE500

PYROPHOSPHATE de TETRAETHYLE (FRENCH) see TCF250

PYROPHOSPHORIC ACID OCTAMETHYLTETRAAMIDE see OCM000

PYROPHOSPHORODITHIOIC ACID, TETRAETHYL ESTER see SOD100

PYROPHOSPHORODITHIOIC ACID, O,O,O,O-TETRAETHYL ESTER see SOD100

PYROPHOSPHORYLTETRAKISDIMETHYLAMIDE see OCM000

PYROSULFUROUS ACID, DIPOTASSIUM SALT see PLR250

PYROSULFURYL CHLORIDE see PPR500

PYROSULPHURIC ACID see SOI520

PYROTARTARIC ACID NITRILE see TLR500

PYROTROPBLAU see CMO250

PYROXYLIC SPIRIT see MGB150

PYROXYLIN see CCU250
PYRROLE see PPS250
PYRROLE-2,5-DIONE see MAM750
PYRROLIDINE see PPS500
l-2-PYRROLIDINECARBOXYLIC ACID see PMH900
PYRROLIDINE, 1-METHYL-2-(3-PYRIDYL)-, SULFATE
see NDR500
PYRROLIDINIUM, 1-CARBOXYMETHYL-1-METHYL-,
IODIDE, METHYL ESTER see CCH300
2-PYRROLIDINONE, 3-((3,5-BIS(1,1-
DIMETHYLETHYL)-4-
HYDROXYPHENYL)METHYLENE)-1-METHOXY- see
MEL100
2-PYRROLIDINONE, 1-CYCLOHEXYL- see CPQ275
2-PYRROLIDINONE, 1-ETHYL- see EPC700
2-PYRROLIDINONE, 1-ISOPROPYL- see IRG100
2-PYRROLIDINONE, 1-(1-METHYLETHYL)-(9CI) see
IRG100
2-PYRROLIDINONE, 1-TALLOW ALKYL derivs. see
TAC200
3-PYRROLINE-2,5-DIONE see MAM750
PYRROLYLENE see BOP500
PYRROL-2-YL KETONE see PPY300
PYSOCOCCINE see SNM500
PYX see PQC525
PYX EXPLOSIVE see PQC525
3ZhP see IGK800
PZh2M see IGK800
PZhO see IGK800
Q-137 see DJC000
QCB see PAV500
QDO see DVR200
QG 100 see SCK600
QIDAMP see AIV500
QIDTET see TBX250
QSAH 7 see PKQ059
QUADRACYCLINE see TBX250
QUAMAQUIL see PGA750
QUAMONIUM see HCQ500
QUANTALAN see CME400
QUANTROVANIL see EQF000
QUARTZ see SCJ500
QUARTZ GLASS see SCK600
QUARTZ SAND see SCK600
QUATERNARIO CPC see CCX000
QUATERNIUM 15 see CEG550
QUATERNOL 1 see DTC600
QUATRACHLOR see BEN000
QUATRESIN see MSB500
QUATREX see TBX250
QUAZO PURO (ITALIAN) see SCJ500
QUECKSILBER (GERMAN) see MCW250
QUECKSILBER CHLORID (GERMAN) see MCY475
QUECKSILBER(I)-CHLORID (GERMAN) see MCW000
QUECKSILBER CHLORUER (GERMAN) see MCW000
QUECKSILBEROXID (GERMAN) see MCT500
QUECKSILBEROXID (GERMAN) see MDF750
QUELLADA see BBQ500
QUEMICETINA see CDP250
QUERCETIN see QCA000
QUERCETINE see QCA000
QUERCETOL see QCA000
QUERCITIN see QCA000
QUERTINE see QCA000
QUERTON 28CL see DTF820
QUESTEX 4 see EIV000
QUESTEX 4H see EIX000
QUESTRAN see CME400

QUETIMID see TEH500
QUETINIL see DCK759
QUIATRIL see DCK759
QUICK see CJJ000
QUICKLIME (DOT) see CAU500
QUICKPHOS see AHE750
QUICKSAN see ABU500
QUICKSET EXTRA see MKA500
QUICKSET SUPER see MKA500
QUICK SILVER see MCW250
QUIDE see PII500
QUIESCIN see RDK000
QUIETIDON see MQU750
QUIETOPLEX see TEH500
QUILONE see LGO100
QUILONUM RETARD see LGZ000
QUINACRINE see ARQ250
QUINACRINE MUSTARD see QDS000
QUINACRINE MUSTARD DIHYDROCHLORIDE see
QDS000
QUINADOME see DNF600
QUINALPHOS see DJY200
QUINAMBICIDE see CHR500
QUINICARDINE see QFS000
QUINIDEX see QFS000
QUINIDINE see QFS000
(+)-QUINIDINE see QFS000
QUININE see QHJ000
(−)-QUININE see QHJ000
β-QUININE see QFS000
QUININE BIMURIATE see QIJ000
QUININE BISULFATE see QMA000
QUININE CHLORIDE see QJS000
QUININE DIHYDROCHLORIDE see QIJ000
(−)-QUININE DIHYDROCHLORIDE see QIJ000
QUININE HYDROCHLORIDE see QJS000
QUININE HYDROGEN SULFATE see QMA000
QUININE MONOHYDROCHLORIDE see QJS000
QUININE MURIATE see QJS000
QUININE SULFATE see QMA000
QUINIZARIN see DMH000
QUINIZARINE GREEN BASE see BLK000
QUINOL see HIH000
8-QUINOL see QPA000
β-QUINOL see HIH000
QUINOLINE see QMJ000
5-QUINOLINESULFONIC ACID, 8,8'-
((HYDROXYSTIBYLENE)BIS(OXY))BIS(7-FORMYL)-,
DISODIUM SALT see AQE320
QUINOLINIC ACID see QOJ200
8-QUINOLINOL see QPA000
8-(QUINOLINOLATO)METHYL MERCURY see
MLH000
(8-QUINOLINOLATO)TRIBUTYLSTANNANE see
TIB000
8-QUINOLINOL, MERCURY COMPLEX see MLH000
QUINOLOR COMPOUND see BDS000
QUINONDO see BLC250
QUINONE see QQS200
o-QUINONE see BDC250
p-QUINONE see QQS200
QUINONE DIOXIME see DVR200
p-QUINONE DIOXIME see DVR200
QUINONE MONOXIME see NLF200
QUINONE OXIME see NLF200
p-QUINONE OXIME see DVR200
QUINOPHENOL see QPA000

6-QUINOXALINAMINE, 5-BROMO-N-(4,5-DIHYDRO-
1H-IMIDAZOL-2-YL)- see BMM575
(2,3-QUINOXALINYLDITHIO)DIMETHYLTIN see
QSJ800
(2-QUINOXALINYLMETHYLENE)-
HYDRAZINECARBOXYLIC ACID METHYL ESTER-
N,N'-DIOXIDE see FOI000
N-(2-QUINOXALINYL)SULFANILAMIDE see QTS000
N¹-2-QUINOXALINYLSULFANILAMIDE see QTS000
N'-2-QUINOXALYLSULFANILAMIDE see QTS000
QUINSORB 010 see DMI600
QUINTAR see DFT000
QUINTAR 540F see DFT000
QUINTESS-N see NCG000
QUINTOCENE see PAX000
QUINTOX see DHB400
QUINTOZEN see PAX000
QUINTOZENE see PAX000
QUINTRATE see PBC250
QUIRVIL see PKQ059
QUOLAC EX-UB see SIB600
QYSA see PKQ059
R 3 see SMQ500
R 8 see MLF250
R 10 see CBY000
R 13 see CLR250
R 14 see PKL750
R 14 see CBY250
R-22 see CFX500
R 23 see CBY750
R 30 see MJP450
R 31 see CHI900
R 40 see MIF765
R-47 see TNF500
R48 see BIF250
R50 see DAD200
88-R see SOP500
R 113 see FOO000
R 114 see FOO509
R 123 see TJY500
R161 see FIB000
R-246 see TND500
5R04 see ARM268
R 561 see PER250
R 717 see AMY500
R 744 see CBU250
Rh 6G see RGW000
R-1303 see TNP250
R 133a see TJY175
R-1492 see MQH750
R 1504 see PHX250
R 1513 see EKN000
R-1608 see EIN500
R 1929 see FLU000
R 2170 see DAP000
R-2498 see TKK500
R 3345 see FHG000
R 3612 see SMQ500
R 40B1 see MHR200
R 4263 see PDW750
R 5240 see PDW750
R7000 see MFB400
R 9985 see MDV500
R-12,564 see LFA020
R12 (DOT) see DFA600
R13 (DOT) see CLR250
R 14487 see DJW890
R14 (DOT) see CBY250

R 16341 see PAP250
R 17635 see MHL000
R 18986 see AIH000
R21 (DOT) see DFL000
R22 (DOT) see CFX500
R 23979 see FPB875
R 35443 see OMG000
R114 (DOT) see DGL600
R500 (DOT) see DFB400
R502 (DOT) see FOO560
R503 (DOT) see FOO562
R12B1 (DOT) see BNA250
R12B2 (DOT) see DKG850
R13B1 (DOT) see TJY100
R142B (DOT) see CFX250
R1132a (DOT) see VPP000
R 8 (fungicide) see MLF250
R 20 (refrigerant) see CHJ500
R 31 (refrigerant) see CHI900
13-RA see VSK955
β-RA see VSK950
RACEMIC LACTIC ACID see LAG000
RACEMIC MANDELIC ACID see MAP000
RACEPHEDRINE HYDROCHLORIDE see EAX500
RACUSAN see DSP400
RADAPON see DGI400
RADAPON see DGI600
RADAZIN see ARQ725
RADDLE see IHC450
RAD-E-CATE see HKC500
RAD-E-CATE 16 see HKC500
RAD-E-CATE 25 see HKC000
RAD-E-CATE 25 see HKC500
RAD-E-CATE 35 see HKC500
RADEDORM see DLY000
RADEPUR see LFK000
RADIASURF 7125 see SKV000
RADIASURF 7155 see SKV100
RADIOSTOL see VSZ100
RADIUM see RAV000
RADIUM F see PJJ750
RADIZINE see ARQ725
RADOCON see BJP000
RADOKOR see BJP000
RADON see RBA000
RADONIL see SNK000
RADONIN see SNN300
RADONNA see CHJ750
RADOSAN see MEO750
RADOX see CFK000
RADOXONE TL see AMY050
RADSTERIN see VSZ100
RAFEX see DUS700
RAFLUOR see SHF500
RALABOL see RBF100
RALGRO see RBF100
RALLY see MRW775
RALONE see RBF100
RAMIK see DVV600
RAMIZOL see AMY050
RAMOR see TEI000
R/AMP see RKP000
RAMPART see PGS000
RAMUCIDE see CJJ000
RANAC see CJJ000
RANCOSIL see SCK600
RANDOX see CFK000

RANEY ALLOY see NCW500
RANEY COPPER see CNI000
RANEY NICKEL see NCW500
RANKOTEX see CIR500
RANTOX T see CFK000
RAPACODIN see DKW800
RAPESEED OIL see RBK200
RAPE SEED OIL see RBK200
RAPHATOX see DUS700
RAPHETAMINE see BBK000
RAPHONE see CIR250
RAPISOL see DJL000
RAS-26 see NIJ500
RASCHIT see CFD990
RASEN-ANICON see CFD990
RASIKAL see SFS000
RASORITE 65 see DXG035
RASPBERRY KETONE see RBU000
RASPBERRY KETONE METHYL ETHER see MFF580
RASTINON see BSQ000
RAT-A-WAY see WAT200
RATBANE 1080 see SHG500
RAT-B-GON see WAT200
RAT-O-CIDE RAT BAIT see RCF000
RAT-GARD see WAT200
RATINDAN 1 see DVV600
RAT & MICE BAIT see WAT200
RAT-NIP see PHP010
RATO see TMO000
RATOMET see CJJ000
RATOX see TEL750
RAT'S END see RCF000
RATS-NO-MORE see WAT200
RATSUL SOLUBLE see WAT220
RATTENGIFTKONSERVE see TEM000
RATTEX see CNV000
RATTRACK see AQN635
RAUCAP see RDK000
RAUDIFORD see RDK000
RAUDIXIN see RDK000
RAUDIXOID see RDK000
RAUGAL see RDK000
RAULEN see RDK000
RAULOYCIN see RDK000
RAULOYDIN see RDK000
RAUMORINE see RDK000
RAUNERVIL see RDK000
RAUNORINE see RDK000
RAUNORMIN "ORZAN" see RDK000
RAUNOVA see RDK000
RAUPASIL see RDK000
RAUPOID see RDK000
RAURINE see RDK000
RAUSAN see RDK000
RAU-SED see RDK000
RAUSEDAN see RDK000
RAUSEDIL see RDK000
RAUSEDYL see RDK000
RAUSERPEN-ALK see RDK000
RAUSERPIN see RDK000
RAUSERPIN-ALK see RDK000
RAUSERPINE see RDK000
RAUSERPOL see RDK000
RAUSINGLE see RDK000
RAUTRIN see RDK000
RAUVILID see RDK000
RAUVLID see RDK000
RAUWASEDIN see RDK000

RAUWILID see RDK000
RAUWILOID see RDK000
RAUWILOID+ see RDK000
RAUWIPUR see RDK000
RAUWOLEAF see RDK000
RAUWOPUR "BYK" see RDK000
RAVEN see CBT750
RAVEN 30 see CBT750
RAVEN 420 see CBT750
RAVEN 500 see CBT750
RAVEN 8000 see CBT750
RAVIAC see CJJ000
RAVINYL see PKQ059
RAVONAL see PBT500
RAVYON see CBM750
RAWILID see RDK000
RAW SHALE OIL see COD750
RAYBAR see BAP000
RAY-GLUCIRON see FBK000
RAYOPHANE see CCU150
RAYOX see TGG760
RAYWEB Q see CCU150
RAZOL DOCK KILLER see CIR250
RAZOXANE see RCA375
RAZOXIN see PIK250
RAZOXIN see RCA375
RB 1509 see CGV250
RBA 777 see CLW000
R-BASE see CBN100
R-C 318 see CPS000
RC 5626 see DTF820
RCA WASTE NUMBER P105 see SFA000
RCA WASTE NUMBER U203 see SAD000
RCA WASTE NUMBER U205 see SBR000
RC 172DBM see AHE250
RC COMONOMER DBM see DED600
RC COMONOMER DOF see DVK600
RCH 1000 see PJS750
RC PLASTICIZER DOP see DVL700
RCRA WASTE NUMBER P001 see WAT200
RCRA WASTE NUMBER P003 see ADR000
RCRA WASTE NUMBER P004 see AFK250
RCRA WASTE NUMBER P005 see AFV500
RCRA WASTE NUMBER P006 see AHE750
RCRA WASTE NUMBER P008 see AMI500
RCRA WASTE NUMBER P009 see ANS500
RCRA WASTE NUMBER P010 see ARB250
RCRA WASTE NUMBER P011 see ARH500
RCRA WASTE NUMBER P012 see ARI750
RCRA WASTE NUMBER P013 see BAK750
RCRA WASTE NUMBER P014 see PFL850
RCRA WASTE NUMBER P015 see BFO750
RCRA WASTE NUMBER P016 see BIK000
RCRA WASTE NUMBER P017 see BNZ000
RCRA WASTE NUMBER P018 see BOL750
RCRA WASTE NUMBER P021 see CAQ500
RCRA WASTE NUMBER P022 see CBV500
RCRA WASTE NUMBER P023 see CDY500
RCRA WASTE NUMBER P024 see CEH680
RCRA WASTE NUMBER P028 see BEE375
RCRA WASTE NUMBER P029 see CNL000
RCRA WASTE NUMBER P030 see COI500
RCRA WASTE NUMBER P031 see COO000
RCRA WASTE NUMBER P033 see COO750
RCRA WASTE NUMBER P036 see DGB600
RCRA WASTE NUMBER P037 see DHB400
RCRA WASTE NUMBER P039 see DXH325
RCRA WASTE NUMBER P042 see VGP000

RCRA WASTE NUMBER P043 see IRF000
RCRA WASTE NUMBER P044 see DSP400
RCRA WASTE NUMBER P046 see DTJ400
RCRA WASTE NUMBER P047 see DUS700
RCRA WASTE NUMBER P048 see DUZ000
RCRA WASTE NUMBER P049 see DXL800
RCRA WASTE NUMBER P050 see EAQ750
RCRA WASTE NUMBER P051 see EAT500
RCRA WASTE NUMBER P054 see EJM900
RCRA WASTE NUMBER P056 see FEZ000
RCRA WASTE NUMBER P057 see FFF000
RCRA WASTE NUMBER P058 see SHG500
RCRA WASTE NUMBER P059 see HAR000
RCRA WASTE NUMBER P060 see IKO000
RCRA WASTE NUMBER P062 see HCY000
RCRA WASTE NUMBER P063 see HHS000
RCRA WASTE NUMBER P064 see MKX250
RCRA WASTE NUMBER P065 see MDC000
RCRA WASTE NUMBER P066 see MDU600
RCRA WASTE NUMBER P067 see PNL400
RCRA WASTE NUMBER P068 see MKN000
RCRA WASTE NUMBER P069 see MLC750
RCRA WASTE NUMBER P070 see CBM500
RCRA WASTE NUMBER P071 see MNH000
RCRA WASTE NUMBER P072 see AQN635
RCRA WASTE NUMBER P073 see NCZ000
RCRA WASTE NUMBER P074 see NDB500
RCRA WASTE NUMBER P075 see NDN000
RCRA WASTE NUMBER P076 see NEG100
RCRA WASTE NUMBER P077 see NEO500
RCRA WASTE NUMBER P078 see NGR500
RCRA WASTE NUMBER P081 see NGY000
RCRA WASTE NUMBER P082 see NKA600
RCRA WASTE NUMBER P084 see NKY000
RCRA WASTE NUMBER P085 see OCM000
RCRA WASTE NUMBER P087 see OKK000
RCRA WASTE NUMBER P088 see EAR000
RCRA WASTE NUMBER P089 see PAK000
RCRA WASTE NUMBER P092 see ABU500
RCRA WASTE NUMBER P093 see PGN250
RCRA WASTE NUMBER P094 see PGS000
RCRA WASTE NUMBER P095 see PGX000
RCRA WASTE NUMBER P096 see PGY000
RCRA WASTE NUMBER P098 see PLC500
RCRA WASTE NUMBER P099 see PLS250
RCRA WASTE NUMBER P101 see PMV750
RCRA WASTE NUMBER P102 see PMN450
RCRA WASTE NUMBER P103 see SBV000
RCRA WASTE NUMBER P104 see SDP000
RCRA WASTE NUMBER P106 see SGA500
RCRA WASTE NUMBER P108 see SMN500
RCRA WASTE NUMBER P109 see SOD100
RCRA WASTE NUMBER P110 see TCF000
RCRA WASTE NUMBER P111 see TCF250
RCRA WASTE NUMBER P112 see TDY250
RCRA WASTE NUMBER P113 see TEL050
RCRA WASTE NUMBER P114 see TEL500
RCRA WASTE NUMBER P115 see TEM000
RCRA WASTE NUMBER P116 see TFQ000
RCRA WASTE NUMBER P118 see PCF300
RCRA WASTE NUMBER P119 see ANY250
RCRA WASTE NUMBER P120 see VDU000
RCRA WASTE NUMBER P120 see VDZ000
RCRA WASTE NUMBER P121 see ZGA000
RCRA WASTE NUMBER P122 see ZLS000
RCRA WASTE NUMBER P123 see CDV100
RCRA WASTE NUMBER P404 see EPC500
RCRA WASTE NUMBER U001 see AAG250

RCRA WASTE NUMBER U002 see ABC750
RCRA WASTE NUMBER U003 see ABE500
RCRA WASTE NUMBER U005 see FDR000
RCRA WASTE NUMBER U006 see ACF750
RCRA WASTE NUMBER U007 see ADS250
RCRA WASTE NUMBER U008 see ADS750
RCRA WASTE NUMBER U009 see ADX500
RCRA WASTE NUMBER U010 see AHK500
RCRA WASTE NUMBER U011 see AMY050
RCRA WASTE NUMBER U014 see IBB000
RCRA WASTE NUMBER U015 see ASA500
RCRA WASTE NUMBER U017 see BAY300
RCRA WASTE NUMBER U018 see BBC250
RCRA WASTE NUMBER U019 see BBL250
RCRA WASTE NUMBER U020 see BBS750
RCRA WASTE NUMBER U021 see BBX000
RCRA WASTE NUMBER U022 see BCS750
RCRA WASTE NUMBER U023 see BFL250
RCRA WASTE NUMBER U024 see BID750
RCRA WASTE NUMBER U025 see DFJ050
RCRA WASTE NUMBER U026 see BIF250
RCRA WASTE NUMBER U027 see BII250
RCRA WASTE NUMBER U028 see DVL700
RCRA WASTE NUMBER U029 see MHR200
RCRA WASTE NUMBER U031 see BPW500
RCRA WASTE NUMBER U032 see CAP500
RCRA WASTE NUMBER U033 see CCA500
RCRA WASTE NUMBER U034 see CDN550
RCRA WASTE NUMBER U035 see CDO500
RCRA WASTE NUMBER U036 see CDR750
RCRA WASTE NUMBER U036 see CDR760
RCRA WASTE NUMBER U037 see CEJ125
RCRA WASTE NUMBER U038 see DER000
RCRA WASTE NUMBER U039 see CFD990
RCRA WASTE NUMBER U041 see EAZ500
RCRA WASTE NUMBER U042 see CHI250
RCRA WASTE NUMBER U043 see VNP000
RCRA WASTE NUMBER U044 see CHJ500
RCRA WASTE NUMBER U045 see MIF765
RCRA WASTE NUMBER U046 see CIO250
RCRA WASTE NUMBER U048 see CJK250
RCRA WASTE NUMBER U049 see CLK235
RCRA WASTE NUMBER U050 see CML810
RCRA WASTE NUMBER U051 see BAT850
RCRA WASTE NUMBER U051 see CMY825
RCRA WASTE NUMBER U052 see CNW500
RCRA WASTE NUMBER U052 see CNW750
RCRA WASTE NUMBER U052 see CNX000
RCRA WASTE NUMBER U052 see CNX250
RCRA WASTE NUMBER U053 see POD000
RCRA WASTE NUMBER U053 see COB250
RCRA WASTE NUMBER U053 see COB260
RCRA WASTE NUMBER U055 see COE750
RCRA WASTE NUMBER U056 see CPB000
RCRA WASTE NUMBER U057 see CPC000
RCRA WASTE NUMBER U058 see CQC650
RCRA WASTE NUMBER U059 see DAC000
RCRA WASTE NUMBER U060 see BIM500
RCRA WASTE NUMBER U061 see DAD200
RCRA WASTE NUMBER U062 see DBI200
RCRA WASTE NUMBER U063 see DCT400
RCRA WASTE NUMBER U064 see BCQ500
RCRA WASTE NUMBER U066 see DDL800
RCRA WASTE NUMBER U067 see EIY500
RCRA WASTE NUMBER U068 see DDP800
RCRA WASTE NUMBER U069 see DEH200
RCRA WASTE NUMBER U070 see DEP800
RCRA WASTE NUMBER U071 see DEP800

RCRA WASTE NUMBER U072 see DEP800
RCRA WASTE NUMBER U073 see DEQ600
RCRA WASTE NUMBER U074 see DEV000
RCRA WASTE NUMBER U075 see DFA600
RCRA WASTE NUMBER U076 see DFF809
RCRA WASTE NUMBER U077 see EIY600
RCRA WASTE NUMBER U078 see VPK000
RCRA WASTE NUMBER U079 see ACK000
RCRA WASTE NUMBER U080 see MJP450
RCRA WASTE NUMBER U081 see DFX800
RCRA WASTE NUMBER U082 see DFY000
RCRA WASTE NUMBER U083 see PNJ400
RCRA WASTE NUMBER U084 see DGG950
RCRA WASTE NUMBER U085 see BGA750
RCRA WASTE NUMBER U086 see DJL400
RCRA WASTE NUMBER U088 see DJX000
RCRA WASTE NUMBER U089 see DKA600
RCRA WASTE NUMBER U090 see DMD600
RCRA WASTE NUMBER U091 see DCJ200
RCRA WASTE NUMBER U092 see DOQ800
RCRA WASTE NUMBER U093 see DOT300
RCRA WASTE NUMBER U094 see DQJ200
RCRA WASTE NUMBER U095 see TGJ750
RCRA WASTE NUMBER U096 see IOB000
RCRA WASTE NUMBER U097 see DQY950
RCRA WASTE NUMBER U098 see DSF400
RCRA WASTE NUMBER U099 see DSF600
RCRA WASTE NUMBER U101 see XKJ500
RCRA WASTE NUMBER U102 see DTR200
RCRA WASTE NUMBER U103 see DUD100
RCRA WASTE NUMBER U105 see DVH000
RCRA WASTE NUMBER U106 see DVH400
RCRA WASTE NUMBER U107 see DVL600
RCRA WASTE NUMBER U108 see DVQ000
RCRA WASTE NUMBER U109 see HHG000
RCRA WASTE NUMBER U110 see DWR000
RCRA WASTE NUMBER U111 see NKB700
RCRA WASTE NUMBER U112 see EFR000
RCRA WASTE NUMBER U113 see EFT000
RCRA WASTE NUMBER U115 see EJN500
RCRA WASTE NUMBER U116 see IAQ000
RCRA WASTE NUMBER U117 see EJU000
RCRA WASTE NUMBER U118 see EMF000
RCRA WASTE NUMBER U119 see EMF500
RCRA WASTE NUMBER U120 see FDF000
RCRA WASTE NUMBER U121 see TIP500
RCRA WASTE NUMBER U122 see FMV000
RCRA WASTE NUMBER U123 see FNA000
RCRA WASTE NUMBER U124 see FPK000
RCRA WASTE NUMBER U125 see FPQ875
RCRA WASTE NUMBER U126 see GGW000
RCRA WASTE NUMBER U127 see HCC500
RCRA WASTE NUMBER U128 see HCD250
RCRA WASTE NUMBER U129 see BBQ500
RCRA WASTE NUMBER U130 see HCE500
RCRA WASTE NUMBER U131 see HCI000
RCRA WASTE NUMBER U132 see HCL000
RCRA WASTE NUMBER U133 see HGS000
RCRA WASTE NUMBER U134 see HHU500
RCRA WASTE NUMBER U135 see HIC500
RCRA WASTE NUMBER U136 see HKC000
RCRA WASTE NUMBER U137 see IBZ000
RCRA WASTE NUMBER U138 see MKW200
RCRA WASTE NUMBER U139 see IGS000
RCRA WASTE NUMBER U140 see IIL000
RCRA WASTE NUMBER U141 see IRZ000
RCRA WASTE NUMBER U142 see KEA000
RCRA WASTE NUMBER U143 see LBG000

RCRA WASTE NUMBER U144 see LCV000
RCRA WASTE NUMBER U146 see LCH000
RCRA WASTE NUMBER U147 see MAM000
RCRA WASTE NUMBER U148 see DMC600
RCRA WASTE NUMBER U149 see MAO250
RCRA WASTE NUMBER U150 see PED750
RCRA WASTE NUMBER U151 see MCW250
RCRA WASTE NUMBER U152 see MGA750
RCRA WASTE NUMBER U153 see MLE650
RCRA WASTE NUMBER U154 see MGB150
RCRA WASTE NUMBER U156 see MIG000
RCRA WASTE NUMBER U157 see MIJ750
RCRA WASTE NUMBER U158 see MJM200
RCRA WASTE NUMBER U159 see MKA400
RCRA WASTE NUMBER U160 see MKA500
RCRA WASTE NUMBER U161 see HFG500
RCRA WASTE NUMBER U162 see MLH750
RCRA WASTE NUMBER U163 see MMP000
RCRA WASTE NUMBER U164 see MPW500
RCRA WASTE NUMBER U165 see NAJ500
RCRA WASTE NUMBER U166 see NBA500
RCRA WASTE NUMBER U167 see NBE700
RCRA WASTE NUMBER U168 see NBE500
RCRA WASTE NUMBER U169 see NEX000
RCRA WASTE NUMBER U170 see NIF000
RCRA WASTE NUMBER U171 see NIY000
RCRA WASTE NUMBER U172 see BRY500
RCRA WASTE NUMBER U173 see NKM000
RCRA WASTE NUMBER U174 see NJW500
RCRA WASTE NUMBER U176 see ENV000
RCRA WASTE NUMBER U177 see MNA750
RCRA WASTE NUMBER U178 see MMX250
RCRA WASTE NUMBER U179 see NLJ500
RCRA WASTE NUMBER U180 see NLP500
RCRA WASTE NUMBER U181 see NMP500
RCRA WASTE NUMBER U182 see PAI250
RCRA WASTE NUMBER U183 see PAV500
RCRA WASTE NUMBER U184 see PAW500
RCRA WASTE NUMBER U185 see PAX000
RCRA WASTE NUMBER U187 see ABG750
RCRA WASTE NUMBER U188 see PDN750
RCRA WASTE NUMBER U189 see PHS000
RCRA WASTE NUMBER U190 see PHW750
RCRA WASTE NUMBER U191 see MOY000
RCRA WASTE NUMBER U193 see PML400
RCRA WASTE NUMBER U194 see PND250
RCRA WASTE NUMBER U196 see POP250
RCRA WASTE NUMBER U197 see QQS200
RCRA WASTE NUMBER U200 see RDK000
RCRA WASTE NUMBER U201 see REA000
RCRA WASTE NUMBER U202 see BCE500
RCRA WASTE NUMBER U204 see SBO000
RCRA WASTE NUMBER U204 see SBQ500
RCRA WASTE NUMBER U206 see SMD000
RCRA WASTE NUMBER U208 see TBQ000
RCRA WASTE NUMBER U209 see TBQ100
RCRA WASTE NUMBER U210 see PCF275
RCRA WASTE NUMBER U211 see CBY000
RCRA WASTE NUMBER U212 see TBT000
RCRA WASTE NUMBER U213 see TCR750
RCRA WASTE NUMBER U214 see TEI250
RCRA WASTE NUMBER U215 see TEJ000
RCRA WASTE NUMBER U216 see TEJ250
RCRA WASTE NUMBER U217 see TEK750
RCRA WASTE NUMBER U218 see TFA000
RCRA WASTE NUMBER U219 see ISR000
RCRA WASTE NUMBER U220 see TGK750
RCRA WASTE NUMBER U221 see TGL500

RCRA WASTE NUMBER U221 see TGL750
RCRA WASTE NUMBER U222 see TGS500
RCRA WASTE NUMBER U223 see TGM740
RCRA WASTE NUMBER U223 see TGM750
RCRA WASTE NUMBER U225 see BNL000
RCRA WASTE NUMBER U226 see MIH275
RCRA WASTE NUMBER U227 see TIN000
RCRA WASTE NUMBER U228 see TIO750
RCRA WASTE NUMBER U230 see TIV750
RCRA WASTE NUMBER U231 see TIW000
RCRA WASTE NUMBER U233 see TIX500
RCRA WASTE NUMBER U234 see TMK500
RCRA WASTE NUMBER U235 see TNC500
RCRA WASTE NUMBER U236 see CMO250
RCRA WASTE NUMBER U237 see BIA250
RCRA WASTE NUMBER U238 see UVA000
RCRA WASTE NUMBER U239 see XGS000
RCRA WASTE NUMBER U242 see PAX250
RCRA WASTE NUMBER U244 see TFS350
RCRA WASTE NUMBER U246 see COO500
RCRA WASTE NUMBER U247 see MEI450
RD 8 see SCK600
RD 120 see SCK600
RD 406 see DGB000
RD 1572 see DBV400
RD 4593 see CIR500
R.D. 13621 see IIU000
RDGE see REF000
RDX see CPR800
RDX and HMX MIXTURES, desensitized with not <10%
phlegmatizer by weight (UN 0391) (DOT) see CPR800
RDX and HMX MIXTURES, wetted with not <15% water
by weight (UN 0391) (DOT) see CPR800
RDX, desensitized (UN 0483) (DOT) see CPR800
RDX, wetted with not <15% water by weight (UN 0072)
(DOT) see CPR800
RE 12420 see DOP600
REACTIVE BLUE 19 see BMM500
REBELATE see DSP400
REBONEX see CBT750
REBRAMIN see VSZ000
RECININE see RJK100
RECIPIN see RDK000
RECOLITE RED LAKE C see CHP500
RECOMBINANT HUMAN TUMOR NECROSIS
FACTOR α see HGL920
RECOMBINANT HUMAN TUMOR NECROSIS
FACTOR-α see HGL920
RECONOX see PDP250
RECTHORMONE OESTRADIOL see EDP000
RECTHORMONE TESTOSTERONE see TBG000
RECTODELT see PLZ000
RECTULES see CDO000
RED #14 see HJF500
RED 104 see ADG250
1695 RED see FMU070
1860 RED see CHP500
11445 RED see BNH500
11554 RED see IHC450
11959 RED see HJF500
11969 RED see ADG250
111440 RED see OHI200
REDAMINA see VSZ000
REDAX see DWI000
RED B see XRA000
RED 2G BASE see NEO500

RED BASE CIBA IX see CLK220
RED BASE CIBA IX see CLK235
RED BASE IRGA IX see CLK220
RED BASE IRGA IX see CLK235
RED BASE NTR see CLK220
RED COPPER OXIDE see CNO000
RED FUMING NITRIC ACID see NEE500
REDIFAL see SNN300
REDI-FLOW see BAP000
RED IRON ORE see HAO875
RED IRON OXIDE see IHC450
REDISOL see VSZ000
RED KB BASE see CLK225
RED LEAD see LDS000
RED LEAD CHROMATE see LCS000
RED LEAD OXIDE see LDS000
RED MERCURIC IODIDE see MDD000
RED No. 5 see XRA000
RED No. 104 see ADG250
RED OCHRE see IHC450
RED OIL see OHU000
RED OXIDE of MERCURY see MCT500
RED PRECIPITATE see MCT500
RED R see FMU070
RED SALT CIBA IX see CLK235
RED SALT IRGA IX see CLK235
RED SCARLET see CHP500
RED-SEAL-9 see ZKA000
REDSKIN see AGJ250
RED 3R SOLUBLE IN GREASE see SBC500
RED SQUILL see RCF000
RED TR BASE see CLK220
RED TRS SALT see CLK235
REDUCED MINERALS WHEY see WBL165
REDUCED-d-PENICILLAMINE see MCR750
REDUCTONE see SHR500
RED ZH see OHI200
REFINED PETROLEUM WAX see PCT600
REFINED SOLVENT NAPHTHA see PCT250
REFORMIN see DJS200
REFRACTORY CERAMIC FIBERS see RCK725
REFRIGERANT 12 see DFA600
REFRIGERANT 14 see CBY250
REFRIGERANT 22 see CFX500
REFRIGERANT 112 see TBP050
REFRIGERANT 113 see FOO000
REFRIGERANT 502 see FOO560
REFRIGERANT 112a see TBP000
REFRIGERANT R 14 see CBY250
REFUSAL see DXH250
REGAL see CBT750
REGAL 99 see CBT750
REGAL 300 see CBT750
REGAL 330 see CBT750
REGAL 600 see CBT750
REGAL 400R see CBT750
REGAL SRF see CBT750
REGENT see CBT750
REGION see SNQ710
REGLON see DWX800
REGLONE see DWX800
REGONOL see GLU000
REGROTON see RDK000
REGULAR BLEACHED SHELLAC see SCC700
REGULOX see DMC600
REGULOX W see DMC600
REGULOX 50 W see DMC600
REGUTOL see DJL000

REHORMIN see DJS200
REICHSTEIN'S F see MIW500
REICHSTEIN'S SUBSTANCE M see CNS750
REIN GUARIN see GLU000
REISE-ENGLETTEN see DYE600
REKAWAN see PLA500
RELACT see DLY000
RELAMINAL see DCK759
RELANIUM see DCK759
RELBAPIRIDINA see PPO000
RELDAN see CMA250
RELEFACT LH-RH see LIU370
RELIBERAN see MDQ250
RELITON YELLOW C see AAQ250
RELON P see PJY500
REMADERM YELLOW HPR see MDM775
REMALAN BRILLIANT BLUE R see BMM500
REMASAN CHLOROBLE M see MAS500
REMAZOL BRILLIANT BLUE R see BMM500
REMICYCLIN see TBX250
REMID see ZVJ000
REMKO see IGK800
REMOL TRF see BGJ250
REMONOL see RDZ900
REMSED see PMI750
REMYLINE Ac see SLJ500
RENACIT 1 see NAP500
RENAFUR see NDY000
RENAGLADIN see VGP000
RENAL AC see ALT250
RENAL EG see ALS990
RENALEPTINE see VGP000
RENALINA see VGP000
RENAL MD see TGL750
RENAL PF see PEY500
RENAL SLA see DBO400
RENARCOL see ARW250
RENARDIN see DMX200
RENARDINE see DMX200
RENBORIN see DCK759
RENESE R see RDK000
RENOFORM see VGP000
RENOLBLAU 3B see CMO250
RENOL MOLYBDATE RED RGS see MRC000
RENOSTYPRICIN see VGP000
RENOSTYPTIN see VGP000
RENOSULFAN see SNN500
REN O-SAL see HMY000
RENSTAMIN see DBM800
REOMOL D 79P see DVL700
REOMOL DOA see AEO000
REOMOL DOP see DVL700
REOMUCIL see CBR675
REPAIRSIN see PKB500
REPARIL SODIUM SALT see EDM000
REPEL see DKC800
REPELLENT 612 see EKV000
REPOC see PJS750
REPPER-DET see DKC800
REPROMIX see MCA000
REPROTEROL HYDROCHLORIDE see DNA600
REPUDIN-SPECIAL see DKC800
REQUTOL see DJL000
R-E-S see RDK000
RESACETOPHENONE see DMG400
β-RESACETOPHENONE see DMG400
RESALTEX see RDK000

RESARIT 4000 see PKB500
RESCUE SQUAD see SHF500
RESCULA see IRR050
RESEDIN see RDK000
RESEDREX see RDK000
RESEDRIL see RDK000
RESE-LAR see RDK000
RESER-AR see RDK000
RESERBAL see RDK000
RESERCAPS see RDK000
RESERCEN see RDK000
RESERCRINE see RDK000
RESERFIA see RDK000
RESERJEN see RDK000
RESERLOR see RDK000
RESERP see RDK000
RESERPAL see RDK000
RESERPAMED see RDK000
RESERPANCA see RDK000
RESERPENE see RDK000
RESERPEX see RDK000
RESERPIDEFE see RDK000
RESERPIL see RDK000
RESERPIN see RDK000
RESERPINA see RDK000
RESERPINE see RDK000
RESERPINUM see RDK000
RESERPKA see RDK000
RESERPOID see RDK000
RESERPUR see RDK000
RESERP "WANDER" see RDK000
RESERSANA see RDK000
RESERUTIN see RDK000
RESIATRIC see RDK000
RESIDINE see RDK000
RESIDUAL(HEAVY) FUEL OIL see FOP200
RESIDUAL OIL SOLVENT EXTRACT see MQV863
RESIDUAL OILS (PETROLEUM), ACID-TREATED
(9CI) see MQV872
RESIDUES (PETROLEUM), THERMAL CRACKED see
RDK100
RESIN (solution) see RDP000
RESIN ACIDS and ROSIN ACIDS, CALCIUM SALTS see
CAW500
RESINE see RDK000
RESINOL ORANGE R see PEJ500
RESINOL RED 2B see SBC500
RESINOL RED G see CMS238
RESINOL YELLOW GR see DOT300
RESIN SCARLET 2R see XRA000
RESIN SOLUTION, flammable (DOT) see RDP000
RESIREN YELLOW TG see AAQ250
RESISTAMINE see TMP750
RESISTOFLEX see PKP750
RESITAN see VBK000
RESLOOM M 75 see HDY000
RESMETHRIN see BEP500
RESMETRINA (PORTUGUESE) see BEP500
RESOACETOPHENONE see DMG400
RESOCALM see RDK000
RESOFORM ORANGE G see PEJ500
RESOFORM ORANGE R see XRA000
RESOFORM RED G see SBC500
RESOFORM YELLOW GGA see DOT300
RESOIDAN see CCK125
RESOMINE see RDK000
RESORCIN see REA000
RESORCIN ACETATE see RDZ900

RESORCINE see REA000
RESORCINE BROWN J see XMA000
RESORCINE BROWN R see XMA000
RESORCIN MONOACETATE see RDZ900
RESORCINOL see REA000
RESORCINOL BIS(2,3-EPOXYPROPYL)ETHER see REF000
RESORCINOL, DIACETATE see REA100
RESORCINOL DIGLYCIDYL ETHER see REF000
RESORCINOL, 2,4-DINITROSO- see DVF300
β-RESORCINOLIC ACID see HOE600
RESORCINOL METHYL ETHER see REF050
RESORCINOL, MONOACETATE see RDZ900
RESORCINOL MONOMETHYL ETHER see REF050
RESORCINOLPHTHALEIN see FEV000
RESORCINOL PHTHALEIN SODIUM see FEW000
RESORCINOL, 2,4,6-TRINITRO-, BARIUM SALT, HYDRATE (2:1:1) see BAO900
RESORCINYL DIGLYCIDYL ETHER see REF000
RESORCITATE see RDZ900
β-RESORCYLIC ACID see HOE600
RESOTROPIN see HEI500
RESOXOL see SNN500
RESPAIRE see ACH000
RESPERIN see RDK000
RESPERINE see RDK000
RESPIFRAL see DMV600
RESPITAL see RDK000
RESPRAMIN see AJD000
RESTAMIN see BBV500
RESTAMINE see BBV500
RESTENIL see MQU750
RESTOVAR see LJE000
RESTRAN see RDK000
RESTROL see DAL600
RESULFON see AHO250
RETACEL see CMF400
RETALON see DAL600
RETARD see DMC600
RETARDER AK see PHW750
RETARDER BA see BCL750
RETARDER ESEN see PHW750
RETARDER J see DWI000
RETARDER PD see PHW750
RETARDER W see SAI000
RETARDEX see BCL750
RETARPEN see BFC750
RETIN-A see VSK950
9-cis-RETINAL see VSK975
9-cis-RETINALDEHYDE see VSK975
RETINOIC ACID see VSK950
13-cis-RETINOIC ACID see VSK955
β-RETINOIC ACID see VSK950
all-trans-RETINOIC ACID see VSK950
RETINOL see VSK600
all-trans RETINOL see VSK600
RETINOL ACETATE see VSK900
RETINOL PALMITATE see VSP000
RETINYL ACETATE see VSK900
all-trans-RETINYL ACETATE see VSK900
RETINYL PALMITATE see VSP000
RETRANGOR see EID200
cis-RETRONECIC ACID ESTER of RETRONECINE see RFP000
cis-RETRONECIC ACID ESTER of RETRONECINE-N-OXIDE see RFU000
RETRORSINE see RFP000

RETRORSINE-N-OXIDE see RFU000
RETROVITAMIN A see VSK600
REUDO see BRF500
REUMACIDE see IDA000
REUMASYL see BRF500
REUMAZOL see BRF500
REUPOLAR see BRF500
REVAC see DJL000
REVENGE see DGI400
REVIDEX see BBK500
REWOMID DLMS see BKE500
REWOPOL HV-9 see PKF000
REWOPOL NLS 30 see SIB600
REXALL 413S see PMP500
REXCEL see CCU150
REXENE see PMP500
REXENE 106 see ADY500
REXOLITE 1422 see SMQ500
REZERPIN see RDK000
REZIFILM see TFS350
RF 10 see CCU250
RFNA see NEE500
RG 600 see ARM268
R-GENE see AQW000
RH see RHF000
RH-787 see PPP750
RH 893 see OFE000
RH-2915 see OQU100
RH 3866 see MRW775
RH-53,866 see MRW775
RHEIC ACID see RHZ700
RHEIN see RHZ700
RHENIUM TRICHLORIDE see RGP000
RHENOCURE CA see DWN800
RHENOSORB C see CAU500
RHENOSORB F see CAU500
RHEOSMIN see RBU000
RHEUMIN TABLETTEN see ADA725
RHINALL see SPC500
RHINATHIOL see CBR675
RHIZOPIN see ICN000
RHODACRYST see VSZ000
RHODAMIN 6G see RGW000
RHODAMINE 6GB see RGW000
RHODAMINE 6G (BIOLOGICAL STAIN) see RGW000
RHODAMINE 6GBN see RGW000
RHODAMINE 6G CHLORIDE see RGW000
RHODAMINE 590 CHLORIDE see RGW000
RHODAMINE 6GCP see RGW000
RHODAMINE 4GD see RGW000
RHODAMINE 6GD see RGW000
RHODAMINE 5GDN see RGW000
RHODAMINE 6Zh-DN see RGW000
RHODAMINE 6GEX ETHYL ESTER see RGW000
RHODAMINE 6G EXTRA see RGW000
RHODAMINE 6G EXTRA BASE see RGW000
RHODAMINE 6G EXTRA BASE see RGW000
RHODAMINE F4G see RGW000
RHODAMINE F5G see RGW000
RHODAMINE F5G CHLORIDE see RGW000
RHODAMINE F 5GL see RGW000
RHODAMINE GDN see RGW000
RHODAMINE 6 GDN see RGW000
RHODAMINE 6 GDN EXTRA see RGW000
RHODAMINE 4GH see RGW000
RHODAMINE 6GH see RGW000
RHODAMINE 6JH see RGW000
RHODAMINE 6ZH see RGW000

RHODAMINE 7JH see RGW000
RHODAMINE J see RGW000
RHODAMINE 5GL see RGW000
RHODAMINE 6G LAKE see RGW000
RHODAMINE LAKE RED 6G see RGW000
RHODAMINE 69DN EXTRA see RGW000
RHODAMINE 6GO see RGW000
RHODAMINE 6GX see RGW000
RHODAMINE Y 20-7425 see RGW000
RHODAMINE ZH see RGW000
RHODANID see ANW750
RHODANIDE see PLV750
RHODANIDE see ANW750
RHODIACHLOR see HAR000
RHODIACID see BJK500
RHODIACIDE see EEH600
RHODIANEBE see MAS500
RHODIA RP 11974 see BDJ250
RHODIASOL see PAK000
RHODIASTAB 83 see PFU300
RHODIATOX see PAK000
RHODIATROX see PAK000
RHODIFAX 16 see CPI250
RHODINE see ADA725
RHODINOL see CMT250
RHODINOL ACETATE see RHA000
RHODINOL (FCC) see DTF400
RHODINYL ACETATE see RHA000
RHODINYL FORMATE see RHA100
RHODIUM see RHF000
RHODIUM CHLORIDE see RHK000
RHODIUM(III) CHLORIDE (1:3) see RHK000
RHODIUM, (1,5-CYCLOOCTADIENE)(2,4-
PENTANEDIONATO)- see CPR840
RHODIUM, ((1,2,5,6-eta)-1,5-CYCLOOCTADIENE)(2,4-
PENTANEDIONATO-O,O')- see CPR840
RHODIUM METAL (OSHA) see RHF000
RHODIUM TRICHLORIDE see RHK000
RHODOCIDE see EEH600
RHODOLNE see SMQ500
RHODOPAS 6000 see AAX175
RHODOPAS M see AAX250
RHODOVIOL see PKP750
RHODOVIOL 4/125 see PKP750
RHODOVIOL 4-125P see PKP750
RHODOVIOL 16/200 see PKP750
RHODOVIOL R 16/20 see PKP750
RHODULINE ORANGE see BJF000
RHOMELLOSE see MIF760
RHOMENE see CIR250
RHONOX see CIR250
RHOPLEX AC-33 (Rohm and Haas) see EMF000
RHOPLEX B 85 see PKB500
RHOTEX GS see SJK000
RHOTHANE see BIM500
RHOTHANE D-3 see BIM500
RHUBARB YELLOW see RHZ700
RHYUNO OIL see SAD000
RIANIL see CLO750
RIBALL see ZVJ000
RIBIPCA see RIK000
RIBODERM see RIK000
RIBOFLAVIN see RIK000
RIBOFLAVINE see RIK000
RIBOFLAVINEQUINONE see RIK000
RIBOFURANOSIDE, GUANINE-9, β-D- see GLS000

RIBOFURANOSIDE, 9H-PURINE-6-THIOL-9 see
MCQ500
9-β-D-RIBOFURANOSYLGUANINE see GLS000
RIBONOSINE see IDE000
β-d-RIBOSYL-6-METHYLTHIOPURINE see MPU000
RIBOSYL-6-THIOPURINE see MCQ500
RICE BRAN WAX see RJF800
RICE STARCH see SLJ500
RICHAMIDE 6310 see BKE500
RICHONATE 1850 see DXW200
RICHONIC ACID B see LBU100
RICHONOL C see SIB600
RICIFON see TIQ250
RICIN (HAITI) see CCP000
RICININ see RJK100
RICININE see RJK100
RICINO (PUERTO RICO) see CCP000
RICINUS COMMUNIS see CCP000
RICINUS OIL see CCP250
RICIRUS OIL see CCP250
RICKETON see CMC750
RICON 100 see SMR000
RICYCLINE see TBX250
RIDOMIL see MDM100
RIDOMIL 2E see MDM100
RIFA see RKP000
RIFADINE see RKP000
RIFAGEN see RKP000
RIFALDAZINE see RKP000
RIFALDIN see RKP000
RIFAMATE see RKP000
RIFAMPICIN see RKP000
RIFAMPICINE (FRENCH) see RKP000
RIFAMPICINUM see RKP000
RIFAMPIN see RKP000
RIFAMYCIN AMP see RKP000
RIFAPRODIN see RKP000
RIFINAH see RKP000
RIFLE POWDER see PLL750
RIFLE POWDER see ERF500
RIFLOC RETARD see CCK125
RIFOBAC see RKP000
RIFOLDIN see RKP000
RIFORAL see RKP000
RIGEDAL see CCK125
RIGETAMIN see EDC500
RIGIDEX see PJS750
RIGIDEX 35 see PJS750
RIGIDEX 50 see PJS750
RIGIDEX TYPE 2 see PJS750
RIGIDIL see BBV500
RIGIDYL see BBV500
RIKEMAL O 71D see GGR200
RIKEMAL OL 100 see GGR200
RIKEMAL S 250 see SKV150
RIKER 601 see TND000
RIMACTAN see RKP000
RIMACTAZID see RKP000
RIMAON see EDW500
RIMIDIN see FAK100
RIMSO-50 see DUD800
RINATIOL see CBR675
RIOL see FLZ050
RIOMITSIN see HOH500
RIPERCOL-L see LFA020
RIRILIM see BBW500
RIRIPEN see BBW500

RISELECT see DGI000
RISERPA see RDK000
RISTAT see HMY000
RITALIN see MNQ000
RITALINE see MNQ000
RITCHER WORKS see MNQ000
RITMENAL see DKQ000
RITOSEPT see HCL000
RITSIFON see TIQ250
RIVADORN see NBU000
RIVANOL see EDW500
RIVASED see RDK000
RIVASIN see RDK000
RIVINOL see EDW500
RIVOMYCIN see CDP250
RL-50 see MRL500
RMC see MAC750
(R-(1α,2β,5α))-5-METHYL-2-(1-METHYLETHYL)-
CYCLOHEXANOL ACETATE (9CI) see MCG750
R(−)-MEZATON see NCL500
RMI9,384A see DLS600
RO 2-9757 see FMM000
RO 4-2130 see SNK000
RO-4-3780 see VSK955
RO 4-5360 see DLY000
RO 4-6316 see DNA200
RO 4-6467 see PME500
RO 5-0360 see DAR400
RO 5-0690 see MDQ250
RO 5-3059 see DLY000
RO 7-1554 see IGH000
RO 7-5050 see MIA250
Ro-5-6901/3 see FMQ000
ROACH SALT see SHF500
ROAD ASPHALT see PCR500
ROAD ASPHALT (DOT) see ARO500
ROAD TAR (DOT) see ARO500
RO-AMPEN see AIV500
RO-AMPEN see AOD125
ROBAMATE see MQU750
ROBANUL see GIC000
ROBENIDINE see RLK890
ROBIMYCIN see EDH500
ROBINUL see GIC000
ROBORAL see PAN100
ROCHIPEL see DIR000
ROCIPEL see DIR000
ROCK CANDY see SNH000
ROCK OIL see PCR250
ROCK SALT see SFT000
ROCORNAL see DIO200
RO-CYCLINE see TBX250
RODANCA see PLV750
RODANIN S-62 (CZECH) see IAQ000
RO-DETH see WAT200
RO-DEX see SMN500
RODEX see FFF000
RODINAL see ALT250
RODINE see RCF000
RODINOL see CMT250
RODIPAL see DIR000
RODOCID see EEH600
ROE 101 see DBB200
ROERIDORM see CHG000
ROGODIAL see DSP400
ROGOR see DSP400
ROGUE see DGI000

ROHYDRA see BAU750
RO-HYDRAZIDE see CFY000
ROIDENIN see IIU000
ROKON see BDF000
ROLAMID CD see BKE500
ROLAZINE see HGP500
ROLL-FRUCT see CDS125
ROLSERP see RDK000
ROMACRYL see PKB500
ROMAN VITRIOL see CNP250
ROMERGAN see DQA400
ROMETIN see CHR500
ROMEZIN see ALF250
ROMICIL see OHM900
ROMOSOL see ART250
ROMPHENIL see CDP250
ROMPUN see DMW000
ROMULGIN O see PKL100
RONDIS R see RNU100
RONILAN see RMA000
RONIN see PPO000
RONNEL see RMA500
ROOTONE see NAK500
ROOTONE see ICP000
ROP 500 F see GIA000
ROPOL see PJS750
ROPOTHENE OB.03-110 see PJS750
ROPTAZOL see NGG500
RORASUL see ASB250
ROSANIL see DGI000
p-ROSANILINE HCL see RMK020
p-ROSANILINE HYDROCHLORIDE see RMK020
ROSCOSULF see SNN300
ROSE BENGAL SODIUM see RMP175
ROSE ETHER see PER000
ROSE GERANIUM OIL ALGERIAN see GDA000
ROSE de GRASSE see RNA000
ROSE de MAI see RNA000
ROSEMARIE OIL see RMU000
ROSEMARY OIL see RMU000
ROSEMIDE see CHJ750
ROSEN OEL (GERMAN) see RNA000
ROSENOL see RNA000
ROSENSTHIEL see MAT250
ROSE OIL see PDD750
ROSE OIL see RNA000
ROSE OIL BULGARIAN see RNA000
ROSE OTTO see RNA000
ROSE QUARTZ see SCI500
ROSE QUARTZ see SCJ500
ROSIN see RNU100
ROSIN CORE SOLDER PYROLYSIS PRODUCTS see
RNU100
ROSIN WW see RNU100
ROSMARIN OIL (GERMAN) see RMU000
RO-SULFIRAM see DXH250
ROTATE see DQM600
ROTAX see BDF000
ROT B see XRA000
ROT C see OHI200
ROTERSEPT see BIM250
ROT G see OHI200
ROT GG FETTLOESLICH see XRA000
ROTHANE see BIM500
ROTOX see MHR200
ROUGE see IHC450
ROUGE CERASINE see OHI200
ROUGE de COCHENILLE A see FMU080

ROUGH & READY MOUSE MIX see WAT200
ROUGH & READY RAT BAIT & RAT PASTE see RCF000
ROUGOXIN see DKN400
ROVRAL see GIA000
ROXARSONE (USDA) see HMY000
ROXEL see RDK000
ROXINOID see RDK000
ROXION U.A. see DSP400
ROXOSUL TABLETS see SNN500
ROXYNOID see RDK000
ROYAL BLUE see DJO000
ROYAL CBTS see CPI250
ROYAL MBTS see BDE750
ROYAL MH-30 see DMC600
ROYAL SLO-GRO see DMC600
ROYAL SPECTRA see CBT750
ROYALTAC see DAI600
ROYAL TMTD see TFS350
ROZOL see CJJ000
ROZTOZOL see CKM000
RP 2275 see AHO250
R.P. 2591 see DFX400
RP 2616 see PPP500
RP 2632 see ALF250
2643-RP see ALF250
RP 2990 see TEX250
3277 R.P. see PMI750
3277 RP see DQA400
RP 3356 see DIR000
3389 R.P. see DQA400
RP 3799 see DIW000
4182 R.P. see DQA400
4753 R.P. see TGD000
6140 RP see PMF500
6909 RP see PIW000
RP7293 see VRF000
RP 8167 see EEH600
RP 8823 see MMN250
RP 8908 see PIW000
RP 10192 see MIJ500
10257 R.P. see TND000
10633 RP see TMK100
RP 13057 see DAC000
13,057 R.P. see DAC000
19583 RP see BDU500
RP 26019 see GIA000
RPA 2 see NAP500
RPA NO. 2 see NAP500
R-PENTINE see CPU500
R(−)-PHENYLEPHRINE see NCL500
4560 RP HYDROCHLORIDE see CKP500
2786 R.P. MALEATE see DBM800
RPR-5 see EIB600
RPR-V see EIB600
R 14 (REFRIGERANT) see CBY250
(R,R,R)-α-TOCOPHEROL see VSZ450
(R,R,R)-α-TOCOPHERYL ACETATE see TGJ050
(2R,4'R,8'R)-α-TOCOPHEROL see VSZ450
(2R,4'R,8'R)-α-TOCOPHERYL ACETATE see TGJ050
RS see CCU250
RS 141 see CJJ250
RS 1280 see CBF250
R SALT see ROF300
R.S. NITROCELLULOSE see CCU250
R105 SODIUM see RMP175

(1R,3S)3[(1'RS)(1',2',2',2'-TETRABROMOETHYL)]-2,2-DIMETHYLCYCLOPROPANECARBOXYLIC ACID (S)-α-CYANO-3-PHENOXYBENZYL ESTER see TBJ600
RTEC (POLISH) see MCW250
RU-4723 see CIR750
RUBATONE see BRF500
"522" RUBBER ACCELERATOR see PIY500
RUBBER CEMENT see CCW250
RUBEANE see DXO200
RUBEANIC ACID see DXO200
RUBENS BROWN see MAT500
RUBESOL see VSZ000
RUBIAZOL A see SNM500
RUBIDIUM see RPA000
RUBIDIUM CHLORIDE see RPF000
RUBIDIUM DICHROMATE see RPK000
RUBIDIUM FLUORIDE see RPP000
RUBIDIUM HYDROXIDE see RPZ000
RUBIDIUM HYDROXIDE SOLUTION (UN 2677) (DOT) see RPZ000
RUBIDIUM HYDROXIDE (UN 2678) (DOT) see RPZ000
RUBIDOMYCIN see DAC000
RUBIDOMYCINE see DAC000
RUBIGAN see FAK100
RUBIGINE see HHU500
RUBIGO see IHC450
RUBINATE 44 see MJP400
RUBINATE M see PKB100
RUBINATE MF 178 see PKB100
RUBINATE MF 182 see PKB100
RUBINATE TDI see TGM740
RUBINATE TDI 80/20 see TGM740
RUBINATE TDI 80/20 see TGM750
RUBITOX see BDJ250
RUBOMYCIN C see DAC000
RUBOMYCIN C 1 see DAC000
RUBRAMIN see VSZ000
RUBRIPCA see VSZ000
RUBROCITOL see VSZ000
RUBRUM SCARLATINUM see SBC500
RUCAINA see DHK400
RUCAINA HYDROCHLORIDE see DHK600
RUCOFLEX PLASTICIZER DOA see AEO000
RUCON B 20 see PKQ059
RUKSEAM see DAD200
RUMESTROL 1 see DKA600
RUMESTROL 2 see DKA600
RUMETAN see ZLS000
RUNA RH20 see TGG760
RUNCATEX see CIR500
RUOCID see AHO250
RUSSIAN COMFREY ROOTS see RRP000
RUTGERS 612 see EKV000
RUTHENIUM see RRU000
RUTHENIUM CHLORIDE see RRZ000
RUTHENIUM COMPOUNDS see RSF000
RUTHENIUM TRICHLORIDE see RRZ000
RUTILE see TGG760
RVK see DXO200
RVM-FG see PAE750
RYLAM see CBM750
RYOMYCIN see HOH500
RYSER see RDK000
RYTHRITOL see PBC250
S51 see BBV500
S115 see NCQ900
S 140 see DAM700

S 151 see EJM500
S 173 see SMQ500
S 202 see DHK600
S 75M see SFO500
S 767 see FAQ800
S-805 see DCS200
S 1065 see MIB750
S 1096 see GGR200
S 1097 see GGR200
S 112A see DSQ000
S 1544 see PFC750
S 2225 see EGV500
S-3151 see AHJ750
S 4068 see THI500
S 5602 see FAR100
S 5660 see DSQ000
S 6900 see DRR200
S 10165 see DGI000
S 1096R see GGR200
S18327 see FGA200
S 9490-3 see PCJ230
S 7481F1 see CQH100
S 65 (polymer) see PKQ059
SA see SAI000
SA 111 see SNJ000
SAATBEIZFUNGIZID (GERMAN) see HCC500
SABARI see HGC500
SACARINA see BCE500
SACCAHARIMIDE see BCE500
SACCHARATED FERRIC OXIDE see IHG000
SACCHARATED IRON see IHG000
SACCHARIN see SJN700
SACCHARINA see BCE500
SACCHARIN ACID see BCE500
SACCHARIN CALCIUM see CAM750
SACCHARINE see BCE500
SACCHARINE SOLUBLE see SJN700
SACCHARINNATRIUM see SJN700
SACCHARINOL see BCE500
SACCHARINOSE see BCE500
SACCHARIN, SODIUM see SJN700
SACCHARIN, SODIUM SALT see SJN700
SACCHARIN SOLUBLE see SJN700
SACCHAROIDUM NATRICUM see SJN700
SACCHAROL see BCE500
SACCHAROSE see SNH000
SACCHARUM see SNH000
SACCHARUM LACTIN see LAR000
SACERIL see DKQ000
SACERIL see DNU000
SACERNO see MKB250
SADH see DQD400
SADOFOS see MAK700
SADOPHOS see MAK700
SADOPLON see TFS350
SADOREUM see IDA000
SAEURE DES PHYTINS (GERMAN) see PIB250
SAEURE FLUORIDE (GERMAN) see FEZ000
SAFARITONE YELLOW G see AAQ250
SAFFRON see SAC100
SAFFRON YELLOW see DUX800
SAFROL see SAD000
SAFROLE see SAD000
SAFROLE MF see SAD000
SAFROTIN see MKA000
SAFSAN see DXE000
SAGATAL see NBU000
SAH 22 see SEM500

SAKOLYSIN (GERMAN) see BHT750
SAL see HMK100
SALACETIN see ADA725
SALACHLOR see SHJ000
SAL AMMONIA see ANE500
SAL AMMONIAC see ANE500
SALAMMONITE see ANE500
SALCETOGEN see ADA725
SAL ENIXUM see PKX750
SALETIN see ADA725
SALICRESIN FLUID see CHW675
SALICYL ALCOHOL see HMK100
SALICYLALDEHYDE DIMETHYL ACETAL
CARBAMATE see SAG100
SALICYLALDEHYDE METHYL ETHER see AOT525
SALICYLIC ACID see SAI000
m-SALICYLIC ACID see HJI100
SALICYLIC ACID, 3,5-DINITRO- see HKE600
SALICYLIC ACID, ISOBUTYL ESTER see IJN000
SALICYLIC ACID, ISOPENTYL ESTER see IME000
SALICYLIC ACID, METHYL ESTER see MPI000
SALICYLIC ACID with PHYSOSTIGMINE (1:1) see
PIA750
SALICYLIC ACID, SODIUM SALT see SJO000
SALICYLIC ACID, p-TOLYL ESTER see THD850
SALIGENIN see HMK100
SALIGENOL see HMK100
SALINE see SFT000
SALISAN see CLH750
SALISOD see SJO000
SALIX see CHJ750
SALMIAC see ANE500
SALOL see PGG750
SALPETERSAEURE (GERMAN) see NED500
SALPETERZUUROPLOSSINGEN (DUTCH) see
NED500
SALSONIN see SJO000
SALT see SFT000
SALT CAKE see SJY000
SALT OF TARTAR see PLA250
SALTPETER see PLL500
SALT of SATURN see LCV000
SALTS of FATTY ACIDS see SAO550
SALUFER see DXE000
SALUNIL see CLH750
SALUPRES see RDK000
SALURETIL see CLH750
SALURIC see CLH750
SALURIN see SIH500
SALUTENSIN see RDK000
SALVACARD see DJS200
SALVACORIN see DJS200
SALVIS see SEP000
SALVO see HKC000
SALVO LIQUID see BCL750
SALVO POWDER see BCL750
SALYRGAN see SIH500
SALYZORON see BBW500
SAMARIUMACETAT (GERMAN) see SAR000
SAMARIUM ACETATE see SAR000
SAMARIUM(III) CHLORIDE see SAR500
SAMARON BRILLIANT VIOLET B see DBY700
SAMARON YELLOW PA3 see AAQ250
SAN see ALQ625
SAN 230 see PHI500
SANASEED see SMN500
SANATRICHOM see MMN250
SANCELER CM-PO see CPI250

SAN-CYAN see COI250
SAND see SCI500
SAND see SCJ500
SAND ACID see SCO500
SANDIMMUN see CQH100
SANDIMMUNE see CQH100
SANDIX see LDS000
SANDOCRYL BLUE BRL see BJI250
SANDOLAN RED N-RS see CMM330
SANDOLIN see DUS700
SANDOPEL BLACK EX see AQP000
SANDORMIN see TEH500
SANDOZ 6538 see DJY200
SANDOZ 52139 see MKA000
SANDRIL see RDK000
SANDRON see RDK000
SANEDRINE see EAW000
SAN-EI BRILLIANT SCARLET 3R see FMU080
SAN-EI TARTRAZINE see FAG140
SANEPIL see DKQ000
SANFURAN see NGE500
SANG gamma see BBQ500
SAN 244 I see DRR200
SAN 6538 I see DJY200
SAN 6913 I see DRR200
SAN 7107 I see DRR200
SAN 52 139 I see MKA000
SANICLOR 30 see PAX000
SANIPIROL see SEP000
SANITIZED SPG see ABU500
SANLOSE SN 20A see SFO500
SANMARTON see FAR100
SANMORIN OT 70 see DJL000
SANOCID see HCC500
SANOCIDE see HCC500
SANODIAZINE see PPP500
SANPRENE LQX 31 see PKP000
SANQUINON see DFT000
SANREX see ADY500
SANSEL ORANGE G see PEJ500
SANSERT see MQP500
SANSPOR see CBF800
SANTAL OIL see OGY220
α-SANTALOL (FCC) see OGY220
SANTALOZONE see DAF150
SANTALYL ACETATE see SAU400
SANTAR see MCT500
SANTAVY'S SUBSTANCE F see MIW500
SANTHEOSE see TEO500
SANTICIZER 160 see BEC500
SANTICIZER 141 (MONSANTO) see DWB800
SANTICIZIER B-16 see BQP750
SANTOBANE see DAD200
SANTOBRITE see PAX250
SANTOBRITE see SJA000
SANTOCEL see SCH002
SANTOCHLOR see DEP800
SANTOCURE see CPI250
SANTOCURE MOR see BDG000
SANTOCURE VULCANIZATION ACCELERATOR see CPI250
SANTOFLEX A see SAV000
SANTOFLEX AW see SAV000
SANTOFLEX IC see PEY500
SANTOMERSE 3 see DXW200
SANTONOX see TFC600
SANTOPHEN see PAX250

SANTOPHEN 20 see PAX250
SANTOQUIN see SAV000
SANTOQUINE see SAV000
SANTOQUIN EMULSION see TMJ000
SANTOQUIN MIXTURE 6 see TMJ000
SANTOTHERM see PJL750
SANTOTHERM FR see PJL750
SANTOWAX see TBC750
SANTOWAX M see TBC620
SANTOWHITE CRYSTALS see TFC600
SANTOX see EBD700
SANWAX 161P see PJS750
SANYO FAST BLUE SALT B see DCJ200
SANYO FAST RED SALT TR see CLK235
SANYO FAST RED TR BASE see CLK220
SANYO LAKE RED C see CHP500
SAOLAN see DSK200
SAPECRON see CDS750
SAPROL see TKL100
SARCELL TEL see SFO500
SARCLEX see DGD600
SARCOCLORIN see BHT750
l-SARCOLYSIN see PED750
dl-SARCOLYSIN see BHT750
p-l-SARCOLYSIN see PED750
dl-SARCOLYSINE see BHT750
SARIN see IPX000
SARIN II see IPX000
SARLACH R see SBC500
SARODORMIN see DYC800
SAROLEX see DCM750
SAROMET see DCK759
SAROTEN see EAI000
SAROTENE see EAI000
SARPAGAN see RDK000
SARPAGEN see RDK000
SARPIFAN HP 1 see AAX175
SAS 2185 see BHL800
SASSAFRAS see SAY900
SASSAFRAS ALBIDUM see SAY900
SASSAFRAS OIL see OHI000
SATIAGEL GS 350 see CCL250
SATIAGUM 3 see CCL250
SATIAGUM STANDARD see CCL250
SATINITE see CAX750
SATIN SPAR see CAX750
SATOX 20WSC see TIQ250
SATURN BROWN LBR see CMO750
SATURN RED see LDS000
SAUTERALGYL see DAM700
SAVENTRINE see DMV600
SAVIT see CBM750
SAVORY OIL (summer variety) see SBA000
SAX see SAI000
SAXIN see SJN700
SAXIN see BCE500
SAXOL see MQV750
SAXOSOZINE see SNN500
SAYTEX 102 see PAU500
SAYTEX 102E see PAU500
SAZZIO see AFL000
SB 475K see SMQ500
SBa 0108E see BAK000
S.B.A. see BPW750
SBO see DJL000
SBP-1382 see BEP500
SBP-1513 see AHJ750
S.B. PENICK 1382 see BEP500

SBS see SMR000
SC 9 see SCQ000
SC-110 see ABU500
SC 201 see SCQ000
SC 2538 see DIR000
SC-4642 see EEH550
SC 9420 see AFJ500
SC 9794 see PII500
SC 10295 see MMN250
SC10363 see VTF000
SC 15090 see TEO500
SC 15983 see AFJ500
SCABANCA see BCM000
SCALDIP see DVX800
SCANBUTAZONE see BRF500
SCANDICAIN see SBB000
SCANDICAIN see CBR250
SCANDICAINE see SBB000
SCANDICANE see SBB000
SCANDISIL see SNN300
SCANDIUM CHLORIDE see SBC000
SCANDIUM(3+) CHLORIDE see SBC000
SCARCLEX see DGD600
SCARLET BASE CIBA II see NMP500
SCARLET BASE IRGA II see NMP500
SCARLET BASE NSP see NMP500
SCARLET B FAT SOLUBLE see OHI200
SCARLET G BASE see NMP500
SCARLET R see FMU070
SCARLET RED see SBC500
SCARLET RED, BIEBRICH see SBC500
SCARLET R (MICHAELIS) see SBC500
SCARLET TR BASE see CLK200
SCATOLE see MKV750
SC 9 (CARBIDE) see SCQ000
SCH 9724 see GCS000
SCHARLACH B see PEJ500
SCHARLACHROT see SBC500
SCHEELES GREEN see CNN500
SCHEELE'S MINERAL see CNN500
SCHERCEMOL 85 see ISC550
SCHERING 36056 see DSO200
SCHERING 36268 see CJJ250
SCHERISOLON see PMA000
SCHEROSON F see CNS750
(−)-SCH 39166 HYDROCHLORIDE see BCP690
SCHRADAN see OCM000
SCHRADANE (FRENCH) see OCM000
SCHULTENITE see LCK000
SCHULTZ No. 39 see HGC000
SCHULTZ No. 95 see FMU070
SCHULTZ No. 770 see FMU059
SCHULTZ No. 1038 see BJI250
SCHULTZ Nr. 208 (GERMAN) see HJF500
SCHULTZ NO. 31 see OHI200
SCHULTZ NO. 541 see SBC500
SCHULTZ NO. 737 see FAG140
SCHULTZ-TAB No. 779 (GERMAN) see RMK020
SCHWEFELDIOXYD (GERMAN) see SOH500
SCHWEFELKOHLENSTOFF (GERMAN) see CBV500
SCHWEFEL-LOST see BIH250
SCHWEFELSAEURELOESUNGEN (GERMAN) see SOI500
SCHWEFELWASSERSTOFF (GERMAN) see HIC500
SCHWEFLIGESAEURE (GERMAN) see SOO500
SCHWEINFURTERGRUEN see COF500
SCHWEINFURT GREEN see COF500

NSC-762 HYDROCHLORIDE see BIE500
SCILLIROSIDE GLYCOSIDE see RCF000
SCINTILLAR see XHS000
SCLAIR 59 see PJS750
SCLAIR 11K see PJS750
SCLAIR 19A see PJS750
SCLAIR 59C see PJS750
SCLAIR 79D see PJS750
SCLAIR 96A see PJS750
SCLAIR 19X6 see PJS750
SCLAIR 2911 see PJS750
SCLAVENTEROL see NGG500
S-COL see SCK600
SCON 5300 see PKQ059
SCONATEX see VPK000
SCONATEX see AAX175
SCOPAMIN see HOT500
SCOPARON see DRS800
SCOPARONE see DRS800
SCOPINE TROPATE see SBG000
SCOPOLAMINE see SBG000
(−)-SCOPOLAMINE see SBG000
SCOPOLAMINE BROMIDE see HOT500
(−)-SCOPOLAMINE BROMIDE see HOT500
SCOPOLAMINE HYDROBROMIDE see HOT500
(−)-SCOPOLAMINE HYDROBROMIDE see HOT500
SCOPOLAMINIUM BROMIDE see HOT500
SCOPOLAMMONIUM BROMIDE see HOT500
SCOPOS see HOT500
SCOTCH PINE NEEDLE OIL see PIH500
SCURENALINE see VGP000
SCUROCAINE see AIT250
SCUTL see ABU500
SCW 1 see SCQ000
SCYAN see SIA500
SD 188 see SMQ500
SD 354 see SMR000
SD 709 see DRM000
SD-1750 see DGP900
SD 1897 see DDL800
SD 2614 see KFK200
SD 3562 see DGQ875
SD 5532 see CDR750
SD 7961 see DGM600
SD 9129 see MRH209
SD 14114 see BLU000
SD 15418 see BLW750
SD 43775 see FAR100
SD ALCOHOL 23-HYDROGEN see EFU000
S DC 200 see SCR400
SDDC see SGM500
SDEH see DJL400
SD-GP 6000 see SCQ000
SD-GP 8000 see SCQ000
SDIC see SGG500
SDM see SNN300
SDMH see DSF600
SDM No. 5 see MAW250
SDM No. 23 see PBC250
SDMO see SNN300
SDP 640 see PJS750
SDPH see DNU000
SEA COAL see CMY760
SEAKEM CARRAGEENIN see CCL250
SEA-LEGS see HGC500
SEARLEQUIN see DNF600
SEA SALT see SFT000
SEATREM see CCL250

SEAWATER MAGNESIA see MAH500
SEAZINA see SNJ000
SEBACIC ACID, DIBUTYL ESTER see DEH600
SEBACIC ACID, DIETHYL ESTER see DJY600
SEBACIL see BAT750
SEBAQUIN see DNF600
SEBIZON see SNQ710
SECAGYN see EDC500
(1-α,3-β,5Z,7E,24R)-9,10-SECOCHOLESTA-5,7,10(19)-
TRIENE-1,3,24-TRIOL see SBL600
9,10-SECOCHOLESTA-5,7,10(19)-TRIENE-1,3,24-
TRIOL, (1-α,3-β,5Z,7E,24R)- see SBL600
9,10-SECOCHOLESTA-5,7,10(19)-TRIEN-3-β-OL see
CMC750
9,10,SECOERGOSTA-5,7,10(19),22-TETRAEN-3-β-OL
see VSZ100
SECONDARY AMMONIUM ARSENATE see DCG800
SECONDARY AMMONIUM PHOSPHATE see ANR500
SECROVIN see DNX500
SECUPAN see EDC500
SECURITY see ARB750
SECURITY see LCK000
SEDABAMATE see MQU750
SEDAFORM see ABD000
SEDALIS SEDI-LAB see TEH500
SEDANTOINAL see MKB250
SEDARAUPIN see RDK000
SEDARAUPINA see RDK000
SEDA-RECIPIN see RDK000
SEDA-SALUREPIN see RDK000
SEDA-TABLINEN see EOK000
SEDERAUPIN see RDK000
SEDESTRAN see DKA600
SEDETINE see OAV000
SEDETOL see EJM500
SEDICAT see EOK000
SEDIMIDE see TEH500
SEDIN see TEH500
SEDIPAM see DCK759
SEDISPERIL see TEH500
SEDIZORIN see EOK000
SEDOFEN see EOK000
SEDOMETIL see DNA800
SEDONAL see EOK000
SEDONEURAL see SFG500
SEDOPHEN see EOK000
SED OSTANTHRENOVA M see CMU475
SEDOVAL see TEH500
SEDRESAN see MEP250
SEDSERP see RDK000
SEDURAL see PDC250
SEDUXEN see DCK759
SEEDRIN see AFK250
SEEDTOX see ABU500
SEEKAY WAX see TIT500
SEFFEIN see CBM750
SEGNALE RED LC see CHP500
S. EGRELTRI ATUNUN (TURKISH) see BML000
SEGURIL see CHJ750
SELECRON see BNA750
SELEKTIN see BKL250
SELEKTON B 2 see EIX500
SELEN (POLISH) see SBO500
1,2,5-SELENADIAZOLE-3-CARBOXAMIDE, 4-AMINO-
see AMN300
1,2,5-SELENADIAZOLE-3-CARBOXAMIDE, 4-AMINO-
N-METHYL- see MGL600

SELENIC ACID see SBN500
SELENIC ACID, liquid (DOT) see SBN500
SELENIC ACID, DIPOTASSIUM SALT see PLR750
SELENIC ACID, DISODIUM SALT, DECAHYDRATE
see SBN505
SELENIDE, DIALLYL- see DBL300
SELENINIC ACID, PROPYL- see PNV760
β-SELENINOPROPIONIC ACID see SBN510
SELENINYL CHLORIDE see SBT500
SELENIOUS ACID see SBO000
SELENIOUS ACID, DIPOTASSIUM SALT (9CI) see
SBO100
SELENIOUS ACID, DISODIUM SALT see SJT500
SELENIOUS ACID DISODIUM SALT
PENTAHYDRATE see SJT600
SELENIOUS ANHYDRIDE see SBQ500
SELENIUM see SBO500
SELENIUM (colloidal) see SBP000
SELENIUM ALLOY see SBO500
SELENIUM BASE see SBO500
SELENIUM(IV) CHLORIDE (1:4) see SBU000
SELENIUM CHLORIDE OXIDE see SBT500
SELENIUM COMPOUNDS see SBP500
SELENIUM CYSTINE see DWY800
SELENIUM, DICHLOROBIS(2-
CHLOROCYCLOHEXYL)- see DEU100
SELENIUM, DICHLOROBIS(2-
ETHOXYCYCLOHEXYL)- see DEU125
SELENIUM DIETHYLDITHIOCARBAMATE see
DJD400
SELENIUM DIMETHYLDITHIOCARBAMATE see
SBQ000
SELENIUM DIOXIDE see SBO000
SELENIUM DIOXIDE see SBQ500
SELENIUM(IV) DIOXIDE (1:2) see SBQ500
SELENIUM DIOXIDE mixed with ARSENIC TRIOXIDE
(1:1) see ARJ000
SELENIUM(IV) DISULFIDE (1:2) see SBR000
SELENIUM DISULFIDE (2.5%) SHAMPOO see SBR500
SELENIUM(IV) DISULFIDE SHAMPOO (2.5%) see
SBR500
SELENIUM DISULPHIDE (DOT) see SBR000
SELENIUM DUST see SBO500
SELENIUM ELEMENTAL see SBO500
SELENIUM FLUORIDE see SBS000
SELENIUM HEXAFLUORIDE see SBS000
SELENIUM HOMOPOLYMER see SBO500
SELENIUM HYDRIDE see HIC000
SELENIUM METAL POWDER, NON-PYROPHORIC
(DOT) see SBO500
SELENIUM MONOSULFIDE see SBT000
SELENIUM OXIDE see SBQ500
SELENIUM OXIDE see SBT200
SELENIUM OXIDE (DOT) see SBT200
SELENIUM OXYCHLORIDE see SBT500
SELENIUM SULFIDE see SBR000
SELENIUM SULFIDE see SBT000
SELENIUM SULPHIDE see SBT000
SELENIUM TETRACHLORIDE see SBU000
2,2'-SELENOBIS(BENZOIC ACID) see SBU150
o,o'-SELENOBIS(BENZOIC ACID) see SBU150
SELENOCYANIC ACID, ESTER WITH α-
HYDROXYACETANILIDE see OOK175
SELENOCYANIC ACID, 2-OXO-2-
(PHENYLAMINO)ETHYL ESTER see OOK175
SELENOCYANIC ACID, POTASSIUM SALT see PLS000
SELENOCYSTINE see DWY800

SELENONIUM, TRIMETHYL- see TME500
SELENONIUM, TRIMETHYL-, CHLORIDE see TME600
SELENOSULFURIC ACID, 2-AMINOETHYL ESTER see AJS950
SELENO-TOLUIDINE BLUE see SBU950
SELENOUREA see SBV000
SELENSULFID (GERMAN) see SBT000
SELEPHOS see PAK000
SELEXOL see DOM100
SELFER see IDE000
SELF ROCK MOSS see CCL250
SELINON see DUS700
SELLA FAST RED RS see CMM330
SELLAITE see MAF500
SELSUN see SBW000
SELSUN BLUE see SBR000
SEL-TOX SSO2 and SS-20 see DXG000
SEM (cytostatic) see TND500
SEMAP see PAP250
SEMBRINA see DNA800
SEMDOXAN see CQC500
SEMDOXAN see CQC650
SEMICARBAZIDE see HGU000
SEMICARBAZIDE HYDROCHLORIDE see SBW500
SEMICARBAZIDE, 1-PHENYL-4-(PHENYLIMINO)-3-THIO- see DWN200
SEMICILLIN see AIV500
SEMOXYDRINE see DBA800
SEMUSTINE see CHD250
SENARMONTITE see AQF000
SENCOR see MQR275
SENCORAL see MQR275
SENCORER see MQR275
SENCOREX see MQR275
SENDOXAN see CQC500
SENDUXAN see CQC500
SENDUXAN see CQC650
SENECA OIL see PCR250
SENEGAL GUM see AQQ500
SENF OEL (GERMAN) see AGJ250
SENKIRKIN see DMX200
SENKIRKINE see DMX200
SENTRY see HOV500
SENTRY GRAIN PRESERVER see PMU750
SEOMINAL see RDK000
SEPPIC MMD see CIR250
SEPTACIL see ALF250
SEPTAMIDE ALBUM see SNM500
SEPTENE see CBM750
SEPTICOL see CDP250
SEPTINAL see SNM500
SEPTIPULMON see PPO000
SEPTISOL see HCL000
SEPTOCHOL see DAQ400
SEPTOFEN see HCL000
SEPTOPLEX see SNM500
SEPTOS see HJL500
SEPTOTAN see PFN000
SEPTRA see SNK000
SEPTRAN see SNK000
SEQ 100 see EIX000
SEQUESTRENE 30A see EIV000
SEQUESTRENE AA see EIX000
SEQUESTRENE Na3 see TNL250
SEQUESTRENE Na 4 see EIV000
SEQUESTRENE SODIUM 2 see EIX500
SEQUESTRENE ST see EIV000

SEQUESTRENE TRISODIUM see TNL250
SEQUESTRENE TRISODIUM SALT see TNL250
SEQUESTRIC ACID see EIX000
SEQUESTROL see EIX000
SEREEN see HOT500
SERENACK see DCK759
SERENIL see CHG000
SERENSIL see CHG000
SEREN VITA see MDQ250
SERENZIN see DCK759
SERFIN see RDK000
SERFOLIA see RDK000
SERIL see MQU750
dl-SERINE see SCA350
l-SERINE DIAZOACETATE see ASA500
l-SERINE DIAZOACETATE (ester) see ASA500
SERINYL BLUE 2G see TBG700
SERINYL HOSIERY YELLOW GD see AAQ250
SERIPLAS YELLOW GD see AAQ250
SERISOL FAST RED 2B see AKE250
SERISOL FAST VIOLET B see DBY700
SERISOL FAST YELLOW GD see AAQ250
SERISOL ORANGE YL see AKP750
SERISTAN BLACK B see AQP000
SERITOX 50 see DGB000
SERMION see NDM000
SEROLFIA see RDK000
SEROTONIN BENZYL ANALOG see BEM750
SERP see RDK000
SERP-AFD see RDK000
SERPALAN see RDK000
SERPALOID see RDK000
SERPANEURONA see RDK000
SERPANRAY see RDK000
SERPASIL see RDK000
SERPASIL APRESOLINE see RDK000
SERPASIL APRESOLINE No. 2 see HGP500
SERPASIL-ESIDREX see RDK000
SERPASIL-ESIDREX K see RDK000
SERPASIL-ESIDREX NO. 1 see RDK000
SERPASIL-ESIDREX NO. 2 see RDK000
SERPASIL PREMIX see RDK000
SERPASOL see RDK000
SERPATE see RDK000
SERPATONE see RDK000
SERPAX see RDK000
SERPAZIL see RDK000
SERPAZOL see RDK000
SERPEDIN see RDK000
SERPEN see RDK000
SERPENA see RDK000
SERPENTIL see RDK000
SERPENTIN see RDK000
SERPENTINA see RDK000
SERPENTINE see ARM250
SERPENTINE see ARM268
SERPENTINE CHRYSOTILE see ARM268
SERPENTINE "PHARBIL" see RDK000
SERPICON see RDK000
SERPIL see RDK000
SERPILOID see RDK000
SERPILUM see RDK000
SERPINE see RDK000
SERPINE (Pharmaceutical) see RDK000
SERPIPUR see RDK000
SERPIVITE see RDK000
SERPLEX K see RDK000
SERPOGEN see RDK000

SERPOID see RDK000
SERPONE see RDK000
SERPRESAN see RDK000
SERPYRIT see RDK000
SERRAL see DKA600
SERTABS see RDK000
SERTAN see DBB200
SERTENS see RDK000
SERTENSIN see RDK000
SERTINA see RDK000
N-SERVE NITROGEN STABILIZER see CLP750
SERVISONE see PLZ000
SESAGARD see BKL250
SESQUIETHYLALUMINUM CHLORIDE see TJP775
SESQUISULFURE de PHOSPHORE (FRENCH) see
PHS500
S-ESTER with O,O-DIMETHYL
PHOSPHOROTHIOATE see MAK700
SET see MDI000
SETACYL DIAZO NAVY R see DCJ200
SETACYL PINK 3B see AKE250
SETACYL YELLOW G see AAQ250
SETACYL YELLOW P-2GL see AAQ250
SETACYL YELLOW 2GN see AAQ250
SETONIL see DCK759
SEVACARB see CBT750
SEVAL see CBT750
SEVENAL see EOK000
SEVICAINE see AIT250
SEVIMOL see CBM750
SEVIN see CBM750
SEVIN 4 see CBM750
SEVIN (OSHA) see CBM750
SEWIN see CBM750
SEXADIEN see DAL600
SEXOCRETIN see DKA600
SEXTONE see CPC000
SEXTONE B see MIQ740
SF 60 see MAK700
S-4068 SF see THI500
S6F HISTYRENE RESIN see SMR000
SF-2052 SULFATE see DAB630
SG-67 see SCH002
SGA see SCK600
SHA CHONG DAN see DXC900
SHA CHONG SHUANG see DXC900
SHALE OIL (DOT) see COD750
SHAMMAH (SAUDI ARABIA) see SED400
SHAMROX see CIR250
SHARSTOP 204 see SGM500
SHAWINIGAN ACETYLENE BLACK see CBT750
SHED-A-LEAF see SFS000
SHED-A-LEAF "L" see SFS000
SHELL 300 see SMQ500
SHELL 4072 see CDS750
SHELL 5520 see PMP500
SHELLAC, BLEACHED see SCC700
SHELL ATRAZINE HERBICIDE see ARQ725
SHELL CARBON see CBT750
SHELL GOLD see GIS000
SHELL MIBK see HFG500
SHELL SD-3562 see DGQ875
SHELL SD-5532 see CDR750
SHELL SD 9129 see MRH209
SHELL SD-14114 see BLU000
SHELL SILVER see SDI500
SHELLSOL 140 see NMX000
SHELL UNDRAUTTED A see AFV500

SHIGATOX see AHO250
SHIGRODIN see BRF500
SHIKIMOLE see SAD000
SHIKISO DIRECT SKY BLUE 5B see CMO500
SHIKISO DIRECT SKY BLUE 6B see CMN750
SHIKISO METANIL YELLOW see MDM775
SHIKOMOL see SAD000
SHIMMEREX see ABU500
SHINKOLITE see PKB500
SHIN-NAITO S see TEH500
SHINNIBROL see TEH500
SHINNIPPON FAST RED GG BASE see NEO500
SHIRAGIKU ROSIN see RNU100
SHIRLAN FLOW see CEX800
SHIRLAN (ZENECA) see CEX800
SHMP see SHM500
SHOALLOMER see PMP500
SHOCK-FEROL see VSZ100
SHOLEX 5003 see PJS750
SHOLEX 5100 see PJS750
SHOLEX 6000 see PJS750
SHOLEX 6002 see PJS750
SHOLEX F 171 see PJS750
SHOLEX F 6050C see PJS750
SHOLEX F 6080C see PJS750
SHOLEX L 131 see PJS750
SHOLEX 4250HM see PJS750
SHOLEX S 6008 see PJS750
SHOLEX SUPER see PJS750
SHOLEX XMO 314 see PJS750
SI see LCF000
SIARKI CHLOREK (POLISH) see SON510
SIARKI DWUTLENEK (POLISH) see SOH500
SIARKOWODOR (POLISH) see HIC500
SIBOL see DKA600
SICO FAT RED BG NEW see CMS238
SICOL 150 see DVL700
SICOL 160 see BEC500
SICOL 250 see AEO000
SICO LAKE RED 2L see CHP500
SICOMET 8400 see EHP700
SICOR ZNP/M see ZJS400
SICOR ZNP/S see ZJS400
SICRON see PKQ059
SIDNOFEN see SPA000
SIENNA see IHC450
SIERRA C-400 see TAB750
SIFERRIT see IHG100
SIGMACELL see CCU150
SIGURAN see HGC500
SILAK M 10 see SCR400
SILANE see SDH575
SILANE 40-43 see DJA330
SILANE 40-47 see TLC600
SILANE, (3-CYANOPROPYL)DIETHOXY(METHYL)-
see COR500
SILANE, (3-CYANOPROPYL)TRIETHOXY- see COR800
SILANE, DIMETHOXYMETHYL- see MJE900
SILANE, ETHOXYDIMETHYL- see EES200
SILANE 48-12 TETRAKIS see OCM100
SILANETETRAMINE, OCTAMETHYL- see OCM100
SILANE, TRICHLOROALLYL- see AGU250
SILANE, TRICHLOROCYCLOHEXYL- see CPR250
SILANE, TRICHLOROETHYL- see EPY500
SILANE, TRICHLOROMETHYL- see MQC500
SILANE, TRIETHOXY(3-CYANOPROPYL)- see COR800
SILANE, TRIETHOXY(3-ISOCYANATOPROPYL)- see
IKG900

SILANE, TRIMETHOXY- see TLB750
SILANE, TRIMETHYLCHLORO- see TLN250
SILANE, VINYL TRICHLORO 1-150 see TIN750
SILANOL, DIPHENYLMETHYL- see DWH550
SILANTIN see DKQ000
SILBER (GERMAN) see SDI500
SILBERNITRAT see SDS000
SILDENAFIL CITRATE see SCF600
SILENE EF see CAW850
SILICA AEROGEL see SCI000
SILICA, AMORPHOS-FUME (ACGIH) see SCH001
SILICA, AMORPHOUS see SCH002
SILICA, AMORPHOUS-DIATOMACEOUS EARTH
(UNCALCINED) (ACGIH) see DCJ800
SILICA, AMORPHOUS FUME see SCH001
SILICA, AMORPHOUS FUMED see SCH002
SILICA, AMORPHOUS-FUSED (ACGIH) see SCK600
SILICA, AMORPHOUS HYDRATED see SCI000
SILICA, CRYSTALLINE see SCI500
SILICA, CRYSTALLINE-CRISTOBALITE see SCJ000
SILICA, CRYSTALLINE-QUARTZ see SCJ500
SILICA, CRYSTALLINE-TRIDYMITE see SCK000
SILICA, CRYSTALLINE-TRIDYMITE (ACGIH, OSHA)
see SCK000
SILICA FLOUR see SCI500
SILICA FLOUR see SCK500
SILICA FLOUR (powdered crystalline silica) see SCJ500
SILICA, FUSED see SCK600
SILICA, FUSED see SCK600
SILICA, FUSED (OSHA) see SCK600
SILICA GEL see SCI000
SILICA GEL see SCL000
SILICA, GEL and AMORPHOUS-PRECIPITATED see
SCL000
SILICANE see SDH575
SILICANE, CHLOROTRIMETHYL- see TLN250
SILICANE, TRICHLOROETHYL- see EPY500
SILICATE D'ETHYLE (FRENCH) see EPF550
SILICATE(2-), HEXAFLUORO-, BARIUM see BAO750
SILICATE(2-), HEXAFLUORO-, BARIUM (1:1) (9CI) see
BAO750
SILICATE(2-), HEXAFLUORO-, CADMIUM (8CI,9CI)
see CAG500
SILICATE(2-), HEXAFLUORO-, CALCIUM (1:1) (9CI)
see CAX250
SILICATE(2-), HEXAFLUORO-, MAGNESIUM (1:1) see
MAF600
SILICATES see SCM500
SILICATE SOAPSTONE see SCN000
SILICA, VITREOUS (9CI) see SCK600
SILICA XEROGEL see SCI000
SILICIC ACID see SCI000
SILICIC ACID see SCL000
SILICIC ACID, BERYLLIUM SALT see SCN500
SILICIC ACID, 2-ETHYLBUTYL ESTER see EHC900
SILICIC ACID, METHYL ESTER of ortho- see MPI750
SILICIC ACID TETRAETHYL ESTER see EPF550
SILICIC ACID, ZIRCONIUM(4+) SALT (1:1) see ZSS000
SILICIC ANHYDRIDE see SCH002
SILICIC ANHYDRIDE see SCJ500
SILICI-CHLOROFORME (FRENCH) see TJD500
SILICIO(TETRACLORURO di) see SCQ500
SILICIUMCHLOROFORM (GERMAN) see TJD500
SILICIUMTETRACHLORID (GERMAN) see SCQ500
SILICIUMTETRACHLORIDE (DUTCH) see SCQ500
SILICIUM(TETRACHLORURE de) (FRENCH) see
SCQ500
SILICOCHLOROFORM see TJD500

SILICOETHANE see DXA000
SILICOFLUORIC ACID see SCO500
SILICON see SCP000
SILICON CARBIDE see SCQ000
SILICON CARBIDE (ACGIH, OSHA) see SCQ000
SILICON CHLORIDE see SCQ500
SILICON CHLORIDE HYDRIDE see DGK300
SILICON DIOXIDE see SCI500
SILICON DIOXIDE see SCK600
SILICON DIOXIDE (FCC) see SCH002
SILICONE 360 see SCR400
SILICONE DC 200 see SCR400
SILICONE DC 360 see SCR400
SILICONE DC 360 FLUID see SCR400
SILICONE DIOXIDE see SCK600
SILICONE RELEASE L 45 see SCR400
SILICONES see SDC000
SILICON FLUORIDE see SDF650
SILICON FLUORIDE BARIUM SALT see BAO750
SILICON MONOCARBIDE see SCQ000
SILICON PHENYL TRICHLORIDE see TJA750
SILICON POWDER, amorphous (DOT) see SCP000
SILICON SODIUM FLUORIDE see DXE000
SILICON TETRACHLORIDE (DOT) see SCQ500
SILICON TETRAFLUORIDE (DOT) see SDF650
SILICON TETRAHYDRIDE see SDH575
SILICON ZINC FLUORIDE see ZIA000
SILIKILL see SCH002
SILIKON ANTIFOAM FD 62 see SCR400
SILMOS T see CAW850
SILMURIN see RCF000
SILOSAN see DIN800
SILOSUPER PINK B see RGW000
SILOTRAS ORANGE TR see PEJ500
SILOTRAS RED T3B see SBC500
SILOTRAS RED TG see CMS238
SILOTRAS SCARLET TB see OHI200
SILOTRAS YELLOW T2G see DOT300
SILOTRAS YELLOW TSG see AAQ250
SILOXANES see SDC000
SILOXANES and SILICONES, DI Me see SCR400
SILTEX see SCK600
SILUNDUM see SCQ000
SILVADENE see PPP500
SILVAN (CZECH) see MKH000
SILVER see SDI500
SILVER (colloidal) see SDI750
SILVER ACETYLIDE see SDJ000
SILVER ACETYLIDE (dry) (DOT) see SDJ000
SILVER AMMONIUM LACTATE see SDL000
SILVER AMMONIUM NITRATE see SDL500
SILVER ARSENITE see SDM100
SILVER ATOM see SDI500
SILVER AZIDE see SDM500
SILVER AZIDE (dry) (DOT) see SDM500
SILVER BOROFLUORIDE see SDN000
SILVER COMPOUNDS see SDO500
SILVER CYANIDE see SDP000
SILVER DIFLUORIDE see SDQ500
SILVER FIR NEEDLE OIL see AAC250
SILVER FIR OIL see AAC250
SILVER(II)·FLUORIDE see SDQ500
SILVER FLUOROBORATE see SDN000
SILVER MATT POWDER see TGB250
SILVER METHANESULFONATE see SDR500
SILVER METHYLSULFONATE see SDR500
SILVER(1+) NITRATE see SDS000
SILVER NITRATE (DOT) see SDS000

SILVER(I) NITRATE (1:1) see SDS000
SILVER(1+) OXIDE see SDU500
SILVER PEROXYCHROMATE see SDW000
SILVER PICRATE (dry) (DOT) see PID200
SILVER PICRATE, wetted with not <30% water, by weight (UN1347) (DOT) see PID200
SILVER PINE OIL see AAC250
SILVER POTASSIUM CYANIDE see PLS250
SILVER TETRAFLUOROBORATE see SDN000
SILVEX (USDA) see TIX500
SILVI-RHAP see TIX500
SILVISAR see HKC500
SILVISAR 510 see HKC000
SILVISAR 550 see MRL750
SILYL BROMIDE see BOE750
SIM see SNK000
SIMANEX see BJP000
SIMAZIN see BJP000
SIMAZINE 80W see BJP000
SIMAZINE (USDA) see BJP000
SIMAZOL see AMY050
SIMPAMINA-D see BBK500
SIMPATEDRIN see BBK000
SIMPLA see SGG500
SINAFID M-48 see MNH000
SINALOST see TNF500
SINCICLAN see DKB000
SINDRENINA see VGP000
SINESALIN COMPOSITION see RDK000
SINITUHO see PAX250
SINNOESTER OGC see GGR200
SINOFLUROL see FLZ050
SINOMIN see SNK000
SINORATOX see DSP400
SINOX see DUS700
SINOX see DUU600
SINTESTROL see DKA600
SINTOMICETINA see CDP250
SINURON see DGD600
SINUTAB see ABG750
SIONIT see SKV200
SIONON see SKV200
SIPEX BOS see TAV750
SIPEX OP see SIB600
SIPOL L8 see OEI000
SIPOL L10 see DAI600
SIPOL L12 see DXV600
SIPOL S see OAX000
SIPOMER DAM see DBK200
SIPONOL S see OAX000
SIPON WD see SIB600
SIPTOX I see MAK700
SIRAN HYDRAZINU (CZECH) see HGW500
SIRLENE see PML000
SIRNIK AMONNY see ANJ750
SIRNIK FOSFORECNY (CZECH) see PHS000
SIROKAL see PLG800
SIRUP see GFG000
SISTOMETRENOL see LJE000
SIXTY-THREE SPECIAL E.C. INSECTICIDE see MNH000
SIXTY-THREE SPECIAL E.C. INSECTICIDE see PAK000
SK 65 see DAB879
SK-100 see TNF500
SK 555 see BHO250
SK-598 see CFA750
SK-106N see NGY000

SK1133 see TND500
SK-3818 see TND250
SK 6882 see TFQ750
SK-15673 see PED750
SK-19849 see BIA250
SK 20501 see CQC650
SK 27702 see BIF750
SK-AMITRIPTYLINE see EAI000
SK-AMPICILLIN see AIV500
SKANE M8 see OFE000
SKATOL see MKV750
SKATOLE see MKV750
ω-SKATOLE CARBOXYLIC ACID see ICN000
SK-CHLORAL HYDRATE see CDO000
SK-CHLOROTHIAZIDE see CLH750
SK-DEXAMETHASONE see SOW000
SK-DIGOXIN see DKN400
SK-DIPHENHYDRAMINE see BAU750
SKEDULE see CBF250
SKEKhG see EAZ500
SKELLYSOLVE F see PCT250
SKELLYSOLVE G see PCT250
SK-ESTROGENS see ECU750
SKF 51 see ILM000
SKF 385 see PET750
SKF 1498 see DQA400
SKF 2538 see DIR000
SKF 5019 see TKK250
SKF 688A see DDG800
SKF 7988 see VRF000
SKF 20,716 see PIW000
SK&F 14287 see DAS000
SK&F 36914 see CLQ500
SKG see CBT500
SKI 24464 see MNA750
SKINO #1 see BSU500
SKINO #2 see EMU500
SK-LYGEN see MDQ250
SK-NIACIN see NCQ900
SK-PHENOBARBITAL see EOK000
SK-PRAMINE see DLH630
SK-PRAMINE HYDROCHLORIDE see DLH630
SK-PREDNISONE see PLZ000
SK-RESERPINE see RDK000
SKS 85 see SMR000
SKT see CBT500
SKT (ADSORBENT) see CBT500
SK-TETRACYCLINE see TBX250
SK-TOLBUTAMIDE see BSQ000
SKY BLUE 4B see CMO500
SKY BLUE 5B see CMO500
1000SL see HKS780
SLAB OIL (OBS.) see MQV875
SLAKED LIME see CAT225
SLEEPAN see TEH500
SLIMICIDE see ADR000
SLIPRO see TEH500
SLO-GRO see DHF200
SLO-GRO see DMC600
SLO-PHYLLIN see TEP000
SLOW-FE see FBN100
SLOW-K see PLA500
SLS see SIB600
SLUG-TOX see TDW500
SMA see SFU500
SMCA see SFU500
SMD 3500 see SMQ500

SMEESANA see AQN635
SMIDAN see PHX250
SMOG see SEB000
SMOKE CONDENSATE, cigarette see SEC000
SMOKELESS TOBACCO see SED400
S MUSTARD see BIH250
SMUT-GO see HCC500
SN 20 see ADY500
S.N. 112 see PPP500
SN 36056 see DSO200
SN 36268 see CJJ250
SN 49537 see TEX600
SNEEZING GAS see CGN000
S.N.G see NGY000
SNIECIOTOX see HCC500
SNOMELT see CAO750
SNOW ALGIN H see SEH000
SNOWGOOSE see TAB750
SNOW TEX see KBB600
SNOW WHITE see ZKA000
SNP see PAK000
SNUFF see SED400
SO see LCF000
SOAMIN see ARA500
SOAP see CNF175
SOAPSTONE see SCN000
SOAP YELLOW F see FEV000
SOBENATE see SFB000
SOBIODOPA see DNA200
SOBITAL see DJL000
SOCAREX see PJS750
SODA ALUM see AHG500
SODA ASH see SFO000
SODA CHLORATE (DOT) see SFS000
SODA LIME see SEE000
SODA LIME with >4% sodium hydroxide (DOT) see
SEE000
SODA LYE see SHS000
SODA LYE see SHS500
SODAMIDE see SEN000
SODA MINT see SFC500
SODANIT see SEY500
SODA NITER see SIO900
SODANTON see DKQ000
SODANTON see DNU000
SODA PHOSPHATE see SJH090
SODASCORBATE see ARN125
SODESTRIN-H see ECU750
SODIO (CLORATO di) (ITALIAN) see SFS000
SODIO (DICROMATO di) (ITALIAN) see SGI000
SODIO, FLUORACETATO di (ITALIAN) see SHG500
SODIO(IDROSSIDO di) (ITALIAN) see SHS000
SODIOPAS see SEP000
SODIO (PERCLORATO DI) (ITALIAN) see PCE750
SODITAL see NBU000
SODIUM see SEE500
SODIUM P-50 see SEQ000
SODIUM (dispersions) see SEF500
SODIUM ACETATE see SEG500
SODIUM ACETATE, anhydrous (FCC) see SEG500
SODIUM ACETYLARSANILATE see AQZ900
SODIUM ACID ARSENATE see ARC000
SODIUM ACID ARSENATE, HEPTAHYDRATE see
ARC250
SODIUM ACID CARBONATE see SFC500
SODIUM ACID METHANEARSONATE see MRL750
SODIUM ACID PHOSPHATE see SJH100
SODIUM ACID PYROPHOSPHATE (FCC) see DXF800

SODIUM ACID SULFATE see SEG800
SODIUM ACID SULFATE (solid) see SEG800
SODIUM ACID SULFITE see SFE000
SODIUM AESCINATE see EDM000
SODIUM ALBAMYCIN see NOB000
SODIUM ALGINATE see SEH000
SODIUM n-ALKYLBENZENE SULFONATE see
SEH500
SODIUM-5-ALLYL-5-(1-METHYLBUTYL)-2-
THIOBARBITURATE see SOX500
SODIUM-dl-5-ALLYL-1-METHYL-5-(1-METHYL-2-
PENTYNYL)BARBITURATE see MDU500
SODIUM ALUMINATE, solid (UN 2812) (DOT) see
AHG000
SODIUM ALUMINATE, solution (UN 1819) (DOT) see
AHG000
SODIUM ALUMINOFLUORIDE see SHF000
SODIUM ALUMINOSILICATE see SEM000
SODIUM ALUMINUM FLUORIDE see SHF000
SODIUM ALUMINUM HYDRIDE (DOT) see SEM500
SODIUM ALUMINUM OXIDE see AHG000
SODIUM ALUMINUM SULFATE see AHG500
SODIUM ALUMINUM TETRAHYDRIDE see SEM500
SODIUM AMIDE see SEN000
SODIUM AMINARSONATE see ARA500
SODIUM p-AMINOBENZENEARSONATE see ARA500
SODIUM d-(−)-α-AMINOBENZYLPENICILLIN see
SEQ000
SODIUM AMINOPHENOL ARSONATE see ARA500
SODIUM p-AMINOPHENYLARSONATE see ARA500
SODIUM AMINOSALICYLATE see SEP000
SODIUM p-AMINOSALICYLATE see SEP000
SODIUM p-AMINOSALICYLIC ACID see SEP000
SODIUM AMINOTRIACETATE see SEP500
SODIUM AMPICILLIN see SEQ000
SODIUM AMYLOBARBITONE see AON750
SODIUM-ANALINE ARSONATE see ARA500
SODIUM ANILARSONATE see ARA500
SODIUM ANTIMONYL TARTRATE see AQI750
SODIUM ANTIMONY TARTRATE see AQI750
SODIUM ARISTOLOCHATE I see SEY050
SODIUM ARSANILATE see ARA500
SODIUM p-ARSANILATE see ARA500
SODIUM ARSANILATE (DOT) see ARA500
SODIUM ARSENATE see SEY100
SODIUM ARSENATE see ARC000
SODIUM ARSENATE see ARD500
SODIUM ARSENATE see ARD600
SODIUM ARSENATE (DOT) see SEY100
SODIUM ARSENATE (DOT) see ARD750
SODIUM ARSENATE DIBASIC, anhydrous see ARC000
SODIUM ARSENATE, DIBASIC, HEPTAHYDRATE see
ARC250
SODIUM ARSENATE HEPTAHYDRATE see ARC250
SODIUM ARSENITE see SEY200
SODIUM ARSENITE see SEY500
SODIUM ARSENITE, solid (DOT) see SEY500
SODIUM ARSENITE, liquid (solution) (DOT) see SEY500
SODIUM ARSENITE, solid (UN 2027) (DOT) see SEY200
SODIUM ARSENITE, aqueous solutions (UN 1686) (DOT)
see SEY200
SODIUM ARSONILATE see ARA500
SODIUM-l-ASCORBATE see ARN125
SODIUM ASCORBATE (FCC) see ARN125
SODIUM AUROTHIOMALATE see GJC000
SODIUM AZIDE see SFA000
SODIUM, AZOTURE de (FRENCH) see SFA000

SODIUM, AZOTURO di (ITALIAN) see SFA000
SODIUM BARBITAL see BAG250
SODIUM BARBITONE see BAG250
SODIUM-1,2 BENZISOTHIAZOLIN-3-ONE-1,1-
DIOXIDE see SJN700
SODIUM BENZOATE see SFB000
SODIUM BENZOIC ACID see SFB000
SODIUM o-BENZOSULFIMIDE see SJN700
SODIUM BENZOSULPHIMIDE see SJN700
SODIUM-2-BENZOSULPHIMIDE see SJN700
SODIUM-o-BENZOSULPHIMIDE see SJN700
SODIUM o-BENZYL-p-CHLOROPHENATE see SFB200
SODIUM o-BENZYL-p-CHLOROPHENOLATE see
SFB200
SODIUM BENZYLPENICILLIN see BFD250
SODIUM BENZYLPENICILLINATE see BFD250
SODIUM BENZYLPENICILLIN G see BFD250
SODIUM BERYLLIUM MALEATE see SFB500
SODIUM BERYLLIUM TARTRATE see SFC000
SODIUM BIBORATE see SFF000
SODIUM BIBORATE see DXG035
SODIUM BIBORATE DECAHYDRATE see SFF000
SODIUM BICARBONATE see SFC500
SODIUM BICHROMATE see SGI000
SODIUM BINOTAL see SEQ000
SODIUM 2-BIPHENYLOLATE see BGJ750
SODIUM (1,1'-BIPHENYL)-2-OLATE see BGJ750
SODIUM, (2-BIPHENYLYLOXY)- see BGJ750
SODIUM BIPHOSPHATE see SJH100
SODIUM BIPHOSPHATE anhydrous see SJH100
SODIUM BIS(2-ETHYLHEXYL) SULFOSUCCINATE
see DJL000
SODIUM BISULFATE, FUSED see SEG800
SODIUM BISULFIDE see SHR000
SODIUM BISULFITE see SFE000
SODIUM BISULFITE see SFE000
SODIUM BISULFITE (1:1) see SFE000
SODIUM BISULFITE, solid (DOT) see SFE000
SODIUM BISULFITE, solution (DOT) see SFE000
SODIUM BORATE DECAHYDRATE see SFF000
SODIUM BOROHYDRIDE see SFF500
SODIUM BROMATE see SFG000
SODIUM BROMIDE see SFG500
SODIUM CACODYLATE (DOT) see HKC500
SODIUM CARBOLATE see SJF000
SODIUM CARBONATE (2:1) see SFO000
SODIUM CARBONATE PEROXIDE see SJB400
SODIUM CARBOXYMETHYL CELLULOSE see SFO500
SODIUM CASEINATE see SFQ000
SODIUM CELLULOSE GLYCOLATE see SFO500
SODIUM CHLORAMINE T see CDP000
SODIUM CHLORATE see SFS000
SODIUM (CHLORATE de) (FRENCH) see SFS000
SODIUM CHLORATE, aqueous solution (DOT) see
SFS000
SODIUM CHLORIDE see SFT000
SODIUM CHLORIDE OXIDE see SHU500
SODIUM CHLORITE see SFT500
SODIUM CHLORITE (UN 1496) (DOT) see SFT500
SODIUM CHLORITE, solution with >5% available
chlorine (UN 1908) (DOT) see SFT500
SODIUM CHLOROACETATE see SFU500
SODIUM 1-(p-CHLOROBENZOYL)-5-METHOXY-2-
METHYLINDOLE-3-ACETATE TRIHYDRATE see
IDA100
SODIUM (4-CHLORO-2-
METHYLPHENOXY)ACETATE see SIL500
SODIUM CHOLATE see SFW000

SODIUM CHOLIC ACID see SFW000
SODIUM CHROMATE see SGI000
SODIUM CHROMATE (VI) see DXC200
SODIUM CHROMATE (DOT) see DXC200
SODIUM CHROMATE DECAHYDRATE see SFW500
SODIUM CITRATE (FCC) see DXC400
SODIUM CLODRONATE see SFX730
SODIUM CLOXACILLIN MONOHYDRATE see SLJ000
SODIUM CMC see SFO500
SODIUM CM-CELLULOSE see SFO500
SODIUM COLISTIMETHATE see SFY500
SODIUM COLISTINEMETHANESULFONATE see
SFY500
SODIUM COLISTIN METHANESULFONATE see
SFY500
SODIUM COMPOUNDS see SFZ000
SODIUM COUMADIN see WAT220
SODIUM CUMENEAZO-β-NAPHTHOL
DISULPHONATE see FAG018
SODIUM CUPROCYANIDE (DOT) see SFZ100
SODIUM CUPROCYANIDE, solid (UN 2316) (DOT) see
SFZ100
SODIUM CUPROCYANIDE, solution (UN 2317) (DOT)
see SFZ100
SODIUM CYANIDE see SGA500
SODIUM CYCLAMATE see SGC000
SODIUM CYCLOHEXANESULFAMATE see SGC000
SODIUM CYCLOHEXANESULPHAMATE see SGC000
SODIUM CYCLOHEXYL AMIDOSULPHATE see
SGC000
SODIUM CYCLOHEXYL SULFAMATE see SGC000
SODIUM CYCLOHEXYL SULFAMIDATE see SGC000
SODIUM CYCLOHEXYL SULPHAMATE see SGC000
SODIUM-2,4-D see SGH500
SODIUM DALAPON see DGI600
SODIUM DBDT see SGF500
SODIUM DECYLBENZENESULFONAMIDE see
DAJ000
SODIUM DECYLBENZENESULFONATE see DAJ000
SODIUM DEDT see SGJ000
SODIUM DEHYDROACETATE (FCC) see SGD000
SODIUM DEHYDROACETIC ACID see SGD000
SODIUM DEOXYCHOLATE see SGE000
SODIUM DEOXYCHOLIC ACID see SGE000
SODIUM DESOXYCHOLATE see SGE000
SODIUM DEXAMETHASONE PHOSPHATE see
DAE525
SODIUM DIACETATE see SGE400
SODIUM DIARSENATE see SEY100
SODIUM DIBUTYLDITHIOCARBAMATE see SGF500
SODIUM DICHLORISOCYANURATE see SGG500
SODIUM DICHLOROCYANURATE see SGG500
SODIUM DICHLOROISOCYANURATE see SGG500
SODIUM-2,4-DICHLOROPHENOXYACETATE see
SGH500
SODIUM-2,2-DICHLOROPROPIONATE see DGI600
SODIUM-α,α-DICHLOROPROPIONATE see DGI600
SODIUM-1,3-DICHLORO-1,3,5-TRIAZINE-2,4-DIONE-
6-OXIDE see SGG500
1-SODIUM-3,5-DICHLORO-s-TRIAZINE-2,4,6-TRIONE
see SGG500
1-SODIUM-3,5-DICHLORO-1,3,5-TRIAZINE-2,4,6-
TRIONE see SGG500
SODIUM DICHLORO-s-TRIAZINETRIONE, dry,
containing more than 39% available chlorine (DOT) see
SGG500
SODIUM DICHROMATE see SGI000

SODIUM DICHROMATE(VI) see SGI000
SODIUM DICHROMATE de (FRENCH) see SGI000
SODIUM DIETHYLBARBITURATE see BAG250
SODIUM-5,5-DIETHYLBARBITURATE see BAG250
SODIUM DIETHYLDITHIOCARBAMATE see SGJ000
SODIUM N,N-DIETHYLDITHIOCARBAMATE see SGJ000
SODIUM DI-(2-ETHYLHEXYL) SULFOSUCCINATE see DJL000
SODIUM DIHYDROGEN ARSENATE see ARD600
SODIUM DIHYDROGEN ORTHOARSENATE see ARD600
SODIUM DIHYDROGEN PHOSPHATE (1:2:1) see SJH100
SODIUM-4-(DIMETHYLAMINO)BENZENEDIAZOSULFONATE see DOU600
SODIUM-p-(DIMETHYLAMINO)BENZENEDIAZOSULFONATE see DOU600
SODIUM-4-(DIMETHYLAMINO)BENZENEDIAZOSULPHONATE see DOU600
SODIUM-p-(DIMETHYLAMINO)BENZENEDIAZOSULPHONATE see DOU600
SODIUM-(4-(DIMETHYLAMINO)PHENYL)DIAZENESULFONATE see DOU600
SODIUM DIMETHYLARSINATE see HKC500
SODIUM DIMETHYLARSONATE see HKC500
SODIUM N,N-DIMETHYLDITHIOCARBAMATE see SGM500
SODIUM DINITRO-o-CRESOLATE, dry or wetted with <15% water, by weight (UN 0234) (DOT) see SGP550
SODIUM DINITRO-o-CRESOLATE, wetted with not <15% water, by weight (UN 1348) (DOT) see SGP550
SODIUM-4,6-DINITRO-o-CRESOXIDE see DUU600
SODIUM DINITRO-o-CRESYLATE see SGP550
SODIUM DIOCTYL SULFOSUCCINATE see DJL000
SODIUM DIOCTYL SULPHOSUCCINATE see DJL000
SODIUM DIOXIDE see SJC500
SODIUM DIPHENYL-4,4'-BIS-AZO-2"-8"-AMINO-1"-NAPHTHOL-3",6 " DISULPHONATE see CMO000
SODIUM DIPHENYLDIAZO-BIS(α-NAPHTHYLAMINESULFONATE) see SGQ500
SODIUM DIPHENYLHYDANTOIN see DNU000
SODIUM DIPHENYL HYDANTOINATE see DNU000
SODIUM-5,5-DIPHENYLHYDANTOINATE see DNU000
SODIUM-5,5-DIPHENYL-2,4-IMIDAZOLIDINEDIONE see DNU000
SODIUM DISULFITE see SII000
SODIUM DITHIONITE (DOT) see SHR500
SODIUM DITOLYLDIAZOBIS-8-AMINO-1-NAPHTHOL-3,6-DISULFONATE see CMO250
SODIUM DITOLYLDIAZOBIS-8-AMINO-1-NAPHTHOL-3,6-DISULPHONATE see CMO250
SODIUM DODECYLBENZENESULFONATE (DOT) see DXW200
SODIUM DODECYLBENZENESULFONATE, dry see DXW200
SODIUM DODECYL SULFATE see SIB600
SODIUM EDETATE see EIV000
SODIUM EDTA see EIV000
SODIUM EOSINATE see BNH500
SODIUM EQUILIN 3-MONOSULFATE see ECW520

SODIUM EQUILIN SULFATE see ECW520
SODIUM ESTRONE SULFATE see EDV600
SODIUM ESTRONE-3-SULFATE see EDV600
SODIUM ETASULFATE see TAV750
SODIUM ETHAMINAL see NBU000
SODIUM ETHASULFATE see TAV750
SODIUM ETHIDRONATE see DXD400
SODIUM ETHYDRONATE see DXD400
SODIUM ETHYLBARBITAL see BAG250
SODIUM ETHYLENEDIAMINETETRAACETATE see EIV000
SODIUM ETHYLENEDIAMINETETRAACETIC ACID see EIV000
SODIUM(2-ETHYLHEXYL)ALCOHOL SULFATE see TAV750
SODIUM-2-ETHYLHEXYL SULFATE see TAV750
SODIUM-2-ETHYLHEXYLSULFOSUCCINATE see DJL000
SODIUM ETHYLISOAMYLBARBITURATE see AON750
SODIUM ETHYLMERCURIC THIOSALICYLATE see MDI000
SODIUM-o-(ETHYLMERCURITHIO)BENZOATE see MDI000
SODIUM ETHYLMERCURITHIOSALICYLATE see MDI000
SODIUM-5-ETHYL-5-(1-METHYLBUTYL)BARBITURATE see NBU000
SODIUM-5-ETHYL-5-(1-METHYLBUTYL)-2-THIOBARBITURATE see PBT500
SODIUM-5-ETHYL-5-PHENYLBARBITURATE see SID000
SODIUM ETIDRONATE see DXD400
SODIUM FERROCYANIDE see SHE350
SODIUM FLUOACETATE see SHG500
SODIUM FLUOACETIC ACID see SHG500
SODIUM FLUOALUMINATE see SHF000
SODIUM FLUORACETATE de (FRENCH) see SHG500
SODIUM FLUORESCEIN see FEW000
SODIUM FLUORESCEINATE see FEW000
SODIUM FLUORIDE see SHF500
SODIUM FLUORIDE, solid and solution (DOT) see SHF500
SODIUM FLUOROACETATE see SHG500
SODIUM FLUOROSILICATE see DXE000
SODIUM FLUORURE (FRENCH) see SHF500
SODIUM FLUOSILICATE see DXE000
SODIUM FORMATE see SHJ000
SODIUM GLUCONATE see SHK800
SODIUM d-GLUCONATE see SHK800
SODIUM GLUTAMATE see MRL500
SODIUM l-GLUTAMATE see MRL500
l(+) SODIUM GLUTAMATE see MRL500
SODIUM GMP see GLS800
SODIUM GUANOSINE-5'-MONOPHOSPHATE see GLS800
SODIUM GUANYLATE see GLS800
SODIUM-5'-GUANYLATE see GLS800
SODIUM HEXADECANOATE see SIZ025
SODIUM HEXAFLUOROALUMINATE see SHF000
SODIUM HEXAFLUOROARSENATE see SHM000
SODIUM HEXAFLUOROSILICATE see DXE000
SODIUM HEXAFLUOSILICATE see DXE000
SODIUM HEXAMETAPHOSPHATE see SHM500
SODIUM HEXAMETAPHOSPHATE see SII500
SODIUM HYDRATE (DOT) see SHS000
SODIUM HYDRATE, solution see SHS500
SODIUM HYDRIDE see SHO500

SODIUM HYDROCORTISONE SUCCINATE see HHR000

SODIUM HYDROCORTISONE-21-SUCCINATE see HHR000

SODIUM HYDROFLUORIDE see SHF500

SODIUM HYDROGEN CARBONATE see SFC500

SODIUM HYDROGEN DIACETATE see SGE400

SODIUM HYDROGEN DIFLUORIDE see SHQ500

SODIUM HYDROGEN FLUORIDE see SHQ500

SODIUM HYDROGEN FLUORIDE see SHQ500

SODIUM HYDROGEN FLUORIDE, solution (DOT) see SHQ500

SODIUM HYDROGEN PHOSPHATE see SJH090

SODIUM HYDROGEN SULFATE, solid (UN 1821) (DOT) see SEG800

SODIUM HYDROGEN SULFATE, solution (UN 2837) (DOT) see SEG800

SODIUM HYDROGEN SULFIDE see SHR000

SODIUM HYDROGEN SULFITE see SFE000

SODIUM HYDROGEN SULFITE, solid (DOT) see SFE000

SODIUM HYDROGEN SULFITE, solution (DOT) see SFE000

SODIUM HYDROSULFIDE see SHR000

SODIUM HYDROSULFIDE, solution (NA 2922) (DOT) see SHR000

SODIUM HYDROSULFIDE, with <25% water of crystallization (UN 2318) (DOT) see SHR000

SODIUM HYDROSULFIDE, with not <25% water of crystallization (UN 2949) (DOT) see SHR000

SODIUM HYDROSULFITE (DOT) see SHR500

SODIUM HYDROSULPHITE see SHR500

SODIUM HYDROXIDE see SHS000

SODIUM HYDROXIDE, dry (DOT) see SHS000

SODIUM HYDROXIDE (liquid) see SHS500

SODIUM HYDROXIDE, bead (DOT) see SHS000

SODIUM HYDROXIDE, flake (DOT) see SHS000

SODIUM HYDROXIDE, solid (DOT) see SHS000

SODIUM HYDROXIDE, granular (DOT) see SHS000

SODIUM HYDROXIDE, solution (FCC) see SHS500

SODIUM-o-HYDROXYBENZOATE see SJO000

SODIUM-4-HYDROXYBUTYRATE see HJS500

SODIUM-γ-HYDROXYBUTYRATE see HJS500

SODIUM (HYDROXYDE de) (FRENCH) see SHS000

SODIUM 2-HYDROXYDIPHENYL see BGJ750

SODIUM p-HYDROXYMERCURIBENZOATE see SHU000

SODIUM o-((3-(HYDROXYMERCURI)-2-METHOXYPROPYL)CARBAMOYL)PHENOXY ACETATE see SIH500

SODIUM HYPOCHLORITE see SHU500

SODIUM HYPOPHOSPHITE see SHV000

SODIUM HYPOSULFITE see SKI000

SODIUM INOSINATE see DXE500

SODIUM-5'-INOSINATE see DXE500

SODIUM IODIDE see SHW000

SODIUM IODINE see SHW000

SODIUM ISOAMYLETHYL BARBITURATE see AON750

SODIUM ISOCYANATE see COI250

SODIUM ISOTHIOCYANATE see SIA500

SODIUM LACTATE see LAM000

SODIUM LAURYLBENZENESULFONATE see DXW200

SODIUM-N-LAURYL SARCOSINE see DXZ000

SODIUM LAURYL SULFATE see SIB600

SODIUM LEVOTHYROXINE see LFG050

SODIUM LUMINAL see SID000

SODIUM MALONYLUREA see BAG250

SODIUM MCPA see SIL500

SODIUM MERCAPTAN see SHR000

SODIUM MERCAPTIDE see SHR000

SODIUM MERCAPTOACETATE see SKH500

SODIUM-2-MERCAPTOBENZOTHIAZOLE see SIG500

SODIUM MERCAPTOMERIN see TFK270

SODIUM MERSALYL see SIH500

SODIUM MERTHIOLATE see MDI000

SODIUM METAARSENATE see ARD500

SODIUM METAARSENATE see ARD750

SODIUM METAARSENITE see SEY500

SODIUM METABISULFITE see SII000

SODIUM METABISULFITE see SII000

SODIUM METABISULPHITE see SII000

SODIUM METAL (DOT) see SEE500

SODIUM, METAL DISPERSION IN ORGANIC SOLVENT see SEF500

SODIUM METAPHOSPHATE see SII500

SODIUM METASILICATE see SJU000

SODIUM METASILICATE, anhydrous see SJU000

SODIUM METAVANADATE see SKP000

SODIUM METHANEARSONATE see MRL750

4-SODIUM METHANESULFONATE METHYLAMINE-ANTIPYRINE see AMK500

SODIUM METHOHEXITAL see MDU500

SODIUM METHOHEXITONE see MDU500

SODIUM METHOXIDE see SIK450

SODIUM dl-1-METHYL-5-ALLYL-5-(1-METHYL-2-PENTYNYL)BARBITURATE see MDU500

SODIUM METHYLAMINOANTIPYRINE METHANESULFONATE see AMK500

SODIUM-4-METHYLAMINO-1,5-DIMETHYL-2-PHENYL-3-PYRAZOLONE 4-METHANESULFONATE see AMK500

SODIUM METHYLATE see SIK450

SODIUM METHYLATE SOLUTIONS in alcohol (UN 1289) (DOT) see SIK450

SODIUM METHYLATE (UN 1431) (DOT) see SIK450

SODIUM (2-METHYL-4-CHLOROPHENOXY)ACETATE see SIL500

SODIUM MOLYBDATE see DXE800

SODIUM MOLYBDATE(VI) see DXE800

SODIUM MONOCHLORACETATE see SFU500

SODIUM MONO- and DIMETHYL NAPHTHALENE SULFONATE see SIM400

SODIUM MONODODECYL SULFATE see SIB600

SODIUM MONOFLUORIDE see SHF500

SODIUM MONOFLUOROACETATE see SHG500

SODIUM MONOHYDROGEN ARSENATE see ARD500

SODIUM MONOHYDROGEN PHOSPHATE (2:1:1) see SJH090

SODIUM MONOIODIDE see SHW000

SODIUM MONOIODOACETATE see SIN000

SODIUM MONOSULFIDE see SJY500

SODIUM MONOXIDE see SIN500

SODIUM MONOXIDE, solid (DOT) see SIN500

SODIUM MYRISTYL SULFATE see SIO000

SODIUM NEMBUTAL see NBU000

SODIUM-22 NEOPRENE ACCELERATOR see IAQ000

SODIUM NITRATE (1:1) see SIO900

SODIUM NITRATE (DOT) see SIO900

SODIUM NITRILOACETATE see SEP500

SODIUM NITRILOTRIACETATE see SEP500

SODIUM NITRILOTRIACETATE see SIP500

SODIUM NITRITE see SIQ500

SODIUM SALT of HEXADECANOIC ACID see SIZ025
SODIUM SALT of HYDROXY-o-CARBOXY-PHENYL-
FLUORONE see FEW000
SODIUM SELENATE see DXG000
SODIUM SELENIDE see SJT000
SODIUM SELENITE see SJT500
SODIUM SELENITE PENTAHYDRATE see SJT600
SODIUM SILICATE see SJU000
SODIUM SILICOALUMINATE see SEM000
SODIUM SILICOFLUORIDE (DOT) see DXE000
SODIUM SORBATE see SJV000
SODIUM SOTRADECOL see SIO000
SODIUM STEARATE see SJV500
SODIUM STEAROYL LACTYLATE see SJV700
SODIUM STIBOGLUCONATE see AQH800
SODIUM SUCARYL see SGC000
SODIUM SULFACHLOROPYRAZINE
MONOHYDRATE see SJW200
SODIUM SULFAMERAZINE see SJW475
SODIUM SULFATE (2:1) see SJY000
SODIUM SULFATEn anhydrous see SJY000
SODIUM SULFHYDRATE see SFE000
SODIUM SULFHYDRATE see SHR000
SODIUM SULFIDE see SJY500
SODIUM SULFIDE, anhydrous (DOT) see SJY500
SODIUM SULFIDE with <30% water of crystallization
(DOT) see SJY500
SODIUM SULFITE (2:1) see SJZ000
SODIUM SULFITE, anhydrous see SJZ000
SODIUM SULFOACETATE derivatives of MONO and
DIGLYCERIDES see SJZ050
SODIUM SULFOCYANATE see SIA500
SODIUM SULFOCYANIDE see SIA500
SODIUM SULFODI-(2-
ETHYLHEXYL)SULFOSUCCINATE see DJL000
SODIUM SULFOXYLATE see SHR500
SODIUM SULPHAMERAZINE see SJW475
SODIUM SULPHATE see SJY000
SODIUM SULPHIDE see SJY500
SODIUM SULPHITE see SJZ000
SODIUM SUPEROXIDE see SJZ100
SODIUM SUPEROXIDE (DOT) see SJZ100
SODIUM l-(+)-TARTRATE see BLC000
SODIUM TARTRATE (FCC) see BLC000
SODIUM TCA SOLUTION see TII250
SODIUM TELLURATE see SKC000
SODIUM TELLURATE(IV) see SKC500
SODIUM TELLURATE, DIHYDRATE see SKC100
SODIUM TELLURATE VI see SKC000
SODIUM TELLURITE see SKC500
SODIUM TETRABORATE see SFF000
SODIUM TETRABORATE see DXG035
SODIUM TETRABORATE DECAHYDRATE see SFF000
SODIUM TETRABORATE ($Na_2B_4O_7$) see DXG035
SODIUM TETRADECYL SULFATE see SIO000
SODIUM TETRAFLUOROBORATE see SKE000
SODIUM TETRAHYDROALUMINATE(1−) see SEM500
SODIUM TETRAHYDROBORATE(1-) see SFF500
SODIUM TETRAPEROXYCHROMATE see SKF000
SODIUM TETRAPOLYPHOSPHATE see SII500
SODIUM THIAMYLAL see SOX500
SODIUM THIOCYANATE see SIA500
SODIUM THIOCYANIDE see SIA500
SODIUM THIOGLYCOLATE see SKH500
SODIUM THIOGLYCOLLATE see SKH500
SODIUM THIOPENTAL see PBT500
SODIUM THIOPENTOBARBITAL see PBT500
SODIUM THIOPENTONE see PBT500

SODIUM THIOSULFATE see SKI000
SODIUM THIOSULFATE, anhydrous see SKI000
SODIUM THYROXIN see LFG050
SODIUM THYROXINATE see LFG050
SODIUM THYROXINE see LFG050
SODIUM l-THYROXINE see LFG050
SODIUM p-TOLUENESULFONYLCHLORAMIDE see
CDP000
SODIUM TOSYLCHLORAMIDE see CDP000
SODIUM TRIMETAPHOSPHATE see SKM500
SODIUM TRIPHOSPHATE see SKN000
SODIUM TRIPOLYPHOSPHATE see SKN000
SODIUM TUNGSTATE see SKN500
SODIUM TUNGSTOPHOSPHATE see SJJ000
SODIUM URATE see SKO575
SODIUM VANADATE see SIY250
SODIUM VANADATE see SKP000
SODIUM VANADIUM OXIDE see SIY250
SODIUM VERONAL see BAG250
SODIUM VERSENATE see EIX500
SODIUM WARFARIN see WAT220
SODIUM WOLFRAMATE see SKN500
SODIZOLE see SNN500
SO-FLO see SHF500
SOFRIL see SOD500
SOFTENIL see TEH500
SOFTENON see TEH500
SOFTIL see DJL000
SOFTRAN see BOM250
SOHNHOFEN STONE see CAO000
SOILBROM-40 see EIY500
SOILBROM-85 see EIY500
SOILFUME see EIY500
SOIL STABILIZER 661 see SMR000
SOK see CBM750
SOL see PKB500
SOLABAR see BAO300
SOLACTOL see LAJ000
SOLADREN see VGP000
SOLAESTHIN see MJP450
SOLAMINE see BEN000
SOLANCARPIDINE see POJ000
SOLANIDINE-S see POJ000
SOLANOSIDE see SKS150
SOLANTIN see DKQ000
SOLANTOIN see DNU000
SOLANTYL see DNU000
SOLAR 40 see DXW200
SOLAR BROWN PL see CMO750
SOLAR LIGHT ORANGE GX see HGC000
SOLAR RUBINE see HJF500
SOLAR VIOLET 5BN see FAG120
SOLAR WINTER BAN see PML000
SOLASKIL see LFA020
SOLASOD-5-EN-3-β-OL see POJ000
SOLASODINE see POJ000
SOLBAR see BAO300
SOLBAR see BAP000
SOLBASE see PJT200
SOLBROL A see HJL000
SOLBROL M see HJL500
SOLCAIN see DHK400
SOLDEP see TIQ250
SOLDESAM see DAE525
SOLESTRO see EDP000
SOLEX see CAW850
SOLFARIN see WAT200

SOLFO BLACK B see DUZ000
SOLFO BLACK BB see DUZ000
SOLFO BLACK G see DUZ000
SOLFO BLACK SB see DUZ000
SOLFO BLACK 2B SUPRA see DUZ000
SOLFO SERPINE see RDK000
SOLFURO di CARBONIO (ITALIAN) see CBV500
SOLGANAL see ART250
SOLGANAL B see ART250
SOLIWAX see DJL000
SOLKA-FIL see CCU150
SOLKA-FLOC see CCU150
SOLKA-FLOC BW see CCU150
SOLKA-FLOC BW 20 see CCU150
SOLKA-FLOC BW 100 see CCU150
SOLKA-FLOC BW 200 see CCU150
SOLKA-FLOC BW 2030 see CCU150
SOLKETAL see DVR600
SOLLICULIN see EDV000
SOLMETHINE see MJP450
SOLOCHROME BLUE FB see HJF500
SOLOSIN see TEP000
SOLOZONE see SJC500
SOL PHENOBARBITAL see SID000
SOL PHENOBARBITONE see SID000
SOLPRENE 300 see SMR000
SOLPYRON see ADA725
SOL SODOWA KWASU
LAURYLOBENZENOSULFONOWEGO (POLISH) see
DXW200
SOLSOL NEEDLES see SIB600
SOLUBLE BARBITAL see BAG250
SOLUBLE FLUORESCEIN see FEW000
SOLUBLE GLUSIDE see SJN700
SOLUBLE GUN COTTON see CCU250
SOLUBLE PENTOBARBITAL see NBU000
SOLUBLE PHENOBARBITAL see SID000
SOLUBLE PHENOBARBITONE see SID000
SOLUBLE PHENYTOIN see DNU000
SOLUBLE SACCHARIN see SJN700
SOLUBLE SULFAMERAZINE see SJW475
SOLUBLE THIOPENTONE see PBT500
SOLUBLE VAN DYKE BROWN see MAT500
SOLUBOND 0-869 see RDP000
SOLUBOND 3520 see RDP000
SOLU-CORTEF see HHR000
SOLU-DECADRON see DAE525
SOLU-GLYC see HHR000
SOLUMEDINE see SJW475
SOLUSOL-75% see DJL000
SOLUSOL-100% see DJL000
SOLUSTIBOSAN see AQH800
SOLUSTIN see AQH800
SOLUSURMIN see AQH800
SOLUTION CONCENTREE T271 see AMY050
SOLUTION GLYCERYL TRINITRATE see NGY000
SOLVANOL see DJX000
SOLVANOM see DTR200
SOLVAR see PKP750
SOLVARONE see DTR200
SOLVAT 14 see BFL000
SOLVENT 111 see MIH275
SOLVENT-DEWAXED HEAVY NAPHTHENIC
DISTILLATE see MQV820
SOLVENT-DEWAXED HEAVY PARAFFINIC
DISTILLATE see MQV825
SOLVENT-DEWAXED LIGHT NAPHTHENIC
DISTILLATE see MQV835

SOLVENT-DEWAXED LIGHT PARAFFINIC
DISTILLATE see MQV840
SOLVENT ETHER see EJU000
SOLVENT ORANGE 15 see BJF000
SOLVENT RED 1 see CMS238
SOLVENT RED 19 see EOJ500
SOLVENT RED 72 see DDO200
SOLVENT-REFINED (mild) HEAVY NAPHTHENIC
DISTILLATE see MQV845
SOLVENT-REFINED (mild) HEAVY PARAFFINIC
DISTILLATE see MQV850
SOLVENT-REFINED (mild) LIGHT NAPHTHENIC
DISTILLATE see MQV852
SOLVENT-REFINED (mild) LIGHT PARAFFINIC
DISTILLATE see MQV855
SOLVENT YELLOW 1 see PEI000
SOLVENT YELLOW 14 see PEJ500
SOLVIC see PKQ059
SOLVIC 523KC see AAX175
SOLVIREX see DXH325
SOLVOSOL see CBR000
SOLYACORD see DJS200
SOLYUSURMIN see AQH800
SOMALIA ORANGE 2R see XRA000
SOMALIA ORANGE A2R see XRA000
SOMALIA ORANGE I see PEJ500
SOMALIA RED III see OHI200
SOMALIA RED IV see SBC500
SOMALIA RED PG see CMS238
SOMALIA YELLOW A see DOT300
SOMALIA YELLOW R see AIC250
SOMBUTOL see EOK000
SOMIO see GFA000
SOMIPRONT see DUD800
SOMNAFAC see MDT250
SOMNASED see DLY000
SOMNIBEL see DLY000
SOMNI SED see CDO000
SOMNITE see DLY000
SOMNOLETTEN see EOK000
SOMNOPENTYL see NBU000
SOMNOS see CDO000
SOMNOSAN see EOK000
SOMNUROL see BNP750
SOMONAL see EOK000
SOMONIL see DSO000
SOMSANIT see HJS500
SONACIDE see GFQ000
SONACON see DCK759
SONAFORM see TDA500
SONATE see ARA500
SONAZINE see CKP500
SONEBON see DLY000
SONISTAN see NBU000
SONNOLIN see DLY000
SONTEC see CDO000
SONTOBARBITAL NABITONE see NBU000
SOOT see SKS750
SOPENTAL see NBU000
SOPHIAMIN see MDQ250
SOPHORETIN see QCA000
SOPHORICOL see GCM350
SOPP see BGJ750
SOPRABEL see LCK000
SOPRABEL see LCK100
SOPRANEBE see MAS500
SOPRATHION see PAK000
SOPRATHION see EEH600

SOPRINAL see BAG250
SOPRINTIN see ABH500
SOPROCIDE see BBQ750
SOPRONTIN see ABH500
SOPROTIN see ABH500
SORBANGIL see CCK125
SORBA-SPRAY Mn see MAU250
SORBESTER P 17 see SKV100
SORBIC ACID see SKU000
SORBIC ACID, POTASSIUM SALT see PLS750
SORBIC ACID, SODIUM SALT see SJV000
SORBIC OIL see PAJ500
SORBICOLAN see SKV200
SORBID see CCK125
SORBIDE NITRATE see CCK125
SORBIDILAT see CCK125
SORBIDINITRATE see CCK125
SORBIMACROGOL LAURATE 300 see PKG000
SORBIMACROGOL OLEATE see PKL100
SORBIMACROGOL TRISTEARATE 300 see SKV195
SORBISLO see CCK125
SORBISTAT see SKU000
SORBISTAT-K see PLS750
SORBISTAT-POTASSIUM see PLS750
SORBITAL O 20 see PKL100
SORBITAN C see SKV150
SORBITAN MONODODECANOATE see SKV000
SORBITAN, MONODODECANOTE, POLY(OXY-1,2-ETHANEDIYL) DERIVATIVES see PKG000
SORBITAN MONOLAURATE see SKV000
SORBITAN MONOOCTADECANOATE see SKV150
SORBITAN, MONOOCTADECANOATE, POLY(OXY-1,2-ETHANEDIYL) DERIVATIVES see PKL030
SORBITAN MONOOLEATE see SKV100
SORBITAN MONOOLEIC ACID ESTER see SKV100
SORBITAN MONOSTEARATE see SKV150
SORBITAN O see SKV100
SORBITAN OLEATE see SKV100
SORBITAN STEARATE see SKV150
SORBITAN, TRIOCTADECANOATE, POLY(OXY-1,2-ETHANEDIYL) derivs. (9CI) see SKV195
SORBITAN, TRISTEARATE, POLYOXYETHYLENE derivatives see SKV195
SORBITE see SKV200
SORBITOL see SKV200
d-SORBITOL see SKV200
SORBITRATE see CCK125
SORBO see SKV200
SORBO-CALCIAN see CAL750
SORBO-CALCION see CAL750
SORBOL see SKV200
SORBONIT see CCK125
SORBON S 60 see SKV150
SORBOSTYL see SKV200
SOREFLON 604 see TAI250
SORETHYTAN (20) MONOOLEATE see PKL100
SORGEN 40 see SKV100
SORGEN 50 see SKV150
SORGHUM GUM see SLJ500
SORGOA see TGB475
SORLATE see PKL100
SORQUAD see CCK125
SORQUAT see CCK125
SORVILANDE see SKV200
SOSIGON see DOQ400
SOTIPOX see TIQ250
SOTYL see NBU000
SOUDAN I see PEJ500

SOUDAN II see XRA000
SOUDAN III see OHI200
SOUFRAMINE see PDP250
SOUP see NGY000
SOUTHERN BENTONITE see BAV750
SOVIET TECHNICAL HERBICIDE 2M-4C see CIR250
SOVIOL see AAX250
SOVOL see PJL750
SOWBUG & CUTWORM BAIT see COF500
SOXINAL PZ see BJK500
SOXINOL CZ see CPI250
SOXINOL PZ see BJK500
SOXISOL see SNN500
SOXYSYMPAMINE see MDT600
SOYBEAN OIL (UNHYDROGENATED) see SKW825
75 SP see DOP600
SP 104 see TCM250
SPA-S-753 see DOS400
SPAN 20 see SKV000
SPAN 55 see SKV150
SPAN 60 see SKV150
SPAN 80 see SKV100
SPANBOLET see SNJ000
SPANISH MARJORAM OIL see MBU500
SPANISH THYME OIL see TFX750
SPANON see CJJ250
SPANONE see CJJ250
SPANTOL see PGA750
SPANTRAN see MQU750
SPARINE HYDROCHLORIDE see PMI500
SPARTOSE OM-22 see CCU150
SPASMEDAL see DAM700
SPASMODOLIN see DAM700
SPASMO-NIT see PAH250
SPATONIN see DIW000
SPAVIT see VSZ450
SP 60 (CHLOROCARBON) see PKQ059
SPEARMINT OIL see SKY000
SPECIAL BLACK 1V & V see CBT750
SPECIAL BLUE X 2137 see EOJ500
SPECIAL SCHWARZ see CBT750
SPECIAL TERMITE FLUID see DEP600
SPECIFEN see EID000
SPECILLINE G see BDY669
SPECTINOMYCIN DIHYDROCHLORIDE see SLI325
SPECTINOMYCIN HYDROCHLORIDE see SLI325
SPECTRACIDE see DCM750
SPECTRAR see INJ000
SPECTRA-SORB UV 9 see MES000
SPECTROLENE BLUE B see DCJ200
SPECTROLENE RED KB see CLK225
SPECTROSIL see SCK600
SPECULAR IRON see IHC450
SPEED see DBA800
"SPEED" see MDQ500
SPENCER 401 see PJY500
SPENCER S-6538 see DJY200
SPENCER S-6900 see DRR200
SPENKEL see PKL500
SPENLITE see PKL500
SPENT SULFURIC ACID (DOT) see SOI500
SPERLOX-S see SOD500
SPERSADOX see DAE525
SPERSUL see SOD500
SPERSUL THIOVIT see SOD500
SP 60 ESTER see AAX250
SPHEROIDINE see FOQ000
SPHERON see CBT750

SPHERON 6 see CBT750
SPHYGMOGENIN see VGP000
SPIKE LAVENDER OIL see SLB500
SPIRESIS see AFJ500
SPIRIDON see AFJ500
SPIRIT see EFU000
SPIRIT of HARTSHORN see AMY500
SPIRIT OF GLONOIN see NGY000
SPIRIT OF GLYCERYL TRINITRATE see NGY000
SPIRIT OF TRINITROGLYCERIN see NGY000
SPIRIT ORANGE see PEJ500
SPIRITS of SALT see HHL000
SPIRITS of TURPENTINE see TOD750
SPIRITS OF WINE see EFU000
SPIRIT of TURPENTINE see TOD750
SPIRIT YELLOW I see PEJ500
SPIRO(1,3-BENZODIOXOLE-2,1'-CYCLOHEXAN)-4-OL, METHYLCARBAMATE see SLD550
SPIROCID see ABX500
SPIROCTANIE see AFJ500
SPIRO(17H-CYCLOPENTA(a)PHENANTHRENE-17,2'-(3'H)-FURAN) see AFJ500
SPIRO(17H-CYCLOPENTA(a)PHENANTHRENE-17,2'(5'H)FURAN), PREGN-4-ENE-21-CARBOXYLIC ACID DERIV. see AFJ500
SPIROFULVIN see GKE000
SPIRO(ISOBENZOFURAN)-1(3H),9'-(9H)XANTHENE-3-ONE, 3',6'-DIHYDROXY-DISODIUM SALT see FEW000
SPIRO(ISOBENZOFURAN-1(3H),9'-(9H)XANTHEN)-3-ONE, 4',5'-DIBROMO-3',6'-DIHYDROXY- (9CI) see DDO200
SPIROLACTONE see AFJ500
SPIROLAKTON see AFJ500
SPIROLANG see AFJ500
SPIRONE see AFJ500
SPIRONOLACTONE see AFJ500
SPIRONOLACTONE A see AFJ500
SPIROZID see ABX500
SPOFADRIZINE see PPP500
SPONGIOFORT see PCU360
SPORILINE see TGB475
SPOR-KIL see ABU500
SPOROSTATIN see GKE000
SPOTRETE see TFS350
SPRACAL see ARB750
SPRAY-DERMIS see NGE500
SPRAY-FORAL see NGE500
SPRAY-HORMITE see SGH500
SPRAYSET MEKP see MKA500
SPRAY-TROL BRANCH RODEN-TROL see WAT200
SPRING-BAK see DXD200
SPRITZ-HORMIT see SGH500
SPROUT/OFF see DMC600
SPROUT-STOP see DMC600
SPS 600 see SMQ500
SPT 50 CPS see EHG100
SPULMAKO-LAX see PDO750
SQ 1489 see TFS350
SQ 3277 see HOC000
SQ 9453 see DUD800
SQ 14055 see TET800
SQ 21977 see MPU000
SQ 22947 see TET800
SQUILL see RCF000
SR 339 see PER250
SR406 see CBG000
SRA 5172 see DTQ400

SRA 7312 see DJY200
SRC-II HEAVY DISTILLATE see CMY750
SRI 859 see MNA750
SRI 1720 see BIF750
SRI 1869 see NKL000
SRI 2200 see CGV250
SRI 2489 see IAN000
SRI 10163-71 see HFT550
SRM 705 see SMQ500
SRM 706 see SMQ500
SRM 1475 see PJS750
SRM 1476 see PJS750
SS see CCU250
SS 578 see DNF600
S.T. 37 see HFV500
ST 90 see SMQ500
ST 155 see CBF250
STABILAN see CMF400
STABILENE see BKA000
STABILISATOR C see DWN800
STABILISATOR SCD see OAT000
STABILIZATOR AR see PFT500
STABILIZER D-22 see DDV600
STABILIZER DLT see TFD500
STABILIZER SCD see OAT000
STABINEX NW 7PS see CAW850
STABLE RED KB BASE see CLK225
STABOXOL 1 see BJE550
STAFAC see VRF000
STAFAST see NAK500
STA-FAST see TIX500
STAFLEN E 650 see PJS750
STAFLEX DBM see DED600
STAFLEX DBP see DEH200
STAFLEX DBS see DEH600
STAFLEX DOP see DVL700
STA-FRESH 615 see SGM500
STAGNO (TETRACLORURO di) (ITALIAN) see TGC250
STALFLEX DOS see BJS250
STAM see DGI000
STAM M-4 see DGI000
STAM F 34 see DGI000
STAM LV 10 see DGI000
STAMPEDE see DGI000
STAMPEDE 3E see DGI000
STAM SUPERNOX see DGI000
STAMYLAN 900 see PJS750
STAMYLAN 1000 see PJS750
STAMYLAN 1700 see PJS750
STAMYLAN 8200 see PJS750
STAMYLAN 8400 see PJS750
STANDACOL CARMOISINE see HJF500
STANDACOL ORANGE G see HGC000
STANDACOL SUNSET YELLOW FCF see FAG150
STANDAMIDD LD see BKE500
STANDAPOL 112 CONC see SIB600
STANDARD LEAD ARSENATE see LCK000
STANEPHRIN see SPC500
STANGEN MALEATE see DBM800
STAN-GUARD 156 see BKO250
STANILO see SLI325
STANNANE, ((p-ACETAMIDOBENZOYL)OXY)TRIPHENYL- see TMV800
STANNANE, ACETOXYDIPHENYLETHYL- see EIM100
STANNANE, ACETOXYTRICYCLOHEXYL- see ABW600

STANNANE, (ACETYLENECARBONYLOXY)TRIPHENYL- see TMW600

STANNANE, (ACETYLOXY)ETHYLDIPHENYL- see EIM100

STANNANE, BROMOTRIETHYL-, compounded with 2-PIPECOLINE (1:1) see TJU850

STANNANE, BUTYLDIETHYLIODO- see BRA550

STANNANE, (BUTYLTHIO)TRIOCTYL- see BSO200

STANNANE, (BUTYLTHIO)TRIPROPYL- see TMY850

STANNANE, CYANATOTRIMETHYL- see TMI100

STANNANE, (CYANOACETOXY)TRIPHENYL- see TMV825

STANNANE, DICYCLOHEXYLOXO- see DGV900

STANNANE, DIMETHYL(2,3-QUINOXALINYLDITHIO)- see QSJ800

STANNANE, ((DIMETHYLTHIOCARBAMOYL)THIO)TRIBUTYL- see TID150

STANNANE, ETHYNYLENEBIS(CARBONYLOXY)BIS(TRIPHENYL- see BLS900

STANNANE, FLUOROTRIS(p-CHLOROPHENYL)- see TNG050

STANNANE, (ISOCYANATO)TRIBUTYL- see THY750

STANNANE, (ISOPROPYLSUCCINYLOXY)TRIBUTYL- see TIE600

STANNANE, (p-NITROPHENOXY)TRIBUTYL- see NII200

STANNANE, TRIBUTYLISOCYANATO- see THY750

STANNANE, TRICHLOROOCTYL- see OGG000

STANNANE, TRIISOPROPYL(UNDECANOYLOXY)- see TKT850

STANNIC BROMIDE see TGB750

STANNIC CHLORIDE, anhydrous (DOT) see TGC250

STANNIC CHLORIDE, pentahydrate (DOT) see TGC282

STANNIC CHLORIDE PENTAHYDRATE see TGC282

STANNIC DICHLORIDE DIIODIDE see TGC280

STANNIC IODIDE see TGD750

STANNIC PHOSPHIDE (DOT) see TGE500

STANN OMF see DVK200

STANNOPLUS see DAE600

STANNORAM see DAE600

STANNOUS CHLORIDE (FCC) see TGC000

STANNOUS FLUORIDE see TGD100

STANNOUS IODIDE see TGD500

STANNOUS OXALATE see TGE250

STANNOUS STEARATE see SLI350

STANNOXYL see TGE300

STANNPLOUS see DAE600

STANOMYCETIN see CDP250

STANSIN see SNN500

STAPHOBRISTOL-250 see SLJ000

STAPHYBIOTIC see SLJ000

STAPHYLOMYCIN see VRF000

STAPYOCINE see VRF000

STAR see GGA000

STARAMIC 747 see SLJ500

STAR ANISE OIL see AOU250

STARCH see SLJ500

α-STARCH see SLJ500

STARCH, CORN see SLJ500

STARCH DUST see SLJ500

STARCH GUM see DBD800

STARCH HYDROXYETHYL ETHER see HLB400

STARCH (OSHA) see SLJ500

STAR DUST see CNE750

STARFOL GMS 450 see OAV000

STARFOL GMS 600 see OAV000

STARFOL GMS 900 see OAV000

STARFOL IPM see IQN000

STARIFEN see EOK000

STARLEX L see CAW850

STARSOL No. 1 see AQQ500

STA-RX 1500 see SLJ500

STATEX see CBT750

STATEX N 550 see CBT750

STATHION see PAK000

STATOMIN MALEATE see DBM800

STATYL see NCN600

STAUFFER N-3049 see EPY000

STAUFFER N-4446 see EMR100

STAUFFER CAPTAN see CBG000

STAUFFER N 521 see DSB200

STAUFFER N 2790 see FMU045

STAUFFER R-1,303 see TNP250

STAUFFER R-1492 see MQH750

STAUFFER R 1504 see PHX250

STAUFFER R 1608 see EIN500

STAUFFER R 14487 see DJW890

STAURODERM see FMQ000

STAVELAN CINATY (CZECH) see TGE250

STAVINOR see LHQ100

STAVINOR 30 see CAX350

STAY-FLO see SHF500

STB see SKE000

STEADFAST see CCP000

STEARALKONIUM CHLORIDE see DTC600

STEARAMINE see OBC000

STEAREX BEADS see SLK000

STEARIC ACID see SLK000

STEARIC ACID, ALUMINIUM SALT see AHA250

STEARIC ACID, BARIUM CADMIUM SALT (4:1:1) see BAI800

STEARIC ACID, CADMIUM SALT see OAT000

STEARIC ACID, LEAD SALT see LDX000

STEARIC ACID, LITHIUM SALT see LHQ100

STEARIC ACID, MONOESTER with ETHYLENE GLYCOL see EJM500

STEARIC ACID, MONOESTER with GLYCEROL see OAV000

STEARIC ACID POTASSIUM SALT. see PLS775

STEARIC ACID, SODIUM SALT see SJV500

STEARIC ACID, TETRADECYL ESTER see TCB100

STEARIC ACID, ZINC SALT see ZMS000

STEARIC MONOGLYCERIDE see OAV000

STEARIX ORANGE see PEJ500

STEARIX RED 4B see SBC500

STEARIX RED 4S see SBC500

STEARIX SCARLET see OHI200

STEAROL see OAX000

STEAROPHANIC ACID see SLK000

STEAROYL PROPLENE GLYCOL HYDROGEN SUCCINATE see SNG600

STEAR YELLOW JB see DOT300

STEARYL ALCOHOL see OAX000

STEARYLAMINE see OBC000

n-STEARYLAMINE see OBC000

STEARYL CITRATE see SLN100

STEARYLDIMETHYLBENZYLAMMONIUM CHLORIDE see DTC600

STEARYL MONOGLYCERIDYL CITRATE see SLN200

STEAWHITE see TAB750

STEBAC see DTC600

STECLIN HYDROCHLORIDE see TBX250

STEINAMID DL 203 S see BKE500
STEINBUHL YELLOW see BAK250
STEINBUHL YELLOW see CAP750
STELADONE see CDS750
STELAZINE see TKK250
STELAZINE DIHYDROCHLORIDE see TKK250
STELLAMINE see DJS200
STELLON PINK see PKB500
STEMETIL see PMF500
STEMPOR see MHC750
STENOLON see MPN500
STENTAL EXTENTABS see EOK000
STEPAN D-50 see IQN000
STEPANOL WAQ see SIB600
STEPHANIA HERNANDIFOLIA Walp., extract see
SLO100
STERAFFINE see OAX000
STERAL see HCL000
STERAMIDE see SNQ710
STERANDRYL see TBG000
STERANE see PMA000
STERASKIN see HCL000
STERAZINE see PPP500
STERCULIA GUM see KBK000
STERICOL see XKA000
STERIDO see BIM250
STERIGMATOCYSTIN see SLP000
STERILIZING GAS ETHYLENE OXIDE 100% see
EJN500
STERISEAL LIQUID #40 see SGM500
STERLING see SFT000
STERLING see CBT750
STERLING N 765 see CBT750
STERLING NS see CBT750
STERLING SO 1 see CBT750
STERLING WAQ-COSMETIC see SIB600
STERNITE 30 see SMQ500
STERNITE ST 30VL see SMQ500
STEROGYL see VSZ100
STEROLAMIDE see TKP500
STEROLONE see PMA000
STERONYL see MPN500
STESOLID see DCK759
STIBANATE see AQH800
STIBANOSE see AQH800
STIBATIN see AQH800
STIBIC ANHYDRIDE see AQF750
2,2',2''-(STIBILIDYNETRIS(THIO))TRIS-
BUTANEDIOIC ACID HEXALITHIUM SALT see
LGU000
STIBILIUM see DKA600
STIBINE see SLQ000
STIBINE OXIDE, TRIPHENYL- see TMQ550
STIBINE SULFIDE, TRIPHENYL- see TMQ600
STIBINE, TRICHLORO- see AQC500
STIBINE, TRIFLUORO- (9CI) see AQE000
STIBINE, TRIS(DODECYLTHIO)- see TNH850
STIBINOL see AQH800
STIBIUM see AQB750
STIBNAL see AQI750
STIBUNAL see AQI750
STICKMONOXYD (GERMAN) see NEG100
STICKSTOFFDIOXID (GERMAN) see NGR500
STICKSTOFFLOST see BIE500
STICKSTOFFWASSERSTOFFSAEURE (GERMAN) see
HHG500
STIGMANOL BROMIDE see POD000
STIGMOSAN BROMIDE see POD000

STIK see NAK500
STIKSTOFDIOXYDE (DUTCH) see NGR500
STIL see DKA600
STILBEN (GERMAN) see SLR000
STILBENE see SLR000
STILBENE 3 see TGE150
α-STILBENECARBONITRILE see DVX600
2,2'-STILBENEDISULFONIC ACID, 4,4'-BIS((4-
ANILINO-6-((2-HYDROXYETHYL)METHYLAMINO)-
s-TRIAZIN- 2-YL)AMINO)-, DISODIUM SALT see
TGE100
2-STILBENESULFONIC ACID, 4-(7-SULFO-2H-
NAPHTHO(1,2-d)TRIAZOL-2-YL)-, DISODIUM SALT
see DXG025
STILBENYL-N,N-DIMETHYLAMINE see DUB800
STILBESTROL see DKA600
STILBESTROL DIETHYL DIPROPIONATE see
DKB000
STILBESTROL DIPROPIONATE see DKB000
STILBESTROL PROPIONATE see DKB000
STILBESTRONATE see DKB000
STILBESTRONE see DKA600
STILBETIN see DKA600
STILBOEFRAL see DKA600
STILBOESTROFORM see DKA600
STILBOESTROL see DKA600
STILBOESTROL DIPROPIONATE see DKB000
STILBOFAX see DKB000
STILBOFOLLIN see DKA600
STILBOL see DKA600
STILKAP see DKA600
STILON see PJY500
STIL-ROL see DKA600
STILRONATE see DKB000
STIMAMIZOL HYDROCHLORIDE see LFA020
STIMINOL see DJS200
STIMULEX see DBA800
STIMULIN see DJS200
STIMULINA see GFO050
STINERVAL see PFC500
STINERVAL see PFC750
STINK DAMP see HIC500
STIPEND see CMA100
STIPINE see SEH000
STIPTANON see EDU500
STIROLO (ITALIAN) see SMQ000
ST. JOHN'S BREAD see LIA000
ST 30UL see SMQ500
STODDARD SOLVENT see SLU500
STOIKON see BCA000
STONE RED see IHC450
STOPAETHYL see DXH250
STOP-DROP see NAK500
STOPETHYL see DXH250
STOPETYL see DXH250
STOPMOLD B see BGJ750
STOP-SCALD see SAV000
STOPSPOT see PFM500
STOPTON ALBUM see SNM500
STOVARSAL see ABX500
STOVARSOL see ABX500
STOVARSOLAN see ABX500
STOXIL see DAS000
STPP see SKN000
STR see SMD000
STRAIGHT-CHAIN ALKYL BENZENE SULFONATE
see LGF825

STRAIGHT-RUN KEROSENE see KEK000
STRATHION see PAK000
STRAWBERRY ALDEHYDE see ENC000
STRAWBERRY RED A GEIGY see FMU080
STRAZINE see ARQ725
STREL see DGI000
STREPAMIDE see SNM500
STREPCEN see SLW500
STREPCIN see SLY500
STREP-GRAN see SLY500
STREPSULFAT see SLY500
STREPTAGOL see SNM500
d-STREPTAMINE, o-6-AMINO-6-DEOXY-α-d-GLUCOPYRANOSYL-(1-4)-o-(3-DEOXY-4-C-METHYL-3-(METHYLAMINO)-β-l-ARABINOPYRANOSYL-(1-6))-N¹)-(3-AMINO-2-HYDROXY-1-OXOPROPYL)-2-DEOXY-, (S)-, SULFATE (1:2) (SALT) see IHN300
STREPTOCLASE see SNM500
STREPTOGRAMIN see VRF000
STREPTOL see SNM500
STREPTOMICINA (ITALIAN) see SLW500
STREPTOMYCES PEUCETIUS see DAC000
STREPTOMYCIN see SLW500
STREPTOMYCIN A see SLW500
STREPTOMYCINE see SLW500
STREPTOMYCIN SESQUISULFATE see SLY500
STREPTOMYCIN SULFATE see SLY500
STREPTOMYCIN SULPHATE B.P. see SLY500
STREPTOMYCINUM see SLW500
STREPTOMYZIN (GERMAN) see SLW500
STREPTOREX see SLY500
STREPTOSIL see SNM500
STREPTOSILTHIAZOLE see TEX250
STREPTOZOCIN see SMD000
STREPTOZONE see SNM500
STREPTOZOTICIN see SMD000
STREPTROCIDE see SNM500
STREPVET see SLY500
STRESNIL see FLU000
STREUNEX see BBQ500
STRICNINA (ITALIAN) see SMN500
STROBANE see MIH275
STROBANE see TBC500
STROBANE-T-90 see CDV100
STRONTIUM ARSENITE see SME500
STRONTIUM ARSENITE, solid (DOT) see SME500
STRONTIUM CHLORATE see SMF500
STRONTIUM CHLORATE, solid or solution (DOT) see SMF500
STRONTIUM CHROMATE (1:1) see SMH000
STRONTIUM CHROMATE (VI) see SMH000
STRONTIUM CHROMATE 12170 see SMH000
STRONTIUM FLUOBORATE see SMI000
STRONTIUM FLUORIDE see SMI500
STRONTIUM MONOSULFIDE see SMM000
STRONTIUM NITRATE (DOT) see SMK000
STRONTIUM(II) NITRATE (1:2) see SMK000
STRONTIUM PEROXIDE see SMK500
STRONTIUM SULFIDE see SMM000
STRONTIUM SULPHIDE see SMM000
STRONTIUM YELLOW see SMH000
STROPHANTHIN G see OKS000
STROPHOPERM see OKS000
STRUMACIL see MPW500
STRUMAZOLE see MCO500
STRYCHININE SULFATE see SMP000
STRYCHNIDIN-10-ONE see SMN500

STRYCHNIDIN-10-ONE, 2,3-DIMETHOXY-(9CI) see BOL750
STRYCHNIDIN-10-ONE, SULFATE (2:1) see SMP000
STRYCHNIN (GERMAN) see SMN500
STRYCHNINE see SMN500
STRYCHNINE, solid and liquid (DOT) see SMN500
STRYCHNINE, 2,3-DIMETHOXY- see BOL750
STRYCHNINE SALT (solid) see SMO500
STRYCHNINE SALTS (DOT) see SMO500
STRYCHNINE SULFATE (2:1) see SMP000
STRYCHNOS see SMN500
STRYON 686 see SMQ500
STRYPTIRENAL see VGP000
STRZ see SMD000
STS see SIO000
STUDAFLUOR see SHF500
STUNTMAN see DMC600
STYPHNIC ACID see SMP500
STYRAFOIL see SMQ500
STYRAGEL see SMQ500
STYRALLYL ALCOHOL see PDE000
STYRALYL ALCOHOL see PDE000
STYREEN (DUTCH) see SMQ000
STYREN (CZECH) see SMQ000
STYREN-ACRYLONITRILEPOLYMER see ADY500
STYRENE see SMQ000
STYRENE-ACRYLONITRILE COPOLYMER see ADY500
STYRENE-BUTADIENE COPOLYMER see SMR000
STYRENE-1,3-BUTADIENE COPOLYMER see SMR000
STYRENE-BUTADIENE POLYMER see SMR000
STYRENE EPOXIDE see EBR000
STYRENE MONOMER (ACGIH) see SMQ000
STYRENE MONOMER, inhibited (DOT) see SMQ000
STYRENE OXIDE see EBR000
STYRENE-7,8-OXIDE see EBR000
STYRENE POLYMER see SMQ500
STYRENE POLYMER with 1,3-BUTADIENE see SMR000
STYRENE POLYMERS see SMQ500
STYREX C see SMQ500
STYROCELL PM see SMQ500
STYROFAN 2D see SMQ500
STYROFLEX see SMQ500
STYROFOAM see SMQ500
STYROL (GERMAN) see SMQ000
STYROLE see SMQ000
STYROLENE see SMQ000
STYROLUX see SMQ500
STYRON see SMQ000
STYRON see SMQ500
STYRON 475 see SMQ500
STYRON 492 see SMQ500
STYRON 666 see SMQ500
STYRON 678 see SMQ500
STYRON 679 see SMQ500
STYRON 683 see SMQ500
STYRON 685 see SMQ500
STYRON 690 see SMQ500
STYRON 440A see SMQ500
STYRON 470A see SMQ500
STYRON 475D see SMQ500
STYRON 69021 see SMQ500
STYRONE see CMQ740
STYRON GP see SMQ500
STYRON 666K27 see SMQ500
STYRON PS 3 see SMQ500
STYRON T 679 see SMQ500

STYRON 666U see SMQ500
STYRON 666V see SMQ500
STYROPIAN see SMQ500
STYROPIAN FH 105 see SMQ500
STYROPOL HT 500 see SMQ500
STYROPOL IBE see SMQ500
STYROPOL JQ 300 see SMQ500
STYROPOL KA see SMQ500
STYROPOR see SMQ000
STYROPOR see SMQ500
STYRYL CARBINOL see CMQ740
STYRYL METHYL KETONE see SMS500
STYRYL OXIDE see EBR000
STYRYLPYRIDINIUM CHLORIDE,
DIETHYLCARBAMAZINE see SMU100
STZ see SMD000
SU 2000 see CBT500
SU 5879 see CFY000
SU-9064 see MQR200
SU-13437 see MCB500
SUAVITIL see BCA000
SUBACETATE LEAD see LCH000
SUBAMYCIN see TBX250
SUBARI see HGC500
SUBCHLORIDE of MERCURY see MCW000
SUBDUE see MDM100
SUBDUE 2E see MDM100
SUBDUE 5SP see MDM100
SUBICARD see PBC250
SUBLIMAT (CZECH) see MCY475
SUBLIMAZE see PDW750
SUBLIMAZE CITRATE see PDW750
SUBLIMED SULFUR see SOD500
SU-BRONTINE see EAG100
SUBSTANCE F see MIW500
SUBTILISIN CARLSBURG see BAC000
SUBTILISIN (9CI, ACGIH) see BAC000
SUBTILISIN NOVO see BAC000
SUBTILOPEPTIDASE A see BAC000
SUBTILOPEPTIDASE B see BAC000
SUBTILOPEPTIDASE BPN' see BAC000
SUBTILOPEPTIDASE C see BAC000
SUBTOSAN see PKQ250
SUCARYL see CPQ625
SUCARYL ACID see CPQ625
SUCARYL CALCIUM see CAR000
SUCARYL SODIUM see SGC000
SUCCARIL see SGC000
SUCCARIL see SJN700
SUCCINBROMIMIDE see BOF500
SUCCINIBROMIMIDE see BOF500
SUCCINIC ACID see SMY000
SUCCINIC ACID ANHYDRIDE see SNC000
SUCCINIC ACID, CADMIUM SALT (1:1) see CAI750
SUCCINIC ACID, DIETHYL ESTER see SNB000
SUCCINIC ACID, 2,3-DIHYDROXY- see TAF750
SUCCINIC ACID-2,2-DIMETHYLHYDRAZIDE see
DQD400
SUCCINIC ACID DINITRILE see SNE000
SUCCINIC ACID, HYDROXY- see MAN000
SUCCINIC ACID, O-ISOPROPYL-O'-
TRIBUTYLSTANNYL ESTER see TIE600
SUCCINIC ACID PEROXIDE (DOT) see SNC500
SUCCINIC ANHYDRIDE see SNC000
SUCCINIC-1,1-DIMETHYL HYDRAZIDE see DQD400
SUCCINIC DINITRILE see SNE000
SUCCINIC PEROXIDE see SNC500
SUCCINODINITRILE see SNE000

SUCCINONITRILE see SNE000
SUCCINYL OXIDE see SNC000
SUCCINYL PEROXIDE see SNC500
SUCCISTEARIN see SNG600
SUCHAR 681 see CBT500
SUCKER-STUFF see DMC600
SUCRA see SJN700
SUCRAPHEN see SPC500
SUCRE EDULCOR see BCE500
SUCRETS see HFV500
SUCRETTE see BCE500
SUCROFER see IHG000
SUCROL see EFE000
SUCROSA see SGC000
SUCROSE see SNH000
SUDAN AX see XRA000
SUDAN G see OHI200
SUDAN G III see OHI200
SUDAN GREEN 4B see BLK000
SUDAN III see OHI200
SUDAN III (G) see OHI200
SUDAN IV see SBC500
SUDAN ORANGE see XRA000
SUDAN ORANGE R see PEJ500
SUDAN ORANGE RPA see XRA000
SUDAN ORANGE RRA see XRA000
SUDAN P see SBC500
SUDAN P III see OHI200
SUDAN R see CMS238
SUDAN RED see XRA000
SUDAN RED 7B see EOJ500
SUDAN RED 290 see CMS238
SUDAN RED 4BA see SBC500
SUDAN RED BB see SBC500
SUDAN RED BBA see SBC500
SUDAN RED G (6CI) see CMS238
SUDAN RED III see OHI200
SUDAN RED IV see SBC500
SUDANROT 7B see EOJ500
SUDAN SCARLET 6G see XRA000
SUDAN X see XRA000
SUDAN YELLOW see DOT300
SUDAN YELLOW R see PEI000
SUDAN YELLOW RRA see AIC250
SUDINE see SNN300
SUESSETTE see SGC000
SUESSTOFF see EFE000
SUESTAMIN see SGC000
SUGAI CHRYSOIDINE see PEK000
SUGAI CONGO RED see SGQ500
SUGAI FAST SCARLET G BASE see NMP500
SUGAI TARTRAZINE see FAG140
SUGANYL see AHO250
SUGAR see SNH000
SUGARIN see SGC000
SUGAR of LEAD see LCV000
SUGARON see SGC000
SU 8842 HYDROCHLORIDE see MQR200
SUICALM see FLU000
SUISYNCHRON see MLJ500
SULADYNE see PDC250
SULAMYD see SNQ710
SUL ANILINOVA (CZECH) see BBL000
SULDIXINE see SNN300
SULEMA (RUSSIAN) see MCY475
SULESTREX see PIK450
SULF-10 see SNQ710
SULFABENZPYRAZINE see QTS000

SULFABROMOMETHAZINE SODIUM see SNH875
SULFACET see SNQ710
SULFACETAMIDE see SNQ710
SULFACETIMIDE see SNQ710
SULFACID LIGHT ORANGE J see HGC000
SULFACTIN see BAD750
SULFACYL see SNQ710
SULFADENE see BDF000
SULFADIAZINE see PPP500
SULFADIMERAZINE see SNJ000
SULFADIMETHOXIN see SNN300
SULFADIMETHOXINE see SNN300
SULFADIMETHOXYDIAZINE see SNN300
SULFADIMETHYLDIAZINE see SNJ000
SULFADIMETHYLISOXAZOLE see SNN500
SULFADIMETHYLPYRIMIDINE see SNJ000
SULFADIMETINE see SNJ000
SULFADIMETOSSINA (ITALIAN) see SNN300
SULFADIMETOXIN see SNN300
SULFADIMEZINE see SNJ000
SULFADIMIDINE see SNJ000
SULFADINE see SNJ000
SULFADSIMESINE see SNJ000
SULFAETHOXYPYRIDAZINE see SNJ100
SULFAFURAZOL see SNN500
SULFAGAN see SNN500
SULFAGUANIDINE see AHO250
SULFAGUINE see AHO250
SULFA-ISODIMERAZINE see SNJ000
SULFAISODIMIDINE see SNJ000
SULFALLATE see CDO250
SULFAMATE see ANU650
SULFAMERADINE see ALF250
SULFAMERAZIN see ALF250
SULFAMERAZINE SODIUM see SJW475
SULFAMETHALAZOLE see SNK000
SULFAMETHIAZINE see SNJ000
SULFAMETHIN see SNJ000
SULFAMETHOXAZOL see SNK000
SULFAMETHOXAZOLE see SNK000
SULFAMETHYLDIAZINE see ALF250
SULFAMETHYLISOXAZOLE see SNK000
SULFAMEZATHINE see SNJ000
SULFAMIC ACID see SNK500
SULFAMIC ACID, MONOAMMONIUM SALT see ANU650
SULFAMIDIC ACID see SNK500
p-SULFAMIDOANILINE see SNM500
SULFAMIDYL see SNM500
SULFAMINSAEURE (GERMAN) see ANU650
N-(5-SULFAMOYL-1,3,4-THIADIAZOL-2-YL)ACETAMIDE see AAI250
SULFAMUL see TEX250
SULFAN see SOR500
SULFANA see SNM500
SULFANALONE see SNM500
SULFANIL see SNM500
SULFANILACETAMIDE see SNQ710
SULFANILAMIDE see SNM500
6-SULFANILAMIDO-2,4-DIMETHOXYPYRIMIDINE see SNN300
5-SULFANILAMIDO-3,4-DIMETHYL-ISOXAZOLE see SNN500
2-SULFANILAMIDO-4,6-DIMETHYLPYRIMIDINE see SNJ000
3-SULFANILAMIDO-5-METHYLISOXAZOLE see SNK000
SULFANILAMIDOPYRIMIDINE see PPP500

2-SULFANILAMIDOQUINOXALINE see QTS000
2-SULFANILAMIDOTHIAZOLE see TEX250
SULFANILGUANIDINE see AHO250
2-SULFANILYL AMINOPYRIDINE see PPO000
2-SULFANILYLAMINOPYRIMIDINE see PPP500
2-(SULFANILYLAMINO)THIAZOLE see TEX250
SULFANILYLGUANIDINE see AHO250
N¹-SULFANILYL-N²-BUTYLCARBAMIDE see BSM000
N¹-SULFANILYL-N²-BUTYLUREA see BSM000
N-SULFANILYL-N'BUTYLUREE (FRENCH) see BSM000
N-SULFANYLACETAMIDE see SNQ710
SULFAPOL see DXW200
SULFAPOLU (POLISH) see DXW200
SULFAPYRIDINE see PPO000
2-SULFAPYRIDINE see PPO000
SULFAPYRIMIDIN (GERMAN) see PPP500
2-SULFAPYRIMIDINE see PPP500
SULFAQUINOXALINE see QTS000
SULFASAN see BKU500
SULFASAN R POWDER see BKU500
SULFASOL see SNN300
SULFASOXAZOLE see SNN500
SULFASTOP see SNN300
SULFATE d'ATROPINE (FRENCH) see ARR500
SULFATE de CUIVRE (FRENCH) see CNP250
SULFATE DIMETHYLIQUE (FRENCH) see DUD100
SULFATE MERCURIQUE (FRENCH) see MDG500
SULFATE de METHYLE (FRENCH) see DUD100
SULFATE de NICOTINE (FRENCH) see NDR500
SULFATEP see SOD100
SULFATE de PLOMB (FRENCH) see LDY000
SULFATES see SNS000
SULFATE de ZINC (FRENCH) see ZNA000
SULFATHIAZOL see TEX250
SULFATHIAZOLE (USDA) see TEX250
SULFATOBIS(DIMETHYLSELENIDE)PLATINUM(II) HYDRATE see SNS100
SULFAZOLE see SNN500
SULFENAMIDE M see BDG000
SULFENAMIDE TS see CPI250
SULFENAX see CPI250
SULFENAX CB see CPI250
SULFENAX CB 30 see CPI250
SULFENAX CB/K see CPI250
SULFERROUS see FBN100
SULFERROUS see FBO000
SULFIDAL see SOD500
SULFIDES see SNT000
SULFIDINE see PPO000
SULFIDOTRICHLORID FOSFORECNY see TFO000
SULFIMEL DOS see DJL000
1,1'-SULFINYLBIS(1,2-DICHLOROETHANE) see SNT100
SULFINYLBIS(METHANE) see DUD800
SULFINYL CHLORIDE see TFL000
SULFISIN see SNN500
SULFISOMEZOLE see SNK000
SULFISOMIDIN see SNJ000
SULFISOMIDINE see SNJ000
SULFISOXAZOLE see SNN500
SULFITE CELLULOSE see CCU150
SULFITE LYE see HIC600
SULFITES see SNT500
SULFIZOLE see SNN500
o-SULFOBENZIMIDE see BCE500
o-SULFOBENZOIC ACID IMIDE see BCE500
SULFOCARBANILIDE see DWN800

SULFOCARBOLIC ACID see HJH500
SULFOCIDINE see SNM500
SULFODIMESIN see SNJ000
SULFODIMEZINE see SNJ000
SULFODOR (CZECH) see EPH000
2-SULFOETHYLAMINE see TAG750
SULFOGUANIDINE see AHO250
SULFOGUENIL see AHO250
N-SULFOMETHYL-POLYMYXIN B SODIUM SALT see SNW600
SULFOMYXIN see SNW600
SULFONA see SOA500
SULFONAMIDE see SNM500
SULFONAMIDE P see SNM500
2-SULFONAMIDOTHIAZOLE see TEX250
2-(4-SULFO-1-NAPHTHYLAZO)-1-NAPHTHOL-4-SULFONIC ACID, DISODIUM SALT see HJF500
SULFONATES see SNY000
o-SULFONBENZOIC ACID IMIDE SODIUM SALT see SJN700
SULFONE-2,4,4',5-TETRACHLORODIPHENYL see CKM000
SULFONIC ACID, MONOCHLORIDE see CLG500
SULFONIMIDE see CBF800
SULFONIUM, (2-((CYCLOHEXYLHYDROXYPHENYLACETYL)OXY)ETHYL)DIMETHYL-, IODIDE see HOJ150
SULFONIUM, (2-HYDROXYETHYL)DIMETHYL-, IODIDE, α-PHENYLCYCLOHEXANEGLYCOLATE see HOJ150
1,1'-SULFONYLBIS(4-AMINOBENZENE) see SOA500
4,4'-SULFONYLBISANILINE see SOA500
p,p-SULFONYLBISBENZAMINE see SOA500
4,4'-SULFONYLBISBENZAMINE see SOA500
p,p-SULFONYLBISBENZENAMINE see SOA500
SULFONYL CHLORIDE see SOT000
SULFONYL CHLORIDE FLUORIDE see SOT500
4,4'-SULFONYLDIANILINE see SOA500
p,p'-SULFONYLDIANILINE see SOA500
N-(SULFONYL-p-METHYLBENZENE)-N'-N-BUTYLUREA see BSQ000
1-p-SULFOPHENYLAZO-2-HYDROXYNAPHTHALENE-6-SULFONATE, DISODIUM SALT see FAG150
1-p-SULFOPHENYLAZO-2-NAPHTHOL-6-SULFONIC ACID, DISODIUM SALT see FAG150
1-(4-SULFOPHENYL)-3-ETHYLCARBOXY-4-(4-SULFONAPHTHYLAZO)-5-HYDROXYPYRAZOLE see OJK325
SULFOPLAN see SNN300
SULFOPON WA 1 see SIB600
SULFOQUANIDINE see AHO250
SULFORON see SOD500
SULFOTEP see SOD100
SULFOTEPP see SOD100
SULFOTEX WALA see SIB600
SULFOXIDE, BIS(1,2-DICHLOROVINYL) see SNT100
SULFOXYL see BDS000
SULFRAMIN 85 see DXW200
SULFRAMIN ACID 1298 see LBU100
SULFRAMIN 40 FLAKES see DXW200
SULFRAMIN 40 GRANULAR see DXW200
SULFRAMIN 1238 SLURRY see DXW200
SULFTECH see SJZ000
SULFUR see SOD500
SULFUR CHLORIDE see SOG500
SULFUR CHLORIDE see SON510

SULFUR CHLORIDE (DI) (DOT) see SON510
SULFUR CHLORIDE (MONO) see SOG500
SULFUR CHLORIDE OXIDE see TFL000
SULFUR DECAFLUORIDE see SOQ450
SULFUR DICHLORIDE see SOG500
SULFUR DIFLUORIDE MONOXIDE see TFL250
SULFUR DIFLUORIDE OXIDE see TFL250
SULFUR DIOXIDE see SOH500
SULFUR DIOXIDE, solution see SOO500
SULFURE de METHYLE (FRENCH) see TFP000
SULFURETED HYDROGEN see HIC500
SULFUR FLOWER (DOT) see SOD500
SULFUR FLUORIDE see SOI000
SULFUR FLUORIDE OXIDE see BLD000
SULFUR HEXAFLUORIDE see SOI000
SULFUR HYDRIDE see HIC500
SULFURIC ACID see SOI500
SULFURIC ACID (mist) see SOI530
SULFURIC ACID, fuming see SOI520
SULFURIC ACID, fuming <30% free sulfur trioxide (DOT) see SOI520
SULFURIC ACID, fuming > or =30% free sulfur trioxide (DOT) see SOI520
SULFURIC ACID, ALUMINUM SALT (3:2) see AHG750
SULFURIC ACID, AMMONIUM NICKEL(2+) SALT (2:2:1) see NCY050
SULFURIC ACID, BARIUM SALT (1:1) see BAP000
SULFURIC ACID, BERYLLIUM SALT (1:1) see BFU250
SULFURIC ACID, BERYLLIUM SALT (1:1), TETRAHYDRATE see BFU500
SULFURIC ACID, BIS(METHYLMERCURY) SALT see BKS810
SULFURIC ACID, CADMIUM(2+) SALT see CAJ000
SULFURIC ACID, CADMIUM SALT, HYDRATE see CAJ250
SULFURIC ACID, CADMIUM SALT, TETRAHYDRATE see CAJ500
SULFURIC ACID, CALCIUM(2+) SALT, DIHYDRATE see CAX750
SULFURIC ACID, CERIUM SALT (2:1) see CDB400
SULFURIC ACID, CHROMIUM (3+) POTASSIUM SALT (2:1:1) see PLB500
SULFURIC ACID, CHROMIUM(3+) SALT (3:2), PENTADECAHYDRATE see CMK425
SULFURIC ACID, COBALT(2+) SALT (1:1) see CNE125
SULFURIC ACID, COPPER(2+) SALT (1:1) see CNP250
SULFURIC ACID, DIAMMONIUM SALT see ANU750
SULFURIC ACID, DIMETHYL ESTER see DUD100
SULFURIC ACID, DIPOTASSIUM SALT see PLT000
SULFURIC ACID, DISODIUM SALT see SJY000
SULFURIC ACID, DITHALLIUM(1+) SALT (8CI, 9CI) see TEM000
SULFURIC ACID, INDIUM SALT see ICJ000
SULFURIC ACID, IRON(2+) SALT (1:1) see FBN100
SULFURIC ACID, IRON (3+) SALT (3:2) see FBA000
SULFURIC ACID, LEAD(2+) SALT (1:1) see LDY000
SULFURIC ACID, MAGNESIUM SALT (1:1), compounded with 2,2'-DITHIOBIS(PYRIDINE) 1,1'-OXIDE see OIU850
SULFURIC ACID, MANGANESE(2+) SALT see MAU250
SULFURIC ACID, MERCURY(2+) SALT (1:1) see MDG500
SULFURIC ACID MIXTURE with SULFUR TRIOXIDE see SOI520
SULFURIC ACID, MONOAMMONIUM SALT see ANJ500
SULFURIC ACID, MONOANHYDRIDE with NITROUS ACID see NMJ000

SULFURIC ACID, MONODODECYL ESTER, SODIUM SALT see SIB600
SULFURIC ACID, MONOETHENYL ESTER, HOMOPOLYMER, POTASSIUM SALT see PKS250
SULFURIC ACID, MONO(2-ETHYLHEXYL)ESTER, SODIUM SALT (8CI) see TAV750
SULFURIC ACID, MONOPOTASSIUM SALT see PKX750
SULFURIC ACID, MONOSODIUM SALT see SEG800
SULFURIC ACID, MONOTETRADECYL ESTER, SODIUM SALT see SIO000
SULFURIC ACID, MYRISTYL ESTER, SODIUM SALT see SIO000
SULFURIC ACID, NICKEL(2$^+$)SALT see NDK500
SULFURIC ACID, NICKEL(2$^+$) SALT (1:1) see NDK500
SULFURIC ACID, NICKEL(2+) SALT, HEXAHYDRATE see NDL000
SULFURIC ACID, THALLIUM SALT see TEL750
SULFURIC ACID, THALLIUM(2+) SALT see TEM050
SULFURIC ACID, THALLIUM(1+) SALT (1:2) see TEM000
SULFURIC ACID, TITANIUM(4+) SALT (2:1) see TGH250
SULFURIC ACID, TITANIUM(4+) SALT (2:1) see TGH252
SULFURIC ACID, ZINC SALT (1:1) see ZNA000
SULFURIC ACID, ZINC SALT (1:1), HEPTAHYDRATE see ZNJ000
SULFURIC ACID, ZIRCONIUM(4+) SALT (2:1) see ZTJ000
SULFURIC AND HYDROFLUORIC ACIDS, MIXTURE (DOT) see HHV000
SULFURIC ANHYDRIDE see SOR500
SULFURIC CHLOROHYDRIN see CLG500
SULFURIC OXIDE see SOR500
SULFURIC OXYCHLORIDE see SOT000
SULFURIC OXYFLUORIDE see SOU500
SULFUR MONOCHLORIDE see SON510
SULFUR MUSTARD see BIH250
SULFUR MUSTARD GAS see BIH250
SULFUROUS ACID see SOO500
SULFUROUS ACID ANHYDRIDE see SOH500
SULFUROUS ACID, 2-(p-tert-BUTYLPHENOXY)CYCLOHEXYL-2-PROPYNYL ESTER see SOP000
SULFUROUS ACID, 2-(p-tert-BUTYLPHENOXY)-1-METHYLETHYL-2-CHLOROETHYL ESTER see SOP500
SULFUROUS ACID, CALCIUM SALT (2:1) (8CI,9CI) see CAN000
SULFUROUS ACID, DIPOTASSIUM SALT see PLT500
SULFUROUS ACID, cyclic ester with 1,4,5,6,7,7-HEXACHLORO-5-NORBORNENE-2,3-DIMETHANOL see EAQ750
SULFUROUS ACID, MONOAMMONIUM SALT see ANB600
SULFUROUS ACID, MONOSODIUM SALT see SFE000
SULFUROUS ACID, SODIUM SALT (1:2) see SJZ000
SULFUROUS ANHYDRIDE see SOH500
SULFUROUS DICHLORIDE see TFL000
SULFUROUS OXIDE see SOH500
SULFUROUS OXYCHLORIDE see TFL000
SULFUROUS OXYFLUORIDE see TFL250
SULFUR OXIDE see SOH500
SULFUR PENTAFLUORIDE see SOQ450
SULFUR PHOSPHIDE see PHS000
SULFUR SELENIDE see SBT000
SULFUR SUBCHLORIDE see SON510

SULFUR TETRAFLUORIDE see SOR000
SULFUR TRIOXIDE see SOR500
SULFUR TRIOXIDE, uninhibited (NA 1829) (DOT) see SOR500
SULFUR TRIOXIDE, inhibited (UN 1829) (DOT) see SOR500
SULFURYL CHLORIDE see SOT000
SULFURYL CHLORIDE FLUORIDE see SOT500
SULFURYL CHLOROFLUORIDE see SOT500
SULFURYL FLUORIDE see SOU500
SULFURYL FLUOROCHLORIDE see SOT500
SULGIN see AHO250
SULKOL see SOD500
SULMET see SNJ000
SULODYNE see PDC250
SULOUREA see ISR000
SULPHABUTIN see BOT250
SULPHACETAMIDE see SNQ710
SULPHADIAZINE see PPP500
SULPHADIMETHOXINE see SNN300
SULPHADIMETHYLISOXAZOLE see SNN500
SULPHADIMETHYLPYRIMIDINE see SNJ000
SULPHADIMIDINE see SNJ000
SULPHADIONE see SOA500
SULPHAFURAZ see SNN500
SULPHAGUANIDINE see AHO250
SULPHAMERAZINE see ALF250
SULPHAMETHALAZOLE see SNK000
SULPHAMETHOXAZOL see SNK000
SULPHAMETHOXAZOLE see SNK000
SULPHAMETHYLISOXAZOLE see SNK000
SULPHAMIC ACID (DOT) see SNK500
SULPHANILAMIDE see SNM500
5-SULPHANILAMIDO-3,4-DIMETHYL-ISOXAZOLE see SNN500
3-SULPHANILAMIDO-5-METHYLISOXAZOLE see SNK000
SULPHASIL see SNQ710
SULPHATHIAZOLE see TEX250
SULPHEIMIDE see CBF800
SULPHISOMEZOLE see SNK000
SULPHISOXAZOL see SNN500
2-SULPHOBENZOIC IMIDE see BCE500
SULPHOBENZOIC IMIDE CALCIUM SALT see CAM750
SULPHOBENZOIC IMIDE, SODIUM SALT see SJN700
SULPHOCARBONIC ANHYDRIDE see CBV500
SULPHOFURAZOLE see SNN500
SULPHON-MERE see SOA500
SULPHONOL FAST RED R see CMM330
SULPHONOL RED R see CMM330
1,1'-SULPHONYLBIS(4-AMINOBENZENE) see SOA500
p,p-SULPHONYLBISBENZAMINE see SOA500
4,4'-SULPHONYLBISBENZAMINE see SOA500
p,p-SULPHONYLBISBENZENAMINE see SOA500
4,4'-SULPHONYLBISBENZENAMINE see SOA500
SULPHONYLDIANILINE see SOA500
p,p-SULPHONYLDIANILINE see SOA500
1-p-SULPHOPHENYLAZO-2-NAPHTHOL-6-SULPHONIC ACID, DISODIUM SALT see FAG150
SULPHOS see PAK000
SULPHUR (DOT) see SOD500
SULPHUR, molten (DOT) see SOD500
SULPHUR, lump or powder (DOT) see SOD500
SULPHUR DIOXIDE, LIQUEFIED (DOT) see SOH500
SULPHURIC ACID see SOI500
SULPHURIC ACID, CADMIUM SALT (1:1) see CAJ000
SULPHUR MUSTARD GAS see BIH250

SULPROFOS see SOU625
SULSOL see SOD500
SULXIN see SNN300
SULZOL see TEX250
SUM 3170 see DCS200
SUMATRA CAMPHOR see BMD000
SUMEDINE see ALF250
SUMICIDIN see FAR100
SUMICURE M see MJQ000
SUMIDUR 44V10 see PKB100
SUMIDUR 44V20 see PKB100
SUMIDUR 44VM see PKB100
SUMIFLY see FAR100
SUMIKATHENE see PJS750
SUMIKATHENE F 702 see PJS750
SUMIKATHENE F 101-1 see PJS750
SUMIKATHENE F 210-3 see PJS750
SUMIKATHENE G 201 see PJS750
SUMIKATHENE G 202 see PJS750
SUMIKATHENE G 701 see PJS750
SUMIKATHENE G 801 see PJS750
SUMIKATHENE G 806 see PJS750
SUMIKATHENE HARD 2052 see PJS750
SUMILIT EXA 13 see PKQ059
SUMILIT PCX see AAX175
SUMINE 2015 see DQP800
SUMINOL MILLING RED RS see CMM330
SUMIPLEX LG see PKB500
SUMIPOWER see FAR100
SUMITEX H 10 see PKP750
SUMITHIAN see DSQ000
SUMITOMO PX 11 see PKQ059
SUMITOX see MAK700
SUMMETRIN see PAG500
SUN-4936 see HGL680
SUNAPTIC ACID B see NAR000
SUNAPTIC ACID C see NAR000
SUNBRELLA see AIH600
SUNFLOWER OIL (UNHYDROGENATED) see
SOU875
SUNFRAL see FLZ050
SUN ORANGE A GEIGY see FAG150
SUNSET YELLOW see FAG150
SUNSET YELLOW BSS see FAG150
SUNSET YELLOW FCF see FAG150
SUNSET YELLOW FCF SUPRA see FAG150
SUNSET YELLOW FU see FAG150
SUNSET YELLOW FU SUPRA see FAG150
SUNSET YELLOW LAKE see FAG150
SUNSOFT O 30B see GGR200
SUNWAX 151 see PJS750
SUN YELLOW see FAG150
SUN YELLOW A-CE see FAG150
SUN YELLOW A-FDC see FAG150
SUN YELLOW EXTRA CONC. A EXPORT see FAG150
SUN YELLOW EXTRA PURE A see FAG150
SUN YELLOW FCF see FAG150
SUPARI, nut extract see BFW000
SUPEOL see GGR200
SUPER 3-1000 see EHP700
SUPERACRYL AE see PKB500
SUPER AMIDE L-9A see BKE500
SUPERBA see CBT750
SUPER-CARBOVAR see CBT750
SUPERCEL 3000 see USS000
SUPER COBALT see CNA250
SUPERCOL see LIA000
SUPERCOL G.F. see GLU000

SUPERCOL U POWDER see GLU000
SUPERCORTIL see PLZ000
SUPER COSAN see SOD500
SUPER CRAB-E-RAD-CALAR see CAM000
SUPER DAL-E-RAD see CAM000
SUPER DAL-E-RAD-CALAR see CAM000
SUPER-DENT see SHF500
SUPER-DE-SPROUT see DMC600
SUPER DYLAN see PJS750
SUPERFLAKE ANHYDROUS see CAO750
SUPERGAN see RDK000
SUPER GLUE see MIQ075
SUPER GLUE see EHP700
SUPERGLYCERINATED FULLY HYDROGENATED
RAPESEED OIL see RBK200
SUPER HARTOLAN see CMD750
SUPERIOR OIL see MQV855
SUPERLYSOFORM see FMV000
SUPERMAN MANEB F see MAS500
SUPERMIKROKALCIT see CAD800
SUPER MOSSTOX see MJM500
SUPERNOX see DGI000
SUPEROL see GGA000
SUPEROL RED C RT-265 see CHP500
SUPEROX see BDS000
SUPEROXOL see HIB050
SUPERPREDNOL see SOW000
SUPER PRODAN see DXE000
SUPER RODIATOX see PAK000
SUPERSEPTIL see SNJ000
SUPERSORBON IV see CBT500
SUPERSORBON S 1 see CBT500
SUPER-SPECTRA see CBT750
SUPER SPROUT STOP see DMC600
SUPER SUCKER-STUFF see DMC600
SUPER SUCKER-STUFF HC see DMC600
SUPERTAH see CMY800
SUPER-TREFLAN see DUV600
SUPER VMP see NAH600
SUPONA see CDS750
SUPONE see CDS750
SUPRA see IHC450
SUPRACAPSULIN see VGP000
SUPRACET BRILLIANT BLUE 2GN see TBG700
SUPRACET BRILLIANT RED 2B see AKE250
SUPRACET DIAZO BLACK A see DPO200
SUPRACET FAST VIOLET B see DBY700
SUPRACET FAST YELLOW G see AAQ250
SUPRACET ORANGE R see AKP750
SUPRADIN see VGP000
SUPRAMIKE see BAP000
SUPRAMYCIN see TBX250
SUPRANEPHRANE see VGP000
SUPRANEPHRINE see VGP000
SUPRANOL see VGP000
SUPRANOL FAST RED 3G see CMM330
SUPRANOL FAST RED GG see CMM330
SUPRANOL RED PBX-CF see CMM330
SUPRANOL RED R see CMM330
SUPRARENIN see VGP000
SUPRASEC 1042 see PKB100
SUPRASEC DC see PKB100
SUPRASIL see SCK600
SUPRASIL W see SCK600
SUPRATHEN see PJS750
SUPRATHEN C 100 see PJS750
SUPREL see VGP000
SUPREMAL see DYE600

SUPREME DENSE see TAB750
SUP'R FLO see MAS500
SUP'R FLO see DXQ500
SUP'R FLO FERBAM FLOWABLE see FAS000
SUPRIMAL see HGC500
SURAUTO see IDW000
SURCHLOR see SHU500
SURCOPUR see DGI000
SUREM see DLY000
SURENINE see VGP000
SURE-SET see CJN000
SURGICAL SIMPLEX see PKB500
SURGI-CEN see HCL000
SURITAL see SOX500
SURITAL SODIUM see SOX500
SURITAL SODIUM (derivative) see SOX500
SURITAL SODIUM SALT see SOX500
SUROFENE see HCL000
SURPLIX see DLH630
SURPRACIDE see DSO000
SURPUR see DGI000
SU SEGURO CARPIDOR see DUV600
SUSPENDOL see ZVJ000
SUSPHRINE see VGP000
SUSTANE see BFW750
SUSTANE see BQI000
SUSTANE see BRM500
SUSTANE 1-F see BQI000
SUSTANONE see TBF500
SUSVIN see MRH209
SUTICIDE see HCQ500
SUXIL see SNE000
SUY-B 2 see IGK800
SUZORITE MICA see MQS250
SUZU see ABX250
SUZU H see HON000
SV-1522 see ABH500
SVC see ABX500
SVO 9 see PKL100
SW 400 see CAW850
SWAT see SOY000
SWEBATE see TAL250
SWEDISH GREEN see CNN500
SWEDISH GREEN see COF500
SWEENEY'S ANT-GO see ARD750
SWEEP see TBQ750
SWEETA see SJN700
SWEET BIRCH OIL see MPI000
SWEET BIRCH OIL see SOY100
SWEET DIPEPTIDE see ARN825
SWEET ORANGE OIL see OGY000
SWEETWOOD BARK OIL see CCO500
SWISS BLUE see BJI250
SY-83 see LAG000
SYDNONE IMINE, 3-(1-METHYL-2-PHENYLETHYL)-,
MONOHYDROCHLORIDE see SPA000
SYDNOPHENE see SPA000
SYDNOPHEN HYDROCHLORIDE see SPA000
SYKOSE see SJN700
SYKOSE see BCE500
SYLANTOIC see DKQ000
SYLANTOIC see DNU000
SYLLIT see DXX400
SYLODEX see ARM268
SYMCLOSEN see TIQ750
SYMCLOSENE see TIQ750
SYMETRYCZNA DWUMETYLOHYDRAZYNA
(POLISH) see DSF600

SYMMETREL see AED250
SYMMETRIC DIMETHYLUREA see DUM200
SYMPAMINA-D see BBK500
SYMPAMINE see BBK000
SYMPATEDRINE see BBK000
SYMPATHIN E see NNO500
SYMPATHIN I see VGP000
m-SYMPATHOL see NCL500
m-SYMPATHOL see SPC500
SYMPATHOLYTIN see DCT050
m-SYMPATOL see NCL500
m-SYMPATOL see SPC500
SYMPHYTUM OFFICINALE L see RRP000
SYMULER EOSIN TONER see BNH500
SYMULER LAKE RED C see CHP500
SYMULON ACID BRILLIANT SCARLET 3R see
FMU080
SYMULON METANIL YELLOW see MDM775
SYMULON SCARLET G BASE see NMP500
SYNALGOS see DQA400
SYNANDRETS see MPN500
SYNANDROL see TBG000
SYNANDROL F see TBF500
SYNANDROTABS see MPN500
SYNAPAUSE see EDU500
SYNASAL see SPC500
SYNASTERON see PAN100
SYNCAINE see AIT250
SYNCAL see BCE500
SYNCELOSE see MIF760
SYNDIOL see EDO000
SYNDIOTACTIC POLYPROPYLENE see PMP500
SYNDROX see MDQ500
SYNDROX see MDT600
SYNELAUDINE see DAM700
m-SYNEPHRINE see NCL500
m-SYNEPHRINE see SPC500
m-SYNEPHRINE HYDROCHLORIDE see SPC500
SYNERGIST 264 see OES000
SYNERONE see TBG000
SYNESTRIN see DKA600
SYNESTRIN see DKB000
SYNESTROL see DAL600
SYNETHENATE see SPC500
SYNFAT 1006 see CAU000
SYNFLORAN see DUV600
SYNGESTERONE see PMH500
SYNGUM D 46D see GLU000
SYNISTAMIN see TAI500
SYNKAY see MMD500
SYNKLOR see CDR750
SYNMIOL see DAS000
SYNOESTRON see DKB000
SYNOTOL L-60 see BKE500
SYNOVEX S see PMH500
SYNOX TBC see BSK000
SYNPENIN see AIV500
SYNPOL 1500 see SMR000
SYNPOR see CCU250
SYNPREN-FISH see PIX250
SYNPRO STEARATE see CAX350
SYNSTIGMIN BROMIDE see POD000
SYNSTIGMINE see DER600
SYNTAR see CMY800
SYNTASE 62 see MES000
SYNTASE 100 see DMI600
SYNTEDRIL see BBV500
SYNTEN YELLOW 2G see AAQ250

SYNTES 12A see EIV000
SYNTESTRIN see DKB000
SYNTESTRINE see DKB000
SYNTEXAN see DUD800
SYNTHETIC 3956 see CDV100
SYNTHETIC EUGENOL see EQR500
SYNTHETIC GLYCERIN see GGA000
SYNTHETIC IRON OXIDE see IHC450
SYNTHETIC LH-RH see LIU370
SYNTHETIC MUSTARD OIL see AGJ250
SYNTHETIC PYRETHRINS see AFR250
SYNTHETIC WINTERGREEN OIL see MPI000
SYNTHOESTRIN see DKA600
SYNTHOFOLIN see DKA600
SYNTHOMYCINE see CDP250
SYNTHOSTIGMINE BROMIDE see POD000
SYNTHRIN see BEP500
SYNTHROID see LFG050
SYNTHROID SODIUM see LFG050
SYNTODRIL see BBV500
SYNTOFOLIN see DKA600
SYNTOLUTAN see PMH500
SYNTON YELLOW 2G see AAQ250
SYNTOPHEROL see VSZ450
SYNTOSTIGMIN see DER600
SYNTOSTIGMIN (tablet) see POD000
SYNTOSTIGMIN BROMIDE see POD000
SYNTOSTIGMINE BROMIDE see POD000
SYNTRON B see EIV000
SYRINGOL see DOJ200
SYS 67ME see SIL500
SYSTAM see OCM000
SYSTANATE MR see PKB100
SYSTANAT MR see PKB100
SYSTHANE see MRW775
SYSTHANE 6 FLO see MRW775
SYSTOPHOS see OCM000
SYTAM see OCM000
SYTLOMYCIN AMINONUCLEOSIDE see ALQ625
SYTOBEX see VSZ000
SZESCIOMETYLENODWUIZOCYJANIAN see DNJ800
SZKLARNIAK see DGP900
T4 see CPR800
T 40 see TGF250
T 72 see PNX000
T 100 see TGM740
T 100 see EHG100
T-144 see IPX000
T-1703 see IRF000
T 1824 see BGT250
T-2002 see BJE750
T-2104 see EIF000
T-2106 see IPX000
α-T see MIH275
β-T see TIN000
TA 12 see TAN750
TAA see TFA000
TABAC (FRENCH) see TGI100
TABACO (SPANISH) see TGI100
TABALGIN see HIM000
TABLE SALT see SFT000
TABOON A see EIF000
TABUN see EIF000
TAC 121 see TGG250
TAC 131 see TGG250
TACALCITOL see SBL600
TACE see CLO750

TACE-FN see CLO750
TACITIN see BCH750
TACOSAL see DKQ000
TACOSAL see DNU000
TACROLIMUS HYDRATE see TAA900
TAENIATOL see MJM500
TAFASAN see MEP250
TAFASAN 6W see MEP250
TAFAZINE see BJP000
TAG see ABU500
TAG-39 see ECU750
TAG 331 see ABU500
TAGAT see SEH000
TAG FUNGICIDE see ABU500
TAG HL 331 see ABU500
TAGN see THN800
TAHMABON see DTQ400
TAK see MAK700
TAKAMINA see VGP000
TAKANARUMIN see ZVJ000
TAKAOKA BRILLIANT SCARLET 3R see FMU080
TAKAOKA METANIL YELLOW see MDM775
TAKATHENE see PJS750
TAKATHENE P 3 see PJS750
TAKATHENE P 12 see PJS750
TAKENATE see XIJ000
TAKENATE 500 see XIJ000
TAKENATE 300C see PKB100
TAKILON see PKQ059
TALARGAN see TEH500
TALBOT see LCK000
TALBOT see LCK100
TALC see TAB750
TALC, containing asbestos fibers see TAB775
TALCORD see AHJ750
TALCUM see TAB750
TALIMOL see TEH500
TALL OIL see TAC000
TALLOL see TAC000
TALLOW BENZYL DIMETHYLAMMONIUM
CHLORIDE see DTC600
N-TALLOWPYRROLIDINONE see TAC200
TALMON see MAO350
TALPHENO see EOK000
TALWAN see DOQ400
TALWIN see DOQ400
TAMARON see DTQ400
TAMAS see BBW500
TAMOFEN see TAD175
TAMOXASTA see TAD175
TAMOXIFEN see NOA600
TAMOXIFEN CITRATE see TAD175
TAMOXIFEN CITRATE see TAD175
TAMPOVAGAN STILBOESTROL see DKA600
TAMPULES see CDP000
TANGANTANGAN OIL see CCP250
TANGELO OIL see TAD250
TANGERINE OIL see TAD500
TANGERINE OIL, COLDPRESSED (FCC) see TAD500
TANGERINE OIL, EXPRESSED (FCC) see TAD500
TANIDIL see DQA400
TANNEX see IDA000
TANNIC ACID see TAD750
TANNIN see TAD750
TANNIN from CHESTNUT see CDM250
TANNIN from MIMOSA see MQV250
TANTALIC ACID ANHYDRIDE see TAF500
TANTALUM see TAE750

TANTALUM-181 see TAE750
TANTALUM CHLORIDE see TAF000
TANTALUM FLUORIDE see TAF250
TANTALUM OXIDE see TAF500
TANTALUM(V) OXIDE see TAF500
TANTALUM PENTACHLORIDE see TAF000
TANTALUM PENTAFLUORIDE see TAF250
TANTALUM PENTAOXIDE see TAF500
TANTALUM PENTOXIDE see TAF500
TANTALUM POTASSIUM FLUORIDE see PLH000
TANTUM see BBW500
TAOMYCIN see HOH500
TAOMYXIN see HOH500
TAP 85 see BBQ500
TA-33MP see TAN750
TAPAR see HIM000
TAPAZOLE see MCO500
TAPHAZINE see BJP000
TAPIOCA see DBD800
TAPIOCA STARCH see SLJ500
TAPIOCA STARCH HYDROXYETHYL ETHER see
HLB400
TAPON see SLJ500
TAP 9VP see DGP900
TAR see CMY800
TAR, from tobacco see CMP800
TARAPACAITE see PLB250
TARAPON K 12 see SIB600
TAR CAMPHOR see NAJ500
TAR, COAL see CMY800
TARDEX 100 see PAU500
TARDIGAL see DKL800
TARDOCILLIN see BFC750
TARFLEN see TAI250
TARGET MSMA see MRL750
TARICHATOXIN see FOQ000
TARIMYL see SOA500
TARLON XB see PJY500
TARNAMID T see PJY500
TAROCTYL see CKP500
TARODYL see GIC000
TARODYN see GIC000
TAR OIL see CMY825
TARRAGON see AFW750
TARRAGON OIL see TAF700
TARTAN see PHK250
TARTAR EMETIC see AQG250
TARTARIC ACID see TAF750
l-(+)-TARTARIC ACID see TAF750
l-TARTARIC ACID, AMMONIUM SALT see DCH000
l-TARTARIC ACID, ANTIMONY POTASSIUM SALT see
AQH000
TARTARIC ACID, DIAMMONIUM SALT see DCH000
TARTARIC ACID, DIBENZOATE see DDE300
TARTARIZED ANTIMONY see AQG250
TARTAR YELLOW FS see FAG140
TARTAR YELLOW N see FAG140
TARTAR YELLOW PF see FAG140
TARTAR YELLOW S see FAG140
TARTRAN YELLOW see FAG140
TARTRAPHENINE see FAG140
TARTRATE ANTIMONIO-POTASSIQUE (FRENCH)
see AQG250
TARTRATED ANTIMONY see AQG250
TARTRATE de NICOTINE (FRENCH) see NDS500
TARTRAZINE see FAG140
TARTRAZINE A EXPO T see FAG140
TARTRAZINE B see FAG140

TARTRAZINE B.P.C. see FAG140
TARTRAZINE EXTRA PURE A see FAG140
TARTRAZINE FD & C YELLOW #5 see FAG140
TARTRAZINE FQ see FAG140
TARTRAZINE G see FAG140
TARTRAZINE LAKE see FAG140
TARTRAZINE LAKE YELLOW N see FAG140
TARTRAZINE M see FAG140
TARTRAZINE MCGL see FAG140
TARTRAZINE N see FAG140
TARTRAZINE NS see FAG140
TARTRAZINE O see FAG140
TARTRAZINE T see FAG140
TARTRAZINE XX see FAG140
TARTRAZINE XXX see FAG140
TARTRAZINE YELLOW see FAG140
TARTRAZOL BPC see FAG140
TARTRAZOL YELLOW see FAG140
TARTRINE YELLOW O see FAG140
TASK see DGP900
TASK TABS see DGP900
TAT CHLOR 4 see CDR750
TATD see DXH250
TATHIONE see GFW000
TATTOO see DQM600
TAUPHON see TAG750
TAURE(o)DON see GJC000
TAURINE see TAG750
TAYSSATO see MEP250
TAZONE see BRF500
TB see CMO250
TBB see THX500
TBBA see BQK500
TBDZ see TEX000
TBE see ACK250
TBEP see BPK250
TBHP-70 see BRM250
TBHQ (FCC) see BRM500
TBOT see BLL750
TBP see TIA250
TBT see BSP500
TBTO see BLL750
2,4,5-TC see TIX500
TC 3-30 see SMQ500
T 1 (Catalyst) see DBF800
TCA see TII250
T-250 CAPSULES see TBX250
TCB see TBO700
TCDBD see TAI000
TCDD see TAI000
2,3,7,8-TCDD see TAI000
TCE see TBQ100
1,1,1-TCE see MIH275
TC HYDROCHLORIDE see TBX250
TCIN see TBQ750
TCM see CHJ500
TCP see TBT000
m-TCPN see TBQ750
TCPP see FQU875
2,4,5-TCPPA see TIX500
TCTH see CQH650
TCTP see TBV750
TD-183 see TBV750
TDA see TGL750
TDBP (CZECH) see TNC500
TDCPP see FQU875
TDE (DOT) see BIM500
p,p'-TDE see BIM500

T-DET see DXY600
TDI see TGM740
2,4-TDI see TGM750
2,6-TDI see TGM800
TDI-80 see TGM740
TDI-80 see TGM750
TDI 80-20 see TGM740
T-82 DIFUMARATE see MDU750
TDI (OSHA) see TGM750
TDOT see TCQ275
TDPA see BHM000
2,5-TDS see TGM400
TEA see TJN750
TEABERRY OIL see MPI000
TEBRAZID see POL500
TEC see TJP750
TECH DDT see DAD200
TECHNETIUM TC 99M SULFUR COLLOID see
SOD500
TECHNICAL BHC see BBQ750
90 TECHNICAL GLYCERINE see GGA000
TECHNICAL HCH see BBQ750
TECHNOPOR see PKQ059
TECH PET F see MQV750
TECOFLEX HR see PKN000
TECQUINOL see HIH000
TECSOL see EFU000
TECTO see TEX000
TECZA see TJR000
TEDIMON 31 see PKB100
TEDION see CKM000
TEDION V-18 see CKM000
TEDP see SOD100
TEDP (OSHA) see SOD100
TEDTP see SOD100
TEF see TND250
TEFAMIN see TEP000
TEFASERPINA see RDK000
TEFILIN see TBX250
TEFLON see TAI250
TEFLON (various) see TAI250
TEFSIEL C see FLZ050
TEG see TJQ000
TEGAFUR see FLZ050
TEGESTER see IQN000
TEGIN see OAV000
TEGIN 503 see OAV000
TEGIN 515 see OAV000
TEGOLAN see CMD750
TEGO-OLEIC 130 see OHU000
TEGOPEN see SLJ000
TEGOPEN see AOC500
TEGOSEPT E see HJL000
TEGOSEPT M see HJL500
TEGOSEPT P see HNU500
TEGO-STEARATE see EJM500
TEGOSTEARIC 254 see SLK000
TEIB see TND000
TEKKAM see NAK500
TEKRESOL see CNW500
TEKWAISA see MNH000
TEL see TCF000
TELAGAN see TEH500
TELARGAN see TEH500
TELARGEAN see TEH500
TELCOTENE see PJS750
TELDRIN see TAI500
TELECOTHENE see PJS750

TELINE see TBX250
TELIPEX see TBG000
TELLOY see TAJ000
TELLUR (POLISH) see TAJ000
TELLURATE see TAI750
TELLURIC ACID see TAI750
TELLURIC(VI) ACID see TAI750
TELLURIC ACID, AMMONIUM SALT see ANV750
TELLURIC ACID, DISODIUM SALT see SKC500
TELLURIC ACID, DISODIUM SALT,
PENTAHYDRATE see TAI800
TELLURIC CHLORIDE see TAJ250
TELLURIUM see TAJ000
TELLURIUM (dust or fume) see TAJ600
TELLURIUM CHLORIDE see TAJ250
TELLURIUM COMPOUNDS see TAJ500
TELLURIUM DIETHYLDITHIOCARBAMATE see
EPJ000
TELLURIUM DIOXIDE see TAJ750
TELLURIUM HEXAFLUORIDE see TAK250
TELLURIUM HYDROXIDE see TAI750
TELLURIUM OXIDE see TAJ750
TELLURIUM TETRACHLORIDE see TAJ250
TELLUROUS ACID, DISODIUM SALT see SKC500
TELMICID see DJT800
TELMID see DJT800
TELMIDE see DJT800
TELMIN see MHL000
TELONE see DGG000
TELONE see DGG950
TELONE II SOIL FUMIGANT see DGG950
TELON FAST BLACK E see AQP000
TELON FAST RED GG see CMM330
TELOTREX see TBX250
TELTOZAN see CAL750
TELVAR see CJX750
TELVAR see DXQ500
TELVAR DIURON WEED KILLER see DXQ500
TELVAR MONURON WEEDKILLER see CJX750
TEMED see TDQ750
TEMEFOS see TAL250
TEMENTIL see PMF500
TEMEPHOS see TAL250
TEMIC see CBM500
TEMIK see CBM500
TEMIK G10 see CBM500
TEMLO see HIM000
TEMOPHOS see TAL250
TEMPANAL see HIM000
TEMPARIN see BJZ000
TEMPLIN OIL see AAC250
TEMPODEX see BBK500
TEMPONITRIN see NGY000
TEMPO-RESERPINA see RDK000
TEMPOSERPINE see RDK000
TEMPRA see HIM000
TEMUR see TDX250
TEN see TJO000
TENAC see DGP900
TENAMENE 2 see DEG200
TENAPLAS see PJS750
TENDOSCEN-COMPR. see RDK000
TENDUST see NDN000
TENIATHANE see MJM500
TENIATOL see MJM500
TENIPOSIDE see EQP000
TENITE 423 see PMP500
TENITE 800 see PJS750

TENITE 1811 see PJS750
TENITE 2910 see PJS750
TENITE 2918 see PJS750
TENITE 3300 see PJS750
TENITE 3340 see PJS750
TENNECETIN see PIF750
TENNECO 1742 see PKQ059
TENN-PLAS see BCL750
TENNUS 0565 see AAX175
TENOX BHA see BQI000
TENOX BHT see BFW750
TENOX HQ see HIH000
TENOX PG see PNM750
TENOX P GRAIN PRESERVATIVE see PMU750
TENOX TBHQ see BRM500
TENSANYL see RDK000
TENSERLIX see RDK000
TENSERPINE "ASSIA" see RDK000
TENSERPINIE see RDK000
TENSINYL see MDQ250
TENSIONAL see RDK000
TENSIONORME see RDK000
TENSIVAL see TEH500
TENSOL 7 see PKB500
TENSOPAM see DCK759
TENTRATE-20 see PBC250
TENURID see DXH250
TENUTEX see DXH250
TEOBROMIN see TEO500
TEODRAMIN see DYE600
TEOFYLLAMIN see TEP000
TEOHARN see MCB000
TEONANACATL see PHU500
TEOS see EPF550
TEP see TJT750
TEPA see TND250
TEPERINE see DLH630
TEPIDONE see SGF500
TEPIDONE RUBBER ACCELERATOR see SGF500
TEPOGEN see SLJ000
TEPP (ACGIH) see TCF250
TEPSERPINE see RDK000
TEQUINOL see HIH000
TERABOL see MHR200
TERAMYCIN HYDROCHLORIDE see HOI000
TERASIL BRILLIANT VIOLET 3B see DBY700
TERASIL YELLOW 2GC see AAQ250
TERASIL YELLOW GBA EXTRA see AAQ250
TERBENZENE see TBD000
TERBIUM CHLORIDE see TAM000
TERBOLAN see RDK000
TERCYL see CBM750
TEREBENTHINE see TOD750
TEREFTALODINITRIL (CZECH) see BBP250
TEREPHTHALIC ACID see TAN750
TEREPHTHALIC ACID CHLORIDE see TAV250
TEREPHTHALIC ACID DICHLORIDE see TAV250
TEREPHTHALIC ACID METHYL ESTER see DUE000
TEREPHTHALIC DICHLORIDE see TAV250
TEREPHTHALONITRILE see BBP250
TEREPHTHALOYL DICHLORIDE see TAV250
TERETON see MFW100
TERFLUZINE see TKK250
TERFLUZINE DIHYDROCHLORIDE see TKK250
TERGEMIST see TAV750
TERGIMIST see TAV750
TERGITOL 4 see SIO000
TERGITOL 08 see TAV750

TERGITOL ANIONIC 08 see TAV750
TERGITOL TMN-10 see TBB775
TERGITOL TP-9 (NONIONIC) see PKF000
TERIMON see TAD175
TERMIL see TBQ750
TERMITKIL see DEP600
TERM-I-TROL see PAX250
TERMOSOLIDO RED LCG see CHP500
TERPENE HYDROCHLORIDE see BMD300
TERPENE POLYCHLORINATES see TBC500
TERPENE RESIN, NATURAL see TBC575
TERPENTINOEL (GERMAN) see PIH750
TERPENTIN OEL (GERMAN) see TOD750
m-TERPHENYL see TBC620
o-TERPHENYL see TBC640
p-TERPHENYL see TBC750
1,3-TERPHENYL see TBC620
TERPHENYLS see TBD000
α-TERPINENE (FCC) see MLA250
γ-TERPINENE (FCC) see MCB750
TERPINEOL see TBD500
α-TERPINEOL ACETATE see TBE250
α-TERPINEOL (FCC) see TBD500
TERPINEOLS see TBD500
TERPINOLENE see TBE000
TERPINYL ACETATE see TBE250
TERPINYL PROPIONATE see TBE600
TERPINYL THIOCYANOACETATE see IHZ000
Δ-1,8-TERPODIENE see MCC250
TERRA ALBA see CAX750
TERRACHLOR see PAX000
TERRACUR P see FAQ800
TERRAFUN see PAX000
TERRAFUNGINE see HOH500
TERRAMITSIN see HOH500
TERRAMYCIN see HOH500
TERRA-SYSTAM see BJE750
TERRA-SYTAM see BJE750
TERRASYTUM see BJE750
TERR-O-GAS 100 see MHR200
TERSAN see TFS350
TERSAN 1991 see BAV575
TERSAN-LSR see MAS500
TERSASEPTIC see HCL000
TERTRACID LIGHT ORANGE G see HGC000
TERTRACID PONCEAU 2R see FMU070
TERTRACID RED CA see HJF500
TERTRACID YELLOW M see MDM775
TERTRAL D see PEY500
TERTRAL EG see ALS990
TERTRAL ERN see NAW500
TERTRAL G see TGL750
TERTRAL P BASE see ALT250
TERTRANESE YELLOW N-2GL see AAQ250
TERTROCHROME BLUE FB see HJF500
TERTRODIRECT BLACK E see AQP000
TERTRODIRECT BLUE 2B see CMO000
TERTRODIRECT BLUE F see CMO500
TERTRODIRECT BLUE FF see CMN750
TERTRODIRECT RED C see SGQ500
TERTROGRAS ORANGE SV see PEJ500
TERTROGRAS RED N see SBC500
TERTROPHENE BRILLIANT GREEN G see BAY750
TERTROPHENE BROWN CG see PEK000
TERTROSULPHUR BLACK PB see DUZ000
TERTROSULPHUR PBR see DUZ000
TERULAN KP 2540 see ADY500

TESCOL see EJC500
TESERENE see DAL600
TESLEN see TBF500
TESPAMINE see TFQ750
TESTAFORM see TBG000
TESTANDRONE see TBF500
TESTAVOL see VSK600
TESTEX see TBG000
TESTHORMONE see MPN500
TESTICULOSTERONE see TBF500
TESTOBASE see TBF500
TESTODET see TBG000
TESTODRIN see TBG000
TESTOGEN see TBG000
TESTONIQUE see TBG000
TESTOPROPON see TBF500
TESTORA see MPN500
TESTORMOL see TBG000
TESTOSTEROID see TBF500
TESTOSTERONE see TBF500
trans-TESTOSTERONE see TBF500
TESTOSTERONE HYDRATE see TBF500
TESTOSTERONE PROPIONATE see TBG000
TESTOSTERONE PROPIONATE see TBG000
TESTOSTERONE-17-PROPIONATE see TBG000
TESTOSTERONE-17-β-PROPIONATE see TBG000
TESTOSTOSTERONE see TBF500
TESTOVIRON see MPN500
TESTOVIRON see TBG000
TESTOVIRON SCHERING see TBF500
TESTOVIRON T see TBF500
TESTOXYL see TBG000
TESTRED see MPN500
TESTREX see TBG000
TESTRONE see TBF500
TESTRYL see TBF500
TESULOID see SOD500
TETA see TJR000
TETD see DXH250
TETIDIS see DXH250
TETLEN see PCF275
O,O,O',O'-TETRAAETHYL-BIS(DITHIOPHOSPHAT) (GERMAN) see EEH600
O,O,O,O-TETRAAETHYL-DIPHOSPHAT, BIS(O,O-DIAETHYLPHOSPHORSAEURE)-ANHYDRID (GERMAN) see TCF250
O,O,O,O-TETRAAETHYL-DITHIONOPYROPHOSPHAT (GERMAN) see SOD100
1,4,5,8-TETRAAMINO-9,10-ANTHRACENEDIONE see TBG700
1,4,5,8-TETRAAMINOANTHRAQUINONE see TBG700
3,3',4,4'-TETRAAMINOBIPHENYL see BGK500
3,3',4,4'-TETRAAMINOBIPHENYL TETRAHYDROCHLORIDE see BGK750
1,3,5,7-TETRAAZAADAMANTANE see HEI500
1,4,7,10-TETRAAZADECANE see TJR000
1,5,8,12-TETRAAZADODECANE see TBI700
TETRABAKAT see TBX250
TETRA-BASE see MJN000
TETRABLET see TBX250
1,1,2,2-TETRABROMAETHAN (GERMAN) see ACK250
TETRABROMIDE METHANE see CBX750
TETRABROMOACETYLENE see ACK250
2,4,5,7-TETRABROMO-9-o-CARBOXYPHENYL-6-HYDROXY-3-ISOXANTHONE, DISODIUM SALT see BNH500
1,1,2,2-TETRABROMOETANO (ITALIAN) see ACK250

S-TETRABROMOETHANE see ACK250
1,1,2,2-TETRABROMOETHANE see ACK250
2,4,5,7-TETRABROMO-3,6-FLUORANDIOL see BNH500
TETRABROMOFLUORESCEIN see BNH500
2',4',5',7'-TETRABROMOFLUORESCEIN DISODIUM SALT see BNH500
TETRABROMOFLUORESCEIN S see BNH500
TETRABROMOFLUORESCEIN SOLUBLE see BNH500
2-(2,4,5,7-TETRABROMO-6-HYDROXY-3-OXO-3H-XANTHENE-9-YL)BENZOIC ACID, DISODIUM SALT see BNH500
TETRABROMOMETHANE see CBX750
TETRABROMO-PLATINUM(2-), DIPOTASSIUM see PLU250
1,1,2,2-TETRABROOMETHAAN (DUTCH) see ACK250
TETRA-n-BUTYLCIN (CZECH) see TBM250
TETRABUTYL DICHLOROSTANNOXANE see TBL500
TETRABUTYLSTANNANE see TBM250
TETRABUTYLTIN see TBM250
TETRABUTYLTITANATE (CZECH) see BSP250
TETRACAP see PCF275
TETRACAPS see TBX250
TETRACARBONYL NICKEL see NCZ000
TETRACARBONYL(TRIFLUOROMETHYLTHIO)MANGANESE dimer see TBN200
TETRACEMATE DISODIUM see EIX500
TETRACEMIN see EIV000
TETRACENE see TEF500
TETRACENE EXPLOSIVE see TEF500
2,4,4',5-TETRACHLOOR-DIFENYL-SULFON (DUTCH) see CKM000
1,1,2,2-TETRACHLOORETHAAN (DUTCH) see TBQ100
TETRACHLOORETHEEN (DUTCH) see PCF275
TETRACHLOORKOOLSTOF (DUTCH) see CBY000
TETRACHLOORMETAAN see CBY000
1,1,2,2-TETRACHLORAETHAN (GERMAN) see TBQ100
TETRACHLORAETHEN (GERMAN) see PCF275
N-(1,1,2,2-TETRACHLORAETHYLTHIO)CYCLOHEX-4-EN-1,4-DIACARBOXIMID (GERMAN) see CBF800
N-(1,1,2,2-TETRACHLORAETHYLTHIO)TETRAHYDROPHTHALAMID (GERMAN) see CBF800
2,4,4',5-TETRACHLOR-DIPHENYL-SULFON (GERMAN) see CKM000
TETRACHLORETHANE see TBQ100
1,1,2,2-TETRACHLORETHANE (FRENCH) see TBQ100
TETRACHLORKOHLENSTOFF, (GERMAN) see CBY000
TETRACHLORMETHAN (GERMAN) see CBY000
1,2,3,4-TETRACHLOROBENZENE see TBN740
TETRACHLOROBENZIDINE see TBO000
2,2',5,5'-TETRACHLOROBENZIDINE see TBO000
3,3',6,6'-TETRACHLOROBENZIDINE see TBO000
3,3',4,4'-TETRACHLOROBIPHENYL see TBO700
3,4,3',4'-TETRACHLOROBIPHENYL see TBO700
2,2',5,5'-TETRACHLORO-(1,1'-BIPHENYL)-4,4'-DIAMINE, (9CI) see TBO000
TETRACHLOROCARBON see CBY000
9-(3',4',5',6'-TETRACHLORO-o-CARBOXYPHENYL)-6-HYDROXY-2,4,5,7-TETRAIODO-3-ISOXANTHONE2Na see RMP175
2,4,5,6-TETRACHLORO-3-CYANOBENZONITRILE see TBQ750
2,2',5,5'-TETRACHLORO-4,4'-DIAMINODIPHENYL see TBO000
2,3,7,8-TETRACHLORODIBENZO(b,e)(1,4)DIOXAN see TAI000

2,3,6,7-TETRACHLORODIBENZO-p-DIOXIN see TAI000

2,3,7,8-TETRACHLORODIBENZO-p-DIOXIN see TAI000

2,3,7,8-TETRACHLORODIBENZO-1,4-DIOXIN see TAI000

1,1,1,2-TETRACHLORO-2,2-DIFLUOROETHANE see TBP000

1,1,1,2-TETRACHLORO-2,2-DIFLUOROETHANE see TBP000

1,1,2,2-TETRACHLORO-1,2-DIFLUOROETHANE see TBP050

TETRACHLORODIPHENYLETHANE see BIM500

TETRACHLORODIPHENYL OXIDE see TBP250

2,4,4',5-TETRACHLORODIPHENYL SULFONE see CKM000

2,4,5,4'-TETRACHLORODIPHENYLSULPHONE see CKM000

TETRACHLOROETHANE see TBP750

sym-TETRACHLOROETHANE see TBQ100

1,1,1,2-TETRACHLOROETHANE see TBQ000

1,1,2,2-TETRACHLOROETHANE see TBQ100

TETRACHLOROETHENE see PCF275

1,1,2,2-TETRACHLOROETHYLENE see PCF275

TETRACHLOROETHYLENE (DOT, ACGIH) see PCF275

N-1,1,2,2-TETRACHLOROETHYLMERCAPTO-4-CYCLOHEXENE-1,2-CARBOXIMIDE see CBF800

N-((1,1,2,2-TETRACHLOROETHYL)SULFENYL)-cis-4-CYCLOHEXENE-1,2-DICARBOXIMIDE see CBF800

N-(1,1,2,2-TETRACHLOROETHYLTHIO)-4-CYCLOHEXENE-1,2-DICARBOXIMIDE see CBF800

TETRACHLOROISOPHTHALONITRILE see TBQ750

TETRACHLOROMETHANE see CBY000

TETRACHLORONAPHTHALENE see TBR000

2,3,4,6-TETRACHLOROPHENOL see TBT000

2,4,5,6-TETRACHLOROPHENOL see TBT000

TETRACHLOROPHENYL ETHER see TBP250

m-TETRACHLOROPHTHALONITRILE see TBQ750

TETRACHLOROSILANE see SCQ500

TETRACHLOROSTANNANE PENTAHYDRATE see TGC282

TETRACHLOROTELLURIUM see TAJ250

3,3,4,4-TETRACHLOROTETRAHYDROTHIOPHENE-1,1-DIOXIDE see TBV300

4,5,6,7-TETRACHLORO-2',4',5',7'-TETRAIODOFLUORESCEIN DISODIUM SALT see RMP175

TETRACHLOROTHIOFENE see TBV750

TETRACHLOROTHIOPHENE see TBV750

2,3,4,5-TETRACHLOROTHIOPHENE see TBV750

TETRACHLOROTHORIUM see TFT000

α-α-α-4-TETRACHLOROTOLUENE see TIR900

p,α-α-α-TETRACHLOROTOLUENE see TIR900

TETRACHLORURE d'ACETYLENE (FRENCH) see TBQ100

TETRACHLORURE de CARBONE (FRENCH) see CBY000

TETRACHLORURE de SILICIUM (FRENCH) see SCQ500

TETRACHLORURE de TITANE (FRENCH) see TGH350

TETRACHLORVINPHOS see TBW100

TETRACICLINA CLORIDRATO (ITALIAN) see TBX250

TETRACID MILLING RED B see CMM330

TETRACID MILLING RED G see CMM330

2,4,4',5-TETRACLORO-DIFENIL-SOLFONE (ITALIAN) see CKM000

1,1,2,2-TETRACLOROETANO (ITALIAN) see TBQ100

TETRACLOROETENE (ITALIAN) see PCF275

TETRACLOROMETANO (ITALIAN) see CBY000

TETRACLORURO di CARBONIO (ITALIAN) see CBY000

TETRACOMPREN see TBX250

TETRACYANONICKELATE(2−) DIPOTASSIUM, HYDRATE see TBW250

TETRACYCLINE CHLORIDE see TBX250

TETRACYCLINE, 7-CHLORO-6-DEMETHYL- see MIJ500

TETRACYCLINE HYDROCHLORIDE see TBX250

TETRACYCLINE, 5-HYDROXY- see HOH500

TETRACYDIN see ABG750

TETRA-D see TBX250

TETRADECANOIC ACID see MSA250

TETRADECANOIC ACID, ISOPROPYL see IQN000

TETRADECANOIC ACID, 1-METHYLETHYL ESTER see IQN000

TETRADECANOL, mixed isomers see TBY500

n-TETRADECOIC ACID see MSA250

TETRADECYL ALCOHOL see TBY500

TETRADECYL DIMETHYL BENZYLAMMONIUM CHLORIDE see TCA500

TETRADECYL OCTADECANOATE see TCB100

TETRADECYL SODIUM SULFATE see SIO000

TETRADECYL STEARATE see TCB100

TETRADECYL SULFATE, SODIUM SALT see SIO000

TETRADICHLONE see CKM000

TETRADIFON see CKM000

TETRADIN see DXH250

TETRADINE see DXH250

TETRADIOXIN see TAI000

TETRADIPHON see CKM000

TETRAETHOXYSILANE see EPF550

O,O,O,O-TETRAETHYL-DIFOSFAAT (DUTCH) see TCF250

TETRAETHYLDITHIODIFOSFAT see SOD100

TETRAETHYL DITHIONOPYROPHOSPHATE see SOD100

TETRAETHYL DITHIOPYROPHOSPHATE see SOD100

O,O,O,O-TETRAETHYL DITHIOPYROPHOSPHATE see SOD100

TETRAETHYL DITHIO PYROPHOSPHATE, liquid (DOT) see SOD100

TETRAETHYL DITHIOPYROPHOSPHATE and GAS MIXTURES, LC50 < or =2000 ppm (DOT) see SOD100

TETRAETHYLENE GLYCOL see TCE250

TETRAETHYLENE GLYCOL DIACRYLATE see ADT050

TETRAETHYLENE GLYCOL DIMETHYL ETHER see PBO500

TETRAETHYLENEPENTAMINE see TCE500

TETRAETHYL LEAD see TCF000

TETRAETHYL LEAD, liquid (DOT) see TCF000

O,O,O',O'-TETRAETHYL S,S'-METHYLENEBISPHOSPHORDITHIOATE see EEH600

O,O,O',O'-TETRAETHYL-S,S'-METHYLENEBISPHOSPHORODITHIOATE see EEH600

TETRAETHYL S,S'-METHYLENE BIS(PHOSPHOROTHIOLOTHIONATE) see EEH600

O,O,O',O'-TETRAETHYL S,S'-METHYLENE DI(PHOSPHORODITHIOATE) see EEH600

TETRAETHYLOLOVO see TCF000

TETRAETHYL ORTHOSILICATE see EPF550

TETRAETHYL ORTHOSILICATE (DOT) see EPF550

TETRAETHYLPLUMBANE see TCF000
TETRAETHYLPLUMBIUM see TCF000
TETRAETHYL PYROFOSFAAT see TCF250
TETRAETHYL PYROPHOSPHATE see TCF250
TETRAETHYL PYROPHOSPHATE, liquid (DOT) see TCF250
TETRAETHYLPYROPHOSPHATE and compressed gas mixtures see TCF260
TETRAETHYL PYROPHOSPHATE and COMPRESSED GAS MIXTURES LC50 < or =200 ppm (UN 1705) (DOT) see TCF260
TETRAETHYL PYROPHOSPHATE and COMPRESSED GAS MIXTURES LC50 >200 ppm, not >5000 ppm (UN 1705) (DOT) see TCF260
TETRAETHYL SILICATE see EPF550
TETRAETHYL SILICATE (DOT) see EPF550
TETRAETHYLSTANNANE see TCF750
TETRAETHYLTHIOPEROXYDICARBONIC DIAMIDE see DXH250
TETRAETHYLTHIRAM DISULPHIDE see DXH250
TETRAETHYLTHIURAM see DXH250
TETRAETHYLTHIURAM DISULFIDE see DXH250
TETRAETHYLTHIURAM DISULPHIDE see DXH250
N,N,N',N'-TETRAETHYLTHIURAM DISULPHIDE see DXH250
TETRAETHYL TIN see TCF750
TETRAETIL see DXH250
O,O,O,O-TETRAETIL-DITIO-PIROFOSFATO (ITALIAN) see SOD100
O,O,O,O-TETRAETIL-PIROFOSFATO (ITALIAN) see TCF250
TETRAFIDON see CKM000
TETRAFINOL see CBY000
TETRAFLUORETHYLENE see TCH500
(T-4)-TETRAFLUOROBERYLLATE(2-) DIAMMONIUM see ANH300
TETRAFLUORO BORATE(1-) LEAD(2+) see LDE000
TETRAFLUOROBORATE-(1−) POTASSIUM see PKY000
TETRAFLUOROBORATE(1-) SODIUM see SKE000
TETRAFLUOROBORATE(1-) STRONTIUM (2:1) see SMI000
TETRAFLUOROBORIC ACID see FDD125
TETRAFLUOROBORIC ACID see HHS600
TETRAFLUOROCARBON see CBY250
TETRAFLUORODICHLOROETHANE see DGL600
1,1,2,2-TETRAFLUORO-1,2-DICHLOROETHANE see FOO509
TETRAFLUOROETHENE see TCH500
TETRAFLUOROETHENE HOMOPOLYMER see TAI250
TETRAFLUOROETHENE POLYMER see TAI250
TETRAFLUOROETHYLENE see TCH500
TETRAFLUOROETHYLENE, inhibited (DOT) see TCH500
TETRAFLUOROETHYLENE HOMOPOLYMER see TAI250
TETRAFLUOROETHYLENE POLYMERS see TAI250
TETRAFLUORO HYDRAZINE see TCI000
TETRAFLUOROMETHANE see CBY250
TETRAFLUOROMETHANE (DOT) see CBY250
2,2,3,3-TETRAFLUORO-4,7-METHANO-2,3,5,6,8,9-HEXAHYDROBENZOSELENOPHENE see TCI100
TETRAFLUORO-m-PHENYLENE DIAMINE DIHYDROCHLORIDE see TCI250
TETRAFLUOROSILANE see SDF650
TETRAFLUOROSULFURANE see SOR000
TETRAFORM see CBY000

TETRAFOSFOR (DUTCH) see PHP010
TETRAGLYME see PBO500
6,7,8,9-TETRAHYDRO-5-AZEPOTETRAZOLE see PBI500
1,2,5,6-TETRAHYDROBENZALDEHYDE see FNK025
1,2,3,6-TETRAHYDROBENZALDEHYDE (DOT) see FNK025
1,2,3,4-TETRAHYDROBENZENE see CPC579
exo-TETRAHYDROBICYCLOPENTADIENE see TLR675
TETRAHYDROBORATE(1−) POTASSIUM see PKY250
Δ^1-TETRAHYDROCANNABINOL see TCM250
Δ^9-TETRAHYDROCANNABINOL see TCM250
(1)-Δ^1-TETRAHYDROCANNABINOL see TCM250
(−)-Δ^9-trans-TETRAHYDROCANNABINOL see TCM250
trans-Δ^9-TETRAHYDROCANNABINOL see TCM250
1-trans-Δ^9-TETRAHYDROCANNABINOL see TCM250
(−)-Δ^1-3,4-trans-TETRAHYDROCANNABINOL see TCM250
exo-TETRAHYDRODI(CYCLOPENTADIENE) see TLR675
7,8,9,10-TETRAHYDRO-N,N-DIETHYL-6H-CYCLOHEPTA(b)QUINOLINE-11-CARBOXAMIDE, see TCO100
TETRAHYDRO-3,5-DIMETHYL-4H,1,3,5-OXADIAZINE-4-THIONE see TCQ275
TETRAHYDRO-2H-3,5-DIMETHYL-1,3,5-THIADIAZINE-2-THIONE see DSB200
TETRAHYDRO-3,5-DIMETHYL-2H-1,3,5-THIADIAZINE-2-THIONE see DSB200
TETRAHYDRO-p-DIOXIN see DVQ000
TETRAHYDRO-1,4-DIOXIN see DVQ000
TETRAHYDRO-2,5-DIOXOFURAN see SNC000
1,2,3,6-TETRAHYDRO-3,6-DIOXOPYRIDAZINE see DMC600
3,4,5,6-TETRAHYDRO-2-(α-ETHOXYBENZYL)-5-ETHYL-5-METHYLPYRIMIDINE see TCQ700
TETRAHYDROFTALANHYDRID (CZECH) see TDB000
TETRAHYDROFURAAN (DUTCH) see TCR750
TETRAHYDROFURAN see TCR750
TETRAHYDROFURANNE (FRENCH) see TCR750
TETRAHYDRO-2-FURANONE see BOV000
1-(TETRAHYDROFURAN-2-YL)-5-FLUOROURACIL see FLZ050
TETRAHYDROFURFURYLAMINE see TCS500
N^1-(2-TETRAHYDROFURYL)-5-FLUOROURACIL see FLZ050
TETRAHYDROGERANIOL see DTE600
1-(TETRAHYDRO-5-HYDROXY-4-METHYL-3-FURANYL)-ETHANONE (9CI) see BMK290
4'-(1,2,3,4-TETRAHYDRO-4-(4-HYDROXY-2-OXO-2H-1-BENZOPYRAN-3-YL)-2-NAPHTHALENYL)(1,1'-BIPHENYL)-4-CARBONITRILE, cis- see TCU110
3a,4,7,7a-TETRAHYDRO-1,3-ISOBENZOFURANDIONE see TDB000
TETRAHYDRO-p-ISOXAZINE see MRP750
TETRAHYDRO-1,4-ISOXAZINE see MRP750
TETRAHYDROLINALOOL see TCU600
3a,4,7,7a-TETRAHYDRO-4,7-METHANOINDENE see DGW000
4a,5,7a,8-TETRAHYDRO-12-METHYL-9H-9,9c-IMINOETHANOPHENANTHRO(4,5-bcd)FURAN-3,5-DIOL see MRO500
1,2,5,6-TETRAHYDRO-1-METHYLNICOTINIC ACID, METHYL ESTER see AQT750

TETRAMETHYL METHYLENE DIAMINE (DOT) see TDR750
p-1',1',4',4'-TETRAMETHYLOKTYLBENZENSULFONAN SODNY (CZECH) see DXW200
TETRAMETHYLOLMETHANE see PBB750
2-(2,3,5,6-TETRAMETHYLPHENOXY)PROPIONIC ACID see BHM000
(2,3,5,6-TETRAMETHYLPHENYL)MERCURY ACETATE see AAS500
TETRAMETHYLPHOSPHORODIAMIDIC FLUORIDE see BJE750
N,N,N,N-TETRAMETHYLPHOSPHORODIAMIDIC FLUORIDE see BJE750
TETRAMETHYLPLUMBANE see TDR500
TETRAMETHYLPYRAZINE see TDU800
2,3,5,6-TETRAMETHYL PYRAZINE (FCC) see TDU800
TETRAMETHYL SILICATE see MPI750
TETRAMETHYLSILIKAT see MPI750
TETRAMETHYLSTANNANE see TDV750
TETRAMETHYLSUCCINONITRILE see TDW250
2,4,6,8-TETRAMETHYL-1,3,5,7-TETROXOCANE see TDW500
TETRAMETHYLTHIOCARBAMOYLDISULPHIDE see TFS350
TETRAMETHYL-O,O'-THIODI-p-PHENYLENE PHOSPHOROTHIOATE see TAL250
O,O,O'O'-TETRAMETHYL-O,O'-THIODI-p-PHENYLENE PHOSPHOROTHIOATE see TAL250
TETRAMETHYLTHIONINE CHLORIDE see BJI250
TETRAMETHYLTHIORAMDISULFIDE (DUTCH) see TFS350
TETRAMETHYLTHIOUREA see TDX000
1,1,3,3-TETRAMETHYLTHIOUREA see TDX000
TETRAMETHYL-THIRAM DISULFID (GERMAN) see TFS350
TETRAMETHYLTHIURAM BISULFIDE see TFS350
TETRAMETHYLTHIURAM DISULFIDE see TFS350
N,N,N',N'-TETRAMETHYLTHIURAM DISULFIDE see TFS350
N,N-TETRAMETHYLTHIURAM DISULPHIDE see TFS350
TETRAMETHYLTHIURAMMONIUM SULFIDE see BJL600
TETRAMETHYLTHIURAM MONOSULFIDE see BJL600
TETRAMETHYLTHIURAM SULFIDE see BJL600
TETRAMETHYL THIURANE DISULFIDE see TFS350
TETRAMETHYL TIN see TDV750
TETRAMETHYLTRITHIO CARBAMIC ANHYDRIDE see BJL600
TETRAMETHYLUREA see TDX250
1,1,3,3-TETRAMETHYLUREA see TDX250
TETRAMETHYLUREE (FRENCH) see TDX250
N,N,N',N'-TETRAMETIL-FOSFORODIAMMIDO-FLUORURO (ITALIAN) see BJE750
TETRAMINE see HOI000
TETRAMINE FAST BROWN BRS see CMO750
TETRAMINE PLATINUM(II) CHLORIDE see ANV800
1,4,5,8-TETRAMINOANTHRAQUINONE see TBG700
1-TETRAMISOLE HYDROCHLORIDE see LFA020
TETRAN see HOH500
TETRANATRIUMPYROPHOSPHAT (GERMAN) see TEE500
TETRAN HYDROCHLORIDE see HOI000
N,2,4,6-TETRANITROANILINE see TDY075
N,2,4,6-TETRANITROANILINE see TDY075
TETRANITROMETHANE see TDY250

N,2,4,5-TETRANITRO-N-METHYLANILINE see TEG250
TETRANITROPENTAERYTHRITE see PBC250
2,3,4,6-TETRANITROPHENOL see TDY600
TETRAN PTFE see TAI250
TETRA OLIVE N2G see APG500
2,4,10,12-TETRAOXA-6,16,17,18-TETRAAZA-3,11-DISTIBATRICYCLO(11.3.1.1^{5,9})OCTADECA-1(17),5,7,9(18),13,15-HEXAENE, 3,11-DIHYDROXY- see AQE305
(±)-(3,5,3',5'-TETRAOXO)-1,2-DIPIPERAZINOPROPANE see PIK250
2,4,5,6-TETRAOXOHEXAHYDROPYRIMIDINE see AFT750
2,4,5,6-TETRAOXOHEXAHYDROPYRIMIDINE HYDRATE see MDL500
TETRAPHENE see BBC250
TETRAPHENYLARSENIUM CHLORIDE see TEA300
TETRAPHENYLARSONIUM CHLORIDE see TEA300
TETRAPHOSPHATE HEXAETHYLIQUE (FRENCH) see HCY000
TETRAPHOSPHOR (GERMAN) see PHP010
TETRAPHOSPHORUS TRISULFIDE see PHS500
TETRAPOM see TFS350
TETRAPOTASSIUM FERROCYANIDE see TEC500
TETRAPOTASSIUM HEXACYANOFERRATE see TEC500
TETRAPOTASSIUM HEXACYANOFERRATE(4-) see TEC500
TETRAPOTASSIUM HEXACYANOFERRATE(II) see TEC500
TETRAPROPYLENE see PMP750
TETRAPROPYLENEPENTAMINE see TED600
TETRAPROPYL LEAD see TED750
TETRASIPTON see TFS350
TETRASODIUM DIARSENATE see SEY100
TETRASODIUM DIPHOSPHATE see TEE500
TETRASODIUM EDTA see EIV000
TETRASODIUM ETHYLENEDIAMINETETRAACETATE see EIV000
TETRASODIUM ETHYLENEDIAMINETETRACETATE see EIV000
TETRASODIUM (ETHYLENEDINITRILO)TETRAACETATE see EIV000
TETRASODIUM PYROPHOSPHATE see TEE500
TETRASODIUM PYROPHOSPHATE, ANHYDROUS see TEE500
TETRASODIUM SALT of EDTA see EIV000
TETRASODIUM SALT of ETHYLENEDIAMINETETRACETIC ACID see EIV000
TETRASOL see CBY000
TETRASTIGMINE see TCF250
TETRASULE see PBC250
TETRA SYTAM see BJE750
TETRATHIURAM DISULFIDE see TFS350
TETRATHIURAM DISULPHIDE see TFS350
TETRATRIMETHYLENEPENTAMINE see TED600
TETRAVEC see PCF275
TETRAVOS see DGP900
TETRA-WEDEL see TBX250
7,8,9,10-TETRAZABICYCLO(5.3.0)-8,10-DECADIENE see PBI500
1,2,3,3A-TETRAZACYCLOHEPTA-8A,2-CYCLOPENTADIENE see PBI500
TETRAZENE see TEF500
1-TETRAZENE, 4-AMIDINO-1-(NITRSOAMINOAMIDINO)-(8CI) see TEF500

TETRAZOBENZENE-β-NAPHTHOL see OHI200
TETRAZO DEEP BLACK G see AQP000
TETRIL see TEG250
TETRINE see EIV000
TETRINE ACID see EIX000
TETRODIRECT BLACK EFD see AQP000
TETRODONTOXIN see FOQ000
TETRODOTOXIN see FOQ000
TETRODOXIN see FOQ000
TETROGUER see PCF275
TETROLE see FPK000
TETRON see TCF250
TETRON-100 see TCF250
TETROPIL see PCF275
TETROSIN OE see BGJ250
TETROSOL see TBX250
TETRYL see TEG250
2,4,6-TETRYL see TEG250
TETRYL FORMATE see IIR000
TETURAM see DXH250
TETURAMIN see DXH250
TEVCOCIN see CDP250
TEVCODYNE see BRF500
TEXACO LEAD APPRECIATOR see BPV100
TEXAN RED TONER D see CHP500
TEXAPON ZHC see SIB600
TEXAS see CBT750
TEXIN 192A see PKO500
TEXIN 445D see PKM250
TEXTILE see SFT500
TEXTONE see SFT500
TEX WET 1001 see DJL000
TFE see TKA350
TFF see TNT500
T-FLUORIDE see SHF500
T-GAS see EJN500
TGD 5161 see SMQ500
β-TGDR see TFJ250
T-GELB BZW, GRUN 1 see QCA000
TH 6040 see CJV250
THACAPZOL see MCO500
THALIDOMIDE see TEH500
THALIN see TEH500
THALINETTE see TEH500
THALLIC OXIDE see TEL050
THALLIUM see TEI000
THALLIUM ACETATE see TEI250
THALLIUM(I) ACETATE see TEI250
THALLIUM(1+) ACETATE see TEI250
THALLIUM BROMIDE see TEI750
THALLIUM(I) CARBONATE (2:1) see TEJ000
THALLIUM CHLORATE see TEJ100
THALLIUM CHLORIDE see TEJ250
THALLIUM(1+) CHLORIDE see TEJ250
THALLIUM(I) FLUORIDE see TEK000
THALLIUM IODIDE see TEK500
THALLIUM(I) IODIDE see TEK500
THALLIUM(1+) IODIDE see TEK500
THALLIUM MONOACETATE see TEI250
THALLIUM MONOCHLORIDE see TEJ250
THALLIUM MONOFLUORIDE see TEK000
THALLIUM MONOIODIDE see TEK500
THALLIUM MONONITRATE see TEK750
THALLIUM MONOSELENIDE see TEL500
THALLIUM NITRATE see TEK750
THALLIUM OXIDE see TEL050
THALLIUM(3+) OXIDE see TEL050

THALLIUM(III) OXIDE see TEL050
THALLIUM PEROXIDE see TEL050
THALLIUM SELENIDE see TEL500
THALLIUM SESQUIOXIDE see TEL050
THALLIUM SULFATE see TEL750
THALLIUM(I) SULFATE (2:1) see TEM000
THALLIUM(II) SULFATE (1:1) see TEM050
THALLIUM SULFATE, solid (DOT) see TEL750
THALLOUS ACETATE see TEI250
THALLOUS CARBONATE see TEJ000
THALLOUS CHLORIDE see TEJ250
THALLOUS FLUORIDE see TEK000
THALLOUS IODIDE see TEK500
THALLOUS NITRATE see TEK750
THALLOUS SULFATE see TEM000
THAM see POC750
THANATE P 210 see PKB100
THANATE P 220 see PKB100
THANATE P 270 see PKB100
THANISOL see IHZ000
THANITE see IHZ000
THBP see TKO250
THC see TCM250
Δ^1-THC see TCM250
Δ^9-THC see TCM250
THEAL TABL. see TEP000
THEELIN see EDV000
THEELOL see EDU500
THEIN see CAK500
THEINE see CAK500
THELESTRIN see EDV000
THELYKININ see EDV000
THENALTON see PAG200
THENARDITE see SJY000
THENARDOL see HIB500
THENOBARBITAL see EOK000
THEOBROMINE see TEO500
THEOCIN see TEP000
THEOFOL see TEP000
THEOGEN see ECU750
THEOGRAD see TEP000
THEOHARN see MCB000
THEOLAIR see TEP000
THEOLIX see TEP000
THEOMINAL see EOK000
THEOPHILCHOLINE see TEH500
THEOPHYL-225 see TEP000
THEOPHYLLIN see TEP000
THEOPHYLLINE see TEP000
THEOPHYLLINE, anhydrous see TEP000
THEOSALVOSE see TEO500
THEOSTENE see TEO500
THERABLOAT see PJK150
THERACANZAN see SNN300
THERADERM see BDS000
THERADIAZINE see PPP500
THERA-FLUR-N see SHF500
THERAPAV see PAH250
THERAPOL see SNM500
THERAZONE see BRF500
THERMA-ATOMIC BLACK see CBT750
THERMACURE see MKA500
THERMAL ACETYLENE BLACK see CBT750
THERMALLY CRACKED RESIDUE see RDK100
THERMALOX see BFT250
THERMATOMIC see CBT750
THERMAX see CBT750

THERMBLACK see CBT750
THERM CHEK 820 see DDV600
THERMINOL FR-1 see PJL750
THERMOASE PC-10 see BAC000
THERMOGUARD B see AQF000
THERMOGUARD S see AQF000
THERMOLITE 813 see DVK200
THERMOLITE 831 see BKK750
THERMOPLASTIC 125 see SMR000
THESAL see TEO500
THESODATE see TEO500
THF see TCR750
THFU see FLZ050
THIABEN see TEX000
THIABENDAZOLE HYDROCHLORIDE see TER500
THIABENDAZOLE (USDA) see TEX000
THIABENZOLE see TEX000
3-THIABUTAN-2-ONE, O-
(METHYLCARBAMOYL)OXIME see MDU600
THIACETAMIDE see TFA000
THIACETIC ACID see TFA500
THIACOCCINE see TEX250
THIACYCLOPENTADIENE see TFM250
THIACYCLOPENTANE see TDC730
THIACYCLOPROPANE see EJP500
(N-1,2,3-THIADIAZOLYL-5)-N'-PHENYLUREA see
TEX600
4-THIAHEPTANEDIOIC ACID see BHM000
THIAMAZOLE see MCO500
THIAMETON see PHI500
THIAMIN see TES750
THIAMINE CHLORIDE see TES750
THIAMINE CHLORIDE HYDROCHLORIDE see
TET300
THIAMINE DICHLORIDE see TET300
THIAMINE HYDROCHLORIDE see TET300
THIAMINE MONOCHLORIDE see TES750
THIAMINE MONONITRATE see TET500
THIAMINE NITRATE see TET500
THIAMIN HYDROCHLORIDE see TET300
THIAMINIUM CHLORIDE HYDROCHLORIDE see
TET300
THIAMUTILIN see TET800
THIAMYLAL SODIUM see SOX500
THIA-4-PENTANAL (DOT) see MPV400
THIA-4-PENTANAL (DOT) see TET900
3-THIAPENTANE see EPH000
THIAPHENE see TFM250
2-THIAPROPANE see TFP000
THIARETIC see CFY000
THIASIN see SNN500
THIATE H see DKC400
THIAZAMIDE see TEX250
THIAZIDE see CLH750
1,3-THIAZOLIDINE, 2-
(METHOXY(METHYLTHIO)PHOSPHINYLIMINO)-3-
METHYL- see MEY200
2-(THIAZOL-4-YL)BENZIMIDAZOLE see TEX000
2-(4-THIAZOLYL)BENZIMIDAZOLE see TEX000
2-(4'-THIAZOLYL)BENZIMIDAZOLE see TEX000
2-(4-THIAZOLYL)-1H-BENZIMIDAZOLE see TEX000
2-(4-THIAZOLYL)-BENZIMIDAZOLE,
HYDROCHLORIDE see TER500
N^1-2-THIAZOLYLSULFANILAMIDE see TEX250
THIAZON see DSB200
THIAZONE see DSB200
THIBENZOLE see TEX000
THIDIAZURON see TEX600

1H-THIENO(3,4-d)IMIDAZOLE-4-PENTANOIC ACID,
HEXAHYDRO-2-OXO-, (3aS-(3a-α-4-β, 6a-α))- see
VSU100
THIERGAN see DQA400
THIFOR see EAQ750
THIIRANE see EJP500
THILANE see TDC730
THILLATE see TFS350
THILOPHENYL see DKQ000
THILOPHENYT see DNU000
THIMECIL see MPW500
THIMER see TFS350
THIMEROSALATE see MDI000
THIMEROSOL see MDI000
THIMET see PGS000
THIMUL see EAQ750
THIOACETAMIDE see TFA000
THIOACETIC ACID see TFA500
THIOALKOFEN BM 4 see TFC600
THIOALLATE see CDO250
THIOANILINE see TFI000
4,4'-THIOANILINE see TFI000
2-THIO-2H-1,3-BENZOXAZINE-2,4(3H)-DIONE see
TFC570
THIOBENZYL ALCOHOL see TGO750
2-THIO-2-BENZYL-PSEUDOUREA
HYDROCHLORIDE see BEU500
4,4'-THIOBIS(ANILINE) see TFI000
4,4'-THIOBISBENZENAMINE see TFI000
4,4'-THIOBIS(6-tert-BUTYL-m-CRESOL) see TFC600
4,4'-THIOBIS(2-tert-BUTYL-5-METHYLPHENOL) see
TFC600
4,4'-THIOBIS(6-tert-BUTYL-3-METHYLPHENOL) see
TFC600
1,1'-THIOBIS(2-CHLOROETHANE) see BIH250
1,1'-THIOBIS(N,N-DIMETHYLTHIO)FORMAMIDE see
BJL600
THIOBIS(DODECYL PROPIONATE) see TFD500
1,1'-THIOBISETHANE see EPH000
4,4'-THIOBIS(3-METHYL-6-tert-BUTYLPHENOL) see
TFC600
1,1'-THIOBIS(2-METHYL-4-HYDROXY-5-tert-
BUTYLBENZENE) see TFC600
THIOCARB see SGJ000
THIOCARBAMATE see ISR000
THIOCARBAMIDE see ISR000
THIOCARBAMISIN see TFD750
THIOCARBAMIZINE see TFD750
THIOCARBAMYLHYDRAZINE see TFQ000
THIOCARBANIL see ISQ000
THIOCARBANILIDE see DWN800
THIOCARBARSONE see CBI250
THIOCARBONIC DICHLORIDE see TFN500
THIOCARBONYL CHLORIDE see TFN500
THIOCARBONYL DICHLORIDE see TFN500
THIOCHLORID FOSFORECNY see TFO000
THIOCHROMAN-4-ONE, OXIME see TEJ000
4-THIOCRESOL see TGP250
o-THIOCRESOL see TGP000
p-THIOCRESOL see TGP250
THIOCYAN see CAY250
THIOCYANATES see TFE500
THIOCYANATE SODIUM see SIA500
THIOCYANATOACETIC ACID ISOBORNYL ESTER
see IHZ000
THIOCYANATOBENZENE see TFF600
THIOCYANATOETHANE see EPP000

α-THIOCYANATOTOLUENE see BFL000
THIOCYANIC ACID, 1,2-ETHANEDIYL ESTER see EJC035
THIOCYANIC ACID, ETHYLENE ESTER see EJC035
THIOCYANIC ACID, ETHYL ESTER see EPP000
THIOCYANIC ACID, MERCURY(2+) SALT see MCU250
THIOCYANIC ACID, PHENYL ESTER see TFF600
1-THIOCYANOBUTANE see BSN500
THIODAN see EAQ750
THIODEMETON see DXH325
p,p-THIODIANILINE see TFI000
4,4'-THIODIANILINE see TFI000
THIODIFENYLAMINE (DUTCH) see PDP250
THIODIGLYCOLIC ACID see MCM750
2,2'-THIODIGLYCOLIC ACID see MCM750
β,β'-THIODIGLYCOLIC ACID see MCM750
THIODIGLYCOLLIC ACID see MCM750
2-THIO-3,5-DIMETHYLTETRAHYDRO-1,3,5-THIADIAZINE see DSB200
THIODIPHENYLAMIN (GERMAN) see PDP250
THIODIPHENYLAMINE see PDP250
O,O'-(THIODI-4,1-PHENYLENE)BIS(O,O-DIMETHYL PHOSPHOROTHIOATE) see TAL250
THIODI-p-PHENYLENEDIAMINE see TFI000
O,O'-(THIODI-p-PHENYLENE)-O,O,O',O'-TETRAMETHYL BIS(PHOSPHOROTHIOATE) see TAL250
THIODIPROPIONIC ACID see BHM000
3,3'-THIODIPROPIONIC ACID see BHM000
β,β'-THIODIPROPIONIC ACID see BHM000
THIODRIL see CBR675
THIOETHANOL see EMB100
2-THIOETHANOL see MCN250
THIOETHANOLAMINE see AJT250
THIOETHYL ALCOHOL see EMB100
THIOETHYL ETHER see EPH000
THIOFACO M-50 see EEC600
THIOFACO T-35 see TKP500
THIOFAN see TDC730
THIOFENOL see PFL850
THIOFIDE see BDE750
THIOFOR see EAQ750
THIOFORMIC ACID, PHENYLAZO-, PHENYLHYDRAZIDE see DWN200
THIOFOSGEN (CZECH) see TFN500
THIOFOZIL see TFQ750
THIOFURAM see TFM250
THIOFURAN see TFM250
THIOFURFURAN see TFM250
(1-THIO-d-GLUCOPYRANOSATO)GOLD see ART250
1-THIO-GLUCOPYRANOSE, MONOGOLD(1+) SALT see ART250
THIOGLUCOSE d'OR (FRENCH) see ART250
THIOGLYCERIN see MRM750
1-THIOGLYCEROL see MRM750
α-THIOGLYCEROL see MRM750
THIOGLYCOL (DOT) see MCN250
THIOGLYCOLANILIDE see MCK000
THIOGLYCOLATESODIUM see SKH500
THIOGLYCOLIC ACID see TFJ100
2-THIOGLYCOLIC ACID see TFJ100
THIOGLYCOLIC ACID ANILIDE see MCK000
THIOGLYCOLIC ACID, SODIUM SALT see SKH500
THIOGLYCOLLIC ACID see TFJ100
THIOGLYCOLLIC ACID, AMMONIUM SALT see ANM500
β-THIOGUANINE DEOXYRIBOSIDE see TFJ250

THIOHEXAM see CPI250
THIOHYPOXANTHINE see POK000
THIOKARBONYLCHLORID (CZECH) see TFN500
THIOLACETIC ACID see TFA500
2-THIOLACTIC ACID see TFK250
THIOLANE see TDC730
THIOLDEMETON see DAP200
2-THIOL-DIHYDROGLYOXALINE see IAQ000
THIOLE see TFM250
THIOLITE see CAX500
THIOL SYSTOX see DAP200
THIOLUX see SOD500
THIOMEBUMAL SODIUM see PBT500
THIOMECIL see MPW500
THIOMERIN SODIUM see TFK270
THIOMERSALATE see MDI000
THIOMETHANOL see MLE650
2-THIO-6-METHYL-1,3-PYRIMIDIN-4-ONE see MPW500
6-THIO-4-METHYLURACIL see MPW500
THIOMETON see PHI500
THIOMIDIL see MPW500
THIOMONOGLYCOL see MCN250
2,5-THIOMORPHOLINEDIONE, 3,3-DIMETHYL-, 2-(o-((METHYLAMINO)CARBONYL)OXIME) see MID860
2,5-THIOMORPHOLINEDIONE, 3-METHYL-, 2-(o-((METHYLAMINO)CARBONYL)OXIME) see MPU600
THIOMUL see EAQ750
THIOMYLAL SODIUM see SOX500
THIONAPHTHOL see NAP500
2-THIONAPHTHOL see NAP500
THIO-β-NAPHTHOL see NAP500
β-THIONAPHTHOL see NAP500
THIONAZIN see EPC500
THIONEMBUTAL see PBT500
THIONEX see BJL600
THIONEX see EAQ750
THIONEX RUBBER ACCELERATOR see BJL600
THIONOACETIC ACID see TFA500
THIONOBENZENEPHOSPHONIC ACID ETHYL-p-NITROPHENYL ESTER see EBD700
THIONOSINE see MCQ500
THIONYL CHLORIDE see TFL000
THIONYL DICHLORIDE see TFL000
THIONYL DIFLUORIDE see TFL250
THIONYL FLUORIDE see TFL250
2-THIO-4-OXO-6-METHYL-1,3-PYRIMIDINE see MPW500
2-THIO-4-OXO-6-PROPYL-1,3-PYRIMIDINE see PNX000
2-THIO-6-OXYPYRIMIDINE see TFR250
THIOPENTAL SODIUM see PBT500
THIOPENTAL SODIUM SALT see PBT500
THIOPENTONE SODIUM see PBT500
THIOPEROXYDICARBONIC ACID DIETHYL ESTER see BJU000
THIOPHAL see TIT250
THIOPHANE see TDC730
THIOPHEN see TFM250
THIOPHENE see TFM250
THIOPHENE, TETRAHYDRO-3,3,4,4-TETRACHLORO-, 1,1-DIOXIDE see TBV300
THIOPHENIT see MNH000
THIOPHENOL (DOT) see PFL850
THIOPHOS see PAK000
THIOPHOSGENE see TFN500
THIOPHOSPHAMIDE see TFQ750

THIOPHOSPHATE de S-N-(1-CYANO-1-
METHYLETHYL)CARBAMOYLMETHYLE et de O,O-
DIETHYLE (FRENCH) see PHK250
THIOPHOSPHATE de O-2,4-DICHLOROPHENYLE et
de O,O-DIETHYLE (FRENCH) see DFK600
THIOPHOSPHATE de O,O-DIETHYLE et de O-(3-
CHLORO-4-METHYL-7-COUMARINYLE) (FRENCH)
see CNU750
THIOPHOSPHATE de O,O-DIETHYLE et de O-(2,5-
DICHLORO-4-BROMO) PHENYLE (FRENCH) see
EGV500
THIOPHOSPHATE de O,O-DIETHYLE et de S-(2-
ETHYLTHIO-ETHYLE) (FRENCH) see DAP200
THIOPHOSPHATE de O,O-DIETHYLE et de o-2-
ISOPROPYL-4-METHYL-6-PYRIMIDYLE (FRENCH)
see DCM750
THIOPHOSPHATE de O,O-DIETHYLE et de O-(4-
NITROPHENYLE) (FRENCH) see PAK000
THIOPHOSPHATE de O,O-DIMETHYLE et de S-2-
ETHYLSULFINYLETHYLE (FRENCH) see DAP000
THIOPHOSPHATE de O,O-DIMETHYLE et de O-2-
ETHYLTHIO-ETHYLE (FRENCH) see DAO800
THIOPHOSPHATE de O,O-DIMETHYLE et de S-2-
ETHYLTHIOETHYLE (FRENCH) see DAP400
THIOPHOSPHATE de O,O-DIMETHYLE et de O-(3-
METHYL-4-NITROPHENYLE) (FRENCH) see DSQ000
THIOPHOSPHATE de O,O-DIMETHYLE et de O-(4-
NITROPHENYLE) (FRENCH) see MNH000
THIOPHOSPHATE de O,O-DIMETHYLE et de O-(2,4,5-
TRICHLOROPHENYLE) (FRENCH) see RMA500
THIOPHOSPHORIC ANHYDRIDE see PHS000
THIOPHOSPHORIC TRICHLORIDE see TFO000
THIOPHOSPHORSAEURE-O,S-
DIMETHYLESTERAMID (GERMAN) see DTQ400
THIOPHOSPHORYL CHLORIDE see TFO000
THIOPHOSPHORYL TRICHLORIDE see TFO000
2-THIOPROPANE see TFP000
2-THIO-6-PROPYL-1,3-PYRIMIDIN-4-ONE see PNX000
6-THIO-4-PROPYLURACIL see PNX000
β-THIOPSEUDOUREA see ISR000
6-THIOPURINE RIBONUCLEOSIDE see MCQ500
6-THIOPURINE RIBOSIDE see MCQ500
2-THIO-1,3-PYRIMIDIN-4-ONE see TFR250
THIORYL see MPW500
THIOSAN see TFS350
THIOSAN see DXH250
THIOSCABIN see DXH250
THIOSEMICARBAZIDE see TFQ000
3-THIOSEMICARBAZIDE see TFQ000
THIOSEPTAL see PPO000
THIOSERINE see CQK000
THIOSTOP N see SGM500
THIOSULFAN see EAQ750
THIOSULFAN TIONEL see EAQ750
THIOSULFIL-A FORTE see PDC250
THIOSULFURIC ACID, S,S'-(2-(DIMETHYLAMINO)-
1,3-PROPANEDIYL) ESTER, DISODIUM SALT see
DXC900
THIOSULFUROUS DICHLORIDE see SON510
THIO-TEP see TFQ750
THIOTEPP see SOD100
THIOTETROLE see TFM250
THIOTEX see TFS350
THIOTHAL SODIUM see PBT500
2-THIO-1-(THIOCARBAMOYL)UREA see DXL800
THIOTHYMIN see MPW500
THIOTHYRON see MPW500

THIOTOX see TFS350
THIOTRIETHYLENEPHOSPHORAMIDE see TFQ750
THIOURACIL see TFR250
2-THIOURACIL see TFR250
6-THIOURACIL see TFR250
2-THIOUREA see ISR000
THIOUREA (DOT) see ISR000
THIOVANIC ACID see TFJ100
THIOVANOL see MRM750
THIOVIT see SOD500
THIOXAMYL see DSP600
6-THIOXOPURINE see POK000
THIOZAMIDE see TEX250
THIPENTAL SODIUM see PBT500
THIRAM see TFS350
THIRAMAD see TFS350
THIRAME (FRENCH) see TFS350
THIRASAN see TFS350
THIRERANIDE see DXH250
THIULIX see TFS350
THIURAD see TFS350
THIURAGYL see PNX000
THIURAM see TFS350
THIURAM E see DXH250
THIURAMIN see TFS350
THIURAMYL see TFS350
THIURANIDE see DXH250
THIURETIC see CFY000
THIURYL see MPW500
THLARETIC see CFY000
THOMPSON-HAYWARD TH6040 see CJV250
THOMPSON'S WOOD FIX see PAX250
THORAZINE see CKP500
THORAZINE HYDROCHLORIDE see CKP500
THORIA see TFT750
THORIUM see TFS750
THORIUM-232 see TFS750
THORIUM CHLORIDE see TFT000
THORIUM DIOXIDE see TFT750
THORIUM HYDRIDE see TFT250
THORIUM METAL, pyrophoric (DOT) see TFS750
THORIUM (4+) NITRATE see TFT500
THORIUM(IV) NITRATE see TFT500
THORIUM OXIDE see TFT750
THORIUM TETRACHLORIDE see TFT000
THORIUM TETRANITRATE see TFT500
THOROTRAST see TFT750
THORTRAST see TFT750
THPA see TDB000
THREARIC ACID see TAF750
THREE ELEPHANT see BMC000
l-THREITOL-1,4-BISMETHANESULFONATE see
TFU500
THREONINE see TFU750
l-THREONINE see TFU750
THRETHYLENE see TIO750
THU see ISR000
(1S,4R,5R)-(−)-3-THUJANONE see TFW000
THUJA OIL see CCQ500
THUJON see TFW000
THUJONE see TFW000
(−)-THUJONE see TFW000
l-THUJONE see TFW000
α-THUJONE see TFW000
THULIUM CHLORIDE see TFW500
THULOL see EDU500
THYALONE see BAG250

THYCAPSOL see MCO500
THYLATE see TFS350
THYLFAR M-50 see MNH000
THYLOGEN MALEATE see DBM800
THYLOQUINONE see MMD500
THYME CAMPHOR see TFX810
THYME OIL see TFX500
THYME OIL RED see TFX750
THYMIAN OEL (GERMAN) see TFX500
THYMIC ACID see TFX810
THYMIDIN see TFX790
THYMIDINE see TFX790
THYMINEDEOXYRIBOSIDE see TFX790
THYMINE-2-DEOXYRIBOSIDE see TFX790
THYM OIL see TFX500
THYMOL see TFX810
m-THYMOL see TFX810
o-THYMOL see CCM000
THYNESTRON see EDV000
THYREONORM see MPW500
THYREOSTAT see MPW500
THYREOSTAT II see PNX000
THYRIL see MPW500
THYROXEVAN see LFG050
l-THYROXINE MONOSODIUM SALT see LFG050
THYROXINE SODIUM SALT see LFG050
l-THYROXINE SODIUM SALT see LFG050
TIAMULIN see TET800
TIAMULINA (ITALIAN) see TET800
TIAZON see DSB200
TIC see TKJ250
TIC MUSTARD see IAN000
TIFOMYCINE see CDP250
TIGLIC ACID see TGA700
TIGLINIC ACID see TGA700
TIKOFURAN see NGG500
TILCAREX see PAX000
TILDIN see BNK000
TILLRAM see DXH250
TIMAZIN see FMM000
TIMET see PGS000
TIMOLET see DLH630
TIMONOX see AQF000
TIMOPED see TGB475
TIMOSIN see MDQ250
TIN see TGB250
TIN (α) see TGB250
TINACTIN see TGB475
TINADERM see TGB475
TIN BIFLUORIDE see TGD100
TIN(IV) BROMIDE (1:4) see TGB750
TIN(II) CHLORIDE (1:2) see TGC000
TIN(IV) CHLORIDE (1:4) see TGC250
TIN CHLORIDE, fuming (DOT) see TGC250
TIN CHLORIDE IODIDE see TGC280
TIN(IV) CHLORIDE, PENTAHYDRATE (1:4:5) see
TGC282
TIN COMPOUNDS see TGC500
TIN DIBUTYL DILAURATE see DDV600
TIN DIBUTYL MERCAPTIDE see DEI200
TIN DICHLORIDE see TGC000
TIN, DIETHYL-, DIIODIDE see DJB000
TIN DIFLUORIDE see TGD100
TINDURIN see TGD000
TINESTAN see ABX250
TINESTAN 60 WP see ABX250
TIN FLAKE see TGB250

TIN FLUORIDE see TGD100
TINIC see NCQ900
TIN(II) IODIDE see TGD500
TIN(IV) IODIDE (1:4) see TGD750
TINMATE see CLU000
TIN, OCTYL-, TRICHLORIDE see OGG000
TIN, OCTYL-, TRIS(ISOOCTYLTHIO GLYCOLLATE)
see OGI000
TINOLITE see CBT750
TINOPAL CBS see TGE150
TINOPAL CBS-X see TGE150
TINOPAL 5BM see TGE100
TINOSTAT see DDV600
TINOX see MIW250
TIN OXALATE see TGE250
TIN(2+) OXALATE see TGE250
TIN(II) OXALATE see TGE250
TIN OXIDE see TGE300
TIN PERBROMIDE see TGB750
TIN PERCHLORIDE (DOT) see TGC250
TIN(IV) PHOSPHIDE see TGE500
TIN POWDER see TGB250
TIN PROTOCHLORIDE see TGC000
TINSET see OMG000
TIN STEARATE see SLI350
TIN TETRABROMIDE see TGB750
TINTETRACHLORIDE (DUTCH) see TGC250
TIN TETRACHLORIDE, anhydrous (DOT) see TGC250
TIN TETRAIODIDE see TGD750
TINTORANE see WAT220
TIN, TRIBUTYL-, ISOTHIOCYANATE see THY750
TIN TRIPHENYL ACETATE see ABX250
TINUVIN 144 see PMK800
TIODIFENILAMINA (ITALIAN) see PDP250
TIOFINE see TGG760
TIOFOS see PAK000
TIOFOSFAMID see TFQ750
TIOFOZIL see TFQ750
TIOINOSINE see MCQ500
TIOMERACIL see MPW500
TIOPENTAL SODIUM see PBT500
TIORALE M see MPW500
TIOTIRON see MPW500
TIOURACYL (POLISH) see TFR250
TIOVEL see EAQ750
TIOXIDE see TGG760
TIOXIDE A-HR see OBU100
TIP-OFF see NAK500
TIRAMPA see TFS350
TISPERSE MB-2X see NBL000
TISPERSE MB-58 see BHA750
TITAANTETRACHLORID (DUTCH) see TGH350
TITANATE see TGF250
TITANDIOXID (SWEDEN) see TGG760
TITANE (TETRACHLORURE de) (FRENCH) see
TGH350
TITANIC ACID, LEAD SALT see LED000
TITANIO TETRACHLORURO di (ITALIAN) see
TGH350
TITANIUM see TGF250
TITANIUM 50A see TGF250
TITANIUM (wet powder) see TGF500
TITANIUM ALLOY see TGF250
TITANIUM compounded with BERYLLIUM (1:12) see
BFR000
TITANIUM CHLORIDE see TGG250
TITANIUM CHLORIDE see TGH350
TITANIUM(III) CHLORIDE see TGG250

TITANIUM COMPOUNDS see TGG500
TITANIUM DIOXIDE see TGG760
TITANIUM DISULFATE see TGH250
TITANIUM DISULFATE see TGH252
TITANIUM FERROCENE see DGW200
TITANIUM METAL POWDER, WET (DOT) see TGF500
TITANIUM NICKEL OXIDE see NDL500
TITANIUM OXIDE see TGG760
TITANIUM POTASSIUM FLUORIDE see PLI000
TITANIUM SULFATE see TGH250
TITANIUM SULFATE SOLUTION see TGH252
TITANIUM SULFATE SOLUTION (DOT) see TGH250
TITANIUM TETRACHLORIDE see TGH350
TITANIUM TRICHLORIDE MIXTURES (UN 2869) (DOT) see TGG250
TITANIUM TRICHLORIDE MIXTURES, pyrophoric (UN 2441) (DOT) see TGG250
TITANIUM TRICHLORIDE, pyrophoric (UN 2441) (DOT) see TGG250
TITANOCENE see DGW200
TITANOCENE, DICHLORIDE see DGW200
TITANOUS CHLORIDE see TGG250
TITANTETRACHLORID (GERMAN) see TGH350
TITRIPLEX see EIX000
TITRIPLEX I see AMT500
TITRIPLEX III see EIX500
TIURAM see DXH250
TIURAM (POLISH) see TFS350
TIURAMYL see TFS350
TIXOTON see BAV750
TJB see DWI000
TK 200 see EHP700
TK 201 see EHP700
TK 1000 see PKQ059
TK 10408 see BRK750
TKB see NKB500
TL4N see DJD600
TL 59 see PER600
TL 69 see DGB600
TL 70 see SOQ450
TL 75 see PHQ500
TL 78 see DNJ800
TL 146 see BIE250
TL 154 see CHF500
TL 161 see BHO250
TL 189 see FOY000
TL 190 see MPI750
TL 207 see CGU199
TL 212 see SOT500
TL 214 see DFH200
TL 229 see HBN600
TL 231 see HFP600
TL 262 see TFO000
TL 266 see NLJ500
TL 294 see DFP200
TL 314 see ADX500
TL 329 see BID250
TL 337 see EJM900
TL 367 see FBS000
TL 373 see EOQ000
TL 389 see DQY950
TL 399 see TIY800
TL 423 see EHK500
TL 466 see IRF000
TL 741 see FIE000
TL 751 see FIH100
TL 792 see BJE750
TL 822 see COO500

TL 869 see SHG500
TL 898 see MCY475
TL 944 see BEU500
TL 1026 see CAG000
TL 1070 see CAG500
TL 1091 see NDC000
TL 1139 see SMI000
TL 1149 see BID250
TL 1163 see TLN250
TL 1182 see CAI000
TL-1380 see PIA750
TL 1450 see MKX250
TL 1483 see EEV200
TL-1504 see MID900
TL 1505 see ABO000
TL 1578 see EIF000
TL 1618 see IPX000
TLA see BPV100
TLD 100 see LHF000
TLP-607 see PAP250
TM see TDK500
TM 30 see CBT750
TM-4049 see MAK700
TMA see TKV000
TMA see TLD500
TMAN see TKV000
TMB see TLM050
7,8,12-TMBA see TLK750
TMDE 6500 see SMQ500
TMEDA see TDQ750
m-TMI see IKG800
TML see TDR500
TMP see TMD250
TMSN see TDW250
TMTD see TFS350
TMTDS see TFS350
TMTM see BJL600
TMTMS see BJL600
TMTU see TDX000
TMU see TDX250
T-2 MYCOTOXIN see FQS000
TNB see TMK500
TNCS 53 see CNP250
TNF-α see HGL920
TNG see NGY000
TNM see TDY250
α-TNT see TMN490
TNT (OSHA) see TMN490
TNT-TOLITE (FRENCH) see TMN490
TNT, dry or wetted with <30% water, by weight (UN 0209) (DOT) see TMN490
TOABOND 40H see AAX250
TOBACCO PLANT see TGI100
TOBACCO REFINED TAR see CMP800
TOBACCO SMOKE CONDENSATE see SEC000
TOBACCO TAR see SEC000
TOBACCO TAR see CMP800
TOCE see DJT400
TOCEN see DJT400
TOCHLORINE see CDP000
TOCOPHEREX see TGJ050
α-TOCOPHEROL see VSZ450
α-TOCOPHEROL ACETATE see TGJ050
(+)-α-TOCOPHEROL ACETATE see TGJ050
d-α-TOCOPHEROL ACETATE see TGJ050
d-α-TOCOPHEROL (FCC) see VSZ450

α-TOLUENETHIOL see TGO750
TOLUENE TRICHLORIDE see BFL250
TOLUENE, VINYL (mixed isomers) see VQK650
o-TOLUENO-AZO-β-NAPHTHOL see TGW000
ar-TOLUENOL see CNW500
α-TOLUENOL see BDX500
α-TOLUIC ACID see PDY850
m-TOLUIC ACID DIETHYLAMIDE see DKC800
α-TOLUIC ACID, ETHYL ESTER see EOH000
α-TOLUIC ACID, α-HYDROXY- see MAP000
α-TOLUIC ALDEHYDE see BBL500
m-TOLUIDIN (CZECH) see TGQ500
o-TOLUIDIN (CZECH) see TGQ750
p-TOLUIDIN (CZECH) see TGR000
TOLUIDINE see CEX800
2-TOLUIDINE see TGQ750
3-TOLUIDINE see TGQ500
4-TOLUIDINE see TGR000
m-TOLUIDINE see TGQ500
o-TOLUIDINE see TGQ750
p-TOLUIDINE see TGR000
o-TOLUIDINE, 6-CHLORO- see CLK227
2-TOLUIDINE HYDROCHLORIDE see TGS500
o-TOLUIDINE HYDROCHLORIDE see TGS500
p-TOLUIDINE HYDROCHLORIDE see TGS750
p-TOLUIDINE, α-(PHENYLTHIO)- see PGN100
p-TOLUIDINIUM CHLORIDE see TGS750
o-TOLUIDYNA (POLISH) see TGQ750
TOLUILENODWUIZOCYJANIAN see TGM750
TOLUINA see BSQ000
TOLUMID see BSQ000
α-TOLUNITRILE see PEA750
m-TOLUOL see CNW750
o-TOLUOL see CNX000
p-TOLUOL see CNX250
TOLUOL (DOT) see TGK750
o-TOLUOL-AZO-o-TOLUIDIN (GERMAN) see AIC250
TOLUOLO (ITALIAN) see TGK750
p-TOLUOLSULFONSAEUREAETHYL ESTER
(GERMAN) see EPW500
p-TOLUOLSULFONSAEURE METHYL ESTER
(GERMAN) see MLL250
p-TOLUQUINONE see MHI250
1,4-TOLUQUINONE see MHI250
TOLU-SOL see TGK750
TOLUVAN see BSQ000
m-TOLUYLENDIAMIN (CZECH) see TGL750
p-TOLUYLENDIAMINE see TGM000
m-TOLUYLENEDIAMINE see TGL750
2,4-TOLUYLENEDIAMINE see TGL750
TOLUYLENE-2,5-DIAMINE see TGM000
2,4-TOLUYLENEDIAMINE (DOT) see TGL750
p-TOLUYLENEDIAMINE SULPHATE see DCE600
TOLUYLENE-2,5-DIAMINE SULPHATE see DCE600
TOLUYLENE-2,4-DIISOCYANATE see TGM750
m-TOLYENEDIAMINE see TGL750
TOLYL ACETATE see MHM100
p-TOLYL ACETATE see MNR250
p-TOLYL ALCOHOL see CNX250
α-TOLYL ALDEHYDE DIMETHYL ACETAL see
PDX000
TOLYLAMINE see TGR000
m-TOLYLAMINE see TGQ500
o-TOLYLAMINE see TGQ750
p-TOLYLAMINE see TGR000
o-TOLYLAMINE HYDROCHLORIDE see TGS500

4-TOLYLARSENOUS ACID see TGV100
5-(o-TOLYLAZO)-2-AMINOTOLUENE see AIC250
1-(o-TOLYLAZO)-2-NAPHTHOL see TGW000
1-(o-TOLYLAZO)-β-NAPHTHOL see TGW000
1-(o-TOLYLAZO)-2-NAPHTHYLAMINE see FAG135
4-(o-TOLYLAZO)-o-TOLUIDINE see AIC250
o-TOLYLAZO-o-TOLYLAZO-2-NAPHTHOL see
SBC500
1-(4-o-TOLYLAZO-o-TOLYLAZO)-2-NAPHTHOL see
SBC500
o-TOLYLAZO-o-TOLYLAZO-β-NAPHTHOL see
SBC500
p-TOLYL BENZOATE see TGX100
TOLYL CHLORIDE see BEE375
o-TOLYL CHLORIDE see CLK100
p-TOLYL CHLORIDE see TGY075
m-TOLYLENEDIAMINE see TGL750
2,4-TOLYLENEDIAMINE see TGL750
4-m-TOLYLENEDIAMINE see TGL750
p,m-TOLYLENEDIAMINE see TGM000
p-TOLYLENEDIAMINE SULPHATE see DCE600
TOLYLENE DIISOCYANATE see TGM740
m-TOLYLENE DIISOCYANATE see TGM750
m-TOLYLENE DIISOCYANATE see TGM800
TOLYLENE-2,4-DIISOCYANATE see TGM750
2,4-TOLYLENEDIISOCYANATE see TGM750
TOLYLENE-2,6-DIISOCYANATE see TGM800
TOLYLENE ISOCYANATE see TGM740
m-TOLYLESTER KYSELINY
METHYLKARBAMINOVE see MIB750
p-TOLYL ETHANOATE see MNR250
p-TOLYL ISOBUTYRATE see THA250
2-TOLYL ISOCYANATE see IKG725
o-TOLYL ISOCYANATE see IKG725
o-TOLYL MERCAPTAN see TGP000
p-TOLYL MERCAPTAN see TGP250
α-TOLYL MERCAPTAN see TGO750
3-TOLYL-N-METHYLCARBAMATE see MIB750
m-TOLYL N-METHYLCARBAMATE see MIB750
p-TOLYL METHYL ETHER see MGP000
o-TOLYL PHOSPHATE see TMO600
4-(p-TOLYL)PIPERAZINYL 3,4,5-
TRIMETHOXYPHENYL KETONE see THF310
p-TOLYL SALICYLATE see THD850
(((p-
TOLYL)SULFAMOYL)IMINO)BIS(METHYLMERCURY
) see MLH100
N-(p-TOLYLSULFONYL)-N'-BUTYLCARBAMIDE see
BSQ000
TOLYLSULFONYLBUTYLUREA see BSQ000
3-(p-TOLYL-4-SULFONYL)-1-BUTYLUREA see BSQ000
p-TOLYLTHIOL see TGP250
1-(p-TOLYL)-4-(3,4,5-
TRIMETHOXYBENZOYL)PIPERAZINE see THF310
TOMATHREL see CDS125
TOMATINE HYDROCHLORIDE see THG300
α-TOMATINE HYDROCHLORIDE see THG300
α-TOMATINE, HYDROCHLORIDE see THG300
TOMATO FIX CONCENTRATE see CJN000
TOMATO HOLD see CJN000
TOMATOTONE see CJN000
TOMIL see DIR000
TOMOFAN see CCU150
TONARIL see TMP750
TONCARINE see MIP750
TONCO-70 see OBG000
TONEDRIN see DBA800

TONEDRON see MDT600
TONER LAKE RED C see CHP500
TONEY RED see OHI200
TONITE see CDN200
TONKA BEAN CAMPHOR see CNV000
TONOCARD see DJS200
TONOCOR see DJS200
TONOFTAL see TGB475
TONOGEN see VGP000
TONOX see MJQ000
TO NTU see BMR750
TONY RED see OHI200
TOPANE see BGJ750
TOPANEL see COB260
TOPANOL see BFW750
TOPAZONE see NGG500
TOPCAINE see EFX000
TOPEX see BDS000
TOP FLAKE see SFT000
TOP FORM WORMER see TEX000
TOPICHLOR 20 see CDR750
TOPICLOR see CDR750
TOPICLOR 20 see CDR750
TOPICYCLINE see TBX250
TOPITOX see CJJ000
TOPITRACIN see BAC250
TOPOKAIN see AIT250
TOPOREX 500 see SMQ500
TOPOREX 830 see SMQ500
TOPOREX 550-02 see SMQ500
TOPOREX 850-51 see SMQ500
TOPOREX 855-51 see SMQ500
TOPSYM see DUD800
TOPZOL see RCF000
TORAK see DBI099
TORAZINA see CKP500
TORBIN see EIN500
TORCH see XPJ000
TORCH BRAND see CBT750
TORDON see PIB900
TORDON 10K see PIB900
TORDON 22K see PIB900
TORDON 101 MIXTURE see PIB900
TORQUE see BLU000
TORSITE see BGJ250
TOSTRIN see TBG000
TOSYL see CDO000
N-TOSYL-l-ALANINE CHLOROMETHYL KETONE see THH360
N-TOSYL-β-ALANINE CHLOROMETHYL KETONE see THH355
N-TOSYL-l-ALANINE DIAZOMETHYL KETONE see THH380
N-TOSYL-β-ALANINE DIAZOMETHYL KETONE see THH375
l-1-TOSYLAMIDO-2-PHENYLETHYL CHLOROMETHYL KETONE see THH450
6-(N-TOSYL)AMINOCAPROIC ACID DIAZOMETHYL KETONE see THH460
TOSYLCHLORAMIDE SODIUM see CDP000
N-TOSYL-d,l-ISOLEUCINE CHLOROMETHYL KETONE see THH470
N-TOSYL-d,l-ISOLEUCINE DIAZOMETHYL KETONE see THH480
N-TOSYL-l-LEUCINE DIAZOMETHYL KETONE see THH490

N-TOSYL-l-PHENYLALANINE CHLOROMETHYL KETONE see THH550
TOSYL-l-PHENYLALANYLCHLOROMETHYL KETONE see THH550
N-TOSYL-l-VALINE CHLOROMETHYL KETONE see THH555
TOTACILLIN see AIV500
TOTALCICLINA see AIV500
TOTAPEN see AIV500
TOTOMYCIN see TBX250
TOTP see TMO600
2,5-TOULENEDIAMINE SULFATE see TGM400
TOX 47 see PAK000
TOXADUST see CDV100
TOXAFEEN (DUTCH) see CDV100
TOXAKIL see CDV100
TOXAN see CBM750
TOXAPHEN (GERMAN) see CDV100
TOXAPHENE see CDV100
TOXICHLOR see CDR750
TOXILIC ACID see MAK900
TOXILIC ANHYDRIDE see MAM000
TOXIN, BLUE GREEN ALGA, MICROCYSTIS AERUGINOSA see AQV990
TOXIN I (MICROCYSTIS AERUGINOSA) see AQV990
TOXIN-LR see AQV990
TOXIN T2 see FQS000
TOXIN T-17 see THI255
TOXIN T 17 (MICROCYSTIS AERUGINOSA) see AQV990
TOXON 63 see CDV100
TOXYLON POMIFERUM see MRN500
TOXYPHEN see CDV100
TOYO EOSINE G see BNH500
TOYOFINE A see CAW850
TOYO OIL ORANGE see PEJ500
TOYO OIL RED BB see SBC500
TOYO OIL YELLOW G see DOT300
TOYO ORIENTAL OIL BLUE G see BLK000
TP see TBG000
2,4,5-TP see TIX500
TPIA see MCB500
TPN (pesticide) see TBQ750
T 100 (POLYSACCHARIDE) see EHG100
TPP see TMT750
d,d-T80-PRALLETHRIN see THI500
TPTA see ABX250
TPTC see CLU000
TPTH see HON000
TPU 2T see PKO500
TPU 10M see PKM250
TPZA see ABX250
TR 201 see SMR000
TRACHOSEPT see DBX400
TRAFARBIOT see AOD125
TRAGACANTH see THJ250
TRAGACANTH GUM see THJ250
TRAGAYA see SEH000
TRA-KILL TRACHEAL MITE KILLER see MCF750
TRAKIPEAL see MDQ250
TRALGON see HIM000
TRAMACIN see AQX500
TRAMETAN see TFS350
TRAMISOL see LFA020
TRAMISOLE see LFA020
TRANILCYPROMINE see PET750
TRANIMUL see DCK759
TRANITE D-LAY see PBC250

TRAN-Q see CJR909
TRANQUIL see PGA750
TRANQUILAN see MQU750
TRANQUILLIN see BCA000
TRANQUIRIT see DCK759
TRANSAMINE see PET750
TRANSANNON see ECU750
TRANSBRONCHIN see CBR675
TRANSERPIN see RDK000
TRANSIT see CHJ750
TRANSPARENT BRONZE SCARLET see CHP500
TRANSPARENT PIGMENT YELLOW O see PIE830
TRANSPARENT YELLOW 2K see PIE800
TRANSPARENT YELLOW O see PIE830
TRANSPLANTONE see NAK500
TRANYLCYPRAMINE see PET750
TRANYLCYPROMINE see PET750
TRANZINE see CKP500
TRAPANAL see PBT500
TRAPANAL SODIUM see PBT500
TRAPEX see ISE000
TRAPEX-40 see ISE000
TRAPEXIDE see ISE000
TRAPIDIL see DIO200
TRAPYMIN see DIO200
TRAQUIZINE see CJR909
TRASAN see CIR250
TRAUMANASE see BMO000
TRAVAD see BAP000
TRAVELIN see DYE600
TRAVELMIN see DYE600
TRAVELON see HGC500
TRAVEX see SFS000
TRAWOTOX see CDO000
TREFANOCIDE see DUV600
TREFICON see DUV600
TREFLAM see DUV600
TREFLAN see DUV600
TREFLANOCIDE ELANCOLAN see DUV600
TRE-HOLD see NAK500
TRELMAR see MQU750
TREMOLITE ASBESTOS see ARM280
TREN see NEI800
TRENAMINE D-200 see OHY000
TRENAMINE D-201 see OHY000
TRENBOLONE see THL600
TREN HP see NEI800
TRENIMON see TND000
TREOMICETINA see CDP250
TREOSULFAN see TFU500
TREPENOL WA see SIB600
TRESPAPHAN see PMP500
TRESULFAN see TFU500
TRETAMINE see TND500
TRETINOIN see VSK950
TRI-4 see DUV600
TRIABARB see EOK000
TRIACETALDEHYDE (FRENCH) see PAI250
TRIACETIN (FCC) see THM500
TRIACETYL GLYCERIN see THM500
TRIAD see TIO750
TRIADIMEFON see CJO250
TRIADIMENOL see MEP250
TRIAETHANOLAMIN-NG see TKP500
TRIAETHYLAMIN (GERMAN) see TJO000
TRIAETHYLENMELAMIN (GERMAN) see TND500
TRIAETHYLENPHOSPHORSAEUREAMID
(GERMAN) see TND250

TRIAETHYLZINNACETAT (GERMAN) see ABW750
TRIAETHYLZINNSULFAT (GERMAN) see BLN500
TRIALLYLAMINE see THN000
TRIALLYL CYANURATE see THN500
TRIAMCINCOLONE ACETONIDE see AQX500
TRIAMCINOLONE ACETONIDE see AQX500
TRIAMCINOLONE-16,17-ACETONIDE see AQX500
TRIAMELIN see TND500
TRIAMIFOS (GERMAN, DUTCH, ITALIAN) see AIX000
TRI(2-AMINOETHYL)AMINE see NEI800
TRIAMINOGUANIDINE NITRATE see THN800
2,4,6-TRIAMINO-s-TRIAZINE see MCB000
2,4,6-TRIAMINO-1,3,5-TRIAZINE see MCB000
2,2',2''-TRIAMINOTRIETHYLAMINE see NEI800
β,β',β''-TRIAMINOTRIETHYLAMINE see NEI800
4,4'4''-TRIAMINOTRIPHENYLMETHAN-
HYDROCHLORID (GERMAN) see RMK020
2,2',2''-TRIAMINOTRIS(ETHYLAMINE) see NEI800
TRIAMIPHOS see AIX000
TRIAMMINEDIPEROXOCHROMIUM(IV) see THP250
TRIAMMONIUM ALUMINUM HEXAFLUORIDE see
THQ500
TRIAMMONIUM AURINTRICARBOXYLATE see
AGW750
TRIAMMONIUM HEXAFLUOROALUMINATE see
THQ500
TRIAMPHOS see AIX000
TRIANGLE see CBT750
TRIANGLE see CNP250
TRI-p-ANISYLCHLOROETHYLENE see CLO750
TRIANOL DIRECT BLUE 3B see CMO250
TRIANON see PPO000
TRIANTINE LIGHT BROWN 3RN see FMU059
TRIANTOIN see MKB250
TRIAPINE see AMI600
TRIASOL see TIO750
TRIATOMIC OXYGEN see ORW000
3,5,7-TRIAZA-1-AZONIAADAMANTANE, 1-(3-
CHLOROALLYL)-, CHLORIDE see CEG550
1,2,3-TRIAZAINDENE see BDH250
1-TRIAZENE, 1,3-DIETHYL-(9CI) see DKD200
TRIAZICHON (GERMAN) see TND000
TRIAZINE A 1294 see ARQ725
1,3,5-TRIAZINE-2,4-DIAMINE, 6-(2-PYRIDINYL)- see
DCD050
s-TRIAZINE, 2,6-DIAMINO-4-BUTYLAMINO- see
BRR800
s-TRIAZINE, 2,6-DIAMINO-4-(2-PYRIDYL)- see
DCD050
s-TRIAZINE-3,5(2H,4H)-DIONE see THR750
1,3,5-TRIAZINE, HEXAHYDRO-1,3,5-TRINITRO-(9CI)
see CPR800
1,3,5-TRIAZINE-2,4,6-TRIAMINE see MCB000
1,3,5-TRIAZINE-2,4,6-TRIAMINE, N-BUTYL- see
BRR800
s-TRIAZINE, 2,4,6-TRIAMINO- see MCB000
s-TRIAZINE TRICHLORIDE see TJD750
s-TRIAZINE-2,4,6(1H,3H,5H)-TRIONE, DICHLORO-,
POTASSIUM DERIV see PLD000
1,3,5-TRIAZINE-2,4,6(1H,3H,5H)-TRIONE, 1,3-
DICHLORO-, POTASSIUM SALT see PLD000
(1,3,5-TRIAZINE-2,4,6-TRIYLTRINITRILO)HEXAKIS
METHANOL see HDY000
(s-TRIAZINE-2,4,6-
TRIYLTRINITRILO)HEXAMETHANOL see HDY000
1,1',1''-s-TRIAZINE-2,4,6-TRIYLTRISAZIRIDINE see
TND500

TRIAZIQUINONE see TND000
TRIAZIQUONE see TND000
TRIAZIRIDINOPHOSPHINE OXIDE see TND250
2,3,5-TRI-(1-AZIRIDINYL)-p-BENZOQUINONE see TND000
TRI(AZIRIDINYL)PHOSPHINE OXIDE see TND250
TRI(-1-AZIRIDINYL)PHOSPHINE OXIDE see TND250
TRIAZIRIDINYLPHOSPHINE SULFIDE see TFQ750
TRIAZIRIDINYL TRIAZINE see TND500
TRIAZOIC ACID see HHG500
TRIAZOLAMINE see AMY050
1H-1,2,4-TRIAZOL-3-AMINE see AMY050
1,2,4-TRIAZOL-5-ONE, 3-NITRO- see NMP620
(1,2,4)TRIAZOLO(4,3-a)PYRIMIDINE see THT280
TRIAZOTION (RUSSIAN) see EKN000
TRIB see SNK000
TRI-BAN see PIH175
TRIBASIC SODIUM PHOSPHATE see SJH200
TRIBASIC ZINC PHOSPHATE see ZJS400
TRIBROMETHANOL see THV000
TRIBROMETHANOL see ARW250
TRIBROMMETHAAN (DUTCH) see BNL000
TRIBROMMETHAN (GERMAN) see BNL000
TRIBROMOALUMINUM see AGX750
TRIBROMOARSINE see ARF250
TRIBROMOETHANOL see THV000
2,2,2-TRIBROMOETHANOL see ARW250
TRIBROMOETHYL ALCOHOL see THV000
TRIBROMOETHYL ALCOHOL see ARW250
2,2,2-TRIBROMOETHYL ALCOHOL see ARW250
2,4,5-TRIBROMOIMIDAZOLE CADMIUM SALT (2:1) see THV500
TRIBROMOMETAN (ITALIAN) see BNL000
TRIBROMOMETHANE see BNL000
TRIBROMONITROMETHANE see NMQ000
TRIBROMOPHOSPHINE see PHT250
1,2,3-TRIBROMOPROPANE see GGG000
sym-TRIBROMOPROPANE see GGG000
TRIBROMOSTIBINE see AQK000
TRIBROMOTRIMETHYLDIALUMINUM see MGC225
TRIBUTILFOSFATO (ITALIAN) see TIA250
TRIBUTOXYBORANE see THX750
TRI-n-BUTOXYBORANE see THX750
TRI(2-BUTOXYETHANOL PHOSPHATE) see BPK250
TRIBUTOXYETHYL PHOSPHATE see BPK250
TRI(2-BUTOXYETHYL) PHOSPHATE see BPK250
TRIBUTYLAMINE see THX250
TRI-n-BUTYLAMINE see THX250
TRI-n-BUTYL BORANE see THX500
TRIBUTYL BORATE see THX750
TRI-n-BUTYL BORATE see THX750
TRIBUTYLBORINE see THX500
TRIBUTYL CELLOSOLVE PHOSPHATE see BPK250
TRIBUTYLCHLOROSTANNANE see CLP500
TRIBUTYLE (PHOSPHATE de) (FRENCH) see TIA250
TRIBUTYLFOSFAAT (DUTCH) see TIA250
TRIBUTYLISOCYANATOSTANNANE see THY750
TRIBUTYL(METHACRYLOXY)STANNANE see THZ000
TRIBUTYL(METHACRYLOYLOXY)STANNANE see THZ000
TRIBUTYL(METHACRYLOYLOXY)-STANNANE POLYMER with METHYL METHACRYLATE (8CI) see OIY000
TRIBUTYL((2-METHYL-1-OXO-2-PROPENYL)OXY)STANNANE see THZ000
TRIBUTYL(NEODECANOYLOXY)STANNANE see TIF250

TRIBUTYLPHOSPHAT (GERMAN) see TIA250
TRIBUTYL PHOSPHATE see TIA250
TRI-n-BUTYL PHOSPHATE see TIA250
TRIBUTYL-PHOSPHINE compounded with NICKELCHLORIDE (2:1) see BLS250
S,S,S-TRIBUTYL PHOSPHOROTRITHIOATE see BSH250
TRIBUTYL PHOSPHOROTRITHIOITE see TIG250
S,S,S-TRIBUTYL PHOSPHOROTRITHIOITE see TIG250
TRIBUTYL(8-QUINOLINOLATO)TIN see TIB000
TRIBUTYLSTANNANE FLUORIDE see FME000
TRI-n-BUTYLSTANNANE HYDRIDE see TIB500
TRI-n-BUTYL-STANNANE OXIDE see BLL750
TRIBUTYLSTANNIC HYDRIDE see TIB500
TRIBUTYLSTANNYL ISOCYANATE see THY750
TRIBUTYLSTANNYL METHACRYLATE see THZ000
TRIBUTYLTIN BENZOATE see BDR750
TRI-n-BUTYLTIN CHLORIDE see CLP500
TRIBUTYLTIN CYCLOHEXANECARBOXYLATE see TID100
TRIBUTYLTIN DIMETHYLDITHIOCARBAMATE see TID150
TRIBUTYLTIN FLUORIDE see FME000
TRIBUTYLTIN HYDRIDE see TIB500
TRI-n-BUTYLTIN HYDRIDE see TIB500
TRIBUTYLTIN ISOCYANATE see THY750
TRI-n-BUTYLTIN ISOCYANATE see THY750
TRIBUTYLTIN ISOPROPYLSUCCINATE see TIE600
TRIBUTYLTIN ISOTHIOCYANATE see THY750
TRIBUTYLTIN METHACRYLATE see THZ000
TRIBUTYLTIN NEODECANOATE see TIF250
TRIBUTYLTIN OXIDE see BLL750
TRIBUTYLTIN SULFIDE see HCA700
S,S,S-TRIBUTYL TRITHIOPHOSPHATE see BSH250
S,S,S-TRIBUTYL TRITHIOPHOSPHITE see TIG250
TRI-N-BUTYL-ZINN BENZOATE (GERMAN) see BDR750
TRI-n-BUTYLZINN-CHLORID (GERMAN) see CLP500
TRIBUTYRIN see TIG750
TRIBUTYROIN see TIG750
TRICADMIUM DINITRIDE see TIH000
TRICALCIUMARSENAT (GERMAN) see ARB750
TRICALCIUM ARSENATE see ARB750
TRICAPRYLIC GLYCERIDE see TMO000
TRICAPRYLIN see TMO000
TRICAPRYLMETHYLAMMONIUM CHLORIDE see MQH000
TRICAPRYLYLMETHYLAMMONIUM CHLORIDE see MQH000
TRICARBALLYLIC ACID, β-ACETOXYTRIBUTYL ESTER see ADD250
TRICARBAMIX Z see BJK500
TRICARBONYL(METHYLCYCLOPENTADIENYL)MANGANESE see MAV750
TRICARNAM see CBM750
TRICESIUM TRICHLORIDE see CDD000
TRICESIUM TRIFLUORIDE see CDD500
TRICESIUM TRIIODIDE see CDE000
TRICHAZOL see MMN250
TRICHLOORAZIJNZUUR (DUTCH) see TII250
1,1,1-TRICHLOOR-2,2-BIS(4-CHLOOR FENYL)-ETHAAN (DUTCH) see DAD200
2,2,2-TRICHLOOR-1,1-BIS(4-CHLOOR FENYL)-ETHANOL (DUTCH) see BIO750
1,1,1-TRICHLOORETHAAN (DUTCH) see MIH275
TRICHLOORETHEEN (DUTCH) see TIO750

1,1'-(2,2,2-TRICHLOROETHYLIDENE)BIS(4-METHOXYBENZENE) see MEI450
1,2-o-(2,2,2-TRICHLOROETHYLIDENE)-α-d-GLUCOFURANOSE see GFA000
TRI(2-CHLOROETHYL)PHOSPHATE see CGO500
TRI-β-CHLOROETHYL PHOSPHATE see CGO500
TRICHLOROETHYLSILANE see EPY500
TRICHLOROETHYLSILICANE see EPY500
TRICHLOROFLUOROMETHANE see TIP500
TRICHLOROFORM see CHJ500
TRICHLOROHEXADECYLSILANE see HCQ000
TRICHLOROHYDRIN see TJB600
(2,2,2-TRICHLORO-1-HYDROXYETHYL)DIMETHYLPHOSPHONATE see TIQ250
2,2,2-TRICHLORO-1-HYDROXYETHYL-PHOSPHONATE, DIMETHYL ESTER see TIQ250
(2,2,2-TRICHLORO-1-HYDROXYETHYL)PHOSPHONIC ACID DIMETHYL ESTER see TIQ250
TRICHLOROISOCYANIC ACID see TIQ750
TRICHLOROISOCYANURIC ACID see TIQ750
1,3,5-TRICHLOROISOCYANURIC ACID see TIQ750
N,N',N"-TRICHLOROISOCYANURIC ACID see TIQ750
TRICHLOROMETAFOS see RMA500
TRICHLOROMETHANE see CHJ500
TRICHLOROMETHANE SULFENYL CHLORIDE see PCF300
TRICHLOROMETHYLBENZENE see BFL250
1-(TRICHLOROMETHYL)BENZENE see BFL250
p-TRICHLOROMETHYLCHLOROBENZENE see TIR900
TRICHLOROMETHYL CHLOROFORMATE see TIR920
TRICHLOROMETHYL CYANIDE see TII750
N-TRICHLOROMETHYLMERCAPTO-4-CYCLOHEXENE-1,2-DICARBOXIMIDE see CBG000
N-(TRICHLOROMETHYLMERCAPTO)PHTHALIMIDE see TIT250
N-(TRICHLOROMETHYLMERCAPTO)-Δ⁴-TETRAHYDROPHTHALIMIDE see CBG000
TRICHLOROMETHYLNITRILE see TII750
1,1,1-TRICHLORO-2-METHYL-2-PROPANOL see ABD000
TRICHLOROMETHYLSTANNANE see MQC750
TRICHLOROMETHYLSULFENYL CHLORIDE see PCF300
TRICHLOROMETHYLSULPHENYL CHLORIDE see PCF300
N-TRICHLOROMETHYLTHIOCYCLOHEX-4-ENE-1,2-DICARBOXIMIDE see CBG000
N-((TRICHLOROMETHYL)THIO)-4-CYCLOHEXENE-1,2-DICARBOXIMIDE see CBG000
2-((TRICHLOROMETHYL)THIO)-1H-ISOINDOLE-1,3(2H)-DIONE see TIT250
N-(TRICHLOROMETHYLTHIO)PHTHALIMIDE see TIT250
TRICHLOROMETHYLTHIO-1,2,5,6-TETRAHYDROPHTHALAMIDE see CBG000
N-((TRICHLOROMETHYL)THIO)TETRAHYDROPHTHALIMIDE see CBG000
N-TRICHLOROMETHYLTHIO-3A,4,7,7A-TETRAHYDROPHTHALIMIDE see CBG000
N-TRICHLOROMETHYLTHIO-cis-Δ⁴-CYCLOHEXENE-1,2-DICARBOXIMIDE see CBG000
TRICHLOROMETHYLTIN see MQC750
TRICHLOROMONOFLUOROMETHANE see TIP500
TRICHLOROMONOSILANE see TJD500

TRICHLORONAPHTHALENE see TIT500
TRICHLORONAT see EPY000
TRICHLORONITROMETHANE see CKN500
TRICHLOROOCTADECYLSILANE see OBI000
TRICHLOROOXOVANADIUM see VDP000
TRICHLOROPHENE see HCL000
2,4,5-TRICHLOROPHENOL see TIV750
2,4,6-TRICHLOROPHENOL see TIW000
2,4,5-TRICHLOROPHENOL, O-ESTER with O,O-DIMETHYL PHOSPHOROTHIOATE see RMA500
2,4,5-TRICHLOROPHENOL-O-ESTER with O-ETHYL ETHYLPHOSPHONOTHIOATE see EPY000
2-(2,4,5-TRICHLOROPHENOXY)PROPIONIC ACID see TIX500
α-(2,4,5-TRICHLOROPHENOXY)PROPIONIC ACID see TIX500
2,4,5-TRICHLOROPHENOXY-α-PROPIONIC ACID see TIX500
2,4,6-TRICHLOROPHENYL CHLOROFORMATE see TIY800
TRICHLOROPHENYLMETHANE see BFL250
TRICHLOROPHENYLSILANE see TJA750
TRICHLOROPHON see TIQ250
TRICHLOROPHOSPHINE SULFIDE see TFO000
1,2,3-TRICHLOROPROPANE see TJB600
TRICHLOROPROPYLSILANE see PNX250
TRICHLOROSILANE see TJD500
TRICHLOROSTIBINE see AQC500
TRICHLOROTITANIUM see TGG250
α,α,α-TRICHLOROTOLUENE see BFL250
ω,ω,ω-TRICHLOROTOLUENE see BFL250
TRICHLORO-s-TRIAZINE see TJD750
1,3,5-TRICHLOROTRIAZINE see TJD750
2,4,6-TRICHLOROTRIAZINE see TJD750
sym-TRICHLOROTRIAZINE see TJD750
2,4,6-TRICHLORO-s-TRIAZINE see TJD750
2,4,6-TRICHLORO-1,3,5-TRIAZINE see TJD750
TRICHLORO-s-TRIAZINETRIONE see TIQ750
1,3,5-TRICHLORO-1,3,5-TRIAZINETRIONE see TIQ750
TRICHLORO-s-TRIAZINE-2,4,6(1H,3H,5H)-TRIONE see TIQ750
2,2',2"-TRICHLOROTRIETHYLAMINE HYDROCHLORIDE see TNF500
TRICHLOROTRIETHYLDIALUMINIUM see TJP775
TRICHLOROTRIETHYLDIALUMINUM see TJP775
TRICHLOROTRIFLUOROETHANE see FOO000
1,1,2-TRICHLORO-1,2,2-TRIFLUOROETHANE (OSHA, ACGIH, MAK) see FOO000
TRICHLOROTRIMETHYLDIALUMINUM see MGC230
1,3,5-TRICHLORO-2,4,6-TRIOXOHEXAHYDRO-s-TRIAZINE see TIQ750
TRICHLORO(VINYL)SILANE see TIN750
TRICHLOROVINYL SILICANE see TIN750
TRICHLORPHENE see TIQ250
2-(2,4,5-TRICHLOR-PHENOXY)-PROPIONSAEURE (GERMAN) see TIX500
O-(2,4,5-TRICHLOR-PHENYL)-O,O-DIMETHYL-MONOTHIOPHOSPHAT (GERMAN) see RMA500
TRICHLORPHON see TIQ250
TRICHLORPHON FN see TIQ250
TRICHLORSILAN (GERMAN) see TJD500
TRICHLOR-TRIAETHYLAMIN-HYDROCHLORID (GERMAN) see TNF500
TRICHLORURE d'ANTIMOINE see AQC500
TRICHLORURE d'ARSENIC (FRENCH) see ARF500
sym-TRICHLOTRIAZIN (CZECH) see TJD750
TRICHLOURACETONITRIL (DUTCH) see TII750

TRICHOCHROMOGENIC FACTOR see AIH600
TRICHOCID see ABY900
TRICHOCIDE see MMN250
TRICHOFURON see NGG500
TRICHOMAN see ABY900
TRICHOMOL see MMN250
TRICHOMONACID "PHARMACHIM" see MMN250
TRICHOPOL see MMN250
TRICHORAD see ABY900
TRICHORAL see ABY900
TRI-CLENE see TIO750
TRI-CLOR see CKN500
TRICLORETENE (ITALIAN) see TIO750
1,1,1-TRICLORO-2,2-BIS(4-CLORO-FENIL)-ETANO
(ITALIAN) see DAD200
1,1,1-TRICLOROETANO (ITALIAN) see MIH275
TRICLOROETILENE (ITALIAN) see TIO750
O-(2,4,5-TRICLORO-FENIL)-O,O-DIMETIL-
MONOTIOFOSFATO (ITALIAN) see RMA500
TRICLOROMETANO (ITALIAN) see CHJ500
TRICLOROMETILBENZENE (ITALIAN) see BFL250
TRICLORO-NITRO-METANO (ITALIAN) see CKN500
TRICLOROSILANO (ITALIAN) see TJD500
TRICLOROTOLUENE (ITALIAN) see BFL250
TRICOFURON see NGG500
TRICOGEN see ABY900
TRICOLAVAL see ABY900
TRICOM see MMN250
TRICON BW see EIX000
TRICORAL see ABY900
12-TRICOSANONE see TJF250
TRICOSTERIL see ABY900
TRICOWAS B see MMN250
TRICRESILFOSFATI (ITALIAN) see TNP500
TRICRESOL see CNW500
TRICRESYLFOSFATEN (DUTCH) see TNP500
TRICRESYL PHOSPHATE see TMO600
TRICRESYL PHOSPHATE see TNP500
TRI-o-CRESYL PHOSPHATE see TMO600
TRICRESYLPHOSPHATE, with more than 3% ortho
isomer (DOT) see TNP500
TRICYANOGEN CHLORIDE see TJD750
TRICYCLOHEXYLHYDROXYSTANNANE see
CQH650
TRICYCLOHEXYLHYDROXYTIN see CQH650
TRICYCLOHEXYLTIN HYDROXIDE see CQH650
TRICYCLOHEXYLZINNHYDROXID (GERMAN) see
CQH650
TRICYCLO(3.3.1.1.$^{(3,7)}$)DECAN-1-AMINE,
HYDROCHLORIDE (9CI) see AED250
exo-TRICYCLO(5.2.1.0^{2,6})DECANE see TLR675
1-TRIDECANAMINE, N,N-DITRIDECYL- see TNN760
1-TRIDECANECARBOXYLIC ACID see MSA250
1-TRIDECANECARBOXYLIC ACID, ISOPROPYL
ESTER see IQN000
2-TRIDECENAL see TJJ400
TRIDESTRIN see EDU500
TRIDEZIBARBITUR see EOK000
TRI-DIGITOXOSIDE (GERMAN) see DKL800
TRI(DIMETHYLAMINO)PHOSPHINE OXIDE see
HEK000
TRIDIMITE (FRENCH) see SCK000
TRIDIPAM see TFS350
TRIDYMITE see SCI500
TRIDYMITE see SCK000
TRIDYMITE 118 see SCK000
α-TRIDYMITE see SCK000

TRIELINA (ITALIAN) see TIO750
TRIEN see TJR000
TRIENBOLONE see THL600
TRI-ENDOTHAL see EAR000
TRIENOLONE see THL600
TRIENTINE see TJR000
TRI-ETHANE see MIH275
N,N',N"-TRI-1,2-ETHANEDIYL PHOSPHORIC
TRIMIDE see TND250
N,N',N"-TRI-1,2-ETHANEDIYLPHOSPHOROTHIOIC
TRIAMIDE see TFQ750
N,N',N"-TRI-1,2-
ETHANEDIYLTHIOPHOSPHORAMIDE see TFQ750
TRIETHANOLAMIN see TKP500
TRIETHANOLAMINE (ACGIH) see TKP500
TRIETHANOMELAMINE see TND500
1,1,3-TRIETHOXYBUTANE see TJL700
TRIETHOXY(3-ISOCYANATOPROPYL)SILANE see
IKG900
TRIETHOXY-3-KYANPROPYLSILAN see COR800
TRIETHOXYMETHANE see ENY500
TRIETHYL ACETYLCITRATE see ADD750
TRIETHYLALUMINUM see TJN750
TRIETHYLALUMINUM see TJN750
TRIETHYLALUMINUM SESQUICHLORIDE see TJP775
TRIETHYLAMINE see TJO000
TRIETHYLAMINE, 2"-CHLORO-1,1'-DIMETHYL-,
HYDROCHLORIDE see CGV600
TRIETHYLBORANE see TJP250
TRIETHYLBORINE see TJP250
TRIETHYLCHLOROPLUMBANE see TJS250
TRIETHYL CITRATE see TJP750
TRIETHYL CITRATE, ACETATE see ADD750
TRIETHYLDIALUMINUM TRICHLORIDE see TJP775
TRIETHYLENE GLYCOL see TJQ000
TRIETHYLENE GLYCOL DIACRYLATE see TJQ100
TRIETHYLENE GLYCOL ETHYL ETHER see EFL000
TRIETHYLENE GLYCOL MONOETHYL ETHER see
EFL000
TRIETHYLENE GLYCOL MONOHEXYL ETHER see
HFT550
2,3,5-TRIETHYLENEIMINO-1,4-BENZOQUINONE see
TND000
TRI(ETHYLENEIMINO)THIOPHOSPHORAMIDE see
TFQ750
2,4,6-TRIETHYLENEIMINO-s-TRIAZINE see TND500
2,4,6-TRI(ETHYLENEIMINO)-1,3,5-TRIAZINE see
TND500
TRIETHYLENEMELAMINE see TND500
N,N',N"-TRIETHYLENEPHOSPHOROTHIOIC
TRIAMIDE see TFQ750
TRIETHYLENEPHOSPHOROTRIAMIDE see TND250
TRIETHYLENETETRAMINE see TJR000
N,N',N"-TRIETHYLENETHIOPHOSPHAMIDE see
TFQ750
N,N',N"-TRIETHYLENETHIOPHOSPHORAMIDE see
TFQ750
TRIETHYLENETHIOPHOSPHOROTRIAMIDE see
TFQ750
TRIETHYLENIMINOBENZOQUINONE see TND000
2,4,6-TRIETHYLENIMINO-s-TRIAZINE see TND500
2,4,6-TRIETHYLENIMINO-1,3,5-TRIAZINE see
TND500
TRIETHYLESTER KYSELINY ACETYLCITRONOVE
see ADD750
TRIETHYLHYDROXY-STANNANE SULFATE (2:1)
(8CI) see BLN500
TRIETHYLHYDROXYTIN SULFATE see BLN500

TRIETHYL LEAD CHLORIDE see TJS250
TRIETHYLOLAMINE see TKP500
TRIETHYL ORTHOFORMATE see ENY500
TRIETHYL PHOSPHATE see TJT750
TRIETHYLPHOSPHINEAUROUS CHLORIDE see CLQ500
TRIETHYL PHOSPHITE see TJT800
TRIETHYLSTANNIUM BROMIDE see BOI750
TRIETHYLTIN ACETATE see ABW750
TRIETHYL TIN BROMIDE see BOI750
TRIETHYLTIN BROMIDE-2-PIPECOLINE see TJU850
TRIETHYLTIN SULPHATE see BLN500
TRIETHYLTRICHLORODIALUMINUM see TJP775
TRIETILAMINA (ITALIAN) see TJO000
TRIFARON see CJM250
TRIFENOXYFOSFIN (CZECH) see TMU250
TRIFENYLFOSFIT (CZECH) see TMU250
TRIFENYLTINACETAAT (DUTCH) see ABX250
TRIFENYL-TINHYDROXYDE (DUTCH) see HON000
TRIFLIC ACID see TKB310
TRIFLORAN see DUV600
TRIFLORPERAZINE DIHYDROCHLORIDE see TKK250
TRIFLUOPERAZINE HYDROCHLORIDE see TKK250
TRIFLUORACETIC ACID see TKA250
TRIFLUORALIN (USDA) see DUV600
3-(5-TRIFLUORMETHYLPHENYL)-,1-DIMETHYLHARNSTOFF (GERMAN) see DUK800
2-(2,2,2-TRIFLUOROACETAMIDO)-4-(5-NITRO-2-FURYL)THIAZOLE see NGN500
TRIFLUOROACETIC ACID (DOT) see TKA250
TRIFLUOROACETYL CHLORIDE see TJX500
TRIFLUOROAMINE OXIDE see NGS500
TRIFLUOROANTIMONY see AQE000
TRIFLUOROARSINE see ARI250
TRIFLUOROBROMOETHYLENE see BOJ000
TRIFLUOROBROMOMETHANE see TJY100
2,2,2-TRIFLUOROCHLOROETHANE see TJY175
1,1,1-TRIFLUORO-2-CHLOROETHANE see TJY175
TRIFLUOROCHLOROETHYLENE (DOT) see CLQ750
1,1,2-TRIFLUORO-2-CHLOROETHYLENE see CLQ750
TRIFLUOROCHLOROMETHANE (DOT) see CLR250
α,α,α-TRIFLUORO-4-CHLOROTOLUENE see CEM825
1,1,1-TRIFLUORO-2,2-DICHLOROETHANE see TJY500
α,α,α-TRIFLUORO-2,6-DINITRO-N,N-DIPROPYL-p-TOLUIDINE see DUV600
TRIFLUOROETHANOIC ACID see TKA250
2,2,2-TRIFLUOROETHANOL see TKA350
(2,2,2-TRIFLUOROETHOXY)ETHENE see TKB250
1,1,1-TRIFLUOROETHYL CHLORIDE see TJY175
2,2,2-TRIFLUOROETHYL VINYL ETHER see TKB250
(3,3,3-TRIFLUORO-2-HYDROXY-2-(TRIFLUOROMETHYL))PROPYL BENZYL KETONE see TKB285
TRIFLUOROMETHANE see CBY750
TRIFLUOROMETHANE SULFONIC ACID see TKB310
TRIFLUOROMETHANE (UN 1984) (DOT) see CBY750
TRIFLUOROMETHANE, refrigerated, liquid (UN 3136) (DOT) see CBY750
3-(TRIFLUOROMETHYL)ANILINE see AID500
m-(TRIFLUOROMETHYL)ANILINE see AID500
3-(TRIFLUOROMETHYL)BENZENAMINE see AID500
(TRIFLUOROMETHYL)BENZENE see BDH500
2-TRIFLUOROMETHYL BENZIMIDAZOLE see TKB800
TRIFLUOROMETHYL CHLORIDE see CLR250

p-(TRIFLUOROMETHYL)CHLOROBENZENE see CEM825
4-(TRIFLUOROMETHYL)-2,6-DINITRO-N,N-DIPROPYLANILINE see DUV600
3-TRIFLUOROMETHYLNITROBENZENE see NFJ500
m-(TRIFLUOROMETHYL)NITROBENZENE see NFJ500
p-TRIFLUOROMETHYLPHENYL CHLORIDE see CEM825
3-(m-TRIFLUOROMETHYLPHENYL)-1,1-DIMETHYLUREA see DUK800
N-(3-TRIFLUOROMETHYLPHENYL)-N'-N'-DIMETHYLUREA see DUK800
N-(m-TRIFLUOROMETHYLPHENYL)-N',N'-DIMETHYLUREA see DUK800
1-(3-TRIFLUOROMETHYLPHENYL)-2-ETHYLAMINOPROPANE HYDROCHLORIDE see PDM250
(TRIFLUOROMETHYL)TRIFLUOROOXIRANE see HDF050
TRIFLUOROMONOBROMOMETHANE see TJY100
TRIFLUOROMONOCHLOROCARBON see CLR250
TRIFLUOROMONOCHLOROETHYLENE see CLQ750
2,2,2-TRIFLUORO-N-(4-(5-NITRO-2-FURYL)-2-THIAZOLYL)ACETAMIDE see NGN500
α,α,α-TRIFLUORO-m-NITROTOLUENE see NFJ500
TRIFLUOROPERAZINE DIHYDROCHLORIDE see TKK250
TRIFLUOROPHOSPHINE see PHQ500
TRIFLUOROPYRAZIN DIHYDROCHLORIDE see TKK250
TRIFLUORO SELENIUM HEXAFLUORO ARSENATE see TKH250
TRIFLUOROSTANNITE HEXADECYLAMINE see TKH300
TRIFLUOROSTANNITE OF HEXADECYLAMINE see TKH300
TRIFLUOROSTIBINE see AQE000
ω-TRIFLUOROTOLUENE see BDH500
α,α,α-TRIFLUOROTOLUENE see BDH500
(α,α,α-TRIFLUORO-m-TOLYL) ISOCYANATE see TKJ250
TRIFLUORO(TRIFLUOROMETHYL)OXIRANE see HDF050
TRIFLUOROVINYLBROMIDE see BOJ000
TRIFLUOROVINYL CHLORIDE see CLQ750
TRIFLUORURE de CHLORE (FRENCH) see CDX750
TRIFLUPERAZINE DIHYDROCHLORIDE see TKK250
TRIFLUPERIDOL see TKK500
TRIFLUPERIDOLO (ITALIAN) see TKK500
TRIFLUPROMAZINE see TKL000
TRIFLURALIN see DUV600
TRIFLURALINA 600 see DUV600
TRIFLURALINE see DUV600
TRIFOCIDE see DUS700
TRIFORINE see TKL100
TRIFORMOL see PAI000
TRIFOSFAMIDE see TNT500
TRIFRINA see DUS700
TRIFTAZIN see TKK250
TRIFUNGOL see FAS000
TRIFUREX see DUV600
TRIGEN see TJQ000
TRIGLYCINE see AMT500
TRIGLYCOL see TJQ000
TRIGLYCOLLAMIC ACID see AMT500
TRIGLYCOL MONOETHYL ETHER see EFL000

TRIMETHOXYFOSFIN see TMD500
TRIMETHOXYMETHANE see TLX600
3,4,5-TRIMETHOXYPHENETHYLAMINE see MDI500
TRIMETHOXYPHOSPHINE see TMD500
3,4,5-TRIMETHOXY-N-(4-
PROPYLCYCLOHEXYL)BENZAMIDE see TLA650
TRIMETHOXY SILANE see TLB750
TRIMETHOXYSILANE (DOT) see TLB750
N-(3-(TRIMETHOXYSILYL)PROPYL)-1,3-
PROPANEDIAMINE see TLC600
TRIMETHYLACETIC ACID see PJA500
TRIMETHYL ACETYL CHLORIDE (DOT) see DTS400
TRIMETHYLAMINE see TLD500
α-α′,α″-TRIMETHYLAMINETRICARBOXYLIC ACID
see AMT500
TRIMETHYLAMINE, anhydrous (UN 1083) (DOT) see
TLD500
TRIMETHYLAMINE, aqueous solutions not >50%
trimethylamine, by weight (UN 1297) (DOT) see TLD500
1,2,4-TRIMETHYL-5-AMINOBENZENE see TLG250
1,2,4-TRIMETHYL-5-AMINOBENZENE
HYDROCHLORIDE see TLG750
TRIMETHYLAMINOMETHANE see BPY250
2,4,5-TRIMETHYLANILIN (CZECH) see TLG250
2,4,5-TRIMETHYLANILINE see TLG250
2,4,6-TRIMETHYLANILINE see TLG500
2,4,5-TRIMETHYLANILINE HYDROCHLORIDE see
TLG750
2,4,6-TRIMETHYLANILINE HYDROCHLORIDE see
TLH000
TRIMETHYLARSINE SELENIDE see TLH250
7,8,12-TRIMETHYLBENZ(a)ANTHRACENE see
TLK750
5:9:10-TRIMETHYL-1:2-BENZANTHRACENE see
TLK750
2,4,5-TRIMETHYLBENZENAMINE see TLG250
2,4,6-TRIMETHYLBENZENAMINE see TLG500
2,4,5-TRIMETHYLBENZENAMINE
HYDROCHLORIDE see TLG750
2,4,6-TRIMETHYLBENZENAMINE
HYDROCHLORIDE see TLH000
TRIMETHYL BENZENE see TLL250
TRIMETHYL BENZENE see TLL500
as-TRIMETHYL BENZENE see TLL750
1,2,3-TRIMETHYL BENZENE see TLL500
1,2,4-TRIMETHYL BENZENE see TLL750
1,2,5-TRIMETHYL BENZENE see TLL750
1,3,5-TRIMETHYL BENZENE see TLM050
sym-TRIMETHYLBENZENE see TLM050
TRIMETHYL BENZENE (ACGIH) see TLM050
TRIMETHYL BENZENE (mixed isomers) see TLL250
TRIMETHYL BENZOL see TLM050
endo-1,7,7-TRIMETHYL-BICYCLO(2.2.1)HEPTAN-2-OL
see BMD000
1,7,7-TRIMETHYLBICYCLO(2.2.1)-2-HEPTANONE see
CBA750
2,6,6-TRIMETHYLBICYCLO(3.1.1)-2-HEPT-2-ENE see
PIH250
4,6,6-TRIMETHYLBICYKLO(3,1,1)HEPT-3-EN see
PIH250
TRIMETHYL BORATE see TLN000
TRIMETHYLBROMOMETHANE see BQM250
β,γ,γ-TRIMETHYLCAPROALDEHYDE
THIOSEMICARBAZONE see TLN125
TRIMETHYLCARBINOL see BPX000
TRIMETHYLCETYLAMMONIUM BROMIDE see
HCQ500

TRIMETHYL-β-CHLORETHYLAMMONIUMCHLORID
see CMF400
TRIMETHYLCHLOROMETHANE see BQR000
TRIMETHYL CHLOROSILANE see TLN250
TRIMETHYLCHLOROSILANE (DOT) see TLN250
TRIMETHYLCHLOROSTANNANE see CLT000
TRIMETHYLCHLOROTIN see CLT000
3,3,5-TRIMETHYL-5-CYANOCYCLOHEXANONE see
IMG500
1,1,3-TRIMETHYL-3-CYCLOHEXENE-5-ONE see
IMF400
3,5,5-TRIMETHYL-2-CYCLOHEXENE-1-ONE see
IMF400
3,5,5-TRIMETHYL-2-CYCLOHEXEN-1-ON (GERMAN,
DUTCH) see IMF400
4-(2,6,6-TRIMETHYL-1-CYCLOHEXEN-1-YL)-3-
BUTEN-2-ONE see IFX000
4-(2,6,6-TRIMETHYL-2-CYCLOHEXEN-1-YL)-3-
BUTEN-2-ONE see IFW000
1-(2,6,6-TRIMETHYL-2-CYCLOHEXEN-1-YL)-1,6-
HEPTADIEN-3-ONE see AGI500
2,2,4-TRIMETHYL-1,2-DIHYDROCHINOLIN see
TLP500
TRIMETHYL-1,2-DIHYDROQUINOLINE see TLP500
2,2,4-TRIMETHYL-1,2-DIHYDROQUINOLINE see
TLP500
1,7,7-TRIMETHYL-o-(3-(2,6-DIMETHYL-4-
MORPHOLINYL)-2-HYDROXYPROPYL)OXIME
(1R,4R)-BICYCLO(2.2.1)HEPTAN-2-ONE see TLP800
1,3,7-TRIMETHYL-2,6-DIOXOPURINE see CAK500
3,7,11-TRIMETHYL-1,6,10-DODECATRIEN-3-YL
ACETATE see NCN800
TRIMETHYLEENTRINITRAMINE (DUTCH) see
CPR800
TRIMETHYLENE see CQD750
TRIMETHYLENE BROMIDE CHLORIDE see BNA825
TRIMETHYLENE CHLOROBROMIDE see BNA825
TRIMETHYLENE CHLOROHYDRIN see CKP600
TRIMETHYLENEDIAMINE see PMK500
TRIMETHYLENE DICHLORIDE see DGF800
1,3-TRIMETHYLENEDINITRILE see TLR500
TRIMETHYLENEETHYLENEDIAMINE see HGI900
exo-TRIMETHYLENENORBORNANE see TLR675
exo-5,6-TRIMETHYLENENORBORNANE see TLR675
TRIMETHYLENE OXIDE see OMW000
TRIMETHYLENETRINITRAMINE see CPR800
sym-TRIMETHYLENETRINITRAMINE see CPR800
TRIMETHYLENOXID (GERMAN) see OMW000
TRIMETHYLESTER KYSELINY BORITE see TLN000
2,2,4-TRIMETHYL-6-ETHOXY-1,2-
DIHYDROQUINOLINE see SAV000
TRIMETHYLFOSFIT see TMD500
TRIMETHYLGLYCIDYLAMMONIUM CHLORIDE see
GGY200
TRIMETHYL GLYCOL see PML000
N,N,N-TRIMETHYL-1-HEXADECANAMINIUM
BROMIDE see HCQ500
TRIMETHYLHEXADECYLAMMONIUM BROMIDE
see HCQ500
1,7,7-TRIMETHYL-o-(3-(HEXAHYDRO-1H-AZEPIN-1-
YL)-2-HYDROXYPROPYL)OXIME (1R,4R)-
BICYCLO(2.2.1)HEPTAN-2-ONE see TLT150
N-1,5-TRIMETHYL-4-HEXENYLAMINE see ILK000
TRIMETHYLHYDRAZINE HYDROCHLORIDE see
TLT750
TRIMETHYLHYDROQUINONE see POG300
2,3,5-TRIMETHYLHYDROQUINONE see POG300

1,7,7-TRIMETHYL-o-(2-HYDROXY-3-(4-MORPHOLINYL)PROPYL)OXIME (1R,4R)-BICYCLO(2.2.1)HEPTAN-2-ONE see TLT757
1,7,7-TRIMETHYL-o-(2-HYDROXY-3-(1-PYRROLIDINYL)PROPYL)OXIME (1R,4R)-BICYCLO(2.2.1)HEPTAN-2-ONE see TLT763
TRIMETHYL-ε-LACTONE (mixed isomers) see TLY000
4,11,11-TRIMETHYL-8-METHYLENE-5-OXATRICYCLO(8.2.0.0(4,6))DODECANE see CCN100
TRIMETHYLNITROSOHARNSTOFF (GERMAN) see TLU750
N-TRIMETHYL-N-NITROSOUREA see TLU750
1,1,3-TRIMETHYL-3-NITROSOUREA see TLU750
1,7,7-TRIMETHYLNORCAMPHOR see CBA750
TRIMETHYLOLPROPANE TRIACETOACETATE see ELJ600
TRIMETHYL ORTHOFORMATE see TLX600
2,4,5-TRIMETHYL Δ-3-OXAZOLINE see TLX800
TRIMETHYL-2-OXEPANONE (mixed isomers) see TLY000
N,N,N-TRIMETHYLOXIRANEMETHANAMINIUM CHLORIDE see GGY200
2,2,4-TRIMETHYLPENTANE see TLY500
2,4,4-TRIMETHYL-2-PENTANETHIOL see MKJ250
2,4,4-TRIMETHYL-2-PENTANETHIOL see OFE030
6,6,9-TRIMETHYL-3-PENTYL-7,8,9,10-TETRAHYDRO-6H-DIBENZO(B,D)PYRAN-1-OL see TCM250
TRIMETHYLPHENYLMETHANE see BQJ250
TRI-2-METHYLPHENYL PHOSPHATE see TMO600
TRIMETHYL PHOSPHATE see TMD250
O,O,O-TRIMETHYL PHOSPHATE see TMD250
TRIMETHYLPHOSPHINE SELENIDE see TMD400
TRIMETHYL PHOSPHITE see TMD500
TRIMETHYL PHOSPHITE see TMD500
TRIMETHYL PHOSPHOROTRITHIOATE see TMD650
TRIMETHYLPYRAZINE see TME270
2,3,5-TRIMETHYLPYRAZINE see TME270
TRIMETHYLSELENONIUM see TME500
TRIMETHYLSELENONIUM CHLORIDE see TME600
TRIMETHYLSELENONIUM ION see TME500
TRIMETHYLSTANNANE SULPHATE see TMI500
TRIMETHYLSTANNYL CHLORIDE see CLT000
TRIMETHYLTHIOUREA see TMH750
1,1,3-TRIMETHYL-2-THIOUREA see TMH750
N,N,N'-TRIMETHYLTHIOUREA see TMH750
TRIMETHYLTIN ACETATE see TMI000
TRIMETHYLTIN CHLORIDE see CLT000
TRIMETHYLTIN CYANATE see TMI100
TRIMETHYLTIN SULPHATE see TMI500
5,7,8-TRIMETHYLTOCOL see VSZ450
α,α,α'-TRIMETHYLTRIMETHYLENE GLYCOL see HFP875
2,4,6-TRIMETHYL-1,3,5-TRIOXAAN (DUTCH) see PAI250
2,4,6-TRIMETHYL-s-TRIOXANE see PAI250
2,4,6-TRIMETHYL-1,3,5-TRIOXANE see PAI250
s-TRIMETHYLTRIOXYMETHYLENE see PAI250
1,3,5-TRIMETHYL-2,4,6-TRIS(3,5-DI-tert-BUTYL-4-HYDROXYBENZYL) BENZENE see TMJ000
1,3,7-TRIMETHYLXANTHINE see CAK500
2-(TRIMETIL-ACETIL)-INDAN-1,3-DIONE (ITALIAN) see PIH175
3,5,5-TRIMETIL-2-CICLOESEN-1-ONE (ITALIAN) see IMF400
2,4,6-TRIMETIL-1,3,5-TRIOSSANO (ITALIAN) see PAI250
TRIMETION see DSP400

TRIMETON see PMS900
TRIMETON see TMJ750
TRIMETOPRIM-SULFA see SNK000
(+)-TRIMIPRAMINE see TMK100
TRIMITAN see TNF500
TRIMUSTINE see TNF500
TRIMUSTINE HYDROCHLORIDE see TNF500
TRINALGON see NGY000
TRINATRIUMPHOSPHAT (GERMAN) see SJH200
TRINEX see TIQ250
TRINITRIN see NGY000
TRINITROACETONITRILE see TMK250
TRINITROANILINE (DOT) see PIC800
TRINITROBENZEEN see TMK500
TRINITROBENZENE see TMK500
1,3,5-TRINITROBENZENE see TMK500
2,4,6-TRINITROBENZENE-1,3-DIOL see SMP500
2,4,6-TRINITRO-1,3-BENZENEDIOL see SMP500
TRINITROBENZENE, dry or wetted with <30% water, by weight (UN 0214) (DOT) see TMK500
TRINITROBENZENE, wetted with not <30% water, by weight (UN 1354) (DOT) see TMK500
TRINITROBENZOIC ACID (dry) see TML000
TRINITROBENZOIC ACID, dry or wetted with <30% water, by weight (UN 0215) (DOT) see TML000
TRINITROBENZOIC ACID, wetted with not <30% water, by weight (UN 1355) (DOT) see TML000
TRINITROBENZOL (GERMAN) see TMK500
4,4,4-TRINITROBUTYRIC ACID see TML100
γ,γ,γ-TRINITROBUTYRIC ACID see TML100
TRINITROCHLOROBENZENE see TML325
TRINITROCYCLOTRIMETHYLENE TRIAMINE see CPR800
2,4,6-TRINITRO-3,5-DIMETHYL-tert-BUTYLBENZENE see TML750
2,4,6-TRINITRO-1,3-DIMETHYL-5-tert-BUTYLBENZENE see TML750
2,2,2-TRINITROETHANOL see TMM000
TRINITROETHANOL (DOT) see TMM000
2,4,6-TRINITROFENOL (DUTCH) see PID000
2,4,6-TRINITROFENOLO (ITALIAN) see PID000
2,4,7-TRINITROFLUOREN-9-ONE see TMM250
2,4,7-TRINITRO-9-FLUORENONE see TMM250
2,4,7-TRINITROFLUORENONE (MAK) see TMM250
TRINITROGLYCERIN see NGY000
TRINITROGLYCEROL see NGY000
TRINITROL see NGY000
TRINITROMETHANE see TMM500
1,3,5-TRINITROPHENOL see PID000
2,4,6-TRINITROPHENOL AMMONIUM SALT see ANS500
TRINITROPHENOL (UN 0154) (DOT) see PID000
TRINITROPHENOL, wetted with not <30% water, by weight (UN 1344) (DOT) see PID000
TRINITROPHENYLMETHYLNITRAMINE see TEG250
2,4,6-TRINITROPHENYLMETHYLNITRAMINE see TEG250
2,4,6-TRINITROPHENYL-N-METHYLNITRAMINE see TEG250
2,4,6-TRINITROPHENYL NITRAMINE (DOT) see TDY075
2,4,6-TRINITROPHENYL (OSHA) see PID000
TRINITROPYRENE see TMN000
1,3,6-TRINITROPYRENE see TMN000
2,4,6-TRINITRORESORCINOL see SMP500
TRINITRORESORCINOL (DOT) see SMP500

TRINITRORESORCINOL, wetted with less than 20% water (DOT) see SMP500
TRINITRORESORCINOL, DRY (DOT) see SMP500
2,4,6-TRINITROTOLUEEN (DUTCH) see TMN490
TRINITROTOLUENE see TMN490
s-TRINITROTOLUENE see TMN490
2,4,6-TRINITROTOLUENE see TMN490
2,4,6-TRINITROTOLUENE see TMN490
sym-TRINITROTOLUENE see TMN490
2,4,6-TRINITROTOLUENE (ACGIH,OSHA) see TMN490
TRINITROTOLUENE (UN 0209) (DOT) see TMN490
TRINITROTOLUENE, wetted with not <30% water, by weight (UN 1356) (DOT) see TMN490
s-TRINITROTOLUOL see TMN490
sym-TRINITROTOLUOL see TMN490
2,4,6-TRINITROTOLUOL (GERMAN) see TMN490
1,3,5-TRINITRO-1,3,5-TRIAZACYCLOHEXANE see CPR800
TRIOCTANOIN see TMO000
TRIOCTANOYLGLYCEROL see TMO000
TRI-n-OCTYL BORATE see TMO550
TRIOCTYL(BUTYLTHIO)STANNANE see BSO200
TRIOCTYLMETHYLAMMONIUM CHLORIDE see MQH000
TRIODURIN see EDU500
TRIOPEPTIN (HUMAN α-COMPONENT) see HGL680
TRIORTHOCRESYL PHOSPHATE see TMO600
TRIOSSIMETELENE (ITALIAN) see TMP000
TRIOVEX see EDU500
2,6,7-TRIOXA-1-ARSABICYCLO(2.2.2)OCTANE, 4-ETHYL- see EQD100
2,6,7-TRIOXA-1-ARSABICYCLO(2.2.2)OCTANE, 4-ISOPROPYL- see IRQ100
2,6,7-TRIOXA-1-ARSABICYCLO(2.2.2)OCTANE, 4-METHYL- see MQH100
TRIOXANE see TMP000
s-TRIOXANE see TMP000
1,3,5-TRIOXANE see TMP000
sym-TRIOXANE see TMP000
5,8,11-TRIOXAPENTADECANE see DDW200
3,6,9-TRIOXAUNDECANE see DIW800
TRIOXIDE(S) see TGG760
2,6,8-TRIOXOPURINE see UVA400
TRIOXYMETHYLEEN (DUTCH) see TMP000
TRIOXYMETHYLEN (GERMAN) see TMP000
TRIOXYMETHYLENE see PAI000
TRIOXYMETHYLENE see TMP000
2,6,8-TRIOXYPURINE see UVA400
TRIPAN BLUE see CMO250
TRI-PCNB see PAX000
TRIPELENAMINE see TMP750
TRIPELENNAMINA (ITALIAN) see TMP750
TRIPELENNAMINE see TMP750
TRIPERIDOL see TKK500
TRIPHACYCLIN see TBX250
TRIPHENYL see TBD000
m-TRIPHENYL see TBC620
p-TRIPHENYL see TBC750
TRIPHENYLACETO STANNANE see ABX250
TRIPHENYLACRYLONITRILE see TMQ250
2,3,3-TRIPHENYLACRYLONITRILE see TMQ250
α,β,β-TRIPHENYLACRYLONITRILE see TMQ250
TRIPHENYLAMINE see TMQ500
TRIPHENYLANTIMONY see TMV250
TRIPHENYLANTIMONY DICHLORIDE see DGO800
TRIPHENYLANTIMONY OXIDE see TMQ550

TRIPHENYL ANTIMONY SULFIDE see TMQ600
TRIPHENYLCHLOROSTANNANE see CLU000
TRIPHENYLCHLOROTIN see CLU000
TRIPHENYLCYANOETHYLENE see TMQ250
TRIPHENYLETHYLENE see TMS250
1,1,2-TRIPHENYLETHYLENE see TMS250
TRIPHENYL PHOSPHATE see TMT750
TRIPHENYLPHOSPHINE see TMU000
TRIPHENYL PHOSPHITE see TMU250
TRIPHENYL STIBINE see TMV250
TRIPHENYLTIN p-ACETAMIDOBENZOATE see TMV800
TRIPHENYLTIN ACETATE see ABX250
TRIPHENYLTIN CHLORIDE see CLU000
TRIPHENYLTIN CYANOACETATE see TMV825
TRIPHENYLTIN HYDROXIDE (USDA) see HON000
TRIPHENYLTIN OXIDE see HON000
TRIPHENYLTIN PROPIOLATE see TMW600
TRIPHENYL-ZINNACETAT (GERMAN) see ABX250
TRIPHENYL-ZINNHYDROXID (GERMAN) see HON000
TRIPHOSPHORIC ACID, SODIUM SALT see SKN000
TRIPHTHAZINE see TKK250
TRIPHTHAZINE DIHYDROCHLORIDE see TKK250
TRIPIPERIDINOPHOSPHINE SELENIDE see TMX350
TRIPLA-ETILO see DBX400
TRIPLEX III see EIX500
TRI-PLUS see TIO750
TRIPOLI see SCI500
TRIPOLY see SKN000
TRIPOLYPHOSPHATE see SKN000
TRIPOMOL see TFS350
TRIPOTASSIUM CITRATE MONOHYDRATE see PLB750
TRIPOTASSIUM HEXACYANOFERRATE see PLF250
TRIPOTASSIUM HEXACYANOMANGANATE(3-) see TMX600
TRIPOTASSIUM TRICHLORIDE see PLA500
TRIPROPARGYL CYANURATE see THN500
TRIPROPYLALUMINUM see TMY100
TRIPROPYLALUMINUM (DOT) see TMY100
TRI-N-PROPYLAMINE see TMY250
TRIPROPYLAMINE (DOT) see TMY250
TRIPROPYL(BUTYLTHIO)STANNANE see TMY850
TRIPROPYLTIN CHLORIDE see CLU250
TRI-n-PROPYLTIN CHLORIDE see CLU250
2,4,6-TRIPROP-2-YNYLOXY-s-TRIAZINE see THN500
TRIPTIDE see GFW000
TRIPTONE see HOT500
TRIS see TNC500
TRIS (flame retardant) see TNC500
TRIS(ACETYLACETONATO)CHROMIUM see TNN250
TRIS(ACETYLACETONATO)CHROMIUM(III) see TNN250
TRISAETHYLENIMINOBENZOCHINON (GERMAN) see TND000
2,4,6-TRIS(ALLYLOXY)TRIAZINE see THN500
TRIS(AMINOETHYL)AMINE see NEI800
TRIS(β-AMINOETHYL)AMINE see NEI800
TRIS(1-AZIRIDINE)PHOSPHINE OXIDE see TND250
2,3,5-TRIS(AZIRIDINO)-1,4-BENZOQUINONE see TND000
2,3,5-TRIS(1-AZIRIDINO)-p-BENZOQUINONE see TND000
TRIS(AZIRIDINYL)-p-BENZOQUINONE see TND000
TRIS(1-AZIRIDINYL)-p-BENZOQUINONE see TND000

TRISODIUM SALT of 3-CARBOXY-5-HYDROXY-1-SULFOPHENYLAZOPYRAZOLE see FAG140
TRISODIUM TRIFLUORIDE see SHF500
TRISODIUM VERSENATE see TNL250
TRIS(2,4-PENTANEDIONATO)CHROMIUM see TNN250
TRIS(2,4-PENTANEDIONATO)CHROMIUM(3+) see TNN250
TRIS(1-PHENYL-1,3-BUTANEDIONATO)CHROMIUM see TNN500
TRIS(1-PHENYL-1,3-BUTANEDIONATO)CHROMIUM(3+) see TNN500
TRIS(1-PHENYL-1,3-BUTANEDIONATO-O,O')CHROMIUM see TNN500
TRIS(1-PHENYL-1,3-BUTANEDIONO)CHROMIUM(III) see TNN500
TRISPHOSPHAMIDE see TNT500
TRISTAR see DUV600
TRIS(TOLYLOXY)PHOSPHINE OXIDE see TNP500
TRIS(o-TOLYL)-PHOSPHATE see TMO600
TRIS(TRIDECYL)AMINE see TNN760
TRISULFON CONGO RED see SGQ500
TRISULFURATED PHOSPHORUS see PHS500
TRITHENE see CLQ750
TRITHEON see ABY900
TRITHION see TNP250
TRITHION MITICIDE see TNP250
TRITICOL see MHC750
TRITISAN see PAX000
TRITOL see TMN490
TRITOLYL PHOSPHATE see TNP500
TRI-2-TOLYL PHOSPHATE see TMO600
TRI-o-TOLYL PHOSPHATE see TMO600
TRITON GR-5 see DJL000
TRITON X 35 see PKF500
TRITON X-40 see DTC600
TRITON X 45 see PKF500
TRITON X 100 see PKF500
TRITON X 102 see PKF500
TRITON X 165 see PKF500
TRITON X 305 see PKF500
TRITON X 405 see PKF500
TRITON X 705 see PKF500
TRITOX see TII750
TRITRIDECYL AMINE see TNN760
TRI-N-TRIDECYLAMINE see TNN760
TRIVAZOL see MMN250
TRI-VC 13 see DFK600
TRIVINYLTIN CHLORIDE see CLU500
TRIVITAN see CMC750
TRIZILIN see DFT800
TRIZINC DIPHOSPHATE see ZJS400
TROCLOSENE see DGN200
TROCLOSENE POTASSIUM see PLD000
TROCOSONE see ECU750
TROFOSFAMID see TNT500
TROFURIT see CHJ750
TROGUM see SLJ500
TROJCHLOREK FOSFORU (POLISH) see PHT275
TROJCHLOROBENZEN see TIK250
TROJCHLOROBENZEN (POLISH) see TIK250
TROJCHLOROETAN(1,1,2) (POLISH) see TIN000
TROJKREZYLU FOSFORAN (POLISH) see TMO600
TROJNITROTOLUEN (POLISH) see TMN490
TROLAMINE see TKP500
TROLEN see RMA500
TROLENE see RMA500
TROLITUL see SMQ500

TROLOVOL see MCR750
TROMASIN see PAG500
TROMBARIN see BKA000
TROMBAVAR see SIO000
TROMBIL see BKA000
TROMBOLYSAN see BKA000
TROMBOSAN see BJZ000
TROMBOVAR see SIO000
TROMETE see SJH200
TROMETHAMINE PROSTAGLANDIN F2-α see POC750
TROMEXAN see BKA000
TROMEXAN ETHYL ACETATE see BKA000
TRONA see SFO000
TRONA see SJY000
TRONA see BMG400
TRONAMANG see MAP750
TRONOX see TGG760
TROPAEOLIN D see DOU600
TROPAEOLIN G see MDM775
2-β-TROPANECARBOXYLIC ACID, 3-β-HYDROXY-, METHYL ESTER, BENZOATE (ESTER) see CNE750
TROPANE, 3-α-((10,11-DIHYDRO-5H-DIBENZO(A,D)CYCLOHEPTEN-5-YL)OXY)-, CITRATE (1:1) see EAG100
3-TROPANYLBENZOATE-2-CARBOXYLIC ACID METHYL ESTER see CNE750
dl-TROPANYL-2-HYDROXY-1-PHENYLPROPIONATE see ARR000
dl-TROPANYL-2-HYDROXY-1-PHENYLPROPIONATE SULFATE see ARR500
TROPEOLIN see BFL000
TROPHICARDYL see IDE000
TROPHOSPHAMID see TNT500
TROPHOSPHAMIDE see TNT500
TROPIC ACID, ESTER with SCOPINE see SBG000
TROPIC ACID, ESTER with TROPINE see ARR000
(−)-TROPIC ACID ESTER with TROPINE see HOU000
TROPIC ACID, 9-METHYL-3-OXA-9-AZATRICYCLO(3.3.1.0^{2,4})NON-7-YL ESTER see SBG000
TROPIC ACID-3-α-TROPANYL ESTER see ARR000
TROPILIDENE see COY000
TROPILIDIN see COY000
TROPINE BENZOHYDRYL ETHER METHANESULFONATE see TNU000
TROPINE TROPATE see ARR000
TROPINTRAN see ARR500
(±)-TROPYL TROPATE see ARR000
dl-TROPYLTROPATE see ARR000
TROTYL see TMN490
TROTYL OIL see TMN490
TROVIDUR see PKQ059
TROVIDUR see VNP000
TROVIDUR PE see PJS750
TROVITHERN HTL see PKQ059
TROYKYD ANTI-SKIN B see EMU500
TROYKYD ANTI-SKIN BTO see BSU500
TROYSAN 142 see DSB200
TROYSAN COPPER 8% see NAS000
TROYSAN ANTI-MILDEW O see TIT250
TRP-P-1 see TNX275
TRP-P-2 see ALD500
TRP-P-1 (ACETATE) see AJR500
TRUE AMMONIUM SULFIDE see ANJ750
TRUFLEX DOA see AEO000
TRUFLEX DOP see DVL700
TRYCITE 1000 see SMQ500

TRYDET SA SERIES see PJV250
TRYPAFLAVINE see DBX400
TRYPANBLAU (GERMAN) see CMO250
TRYPAN BLUE see CMO250
TRYPAN BLUE SODIUM SALT see CMO250
TRYPARSAMIDE see CBJ750
TRYPOXYL see ARA500
TRYPTAZINE DIHYDROCHLORIDE see TKK250
TRYPTIZOL see EAI000
TRYPTIZOL HYDROCHLORIDE see EAI000
(−)-TRYPTOPHAN see TNX000
l-TRYPTOPHAN see TNX000
dl-TRYPTOPHAN see TNW500
dl-TRYPTOPHAN, pyrolyzate 1 see TNX275
TRYPTOPHANE see TNX000
l-TRYPTOPHAN (FCC) see TNX000
l-TRYPTOPHAN, 5-HYDROXY-, (9CI) see HOA600
TRYPTOPHAN P1 see TNX275
TRYPTOPHAN P2 see ALD500
TS-160 see TNF500
TSAPOLAK 964 see CCU250
TSC see TFQ000
T-SERP see RDK000
TSIKLAMID see ABB000
TSIMAT see BJK500
TSIRAM (RUSSIAN) see BJK500
TSITREX see DXX400
TSIZP 34 see ISR000
TSP see SJH200
TSPA see TFQ750
TSPP see TEE500
TST see EIV000
T-STUFF see HIB050
TSUKUBAENOLIDE HYDRATE see TAA900
TSUMACIDE see MIB750
TSUMAUNKA see MIB750
TTD see TFS350
TTD see DXH250
T-17 TOXIN see THI255
T²-TRICHOTHECENE see FQS000
TTS see DXH250
TTX see FOQ000
TU see TFR250
2-TU see TFR250
TUADS see TFS350
TUAZOLE see MDT250
TUBADIL see TOA000
TUBARINE see TOA000
TUBERIT see CBM000
TUBERITE see CBM000
TUBERSAN see SEP000
TUBOCIN see RKP000
TUBOCURARINE CHLORIDE see TOA000
(+)-TUBOCURARINE CHLORIDE see TOA000
d-TUBOCURARINE CHLORIDE see TOA000
TUBOCURARINE, CHLORIDE, HYDROCHLORIDE, (+)- (8CI) see TOA000
d-TUBOCURARINE DICHLORIDE see TOA000
TUBOCURARINE HYDROCHLORIDE see TOA000
(+)-TUBOCURARINE HYDROCHLORIDE see TOA000
d-TUBOCURARINE HYDROCHLORIDE see TOA000
TUBOTHANE see MAS500
TUBOTIN see ABX250
TUBOTIN see HON000
TUEX see TFS350
TUFF-LITE see PMP500
TUGON see TIQ250
TUGON FLY BAIT see TIQ250

TUGON STABLE SPRAY see TIQ250
TULABASE FAST GARNET GB see AIC250
TULABASE FAST GARNET GBC see AIC250
TULABASE FAST RED TR see CLK220
TULADISPERSE FAST YELLOW 2G see AAQ250
TULISAN see TFS350
TULUYLENDIISOCYANAT see TGM750
TUMBLEAF see SFS000
TUMESCAL OPE see BGJ250
TUMEX see QPA000
TUMOR NECROSIS FACTOR-α see HGL920
TUNG NUT MEALS see TOA500
TUNG NUT OIL see TOA510
TUNGSTEN see TOA750
TUNGSTEN BLUE see TOC750
TUNGSTEN CARBIDE see TOB500
TUNGSTEN COMPOUNDS see TOC500
TUNGSTEN FLUORIDE see TOC550
TUNGSTEN HEXAFLUORIDE see TOC550
TUNGSTEN OXIDE see TOC750
TUNGSTEN TRIOXIDE see TOC750
TUNGSTIC ANHYDRIDE see TOC750
TUNGSTIC OXIDE see TOC750
TUNGSTOPHOSPHORIC ACID (8CI) see PHU750
TUNGSTOPHOSPHORIC ACID, SODIUM SALT see SJJ000
TUNICIN see CCU150
TUPHETAMINE see BBK500
TUR see CMF400
TURBINAIRE see DAE525
TURCAM see DQM600
TURF-CAL see ARB750
TURGEX see HCL000
TURISYNCHRON see MLJ500
TURMERIC see TOD625
TURMERIC OIL see COG000
TURMERIC OLEORESIN see COG000
TURPENTINE see TOD750
TURPENTINE CAMPHOR see BMD300
TURPENTINE OIL see TOD750
TURPENTINE OIL, RECTIFIER see TOD750
TURPENTINE STEAM DISTILLED see TOD750
TURPENTINE SUBSTITUTE (UN 1300) (DOT) see TOD750
TURPENTINE (UN 1299) (DOT) see TOD750
TURPINAL SL see HKS780
TUSSAPAP see HIM000
TUTANE see BPY000
TV 02 see SBL600
TVS 8105 see DVK200
TWEEN 20 see PKG000
TWEEN 60 see PKL030
TWEEN 65 see SKV195
TWEEN 80 see PKL100
TWINKLING STAR see AQF000
TWIN LIGHT RAT AWAY see WAT200
TX 100 see PKF500
TYCLAROSOL see EIV000
TYDEX see BBK500
TYGON see AAX175
TYLAN see TOE600
TYLANDRIL see RDK000
TYLENOL see HIM000
TYLON see TOE600
TYLOSE 444 see MIF760
TYLOSE 666 see SFO500
TYLOSE A4S see MIF760

TYLOSE MF see MIF760
TYLOSE MH see MIF760
TYLOSE MH20 see MIF760
TYLOSE MH50 see MIF760
TYLOSE MH300 see MIF760
TYLOSE MH1000 see MIF760
TYLOSE MH2000 see MIF760
TYLOSE MH300P see MIF760
TYLOSE MH4000 see MIF760
TYLOSE SAP see MIF760
TYLOSE SL see MIF760
TYLOSE SL 100 see MIF760
TYLOSE SL 400 see MIF760
TYLOSE SL 600 see MIF760
TYLOSE TWA see MIF760
TYLOSIN see TOE600
TYLOSIN HYDROCHLORIDE see TOE750
TYLOSIN and SULFAMETHAZINE see TOE810
TYLOSTERONE see DKA600
TYOX A see BHM000
TYOX B see TFD500
TYPOGEN CARMINE see EOJ500
TYRANTON see DBF750
TYRIL see ADY500
TYRIN see PJS750
TYROSINE see TOG300
l-TYROSINE see TOG300
p-TYROSINE see TOG300
l-p-TYROSINE see TOG300
TYVEK see PJS750
U 02 see CBT500
U-14 see POC750
U 46 see CIR500
U46 see DGB000
475U see SMQ500
U625 see SMQ500
666U see SMQ500
U-1149 see FOU000
U 1363 see DVV600
U-1434 see EEI000
U-3886 see SFA000
U-4224 see DSB000
U-4748 see POK000
U 4905 see HHR000
U-5227 see AQN635
U-5897 see CDT750
U-5954 see DOO800
U-5965 see TBX250
U 6020 see PLZ000
U-6062 see CDP250
U-6233 see BDH250
U 6324 see PGN250
U-6421 see NGE500
U-6591 see NOB000
U 6987 see BSM000
U-7743 see CHW675
U-8344 see BIA250
U-8953 see FMM000
U-9889 see SMD000
U-10149 see LGD000
U 10997 see MQS225
U-12062 see DVJ200
U-14583 see POC500
U 15030 see DKC400
U 18496 see ARY000
U 25,352 see CHS250
U 25,354 see DGG400
U 27,151 see BST900

U-32.104 see MHC750
U 1 (polymer) see PKQ059
UA 1 see SCQ000
UA 2 see SCQ000
UA 3 see SCQ000
UA 4 see SCQ000
UBATOL U 2001 see SMQ500
UC 7744 see CBM750
UC-21149 see CBM500
UCAR 17 see EJC500
UCAR 130 see AAX250
UCAR BUTYLPHENOL 4-T see BSE500
UCAR SOLVENT LM (OBS.) see PNL250
UCAR SOLVENT 2LM see DWT200
UCB 170 see HGC500
UCB 492 see CJR909
UCB 1414 see HHK050
UCB 4445 see BOM250
U.CB 4492 see CJR909
UCC 974 see DSB200
UCC 6863 see SMQ500
UCET see CBT750
UCET TEXTILE FINISH 11-74 (OBS.) see VOA000
UC LIQUID G see SCR400
U-COMPOUND see UVA000
UCON 12 see DFA600
UCON 22 see CFX500
UCON 112 see TBP050
UCON 113 see FOO000
UCON 114 see FOO509
UCON FLUOROCARBON 113 see FOO000
UCON 12/HALOCARBON 12 see DFA600
UCON 22/HALOCARBON 22 see CFX500
UCON 113/HALOCARBON 113 see FOO000
UCON 500/HALOCARBON 500 see DFB400
UCON REFRIGERANT 11 see TIP500
UDMH (DOT) see DSF400
UDOLAC see SOA500
U46 DP-FLUID see DGB000
UF 1 see DMI600
UF 15 see SCQ000
UF 021 see IRR050
UJOVIRIDIN see CCK000
UK 14304 see BMM575
UK14304 see BMM575
UK 92480-10 see SCF600
UKARB see CBT750
UKOPEN see AOD125
U 46 KV-ESTER see CIR500
U 46 KV-FLUID see CIR500
ULACORT see PMA000
ULCERFEN see TEH500
ULSTRON see PMP500
ULTRABION see AIV500
ULTRA BRILLIANT BLUE P see DSY600
ULTRABRON see AIV500
ULTRACIDE see DSO000
ULTRACORTEN see PLZ000
ULTRACORTENE-H see PMA000
ULTRAMARINE GREEN see CMJ900
ULTRAMARINE YELLOW see BAK250
ULTRAMID BMK see PJY500
ULTRA SULFATE SL-1 see SIB600
ULTRAWET K see DXW200
ULTRON see PKQ059
ULUP see FMM000
ULVAIR see MRH209
UMBETHION see CNU750

UMBRATHOR see TFT750
UMBRIUM see DCK759
U 46 M-FLUID see CIR250
UNADS see BJL600
UNAMIDE J-56 see BKE500
1,2,3,4,5,5,6,7,9,10,10-
UNDECACHLOROPENTACYCLO(5.3.0.O^{2,6}.O^{3,9}.O^{4,8})D
ECANE see MRI750
5,9-UNDECADIEN-2-ONE, 6,10-DIMETHYL- see
GDE400
UNDECANAL see UJJ000
1-UNDECANAL see UJJ000
n-UNDECANAL see UJJ000
UNDECANALDEHYDE see UJJ000
UNDECANE see UJS000
n-UNDECANE see UJS000
1-UNDECANECARBOXYLIC ACID see LBL000
UNDECANOIC ACID, TRIISOPROPYLSTANNYL
ESTER see TKT850
n-UNDECANOL see UNA000
2-UNDECANONE see UKS000
6-UNDECANONE see ULA000
UNDECAN-6-ONE see ULA000
10-UNDECENAL see ULJ000
1-UNDECEN-10-AL see ULJ000
10-UNDECENAL DIGERANYL ACETAL see UNA100
1-UNDECENE, 11,11-BIS((3,7-DIMETHYL-2,6-
OCTADIENYL)OXY)- see UNA100
10-UNDECENOIC ACID, BUTYL ESTER see BSS100
9-UNDECENOIC ACID, METHYL ESTER see ULS400
2-UNDECENOL see ULS875
UNDECYL ALCOHOL see UNA000
UNDECYL ALDEHYDE see UJJ000
n-UNDECYL ALDEHYDE see UJJ000
UNDECYLENALDEHYDE see ULJ000
10-UNDECYLENEALDEHYDE see ULJ000
UNDECYLENIC ALDEHYDE see ULJ000
UNDECYLENIC ALDEHYDE DIGERANYL ACETAL
see UNA100
UNDECYLIC ALDEHYDE see UJJ000
UNDEN see EDV000
UNFINISHED LUBRICATING OIL see COD750
UNIBARYT see BAP000
UNICEL-ND see DVF400
UNICEL NDX see DVF400
UNICHEM see PKQ059
UNICHLOR see PAH780
UNICHLOR 50 see PAH780
UNICIN see TBX250
UNICROP MANEB see MAS500
UNIDERM WGO see WBJ700
UNIDIGIN see DKL800
UNIDRON see DXQ500
UNIFLEX DOS see BJS250
UNIFOS DYOB S see PJS750
UNIFOS EFD 0118 see PJS750
UNIFUME see EIY500
UNI-GUAR see GLU000
UNILORD see RDK000
UNIMATE GMS see OAV000
UNIMATE IPM see IQN000
UNIMOLL BB see BEC500
UNIMYCETIN see CDP250
UNIMYCIN see TBX250
UNION BLACK EM see AQP000
UNION CARBIDE 7,744 see CBM750
UNION CARBIDE A-150 see TIN750
UNION CARBIDE LIQUID G see SCR400

UNIPINE see PIH750
UNIPON see DGI400
UNIPON see DGI600
UNIROYAL see DFT000
UNIROYAL D014 see SOP000
UNISEDIL see DCK759
UNISOL RH see SFO500
UNISOMNIA see DLY000
UNISTRADIOL see EDP000
UNISULF see SNN500
UNITANE O-110 see TGG760
UNITED see CBT750
UNITED CHEMICAL DEFOLIANT No. 1 see SFS000
UNITENE see MCC250
UNITENSEN see RDK000
UNITENSEN see CKP500
UNITERTRACID LIGHT ORANGE G see HGC000
UNITERTRACID YELLOW TE see FAG140
UNITESTON see TBG000
UNITOX see CDS750
UNIVERM see CBY000
UNIVOL U 316S see MSA250
UNLEADED GASOLINE see GCE100
UNLEADED MOTOR GASOLINE see GCE100
UNON P see TAI250
UNOPROSTONE ISOPROPYL ESTER see IRR050
UNOXAT EPOXIDE 269 see LFV000
UNOX EPOXIDE 206 see VOA000
UP 1 see SMQ500
UP 2 see SMQ500
UP 1E see SMR000
UP 27 see SMQ500
UPIOL see BNP750
UPM see SMQ500
UPM703 see SMQ500
UPM508L see SMQ500
URACIL see UNJ800
URACILLOST see BIA250
URACILMOSTAZA see BIA250
URACIL MUSTARD see BIA250
URACTONE see AFJ500
URADAL see BNK000
URAGAN see SIH500
URAGAN see BMM650
URAGON see BMM650
URALGIN see EID000
URAMINE T 80 see HLU500
URAMUSTIN see BIA250
URAMUSTINE see BIA250
URANIN see FEW000
URANINE A EXTRA see FEW000
URANINE USP XII see FEW000
URANINE YELLOW see FEW000
URANIUM see UNS000
URANIUM ACETATE see UPS000
URANIUM(IV) CHLORIDE see UQJ000
URANIUM FLUORIDE (fissile) see UOJ000
URANIUM FLUORIDE OXIDE see UQA000
URANIUM HEXAFLUORIDE, fissile excepted or non-
fissile (UN 2978) (DOT) see UOJ000
URANIUM HEXAFLUORIDE, fissile (containing >1% U-
235) (UN 2977) (DOT) see UOJ000
URANIUM(III) HYDRIDE see UPA000
URANIUM METAL, pyrophoric (DOT) see UNS000
URANIUM OXYACETATE see UPS000
URANIUM OXYFLUORIDE see UQA000
URANIUM TETRACHLORIDE see UQJ000
URANYL ACETATE see UPS000

URANYL CHLORIDE see URA000
URANYL FLUORIDE see UQA000
URANYL NITRATE (solid) see URA200
URANYL NITRATE HEXAHYDRATE see URS000
URANYL NITRATE HEXAHYDRATE, solution (DOT) see URS000
URAPRINT 62-126 see CAW500
URAZIUM see PDC250
URBANYL see CIR750
URBASON see MOR500
URBASONE see MOR500
URBIL see MQU750
URBOL see ZVJ000
UREA see USS000
UREA DIOXIDE see HIB500
UREA, ETHYL- see EQD875
UREA, 1-ETHYL- see EQD875
UREA HYDROGEN PEROXIDE (DOT) see HIB500
UREA HYDROGEN PEROXIDE SALT see HIB500
UREA HYDROPEROXIDE see HIB500
UREA, MONONITRATE (8CI,9CI) see UTJ000
UREA NITRATE see UTJ000
UREA NITRATE (wet) see UTJ000
UREA NITRATE, dry or wetted with <20% water, by weight (UN 0220) (DOT) see UTJ000
UREA NITRATE, wetted with not <20% water, by weight (UN 1357) (DOT) see UTJ000
UREA PEROXIDE (DOT) see HIB500
UREAPHIL see USS000
p-UREIDOBENZENEARSONIC ACID see CBJ000
(p-UREIDOBENZENEARSYLENEDITHIO)DI-o-BENZOIC ACID see TFD750
4-UREIDO-1-PHENYLARSONIC ACID see CBJ000
(p-UREIDOPHENYLARSYLENEDITHIO)DIACETIC ACID see CBI250
(p-UREIDOPHENYLARSYLENEDITHIO)DI-o-BENZOIC ACID see TFD750
UREOPHIL see USS000
URETAN ETYLOWY (POLISH) see UVA000
URETHAN see UVA000
URETHANE see UVA000
URETHANE POLYMERS see PKL500
URETHYLANE see MHZ000
UREVERT see USS000
UREX see CHJ750
URGENEA MARITIMA see RCF000
URI see SPC500
URIBEN see EID000
URIC ACID see UVA400
URIC ACID, MONOSODIUM SALT see SKO575
URICEMIL see ZVJ000
URICOVAC see DDP200
URIDINAL see PDC250
URINEX see CLH750
URIPLEX see PDC250
URISOXIN see SNN500
URITAS see ZVJ000
URITONE see HEI500
URITRISIN see SNN500
URIZEPT see NGE000
URNER'S LIQUID see DEL000
UROBENYL see ZVJ000
UROBIOTIC-250 see PDC250
URODIAZIN see CFY000
URODINE see PDC250
URODIXIN see EID000
UROFEEN see PDC250
UROGAN see SNN500

UROKINASE see UVS500
UROKINASE (ENZYME-ACTIVATING) see UVS500
UROMALINE see MAP000
UROMAN see EID000
UROMIDE see PDC250
UROMYCINE see GCO000
URONEG see EID000
UROPHENYL see PDC250
UROPYRIDIN see PDC250
UROPYRINE see PDC250
UROSEMIDE see CHJ750
UROSIN see ZVJ000
UROSULFON see SNQ710
UROSULFONE see SNQ710
UROTROPIN see HEI500
UROTROPINE see HEI500
UROX B WATER SOLUBLE CONCENTRATE WEED KILLER see BMM650
UROX HX GRANULAR WEED KILLER see BMM650
URSOFERRAN see IGS000
URSOL BROWN O see CEG600
URSOL BROWN RR see ALL750
URSOL D see PEY500
URSOL EG see ALS990
URSOL ERN see NAW500
URSOL OLIVE 6G see CFK125
URSOL P see ALT250
URSOL P BASE see ALT250
URSOL SLA see DBO400
USACERT FD & C YELLOW NO. 6 see FAG150
USACERT RED No. 1 see FAG018
USACERT YELLOW NO. 5 see FAG140
USACERT YELLOW NO. 6 see FAG150
USAF D-1 see IDW000
USAF D-3 see CHU500
USAF D-5 see BSO500
USAF D-9 see CBM000
USAF M-4 see ISQ000
USAF M-7 see CEI500
USAF P-2 see BJK500
USAF P-5 see TFS350
USAF P-7 see DXQ500
USAF P-8 see CJX750
USAF P-220 see QQS200
USAF A-233 see AIR250
USAF A-4600 see MAO250
USAF A-6598 see DXP200
USAF A-8564 see BIQ500
USAF A-8565 see HIM500
USAF A-9442 see SNE000
USAF A-9789 see DVX600
USAF A-14980 see MIV300
USAF AB-315 see DXJ800
USAF AM-1 see DJI400
USAF AM-3 see EMU500
USAF AM-5 see AAH250
USAF AM-6 see BSU500
USAF AN-7 see MFC700
USAF AN-8 see TDK000
USAF AN-11 see CAM000
USAF B-15 see TFC600
USAF B-17 see BKU500
USAF B-19 see DWC600
USAF B-24 see SAV000
USAF B-30 see TFS350
USAF B-32 see BJL600
USAF B-33 see BDE750
USAF B-33 see DXH250

USAF B-35 see SGF500
USAF B-40 see MRM750
USAF B-44 see DXL800
USAF B-58 see FPM000
USAF B-100 see DJY800
USAF B-121 see MAL250
USAF CB-7 see BAC250
USAF CB-11 see GLS000
USAF CB-13 see FMT000
USAF CB-18 see AEH000
USAF CB-19 see NCG000
USAF CB-20 see TET300
USAF CB-21 see TFA000
USAF CB-22 see NBE500
USAF CB-27 see RDK000
USAF CB-29 see ICW000
USAF CB-30 see THR750
USAF CB-34 see CQJ750
USAF CB-35 see TFJ100
USAF CB-36 see MCM750
USAF CB-37 see MRM750
USAF CB-96 see HOO100
USAF CS-6 see NNM000
USAF CY-2 see CAQ250
USAF CY-4 see NAP500
USAF CY-5 see BDE750
USAF CY-6 see MJN250
USAF CY-7 see BDG000
USAF CY-9 see MES000
USAF CY-10 see NBA500
USAF DO-1 see CEB250
USAF DO-12 see HHR500
USAF DO-21 see EKR500
USAF DO-28 see DMI600
USAF DO-29 see CDY850
USAF DO-36 see DVF200
USAF DO-41 see TNC500
USAF DO-44 see EMU500
USAF DO-45 see AAG000
USAF DO-46 see AKB000
USAF DO-50 see MGG000
USAF DO-54 see MHF750
USAF DO-59 see PGH000
USAF DO-68 see DGK200
USAF E-2 see MCM750
USAF EA-1 see NGG500
USAF EA-2 see NGE000
USAF EA-4 see NGE500
USAF EK see TKO250
USAF EK-3 see AAQ500
USAF EK-206 see PEY750
USAF EK-218 see QMJ000
USAF EK-245 see DWN800
USAF EK-338 see DOT300
USAF EK-356 see HIH000
USAF EK-394 see PEY500
USAF EK-442 see BBL000
USAF EK-488 see ABE500
USAF EK-496 see ABH000
USAF EK-497 see ISR000
USAF EK-510 see TGP250
USAF EK-600 see CBN000
USAF EK-631 see AHR240
USAF EK-704 see ASL250
USAF EK-794 see QPA000
USAF EK-906 see EMU500
USAF EK-1047 see DJC400
USAF EK-1235 see EOL500

USAF EK-1239 see AAY000
USAF EK-1270 see DWC600
USAF EK-1275 see TFQ000
USAF EK-1375 see PEI000
USAF EK-1509 see TGO750
USAF EK-1569 see PGN250
USAF EK-1597 see EEC600
USAF EK-1651 see DXP600
USAF EK-1719 see TFA000
USAF EK-1803 see DKC400
USAF EK-1860 see TFM250
USAF EK-1995 see COH500
USAF EK-2089 see TFS350
USAF EK-2122 see HBD500
USAF EK-2124 see BEU500
USAF EK-2219 see BGJ250
USAF EK-2596 see SGJ000
USAF EK-2676 see TGP000
USAF EK-3092 see DWN200
USAF EK-3302 see ELL500
USAF EK-4037 see AIG000
USAF EK-4196 see MCN250
USAF EK-4376 see AIF500
USAF EK-4394 see DXO200
USAF EK-4628 see HES000
USAF EK-5185 see MMD500
USAF EK-5199 see SKH500
USAF EK-5296 see CKT500
USAF EK-5432 see BDE750
USAF EK-6279 see SIN000
USAF EK-6454 see MPW500
USAF EK-6540 see BCC500
USAF EK-6561 see ALQ000
USAF EK-6583 see MCK000
USAF EK-P-433 see ANW750
USAF EK-P-583 see FOU000
USAF EK-P-737 see TFA500
USAF EK-T-434 see SIA500
USAF EK-P-5976 see AQN635
USAF EK-P-6255 see BJL600
USAF EK-P-6281 see DXL800
USAF EK-T-6645 see BKU500
USAF EL-30 see MCO500
USAF EL-42 see EEH000
USAF EL-62 see IAQ000
USAF EL-101 see BQP250
USAF FO-1 see BQP250
USAF GE-1 see MNV750
USAF GY-2 see BLE500
USAF GY-3 see BDF000
USAF GY-5 see BIX000
USAF GY-7 see BHA750
USAF KE-7 see OAV000
USAF KE-8 see HKJ000
USAF KE-11 see EJM500
USAF KE-20 see HIN500
USAF KF-3 see PES750
USAF KF-5 see CDN500
USAF KF-11 see CEQ600
USAF KF-13 see DVX200
USAF KF-21 see PEA750
USAF KF-25 see EHP500
USAF MA-2 see BDH000
USAF MA-4 see AID500
USAF MA-5 see NFJ500
USAF MA-16 see BDH500
USAF ME-1 see BAD750
USAF MK-6 see DXO200

USAF MO-2 see ANM500
USAF ND-09 see PHY000
USAF ND-54 see DMI600
USAF ND-59 see DMH400
USAF PD-20 see DOT800
USAF Q-1 see FPW000
USAF Q-2 see TCS500
USAF RH-1 see MDN500
USAF RH-3 see DPG600
USAF RH-6 see OFK000
USAF RH-7 see HGP000
USAF RH-8 see MLC750
USAF SC-2 see GKE000
USAF SE-3 see TML100
USAF SN-9 see TEX250
USAF ST-40 see MGA750
USAF SZ-1 see BNM250
USAF TH-9 see DXM600
USAF UCTL-1856 see AMC000
USAF XF-21 see BCC500
USAF XR-19 see PFL850
USAF XR-22 see AMY050
USAF XR-29 see BDF000
USAF XR-41 see CJX750
USAF XR-42 see DXQ500
USALAKE FD & C YELLOW NO. 6 LAKE see FAG150
U.S. BLENDED LIGHT TOBACCO CIGARETTE REFINED TAR see CMP800
U.S.P. MENTHOL see MCG250
USP METHYLCELLULOSE see MIF760
USP SODIUM CHLORIDE see SFT000
USPULUM see CHW675
USP XIII STEARYL ALCOHOL see OAX000
USR 604 see DFT000
U.S. RUBBER 604 see DFT000
U.S. RUBBER D-014 see SOP000
USTINEX see CIR250
UTOSTAN see PDC250
UVALERAL see BNP750
UV CHEK AM 104 see BIW750
UVINUL 400 see DMI600
UVINUL M 40 see MES000
UZONE see BRF500
V-18 see CKM000
825TV see SMQ500
VA 0112 see AAX250
VABROCID see NGE500
VAC see VLU250
VACATE see CIR250
VACOR see PPP750
VACUUM RESIDUUM see MQV755
VADROCID see NGE500
VAFLOL see VSK600
VAGD see AAX175
VAGESTROL see DKA600
VAGILEN see MMN250
VAGILIA see SNN500
VAGIMID see MMN250
VAGISEPT see ABX500
VAGOFLOR see ABX500
VAGOSTIGMIN see DER600
VAGOSTIGMINE BROMIDE see POD000
VALADOL see HIM000
VALAMINA see EEH000
VALAMINE see IIM000
VALAMINETTEN see EEH000
VALDRENE see BAU750
VAL-DROP see SFS000

VALENTINITE see AQF000
VALEO see DCK759
VALERAL see VAG000
n-VALERALDEHYDE see VAG000
VALERAN DI-n-BUTYLCINICITY (CZECH) see DEA600
VALERIANIC ACID see VAQ000
VALERIANIC ALDEHYDE see VAG000
VALERIC ACID see VAQ000
n-VALERIC ACID see VAQ000
VALERIC ACID ALDEHYDE see VAG000
VALERIC ACID, 2-AMINO-3-METHYL- see IKX000
VALERIC ACID, 2-AMINO-4-METHYL- see LES000
VALERIC ALDEHYDE see VAG000
4-VALEROLACTONE see VAV000
γ-VALEROLACTONE (FCC) see VAV000
VALERON see PJS750
VALERONE see DNI800
Δ-VALEROSULTONE see BOU250
VALERYLALDEHYDE see VAG000
VALERYL CHLORIDE see VBA000
4-VALERYLPYRIDINE see VBA100
VALETHAMATE see VBK000
VALETHAMATE BROMIDE see VBK000
VALEXONE see BAT750
VALFLON see TAI250
VALGESIC see HIM000
VALGIS see TEH500
VALGRAINE see TEH500
VALI FAST RED 1308 see RGW000
VALINE see VBP000
VALINE ALDEHDYE see IJS000
l-VALINE (FCC) see VBP000
VALITRAN see DCK759
VALIUM see DCK759
VALKACIT CA see DWN800
VALLADAN see BCA000
VALLAROSOLANOSIDE see VCA100
VALLERGINE see DQA400
VALMID see EEH000
VALMIDATE see EEH000
VALSPEX 155-53 see PJS750
VALZIN see EFE000
VANADIC ACID, AMMONIUM SALT see ANY250
VANADIC ACID, MONOSODIUM SALT see SKP000
VANADIC ACID, POTASSIUM SALT see PLK810
VANADIC(II) ACID, TRISODIUM SALT see SIY250
VANADIC ANHYDRIDE see VDU000
VANADIC OXIDE see VEA000
VANADIO, PENTOSSIDO di (ITALIAN) see VDU000
VANADIUM see VCP000
VANADIUM compounded with BERYLLIUM (1:12) see BFR250
VANADIUM BROMIDE see VEK000
VANADIUM CHLORIDE see VEF000
VANADIUM(III) CHLORIDE see VEP000
VANADIUM COMPOUNDS see VCZ000
VANADIUM DUST and FUME (ACGIH) see VDU000
VANADIUM DUST and FUME (ACGIH) see VDZ000
VANADIUM (OSHA) see VDZ000
VANADIUM OXIDE see VEA000
VANADIUM(V) OXIDE see VDU000
VANADIUM OXYTRICHLORIDE see VDP000
VANADIUM PENTAOXIDE see VDU000
VANADIUMPENTOXID (GERMAN) see VDU000
VANADIUM PENTOXIDE (dust) see VDU000
VANADIUM PENTOXIDE (fume) see VDZ000

VANADIUM PENTOXIDE, non-fused form (DOT) see VDU000
VANADIUM PENTOXIDE, nonfused form (DOT) see VDZ000
VANADIUMPENTOXYDE (DUTCH) see VDU000
VANADIUM, PENTOXYDE de (FRENCH) see VDU000
VANADIUM SESQUIOXIDE see VEA000
VANADIUM TETRACHLORIDE see VEF000
VANADIUM TRIBROMIDE see VEK000
VANADIUM TRICHLORIDE see VEP000
VANADIUM TRICHLORIDE OXIDE see VDP000
VANADIUM TRIOXIDE see VEA000
VANADYL SULFATE see VEZ000
VANADYL TRICHLORIDE see VDP000
VANANOTE see MDW750
VANAY see THM500
VANCIDA TM-95 see TFS350
VANCIDE 89 see CBG000
VANCIDE FE95 see FAS000
VANCIDE KS see HON000
VANCIDE MANEB 80 see MAS500
VANCIDE MZ-96 see BJK500
VANCIDE P see ZMJ000
VANCIDE TM see TFS350
VANDEX see SBO500
VAN DYK 264 see OES000
VAN DYKE BROWN see MAT500
VANGARD K see CBG000
VANGUARD N see BIW750
VANICIDE see CBG000
VANILLA see VFK000
VANILLAL see EQF000
VANILLALDEHYDE see VFK000
VANILLIC ALDEHYDE see VFK000
VANILLIN see VFK000
p-VANILLIN see VFK000
VANILLYL ACETONE see VFP100
VANIROM see EQF000
VANITROPE see IRY000
VANLUBE PCX see BFW750
VANOBID see LFF000
VANOXIDE see BDS000
VANZOATE see BCM000
VAPO-N-ISO see DMV600
VAPONA see DGP900
VAPONEFRIN see VGP000
VAPONITE see DGP900
VAPOPHOS see PAK000
VAPOROLE see IMB000
VAPORPAC see ILM000
VAPOTONE see TCF250
VARAMID ML 1 see BKE500
VARDHAK see NAK500
VARFINE see WAT220
VARIOFORM I see ANN000
VARIOFORM II see USS000
VARISOFT SDC see DTC600
VARITOX see TII250
VARNISH MARKER'S NAPHTHA see PCT250
VARNOLINE see SLU500
VAROX see DRJ800
VAROX DCP-R see DGR600
VAROX DCP-T see DGR600
VASAL see PAH250
VASAZOL see DJS200
VASCARDIN see CCK125
VASCUALS see VSZ450
VASITOL see PBC250

VASOCONSTRICTINE see VGP000
VASOCONSTRICTOR see VGP000
VASODIATOL see PBC250
VASODRINE see VGP000
VASOGLYN see NGY000
VASORBATE see CCK125
VASOSPAN see PAH250
VASOTON see VGP000
VASOTONIN see VGP000
VASOTRATE see CCK125
VASO-80 UNICELLES see PBC250
VASOXINE see MDW000
VASOXINE HYDROCHLORIDE see MDW000
VASOXYL HYDROCHLORIDE see MDW000
VATERITE see CAO000
VAT GRAY S see CMU475
VAT GREY S see CMU475
VATRAN see DCK759
VATROLITE see SHR500
VATSOL OT see DJL000
VAZO 64 see ASL750
V-BRITE see SHR500
VC see VNP000
V-C CHEMICAL V-C 9-104 see EIN000
VCM see VNP000
VCN see ADX500
VC13 NEMACIDE see DFK600
VCR see LEY000
VCR SULFATE see LEZ000
VDC see VPK000
VDF see VPP000
VECTAL see ARQ725
VECTAL SC see ARQ725
VEDRIL see PKB500
VEE GEE GELATIN see PCU360
VEGADEX see CDO250
VEGADEX SUPER see CDO250
VEGETABLE GUM see DBD800
VEGETABLE OIL see VGU200
VEGETABLE OIL MIST (OSHA) see VGU200
VEGETABLE PEPSIN see PAG500
VEGETABLE (SOYBEAN) OIL, brominated see BMO825
VEGFRU see PGS000
VEGFRU FOSMITE see EEH600
VEGFRU MALATOX see MAK700
VEHAM-SANDOZ see EQP000
VEHEM see EQP000
VEL 4283 see MKA000
VEL 4284 see DRR200
VELARDON see PAG500
VELBAN see VLA000
VELBE see VLA000
VELDOPA see DNA200
VELFLON see TAI250
VELMOL see DJL000
VELPAR see HFA300
VELPAR WEED KILLER see HFA300
VELSICOL 104 see HAR000
VELSICOL 506 see LEN000
VELSICOL 1068 see CDR750
VELSICOL COMPOUND "R" see MEL500
VELSICOL 53-CS-17 see EBW500
VELSICOL 58-CS-11 see MEL500
VELSICOL VCS 506 see LEN000
VELUSTRAL KPA see PJS750
VELVETEX see CBT750
VENA see BBV500
VENCIPON see EAW000

VENDACID LIGHT ORANGE 2G see HGC000
VENDEX see BLU000
VENETIAN RED see IHC450
VENTOX see ADX500
VENTRAMINE see DJS200
VENTUROL see DXX400
VENZONATE see BCM000
VEPESID see EAV500
VERATROLE METHYL ETHER see AGE250
VERAX see BBW500
VERAZINC see ZNA000
VERCIDON see DJT800
VERDANTIOL see LFT100
VERDICAN see DGP900
VERDIPOR see DGP900
VERDONE see CIR250
VERESENE DISODIUM SALT see EIX500
VERGFRU FORATOX see PGS000
VERILOID see RDK000
3427 VERI PUR PINK see ADG250
VERMICID see AOR500
VERMICIDE BAYER 2349 see TIQ250
VERMICIDIN see MHL000
VERMIRAX see MHL000
VERMITIN see PDP250
VERMITIN see DFV400
VERMIZYM see PAG500
VERMOESTRICID see CBY000
VERMOX see MHL000
VERNAMYCIN see VRF000
VERNINE see GLS000
VERONAL SODIUM see BAG250
VERON P 130/1 see PKQ059
VEROSPIRON see AFJ500
VEROSPIRONE see AFJ500
VERPANYL see MHL000
VERROL see VSZ450
VERSALIDE see ACL750
VERSENE 9 see TNL250
VERSENE 100 see EIV000
VERSENE ACID see EIX000
VERSENE NTA ACID see AMT500
VERSENE POWDER see EIV000
VERSENE SODIUM 2 see EIX500
VERSNELLER NL 63/10 see DQF800
VERSOMNAL see EOK000
VERSOTRANE see PFN000
TRANS-VERT see MRL750
VERTAC see BQI000
VERTAC see DGI000
VERTAC 90% see CDV100
VERTAC METHYL PARATHION TECHNISCH 80% see MNH000
VERTAC TOXAPHENE 90 see CDV100
VERTHION see DSQ000
VERTIMEC see ARW200
VERTISAL see MMN250
VERTOLAN see SNJ000
VESAKONTUHO MCPA see CIR250
VESPARAX-WIRKSTOFF see HHK050
VESPARAZ-WIRKSTOFF see CJR909
VESPRIN see TKL000
VESTIN see PDC250
VESTINOL AH see DVL700
VESTINOL OA see AEO000
VESTOLEN see PJS750
VESTOLEN A 616 see PJS750
VESTOLEN A 6016 see PJS750

VESTOLEN P 5232G see SMQ500
VESTOLIT B 7021 see PKQ059
VESTROL see TIO750
VESTYRON see SMQ500
VESTYRON 512 see SMQ500
VESTYRON 114-12 see SMQ500
VESTYRON HI see SMR000
VESTYRON MB see SMQ500
VESTYRON N see SMQ500
VETAFLAVIN see DBX400
VETALAR see CKD750
VETALOG see AQX500
VETA-MERAZINE see ALF250
VETAMOX see AAI250
VETARCILLIN see BFC750
VETARSENOBILLON see NCJ500
VETBUTAL see NBU000
VETERINARY NITROFURAZONE see NGE500
VETICILLIN see BFD250
VETICOL see CDP250
VETIDREX see CFY000
VETIOL see MAK700
VETOL see MAO350
VETOX see CBM750
VETQUAMYCIN-324 see TBX250
VETRANQUIL see ABH500
VETSIN see MRL500
VETSTREP see SLY500
VIAGRA see SCF600
VI-ALPHA see VSK600
VIANIN see AOR500
VIANSIN see MDQ250
VIBALT see VSZ000
VIBATEX S see PKP750
VIBAZINE see BOM250
VIBAZINE see HGC500
VIBISONE see VSZ000
VI-CAD see CAE250
VICCILLIN see AIV500
VICCILLIN S see AIV500
VICILLIN see AIV500
VICIN see BFC750
VICKNITE see PLL500
VICTORIA ORANGE see DUT600
VICTORIA SCARLET 3R see FMU080
VICTORIA YELLOW see DUT600
VICTOR TSPP see TEE500
VIDARABIN see AQQ900
VIDARABINE see AQQ900
VIDDEN D see DGG000
VIDDEN D see DGG950
VIDLON see PJY500
VIDOPEN see AOD125
VIENNA GREEN see COF500
VIGANTOL see VSZ100
VIGORSAN see CMC750
VIKANE see SOU500
VIKANE FUMIGANT see SOU500
VILLIAUMITE see SHF000
VILLIAUMITE see SHF500
VINAC B 7 see AAX250
VINACOL MH see PKP750
VINALAK see PKP750
VINAMAR see EQF500
VINAMUL N 710 see SMQ500
VINAMUL N 7700 see SMQ500
VINAROL see PKP750
VINAROL DT see PKP750

VINAROLE see PKP750
VINAROL ST see PKP750
VINAVILOL 2-98 see PKP750
VINBLASTINE SULFATE see VLA000
VINCALEUKOBLASTINE SULFATE see VLA000
VINCALEUKOBLASTINE SULFATE (1:1) (SALT) see VLA000
VINCLOZOLIN (GERMAN) see RMA000
VINCRISTINE see LEY000
VINCRISTINE SULFATE ONCORIN see LEZ000
VINCRISTINSULFAT (GERMAN) see LEZ000
VINCRISUL see LEZ000
VINCRYSTINE see LEY000
VINEGAR ACID see AAT250
VINEGAR NAPHTHA see EFR000
VINEGAR SALTS see CAL750
VINESTHENE see VOP000
VINESTHESIN see VOP000
VINETHEN see VOP000
VINETHENE see VOP000
VINETHER see VOP000
VINICIZER 80 see DVL700
VINICIZER 85 see DVL600
VI-NICOTYL see NCR000
VI-NICTYL see NCR000
VINIDYL see VOP000
VINIKA KR 600 see PKQ059
VINIKULON see PKQ059
VINILE (ACETATO di) (ITALIAN) see VLU250
VINILE (BROMURO di) (ITALIAN) see VMP000
VINILE (CLORURO di) (ITALIAN) see VNP000
VINIPLAST see PKQ059
VINIPLEN P 73 see PKQ059
VINISIL see PKQ250
VINKEIL 100 see EIX000
VINKRISTIN see LEY000
VINNAROL see PKP750
VINNOL E 75 see PKQ059
VINNOL H 10/60 see AAX175
VINOFLEX see PKQ059
VINOFLEX MO 400* see IJQ000
VINOL see PKP750
VINOL 125 see PKP750
VINOL 205 see PKP750
VINOL 351 see PKP750
VINOL 523 see PKP750
VINOL UNISIZE see PKP750
VINOTHIAM see TES750
VINSTOP see SGM500
VINTHIONINE see VLU200
VINYDAN see VOP000
VINYLACETAAT (DUTCH) see VLU250
VINYLACETAT (GERMAN) see VLU250
VINYL ACETATE see VLU250
VINYL ACETATE, inhibited (DOT) see VLU250
VINYL ACETATE HOMOPOLYMER see AAX250
VINYL ACETATE H.Q. see VLU250
VINYL ACETATE POLYMER see AAX250
VINYL ACETATE RESIN see AAX250
VINYL ACETATE–VINYL CHLORIDE COPOLYMER see AAX175
VINYL ACETATE–VINYL CHLORIDE POLYMER see AAX175
VINYL ALCOHOL POLYMER see PKP750
VINYL AMIDE see ADS250
VINYL A MONOMER see VLU250
VINYLBENZEN (CZECH) see SMQ000

VINYLBENZENE see SMQ000
VINYLBENZENE POLYMER see SMQ500
VINYLBENZOL see SMQ000
VINYLBROMID (GERMAN) see VMP000
VINYL BROMIDE see VMP000
VINYL BROMIDE, inhibited (DOT) see VMP000
VINYL BUTYL ETHER see VMZ000
VINYL-n-BUTYL ETHER see VMZ000
VINYL BUTYRATE see VNF000
VINYL BUTYRATE, INHIBITED (DOT) see VNF000
VINYLBUTYROLACTAM see EEG000
N-VINYLBUTYROLACTAM POLYMER see PKQ250
VINYL CARBAMATE see VNK000
VINYLCARBINOL see AFV500
VINYL CARBINYL CINNAMATE see AGC000
VINYLCHLON 4000LL see PKQ059
VINYLCHLORID (GERMAN) see VNP000
VINYL CHLORIDE see VNP000
VINYL CHLORIDE HOMOPOLYMER see PKQ059
VINYL CHLORIDE MONOMER see VNP000
VINYL CHLORIDE POLYMER see PKQ059
VINYL CHLORIDE–VINYL ACETATE POLYMER see AAX175
VINYL-2-CHLOROETHYL ETHER see CHI250
VINYL-β-CHLOROETHYL ETHER see CHI250
VINYL C MONOMER see VNP000
VINYL CYANIDE see ADX500
4-VINYLCYCLOHEXENE see CPD750
1-VINYLCYCLOHEXENE-3 see CPD750
1-VINYLCYCLOHEX-3-ENE see CPD750
4-VINYLCYCLOHEXENE-1 see CPD750
4-VINYL-1-CYCLOHEXENE see CPD750
VINYL CYCLOHEXENE DIEPOXIDE see VOA000
4-VINYLCYCLOHEXENE DIEPOXIDE see VOA000
4-VINYL-1-CYCLOHEXENE DIEPOXIDE see VOA000
4-VINYL-1,2-CYCLOHEXENE DIEPOXIDE see VOA000
VINYL CYCLOHEXENE DIOXIDE see VOA000
4-VINYLCYCLOHEXENE DIOXIDE see VOA000
1-VINYL-3-CYCLOHEXENE DIOXIDE see VOA000
4-VINYL-1-CYCLOHEXENE DIOXIDE (MAK) see VOA000
VINYLE (ACETATE de) (FRENCH) see VLU250
VINYLE (BROMURE de) (FRENCH) see VMP000
VINYLE (CHLORURE de) (FRENCH) see VNP000
VINYLESTER KYSELINY MASELNE see VNF000
VINYLESTER KYSELINY OCTOVE see VLU250
VINYL ETHANOATE see VLU250
VINYL ETHER see VOP000
VINYLETHYLENE see BOP500
VINYL ETHYL ETHER see EQF500
VINYL ETHYL ETHER, inhibited (DOT) see EQF500
VINYLETHYLNITROSAMIN (GERMAN) see NKF000
VINYLETHYLNITROSAMINE see NKF000
VINYL FLUORIDE see VPA000
VINYL FLUORIDE, inhibited (DOT) see VPA000
VINYLFORMIC ACID see ADS750
S-VINYL-dl-HOMOCYSTEINE see VLU200
VINYLIDENE CHLORIDE see VPK000
VINYLIDENE CHLORIDE (II) see VPK000
VINYLIDENE DICHLORIDE see VPK000
VINYLIDENE DIFLUORIDE see VPP000
VINYLIDENE FLUORIDE see VPP000
VINYLIDINE CHLORIDE see VPK000
1-VINYLIMIDAZOLE HOMOPOLYMER see VPP100
N-VINYLIMIDAZOLE HOMOPOLYMER see VPP100
N-VINYLIMIDAZOLE POLYMER see VPP100

VINYL ISOBUTYL ETHER (DOT) see IJQ000
VINYL ISOBUTYL ETHER, inhibited (DOT) see IJQ000
VINYLITE VYDR 21 see AAX175
VINYLKYANID see ADX500
VINYL METHYL ETHER (DOT) see MQL750
VINYL METHYL KETONE see BOY500
VINYLOFOS see DGP900
VINYLON FILM 2000 see PKP750
VINYLOPHOS see DGP900
VINYLOXIRANE see EBJ500
VINYLPHATE see CDS750
α-VINYLPIPERONYL ALCOHOL see BCJ000
VINYL PRODUCTS R 3612 see SMQ500
VINYL PRODUCTS R 10688 see AAX250
VINYL PROPIONATE see VQK000
N-VINYLPYRROLIDINONE see EEG000
1-VINYL-2-PYRROLIDINONE see EEG000
N-VINYL-2-PYRROLIDINONE see EEG000
VINYLPYRROLIDONE see EEG000
N-VINYLPYRROLIDONE see EEG000
1-VINYL-2-PYRROLIDONE see EEG000
N-VINYL-2-PYRROLIDONE (ACGIH) see EEG000
1-VINYL-2-PYRROLIDONE CROSSLINKED
INSOLUBLE POLYMER see PKQ150
N-VINYLPYRROLIDONE POLYMER see PKQ250
VINYLSILICON TRICHLORIDE see TIN750
VINYLSTYRENE see DXQ740
m-VINYLSTYRENE see DXQ745
VINYL TOLUENE see VQK650
3- and 4-VINYL TOLUENE (mixed isomers) see VQK650
VINYL TOLUENE, inhibited mixed isomers (DOT) see
VQK650
VINYL TRICHLORIDE see TIN000
VINYL TRICHLOROSILANE (DOT) see TIN750
VINYL TRICHLOROSILANE, INHIBITED (DOT) see
TIN750
VIOCID see AOR500
VIOFORM see CHR500
VIOFORM N.N.R. see CHR500
VIOFURAGYN see NGG500
VIOLET 3 see XGS000
VIOLET 2S see DBY700
12416 VIOLET see AOR500
VIOLET CP see AOR500
VIOLET GENCIANOVA see AOR500
VIOLET KRYSTALOVA see AOR500
VIOLET LEAF ALCOHOL see NMV780
VIOLET LEAF ALDEHYDE see NMV760
VIOLET 6BN see AOR500
VIOLET 5BO see AOR500
VIOLET XXIII see AOR500
VIOLET ZASADITA 3 see AOR500
VIOPSICOL see MDQ250
VIOPSICOL see LFK000
VIO-SERPINE see RDK000
VIOSTEROL see VSZ100
VIOXAN see CBM750
VIOZENE see RMA500
VI-PAR see CIR500
VI-PEX see CIR500
VIRCHEM see SHR500
VIRCHEM 931 see ZJS400
VIRGIMYCIN see VRF000
VIRGINIA CAROLINA VC 9-104 see EIN000
VIRGINIAMYCIN see VRF000
VIRIDINE see PDX000
VIROFRAL see AED250

VIRORMONE see TBF500
VIROSTERONE see TBF500
VIRSET 656-4 see MCB000
VIRTEX CC see SHR500
VIRTEX D see SHR500
VIRTEX L see SHR500
VIRTEX RD see SHR500
VIRUBRA see VSZ000
VISADRON see NCL500
VISADRON see SPC500
VISCARIN see CCL250
VISCARIN 402 see CAO250
VISCOL see MIF760
VISCOL 350P see PMP500
VISCOLEO OIL see VGU200
VISCONTRAN L52 see MIF760
VISCOSOL see MIF760
VISKING CELLOPHANE see CCT250
VISKO-RHAP see DGB000
VISTABAMATE see MQU750
VITACIN see ARN000
VITAMIN A see VSK600
VITAMIN A1 see VSK600
VITAMIN A ACETATE see VSK900
trans-VITAMIN A ACETATE see VSK900
VITAMIN A ACID see VSK950
13-cis-VITAMIN A ACID see VSK955
VITAMIN A1 ALCOHOL see VSK600
all-trans-VITAMIN A ALCOHOL see VSK600
VITAMIN A ALCOHOL ACETATE see VSK900
9-cis-VITAMIN A ALDEHYDE see VSK975
VITAMIN A PALMITATE see VSP000
VITAMIN B1 see TES750
VITAMIN B2 see RIK000
VITAMIN B3 see NCR000
VITAMIN B4 see AEH000
VITAMIN B-5 see CAU750
VITAMIN B6 see PPK250
VITAMIN B7 see VSU100
VITAMIN B7 see BGD100
VITAMIN Bc see FMT000
VITAMIN B12 (FCC) see VSZ000
VITAMIN B HYDROCHLORIDE see TET300
VITAMIN B6-HYDROCHLORIDE see PPK500
VITAMIN B1 MONONITRATE see TET500
VITAMIN B1 NITRATE see TET500
VITAMIN BX see AIH600
VITAMIN B1 see TET300
VITAMIN B12 COMPLEX see VSZ000
VITAMIN B6 HYDROCHLORIDE see VSU000
VITAMIN C see ARN000
VITAMIN C see ARN125
VITAMIN C SODIUM see ARN125
VITAMIN D2 see VSZ100
VITAMIN D3 see CMC750
VITAMIN E see VSZ450
VITAMIN E ACETATE see TGJ050
VITAMIN G see RIK000
VITAMIN H see VSU100
VITAMIN H see AIH600
VITAMIN H see BGD100
VITAMIN K see VSZ500
VITAMIN K3 see MMD500
VITAMIN K2(O) see MMD500
VITAMIN L see API500
VITAMIN M see FMT000
VITAMIN PP see NCR000
VITAMISIN see ARN000

VITANEURON see TES750
VITAPLEX E see VSZ450
VITAPLEX N see NCQ900
VITARUBIN see VSZ000
VITA-RUBRA see VSZ000
VITASCORBOL see ARN000
VITAVAX see CCC500
VITAVEL-A see VSK600
VITAVEL-D see VSZ100
VITAYONON see VSZ450
VITEOLIN see VSZ450
VITINC DAN-DEE-3 see CMC750
VITON see BBQ500
VITPEX see VSK600
VITRAL see VSZ000
VITRAN see TIO750
VITREOSIL IR see SCK600
VITREOUS QUARTZ see SCK600
VITREOUS SILICA see SCK600
VITREX see PAK000
VITRIFIED SILICA see SCK600
VITRIOL BROWN OIL see SOI500
VITRIOL, OIL OF (DOT) see SOI500
VITRIOL RED see IHC450
VI-TWEL see VSZ000
VIVAL see DCK759
VLB MONOSULFATE see VLA000
VLVF see AAX175
VM-26 see EQP000
VMCC see AAX175
VMI 10-3 see AMI500
VM & P NAPHTHA see PCT250
VM&P NAPHTHA see PCT250
VM and P NAPHTHA see PCT250
VM & P NAPHTHA (ACGIH,OSHA) see PCT250
VN 1 see NDZ000
VOFATOX see MNH000
VOGAN see VSK600
VOGAN-NEU see VSK600
VOGEL'S IRON RED see IHC450
VOLAMIN see EEH000
VOLATILE OIL OF MUSTARD see AGJ250
VOLATON see BAT750
VOLCLAY see BAV750
VOLCLAY BENTONITE BC see BAV750
VOLFARTOL see TIQ250
VOLIDAN see VTF000
VOMEX A see DYE600
VOMISSELS see HGC500
VONAMYCIN POWDER V see NCE000
VONDACEL BLACK N see AQP000
VONDACEL BLUE 2B see CMO000
VONDACEL BLUE FF see CMN750
VONDACEL BLUE HH see CMO500
VONDACEL RED CL see SGQ500
VONDACID METANIL YELLOW G see MDM775
VONDACID TARTRAZINE see FAG140
VONDALHYDE see DMC600
VONDAMOL FAST RED RS see CMM330
VONDCAPTAN see CBG000
VONDODINE see DXX400
VONDRAX see DMC600
VONDURON see DXQ500
VONEDRINE see DBA800
VONTERYL VIOLET 2B see DBY700
VONTERYL YELLOW G see AAQ250
VONTERYL YELLOW R see AAQ250
VOPCOLENE 27 see OHU000

VORLEX see ISE000
VOROX see AMY050
VOROX AA see AMY050
VOROX AS see AMY050
VORTEX see ISE000
VOTEXIT see TIQ250
VP 1940 see ABX250
VP 16213 see EAV500
825TV-PS see SMQ500
V-PYROL see EEG000
V-SERP see RDK000
VT 1 see TGF250
VUAGT 1-4 see DXH325
VUAGT 1964 see DXH325
VUAGT-I-4 see TFS350
VUCINE see EDW500
VULCACEL B-40 see DVF400
VULCACEL BN see DVF400
VULCACID D see DWC600
VULCACURE see SGF500
VULCACURE see BIX000
VULCACURE see BJC000
VULCAFIX SCARLET R see CHP500
VULCAFOR BSM see BDG000
VULCAFOR CBS see CPI250
VULCAFOR HBS see CPI250
VULCAFOR TMTD see TFS350
VULCALENT A see DWI000
VULCAMEL TBN see NAP500
VULCAN see CBT250
VULCAN RED LC see CHP500
VULCATARD see DWI000
VULCOL FAST RED L see CHP500
VUL-CUP see BHL100
VUL-CUP 40KE see BHL100
VUL-CUP R see BHL100
VULKACIT C see CPI250
VULKACIT CZ see CPI250
VULKACIT CZ/C see CPI250
VULKACIT CZ/K see CPI250
VULKACIT D/C see DWC600
VULKACIT DM see BDE750
VULKACIT DM/MGC see BDE750
VULKACIT DOTG/C see DXP200
VULKACIT LDA see BJC000
VULKACIT LDB/C see BIX000
VULKACIT MERCAPTO see BDF000
VULKACIT MTIC see TFS350
VULKACIT NPV/C2 see IAQ000
VULKACIT THIURAM see TFS350
VULKACIT THIURAM/C see TFS350
VULKACIT THIURAM MS/C see BJL600
VULKACIT ZM see BHA750
VULKALENT A (CZECH) see DWI000
VULKANOX 4020 see PEY500
VULKANOX HS/LG see TLP500
VULKANOX HS/POWDER see TLP500
VULKASIL see SCH002
VULKAZIT see DWC600
VULNOPOL NM see SGM500
VULTROL see DWI000
VULVAN see TBG000
VUMON see EQP000
VX see EIG000
VYAC see VLU250
VYDATE see DSP600
VYDATE L INSECTICIDE/NEMATICIDE see DSP600

VYDATE L OXAMYL INSECTICIDE/NEMATOCIDE see DSP600
VYGEN 85 see PKQ059
VYNAMON ORANGE CR see LCS000
VYNAMON SCARLET BY see MRC000
VYNW see AAX175
W 101 see PMP500
W 483 see DIR000
W 524 see TKL100
W 1544 see PFC500
W 1655 see PDC250
W 1929 see SFY500
W-2946M see DNA600
WACHOLDERBEER OEL (GERMAN) see JEA000
WACKER S 14/10 see BJE750
WALNUT STAIN see MAT500
WALSRODER MC 20000S see MIF760
WAMPOCAP see NCQ900
WANADU PIECIOTLENEK (POLISH) see VDU000
WAPNIOWY TLENEK (POLISH) see CAU500
WARAN see WAT220
WARCOUMIN see WAT220
WARDAMATE see MQU750
WARDUZIDE see CLH750
WARECURE C see IAQ000
WARFARIN see WAT200
WARFARINE (FRENCH) see WAT200
WARFARIN SODIUM see WAT220
WARFILONE see WAT220
WARKEELATE ACID see EIX000
WARKEELATE PS-43 see EIV000
WASH OIL see CMY825
WASSERSTOFFPEROXID (GERMAN) see HIB050
WATANIDIPINE HYDROCHLORIDE see WAT230
(S)-(+)-WATANIDIPINE HYDROCHLORIDE see NDY550
WATANIDIPINE HYDROCHLORIDE (S)-ENANTIOMER see NDY550
WATER-d2 (9CI) see HAK000
WATERCARB see CBT500
WATER GLASS see SJU000
WATERSTOFPEROXYDE (DUTCH) see HIB050
WATER²-H2 see HAK000
WATTLE GUM see AQQ500
WAXAKOL ORANGE GL see PEJ500
WAXAKOL RED BL see SBC500
WAXAKOL VERMILION L see XRA000
WAXAKOL YELLOW NL see AIC250
WAX LE see PJS750
WAXOLINE GREEN see BLK000
WAXOLINE ORANGE A see BJF000
WAXOLINE RED O see SBC500
WAXOLINE RED OM see SBC500
WAXOLINE RED OS see SBC500
WAXOLINE YELLOW AD see DOT300
WAXOLINE YELLOW I see PEJ500
WAXOLINE YELLOW O see IBB000
WAXSOL see DJL000
WAYNE RED X-2486 see CHP500
WEATHER COAT 1000 see ZJS400
WEC 50 see TIQ250
WECKAMINE see BBK000
WECOLINE 1295 see LBL000
WECOLINE OO see OHU000
WEDDING BELLS see DJO000
WEECON see COI250
WEED 108 see MRL750
WEEDANOL CYANOL see PLC250

WEEDAR ADS see AMY050
WEEDAR AT see AMY050
WEEDAR MCPA CONCENTRATE see CIR250
WEEDAZIN see AMY050
WEEDAZIN ARGINIT see AMY050
WEEDAZOL see AMY050
WEEDAZOL GP2 see AMY050
WEEDAZOL SUPER see AMY050
WEEDAZOL T see AMY050
WEEDAZOL TL see AMY050
WEEDAZOL TL see ANW750
WEEDBEADS see SJA000
WEED-B-GON see TIX500
WEED DRENCH see AFV500
WEED-E-RAD see MRL750
WEEDEX A see ARQ725
WEEDEX GRANULAT see AMY050
WEED-HOE see MRL750
WEEDOCLOR see AMY050
WEEDONE see PAX250
WEEDONE 128 see IOY000
WEEDONE 170 see DGB000
WEEDONE CRAB GRASS KILLER see PLC250
WEEDONE DP see DGB000
WEEDONE MCPA ESTER see CIR250
WEED-RHAP see CIR250
WEEDTRINE-D see DWX800
WEEVILTOX see CBV500
WEGLA DWUSIARCZEK (POLISH) see CBV500
WEGLA TLENEK (POLISH) see CBW750
WEHYDRYL see BAU750
WEISS PHOSPHOR (GERMAN) see PHP010
WEISSPIESSGLANZ see AQF000
WELDING FUMES see WBJ000
WELFURIN see NGE000
WELVIC G 2/5 see PKQ059
WEPSIN see AIX000
WEPSYN see AIX000
WEPSYN 155 see AIX000
WESCOZONE see BRF500
WESPURIL see MJM500
WEST INDIAN BAY OIL see LBK000
WEST INDIAN LEMONGRASS OIL see LEH000
WESTOCAINE see AIT250
WESTRON see TBQ100
WESTROSOL see TIO750
WETAID SR see DJL000
WET-TONE B see MSB500
WEX 1242 see PMP500
W-GUM see SLJ500
WH 7286 see DMW000
WH 7508 see PIW000
WHATMAN CC-31 see CCU150
WHEAT GERM OIL see WBJ700
WHEAT GLUTEN see WBL100
WHEAT HUSK OIL see WBJ700
WHEY, PROTEIN CONCENTRATE see WBL155
WHEY, REDUCED MINERALS see WBL165
WHISKEY see WBS000
1700 WHITE see TGG760
WHITE ARSENIC see ARI750
WHITE ASBESTOS see ARM268
WHITE ASBESTOS (chrysotile, actinolite, anthophyllite, tremolite) (DOT) see ARM268
WHITE CAMPHOR OIL see CBB500
WHITE CAUSTIC see SHS000
WHITE CAUSTIC, solution see SHS500
WHITE CEDAR OIL see CCQ500

WHITE COPPERAS see ZNA000
WHITE CRYSTAL see SFT000
WHITE LEAD see LCP000
WHITE MERCURY PRECIPITATED see MCW500
WHITE MINERAL OIL see MQV750
WHITE MINERAL OIL see MQV875
WHITE OIL OF CAMPHOR see CBB500
WHITE PHOSPHORUS see PHP010
WHITE PRECIPITATE see MCW500
WHITE SEAL-7 see ZKA000
WHITE SHELLAC see SCC700
WHITE SPIRITS see SLU500
WHITE STAR see AQF000
WHITE STREPTOCIDE see SNM500
WHITE TAR see NAJ500
WHITE VITRIOL see ZNA000
WHITE VITRIOL see ZNJ000
WICKENOL 101 see IQN000
WICKENOL 155 see OFG100
WICKENOL 158 see AEO000
WICKENOL 303 see AHA000
WICKENOL 321 see AHA000
WICKENOL 323 see AHA000
WICKENOL 324 see AHA000
WIDLON see PJY500
WIE OBEN see DJY200
WIJS' CHLORIDE see IDS000
WILKINITE see BAV750
WILLBUTAMIDE see BSQ000
WILLESTROL see DKB000
WILLNESTROL see DAL600
WILLOSETTEN see SJN700
WILPO see DTJ400
WILT PRUF see PKQ059
WILTZ-65 see NAS000
WIN 1539 see KFK000
WIN 4369 see PBS000
WIN 5162 see DMV600
WIN 5606 see BCA000
WIN 18,320 see EID000
WIN 20228 see DOQ400
WIN 22005 see UVS500
WIN 24933 see LIM000
WINACET D see AAX250
WINE see WCA000
WINE ETHER see ENW000
WING STOP B see SGM500
WINIDUR see PKQ059
WIN-KINASE see UVS500
WINTERGREEN OIL (FCC) see MPI000
WINTERGREEN OIL, SYNTHETIC see MPI000
WINTERWASH see DUS700
WINTOMYLON see EID000
WINYLU CHLOREK (POLISH) see VNP000
WIRKSTOFF 37289 see EPY000
WITAMINA PP see NCR000
WITAMOL 320 see AEO000
WITCARB 940 see CBT500
WITCIZER 300 see DEH200
WITCIZER 312 see DVL700
WITCO see CBT750
WITCOBLAK NO. 100 see CBT750
WITCO G 339S see CAX350
WITCONOL MS see OAV000
WITCONOL MST see OAV000
WITCO 1298 SULFONIC ACID see LBU100
WITTOX C see NAS000
WJG 11 see PJS750

WL-5792 see DGM600
WL 18236 see MDU600
WL 19805 see BLW750
WL 43479 see AHJ750
WL 43775 see FAR100
WN 12 see ISE000
WNF 15 see PJS750
WNYESTRON see EDV000
WOCHEM No. 320 see OHU000
WOFATOS see MNH000
WOFATOX see MNH000
WOFAVERDIN see CCK000
WOFOTOX see MNH000
WOJTAB see PLZ000
WOLFRAM see TOA750
WOLFRAMITE see TOC750
WONDER TREE see CCP000
WONUK see ARQ725
WOOD ALCOHOL (DOT) see MGB150
WOOD DUST see WCJ100
WOOD ETHER see MJW500
WOOD NAPHTHA see MGB150
WOOD SPIRIT see MGB150
WOOL FAT see LAU550
WOOLLEY'S ANTISEROTONIN see BEM750
WOOL MORDANT see CMK425
WOOL ORANGE 2G see HGC000
WOOL VIOLET see FAG120
WOOL YELLOW see FAG140
WORM-AGEN see HFV500
WORM-CHEK see LFA020
WORMWOOD see SMY000
WORMWOOD ACID see SMY000
WOTEXIT see TIQ250
WP 155 see AIX000
WR 448 see SOA500
WR 2978 see TGD000
W-13 STABILIZER see SLJ500
WURM-THIONAL see PDP250
WVG 23 see PJS750
WY 509 see DQA400
WY 554 see DAM700
WY-1172 see ABH500
WY-3478 see HJS500
WY-5103 see AIV500
WY-14,643 see CLW250
WY-42956 see BCP250
WYACORT see MOR500
WYCILLINA see BFC750
WYEX see CBT750
WYSEALS see MQU750
X 119 see DTP000
X 149 see BOT250
X 250 see PAE750
X 600 see SMQ500
XA 2 see CFA750
XA 41 see XAA500
X-AB see PKQ059
XALATAN see XAA500
X-ALL Liquid see AMY050
XAMAMINA see DYE600
XANTHACRIDINUM see DBX400
XANTHAN GUM see XAK800
XANTHAURINE see QCA000
XANTHINE-7-N-OXIDE see HOP259
XANTHOTOXIN see XDJ000

XANTHYLIUM, 9-(2-(ETHOXYCARBONYL)PHENYL)-3,6-BIS(ETHYLAMINO)-2,7-DIMETHYL-, CHLORIDE see RGW000
XANTURAT see ZVJ000
XARIL see ABG750
XAXA see ADA725
m-XDI see XIJ000
XE 340 see CBT500
XENENE see BGE000
o-XENOL see BGJ250
XENON see XDS000
XENON (UN 0236) (DOT) see XDS000
XENON, refrigerated liquid (cryogenic liquids) (UN 2591) (DOT) see XDS000
XENYLAMIN (CZECH) see AJS100
XENYLAMINE see AJS100
XENYTROPIUM BROMIDE see PEM750
XERAC see BDS000
XF-13-563 see SCR400
XF 4175L see CBT500
XILENOLI (ITALIAN) see XKA000
XILIDINE (ITALIAN) see XMA000
XILOCAINA (ITALIAN) see DHK400
XILOLI (ITALIAN) see XGS000
XITIX see ARN000
XL-50 see PDP250
XL 1246 see PJS750
XL 335-1 see PJS750
XL ALL INSECTICIDE see NDN000
XNM 68 see PJS750
XO 440 see PJS750
XYCAINE HYDROCHLORIDE see DHK600
XYCIANE see DHK400
XYLAZINE (USDA) see DMW000
XYLENE see XGS000
m-XYLENE see XHA000
o-XYLENE see XHJ000
p-XYLENE see XHS000
1,2-XYLENE see XHJ000
1,3-XYLENE see XHA000
1,4-XYLENE see XHS000
XYLENE BLUE VSG see FMU059
m-XYLENE, 5-tert-BUTYL-2,4,6-TRINITRO- see TML750
m-XYLENE-α,α'-DIAMINE see XHS800
m-XYLENE DIISOCYANATE see XIJ000
XYLENE FAST ORANGE G see HGC000
XYLENE FAST YELLOW GT see FAG140
XYLENE MUSK see TML750
XYLENEN (DUTCH) see XGS000
m-XYLENESULFONIC ACID see XJJ000
2,4-XYLENESULFONIC ACID see XJJ000
m-XYLENE-4-SULFONIC ACID see XJJ000
XYLENOL see XKA000
m-XYLENOL see XKJ500
o-XYLENOL see XKJ000
p-XYLENOL see XKS000
2,3-XYLENOL see XKJ000
2,4-XYLENOL see XKJ500
2,5-XYLENOL see XKS000
3,5-XYLENOL see XLS000
1,2,5-XYLENOL see XKS000
1,3,5-XYLENOL see XLS000
XYLENOLEN (DUTCH) see XKA000
XYLENOLS (DOT) see XKA000
XYLESTESIN see DHK400
XYLESTESIN HYDROCHLORIDE see DHK600
m-XYLIDENE DIISOCYANATE see XIJ000

2,4-XYLIDENE (MAK) see XMS000
XYLIDINE see XMA000
m-XYLIDINE see XMS000
o-XYLIDINE see XMJ000
o-XYLIDINE see XNJ000
2,3-XYLIDINE see XMJ000
2,4-XYLIDINE see XMS000
2,5-XYLIDINE see XNA000
2,6-XYLIDINE see XNJ000
3,4-XYLIDINE see XNS000
3,5-XYLIDINE see XOA000
m-4-XYLIDINE see XMS000
p-XYLIDINE (DOT) see XNA000
XYLIDINEN (DUTCH) see XMA000
3,4-XYLIDINE, N-(2-OXAZOLIN-2-YL)- see XOS500
XYLIDINE PONCEAU see FMU070
3,4-XYLIDINO-2-OXAZOLINE see XOS500
XYLITE see XPJ000
XYLITE (SUGAR) see XPJ000
XYLITOL see XPJ000
XYLITON see XPJ000
l-XYLOASCORBIC ACID see ARN000
XYLOCAIN see DHK400
XYLOCAINE HYDROCHLORIDE see DHK600
XYLOCARD see DHK600
XYLOCITIN see DHK400
XYLOCITIN HYDROCHLORIDE see DHK600
XYLOIDIN see CCU250
o-XYLOL see XHJ000
XYLOL (DOT) see XGS000
m-XYLOL (DOT) see XHA000
p-XYLOL (DOT) see XHS000
XYLOLE (GERMAN) see XGS000
XYLONEURAL see DHK600
XYLOTOX see DHK400
XYLOTOX HYDROCHLORIDE see DHK600
2,3-XYLYLAMINE see XMJ000
2,6-XYLYLAMINE see XNJ000
3,4-XYLYLAMINE see XNS000
3,5-XYLYLAMINE see XOA000
1-XYLYLAZO-2-NAPHTHOL see XRA000
1-(o-XYLYLAZO)-2-NAPHTHOL see XRA000
1-(2,4-XYLYLAZO)-2-NAPHTHOL see XRA000
1-XYLYLAZO-2-NAPHTHOL-3,6-DISULFONIC ACID, DISODIUM SALT see FMU070
1-XYLYLAZO-2-NAPHTHOL-3,6-DISULPHONIC ACID, DISODIUM SALT see FMU070
1-(2,4-XYLYLAZO)-2-NAPHTHOL-3,6-DISULPHONIC ACID, DISODIUM SALT see FMU070
XYLYL BROMIDE see XRS000
m-XYLYLENDIAMIN (CZECH) see XHS800
m-XYLYLENDIISOKYANAT see XIJ000
XYLYLENDIISOKYANAT (CZECH) see XIJ000
o-XYLYLENE DICHLORIDE see DGP400
m-XYLYLENE DIISOCYANATE see XIJ000
p-XYLYLENE ISOCYANATE see XSS260
((2,6-XYLYL)SULFINYL)METHYLCARBAMIC ACID-2,3-DIHYDRO-2,2-DIMETHYL-7-BENZOFURANYL ESTER see XWS100
XYLZIN see DMW000
Y 2 see CBM000
Y-4 see FAG140
Y 40 see SCK600
Y 195 see PKL750
Y 218 see PKM250
Y 221 see PKM500
Y-223 see PKN000
Y-238 see CDV625

Y 302 see PKP000
Y 9030 see IKG900
YADALAN see CLH750
YALTOX see CBS275
YAMAMOTO METHYLENE BLUE B see BJI250
YANOCK see FFF000
YARMOR see PIH750
YARMOR PINE OIL see PIH750
YASOKNOCK see SHG500
YATROCIN see NGE500
YE 5626 see SCQ000
1310 YELLOW see FAG140
1351 YELLOW see FAG150
1409 YELLOW see FAG140
1899 YELLOW see FAG150
11363 YELLOW see MDM775
11712 YELLOW see FEV000
11824 YELLOW see FEW000
12417 YELLOW see FEW000
YELLOW AB see FAG130
YELLOW CROSS LIQUID see BIH250
YELLOW CUPROCIDE see CNO000
YELLOW FERRIC OXIDE see IHC450
YELLOW G SOLUBLE in GREASE see DOT300
YELLOW LAKE 69 see FAG140
YELLOW LEAD OCHER see LDN000
YELLOW MERCURIC OXIDE see MCT500
YELLOW MERCURY IODIDE see MDC750
YELLOW No. 2 see FAG130
YELLOW NO. 5 see FAG140
YELLOW NO. 6 see FAG150
YELLOW NO. 5 FDC see FAG140
YELLOW OB see FAG135
YELLOW ORANGE S see FAG150
YELLOW ORANGE SPECIALLY PURE 85 see FAG150
YELLOW ORANGE S SPECIALLY PURE see FAG150
YELLOW OXIDE of IRON see IHC450
YELLOW OXIDE of MERCURY see MCT500
YELLOW PHOSPHORUS see PHP010
1903 YELLOW PINK see BNH500
YELLOW PRECIPITATE see MCT500
YELLOW PRUSSIATE of SODA see SHE350
YELLOW PYOCTANINE see IBB000
YELLOW RELITON G see AAQ250
YELLOW RESIN see RNU100
YELLOW SF FOR FOOD see FAG150
YELLOW SUN see FAG150
YELLOW SY FOR FOOD see FAG150
YELLOW TRANSPARENT 2K see PIE800
YELLOW TRANSPARENT O see PIE830
YELLOW TRANSPARENT PIGMENT K see PIE800
YELLOW TRANSPARENT PIGMENT O see PIE830
YELLOW ULTRAMARINE see CAP500
YELLOW Z see AAQ250
YH 9030 see IKG900
YM-31636 see IAT275
YODOCHROME METANIL YELLOW see MDM775
YODOMIN see TEH500
YODOXIN see DNF600
YOHIMBAN-16-CARBOXYLIC ACID derivative of
BENZ(g)INDOLO(2,3-a)QUINOLIZINE see RDK000
YOHIMBAN-16-CARBOXYLIC ACID, 17-HYDROXY-,
METHYL ESTER, MONOHYDROCHLORIDE,
(16α,17β)- see YBS500
β-YOHIMBINE HYDROCHLORIDE see YBS500
β-YOHIMBIN HYDROCHLORIDE see YBS500
YOMESAN see DFV400

YOSHINOX 425 see MJN250
YOSHINOX S see TFC600
YPERITE see BIH250
YTTERBIUM see YDA000
YTTERBIUM CHLORIDE see YDJ000
YTTERBIUM NITRATE see YDS800
YTTERBIUM TRICHLORIDE see YDJ000
YTTRIA see YGA000
YTTRIUM see YEJ000
YTTRIUM-89 see YEJ000
YTTRIUM CHLORIDE see YES000
YTTRIUM(III) NITRATE (1:3) see YFJ000
YTTRIUM OXIDE see YGA000
YTTRIUM TRICHLORIDE see YES000
YUGOVINYL see PKQ059
YUKALON EH 30 see PJS750
YUKALON HE 60 see PJS750
YUKALON K 3212 see PJS750
YUKALON LK 30 see PJS750
YUKALON MS 30 see PJS750
YUKALON PS 30 see PJS750
YUKALON YK 30 see PJS750
Z 4828 see TNT500
Z 4942 see IMH000
ZACLONDISCOIDS see HHS000
ZACTIRIN COMPOUND see ABG750
ZACTRAN see DOS000
ZADONAL see EOK000
ZAFFRE see CND125
ZAGREB see BIE500
ZAHARINA see BCE500
ZAHLREICHE BEZEICHNUNGEN (GERMAN) see
DUS700
ZAMINE see BBK500
ZANOSAR see SMD000
ZAPRAWA NASIENNA PLYNNA see MLF250
ZARDA (INDIA) see SED400
ZASSOL see COI250
ZEAPUR see BJP000
ZEARALANOL see RBF100
ZEARANOL see RBF100
ZEAZIN see ARQ725
ZEAZINE see ARQ725
ZECTANE see DOS000
ZECTRAN see DOS000
ZEIDANE see DAD200
ZEIN see ZAT100
ZELAN see CIR250
ZELEN OSTANTHRENOVA BRILANTNI FFB
(CZECH) see JAT000
ZELIO see TEL750
ZENADRID (VETERINARY) see PLZ000
ZENALOSYN see PAN100
ZENITE see BHA750
ZENITE SPECIAL see BHA750
ZENTINIC see PAG200
ZENTRONAL see DKQ000
ZENTROPIL see DKQ000
ZENTROPIL see DNU000
ZEOGEL see PAE750
ZEPHROL see EAW000
ZEPHYR see ARW200
ZERANOL (USDA) see RBF100
ZERDANE see DAD200
ZERLATE see BJK500
ZERTELL see CMA250
ZESET T see VLU250
ZEST see MRL500

ZESTE see ECU750
ZETAR see CMY800
ZETAX see BHA750
ZETIFEX ZN see TNC500
ZEXTRAN see DOS000
ZF 36 see PJS750
Z-GLY see CBR125
ZIARNIK see ABU500
ZIDE see CFY000
ZIMATE see BJK500
ZIMATE METHYL see BJK500
ZIMCO see VFK000
ZIMTALDEHYDE see CMP969
ZIMTSAEURE (GERMAN) see CMP975
ZINC see ZBJ000
ZINC ACETATE plus MANEB (1:50) see EIQ500
ZINC ACID PHOSPHATE see ZJS400
ZINC ARSENATE see ZDJ000
ZINC ARSENATE, BASIC see ZDJ000
ZINC-m-ARSENITE see ZDS000
ZINC ARSENITE, solid (DOT) see ZDS000
ZINC ASHES (UN 1435) (DOT) see ZBJ000
ZINC-2-BENZOTHIAZOLETHIOLATE see BHA750
ZINC BENZOTHIAZOLYL MERCAPTIDE see BHA750
ZINC BENZOTHIAZOL-2-YLTHIOLATE see BHA750
ZINC BENZOTHIAZYL-2-MERCAPTIDE see BHA750
ZINC BERYLLIUM SILICATE see BFV250
ZINC-BIBUTYLDITHIOCARBAMATE see BIX000
ZINC BIS(DIMETHYLDITHIOCARBAMATE) see
BJK500
ZINC
BIS(DIMETHYLDITHIOCARBAMOYL)DISULPHIDE
see BJK500
ZINC,((BIS(d-GLUCONATO-O^1),O^2))- (9CI) see ZIA750
ZINC BUTTER see ZFA000
ZINC CARBONATE (1:1) see ZEJ050
ZINC CHLORATE see ZES000
ZINC CHLORIDE see ZFA000
ZINC CHLORIDE (ACGIH,OSHA) see ZFA000
ZINC CHLORIDE, anhydrous (UN 2331) (DOT) see
ZFA000
ZINC CHLORIDE, solution (UN 1840) (DOT) see
ZFA000
ZINC (CHLORURE de) (FRENCH) see ZFA000
ZINC CHROMATE see ZFJ100
ZINC CHROMATE HYDROXIDE see CMK500
ZINC CHROMATE(VI) HYDROXIDE see ZFJ100
ZINC CHROMATE(VI) HYDROXIDE see CMK500
ZINC CHROMATE, POTASSIUM DICHROMATE, and
ZINC HYDROXIDE (3:1:1) see ZFJ122
ZINC CHROMATE with ZINC HYDROXIDE and
CHROMIUM OXIDE (9:1) see ZFJ125
ZINC CHROME see PLW500
ZINC CHROME YELLOW see ZFJ100
ZINC CHROMIUM OXIDE see ZFJ100
ZINC CITRATE see ZFJ250
ZINC COMPOUNDS see ZFS000
ZINC CYANIDE see ZGA000
ZINC-DIBUTYLDITHIOCARBAMATE see BIX000
ZINC-N,N-DIBUTYLDITHIOCARBAMATE see BIX000
ZINC DICHLORIDE see ZFA000
ZINC DICYANIDE see ZGA000
ZINC DIETHYLDITHIOCARBAMATE see BJC000
ZINC-N,N-DIETHYLDITHIOCARBAMATE see BJC000
ZINC DIMETHYLDITHIOCARBAMATE see BJK500
ZINC N,N-DIMETHYLDITHIOCARBAMATE see
BJK500
ZINC DISTEARATE see ZMS000

ZINC DITHIONITE see ZGJ100
ZINC DITHIONITE see ZIJ100
ZINC DITHIONITE (DOT) see ZIJ100
ZINC DUST see ZBJ000
ZINC DUST (DOT) see ZBJ000
ZINC ETHIDE see DKE600
ZINC ETHYL (DOT) see DKE600
ZINC FLUORIDE see ZHS000
ZINC FLUOROSILICATE see ZIA000
ZINC FLUOROSILICATE (DOT) see ZIA000
ZINC FLUORURE (FRENCH) see ZHS000
ZINC FLUOSILICATE see ZIA000
ZINC GLUCONATE see ZIA750
ZINC HEXAFLUOROSILICATE see ZIA000
ZINC HYDROSULFITE see ZGJ100
ZINC HYDROSULFITE see ZIJ100
ZINC HYDROSULFITE see ZIJ100
ZINC HYDROSULFITE (DOT) see ZGJ100
ZINC HYDROXYCHROMATE see ZFJ100
ZINC HYDROXYCHROMATE see CMK500
ZINCITE see ZKA000
ZINC MANGANESE BERYLLIUM SILICATE see
BFS750
ZINCMATE see BJK500
ZINC MERCAPTOBENZOTHIAZOLATE see BHA750
ZINC-2-MERCAPTOBENZOTHIAZOLE see BHA750
ZINC MERCAPTOBENZOTHIAZOLE SALT see
BHA750
ZINC MERCURY CHROMATE COMPLEX see ZJA000
ZINC METAARSENITE see ZDS000
ZINC METHARSENITE see ZDS000
ZINC MURIATE, solution (DOT) see ZFA000
ZINC NITRATE see ZJJ000
ZINCO (CLORURO di) (ITALIAN) see ZFA000
ZINC OCTADECANOATE see ZMS000
ZINCO (FOSFURO di) (ITALIAN) see ZLS000
ZINCOID see ZKA000
ZINC OMADINE see ZMJ000
ZINC ORTHOPHOSPHATE see ZJS400
ZINC OXIDE see ZKA000
ZINC OXIDE FUME (MAK) see ZKA000
ZINC PERCHLORATE HEXAHYDRATE see ZKS100
ZINC PERMANGANATE see ZLA000
ZINC PEROXIDE see ZLJ000
ZINC PHOSPHATE (3:2) see ZJS400
ZINC PHOSPHIDE see ZLS000
ZINC (PHOSPHURE de) (FRENCH) see ZLS000
ZINCPOLYANEMINE see ZMJ000
ZINC POTASSIUM CHROMATE see CMK400
ZINC POWDER see ZBJ000
ZINC POWDER (DOT) see ZBJ000
ZINC PT see ZMJ000
ZINC PYRIDINE-2-THIOL-1-OXIDE see ZMJ000
ZINC PYRIDINETHIONE see ZMJ000
ZINC PYRION see ZMJ000
ZINC PYRITHIONE see ZMJ000
ZINC STEARATE see ZMS000
ZINC SULFATE see ZNA000
ZINC SULFATE see ZNJ000
ZINC SULFATE (1:1) HEPTAHYDRATE see ZNJ000
ZINC SULFATE HEPTAHYDRATE (1:1:7) see ZNJ000
ZINC SULPHATE see ZNA000
ZINC SUPEROXIDE see ZLJ000
ZINC TETRAOXYCHROMATE 76A see ZFJ100
ZINC-TOX see ZLS000
ZINC VITRIOL see ZNA000
ZINC VITRIOL see ZNJ000
ZINC WHITE see ZKA000

ZINC YELLOW see PLW500
ZINC YELLOW see ZFJ100
ZINC YELLOW see ZFJ125
ZINC YELLOW see CMK500
ZINGERONE see VFP100
ZINGIBERONE see VFP100
ZINK-BIS(N,N-DIMETHYL-DITHIOCARBAMAAT)
(DUTCH) see BJK500
ZINK-BIS(N,N-DIMETHYL-DITHIOCARBAMAT)
(GERMAN) see BJK500
ZINKCARBAMATE see BJK500
ZINKCHLORID (GERMAN) see ZFA000
ZINKCHLORIDE (DUTCH) see ZFA000
ZINK-(N,N-DIMETHYL-DITHIOCARBAMAT)
(GERMAN) see BJK500
ZINKFOSFIDE (DUTCH) see ZLS000
ZINKOSITE see ZNA000
ZINKPHOSPHID (GERMAN) see ZLS000
ZINN (GERMAN) see TGB250
ZINNTETRACHLORID (GERMAN) see TGC250
ZINOPHOS see EPC500
ZINPOL see PJS750
ZIPAN see DCK759
ZIRAM see BJK500
ZIRAMVIS see BJK500
ZIRASAN see BJK500
ZIRBERK see BJK500
ZIRCAT see ZOA000
ZIRCON see ZSS000
ZIRCONATE, BARIUM (1:1) see BAP750
ZIRCONIUM see ZOA000
ZIRCONIUM (ACGIH,OSHA) see ZOA000
ZIRCONIUM CHLORIDE see ZPA000
ZIRCONIUM(IV) CHLORIDE (1:4) see ZPA000
ZIRCONIUM CHLORIDE HYDROXIDE see ZPJ000
ZIRCONIUM CHLOROHYDRATE see ZPJ000
ZIRCONIUM COMPOUNDS see ZQA000
ZIRCONIUM, DICHLORO-DI-pi-
CYCLOPENTADIENYL- see ZTK400
ZIRCONIUM FLUORIDE see ZQS000
ZIRCONIUM GLUCONATE see ZQS100
ZIRCONIUM HYDRIDE see ZRA000
ZIRCONIUM HYDROXYCHLORIDE see ZPJ000
ZIRCONIUM METAL, dry, chemically produced, finer than
20 mesh particle size (UN 2008) see ZOA000
ZIRCONIUM NITRATE see ZSA000
ZIRCONIUM OXYCHLORIDE see ZSJ000
ZIRCONIUM PICRAMATE, dry or wetted with <20%
water, by weight (UN 0236) (DOT) see PIC750
ZIRCONIUM PICRAMATE, wetted with not <20% water,
by weight (UN 1517) (DOT) see PIC750
ZIRCONIUM POTASSIUM FLUORIDE see PLG500
ZIRCONIUM POWDER, dry (UN 2008) (DOT) see
ZOA000
ZIRCONIUM POWDER, wetted with not <25% water
(UN 1358) (DOT) see ZOA000
ZIRCONIUM SCRAP (UN 1932) (DOT) see ZOA000
ZIRCONIUM(IV) SILICATE (1:1) see ZSS000
ZIRCONIUM(IV) SULFATE (1:2) see ZTJ000
ZIRCONIUM TETRACHLORIDE (DOT) see ZPA000
ZIRCONIUM TETRACHLORIDE, solid (DOT) see
ZPA000
ZIRCONIUM TETRAFLUORIDE see ZQS000
ZIRCONIUM, dry, coiled wire, finished metal sheets, strip
(UN 2858) (DOT) see ZOA000
ZIRCONIUM, dry, finished sheets, strip or coiled wire (UN
2009) (DOT) see ZOA000
ZIRCONOCENE, DICHLORIDE see ZTK400

ZIRCONYL CHLORIDE see ZSJ000
ZIRCONYL HYDROXYCHLORIDE see ZPJ000
ZIRCONYL NITRATE see BLA000
ZIRCONYL SULFATE see ZTJ000
ZIREX 90 see BJK500
ZIRIDE see BJK500
ZIRPON see MQU750
ZIRTHANE see BJK500
ZITEX H 662-124 see TAI250
ZITHIOL see MAK700
ZITOX see BJK500
ZITRONEN OEL (GERMAN) see LEI000
ZLUT DISPERZNI 3 see AAQ250
ZLUT KYSELA 23 see FAG140
ZLUT MARCIOVA see DUX800
ZLUT MASELNA (CZECH) see DOT300
ZLUT MASELNA OB see FAG135
ZLUT NAFTOLOVA see DUX800
ZLUT PIGMENT 100 see FAG140
ZLUT POTRAVINARSKA 3 see FAG150
ZLUT POTRAVINARSKA 4 see FAG140
ZLUT PRIRODNI 8 see MRN500
ZLUT PRIRODNI 11 see MRN500
ZLUT ROZPOUSTEDLOVA 6 see FAG135
ZLUT ROZPOUSTEDLOVA 77 see AAQ250
ZMA see ZDS000
ZMBT see BHA750
Z-METHYLPROPYL ACRYLATE see IIK000
ZnMB see BHA750
ZOALENE see DUP300
ZOAMIX see DUP300
ZOAQUIN see DNF600
ZOBA BLACK D see PEY500
ZOBA BROWN P BASE see ALT250
ZOBA BROWN RR see ALL750
ZOBA EG see ALS990
ZOBA ERN see NAW500
ZOBA 3GA see ALT000
ZOBA GKE see TGL750
ZOBA SLE see DBO400
ZOGEN DEVELOPER H see TGL750
ZOLAPHEN see BRF500
ZOLIDINUM see BRF500
ZOLON see BDJ250
ZOLONE see BDJ250
ZOLONE PM see BDJ250
ZOOCOUMARIN (RUSSIAN) see WAT200
ZOOFURIN see NGE000
ZOOLON see BDJ250
ZOPAQUE see TGG760
ZORANE see BRF500
ZORIFLAVIN see DBX400
ZOTOX see ARB250
ZOTOX see ARH500
ZOTOX CRAB GRASS KILLER see ARB250
ZP see ZLS000
ZPF see ZJS400
ZP-SB see ZJS400
ZUTRACIN see BAC250
ZWAVELWATERSTOF (DUTCH) see HIC500
ZWAVELZUUROPLOSSINGEN (DUTCH) see SOI500
ZWITSALAX see DMH400
ZYGOSPORIN A see ZUS000
ZYKLOHEXANDIAMINETETRAESSIGSAEURE
KADMIUMKOMPLEXE see CAD900
ZYKLOPHOSPHAMID (GERMAN) see CQC650
ZYLOPRIM see ZVJ000
ZYLORIC see ZVJ000

ZYNOPLEX see TAD175

ZYTEL 211 see PJY500
ZYTOX see MHR200

DOT Guide Cross-Index

NA 0349 see GJU600	NA 3018 see MNH000	UN 0411 see PBC250
NA 0473 see BAO900	NA 3018 see TCF250	UN 0452 see GJU600
NA 0473 see LDP000	NA 9163 see ZTJ000	UN 0483 see CPR800
NA 0473 see MMP000	NA 9202 see CBW750	UN 0484 see CQH250
NA 1051 see HHS000	NA 9206 see MOB399	UN 0490 see NMP620
NA 1247 see MLH750	NA 9260 see AGX000	UN 1001 see ACI750
NA 1361 see CDI000	NA 9269 see TLB750	UN 1002 see AFG250
NA 1361 see CDI250	UN 0004 see ANS500	UN 1003 see AFG250
NA 1361 see CDJ000	UN 0027 see ERF500	UN 1005 see AMY500
NA 1463 see CMK000	UN 0027 see PLL750	UN 1006 see AQW250
NA 1499 see SIT750	UN 0028 see ERF500	UN 1951 see AQW250
NA 1549 see AQE000	UN 0028 see PLL750	UN 1008 see BMG700
NA 1549 see AQK000	UN 0072 see CPR800	UN 1009 see TJY100
NA 1556 see DFP200	UN 0074 see DUR800	UN 1010 see BOP100
NA 1557 see ARI000	UN 0075 see DJE400	UN 1011 see BOR500
NA 1557 see ARJ100	UN 0076 see DUY600	UN 1013 see CBU250
NA 1574 see CAM300	UN 0079 see HET500	UN 1014 see CBV250
NA 1574 see CAM500	UN 0081 see DYG000	UN 1015 see CBV000
NA 1649 see TCF000	UN 0110 see GJU600	UN 1016 see CBW750
NA 1665 see NMS000	UN 0114 see TEF500	UN 1017 see CDV750
NA 1707 see TEL750	UN 0118 see CPR800	UN 1018 see CFX500
NA 1760 see TGH250	UN 0129 see LCM000	UN 1020 see CJI500
NA 1760 see TGH252	UN 0130 see LEE000	UN 1022 see CLR250
NA 1778 see SCO500	UN 0133 see MAW250	UN 1023 see HHJ500
NA 1807 see PHS250	UN 0135 see MDC000	UN 1026 see COO000
NA 1811 see PKU250	UN 0143 see NGY000	UN 1027 see CQD750
NA 1829 see SOR500	UN 0144 see NGY000	UN 1028 see DFA600
NA 1831 see SOI520	UN 0146 see NMB000	UN 1029 see DFL000
NA 1911 see DDI450	UN 0147 see NMQ500	UN 1032 see DOQ800
NA 1956 see HDF050	UN 0151 see PBT050	UN 1033 see MJW500
NA 1967 see PAK230	UN 0153 see PIC800	UN 1035 see EDZ000
UN 1380 see PAT750	UN 0154 see PID000	UN 1036 see EFU400
NA 1986 see AFJ000	UN 0155 see TML325	UN 1037 see EHH000
NA 1987 see AFJ000	UN 0208 see TEG250	UN 1038 see EIO000
NA 1993 see DHE800	UN 0209 see TMN490	UN 1039 see EMT000
NA 1993 see DHE900	UN 0214 see TMK500	UN 1040 see EJN500
NA 1993 see FOP000	UN 0215 see TML000	UN 1041 see EJO000
NA 1999 see ARO500	UN 0219 see SMP500	UN 1045 see FEZ000
NA 2212 see ARM250	UN 0220 see UTJ000	UN 1046 see HAM500
NA 2212 see ARM260	UN 0222 see ANN000	UN 1048 see HHJ000
NA 2212 see ARM268	UN 0224 see BAI000	UN 1049 see HHW500
NA 2212 see ARM275	UN 0226 see CQH250	UN 1050 see HHL000
NA 2212 see ARM280	UN 0234 see SGP550	UN 1052 see HHU500
NA 2215 see MAK900	UN 0235 see PIC500	UN 1053 see HIC500
NA 2672 see ANK250	UN 0236 see PIC750	UN 1055 see IIC000
NA 2762 see AFK250	UN 0282 see NHA500	UN 1060 see MFX600
NA 2783 see DXH325	UN 0284 see GJU600	UN 1061 see MGC250
NA 2783 see MNH000	UN 0285 see GJU600	UN 1063 see MIF765
NA 2783 see PAK000	UN 0292 see GJU600	UN 1064 see MLE650
NA 2783 see TCF250	UN 0293 see GJU600	UN 1065 see NCG500
NA 2783 see TIQ250	UN 0318 see GJU600	UN 1066 see NGP500
NA 2809 see MCW250	UN 0340 see CCU250	UN 1067 see NGU500
NA 2811 see SBT200	UN 0341 see CCU250	UN 1069 see NMH000
NA 2821 see PDN750	UN 0342 see CCU250	UN 1070 see NGU000
NA 2845 see MOC000	UN 0343 see CCU250	UN 1072 see OQW000
NA 2920 see DEV200	UN 0372 see GJU600	UN 1073 see OQW000
NA 2922 see SHR000	UN 0385 see NFJ000	UN 1075 see LGM000
NA 2927 see EOP600	UN 1578 see NFS525	UN 1076 see PGX000
NA 2927 see EOR000	UN 0391 see CPR800	UN 1077 see PMO500
NA 2949 see SHR000	UN 0394 see SMP500	UN 1079 see SOH500
	UN 0402 see PCD500	UN 1080 see SOI000

UN 1081 see TCH500	UN 1181 see EHG500	UN 1264 see PAI250
UN 1082 see CLQ750	UN 1182 see EHK500	UN 1265 see EIK000
UN 1083 see TLD500	UN 1862 see EHO200	UN 1265 see PBK250
UN 1085 see VMP000	UN 1183 see DFK000	UN 1267 see PCR250
UN 1086 see VNP000	UN 1184 see EIY600	UN 1268 see PCS250
UN 1087 see MQL750	UN 1185 see EJM900	UN 1270 see NAH600
UN 1088 see AAG000	UN 1188 see EJH500	UN 1271 see PCT250
UN 1089 see AAG250	UN 1189 see EJJ500	UN 1272 see PIH750
UN 1090 see ABC750	UN 1190 see EKL000	UN 1274 see PND000
UN 1091 see ABC750	UN 1192 see LAJ000	UN 1275 see PMT750
UN 1092 see ADR000	UN 1193 see MKA400	UN 1276 see PNC250
UN 1093 see ADX500	UN 1194 see ENN000	UN 1277 see PND250
UN 1098 see AFV500	UN 1195 see EPB500	UN 1278 see CKP750
UN 1099 see AFY000	UN 1196 see EPY500	UN 1279 see PNJ400
UN 1100 see AGB250	UN 1198 see FMV000	UN 1280 see PNL600
UN 1104 see AOD725	UN 1199 see FPQ875	UN 1281 see PNM500
UN 1104 see AOD735	UN 1201 see FQT000	UN 1282 see POP250
UN 1106 see AOJ000	UN 1203 see GBY000	UN 1288 see COD750
UN 1106 see PBV500	UN 1204 see NGY000	UN 1289 see SIK450
UN 1108 see AOI800	UN 1206 see HBC500	UN 1292 see EPF550
UN 1110 see MGN500	UN 1207 see HEM000	UN 1294 see TGK750
UN 1112 see AOL250	UN 1208 see HEN000	UN 1295 see TJD500
UN 1114 see BBL250	UN 1212 see IIL000	UN 1296 see TJO000
UN 1123 see BPU750	UN 1213 see IIJ000	UN 1297 see TLD500
UN 1123 see BPV000	UN 1214 see IIM000	UN 1298 see TLN250
UN 1123 see BPV100	UN 1218 see IMS000	UN 1299 see TOD750
UN 1125 see BPX750	UN 1219 see INJ000	UN 1300 see TOD750
UN 1126 see BMX500	UN 1220 see INE100	UN 1301 see VLU250
UN 1128 see BRK000	UN 1221 see INK000	UN 1302 see EQF500
UN 1129 see BSU250	UN 1222 see IQP000	UN 1303 see VPK000
UN 1130 see CBB500	UN 1223 see KEK000	UN 1304 see IJQ000
UN 1131 see CBV500	UN 1228 see CPW300	UN 1305 see TIN750
UN 1134 see CEJ125	UN 1228 see FPM000	UN 1307 see XGS000
UN 1135 see EIU800	UN 1228 see HBD500	UN 1309 see AGX000
UN 1136 see CMY900	UN 1228 see HES000	UN 1310 see ANS500
UN 1143 see COB250	UN 1228 see IMU000	UN 1312 see BMD000
UN 1144 see COC500	UN 1228 see LBX000	UN 1313 see CAW500
UN 1145 see CPB000	UN 1228 see PBM000	UN 1314 see CAW500
UN 1146 see CPV750	UN 1228 see PML500	UN 1318 see CNE000
UN 1147 see DAE800	UN 1228 see TGO750	UN 1320 see DUY600
UN 1148 see DBF750	UN 1228 see TGP000	UN 1328 see HEI500
UN 1149 see BRH750	UN 1228 see TGP250	UN 1332 see TDW500
UN 1150 see DFH800	UN 1229 see MDJ750	UN 1334 see NAJ500
UN 1153 see EJE500	UN 1230 see MGB150	UN 1336 see NHA500
UN 1154 see DHJ200	UN 1231 see MFW100	UN 1337 see NMB000
UN 1155 see EJU000	UN 1233 see HFJ000	UN 1338 see PHO500
UN 1156 see DJN750	UN 1234 see MGA850	UN 1339 see PHQ750
UN 1157 see DNI800	UN 1235 see MGC250	UN 1340 see PHS000
UN 1158 see DNM200	UN 1237 see MHY000	UN 1341 see PHS500
UN 1159 see IOZ750	UN 1238 see MIG000	UN 1343 see PHT750
UN 1160 see DOQ800	UN 1239 see CIO250	UN 1344 see PID000
UN 1161 see MIF000	UN 1242 see DFS000	UN 1346 see SCP000
UN 1162 see DFE259	UN 1243 see MKG750	UN 1347 see PID200
UN 1163 see DSF400	UN 1244 see MKN000	UN 1348 see SGP550
UN 1164 see TFP000	UN 1245 see HFG500	UN 1349 see PIC500
UN 1165 see DVQ000	UN 1246 see MKY500	UN 1350 see SOD500
UN 1166 see DVR800	UN 2397 see MLA750	UN 1354 see TMK500
UN 1167 see VOP000	UN 1248 see MOT000	UN 1355 see TML000
UN 1170 see EFU000	UN 1249 see PBN250	UN 1356 see TMN490
UN 1171 see EES350	UN 1250 see MQC500	UN 1357 see UTJ000
UN 1172 see EES400	UN 1251 see BOY500	UN 1358 see ZOA000
UN 1173 see EFR000	UN 1255 see NAH600	UN 1360 see CAW250
UN 1175 see EGP500	UN 1256 see NAH600	UN 1361 see CBT500
UN 1176 see BMC250	UN 1257 see GBY000	UN 1362 see CBT500
UN 1178 see DHI000	UN 1259 see NCZ000	UN 1363 see CNR000
UN 1179 see EHA500	UN 1261 see NHM500	UN 1366 see DKE600
UN 1180 see EHE000	UN 1262 see OCU000	UN 1369 see DSY600

UN 1376 see IHG100	UN 1493 see SDS000	UN 1604 see EEA500
UN 1381 see PHP010	UN 1494 see SFG000	UN 1605 see EIY500
UN 2447 see PHP010	UN 1495 see SFS000	UN 1606 see IGN000
UN 1382 see PLT250	UN 1496 see SFT500	UN 1607 see IGO000
UN 1384 see SHR500	UN 1498 see SIO900	UN 1611 see HCY000
UN 1385 see SJY500	UN 1500 see SIQ500	UN 1613 see HHS000
UN 1396 see AGX000	UN 1502 see PCE750	UN 1614 see HHS000
UN 1397 see AHE750	UN 1503 see SJC000	UN 1616 see LCV000
UN 1400 see BAH250	UN 1504 see SJC500	UN 1617 see ARC750
UN 1402 see CAN750	UN 1505 see SJE000	UN 1617 see LCK000
UN 1403 see CAQ250	UN 1506 see SMF500	UN 1617 see LCK100
UN 1407 see CDC000	UN 1507 see SMK000	UN 1618 see LCL000
UN 1408 see FBG000	UN 1509 see SMK500	UN 1620 see LCU000
UN 1410 see LHS000	UN 1510 see TDY250	UN 1621 see LIC000
UN 1411 see LHS000	UN 1511 see HIB500	UN 1622 see ARD000
UN 1413 see LHT000	UN 1513 see ZES000	UN 1623 see MDF350
UN 1414 see LHH000	UN 1514 see ZJJ000	UN 1624 see MCY475
UN 1415 see LGO000	UN 1515 see ZLA000	UN 1625 see MDF000
UN 1417 see LHP000	UN 1516 see ZLJ000	UN 1626 see PLU500
UN 1418 see MAC750	UN 1517 see PIC750	UN 1627 see MDE750
UN 1419 see AHD250	UN 1541 see MLC750	UN 1629 see MCS750
UN 1420 see PKT250	UN 1545 see AGJ250	UN 1629 see MDE250
UN 1423 see RPA000	UN 1546 see DCG800	UN 1630 see MCW500
UN 1426 see SFF500	UN 1547 see AOQ000	UN 1631 see MCX500
UN 1427 see SHO500	UN 1548 see BBL000	UN 1636 see MDA250
UN 1428 see SEE500	UN 1550 see AQE250	UN 1642 see MDA500
UN 1431 see SIK450	UN 1551 see AQG250	UN 1637 see MDC500
UN 1432 see SJI500	UN 1553 see ARB250	UN 1638 see MDC750
UN 1435 see ZBJ000	UN 1554 see ARB250	UN 1639 see MCV250
UN 1436 see ZBJ000	UN 1555 see ARF250	UN 1640 see MDF250
UN 1437 see ZRA000	UN 1558 see ARA750	UN 1641 see MCT500
UN 1438 see AHD750	UN 1559 see ARH500	UN 1643 see NCP500
UN 1439 see ANB500	UN 1560 see ARF500	UN 1644 see MCU000
UN 1442 see PCD500	UN 1561 see ARI750	UN 1646 see MCU250
UN 1444 see ANR000	UN 1562 see ARE500	UN 1647 see BNM750
UN 1445 see BAJ500	UN 1562 see ARE750	UN 1648 see ABE500
UN 1446 see BAN250	UN 1565 see BAK750	UN 1650 see NBE500
UN 1447 see PCD750	UN 1567 see BFO750	UN 1651 see AQN635
UN 1448 see PCK000	UN 1569 see BNZ000	UN 1653 see NDB500
UN 1449 see BAO250	UN 1570 see BOL750	UN 1654 see NDN000
UN 1451 see CDE250	UN 1571 see BAI000	UN 1656 see NDP400
UN 1452 see CAO500	UN 1572 see HKC000	UN 1657 see NDR000
UN 1453 see CAP000	UN 1573 see ARB750	UN 1658 see NDR500
UN 1454 see CAU000	UN 1575 see CAQ500	UN 1659 see NDS500
UN 1456 see CAV250	UN 1577 see CGL750	UN 1660 see NEG100
UN 1457 see CAV500	UN 1578 see CJA950	UN 1661 see NEN500
UN 1463 see CMK000	UN 1578 see CJB250	UN 1661 see NEO000
UN 1466 see FAY200	UN 1578 see CJB750	UN 1661 see NEO500
UN 1466 see IHB900	UN 1579 see CLK235	UN 1662 see NEX000
UN 1467 see GLA000	UN 1580 see CKN500	UN 1663 see NIF000
UN 1469 see LDO000	UN 1583 see CKN500	UN 1663 see NIF010
UN 1470 see LDS499	UN 1584 see PIE500	UN 1664 see NMO500
UN 1471 see LHJ000	UN 1586 see CNN500	UN 1664 see NMO525
UN 1472 see LHO000	UN 1587 see CNL000	UN 1664 see NMO550
UN 1474 see MAH000	UN 1589 see COO750	UN 1669 see PAW500
UN 1475 see PCE000	UN 1591 see DEP600	UN 1670 see PCF300
UN 1476 see MAH750	UN 1592 see DEP800	UN 1671 see PDN750
UN 1484 see PKY300	UN 1593 see MJP450	UN 1672 see PFJ400
UN 1485 see PLA250	UN 1594 see DKB110	UN 1673 see PEY000
UN 1486 see PLL500	UN 1595 see DUD100	UN 1673 see PEY250
UN 1487 see PLM000	UN 1597 see DUQ180	UN 1673 see PEY500
UN 1488 see PLM500	UN 1597 see DUQ200	UN 1674 see ABU500
UN 1489 see PLO500	UN 1597 see DUQ400	UN 1677 see ARD250
UN 1490 see PLP000	UN 1597 see DUQ600	UN 1678 see PKV500
UN 1491 see PLP250	UN 1599 see DUY600	UN 1679 see PLC175
UN 1422 see PLS500	UN 1600 see PGN100	UN 1680 see PLC500
UN 1492 see DWQ000	UN 1603 see EGV000	UN 1683 see SDM100

UN 1684 see SDP000	UN 1760 see FBI000	UN 1837 see TFO000
UN 1685 see ARD750	UN 1761 see DBU800	UN 1838 see TGH350
UN 1685 see SEY100	UN 1762 see CPE500	UN 1839 see TII250
UN 1686 see SEY200	UN 1764 see DEL000	UN 1840 see ZFA000
UN 1686 see SEY500	UN 1765 see DEN400	UN 1841 see AAG500
UN 1687 see SFA000	UN 1766 see DGF200	UN 1845 see CBU250
UN 1688 see HKC500	UN 1767 see DEY800	UN 1846 see CBY000
UN 1689 see SGA500	UN 1768 see PHF250	UN 1847 see PLT250
UN 1690 see SHF500	UN 1769 see DFF000	UN 1848 see PMU750
UN 1691 see SME500	UN 1770 see BNG750	UN 1859 see SDF650
UN 1692 see SMN500	UN 1770 see BNH000	UN 1860 see VPA000
UN 1692 see SMO500	UN 1771 see DYA800	UN 1862 see COB750
UN 1695 see CDN200	UN 1773 see FAU000	UN 1865 see PNQ500
UN 1697 see CEA750	UN 1775 see FDD125	UN 1866 see RDP000
UN 1698 see PDB000	UN 1775 see HHS600	UN 1868 see DAE400
UN 1699 see CGN000	UN 1776 see PHJ250	UN 1869 see MAC750
UN 1701 see XRS000	UN 1777 see FLZ000	UN 1870 see PKY250
UN 1702 see TBP750	UN 1779 see FNA000	UN 1872 see LCX000
UN 1704 see SOD100	UN 1780 see FOY000	UN 1873 see PCD250
UN 1705 see TCF260	UN 1781 see HCQ000	UN 1884 see BAO000
UN 1708 see TGQ500	UN 1782 see HDE000	UN 1885 see BBX000
UN 1708 see TGQ750	UN 1858 see HDF000	UN 1886 see BAY300
UN 1708 see TGR000	UN 1783 see HEO000	UN 1887 see CES650
UN 1709 see TGL750	UN 1784 see HFX500	UN 1888 see CHJ500
UN 1710 see TIO750	UN 1786 see HHV000	UN 1889 see COO500
UN 1712 see ZDS000	UN 1787 see HHI500	UN 1891 see EGV400
UN 1713 see ZGA000	UN 1788 see HHJ000	UN 1892 see DFH200
UN 1714 see ZLS000	UN 1789 see HHL000	UN 1894 see PFN100
UN 1715 see AAX500	UN 1790 see HHU500	UN 1895 see MCU750
UN 1716 see ACD750	UN 1791 see SHU500	UN 1897 see PCF275
UN 1717 see ACF750	UN 1792 see IDS000	UN 1898 see ACO500
UN 1718 see ADF250	UN 1793 see PHE500	UN 1902 see DNK800
UN 1722 see AGB500	UN 1794 see LDY000	UN 1905 see SBN000
UN 1723 see AGI250	UN 1798 see HHM000	UN 1907 see SEE000
UN 1724 see AGU250	UN 1799 see NNE000	UN 1908 see SFT500
UN 1725 see AGX750	UN 1800 see OBI000	UN 1910 see CAU500
UN 1726 see AGY750	UN 1801 see OGE000	UN 1911 see DDI450
UN 1727 see ANJ000	UN 1802 see PCD250	UN 1913 see NCG500
UN 1728 see PBY750	UN 1803 see HJH500	UN 1914 see BSJ500
UN 1729 see AOY250	UN 1804 see TJA750	UN 1915 see CPC000
UN 1730 see AQD000	UN 1805 see PHB250	UN 1916 see DFJ050
UN 1731 see AQD000	UN 1806 see PHR500	UN 1917 see EFT000
UN 1732 see AQF250	UN 1808 see PHT250	UN 1918 see COE750
UN 1733 see AQC500	UN 1809 see PHT275	UN 1919 see MGA500
UN 1736 see BDM500	UN 1810 see PHQ800	UN 1921 see PNL400
UN 1737 see BEC000	UN 1812 see PLF500	UN 1922 see PPS500
UN 1738 see BEE375	UN 1813 see PLJ500	UN 1923 see CAN000
UN 1739 see BEF500	UN 1814 see PLJ500	UN 1928 see MLE000
UN 1741 see BMG500	UN 1816 see PNX250	UN 1931 see ZGJ100
UN 1742 see BMG750	UN 1817 see PPR500	UN 1931 see ZIJ100
UN 1744 see BMP000	UN 1818 see SCQ500	UN 1932 see ZOA000
UN 1745 see BMQ000	UN 1819 see AHG000	UN 1935 see COI500
UN 1746 see BMQ325	UN 1821 see SEG800	UN 1938 see BMR750
UN 1747 see BSR000	UN 2837 see SEG800	UN 1939 see PHU000
UN 2716 see BST500	UN 1823 see SHS000	UN 1940 see TFJ100
UN 1748 see HOV500	UN 1823 see SHS500	UN 1941 see DKG850
UN 1749 see CDX750	UN 1824 see SHS000	UN 1942 see ANN000
UN 1750 see CEA000	UN 1824 see SHS500	UN 1952 see EJO000
UN 1751 see CEA000	UN 1825 see SIN500	UN 1954 see DFB400
UN 1752 see CEC250	UN 1827 see TGC250	UN 1958 see DGL600
UN 1753 see CKM250	UN 1829 see SOR500	UN 1959 see VPP000
UN 1754 see CLG500	UN 1830 see SOI500	UN 1961 see EDZ000
UN 1755 see CMK000	UN 1832 see SOI500	UN 1962 see EIO000
UN 1756 see CMJ560	UN 1833 see SOO500	UN 1963 see HAM500
UN 1757 see CMJ560	UN 1834 see SOT000	UN 1964 see HHJ500
UN 1758 see CML125	UN 1835 see TDK500	UN 1965 see HHJ500
UN 1759 see FBI000	UN 1836 see TFL000	UN 1966 see BFO750

UN 1966 see HHW500	UN 2196 see TOC550	UN 2257 see PKT250
UN 1969 see MOR750	UN 2197 see HHI500	UN 2258 see PMK250
UN 1971 see MDQ750	UN 2198 see PHR750	UN 2259 see TJR000
UN 1972 see MDQ750	UN 2199 see PGY000	UN 2260 see TMY250
UN 1973 see FOO560	UN 2201 see NGU000	UN 2261 see XKA000
UN 1974 see BNA250	UN 2202 see HIC000	UN 2262 see DQY950
UN 1975 see NGT500	UN 2203 see SDH575	UN 2264 see DRF709
UN 1976 see CPS000	UN 2204 see CCC000	UN 2269 see AIX250
UN 1977 see NGP500	UN 2205 see AER250	UN 2270 see EFU400
UN 1978 see PMJ750	UN 2206 see AGJ000	UN 2271 see EGI750
UN 1982 see CBY250	UN 2206 see BMT150	UN 2271 see ODI000
UN 1984 see CBY750	UN 2206 see CHL250	UN 2272 see EGK000
UN 1986 see EFU000	UN 2206 see CKA750	UN 2273 see EGK500
UN 1987 see EFU000	UN 2206 see CKB000	UN 2275 see EGW000
UN 1991 see NCI500	UN 2206 see FHC200	UN 2276 see EKS500
UN 1994 see IHG500	UN 2206 see IKH000	UN 2277 see EMF000
UN 2001 see NAR500	UN 2206 see IKH099	UN 2278 see HBJ000
UN 2008 see ZOA000	UN 2206 see TGM750	UN 2279 see HCD250
UN 2009 see ZOA000	UN 2206 see THY750	UN 2280 see HEO000
UN 2011 see MAI000	UN 2206 see TKJ250	UN 2281 see DNJ800
UN 2014 see HIB010	UN 2206 see XSS260	UN 2283 see IIY000
UN 2014 see HIB050	UN 2207 see AGJ000	UN 2284 see IJX000
UN 2020 see CJK750	UN 2207 see BMT150	UN 2290 see IMG000
UN 2021 see CJK750	UN 2207 see CHL250	UN 2293 see MEX250
UN 2022 see CNW500	UN 2207 see CKA750	UN 2294 see MGN750
UN 2023 see EAZ500	UN 2207 see CKB000	UN 2295 see MIF775
UN 2027 see SEY200	UN 2207 see COI250	UN 2296 see MIQ740
UN 2027 see SEY500	UN 2207 see FHC200	UN 2297 see MIR250
UN 2029 see HGS000	UN 2207 see FLR100	UN 2298 see MIU500
UN 2030 see HGU500	UN 2207 see IKG800	UN 2299 see DEM800
UN 2031 see NED500	UN 2207 see IKH000	UN 2300 see EOS000
UN 2032 see NEE500	UN 2207 see IKH099	UN 2301 see MKH000
UN 2036 see XDS000	UN 2207 see OBG000	UN 2302 see MKW450
UN 2038 see DVG600	UN 2207 see TGM750	UN 2303 see MPK250
UN 2044 see NCH000	UN 2207 see TGM800	UN 2304 see NAJ500
UN 2045 see IJS000	UN 2207 see THY750	UN 2307 see NFS700
UN 2047 see DGG700	UN 2207 see TKJ250	UN 2310 see ABX750
UN 2047 see DGG950	UN 2207 see XIJ000	UN 2311 see EEL100
UN 2047 see DGI700	UN 2207 see XSS260	UN 2312 see PDN750
UN 2048 see DGW000	UN 2209 see FMV000	UN 2315 see PJL750
UN 2049 see DIU000	UN 2210 see MAS500	UN 2316 see SFZ100
UN 2051 see DOY800	UN 2213 see PAI000	UN 2317 see SFZ100
UN 2052 see MCC250	UN 2214 see PHW750	UN 2318 see SHR000
UN 2053 see MKW600	UN 2215 see MAM000	UN 2320 see TCE500
UN 2054 see MRP750	UN 2218 see ADS750	UN 2323 see TJT800
UN 2055 see SMQ000	UN 2219 see AGH150	UN 2325 see TLM050
UN 2056 see TCR750	UN 2222 see AOX750	UN 2416 see TLN000
UN 2058 see VAG000	UN 2224 see BCQ250	UN 2329 see TMD500
UN 2059 see CCU250	UN 2225 see BBS750	UN 2330 see UJS000
UN 2074 see ADS250	UN 2226 see BFL250	UN 2331 see ZFA000
UN 2075 see CDN550	UN 2227 see MHU750	UN 2332 see AAH250
UN 2076 see CNW750	UN 2232 see CDY500	UN 2333 see AFU750
UN 2076 see CNX000	UN 2238 see CLK100	UN 2334 see AFW000
UN 2076 see CNX250	UN 2238 see CLK130	UN 2336 see AGH000
UN 2077 see NBE700	UN 2238 see TGY075	UN 2337 see PFL850
UN 2078 see TGM740	UN 2240 see CLG500	UN 2338 see BDH500
UN 2079 see DJG600	UN 2243 see CPF000	UN 2339 see BMX750
UN 2130 see NNA100	UN 2245 see CPW500	UN 2341 see BNP250
UN 2171 see DNS000	UN 2246 see CPX750	UN 2343 see BNU500
UN 2186 see HHL000	UN 2247 see DAG400	UN 2344 see BNY000
UN 2187 see CBU250	UN 2248 see DDT800	UN 2345 see PMN500
UN 2188 see ARK250	UN 2249 see BIK000	UN 2346 see BOT500
UN 2190 see ORA000	UN 2649 see BIK250	UN 2347 see BRR900
UN 2191 see SOU500	UN 2251 see NNG000	UN 2348 see BPW100
UN 2192 see GEI100	UN 2252 see DOE600	UN 2350 see BRU780
UN 2194 see SBS000	UN 2253 see DQF800	UN 2352 see VMZ000
UN 2195 see TAK250	UN 2256 see CPC579	UN 2356 see CKQ000

UN 2357 see CPF500	UN 2457 see DQT400	UN 2530 see IJW000
UN 2359 see DBI600	UN 2458 see HCQ600	UN 2531 see MDN250
UN 2360 see DBK000	UN 2463 see AHB500	UN 2535 see MMA250
UN 2361 see DNH400	UN 2464 see BFT000	UN 2538 see NHP990
UN 2362 see DFF809	UN 2465 see DGN200	UN 2541 see TBE000
UN 2363 see EMB100	UN 2466 see PLE260	UN 2542 see THX250
UN 2364 see IKG000	UN 2468 see TIQ750	UN 2547 see SJZ100
UN 2366 see DIX200	UN 2470 see PEA750	UN 2548 see CDX250
UN 2368 see PIH250	UN 2471 see OKK000	UN 2552 see HDA000
UN 2369 see BPJ850	UN 2473 see ARA500	UN 2553 see NAH600
UN 2371 see IHR220	UN 2474 see TFN500	UN 2554 see CIU750
UN 2372 see TDQ750	UN 2475 see VEP000	UN 2555 see CCU250
UN 2373 see EFT500	UN 2477 see ISE000	UN 2556 see CCU250
UN 2374 see DHH800	UN 2478 see AGJ000	UN 2557 see CCU250
UN 2375 see EPH000	UN 2478 see BMT150	UN 2558 see BNI000
UN 2377 see DOO600	UN 2478 see CHL250	UN 2564 see TII250
UN 2378 see DOS200	UN 2478 see CKA750	UN 2565 see DGT600
UN 2379 see DQU600	UN 2478 see CKB000	UN 2567 see SJA000
UN 2380 see DHG000	UN 2478 see COI250	UN 2572 see PFI000
UN 2381 see DRQ400	UN 2478 see FHC200	UN 2573 see TEJ100
UN 2265 see DSB000	UN 2478 see FLR100	UN 2574 see TNP500
UN 2382 see DSF600	UN 2478 see IKG800	UN 2576 see PHU000
UN 2383 see DWR000	UN 2478 see IKH000	UN 2579 see PIJ000
UN 2384 see PNM000	UN 2478 see IKH099	UN 2580 see AGX750
UN 2385 see ELS000	UN 2478 see OBG000	UN 2581 see AGY750
UN 2386 see EOS500	UN 2478 see TGM750	UN 2582 see FAU000
UN 2387 see FGA000	UN 2478 see THY750	UN 2587 see QQS200
UN 2389 see FPK000	UN 2478 see TKJ250	UN 2591 see XDS000
UN 2390 see IEH000	UN 2478 see XSS260	UN 2599 see FOO562
UN 1608 see IGM000	UN 2480 see MKX250	UN 2603 see COY000
UN 2393 see IIR000	UN 2481 see ELS500	UN 2606 see MPI750
UN 2394 see PMV250	UN 2482 see PNP000	UN 2607 see ADR500
UN 2396 see MGA250	UN 2485 see BRQ500	UN 2608 see NIY000
UN 2398 see MHV859	UN 2487 see PFK250	UN 2610 see THN000
UN 2399 see MOG500	UN 2488 see CPN500	UN 2611 see CKR500
UN 2400 see ITC000	UN 1763 see CPR250	UN 2456 see CKS000
UN 2401 see PIL500	UN 2489 see MJP400	UN 2612 see MOU830
UN 2403 see MQK750	UN 2490 see BII250	UN 2614 see IMW000
UN 2404 see PMV750	UN 2491 see EEC600	UN 2615 see EPC125
UN 2407 see IOL000	UN 2493 see HDG000	UN 2616 see IOI000
UN 2411 see BSX250	UN 2495 see IDT000	UN 2617 see MIQ745
UN 2412 see TDC730	UN 2496 see PMV500	UN 2618 see VQK650
UN 2414 see TFM250	UN 2497 see SJF000	UN 2619 see DQP800
UN 2417 see CCA500	UN 2498 see FNK025	UN 2621 see ABB500
UN 2418 see SOR000	UN 2501 see TND250	UN 2622 see GGW000
UN 2420 see HCZ000	UN 2502 see VBA000	UN 2626 see CDU000
UN 2422 see OBO000	UN 2503 see ZPA000	UN 2628 see PLG000
UN 2426 see ANN000	UN 2505 see ANH250	UN 2629 see SHG500
UN 2427 see PLA250	UN 2506 see ANJ500	UN 2630 see SJT500
UN 2428 see SFS000	UN 2507 see CKO750	UN 2642 see FIC000
UN 2429 see CAO500	UN 2508 see MRD500	UN 2206 see FLR100
UN 2431 see AOV900	UN 2509 see PKX750	UN 2643 see MHR250
UN 2432 see DIS700	UN 2511 see CKS750	UN 2644 see MKW000
UN 2435 see DFJ800	UN 2512 see ALT000	UN 2646 see HCE500
UN 2436 see TFA500	UN 2512 see ALT250	UN 2647 see MAO250
UN 2437 see DFQ800	UN 2514 see PEO500	UN 2648 see NIE600
UN 2438 see DTS400	UN 2515 see BNL000	UN 2650 see DFU000
UN 2439 see SHQ500	UN 2516 see CBX750	UN 1152 see DFX000
UN 2440 see TGC282	UN 2517 see CFX250	UN 2651 see MJQ000
UN 1433 see TGE500	UN 2521 see KFA000	UN 2655 see PLH750
UN 2441 see TGG250	UN 2522 see DPG600	UN 2656 see QMJ000
UN 2442 see TIJ150	UN 2524 see ENY500	UN 2657 see SBR000
UN 2443 see VDP000	UN 2525 see DJT200	UN 2658 see SBO500
UN 2444 see VEF000	UN 2526 see FPW000	UN 2659 see SFU500
UN 2448 see SOD500	UN 2527 see IIK000	UN 2661 see HCL500
UN 2451 see NGW000	UN 2528 see IIW000	UN 2662 see HIH000
UN 2453 see FIB000	UN 2529 see IJU000	UN 2664 see DDP800

UN 2666 see EHP500	UN 2761 see DHB400	UN 2950 see MAC750
UN 2668 see CDN500	UN 2785 see MPV400	UN 2951 see OPE000
UN 2670 see TJD750	UN 2785 see TET900	UN 2952 see ASL750
UN 2671 see AMI000	UN 2789 see AAT250	UN 2956 see TML750
UN 2671 see AMI250	UN 2790 see AAT250	UN 2965 see BMH000
UN 2671 see AMI500	UN 2798 see DGE400	UN 2966 see MCN250
UN 2673 see CEH250	UN 2799 see PFW200	UN 2967 see SNK500
UN 2674 see DXE000	UN 2799 see PFW210	UN 2968 see MAS500
UN 2676 see SLQ000	UN 2802 see CNJ950	UN 2969 see CCP000
UN 2677 see RPZ000	UN 2802 see CNK500	UN 2970 see BBS300
UN 2678 see RPZ000	UN 2805 see LHH000	UN 2972 see DVF400
UN 2679 see LHI100	UN 2806 see LHM000	UN 2973 see DRO400
UN 2680 see LHI100	UN 2812 see AHG000	UN 2975 see TFS750
UN 2681 see CDD750	UN 2815 see AKB000	UN 2976 see TFT500
UN 2682 see CDD750	UN 2817 see ANJ000	UN 2977 see UOJ000
UN 2684 see DIY800	UN 2818 see ANT000	UN 2978 see UOJ000
UN 2685 see DJI400	UN 2819 see PBW750	UN 2979 see UNS000
UN 2686 see DHO500	UN 2820 see BSW000	UN 2980 see URS000
UN 2687 see DGU200	UN 2822 see CKW000	UN 2981 see URA200
UN 2688 see BNA825	UN 2823 see COB500	UN 2984 see HIB005
UN 2689 see CDT750	UN 2826 see CLJ750	UN 2984 see HIB010
UN 2692 see BMG400	UN 2829 see HEU000	UN 2989 see LCV100
UN 2698 see TDB000	UN 2830 see LHK000	UN 3022 see BOX750
UN 2699 see TKA250	UN 2831 see MIH275	UN 3023 see MKJ250
UN 2708 see MHV750	UN 2835 see SEM500	UN 3023 see OFE030
UN 2710 see DWT600	UN 2838 see VNF000	UN 3054 see CPB625
UN 2711 see DDJ900	UN 2839 see AAH750	UN 3055 see AJU250
UN 2713 see ADJ500	UN 2840 see BSU500	UN 3056 see HBB500
UN 2717 see CBA750	UN 2841 see DCH200	UN 3057 see TJX500
UN 2719 see BAI750	UN 2842 see NFY500	UN 3064 see NGY000
UN 2720 see CMJ600	UN 2845 see EOQ000	UN 3071 see CPW300
UN 2721 see CNJ900	UN 2849 see CKP600	UN 3071 see FPM000
UN 2724 see MAS900	UN 2851 see BMG800	UN 3071 see HBD500
UN 2725 see NDG000	UN 2853 see MAF600	UN 3071 see HES000
UN 2726 see NDG550	UN 2854 see COE000	UN 3071 see IMU000
UN 2727 see TEK750	UN 2855 see ZIA000	UN 3071 see LBX000
UN 2728 see ZSA000	UN 2858 see ZOA000	UN 3071 see PBM000
UN 2729 see HCC500	UN 2859 see ANY250	UN 3071 see PML500
UN 2733 see AOE200	UN 2862 see VDU000	UN 3071 see TGO750
UN 2733 see BPY000	UN 2864 see PLK810	UN 3071 see TGP000
UN 2733 see BPY250	UN 2865 see OLS000	UN 3071 see TGP250
UN 2733 see DAG600	UN 2869 see TGG250	UN 3080 see AGJ000
UN 2733 see HBL600	UN 2870 see AHG875	UN 3080 see BMT150
UN 2733 see HFK000	UN 2871 see AQB750	UN 3080 see CHL250
UN 2733 see OEK010	UN 2872 see DDL800	UN 3080 see CKA750
UN 2733 see PBV505	UN 2873 see DDU600	UN 3080 see CKB000
UN 2734 see AOE200	UN 2874 see FPU000	UN 3080 see COI250
UN 2734 see BPY000	UN 2875 see HCL000	UN 3080 see FHC200
UN 2734 see BPY250	UN 2876 see REA000	UN 3080 see FLR100
UN 2734 see DAG600	UN 2877 see ISR000	UN 3080 see IKG800
UN 2734 see HBL600	UN 2879 see SBT500	UN 3080 see IKH000
UN 2734 see HFK000	UN 2906 see IMG000	UN 3080 see IKH099
UN 2734 see PBV505	UN 2907 see CCK125	UN 3080 see OBG000
UN 2738 see BQH850	UN 2931 see VEZ000	UN 3080 see TGM750
UN 2739 see BSW550	UN 2933 see CKT000	UN 3080 see THY750
UN 2740 see PNH000	UN 2936 see TFK250	UN 3080 see TKJ250
UN 2746 see CBX109	UN 2937 see PDE000	UN 3080 see XIJ000
UN 2750 see DGG400	UN 2938 see MHA750	UN 3080 see XSS260
UN 2751 see DJW600	UN 2943 see TCS500	UN 3083 see PCF750
UN 2752 see EBQ700	UN 2945 see MHV000	UN 3136 see CBY750
UN 2752 see EKM200	UN 2946 see ALS990	UN 3149 see PCL500
UN 2761 see AFK250	UN 2948 see AID500	